FUNDAMENTAL VIROLOGY

Third Edition

FUNDAMENTAL VIROLOGY

Third Edition

Editors-in-Chief

Bernard N. Fields, M.D.
Department of Microbiology and Molecular Genetics
Harvard Medical School
Boston, Massachusetts

David M. Knipe, Ph.D.
Department of Microbiology and Molecular Genetics
Harvard Medical School
Boston, Massachusetts

Peter M. Howley, M.D.
Department of Pathology
Harvard Medical School
Boston, Massachusetts

Associate Editors

Robert M. Chanock, M.D.
Laboratory of Infectious Diseases
National Institute of Allergy and Infectious Disease
National Institutes of Health
Bethesda, Maryland

Joseph L. Melnick, Ph.D., Sc.D.
Department of Virology and Epidemiology
Baylor College of Medicine
Texas Medical Center
Houston, Texas

Thomas P. Monath, M.D.
Ora Vax, Inc.
Cambridge, Massachusetts

Bernard Roizman, Sc.D.
Department of Molecular Genetics and Cell Biology
University of Chicago
Chicago, Illinois

Stephen E. Straus, M.D.
Laboratory of Clinical Investigation
National Institutes of Health
Bethesda, Maryland

Lippincott-Raven Publishers, Philadelphia, 227 East Washington Square, Philadelphia, PA 19106

Made in the United States of America

Library of Congress Cataloging-in-Publication Data

Fundamental virology / editors-in-chief, Bernard N. Fields, David M. Knipe, Peter M. Howley ; associate editors, Robert M. Chanock . . . [et al.]—3rd ed.
p. cm.
Consists of selected chapters reprinted from: Fields virology. 3rd ed. 1996
Includes bibliographical references and index.
ISBN 0-7817-0284-4
1. Virology. I. Fields, Bernard N. II. Knipe, David M. (David Mahan). III. Howley, Peter M. IV. Fields virology.
[DNLM: 1. Viruses. 2. Virus Replication. QW 160 F981 1995]
QR360.F847 1995
616′.0194—dc20
DNLM/DLC
for Library of Congress 95-31493

The material contained herein appeared in *Fields Virology*, Third Edition, edited by Fields, Knipe, Howley, et al. Lippincott-Raven Publishers, Philadelphia © 1996.

Great care has been taken to maintain the accuracy of the information contained in the volume. However, neither Lippincott-Raven Publishers nor the editors can be held responsible for errors or for any consequences arising from the use of the information contained herein.

Materials appearing in this book prepared by individuals as part of their official duties as U.S. Government employees are not covered by the above-mentioned copyright.

9 8 7 6 5 4 3 2

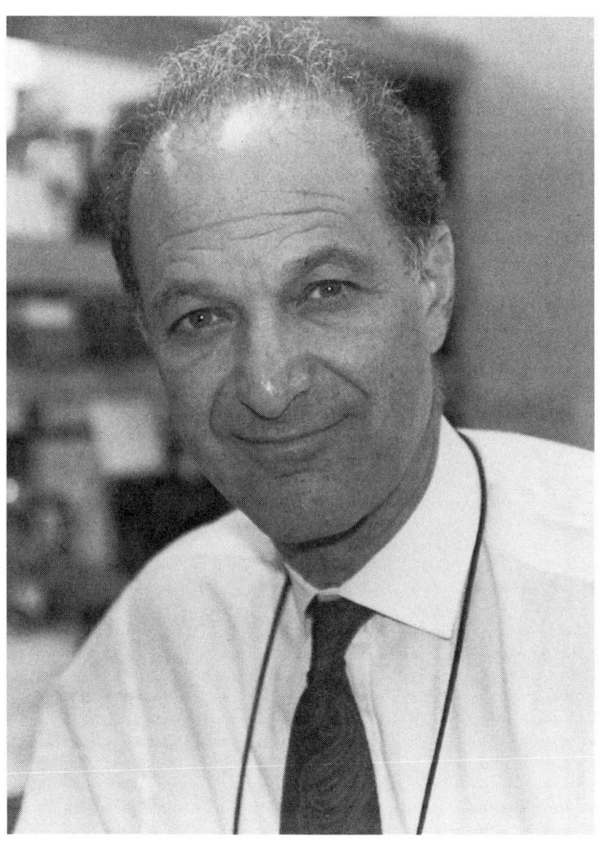

Bernard N. Fields, 1938–1995

Bernard N. Fields transmitted enthusiasm and warmth, love of life, and love of people to everyone who had the privilege of knowing him. In his scientific work he combined his training in medicine and infectious disease with training in basic virology to give a unique view of the biology of viruses. Through his studies of reoviruses he demonstrated that the stages of viral infection and pathogenesis in the host organism could be broken into discrete steps and the functions of individual viral gene products could be defined through genetic analysis. This laid the groundwork for the current field of molecular pathogenesis of viruses. Bernie's unique perspective on virology also led to the conception of this book, for which he brought together molecular and medical virologists. After he was diagnosed with pancreatic cancer in 1992, Bernie decided that one of his priorities was to publish the Third Edition of *Fields Virology.* He worked toward that goal with a sense of urgency over the following two years. It was a source of great satisfaction to Bernie when the final chapters were delivered to Raven Press in December of 1994. With his death on January 31, 1995, we have lost a great friend and statesman of biomedical science. On behalf of everyone who contributed to the Third Edition of *Fields Virology,* we dedicate this book to the memory of Bernard N. Fields.

Contents

Unclassified Agents

Contributing Authors

Rafi Ahmed, Ph.D.
Department of Microbiology and Immunology
UCLA School of Medicine
10833 Le Conte Avenue
Los Angeles, California 90024-1747

Kenneth I. Berns, M.D., Ph.D.
Department of Microbiology
New York Hospital-Cornell Medical Center
525 East 68th Street
New York, New York 10021

Allan M. Campbell, Ph.D.
Department of Biological Science
Stanford University
Stanford, California 94305

Robert M. Chanock, M.D.
Laboratory of Infectious Diseases
NIAID, Building 7
9000 Rockville Pike
National Institutes of Health
Bethesda, Maryland 20892

Donald M. Coen, Ph.D.
Department of Biological Chemistry and Molecular Pharmacology
Harvard Medical School
1250 Longwood Avenue
Boston, Massachusetts 02115

John M. Coffin, Ph.D.
Department of Molecular Biology and Microbiology
Tufts University School of Medicine
136 Harrison Avenue
Boston, Massachusetts 02111

Charles N. Cole, Ph.D.
Department of Biochemistry
Dartmouth Medical School
Hanover, New Hampshire 03756

Mary K. Estes, M.D.
Division of Molecular Virology
One Baylor Plaza
Baylor College of Medicine
Houston, Texas 77030

Bernard N. Fields, M.D.
Department of Microbiology and Molecular Genetics
Harvard Medical School
200 Longwood Avenue
Boston, Massachusetts 02115

Donald Ganem, M.D.
Department of Microbiology
University of California
San Francisco, California 94143

Stephen Harrison, Ph.D.
Department of Biochemistry and Molecular Biology
Howard Hughes Medical Institute
Harvard University
7 Divinity Avenue
Cambridge, Massachusetts 02138

Kathryn V. Holmes, Ph.D.
Department of Microbiology
Box B-175
University of Colorado
Health Sciences Center
4200 East 9th Avenue
Denver, Colorado 80262

Peter M. Howley, M.D.
Department of Pathology
Harvard Medical School
200 Longwood Avenue
Boston, Massachusetts 02115

Elliott Kieff, M.D., Ph.D.
Brigham and Women's Hospital
Thorn Building 12-S
75 Francis Street
Boston, Massachusetts 02115

David M. Knipe, Ph.D.
Department of Microbiology and Molecular Genetics
Harvard Medical School
200 Longwood Avenue
Boston, Massachusetts 02115

Daniel Kolakofsky, Ph.D.
Department of Genetics and Microbiology
University of Geneva School of Medicine
9, Avenue de Champel
CH-1211 Geneva
Switzerland

Michael M. C. Lai, M.D., Ph.D.
Department of Microbiology
University of Southern California
School of Medicine
2011 Zonal Avenue, HMR 401
Los Angeles, California 90033-1054

Robert A. Lamb, Ph.D., Sc.D.
Department of Biochemistry, Molecular Biology and Cell Biology
Northwestern University
2153 Sheridan Road
Evanston, Illinois 60208-3500

Arnold J. Levine, Ph.D.
Chairman, Department of Molecular Biology
Lewis Thomas Laboratory
Princeton University
Princeton, New Jersey 08541

Paul A. Luciw, M.D.
Department of Medical Pathology
School of Medicine
University of California
Davis, California 95616

Joseph L. Melnick, M.D.
Department of Virology and Epidemiology
Baylor College of Medicine
Texas Medical Center
One Baylor Plaza
Houston, Texas 77030

Lois K. Miller, Ph.D.
Department of Entomology
University of Georgia
Athens, Georgia 30602

Thomas P. Monath, M.D.
Vice President, Research and Medical Affairs
OraVax, Inc.
230 Albany Street
Cambridge, Massachusetts 02139

Lynda A. Morrison, Ph.D.
Department of Microbiology and Molecular Genetics
Harvard Medical School
200 Longwood Avenue
Boston, Massachusetts 02115

Bernard Moss, M.D., Ph.D.
National Institute of Allergy and Infectious Diseases
Building 5, Room 324
National Institutes of Health
9000 Rockville Pike
Bethesda, Maryland 20014

Frederick A. Murphy, D.V.M., Ph.D.
School of Veterinary Medicine
Office of the Dean
University of California
Davis, California 95616-8734

Joseph R. Nevins, Ph.D.
Howard Hughes Medical Institute
Department of Genetics
Duke University Medical Center
P.O. Box 3054
Durham, North Carolina 27710

Max L. Nibert, M.D., Ph.D.
Institute for Molecular Virology
University of Wisconsin - Madison
1525 Linden Drive
635 Bock Laboratories
Madison, Wisconsin 53706-1596

Michael B. A. Oldstone. M.D.
Scripps Clinic and Research Foundation
10666 North Torrey Pines Road
La Jolla, California 92037

Stanley B. Prusiner, M.D.
Department of Neurology
University of California
San Francisco, California 94143-0518

Robert F. Ramig, Ph.D.
Department of Virology and Epidemiology
Baylor College of Medicine
One Baylor Plaza
Texas Medical Center
Houston, Texas 77030

Bernard Roizman, Sc.D.
Department of Molecular Genetics and Cell Biology
University of Chicago
910 East 58th Street
Chicago, Illinois 60627

John K. Rose, Ph.D.
Department of Pathology
Yale University School of Medicine
P.O. Box 3333
New Haven, Connecticut 06510

Roland R. Rueckert, Ph.D.
Department of Biochemistry
Biophysics Laboratory
University of Wisconsin
1525 Linden Drive
Madison, Wisconsin 53706

Leslie A. Schiff, Ph.D.
Department of Microbiology
University of Minnesota Medical School
Box 196 UMH
420 Delaware Street SE
Minneapolis, Minnesota 55455

Milton J. Schlesinger, Ph.D.
Department of Microbiology and Immunology
Washington University School of Medicine
P.O. Box 8093
660 South Euclid Avenue
St. Louis, Missouri 63110

Sondra Schlesinger, Ph.D.
Department of Microbiology and Immunology
Washington University School of Medicine
P.O. Box 8230
660 South Euclid Avenue
St. Louis, Missouri 63110

Connie S. Schmaljohn, Ph.D.
Virology Division
USAMRIID
Fort Detrick
Frederick, Maryland 21701-5011

Amy E. Sears, Ph.D.
Department of Microbiology and Immunology
Emory University School of Medicine
1510 Clifton Road
Atlanta, Georgia 30322

Ganes C. Sen, Ph.D.
Department of Molecular Biology
The Cleveland Clinic Foundation
9500 Euclid Avenue
Cleveland, Ohio 44195

John G. Shaw, Ph.D.
University of Kentucky
College of Agriculture
Department of Plant Pathology
Lexington, Kentucky 40546

Thomas Shenk, Ph.D.
Howard Huges Medical Institute
Department of Molecular Biology
Princeton University
Princeton, New Jersey 08544-1014

John J. Skehel, Ph.D., F.R.S.
National Institute of Medical Research
Mill Hill, The Ridgeway
London, NW7, IAA
United Kingdom

Peter Southern, Ph.D.
Department of Microbiology
University of Minnesota
420 Delaware Street SE
Minneapolis, Minnesota 55455-0380

Ellen G. Strauss, Ph.D.
Division of Biology 156-29
California Institute of Technology
Pasadena, California 91125

James H. Strauss, Ph.D.
Division of Biology 156-29
California Institute of Technology
Pasadena, California 91125

Stephen E. Straus, M.D.
Laboratory of Clinical Investigation
National Institute of Allergy and Infectious Diseases
National Institutes of Health
Bldg. 10, Room 11N113
Bethesda, Maryland 20892

John M. Taylor, Ph.D.
Fox Chase Cancer Center
Institute for Cancer Research
7701 Burholme Avenue
Philadelphia, Pennsylvania 19111

Kenneth L. Tyler
Neurology Service (127)
Veterans Administration Medical Center
1055 Clermont Street
Denver, Colorado 80220

Jan Vilček, M.D.
New York University Medical Center
550 First Avenue
New York, New York 10016

Peter K. Vogt, Ph.D.
SBR-7 Department of Molecular and Experimental Medicine
The Scripps Research Institute
10666 North Torrey Pines Road
La Jolla, California 92037

Robert R. Wagner, M.D.
Microbiology Department
University of Virginia School of Medicine
Box 441
Charlottesville, Virgina 22908

J. Lindsay Whitton, M.D., Ph.D., M.R.C.
Neuropharmacology, CVN9
Scripps Clinic and Research Foundation
10666 North Torrey Pines Road
La Jolla, California 92037

Reed B. Wickner, M.D.
National Institute of Diabetes, Digestive, and Kidney Diseases
National Institutes of Health
Bldg. 8, Room 207
Bethesda, Maryland 20892-0001

Donald C. Wiley, Ph.D.
Department of Biochemistry and Molecular Biology
Howard Hughes Medical Institute
Harvard University
7 Divinity Avenue
Cambridge, Massachusetts 02138

Preface

Despite important medical advances over the past 25 years, viruses remain as major pathogens in humans. Viruses have also been important tools that have contributed to significant discoveries in biology over the past half century. Because of their simplicity, bacterial viruses played a focal role in many of the important developments in molecular biology. More recently, animal viruses have had an equally powerful impact on the study of eukaryotic molecular genetics. In conjunction with these fundamental discoveries, the natural history of infectious diseases has seen equally remarkable changes. Smallpox has disappeared and AIDS has appeared. With the increasing use of immunosuppression, many indigenous or "latent" viruses have taken on increasing importance. In addition, many "classic" viral infections have been controlled, in part, by effective vaccines (polio, measles, rubella) while others have resisted effective preventatives (respiratory syncytial virus). With the striking new insights in molecular biology, some important new insights into fundamental features of viruses as infectious agents have resulted. There are now new ways to make vaccines, and the biochemistry of viruses has begun to answer classic questions about epidemiology and pathogenesis. The goal of *Fields Virology,* from which the present volume is derived, is to bring together basic and medical aspects of virology in a more unified, comprehensive presentation than that provided by general textbooks. Thus, *Fields Virology* is a reference and textbook for medical and graduate students as well as scientists, physicians, and investigators interested in viruses as they are represented in the biological sciences.

This book, *Fundamental Virology Third Edition,* consists of a set of chapters reprinted from *Fields Virology Third Edition.* Its goal is to provide a text for graduate and upper-level undergraduate students, researchers, scientists, and investigators whose primary interests are directed toward basic aspects of virology rather than its more clinical or applied features. Like previous editions, this Third Edition is divided into two parts. The first part, Chapters 1 through 15, presents the basic concepts of general virology while the second part, Chapters 16 through 37, describes the biochemistry, molecular biology, and cellular aspects of replication of viruses of the different groups.

An enormous explosion in the information about viruses has occurred since the last edition. In addition to updating all chapters, we have included new chapters on viruses infecting (i) plants; (ii) insects; (iii) yeasts, fungi, and parasites; and (iv) bacteria. This text will be very useful in courses in general or molecular virology. We hope course instructors will supplement the material in this book with material on pathogenesis of specific viruses from *Fields Virology Third Edition* to further unify the fields of molecular virology and pathogenesis.

Many people have contributed to the preparation of this book; however, we want to specifically thank Marcia Kazmierczak here at Harvard Medical School for her excellent and invaluable assistance to us and the authors in the preparation of the book.

Bernard N. Fields, M.D.
David M. Knipe, Ph.D.
Peter M. Howley, M.D.

FUNDAMENTAL VIROLOGY

Third Edition

Fundamental Virology, Third Edition
edited by B.N. Fields, D.M. Knipe, P.M. Howley, et al.
Lippincott - Raven Publishers, Philadelphia © 1996

CHAPTER 1

The Origins of Virology

Arnold J. Levine

Virology, as a subject matter, has had a remarkable history. Viruses, because of their predatory nature, have shaped the history and evolution of their hosts. Virtually all living organisms, when studied carefully, have viral parasites, and so these smallest of living entities exert significant forces upon all life forms, including themselves. The medical consequences of viral infections of humans have altered our history and have resulted in extraordinary efforts on the part of virologists to study, understand, and eradicate these agents. These virologists have elucidated new principles of life processes and taken major new directions in science. Many of the concepts and tools of molecular biology have been derived from the study of viruses and their host cells. This chapter is an attempt to review selected portions of this history as it relates to the development of new concepts in virology (50, 51).

THE DEVELOPMENT OF THE CONCEPT OF VIRUSES

The Early Period: The 19th Century

By the last half of the 19th century, the existence of a diverse microbial world of bacteria, fungi, and protozoa was well established. As early as 1840, the noted German anatomist, Jacob Henle of Gottingen (the discoverer of Henle's loop and the grandfather of 20th-century virologist Werner Henle), hypothesized the existence of infectious agents that were too small to be observed with the light microscope and that were able to cause specific diseases. In the absence of any direct evidence for such entities, however, his ideas failed to be accepted. It was at this time that three major advances in microbiology came together to set the stage for the development of the concept of a submicroscopic agent that would come to be called a *virus*.

The first of these ideas was the demonstration that the spontaneous generation of organisms did not occur. This notion had a long history, with experiments both supporting and refuting it. The credit, however, for finally disproving this hypothesis is commonly given to Louis Pasteur (1822–95), who employed his swan-neck flasks to strike a mortal blow to the concept of spontaneous generation. Pasteur went on to study fermentation by different microbial agents. During these studies, he made it clear that "different kinds of microbes are associated with different kinds of fermentations" and he extended this concept to disease processes. Building upon this, Robert Koch (1843–1910), a student of Jacob Henle and a country doctor in a small German village, demonstrated that the anthrax bacillus was the cause of this disease (1876) and that the tubercle bacillus was the cause of tuberculosis in humans (1882). Little of this would have been possible without the third major contribution by Joseph Lister (1827–1912). Once it was clear that organisms reproduce new organisms, the importance of a sterile field, whether in surgery or for the isolation of new organisms, became

A.J. Levine: Department of Molecular Biology, Princeton University, Princeton, New Jersey 08544-1014.

clear. Lister contributed the techniques of limiting dilution to obtain pure cultures of organisms, while Koch developed solid media, the isolation of separate individual colonies of bacteria to obtain pure cultures, and the use of stains to visualize these microorganisms. While many scientists of that day contributed to these tools and concepts, it was principally Pasteur, Lister, and Koch who put together a new experimental approach for medical science.

These studies formalized some of Jacob Henle's original ideas in what are now termed *Koch's postulates* for defining whether an organism was indeed the causative agent of a disease. These postulates state that (a) the organism must be regularly found in the lesions of the disease, (b) the organism must be isolated in pure culture, (c) inoculation of such a pure culture of organisms into a host should initiate the disease, and (d) the organism must be recovered once again from the lesions of the host. By the end of the 19th century, these concepts became the dominant paradigm of medical microbiology. They outlined an experimental method to be used in all situations. It was only when these rules broke down and failed to yield a causative agent that the concept of a virus was born.

The Discovery Period: 1886–1903

Adolf Mayer (1843–1942), a German scientist trained in the field of chemical technology (who had studied fermentation and plant nutrition), became the director of the Agricultural Experimental Station at Wageningen, Holland, in 1876. A few years later (1879), he began his research on diseases of tobacco and, although he was not the first to describe such diseases, he named the disease *tobacco mosaic disease* after the dark and light spots on infected leaves. In one of Mayer's experiments, he inoculated healthy plants with the juice extracted from diseased plants by grinding up the infected leaves in water. This was the first experimental transmission of a viral disease of plants, and Mayer reported that, "in nine cases out of ten (of inoculated plants), one will be successful in making the healthy plant . . . heavily diseased" (109). Although these studies established the infectious nature of the disease, neither a bacterial nor a fungal agent could be consistently cultured or detected in these extracts and so Koch's postulates could not be satisfied. In a preliminary communication in 1882 (108), he speculated that the cause could be a "soluble, possibly enzyme-like contagium, although almost any analogy for such a supposition is failing in science." However, four years later, in his definitive paper on this subject, Mayer concluded that the mosaic disease "is bacterial, but that the infectious forms have not yet been isolated, nor are their forms and mode of life known" (109).

The next step was taken by Dimitri Ivanofsky (1864–1920), a Russian scientist working in St. Petersburg. In 1887 and again in 1890, Ivanofsky was commissioned by the Russian Department of Agriculture to investigate the cause of a tobacco disease on plantations in Bassarabia, Ukraine, and the Crimea. Ivanofsky rapidly repeated Mayer's observations, demonstrating that the sap of infected plants contained an agent able to transmit the disease to healthy plants, and he added one additional step. He passed the infected sap through a filter that blocked the passage of bacteria, the Chamberland filter, made of unglazed porcelain. The Chamberland filter, perfected to purify water by Charles Chamberland, one of Pasteur's collaborators, contained pores small enough to retard most bacteria. On February 12, 1892, Ivanofsky reported to the Academy of Sciences of St. Petersburg that "the sap of leaves infected with tobacco mosaic disease retains its infectious properties even after filtration through Chamberland filter candles" (79). The importance of this experiment is that it provided an operational definition of viruses, an experimental technique by which an agent could qualify as a virus.

Ivanofsky, like Mayer before him, failed to culture an organism from the filtered sap and failed to satisfy Koch's postulates. He too was bound by the paradigm of the times suggesting that the filter might be defective or something might be wrong with his methods. He even suggested the possibility that a toxin (not a living-reproducing substance) might pass through the filter and cause the disease. As late as 1903, when Ivanofsky published his thesis (80), he could not depart from the possibility that bacteria caused this disease and that he and others had somehow failed to culture them. The dogma of the times and the obvious success of Koch's postulates kept most scientists from interpreting their data in a different way. It is equally curious that, at this time (1885), Pasteur was working with viruses and developing the rabies vaccine (117) but never investigated the unique nature of that infectious agent.

The third scientist to play a key role in the development of the concept of viruses was Martinus Beijerinck (1851–1931), a Dutch soil microbiologist who collaborated with Adolf Mayer at Wageningen. Beijerinck also showed that the sap of infected tobacco plants could retain its infectivity after filtration through a Chamberland candle filter. [He was unaware of Ivanofsky's work at the time (1898).] He then extended these studies by showing that the filtered sap could be diluted and then regain its "strength" after replication in living, growing tissue of the plant. The agent could reproduce itself (which meant that it was not a toxin) but only in living tissue, not in the cell-free sap of the plant. This explained the failure to culture the pathogen outside of its host and set the stage for an organism, smaller than bacteria (a filterable agent), not observable in the light microscope, and able to reproduce itself only in living cells or tissue. Beijerinck called this agent a "contagium vivum fluidum" (9), or a contagious living liquid. This concept began a 25-year debate about the nature of viruses; were they liquids or particles? This conflict was laid to rest when d'Herelle developed the plaque assay in 1917 (33) and when the first electron micrographs were taken of tobacco mosaic virus in 1939 (88).

Thus, Mayer, Ivanofsky, and Beijerinck each contributed to the development of a new concept: a filterable agent too small to observe in the light microscope but able to cause disease by multiplying in living cells. Loeffler and Frosch (101) rapidly described and isolated the first filterable agent from animals, the foot-and-mouth disease virus, and Walter Reed and his team in Cuba (1901) recognized the first human filterable virus, yellow fever virus (122). The term *virus* [from the Latin for slimy liquid or poison (76)] was at that time used interchangeably for any infectious agent and so was applied to tobacco mosaic virus. The literature of the first decades of the 20th century most often referred to these infectious entities as *filterable agents,* and this was indeed the operational definition of viruses. Sometime later the term *virus* became restricted in use for those agents that fulfilled the criteria developed by Mayer, Ivanofsky, and Beijerinck and that were the first agents to cause a disease that could not be proven by using Koch's postulates.

The Plant Viruses and the Chemical Period: 1929–1956

Tobacco mosaic virus (TMV) continued to play a central role in exploring the nature and properties of viruses. The early decades of the 20th century saw the development of techniques to purify enzymes (proteins). The notion that viruses were proteins and so could be purified in the same way was first appreciated and applied by Vinson and Petre (1927–31) at the Boyce Thompson Institute in Philadelphia. They precipitated the infectious TMV agent [using an infectivity assay developed by Holmes (75)] from the crude sap of infected plants using selected salts, acetone, or ethyl alcohol (144). They showed that the infectious virus could move in an electric field, in a manner just like proteins. At the same time, H. A. Purdy-Beale, also at the Boyce Thompson Institute, produced in rabbits antibodies directed against TMV that could neutralize the infectivity of this agent (121). At the time, this was taken as further proof of the protein nature of viruses. With the advent of purification procedures for viruses, both physical and chemical measurements of the virus became possible. The strong flow birefringence of purified preparations of TMV were interpreted (correctly) to show an asymmetric particle or rod-shaped particle (136). Max Schlesinger (129), working on purified preparations of bacteriophages in Frankfurt, Germany, showed that they were composed of proteins and contained phosphorus and deoxyribonucleic acid. This led to the first suggestion that viruses were composed of nucleoproteins. The crystallization of TMV in 1935 by Wendell Stanley (132), working at the Rockefeller Institute branch in Princeton, New Jersey, brought this infectious agent into the world of the chemists. Within a year, Bawden and Pirie (7,8) had demonstrated that crystals of TMV contained 0.5% phosphorus and 5% RNA. The first "view" of a virus came from x-ray crystallography using these crystals to show rods of a constant diameter aligned in hexagonal arrays containing RNA and protein (14). The first electron micrographs of any virus, TMV, were taken with a microscope built in Germany and confirmed the rod shape of the virus particle (88).

The x-ray diffraction patterns (14) suggested that TMV was built up from repeating subunits. These data and other considerations led Crick and Watson (29) to realize that most simple viruses had to consist of one or a few species of identical protein subunits. By 1954–55, several techniques had been developed to dissociate TMV protein subunits and TMV could be reconstituted from its RNA and protein subunits (57) to produce infectious virus. The principles of virus self-assembly (20) were appreciated using TMV, and in 1962 Caspar and Klug (23) elegantly described the geometric principles of icosahedral virus structure for many of the isometric viruses. Thus, from 1929 to 1962, both the structures and the chemical compositions of viruses were elucidated.

As early as 1926, H. H. McKinney (110) reported the isolation of "variants" of TMV with a different plaque morphology that bred true and could be isolated from several geographic locations (111). In 1933, Jensen (85) confirmed McKinney's observations and demonstrated reversion mutations for this phenotype. Thus, viruses, like all living and replicating entities, could mutate and therefore had genetic information. That the genetic information resides in the RNA of TMV was shown 30 years later by an infectious RNA assay (56,65). This was the first demonstration that RNA could be a genetic material. This followed Avery's DNA transformation experiments with pneumococcus (4) and the Hershey-Chase experiment with bacteriophages (73), both of which demonstrated that DNA was the more common genetic material. TMV RNA and its nucleotide sequence helped to confirm codon assignments for the genetic code, added clear evidence for the universality of the genetic code and helped to elucidate the mechanisms of mutation by several diverse mutagens (55). TMV and its related plant viruses have contributed significantly to both the origins and the development of virology.

The Bacteriophages

The Early Period: 1915–1940

In 1915, Frederick W. Twort was the superintendent of the Brown Institution in London. In his research, Twort was looking for variants of vaccinia virus, used in the smallpox vaccine, that would replicate in simple defined media outside of living cells. In one of his experiments, he inoculated a culture dish of nutrient agar with an aliquot of the smallpox vaccine and, although the virus failed to replicate, bacterial contaminants grew in the agar dish very readily. As Twort continued to incubate his cultures, he noticed that some bacterial colonies underwent a visible change and became "watery looking" (more transparent). Such colonies were no longer able to replicate when subcultured (the bacteria had been killed). Twort called this phenomenon *glassy transformation,* and he went on to show that infecting a normal colony of bacteria with the glassy transforming principle would kill the bacteria. The glassy entity readily passed through a porcelain filter, could be diluted one million fold, and when placed upon fresh bacteria would regain its strength or titer (140–142).

Twort (140) published a short note describing all this and suggested that one explanation of his observation was a virus of bacteria. Twort's research was interrupted by World War I, in which he served. When he returned to London, he did not continue this line of research and made no further contributions in this area.

At the same time, Felix d'Herelle, a Canadian medical bacteriologist, was working at the Pasteur Institute in Paris. In August 1915, a cavalry squadron of French soldiers were quartered in Maisons-Lafitte, just outside Paris, and a rampant *Shigella* dysentery infection was devastating the entire outfit. d'Herelle readily isolated the dysentery bacillus from filtered emulsions of the feces of sick men and cultured them. As the bacteria grew and covered the surface of the Petri dish, d'Herelle occasionally observed clear circular spots where no bacteria grew, and he called these *taches vierges* or *plaques.* d'Herelle (33,34) was able to follow the course of an infection in a single patient, noting when the bacteria were most plentiful and when the plaques appeared. He was able to demonstrate that the plaques appeared on the fourth day after infection and killed the bacteria in the culture dish; interestingly, the patient began to improve on the fourth day after infection. d'Herelle named these viruses *bacteriophages,* and he went on to develop techniques utilized to this day in virology. He developed the use of limiting dilutions with the plaque assay to titer the virus preparation. He reasoned that the appearance of plaques showed that the virus was particulate or "corpuscular." d'Herelle also demonstrated that the first step of a virus infection was for the agent to attach (adsorption) to the host cell, which he showed by the co-sedimentation of virus and host after mixing the two. (He showed that the virus was lost from the supernatant fluid.) The attachment of a virus occurred only when bacteria sensitive to the virus were mixed with it, demonstrating the host range specificity of a virus at the adsorption step. He described cell lysis and the release of infectious virus in clear and modern terms. d'Herelle (34,35) was in many ways one of the founders of the principles of modern virology.

By 1921, an increasing number of lysogenic bacterial strains had been isolated, and it became impossible to separate the virus from its host in some experiments. This led Jules Bordet (18) of the Pasteur Institute in Brussels to suggest that the transmissible agent described by d'Herelle was nothing more than a bacterial enzyme that stimulates its own production. Although that was an incorrect conclusion, it is remarkably close to the present ideas of prion structure and replication (see Chapter 91 in *Fields Virology,* 3rd ed.).

Throughout the decades of the 1920s and 1930s, d'Herelle focused his efforts upon the potential medical applications of his research, but it never bore fruit. What basic research was being carried out at this time was often confused by the interpretations due to the strong personalities of individual scientists in the field. Although it was clear that there were many diverse bacteriophages and that some were lytic while some were lysogenic, their interrelationships remained ill-defined. The highlight of this period was the demonstration by Max Schlesinger (128,130) that purified phages had a maximum linear dimension of 0.1 micron and a mass of about 4×10^{-16} grams and that they were composed of protein and DNA in roughly equal proportions. In 1936, no one quite knew what to make of that observation, but it would begin to make a great deal of sense over the next 20 years.

The Modern Period: 1938–1970

Max Delbrück was trained as a physicist at the University of Gïttingen, and his first position was at the Kaiser Wilhelm Institute for Chemistry in Berlin. There he joined a diverse group of individuals actively discussing how quantum physics related to an understanding of heredity. Delbrück's interest in this area led him to develop a quantum mechanical model of the gene, and in 1937 he applied for and obtained a fellowship to study at Caltech. Once at Caltech, he teamed up with another research fellow, Emory Ellis (44), who was working with a group of bacteriophages, T2, T4, T6 (the T-even phages). Delbrück soon appreciated that these viruses were ideal for the study of virus replication. These phages represented a way to probe how genetic information could determine the structure and function of an organism. From the beginning, these viruses were viewed as model systems for understanding cancer viruses or even for understanding how a sperm fertilizes an egg and a new organism develops. Ellis and Delbrück (45) designed the one-step growth curve experiment, where an infected bacterium liberates hundreds of phages synchronously after a one-half hour latent or eclipse period. This experiment defined the latent period, where viral infectivity was lost. This became the experimental paradigm of the phage group.

At the outbreak of World War II, Delbrück remained in the United States (at Vanderbilt University) and met an Italian refugee, Salvador E. Luria, who had fled to America and was working at Columbia University (with phages T1 and T2). They met at a meeting in Philadelphia on December 28, 1940, and spent the next two days planning experiments at Columbia University in New York. These two scientists were to recruit and lead a growing group of researchers focused upon bacterial viruses as a model for understanding life processes. Central to their success was an invitation to spend the summer of 1941 at Cold Spring Harbor Laboratory doing experiments. The result was that a German physicist and an Italian geneticist joined forces throughout the years of World War II to travel throughout the United States and recruit a new generation of biologists that came to be known as *the phage group.*

Shortly thereafter, Tom Anderson, an electron microscopist at the RCA Laboratories in Princeton, New Jersey, met Delbrück, and by March 1942 the first clear pictures of bacteriophages had been obtained (104). About the same time, the first mutants of these phages were isolated and characterized (102). By 1946, the first phage course was being taught at Cold Spring Harbor and, in March 1947, the first phage meeting attracted eight people. From these humble beginnings grew the field of molecular biology, which focused upon the bacterial host and its viruses.

The next 25 years (1950–75) was an intensely productive period of virus research with bacteriophages. Hundreds of virologists produced thousands of publications that covered three major areas: (i) lytic infection of *Escherichia coli* with the T-even phages; (ii) the nature of lysogeny, using lambda phage; and (iii) the replication and properties of several unique phages such as ϕX174 (single-stranded circular DNA), the RNA phages, T7, and so forth. It is simply not possible here to review all of this literature, which laid the foundations for modern molecular virology and biology, so only selected highlights of this era will be mentioned.

By 1947–48, the idea of examining, at the biochemical level, the events occurring in phage-infected cells during the latent period had come into its own. Seymour Cohen [who had trained first with Erwin Chargaff at Columbia University, studying lipids and nucleic acids, and then with Wendell Stanley working on TMV RNA and who had taken Delbrück's phage course (1946) at Cold Spring Harbor] examined the effects of phage infection upon DNA and RNA levels in infected cells using a colorimetric analysis (26,27). These studies showed a dramatic alteration of macromolecular synthesis in phage-infected cells: (a) The net accumulation of RNA stopped in these cells. [Later this would be the basis for detecting a rapidly turning over species of RNA and the first demonstration of messenger RNA (3)]. (b) There was a cessation of DNA synthesis for seven minutes followed by a resumption of DNA synthesis at a 5- to 10-fold increased rate. (c) At the same time, Monod and Wollman (112) showed that the synthesis of a cellular enzyme, the inducible β-galactosidase, was inhibited after phage infection. These experiments divided the viral latent period into early (prior to DNA synthesis) and late times. These results, more importantly, made the clear point that a virus could redirect cellular macromolecular synthetic processes in infected cells (28).

By the end of 1952, two experiments had a critical effect upon this field. First, Hershey and Chase (73) utilized differentially labeled viral proteins ($^{35}SO_4$) and nucleic acids ($^{32}PO_4$) to follow phage attachment to bacteria. They were able to shear the viral protein coats from the bacteria using a Waring blender and thus leave only the DNA associated with the infected cells. This enabled them to prove that the DNA had all the information needed to reproduce 100 new viruses. The Hershey-Chase experiment came at the right time to be appreciated in light of the novel structure of DNA elucidated by Watson and Crick one year later (145). Together, these experiments formed a cornerstone of the molecular biology revolution (21). The second experiment in virology carried out in 1953, which was to have an important influence, was done by by G. R. Wyatt and S. S. Cohen (155). They identified a new base, hydroxymethylcytosine, in T-even phage DNA; this new base took the place of cytosine, which was present in bacterial DNA. This began a ten-year study of how deoxyribonucleotides were synthesized in bacteria and phage infected cells, and lead to the critical observation that the virus introduces genetic information for a new enzyme into the infected cell (53). By 1964, Mathews et al. (28) had proved that hydroxymethylase does not exist in uninfected cells and must be encoded for by the virus. These experiments introduced the concept of early enzymes, utilized in deoxypyrimidine biosynthesis and DNA replication (90), and provided clear biochemical proof that the virus encoded new information expressed as proteins in an infected cell. A detailed genetic analysis of these phages identified and mapped the genes encoding these phage proteins and added to this concept. Indeed, the genetic analysis of the rII A and B cistrons of T-even phages became one of the best-studied examples of "genetic fine structure" (10,11). The replication of viral DNA using phage mutants and extracts to complement and purify enzyme activities in vitro contributed a great deal to our modern understanding of how DNA duplicates itself (1). Finally, a detailed genetic analysis of phage assembly, utilizing the complementation of phage assembly mutants in vitro, is a lucid example of how complex structures are built by living organisms using the principles of self-assembly (41). The genetic and biochemical analysis of phage lysozyme helped to elucidate the molecular nature of mutations (135), and phage (amber) mutations provided a clear way to study second-site suppressor mutations at the molecular level (12). The circular genetic map of the T-even phages (134) was explained by the circularly permuted, terminally redundant (heterozygotes of phages) conformation of these DNAs (139).

The remarkable reprogramming of viral and cellular protein synthesis in phage-infected cells was dramatically revealed by an early use of sodium dodecyl sulfate (SDS)-polyacrylamide gels (93), showing that viral proteins are made in a specific sequence of events ever subdividing the early and late proteins. The underlying mechanism of this temporal regulation led to the discovery of sigma factors modifying RNA polymerase and conferring gene specificity (67). The study of gene regulation at almost every level (transcription, RNA stability, protein synthesis, protein processing) was revealed from a set of original contributions derived from an analysis of phage infections.

Although this remarkable progress had begun with the lytic phages, no one knew quite what to make of the lysogenic phages. This changed in 1949 when Andre Lwoff at the Pasteur Institute began his studies with *Bacillus megaterium* and its lysogenic phages. By use of a micromanipulator, single bacteria were shown to divide up to 19 times, never liberating a virus. When lysogenic bacteria were lysed from without, no virus was detected. But from time to time a bacterium spontaneously lysed and produced many viruses (106). The influence of ultraviolet light in inducing the release of these viruses was a key observation that began to outline this curious relationship between a virus and its host (107). By 1954, Jacob and Wollman (82,83) at the Pasteur Institute had made the important observation that a genetic cross between a lysogenic bacterial strain (Hfr, lambda) and a nonlysogenic recipient resulted in the induction of the virus after conjugation, a process they called *zygotic induction.* In fact, the position of the lysogenic phage or prophage in the chromosome of its host *E. coli* could be mapped by the standard interrupted mating experiment after a genetic cross (83). This was one of the most critical experiments in the conceptual understanding of lysogenic viruses for several reasons: (a) it showed that a virus behaved like a bacterial gene on a chromosome in a bacteria; (b) it was one of the first experimental results to suggest that the viral genetic material was kept quiescent in bacteria by negative regulation, which was lost as the chromosome passed from the lysogenic donor bacteria to the nonlysogenic recipient host; and (c) this helped Jacob and Monod to realize as early as 1954 that the "induction of enzyme synthesis and of phage development are the expression of one and the same phenomenon" (106). These experiments laid the foundation for the operon model and the nature of coordinate gene regulation. Although the structure of DNA was elucidated in 1953 (145) and zygotic induction was described in 1954, the relationship between the bacterial chromosome and the viral chromosome in lysogeny was still referred to as the *attachment site* and literally thought of in those terms. It was not until Campbell (22) proposed the model for lambda integration of DNA into the bacterial chromosome, based upon the fact that the sequence of phage markers was different in the integrated state than in the replicative or vegetative state, that the truly close relationship between a virus and its host was appreciated. This led to the isolation of the negative regulator or repressor of lambda, a clear understanding of immunity in lysogens, and one of the early examples of how genes are regulated coordinately (120). The genetic analysis of the lambda bacteriophage life cycle is one of the great intellectual adventures in microbial genetics (72). It deserves to be reviewed in detail by all students of molecular virology and biology.

The lysogenic phages such as P22 of *Salmonella typhimurium* provided the first example of generalized transduction (157), while lambda provided the first example of specialized transduction (114). That viruses could carry within them cellular genes and transfer such genes from one cell to another provided not only a method for fine genetic mapping, but also a new concept in virology. As the genetic elements of bacteria were studied in more detail, it became clear that there was a remarkable continuum from lysogenic phages to episomes, transposons and retrotransposons, insertion elements, retroviruses, hepadnoviruses, viroids, virosoids (in plants), and prions. Genetic information moves between viruses and their hosts to the point where definitions and classifications begin to blur.

The genetic and biochemical concepts that derive from the study of bacteriophages made the next phase of virology possible. The lessons of the lytic and lysogenic phages were often relearned and modified as the animal viruses were studied.

ANIMAL VIRUSES

The Early Period (1898–1965): Discovery and Cell Culture

Some of the highlights of the early period of animal virology (1898–1965) are summarized in Table 1. Once the concept of a filterable virus took hold, this experimental procedure was applied to many diseased tissues. Filterable agents, unable to be seen in a light microscope, that replicate only in living animal tissue were found. There were truly some surprises, such as a virus—yellow fever virus—transmitted by a mosquito vector (122), specific visible pathological inclusion bodies (viruses) in infected tissue (80,116), and even viral agents that can "cause cancer" (43,123). Throughout this early time period (1900–30), a wide variety of viruses were found (Table 1) and characterized for their size (using the different pore sizes of filters), resistance to chemical or physical agents (alcohol, ether, etc.), and pathogenic effects. Just based upon these properties, it became clear that viruses were a very diverse group of agents. Some were even observable in the light microscope (vaccinia in dark-field optics). Some were inactivated by ether, while others were not. The range of viral diseases affected every tissue type. Viruses gave rise to chronic or acute disease; they were persistent agents or recurred in a periodic fashion. Viruses might cause cellular destruction or induce cellular proliferation. For the early virologists, unable to see their agents in a light microscope and often confused by this great diversity, there had to be an element of faith in their studies. In 1912, S. B. Wolbach, an American pathologist, remarked that "it is quite possible that when our knowledge of filterable viruses is more complete, our conception of living matter will change considerably, and that we shall cease to attempt to classify the filterable viruses as animal or plant" (152).

TABLE 1. *Landmarks in animal virus research: the early period (1898–1965)*

Date	Virologist	Discovery
1898	F. Loeffler & P. Frosch	First demonstration of a filterable animal virus; foot-and-mouth disease virus
1898	G. Sanarelli	Myxomatous virus
1901	W. Reed et al.	First human virus; yellow fever virus
1901	A. Lode & J. Gruber	Fowl plague virus
1903	P. Remlinger & Riffat-Bay	Rabies virus
1903	A. Negri	The inclusion bodies of rabies virus
1908	V. Ellerman & O. Bang	First demonstration of a leukemia-causing virus
1909	K. Landsteiner & E. Popper	Poliovirus
1911	P. Rous	First demonstration of a solid tumor virus; Rous sarcoma virus
1912	A. Carrel	Tissue culture of chick embryo explant
1913	E. Steinhardt et al.	An early example of virus propagation in tissue culture
1931	J. Furth	Use of mice as a host for viruses
1931	A. Woodruff & E. Goodpasture	Use of the embryonated hen's egg as a host for viruses
1931	R. Shope	Swine influenza virus
1933	W. Smith et al.	Human influenza virus
1933	R. Shope	Rabbit papilloma virus (first DNA tumor virus)
1933	Staff of the Jackson Memorial Laboratory	Mouse mammary tumors
1936	P. Rous & J. Beard	Rabbit papilloma virus induces carcinomas in a different species
1941–50	G. Hirst	Influenza virus hemagglutination (HA), HA inhibition by antibody, receptor-destroying enzyme (neuraminidase—the first virus-associated enzyme)
1948–55	Sanford et al.	Culture of single animal cells
	J. Enders et al.	Nonneural tissue (human) supports poliovirus replication in culture
	G. Gey et al.	Single cell cultures, HeLa cervical carcinoma
	H. Eagle	The optimal medium for growing cells
1952–65	Dulbecco, Darnell, Baltimore, et al.	Poliovirus—plaque assay and origins of the molecular biology of animal viruses using poliovirus

The way out of this early confusion was led by the plant virologists and the development of techniques to purify viruses and characterize both the chemical and physical properties of these agents (see "The Plant Viruses and the Chemical Period: 1929–1956"). The second path out of this problem came from the studies with bacteriophages, where single cells infected with viruses in culture were much more amenable to experimental manipulation and clear answers than were virus infections of whole animals. While the plant virologists of that day were tethered to their greenhouses and the animal virologists were bound to their animal facilities, the viruses of bacteria were studied in Petri dishes and test tubes. Progress in simplifying the experimental system under study came one step at a time; from studying animals in the wild, to laboratory animals, such as the mouse (59), or the hen's egg (153), to the culture of tissue, and then to single cells in culture. Between 1948 and 1955, a critical transition converting animal virology into a laboratory science came in four important steps: (i) Sanford et al. (126) at the National Institutes of Health (NIH) overcame the difficulty of culturing single cells, (ii) George Gey (64) and his colleagues at Johns Hopkins Medical School cultured and passaged human cells for the first time and developed a line of cells (HeLa) from a cervical carcinoma, (iii) Harry Eagle (40) at the NIH developed an optimal medium for the culture of single cells, and (iv) in a demonstration of the utility of all this, Enders and his colleagues (46) showed that poliovirus could replicate in a nonneuronal human explant of embryonic tissues. These ideas, technical achievements, and experimental materials had two immediate effects in virology. First, they led to the development of the polio vaccine, as the first vaccine produced in cell culture. From 1798 to 1949, all of the vaccines in use (smallpox, rabies, yellow fever, influenza) had been grown in animals or embryonated hens' eggs. Poliovirus was grown in monkey kidney cells (74,98) incubated in flasks. Second, the exploitation of cell culture for the study of viruses began the modern era of molecular virology. The first plaque assay for an animal virus in culture was with poliovirus (36), and it led to an analysis of poliovirus every bit as detailed and important as the contemporary work with bacteriophages. The simplest way to document this statement is for the reader to compare the first edition of *General Virology* by S. E. Luria in 1953 (103) to the second edition by S. E. Luria and J. E. Darnell in 1967 (105) and to examine the experimental descriptions of poliovirus infection of cells. The present era of virology had arrived, and it would continue to be full of surprises.

The Modern Period: 1960 to the Present

In this chapter, information has been presented chronologically or in separate virus groups (plant viruses, bacteriophages, and animal viruses), which reflects the historical separation of these fields. In this section, the format changes because the motivation for studying viruses began to change during this period. Virologists began to use viruses to probe questions central to understanding all life processes. Because viruses replicate in and are dependent upon their host cells, they must use the rules, signals, and regulatory pathways of the host. Virologists began to make contributions to all facets of biology. These ideas began with the phage group and were continued by the animal virologists. Second, during this period (1970 to the present), the recombinant DNA revolution began, and both bacteriophages and animal viruses played a critical and central role in this revolution. Viruses were used to probe many diverse questions in biology. For these reasons, the organization of this section focuses upon the accomplishments of cellular and molecular biology, where viruses were used.

The Role of Animal Viruses in Understanding the Basic Outlines of Eukaryotic Gene Regulation

The closed circular and superhelical nature of polyoma virus DNA was first elucidated by Dulbecco and Vogt (37) and Weil and Vinograd (147). The underlying reason for this structure of DNA was first shown to be the packaging of the DNA of simian vacuolating virus 40 (SV40), wound around nucleosomes (63), which produced a superhelix when the histones were removed. These observations proved to be excellent models for the *E. coli* genome (154) and the mammalian chromosome (94). The unique genomes of viruses, with single-stranded DNA (131), plus or minus strand RNA, or double-stranded RNA, are the only life forms that have adopted these modes of information storage.

Important elements of the eukaryotic transcription machinery have been elucidated with viruses. The first transcriptional enhancer element (acts in an orientation- and distance-independent fashion) was described in the SV40 genome (68), as was a distance- and orientation-dependent promoter element observed with the same virus. The transcription factors that bind to the promoter, SP-1 (38), or the enhancer element, such as AP-1, AP-2 (96), and are essential to promote transcription along with the basal factors, were first described with SV40. AP-1 is composed of *fos* and *jun* family member proteins, demonstrating the role of transcription factors as oncogenes (17). Indeed, the great majority of experimental data obtained for basal and accessory transcription factors comes from in vitro transcription systems using the adenovirus major late promoter or the SV40 early enhancer-promoter (146). Our present day understanding of RNA polymerase III promoter recognition comes, in part, from an analysis of the adenovirus VA gene transcribed by this polymerase (54).

Almost everything we know about the steps of messenger RNA (mRNA) processing began with observations made with viruses. RNA splicing of new transcripts was first described with the adenoviruses (13,25). Polyadenylation of mRNA was first observed with poxviruses (86) using a system where the first DNA-dependent RNA polymerase in a virion was discovered (87). The signal for polyadenylation in the mRNA was first found using SV40 (52). The cap and methylation of bases at the 5'-end of mRNA was first detected using reoviruses (60). What little is known about the process of RNA transport out of the nucleus has shown a remarkable discrimination of viral and cellular mRNAs by the adenovirus E1B-55Kd protein (118).

Translational regulation has been profitably studied using poliovirus RNA (or TMV RNA) where internally regulated initiation entry sites (IRES) have been described (84). The discovery of the role of interferon, in inducing a set of gene products that act upon translational regulatory events, owes its origins to virology (77,78). Similarly, the viral defenses against interferon by the adenovirus VA RNA has provided unique insight to the role of eIF-2 phosphorylation events (89). Posttranslational processing of proteins by proteases, carbohydrate addition to proteins in the Golgi apparatus, phosphorylation by a wide variety of important cellular protein kinases, or the addition of fatty acids to membrane-associated proteins have all been profitably studied using viruses. Indeed, a good deal of our present-day knowledge in cell biology of how protein trafficking occurs and is regulated in cells comes from the use of virus-infected cell systems. Clearly, the field of gene regulation has relied upon virology for many of its central tenants.

The Role of Animal Viruses in the Recombinant DNA Revolution

The discovery of the enzyme reverse transcriptase in retroviruses (5,138) not only helped to prove how retroviruses replicate, but also provided an essential tool to produce complementary DNAs (cDNAs). The first restriction enzyme map of a chromosome, HindII plus III, was with SV40 DNA (30,31), and the first DNA to show the specificity of a restriction enzyme was SV40 DNA with EcoRI (113,115). Some of the earliest DNA cloning experiments used SV40 DNA into lambda, or human β-hemoglobin genes into SV40 DNA, to construct the first mammalian expression vectors (81). Indeed, a debate about these very experiments led to a temporary moratorium on all such recombinant experiments. From the beginning, several animal viruses had been developed into expression vectors for foreign genes, including SV40 (66), the retroviruses (148), the adenoviruses (62,70), and adeno-associated virus (125), which has the remarkable property of site preferential integration (91). Modern day strategies of gene therapy will surely rely upon some of these recombinant viruses. The first cDNA cloning of hemoglobin sequences utilized lambda vectors for the cloning and replication of these mRNA copies. In a nice twist of events, the long-elusive hepatitis virus C (non-A, non-B) was cloned from serum using recombinant DNA techniques, reverse transcriptase, and lambda phage vectors (24).

The Role of Animal Virology in Oncology

It is not too strong a statement to say that we owe a great proportion of our present understanding of the origins of human cancers to two major groups of animal viruses, the retroviruses and DNA tumor viruses. The oncogenes were first discovered and proven to exist in a virus and then in the host cell genome using Rous sarcoma virus (133). A wide variety of retroviruses have captured, altered, and delivered oncogenes to the virologists (see Chapter 9). The insertion of retroviruses into the genomes of cancerous cells also helped to locate additional oncogenes (69). The second group of genes that contribute to the origins of human cancers, the tumor suppressor genes (97), has been shown to be intimately associated with the DNA tumor viruses. Genetic alterations at the p53 locus are the single most common mutations known to occur in human cancers (60% to 65% of the time) (99). The p53 protein was first discovered in association with the SV40 large T-antigen (95,100). SV40, the human adenoviruses, and the human papillomaviruses all encode oncogenes that produce proteins that interact with and inactivate the functions of two tumor suppressor gene products, the retinoblastoma susceptibility gene product (Rb) and p53 (32,39,95,100,127,149,151). The cellular oncogenes and the tumor suppressor genes in human cancers have been studied and understood most profitably using these viruses.

The viruses that cause cancers have provided some of the most extraordinary episodes in modern animal virology. The recognition of a new disease and the unique geographic distribution of Burkitt's lymphomas in Africa (19) set off a search for viral agents that cause cancers in humans. From D. Burkitt (19) to Epstein, Achong, and Barr (47) to W. Henle and G. Henle (71), the story of the Epstein-Barr virus and its role in several cancers, as well as infectious mononucleosis, provides us with the best in detective story science. The story is not yet complete and many mysteries remain. Similarly, the identification of a new pathological disease, adult T-cell leukemia, in Japan by K. Takatsuki (137,143) led to the isolation of a virus that causes the disease by I. Miyoshi and Y. Hinuma (156) and the realization that this virus (HTLV-1) had been found previously (119) by Gallo and his colleagues. Although this discovery provided the virus, there is yet to be a satisfactory explanation of how this virus contributes to adult T-cell leukemia.

Equally interesting is the road to the hepatitis B virus and hepatocellular carcinomas. By 1967, S. Krugman and his colleagues (92) had good evidence distinguishing between hepatitis A and B viruses, and in the same year B. Blumberg et al. (16) detected the Australia antigen. Through a tortured path it eventually became clear that the Australia antigen was a diagnostic marker (a coat protein) for hepatitis B. While this freed the blood supply of this dangerous virus, Hilleman at Merck, Sharp and Dohme and the Chiron Corporation (which later isolated the hepatitis C virus) went on to produce the first human vaccine that prevents hepatitis B infections and very likely hepatocellular carcinomas associated with chronic virus infections (see Chapter 36). The idea of a vaccine that can prevent cancer [first proven with the Mareck's disease virus and T-cell lymphomas in chickens (15,42)] comes some 82 to 85 years after the first discoveries of tumor viruses by Ellerman, Bang, and Rous. At present, an experiment is under way in Taiwan, where 63,500 newborn infants have been inoculated to prevent hepatitis B infections. Based upon the epidemiological predictions, this vaccination program should result in 8,300 fewer cases of liver cancer in that population some 40 to 50 years from now.

Vaccines

The Salk and then Sabin poliovirus vaccines were the first beneficial products of the cell culture revolution. In the early 1950s in the United States, just before the introduction of the Salk vaccine, about 21,000 cases of poliomyelitis were reported annually. Today, the number is fewer than ten (see Chapter 16) (98).

Among the most remarkable achievements of our century is the complete eradication of smallpox, a disease with a history of over 2,000 years (150). In 1966, the World Health Organization began a program to immunize all individuals who had come into contact with an infected person. This strategy, as opposed to trying to immunize an entire population (which simply was not possible), worked and, in October 1977, Ali Maolin of Somalia was the last person in the world to have a naturally occurring case of smallpox (barring laboratory accidents). Because smallpox has no animal reservoir and requires people-to-people contact for its spread, most scientists agree that we are free of this disease (150). What most scientists do not agree upon is whether we should store smallpox virus samples as a reference for the future (2).

The viral vaccines used in the past have included live attenuated vaccines, killed virus vaccines, and subunit vaccines. Both the killed virus vaccine (Salk) and the recombinant subunit vaccine (hepatitis B, S antigen) were new to the modern era of virology. In the future, we will see one virus (e.g., vaccinia virus) presenting the antigens of a different virus, the injection of DNA encoding viral antigens, and the use of specific interleukins or hormones with vaccines to stimulate immunity at specific locations in the host and elucidate specific immunoglobulin classes. Considering that the first vaccines (for smallpox) were reported in the Chinese literature of the tenth century (48), these ideas can be traced back in time to the origins of virology.

While vaccines have been extraordinarily successful in preventing specific diseases, there had been very few natural products or chemotherapeutic agents that cured or reduced the symptoms of virus infections. That changed dramatically with the development of acyclovir by Burrows-Wellcome, which requires both a viral enzyme (thymidine kinase) to activate it (phosphorylate) and a second enzyme to incorporate it into viral DNA, employing its specificity for the viral encoded DNA polymerase. This drug blocks HSV-2 reactivation from latency and stopped a growing epidemic in the 1970s and 1980s (Chapter 15 in *Fields Virology,* 3rd ed.). Similarly, the interferons (Chapter 15 in *Fields Virology,* 3rd ed.) have come into use clinically for hepatitis B and C infections, cancer therapy, and multiple sclerosis. The interferons, found in the course of studying virus interference (77,78), will modulate the immune response and have an increasing role to play in treatments of many clinical syndromes.

Epidemiological Adventures in Virology

The field of the epidemiology of viruses (see Chapter 9 in *Fields Virology,* 3rd ed.) got a terrific boost in the last half of the 20th century with the advent of specific molecular tools (antibodies, polymerase chain reaction, rapid diagnostic tests) to detect viruses in body fluids or tissue samples, compare and classify these agents rapidly, and determine the relationships between virus strains. The marriage of behavioral, geographic, and molecular epidemiology made this a most powerful science. The previous section reviewing advances in oncology pays tribute to this strategy by D. Burkitt and K. Takatsuki leading to the identification of Epstein-Barr virus (EBV) and human T-cell leukemia virus (HTLV-1). Similarly, the recombinant DNA revolution overcame the problems of propagating the human papillomaviruses, permitting the isolation of new virus serotypes and setting off epidemiological correlations for high- or low-risk cancer viruses (158). The human papillomaviruses (see Chapter 30 this volume and Chapter 66 in *Fields Virology,* 3rd ed.) differ in transmission, location on the body, their nature of pathogenesis, persistence, and so forth. Similarly, the first recognition of the so-called slow viruses, which cause spongiform encephalopathies (Chapters 90 and 91 in *Fields Virology,* 3rd ed.), came from epidemiological studies with humans and animals.

With the evolutionary development of a sophisticated immune system in vertebrates, the viruses of these hosts faced a new challenge in that they could no longer productively infect their hosts more than once. For their part, these viruses responded with the establishment of latency (herpesviruses), attacking directly the cells involved in the immune response (human immunodeficiency virus), setting up persistent or chronic infections, or devising rapid methods for antigenic changes (influenza A virus). This ability of influenza A to undergo antigenic drift and shift resulted in the development of a lifestyle that uses Darwinian principles in a time frame shorter than that of any other organism. G. Hirst and his colleagues (1941–50) developed the diagnostic tools to follow this. This permitted both the typing of the hemagglutinin of influenza A strains as well as monitoring of the antibody response to it in patients (see Chapter 22 this volume and Chapter 46 in *Fields Virology,* 3rd ed.). This has been used, with more and more sophisticated molecular approaches, to prove the presence of animal reservoirs for this virus, the reassortment of chromosomes between human and animal virus strains (antigenic shift), and a high rate of mutation (antigenic drift) due to RNA-dependent RNA synthesis with no known RNA editing or corrective mechanisms. How these molecular events led to episodic local epidemics and worldwide pandemics has now been explained in broad outline. Although most viruses tend to come to equilibrium with their hosts and become endemic [see, for example, the introduction of the North American myxoma-fibroma virus into the wild European rabbit population in Australia (49)], influenza A virus remains epidemic, changing from local to pandemic via an antigenic shift of its HA subunit gene. These studies (Chapter 22 this volume and Chapter 46 in *Fields Virology,* 3rd ed.) have revealed an extraordinary lifestyle. Similarly, the study of the mechanisms of viral pathogenesis and the role of the immune system in this process have led to new insights in the virus-host relationship.

An overview of the modern period of animal virology would not be complete without mentioning the appearance of an apparently new virus during the decades of the 1970s and 1980s, the human immunodeficiency virus. It was first recognized as a new disease entity by clinicians and epidemiologists, and they rapidly tracked down the venereal mode of virus transmission. Blood products and transplants provided the fluid and tissue samples for virologists to detect this virus. The first published report of acquired immunodeficiency syndrome (AIDS) was in June 1981. The first publications describing a candidate virus for this disease were in 1983 (6) and then 1984 (61). Had this virus and disease begun in 1961 instead of 1981, neither the nature of retroviruses nor the existence of its host cell (CD-4 helper T-cell) would have been understood. The rapid development of a diagnostic test has helped to remove the virus from the blood supply and to test individuals for the virus. A treatment that controls or eliminates this virus remains elusive (see Chapters 61 and 62 in *Fields Virology,* 3rd ed.).

AFTERWORD: D'HERELLE'S DREAM AND KOCH'S POSTULATES

It was d'Herelle's dream to use bacteriophages as his "magic bullet" to kill bacteria and cure diseases caused by these agents. The ultimate historical irony of that dream was played out by some of the participants described in this chapter over the same time frame of this chapter.

The story begins in 1884, when Friedrich Johannes Loeffler in Germany used Koch's postulates to isolate and identify the bacteria that causes diphtheria. [Fourteen years later, he and Paul Frosch (both had been trained by R. Koch) would isolate the foot-and-mouth disease virus, the first animal virus to be isolated.] Loeffler was surprised to note, however, that when he inoculated the bacteria into an animal, the bacilli were detected only at the site of the local injection, whereas the abnormalities responsible for the disease were visible in the heart, liver, kidneys, and so forth. Loeffler hypothesized that the bacilli produced a toxin that caused the disease at remote sites in the body with no detectable bacteria. In the next step (in 1888), Emile Roux and A. Yersin (at the Pasteur Institute) demonstrated a heat-labile soluble toxin in the fluid phase above the diphtheria bacillus cultures. An injection of the toxin (which could be filtered through a Chamberland filter) into animals reproduced the symptoms. Emile Roux was a student of Pasteur's who worked with Chamberland on his filters. When Ivanofsky came up with the idea that the filterable agent in his TMV preparations was a toxin, he credited E. Roux for this idea and his work on diphtheria toxin as precedence for this notion (76). In 1898, E.I.E. Nocard and E. Roux isolated from cows a pleuropneumonia organism that could pass a Chamberland filter, could be seen as microscopic dots in the light microscope, and could be replicated in collodion sacs in a cell-free meat infusion in the peritoneal cavity of a rabbit. In this way they discovered mycoplasma, a bacteria that violated the "filterable agent" concept of a virus or a liquid that could only replicate in an intracellular environment. In fact, in 1903, Roux challenged Beijerinck's idea of a fluid contagium by calling it "very original" but not distinguishable from his tiny mycoplasma spores, which were microorganisms. In 1903, Emile Roux (124) wrote the first review article on viruses, entitled "Sur les microbes dits 'invisible'."

By 1890 in Germany, von Behring and Kitasato treated the diphtheria toxin with chemical agents to inactivate it and immunized animals with this. They demonstrated that serum from immunized animals protected other animals from the toxin and, on Christmas night in 1891 in Berlin, this antitoxin was first given to a child with diphtheria. Shibasaburo Kitasato had also developed a microorganism filter with somewhat finer pores than the Chamberland filter, and he even sold it commercially. Loeffler, in 1898, used both the Chamberland and the Kitasato filters to test whether foot-and-mouth disease virus was retained by or passed these two filters. This virus readily passed through the Chamberland filter but lost virulence by repeated filtration through a Kitasato filter. Loeffler used these data to claim that the virus was particulate or corpuscular, partly stopped by the Kitasato filter, and he thus challenged Beijerinck's concept of a fluid contagium (a liquid virus). Beijerinck responded that he felt that the foot-and-mouth disease virus was adsorbed to the Kitasato filter and was not stopped by its smaller pore size. The scientific arguments concerning the nature of viruses continued.

In 1923, G. Ramon introduced a formalin-treated diphtheria toxin as an immunizing agent, and an effective vaccine was in hand that all but eliminated this disease from countries that had an active vaccination program. Although that should have been the end of a good story, it was not. In 1951, V. J. Freeman (58) made the unexpected observation that all virulent strains of *C. diphtheriae* are lysogenic with a phage called *beta.* If the bacteria are cured of this prophage, the bacteria fail to produce toxin and are avirulent. Indeed, the gene for this toxin is encoded by the phage genome and is regulated by the metabolic state of the bacteria. The lysogenic virus causes this disease. Koch's postulates had been circumvented by the intimate association of a virus and its host as first described by A. Lwoff and his colleagues. d'Herelle's dream of a phage curing a disease was given a cruel twist of fate in this case. These filterable agents, be they toxins, viruses, or even viruses that produce toxins when lysogenic in their host and so cannot pass a filter, have all the power to present a most confusing or at the very least a complex picture to the observer.

It is a tribute to the scientists of those times that they came so close to describing reality, while inferring the existence of organisms they could not see except for their effect upon a host. These 100 years of virology have forged new concepts and provided novel insights into the processes of life.

REFERENCES

1. Alberts BM, Bedinger BP, Formosa T, Jongeneel CV, Kreuzer KN. Studies on DNA replication in the bacteriophage T4 in vitro systems. *Cold Spring Harb Symp Quant Biol* 1982;47:655–668.
2. Altman LK. Smallpox virus, frozen in 2 labs, escapes a scalding end for now. *The New York Times,* Natl Ed. 1993 Dec 25:1,8.
3. Astrachan L, Volkin E. Properties of ribonucleic acid turnover in T2-infected *Escherichia coli. Biochem Biophys Acta* 1958;29:536–539.
4. Avery OT, MacLeod CM, McCarty M. Studies on the chemical nature of the substance inducing transformation of pneumococcal types: induction of transformation by a deoxyribonucleic acid fraction isolated from pneumococcus type III. *J Exp Med* 1944;79:137–158.
5. Baltimore C. RNA-dependent DNA polymerase in virions of RNA tumour viruses. *Nature* 1970;226:1209–1222.
6. Barre-Sinoussi F, Chermann JC, Rey F, et al. Isolation of a T-lymphotropic retrovirus from a patient at risk for acquired immune deficiency syndrome (AIDS). *Science* 1983;110:868–871.
7. Bawden FC, Pirie NW. The isolation and some properties of liquid crystalline substances from solanaceous plants infected with three strains of tobacco mosaic virus. *Proc R Soc* 1937;123:274–320.
8. Bawden FC, Pirie NW, Bernal JD, Fankuchen I. Liquid crystalline substances from virus infected plants. *Nature* 1936;138:1051–1052.
9. Beijerinck MW. Concerning a contagium vivum fluidum as a cause of the spot-disease of tobacco leaves. *Verh Akad Wetensch, Amsterdam, II* 1898;6:3–21.
10. Benzer S. Fine structure of a genetic region in bacteriophage. *Proc Natl Acad Sci USA* 1955;41:344–354.
11. Benzer S. Genetic fine structure. In: *Harvey lectures, vol 56.* New York: Academic Press; 1961:1.
12. Benzer S, Champe SP. Ambivalent rII mutants of phage T4. *Proc Natl Acad Sci USA* 1961;47:1025–1038.
13. Berget SM, Moore C, Sharp PA. Spliced segments at the 5' terminus of adenovirus 2 late mRNA. *Proc Natl Acad Sci USA* 1977;74:3171–3175.

14. Bernal JD, Fankuchen I. X-ray and crystallographic studies of plant virus preparations. *J Gen Physiol* 1941;25:111–165.
15. Biggs PM, Payne LN, Milne BS, et al. Field trials with an attenuated cell-associated vaccine for Mareck's disease. *Vet Rec* 1970;87:704–709.
16. Blumberg BS, Gerstley BJS, Hungerford DA, London WT, Sutneck AI. A serum antigen (Australia antigen) in Down's syndrome, leukemia and hepatitis. *Ann Intern Med* 1967;66:924–931.
17. Bohmann D, Bos TJ, Admon A, Nishimura T, Vogt PL, Tjian R. Human proto-oncogene *c-jun* encodes a DNA binding protein with structural and functional properties of transcription factor AP-1. *Science* 1987;328:1386–1392.
18. Bordet J. Concerning the theories of the so-called "bacteriophage." *Br Med J* 1922;2:296.
19. Burkitt D. A children's cancer dependent on climatic factors. *Nature* 1962;194:232–234.
20. Butler PJG, Klug A. Assembly of the particle of TMV from RNA and disk of protein. *Nature New Biol* 1971;229:47.
21. Cairns J. The autoradiography. In: Cairns J, Stent GS, Watson JD, eds. *Phage and the origins of molecular biology.* Cold Spring Harbor, NY: Cold Spring Harbor Laboratory Press; 1966:252–257.
22. Campbell AM. Episomes. *Adv Genet* 1962;11:101–145.
23. Caspar DLD, Klug A. Physical principles in the construction of regular viruses. *Cold Spring Harb Symp Quant Biol* 1962;27:1–32.
24. Chov Q-L, Kuo G, Weiner AJ, Overby LR, Bradley DW, Houghton M. Isolation of a cDNA clone derived from a blood borne non-A, non-B viral hepatitis genome. *Science* 1989;88:359–362.
25. Chow LT, Gelinas RE, Broker TR, Roberts RJ. An amazing sequence arrangement at the 5' ends of adenovirus 2 messenger RNAs. *Cell* 1977;12:1–8.
26. Cohen SS. The synthesis of bacterial viruses in infected cells. *Cold Spring Harb Symp Quant Biol* 1947;12:35–49.
27. Cohen SS. Synthesis of bacterial viruses; synthesis of nucleic acid and proteins in *Escherichia coli* infected with T2r$^+$ bacteriophage. *J Biol Chem* 1948;174:281–295.
28. Cohen SS. *Virus-induced enzymes.* New York: Columbia University Press 1968.
29. Crick FHC, Watson JD. The structure of small viruses. *Nature* 1956;177:473–475.
30. Danna K, Nathans D. Specific cleavage of SV40 DNA by restriction endoculease of *H. influenzae. Proc Natl Acad Sci USA* 1971;68:2913–2918.
31. Danna KJ, Sack GH Jr, Nathans D. Studies of simian virus 40 DNA. VII. A cleavage map of the SV40 genome. *J Mol Biol* 1973;78:363–376.
32. DeCaprio JA, Ludlow JW, Figge J, et al. SV40 large tumor antigen forms a specific complex with the product of the retinoblastoma susceptibility gene. *Cell* 1988;54:275–283.
33. d'Herelle FH. Sur un microbe invisible antagoniste dis bacilles dysentériques. *C R Hebd Seanc Acad Sci Paris* 1917;165:373–390.
34. d'Herelle F. Le microbe bactériophage, agent d'immunité dans la peste et le barbone. *C R Hebd Seanc Acad Sci Paris* 1921;172:99.
35. d'Herelle F. *The bacteriophage and its behavior.* Baltimore: Williams & Wilkins, 1926.
36. Dulbecco R, Vogt M. Some problems of animal virology as studied by the plaque technique. *Cold Spring Harb Symp Quant Biol* 1953; 18:273–279.
37. Dulbecco R, Vogt M. Evidence for a ring structure of polyoma virus DNA. *Proc Natl Acad Sci USA* 1963;50:236–243.
38. Dynan WS, Tjian R. The promoter specific transcription factor Sp1 binds to upstream sequences in the SV40 early promoter. *Cell* 1983;35:79–87.
39. Dyson N, Howley PM, Munger K, Harlow E. The human papillomavirus-16 E7 oncoprotein is able to bind to the retinoblastoma gene product. *Science* 1989;243:934–937.
40. Eagle H. The specific amino acid requirements of a human carcinoma cell strain HeLa in tissue culture. *J Exp Med* 1955;102:37–48.
41. Edgar RS, Wood WB. Morphogenesis of bacteriophage T4 in extracts of mutant-infected cells. *Proc Natl Acad Sci USA* 1966;55:498–505.
42. Eidson CS, Kleven SH, Anderson DP. Vaccination against Marek's disease. In: *Oncogenesis and herpesvirus.* Lyon: International Ongeny Research on Cancer; 1972:147.
43. Ellermann V, Bang O. Experimentelle Leukamie bei Huhnern. *Zentralbl Bakteriol Alet I* 1908;46:595–597.
44. Ellis EL. Bacteriophage: one-step growth. In: Cairns J, Stent GS, Watson JD, eds. *Phage and the origins of molecular biology.* Cold Spring Harbor, NY: Cold Spring Harbor Laboratory Press; 1966:53–62.
45. Ellis EL, Delbrück M. The growth of bacteriophage. *J Gen Physiol* 1939;22:365–384.
46. Enders JF, Weller TH, Robbins FC. Cultivation of the Lansing strain of poliomyelitis virus in cultures of various human embryonic tissues. *Science* 1949;109:85–87.
47. Epstein MA, Achong BG, Barr YM. Virus particles in cultured lymphoblasts from Burkitt's lymphoma. *Lancet* 1964;1:702–703.
48. Fenner F, Nakano JH. Poxviridae: The poxviruses. In: Lennette EH, Halonen P, Murphy FA, eds. *The laboratory diagnosis of infectious diseases: principles and practice.* Volume 2: *Viral, rickettsial and chlamydial diseases.* New York: Springer-Verlag; 1988:177–210.
49. Fenner F, Woodroofe GM. Changes in the virulence and antigenic structure of strains of myxomavirus recovered from Australian wild rabbits between 1950 and 1964. *Aust J Exp Biol Med Sci* 1965;43:359–374.
50. Fields BN, Knipe DM, Chanock RM, Hirsch MS, Melnick JL, Monath TP, Roizman B, eds. *Fields virology.* 2nd ed. New York: Raven Press, 1990.
51. Fields BN, Knipe DM, Chanock RM, Melnick J, Roizman B, Shope R, eds. *Fields virology.* 1st ed. New York: Raven Press, 1985.
52. Fitzgerald M, Shenk T. The sequence 5'-AAUAAA-3' forms part of the recognition site for polyadenylation of late SV40 mRNAs. *Cell* 1981;24:251–260.
53. Flaks JG, Cohen SS. Virus-induced acquisition of metabolic function. I. Enzymatic formation of 5'-hydroxymethyldeoxycytidylate. *J Biol Chem* 1959;234:1501–1506.
54. Fowlkes DM, Shenk T. Transcriptional control regions of the adenovirus VAI RAN gene. *Cell* 1980;22:405–413.
55. Fraenkel-Conrat H, Singer B. The chemical basis for the mutagenicity of hydroxylamine and methoxyamine. *Biochim Biophys Acta* 1972;262:264.
56. Fraenkel-Conrat H, Singer B, Williams RC. Infectivity of viral nucleic acid. *Biochim Biophys Acta* 1957;25:87–96.
57. Fraenkel-Conrat H, Williams RC. Reconstitution of active tobacco mosaic virus from its inactive protein and nucleic acid components. *Proc Natl Acad Sci USA* 1955;41:690–698.
58. Freeman VJ. Studies on the virulence of bacteriophage-infected strains of *Corynbacterium diptheriae. J Bacteriol* 1951;61:675–688.
59. Furth J, Strumia M. Studies on transmissible lymphoid leukemia of mice. *J Exp Med* 1931;53:715–726.
60. Furuichi Y, Morgan M, Muthukrishnan S, Shatkin AJ. Reovirus mRNA contains a methylated, blocked 5'-terminal structure: m^{7}G(5')ppp(5')G^{m}pCp. *Proc Natl Acad Sci USA* 1975;72:362–366.
61. Gallo RC, Salahuddin SZ, Popovic M, et al. Frequent detection and isolation of cytopathic retroviruses (HTLV-III) from patients with AIDS and at risk for AIDS. *Science* 1984;224:497–500.
62. Gaynor RB, Hillman D, Berk AJ. Adenovirus early region 1A protein activates transcription of a nonviral gene introduced into mammalian cells by infection or transfection. *Proc Natl Acad Sci USA* 1984; 81:1193–1197.
63. Germond JE, Hirt B, Oudet P, Gross-Bellard M, Chambon P. Folding of the DNA double helix in chromatin-like structures from simian virus 40. *Proc Natl Acad Sci USA* 1975;72:1843–1847.
64. Gey GO, Coffman WD, Kubicek MT. Tissue culture studies of the proliferative capacity of cervical carcinoma and normal epithelium. *Cancer Res* 1952;12:264–265.
65. Gierer A, Schramm G. Infectivity of ribonucleic acid from tobacco mosaic virus. *Nature* 1956;177:702–703.
66. Goff SP, Berg P. Construction of hybrid viruses containing SV40 and lambda phage DNA segments and their propagation in cultured monkey cells. *Cell* 1976;9:695–705.
67. Gribskov M, Burgess RR. Sigma factors from *E. coli,* B. *subtilis,* phage SP01 and phage T4 are homologous proteins. *Nucleic Acids Res* 1986;14:6745–6763.
68. Gruss P, Dhar R, Khoury G. Simian virus 40 tandem repeated sequences as an element of the early promoter. *Proc Natl Acad Sci USA* 1981;78:943–947.
69. Hayward WS, Neel BG, Astrin SM. Activation of a cellular onc gene by promoter insertion in ALV-induced lymphoid leukosis. *Nature* 1981;290:475–480.
70. Hearing P, Shenk T. Sequence-independent autoregulation of the adenovirus type 5 E1A transcription unit. *Mol Cell Biol* 1985; 5:3214–3221.
71. Henle G, Henle W, Diehl V. Relation of Burkitt's tumor-associated herpes-type virus to infectious mononucleosis. *Proc Natl Acad Sci USA* 1968;59:94–101.

72. Hershey AD, ed. *The bacteriophage lambda*. New York: Cold Spring Harbor Laboratory Press; 1971.
73. Hershey AD, Chase M. Independent functions of viral protein and nucleic acid in growth of bacteriophage. *J Gen Physiol* 1952; 36:39–56.
74. Hilleman MR. Historical and contemporary perspectives in vaccine developments: from the vantage of cancer. In: Melnick JL, ed. *Progress in medical virology*, vol 39. Switzerland: S Karger Publishers; 1992; 1–18.
75. Holmes FA. Local lesions in tobacco mosaic. *Bot Gaz* 1929;87:39–55.
76. Hughes SS. *The virus, a history of the concept*. London: Heinemann Education Books; 1977.
77. Isaacs A, Lindenmann J. Virus interference I. *Proc R Soc Lond* 1957;147B:258–267.
78. Isaacs A, Lindenmann J. The interferon II. Some properties of interferon. *Proc R Soc Lond* 1957;147B:268–273.
79. Ivanofsky D. Concerning the mosaic disease of the tobacco plant. *St Petersburg Acad Imp Sci Bul* 1892;35:67–70.
80. Ivanofsky D. On the mosaic disease of tobacco. *Z Pfanzenkr* 1903; 13:1–41.
81. Jackson DA, Symons RH, Berg P. Biochemical method for inserting new genetic information into DNA of simian virus 40: circular SV40 DNA molecules containing lambda phage genes and the galactose operon of *Escherichia coli*. *Proc Natl Acad Sci USA* 1972;69:2904–2909.
82. Jacob F, Wollman EL. Etude génétique d'un bactériophage tempéré d'*Escherichia coli*. I. Le système génétique du bactériophage λ. *Ann Inst Pasteur* 1954;87:653–673.
83. Jacob F, Wollman EL. *Sexuality and the genetics of bacteria*. New York: Academic Press, 1961.
84. Jan SK, Davies MV, Kaufman RJ, Wimmer E. Initiation of protein synthesis by internal entry of ribosomes into the 5' nontranslated region of encephalomyocarditis virus RNA in vivo. *J Virol* 1989;63:1651–1660.
85. Jensen JH. Isolation of yellow-mosaic virus from plants infected with tobacco mosaic. *Phytopathology* 1933;23:964–974.
86. Kates J, Beeson J. Ribonucleic acid synthesis in vaccinia virus. II. Synthesis of polyriboadenylic acid. *J Mol Biol* 1970;50:19–23.
87. Kates JR, McAuslan BR. Poxvirus DNA-dependent RNA polymerase. *Proc Natl Acad Sci USA* 1967;58:134–141.
88. Kausche GA, Ankuch PF, Ruska H. Die sichtbarmachung von PF lanzlichem virus in ubermikroskop. *Naturwissenschaften* 1939;27:292–299.
89. Kitajewski J, Schneider RJ, Safer B, Shenk T. An adenovirus mutant unable to express VAI RNA displays different growth responses and sensitivity to interferon in various host cell lines. *Mol Cell Biol* 1986;6:4493–4498.
90. Kornberg A. Biological synthesis of deoxyribonucleic acid. *Science* 1960;131:1503–1508.
91. Kotin RM, Siniscalco RJ, Samulski RJ, et al. Site-specific integration by adenovirus-associated virus. *Proc Natl Acad Sci USA* 1990;87:2211–2215.
92. Krugman S, Giles JP, Hammond J. Infectious hepatitis: evidence for two distinctive clinical, epidemiological and immunological types of infection. *JAMA* 1967;200:365–373.
93. Laemmli UK. Cleavage of structural proteins during the assembly of the head of bacteriophage T4. *Nature* 1970;227:680–685.
94. Laemmli UK, Cheng SM, Adolph KW, Paulson JR, Brown JA, Baumbach WR. Metaphase chromosome structure: the role of nonhistone proteins. *Cold Spring Harb Symp Quant Biol* 1978;42:351–360.
95. Lane DP, Crawford LV. T antigen is bound to a host protein in SV40-transformed cells. *Nature* 1979;278:261–263.
96. Lee W, Haslinger A, Karin M, Tjian R. Activation of transcription by two factors that bind promoter and enhancer sequences of the human metallothionein gene and SV40. *Nature* 1987;325:368–372.
97. Levine AJ. The tumor suppressor genes. *Ann Rev Biochem* 1993;62:623–651.
98. Levine AJ. The origins of the small DNA tumor viruses. *Adv Cancer Res* 1994;65:141–168.
99. Levine AJ, Momand J, Finlay CA. The p53 tumor suppressor gene. *Nature* 1991;351:453–456.
100. Linzer DIH, Levine AJ. Characterization of a 54K dalton cellular SV40 tumor antigen in SV40 transformed cells. *Cell* 1979;17:43–52.
101. Loeffler F, Frosch P. *Zentralbl Bakt 1 Orig* 1898;28:371.
102. Luria SE. Mutations of bacterial viruses affecting their host range. *Genetics* 1945;30:84–99.
103. Luria SE, ed. *General virology*. 1st ed. New York: Wiley; 1953.
104. Luria SE, Anderson TF. Identification and characterization of bacteriophages with the electron microscope. *Proc Natl Acad Sci USA* 1942;28:127–130.
105. Luria SE, Darnell JE, eds. *General virology*. 2nd ed. New York: J Wiley and Sons; 1967.
106. Lwoff A. The prophage and I. In: Cairns J, Stent GS, Watson JD, eds. *Phage and the origins of molecular biology*. Cold Spring Harbor, NY: Cold Spring Harbor Laboratory Press; 1961:88–99.
107. Lwoff A, Siminovitch L, Kjeldgaard N. Induction de la lyse bactériophagique de la totalité d'une population microbienne lysogène. *C R Acad Sci Paris* 1950;231:190–191.
108. Mayer A. On the mosaic disease of tobacco; preliminary communication. *Tijdschr Landbouwk* 1882;2:359–364.
109. Mayer A. On the mosaic disease of tobacco. *Landwn VersStnen* 1886;32:451–467.
110. McKinney HH. Factors affecting the properties of a virus. *Phytopathology* 1926;16:753–758.
111. McKinney HH. Mosaic diseases in the Canary Islands, West Africa, and Gibraltar. *J Agric Res* 1929;39:557–578.
112. Monod J, Wollman EL. L'inhibition de la croissance et de l'adaption enzymatique chez les bactéries infectées par le bactériophage. *Ann Inst Pasteur* 1047;73:937–957.
113. Morrow JF, Berg P. Cleavage of simian virus 40 DNA at a unique site by a bacterial restriction enzyme. *Proc Natl Acad Sci USA* 1972;69:3365–3369.
114. Morse ML, Lederberg EM, Lederberg J. Transduction in *Escherichia coli* K12. *Genetics* 1956;41:142–156.
115. Mulder C, Delius H. Specificity of the break produced by restricting endonuclease R_I in SV40 DNA as revealed by partial denaturation. *Proc Natl Acad Sci USA* 1972;69:3215.
116. Negri A. Beitrag zum Stadium der Aetiologie der Tollwuth. *Z Hyg Infektkrankh* 1903;43:507–528.
117. Pasteur L. Méthode pour prévenir la rage après morsure. *C R Acad Sci* 1885;101:765–772.
118. Pilder S, Moore M, Logan J, Shenk T. The adenovirus E1B-55Kd transforming polypeptide modulates transport or cytoplasmic stabilization of viral and host cell mRNA. *Mol Cell Biol* 1986;6:470–476.
119. Poiesz BJ, Ruscetti FW, Gazdar AF, et al. Detection and isolation of type C retrovirus particles from fresh and cultured lymphocytes of a patient with cutaneous T-cell lymphoma. *Proc Natl Acad Sci USA* 1980;77:7415–7419.
120. Ptashne M. *A genetic switch, gene control and phage lambda*. Palo Alto, CA: Blackwell Science Publishers; 1987.
121. Purdy-Beale HA. Immunologic reactions with tobacco mosaic virus. *J Exp Med* 1929;49:919–935.
122. Reed W, Carroll J, Agramonte A, Lazear J. *Senate Documents* 1901;66(822):156.
123. Rous P. A sarcoma of the fowl transmissible by an agent separable from the tumor cells. *J Exp Med* 1911;13:397–399.
124. Roux E. Sur les microbes dits invisible. *Bull Inst Pasteur Paris* 1903;1:7–12,49–56.
125. Samulski RJ, Chang L-S, Shenk T. Helper-free stocks of recombinant adeno-associated viruses: normal integration does not require viral gene expression. *J Virol* 1989;63:3822–3828.
126. Sanford KK, Earle WR, Likely GD. The growth in vitro of single isolated tissue cells. *J Natl Cancer Inst* 1948;23:1035–1069.
127. Sarnow P, Ho YS, Williams J, Levine AJ. Adenovirus E1B-58Kd tumor antigen and SV40 large tumor antigen are physically associated with the same 54Kd cellular protein in transformed cells. *Cell* 1982;28:387–394.
128. Schlesinger M. Die bestimmung von teilchengrïsse und spezifischem gewicht des bakteriophagen durch zentrifugierversuche. *Z Hyg Infektionskrankh* 1932;114:161.
129. Schlesinger M. Zur Frage der chemischen Zusammensetzung des Bakteriophagen. *Biochem Z* 1934;273:306–311.
130. Schlesinger M. The fuelgen reaction of the bacteriophage substance. *Nature* 1936;138:508–509.
131. Sinsheimer RL. A single stranded DNA from bacteriophage Ø174. *Brookhaven Symp Biol* 1959;12:27–34.
132. Stanley W. Isolation of a crystaline protein possessing the properties of tobacco-mosaic virus. *Science* 1935;81:644–645.
133. Stehelin D, Varmus HE, Bishop JM, Vogt PK. DNA related to transforming gene(s) of avian sarcoma viruses is present in normal avian DNA. *Nature* 1976;260:170–173.

134. Streisinger G, Edgar RS, Denhardt GH. Chromosome structure in phage T4. I. Circularity of the linkage map. *Proc Natl Acad Sci USA* 1964;51:775–779.
135. Streisinger G, Mukai F, Dreyer WJ, Miller B, Horiuchi S. Mutations affecting the lysozyme of phage T4. *Cold Spring Harb Symp Quant Biol* 1961;26:25–30.
136. Takahashi WN, Rawlins RE. Method for determining shape of colloidal particles; application in study of tobacco mosaic virus. *Proc Natl Acad Sci USA* 1932;30:155–157.
137. Takatsuki K, Uchuyama T, Ueshima Y, et al. Adult T-cell leukemia: proposal as a new disease and cytogenetic, phenotypic and function studies of leukemic cells. *Gann Monogr* 1982;28:13–21.
138. Temin HM, Mizutani S. RNA-dependent DNA polymerase in virions of Rous sarcoma virus. *Nature* 1970;226:1211–1213.
139. Thomas CA Jr. The arrangement of information in DNA molecules. *J Gen Physiol* 1966;49:143–169.
140. Twort FW. An investigation on the nature of the ultramicroscopic viruses. *Lancet* 1915;189:1241–1243.
141. Twort FW. The bacteriophage: the breaking down of bacteria by associated filter-passing lysins. *Br Med J* 1922;2:293.
142. Twort FW. The discovery of the bacteriophage. *Sci News* 1949; 14:33.
143. Uchiyama T, Yodoi J, Sagawa K, Takatsuki K, Uchino H. Adult T-cell leukemia: clinical and hematologic features of 16 cases. *Blood* 1977;50:481–492.
144. Vinson CG, Petre AW. Mosaic disease of tobacco. *Botan Gaz* 1929;87:14–38.
145. Watson JD, Crick FHC. A structure for deoxyribonucleic acid. *Nature* 1953;171:737–738.
146. Weil PA, Luse DS, Segall J, Roeder RG. Selective and accurate initiation of transcription at the Ad2 major late promoter in a soluble system dependent on purified RNA polymerase II and DNA. *Cell* 1979;18:469–484.
147. Weil R, Vinograd J. The cyclic helix and cyclic coil forms of polyoma viral DNA. *Proc Natl Acad Sci USA* 1963;50:730–736.
148. Weiss R, Teich N, Varmus H, Coffin J, eds. *RNA tumor viruses*. New York: Cold Spring Harbor Laboratory Press, 1982.
149. Werness BA, Levine AJ, Howley PM. Association of human papillomavirus types 16 and 18 E6 proteins with p53. *Science* 1990; 248:76–79.
150. World Health Organization. *The global eradication of smallpox*. Final report of the Global Commission for the Certification of Smallpox Eradication (History of International Public Health, no 4). Geneva: World Health Organization; 1980.
151. Whyte P, Buchkovich KJ, Horowitz JM, Friend SH, Raybuck M, Weinberg RA, Harlow E. Association between an oncogene and an anti-oncogene; the adenovirus E1a proteins bind to the retinoblastoma gene product. *Nature* 1988;334:124–129.
152. Wolbach SB. The filterable viruses, a summary. *J Med Res* 1912; 27:1–25.
153. Woodruff AM, Goodpasture EW. The susceptibility of the chorio-allantoic membrane of chick embryos to infection with the fowl-pox virus. *Am J Pathol* 1931;7:209–222.
154. Worcel A, Burgi E. On the structure of the folded chromosome of *E. coli*. *J Mol Biol* 1972;71:127–139.
155. Wyatt GR, Cohen SS. The basis of the nucleic acids of some bacterial and animal viruses: the occurence of 5-hydroxymethylcytosine. *Biochem J* 1952;55:774–782.
156. Yoshida M, Miyoshi I, Hinuma Y. Isolation and characterization of retrovirus from cell lines of human adult T-cell leukemia and its implications in the disease. *Proc Natl Acad Sci USA* 1982; 79:2031–2035.
157. Zinder ND, Lederberg J. Genetic exchange in *Salmonella*. *J Bacteriol* 1952;64:679–699.
158. zur Hausen H. Viruses in human cancers. *Science* 1991; 254:1167–1172.

Fundamental Virology, Third Edition
edited by B.N. Fields, D.M. Knipe, P.M. Howley, et al.
Lippincott - Raven Publishers, Philadelphia © 1996

CHAPTER 2

Virus Taxonomy

Frederick A. Murphy

F. A. Murphy: School of Veterinary Medicine, Office of the Dean, University of California, Davis, California 95616.

> The world, said Paul Valery, is equally threatened with two catastrophes: order and disorder. So is virology.
>
> Lwoff, Horne, and Tournier (81)

The earliest experiments involving viruses were designed to separate them from microbes that could be seen in the light microscope and that usually could be cultivated on rather simple media. In the experiments that led to the first discoveries of viruses, by Beijerinck and Ivanovski (tobacco mosaic virus), Loeffler and Frosch (foot-and-mouth disease virus), and Reed and Carroll (yellow fever virus) at the turn of the century, one single physicochemical characteristic was measured, that being their small size as assessed by filterability (134). No other physicochemical measurements were made at that time, and most studies of viruses centered on their ability to cause infections and diseases. The earliest efforts to classify viruses, therefore, were based on perceived common pathogenic properties, common organ tropisms, and common ecological and transmission characteristics. For example, viruses that share the pathogenic property of causing hepatitis (e.g., hepatitis A virus, hepatitis B virus, hepatitis C virus, yellow fever virus, and Rift Valley fever virus) would have been brought together as "the hepatitis viruses," and plant viruses causing mosaics (e.g., cauliflower mosaic virus, ryegrass mosaic virus, brome mosaic virus, alfalfa mosaic virus, and tobacco mosaic virus) would have been brought together as "the mosaic viruses."

Although the first studies of viruses were begun at the turn of the century, it was not until the 1930s that evidence of the structure and composition of virions started to emerge. This prompted Bawden (5,6) to propose for the first time that viruses be grouped on the basis of shared virion properties. Among the first taxonomic groups constructed on this basis were the herpesvirus group (2), the myxovirus group (3), the poxvirus group (45), and several groups of plant viruses with rod-shaped or filamentous virions (11). In the 1950s and 1960s, there was an explosion in the discovery of new viruses. Prompted by a rapidly growing mass of data, several individuals and committees independently advanced classification schemes. The result was confusion over competing, conflicting schemes, and for the first but not the last time it became clear that virus classification and nomenclature are topics that give rise to very strongly held opinions.

THE INTERNATIONAL COMMITTEE ON TAXONOMY OF VIRUSES

Against this background, in 1966 the International Committee on Nomenclature of Viruses (ICNV)[1] was established at the International Congress of Microbiology in Moscow. At that time, virologists already sensed a need for a single, universal taxonomic scheme. There was little dispute that the hundreds of viruses being isolated from humans, animals, plants, invertebrates, and bacteria should be classified in a single system and that this system should separate the viruses from all other biological entities. Nevertheless, there was much dispute over the taxonomic system to be used. Lwoff et al. (81) argued for the adoption of an all-embracing scheme for the classification of viruses into subphyla, classes, orders, suborders, and families. Descending hierarchical divisions were to be based, arbitrarily and monothetically, on nucleic acid type, capsid symmetry, presence or absence of an envelope, and so forth. Opposition to this scheme was based on its arbitrariness in deciding the relative importance of virion characteristics to be used and on the argument that not enough was known about the characteristics of most viruses to warrant an elaborate hierarchy. An alternative proposal was set forth in 1966 by Gibbs, Harrison, Watson, and Wildy (52,53); in this system, divisions were based on multiple criteria (polythetic criteria) and categorized by "cryptograms" (coded notations of eight virus characters). These early efforts succeeded well in stimulating interest in the development of the universal taxonomy system that evolved in the 1970s and has been built on ever since (14,17,85,139).

In the universal scheme developed by the ICTV, virion characteristics are considered and weighted as criteria for making divisions into families, in some cases subfamilies, and genera (until recently, the scheme did not use any hierarchical level higher than that of family, but now one order, the order *Mononegavirales*, has been approved). In each case, the relative hierarchy and weight assigned to each characteristic used in defining taxa is set arbitrarily and still influenced by prejudgments of relationships that "we would like to believe (from an evolutionary standpoint), but are unable to prove" (44). At its meeting in Mexico City in 1970, the ICTV approved the first two families and 24 floating genera (85,139). At that time, 16 plant virus groups were also designated (61). Since then, the ICTV has published six reports entitled "The Classification and Nomenclature of Viruses" (43,48,83,84,96,139). The sixth report of the ICTV, published in 1995, records a universal taxonomy scheme comprising one order, 71 families, 11 subfamilies, and 164 genera, including many floating genera, and more than 4,000 member viruses. The system still contains hundreds of unassigned viruses, largely because of a lack of data.

THE UNIVERSAL SYSTEM OF VIRUS TAXONOMY

Today there is a sense that a significant fraction of all existing viruses of humans, domestic animals, and economically important plants have already been isolated and entered into the taxonomic system. This sense is based on the infrequency in recent years of discoveries of viruses that do not fit into present taxa. Of course, this sense does

[1] The International Committee on Nomenclature of Viruses (ICNV) became the International Committee on Taxonomy of Viruses (ICTV) in 1973. Today, the ICTV operates under the auspices of the Virology Division of the International Union of Microbiological Societies. The ICTV has six Subcommittees, 45 Study Groups, and over 400 participating virologists.

not extend to the viruses infecting the myriad other species populating the Earth. This present sense of the diversity of the viruses, however imperfect, does point once again to the need for a universal, usable taxonomic system—a system to keep track of the large number of viruses being isolated and studied throughout the world, a system to tie viral characteristics to virus names. The present universal system of virus taxonomy is useful and usable. It is set arbitrarily at hierarchical levels of order, family, subfamily, genus, and species. Lower hierarchical levels, such as subspecies, strain, variant, and so forth, are established by international specialty groups and by culture collections.

Virus Orders

Virus orders represent groupings of families of viruses that share common characteristics and are distinct from other orders and families. Virus orders are designated by names with the suffix *-virales*. To date, one order has been approved by the ICTV, the order *Mononegavirales*, comprising the families *Paramyxoviridae*, *Rhabdoviridae*, and *Filoviridae*. It is the ICTV's intention to move slowly in the approval of orders, limiting use to those instances where there is good evidence of phylogenetic relationship among the viruses of member families.

Virus Families and Subfamilies

Virus families represent groupings of genera of viruses that share common characteristics and are distinct from the member viruses of other families. Virus families are designated by names with the suffix *-viridae*. Despite concerns about the arbitrariness of early criteria for creating these taxa, most of the original families have stood the test of time and are still intact. This level in the taxonomic hierarchy now seems stable; indeed, it is the benchmark of the entire universal taxonomy system. Most of the families of viruses have distinct virion morphology, genome structure, and/or strategies of replication, indicating phylogenetic independence or great phylogenetic separation. At the same time, the virus family is being recognized as a taxon uniting viruses with a common, even if distant, phylogeny. In four families, namely the families *Poxviridae*, *Herpesviridae*, *Parvoviridae*, and *Paramyxoviridae*, subfamilies have been introduced to allow for a more complex hierarchy of taxa, in keeping with the apparent intrinsic complexity of the relationships among member viruses. Subfamilies are designated by terms with the suffix *-virinae*.

Virus Genera

Virus genera represent groupings of species of viruses that share common characteristics and are distinct from the member viruses of other genera. Virus genera are designated by terms with the suffix *-virus*. This level in the hierarchy of taxa also seems stable and in many cases may be considered a benchmark for setting definitions of other taxa, especially species. The criteria used for creating genera differ from family to family. As more viruses are discovered and studied, there is pressure in many families to use smaller and smaller genetic, structural, or other differences to create new genera. Because evidence of common phylogeny has entered the definition of many families, it is logical that even more such evidence will become the basis for defining genera. In fact, it might be said that in the future genera will not stand where evidence is obtained of distinct phylogenies among member species.

Virus Species

The species taxon has always been regarded as the most important hierarchical level in classification, but with the viruses it has proved to be the most difficult to address. After years of controversy, in 1991 the ICTV accepted the definition of a virus species proposed by van Regenmortel (130) as follows: "A virus species is defined as a polythetic class of viruses that constitutes a replicating lineage and occupies a particular ecological niche." Members of a polythetic class are defined by more than one property, and no single property is essential or necessary. One major advantage in this definition is that it can accommodate the inherent variability of viruses and does not depend on the existence of a single unique characteristic. Similarly, it can accommodate the different traditions of virologists working in different areas of virology, in some cases accommodating "the lumpers" and in others "the splitters."

The ICTV study groups are now determining the specific properties to be used to define species in the taxon for which they are responsible. It seems clear that the term species eventually will be defined somewhat similarly to the term virus, although in many cases the term virus matches best with subspecies, strain, or even variant. Just as the term virus is defined differently in different virus families, so will species be defined differently, in some cases with emphasis on genome properties and in others on structural, physicochemical, or serological properties. Some viruses have already been designated as species, for example, Sindbis virus, Newcastle disease virus, poliovirus 1, vaccinia virus, Fiji disease virus, and tomato spotted wilt virus. However, these examples do not reflect the difficulty that is being encountered in deciding whether a particular virus should be designated as a species or as a strain.

VIRUS NOMENCLATURE

The Usage of Formal Taxonomic Nomenclature

In formal taxonomic usage, the first letters of virus family, subfamily, and genus names are capitalized; the genus and species names are printed in italics (underlined when typewritten). Species designations are not capitalized (unless they are derived from a place name or a host family or genus name). In formal usage, the name of the taxon should precede the term for the taxonomic unit, for example, the family *Paramyxoviridae*, the genus *Morbillivirus*." Furthermore, it was decided years ago that virus nomenclature would not involve the use of Latinized binomial terms. For example, terms such as *Flavivirus fabricis*, *Orthopoxvirus variolae*, and *Herpesvirus varicellae*, which were used at one time, have been abandoned. The following represent examples of full formal taxonomic terminology:

1. Family *Poxviridae*, subfamily *Chordopoxvirinae*, genus *Orthopoxvirus,* vaccinia virus.
2. Family *Herpesviridae*, subfamily *Alphaherpesvirinae*, genus *Simplexvirus,* human herpesvirus 2 (herpes simplex virus 2).
3. Family *Picornaviridae*, genus *Enterovirus,* poliovirus 1.
4. Order *Mononegavirales*, Family *Rhabdoviridae*, genus *Lyssavirus,* rabies virus.
5. Family *Bunyaviridae*, genus *Tospovirus,* tomato spotted wilt virus.
6. Family *Bromoviridae*, genus *Bromovirus,* brome mosaic virus.
7. Genus *Sobemovirus,* southern bean mosaic virus.
8. Family *Totiviridae*, genus *Totivirus,* Saccaromyces cerevisae virus L-A.
9. Family *Tectiviridae*, genus *Tectivirus,* enterobacteria phage PRD 1
10. Family *Plasmaviridae*, genus *Plasmavirus,* or Acholeplasma L2

The Vernacular Usage of Virus Nomenclature

In informal vernacular usage, virus family, subfamily, genus, and species names are written in lower case Roman script; they are not capitalized, nor are they printed in italics or underlined. In informal usage, the name of the taxon should not include the formal suffix, and the name of the taxon should follow the term for the taxonomic unit; for example, ". . . the picornavirus family . . . the enterovirus genus."

The use of vernacular terms for virus taxonomic units and virus names should not lead to unnecessary ambiguity or loss of precision in virus identification. The formal family, subfamily, and genus terms and standard ICTV vernacular species terms, rather than any synonyms or transliterations, should be used as the basis for choosing vernacular terms.

One particular source of ambiguity in vernacular nomenclature lies in the common use of the same root terms in formal family and genus names. Imprecision stems from not being able to easily identify in vernacular usage which hierarchical level is being cited. For example, the vernacular name "paramyxovirus" might refer to the family *Paramyxoviridae*, the genus *Paramyxovirus,* or one of the species in the genus *Paramyxovirus,* such as one of the human parainfluenza viruses. Some virologists have suggested that this problem be solved by renaming taxa so that the same root term is never used at multiple hierarchical levels. However, there is no consensus for this proposal. In fact, as plant virus taxonomy switches away from groups and toward families and genera, this problem will be exacerbated. The solution in vernacular usage is to avoid "jumping" hierarchical levels and to add taxon identification wherever needed. For example, when citing the taxonomic placement of human parainfluenza virus 1, the term "paramyxovirus" should refer firstly to the genus, not the subfamily or family, and taxon identification should always be added: "human parainfluenza virus 1 is a member of the paramyxovirus genus," rather than " human parainfluenza virus 1 is a paramyxovirus." Most examples like this exemplify the advantage of switching, where necessary, into formal nomenclature usage: "human parainfluenza virus 1 is a species in the genus *Paramyxovirus,* family *Paramyxoviridae*." In this example, as is usually the case, adding the information that this virus is also a member of the subfamily *Paramyxovirinae* and the order *Mononegavirales* is unnecessary.

STRUCTURAL, GENOMIC, PHYSICOCHEMICAL, AND REPLICATIVE PROPERTIES OF VIRUSES USED IN TAXONOMY

The way by which viruses are characterized, for taxonomic and other purposes, is changing rapidly. In the past, laboratory techniques have included characterizations of virion morphology (by electron microscopy), virion stability (by varying pH and temperature, adding lipid solvents and detergents, etc.), virion size (by filtration through fibrous and porous microfilters), and virion antigenicity (by many different serologic methods). These means worked because after large numbers of viruses had been studied and their characteristics placed into the universal taxonomic scheme, it was necessary in most cases to only measure a few characteristics to place a new virus, especially a new variant from a well-studied source, in its proper taxonomic niche. For example, a new adenovirus, isolated from the human respiratory tract and identified by serologic means, was easy to place in its niche in the family *Adenoviridae*, genus *Mastadenovirus.* The exceptions occurred when a new virus was found that did not have a familiar set of properties. Such a virus became a candidate prototype for a new taxon, generally a new family or genus. In such cases, comprehensive characterization of all virion properties was deemed necessary.

One particularly important technological advance underpinning the development of modern virus taxonomy was the invention by Brenner and Horne (12) of the negative staining technique for electron microscopic examination of virions. The impact of this technique was immediate: (a) virions could be characterized with respect to size, shape, surface structure, presence or absence of an envelope, and, often, symmetry; (b) the method could be applied simply and universally; and (c) virions could be characterized in unpurified material, including diagnostic specimens. Negative staining has facilitated the rapid accumulation of data about the physical properties of many viruses. Thin-section electron microscopy of virus-infected cell cultures and tissues of infected humans, animals (including experimental animals), and plants has provided complementary data on virion morphology, mode and site of virion morphogenesis (e.g., site of budding), and so forth. Thus, in many cases viruses were placed in their appropriate family, and in some instances in their appropriate genus, after simple visualization and measurement by negative-stain and/or thin-section electron microscopy (93).

The fundamental molecular bases for many of the empirical virion property measurements that were originally used to construct virus families and genera are now well understood. Many of the characteristics that have been used in deciding taxonomic constructions are listed in Table 1. Moreover, through the use of monoclonal antibodies, synthesized peptides, and epitope mapping, there is new understanding of the molecular bases for those serological reactions that were originally used to construct families and genera. Today, genome sequencing, or partial sequencing, is often performed early in virus identification, and even in diagnostic activities. For comparison, genome sequences are available in readily accessible databases for the prototype viruses of nearly all taxa. Sequence data is even driving consideration of the construction of new families and new genera before other data is available. It is likely because of their absolute nature that genome sequence data will become the base for further refinement and expansion of the universal taxonomic scheme. Genome sequence data are also the key to the advance of phylogenetic taxonomy. In addition, the derivatives of sequencing are advancing as taxonomic criteria; for example, genome organization, gene order, strategy of replication and other genetic considerations have been added to the taxonomic decision process (94–96,98).

TABLE 1. *Some properties of viruses used in taxonomy*

- Virion properties
  - Morphology
    - Virion size
    - Virion shape
    - Presence or absence and nature of peplomers
    - Presence or absence of an envelope
    - Capsid symmetry and structure
  - Physicochemical and physical properties
    - Virion molecular mass (Mr)
    - Virion buoyant density (in CsCl, sucrose, etc.)
    - Virion sedimentation coefficient
    - pH stability
    - Thermal stability
    - Cation stability (Mg^{2+}, Mn^{2+})
    - Solvent stability
    - Detergent stability
    - Irradiation stability
  - Genome
    - Type of nucleic acid (DNA or RNA)
    - Size of genome in kb/kbp
    - Strandedness: single-stranded or double-stranded
    - Linear or circular
    - Sense (positive-sense, negative-sense, ambisense)
    - Number and size of segments
    - Nucleotide sequence
    - Presence of repetitive sequence elements
    - Presence of isomerization
    - G&C content ratio
    - Presence or absence and type of 5′ terminal cap
    - Presence or absence of 5′ terminal covalently linked protein
    - Presence or absence of 3′ terminal poly (A) tract
  - Proteins
    - Number, size, and functional activities of structural proteins
    - Number, size, and functional activities of nonstructural proteins
    - Details of special functional activities of proteins, especially transcriptase, reverse transcriptase, hemagglutinin, neuraminidase, and fusion activities
    - Amino acid sequence or partial sequence
    - Glycosylation, phosphorylation, myristylation property of proteins
    - Epitope mapping
  - Lipids
    - Content, character, etc.
  - Carbohydrates
    - Content, character, etc.
  - Genome organization and replication
    - Genome organization
    - Strategy of replication
    - Number and position of open reading frames
    - Transcriptional characteristics
    - Translational characteristics
    - Site of accumulation of virion proteins
    - Site of virion assembly
    - Site and nature of virion maturation and release
- Antigenic properties
  - Serologic relationships, especially as obtained in reference centers
- Biologic properties
  - Natural host range
  - Mode of transmission in nature
  - Vector relationships
  - Geographic distribution
  - Pathogenicity, association with disease
  - Tissue tropisms, pathology, histopathology

THE FUTURE OF VIRUS TAXONOMY

The Advance of Phylogenetic Taxonomy and the Use of Higher Taxa

Until recently, one of the rules of virus taxonomy stated that the system was not meant to imply any phylogenetic relationships. Because viruses leave no fossils (except perhaps within arthropods and other creatures embedded in amber!), it was presumed that there never would be enough evidence to prove whether different taxa had common evolutionary roots. In fact, because the prevailing concept was that each kind of virus was derived separately from its host, it was considered foolish to consider any idea of a single evolutionary "tree" for the viruses (51). The generally very different morphological and physicochemical characteristics of the member viruses of many different taxa supported this view.

Now, as genome sequencing of many viruses, and many organisms from archebacteria to humans, is revealing many conserved functional or "fossil" domains of ancient lineage, for the first time the archeology of the viruses is being explored from the perspective of data, not just "armchair theory." We now know that many viruses have gained some functional genes from their hosts (and hosts have gained some genes from viruses) and have gained other genes from other viruses; that is, viral genomes seem to represent more or less ancient "grab-bags" of genes, fine-tuned by the Darwinian forces of selection into replicative machines with extraordinary functional economy. We now know that the genomes of viruses in different families, in most cases, are extremely different from each other, but we also know that in some cases seemingly unrelated viruses (and taxa) are similar—similar in gene order and arrangement, fine points of strategy of replication, and even in conserved sequence domains encoding similarly functioning proteins. Overall, the differences between most taxa are so great that it still seems foolish to think of building a monophylogenous tree uniting all the viruses. On the other hand, the unexpected similarities have prompted some consideration of a partial phylogenetic taxonomy (51,54–56,70).

As this evidence of phylogenetic relationships between families was studied, a desire emerged to reflect these relationships in the universal taxonomic scheme. However, there has been no wish to combine families exhibiting distant phylogenetic relationships and thereby have them lose their practical identities. As one virologist stated, "The family is the fixed point, the benchmark, in virus taxonomy, so let's not do anything to change this." Instead, there has been increasing interest in capturing these relationships by uniting distantly related families in higher taxa, namely orders. The order *Mononegavirales*, comprising the families *Paramyxoviridae*, *Rhabdoviridae*, and *Filoviridae*, was formed in recognition that the member viruses had common sequences in their nucleocapsid genes and similar gene arrangements and gene products (104). At this point, ICTV is committed to reserving the hierarchical level of order solely for recognizing phylogenetic relationships.

There is a further occasion for considering the grouping of families together; this involves many of the positive-sense, single-stranded RNA viruses. Similarities in genome organization, gene arrangement, and genome sequence in particular domains between viruses in several taxa, representing diverse vertebrate, invertebrate, plant, and bacterial viruses, have been observed in the past 10 years. Kamer and Argos (67) first aligned RNA-dependent RNA polymerase gene sequences of several plant, animal, and bacterial viruses. Goldbach (54,55) greatly broadened this approach and used the data to explore the possible paths of evolution of the many positive-sense RNA viruses. He proposed the formation of several "supergroups" ("superfamilies") to formalize the recognition of similarities. Gibbs (51) and Strauss, Strauss and Levine (124,125) have explored the possible mechanisms underpinning such evolutionary relationships. Today, work is centered on assessment of many characteristics of the RNA viruses: (a) genome organization and gene order; (b) presence or absence of a 5'-terminal covalently linked protein, a 3'-terminal poly (A) tract, or a 5'-terminal cap; (c) presence or absence of subgenomic RNA; (d) polyprotein processing and enzymology; and so forth. At the heart of the matter is the assessment of conserved sequences in genes encoding the RNA-dependent RNA polymerases, helicases, and proteases. These characteristics are being used as the basis for several different ideas for constructing higher taxa (57,58,73–75,82,124, 125,133).

In their most recent proposals, Goldbach and de Haan (58) and Koonin and Dolja (74) have proposed three major clusters of positive-sense, single-stranded RNA viruses, one for the picornalike viruses, one for the togalike viruses, and one for the flavilike viruses (although Goldbach calls attention to the major differences among the viruses in the latter cluster). Goldbach and de Haan (58) continue to use the terms supergroups and superfamilies for these clusters, but Koonin and Dolja (74) have taken this a step further in using the taxonomic hierarchical levels of class and order to denote the same groupings. It was thought for a time that the approach also could be extended to the double-stranded RNA viruses, but recent evidence suggests a polyphyletic origin of double-stranded RNA viruses from different groups of positive-sense RNA viruses (16,74). This matter will be debated by the ICTV over the next few years, but already it is clear that most virologists do not wish to abandon the present system, which is based on assessment and weighting of multiple virion characteristics. As one virologist stated, "I am alarmed by the idea of erecting higher taxa upon a scheme that assumes that the polymerase is the virus." Clearly, there will be interest in melding phylogenetic considerations and traditional approaches into a unifying system.

Within the subject of phylogenetic taxonomy, one of the most interesting debates centers on which characteristics of an organism are most ancient, which are most recent, which are most stable, and which are most changing. With increasing knowledge of which characteristics are conserved through evolutionary divergence, arbitrary cladistic taxonomy seemingly must be melded with phylogenetic taxonomy. Many virologists over the years have considered that virion structural elements were most ancient; after all, cumulative mutations in genes encoding proteins for icosahedral capsids could only lead to lethal instability. Similarly, many virologists have considered that viral genome expression strategies were most ancient; again, even if a virus figured out how to reinvent its capsid, changes in integrated multigenic replication steps would certainly be lethal. At present, however, the finding of conserved sequence domains in polymerases, helicases, and proteases—but not in structural or other genes of the positive-sense, single-stranded RNA viruses—suggests that the whole subject must be revisited. Of course, horizontal gene transfer between viruses and dynamic gene acquisition from host cells adds to the sense of phylogenetic complexity. It is unfortunate that viruses have such small genomes; there will not be an opportunity to find confirmatory evidence of relationships by analyzing additional genes (63).

The ICTV Database

The number of viruses occupying geographic and/or host niches as pathogens or silent passengers of humans, animals, plants, invertebrates, protozoa, fungi, and bacteria is very large. Our lists are increasing regularly as we search in new niches and as the sensitivity and specificity of our techniques for detection get better and better. Today, the ICTV recognizes more than 4,000 viruses. Specialty groups keep track of far more viruses, virus strains, and subtypes, each having particular health or economic distinction and importance. It has been estimated that more that 30,000 viruses, virus strains, and subtypes are being tracked in various specialty laboratories, reference centers, and culture collections communicating with the World Health Organization (WHO), Food and Agriculture Organization (FAO), and other international agencies. Furthermore, the development of the viral quasispecies concept, with its prediction of rapid evolution of variants that may become fixed in nature as new species, portends future needs to track even more viral entities (34,38,64,143). It has been estimated that to describe a virus comprehensively, approximately 500 to 700 characters must be determined (4,9). This means that to comprehensively describe all the known viruses of the world we must "fill in the blanks" for 3 to 21 million data points (of course, many of the data are the same when entering related viruses). This situation is even more complex; as we add more and more genome sequence information, our data systems will become truly enormous.

A major goal of the ICTV is to design, build, and make available to all virologists, worldwide, a universal virus database, the ICTV database (ICTVdB). This database will encompass data that are now used in developing and managing the universal system for virus taxonomy. The ICTVdB is designed to describe viruses down to the species level, in keeping with the level of responsibility of the ICTV, and will interface with the databases of international specialty groups that are cataloguing data down to strain, variant, and isolate levels, that is, levels important in medicine, agriculture, and other scholarly fields. The ICTV's goal is to design the database to feed directly into user-friendly programs that will be directly accessible to users (101). Particular products, tailored to particular users, will be compressed to fit into equipment and software that can be readily accessed around the world.

Several virus databases are in operation in the world that will be interfaced with the ICTVdB, including (a) the plant virus database operated at the Australian National University in Canberra, Australia [the Virus Identification Data Exchange (VIDE) Project]; (b) the veterinary virus database operated by the CSIRO/Australian Animal Health Laboratory in Geelong, Australia (the VIREF Project) (AJ Della Porta, personal communication, 1993); and (c) an arbovirus database operated for the American Committee on Arthropod-borne Viruses (ACAV) by the Centers for Disease Control in Ft. Collins, Colorado (68). The most advanced of these databases is the VIDE Project on plant viruses. This project involves a worldwide network of more than 200 collaborating plant virologists and a database that now contains 569 characters for more than 890 plant virus species in 55 genera (9,18; AJ Gibbs, personal communication, 1993). Using these databases as a foundation, the ICTV has laid out a plan to develop the universal virus database, the ICTVdB, over the next 10 years.

TAXONOMY AND UNAMBIGUOUS VIRUS IDENTIFICATION

Unambiguous virus identification is a major virtue of the universal system of taxonomy (92), and of particular value when the editor of a journal requires precise naming of viruses cited in a publication. At a minimum, precise naming avoids problems caused by synonyms, transliterated vernacular names, and local laboratory jargon. Precise virus identification includes taxonomic status, such as name of family, genus, and species, as well as strain designation terms. The matter of deciding how type species and strains are chosen and designated remains the responsibility of international specialty groups, some of which operate under the auspices of the WHO and other agencies.

One of the best models for the kind of description necessary to avoid ambiguity in virus strain identification is that of the American Type Culture Collection in its frequently updated catalogue of animal and plant viruses (1). For example, St. Louis encephalitis virus is listed as follows:

St. Louis encephalitis virus class III ATCC VR-80

Strain: Hubbard. Original source: Brain of patient, Missouri, 1937. Reference: McCordock HA et al., *Proc Soc Exp Biol Med* 1937;37:288. Preparation: 20% SMB in 50% NIRS infusion broth; supernatant of low-speed centrifugation. Host of choice: sM (i.c.); M (i.c.). Incubation: 3–4 days. Effect: Death. Host range: M, Ha, CE, HaK, CE cells. Special characteristics: Infected brain tissue will have a titer of ~10^7. Agglutinates goose and chicken red blood cells. Cross-reacts with many or all members of group B arboviruses. Deposited by: W.McD. Hammon.

Taxonomy and the Adequate Description of New Viruses

Because thousands of viruses have been isolated from human and animal specimens, there have been many errors and duplications over the years; that is, viruses isolated in different laboratories have been given different names and the chance for coexistence in various virus lists (92). Viruses also have been placed in the wrong lists, the most notable instance involving the emergence of the family *Reoviridae* (genus *Orthoreovirus*) from the initial placement of its prototype viruses in the list of human enteroviruses. Recently, several named but serologically "ungrouped" viruses listed in the *International Catalogue of Arboviruses* (68) were found to actually be unrecognized isolates of Lassa virus and Rift Valley fever virus, viruses that must be handled under maximum containment conditions. The reasons for these and similar problems have been inadequate characterization and description of viruses by those who isolate them and inadequate review of data by international specialty groups. Such problems seem to be declining in frequency, but continuous attention is warranted because consequences can be serious. Assuring the adequacy of characterization and description of new viruses is a particular responsibility of reference laboratories, international reference centers, international specialty groups, and culture collections.

When an "unknown" is first studied in a laboratory, its initial characterization may involve only standardized protocols. That is, only a few characteristics may be determined before the application of specific identification procedures. Only when an unknown fails to yield to routine procedures is there call for more extensive study. One key to simplifying and rationalizing such study is to set useful techniques into a proper sequence based on taxonomic characteristics. This sequence of procedures should include logical short-cuts to avoid extra effort and expense. For example, negative-stain electron microscopy represents a logical short-cut for the initial placement of unknowns that emerge from characterization protocols. If an unknown is shown by electron microscopy to be a rhabdovirus, there is little value in checking whether its genome is DNA or RNA. Likewise, there is little value in performing serology against other than the known rhabdoviruses (except perhaps in testing for the presence of contaminant viruses). Comprehensive characterization is the key to discovery of novel viruses, but fully characterizing usual isolates rarely contributes to such discovery.

Taxonomy in Diagnostic Virology

A universal system for taxonomy and nomenclature of viruses is a practical necessity whenever large numbers of distinct isolates are being analyzed, as in a reference diagnostic laboratory. The clinician usually makes a preliminary diagnosis of a viral disease on the basis of four kinds of evidence: (a) clinical features, which allow recognition with varying certainty in typical cases of many viral diseases (e.g., varicella exanthem, herpes simplex gingivostomatitis); (b) epidemic behavior, which in a typical population may allow recognition (e.g., epidemic influenza, arbovirus diseases such as dengue and yellow fever, enterovirus exanthems); (c) circumstances of occurrence, which may indicate probable etiology (e.g., respiratory syncytial virus as the primary cause of croup and bronchiolitis in infants, hepatitis B or hepatitis C as the likely cause of hepatitis after blood transfusion); and (d) organ involvement, which may suggest a probable etiology (e.g., mumps virus as the cause of parotitis, viruses in general as the cause of 80–90% of acute respiratory infections). Shortcomings in the predictive value of these kinds of evidence suggest that the laboratory diagnostician as well as the clinician must appreciate the range of possible etiologic agents in particular disease syndromes. In many circumstances, there is value in initially assembling an inclusive "long list" of possible etiologic agents so that no candidate agent is overlooked. In most instances this is done informally, and the process is adjusted to the complexity of the case.

The universal system of virus taxonomy may be used as the source of the long list of candidate etiologic agents. The system serves to organize the long list logically, and because the system is comprehensive, it is unlikely that known viruses will be overlooked. The observations that serve to place an etiologic agent in its proper family and genus should also play a major role in shortening the list and should in most cases provide etiologic information needed to select immunologic (serologic) identification techniques. For example, the long list of possible etiologies for a slowly progressive central nervous system disease would include many viruses that are difficult or impossible to cultivate. However, the identification of spherical, 45-nm, nonenveloped virions in the nuclei of cells in a brain biopsy from a patient with such a disease would contribute to shortening the list to the family *Papovaviridae*, genus *Polyomavirus*, thereby suggesting a diagnosis of JC or SV 40 virus-induced progressive multifocal leukoencephalopathy. In this case, many viruses known to invade the brain and cause slowly progressive neurologic disease would be eliminated from the differential diagnostic list; e.g., member viruses of the families *Picornaviridae*, *Togaviridae*, *Flaviviridae*, *Paramyxoviridae*, *Rhabdoviridae*, *Bunyaviridae*, *Arenaviridae*, *Retroviridae*, *Adenoviridae*, and *Herpesviridae*.

Classification of Viruses on the Basis of Epidemiologic Criteria

Separate from the formal universal taxonomic system and the formal and vernacular nomenclature that stems from it are other classifications of viruses useful in instructional, clinical, epidemiological, and diagnostic settings. These classifications are based on virus tropisms and modes of transmission. Most viruses of humans and ani-

TABLE 2. *Criteria for disease causation: a unified concept appropriate for viruses as causative agents of disease, based on the Henle-Koch postulates, developed by A. S. Evans (41)*

1. Prevalence of the disease is significantly higher in subjects exposed to the putative virus than in those not so exposed.
2. Incidence of the disease is significantly higher in subjects exposed to the putative virus than in those not so exposed (prospective studies).
3. Evidence of exposure to the putative virus is present more commonly in subjects with the disease than in those without the disease.
4. Temporally, the onset of disease follows exposure to the putative virus and an incubation period that follows a normal pattern.
5. A spectrum of signs and symptoms follows exposure to the putative virus, presenting pattern of response, from mild to severe.
6. A measurable host response, such as an antibody response and/or a cell-mediated immune response, follows exposure to the putative virus. In those individuals lacking prior experience, the response appears regularly, and in those individuals with prior experience, the response is anamnestic.
7. Experimental reproduction of the disease follows deliberate exposure of animals or humans to the putative virus, but nonexposed control subjects remain disease free. Deliberate exposure may be in the laboratory or in the field, as with sentinel animals.
8. Elimination of the putative virus and/or its vector decreases the incidence of the disease.
9. Prevention or modification of infection, via immunization or drugs, decreases the incidence of the disease.
10. "The whole thing should make biologic and epidemiologic-sense."

mals are transmitted congenitally or by inhalation, ingestion, injection (including via arthropod bites), or close contact (including sexual contact) (137).

Enteric viruses are usually acquired by ingestion (fecal-oral transmission) and replicate primarily in the intestinal tract. The term is usually restricted to viruses that remain localized in the intestinal tract, rather than causing generalized infections. Enteric viruses are included in the families *Picornaviridae* (genus *Enterovirus*), *Caliciviridae*, *Astroviridae*, *Coronaviridae*, *Reoviridae* (genera *Rotavirus* and *Reovirus*), *Parvoviridae*, and *Adenoviridae*.

Respiratory viruses are usually acquired by inhalation (respiratory transmission) or by fomites (hand-to-nose/mouth/eye transmission) and replicate primarily in the respiratory tract. The term is usually restricted to viruses that remain localized in the respiratory tract, rather than causing generalized infections. Respiratory viruses are included in the families *Picornaviridae* (genus *Rhinovirus*), *Caliciviridae*, *Coronaviridae* (genus *Coronavirus*), *Paramyxoviridae* (genera *Paramyxovirus* and *Pneumovirus*), *Orthomyxoviridae*, and *Adenoviridae*.

Arboviruses (from "*ar*thropod-*bo*rne viruses") replicate in their hematophagous (blood-feeding) arthropod hosts and are then transmitted by bite to vertebrate hosts, wherein virus replication produces viremia of sufficient magnitude to infect other blood-feeding arthropods. Thus, the cycle is perpetuated. Part of the arthropod cycle can be bypassed by some viruses via vertical transmission, wherein transovarial infection passes virus directly from one arthropod generation to the next. Arboviruses are included in the families *Togaviridae*, *Flaviviridae*, *Rhabdoviridae*, *Bunyaviridae*, *Reoviridae* (genera *Orbivirus* and *Coltivirus*), and the unnamed family containing African swine fever virus.

Oncogenic viruses are acquired by close contact (including sexual contact), injection, and unknown means. The viruses usually infect only specific target tissues, where they usually become persistent and may evoke transformation of host cells, which may in turn progress to malignancy. Viruses that have demonstrated the capacity to be oncogenic, in experimental animals or in nature (but not necessarily in humans or domestic animals), are included in the families *Retroviridae*, *Hepadnaviridae*, *Papovaviridae*, *Adenoviridae*, and *Herpesviridae*.

Taxonomy and the Causal Relationship between Virus and Disease

One of the landmarks in the study of infectious diseases was the development of the Henle-Koch postulates of causation. They were originally drawn up for bacteria and protozoa, but were revised in 1937 by Rivers and again in 1982 by Evans in attempts to accommodate the special problem of proving disease causation by viruses (41) (Table 2). The problem is still difficult, especially when viruses are considered as causative of chronic diseases (including several of the hepatitides), neoplastic diseases, and slowly progressive neurological diseases. Because most such diseases cannot be reproduced in experimental animals, virologists have had to evaluate causation indirectly via "guilt by association," an approach that relies to a large degree on epidemiologic data and patterns of serologic reactions in populations. The framework of virus taxonomy, again, plays a role, this time in helping to evaluate some of the criteria for causation. This is especially the case in evaluating the likelihood that particular kinds of viruses might be etiologically, rather than

TABLE 3. *Families containing human and animal viruses*

VIRUS FAMILY	CHARACTERISTICS
RNA viruses	
Picornaviridae *Caliciviridae* *Astroviridae*	Single-stranded RNA positive-sense nonsegmented nonenveloped
Togaviridae *Flaviviridae*	Single-stranded RNA positive-sense nonsegmented enveloped
Coronaviridae Floating genus: *Arterivirus*	Single-stranded RNA positive-sense nonsegmented enveloped nested set transcription
Order: *Mononegavirales* *Paramyxoviridae* *Rhabdoviridae* *Filoviridae*	Single-stranded RNA negative-sense nonsegmented enveloped
Orthomyxoviridae *Bunyaviridae* *Arenaviridae*	Single-stranded RNA negative-sense (some ambisense genes) segmented enveloped
Reoviridae *Birnaviridae*	Double-stranded RNA positive-sense segmented nonenveloped
Retroviridae	Single-stranded RNA positive-sense retroid DNA step in replication
DNA viruses	
Hepadnaviridae	Double-stranded/ single-stranded DNA enveloped retroid DNA step in replication
Circoviridae *Parvoviridae*	Single-stranded DNA nonenveloped
Papovaviridae *Adenoviridae*	Double-stranded DNA nonenveloped
Herpesviridae *Poxviridae* *Iridoviridae* Family: unnamed, African swine fever virus	Double-stranded DNA enveloped
Subviral agents: Satellites, Viroids, and Agents of Spongiform Encephalopathies (Prions)	
floating genus: *Deltavirus*	Single-stranded RNA negative-sense nonsegmented defective, satellite
taxon: undefined, unnamed agents of spongiform encephalopathies (prions)	No known nucleic acid prions, "self- replicating" proteins

coincidentally or opportunistically, associated with a given disease. For example, very early in the investigation of the acquired immunodeficiency syndrome (AIDS), when many kinds of viruses were being isolated from patients and many candidate etiologic agents were being advanced publicly, several virologists working on the disease predicted that the etiologic agent would turn out to be a member of the family *Retroviridae*. This prediction was based on review of the biological and pathogenetic properties of member viruses of each family, the ruling out of most, and the recognition of similarities between characteristics of the diseases caused by known animal retroviruses and the disease AIDS. This prediction guided some of the early experimental approaches toward finding the etiologic agent. Later, after the discovery of the etiologic agent, human immunodeficiency virus (HIV), taxonomic considerations predicted its placement in the genus *Lentivirus* of the family *Retroviridae*. This taxonomic prediction, in turn, has been guiding experimental design in many areas, including cell biology, diagnostics, and epidemiology.

A TAXONOMIC DESCRIPTION OF THE FAMILIES OF VIRUSES CONTAINING HUMAN AND ANIMAL PATHOGENS

Of the 71 families of viruses recognized by the ICTV, 24 contain viruses that infect humans and animals (Table 3) (96). There are also two "floating" genera containing human or animal viruses (the genera *Arterivirus* and *Deltavirus*). An unresolved issue concerns the order of listing the virus families; this issue again points out the arbitrariness of weighting virion characteristics for taxonomic purposes (96). The listing of the families in Table 3 is in an order approximating the order of the chapters in later sections of this book. The order is set by nucleic acid type and strandedness (RNA or DNA, single- or double-stranded); genome replication strategy; genome sense (positive, negative, or ambisense); genome segmentation; and the presence or absence of an envelope. An alphabetic listing of some of the more common pathogenic viruses of humans (Table 4) and animals (Table 5) is also provided for cross-referencing purposes.

The Family *Picornaviridae*

Family: *Picornaviridae* (the picornaviruses) (Fig. 1)
Genus: *Enterovirus* (enteroviruses)
Genus: *Cardiovirus* (cardioviruses)
Genus: *Rhinovirus* (rhinoviruses)
Genus: *Aphthovirus* (foot-and-mouth disease viruses)
Genus: *Hepatovirus* (hepatitis A virus)

Characteristics (*29*,60,100,123)

Virions are 28–30 nm in diameter and are composed of 60 protein subunits (protomers) arranged in icosahedral symmetry (T = 1, pseudo T = 3). The atomic structure of representative viruses of four of the five genera has been solved. Virion molecular weight (Mr) is 8–9 × 10^6; buoyant density is 1.33–1.45 g/cm^3 in CsCl, varying with the genus. The genome consists of a single molecule of linear, positive-sense, single-stranded RNA, 7.2–8.4 kb in size. The genome has a 5'-untranslated sequence of variable length, an open reading frame encoding the polyprotein precursor to the structural and nonstructural proteins, a 3'-short noncoding sequence, and a 3'-poly (A) tract of variable length. A small protein, VPg, is linked covalently to the 5'-terminus. Virions are constructed from 60 copies each of four capsid proteins (e.g., poliovirus VP2, VP3, VP1 with Mr 24,000–41,000, VP4 with Mr 5,500–13,500) and a single copy of the genome linked protein (VPg, Mr ~2,400). One or two of the nonstructural proteins have proteolytic activity, another nonstructural protein is an RNA-dependent RNA polymerase. RNA replication involves the synthesis of a complementary RNA that serves as a template for genome RNA synthesis. Genome RNA also serves as messenger RNA (mRNA); translation yields a polyprotein that is cleaved into all the viral proteins, including those proteins that serve as enzymes for specific cleavages. Replication and assembly take place in cytoplasm, and virus is released via cell destruction. Infection is generally acute and cytolytic, but persistent infections occur with some viruses. The viruses have narrow host ranges. Transmission is horizontal, mainly by contact, fecal/oral, or airborne routes. Some viruses are unstable below pH 7 and some are stabilized by divalent cations. Virions are insensitive to ether, chloroform, and nonionic detergents.

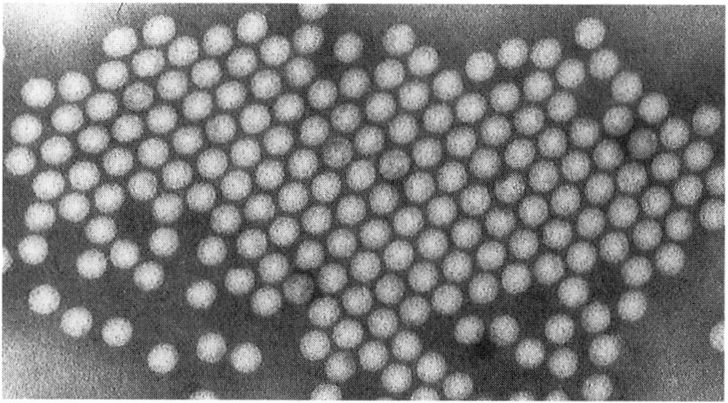

FIG. 1. Family *Picornaviridae*, genus *Enterovirus*, poliovirus 1 (×140,000). Micrographs provided courtesy of C. Goldsmith, A.K. Harrison, S.G. Whitfield, E.L. Palmer, M.L. Martin, L.D. Pearson, T. Baker, E.H. Cook, D. Bradley, C-H. von Bonsdorff, J. Esposito, C. Smale, M.V. Nermut, D. Hockley, C. Grief, C. Mebus, V.I. Heine, F.X. Heinz, D. Cubitt, R. Glass, S. Monroe, M. Weiss, M.C. Horzinek, S. Nakata, D.O. Matson, J.M. Taylor, J.L. Gerin, R. Purcell, H. Gelderblom, M. Szymanski, and M. Wurtz.

Human Pathogens

Genus *Enterovirus*: polioviruses 1, 2, 3; coxsackieviruses A1–22,[2] A24; coxsackieviruses B1–6; human echoviruses 1–7, 9, 11–27, 29–33; human enteroviruses 68–71. Genus *Rhinovirus*: human rhinoviruses 1–100, 1A, 1B, and Hanks. Genus *Hepatovirus*: hepatitis A virus.

Animal Pathogens

Genus *Enterovirus*: swine vesicular disease virus; Theiler's murine encephalomyelitis virus; porcine enteroviruses 1–8; bovine enteroviruses 1–7; simian enteroviruses 1–18. Genus *Cardiovirus*: encephalomyocarditis (EMC) virus. Genus *Rhinovirus*: bovine rhinoviruses 1–3; equine rhinoviruses 1 and 2. Genus *Aphthovirus*: foot-and-mouth disease viruses O, A, C, SAT 1–3, and ASIA 1. Genus *Hepatovirus*: simian hepatitis A virus.

The Family *Caliciviridae*

Family: *Caliciviridae* (the caliciviruses) (Fig. 2)
Genus: *Calicivirus* (caliciviruses)

Characteristics (21,66,87,118)

Virions are 30–38 nm in diameter with 32 cup-shaped surface depressions arranged in icosahedral symmetry; virions are composed of 180 protein molecules arranged in dimers. Virion Mr is ~15 × 10^6; buoyant density is 1.33 to 1.40 g/cm^3 in CsCl. The genome consists of a single molecule of linear, positive-sense, single-stranded RNA, 7.4–7.7 kb in size. 3'-coterminal subgenomic RNAs (2.2–2.4 kb)

[2]As a result of characterization done after number series were completed, there are now some gaps and shifts: coxsackievirus A23 = echovirus 9; echovirus 10 = reovirus 1; echovirus 28 = human rhinovirus 1A; human enterovirus 72 = hepatitis A virus; and echovirus 22 = a unique picornavirus that has not yet been classified.

TABLE 4. *Some pathogenic viruses of humans*

Virus[a]	Family	Genus
Adenovirus 1–49	*Adenoviridae*	*Mastadenovirus*
Astrovirus 1, 2	*Astroviridae*	*Astrovirus*
B virus (Cercopithecus herpesvirus)	*Herpesviridae*	*Simplexvirus*
BK virus	*Papovaviridae*	*Polyomavirus*
Bunyamwera virus	*Bunyaviridae*	*Bunyavirus*
California encephalitis virus	*Bunyaviridae*	*Bunyavirus*
Central European encephalitis virus	*Flaviviridae*	*Flavivirus*
Chikungunya virus	*Togaviridae*	*Alphavirus*
Colorado tick fever virus	*Reoviridae*	*Coltivirus*
Congo-Crimean hemorrhagic fever virus	*Bunyaviridae*	*Nairovirus*
Cowpox virus	*Poxviridae*	*Orthopoxvirus*
Coxsackieviruses A 1–21 and A 24	*Picornaviridae*	*Enterovirus*
Coxsackieviruses B 1–6	*Picornaviridae*	*Enterovirus*
Creutzfeldt-Jakob disease agent	Unclassified	Unnamed, prions
Dengue viruses 1–4	*Flaviviridae*	*Flavivirus*
Duvenhage virus	*Rhabdoviridae*	*Lyssavirus*
Eastern equine encephalitis virus	*Togaviridae*	*Alphavirus*
Ebola virus	*Filoviridae*	*Filovirus*
Echoviruses 1–9 and 11–27 and 29–34	*Picornaviridae*	*Enterovirus*
Enteroviruses 68–71	*Picornaviridae*	*Enterovirus*
Epstein-Barr virus (human herpesvirus 4)	*Herpesviridae*	*Lymphocryptovirus*
Hantaan virus	*Bunyaviridae*	*Hantavirus*
Hepatitis A virus	*Picornaviridae*	*Hepatovirus*
Hepatitis B virus	*Hepadnaviridae*	*Orthohepadnavirus*
Hepatitis C virus	*Flaviviridae*	Unnamed
Hepatitis delta virus		*Deltavirus*
Hepatitis E virus	Unclassified	Unclassified
Herpes simplex viruses 1 and 2 (human herpesviruses 1 and 2)	*Herpesviridae*	*Simplexvirus*
Human enteric coronavirus	*Coronaviridae*	*Coronavirus*
Human cytomegalovirus (human herpesvirus 5)	*Herpesviridae*	*Cytomegalovirus*
Human herpesviruses 6A, 6B, and 7	*Herpesviridae*	*Roseolovirus*
Human immunodeficiency viruses 1 and 2	*Retroviridae*	*Lentivirus*
Human respiratory coronaviruses 229E and OC43	*Coronaviridae*	*Coronavirus*
Human rotaviruses	*Reoviridae*	*Rotavirus*
Human T-lymphotropic viruses 1 and 2	*Retroviridae*	Unnamed, HTLV/BLV viruses
Influenza viruses A and B	*Orthomyxoviridae*	*Influenzavirus A, B*
Japanese encephalitis virus	*Flaviviridae*	*Flavivirus*
JC virus	*Papovaviridae*	*Polyomavirus*
Junin virus (Argentine hemorrhagic fever virus)	*Arenaviridae*	*Arenavirus*
Kuru agent	Unclassified	Unnamed, prions
Kyasanur forest virus	*Flaviviridae*	*Flavivirus*
La Crosse virus	*Bunyaviridae*	*Bunyavirus*
Lassa virus	*Arenaviridae*	*Arenavirus*
Lymphocytic choriomeningitis virus	*Arenaviridae*	*Arenavirus*
Machupo virus (Bolivian hemorrhagic fever virus)	*Arenaviridae*	*Arenavirus*
Marburg virus	*Filoviridae*	*Filovirus*
Mayaro virus	*Togaviridae*	*Alphavirus*
Measles virus	*Paramyxoviridae*	*Morbillivirus*
Mokola virus	*Rhabdoviridae*	*Lyssavirus*
Molluscum contagiosum virus	*Poxviridae*	*Moluscipoxvirus*
Monkeypox virus	*Poxviridae*	*Orthopoxvirus*
Muerto Canyon virus	*Bunyaviridae*	*Hantavirus*
Mumps virus	*Paramyxoviridae*	*Rubulavirus*
Murray Valley encephalitis virus	*Flaviviridae*	*Flavivirus*
Norwalk virus (and related viruses)	*Caliciviridae*	*Calicivirus*
O'nyong-nyong virus	*Togaviridae*	*Alphavirus*
Omsk hemorrhagic fever virus	*Flaviviridae*	*Flavivirus*
Orf virus (contagious pustular dermatitis virus)	*Poxviridae*	*Parapoxvirus*
Oropouche virus	*Bunyaviridae*	*Bunyavirus*
Papillomaviruses 1–60	*Papovaviridae*	*Papillomavirus*
Parainfluenza viruses 1 and 3	*Paramyxoviridae*	*Paramyxovirus*
Parainfluenza viruses 2 and 4	*Paramyxoviridae*	*Rubulavirus*
Parvovirus B-19	*Parvoviridae*	*Parvovirus*
Polioviruses 1–3	*Picornaviridae*	*Enterovirus*
Pseudocowpox virus (milker's nodule virus)	*Poxviridae*	*Parapoxvirus*
RA-1 virus	*Parvoviridae*	*Parvovirus*
Rabies virus	*Rhabdoviridae*	*Lyssavirus*
Respiratory syncytial virus	*Paramyxoviridae*	*Pneumovirus*
Rhinoviruses 1–113	*Picornaviridae*	*Rhinovirus*
Rift Valley fever virus	*Bunyaviridae*	*Phlebovirus*

TABLE 4. *Some pathogenic viruses of humans (continued)*

Virus[a]	Family	Genus
Rocio virus	*Flaviviridae*	*Flavivirus*
Ross River virus	*Togaviridae*	*Alphavirus*
Rubella virus	*Togaviridae*	*Rubivirus*
Russian spring-summer encephalitis virus	*Flaviviridae*	*Flavivirus*
Sandfly fever-Naples virus	*Bunyaviridae*	*Phlebovirus*
Sandfly fever-Sicilian virus	*Bunyaviridae*	*Phlebovirus*
St. Louis encephalitis virus	*Flaviviridae*	*Flavivirus*
SV 40 virus	*Papovaviridae*	*Polyomavirus*
Tahyna virus	*Bunyaviridae*	*Bunyavirus*
Vaccinia virus	*Poxviridae*	*Orthopoxvirus*
Varicella-zoster virus (human herpesvirus 3)	*Herpesviridae*	*Varicellovirus*
Variola virus	*Poxviridae*	*Orthopoxvirus*
Venezuelan equine encephalitis virus	*Togaviridae*	*Alphavirus*
Vesicular stomatitis viruses	*Rhabdoviridae*	*Stomatovirus*
West Nile virus	*Flaviviridae*	*Flavivirus*
Eastern equine encephalitis virus	*Togaviridae*	*Alphavirus*
Yellow fever virus	*Flaviviridae*	*Flavivirus*

[a]Arthropod-borne viruses are listed according to their vertebrate host.

are synthesized intracellularly and may also be encapsidated by some viruses (e.g., rabbit hemorrhagic disease virus). Nonstructural proteins (a putative helicase, a cysteine protease, and an RNA-dependent RNA polymerase) are encoded at the 5'-end of the RNA, structural proteins at the 3'-end. Virions are constructed from one major species of protein (Mr 59,000–71,000) whose N-terminus is usually blocked. A protein (VPg, Mr 10,000–15,000) is covalently attached to the 5'-end of the RNA of most viruses. A minor protein (Mr 28,000–30,000) has been found in some viruses. Replication and assembly take place in cytoplasm, and virus is released via cell destruction. The viruses have narrow host ranges. Virions are insensitive to ether, chloroform, and detergents. Some viruses are inactivated, others are enhanced by trypsin.

Human Pathogens

Genus *Calicivirus*: Norwalk and similar viruses, such as Southampton, Snow Mountain, Hawaii, and Taunton viruses, all of which are proven causes of gastroenteritis; hepatitis E virus.[3]

Animal Pathogens

Genus *Calicivirus:* vesicular exanthema viruses 1–12 of swine; San Miguel sea lion viruses 1–8; feline caliciviruses; rabbit hemorrhagic disease virus; European brown hare disease virus; canine calicivirus; bovine enteric calicivirus; porcine enteric calicivirus; chimpanzee calicivirus (Pan-1); mink calicivirus; reptile calicivirus; walrus calicivirus; and fowl calicivirus.

The Family *Astroviridae*

Family: *Astroviridae* (the astroviruses) (Fig. 3)
Genus: *Astrovirus* (astroviruses)

Characteristics (89,140)

Virions are 28–30 nm in diameter and spherical in shape. A distinctive five- or six-pointed star is discernible on the surface of some virions in some preparations. Virion Mr is ~8×10^6; buoyant density is 1.36–1.39 g/cm^3 in CsCl. The genome consists of a single molecule of linear, positive-sense, single-stranded RNA, 7.2–7.9 kb in size; it has a 3'-poly (A) tract of variable length. The precise protein composition of the viruses remains unclear; however, all viruses have at least two, possibly three, major proteins (Mr 29,000–39,000); several viruses also contain smaller proteins (Mr 13,000–36,000). Replication involves the synthesis of a 3'-0subgenomic RNA (2.8 kb in size), but has not otherwise been determined. The putative RNA-depen-

[3]Hepatitis E virus has been classified as a calicivirus, but evidence has been accumulating that it is so different that it may be placed in a separate taxon. Virions are 30–32 nm in diameter, exhibit icosahedral symmetry of unknown detail, and have surface features (indentations and projections) that are more defined than on picornaviruses and less than on caliciviruses. Virion buoyant density is 1.39–1.40 g/cm^3 in CsCl and 1.29 g/cm^3 in potassium tartrate/glycerol. The genome consists of a single molecule of linear, positive-sense, single-stranded RNA, 7.6 kb in size; it has a 3' poly (A) tract about 300 bases in length. The genome contains two (or three) noncontiguous large open reading frames that span nearly the whole genome. The 5' open reading frame, encoding nonstructural genes, spans 5 kb; the 3' open reading frame(s), encoding structural genes, spans 2 kb (deduced gene order is 5'-methyltransferase, protease, x-gene, helicase, polymerase, capsid-3'). This combination of putative nonstructural genes, the gene order, and the paucity of sequence similarities with picornaviruses, caliciviruses, or astroviruses are the basis for a proposal to place the virus in a separate taxon (74,126). Replication and assembly take place in cytoplasm, and virus is released via cell destruction. The virus naturally infects only humans; transmission is by the fecal-oral route and often involves water supplies. The virus is the cause of large epidemics of hepatitis, mostly in developing countries, in which there is a high mortality rate in women infected during pregnancy. Virions are rather insensitive to solvents and detergents and are labile to freeze-thawing and high concentrations of salts.

TABLE 5. *Some pathogenic viruses of animals*

Virus	Family	Genus
African horsesickness viruses 1–9	*Reoviridae*	*Orbivirus*
African swine fever virus	Unnamed	Unnamed
Aleutian mink disease virus	*Parvoviridae*	*Parvovirus*
Avian reticuloendotheliosis virus	*Retroviridae*	Unnamed, avian type C retroviruses
Avian sarcoma and leukosis viruses	*Retroviridae*	Unnamed, avian type C retroviruses
B virus (Cercopithecus herpesvirus)	*Herpesviridae*	*Simplexvirus*
Berne virus (horses)	*Coronaviridae*	*Torovirus*
Bluetongue viruses 1–25	*Reoviridae*	*Orbivirus*
Border disease virus (sheep)	*Togaviridae*	*Pestivirus*
Borna disease virus (horses)	Unclassified	Unclassified
Bovine enteroviruses 1–7	*Picornaviridae*	*Enterovirus*
Bovine ephemeral fever virus	*Rhabdoviridae*	*Ephemerovirus*
Bovine immunodeficiency virus	*Retroviridae*	*Lentivirus*
Bovine leukemia virus	*Retroviridae*	Unnamed HTLV/BLV viruses
Bovine mamillitis virus	*Herpesviridae*	*Simplexvirus*
Bovine papillomaviruses	*Papovaviridae*	*Papillomavirus*
Bovine papular stomatitis virus	*Poxviridae*	*Parapoxvirus*
Bovine respiratory syncytial virus	*Paramyxoviridae*	*Pneumovirus*
Bovine virus diarrhea virus	*Togaviridae*	*Pestivirus*
Breda virus (calves)	Coronaviridae	*Torovirus*
Canine adenovirus 2	*Adenoviridae*	*Mastadenovirus*
Canine distemper virus	*Paramyxoviridae*	*Morbillivirus*
Canine parvovirus	*Parvoviridae*	*Parvovirus*
Caprine arthritis-encephalitis virus	*Retroviridae*	*Lentivirus*
Cowpox virus	*Poxviridae*	*Orthopoxvirus*
Eastern equine encephalitis virus	*Togaviridae*	*Alphavirus*
Ebola virus	*Filoviridae*	*Filovirus*
Ectromelia virus (mousepox virus)	*Poxviridae*	*Orthopoxvirus*
Encephalomyocarditis virus	*Picornaviridae*	*Cardiovirus*
Epizootic hemorrhagic disease viruses (deer)	*Reoviridae*	*Orbivirus*
Equine abortion virus (EHV1)	*Herpesviridae*	*Varicellovirus*
Equine adenoviruses	*Adenoviridae*	*Mastadenovirus*
Equine arteritis virus		*Arterivirus*
Equine coital exanthema virus (EHV3)	*Herpesviridae*	*Varicellovirus*
Equine infectious anemia virus	*Retroviridae*	*Lentivirus*
Equine rhinopneumonitis virus (EHV4)	*Herpesviridae*	*Varicellovirus*
Feline calicivirus	*Caliciviridae*	*Calicivirus*
Feline immunodeficiency virus	*Retroviridae*	*Lentivirus*
Feline infectious peritonitis virus	*Coronaviridae*	*Coronavirus*
Feline panleukopenia virus	*Parvoviridae*	*Parvovirus*
Feline sarcoma and leukemia viruses	*Retroviridae*	Unnamed, mammalian type C retroviruses
Fibroma viruses of rabbits and hares and squirrels	*Poxviridae*	*Myxomavirus*
Foot-and-mouth disease viruses	*Picornaviridae*	*Aphthovirus*
Fowlpox virus	*Poxviridae*	*Avipoxvirus*
Hemagglutinating encephalomyelitis virus (swine)	*Coronaviridae*	*Coronavirus*
Hog cholera virus	*Togaviridae*	*Pestivirus*
Infectious bovine rhinotracheitis virus	*Herpesviridae*	*Simplexvirus*
Infectious bronchitis virus (fowl)	*Coronaviridae*	*Coronavirus*
Infectious bursal disease virus (fowl)	*Birnaviridae*	*Avibirnavirus*
Infectious canine hepatitis virus	*Adenoviridae*	*Mastadenovirus*
Infectious hematopoietic necrosis virus (fish)	*Rhabdoviridae*	Unclassified
Infectious laryngotracheitis virus (fowl)	*Herpesviridae*	Unclassified
Infectious pancreatic necrosis virus (fish)	*Birnaviridae*	*Aquabirnavirus*
Influenza viruses of swine, horses, seals, and fowl	*Orthomyxoviridae*	*Influenzavirus A, B*
Japanese encephalitis virus	*Flaviviridae*	*Flavivirus*
Lactic dehydrogenase virus (mice)		*Arterivirus*
Lymphocytic choriomeningitis virus	*Arenaviridae*	*Arenavirus*
Maedi/visna virus (sheep)	*Retroviridae*	*Lentivirus*
Marburg virus	*Filoviridae*	*Filovirus*

TABLE 5. *Some pathogenic viruses of animals (continued)*

Virus	Family	Genus
Marek's disease virus (fowl)	*Herpesviridae*	Unclassified
Mink enteritis virus	*Parvoviridae*	*Parvovirus*
Minute virus of mice	*Parvoviridae*	*Parvovirus*
Mouse hepatitis viruses	*Coronaviridae*	*Coronavirus*
Mouse mammary tumor virus	*Retroviridae*	Unnamed, mammalian type B retroviruses
Mouse poliomyelitis virus (Theiler's virus)	*Picornaviridae*	*Cardiovirus*
Mucosal disease virus (cattle)	*Togaviridae*	*Pestivirus*
Myxoma virus	*Poxviridae*	*Myxomavirus*
Nairobi sheep disease virus	*Bunyaviridae*	*Nairovirus*
Newcastle disease virus (fowl)	*Paramyxoviridae*	*Paramyxovirus*
Orf virus (contagious pustular dermatitis virus)	*Poxviridae*	*Parapoxvirus*
Parainfluenza virus 3	*Paramyxoviridae*	*Rubulavirus*
Parainfluenza virus 1 (Sendai virus)	*Paramyxoviridae*	*Paramyxovirus*
Peste-des-petits-ruminants virus (sheep and goats)	*Paramyxoviridae*	*Morbillivirus*
Pneumonia virus of mice	*Paramyxoviridae*	*Pneumovirus*
Progressive pneumonia virus of sheep	*Retroviridae*	*Lentivirus*
Pseudocowpox virus (milker's nodule virus)	*Poxviridae*	*Parapoxvirus*
Pseudorabies virus	*Herpesviridae*	*Varicellovirus*
Rabbit hemorrhagic disease virus	*Caliciviridae*	*Calicivirus*
Rabies virus	*Rhabdoviridae*	*Lyssavirus*
Reoviruses 1–3	*Reoviridae*	*Orthoreovirus*
Rift Valley fever virus	*Bunyaviridae*	*Phlebovirus*
Rinderpest virus	*Paramyxoviridae*	*Morbillivirus*
Rotaviruses of many species	*Reoviridae*	*Rotavirus*
Scrapie agent (sheep and goats)	Unclassified	Unnamed, prions
Sheeppox virus	*Poxviridae*	*Sheeppoxvirus*
Shope papillomavirus	*Papovaviridae*	*Papillomavirus*
Simian immunodeficiency viruses	*Retroviridae*	*Lentivirus*
Swine vesicular disease virus	*Picornaviridae*	*Enterovirus*
Swinepox virus	*Poxviridae*	*Swinepox*
Tick-borne encephalitis viruses	*Flaviviridae*	*Flavivirus*
Transmissible gastroenteritis virus (swine)	*Coronaviridae*	*Coronavirus*
Turkey bluecomb virus	*Coronaviridae*	*Coronavirus*
Venezuelan equine encephalitis virus	*Togaviridae*	*Alphavirus*
Vesicular exanthema virus (swine)	*Caliciviridae*	*Calicivirus*
Vesicular stomatitis viruses	*Rhabdoviridae*	*Stomatovirus*
Wasting disease of deer and elk	Unclassified	Unnamed, prions
Wesselsbron virus	*Flaviviridae*	*Flavivirus*
Western equine encephalitis virus	*Togaviridae*	*Alphavirus*

dent RNA polymerase seems to be encoded in a different reading frame from other proteins (putative helicase and protease), implying that translation involves ribosomal frameshifting. Mature virions are often seen in crystalline arrays in the cytoplasm of infected cells. The viruses have narrow host ranges; transmission is by the fecal/oral route. The viruses are distributed worldwide. Virions are rather thermostable and are resistant to pH 3, ether, chloroform, lipid solvents, and nonionic, anionic, and zwitterionic detergents. Tryspin is required in cell culture growth medium for serial propagation of several of the viruses.

Human Pathogens

Genus *Astrovirus*: human astroviruses 1–5.

Animal Pathogens

Genus *Astrovirus*: bovine astroviruses 1 and 2; ovine astrovirus; porcine astrovirus; canine astrovirus; and duck astrovirus.

The Family *Togaviridae*

Family: *Togaviridae* (the togaviruses) (Fig. 4)
Genus: *Alphavirus* (alphaviruses)
Genus: *Rubivirus* (rubella virus)

Characteristics (35,68,88,103)

Virions are 70 nm in diameter and spherical, with a lipid envelope and peplomers composed of a heterodimer of two

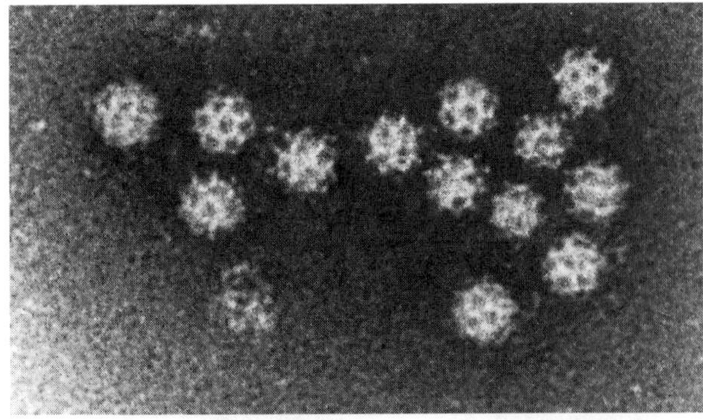

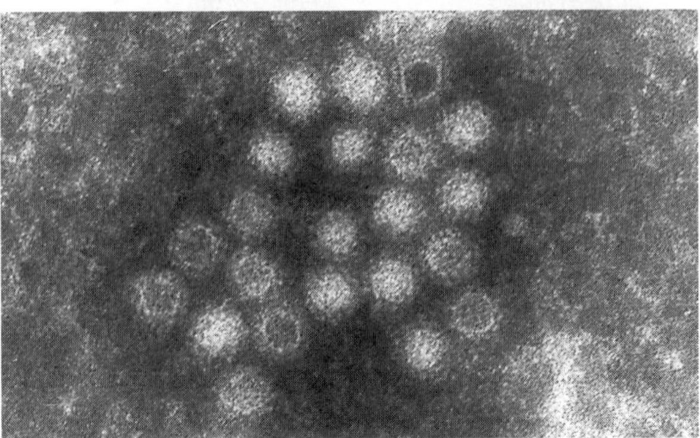

FIG. 2. Family *Caliciviridae*, genus *Calicivirus,* human calicivirus, Sapporo strain (top ×144,0000); Norwalk virus (bottom ×140,000).

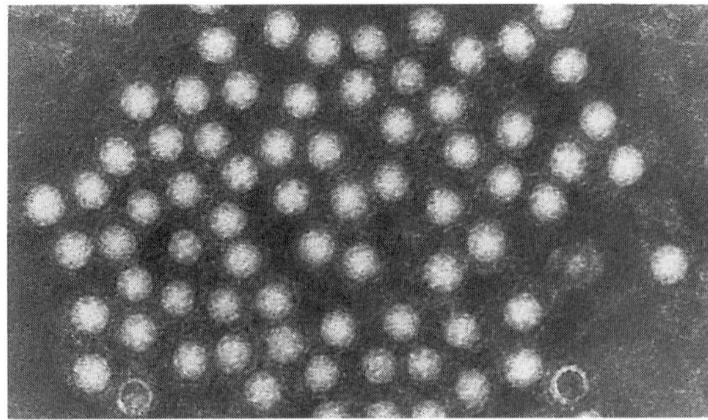

FIG. 3. Family *Astroviridae*, genus *Astrovirus,* human astrovirus 2 (×120,000).

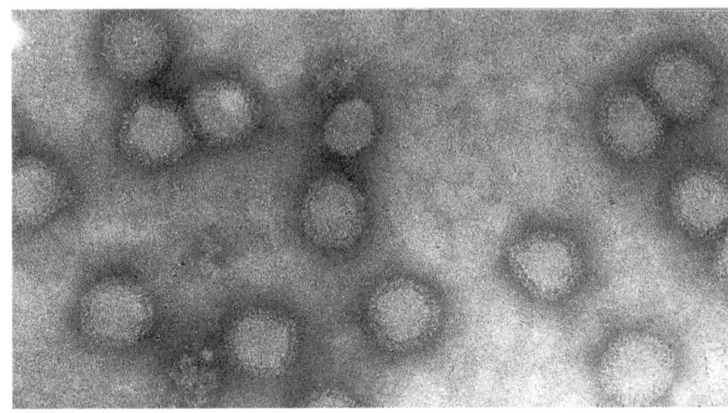

FIG. 4. Family *Togaviridae*, genus *Rubivirus,* rubella virus (×105,000).

glycoproteins. The peplomers are placed in a T = 4 icosahedral lattice consisting of 80 trimers (alphaviruses). The surface of rubella virions appears pleomorphic. The envelope is tightly organized around an icosahedral nucleocapsid that is 40 nm in diameter and exhibits T = 4 icosahedral symmetry. Virion Mr is ~52 × 10^6; buoyant density is 1.20 g/cm^3 (alphaviruses) or 1.18–1.19 g/cm^3 (rubella virus) in sucrose. The genome consists of a single molecule of linear, positive-sense, single-stranded RNA, 9.7–11.8 kb in size. The RNA is capped at the 5'-terminus and polyadenylated at the 3'-end. The structural proteins include a basic capsid protein (C, Mr 30,000–33,000) and two envelope glycoproteins (E1 and E2, Mr 45,000–58,000). Some alphaviruses have a third envelope protein, E3 (Mr 10,000). Virions contain lipids that are derived from host cell membranes. Virion glycoproteins contain high mannose and complex N-linked glycans on the envelope glycoproteins (rubella virus E2 protein also contains O-linked glycans). RNA replication involves the synthesis of complementary RNA, which serves as a template for genomic RNA synthesis. A subgenomic mRNA (26*S* in alphaviruses and 24*S* in rubella virus) is synthesized in excess, from which a viral polyprotein is translated and cleaved into the structural proteins (gene order in the 26*S* mRNA is C-E3-E2-6K-E1). Genome RNA serves as a template for mRNA for the synthesis of a polyprotein precursor for the nonstructural proteins; the polyprotein is cleaved into four final products (nsp1, 2, 3, 4). Replication takes place in cytoplasm, and assembly involves budding through host cell membranes. The infection of vertebrate cells by alphaviruses is cytolytic and involves the shutdown of host cell macromolecular synthesis. In mosquito cells, alphaviruses usually establish a noncytolytic persistent infection. Alphaviruses are transmitted between vertebrates by mosquitoes and certain other hematophagous arthropods; they have wide host ranges and worldwide distribution. Rubella virus occurs worldwide but infects only humans and is transmitted directly by contact and aerosols. The viruses are stable at pH 7–8, but are rapidly inactivated at acidic pH; the viruses are thermolabile and sensitive to organic solvents and detergents. Most of the viruses exhibit pH-dependent hemagglutinating activity.

Human Pathogens

Genus *Alphavirus*: eastern equine encephalitis virus; western equine encephalitis virus; Venezuelan equine encephalitis virus; Sindbis virus (and variants Ockelbo and Babanki viruses); chikungunya virus; o'nyong-nyong virus; Igbo Ora virus; Ross River virus; Mayaro virus; and Barmah Forest virus. Genus *Rubivirus*: rubella virus.

Animal Pathogens

Genus *Alphavirus*: eastern equine encephalitis virus; western equine encephalitis virus; Venezuelan equine encephalitis virus; and Getah virus.

The Family *Flaviviridae*

Family: *Flaviviridae* (the flaviviruses) (Fig. 5)
Genus: *Flavivirus* (flaviviruses)
Genus: *Pestivirus* (mucosal disease viruses)
Genus: (unnamed, hepatitis C virus)

Characteristics (19,23,26,136)

Virions are 45–60 nm in diameter and spherical, with a lipid envelope and fine peplomers that do not show any structure or symmetrical placement. The viral core is spherical, but its structure is unknown. Virion Mr is ~60 $\times$ 10^6 (flaviviruses); buoyant density is 1.20–1.23 g/cm^3 in sucrose. The genome consists of a single molecule of linear, positive-sense, single-stranded RNA, 10.7 kb (flaviviruses), 12.5 kb (pestiviruses), or 9.5 kb (hepatitis C virus) in size. The 5'-end structure of the RNA of flaviviruses consists of a type 1 cap. Except for some tick-borne flaviviruses, the RNA does not contain a 3'-terminal poly (A) tract. Virions contain two or three membrane-associated proteins and a core protein. Virions contain lipids that are derived from host cell membranes. Virions contain carbohydrates in the form of glycolipids and glycoproteins. RNA replication involves the synthesis of a complementary RNA, which serves as template for genomic RNA synthesis. A single, long, open reading frame encodes a polyprotein that is proteolytically cleaved into all the viral proteins. The structural proteins are encoded at the 5'-end, the nonstructural proteins including a protease, helicase, and polymerase, at the 3'-end of the RNA (gene order in flaviviruses is 5'-C-preM-E-NS1-NS2A-NS2B-NS3-NS4A-NS4B-NS5-3'). Replication takes place in cytoplasm, and assembly involves passage through and envelopment by internal host cell membranes (true budding, i.e., progressive membrane deformation, is not seen). Replication commonly is accompanied by a characteristic proliferation of intracellular membranes. The infection of cells of vertebrate origin by flaviviruses is cytolytic. Flaviviruses are transmitted between vertebrates by mosquitoes and ticks; they may be transmitted transovarially (mosquitoes, ticks) and transstadially (ticks) (no arthropod vector is involved in the transmission of some bat-associated flaviviruses). Some viruses have a limited vertebrate host range, others have a wide host range and worldwide distribution. Pestiviruses infect only certain animals and are transmitted by direct and indirect contact (e.g., fecally contaminated food, urine, or nasal secretions, etc.); all pestiviruses are also transmitted transplacentally and congenitally. Hepatitis C virus infects only humans and is transmitted by close contact, sexual contact, and blood transfusion. The viruses are stable at pH 7–8, but are inactivated at acidic pH; the viruses are thermolabile and sensitive to organic solvents and detergents. Many of the flaviviruses exhibit pH-dependent hemagglutinating activity.

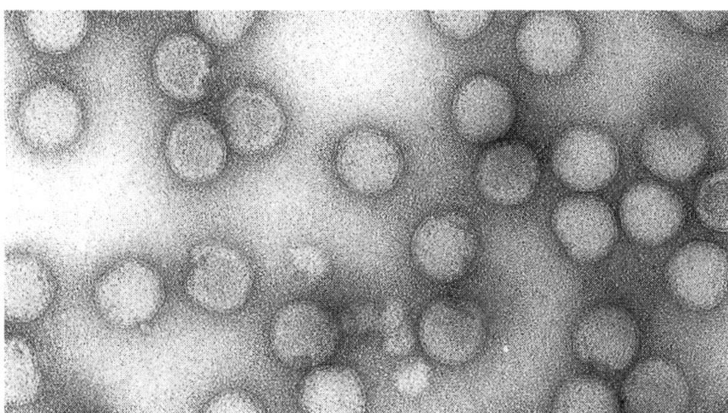

FIG. 5. Family *Flaviviridae*, genus *Flavivirus*, European tick-borne encephalitis virus (×126,000).

Human Pathogens

Genus *Flavivirus*: yellow fever virus; dengue viruses 1–4; West Nile virus; St. Louis encephalitis virus; Japanese encephalitis virus; Murray Valley encephalitis virus; Rocio virus; tick-borne encephalitis viruses (including European and far eastern tick-borne encephalitis viruses, Russian Spring-Summer encephalitis virus, Kyasanur forest disease virus, Omsk hemorrhagic fever virus, louping ill virus, and Powassan virus); and others. Genus (unnamed): hepatitis C virus.

Animal Pathogens

Genus *Flavivirus:* yellow fever virus; Japanese encephalitis virus; tick-borne encephalitis viruses; Wesselsbron virus; louping ill virus; and Israel turkey meningoencephalitis virus. Genus *Pestivirus*: bovine virus diarrhea virus (mucosal disease virus); hog cholera virus; and border disease virus of sheep.

The Family *Coronaviridae*

Family: *Coronaviridae* (the coronaviruses) (Fig. 6)
Genus: *Coronavirus* (coronaviruses)
Genus: *Torovirus* (toroviruses)

Characteristics (22,76,120,122)

Virions are 80–220 nm in diameter and pleomorphic but roughly spherical in shape (coronaviruses) or 120–140 nm in diameter and disk, kidney, or rod shaped (toroviruses).

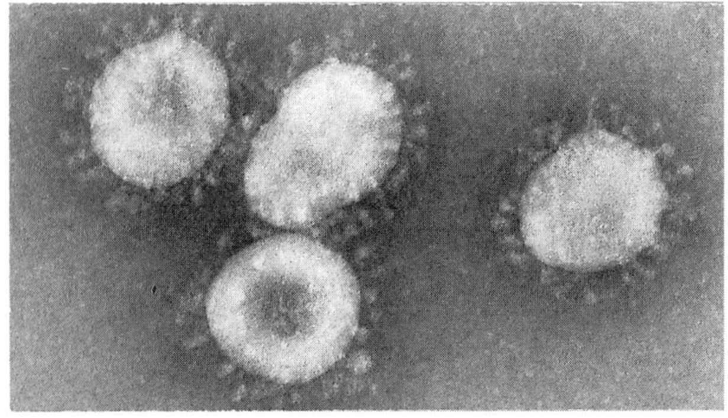

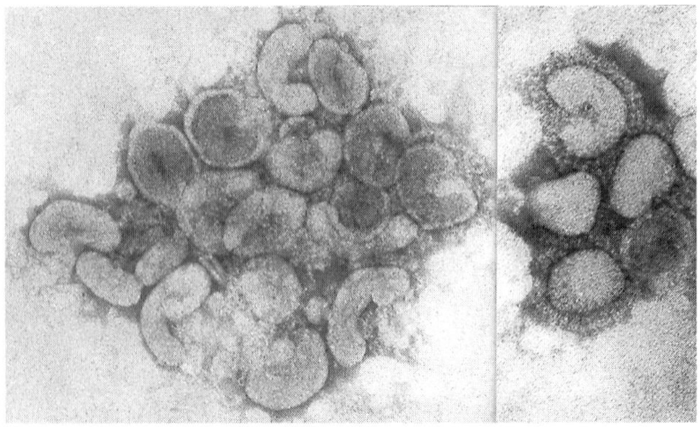

FIG. 6. Family *Coronaviridae*, genus *Coronavirus,* human coronavirus OC 43 (top ×140,000); genus *Torovirus,* Berne virus (bottom; composite ×93,600).

Virions are enveloped and have large club-shaped peplomers about 20 nm in length. The nucleocapsid is helical with a diameter of 10–20 nm (coronaviruses) or tightly coiled and doughnut shaped (toroviruses). Virion Mr is ~400×10^6; buoyant density is 1.15–1.19 g/cm^3 in sucrose and 1.23–1.24 g/cm^3 in CsCl (coronaviruses). The genome consists of a single molecule of linear, positive-sense, single-stranded RNA, 20–30 kb in size. The RNA has a 5'-terminal cap and a 3'-terminal poly (A) tract. Virions contain a large surface glycoprotein (S, Mr 180,000–220,000), an intermembrane protein (M, Mr 27,000–35,000), and a nucleocapsid protein (N, Mr 19,000–60,000). In addition, some coronaviruses contain a hemagglutinin-esterase protein (HE, Mr 65,000) that forms short surface projections and a small membrane protein (SM, Mr 10,000–12,000). Virions contain lipid in their envelopes, and some of their surface proteins are heavily glycosylated. The genomic RNA serves as the mRNA for the RNA polymerases (Pol 1a, Mr 440,000–500,000; Pol 1b, Mr 300,000–308,000); when translated, the polymerase components are responsible for the formation of full-length complementary and progeny RNA species and for the production of subgenomic mRNAs. One species of genome length complementary RNA acts as a template for the synthesis of a 3'-coterminal nested set of subgenomic mRNAs that are 5'-capped and 3'-polyadenylated. Synthesis of mRNA species from this template involves a process of discontinuous transcription, probably by a leader-priming mechanism. The viruses synthesize five to seven major subgenomic mRNAs (varies with the virus). Only the 5'-unique regions of the mRNAs are thought to be translationally active. Virions mature in the cytoplasm by budding through the endoplasmic reticulum and Golgi membranes. The viruses have narrow host ranges. Aerosol, fecal/oral, and fomite transmission are common. Virions are sensitive to heat, lipid solvents, nonionic detergents, formaldehyde, and oxidizing agents. Some viruses are stable at pH 3.0.

Human Pathogens

Genus *Coronavirus*: human coronaviruses 229-E and OC43 and others (causing the common cold, upper respiratory tract infection, probably pneumonia, and possibly gastroenteritis). Genus *Torovirus*: recent evidence indicates that toroviruses infect humans and cause enteric and respiratory disease.

Animal Pathogens

Genus *Coronavirus:* infectious bronchitis virus of fowl; turkey bluecomb virus; transmissible gastroenteritis virus of swine; hemagglutinating encephalomyelitis virus of swine; porcine epidemic diarrhea virus; calf coronavirus; feline infectious peritonitis virus; feline enteric coronavirus; canine coronavirus; mouse hepatitis viruses; rat coronavirus (sialodacryoadenitis virus); and rabbit coronavirus. Genus *Torovirus:* Berne virus (horses); Breda virus (calves); bovine respiratory torovirus; porcine torovirus; and feline torovirus.

The Floating Genus *Arterivirus*

Genus: *Arterivirus* (arteriviruses) (Fig. 7)

Characteristics (25,30)

Virions are 50–70 nm in diameter and consist of an isometric nucleocapsid ~35 nm in diameter surrounded by a lipid envelope decorated with 12- to 15-nm ringlike surface structures. Virion buoyant density is 1.13–1.17 g/cm^3 in sucrose and 1.17–1.20 g/cm^3 in CsCl. The genome consists of a single molecule of linear, positive-sense, single-stranded RNA, 15 kb in size. The RNA has a 5'-terminal cap and a 3'-terminal poly (A) tract. Virions contain a nucleocapsid protein (N, Mr 14,000), a nonglycosylated membrane-spanning protein (M, Mr 16,000), and several glycosylated surface proteins (GS, Mr 25,000; GL, Mr 30,000–42,000). The virions contain lipid in their envelopes and carbohydrates as part of their glycoproteins. The genomic RNA serves as a template for the synthesis of progeny RNA via a complementary-sense intermediate. Genome length negative-sense RNA acts as a template for

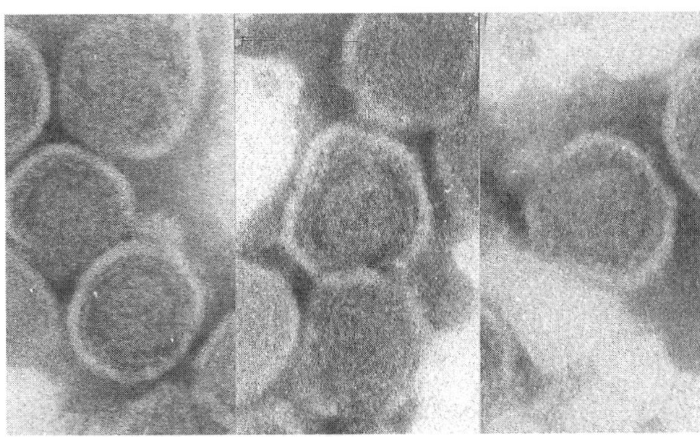

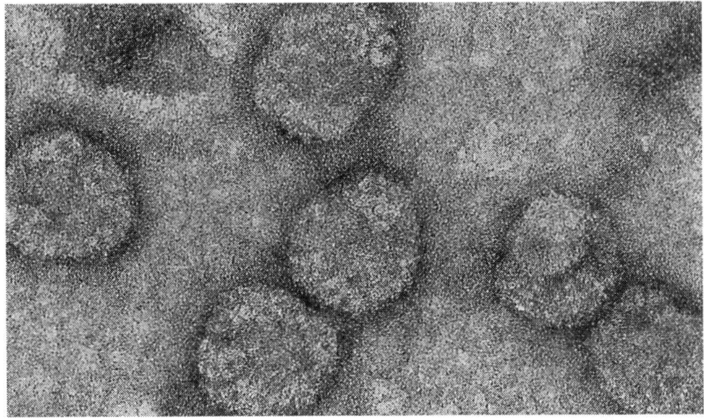

FIG. 7. Genus *Arterivirus*, lactate dehydrogenase-elevating virus (top; virions penetrated by stain; composite ×152,000); equine arteritis virus (bottom; virion surface ×152,000).

the synthesis of a 3'-coterminal nested set of six subgenomic mRNAs. Synthesis of mRNA species involves a process of discontinuous transcription, probably by a leader-priming mechanism, and, like the coronaviruses, frameshifting in the transcription of the polymerase gene. Primary host cells are macrophages. Persistent infections are established regularly. Transmission is by contact (including sexual contact), fomites, and aerosol.

Human Pathogens

Genus *Arterivirus*: none known.

Animal Pathogens

Genus *Arterivirus*: equine arteritis virus; lactate dehydrogenase-elevating virus (mice); simian hemorrhagic fever virus; Lelystad virus (porcine reproductive and respiratory syndrome virus); and VR2332 virus (swine).

The Family *Paramyxoviridae*

Order: *Mononegavirales* (the nonsegmented, negative-sense, single-stranded RNA viruses)
Family: *Paramyxoviridae* (the paramyxoviruses) (Fig. 8)
Subfamily: *Paramyxovirinae* (the paramyxo- and morbilliviruses)
Genus: *Paramyxovirus* (paramyxoviruses)
Genus: *Rubulavirus* (rubulaviruses)
Genus: *Morbillivirus* (morbilliviruses)
Subfamily: *Pneumovirinae* (the pneumoviruses)
Genus: *Pneumovirus* (pneumoviruses)

Characteristics (71,72,105)

Virions are 150–300 nm in diameter and pleomorphic, but usually spherical in shape (filamentous forms also occur), and consist of a lipid-containing envelope, with large peplomers 8–12 nm in length, spaced 7–10 nm apart, within which is coiled a helical nucleocapsid. The nucleocapsid is 13–18 nm in diameter, up to 1,000 nm in length, with a 5.5- to 7-nm pitch (varying with the genus). Virion Mr is >500 × 10^6, and much more for multiploid virions; buoyant density is 1.18–1.20 g/cm^3 in sucrose. The genome consists of a single molecule of linear, negative-sense, single-stranded RNA 16–20 kb in size. Intracellularly, genome-size RNA occurs only as nucleocapsids. Virions contain 10–12 proteins (Mr 5,000–250,000); proteins common to all genera include three nucleocapsid-associated proteins (N or NP, an RNA-binding protein; P, a phosphoprotein; and L, a large putative polymerase protein), an unglycosylated envelope protein (M), and two glycosylated envelope proteins, comprising a fusion protein (F) and an attachment protein (G, H, or HN). The surface glycoproteins of some viruses have neuraminidase activity (e.g., genus *Paramyxovirus*), hemagglutinating activity (genera *Paramyxovirus* and *Morbillivirus*), and fusion activity (all genera). Hemagglutinating and fusion activities are based in separate surface proteins (HN or HA, and F proteins). Virions contain lipids that are derived from the host cell plasma membrane. Virions contain carbohydrates as N-linked and O-linked glycans on surface proteins. RNA replication first involves mRNA transcription from the genomic RNA via the virion transcriptase; later, using the protein products of this transcription there is production of full-length positive-stranded template, which in turn is used for the synthesis of genomic RNA. The genome is transcribed

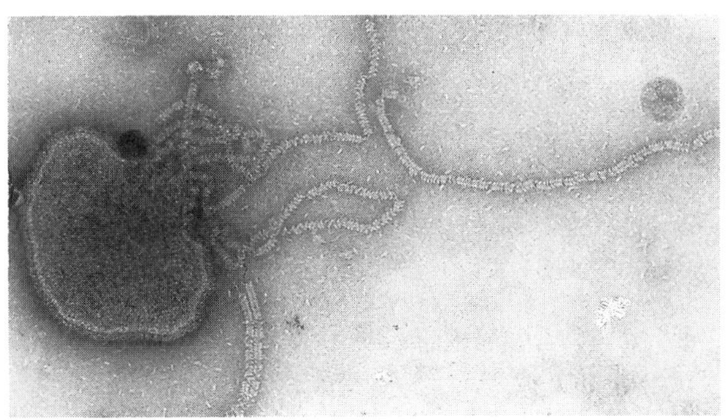

FIG. 8. Family *Paramyxoviridae*, genus *Paramyxovirus*, human parainfluenza virus 1 (×56,000).

processively from the 3'-end by virion-associated enzymes into six to 10 separate, subgenomic, viral-complementary mRNAs. The mRNAs are capped and have a 3'-poly (A) tract. Replication takes place in cytoplasm, and assembly occurs via budding on plasma membranes. The viruses have narrow host ranges; they have only been conclusively identified in vertebrates, mostly in mammals and birds. Transmission is horizontal mainly by aerosols and droplets. Morbilliviruses may cause persistent infections. Virions are sensitive to heat, lipid solvents, nonionic detergents, formaldehyde, and oxidizing agents.

Human Pathogens

Genus *Paramyxovirus*: human parainfluenza viruses 1 and 3. Genus *Rubulavirus*: mumps virus; human parainfluenza viruses 2, 4a, and 4b. Genus *Morbillivirus*: measles virus. Genus *Pneumovirus*: human respiratory syncytial virus.

Animal Pathogens

Genus *Paramyxovirus*: bovine parainfluenza virus 3; mouse parainfluenza virus 1 (Sendai virus); and simian parainfluenza virus 10. Genus *Rubulavirus*: Newcastle disease virus (avian paramyxovirus 1); avian paramyxovirus 2 (Yucaipa virus); avian paramyxoviruses 3, 4, 5 (Kunitachi virus), 6, 7, 8, and 9; porcine rubulavirus (la-Piedad-Michoacan-Mexico virus); and simian parainfluenza viruses 5 and 41. Genus *Morbillivirus*: canine distemper virus; rinderpest virus; peste-des-petits-ruminants virus (goats and sheep); equine morbillivirus (Australia), dolphin distemper virus; porpoise distemper virus; and phocine (seal) distemper virus. Genus *Pneumovirus*: bovine respiratory syncytial virus; pneumonia virus of mice.

The Family *Rhabdoviridae*

Order: *Mononegavirales* (the nonsegmented, negative-sense, single-stranded RNA viruses)
Family: *Rhabdoviridae* (the rhabdoviruses) (Fig. 9)
Genus: *Vesiculovirus* (vesiculoviruses)
Genus: *Lyssavirus* (lyssaviruses)
Genus: *Ephemerovirus* (ephemeroviruses)
Genus: *Cytorhabdovirus* (plant rhabdovirus group A)
Genus: *Nucleorhabdovirus* (plant rhabdovirus group B)

Characteristics (15,19,49,141)

Virions are bullet shaped (plant viruses are often bacilliform), 70–85 nm in diameter and 130–380 nm long (the most common length is 180 nm). Virions consist of an envelope, with large peplomers (8–10 nm long, 3 nm in diameter, consisting of trimers of the virus glycoprotein), within which is a helically coiled cylindrical nucleocapsid

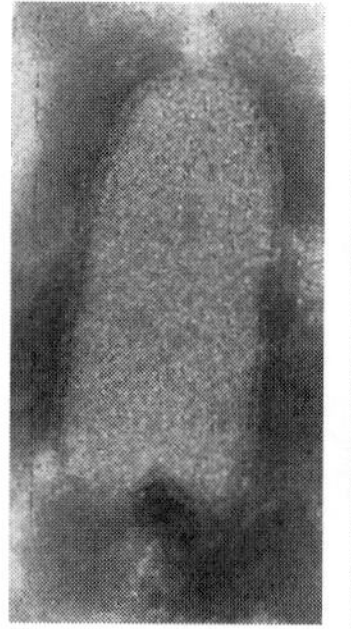
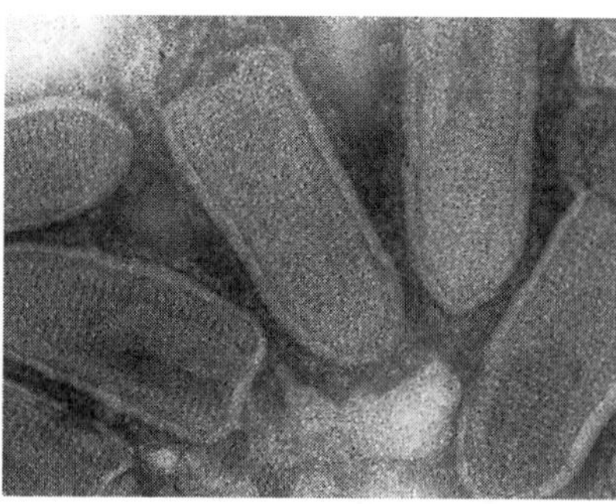

FIG. 9. Family *Rhabdoviridae*, genus *Lyssavirus*, rabies virus (left); genus *Vesiculovirus*, vesicular stomatitis virus-Indiana (right) (×156,000).

(spacing 4.5–5 nm). Virion Mr is 300–1,000 × 10^6; buoyant density is 1.19–1.20 g/cm^3 in CsCl and 1.17–1.19 g/cm^3 in sucrose. The genome consists of a single molecule of linear, negative-sense, single-stranded RNA, 13–16 kb in size. The RNA has a 5'-terminal triphosphate and is not polyadenylated; its ends have inverted complementary sequences. The viruses generally have five proteins: L, Mr 220,000–240,000, the RNA-dependent RNA polymerase that functions in transcription, replication, mRNA 5'-capping, and 3'-poly (A) tract formation; G, Mr 65,000–90,000, the glycoprotein that forms the peplomers; N, Mr 47,000–62,000, the major component of the viral nucleocapsid; NS (or P or M1 for rabies virus), Mr 20,000–30,000, a component of the viral polymerase; and M (or M2 for rabies virus), Mr 20,000–30,000, a basic protein that facilitates budding by binding to the nucleocapsid and to the cytoplasmic domain of the glycoprotein. Virions contain lipids, their composition reflecting the composition of host cell membranes. Carbohydrates are present as N-linked glycan chains on the glycoprotein and as glycolipids. Replication first involves mRNA transcription from the genomic RNA via the virion polymerase; later, using the protein products of this transcription, there is production of full-length, positive-stranded templates, which in turn are used for the synthesis of genomic RNA. The genome contains at least five open reading frames in the order 3'-N-NS-M-G-L-5'; some viruses have additional genes interposed. Genes are transcribed as 5'-capped, 3'-polyadenylated and generally monocistronic mRNAs. Replication takes place in cytoplasm, and assembly occurs via budding upon plasma (vesiculoviruses) or intracytoplasmic (lyssaviruses) membranes. Rabies virus produces prominent cytoplasmic inclusion bodies (Negri bodies) in infected cells. Certain plant viruses replicate in the nucleus. The viruses have broad host ranges; many replicate in and are transmitted by arthropods, others infect fish, plants, and invertebrates. Rabies virus is transmitted between mammals by bite. Virus infectivity is stable in the range pH 5–10, but viruses are thermolabile and sensitive to ultraviolet (UV) and X-irradiation, lipid solvents, and detergents. Some viruses exhibit hemagglutinating activity.

Human Pathogens

Genus *Vesiculovirus*: vesicular stomatitis–New Jersey virus; vesicular stomatitis–Indiana virus (VSV-I type 1), Cocal virus (VSV-I type 2); Alagoas virus (VSV-I type 3); Maraba virus (VSV-I type 4); Chandipura virus; Piry virus; and Isfahan virus. Genus *Lyssavirus*: rabies virus; European bat viruses 1 and 2; Mokola virus; Duvenhage virus; and Lagos bat virus.

Animal Pathogens

Genus *Vesiculovirus*: vesicular stomatitis–New Jersey virus; vesicular stomatitis–Indiana virus (VSV-I type 1), Cocal virus (VSV-I type 2); Alagoas virus (VSV-I type 3); Maraba virus (VSV-I type 4); Chandipura virus; Piry virus; Isfahan virus; pike fry rhabdovirus (grass carp rhabdovirus); and spring viremia of carp virus. Genus *Lyssavirus*: rabies virus; European bat viruses 1 and 2; Mokola virus; Duvenhage virus; Lagos bat virus; Kotonkan virus; and Obodhiang virus. Genus *Ephemerovirus*: bovine ephemeral fever virus. Ungrouped: viral hemorrhagic septicemia virus of salmon (Egtved virus) and infectious hematopoietic necrosis virus of fish.

The Family *Filoviridae*

Order: *Mononegavirales* (the nonsegmented, negative-sense, single-stranded RNA viruses)
Family: *Filoviridae* (the filoviruses) (Fig. 10)
Genus: *Filovirus* (filoviruses)

Characteristics (42,69,90,117)

Virions are enveloped and pleomorphic and appear as long filamentous forms, sometimes with extensive branching and sometimes as U-shaped, 6-shaped, or circular forms. Virions have a uniform diameter of 80 nm and vary greatly in length (up to 14,000 nm); virions recovered by gradient centrifugation are generally bacilliform and uniform in length (Ebola virus, ~1,000 nm; Marburg virus, ~800 nm). Virions consist of a lipid-containing envelope, with large peplomers (7–10 nm in length, spaced at 10 nm intervals), surrounding a rigid helically symmetrical nucleocapsid (50 nm in diameter) with cross-striations (periodicity 5 nm). Virion Mr is >500 $\times$ 10^6; buoyant density is 1.14 g/cm^3 in potassium tartrate. The genome consists of a single molecule of linear, negative-sense, single-stranded RNA, 19.1 kb in size, with complementary end sequences. Virions contain seven major proteins (sizes for Marburg virus): L, Mr 180,000, the RNA-dependent RNA polymerase; GP, Mr 170,000, the peplomer glycoprotein (peplomers are formed as trimers); NP, Mr 96,000, the nucleoprotein; M, Mr 38,000, the matrix or membrane-associated protein; P, Mr 32,000, possibly a polymerase component; an Mr 28,000 minor nucleoprotein; and an Mr 24,000 second matrix or membrane-associated protein. Except for the structural proteins, no other viral proteins have been detected in infected cells. Virions contain lipids, their composition reflecting the composition of host cell membranes. Virions contain carbohydrates as N-linked glycans of the complex, hybrid, and oligomannosidic type and O-linked glycans of the neutral mucin type. The glycans constitute ~50% of the mass of the glycoprotein. Replication occurs via synthesis of a complete complementary RNA species. RNAs are transcribed by sequential interrupted synthesis; at the gene boundaries there are conserved transcriptional stop and start signals. In addition, there are relatively long 3'- and 5'-noncoding regions. Replication takes place in the cytoplasm, and assembly involves envelopment via budding of preformed nucleocapsids. Nucleocapsids accumulate in the cytoplasm, forming prominent inclusion bodies. The natural history of the viruses remains unknown. The viruses are sensitive to lipid solvents, detergents, hypochlorite and phenolic disinfectants, and UV- and -irradiation. The viruses are Biosafety Level 4 pathogens; they must be handled in the laboratory under maximum containment conditions.

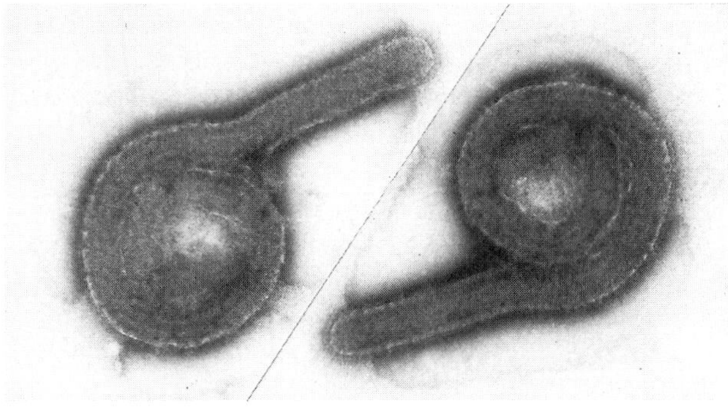

FIG. 10. Family *Filoviridae*, genus *Filovirus*, Ebola virus (composite) ($\times$46,200).

Human Pathogens

Genus *Filovirus*: Marburg virus; Ebola virus, subtype Zaire; and Ebola virus, subtype Sudan.

Animal Pathogens

Genus *Filovirus*: Marburg virus; Ebola virus, subtype Zaire; Ebola virus, subtype Sudan; and Ebola virus, subtype Reston (Marburg and Ebola virus, subtype Reston have been the cause of natural infections in monkeys; all of the viruses have elicited experimental disease in monkeys, guinea pigs, hamsters, and mice).

The Family *Orthomyxoviridae*

Family: *Orthomyxoviridae* (the influenza viruses) (Fig. 11)
Genus: *Influenzavirus A, B* (influenza A and B viruses)
Genus: *Influenzavirus C* (influenza C virus)
Genus: (unnamed, Thogoto-like viruses)

Characteristics (37,77)

Virions are pleomorphic, often spherical, and 80–120 nm in diameter. Filamentous forms several micrometers in length also occur. Virions consist of a lipid-containing envelope, with large peplomers, 10–14 nm in length and 4–6 nm in diameter, representing trimeric hemagglutinin and tetrameric neuraminidase structures. Within the envelope there are helically symmetrical nucleocapsids of different size classes, 150–130 nm in length with a loop at one end. Virion Mr is 250×10^6; buoyant density is 1.19 g/cm^3 in sucrose. The genome consists of eight (influenza viruses A and B), seven (influenza virus C and Dhori virus), or six (Thogoto virus) molecules of linear, negative-sense, single-stranded RNA, 10–13.6 kb in overall size. Segment lengths range from 900 to 2,350 nucleotides. Genome segments have conserved and partially complementary 5'- and 3'-end sequences. Virions contain seven to nine major proteins; structural proteins common to all genera include three polymerase proteins (PA, PB1, and PB2 in influenza A virus), a nucleocapsid protein (NP), a hemagglutinin (HA), and a nonglycosylated membrane or matrix protein (M1 or M). In addition, depending on the genus, viruses may encode two nonstructural proteins (NS1, NS2). Virion enzymes include a transcriptase (PB1 in influenza A virus), an endonuclease (PB2 in influenza A virus), and a receptor-destroying enzyme (neuraminidase in influenza A virus, HEF protein in influenza C virus). Virions contain lipids, reflecting the composition of the host cell plasma membrane. Virions contain carbohydrates in the form of glycoproteins and glycolipids (N-glycosidic side chains of glycoproteins, glycolipids, and mucopolysaccharides); their composition is host and virus dependent. RNA replication involves primary mRNA transcription from each segment of the genomic RNA via a virion polymerase, primed by a capped oligonucleotide (eight to 15 nucleotides in length) appropriated from host cell mRNA; later, using the protein products of this transcription, full-length complementary RNA (neither capped nor polyadenylated) is synthesized from each segment and used as a template for the synthesis of genomic RNA segments. Replication takes place in the nucleus (RNA synthesis, N, M, and NS1 protein accumulation) and cytoplasm, and assembly occurs via budding upon plasma membranes. Closely related viruses can reassort genes during mixed infections. Particular influenza A viruses infect humans and other mammalian and avian species; interspecies transmission is believed to account for gene reassortment and the emergence of new human pandemic strains. Influenza B viruses naturally infect only humans. Transmission among humans is by aerosols and droplets and is water-borne among ducks. Thogoto and Dhori viruses are transmitted by ticks and replicate in both ticks and mammals. Virions are sensitive to heat, lipid solvents, nonionic detergents, formaldehyde, irradiation, and oxidizing agents. The viruses exhibit hemagglutinating activity.

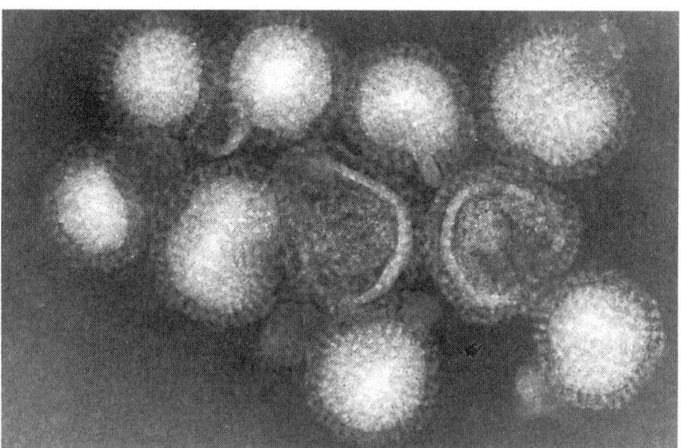

FIG. 11. Family *Orthomyxoviridae*, genus *Influenzavirus A, B,* influenza A virus [A/HK/8/68 (H3N2)] (×150,000).

Human Pathogens

Genus *Influenzavirus A, B*: influenza A viruses and influenza B viruses. Genus *Influenzavirus C*: influenza C viruses.

Animal Pathogens

Genus *Influenzavirus A, B*: influenza A viruses infect swine, horses, seals, fowl, ducks and many other species of birds. Genus *Influenzavirus C*: influenza C viruses infect swine.

The Family *Bunyaviridae*

Family: *Bunyaviridae* (the bunyaviruses) (Fig. 12)
Genus: *Bunyavirus* (bunyaviruses)
Genus: *Phlebovirus* (phleboviruses)
Genus: *Nairovirus* (nairoviruses)
Genus: *Hantavirus* (hantaviruses)
Genus: *Tospovirus* (tospoviruses)

Characteristics (7,10,39,97)

Virions are spherical or pleomorphic, 80–120 nm in diameter, and consist of a lipid-containing envelope, with

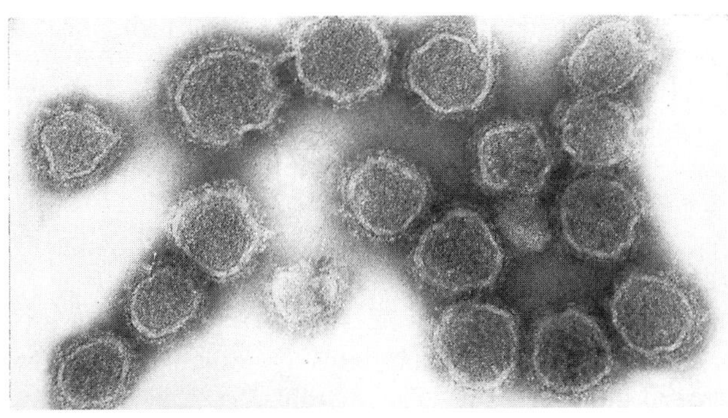

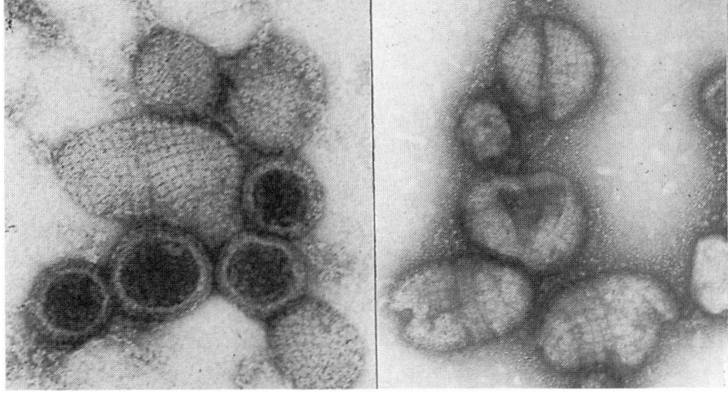

FIG. 12. Family *Bunyaviridae*, genus *Bunyavirus*, LaCrosse virus (top ×112,000); genus *Hantavirus*, Hantaan virus (bottom ×140,400).

fine peplomers 5–10 nm in length, within which are three circular, helically symmetrical nucleocapsids 2–2.5 nm in diameter and 200–3,000 nm in length. Virion Mr is 300–400 × 10^6; buoyant density is 1.16–1.18 g/cm^3 in sucrose and 1.20–1.21 g/cm^3 in CsCl. The genome consists of three molecules (L, M, S) of "circular" (terminal nucleotides of each viral RNA species are base-paired forming noncovalently closed, circular RNAs and ribonucleocapsids), negative- or ambisense, single-stranded RNA, 11–21 kb in overall size. Terminal sequences of gene segments are conserved among viruses in each genus but are different among genera. The L segment encodes the viral transcriptase-replicase, the M segment the envelope glycoproteins, and the S segment the nucleocapsid protein. Phleboviruses and tospoviruses have an ambisense S RNA; they encode NSS proteins in the 5'-half of virion S RNA. The viruses have four structural proteins, two external glycoproteins (G1 and G2, Mr 55,000–108,000 and 29,000–70,000), a nucleocapsid protein (N, Mr 19,000–54,000), and a large transcriptase protein (L, Mr 250,000–330,000). Virions contain lipids (phospholipids, sterols, fatty acids, and glycolipids) that are derived from host cell (Golgi) membranes. Virions contain carbohydrates as N-linked high mannose glycans on the G1 and G2 proteins. RNA replication involves a primary transcription of mRNA from each segment of the genomic RNA via a virion transcriptase; later, using the protein products of this transcription, there is production of full-length complementary RNA for each segment, each of which in turn is used as template for the synthesis of genomic RNA segments. Replication takes place in the cytoplasm, and assembly occurs via budding usually upon Golgi membranes. Closely related viruses can reassort gene segments during mixed infections. The viruses (except Hantaviruses) replicate in vertebrates and arthropods; various viruses are transmitted by mosquitoes, ticks, phlebotomine flies, thrips, and other arthropods. Transovarial and venereal transmission occurs in some vector mosquito species. The viruses are generally cytolyic in their vertebrate hosts, but not in their invertebrate hosts. Hantaviruses are transmitted by persistently infected rodents via aerosolization of urine, saliva, and feces. Some viruses have narrow host ranges, others have wide host ranges and occur worldwide. The viruses are sensitive to heat, detergents, and formaldehyde; some exhibit hemagglutinating activity.

Human Pathogens

Genus *Bunyavirus*: Bunyamwera virus; Bwamba virus; Oriboca virus; Oropouche virus; Guama virus; LaCrosse virus; Jamestown Canyon virus; California encephalitis virus; snowshoe hare virus; and Tahyna virus. Genus *Phlebovirus*: sandfly fever-Naples virus; sandfly fever-Sicilian virus; Rift Valley fever virus. Genus *Nairovirus*: Crimean-Congo hemorrhagic fever virus. Genus *Hantavirus*: Hantaan virus, Seoul virus (hemorrhagic fever with renal syndrome); Sin Nombre virus (acute respiratory distress syndrome); Puumala virus (nephropathia epidemica).

Animal Pathogens

Genus *Bunyavirus*: Akabane virus and Aino virus. Genus *Phlebovirus*: Rift Valley fever virus. Genus *Nairovirus*: Nairobi sheep disease virus.

The Family *Arenaviridae*

Family: *Arenaviridae* (the arenaviruses)
Genus: *Arenavirus* (arenaviruses) (Fig. 13)

Characteristics (91,102,114)

Virions are spherical to pleomorphic, 50–300 nm in diameter (mean 110–130 nm), with a dense lipid-containing envelope and large club-shaped peplomers, 8–10 nm in length, within which are two loosely helical, circular, nucleocapsid structures and a variable number of ribosomes. Virion Mr has not been determined; buoyant density is 1.17–1.18 g/cm^3 in sucrose and 1.19–1.20 g/cm^3 in CsCl. The genome consists of two molecules (L and S, Mr 2.2–2.8 × 10^6 and 1.1 × 10^6) of circular (ends hydrogen-bonded), negative- and ambisense, single-stranded RNA,

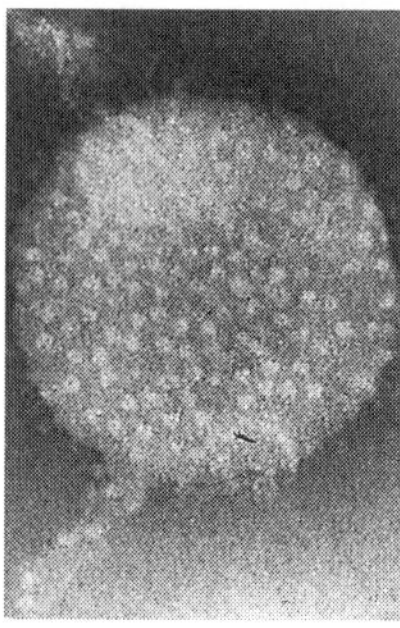
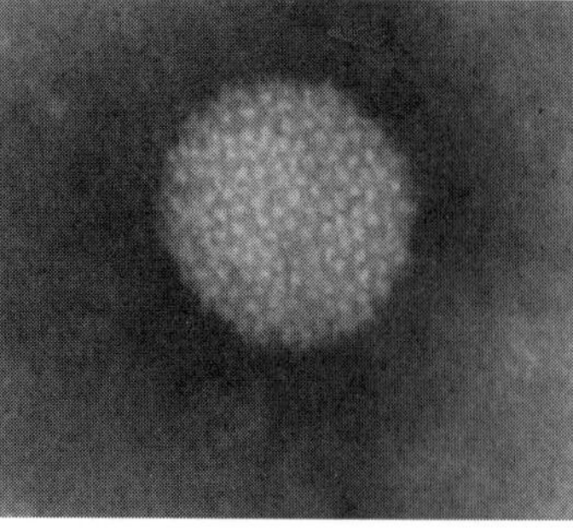

FIG. 13. Family *Arenaviridae*, genus *Arenavirus*, lymphocytic choriomeningitis virus (left) and Tacaribe virus (right) (×162,500).

10–14 kb in overall size. The 3'-terminal sequences (19–30 nucleotides) are similar between the two RNAs and conserved among the viruses. Preparations of purified virus contain 28S, 18S, and 4-6S RNAs attributed to encapsidated host cell ribosomes. The viruses have four major proteins: a nucleoprotein (N or NP, Mr 63,000–72,000), an RNA-dependent RNA polymerase (L, Mr ~200,000), and two glycoproteins (G1, G2; Mr 34,000–44,000). The viruses also contain a putative zinc binding protein (Z or p11; Mr 10,000–14,000). Virions contain lipids similar in composition to those of the host cell plasma membrane. Carbohydrates are in the form of complex glycans on both glycoproteins. The L and S RNAs of arenaviruses each have an ambisense coding arrangement: N is encoded in the viral-complementary sequence corresponding to the 3'-half of the S segment, whereas the viral glycoprotein precursor is encoded in the viral-sense sequence corresponding to the 5'-half of the S segment. The S intergenic region contains nucleotide sequences with the potential of forming hairpin configurations, which may function to terminate mRNA transcription from the viral and viral-complementary S RNAs. The ambisense viral L RNA encodes in its viral-complementary sequence the L protein and in the viral-sense 5'-end sequence the Z protein. Because of the ambisense coding arrangement, replication involves a primary transcription of only N and L mRNAs that can be synthesized from the genomic RNAs by the virion polymerase before translation. The products of these mRNAs are presumed to be involved in the synthesis of full-length viral complementary species, which serve as templates for the synthesis of the glycoprotein mRNA and the synthesis of full-length viral RNAs. The viruses can form reassortant progeny. Replication takes place in cytoplasm, and assembly occurs via budding on plasma membrane. The viruses are sensitive to pH <5.5 and >8.5, heat, detergents, organic solvents, and UV- and γ-irradiation. The human pathogens, Lassa, Machupo, Junin, and Guanarito viruses, are Biosafety Level 4 pathogens and can be handled in the laboratory only under maximum containment conditions.

Human Pathogens

Genus *Arenavirus*: lymphocytic choriomeningitis (LCM) virus; Lassa virus; Machupo virus (Bolivian hemorrhagic fever); Junin virus (Argentine hemorrhagic fever); and Guanarito virus (Venezuelan hemorrhagic fever).

Animal Pathogens

Genus *Arenavirus*: lymphocytic choriomeningitis (LCM) virus of mice and hamsters.

The Family *Reoviridae*

Family: *Reoviridae* (the reoviruses) (Fig. 14)
Genus: *Orthoreovirus* (reoviruses)
Genus: *Orbivirus* (orbiviruses)
Genus: *Rotavirus* (rotaviruses)
Genus: *Coltivirus* (coltiviruses)
Genus: *Aquareovirus* (golden shiner virus)
Genus: *Cypovirus* (cytoplasmic polyhedrosis viruses)
Genus: *Phytoreovirus* (plant reovirus group 1)
Genus: *Fijivirus* (plant reovirus group 2)
Genus: *Oryzavirus* (plant reovirus group 3)

Characteristics (96,111,113)

Virions appear nearly spherical and have a diameter of 60–80 nm; they are nonenveloped and have two or three shells (each with icosahedral symmetry). The viruses of each genus differ in morphologic and physicochemical details. Outer shells differ most between genera; the orthoreoviruses have a well-defined outer capsid composed of hexagonal and pentagonal units, whereas the orbiviruses have a diffuse outer shell and the rotaviruses have a complex outer shell structure with T = 13 icosahedral symmetry and 132 channels superimposed and extending from the surface to the core and 60 short spikes extending from the surface of the virion. The inner shell of all viruses is similar in size (50–65 nm) with icosahedral symmetry and discernible subunits. Virion Mr is ~80–130 × 10^6; buoyant density is 1.36–1.39 g/cm^3 in CsCl. The genome consists of linear, double-stranded RNA divided into 10 (genus *Reovirus* and *Orbivirus*), 11 (genus *Rotavirus*), or 12 (genus *Coltivirus*) segments, and 23 (genus *Reovirus*), 18 (genus *Orbivirus*), 16–21 (genus *Rotavirus*), or 27 (genus *Coltivirus*) kbp in overall genome size. The individual RNA segments vary in size from 680 bp (the smallest rotavirus segment) to 3,900 bp (the largest orthoreovirus segment). The positive strands of each duplex have 5'-terminal caps (type 1 structure), and the negative strands have phosphorylated 5'-termini. The RNAs lack 3'-terminal poly (A) tracts. The viruses have 10–12 structural proteins; there are at least three major capsid proteins and at least three proteins constituting the virion RNA poly-

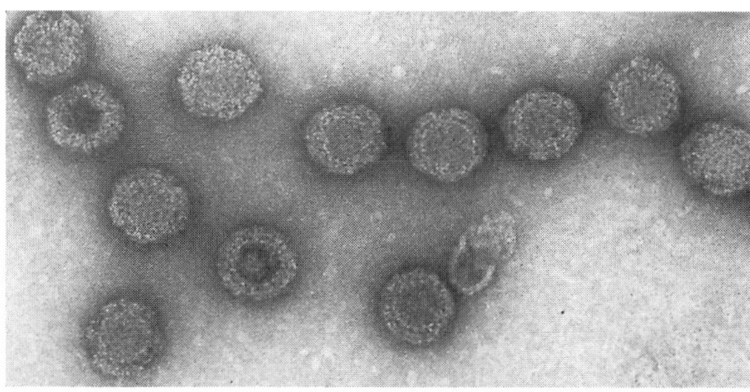

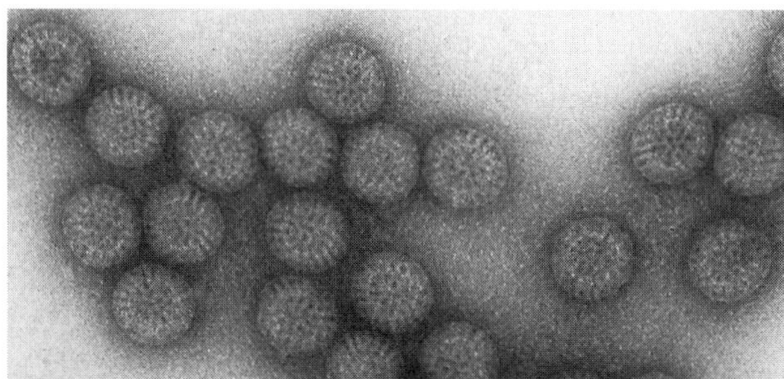

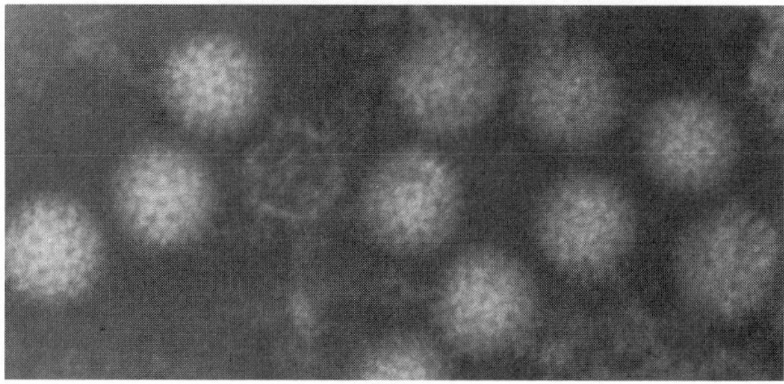

FIG. 14. Family *Reoviridae*, genus *Orthoreovirus*, reovirus 2 (top); genus *Rotavirus*, human rotavirus, type unknown (middle); genus *Orbivirus*, bluetongue virus 11 (bottom ×105,000; top and middle ×126,000 bottom).

merase complex and associated enzymes involved in mRNA synthesis (including elongation, capping, and methylation). The proteins range in size from Mr 15,000–155,000. Mature virions do not contain lipid; depending on the genus, a myristyl residue may be covalently attached to one of the virion proteins. In some genera one of the outer virion proteins may be glycosylated with high mannose glycans, or O-linked N-acetylglucosamine. Repetitive asymmetric transcription of full-length mRNAs from each double-stranded RNA segment occurs within transcriptionally active virus particles throughout the infection course, and the mRNA products are extruded from these particles. The process of RNA replication is not completely known; there is some evidence that sets of capped mRNAs and certain proteins are incorporated into assortment complexes that are the precursors of progeny virus particles. These complexes, together with structural proteins, are encapsidated into subviral particles where they are transcribed to form double-stranded progeny RNA species. Replication and assembly take place in cytoplasm, often in association with granular or fibrillar inclusion bodies. Genome segment reassortment occurs readily in cells coinfected with closely related viruses. Rotaviruses and reoviruses are spread by direct contact and indirectly by fomites (fecal-oral transmission); orbiviruses and Coltiviruses are transmitted via arthropods (e.g., gnats, mosquitoes, or ticks). Reovirus distribution is ubiquitous and worldwide; the distribution of some other viruses is geographically limited. Virus infectivity is moderately resistant to heat, organic solvents, detergents, and irradiation. pH stability varies between genera.

Human Pathogens

Genus *Reovirus*: reoviruses 1, 2, and 3 have been isolated from humans, but a causal relationship to illness is rare. Genus *Orbivirus*: Orungo virus (febrile illness in Nigeria and Uganda); Kemerovo virus (febrile illness in Russia and Egypt). Genus *Rotavirus*: group A and B human rotaviruses. Genus *Coltivirus*: Colorado tick fever virus; Eyach virus.

Animal Pathogens

Genus *Orthoreovirus*: reoviruses l, 2, and 3 (mice and other species); avian reoviruses. Genus *Orbivirus*: bluetongue viruses 1–25; African horsesickness viruses 1–9; epizootic hemorrhagic disease of deer viruses 1–2; Ibaraki virus; equine encephalosis viruses; and other viruses affecting various animal species. Genus *Rotavirus*: group A rotaviruses (including simian rotavirus SA11); group C rotaviruses (porcine); group D rotaviruses (fowl); group E rotaviruses (porcine); and group F rotaviruses (avian), all of which are important causes of diarrhea. Genus *Aquareovirus:* golden shiner virus; several salmon reoviruses; grass carp reovirus; chub reovirus; and channel catfish reovirus.

The Family *Birnaviridae*

Family: *Birnaviridae* (the birnaviruses) (Fig. 15)
Genus: *Aquabirnavirus* (fish birnaviruses)
Genus: *Avibirnavirus* (fowl birnaviruses)
Genus: *Entomobirnavirus* (insect birnaviruses)

Characteristics (32,33)

Virions are nonenveloped, have icosahedral symmetry, and are 60 nm in diameter. Virions have 132 morphological subunits in a T = 9 arrangement. Virion Mr is 55 × 10^6; buoyant density is 1.33 g/cm^3 in CsCl. The genome consists of two molecules (A, B) of linear, double-stranded

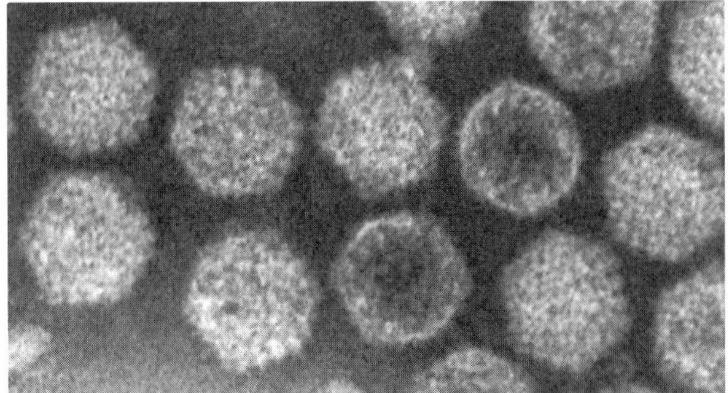

FIG. 15. Family *Birnaviridae*, genus *Birnavirus*, infectious bursal disease virus (×189,000).

RNA, 5.7–5.9 kbp in overall size (infectious pancreatic necrosis virus: A segment 3,092 bp; B segment 2,784 bp). Both genome segments contain a 5'-genome-linked protein (VPg). There are no poly (A) tracts at the 3'-ends of the RNA segments. Virions contain five proteins: VP1 (Mr 94,000), the RNA-dependent RNA polymerase; pre-VP2 (Mr 62,000) and VP2 (Mr 54,000), the major capsid proteins; VP3 (Mr 30,000), an internal capsid protein; and VPg (Mr 94,000), the genome-linked protein. Infected cells also contain NS or VP4 (Mr 29,000), a virus-coded protease, and a positively charged minor protein (Mr 17,000). Virions contain no lipid. The VP2 capsid protein may be glycosylated. Replication involves the synthesis by the virion RNA-dependent RNA polymerase of two genome length mRNAs, one from each of the genome segments. Virus RNA is transcribed by a semi-conservative strand displacement mechanism. Segment A mRNA is translated to a Mr 10^6 polyprotein that is cleaved to form (5' to 3') the pre-VP2, VP4, and VP3 proteins. Pre-VP2 is later processed by a slow maturation cleavage to produce VP2. The mRNA from segment B is translated to form VP1. Virus particles assemble and accumulate in the cytoplasm. Natural hosts include salmonid fish, other freshwater and marine fishes, bivalve mollusks, chickens, ducks, turkeys, and other domestic fowl. The viruses are transmitted both vertically and horizontally; geographic distribution is worldwide. The viruses are sensitive to acid, heat, detergents, organic solvents, and UV- and γ-irradiation.

Human Pathogens

Genus *Birnavirus*: none known.

Animal Pathogens

Genus *Birnavirus*: infectious bursal disease virus of chickens, ducks, and turkeys; infectious pancreatic necrosis virus of fish.

The Family *Retroviridae*

Family: *Retroviridae* (the retroviruses)
Genus: unnamed, mammalian type B retroviruses
Genus: unnamed, mammalian and reptilian type C retroviruses
Genus: unnamed, avian type C retroviruses
Genus: unnamed, type D retroviruses
Genus: unnamed, HTLV/BLV viruses
Genus: *Lentivirus* (lentiviruses) (Fig. 16)
Genus: *Spumavirus* (spumaviruses)

Characteristics (28,36,132)

Virions are spherical, enveloped, and 80–100 nm in diameter; they consist of a lipid-containing envelope, with peplomers 8 nm in length, surrounding an icosahedral capsid, which in turn contains a helical nucleocapsid. The nucleocapsid is eccentric in type B virions, concentric in type C, HTLV/BLV, and spumavirus virions, and rod or cone shaped in lentivirus virions. Virion buoyant density is 1.16–1.18 g/cm^3 in sucrose. The genome consists of a homodimer of linear, positive-sense, single-stranded RNA; each monomer is 7–11 kb in size. The monomers are held together noncovalently by hydrogen bonds and have a 3'-poly (A) tract, with a 5'-cap structure (type 1). Most of the viruses have four main genes coding for the virion proteins in the order 5'-*gag, pro, pol, env*-3'. Many of the viruses contain other genes encoding nonstructural proteins important for the regulation of gene expression

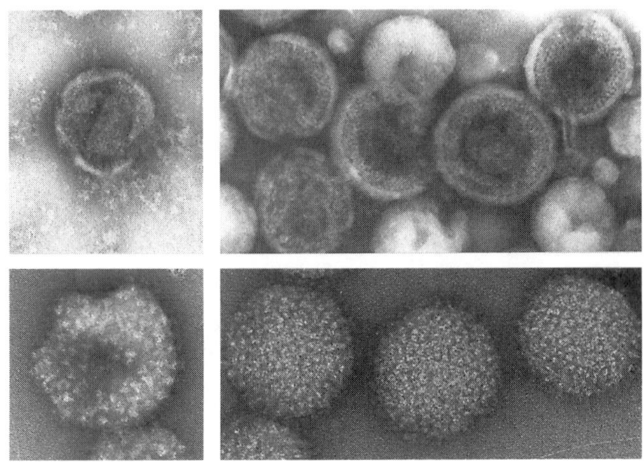

FIG. 16. Family *Retroviridae*, genus *Lentivirus*, human immunodeficiency virus 1, displaying a typical conical core (top left ×104,000); human immunodeficiency virus 1, immature virions displaying structured protein layer beneath the envelope (top, right × 104,000); simian immunodeficiency virus, displaying triangular peplomers (bottom left × 126,000); unnamed genus (mammalian type B retroviruses), mouse mammary tumor virus, displaying closely spaced peplomers (bottom right × 112,000).

and virus replication; others carry (as inserts or as substitutions) cell-derived sequences that are important in pathogenesis. Virion proteins include two envelope proteins encoded by the viral *env* gene, three to six internal nonglycosylated structural proteins encoded by the *gag* gene, and, in order from the amino terminus, a matrix, capsid, and nucleocapsid protein. Other proteins are a protease encoded by the *pro* gene and a reverse transcriptase and integrase encoded by the *pol* gene. Virions contain lipids derived from the plasma membrane of the host cell. Virions contain carbohydrates as glycosylated envelope surface proteins. RNA replication is unique: replication starts with reverse transcription of virion RNA into cDNA. cDNA synthesis involves the concomitant digestion of the viral RNA (RNAse of the reverse transcriptase) with the products of this digestion serving to prime positive-sense cDNA synthesis on the negative-sense DNA transcripts. In final form, the linear double-stranded DNA transcripts derived from the viral genome contain long terminal repeats; this DNA is circularized, integrated into the host chromosomal DNA, and then used for transcription, including transcription of full-length genomic RNA species. There are several classes of mRNA varying with the virus. An mRNA comprising the whole genome is a template for the translation of the *gag, pro,* and *pol* genes; translation yields polyprotein precursors that are cleaved to yield the structural proteins, protease, reverse transcriptase, and integrase. Virion assembly occurs via budding on plasma membranes. The viruses are widely distributed in vertebrates; endogenous proviruses (products of ancient infections of germ line cells, inherited as Mendelian genes) also occur widely in vertebrates. The viruses are associated with many different diseases, including leukemias, lymphomas, sarcomas, carcinomas, immunodeficiencies, autoimmune diseases, lower motor neuron diseases, and several acute diseases involving tissue damage. Virions are sensitive to heat, detergents, and formaldehyde, but relatively resistant to UV light.

Human Pathogens

Genus (unnamed, HTLV/BLV viruses): human T-cell lymphotropic viruses 1 and 2. Genus *Lentivirus*: human immunodeficiency viruses 1 and 2.

Animal Pathogens

Genus (unnamed, mammalian type B retroviruses): mouse mammary tumor virus. Genus (unnamed, mammalian and reptilian type C retroviruses): many murine leukemia viruses [e.g., Abelson, AKR (endogenous), Friend, Maloney murine leukemia viruses]; many murine sarcoma viruses (e.g., Harvey, Kirsten, Moloney, Finkel-Biskis-Jinkins murine sarcoma viruses); feline leukemia virus; feline sarcoma viruses; gibbon ape leukemia virus; woolly monkey sarcoma virus; porcine type C virus; guinea pig type C virus; and viper type C virus. Genus (unnamed, avian type C retroviruses): Rous sarcoma virus; avian carcinoma viruses; avian sarcoma viruses; avian leukosis viruses; avian myeloblastosis viruses; avian reticuloendotheliosis viruses; and duck spleen necrosis virus. Genus (unnamed, type D retroviruses): Mason-Pfizer monkey virus; simian type D virus 1; Langur type D virus; squirrel monkey type D virus; and ovine pulmonary adenocarcinoma virus (Jaagsiekte). Genus (unnamed, HTLV/BLV viruses): simian T-cell lymphotropic viruses; bovine leukemia virus. Genus *Lentivirus*: simian immunodeficiency viruses (African green monkey, sooty mangabey, stump-tailed macaque, pig-tailed macaque, Rhesus, chimpanzee, and mandrill viruses); visna/maedi virus; caprine arthritis-encephalitis virus; equine infectious anemia virus; feline immunodeficiency virus; and bovine immunodeficiency virus.

The Family *Hepadnaviridae*

Family: *Hepadnaviridae* (the hepadnaviruses) (Fig. 17)
Genus: *Orthohepadnavirus* (hepadnaviruses of mammals)
Genus: *Avihepadnavirus* (hepadnaviruses of birds)

Characteristics (50,59,65)

Virions are spherical and occasionally pleomorphic, 40–48 nm in diameter, with no evident surface projections. The envelope contains the surface antigens (HBsAg) and surrounds an icosahedral, 27–35 nm diameter nucleocapsid core constructed from 180 capsomers (core antigen, HBcAg) arranged in T = 3 symmetry. Buoyant density is 1.25 g/cm^3 in CsCl. The genome consists of a single molecule of DNA, which is circular (maintained by base-pairing of cohesive ends), nicked, and mainly

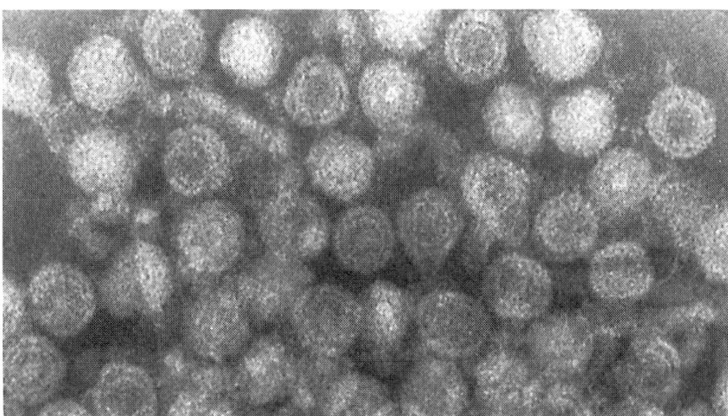

FIG. 17. Family *Hepadnaviridae*, genus *Hepadnavirus,* hepatitis B virus (×175,000).

double-stranded (but with a large single-stranded gap), 3.2 kbp in size when fully double-stranded. One strand (negative-sense) is full-length, the other varies in length. The 5'-end of the negative-sense DNA has a covalently attached terminal protein; the 5'-end of the positive-sense DNA has a covalently attached 19-nucleotide, 5'-capped oligoribonucleotide primer. Lipids are present in the envelope of virions as phospholipids, sterols, and fatty acids. Carbohydrates are present in virions as N-linked glycans of complex and high mannose types. Virion envelope proteins consist of three antigenically distinct species: S-proteins (p24, GP27), M-proteins (P33, GP36), and L-proteins (P39, GP42). The 20- to 25-nm particles (HBsAg) contain predominantly S-proteins and occasionally M-proteins. Filamentous forms contain these proteins and L-proteins. The virion core is composed principally of core antigen (HBcAg). Enzymes associated with virions include a protein kinase and reverse transcriptase with RNA- and DNA-dependent DNA polymerase and RNAse H activities. DNA replication involves repair of the single-stranded gap, conversion to a supercoiled helix, and transcription of two classes of RNA (mRNA for protein synthesis and genomic RNA, which is transcribed in turn by the virus-specific reverse transcriptase to make genomic DNA). Replication in hepatocytes takes place in the nucleus; HBsAg production occurs in the cytoplasm in massive amounts and is formed into 22-nm HBsAg particles that are shed into the bloodstream, producing antigenemia. The viruses are host specific, and persistence is associated with chronic disease and neoplasia. Although integration of viral DNA into the host genome is not required for replication, this does occur in hepatitis B-associated hepatocellular carcinoma. The viruses are sensitive to acid, heat, detergents, organic solvents, and UV- and γ-irradiation.

Human Pathogens

Genus *Orthohepadnavirus*: hepatitis B virus.

Animal Pathogens

Genus *Orthohepadnavirus*: woodchuck (*Marmota monax*) hepatitis virus; ground squirrel (*Spermophilus beecheyi*) hepatitis virus; and red-bellied squirrel (*Callosciurus erythracus*) hepatitis virus. Genus *Avihepadnavirus*: Pekin duck (*Anas domesticus*) hepatitis virus; heron hepatitis virus.

The Family *Circoviridae*

Family: *Circoviridae* (the circoviruses) (Fig. 18)
Genus: *Circovirus* (circoviruses)

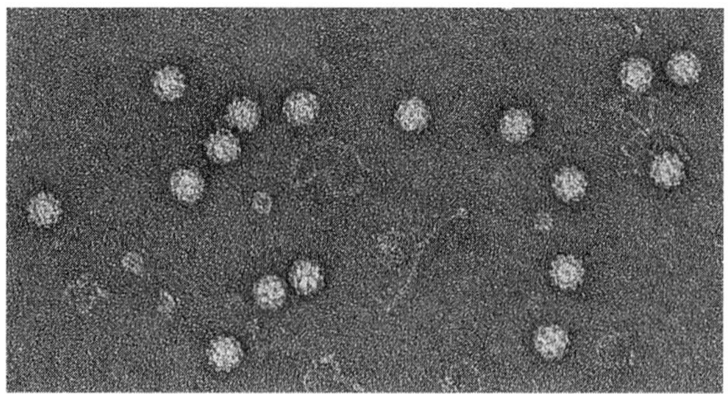

FIG. 18. Family *Circoviridae*, genus *Circovirus*, chicken anemia virus (× 150,000).

Characteristics (99,129)

Virions are nonenveloped, have icosahedral symmetry, and range from 15 to 22 nm in diameter. Virion buoyant density is 1.33–1.37 g/cm^3 in CsCl. The genome consists of a single molecule of circular, negative-sense, single-stranded DNA, 1.7 to 2.3 kb in size. Virion proteins vary in number and size between viruses: one protein (Mr 36,000–50,000) has been described in some viruses, three (Mr 16,000–26,000) in others. Virions do not contain lipids or carbohydrates. Details of viral replication are not known, but circular double-stranded DNA replicative intermediates have been detected in infected cells. The genome of chicken anemia virus contains three partially overlapping open reading frames. The viruses replicate in the nucleus and, like parvoviruses, probably depend on cellular proteins produced during the S phase of the cell cycle. The viruses appear to be specific for the species of origin and worldwide in distribution; their mode of transmission is not known. The avian viruses are important pathogens. The viruses resist heating at 60°C for 30 min.

Human Pathogens

Genus *Circovirus*: none known.

Animal Pathogens

Genus *Circovirus*: chicken anemia virus; psittacine beak and feather disease virus; and porcine circovirus.

The Family *Parvoviridae*

Family: *Parvoviridae* (the parvoviruses)
Subfamily: *Chordoparvovirinae* (the parvoviruses of vertebrates)
Genus: *Parvovirus* (parvoviruses of mammals and birds) (Fig. 19)

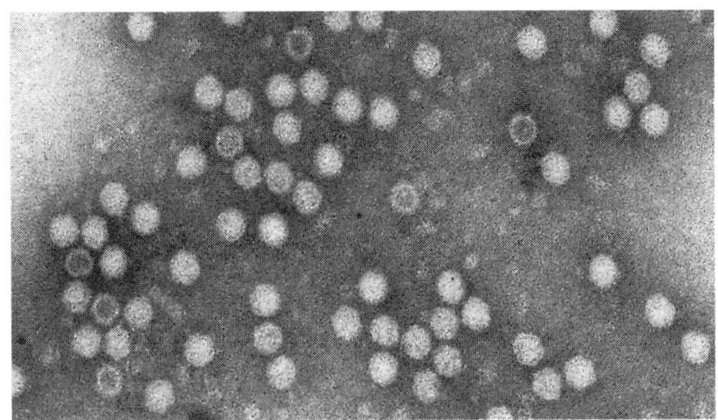

FIG. 19. Family *Parvoviridae*, genus *Parvovirus*, H-1 virus (×133,000).

Genus: *Erythrovirus* (human parvovirus B 19-like viruses)
Genus: *Dependovirus* (adeno-associated viruses)
Subfamily: *Entomoparvovirinae* (the parvoviruses of invertebrates)
Genus: *Densovirus* (insect parvovirus group A)
Genus: *Iteravirus* (insect parvovirus group B)
Genus: *Contravirus* (insect parvovirus group C)

Characteristics (96,121)

Virions are nonenveloped, have icosahedral symmetry (32 capsomeres in a T = 3 arrangement), and are 18–26 nm in diameter. Virion Mr is 5.5–6.2 × 10^6; buoyant density is 1.39–1.42 g/cm^3 in CsCl. The genome consists of a single molecule of single-stranded DNA, ~5 kb in size. Viruses of the genus *Parvovirus* preferentially encapsidate negative-stranded DNA, whereas members of the other genera encapsidate positive- and negative-stranded DNA in various proportions. The viruses have three to four major proteins: the Mr of VP1 is 80,000–96,000, VP2 is 64–85,000, VP3 is 60–75,000, and VP4 is 49–52,000. The viral proteins represent alternative forms of the same gene product. Virions lack lipids and carbohydrates. DNA replication is complex, involving the formation of hairpin structures, extension to form a complete double-stranded intermediate, endonuclease cleavage, and repetition of these cycles. The viruses have two major genes: the REP (or NS) gene that encodes functions required for transcription and DNA replication, and the CAP (or S) gene that encodes the coat proteins. Both genes are encoded in the same DNA strand in the vertebrate viruses (for some viruses alternative splicing allows different forms of REP gene products to be produced). The CAP gene produces up to three proteins. Replication and assembly take place in the nucleus; replication requires host cell functions of the late S phase of the cell division cycle, indicating a close association between host and viral DNA synthesis, probably involving host DNA polymerase(s). The viral genome must be converted into double-stranded DNA before mRNA transcription occurs. DNA synthesis derives from a self-priming mechanism and involves palindromic sequences. Some of the viruses benefit from coinfection with other viruses, such as adenoviruses or herpesviruses, or from the effects of chemical or other treatments of the host cell (affecting host cell rather than virus replication functions). The viruses have narrow host ranges; are resistant to acid, heat, detergents, organic solvents, and exposure to pH 3–9; and are sensitive to treatment with formaldehyde, β-propiolactone, hydroxylamine, oxidizing agents, and UV- and γ-irradiation.

Human Pathogens

Genus *Parvovirus*: RA-1 virus has been associated with rheumatoid arthritis. Genus *Erythrovirus:* human parvovirus B-19 (the cause of aplastic crisis in hemolytic anemias and in sickle-cell disease, erythema infectiosum or fifth disease, and spontaneous abortion, fetal death, and hydrops fetalis).

Animal Pathogens

Genus *Parvovirus*: feline panleukopenia virus; canine parvovirus; minute virus of canines; mink enteritis virus; Aleutian mink disease virus; bovine parvovirus; porcine parvovirus; goose parvovirus; and minute virus of mice and other murine parvoviruses (e.g., HB, H-1, Kilham rat virus).

The Family *Papovaviridae*

Family: *Papovaviridae* (the papovaviruses)
Genus: *Papillomavirus* (papillomaviruses) (Fig. 20)
Genus: *Polyomavirus* (polyomaviruses)

Characteristics (24,79,86,115,116)

Virions are nonenveloped, have icosahedral symmetry, and are 45 nm (polyomaviruses) or 55 nm (papillo-

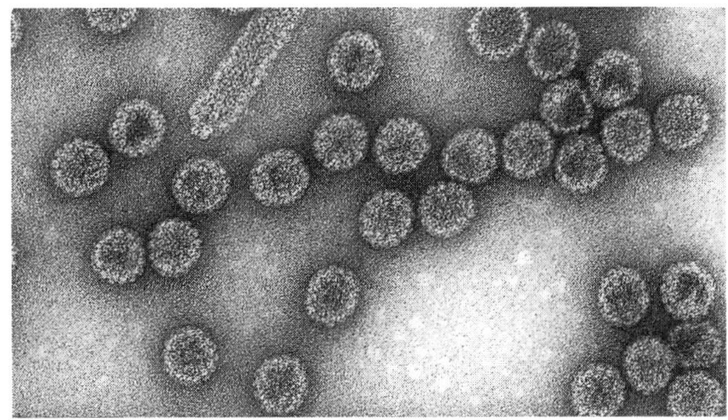

FIG. 20. Family *Papovaviridae*, genus *Papillomavirus*, human papillomavirus (untyped) (×105,000).

maviruses) in diameter. Virions have 72 capsomeres (360 VP1 subunits arranged in 72 pentavalent capsomeres, shown for polyoma and SV40 viruses) in a skewed (T = 7) arrangement. Filamentous and tubular particles result from aberrant assembly. Virion Mr is 25 × 10^6 (polyomaviruses) or 47 × 10^6 (papillomaviruses); buoyant density is 1.20 g/cm^3 in sucrose and 1.34–1.35 g/cm^3 in CsCl. The genome consists of a single molecule of circular, double-stranded DNA, 5 kbp (polyomaviruses) or 8 kbp (papillomaviruses) in size. The genome may persist in infected cells in an integrated form (polyomaviruses) or in an episomal form (papillomaviruses). The virus genomes encode five to 10 proteins (Mr 3,000–88,000), including three capsid proteins (VP1, VP2, and VP3), and in some viruses a nonstructural protein that may facilitate capsid assembly. Virions contain no lipids or carbohydrates. Transcription is complex, divided into early and late stages, each controlled by separate promoters, each occurring on opposite DNA strands (polyomaviruses) or on the same strand (papillomaviruses). Precursor mRNAs undergo posttranscriptional processing that includes capping and polyadenylation of 5'- and 3'-termini, respectively, as well as differential splicing of messages and use of overlapping open reading frames. Posttranslational modifications of some early and late viral proteins include phosphorylation, N-acetylation, fatty acid acylation, ADP-ribosylation, methylamination, adenylation, glycosylation, and sulphation. DNA replication starts at a fixed point on the genome, which remains in circular configuration throughout the process; it involves a unique origin of replication and the action of host DNA polymerase(s). Replication proceeds bidirectionally; late in the replication cycle, rolling circle-type molecules have been identified. Replication and assembly occur in the nucleus, and virion release is via cell destruction. Most of the viruses have narrow host ranges; they are transmitted by contact (including sexual contact) and by aerosol and are distributed worldwide. The papillomaviruses cause tumors (warts, papillomas, carcinomas); the polyomaviruses have been associated with urinary tract infection and central nervous system demyelinating disease. Virions are resistant to ether and acid and are rather resistant to heat.

Human Pathogens

Genus *Polyomavirus*: BK virus (kidney infection and mild respiratory disease); JC virus; and simian virus 40 (all three viruses can cause progressive multifocal leukoencephalopathy). Genus *Papillomavirus*: human papillomaviruses 1–60 (some viruses are associated with benign genital condylomas, others are associated with genital and oral/pharyngeal carcinomas).

Animal Pathogens

Genus *Polyomavirus*: mouse polyomavirus; African green monkey B-lymphotropic polyomavirus; baboon polyomavirus 2; simian viruses 12 and 40; fetal rhesus kidney virus; bovine polyomavirus; budgerigar fledgling disease virus; hamster polyomavirus; K virus (rabbits); and rabbit kidney vacuolating virus. Genus *Papillomavirus*: cottontail rabbit papillomavirus (Shope papillomavirus); bovine papillomaviruses 1, 2, and 4; canine oral papillomavirus; deer papillomavirus (deer fibroma virus); elephant papillomavirus; equine papillomavirus; and ovine papillomavirus.

The Family *Adenoviridae*

Family: *Adenoviridae* (the adenoviruses)
Genus: *Mastadenovirus* (mammalian adenoviruses) (Fig. 21)
Genus: *Aviadenovirus* (avian adenoviruses)

Characteristics (27,62,112,138)

Virions are nonenveloped, appear hexagonal in outline, have icosahedral symmetry, and are 80–110 nm in diameter. Virions have 240 nonvertex capsomers (hexons) 8–10 nm in diameter and 12 vertex capsomers (pentons) with fibers that protrude 9–35 nm from the virion surface. Virion Mr is 150–180 × 10^6; buoyant density is 1.32–1.35 g/cm^3 in CsCl. The genome consists of a single molecule of linear, double-stranded DNA, 36–38 kbp in size. A virus-coded terminal protein is covalently linked to the 5'-end of each DNA strand, and there are inverted terminal repeats of 50–200 bp. Virions have at least 12 structural proteins and many nonstructural proteins. Virions do not contain lipid; fibers and some of the nonstructural proteins are glycosylated. DNA replication is complex and distinct from that of other viruses; it involves circularization and strand displacement using a protein-priming mechanism (terminal protein) together with a virus-coded DNA polymerase and DNA binding proteins in concert with cellular factors. Transcription involves early and late genes distributed along both strands of the DNA. Transcription by host RNA polymerase II involves both DNA strands and initiates from

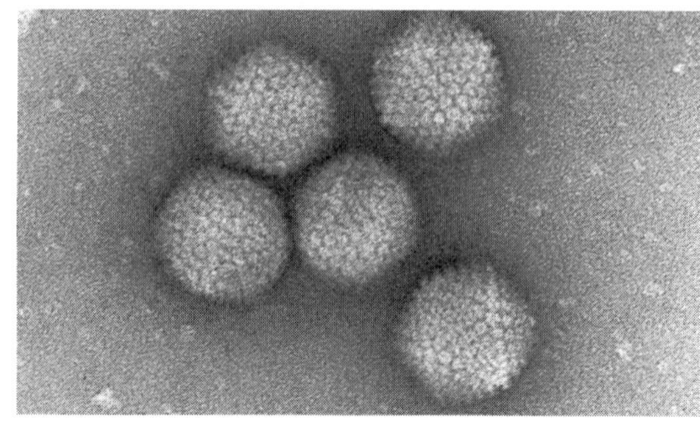

FIG. 21. Family *Adenoviridae*, genus *Mastadenovirus,* human adenovirus type 2 (×120,000).

four early, two intermediate, and one major late promoter. All primary transcripts are capped and polyadenylated. There are complex splicing patterns to produce families of mRNAs. Virions are assembled in the nucleus, often in paracrystalline arrays, and are released by cell destruction. Most viruses have a narrow host range. Experimentally, several of the viruses cause tumors in newborns of heterologous species. The viruses are stable under mild acidic conditions and are insensitive to lipid solvents.

Human Pathogens

Genus: *Mastadenovirus*: human adenoviruses 1–49.

Animal Pathogens

Genus *Mastadenovirus:* simian adenoviruses 1–27; bovine adenoviruses 1–9; porcine adenoviruses 1–4; ovine adenoviruses 1–6; equine adenoviruses 1 and 2; infectious canine hepatitis virus (canine adenovirus 1); canine adenovirus 2; caprine adenovirus 1; and murine adenoviruses 1 and 2. Genus *Aviadenovirus*: fowl adenoviruses 1–12; turkey adenoviruses 1–3; goose adenoviruses 1–3; pheasant adenovirus 1; and duck adenoviruses 1 and 2.

The Family *Herpesviridae*

Family: *Herpesviridae* (the herpesviruses)
Subfamily: *Alphaherpesvirinae* (the herpes simplex–like viruses)
Genus: *Simplexvirus* (simplexviruses) (Fig. 22)
Genus: *Varicellovirus* (varicelloviruses)
Subfamily: *Betaherpesvirinae* (the cytomegaloviruses)
Genus: *Cytomegalovirus* (cytomegaloviruses)
Genus: *Muromegalovirus* (murine cytomegaloviruses)
Genus: *Roseolovirus* (human herpesvirus 6)
Subfamily: *Gammaherpesvirinae* (the lymphocyte-associated herpesviruses)
Genus: *Lymphocryptovirus* (Epstein-Barr–like viruses)
Genus: *Rhadinovirus* (saimiri-ateles–like herpesviruses)

Characteristics (109,110)

Virions are spherical or pleomorphic, have an overall diameter of 150–200 nm, and consist of (a) an envelope with surface projections, (b) a tegument consisting of amorphous material, (c) an icosahedral nucleocapsid 100 nm in diameter with 162 (150 hexameric, 12 pentameric) prismatic capsomeres, and (d) a core consisting of a fibrillar spool on which the DNA is wrapped. Virion buoyant density is 1.20–1.29 g/cm^3 in CsCl. The genome consists of a single molecule of linear, double-stranded DNA 124–235 kbp in size. Genomic DNA sometimes contains terminal and internal reiterated sequences, sometimes forming two covalently linked components (L and S) that result in the formation of two or four isomeric forms oriented differently in the various subfamilies and genera. The viruses have been divided into six groups depending on the presence or absence and arrangement of the reiterated sequences. Virions have more than 30 structural proteins, and at least as many nonstructural proteins. Among the proteins specified by all herpesviruses are a DNA polymerase, DNA binding proteins, and a protease; there are many additional proteins with enzymatic activities, including as many as three protein kinases. Human herpesvirus 1 (herpes simplex virus 1) contains 11 glycosylated and at least two nonglycosylated proteins in its envelope, including an Fc receptor. Virions contain lipids, their composition matching host cell nuclear or other membranes. Carbohydrates associated with the viral envelope glycoproteins are of the complex high mannose type. Replication starts with circularization of viral DNA; transcription and translation are coordinately regulated and sequentially ordered in a cascade with three major stages. Immediate early genes are transcribed by nuclear enzymes, and mRNAs are transported to the cytoplasm and translated; proteins are then transported to the nucleus and are involved in the synthesis of additional mRNAs. Early proteins are involved in the replication of the viral DNA by a rolling circle mechanism. Late mRNAs are translated mostly into the structural proteins. Replication takes place in the nucleus, and capsids acquire their envelopes via budding through the inner lamella of the nuclear envelope. Virions are released via transport across the cytoplasm in membranous vesicles, which then fuse with the plasma membrane. The viruses generally have narrow host ranges. Transmission is usually by contact (infected cells in saliva, urogenital excretions, or free virus in aerosols). Some viruses induce neoplasia and most persist for the lifetime of their host. The viruses are sensitive to acid, heat, detergents, organic solvents, and UV- and γ-irradiation.

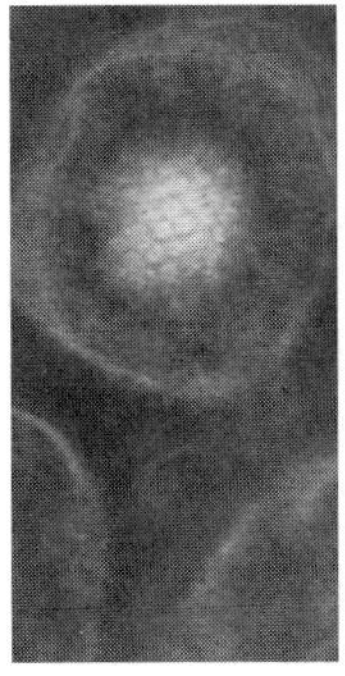
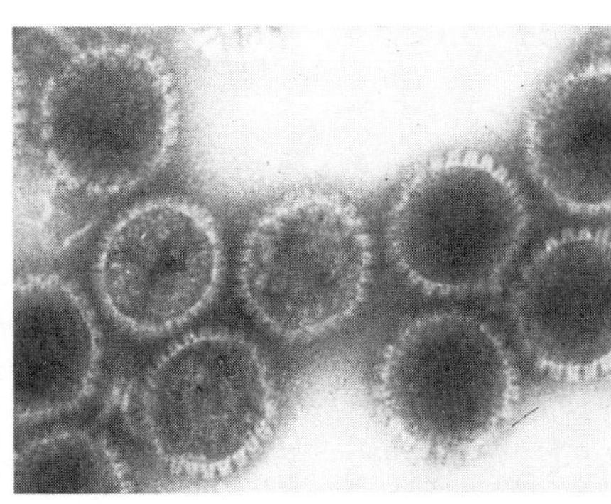

FIG. 22. Family *Herpesviridae*, genus *Simplexvirus,* human herpesvirus 1 (herpes simplex virus 1), enveloped (left), and naked capsids (right) ($\times$147,000).

Human Pathogens

Subfamily *Alphaherpesvirinae*/Genus *Simplexvirus:* human herpesviruses 1 and 2 (herpes simplex viruses 1 and 2); herpesvirus B (cercopithecine herpesvirus 1). Genus *Varicellovirus*: human herpesvirus 3 (varicella-zoster virus). Subfamily *Betaherpesvirinae*/Genus *Cytomegalovirus:* human herpesvirus 5 (human cytomegalovirus). Genus *Roseolovirus:* human herpesviruses 6A, 6B, and 7. Subfamily *Gammaherpesvirinae*/Genus *Lymphocryptovirus:* human herpesvirus 4 (Epstein-Barr virus).

Animal Pathogens

Subfamily *Alphaherpesvirinae*/Genus *Simplexvirus:* herpesvirus B (cercopithecine herpesvirus 1); bovine herpesvirus 2 (bovine mamillitis virus, pseudo-lumpy skin disease virus). Genus *Varicellovirus*: equid herpesvirus 1 (equine abortion virus); equid herpesvirus 4 (equine rhinopneumonitis virus); bovine herpesvirus 1 (infectious bovine rhinotracheitis virus); and pseudorabies virus (suid herpesvirus 1, Aujeszky's disease virus). Unclassified in the Subfamily: many simian herpesviruses [ateline herpesvirus 1 (spider monkey herpesvirus); cercopithecine herpesviruses 2 (SA8), 6, 7, and 9; saimiriine herpesvirus 1 (marmoset herpesvirus)]; equid herpesvirus 3 (equine coital exanthema virus); asinine herpesviruses 1 and 3; bovine herpesvirus 5 (bovine encephalitis herpesvirus); canid herpesvirus 1 (canine herpesvirus); caprine herpesvirus 1 (goat herpesvirus); cervid herpesviruses 1 and 2; felid herpesvirus 1 (feline rhinotracheitis virus); gallid herpesvirus 1 (infectious laryngotracheitis virus); and anatid herpesvirus 1 (duck plague herpesvirus). Subfamily *Betaherpesvirinae*/Genus *Muromegalovirus:* mouse cytomegalovirus 1. Unclassified in the Subfamily: many simian/primate herpesviruses [aotine herpesviruses 1 and 3 (aotus cytomegaloviruses 1 and 3); callitrichine herpesvirus 2 (marmoset cytomegalovirus 2); pongine herpesviruses 1, 2, and 3 (chimpanzee, orangutan, and gorilla herpesviruses); cebine herpesviruses 1 and 2 (capuchin herpesviruses); cercopithecine herpesviruses 3, 4, 5, and 8 (simian cytomegaloviruses); cercopithecine herpesvirus 12 (baboon herpesvirus)]; equid herpesviruses 2 and 5 (equine cytomegaloviruses); suid herpesvirus 2 (swine cytomegalovirus, inclusion-body rhinitis virus); caviid herpesvirus 2 (guinea pig cytomegalovirus); murid herpesvirus 2 (rat cytomegalovirus). Subfamily *Gammaherpes*virinae/Unclassified in the Subfamily: alcelaphine herpesvirus 1 (malignant catarrhal fever virus of cattle, wildebeest herpesvirus); ovine herpesvirus 2 (sheep-associated malignant catarrhal fever of cattle herpesvirus); bovine herpesvirus 4 (Movar herpesvirus); equid herpesviruses 2 and 5; caviid herpesvirus 1 (guinea pig herpesvirus 1); and herpesvirus saimiri 2 (saimiriine herpesvirus 2, squirrel monkey herpesvirus). Unclassified in the Family: many herpesviruses of rodents; channel catfish herpesvirus and other herpesviruses of fish; chelonid herpesvirus 1 (gray patch disease agent of green sea turtle) and other herpesviruses of turtles, lizards, and amphibia; several herpesviruses of snakes; gallid herpesviruses 2 and 3 (Marek's disease viruses); and psittacid herpesvirus 1 (parrot herpesvirus, Pacheco's disease) and many herpesviruses of other birds.

The Family *Poxviridae*

Family: *Poxviridae* (the poxviruses) (Fig. 23)
Subfamily: *Chordopoxvirinae* (the poxviruses of vertebrates)
Genus: *Orthopoxvirus* (orthopoxviruses)
Genus: *Parapoxvirus* (parapoxviruses)
Genus: *Avipoxvirus* (fowlpoxviruses)
Genus: *Capripoxvirus* (sheeppoxlike viruses)
Genus: *Leporipoxvirus* (myxomaviruses)
Genus: *Suipoxvirus* (swinepoxviruses)
Genus: *Molluscipoxvirus* (molluscum contagiosum viruses)
Genus: *Yatapoxvirus* (yabapox and tanapox viruses)
Subfamily: *Entomopoxvirinae* (the poxviruses of insects)
Genus: *Entomopoxvirus A* (insect poxvirus group A)
Genus: *Entomopoxvirus B* (insect poxvirus group B)
Genus: *Entomopoxvirus C* (insect poxvirus group C)

Characteristics (40,46,47)

Virions are large and brick shaped, 220–450 nm long × 140–260 nm wide × 140–260 nm thick (or ovoid in the case of the genus *Parapoxvirus,* 250–300 nm long × 160–190 nm wide), and consist of (a) an external envelope derived from the cell but containing virus-specified proteins, (b) a complex coat of regular spiral filaments 10–20 nm in diameter, and (c) an internal biconcave or cylindrical core that contains the genome DNA and proteins organized in a nucleoprotein complex and one or two lateral bodies. Virion buoyant density is 1.16 g/cm^3 in sucrose and 1.25 g/cm^3 in CsCl. The genome consists of a single molecule of linear, covalently closed, double-stranded DNA

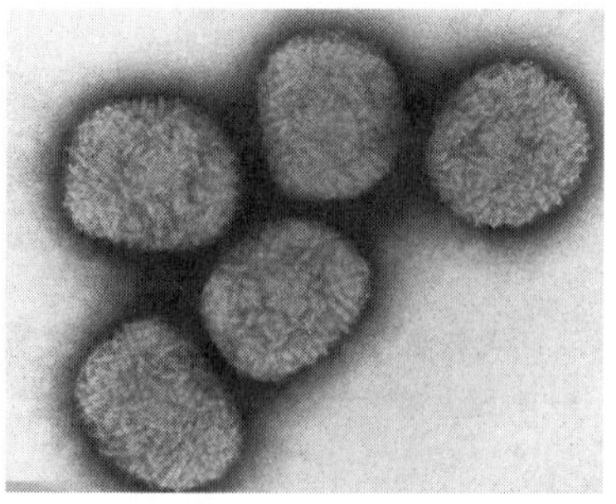

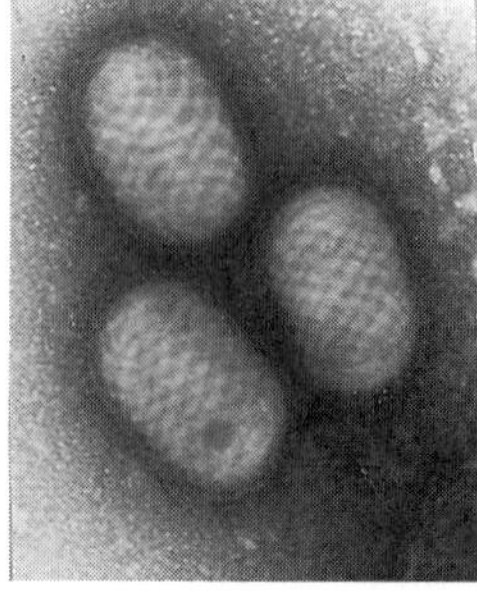

FIG. 23. Family *Poxviridae,* genus *Orthopoxvirus,* vaccinia virus (left) (×20900); genus *Parapoxvirus,* orf virus (right) (×33,000).

130–375 kbp in size. Variably sized tandem repeat sequence arrays are often present near the ends of the DNA. Genomes encode 150–300 proteins varying with the virus; about 100 proteins are present in virions (including a DNA-dependent DNA transcriptase and a DNA-dependent RNA polymerase). Enveloped virions contain lipids, including glycolipids, derived from cellular lipids and synthesized *de novo* during infection; virions contain carbohydrates as glycoproteins. The genome contains closely spaced open reading frames preceded by promoters that temporally regulate transcription of three classes of genes: early genes that encode many nonstructural proteins, including enzymes involved in replicating the genome and modifying DNA, RNA, and proteins; intermediate genes, expressed during the period of DNA replication and regulating transcription; and late genes that mainly encode virion structural proteins. The mRNAs are capped, polyadenylated at their 3'-termini, and not spliced. DNA replication is directed mainly by viral enzymes; although incompletely understood, it seems to involve a self-priming, unidirectional, strand displacement mechanism in which concatemerized replicative intermediates serve as templates for the synthesis of genomic DNA. Virion morphogenesis proceeds via coalescence of DNA within crescent-shaped, lipoprotein bilayers that are coated with spicules; lipoprotein structures enclose the genome to form immature particles that then mature by the addition of coat layers. Replication and assembly occur in cytoplasm in viroplasm ("viral factories"), and virions are released by budding (enveloped virions) or by cell destruction (nonenveloped virions). Transmission is by direct contact, by fomites, by aerosol, or mechanically by arthropods. The viruses generally have narrow host ranges. Some viruses are insensitive to ether and heat stable; generally, they are sensitive to common detergents, formaldehyde, and oxidizing agents.

Human Pathogens

Genus *Orthopoxvirus*: variola (smallpox) virus; vaccinia virus; monkeypox virus; and cowpox virus. Genus *Parapoxvirus*: orf (contagious pustular dermatitis) virus; pseudocowpox (milker's nodule) virus. Genus *Yatapoxvirus*: yabapox virus; tanapox virus. Genus *Molluscipoxvirus*: molluscum contagiosum virus.

Animal Pathogens

Genus *Orthopoxvirus*: vaccinia virus (seen as rabbitpox in colonized rabbits and as buffalopox in buffaloes and cattle); cowpox virus; ectromelia (mousepox) virus; rabbitpox virus; monkeypox virus; camelpox virus; and raccoonpox virus. Genus *Parapoxvirus*: orf (contagious pustular dermatitis) virus; pseudocowpox (milker's nodule) virus; bovine papular stomatitis virus; and parapox virus of red deer. Genus *Avipoxvirus*: many specific bird poxviruses, including fowlpox virus; canarypox virus; and pigeonpox virus. Genus *Capripoxvirus*: sheeppox virus; goatpox virus; and lumpy skin disease virus (cattle). Genus *Leporipoxvirus*: myxoma virus (rabbits); rabbit fibroma virus (Shope fibroma virus); hare fibroma virus; and squirrel fibroma virus. Genus *Suipoxvirus*: swinepox virus. Genus *Mosuscipoxvirus*: molluscum contagiosum virus. Genus *Yatapoxvirus*: tanapox virus and yabapox virus (monkeys).

The Unnamed Family for African Swine Fever Virus and the Family *Iridoviridae*

Family: unnamed, African swine fever-like viruses (Fig. 24)
Genus: unnamed, African swine fever-like viruses
Family: *Iridoviridae* (the iridoviruses)
Genus: *Iridovirus* (small iridescent insect viruses)
Genus: *Chloriridovirus* (large iridescent insect viruses)
Genus: *Ranavirus* (frog iridoviruses)
Genus: *Lymphocystivirus* (lymphocystis viruses of fish)
Genus: unnamed, goldfish iridoviruses

Characteristics (31,131)

African swine fever virus and iridovirus virions consist of a core structure 70–100 nm diameter surrounded by a nucleocapsid 172–191 nm in diameter and a lipid-containing envelope (missing on some iridoviruses). The nucleocapsid exhibits icosahedral symmetry (T = 189–217) with 1,892–2,172 capsomers (capsomers are 13 nm in diameter and appear as hexagonal prisms with a central hole). African swine fever virus has an overall diameter of 175–215 nm (iridoviruses have diameters of 125–300 nm). Virion buoyant density is 1.19–1.24 g/cm^3 in CsCl. The genome of African swine fever virus consists of a single molecule of linear, covalently close-ended, double-stranded DNA, 170–190 kbp in size (varying among iso-

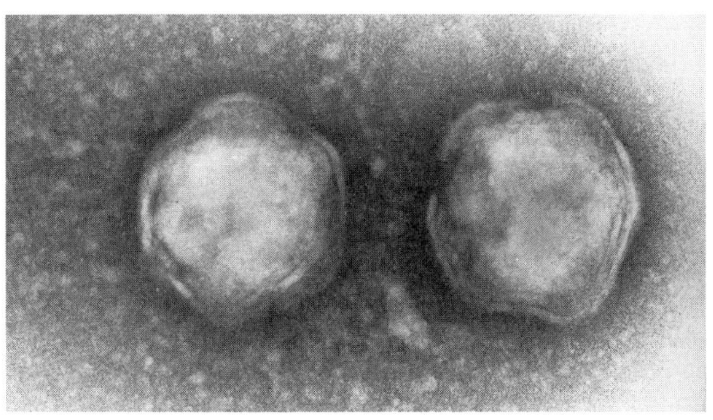

FIG. 24. Unnamed family, African swine fever virus ×75,000).

lates). Iridoviruses have genome sizes of 150–350 kbp. Terminal inverted sequences consist of arrays of tandem repeats adjacent to the ends. Genes are encoded on both DNA strands and are generally closely spaced. At several locations short tandem repeat arrays are found in intergenic regions. Virions contain more than 54 proteins and several virion-associated enzymes required for transcription and posttranscriptional modification of mRNA, including RNA polymerase, poly (A) polymerase, and mRNA capping enzymes. Synthesis of more than 100 virus-induced proteins has been detected in infected cells. Virions contain lipids, including glycolipids; carbohydrates have been demonstrated in virions only in the form of glycolipids. Early mRNA synthesis and processing begins immediately after virus infection. Transcripts are 3'-polyadenylated and 5'-capped. DNA replication proceeds by a self-priming mechanism. Replication occurs in the cytoplasm (although the nucleus is needed for viral DNA synthesis), and virions are released by budding or cell destruction. The only animal species naturally infected by African swine fever virus are domestic and wild swine, warthogs, bushpigs, and giant forest hogs. Virus strains differ in virulence, from apathogenic to causing severe disease and near 100% mortality. African swine fever virus transmission occurs via direct contact, infected meat, fomites, and biting flies, and biologically by soft ticks of the genus *Ornithodoros.* The virus can be transmitted in ticks transstadially, transovarially, and sexually. Virions are sensitive to heat, ether, chloroform, and detergents but are stable over a wide range of pH.

Human Pathogens

The Unnamed Family/Genus for African Swine Fever Virus: none known. Family *Iridoviridae*: none known.

Animal Pathogens

The Unnamed Family/Genus for African Swine Fever Virus: African swine fever virus. Family *Iridoviridae*, genus *Ranavirus:* frog viruses 1–3, 5–24, L2, L4, and L5. Genus *Lymphocystivirus*: flounder iridescent virus; lymphocystis disease virus of fish. Unnamed genus: goldfish viruses 1 and 2.

SUBVIRAL AGENTS: DEFECTIVE VIRUSES, SATELLITES, VIROIDS, AND AGENTS OF SPONGIFORM ENCEPHALOPATHIES (PRIONS)

The Floating Genus *Deltavirus*

Genus: *Deltavirus* (hepatitis delta virus) (Fig. 25)

Characteristics (78,127,128)

Hepatitis delta virus is defective and requires certain helper functions for replication; such functions can be supplied by hepatitis B virus or woodchuck hepatitis virus. Virions are spherical, ~36–43 nm in diameter with no surface projections. The envelope is acquired from the helper virus (HBsAg, when the helper is hepatitis B virus); within is a stable ribonucleoprotein complex forming a spherical core structure 18 nm in diameter. The genome consists of a single molecule of circular, negative-sense, single-stranded RNA, ~1,700 nucleotides in size; it can fold into an unbranched, rod-shaped structure formed by intramolecular base-pairing. Genome replication involves RNA-directed RNA synthesis via a rolling circle mechansim that generates complementary oligomeric forms and involves site-specific autocatalytic cleavage and ligation to generate monomers. The complementary intermediate form is referred to as the antigenome. Only one hepatitis delta virus mRNA is found in infected liver; it directs the synthesis of the single virus protein, hepatitis delta antigen (HDAg, Mr 22,000–24,000); this protein exists in two forms that differ by a 19-amino acid carboxy-terminal extension. The smaller form is needed for genome replication, the larger for particle assembly. The genome structure and catalytic activities of hepatitis delta virus closely resemble those of some viroids and satellite viruses found in certain plants. The translation of hepatitis delta antigen and the dependency on hepadnavirus replication distinguish hepatitis delta virus from plant associated agents.

Human Pathogens

Genus *Deltavirus*: hepatitis delta virus (found in nature only in humans infected with hepatitis B virus).

Animal Pathogens

Genus *Deltavirus*: none known; hepatitis delta virus has been experimentally transmitted to the woodchuck in the presence of woodchuck hepatitis virus.

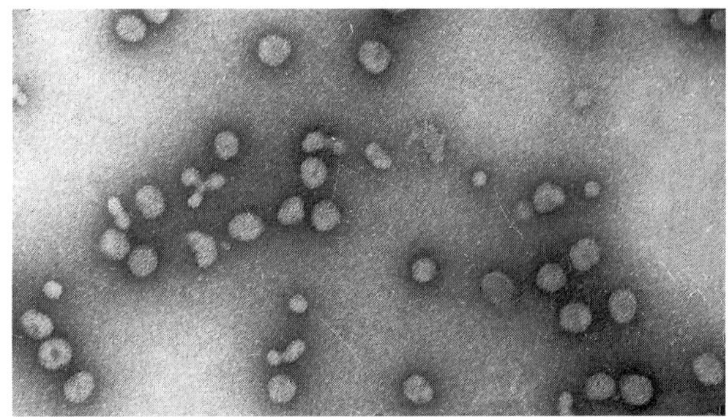

FIG. 25. Genus *Deltavirus,* hepatitis delta virus. Virions (36–43 nm in diameter) purified from serum of an experimentally infected chimpanzee (there are also a few contaminating HBsAg

The Undefined, Unnamed Taxon for the Agents of Spongiform Encephalopathies (Prions)

Characteristics (106,107,135)

Prions are small, proteinaceous infectious particles that contain no detectable nucleic acid of any form but are transmissible among selected species where they induce fatal neurologic disease. Microsomal fractions of brain material enriched for prion infectivity contain numerous microsomes, which when subjected to limited proteolysis generate smooth, ribbonlike, rod-shaped particles (diameter 11 nm; length 165 nm, range 25–550 nm). These rods are composed largely of a protein designated the scrapie isoform of the prion protein, or PrP^{Sc} (Mr 33,000–35,000). A posttranslational process, as yet undefined, generates PrP^{Sc} from the normal cellular isoform of the protein, PrP^{C}. Both PrP^{Sc} and PrP^{C} are encoded by a single-copy chromosomal gene. Although the inoculation into susceptible animals of prion material initiates the production of PrP^{Sc}, the synthesis of PrP^{Sc} originates from the host PrP gene. Prions aggregate into particles of nonuniform size that cannot be solubilized by detergents but can by phospholipids. The multiplication of prion infectivity involves the posttranslational conversion of PrP^{C}, or another precursor, to PrP^{Sc}. Infecting PrP^{Sc} molecules are presumed to combine with homologous host-encoded PrP^{C} molecules giving rise to new PrP^{Sc} molecules. The mechanism of natural transmission of scrapie among sheep and goats is unknown; transmission of the other agents of animals involves consumption of feed supplements containing tissues of infected animals. Similarly, kuru of the New Guinea Fore people is thought to have resulted from the consumption of brain tissue during ritualistic cannibalism. Some transmission of Creutzfeldt-Jakob disease is iatrogenic, but most is of unknown origin. The familial diseases of humans all occur as dominant inherited mutations in the PrP gene. Prions resist inactivation by nucleases, UV irradiation at 254 nm, treatment with psoralens, divalent cations, metal ion chelators, acid (pH 3–7), hydroxylamine, formaldehyde, boiling, or proteases. Prion infectivity is diminished by prolonged digestion with proteases or by treatments with urea, boiling in detergents, alkali (pH >10), autoclaving at 132°C for more than 2 h, denaturing organic solvents (e.g., phenol), or chaotropic agents such as guanidine isocyanate.

Human Pathogens

Undefined, unnamed taxon: kuru agent; Creutzfeldt-Jakob disease agent; Gerstmann-Strüssler-Scheinker syndrome agent; fatal familial insomnia agent.

Animal Pathogens

Undefined, unnamed taxon: scrapie agent of sheep and goats; transmissible mink encephalopathy agent; chronic wasting disease agent of mule deer and elk; bovine spongiform encephalopathy agent; feline spongiform encephalopathy agent; exotic ungulate encephalopathy agent of nyala and greater kudu.

UNCLASSIFIED HUMAN AND ANIMAL VIRUSES

Borna Disease Virus

The virus causes a meningoencephalomyelitis in a wide variety of vertebrate hosts, including horses, sheep, cats, and birds, and has been experimentally transmitted to rodents, rabbits, and primates (macaques) (80,108). Serologic evidence suggests that the virus can infect humans and cause neuropsychiatric disorders (8). The virus has never been seen by electron microscopy; however, in recent studies it has been characterized as a negative-sense, single-stranded RNA virus, seemingly related to the member viruses of the order *Mononegavirales* (families *Paramyxoviridae*, *Rhabdoviridae*, and *Filoviridae*) (13). In contrast to the member viruses of these established taxa, however, the 8.9-kb genome of Borna disease virus is transcribed in the host cell nucleus into subgenomic RNAs and produces high levels of polycistronic messages. The genome contains three transcriptional units coding for five proteins through polymerase readthrough and posttranscriptional RNA splicing (119). Deduced amino acid sequences have shown only very limited homology with known viruses (108). These observations suggest that this virus will require construction of a new taxon.

Hepatitis X virus

There are patients with hepatitis, parenterally acquired by blood transfusion, sexual contact, and so forth, where tests for all known hepatitis viruses are negative. Blood and liver tissue from some of these patients have caused liver pathology in primates, and this evidence of infection has been found when primate blood and tissue have been serially passaged. One passage series seems to have involved a small, solvent-resistant agent, but in this series, as in all others, no virus has been observed by electron microscopy or characterized further (D Bradley, personal communication, 1993).

Unclassified Arboviruses

Several viruses that have been isolated in the course of arbovirus investigations have been only partially characterized; in all cases they are available in reference collections. All have been passaged in mammalian cell cultures and in some cases in experimental animals. Some physicochemical data are available as well, but not enough to place them in established taxa. This is a dynamic list; at present it includes the following:

1. Araguari virus (BeAn 174214): isolated from *Philander opossum* (marsupial) in Brazil, initially in 1969; lipid solvent sensitive, RNA and protein profiles, hemagglutinin, electron microscopy (142).
2. Aride virus (EgArT 3088): isolated from *Amblyomma loculosum* (tick) in the Seychelles, initially in 1973; lipid solvent sensitive.
3. Jembrana virus: isolated from *Bos javanicus* (Bali cattle) in the course of an epidemic marked by 26,000 to 70,000 deaths in 1964; lipid solvent sensitive, membrane filtration data, electron microscopy (142).
4. Johnston Atoll virus (original): isolated from *Ornithodoros capensis* (tick) in Johnston Atoll and Hawaii, United States, Australia, and New Zealand, initially in 1964; lipid solvent sensitive, RNA and protein profiles, electron microscopy (142).
5. Midway virus (RML 47153): isolated from *Ornithodoros capensis* and *O. denmarki* (ticks) on Midway Island, United States (also isolated on other islands in Hawaii and in Japan), initially in 1966; lipid solvent sensitive, RNA (via BUdR and actinomycin D), membrane filtration data, electron microscopy (142).
6. Nyamanini virus (SAAn 12526): isolated from *Bubulcus ibis* (cattle egret) in Natal, South Africa (also isolated in Egypt, Nigeria and India), initially in 1-957; lipid solvent sensitive, electron microscopy (142).
7. Oubangui virus (ArB 3816): isolated from *Culex guiarti* (mosquito) in Central African Republic in 1972; lipid solvent sensitive.
8. Oyo virus (2898): isolated from *Sus scrofa* (domestic pig) in Nigeria in 1964; lipid solvent sensitive.
9. Quaranfil virus (Ar 1113): isolated from *Argas arboreus* (tick), *Bubulcus ibis* (cattle egret) and *Columba livia* (pigeon) in Egypt (isolates have also been obtained in South Africa, Nigeria, Afghanistan, Kuwait, Iraq, and Yemen), initially in 1953; lipid solvent sensitive, RNA and protein profiles, electron microscopy (142).
10. Salanga virus (AnB 904a): isolated from *Aethomys medicatus* (vertebrate) in Central African Republic, in 1971; lipid solvent sensitive.
11. Sebokele virus (AnB 1227b): isolated from *Hylomyscus* and *Praomys* species (rodents) in Central African Republic, initially in 1972; not lipid solvent sensitive (142).
12. Sembalam virus (CMC 8518): isolated from *Nycticorax nycticorax* (night heron) and *Ardea cinerea* (bird) in India, initially in 1963; lipid solvent sensitive.
13. Slovakia virus (265): isolated from *Argas* persicus (tick) in Slovakia, in 1976; not lipid solvent sensitive (142).
14. Somone virus (DakArD 4499): isolated from *Amblyomma variegatum* and *Boophilus decoloratus* (ticks) in Senegal, initially in 1968; lipid solvent sensitive (142).
15. Toure virus (DakAnD 4611): isolated from *Tatera kempi* (rodent) in Senegal, initially in 1968; lipid solvent sensitive (142).
16. Venkatapuram virus (IG 2464): isolated from *Culex vishnui* (mosquito) in India, in 1955; lipid solvent sensitive.

AN OVERVIEW OF THE FAMILIES OF VIRUSES CONTAINING PLANT, INVERTEBRATE, FUNGAL, AND BACTERIAL VIRUSES

In the Sixth Report of the ICTV (Virus Taxonomy, The Classification and Nomenclature of Viruses, 1995) (96), the taxa encompassing all vertebrate, invertebrate, plant, fungal, and bacterial viruses are presented in an order somewhat different than that used in this book. The listing of taxa presented in Table 6 follows the order used by the ICTV. The listing includes one order, 71 families, and 164 genera. The names of orders, families, and genera approved by the ICTV are printed in italics. Names that have not yet been approved are printed in parentheses in standard type. Species names, whether approved or not, are printed in standard type. In most cases, the examples given of member viruses are the type species of the taxon. Taxa for satellites, viroids, and prions (the agents of spongiform encephalopathies) are appended, but are not as yet complete or approved by the ICTV.

ENDNOTE

The author wishes to express his appreciation to the members of the International Committee on Taxonomy of Viruses (ICTV) and to the more than 400 virologists who have participated in its recent proceedings. This chapter follows the taxonomic system of the ICTV, differing only in focusing mostly on the viruses of vertebrates; for up-to-date information on the continuing proceedings of the Committee, the reader is referred to the most current Report of the ICTV (at this writing, this is *Virus Taxonomy–The Sixth Report of the International Committee on Taxonomy of Viruses.* Springer-Verlag, Vienna, 1995).

TABLE 6. *The families of viruses containing plant, invertebrate, fungal, bacterial, and vertebrate viruses*

Order	Family	Subfamily	Genus	Host category: type species or example
			Double-stranded DNA viruses	
	Myoviridae		"Unnamed, the T4-like phages"[a]	Bacteria: coliphage T4
	Siphoviridae		"Unnamed, the λ-like phages"	Bacteria: coliphage λ
	Podoviridae		"Unnamed, the T7-like phages"	Bacteria: coliphage T7
	Tectiviridae		*Tectivirus*	Bacteria: enterobacteria phage PRD 1
	Corticoviridae		*Corticovirus*	Bacteria: Alteromonas phage PM2
	Plasmaviridae		*Plasmavirus*	Mycoplasma: Acholeplasma phage L2
	Lipothrixviridae		*Lipothrixvirus*	Archaebacteria: Thermoproteus phage 1
	Fuselloviridae		*Fusellovirus*	Archaebacteria: Sulfolobus phage 1
	Poxviridae	Chordopoxvirinae	*Orthopoxvirus*	Vertebrates: vaccinia virus
			Parapoxvirus	Vertebrates: orf virus
			Avipoxvirus	Vertebrates: fowlpox virus
			Capripoxvirus	Vertebrates: sheeppox virus
			Leporipoxvirus	Vertebrates: myxoma virus
			Suipoxvirus	Vertebrates: swinepox virus
			Molluscipoxvirus	Vertebrates: molluscum contagiosum virus
			Yatapoxvirus	Vertebrates: Yaba monkeypox virus
		Entomopoxvirinae	*Entomopoxvirus A*	Invertebrates: Melolontha melolontha virus
			Entomopoxvirus B	Invertebrates: Amsacta moorei virus
			Entomopoxvirus C	Invertebrates: Chironomus luridus virus
	"Unnamed, African swine fever-like viruses"		"Unnamed, African swine fever-like viruses"	Vertebrates[b]: African swine fever virus
	Iridoviridae		*Iridovirus*	Invertebrates: Chilo iridescent virus
			Chloriridovirus	Invertebrates: mosquito iridescent virus
			Ranavirus	Vertebrates: frog virus 3
			Lymphocystivirus	Vertebrates: flounder iridescent virus
			"Unnamed, goldfish virus 1-like viruses"	Vertebrates: goldfish virus 1
	Phycodnaviridae		*Phycodnavirus*	Green algae: Paramecium bursaria Chlorella virus 1
	Baculoviridae		*Nucleopolyhedrovirus*	Invertebrates: Autographa californica nuclear polyhedrosis virus
			Granulovirus	Invertebrates: Plodia interpunctella virus
	Herpesviridae	Alphaherpesvirinae	*Simplexvirus*	Vertebrates: human herpesvirus 1 (herpes simplex virus 1)
			Varicellovirus	Vertebrates: human herpesvirus 3 (varicella-zoster virus)
		Betaherpesvirinae	*Cytomegalovirus*	Vertebrates: human herpesvirus 5 (human cytomegalovirus)
			Muromegalovirus	Vertebrates: mouse cytomegalovirus 1
			Roseolovirus	Vertebrates: human herpesvirus 6B
		Gammaherpesvirinae	*Lymphocryptovirus*	Vertebrates: human herpesvirus 4 (Epstein-Barr virus)
			Rhadinovirus	Vertebrates: ateline herpesvirus 2
	Adenoviridae		*Mastadenovirus*	Vertebrates: human adenovirus 2
			Aviadenovirus	Vertebrates: fowl adenovirus 1
			Rhizidovirus	Fungi: Rhizidomyces virus
	Papovaviridae		*0Papillomavirus*	Vertebrates: cottontail rabbit papillomavirus (Shope)
			Polyomavirus	Vertebrates: polyomavirus
	Polydnaviridae		*Ichnovirus*	Invertebrates: Campoletis sonorensis virus

TABLE 6. *Continued.*

Order	Family	Subfamily	Genus	Host category: type species or example
			Bracovirus	Invertebrates: Cotesia melanoscela virus
		Single-stranded DNA viruses		
	Inoviridae		*Inovirus*	Bacteria: coliphage fd
			Plectrovirus	Mycoplasma: Acholeplasma phage L51
	Microviridae		*Microvirus*	Bacteria: coliphage ÌX174
			Spiromicrovirus	Spiroplasma: Spiroplasma phage SpV4
			Bdellomicrovirus	Bacteria: Bdellovibrio phage MAC 1
			Chlamydiamicrovirus	Chlamydia: Chlamydia phage Chp 1
	Geminiviridae		"Unnamed, Subgroup I viruses"	Plants: maize streak virus
			"Unnamed, Subgroup II viruses"	Plants: beet curly top virus
			"Unnamed, Subgroup III viruses"	Plants: tomato golden mosaic virus
	Circoviridae		*Circovirus*	Vertebrates: chicken anemia virus
	Parvoviridae	Chordoparvovirinae	*Parvovirus*	Vertebrates: minute virus of mice
			Dependovirus	Vertebrates: adeno-associated virus 2
			Erythrovirus	Vertebrates: human parvovirus B 19
		Entomoparvovirinae	*Densovirus*	Invertebrates: Junonia coenia virus
			Iteravirus	Invertebrates: Bombyx mori virus
			Contravirus	Invertebrates: Aedes aegypti virus
		DNA and RNA reverse transcribing viruses		
			Badnavirus	Plants: commelina yellow mottle virus
			Caulimovirus	Plants: cauliflower mosaic virus
	Hepadnaviridae		*Orthohepadnavirus*	Vertebrates: hepatitis B virus
			Avihepadnavirus	*Vertebrates: duck hepatitis virus*
	Retroviridae		"Unnamed, mammalian type B retroviruses"	Vertebrates: mouse mammary tumor virus
			"Unnamed, mammalian type C retroviruses"	Vertebrates: murine leukemia virus
			"Unnamed, avain type C retroviruses"	Vertebrates: avian leukosis virus
			"Unnamed, mammalian type D retroviruses"	Vertebrates: Mason-Pfizer monkey virus
			"Unnamed, HTLV/BLV viruses"	Vertebrates: bovine leukemia virus
			Lentivirus	Vertebrates: human immunodeficiency virus 1
			Spumavirus	Vertebrates: human foamy virus 1
		Double-stranded RNA viruses		
	Cystoviridae		*Cystovirus*	Bacteria: Pseudomonas phage Ì6
	Reoviridae		*Orthoreovirus*	Vertebrates: reovirus 3
			Orbivirus	Vertebrates: bluetongue virus 1
			Coltivirus	Vertebrates: Colorado tick fever virus
			Rotavirus	Vertebrates: simian rotavirus SA11
			Aquareovirus	Vertebrates: golden shiner virus
			Cypovirus	Invertebrates: Bombyx mori cytoplasmic polyhedrosis virus 1
			Phytoreovirus	Plants: wound tumor virus
			Fijivirus	Plants: Fiji disease virus
			Oryzavirus	Plants: rice ragged stunt virus
	Birnaviridae		*Aquabirnavirus*	Vertebrates: infectious pancreatic necrosis virus
			Avibirnavirus	Vertebrates: infectious bursal disease virus
			Entomobirnavirus	Invertebrates: Drosophila X virus

TABLE 6. *Continued.*

Order	Family	Subfamily	Genus	Host category: type species or example
	Totiviridae		*Totivirus*	Fungi: Saccharomyces cerevisiae virus L-A
			Giardiavirus	Protozoa: Giardia lamblia virus
			Leishmaniavirus	Protozoa: Leishmania brasiliensis virus 1-1
	Partitiviridae		*Partitivirus*	Fungi: Gaeumannomyces graminis virus 019/6A
			Chrysovirus	Fungi: Penicillium chrysogenum virus
			Alphacryptovirus	Plants: White clover cryptic virus I
			Betacryptovirus	Plants: White clover cryptic virus II
	Hypoviridae		*Hypovirus*	Fungi: Cryphonectria parasitica virus 1-EP713
		Negative-sense, single-stranded RNA viruses		
Mono-negavirales				
	Paramyxoviridae	Paramyxovirinae	*Paramyxovirus*	Vertebrates: parainfluenza virus 1
			Morbillivirus	Vertebrates: measles virus
			Rubulavirus	Vertebrates: mumps virus
		Pneumovirinae	*Pneumovirus*	Vertebrates: respiratory syncytial virus
	Rhabdoviridae		*Lyssavirus*	Vertebrates: rabies virus
			Vesiculovirus	Vertebrates: vesicular stomatitis Indiana virus 1
			Ephemerovirus	Vertebrates: bovine ephemeral fever virus
			Cytorhabdovirus	Plants: lettuce necrotic yellows virus
			Nucleorhabdovirus	Plants: potato yellow dwarf virus
	Filoviridae		*Filovirus*	Vertebrates: Marburg virus
	Orthomyxoviridae		*Influenzavirus A, B*	Vertebrates: influenza A virus A/PR/8/34(H1N1)
			Influenzavirus C	Vertebrates: influenza C virus
			"Unnamed, Thogoto-like viruses"	Vertebrates: Thogoto virus
	Bunyaviridae		*Bunyavirus*	Vertebrates: Bunyamwera virus
			Nairovirus	*Vertebrates:* Nairobi sheep disease virus
			Phlebovirus	Vertebrates: sandfly fever Sicilian virus
			Hantavirus	Vertebrates: Hantaan virus
			Tospovirus	Plants: tomato spotted wilt virus
	Arenaviridae		*Arenavirus*	Vertebrates: lymphocytic choriomeningitis virus
			Tenuivirus	Plants: rice stripe virus
		Positive-sense, single-stranded RNA viruses		
	Leviviridae		*Levivirus*	Bacteria: coliphage MS2
			Allolevirus	Bacteria: coliphage Q·
	Picornaviridae		*Enterovirus*	Vertebrates: poliovirus 1
			Aphthovirus	Vertebrates: foot-and-mouth disease virus O
			Cardiovirus	Vertebrates: encephalomyocarditis virus
			Hepatovirus	Vertebrates: hepatitis A virus
			Rhinovirus	Vertebrates: human rhinovirus 1A
	Sequiviridae		*Sequivirus*	Plants: parsnip yellow fleck virus
			Waikavirus	Plants: rice tungro spherical virus
	Comoviridae		*Comovirus*	Plants: cowpea mosaic virus
			Nepovirus	Plants: tobacco ringspot virus
			Fabavirus	Plants: broad bean wilt virus 1
	Potyviridae		*Potyvirus*	Plants: potato virus Y
			Bymovirus	Plants: barley yellow mosaic virus
			Rymovirus	Plants: ryegrass mosaic virus
	Caliciviridae		*Calicivirus*	Vertebrates: vesicular exanthema of swine virus
	Astroviridae		*Astrovirus*	Vertebrates: human astrovirus 1
	Nodaviridae		*Nodavirus*	Invertebrates: Nodamura virus

TABLE 6. *Continued.*

Order	Family	Subfamily	Genus	Host category: type species or example
	Tetraviridae		"Unnamed, Nudaurelia β capensis-like viruses" "Unnamed, Nudaurelia ω capensis-like viruses"	Invertebrates: Nudaurelia β capensis virus Invertebrates: Nudaurelia ω capensis virus
			Sobemovirus	Plants: southern bean mosaic virus
			Luteovirus	Plants: barley yellow dwarf virus
			Enamovirus	Plants: pea enation mosaic virus
			Umbravirus	Plants: carrot mottle virus
	Tombusviridae		*Tombusvirus* *Carmovirus*	Plants: tomato bushy stunt virus Plants: carnation mottle virus
			Necrovirus	Plants: tobacco necrosis virus
			Dianthovirus	Plants: carnation ringspot virus
			Machlomovirus	Plants: maize chlorotic mottle virus
	Coronaviridae		*Coronavirus* *Torovirus*	Vertebrates: avian infectious bronchitis virus Vertebrates: Berne virus
			Arterivirus	Vertebrates: equine arteritis virus
	Flaviviridae		*Flavivirus* *Pestivirus* "Unnamed, hepatitis C-like viruses"	Vertebrates: yellow fever virus Vertebrates: bovine virus diarrhea virus Vertebrates: hepatitis C virus
	Togaviridae		*Alphavirus* *Rubivirus*	Vertebrates: Sindbis virus Vertebrates: rubella virus
			Tobamovirus	Plants: tobacco mosaic virus
			Tobravirus	Plants: tobacco rattle virus
			Hordeivirus	Plants: barley stripe mosaic virus
			Furovirus	Plants: soil-borne wheat mosaic virus
	Bromoviridae		*Bromovirus* *Cucumovirus* *Ilarvirus* *Alfamovirus*	Plants: brome mosaic virus Plants: cucumber mosaic virus Plants: tobacco streak virus Plants: alfalfa mosaic virus
			Idaeovirus	Plants: rasberry bushy dwarf virus
			Closterovirus	Plants: beet yellows virus
			Capillovirus	Plants: apple stem grooving virus
			Trichovirus	Plants: apple chlorotic leaf spot virus
			Tymovirus	Plants: turnip yellow mosaic virus
			Carlavirus	Plants: carnation latent virus
			Potexvirus	Plants: potato X virus
	Barnaviridae		*Barnavirus*	Fungi: mushroom bacilliform virus
			Marafivirus	Plants: maize rayado fino virus
		Subviral agents: satellites, viroids, and prions		
Taxon: undefined, unnamed			Satellites	Plants: cucumber mosaic virus satellite
Genus			*Deltavirus*	Vertebrates: hepatitis delta virus
Taxon: undefined, unnamed			Viroids	Plants: potato spindle tuber viroid
Taxon: undefined, unnamed			Prions	Vertebrates: scrapie agent

[a]Quotation marks are used to denote that the taxon has not been named or that the taxon name has not been approved by the ICTV.
[b]Vertebrate arthropod-borne viruses are listed according to their vertebrate hosts.

REFERENCES

1. American Type Culture Collection. *Catalogue of animal viruses and antisera, Chlamydiae and Rickettsiae.* 6th ed. Rockville, MD: American Type Culture Collection; 1990.
2. Andrewes CH. Nomenclature of viruses. *Nature* 1954;173: 260–261.
3. Andrewes CH, Bang FB, Burnet FM. A short description of the Myxovirus group (influenza and related viruses). *Virology* 1955;1:176–180.
4. Atherton JG, Holmes IR, Jobbins EH. ICTV code for the description of virus characters. *Monogr Virol* 1983;14:1–154.
5. Bawden FC. *Plant viruses and virus diseases.* 1st ed. Waltham, MA: Chronica Botanica Company; 1941.
6. Bawden FC. *Plant viruses and virus diseases.* 3rd ed. Waltham, MA: Chronica Botanica Company; 1950.
7. Bishop DHL, Calisher CH, Casals J, et al. Bunyaviridae. *Intervirology* 1980;14:125–143.
8. Bode L, Ferszt R, Czech G. Borna disease virus infection and affective disorders in man. *Archives of Virology Supplementum*, 1993;7:159–167.
9. Boswell KF, Dallwitz MJ, Gibbs AJ, Watson L. The VIDE (Virus Identification Data Exchange) project: a data bank for plant viruses. *Rev Plant Pathol* 1986;65:221–231.
10. Bouloy M. Bunyaviridae: genome organization and replication strategies. *Adv Virus Res* 1991;40:235–66.
11. Brandes J, Wetter C. Classification of elongated plant viruses on the basis of particle morphology. *Virology* 1959;8:99–109.
12. Brenner S, Horne RW. A negative staining method for high resolution electron microscopy of viruses. *Biochim Biophys Acta* 1959;34:103–110.
13. Briese T, Schneemann A, Lewis AJ, Park Y-S, Kim S, Lipkin WI. Genomic organization of Borna disease virus. *Proc Natl Acad Sci USA* 1994;91:4362–4366.
14. Brown F. The classification and nomenclature of viruses: summary of results of meetings of the ICTV in Sendai, September 1984. *Intervirology* 1986;25:141–143.
15. Brown F, Bishop DHL, Crick J, et al. Rhabdoviridae. *Intervirology* 1979;12:1–17.
16. Bruenn JA. Relationship among the positive strand and double-strand RNA viruses as viewed through their RNA-dependent RNA polymerases. *Nucl Acids Res* 1991;19:217–225.
17. Brunt A, Crabtree K, Gibbs AJ, Watson L, eds. *Viruses of plants.* 2 volumes. London: C.A.B. International; 1992.
18. Buchen-Osmond C, Blaine LD, Gibbs AJ. Towards a comprehensive virus data base. In: *Proceedings of the Ninth International Congress of Virology.* Glasgow: W76-1, 1993:116.
19. Calisher CH, Karabatos N, Dalrymple JM, et al. Antigenic relationships between flaviviruses as determined by cross-neutralization tests with polyclonal antisera. *J Gen Virol* 1989;70:37–43.
20. Calisher CH, Karabatsos N, Zeller H, et al. Antigenic relationships among rhabdoviruses from vertebrates and hematophagous arthropods. *Intervirology* 1989;30:241–257.
21. Carter MJ, Milton ID, Meanger J, Bennett MJ, Gaskell RM, Turner PC. The complete nucleotide sequence of a feline calicivirus. *Virology* 1992;190:443–448.
22. Cavanagh D, Brian DA, Enjuanes L, et al. Recommendations of the coronavirus study group for the nomenclature of the structural proteins, mRNAs and genes of coronaviruses. *Virology* 1990;176:306–307.
23. Chambers TJ, Hahn CS, Galler R, Rice CM. Flavivirus genome organization, expression, and replication. *Annu Rev Microbiol* 1990;44:649–688.
24. Chan S-Y, Bernard H-U, Ong C-K, Chan S-P, Hofmann B, Delius H. Phylogenetic analysis of 48 papillomavirus types and 28 subtypes and variants: a showcase for the molecular evolution of DNA viruses. *J Virol* 1992;66:5714–5725.
25. Chirnside ED. Equine arteritis virus: an overview. *Br Vet J* 1992; 148:181–198.
26. Choo QL, Richman KH, Han JH, et al. Genetic organization and diversity of the hepatitis C virus. *Proc Natl Acad Sci USA* 1991; 88:2451–2455.
27. Chroboczek J, Bieber F, Jacrot B. The sequence of the genome of adenovirus type 5 and its comparison with the genome of adenovirus type 2. *Virology* 1992;186:280–285.
28. Coffin JM. Structure and classification of retroviruses. In: Levy J, ed. *The retroviridae.* Vol 1. New York: Plenum; 1993.
29. Cooper PD, Agol VI, Bachrach HL, et al. Picornaviridae: second report. *Intervirology* 1978;10:165–180.
30. Den Boon JA, Snijder EJ, Chirnside ED, de Vries AAF, Horzinek MC, Spaan WJM. Equine arteritis virus is not a togavirus but belongs to the coronavirus-like superfamily. *J Virol* 1991;65:2910–2920.
31. Dixon LK, Wilkinson PJ, Sumpton KJ, Ekue F. Diversity of the African swine fever virus genome. In: Darai G, ed. *Molecular biology of iridoviruses.* Boston: Kluwer Academic; 1990:271–296.
32. Dobos P, Hill BJ, Hallett R, Kells DTC, Becht H, Teninges D. Biophysical and biochemical characterization of five animal viruses with bisegmented double-stranded RNA genomes. *J Virol* 1979;32: 593–605.
33. Dobos P, Roberts TE. The molecular biology of infectious pancreatic necrosis virus; a review. *Can J Microbiol* 1983;29:377–384.
34. Dolja VV, Carrington JC. Evolution of positive-strand RNA viruses. *Semin Virol* 1992;3:315–326.
35. Dominguez GD, Wang C-Y, Frey TK. Sequence of the genome RNA of rubella virus: evidence for genetic rearrangement during togavirus evolution. *Virology* 1990;177:225–238.
36. Doolittle RF, Feng D-F, Johnson MS, McClure MA. Origins and evolutionary relationships of retroviruses. *Q Rev Biol* 1989;64:1–30.
37. Dowdle WR, Davenport FM, Fukumi H, et al. *Orthomyxoviridae. Intervirology* 1975;5:245–251.
38. Eigen M. Viral quasispecies. *Scientific American* 1993;269(1): 42–49.
39. Elliott RM, Schmaljohn CS, Collett MS. Bunyaviridae genome structure and gene expression. *Current topics in microbiology and immunology.* Berlin: Springer-Verlag; 1991:91–142.
40. Esposito JJ, Fenner F. Poxvirus infections in humans. In: Balows A, editor-in-chief. *Manual of clinical microbiology*, 5th ed. Washington, DC:American Society for Microbiology;1991:858–867.
41. Evans AS. *Viral Infections of Humans. Epidemiology and Control.* 2nd ed. New York: Plenum; 1982.
42. Feldmann H, Muhlberger E, Randolf A, et al. Marburg virus, a filovirus: messenger RNAs, gene order, and regulatory elements of the replication cycle. *Virus Res* 1992;24:1–19.
43. Fenner F. The classification and nomenclature of viruses: second report of the International Committee on Taxonomy of Viruses. *Intervirology* 1976;7:1–115.
44. Fenner F. The classification of viruses; why, when and how. *Aust J Exp Biol Med Sci* 1974;52:223–231.
45. Fenner F, Burnet FM. A short description of the poxvirus group (vaccinia and related viruses). *Virology* 1957;4:305–310.
46. Fenner F, Nakano JH. Poxviridae: the poxviruses. In: Lennette EH, Halonen P, Murphy FA, eds. *Laboratory diagnosis of infectious diseases: principles and practice.* Volume 2: viral, rickettsial and chlamydial diseases. New York: Springer-Verlag; 1988;177–207.
47. Fenner F, Wittek R, Dumbell KR. *The orthopoxviruses.* New York: Academic; 1989.
48. Francki RIB, Fauquet CM, Knudson DL, Brown F. *The classification and nomenclature of viruses: fifth report of the International Committee on Taxonomy of Viruses.* Vienna: Springer-Verlag; 1991.
49. Frerichs GN. Rhabdoviruses of fishes. In: Ahne W, Kurstak E, eds. *Viruses of lower vertebrates.* New York: Springer-Verlag; 1989:316–332.
50. Ganem D, Varmus HE. The molecular biology of the hepatitis B viruses. *Annu Rev Biochem* 1987;56:651–693.
51. Gibbs AJ. Molecular evolution of viruses: "trees, clocks and modules." *J Cell Sci* 1987;7(suppl):319–337.
52. Gibbs AJ, Harrison BD. Realistic approach to virus classification and nomenclature. *Nature* 1966;218:927–929.
53. Gibbs AJ, Harrison BD, Watson DH, Wildy P. What's in a virus name? *Nature* 1966;209:450–454.
54. Goldbach RW. Genome similarities between plant and animal RNA viruses. *Microbiol Sci* 1987;4:197–202.
55. Goldbach RW. Molecular evolution of plant RNA viruses. *Annu Rev Phytopathol* 1986;24:289–310.
56. Goldbach RW, Wellink J. Evolution of plus-strand RNA viruses. *Intervirology* 1988;29:260–268.
57. Goldbach RW, Le Gall O, Wellink J. Alpha-like viruses in plants. *Semin Virol* 1991;2:19–25.
58. Goldbach RW, de Haan P. RNA viral supergroups and the evolution of RNA viruses. In: Morse SS, ed. *The evolutionary biology of viruses.* New York: Raven; 1993:105–119.

59. Gust ID, Burrell CJ, Coulepis AG, Robinson WS, Zuckerman AJ. Taxonomic classification of human hepatitis B virus. *Intervirology* 1986;25:14–29.
60. Gust ID, Coulepis AG, Feinstone SM, et al. Taxonomic classification of hepatitis A virus. *Intervirology* 1983;20:1–7.
61. Harrison BD, Finch JT, Gibbs AJ, et al. Sixteen groups of plant viruses. *Virology* 1966;45:356–363.
62. Hierholzer JC, Wigand R, Anderson LJ, Adrian T, Gold JWM. Adenoviruses from patients with AIDS; a plethora of serotypes and a description of five new serotypes of subgenus D (types 43–47). *J Infect Dis* 1988;158:804–813.
63. Holland JJ, Spindler K, Horodyski F, Grabau E, Nichol S, Van de Pol S. Rapid evolution of RNA genomes. *Science* 1982;215: 1577–1585.
64. Holland JJ, de la Torre JC, Steinhauer DA. RNA virus populations as quasispecies. *Curr Top Microbiol Immunol* 1992;176:1–20.
65. Howard C, Melnick J. Classification and taxonomy of hepatitis viruses. In: Hollinger F, Lemon S, Margolis H, eds. *Viral hepatitis and liver disease.* Baltimore: Williams & Williams; 1991:890–892.
66. Jiang X, Graham DY, Wang K, Estes MK. Norwalk virus genome cloning and characterization. *Science* 1991;250:1580–1583.
67. Kamer G, Argos P. Primary structural comparison of RNA-dependent polymerases from plant, animal and bacterial viruses. *Nucl Acids Res* 1984;12:7269–7282.
68. Karabatsos N, ed. *International catalogue of arboviruses.* 3rd ed. San Antonio: American Society of Tropical Medicine and Hygiene; 1985.
69. Kiley MP, Bowen ETW, Eddy GA, et al. Filoviridae: a taxonomic home for Marburg and Ebola viruses? *Intervirology* 1982;18:24–32.
70. Kingsbury DW. Biological concepts in virus classification. *Intervirology* 1988;29:242–253.
71. Kingsbury DW. *The paramyxoviruses.* New York: Plenum; 1991.
72. Kingsbury DW, Bratt MA, Choppin PW, et al. Paramyxoviridae. *Intervirology* 1978;10:137–152.
73. Koonin EV. The phylogeny of RNA-dependent RNA polymerases of positive-strand RNA viruses. *J Gen Virol* 1991;72:2197–2206.
74. Koonin EV, Dolja VV. Evolution and taxonomy of positive-strand RNA viruses: implications of comparative analysis of amino acid sequences. *Crit Rev Biochem Mol Biol* 1993;28:375–430.
75. Koonin EV, Gorbalenya AE. Evolution of RNA genomes: does the high mutation rate necessitate a high rate of evolution of viral proteins? *J Mol Evolution* 1989;28:524–527.
76. Koopmans M, Horzinek MC. *Toroviruses* of animals and humans. A Review. *Adv Virus Res* 1994;43:233–273.
77. Krug RM. The influenza viruses In: Fraenkel-Conrat H, Wagner RR, eds. *The viruses.* New York: Plenum; 1990.
78. Lai MMC, Chao Y-C, Chang M-F, Lin J-H, Gust I. Functional studies of hepatitis delta antigen and delta virus RNA. In: Gerin JL, Purcell RH, Rizzeto M, eds. *The hepatitis delta virus.* New York: Alan R. Liss; 1991:283–292.
79. Lambert PF. Papillomavirus DNA replication. *J Virol* 1991;65:3417–3420.
80. Ludwig H, Bode HL, Gosztonyi G. Borna disease: a persistent virus infection of the central nervous system. *Prog Med Virol* 1988;35:107–151.
81. Lwoff A, Horne R, Tournier P. A system of viruses. *Cold Spring Harb Symp Quant Biol* 1962;27:51–55.
82. Mahy BWJ. Related viruses of the plant and animal kingdoms. *Semin Virol* 1991;2:1–77.
83. Matthews REF. Classification and nomenclature of viruses: fourth report of the International Committee on Taxonomy of Viruses. *Intervirology* 1982;17:1–199.
84. Matthews REF. Classification and nomenclature of viruses: third report of the International Committee on Taxonomy of Viruses. *Intervirology* 1979;12:132–296.
85. Matthews REF, ed. *A critical appraisal of viral taxonomy.* Boca Raton, Florida: CRC Press; 1983.
86. Melnick JL, Allison AC, Butel JS, et al. Papovaviridae. *Intervirology* 1974;3:106–120.
87. Meyers G, Wirblich C, Thiel HJ. Rabbit haemorrhagic disease virus—molecular cloning and nucleotide sequencing of a calicivirus genome. *Virology* 1991;184:664–676.
88. Monath TP, ed. *The arboviruses: epidemiology and ecology.* 5 volumes. Boca Raton, Florida: CRC Press; 1988.
89. Monroe SS, Stine SE, Gorelkin L, Herrmann JE, Blacklow NR, Glass RI. Temporal synthesis of proteins and RNAs during human astrovirus infection of cultured cells. *J Virol* 1991;65:641–648.
90. Muhlerger E, Sanchez A, Randolf A, et al. The nucleotide sequence of the L gene of Marburg virus, a filovirus: homologies with paramyxoviruses and rhabdoviruses. *Virology* 1992;187:534–547.
91. Murphy FA. Arenavirus taxonomy: a review. In: Monath TP, ed. Arenaviral infections of public health importance. *Bull World Health Organ* 1975;52:389–392.
92. Murphy FA. Current problems in vertebrate virus taxonomy. In: Matthews REF, ed. *A critical appraisal of viral taxonomy.* Boca Raton, Florida: CRC Press; 1983:37–62.
93. Murphy FA. Taxonomy of animal viruses. In: Nermut MV, Steven AC, eds. *Animal virus structure.* Amsterdam: Elsevier; 1987.
94. Murphy FA. Virus taxonomy. In: Fields BN, Knipe DM, Chanock RM, et al., eds. *Virology.* 1st ed. New York: Raven; 1985:7–26.
95. Murphy FA. Virus taxonomy and nomenclature. In: Lennette EH, Halonen P, Murphy FA, eds. *Laboratory diagnosis of infectious diseases, principles and practices.* Volume II: viral, rickettsial and chlamydial diseases. New York: Springer-Verlag; 1988:153–176.
96. Murphy FA, Fauquet CM, Bishop DHL, et al. *Virus taxonomy—the classification and nomenclature of viruses: sixth report of the International Committee on Taxonomy of Viruses.* Vienna: Springer-Verlag; 1995.
97. Murphy FA, Harrison AK, Whitfield SG. Bunyaviruses. Morphologic and morphogenetic similarities of Bunyamwera serologic supergroup viruses and several other arthropod-borne viruses. *Intervirology* 1973;l:297–316.
98. Murphy FA, Kingsbury D. Virus taxonomy. In: Fields BN, Knipe DM, Chanock RM, et al., eds. *Virology.* 2nd ed. New York: Raven; 1990:9–36.
99. Noteborn N, De Boer G, van Roozelaar D, et al. Characterization of cloned chicken anermia virus DNA that contains all elements for the infectious replication cycle. *J Virol* 1991;65:3131–3139.
100. Palmenberg AC. Sequence alignments of picornaviral capsid proteins. In: Selmer BL, Ehrenfeld E, eds. *Molecular aspects of picornavirus infection and detection.* Orlando: Academic; 1989;211–230.
101. Pankhurst RJ, Aitchison RR. An on-line identification program. In: Pankhurst RJ, ed. *Biological identification with computers.* London: Academic; 1975;181–185.
102. Pfau CJ, Bergold GH, Casals J, et al. Arenaviridae. *Intervirology* 1974;4:207–218.
103. Porterfield JS, Casals J, Chumakov MP, et al. Togaviridae. *Intervirology* 1978;9:129–148.
104. Pringle CR. The order Mononegavirales. *Arch Virol* 1991;117:137–140.
105. Pringle CR. Paramyxoviruses and disease. In: Russell WC, Almond JW, eds. *Society for General Microbiology Symposium 40: the molecular basis of virus disease.* Cambridge: Cambridge University Press; 1987;51–90.
106. Prusiner SB. Molecular biology of prion diseases. *Science* 1991;252: 1515–1522.
107. Prusiner SB, Collinge J, Powell J, Anderton B, eds. *Prion diseases of humans and animals.* London: Ellis Horwood; 1993.
108. Richt JA, Herzog S, Pyper J, et al. Borna disease virus: Nature of the etiologic agent and significance of infection in man. *Archives of Virology Supplementum*, 1993;7:101–109.
109. Roizman B, Carmichael LE, Deinhardt F, et al. Herpesviridae: definition, provisional nomenclature and taxonomy. *Intervirology* 1982; 16:201–217.
110. Roizman B, Desrosiers RC, Fleckenstein B, Lopez C, Minson AC, Studdert MJ. The family Herpesviridae: an update. *Arch Virol* 1992;123:425–449.
111. Roy P, Gorman GM, eds. Bluetongue viruses. *Curr Top Microbiol Immunol* 1990;162:1–200.
112. Russell WC, Bartha A, deJong JC, et al. Adenoviruses: fifth report. *Arch Virol* 1991;(suppl 2):140–144.
113. Saif LJ. Nongroup A rotaviruses. In: Saif LJ, Theil KW, eds. *Viral diarrhoeas of man and animals.* Boca Raton, Florida: CRC Press; 1990:73–95.
114. Salvato M, ed. *The arenaviridae.* New York: Plenum; 1993.
115. Salzman NP. Volume 1: the polyomaviruses. *The papovaviridae.* New York: Plenum; 1986.
116. Salzman NP, Howley PM. Volume 2: the papillomaviruses. *The papovaviridae.* New York: Plenum; 1987.

117. Sanchez A, Kiley MP, Klenk H-D, Feldman H. Sequence analysis of the Marburg virus nucleoprotein gene: comparison to Ebola virus and other non-segmented negative strand RNA viruses. *J Gen Virol* 1992;73:347–357.
118. Schaffer FL, Bachrach HL, Brown F, et al. Caliciviridae. *Intervirology* 1980;14:1–6.
119. Schneider PA, Schneemann A, Lipkin WI. RNA splicing in Borna disease virus, a nonsegmented, negative-strand RNA virus. *J Virol* 1994;68:5007–5012.
120. Siddell S, Anderson R, Cavanaugh D, et al. *Coronaviridae*. *Intervirology* 1983;20:181–190.
121. Siegl G, Bates RC, Berns KI, et al. Characteristics and taxonomy of the family Parvoviridae. *Intervirology* 1985;23:61–73.
122. Snijder EJ, Horzinek MC. *Toroviruses*: replication, evolution and comparison with other members of the coronavirus-like superfamily. *J Gen Virol* 1993;74:2305–2316.
123. Stanway G. Structure, function and evolution of picornaviruses. *J Gen Virol* 1990;71:2483–2501.
124. Strauss JH, Strauss EG. Evolution of RNA viruses. *Annu Rev Microbiol* 1988;42:657–683.
125. Strauss JH, Strauss EG, Levine AJ. Virus evolution. In: Fields BN, Knipe DM, eds. *Fundamental virology*. New York: Raven; 1990: 167–190.
126. Tam AW, Smith MW, Guerra ME, et al. Hepatitis E virus molecular cloning and sequencing of the full length viral genome. *Virology* 1991;185:120–131.
127. Taylor JM. Human hepatitis delta virus. *Curr Top Microbiol Immunol* 1991;168:141–166.
128. Taylor JM. Structure and replication of hepatitis delta virus. In: Hollinger FB, Lemon SM, Margolis H, eds. *Viral hepatitis and liver disease*. Baltimore: Williams & Wilkins; 1991;460–463.
129. Todd D, Niagro F, Ritchie B, et al. Comparison of three animal viruses with circular single-stranded DNA genomes. *Arch Virol* 1991;117: 129–135.
130. van Regenmortel MHV. Virus species, a much overlooked but essential concept in virus classification. *Intervirology* 1990; 31:241–254.
131. Vinuela E. African swine fever. *Curr Top Microbiol Immunol* 1985;116:151–170.
132. Vogt PK. The oncovirinae—a definition of the group. In: *Report No. 1 of WHO Collaborating Centre for Collection and Evaluation of Data on Comparative Virology*. Munich: UNI-Druck; 1976:327–339.
133. Ward CW. Progress towards a higher taxonomy of viruses. *Res Virol* 1993;144:419–453.
134. Waterson AP, Wilkinson L. *An introduction to the history of virology*. London: Cambridge University Press; 1978.
135. Weissmann CA. "Unified theory" of prion propagation. *Nature* 1991; 352:679–683.
136. Westaway EG, Brinton MA, Gaidamovich SY, et al. Flaviviridae. *Intervirology* 1985;24:183–192.
137. White DO, Fenner F. *Medical virology*. 4th ed. Orlando: Academic, 1994.
138. Wigand R, Bartha A, Dreizin RS, et al. Adenoviridae: second report. *Intervirology* 1982;18:169–176.
139. Wildy P. Classification and nomenclature of viruses: first report of the International Committee on Nomenclature of Viruses. *Monogr Virol* 1971;5:1–181.
140. Willcocks MM, Carter MJ, Madeley CR. Astroviruses. *Rev Med Virol* 1992;2:97–106.
141. Wunner WH. The chemical composition and molecular structure of rabies viruses. In: Baer GE, ed. *The natural history of rabies*. Boca Raton, Florida: CRC Press; 1990:31–67.
142. Zeller HG, Karabatsos N, Calisher CH, Digoutte JP, Murphy FA, Shope RE. Electron microscopic and antigenic studies of uncharacterized viruses. I. Evidence suggesting the placement of viruses in the families Arenaviridae, Paramyxoviridae, and Poxviridae. *Arch Virol* 1989;108:191–209.
143. Zimmern D. Evolution of RNA viruses. In: Holland JJ, Domingo E, Alquist P, eds. *RNA genetics*. Boca Raton, Florida: CRC Press; 1988:211–240.

Fundamental Virology, Third Edition
edited by B.N. Fields, D.M. Knipe, P.M. Howley, et al.
Lippincott - Raven Publishers, Philadelphia © 1996

CHAPTER 3

Virus Structure

Stephen C. Harrison, John J. Skehel, and Don C. Wiley

A virus particle is a structure that has evolved to transfer nucleic acid from one cell to another. It is, in effect, an extracellular organelle. The nucleic acid may be either RNA or DNA, and in both cases particles of varying complexity are found. Structural relationships among viruses are conveniently discussed by comparing viruses having similar replication strategies. There is a particularly important distinction between enveloped viruses (those with a lipid-bilayer membrane) and nonenveloped viruses. The distinction corresponds to a difference in the way the virus leaves and enters a cell. This chapter deals with the structural organization of virus particles, both nonenveloped and enveloped. A number of specific examples need to be described in sufficient detail to make the principles clear, but the reader should also consult chapters on individual virus groups for further discussion of the relationships between structural features and host–cell interactions.

An extensive nomenclature has developed to describe virus structures. It is useful at the outset to define the following terms, in a manner corresponding to common biochemical usage. Other terms, such as those describing virus symmetry, are defined at appropriate points in the text.

Subunit or protein subunit: A single folded polypeptide chain [e.g., the 17,000-MW protein of tobacco mosaic virus (TMV) or VP1 of poliovirus].

Structure unit: A collection of one or more nonidentical protein subunits that together form the chemical building block of a larger assembly (e.g., VP1, VP2, VP3, and VP4 in poliovirus or E1, E2, and E3 in the outer shell of Semliki Forest virus). The word *protomer* is often used for this same purpose.

S. C. Harrison and D. C. Wiley: Howard Hughes Medical Institute, Harvard University, Department of Molecular and Cellular Biology, Cambridge, Massachusetts 02138.

J. J. Skehel: National Institute for Medical Research, London, NW7 1AA, United Kingdom.

Assembly unit: A (usually symmetrical) set of subunits or structure units that is an important intermediate or subassembly in the formation of a larger structure (e.g., the VP1 pentamer in polyomavirus and SV40).

Morphologic unit: The apparent lumps or clusters seen on the surface of a particle by electron microscopy. They generally correspond to projecting parts of the protein subunits, clustered about particular local axes of symmetry. Because morphologic units do not necessarily correspond to chemically unique entities, this is a term of convenience only, for use in describing electron micrographs. The word *capsomere* has also been used to describe such clusters, particularly when they do correspond to a chemically definable oligomer of one or more viral proteins.

Capsid: The protein shell directly surrounding viral nucleic acid. The word *coat* or *shell* is often just as clear in context.

Nucleocapsid: The complete protein–nucleic acid complex that is the packaged form of the genome in a virus particle (e.g., the core of Sindbis virus or the filamentous complexes of N protein and RNA in VSV, influenza, and so on). This is a useful term principally in cases where the nucleocapsid is a definite substructure of a more complex virus particle.

Envelope: The lipid bilayer and associated glycoproteins that surround many types of virus particle.

Virion: The entire virus particle.

HOW VIRUS STRUCTURES ARE STUDIED

Electron microscopy is the most useful way to determine the general morphology of a virus particle. Chapter 2 contains electron micrographs of many of the important types of virus. For examining infected cells and larger isolated particles, the traditional thin-sectioning methods can be used. The thickness of a section and the coarseness of staining methods limit resolution to about 50–75 Å, even in the best cases. (*Resolution* means "the approximate minimum size of a substructure that can be detected." Recall that one atomic diameter is 2–3 Å; an α-helix, 10 Å; and a DNA double helix, 20 Å.) Negative staining, using uranyl acetate, potassium phosphotungstate, or related electron-dense compounds, gives somewhat more detailed images of isolated and purified virus particles (see figures in Chapter 2). Viruses embedded in negative stain are often relatively well preserved. But since electron micrographs usually reveal contrast from both the top and bottom surfaces of embedded particles, visual interpretation of finer aspects of the image can be difficult.

Quantitative methods for image analysis can be applied to micrographs of particularly well-preserved particles, especially those with high symmetry. An example of a computed three-dimensional-image reconstruction from electron micrographs of negatively stained virus particles is shown in Fig. 1. The typical resolution for such an analysis is 30–40 Å. Methods for preserving viruses and other macromolecular assemblies for electron microscopy, by rapid freezing to liquid nitrogen temperatures, permit visualization of contrast from the structures in the particle itself and not just from the "envelope" created by a surrounding of negative stain. Examples of structure computed from micrographs of such frozen-hydrated specimens are shown in Fig. 2 (See also colorplate 1.).

The detail obtained from even the most elegant of electron microscopy methods falls well short of the level required to understand molecular interactions. A molecular-resolution picture can be obtained by x-ray diffraction methods, if single crystals of the relevant structure can be obtained. It has been known since the 1930s that simple plant viruses, such as tomato bushy stunt virus (TBSV), can be crystallized (8), and the first x-ray studies of such crystals were carried out as early as 1938 (10). Crystallization of poliovirus (141) and other important animal viruses showed that the approach could be extended to human pathogens. The first complete high-resolution structure of a crystalline virus was obtained from TBSV in 1978 (74), and since then the structures of a number of animal pathogens have been determined, including human rhinovirus (123), poliovirus (82), foot-and-mouth disease virus (2), and canine parvovirus (151). Only very regular structures can form single crystals, and

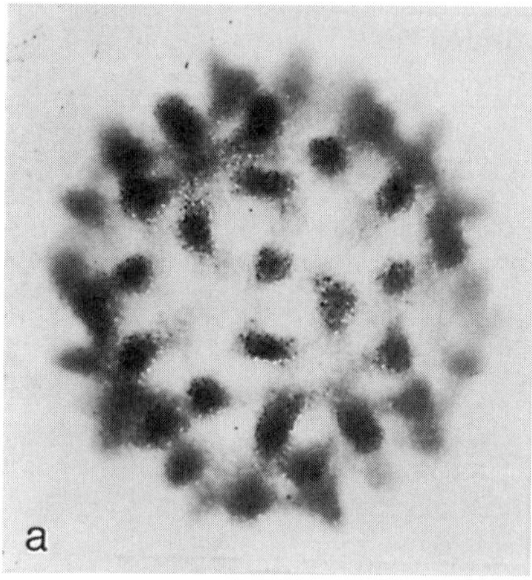

FIG. 1. Three-dimensional image reconstruction from electron micrographs of negatively stained tomato bushy stunt virus (TBSV). Virus particle is represented in *black*; the surrounding stain is *light*. The view is nearly along an icosahedral twofold axis of symmetry (see Figs. 10 and 11B). The particle has 90 protruding knobs. They are the major source of contrast with the uranyl acetate negative stain in which the particle is embedded. As the more detailed results of x-ray crystallography show (see Fig. 3), these knobs are the pairwise-clustered P domains that project from each of the 180 protein subunits. (Courtesy of RA Crowther and J Finch.)

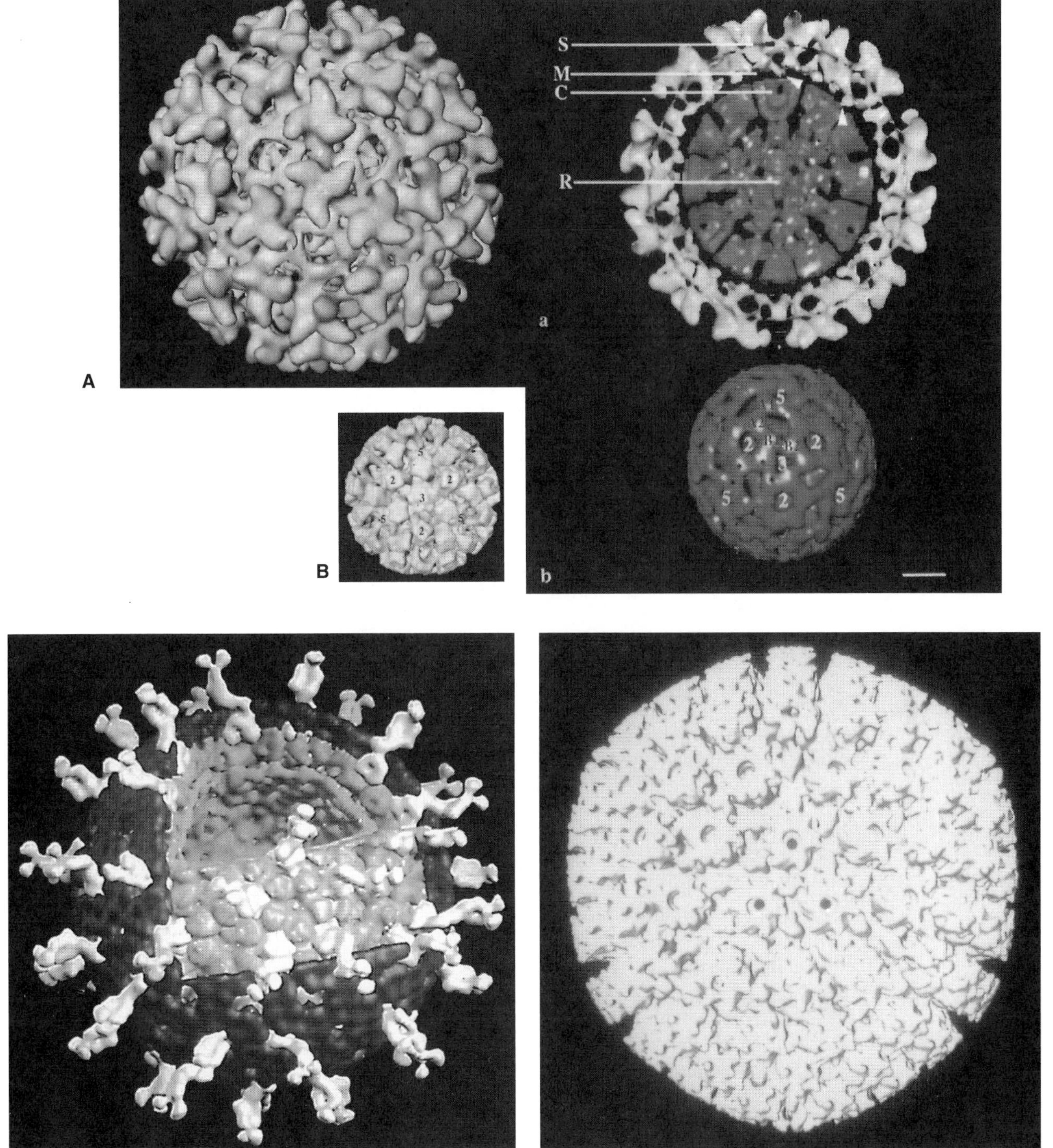

FIG. 2. Three-dimensional reconstructions of icosahedrally symmetric virus particles from cryoelectron micrographs. The contrast comes from the viral components themselves—protein, nucleic acid, and lipid—not from stain, as in Fig. 1. Therefore, meaningful detail is present at all radii in the image. **(A)** Sindbis virus, an enveloped, positive-strand RNA virus. Surface view of virion (left), showing T = 4 arrangement of viral glycoprotein; cross-sectional view (top right), showing glycoprotein (light) and core (dark), with the lipid bilayer as a low-density gap between them; surface view of core (bottom right). (Courtesy of BVV Prasad. See also Fig. 19.) **(B)** The capsid of a calicivirus, a T = 3 particle. The view is along an icosahedral threefold axis, which has quasi-sixfold character in a T = 3 structure. The projections are dimer-clustered domains of the protein subunit, and the surface topography strongly resembles that of the TBSV particle shown in Figs. 1 and 3. (Courtesy of BVV Prasad.) **(C)** Rotavirus, a nonenveloped, double-stranded RNA virus. The particle has three protein layers: the outer shell (VP7), the inner shell (VP6) and the core (VP2). The spikes represent the hemagglutinin (VP4)— the protein through which the virus is believed to recognize its receptor. The outer and inner shells conform to a T = 13 icosahedral lattice. (Courtesy of M Yeager.) **(D)** Nucleocapsid of herpes simplex virus, a double-stranded DNA virus. The surface contains a principal structural protein (VP5), arranged in hexameric and pentameric clusters in a T = 16 icosahedral lattice. (Courtesy of BVV Prasad.)

to study the molecular details of larger and more complex virus particles, it is necessary to "dissect" them into well-defined subunits or substructures. The influenza virus hemagglutinin (164) and adenovirus hexon (121) are examples of important molecular components studied in this way.

It should be emphasized that no study of virion structure can be fully informative without a detailed catalog of macromolecular components and their relative proportions. Moreover, modern cloning approaches ensure that complete protein sequences are almost always available. As more and more proteins are being studied by x-ray crystallographic methods, it is becoming possible to deduce certain features of three-dimensional conformation from the amino acid sequence of protein by identifying homology with proteins of known structure. For example, from the structures of picornaviruses already determined, we can predict many molecular details of other picornaviruses.

BACKGROUND ON PROTEIN STRUCTURE AND MACROMOLECULAR ASSEMBLIES

Domains, Hinges, and Arms in Folded Proteins

The three-dimensional structure of a protein is stabilized by noncovalent interactions among its amino acid residues. Polypeptide chains of viral structural proteins usually contain one or more regions that form globular domains (Fig. 3). Such folded domains are fairly rigid, stable structures, and their stability does not depend on interactions with other parts of the subunit or with other subunits. The polypeptide chains may also contain regions that are extended and flexible in the free protein and that form defined structures only when interacting with other chains in an assembled virion. The term *domain* is reserved for the independently folded regions, generally about 100–200 amino acid residues in size; words like *arm* and *hinge* are used to refer to more flexible extensions and connections. The organization of a protein domain is usually defined by the way elements of secondary structure—α-helices and β-sheets—pack against one another.

The amino acid sequence of a polypeptide chain determines how it will fold up in solution. It is important to emphasize, however, that our present capacity to compute the stability of possible structures does not permit adequate predictions from knowledge of sequence alone. Even the well-known methods for predicting secondary structure have too many inherent ambiguities to be genuinely useful. Strong predictions can only be made when the sequence of amino acids in some part of a protein is recognizably similar to the sequence of a segment of known structure in another protein. Such comparisons become even more powerful when many sequences can be aligned and when more than one structure has been determined. Fortunately, proteins of similar function often have similar structure. When a few related three-dimensional structures have been worked out by crystallographic methods, useful models for other members of the class can often be proposed.

Assemblies

The noncovalent bonds between subunits in an assembly such as the shell of a virus are of precisely the same sort that stabilize a folded protein domain. The interface between two subunits can therefore look very much like the interior of a single domain, with amino acid side chains tightly packed against one another. Contacts between polar groups involve hydrogen bonds and salt bridges. Contacts between nonpolar groups contribute van der Waals forces and hydrophobic interactions. A remarkable feature of most of the virus structures so far determined is the way in which the polypeptide chain from one subunit can extend under or over domains of neighboring subunits. Often these extended polypeptide arms intertwine with others. Contacts between individual subunits determine the overall structure of an assembly. If a number of identical units are involved, repeated contacts of the same kind must occur, and the resulting structure exhibits *symmetry*. Some simple viruses form spontaneously from their dissociated components. This *self-assembly* process is driven by the stability of the interactions between protein subunits under conditions favoring association. More complex structures may require enzyme-catalyzed modifications of the subunits to trigger or to stabilize assembly. In some cases, scaffold proteins—species that participate in assembly but do not remain in the final structure—can be involved (21).

Membrane Proteins

Membrane-associated proteins can be anchored to the lipid bilayer by a variety of means (134). Proteins that actually traverse the bilayer may do so only once, with a simple hydrophobic connecting segment joining the partitions on either side, or several times, with multiple membrane spanning segments. Covalently attached lipids can anchor proteins that reside entirely on one side of the membrane. Some membrane-associated proteins are attached only through interactions with other proteins that are anchored in one of the ways just described.

There are significant differences in the structural features of proteins (or portions of proteins) found on the extracellular and intracellular sides of a cell membrane. Because of a gradient in redox potential across the membrane, extracellular domains often contain disulfide bonds and rarely have free cysteines, whereas intracellular do-

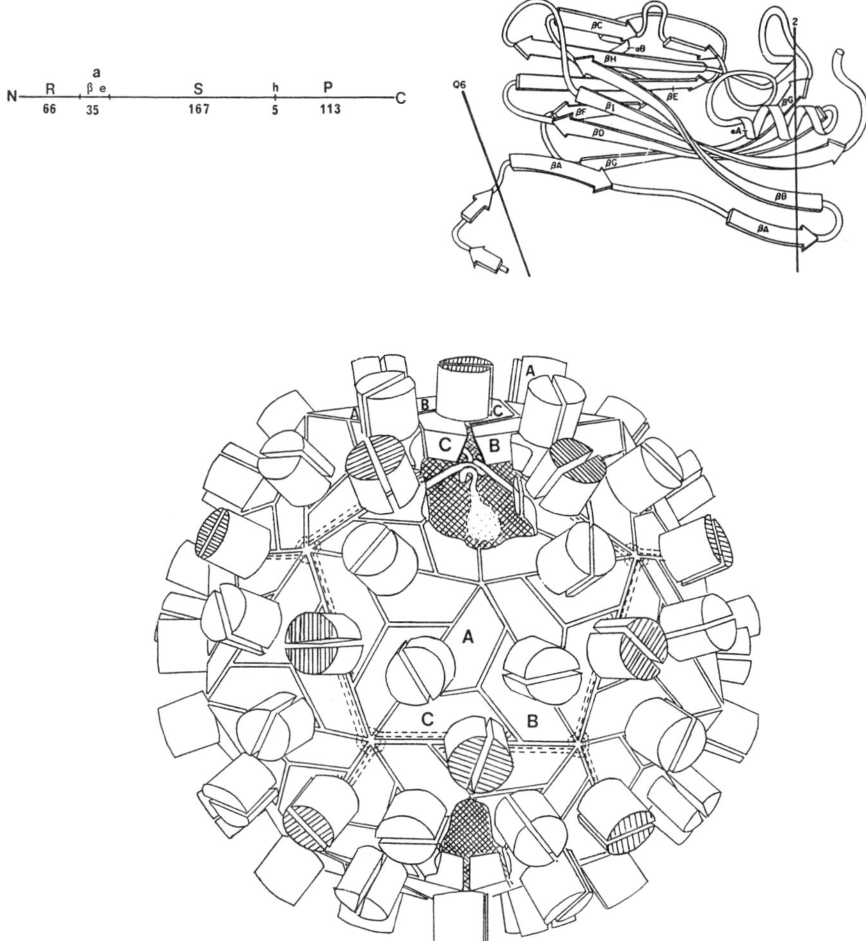

FIG. 3. Molecular architecture of tomato bushy stunt virus (TBSV), illustrating some of the principles of domain organization described in the text. TBSV is a simple, icosahedrally symmetric, RNA-containing particle with 180 identical subunits. The 40-kD coat protein is organized into four regions: a positively charged R segment, a connecting arm (a), and two compactly folded domains— S ("shell") and P ("projecting"). There is a hinge (h) between the S and P domains. The R segment is probably irregularly folded in the particle interior; it interacts with viral RNA. The arm is folded against the S domain on 60 of the subunits (the shaded units in positions symmetrically equivalent to "C"), but it extends inward in a more disordered way on the remaining 120 (positions equivalent to "A" and "B"). The folded arms form an internal framework (dashed lines in bottom diagram), with three arms meeting and interdigitating at each of the icosahedral threefold positions (cutaway). The relationship of the domains and connecting segments to the sequence (N terminus to C terminus) is shown at the top left; the folded structure of the S domain, at the top right; the packing of units in the viral shell, at the bottom. The fold of the S domain is a sandwich of two β sheets (compare Figs. 14 and 17).

mains do not contain disulfides. The routes of protein biosynthesis also ensure that only extracellular domains (or secreted proteins) are glycosylated and that different covalent lipid modifications are used for anchoring extracellular and intracellular proteins.

Biosynthesis

Proteins destined to be secreted across membranes and many destined to become integrated into membranes (including all the viral glycoproteins) are synthesized on membrane-bound polyribosomes (106). Most of these proteins are translated with an amino-terminal peptide extension termed a *signal peptide*, or *leader*, sequence (12,13,98). This signal peptide directs the translocation of the nascent polypeptide chain across the membrane of the endoplasmic reticulum (ER) into the lumen. The signal sequences found in viral glycoproteins all share a central region of uncharged, mainly hydrophobic residues often bounded by a positively charged residue on the amino-terminal end. There does not appear to be any specific sequence requirement other than that the largely hydrophobic stretch be at least 11 amino acids long (156).

In most cases the signal peptide is cleaved from the newly synthesized protein by a "signal peptidase" during the translocation (169). This cleavage occurs on the lumenal side of the ER membrane.

Simultaneous with the translocation that accompanies translation, oligosaccharide chains are attached to the protein as it emerges into the lumen of the rough ER (88). A precursor oligosaccharide is transferred *en bloc* from a dolichol lipid carrier to an N-glycosidic linkage with the asparagine in Asn-X-Ser/Thr sequences. Some processing of this precursor, to a mannose-rich structure, also occurs in the ER. Cell fractionation and electron microscopy have demonstrated that viral glycoproteins then pass from the ER to the Golgi complex *en route* to the cell surface (63,77,87,150). Many oligosaccharides are processed during passage through the Golgi complex by mannosyl glycosidases, which trim off mannose residues, and glycosyl transferases, which add galactose, fucose, N-acetyl glucosamine, and neuraminic acid, yielding "complex" oligosaccharides (Fig. 4) (88). Addition of sugars to certain serine or threonine side chains may also occur in the Golgi complex.

The path that the processed glycoproteins take to the cell surface is incompletely understood. Certain epithelial cells form monolayers in which tight junctions divide the cell membrane into two regions—an apical surface and a basolateral surface. Several enveloped viruses "bud" exclusively from one surface or the other. Influenza and parainfluenza viruses, for example, are assembled at the apical surface, whereas VSV is formed exclusively at the basolateral surface (319,324).

Structural Classes

Viral membrane proteins belong to at least four structural classes (Fig. 5). Members of the largest class, which includes most viral surface glycoproteins, are held in the lipid bilayer by a single transmembrane anchoring peptide. A number of these proteins are further linked to the bilayer by palmityl groups, which form thioesters with cysteine residues near the position where the polypeptide chain emerges on the cytoplasmic side of the membrane. The bulk of the mass of viral proteins with a single transmembrane anchor is usually on the external side of the membrane, with a smaller domain on the inner (cytoplasmic) side. The commonest orientation places the C-terminus inside the membrane but the opposite orientation is also found. In the former case, membrane insertion is generally determined by a cleaved N-terminal signal

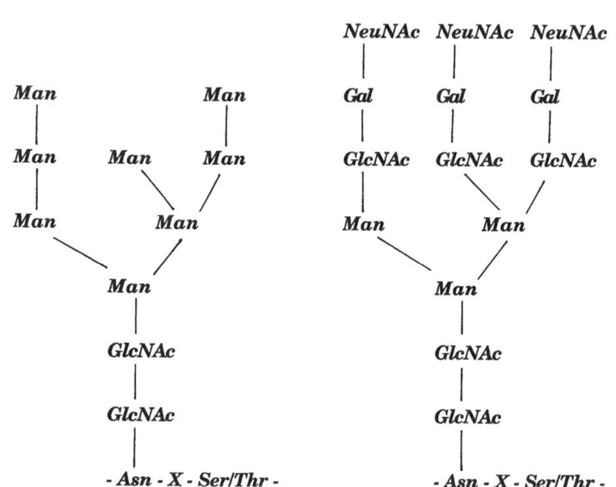

FIG. 4. Mannose-rich (left) and complex (right) N-linked oligosaccharides. The mannose-rich oligosaccharides are composed of *N*-acetyl-glucosamine (GlcNAc) and mannose (Man). Carbohydrate processing enzymes in the Golgi remove the outer mannoses and add fucose, *N*-acetyl glucosamine, galactose, and terminal *N*-acetyl neuraminic acid (NeuNAc, sialic acid). Not all oligosaccharide chains are identically processed. The structure shown here is only one example. Moreover, some chains remain as mannose-rich structures, even on mature glycoproteins.

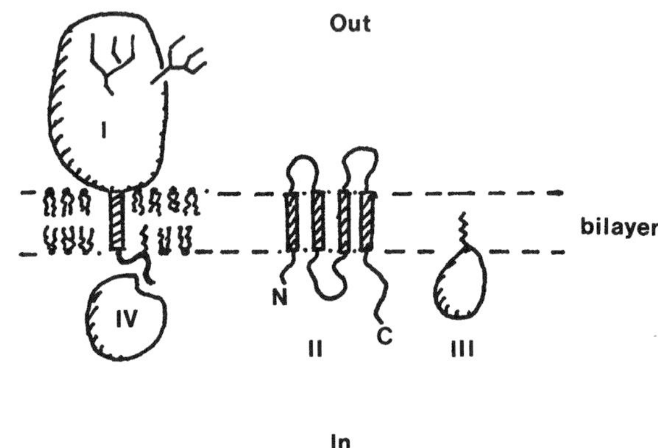

FIG. 5. Four classes of viral membrane protein. In I, there is a single transmembrane segment. The extracellular domain is glycosylated. Cysteine residues near the cytoplasmic face of the membrane are sometimes palmitylated. In this illustration, the cytoplasmic domain is relatively small, corresponding to many viral glycoproteins. In II, the polypeptide chain passes several times back and forth across the bilayer. The loops between these segments may be of various sizes—even entire folded domains. III is essentially a cytoplasmic protein, anchored to the membrane by a covalently attached lipid— either a myristyl group at the N-terminus or a prenyl group at the C-terminus. IV is also a cytoplasmic protein, bound to the membrane by association with another, membrane-embedded protein. The illustration shows a protein associated with the cytoplasmic tail of class I glycoprotein.

sequence, whereas in the latter, the transmembrane segment serves as the insertion signal, and there is no cleavage. In cases where two or more membrane proteins are derived from a polyprotein precursor (e.g., in enveloped, positive-strand RNA viruses), the transmembrane anchor of one protein is followed by a signal for the next. Cleavage by signal peptidase (on the lumenal side of the membrane) separates the two proteins, but the N-terminal partner retains both hydrophobic segments and thus terminates in a transmembrane "hairpin." Members of the second class of membrane proteins can form channels and pumps; these proteins are more extensively embedded in the lipid. The polypeptide chain passes alternately back and forth across the bilayer, in many cases as α-helices. Members of the third class lie on the inner (cytoplasmic) surface of the membrane and have a lipid anchor. An N-terminal myristyl group is the only such anchor found so far in virions, but a number of nonstructural proteins (such as the viral oncoprotein Ras) are anchored by a C-terminal prenyl group. Members of a fourth class associate with the viral membrane by virtue of their interactions with the cytoplasmic "tail" of a surface glycoprotein but are not otherwise anchored.

In the viral glycoproteins, the transmembrane anchor peptide comprises 18–27 primarily hydrophobic amino acid residues. The hydrophobic section of the anchor probably forms an α-helix. In an α-helix the hydrogen-bonding capacities of the carbonyl oxygens and amide nitrogens in the main chain are satisfied through internal hydrogen bonds, thus avoiding unpaired polar atoms in the hydrophobic environment of the lipid bilayer. The 18–27 uncharged amino acids are just sufficient to traverse the bilayer in an α-helical conformation.

Isolation and Hydrophobic Properties of Viral Glycoproteins

Viral membrane glycoproteins can be isolated intact from the lipid bilayer of membranes only by solubilizing the membranes in detergent. In the first stage of the process, detergent binding to the membrane causes it to rupture, releasing the nucleocapsid and other internal components (81). Subsequent steps are illustrated in Fig. 6. Nonionic detergents solubilize membrane proteins by interacting with the hydrophobic parts of the proteins and lipids. In excess detergent, each membrane protein and lipid molecule becomes inserted into its own detergent micelle. If the lipids are removed, what remains is a monodisperse solution of protein–detergent complexes. Since most nonionic detergents do not denature proteins, these protein–detergent complexes are stable and active (80). If, in the absence of lipid, the detergent is removed, the hydrophobic anchor peptides of the glycoprotein aggregate with one another to form a protein/protein micelle (Fig. 6). If the detergent is removed in the presence of lipid, the bilayer lipid vesicles form with the glycoproteins reinserted into them (Fig. 6C). Thus, the hydrophobic anchor peptides of glycoproteins are never found in the aqueous phase but always either in a hydrophobic lipid bilayer or surrounded by the hydrophobic portion of a detergent molecule. Figure 7 shows an electron micrograph of protein–protein micelles of the influenza hemagglutinin, obtained by removing detergent from a monodisperse solution of protein–detergent complexes (90). Because of their multidentate character, protein–protein micelles of virus glycoproteins have been used in vaccine preparations and in biochemical studies.

To study the external, hydrophilic domain of membrane proteins, it has been possible in a number of cases to remove the hydrophobic anchor peptide by proteolytic digestion of membranes or protein–detergent complexes (Fig. 6D) (17,28,70,139,145). This treatment releases the hydrophilic portion of the molecule from the membrane, often as a soluble and active protein. Because parts of a viral glycoprotein exist in three distinct environments—the extracellular space, the lipid bilayer, and the cytoplasm—it is reasonable to expect that the structure will be divided into regions that are folded independently in order to be stable in the distinct environments. The observation that the anchor peptides of glycoproteins can be removed, releasing an active external domain, is one experimental confirmation of this view.

TYPES OF VIRUS STRUCTURE AND THEIR CHARACTERISTICS

Viruses come in a great variety of shapes and sizes, as can be seen from the examples described in detail later. Structural features are determined by requirements for assembly, exit, transmission, attachment, penetration, and uncoating. These are, in effect, the functions of the virion. The best way to describe the structural features and to outline the varieties and similarities of molecular architecture so far detected is through descriptions of particular viruses. Most of the chapter is devoted to such specifics. This section contains some more general remarks.

Helical and Icosahedral Symmetry

Viral substructures assemble from protein subunits that are specified by the viral genome. Genetic economy then dictates that these substructures be made of many identical copies of one or a few kinds of protein (36). Repeated occurrence of similar protein–protein interfaces leads to a symmetrical arrangement of the subunits. In the simple example shown in Fig. 8, two commas face each other and interact through complementary surfaces h (at the head) and t (at the tail). In a symmetrical arrangement (Fig. 8A), the two h–t contacts are the same; in a nonsymmetrical arrangement (Fig. 8B), they are differ-

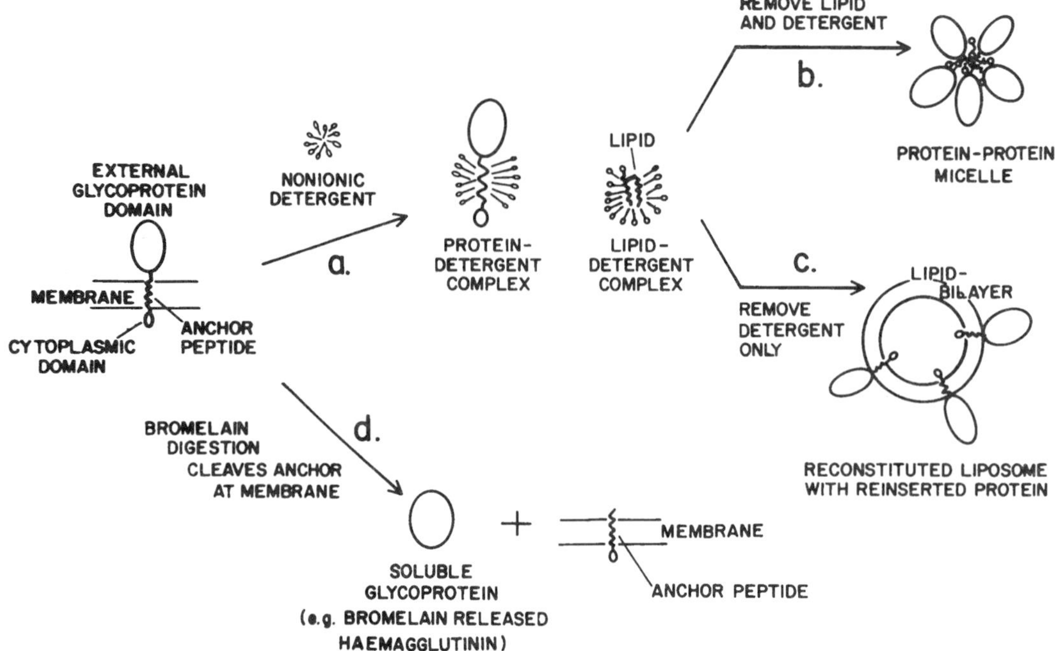

FIG. 6. The isolation of viral membrane glycoproteins, illustrated by the influenza virus hemagglutinin as an example. **(A)** Excess nonionic detergent solubilizes membrane proteins by interacting with the hydrophobic (*wavy line*) parts of the proteins and lipids. **(B)** Removal of the detergent in the absence of lipid results in protein/protein micelles, formed by association of the hydrophobic anchor peptides. **(C)** Removal of the detergent in the presence of lipid results in the reconstitution of lipid-bilayer vesicles and the reinsertion of the proteins into the artificial membrane. **(D)** In some cases, treatment of membrane proteins with proteases removes the hydrophobic anchor peptide near the membrane interface, releasing a soluble and active fragment of the glycoprotein.

ent. If there is a preferred pattern of noncovalent interactions between amino acid side chains at h and t (a free-energy minimum), then a symmetrical arrangement ensures that optimal contacts occur at both head–tail interfaces. An asymmetric arrangement can at best lead to one optimal and one nonoptimal interaction. The symmetry that describes the pair of commas in Fig. 8A is a very simple one—a twofold rotation axis. More complex assemblies can have more elaborate symmetries, which specify the spatial pattern of repeated protein–protein interactions. Rodlike structures, such as the filamentous nucleocapsids of many enveloped viruses and some of the plant and bacterial viruses, have helical symmetry. Many spherical viruses have icosahedral symmetry.

Helical Symmetry

Helical symmetry is conveniently described by the number of units per turn, u (not necessarily integral), and the axial rise per unit, p. The pitch of the helix, P, is equal to $u \times p$. Helical structures can have a rotation axis coincident with the helix axis (e.g., the T4 phage tail, with a sixfold axis) or an array of twofold axes perpendicular to the helix axis (as in DNA). The diagram of TMV in Fig.

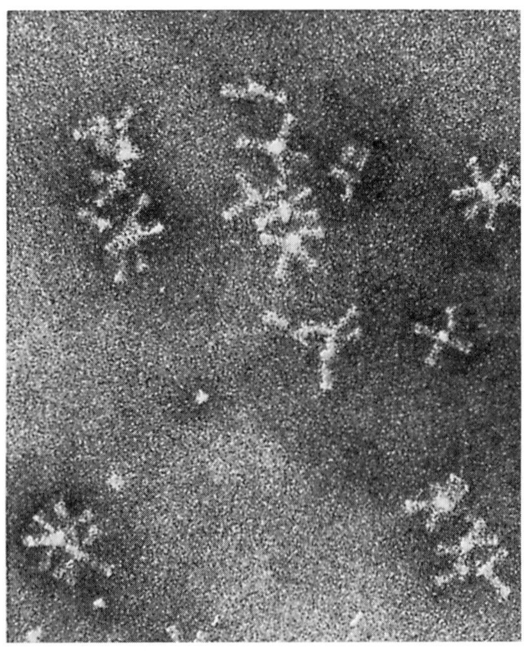

FIG. 7. Protein/protein micelles of the influenza hemagglutinin prepared by removing detergent from the detergent-purified protein (6).

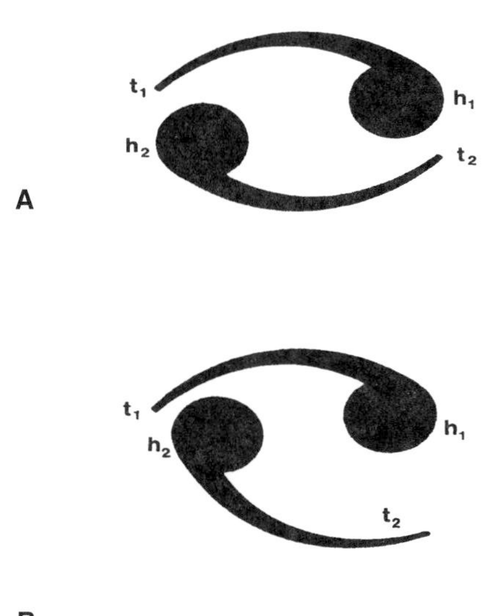

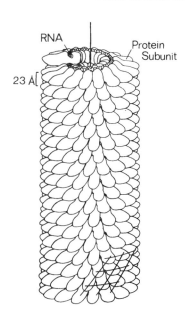

FIG. 8. **(A)** Two commas, symmetrically related to each other by a twofold rotation axis. If these commas represent protein subunits, then amino acid side chains at the head surface of one subunit, h_1, interact with complementary side chains at the tail surface of the other, t_2, in the same way that those at h_2 interact with t_1. **(B)** In a nonsymmetrical pair of commas, the h_1 t_2 contact is different from h_2 t_1. Only one of the two can be optimal. Thus, a stable and unique assembly is often one with subunits that can interact favorably in a symmetric structure.

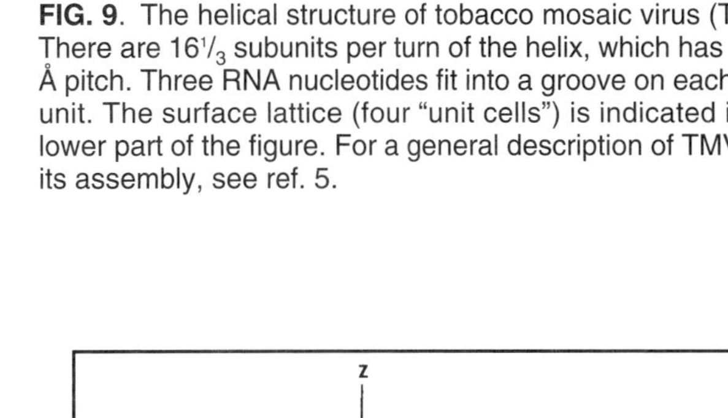

FIG. 9. The helical structure of tobacco mosaic virus (TMV). There are $16^1/_3$ subunits per turn of the helix, which has a 23-Å pitch. Three RNA nucleotides fit into a groove on each subunit. The surface lattice (four "unit cells") is indicated in the lower part of the figure. For a general description of TMV and its assembly, see ref. 5.

9 illustrates that a structure with helical symmetry (in this case having $u = 16^1/_3$, $p = 1.4$ Å) can also be described by reference to a *surface lattice*—a network connecting equivalent units in the curved surface. The lattice lines correspond to different helical paths, as shown in Fig. 9.

Icosahedral Symmetry

Most closed-shell virus particles have structures based on icosahedral symmetry. The icosahedron is a figure with 20 triangular faces, characterized by a symmetry that involves a collection of rotation axes (Fig. 10). An object with icosahedral symmetry need not have the shape of an icosahedral solid, but it must have the appropriate rotation axes. Thus, the TBSV particle shown in Fig. 1 has a bumpy, nearly spherical outer aspect, but its symmetry corresponds precisely to that of the solid shown in Fig. 10. This symmetry is the most efficient of possible arrangements for subunits in a closed shell, in the sense that it uses the smallest unit to build a shell of fixed size (22). There are exactly 60 identical elements in the surface of any icosahedrally symmetric structure, related to each other by twofold, threefold, and fivefold rotation axes. (Presence of a twofold rotation axis means that when

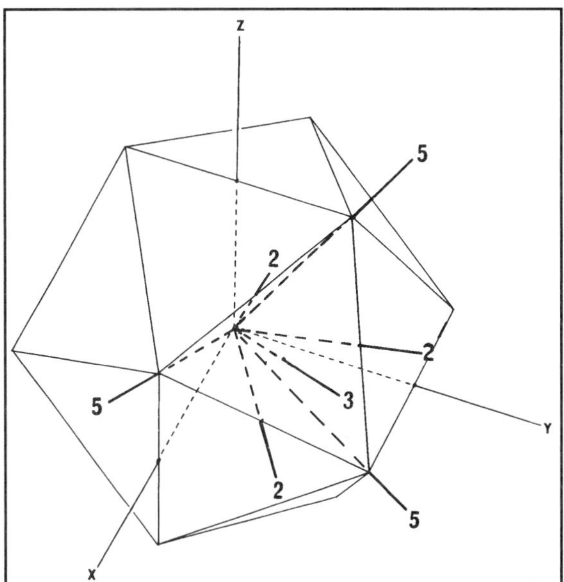

FIG. 10. Icosahedron, showing some of the elements of icosahedral symmetry. There are twofold, threefold, and fivefold axes of rotational symmetry. Note the important difference between *symmetry* and *shape*. Symmetry is the key feature of a structure. It may be defined as the collection of all operations (e.g., rotation axes) that bring the object into self-coincidence. Shape is the overall physical envelope of the object. Many objects that have icosahedral symmetry do not have icosahedral shape; for example, the pentagonal dodecahedron. Indeed, most viruses that have icosahedral symmetry do *not* have icosahedral shape (see Figs. 1 and 2).

the object is rotated by 180°, its appearance is identical to the way it looked in the starting position; with a threefold axis, both 120° and 240° rotations bring the object into self-identity; and so on.)

The way in which physical units can pack with this symmetry is shown in Fig. 11A. Commas related by twofold axes make contact head to head; those related by threefold axes make contact neck to neck; those related by fivefold axes make contact tail to tail. Closed-shell symmetries of higher order than icosahedral are, in a strict mathematical sense, not possible. Most viruses have more nucleic acid than can be packed within a shell defined by 60 subunits of reasonable molecular weight, however, and in these particles each of the 60 equivalent structural elements is composed of a number of protein chains. In some cases (e.g., poliovirus; see Chapter 16) these chains are chemically distinct. In other cases, however, they are genetically and chemically the same, and the viral shell is composed of some multiple of 60 identical units. Since not all the subunits of a given kind are symmetrically related to each other, a particular unit must be "switched" into one of several possible modes of interaction with its neighbors, depending on its location in the shell. This multiplicity raises two important problems for assembly. The first problem is one of protein architecture: How can the potential for alternative bonding modes be built into a subunit? The second problem is one of regulated assembly: How can the switching among these modes be controlled precisely in order to prevent irregularity in the shell and incorrect closure? The structures described later in this chapter show various molecular solutions to these problems.

One class of solutions to the first problem (22) depends on a small degree of flexibility in the intersubunit contacts and on the characteristics of an icosahedral surface. An example is shown in Fig. 11B, where 180 commas are arranged with icosahedral symmetry. All the commas make similar contacts, but they fall into three classes, denoted A, B, and C. One A comma, one B comma, and one C comma can be taken as the fundamental "asymmetric unit," which generates the entire structure when replicated by the 60-fold icosahedral symmetry. As in the simple 60-unit structure of Fig. 11A, the contacts that hold the structure together involve head-to-head bonds relating a pair of units, back-to-back bonds relating rings of three, and tail-to-tail bonds relating rings of five or six. Thus, although distinct in details of their packing, A, B, and C units in Fig. 11B are in fact very similar in the way they interact. All subunits form tail-to-tail bonds—those of the A-type in rings of five, as in the simple 60-subunit assembly, and those of the B and C types alternating in rings of six. Homologous parts of subunits make similar contacts in all cases. This sort of bonding was called "quasi-equivalent" (22), to emphasize that similar though not identical contacts are made by all the subunits. Packing of this kind occurs in a number of structures of positive-strand RNA viruses, but with an important additional feature. A flexible arm on each subunit can be ordered or disordered, helping to determine which of the A, B, or C modes of interaction the subunit adopts (71).

Icosahedral packings based on more than 180 subunits with closely related interactions can also be generated, but only for certain multiples of 60 (22). It can be shown that, in general, $T = h^2 + hk + k^2$ (h and k are integers) gives the permitted multiples of 60 units. For T = 7 (and many higher triangulations), there are two enantiomorphic possibilities. The number T giving the multiple of 60 units in the structure is called the triangulation number because of the way in which the icosahedral surface can be thought to be subtriangulated to accommodate the larger numbers of subunits. Structures have been identified corresponding to T = 3 (caliciviruses and many plant and insect RNA viruses), T = 4 (the cores and glycoprotein shells of alphaviruses), T = 7 (heads of bacteriophage P22), and T = 13 (rotaviruses and orbiviruses). Viruses with still larger shells generally use different kinds of subunits for fivefold vertices and have other complexities. Such structures need an organized assembly pathway, with one set of protein subunits acting as a framework or adapter for positioning another.

Observe that an icosahedral shell may be described by a surface lattice, just as in the helical case. An icosahedral surface lattice is necessarily related to a hexagonal network of equilateral triangles. Figures 11C and D show how a curved surface may be generated by introducing fivefold vertices into a planar hexagonal lattice consisting initially of only sixfold vertices. Precisely 12 fivefold vertices are needed to complete a closed shell. If the spacing of the 12 fivefold vertices is regular, the overall structure has icosahedral symmetry. The spacing of fivefold vertices in the network (nearest neighbors, second-nearest neighbors, etc.) determines the triangulation number (T = 1, T = 3, etc.). Figure 11E shows a T = 4 structure, produced from a hexagonal lattice of commas by introducing curvature in this way. The outer shells of a number of larger virus particles are formed by assembly units packed at the sixfold and fivefold positions of an icosahedral surface lattice, but the units do not necessarily have the symmetry of the position they occupy. For example, the capsids of SV40 and of polyomavirus are composed of 72 VP1 pentamers at all the vertices (12 fivefold and 60 sixfold) of a T = 7 icosahedral surface lattice (see the descriptions of polyoma and SV40 later) (93,117). The subunit interactions cannot be quasi-equivalent, because 60 of the pentamers are surrounded by *six* other pentamers, rather than five. The surface lattice does, however, represent a way of obtaining *close packing*, even when the objects do not have sixfold symmetry and when the interactions of these objects with their six neighbors are not all the same. Protein oligomers packed in this way make a good protective shell, and this probably explains why adenovirus particles appear to be T = 25 structures, or papovaviruses T = 7, even when the actual local sym-

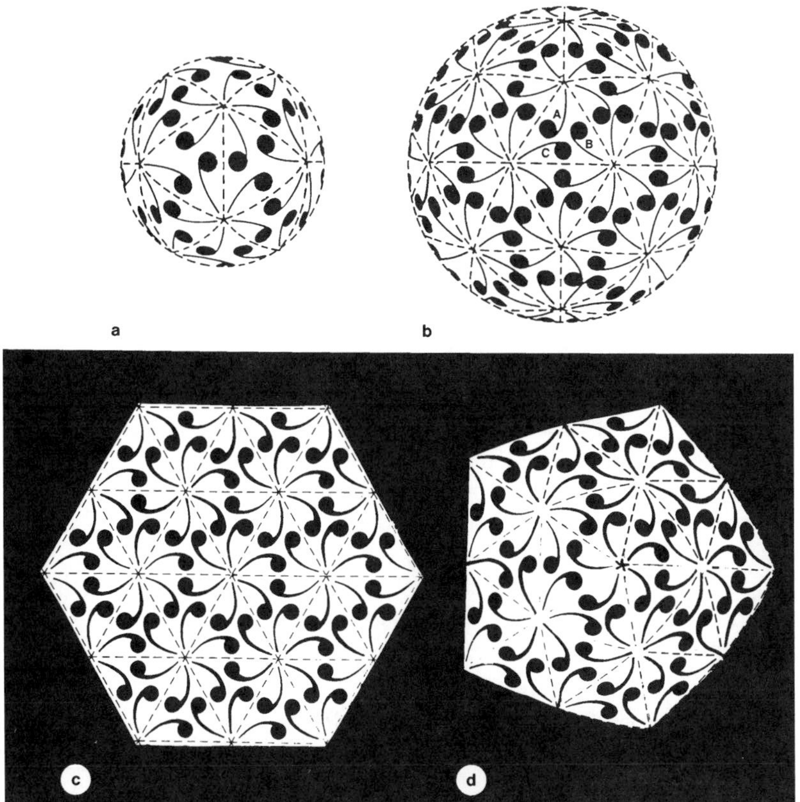

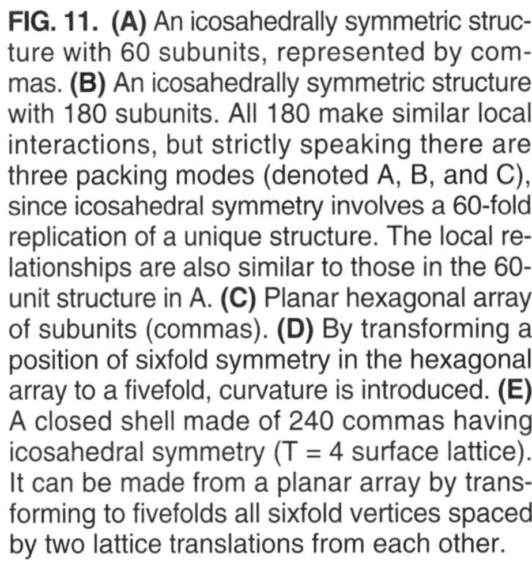

FIG. 11. **(A)** An icosahedrally symmetric structure with 60 subunits, represented by commas. **(B)** An icosahedrally symmetric structure with 180 subunits. All 180 make similar local interactions, but strictly speaking there are three packing modes (denoted A, B, and C), since icosahedral symmetry involves a 60-fold replication of a unique structure. The local relationships are also similar to those in the 60-unit structure in A. **(C)** Planar hexagonal array of subunits (commas). **(D)** By transforming a position of sixfold symmetry in the hexagonal array to a fivefold, curvature is introduced. **(E)** A closed shell made of 240 commas having icosahedral symmetry (T = 4 surface lattice). It can be made from a planar array by transforming to fivefolds all sixfold vertices spaced by two lattice translations from each other.

metry of the surface lattice does not turn out to correspond to a folded hexagonal net.

Elongated, closed-shell structures (such as the head of T4 bacteriophage) can be constructed from hemispherical "caps" having an icosahedral surface lattice and tubular "sides" with a matching helical surface lattice (21). Assembly units of icosahedral viruses can sometimes be found in aberrant tubular structures with helical symmetry: Since the chemical interactions between units are likely to be similar in the two cases, it is not surprising that the helical surface lattice of the tubes is related to the icosahedral surface lattice of the normal shell.

Not all virus particles show clear external symmetry. Long, filamentous structures may appear flexible even though their subunits have a locally regular helical arrangement. In large enveloped viruses, the membrane itself can hold external proteins together if they are anchored in the lipid bilayer. Thus, close and specific lateral interactions between these proteins are not required, and a symmetric surface arrangement is not necessarily expected. It is clear in some cases (e.g., alphaviruses and bunyaviruses) that tight lateral interactions do occur, although the precise protein arrangement in the alphavirus surface is also dictated by contacts between the icosahedrally symmetric core and the inward-projecting "tails" of the membrane glycoprotein. In other cases (e.g., retroviruses), it is not known whether there is even any local symmetry in the arrangement of envelope proteins. Distortions of the particles when subjected to conventional electron microscopy have hindered direct analysis.

Assembly Pathways

Only the simplest virus particles appear to self-assemble directly from the primary products of biosynthesis (e.g., protein subunits and genomic RNA). In most cases, various additional steps intervene, involving processes such as subcellular localization, protein modification, or cleavage of protein precursors. A significant property of larger and more complex virus structures is that particles are constructed from distinct *subassemblies*. A dramatic illustration is found in bacteriophages such as T4 (21). Heads, tails, and tail fibers assemble independently in pathways that have defined, sequential character. Within a pathway, a particular intermediate serves to nucleate addition of the next component. For example, tail core subunits in T4 phage do not associate with each other unless assembly is initiated on a baseplate. Likewise, budding of Sindbis or Semliki Forest virus occurs only around preassembled cores (see later). Pathways involving subassemblies contain a built-in accuracy check point, because incorrectly formed subassemblies are not likely to be able to incorporate into the larger structure.

Nucleic Acid Packaging

Incorporation of viral nucleic acid must be specific but independent of most of its base sequence. Thus, viral genomes generally contain a "packaging signal"—a short sequence or set of sequences that directs encapsidation. In a number of cases, such as large sDNA viruses, genome replication and packaging are coordinated, and initiation of packaging involves the replication machinery.

In single-strand RNA viruses it appears that no definite overall secondary or tertiary fold is needed for the genomic RNA, aside from the restriction that it fit within the shell. Positively charged, inward-projecting arms of coat protein subunits or distinct, positively charged internal proteins are often present to neutralize the negatively charged DNA or RNA phosphate groups. In some cases polyamines serve the same function. Charge neutralization as such does not determine specificity for viral nucleic acid. Specific incorporation generally requires a defined "packaging sequence," or a set of such sequences (129). Single-strand RNA packaging sequences appear to involve formation of a three-dimensional RNA structure (e.g., a stem-loop) recognized by one or more coat-protein subunits (Fig. 12). Coordinated base substitutions can be tolerated as long as they are consistent with the required RNA fold, thereby restricting only a very few bases that make direct contacts with protein. Examples of this sort of protein/RNA recognition are found in the assembly of phage R17 (Fig. 12A) and of TMV (Fig. 12B) and in the specific binding of the HIV *tat* and *rev* gene products to their RNA sites (TAR and RRE, respectively). TMV assembly begins at an internal origin sequence, about 1 kb from the 3' end of the genome (168). A 75-base sequence containing a presumed stem-loop structure is sufficient to initiate specific encapsidation (152). In the cases of Tat and Rev, a small, highly basic part of the protein determines much of the specificity, and a defined RNA structure is essential for the interaction (116). The detailed three-dimensional structure of a viral RNA packaging signal complexed with its specific protein recognition element has yet to be determined. For a few icosahedral RNA and ssDNA viruses, crystal structures have shown segments of ordered nucleic acid bound to sites on the inside surfaces of the coat proteins. These repeated interactions probably result from tight packing of the nucleic acid within the virion, and there is no evidence that they have any sequence specificity. In picornaviruses, where encapsidation and replication occur together, specific packaging might involve some coordination of these processes (see Chapter 21). In hepadnaviruses, which package viral RNA and then copy it to DNA, binding of reverse transcriptase to a stem-loop structure is necessary for RNA encapsidation (Fig. 12C) (6,112).

Double-stranded DNA packaging signals might, on general grounds, be expected to involve direct recognition of a sequence of base pairs, as in recognition by transcriptional control proteins. A series of repeated AT-rich sequences near the left-hand end of the adenovirus type 5 genome is required for efficient packaging, but a protein that binds to them has not yet been identified (61).

The mechanism of overall condensation or encapsidation of a viral genome is distinct from the specific recognition just discussed. In helical assemblies such as TMV or the nucleocapsid of VSV, each subunit interacts with a definite number of nucleotides and fully unwinds any local RNA secondary structure. In icosahedral viruses, the nucleic acid secondary structure is in general maintained, and there is less regular contact between protein and RNA or DNA. Various strategies are seen. Many small RNA viruses (e.g., the T = 3 plant viruses, the cores of alphaviruses) have positively charged, inward-projecting arms from the coat protein subunits. These arms help to neutralize the RNA charge. Some part of the arm on one or a few subunits may also mediate specific recognition. In the picornaviruses, polyamines are incorporated to achieve charge neutralization. Double-stranded DNA viruses use either cellular histones to condense the viral chromosome (polyomaviruses) or basic, virus-encoded proteins (adenoviruses, herpesviruses). The shells of small RNA viruses and of the polyomaviruses probably assemble around the viral genome, helping to condense it as assembly proceeds. The shells of adenoviruses and herpesviruses are believed to preassemble, and the viral DNA is thought to be incorporated into them by an active mechanism (37,43). The DNA bacteriophages package their genomes by such a process. Phage lambda, for example, cleaves its DNA from the concatenated product of replication and "pumps" it into a head precursor (9).

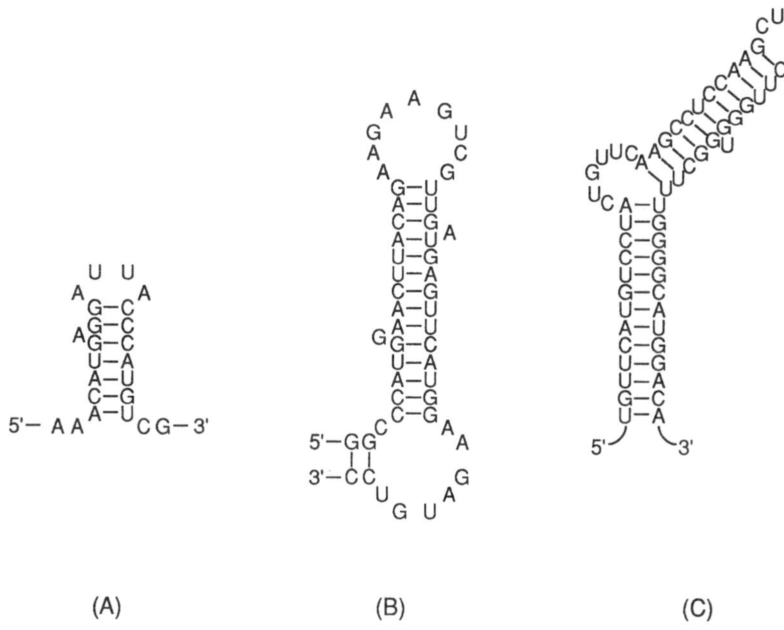

FIG. 12. Sequences (and proposed secondary structures) of RNA segments that direct specific assembly. **(A)** The site of coat-protein binding for the RNA phage R17 (18b). **(B)** The origin of assembly for tobacco mosaic virus (168). **(C)** The stem-loop structure for binding of hepadnaviral reverse transcriptase, an event required for RNA encapsidation (112).

Viral Membranes

Most enveloped viruses (except for the large and very complex poxviruses) acquire their membrane or "envelope," a lipid bilayer and associated proteins, by budding through an appropriate cellular membrane—the plasma membrane in many cases, the ER, Golgi, or nuclear membranes in others (Fig. 13). Using the cell's compartmentalization mechanisms, these viruses direct the insertion of their surface glycoproteins into the cell membrane. Subsequent events are only incompletely understood, but in some enveloped viruses there appears to be a transmembrane interaction of the membrane glycoproteins and the components of the virus in the cytoplasm, followed by a pinching off from the cell surface or into the lumen of the ER or Golgi. The lipids in the resulting bilayer derive from the cell, whereas the proteins are virally encoded (29).

Figure 13 illustrates features of the budding process in two examples—an alphavirus (Sindbis) and an orthomyxovirus (influenza). The structure of the nucleocapsid varies with virus type. It is a compact, spherical particle in the alphavirus (72), a filamentous helical nucleocapsid in paramyxoviruses and rhabdoviruses (47), and a multisegmented helical nucleocapsid in the orthomyxoviruses (131). The viral glycoproteins are anchored in the cellular membrane by a transmembrane hydrophobic peptide, which terminates in a small cytoplasmic domain. In the alphaviruses, a core particle (nucleocapsid) assembles independently in the cytoplasm, in a manner similar to the formation of a nonenveloped virion. Interactions between the core and the cytoplasmic tail of the glycoprotein then determine the location of budding. In orthomyxoviruses such as influenza, the M ("matrix") protein associates with the nucleocapsid segments and with the underside of the membrane, presumably by interaction with the cytoplasmic domains of the glycoproteins. Specific interactions both laterally among surface glycoproteins and between glycoproteins and the underlying core or M proteins drive the budding process, as illustrated. Absolute specificity is sometimes violated, as in cases of phenotypic mixing, where, for example, SV5 glycoproteins can be found in the membrane of VSV (27). In certain retroviruses, cleavage of core proteins after budding appears to lead to major internal rearrangements (see later and Chapter 26).

In viral cell entry and replication, virus membrane proteins are associated with at least four activities: receptor binding, membrane fusion, secondary uncoating or transcriptase activation, and receptor destruction. The first two, receptor binding and membrane fusion, appear to be obligatory functions. For some viruses (e.g., rhabdoviruses), the single membrane glycoprotein is responsible for both. Combination of functions within a single molecule also occurs in paramyxoviruses, in which the receptor-binding and receptor-destroying activities reside in the HN glycoprotein and fusion is a property of the F glycoprotein. Influenza C viruses have a single type of glycoprotein that mediates three distinct processes—receptor binding, membrane fusion, and receptor destruction. Influenza A viruses have a hemagglutinin (HA), which carries out binding and fusion, and a separate neu-

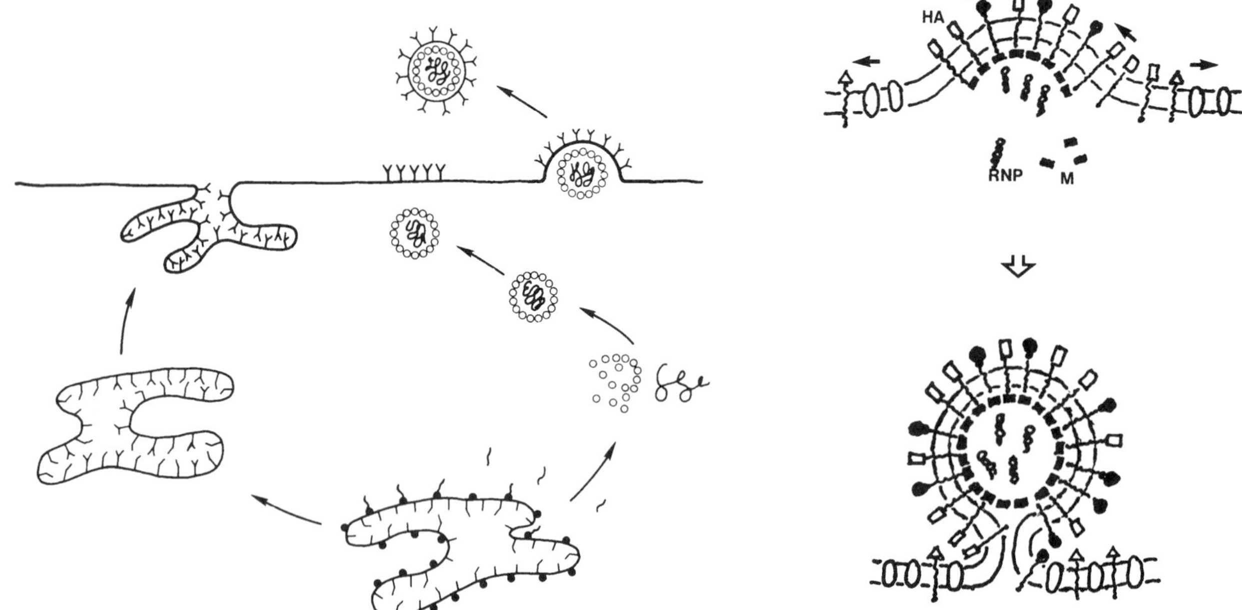

FIG. 13. Budding of enveloped viruses. **(A)** Diagram of Sindbis virus biosynthesis and assembly. The viral structural proteins are synthesized as a polyprotein from a single message. The core protein is cleaved from the precursor at an early stage. The polysomes then associate with ER membranes to complete synthesis of the glycoproteins, which are exported to the cell surface by the standard constitutive route through the Golgi compartments. Core protein and RNA assemble to form nucleocapsids, which associate with glycoprotein patches and initiate budding. **(B)** Influenza virus budding. Nucleocapsid protein (N), matrix protein (M), and glycoproteins (HA, NA) are synthesized from independent messages. Glycoproteins arrive at the cell surface by the usual route. Budding is a co-assembly at the cell surface of the glycoproteins with M and with RNP segments. Host cell proteins are excluded (*arrows*).

raminidase (NA), which destroys the sialic acid receptor. In addition, they contain an ion channel protein, M_2, which assists uncoating and transcriptase activation. Thus, only these viruses are at present known to have all four activities.

Not all viruses with membranes seem to have all four functions. In particular, receptor destruction may only be required for viruses with abundant and ubiquitous receptors or with receptors that would otherwise be components of the virion. For example, terminal sialic acid residues, if not removed from myxovirus glycoproteins by viral neuraminidase, would serve as receptors to aggregate and decrease the number of infectious virions.

The lipid composition of viral membranes has been shown to reflect that of the cellular membrane through which the particle buds (29). Thus, viruses such as influenza and VSV, which emerge through the plasma membrane, contain phospholipids and cholesterol in characteristic proportions, whereas flaviviruses, which emerge into the lumen of the ER, contain almost no cholesterol. The presence of cholesterol tends to increase the thickness of a bilayer, by restricting free rotation about single bonds in the fatty-acid chains of adjacent phospholipids. The lengths of the α-helical transmembrane segments in various viral glycoproteins vary accordingly: from about 26 residues in flu to 18–20 in yellow fever.

NONENVELOPED POSITIVE-STRAND RNA VIRUSES

High-resolution, three-dimensional structures for a number of vertebrate, plant, and insect positive-strand RNA viruses have been determined, because the small, nonenveloped, icosahedral particles are in general very stable and form good crystals. An unexpected discovery to emerge from these structures is that most are variations on a common theme (73). This common theme is schematically represented by Fig. 11B, showing 180 major "building blocks" of similar shape, packed in an icosahedral array. The blocks are protein domains based on an all-β-sheet organization, known variously as a "jelly-roll β-barrel," a "Swiss-roll β-barrel," or an "eight-strand, antiparallel β-barrel." The topology of the folded polypeptide chain characteristic of this barrel is present in a number of proteins, but the size, the shape, and the disposition of loops are specifically typical of the domains in virus capsids. This structure, schematized in Fig. 14, can therefore be called a "viral capsid β-barrel." The fold is based on a framework of two apposed β-sheets, each with four antiparallel strands. The strands are denoted by capital letters B–I. (A is missing for historical reasons.) Strands B, I, D, and G make up one sheet, which twists sharply

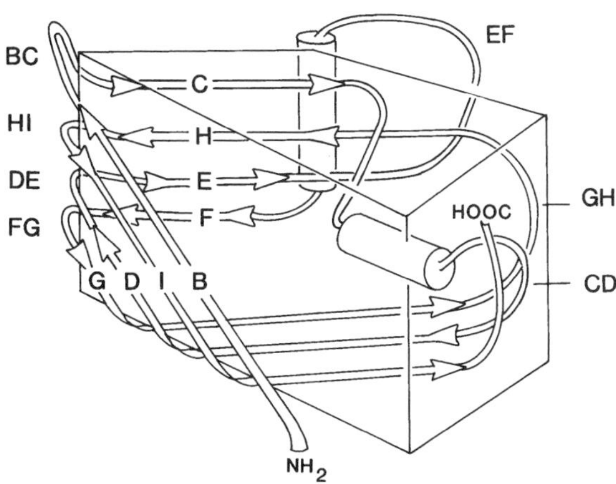

FIG. 14. Diagram of the way in which a polypeptide chain is folded to make a viral capsid β-barrel domain. The overall aspect is wedgelike. Capital letters B–I denote β-strands in the framework of the domain. They form two sheets (B-I-D-G and C-H-E-F). Loops between these strands are denoted by BC, CD, and so on. These loops and the N- and C-terminal extensions can vary greatly in length and conformation (see Fig. 17). The diagram is oriented so that the top part of the domain faces the outside of the virion.

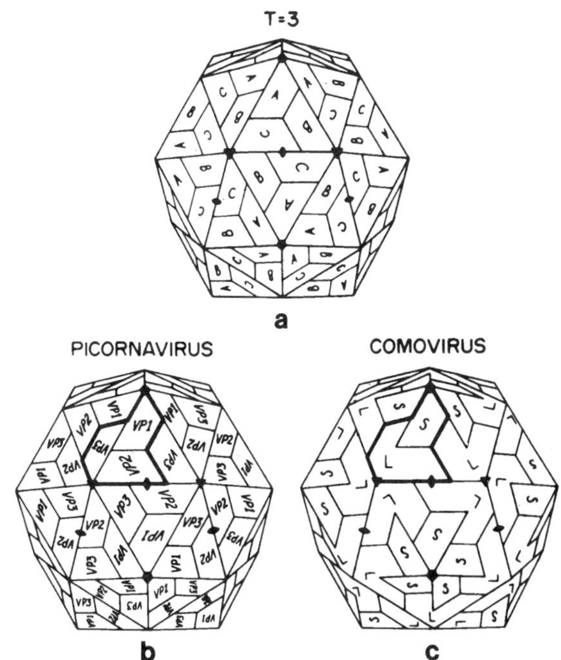

FIG. 15. Comparison of the packing of the β-barrel domains in **(A)** $T = 3$ plant and insect viruses; **(B)** vertebrate picornaviruses; and **(C)** plant comoviruses. In $T = 3$ structures, A, B, and C subunits are chemically identical. In picornaviruses, VP1, VP2, and VP3 are different, but similar in the way the polypeptide chain folds (see Fig. 17). In comoviruses, the large subunit contains two β-barrel domains, in positions corresponding to C and B in $T = 3$ structures or VP2 and VP3 in picornaviruses, and the small subunit contains one such domain, in a position corresponding to A or VP1. (Courtesy of J. Johnson.)

and thus makes up both the inner surface and one "side wall" of the wedge-shaped structure. Strands C, H, E, and F make up the second, smaller sheet, which forms the other side wall of the wedge. Viral capsid β-barrel domains vary somewhat in size because the loops, especially the CD, EF, and GH loops at the "back" of the barrel, can vary substantially in length and structure. Most known domains of this kind contain 150–200 amino acid residues. The wedgelike shape of the barrel allows the domains to pack tightly around symmetry axes (Fig. 15). As described in the previous section, the 60-fold character of icosahedral symmetry implies that the 180 β-barrel domains in a capsid must be grouped into 60 identical sets. In many plant and insect viruses, there is only one kind of subunit, and the three domains in a set are chemically identical. The packing in Fig. 15A allows quasi-equivalent interactions at corresponding interfaces (see descriptions of TBSV and BBV later). These viruses are therefore said to have T = 3 structures. In the vertebrate picornaviruses, the three domains in a set belong to independent chains cleaved from a common precursor. Nonetheless, all three have central regions with similar folds, and the packing of the domains in the viral shell conserves the approximate relative orientations found in the T = 3 structures (Fig. 15B). It has been suggested that the picornavirus capsid proteins could have evolved from a T = 3 capsid protein by gene duplication and divergence, with the addition of cleavage steps. Indeed, in at least one class of plant viruses, an intermediate situation occurs: The three β-barrel domains are different, but two (in B and C locations) remain covalently connected (see Fig. 15C). Whatever the evolutionary history, the structural similarities suggest that the entire set of icosahedral, positive-strand RNA viruses represents a common solution to the problem of how to design a vehicle for delivery of mRNA to the ribosomes of a host cell.

Picornaviruses

High-resolution structures of human rhinovirus 14 (HRV 14) (123), poliovirus (types 1 and 3) (82), mengovirus (94), and foot-and-mouth disease virus (FMDV)

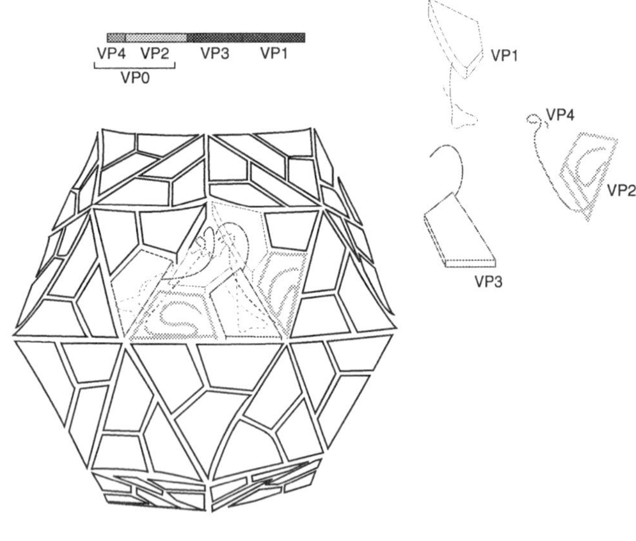

A

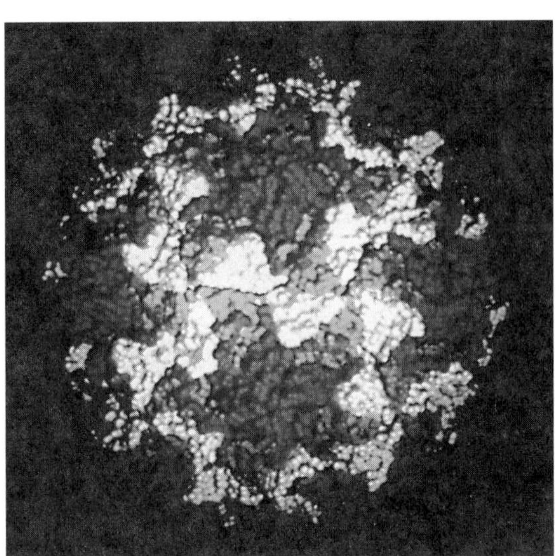

B

FIG. 16. (A) Overview of the packing of subunits in picornaviruses. The proteins are cleaved from a precursor, as shown. VP1–VP3 are represented by wedge-shaped blocks (the viral capsid β-barrel domains) with N- and C-terminal extensions. The N-terminal extensions interdigitate to form an internal framework. VP4 is, in effect, part of the N-terminal extension of VP0. In polio and rhinovirus, the prominent GH of VP1 loop lies across VP2 and VP3 as shown. In FMDV it forms an even larger, disordered projection. An "exploded" view of one protomer is shown at the right. **(B)** Surface view of poliovirus, in the same orientation as in (A).

(2) have been determined. These represent the four major groups of vertebrate picornaviruses. The general architecture is the same for all (Fig. 16) (see also colorplate 2). The particles have an outer diameter of about 300 Å. The three capsid proteins—VP0, VP1, and VP3—are cleaved by a viral protease from a polyprotein precursor (see Chapter 16). Each contains a central domain with a β-barrel fold and N- and C-terminal extensions. VP0 is cleaved during or after assembly into VP4 (initially part of its N-terminal extension) and VP2 (the entire remainder). VP4 bears a myristyl group at its N-terminus. The β-barrel domains of VP1, VP2, and VP3 from poliovirus are compared in Fig. 17. The framework, β-strand elements are relatively similar, but the loops that join them vary significantly. These loops, together with the C-terminal extensions, form the outer surface of the virion and impart its distinctive appearance to the outside world—that is, its antigenic character and its receptor complementarity. The common fold for the central domain of all three subunits is not predictable from any sequence alignment, but the shape of the domain is nearly invariant, consistent with its primary role in packing to create a rigid and well-defined viral shell. The N-terminal extensions of the subunits decorate the inner surface of this shell, forming an elaborately interdigitating network that directs and stabilizes the overall assembly. Parts of this network—VP4 and the arm of VP1—probably also help effect cell entry after an overall conformational change in the virion (see later).

The loops between β-roll framework elements are of variable size and structure, not only among the three kinds of subunits in any one virus, but also between corresponding subunits of different classes of picornavirus. In poliovirus, VP1 has particularly prominent BC and GH loops, creating a deep surface groove between them (see Fig. 17B). When five VP1 units are packed together in the virus, the confluence of these grooves creates a depression circulating around the fivefold axis in the capsid surface (Fig. 16B). HRV 14 has similar BC and GH loops and a similar depression described as a "canyon" (123). The canyon has been shown by mutational analysis and by direct electron-microscopic visualization to be the site of the major-group human rhinovirus/receptor interaction (34,105). Viruses that are subject to neutralization by host antibodies can escape the host immune system by changing surface residues. It has been suggested that sites for receptor binding, which may not change in this way without loss of affinity, are so structured that they cannot be the sole determinant of a neutralization epitope (32). The rhinovirus canyon, where the key receptor contacts appear to be at the bottom of a cavity that is too small to admit the tip of an Fab fragment, illustrates

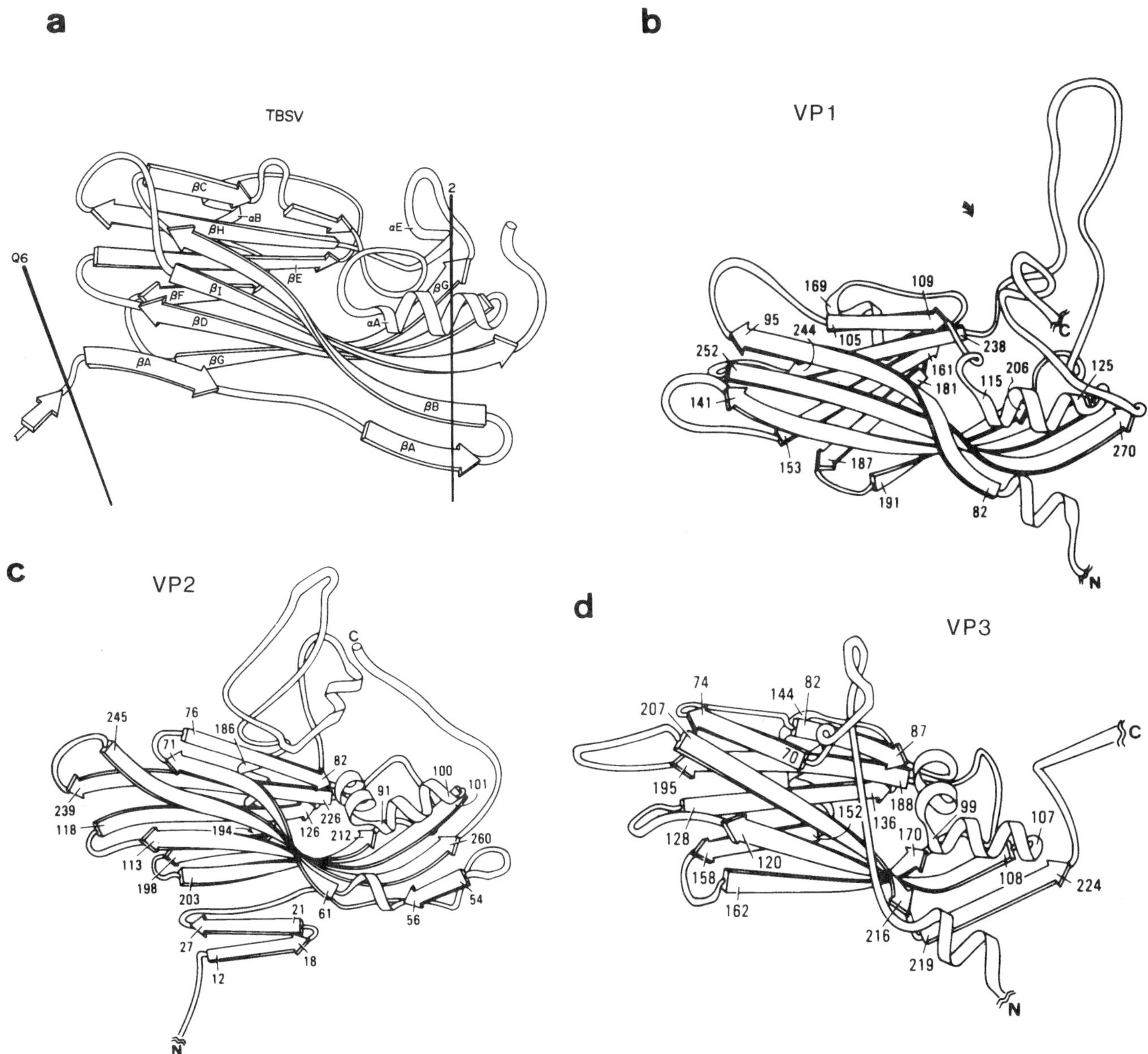

FIG. 17. Comparison of β-barrel domains in **(A)** TBSV (the S domain) (40), **(B–D)** poliovirus VP1, VP2, and VP3 (42). The *arrow* on the diagram of VP1 shows how the prominence of the BC (96–104) and GH (207–237) loops effectively creates a depression in the subunit surface. This is the so-called canyon in HRV 14 and the opening of its drug-binding site.

one possible mechanism. The major group rhinovirus receptor is ICAM-1, a monomeric adhesion molecule containing five concatenated domains with immunoglobulin (Ig) homology (65,140). The tip of a single Ig domain can fit into the rhinovirus binding site, but the heterodimeric antigen-recognition region of an antibody molecule cannot. It is indeed the tip of ICAM-1 that binds the virus (105). Another mechanism for avoiding neutralization is illustrated by foot-and-mouth disease virus (FMDV), which binds to receptors that recognize the fibronectin Arg-Gly-Asp (RGD) sequence (50). There is an RGD peptide on the GH-loop of FMDV VP1 (2). This very extended loop projects in a disordered way from the viral surface, creating a conformationally as well as a mutationally variable surround for the binding peptide.

The hydrophobic core of the β-barrel in VP1 of HRV14 and other human rhinoviruses contains a cavity that can accommodate certain drugs known to inhibit viral growth (3,137). The corresponding site in the poliovirus structure is occupied by an extended bit of hydrocarbon-like density of still unknown chemical identity (82). Thus, VP1 seems to have a degree of flexibility that can be suppressed by inserting an appropriate small molecule.

T = 3 Plant and Insect Viruses

The first viruses to be seen at high resolution were T = 3 plant viruses—TBSV (74) and SBMV (1). The organization of these viruses (shown in Fig. 3) resembles that of the picornaviruses, but there is only one kind of subunit. It consists of a central jelly-roll domain ("S domain" to indicate "shell") with N-terminal and C-terminal extensions. In TBSV, the C-terminal extension forms a complete additional domain (the "P domain," to indicate its projecting character), but in some T = 3 structures there is hardly any C-terminal extension at all. The packing of S domains relates quite closely to the packing of jelly-roll domains in picornaviruses. The A location corresponds to VP1; B, to VP3; C, to VP2. The overall diameter of the S domain shell is about 300 Å, very similar to that of the picornavirus particles. The N-terminal extensions in TBSV and SBMV have two parts—a positively charged, disordered "R domain," for neutralizing RNA, and a connecting arm between it and the S domain. The arm is ordered only on the 60 subunits in C locations, where it folds along the base of the S domain and interdigitates with two others around the icosahedral threefold axis (Fig. 3). The 60 ordered arms form an interconnected, internal framework. The viruses assemble from subunit dimers. The complete T = 3 particle may be described as an assembly of 60 A/B dimers and 30 C/C dimers. The C/C dimers create "flatter" areas in the surface than the A/B dimers, since their S domains must spread apart to accommodate the folded arms (see Fig. 3). Coordination of assembly involves sequential formation of the internal framework, which in turn determines whether a dimer adopts an A/B conformation, with disordered arms and sharper curvature, or a C/C conformation, with ordered arms propagating the framework and flatter curvature (138).

The particles of black beetle virus (BBV) and other nodaviruses, which infect vertebrates as well as insects, are similar in many of their structural features to the plant virus particles just described (84). Their subunits contain jelly-roll β-barrel domains that pack like the TBSV S domains in Fig. 3. As in TBSV, an amino terminal arm, ordered on the C subunits, serves as a conformational switch. In addition, the nodavirus particles have a duplex stem of ordered RNA (about 10 bp) lying across the twofold axis and along the C/C interface, beneath the amino-terminal arm (48). Together, the RNA and the arm act like a "wedge" to stabilize the flatter curvature of a C/C dimer. The nodavirus subunit, like the picornavirus VP0, undergoes an autolytic cleavage after assembly (55). The cleavage generates a carboxy-terminal peptide, partly helical and partly disordered, on the interior of the particle. The 15-residue helical segment is amphipathic, and it may play a role during entry analogous to that postulated for VP4 and the amino terminus of VP1 in poliovirus (see the next section).

Cell-Induced Conformational Changes

Polioviruses and rhinoviruses undergo a conformational change upon interaction with the cell surface (67). The critical event must involve an expansion, since VP4 is lost from the interior of the virion. It is triggered by interaction with the cellular receptor (59). That is, the receptor serves not only to bind the virus to a cell but also to initiate uncoating. The altered virus particles are protease sensitive and bind to liposomes. In poliovirus, the structural change is accompanied by exposure of the hydrophobic amino terminus of VP1 (52). This arm appears to be associated with VP4 in the particle interior, and its externalization could easily accompany loss of VP4. There is evidence that VP4 and the amino-terminal region of VP1 both play a role in penetration (52,101). The expansion of plant viruses, such as tomato bushy stunt virus and turnip crinkle virus, provides a possible structural model (122). In these viruses, an interface containing bound Ca^{2+} is destabilized by removal of divalent cations at pH > 7, resulting in expansion of the particle and extrusion of amino-terminal arms from A and B subunits through fenestrations in its expanded surface (75). The expanded virion is believed to be an uncoating intermediate, produced by the low cytoplasmic Ca^{2+} concentration after viral entry through abrasions in the cell surface. Animal viruses require a built-in membrane penetration mechanism. Those with lipid envelopes have fusion glycoproteins, activated in many cases by the low endosomal pH. The externalization of a hydrophobic and a myristylated peptide may represent an analogous property. One circumstantial correlation that supports such a role is the absence of a VP4 homolog in plant "picorna" viruses such as CpMV, which presumably do not enter by endocytosis, and the presence of a cleaved carboxy-terminal arm in insect nodaviruses, which presumably do.

The VP1 amino-terminal arm is shorter in cardioviruses (such as ME and mengovirus) and in aphthoviruses (such as FMDV) than it is in polioviruses and rhinoviruses and does not contain the putative amphipathic helix. The receptor-bound forms of these viruses appear to dissociate, rather than merely to expand, especially at low pH (69).

Viral expansion and dissociation are cooperative processes. Only a single or a few bound receptors per virion would presumably be necessary to trigger the change in a picornavirus. Likewise, one or a few "cross-links" might be sufficient to inhibit it. Neutralization of poliovirus can be accomplished by only a few bound antibodies and cannot be effected by Fab fragments. Indeed,

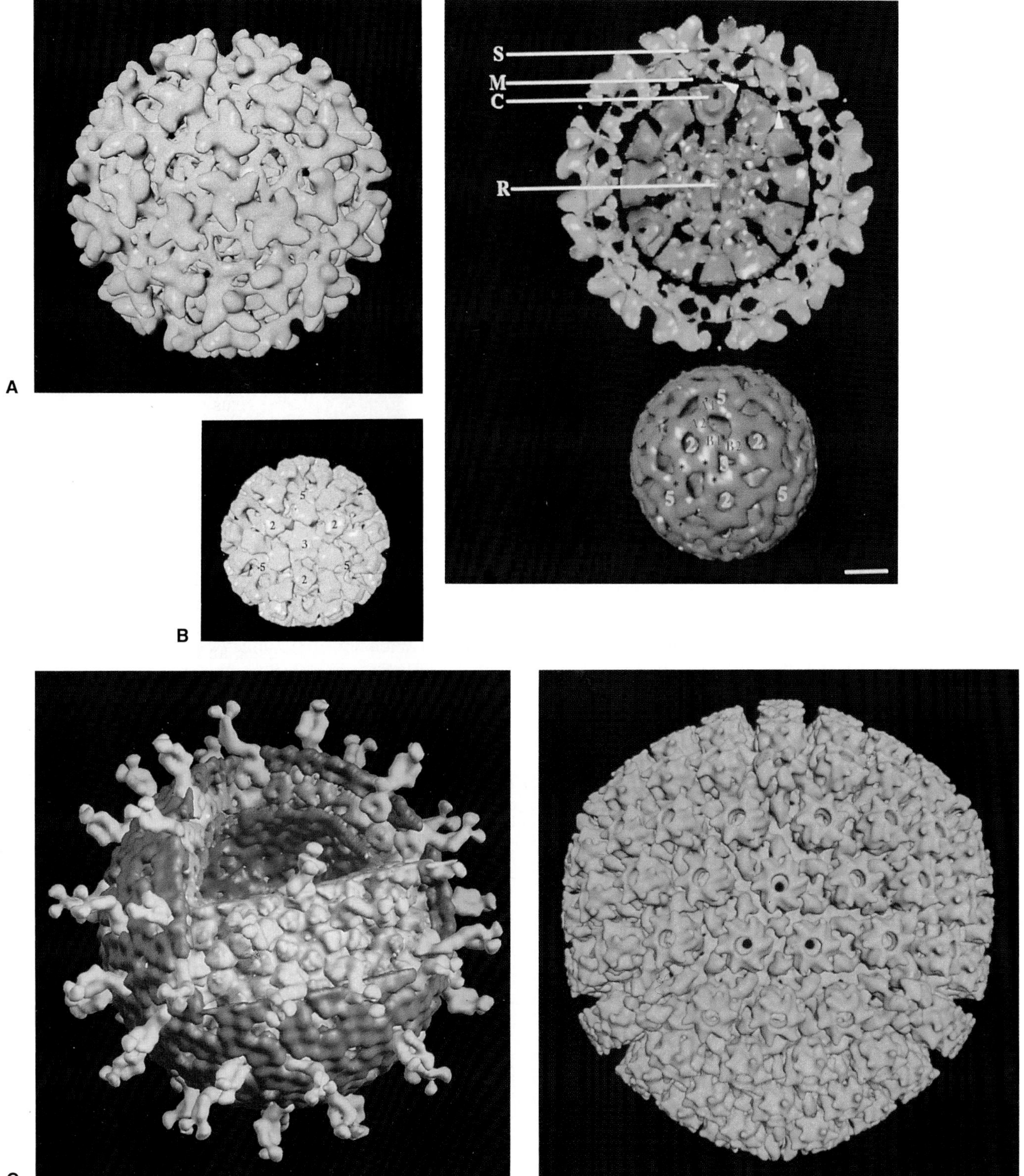

Colorplate 1. (Figure 2) Three-dimensional reconstructions of icosahedrally symmetric virus particles from cryoelectron micrographs. The contrast comes from the viral components themselves—protein, nucleic acid, and lipid—not from stain, as in Fig. 1. Therefore, meaningful detail is present at all radii in the image. **(A)** Sindbis virus, an enveloped, positive-strand RNA virus. Surface view of virion (left), showing T = 4 arrangement of viral glycoprotein; cross-sectional view (top right), showing glycoprotein (yellow) and core (blue), with the lipid bilayer as a low-density gap between them; surface view of core (bottom right). (Courtesy of BVV Prasad. See also Fig. 19.) **(B)** The capsid of a calicivirus, a T = 3 particle. The view is along an icosahedral threefold axis, which has quasi-sixfold character in a T = 3 structure. The projections are dimer-clustered domains of the protein subunit, and the surface topography strongly resembles that of the TBSV particle shown in Figs. 1 and 3. (Courtesy of BVV Prasad.) **(C)** Rotavirus, a nonenveloped, double-stranded RNA virus. The particle has three protein layers, color-coded blue: the outer shell (VP7), violet: the inner shell (VP6) and red: the core (VP2). The yellow spikes represent the hemagglutinin (VP4)—the protein through which the virus is believed to recognize its receptor. The outer and inner shells conform to a T = 13 icosahedral lattice. (Courtesy of M Yeager.) **(D)** Nucleocapsid of herpes simplex virus, a double-stranded DNA virus. The surface contains a principal structural protein (VP5), arranged in hexameric and pentameric clusters in a T = 16 icosahedral lattice. (Courtesy of BVV Prasad.)

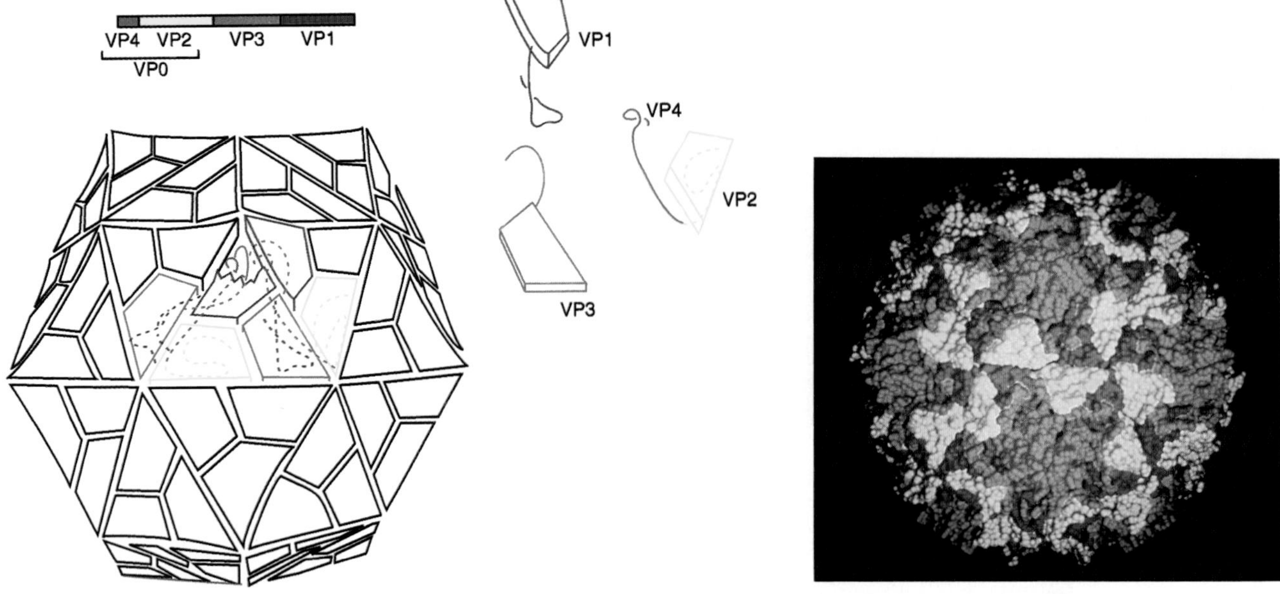

Colorplate 2. (Figure 16) **(A)** Overview of the packing of subunits in picornaviruses. The proteins are cleaved from a precursor, as shown. VP1–VP3 are represented by wedge-shaped blocks (the viral capsid β-barrel domains) with N- and C-terminal extensions. The N-terminal extensions interdigitate to form an internal framework. VP4 is, in effect, part of the N-terminal extension of VP0. In polio and rhinovirus, the prominent GH of VP1 loop lies across VP2 and VP3 as shown. In FMDV it forms an even larger, disordered projection. An "exploded" view of one protomer is shown at the right. **(B)** Surface view of poliovirus, with the same shading as in (A).

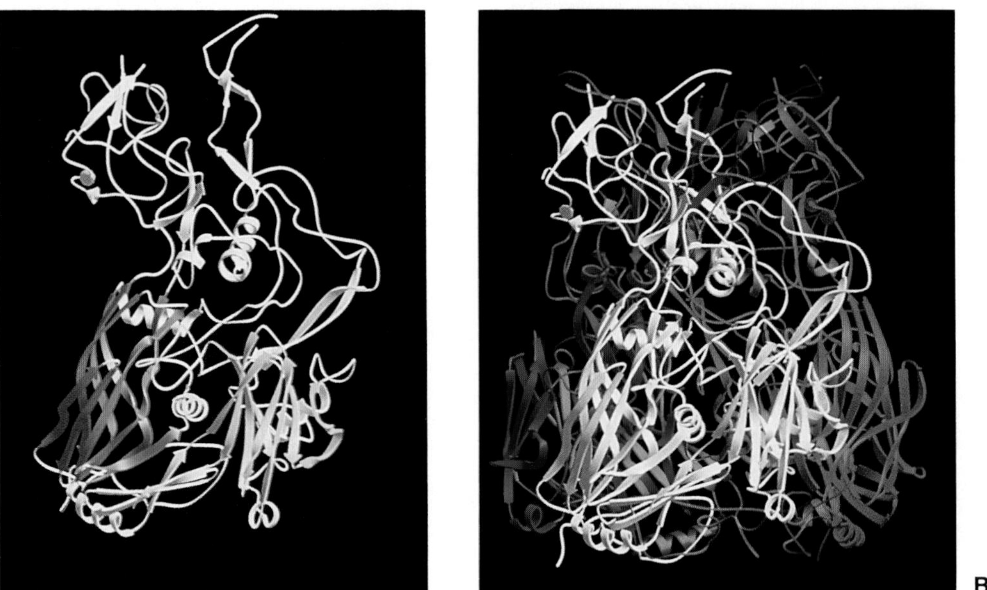

Colorplate 3. (Figure 21) **(A)** The polypeptide chain of the adenovirus hexon. The direction of view is perpendicular to the threefold axis of the hexon. The two viral capsid β-barrel domains are shown in color; the loops between strands in each domain and between domains are shown in white. Note that these domains are oriented so that their β-strands would run radially in the virus particle, rather than tangentially, as in the picornaviruses. **(B)** The hexon trimer. The white subunit, in the foreground, is in the same orientation as the single subunit in (A). The loops of the individual chains fold extensively with each other. These loops would probably not be ordered in a monomeric subunit, and it is likely that subunit folding and trimer assembly occur coordinately.

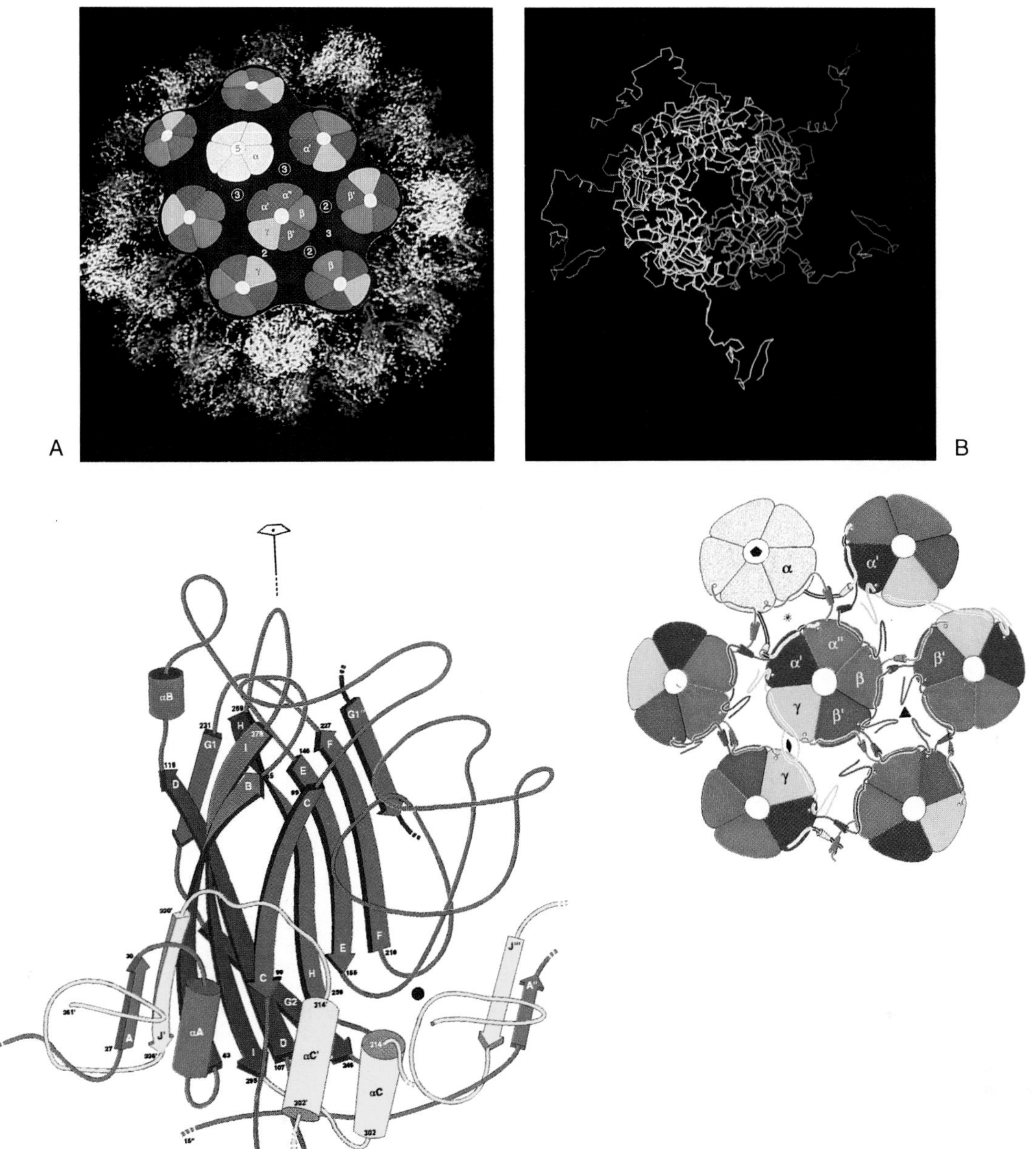

Colorplate 4. (Figure 19) Organization of the polyomaviruses, as illustrated by SV40. **(A)** Overview of the particle, showing the packing of VP1 pentamers. The diagram is based on a molecular graphics representation in which the polypeptide chain of each subunit is shown as a folded, colored line. The subunits of pentamers on fivefold positions are shown in white; those of pentamers in general, six-coordinated positions are shown in colors. Overlaid on the central part of the image are block representations. The six colors indicate six quite different environments for the VP1 subunit. **(B)** A six-coordinated VP1 pentamer, "extracted" from the model in (A). The extended carboxy-terminal arms are shown in the conformations they adopt in the assembled particle; in the free pentamer they are disordered and flexible. **(C)** The SV40 subunit, viewed normal to the pentamer axis. Strands are represented as ribbons; helices, as cylinders; loops as narrow tubes. The two sheets of the β-barrel domain are colored blue and red; the intervening loops are violet; the carboxy-terminal arms are yellow; the amino-terminal arms are green. The complete yellow carboxy-terminal arm seen here actually emanates from a subunit in another pentamer; only the initial, α-helical segment of the arm from this subunit is shown, since it then extends out of the page. Note how the green, amino-terminal segment clamps the invading arm in place. The small ball represents the site of a Ca^{2+} ion. **(D)** Pattern of interchange of arms in the virion. The shading scheme corresponds to (A) and (B). The body of a pentamer is shown as a five-petalled flower; the arms are cylinders (α-helices) and lines. The small open circles mark Ca^{2+} sites.

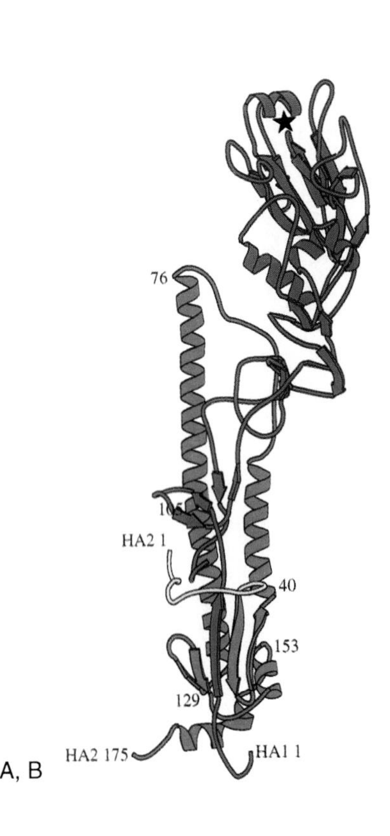

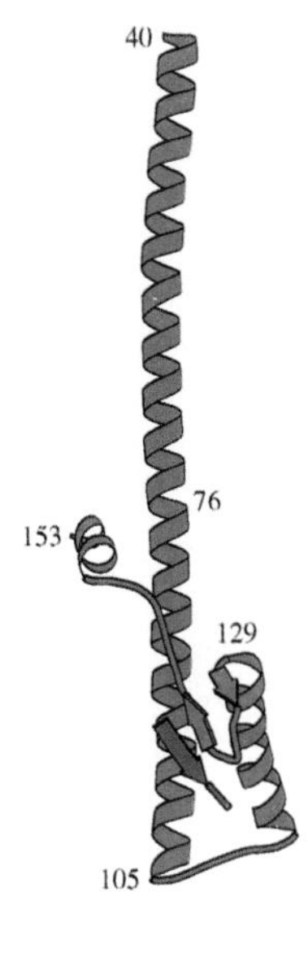

A, B

Colorplate 5. (Figure 27) The folded structure of the influenza virus hemagglutinin (HA) and its rearrangement when exposed to low pH. **(A)** The HA monomer: HA_1 is in blue; HA_2 is in red; the fusion peptide at the N-terminus of HA_2 is in white. Residue numbers in HA_2 are shown at several key positions, in order to assist in visualizing the conformational change that occurs on fusion activation. The receptor-binding pocket in the β-barrel "top" domain of HA_1 is indicated with a star. The viral membrane would be at the bottom of this figure. The transmembrane segment, which follows HA_2 175 is not shown. **(B)** The HA_2 monomer in the fusion-active form. The fragment shown is produced as the result of digesting the activated HA with thermolysin, removing most of HA_1 and the fusion peptide of HA_2. A short segment of HA_1 (blue) remains bound to the rearranged HA_2, as shown. Note the dramatic conformational change, in which residues 40–105 become a continuous α-helix. The introduction of a kink at residue 105 also causes the "bottom" part of the molecule to rotate upward. **(C)** The HA trimer. The subunit toward the right in this illustration is viewed in essentially the same orientation as the monomer in (A). Parts of HA_2 are colored: the short helix and loop (residues 40–75), in red; the long helix from its N-terminus to the point where the fusion peptide inserts along the threefold (residues 76–105) and the remainder of the long helix, in violet; and the β-sheet that follows (residues 106–140), green; one of the C-terminal helices of HA_2 (residues 141–162), yellow. The color scheme is designed to assist in visualizing the conformational change. Exposure to low pH initiates this change, causing the top domains of HA_1 to move apart (*arrow*). **(D)** The fusion-active form of HA_2. The tops of HA_1 may be imagined to have spread to either side. The shaded segments of HA_2 correspond to the similarly colored segments in (C). (Courtesy of F. Hughson.)

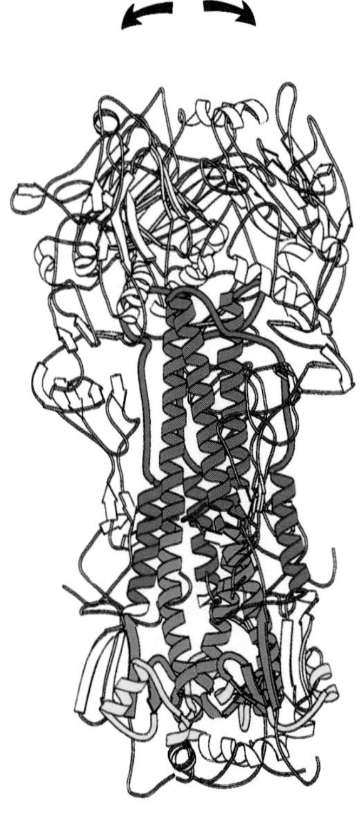

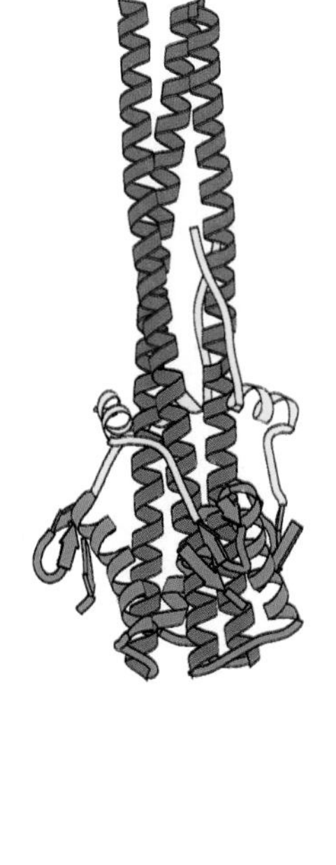

C, D

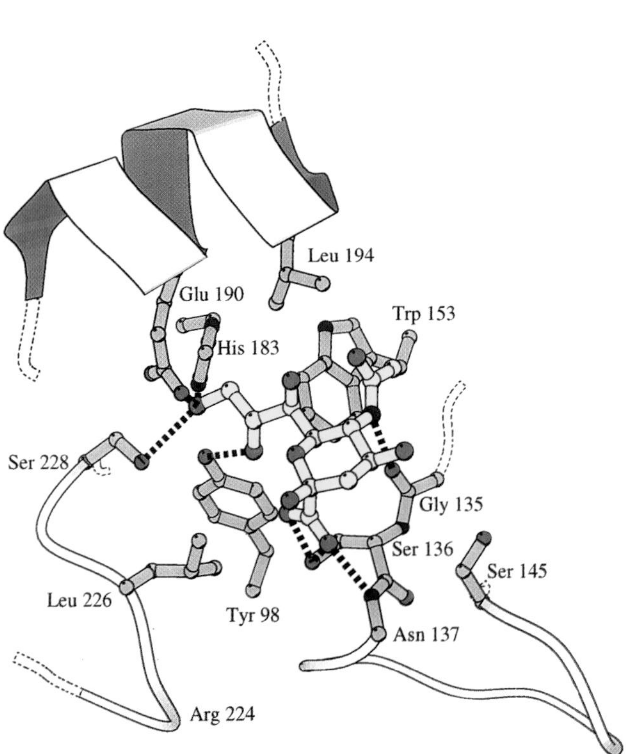

Colorplate 6. (Figure 28) A detailed view of the receptor binding site on influenza virus HA_1, showing bound sialic acid (yellow). Side chains from HA_1 that contact the sialic acid are shown. The view is essentially a blow-up of the region indicated by a star in Fig. 27A.

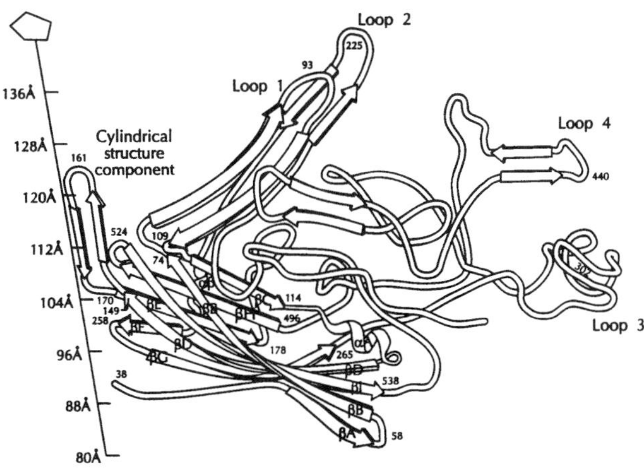

FIG. 18. The canine parvovirus subunit, shown in the same view as the viral subunits in Fig. 17. Note the very extensive BC, EF, and GH loops and the β-ribbon insertion in the DE loop.

it has been proposed that antibodies must cross-link two protomers in order to neutralize the virus (45).

SINGLE-STRANDED DNA VIRUSES

The Parvoviruses are, as their name suggests, very small particles that contain a 5-kb single-stranded DNA genome (see Chapter 31 and ref. 118). In canine parvovirus (CPV) and its relatives, there are 60 copies of the coat protein, which comes in three different lengths. VP1 (85 kD) and VP2 (65 kD) are generated by in-frame alternative splicing, so that VP1 contains about 200 additional residues at its amino terminus. VP3 is produced by proteolytic cleavage of 20 residues from the amino terminus of VP2. Each virion contains a small number of VP1 subunits, with VP2 and VP3 being the major species. The proportion of VP2 cleaved to VP3 can be enhanced by treatment with protease.

The structure of CPV (151) shows that the subunit contains a viral capsid β-barrel domain (Fig. 18), similar to the one found in RNA viruses, arranged in a T=1 icosahedral lattice. There are substantial insertions in the BC, EF, and GH loops (the last containing more than 200 residues), a short β-ribbon insertion in the DE loop, and a 40-residue carboxy-terminal extension. The tapered end of the wedge-shaped β-roll packs around an icosahedral fivefold axis, similar to the fivefold packing of VP1 in picornaviruses and the A subunits in T = 3 plant and insect viruses. Threefold contacts are mediated largely by the GH loop. The DE insertion forms a hollow, cylindrical protrusion along the fivefold axis. The amino-terminal arm of one of the five VP2 subunits may project outward along the fivefold, and the cleavage to VP3 may occur because of the accessibility of this arm to protease (151).

DOUBLE-STRANDED DNA VIRUSES

The double-stranded DNA (dsDNA) viruses include both relatively simple particles, such as simian virus 40 (SV40), with a 5-kbp genome, and very complex ones, such as the herpesviruses, with 150-kbp genomes. Except for the extremely complex poxvirus, which will not be considered here, their capsids are icosahedrally symmetric, but there are a variety of strategies for accurate assembly of larger shells.

SV40 and Murine Polyomavirus

These viruses are constructed from 72 pentamers of VP1 (40–45 kD), and each pentamer appears to incorporate one copy of an internal protein (VP2 or VP3). The shell is about 500 Å in external diameter. It packages a so-called minichromosome—the closed, circular viral DNA genome complexed with cellular core histones into about 25 nucleosomes (30). The organization of SV40 and polyoma appears to present a structural puzzle. As described earlier (Fig. 10), an icosahedral structure has 12 fivefold axes, and hence just 12 of the VP1 pentamers occupy positions of fivefold symmetry. The remaining 60, which lie at the six-coordinated positions of a T = 7 lattice, are surrounded by six other pentamers (Fig. 19A) (See also colorplate 4.). How can a pentamer fit into hexavalent surround? Why does the structure not follow the principle of quasi-equivalence, with *hexamers* of VP1 occupying the six-coordinated positions?

The answers to these questions can be seen in the crystallographically determined structures of SV40 and polyoma (93,117,142). The VP1 pentamers are extremely stable and robust subassemblies, and the structure is based on use of this single type of building block. The pentamers are joined by extended polypeptide arms (the C-termini of VP1) rather than by contacts across complementary surfaces. This type of linkage allows flexibility in the way a given subunit can see the subunits of neighboring pentamers, while still tying onto them in a specific manner. VP1 folds into an N-terminal arm, a central β-barrel domain, and a long C-terminal arm (Fig. 19C). The β-barrel has the same sort of polypeptide chain fold as in the RNA-virus and parvovirus subunits, but its shape is somewhat different because the sheets pack rather differently against each other. The β-strands run approximately parallel to the fivefold axis, rather than nearly perpendicular to it. The structure of the domain is the same in all VP1 subunits, independent of pentavalent or hexavalent packing, except for a small shift in the CD loop. The only major structural variation in the subunit occurs in the carboxy-terminal arms, which extend away from their subunit of origin and invade an adjacent pentamer. Each arm donates a strand to the edge of a β-sheet, and this strand is clamped in place by an additional strand from the amino-terminal part of the target subunit (Fig. 19C). Almost all interpentamer contacts involve the carboxy-terminal arms, which are likely to be flexible and unstructured on free pentamers. The amino terminus of each subunit is also likely to be unstructured because it can only clamp into place when the C-terminal arm from another pentamer is present. Specificity of assembly is ensured by the sequence of amino acid residues in the inserted strand, which make many contacts in the target subunit, but restrictions on a unique relative orientation of two pentamers are relieved by not needing to match two preformed complementary surfaces. A helical segment at the beginning of the arm appears to give rigidity to the assembled structure by packing between one or two other such helices and the body of the target subunit (Fig. 19C,D).

In addition to the clamp from the amino-terminal segment, there is an ion-mediated contact that locks the arm against its target. The site shown in Fig. 19C probably binds Ca^{2+}; the liganding residues include an aspartic acid in the arm and two glutamic acid residues in the target pentamer. There is evidence that withdrawal of Ca^{2+} plays a role in viral uncoating (30). Reduction of disulfide bonds is also believed to be required, and there are indeed in SV40 and polyomavirus disulfides that further clamp the C-terminal arm. The disulfides probably form after the virus escapes from an infected cell. Their reduction, along with abstraction of the Ca^{2+}, is likely to occur on uptake into a target cell, because of the reducing character of the intracellular environment.

Polyoma VP1 expressed in *E. coli* is isolated as pentamers (92). These pentamers can be made to assemble into capsidlike structures under various conditions (126). Thus, the posttranslational modifications detected on VP1 derived from virions (phosphorylation and acetylation) are not essential for directing the alternative five- and six-coordinated packings. The *in vitro* assembled capsids are somewhat variable in size, however, suggesting that additional mechanisms contribute to the accuracy of virus assembly (127). Capsidlike particles, isolated from nuclei of insect or mouse cells transfected with VP1 expression vectors, are much more perfect (49,100). Nuclei contain "chaperone" proteins that may help disassemble incorrect structures and increase the yield of perfect capsids. The internal proteins, VP2 and VP3, are evidently not required for the observed accuracy of particle formation.

The receptors for initial binding of polyomavirus to the cell surface are sialic-acid-bearing oligosaccharides attached to membrane glycoproteins (108). The binding site on VP1 for these sialo-oligosaccharides is a shallow groove on the outward facing rim of the subunit (142). There is a pocket at one end of the groove for sialic acid. The overall shape of the oligosaccharide, as determined by the stereochemistry of the glycosidic linkages and by internal hydrogen bonding, is particularly important for a proper fit into the site on VP1.

The internal proteins, VP2 and VP3, overlap in sequence: VP2 has about 100 additional residues at its amino terminus and bears an N-myristyl group. The common carboxy terminus serves to tether VP2 and VP3 to VP1, by inserting into the inward-facing conical hollow along the axis of the pentamer (66,93). In polyoma, about 40 residues near the C-terminus of VP2/3 are sufficient for binding to VP1 (5). VP2 appears to be important for entry or uncoating (125). How it functions in this way is not known. It is interesting to note that another N-myristylated, internal protein—VP4 of picornaviruses—also has a role in viral entry (101). One possibility, suggested by the membrane-targeting characteristics of the myristyl group, is that VP2 helps disrupt the vesicular membrane that surrounds the virion after uptake into a cell. A function of this sort would require that VP2 emerge from the viral interior during entry, but there is at present no evidence for such a step. It is not clear why both VP2 and VP3 are needed in the virion, since the sequence of the former completely contains that of the latter. It is possible that there is insufficient space in the particle interior for 72 copies of VP2 and that only a modest "dose" of VP2 is needed to fulfill its entry function. The truncated version (VP3) may then have evolved to serve whatever additional function is required on every pentamer, without wasting internal space.

The papillomaviruses have an all-pentamer architecture like the polyomaviruses, but the virions are somewhat larger (4). It is possible that the detailed molecular organization is also similar, despite the lack of any obvious sequence relations between L1 (papillomaviruses) and VP1 (polyomaviruses). The papillomaviruses have a single species of internal protein (L2).

Adenoviruses

The adenovirus particle is more complex than those described so far. It illustrates how large structures are constructed from subassemblies. The diagrams in Fig. 20 show the adenovirus particle and the substructures into which it can be resolved (46,56). The outer shell has a somewhat rounded icosahedral shape. At first glance it appears to have a triangulation number of 25. The structures at the fivefold positions ("pentons") are different from the rest ("hexons"), however, and the hexons are chemically trimers rather than hexamers. Thus, the structure really does not correspond to a simple subtriangulated icosahedral design.

The hexons are trimers of a 110K polypeptide (conventionally denoted II). The high-resolution structure of the hexon (Fig. 21A and B, colorplate 3) (121) shows that each subunit contains two rather similar domains, each of which roughly resembles the β-barrel part of VP1 from SV40 and polyoma. That is, the β-strands in the hexon domains are connected sequentially, as in the usual jelly

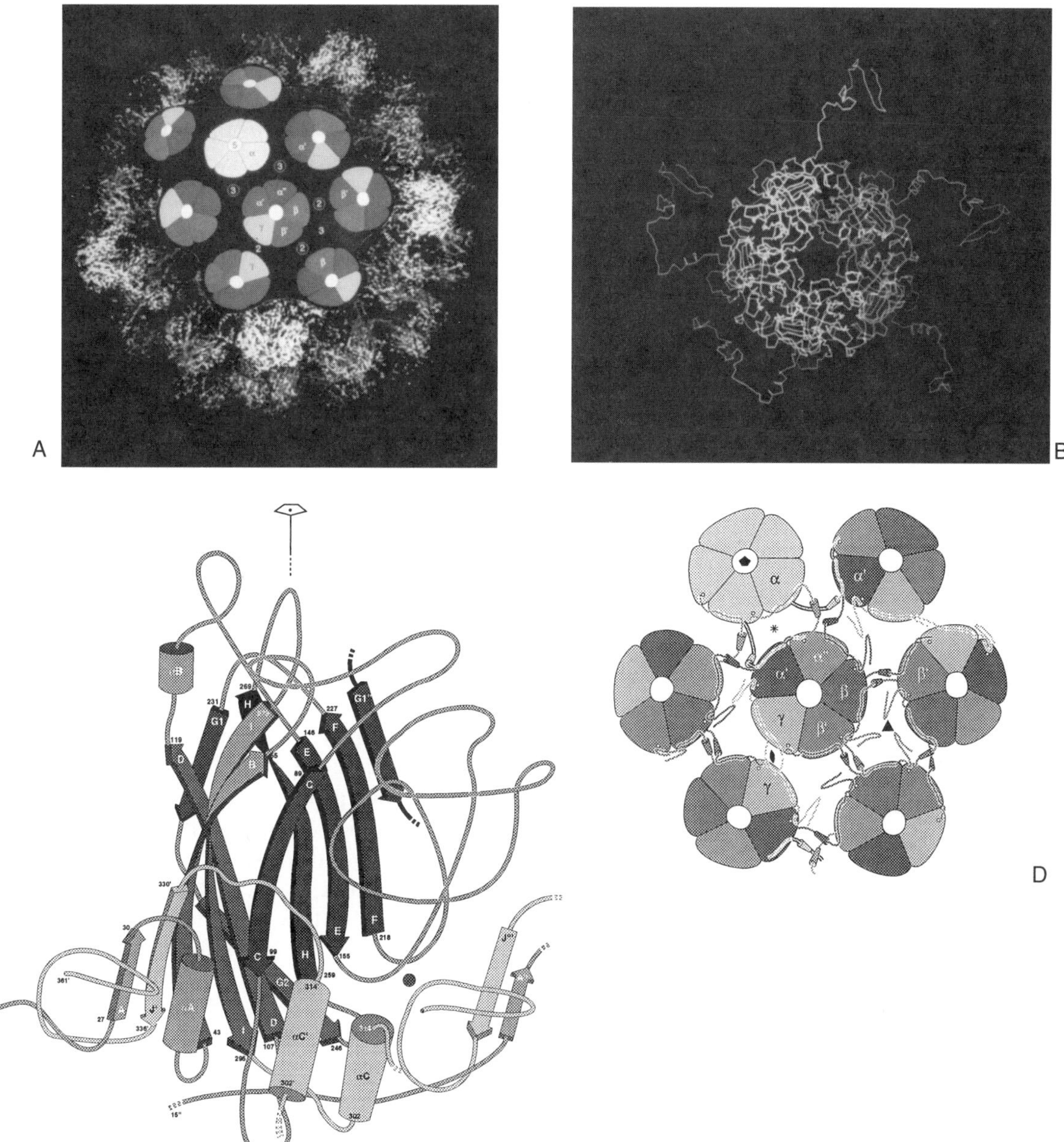

FIG. 19. Organization of the polyomaviruses, as illustrated by SV40. **(A)** Overview of the particle, showing the packing of VP1 pentamers. The diagram is based on a molecular graphics representation in which the polypeptide chain of each subunit is shown as a folded line. The subunits of pentamers on fivefold positions are shown in white; those of pentamers in general, six-coordinated positions are shown in darker shades. Overlaid on the central part of the image are block representations. The six shades indicate six quite different environments for the VP1 subunit. **(B)** A six-coordinated VP1 pentamer, "extracted" from the model in (A). The extended carboxy-terminal arms are shown in the conformations they adopt in the assembled particle; in the free pentamer they are disordered and flexible. **(C)** The SV40 subunit, viewed normal to the pentamer axis. Strands are represented as ribbons; helices, as cylinders; loops as narrow tubes. The two sheets of the β-barrel domain are in different dark shades; the carboxy-terminal arms are pale; the amino-terminal arms are somewhat darker. The complete carboxy-terminal arm seen here actually emanates from a subunit in another pentamer; only the initial, α-helical segment of the arm from this subunit is shown, since it then extends out of the page. Note how the amino-terminal segment clamps the invading arm in place. The small ball represents the site of a Ca^{2+} ion. **(D)** Pattern of interchange of arms in the virion. The shading scheme corresponds to (A) and (B). The body of a pentamer is shown as a five-petalled flower; the arms are cylinders (α-helices) and lines. The small open circles mark Ca^{2+} sites.

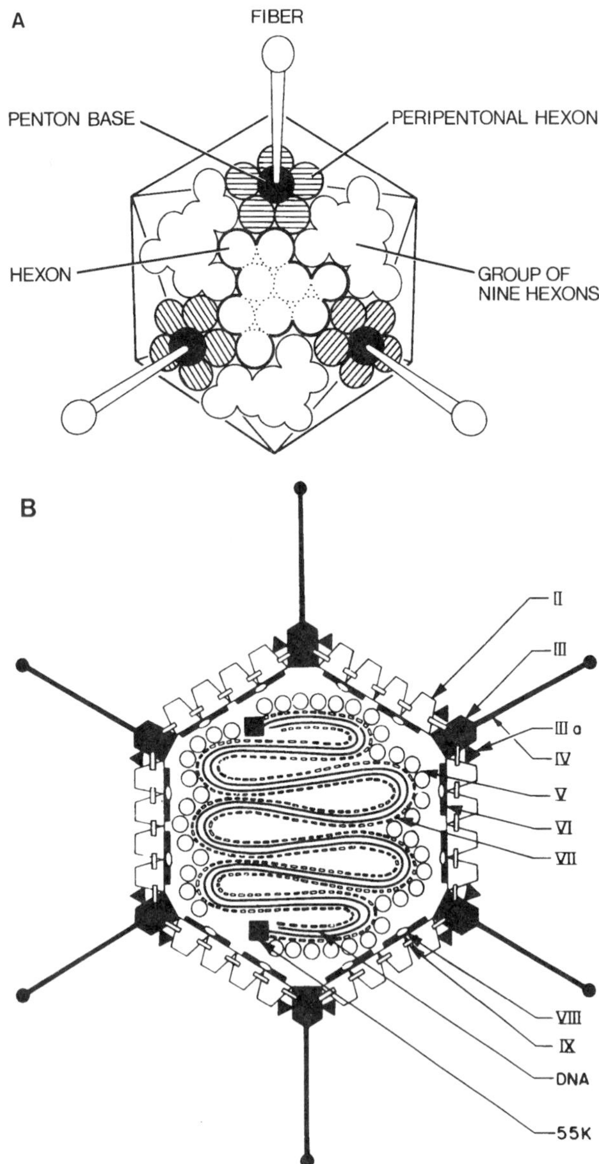

FIG. 20. (A) Diagram of the adenovirus particle, showing the location of major protein subassemblies. The nine hexons drawn as a group and the "peripentonal" hexons are trimers of the same polypeptide; they are distinguished only by their location in the structure. (Illustration by John Mack. From ref. 18a, with permission.) **(B)** Cross section, showing the probable location of the principal polypeptide components and the viral DNA. The assignment for proteins other than II (hexon), III (penton base), and IV (penton fiber) is based on studies of stepwise dissociation and on difference cryoelectron microscopy (see Fig. 21C). (Redrawn from ref. 56 with permission.)

roll framework, and they run radially, as in the polyomavirus pentamer. The major loops in each domain are between strands D and E and strands F and G. These loops project outward, interacting tightly with corresponding loops from other subunits in the trimer. Their conformation therefore depends on trimer formation. The loops bear the principal type-specific antigenic determinants. Assembly of the hexon from newly synthesized protein *in vivo* appears to require a factor, the "100K protein," which is also encoded by the virus. It serves as a chaperone for assembly and nuclear transport. This protein does not form part of the final structure (23).

The penton comprises a pentameric base (polypeptide III) and a trimeric fiber (IV). A three-dimensional image reconstruction of adenovirus type 2 from cryoelectron micrographs shows that the penton base fits into the virion shell at the same level as the hexons (143). Its immediate contacts are the so-called peripentonal hexons (Fig. 20). Polypeptide VI, identified by "subtracting" the known hexon x-ray structure from the image reconstruction of the virion, lies inside the hexon shell and appears to hold the peripentonal hexons in place (144). The amino acid sequence of the penton fiber reveals 22 approximate repeats that form the shaft and a C-terminal 181-residue domain that forms a knob at the tip (64). A short N-terminal segment of 44 residues probably anchors the fiber in the base. Both the fiber and the base are believed to participate in contacts important for cell entry. In adenovirus type 2, the fiber knob mediates initial attachment, and the base has been shown to make a (presumably subsequent) interaction through an Arg-Gly-Asp (RGD) sequence with appropriate cell-surface integrins.

Dissociation of adenovirus particles by various methods yields groups of nine hexons, as shown in Fig. 20. They are derived from virions as indicated, and they include all but the peripentonal hexons. The hexons in the groups of nine are held together by viral polypeptide IX, which copurifies with these structures. Its location has been determined by scanning electron microscopy of groups of nine and by difference cryoelectron microscopy of virions (53,144). Another "minor" component, polypeptide IIIa, appears to hold groups of nine together in the virion shell (144). Thus, different proteins are responsible (1) for "cementing" the threefold lattice within a face of the icosahedral shell (polypeptide IX), (2) for stabilizing the hexon/hexon interactions across an edge of the icosahedral surface (IIIa), and (3) for anchoring the peripentonal hexons (VI). Of course, each of these components could have additional functions.

The organization of the adenovirus core is still not well understood. Two rather basic proteins, V and VII, are associated with the viral DNA, which is loaded into empty precursor shells. Insertion of DNA into preassembled heads is a common feature in bacteriophage assembly, and it may be a particularly suitable way to package larger dsDNA genomes.

Herpesviruses

Herpesvirus particles are still more complex: a nucleocapsid 1,200 Å in diameter is surrounded by a thick layer of additional protein (the "tegument") and an outer en-

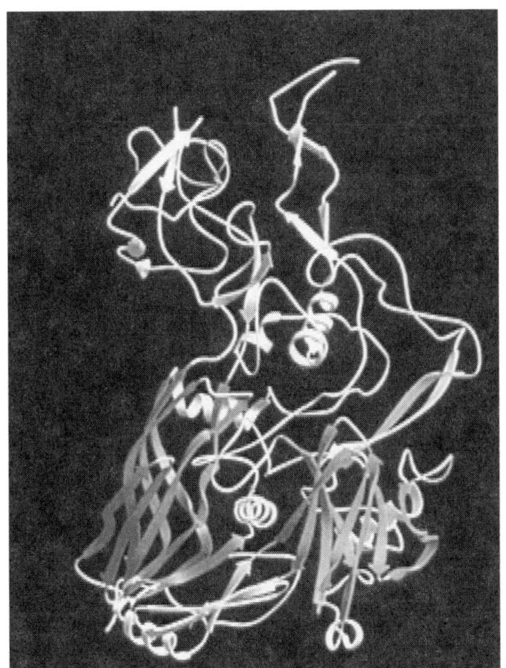

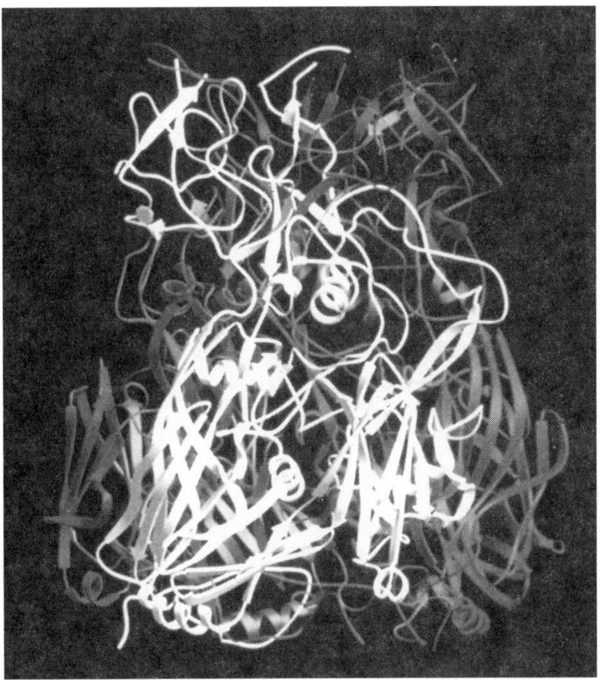

FIG. 21. **(A)** The polypeptide chain of the adenovirus hexon. The direction of view is perpendicular to the threefold axis of the hexon. The two viral capsid β-barrel domains are shown in gray; the loops between strands in each domain and between domains are shown in a lighter shade. Note that these domains are oriented so that their β-strands would run radially in the virus particle, rather than tangentially, as in the picornaviruses. **(B)** The hexon trimer. The white subunit, in the foreground, is in the same orientation as the single subunit in (A). The loops of the individual chains fold extensively with each other. These loops would probably not be ordered in a monomeric subunit, and it is likely that subunit folding and trimer assembly occur coordinately.

velope with spikelike glycoproteins. The nucleocapsid assembles in the nucleus of an infected cell, and subsequent maturation involves a complex sequence of events through which the tegument and outer membrane are acquired (see Chapter 71 in *Fields Virology,* 3rd ed.). Analysis of electron micrographs (see Fig. 2) shows that the nucleocapsid shell of herpes simplex virus is composed of hexameric and pentameric units of the principal structural protein, VP5 (155 kD), located at the vertices of a T = 16 icosahedral lattice (133,167). This shell is about 175 Å thick (radius 425–600 Å). Proteins known as VP23 (36 kD) and VP19c (57 kD) probably form connections among the VP5 hexamers and pentamers. An inner shell, not present in "empty" capsids, occupies radii between about 325 and 425 Å (133).

DOUBLE-STRANDED RNA VIRUSES

The reoviruses, rotaviruses, and orbiviruses are related in the organization of their segmented dsRNA genomes. They differ strikingly, however, in key aspects of their maturation strategy with important consequences for virion organization (see Chapter 51 in *Fields Virology,* 3rd ed.). Reoviruses assemble entirely in the cytoplasm. Their organization can be described roughly as follows: An icosahedrally symmetric "core" is surrounded by an outer protein shell, including hemagglutinin oligomers that insert on the fivefold vertices. Rotaviruses assemble in two separate cellular compartments. The internal structure of rotaviruses [generally called a *single-shelled particle* (SSP) in the literature] assembles in the cytoplasm but then buds through the membrane of the rough ER, acquiring a temporary envelope. In the mature virion, an outer glycoprotein shell replaces the membrane.

Reoviruses

The locations of the major reovirus proteins are shown diagrammatically in Fig. 22A. When the virus is treated with protease, σ3 is released, μ1/μ1c is cleaved, and σ1 undergoes a conformational change to form a thin, fibrous structure projecting from the fivefold positions (54,103). The modified virion, known as an "intermediate subviral particle" (ISVP), is infectious. Indeed, it is believed to be an obligate intermediate, generated either extracellularly or in a lysosome (15,147). Cell entry and penetration result in further disassembly to cores, which are transcriptionally active complexes (136). The structures of virions, ISVPs, and cores have been studied comparatively by cryoelectron microscopy at about 30 Å res-

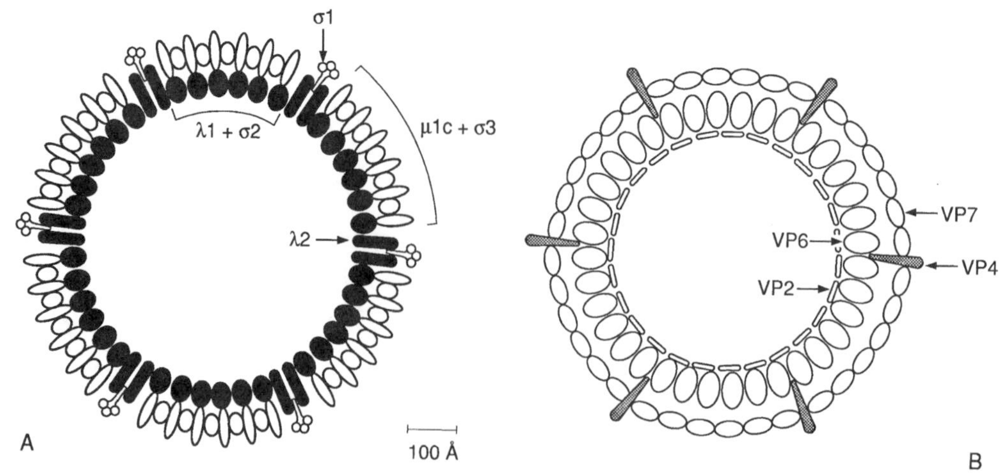

FIG. 22. Diagram showing the locations of principal structural proteins in dsRNA viruses. **(A)** Reovirus; **(B)** rotavirus.

olution (Fig. 23) (42). The core appears to serve as a scaffold for the organization of the outer shell and for insertion of σ1. Its most prominent features are the broad fivefold spikes, formed by λ2 pentamers, which project about 60 Å from the surface. The precise arrangement of the other core structural proteins (λ1 and σ2) cannot be inferred from the present analysis; there are probably 120 copies of each in the particle (35). The proteins of the outer layer, μ1 and σ3, "fill in" between the λ2 spikes. Their arrangement conforms to a T = 13*ℓ* icosahedral lattice, but 15 copies (five trimers) are missing around each fivefold, because of the bulk of the l2 subunits (42). When ISVPs are generated, σ1 undergoes a striking change from a compact form that cannot readily be identified in the reconstructed images to an extended and somewhat flexible fiber (54). Studies of isolated σ1 show that it forms a trimer (146), and the sequence indicates that the amino-terminal end, inserted into the λ2 spike, is an α-helical coiled coil (102). The carboxy-terminal tip contains the cell attachment site; the conformational change presents this site in a particularly exposed way. Steps in cell entry that follow attachment appear to involve μ1. The μ1 protein is present in virions largely as a product (μ1N/μ1c) with a cleaved peptide bond about 40 residues from the amino terminus. It bears an N-terminal myristyl group (104). Formation of ISVPs involves a second cleavage of μ1c (103). The analogy with VP0 (VP4/VP2) of the picornaviruses and with VP2 of the polyomaviruses is noteworthy. In all three cases, a myristylated protein, covered in the intact virion, appears to mediate penetration. In the reovirus/picornavirus comparison, there is the further parallel of a maturation cleavage that generates a myristylated peptide retained in the virion. The final product of reovirus entry is the core, which serves as a cytoplasmic transcription complex for mRNA. Transition from ISVP to core involves not only the loss of μ1 and σ1, but also conformational changes in the outer part of the λ2 spikes. The spikes are believed to serve as channels for entry of nucleotide triphosphates and for exit of nascent mRNA chains (7). In the cores, they have an 80-Å central channel, opened by the exit of σ1 and the conformational change at the surface (42).

Rotaviruses

Rotavirus architecture is also based on a T = 13*ℓ* lattice. The locations of the major proteins in nested shells are diagrammed in Fig. 22B. Cryoelectron microscopy of virions and SSPs shows that both outer and inner shells conform to this symmetry (Fig. 2) (114,166). The outer shell contains 780 subunits of glycoprotein VP7 arranged with trimer clustering in a rather open, fenestrated lattice. Sixty dimeric spikes of the hemagglutinin, VP4, project from this lattice through openings on one set of local sixfold positions (132,165). These spikes are anchored to the inner shell at corresponding positions. The inner shell contains 780 copies of VP6 assembled into trimers. The outer parts of these trimers form 260 "pillars" supporting the VP7 shell; inner parts spread out to form a more continuous layer. A so-called core underlies the VP6 shell. Its principal structural protein is VP2. Cores can be obtained as distinct particles by controlled disruption of SSPs.

Orbiviruses

Orbiviruses resemble rotaviruses in their general organization, but their outer shell is less clearly defined in images obtained by negative-stain electron microscopy. Cryoelectron micrographs of bluetongue virus cores show that they are direct counterparts of rotavirus SSPs (113). The position, dimensions, and shape of the 260 trimers

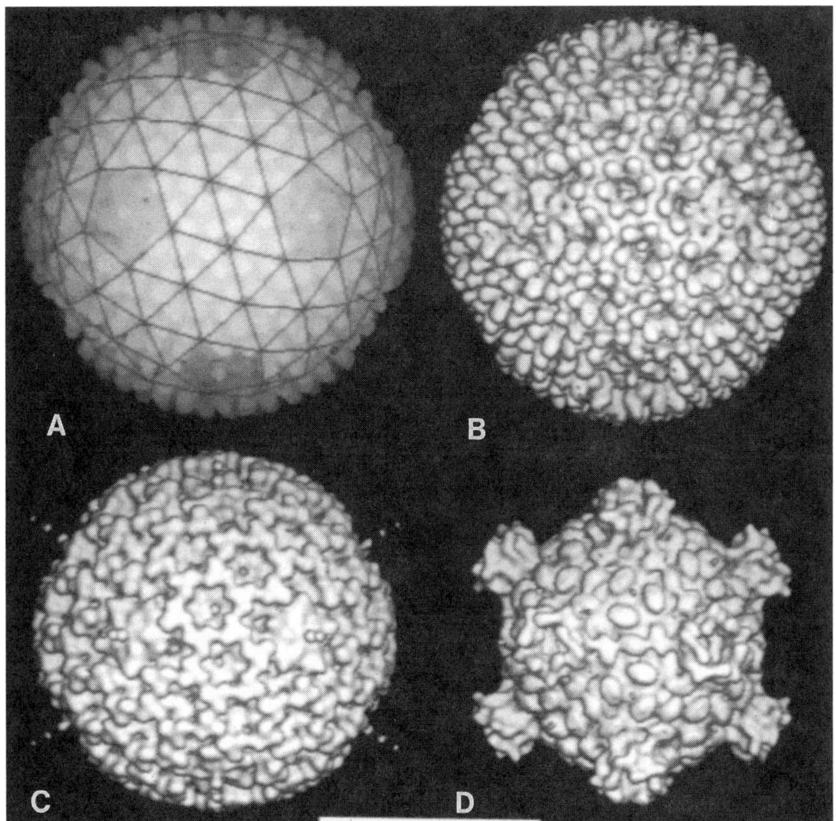

FIG. 23. Three-dimensional image reconstructions of reovirions and subviral particles, based on cryoelectron microscopy. **(A)** Icosahedral net, showing T = 13l lattice. The pentameric cutouts around each fivefold show where $\sigma3/\mu1$ trimers are "missing" because of the 12 spikes that emerge from the core. **(B)** Complete virus particle. The surface bumps are from protein $\sigma3$. **(C)** ISVP. Because $\sigma3$ has dissociated, the surface bumps have disappeared, revealing trimeric clusters of $\mu1c$ just beneath. Note also the conformational change in the pentameric $\lambda2$ spike (for location, refer to A) and the $\sigma1$ fibers projecting along the fivefold axes. **(D)** Core particle. Loss of $\mu1c$ and of $\sigma1$ leaves the 12 spikes projecting prominently from the surface and exposes the $\sigma2$ protein between these spikes (42). Bar = 500 Å. (Courtesy of K Dryden.)

of bluetongue VP7 correspond closely to those of rotavirus VP6. There is also an inner lattice, composed of VP3, that corresponds to the rotavirus core.

ENVELOPED POSITIVE-STRAND RNA VIRUSES

The alphaviruses, such as Sindbis, and the flaviviruses, such as yellow fever, have the simplest of any enveloped virion. The organization of the alphavirus particle is reasonably well understood (Fig. 24A), and there is a high-resolution structure for the Sindbis core subunit.

In the alphaviruses, two protein shells—an outer glycoprotein layer and an inner core (or nucleocapsid)—are separated by a lipid bilayer (72). Both the core and the outer glycoprotein shell conform to T = 4 icosahedral lattices (107,154). The core (C) subunits assemble in the cytoplasm, and the particle buds through the plasma membrane, acquiring in the process its envelope with 240 copies each of the glycoproteins E1 and E2. In some cases, an additional glycosylated species, E3, is present. It derives from the precursor, E2, which is cleaved to generate E3 + E2. It is lost in some viruses, such as Sindbis, and retained in others, such as SFV. Both E1 and E2 have hydrophobic, transmembrane segments (presumably α-helical) at their C-termini, but only E1 has a significant "cytoplasmic tail." The tail determines specific recognition of the nucleocapsid during budding. E1 and E2 are associated as heterodimers, which in turn form 80 trimeric clusters. These clusters give the particle its characteristic appearance in the electron microscope (Fig. 2A). The T = 4 structure of the core probably plays an important role in organizing the glycoprotein lattice, but strong lateral glycoprotein interactions can also help exclude host-cell proteins and drive the budding process (153). These lateral interactions are sufficient to direct spontaneous formation of a hexagonal array when

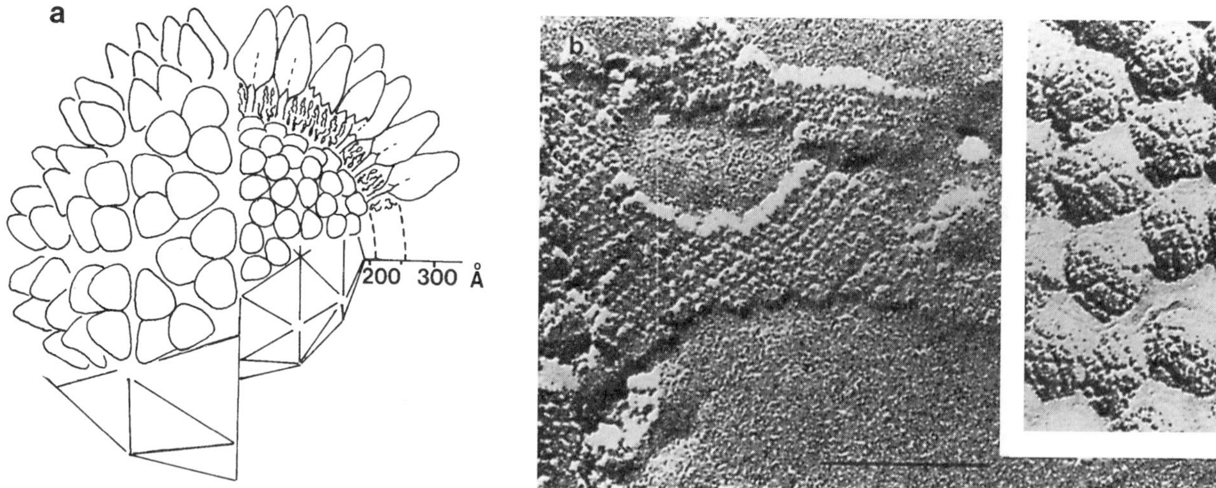

FIG. 24. (A) Overview of the organization of the Sindbis virus particle. See also Fig. 2A. The pear-shaped objects represent the external parts of E1 and E2, closely associated as heterodimers and further clustered into trimers in the T = 4 icosahedral lattice that describes the surface organization (11). There are 240 E1:E2 heterodimers in the complete particle. Hydrophobic transmembrane segments of E1 and E2 penetrate the lipid bilayer. A 33-residue internal domain on E1 is believed to mediate interactions with the core. The core also has a T = 4 icosahedral structure (7). **(B)** Rotary shadowed freeze-etch micrographs of Sindbis virions and of hexagonal arrays that are formed by the E1 and E2 glycoproteins when they are released into lipid-detergent vesicles by treatment of virions with low concentrations of nonionic detergent (10) Bar = 1,000 Å.

Sindbis glycoproteins are isolated at high concentrations in membrane/detergent vesicles (Fig. 24B).

The core subunit contains a flexible, roughly 110-residue, N-terminal arm, rich in positively charged amino acids, and a 150-residue C-terminal domain with a chymotrypsin-like fold (26). This domain indeed must function as a protease, to excise the core protein itself from a polyprotein precursor (68,96). The residues of the catalytic triad have the same spatial relationships that they have in the serine proteases, and the C-terminus of the protein lies in a specificity pocket, as one would expect if it were the immediate product of a cleavage (26). The protein functions as an enzyme for this one reaction only, since its C-terminal residue continues to block access to the active site. The presumed interaction of the E1 tail with the core has not yet been seen, nor are the detailed structures of the togavirus glycoproteins known. These proteins undergo a significant rearrangement when exposed to low pH, thereby revealing a surface that can mediate fusion of viral and endosomal membranes during entry. The surface lattice formed by the glycoproteins appears to be quite perfect and sufficiently rigid that intact Sindbis particles can be crystallized (76).

The flavivirus particles also appear to be based on an icosahedral design. The core protein is small and rich in positively charged amino acid residues. The protease activity required to liberate it from the precursor is provided by the cellular signal protease and by another viral enzyme, so that it does not need to bear a chymotrypsin-like domain (25). The core buds into the endoplasmic reticulum, where it forms an immature particle containing two envelope glycoproteins, known as prM and E; cleavage of prM to M, which occurs during passage of the virus out of the cell, removes most of the external domain of the preM and leaves E as the major surface structure (25). E is a dimer on the surface of virions not exposed to low pH, and the structure of this dimer from tick-borne encephalitis virus (cleaved from its membrane anchor by trypsin) has recently been determined (Fig. 25) (119a). It is a very elongated molecule, and its long dimension must lie parallel to the viral surface. The E protein is therefore likely to form a latticelike network on the outside of the membrane bilayer. Trimers of E are recovered when the virus is exposed to low pH, indicating that the fusion-activating conformational transition involves a rearrangement of the E-protein network, with trimer contacts now predominating. It is therefore probable that large conformational changes are involved, reminiscent of those known to occur when influenza virus hemagglutinin is exposed to low pH (see later). The E protein does not have

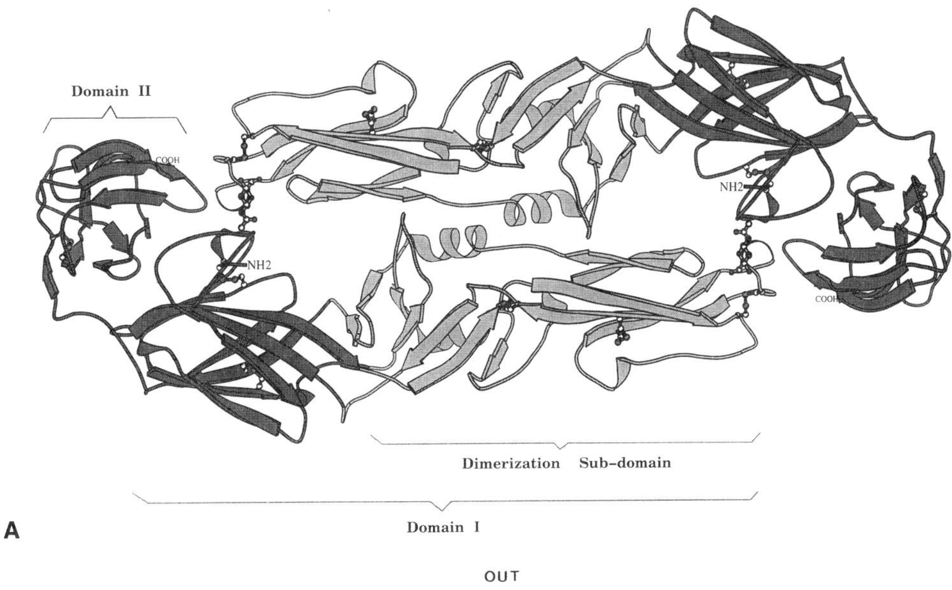

OUT

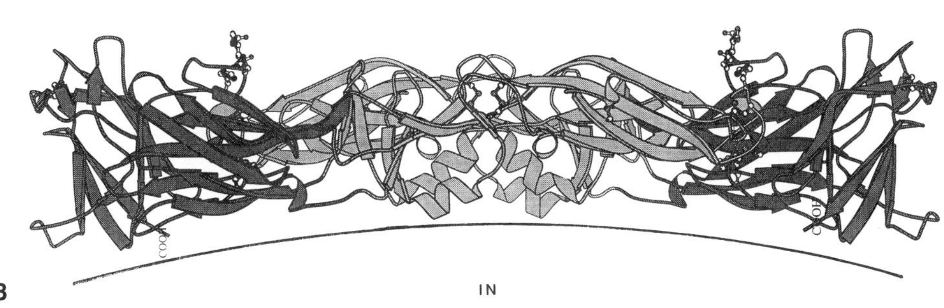

IN

FIG. 25. Diagram showing the polypeptide chain of a flavivirus envelope glycoprotein. The chain trace illustrated here is from the structure of the dimeric external portion of the tick-borne encephalitis virus envelope protein (119a). The fragment that was crystallized included all but about 45 residues N-terminal to the transmembrane segment. Sequences of other flavivirus glycoproteins show extensive conservation, indicating that all must have essentially the same folded structure. **(A)** View looking onto the face of the molecule that would face outward when the protein lies on the virion surface. **(B)** "Side" view, showing approximate curvature of the viral membrane bilayer.

a "fusion sequence" at its amino terminus, however, and the detailed molecular mechanism by which it adopts a fusogenic configuration must differ in important respects from the one that has been seen in studies of HA.

NEGATIVE-STRAND RNA VIRUSES

The enveloped negative-strand RNA viruses are larger than the togaviruses and flaviviruses. Their genomes, whether segmented or unique, are packaged into filamentous RNP structures by association with a protein generally denoted N. The intact helical assembly of N with RNA serves as the template for positive-strand transcription. Electron micrographs of negatively stained, disrupted virions suggest similarities in the structures of these nucleocapsids, but detailed analyses are still lacking. The viral membranes contain one or two glycoproteins, anchored by a presumably α-helical transmembrane stem. In orthomyxoviruses, paramyxoviruses, and rhabdoviruses, there is a major internal protein denoted M, which may help to organize the envelope by contacting the tails of the glycoprotein anchors. The only proteins from these viruses that have been studied at high resolution are the hemagglutinin and neuraminidase of influenza. The hemagglutinin provides the most detailed picture available for the binding, fusion, and entry steps of any enveloped virus. In bunyaviruses, there is no M protein, and nucleocapsids are thought to interact with glycoprotein tails directly. The structures of influenza virus, of vesicular stomatitis virus (VSV), and of a bunyavirus are described as examples in the following sections.

Influenza Virus

The structure of the influenza A viral membrane is discussed in some detail here, because the full three-di-

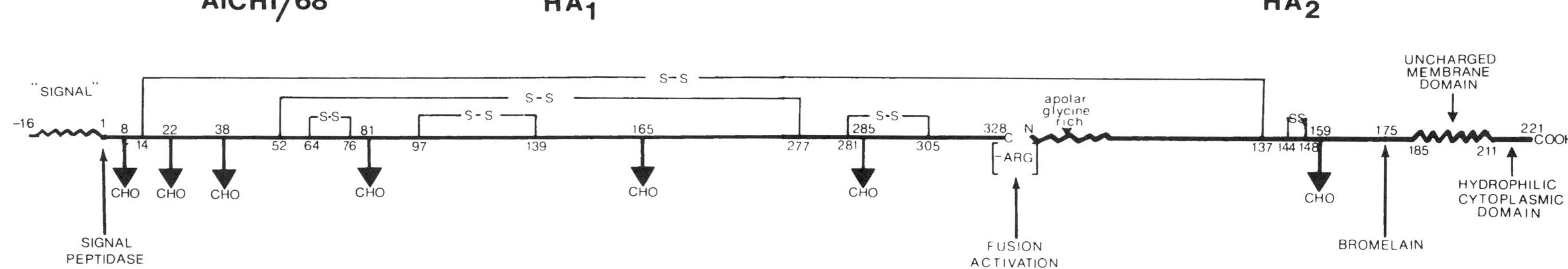

FIG. 26. Schematic drawing of the primary structure of the 1968 Hong Kong influenza virus hemagglutinin, showing the external domain (HA plus HA_2 1–185), the membrane anchor (HA_2 185–211), and the cytoplasmic domain (HA_2 212–221). The cleavage site between HA_1 and HA_2 is labeled "fusion activation." The signal sequence (removed by signalase), S–S bridges, carbohydrate (CHO) attachment sites, and the fusion peptide (N-terminus of HA_2) are shown.

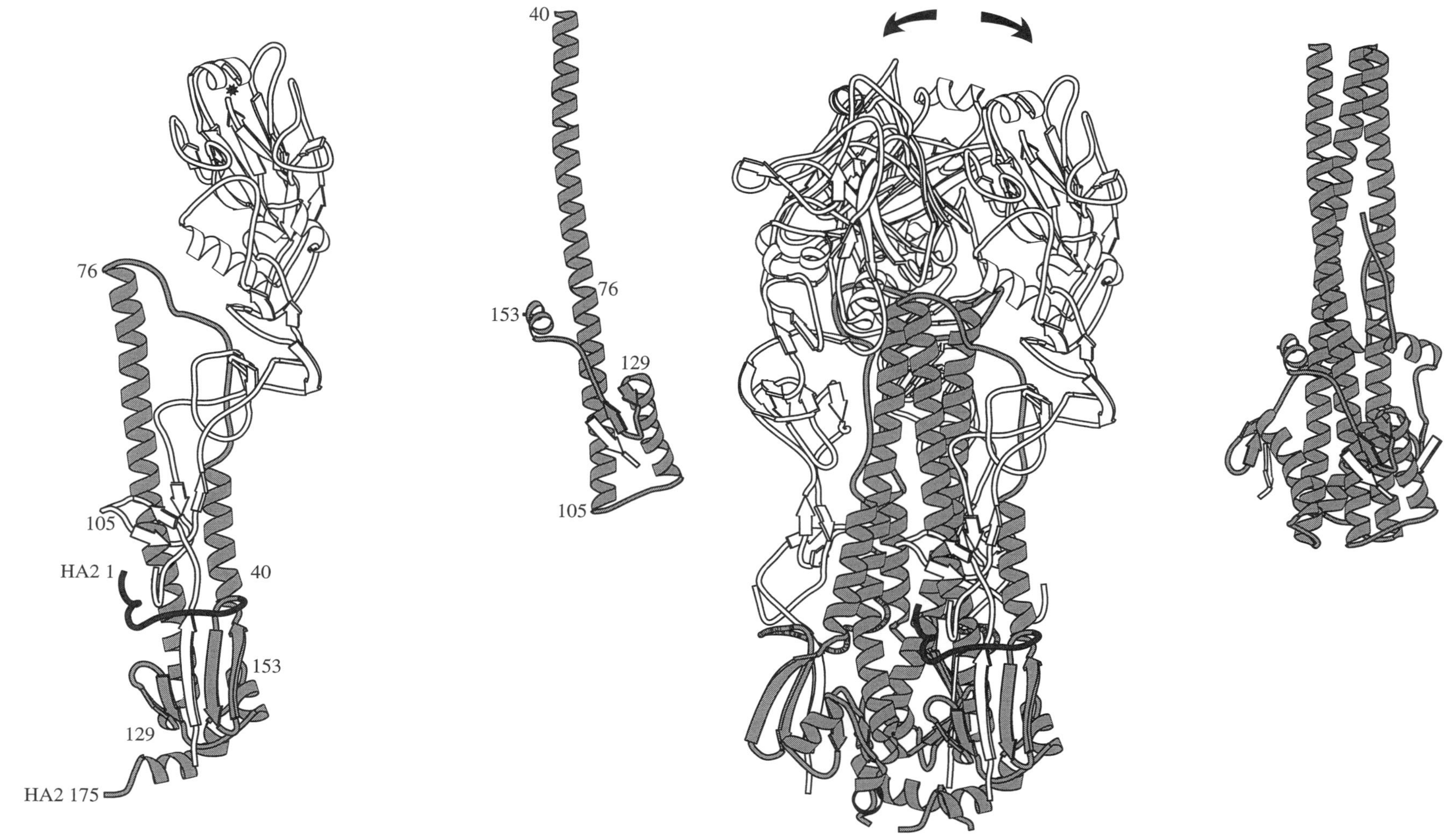

FIG. 27. The folded structure of the influenza virus hemagglutinin (HA) and its rearrangement when exposed to low pH. **(A)** The HA monomer: HA_1 is light; HA_2 is dark; the fusion peptide at the N-terminus of HA_2 is black. Residue numbers in HA_2 are shown at several key positions, in order to assist in visualizing the conformational change that occurs on fusion activation. The receptor-binding pocket in the β-barrel "top" domain of HA_1 is indicated with a star. The viral membrane would be at the bottom of this figure. The transmembrane segment, which follows HA_2 175 is not shown. **(B)** The HA_2 monomer in the fusion-active form. The fragment shown is produced as the result of digesting the activated HA with thermolysin, removing most of HA_1 and the fusion peptide of HA_2. A short segment of HA_1 (white) remains bound to the rearranged HA_2, as shown. Note the dramatic conformational change, in which residues 40–105 become a continuous α-helix. The introduction of a kink at residue 105 also causes the "bottom" part of the molecule to rotate upward. **(C)** The HA trimer. The subunit toward the right in this illustration is viewed in essentially the same orientation as the monomer in (A). HA_2 is shaded: the short helix and loop (residues 40–75); the long helix from its N-terminus to the point where the fusion peptide inserts along the threefold (residues 76–105) and the remainder of the long helix and the β-sheet that follows (residues 106–140); one of the C-terminal helices of HA_2 (residues 141–162). Exposure to low pH initiates a conformational change, causing the top domains of HA_1 to move apart (*arrow*). **(D)** The fusion-active form of HA_2. The tops of HA_1 may be imagined to have spread to either side. (Courtesy of F. Hughson.)

mensional structures of its two glycoproteins, HA and NA, are known. In addition, as mentioned in an earlier section, influenza A is the only virus for which proteins that carry out all four membrane-associated activities (binding, fusion, receptor destruction, secondary uncoating) have been identified. It therefore serves at present as a general paradigm for viral membrane function. The only other viral membrane protein of known structure is the flavivirus glycoprotein described earlier.

The Hemagglutinin

The Hong Kong influenza virus HA is used here as the prototype (Fig. 26). The primary translation product of its mRNA is a 567-residue polypeptide, N-glycosylated at seven asparagine residues in the canonical Asn-X-Ser/Thr sequences (16). It has an NH_2-terminal signal sequence, required for transfer of the nascent polypeptide into the endoplasmic reticulum and removed proteolytically in a subsequent step. There is also a COOH-terminal hydrophobic region, which serves to anchor the completed molecule in cellular and viral membranes. During infection this primary translation product, HA_0, is cleaved into two glycopolypeptides, HA_1 and HA_2, which remain disulfide linked. Cleavage generates the COOH terminus of HA_1 and the NH_2 terminus of HA_2. It is carried out intracellularly by a furinlike protease or, more frequently, extracellularly by factor X–like enzymes (86). The sequence in HA recognized intracellular by furin consists of several basic residues. The recognition sequence of the enzymes that cleave extracellularly is a single arginine residue. Intracellular cleavage at polybasic sequences is characteristic of highly pathogenic influenza viruses (86). Cleavage is an absolute requirement for infectivity, since it is necessary for the membrane fusion activity mediated by HA. The sequence that follows the newly formed NH_2-terminus of HA_2, known as the "fusion peptide," is highly conserved in all HAs. Twenty-residue peptide analogs of this NH_2-terminal region can cause membranes to fuse *in vitro* (91,161).

The assembled HA molecule is a trimer of this disulfide-linked HA_1/HA_2 subunit (162). Its three-dimensional structure (Fig. 27, colorplate 5) is known from x-ray studies of crystals of a soluble fragment released from viral membranes by digestion with the protease bromelain (164). The HA_1 chain of each subunit extends outward about 140 Å from its NH_2 terminus near the membrane, to which it was initially attached via the signal sequence. Most of HA_1 folds into a globular domain with a β-jelly-roll structure. This domain forms the "top" of the molecule, and it bears the site for attachment to sialic acid, the cell surface receptor for influenza. The COOH-terminal segment of HA_1 loops back toward the virus membrane, and the site of HA_0 precursor cleavage is approximately 30 Å from the bilayer surface. The NH_2-terminal fusion-peptide of HA_2 is buried in the center of the trimer. The HA_2 chain has a central, hairpinlike structure made of two α-helices linked by a region of extended chain. There is also a membrane-proximal β-sheet, composed of two strands from near the NA_2 terminus of HA_2, two from near its COOH terminus, and a central strand from the NH_2 terminus of HA1. The longer of the two α-helices in the HA_2 hairpin (52 residues) forms the central trimer contact by associating into a three-chain coiled coil. The coiled coil splays apart at its base, to allow insertion of the fusion peptides. The shorter helix of the hairpin (19 residues), which is packed against two long helices, forms the outer part of the HA stalk. The structures of the transmembrane and internal segments of HA are not known, because the crystallizable fragment has been cleaved with bromelain about 10 residues to the NH_2-terminal side of the membrane insertion point.

Membrane Fusion

Following binding to the cell surface, many enveloped viruses, including influenza, are taken in to endosomes and thus exposed to the low pH of these cellular compartments (95). At endosomal pH, fusion glycoproteins are activated. They undergo a conformational change that permits them to mediate fusion of viral and endosomal membranes. This event allows transfer of the viral genome into the cell and initiation of viral replication. For influenza, the membrane fusion potential of the HA, primed by cleavage of HA_0 into HA_1 and HA_2 in the previous infectious cycle, is triggered between pH 5.0 and pH 6.0, depending on the particular HA. There are dramatic pH-induced changes in its structure (135). These changes have been explored by electron microscopic and antigenic analyses of HA after incubation at the pH of fusion (38) and by studies of mutant HAs that fuse membranes at a different pH or that contain novel disulfide bonds requiring reduction before low pH activation (39,58). The results of such studies show that conformational changes are necessary for membrane fusion activity. The nature and extent of the changes have been demonstrated by x-ray analysis of a crystalline fragment of HA_2 in the fusion pH conformation (18). There are large differences from the native molecule at both ends of HA_2 (Fig. 27B,D). The HA_2 hairpin straightens out, and the central segment (previously a doubled-over strand) adopts an α-helical conformation (19). A three-chain, α-helical coiled coil now extends from near the NH_2 terminus of HA_2 at the "top" of the molecule to the point where the three long helices splay apart in the neutral pH form. At this position, the HA_2 chain folds back, and the C-terminal part of the former long helix now packs against the N-terminal part. The NH_2-terminal fusion peptide is thus displaced more than 100 Å from its central, buried location to the top of a long, triple-stranded α-helical coiled coil (Fig. 27B,D). These changes must clearly be accompa-

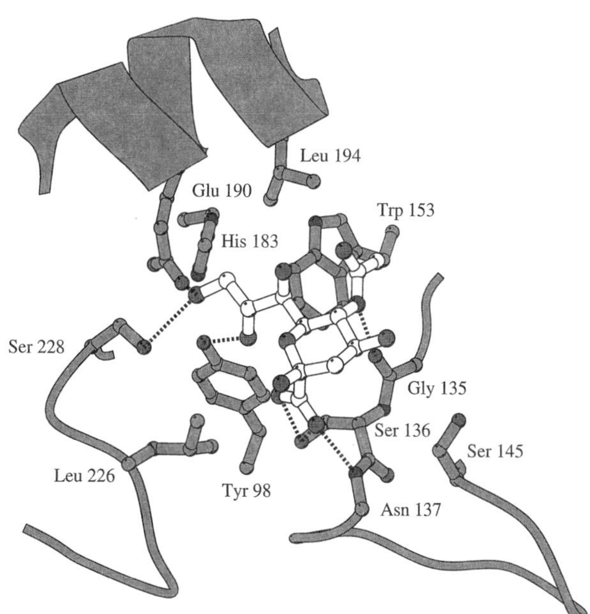

FIG. 28. A detailed view of the receptor binding site on influenza virus HA_1, showing bound sialic acid (center). Side chains from HA_1 that contact the sialic acid are shown. The view is essentially a blow-up of the region indicated by a star in Fig. 27A.

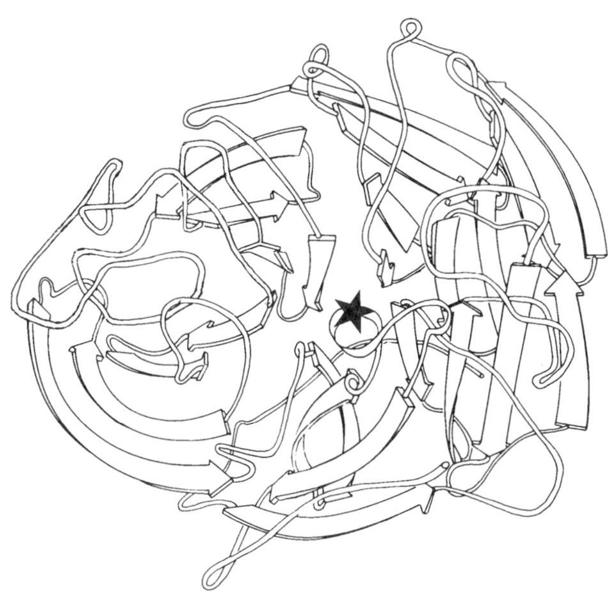

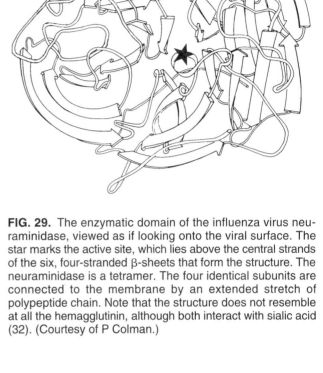

FIG. 29. The enzymatic domain of the influenza virus neuraminidase, viewed as if looking onto the viral surface. The star marks the active site, which lies above the central strands of the six, four-stranded β-sheets that form the structure. The neuraminidase is a tetramer. The four identical subunits are connected to the membrane by an extended stretch of polypeptide chain. Note that the structure does not resemble at all the hemagglutinin, although both interact with sialic acid (32). (Courtesy of P Colman.)

nied by detrimerization of the membrane distal globular domains of HA_1, to make way for the extension of the coiled coil. Although the mechanism of fusion itself is not known, it seems most likely that these elaborate conformational rearrangements deliver the extruded fusion peptide to interact with the endosomal membrane, bridging the virus and endosomal membranes and possibly shortening the distance between them to facilitate their union. Similarities in the primary structure of other fusion glycoproteins, in particular those of the paramyxoviruses and retroviruses, allow the prediction of helical structures adjacent to analogous fusion peptides and suggest that they may participate in membrane fusion in a similar way (24). Not all such fusion glycoproteins appear to require low pH activation, however; a number fuse the virus membranes directly with the cell surface membrane. In these cases the membrane fusion potential of the glycoproteins, which, like HA, are primed by cleavage of a biosynthetic precursor glycopolypeptide, may be activated by the process of receptor binding.

Receptor Binding

HA binds viruses to cells by recognizing the terminal sialic acid residues of the carbohydrate side chains on cell surface glycoproteins or glycolipids. Sialidase treatment of cells prevents binding; resialylation restores it (109). The receptor binding site is a shallow pocket at the membrane distal tip of each HA subunit. It has been identified both by studies of the sequences of mutant HAs that preferentially bind to sialic acid in either α-2,6-linkage or α-2,3-linkage to the penultimate residues of carbohydrate side-chains and by crystallographic analysis of HA complexed with sialoside receptor analogs (115,160). The receptor pocket is formed by a cluster of amino acid residues, conserved in all HAs of type A influenza viruses, noncontiguous in the amino acid sequence but brought together by the folding of the HA_1 glycopolypeptide chain (Fig. 28. colorplate 6). Binding is specific for α-anomeric sialosides. Important for this specificity are hydrogen bonds from conserved donor groups on HA to the axial carboxylate. Additional hydrogen bonds link the 6-glyceryl and 5-*N*-acetyl substituents to other conserved groups on the protein. The affinity of the site for sialosides is low, with a dissociation constant of about 2×10^{-3}. This weak binding contrasts with the interaction between the HIV envelope glycoprotein gp120 and its receptor, CD4, which has a K_D of about 10^{-9} M (99). Effective binding of influenza virus to cell membranes must therefore involve multiple, cooperative interactions with sialylated carbohydrate chains. The density of such glycosyl groups on a cell surface makes such an oligovalent attachment possible.

The Neuraminidase

The neuraminidase, NA, is a tetramer of identical subunits. It has a molecular weight of approximately 240,000, and it appears in the electron microscope to have a box-shaped head on a thin stalk that is anchored in the viral membrane. Both head and stalk are N-glycosylated at two to five sites, depending on the particular NA. The molecule projects form the viral surface by about 120 Å; it is therefore similar in length to the 135-Å HA. NA is anchored in the membrane by an NH_2-terminal hydrophobic region that also serves as its signal sequence. Tetrameric heads, enzymically and antigenically indistinguishable from the intact molecule, can be released from the stalks by proteolytic digestion, and their three-dimensional structures have been determined by x-ray crystallography (32). These analyses show that each subunit consists of six four-stranded, W-shaped β-sheets in which the strands are connected by reverse turns. The fourth strand of each sheet is connected across the upper face to the first strand of the following sheet. The first strands of the sheets nearest the center of the subunit are almost parallel with the fourfold axis of symmetry of the complete tetramer, and the sheets gradually twist so that the fourth strands are almost at right angles to the first. This arrangement gives the subunit the shape of a six-bladed propeller (Fig. 29).

NA catalyzes the cleavage of α-(2,3) and α-(2,6) glycosidic linkages between terminal sialic acids and the penultimate saccharide residues of glycoprotein and glycolipid carbohydrate side chains (33). As a result, it removes the sialic acid residues recognized by HA and destroys the influenza receptor activity of glycoconjugates. During viral replication, NA activity ensures efficient release of newly assembled viruses from the desialylated infected cell surface, and the production of desialylated viral glycoproteins prevents aggregation of virus particles by inappropriate HA/receptor interactions. These essential roles make NA a target in antiviral drug design programs based on the structure of the enzyme active site and its interactions with sialic acid.

X-ray studies of crystalline NA/sialic acid complexes show that the enzymatic active site is structurally quite different from the sialic acid receptor recognition site in the HA (157). The active site is located on the membrane distal surface near the axis relating the six β-sheets. It contains seven charged residues, which are completely conserved in all influenza NAs. The carboxyl group of the α-anomeric sialic acid reaction product interacts in an equatorial orientation with three conserved arginine residues, and the 5-*N*-acetyl and 6-glyceryl groups also interact with conserved amino acid side chains. Detailed examination of the site and computer predictions of modifications to bound sialic acid that might increase interactions with the enzyme showed that the 4-hydroxyl group, which is directed toward conserved glutamic acid 119, was a candidate for modification. Introduction of a guanidinyl group at this position in a previously characterized inhibitory sialic acid analog has led to the production of a potent and specific neuraminidase inhibitor and anti-influenza compound (157).

Antigenic Variation

Analyses of the structures of NA and HA have been directed in large measure toward understanding the molecular basis of antigenic variation—a notorious property of influenza virus. As surface glycoproteins, HA and NA are exposed to interactions with antibodies that block infectivity. During an influenza pandemic period they both vary considerably, and to about the same extent. The regions of the molecules subject to variation can be identified by locating the positions of changes in amino acid sequence on the three-dimensional structures of HA and NA. Such studies indicate that the residues primarily involved in both glycoproteins are components of their membrane distal surfaces. The antigenic significance of the changes can be inferred from similar analyses of the sequence changes in antigenic variant glycoproteins, selected by growing virus in the presence of monoclonal antibodies that block infectivity. There is an excellent correspondence between the locations of changes in natural variants and those selected *in vitro* (163). X-ray studies of monoclonal antibody selected variants of both HA and NA glycoproteins have also shown that the site of the amino acid substitution in the variant defines the site of antibody binding. That is, the structural differences between wild type and variant HAs and NAs are confined to the location of the substituted amino acid. The most direct formal proof of this conclusion has come from x-ray studies of NA-monoclonal Fab complexes (31). They indicate that about 700$Å^2$ of the membrane-distal NA surface are buried in the bound antibody fragment, including approximately 17 amino acid residues in five discontinuous loops. All the amino acids substituted in different variant NAs selected by the antibody are components of the buried surface, directly proving the conclusion that such substitutions define the location of the epitope. Structural changes detected in the bound NA, although small, may be important for the strength of the antibody/antigen interaction.

The Ion Channel Protein, M2

The influenza A virus ion channel, a relatively minor component of the viral membrane, is formed by the nonglycosylated transmembrane polypeptide, M_2 (79). This protein is the smaller product of the gene that encodes the matrix protein, M_1. The channel is a tetramer that contains two noncovalently associated dimers, each with an internal disulfide link (83,148). The polypeptide chain is oriented so that, of the 97 amino acids, the NH_2-terminal

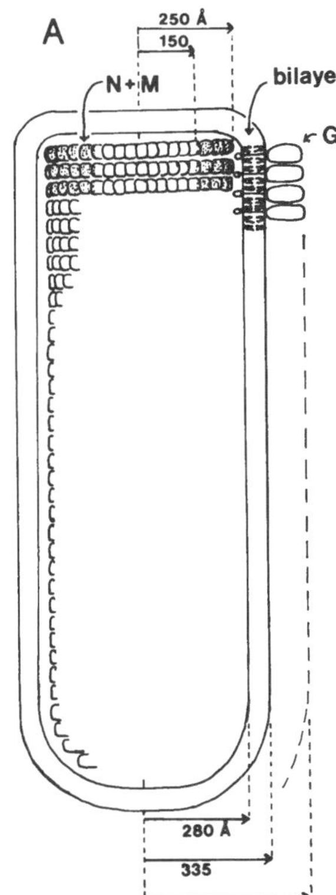

FIG. 30. **(A)** Organization of the VSV particle, showing the relationship of nucleocapsid (N), "matrix" protein (M), and glycoprotein (G) to the lipid bilayer. The dimensions are based on electron microscopy (11) and on small-angle x-ray diffraction (DC Wiley, unpublished observations). **(B)** Electron micrograph of intact and disrupted VSV. The tubular structures formed by M and nucleocapsid (NC) appear to give overall coherence to the particle. (Courtesy of C-H von Bonsdorff.) **(C)** Freeze-etch micrograph of a partially disrupted VSV particle, showing the core structure, some extruded nucleocapsid, and cross-fractured membrane. (Courtesy of J Heuser.)

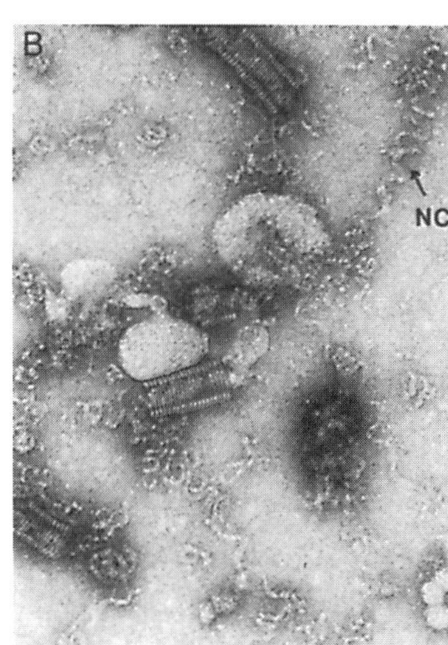

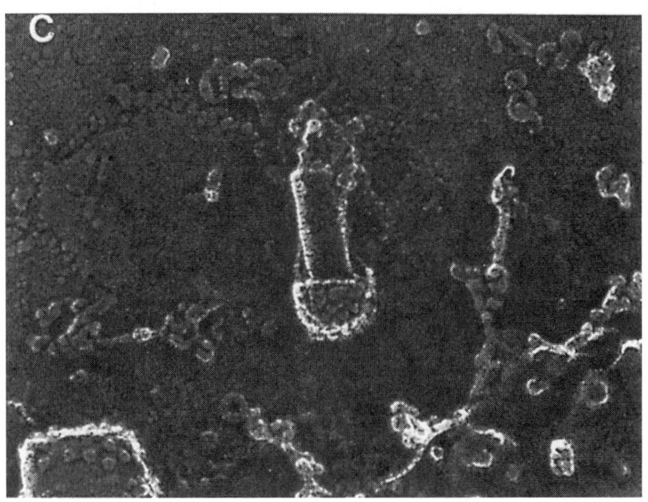

24 residues are external to the virus membrane, 19 residues form a hydrophobic membrane spanning region, and 54 at the COOH terminus are inside the virus (89). It has been suggested that the transmembrane regions are amphipathic helices, associated in the tetramer to form a hydrophilic channel (148). Electrophysiologic studies of Xenopus oocytes expressing M_2 have shown a channel permeable to sodium ions, activated at low pH, and influenced in its activity and ion selectivity by sequence changes in the hydrophobic transmembrane region (111). In infected cells, its function seems to involve proton transport both into virus particles in endosomes and out of Golgi vesicles. These conclusions are drawn primarily from experiments on the mechanism of action of an anti-influenza drug, amino adamantine, resistance to which maps genetically to M_2. Resistance mutations specifically involve amino acid substitutions in the hydrophobic transmembrane region of the M_2 polypeptide (79). The drug appears to have two significant effects: It inhibits dissociation of the virion genome/transcriptase complex from the matrix protein, M (a process that requires acidification of the infecting particle), and it pre-

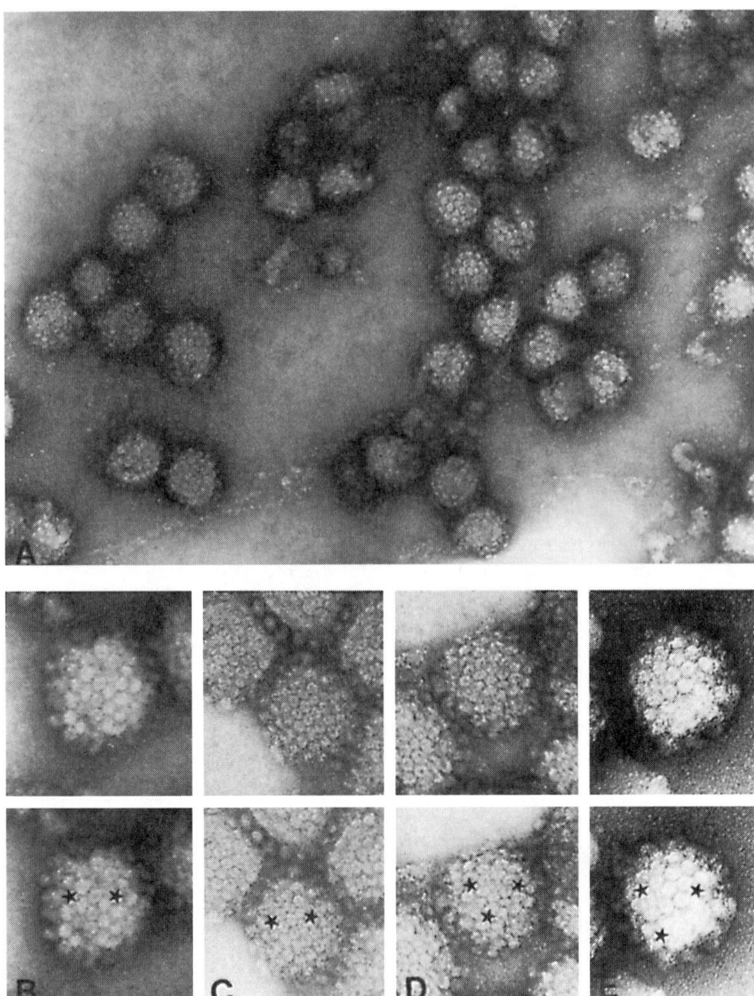

FIG. 31. Uukuniemi virus particles fixed with glutaraldehyde before purification and negatively stained with uranyl acetate. **(A)** A field of slightly flattened particles in which the surface structure is clearly resolved. **(B–E)** Selected particles showing the hexagonal array of subunits, with the positions of pentamers marked by stars. In all particles shown, the arrangement of glycoproteins corresponds to an icosahedral surface lattice of T = 12. (From ref. 12, with permission.)

maturely triggers the irreversible structural changes in HA when the hemagglutinin precursor is cleaved intracellularly into HA_1 and HA_2. Both effects of the drug are interpreted to result from inhibition of a proton channel activity of M_2. In the first case, acidification of virus in endosomes by the activity of virion M_2 channels is prevented. In the second, the M_2-mediated efflux of protons from acidic trans-Golgi vesicles is blocked, and cleaved HA is therefore exposed to low pH. Single amino acid substitutions in the M_2 transmembrane region similarly influence both steps (78). Thus, a common activity mediates both functions by allowing a common direction for proton flux—from NH_2 to COOH terminus of the subunit; from the lumen of the Golgi to the cytoplasm; and from the lumen of the endosome to the interior of the virion.

Vesicular Stomatitis Virus

The virus particle contains three proteins in large quantity (G, N, and M) and two minor components (NS and L). The glycoprotein (G, 65K), the nucleocapsid protein (N, 50K), and the "matrix" protein (M, 29K) are in the approximate molar ratio of 1:2:3 or 1:2:4 (20). They are organized as shown in Fig. 30. The overall shape and coherence of the particle probably depends on M, which forms tubular structures in conjunction with the nucleocapsid. These tubes appear to be a shallow helix with about 40 subunits per turn and about 35 turns per particle (Fig. 30), but the precise structure is still uncertain. Free nucleocapsid appears to be a more tightly coiled structure, with the aspect of a beaded strand (presumably subunits of N bound to RNA) wrapped into a 150-Å-diameter helix. It can often be seen as if extruded from a partly disrupted virion (Fig. 30). When extended on the carbon film of an electron microscope grid, it can appear sinusoidal but with a local radius of curvature similar to the radius of the coil. Some experiments have suggested that the N protein preferentially assembles on the sequence at the 5' end of the genome (14). The glycoprotein (G) communicates with internal structures across the lipid bilayer via a hydrophobic membrane anchor and a small internal domain (see Chapter 19). It is generally supposed that the internal domain makes contact with M, but di-

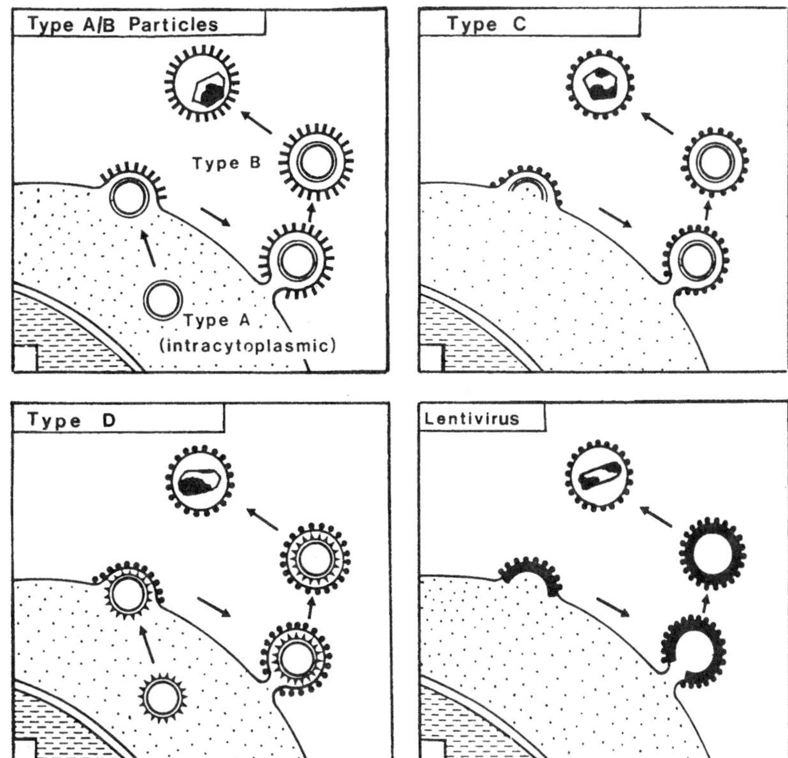

FIG. 32. Diagrammatic representation of the organization of different classes of retrovirus. (Redrawn from ref. 3.)

rect proof is still lacking. The phenomenon of phenotypic mixing and the detection of pseudotypes (see later chapters) indicate that there may be important flexibility in the interaction between internal structures and glycoprotein tails.

Uukuniemi Virus

Bunyaviruses are spherical particles, about 950 Å in diameter, containing three principal structural proteins, a less abundant protein (the polymerase), and three segments of RNA (see Chapter 22). Two of the proteins (G1 and G2) are glycosylated. Considering its size, the architecture of the particle is remarkably simple. Electron micrographs of Uukuniemi virus show morphologic units, 100–120 Å in diameter, arranged in the viral membrane in a T = 12 icosahedral lattice (Fig. 31) (155). The simplest interpretation is that these units correspond to hexameric or pentameric clusters of G1:G2 heterodimers, since G1 and G2 are present in roughly equimolar amounts. There is no internal protein to form an icosahedral shell, so lateral interactions in the glycoprotein lattice must determine the size and shape of the particle. These interactions must also drive budding of the virus through the membrane of the Golgi complex, where it matures. The nucleocapsids are presumably incorporated by interactions of N protein subunits with cytoplasmic tails of G1 or G2. Each RNA segment circularizes through inverted complementary sequences at the 3' and 5' ends, and the nucleocapsids appear to be helically coiled complexes of N with circular RNA (see Chapter 22).

RETROVIRUSES

Retroviruses are enveloped particles, about 1,000 Å in diameter, containing a single-envelope glycoprotein, several internal proteins derived during or after assembly by cleavage of a precursor polyprotein, and two copies of the genomic RNA. The traditional classification is by the appearance of the particles and of their precursors in thin sections of infected cells (11,51,85). Classes differ in the appearance of the internal structures and glycoprotein knobs or spikes and with respect to whether the core assembles in the cytoplasm or at the plasma membrane during budding (Fig. 32). In all cases, the internal structure of the particle rearranges after budding, presumably reflecting cleavage of the gag protein precursor (40). In the most common pattern of maturation, exemplified by the type C avian and mammalian leukemia/sarcoma viruses such as RSV and MuLV as well as by HTLV-1 and HIV, there is co-assembly of core and envelope at the plasma membrane during the budding process. Freshly budded, "immature" particles show a ring of densely staining material just beneath the circular outline of the membrane and a relatively lighter center. The appearance of the core structures changes during maturation, with a shift in the

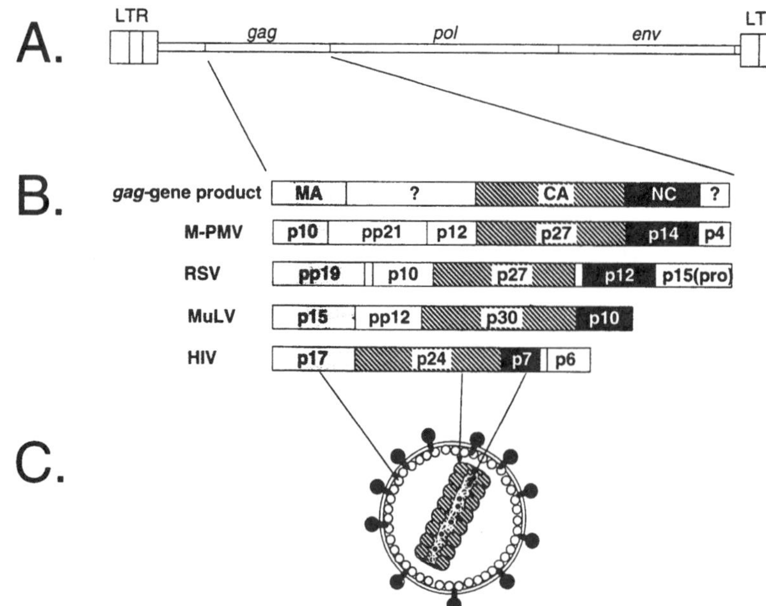

FIG. 33: Organization of the *gag* region of retroviral genomes and its relationship to the organization of a mature virus particle. **(A)** Overall arrangement of retroviral genomes. **(B)** The subunit products of cleavage of the *gag* precursor and their specific names in the case of Mason-Pfizer monkey virus (M-PMV), Rous sarcoma virus (RSV), murine leukemia virus (MuLV), and human immunodeficiency virus (HIV). **(C)** Representation of a virion, showing locations of the MA, CA, and NC gene products. [From (85) with permission.]

strongly staining material toward the center of the particle. In the case of the lentiviruses, the rearranged core has a tubular aspect. In the alternative pattern of maturation, exemplified by MMTV and M-PMV, formation of a cytoplasmically assembled, spherical core (known as an A-type particle) precedes budding. In the case of B-type particles, such as MMTV, subsequent maturation yields an eccentric, strongly staining internal structure. In D-type particles, such as M-PMV, the rearranged internal structure is tubular.

During budding, two or three polyproteins form the core of the immature particle. All are translated from unspliced viral mRNA. The most abundant is the Gag precursor, initiated near the 5' end of the message. The others contain most of the Gag sequences fused to additional domains, either by ribosomal frameshift or by suppressed termination (40). The efficiency of these processes regulates the level of the additional domains in the virion, generally at around 5% of the normal Gag product. The fusion products are known as Gag-Pol or in some cases Gag-Pro and Gag-Pro-Pol. Their domains are proenzymes, which rearrange to form active oligomers after cleavage from the precursor. The associated enzymatic activities include a protease, which is a domain of either the Gag or Gag-Pol product in most retroviruses, but the product of a separate reading frame (Pro) in some others; a reverse transcriptase (RT); a ribonuclease H; and an integrase. The presence of Gag domains on all these polyproteins presumably ensures their co-assembly into the virus particle.

Properties of the domains of the gag protein appear to determine the morphogenetic events that characterize retroviral budding and maturation. Indeed, mutations in *gag* can change the morphogenetic pathway so that core assembly accompanies budding, as in C-type particles (120). The processed segments of the *gag* gene product differ somewhat from one group of viruses to another, but at least three components are found in all retroviruses. These are denoted MA (matrix), CA (capsid), and NC (nucleocapsid), and they appear in that order in the unprocessed precursor (Fig. 33). Cleavage by the viral protease, which is present within the virion because of its inclusion as a domain of one of the polyproteins, yields distinct subunits corresponding to the Gag segments. The N terminus of Gag is myristylated in all mammalian retroviruses, and the myristyl group is retained on MA. In most cases that have been studied, myristylation is essential for directing the precursor to the membrane and for budding of the particle (119,130). Contacts between internal proteins and tails of the envelope glycoprotein are apparently not required for budding, however, since efficient extrusion of "bald" particles has been observed in *env* deletion mutants (119). Thus, viral budding can be driven by the core alone, targeted to the plasma membrane by covalently attached lipid. This mechanism can be contrasted to the joint importance of core structure and glycoprotein lateral contacts in assembly of the alphaviruses and to budding driven by formation of a glycoprotein lattice in the bunyaviruses.

The myristylated, N-terminal MA domain anchors the Gag polyprotein to the inner face of the viral membrane. After cleavage, the other domains lie in roughly concentric structures, in the order in which they are found in the precursor. The CA domain forms a capsid shell that encloses the genomic RNA, the NC protein tightly bound to the RNA, and the RT and integrase enzymes. After infection, reverse transcription occurs within this shell in the cytoplasm of the new host cell. The "maturation" step—cleavage of the gag precursor and rearrangement of the MA and NC components—thus generates a subviral particle that functions directly in the next infectious cycle. The molecular details of the MA and CA domains and their lateral associations are not yet known. The organized structure of the capsid shell clearly differs from retrovirus to retrovirus, being quasi-spherical in some cases and roughly tubular in others. The NC subunit of nearly all retroviruses contains one or two elements with the sequence Cys-X_2-Cys-X_4-His-X_4-Cys. Each such element binds a Zn^{2+} ion, which stabilizes a tightly folded structure (149). Mutational evidence indicates that these elements play a role in specific recognition of the packaging sequences in viral genomic RNA (60,97).

Retroviral envelope glycoproteins have a number of properties in common with the influenza HA. They are synthesized as precursors, which must be cleaved in order to produce fusion-competent species. Moreover, the amino acid sequence at the new N-terminus generated by the cleavage of retroviral glycoproteins resembles that of the fusion peptide at the N-terminus of influenza HA_2. HIV-*env* is an oligomer—either trimer or tetramer (41,110,128). The envelope glycoprotein of Rous sarcoma virus is a trimer (44). One important difference between the fusion activities of influenza and HIV glycoproteins is in the requirement for low pH. HIV enters directly at the cell surface, rather than through endosomes, and it does not require low pH for fusion. That is, proton binding is not needed to trigger the change to a fusion-active conformation. Instead, receptor binding appears to induce such a change directly.

PERSPECTIVES

The 6 years since the last edition of this text have witnessed major extensions in our understanding of virus structure. The most important single theme to emerge is that virus particles are not just packages. They do things, often through dramatic conformational changes. Crystallography and electron microscopy show how these changes occur—for example, in the low-pH fusion activation of flu HA (18), the expansion of TBSV (122), and the formation of ISVPs from reovirions (42). Conformational rearrangements in viral structural proteins are particularly crucial for viral entry into a cell. Most animal virus particles are, in their mature forms, actually metastable structures, although often very robust ones. They are *primed* for conformational rearrangement in response to some *trigger*, associated with early steps in cell binding or uptake. The priming frequently involves a proteolytic cleavage. The trigger can be an event such as receptor binding at the cell surface or proton binding in an endosome. The envelope glycoproteins of influenza viruses and of retroviruses are good examples. Cleavage of flu HA_0 to HA_1 and HA_2 generates a structure that can rearrange (as shown in Fig. 27) on exposure to low pH. The change is irreversible; the rearranged structure is actually stable at neutral pH, but proton binding is needed to catalyze the conformational transition. Cleavage of HIV gp160 primes it for a conformational change triggered by CD4. Polio and other picornaviruses are likewise primed by the VP0 $\rightarrow$ VP4 + VP2 cleavage to undergo major structural reorganization when triggered by receptor binding. The receptor is actually a catalyst of the change, since it does not bind to the cell-altered form of the virion. In flaviviruses, the priming of the envelope glycoprotein to undergo its low-pH-triggered transition involves cleavage not of the E protein itself, but of its "chaperone," pre-M, and consequent loss of most of the pre-M ectodomain. In reoviruses, proteolytic cleavage leads to even more extensive reorganization, as comparison of virions and ISVPs has revealed (see Fig. 23).

Are there prospects for progress in antiviral drug discovery or design, or for vaccine design, based on progress in understanding structural rearrangements in virion proteins and activities of virion-associated enzymes? Studies of antipicornaviral drugs bound to rhinovirus (137) (and reviewed in ref. 57), and poliovirus (62) and design of influenza neuraminidase inhibitors (157) indicate that structure-based approaches are now possible. Likewise, structures of viral receptors (124,158) and receptor/virus complexes (105,142,159,160) suggest strategies for modeling inhibitors of viral attachment. Moreover, analysis of the structural correlates of virulence may facilitate the engineering of attenuated vaccine strains [e.g., for flaviviruses (119a)]. These first steps also reveal the problems and pitfalls. It is clear, for example, that a more complete understanding of the receptor-induced conformational rearrangement or dissociation of picornaviruses will be necessary in order to rationalize the structure/function properties of drugs that bind in the VP1 pocket. Soluble mimics of viral receptors may turn out in many cases to be poor inhibitors, because productive attachment to the cell is polyvalent, and monomeric receptor viral mimics therefore bind only weakly. The importance of a triggered rearrangement for cell entry in many types of virus suggests that a broadly useful drug discovery strategy might be to seek compounds capable of triggering the conformational change before the virus has reached the appropriate site of infection. Such issues are likely to be the focus of structural virology in the coming years.

REFERENCES

1. Abad-Zapatero C, Abdel-Meguid SS, Johnson JE, et al. Structure of southern bean mosaic virus at 2.8 Å resolution. *Nature* 1980;286:33–39.
2. Acharya R, Fry E, Stuart D, Fox G, Rowlands D, Brown F. The three-dimensional structure of foot-and-mouth disease virus at 2.0 Å resolutions. *Nature* 1989;337:709–716.

3. Badger J, Minor I, Kremer MJ, et al. Structural analysis of a series of antiviral agents complexed with human rhinovirus 14. *Proc Natl Acad Sci USA* 1988;85:3304–3308.
4. Baker TS, Newcomb WW, Olson NH, Cowsert LM, Olson C, Brown JC. Structures of bovine and human papillomaviruses. Analysis by cryoelectron microscopy and three-dimensional reconstruction. *Biophys J* 1991;60:1445–1456.
5. Barouch DH, Harrison SC. Interactions among the major and minor coat proteins of polyomavirus. *J Virol* 1994;68:3982–3989.
6. Bartenschlager R, Junker-Niepmann M, Schaller H. The P-gene product of hepatitis B virus is required as a structural component for genomic RNA encapsidation. *J Virol* 1990;64:5324–5332.
7. Bartlett NM, Gillies SC, Bullivant S, Bellamy AR. Electron microscopy study of reovirus reaction cores. *J Virol* 1974;14:315–326.
8. Bawden FC, Pirie NW. Crystalline preparations of tomato bushy stunt virus. *Br J Exp Pathol* 1938;29:251–263.
9. Becker A, Murialdo H. Bacteriophage lambda DNA: the beginning of the end. *J Bacteriol* 1990;172:2819–2824.
10. Bernal JD, Fankuchen I. Structure types of protein "crystals" from virus-infected plants. *Nature* 1939;139:923–924.
11. Bernhard W. Electron microscopy of tumor cells and tumor viruses: a review. *Cancer Res* 1958;18:491–509.
12. Blobel G. Control of intracellular protein traffic. *Methods Enzymol* 1983;96:663–682.
13. Blobel G, Dobberstein B. Transfer of proteins across membranes. *J Cell Biol* 1975;67:835–851.
14. Blumberg BM, Giorgi C, Kolakofsky D. N protein of vesicular stomatitis virus selectively encapsidates leader RNA in vitro. *Cell* 1983;32:559–567.
15. Bodkin DK, Fields BN. Growth and survival of reovirus in intestinal tissue: role of the L2 and S1 genes. *J Virol* 1989;63:1188–1193.
16. Both GW, Sleigh MS. Complete nucleotide sequence at the haemagglutinin gene from a human influenza virus of the Hong Kong subtype. *Nucleic Acids Res* 1980;8:2561–2575.
17. Brand CM, Skehel JJ. Crystalline antigen from the influenza virus envelope. *Nature* 1972;238:145–147.
18. Bullough PA, Hughson FM, Skehel JJ, Wiley DC. The structure of influenza haemagglutinin at the pH of membrane fusion. *Nature* 1994;371:37–43.
18a. Burnett RM. Structural investigations on hexon, the major coat protein of adenovirus. In: McPherson A, Jurnak FA, eds. *Biological macromolecules and assemblies*, vol. I, *The viruses*. New York: Wiley; 1984:337–385.
18b. Carey J, Cameron V, de Haseth PL, Uhlenbeck OC. Sequence-specific interaction of R17 coat protein with its ribonucleic acid binding site. *Biochemistry* 1993;22:2601–2610.
19. Carr CM, Kim PS. A spring-loaded mechanism for the conformational change of influenza hemagglutinin. *Cell* 1993;73:823–832.
20. Cartwright B, Smale CJ, Brown F, Hull R. Model for vesicular stomatitis virus. *J Virol* 1972;10:256–260.
21. Casjens S, King J. Virus assembly. *Annu Rev Biochem* 1975;44:555–611.
22. Caspar DLD, Klug A. Physical principles in the construction of regular viruses. *Cold Spring Harbor Symp Quant Biol* 1962;27:1–22.
23. Cepko C, Sharp P. Assembly of adenovirus major capsid protein is mediated by a nonvirion protein. *Cell* 1982;31:407–415.
24. Chambers P, Pringle CR, Easton AJ. Heptad repeat sequences are located adjacent to hydrophobic regions in several types of virus fusion glycoproteins. *J Gen Virol* 1990;71:3075–3080.
25. Chambers TJ, Hanh CS, Galler R, Rice CM. Flavivirus genome organization, expression, and replication. *Annu Rev Microbiol* 1990;44:649–688.
26. Choi H-K, Tong L, Minor W, Dumas PB U., Rossmann M, Wengler G. Structure of sindbis virus core protein reveals a chymotrypsin-like serine proteinase and the organization of the virion. *Nature* 1991;354:37–43.
27. Choppin PW, Compans RW. Phenotypic mixing of envelope proteins of the parainfluenza virus SV5 and vesicular stomatitis virus. *Virology* 1970;5:609–616.
28. Choppin PW, Compans RW. Replication of paramyxoviruses. *Compr Virol* 1975;4:95.
29. Choppin PW, Klenk HD, Compans RW, Caliguiri A, ed. The parainfluenza virus SV5 and its relationship to the cell membrane. *Perspectives in virology*. New York: Academic Press; 1971.
30. Christiansen G, Landers T, Griffith J, Berg PJ. Characterization of components released by alkali disruptions of simian virus 40. *J Virol* 1977;21:1079–1084.
31. Colman PM, Laver WG, Varghese JN, et al. Three-dimensional structure of a complex of antibody with influenza virus neuraminidase. *Nature* 1987;326:358–363.
32. Colman PM, Varghese JN, Laver WG. Structure of the catalytic and antigenic sites in influenza virus neuraminidase. *Nature* 1983;303:41–47.
33. Colman PM, Ward CW. Structure and diversity of influenza virus neuraminidase. *Curr Top Microbiol Immunol* 1985;114:177–255.
34. Colonno RJ, Conda JH, Mizutani S, Callahan P, Davies M, Murchko MA. Evidence for the direct involvement of the rhinovirus canyon in receptor binding. *Proc Natl Acad Sci USA* 1988;85:5449–5453.
35. Coombs KM, Fields BN, Harrison SC. Crystallization of the reovirus type 3 Dearing core: crystal packing is determined by the λ2 protein. *J Mol Biol* 1990;215:1–5.
36. Crick FHC, Watson JD. Structure of small viruses. *Nature* 1956;177:473–475.
37. D'Halluin JC, Milleville M, Boulanger PA, Martin GR. Temperature-sensitive mutants of adenovirus type 2 blocked in virion assembly: accumulation of light intermediate particles. *J Virol* 1978;26:344–356.
38. Daniels RS, Douglas AR, Skehel JJ, Wiley DC. Analyses of the antigenicity of influenza haemagglutinin at the pH optimum for virus-mediated membrane fusion. *J Gen Virol* 1983;64:1657–1662.
39. Daniels RS, Downie JC, Hay AJ, et al. Fusion mutants of the influenza virus hemagglutinin glycoprotein. *Cell* 1985;40:431–439.
40. Dickson C, Eisenman R, Fan H, Hunter E, Teich N. Protein biosynthesis and assembly. In: Weiss R, Teich N, Varmus H, Coffin J, eds. *RNA tumor viruses*. Cold Spring Harbor, NY: Cold Spring Harbor Laboratory; 1984:513–648.
41. Doms RW, Earl PL, Chakrabarti S, Moss B. Human immunodeficiency virus types 1 and 2 and simian immunodeficiency virus *env* proteins possess a functionally conserved assembly domain. *J Virol* 1990;64:3537.
42. Dryden KA, Wang G, Yeager M, et al. Early steps in reovirus infection are associated with dramatic changes in supramolecular structure and protein conformation: analysis of virions and subviral particles by cryoelectron microscopy and image reconstruction. *J Cell Biol* 1993;122:1023–1041.
43. Edvardsson B, Ustacelebi S, Williams J, Philipson L. Assembly intermediates among adenovirus type 5 temperature-sensitive mutants. *J Virol* 1978;25:641–651.
44. Einfeld D, Hunter E. Oligomeric structure of a prototype retrovirus glycoprotein. *Proc Natl Acad Sci USA* 1988;85:8688–8692.
45. Emini EA, Ostapchuk P, Wimmer E. Bivalent attachment of antibody onto poliovirus leads to conformational alteration and neutralization. *J Virol* 1983;48:547–550.
46. Everitt E, Lutter L, Philipson L. Structural proteins of adenovirus, XII. Location and neighbor relationship among proteins of adenovirion type 2 as revealed by enzymatic iodination, immunoprecipitation and chemical cross-linking. *Virology* 1975;67:197–208.
47. Finch JT, Gibbs AJ. The structure of the nucleocapsid filaments of the paramyxo-viruses. In: Barry RD, Mahy BWJ, eds. *The biology of large RNA viruses*. New York: Academic Press; 1970:115–132.
48. Fisher AJ, Johnson JE. Ordered duplex RNA controls capsid architecture in an icosahedral animal virus. *Nature* 1993;361:176–179.
49. Forstova J, Krauzewicz N, Wallace S, et al. Cooperation of structural proteins during late events in the life cycle of polyomavirus. *J Virol* 1993;67:1405–1413.
50. Fox G, Parry NR, Barnett PV, McGinn B, Rowlands DJ, Brown F. The cell attachment site on foot-and-mouth disease virus includes the amino acid sequence RGD (arginine-glycine-aspartic acid). *J Gen Virol* 1989;70:625–637.
51. Frank H. Retroviridae. In: Nermut MV, Steven AC, ed. *Animal virus structure*. Amsterdam: Elsevier; 1987:253–256.
52. Fricks CE, Hogle JM. Cell-induced conformational change in poliovirus: externalization of the amino terminus of VP1 is responsible for liposome binding. *J Virol* 1990;64:1934–1945.
53. Furciniti PS, Van Oostrum J, Burnett RM. Adenovirus polypeptide IX revealed as capsid cement by difference images from electron microscopy and crystallography. *EMBO J* 1989;8:3563–3570.
54. Furlong DB, Nibert ML, Fields BN. Sigma 1 protein of mammalian reoviruses extends from the surfaces of viral particles. *J Virol* 1988;62:246–256.

55. Gallagher TM, Rueckert RR. Assembly-dependent maturation cleavage in provirions of a small icosahedral insect ribovirus. *J Virol* 1988;62:3399–3406.
56. Ginsberg HS. Adenovirus structural proteins. In: Fraenkel-Conrat H, Wagner RR, eds. *Comprehensive Virology*. New York: Plenum; 1979:409–457.
57. Giranda VL. Structure-based drug design of antirhinoviral compounds. *Structure* 1994;2:695–698.
58. Godley L, Pfeifer J, Steinhauer D, et al. Introduction of intersubunit disulfide bonds in the membrane-distal region of the influenza hemagglutinin abolishes membrane fusion activity. *Cell* 1992;68:635–645.
59. Gomez Yatal A, Kaplan G, Racaniello VR, Hogle JM. Characterization of poliovirus conformational alteration mediated by soluble cell receptors. *Virology* 1993;197:501–505.
60. Gorelick RJ, Henderson LE, Hansen JP, Rein A. Point mutants of Moloney murine leukemia virus that fail to package viral RNA: evidence for specific RNA recognition by a "zinc-finger-like" protein sequence. *Proc Natl Acad Sci USA* 1988;85:8420–8424.
61. Grable M, Hearing P. *cis-* and *trans* requirements for the selective packaging of adenovirus type 5 DNA. *J Virol* 1992;66:723–731.
62. Grant RA, Hiremath C, Filman DJ, Syed R, Andries K, Hogle JM. Structures of polioviruses complexed with antiviral drugs: implications for viral stability and drug design. *Curr Biol* 1994;4:784–797.
63. Green J, Griffiths G, Louvard D, Quinn P, Warren G. Passage of viral membrane proteins through the Golgi complex. *J Mol Biol* 1981;152:663–698.
64. Green NM, Wrigley NG, Russel WC, Martin SR, McLachlan AD. Evidence for a repeating β-sheet structure in the adenovirus fibre. *EMBO J* 1983;8:1357–1365.
65. Greve JM, Davis G, Meyer AM, et al. The major human rhinovirus receptor is ICAM-1. *Cell* 1989;56:839–847.
66. Griffith JP, Griffith DL, Rayment I, Murakami T, Caspar DLD. Inside polyomavirus at 25-Å resolution. *Nature* 1992;355:652–654.
67. Guttman N, Baltimore D. A plasma membrane component able to bind and alter virions of poliovirus type 1: studies on cell-free alteration using a simplified assay. *Virology* 1977;82:25–36.
68. Hahn CS, Strauss JH. Site-directed mutagenesis of the proposed catalytic amino acids of the sindbis virus capsid protein autoprotease. *J Virol* 1990;64:3069–3073.
69. Hall L, Rueckert RR. Infection of mouse fibroblasts by cardioviruses: premature uncoating and its prevention by elevated pH and magnesium chloride. *Virology* 1971;43:152–165.
70. Hall W, Martin SJ. The biochemical and biological characteristics of the surface components of measles virus. *J Gen Virol* 1974;22:63.
71. Harrison SC. Multiple modes of subunit association in the structures of simple spherical viruses. *Trends Biochem Sci* 1984;9:345–351.
72. Harrison SC. Alphavirus structure. In: Schlesinger S, Schlesinger M, ed. *The Togaviridae and Flaviviridae*. New York: Plenum Press; 1986:21–34.
73. Harrison SC. Common features in the design of small RNA viruses. In: Oldstone MFA, Notkins A, ed. *Concepts in viral pathogenesis*, Vol. III. New York: Springer-Verlag; 1989:3–19.
74. Harrison SC, Olson A, Schutt CE, Winkler FK, Bricogne G. Tomato bushy stunt virus at 2.9 Å resolution. *Nature* 1978;276:368–373.
75. Harrison SC, Sorger PK, Stockley PG, Hogle JM, Altman R, Strong RK. *Positive strand RNA viruses*. New York: Alan R. Liss; 1987.
76. Harrison SC, Strong RK, Schlesinger S, Schlesinger M. Crystallization of Sindbis virus and its nucleocapsid. *J Mol Biol* 1992;226:277–280.
77. Hay A. Studies on the formation of the influenza virus envelope. *Virology* 1974;60:398–418.
78. Hay AJ. The mechanism of action of amantadine and rimantadine against Influenza viruses. In: Notkins AL, Oldstone MBA, ed. *Concepts in viral pathogenesis*. New York: Springer-Verlag; 1989:361–367.
79. Hay AJ. The action of adamantanamines against influenza A viruses: inhibition of the M_2 ion channel protein. *Semin Virol* 1992;3:21–30.
80. Helenius A, McCaslin DR, Fries E, Tanford C. Properties of detergents. *Methods Enzymol* 1979;56:734–749.
81. Helenius S, Soderlund H. Stepwise dissociation of the Semliki Forest virus membrane with Triton X-100. *Biochem Biophys Acta* 1973;307:287–300.
82. Hogle JM, Chow M, Filman DJ. Three-dimensional structure of poliovirus at 2.9 Å resolution. *Science* 1985;229:1358–1365.
83. Holsinger LJ, Lamb RA. Influenza virus M_2 integral membrane protein is a homotetramer stabilized by formation of disulfide bonds. *Virology* 1991;183:32–43.
84. Hosur MV, Schmidt T, Tucker RC, et al. Structure of an insect virus at 3.0 Å resolution. *Proteins* 1987;2:167–176.
85. Hunter E. Macromolecular interactions in the assembly of HIV and other retroviruses. *Semin Virol* 1994;5:71–83.
86. Klenk H-D, Rott R. The molecular biology of influenza virus pathogenicity. *Adv Virus Res* 1988;34:247–281.
87. Klenk HD, Wollert W, Rott R, Scholtissek C. Association of influenza virus proteins with cytoplasmic fractions. *Virology* 1974;57:28–41.
88. Kornfeld R, Kornfeld S. Assembly of asparagine-linked oligosaccharides. *Annu Rev Biochem* 1985;54:631–661.
89. Lamb RA, Zebedee SL, Richardson CD. Influenza virus M_2 protein is an integral membrane protein expressed on the infected-cell surface. *Cell* 1985;40:627–633.
90. Laver WG, Valentine RC. Morphology of the isolated haemagglutinin and neuraminidase subunits of influenza virus. *Virology* 1969;38:105.
91. Lear JD, DeGrado WF. Membrane binding and conformational properties of peptides representing the NH_2 terminus of influenza HA-2. *J Biol Chem* 1987;262:6500–6505.
92. Leavitt AD, Roberts TM, Garcea RL. Polyoma virus major capsid protein. VP_1: purification after high-level expression in *Escherichia coli*. *J Biol Chem* 1985;260:12803–12809.
93. Liddington RC, Yan Y, Zhao HC, Sahli R, Benjamin TL, Harrison SC. Structure of simian virus 40 at 3.8 Å resolution. *Nature* 1991; 354:278–284.
94. Luo M, Vriend G, Kamer G, et al. The atomic structure of Mengo virus at 3.0 Å resolution. *Science* 1987;235:182–191.
95. Marsh M, Helenius A. Virus entry into animal cells. *Adv Virus Res* 1989;36:107–151.
96. Melancon P, Garoff H. Processing of the Semliki forest virus structural polyprotein: role of the capsid protease. *J Virol* 1987;61:1301–1309.
97. Meric C, Gouilloud E, Spahr P-F. Mutations in Rous sarcoma virus nucleocapsid protein p12 (NC): deletions of Cys-His boxes. *J Virol* 1988;62:3328–3333.
98. Milstein C, Brownlee G. A possible precursor of immunoglobulin light chains. *Nature New Biol* 1972;239:117–120.
99. Moebius U, Clayton LK, Abraham S, Harrison SC, Reinherz EL. The HIV gp120 binding site on CD4: Delineation by quantitative equilibrium and kinetic binding studies of mutants in conjunction with a high resolution CD4 atomic structure. *J Exp Med* 1992;176:507–517.
100. Montross L, Watkins S, Moreland RB, Mamon H, Caspar DLD, Garcea RL. Nuclear assembly of polyomavirus capsids in insect cells expressing the major capsid protein VP1. *J Virol* 1991;65:4991–4998.
101. Moscufo N, Yafal AG, Rogove A, Hogle J, Chow M. A mutation in VP4 defines a new step in the late stages of cell entry by poliovirus. *J Virol* 1993;67:5075–5078.
102. Nibert ML, Dermody TS, Fields BN. Structure of the reovirus cell-attachment protein: a model for the domain organization of s1. *J Virol* 1990;64:2976–2989.
103. Nibert ML, Fields BN. A carboxy-terminal fragment of protein ml/m1C is present in infectious subvirion particles of mammalian reoviruses and is proposed to have a role in penetration. *J Virol* 1992;66:6408–6418.
104. Nibert ML, Schiff LA, Fields BN. Mammalian reoviruses contain a myristylated structural protein. *J Virol* 1991;65:2372–2380.
105. Olson NH, Kolatkar PR, Oliveira MA, et al. Structure of a human rhinovirus complexed with its receptor molecule. *Proc Natl Acad Sci USA* 1993;90:507–511.
106. Palade G. Intracellular aspects of the process of protein synthesis. *Science* 1975;189:347–358.
107. Paredes AM, Brown DT, Rothnagel R, et al. Three-dimensional structure of a membrane-containing virus. *Proc Natl Acad Sci USA* 1993;90:9095–9099.
108. Paulson JC. Interactions of animal viruses with cell surface receptors. In: Conn PM, ed. *The receptors*. Orlando: Academic Press; 1985:131–219.
109. Paulson JC, Sadler JE, Hill RL. Restoration of specific myxovirus receptors to asialoerythrocytes by incorporation of sialic acid with pure sialyltransferases. *J Biol Chem* 1979;254:2120–2124.
110. Pinter A, Honnen WJ, Tilley SA, et al. Oligomeric structure of gp41, the transmembrane protein of human immunodeficiency virus type 1. *J Virol* 1989;63:2674.

111. Pinto LH, Holsinger LJ, Lamb RA. Influenza virus M2 protein has ion channel activity. *Cell* 1992;69:517–528.
112. Pollack JR, Ganem D. An RNA stem-loop structure directs hepatitis B virus genomic RNA encapsidation. *J Virol* 1993;67:3254–3263.
113. Prasad BV, Yamaguchi S, Roy P. Three dimensional structure of single-shelled bluetongue virus. *J Virol* 1992;66:2135–2142.
114. Prasad BVV, Wang GJ, Clerx JPM, Chiu W. Three-dimensional structure of rotavirus. *J Mol Biol* 1988;199:269–275.
115. Pritchett TJ, Brossmer R, Rose U, Paulson JC. Receptor determinants of human and animal influenza virus isolates: differences in receptor specificity of the H3 hemagglutinin based on species of origin. *Virology* 1983;127:361–373.
116. Puglisi JD, Tan R, Calnan BJ, Frankel AD, Williamson JR. Conformation of the TAR RNA-arginine complex by NMR spectroscopy. *Science* 1992;76:76–80.
117. Rayment I, Baker TS, Caspar DLD, Murakami WT. Polyoma virus capsid structure at 22.5 Å resolution. *Nature* 1982;295:110–115.
118. Reed AP, Jones EV, Miller TJ. Nucleotide sequence and genome organization of canine parvovirus. *J Virol* 1988;62:266–276.
119. Rein A, McClure MR, Rice NR, Luftig RB, Schultz AM. Assembly of Moloney murine leukemia virus: requirement for myristylation site in $PR65^{gag}$. In: Brinton MA, Rueckert RR, ed. *Positive-strand RNA viruses.* New York: Alan R. Liss; 1987:227–237.
119a. Rey FA, Heinz FX, Mandl C, Kunz C, Harrison SC. The envelope glycoprotein from tick-borne encephalitis virus. *Nature* 1995;375:291–298.
120. Rhee SS, Hunter E. A single amino-acid substitution within the matrix protein of a type D retrovirus converts its morphogenens to that of a type C retrovirus. *Cell* 1990;63:77–86.
121. Roberts MM, White JL, Grutter MG, Burnett RM. Three-dimensional structure of the adenovirus major coat protein hexon. *Science* 1986;232:1148–1151.
122. Robinson IK, Harrison SC. Structure of the expanded state of tomato bushy stunt virus. *Nature* 1982;297:563–568.
123. Rossmann MG, Arnold E, Erickson JW. Structure of a human common cold virus and functional relationship to other picornaviruses. *Nature* 1985;317:145–153.
124. Ryu S-E, Kwong PD, Truneh A, et al. Crystal structure of HIV-binding recombinant fragment of CD4. *Nature* 1990;348:419–423.
125. Sahli R, Freund R, Dubensky T, Garcea R, Bronson R, Benjamin T. Defect in entry and altered pathogenicity of a polyoma virus mutant blocked in VP2 myristylation. *Virology* 1993;192:142–153.
126. Salunke D, Caspar DLD, Garcea RL. Self-assembly of purified polyomavirus capsid VP1. *Cell* 1986;46:895–904.
127. Salunke DM, Caspar DLD, Garcea RL. Polymorphism in the assembly of polyomavirus capsid protein VP1. *Biophys J* 1989;56:887–900.
128. Schawaller M, Smith GE, Skehel JJ, Wiley DC. Studies with crosslinking reagents on the oligomeric structure of the *env* glycoprotein of HIV. *Virology* 1989;172:367.
129. Schlesinger S, Makino S, Lineal ML. *cis*-acting genomic elements and *trans*-acting proteins involved in the assembly of RNA viruses. *Semin Virol* 1994;5:39–49.
130. Schultz AM, Rein A. Unmyristylated Moloney murine leukemia virus Pr65gag is excluded from virus assembly and maturation events. *J Virol* 1989;63:2370–2373.
131. Schulze IT. The structure of influenza virus: 1. The polypeptides of the virion. *Virology* 1970;42:890.
132. Shaw AL, Rothnagel R, Chen D, Ramig RF, Chiu W, Prasad BV. Three dimensional visualization of the rotavirus hemagglutinin structure. *Cell* 1993;74:693–701.
133. Shrag JD, Prasad BVV, Rixon FJ, Chiu W. Three-dimensional structure of the HSV1 nucleocapsid. *Cell* 1989;56:651–660.
134. Singer SJ. The structure and insertion of integral proteins in membranes. *Annu Rev Cell Biol* 1990;6:247–296.
135. Skehel JJ, Bayley PM, Brown EB, et al. Changes in the conformation of influenza virus haemagglutinin at the pH optimum of virus-mediated membrane fusion. *Proc Natl Acad Sci USA* 1982;79: 968–972.
136. Skehel JJ, Joklik WK. Studies on the *in vitro* transcription of reovirus RNA catalyzed by reovirus cores. *Virology* 1969;39:822–831.
137. Smith TJ, Kremer MJ, Luo M, et al. The site of attachment in human rhinovirus 14 for antiviral agents that inhibit uncoating. *Science* 1986;233:1286–1293.
138. Sorger PK, Stockley PG, Harrison SC. Structure and assembly of turnip crinkle virus, II. Mechanism of reassembly *in vitro*. *J Mol Biol* 1986;191:639–658.
139. Springer T, Strominger J, Mann D. Partial purification of detergent-soluble HL-A antigen and its cleavage by papain. *Proc Natl Acad Sci USA* 1974;71:1539–1543.
140. Staunton DE, Merluzzi VJ, Rothlein R, Barton R, Marlin SD, Springer TA. A cell adhesion molecule, ICAM-1, is the major surface receptor for rhinoviruses. *Cell* 1989;56:849–853.
141. Steere RL, Schaffer FL. The structure of crystals of purified Mahoney poliovirus. *Biochem Biophys Acta* 1958;28:241.
142. Stehle T, Yan Y, Benjamin TL, Harrison SC. Structure of murine polyomavirus complexed with an oligosaccharide receptor fragment. *Nature* 1994;369:160–163.
143. Stewart PL, Burnett RM, Cyrklaff M, Fuller SD. Image reconstruction reveals the complex molecular organization of adenovirus. *Cell* 1991;67:145–154.
144. Stewart PL, Fuller SD, Burnett RM. Difference imaging of adenovirus: bridging the resolution gap between X-ray crystallography and electron microscopy. *EMBO J* 1993;12:2589–2599.
145. Strittmater P, Velick SF. The isolation and properties of microsomal cytochrome. *J Biol Chem* 1956:221–253.
146. Strong JE, Leone G, Duncan R, Sharma RK, Lee RWK. Biochemical and biophysical characterization of the reovirus cell attachment protein s1: evidence that it is a homotrimer. *Virology* 1991;184:23–32.
147. Sturzenbecker LJ, Nibert ML, Furlong D, Fields BN. Intracellular digestion of reovirus particles requires a low pH and is an essential step in the viral infectious cycle. *J Virol* 1987;61:2351–2361.
148. Sugrue RJ, Hay AJ. Structural characteristics of the M_2 protein of influenza A viruses: evidence that it forms a tetrameric channel. *Virology* 1991;180:617–624.
149. Summers MF, South TL, Kim B, Hare DR. High-resolution structure of an HIV zinc fingerlike domain via a new NMR-based distance geometry approach. *Biochemistry* 1990;29:329–340.
150. Tokuyasu K, Singer SJ. Passage of an integral membrane protein, the vesicular stomatitis virus glycoprotein, through the Golgi apparatus en route to the plasma membrane. *Proc Natl Acad Sci USA* 1981; 78:1746–1750.
151. Tsao J, Chapman MS, Agbandge M, et al. The three-dimensional structure of canine parvovirus and its functional implications. *Science* 1991;251:1456–1464.
152. Turner DR, Joyce LE, Butler PJG. The tobacco mosaic virus assembly origin RNA. *J Mol Biol* 1988;203:531–547.
153. Von Bonsdorff C-H, Harrison SC. Hexagonal glycoprotein arrays from Sindbis virus membranes. *J Virol* 1978;28:578–583.
154. Von Bonsdorff CH, Harrison SC. Sindbis virus glycoproteins form a regular icosahedral surface lattice. *J Virol* 1975;16:141–145.
155. Von Bonsdorff CH, Pettersson RF. Surface structure of Uukuniemi virus. *J Virol* 1975;16:1296–1307.
156. von Heijne G. Signal sequences: the limits of variation. *J Mol Biol* 1985;184:99–105.
157. von Itzstein M, Wu W-Y, Kok GB, et al. Rational design of potent sialidase-based inhibitors of influenza virus replication. *Nature* 1993;363:418–423.
158. Wang J, Yan Y, Garrett TP, et al. Atomic structure of a fragment of human CD4 containing two immunoglobulin-like domains. *Nature* 1990;348:411–419.
159. Watowich SJ, Skehel JJ, Wiley DC. Crystal structures of influenza virus hemagglutinin in complex with high-affinity receptor analogs. *Structure* 1994;2:719–731.
160. Weis W, Brown J, Cusack S, Paulson JE, Skehel JJ, Wiley DC. The structure of the influenza virus haemagglutinin complexed with its receptor, sialic acid. *Nature* 1988;333:426–431.
161. Wharton SA, Martin SR, Ruigrok RWH, Skehel JJ, Wiley DC. Membrane fusion by peptide analogues of influenza virus haemagglutin. *J Gen Virol* 1988;69:1847–1857.
162. Wiley DC, Skehel JJ. The structure and function of the hemagglutinin membrane glycoprotein of influenza virus. *Annu Rev Biochem* 1987; 56:365–394.
163. Wiley DC, Wilson IA, Skehel JJ. Structural identification of the antibody-binding sites of Hong Kong influenza haemagglutinin and their involvement in antigenic variation. *Nature* 1981;289:373–378.
164. Wilson IA, Skehel JJ, Wiley DC. Structure of the haemagglutinin membrane glycoprotein of influenza virus at 3 Å resolution. *Nature* 1981;289:366–373.
165. Yeager M, Berriman JA, Baker TS, Bellamy AR. Three-dimensional structure of the rotavirus haemagglutinin VP4 by cryo-electron mi-

croscopy and difference map analysis. *EMBO J* 1994;13:1011–1018.
166. Yeager M, Dryden A, Olson NH, Greenberg HB, Baker TS. Three-dimensional structure of rhesus rotavirus by cryoelectron microscopy and image reconstruction. *J Cell Biol* 1990;110:2133–2144.
167. Zhou ZH, Prasad BVV, Jakana J, Rixon FJ, Chiu W. Protein subunit structures in the herpes simplex virus A-capsid determined form 400 kV spot-scan electron cryomicroscopy. *J Mol Biol* 1994;in press:
168. Zimmern D, Wilson TMA. Location of the origin for viral reassembly on tobacco mosaic virus RNA and its relation to stable fragment. *FEBS Lett* 1976;71:294–298.
169. Zwizinski C, Wickner W. Purification and characterization of leader (signal) peptidase from *Escherichia coli. J Biol Chem* 1980; 255: 7973–7977.

Fundamental Virology, Third Edition
edited by B.N. Fields, D.M. Knipe, P.M. Howley, et al.
Lippincott - Raven Publishers, Philadelphia © 1996

CHAPTER 4

Viral Genetics

Donald M. Coen and Robert F. Ramig

Animal virus genetics, the ultimate goal of which is to gain a detailed understanding of the structure and function of the viral genome and each of the viral gene products, can be considered to have begun with the development of plaque assays for cytolytic viruses by Dulbecco (74,75). The use of plaque assays allowed precise enumeration of progeny, the production of pure clonal strains of virus, and a system suitable for use with conditional-lethal mutations. Similarly, the development of a focus assay (347) was a key event in the development of genetic systems for the noncytolytic, transforming viruses.

A number of other techniques and methods have been important in the development of the genetics of specific virus groups. The development of high-resolution electrophoresis systems for proteins and nucleic acids allowed the use of polymorphisms of electrophoretic mobility of RNA or protein to be used as segregating genetic markers (250,280). These, combined with restriction endonucleases, opened DNA viruses to similar analyses using migrational polymorphisms of DNA fragments (122,189,235). The advent of techniques for transfection, molecular cloning, sequencing, fine mapping transcripts, expressing proteins, and site-directed mutagenesis has greatly increased the power and refinement of genetic analysis.

This chapter, which draws upon chapters in earlier editions of this textbook by D.M. Knipe and by the present authors (41,181, 278,279), will begin with a discussion of the fundamental genetic reagent—mutation—and how various kinds of viral mutations are generated. Molecular techniques make it possible to introduce nearly any mutation into many viral genomes. Different kinds of mutations have particular uses for molecular analysis. Interactions among viruses and between the virus and host cell can affect the phenotypic expression of mutant and wild-type alleles; certain of these interactions, such as those of defective viruses, will be emphasized. The chapter will then cover how genes and mutations are mapped using genetic techniques, which mainly entail recombination mechanisms, and mol-

D. M. Coen: Department of Biological Chemistry and Molecular Pharmacology, Harvard Medical School, Boston, Massachusetts 02115.
R. F. Ramig: Division of Molecular Virology, Baylor College of Medicine, Houston, Texas 77030.

ecular methods including DNA sequencing. Genetic approaches to function, including classical complementation analysis and molecular methods for studying genes outside the viral genome, such as the use of viral vectors, will then be described. Finally, the use of interfering gene products for approaching gene function will be summarized.

Due to the constraints of space, the consideration of any one genetic system, genetic phenomenon, technique, or application of that technique cannot be comprehensive. Instead, a limited number of examples, whose choices are necessarily subjective and arbitrary, are presented. There are, however, a significant number of review articles that treat the genetics of individual viruses in a comprehensive manner, and the reader is referred to these reviews for details necessarily omitted here. The following is a listing (incomplete) of reviews on systems that are considered in this chapter: SV40 and polyoma (76,169,345), adenoviruses (107,108,391), adenovirus-SV40 hybrids (256), herpesviruses (218,299,300,316,340), poxviruses (48,162,237), picornaviruses (23,53,383), togaviruses (35,263,295,336), rhabdoviruses (269–271), retroviruses (19,43,110,167, 200, 365), reoviruses (79,86,111,284,286,307), influenza viruses (171,191,205,250,252,374), bunyaviruses (16,21), paramyxoviruses (24), and coronavirus (190). In addition, a number of useful reviews of specific genetic phenomena have been written: mutation (66,68,69,139,145,306,329,337), reversion and suppression (130,283), infectious clones of RNA viruses (23), complementation (89), recombination (70,92,157,251,277,331,377,382), and interference (143,149).

MUTATION

Terminology

The terms *strain*, *type*, *variant*, and *mutant* have all been used extensively and without discrimination. All these terms generally have been used to designate a virus that differs in some heritable way from a parental or wild-type virus. *Wild type* has a narrow connotation among molecular geneticists, who use it to designate the original, usually laboratory-adapted, strain of virus from which mutants are selected and to which those mutants are compared. Thus, *wild type* is an arbitrary designation that is attached to a specific virus population and generally is applied to it only with respect to the type of mutation under study, such as temperature-sensitive (ts) mutations. However, a wild-type or ts+ virus may contain mutations that are not of the ts class with respect to which the wild type is defined. It is now recognized that the form of a virus or gene designated wild type may not accurately reflect the nature of the virus or gene as it is isolated from nature. This aspect of the arbitrary designation of wild type is acknowledged by differentiating carefully between laboratory wild types and new virus isolates from the natural host, which have been referred to as *field isolates.*

Through use, specific connotations have also come to be associated with the terms *strain*, *type*, and *variant*. *Strain* is often used to designate different wild types of the same virus, for example, vesicular stomatitis virus strain Orsay and strain New Jersey. *Type* has come to be synonymous with serotype as determined by neutralization of infectivity, for example, reovirus serotypes 1, 2, and 3. *Variant* is generally used to indicate a virus that is phenotypically different from the wild type but for which the genotypic basis of the variation is often not known, for example, neutralization variants.

Spontaneous Mutation

Some viruses yield a high proportion of mutants on passage in the absence of any known mutagen. These spontaneous mutations accumulate in the genomes of viruses and introduce the variation in phenotype that is subjected to selection pressure during the evolution of a virus. The rates of spontaneous mutation can be as low as 10^{-8} to 10^{-11} per incorporated nucleotide in DNA genomes. In viruses with RNA genomes, much higher rates of spontaneous mutation have been measured, on the order of 10^{-3} to 10^{-4} per incorporated nucleotide. The difference in rates of mutation has been ascribed to the low fidelity of RNA genome replication, presumably due in part to the lack of proofreading activities found in RNA replication enzymes (139). Thus, while the genomes of DNA viruses are relatively stable, the high rates of spontaneous mutation in RNA genomes suggest that the concept of wild type is very fleeting in RNA genome systems (66,145). However, data from the RNA phage Qβ indicated that at equilibrium a wild type dominated the population because it replicated more rapidly than the spontaneously arising mutants (65). Unfortunately for the geneticist, there are a number of factors that promote disequilibrium in genome populations (139), and these factors often favor the accumulation of mutants in a virus population. The end result of spontaneous mutation is difficulty in maintaining a homogeneous population of virus, whether it be wild type or mutant. Viruses are standardly subcloned at intervals to combat this problem, but mutation often occurs during plaque formation or virus growth so that it is often difficult to produce genetically homogeneous high-titer virus stocks (28,371).

The techniques of molecular biology have recently allowed more direct examination of spontaneous mutation rates, especially in RNA genome viruses. These studies (255,329,332,373) confirmed that the spontaneous mutation rates of RNA viruses were high but that the rates varied over several orders of magnitude (10^{-3} to 10^{-6}) depending on the virus.

Induced Mutation

Many mutations isolated in studies of animal viruses have been derived from populations of the wild type treated with mutagens to increase the frequency of mutants in the population. The major problem associated with the use of mutagens is selecting an appropriate mutagen dose such that the frequency of mutations with the desired phenotype is increased but the probability of multiple mutations in an individual genome is minimized. Since most mutagens result in a loss of infectivity or a reduced yield of virus from infected cells, a dose-response curve can be generated for each virus-cell—mutagen combination used. With the dose-response curve, the number of lethal "hits" per average genome can be calculated using the Poisson distribution. However, since many mutagenic events are not lethal, the number of lethal hits at a given mutagen dose represents the absolute minimum number of mutagenic events induced by the mutagen at that dose. In practice, to maximize the probability of isolating mutants containing single mutations, mutants are picked from the lowest mutagen dose giving a reasonable frequency of mutations with the desired phenotype.

A large number of different mutagens have been used in animal virus systems, but these fall into a small number of classes defined by the mechanism of mutagenesis. These mechanisms have been thoroughly reviewed (68,69,95) and are considered only briefly here. One class, often called *in vitro* mutagens, acts by chemical modification of the resting nucleic acid of the virus particle, causing changes in base pairing at subsequent replication and resulting in transitions and transversions. Among the *in vitro* mutagens are nitrous acid, hydroxylamine, and the alkylating agents. A second major class is the *in vivo* mutagens that require metabolically active nucleic acid for their action. One group of *in vivo* mutagens contains the base analogs that are incorporated into the nucleic acid following normal base pairing rules. Once incorporated these analogs are capable of undergoing tautomeric shifts that cause them to pair with different bases at subsequent replications, resulting in transitions and transversions. Another group of *in vivo* mutagens contains the intercalating agents. These agents have a planar structure with hydrophobic surfaces and are intercalated into the stack of bases. On subsequent replication of the nucleic acid, distortion of the stack causes the formation of base insertions or deletions. Ultraviolet (UV) light is occasionally used as a mutagen. The major products of UV-irradiation are pyrimidine dimers. In DNA, the dimers are excised and the duplex is repaired with mutation thought to result from base misincorporation that occurs during repair. In RNA, the mechanism of UV mutagenesis is not known.

A property associated with most mutations is the ability to revert to the wild type, and each mutation has a characteristic frequency of reversion that can be measured accurately. Frequencies of reversion have usually been measured more accurately than the forward mutation rate because of the strong selective pressures that can be used against the mutant, revealing rare revertants.

Engineered Mutations

Molecular techniques now make it possible to introduce nearly any type of mutation into many viruses. The first step for engineering a mutation into a virus is to clone the sequence of interest as a recombinant DNA molecule. The second step is to demonstrate the authenticity of the cloned sequence, preferably by its biological activity. These two first steps are invaluable not only for mutagenesis but for other molecular analyses. The third step is to subject the cloned sequence to *in vitro* mutagenesis. The fourth is to introduce the mutated sequence into the viral genome. This permits testing the phenotypes of the mutant virus. A crucial fifth step is to demonstrate that any mutant phenotype observed is due to the engineered mutation.

Recombinant DNA Cloning of Viral Sequences

Recombinant DNA cloning of viral sequences is usually an obligatory first step for molecular genetic analysis of animal viruses. For DNA viruses that are difficult to grow, such as hepatitis B or papillomaviruses, or for retroviruses in which proviral DNA molecules are rare, molecular cloning allows the isolation of large amounts of a given DNA segment for sequencing, transcript mapping, mutagenesis, and studies of gene function. For those DNA viruses that can be prepared in quantity, some of these analyses can be performed with virion DNA. However, recombinant DNA technology can guarantee the purity of the DNA sequences, as implied by the term *cloning*. For RNA viruses, the cloning of cDNA copies enables the virologist to perform biochemical and genetic manipulations currently impossible with RNA genomes. Manuals detailing the different kinds of vectors that can be used, procedures for molecular cloning of DNA molecules, methods for identifying recombinant clones, etc. have been extensively merchandised (e.g.,5). Approaches pertinent to specific viruses will be described here.

Cloning Genomes of Double-Stranded DNA Viruses

Probably the easiest virus genomes to clone are small circular double-stranded DNA molecules, such as those of papovaviruses. These can be cleaved with a restriction enzyme that cuts only once and ligated to a similarly cleaved vector (259). A similar strategy has been used to clone the linear DNA genome of adenovirus after circularizing virion DNA (106). Larger double-stranded DNA viral genomes are generally not cloned in one piece, but rather as a library of overlapping subgenomic fragments. Linear DNA virus genomes pose the additional problem of yielding terminal fragments with one end (one of the original termini of the genome) not corresponding to any restriction enzyme cleavage site. This problem can be overcome by strategies such as adding a synthetic linker containing a restriction enzyme recognition site (233) to the end of the viral DNA molecule. With the development of vectors that can accommodate very large inserts (27,244), it should now be possible to clone large DNA virus genomes intact as a single insert.

Cloning Genomes of Single-Stranded DNA Viruses and RNA Viruses

Single-stranded DNA virus genomes can be cloned following isolation of double-stranded replicative forms (293) or annealing of complementary virion DNAs (312). RNA virus genomes can be cloned using the standard cDNA cloning procedures used to make DNA copies of mRNAs. An especially dramatic example of this approach was the isolation, using random primers, of cDNA clones for a non-A, non-B hepatitis agent, which had been virtually uncharacterized previously (39). Indeed, molecular cloning permitted the identification and characterization of this agent, now known as hepatitis C. In the case of retroviruses, one usually allows the viruses to make the cDNA copies and then clones either the unintegrated viral DNA (124) or the integrated provirus (361).

Biological Activity of Recombinant DNA Copies of Viral Genomes

Infectivity

Not all recombinant DNA copies of a viral genome or viral gene are faithful. For valid functional analyses of molecular clones, it is essential to test them for biological activity. When it is possible to clone entire DNA virus genomes intact as a single insert, biological activity can be demonstrated by introducing the DNA into permissive cells (*transfection*) and recovering infectious virus. Depending on the virus, it may or may not be necessary to remove vector sequences, to recircularize the inserted viral DNA, or to use plasmids containing tandem repeats (106,117,203, 204,259,312,323,380,386). For those larger DNA viruses that have not yet been cloned as a single insert, it is often possible to test cloned subfragments individually for function by complementation (as will be discussed later in the chapter). A more comprehensive approach is to transfect cells with a mixture of overlapping cloned DNA fragments representing the entire virus genome, which can recombine and give rise to infectious virus (360). Unfortunately, it is not possible to use infectivity to test the biological activity of DNA cloned from viruses whose DNAs are not infectious, such as poxviruses.

Molecular clones of positive strand RNA viruses can be tested for infectivity either by transfection of complete cloned cDNA copies (276) or by transfection of *in vitro* transcripts of cDNA clones (230). Specific issues relating to these methods have been reviewed (23). One particularly striking version of these approaches is the generation of infectious poliovirus from cloned cDNA using *in vitro* transcription followed by translation in a cell-free extract of uninfected human cells (233).

As this chapter went to press, the very first publications on the generation of a nondefective, infectious virus solely from cloned cDNA copies of negative-strand RNA viruses began to appear (318a). This accomplishment was preceded by a series of technical advances, as reviewed in García-Sastre and Palese (101). These will be discussed later in the section on introduction of mutations into viruses. Presently, no one has yet published on incorporating sequences from cloned cDNA into a double-stranded RNA virus, but it has been possible to amplify and express a synthetic analog of a rotavirus gene (113). Rapid progress with both negative-strand and double-stranded RNA viruses is anticipated in generation of nondefective infectious virus, testing of portions of cloned genomes for biological activity, and incorporation of mutations into viruses (see later).

Biological Activity of Cloned Fragments

Individual cloned fragments of viral genomes can be tested for biological activity in several ways. One way to do this is to ligate the cloned fragment to an incomplete viral DNA lacking that fragment. For example, in the adenovirus system, a cloned fragment from one genomic terminus has been tested for activity by ligation to virion DNA from which that fragment had been cleaved (Fig. 1) (333,335). A variation on this theme is provided in the negative-strand virus systems (see later), in which a cloned sequence can be transcribed and the transcript transfected into cells expressing viral proteins by means of homologous or heterologous helper viruses to give rise to infectious progeny. This indicates that the cloned viral sequences are functional.

A second way to test for biological activity is to assay whether a cloned DNA fragment can complement a virus mutant in *trans*. These procedures will be discussed in more detail in the section on expression of viral genes in mammalian cells. A potential disadvantage of complementation assays is that care must be taken to ensure that all genes within a cloned fragment are tested. This problem is not trivial considering the propensity of viruses to contain overlapping genes. A second potential problem is intragenic complementation (see section on complementation). Thus, the complementation of a virus mutation by one gene within a cloned

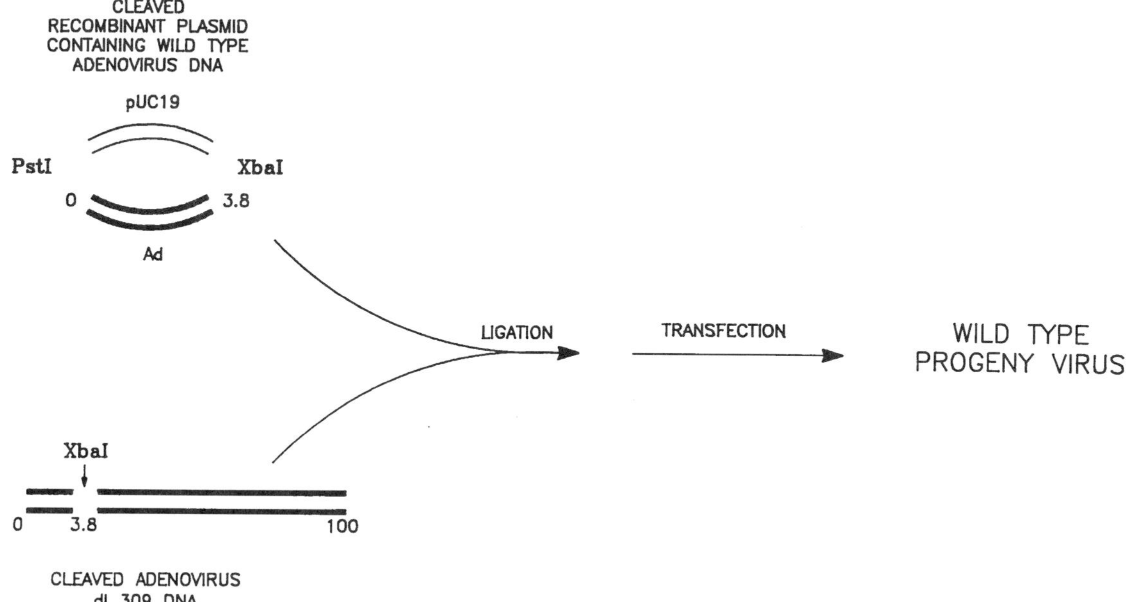

FIG. 1. Reconstruction of a wild-type adenovirus from restriction enzyme cleaved viral DNA and a recombinant DNA plasmid. The recombinant plasmid contains adenovirus DNA (Ad) from the left-hand terminus (map unit 0) to the leftmost *Xba*I site (map unit 3.8), inserted into plasmid pUC19 between its *Pst*I and *Xba*I sites. (As the left-hand terminus of adenovirus DNA does not encode a *Pst*I site, insertion into the plasmid's *Pst*I site entailed deoxyguanosine-deoxycytosine tailing.) The recombinant plasmid is cleaved with *Pst*I and *Xba*I to release the viral DNA fragment, which is ligated to the large fragment formed when adenovirus mutant dl 309 DNA is cleaved with *Xba*I. The ligation mixture is transfected into cells, giving rise to virus that can grow on nontransformed human cells, thereby distinguishing it from its dl 309 parent. Therefore, the cloned DNA sequences are biologically active. [Interestingly, the virus progeny lose the deoxyguanosine-deoxycytosine tails introduced during cloning (simplified from experiments described in refs. 333 and 335)].

fragment might not guarantee that the entire cloned fragment would permit infectivity if contained within viral DNA.

Activities of cloned viral DNA fragments can be assessed in many other ways, such as by testing the ability of viral promoters to confer regulated gene expression on reporter genes or by testing the ability of cloned viral genes to express active enzymes in heterologous expression systems. These methods are valuable and will be discussed later in the chapter. However, they cannot guarantee the authenticity of a cloned segment as completely as methods that measure infectivity or complementation. Additionally, it is possible to compare the sequence of cloned DNA with that derived by directly sequencing viral nucleic acids. However, there is an appreciable error rate in sequencing, even in the best laboratories.

In Vitro *Mutagenesis of Viral DNA Sequences*

Recombinant DNA technology has made it possible to introduce different kinds of mutations into cloned DNA at sites predetermined with a high degree of precision. Three kinds will be discussed here—deletion mutations, linker mutations, and point mutations—as will their use in functional analysis of specific coding sequences and *cis*-acting elements.

Deletion Mutations

Deletion mutations are among the easiest to construct, and their use predates the availability of cloned viral genomes or sequence information. The simplest strategy for creating deletions involves digestion of a circular DNA with restriction enzymes and recircularization to create smaller circles. The deletions that can be created by this method depend upon the availability of convenient restriction enzyme sites. More systematic deletion mutagenesis can be performed by linearizing a circular DNA with a restriction enzyme and then using exonucleases whose extent of digestion can be readily controlled. After recircularization, a population of molecules with different length deletions is obtained.

Specifically targeted deletions and other mutations can be constructed using oligonucleotide-directed mutagenesis, or by use of the polymerase chain reaction (PCR). The im-

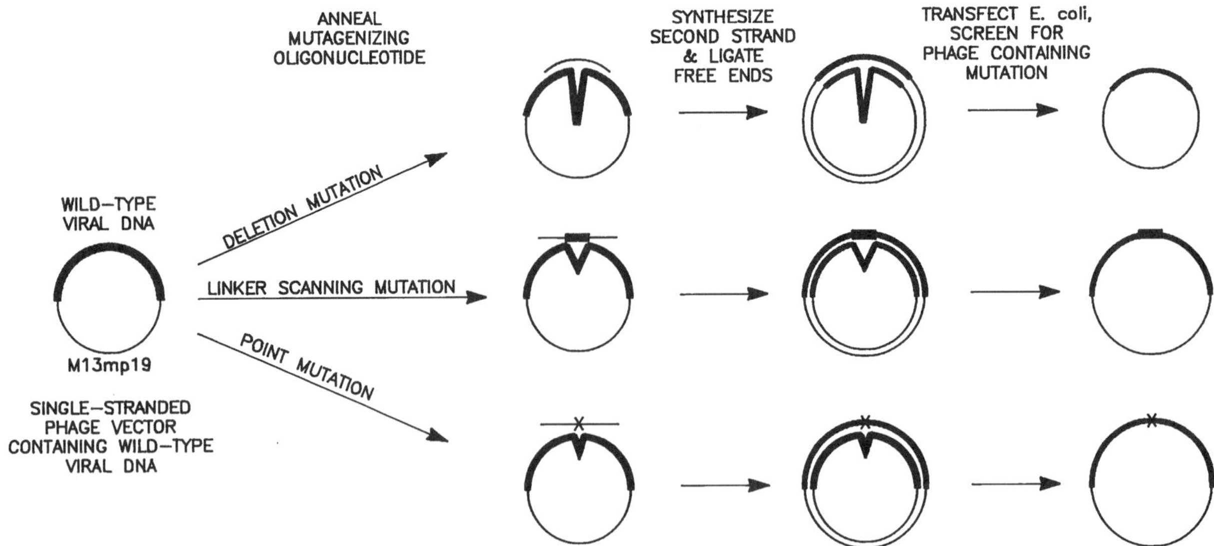

FIG. 2. The use of oligonucleotide site-directed mutagenesis to create deletion, linker scanning, or point mutations in viral DNA. Wild-type viral DNA is cloned into single-stranded bacteriophage vectors such as M13mpl9. Single-stranded phage DNA is isolated and annealed with the appropriate mutagenizing oligonucleotide. For a deletion mutation, the oligonucleotide anneals to two noncontiguous regions of the DNA, looping out the viral DNA to be deleted. For a linker scanning mutation, the oligonucleotide contains the same number of bases as the corresponding region of cloned viral DNA, but some of the internal bases are noncomplementary so as to comprise a restriction enzyme recognition site (indicated by a box). For a point mutation, an oligonucleotide with a single internal base mismatch is annealed. The annealed oligonucleotides serve as primers for extension *in vitro* by DNA polymerase, and the ensuing double-stranded circles are closed with DNA ligase. The circles are transfected into *E. coli* in which bacteriophage that contain the desired mutation are formed via host repair and bacteriophage replication processes. Various strategies are available to increase the efficiency of mutagenesis and for detection of mutagenized phage.

portance of these techniques was emphasized in 1993 with the awarding of the Nobel Prize in Chemistry to their inventors. Both methods require knowledge of at least part of the DNA sequence. In the oligonucleotide-directed technique (Fig. 2), cloned single-stranded DNA in a single-strand DNA bacteriophage vector is hybridized to an oligonucleotide that is complementary to sequences on either side of the region to be deleted. After DNA synthesis *in vitro*, the resulting heteroduplex, with wild-type sequences on one strand and the deletion on the other, is transfected into bacteria and the resulting phage plaques are screened for the presence of the deletion. A number of methods have been developed to increase the frequency of phage that contain the mutation (187,346). PCR methods have the advantage of not requiring single-strand phage. One PCR method for creating deletions is outlined in Fig. 3 and is a version of *inverse PCR* (246). In this method, circular double-stranded DNA is denatured and two primers are annealed to opposite strands, with their 3' ends pointing away from the segment to be deleted. These are then used to amplify the segment between the 3' ends of the primer, giving rise to a linear molecule that lacks the segment to be deleted. This can then be recircularized and cloned. Both oligonucleotide and PCR methods can give rise to adventitious mutations, which can be checked by DNA sequencing.

Deletion mutations have two major uses. First, they can be used to construct null mutations, as discussed later in the chapter. Second, deletion mutations can be used to assign functions to specific parts of a protein molecule or *cis*-acting sequence. For example, deletion analysis of the vesicular stomatitis virus glycoprotein gene showed that the C terminus of the protein serves to anchor the glycoprotein in lipid membranes (303). Similarly, a detailed deletion analysis on the HSV thymidine kinase (*tk*) gene defined a transcriptional control region required both for wild-type levels of *tk* mRNA and for accurate transcription initiation (219). Potential problems with deletion mutations include an inability to address the roles of specific amino acids or base pairs, the potential for nonspecific effects such as protein denaturation or instability, and inhibition or unmasking of function by altering protein tertiary structure or changing spatial relationships in proteins or *cis*-acting elements. For these reasons, more detailed mutational analysis is often undertaken.

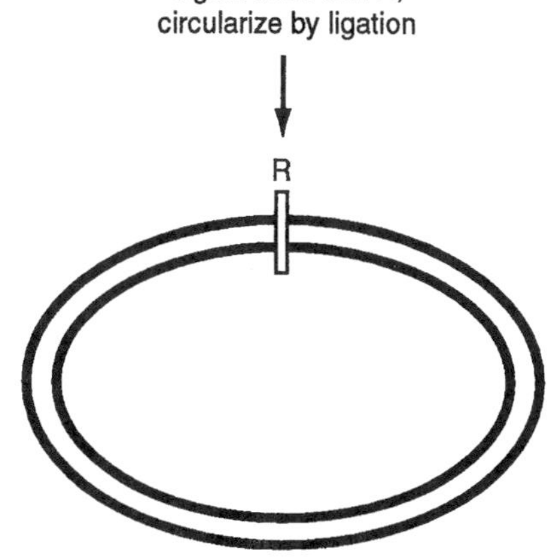

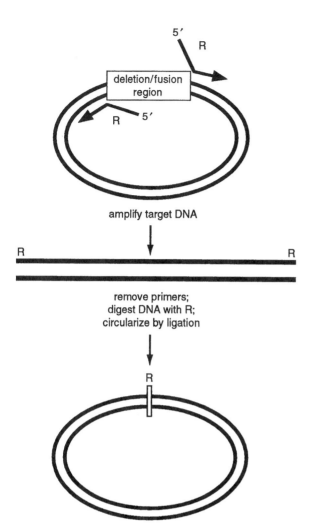

FIG. 3. Construction of deletion mutations by inverse PCR (246). Wild-type DNA is cloned into a circular plasmid. Primers complementary to sequences flanking the region to be deleted are synthesized. Each primer contains at its 5' end a sequence corresponding to a restriction enzyme site (R) while the 3' ends point away from the region to be deleted (arrowheads). The plasmid DNA is denatured, the primers annealed, and PCR is performed to yield a linear molecule lacking the region to be deleted, with restriction enzyme sites (R) at each end. The primers are removed (usually by gel purification of the linear PCR product), the ends of the PCR product are digested with the restriction enzyme to create two enzyme half-sites, and the molecule is circularized using DNA ligase. The circles can be introduced into *E. coli* and maintained as a plasmid there, providing the deletion did not remove any sequences important for that. From ref. 5, with permission.

Linker Mutagenesis

Mutagenesis using restriction enzyme linkers affords a more detailed analysis than does deletion mutagenesis, while allowing ready screening for the mutation by the presence of a new restriction site. The technique of linker scanning mutagenesis was devised by McKnight and Kingsbury (220) both to allow detection of short promoter elements in the *tk* promoter and to overcome the spacing problems encountered in deletion mutagenesis. The technique essentially replaces wild-type sequences with a restriction enzyme linker. Linker scanning mutations can be created by joining deletion mutants with the linker or by using oligonucleotide-directed or PCR mutagenesis techniques, in which the mutation is incorporated into an oligonucleotide used to prime DNA synthesis (see Figs. 2 and 3). The result is a cluster of point mutations at positions where the linker differs from the wild-type sequence. Assays of the set of linker scanning *tk* mutations defined several promoter domains that correspond to binding sites for various cellular DNA binding proteins (42,77,165,220).

Although linker scanning mutagenesis has been applied to protein coding sequences, linker insertion mutagenesis has been more popular in this context. In one version of this method, linkers corresponding to a small number of codons (e.g., multiples of three base pairs) are inserted at various restriction sites throughout a protein-coding gene. Many sites can be mutagenized, especially using limited digestion by frequently cutting restriction enzymes. Problems caused by changes in spacing are limited as the insertions are short and the method is somewhat easier than linker scanning mutagenesis. A sophisticated form of this procedure (202) has been used by Goff and his colleagues to separate the domain encoding the reverse transcriptase activity of Moloney murine leukemia virus from that encoding RNase H; deletion mutagenesis had failed to separate these two activities (344).

A second version of linker insertion mutagenesis entails insertion of linkers that cause polypeptide chain termination in all three possible reading frames (nonsense mutations). This leads to polypeptides of various lengths, which can exhibit subsets of the activities of the original protein. Both this method and the creation of shifts in the translational frame by insertion or deletion of any number of base pairs that is not divisible by three can, in principle, be used to construct null mutants or to assign functions to specific parts of protein molecules. Many of the limitations of deletion mutations apply to these mutations as well.

Point Mutations

Point mutations can identify specific roles of individual base pairs in *cis*-acting regulatory elements or of individual amino acids in proteins. Single base or amino acid alterations can have enormous phenotypic effects, pinpointing the importance of these residues. On the other hand, any mutation, but especially a point mutation, can exert no discernible effect yet alter an important residue if, for example, that residue is one of several that have the same activities. For example, promoters can have repeated elements so that the inactivation of one will produce little effect. For this reason, it is often valuable to combine point mutations with lower resolution mutagenesis procedures. There are two basic procedures for creating point mutations in cloned viral DNAs—random mutagenesis of specific fragments or site-directed mutagenesis. Cloned DNA fragments can be mutagenized randomly by exposure to chemical mutagens such as nitrous acid or hydroxylamine (40,49,313) or by replication, for example, during PCR, with a mutagenic DNA polymerase (196). This approach allows one to create point mutations in relatively large regions of a genome without any predetermined bias as to which mutations will prove interesting. Only those mutations that give rise to a distinct phenotype are then chosen for further study. This has been especially useful in introducing mutations into regions of the genome in which no mutations had previously been found. Chu et al. (40) mutagenized such a region of HSV DNA with hydroxylamine, incorporated the mutagenized fragment into HSV by marker rescue methods, and recovered three new ts mutants with lesions within this region. One of these mutants subsequently was shown to be negative for DNA synthesis and likely to affect a double-stranded DNA binding protein, indicating an essential role for this protein in viral DNA synthesis (213), as predicted from transient expression assays of HSV DNA synthesis (385).

Site-directed point mutations allow one to address more precise questions of function. There have been a number of methods developed for site-directed mutagenesis, but the most popular include oligonucleotide-directed mutagenesis and PCR-based methods. These methods have already been described in the section on deletion mutations (see Figs. 2 and 3). In each case the mutation is introduced into an oligonucleotide used to prime DNA synthesis. Using these procedures, a short region of the genome can be saturated with every possible point mutation by using oligonucleotides that contain some proportion of each of the four bases at each position.

Site-directed mutagenesis can help test hypotheses regarding *cis*-acting elements such as promoters and to elucidate protein structure-function relationships by testing the importance of specific residues. It can be especially valuable in viral systems for testing the various roles of overlapping reading frames. One example was the construction of polyomavirus containing a mutation that terminates translation of the middle-T antigen while causing a conservative alanine to valine change in the overlapping small-T antigen reading frame (31). The mutant lacks 37 amino acids at the carboxyl terminus of middle-T antigen and is defective for cell transformation, membrane attachment, and associated protein kinase activity. These data were among those that dissected the roles for the various polyoma T-antigens in oncogenic transformation and correlated biochemical properties of middle-T antigen with this process.

Introduction of Mutated Sequences into Viral Genomes

As discussed in the section on biological activity of recombinant DNA copies of viral genomes, to ensure that a cloned DNA fragment is a faithful, wild-type sequence, one tests the fragment for its ability to produce wild-type infectious virus. Similarly, to test the effect of a mutation on a viral function in its authentic context, one incorporates the mutation into the viral genome. Methods used to do this are presented later along with examples of how the technique has been used to explore viral function.

For all the methods presented, there are certain important precautions. It is important to ascertain that the virus contains the mutation created *in vitro* by restriction enzyme and/or sequencing analysis. If the mutated virus expresses a particular phenotype, it is then important to ascertain that the phenotype is due only to the created mutation, and not to some adventitious mutation, using methods such as those described in the section on mapping mutations later in the chapter.

Direct Mutagenesis of Infectious Cloned DNAs

For those viral genomes that can be cloned as intact, infectious molecules or that can be cloned and then transcribed into infectious RNA, introducing mutations into virus is straightforward. One simply creates the desired mutation in the cloned DNA, either transfects the DNA or transcribes it and transfects the RNA into susceptible cells, and recovers virus. To recover potentially lethal mutations in a *trans*-acting gene, it is best to transfect cells that are transformed with the gene, which can complement the mutations, or to propagate the mutants with a helper virus. The construction of human immunodeficiency virus (HIV) mutants containing mutations in the *tat* gene is an example of this approach. Infectious proviral clones containing large deletions in the *tat* gene that failed to *trans*-activate expression of a reporter gene linked to the HIV LTRs failed to produce virus or viral structural proteins upon transfection into susceptible cells. This defect could be complemented either transiently by cotransfection with a plasmid expressing a *tat* cDNA or in cell lines stably expressing *tat* from a retroviral vector (60,90). These experiments demonstrated that *tat* function was essential for HIV replication and that its activity in the context of the virus genome resembles that in isolation.

For *cis*-acting mutations, failure to recover a virus with an engineered mutation is taken as evidence, albeit negative, that the mutations are lethal. Often, less penetrant mutations can be informative. An example of this is the 5' untranslated region of poliovirus. Several different kinds of mutations have been engineered into infectious cloned DNA copies of the poliovirus genome and then introduced into cells either by transfection of the DNA or of RNA transcribed from the DNA (62,161,355). A number of the mutated DNAs did not give rise to virus, while among those that did, several exhibited slow and/or temperature-sensitive growth. The results taken together indicate that this region is required for viral replication. Studies of this genetic element in isolation suggest that it functions at least in part to mediate cap-independent translation (260) as discussed later in the chapter.

Ligation of Mutated Fragments to Reconstruct DNA Viruses

As discussed previously, one strategy for assessing the biological activity of cloned DNA fragments of certain viruses including adenoviruses is to ligate the fragment to the virion DNA from which that fragment had been cleaved (Fig. 1). When the cloned fragment contains a mutation, this method can be used to generate mutant viruses. This approach has been used extensively to study adenovirus functions involved in gene expression and in oncogenic transformation. An example is the construction of mutants containing mutations in either of the two genes encoding the small virus-associated (VA) RNAs. Deletion of one of the genes, but not the other, leads to poor growth due to inefficient translation of viral mRNAs at late times in the infectious cycle (349). These results have led to studies implicating VA RNAs in translational control at the level of phosphorylation of initiation factor eIF-2, in which the mutant viruses have played important roles (318).

Construction of Mutations Using Homologous Recombination

In many DNA and RNA virus systems, approaches employing homologous recombination, usually *in vivo*, have been necessary to construct mutant viruses. The mechanisms of recombination are similar to those during mixed infection and are discussed later in the chapter.

Marker Transfer in DNA Viruses

For large DNA viruses, and even for some smaller ones, mutant viruses have been constructed using the methods outlined in Fig. 4, which shows how the method can be used to construct a drug-resistant mutant. These techniques have been termed *marker transfer* or *allele replacement* in that the wild-type allele is replaced with a mutation that is transferred from a mutant fragment. Wild-type sequences flanking the mutation are required to permit homologous recombination. One can cotransfect infectious viral DNA with a cloned DNA fragment containing the mutation of interest, as is common with herpesviruses, or one can transfect the mutagenized fragment and infect with virus, as is common with poxviruses. In both cases, if the mutation causes a specific predictable phenotype, one can select (e.g., drug resistance) or screen (e.g., temperature sensi-

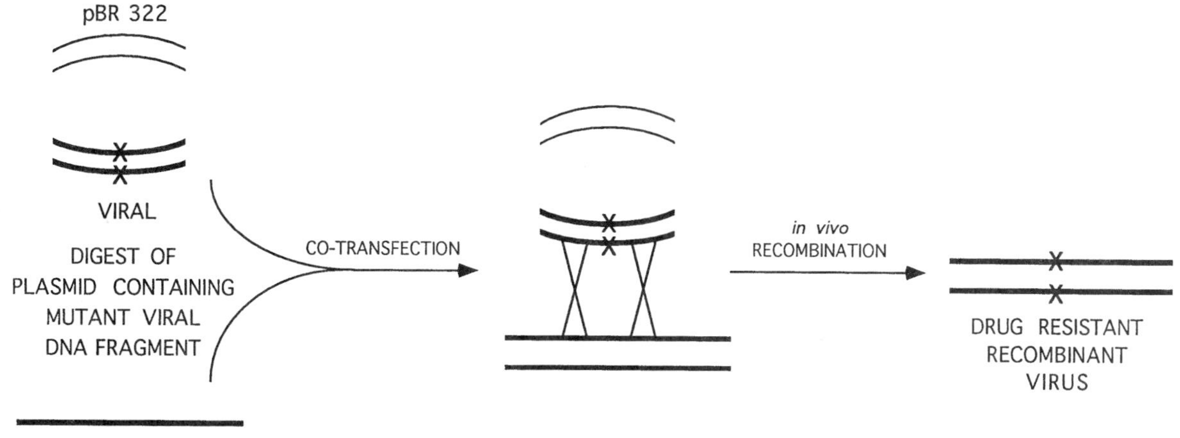

FIG. 4. Marker transfer by cotransfection of cloned mutant DNA and infectious wild-type DNA. The recombinant DNA plasmid contains mutant viral sequences (-X-), which should confer a drug-resistance phenotype when incorporated into the viral genome. The restriction enzyme digested plasmid is mixed with intact infectious wild-type viral DNA and the mixture is transfected into cells. Homologous recombination between the cloned mutant sequences and the wild-type genome occurs in the transfected cells. Drug-resistant viruses can be selected and the percentage of drug-resistant progeny quantified. When the percentage of drug-resistant progeny arising from the transfection containing the mutant plasmid is substantially higher than that arising from transfected infectious wild-type viral DNA alone, then the mutation has likely been transferred from the plasmid to the virus.

tivity) for that phenotype. One can also screen plaques for the presence of the mutation (genotype) by methods such as plaque hybridization. Generally, however, only a small fraction of the progeny arising from *in vivo* recombination contains the mutation of interest, especially when the mutant recombinant is at a growth disadvantage compared to the wild-type parent. This is much less efficient than the methods outlined previously involving direct transfection or ligation. As a result, a number of techniques have been developed to increase the proportion of mutants.

Marker Transfer Linked to a Viral Gene

Some viral genes permit selection either for or against expression. For example, mutations affecting the thymidine kinase (*tk*) genes found in most herpesviruses and poxviruses can be transferred into wild-type genomes and selected for resistance to drugs that are phosphorylated by *tk*. A mutant *tk* gene was used to transfer the linker scanning mutations (see the section on linker mutagenesis) in the *tk* promoter into the HSV genome (42), as diagrammed in Fig. 5. Plasmids were constructed in which the linker scanning mutations were linked to *tk* protein-coding sequences that conferred resistance at 39° to the drug acyclovir. These plasmids were cotransfected with infectious DNA from *tk*$^+$ HSV. The progeny virus were selected for resistance to acyclovir at 39°. In this way, the linker scanning mutations were introduced into the genome as an unselected marker linked to the drug resistance mutation. Using this procedure, as many as 40% of the selected progeny can contain the engineered mutation (151). An advantage of this procedure was that it permitted the isolation of the HSV linker scanning mutants regardless of their effect on *tk* expression. Studies of these mutants supported the hypothesis that *tk* transcription was induced by viral regulatory proteins interacting with cellular transcription factors without binding to a specific site in the *tk* promoter (42).

Procedures Using Foreign Marker Genes

Related approaches have been developed for constructing specific mutations in *trans*-acting functions. One approach was first developed to aid in construction of vaccinia virus vectors expressing foreign genes (253) and uses the *E. coli lacZ* gene under the control of appropriate gene expression signals as an insertional mutagen. Insertion of this "cassette" by marker transfer results in a mutant virus that expresses β-galactosidase and will form blue plaques when plated in the presence of the chromogenic substrate,

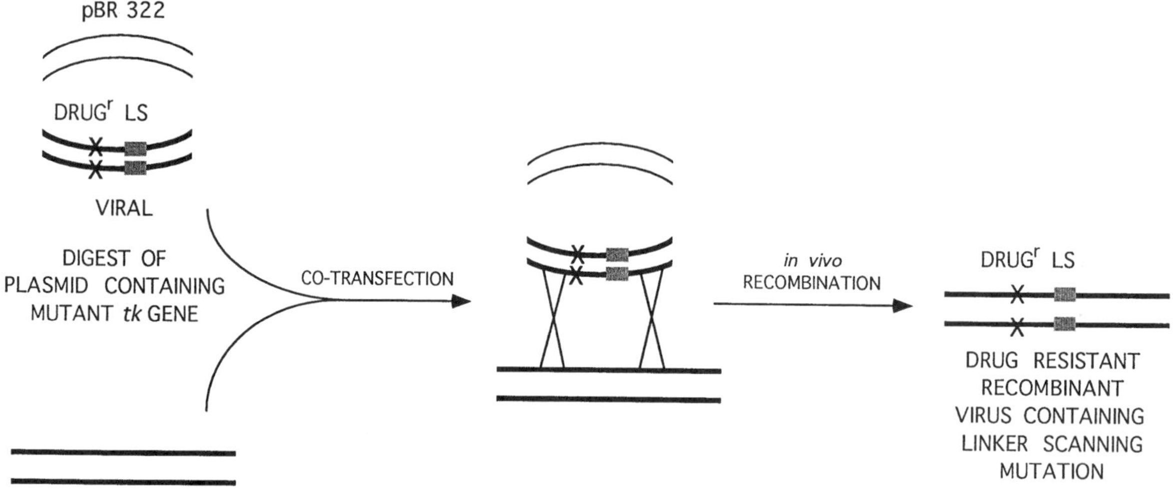

FIG. 5. Linked marker transfer of a linker scanning mutation and a drug resistance marker. A mutant herpes simplex virus thymidine kinase (*tk*) gene containing a drug-resistance mutation (-X-) in protein coding sequences and a linker scanning (LS) mutation (shaded box) in sequences upstream of the gene is cloned into pBR322. The recombinant DNA plasmid is cleaved with a restriction enzyme, mixed with intact infectious wild-type viral DNA, and the mixture is transfected into cells. Recombination between the cloned mutant sequences and the wild-type genome occurs in the transfected cells. When the LS mutation and the drug-resistance mutation are not too far apart (several kilobase pairs), selection for the drug-resistance mutation results frequently in the isolation of recombinant virus carrying the LS mutation. Thus, the LS mutation is introduced into the genome as an unselected marker by linkage to the drug resistance mutation (see ref. 42). Several other similar strategies for enriching for viruses containing engineered mutations are available (e.g., those discussed in the text or linkage of mutations to wild-type sequences that rescue *ts* mutations).

X-gal. Using such a blue plaque mutant as a source of infectious DNA and marker rescue methods, one can then replace the *lacZ* mutation with any desired mutation or foreign gene, screening for white plaques. This obviates the necessity of linkage to a previously established marker in a viral gene.

Procedures to Increase or Select for Recombination

A method developed originally in the pseudorabies virus system takes advantage of the Cre-*lox* site–specific recombination machinery of bacteriophage P1. In this approach, the specific *loxP* recombination site is introduced into viral DNA by marker transfer. The infectious DNA is then recombined *in vitro* with plasmids that also contain the *loxP* site, using purified Cre protein. The infectious DNA is then recovered and transfected into cells. Using this method, as many as 20% of the progeny have been mutant (100,315).

Transfection of a library of cosmids or plasmids containing large overlapping inserts, which taken together contain the entire viral genome, can generate infectious virus (44,55,61,360). When one of the cosmids contains a mutation, nearly all of the infectious virus generated is mutant. This approach, originally developed in pseudorabies virus, should be especially valuable for situations where it is difficult to plaque purify virus and/or where there is no ready selection for the mutant phenotype. A variation on this method has been very useful in the baculovirus system, where virion DNA is circular. A restriction site that permits linearization has been engineered into baculovirus DNA using marker transfer. Cotransfection of the linear DNA with mutant plasmid then permits recircularization by homologous recombination and thus restores the infectivity of the DNA so that the majority of progeny virus contain the mutation of interest, or for overexpression studies, the foreign gene of interest (178).

Insertion of Viral DNA Sequences at New Sites in a Viral Genome

Naturally arising rearranged viral genomes, such as those that are found in defective interfering particles, have proven valuable, especially in identifying *cis*-acting signals involved in virus packaging and replication (see section on defectiveness). Molecular genetic approaches make it possible to construct viruses in which sequences normally found at one site are now found at another. In the herpes simplex virus system, the *tk* gene has proven valuable for such studies. For example, insertion of HSV sequences from tens of kilobases away into the *tk* gene has allowed identification of sequences involved in cleaving the termini of HSV DNA and insight into the mechanism of cleavage (232,363).

Targeted Recombination of Short Transcripts

In several RNA virus systems, it is possible to engineer mutations into virus using RNA recombination mechanisms. This has proven especially valuable in the positive-strand coronaviruses, in which progress towards construction of full-length cDNA clones has been slow because of the large size of their RNA genomes (>25 kb). The procedures, which are similar to the marker transfer approaches used for large DNA viruses outlined earlier, entail the cotransfection of synthetic transcripts with genomic RNA or the expression of such transcripts in coronavirus-infected cells using vaccinia virus vectors (182,199,359). RNA recombination ensues between the synthetic transcript and the coronavirus genome. Recombinant mutant virus progeny can be detected physically using methods such as PCR and recovered. Two groups (182,359) used an approach akin to linked marker transfer to incorporate sequence changes into the mouse hepatitis virus (MHV) *N* gene by rescuing a *ts* mutant whose thermolability was associated with a deletion in the *N* gene. The deleted sequence was restored and the engineered mutation incorporated in progeny virus that survived high temperatures. These techniques, which have been termed *targeted RNA recombination,* may permit detailed mutational analyses of this virus group.

Introduction of Mutations into Negative-Strand RNA Viruses

During the late 1980s and early 1990s, the first successes were achieved in introducing engineered mutations into negative-strand RNA viruses. Several different approaches have been taken. In the influenza system, advantage has been taken of the segmented nature of the viral genome. A cloned copy of one segment of the viral genome is made mutant by any of the techniques described above. In the original successful strategy (207), it is then transcribed *in vitro*, the transcript is reconstituted with purified influenza polymerase proteins, and the ribonucleoprotein is transfected into cells infected with a helper virus. If the mutant transcript contains the proper replication and packaging signals, it can be replicated alongside the helper virus transcripts and then, when the transcripts are packaged, sorted into virions. At present, the efficiency of recombination is low, requiring techniques similar to those outlined earlier for DNA viruses to aid recovery of mutants. Various versions of this approach are now being used widely to investigate influenza virus biology.

With viruses that contain a single negative strand of RNA, workers have succeeded in amplifying, expressing, and packaging foreign genes into infectious, but defective, viruses by transfection of naked *in vitro* transcripts into helper-virus infected cells (47,63,254). The recombinant viruses were detected by their ability to infect new cells and express reporter genes. A different approach has succeeded in generating infectious defective particles of vesicular stomatitis virus (VSV) (258). This strategy uses a plasmid encoding a defective-interfering (DI) RNA (see section on defectiveness) driven by a promoter for a bacteriophage polymerase with sequences encoding a ribozyme immediately downstream of the DI sequences. This plasmid was cotransfected along with plasmids encoding each of the five individual viral proteins, also under the control of the bacteriophage promoter, into cells that had previously been infected with a vaccinia virus expressing the bacteriophage RNA polymerase. This polymerase transcribed the DI-ribozyme fusion RNA. The ribozyme generated the correct 3' of the DI RNA by autolytic cleavage. At the same time, the bacteriophage RNA polymerase transcribed the VSV protein-coding genes, which were translated into proteins that replicated and packaged the DI RNA. These still evolving methods should lead to the introduction of a variety of types of mutations into negative-strand RNA viruses.

Demonstration that Mutant Phenotypes Are Due to Specific Mutations

A final critical step in engineering a virus mutant is to ensure that any phenotypes observed are due to the engineered mutation and not some adventitious mutation. This is much more than a hypothetical concern as many of the methods employed to engineer mutations and the transfection methods employed to introduce the mutations into virus give rise to changes at other sites at appreciable frequency. There are several strategies that can be employed to reduce these problems. It is best to mutagenize cloned DNAs that are short enough to permit sequencing of the entire segment to ascertain that there are no other mutations. Once that has been done, if it can be shown that the mutant DNA substantially increases the frequency of the mutant phenotype following transfection, that is strong evidence that the mutation confers the phenotype. It is often wise to obtain independent viral isolates (i.e., separated before or immediately following transfection) from the same mutant plasmid. If the independent isolates exhibit the same phenotype, then it is very likely due to the engineered mutation. None of these strategies is as convincing, however, as using molecular mapping methods to demonstrate that the mutation conferring the phenotype lies within a specific nucleic acid sequence and then sequencing that region to show that the only mutation is the engineered mutation. Methods for mapping mutations are outlined later in this chapter.

Types of Mutation and their Use for Molecular Analysis

A number of different types of mutants have been selected from animal viruses. Good mutants for genetic studies are easily scored, are reasonably stable, and should result from a single mutation (84). A few of these are discussed here.

Null Mutations

Null mutations completely inactivate the function of a gene. In principle, any kind of nucleic acid change can yield the null phenotype. However, not even large deletions nor insertions, nonsense, nor frameshift mutations will necessarily cause the null phenotype. Some of these mutations leave intact protein fragments that retain function. Moreover, insertion, nonsense, and frameshift mutations can sometimes be bypassed by translational mechanisms such as nonsense suppression, reinitiation, and ribosomal frameshifting. Indeed, if one is engineering a null mutant it is safest to delete the entire gene. This requires that the mutation be in a nonessential gene or that a method exists for propagating the otherwise defective virus.

A null mutant can be critical for many analyses, especially to determine if a gene is essential for some process. There are many cases in which a very small amount of function retained by a mutant gene is sufficient, for example, for viral replication in a particular cell type or some manifestation of disease. Along the same lines, a null mutant can be critical if one wishes to determine that a gene is *not* important for some process.

Temperature-Sensitive Mutations

Temperature-sensitive (*ts*) mutations have proved to be highly useful in animal viruses, primarily because of their conditional-lethal phenotype. Temperature-sensitive mutants are usually produced by missense mutations that alter the nucleotide sequence of a gene so that the resulting protein product of that gene is unable to assume or maintain its functional configuration at the nonpermissive (high) temperature. The protein, however, is able to assume a functional configuration at permissive (low) temperature, a property that allows the mutant to be propagated. The *ts* mutation can help identify the mutant protein or show that the protein is structural because the protein or the virions produced at the permissive temperature may be thermolabile. However, the ts phenotype can be due to temperature-sensitive defects in synthesis or assembly of a protein or due to the requirement for a gene product at high temperature but not low. Since *ts* mutations are conditional-lethal, one can theoretically select *ts* mutations in any viral gene encoding a required gene product. Indeed, in some viruses *ts* mutations have been isolated that identify every viral gene (284,319), and *ts* mutants have been isolated in many genes of other viruses.

Several properties of *ts* mutants are worthy of note. Generally, *ts* mutants produce full-size proteins that usually have the same immunologic specificity as the wild-type protein. Certain *ts* mutants are "leaky"; that is, some functional activity is retained at the nonpermissive temperature, although function is so impaired that very little infectious virus is produced. On the other hand, *ts* mutants rarely function as effectively as wild type at permissive temperature. This often makes it difficult to produce high-titer stocks since revertants to wild type, although rare, have a significant growth advantage even at permissive temperature. Because temperature can be readily manipulated up or down, *ts* mutations have been valuable in so-called temperature shift experiments for assessing the time during which a gene product is required for some process. Within a given gene, *ts* mutations can produce different phenotypes, helping to identify specific functional sites and domains.

Cold-Sensitive Mutations

Cold-sensitive (*cs*) mutations are conditionally lethal at low (nonpermissive) temperature and grow nearly as well as wild type at high (permissive) temperature. They can have many of the same uses as *ts* mutations. However, the necessity of using significantly suboptimal nonpermissive temperatures with these mutants makes their general use in eukaryotic cells impractical. Despite this limitation, *cs* mutations have provided useful genetic tools (228a). They also are being explored as attenuating mutations for vaccines (208).

Plaque Morphology Mutations

Plaque morphology mutants have altered morphology due to one of a variety of metabolic differences between mutant and wild type. In adenovirus, large plaque mutants have been shown to release virus from the host cell more rapidly than wild type (179). Syncytial (*syn*) mutants of viruses like herpes simplex (298) cause neighboring cells to fuse rather than undergo typical cytolytic changes. These phenotypes can provide useful genetic markers and the mutants can shed light on processes such as membrane fusion and viral egress.

Host Range Mutations

Host range mutations have been selected from a number of viruses. Mutants of adenovirus type 5 have been isolated that can grow in human but not in hamster cells (342), and similar host range mutants have been isolated from numerous other viruses. Benjamin (9) selected host range mutants of polyoma virus that were able to grow in polyoma-transformed cells but not in normal mouse cells. These were invaluable in defining the viral oncogene, middle T antigen. This approach has been extended using transfection methods, as described later in this chapter, to develop cell lines that can complement otherwise lethal virus mutants. Another type of host-range system that has been central in bacterial virus genetics are cells that can suppress nonsense mutations. There have been many efforts to develop these for eukaryotic viruses (322); however, none has yet come into general use.

Drug Resistance Mutations

Drug resistance mutations in herpesvirus and poxvirus *tk* genes, and their use in construction of viral mutants and expression of viral genes, was discussed earlier. Drug resistance has also been used to assign functions to specific genes or to specific portions of proteins. For example, mutations conferring resistance to therapeutic doses of amantadine map within the influenza virus *M2* gene (132) and the altered residues helped define an ion channel formed by the M2 protein (264). Drug resistance mutations have been instrumental in elucidating basic genetic mechanisms, for example, guanidine resistance was key in the demonstration that RNA recombination in picornaviruses occurred by a mechanism of copy-choice (177). Finally, when drugs are used to inhibit viral processes such as DNA replication, drug-resistant mutants provide useful controls for the specificity of inhibition.

Antibody Escape/Resistance Mutations

Virus mutants that specify altered surface proteins can be selected by the resistance of the virus to neutralizing antisera against that protein. The resulting neutralization escape mutants can then be analyzed by DNA sequencing for the positions of mutations and by reactivity with a panel of monoclonal antibodies for effects on specific epitopes (103). Variations on this theme have been developed that allow selection of monoclonal antibody–resistant mutants (144, 183). Monoclonal antibody–resistant mutants can be isolated with non-neutralizing antibodies, whereas neutralization escape mutants require antibodies with neutralizing activity. Mutants of these types have been particularly useful in studies of viral surface proteins, particularly those that are neutralizing antigens. For example, neutralization escape mutants have been examined with reference to the three-dimensional structure of the influenza virus hemagglutinin, permitting mapping of the antigenic sites of the molecule (see Chapter 3).

Revertants, Suppressors, and Heterozygosis

Reversion of a viral mutant to wild-type phenotype can occur by one of three pathways: (a) reversion by back-mutation at the nucleotide of the original mutation can restore the wild-type nucleotide sequence (true reversion); (b) a mutation can occur at a second site in the same gene as the original mutation, restoring the wild-type phenotype (intragenic suppression); or (c) reversion can be mediated by a second mutation in a gene different from the gene containing the original mutation (extragenic suppression). The latter two pathways constitute genetic suppression. The suppressor mutation bypasses the defect of the original mutation, resulting in a virus with a mutant genotype but a wild-type phenotype (called a pseudorevertant to distinguish it from a true revertant). Suppression is traditionally demonstrated by a backcross of a suspected pseudorevertant to wild type (both are phenotypically wild type), yielding progeny of the cross that exhibit the mutant phenotype. A variety of molecular techniques can now be used, such as sequencing, to demonstrate the presence of the original mutant genotype. Suppression of the ts phenotype of a *ts* mutation was first noted in reovirus (281), where backcross of a suspected pseudorevertant (ts^+) of a *ts* mutant to wild type (ts^+) yielded a high frequency of progeny with the ts phenotype. The rescued *ts* mutants were genetically identical to the *ts* mutant from which the pseudorevertant was originally isolated. A systematic survey of revertants of reovirus *ts* mutants from all mutant groups showed that reversion via extragenic suppression was the primary means of generating revertant phenotype since greater than 90% of the revertants contained extragenically suppressed *ts* mutations (282). In these studies, no phenotype other than suppressor activity could be associated with suppressor mutations and, as a result, relatively little is known about the mechanism of suppression in reovirus.

A powerful use of extragenic suppression is in detecting and analyzing protein-protein and protein-DNA interactions. This rests on mechanisms wherein the mutant phenotype of a *cis*-acting element or protein is suppressed through physical interaction with a second mutant (suppressor) protein (223,242,328a). This requires that suppression be allele specific; that is, not every mutation in the first gene should be suppressed by every mutation in the second gene. A particularly elegant use of this approach was the isolation of suppressors of *cs* mutations in the SV40 origin of DNA replication. These suppressor mutations, which were allele specific, mapped within T antigen, demonstrating that T antigen interacts with the origin *in vivo* and suggesting a path for the evolution of new viral replicons (328a).

Suppression of mutant phenotype may be a phenomenon of some biological significance in animal viruses. The high frequency of extragenic suppression observed with reovirus suggests that suppression may represent a general mechanism by which RNA viruses lacking DNA intermediates in their life cycles can overcome the effects of deleterious mutations (282,283). This mechanism would bypass the effects of deleterious mutations that accumulate spontaneously and cannot be removed (or are removed very inefficiently) by recombination in RNA genomes. An influenza virus vaccine that was attenuated by the presence of *ts* mutations was found to generate a suppressed pseudorevertant during a vaccine trial (242). Tests in volunteers revealed that the pseudorevertant had regained the virulent phenotype, apparently as a result of suppression of the attenuating mutations (350).

Heterozygosis in viruses is similar to suppression in that the phenotype of a mutant allele becomes masked. However, in this case, this is caused by the presence of a wild-type copy of the gene in the same virion rather than by a second mutation. In the genetics of higher organisms, heterozygosis implies that diploid chromosomes differ in allelic markers at one or more loci. Among animal viruses, heterozygosity of this sort is not generally seen because, with a few exceptions, the genomes of these viruses are haploid. However, the genomes of the retroviruses are fully diploid, consisting of two copies of the genomic RNA molecule. Retroviruses can be heterozygous for all markers, and heterozygotes have been postulated to be an intermediate in the recombination process (376). In DNA viruses that contain reiterated sequences, and thus are partially diploid, heterozygosity may also occur. For example, the genome of herpes simplex virus contains terminal and internal repetitive sequences, which allow some alleles to be present in two or more copies (325,369). Although the alleles in the repeated sequences are generally identical, some herpes simplex virus type 1/type 2 recombinants have been isolated that contain heterozygous terminal repetitious sequences. This heterozygosity affects the isomerization of the genome that normally occurs with herpesviruses (58,59).

Heterozygosis has been observed in some enveloped viruses with haploid genomes; this heterozygosity appears to arise from aberrant packaging of genomes at maturation, resulting in multiploid particles. Complementation between genomes in heteropolyploid particles has been misinterpreted as recombination (56). Multiploid particles have not been documented in nonenveloped viruses, in which the structural constraints of packaging make it unlikely that two genomes could be packaged into a single capsid. Correspondingly, heterozygosis has not been reported in these viruses.

Other Types of Mutations

Differences in virulence or specific pathogenic properties of different virus strains have recently come to be used as genetic markers in studies of the molecular basis of these phenomena in a variety of viruses (324). Interesting examples of this are mutants of lytic viruses selected for their ability to establish a persistent infection in a susceptible host (142,393). Variations in the electrophoretic migration rates of nucleic acids and proteins have also served as useful genetic markers in a number of virus systems.

VIRUS-VIRUS AND VIRUS-HOST INTERACTIONS THAT AFFECT PHENOTYPE

The interactions between different viruses in mixed infections form the foundation for the genetic analyses of recombination and complementation. These methods will be discussed subsequently. First, a number of other interactions between viruses and between virus and host cell will be covered. These can affect the phenotypic expression of wild-type or mutant alleles and often result in phenotypic masking of the true viral genotype. They include phenotypic mixing, interference, and defectiveness among viruses, and integration and persistence between virus and host cell.

Phenotypic Mixing

Phenotypic mixing refers to the process by which individual progeny of a mixed infection contain structural proteins (capsid or envelope) derived from both parental viruses. In the extreme case, a progeny genome genotypically identical to one parent is enclosed within a capsid or envelope entirely specified by the other parent. This mixing of genomes and structural proteins results in virus particles in which the phenotypic properties of the virions do not reflect the phenotypic potential of the enclosed genome. However, on subsequent infection, expression of the genome results in progeny with congruent genotype and phenotype. Thus, phenotypic mixing is a transient phenomenon.

Phenotypic mixing is a relatively common occurrence in naked capsid viruses that are closely related. The ability of capsid proteins to intermix and yield infectious virions implies that the mixed proteins must perform very similar structural roles in the virion. Phenotypic mixing of structural antigens has been observed between closely related viruses such as poliovirus serotypes 1 and 2 (195) and between more distantly related viruses such as echovirus 7 and coxsackievirus A9 (152). In some cases of phenotypic mixing, most of the progeny have completely heterologous capsids (transcapsidation) (286). Pathogenic viruses such as poliovirus and coxsackievirus (140) can transcapsidate *in vitro*, a finding that may have epidemiologic consequences if transcapsidation can also occur *in vivo*.

Phenotypic mixing also readily occurs with some of the enveloped RNA and DNA viruses. In these viruses, mixing often occurs when nucleocapsids of one virus are contained within an envelope specified by the other, a phenomenon often referred to as pseudotype formation. The complementation seen between the replication-defective RNA tumor viruses and their helper viruses (see later) results in pseudotypes, as the envelope glycoproteins of the virions are generally encoded by the helper virus. Vesicular stomatitis virus has been found to form pseudotypes with a number of other enveloped viruses including fowl plague virus (394), avian and murine RNA tumor virus complexes (395), and herpes simplex virus (148). However, there are constraints on the ability of the envelope of one virus to enclose the nucleocapsids of another virus as evidenced by the failure to detect pseudotype formation between vesicular stomatitis virus and Sindbis virus (26). Not all phenotypically mixed virions obtained from mixed infection of enveloped viruses are pseudotypes. Mixing of envelope proteins can also occur and often results in virions that can be neutralized by antisera to both parents (394).

Interference

Several types of interference have been noted with animal viruses (84). Of greatest interest to the geneticist is homologous interference, which occurs within the cell and is exhibited against only homologous virus or closely related viruses.

One well-documented form of homologous interference is that which occurs in viruses that have been serially passaged at high multiplicity (149). In this situation the total yield of virus particles remains relatively constant, but the yield of infectious virus drops with increasing passage. Thus, there appears to be an interference with the growth of the infectious portion of the virus population. Examination of these interfering virus populations has shown that a large proportion of the viral genomes contains deletions or has deleted genome segments (147,367). It has been proposed that the deletion mutants interfere with the growth of complete virus by effectively competing for components of the replication apparatus [e.g., polymerase (149)], giving rise to an ever-increasing proportion of defective interfering (DI) virus. However, it has been suggested that, while DI viruses may compete with complete viruses for components of the replication apparatus, they do not specifically interfere with the growth of complete virus. Ahmed and Fields (3) found that *ts* mutants could be rescued from DI stocks of reovirus and, furthermore, these *ts* mutants all had a strong interfering phenotype. Thus, an interfering phenotype may be determined not by the gene(s) deleted but by mutations in the genes that are present or by both.

Homologous interference is also seen with mutants other than deletions. Certain *ts* mutations have been found to interfere with the growth of wild-type virus at both permissive and nonpermissive temperatures in a number of virus systems (34,170,392). Some *ts* mutants are also capable of interfering with the growth of other *ts* mutants (50), a result that can affect the yields of mixed infection in complementation tests. Interference with the growth of wild-type virus also can reduce the recombinant yields observed in recombination tests with *ts* mutants, which may affect the recombination frequencies obtained (34). The use of interfering viral gene products to investigate gene function is discussed later in this chapter.

Defectiveness

Defective viruses have genomes lacking adequate function in one or more of the essential genes required for autonomous viral replication. Defective viruses require helper activity from another virus genome or virus gene(s) for replication and/or maturation. Here we are concerned with classes of defective viral genomes that are biologically active.

Integrated Defective Genomes

There are a large number of human and animal virus-cell interactions in which the integrated DNA of viral genomes can cause profound and lasting modification of normal cell functions. Integrated defective (or nondefective) viral genomes can cause phenotypic changes in host cells by transduction, *trans*-activation, *cis*-activation, inactivation by integration, genetic conversion by expression of viral integrated viral genes, chromosome breakage and rearrangements at or near sites of integration, and superinfection immunity. The transforming genomes of defective tumor viruses provide an excellent example of tranduction resulting in aberrant expression of cellular oncogenes (see Chapter 9).

Satellite Viruses

Satellite viruses are defective viruses that require the gene products produced by other, often unrelated, viruses. Some share genome sequence homology with their helper virus, but many have unique genome sequences. Two groups of satellites of human viruses have been characterized. The adeno-associated viruses (AAV) are small icosahedral viruses containing a single strand of DNA that encodes three capsid proteins. AAV can replicate only in the nucleus of cells simultaneously infected with adenovirus (or herpesvirus), and they interfere with the replication of the helper adenovirus and with adenovirus transformation of cells (see Chapter 31 in this volume and Chapter 70 in *Fields Virology,* 3rd ed.). A satellite virus called hepatitis delta requires coinfection with hepatitis B virus helper for replication. Hepatitis delta encodes the delta antigen in its RNA genome but requires replicative functions and part of the viral protein envelope from the hepatitis B helper. The delta agent appears to be associated with the most severe forms of acute and chronic hepatitis (see Chapter 36 in this volume and Chapter 88 in *Fields Virology,* 3rd ed.).

Helper-Dependent Defective Interfering Viruses

The defective interfering (DI) populations, which were briefly discussed earlier in terms of their interference, constitute the best studied of the defective viruses. DI particles of viruses accumulate in stocks of virus that are passaged serially at high multiplicity. DI particles have three major characteristics: defectiveness, interference, and enrichment (149). Defectiveness implies that DI viruses have lesions in required genes. For most DI particles these mutations are deletions, but point mutations have been described and *ts* mutants can be considered to be conditionally defective. Interference indicates that the DI particles interfere with the growth of the complete helper virus and with other homologous viruses. The third property of DI virus is the ability to enrich itself at the expense of standard virus. DI particles have been observed in many viruses, but those of vesicular stomatitis virus have been studied in detail because the particles containing deleted genomes are shorter than standard virus and can be purified by rate zonal centrifugation (reviewed in 143, 146, and 147). The DI particles isolated from many other viruses also have the properties of defectiveness, interference, and enrichment and share many of the properties elucidated for DI particles of vesicular stomatitis virus.

DI particles may have biological significance. They are formed during *in vivo* infections (141) and have been postulated to have a role in the limitation of the disease process. DI virus has also been shown to play a role in the establishment and maintenance of persistent infections in tissue culture (142) and *in vivo* (141,143,164). Finally, large concentrations of DI particles are found in live, attenuated vaccines (222) and may have a role in the efficacy of those vaccines.

Many other defective viruses have been described. One of the more interesting forms of defective virus is the large number of noninfectious virus particles that are present in nearly all stocks of animal viruses. In some cases, a large fraction of the noninfectious particles can be rendered infectious (78) by treatments of various kinds, indicating that they are potentially infectious. The nature of the defect that renders these particles noninfectious or potentially infectious is often not known. However, it may reflect the use of suboptimal assay systems in which the efficiency of plaque formation by infectious particles is less than 100%.

Virus-Host Interactions

In addition to the interactions that occur between viruses during mixed infection, there are interactions that occur between the virus and the host cell that can influence viral phenotype. Many of these interactions occur in any host cell, but some of the interactions require specific types of host cell with certain host genotypes. Two of these virus-host interactions are briefly discussed here in terms of genetic phenomena: integration and persistence. The normal parasite-host interactions of viral infection are considered in Chapter 8. The particular virus-host interaction called transformation is considered in Chapter 9 and in chapters on specific oncogenic viruses. Persistence is considered in Chapters 6 and 7 and in chapters on specific viruses.

Integration

The genomes of transforming viruses are generally found to be stably associated with the transformed host cell. For the viral geneticist, this means that the resident viral genomes can influence the activity of wild-type and mutant viruses that infect these cells. In some cases this can lead to interference, as in the resistance to superinfection exhibited by cells infected with retroviruses (see Chapter 26), and in others it can lead to complementation of mutant viruses, as discussed earlier under host-range mutants. In many cases, transformation entails integration of the virus, which is recombination between viral and host genomes. This aspect of transforming viruses is discussed briefly later. Additional information can be found in Chapter 9 and in chapters on the relevant viruses.

Among the DNA tumor viruses, SV40 integration has been the most extensively studied. In SV40, viral DNA sequences are found linked covalently with the host-cell DNA. The amount of DNA integrated varies between cell lines and has been determined to range from <1 to >10 genome equivalents per diploid host genome (20). Transformation can be mediated by subgenomic fragments of the SV40 genome, and these fragments are also found integrated into the cellular DNA. The site of integration is not constant in either the viral or cellular genomes (311). Two mechanisms have been proposed for integration. It could be mediated by legitimate recombination between partially homologous viral and host sequences, or it could occur by illegitimate recombination between nonhomologous sequences. Available data favor the second hypothesis (352). One should note that some DNA viruses transform without integrating (e.g., papillomaviruses), although the presence and expression of episomal DNA is required for initiation and maintenance of the transformed state (356).

The genomes of retroviruses are found in the integrated proviral state in infected and transformed cells. The RNA genome of a retrovirus is reverse transcribed into DNA as one of the initial events of the replicative cycle, and the resultant DNA provirus is then integrated into the host-cell genome. The mechanism of retrovirus integration has three distinguishing features (362): (a) The site of integration into the host chromosome is relatively nonspecific and shares no homology with viral sequences, indicating that integration is by illegitimate recombination. (b) A short host sequence at the integration site is duplicated, giving rise to direct repeats of a few host nucleotides flanking the provirus. (c) The site of integration in the viral sequence is rigidly determined. Thus, retrovirus integration occurs at specific viral sequences but at nonspecific host-cell sequences. This integration behavior, as well as the structure of the provirus, is similar to the structure and integration of the transposable genetic elements found in prokaryotic and eukaryotic genomes and has led to the proposal that retroviruses evolved from these transposable elements (348). The ability of retroviruses to "capture" cellular genes and integrate into the host chromosome presents the interesting possibility that they may act as transducing agents, moving host genetic information in a manner analogous to the transducing phages.

Persistent Infection

Persistent infections have been established with a wide variety of animal viruses. When a lytic virus establishes a persistent infection in a susceptible host, strong selective pressures are exerted on both the virus and the host cell. Either the virus must change so that it no longer injures the host, or the cell must change so that it can support virus replication without being lysed (2). Considerable attention has focused on the first alternative (reviewed in 143 and 393), and mention is made of these kinds of mutations in the section on mutation earlier in this chapter, but very little notice has been taken of the second alternative. Cells "cured" of their persistent infection have been selected from populations of persistently infected cells. These cells are often resistant to reinfection with wild-type virus (364). Spontaneously appearing resistant variants have been selected from populations of uninfected cells (381). These findings suggest that cellular genes determine the susceptibility to viral infection and that mutations in these genes occur spontaneously and are selected during persistent infection. Evidence to support the notion of cellular mutation during persistent infection has come from a careful examination of cells persistently infected with reovirus (2). "Cured" subcultures of several cell lines persistently infected with reovirus were obtained, and they exhibited no evidence of reovirus infection. These "cured" cell lines showed a uniform reduction in their capacity to support the replication of wild-type reovirus, although virus from their persistently infected parental cell lines replicated normally in the "cured" cells. Furthermore, when these cells were infected with the wild type, significant numbers of cells became persistently infected once again, behavior atypical of infection with the wild type. Thus, the persistently infected cells and the virus present in them appeared to have coevolved, producing variants of both virus and host that were better adapted for the persistent state, apparently through regulation of the lytic potential of the infection. Specific aspects of persistent infection with the various viruses are considered in the chapters on the relevant viruses.

MAPPING MUTATIONS AND GENES

The ability to map mutations is valuable for many reasons, not the least of which is the assignment of mutant phenotypes to specific genes, which in fact can identify genes. In viral genetics, mutations were originally mapped using recombination methods. In systems where molecular techniques are not yet available, these methods are still seeing considerable use. Indeed, with segmented genomes, reassortment mapping remains a powerful technique. This section will first briefly review recombination mapping, then reassortment mapping, and will then discuss mechanisms of recombination. It will conclude with a discussion of molecular techniques for mapping mutations and methods such as sequencing and transcript mapping to map genes.

Recombination Mapping

Recombination is a physical interaction of viral genomes in the mixed infected cell. The results of recombination are progeny genomes that contain genetic information in nonparental combinations. In animal viruses this interaction can occur by three distinct mechanisms, with the mechanism depending on the physical organization of the viral genome: breakage-reunion, copy-choice, and reassortment. (These mechanisms are reviewed later after discussions of the uses of recombination.) In systems with unimolecular genomes in which recombination is via a mechanism of breakage-reunion or copy-choice, the probability that a recombination event will occur between two mutations should be proportional to the physical distance separating those mutations on the chromosome. Thus, mutant pairs show a continuous range of recombination frequencies, up to a theoretical maximum of 50%. This range allows the markers to be arrayed onto linear maps that assign a precise location for each mutation relative to the other mutations on the chromosome. The 50% theoretical maximum recombination frequency follows from the multiplicity of genetic exchanges that can occur between distant markers. Only progeny chromosomes having an odd number of exchanges between two markers will have recombinant configuration of those markers, and if many exchanges occur only 50% will have the required odd number of exchanges.

In the viruses with segmented genomes, a different behavior is observed. When pairwise crosses of mutants are made, the frequency of recombinant progeny is either very high or undetectably low. This all-or-none pattern of recombination is interpreted as follows. In the case of no recombination, both mutants are assumed to have mutations in the same genome segment, so that it is impossible to reassort the segments into progeny having a nonparental phenotype (segments not marked with mutations may have reassorted into nonparental combinations but cannot be detected). In the pairs of mutants for which high frequencies of recombinant progeny are observed, the parental mutants are assumed to have mutations in different segments so that reassortment can generate progeny with nonparental configurations of the marked segments (segments not marked can also reassort). No consistent gradient of recombination frequencies is observed, indicating that recombination within segments by the breakage-reunion or copy-choice mechanisms does not occur at detectable levels. In systems that reassort markers, a 50% recombination frequency is expected. However, this frequency is rarely obtained as a number of poorly understood factors can affect the reassortment of genome segments.

Recombination analyses are most easily performed with mutants of the conditional-lethal type. To perform recombination tests, cells are mixedly infected with two mutants and single infections are performed with each mutant parent. The infected cells are incubated at the permissive condition (low temperature for *ts* mutants) for a time sufficient to allow a complete virus replication cycle. The yields of the mixedly infected and singly infected cells are determined at the nonpermissive condition (high temperature for *ts* mutants) to quantify recombinants having the wild-type phenotype, or revertants in the case of the single infections. The mixed infection is also assayed at the permissive condition to determine the total virus yield of the infection. A recombination frequency or recombination index can then be calculated.

At this writing, recombination other than reassortment mapping is rarely used for mapping mutations, as it has been supplanted by more precise and reliable molecular methods. However, it can still be used to test for whether two mutations are likely to be in the same gene (allelic) by virtue of a very low recombination frequency and for rapid construction or resolution of double mutants when there is a convenient selection or screen for the desired progeny.

Reassortant Mapping

The all-or-none nature of reassortment and its use for mapping mutations are illustrated by recombination data from reovirus (284) shown in Table 1. The frequencies of ts^+ recombinants for the 10 mutants tested varied from 0.00% to 10.99%. Although a gradient of reassortment frequencies is observed, the mutants cannot be ordered onto a self-consistent, additive linear map. Furthermore, a statistical analysis of replicate crosses with some of these mutants showed that no linkage between mutants could be detected (85). The absence of linkage is expected for recombination by the mechanism of reassortment. *ts*192, a mutant being mapped, failed to yield *ts*+ reassortants when crossed with the prototype A mutant, *tsA*(350) but did yield ts^+ reassortants when crossed with the prototype mutants of groups B–J. This reassortment result indicates that *ts*192 and *tsA*(350) have mutations in cognate genome segments, and *ts*192 was assigned to reassortment group A. The prototype mutants of groups A–J yielded ts^+ reassortants in all pairwise crosses, confirming the prior assignment of these mutants to noncognate genome segments. In none of the crosses did the frequencies of ts^+ recombinants reach the 25% theoretically expected in a randomly reassorting system (note that reassortment frequencies are reported as $\%ts^+$ reassortants and are uncorrected for the expected double mutant reassortant class). The low fre-

TABLE 1. *Reassortment between reovirus* ts *mutants*

Mutant	Percentage of *ts*⁺ reassortants[a] when crossed with									
	tsA(201)	tsB(352)	tsC(447)	tsD(357)	tsE(320)	tsF(556)	tsG(453)	tsH(26/8)	tsI(138)	tsJ(128)
ts192	0.00	5.29	2.79	4.24	2.35	1.36	2.11	2.27	1.24	1.42
tsA(201)		3.06	3.81	4.43	4.13	2.04	5.90	2.31	0.95	3.42
tsB(352)			3.73	2.08	5.01	2.37	1.89	6.91	3.42	2.30
tsC(447)				7.93	8.31	1.52	8.61	10.99	9.00	2.67
tsD(357)					9.26	3.75	6.37	10.93	5.30	7.01
tsE(320)						5.61	7.80	5.39	2.43	3.31
tsF(556)							1.59	2.28	0.93	3.53
tsG(453)								6.35	3.64	6.78
tsH(26/8)									3.57	3.71
tsI(138)										2.65
tsJ(128)										

[a]Recombination frequencies reported as the percentage of ts^+ reassortants. Uncorrected for unscored double-mutant reassortants.

Modified from Ramig et al. (284), with permission.

quencies of recombinants observed in reassorting viruses are not understood but may be related to other properties of the mutants (see above).

Intertypic Crosses

Genetic studies have often made use of different serotypes or strains of the same virus as the two parents in crosses. Although these crosses generally make use of *ts* mutations in one or both of the parents, the serotypic or strain differences of the two parents introduce a large number of additional useful markers into the cross.

The most frequently used markers are serotype- or strain-specific differences in the electrophoretic migration of nucleic acid. In the segmented genome viruses there are natural polymorphisms that affect the migration rates of genome segments of viruses that can be used as parents. Figure 6 illustrates the use of nucleic acid fragment size or segment migration rates as markers in conjunction with *ts* mutations for segmented genome RNA viruses. Two *ts* mutants are crossed and ts^+ recombinant progeny are selected. The RNAs of the ts^+ recombinants are compared to the RNAs of the *ts* parental strains by electrophoresis. In both cases, the portion of the genome contributed to the recombinant by each parent can be determined. The restriction fragments or genome segments containing the counterselected *ts* lesions can be identified because they are excluded from the ts^+ recombinants. Analysis of the

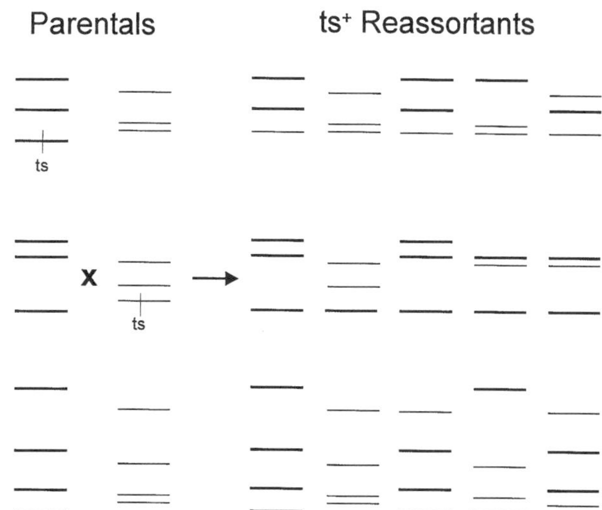

FIG. 6. Diagrammatic illustration of an intertypic cross between *ts* mutants of reovirus. The genome segments are displayed as resolved by polyacrylamide gel electrophoresis, in which each parental serotype has a unique pattern of migration rates for its segments (also indicated by line weight). The parental origin of each genome segment of *ts*+ reassortants is determined by electrophoretic migration. The nonrandom segregation of some genome segments allows deduction of the segment in each parental virus that contains the *ts* lesion. Reassortants with defined constellations of genome segments can be nonselectively isolated and subsequently tested for expression of nonselectable phenotypes, a method useful in identification of genes controlling pathogenesis and other phenotypes.

segregation patterns of fragments or segments of a large number of ts^+ recombinants allows the location of the counterselected mutation in each parent to be determined accurately. This method has provided the main means of mapping *ts* mutations onto genome segments in segmented genome viruses.

Serotype- or strain-specific polymorphisms in the electrophoretic migration rates of viral proteins have also provided markers. Once the parental origin of each genome segment of a recombinant is established, the parental origin of each of the proteins synthesized by that clone is determined. Analysis of a number of recombinant clones shows that the presence of a protein from one parent always segregates with a certain restriction fragment or genome segment from that parent. Thus, the protein in question cosegregates with the fragment or segment in which it is encoded. The physical locations of genes encoding specific viral proteins have been mapped by this method in a large number of viruses.

In general, any property that differs between two parental strains of virus can be mapped to a specific gene(s) through the use of intertypic crosses. The only requirement is that the parental genomes have electrophoretically distinguishable restriction fragments or genome segments so that segregation of the property being mapped can be correlated with segregation of fragments or segments. The intertypic multifactor cross has proved particularly useful in studies of the genetics of virus pathogenesis in both DNA viruses (127,158) and RNA viruses (247,319,324).

Genetic Reactivation

Genetic reactivation represents a special case of recombination. In this case, one or both of the parental viruses are not infectious, but in mixed infection, infectious progeny are produced that contain markers from both parents. These infectious progeny represent recombinants or reassortants in which the inactivating lesion(s) of the noninfectious parent(s) has been removed. Reactivation between an infectious parent and an inactivated parent is called cross-reactivation or marker rescue. Reactivation in the situation in which both parents have been inactivated is called multiplicity reactivation. Multiplicity reactivation is observed only if the inactivating lesions of the two parents are at different sites on the genome, so that recombination or reassortment can produce a viable progeny genome.

Recombination Mechanisms

The mechanisms of recombination have not been investigated extensively in animal virus systems. Although some viruses encode enzymes that could function in a recombination system, mutants with specific recombination defects have not been isolated in animal viruses, which suggests either that recombination is such an essential viral function that one cannot obtain conditional-lethal mutants or, more likely, recombination is mediated by host-encoded enzymes.

"Intramolecular Recombination"

In viruses that have a single genome molecule, including all the DNA viruses and RNA viruses that replicate via DNA intermediates, recombination involves the breakage and reformation of covalent bonds within the nucleic acid, that is, a breakage-reunion mechanism of generating nonparental progeny genomes (referred to here as *intramolecular recombination*) elucidated for bacteriophages and bacteria (226). The various models for recombination in prokaryotic and eukaryotic cells have been discussed thoroughly in a number of reviews and books (70,92,251,277, 331,377,382) and are not considered here. Attention has begun to focus on the DNA of animal viruses as a model system for the study of recombination in animal cells (305,370). The features of intramolecular recombination in DNA animal viruses that have been examined are consistent with the breakage-reunion mechanisms of several different recombination models (22,240,241,366,384,389, 390).

Copy-Choice Recombination

In some virus groups that have a single chromosome composed of RNA, recombination has long been recognized [poliovirus (51,52)] or more recently recognized [aphthovirus (174,175); coronavirus (168,211); alphavirus (125,337)]. Only recently have studies with poliovirus, foot-and-mouth disease virus, and coronavirus provided genetic and biochemical proof of intramolecular recombination (168,174,175,177,211,351). These studies demonstrated recombinant configurations of biochemical markers and RNA sequences, as well as *ts* markers, in putative recombinant progeny. The precise mode of recombination in these viruses is not understood, but the best evidence to date suggests that the polymerase switches template strands during RNA synthesis. This *copy-choice* mechanism does not involve the breakage of covalent bonds. The details of the copy-choice mechanism have not been elucidated, but several studies have suggested that specific sequence homologies or secondary structures are involved in directing the strand-switch by the polymerase (88,301). However, other studies have suggested that the strand-switch is not sequence or structure directed and can occur at many sites on the chromosome (1,168,177). One study (177) indicated that the strand-switch occurred during synthesis of negative-sense RNA in poliovirus; similar information is lacking in other systems.

Reassortment

In viruses with segmented genomes, including influenza viruses, reoviruses, bunyaviruses, and arenaviruses, no covalent bonds are broken in the recombination process. Instead, the genome segments are randomly reassorted in the progeny, a mechanism referred to as *reassortment*. The segmented genome viruses characteristically have particle to plaque-forming-unit ratios too low to be consistent with a random selection of genome segments from the intracellular pool. Thus, there must be a mechanism active in the infected cell to ensure that each particle receives one copy of each genome segment. Reassortment most logically could occur at the stage of morphogenesis at which segments are selected from the intracellular pool for packaging, and thus the reassortment mechanism most likely is intimately associated with morphogenesis. Such an association might explain the failure to isolate mutants of segmented viruses with recombination defects, since morphogenesis would most likely also be affected. The mechanism by which segmented genome viruses reassort segments remains one of the interesting and unanswered questions of animal virology.

Molecular Methods for Mapping Mutations

Most of the molecular methods used for mapping mutations are essentially identical to those used to introduce mutations outlined previously in the chapter, except that one introduces wild-type sequences into a mutant virus rather than mutant sequences into a wild-type virus. These methods are formally equivalent to genetic reactivation discussed earlier in which a fragment of wild-type DNA can be considered as an inactivated virus and an infectious mutant viral genome can be considered as an active virus. For this reason, restoration of a wild-type phenotype using cloned DNA fragments is generally given the term from genetic reactivation, *marker rescue*. An example of marker rescue for a genome that can be cloned as an infectious molecule would be a mutant papova virus. In this case, one can clone the mutant viral DNA, replace a small segment that contain the mutation with wild-type sequences, and transfect the new construct to give rise to a restored virus, which should be wild type for the relevant phenotype. Similarly, for large DNA viruses one can map the mutation by homologous recombination, for coronaviruses by targeted recombination, etc. For mutations in *trans*-acting genes in many virus systems, one can also use complementation by cloned DNA (see later) to show that the phenotype is due to a mutation in a specific gene. Regardless of the mapping method used, the final step is to sequence the mutant DNA corresponding to that replaced in the restored virus or in the complementing DNA to determine the location of the mutation to the nucleotide. Of course, these techniques can be used to map engineered or nonengineered mutations, and, many of these methods were first developed before methods for engineering mutations.

Molecular Approaches to Gene Mapping

Previous sections described classical genetic and molecular genetic methods for mapping mutations, which can be used to locate genes. The next section will cover molecular approaches to mapping genes, which have proved to be very useful. The first approach to be discussed will be DNA sequencing. Then, methods for mapping transcripts and polypeptides will be reviewed. Other mapping approaches, which will not be covered in detail, include the use of UV irradiation (6,46,99) or protein synthesis inhibitors (292) to map gene order.

Sequence Analysis

A prerequisite for detailed analyses of viral functions is the availability of sequence information. Complete sequences have been deciphered for many viral genomes, the largest being the 230 kilobase pair (kbp) human cytomegalovirus sequence (36). Much information can be gained from DNA sequencing:(a) Long translational reading frames uninterrupted by stop codons predicted from the DNA sequence (open reading frames, orfs) suggest the presence of protein-coding sequences. The likelihood that an orf actually encodes protein increases if the codons for given amino acids occur with the frequency used by authentic protein-coding sequences of the virus. Even short orfs, which can overlap other protein-coding sequences, can encode protein, particularly via mRNA splicing. A virus that makes surprisingly extensive use of short overlapping orfs in this way is human immunodeficiency virus (38). (b) Hypotheses about the function of predicted protein sequences can be formulated when the sequence is similar to that of a protein with known function. For example, sequence analysis of vaccinia virus revealed an orf with similarity to known growth factors (17,25,291). This predicted that vaccinia virus would encode a growth factor, which was borne out in subsequent studies (176,338,357). Similarly, it is possible to predict that specific portions of a predicted protein have particular functions, such as signal sequences for targeting proteins to the endoplasmic reticulum, transmembrane sequences, glycosylation sites, nucleotide binding sites, etc. (c) It is possible to identify short sequence motifs that serve as signals in gene expression. Methionine codons likely to be used to initiate translation are found at the beginnings of orfs, embedded in a consensus sequence GCCGCCA/GCCAUGG (184–186). Sites of mRNA cleavage and polyadenylation are almost always signaled by AAUAAA followed by certain less obvious signals (91,272). The starts of transcription by RNA polymerase II are usually about 25 bp downstream of canonical "TATA boxes" (12). Other *cis*-acting elements, such as promoter and enhancer elements or origins of replication, can also sometimes be predicted from their sequence similarity to elements with known functions.

Transcript and Polypeptide Mapping

All of the methods for mapping transcripts and polypeptides onto cloned DNA segments were first developed using virus systems. Details of performing these procedures can be found in various laboratory manuals (e.g.,5). The information gained from transcript and polypeptide mapping can be invaluable in interpreting sequence data. The locations of mRNA 5' ends, splice sites, etc. can often be crucial in elucidating viral regulatory mechanisms. To determine whether certain open reading frames are spliced together to encode a virus protein usually requires S-1 mapping, sequencing of cDNA clones, or similar methods. Mapping the 5' end of mRNAs is crucial to determining if all of a long open reading frame will contribute to the final protein product. *In vitro* translation methods, such as hybrid-arrested and hybrid-selected translation (257,294) or *in vitro* transcription-translation (see below under heterologous expression), can be used to map proteins found in infected cells to specific regions of viral genomes.

For all their power, however, DNA sequencing and transcript and polypeptide mapping techniques can only lead to hypotheses regarding function or regulation. They can, nevertheless, greatly aid in the design of experiments to test these hypotheses.

APPROACHES TO GENE FUNCTION

There are many ways to analyze gene function. The traditional approach is classical complementation analysis, which will be the subject of the next section of this chapter. This will then be followed by molecular methods for studying viral genes outside the viral genome, which can offer a second route to complementation analysis, and can permit the detailed analysis of viral *cis*- and *trans*-acting functions. Finally, molecular approaches to gene function using interfering gene products will be covered.

Complementation

Complementation refers to the interaction of viral gene products in mixedly infected cells that results in the yield of one or both parental mutants being enhanced while their genotypes remain unchanged. This definition reflects the fact that one or both of the viruses (mutants) provides a gene product in which the other is defective, allowing one or both of the mutants to grow in the mixedly infected cell. It follows that if both viruses are defective in the same gene product (function), neither parent will be able to provide the missing function and no complementation will occur. Thus, the complementation test can be used to place viral mutants into functional groups. Two mutants that are unable to complement are generally considered to have defects in the same gene and gene product, whereas two mutants able to complement are considered to be defective in different genes and gene products. Division of mutants into complementation groups can be done without any knowledge of the specific biochemical defects of those mutants. Theoretically, there can be as many complementation groups as genes. However, the absolute lethality of some mutations, or the nonessential function of others, often results in fewer complementation groups than genes.

Complementation tests are most easily performed with mutants of the conditional-lethal type. To perform complementation tests, cells are mixedly infected with two mutants and, as controls, cells are infected with each of the mutants alone. The infected cells are incubated at the nonpermissive condition (high temperature in the case of *ts* mutations). After sufficient time for viral growth, the total yields of the mixed and control infections are determined at the permissive condition (low temperature in the case of *ts* mutations). A complementation index is calculated as the ratio of the yield of the mixedly infected cells to the sum of the yields of the single infection controls. Complementation is measured as an enhanced yield from the mixedly infected cells, and any complementation index greater than 1.0 indicates complementation. To discount the effects of multiplicity and other factors on the yield of the mixed infection, complementation indices of 2.0 or greater are generally used as the cut-off point between a positive and a negative complementation test. Since virus growth occurs in the complementing situation, recombination can occur, giving rise to wild-type recombinants in the yield of the mixed infection. Terms are often incorporated into the equations for calculation of complementation indices to correct for the contribution of recombination to the enhanced yield.

Two types of complementation have been observed. The most typical is nonallelic (intergenic) complementation in which mutants defective in different functions assist each other in multiplication by supplying the function defective in the other virus. Nonallelic complementation is useful for determination of the number of different genes identified by mutation in a given mutant collection and has been used extensively in genetic characterization of animal virus mutants. Allelic (intragenic) complementation has been observed much less frequently and is usually thought to occur when the gene product defective in both parents is a multimeric protein (89). In this case, the two parents have defects in different domains of the same protein. If the multimeric functional form of the protein is composed of subunits from a single parent, the multimer is nonfunctional. However, if the multimer is composed of subunits from both parents, the multimer is able to assume a functional configuration and complementation is observed. Allelic complementation has been observed occasionally in animal virus mutants (e.g. 216,327).

Complementation between mutants can be asymmetric. In this case, the complementation is one way and only one of the parental mutants is represented among the progeny. Asymmetric complementation has been observed between mutants of poliovirus (50). In some virus systems complementation has been observed to be very inefficient (34), although the failure to complement may be secondary to other properties of the mutants (see earlier). In other virus systems complementation is extremely efficient, so efficient that measurement of other viral interactions is difficult against the background due to complementation (72).

When viral mutants are collected, complementation is often the first genetic test carried out with the mutants. Complementation tests are easy to perform and rapidly divide the mutants into functional groups that can be studied further by other means. Although complementation analyses have been carried out with virtually every mutant collection, they have been exceedingly useful in virus groups in which recombination analysis is not possible due to the absence of significant recombination. Thus, complementation analyses have been particularly useful with mutants of togaviruses (263), rhabdoviruses (269,270), and paramyxoviruses (24). Table 2 contains complementation data derived from standard complementation tests with Sindbis virus (263). On the basis of these data, the mutants tested were divided into six complementation groups. Several of the problems frequently encountered in complementation analysis are illustrated by these data. For example, *ts*11 is assigned to a complementation group designated A'. This is due to the variable results obtained with this mutant. It was originally assigned to an independent group, then was included with group A, and finally returned to an independent group. Thus, the status of *ts*11 and complementation group A should be regarded as unsettled. The group B mutant, *ts*6, complements all other mutants to a much greater degree than is seen with other complementing pairs. Among the group A mutants the degree of complementation is quite variable. These last two points illustrate the problems of defining a significant degree of complementation and distinguishing that level from an absence of complementation.

Complementation is observed in a number of natural situations in animal virus systems. As discussed previously in this chapter, defective viruses that arise during passage of viruses are maintained in the virus population by complementation with the few viruses with complete genomes that are also present in the population (helper virus). Another situation involves the acute transforming retroviruses, most of which are replication defective due to deletions of genes encoding replication functions (200). Stocks of these viruses contain, in addition to the replication-defective transforming virus, a related virus that is replication competent and transformation defective. This second virus rescues the replication-defective virus by complementa-

TABLE 2. *Complementation between* ts *mutants of Sindbis virus*

		Complementation level[a] in mixed infection with											
Complementation group	Mutant	ts21	ts19	ts17	ts4	ts24	ts11	ts6	ts2	ts5	ts10	ts23	ts20
A	ts21		0.2	0.1	0.1	1.5	17.6	200	43	27	24	nd	nd
A	ts19			0.4	0.5	3.2	120	485	nd	16	19	43	12
A	ts17				nd	4.5	30	393	25	31	17	19	23
A	ts4					0.4	9	127	19	nd	27	nd	8
A	ts24						4.2	364	22	29	18	16	5
A'	ts11							280	1.8	50	43	nd	12
B	ts6								62	93	64	68	21
C	ts2									1	36	27	5.2
C	ts5										345	61	8
D	ts10											0.7	16.7
D	ts23												8.5
E	ts20												

[a]The complementation level is the ratio of the viral titer produced by a mixed infection of two *ts* mutants at the restrictive temperature to the sum of the viral titers produced in parallel single infections.

nd not done.

From Pfefferkorn (263), with permission.

tion, keeping it present within the virus population. In some cases, unrelated viruses are able to complement. For example, human adenoviruses are unable to complete the replication cycle in certain monkey kidney cells. However, in cells coinfected with SV40 and adenovirus, yields of both SV40 and adenovirus are synthesized (245), indicating that these two unrelated viruses can complement with respect to their host restriction defects.

Studying Viral Genes Outside the Viral Genome

Thus far, this chapter has focused on studying viral genes in the context of the viral genome. Before the advent of recombinant DNA methods, it was often impossible to test the function of certain viral proteins or *cis*-acting elements in the absence of other viral or cellular proteins. In many cases, such studies of viral genetic elements in isolation is not only easier but is necessary for elucidating mechanism. This is especially true for essential *cis*-acting elements and proteins in which many mutations in the context of the viral genome simply ablate viral replication, without shedding light on why.

The next several sections will focus on methods for studying viral genes outside the viral genome, first in mammalian cells. This permits assays of complementation by cloned genes as well as mechanistic studies. These methods also frequently make use of viruses as vectors for efficient expression, and several viral vectors will be covered. These vectors are also of interest in terms of their potential utility as vaccines and delivery of gene therapy, and examples of these uses will be provided. Finally, methods for studying viral genes in heterologous systems will be addressed.

Expression of Viral Genes in Mammalian Cells

Mammalian cells are the normal hosts for most of the viruses covered in this textbook. The introduction of isolated viral genetic elements into mammalian cells can entail either transient expression or expression after the genetic material is stably associated with the cell. Either transient or stable expression can be achieved by direct introduction of the genetic material into cells or by use of heterologous virus genomes as vectors. An outstanding use of these methods is for complementation analysis. This will be discussed first in general terms. Then, transfection methods will be covered, along with examples of their use for mechanistic studies. Then, viral vectors will be surveyed, with examples of their use not only for basic research, but also their potential for clinical application.

Complementation Using Cloned Sequences

Classical viral complementation analyses, described earlier, where cells are infected with a pair of virus mutants under nonpermissive conditions, are invaluable. However, more sophisticated analyses have been permitted by the advent of methods for introducing cloned segments of viral genomes into cells by transfection or by infection with a heterologous virus. This approach can be formally considered in terms of a classical complementation assay with the cloned segment taking the place of one of the paired mutants. In this case, the cloned segment mimics a mutant in which all viral sequences are deleted except the cloned segment. Indeed, the major difference between this approach and classical complementation is that rather than mixedly infecting cells with the two mutants, one introduces one mutant genome by normal infection and the other by transfection or a viral vector. A major advantage of the newer approach is that when a cloned segment can complement a virus mutant, that permits assignment of the mutant function to the cloned segment. A second major advantage is that this approach permits one to propagate viruses with otherwise lethal mutations without the use of helper viruses, for example, on cell lines stably containing the cloned segment of viral DNA.

Additionally, methods such as those described under the heading of engineered mutations can be used to construct homologous viruses that permit complementation analyses and assignment of functions to specific cloned segments. One method is to engineer viruses that themselves carry large deletions of viral material and to see if these can complement specific mutants, in effect as helper viruses. A second approach is to engineer viruses with a segment introduced into a new site in the viral genome. In this case, a partially diploid virus is created such that if one segment is mutant, the other segment can complement the defect.

Transfection

The concept of transfection—the initiation of infection by introduction of viral nucleic acids into cells—was presented in the section on biological activity of recombinant DNA copies of viral genomes earlier in this chapter. The term *transfection* is now widely used to describe the introduction of any nucleic acid into animal cells. Naked nucleic acids enter cells and exhibit biological activity with very poor efficiency. Several procedures have been developed to enhance this process. The most common procedure used involves forming a coprecipitate of nucleic acid with calcium phosphate, which cells apparently take up more readily than nucleic acids alone (115). Other methods include using polycationic compounds, such as DEAE-dextran and poly-L-ornithine, sometimes in combination with compounds such as chloroquine, which evidently inhibit lysosomal degradation of nucleic acids (206); natural or artificial lipid membranes to fuse with cells (83,93,317,378); microinjection using pressure (4,30) or iontophoresis; electroporation (243,267), in which nucleic acids evidently enter cells through holes created in membranes by electric discharge; scrape loading and sonication loading, in which mechanical shear permits DNAs to enter cells; and high-velocity microprojectiles (82,180)

Transient Expression After Transfection

Given the proper regulatory signals, virtually any gene can be expressed efficiently for a short time after transfection. This expression, which usually peaks at 48–72 hours post-transfection, can be monitored in what are called *transient expression* assays by complementation (274), immunofluorescence (13), immunoprecipitation (275), enzyme assays (112), or by hybridization techniques used to detect transcripts (123). Transient expression assays combined with immunofluorescence or cell-fractionation methods can be used to test the ability of viral proteins to achieve their correct subcellular localization (273). Although expression often occurs in only a small fraction of the cells, it can be efficient enough to measure substantial increases or decreases in gene expression. This approach has been especially useful in studying the effects of *cis*-acting elements and *trans*-acting factors from viral systems.

Analysis of Cis-*Acting Elements*

Transient expression was used to help elucidate how poliovirus RNA is able to initiate translation efficiently despite the lack of a 7-methylguanosine cap on its 5' end. Pelletier and Sonenberg (260) constructed a plasmid in which the 5' untranslated region (UTR) of poliovirus RNA was inserted upstream of the reporter gene encoding bacterial chloramphenicol acetyltransferase (CAT) and downstream of another reporter gene encoding herpes simplex virus thymidine kinase (*tk*). In a control plasmid, the 5' UTR was omitted. These constructs were transfected separately into cells in which they both expressed *tk* and CAT 2 days post-transfection. However, when these transfected cells were infected with poliovirus, only CAT was detected and only from the plasmid containing the 5' UTR. These results and others led to the proposal that the poliovirus 5' UTR directs internal initiation of translation, unlike most eukaryotic mRNAs. They also illustrate the utility of experiments in which viral elements are removed from the parent virus and tested for function. Performing a similar experiment with poliovirus itself would have been much more difficult because a mutant virus lacking the 5' UTR would simply not grow.

Analysis of Trans-*Acting Factors*

Transient expression has been used to show that virus gene products can regulate viral gene expression independent of other viral genes. An early example of this was the demonstration that the adenovirus ElA gene increased the expression of the adenovirus 72 kD DNA-binding protein gene when plasmids containing the genes were cotransfected into cells (150). In that case, a body of earlier work, including analysis of adenovirus mutants, had implicated ElA as a positive regulatory factor. Transient expression approaches have subsequently identified viral regulatory genes in the absence of prior standard genetic analysis. It has been critical, however, in several of these cases, to ensure that these genes function in the authentic context of the viral genome as regulators.

Stable Expression After Transfection: Cell Transformation

The ability of cells to maintain and express transfected DNAs in a stable manner, evidently through integration of the transfected DNA into cell chromosomes, was first demonstrated when transfection of papova and adenovirus DNA fragments led to changes in the growth properties of cells (115). A few years later, thymidine kinase-negative (tk^-) cells, which cannot grow in HAT medium (growth medium containing hypoxanthine, aminopterin, and thymidine) (201), were converted to HAT resistance by transfection with herpes simplex virus (HSV) DNA, which contains a functional *tk* gene (6,210,379). Both of these experiments are paradigms of *cell transformation*, in which acquisition of new DNA sequences changes the phenotype of a cell (261), which should be distinguished from changes in cell growth characteristic of *growth transformation* or *oncogenic transformation*. The latter arise from a variety of mechanisms, only one of which is the acquisition of new DNA sequences.

Cotransfection with Selectable Markers

Most viral genes are not known to change the phenotype of cells in a way that permits selection. To obtain cell lines that acquire and express these genes stably, it is necessary to introduce them with a selectable marker. Early studies using calcium phosphate–mediated transfection of tk^- cells with the HSV *tk* gene and another DNA demonstrated that selection for retention and expression of the *tk* gene leads to the isolation of cell lines that also contain the other DNA. Even though the *tk* gene and the other DNA initially are not covalently joined, during the transfection process they become physically linked (262). This procedure has been extended to cells that do not have an unusual genotype, such as tk^-, by using dominant selectable markers. These markers include the prokaryotic xanthine guanine phosphoribosyl transferase (*gpt*), tryptophan synthase (*trp*B) and histidinol dehydrogenase (*his*D) genes (131,239); prokaryotic and eukaryotic dihydrofolate reductase (*dhfr*) genes; and prokaryotic aminoglycoside phosphotransferase genes that can inactivate antibiotics such as G418 and hygromycin, which block eukaryotic protein synthesis (45,160,314). This last class has seen the most widespread use thus far.

Cell lines containing viral genes constructed by cotransfection with a selectable marker can vary considerably in their constitutive expression, from high levels to undetectable levels. Low levels of expression can be desirable if the viral gene product is toxic to the cell or interferes with virus replication. In such cases, inducible promoters such as the mouse mammary tumor virus (MMTV) long terminal repeat, which is glucocorticoid inducible, or linked amplifiable genes such as *dhfr* can be used to obtain appropriate levels of expression in given situations. Frequently, the best inducible promoter for a viral gene is the gene's own promoter, which contains signals that respond to viral *trans*-acting factors so that it is induced upon virus infection.

Uses of Cells Transformed with Viral Genes

There are many uses for cell transformation. A major use is to create host cells for the isolation and propagation of mutant viruses defective in specific viral genes. The 293 cell line (116), which was oncogenically transformed with adenovirus DNA, has been indispensable in genetic studies of adenovirus mutants with lesions in the ElA and ElB genes (14,166). Cell lines that express viral functions by cotransforming cells with viral DNA fragments and selectable markers are now widely used for genetic studies and have permitted the isolation of viruses with lesions in genes not identified by studies of conventional markers such as temperature-sensitive mutations, as described in the section on engineering viral mutants. Although a potential problem with this approach is regeneration of wild-type virus due to recombination of the viral mutant with the viral DNA carried by the host cell, various strategies can be employed to minimize or eliminate such recombination.

Mammalian Viral Vectors: General Considerations

Viruses provide an excellent way to introduce viral or cellular genes into mammalian cells. They offer the advantage of being able to introduce multiple copies of the gene of interest into nearly every cell in a culture. Depending upon the virus used as a vector, the gene can be expressed transiently or stably. As reviewed earlier, viral vectors can be used, much as transfection methods, for complementation analysis. They can also be used to express large amounts of a protein product; indeed, with certain systems such as vaccinia virus, expression levels can be nearly as high as the best nonmammalian expression systems. Viral vectors are especially useful, however, for their ability to infect mammalian hosts. This has been used in two major areas: (a) to study immune function with applications to the development of vaccines and (b) to deliver genes to specific mammalian cells with application to the development of therapies. Examples of each of these uses of viral vectors will be provided later.

Viral Genomes as Vectors: Transient Expression

Viruses commonly used for transient expression have been derived mainly from papovavirus, adenovirus, herpesvirus, and poxvirus.

Papovavirus Vectors

The earliest virus vectors were derived from the papovavirus, SV40, and as this virus undergoes lytic replication, the vectors express their genes transiently (128,238). These vectors suffer several disadvantages. Foreign DNA is inserted into essential genes so that propagation of the vectors requires helper virus. The maximal size of DNA that can be packaged by SV40 is not much greater than the 5,243 base pairs that normally constitute the SV40 genome, constraining the size of the foreign DNA that can be inserted. Nevertheless, SV40 vectors have been highly valuable, in part because of the ease of mutagenesis in this system (see section on introduction of mutated sequences into viral genomes). One example is the use of these vectors to examine the influenza hemagglutinin. A cDNA encoding the hemagglutinin was inserted in place of SV40 late genes and the recombinant virus was propagated with a helper virus deleted in the early region. Cells infected with the recombinant virus stock expressed hemagglutinin efficiently on the cell surface in a manner indistinguishable from authentic hemagglutinin (104). Studies of a variety of mutated hemagglutinin genes expressed from these vectors have dissected the roles of various parts of the molecule in the biosynthesis, assembly, maturation, intracellular transport, and function of this protein (67,105), which have helped make it a paradigm in studies of glycoproteins.

An alternate use of the SV40 system has been a set of cell lines, called COS cells, which contain SV40 genomes that lack a functional origin of replication (109). These cells express the SV40 T antigen and thus can amplify transfected molecules containing a functional SV40 origin. They thus serve as an efficient host cell for transient expression.

Large DNA Virus Vectors

Vectors have also been developed from the larger DNA viruses of the adenovirus, herpesvirus, and poxvirus families. An advantage of these viruses as expression vectors is the amount of foreign DNA that can be packaged. This is because they can package genomes that contain substantial insertions of DNA and because genes that are not essential for virus growth in cultured cells can be deleted, providing even more room for foreign genes. The deletion of such genes is also a strategy for attenuation of pathogenicity.

Adenovirus Vectors

Most adenovirus vectors have made use of 293 cells, which permit one to substitute the viral E1 region with the foreign DNA of interest yet propagate the recombinant virus to high titers in the absence of helper. The use of the viral late promoter and tripartite leader allow the foreign gene to be expressed at the high levels typical of the viral late structural proteins (15). Adenovirus vectors are prepared either using a strategy similar to that outlined in Fig. 1 or through the use of infectious bacterial plasmids (106). As adenoviruses infect many tissues in mammals, including nondividing cells, they have been explored as vectors for gene therapy. A potentially exciting use takes advantage of adenoviruses' tropism for the respiratory tract to deliver the cystic fibrosis transmembrane conductance regulator gene to airway epithelium in rats (304), which may lead to a therapy for cystic fibrosis in humans. Replication-deficient adenovirus vectors have also been used for stable expression (see later).

Herpesvirus Vectors

Herpesvirus vectors can be readily constructed by the same strategies used to introduce mutations into these viruses. A second strategy for using herpesviruses as a vector has been to construct subgenomic replicons (330,334), which can be propagated both in mammalian cells with the aid of helper virus and in *E. coli*. Both strategies have been used with herpes simplex virus (HSV) to deliver foreign genes to cells *in vitro* and in mice (102,138,249,326). A potential clinical use for herpesviruses takes advantage of certain virus mutants to be attenuated for pathogenicity in the nervous system yet still replicate in and kill dividing cells. This has led to the successful therapy of nervous system tumors in experimental animals (215). Another potential use of herpesviruses is as polyvalent vaccine vectors, stemming from their medical and agricultural importance. Herpesvirus vectors also have potential for stable expression in neuronal cells (see later).

Poxvirus Vectors

The prototype poxvirus, vaccinia virus, has become a very popular vector (209,236). Recombinants are readily constructed by marker rescue procedures, which do not require the preparation of infectious DNA, and several schemes have been developed to facilitate the selection and screening of recombinant viruses (81,94,253). Since vaccinia virus uses its own transcriptional machinery, it is necessary to link foreign genes to vaccinia virus transcriptional signals. An interesting alternative strategy has been to create viral recombinants in which a bacteriophage RNA polymerase is expressed under the control of vaccinia virus transcriptional signals while the foreign gene of interest is placed under the control of the cognate bacteriophage promoter (98).

Like several other vectors, vaccinia virus vectors have proven useful for overexpression of heterologous viral proteins (339), for which they are probably the best suited of the mammalian viruses, and for transient expression to permit complementation of virus mutants, especially those of negative-strand RNA viruses. One example was provided earlier in the discussion of the expression of cloned VSV genes that permit the generation of infectious defective-interfering particles (258). They are also particularly well suited as tools for studies of cellular immunity against other viruses. Vaccinia virus recombinants can infect a wide range of cell types and express foreign viral antigens on the cell surface in proper association with histocompatibility antigens. This permits the generation of a library of target cells for recognition and lysis by autologous cytotoxic T lymphocytes (CTLs). Moreover, the same recombinants can be used to vaccinate animals in an effort to induce cell-mediated immunity *in vivo*. The specificity of the induced CTLs can then be tested on various target cells infected either with various recombinants or different strains of the appropriate virus. For example, vaccinia virus recombinants bearing influenza virus proteins have been used to demonstrate that influenza hemagluttinin induces CTLs that distinguish among various influenza strains and that the influenza virus nucleoprotein is a major target for cross-reactive antiinfluenza CTLs (10,11,388). The ability to induce high levels of both cellular and humoral immunity, combined with the successful history of vaccinia virus as a vaccine against smallpox, have excited interest in vaccinia vectors as polyvalent vaccines for clinical use, although concerns regarding safety remain. It is likely that the applications of vaccinia virus vectors will continue to expand.

RNA Virus Vectors

With advances in molecular methods to engineer mutations into RNA viruses have come advances in the development of vectors derived from RNA viruses. One early example was the use of defective Sindbis virus genomes with viral sequences replaced by foreign genes, which can be propagated with the aid of helper virus (197). Subsequently, nondefective vectors have been developed. These and other togavirus vectors have several advantages, including ease of manipulation, the rapid generation of high-titered stocks following transfection, robust protein expression, and wide host range. One promising use of these is to study cellular immunity, providing an alternative to vaccinia virus (126). Indeed, several RNA viruses have been explored as potential vaccine vectors based both on their ability to express foreign epitopes and clinical experience with live or killed vaccines derived from these agents (80,198).

Viral Genomes as Vectors: Stable Expression

Whereas the expression vectors discussed earlier are derived from lytic viruses that associate only transiently with cells before killing them, another class of expression vectors has been derived from nonlytic viruses that associate stably with cells. Although these latter vectors have not been used extensively for overexpression of foreign proteins, they permit long-term studies that cannot be performed with those derived from lytic viruses.

Bovine Papillomavirus and Epstein-Barr Virus Vectors

One such stable expression vector is derived from bovine papillomavirus (BPV). BPV contains a small circular DNA genome that will oncogenically transform cells, providing a phenotypic selection. Only a fraction of the genome is required for oncogenic transformation of cells; the remainder can be replaced with the foreign gene of interest and an antibiotic resistance marker and an *E. coli* origin of replication, permitting propagation in *E. coli*. As circular episomes, BPV vectors are "shuttle vectors," which can be easily isolated away from mammalian chromosomal DNA and recovered in *E. coli* (193,203,231). Because of the episomal character, there are no differences in chromosomal flanking sequences to affect levels of expression, in contrast with stable transformation after transfection. However, other factors can affect levels of expression from BPV recombinants (248). Another drawback to BPV vectors is that currently they are not packaged into infectious virus and must be introduced into cells by transfection methods. A stable virus expression vector with similar properties is based upon the Epstein-Barr virus (EBV) replicon that permits episomal replication in cells. The replicon consists of a *cis*-acting element, oriP, and a *trans*-acting factor, EBNA-1 (387). An EBV vector has been developed for cloning and expressing cDNAs (214). EBV and BPV vectors have been useful in studying mutagenesis (71,73,133).

Retrovirus Vectors

Expression vectors derived from retroviruses have been used extensively to create cell lines expressing particular proteins. Retrovirus vectors can be packaged efficiently so that every cell in a culture is infected. In retroviral vectors, the virus protein-coding sequences are replaced by foreign genes, while *cis*-acting sequences for priming reverse transcription, for transcription, and for packaging are retained. This limits the amount of foreign sequences that can be packaged to <10 kilobase pairs. Packaging cell lines that contain mutant retroviruses with defects in *cis*-acting elements have been constructed. These cell lines express all the necessary *trans*-acting replication and packaging functions but produce little or no infectious virus (212). Transfecting packaging lines with bacterial plasmids that contain the retrovirus expression vector with the foreign gene of interest yields stocks of replication-defective viruses that will transduce the gene to new cells. The host ranges of the vectors are dependent upon the virus envelope genes resident in the packaging cell lines. Several clever variations have been incorporated into this strategy to lower the background of infectious virus arising by reversion or recombination; to permit easy selection or screening; and to permit recovery as a shuttle vector (57,328,375). When genes that are normally spliced are inserted into a retrovirus vector, cDNA copies are formed during generation of the virus. This strategy was used to obtain cDNA copies of three different transcripts that arise from the adenovirus ElA gene (33).

Given the natural ability of retroviruses to transduce oncogenes, it is not surprising that retroviral vectors are particularly well suited for studies of the mechanisms of oncogenic transformation, especially when used to deliver mutagenized oncogenes (32,173,265). Retroviral vectors are also being used extensively in *in vivo* experiments to study cell lineages during mammalian development (268) and, because of their high efficiency, stability, and lack of spread, in the development of human gene therapies such as gene replacement (227). An additional therapeutic use exploits the fact that many retroviruses can only productively infect dividing cells. This has led to strategies in which tumors have been transduced with retroviral vectors that express the HSV thymidine kinase, which can activate the cytotoxic drug, ganciclovir. This appears to kill the transduced tumor cells and, by a "bystander" mechanism, neighboring dividing cells, while sparing normal nondividing tissue (54,343).

Parvovirus Vectors

There has been growing interest in parvoviruses as vectors, especially for gene therapy. Some of this effort has involved autonomous parvoviruses, especially because of their preference for neoplastically transformed cells, which may form the basis for tumor therapy (309). Much effort has focused on adeno-associated virus (AAV), which integrates at specific sites to establish a latent state with high frequency, is nonpathogenic, and infects a broad range of cells and has several other appealing properties (see Chapter 31). An early report on the use of AAV showed that it could transduce an antibiotic resistance marker into cells in culture (137). Syndromes under investigation for gene therapy include hemoglobinopathies, in which an AAV vector has permitted regulated high level of expression of a human γ-globin gene in an erythroid cell, which has proven difficult with retroviral vectors (372).

Herpesviruses and Adenoviruses

The ability of herpesviruses and adenoviruses to infect nondividing cells, and especially the ability of α-herpesviruses to establish stable latent infections in the nervous system, has engendered excitement about their potential for basic studies of neurophysiology and for gene therapy. A number of efforts have been made in this direction using attenuated herpesvirus recombinant and amplicon vectors and replication-deficient adenovirus vectors (64,102,138, 194). At this writing, numerous practical aspects with these vectors remain to be resolved.

Expression of Viral Genes in Heterologous Systems

It is often difficult or impossible to test the functions of certain viral proteins in the absence of other viral or cellular proteins or to prepare monospecific antisera against them because their low abundance in virus-infected cells prevents their purification. Moreover, detailed biochemical and biophysical studies require milligrams of protein. Recombinant DNA technologies have permitted virologists to overcome these problems via the ability to express cloned viral genes efficiently in heterologous systems. Rather than enumerating all such expression systems, brief discussions of three popular nonmammalian systems and certain *in vitro* methods will be presented. Certain mammalian systems, such as vaccinia virus vectors, can be used in a similar fashion and have been discussed already.

Expression in Escherichia coli

Ease of manipulation, availability of vectors that permit high levels of expression, large capacity bacterial fermenters, and a wealth of genetic and biochemical information have made *E. coli* the first heterologous expression system to which most virologists turn. It is possible to express viral proteins in *E. coli* as fusion proteins, in which viral coding sequences are fused to a prokaryotic gene, such as the *lac*Z gene, which encodes β-galactosidase, or to express authentic viral proteins by fusing viral coding sequences to prokaryotic control sequences.

Gene Fusions

Gene fusions can be constructed quickly. Vectors that allow screening for insertion and expression of random orfs as fusion proteins by restoration of the reading frame of the *lacZ* gene have been developed (118). Fusion proteins are sometimes more stable than when the viral protein is expressed alone. They are particularly useful for raising monospecific antisera that can then be used to identify and study viral proteins in infected or transformed cells. Moreover, they often retain activities expected of authentic proteins.

This approach is illustrated by studies of one of the protein components of the Epstein-Barr virus nuclear antigen (EBNA), EBNA-1. EBNA was originally identified in EBV-immortalized human lymphocytes by its reaction with human sera in immunofluorescence experiments (289). Attempts to purify EBNA proteins were hampered by the scarcity of these proteins and by the lack of sensitive functional assays (321). Early molecular genetic approaches included the identification and mapping of EBV transcripts expressed coincident with EBNA (134). Orfs contained within the regions to which the transcripts mapped were expressed in *E. coli* as fusion proteins (136). Antisera raised in rabbits against these fusion proteins identified various proteins in EBV-immortalized lymphocytes (135,136). One of these proteins, EBNA-1, varies in apparent molecular weight from 78 to 85 kD, depending on the EBV strain used to immortalize the lymphocytes; this size difference correlates with differences in numbers of DNA repeats within the relevant region of each strain's genome (135,136). These data effectively mapped part of the EBNA-1 orf on EBV DNA. Subsequently, it was found that EBNA–β-galactosidase fusion proteins can bind specifically to regions of EBV DNA (288), corresponding to portions of an origin of replication whose activity is dependent upon EBNA-1 (387). This subsequently led to the discovery of similar activity in authentic EBNA-1 purified from EBV-immortalized lymphocytes (163,290). Thus, the expression of EBNA-1 fusion proteins in *E. coli* provided a powerful approach to viral protein identification and function.

Expression of Authentic Proteins

Clearly, to study all the properties of a viral protein and to obviate potential artifacts, it is better to express the authentic protein rather than a fusion protein. For the expression of viral proteases, it has been possible to obtain authentic enzyme from fusion proteins by autocatalytic excision (129). More often, the initiation codon of the viral protein is fused appropriate distances downstream of a strong, inducible bacterial promoter and a Shine–Dalgarno sequence that is part of a ribosome binding site (297). Inducible promoters are valuable because certain viral proteins such as the vesicular stomatitis virus G protein (302) are deleterious to the growth of *E. coli* so that the bacteria survive only when expression of the viral protein is repressed. A separate problem is that viral proteins are sometimes unstable in *E. coli*; form insoluble, denatured aggregates; or simply are expressed poorly. Despite these potential pitfalls, there are many examples of viral proteins retaining important activities. For example, the human immunodeficiency virus (HIV) reverse transcriptase has been expressed as active enzyme at high levels in *E. coli* by several groups. In one case (192), since the enzyme is normally synthesized as part of a polyprotein and therefore lacks the necessary translation signals, its coding sequence was first cloned as part of a larger HIV segment. The flanking sequences were then removed by deletion mutagenesis and a stop codon was inserted to generate the authentic carboxy terminus. A protein with the same primary sequence as the authentic 66 kD viral protein, except for two or three additional N-terminal residues, was produced in *E. coli*, and this was processed to yield heterodimers of 66 and 51 kD that are physically and kinetically similar to the native enzyme. This preparation of recombinant enzyme and those from other groups have permitted x-ray crystallographic and detailed enzymological studies of this enzyme, which, along with screens for compounds that inhibit enzyme activity, may facilitate the development of compounds active against HIV.

Expression in Yeast

Like *E. coli*, yeast cells can be grown in large quantity and are amenable to genetic manipulation. Expression vectors with relatively strong promoters are available in yeast, and signals can be added that direct secretion of the expressed protein. A successful application of a yeast expression system has been expression of the hepatitis B surface antigen (229,358). In this case, not only is expression efficient, but the expressed protein assembles into particles similar to those found in the blood of hepatitis patients (217), which are immunogenic and form the basis for a safe and efficacious vaccine (159,217,320). Interestingly, assembly and immunogenicity occur even though the protein expressed in yeast is unglycosylated, while the authentic viral protein is glycosylated.

Other studies have taken advantage of the wealth of information regarding yeast transcription, which has permitted sensitive assays for regions of proteins that activate transcription or that participate in protein-protein interactions (87,396). Yeast genetics might be brought to bear even more directly on viral processes, as illustrated by the successful replication and transcription of brome mosaic virus RNA replicons in yeast (156).

Expression in Baculovirus-Infected Insect Cells

Vectors derived from baculoviruses are very popular (228,341). The prototype virus of this family is *Autographa* californica nuclear polyhedrosis virus (AcNPV). AcMNPV expresses large amounts of a polyhedrin protein, which is not essential for growth in *Spodoptera* frugiperda cells in culture. Most baculovirus expression vectors make use of these features by replacing the polyhedrin gene with the foreign gene of interest. This places the foreign gene under the control of polyhedrin gene expression signals. Construction of these vectors entail marker transfer procedures reviewed earlier in this chapter. Recombinants can be recovered at high frequency and readily identified by alterations in plaque morphology due to loss of polyhedrin expression or expression of β-galactosidase from a *lac*Z gene incorporated into the marker transfer protocol.

Although baculovirus vectors are not as easy to manipulate as *E. coli* or yeast vectors, they are capable of producing large amounts of protein. Because insect cells utilize many of the same signals for post-translational modification and secretion as mammalian cells, the protein produced is often indistinguishable from the authentic viral protein. Baculovirus vectors should be generally applicable and have already proved especially useful in examining macromolecular complexes because it is easy to infect insect cells with more than one recombinant baculovirus. An example of this has been the assembly of particles of bluetongue virus by simultaneous expression of four viral structural proteins (96).

Expression Using In Vitro *Methods*

A number of investigators have used *in vitro* methods to express viral genes. In many of these systems, DNA is transcribed with an appropriate RNA polymerase into RNA and the RNA can then be translated into polypeptides with a cell-free translation extract. Although these transcription-translation systems do not produce large amounts of protein [although the amounts can be appreciable (330a)], they often yield enough to detect enzymatic or other activities and can be used to radiolabel gene products to high specific activity. Additionally, the effects of engineered mutations can be assayed quickly.

Early versions of such systems employed *E. coli* RNA polymerase and *E. coli*, wheat germ, or rabbit reticulocyte translation machinery and succeeded in identifying and mapping papovavirus genes (53,296,308). More recent versions make use of bacteriophage RNA polymerases that initiate with high specificity and efficiency at cognate promoter sequences and then elongate processively so that the majority of their products are full-length transcripts. The cell-free transcripts can be useful as a source of infectious RNA (230), as a labeled substrate for studies of splicing or polyadenylation (121,234), and as hybridization probes (225). The transcripts can be capped with the 7-methylguanosine moiety found at the 5' end of most eukaryotic mRNAs, either by using a capping enzyme or by incorporation of cap analogs in the transcription reaction. Either capped or uncapped transcripts can be translated in appropriate translation extracts. For example, such systems have been used to identify unusual translation strategies in picornavirus and retrovirus RNAs (155,260) and to show that the herpes simplex virus *trans*-activating protein VP16 can form complexes with cellular DNA-binding proteins (221).

Chemical Synthesis of Peptides

Investigators have also succeeded in chemical synthesis of viral proteins. These have not only been useful for biochemical studies, but in some cases the peptides have been active when added to cells. For example, chemically synthesized peptides encoded within the adenovirus Ela and human immunodeficiency virus *tat* genes, when introduced into cells, *trans*-activate gene expression, providing evidence for regions of the proteins important for biological activity (119,120). However, as there have been discrepancies between the results of these studies and mutational analyses (310), the mechanisms of *trans*-activation may differ from those ordinarily used by the viruses.

Interfering Gene Products for Analysis of Gene Function

Although the construction of specific mutations is still the most widely used method to analyze specific viral functions, new methods entailing expression of interfering gene products offer promising alternatives.

Antisense RNA and Oligonucleotides

One way to interfere with specific viral gene products is to use nucleic acid complementary to a given mRNA to interfere with its expression. These nucleic acids can be introduced into the cell either by expression of RNA following transfection of cloned DNA or from a viral vector or via direct uptake by cells of oligonucleotides, especially if they have been modified to resist nuclease digestion. This approach, known as *antisense,* has met with some success and holds special promise in systems in which it is difficult to construct specific mutations. The mechanisms of antisense inhibition of gene expression in mammalian cells appear complex and will not be discussed here.

The initial reports of the use of antisense RNA in mammalian cells used the herpes simplex virus (HSV) *tk* gene as the test gene (153,154,172). Various plasmid constructs expressing sequences complementary to *tk* mRNA were transfected into mammalian cells and succeeded in decreasing *tk* expression, either from stably expressed *tk* genes or from cotransfected *tk* genes, but did not decrease the expression of various control genes. Since then there have been many examples of the use of antisense RNA to study viral functions. In one example, human leukocyte cell lines were infected with retrovirus vectors expressing RNA complementary to either the 5' end of human T-cell leukemia virus type I (HTLV-1) mRNA or to mRNA encoding the HTLV-1 *tax* gene (368). HTLV-1 replication and immortalization were inhibited in cells harboring either retrovirus vector. These results are consistent with results showing that the *tax* gene is essential for virus replication (37). Interestingly, one of the retrovirus vector constructs also inhibited cell proliferation, emphasizing that care must be taken to ensure the specificity of antisense inhibition.

Interfering Proteins

A second way to interfere with specific viral gene products is by overexpression of a wild-type or mutant form of a virus protein. There are several examples of overexpression of wild-type viral proteins leading to inhibition of virus replication. In one case, COS cells transfected with plasmids with the vesicular stomatitis virus (VSV) L gene expressed L protein (part of the virion RNA polymerase) in increasing amounts through 3 days post-transfection, yet complementation of VSV L gene mutants peaked at only 12 hr post-transfection and then declined (224). Cells with high levels of L protein also inhibited wild-type VSV replication but not measles virus replication; inhibition depended upon the integrity of L protein–coding sequences. The results suggest that high expression of L protein leads to depletion of other factors required for virus replication.

An early example of inhibition of virus replication by a mutant form of a virus protein was provided by studies of the HSV protein, VP16. HSV contains a virion factor, consisting at least in part of VP16, that *trans*-activates HSV immediate early genes (29,266). Transient expression of a mutant form of VP16 lacking its C-terminal 78 amino acids blocks this *trans*-activation in a dominant-negative manner, presumably because it binds to cellular factors, preventing their association with wild-type VP16 (354). Cell lines stably transformed with this mutant form of VP16 exhibited decreases in HSV yield and immediate early gene expression, while not interfering with the growth of pseudorabies virus (97). These results provided evidence that VP16 *trans*-activation plays an important, although not necessarily essential, role in HSV replication.

It has not escaped notice that antisense and protein inhibition of virus replication could provide ways of preventing or treating virus diseases. This has led to active investigations of these possibilities.

SUMMARY AND PERSPECTIVES

It should be clear from this chapter that viral genetics is a discipline grounded in the rigorous and formal considerations of classical genetics but amenable to the rapid and precise techniques of molecular biology. One way to summarize this chapter is to outline the studies that virologists would perform to characterize a hypothetical newly recognized virus. Different strains of the virus would be isolated and distinguished from related viruses. A DNA copy of the genome would be cloned. Efforts to demonstrate biological activity of the cloned DNA would begin, as would sequence and transcript mapping analyses. These latter analyses would predict potential regulatory elements and viral proteins, and some of their likely functions, such as enzymatic activities. Various kinds of mutations would be engineered in these genetic elements, tested for their effects on virus biology, and mapped. The cloned genes could be used to complement viral mutants. *Cis*-acting regulatory elements and potential *trans*-acting factors controlling gene expression would be tested in stable and transient expression assays in mammalian cells. The open reading frames would be expressed in heterologous systems such as *E. coli* to help generate antisera to identify the proteins in infected cells and to identify infected hosts. The proteins would be tested for ligand-binding and enzymatic activities and characterized by biochemical and biophysical methods. Possible interesting translational strategies might be tested by *in vitro* transcription and translation. If the virus were not too large, more than enough would be known about each virus gene in a few years to name the genes according to their function.

This hypothetical chain of events is not dissimilar to what ensued when human immunodeficiency virus was recognized. The fast pace of progress in understanding this virus is testimony to the value of viral genetics. Much of the rest of this textbook provides examples of the use of the genetic approaches outlined in this chapter, as well as other approaches for which there was no space. Viral genetics has also entered the clinic in the diagnosis, prevention, and treatment of viral infections. It is hoped that the examples discussed in this chapter provide a sense of the power of these approaches, as well as their limitations.

REFERENCES

1. Agut H, Kean KM, Bellocq C, Fichot O, Girard M. Intratypic recombination of polioviruses: evidence for multiple crossing-over sites on the viral chromosome. *J Virol* 1987;61:1722–1725.
2. Ahmed R, Canning WM, Kauffman RS, et al. Role of the host cell in persistent viral infection: coevolution of L cells and reovirus during persistent infection. *Cell* 1981;25:325–332.
3. Ahmed R, Fields BN. Reassortment of genome segments between reovirus defective interfering particles and infectious virus: construction of temperature-sensitive and attenuated viruses by rescue of mutations from DI particles. *Virology* 1981;111:351–363.
4. Anderson W, Killos L, Sander-Haigh L, Kretschmer PJ, Diacumakos E.G. Replication and expression of thymidine kinase and human globin genes microinjected into mouse fibroblast. *Proc Natl Acad Sci USA* 1980;5399–5403.
5. Ausubel FM, Brent R, Kingston RE, Moore DD, Seidman JG, Smith JA, Struhl K. *Current protocols in molecular biology*. New York: Wiley Interscience; 1994.
6. Bacchetti S, Graham FL. Transfer of the gene for thymidine kinase to thymidine-kinase deficient human cells by purified herpes simplex viral DNA. *Proc Natl Acad Sci USA* 1977;74:1590–1594.
7. Ball LA, White CN. Order of transcription of genes of vesicular stomatitis virus. *Proc Natl Acad Sci USA* 1976;73:442–446.
8. Banerjee AK, Barik S. Gene expression of virsicular stomatitis virus genome RNA. *Virology* 1992;188:417–428.
9. Benjamin TL. Host range mutants of polyoma virus. *Proc Natl Acad Sci USA* 1970;67:394–399.
10. Bennink JR, Yewdell JW, Smith GL, Moller C, Moss B. Recombinant vaccina virus primes and stimulates influenza virus haemagglutinin-specific cytotoxic T lymphocytes. *Nature (Lond)* 1984; 311:578–579.
11. Bennink JR, Yewdell JW, Smith GL, Moss B. Recognition of cloned influenza hemagglutinin gene products by cytotoxic T lymphocytes. *J Virol* 1986;57:786–791.
12. Benoist C, O'Hare K, Breathnach R, Chambon P. The ovalbumin gene-sequence of putative control genes *Nucleic Acids Res* 1980;8:127–142.
13. Benoist C, Chambon P. *In vivo* sequence requirements of the SV40 early promoter region. *Nature (Lond)* 1981;290:304–310.
14. Berk AJ, Lee F, Harrison T, Williams J, Sharp PA. Pre-early adenovirus 5 gene product regulates synthesis of early viral messenger RNAs. *Cell* 1979;17:935–944.
15. Berkner KL, Schaffhausen BS, Roberts TM, Sharp PA. Abundant expression of polyomavirus middle T antigen and dihydrofolate reductase in an adenovirus recombinant. *J Virol* 1987 61:1213–1220.
16. Bishop DHL. Genetic potential of bunyavirus. *Curr Top Microbiol Immunol* 1978;86:1–33.
17. Blomquist MC, Hunt LT, Barker WC. Vaccinia virus 19-kilodalton protein: relationship to several mammalian proteins, including two growth factors. *Proc Natl Acad Sci USA* 1984;81:7363–7367.
18. Bond VC, Wold B. Poly-L-ornithine-mediated transformation of mammalian cells. *Mol Cell Biol* 1987;7:2286–2293.
19. Boris-Lawrie KA, Temin HM. Recent advances in retrovirus vector technology. *Curr Opin Genet Dev* 1993;3:102–109.
20. Botchan M, McKenna G, Sharp PA. Cleavage of integrated SV40 by RI restriction endonuclease. *Cold Spring Harbor Symp Quant Biol* 1974;38:383–395.
21. Bouloy M. Bunyaviridae: genome organization and replication strategies. *Adv Virus Res* 1991;40:235–275.
22. Boursnell MEG, Mautner V. Recombination in adenovirus: crossover sites in intertypic recombinants are located in regions of homology. *Virology* 1981;112:198–209.
23. Boyer JC, Haenni AL. Infectious transcripts and cDNA clones of RNA viruses. *Virology* 1993;198:415–426.
24. Bratt MA, Hightower LE. Genetic and paragenetic phenomena in paramyxoviruses. In: Fraenkel-Conrat H, Wagner RR, eds. *Comprehensive virology*, vol. 9. New York: Plenum Press; 1977:457–533.
25. Brown JP, Twardzik DR, Marquardt M, Todaro GJ. Vaccinia virus encodes a polypeptide homologous to epidermal growth factor and transforming growth factor. *Nature (Lond)* 1985;313:491–492.
26. Burge BW, Pfefferkorn ER. Phenotypic mixing between group A arboviruses. *Nature (Lond)* 1966;210:1397–1399.
27. Burks DT, Carle GF, Olson MV. Cloning of large fragments of exogenous DNA into yeast by means of artificial chromosome vectors. *Science* 1987;236:806–812.
28. Burnet FM. *Principles of animal virology*. 2nd ed. New York: Academic Press; 1960.
29. Campbell MEM, Palfreyman JW, Preston CM. Identification of herpes simplex virus DNA sequences which encode a *trans*-acting polypeptide responsible for stimulation of immediate early transcription. *J Mol Biol* 1984;180:1–19.
30. Capecchi M. High efficiency transformation by direct microinjection of DNA into cultured mammalian cells. *Cell* 1980;22:479–488.
31. Carmichael GG, Schaffhausen BS, Dorsky DI, Oliver DB, Benjamin TL. Carboxy terminus of polyoma middle-sized tumor antigen is required for attachment to membranes, associated protein kinase activities and cell transformation. *Proc Natl Acad Sci USA* 1982;79: 3579–3583.
32. Carwright CA, Eckhart W, Simon S, Kaplan PL. Cell transformation by pp60C src mutated in the carboxy-terminal regulatory domain. *Cell* 1987;49:83–91 .
33. Cepko CL, Roberts BE, Mulligan RC. Construction and applications of a highly transmissible murine retrovirus shuttle vector. *Cell* 1984;37:1053–1062.
34. Chakraborty PR, Ahmed R, Fields BN. Genetics of reovirus: the relationship of interference to complementation and reassortment of temperature-sensitive mutants at nonpermissive temperature. *Virology* 1979;94:119–127.
35. Chambers TJ, Hahn CS, Galler R, Rice CM. Flavivirus genome organization, expression, and replication. *Ann Rev Microbiol* 1990;44: 649–688.
36. Chee MS, Bankier AT, Beck S, et al. Analysis of the protein-coding content of the sequnce of human cytomegalovirus strain AD169. *Curr Topics Microbiol Immunol* 1990;154:125–170.
37. Chen ISY, Slamon DJ, Rosenblatt JD, Shah NP, Quan SG, Wachsman W. The X gene is essential for HTLV-1 replication. *Science* 1985;22: 54–58.
38. Chen ISY. Regulation of AIDS virus expression. *Cell* 1986;47:1–2.
39. Choo Q-L, Kuo G, Wiener AJ, Overby LR, Bradley DW, Houghton M. Isolation of cDNA clone derived from a blood-borne non-A, non-B viral hepatitis genome. *Science* 1989;361:359–362.

40. Chu C-T, Parris DS, Dixion RAF, Farber FE, Schaffer PA. Hydroxylamine mutagenesis of HSV DNA and DNA fragments: introduction of mutations into selected regions of the viral genome. *Virology* 1979;98:168–181.
41. Coen DM. Molecular genetics of animal viruses. In: Fields BN, Knipe DM, Chaock, RM, Hirsch MS, Melnick JL, Monath TP, Roizman B, eds. *Fields virology,* 2nd ed. New York: Raven Press; 1990:123–150.
42. Coen DM, Weinheimer SP, McKnight SL. A genetic approach to promoter recognition during trans induction of viral gene expression. *Science* 1986;234:53–59.
43. Coffin JM, Stoye JP, Frankel WN. Genetics of endogenous murine leukemia viruses. *Ann NY Acad Sci* 1989;567:39–49.
44. Cohen J, Seidel KE. Generation of a varicella-zoster virus (VZV) and viral mutants from cosmid DNAs: VZV thymidylate synthetase is not essential for replication *in vitro. Proc Natl Acad Sci USA* 1993; 90:7376–7380.
45. Colbere-Garapin F, Horodniceanu F, Kourilsky P, Garapin A-C. A new dominant hybrid selective marker for higher eukaryotic cells. *J Mol Biol* 1981;150:1–14.
46. Collins PL, Hightower LE, Ball LA. Transcriptional map for Newcastle disease virus. *J Virol* 1980;35:682–693.
47. Collins PL, Mink MA, Stec DS. Rescue of synthetic analogs of respiratory syncytial virus genomic RNA and effect of truncations and mutation on the expression of a foreign reporter gene. *Proc Natl Acad Sci USA* 1991;88:9663–9667.
48. Condit RC, Niles EG Orthopoxvirus genetics. *Curr Top Microbiol Immunol* 1990;163:1–39.
49. Conley AJ, Knipe DM, Jones PC, Roizman B. Molecular genetics of herpes simplex virus. VII. Characterization of a temperature-sensitive mutant produced by *in vitro* mutagenesis and defective in DNA synthesis and accumulation of polypeptides. *J Virol* 1981;37:191–206.
50. Cooper PD. Rescue of one phenotype in mixed infections with heat-defective mutants of poliovirus type 1. *Virology* 1965;25:431–438.
51. Cooper PD. The genetic analysis of poliovirus. In: Levy HB, ed. *The biochemistry of viruses*. New York: Marcel-Dekker; 1969:177–218.
52. Cooper PD. Genetics of picornaviruses. In: Fraenkel-Conrat H, Wagner RR, eds. *Comprehensive virology*, vol. 9. New York: Plenum Press; 1977:133–207.
53. Crawford LV, Gesteland RF. Synthesis of polyoma proteins *in vitro*. *J Mol Biol* 1973;74:627–634.
54. Culver KW, Ram Z, Wallbridge S, Ishii H, Oldfield EH, Blaese, RM. In vivo gene transfer with retroviral vector-producer cells for treatment of experimental brain tumors. *Science* 1992;256:1550–1552.
55. Cunningham C, Davison AJ. A cosmid-based system for constructing mutants of herpes simplex virus type 1.*Virology* 3;197:116–127.
56. Dahlberg JE, Simon EH. Recombination in Newcastle disease virus (NDV): the problem of complementing heterozygotes. *Virology* 1969; 38:490–493.
57. Danos O, Mulligan RC. Safe and efficient generation of recombinant retroviruses with amphotropic and ecotropic host ranges. *Proc Natl Acad Sci USA* 1988;85:6460–6464.
58. Davison AJ, Wilkie NM. Inversion of the two segments of the herpes simplex virus genome in intertypic recombinants. *J Gen Virol* 1983; 64:1–18.
59. Davison AJ, Wilkie NM. Either orientation of the L segment of the herpes simplex virus genome may participate in production of viable intertypic recombinants. *J Gen Virol* 1983;64:247–250.
60. Dayton AI, Sodroski JG, Rosen CA, Goh WC, Haseltine WA. The *trans*-activator gene of the human T cell lymphotropic virus type III is required for replication. *Cell* 1986;44:941–947.
61. deWind N, Zijderveld A, Glazenburg K, Gielkens A, Berns A. Linker insertion mutagenesis of herpesviruses: inactivation of single genes within the Us region of pseudorabies virus. *J Virol* 1990; 64:4691–4696.
62. Dildine SL, Semler BL. The deletion of 41 proximal nucleotides reverts a poliovirus mutant containing a temperature-sensitive lesion in the 5' noncoding region of genomic RNA. *J Virol* 1989;63:847–862.
63. Dimock K, Collins PL. Rescue of synthetic analogs of genomic RNA and replicative-intermediate RNA of human parainfluenza virus type 3 *J Virol* 1993;67:2772–2778.
64. Dobson AT, Margolis TP, Searati F, Stevens JG, Feldman LT. A latent, non-pathogenic HSV-1 derived vector stably expresses β-galactosidase in mouse neurons. *Neuron* 1990;5:353–360.
65. Domingo E, Sabo D, Taniguchi T, Weissman C. Nucleotide sequence heterogeneity of an RNA phage population. *Cell* 1978;13:735–744.
66. Domingo E, Martinez-Salas E, Sobrino F, et al. The quasispecies (extremely heterogeneous nature of viral RNA genome populations): biological relevance—a review. *Gene* 1985;40:1–8.
67. Doyle C, Roth MG, Sambrook J, Gething M-J. Mutations in the cytoplasmic domain of the influenza hemagglutinin affect different stages of intracellular transport. *J Cell Biol* 1985;100:704–714.
68. Drake JW. Mutagenic mechanisms. *Annu Rev Genetics* 1969;3:247–268.
69. Drake JW. The biochemistry of mutagenesis. *Annu Rev Biochem* 1976;45:11–37.
70. Dressler D, Potter H. Molecular mechanisms in genetic recombination. *Annu Rev Biochem* 1982;51:727–761.
71. Drinkwater NR, Klinedinst DK. Chemically induced mutagenesis in a shuttle vector with a low-background mutant frequency. *Proc Natl Acad Sci USA* 1986;83:3402–3406.
72. Dubbs DR, Rachmeler M, Kit S. Recombination between temperature-sensitive mutants of simian virus 40. *Virology* 1974;57:161–174.
73. DuBridge RB, Tang P, Hsia H-C, Leong P-M, Miller JH, Calos MP. Analysis of mutation in human cells by using an Epstein-Barr virus shuttle vector. *Mol Cell Biol* 1987;7:379–387.
74. Dulbecco R, Vogt M. Plaque formation and the isolation of pure lines with poliomyelitis viruses. *J Exp Med* 1954;99:167–182.
75. Dulbecco R. Production of plaques in monolayer tissue culture by single particles of an animal virus. *Proc Natl Acad Sci USA* 1952; 38:747–754.
76. Eckhart W. Genetics of polyoma virus and simian virus 40. In: Fraenkel-Conrat H, Wagner RR, eds. *Comprehensive virology*, vol. 9. New York: Plenum Press; 1977:1–35.
77. Eisenberg SP, Coen DM, McKnight SL. Promoter domains required for expression of plasmid-borne copies of the herpes simplex virus thymidine kinase gene in virus-infected mouse fibroblasts and microinjected frog oocytes. *Mol Cell Biol* 1985;5:1940–1947.
77a. Elion E. Constructing recombinant DNA molecules by the polymerase chain reaction. In: Ausubel FM, Brent R, Kingston RE, Moore DD, Seidman JG, Smith JA, Struhl, K, eds. *Current protocols in molecular biology*. New York: Wiley Interscience: 1994:3.17.1,10.
78. Estes MK, Graham DY, Mason BB. Proteolytic enhancement of rotavirus infectivity: molecular mechanisms. *J Virol* 1981;39:879–888.
79. Estes MK, Cohen J. Rotavirus gene structure and function. *Microbiol Rev* 1989;53:410–449.
80. Evans DJ, Mckeating J, Meredith JM, et al. An engineered poliovirus chimaera elicits broadly reactive HIV-1 neutralizing antibodies. *Nature (Lond)* 1989;339:385–388.
81. Falkner FG, Moss B. *Escherichia coli gpt* gene provides dominant selection for vaccinia virus open reading frame expression vectors. *J Virol* 1988;62:1849–1854.
82. Fechheimer M, Boylan JF, Parker S, Sisken JE, Patel GL, Zimmer S. Transfection of mammalian cells with plasmid DNA by scrape loading and sonication loading. *Proc Natl Acad Sci USA* 1987;84: 8463–8467.
83. Felgner PL, Gadek TR, Holm M, et al. Lipofection: a highly efficient, lipid-mediated DNA-transfection procedure. *Proc Natl Acad Sci USA* 1987;84:7413–7417.
84. Fenner F, McAuslan BR, Mims CA, et al. *The biology of animal viruses*. 2nd ed. New York: Academic Press; 1974:281.
85. Fields BN. Temperature-sensitive mutants of reovirus type 3: features of genetic recombination. *Virology* 1971;46:142–148.
86. Fields BN. Genetics of reovirus. *Curr Top Microbiol Immunol* 1981;91:1–24.
87. Fields S, Song, O-K. A novel genetic system to detect protein-protein interactions. *Nature (Lond)* 1989;340:245–256.
88. Fields S, Winter G. Nucleotide sequence of influenza virus segments 1 and 3 reveal mosaic structure of a small viral RNA segment. *Cell* 1982;28:303–313.
89. Fincham RS. *Genetic complementation*. New York: WA Benjamin; 1966.
90. Fisher AG, Feinberg MB, Josephs SF, et al. The *trans*-activator gene of HTLV-III is essential for virus replication. *Nature (Lond)* 1986; 320:367–371.
91. Fitzgerald M, Shenk T. The sequence 5'-AAUAAA-3' forms part of the recognition site for polyadenylation of late SV40 mRNAs. *Cell* 1981;24:251–260.
92. Fox MS. Some features of genetic recombination in prokaryotes. *Annu Rev Genet* 1978;12:47–68.
93. Fraley RT, Fornari CS, Kaplan S. Entrapment of a bacterial plasmid in phospholipid vesicles: potential for gene transfer. *Proc Natl Acad Sci USA* 1979;76:3348–3352.
94. Franke CA, Rice CM, Strauss JH, Hruby DE. Neomycin resistance as

a dominant selectable marker for selection and isolation of vaccinia virus recombinants. *Mol Cell Biol* 1985;5:1918–1924.
95. Freese E. Molecular mechanisms of mutations. In: Taylor JH, ed. *Molecular genetics*, vol. 1. New York: Academic Press; 1963:207.
96. French TJ, Marshall JJA, Roy P. Assembly of double-shelled, virus-like particles of bluetongue virus by the simultaneous expression of four structural proteins. *J Virol* 1990;64:5695–5700.
97. Friedman AD, Triezenberg SJ, McKnight SL. Expression of a truncated viral *trans*-activator selectively impedes lytic infection by its cognate virus. *Nature (Lond)* 1988;335:452–454.
98. Fuerst TR, Earl P, Moss B. Use of a hybrid vaccinia virus-T7 RNA polymerase system for expression of target genes. *Mol Cell Biol* 1987;7:2538–2544.
99. Fuller FJ, Marcus PI. Sindbis virus. I. Gene order of translation *in vivo*. *Virology* 1980;107:441–451.
100. Gage PP, Sauer B, Levine M, Glorioso J. A cell-free recombination system for site-specific integration of multigenic shuttle plasmids into the herpes simplex virus type 1 genome. *J Virol* 1992;66:5509–5515.
101. García-Sastre A, Palese P. Genetic manipulation of negative-strand RNA viruses genomes. *Ann Rev Microbiol* 1993;47:765–790.
102. Geller AI, Breakefield XO. A defective HSV-1 vector expresses *Escherichia coli* β-galactosidase in cultured peripheral neurons. *Science* 1988;241:1667–1669.
103. Gerhard, W, Webster, RG. Antigenic drift in influenza viruses. I. Selection and characterization of antigenic variants of A/Pr/8/34 (H0N1) influenza virus with monoclonal antibodies. *J Exp Med* 1978;148: 383–392.
104. Gething M-J, Sambrook J. Cell-surface expression of influenza haemagglutinin from a cloned DNA copy of the RNA gene. *Nature (Lond)* 1981;293:620–625.
105. Gething M-J, McCammon K, Sambrook, J. Expression of wild-type and mutant forms of influenza hemagglutinin: the role of folding in intracellular transport. *Cell* 1986;46:939–950.
106. Ghosh-Choudhoury G, Haj-Ahmad Y, Brinkley P, Rudy J, Graham FL. Human adenovirus cloning vectors based on infectious bacterial plasmids. *Gene* 1986;50:161–171.
107. Ginsberg HS, Young CSH. Genetics of adenoviruses. *Adv Cancer Res* 1976;23:91–130.
108. Ginsberg HS, Young CSH. Genetics of adenoviruses. In: Fraenkel-Conrat H, Wagner RR, eds. *Comprehensive virology*, vol. 9. New York: Plenum Press; 1977:27–88.
109. Gluzman Y, Sambrook J, Frisque RJ. Expression of early genes of origin-defective mutants of simian virus 40. *Proc Natl Acad Sci USA* 1980;77:3898–3902.
110. Goff SP. Genetics of retroviral integration. *Ann Rev Genet* 1992; 26:227–244.
111. Gombold JL, Ramig RF. Rotavirus genetics. *Curr Top Microbiol Immunol* 1994;185, in press.
112. Gorman CM, Moffat LF, Howard BH. Recombinant genomes which express chloramphenicol acetyltransferase in mammalian cells. *Mol Cell Biol* 1982;2:1044–1051.
113. Gorziglia MI, Collins PL. Intracellular amplification and expression of synthetic analog of genomic RNA bearing a foreign marker gene: mapping *cis*-acting nucleotides in the 3'noncoding region. *Proc Natl Acad Sci USA* 1992;89:5784–5788.
114. Graham FL, van der Eb AJ. A new technique for the assay of infectivity of human adenovirus 5 DNA. *Virology* 1973;52:456–467.
115. Graham FL, Abrahams PJ, Mulder C, et al. Studies on *in vitro* transformation by DNA and DNA fragments of human adenoviruses and simian virus 40. *Cold Spring Harbor Symp Quant Biol* 1974;39: 637–650.
116. Graham FL, Smiley J, Russell WC, Nairn R. Characteristics of a human cell line transformed by DNA from human adenovirus type 5. *J Gen Virol* 1977;36:59–72.
117. Graham FL. Covalently closed circles of human adeonvirus DNA are infectious. *EMBO J* 1984;3:2917–2922.
118. Gray MR, Colot HV, Guarente L, Rosbash M. Open reading frame cloning: identification, cloning and expression of open reading frame DNA. *Proc Natl Acad Sci USA* 1982;79:6598–6602.
119. Green M, Loewenstein PM. Autonomous functional domains of chemically synthesized human immunodeficiency virus *Tat trans*-activator protein. *Cell* 1988;55:1179–1188.
120. Green M, Loewenstein PM, Pusztai R, Symington JS. An adeonvirus ElA protein domain activates transcription *in vivo* and *in vitro* in the absence of protein synthesis. *Cell* 1988;53:921–926.
121. Green MR, Maniatis T, Melton DA. Human β-globin pre-mRNA synthesized *in vitro* is accurately spliced in Xenopus oocyte nuclei. *Cell* 1983;32:681–694.
122. Grodzicker T, Williams J, Sharp P, Sambrook J. Physical mapping of temperature-sensitive mutations of adenovirus. *Cold Spring Harbor Symp Quant Biol* 1974;39:439–446.
123. Gruss P, Dhar R, Khoury G. Simian virus 40 tandem repeated sequences as an element of the early promoter. *Proc Natl Acad Sci USA* 1981; 76:4317–4321.
124. Hager GL, Chang EH, Chan HW, et al. Molecular cloning of the Harvey sarcoma virus closed circular DNA intermediates: initial structural and biological characterization. *J Virol* 1979;31:795–809.
125. Hahn CS, Lustig S, Strauss EG, Strauss JH. Western equine encephalitis virus is a recombinant virus. *Proc Natl Acad Sci USA* 1988;85: 5997–6001.
126. Hahn CS, Hahn YS, Braciale TJ, Rice CM. Infectious Sindbis virus transient expression vectors for studying antigen processing and presentation. *Proc Natl Acad Sci USA* 1992;89:2679–2683.
127. Halliburton IW, Honess RW, Killington RA. Virulence is not conserved in recombinants between herpesvirus types 1 and 2. *J Gen Virol* 1986; 68:1435–1444.
128. Hamer DH, Leder P. Expression of the chromosomal mouse β-globin gene cloned in SV40. *Nature (Lond)* 1979;281:35–40.
129. Hanecak R, Semler BL, Ariga H, Anderson CW, Wimmer E. Expression of a cloned gene segment of poliovirus in *E. coli*: evidence for autocatalyic production of the viral proteinase. *Cell* 1984; 37:1063–1073.
130. Hartman PE, Roth JR. Mechanisms of suppression. *Adv Genet* 1973;17:1–105.
131. Hartman SC, Mulligan RC. Two dominant-acting selectable markers for gene transfer studies in mammalian cells. *Proc Natl Acad Sci USA* 1988;85:8047–8051.
132. Hay AJ, Wolstenholme AJ, Skehel JJ, Smith MH. The molecular basis of the specific anti-influenza action of amantadine. *EMBO J* 1985;4:3021–3024.
133. Heinzel SS, Krysan PJ, Calos MP, DuBridge RB. Use of simian virus 40 replication to amplify Epstein-Barr virus shuttle vectors in human cells. *J Virol* 1988;62:3738–3746.
134. Heller M, van Santen V, Kieff E. Simple repeat sequence in Epstein-Barr virus DNA is transcribed in latent and productive infections. *J Virol* 1982;44:311–320.
135. Hennesey K, Heller M, van Santen V, Kieff E. Simple repeat array in Epstein-Barr virus DNA encodes part of the Epstein-Barr nuclear antigen. *Science* 1983;220:1396–1398.
136. Hennessey K, Kieff E. One of two Epstein-Barr virus nuclear antigens contains a glycine-alanine copolymer domain. *Proc Natl Acad Sci USA* 1983;80:5665–5669.
137. Hermonat PL, Muzyczka N. Use of adeno-associated virus as a mammalian DNA cloning vector: transduction of neomycin resistance into mammalian tissue culture cells. *Proc Natl Acad Sci USA* 1984; 81:6466.
138. Ho DY, Mocarski ES. β-galactosidase as a marker in the peripheral and neural tissues of the herpes simplex virus-infected mouse. *Virology* 1988;167:279–283.
139. Holland J, Spindler K, Horodyski F, Grabau E, Nichol S, Van de Pol S. Rapid evolution of RNA genomes. *Science* 1982;215:1577–1585.
140. Holland JJ, Cords CE. Maturation of poliovirus RNA with capsid protein coded by heterologous enteroviruses. *Proc Natl Acad Sci USA* 1964;51:1082–1085.
141. Holland JJ, Villareal LP. Purification of defective interfering T particles of vesicular stomatitis and rabies viruses generated in vivo in the brains of newborn mice. *Virology* 1975;67:438–449.
142. Holland JJ, Grabau E, Jones CL, Semler BL. Evolution of multiple genome mutations during long term persistent infection with vesicular stomatitis virus. *Cell* 1979;16:495–504.
143. Holland JJ, Kennedy SI, Semler BL, Jones L, Roux L, Grabau E. Defective interfering RNA viruses and the host cell response. In: Fraenkel-Conrat H, Wagner RR, eds. *Comprehensive virology,* vol. 16. New York: Plenum Press; 1980:137–192.
144. Holland TC, Sandri-Goldin RM, Holland LE, Martin SD, Levine M, Glorioso JC. Physical mapping of the mutation in an antigenic variant of herpes simplex virus type 1 by use of immunoreactive plaque assay. *J Virol* 1983;46:649–652.
145. Holland JJ. Continuum of change in RNA virus genomes. In: Notkins

AL, Oldstone MBA, eds. *Concepts in viral pathogenesis*. New York: Springer-Verlag; 1984:137–143.
146. Holland JJ. Defective interfering rhabdoviruses. In: Wagner RR, ed. *The rhabdoviruses*. New York: Plenum Press; 1987:297–360.
147. Huang AS, Wagner RR. Comparative sedimentation coefficients of RNA extracted from plaque-forming and defective particles of vesicular stomatitis virus. *J Mol Biol* 1966;22:381–384.
148. Huang AS, Palma EL, Hewlett N, Roizman B. Pseudotype formation between enveloped RNA and DNA viruses. *Nature (Lond)* 1974;252: 743–745.
149. Huang AS, Baltimore D. Defective interfering animal viruses. In: Fraenkel-Conrat H, Wagner RR, eds. *Comprehensive virology*, vol. 10. New York: Plenum Press; 1977:73–116.
150. Imperiale MJ, Feldman LT, Nevins JR. Activation of gene expression by adenovirus and herpes virus regulatory genes acting in *trans* and by a *cis*-acting adenovirus enhancer element. *Cell* 1983;35:127–136.
151. Irmiere AF, Manos MM, Jacobson JG, Gibbs JS, Coen DM. Effect of an amber mutation in the herpes simplex virus thymidine kinase gene on polypeptide synthesis and stability. *Virology* 1989;168:210–220.
152. Itoh H, Melnick JL. Double infections of single cells with ECHO 7 and Coxsackie A9 viruses. *J Exp Med* 1959;109:393–406.
153. Izant JG, Weintraub, H. Inhibition of thymidine kinase gene expression by anti-sense RNA: a molecular approach to genetic analysis. *Cell* 1984;36:1007–1015.
154. Izant JG, Weintraub, H. Constitutive and conditional suppression of exogenous and endogenous genes by anti-sense RNA. *Science* 1985;229:345–352.
155. Jacks T, Varmus HE. Expression of the Rous sarcoma virus *pol* gene by ribosomal frameshifting. *Science* 1985;230:1237–1242.
156. Janda M, Ahlquist P. RNA-dependent replication, transcription, and persistence of brome mosaic virus RNA replicons in *S. cerevisiae*. *Cell* 1993;72:961–970.
157. Jarvis, TC, Kirkegaard, K. The polymerase in its labyrinth: mechanisms and implications of RNA recombination. *Trends Genet* 1991; 7:186–191.
158. Javier RT, Sedarati F, Stevens J. Two avirulent herpes simplex viruses generate lethal recombinants *in vivo*. *Science* 1986;234:746–748.
159. Jilg W, Lorbeer B, Schmidt M, Wilske B, Zoulek G, Deinhardt, F. Clinical evaluation of a recombinant hepatitis B vaccine. *Lancet* 1984; 2:1174–1175.
160. Jimenez A, Davies, J. Expression of a transposable antibiotic resistance element in *Saccharomyces*. *Nature (Lond)* 1980;287:869–871.
161. Johnson VH, Semler BL. Defined recombinants of poliovirus and coxsackievirus: sequence-specific deletions and functional substitutions in the 5'-noncoding regions of viral RNAs. *Virology* 1988;162:47–57.
162. Johnson GP, Goebel SJ, Paoletti E. An update on the vaccinia virus genome. *Virology* 1993;196:381–401.
163. Jones CH, Hayward SD, Rawlins. Interaction of the lymphocyte-derived Epstein-Barr Virus Nuclear Antigen EBNA-1 with its DNA-binding sites. *J Virol* 1989;63:101–110.
164. Jones CL, Holland JJ. Requirements for DI particle prophylaxis against VSV infection *in vivo*. *J Gen Virol* 1980;49:215–220.
165. Jones KA, Yamamoto KR, Tjian R. Two distinct transcription factors bind to the HSV thymidine kinase promoter *in vitro*. *Cell* 1985;559–572.
166. Jones N, Shenk T. An adenovirus type 5 early gene function regulates expression of other viral genes. *Proc Natl Acad Sci USA* 1979;76: 3665–3669.
167. Katz RA, Skalka AM. Generation of diversity in retroviruses. *Ann Rev Genet* 1990;24:409–445.
168. Keck JG, Stohlman SA, Soe LH, Makino S, Lai MMC. Multiple recombination sites at the 5'-end of murine coronavirus RNA. *Virology* 1987;156:331–341.
169. Kelly TJ, Nathans D. The genome of simian virus 40. *Adv Virus Res* 1977;21:85–120.
170. Keranen S. Interference of wild type virus replication by an RNA negative temperature sensitive mutant of Semliki Forest virus. *Virology* 1977;80:1–11.
171. Kilbourne ED. *Influenza*. New York: Academic Press; 1987:111–154.
172. Kim SK, Wold BS. Stable reduction of thymidine kinase activity in cells expressing high levels of anti-sense RNA. *Cell* 1985;42:129–138.
173. Kimelman D. A novel general approach to eucaryotic mutagenesis functionally identifies conserved regions within the adenovirus 13S ElA polypeptide. *Mol Cell Biol* 1986;6:1487–1496.
174. King AM, Slade WR, Neuman JWI, McCahon D. Temperature sensitive mutants of foot-and-mouth disease virus with altered structural polypeptides. II. Comparisons of recombination and biochemical maps. *J Virol* 1980;34:67–72.
175. King AM, McCahon D, Slade WR, Neuman JWI. Recombination in RNA. *Cell* 1982;29:921–928.
176. King CS, Cooper JA, Moss B, Twardizik DR. Vaccinia virus growth factor stimulates tyrosine protein kinase activity of A431 epidermal growth factor receptors. *Mol Cell Biol* 1986;6:332–336.
177. Kirkegaard K, Baltimore D. The mechanism of RNA recombination of poliovirus. *Cell* 1986;47:433–441.
178. Kitts PA, Ayres MD, Possee RD. Linearization of baculovirus DNA enhances the recovery of recombinant virus expression vectors. *Nucleic Acids Res* 1990;18:5667–5672.
179. Kjellen LE. A variant of adenovirus type 5. *Arch Ges Virusforsch* 1963;13:482–488.
180. Klein TM, Gradziel T, Fromm ME, Sanford JC. High-velocity microprojectiles for delivering nucleic acids into living cells. *Nature (Lond)* 1987;327:70–73.
181. Knipe DM. Molecular genetics of animal viruses. In: Fields BN, Knipe DM, Chanock RM, Melnick JL, Roizman B, Shope RE, eds. *Virology*. New York: Raven Press; 1985:129–143.
182. Koetzner CA, Parker MM, Ricard CS, Sturman LS, Masters PS. Repair and mutagenesis of the genome of a deletion mutant of the coronavirus mouse hepatitis virus by targeted RNA recombination. *J Virol* 1992;66:1841–1848.
183. Kousoulas KG, Pellet PE, Pereira L, Roizman B. Mutations affecting conformation or sequence of neutralizing epitopes identified by viable plaques segregated from syn and ts domains of HSV-1 (F) gB gene. *Virology* 1984;135:379–394.
184. Kozak M. Point mutations define a sequence flanking the AUG initiator condon that modulates translation by eukaryotic ribosomes. *Cell* 1986;44:283–292.
185. Kozak M. At least six nucleotides preceding the AUG initiator codon enhance translation in mammalian cells. *J Mol Biol* 1987;196: 947–950.
186. Kozak M. An analysis of 5'-noncoding sequences from 699 vertebrate messenger RNAs. *Nucleic Acids Res* 1987;15:8125–8148.
187. Kunkel TA. Rapid and efficient site-specific mutagenesis without phenotypic selection. *Proc Natl Acad Sci USA* 1985;82:488–492.
188. Lai C-J, Nathans D. Deletion mutants of simian virus 40 generated by enzymatic excision of DNA segments from the viral genome. *J Mol Biol* 1974;89:179–193.
189. Lai C-J, Nathans D. Mapping the genes of simian virus 40. *Cold Spring Harbor Symp Quant Biol* 1974;39:53–60.
190. Lai MM. Coronavirus: organization, replication and expression of its genome. *Ann Rev Microbiol* 1990;44:303–333.
191. Lamb RA, Horvath CM. Diversity of coding strategies in influenza viruses. *Trends Genet* 1991;7:261–266.
192. Larder B, Purifoy D, Powell K, Darby G. AIDS virus reverse transcriptase defined by high level expression in *Escherichia coli*. *EMBO J* 1987;6:3133–3137.
193. Law M-F, Lowy DR, Dvoretzky I, Howley, PM. Mouse cells transformed by bovine papillomavirus contain only extrachromosomal viral DNA sequences. *Proc Natl Acad Sci USA* 1981;78:2727–2731.
194. Le Gal La Salle G, Robert JJ, Berrard S, Ridoux V, Stratford-Perricaudet LD, Perricaudet M, Mallet J. An adenovirus vector for gene transfer into neurons and glia in the brain. *Science* 1993;259:988–990.
195. Ledinko N, Hirst GK. Mixed infection of cells with poliovirus types 1 and 2. *Virology* 1961;14:207–219.
196. Leung DW, Chen E, Goeddel DV. A method for random mutagenesis of a defined DNA segment using a modified polymerase chain reaction. *Technique* 1989;1:11–15.
197. Levis R, Huang H, Schlesinger S. Engineered defective interfering RNAs of Sindbis virus express bacterial chloramphenicol acetyltransferase in avian cells. *Proc Natl Acad Sci USA* 1987;84:4811–4815.
198. Li S, Schulman JL, Moran T, Bona C, Palese P. Influenza A virus transfectants with chimeric hemagglutinins containing epitopes from different subtypes. *J Virol* 1992;66:399–404.
199. Liao C-L, Lai MMC. RNA recombination in a coronavirus: recombination between viral genomic RNA and transfected RNA fragments. *J Virol* 1992;66:6117–6124.
200. Linial M, Blair D. Genetics of retroviruses. In: Weiss R, Teich N, Varmus H, Coffin J, eds. *RNA tumor viruses*. Cold Spring Harbor, NY: Cold Spring Harbor Laboratory; 1982:649–784.

201. Littlefield J. Selection of hybrids from matings of fibroblasts *in vitro* and their presumed recombinants. *Science* 1964;145:709–710.
202. Lobel LI, Goff SP. Construction of mutants of Moloney murine leukemia virus by suppressor-linker insertional mutagenesis: positions of viable insertion mutants. *Proc Natl Acad Sci USA* 1984;81: 4149–4153.
203. Lowy DR, Dvoretzky I, Shober R, Law M-F, Engel L, Howley PM. *In vitro* tumorigenic transformation by a defined sub-genomic fragment of bovine papilloma virus. *Nature (Lond)* 1980;287:72–74.
204. Lowy DR, Rands E, Chattopadhyay SK, Garon CF, Hager GL. Molecular cloning of infectious integrated murine leukemia virus DNA from infected mouse cells. *Proc Natl Acad Sci USA* 1980;77:614–618.
205. Luo, G, Palese, P. Genetic analysis of influenza virus. *Curr Opin Genet Dev* 1992:2:77–81.
206. Luthman H, Magnusson G. High efficiency polyoma DNA transfection of chloroquine treated cells. *Nucleic Acids Res* 1983;11:1295–1307.
207. Luytjes W, Krystal M, Enami M, Parvin J, Palese P. Amplification, expression, packaging of a foreign gene by influenza virus. *Cell* 1989; 59:1107–1113.
208. Maassab HF, LaMontagne JR, DeBorde DC. Live Influenza virus vaccines. In: Plotkin SA, Mortimer EA, eds. *Vaccines*. Philadelphia: WB Saunders; 1988:435–437.
209. Mackett M, Smith GL. Vaccinia virus expression vectors. *J Gen Virol* 1986;67:2067–2082.
210. Maitland NJ, McDougall JK. Biochemical transformation of mouse cells by fragments of herpes simplex virus DNA. *Cell* 1977;11:233–241.
211. Makino S, Keck JG, Stohlman SA, Lai MMC. High-frequency RNA recombination of murine coronaviruses. *J Virol* 1986;57:729 –737.
212. Mann R, Mulligan RC, Baltimore D. Construction of a retrovirus packaging mutant and its use to produce helper-free defective retrovirus. *Cell* 1983;33:153–159.
213. Marchetti ME, Smith CA, Schaffer PA. A temperature-sensitive mutation in a herpes simplex virus type 1 gene required for viral DNA synthesis maps to coordinates 0.609 through 0.614 in UL. *J Virol* 1988;62:715–721.
214. Margolskee RF, Kavathas P, Berg P. Epstein-Barr virus shuttle vector for stable episomal replication of cDNA expression libraries in human cells. *Mol Cell Biol* 1988;8:2837–2847.
215. Martuza RL, Malick A, Markert JM, Ruffner KL, Coen DM. Experimental therapy of human glioma by means of a genetically engineered virus mutant. *Science*.1991;252:854–856.
216. Massicot JG, VanWyke K, Chanock RM, Murphy BR. Evidence for intrasegmental complementation between two influenza A viruses having *ts* mutants in their P1 genes. *Virology* 1982;117:496–500.
217. McAleer WJ, Buynak EB, Maigetter RZ, Wampler DE, Miller WJ, Hilleman MR. Human hepatitis B vaccine from recombinant yeast. *Nature (Lond)* 1984;307:178–180.
218. McGeoch DJ. The genomes of the human herpesviruses: contents, relationships, and evolution. *Ann Rev Microbiol* 1989;43:235–265.
219. McKnight SL, Gavis ER, Kingsbury R, Axel R. Analysis of transcriptional regulatory signals of the HSV thymidine kinase gene: identification of an upstream control region. *Cell* 1981;25:385–398.
220. McKnight SL, Kingsbury R. Transcriptional control signals of a eukaryotic protein-coding gene. *Science* 1982;217:316–324.
221. McKnight JL, Kristie TM, Roizman B. Binding of the virion protein mediating α gene induction in herpes simplex virus 1-infected cells to its *cis* site requires cellular proteins. *Proc Natl Acad Sci USA* 1987;84:7061–7065.
222. McLaren LC, Holland JJ. Defective interfering particles from poliovirus vaccine and vaccine reference strains. *Virology* 1974;60: 579–583.
223. McPhillips TH, Ramig RF. Extragenic suppression of temperature-sensitive mutants in reovirus: mapping suppressor mutations. *Virology* 1984;135:428–439.
224. Meier E, Harmision GG, Schubert M. Homotypic and heterotypic exclusin of vesicular stomatitis virus replication by high levels of recombinant polymerase protein L. *J Virol* 1987;61:3133–3142.
225. Melton DA, Krieg PA, Rebagliati MR, Maniatis T, Zinn K, Green MR. Efficient *in vitro* synthesis of biologically active RNA and RNA hybridization probes from plasmids containing a bacteriophage SP6 promoter. *Nucleic Acids Res* 1984;12:7035–7056.
226. Meselson M, Weigle JJ. Chromosome breakage accompanying genetic recombination in bacteriophage. *Proc Natl Acad Sci USA* 1961;47: 857–868.
227. Miller AD. Human gene therapy comes of age. *Nature (Lond)* 357: 455–460.
228. Miller LK. Baculoviruses as gene expression vectors. *Annu Rev Microbiol* 1988;42:177–179.
229. Miyanohara A, Toh-e A, Nazaki C, Hamada F, Ohtomo N, Matsubara K. Expression of hepatitis B surface antigen in yeast. *Proc Natl Acad Sci USA* 1983;80:1–5.
230. Mizutani S, Colonno RJ. *In vitro* synthesis of an infectious RNA from cDNA clones of human rhinovirus type 14. *J Virol* 1985;56:628–632.
231. Moar MH, Campo MS, Laird AM, Jarrett WFH. Persistence of non-intergrated viral DNA in bovine cells transformed *in vitro* by bovine pappilloma virus type 2. *Nature (Lond)* 1981;293:749–751.
232. Mocarski ES, Roizman B. Structure and role of the herpes simplex virus DNA termini in inversion, circularization and generation of virion DNA. *Cell* 1982;31:89–97.
233. Molla A, Paul AV, Wimmer E. Cell-free, de novo synthesis of poliovirus. *Science* 1991;254:1647–1651.
234. Moore CL, Sharp PA. Site-specific polyadenylation in a cell-free reaction. *Cell* 984; 36:581–591 .
235. Morse LS, Buchman TG, Roizman B, Schaffer PA. Anatomy of herpes simplex virus DNA. IX. Apparent exclusion of some parental arrangements in the generation of intertypic (HSV-1 X HSV-2) recombinants. *J Virol* 1977;24:231–248.
236. Moss B, Flexner C. Vaccinia virus expression vectors. *Annu Rev Immunol* 1987;5:305–324.
237. Moss B. Poxvirus vectors: cytoplasmic expression of transferred genes. *Curr Opin Genet Dev* 1993;3:86–90.
238. Mulligan RC, Howard BH, Berg P. Synthesis of rabbit β-globin in cultured monkey kidney cells following infection with a SV40 β-globin recombinant genome. *Nature (Lond)* 1979;277:108–114.
239. Mulligan RC, Berg P. Expression of a bacterial gene in mammalian cells. *Science* 1980;209:1422–1427.
240. Munz PL, Young C, Young CSH. The genetic analysis of adenovirus recombination in triparental and superinfection crosses. *Virology* 1983; 126:576–586.
241. Munz PL, Young CSH. Polarity in adenovirus recombination. *Virology* 1984;135:503–514.
242. Murphy BR, Tolpin MD, Massicot JG, Kim HY, Parrott RH, Chanock RM. Escape of a highly defective influenza A virus mutant from its temperature sensitive phenotype by extragenic suppression and other types of mutation. *Ann NY Acad Sci* 1980;354:172–182.
243. Neumann E, Schaefer-Ridder M, Wang Y, Hofschneider PH. Gene transfer into mouse myeloma cells by electroporation in high electric fields. *EMBO J* 1982;1:841–845.
244. O'Connor M, Peifer M, Bender W. Construction of large DNA segments in *E. coli*. *Science* 1989;244:1307–1312.
245. O'Conor GT, Rabson AS, Berezesky IK, Paul PJ. Mixed infection with simian virus 40 and adenovirus 12. *J Natl Cancer Inst* 1963;31: 903–915.
246. Ochman H, Gerber AS, Hartl DL. Genetic applications of an inverse polymerase chain reaction. *Genetics* 1988;120:621–625.
247. Offit PA, Blavat G, Greenberg HB, Clark HF. Molecular basis of rotavirus virulence: Role of gene segment 4. *J Virol* 1986;57:46–49.
248. Ostrowski MC, Richard-Foy H, Wolford RG, Bivard DS, Hager GL. Glucocorticoid regulation of transcription at an amplified episomal promoter. *Mol Cell Biol* 1983;3:2045–2057.
249. Palella TD, Silverman LJ, Schroll CT, Homa FL, Levine M, Kelley WN. Herpes simplex virus-mediated human hypoxanthine-guanine phosphoribosyltransferase gene transfer into neuronal cells. *Mol Cell Biol* 1988;8:457–460.
250. Palese P. The genes of influenza virus. *Cell* 1977;10:1–10.
251. Palese P. Reassortment continuum. In: Notkins AL, Oldstone MBA, eds. *Concepts in viral pathogenesis*. New York: Springer-Verlag; 1984:144–151.
252. Palese P, Garcia-Sastre A. Genetic manipulation of negative-strand RNA virus genomes. *Ann Rev Microbiol* 1993;47.
253. Panicali D, Grzelecki A, Huang C. Vaccinia virus vectors utilizing the β-galactosidase assay for rapid selection of recombinant viruses and measurement of gene expression. *Gene* 1986;47:193–199.
254. Park KH, Huang T, Correia FF, Krystal M. Rescue of a foreign gene by Sendai virus. *Proc Natl Acad Sci USA* 1991;88:5537–5541.
255. Parvin JD, Moscona A, Pan WT, et al. Measurement of the mutation rates of animal viruses: influenza A virus and poliovirus type 1. *J Virol* 1986;59:377–383.
256. Patch CT, Levine AS, Lewis AM. The adenovirus-SV40 hybrid viruses. In: Fraenkel-Conrat H, Wagner RR, eds. *Comprehensive virology*, vol. 13. New York: Plenum Press; 1979:459–487.

257. Paterson BM, Roberts BE, Kuff EL. Structural gene identification and mapping by DNA-mRNA hybrid-arrested cell-free translation. *Proc Natl Acad Sci USA* 1977;74:4370–4374.
258. Pattnaik AK, Ball LA, LeGrone AW, Wertz GW. Infectious defective interfering particle of VSV from transcripts of a cDNA clone. *Cell* 1992;69:1011–1020.
259. Peden KWC, Pipas JM, Pearson-White S, Nathans D. Isolation of mutants of an animal virus in bacteria. *Science* 1980;209:1392–1396.
260. Pelletier J, Sonenberg N. Internal initiation of translation of eukaryotic mRNA directed by a sequence derived from poliovirus RNA. *Nature (Lond)* 1988;334:320–325.
261. Pellicer A, Robins D, Wold B, et al. Altering genotype and phenotype by DNA-mediated gene transfer. *Science* 1980;209:1414–1422.
262. Perucho M, Hanahan D, Wigler M. Genetic and physical linkage of exogenous sequences in transformed cells. *Cell* 1980;22:309–317.
263. Pfefferkorn ER. Genetics of togaviruses. In: Fraenkel-Conrat H, Wagner RR, eds. *Comprehensive virology*, vol. 9. New York: Plenum Press; 1977:209–238.
264. Pinto LH, Holsinger LJ, Lamb RA. Influenza virus M2 protein has ion channel activity. *Cell* 1992;69:517–528.
265. Piwnica-Worms H, Saunders KB, Roberts TM, Smith AE, Cheng SH. Tyrosine phosphorylation regulates the biochemical and biological properties of pp60C src. *Cell* 1987;49:75–82.
266. Post LE, Mackem S, Roizman B. Regulation of alpha genes of herpes simplex virus: expression of chimeric genes produced by fusion of thymidine kinase with alpha gene promoters. *Cell* 1981;24:555–556.
267. Potter H, Weir L, Leder P. Enhancer-dependent expression of human k immunoglobulin genes introduced into mouse pre-B lymphocytes by electroporation. *Proc Natl Acad Sci USA* 1984;81:7161–7165.
268. Price J, Turner D, Cepko C. Lineage analysis in the vertebrate nervous system by retrovirus-mediated gene transfer. *Proc Natl Acad Sci USA* 1987;84:156–160.
269. Pringle CR. Conditional-lethal mutants of vesicular stomatitis virus. *Curr Top Microbiol Immunol* 1975;69:85–116.
270. Pringle CR. Genetics of rhabdoviruses. In: Fraenkel-Conrat H, Wagner RR, eds. *Comprehensive virology*, vol. 9. New York: Plenum Press; 1977:239–289.
271. Pringle CR. Rhabdovirus genetics. In: Wagner RR, ed. *The rhabdoviruses*. New York: Plenum Press; 1987:167–243.
272. Proudfoot N, Brownlee G. 3' non-coding region sequences in eukaryotic mRNA. *Nature (Lond)* 1976;263:211–214.
273. Quinlan MP, Chen LB, Knipe DM. The intranuclear location of a herpes simplex virus DNA-binding protein is determined by the status of viral DNA replication. *Cell* 1984;36:857–868.
274. Quinlan MP, Knipe DM. A genetic test for expression of a functional herpes simplex virus DNA-binding protein from a transfected plasmid. *J Virol* 1985;54:619–622.
275. Quinlan MP, Knipe DM. Stimulation of expression of a herpes simplex virus DNA-binding protein by two viral functions. *Mol Cell Biol* 1985;5:957–963.
276. Racaniello VR, Baltimore D. Molecular cloning polio virus DNA and determination of the complete nucleotide sequence of the viral genome. *Proc Natl Acad Sci USA* 1981;78:4887–4891.
277. Radding CM. Genetic recombination: strand transfer and mismatch repair. *Annu Rev Biochem* 1978;47:847–880.
278. Ramig RF. Principles of animal virus genetics. In: Fields BN, Knipe DM, Chanock RM, Melnick JL, Roizman B, Shope RE, eds. *Virology*. New York: Raven Press; 1985:101–127.
279. Ramig RF. Principles of animal virus genetics. In: Fields BN, Knipe DM, Chanock RM, Hirsch MS, Melnick JL, Monath TP, Roizman B, eds. *Fields virology*, 2nd ed. New York: Raven Press; 1990:95–122.
280. Ramig RF, Cross RK, Fields BN. Genome RNAs and polypeptides of reovirus serotypes 1, 2 and 3. *J Virol* 1977;22:726–733.
281. Ramig RF, White RM, Fields BN. Suppression of the temperature sensitive phenotype of a mutant of reovirus type 3. *Science* 1977; 195:406–407.
282. Ramig RF, Fields BN. Revertants of temperature sensitive mutants of reovirus: evidence for frequent extragenic suppression. *Virology* 1979;92:155–167.
283. Ramig RF. Suppression of temperature-sensitive phenotype in reovirus: An alternative pathway from *ts* to *ts*[+] phenotype. In: Fields BN, Jaenisch R, Fox CF, eds. *Animal virus genetics*. New York: Academic Press; 1980:633–642.
284. Ramig RF, Fields BN. Genetics of reoviruses. In: Joklik WK, ed. *The reoviridae*. New York: Plenum Press; 1983:197–228.
285. Ramig RF, Ahmed R, Fields BN. A genetic map of reovirus: assignment of the newly defined mutant groups H, I and J to genome segments. *Virology* 1983;125:299–313.
286. Ramig RF, Ward RL. Genomic segment reassortment in rotaviruses and other reoviridae. *Adv Virus Res* 1991;39:164–207.
287. Rapp F, Butel JS, Melnick JL. SV40-adenovirus hybrid populations: transfer of SV40 determinants from one type of adenovirus to another. *Proc Natl Acad Sci USA* 1965;54:717–724.
288. Rawlins DR, Milman G, Hayward SD. Sequence-specific DNA binding of the Epstein-Barr virus nuclear antigen (EBNA-1) to clustered sites in the plasmid maintenance region. *Cell* 1985;42:659–668.
289. Reedman BM, Klein G. Cellular localization of an Epstein-Barr virus (EBV)-associated complement-fixing antigen in producer and non-producer lymphoblastoid cell lines. *Int J Cancer* 1973; 11:499–520.
290. Reischig J, Barsch D, Polack A, Vonka V, Hirsch I. Electron microscopy of binding of Epstein-Barr virus (EBV) nuclear antigen (EBNA-1) to EBV DNA. *Virology* 1987;160:498–501.
291. Reisner AH. Similarity between the vaccinia virus 19K early protein and epidermal growth factor. *Nature (Lond)* 1985;313:801–803.
292. Rekosh D. Gene order of the poliovirus capsid proteins. *J Virol* 1972;9:479–487.
293. Rhode SL III, Klaassen, B. DNA sequence of the 5' terminus containing the replication origin of parvovirus replicative form DNA. *J Virol* 1982;41:990–999.
294. Ricciardi RP, Miller JS, Roberts BE. Purification and mapping of specific mRNAs by hybridization selection and cell free translation. *Proc Natl Acad Sci USA* 1979;76:4927–4931.
295. Rice CM, Strauss EG, Strauss JH. Structure of the flavivirus genome. In: Schlesinger S, Schlesinger MJ, eds. *The togaviridae and flaviridiridae*. New York: Plenum Press; 1986:279–326.
296. Roberts BE, Gorecki M, Mulligan RC, Danna KJ, Rozenblatt S, Rich A. SV40 DNA directs the synthesis of authentic viral polypeptides in a linked transcription-translation cell free system. *Proc Natl Acad Sci USA* 1975;72:1922–1926.
297. Roberts TM, Kacich R, Ptashne M. A general method for maximizing the expression of a cloned gene. *Proc Natl Acad Sci USA* 1979; 76:760–764.
298. Roizman B. Polykaryosis: results from fusion of mononucleated cells. *Cold Spring Harbor Symp Quant Biol* 1962;27:327–342.
299. Roizman B. The organization of the herpes simplex virus genomes. *Annu Rev Genet* 1979;13:25–57.
300. Roizman B. The structure and isomerization of herpes simplex virus genomes. *Cell* 1979;16:481–494.
301. Romanova LI, Blinov VM, Tolskaya EA, et al. The primary structure of crossover regions of intertypic poliovirus recombinants: a model of recombination between RNA genomes. *Virology* 1986; 155:202–213.
302. Rose JK, Shafferman A. Conditional expression of the vesicular stomatitis virus glycoprotein gene in *Escherichia coli*. *Proc Natl Acad Sci USA* 1981;78:6670–6674.
303. Rose JK, Bergmann JE, Expression from cloned cDNA of cell-surface and secreted forms of vesicular stomatitis virus in eukaryotic cells. *Cell* 1982;30:753–762.
304. Rosenfeld MA, Yoshimura K, Trapnell BC, et al. *In vivo* transfer of the human cystic fibrosis transmembrane conductance regulator gene to the airway epithelium. *Cell* 1992;68:143–155.
305. Roth DB, Porter TN, Wilson JH. Mechanisms of nonhomologous recombination in mammalian cells. *Mol Cell Biol* 1985;5:2599–2608.
306. Roth JR. Frameshift mutations. *Annu Rev Genet* 1974;43:319–346.
307. Roy P, Marshall JJ, Frenc, TJ. Structure of the bluetongue virus genome and its encoded proteins. *Curr Top Microbiol Immunol* 1990;162:43–87.
308. Rozenblatt S, Mulligan RC, Gorecki M, Roberts BE, Rich A. Direct biochemical mapping of eukaryotic viral DNA by means of a linked transcription-translation cell-free system. *Proc Natl Acad Sci USA* 1976;73:2747–2751.
309. Russell S, Brandenburger A, Flemming CL, Collins MKL, Rommelaere J. Transformation-dependent expression of interleukin genes delivered by a recombinant parvovirus. *J Virol* 1992;66:2821–2828.
310. Sadaie MR, Rappaport J, Benter T, Josephs SF, Willis R, Wong-Staal F. Missense mutations in an infectious human immunodeficiency viral genome: functional mapping of *tat* and identification of the rev splice acceptor. *Proc Natl Acad Sci USA* 1988;85:9224–9228.
311. Sambrook JF, Greene R, Stringer J, Mitchison T, Hu SL, Botchan M. Analysis of the sites of integration of viral DNA sequences in rat cells

transformed by adenovirus 2 or SV40. *Cold Spring Harbor Symp Quant Biol* 1979;44:569–576.
312. Samulski RJ, Berns KI, Tan M, Muzyczka N. Cloning of AAV into pBR322: rescue of intact virus from the recombinant plasmid in human cells. *Proc Natl Acad Sci USA* 1982;79:2077–2081.
313. Sandri-Goldin RM, Levine M, Gloriso JC. Methods for induction of mutations in physically defined regions of the herpes simplex virus genome. *J Virol* 1981;38:41–49.
314. Santerre RF, Allen NE, Hobbs JN Jr., Rao RN, Schmidt RJ. Expression of prokaryotic genes for hygromycin B and G418 resistance as dominant-selection markers in mouse L cells. *Gene* 1984;30:147–156.
315. Sauer B, Whealy M, Robbins A, Enquist L. Site-specific insertion of DNA into a pseudorabies virus vector. *Proc Natl Acad Sci USA* 1987;84:9108–9112.
316. Schaffer PA. Temperature sensitive mutants of herpesviruses. *Curr Top Microbiol Immunol* 1975;70:51–100.
317. Schaffner W. Direct transfer of cloned genes from bacteria to mammalian cells. *Proc Natl Acad Sci USA* 1980;77:2163–2167.
318. Schneider RJ, Shenk T. Impact of virus infection on host cell protein synthesis. *Ann Rev Biochem* 1987;56:317–332.
318a. Schnell MJ, Mebatsion T, Conzelmann K-K. Infectious rabies viruses from cloned cDNA. *EMBO J* 1994;13:4195–4203.
319. Scholtissek C. Influenza virus genetics. *Adv Genet* 1979;20:1–36.
320. Scolnick EM, McLean AA, West DJ, McAleer WJ, Miller WJ, Buynak EB. Clinical evaluation in healthy adults of a hepatitis B vaccine made by recombinant DNA. *JAMA* 1984;251:2812–2815.
321. Sculley TB, Kreofsky T, Pearson GR, Spelsberg TC. Partial purification of the Epstein-Barr virus nuclear antigen(s). *J Biol Chem* 1983;258:3974–3982.
322. Sedivy JM, Capone JP, Raj Bhandary UL, Sharp PA. An inducible mammalian amber suppressor: propagation of a poliovirus mutant. *Cell* 1987;50:379–389.
323. Sells MA, Chen M-L, Acs G. Production of hepatitis B virus particles in Hep G2 cells transfected with cloned hepatitis B virus DNA. *Proc Natl Acad Sci USA* 1987;84:1005–1009.
324. Sharpe AH, Fields BN. Pathogenesis of reovirus. In: Joklik WK, ed. *The reoviridae*. New York: Plenum Press; 1983:229–285.
325. Sheldrick P, Berthelot N. Inverted repetitions in the chromosome of herpes simplex virus. *Cold Spring Harbor Symp Quant Biol* 1974;39: 667–678.
326. Shih MF, Arsenakis M, Tiollais P, Roizman B. Expression of hepatitis B virus S gene by herpes simplex virus type 1 vectors carrying alpha- and beta-regulated gene chimeras. *Proc Natl Acad Sci USA* 1984;81: 5867–5870.
327. Shimizu K, Mullinix MG, Chanock RM, Murphy BR. Temperature-sensitive mutants of influenza A/Udorn/72 (H3N2) virus. II. Genetic analysis and demonstration of intrasegmental complementation. *Virology* 1982;117:45–61.
328. Shimotohno K, Temin HM. Formation of infectious progeny virus after insertion of herpes simplex thymidine kinase gene into DNA of an avian retrovirus. *Cell* 1981;26:67–77.
328a. Shortle DR, Margolskee RF, Nathans D. Mutational analysis of the simian virus 40 replicon: pseudorevertants of mutants with a defective replication origin. *Proc Natl Acad Sci USA* 1979;76:6128–6131.
329. Smith DB, Inglis SC. The mutation rate and variability of eukaryotic viruses: an analytical review. *J Gen Virol* 1987;68:2729–2740.
330. Spaete RR, Frenkel N. The herpes simplex virus amplicon: a new eukaryotic defective-virus cloning-amplifying vector. *Cell* 1982;30: 295–314.
330a. Spirin AS, Baranov YI, Ryabova LA, Ovodov SY, Alakhov YB. A continuous cell-free translation system capable of producing polypeptides in high yield. *Science* 1988;242:1162–1164.
331. Stahl FW. *Genetic recombination: thinking about it in phage and fungi.* San Francisco: WH Freeman; 1979.
332. Steinhauer DA, Holland JJ. Direct method for quantitation of extreme polymerase error frequencies at selected single base sites in viral RNA. *J Virol* 1986;57:219–228.
333. Stow ND. Cloning of a DNA fragment from the left-hand terminus in the adenovrius type 2 genome and its use in site-directed mutagenesis. *J Virol* 1981;37:171–180.
334. Stow ND. Localization of an origin of DNA replication within the TRS/IRs repeated region of the herpes simplex virus type 1 genome. *EMBO J* 1982;1:863–867.
335. Stow ND. The infectivity of adenovirus genomes lacking DNA sequences from their left-hand termini. *Nucleic Acids Res* 1982;10: 5105–5119.
336. Strauss EG, Strauss JH. Structure and replication of the alphavirus genome. In: Schlesinger S, Schlesinger MJ, eds. *The togaviridae and flavidiridae*. New York: Plenum Press; 1986:35–90.
337. Strauss JH, Strauss EG. Evolution of RNA viruses. *Annu Rev Microbiol* 1988;42:657–683.
338. Stroobant P, Rice AP, Gillick WJ, Chen IM, Waterfield MD. Purification and characterization of vaccinia virus growth factor. *Cell* 1985;42:383–393.
339. Stunnenberg HG, Lange H, Philippson L, van Miltenberg RT, van der Vliet PC. High expression of functional adenovirus DNA polymerase and precursor terminal protein using recombinant vaccinia virus. *Nucleic Acids Res* 1988;16:2431–2444.
340. Subak-Sharpe JH, Timbury MC. Genetics of herpesviruses. In: Fraenkel-Conrat H, Wagner RR, eds. *Comprehensive virology*, vol. 9. New York: Plenum Press; 1977:89–131.
341. Summers MD, Smith GE. *A manual of methods for baculovirus vectors and insect cell culture procedures.* College Station, TX: Texas Agricultural Experiment Station; 1987.
342. Takahashi M. Isolation of conditional lethal mutants (temperature sensitive and host dependent mutations) of adenovirus type 5. *Virology* 1972;49:815–817.
343. Takamiya Y, Short MP, Ezzeddine ZD, Moolten FL, Breakefield XO, Martuza RL. Gene therapy of malignant brain tumors: a rat glioma line bearing the herpes simplex virus type 1-thymidine kinase gene and wild type retrovirus kills other tumor cells. *J Neurosci Res* 1992; 33:493–503.
344. Tanese N, Goff SP. Domain structure of the Moloney murine leukemia virus reverse transcriptase: mutational analysis and separate expression of the DNA polymerase and RNase H activities. *Proc Natl Acad Sci USA* 1988;85:1777–1781.
345. Tegtmeyer P. Genetics of SV40 and polyoma virus. In: Tooze J, ed. *DNA tumor viruses.* Cold Spring Harbor, NY: Cold Spring Harbor Laboratory; 1980:297–338.
346. Taylor JW, Ott J, Eckstein F. The rapid generation of oligonucleotide-directed mutations at high frequency using phosphorothioate-modified DNA. *Nucleic Acids Res* 1985;13:8765–8785.
347. Temin HM, Rubin H. Characteristics of an assay for Rous sarcoma virus and Rous sarcoma cells in tissue culture. *Virology* 1958;6:669–688.
348. Temin HM. Origin of retroviruses from cellular moveable genetic elements. *Cell* 1980;21:599–600.
349. Thimmappaya B, Weinberger C, Schneider RJ, Shenk T. Adenovirus VAI RNA is required for efficient translation of viral mRNA at late times after infection. *Cell* 1982;31:543–551.
350. Tolpin MD, Clements ML, Levine MM, et al. Evaluation of a phenotypic revertant of the A/Alaska/77-ts-1A2 reassortant virus in hamsters and in seronegative adult volunteers: further evidence that the temperature sensitive phenotype is responsible for attenuation of ts-1A2 reassortant viruses. *Infect Immun* 1982;36:645–650.
351. Tolskaya EA, Romanova LA, Kolesnikova MS, Agol VI. Intertypic recombination in poliovirus: genetic and biochemical studies. *Virology* 1983;124:121–132.
352. Topp WC, Lane D, Pollack R. Transformation by SV40 and polyoma viruses. In: Tooze J, ed. *DNA tumor viruses*. Cold Spring Harbor, NY: Cold Spring Harbor Laboratory; 1980:205–296.
353. Traktman P. Poxviruses: an emerging portrait of biological strategy. *Cell* 1990;62:621–626.
354. Triezenberg SJ, Kingsbury RC, McKnight SL. Functional dissection of VP16, the *trans*-activator of herpes simplex virus immediate early gene expression. *Genes Dev* 1988;2:718–729.
355. Trono D, Andino R, Baltimore D. An RNA sequence of hundreds of nucleotides at the 5' end of poliovirus RNA is involved in allowing viral protein synthesis. *J Virol* 1988;62:2291–2299.
356. Turek LP, Byrne JC, Lowy DR, Dvoretzky I, Friedman RM, Howley PM. Interferon induces morphologic reversion with elimination of extrachromosomal viral genomes in bovine papilloma virus-transformed mouse cells. *Proc Natl Acad Sci USA* 1982;79: 7914–7918.
357. Twardzik DR, Brown JP, Ranchalis JE, Todaro GJ, Moss B. Vaccinia virus-infected cells release a novel polypeptide functionally related to transforming and epidermal growth factors. *Proc Natl Acad Sci* 1985;82:5300–5304.
358. Valuenzuela P, Medina A, Rutter WJ, Ammerer G, Hall BD. Synthe-

sis and assembly of hepatitis B virus surface antigen particles in yeast. *Nature (Lond)* 1982;298:347–350.
359. van der Most RG, Heijnen L, Spaan WJ, de Groot RJ. Homologous RNA recombination allows efficient introduction of site-specific mutations into the genome of coronavirus MHV-A59 via synthetic co-replicating RNAs. *Nucleic Acids Res* 1992;20:3375–3381.
360. van Zijl M, Quint W, Briaire J, de Rover T, Gielkens A, Berns A. Regeneration of herpesviruses from molecularly cloned subgenomic fragments. *J Virol* 1988;62:2191–2195.
361. Vande Woude GF, Oskarsson M, Enquist LW, Nomura S, Sullivan M, Fischinger PJ. Cloning of integrated Moloney sarcoma proviral DNA sequences in bacteriophage. *Proc Natl Acad Sci USA* 1979;76: 4464–4468.
362. Varmus HE. Form and function of retroviral proviruses. *Science* 1982;216:812–820.
363. Varmuza SL, Smiley JR. Signals for site-specific cleavage of HSV DNA: maturation involves two separate cleavage events at sites distal to the recognition sequences. *Cell* 1985;1:793–802.
364. Vogt M, Dulbecco R. Properties of HeLa cell culture with increased resistance to poliomyelitis virus. *Virology* 1958;5:425–434.
365. Vogt PK. Genetics of RNA tumor viruses. In: Fraenkel-Conrat H, Wagner RR, eds. *Comprehensive virology*, vol. 9. New York: Plenum Press; 1977:341–455.
366. Volkert FC, Young CSH. The genetic analysis of recombination using adenovirus overlapping terminal DNA fragments. *Virology* 1983; 125:175–193.
367. Von Magnus P. Incomplete forms of Influenza virus. *Adv Virus Res* 1954;2:59–86.
368. von Ruden T, Gilboa E. Inhibition of human T-cell leukemia virus type I replication in primary human T cells that express antisense RNA. *J Virol* 1989;63:677–682.
369. Wadsworth S, Jacob J, Roizman B. Anatomy of herpes simplex virus DNA. II. Size, composition and arrangement of inverted terminal repetitions. *J Virol* 1975;15:1487–1497.
370. Wake CT, Wilson JH. Defined oligomeric SV40 DNA: a sensitive probe of general recombination in somatic cells. *Cell* 1980;21:141–148.
371. Walen KH. Demonstration of inapparent heterogeneity in a population of an animal virus by single burst analysis. *Virology* 1963;20:230–234.
372. Walsh CE, Liu JM, Xiao X, Young NS, Nienhuis AW, Samulski RJ. Regulated high level expression of a human gamma-globin gene introduced into erythroid cells by an adeno-associated virus vector. *Proc Natl Acad Sci USA* 1992;89:7257–7261.
373. Ward CD, Stokes MAM, Flanegan JB. Direct measurement of the poliovirus RNA polymerase error frequency *in vitro*. *J Virol* 1988; 62:558–562.
374. Webster RG, Laver WG, Air GM, Schild GC. Molecular mechanisms of variation in influenza viruses. *Nature (Lond)* 1982;296:115–121.
375. Wei C-M, Gibson M, Spear P, Scolnick, EM. Construction and isolation of a transmissible retrovirus containing the src gene of Harvey murine sarcoma virus and the thymidine kinase gene of herpes simplex virus type 1. *J Virol* 1981;39:935–944.
376. Weiss RA, Mason WS, Vogt PK. Genetic recombinants and heterozygotes derived from endogenous and exogenous avian RNA tumor viruses. *Virology* 1973;52:535–552.
377. Whitehouse HLK. *Genetic recombination: understanding the mechanisms*. New York: Wiley; 1982.
378. Wiberg FC, Sunnerhagen P, Kaltoft K, Zeuthen J, Bjursell G. Replication and expression in mammalian cells of transfected DNA: description of an improved erythrocyte ghost fusion technique. *Nucleic Acids Res* 1983;11:7287–7302.
379. Wigler M, Silverstein S, Lee L-S, Pellicer A, Cheng Y-C, Axel R. Transfer of purified herpes virus thymidine kinase gene to cultured mouse cells. *Cell* 1977;11:223–232.
380. Will H, Cattaneo R, Koch HG, Darai G, Schaller H. Cloned HBV DNA causes hepatitis in chimpanzees. *Nature* 1982;299:740–742.
381. Wilson JH, DePamphilis M, Berg P. Simian virus 40 permissive cell interactions: selection and characterization of spontaneously arising monkey cells that are resistant to simian virus 40 infection. *J Virol* 1976;20:391–399.
382. Wilson JH. *Genetic recombination*. Menlo Park, CA: Benjamin-Cummings; 1985.
383. Wimmer E, Hellen CUT, Cao X. Genetics of polio virus. *Ann Rev Genet* 1993;27:353–436.
384. Wolgemuth DJ, Hsu MT. Visualization of genetic recombination intermediates of human adenovirus type 2 DNA from infected HeLa cells. *Nature (Lond)* 1980;287:168–170.
385. Wu CA, Nelson NJ, McGeoch DJ, Challberg MD. Identification of herpes simplex virus type 1 genes required for origin-dependent DNA synthesis. *J Virol* 1988;62:435–443.
386. Yaginuma K, Shirakata Y, Kobayashi M, Koike K. Hepatitis B virus (HBV) particles are produced in a cell culture system by transient expression of transfected HBV DNA. *Proc Natl Acad Sci USA* 1987;84:2678–2682.
387. Yates JL, Warren N, Sugden B. Stable replication of plasmids derived from Epstein-Barr virus in various mammalian cells. *Nature (Lond)* 1985;313:812–815.
388. Yewdell JW, Bennink JR, Smith GL, Moss B. Influenza A virus nucleoprotein is a major target antigen for cross-reactive anti-influenza virus cytotoxic T lymphocytes. *Proc Natl Acad Sci USA* 1985;82: 1785–1789.
389. Young CSH, Silverstein SJ. The kinetics of adenovirus recombination in homotypic and heterotypic genetic crosses. *Virology* 1980; 101:503–515.
390. Young CSH, Cachianes G, Munz PL, Silverstein S. Replication and recombination in adenovirus-infected cells are temporally and functionally related. *J Virol* 1984;51:571–577.
391. Young CSH, Shenk T, Ginsberg HS. The genetic system. In: Ginsberg HS, ed. *The adenoviruses*. New York: Plenum Press; 1984:125–172.
392. Youngner JS, Quagliani DO. Temperature sensitive mutants of VSV are conditionally defective particles that interfere with and are rescued by wild type virus. *J Virol* 1978;19:102–107.
393. Youngner JS, Preble OT. Viral persistence: evolution of viral populations. In: Fraenkel-Conrat H, Wagner RR, eds. *Comprehensive virology*, vol. 16. New York: Plenum Press; 1980:73–155.
394. Zavada J, Rosenburgova M. Phenotypic mixing of vesicular stomatitis virus with fowl plague virus. *Acta Virol* 1972;16:103–114.
395. Zavada J. Pseudotypes of vesicular stomatitis virus with the coat of murine leukemia and avian myeloblastosis viruses. *J Gen Virol* 1972;15:183–191.
396. Zervos AS, Gyuris J, Brent R. Mxi1, a protein that specifically interacts with Max to bind Myc-Max recognition sites. *Cell* 1993;72: 223–232.

Fundamental Virology, Third Edition
edited by B.N. Fields, D.M. Knipe, P.M. Howley, et al.
Lippincott - Raven Publishers, Philadelphia © 1996

CHAPTER 5

Virus Evolution

Ellen G. Strauss, James H. Strauss, and Arnold J. Levine

It is clear to any virologist that viruses are highly evolved organisms and that they are continuing to evolve today. As obligate intracellular parasites with limited genome size, they are among the most efficient and economical of life forms. Constrained variously by geometric limitations on size and shape and (in the case of RNA viruses) by the inherent error frequencies in their replication processes, viruses encode primarily those functions that they cannot adapt from their hosts and waste few nucleotides on nonfunctional nucleic acid. Indeed, some viruses have evolved ways to increase their coding capacity by utilizing more than one reading frame to encode unrelated proteins by alternative initiation during translation (34,45), by differential splicing of mRNAs (61), or by frameshifting by the insertion of nontemplated nucleotides within an open reading frame (ORF) during transcription (80). An additional constraint on certain viruses is the necessity to replicate alternately in two very different hosts (e.g., the replication of yellow fever virus in mosquitoes and humans). Finally, all viruses face two competing structural requirements; extracellular virion particles must be not only sufficiently stable to survive outside of a cell for minutes or days, but also capable of rapid disassembly upon entry into a susceptible host cell.

What, then, does the subject of virus evolution entail? Viruses have left no fossil record; moreover, even the symptoms that we recognize today as virus-caused can be tracked back only a few millennia to the beginnings of recorded human history. For comparative studies, we are in possession of virus isolates not more than 80 years old. From a detailed analysis of the properties of the viruses extant today, we can hope to extrapolate backward toward the origin of viruses and to project forward toward future patterns of change and evolution, especially among human pathogens. To intervene and protect humans and animals from viral outbreaks, we need to know how viruses change in response to changing environmental factors, how rapidly they change, and to what type of selection pressures they are responding.

Viruses are very diverse in size, structure, and genome organization. For our purposes, the essential characteristics that define a virus are that it is an obligate intracellular parasite whose DNA or RNA is encapsidated in protein encoded in its genome, that has an evolutionary history independent of that of its host. Beyond this, there is virtual-

E. G. Strauss and J. H. Strauss: Division of Biology, California Institute of Technology, Pasadena, California 91125.
A. J. Levine: Department of Molecular Biology, Lewis Thomas Laboratory, Princeton University, Princeton, New Jersey 08544.

ly a continuum of viruses, virusoids, plasmids, transposable elements, insertion elements, and so forth. Current comparisons of viruses at the level of their replication and translation strategies, the nucleotide sequences of their genomes, the amino acid sequences of their encoded proteins, and detailed structural studies at atomic resolution are all contributing to a much clearer picture of the interrelationships among viruses and among virus families. All viruses can be divided into three main groups: those viruses with DNA as their genetic material, those with both DNA and RNA as their genetic material at different stages of their life cycle (retroviruses, caulimoviruses, hepadnaviruses), and those with RNA as their genetic material. The differences between these three groups are profound and may indicate independent origins as well as differences in their evolution.

THEORIES OF THE ORIGIN OF VIRUSES

To discuss theories of the origin of viruses, we need to define what we mean by *originate* and how we propose to recognize an origin event. Here we have defined the origin of a virus or genetic element as that time when its replication and evolution become independent of the macromolecules from which it was derived. When a virus acquires the genetic information to determine its own propagation and destiny, then it has achieved the status of a new and independent genetic element.

Three different theories have been proposed to explain the origins of viruses: (i) the regressive theory of virus origins (12,38,69,70), (ii) the theory that viruses originated from host cell RNA (14,108) and/or DNA (49) components, and (iii) the theory that viruses originated and evolved along with the most primitive molecules containing self-replicating abilities (75,108). The first two theories assume that viruses originated after their hosts were fully developed, whereas the third views viruses as having co-evolved with cells from the origin of life itself.

Regressive Theory of Virus Origins

The regressive theory of virus origins proposes that viruses are degenerate forms of intracellular parasites. Such intracellular parasitism could result when a microorganism becomes critically dependent upon a metabolite that cannot pass through a cell membrane. Once inside, such a parasite could progressively lose other biosynthetic abilities without affecting its selection. The essential functions a parasite must retain would be an origin of DNA replication (a *cis*-acting site), a way to regulate replication (a *trans*-acting DNA replication initiation protein), and a way to interact with the host biosynthetic and replicative machinery (both *cis*- and *trans*-acting functions). The end result of selection would be a small DNA molecule or plasmid that replicates as an obligate intracellular parasite (49).

At present, there are some problems with this type of explanation for the origin of viruses (68), especially if the goal is a single theory to explain the origin of all viruses, both RNA- and DNA-containing. Present-day examples of intracellular parasitism could be viewed as modern-day "intermediates" in such an evolutionary process (19). Parasitic bacteria such as the leprosy bacillus have apparently evolved in this degenerative direction. Rickettsiae, which are small gram-negative bacteria that multiply by fission and that have lost some biosynthetic activities essential for free living forms, have been considered to be possible intermediates between bacteria and viruses. Even closer to viruses are the chlamydiae, which are also prokaryotic parasites with a more extensive loss of biosynthetic abilities. They have no cell wall but do contain both DNA and RNA macromolecules and multiply by binary fission, retaining their cellular structure. Indeed, for a long time these agents, the lymphogranuloma-psittacosis-trachoma group, were classified as viruses until it became clear that they are degenerative bacteria.

A similar regressive theory has been proposed to explain the origin of cellular organelles such as mitochondria and chloroplasts (92,113). These organelles are about the same size as bacteria and contain circular DNA genomes (like bacteria but much smaller) that encode the information to make ribosomal RNA and transfer RNA for an organelle-specific translation apparatus. These organelles also contain genes that produce proteins that are synthesized and retained within the organelle.

Viruses, on the other hand, never encode ribosomal RNA subunits, nor do they set up their own translational machinery. Most viruses complete their life cycle by packaging their genetic information into a coat composed, in the main, of viral encoded proteins. Viral coats may be thought of as being analogous to some subcellular particulate structures (bacterial flagella, actin filaments) that self-assemble from protein subunits. The regressive theory would explain virus origins in two stages. First, selection for an organelle (92,113) followed by the production of a free DNA plasmid in the cell (49). Second, the formation of viral capsid proteins via random mutations in the genes encoding the subunits of subcellular structures. The newly acquired trait of transmission from cell to cell would be favorably selected for over evolutionary time scales (70). The example of virusoids evolving from viroids (discussed below) is sometimes cited as an example of the regressive theory.

Origin from Cellular RNA and/or DNA Components

In its simplest form, this theory postulates that viruses arose from normal cellular components that gained the ability to replicate autonomously and thereby evolve independently. This is an attractive proposal, since it could account for the origins of all viruses—DNA viruses arising from plasmids or transposable elements, retroviruses from retrotransposons, and RNA viruses from self-replicating mRNAs. The requirements for autonomy are twofold; all viruses need an origin of replication and a source of a protective (protein) coat for the extracellular phase of their existence. In addition, most RNA viruses need an RNA replicase enzyme.

Origins of Replication

Autonomously replicating molecules, be they RNA or DNA, require some kind of origin of replication or start site; by definition, even viroids must have such a site. For RNA viruses, origins of replication are usually the ends of linear molecules. In some cases the end of the RNA is capable of folding into a stable three-dimensional structure; in others it has a protein bound covalently to it. In still other cases no special feature other than the linear sequence distinguishes the start site. In DNA viruses the origin is usually internal and contains elements that are discussed in more detail below. Events at the origin of replication, such as providing a protein or nucleic acid primer, commonly require (*trans*-acting) sequence recognition proteins that regulate each round of replication. Because viruses rarely encode all of the proteins required for viral RNA or DNA replication, viral proteins often act in concert with host cell-encoded gene products to initiate replication.

Many DNA viruses also encode their own DNA polymerases (T4, T7, adenoviruses, herpesviruses, poxviruses) and other replicative enzymes (primases, helicases, ligases, single-stranded binding proteins) (59). RNA viruses encode RNA-dependent RNA polymerases, capping enzymes, polyadenylation enzymes, and in some cases proteases to process primary translation products (68). Any theory of virus origin and evolution must be able to accommodate these diverse functions and their complex interactions.

Viroids and Virusoids

Viroids and virusoids are the smallest known autonomously replicating molecules (reviewed in references 20 and 100). *Viroids* are plant pathogens that are composed of single-stranded covalently circular RNA molecules 240 to 375 residues long. *Virusoids* are satellite RNAs of the same size, which are encapsulated by proteins encoded by helper viruses. In both cases the RNA is extensively self-complementary and base-paired throughout its length. Viroids replicate primarily in the nucleolus, where the plus strand infectious RNA is copied into multimeric minus strand RNAs by DNA-dependent RNA polymerase II, and the replication of this template into multimeric plus strands is performed by DNA-dependent RNA polymerase I (20). On the other hand, virusoids replicate wholly in the cytoplasm, using RNA-dependent RNA polymerases either from their host cells or from their helper viruses. Both viroids and virusoids seem to replicate by a rolling circle model where the plus strand acts as a template to produce many linear minus strands in a tandem array. In some virusoids and viroids, both plus strands and minus strands self-cleave to generate monomeric units. In other viroids and virusoids, the minus strand is not cleaved and the only monomeric forms are plus strands (20). Although virusoids seem not to multiply in the absence of their helpers, it is unclear whether this is because they lack replication functions or lack the genes encoding the encapsidating proteins (20). It is clear from a comparison of several viroids that their genomes are composed of regions or domains with significant sequence homologies and other regions that have high sequence variability (100). Viroids could have arisen from cellular RNAs that acquired an origin of replication. Neither viroids or virusoids encode any protein species (20), and they are completely dependent upon host functions for replication, unlike other RNA viruses. Unlike animal cells, many types of plant cells contain RNA-dependent RNA polymerase activities capable of replicating virusoids (15). It is unclear whether these host cell polymerases are the "fossil footprints" of unidentified plant viruses (which now either are integrated into the genome or are degenerate or cryptic) or whether these enzymes have essential roles in the uninfected plant.

The concept that viruses have their origin in such autonomously replicating and self-splicing RNA molecules has some attractive possibilities, although it might be a very long evolutionary route to complex present-day viruses. However, it is also possible that, rather than progenitors or models for viral origins, these small RNAs may be recent and sophisticated parasites, which could have arisen de novo from cellular elements or by degenerative evolution of more complex self-replicating RNA viruses, retaining only their capacity for self-splicing and the ability to cause disease in plants.

Plasmids, IS Elements, and Transposons

There are a number of cellular DNA molecules, such as the cellular plasmids and transposons, whose origin and evolution may be related to those of viruses. Bacterial episomes are circular DNA molecules that can exist either in an integrated state within the host chromosome or as autonomous plasmids that replicate once per generation (42,49,68). The minimal requirements for plasmid duplication are the same as for viruses but, in addition, those molecules must be segregated efficiently to the daughter cells after cell division. Plasmids can also contain additional genes that encode products useful to the host cell, such as genes for antibiotic resistance or genes to promote genetic exchange and conjugation of plasmids and host DNA (49,91). Some restriction enzyme modification systems are encoded by plasmid genes as well as the genetic information for some biodegradative enzyme activities (1). The origin and evolution of plasmids may tell us something about the origins of DNA viruses.

A careful analysis of plasmid sequences has shown that many of the banks of genes in plasmids (e.g., for antibiotic resistance) are carried into the plasmid as a genetic unit on a movable genetic element called a transposon (1,91). Transposons are DNA elements (750 bp to 40,000 bp in size) that can move from one site in the DNA to another. In most and probably all transposons, the movement is mediated by two insertion elements (IS elements) on either end of the transposon that encode the functions required for their own movement (16,56). All IS elements contain *cis*-acting terminal inverted repeated DNA sequences (20 to 40 base pairs long) flanking the gene encoding a *transposase* and its regulatory sequences (1,108). The transposase acts by binding to the inverted terminal repeated sequences and to adjacent nucleotide sequences (3 to 12 base pairs) at the integration site in chromosomal DNA. This enzyme then transposes the element to a new chromosomal location. Some IS elements also encode a second enzyme called *resolvase,* which mediates a second stage of transposition (108).

There are basically three types of transposition events: (i) Simple transposition employs the transposase only to excise the element and move it to a different site. (ii) Replicative transposition involves the transposase (and host enzymes) duplicating the DNA element, forming a tandem integration intermediate that is acted upon by resolvase to produce a free copy that moves to a new location. The transposon Tn3 is an example of such an element; it contains terminal repeated DNA sequences and genes encoding both a transposase and a resolvase, and it carries out replicative transposition (56). (iii) Complex replicative transposition involves an RNA intermediate. The DNA element, known as a *retrotransposon,* is first copied into RNA. This RNA encodes a reverse transcriptase activity, which copies it back into DNA. The DNA copy is then integrated by site-specific cleavage into a new location by a specific integrase activity such that it can be transcribed again into an identical linear RNA copy. Class 1 retrotransposons, which include Ty-1 in yeast and Copia and 412 in *Drosophila* (91), share significant sequence homology with retroviruses (24).

These remarkable movable elements accelerate the rate of evolution by moving units or blocks of genes into and out of cellular plasmids and viral chromosomes. This has been called *modular evolution,* and there are many examples where viral infection results in the insertion of transposons and associated genes from the host cells into progeny viruses (108). Thus, transposons carried by viruses can move between hosts as infectious units accelerating the rate of viral and host cell evolution by adding new gene combinations to the pool. IS elements can also have a marked effect upon viral evolution by inserting into the viral genome in such a way that they inactivate a viral gene (9,91). Conversely, some IS elements encode enhancer or promoter signals that can activate or positively regulate the transcription of genes adjacent to the point of insertion. This has been observed with IS 2 and 3 elements in *E. coli* and the Ty-1 element in yeast (1,91). Furthermore, when retrovirus DNA derived from an avian leukosis virus is inserted into the host chromosome near the oncogene C-*myc*, the insertion results in increased transcription of that gene and the development of a bursal lymphoma in the infected chickens (43). Similar results have been reported in the mouse, where retrovirus insertion sites have been used to identify adjacent DNA sequences as oncogenes (77).

Many DNA viruses share suggestive similarities in genome organization with the transposons. Like transposons, the genomes of herpesviruses, adenoviruses, and parvoviruses contain inverted terminal repeated sequences; like transposons, the viruses contain genes between these terminal repeated sequences that facilitate virus replication and movement from one cell to another. Whether these structural or sequence organizational motifs found in viruses and transposable elements are the result of evolution from common ancestors or from convergent evolution to find common solutions to replicative problems remains unclear. Interestingly, there is a bacterial virus, Mu, that can clearly be classified as both a virus and a transposable element. Unlike all other lysogenic bacteriophages, Mu randomly integrates into the host chromosome in both the lytic and the lysogenic state. During the lytic cycle, the virus transposes to a new position in the host chromosome, and the productive cycle of this virus seems to require these transposition events (9,91).

Self-replicating mRNAs

RNA viruses, which in general are smaller and have more limited coding capacity than do DNA viruses, may have had an independent origin. One possibility for the origin of RNA viruses is that a cellular mRNA (e.g., the mRNA encoding a DNA-dependent RNA polymerase) acquired the ability to replicate itself by obtaining an origin of replication. If such a self-replicating mRNA could borrow a gene for a protein coat, then a very simple RNA virus would be born.

Origin from Self-Replicating RNA Molecules

A reasonable case has been made that polymers of ribonucleotides can contain both the information required and the functional capacity needed to permit a self-replicating system to originate (50). The fact that RNA molecules can catalyze several types of chemical reactions has stimulated interest in these mechanisms playing a role in the origin of life processes and the evolution of viruses (13,108). At least three related enzymatic activities can be carried out by small and simple RNA sequences: (i) a ribonuclease activity or

(RNase P) of *E. coli*, (ii) self-splicing by RNA molecules (ribozymes) to delete internal nucleotide sequences, and (iii) a template-dependent synthesis of polycytidylic acid using an RNA primer (13,14). Thus, known RNA molecules can carry out three fundamental reactions that are necessary for replication and evolution. These observations argue in favor of an RNA-based evolving system that preceded our current DNA-RNA-protein lifestyle (50). These original RNA information-duplication systems are then postulated to have evolved into an RNA-protein–mediated set of reactions. In a third step, DNA, which is more stable than RNA, becomes the information storage system of life. The very reactivity of RNA, which favored its use in catalysis, mediates against its use as a stable molecule for the storage of information (75). As these molecules were packaged into cells or organisms, some of the genetic information was combined into a unit called the *host cell.* This theory suggests that other self-replicating RNAs co-evolved and parasitized these hosts, eventually becoming the RNA viruses. Modern-day viroids and virusoids, which retain some of these RNA enzymatic capabilities, are thought by some to be "fossils" of the prebiotic RNA world.

FACTORS AFFECTING DNA VIRUS EVOLUTION

Mutation

The two major forces acting upon viral genomes to generate diversity that can be tested for environmental survival and replicative fitness are mutation and recombination. Some viruses have a good deal of control over their own rates of mutation and even the frequency of recombination. They exert control by encoding viral enzymes for DNA polymerization and other replicative functions and even recombinational functions. Mutations in the T4 bacteriophage DNA polymerase or the DNA replicative single-strand DNA binding protein (gene 32 protein) can alter the rate of mutation of the T4 virus genome (31). The adenoviruses, herpesviruses, and poxviruses all encode their own DNA polymerases. Although the host cell has evolved an elaborate system of testing for mismatched base pairs at or near a replication fork (proofreading), it is not clear whether these viral enzymes have similar editing functions, as the rate of mutation has not been determined for many DNA viruses. On the other hand, the mutation rate (in a single cycle of replication) for a chicken retrovirus genome (RNA to DNA to RNA) seems to be about 1 in 10^5 to 1 in 10^4 nucleotides per replication cycle (25,62), which is higher (by more than 4 orders of magnitude) than similar estimates for the host cell genome (6). This suggests that retroviruses, like other RNA viruses, are evolving much more rapidly than are their hosts.

In addition to high rates of mutation per generation, viruses live through many generations in a short time span. For example, a single adenovirus particle can infect a cell and produce 250,000 DNA molecules during a single replication cycle, which is the equivalent of 17 generations (22). A third reason why viruses may evolve rapidly is that infection at high multiplicity permits the replication or duplication of defective viruses by genetic complementation, which can then be passed on to new hosts. Such "evolutionary variants" have been produced in cell culture and studied with SV40 (7,74) and polyoma viruses (66). The viral genomes contain deletions and duplications of viral sequences, insertions and substitutions of cellular DNA, and rearrangements of viral genes and viral *cis*-acting sites. Both highly reiterated and unique copies of cellular DNAs are found in the viral genome of defective variants, and this mating with the host cell chromosome seems to be fairly random and not to occur by homologous recombination. All evolutionary variants retain an origin of DNA replication, which is commonly reiterated one or several times. Clearly, the growth of variants at high multiplicity permits the loss of *trans*-acting functions that can be complemented, but *cis*-acting gene functions, like an origin of replication and specific encapsidation sequences, are retained (103).

In some cases, relatively simple mutations in certain *cis*-acting signals can have large effects upon the biology of the virus. Polyoma is a virus that replicates in many different tissues of the mouse and is able to form tumors in many tissues (therefore the name poly-oma) (103). The virus fails to replicate efficiently in embryonic mouse cells or in cells derived from a tumor of such embryonic cells (murine embryonal carcinoma cells) (63). However, mutants of polyoma virus have been selected for growth in various embryonal carcinoma cell lines (Fig. 1) (30,52,90,106) and, in all cases, the mutations map in the viral *cis*-acting enhancer element that regulates transcription and DNA replication (63). Furthermore, different cell lines of embryonal carcinoma cells select for different mutations (90,106). This is an example of a relatively simple set of mutations that change the host range of a virus. In a new host new selection pressures predominate and force the virus to evolve along a new pathway. Mutations that alter the host range of viruses may be in coat protein genes (111), in gene products that mediate the penetration of a virus into a cell (89), in uncoating gene functions (93), or in gene products involved in transcriptional regulation and splicing (57), as well as the *cis*-acting mutants that regulate viral enhancer functions, discussed above. Simple mutations that affect the interaction of the virus and its host and change the host range can lead ultimately to significant alterations in the virus.

Recombination

Clustering of cis*-acting Elements*

Some viruses contain the genetic information to encode proteins that enhance recombination between viral genomes

1) Growth of Polyoma Host Range Mutants

Virus	Virus Titer in PFU/ml	
	Mouse cells (3T3)	Embryonal carcinoma cells (F9)
Wild type	2.0×10^7	$<10^2$
Mutant 1	1.2×10^7	9×10^4
Mutant 2	2.3×10^7	2.2×10^5

2) Sequence Alterations in the Mutants

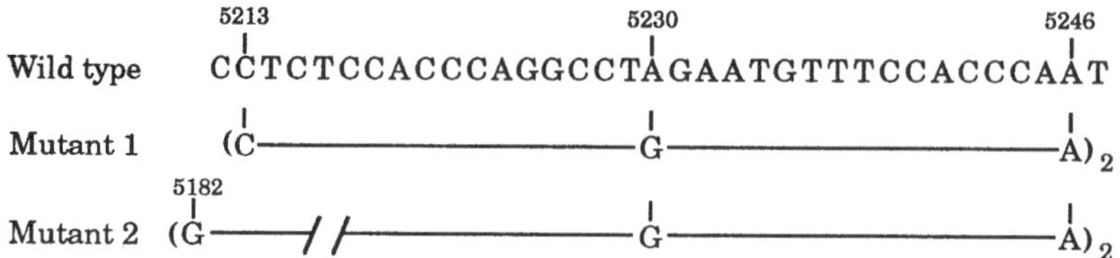

3) Polyoma Transcription and Replication Control Region

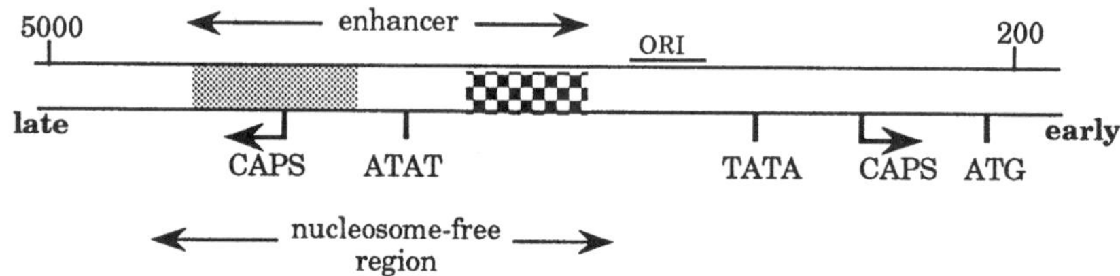

FIG. 1. Polyoma virus host range mutations. **1:** Growth of wild-type polyoma virus and two host range mutants in mouse 3T3 cells and F9 murine embryonal carcinoma cells. **2:** Determination of host range mutations. Each mutant contained a single A→G substitution at nucleotide 5230 and a tandem duplication of the sequence shown between the parentheses. **3:** A map of the enhancer region of the polyoma virus genome, showing the location of these mutations *(checkerboard box)* (90) and others that permit growth in another cell line nonpermissive for wild-type polyoma (*shaded box*) (106). The enhancer is a *cis*-acting element that regulates transcription and DNA replication of the virus, and the DNA for these signals is localized in a nucleosome-free region of the viral genome. Many *cis*-acting sites are clustered in this region, including an origin of DNA replication (ori) and transcriptional signals (TATA and CAPS), as well as the ATG start of translation. *Early* and *late* refer to two sets of genes expressed at different times in the replication cycle.

(i.e., the lambda virus *exo* and *bet* gene products) (108) and thus speed up their own evolution. *Cis*-acting sites on viral genomes usually interact with (bind to) a viral or cellular *trans*-acting protein (repressor-operator). Recombination between related viruses might produce a genome containing a *trans*-acting gene product trying unsuccessfully to interact with a *cis*-acting element derived from another virus. To avoid these difficulties, viral genomes are commonly organized with *cis*-acting elements located near the genes encoding the proteins that bind them (see Fig. 1). The lambda virus C_I and Cro protein genes are located adjacent to their operator binding sites (108). In bacteriophage T7, seven different *cis*-acting signals are clustered into 193 base pairs including transcriptional signals, the origin of DNA replication, mRNA processing sites, and a ribosomal binding site (28). Such clustering ensures that this integrated unit of *cis*-acting signals will be inherited together during recombination.

A good example of the stringent controls on genome organization in viruses is the organization of the late genes that encode the structural proteins of adenoviruses. The genes for these viral proteins are organized into five families (L1 to L5) that span a length of 30,000 base pairs or about 80% of the genome (22,103). Transcription starts at a single site (16.3% in from one end of the linear genome) and covers the entire 30-kb DNA. This long primary transcript is processed by splicing and polyadenylated to produce each of the five families of late RNAs. Each of these five RNAs then serves as the precursor of a set of spliced mRNAs (3 to 5 per family) that provides the 20 different mRNAs for the synthesis of the late proteins. The 20 different structural proteins and late mRNAs are all made in different quantities, which is regulated by the splicing patterns of the primary transcript, which was, in turn, determined by the position of the gene within the major late transcript (114).

Generating Diversity by Recombination

From the discussion above of transposons and of nonhomologous recombination between the SV40 and polyoma variants and their hosts, it is clear that recombination plays a large role in evolutionary change among DNA viruses. Several additional examples will serve to reinforce this point. Bacteriophages T3 and T7 are closely related viruses, based on genome organization and sequence homologies. In both viruses, the gene named 0.3 serves the same function [i.e., to block a host cell restriction enzyme from cleaving the viral DNA (28,72)] and is located in the same relative position in the genome. However, the genes and their products are completely different. The T7 antirestriction protein binds to and inactivates the *E. coli* restriction enzymes *Eco*B and *Eco*K, preventing them from cutting the viral DNA. The T3 analog of this gene produces an enzyme that hydrolyzes the co-factor S-adenosylmethionine required for restriction enzyme function. One 0.3 gene could not have evolved from the other by an accumulation of point mutations, and, clearly, these two closely related viruses have acquired their antirestriction enzyme protection from different sources, most likely via recombination.

Yet another example of an unusual acquisition of genetic information can be seen in some of the bacteriophage T4 tRNA genes. This virus contains three different genes for tRNA molecules that each produce a primary transcript, which is then required to undergo a self-splicing reaction to eliminate an intron before a functional tRNA is produced (37). Such tRNA genes have been observed in eukaryotic cells and organisms but never in prokaryotes, such as *E. coli*, which is the natural host of T4. The three tRNA genes are located on the T4 chromosome at three different sites, suggesting that a single tRNA original gene has been duplicated and transposed. But how did T4 acquire a self-splicing tRNA gene in a prokaryotic environment? There are, of course, two possibilities: it evolved in the T4 genome by mutation, trial and error, and selection, or it was acquired from a eukaryotic host. *E. coli,* the host of T4, can live in the intestinal tract of many vertebrates, and it is quite possible that at some time in the evolutionary past T4 acquired eukaryotic DNA.

The role of recombination in the constant change of viruses can be seen in a study of human adenoviruses derived from their natural host. Over 40 isolates of human adenoviruses have been classified into seven subgroups on the basis of plaque neutralization tests, hemagglutination, tumorigenesis in animals, restriction endonuclease digestion, and sequence homologies (22). With the use of these data and the size of the structural proteins from different serotypes, attempts have been made to reconstruct evolutionary lineages starting with a first human progenitor virus and leading to the diversity of the present day viruses (88,107). What is interesting from these studies is that new clinical isolates of human adenoviruses arise from recombination of existing serotypes as well as from mutational changes.

Evolutionary Constraints Imposed by Limited Genome Size

Gene Duplication

One constraint on the evolution of viruses may be the limitations imposed on the size of a genome by the volume available to be packaged, in particular in the viruses with isometric heads or capsids. Viral genomes are thought of as having very little room for "noncoding DNA" and, indeed, in the 40,000 bp of the phage T7 genome there are very few nucleotides that are neither coding nor part of *cis*-acting elements (28). Many gene families have arisen and evolved through gene duplication and genetic drift, but this mechanism of gaining diversity is lost if the virus has no room in its genome for duplications.

Not all viruses are constrained in this way, however. Certain herpesviruses, such as Epstein-Barr virus, contain a large number of repetitive DNA sequences amplified or duplicated from a simple core sequence (4). Interestingly, some of these repetitive sequences have been incorporated into essential genes (44) and into *cis*-acting elements required for virus replication (84) (i.e., they have evolved into functional units). In the latent state, the EBV genome contains an origin of replication composed of two *cis*-acting elements: (i) a 30-bp motif that is repeated about 20 times in tandem and (ii) an A+T-rich element containing a 65-bp sequence repeated in a dyad symmetry (84). An examination of these two elements indicates how duplications and mutations produced these sequences in discrete steps (Fig. 2). Within these sequences, which are required for EBV DNA replication, is the motif TAGCATATGCTA, the specific *cis*-acting binding site for an essential protein for DNA replication, the EBV nuclear antigen-1 (EBNA-1) (83). The EBNA-1 protein itself is composed of two domains separated by a repeating glycine-alanine polypeptide that is encoded by a highly repeated sequence, GGA-GCA (44). The carboxy-terminal (C-terminal) domain of EBNA-1 binds to the DNA replication origin and presumably the amino-terminal (N-terminal) domain interacts with the host cell gene products required for viral DNA replication.

The EBNA-2 protein is also constructed from N-terminal and C-terminal domains linked together by a repetitive sequence, CCC-CCA-CCA, which encodes a polyproline linker connecting these two domains (4). In herpesviruses, new recombinant proteins seem to be generated by linking together functional units from diverse genes with amplified repetitive DNA sequences that are translated into

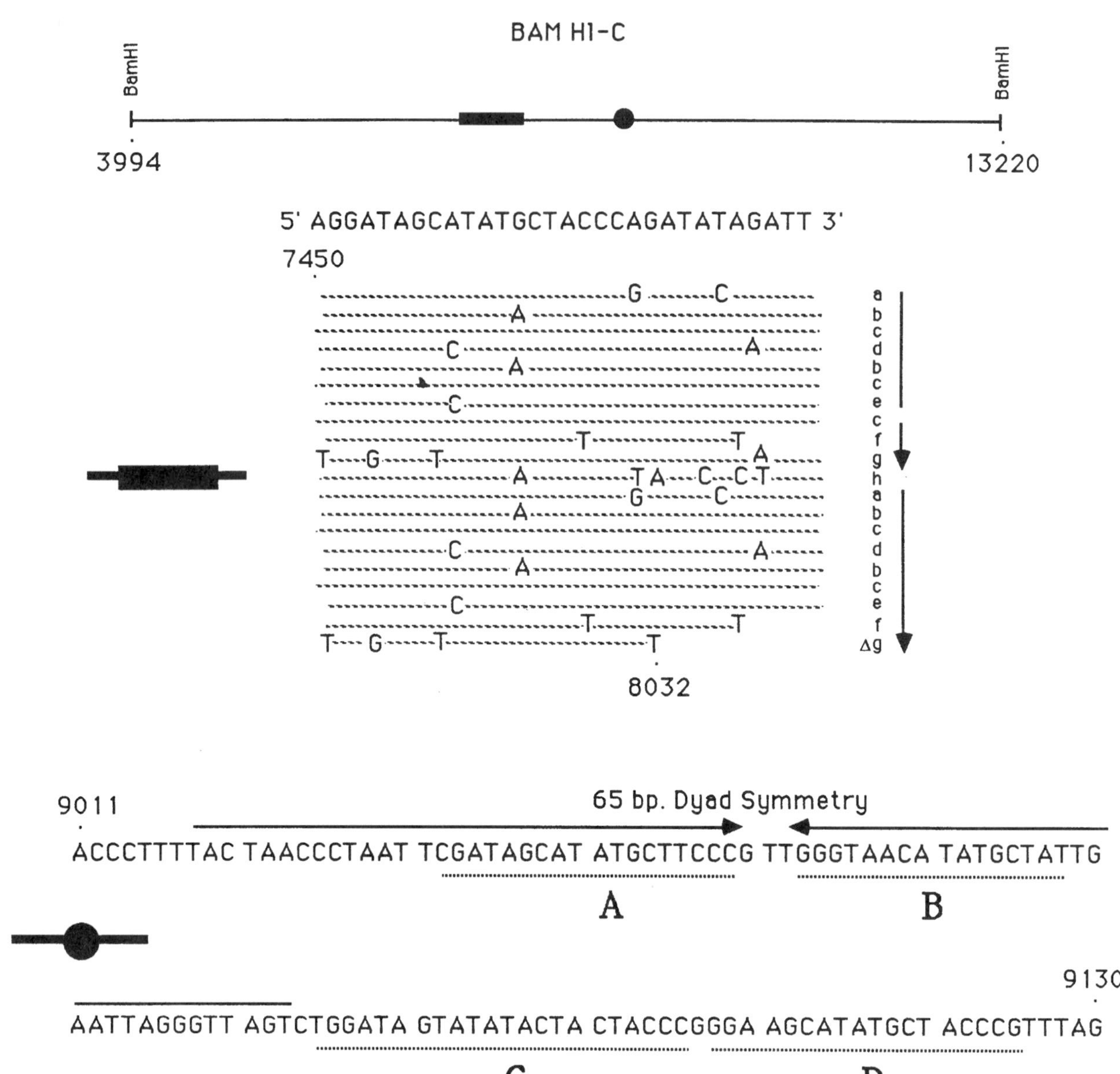

FIG. 2. The sequence of the origin of DNA replication (in the latent state) of Epstein-Barr virus. The origin of DNA replication of Epstein-Barr virus is a good example of an amplified simple repetitive sequence that has evolved into a functional element. The latent origin of DNA replication is composed of two *cis*-acting elements, a 30–base pair sequence repeated 20 times (*rectangle*) and a 65-base pair sequence repeated in a dyad symmetry (*circle*). The EBNA-1 protein binds at these two sites at TAGCATATGCTA and is required to initiate each round of DNA replication. The AT-rich region with dyad symmetry (*circle*) is typical of the structures found at viral origins of DNA replication. (Adapted from Lupton and Levine, ref. 67, with permission.)

polypeptide connectors. Clearly, these viral genomes have the luxury of increasing their genome size or complexity by duplications, but others do not.

Nonessential Genes

The discussion of how much space a viral genome should leave for evolutionary growth and enhancement of its ability to adapt to new situations brings up the question of whether viral genomes contain genes that are not essential for their replication. The definition of nonessential viral genes is complicated by the fact that a gene product may be essential under one condition but not under a different set of circumstances. For example, the thymidine kinase gene is not required for optimal growth of herpes simplex type 1 virus in cell culture (55) but is required for optimal growth in an animal (101). Thymidine kinase is an enzyme that permits exogenous thymidine to be incorporated into DNA but is not part of the usual DNA biosynthetic pathway; instead, it is on a scavenging pathway. Thus, the need for this enzyme by the virus depends upon the activity of the normal biosynthetic pathway, which is high in cells in culture, making thymidine kinase dispensable, but low in animals.

Another example of a "nonessential gene" is the adenovirus E3 glycoprotein gene. The gene for this protein can be deleted, and the mutant virus replicates normally in cell culture (85); when placed back into the host animal, however, it is more virulent than the wild-type virus. The E3 glycoprotein acts at the endoplasmic reticulum to block the proper maturation of the class I major histocompatibility antigens on the surface of infected cells (11). In wild-type virus-infected cells, the immune response by cytotoxic T cells is attenuated, resulting in, paradoxically, higher virus titers but less abnormality. Mutants that lack the E3 glycoprotein produce lower titers in vivo but more severe disease brought on by immune cell killing. Here the virus has evolved a way to simultaneously attenuate the immune response, enhance virus production, and allow for long-term chronic or persistent growth (22). In this example, viral gene products that interact with the immune system of the host are "nonessential" for replication in cells in culture. Clearly, viral genes can be designated as essential or nonessential only within a particular context.

G+C Content and Codon Usage in DNA Viruses

DNA viruses can have a very broad distribution of G+C/A+T ratios in their genomes, even among members of a single virus family. For example, the human herpesviruses can range from 46% G+C content in varicella zoster to 69% G+C content in herpes simplex type 2 (103). This is undoubtedly a reflection of the codon usage and of duplication of repetitive sequences of high or low G+C content (4). The degeneracy of the genetic code permits wide variations (about 33%) in the G+C content of the DNA without altering the proteins encoded, but virus genomes may be less efficiently translated when the G+C content of the virus is very different from that of the host cell. For example, bacteriophage T4 has a G+C content of 35% and replicates in a bacterium (*E. coli*) with a 50% G+C content. Similarly, pseudorabies virus with 74% G+C content replicates in mammalian cells with 42% G+C content. This means that tRNA species for codons that are rare in the host cell and common in the virus could become limiting during translation of viral mRNAs. Some viruses, like T4, encode their own genes for tRNAs (108), presumably to overcome this problem. Other viruses, like herpesviruses, do not use this strategy, and it is not clear whether they have developed alternative mechanisms to overcome a low abundance of cellular tRNA for common virus codons.

An equally interesting question is why some viruses retain a G+C content and pattern of codon usage that is different from that of their host cells. Many of the herpesviruses maintain a stable G+C content some 25% higher than that of their host cells (103). With the degeneracy of the genetic code and a mutation rate of 10^{-5} to 10^{-7} bases per replication cycle, a viral genome could readily approach the G+C content of its host cell in a relatively short time evolutionarily. Furthermore, it could do this without altering the amino acid sequence of the viral proteins because of the extensive third position degeneracy of the genetic code. Thus, there must be selective pressures that maintain the G+C content of the virus and keep it so different from that of its hosts.

One possible source of selective pressure is illustrated by the low usage of CpG doublets in vertebrate DNA and their viruses. The CpG doublet is a signal to methylate the C-residue, and methylated C-residues may lead to lower levels of transcription, altered chromatin structure, and even higher mutation rates than unmethylated C-residues (18,21). Thus, A, T, or C may be favored over G in the third position of a codon following a C in the second position when XCA, XCT, XCC, and XCG all encode the same amino acid.

The G+C content of a DNA virus may also be influenced by the DNA polymerase and the viral DNA replication machinery. Many DNA viruses encode their own DNA polymerases, which have a characteristic error frequency; certain polymerases even systematically make a particular kind of error (A+T to G+C, for example) (5,31). Furthermore, duplications of short sequences can occur at replication forks (59). Homopolymer runs of bases in DNA (e.g., CCCCCCCC) or palindromic sequences (e.g., CCCCGGGG) seem to be very good targets for duplication. When a polymerase or a replication fork, composed of viral proteins, favors G+C-rich sequences to duplicate, the overall G+C content of the virus will increase. The repeated linker or connector sequences found in the EBNA-1 and -2 genes (GGA-GCA and CCC-CCA-CCA) suggest that this probably occurs in EBV, which has a G+C content of 59% (4). Thus, the viral replication machinery may control the G+C content of a virus, which in turn affects codon usage and mutation rates. Even if the altered G+C content of many viruses arises from biases introduced by their own polymerases during replication, however, it must in addition confer a selective advantage, or it would not have been maintained throughout evolution.

The replication machinery argument predicts that the simpler viruses, with smaller genomes and fewer viral genes involved in DNA replication, would have G+C contents closer to those of their host cells, and this seems to be the case. The two classes of DNA viruses that do not encode their own DNA polymerase (but do have their own DNA initiation proteins) are the parvoviruses and the papovaviruses. The G+C content of parvoviruses is between 41% and 53%, and that of papovaviruses is between 41% and 49%; these are comparable to the 42% G+C content of their mammalian host.

Measuring Evolutionary Relatedness Using DNA or Protein Sequences

The genomes from a large number of viruses have now been cloned, and their complete nucleotide sequences determined. These sequences provide a database with which to study the comparative genome organizations, nucleotide sequences, and protein sequences from homologous genes found in different viruses. In some cases, these analyses have taken the form of comparing families of viruses, such as the herpesviruses (51). In yet another study a single virus species (human papilloma virus 16 or 18), isolated from different geographic locations from hosts with distinct ethnic or racial backgrounds, has yielded useful in-

TABLE 1. *Characteristics of RNA virus superfamilies*

Group	Lineage	Virus groups	Helicase type	Common features	Divergent characteristics: Morphology	Divergent characteristics: Hosts and vectors
Super 1 (POL 1)	Picorna-like	Picornaviridae	III		Icosahedral	Mammals
		Comovirus			Separate	Plants
				5'-VPg	encapsidation	
		Nepovirus		3'-poly(A)		Mammals
		Calicivirus		No subgenomic RNAs		
				Polyprotein processing		
	Poty-like	Potyvirus	II	No overlapping ORFs	Rod-shaped	Plants
		Bymovirus				
	Sobemo-like	Sobemovirus	None		Isometric	Plants
		Luteovirus				
		Nodavirus				Insect
	Arteri-like	Corona virus	I	5'-cap		Mammals
		Arterivirus		3'-poly(A)		
		Torovirus		Nested set of mRNAs		
				Enveloped		
Super 2 (POL 2)	Phage	RNA coliphages	None			Bacteria
	Flavi-like	Flavivirus	II			Humans
				One ORF	Enveloped	
	Pesti-like	Pestivirus	II	No 3'-poly(A)		Mammals
	Carmo-like	Carmovirus	None		Icosahedral	Plants
		Tombusvirus				
Super 3 (POL 3)	Tymo-like	Tymovirus	I		Icosahedral	Plants
		Carlavirus			Filamentous	
		Potexvirus				
		Capillovirus				
	Rubi-like	Rubella	I		Enveloped	Mammals
		Hepatitis E		5'-caps		Birds, Humans
		Alphaviruses		Subgenomic mRNAs		Insects
				No overlapping ORFs		
				Readthrough (most)		
	Tobamo-like	Tobamovirus	I		Rod-shaped	Plants
		Tricornavirus				Icosahedral
		Hordeivirus				Bacilliform
		Tobravirus				Filamentous
		Closterovirus				
Minus strand RNA						Mammals
Paramyxoviridae				Enveloped	Some segmented genomes	Birds
						Fish
Rhabdoviridae				Self-complementary termini	Some with M protein	Insects
						Plants
Orthomyxoviridae					Pleiomorphic	
Bunyaviridae				Helical capsid		
Arenaviridae				Overlapping ORFs		
Filoviridae					Enveloped rod	
Double-strand RNA						
Reoviridae				Segmented genome		Vertebrates
Birnaviridae				5'-cap		Plants
				3'-OH		Arthropods
				SS RNA intermediates		Mollusks

VPg, genome-linked protein; ORF, open reading frame.
Data for this figure came from refs. 58, 98, and 103.

formation (46,78). In a third example of this approach, a single virus, hepatitis C virus, replicating and evolving in a single patient before and after a liver transplant, was studied, and its changing nucleotide sequence was analyzed (73). Similarly, there is a growing database of sequences derived from the variable domains (the V3 loop) of the envelope protein of the human immunodeficiency virus (see Chapter 27), which has elucidated sets and subsets of virus variants critical to an understanding of the evolution of this virus and its immunological diversity over just the past ten years.

Using amino acid sequences derived from proteins encoded by homologous genes in two related viruses has some limitations for evolutionary comparisons. For example, the 0.3 genes from bacteriophages T3 and T7 have completely different nucleotide sequences encoding different proteins. The remainder of these viral genomes are closely related. Thus, a comparison of different genes from one virus with their homologous genes in a related virus often provides different degrees of relatedness. This is because of the acquisition of single genes or sets of genes by genetic recombination between diverse genomes. In addition, selection pressures, which fix mutations in a gene, can act differently on two different genes.

For these reasons, individual gene comparisons have been of limited value in evolutionary studies. Rather, genomic nucleotide sequences and, in particular, the relative abundances of di-, tri-, and tetranucleotides have been profitably used for evolutionary analysis (51). Quite clearly, some regions of a viral genome evolve more rapidly than do others. Such sequences have been utilized to track the rate of evolutionary changes in a single virus species as it spreads into new geographic locations (46). Indeed, an analysis of this type has even been a useful tool to follow migrations of the host, in this case humans, over a 200,000- to 1,000,000-year period (46,78). The close relationship between a virus and its host often results in co-evolution and a shared fate.

FACTORS AFFECTING RNA VIRUS EVOLUTION

RNA Virus Superfamilies

RNA viruses are a diverse group; they infect prokaryotes (the bacteriophages) as well as many eukaryotes, both plants and animals. Most RNA viruses have single-stranded RNA as their genetic material, although one family has double-stranded RNA. RNA viruses seem to be a special class of mRNA molecules (introns are conspicuously lacking among the RNA viruses) that have evolved mechanisms for replication, usually in the cell cytoplasm, as well as packaging strategies that allow them to survive in the extracellular environment. Eukaryotic cells have a wide arsenal of activities to control the half-lives of mRNAs, and these nucleases have made it difficult to isolate intact RNA viral genomes from cells. The study of RNA viruses has also been hampered by the extreme virulence of many isolates (which necessitates stringent containment for the preparation of amounts of material sufficient for biochemical study), the difficulties inherent in direct sequencing of RNA, and the failure of most RNA viruses to undergo ready recombination. The advent of cDNA cloning and of the sequencing of RNA viral genomes as plasmid copies has obviated many of these problems and enormously increased the potential for learning about RNA viruses. The recent development of cDNA clones from which infectious virus can be recovered has made it possible to engineer specific mutations at will at the DNA level and rescue RNA virus containing these changes, and to make recombinant viruses, and will in turn usher in a whole new era of RNA virology.

Conventional RNA viruses have traditionally been divided into three main groups: the plus stranded viruses (those in which the input genome is translated into protein and whose deproteinized nucleic acid is sufficient to initiate infection), the minus stranded viruses (in which the input genome is complementary to the message sense and must be transcribed by virion-associated enzymes before translation can occur), and the double-stranded viruses. These major groups have, in turn, been subdivided into virus families on the basis of virus structure, hosts, and epidemiology. Determination of the nucleotide sequences of many viruses has revealed striking homologies among otherwise disparate groups (reviewed in references 36, 59, 99, and 104); from such comparisons the RNA viruses have been grouped into superfamilies, as shown in Table 1. A single superfamily may embrace a wide variety of virion morphologies [e.g., members of superfamily 3 can be enveloped, isometric nonenveloped, rod-shaped (helical), or bacilliform] and a wide phylogenetic range of hosts (most superfamilies have members that infect both plants and animals). However, in each case it is believed that all members of the superfamily have diverged from a common ancestor. That all of the RNA viruses can be classified into so few superfamilies suggests that RNA viruses probably arose very few times in the evolutionary history of life on earth. In fact, we believe that RNA viruses may have arisen only once and that the many types of extant viruses were derived from a single protovirus by linear divergence, recombination, and gene duplication.

Another group of RNA-containing viruses, not included in Table 1, are the retroviruses. Retroviruses, with an obligate DNA intermediate in their replication, have more in common with retrotransposons than with other RNA viruses. Like retroposons, the retroviruses contain long terminal repeats and encode a reverse transcriptase activity. Retroviruses share amino acid sequence homologies in the reverse transcriptase with that of class 1 retrotransposons, like Ty-1 and Copia. In addition, there are sequence similarities in the terminal repeats, in the transcription signals, and in the primer binding site for initiation of replication (94). Like other RNA viruses they can mutate very quickly because of errors during RNA replication, and like DNA viruses they can recombine during their DNA phase and often acquire cellular genes. Retroviruses, such as human immunodeficiency virus (HIV), have become medically very important, but their evolution is probably completely distinct from that of the other RNA viruses.

Rates of Divergence

The rate of divergence of RNA viruses is very rapid, so rapid, in fact, that it is somewhat surprising that sequence homology is detectable between, for example, the plant and animal viruses. Numerous authors (47,95,98) have explored this concept in detail and have concluded that the rate of divergence of RNA viral genomes at the nucleotide level is 0.03% to 2.0% per nucleotide per year, a rate about one million times the rate for eukaryotic DNA genomes (6). Other authors have concluded that certain groups, like the New World equine encephalitis viruses, have diverged even more slowly, on the order of 0.014% per nucleotide per year (109). Some of these estimates are based on systems that may not be representative of the rate of divergence in nature, such as the rate of acquisition of mutations by viruses passed in tissue culture (96) or the error frequency of RNA replicase enzymes (reviewed in refs. 47 and 98). Other estimates are based upon natural divergence, such as the rate of change in influenza isolates (10,87). However, certain counterexamples support a slower rate of RNA virus divergence; for example, a strain of turnip yellow mosaic virus is thought to have changed by only 1% during the last 10,000 years (33). Moreover, certain residues and/or domains seem to change much more slowly, as shown by the rate of reversion of selected mutations in poliovirus (79) and Sindbis virus (29). However, it seems clear that the overall rate of divergence of RNA viruses is so rapid that two currently extant viruses, one infecting plants and one infecting animals, which retain sequence similarities within their replicase proteins (2,41), cannot possibly have diverged from one another when the plants diverged from the animals. Indeed, most viruses we know today have probably evolved since the last ice age. This conclusion is unsettling to certain theorists, who have proposed various arguments to resolve this "dilemma" (24,33). That is not to say that RNA viruses have not existed for hundreds of millions of years (and most virologists believe implicitly that viruses have been present for a very long time), but we may never know what the earlier viruses were like or even be able to prove unambiguously that they existed. Another correlate of this result is that horizontal radiation of RNA viruses must be widespread and that acquisition of new hosts must be an important aspect of RNA virus evolution. This acquisition requires the ability to enter target cells of the new hosts through appropriate receptors and the ability to be transmitted efficiently from one host to another. Indeed, once an RNA virus has passed the barriers of transmissibility, adsorption, and penetration, the interior of one eukaryotic cell must look very like another for the purposes of replication.

Mechanisms for Creating Divergence

Mutation

One of the most important mechanisms for producing new types of RNA viruses (and generating divergence within existing strains) is mutation, mostly in the form of point mutation but also including deletion and rearrangement. Most RNA polymerases (replicases) seem to lack proofreading activities (but see ref. 48), and their inherent error frequency is on the order of 10^{-4} (47,97,98). This, together with the large number of replication cycles that a virus can undergo in a year, results in the high rate of divergence noted above. In addition, even the virus yield of a single cell will be a population of genomes each with one or more changes from the average sequence (23). Most replicase errors are either neutral (e.g., they contribute to the randomization of codons by inserting silent mutations without changing coding potential) or deleterious, in which case they are quickly removed from the competitive world of viral reproduction. This type of negative selection reduces the overall rate of mutation to less than the error frequency of the polymerase. However, such a preexisting catalog of altered genomes enables a virus population to respond very quickly to changing environmental situations.

It is relatively easy to isolate virus mutants in the laboratory, and the rates of mutation and reversion have been used to estimate the rate of RNA virus divergence (reviewed in ref. 95). For example, a large number of antibody-escape variants have been isolated in different systems (i.e., viruses that grow in the presence of neutralizing antibodies). These variants in turn have been used to map the locations on surface proteins of the relevant antigenic epitopes. Such mutants mimic in the laboratory the situation in an animal host, where the virus must shift its antigenic characteristics to escape neutralization by the host immune system.

Both naturally occurring and laboratory selected variants illustrate the relative plasticity of RNA virus genomes. One particular type of mutation that has been identified, in the case of vesicular stomatitis virus, is a mutation in the polymerase gene which appears to increase the frequency of mutations occurring in other parts of the genome, presumably by increasing the error frequency of the enzyme (81). Whether or not such "mutator" genes are widespread among RNA viruses is unclear.

Reassortment

For viruses in which the genome is present in several discrete genome segments, there is a well-known mechanism for generating diversity, which has been termed *reassortment.* Two viruses with segmented genomes produce progeny with various combinations of these segments, much as higher organisms reassort chromosomes. This strategy provides a rapid method for the production of viruses with totally new potentials and is thought to be the basis for the dramatic antigenic shifts in influenza virus, for example, which lead to global pandemics (110). In such a reassortment, a human influenza virus reassorts with one from another animal, and the result is a new strain with the genes essential for replication in humans combined with wholly new surface antigens against which the human population has no immunity. It can even be argued that the segmented genome evolved in influenza viruses because of this capability. Uncontrolled reassortment may also carry penalties. Among the Bunyaviridae, reassortment occurs only between members of the same genus and is prevented between members of different genera (104). It is of interest in this regard that many of the plant virus members of superfamilies 1 and 3 have bipartite or tripartite genomes, some of which are even separately encapsidated, whereas other plant virus members and all of the animal virus representatives are monopartite. This suggests that during evolution viruses can either break apart to form segmented genomes or unite segments to form monopartite genomes, depending upon conditions. The selective advantage of a segmented genome is the ease of genetic exchange via reassortment, whereas the advantage of a monopartite genome lies in the ease of packaging and transmission of a single RNA molecule containing the entire genetic complement of a virus. Among plant viruses, whose transmission is primarily mechanical, there seems to be little selective disadvantage to separate encapsidation of genome segments.

Recombination

Recombination, in which a single polynucleotide contains sequences covalently joined which have originated from two parental types, is only infrequently utilized among the RNA viruses. Eukaryotic cells in general do not contain RNA "recombinases," and RNA viruses do not encode enzymes with the capability for efficient RNA recombination. That is not to say that recombination does not exist. It has been shown in the laboratory that true recombination occurs with a significant frequency among the picornaviruses (17,102) and among the coronaviruses (71) and that the mechanism involves copy-choice, rather than breakage and relinkage (54). In addition, it has been possible to demonstrate recombination in certain plant virus systems (8) and in alphaviruses. For alphaviruses a bipartite Sindbis virus genome was constructed in which the nonstructural proteins were encoded in a genomic RNA from which the structural proteins had been deleted, and the structural proteins were produced from a defective interfering (DI) RNA (32). Co-infection with these two RNAs produced infectious virions that contained copies of both RNAs. However, these two RNAs also underwent recombination to produce a nondefective RNA capable of a complete replication cycle (112).

Recombination has also been shown to occur in animal hosts, as evidenced by the recovery of recombinant polioviruses from vaccinees (53). It is difficult to extrapolate from this case to obtain an accurate estimate of the frequency of such recombination events in nature, however. Vaccinees received high doses of three different strains of poliovirus simultaneously, which is unusual in natural infection. Now that a large number of viral RNA genomes have been sequenced, however, it is becoming clear that recombination among RNA viruses has been a major force in generating diversity among this group (see below) and that recombinant viruses that can compete successfully in nature have arisen with a finite (but probably low) frequency. One widespread and "successful" virus, Western equine encephalitis virus, was shown to be a recombinant between Eastern equine encephalitis virus and a New World relative of Sindbis virus (39). Among the various families of plant viruses, so much recombination has apparently occurred that it is difficult to construct a linear phylogenetic tree (58).

Evolution of Plus Stranded Viruses

Conservation of Replication Enzymes

During the last decade, the complete nucleotide sequences of a large number of RNA virus genomes have been obtained, and the deduced amino acid sequences of the encoded proteins have been subjected to computer-assisted alignment and analysis. The majority of these sequences are of plus strand RNA viruses, primarily plant viruses. Of the virus-encoded proteins, the replicases are the most highly conserved, and all plus strand RNA virus replicases can be grouped unambiguously into one of three types. Other conserved polypeptides include those for helicases (unwinding activities), which are found in almost all viruses with genomes larger than 6 kb and which also separate into three types, genes for methyltransferases (to provide 5'-terminal methylated caps), genes for genome-linked proteins for uncapped genomes, and genes for proteinases to process polyprotein precursors. The proteinases are of two distinct varieties, those that resemble chymotrypsin (although some have cysteine in place of serine in the active site) and those that resemble papain. Many of these proteinases have substantially diverged from their cellular counterparts and retain only small blocks of identifiable protein sequence homology around the catalytic residues. Outside of these replicase proteins, there are only a few other homologous proteins of widespread distribution; one is a plant protein that promotes the spread of virus from cell to cell in plants, and another is a conserved animal virus protein (X) of unknown function (58).

In Table 1, plus stranded viruses were assigned to superfamily 1, 2, or 3 on the basis of sequence alignments of their RNA-dependent RNA polymerases (58). In Fig. 3 the genomes of most of the virus groups described in Table 1 are shown schematically to illustrate the genome organization of their nonstructural genes. Superfamily 1 contains four "lineages" that are distinguished from one another by their helicases or lack thereof. Presumably, starting with a polymerase module, each lineage acquired its helicase independently. Superfamily 2 has fewer members but includes viruses of bacteria, of mammals, and of plants. Superfamily 3 is again divided into lineages, but in this case all lineages have the same type of helicase.

Even though it appears in Fig. 3 that almost any organization is possible, there is a general overall order to the replicase protein array. Methyltransferases, if present, are generally at the 5'-end of the array, and polymerases are at the 3'-end. Helicases are generally upstream of polymerases, and papain-like proteinases are generally upstream of helicases. Within a group (e.g., flaviviruses), all members have the same genes in the same order. If the structural proteins are translated as part of a long ORF that also encodes the replicase array, they are almost always at the 5'-end; any attenuation of translation due to premature termination would then favor an excess of structural proteins over replicase. When structural proteins are encoded at the 3'-end of the genome, they are usually translated from a subgenomic RNA.

In Fig. 4 a subset of superfamily 3 is shown in more detail. In this comparison tobacco mosaic virus (a rod-shaped plant virus), brome mosaic virus (BMV) (a plant virus with three separately encapsulated RNA segments), and Sindbis virus (SIN) (an enveloped animal virus) are shown, with different shading to indicate domains of homology. Within these conserved domains, the amino acid identity may not be more than 20% but the sequence similarities

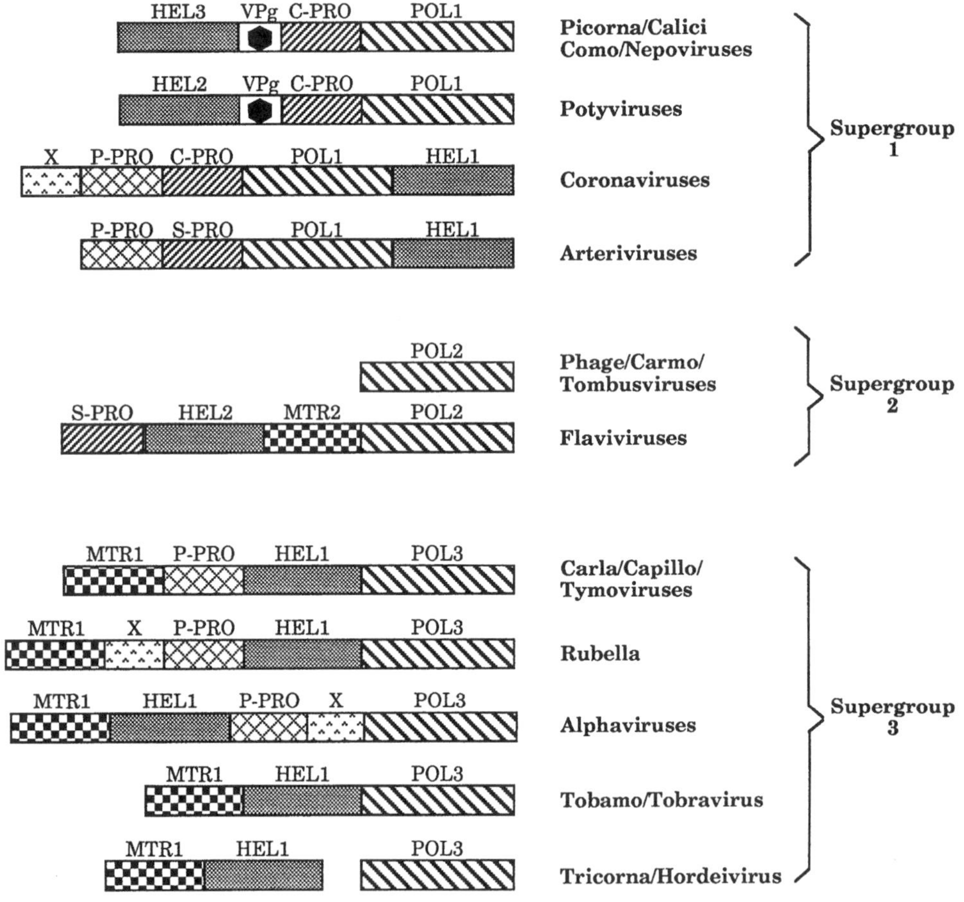

FIG. 3. Organization of plus strand RNA genomes. The genomes of plus strand RNA viruses are made up of a small number of genes for replicative functions which have been mixed and matched in a number of combinations. These include a helicase (*dark gray*), a genome-linked protein (*solid hexagon*), a chymotrypsin-like proteinase (*fine diagonals*), a polymerase (*bold diagonals*), a papain-like proteinase (*crosshatch*), a methyltransferase (*checkerboard*), and a region of unknown function X (*stippled*). Some of these units come in more than one type (like POL1, POL2, and POL3) as defined by amino acid sequence alignments. In this schematic representation, the units are not shown to scale, and the structural proteins have been omitted for simplicity. This figure is reprinted from Strauss and Strauss, ref. 100a, with permission.

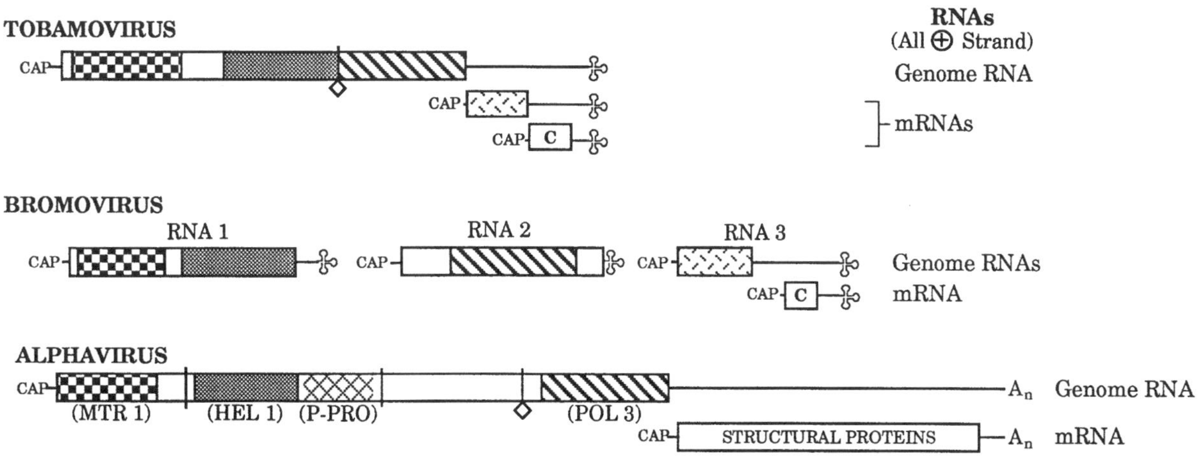

FIG. 4. Genome organization and amino acid homologies between Sindbis virus and two families of plant viruses. Untranslated regions are shown as *lines,* translated regions as *open boxes,* on both genomic and subgenomic RNAs. ◊, termination codons readthrough to produce downstream products. Within the translated regions of the genome RNAs, there are three domains of homology: methytransferase (MTR1) is *checkerboard,* helicase (HEL 1) is *gray,* and polymerase (POL3) is shown in *diagonal lines.* The *box* marked *C* is the gene for the capsid protein; P-PRO is a papain-like proteinase found in alphaviruses. Another domain believed to contribute to spread within plants is stippled on certain subgenomic RNAs. This figure is reprinted from Strauss and Strauss, ref. 100a, with permission.

extend over hundreds of residues. This also illustrates that two methods seem to be used to regulate the production of the POL 3 domain; in SIN and TMV this domain is downstream of a (leaky) termination codon, whereas in BMV it is encoded on a different RNA segment.

Transcriptional and Translational Control Signals

Studies on the nature and function of control elements for replication, transcription, and translation of RNA viruses have lagged behind the corresponding studies for DNA viruses. From comparative sequencing studies it was proposed that certain conserved sequence elements in viral RNAs serve as replication and transcription signals (similar to the origins of replication in DNA viruses) (reviewed in ref. 99). Replication and transcription enzymes as well as cellular factors must bind to these elements to initiate transcription and replication. Specificity of binding may reside either in the linear core sequence (usually 10 to 20 nucleotides in length) or in the specific secondary structure of the element. Most recent studies have involved site-specific mutagenesis and deletion analysis of full-length cDNA clones of the viruses, from which infectious RNA can be transcribed. In such a system the promoter sequences required for replication of brome mosaic virus, which are present in the 3'-terminal tRNA-like structure present on each of the three RNA genome segments, have been elegantly characterized (26,27). Debilitating site-specific mutations introduced into this structure on one RNA segment were repaired in viable revertants both by direct reversion (polymerase error) and by recombination with another RNA segment during infection of intact plants (82).

In alphaviruses, three domains of conserved nucleotide sequence thought to be control elements have been intensively studied: a stem-loop structure at the 5'-terminus of the genome; the 3'-noncoding region, which ends in a highly conserved nucleotide sequence adjacent to the poly(A); and the promoter for subgenomic mRNA transcription. Studies with DI RNAs of Sindbis virus have determined the nature of sequence elements required for replication and packaging of DI RNA (65). The promoter for subgenomic RNA synthesis consists of 24 nucleotides, and this promoter is functional when placed in other locations in the genome and also can serve as a promoter for a nonviral heterologous gene (64). Mutagenesis and deletion analysis of the 5'- and 3'-terminal conserved elements of SIN have been performed using a full-length infectious clone of the virus (60,76). Surprisingly, many deletions and substitutions in these elements were tolerated, to the extent of producing viable progeny, but in most cases the viruses were crippled relative to the wild type. Many mutations were host-specific in their effects, indicating that host factors probably bind to these sequences and suggesting that the highly conserved sequences found in natural isolates represent a consensus best-fit sequence (i.e., a compromise between the optimal promoter in vertebrate cells and the optimal promoter in mosquito cells).

Three-Dimensional Conformation of Capsid Proteins

One final level at which there seem to be detectable remnants of common ancestry among a large number of otherwise unrelated viruses is in the three-dimensional conformations of their capsid proteins. There are pronounced similarities among the capsids of southern bean mosaic virus, other icosahedral plant viruses, and the picornavirus capsid at atomic resolution (reviewed in ref. 86). In these viruses, the capsid proteins are folded into an "eight-fold β-barrel" conformation, with the loops of less structured amino acid sequence located between the β-sheets. Although detailed x-ray data are not available for most viruses, it has been possible to fold the linear amino acid sequences of the capsid protein of hepatitis B virus into the known conformation of the picornavirus capsid proteins; when this is done, the predicted structure is in accord with known properties of the capsid. Thus, the location of known antigenic sites in the hepatitis B core protein are consistent with this predicted structure (3). However, when the complete structure of the Sindbis capsid protein was solved, it was found to have a completely different configuration that is very similar to that of chymotrypsin, consistent with its autoproteolytic activity (40). Thus, it seems that viruses can "borrow" a variety of cellular components, some with enzymatic activities, and adapt them for use as structural proteins.

Evolution of Minus Strand Viruses

The minus stranded viruses form a fairly homogeneous group structurally and share some common features in their genome organization and replication strategy (Fig. 5). In this figure the genomes of representatives of five of the six families of minus strand viruses have been aligned. It is clear that minus strand viruses are made up of a small number of genes, regardless of whether the genome is a single

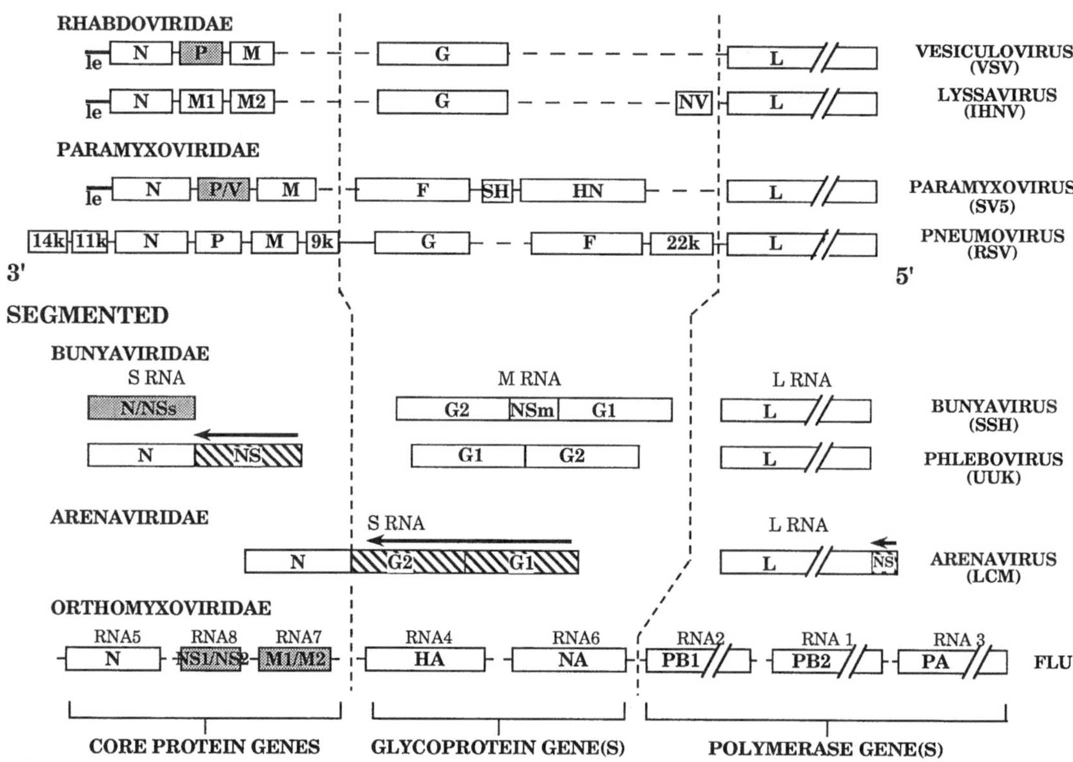

FIG. 5. Genome organization among the minus strand RNA viruses. The maps of the genes of the Rhabdoviridae, the Paramyxoviridae, the Bunyaviridae, and the Arenaviridae have been aligned to maximize functional similarity between gene products. The individual gene segments of the Orthomyxoviridae have been arranged according to similarity of function with the two groups above. Abbreviations of virus names are as follows: *VSV,* vesicular stomatitis virus; *IHNV,* infectious hematopoietic necrosis virus; *RSV,* respiratory syncytial virus; *SV5,* simian virus 5; *SSH,* snowshoe hare virus; *UUK,* Uukuniemi virus; *LCM,* lymphocytic choriomeningitis virus; *FLU,* influenza A. *le* is a nontranslated leader sequence. The gene products are abbreviated as follows: *N,* nucleoprotein; *P,* phosphoprotein; *M* (*M1, M2*), matrix proteins; *G* (*G1, G2*), membrane glycoproteins; *F,* fusion glycoprotein; *HN,* hemagglutinin-neuraminidase glycoprotein; *L,* replicase; *NA,* neuraminidase glycoprotein; *HA,* hemagglutinin glycoprotein; *NS* (*NV, SH,* NS_s, NS_m), nonstructural proteins; *PB1, PB2,* and *PA,* components of the influenza replicase. Within a given genome, the genes are drawn roughly to scale. *Shaded genes* are those in which multiple proteins are produced from different ORFs; *diagonal lines* are genes transcribed in the ambisense strategy. (Data for this figure are from Tordo et al., ref. 104.)

RNA molecule or a number of segments, and that these are different from the modules making up the plus strand viruses. Although extensive sequence information is available, only limited homologies among the various virus-encoded proteins have been detected. The most extensive of these is among the polymerase proteins of the nonsegmented minus stranded viruses (reviewed in ref. 105). In terms of their replication and morphology, the minus strand viruses are all very similar, and presumably they are derived from a single ancestor. All of them are enveloped, and all have a helical nucleocapsid. All of them have short self-complementary sequences at the ends of their genomes or genome segments, suggesting that the replicase binding site to promote both plus and minus strand synthesis is the same. These terminal sequences, which may also include encapsidation sequences, are missing from their plus strand mRNAs, which in no case are exact complements of the genome.

The minus strand viruses have also evolved mechanisms to increase the coding capacity of their genomes. Certain viruses use more than one initiation codon to start translation in two overlapping ORFs (34); others, like influenza, translate two different proteins from spliced and unspliced forms of a single mRNA (61). Paramyxoviruses insert nontemplated G residues during transcription to shift the frame downstream of the insertions (80).

CONCLUDING REMARKS

In attempting to describe the evolution of viruses, we have illustrated that there are obvious differences between the DNA viruses, the retroviruses, and the RNA viruses. It is unclear whether all viruses had a single origin or whether there were multiple origins. However, the different epidemiological histories of these three main types of viruses and the relative importance of genetic interchange among them or between the viruses and their hosts lead to different concepts of the current evolution of extant viruses and different predictions for future viruses. However, for RNA viruses, it is possible to enumerate some general evolutionary principles.

1. RNA viruses evolve so rapidly that only the most important functional motifs and capacities are retained between widely divergent viruses. These functions are almost all involved with RNA replication.
2. RNA virus genomes are made up of only a small number of genes, almost all of which are essential.
3. RNA virus genome organization is the result of two contradictory evolutionary forces, conservation of the array of nonstructural genes and rearrangement due to recombination.
4. Sufficient modular evolution and gene shuffling via recombination has taken place that it is not possible to draw a single phylogenetic tree for the plus strand RNA viruses, let alone one for all RNA-containing viruses.

With the exception of the poxviruses (now virtually vanquished as human pathogens by classical vaccination), which differ from other DNA viruses in morphology and their wholly cytoplasmic replication and which may in fact have a separate origin, the DNA viruses do not cause the rapidly moving worldwide pandemics of human disease that are the rule for RNA viruses. Indeed, it is within these pandemics that much of the information on the variability and divergence of viruses over time has been accumulated. The antigenic shifts of influenza during an epidemic within a previously immunized population provide another setting in which to study virus evolution. In contrast to RNA viruses (which include such scourges as yellow fever, dengue, Rift valley fever, encephalitic viruses, measles, mumps, and rubella), the majority of DNA viruses set up persistent or latent infections within their hosts, which may lead to malignant transformation and chronic disease. Sequestered within their hosts, these DNA viruses may be quiescent for decades and then reemerge with their genomes virtually unchanged. Thus, their apparent rate of divergence may be considerably less than that for cytolytic viruses, regardless of the inherent error frequency of their replicases.

REFERENCES

1. Movable genetic elements. *Cold Spring Harbor Symp Quant Biol* 1980;1:1–445.
2. Ahlquist P, Strauss EG, Rice CM, et al. Sindbis virus proteins nsP1 and nsP2 contain homology to nonstructural proteins from several RNA plant viruses. *J Virol* 1985;53:536–42.
3. Argos P, Fuller SD. A model for the hepatitis B virus core protein: prediction of antigenic sites and relationship to RNA virus capsid proteins. *EMBO J* 1988;7:819–24.
4. Baer R, Bankier AT, Biggen MD, et al. DNA sequence and expression of the B95-8 Epstein-Barr virus genome. *Nature* 1984;310:207–21.
5. Bessman MJ, Muzyczka N, Goodman MF, Schnaar RL. Studies on the biochemical basis of spontaneous mutation: II. The incorporation of base and its analogue into DNA by wild-type, mutator and antimutator DNA polymerases. *J Mol Biol* 1974;88:409–21.
6. Britten RJ. Rates of DNA sequence evolution differ between taxonomic groups. *Science* 1986;231:1393–8.
7. Brockman WW, Lee TNH, Nathans D. Characterization of cloned evolutionary variants of simian virus 40. *Cold Spring Harbor Symp Quant Biol* 1975;39:119–24.
8. Bujarski JJ, Kaesberg P. Genetic recombination between RNA components of a multipartite plant virus. *Nature* 1986;321:528–31.
9. Bukhari A, Shapiro JA, Adhya SL. *DNA insertion elements, plasmids and episomes*. New York: Cold Spring Harbor Laboratory; 1977.
10. Buonagurio DA, Nakada S, Parvin S, et al. Evolution of human influenza A virus over 50 years: rapid uniform rate of change in NS gene. *Science* 1986;232:980–2.
11. Burgert HG, Kvist S. An adenovirus type 2 glycoprotein blocks cell surface expression of human histocompatibility class 1 antigens. *Cell* 1985;41:987–97.
12. Burnet FM. *Virus as an organism*. Cambridge: Harvard University Press; 1945.
13. Cech TR. A model for the RNA catalyzed replication of RNA. *Proc Natl Acad Sci USA* 1986;83:4360–3.
14. Cech TR, Bass BL. Biological catalysis by RNA. *Annu Rev Biochem* 1986;55:599–629.
15. Chifflot S, Sommer P, Hartmann D, Stussi-Garaud C, Hirth L. Replication of alfalfa mosaic virus RNA: evidence for a soluble replicase in healthy and infected tobacco leaves. *Virology* 1980;100:91–100.

16. Cohen S, Shapiro J. Transposable genetic elements. *Sci Am* 1980; 242:40–50.
17. Cooper PD. Genetics of picornaviruses. In: H. Fraenkel-Conrat and R. R. Wagner, eds. *Regulation and genetics*. New York: Plenum Press; 1977:133–207.
18. Coulondre C, Miller J, Farabaugh P, Gilbert W. Molecular basis of base substitution hotspots in *E. coli*. *Nature* 1978;274:776–80.
19. Davis BD, Dulbecco R, Eisen HN, Ginsberg HS. *Microbiology*. New York: Harper and Row; 1980.
20. Diener TO, ed. The viroids. In: *The viruses*. New York: Plenum Press; 1987:344.
21. Doerfler W. DNA methylation and gene activity. *Annu Rev Biochem* 1983;52:93–124.
22. Doerfler W, ed. The molecular biology of adenoviruses, 30 years of research 1953–1983. *Curr Top Microbiol Immunology* 1983;1–3.
23. Domingo E, Martinez-Salas E, Sobrino F, et al. The quasi-species (extremely heterogeneous) nature of viral RNA genome populations: biological relevance—a review. *Gene* 1985;40:1–8.
24. Doolittle RF, Feng D-F, Johnson MS, McClure MA. Origins and evolutionary relationships of retroviruses. *Q Rev Biol* 1989;64:1–30.
25. Dougherty J, Temin HM. Determination of the rate of base pair substitution and insertion mutations in retrovirus replication. *J Virol* 1988;62:2817–22.
26. Dreher TW, Hall TC. Mutational analysis of the sequence and structural requirements in brome mosaic virus RNA for minus strand promoter activity. *J Mol Biol* 1988;201:31–40.
27. Dreher TW, Hall TC. Mutational analysis of the tRNA mimicry of brome mosaic virus RNA. Sequence and structural requirements for aminoacylation and 3'-adenylation. *J Mol Biol* 1988;201:41–55.
28. Dunn JJ, Studier FW. Complete nucleotide sequence of bacteriophage T7 DNA and locations of T7 genetic elements. *J Mol Biol* 1983;166: 477–535.
29. Durbin RK, Stollar V. Sequence analysis of the E2 gene of a hyperglycosylated, host restricted mutant of Sindbis virus and estimation of mutation rate from frequency of revertants. *Virology* 1986;154:135–43.
30. Fujimura FK, Silbert P, Eckhart W, Linney E. Polyoma virus infection of retinoic acid induced differentiated teratocarcinoma cells. *J Virol* 1981;39:306–12.
31. Gallin FD, Nossal NG. Control of mutation frequency by bacteriophage T4 DNA polymerase II. Accuracy of nucleotide selection by the L88 mutation, CB120 antimutator and wild-type T4 DNA polymerase. *J Biol Chem* 1976;251:5225–31.
32. Geigenmuller-Gnirke U, Weiss B, Wright R, Schlesinger S. Complementation between Sindbis viral RNAs produces infectious particles with a bipartite genome. *Proc Natl Acad Sci USA* 1991;88:3253–57.
33. Gibbs A. Molecular evolution of viruses: "trees," "clocks," and "modules." *J Cell Sci Suppl* 1987;7:319–37.
34. Giorgi C, Blumberg BM, Kolakovsky D. Sendai virus contains overlapping genes expressed from a single mRNA. *Cell* 1983;35:829–36.
35. Goldbach R, LeGall O, Wellink J. Alpha-like viruses in plants. *Semin Virol* 1991;2:19–25.
36. Goldbach RW. Molecular evolution of plant RNA viruses. *Annu Rev Phytopathol* 1986;24:289–310.
37. Gott JM, Shub DA, Belfort M. Multiple self-splicing introns in bacteriophage T4: evidence from autocatalytic GTP labeling of RNA in vitro. *Cell* 1986;47:81–87.
38. Green RB. On the nature of filterable viruses. *Science* 1935;82:443–5.
39. Hahn CS, Lustig S, Strauss EG, Strauss JH. Western equine encephalitis virus is a recombinant virus. *Proc Natl Acad Sci USA* 1988;85: 5997–6001.
40. Hahn CS, Strauss EG, Strauss JH. Sequence analysis of three Sindbis virus mutants temperature-sensitive in the capsid autoprotease. *Proc Natl Acad Sci USA* 1985;82:4648–52.
41. Haseloff J, Goelet P, Zimmern D, et al. Striking similarities in amino acid sequence among nonstructural proteins encoded by RNA viruses that have dissimilar genomic organization. *Proc Natl Acad Sci USA* 1984;81:4358–62.
42. Hayes W. *The genetics of bacteria and their viruses*. New York: Wiley and Son; 1986.
43. Hayward WS, Neel BG, Astrin SR. Activation of a cellular one gene by promoter insertion in ALV-induced lymphoid leukosis. *Nature* 1981;290:475–80.
44. Hearing JC, Nicolas JC, Levine AJ. Identification of Epstein-Barr virus sequences that encode a nuclear antigen expressed in latently infected lymphocytes. *Proc Natl Acad Sci USA* 1984;81:4373–8.
45. Herman RC. Internal initiation of translation on the vesicular stomatitis virus phosphoprotein mRNA yields a second protein. *J Virol* 1986;58:797–804.
46. Ho L, Chan S-Y, Burk RD, Das BC, Fujinaga K, Icenogle JP, Kahn T, et al. The genetic drift of human papillomavirus type 16 is a means of reconstructing prehistoric viral spread and the movement of ancient human populations. *J Virol* 1993;67:6413–23.
47. Holland J, Spindler K, Horodyski F, et al. Rapid evolution of RNA genomes. *Science* 1982;215:1577–85.
48. Ishihama A, Mizumoto K, Kawakami K, Kato A, Honda A. Proofreading function associated with the RNA-dependent RNA polymerase from influenza virus. *J Biol Chem* 1986;261:10417–21.
49. Jacob F, Wollman EL. *Sexuality and the genetics of bacteria*. New York: Academic Press; 1961:327–32.
50. Joyce GF. The rise and fall of the RNA world. *New Biol* 1991;3: 399–407.
51. Karlin S, Mocarski ES, Schachtel GA. Molecular evolution of herpesviruses: genomic and protein sequence comparisons. *J Virol* 1994;68:1886–1902.
52. Katinka M, Vasseur M, Montreau N, Yaniv M, Blangy D. Polyoma DNA sequences involved in control of viral gene expression in murine embryonal carcinoma cells. *Nature* 1981;290:720–2.
53. Kew OM, Nottay BK. Evolution of the oral polio vaccine strains in humans occurs by both mutation and intramolecular recombination. In: R. M. Chanock, ed. *Modern approaches to vaccines: molecular and chemical basis of virus virulence*. Cold Spring Harbor, NY: Cold Spring Harbor Laboratories; 1984:357–62.
54. Kirkegaard K, Baltimore D. The mechanism of RNA recombination in poliovirus. *Cell* 1986;47:433–43.
55. Kit S, Dubbs DR. Acquisition of thymidine kinase activity by herpes simplex infection mouse fibroblast cells. *Biochem Biophys Res Commun* 1963;11:55–9.
56. Kleckner N. Transposable elements in prokaryotes. *Annu Rev Genet* 1981;15:341–404.
57. Klessig DF, Grodzicker T. Mutations that allow human Ad2 and Ad5 to express late genes in monkey cells map in the viral gene encoding the 72 Kd DNA binding protein. *Cell* 1979;17:957–63.
58. Koonin EV, Dolja VV. Evolution and taxomony of positive-strand RNA viruses: implications of comparative analysis of amino acid sequences. *Crit Rev Biochem Mol Biol* 1993;28:375–430.
59. Kornberg A. *DNA replication*. San Francisco: Freeman; 1980.
60. Kuhn RJ, Hong Z, Strauss JH. Mutagenesis of the 3' nontranslated region of Sindbis virus RNA. *J Virol* 1990;64:1465–76.
61. Lamb RA, Choppin PW. The gene structure and replication of influenza virus. *Annu Rev Biochem* 1983;52:467–506.
62. Leider JM, Palese P, Smith FI. Determination of the mutation rate of a retrovirus. *J Virol* 1988;62:3084–91.
63. Levine AJ. The nature of the host range restriction of SV40 and polyoma viruses in embryonal carcinoma cells. *Curr Top Microbiol Immunol* 1982;101:1–30.
64. Levis R, Schlesinger S, Huang HV. The promoter for Sindbis virus RNA-dependent subgenomic RNA transcription. *J Virol* 1990;64: 1726–33.
65. Levis R, Weiss BG, Tsiang M, Huang H, Schlesinger S. Deletion mapping of Sindbis virus DI RNAs derived from cDNAs defines the sequences essential for replication and packaging. *Cell* 1986;44:137–45.
66. Lund E, Fried M, Griffin BE. Polyoma virus defective DNAs. I. Physical maps of a related set of defective molecules (D76, D91, D92). *J Virol* 1977;117:473–85.
67. Lupton S, Levine AJ. Mapping genetic elements of Epstein-Barr virus that facilitate extrachromosomal persistence of Epstein-Barr virus derived plasmids in human cells. *Mol Cell Biol* 1985;5:2533–9.
68. Luria SE, Darnell JE, Baltimore D, Campbell A. *General virology*. New York: John Wiley and Sons; 1978.
69. Lwoff A. *L'evolution physiologique: Çtudes des pertes de fonctions chez les microorganisms*. Paris: Hermann Press; 1943.
70. Lwoff A. Factors influencing the evolution of viral diseases at the cellular level and in the organism. *Bacteriol Rev* 1959;23:109–24.
71. Makino S, Keck JG, Stohlman ST, Lai MMC. High-frequency RNA recombination of murine coronaviruses. *J Virol* 1986;57:729–37.
72. Mark K, Studier FW. Purification of the gene 0.3 protein of bacteriophage T7, an inhibitor of the DNA restriction system. *J Biol Chem* 1981;256:2573–8.
73. Martell M, Esteban JE, Quer J, Vargas V, Esteban R, Guardia J, Gomes Z. Dynamic behavior of hepatitis C virus (HCV) quasispecies in pa-

tients undergoing orthotopic liver transplantation *J Virol* 1994; 68:3425–36.
74. Mertz JE, Carbon J, Herzberg M, Davis R, Berg P. Isolation and characterization of individual clones of simian virus 40 mutants containing deletions, duplications and insertions in their DNA. *Cold Spring Harbor Symp Quant Biol* 1975;39:69–75.
75. Miller SL, Orgel LE. *The origins of life on earth*. Englewood Cliffs, NJ: Prentice Hall; 1974.
76. Niesters HGM, Strauss JH. Defined mutations in the 5' nontranslated sequence of Sindbis virus RNA. *J Virol* 1990;64:4162–8.
77. Nusse R, Van Ooyen A, Cox D, Fung YKT, Varmus H. Mode of proviral activation of a putative mammary oncogene (Int-1) on mouse chromosome 15. *Nature* 1984;307:131–6.
78. Ong C-K, Chan S-Y, Campo MS, et al. Evaluation of human papillomavirus type 18. An ancient phylogenetic root in Africa and intratype diversity reflect coevolution with human ethnic groups. *J Virol* 1993;67:6424–31.
79. Parvin JD, Moscona A, Pan WT, Leider JM, Palese P. Measurement of the mutation rates of animal viruses: influenza A virus and poliovirus type 1. *J Virol* 1986;59:377–83.
80. Pelet T, Curran J, Kolakovsky D. The P gene of bovine parainfluenza virus 3 expresses all three reading frames from a single mRNA editing site. *EMBO J* 1991;10:443–8.
81. Pringle CR, Devine V, Wilkie M, et al. Enhanced mutability associated with a temperature-sensitive mutant of vesicular stomatitis virus. *J Virol* 1981;39:377–89.
82. Rao ALN, Hall TC. Recombination and polymerase error facilitate restoration of infectivity in brome mosaic virus. *J Virol* 1993; 67:969–79.
83. Rawlins DR, Milman G, Hayward SD, Hayward G. Sequence specific DNA binding of Epstein-Barr virus nuclear antigen (EBNA-1) to clustered sites in the plasmid maintenance region. *Cell* 1985;42:859–68.
84. Reisman D, Yattes J, Sugden BA. A putative origin of replication of plasmids derived from Epstein-Barr virus is composed of two *cis*-acting components. *Mol Cell Biol* 1985;5:1822–32.
85. Ross S, Levine AJ. The genomic map position of the adenovirus type 1 glycoprotein. *Virology* 1979;99:427–30.
86. Rossman MG, Rueckert RR. What does the molecular structure of viruses tell us about viral functions? *Microbiol Sci* 1987;4:206–14.
87. Saitou N, Nei M. Polymorphism and evolution of influenza A virus genes. *Mol Biol Evol* 1986;3:57–74.
88. Sambrook J, Sleigh M, Engler JA, Broker TR. The evolution of the adenoviral genome. *Ann NY Acad Sci* 1980;354:426–52.
89. Scheid A, Choppin PW. Protease activation mutants of Sendai virus: activation of biological properties by specific proteases. *Virology* 1976;69:265–77.
90. Sekikawa K, Levine AJ. Isolation and characterization of polyoma host range mutants that replicate in nullipotent embryonal carcinoma cells. *Proc Natl Acad Sci USA* 1981;78:1100–4.
91. Shapiro JA, ed. *Mobile genetic elements*. New York: Academic Press; 1983.
92. Shinozaki K, Ohme M, Tanaka M, et al. The complete nucleotide sequence of the tobacco chloroplast genome: its gene organization and expression. *EMBO J* 1986;5:2043–9.
93. Simons K, Garoff H, Helenius A. How an animal virus gets into and out of its host cell. *Sci Am* 1982;246:58–66.
94. Singer M, Berg P. *Genes & genomes*. Mill Valley, CA: University Science Books; 1991.
95. Smith DB, Inglis SC. The mutation rate and variability of eukaryotic viruses: an analytical review. *J Gen Virol* 1987;68:2727–40.
96. Sobrino F, Davila M, Ortin J, Domingo E. Multiple genetic variants arise in the course of replication of foot-and-mouth disease virus in cell culture. *Virology* 1983;128:310–8.
97. Steinhauer DA, Holland JJ. Direct method for quantitation of extreme polymerase error frequencies at selected single base sites in viral RNA. *J Virol* 1986;59:545–50.
98. Steinhauer DA, Holland JJ. Rapid evolution of RNA viruses. *Annu Rev Microbiol* 1987;41:409–33.
99. Strauss EG, Strauss JH. Replication strategies of the single-stranded RNA viruses of eukaryotes. *Curr Top Microbiol Immunol* 1983;105:1–98.
100. Symons RH, ed. Viroids and related pathogenic RNAs. In: *Seminars in Virology*, 1990;1:75–162.
100a. Strauss JH, Strauss EG. The alphaviruses: gene expression, replication, and evolution. *Micro Revs*. 1994;58:491–562.
101. Tenser RB, Dunstan ME. Herpes simplex virus thymidine kinase expression in infection of the trigeminal ganglion. *Virology* 1979;99: 417–22.
102. Tolskaya EA, Romanov LA, Kolesnikova MS, Agol VI. Intertypic recombination in poliovirus: genetic and biochemical studies. *Virology* 1983;124:121–32.
103. Tooze J, ed. The DNA tumor viruses. In: *Molecular biology of tumor viruses* Part 2, 2nd ed. Cold Spring Harbor, NY: Cold Spring Harbor Laboratory, 1980.
104. Tordo N, de Haan P, Goldbach R, Poch O. Evolution of negative-stranded RNA genomes. *Semin Virol* 1992;3:341–57.
105. Tordo N, Poch O, Ermine A, Keith G, Rougeons F. Completion of the rabies virus genome sequence determination: highly conserved domains among the L (polymerase) proteins of unsegmented negative-strand RNA viruses. *Virology* 1988;165:565–76.
106. Vasseur M, Kress C, Montreau N, Blangy D. Isolation and characterization of polyoma virus mutants able to develop in embryonal carcinoma cells. *Proc Natl Acad Sci USA* 1980;77:1069–72.
107. Wadell G. Molecular epidemiology of human adenoviruses. *Curr Top Microbiol Immunol* 1984;110:191–220.
108. Watson JD, Hopkins N, Roberts JW, Steitz JA, Weiner A. *The molecular biology of the gene*. Menlo Park, CA: Benjamin-Cummings Publishing; 1987.
109. Weaver SC, Rico-Hess R, Scott TW. Genetic diversity and slow rates of evolution in New World alphaviruses. *Curr Top Microbiol Immunol* 1992;176:99–117.
110. Webster RG, Laver WG, Air GM, Schild GC. Molecular mechanisms of variation in influenza viruses. *Nature* 1982;296:115–21.
111. Weiner HL, Powers ML, Fields BN. Absolute linkage of virulence and central nervous system tropism of reoviruses to viral hemagglutinin. *J Infect Dis* 1980;141:609–14.
112. Weiss BG, Schlesinger S. Recombination between Sindbis virus RNAs. *J Virol* 1991;65:4017–25.
113. Yang D, Oyaizu Y, Oyaizu H, Olsen GJ, Woese CR. Mitochondrial origins. *Proc Natl Acad Sci USA* 1985;82:4443–7.
114. Ziff EB, Evans RM. Coincidence of the promoter and capped 5' terminus of RNA from adenovirus 2 late transcription unit. *Cell* 1978;15:1463–76.

Fundamental Virology, Third Edition
edited by B.N. Fields, D.M. Knipe, P.M. Howley, et al.
Lippincott - Raven Publishers, Philadelphia © 1996

CHAPTER 6

Pathogenesis of Viral Infections

Kenneth L. Tyler and Bernard N. Fields

Viral pathogenesis can be defined as the method by which viruses produce disease in the host (84,194,453). In this chapter we focus almost entirely on the pathogenesis of acute viral infections as they occur in humans and other mammalian hosts. Much of our existing knowledge of viral pathogenesis is based on experimental studies using animal models of natural human infections. It is important to recognize that these are at best imperfect analogies to human infections. Nonetheless, it is possible to make observations and perform experimental manipulations in animals that are not possible in naturally occurring infections of humans. The knowledge obtained in this fashion has been invaluable in understanding how viruses produce human diseases.

Pathogenesis concerns itself with the mechanisms by which viruses injure discrete populations of cells in different organs to produce the signs and symptoms of disease in a particular host. Yet it is vitally important to recognize that the production of disease is in fact a relatively unusual outcome of viral infection of a host (Fig. 1). The capacity of a virus, compared to other closely related viruses, to produce disease in a host is referred to as its *virulence.* Since virulence depends on a variety of viral and host factors, it must be defined in terms of specific viral factors (e.g., dose of virus, route of entry) as well as specific host factors (e.g., age, immune status, and species of host). With many "viru-

K. L. Tyler: Departments of Neurology, Microbiology, Immunology and Medicine, University of Colorado Health Sciences Center, and Neurology Service, Denver VA Medical Center, Denver, Colorado 80262.

B. N. Fields: Departments of Microbiology and Molecular Genetics, and Medicine and the Shipley Institute of Medicine, Harvard Medical School, Boston, Massachusetts 02115.

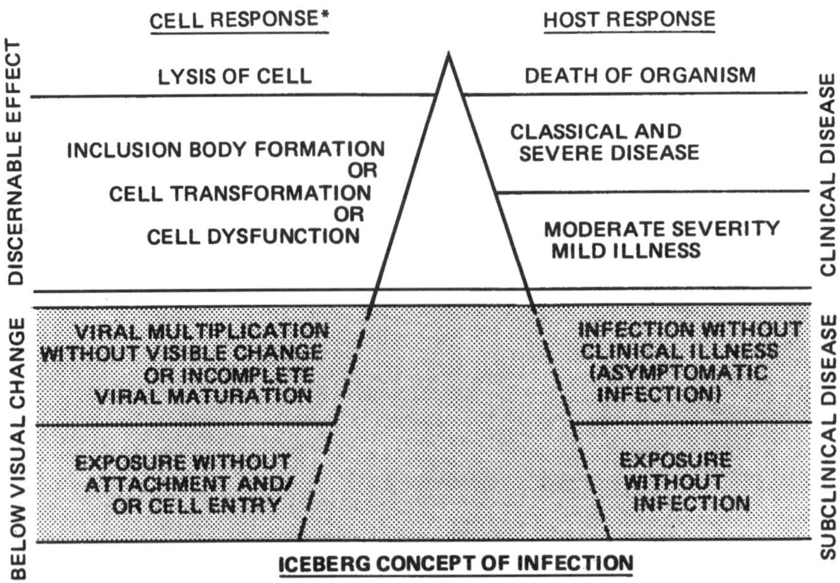

FIG. 1. Varieties of host and cellular responses to virus infection. (From Evans and Brachman, ref. 1, with permission.)

lent" viruses, subclinical or inapparent infections often outnumber cases of symptomatic illness. This has been repeatedly demonstrated, for example, during epidemics of mosquito-borne encephalitis, such as that caused by Western equine encephalitis virus, where the ratio of serologically documented infections may exceed the cases of acute encephalitis by several hundredfold. However, for certain classic viral infections, such as measles, smallpox, rabies, and influenza, almost all infected individuals develop disease.

In addition to acute infection, the interaction between a virus and a host may lead to a variety of other outcomes, including the development of persistent or latent infection (see Chapter 7) and oncogenesis (see Chapter 9). These aspects of viral infection are reviewed in other chapters and are not considered further here. As viral infection proceeds, the virus encounters a constant series of obstacles within the host. Among the most important of these are the various components of the host's immune system (see Chapter 10) and the production by the host of a variety of protein mediators, including cytokines (see Chapter 11). Since these aspects of viral infection are also reviewed in individual chapters, they are not dealt with further in this chapter. Finally, the details of how viral infection results in alteration of host cell function and cytopathology are also reviewed separately (see Chapter 8).

Implicit to the study of pathogenesis is the concept that, despite the multitude of different types of viruses and hosts, there are common themes and common strategies that have evolved in viral-host interactions. These can be considered conveniently in terms of a series of stages that characterize virus-host interactions (Table 1).

A virus must be sufficiently stable in the environment to survive until it encounters a susceptible host. It must then enter the host, through one of a number of available portals (e.g., the respiratory or the gastrointestinal tract). Once inside the host, the virus must reach a primary site, typically in proximity to the site of entry, where it can undergo initial (primary) replication. In the case of viruses that primarily produce localized infections, they may be released from surface mucosal epithelium or analogous sites and be directly shed from the host (e.g., in stool or other bodily fluids or through coughing and sneezing). After entering the environment, these viruses may then reinitiate the cycle of virus-host interaction by being introduced into a new host.

Viruses that produce systemic infections rather than localized infection must cross mucosal barriers and spread to distant sites in the host using defined pathways of spread such as the bloodstream, lymphatics, or nerves. After spreading or at the site of local infection, viruses are targeted to specific populations of cells within individual organs or tissues. The capacity of viruses to infect and productively multiply in discrete tissues or populations of cells within these tissues is referred to as *tropism.* Each of these stages in viral pathogenesis is discussed in more detail in the following sections.

The study of viral pathogenesis can be conceptualized at a number of levels. At the level of the whole organism,

TABLE 1. *Stages in viral pathogenesis*

Animal	Cell
Entry into the host	Adsorption
Primary replication	Penetration
Spread through the host	Uncoating
Cell and tissue tropism	Transcription
Host immune responses	Translation
Secondary replication	Replication
Cell injury	Assembly
Persistence	Release

pathogenesis involves the investigation of such questions as: How does a virus enter the host? Where does it undergo initial replication? How does it spread in the host? What organs and tissues does it infect? How is it transmitted? The next level of inquiry is at the cellular level. What is the nature of the viral receptor on host cells? What is the mechanism by which the virus enters these cells? How does viral infection lead to alteration of host cell function? What are the pathways and mechanism(s) by which virus is released from infected cells? Finally, advances in molecular biological techniques and in fields such as x-ray crystallography have allowed inquiry to proceed to the level of atomic structure and the sequence of proteins and nucleic acids. In subsequent sections we have tried to discuss viral pathogenesis in terms of current knowledge as it exists at each of these levels.

In somewhat simplistic terms, each stage or level in viral pathogenesis can be seen as posing a series of obstacles that the virus may overcome to proceed to the subsequent level. The successful completion of all of the stages in viral pathogenesis results in disease, whereas failure to complete these stages results in either abortive or nonproductive infection or total failure to infect. The structural design of a virus can be thought of in terms of the particular functions that the virus must perform to overcome the existing obstacles successfully at a particular point in the pathogenesis cycle. For example, the structure of the virus must enable it to remain infectious after being exposed to conditions such as temperature and humidity that are present in the ambient environment. For many viruses, this "stabilizing" function, analogous to that of a bacterial spore, is played by the envelope or, in the case of nonenveloped viruses, by the capsid proteins. Once a virus has entered the host, the parts of the viral structure that function to ensure environmental stability or to promote transmission become nonessential and are often removed or discarded as the virus proceeds through subsequent stages in pathogenesis. For example, the proteolytic removal of capsid proteins may, at this stage, actually enhance infectivity rather than attenuate it (491).

In the future, studies of viral pathogenesis will undoubtedly continue to enhance current knowledge concerning the role played by individual viral genes in determining particular viral functions. This in turn will require better definition and understanding of the events that occur at particular stages in viral pathogenesis. The availability of molecular biological techniques and tools such as x-ray crystallography will undoubtedly provide increasing insight into the structure of viruses and their component parts. This will certainly lead to a better elucidation of the relationship between viral structure and the functions required of viruses at specific stages in their pathogenesis. Indeed, a major impetus for understanding viral structure is the hope that this understanding of structure at atomic levels will provide new insights into viral function. As understanding of the mechanisms by which viruses produce disease increases, this should provide more rational strategies for the control, prevention, and therapy of viral diseases. The goal of this chapter is to provide an overview of the current state of the art.

ENTRY

Viruses that infect humans to produce either systemic disease or localized infections enter the host through a variety of routes, including direct inoculation, or through the respiratory, gastrointestinal, or genitourinary route (Fig. 2).

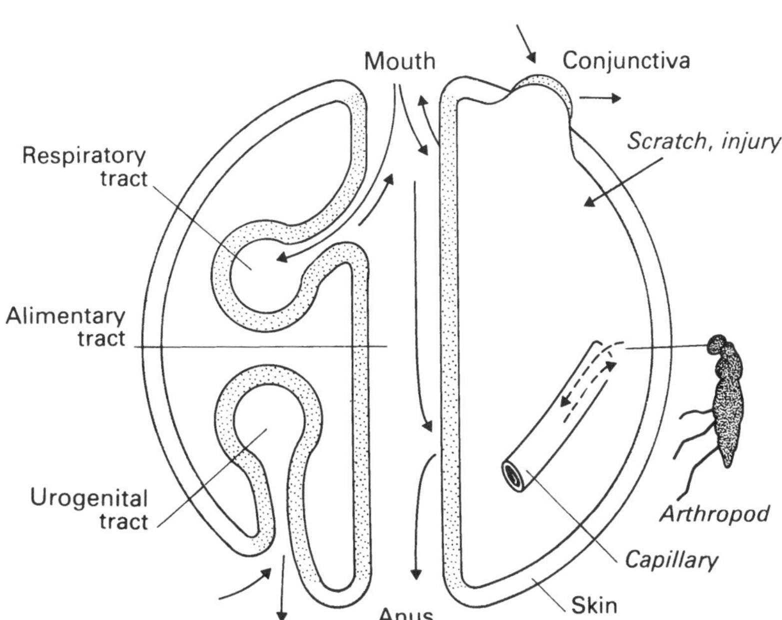

FIG. 2. Sites of entry of viruses into the host. (From Mims and White, ref. 453, with permission.)

Skin

Entry can occur through either induced or acquired breeches in the integrity of the skin or via the mucosal surfaces of the respiratory, gastrointestinal, or genitourinary tracts. The outer layer of the epidermis is composed of the dead keratinized cells of the stratum corneum. These dead cells obviously cannot support viral replication (194). Underneath these cells are the living cells of the stratum Malpighii. Some viruses, including the papillomaviruses responsible for human wart infections, enter the skin at sites where loss in the integrity of the epithelium has resulted from accidental scrapes or abrasions. In the process of vaccination or variolation, virus is deliberately inoculated into the skin through iatrogenic abrasions. Certain animal papillomaviruses are mechanically transmitted by arthropod vectors (137,138). After entry, these viruses replicate in epithelial cells at the site of inoculation (8,496,627,649). Since the epidermal layer is devoid of blood vessels, lymphatics, and nerve fibers, viruses that initiate epidermal infection are typically restricted to the site of entry and only rarely disseminate to produce systemic infections.

Certain of the poxviruses, including mousepox (ectromelia), cowpox, and rabbitpox (myxoma), also commonly initiate skin infection through traumatic openings in the skin, but the initial infectious process seems to begin in the dermis (193,570). Studies with mousepox have indicated that, for poxviruses, replication in fibroblasts and macrophages within the dermis is an important part of the infectious cycle (570,571).

Another route of dermal infection is via the bites of arthropod vectors such as mosquitoes, mites, ticks, and sandflies (see Chapters 29 and 32 in *Fields Virology,* 3rd ed.). Transmission of virus may be purely mechanical (i.e., via the mouthparts of the biting insect), as illustrated by myxoma virus infections. Alternatively, replication of the virus in the arthropod may be an essential part of the viral life cycle (105,262,474,480). Most of the arthropod-borne toga- and flaviviruses, which include the agents of epidemic encephalitis and yellow fever, seem to replicate in the gut of the insect vector with subsequent spread to salivary tissue. Virus is then inoculated into the host during feeding. Once a virus has reached the dermis, it has access to blood vessels, lymphatics, and dendritic cells, including macrophages present in the skin.

Deep inoculation, through the dermis and into the subcutaneous tissue and muscle, can follow hypodermic needle injections, acupuncture, ear piercing, tattooing, or animal bites. Hepatitis B virus may gain access to the host through needle inoculation of various types (286,315,397, 408) and even in odd cases via human bites (422). Infection with both rabies and herpesvirus simiae is characteristically initiated by the bites of infected animals (22,470).

Respiratory Tract

Inhibition of viral entry through the skin depends primarily on the mechanical integrity of the epidermal barrier. By contrast, more complex host defense systems are involved in inhibiting viral entry through mucosal surfaces (see ref. 446 for review). In the respiratory tract goblet cells secrete mucus which, propelled by the action of ciliated epithelial cells, clears foreign material by its continual upward movement through the respiratory tract. Both humoral and cellular immune mechanisms also act to protect the host against mucosal infection. Immunoglobulin A (IgA) is present in respiratory secretions, and collections of lymphoid cells occur in the tonsils and adenoids as well as in the subepithelial tissue of the nasopharynx and trachea (488). The alveoli also contain macrophages and other phagocytic cells. An additional factor in host defense is the cool temperature of the upper respiratory tract. The cooler temperature of the upper airways acts to inhibit the replication of many viruses; however, this temperature is optimum for the replication of rhinoviruses, a major cause of upper respiratory infections (see Chapter 23 in *Fields Virology,* 3rd ed.). Despite the nature of host barriers to respiratory infection, a number of viruses, including herpesviruses, adenoviruses, myxoviruses, paramyxoviruses, and rhinoviruses, can initiate infection via this route.

Viruses that enter the respiratory tract do so primarily in the form of either aerosolized droplets or saliva. Coughing and sneezing can generate enormous numbers of small aerosolized particles (222). Particles smaller than 5 μm in diameter can remain airborne for long periods, a property that can be dramatically enhanced by the presence of air currents. The distribution of inhaled aerosol particles within the respiratory tract is largely a function of particle size (352). Large particles rarely escape the filtering action of the nasal turbinates, whereas small particles (<5 μm diameter) are capable of reaching the alveolar spaces (399). Environmental factors including air temperature and humidity also influence the stability of aerosolized viral particles (149–151,352). The physicochemical properties and structure of virion particles can also influence their stability as aerosols. For example, enveloped viruses seem to be less susceptible to desiccation in arid (low-humidity) conditions than are their nonenveloped counterparts (149,150,268). This may provide one explanation for the fact that enveloped viruses (e.g., influenza, parainfluenza, respiratory syncytial virus) are more common causes of lower respiratory tract infections than are their nonenveloped counterparts (e.g., adenovirus, rhinovirus).

The proper functioning of the mucociliary transport system seems to be an important determinant in resistance to viral infection of the respiratory tract. Drugs that inhibit this transport process increase host susceptibility to a number of viral infections (31). Susceptibility to Sendai virus infection in mice has been linked to a genetic polymorphism in mucociliary transport function (78), providing additional evidence for the importance of this system in resistance to viral infection.

Gastrointestinal Tract

Many viruses enter the host via the gastrointestinal (GI) tract, where they may either initiate local infection (e.g., rotaviruses, coronaviruses, adenovirus, Norwalk agent) or invade the host to produce systemic illness (e.g., enteroviruses, hepatitis A). Local infections can be defined as those confined to epithelial cells adjacent to the intestinal lumen, whereas systemic infection occurs when viruses cross the mucosal layer to invade underlying tissues and spread within the host.

The physicochemical environment in the upper GI tract seems to be extremely inhospitable to invading microorganisms. The secretion of acid by gastric parietal cells may reduce the intraluminal pH to 2.0 or lower. In addition, a variety of proteases are secreted by gastric and pancreatic cells, and bile salts enter the duodenum from the biliary tract. Mucus is secreted by both gastric and intestinal cells and may contain both specific (e.g., IgA) and nonspecific inhibitors of viral infection. To initiate infection via the GI route, a virus would ideally have the properties of acid stability, resistance to loss of infectivity in the presence of bile salts, and resistance to inactivation by proteolytic enzymes. In fact, the majority of viral pathogens that enter via the GI tract generally exhibit all of these properties.

Almost all picornaviruses are capable of producing enteric infections in humans. A notable exception to this rule is the rhinoviruses. In distinction to other picornaviruses and to rotaviruses, rhinoviruses are extremely acid labile and undergo complete inactivation with loss of infectivity at low pH (489,651). pH-induced inactivation seems to result from the loss of the VP4 surface protein and resulting leakage of the viral RNA, which produces noninfectious "empty capsids" (651). Interestingly, many of the enteroviruses undergo an analogous type of degradation under alkaline conditions (334). Similar investigations with reoviruses suggest that under certain circumstances exposure to alkaline pH can result in the loss of the viral cell attachment protein from the viral outer capsid (163,164). Changes in the outer capsid structure of rotaviruses have also been observed at both extremely acidic or alkaline pH (Chapter 56 in *Fields Virology,* 3rd ed.).

The effect of proteolytic enzymes, such as those present in the gastrointestinal tract, on the infectivity of enteric viruses has also been investigated. Trypsin, pancreatin, and elastase have all been shown to enhance rotavirus infectivity (124,183,184) and are in fact routinely incorporated into procedures to isolate and cultivate these viruses. Conversely, inhibition of proteolysis may result in attenuation of rotavirus infection (711). The importance of proteolytic processing of virions *in vivo* has recently been demonstrated for reoviruses (38,62). Reovirus particles present in the intestinal lumen are converted into infectious subviral particles (ISVPs) by proteolytic processing of their $\sigma 3$ and $\mu 1C$ outer capsid proteins. ISVPs seem to be the form of the virion responsible for the subsequent initiation of intestinal infection. Inhibition of intraluminal proteolytic digestion dramatically inhibits growth of reovirus in the small intestine (38).

Studies with viruses indicate that host proteases act primarily to alter the viral outer capsid or envelope proteins. For example, proteolytic digestion of rotaviruses with trypsin, pancreatin, or elastase results in cleavage of the outer capsid protein VP4 and the appearance of two cleavage products VP8 and VP5 (183,184). Enhancement of coronavirus infectivity also occurs in the presence of proteolytic enzymes (650) and may be due to the cleavage of the E2 peplomer glycoprotein (the viral cell attachment protein) from an inactive precursor into its active form. Genetic studies with reovirus reassortants, derived from parental viruses, which differ in their susceptibility to proteolytic digestion, indicate that the reovirus major outer capsid protein $\mu 1C$ is the major determinant of protease sensitivity (593).

Detailed studies of the mechanism of action at a molecular level of proteolytic digestion of viral proteins have been made with influenza viruses and Sendai virus (218, 223,374,383,564,608,742). These studies provide insights into how proteolytic processing of viral proteins can influence viral pathogenicity. Cellular exopeptidases cleave the hemagglutinin (HA) of influenza into two disulfide-bonded subunits (HA_1 and HA_2). After virus enters target cells through the process of receptor-mediated endocytosis, the amino-terminus of the HA_2 protein undergoes a pH-dependent conformational change that facilitates the subsequent fusion of the influenza virus envelope with the inner membrane of the endocytic vesicle (630). This in turn results in the escape of the viral nucleocapsid into the target cell's cytoplasm, where replication occurs. If cleavage of the HA does not occur, then influenza virus still binds to cell surface receptors but is unable to initiate a productive infection. Inhibition of HA cleavage by administration of protease inhibitors results in attenuation of influenza virus infection *in vivo* (665,742,770,771).

Thus, many viruses have evolved so that proteolytic processing actually facilitates rather than inhibits viral infection (491). The mechanisms by which this facilitation occurs are gradually becoming better understood. As a general principle, proteolytic processing of these viruses usually results in partial cleavage of specific viral surface proteins, which in turn facilitates specific events, such as membrane fusion or transcriptional activation, in the viral replicative cycle. Conversely, viruses that are easily inactivated by proteolytic processing are unlikely to initiate infection via sites such as the GI tract, where proteolytic enzymes are a prominent part of the internal milieu.

The presence of bile salts in the intestinal lumen may also play an important role in determining the types of virus that are capable of initiating GI infection. Viral envelopes are derived from host cell lipoprotein bilayer membranes and are particularly susceptible to dissociation by bile salts. Conversely, the protein capsids of nonenveloped viruses are quite resistant to the action of bile salts. This may explain why, with the exception of coronaviruses, enveloped viruses do not initiate infection via the upper gastrointestinal tract. Coronaviruses are susceptible to inactivation by bile salts *in vitro,* and the factors that enable them to survive inactivation *in vivo* remain unknown.

Some viruses, such as the rotaviruses, produce enteric infections that are typically limited to the intestinal epithelial cells (402). Local infection of intestinal epithelial cells also occurs during adenovirus, astrovirus, calicivirus, coronavirus, and parvovirus infection (603). The factors that determine whether viral infection remains limited to the GI tract or spreads to produce systemic illness are largely unknown. Some enteric viruses, including reoviruses and poliovirus, spread from the lumen of the small intestine to collections of submucosal lymphoid tissue known as Peyer's patches (59,60,465,602,749,750).

The early events in the spread of reoviruses after intestinal infection have been investigated extensively using electron microscopy. Virus initially adheres to the surface of specialized intestinal epithelial cells (M cells) that overlie the surface of Peyer's patches. Reovirions can be sequentially visualized at the surface of M cells, being transported in vesicles through these cells, and subsequently free in the extracellular space of Peyer's patches (40,749,750) (Fig. 3). The M cell pathway is also utilized by poliovirus to penetrate the intestinal epithelial barrier (629), suggesting that this may be a general pathway for the entry of enteric viruses that spread beyond the mucosal surface.

The recognition that anal intercourse is a major risk factor for the transmission of HIV infection has also led to the recognition that some viruses that produce systemic disease may gain entry to the host through the lower GI tract. HIV has been detected in bowel epithelium and enterochromaffin cells of patients with HIV and can infect human colonic cell lines in culture (395,434,486,507). The mechanisms by which HIV infection is initiated through the lower GI tract remain to be established.

Genitourinary Tract

In addition to the respiratory and gastrointestinal tracts, other mucosal surfaces may provide sites for viral entry

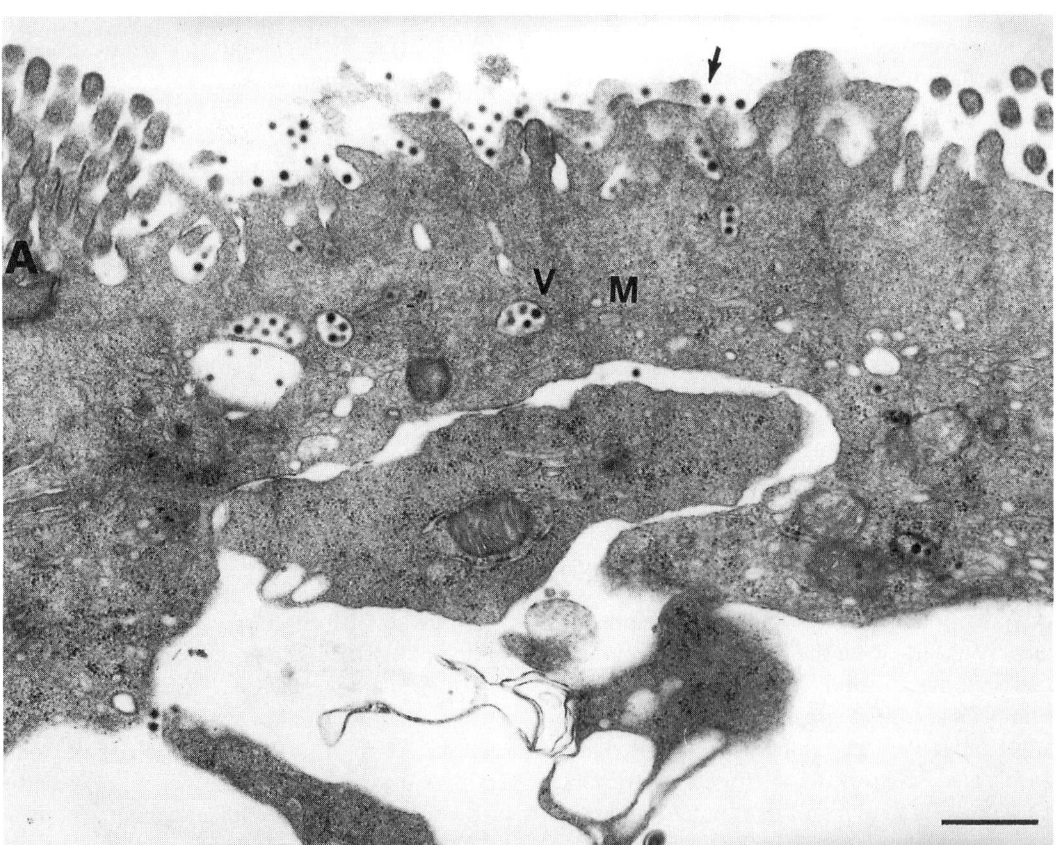

FIG. 3. Reovirus serotype 1 strain Lang virions on the surface (*arrow*) and within vesicles (*V*) of an M cell in the ileal epithelium overlying a Peyer's patch. (From Wolf et al., ref. 750, with permission.)

into the host. Sexual activity may produce tears or abrasions in the vaginal epithelium or trauma to the urethra allowing viral entry via the genitourinary route. Sexually transmitted viruses in humans include HIV, herpes simplex virus (HSV), and human papillomaviruses 11, 16, and 18 (the agents responsible for genital warts or condyloma acuminata). In the case of HIV, infection may be facilitated by breaks in the genital epithelial barrier, but the virus may also productively infect cervical epithelial cells and dendritic (Langerhans) cells within the vaginal epithelium (378,535,661). HHV-6 (114) and HPV 16 and 18 (754) have also been shown to infect cervical epithelial cells in culture.

Acute hemorrhagic cystitis in young boys has been associated with adenoviruses 11 and 21 (469,494); however, it is not known whether these viruses reach the bladder directly through the urethra or indirectly via another site of entry. Other viruses, including polyomaviruses, can produce viruria, but this is generally not associated with symptomatic disease. In some of these cases, virus seems to reach the kidney, presumably through the bloodstream, and then enter the urine (437). Little is known about the nature of host factors that influence viral invasion by these routes. The nature of the cervical mucus, the pH of vaginal secretions, and the chemical composition of the urine may all play a role in host defense against infection.

Conjunctiva

The conjunctiva may also provide a route for the entry of viruses that either produce local disease (conjunctivitis) or more rarely disseminate from this site to produce systemic infection (e.g., paralytic illness with enterovirus 70). Direct inoculation of virus onto the conjunctiva during ophthalmological procedures (e.g., foreign body removal, tonometry) can produce keratoconjunctivitis, and many cases have been attributed to adenovirus 8, 19, or 37. Similarly, the conjunctivitis that follows swimming in public areas (swimming pool conjunctivitis) seems frequently to be due to direct infection most often with adenoviruses (142,253,340). Conjunctivitis can also occur in association with enteroviral infections (e.g., echo 7,11; coxsackie A24,B2). In most cases it appears that infection is due to direct inoculation of virus into the eye. Systemic spread is exceedingly rare, although, as noted, it is an important feature of the acute hemorrhagic conjunctivitis produced by enterovirus 70. This virus can spread from the conjunctiva to the central nervous system, where it can produce poliomyelitis, radiculomyelopathy, or cranial nerve palsies (275,313,355,714). Conjunctivitis may also occur as part of a systemic viral infection in which virus reaches the conjunctivae after the induction of viremia rather then directly from the external surface. This type of process seems to account for the conjunctivitis commonly associated with acute measles infection.

SPREAD IN THE HOST

Localized Versus Systemic Infection

Some viruses produce the brunt of their tissue injury in close proximity to their site of entry into the host. This pattern of infection characterizes the upper respiratory infections of influenza, parainfluenza, rhinoviruses, and coronaviruses; the gastrointestinal infections caused by rotaviruses; and the dermatological infections of the papillomaviruses. In these cases spread of virus occurs primarily as a result of contiguous infection of adjacent cells. Virus rarely spreads beyond the epithelial cell layer, although in some cases involvement of regional lymph nodes and even systemic invasion can occur. The factors that act to confine the scope of these infections and that act to impede systemic dissemination are largely unknown.

Polarized Infection of Epithelial Cells

Certain viruses preferentially bud from either the apical or basal surfaces of polarized epithelial cells (see refs. 127 and 687 for review) (Fig. 4). Preferential release of virus from a particular surface of polarized epithelial cells may influence whether the subsequent pattern of disease is localized or becomes systemic. If infection of an epithelial cell layer results exclusively in release of virus toward the luminal surface, then this favors the development of localized infection. Conversely, release of virus toward subepithelial tissues may facilitate mucosal invasion and the development of systemic infection. In the case of Sendai virus, for example, wild-type virus buds from the apical surface of polarized bronchial epithelial cells and produces bronchial infection. Sendai mutants that have the capacity to bud from either the apical or the basolateral surface of epithelial cells have extended tissue tropism and enhanced virulence *in vivo* (666).

Examples of viruses with polarized patterns of release from epithelial cells include vaccinia, vesicular stomatitis virus (VSV), and certain retroviruses, which are assembled at and bud from the basolateral cell membrane, and influenza, parainfluenza, SV40, polio, and Sendai viruses, which use the apical surface (125,216,565,581,586,688). Not all viruses exhibit polarized patterns of infection and release, and some, such as poliovirus, are capable of bidirectional entry into polarized epithelial cells (689). In cases of polarized infection by enveloped viruses, envelope glycoproteins are inserted into the appropriate region of the cell membrane before viral budding. This suggests that it is the initial targeting and localization of these proteins to specific regions of the cell membrane that determine the subsequent polarity of viral budding (514,580,584,586). The information required for appropriate targeting of these proteins to the cell surface is contained in specific regions of the targeted proteins, and even single amino acid sub-

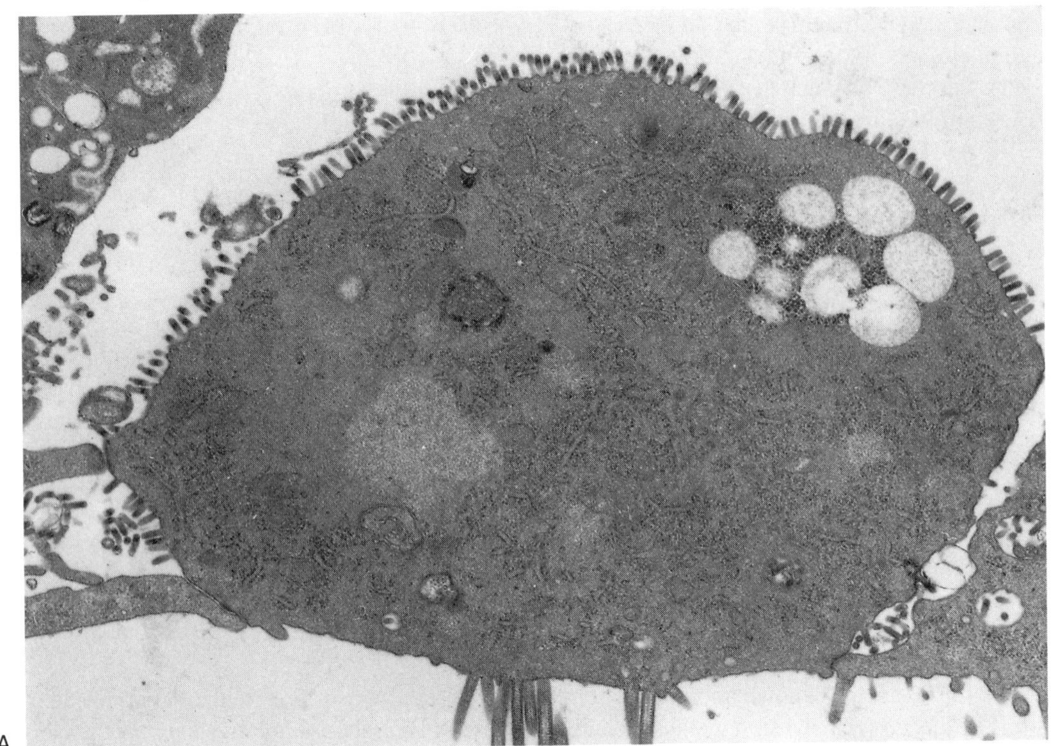

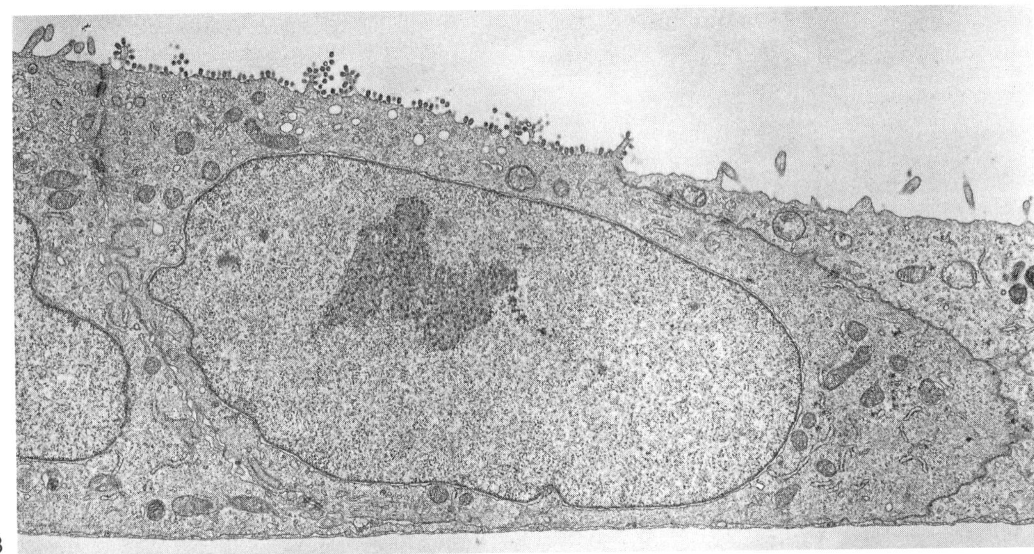

FIG. 4. A: Polarized budding of SV5 and VSV virions from different surfaces of the same MDCK cell. Filamentous SV5 buds from the apical surface, while bullet-shaped VSV particles assemble at the basolateral domain of the cell. (From Rindler et al., ref. 565, with permission.) **B:** Budding of influenza virus from the apical surface of an MDCK cell. (From Rodriguez-Boulan and Sabatini, ref. 581, with permission.)

stitutions in the appropriate domain can alter polarized delivery (69,128,483,584–586,671).

Hematogenous Spread

Viruses that produce systemic disease must spread from their site of entry into the host to their ultimate target tissues. The two major pathways for this spread, via the bloodstream and via nerves, are discussed separately. Direct inoculation of virus into the bloodstream may occur with the bite of an arthropod vector, through iatrogenic inoculation with a contaminated needle, or by the transfusion of contaminated blood products (passive viremia). In the case of experimental bunyavirus infection, if the inoculum size is sufficient, passive viremia may be adequate to deliver virus to the central nervous system (CNS) and initiate a lethal encephalitis (Neal Nathanson, *personal communication, 1994*). A more com-

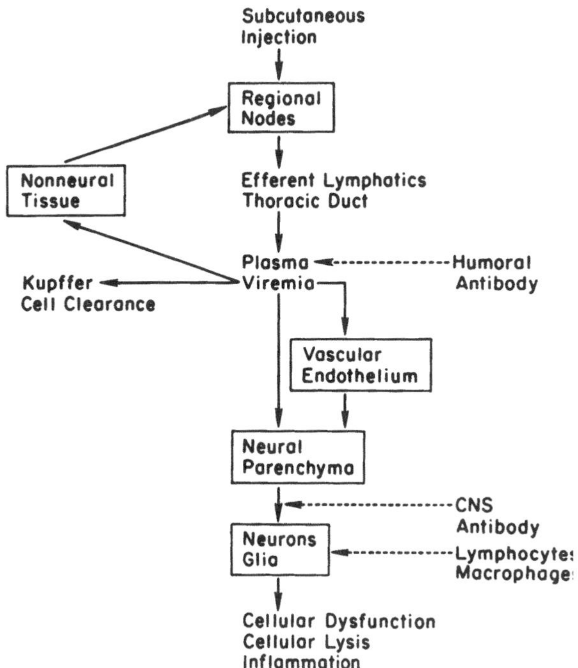

FIG. 5. Stages in the pathogenesis of flavivirus infection. (From Nathanson, ref. 480, with permission.)

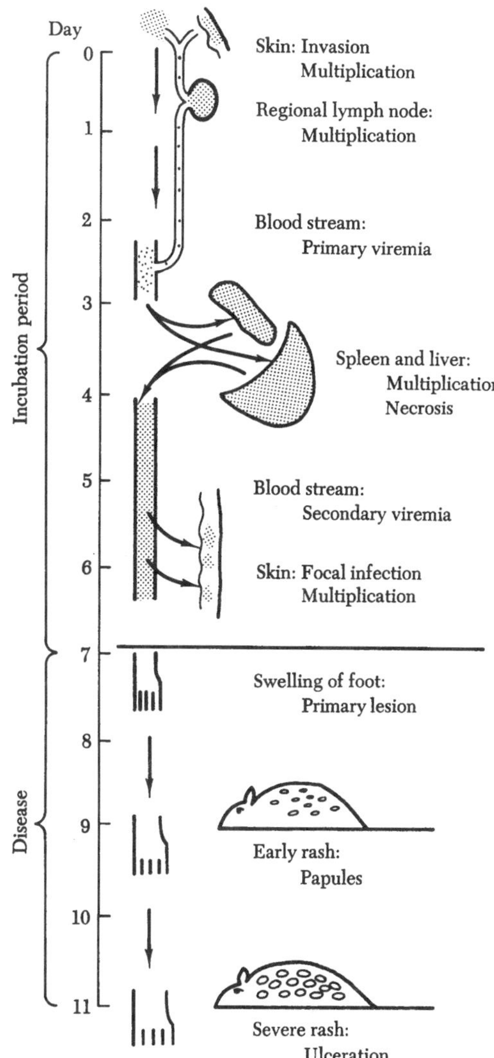

FIG. 6. The pathogenesis of mousepox (ectromelia). (From Fenner, ref. 193, with permission.)

mon sequence of events, exemplified by viruses such as flaviviruses, measles, and polio, is for primary replication at the site of entry followed by spread to regional lymph nodes (6,58,425–427,480,602). From the regional lymph nodes, virus can gain access to efferent lymphatics, the thoracic duct, and the systemic circulation. Ectromelia (mousepox) may replicate in lymphatic endothelial cells, further increasing its chances of reaching the bloodstream (570).

In almost all cases, some degree of replication at a primary site near the site of entry in the host seems to precede the onset of viremia. In some cases primary replication is followed by a transient low-titer viremia (primary viremia), which serves to spread virus to sites, such as the reticuloendothelial system, where additional replication leads to a more sustained and higher titer (secondary viremia) (Fig. 5). These events were elegantly defined by Fenner (193) in a series of classic studies of the pathogenesis of ectromelia (mousepox) (Fig. 6). Although viremia occurs in many human infections (see below), it is often difficult to demonstrate discrete phases of primary and secondary viremia.

Once virus has reached the circulation, it can travel either free in the plasma or in association with cellular elements (Table 2). It is important to recognize that not all the associations between viruses and cellular elements listed in the table are known to be of pathogenetic significance.

The magnitude and duration of viremia reflect the outcome of a series of interactions between the virus and the host. Viral replication at the primary site and subsequently in a variety of other tissues provides a continuing source of virus input into the circulation. Important sites of primary replication for viruses that spread through the bloodstream include skeletal muscle for togaviruses (245,318, 473,480); connective tissue, muscle, endothelial cells, and reticuloendothelial organs for flaviviruses; and brown fat for enteroviruses (58,60). At the same time as virus is entering the blood compartment from these sites, a variety of clearance mechanisms are actively removing virus from the circulation. The magnitude of viremia varies as a result of the dynamic interrelationship between the amount of virus entering the blood compartment and the efficiency with which it is removed. The extent to which virus input into the bloodstream exceeds the efficiency of host clearance determines the magnitude of viremia. Viremia persists for as long as virus input into the bloodstream continues and exceeds the host's capacity to clear virus from the circulation. The mean duration for which infectious virus remains in the bloodstream varies for different viruses (Neal Nathanson, *personal communication, 1994*).

TABLE 2. *Methods of viral spread within cells and through the bloodstream*

Free in plasma	Togaviruses, picornaviruses
Red cell associated	CTFV
Platelet associated	HSV, LCMV, retroviruses
Lymphocyte associated	EBV, CMV, LCMV, mumps, measles, rubella, HBV, JC virus, and BK
Monocyte/macrophage associated	Polio, CMV, HIV, lentiviruses, LCMV, African swine fever, measles
Neutrophil associated	Influenza

The listing of a virus family indicates that one or more member of the family may spread in association with these cellular elements. See text and refs. 98, 160, 176, 182, 206, 280, 298,421,435,438,554,654,669). CMV, cytomegalovirus; CTFV, Colorado tick fever virus; EBV, Epstein-Barr virus; HBV, hepatitis B virus; HIV, human immunodeficiency virus; HSV, herpes simplex virus; LCMV, lymphocytic choriomeningitis virus.

Phagocytic cells in the reticuloendothelial system (RES) and serum factors, including complement and antibody, act to aid in clearance of virus from the bloodstream (453). One of the most important determinants of clearance is particle size (79,453). Rapid clearance (>99.99% of an intravenous inoculum cleared within an hour) is seen after inoculation of a large virus such as vaccinia into the bloodstream of a mouse. A similar pattern of clearance is seen with small viruses opsonized by antibody or complement. The coating of virus with antibody or complement may facilitate clearance by phagocytic cells such as macrophages because of the presence of receptors on these cells for the Fc portion of antibody molecules and for complement (C_3 receptor). The net charge on a viral particle may also influence its clearance. In the case of Venezuelan equine encephalitis (VEE) virus, particles with high net charge are rapidly cleared from the bloodstream, whereas those with low charge show delayed clearance (307).

The nature of the protein(s) on the outer surface of the virus may also be an important determinant of the pattern of clearance. For example, the sialic acid content of the envelope of Sindbis virus seems to influence the degree of viral clearance (279). In the case of reoviruses, the σ_1 outer capsid protein determines the organ-specific pattern of viral clearance from the bloodstream after an intravenous inoculation. This pattern can be altered by the presence of antibody on the viral surface (708).

The nature of the interaction between virus and the macrophages of the RES seems to be an important determinant of the development of viremia (see ref. 453 for review). Obviously, if a virus can avoid phagocytosis by macrophages, this would facilitate the maintenance of viremia. For some viruses, uptake by macrophages results in inactivation and is a major factor in terminating viremia. In this type of infection, factors that inhibit the phagocytic capacity of the macrophages can serve to amplify viremia and potentiate the severity of viral infection. Experimental inhibition of macrophage clearance can be achieved with agents such as Thorotrast and silica. This type of blockade markedly enhances viremia induced by yellow fever virus (YFV) (773,774) and Semliki Forest virus (SFV) (453) and potentiates infection due to HSV (462). Under experimental circumstances, extremely high-titer viral inoculums can overwhelm the phagocytic capacity of the RES and allow for potentiation of viremia (293,524).

Certain viruses may replicate in macrophages, which serves to further amplify viremia rather then terminating it. A number of viruses including togaviruses (e.g., lactate dehydrogenase–elevating virus, Sendai), poxviruses (e.g., ectromelia), lentiviruses (e.g., HIV, visna, equine infectious anemia virus, Aleutian disease virus), coronaviruses (e.g., mouse hepatitis virus), arenaviruses (e.g., lymphocytic choriomeningitis virus), and reoviruses have been shown to replicate in macrophages (453). Because macrophages have receptors for the Fc portion of antibody, viruses coated with antibody may be more readily taken up by these cells. In some cases this facilitated uptake actually leads to an enhanced replication of these viruses in macrophages (antibody-mediated enhancement) (86,525). Antibody-mediated enhancement may play an important role in the pathogenesis of viral infections such as dengue fever (258,540).

Differences in the susceptibility of macrophages to viral infection may account for variation in the age-related susceptibility of animals to certain types of viral infection (278,316) (see "Host Factors"). For example, alveolar macrophages derived from young mice will support the replication of herpes simplex virus. Transfer of adult macrophages to young mice increases their resistance to HSV infection (278,316,461). Viral virulence may also be related to the capacity of a particular strain of virus to grow in macrophages. For example, virulent strains of ectromelia grow to high titer in hepatic macrophages, whereas avirulent strains do not (453,571,572). Finally, the genetically conferred resistance of certain strains of inbred mice to certain viral infections correlates well with the susceptibility of their cultured macrophages to infection *in vitro*. This phenomenon has been particularly well exemplified by certain strains of mouse hepatitis virus (34).

The vascular endothelial cells also seem to play an important role in the maintenance of viremia for certain viruses (326,453). The capacity of viruses to infect endothelial cells has been studied both *in vitro* and *in vivo* for a large number of viruses (131,199,208,323,379,433, 519,530,707,717,776) and is discussed more fully in subsequent sections.

There is a general correlation between the capacity of blood-borne neurotropic viruses to generate a high-titer viremia and their neuroinvasiveness (292,293,480,557). Similarly, attenuated virulence may reflect an incapacity to generate sufficient viremia, even for viruses that are fully virulent after direct CNS inoculation (228,229,310).

For example, mutants of Semliki Forest virus generate high-titer viremia but, unlike their wild-type counterparts, are unable to invade the CNS (17,18). Thus, the capacity to generate a viremia of adequate magnitude and duration may be necessary for neuroinvasiveness but is not always sufficient.

Tissue Invasion

The factors that determine the capacity of blood-borne virus to escape the vascular compartment and invade specific organs are poorly understood. The problem has probably been most extensively investigated as it applies to invasion of the central nervous system by blood-borne viruses (Fig. 7) (321,326,453). In most regions of the CNS the capillary endothelial cells of the cerebral microvasculature are joined by tight junctions (zona occludens) and there is an underlying dense basement membrane. A notable exception to this pattern occurs in the choroid plexus, where the capillary endothelium is fenestrated and the basement membrane is generally sparse. Some viruses invade the CNS from the bloodstream by entering the stroma of the choroid plexus through the fenestrated capillary endothelium and then either infecting or being passively transported across the choroid plexus epithelial cells into the cerebrospinal fluid (CSF). Once in the CSF, virus can infect the ependymal cells lining the walls of the ventricles and can then invade the underlying brain tissue. This pattern of infection has been shown for a number of viruses, including mumps (270,751), arboviruses (404), and rat virus (401). Visna, a lentivirus, can be shown to infect cultured choroid plexus epithelial cells *in vitro* (479), which suggests that this pathway may be used by this virus *in vivo* as well. In some cases direct infection of choroid plexus may not be demonstrable, although the sequential appearance of virus in CSF and ependymal cells suggests the possibility that virus has used this route initially to reach the CSF (741).

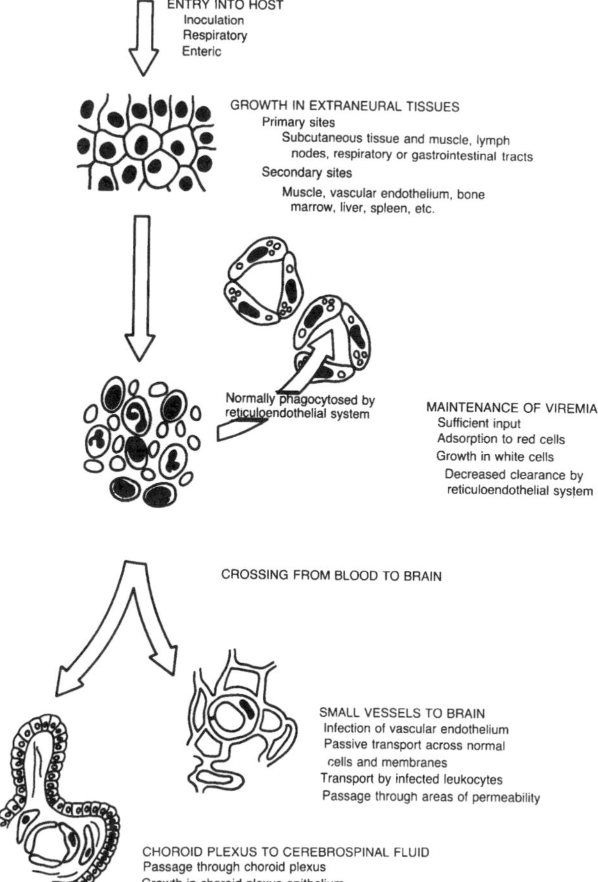

FIG. 7. Pathways of entry of blood-borne viruses into the central nervous system. (From Johnson, ref. 321, with permission.)

A number of viruses that invade the CNS do so by either directly infecting or being passively transported across capillary endothelial cells (33). Infection of endothelial cells has been proposed as a mechanism for CNS invasion of such disparate groups of viruses as picornaviruses (131,776), togaviruses (318,480,521), bunyaviruses (322), parvoviruses, and murine retroviruses (433,530) and has been demonstrated *in vitro* for many viruses (131,208,379,707,717). In the case of some murine retroviruses, there is a close association between neuropathogenicity and the capacity to bind to brain endothelial capillary cells *in vitro* (433).

Transendothelial transport of viruses including measles, mumps, canine distemper virus, togaviruses, and lentiviruses may provide another route of penetration of the endothelium and may occur via the diapedesis of infected monocytes, leukocytes, and lymphocytes (204,254,521,655,751). This process may be facilitated by viral modulation of the activity of LFA-1 and other cell surface adhesion molecules (19,276,583). Finally, factors that increase vascular permeability, such as vasogenic amines, may also facilitate viral invasion of sites such as the CNS (616).

Neural Spread

Another important route for viral spread is through nerves (321,326,361) (Fig. 8). Although this was first postulated as a mechanism for the spread of rabies as early as the 18th century (Morgagni, cited in ref. 321), it was not clearly demonstrated until the elegant work by Pasteur and his contemporaries (492,589). During the first half of the 20th century, experiments by Goodpasture, Hurst, Sabin, Bodian, and others established that herpesviruses, polioviruses, and certain arboviruses were also capable of neural spread (187,235,289,598,601). These early observations have been repeatedly confirmed in more modern studies (225,367,372,470–472,562,680,681,695), and the list of neurally spreading viruses has been expanded to include reoviruses (199,466,691), pseudorabies virus (94,95,695), coronaviruses (36,527,528), and Borna disease virus (93,458).

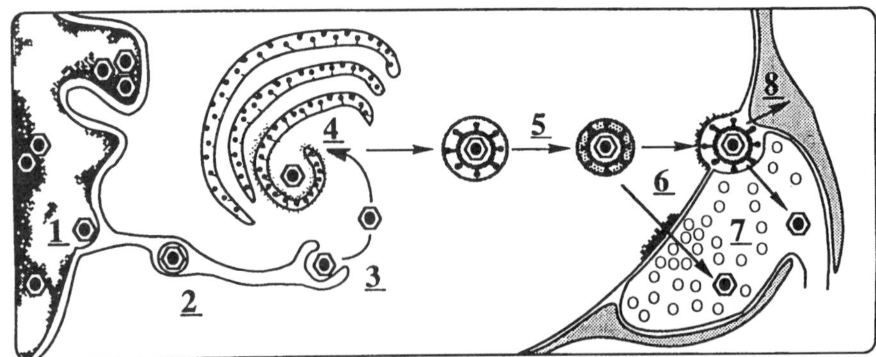

FIG. 8. Significant events that lead to replication and transneuronal passage of pseudorabies virus. Virions replicate in the cell nucleus and acquire a membrane envelope as they bud from the nuclear membrane (*1*), traverse the endoplasmic reticulum (*ER*) (*2*), and are subsequently released into the cytoplasm after a fusion event between the virion membrane and ER (*3*). Virions acquire an envelope at the Golgi apparatus (*4*), which consolidates and becomes denser (*5*). Transneuronal passage of virus occurs after fusion of the virion envelope with the postsynaptic membrane (*6*). The nucleocapsid can subsequently undergo retrograde axonal transport toward the neuronal cell body (*7*). (From Card et al., ref. 94, with permission.)

Classic experiments by Howe and Bodian (289) on the spread of poliovirus through the sciatic nerve of the monkey established several important principles of neural spread that have been repeatedly confirmed in other systems (23). They demonstrated that primary replication of virus in extraneural tissues was not an absolute prerequisite for neural spread by showing that poliomyelitis could develop in monkeys after immersion of the cut distal end of the sciatic nerve into a solution containing virus. They also established that the incidence of subsequent neurological disease, which reflected the efficiency of neural spread, depended on a variety of factors, including the amount of nerve exposed to virus, the duration of this exposure, and the concentration of virus present (289). Under circumstances of natural infection, direct exposure of a nerve to virus is presumably unusual, and primary replication of virus at peripheral sites, although not essential for subsequent neural spread (133,619), may facilitate this process by amplifying the size of the initial inoculum.

The advent of techniques for localizing viral antigens in tissues allowed the sequential progression of viral infection along neural pathways to be clearly documented (111,317,319,432,470–472). The direct visualization of herpesvirus and rabies virus particles within axons provided further evidence for intraaxonal transport (111,277, 303,304,311,362,470,472,548). A natural corollary of these observations was the demonstration that axonal transport also occurs in cultured neurons *in vitro* (414,415,682, 683,772) and that inhibitors of axonal transport also inhibited the spread of neurally spreading viruses *in vivo* and *in vitro* (53,100,101,363,364,415,680,691). Both kinetic studies and experiments using pharmacological inhibitors indicate that most viruses that spread neurally do so via fast axonal transport (361,691). *In vivo* studies, in which the rate of viral transport is generally measured by dividing the distance traveled by the elapsed time between inoculation and detection at a distant site, have resulted in calculated transport values of 2 to 16 mm/day for viruses as diverse as polio (289), rabies (304), herpes simplex (364), and reovirus T3 Dearing (691). These rates are faster than those associated with slow axonal transport but far slower than the rate of fast axonal transport (>100 mm/day). The discrepancy may be due to the fact that viral transit times also reflect the time required for a variety of events in addition to transport (e.g., adsorption, penetration and uncoating, and in some cases replication). The rate of viral transport *in vitro* is often (682,683,772) but not invariably (415) considerably faster than that calculated from experiments *in vivo.*

Further confirmation of the use of fast axonal transport by viruses comes from pharmacological studies. For example, selective inhibition of fast axonal transport with colchicine, but not the inhibition of slow axonal transport with imidodiproprionitrile (IDPN), inhibits the neural spread of reovirus T3 (691). Colchicine has also been shown repeatedly to inhibit the neural spread of both rabies virus and HSV both to the CNS (53,101,364,680) and within the CNS (100). Similarly, taxol, nocodazole, and vinblastine, agents that disrupt the microtubules that serve as the structural guides for fast axonal transport, can inhibit the neural spread of viruses *in vivo* and *in vitro* (361,415). A selective inhibitor of retrograde axonal transport, erythro-9,3-(2-hydroxynonyl)adenine, has also been shown to inhibit transport of herpes simplex virus (363), as has an inhibitor of axonal transport in sensory neurons (capsaicin) (641).

The exact mechanism(s) by which axonal transport of viruses occurs has not been established. In the case of enveloped viruses, such as rabies and herpes simplex, both mature virions and naked nucleocapsids have been visualized within axons (111,303,304,311,362,472). Fast axonal transport typically involves the movement of material contained in vesicles along microtubules, yet it remains unclear whether virus is transported in this fashion. Some viruses, including reoviruses and adenoviruses, are capable of binding to isolated purified microtubules (21), suggesting that the movement of extravesicular virus could also occur. It is easy to see how viruses that initially enter nerve cells via receptor-mediated endocytosis, and thus appear in vesicles, could be transported to the perikaryon in a vesicle-associated fashion. In some cases, virus might get inserted into vesicles via the Golgi apparatus, in a fashion akin to the sorting of certain cellular macromolecules. Defining the nature of neural transport of viruses at a cellular and molecular level remains an unsolved problem in viral pathogenesis.

In vivo studies of viral spread clearly indicate that viruses can cross the neuromuscular junction and travel transneuronally at both synaptic and nonsynaptic sites (25,95,361, 362,365,445,642,652,694–696), with the predominant route depending on the virus. In some cases transneuronal spread may be facilitated by the pattern of viral release from the nerve cell or myocytes. Polarized sorting of viral glycoproteins to the axon and dendrites of cultured neurons (161), a process analogous to that seen in epithelial cells (see above), undoubtedly plays a role in the patterns of viral release and transport in neurons. Rabies virus accumulates in myocytes at the neuromuscular junction (470,471) and may bind to the nicotinic acetylcholine receptor at the neuromuscular junction (85,260,388,389,722). The synaptic terminals of nerve cells may contain increased numbers of receptors for certain viruses, including herpes simplex and rabies (414,415,700,701,772), perhaps explaining how entry at the distal axon terminal is facilitated. Rabies virions also bud from peripheral nerve cells along their axons at the nodes of Ranvier, at the boundary between apposing myelin-producing Schwann cells (311,470). A related virus, vesicular stomatitis virus, will preferentially bud from synaptic sites in the presence of antiviral antibody, suggesting that the synaptic region of the axon terminal may have unique structural features that facilitate the exocytosis of some viruses (170).

The capacity of viruses to spread along discrete neuronal pathways and to spread transneuronally across synapses has made them valuable tools for defining neuroanatomic pathways. Viruses have been used to map synaptically linked interneuronal circuits both within the central nervous system and between the peripheral and central nervous systems. Injection of virus into a particular target tissue and sequential study of its central transit may allow the peripheral and central innervation of these structures to be mapped with exquisite fidelity. Herpes simplex virus and pseudorabies virus have both been used extensively for this purpose (94,95,285,371,642,652,694–696).

It is important to recognize that spreading via the bloodstream and spreading through nerves are not mutually exclusive processes. Some viruses may use different pathways to reach different organs or different pathways at different stages in pathogenesis. For example, it has been suggested that certain flaviviruses may spread via the bloodstream to reach the neurons of the olfactory bulb and then travel retrogradely via neural spread within these specialized neurons to reach the CNS (456). Varicella-zoster virus (VZV) disseminates to the skin via the bloodstream to produce the classic exanthemal rash of chickenpox but then travels up the axons of sensory neurons to reach sensory ganglia, where the virus remains latent. Reactivation of VZV is associated with spread of virus via nerves to reach the skin and the appearance of the dermatomal lesions of shingles (248).

In some cases, different strains of the same virus seem to spread preferentially via different pathways. Under experimental conditions, some strains of poliovirus spread to the CNS via nerves (562), whereas others do so via the bloodstream. For a long time controversy existed concerning the route of spread used by poliovirus to reach the CNS after enteric infection in humans (58,60,186,289,330,481,602). The conventional view is that spread of poliovirus to the CNS is via the bloodstream, although the neural route has never been definitively excluded (321). Recent studies in the mouse model have confirmed the importance of neural pathways as the route of spread of virus from muscle to CNS (562), and studies with other enteric viruses have led to renewed recognition of the importance of neural pathways as a potential route of spread from the intestine to the CNS (466), which should provide renewed impetus to reinvestigate the importance of this pathway in poliovirus pathogenesis (465).

Molecular and Genetic Determinants of Viral Spread

Although the capacity of viruses to spread via different routes in the infected host has been recognized for some time, it has only been recently that the nature of specific viral genes involved in determining this process has been identified. After intramuscular inoculation, reovirus T1 Lang (T1L) spreads to the CNS via the bloodstream, whereas T3 Dearing (T3D) spreads via nerves. The viral S1 gene, which encodes the outer capsid protein σ_1, determines the capacity of T1L × T3D reassortants to spread via either the bloodstream or nerves to reach the CNS (691). T1 and T3 also show different patterns of spread after oral inoculation into mice. T3 does not disseminate to extraintestinal organs, but T1 does. The pattern of spread of T1 suggests that virus initially spreads via local lymphatics and then via the bloodstream. Studies with T1L × T3D reassortants indicate that the capacity of T1 to spread extraintestinally is also determined by the S1 gene (336). Recent studies indicate that the pattern of clearance of T1 and T3 from the bloodstream after intravenous inoculation differs, and these differences are determined by the S1 gene (708). Thus, several studies with reoviruses indicate that their pattern of spread in the infected host is determined by the nature of the viral S1 gene and the outer capsid protein (σ1) that it encodes.

In a series of recent studies, genetic factors that influence the capacity of herpes simplex virus to spread to and invade the CNS have also been investigated. There seem to be numerous factors that influence HSV neuroinvasiveness, neurovirulence, and latency (see ref. 647 for review).

The region of the HSV genome between 0.25 and 0.53 map units (m.u.) seems to contain genes that are critical in allowing HSV-1 to spread to the CNS after footpad inoculation in mice (neuroinvasiveness) (672). An area contained within this region (0.31 to 0.44 m.u.) has also been shown to influence the capacity of intertypic HSV-1 × HSV-2 recombinants to spread to the CNS via nerves after corneal inoculation in mice (499). A number of viral proteins, including the DNA polymerase, thymidine kinase, the nucleoprotein p40, and the gB envelope glycoprotein, are encoded by genes contained in these regions. Recent evidence suggests that mutations in gB may be particularly important in influencing the neuroinvasive capacity of certain HSV strains, including KOS and ANG (765). Another genetic mapping study has implicated the DNA polymerase (0.413 to 0.426 m.u.) (145). The mechanism by which alterations in the polymerase gene enhance or attenuate neural spread remain speculative. It seems likely that the effect is an indirect one having to do with replication competence rather than a direct effect on axon transport of virus, since it can be shown that recombinants with differences in this region of the genome vary dramatically in their capacity to replicate in peripheral nervous system tissue (145,499,672).

Mutations within the gD envelope glycoprotein can also influence HSV neuroinvasiveness. For example, the neuroinvasive HSV 1 ANG-path strain differs from its nonneuroinvasive parent by only a single amino acid substitution within gD (305).

These studies suggest that, in the case of HSV, neuroinvasiveness may be affected by mutations involving a number of steps in pathogenesis. Mutations in envelope glycoproteins such as gB and gD may affect viral attachment or penetration. Polymerase mutations and mutations in viral immediate early proteins may hamper viral replication.

Strain-specific patterns in the transport of herpesviruses have been investigated for both herpes simplex virus and pseudorabies virus (PRV) (97). For example, the PRV-Becker, PRV-Bartha, and PRV-91 strains all differ in their pattern of neuronal uptake and subsequent neural spread (96,97,642). The PRV-Becker and PRV-91 strains differ only in the deletion of the gene encoding the PRV gI and gp63 glycoproteins, suggesting that the complex formed by these proteins may be critical in determining differences in neuronal uptake and spread (96,97,347,642,735).

Studies in the bunyavirus system have identified viral genes that are important in the capacity of these viruses to spread from the periphery to the CNS (neuroinvasiveness). Comparisons between nonneuroinvasive and neuroinvasive strains indicate that the viral M genome segment, which encodes the two envelope glycoproteins (G1, G2) is the major determinant of the neuroinvasive phenotype (228,230,309). Subsequent studies with monoclonal antibody–resistant bunyavirus variants suggest that the viral G1 glycoprotein may be the important determinant.

Studies with monoclonal antibody–resistant mutants of the flavivirus, Japanese encephalitis virus (JEV), indicate that mutations in the E envelope glycoprotein can reduce the neuroinvasiveness of these variants without altering their virulence after intracranial inoculation (102). Thus, studies in viral systems including reoviruses, togaviruses, herpesviruses, and bunyaviruses indicate that viral surface glycoproteins may be critical determinants of the capacity of these viruses to reach the CNS from peripheral sites.

TROPISM

The pattern of systemic illness produced during an acute viral infection depends in large part on the specific host organs infected and in many cases on the capacity of viruses to infect discrete populations of cells within these organs (tropism). The tissue tropism of a virus is influenced by the interaction between a variety of host and viral factors. Although a great deal of attention has focused on the importance of viral cell attachment proteins and specific viral receptors on target cells (2,134,265,407,439), it has also become clear that a variety of other virus-host interactions can play an important role in determining the ultimate tropism of a virus. This fact is clearly indicated by experiments that demonstrate that the mere presence of a functional viral receptor may be insufficient to allow viral infection of target cells, as has been shown with HIV, poliovirus, rotaviruses, and mouse hepatitis virus (MHV) (e.g., see refs. 37, 116, 117, 123, 263, 423, 563, and 762). For example, intracellular factors and the route of viral entry may be important determinants of tropism (see below). Both poliovirus and MHV seem to require additional cellular factors after receptor binding, although the mechanism by which they act remains unclear (622,762). Similarly, although CD4 is clearly an HIV receptor, many cells expressing CD4 remain resistant to HIV infection (116,123,263,423). In many cases these cell lines still bind virus, but they often do not allow subsequent viral fusion and entry (see below).

Binding to Target Cells and Penetration of Cell Membranes

The importance of viral receptors in determining tissue tropism was initially emphasized by Holland and McLaren for polioviruses (282,283). Since that time there has been an explosion of knowledge concerning the nature of viral cell recognition proteins and their target cell receptors. A full discussion of each viral system and its target cell receptor is beyond the scope of this chapter, and readers are directed to the chapters dealing with individual viral families for more detailed information and more complete references. A partial list of virus families and examples of possible associated host cell receptor(s) can be found in Table 3.

As the list of potential receptors for specific viruses becomes more and more complete, it is possible to identify certain common themes. Different viruses may utilize the same cellular structures as a receptor. For example, sialic acid residues are important components of the receptor for certain coronaviruses (human coronavirus OC43, bovine coronavirus), orthomyxoviruses (influenza A and B), and reoviruses (T3 Dearing). Similarly, gangliosides have been implicated as potential receptors for paramyxoviruses (Sendai) and rhabdoviruses (rabies). Sometimes several different viruses belonging to the same family utilize the same receptor, as seems to be the case with heparan sulfate molecules, which are used by HSV-1, human cytomegalovirus, pseudorabies, and bovine herpesvirus-1 for initial attachment. In some cases, even though the receptor molecules are not identical, they may share enough structural similarity to merit inclusion together as part of a superfamily. For example, many rhinoviruses, poliovirus, and HIV-1 bind to molecules that are members of the immunoglobulin superfamily (739). Certain types of cell surface molecules seem to be utilized as viral receptors far more frequently than would be expected by chance. Among this group are the human membrane cofactor proteins, which include regulators of complement activation (such as the measles receptor), CD46 (which binds C3b and C4b), and the Epstein-Barr virus receptor, CD21 (which serves as a C3d receptor). HIV-1 and certain other retroviruses bind to CD4. Major histocompatibility complex antigens

TABLE 3. *Putative viral receptors*

Family	Virus	Possible receptor	Refs.
Adenoviridae	Adenovirus type 2	Integrin $a_v\beta_3$ and $a_v\beta_5$	(740)
Coronaviridae	MHV	Carcinoembryonic antigens	(174,743,761)
	TGEV, 229E	Aminopeptidase N	(759)
	OC43, bovine	Sialic acid residues	(709)
Herpesviridae	EBV	CD21 (CR2 receptor)	(197,205,440,487)
	HSV	Heparan sulfate	(624,634,755)
	CMV	Heparan sulfate	(129)
		β_2Microglobulin/MHC I	(250,757)
	BHV-1, BHV-4	Heparan sulfate	(504,704)
	Pseudorabies	Heparin sulfate	(332)
Myxoviridae	Influenza A, B	Sialic acid residues	(274,731)
	Sendai	Gangliosides	(429)
	Measles	CD46	(478)
		Moesin	(172)
Parvoviridae	B19	Erythrocyte P antigen (globoside)	(77)
Picornaviridae	Polio	IgG superfamily (PVR)	(447)
	Rhinoviruses (major group)	ICAM-1	(241,644,676)
	Echovirus 1, 8	Integrin VLA-2 ($a_2\beta_1$)	(49,50)
Poxviridae	Vaccinia	EGF receptor	(180)
Reoviridae	Reovirus T3 (Dearing)	Sialic acid residues	(221,522,523)
		EGF receptor	(662)
	Rotavirus SA11	Gangliosides	(656)
		Sialic acids	(39)
Retroviridae	HIV-1	CD4	(136,350,423,441,606,732)
		Galactosylceramide	(52,189)
	MuLV	y+ amino acid transporter	(345,719)
	GLV	Phosphate permease	(312)
Rhabdoviridae	Rabies	Acetylcholine receptor	(85,260,388,389)
		Gangliosides	(657)
		Phospholipids	(756)
	VSV	Phosphatidyl serine	(610)
Togaviridae	Sindbis	Laminin receptor	(720)
	Semliki Forest	HLA H2-K, H2-D	(267,506)
	Lactate dehydrogenase	Ia	(301,358)

Refer to individual viral chapters for full details of the evidence associated with particular viruses and receptors. Some putative receptors listed remain the subject of ongoing debate and controversy. BHV, bovine herpesvirus; EGF, epidermal growth factor; GLV, Gross leukemia virus; IgG, immunoglobulin G; MHC, major histocompatibility complex; MHV, mouse hepatitis virus; MuLV, murine leukemia virus; PVR, poliovirus receptor; TGEV, transmissable gastroenteritis virus; VSV, vesicular stomatitis virus.

have been implicated as putative receptors for cytomegalovirus (CMV), Semliki Forest virus, and lactate dehydrogenase–elevating virus (LDEV), although in some cases the assignment remains controversial (506). Other cell surface antigens, including the members of the carcinoembryonic antigen family (coronaviruses) and the erythrocyte P antigen (parvovirus B19), also serve as viral receptors. Receptors for neurotransmitters, growth factors, and laminin, including the acetylcholine receptor (rabies), the EGF receptor (vaccinia, reovirus T3 Dearing), and the laminin receptor (Sindbis) may be involved in virus-cell recognition, although controversy again exists about the validity of certain of these assignments (556). Among the new categories of proteins that have been implicated as potential receptors are permeases (Gibbon leukemia virus), aminopeptidases (coronaviruses), amino acid transporter proteins (Mahoney leukemia virus), and membrane organization proteins such as moesin (measles).

General principles have emerged from studies of virus-receptor interactions (see ref. 265 for recent review). Many viruses seem capable of interacting with more than one receptor molecule (see Table 3). Binding to different receptors by a single virus may be mediated either by a single cell attachment protein or by different proteins. For example, HIV-1 uses the gp120 envelope glycoprotein (52,189, 441) to bind to both CD4 (136,350,423,441,606) and galactosylceramide (52,189), although different domains of the protein seem to be involved. Similarly, measles virus seems to bind via its H glycoprotein to both membrane cofactor protein CD46 (478) and the membrane-organizing protein, moesin (172). Adenovirus 2 binds to certain integrin molecules through the penton base, which contains an Arg-Gly-Asp (RGD) sequence of the type recognized by cell adhesion receptors of the integrin family (24,740). For initial entry into target cells, adenovirus also attaches with high affinity to cell surface receptors via the fiber protein (see ref. 237). Similarly, Sindbis virus binds to the high-affinity laminin receptor on many mammalian cells (720) but seems to use a distinct receptor for binding to mouse neuronal cells (692). Single amino acid substitutions in the Sindbis E2 protein can alter binding to neuroblastoma cells but not BHK cells (692), suggesting that distinct receptor-binding domains may exist on the E2 spike.

The view that attachment is a monophasic interaction between a single viral cell recognition protein and a single host receptor is inadequate to explain many types of virus-cell interactions. In the case of the alphaherpesviruses, it seems that internalization is better understood in terms of a cascade of events that involve different glycoproteins and different cell surface molecules at different stages (215,634). Different cell surface proteins may be utilized for initial attachment and entry into target cells and for cell-to-cell spread across closely apposed populations of cells (159). An initial HSV attachment step seems to involve the interaction between gC (and possibly gB) and heparan sulfate molecules (271,368,755). This is followed by a stable attachment phase that involves gD, and possibly gH, and occurs independently of heparan sulfate binding (215). Other glycoproteins, including gE and gI, may facilitate cell-to-cell spread across contiguous cell layers (159). An analogous series of events has been postulated to occur with other herpesviruses, including pseudorabies (332), human cytomegalovirus (129), and bovine herpesvirus (BHV-1) (504). Even for viruses that may utilize only a single receptor, a progression from weak initial binding to secondary stable attachment (adhesion strengthening) may occur as a result of factors such as temperature shifts or recruitment of receptors to allow multivalent binding (265).

For all viruses, initial attachment serves as the first in a series of steps that ultimately deliver the genome to a site in the cell where it can begin replication. For enveloped viruses, penetration of the cytoplasmic membrane takes place after fusion between the viral envelope and the cell membrane. For nonenveloped viruses, penetration of the cytoplasmic membrane is still not completely understood, but it involves different strategies (see refs. 281, 430, 737, and 738 for review). In some cases, as discussed above for alphaherpesviruses, initial attachment and subsequent fusion may be mediated, entirely or in part, by different viral proteins. In other cases, the interaction of a virion cell attachment protein (VAP) with its target cell receptor may initiate conformational or other changes in other virion proteins that may in turn be essential for subsequent steps in virus-cell interaction, including fusion. In some cases, an envelope glycoprotein, such as the vaccinia HA, may have fusion-inhibitory activity, with removal of the protein activating the fusion protein(s) (503). As another example, binding of the alphaviruses Sindbis and Semliki Forest to host cells results in structural reorganization of the envelope glycoproteins on the virion surface, which exposes previously hidden hydrophobic fusogenic domains (74,202,716). In the case of SFV, dissociation of the E1/E2 heteromer is followed by the appearance of E1 homotrimers that initiate subsequent fusion events (715). In the case of HIV, a conformational change in the gp120 envelope glycoprotein, which serves as the cell recognition protein, may increase exposure of the N-terminus of the transmembrane (TM) protein, which contains a stretch of nonpolar amino acids that are essential for fusion activity (see ref. 442 for review). Conformational changes in the envelope glycoproteins required for fusion may also occur after initial receptor binding has resulted in internalization of the virion into acidified vesicles, as occurs with influenza (545,645, 737,738) and lymphocytic choriomeningitis virus (LCMV) (155). Given these results, it is not surprising that alterations in fusion proteins can dramatically alter viral host range and cell tropism, independent of changes in viral cell recognition proteins (214,327).

Fusion events may occur under pH-independent or pH-dependent conditions (281,374,430,737,738,742). The fusions of the envelopes of herpesviruses, coronaviruses, paramyxoviruses, and certain retroviruses with the plasma membranes of target cells are all examples of pH-independent fusion. Conversely, fusion activity may occur only after endocytosis has delivered virions into the acidic environment of endosomal vesicles. This type of pH-dependent fusion is characteristic of orthomyxoviruses, togaviruses, rhabdoviruses, bunyaviruses, and arenaviruses (63,737,738). It is important to recognize that the type of entry pathway utilized by a virus may depend on the cell type. For example, EBV entry into lymphoblastoid and epithelial cells seems to occur via direct fusion of the viral envelope with the target cell plasma membrane, whereas entry into normal B cells involves initial internalization into intracellular vesicles (452). In some cases, a single protein serves both as the viral cell recognition and fusion protein (e.g., the HA of influenza A), whereas in other viruses these activities are separated (e.g., paramyxoviruses, HIV).

Some general features of virion fusion proteins have been identified (737,738). For enveloped viruses, typical fusion proteins have external N-termini and internal C-termini, with most of their mass lying outside the viral envelope. Many of these proteins exist as oligomers, typically trimers or less frequently tetramers. This is perhaps best exemplified by the influenza hemagglutinin, which is a trimer composed of three identical disulfide-linked subunits, each of which is involved in fusion (742). Many fusion proteins contain a distinctive stretch of apolar amino acids referred to as a fusion sequence, although in some cases proteins capable of fusion do not have obvious "fusion sequences." Fusion sequences vary from short stretches 3 to 6 amino acids to longer sequences 24 to 36 amino acids in length and seem to often have the predicted secondary structure either of amphipathic helices with hydrophobic amino acids on one helical face and apolar residues on the other or of stretches of apolar amino acids. Since fusion proteins are typically inserted into viral envelopes via apolar transmembrane-anchoring segments, the presence of a second apolar region (the fusion peptide sequence) presumably facilitates their simultaneous interaction with the target cell membrane. The fusion sequence may be located at the N-terminal of the fusion protein, as is the case with orthomyxovirus HAs, paramyxovirus F proteins, and many retroviral envelope fusion proteins. In the case of SIV the fusion sequence may be located within the intracytoplasmic domain of the TM protein (566).

Nonenveloped viruses, like their enveloped counterparts, undergo conformational changes consequent to receptor binding. Intracellular delivery of the viral capsid may also be a complex event involving several viral proteins. For example, in the case of reoviruses, the σ_1 outer capsid protein serves as the viral cell recognition protein (384). Proteolytic removal of other outer capsid proteins, as occurs during natural infection in the lumen of the intestine (38,62), is associated with a change in the σ_1 cell recognition protein from a compact to an extended fibrous form (167). This transition may facilitate subsequent receptor binding. Proteolytic digestion also exposes a fragment on another outer capsid protein, $\mu 1$, that is proposed to have a role in membrane penetration (412,490).

During the entry of polioviruses into target cells, conformational transitions also occur that are dependent upon binding of the virion to its cell surface receptor (467). The VP4 capsid protein and the amino-terminus of VP1 are extruded from the virion interior (207,467). These proteins seem to play a role in membrane-binding interactions that are important postattachment steps in poliovirus entry.

Viral Gene Expression

For some viruses the interaction between the viral cell attachment protein and host cell receptors is the principal determinant of tropism (see Table 3). However, as more cellular receptors become identified, it has become apparent that, although the interaction between a virus and its receptor is often a major determinant of viral tropism, it is not the only factor. Other important factors, especially for retroviruses and papovaviruses, are the elements in the viral genome that regulate the transcription of viral genes in a cell-, tissue-, or disease-specific fashion, including enhancers and transcriptional activators (11,428,444,517,577,579,617,721,745). A full review of the many examples of viral transcriptional regulatory elements is beyond the scope of this chapter, and readers should refer to the individual chapters dealing with specific viruses for more complete discussion and references.

Both SV40 and polyoma have enhancer elements that show some degree of cell-type specificity (11,272,509, 510,578,579,721,745), although they also behave in a "promiscuous" fashion and stimulate viral gene expression to some degree in many varieties of differentiated cells. When the SV40 enhancer and early promoter are linked to a marker gene [chloramphenicol acetyltransferase (CAT)], gene expression is about fivefold greater in monkey kidney cells compared to mouse cells (373). The SV40 enhancer consists of tandemly repeated sequences that contain functionally independent elements that have distinct patterns of cell-type specificity (272), which can be mimicked by synthetic oligonucleotides corresponding to the nucleotide sequences of the individual elements (509,510).

Wild-type polyomavirus does not replicate in undifferentiated mouse embryonal carcinoma (EC) cells, and there is no early gene expression (212). Polyoma virus mutants that can infect EC cells have point mutations and duplications within the polyoma enhancer (212,679). Polyoma enhancer sequence alterations can alter enhancer-specific viral DNA replication in an organ-specific way and affect acute and persistent infections differently (578). It seems that certain types of EC cells make a negative regulatory factor that interacts with the wild-type polyomavirus enhancer, but not with the mutant enhancer, to repress its activity (511,758). Aging and age-specific transcription factors also affect organ-specific transcription in polyoma virus-infected mice (11,745).

Another striking example of enhancer-related cell specificity occurs with JC papovavirus, the agent responsible for the neurological illness progressive multifocal leukoencephalopathy (PML). When the JC virus (JCV) enhancer is coupled to the gene for CAT and this construct is tested for activity in human fetal glial cells and HeLa cells, the enhancer is active only in the glial cells (341). This corresponds well with the fact that JCV permissively infects only oligodendroglia within the central nervous system.

Further information on the tissue specificity of papovavirus enhancers has come from studies using transgenic mice (70,632). The offspring of transgenic mice made with early region genes (promoter, enhancer, and T-antigen genes) from JCV develop a neurological disorder characterized by tremor and seizures. Pathologically, they have abnormalities in CNS myelin due to dysfunction of the myelin-producing oligodendrocytes. Oligodendrocyte dysfunction correlates with the expression of JCV DNA and the presence of high levels of JCV-specified mRNA in these cells. It has been suggested that the cell specificity of JCV gene expression is due to the viral enhancer and that cell dysfunction itself is due to expression of virally encoded T antigen (632).

Transgenic mice made using the early control region and early genes of SV40 virus often develop choroid plexus epithelial cell tumors (70). Deletion of the SV40 enhancer region from the microinjected DNA results in a dramatic decrease in the incidence of these tumors, suggesting that the enhancer region may play an important role in the choroid plexus specificity of tumor development, with large T-antigen expression accounting for tumor development (70,515,702). Further studies with polyoma virus indicate that enhancer requirements determine differences in cell-specific control of DNA replication in normal versus transformed pancreas cells (579).

Papillomaviruses also contain enhancers with both inducible and constitutive elements (226). In the case of human papillomavirus 11, the enhancer may play an important role in tissue tropism, as it can be shown that the enhancer is specifically active in keratinocytes (444) and even in specific types of keratinocytes (646). Replication and gene expression of papillomaviruses increases as skin cells differentiate from basal cells to mature keratinocytes (46).

The role of enhancer elements in tissue and disease specificity has also been investigated extensively with avian and murine retroviruses (143,513,723) and HIV (see ref. 517 for review). Enhancer elements within the long terminal repeats (LTRs) play a key role in the capacity of avian and murine retroviruses to produce both neoplastic and nonneoplastic diseases (see ref. 188 for review). For example, certain strains of avian leukosis virus induce lymphomas, whereas other strains produce osteopetrosis in avian hosts. The capacity of these viruses to produce these different disease patterns is determined by enhancer regions with the LTR region of the viral genome, although non-LTR gene sequences may also play a role (75, 574–576,618).

Enhancer regions also determine the disease specificity of murine leukemia viruses. After inoculation into mice, Friend murine leukemia virus (F-MuLV) produces erythroleukemia and Moloney murine leukemia virus (M-MuLV) produces T-lineage lymphoid cell neoplasms (103,112,113,152,153,763). When recombinant viruses containing the enhancer regions of either F-MuLV or M-MuLV are made, the pattern of disease production correlates with the enhancer type (152,153). Similarly, replacement of the M-MuLV enhancer sequences in the LTR with the enhancer sequences from SV40 results in loss of the capacity of recombinant viruses to produce T-cell lymphomas and the appearance instead of B-lineage lymphomas and acute myeloid leukemia in infected mice (259). Recombinant polyoma virus that has its B enhancer element replaced with the 72-base pair repeat enhancer from Moloney leukemia virus genome has strong specificity for the pancreas. The LTRs also seem to play a role in the induction of neurodegenerative disease by certain murine retroviruses (518). The role of the LTRs in determining tropism and disease pattern in HIV is less clear than the role played by the envelope glycoproteins (536). However, a variety of cellular factors regulate HIV gene expression and may influence viral load and disease progression (517). These findings indicate that tissue specificity can be achieved by apparently modular elements within viral genomes (577).

Enhancer regions are also present in the genomes of herpesviruses (65,118,377). The EBV virus enhancer contains at least two domains. The A domain is constitutively active in fibroblasts, epithelial cells, and certain myeloid lineage cells but not in B lymphocytes. The B domain is active in B lymphocytes but requires the presence of a *trans*-activating factor encoded by the EBV genome for activity (118).

Viral enhancers have also been found within the hepatitis B virus (HBV) genome (110,308,551,620,677,705,760). Some HBV enhancer elements are more active in liver-derived cells (14,110,308,551,760) compared to nonhepatic cell lines. Transgenic mice given injections of the entire HBV genome (20,83) also show a tissue-specific pattern of HB_sAg expression, although antigen expression is not exclusively confined to liver tissue.

Site of Entry and Pathway of Spread

The tropism of a virus may depend on both its site of entry and its pathway of spread in the infected host. This principle has clearly been illustrated for some neurotropic viruses. For example, in a series of classic studies Howe and Bodian (289) inoculated the Rockefeller MV strain of poliomyelitis into rhesus monkeys via various routes and then studied the distribution of lesions within the CNS. Some areas of the brain, including the sensorimotor cortex and certain thalamic nuclei, were always infected, regardless of the route of viral inoculation. Other areas, such as the hippocampus and striate cortex, were never involved. Finally, there were areas that were selectively involved only after viral inoculation via specific routes (e.g., the septal nuclei and olfactory bulbs after intranasal inoculation). Similar results have been obtained with a variety of viruses that depend exclusively on neural spread to travel from the site of inoculation to distant target tissues, including vesicular stomatitis (600), herpes simplex (12), pseudorabies (432), and rabies (22) viruses (Fig. 9) (see "Neural Spread").

The dependence of neurotropism on the site of viral inoculation is also illustrated by the appearance of neurally spreading viruses such as herpes simplex, rabies, and reovirus T3 Dearing in the region of the spinal cord innervating the skin and muscles at the site of inoculation. The same principle is illustrated when neurally spreading viruses are inoculated at different sites. Their ultimate pattern of tropism then depends on the neural pathways innervating the site of inoculation. Inoculation at a specific site produces a unique pattern of viral distribution (22,95,133, 199,289,562,691,696).

The effect of site of entry on the ultimate tropism of viruses has not been extensively investigated for hematogenously spreading viruses. For polyomaviruses, the route of inoculation seems to determine the site of primary replication and the eventual site of persistent replication. The mechanism for this effect has not been clearly established, although the viral VP1 protein type seems to be involved in this effect (168,169).

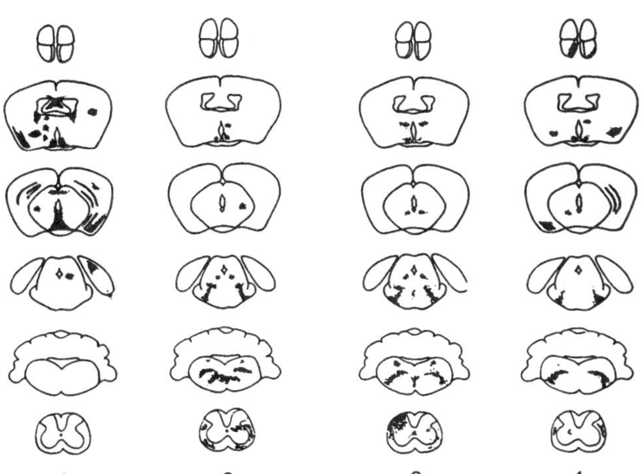

FIG. 9. The effect of route of inoculation of HSV 1 on the distribution of viral antigen in the central nervous system. Diagrams of coronal sections of mouse brain and spinal cord show the location of viral antigen (*black dots*) after inoculation through the following routes: (*1*) intracerebral, (*2*) intravenous, (*3*) sciatic nerve, and (*4*) intranasal. (From Anderson and Field, ref. 12, with permission.)

For viruses whose ultimate tissue tropism depends on the generation of viremia, it is difficult to understand how route of entry should alter tropism unless inoculation at different sites results in differences in the magnitude or duration of viremia. An interesting and unexplained exception to this rule may be the "provoking effect" described in studies with poliomyelitis. It was shown that, if intracardiac injection of poliovirus serotype 1 into a monkey was preceded by an intramuscular injection, paralysis tended to occur more frequently in the injected limb (hence the so-called provoking effect of trauma or intramuscular injection on the subsequent development of paralysis) (57). It was suggested that local trauma could selectively increase the permeability of blood vessels in the region of the spinal cord innervating the traumatized site. This in turn would allow blood-borne virus to localize specifically in that region of the spinal cord and produce preferential paralysis of the traumatized limb. Direct evidence for this sequence of events is lacking, and the mechanism of the provoking effect must be considered as still unsolved.

The provoking effect has been suggested as a possible mechanism to explain the high incidence of paralysis in the inoculated limb of individuals injected with incompletely inactivated lots of poliovirus vaccine (482) and might represent an example of dependence of the tropism of a hematogenously spreading virus on the site of viral inoculation. However, the mechanism of the provoking effect remains obscure, and it remains unclear how intramuscularly inoculated poliovirus in humans reaches the CNS. If virus spreads from the site of inoculation to the CNS via nerves, a similar preferential distribution of paralysis might also be expected to occur.

For some viruses the pathway of spread may vary depending on the site of inoculation, and this difference may in turn influence tropism. This possibility is illustrated by studies in mice with the neurotropic NWS strain of influenza (558). After intraperitoneal inoculation, virus spreads through the bloodstream, and viral antigen is found in the meninges, choroid plexus, and ependymal cells and in perivascular locations in the brain parenchyma. After intranasal inoculation, virus spreads through nerves and antigen is noted primarily in the olfactory bulbs, trigeminal ganglia, and brainstem. Thus, in this case the site of inoculation influences the route of viral spread, and the route of spread affects the ultimate tissue tropism.

MOLECULAR AND GENETIC DETERMINANTS OF VIRULENCE

The molecular and genetic determinants of viral virulence are complex. The following sections provide selected examples of how viral envelope and capsid proteins; core, matrix, and nonstructural proteins; and noncoding regions of the viral genome can all influence pathogenesis and virulence. A full discussion of these issues for every virus is beyond the scope of this chapter, and interested readers are referred to the chapters dealing with individual viral families for additional details and more complete references.

Envelope Glycoproteins and Capsid Proteins

As discussed earlier, envelope glycoproteins and capsid surface proteins frequently function as viral cell recognition proteins. It is not surprising that alterations in these proteins can alter viral tropism and pathogenicity both *in vitro* and *in vivo*, although it is important to recognize that this may be mediated by mechanisms other than effects on target cell binding and receptor recognition.

Naked Icosahedral Viruses: Picornaviruses and Reoviruses

The availability of the complete nucleotide sequence of virulent poliovirus strains, their attenuated vaccine derivatives, and virulent revertants (see refs. 9 and 549 for review) has greatly facilitated analysis of the molecular and genetic determinants of picornavirus virulence. Mouse models of poliovirus infection (330,375) and, more recently, transgenic mice expressing the human poliovirus receptor (287,560,563) have also been extremely useful tools for analyzing poliovirus virulence and pathogenesis. The importance of mutations in the poliovirus 5'-noncoding region and of RNA polymerase in poliovirus tropism and neurovirulence is discussed in subsequent sections.

Mutations in poliovirus outer capsid proteins including VP1 (e.g., AA position 143) (420,667,734) are determinants of attenuation *in vitro* and *in vivo.* In the mouse model of poliomyelitis, both 5'-noncoding mutations and mutations in VP1 (AA#143 and AA#22) and VP2 (AA#31) are important in virulence (132,468,561). Normal mice are susceptible to infection with mouse-adapted strains of P2/Lansing but not to infection with P1/Mahoney. Replacement of eight amino acids composing the BC loop of VP1 from Mahoney with the corresponding Lansing sequences restores mouse susceptibility, indicating the critical role of this VP1 region in determining poliovirus host range (475). The importance of the VP1-143 change *in vivo* has been confirmed by sequence analysis of eight isolates from cases of vaccine-associated (P2/Sabin) poliomyelitis. In two studies, all vaccine-associated cases of polio had lost the VP1-143 Ile residue associated with attenuation (181,420).

Outer capsid proteins are also determinants of virulence for other picornaviruses. For coxsackievirus B4, the VP1 capsid protein seems to be an important determinant of virulence, with a threonine residue at position 129 playing a major role (87). The VP1 protein also plays a critical role in the pathogenesis of infection with Theiler's murine encephalomyelitis virus (TMEV). Amino acid substitutions at position 101 have been shown to be particularly critical in modulating TMEV-induced neurological disease (713,777,778). However, TMEV neurovirulence is clearly a complex process that can be influenced both by regions of the genome encoding capsid proteins and by noncoding regions (see below) (89,211,768).

The importance of capsid proteins in reovirus pathogenesis has been demonstrated in a variety of experimental models. Reovirus T1 Lang and T3 Dearing differ in their CNS tropism, with the former infecting ependymal cells and the latter neurons. Studies with T1L × T3D reassortants indicate that the viral S1 gene, which encodes the reovirus cell recognition protein $\sigma 1$, determines this pattern (727,728). In this case, alterations in virulence are due to differences in tropism, with neuronal infection (encephalitis) being associated with enhanced mortality compared to ependymal cell infection. The S1 gene is also a key determinant of tropism in the pituitary and retina. Monoclonal antibody–resistant variants of T3D, which contain single amino acid substitutions in $\sigma 1$, have attenuated CNS growth, a restricted pattern of neurotropism, and reduced lethality (41,42,636).

The virulence of different reovirus strains belonging to the T3 serotype, and hence producing the encephalitic pattern of disease, has also been compared. Reassortants between the virulent T3D strain and the attenuated T3C8 strain indicate that the viral M2 gene, which encodes the outer capsid protein $\mu 1$, is an important determinant of neurovirulence for T3 strains (290). Recent studies suggest that proteolytic derivatives of the $\mu 1$ protein play an important role in membrane penetration by reoviruses (412,490). The fact that alterations in the capacity of reoviruses to penetrate membranes plays a role in reovirus virulence resembles findings with penetration mutants of Sindbis and VEE (see below).

Enveloped Viruses

In the case of the alpha togaviruses, as exemplified by Sindbis, mutations in either the E1 or E2 envelope glycoprotein result in altered virulence in mice (144, 244,314,390,413,533,685,712), and mutations in both proteins result in highly attenuated viral strains (534). E1 and E2 form a stable, noncovalently linked heterodimer that serves as the virion cell recognition protein. It has recently been suggested that an amino acid change in E2 (residue 172) may decrease the binding of virus to neural but not nonneural cells. More virulent isolates seem to bind more efficiently to neural cells, induce earlier viral RNA synthesis in neural cells, and replicate more rapidly in neurons (685).

In both Sindbis and Venezuelan encephalitis virus, mutations in E2 (e.g., AA#114) coordinately affect both the capacity of virions to penetrate cell membranes and their virulence. Interestingly, rapid rather than delayed membrane penetration seems to correlate with attenuation (144,328,526,533,597). These results, and those in other viral systems (see below), suggest that altered interaction of viruses with host cell membranes, as well as with target receptors, may be important determinants of tropism and virulence.

Differences in flavivirus virulence have also been related to changes in the envelope (E) protein (405). Amino acid sequence comparison of the attenuated vaccine strain of yellow fever virus (17D) with its virulent Asibi parent shows a series of mutations, the majority of which are clustered in the E protein gene (255). The E protein is also an important determinant of the virulence of Murray Valley encephalitis virus. In this case mutations frequently involve a portion of the E protein that contains an R-G-D sequence (405). This type of protein sequence has been shown to be an important feature of proteins binding to the integrin family on cell membrane receptors. Similar or related sequences have been found in other flaviviruses, and they may be important determinants of cell attachment. Mutations within the E protein may also affect the neuroinvasiveness and neurovirulence of flaviviruses (102,339).

Mutations in the E2 envelope glycoprotein are important determinants of coronavirus virulence. Antigenic variants of MHV-4 and JHM virus, two neurotropic coronaviruses, with alterations in the E2 protein, have attenuated neurovirulence and altered patterns of CNS disease (139,200,726). Deletions in the envelope gene result in mutants of MHV-4 with decreased capacity to spread within the CNS and attenuated virulence (190).

The envelope glycoprotein (G protein) is an important determinant of rabies virulence. Monoclonal antibodies have been used to select mutant viruses with changes in antigenic site III (residues 330–338) of the G protein. Site III mutants with an arginine or lysine residue at position 333 remain virulent, whereas those with a variety of amino acid substitutions at this position are attenuated (158,615,690). Rabies variants with genes in antigenic site II of the G protein also frequently have attenuated neurovirulence (541), although not to the extent seen with site III mutants. This site is a nonlinear epitope that encompasses residues 34 to 42 and 198 to 200. The mechanism of attenuation of site III G protein mutants has been investigated. These mutants spread more slowly within the CNS than do their virulent counterparts (157) and seem to have lost the capacity to invade certain types of nerve fibers (367,372).

The genetic determinants of influenza virus virulence have been extensively investigated. Cleavage of the viral hemagglutinin is required for initiation of influenza infection *in vitro* and acts as a determinant of virulence *in vivo* (612,725). HA cleavage is affected by its glycosylation (154) and is also influenced by the neuraminidase protein. It has been suggested that removal of sialic acid residues on HA by neuraminidase may facilitate subsequent cleavage of HA by endogenous proteases (see ref. 396). Changes in the viral HA and the neuraminidase proteins have both been clearly shown to influence influenza virus pathogenicity (154,396,476,653,725). The hemagglutinin-neurominidase (HN) protein of mumps is also a major determinant of its pathogenicity (410).

Reassortants between virulent and avirulent strains and monoclonal antibody–resistant variants have been used to study the molecular mechanisms of bunyavirus virulence. As discussed earlier, the viral M gene, which encodes the two viral glycoproteins (G1 and G2) is an important determinant of the capacity of bunyaviruses to spread from the periphery to the CNS (neuroinvasiveness) (228,229,246,309,628). The G1 protein in particular plays an important role in this process (230). However, neuroinvasiveness is clearly under polygenic control as mutations affecting the viral L gene, encoding the polymerase, can also influence this phenotype (246).

Mutations in the LCMV glycoprotein affect an early step in infection, possibly viral attachment-penetration, and may be a determinant, along with polymerase gene mutations, in the macrophage tropism of certain LCMV strains (435). The capacity of certain LCMV strains to infect the growth hormone–producing cells of the pituitary maps to the viral S gene segment that encodes the viral nucleoprotein and the two envelope glycoproteins (568). Changes in the gp120 SU protein of HIV-1 can dramatically alter the tropism and the virulence of these viruses. Some HIV isolates infect only primary cells (e.g., CD4+ T cells, macrophages, and dendritic cells), whereas others can also infect immortalized CD4+ T-cell lines. In many cases the block in viral infection seems to be at the level of viral entry (346,497). However, in some cases alterations in the envelope glycoproteins seem to affect replication without restricting viral entry (463,611). Single amino acid substitutions in restricted regions of gp120, including both the CD4 binding domain and other areas, have been shown to alter HIV-1 tropism (66,92,130,297,464,626,660,733). The receptor choice determinants of other retroviruses, including murine leukemia viruses, have also been localized to specific regions of their SU glycoproteins (43,460,512).

Another mechanism of envelope protein effects on virulence has been identified in certain Moloney murine leukemia virus strains. Neuropathogenicity correlates with single amino acid substitutions within the envelope precursor protein (gPr80^{env}). This mutation results in inefficient transport of envelope proteins from the endoplasmic reticulum of infected cells and cytopathicity due to accumulation of this protein and aberrant virus particles within cells (625,659). Isolated HIV gp120 protein has also been shown to be directly toxic to nerve cells in culture. The mechanism of this toxicity remains to be established, but among the possible actions of gp120 are mediation of changes in intracellular calcium levels and inhibition of the action of essential neurotrophic factors (see below).

The TM protein of HIV-1 also contains a region referred to as the lentivirus lytic peptide (LLP-1), which shares predicted structural similarity with a variety of cytolytic peptides and may be involved in HIV-induced cytopathogenicity (see below).

The importance of envelope glycoproteins in the cell attachment and membrane fusion of herpesviruses has already been discussed, as has their potential role in the pattern of spread of these viruses. Recent studies with pseudorabies virus indicate that the gI and gp63 proteins, which are homologs of the HSV envelope glycoproteins gE and gI, can affect the neurotropism of PRV as well as its pattern of neuronal uptake and axonal transport (95–97,735).

Viral Polymerases and Core, Matrix, and Nonstructural Proteins

Recent studies in viral systems have identified genes encoding viral polymerases and other core proteins as potentially important determinants of virulence. It is important to recognize that mutations that inhibit viral replication will obviously result in attenuation. Mutations resulting in impaired replication in all cells should be differentiated from those mutations that selectively impair the capacity of a virus to replicate in some cells but not others. These types of "host range" mutations may provide important information about virulence and attenuation.

There is evidence that mutations in the RNA polymerase ($3D^{pol}$) of poliovirus 1 may be a determinant of neuroattenuation in both monkeys and mice (663), although the effect is less pronounced than that of attenuating mutations in the 5'-noncoding region (see below). Mutations in the 2B and 2C region of hepatitis A virus dramatically affect the kinetics of viral growth in cultured cells and may also influence host range (177,178,217). Mutations in the 2B region have also been found in the attenuated P1/Sabin vaccine strain (495). The functions of the 2B and 2C proteins of picornaviruses are poorly understood, although they seem to play a role in replication and viral RNA synthesis (537).

Mutations in 2A and 2B have been found in the attenuated P1/Sabin vaccine strain (495). The product of 2A, $2A^{pro}$, seems to be a trypsin-like serine protease that is involved in the proteolytic process of the picornavirus capsid precursor region (537). The specific role of these mutations compared to those in the capsid proteins and noncoding regions in the attenuated phenotype have not been established. The genomes of flaviviruses also encode a trypsin-like serine proteinase (NS3). Mutations in NS3, or in an associated protein, NS2B, alter the proteolytic processing of the flavivirus polyprotein and attenuate viral replication (106).

Matrix proteins, which are components of many enveloped viruses, may also play a role in cytopathicity. VSV matrix (M) protein binds the ribonucleoprotein core of VSV to the intracellular plasma membrane of infected cells as a prelude to viral budding. M protein is also responsible for characteristic cytopathogenic changes induced in VZV-infected cells (56) and that this may be due to its inhibition of host cell–directed gene transcription (54). The role of the M protein in influencing pathogenesis *in vivo* remains to be determined.

The role of the hemmagglutinin (HA) and neuraminidase (NM) proteins in influencing the tropism and pathogenesis of influenza have been discussed. Although the results from influenza reassortant studies are frequently complex, the influenza virus polymerase gene has been implicated repeatedly as a determinant of virulence (e.g., see ref. 633).

While the G1 glycoprotein of bunyaviruses is an important determinant of the capacity of these viruses to spread from peripheral sites of replication to the CNS (see above), recent studies suggest that the capacity to cause disease after direct CNS inoculation of virus (neurovirulence) may be determined by the L RNA segment that encodes the virion polymerase (179). These results are in accord with earlier studies suggesting that both the L and S gene segments had a modulating effect on the capacity of the M segment to determine both virulence (309,590) and the neuroinvasive phenotype (246). La Crosse bunyavirus polymerase mutants replicate poorly in mouse CNS but normally in BHK cells, suggesting that the polymerase mutation results in a "host range" alteration rather than in generalized viral attenuation.

Comparison of virulent and avirulent strains of LCMV in the guinea pig also suggest that the L RNA segment, which encodes the arenavirus polymerase, is an important determinant of LCMV virulence (246,567). Mutations in the L segment are also involved in the generation of persistent infection in adult mice (3). Recent studies suggest that mutations in the L segment are correlated with the macrophage tropism of LCMV isolates. A single amino acid change in the viral polymerase (residue 1079) is a major determinant of infectious virus production in macrophages (435). Reversion of this mutation results in loss of macrophage tropism.

The importance of viral core proteins in influencing the pattern of disease is clearly exemplified by reoviruses. Genetic studies using reassortant viruses derived from nonmyocarditic and efficiently myocarditic reovirus strains indicate that the myocarditic phenotype is associated with the viral M1 gene in all strains tested and additionally with the L1 and L2 genes in specific viral strains (623). These genes encode a viral core protein, the virion polymerase, and the guanyl transferase, respectively. The viral M1 and L1 genes influence viral growth in cultured cardiac myocytes (436), although growth capacity alone does not directly correlate with induction of myocarditis *in vivo*. However, the capacity of viruses to induce a cytopathic effect (CPE) in cardiac myocytes *in vitro* does correlate with myocarditic potential *in vivo* (44).

For another member of the family Reoviridae, murine rotaviruses, recent evidence suggests that the gene 5 product, NS53, a protein postulated to be involved in replication, may also be a determinant of their virulence and capacity to spread in the infected mouse (73). Another rotavirus nonstructural protein may influence viral cytopathicity through its effects on cytosolic calcium levels (see below).

Although the viral envelope glycoproteins seem to be major determinants of the tropism of HIV and other retroviruses, there is abundant evidence that nonenvelope proteins can contribute to both tropism and cytopathicity. Among the HIV genes and cis-acting sequences that have been suggested to have a role in cytopathicity are *tat* (115), *sor* (198), *vif* (604), and *nef*, the RRE (147), and portions of the LTRs (536). The roles of these proteins and sequences in HIV replication and pathogenesis are discussed in Chapters 61 and 62.

Parvovirus cytotoxicity is also mediated in part by nonstructural proteins (68,88,386). It has recently been shown that the NS proteins activate the promoter of the human c-*erbA1* gene, which encodes a thyroid hormone receptor (αT3R), and that thyroid hormone (T3) can enhance the susceptibility of cells to parvovirus-induced CPE (703). These results suggest a possible mechanism by which the NS proteins might influence parvovirus cytopathicity.

The molecular determinants of HSV virulence are complex (see ref. 647 for review). Mutations involving viral envelope glycoproteins, DNA polymerase, and dUTPase have all been shown to affect neurovirulence. Intertypic reassortants between HSV-1 × HSV-2 indicate that the enhanced capacity of HSV-2 to spread to the CNS from the cornea is due to the DNA polymerase (145). The same nucleotide sequences within the polymerase gene that influence the pattern of viral spread also seem to determine the capacity of HSV 2 to grow in mouse tissues and human lymphocytes (382). Recent studies suggest that ICP34.5, a protein encoded by a gene in the HSV inverted repeat, is an additional determinant of HSV neurovirulence (120). This protein may also be an inhibitor of apoptotic cell death in neurons (121), although whether this contributes to its effects on pathogenesis *in vivo* is unknown.

Noncoding Regions of the Viral Genome

Recent evidence from viral systems suggests that noncoding regions of viral genomes may be important determinants of virulence and pathogenicity, presumably through their effects on gene replication and transcription.

The role of the 5'-noncoding region in picornavirus virulence has been extensively studied. The poliovirus 5'-noncoding region is nearly 750 bases long and accounts for over 10% of the entire viral genome. This region contains extensive RNA secondary structure, including multiple stem-loop and cloverleaf structures (631). One region of the 5'-noncoding region, referred to as the ribosome landing pad or internal ribosome entry site, directs cap-independent initiation of poliovirus protein synthesis. It is assumed that nucleotide changes in these regions alter RNA secondary structure and thereby affect translational efficiency (375,631). One recent study suggests that an attenuating (U) substitution at 5'-noncoding position 472 in polio strains results in weaker interaction between polio RNA and the host cell translational initiation factor eIF-2 (658).

Evidence for the importance of the 5'-noncoding region in poliovirus virulence has come from comparison of the nucleotide sequences of virulent and attenuated (e.g., vaccine strains) of polio and their revertants (9,185,495,549,643). The P/Leon vaccine strain contains a cytosine (C) to uracil (U) change at position 472 that seems to be critical for its attenuated phenotype (185,375). Virulent revertants from the vaccine strain have reversion of U to C at this position (185). Interestingly, passage of the P/Sabin vaccine strain through the human gastrointestinal tract rapidly selects for reversion from C to U (9,185,454,549). There are also several changes in the 5'-noncoding region between the virulent P1/Mahoney and its attenuated P1/Sabin vaccine derivative (495). One of these alterations, a G to A change at position 480, seems to be particularly important for the attenuated phenotype (337,508,549). A nearby position (481) has also been shown to be a critical determinant of virulence in studies of the mouse adapted P2/Lansing strain (375,376) and the type 2 poliovirus vaccine strain (419,420,561).

The importance of the 5'-noncoding region in virulence has also been established for other picornaviruses including mengovirus, Theiler's murine encephalomyelitis virus, and hepatitis A virus. In the case of mengovirus, changes in the length of the long homopolymeric (C) tract within the 5'-noncoding region can dramatically attenuate viruses *in vivo* without affecting their growth in cell culture (171). TMEV, like poliovirus, has a long (>1,000 bp) untranslated 5' region, which contains putative internal ribosome entry sites (30). Deletions of even a single nucleotide base or insertions as small as three bases within the 5'-noncoding region can dramatically attenuate TMEV neurovirulence (29,543). Similarly, differences between neurovirulent and nonneurovirulent TMEV strains seem to be due in part to differences in the 5'-noncoding region (400,768), although the genetic determinants of TMEV neurovirulence are clearly complex and influenced by multiple genomic regions (768). In the case of hepatitis A virus, critical mutations in the 5'-noncoding region affect viral virulence and adaptation to cell culture (146,178,217).

The effects of deletion mutants in the 3'- and 5'-noncoding region of the genome have also been studied for the alpha togaviruses including Sindbis (369) and VEE (348). Sindbis virus 5'-noncoding mutants typically show defects in RNA synthesis and altered rates of growth in cultured cells. Effects of viral mutations on growth rate were dependent on the host species of the infected cell, suggesting that the interaction of host factors with the virion RNA were critical in determining these effects. These viruses also differed in their capacity to produce disease in mice, suggesting that the noncoding region is an important determinant of virulence *in vivo* as well as *in vitro.*

Mouse challenge experiments with VEE viruses containing sequences corresponding to those of the virulent Trinidad donkey (TD) strain or its attenuated vaccine derivative (TC-83) also indicate that attenuation is determined in part by mutations within the 5'- noncoding region, although changes in the E2 envelope glycoprotein also play a role (348).

The 3'- and 5'-noncoding regions of the NM gene of influenza virus have also been shown to affect influenza virus virulence in mice (476). This effect presumably occurs by altering transcription of the NM gene, whose protein is known to be a determinant of influenza virulence (see above).

Studies with murine retroviruses suggest that, although the envelope glycoproteins are important determinants of viral tropism and pathogenicity, sequences in the 5'-untranslated region of the genome may influence viral replication kinetics and the development of disease (539). Whether untranslated regions play a role in determining the pathogenesis of other retroviruses remains to be established.

SELECTED MECHANISMS OF VIRAL VIRULENCE AND CYTOPATHOGENICITY

It is obviously impossible to review the wide variety of mechanisms that viruses have evolved for inducing cytopathicity and for avoiding host antiviral defenses. Virus-host cell interactions are discussed more fully in Chapter 8. Special types of virus-host interactions, including those leading to viral persistence and cell transformation by viruses, are also discussed in individual chapters (Chapters 7 and 9). The host's immune responses to viruses are covered in Chapters 10 and 11, although some of the methods adop ted by viruses to avoid host immune surveillance are also covered in this section.

Virally Encoded Inhibitors of Host Immune Responses

Many viruses have adopted complex strategies to avoid discrete components of the host immune response, including cell-mediated immune responses and antiviral mediators (reviewed in ref. 234). An adenovirus E3 protein, gp19K, localizes to the membrane of target cell endoplasmic reticulum, where it binds newly synthesized MHC class I antigens and prevents their transport to the cell surface (13,82). The resulting down-regulation of cell surface MHC class I may reduce recognition of adenovirus-infected cells by $CD8^{\pm}T_{CTL}$, leading to persistent infection (234,748). The magnitude of the E3 effect depends on the target cell infected and may be influenced by expression of adenovirus E1A protein (588). Two additional E3 proteins (14.7K, 10.4K) and the E1B 19K protein can protect adenovirus-infected cells from cytolysis by tumor necrosis factor (TNF) (232,233,552). The exact mechanism has not yet been established but may involve protection from TNF-induced apoptosis (553).

Herpesviruses, including EBV and CMV, have evolved strategies that result in reduced immunogenic presentation of certain antigens, including transcription factors (e.g., CMV IE protein, EBV EBNA1), resulting in absent or diminished T_{CTL} recognition of these antigens (224,231). In some cases this may be due to selective interference with processing of these proteins that precedes their entry into the endoplasmic reticulum and subsequent binding with class I major histocompatibility complex (MHC).

Many herpesviruses including VZV and HSV also encode Fc receptors that can bind immunoglobulins (47,403,484). For HSV, gE and gI function as an FcR (47). The function of these Fc binding sites remains unclear, although deletion of these proteins may be associated with enhanced susceptibility to antibody and complement-mediated lysis of virus or virus-infected cells.

Herpesviruses and vaccinia also encode proteins that function as receptors for complement components including C3b (209). In the case of HSV, gC functions as a complement-binding protein (209) and inhibits complement-mediated cell lysis and virus neutralization (210,264). Herpesvirus saimiri (HVS) has a gene encoding a protein with sequence similarity to the human terminal complement regulatory protein CD59 (4,5,587). Cells expressing CD59 are protected from complement-mediated cell lysis (587), suggesting that this protein plays a role in immune evasion by HVS. The major secretory protein of vaccinia virus is also a complement-control protein (357) that inhibits the complement cascade (356,443) and thereby prevents antibody-dependent complement-enhanced neutralizat ion of vaccinia and enhances vaccinia virulence (302).

Influenza virus infection of target cells can inhibit the cytolytic action of interferon (IFN) (335). Influenza virus prevents activation of the IFN-induced double-stranded RNA (dsRNA)–activated protein kinase DAI (191,335,385). IFN-induced activation of double-stranded RNA-activated inhibitor of translation (DAI) kinase inhibits protein synthesis by phosphorylating, and thereby inactivating, the α subunit of translational initiation factor e IF-2 (335). Preventing the activation of DAI kinase by IFN allows continued synthesis of influenza virus proteins in infected cells.

Poxviruses encode a variety of proteins that may interfere with components of the IFN pathway. These include genes encoding an eIF-2 homolog (45), a dsRNA binding protein, and an inhibitor of the IFN-induced dsRNA-dependent protein kinase (109) and an IFN-gamma receptor homolog (697).

Cell surface adhesion molecules including integrins are critical mediators of a variety of cell-cell interactions, including a variety of immune responses and the adherence to and subsequent migration of activated leukocytes across endothelial cells. It has recently been recognized that some viruses, including HIV-1 (276,583), measles (19), and EBV (240), can modulate the expression of integrins including LFA-1, LFA-3, and ICAM-1, which may facilitate the subsequent dissemination of virus or help these viruses avoid immune surveillance.

A number of viruses have now been shown to encode proteins that function as "superantigens" (227,294,300,380). Superantigens react with populations of T cells expressing particular V receptor subsets. This interaction can result in selective proliferation and/or deletion of these particular T-cell populations. Changes in specific T-cell populations in turn can dramatically affect the host response to the infecting pathogen and the subsequent response to other organisms.

A region of the HIV-1 TM protein (residues 576–590) encodes an immunosuppressive domain. Peptides homologous to this region can block lymphoproliferation, perhaps by interfering with plasma membrane-associated protein kinase C (331,596).

Growth Factor, Cytokine, and Transmitter Modulation

Peptide growth factors have multifaceted effects on target cells that seem to be critical in the regulation of cell growth, proliferation, and differentiation. These actions, which occur subsequent to interaction of peptide growth factors with their cell surface receptors, trigger a variety of intracellular signaling pathways. Stimulation of peptide growth factor receptors may facilitate viral infection by altering the proliferative or differentiated state of target cells. For example, vaccinia virus encodes homologs of both epidermal growth factor (EGF) (55,76) and endothelial growth factor (417). These homologs can be shown to be biologically active *in vitro* and *in vivo* and are important in viral pathogenesis. For example, deletion of the vaccinia EGF homolog reduces viral virulence (81), and blockade of the EGF receptor on target cells inhibits vaccinia infection (180).

In some cases, even when viruses do not encode growth factor homologs, they may be capable of interacting with growth factor recepto rs in a manner that facilitates infection. Reovirus type 3 Dearing grows poorly in 3T3 fibroblasts lacking EGF receptor. Transfection of these cells with a functional, but not with a mutated, nonfunctional EGF receptor results in markedly enhanced viral growth and cytopathogenicity (662). In adenovirus-infected cells, the 10.4K and 14.5K E3 encoded proteins form a complex that downregulate s EGF receptor expression, although the effects this has on cytopathicity have not been determined (675).

The envelope glycoprotein of the retrovirus Friend spleen focus-forming virus (SFFV) interacts with members of the cytokine receptor superfamily, including the erythropoietin and interleukin 2 (IL-2) receptors (99,196,360,684,775). Stimulation of erythropoiesis and erythroblastosis by this virus may be an important determinant for the subsequent production of erythroleukemia in infected animals.

Stimulation of cells by cytokines including IL-6, TNF-α, and GM-CSF may enhance the replication of certain viruses, such as HIV and SIV, in specific cells, including macrophages and lymphocytes (203,359,531,532). These viruses, in turn, can induce the production of these factors from infected cells (448,477).

The EBV-encoded BCRF1 protein is a homolog of IL-10, with >90% amino acid sequence identity (291,457). IL-10 is normally secre ted by T helper cells and acts to inhibit some of the actions of IFN-gamma. It has been suggested that this protein may facilitate EBV infection by inhibiting early natural killer (NK) and T_{CTL} responses to EBV (370,648).

It has recently been shown that parvovirus nonstructural proteins can interact with the promoter of the human c-erbA1 gene and there by induce cell surface expression of a thyroid hormone receptor (αT3R) (703). Cells expressing αT3R are sensitized to parvovirus cytopathicity when subsequently exposed to thyroid (T3) hormone, suggesting that receptor expression may have direct functional consequences.

Infection of neuronal cells by a variety of viruses can depress or alter levels of neurotransmitters and their synthesizing enzymes in the absence of histological evidence of cell injury (398,591,592). Similarly, infection of neuroendocrine cells, such as the growth hormone–producing cells of the pituitary, can depress hormone production without producing gross cytopathology. In the case of LCMV infection of these cells, altered growth hormone production is directly responsible for clinical disease (505,568).

The HIV gp120 protein has a region of amino acid similarity with neuroleukin. Neuroleukin acts as a nerve growth factor and enhances the survival of cultured spinal cord and dorsal root ganglion neurons. In some reports, gp120 seems to inhibit the action of neuroleukin and decreases the number of surviving neurons in culture (252). Whether the capacity of gp120 to inhibit neuroleukin plays a role in HIV pathogenesis *in vivo* remains to be established.

Apoptosis

Viruses may trigger an internally programmed pathway of cell death (apoptosis) in infected target cells. The morphological hallma rk of apoptotic cell death is the appearance of condensed fragments of nuclear chromatin, which can be visualized by electron micros copy of cells. Chromatin fragmentation results from activation of a calcium-dependent endonuclease that fragments cellular genomic DNA into oligomers of 180- to 200-bp multiples. Apoptotic cell death is not associated with a host cell inflammatory response and may in part represent a mechanism for eliminating cells when associated inflammation might be deleterious. Apoptotic cell death seems to be a normal part of the ontogeny of both the lymphoid and nervous systems, and it is not surprising that viral apoptosis has been best studied in neural and lymphoid cells.

For viruses that normally produce acute lytic infection associated with the apoptotic death of infected cells, inhibition of this process may be an important mechanism for the generation of nonlytic persistent and latent infections (239,269,391) (see Chapter 7). Apoptosis seems to be an important mechanism for the depletion of CD4+ cells both in cell culture and in HIV-infected individuals and may contribute to virus-induced immune deficiency (354,381, 670). In the case of HIV, expression of the gp160 envelope glycopr otein in susceptible cells seems to be sufficient to induce apoptosis (411). Direct apoptotic cell death has also been seen in ly mphocytes after EBV infection (338,693). Similarly, expression of the adenovirus E1A proteins may induce apoptosis in susceptible cells, and this can be inhibited by other adenovirus proteins including the 19kD E1B protein (553,736). Expression of the parvov irus nonstructural proteins has also been reported to induce programmed cell killing (88).

In addition to direct induction of apoptotic cell death, viruses may also prime infected cells to undergo apoptosis upon subsequent stimulation. For example, stimulation of the TcR-CD3 complex on T lymphocytes, as occurs during antigenic challenge, may induce apoptosis of HIV- and LCMV-infected cells (236,555).

In addition to being able to induce apoptosis, many viruses contain proteins that function as inhibitors of apoptosis. Among the vir al inhibitors of apoptosis that have been identified are the HSV-1 ICP34.5 (121), EBV latent membrane protein 1 (269), and the adenovirus E1B 19kD protein (553). An African swine fever virus gene with homology to the apoptosis-inhibiting proto-oncogene *bcl-2* and to the EBV BHRF1 gene has also recently been reported (485). As noted above, the resultant inhibition of cyto lysis may facilitate the development of persistent or latent infections. In addition, inhibition of apoptosis may also blunt the antiviral effects of cytokines or other mediators (e.g., TNF), which may trigger programmed cell death.

Disruption of Intracellular Calcium Homeostasis

Disruption of intracellular calcium homeostasis seems to be a final common pathway for the development of many types of irreversi ble cell injury. A number of viruses including rotavirus (450,674), CMV (493), and HIV (166) are known to produce increases in intracellular calcium, and this may be an important mechanism for viral CPE. Blocking this effect by reducing extracellular calc ium levels *in vitro* or through pharmacological treatment of infected cells with calcium channel antagonists may block CPE (166). The molecular mechanism(s) behind virally induced changes in cellular calcium levels are unknown but could conceivably involve disruption of plasma membrane calcium translocases, membrane calcium channels, calcium pumps, or intracellular compartmentalization of calcium. In the case of HIV, changes in intracellular calcium may be mediated by binding of the gp120 envelope glycoprotein. Recent st udies with rotaviruses suggest that a nonstructural glycoprotein (NSP4) plays a key role in affecting intracellular calcium levels (674). The HIV-1 transmembrane protein contains a linear region of 28 amino acids (residues 828–855), referred to as the *lentivirus lytic peptide* (LLP-1) (451). This region has been proposed to form a structure similar to certain cytolytic peptides and cation channel proteins. This suggests that TM-induced disruption of target cell membranes, with resulting alteration in ion fluxes, may play a role in HIV cytopathogenicity.

Reactive Oxygen and Nitrogen Intermediates

Inducible reactive oxygen and nitrogen intermediates produced by macrophages and other cells, either spontaneously or after stimu lation by cytokines or other mediators, may play an important role in certain types of cell destruction. Nitric oxide synthase (NOS) is present in macrophages, and its induction leads to the production of reactive oxygen and nitrogen intermediates, including nitric oxide (NO). Nitric oxide and superoxides are highly cytotoxic. NO may also serve as a neurotransmitter within the brain, and its action has been shown to facilitate excitotoxic cell injury by glutamate in neuronal cultures.

Neuronal destruction in Borna disease virus (BDV)–infected rats is due to an immunopathological response rather than to dir ect viral injury of infected cells. Cytokines including Il-1, Il-6, and TNF-α reach peak levels coincident with the peak in inflamma tory responses in brains of BDV-infected animals. Cytokines may mediate BDV neuronal injury by stimulating inflammatory cells to pro duce superoxide and NO by inducing NOS activity (769). The severity of neurological signs and the degree and locale of CNS abnormality correlate with the induction of NOS (769). Oxygen radicals may also be involved in the tissue injury produced during infection with influenza virus (500). It remains to be established whether the induction of reactive oxygen and nitrogen intermediates will prove to be a common mechanism of cytotoxicity.

Deoxynucleoside Triphosphate Pools

Deoxynucleoside triphosphates are precursors necessary for the replication of many viral genomes. The concentrations of these precursors can vary extensively at different stages of the cell cycle and between different cell types. Many viruses contain an enzymatic ribonucleotide reductase activity that catalyzes the reduction of ribonucleoside diphosphates to their deoxyribonucleos ide counterparts. Evidence from viral systems including herpesviruses, vaccinia, HIV-1 and other retroviruses suggests that inhibition or modulation of this activity may attenuate viral virulence and alter the kinetics of viral replication (90,119,306,449).

A number of viruses including herpesviruses, poxviruses, and certain retroviruses are known to encode dUTPases. dUTPases are importa nt in nucleotide biosynthesis and catalyze the hydrolysis of dUTP to dUMP and PP_i. dUMP in turn is used as a substrate for thymidylate synthetases in the generation of TTP. HSV-1 mutants deficient in dUTPase activity are attenuated for neurovirulence, neuroi nvasiveness, and reactivation from latency (546). Deletion of a putative dUTPase domain from the *pol* gene of equine infectious anemia virus, a retrovirus belonging to the lentivirus subfamily, results in markedly reduced replication of virus in nondividing cell s (e.g., macrophages), but not in dividing cells (673). These results suggest that virally encoded dUTPase activity may be an imp ortant determinant of virulence for some viruses in nonreplicating, terminally differentiated cells, especially in the nervous and immune systems.

TRANSMISSION OF VIRAL INFECTIONS

After infecting a susceptible host, viruses must leave this host and enter the environment. The infectious cycle then begins anew with infection of a susceptible host (Fig. 10). Many of the basic issues and concepts in the transmission of viral infection are reviewed in Chapter 9 in *Fields Virology,* 3rd ed. and are only briefly covered here. Virus transmission typically begins with shedding from the infected host through respiratory, enteric, or genitourinary secretions. However, there are basic exceptions to this general pattern. Arbovirus infections typically involve the ingestion by an arthropod vector of a blood meal from a viremic host, and transmission occurs when the vector feeds on another susceptible host. The transmitting fluid, blood, is never really "shed" from the body. A similar pattern of transmission can occur when contaminated tissues or blood products are removed from an infected individual and transplanted, transfused, or inoculated into a susceptible host. This pattern is exemplified in the transmission of HIV and hepatitis B virus infection between drug addicts sharing blood-contaminated needles. Inadvertent transfusion of contaminated blood or blood products may result in infection with HBV, hepatitis C, CMV, EBV, HIV and HTLV-1 (195,542).

In cases involving blood-borne transmission of viral infection, a number of factors influence the subsequent likelihood of infection. These include the titer of virus in the blood, the duration of the viremic state, the amount of material transmitted, and the route of transmission. Before the advent of adequate tests for screening donated blood, HBV infection was a serious potential complicati on of transfusion. This undoubtedly reflected the incidence of infection in certain populations (e.g., professional blood donors), the fact that a chronic carrier state with persisting viremia could occur, and the fact that up to 10^7 infectious doses of virus could be present in a milliliter of blood (35). By contrast, posttransfusion infection with hepatitis A virus is an extremely rare event. This undoubtedly reflects the short duration of the viremic stage and the low concentration of hepatitis A in the blood (284).

Viral infection of the skin is a hallmark of many types of viral infection, including those produced by measles, rubella, certain enteroviruses, herpes simplex, varicella-zoster, various poxviruses, and the human papillomaviruses (453). In many of these cases, even though infectious virus can be demonstrated in skin lesions, this does not appear to be a significant source of viral transmission. Exceptions to this rule include the spread of genital herpes simplex and occasionally herpes labialis from infected skin lesion s (140), rare cases of chickenpox contacted after exposure

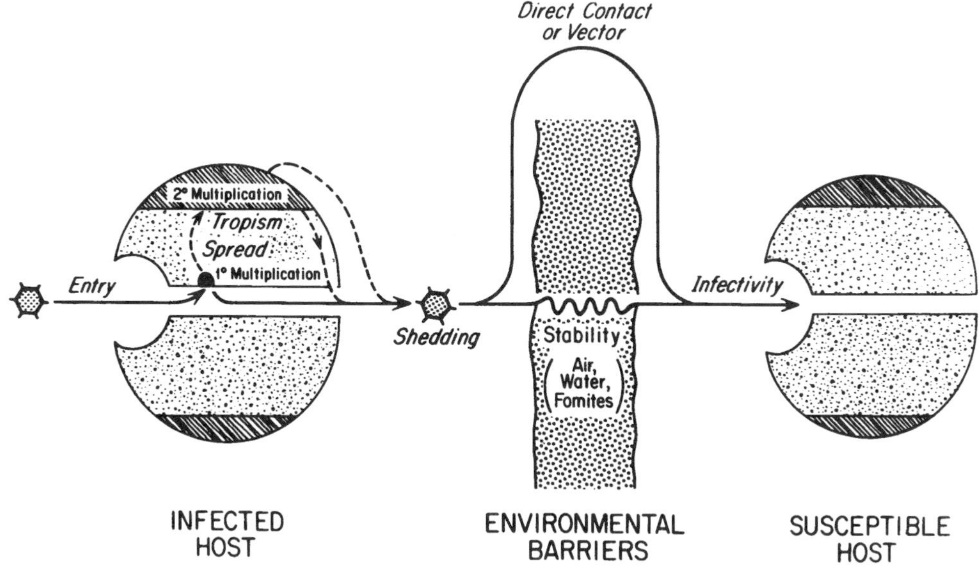

FIG. 10. Transmission of viruses from the infected host into the environment and to novel susceptible hosts. (From Keroack et al., ref. 342, with permission.)

to shingles (80), and the mechanical transmission of human papillomaviruses. In the case of certain poxviruses and papillomaviruses of animals, skin lesions provide a source of virus that can infect biting arthropods, which in turn transmit the virus to new hosts (194). Smallpox virus remained infectious in dried skin crusts for extensive periods, and fomites provided an important route of transmission for this disease before its eradication.

Viral transmission from the upper and lower respiratory tracts depends on aerosols generated during coughing or sneezing or in infected saliva. Some of the factors that influence the stability of viruses in aerosols are reviewed in the section dealing with viral entry. Transmission via infected secretions, which can contain large amounts of infectious virus (516), is important in the genesis of upper respiratory infections, as well as systemic viral infections (e.g., measles, varicella) that enter the host via the respiratory route.

Infected saliva plays an important role in the transmission of Epstein-Barr virus, either through intimate oral contact or possibly by salivary residues left on cups, food, toys, or other objects. Mumps virus is shed in the saliva and can be transmitted after nasal or buccal mucosal inoculation of virus. Salivary excretion of CMV can frequently be demonstrated, but the importance of oral secretions in disease transmission remains unknown. HBV has been isolated from saliva (10,28,266,614), but in experimental transmission studies saliva seems to be infectious only after parenteral inoculation (10,28,614). This would seem to suggest that, although saliva may contain HBV, transmission via this route is unusual unless there is associated viral inoculation (e.g., biting) (422). Saliva has also been shown to contain HIV and HHV-6 (392), although salivary secretions have never been shown to play a significant role in viral transmission. Certain murine retroviruses can be isolated from saliva, which may play a role in male-to-male viral transmission when infected saliva is parenterally inoculated during fighting (538). Perhaps the most widely recognized example of salivary transmission of virus occurs with rabies (192,706). In the case of rabies infection, inoculation of infected saliva through the bite of a rabid animal seems to be the predominant mode of viral entry.

For many viruses shedding occurs by way of the stool. This route of transmission is central to the pathogenesis of enterovirus infection (e.g., poliovirus) (58) and also occurs with rotavirus and hepatitis A (156,550). Infection of susceptible hosts typically occurs by way of fecal-oral transmission. Virus-infected stool can serve to transmit virus directly under circumstances in which proper hygienic measures are not used or are difficult to practice (e.g., with young infants or institutionalized patients). Infected stool can also contaminate water supplies when sewage and waste disposal conditions are substandard.

A number of viruses including CMV, hepatitis A, HBV, mumps, polyomaviruses, and HIV have been isolated from urine (15,349,453,698,699). However, in none of these cases does urine seem to be a major source for viral transmission. By contrast, urinary excretion is central to zoonotic transmission of many arenaviruses and hantaviruses. Human exposure to infected rodent urine or aerosolized urine-contaminated material may initiate viral infection.

Viral infection of semen occurs with HIV, murine retroviruses (393,394,538,746), CMV, and HBV (10,266,614). In HIV infection and, to a lesser extent, for CMV and HBV, infected semen seems to be important in disease transmission. HIV has been found both in cell-free seminal fluid (64) and in lymphoid cells cultured from semen of HIV+ men (766,767). Infection may be facilitated by breaks in the cervical, vaginal, or urethral epithelium, although infection of intact epithelial cells may also occur. Infected semen has also been implicated in the horizontal transmission of murine retroviruses (393,538).

Viruses including HIV have also been isolated from the genital secretions and cervix of infected women (535,710, 747). These secretions may play a role in female-to-male heterosexual transmission of HIV infection. It is worth emphasizing that transmission of viral, as well as nonviral agents, via this route in humans can be almost completely eliminated through the use of condoms.

Several viruses including CMV, mumps, rubella, caprine arthritis-encephalitis virus, certain flaviviruses, minute virus of mice (MVM), and mouse mammary tumor virus have been isolated from human or animal milk or colostrum (453). Infected breast milk provides a source of transmission of CMV from mother to child during the perinatal period (175,640). Hepatitis B virus infection can also be transmitted from mother to child during the perinatal period. Although this type of transmission is well documented, and virus can be found in the milk of infected mothers, it is unclear what role infected milk plays in disease transmission. Maternal milk has also been implicated in the vertical transmission of visna maedi and may play a role in human transmission of HTLV-I. Although mumps virus can be found in human milk (344) and rare cases of perinatal mumps infection occur (329), infected milk is not important in viral transmission.

Investigations of the role played by specific viral genes and the proteins they encode in determining the transmissibility of viral infections are extremely limited (343,613). In the case of reoviruses, the capacity of certain strains of virus to spread from an infected mouse to its uninfected littermates is related to both the magnitude of viral growth in intestinal tissues and the amount of virus shed in the stool (343). Studies with reassortant viruses derived from low- and high-transmission reovirus strains indicated that the viral L2 gene, which encodes the core spike protein λ2, is the major determinant of transmission (343). Since this gene is also a major determinant of the capacity of reoviruses to grow in intestinal tissue after peroral inoculation (61), it may facilitate transmission by increasing viral growth and shedding.

Studies have also been made of the aerosol transmission of influenza virus between mice (613). Although a highly transmissible influenza strain (Jap/305) and a poorly transmissible one (Ao/NWS) grow to equivalent titer in the lungs, the amount of Jap/305 is considerably higher in bronchial secretions and in exhaled air. It was suggested that the neuraminidase protein might account for diffe rences in the amount of virus released from respiratory epithelial cells into bronchial secretions and expelled air, although direct evidence for this is lacking.

HOST FACTORS

It is important to emphasize that the outcome of viral infection of a particular host depends not only on viral factors, but also on a variety of host factors. The role played by host factors in infection has been comprehensively reviewed (194,453) and interested readers are referred to these sources for more detailed information. Aspects of the host that play an important role in determ ining the epidemiology of viral infection are also discussed in Chapter 9 in *Fields Virology,* 3rd ed.

Experience with human infections has repeatedly shown that, when a large population is exposed to the same viral pathogen, such as occurs, for example, during epidemics of encephalitis, the result is a range of outcomes varying from asymptomatic infection to significant and even fatal disease. The same pattern has been reproduced in inadvertent natural experiments such as the inoculation of HBV-contaminated lots of yellow fever virus vaccine into 45,000 military personnel (607). Clinical hepatitis occurred in only 2% of those vaccinated (914 cases), and of this group only a small minority (4%) developed severe disease. The variation in outcomes seems to have been due to differences in host susceptibility, since the amount of HBV inoculated can be presumed to have been fairly uni form among the vaccinees.

In 1955 nearly 120,000 grade-school children were vaccinated with improperly inactivated lots of poliovirus type 1 prepared by Cutter Laboratories (482) ("Cutter incident"). It was subsequently estimated that about 50% of those vaccinated were susceptible to in fection (the remainder had preexisting antibody), and of this group at least 10% to 25% became infected (as estimated by the presenc e of minor illness or fecal excretion of virus). At least 60 cases of paralytic poliomyelitis were ultimately documented among these vaccinees. Thus, the Cutter incident again emphasizes the fact that, when a large population of individuals are inoculated with approximately similar doses of virus, there is a wide range of possible outcomes, which presumably depend on a variety of factors specific to the host.

The importance of host factors in determining the outcome of viral infection has been repeatedly demonstrated in animal models. Classic studies with VSV infection of mice clearly illustrated the importance of host factors, including immune status, genetic background, age, and nutrition, in determining the outcome of infection (600). The role of the host immune response and mediators such as interferon in influencing viral infection are discussed separately (see Chapters 10 and 11).

The genetic constitution of the host is undoubtedly one of the most important factors influencing the outcome of viral infection and has been extensively investigated using inbred strains of mice (32,71,72,582). It was recognized as early as 1937, by Webster, that it was possible to breed strains of mice selectively to differ in their resistance to flaviviruses, including St. Louis encephalitis virus and louping-ill virus (724). This observation was later extended to other flaviviruses (141,480,605), including yellow fever virus (599) and West Nile virus (247). In general, susceptibility correlated with increased levels of viral replication in tissues. In some cases resistant mice seemed to be more susceptible to the antiviral action of interferon (261), although at least one resistance gene (Flv^r) confers flavivirus-specific resistance by an interferon-independent mechanism (71).

Since the early studies with flaviviruses, genetic factors influencing susceptibility to infection have been found involving herpesviruses (7,27,104,249,409,547,621,635), poxviruses (498,609), papovaviruses (72), rhabdoviruses (406), coronaviruses (34, 67,353), retroviruses (71,544), myxoviruses (256,257,638), arenaviruses (505,678), and Borna disease virus (273,595). In almost all cases the genes determining resistance to different types of virus seem to segregate independently, suggesting that a variety of mechanisms are involved rather than a single global type of resistance or susceptibility. Although it would seem logical that genetic resistance to infection would be linked to host genes controlling the immune response (e.g., H-2 in the mouse), this is in fact surprisingly infrequent (71,72). Susceptibility may also be associated with the presence or absence of the appropriate host cell receptors on target cells. This mechanism may explain the susceptibility of some strains of mice to intestinal infection with MHV (67). However, some strains of mice remain resistant to MHV infection despite the presence of functional virus receptors (762), indicating that other factors are also involved.

The genetic basis for resistance of certain strains of mice to infection with influenza viruses has been extensively investigated. A2G mice are resistant to infection with influenza A. Resistance is inherited as an autosomal dominant trait that is associated with the *Mx* allele on mouse chromosome 16 (639). Resistance is mediated by the action of IFN-γ and IFN-ß, which induce a 72-kd protein (Mx protein) in cells derived from resistant mice. The Mx protein accumulates in the nucleus of cells and inhibits viral replication by inhibiting viral mRNA synthesis (165,366). Treatment of A2G mice with antibody to interferon renders them susceptible to infection (256,257). Transfection of cells derived from susceptible mice with cDNA encoding of the Mx protein confers protection against influenza A infection *in vitro* (639). Mice that are susceptible to infection seem to have either deletions or nonsense mutations in the *Mx* gene (637).

Human viral infections provide numerous examples of the correlation between the age of the host and the severity of viral infection. Some viruses tend to produce less severe infection in infants (e.g., varicella, mumps, polio, EBV, hepatitis A), whereas others are more severe (e.g., rotaviruses, respiratory syncytial virus (RSV)) (see, for example, refs. 462 and 730). Age-related susceptibility to virus infection also occurs in a variety of experiment al viral infections in mice (48, 201, 238, 242, 316, 324, 325, 330,387,418,559,729).

The basis for the age dependence of viral infection is poorly understood. Some types of enhanced resistance to viral infection with increasing age may reflect the maturation of both specific and nonspecific components of the host's immune system, including phagocytosis, NK cell activity, cell-mediated cytotoxicity, and antibody production. Organ-specific differences in the immune response may explain why some tissues in adult animals become resistant to viral infection, whereas others do not (745).

Immunological factors do not seem to be the key determinants of age-related susceptibility to infection with alphaviruses, such as Sindbis (243,559). In the case of encephalitis, the state of neuronal maturity seems to be an important factor in determining age- related susceptibility (501). It has recently been shown that changes in the Sindbis E2 glycoprotein can compensate for age-dependent neuronal restriction of replication (686). Dependence of viral infection on the state of cellular differentiation has also b een described with coronaviruses (51,744), herpesviruses (173), polyomaviruses (46,213,424), parvoviruses (251,668), retroviruses (135,220,416), and arenaviruses (148).

Some viruses preferentially infect mitotically active cells, and differences in the organ distribution of these cells during develop ment and maturation may influence the pattern of disease. For example, prenatal infection of cats with feline parvovirus results in destruction of the germinal cells of the cerebellum and subsequent cerebellar aplasia. Conversely, infection of adult cats involves primarily mitotically active bone marrow and intestinal epithelial cells and does not result in neurological illness (320). Induction of cellular damage or injury may also trigger responses that make previously nonpermissive cells able to support viral replicat ion (16,529,594).

Hormones can also influence the outcome of viral infection. This may provide one explanation for the frequent observation that male mice are more susceptible to a variety of viral infectious than are their female counterparts (406,752). A number of viral infect ions, including those with polio, hepatitis A and B, and smallpox viruses, are commonly more severe during pregnancy (122,453), although it is not known whether this reflects hormonal alterations or other factors. Reactivation of polyomavirus infection also occurs more commonly during pregnancy, although this typically occurs without associated clinical disease (126). Pregnant mice are more susceptible to intravaginal infection with herpes simplex (26) but not to intraperitoneal or intranasal infection (764). It has been suggested that this may be due to local hormonal effects (26). In experimental studies, the administration of both steroid and thyroid hormones can be shown to affect adversely the outcome of certain types of viral infection (295,299).

The nutritional state of the host can also exert a marked influence on the outcome of viral infection (107,108). For example, pro tein malnutrition dramatically exacerbates the severity of measles infection, perhaps by depressing host cellular immunity (108,162, 459). A similar increase in susceptibility to coxsackievirus and flavivirus infection occurs in experimental infection of malnourished mice (292,753). At the opposite extreme, hyperalimentation and the induction of hypercholesterolemia may also be associated with increased severity of viral infection (91).

Personal habits may also play a role in altering the severity of viral infection, as well as exerting their more obvious effect on risk of exposure. For example, the severity of influenza is increased in chronic smokers (333), which is related to the deterioration in mucociliary clearance. It has long been recognized that preceding vigorous exercise may worsen the severity of a subsequent bout of poliomyelitis (288), although the mechanism for this has never been clearly established. Experimentally, exercise can also be shown to increase the severity of coxsackievirus infection in mice (219).

The role of host responses such as fever and inflammation in combatting viral infection has been examined (573). Classic studies of myxoma infection in rabbits clearly demonstrated that increasing body temperature increased protection against infection and that decreasing temperature increased the severity of infection (431,520). Similar results have been found with ectromelia and coxsac kievirus infections in mice (569,718). Blocking the development of fever with drugs (e.g., salicylates) increases the mortality of vaccinia infection in rabbits and increases the shedding of influenza by ferrets (296), providing additional evidence for an antiviral effect of fever.

The importance of host cell enzymes in influencing the outcome of viral infection was discussed earlier in this chapter. It is well exemplified in the role played by host cell enzymes in the pathogenesis of myxovirus infection (see ref. 608 for review), although other examples exist. Influenza requires cleavage of the viral hemagglutinin from an inactive precursor form into two disulfide-bonded subunits to become infectious. This cleavage is mediated by host cell trypsin-like proteases, and without this cleavage virus is avirulent (351,608). The HA proteins of mammalian influenza viruses and nonpathogenic avian viruses are susceptible to proteolytic cleavage in a limited number of cell types, whereas the HAs of many pathogenic avian influenza viruses are cleaved by proteases present in many types of host cells (351). Differences in the susceptibilities of these HAs to cleavage are due to the presence of only a single arginine residue at the cleavage site in the mammalian and nonpathogenic avian viruses, whereas the pathogenic avian viruses typically have several basic amino acids present at the cleavage site (351). Mutations in the influenza HA near the cleavage site can dramatically alter both the host range and the pathogenicity of influenza viruses (154,502).

A host cell protease-mediated cleavage similar to that required for influenza virus is required for activation of the fusion activity of paramyxoviruses such as Sendai virus. Virus containing an uncleaved fusion (F_O) protein does not replicate after inoculation into mice. However, virus will replicate if the F protein has previously been cleaved *in vitro* into the disulfide-bonded F_1 and F_2 subunits by trypsin. Mutant viruses whose F protein is cleavable by chymotrypsin but not by trypsin also do not produce disease unless the protein is cleaved *in vitro* (455,664).

Host cell enzymes may also play an indirect role in susceptibility to Sindbis virus infection. The envelope of the virus contains sialic acid residues derived as the virus buds from host cell plasma membranes. If the membranes are rich in sialic acid, virus is a more potent activator of complement. Complement activation and subsequent opsonization of virus facilitate viral clearance and thus decrease viral virulence (279). Thus, in this case, the biochemical composition of host cell membranes alters the subsequent virulence of virus.

REFERENCES

1. Evans AS, Brachman PS, eds. *Bacterial infections of humans.* 2nd ed. New York: Plenum Press; 1991:31.
2. *Virus attachment and entry into cells.* Washington: American Society for Microbiology; 1986.
3. Ahmed R, Simon R, Matloubian M, et al. Genetic analysis of *in vivo* selected viral variants causing chronic infection: importance of mutation in the L RNA segment of lymphocytic choriomeningitis virus. *J Virol* 1988;62:3301–3308.
4. Albrecht J-C, Fleckenstein B. New member of the multigene family of complement control proteins in *Herpesvirus saimiri. J Virol* 1992; 66:3937–3940.
5. Albrecht J-C, Nicholas J, Cameron KR, et al. *Herpesvirus saimiri* has a gene specifying a homologue of the cellular membrane glycoprotein CD59. *Virology* 1992;190:527–530.
6. Albrecht P. Pathogenesis of neurotropic arbovirus infection. *Curr Top Microbiol Immunol* 1968;43:44–91.
7. Allan JE, Shellam GR. Genetic control of murine cytomegalovirus infection: viral titres in resistant and susceptible strains of mice. *Arch Virol* 1984;81:139–150.
8. Almeida JD, Howatson AF, Williams MG. Electron microscopic study of human warts, sites of virus production and the nature of the inclusion bodies. *J Invest Dermatol* 1962;38:337–345.
9. Almond JW. The attenuation of poliovirus neurovirulence. *Annu Rev Microbiol* 1987;41:153–180.
10. Alter HJ, Purcell RH, Gerin JL, et al. Transmission of hepatitis B surface antigen positive saliva and semen. *Infect Immun* 1977;16:928–933.
11. Amalfitano A, Martin LG, Fluck MM. Different roles for two enhancer domains in the organ- and age-specific pattern of polyomavirus replication in the mouse. *Mol Cell Biol* 1992;12:3628–3635.
12. Anderson J, Field H. The distribution of herpes simplex virus type 1 antigen in the mouse central nervous system after different routes of inoculation. *J Neurol Sci* 1983;60:181–195.
13. Andersson M, Paabo S, Nilsson T, et al. Impaired intracellular transport of class I MHC antigens as a possible means for adenoviruses to evade immune surveillance. *Cell* 1985;43:215–222.
14. Antonucci TK, Rutter WJ. Hepatitis B virus (HBV) promoters are regulated by the HBV enhancer in a tissue-specific manner. *J Virol* 1989;63:579–583.
15. Arthur RR, Shah KV. Occurrence and significance of papovavirus BK and JC in the urine. *Prog Med Virol* 1989;36:42–61.
16. Atencio IA, Shadan FF, Zhou XJ, et al. Adult mouse kidneys become permissive to acute polyomavirus infection and reactivate persistent infections in response to cellular damage and regeneration. *J Virol* 1993;67:1424–1432.
17. Atkins GJ, Sheahan BJ, Dimmock NJ. Semliki Forest virus infection of mice: a model for genetic and molecular analysis of viral pathogenicity. *J Gen Virol* 1985;66:395–408.
18. Atkins GJ, Sheahan BJ. Semliki Forest virus neurovirulence mutants have altered cytopathogenicity for central nervous system cells. *Infect Immun* 1982;36:333–341.
19. Attibele N, Wyde PR, Trial J, et al. Measles virus-induced changes in leukocyte function antigen 1 expression and leukocyte aggregation: possible role in measles virus pathogenesis. *J Virol* 1993;67:1075–1079.
20. Babinet CH, Farza H, Morello D, et al. Specific expression of hepatitis B surface antigen (HBsAg) in transgenic mice. *Science* 1985; 230:1160–1163.
21. Babiss L, Luftig R, Weatherbee J, et al. Reovirus serotypes 1 and 3 differ in their *in vitro* association with microtubules. *J Virol* 1979; 30:863–874.
22. Baer GM: Pathogenesis to the central nervous system. In: Baer GM, ed. *The natural history of rabies.* Orlando: Academic Press; 1975:181–198.
23. Baer GM. Animal models in the pathogenesis and treatment of rabies. *Rev Infect Dis* 1988;10(Suppl 4):S739–S750.
24. Bai M, Harfe B, Freimuth P. Mutations that alter an Arg-Gly-Asp (RGD) sequence in the adenovirus type 2 penton base protein abolish its cell-rounding activity and delay virus reproduction in flat cells. *J Virol* 1993;67:5198–205.
25. Bak I, Markham C, Cook M, et al. Intra-axonal transport of herpes simplex virus in the rat central nervous system. *Brain Res* 1977;136:415–429.
26. Baker D, Plotkin S. Enhancement of vaginal infection in mice by herpes simplex virus type II with progesterone. *Proc Soc Exp Biol Med* 1978;158:131–134.
27. Bancroft G, Shellam G, Chalmer J. Genetic influences on the augmentation of natural killer (NK) cells during murine cytomegalovirus infection: correlation with pattern of resistance. *J Immunol* 1981; 126:988–994.
28. Bancroft WH, Snitbhan R, Scott RM, et al. Transmission of hepatitis B virus to gibbons by exposure to human saliva containing hepatitis B surface antigen. *J Infect Dis* 1977;135:79–85.
29. Bandyopadhyay PK, Pritchard AE, Jensen K, et al. A three-nucleotide insertion in the H stem-loop of the 5' untranslated region of Theiler's virus attenuates neurovirulence. *J Virol* 1993;67:3691–3695.
30. Bandyopadhyay PK, Wang C, Lipton HL. Cap-independent translation by the 5' untranslated region of Theiler's murine encephalomyelitis virus. *J Virol* 1992;66:6249–6256.
31. Bang F, Bang B, Foard M. Responses of upper respiratory mucosa to drugs and viral infections. *Am Rev Respir Dis* 1966;93(suppl): 5142–5149.
32. Bang F. Genetics of resistance of animal to viruses. I. Introduction and studies in mice. *Adv Virus Res* 1978;23:269–348.
33. Bang F, Luttrell C. Factors in the pathogenesis of virus diseases. *Adv Virus Res* 1961;8:199–244.
34. Bang F, Warwick A. Mouse macrophages as host cells for the mouse hepatitis virus and the genetic basis of their susceptibility. *Proc Natl Acad Sci USA* 1960;46:1065–1075.
35. Barker LF, Murray R. Relationship of virus dose to incubation time of clinical hepatitis and time of appearance of hepatitis-associated antigen. *Am J Med Sci* 1972;263:27–33.
36. Barnett EM, Perlman S. The olfactory nerve and not the trigeminal nerve is the major site of CNS entry for mouse hepatitis virus, strain JHM. *Virology* 1993;194:185–191.
37. Bass DM, Baylor MR, Chen C, et al. Liposome mediated transfection

of intact viral particles reveals that plasma membrane determines permissivity of tissue culture cells to rotavirus. *J Clin Invest* 1992; 90:2313–2320.
38. Bass DM, Bodkin D, Dambrauskas R, et al. Intraluminal proteolytic activation plays an important role in replication of type 1 reovirus in the intestines of neonatal mice. *J Virol* 1990;64:1830–1833.
39. Bass DM, Mackow ER, Greenberg HB. Identification and partial characterization of a rhesus rotavirus binding glycoprotein on murine enterocytes. *Virology* 1991;183:602–610.
40. Bass DM, Trier JS, Dambrauskas R, et al. Reovirus type 1 infection of small intestinal epithelium in suckling mice and its effect on M cells. *Lab Invest* 1988;58:226–235.
41. Bassel-Duby R, Jayasuriya A, Chatterjee D, et al. Sequence of reovirus haemagglutinin predicts a coiled-coil structure. *Nature* 1985;315:421–423.
42. Bassel-Duby R, Spriggs DR, Tyler KL, et al. Identification of attenuating mutations on the reovirus type 3 S1 double-stranded RNA segment with a rapid sequencing technique. *J Virol* 1986;60:64–67.
43. Battini J, Heard JM, Danos O. Receptor choice determinants in the envelope glycoprotein of amphotropic, xenotropic, and polytropic murine leukemia viruses. *J Virol* 1992;66:1468–1475.
44. Baty CJ, Sherry B. Cytopathogenic effect in cardiac myocytes, but not in cardiac fibroblasts, is correlated with reovirus-induced acute myocarditis. *J Virol* 1993;67:6295–6298.
45. Beattie E, Tartaglia J, Paoletti E. Vaccinia virus encoded eIF-2 alpha homolog abrogates the antiviral effect of interferon. *Virology* 1991; 183:419–422.
46. Bedell MA, Hudson JB, Golub TR, et al. Amplification of human papillomavirus genomes *in vitro* is dependent on epithelial differentiation. *J Virol* 1991;65:2254–2260.
47. Bell S, Cranage M, Borysiewicz LK, et al. Induction of immunoglobulin G by Fc receptors by recombinant vaccinia virus expressing glycoproteins E and I of herpes simplex virus type 1. *J Virol* 1990; 64:2181–2186.
48. Ben-Hur T, Hadar J, Shtram Y, et al. Neurovirulence of herpes simplex virus type 1 depends on age in mice and thymidine kinase expression. *Arch Virol* 1983;78:307–308.
49. Bergelson JM, Shepley MP, Chan BMC, et al. Identification of the integrin VLA-2 as a receptor for echovirus. *Science* 1992;255:1718–1720.
50. Bergelson JM, St. John N, Kawaguchi S, et al. Infection by echoviruses 1 and 8 depends on the alpha2 subunit of human VLA-2. *J Virol* 1993;67:6847–6852.
51. Beushausen S, Dales S. *In vivo* and *in vitro* models of demyelinating disease. XXI. Relationship between differentiation of rat oligodendrocytes and control of JHMV replication. *Adv Exp Med Biol* 1987; 218:239–254.
52. Bhat S, Spitalnik SL, Gonzalez-Scarano F, et al. Galactosyl ceramide or a derivative is an essential component of the neural receptor for human immunodeficiency virus type 1 envelope glycoprotein gp120. *Proc Natl Acad Sci USA* 1991;88:7131–7134.
53. Bijlenga G, Heaney T. Post-exposure local treatment of mice infected with rabies with two axonal flow inhibitors, colchicine and vinblastine. *J Gen Virol* 1978;39:381–385.
54. Black BL, Rhodes RB, McKenzie M, et al. The role of vesicular stomatits virus matrix protein in inhibition of host-directed gene expression is genetically separable from its function in virus assembly. *J Virol* 1993;67:4814–4821.
55. Blomquist MC, Hunt LT, Barker WC. Vaccinia virus 19-kilodalton protein: relationship to several mammalian proteins, including two growth factors. *Proc Natl Acad Sci USA* 1984;81:7363–7367.
56. Blondel D, Harmison GG, Schubert M. Role of matrix protein in cytopathogenesis of vesicular stomatitis virus. *J Virol* 1990;64:1716–1725.
57. Bodian D. Viremia in experimental poliomyelitis. II. Viremia and the mechanism of the "provoking" effect of injections or trauma. *Am J Hyg* 1954;60:358–370.
58. Bodian D. Emerging concepts of poliomyelitis infection. *Science* 1955; 122:105–108.
59. Bodian D. Poliovirus in chimpanzee tissues after virus feeding. *Am J Hyg* 1956;64:181–197.
60. Bodian D: Poliomyelitis: pathogenesis and histopathology. In: Rivers TM, Horsfall FL Jr, eds. *Viral and rickettsial infections of man.* Philadelphia: Lippincott; 1959:479–498.
61. Bodkin D, Fields BN. Growth and survival of reovirus in intestinal tissue: role of the L2 and S1 genes. *J Virol* 1989;63:1188–1193.
62. Bodkin D, Nibert ML, Fields BN. Proteolytic digestion of reovirus serotype 1 in the intestinal lumen of neonatal mice. *J Virol* 1989;63:4676–4681.
63. Borrow P, Oldstone MBA. Mechanism of lymphocytic choriomeningitis virus entry into cells. *Virology* 1994;198:1–9.
64. Borzy JS, Connell RS, Kiessling AA. Detection of human immunodeficiency virus in cell-free seminal fluid. *J Acquir Immune Defic Syndr* 1988;1:419–424.
65. Boshart M, Werber F, John G, et al. A very strong enhancer is located upstream of an immediate early gene of human cytomegalovirus. *Cell* 1985;41:521–530.
66. Boyd MT, Simpson GR, Cann AJ, et al. A single amino acid substitution in the V1 loop of human immunodeficiency virus type 1 gp120 alters cellular tropism. *J Virol* 1993;67:3649–3652.
67. Boyle JF, Weismiller DG, Holmes KV. Genetic resistance to mouse hepatitis virus correlates with the absence of virus-binding activity on target tissues. *J Virol* 1987;61:185–189.
68. Brandenburger A, Legendre D, Avalosse B, et al. NS-1 and NS-2 proteins may act synergistically in the cytopathogenicity of parvovirus MVMp. *Virology* 1990;174:576–584.
69. Brewer CB, Roth MG. A single amino acid change in the cytoplasmic domain alters the polarized delivery of influenza virus hemagglutinin. *J Cell Biol* 1991;114:413–421.
70. Brinster R, Chen H, Messing A, et al. Transgenic mice harboring SV40 T-antigen genes develop characteristic brain tumors. *Cell* 1984; 37:367–379.
71. Brinton M, Blank K, Nathanson N: Host genes that influence susceptibility to viral diseases. In: Notkins AL, Oldstone MBA, eds. *Concepts in viral pathogenesis.* New York: Springer-Verlag; 1984:71–78.
72. Brinton M, Nathanson N. Genetic determinants of virus susceptibility: epidemiologic implications of murine models. *Epidemiol Rev* 1981;3:115–139.
73. Broome RL, Vo PT, Ward RL, et al. Murine rotavirus genes encoding outer capsid proteins VP4 and VP7 are not major determinants of host range restriction and virulence. *J Virol* 1993;67:2448–2455.
74. Brown DT, Edwards J. Structural changes in alphaviruses accompanying the process of membrane penetration. *Semin Virol* 1992;3:519–527.
75. Brown DW, Blais BP, Robinson HL. Long terminal repeat (LTR) sequences, env, and a region near 5'LTR influence the pathogenic potential of recombinants between Rous-associated virus types 0 and 1. *J Virol* 1988;62:3431–3437.
76. Brown JP, Twardzik DR, Marquardt H, et al. Vaccinia virus encodes a polypeptide homologous to epidermal growth factor and transforming growth factor. *Nature* 1985;313:491–492.
77. Brown KE, Anderson SM, Young NS. Erythrocyte P antigen: cellular receptor for B19 parvovirus. *Science* 1993;262:114–117.
78. Brownstein D. Resistance/susceptibility to lethal Sendai virus infection genetically linked to a mucociliary transport polymorphism. *J Virol* 1987;61:1670–1671.
79. Brunner KT, Hurez D, McCluskey RT, et al. Blood clearance of P32-labeled vesicular stomatitis and Newcastle disease viruses by the reticuloendothelial system in mice. *J Immunol* 1960;85:99–104.
80. Bruusgaard E. The mutual relation between zoster and varicella. *Br J Dermatol* 1932;44:1–24.
81. Buller RM, Chakrabati S, Cooper JA, et al. Deletion of the vaccinia virus growth factor gene reduces virus virulence. *J Virol* 1988; 62:866–874.
82. Burgert HG, Kvist S. An adenovirus type 2 glycoprotein blocks cell surface expression of human histocompatibility class I antigen. *Cell* 1985;41:987–997.
83. Burk RD, DeLoia JA, Elawady MK, et al. Tissue preferential expression of the hepatitis B virus (HBV): surface antigen gene in two lines of HBV transgenic mice. *J Virol* 1988;62:649–654.
84. Burnet FM. *Principles of animal virology.* Orlando: Academic Press; 1960.
85. Burrage TG, Tignor GH, Smith AL. Rabies virus binding at neuromuscular junctions. *Virus Res* 1985;2:273–289.
86. Burstin SJ, Brandriss MW, Schlesinger JJ. Infection of a macrophage-like cell line P388D1 with reovirus: effects of immune ascitic fluids and monoclonal antibodies on neutralization and enhancement of viral growth. *J Immunol* 1983;130:2915–2919.
87. Caggana M, Chan P, Ramsingh A. Identification of a single amino acid residue in the capsid protein VP1 of coxsackievirus B4 that determines

the virulent phenotype. *J Virol* 1993;67:4797–4803.
88. Caillet-Fauquet P, Perros M, Brandenburger A, et al. Programmed killing of human cells by means of an inducible clone of parvoviral genes encoding the non-structural proteins. *EMBO J* 1990;9:2989–2995.
89. Calenoff MA, Faaberg KS, Lipton HL. Genomic regions of neurovirulence and attenuation in Theiler's murine encephalomyelitis virus. *Proc Natl Acad Sci USA* 1990;87:987–992.
90. Cameron J, McDougall I, Marsden HS, et al. Ribonucleotide reductase encoded by herpes simplex virus is a determinant of the pathogenicity of the virus in mice and a valid antiviral target. *J Gen Virol* 1988;69:2607–2612.
91. Campbell A, Lorio R, Madge G, et al. Dietary hepatic cholesterol elevation: effects on coxsackie B5 infection and inflammation. *Infect Immun* 1982;37:307–317.
92. Cann AJ, Churcher MJ, Boyd M, et al. The region of the envelope glycoprotein gene of human immunodeficiency virus type 1 responsible for determination of cell tropism. *J Virol* 1992;66:305–309.
93. Carbone KM, Duchala CS, Griffin JW, et al. Pathogenesis of Borna disease virus in rats: evidence that intra-axonal spread is the major route for dissemination and determinant for disease incubation. *J Virol* 1987;61:3431–3440.
94. Card JP, Rinaman L, Lynn RB, et al. Pseudorabies virus infection of the rat central nervous system: ultrastructural characterization of viral replication, transport, and pathogenesis. *J Neurosci* 1993;13:2515–2539.
95. Card JP, Rinaman L, Schwaber JS, et al. Neurotropic properties of pseudorabies virus: uptake and transneuronal passage in the rat central nervous system. *J Neurosci* 1990;10:1974–1994.
96. Card JP, Whealy ME, Robbins AK, et al. Pseudorabies virus envelope glycoprotein gI influences both neurotropism and virulence during infection of the rat visual system. *J Virol* 1992;66:3032–3041.
97. Card JP, Whealy ME, Robbins AK, et al. Two alpha-herpesvirus strains are transported differentially in the rodent visual system. *Neuron* 1991;6:957–969.
98. Carillo C, Borca MV, Alfonso CL, et al. Long-term persistent infection of swine monocytes/macrophages with African swine fever virus. *J Virol* 1994;68:580–583.
99. Casadevall N, Lacombe C, Muller O, et al. Multimeric structure of the membrane erythropoietin receptor of murine erythroleukemia cells (Friend cells): cross-linking of erythropoietin with spleen focus-forming virus envelope protein. *J Biol Chem* 1991;266:16015–16020.
100. Ceccaldi PE, Ermine A, Tsiang H. Continuous delivery of colchicine in the rat brain with osmotic pumps for inhibition of rabies virus transport. *J Virol Methods* 1990;28:79–84.
101. Ceccaldi PE, Gillet JP, Tsiang H. Inhibition of the transport of rabies virus in the central nervous system. *J Neuropathol Exp Neurol* 1989; 48:620–630.
102. Cecilia D, Gould EA. Nucleotide changes responsible for loss of neuroinvasiveness in Japanese encephalitis virus neutralization-resistant mutants. *Virology* 1991;181:70–77.
103. Celander D, Haseltine WA. Tissue specific transcription preference as a determinant of cell tropism and leukemogenic potential of murine retroviruses. *Nature* 1984;312:159–163.
104. Chalmer J, MacKenzie J, Stanley NF. Resistance to murine cytomegalovirus linked to the major histocompatability complex of the mouse. *J Gen Virol* 1977;37:107–114.
105. Chamberlain RW, Sudia WD. Mechanisms of transmission of viruses by mosquitoes. *Annu Rev Entomol* 1961;6:371–390.
106. Chambers TJ, Nestorowicz A, Amberg SM, et al. Mutagenesis of the yellow fever virus NS2B protein: effects on proteolytic processing, NS2B-NS3 complex formation, and viral replication. *J Virol* 1993; 67:6797–6807.
107. Chandra R. Nutritional deficiency and susceptibility to infection. *Bull WHO* 1979;57:167–177.
108. Chandra R. Nutrition, immunity, and infection: present knowledge and future directions. *Lancet* 1983;1:688–691.
109. Chang H-W, Watson JC, Jacobs BL. The E3L gene of vaccinia virus encodes an inhibitor of the interferon-induced, double-stranded RNA-dependent protein kinase. *Proc Natl Acad Sci USA* 1992;89:4825–4829.
110. Chang HK, Wang BY, Yuh CH, et al. A liver specific nuclear factor interacts with the promoter region of the large surface protein gene of human hepatitis B virus. *Mol Cell Biol* 1989;9:5189–5197.
111. Charlton K, Casey G. Experimental rabies in skunks: immunofluorescent, light, and electron microscopic studies. *Lab Invest* 1979;41:36–44.
112. Chatis P, Holland C, Hartley J, et al. Role of the 3' end of the genome in determining the disease specificity of Friend and Moloney murine leukemia viruses. *Proc Natl Acad Sci USA* 1983;80:4408–4441.
113. Chatis P, Holland C, Silver J, et al. A 3' end fragment encompassing the transcriptional enhancers of non-defective Friend virus confers erythroleukemogenicity on Moloney leukemia virus. *J Virol* 1984; 52:248–254.
114. Chen M, Popescu N, Woodworth C, et al. Human herpesvirus 6 infects cervical epithelial cells and transactivates human papillomavirus gene expression. *J Virol* 1994;68:1173–1178.
115. Cheng-Mayer C, Shioda T, Levy JA. Host range, replicative, and cytopathic properties of human immunodeficiency virus type 1 are determined by very few amino acid changes in Tat and gp120. *J Virol* 1991;65:6931–6941.
116. Chesebro B, Buller R, Portis J, et al. Failure of human immunodeficiency virus entry and infection in CD4+ human brain and skin cells. *J Virol* 1990;64:215–221.
117. See ref. 116.
118. Chevallier-Greco A, Groffat H, Manet E, et al. The Epstein-Barr virus (EBV) DR enhancer contains two functionally different domains: domain A is constitutive and cell specific, domain B is transactivated by the EBV early promoter R. *J Virol* 1989;63:615–623.
119. Child SJ, Palumbo GJ, Buller RM. Insertional inactivation of the large subunit of ribonuclease reductase encoded by vaccinia virus is associated with reduced virulence *in vivo*. *Virology* 1990;174:625–629.
120. Chou J, Kern ER, Whitley RJ, et al. Mapping of herpes simplex virus-1 neurovirulence to γ1.34.5, a gene nonessential for growth in culture. *Science* 1990;250:1262–1266.
121. Chou J, Roizman B. The γ1.34.5 gene of herpes simplex virus 1 precludes neuroblastoma cells from triggering total shutoff of protein synthesis characteristic of programmed cell death in neuronal cells. *Proc Natl Acad Sci USA* 1992;89:3266–3270.
122. Christie A, Allan A, Aref M, et al. Pregnancy hepatitis in Libya. *Lancet* 1976;2:827–829.
123. Clapham PR, Blanc D, Weiss RA. Specific cell surface requirements for the infection of CD4+ cells by human immunodeficiency virus types 1 and 2 and by simian immunodeficiency virus. *Virology* 1991;181: 703–715.
124. Clark SM, Roth JR, Clark ML, et al. Tryptic enhancement of rotavirus infectivity: mechanism of enhancement. *J Virol* 1981;39:816–822.
125. Clayson E, Brando VJ, Compans RW. Release of simian virus 40 virions from epithelial cells is polarized and occurs without cell lysis. *J Virol* 1989;63:2278–2288.
126. Coleman DV, Wolfendale MR, Daniel RA, et al. A prospective study of human polyomavirus infection in pregnancy. *J Infect Dis* 1980; 142:1–8.
127. Compans RW, Srinivas RV. Protein sorting in polarized epithelial cells. *Curr Top Microbiol Immunol* 1991;170:141–181.
128. Compton T, Ivanov IE, Gottlieb T, et al. A sorting signal for the basolateral delivery of vesicular stomatitis virus (VSV) G protein lies in its luminal domain: analysis of the targeting of VSV G-influenza hemagglutinin chimeras. *Proc Natl Acad Sci USA* 1989;86:4112–4116.
129. Compton T, Nowlin DM, Cooper NR. Initiation of human cytomegalovirus infection requires initial interaction with cell surface heparan sulfate. *Virology* 1993;193:834–841.
130. Cordonnier AL, Montagnier L, Emerman M. Single amino acid changes in HIV envelope affect viral tropism and receptor binding. *Nature* 1989;340:571–574.
131. Couderc T, Barzu T, Horaud F, et al. Poliovirus permissivity and specific receptor expression on human endothelial cells. *Virology* 1990;174:95–102.
132. Couderc T, Hogle J, Le Blay H, et al. Molecular characterization of mouse-virulent poliovirus type 1 Mahoney mutants: involvement of residues of polypeptides VP1 and VP2 located on the inner surface of the capsid protein shell. *J Virol* 1993;67:3808–3817.
133. Coulon P, Derbin C, Kucera P, et al. Invasion of the peripheral nervous systems of adult mice by the CVS strain of rabies virus and its avirulent derivative Av01. *J Virol* 1989;63:3550–3554.
134. Crowell RL, Hsu K-HL. Isolation of cellular receptors for viruses. In: Notkins AL, Oldstone MBA, eds. *Concepts in viral pathogenesis II.* New York: Springer-Verlag; 1986:117–125.
135. Czub M, Czub S, McAtee F, et al. Age-dependent resistance to murine retrovirus-induced spongiform neurodegeneration results from central nervous system–specific restriction of viral replication. *J Virol* 1991;65:2539–2544.

136. Dalgleish AG, Beverly PCL, Clapham PR, et al. The CD4 (T4) antigen is an essential component of the receptor for the AIDS retrovirus. *Nature* 1985;312:763–767.
137. Dalmat H. Arthropod transmission of rabbit papillomatosis. *J Exp Med* 1958;108:9–20.
138. Dalmat H. Arthropod transmission of rabbit fibromatosis (Shope). *J Hyg* 1959;57:1–30.
139. Dalziel R, Lampert PW, Talbot PJ, et al. Site-specific alteration of murine hepatitis virus type 4 peplomer glycoprotein E2 results in reduced neurovirulence. *J Virol* 1986;59:463–471.
140. Daniels CA, LeGoff SG, Notkins AL. Shedding of infectious virus-antibody complexes from vesicular lesions of patients with recurrent herpes labialis. *Lancet* 1975;2:524–528.
141. Darnell M, Koprowski H. Genetically determined resistance to infection with group B arboviruses. II. Increased production of interfering particles in cell cultures from resistant mice. *J Infect Dis* 1974; 129:248–256.
142. Darougar S, Quinlan MP, Gibson JA, et al. Epidemic keratoconjunctivitis and chronic papillary conjunctivitis in London due to adenovirus type 19. *Br J Ophthal* 1977;61:76–85.
143. Davis B, Linney E, Fan H. Suppression of leukemia virus pathogenicity by polyoma virus enhancers. *Nature* 1985;314:550–553.
144. Davis NL, Fuller FJ, Dougherty WG, et al. A single nucleotide change in the E2 glycoprotein gene of Sindbis virus affects penetration rate in cell culture and virulence in neonatal mice. *Proc Natl Acad Sci USA* 1986;83:6771–6775.
145. Day S, Lausch R, Oakes J. Evidence that the gene for herpes simplex virus type 1 DNA polymerase accounts for the capacity of an intertypic recombinant to spread from eye to central nervous system. *Virology* 1988;163:166–173.
146. Day SP, Murphy P, Brown EA, et al. Mutations within the 5' nontranslated region of hepatitis A virus RNA which enhance replication in BS-C-1 cells. *J Virol* 1992;66:6533–6540.
147. Dayton ET, Konings DAM, Lim SY, et al. The RRE of human immunodeficiency virus type 1 contributes to cell-type-specific viral tropism. *J Virol* 1993;67:2871–2878.
148. De La Torre JC, Rall G, Oldstone C, et al. Replication of lymphocytic choriomeningitis virus is restricted in terminally differentiated neurons. *J Virol* 1993;67:7350–7359.
149. DeJong JG. The survival of measles virus in air in relation to the epidemiology of measles. *Arch Ges Virus Forsch* 1965;16:97–102.
150. DeJong JG, Harmsen M, Platinga AD, et al. Inactivation of Semliki Forest virus in aerosols. *Appl Environ Microbiol* 1976;32:315–319.
151. DeJong JG, Winkler KC. The inactivation of poliovirus in aerosols. *J Hyg* 1968;66:557–565.
152. DesGroseillers L, Jolicoeur P. Mapping the viral sequences conferring leukemogenicity and disease specificity in Moloney and amphotropic murine leukemia viruses. *J Virol* 1984;52:448–456.
153. DesGroseillers L, Rassart E, Jolicoeur P. Thymotropism of murine leukemia virus is conferred by its long terminal repeat. *Proc Natl Acad Sci USA* 1983;80:4203–4207.
154. Deshpande K, Fried V, Ando M, et al. Glycosylation affects cleavage of an H5N2 influenza virus hemagglutinin and regulates virulence. *Proc Natl Acad Sci USA* 1987;84:36–40.
155. Di Simone C, Zandonatti MA, Buchmeier MJ. Acidic pH triggers LCMV membrane fusion activity and conformational change in the glycoprotein spike. *Virology* 1994;189:455–465.
156. Dienstag JL, Feinstone SM, Kapikian AZ, et al. Fecal shedding of hepatitis-A antigen. *Lancet* 1975;1:765–767.
157. Dietzschold B, Wiktor T, Trojanowski J, et al. Differences in cell-to-cell spread of pathogenic and apathogenic rabies virus *in vivo* and *in vitro*. *J Virol* 1985;56:12–18.
158. Dietzschold B, Wunner WH, Wiktor TJ, et al. Characterization of an antigenic determinant of the glycoprotein that correlates with pathogenicity of rabies virus. *Proc Natl Acad Sci USA* 1983;80:70–74.
159. Dingwell KS, Brunetti CR, Hendricks RL, et al. Herpes simplex virus glycoproteins E and I facilitate cell-to-cell spread *in vivo* and across junctions of cultured cells. *J Virol* 1994;68:834–845.
160. Dorries K, Vogel E, Gunther S, et al. Infection of human polyomaviruses JC and BK in peripheral blood leukocytes from immunocompetent individuals. *Virology* 1994;198:59–70.
161. Dotti CG, Simons K. Polarized sorting of viral glycoproteins to the axon and dendrites of hippocampal neurons in culture. *Cell* 1990;62:63–72.
162. Dover A, Escobar J, Duenas A, et al. Pneumonia associated with measles. *JAMA* 1975;234:612–614.
163. Drayna D, Fields BN. Biochemical studies on the mechanisms of chemical and physical inactivation of reovirus. *J Gen Virol* 1982;63:161–170.
164. Drayna D, Fields BN. Genetic studies on the mechanism of chemical and physical inactivation of reovirus. *J Gen Virol* 1982;63:149–159.
165. Dreiding P, Staeheli P, Haller O. Interferon-induced protein Mx accumulates in nuclei of mouse cells expressing resistance to influenza viruses. *Virology* 1985;140:192–196.
166. Dreyer EB, Kaiser PK, Oggerman JT, et al. HIV-1 coat protein neurotoxicity prevented by calcium channel antagonists. *Science* 1990; 248:1419–1421.
167. Dryden KA, Wang G, Yeager M, et al. Early steps in reovirus infection are associated with dramatic changes in supramolecular structure and protein conformation: analysis of virions and subviral particles by cryoelectron microscopy and image reconstruction. *J Cell Biol* 1993; 122:1023–1041.
168. Dubensky TW, Freund R, Dawe CJ, et al. Polyomavirus replication in mice: influences of VP1 type and route of inoculation. *J Virol* 1991; 65:342–349.
169. Dubensky TW, Villarreal LP. The primary site of replication alters the eventual site of persistent infection by polyomavirus in mice. *J Virol* 1984;50:541–546.
170. Dubois-Dalcq M, Hooghe-Peters E, Lazzarini R. Antibody induced modulation of rhabdovirus infection of neurons in vitro. *J Neuropathol Exp Neurol* 1980;39:507–522.
171. Duke GM, Osorio JE, Palmenberg AC. Attenuation of Mengo virus through genetic engineering of the 5' noncoding poly(c) tract. *Nature* 1990;343:474–476.
172. Dunster LM, Schneider-Schaulies J, Loffler S, et al. Moesin: a cell membrane protein linked with susceptibility to measles virus infection. *Virology* 1994;198:265–274.
173. Dutko F, Oldstone MBA. Cytomegalovirus causes a latent infection in undifferentiated cells and is activated by induction of a cell differentiation. *J Exp Med* 1981;154:1636–1651.
174. Dveskler GS, Dieffenbach CW, Cardellichio CB, et al. Several members of the carcinoembryonic antigen-related glycoprotein family are functional receptors for the coronavirus mouse hepatitis virus A59. *J Virol* 1993;67:1–8.
175. Dworsky ME, Yow M, Stagno S, et al. Cytomegalovirus infection of breast milk and transmission in infancy. *Pediatrics* 1983;72:295–299.
176. Embretson J, Zupancic M, Ribas JL, et al. Massive covert infection of helper T lymphocytes and macrophages by HIV during the incubation period of AIDS. *Nature* 1993;362:359–362.
177. Emerson SU, Huang YK, Purcell RH. 2B and 2C mutations are pivotal but mutations throughout the genome of HAV contribute to adaptation to cell culture. *Virology* 1993;194:475–480.
178. Emerson SU, McCrill C, Rosenblum B, et al. Mutations responsible for adaptation of hepatitis A virus to efficient growth in cell culture. *J Virol* 1991;65:4882–4886.
179. Endres MJ, Griot C, Gonzalez-Scarano F, et al. Neuroattenuation of an avirulent *Bunyavirus* variant maps to the L RNA segment. *J Virol* 1991;65:5465–5470.
180. Eppstein DA, Marsh YV, Schreiber AB, et al. Epidermal growth factor receptor occupancy inhibits vaccinia virus infection. *Nature* 1985;318:663–665.
181. Equestre M, Genovese D, Cavalieri F, et al. Identification of a consistent pattern of mutations in neurovirulent variants derived from the Sabin vaccine strains of poliovirus type 2. *J Virol* 1991;65:2707–2710.
182. Esolen LM, Ward BJ, Moench TR, et al. Infection of monocytes during measles. *J Infect Dis* 1993;168:47–52.
183. Espejo RT, Lopez S, Arias C. Structural polypeptides of simian rotavirus SA11 and the effect of trypsin. *J Virol* 1981;37:156–160.
184. Estes MK, Graham DY, Mason BB. Proteolytic enhancement of rotavirus infectivity: molecular mechanisms. *J Virol* 1981;39:879–888.
185. Evans D, Dunn G, Minor P. Increased neurovirulence associated with a single nucleotide change in a non-coding region of the Sabin type 3 poliovaccine genome. *Nature* 1985;314:548–550.
186. Faber H. *The pathogenesis of poliomyelitis*. Springfield, IL: Charles C Thomas; 1955.
187. Fairbrother R, Hurst E. The pathogenesis of and propagation of the virus in experimental poliomyelitis. *J Pathol Bacteriol* 1930;33:17–45.
188. Fan H. Influences of the long terminal repeats on retrovirus pathogenicity. *Semin Virol* 1990;1:165–174.

189. Fantini J, Cook DG, Nathanson N, et al. Infection of colonic epithelial cell lines by type 1 human immunodeficiency virus is associated with cell surface expression of galactosylceramide, a potential alternative gp120 receptor. *Proc Natl Acad Sci USA* 1993;90:2700–2704.
190. Fazakerley JK, Parker SE, Bloom F, et al. The V5A13.1 envelope glycoprotein deletion mutant of mouse hepatitis virus type 4 is neuroattenuated by its reduced rate of spread in the central nervous system. *Virology* 1992;187:178–188.
191. Feigenblum D, Schneider RJ. Modification of eukaryotic initiation factor 4F during infection by influenza virus. *J Virol* 1993;67:3027–3035.
192. Fekadu M, Shaddock JH, Baer GM. Excretion of rabies virus in the saliva of dogs. *J Infect Dis* 1982;145:715–719.
193. Fenner F. Mousepox (infectious ectromelia of mice): a review. *J Immunol* 1949;63:341–373.
194. Fenner F. *The biology of animal viruses.* Orlando: Academic Press; 1968.
195. Feorina PM, Jaffe HW, Palmer E, et al. Transfusion-associated acquired immunodeficiency syndrome: evidence for persistence in blood donors. *N Engl J Med* 1985;312:1293–1296.
196. Ferro FE Jr, Kozak SL, Hoatlin ME, et al. Cell surface site for mitogenic interaction of erythropoietin receptors with the membrane glycoprotein encoded by Friend erythroleukemia virus. *J Biol Chem* 1992;268:5741–5747.
197. Fingeroth JD, Clabby ML, Strominger JD. Characterization of a T-lymphocyte Epstein-Barr virus/C3d receptor (CD21). *J Virol* 1988; 62:1442–1447.
198. Fisher AG, Ensoli B, Ivanoff L, et al. The sor gene of HIV-1 is required for efficient viral transmission in vitro. *Science* 1987;237:888–93.
199. Flamand A, Gagner JP, Morrison LA, et al. Penetration of the nervous system of suckling mice by mammalian reoviruses. *J Virol* 1991; 65:123–131.
200. Fleming J, Trousdale M, El-Zaatari F, et al. Pathogenicity of antigenic variants of murine coronavirus JHM selected with monoclonal antibodies. *J Virol* 1986;58:869–875.
201. Fleming P. Age-dependent and strain-related differences of virulence of Semliki Forest virus in mice. *J Gen Virol* 1977;37:93–105.
202. Flynn DC, Meyer WJ, MacKenzie JM, et al. A conformational change in Sindbis virus glycoproteins E1 and E2 is detected at the plasma membrane as a consequence of virus-cell interaction. *J Virol* 1990; 64:3643–3653.
203. Folks TM, Clouse KA, Justement J, et al. Tumor necrosis factor alpha induces expression of human immunodeficiency virus in a chronically infected T-cell clone. *Proc Natl Acad Sci USA* 1989;86:2365–2368.
204. Fournier J-G, Tardieu M, Lebon P, et al. Detection of measles virus RNA in lymphocytes from peripheral blood and brain perivascular infiltrates of patients with subacute sclerosing panencephalitis. *N Engl J Med* 1985;313:910–915.
205. Frade R, Barel M, Ehlin-Henriksson B, et al. gp140, the C3d receptor of human lymphocytes, is also the Epstein-Barr virus receptor. *Proc Natl Acad Sci USA* 1985;82:1490–1493.
206. Freistadt MS, Fleit HB, Wimmer E. Poliovirus receptor on human blood cells: a possible extraneural site of poliovirus replication. *Virology* 1993;195:798–803.
207. Fricks CE, Hogle JM. Cell-induced conformational change in poliovirus: externalization of the amino terminus of VP1 is responsible for liposome binding. *J Virol* 1990;64:1934–1945.
208. Friedman H, Macarek E, MacGregor RA, et al. Virus infection of endothelial cells. *J Infect Dis* 1981;143:266–273.
209. Friedman HM, Cohen GH, Eisenberg RJ, et al. Glycoprotein C of herpes simplex virus 1 acts as a receptor for the C3b complement component on infected cells. *Nature* 1984;309:633–635.
210. Fries LF, Friedman HM, Cohen GH, et al. Glycoprotein C of herpes simplex virus 1 is an inhibitor of the complement cascade. *J Immunol* 1986;137:1636–1641.
211. Fu J, Stein S, Rosenstein L, et al. Neurovirulence determinants of genetically engineered Theiler viruses. *Proc Natl Acad Sci USA* 1990; 87:4125–4129.
212. Fujimura FK, Deinuger PL, Friedmann T, et al. Mutation near the polyoma DNA replication origin permits productive infection of F9 embryonal carcinoma cells. *Cell* 1981;23:809–814.
213. Fujimura FK, Silbert PE, Eckhart W, et al. Polyoma virus infection of retinoic acid-induced differentiated teratocarcinoma cells. *J Virol* 1981;39:306–312.
214. Fujita K, Silver J, Peden K. Changes in both gp120 and gp41 can account for increased growth potential and expanded host range of human immunodeficiency virus type 1. *J Virol* 1992;66:4445–4451.
215. Fuller AO, Lee W-C. Herpes simplex virus type 1 entry through a cascade of virus-cell interactions requires different roles of gD and gH in penetration. *J Virol* 1992;66:5002–5012.
216. Fuller SD, von Bonsdorf C-H, Simons K. Vesicular stomatitis virus infects and matures only through the basolateral surface of the polarized epithelial cell line, MDCK. *Cell* 1984;38:65–77.
217. Funkhouser AW, Purcell RH, D'Hondt E, et al. Attenuated hepatitis A virus: genetic determinants of adaptation to growth in MRC-5 cells. *J Virol* 1994;68:148–157.
218. Garten W, Bosch FX, Linder D, et al. Proteolytic activation of the human influenza virus hemagglutinins: the structure of the cleavage site and the enzyme involved in cleavage. *Virology* 1981;115:361–374.
219. Gatmaitan B, Chason J, Lerner A. Augmentation of the virulence of murine Coxsackie virus B-3 myocardiopathy by exercise. *J Exp Med* 1970;131:1121–1136.
220. Gendelman HE, Narayan O, Kenedy-Stoskopf S, et al. Tropism of sheep lentiviruses for monocytes: susceptibility to infection and virus gene expression increase during maturation of monocytes to macrophages. *J Virol* 1986;58:67–74.
221. Gentsch J, Pacitti A. Differential interaction of reovirus type 3 with sialylated components on animal cells. *Virology* 1987;161:245–248.
222. Gerone PJ, Couch RB, Keeter GV, et al. Assessment of experimental and natural viral aerosols. *Bacteriol Rev* 1966;30:576–584.
223. Gething MJ, White JM, Waterfield MD. Purification of the fusion protein of Sendai virus: analysis of the NH_2-terminal sequence generated during precursor activation. *Proc Natl Acad Sci USA* 1978;75:2737–2740.
224. Gilbert MJ, Riddell SR, Li C-R, et al. Selective interference with class I major histocompatability complex presentation of the immediate-early protein following infection with human cytomegalovirus. *J Virol* 1993;67:3461–3469.
225. Gillet JP, Derer P, Tsiang H. Axonal transport of rabies virus in the central nervous system of the rat. *J Neuropathol Exp Neurol* 1986; 45:619–634.
226. Gius D, Grossman S, Bedell M, et al. Inducible and constitutive enhancer domains in the non-coding region of human papillomavirus type 18. *J Virol* 1988;62:665–672.
227. Golovkina TV, Chernonsky A, Dudley JP, et al. Transgenic mouse mammary tumor virus superantigen expression prevents viral infection. *Cell* 1992;69:637–645.
228. Gonzalez-Scarano F, Beaty FB, Sundin D, et al. Genetic determinants of the virulence and infectivity of La Crosse virus. *Microb Pathog* 1988;4:1–7.
229. Gonzalez-Scarano F, Jacoby D, Griot C, et al. Genetics, infectivity and virulence of California serogroup viruses. *Virus Res* 1992;24:123–135.
230. Gonzalez-Scarano F, Janssen R, Najjar J, et al. An avirulent G1 glycoprotein variant of LaCrosse bunyavirus with defective fusion function. *J Virol* 1985;54:757–763.
231. Gooding LR. Virus proteins that counteract host immune defenses. *Cell* 1992;71:5–7.
232. Gooding LR, Aquino L, Duerksen-Hughes P, et al. The E1B 19,000-molecular-weight protein of group C adenoviruses prevents tumor necrosis factor cytolysis of human cells but not of mouse cells. *J Virol* 1991;65:3083–3094.
233. Gooding LR, Ranheim TS, Tollefson AE, et al. The 10,400- and 14,500-dalton proteins encoded by region E3 of adenovirus function together to protect many but not all mouse cell lines against lysis by tumor necrosis factor. *J Virol* 1991;65:4114–4123.
234. Gooding LR, Wold WSM. Molecular mechanisms by which adenoviruses counteract antiviral immune defenses. *Crit Rev Immunol* 1990;10:53–71.
235. Goodpasture E. The axis-cylinders of peripheral nerves as portals of entry for the virus of herpes simplex in experimentally infected rabbits. *Am J Pathol* 1925;1:11–28.
236. Gougeon ML, Laurent-Crawford AG, Hovanessian AG, et al. Direct and indirect mechanisms mediating apoptosis during HIV infection: contribution to *in vivo* CD4 T cell depletion. *Semin Immunol* 1993;5:187–194.
237. Greber UF, Willetts M, Webster P, et al. Stepwise dismantling of adenovirus 2 during entry into cells. *Cell* 1993;75:477–486.
238. Greenlee J. Effect of host age on experimental K virus infection in mice. *Infect Immun* 1981;33:297–303.

239. Gregory CD, Dive C, Henderson S, et al. Activation of Epstein-Barr virus latent genes protects human B cells from death by apoptosis. *Nature* 1991;349:612–614.
240. Gregory CD, Murray RJ, Edwards CF, et al. Down regulation of cell adhesion molecules LFA-3 and ICAM-1 in Epstein-Barr virus positive Burkitt's lymphoma underlies tumor escape from virus-specific T cell surveillance. *J Exp Med* 1988;167:1811–1824.
241. Greve J, Davis G, Meyer A, et al. The major human rhinovirus receptor is ICAM-1. *Cell* 1989;56:839–847.
242. Griffin D, Mullinix J, Narayan O, et al. Age dependence of viral expression: comparative pathogenesis of two rodent-adapted strains of measles virus in mice. *Infect Immun* 1974;9:690–695.
243. Griffin DE. Role of the immune response in age-dependent resistance of mice to encephalitis due to Sindbis virus. *J Infect Dis* 1976; 133:456–464.
244. Griffin DE. Molecular pathogenesis of Sindbis virus encephalitis in experimental animals. *Adv Virus Res* 1989;36:255–271.
245. Grimley P, Friedman R. Arboviral infection of voluntary striated muscles. *J Infect Dis* 1970;122:45–52.
246. Griot C, Pekosz A, Lukac D, et al. Polygenic control of neuroinvasiveness in California serogroup bunyaviruses. *J Virol* 1993;67:3861–3867.
247. Groschel D, Koprowski H. Development of a virus-resistant inbred mouse strain for the study of innate resistance to arbo B viruses. *Arch Ges Virusforsch* 1965;17:379–391.
248. Grose C. Varicella-zoster virus infections, chickenpox (varicella), and shingles (zoster). In Glaser R, Gotlieb-Stematsky T, eds. *Human herpesvirus infections: clinical aspects*. New York: Dekker; 1982:85–150.
249. Grundy J, MacKenzie J, Stanley N. Influence of H-2 and non H-2 linked genes on resistance to murine cytomegalovirus infection. *Infect Immun* 1981;32:277–286.
250. Grundy JE, McKeating JA, Ward PJ, et al. Beta 2-microglobulin enhances the infectivity of cytomegalovirus and when bound to virus enables class I HLA molecules to be used as a virus receptor. *J Gen Virol* 1987;68:793–803.
251. Guetta E, Ron D, Tal J. Developmental-dependent replication of minute virus of mice in differentiated mouse testicular lines. *J Gen Virol* 1986;67:2549–2554.
252. Gurney M, Lee M, Ho D, et al. Functional interaction and partial homology between human immunodeficiency virus and neuroleukin. *Science* 1987;237:1047–1051.
253. Guyer B, O'Day DM, Hierholzer JC, et al. Epidemic keratoconjunctivitis: a community outbreak of mixed adenovirus type 8 and type 19 infection. *J Infect Dis* 1975;132:142–150.
254. Haase AT. Pathogenesis of lentiviruses infections. *Nature* 1986; 322:130–136.
255. Hahn CS, Dalrymple J, Strauss J, et al. Comparison of the virulent Asibi strain of yellow fever virus with the 17D vaccine strain derived from it. *Proc Natl Acad Sci USA* 1987;84:2019–2023.
256. Haller O, Arnheiter H, Gresser I, et al. Genetically determined, interferon-dependent resistance to influenza virus in mice. *J Exp Med* 1979;149:601–612.
257. Haller O, Arnheiter H, Lindenmann J, et al. Host gene influences sensitivity to interferon action selectively for influenza virus. *Nature* 1980;283:660–662.
258. Halstead SB. *In vivo* enhancement of dengue virus infection in rhesus monkeys by passively transferred antibody. *J Infect Dis* 1979; 140:527–533.
259. Hanecak R, Pattengale P, Fan H. Addition or substitution of simian virus 40 enhancer sequences into the Moloney murine leukemia virus (M-MuLV) long terminal repeat yields infectious M-MuLV with altered biological properties. *J Virol* 1988;62:2427–2436.
260. Hanham CA, Zhao F, Tignor GH. Evidence from the anti-idiotype network that the acetylcholine receptor is a rabies virus receptor. *J Virol* 1993;67:530–542.
261. Hanson B, Koprowski H, Baron S, et al. Interferon-mediated natural resistance of mice to arbo B virus infection. *Microbios* 1969;1B:51–68.
262. Hardy JL, Houk EJ, Kramer LD, et al. Intrinsic factors affecting vector competence of mosquitoes for arboviruses. *Annu Rev Entomol* 1983;28:229–262.
263. Harrington RD, Geballe AP. Cofactor requirement for human immunodeficiency virus type 1 entry into a CD4-expressing human cell line. *J Virol* 1993;67:5939–5947.
264. Harris SL, Frank I, Yee A, et al. Glycoprotein C of herpes simplex virus type 1 prevents complement-mediated cell lysis and virus neutralization. *J Infect Dis* 1990;162:331–337.
265. Haywood AM. Virus receptors: binding, adhesion strengthening, and changes in viral structure. *J Virol* 1994;68:1–5.
266. Heathcote J, Jenny M, Cameron C, et al. Hepatitis-B antigen in saliva and semen. *Lancet* 1974;1:71–73.
267. Helenius A, Morein B, Fries E, et al. Human (HLA-A and HLA-B) and murine (H2K and H2D) histocompatability antigens are cell surface receptors for Semiliki Forest virus. *Proc Natl Acad Sci USA* 1978;75: 3846–3850.
268. Hemmes HH, Winklerk KC, Kool SM. Virus survival as a seasonal factor in influenza and poliomyelitis. *Nature* 1960;188:430–431.
269. Henderson S, Rowe M, Gregory C, et al. Induction of bcl-2 expression by Epstein-Barr virus latent membrane protein 1 protects infected B cells from programmed cell death. *Cell* 1991;65:1107–1115.
270. Herndon RM, Johnson RT, Davis LE, et al. Ependymitis in mumps virus meningitis: electron microscopic studies of cerebrospinal fluid. *Arch Neurol* 1974;30:475–479.
271. Herold BC, WuDunn D, Soltys N, et al. Glycoprotein C of herpes simplex virus type 1 plays a principal role in the adsorption of virus to cells and infectivity. *J Virol* 1991;65:1090–1098.
272. Herr W, Clarke J. The SV40 enhancer is composed of multiple independent elements that can functionally compensate for one another. *Cell* 1986;45:461–470.
273. Herzog S, Frese K, Rott R. Studies on the genetic control of resistance of black hooded rats to Borna disease. *J Gen Virol* 1991;72:535–540.
274. Higa HH, Rogers GN, Paulson JC. Influenza virus hemagglutinins differentiate between receptor determinants bearing N-acetyl-N-glycollyl- and N-O-diacetylneuraminic acid groups. *Virology* 1985; 144:279–282.
275. Higgins PG. Enteroviral conjunctivitis and its neurological complications. *Arch Virol* 1982;73:91–101.
276. Hildreth JEK, Orentas R. Involvement of a leukocyte adhesion receptor (LFA-1) in HIV induced syncytium formation. *Science* 1989; 244:1075.
277. Hill T, Field H, Roome A. Intra-axonal location of herpes simplex virus particles. *J Gen Virol* 1972;15:253–255.
278. Hirsch MS, Zisman B, Allison AC. Macrophages and age-dependent resistance to herpes simplex virus in mice. *J Immunol* 1970;104: 1160–1165.
279. Hirsch RL, Griffin DE, Winkelstein JA. Natural immunity to Sindbis virus is influenced by host tissue sialic acid content. *Proc Natl Acad Sci USA* 1983;80:548–550.
280. Ho DD, Rota TR, Hirsch M. Infection of monocyte/macrophages by human T lymphotropic virus type III. *J Clin Invest* 1986;77:1712–1715.
281. Hoekstra D, Kok JW. Entry mechanisms of enveloped viruses: implications for fusion of intracellular membranes. *Biosci Rep* 1989;9:273–305.
282. Holland JJ. Receptor affinities as major determinants of enterovirus tissue tropism in humans. *Virology* 1961;15:312–326.
283. Holland JJ, McLaren LC, Syverton JJ. The mammalian cell-virus relationship. IV. Infection of naturally insusceptible cells with enterovirus nucleic acid. *J Exp Med* 1959;110:65–80.
284. Hollinger FB, Khan NC, Oefinger PE, et al. Post-transfusion hepatitis A. *JAMA* 1983;250:2313–2317.
285. Hoover JE, Strick PL. Multiple output channels in the basal ganglia. *Science* 1993;259:819–821.
286. Hopkins GB, Gostling JV, Hill I, et al. Hepatitis after tattooing: a fatal case. *Br Med J* 1973;3:210–211.
287. Horie H, Koike S, Kurata T, et al. Transgenic mice carrying the human poliovirus receptor: new animal model for study of poliovirus neurovirulence. *J Virol* 1994;68:681–688.
288. Horstmann D. Acute poliomyelitis: relation of physical activity at the time of onset to the course of the disease. *JAMA* 1950;142:236–241.
289. Howe H, Bodian D: *Neural mechanisms in poliomyelitis.* New York: Commonwealth Fund; 1942.
290. Hrdy D, Rubin D, Fields B. Molecular basis of reovirus neurovirulence: role of the M2 gene in avirulence. *Proc Natl Acad Sci USA* 1982;79: 1298–1302.
291. Hsu D-H, Malefyt RD, Fiorentino DF, et al. Expression of interleukin-10 activity by Epstein-Barr virus protein BCRF1. *Science* 1990;250: 830–832.
292. Huang C. Studies of virus factors as causes of inapparent infection in Japanese B encephalitis: virus strains, viremia, stability to heat and infective dosage. *Acta Virol* 1957;1:36–45.

293. Huang CH, Wong C. Relation of the peripheral multiplication of Japanese B encephalitis virus to the pathogenesis of infection in mice. *Acta Virol* 1963;7:322–330.
294. Hugin AW, Vacchio MS, Morse HC III. A virus-encoded "superantigen" in a retrovirus-induced immunodeficiency syndrome. *Science* 1991;252:427.
295. Hurst E, Melvin P, Thorpe J. The influence of cortisone, ACTH, thyroxine, and thiouracil on equine encephalomyelitis in the mouse and on its treatment with mapacrine. *J Comp Pathol* 1960;70:361–373.
296. Husseini R, Sweet C, Collie M, et al. Elevation of nasal viral levels by suppression of fever in ferrets infected with influenza viruses of differing virulence. *J Infect Dis* 1982;145:520–524.
297. Hwang SS, Boyle TJ, Lyerly HK, et al. Identification of the envelope V3 loop as the primary determinant of cell tropism in HIV-1. *Science* 1991;253:71–74.
298. Ibanez CE, Schrier R, Ghazal P, et al. Human cytomegalovirus productively infects differentiated macrophages. *J Virol* 1991;65: 6581–6588.
299. Imam I, Hammon W. Susceptibility of hamsters to peripherally inoculated Japanese B and St. Louis viruses following cortisone, x-ray, trauma. *Proc Soc Exp Biol Med* 1957;95:6–11.
300. Imberti L, Sottini A, Bettinardi A, et al. Selective depletion in HIV infection of T cells that bear specific V beta sequences. *Science* 1991;254:860–862.
301. Inada T, Mims CA. Mouse Ia antigens are receptors for lactate dehydrogenase virus. *Nature* 1984;309:59–61.
302. Isaacs SN, Kotwal GJ, Moss B. Vaccinia virus complement-control protein prevents antibody-dependent complement enhanced neutralization of infectivity and contributes to virulence. *Proc Natl Acad Sci USA* 1992;89:628–632.
303. Iwasaki Y, Clark H. Cell to cell transmission of virus in the central nervous system. II. Experimental rabies in the mouse. *Lab Invest* 1975;33:391–399.
304. Iwasaki Y, Liu D, Yamamoto T, et al. On the replication and spread of rabies virus in the human central nervous system. *J Neuropathol Exp Neurol* 1985;44:185–195.
305. Izumi KM, Stevens JG. Molecular and biological characterization of a herpes simplex virus type 1 (HSV-1) neuroinvasiveness gene. *J Exp Med* 1990;172:487–496.
306. Jacobson JG, Leib DA, Goldstein DJ, et al. A herpes simplex virus ribonucleotide reductase deletion mutant is defective for productive acute and reactivatable latent infections of mice and for replication in mouse cells. *Virology* 1989;173:276–283.
307. Jahrling PB, Eddy GA. Comparisons among members of the Venezuelan encephalitis virus complex using hydroxyapatite column chromatography. *Am J Epidemiol* 1977;106:408–417.
308. Jameel S, Siddiqui A. The human hepatitis B virus enhancer requires *trans*-acting cellular factors for activity. *Mol Cell Biol* 1986;6:710–715.
309. Janssen R, Nathanson N, Endres M, et al. Virulence of La Crosse virus is under polygenic control. *J Virol* 1986;59:1–7.
310. Janssen RS, Gonzalez-Scarano F, Nathanson N. Mechanisms of bunyavirus virulence. *Lab Invest* 1984;50:447–455.
311. Jenson A, Rabin E, Bentinck D, et al. Rabiesvirus neuronitis. *J Virol* 1969;3:265–269.
312. Johann SV, Gibbons JJ, O'Hara B. GLVR1, a receptor for gibbon ape leukemia virus, is homologous to a phosphate permease of *Neurospora crassa* and is expressed at high levels in brain and thymus. *J Virol* 1992;66:1635–1640.
313. John TJ, Christopher S, Abraham J. Neurological manifestitations of acute hemorrhagic conjunctivitis due to enterovirus. *Lancet* 1981;2: 1283–1284.
314. Johnson BJB, Kinney RM, Kost CL, et al. Molecular determinants of alphavirus neurovirulence: nucleotide and deduced protein sequence changes during attenuation of Venezuelan equine encephalitis virus. *J Gen Virol* 1986;67:1951–1960.
315. Johnson CJ, Anderson H, Spearman J, et al. Ear piercing and hepatitis: nonsterile instruments for ear piercing and the subsequent onset of viral hepatitis. *JAMA* 1974;227:1165.
316. Johnson RT. The pathogenesis of herpes virus encephalitis. II. A cellular basis for the development of resistance with age. *J Exp Med* 1964;120:359–374.
317. Johnson RT. The pathogenesis of herpes encephalitis. I. Virus pathway to the nervous system of suckling mice demonstrated by fluorescent antibody staining. *J Exp Med* 1964;119:343–356.
318. Johnson RT. Virus invasion of the central nervous system: a study of Sindbis virus infection in the mouse using fluorescent antibody. *Am J Pathol* 1965;46:929–943.
319. Johnson RT. Experimental rabies: studies of vulnerability and pathogenesis using fluorescent antibody staining. *J Neuropathol Exp Neurol* 1965;24:662–675.
320. Johnson RT. Selective vulnerability of neural cells to viral infection. *Brain* 1980;103:447–472.
321. Johnson RT. *Viral infections of the nervous system.* New York: Raven Press; 1982.
322. Johnson RT. Pathogenesis of La Crosse virus in mice. In: Calisher C, Thompson W, eds. *California serogroup viruses.* New York: Alan R Liss; 1983:139–144.
323. Johnson RT, Johnson KP. California encephalitis. II. Studies of experimental infection in the mouse. *J Neuropathol Exp Neurol* 1968;27: 390–400.
324. Johnson RT, McFarland HF, Levy SE. Age-dependent resistance to viral encephalitis: studies of infections due to Sindbis virus in mice. *J Infect Dis* 1972;125:257–262.
325. See ref. 324.
326. Johnson RT, Mims CA. Pathogenesis of viral infections of the nervous system. *N Engl J Med* 1968;278:23–30,84–92.
327. Johnston PB, Dubay JW, Hunter E. Truncations of the simian immunodeficiency virus transmembrane protein confer expanded virus host range by removing a block to virus entry into cells. *J Virol* 1993;67: 3077–3086.
328. Johnston RE, Smith JF. Selection for accelerated penetration in cell culture coselects for attenuated mutants of Venezuelan equine encephalitis virus. *Virology* 1988;162:437–443.
329. Jones JF, Fulginiti VA. Perinatal mumps infection. *J Pediatr* 1980;96:912–914.
330. Jubelt B, Narayan O, Johnson RT. Pathogenesis of human poliovirus infection in mice. II. Age-dependency of paralysis. *J Neuropathol Exp Neurol* 1980;39:149–159.
331. Kadota J, Cianciolo G, Snyderman R. A synthetic peptide homologous to retroviral transmembrane envelope protein depresses protein kinase C: a potential mechanism for immunosuppression. *Microbiol Immunol* 1991;35:443–459.
332. Karger A, Mettenleiter TC. Glycoproteins gII and gp50 play dominant roles in the biphasic attachment of pseudorabies virus. *Virology* 1993;194:654–663.
333. Kark J, Lubiush M, Rannon L. Cigarette smoking as a risk factor for epidemic A (H1N1) influenza in young men. *N Engl J Med* 1982;307:1042–1046.
334. Katagiri S, Aikawa S, Hinuma Y. Stepwise degradation of poliovirus capsid by alkaline treatment. *J Gen Virol* 1971;13:101–109.
335. Katze MG, Krug RM. Translational control in influenza virus-infected cells. *Enzyme* 1990;44:265–277.
336. Kauffman R, Wolf J, Finberg R, et al. The sigma 1 protein determines the extent of spread of reovirus from the gastrointestinal tract of mice. *Virology* 1983;124:403–410.
337. Kawamura N, Kohara M, Abe S, et al. Determinants in the 5'-noncoding region of poliovirus Sabin 1 RNA that influence the attenuation phenotype. *J Virol* 1989;63:1302–1309.
338. Kawanishi M. Epstein-Barr virus induces fragmentation of chromosomal DNA during lytic infection. *J Virol* 1994;67:7654–7658.
339. Kawano H, Rostapshov V, Rosen L, et al. Genetic determinants of dengue type 4 virus neurovirulence for mice. *J Virol* 1993;67:6567–6575.
340. Kemp MC, Hierholzer JC, Cabradilla CP, et al. The changing etiology of epidemic keratoconjunctivitis: antigenic and restriction enzyme analysis of adenovirus types 19 and 37 isolated over a 10-year period. *J Infect Dis* 1983;148:29–33.
341. Kenney S, Natarajan V, Strike D, et al. JC virus enhancer-promoter active in human brain cells. *Science* 1984;226:1337–1339.
342. Keroack M, Bassel-Duby R, Fields B. Genetic alterations in reovirus and their impact on host and environment. In: Fields BN, Martin M, Kamely D, eds. *Genetically altered viruses and the environment.* Cold Spring Harbor, NY: Cold Spring Harbor Laboratory Press; 1985:165–179.
343. Keroack M, Fields B. Viral shedding and transmission between hosts determined by reovirus L2 gene. *Science* 1986;232:1635–1638.
344. Kilham L. Mumps virus in human milk and in milk of an infected monkey. *JAMA* 1951;146:1231.

345. Kim JW, Closs EI, Albritton LM, et al. Transport of cationic amino acids by the mouse ecotropic retrovirus receptor. *Nature* 1991;352:725–728.
346. Kim S, Ikeuchi K, Groopman J, et al. Factors affecting cellular tropism of human immunodeficiency virus. *J Virol* 1990;64:5600–5604.
347. Kinman TG, de Wind N, Oei-Lie N, et al. Contribution of single genes within the unique short region of Aujesky's disease virus (suid herpesvirus type 1) to virulence, pathogenesis and immunogenicity. *J Gen Virol* 1992;73:243–251.
348. Kinney RM, Chang G-J, Tsuchiya KR, et al. Attenuation of Venezuelan equine encephalitis virus strain TC-83 is encoded by the 5'-noncoding region and the E2 envelope glycoprotein. *J Virol* 1993;67:1269–1277.
349. Kitamura T, Aso Y, Kuniyoshi N, et al. High incidence of urinary JC virus in nonimmunocompromised older patients. *J Infect Dis* 1990;161:1128–1133.
350. Klatzman D, Champagne E, Charmaret S, et al. T-lymphocyte T4 molecule behaves as the receptor for human retrovirus LAV. *Nature* 1984;312:767–768.
351. Klenk H-D, Rott R. Biology of influenza virus pathogenicity. *Adv Virus Res* 1988;34:247–281.
352. Knight V, Gilbert BE, Wilson SL. Airborne transmission of virus infections. In Fields B, Martin M, Kamely K, eds. *Genetically altered viruses and the environment.* Cold Spring Harbor, NY: Cold Spring Harbor Laboratory; 1985:73–94.
353. Knobler RL, Taylor BA, Woddell MK, et al. Host genetic control of mouse hepatitis virus type-4 (JHM strain) replication. *Exp Clin Immunogenet* 1984;1:217–222.
354. Koga Y, Sasaki M, Yoshida H, et al. Cytopathic effect determined by the amount of CD4 molecules in human cell lines expressing envelope glycoprotein of HIV. *J Immunol* 1990;144:94–102.
355. Kono R, Miyamura K, Tajiri E, et al. Neurological complications associated with acute hemorrhagic conjunctivitis virus infection and its serological confirmation. *J Infect Dis* 1974;129:590–593.
356. Kotwal GJ, Isaacs SN, Mckenzie R, et al. Inhibition of the complement cascade by the major secretory protein of vaccinia virus. *Science* 1990;250:827–830.
357. Kotwal GJ, Moss B. Vaccinia virus encodes a secretory polypeptide structurally related to complement control proteins. *Nature* 1988;335:176–181.
358. Kowalchyk K, Plagemann PGW. Cell surface receptors for lactate dehydrogenase elevating virus on subpopulation of macrophages. *Virus Res* 1985;2:211–229.
359. Koyanagi Y, O'Brien WA, Zhao JQ, et al. Cytokines alter production of HIV-1 from primary mononuclear phagocytes. *Science* 1988;241:1673–1675.
360. Kozak SL, Hoatlin ME, Ferro FJ Jr, et al. A Friend virus mutant that overcomes Fv-2rr host resistance encodes a small glycoprotein that dimerizes, is processed to cell surfaces, and specifically activates erythropoietin receptors. *J Virol* 1993;67:2611–2620.
361. Kristensson K. Implications of axoplasmic transport for the spread of viruses in the nervous system. In Weiss DG, Gorio A, eds. *Axoplasmic transport in physiology and pathology.* New York: Springer-Verlag; 1982:153–158.
362. Kristensson K, Ghetti B, Wisniewski H. Study on the propagation of herpes simplex virus (type 2) into the brain after intraocular injection. *Brain Res* 1974;69:189–201.
363. Kristensson K, Lycke E, Ryotta M, et al. Neuritic transport of herpes simples virus in rat sensory neurons *in vitro.* Effects of substances interacting with microtubular function and axonal flow (nocodazole, taxol and erythro-9-3(2-hydroxynonyl)adenine). *J Gen Virol* 1986;67:2023–2028.
364. Kristensson K, Lycke E, Sjostand J. Spread of herpes simplex virus in peripheral nerves. *Acta Neuropathol* 1971;17:44–53.
365. Kristensson K, Nennesmo I, Persson L, et al. Neuron to neuron transmission of herpes simplex virus. *J Neurol Sci* 1982;54:149–156.
366. Krug RM, Shaw M, Broni B, et al. Inhibition of influenza viral mRNA synthesis in cells expressing the interferon-induced Mx gene product. *J Virol* 1985;56:201–206.
367. Kucera P, Dolivo M, Coulon P, et al. Pathways of the early propagation of virulent and avirulent rabies strains from the eye to the brain. *J Virol* 1985;55:158–162.
368. Kuhn JE, Kramer MD, Willenbacher W, et al. Identification of herpes simplex virus type 1 glycoproteins interacting with the cell surface. *J Virol* 1990;64:2491–2497.
369. Kuhn RJ, Griffin DE, Zhang H, et al. Attenuation of Sindbis virus neurovirulence by using defined mutations in the nontranslated regions of the genome RNA. *J Virol* 1992;66:7121–7127.
370. Kurilla MG, Swaminathan S, Welsh RM, et al. Effects of virally expressed interleukin-10 on vaccinia virus infection in mice. *J Virol* 1993;67:7623–7628.
371. Kuypers HGJM, Ugolini G. Viruses as transneuronal tracers. *Trends Neurosci* 1990;13:72–75.
372. Lafay F, Coulon P, Astic L, et al. Spread of the CVS strain of rabies virus and of the avirulent mutant Av01 along the olfactory pathways of the mouse after intranasal inoculation. *Virology* 1991;183:320–330.
373. Laimins L, Khoury G, Gorman C, et al. Host-specific activation of gene expression by 72 base pair repeats of simian virus 40 and Moloney murine leukemia virus. *Proc Natl Acad Sci USA* 1982;79:6453–6457.
374. Lamb RA. Paramyxovirus fusion: a hypothesis for changes. *Virology* 1993;197:1–11.
375. Lamonica N, Almond JW, Racaniello VR. A mouse model for poliovirus neurovirulence identifies mutations that attenuate the virus for humans. *J Virol* 1987;61:2917–2920.
376. Lamonica N, Meriam C, Racaniello V. Mapping of sequences required for mouse neurovirulence of poliovirus type 2 Lansing. *J Virol* 1986;57:515–525.
377. Lang JC, Spandidos DA, Wilkie NM. Transcriptional regulation of a herpes simplex virus immediate early gene is mediated through an enhancer-type sequence. *EMBO J* 1984;3:389–395.
378. Langhoff E, Terwilliger EF, Bos J, et al. Replication of human immunodeficiency virus type 1 in primary dendritic cell cultures. *Proc Natl Acad Sci USA* 1991;88:7998–8002.
379. Lathey JL, Wiley CA, Verity MA, et al. Cultured human brain capillary endothelial cells are permissive for infection by human cytomegalovirus. *Virology* 1990;176:266–273.
380. Laurence J, Hodtsev AS, Posnett DP. Superantigen implicated in dependence of HIV-1 replication in T cells on TCR V beta expression. *Nature* 1992;358:2550–2559.
381. Laurent-Crawford AG, Krust GB, Muller S, et al. The cytopathic effect of HIV is associated with apoptosis. *Virology* 1991;185:829–839.
382. Lausch RN, Yeung KC, Miller JZ, et al. Nucleotide sequences responsible for the inability of a herpes simplex virus type 2 strain to grow in human lymphocytes are identical to those responsible for its inability to grow in mouse tissues following ocular infection. *Virology* 1990;176:319–328.
383. Lazarowitz SG, Choppin PW. Enhancement of the infectivity of influenza A and B viruses by proteolytic cleavage of the hemagglutinin polypeptide. *Virology* 1975;68:440–454.
384. Lee PWK, Hayes EC, Joklik WK. Protein sigma 1 is the reovirus cell attachment protein. *Virology* 1981;108:156–163.
385. Lee TG, Hovanessian A, Katze MG. Purification and partial characterization of a cellular inhibitor of the interferon induced 68,000 Mr protein kinase from influenza virus infected cells. *Proc Natl Acad Sci USA* 1990;87:6208–6212.
386. Legendre D, Rommelaere J. Terminal regions of the NS-1 protein of the parvovirus minute virus of mice are involved in cytotoxicity and promoter trans inhibition. *J Virol* 1992;66:5705–5713.
387. Lennette EH, Koprowski H. Influence of age on the susceptibility of mice to infection with certain neurotropic viruses. *J Immunol* 1944;49:175.
388. Lentz T. Rabies virus receptors. *Trends Neurosci* 1985;8:360–364.
389. Lentz TL, Burrage TG, Smith AL, et al. Is the acetylcholine receptor a rabies virus receptor? *Science* 1982;215:182–184.
390. Levine B, Griffin DE. Molecular analysis of neurovirulent strains of Sindbis virus that evolve during persistent infection of scid mice. *J Virol* 1993;67:6872–6875.
391. Levine B, Huang Q, Isaacs JT, et al. Conversion of lytic to persistent alphavirus infection by the bcl-2 cellular oncogene. *Nature* 1993;361:739–742.
392. Levy J, Greenspan D, Ferro F, et al. Frequent isolation of HHV-6 from saliva and high seroprevalence of the virus in the population. *Lancet* 1990;335:1047–1050.
393. Levy J, Joyner J, Borenfreund E. Mouse sperm can horizontally transmit type C viruses. *J Gen Virol* 1980;51:439–443.
394. Levy J, Shimabukuro J. Recovery of AIDS-associated retroviruses from patients with AIDS, related conditions, and clinically healthy individuals. *J Infect Dis* 1985;152:734–738.
395. Levy JA, Margaretten W, Nelson J. Detection of HIV in enterochro-

maffin cells in the rectal mucosa of an AIDS patient. *Am J Gastroenterol* 1989;84:787–789.
396. Li S, Schulman J, Itamura S, et al. Glycosylation of neuraminidase determines the neurovirulence of influenza A/WSN/33 virus. *J Virol* 1993;67:6667–6673.
397. Limentani AE, Elliot LM, Noah ND, et al. An outbreak of hepatitis B from tattooing. *Lancet* 1979;2:86–88.
398. Lipkin WI, Battenberg LF, Bloom FE, et al. Viral infection of neurons can depress neurotransmitter mRNA levels without histologic injury. *Brain Res* 1988;451:333–339.
399. Lippmann M, Yeates DB, Albert RE. Deposition, retention, and clearance of inhaled particles. *Br J Indust Med* 1980;37:337–362.
400. Lipton HL, Calenoff M, Bandyopadhyay PK, et al. The 5'-noncoding sequences from a less virulent Theiler's virus dramatically attenuate GDVII neurovirulence. *J Virol* 1991;65:4370–4377.
401. Lipton HL, Johnson RT. The pathogenesis of rat virus infections in the newborn hamster. *Lab Invest* 1972;27:508–513.
402. Little LM, Shadduck JA. Pathogenesis of rotavirus infection in mice. *Infect Immun* 1982;38:755–763.
403. Litwin V, Sandor M, Grose C. Cell surface expression of the varicella-zoster virus glycoproteins and Fc receptor. *Virology* 1990;178:263–272.
404. Liu C, Voth D, Rodina P, et al. A comparative study of the pathogenesis of western equine and eastern equine encephalomyelitis virus infections in mice by intracerebral and subcutaneous inoculations. *J Infect Dis* 1970;122:53–63.
405. Lobigs M, Usha R, Nestorowicz A, et al. Host cell selection of Murray Valley encephalitis virus variants altered at an RGD sequence in the envelope protein and in mouse virulence. *Virology* 1990;176:587–595.
406. Lodmell D. Genetic control of resistance to street rabies virus in mice. *J Exp Med* 1983;157:451–460.
407. Lonberg-Holm K. Attachment of animal viruses to cells: an introduction. In: Lonberg-Holm K, Philpson L, eds. *Virus receptors (part 2)*. London: Chapman and Hall; 1981.
408. Long GE, Rickman LS. Infectious complications of tattoos. *Clin Infect Dis* 1994;18:610–619.
409. Lopez C. Resistance to HSV-1 in the mouse is governed by two major independently segregating, non H-2 loci. *Immunogenetics* 1980;11:87–92.
410. Love A, Rydbeck R, Kristensson K, et al. Hemagglutinin-neuraminidase glycoprotein as a determinant of pathogenicity in mumps virus hamster encephalitis: analysis of mutants selected with monoclonal antibodies. *J Virol* 1985;53:67–74.
411. Lu Y-Y, Koga Y, Tanaka K, et al. Apoptosis induced in CD4-2D cells expressing gp160 of human immunodeficiency virus type 1. *J Virol* 1994; 68:390–399.
412. Lucia-Jandris P, Hooper JW, Fields BN. Reovirus M2 gene is associated with chromium release from mouse L cells. *J Virol* 1993;67: 5339–5345.
413. Lustig S, Jackson AC, Hahn CS, et al. Molecular basis of Sindbis virus neurovirulence in mice. *J Virol* 1988;62:2329–2336.
414. Lycke E, Kristensson K, Svennerhold B, et al. Uptake and transport of herpes simplex virus in neurites of rat dorsal root ganglia cells in culture. *J Gen Virol* 1984;65:55–64.
415. Lycke E, Tsiang H. Rabies virus infection of cultured rat sensory neurons. *J Virol* 1987;61:2733–2741.
416. Lynch WP, Portis JL. Murine retrovirus-induced spongiform encephalopathy: disease expression is dependent on postnatal development of the central nervous system. *J Virol* 1993;67:2601–2610.
417. Lyttle DJ, Fraser KM, Fleming SB, et al. Homologs of vascular endothelial growth factor are encoded by the poxvirus Orf virus. *J Virol* 1994;68:84–92.
418. Maas HJL, DeBoer GF, Groenendal JE. Age-related resistance to avian leukosis virus. 3. Infectious virus, neutralizing antibody, and tumors in chickens inoculated at various ages. *Avian Pathol* 1982;11: 309–327.
419. Macadam AJ, Pollard SR, Ferguson G, et al. The 5' noncoding region of the type 2 poliovirus vaccine strain contains determinants of attenuation and temperature sensitivity. *Virology* 1991;181:451–458.
420. Macadam AJ, Pollard SR, Ferguson G, et al. Genetic basis of attenuation of the Sabin type 2 vaccine strain of poliovirus in primates. *Virology* 1993;192:18–26.
421. Maciejewski JP, Bruening EE, Donahue RE, et al. Infection of mononuclear phagocytes with human cytomegalovirus. *Virology* 1993;195: 327–336.
422. MacQuarrie MB, Forghani B, Wolochow D. Hepatitis B transmitted by a human bite. *JAMA* 1974;230:723–724.
423. Maddon P, Dalgleish A, McDougal J, et al. The T4 gene encodes the AIDS virus receptor and is expressed in the immune system and brain. *Cell* 1986;47:333–348.
424. Maione R, Felsani A, Pozzi L, et al. Polyomavirus genome and polyomavirus enhancer-driven gene expression during myogenesis. *J Virol* 1989;63:4890–4897.
425. Malkova D. The role of the lymphatic system in experimental infection with tick-borne encephalitis. *Acta Virol* 1960;4:233–240.
426. Malkova D. Role of free cells in the lymph and blood vessels during viremia in animals experimentally infected with tickborne encephalitis virus. II. Virus bound in vivo to the cellular blood component in mice. *Acta Virol* 1967;11:317–320.
427. Malkova D, Frankova V. The lymphatic system in the development of experimental tick-borne encephalitis in mice. *Acta Virol* 1959;3:210–214.
428. Maniatis T, Goodbourn S, Fischer JA. Regulation of inducible and tissue-specific gene expression. *Science* 1987;236:1237–1245.
429. Markwell MAK, Svennerholm L, Paulson JC. Specific gangliosides function as host cell receptors for Sendai virus. *Proc Natl Acad Sci USA* 1981;78:5406–5410.
430. Marsh M, Helenius A. Virus entry into animal cells. *Adv Virus Res* 1989;36:107–151.
431. Marshall I. The influence of ambient temperature on the course of myxomatosis in rabbits. *J Hyg* 1959;57:484–497.
432. Martin X, Dolivo M. Neuronal and transneuronal tracing in the trigeminal system of the rat using the herpes virus suis. *Brain Res* 1983; 273:253–276.
433. Masuda M, Hoffman PM, Ruscetti SK. Viral determinants that control the neuropathogenicity of PVC-211 murine leukemia virus *in vivo* determine brain capillary endothelial cell tropism *in vitro*. *J Virol* 1993; 67:4580–4587.
434. Mathjis JM, Hing MC, Grierson J, et al. HIV infection of rectal mucosa. *Lancet* 1988;1:1111.
435. Matloubian M, Kolhekar SR, Somasundaram T, et al. Molecular determinants of macrophage tropism and viral persistence: importance of single amino acid changes in the polymerase and glycoprotein of lymphocytic choriomeningitis virus. *J Virol* 1993;67:7340–7349.
436. Matoba Y, Sherry BN, Fields BN, et al. Identification of the viral genes responsible for growth of strains of reovirus in cultured mouse heart cells. *J Clin Invest* 1991;87:1628–1633.
437. McCance D. Persistence of animal and human papovaviruses in renal and nervous tissues. In: Sever JL, Madden DL, eds. *Polyomaviruses and human neurological diseases*. New York: Alan R Liss; 1983.
438. McChesney MB, Fuginami RS, Lerchae NW, et al. Virus induced immunosuppression: infection of peripheral blood mononuclear cells and suppression of immunoglobulin synthesis during natural infection of rhesus monkeys. *J Infect Dis* 1989;159:757–760.
439. McClintock PR, Notkins AL. Viral receptors: expression, regulation and relationship to infectivity. In: Notkins AL, Oldstone MBA, eds. *Concepts in viral pathogenesis*. New York: Springer-Verlag; 1984: 97–101.
440. McClure JE. Cellular receptor for Epstein-Barr virus. *Prog Med Virol* 1992;39:116–138.
441. McDougal JS, Kennedy MS, Sligh JM, et al. Binding of HTLV III/LAV to T4+ T cells by a complex of the 110K viral protein and the T4 molecule. *Science* 1986;231:382–385.
442. McKeating JA, Wiley RL. Structure and function of the HIV envelope. *AIDS* 1989;3(Suppl 1):S35–S41.
443. Mckenzie R, Kotwal GJ, Moss B, et al. Regulation of complement activity by vaccinia virus complement-control protein. *J Infect Dis* 1992;166:1245–1250.
444. McKnight S, Tijan R. Transcriptional selectivity of viral genes in mammalian cells. *Cell* 1986;46:795–805.
445. McLean JH, Shipley MT, Bernstein DI. Golgi-like transneuronal retrograde labeling with CNS injections of herpes simplex virus type 1. *Brain Res Bull* 1989;22:867–881.
446. McNabb PC, Tomasi TB. Host defense mechanisms at mucosal surfaces. *Annu Rev Microbiol* 1981;35:477–496.
447. Mendelsohn C, Wimmer E, Racaniello V. Cellular receptor for poliovirus: molecular cloning, nucleotide sequence, and expression of a new member of the immunoglobulin superfamily. *Cell* 1989;56: 855–865.

448. Merill JE, Koyanagi Y, Chen ISY. Interleukin 1 and tumor necrosis factor alpha can be induced from mononuclear phagocytes by human immunodeficiency virus type 1 binding to the CD4 receptor. *J Virol* 1989;63:4404–4408.
449. Meyerhans A, Vartanian J-P, Hultgren C, et al. Restriction and enhancement of human immunodeficiency virus type 1 replication by modulation of intracellular deoxynucleoside triphosphate pools. *J Virol* 1994;68:535–540.
450. Micehelangeli F, Ruiz M-C, del Castillo JR, et al. Effect of rotavirus on intracellular calcium homeostasis in cultured cells. *Virology* 1991; 181:520–527.
451. Miller MA, Cloyd MW, Liebmann J, et al. Alterations in cell membrane permeability by the lentivirus lytic peptide (LLP-1) of HIV-1 transmembrane protein. *Virology* 1993;196:89–100.
452. Miller N, Hutt-Fletcher LM. Epstein-Barr virus enters B cells and epithelial cells by different routes. *J Virol* 1992;66:3409–3414.
453. Mims CA, White DO. *Viral pathogenesis and immunology*. Oxford: Blackwell Scientific Publications; 1984.
454. Minor PD, Dunn G. The effect of sequences in the 5'-non-coding region of poliovirus in the human gut. *J Gen Virol* 1988;69:1091–1096.
455. Mochizuki Y, Tashiro M, Homma M. Pneumopathogenicity in mice of a Sendai virus mutant, tsrev-58, is accompanied by *in vitro* activation with trypsin. *J Virol* 1988;62:3040–3042.
456. Monath TP, Cropp C, Harrison A. Mode of entry of a neurotropic arbovirus into the central nervous system. *Lab Invest* 1983;48:399–410.
457. Moore KW, Viera P, Fiorentino DF, et al. Homology of cytokine synthesis inhibitory factor (IL-10) to Epstein-Barr virus gene BCRF1. *Science* 1990;248:1230–1234.
458. Morales JA, Herzog S, Kompter C, et al. Axonal transport of Borna virus along olfactory pathways in spontaneously and experimentally infected rats. *Med Microbiol Immunol* 1988;177:51–68.
459. Morely D. The severe measles of West Africa. *Proc R Soc Med* 1969;57:846–849.
460. Morgan RA, Nussbaum O, Muenchau DD, et al. Analysis of the functional and host-range determining regions of the murine ecotropic and amphotropic retrovirus envelope proteins. *J Virol* 1993;67:4712–4721.
461. Mogensen S. Macrophages and age-dependent resistance to hepatitis induced by herpes simplex virus type 2. *Infect Immun* 1978;19:46–50.
462. Mogensen S. Role of macrophages in natural resistance to virus infection. *Microbiol Rev* 1979;43:1–26.
463. Mori K, Ringler DJ, Desrosiers RC. Restricted replication of simian immunodeficiency virus strain 239 in macrophages is determined by env but is not due to restricted entry. *J Virol* 1993;67:2807–2814.
464. Mori K, Ringler DJ, Kodama T, et al. Complex determinants of macrophage tropism in env of simian immunodeficiency virus. *J Virol* 1992;66:2067–2075.
465. Morrison LA, Fields BN. Parallel mechanisms in neuropathogenesis of enteric virus infections. *J Virol* 1991;65:2767–2772.
466. Morrison LA, Sidman RL, Fields BN. Direct spread of reoviruses from the intestinal lumen to the central nervous system through vagal autonomic fibers. *Proc Natl Acad Sci USA* 1991;88:3852–3856.
467. Moscufo N, Yafal AG, Rogove A, et al. A mutation in VP4 defines a new step in the late stages of cell entry by poliovirus. *J Virol* 1003; 67:5075–5078.
468. Moss EG, Racaniello VR. Host range determinants located on the interior of the poliovirus capsid. *EMBO J* 1991;10:1067–1074.
469. Mufson MA, Belshe RB, Horrigan TJ, et al. Cause of acute hemorrhagic cystitis in children. *Am J Dis Child* 1973;126:605–609.
470. Murphy FA. Rabies pathogenesis: brief review. *Arch Virol* 1977;54: 279–297.
471. Murphy FA, Bauer SP. Early street rabies virus infection in striated muscle and later progression to the central nervous system. *Intervirology* 1974;3:256–268.
472. Murphy FA, Harrison AK, Winn WC Jr, et al. Comparative pathogenesis of rabies and rabies-like viruses: viral infection and transit from inoculation site to the central nervous system. *Lab Invest* 1973;29:1–16.
473. Murphy FA, Taylor W, Mims C, et al. Pathogenesis of Ross River virus infection in mice. II. Muscle, heart, and brown fat lesions. *J Infect Dis* 1973;127:129–138.
474. Murphy FA, Whitfield SG, Sudia WD, et al. Interactions of vector with vertebrate pathogenic viruses. In: Maramorsch K, Shope RE, eds. *Invertebrate immunity*. Orlando: Academic Press; 1979:25.
475. Murray MG, Bradley J, Yang X-F, et al. Poliovirus host range is determined by a short amino acid sequence in neutralization antigenic site I. *Science* 1988;241:213–215.
476. Muster T, Subbarao EK, Enami M, et al. An influenza virus containing influenza B virus 5'- and 3'-noncoding regions on the neuraminidase gene is attenuated in mice. *Proc Natl Acad Sci USA* 1991;88:5177–5181.
477. Nakajima K, Martinez-Maza O, Hirano T, et al. Induction of Il-6 (B cell stimulatory factor-2/IFN-beta 2) production by HIV. *J Immunol* 1989;142:531–536.
478. Naniche D, Varior-Krisnan G, Cervoni F, et al. Human membrane cofactor protein (CD46) acts as a cellular receptor for measles virus. *J Virol* 1993;67:6025–6032.
479. Narayan O, Griffin D, Silverstein A. Slow virus infection: replication and mechanisms of persistence of Visna virus in sheep. *J Infect Dis* 1977;135:800–806.
480. Nathanson N. Pathogenesis. In: Monath TP, ed. *St. Louis encephalitis*. Washington, DC: American Public Health Association; 1980:201–236.
481. Nathanson N, Bodian D. Experimental poliomyelitis following intramuscular virus injection. I. The effect of neural block on a neurotropic and a pantropic strain. *Bull Johns Hopkins Hosp* 1961;108:308–319.
482. Nathanson N, Langmuir A. The Cutter incident. Poliomyelitis following formaldehyde-inactivated poliovirus vaccination in the United States during the spring of 1955. *Am J Hyg* 1963;78:16–81.
483. Nayak DP, Jabbar MA. Structural domains and organizational conformation involved in sorting and transport of influenza virus transmembrane proteins. *Annu Rev Microbiol* 1989;43:465–501.
484. Neidhardt H, Schroder CH, Kaerner HC. Herpes simplex virus type 1 glycoprotein E is not indispensible for viral infection. *J Virol* 1987; 61:600–603.
485. Neilan JG, Lu Z, Afonso CL, et al. An African Swine fever virus gene with similarity to the proto-oncogene bcl-2 and the Epstein-Barr virus gene BHRF1. *J Virol* 1993;67:4391–4394.
486. Nelson JA, Wiley CA, Reynolds-Kohler C, et al. Human immunodeficiency virus detected in bowel epithelium from patients with gastrointestinal symptoms. *Lancet* 1988;1:259–262.
487. Nemerow GR, Wolfert R, McNaughton ME, et al. Identification and characterization of the Epstein-Barr virus receptor on human B lymphocytes and its relationship to the C3d complement receptor (CR2). *J Virol* 1985;55:347–351.
488. Newhouse M, Sanchis J, Blenenstock J. Lung defense mechanisms. *N Engl J Med* 1976;295:990–8,1045–1052.
489. Newman JF, Rowlands DJ, Brown F. A physiochemical subgrouping of the mammalian picornaviruses. *J Gen Virol* 1973;18:171–180.
490. Nibert ML, Fields BN. A carboxy-terminal fragment of protein u1/u1C is present in infectious subviral particles of mammalian reoviruses and is proposed to have a role in penetration. *J Virol* 1992;66:6408–6418.
491. Nibert ML, Furlong DB, Fields BN. Mechanisms of viral pathogenesis. Distinct forms of reoviruses and their roles during replication in cells and host. *J Clin Invest* 1991;88:727–734.
492. Nicolau S, Mateiesco E. Septinevrites a virus rabique des rues. Preuves de la marche centrifuge du virus dans les nerfs peripheriques des lapins. *C R Acad Sci* 1928;186:1072–1074.
493. Nokta M, Easton D, Steinsland S, et al. Ca2+ responses in cytomegalovirus-infected fibroblasts of human origin. *Virology* 1574; 259:267.
494. Nomazaki Y, Kumaska T, Yano N, et al. Further study of acute hemorrhagic cystitis due to adenovirus type II. *N Engl J Med* 1973;289: 344–347.
495. Nomoto A, Omata T, Toyoda H, et al. Complete nucleotide sequence of the attenuated poliovirus Sabin 1 strain genome. *Proc Natl Acad Sci USA* 1982;79:5793–5797.
496. Noyes WF. Studies on the Shope rabbit papilloma virus. II. The location of infective virus in papillomas of the cottontail rabbit. *J Exp Med* 1959;109:423–428.
497. O'Brien WA, Koyanagi Y, Namazie A, et al. HIV-1 tropism for mononuclear phagocytes can be determined by regions of gp120 outside the CD4-binding domain. *Nature* 1990;348:69–73.
498. O'Neill H, Blanden R. Mechanisms determining innate resistance to ectromelia virus infection in C57BL mice. *Infect Immun* 1983;41:1391.
499. Oakes J, Gray W, Lausch R. Herpes simplex virus type 1 DNA sequences which direct spread of virus from the cornea to the central nervous system. *Virology* 1986;150:513–517.
500. Oda T, Akaike T, Hamamoto T, et al. Oxygen radicals in influenza-induced pathogenesis and treatment with polymer-conjugated SOD. *Science* 1989;244:974–976.
501. Ogata A, Nagashima K, Hall WW, et al. Japanese encephalitis virus

neurotropism is dependent on the degree of neuronal maturity. *J Virol* 1991;65:880–886.
502. Ohuchi M, Orlich M, Ohuchi R, et al. Mutations at the cleavage site of the hemagglutinin alter the pathogenicity of influenza virus A/chick/Penn/83 (H5N2). *Virology* 1989;168:274–280.
503. Oie M, Shida H, Ichihashi Y. The function of the vaccinia hemagglutinin in the proteolytic activation of infectivity. *Virology* 1990;176: 494–504.
504. Okazaki K, Matsuzaki T, Sugahara Y, et al. BHV-1 adsorption is mediated by the interaction of glycoprotein gIII with heparinlike moiety on the cell surface. *Virology* 1991;181:666–670.
505. Oldstone MBA, Ahmed R, Buchmeier M, et al. Perturbation of differentiated functions during viral infection *in vivo*. I. Relationship of lymphocytic choriomeningitis virus and host strains to growth hormone deficiency. *Virology* 1985;142:158–174.
506. Oldstone MBA, Tishon A, Dutko FJ, et al. Does the major histocompatability complex serve as a specific receptor for Semliki Forest virus? *J Virol* 1980;34:256–265.
507. Omary MB, Brenner DA, de Grandpre LY, et al. HIV-1 infection and expression in human colonic cells: infection and expression in CD4+ and CD4 cell lines. *AIDS* 1991;5:275–281.
508. Omata T, Kohara M, Kuge S, et al. Genetic analysis of the attenuation phenotype of poliovirus types. *J Virol* 1986;58:348–358.
509. Ondek B, Gloss L, Herr W. The SV40 enhancer contains two distinct levels of organisation. *Nature* 1988;333:40–45.
510. Ondek B, Shepard A, Herr W. Discrete elements within the SV40 enhancer region display different cell-specific enhancer activities. *EMBO J* 1987;6:1017–1025.
511. Ostapchuk P, Diffley JFX, Bruder JT, et al. Interaction of a nuclear factor with the polyoma virus enhancer region. *Proc Natl Acad Sci USA* 1986;83:8550–8554.
512. Ott D, Rein A. Basis for receptor specificity of nonecotropic murine leukemia virus surface glycoprotein gp70 SU. *J Virol* 1992;66: 4632–4638.
513. Overbeck P, Lai S-P, Van Quill K, et al. Tissue-specific expression in transgenic mice of a fused gene containing RSV terminal sequences. *Science* 1986;231:1574–1577.
514. Owens RJ, Dubay JW, Hurter G, et al. Human immunodeficiency virus envelope protein determines the site of virus release in polarized epithelial cells. *Proc Natl Acad Sci USA* 1991;88:3987–3991.
515. Palmiter R, Chen H, Messing A, et al. SV40 enhancer and large-T antigen are instrumental in development of choroid plexus tumors in transgenic mice. *Nature* 1985;316:457–460.
516. Pancic F, Carpenter DC, Caine PE. Role of infectious secretions in the transmission of rhinoviruses. *J Clin Microbiol* 1980;12:567.
517. Pantaleo G, Graziosi C, Fauci AS. The immunopathogenesis of human immunodeficiency virus infection. *N Engl J Med* 1993;328:327–335.
518. Paquette Y, Kay DG, Rassart E, et al. Substitution of the U3 long terminal repeat region of the neurotropic CAS-Br-E retrovirus affects its disease inducing potential. *J Virol* 1990;64:3742–3752.
519. Park BH, Lavi E, Blank KJ, et al. Intracerebral hemorrhages and syncytium formation induced by endothelial infection with a murine leukemia virus. *J Virol* 1993;67:6015–6024.
520. Parker R, Thompson R. The effect of external temperature on the course of infectious myxomatosis of rabbits. *J Exp Med* 1942;75:567–573.
521. Pathak S, Webb HE. Possible mechanisms for the transport of Semliki Forest virus into and within mouse brain: an electron microscopic study. *J Neurol Sci* 1974;23:175–184.
522. Paul R, Lee P. Glycophorin in the reovirus receptor on human erythrocytes. *Virology* 1987;159:94–101.
523. Paul RW, Choi AH, Lee PWK. The alpha-anomeric form of sialic acid is the minimal receptor determinant recognized by reovirus. *Virology* 1989;172:382–385.
524. Peck JL Jr, Sabin AB. Multiplication and spread of the virus of St. Louis encephalitis in mice with special emphasis on its fate in the alimentary tract. *J Exp Med* 1947;85:647–662.
525. Peiris JS, Porterfield JS. Antibody-dependent enhancement of plaque formation on cell lines of macrophage origin-a sensitive assay for antiviral antibody. *J Gen Virol* 1981;57:119–125.
526. Pence DF, Davis NL, Johnston RE. Antigenic and genetic characterization of Sindbis virus monoclonal antibody escape mutants which define a pathogenesis domain on glycoprotein E2. *Virology* 1990;175: 41–49.
527. Perlman S, Evans G, Afifi A. Effect of olfactory bulb ablation on spread of a neurotropic coronavirus into the mouse brain. *J Exp Med* 1990;172:1127–1132.
528. Perlman S, Jacobsen G, Afifi A. Spread of a neurotropic coronavirus into the CNS via the trigeminal and olfactory nerves. *Virology* 1989; 170:556–560.
529. Piccoli DA, Witzleben CL, Guico CJ, et al. Synergism between hepatic injuries and a nonhepatotropic reovirus in mice. Enhanced hepatic infection and cell death. *J Clin Invest* 1990;86:1038–1045.
530. Pitts O, Powers M, Bilello J, et al. Ultrastructural changes associated with retroviral replication in central nervous system capillary endothelial cells. *Lab Invest* 1987;56:401–409.
531. Poli G, Bressler P, Kinter A, et al. Interleukin 6 induces human immunodeficiency virus expression in infected monocytic cells alone and in synergy with tumor necrosis factor alpha by transcriptional and post-transcriptional mechanisms. *J Exp Med* 1990;172:151–158.
532. Poli G, Kinter A, Justement JS, et al. Tumor necrosis factor alpha functions in an autocrine manner in the induction of human immunodeficiency virus expression. *Proc Natl Acad Sci USA* 1990;87:782–785.
533. Polo JM, Davis NL, Rice CM, et al. Molecular analysis of Sindbis virus pathogenesis in neonatal mice using recombinants constructed *in vitro*. *J Virol* 1988;62:2124–2133.
534. Polo JM, Johnston RE. Attenuating mutations in glycoproteins E1 and E2 of Sindbis virus produce a highly attenuated strain when combined *in vitro*. *J Virol* 1990;64:4438–4444.
535. Pomerantz RJ, de la Monte SM, Donnegan SP, et al. Human immunodeficiency virus infection of the human cervix. *Ann Intern Med* 1988;108:321–327.
536. Pomerantz RJ, Feinberg MG, Andino R, et al. The long terminal repeat is not a major determinant of the cellular tropism of human immunodeficiency virus type 1. *J Virol* 1991;65:1041–1045.
537. Porter AG. Picornavirus nonstructural proteins: emerging roles in virus replication and inhibition of host cell functions. *J Virol* 1994;67: 6917–6921.
538. Portis JL, McAtee F, Hayes S. Horizontal transmission of murine retroviruses. *J Virol* 1987;61:1037–1044.
539. Portis JL, Perryman S, McAtee FJ. The R-U5-5' leader sequence of neurovirulent wild mouse retrovirus contains an element controlling the incubation period of neurodegenerative disease. *J Virol* 1991;65: 1877–1883.
540. Poterfield JS. Immunological enhancement and the pathogenesis of dengue hemorrhagic fever. *J Hyg* 1982;89:355–364.
541. Prehaud C, Coulon P, Lafay F, et al. Antigenic site II of the rabies virus glycoprotein: structure and role in viral virulence. *J Virol* 1988;62:1–7.
542. Prince AM, Szmuness W, Milian SJ, et al. A serologic study of cytomegalovirus infections associated with blood transfusions. *N Engl J Med* 1971;284:1125–1131.
543. Pritchard AE, Calenoff MA, Simpson S, et al. A single base deletion in the 5' noncoding region of Theiler's virus attenuates neurovirulence. *J Virol* 1992;66:1951–1958.
544. Purchase HG, Gilmour DG, Romero CH, et al. Postinfection genetic resistance to avian lymphoid leukosis resides in B target cell. *Nature* 1977;270:61–62.
545. Puri A, Booy FP, Doms RW, et al. Conformational changes and fusion activity of influenza virus hemagglutinin of the H2 and H3 subtypes. Effects of acid pretreatment. *J Virol* 1990;64:3824–3832.
546. Pyles RB, Sawtell NM, Thompson RL. Herpes simplex virus type 1 dUTPase mutants are attenuated for neurovirulence, neuroinvasiveness, and reactivation from latency. *J Virol* 1992;66:6706–6713.
547. Quinnan GV Jr, Manischewitz JF. Genetically determined resistance to lethal murine cytomegalovirus infection is mediated by interferon-dependent and -independent restriction of virus replication. *J Virol* 1987;61:1875–1881.
548. Rabin E, Jenson A, Melnick J. Herpes simplex virus in mice: electron microscopy of neural spread. *Science* 1968;162:126–129.
549. Racaniello VR. Poliovirus neurovirulence. *Adv Virus Res* 1988;34: 217–246.
550. Rakela J, Mosley JW. Fecal excretion of hepatitis A virus in humans. *J Infect Dis* 1977;135:933–938.
551. Raney AK, Milich DR, Easton AJ, et al. Differentiation-specific transcriptional regulation of the hepatitis B virus large surface antigen gene in human hepatoma cell lines. *J Virol* 1990;64:2360–2368.
552. Ranheim TS, Shisler J, Horton TM, et al. Characterization of mutants within the gene for adenovirus E3 14.7-kilodalton protein which prevents cytolysis by tumor necrosis factor. *J Virol* 1993;67:2159–2167.

553. Rao L, Debbas M, Sabbatini P, et al. The adenovirus E1A proteins induce apoptosis, which is inhibited by the E1B 19-kDa and Bcl-2 proteins. *Proc Natl Acad Sci USA* 1992;89:7742–7746.
554. Ratcliffe D, Migliorisi G, Cramer E. Translocation of influenza by migrating neutrophils. *Cell Mol Biol* 1992;38:63–70.
555. Razvi ES, Welsh RM. Programmed cell death of T lymphocytes during acute viral infection: a mechanism for virus-induced immune deficiency. *J Virol* 1993;67:5754–5765.
556. Reagan KJ, Wunner WH. Rabies virus interaction with various cell lines is independent of the acetylcholine receptor. *Arch Virol* 1985;84:277–282.
557. Reid H, Doherty P. Louping-ill encephalomyelitis in the sheep. I. The relationship of viremia and the antibody response to susceptibility. *J Comp Pathol* 1971;81:521–529.
558. Reinacher M, Bonin J, Narayan O, et al. Pathogenesis of neurovirulent influenza A virus infection in mice. *Lab Invest* 1983;49:686–692.
559. Reinarz AB, Broome MG, Sagik BP. Age resistance of mice to Sindbis virus infection: viral replication as a function of host age. *Infect Immun* 1971;3:268–273.
560. Ren R, Costantini F, Gorgacz EJ, et al. Transgenic mice expressing a human poliovirus receptor: a new model for poliomyelitis. *Cell* 1991;63:353–362.
561. Ren R, Moss EG, Racaniello VR. Identification of two determinants that attenuate vaccine-related type 2 poliovirus. *J Virol* 1991;65:1377–1382.
562. Ren R, Racaniello VR. Poliovirus spreads from muscle to the central nervous system by neural pathways. *J Infect Dis* 1992;166:747–752.
563. Ren R, Racaniello VR. Human poliovirus receptor gene expression and poliovirus tissue tropism in transgenic mice. *J Virol* 1992;66: 296–304.
564. Richardson CD, Scheid A, Choppin PW. Specific inhibition of paramyxovirus and myxovirus replication by oligopeptides with amino acid sequences similar to those at the N-termini of the F-2 or HA2 viral polypeptides. *Virology* 1980;105:205–222.
565. Rindler MJ, Ivanov IE, Plesken H, et al. Viral glycoproteins destined for apical or basolateral plasma membrane domains traverse the same Golgi apparatus during their intracellular transport in doubly infected Madin-Darby canine kidney cells. *J Cell Biol* 1984;98:1304–1319.
566. Ritter CD, Mulligan MJ, Lydy SL, et al. Cell fusion activity of the simian immunodeficiency virus envelope protein is modulated by the intracytoplasmic domain. *Virology* 1993;197:255–264.
567. Riviere Y, Ahmed R, Southern P, et al. Genetic mapping of lymphocytic choriomeningitis virus pathogenicity: virulence in guinea pigs is associated with the L RNA segment. *J Virol* 1985;55:704–709.
568. Riviere Y, Ahmed R, Southern P, et al. Perturbation of differentiated functions during viral infection *in vivo*. II. Viral reassortants map growth hormone defect to the S RNA of the lymphocytic choriomeningitis virus genome. *Virology* 1985;142:175–182.
569. Roberts J. Enhancement of the virulence of attenuated ectromelia in mice maintained in a cold environment. *Aust J Exp Biol Med Sci* 1964;42:657–666.
570. Roberts JA. Histopathogenesis of mousepox. II. Cutaneous infection. *Br J Exp Pathol* 1962;43:462–468.
571. Roberts JA. Growth of virulent and attenuated ectromelia virus in cultured macrophages from normal and ectromelia-immune mice. *J Immunol* 1964;92:837–842.
572. Roberts JA. Histopathogenesis of mousepox. III. Ectromelia virulence. *Br J Exp Pathol* 1964;44:465–472.
573. Roberts N. Temperature and host defense. *Microbiol Rev* 1979;43:241–259.
574. Robinson HL, Blais BM, Tsichlis PN, et al. At least two regions of the viral genome determine the oncogenic potential of avian leukosis viruses. *Proc Natl Acad Sci USA* 1982;79:1225–1229.
575. Robinson HL, Jensen L, Coffin JM. Sequences outside of the long terminal repeat determine the lymphomogenic potential of Rous-associated virus type 1. *J Virol* 1985;55:752–759.
576. Robinson HL, Reinsch SS, Shank PR. Sequences near the 5'-long terminal repeat of avian leukosis viruses determine the ability to induce osteopetrosis. *J Virol* 1986;59:45–49.
577. Rochford R, Campbell BA, Villarreal LP. A pancreas specificity results from the combination of polyomavirus and Moloney murine leukemia virus enhancer. *Proc Natl Acad Sci USA* 1987;84:449–453.
578. Rochford R, Moreno JP, Peake ML, et al. Enhancer dependence of polyomavirus persistence in mouse kidneys. *J Virol* 1992;66:3287–3297.
579. Rochford R, Villarreal LP. Polyomavirus DNA replication in the pancreas and in a transformed pancreas cell line has distinct enhancer requirements. *J Virol* 1991;65:2108–2112.
580. Rodriguez-Boulan E, Pendergast M. Polarized distribution of viral envelope proteins in the plasma membrane of infected epithelial cells. *Cell* 1980;20:45–54.
581. Rodriguez-Boulan E, Sabatini DD. Asymmetric building of viruses in epithelial monolayers: a model system for study of epithelial polarity. *Proc Natl Acad Sci USA* 1978;75:5071–5075.
582. Rosenstreich D, Weinblatt A, O'Brien A. Genetic control of resistance to infection in mice. *Crit Rev Immunol* 1982;3:263–330.
583. Rossen DR, Smith CW, Laughter AH, et al. HIV-1 stimulated expression of CD11/CD18 integrins and ICAM-1: a possible mechanism for the extravascular dissemination of HIV-1 infected cells. *Trans Assoc Am Physicians* 1989;102:117–130.
584. Roth MG, Compans RW, Giusti L, et al. Influenza virus hemagglutinin expression is polarized in cells infected with recombinant SV40 viruses carrying cloned hemagglutinin DNA. *Cell* 1983;33:435–442.
585. Roth MG, Gunderson GD, Patil N, et al. The large external domain is sufficient for the correct sorting of secreted chimeric influenza virus hemagglutinins in polarized monkey kidney cells. *J Cell Biol* 1987;104:848–860.
586. Roth MG, Srinivas RV, Compans RW. Basolateral maturation of retroviruses in polarized epithelial cells. *J Virol* 1983;45:1065–1073.
587. Rother RP, Rollins SA, Fodor WL, et al. Inhibitor of complement-mediated cytolysis by the terminal complement inhibitor of *Herpesvirus saimiri*. *J Virol* 1994;68:730–737.
588. Routes JM, Metz BA, Cook JL. Endogenous expression of E1A in human cells enhances the effect of adenovirus E3 on class I major histocompatibility complex antigen expression. *J Virol* 1993;67:3176–3181.
589. Roux E. Notes de laboratoire sur la presence du virus rabique dans les nerfs. *Ann Inst Pasteur* 1888;2:18–27.
590. Rozhon E, Gensemer P, Shope R, et al. Attenuation of virulence of a bunyavirus involving an L RNA defect and isolation of LAC/SSH/LAC and LAC/SSH/SSH reassortants. *Virology* 1981;111:125–138.
591. Rubenstein R, Price RW. Early inhibition of acetylcholinesterase and choline acetyltransferase activity in herpes simplex virus type 1 infection of PC12 cells. *J Neurochem* 1984;42:142–150.
592. Rubenstein R, Price RW, Joh T. Alteration of tyrosine hydroxylase activity in PC12 cells infected with herpes simplex virus type 1. *Arch Virol* 1985;83:65–82.
593. Rubin DH, Fields BN. Molecular basis of reovirus virulence: role of the M2 gene. *J Exp Med* 1980;152:853–868.
594. Rubin DH, Morrison AH, Witzelben CL, et al. Site of reovirus replication in the liver is determined by the type of hepatocellular insult. *J Virol* 1990;64:4593–4597.
595. Rubin SA, Waltrip RW, Bautista JR, et al. Borna disease virus in mice: host-specific differences in disease expression. *J Virol* 1993;67:548–552.
596. Ruegg CL, Monell CR, Strand M. Inhibition of lymphoproliferation by a synthetic peptide with sequence identity to gp41 human immunodeficiency virus type 1. *J Virol* 1989;63:3257–3260.
597. Russell DL, Dalrymple JM, Johnston RE. Sindbis virus mutations which coordinately affect glycoprotein processing, penetration, and virulence in mice. *J Virol* 1989;53:1619–1629.
598. Sabin A. The nature and rate of centripetal progression of certain neurotropic viruses along peripheral nerves. *Am J Pathol* 1937;13:615–617.
599. Sabin A. Genetic factors affecting susceptibility and resistance to virus diseases of the nervous system. *Res Publ Assoc Res Nerv Ment Dis* 1954;33:57.
600. Sabin A, Olitsky P. Influence of host factors on neuroinvasiveness of vesicular stomatitis virus. *J Exp Med* 1937;66:15–34,35–57;67:201–208, 229–249.
601. Sabin A, Olitsky P. Pathological evidence of axonal and transsynaptic progression of vesicular stomatitis and eastern equine encephalomyelitis viruses. *Am J Pathol* 1937;13:615.
602. Sabin AB. Pathogenesis of poliomyelitis. Reappraisal in the light of new data. *Science* 1956;123:1151–1157.
603. Saif LJ. Comparative aspects of enteric viral infections. In Saif LJ, Theil KW, eds. *Viral diarrheas of man and animal.* Boca Raton: CRC Press; 1990:9–31.
604. Sakai K, Xiaoyue MA, Gordienko I, et al. Recombinational analysis of a natural noncytopathic human immunodeficiency virus type 1 (HIV-1) isolate: role of the *vif* gene in HIV-1 infection kinetics and cytopathicity. *J Virol* 1991;65:5765–5773.

605. Sangster MY, Shellam GR. Genetically controlled resistance to flaviviruses within the house mouse complex of species. *Curr Top Microbiol Immunol* 1986;127:313–318.
606. Sattentau QJ, Weiss RA. The CD4 antigen: physiologic ligand and HIV receptor. *Cell* 1988;52:631–633.
607. Sawyer W, Meyer K, Eaton M, et al. Jaundice in army personnel in the western region of the United States and its relation to vaccination against yellow fever. *Am J Hyg* 1944;39:337–430.
608. Scheid A, Choppin PW. Proteolytic cleavage and viral pathogenesis. In Notkins AL, Oldstone MBA, eds. *Concepts in viral pathogenesis.* New York: Springer-Verlag; 1984:26–31.
609. Schell K. Studies in the innate resistance of mice to infection with mousepox. I. Resistance and antibody protection. *Aust J Exp Biol Med Sci* 1960;38:271–288.
610. Schlegel R, Tralka S, Willingham MC, et al. Inhibition of VSV binding and infectivity of phosphatidylserine. Is phosphatidylserine a VSV binding site? *Cell* 1983;32:639–646.
611. Schmidtmayerova H, Bolmont C, Baghdiguian S, et al. Distinctive pattern of infection and replication of HIV 1 strains in blood-derived macrophages. *Virology* 1992;190:124–133.
612. Scholtissek C, Rott R, Orlich M, et al. Correlation of pathogenicity and gene constellation of an influenza A virus (fowl plague): I. Exchange of a single gene. *Virology* 1977;81:74–80.
613. Schulman JL. Experimental transmission of influenza virus infection in mice. IV. Relationship of transmissibility of different strains of virus and recovery of airborne virus in the environment of effector mice. *J Exp Med* 1967;125:479–488.
614. Scott RM, Snitbhan R, Bancroft WH, et al. Experimental transmission of hepatitis B virus by semen and saliva. *J Infect Dis* 1980;142:67–71.
615. Seif I, Coulon P, Rollin PE, et al. Rabies virulence: effect on pathogenicity and sequence characterization of rabies virus mutations affecting antigenic site III of the glycoprotein. *J Virol* 1985;52:926–934.
616. Sellers M. Studies on the entry of viruses into the central nervous system of mice via the circulation. Differential effects of vasoactive amines and CO on virus infectivity. *J Exp Med* 1969;129:719–746.
617. Serfling E, Jasin M, Shaffner W. Enhancers and eukaryotic gene transcription. *Trends Genet* 1985;1:224–230.
618. Shank PR, Schatz PJ, Jensen LM, et al. Sequences in the gag-pol-5' env region of avian leukosis viruses confer the ability to induce osteopetrosis. *Virology* 1985;145:94–104.
619. Shankar V, Dietzschold B, Koprowski H. Direct entry of rabies virus into the central nervous system without prior local replication. *J Virol* 1991;65:2736–2738.
620. Shaul Y, Rutter WJ, Laub O. A human hepatitis B viral enhancer element. *EMBO J* 1985;5:1967–1971.
621. Shellam GR, Flexman JP. Genetically determined resistance to murine cytomegalovirus and herpes simplex virus in newborn mice. *J Virol* 1986;58:152–156.
622. Shepley M, Sherry B, Weiner HL. Monoclonal antibody identification of a 100kDa membrane protein in HeLa cells and human spinal cord involved in poliovirus attachment. *Proc Natl Acad Sci USA* 1988;85:7743–7747.
623. Sherry B, Fields BN. The reovirus M1 gene, encoding a viral core protein, is associated with the myocarditic phenotype of a reovirus variant. *J Virol* 1989;63:4850–4856.
624. Shieh MT, WuDunn D, Montgomery RI, et al. Cell surface receptors for herpes simplex virus are heparin sulfate proteoglycans. *J Cell Biol* 1992;116:1273–1281.
625. Shikova E, Lin Y-C, Saha K, et al. Correlation of specific virus-astrocyte interactions and cytopathic effects induced by ts1, a neurovirulent mutant of Moloney murine leukemia virus. *J Virol* 1993;67: 1137–1147.
626. Shioda T, Levy JA, Cheng-Meyer C. Macrophage and Tcell-line tropisms of HIV-1 are determined by specific regions of the envelope gp120 gene. *Nature* 1991;349:167–169.
627. Shope RE, Hurst EW. Infectious papillomatosis of rabbits. *J Exp Med* 1933;58:607–624.
628. Shope RE, Rozhon E, Bishop D. Role of the middle-sized bunyavirus RNA segment in mouse virulence. *Virology* 1981;114:273–276.
629. Sicinski P, Rowinski J, Warchol JB, et al. Poliovirus type 1 enters the human host through intestinal M cells. *Gastroenterology* 1990;98:56–58.
630. Skehel JJ, Bayley PM, Brown EB, et al. Changes in the conformation of influenza virus hemagglutinin at the pH optimum of virus mediated membrane fusion. *Proc Natl Acad Sci USA* 1982;79:968–972.
631. Skinner MA, Racaniello VR, Dunn G, et al. New model for the secondary structure of the 5' noncoding RNA of poliovirus is supported by biochemical and genetic data that also show that RNA secondary structure is important in neurovirulence. *J Mol Biol* 1989;207:379–392.
632. Small J, Scangos G, Cork L, et al. The early region of human papovavirus JC induces dysmyelination in transgenic mice. *Cell* 1986;46:13–18.
633. Snyder MH, Buckler-White AJ, London WT, et al. The avian influenza nucleoprotein gene and a specific constellation of avian and human virus polymerase genes each specify attenuation of avian-human influenza A/Pintail/79 reassortant viruses for monkeys. *J Virol* 1987;61: 2857–2863.
634. Spear PG. Entry of alphaherpesviruses into cells. *Semin Virol* 1993;4: 167–180.
635. Sprecher E, Becker Y. Herpes simplex virus type 1 pathogenicity in footpad and ear skin depends on Langerhans cell density, mouse genetics, and virus strain. *J Virol* 1987;61:2515–2522.
636. Spriggs DR, Bronson RT, Fields BN. Hemagglutinin variants of reovirus type 3 have altered central nervous system tropism. *Science* 1983;220: 505–507.
637. Staeheli P, Grob R, Meier E, et al. Influenza virus-susceptible mice carry MX genes with a large deletion or a nonsense mutation. *Mol Cell Biol* 1988;8:4518–4523.
638. Staeheli P, Haller O, Boll W, et al. Mx protein: constitutive expression in 3T3 cells transformed with cloned Mx cDNA confers selective resistance to influenza virus. *Cell* 1986;44:147–158.
639. Staeheli P, Pravtcheva D, Lundin LG, et al. Interferon-regulated influenza virus resistance gene *Mx* is localized on mouse chromosome 16. *J Virol* 1986;58:967–969.
640. Stagno S, Reynolds DW, Pass RF, et al. Breast milk and the risk of cytomegalovirus infection. *N Engl J Med* 1980;302:1073–1076.
641. Stanberry LR. Capsaicin interferes with the centrifugal spread of virus in primary and recurrent genital herpes simplex virus infections. *J Infect Dis* 1990;162:29–34.
642. Standish A, Enquist LW, Schwaber JS. Innervation of the heart and its central medullary origin defined by viral tracing. *Science* 1994;263: 232–234.
643. Stanway G, Hughes PJ, Mountford RC, et al. Comparison of the complete nucleotide sequences of the genomes of the neurovirulent poliovirus P3/Leon/37 and its attenuated Sabin vaccine derivative P3/Leon/12a1b. *Proc Natl Acad Sci USA* 1984;81:1539–1543.
644. Staunton DE, Merluzzi VJ, Rothlein R, et al. A cell adhesion molecule, ICAM-1, is the major surface receptor for rhinoviruses. *Cell* 1989; 56:849–853.
645. Stegmann T, White J, Helenius A. Intermediates in influenza induced membrane fusion. *EMBO J* 1990;13:4231–4241.
646. Steinberg BM, Auborn KJ, Bransma JL, et al. Tissue site-specific enhancer function of the upstream regulatory region of human papillomavirus type II in cultured keratinocytes. *J Virol* 1989;63:957–960.
647. Stevens JG. Herpes simplex virus: neuroinvasiveness, neurovirulence, and latency. *Semin Neurosci* 1991;3:141–147.
648. Stewart JP, Rooney CM. The interleukin-10 homolog encoded by Epstein-Barr virus enhances the reactivation of virus-specific cytotoxic T cell and HLA-unrestricted killer cell responses. *Virology* 1992; 191:773–782.
649. Stone RS, Shope RE, Moore DH. Electron microscope study of the development of the papilloma virus in the skin of the rabbit. *J Exp Med* 1959;10:543–546.
650. Storz J, Rott R, Kaluza G. Enhancement of plaque formation and cell fusion of an enteropathogenic coronavirus by trypsin treatment. *Infect Immun* 1981;31:1214–1222.
651. Stott EJ, Killington RA. Rhinoviruses. *Annu Rev Microbiol* 1972;26: 503–524.
652. Strack AM, Loewy AD. Pseudorabies virus: a highly specific transneuronal cell body marker in the sympathetic nervous system. *J Neurosci* 1990;10:2139–2147.
653. Sugiura A, Ueda M. Neurovirulence of influenza virus in mice. I. Neurovirulence of recombinants between virulent and avirulent strains. *Virology* 1979;101:440–449.
654. Sullivan JL, Barry DW, Lucas SJ, et al. Measles virus infection of human mononuclear cells. Acute infection of peripheral blood lymphocytes and monocytes. *J Exp Med* 1975;142:773–784.
655. Summer BA, Griesen HA, Apper MG. Possible initiation of viral encephalomyelitis in dogs by migrating lymphocytes infected with distemper. *Lancet* 1978;1:187–189.
656. Superti F, Donelli G. Gangliosides as binding sites in SA-11 rotavirus

infection of LLC-MK2 cells. *J Gen Virol* 1991;72:2467–2474.
657. Superti F, Hauttecoeur B, Morelec MJ, et al. Involvement of gangliosides in rabies virus infection. *J Gen Virol* 1986;67:47–56.
658. Svitkin Y, Pestova T, Maslova S, et al. Point mutations modify the response of poliovirus RNA to a translation initiation factor: a comparison of neurovirulent and attenuated strains. *Virology* 1988;166:394–404.
659. Szurek PF, Floyd E, Yuen PH, et al. Site-directed mutagenesis of the codon for Ile-25 in gPr80 alters the neurovirulence of ts1, a mutant of Moloney murine leukemia virus TB. *J Virol* 1990;64:5241–5249.
660. Takeuchi Y, Akutsu M, Murayama K, et al. Host range mutant of human immunodeficiency virus type 1: modification of cell tropism by a single point mutation at the neutralization epitope in the env gene. *J Virol* 1991;65:1710–1718.
661. Tan X, Pearce-Pratt R, Phillips DM. Productive infection of a cervical epithelial cell line with human immunodeficiency virus: implications for sexual transmission. *J Virol* 1993;67:6447–6452.
662. Tang D, Strong JE, Lee PWK. Recognition of the epidermal growth factor receptor by reovirus. *Virology* 1993;197:412–414.
663. Tardy-Panit M, Blondel B, Martin A, et al. A mutation in the RNA polymerase of poliovirus contributes to attenuation in mice. *J Virol* 1993;67:4630–4638.
664. Tashiro M, Homma M. Pneumotropism of Sendai virus in relation to protease-mediated activation in mouse lungs. *Infect Immun* 1983;39:879–888.
665. Tashiro M, Klenk H-D, Rott R. Inhibitory effect of a protease inhibitor, leupeptin, on the development of influenza pneumonia, mediated by a concomitant bacteria. *J Gen Virol* 1987;68:2039–2041.
666. Tashiro MJ, Seto JT, Choosakul S, et al. Budding site of Sendai virus in polarized epithelial cells is one of the determinants for tropism and pathogenicity in mice. *Virology* 1992;187:413–422.
667. Tatem JM, Weeks-Levy C, Georgiu A, et al. A mutation present in the amino terminus of Sabin 3 poliovirus VP1 protein is attenuating. *J Virol* 1992;66:3194–3197.
668. Tattersall P. Replication of the parvovirus MVM. I. Dependence of virus multiplication and plaque formation on cell growth. *J Virol* 1972;10:586–590.
669. Taylor-Wiedeman J, Sissons JGP, Borysiewicz LK, et al. Monocytes are a major site of persistence of human cytomegalovirus in peripheral blood mononuclear cells. *J Gen Virol* 1991;72:2059–2064.
670. Terai C, Kornbluth RS, Pauza CD, et al. Apoptosis as a mechanism of cell death in cultured T lymhoblasts acutely infected with HIV-1. *J Clin Invest* 1991;87:1710–1715.
671. Thomas DC, Brewer CB, Roth MG. Vesicular stomatitis virus glycoprotein contains a dominant cytoplasmic sorting signal critically dependent upon a tyrosine. *J Biol Chem* 1993;268:3313–3320.
672. Thompson RL, Cook M, Devi-Rao G, et al. Functional and molecular analysis of the avirulent wild-type herpes simplex virus type 1 strain KOS. *J Virol* 1986;58:203–211.
673. Threadgill DS, Steagall WK, Flaherty MT, et al. Characterization of equine infectious anemia virus dUTPase: growth properties of dUTPase-deficient mutant. *J Virol* 1993;67:2592–2600.
674. Tian P, Hu Y, Schilling WP, et al. The nonstructural glycoprotein of rotavirus affects intracellular calcium levels. *J Virol* 1994;68:251–257.
675. Tollefson AE, Stewart AR, Yei S, et al. The 10,400 and 14,500-dalton proteins encoded by region E3 of adenovirus form a complex and function together to down-regulate the epidermal growth factor receptor. *J Virol* 1991;65:3095–3105.
676. Tomassini JE, Graham D, Dewitt CM, et al. cDNA cloning reveals that the major group rhinovirus receptor on HeLa cells is intercellular adhesion molecule 1. *Proc Natl Acad Sci USA* 1989;86:4907–4911.
677. Tongoni A, Cattaneo R, Serfling E, et al. A novel expression selection approach allows precise mapping of the hepatitis B virus enhancer. *Nucleic Acids Res* 1985;13:7457–7472.
678. Tosolini F, Mims C. Effect of murine strain and viral strain on the pathogenesis of lymphocytic choriomeningitis infection and a study of footpad responses. *J Infect Dis* 1971;123:134–144.
679. Tseng RW, Fujimura FK. Multiple domains in the polyomavirus B enhancer are required for productive infection of F9 embryonal carcinoma cells. *J Virol* 1988;62:2890–2895.
680. Tsiang H. Evidence for intraaxonal transport of fixed and street rabies virus. *J Neuropathol Exp Neurol* 1979;38:286–299.
681. Tsiang H. Pathophysiology of rabies virus infection of the nervous system. *Adv Virus Res* 1993;42:375–412.
682. Tsiang H, Ceccaldi PE, Lycke E. Rabies virus infection and transport in human sensory dorsal root ganglia neurons. *J Gen Virol* 1991;72:1191–1194.
683. Tsiang H, Lycke E, Ceccaldi PE, et al. The anterograde transport of rabies virus in rat sensory dorsal root ganglia neurons. *J Gen Virol* 1989;70:2075–2085.
684. Tsichlis PN, Bear SE. Infection by mink cell focus-forming viruses confers interleukin 2 (IL-2) independence on an IL-2 dependent rat T-cell lymphoma line. *Proc Natl Acad Sci USA* 1991;88:4611–4615.
685. Tucker PC, Griffin DE. The mechanism of altered Sindbis virus neurovirulence associated with a single-amino-acid change in the E2 glycoprotein. *J Virol* 1991;65:1551–1557.
686. Tucker PC, Strauss EG, Kuhn RJ, et al. Viral determinants of age-dependent virulence of Sindbis virus for mice. *J Virol* 1993;67:4605–4610.
687. Tucker SP, Compans RW. Virus infection of polarized epithelial cells. *Adv Virus Res* 1992;42:187–247.
688. Tucker SP, Thornton CL, Wimmer E, et al. Vectorial release of poliovirus from polarized human intestinal epithelial cells. *J Virol* 1993;67:4274–4282.
689. Tucker SP, Thornton CL, Wimmer E, et al. Bidirectional entry of poliovirus into polarized epithelial cells. *J Virol* 1993;67:29–38.
690. Tuffereau C, Leblois H, Benejean J, et al. Arginine or lysine in position 333 of ERA and CVS glycoprotein is necessary for rabies virulence in adult mice. *Virology* 1989;172:206–212.
691. Tyler K, McPhee D, Fields B. Distinct pathways of viral spread in the host determined by reovirus S1 gene segment. *Science* 1986;233:770–774.
692. Ubol S, Griffin DE. Identification of a putative alphavirus receptor on mouse neural cells. *J Virol* 1991;65:6913–6921.
693. Uehara T, Miyawaki T, Ohta K, et al. Apoptotic death of primed CD45RO+ T lymphocytes in Epstein-Barr virus-induced infectious mononucleosis. *Blood* 1992;80:452–458.
694. Ugolini G. Transneuronal transfer of herpes simplex virus type 1 (HSV 1) from mixed limb nerves to the CNS. I. Sequence of transfer from sensory, motor and sympathetic nerve fibres to the spinal cord. *J Comp Neurol* 1992;326:527–548.
695. Ugolini G, Kuypers HGJM, Simmons A. Retrograde transneuronal transfer of herpes simplex virus type 1 (HSV 1) from motoneurons. *Brain Res* 1987;422:242–256.
696. Ugolini G, Kuypers HGJM, Strick PL. Transneuronal transfer of herpes virus from peripheral nerves to cortex and brainstem. *Science* 1989;243:89–91.
697. Upton C, Mossman K, McFadden G. Encoding of a homolog of the IFN gamma receptor by myxovirus. *Science* 1992;258:1369–1372.
698. Utz JP. Viruria in man. *Prog Med Virol* 1964;6:71–81.
699. Utz JP, Houk VN, Alling DW. Clinical and laboratory studies of mumps. IV. Viruria and abnormal renal function. *N Engl J Med* 1964;270:1283–1286.
700. Vahlne A, Nystrom B, Sandberg M, et al. Attachment of herpes simplex virus to neurons and glial cells. *J Gen Virol* 1978;40:359–371.
701. Vahlne A, Sveanerholm B, Lycke E. Evidence of herpes simplex virus type-selective receptors on cellular plasma membranes. *J Gen Virol* 1979;44:217–225.
702. Van Dyke T, Finlay C, Miller D, et al. Relationship between simian virus 40 large tumor antigen expression and tumor formation in transgenic mice. *J Virol* 1987;61:2029–2032.
703. Vanacker J-M, Laudet V, Adelmant G, et al. Interconnection between thyroid hormone signalling pathways and parvovirus cytotoxic functions. *J Virol* 1993;67:7668–7672.
704. Vanderplasschen A, Bublot M, Dubuisson J, et al. Attachment of the gammaherpesvirus bovine herpesvirus 4 is mediated by the interaction of gp8 glycoprotein with heparinlike moieties on the cell surface. *Virology* 1993;196:232–240.
705. Vannice JL, Levinson AD. Properties of the human hepatitis B virus enhancer: position effects and cell type non-specificity. *J Virol* 1988;62:1305–1313.
706. Vaughn JB, Gerhardt P, Newell KW. Excretion of street rabies virus in the saliva of dogs. *JAMA* 1965;193:363–368.
707. Verdin EM, King GL, Maratos-Flier E. Characterization of a common high-affinity receptor for reovirus serotypes 1 and 3 on endothelial cells. *J Virol* 1989;63:1318–1325.
708. Verdin EM, Lynn SP, Fields BN, et al. Uptake of reovirus serotype 1 by the lungs from the bloodstream is mediated by viral hemagglutinin. *J Virol* 1988;62:545–551.

709. Vlasak R, Luytjes W, Spaan W, et al. Human and bovine coronaviruses recognize sialic acid-containing receptors similar to those of influenza C virus. *Proc Natl Acad Sci USA* 1988;85:4526–4529.
710. Vogt MW, Will DJ, Craven DE, et al. Isolation of HTLVIII/LAV from cervical secretions of women at risk for AIDS. *Lancet* 1986;1:525–527.
711. Vonderfecht S, Miskuff R, Wee S, et al. Protease inhibitors suppress the *in vitro* and *in vivo* replication of rotaviruses. *J Clin Invest* 1988;82:2011–2016.
712. Vrati S, Faragher SG, Weir RC, et al. Ross River virus mutant with the deletion in the E2 gene: properties of the virion, virus-specific macromolecular synthesis, and attenuation of virulence for mice. *Virology* 1986;151:222–232.
713. Wada Y, Pierce ML, Fujinami RS. Importance of amino acid 101 within capsid protein VP1 for modulation of Theiler's virus-induced disease. *J Virol* 1994;68:1219–1223.
714. Wadia HH, Irani PF, Katrak SM. Lumbosacral radiculomyelitis associated with pandemic acute hemorrhagic conjunctivitis. *Lancet* 1973;1:350–352.
715. Wahlberg JM, Bron R, Wilschut J, et al. Membrane fusion of Semliki Forest virus involves homotrimers of the fusion protein. *J Virol* 1992;66:7309–7318.
716. Wahlberg JM, Garoff H. Membrane fusion process of Semliki Forest virus. I. Low pH-induced rearrangement in spike protein quaternary structure precedes virus penetration into cells. *J Cell Biol* 1992; 116:339–348.
717. Waldman WJ, Roberts WH, Davis DH, et al. Preservation of natural cytopathogenicity of cytomegalovirus by propagation in endothelial cells. *Arch Virol* 1991;117:143–164.
718. Walker D, Boring W. Factors influencing host-virus interactions. III. Further studies on the alterations of coxsackie virus infection in adult mice by environmental temperature. *J Immunol* 1958;80:39–44.
719. Wang H, Kavanaugh MP, North RA, et al. Cell-surface receptor for ecotropic murine retroviruses is a basic amino-acid transporter. *Nature* 1991;352:729–731.
720. Wang K-S, Kuhn RJ, Strauss EG, et al. High-affinity laminin receptor is a receptor for Sindbis virus in mammalian cells. *J Virol* 1992;66: 4992–5001.
721. Wasylyk B, Imler JL, Chatton B, et al. Negative and positive factors determine the activity of the polyoma virus enhancer alpha domain in undifferentiated and differentiated cell types. *Proc Natl Acad Sci USA* 1988;85:7952–7956.
722. Watson H, Tignor G, Smith A. Entry of rabies virus into the peripheral nerves of mice. *J Gen Virol* 1981;56:371–382.
723. Weber F, Schaffner W. Enhancer activity correlates with the oncogenic potential of avian retroviruses. *EMBO J* 1985;4:949–956.
724. Webster L. Inheritance of resistance of mice to enteric bacterial and neurotropic virus infections. *J Exp Med* 1937;65:261–286.
725. Webster RG, Rott R. Influenza virus A pathogenicity: the pivotal role of hemagglutinin. *Cell* 1987;50:665–666.
726. Wege H, Winter J, Meyermann R. The peplomer protein E2 of coronavirus JHM as a determinant of neurovirulence: definition of critical epitopes by variant analysis. *J Gen Virol* 1988;69:87–98.
727. Weiner HL, Drayna D, Averill DR, et al. Molecular basis of reovirus virulence: the role of the S1 gene. *Proc Natl Acad Sci USA* 1977;74: 5744–5748.
728. Weiner HL, Powers ML, Fields BN. Absolute linkage of virulence with central nervous system cell tropism of reovirus to hemagglutinin. *J Infect Dis* 1980;141:509–616.
729. Weiner LP, Cole GA, Nathanson N. Experimental encephalitis following peripheral inoculation of West Nile virus in mice of different ages. *J Hyg* 1970;68:435–446.
730. Weinstein L. Influence of age and sex on susceptibility and clinical manifestation in poliomyelitis. *N Engl J Med* 1957;257:47–52.
731. Weis W, Brown J, Cusack S, et al. Structure of the influenza virus hemagglutinin complexed with its receptor, sialic acid. *Nature* 1988;333:426–431.
732. Weiss RA. Human immunodeficiency virus receptors. *Semin Virol* 1992;3:79–84.
733. Westervelt P, Trowbridge DB, Epstein LG, et al. Macrophage tropism determinants of human immunodeficiency virus type 1 in vivo. *J Virol* 1992;66:2577–2582.
734. Westrop G, Evans D, Dunn G, et al. Genetic basis of attenuation of the Sabin type 3 oral poliovaccine. *J Virol* 1989;63:1338–1343.
735. Whealy ME, Card JP, Robbins AK, et al. Specific pseudorabies virus infection of the rat visual system requires both gI and gp63 glycoproteins. *J Virol* 1993;67:3786–3797.
736. White E, Cipriani R, Sabbatini P, et al. Adenovirus E1B 19-kilodalton protein overcomes the cytotoxicity of E1A proteins. *J Virol* 1991;65: 2968–2978.
737. White JM. Viral and cellular membrane fusion proteins. *Annu Rev Physiol* 1990;52:675–697.
738. White JM. Membrane fusion. *Science* 1992;258:917–924.
739. White JM, Littman DR. Viral receptors of the immunoglobulin superfamily. *Cell* 1989;56:725–728.
740. Wickham TJ, Mathias P, Cheresh DA, et al. Integrins $alpha_v beta_3$ and $alpha_v beta_5$ promote adenovirus internalization but not virus attachment. *Cell* 1993;73:309–319.
741. Wiley C, Schrier R, Denaro F, et al. Localization of cytomegalovirus proteins and genome during fulminant central nervous system infection in an AIDS patient. *J Neuropathol Exp Neurol* 1986;45:127–139.
742. Wiley DC, Skehel JJ. The structure and function of the hemagglutinin membrane glycoprotein of influenza virus. *Annu Rev Biochem* 1987;56:365–394.
743. Williams RK, Jiang G-S, Holmes KV. Receptor for mouse hepatitis virus is a member of the carcinoembryonic antigen family of glycoproteins. *Proc Natl Acad Sci USA* 1991;88:5533–5536.
744. Wilson G, Beushausen S, Dales S. *In vivo* and *in vitro* models of demyelinating diseases. XV. Differentiation influences the regulation of coronavirus infection in primary explants of mouse CNS. *Virology* 1986;151:253–264.
745. Wirth JJ, Amalfitano A, Gross R, et al. Organ- and age-specific replication of polyomavirus in mice. *J Virol* 1992;66:3278–3286.
746. Wofsy C, Cohen J, Haver L, et al. Isolation of the AIDS-associated retrovirus from vaginal and cervical secretions of women with antibodies to the virus. *Lancet* 1986;1:527–529.
747. See ref. 746.
748. Wold WSM, Gooding LR. Region E3 of adenovirus: a cassette of genes involved in host immunosurveillance and virus-cell interactions. *Virology* 1991;184:1–8.
749. Wolf JL, Kauffman RS, Finberg R, et al. Determinants of reovirus interaction with intestinal M cells and absorptive cells of murine intestine. *Gastroenterology* 1983;85:291–300.
750. Wolf JL, Rubin D, Finberg R, et al. Intestinal M cells: a pathway for entry of reovirus into the host. *Science* 1981;212:471–472.
751. Wolinsky JS, Klassen T, Baringer JR. Persistence of neuroadapted mumps virus in brains of newborn hamsters after intraperitoneal inoculation. *J Infect Dis* 1976;133:260–267.
752. Wong C, Woodruff JJ, Woodruff JF. Generation of cytotoxic T lymphocytes during coxsackievirus B3 infection. III. Role of sex. *J Immunol* 1977;119:591–597.
753. Woodruff J. The influence of quantitated post-weaning undernutrition on coxsackievirus B3 infection of adult mice. *J Infect Dis* 1970;121: 164–181.
754. Woodworth CD, Bowden PE, Doninger J, et al. Characterization of normal human exocervical epithelial cells immortalized in vitro by papillomavirus types 16 and 18 DNA. *Cancer Res* 1988;48:4620–4628.
755. WuDunn DW, Spear PG. Initial interaction of herpes simplex virus with cells is binding to heparan sulfate. *J Virol* 1989;63:52–58.
756. Wunner WH, Reagan KJ, Koprowski H. Characterization of saturable binding sites for rabies virus. *J Virol* 1984;50:691–697.
757. Wykes MN, Shellam GR, McCluskey J, et al. Murine cytomegalovirus interacts with major histocompatability complex class I molecules to establish cellular infection. *J Virol* 1993;67:4182–4189.
758. Xiao JH, Davidson I, Ferrandon D, et al. One cell-specific and three ubiquitous nuclear proteins bind *in vitro* to overlapping motifs in the domain B1 of the SV40 enhancer. *EMBO J* 1987;6:3005–3013.
759. Yeager CL, Ashmun RA, Williams RK, et al. Human aminopeptidase N is a receptor for human coronavirus 229E. *Nature* 1992;357:420–422.
760. Yee JK. A liver-specific enhancer in the core promoter region of the human hepatitis B virus. *Science* 1989;246:658–661.
761. Yokomori K, Lai MC. Mouse hepatitis virus utilizes two carcinoembryonic antigens as alternative receptors. *J Virol* 1992;66:6194–6199.
762. Yokomori K, Lai MC. The receptor for mouse hepatitis virus in the resistant mouse strain SJL is functional: implications for the requirement of a second factor for viral infection. *J Virol* 1992;66:6931–6938.
763. Yoshimura FK, Davison B, Chaffin K. Murine leukemia virus long terminal repeat sequences can enhance gene activity in a cell type–specific manner. *Mol Cell Biol* 1985;5:2832–2835.

764. Young E, Gomez C. Enhancement of herpes virus type 2 infections in pregnant mice. *Proc Soc Exp Biol Med* 1979;160:416–420.
765. Yuhasz SA, Stevens JG. Glycoprotein B is a specific determinant of herpes simplex virus type 1 neuroinvasiveness. *J Virol* 1993;67:5948–5954.
766. Zagury D, Bernard J, Leibowich J, et al. HTLV-III in cells cultured from semen of two patients with AIDS. *Science* 1984;226:449–451.
767. Zagury D, Fouchard M, Cheynier M, et al. Evidence for HTLV-III in T cells from semen of AIDS patients. *Cancer Res* 1985;45(Suppl 1):4595–4597.
768. Zhang L, Senkowski A, Shim B, et al. Chimeric cDNA studies of Theiler's murine encephalomyelitis virus neurovirulence. *J Virol* 1993;67:4404–4408.
769. Zheng YM, Schafer MK-H, Weihe E, et al. Severity of neurological signs and degree of inflammatory lesions in the brain of rats with Borna disease correlate with the induction of nitric oxide synthase. *J Virol* 1993;67:5786–5791.
770. Zhirnov O, Ovcharenko A, Bukrinskaya A. Protective effect of protease inhibitors in influenza infected animals. *Arch Virol* 1982;73:263–272.
771. Zhirnov O, Ovcharenko A, Bukrinskaya A. Suppression of influenza virus replication in infected mice by protease inhibitors. *J Gen Virol* 1984;65:191–196.
772. Ziegler R, Herman R. Peripheral infection in culture of rat sensory neurons by herpes simplex virus. *Infect Immun* 1980;28:620–623.
773. Zisman B, Hirsch MS, Allison AC. Selective effects of antimacrophage serum, silica and anti-lymphocyte serum on pathogenesis of herpes virus infection of young adult mice. *J Immunol* 1970;104:1155–1159.
774. Zisman B, Wheelock EF, Allison AC. Role of macrophages and antibody in resistance of mice against yellow fever virus. *J Immunol* 1971;107:236–243.
775. Zon LI, Moreau J-F, Koo J-W, et al. The erythropoietin receptor transmembrane region is necessary for activation by the Friend spleen focus-forming virus gp55 glycoprotein. *Mol Cell Biol* 1992;12:2949–2957.
776. Zurbriggen A, Fujinami R. Theiler's virus infection in nude mice: viral RNA in vascular endothelial cells. *J Virol* 1988;62:3589–3596.
777. Zurbriggen A, Fujinami RS. A neutralization-resistant Theiler's virus variant produces an altered disease pattern in the mouse central nervous system. *J Virol* 1989;63:1505–1513.
778. Zurbriggen A, Thomas C, Yamada M, et al. Direct evidence of a role for amino acid 101 of VP-1 in central nervous system disease in Theiler's murine encephalomyelitis virus infection. *J Virol* 1991;65:1929–1937.

Fundamental Virology, Third Edition
edited by B.N. Fields, D.M. Knipe, P.M. Howley, et al.
Lippincott - Raven Publishers, Philadelphia © 1996

CHAPTER 7

Persistence of Viruses

Rafi Ahmed, Lynda A. Morrison, and David M. Knipe

Virus survival in nature requires continuous infection of susceptible individuals. Within an infected host, viruses can either cause acute infections or can establish long-term persistence. During acute infections, virus is cleared by the host immune response, necessitating rapid transmission or capacity for extra-organismal survival. Some acute viruses, such as measles or mumps, survive by constant infection of human populations only, whereas others, such as influenza, yellow fever, or rabies viruses, circulate in more than one species. Still other viruses possess structural features that permit them to survive the rigors of an extra-organismal environment until contact with a susceptible host occurs. For example, poxviruses are stable in a dried form, whereas enteric viruses, such as poliovirus or rotaviruses, can survive in water supplies until ingested by susceptible individuals. The parameters that define these types of viral spread and survival during acute infections are described in Chapter 9. Alternatively, viruses may persist within an individual host organism for extended periods of time. These virus infections begin as acute infections but progress to latent or chronic infections during which the virus is transmitted periodically to new host organisms. A list of viruses that can persist in humans is given in Table 1. As can be seen from this table, the ability to persist *in vivo* is not confined to a particular virus group, and a variety of DNA- and RNA-containing viruses can establish long-term infections.

Persistent viruses cause an increasing proportion of the viral disease burden borne by humans—for example, AIDS caused by human immunodeficiency virus (HIV), chronic hepatitis and hepatocellular carcinoma caused by infection with hepatitis B virus, anogenital cancer associated with human papilloma viruses, disseminated herpes simplex virus 2 infection in the newborn, etc. (see Table 1). It is likely that sensitive techniques, such as polymerase chain reaction (PCR) will uncover even more evidence of low-level viral persistence. Thus, additional disease may be found to be caused by viruses persisting at low levels. As our knowledge of viral replication and immunologic clearance mechanisms in acute viral infections has increased, there is now a greater urgency to understand the mechanisms of viral persistence.

During acute infection of the host, many viruses inhibit the metabolism of host cells they productively infect so that cytopathic effects or cell death ultimately result. Obvious-

R. Ahmed: Department of Microbiology and Immunology, UCLA School of Medicine, Los Angeles, California 90024.
L. A. Morrison and D. M. Knipe: Department of Microbiology and Molecular Genetics, Harvard Medical School, Boston, Massachusetts 02115.

TABLE 1. *Viruses that persist in humans*

Virus	Site of persistence	Consequence
DNA viruses		
Adenovirus	Adenoids, tonsils, lymphocytes	None known
Cytomegalovirus	Kidney, salivary glands, lymphocytes? macrophages? stromal cells?	Pneumonia, retinitis
Epstein–Barr virus	Pharyngeal epithelial cells, B cells	Infectious mononucleosis, Burkitt's lymphoma, nasopharyngeal carcinoma, non-Hodgkin's lymphoma, oral hairy leukoplakia
Herpes simplex virus 1 and 2	Sensory ganglia neurons	Cold sores, genital herpes, encephalitis, keratitis
Human herpesvirus 6	Lymphocytes	Exanthem subitum
Varicella–zoster virus	Sensory ganglia neurons and/or satellite cells	Varicella, zoster
Hepatitis B virus	Hepatocytes, lymphocytes? macrophages?	Hepatitis, hepatocellular carcinoma
Hepatitis D virus	Hepatocytes	Exacerbation of chronic HBV infection
Papilloma virus	Epithelial skin cells	Papilloma, carcinoma
Parvovirus B19	Erythroid progenitor cells in bone marrow	Aplastic crisis in hemolytic anemia, chronic bone marrow deficiency
Polyomavirus BK	Kidney	Hemorrhagic cystitis
Polyomavirus JC	Kidney, oligodendrocytes in CNS	Progressive multifocal leukoencephalopathy
RNA Viruses		
Hepatitis C virus	Hepatocytes, lymphocytes? macrophages?	Hepatitis, hepatocellular carcinoma
Measles virus[a]	Neurons and supporting cells in CNS	Subacute sclerosing panencephalitis, measles inclusion body encephalitis
Rubella virus[a]	CNS	Progressive rubella panencephalitis, insulin-dependent diabetes? juvenile arthritis?
Retroviruses		
Human immunodeficiency virus	$CD4^+$ T cells, monocytes macrophages, microglia	AIDS
Human T-cell leukemia virus I	T cells	T cell leukemia, tropical spastic paraparesis, polymyositis
Human T-cell leukemia virus II	T cells	None known

[a]Measles and rubella viruses typically cause acute infections. However, in rare instances these viruses have been shown to persist in the CNS.

ly, if a cytolytic virus is going to establish a persistent infection, alternative virus-host cell interactions must occur to limit the cytopathic effects of the virus. Much of the story of persistent infections told here describes the mechanisms by which the effects of a virus on the host cell are attenuated so that the cell can survive and the virus can persist within it. In many ways this is the classic story of a parasite, needing to survive and persist within a host without harming it, while nonetheless resisting its defenses.

To present the subject of viral persistence, we have divided this chapter into three sections. The first section defines the different patterns of viral infection seen in a host organism. The second part considers, in general terms, the possible mechanisms and strategies that viruses employ to establish and maintain a persistent infection. This section draws on examples from human viral infections as well as from animal model infections to illustrate specific concepts and principles. The third section examines in greater detail the mechanisms by which individual viruses persist within a human host. In this section we have not attempted to cover all viruses known to persist in humans but have selected representative examples to illustrate the various types of virus-host interactions that occur during persistent infections. Other chapters within this book examine the implications of viral persistence within the context of viral pathogenesis (Chapter 6), virus-host cell interactions (Chapter 8), viral evolution (Chapter 5), and epidemiology (Chapter 9 in *Fields Virology,* 3rd ed.).

I. Acute viral infection followed by viral clearance

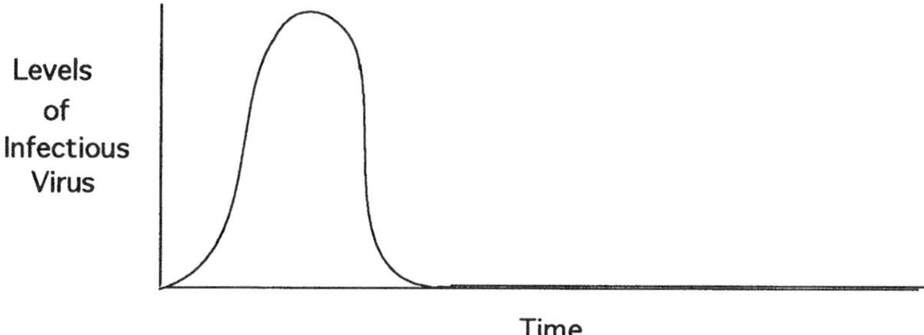

II. Acute viral infection followed by latent infection and periodic reactivation

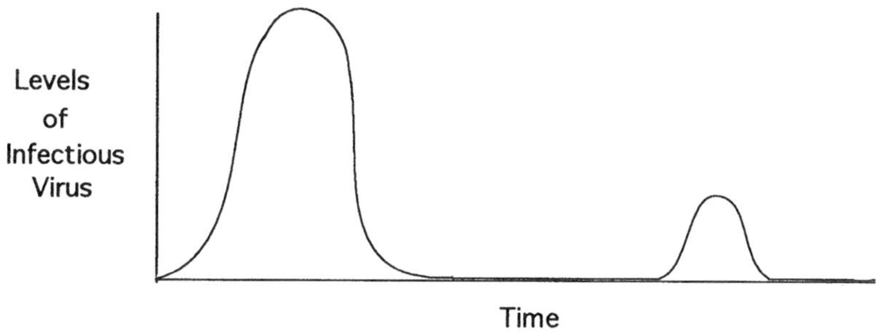

III. Acute viral infection followed by chronic infection

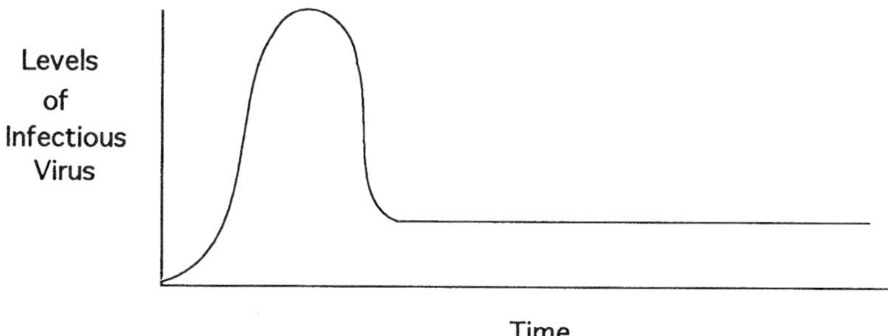

IV. Slow chronic infections

Levels of Infectious Virus

Time

FIG. 1. General patterns of viral infection in the host organism.

PATTERNS OF VIRAL INFECTION

Viral infections of a host organism can be divided into several general categories, based on the patterns and levels of infectious virus detectable in the organism at various times after infection (Fig. 1). The four types of infection illustrated in Fig. 1 are as follows:

1. Acute infection followed by viral clearance by the host immune response.
2. Acute infection followed by latent infection in which viruses persist in a noninfectious form with intermittent periods of viral reactivation and shedding. These viruses must be capable of causing productive infection in certain cells or under certain conditions and

TABLE 2. *Examples of latent infection followed by periodic reactivation*

Example	Mechanism of establishment	Maintenance	Stimulatory mechanism for reactivation
Herpes simplex virus in sensory ganglion neurons	Limited transcription and possibly replication of genome	Non-replicating episome	Neuronal activation or damage
Epstein-Barr virus in B lymphocytes	Limited transcription of viral genome	Replication of viral DNA as episome	Antigen activation or other stimuli of B cells
Papilloma virus in basal cells of epidermis	Limited transcription of viral genome	Replication of viral DNA as episome	Differentiation of basal cells

nonpermissive infection in other cells (Table 2). The important issues here are the mechanisms by which the cytopathic potential of these viruses is limited so they can establish latency, how they are maintained in a latent form, and how they reactivate.

3. Acute infection followed by a chronic infection in which infectious virus is continuously shed from or is present in infected tissues. Chronic infections can be established if the host immune response cannot eliminate virus generated during an acute infection. Productive infection of host cells may be followed by spread to cells that are less permissive, or by evolution of an immune response that dampens viral replication but cannot completely clear the virus from the host.
4. Slow, chronic infection without acute infection. This pattern is observed only with unconventional agents such as those causing spongiform encephalopathies. Because such infections are not caused by any of the known viruses, we will not discuss this pattern further, but the reader is referred to Chapter 37 in this volume and Chapters 90–91 in *Fields Virology,* 3rd ed. on these unconventional agents.

We have defined idealized patterns of viral infections of the host, but in fact many viruses show different patterns of infection in different tissues or cell types or combine these general patterns so that their infection of the host does not fit cleanly within these definitions. Thus, these patterns should be viewed as themes with variations and not as rigid definitions.

MECHANISMS OF VIRAL PERSISTENCE

Three general conditions must be fulfilled for a virus to persist over a long term in a host. First, the virus must be able to infect host cells without being cytopathic. Second, there must be mechanisms for long-term maintenance of the viral genome in host cells. Third, the virus has to avoid detection and elimination by the host's immune system. Viruses have devised various strategies to accomplish these goals, and we now consider general mechanisms. Specific examples illustrating the various concepts are given later in "Selected Examples of Viruses that Persist in Humans."

Restriction of Viral Cytolytic Effect

Survival of critical numbers of infected cells is a basic requirement for viral persistence. Therefore, a virus can persist in a host cell only if it does not kill the cell or produce excessive damage. Viruses that cause nonlytic productive infections are well suited for persistence, and these viruses are the ones most likely to establish a chronic infection. For example, most arenaviruses can replicate in a cell without killing it or affecting the growth rate of the cell (35,179). This nonlytic phenotype permits the various arenaviruses to survive in nature as lifelong chronic infections of their respective rodent hosts (139,180,258). Another example of a nonlytic virus-host cell interaction is chronic infection of liver cells with hepatitis B virus (HBV), which productively infects hepatocytes with minimal to no cell injury (57). However, for viruses that usually inhibit host cell metabolism and lyse the cell, the lytic phenotype is a potential barrier to the successful establishment of a persistent infection. Conditions under which the cytopathic effect of lytic viruses is limited include (i) infection of nonpermissive cell types or of cells in a nonpermissive environment, or (ii) evolution of viral variants that either are less cytolytic or interfere with replication of wild-type virus. Another possible mechanism for limitation of viral cytopathic effect would be for the virus to evolve complex regulatory controls that actively shut down viral gene expression in certain cell types, but there are no known examples of this scenario. Instead, certain cell types are nonpermissive, imposing a down-regulatory effect on viral gene expression, and it appears that viruses may have merely evolved to persist in this restrictive environment.

It is also worth noting that a virus may be lytic for certain cell types but not for others. This is particularly relevant for persistence *in vivo*, since many different cell types are present in the whole animal. For instance, there is evidence suggesting that HIV is more lytic for T cells than for monocytes/macrophages. Productive infection of T cells can result in cell death, whereas HIV-infected mononuclear phagocytes produce virus for considerable periods of time without cell lysis (106,131,162,176,278). Similarly, Sindbis virus, the prototype alphavirus, causes a lytic infection in neuroblastoma cells or freshly explanted dorsal root ganglion neurons but establishes a persistent infection in neurons that have differentiated in the presence of nerve growth factor (187). Although persistence of alphaviruses in humans has not been documented, recent studies have shown that Sindbis virus can persist in neurons of adult mice without causing overt damage (185,186,188).

Infection of Nonpermissive Cells

The best examples of restricted viral gene expression involve infection of non- or semipermissive cells. For example, sensory neurons are nonpermissive for the normally cytolytic herpes simplex virus (HSV), allowing little or no viral gene expression. In these cells a latent infection is established. Similarly, B lymphocytes are nonpermissive for Epstein-Barr virus (EBV). In this case, selective expression of EBV genes not involved in lytic infection permits viral DNA replication to occur when the host cell divides, while nonetheless maintaining the noncytolytic state. In addition, basal layer skin cells are nonpermissive for papillomaviruses, and latent infections are established in these cells. These viruses remain in the latent state until induced to reactivate by perturbations of the host cell environment, such as cell injury or cellular differentiation, which are thought to at least temporarily convert a nonpermissive cell type to one permissive for viral replication (Table 2). Latent herpesvirus and papillomavirus infections are discussed in more detail in a later section of this chapter.

Another example of restricted gene expression by a normally cytolytic virus comes from *in vitro* studies on infection of teratocarcinoma cells by cytomegalovirus (CMV) (82,112). CMV causes a productive infection of differentiated teratocarcinoma cells, but infection of undifferentiated cells results in incomplete viral transcription. If these nonproductively infected cells are then driven to differentiate by treatment with retinoic acid (or by any other signals leading to differentiation), multiple transcripts associated with lytic infection are produced, viral proteins are synthesized, and infectious progeny are released. *In vivo*, CMV causes a nonproductive infection of human peripheral blood mononuclear cells (PBMC) (87,156,216,261). Infection is confined predominantly to monocytes, and viral gene expression appears to be limited to the immediate early gene products. There is now some evidence that differentiation of monocytes into macrophages can convert this nonproductive infection into a productive one (326).

"Smoldering" Infection of Low Numbers of Permissive Cells

Studies on persistent virus infections of cultured cells have shown that lytic viruses can persist *in vitro* under conditions where only a small fraction of the total cells is infected at any one time. In such carrier cultures the few infected cells release viruses and are killed, but the progeny viruses again infect only a small number of the total cells, while the majority of the cells remains uninfected. This restriction could be due to the availability of only a limited number of susceptible cells at a given time or due to the presence of soluble inhibitors such as interferon in the culture medium (342,343,359). This type of "smoldering" or "cycling" infection has been implicated in the persistence of lactic dehydrogenase virus in mice and adenoviruses in humans (5,197,253). It is possible that such "smoldering" infections are more common *in vivo* than has been appreciated.

Evolution of Viral Variants

There is considerable evidence from *in vitro* studies that viral mutants generated from normally cytocidal wild-type viruses possess modified lytic potential (136,197,347,359). Several types of viral variants, such as temperature-sensitive and small plaque mutants and defective-interfering particles (Chapter 4) have been implicated in the establishment of persistent infections *in vitro*. In some systems, the genetic basis of attenuation is known, and the viral gene(s) that plays a role in persistence has been identified. For example, studies on reovirus persistence have shown that mutations in specific viral genes are crucial for the establishment and maintenance of persistent infection in L-929 cells (6,8,76,160). Also, mutational alterations of the glycoproteins of Sindbis virus have been shown to modify the lytic potential of Sindbis virus for neurons (186). These studies illustrate the concept that a few amino acid changes in selected viral genes can convert a lytic infection into a persistent one.

In addition to selection of viral variants, there is also the possibility of obtaining variants of the host cell. There have been several reports describing the selection of mutant cells during persistent infection of cell lines in culture. For example, HeLa cells with increased resistance to poliovirus or coxsackie A9 virus were selected in carrier cultures (2,324,337). The increased resistance of HeLa cell variants resulted from blocks at different steps of the poliovirus life cycle: Some cell lines were nonpermissive for poliovirus infections because of decreased levels of expression of the poliovirus receptor, whereas others were nonpermissive for poliovirus replication at intracellular steps (56,157). Similarly, African green monkey kidney cells partially resistant to simian virus 40 were isolated from persistently infected cell lines (260,352). A persistently infected cell culture provides a dynamic environment, and coevolution of both host cells and virus has been described during persistence of reovirus in L cells (7,76). It was shown that both the original L cells and the parental reovirus had changed, followed by a coselection of variant cells and virus. During the course of the persistent infection, mutant L cells that were highly resistant to lysis by reovirus were selected, and this was followed by selection of an altered reovirus that grew much better in these mutant cells. These observations on host-virus coevolution described for reovirus have also been shown to occur during persistent infection of BHK cells with foot-and-mouth disease virus and in L cells infected with minute virus of mice (72–74,270). Although these studies have shown that host cells in culture can evolve to a nonpermissive state allowing persistent infection, it is likely that cells in a host are constrained in their potential for this type of evolution.

Mechanisms of Viral Genome Maintenance

The second requirement for viral persistence is a mechanism for maintenance of the viral genome in the persistently infected cell. If the host cell is dividing, there must be replication of the viral genome so that it is not diluted among daughter cells. For retroviruses, the mechanism to achieve this result is a part of the viral replicative process. The DNA copy of the viral RNA genome is integrated into the host chromosome so that the viral genome is propagated with the host chromosomes. Similarly, the genome of parvoviruses is integrated into the host chromosome. The genomes of some DNA viruses, such as papillomaviruses and Epstein-Barr virus, are maintained as episomal circular molecules with their replication tied to the host cell cycle but promoted by viral proteins. In contrast, in nondividing neurons the DNA genome of herpes simplex virus is maintained without replication, probably as a circular episomal form. True latent infections are more common with DNA viruses, probably because the host cell carries out processes that can retain the viral DNA genome in a quiescent form, either integrated or as an extrachromosomal plasmid. In contrast, there are few, if any, mechanisms for long-term propagation of RNA molecules in a host cell. Therefore, RNA viruses can persist only by a low-grade, chronic infection in which the viral RNA is replicated continuously at a low rate, usually by the viral replication apparatus. The section entitled "Selected Examples of Viruses that Persist in Humans" examines the specifics of the mechanisms by which genomes are maintained within the persistently infected cell for several viruses of clinical importance.

Evasion of Host Immunity

The primary function of the host's immune system is recognition and elimination of foreign materials. Thus, a major component of viral persistence is evasion of immunologic surveillance. Successful resolution of a virus infection depends on the appropriate immune responses being sufficient to inhibit viral growth and dissemination. Viruses have developed various strategies to shift the balance so as to avoid elimination by the host's immune system (5,113,114,199,215,305).

Antiviral Immunity

Antibodies and T cells are the two main antigen-specific effector systems for resolving viral infections (81). For a detailed description of immunity to viruses, the reader is referred to Chapter 10. However, before discussing the various escape mechanisms that viruses have evolved, it is useful to review how T and B cells control viral infections and to define the critical molecules involved in recognition of viral materials by T cells and antibody (Table 3). Antibodies can recognize either free virus or virus-infected cells. They control virus infections by neutralizing virus particles and by killing infected cells through complement-mediated cytotoxicity or antibody-dependent cell-mediated cytotoxicity (ADCC). The critical viral proteins recognized in these processes are surface glycoproteins or outer capsid proteins, and although antibodies against internal and nonstructural viral proteins are also made, these do not participate in viral neutralization or antibody-mediated killing of infected cells. Antibody binding to viral glycoproteins at the cell surface can also down-regulate the expression of viral genes inside the infected cell (105), but the mechanism by which this effect occurs is not understood. This interesting phenomenon, first described for measles virus-infected cells in culture (103,104), may also operate *in vivo* in controlling Sindbis virus infection in neurons (188). Similar observations have been made using antibodies to the surface glycoproteins of rabies virus and Sendai virus (79,208). These studies have also suggested that antibody can neutralize virus intracellularly (79,188,208). Thus, antibody can control viral infections by several different mechanisms, and the critical recognition molecules appear to be viral surface glycoproteins or outer-capsid proteins. Obviously, then, changes in either the structure or expression of these viral outer-capsid or surface glycoproteins could be important mechanisms by which viruses can avoid elimination by antibody. Alterations in other viral proteins (internal and nonstructural) are probably not critical in escape from humoral immunity. However, it should be noted that there are some reports of antibody to internal or core proteins also playing a role in protection (192,313).

In contrast to antibodies, which recognize viral proteins by themselves, T cells only see viral antigen in association with host major histocompatibility complex (MHC) molecules (108,362). The antigen-specific T-cell receptor recognizes short viral peptides bound to cellular MHC molecules (70,212). An important consequence of this mode of recognition is that T cells cannot recognize free virus, and their antiviral activities are confined to infected cells. Thus, the T-cell arm of the immune system has evolved for surveillance of infected cells, whereas antibody serves as the primary defense against free virus. T cells are further subdivided into two subsets, $CD4^+$ and $CD8^+$ T cells (108,191). $CD4^+$ T cells recognize viral peptides in association with MHC class II antigens, whereas $CD8^+$ T cells recognize viral peptides bound to MHC class I molecules (31,34,195). These peptide fragments can be derived from any viral protein, structural (either surface or internal) or nonstructural. Thus, all viral proteins can be potential targets for T-cell recognition. The limiting factors are the intracellular pro-

TABLE 3. *Anti-viral T and B cell immunity*

Effector system	Recognition molecule	Mechanism of viral control
Antibody	Surface glycoproteins or outer-capsid proteins of virus particles	(i) Neutralization of virus (ii) Opsonization of virus particles
	Viral glycoproteins expressed on membranes of infected cells	(i) Antibody-complement mediated and antibody-dependent cell-mediated cytotoxicity (ADCC) of virus-infected cells (ii) Down-regulation of intracellular viral gene expression
$CD4^+$ T cells	Viral peptides (10–20 mers) presented by MHC class II molecules	(i) Release of antiviral cytokines (interferon-γ, TNF) (ii) Activation/recruitment of macrophages (iii) Help for antiviral antibody production (iv) Help for $CD8^+$ CTL responses (v) Killing of virus-infected cells?
$CD8^+$ T cells	Viral peptides (8–10 mers) presented by MHC class I molecules of infected cells	(i) Killing of virus-infected cells (ii) Release of antiviral cytokines (interferon-γ, TNF) (iii) Activation/recruitment of macrophages

cessing of the protein and the capability of the peptides to bind to MHC molecules (i.e., their affinity for various MHC molecules). The efficient interaction of virus-specific T cells with virus-infected cells depends not only on binding of the antigen-specific T-cell receptor (TCR) with the MHC/peptide complex, but also on several other accessory molecules that serve to increase adherence between T cells and their infected target cells (306). Some of the molecules involved in adhesion include LFA-1, CD2, CD4, and CD8 on T cells and their corresponding ligands on target cells, ICAM-1, LFA-3, and the MHC class II and I molecules, respectively. Thus, in contrast to antibody-virus interaction that primarily involves only two "players," recognition of virus-infected cells by T cells depends on a large number of molecules. Alterations in the structure or expression of any of these proteins can interfere with the effector functions of T cells and provide a possible mechanism of escape from T-cell-mediated immunity.

How do T cells control virus infections? The primary mechanism employed by $CD8^+$ T cells is killing of virus-infected cells. These $CD8^+$ cytotoxic T lymphocytes (CTL) are highly efficient in this process and can kill infected cells by two distinct mechanisms: (i) a secretory and membranolytic pathway involving perforin and granzymes, and (ii) a nonsecretory receptor-mediated pathway involving fas, a protein of the tumor necrosis factor (TNF) receptor family (24,154,344). Two recent studies using perforin gene knockout mice have shown that the perforin pathway is essential for clearing lymphocytic choriomeningitis virus (LCMV) infection *in vivo* (155,344). In addition to their killing function, $CD8^+$ T cells also control virus growth by producing antiviral cytokines such as interferon-γ and TNF, which interfere intracellularly with virus replication (29,81). $CD4^+$ T cells contribute to antiviral immunity in many different ways; they produce antiviral cytokines, are involved in activation and recruitment of macrophages, and provide help for both antibody production and $CD8^+$ CTL responses (29,81,204,223,245). Thus, $CD4^+$ T cells play a central role in antiviral immunity. Virus-specific $CD4^+$ CTL have also been described in several systems but the contribution of killing per se by $CD4^+$ T cells to control of virus infections *in vivo* is not clear (81,178,224).

Viral Evasion Strategies

Having reviewed the basics of antiviral immunity, let us now consider the different strategies viruses use to evade the immune system.

Restricted Expression of Viral Genes

Restricted viral gene expression not only reduces the lytic potential of viruses, as discussed earlier, but also provides a simple and highly effective mechanism by which infected cells escape detection by the host's immune response. This strategy is employed to varying degrees by nearly all the viruses that cause persistent infections. However, it is best exemplified by the herpesviruses and retroviruses. The most extensively analyzed system is latent infection of neurons by HSV, in which viral gene expression is completely turned off except for transcription from one region of the genome, and there appear to be no viral proteins expressed in the infected neurons (316,317). Under such conditions the virus essentially becomes invisible to the immune system, since immunity is directed against foreign proteins and is not programmed to distinguish between "self" and "foreign" nucleic acid. Thus, absolute latency is the ideal way of evading the immune system. However, this is not always feasible from the virus's point of view because certain viral proteins are often essential for replicating viral DNA during latency. For example, during latent infection of B cells by EBV, expression of one of the EBV proteins, called EBNA-1, is necessary for propagating the episomal EBV genome through cycles of cell division. A permanent state of latency is also not in the best interest of the virus in terms of transmission. Without production of infectious virus there can be no horizontal spread—an important mode of viral transmission. Thus, it is not surprising that even viruses that are highly efficient in establishing latent infections go through intermittent phases of productive infection.

Infection of Immunologically Privileged Sites

Another strategy employed by viruses is infection of tissues and cell types that are not readily accessible to the immune system (19). A site of persistence favored by many viruses (Tables 1 and 4) is the central nervous system (CNS). At least two factors favor viral persistence in the CNS: first, the presence of the blood/brain barrier, which limits lymphocyte trafficking through the CNS, and second, the presence of specialized cells such as neurons, which express neither MHC class I nor class II molecules and hence cannot be directly recognized by T cells (9,146,149,240). The kidney is another organ where viruses tend to persist (9,42,147). The human polyomaviruses BK and JC replicate in the kidney with almost lifelong shedding into the urine. CMV is also found in the kidney and is shed for long periods of time. It is not fully understood why the immune system is less effective in eliminating microbes from the kidney because no blood/kidney barrier exists, and extensive trafficking of lymphocytes through the kidney occurs. However, a recent study has shown that although T cells infiltrate the kidney, they have limited access to infected epithelial cells. These cells are protected by an intact basement membrane and microvascular endothelium, and this barrier has to be broken before T cells can have direct contact with their target cells (13). The scenario of limited T-cell access would also apply to epithelial surfaces of other secretory or excretory glands, and it is worth noting that the salivary gland, a secretory tissue, is a favored site of persistence for viruses such as CMV and EBV. Replication of papillomaviruses in the epidermis provides another example of persistence at an immunologically privileged site. The productive cycle occurs only in differentiating keratinocytes, and as a result, the infected cell and the viral particles are physically separated from the host's immune response by a basement membrane.

Antigenic Variation

The emergence of viral variants during persistence is a well-documented phenomenon (135,310,359). Viruses, especially those with RNA genomes, can undergo mutation at high frequencies, and under the appropriate selective pressure variants can arise rapidly (137). This ability to mutate quickly can provide a means of evading both T- and B-cell immunity.

Viral Escape from Antibody Recognition. There are many examples of antibody-resistant variants, and mutational alteration of viral proteins at sites critical for antibody recognition is a highly effective means of escape from neutralizing antibody. The classic example of this is the antigenic "shift" and "drift" seen among influenza viruses (242,346). These antigenic changes are due to alterations in the two surface glycoproteins of influenza virus, the hemagglutinin and the neuraminidase. Influenza virus does not establish a persistent infection in individuals, but the emergence of antigenic variants does contribute to the persistence of influenza virus at the population level (see Chapter 46 in *Fields Virology,* 3rd ed. for details on this subject). The best example of antigenic variation in a given individual during a persistent infection comes from the lentiviruses. Kono and colleagues (166,167) were the first to report antigenic drift of equine infectious anemia virus (EIAV) in chronically infected horses. They showed that sera taken at various times from an infected animal were able to neutralize virus isolates from previous clinical episodes but failed to neutralize subsequent EIAV isolates effectively. Additional studies have shown that these serologic changes correlate with alterations in the glycoproteins of EIAV (218,276). Narayan and colleagues have demonstrated the presence of antigenic variants in sheep chronically infected with visna virus (60,225–228). The altered neutralization properties of these variants have been correlated with genetic changes in the visna virus glycoproteins. Antigenic variants have also been shown to emerge during persistence of caprine arthritis/encephalitis virus (CAEV) in goats, simian immunodeficiency virus (SIV) in monkeys, and HIV in humans (40,41,93,225,259, 265,351). Thus, the appearance of serologically different variants during chronic infection is a common feature of all lentiviruses. However, the biological importance of antigenic variation in these viruses is not fully understood. With the possible exception of EIAV, where the recurrent episodes of disease (clinical symptoms associated with bursts of viremia) correlate with the emergence of antibody escape variants, the phenomenon of antigenic variation among lentiviruses does not seem to be essential for either persistence or induction of disease (93,194, 218,226,276,332). In many cases the "parental" virus continues to persist along with the variants in the infected host in spite of the presence of neutralizing antibodies. Thus, although neutralizing antibodies select antigenic variants, they are unable to eliminate the "parental wild-type" virus completely, and both variant and wild-type virus often coexist in the persistently infected host.

Viral Escape from T-Cell Recognition. Antigenic variation in viral peptides at residues involved in the processing and binding of the peptides to MHC molecules or alterations in viral peptides at sequences that directly contact the TCR can result in loss of recognition by the appropriate T cell. Theoretically, such alterations can occur in viral peptides presented by either MHC class I or class II molecules, resulting in escape from CD8$^+$ and CD4$^+$ T cells, respectively. To date, few, if any, studies have systematically analyzed antigenic variation within viral epitopes recognized by CD4$^+$ T cells. In large part this is due to the paucity of information on viral peptides seen by CD4$^+$ T cells. In contrast, more information on CD8$^+$ T-cell epitopes is available, and there are several well-documented examples of antigenic variation resulting in viral escape from CTL recognition (175). CTL escape variants were first demonstrated using the LCMV system (251). In this study it was shown that a single amino acid change can abrogate CTL recognition, leading to persistence of the mutant virus *in vivo*. It should be noted, however, that this profound biological effect of the CTL escape variant was seen in transgenic mice that expressed a single LCMV-specific TCR and could only make a monoclonal CTL response (251). It remains to be seen whether CTL escape variants will have a similar advantage under more physiologic conditions when antiviral CTL responses are likely to be polyclonal and multispecific.

CTL escape variants have been shown to occur in several viruses infecting humans, including HIV, EBV, and HBV (63,69,71,88,175,190,249). There is substantial evidence that CTL play a role in controlling, at least to some degree, infections by all three of these viruses (i.e., HIV, EBV, and HBV). Thus, it is tempting to propose that in these cases generation of CTL escape variants is critical for maintenance of the persistent infection. However, in most of the studies where CTL escape variants were identified, these variants did not go on to become the predominant viral species (175). Also, in a longitudinal study of HIV-infected individuals, no escape variants were detected over a 14-month period, demonstrating that the presence of such variants is not essential for HIV persistence (213). Similar observations have been made with SIV (54). Although these studies question the role of CTL escape variants in viral persistence, escape from CTL can also occur by several other mechanisms (discussed later). CTL lysis may also be avoided if the density of viral peptides present on the cell surface is too low to trigger efficient lysis of the target cells. Tsomides and colleagues (334) have provided evidence for this by showing that the low density of certain HIV peptides on naturally infected cells limits the effectiveness of HIV-specific CTL.

Two recent studies on HIV and HBV have provided evidence for a novel strategy of foiling the CTL response (26,163). These studies have identified mutants of HIV and HBV that can interfere with CTL function *in vitro*. These variants have alterations in epitopes recognized by CTL. However, unlike CTL escape variants that are not recognized by CTL (i.e., are invisible to the TCR), these mutant epitopes can still interact with the TCR, but instead of giving a stimulatory signal, they act as antagonists and render the CTL nonfunctional. This phenomenon has been termed *TCR antagonism*, and although the mechanism is not understood, it provides a highly effective means of evading the CTL response (145,301). An important aspect of this strategy is that the "antagonistic" variants can block CTL-mediated lysis of cells that are coinfected with viruses containing wild-type sequences. This would provide a possible explanation for survival of wild-type virus in the face of an ongoing CTL response. The one caveat to these exciting observations is that, at present, we do not know whether TCR antagonism is effective *in vivo*.

Suppression of Cell Surface Molecules Required for T-Cell Recognition

Viruses can escape T-cell recognition not only by mutation of the sequences encoding the epitope (peptide) seen by the TCR but also by down-regulating the expression of any one of the several host molecules that are necessary for efficient T-cell recognition of virus-infected cells. These include host MHC class I (for CD8$^+$ cells) or class II molecules (for CD4$^+$ cells), as well as several adhesion molecules such as ICAM-1 and LFA-3. Viruses have developed strategies for disruption of this multimeric interaction by selectively inhibiting the expression of these critical host cell molecules.

Reduction of MHC class I antigen expression on host cells as a consequence of viral infection has been reported for several viruses (148,201,256,262,287,357). This is an effective means of avoiding recognition by CD8$^+$ T cells. The best-documented example of virus-mediated suppression of MHC class I antigens is by the adenoviruses (12,38,39,85,125,241,287,295,353). Human adenoviruses can down-regulate MHC expression by two distinct mechanisms. One of the early proteins of adenovirus type 2, the E3 19-kd protein, termed E3/19K, can bind to and form a complex with MHC antigens. The formation of this E3/MHC class I complex prevents the MHC antigens from being correctly processed by inhibiting their terminal glycosylation. This results in reduced cell surface expression of the class I antigen. Adenovirus type 12 (an oncogenic serotype) prevents MHC class I expression by a different mechanism, which involves either a transcriptional block or lack of export of mRNA from the nucleus. Another early protein, the E1a protein of adenovirus type 12, has been implicated in this inhibition. The biological significance of this phenomenon was demonstrated by showing that adenovirus 12-induced tumors that failed to express their endogenous MHC antigens escaped immunologic surveillance and continued to grow upon transfer into an immunocompetent host (25). When these MHC-negative, adenovirus 12-induced tumors were transfected with an exogenous MHC class I gene (whose expression was not suppressed) and then transferred into immunocompetent animals, the tumor was rejected (325). Adenovirus down-regulation of MHC class I antigens may also operate *in vivo* to promote persistent infection. However, despite the many elegant studies on suppression of MHC class I by adenoviruses, there is still no evidence that persistence of human adenoviruses during the natural infection is related to down-regulation of MHC class I molecules.

Several reports have documented virus-mediated suppression of MHC class II antigens (36,152,184,205–207,248,262,286,293). Among the viruses infecting humans, CMV, HIV, and measles virus have been shown to interfere with MHC class II expression. The mechanism of viral interference is currently not understood, but in all the cases examined, viral infection did not affect the basal level expression of MHC class II molecules. Instead the viral effect was directed toward inhibiting the interferon-γ-mediated up-regulation of MHC class II mRNA transcription. Increased expression of MHC class II by interferon-γ is likely to play a key role in antigen presentation, and interference with this step may prevent the generation of an effective immune response against the virus.

Studies by Rickinson and colleagues (117) have shown that down-regulation of the cell surface adhesion molecules LFA-3 and ICAM-1 is involved in the escape of EBV-positive Burkitt's lymphoma (BL) cell lines from EBV-specific CTL. Certain EBV-positive BL cell lines were not killed by MHC-matched, virus-specific CTL in assays where EBV-transformed B lymphoblastoid cells derived from the normal B cells of the same patients were readily lysed. The resistance of these BL cells to CTL killing was not due to altered expression of MHC class I genes or lack of viral gene expression but instead was correlated with a reduced level of the adhesion molecules, LFA-3 and ICAM-1, on the tumor cell surface. The mechanism involved in selective suppression of these adhesion molecules is not known, but these studies nicely illustrate yet another viral strategy of circumventing the immune response.

Interference with Cytokine Function

Cytokines are an important part of antiviral immunity (29,223). Recent studies have described several viral proteins that act as "defense molecules" by interfering with cytokine function (114,199,305). Three adenovirus early proteins, E3-14.7K, E3-10.4K/14.5K, and E1B-19K, can protect virus-infected cells from lysis by TNF. The mechanism by which these adenovirus proteins counteract TNF is not known currently. The poxviruses also encode a protein, T2, that inhibits the action of TNF. The poxvirus T2 protein is a homolog of the cellular receptor for TNF and is released from infected cells, thus serving as a decoy. T2 binds TNF and prevents TNF from binding its true cellular receptor and destroying virus-infected cells. Other examples of viral defense molecules include the EBV protein BCRF1, which is a homolog of IL-10 and can block synthesis of IL-2 and interferon-γ, and a secretory protein encoded by myxoma virus that binds interferon-γ (219,335). In some instances viral RNA itself can function as a defense molecule. The adenovirus VA RNA, the HIV TAR RNA, and the EBV EBER RNA can inhibit the antiviral effects of interferon-α/β. Interferons induce the synthesis of a phosphoprotein called DAI that, in the presence of double-stranded RNA, phosphorylates initiation factor eIF-2 and prevents initiation of translation. The extensive secondary structure of these viral RNAs allows them to bind to DAI, inhibit dsRNA-binding, and thereby interfere with the action of interferon (114).

Several viral "defense" molecules have been identified in many different viruses (114,305). This rapidly expanding area of research is exciting and provocative, but most of the findings are based on *in vitro* studies, and it is essential that these observations be put to the test *in vivo*. In fact, in one instance we already know that the presence of viral defense molecules is not sufficient to evade the host's immune response completely. Several molecules that interfere with various cytokines have been found in vaccinia virus, and yet vaccinia virus does not cause a persistent infection and is efficiently eliminated by the immune system. Although these proteins are likely to play a critical role in the pathogenesis of vaccinia virus, it appears that at least in this instance, they are not sufficient to provide long-term persistence within an individual host.

Immunologic Tolerance

Finally, perhaps the most efficient means of establishing and maintaining a chronic infection is to selectively silence the effector system responsible for clearing the virus. The classic example of this is the suppression of LCMV-specific CTL responses in congenitally infected carrier mice (35,179). Adult mice infected with LCMV mount a vigorous cellular and humoral response against the virus and clear the infection within 2 weeks. This clearance is mediated primarily by LCMV-specific $CD8^+$CTL (10,43,202,221,362, 363). In contrast to the acute infection seen in adults, mice infected with LCMV at birth or *in utero* become chronically infected, showing lifelong viremia with high levels of infectious virus and viral antigen in most of their major organs. The persistence of LCMV in these carrier mice is accompanied by lack of T-cell response to the virus. This is a highly specific defect. Such persistently infected mice exhibit no generalized immune suppression and respond normally to other antigens, but they show no detectable T-cell responses against LCMV because the virus-specific T cells have been clonally deleted within the thymus upon seeing viral antigen (11,35,147,222,252). The inability of the carrier mice to eliminate virus is primarily due to this T-cell defect. The best evidence for this comes from experiments showing that adoptive transfer of LCMV-specific $CD8^+$ T cells results in clearance of virus from carrier mice (9,110,146,240,338). Although other viral systems have not been studied as extensively, it is likely that decreased virus-specific T-cell responses are an important factor in the establishment and maintenance of many persistent viral infections.

Immunologic tolerance plays a role in the establishment of chronic HBV infections as well (57). Most children born to HBV-infected mothers go on to become HBV carriers. It is likely that in these neonatally infected carriers at least some of the HBV-specific T cells undergo clonal deletion either within the thymus or in the periphery. Even during adult onset of HBV infection, it is possible that some of the virus-specific T cells may be deleted in the periphery through overstimulation by high doses of viral antigen, as has been documented during infection of adult mice with macrophage-tropic and invasive strains of LCMV that rapidly produce a high antigen load in many tissues (11,203,222). Under these conditions, virus-specific $CD8^+$ CTL are overstimulated and driven to clonal exhaustion (deletion) in the periphery. Recent studies have shown that $CD4^+$ T cells play a critical role in sustaining $CD8^+$ CTL responses during chronic infection; under conditions of $CD4^+$ T-cell deficiency there is rapid exhaustion of LCMV-specific CTL in the periphery (22,202). These findings have implications for chronic viral infections in general and may provide a possible explanation for the loss of HIV-specific CTL activity that is seen during the late stages of AIDS, when $CD4^+$ T cells become limiting and their numbers fall below a critical threshold necessary for maintaining CTL function.

The various viral evasion mechanisms that we have discussed are summarized in Table 4. Although we have cited specific examples to illustrate the different strategies, it is unlikely that a single mechanism can account for the persistence of a given virus. It is more likely that a combination of these strategies, plus other mechanisms of which we are currently unaware, contributes to the persistence of virus in an otherwise immunocompetent host.

SELECTED EXAMPLES OF VIRUSES THAT PERSIST IN HUMANS

This section examines the mechanisms by which several viruses establish persistent infections. These are some of the best-studied examples of persistent infections, and space limitations prevent us from discussing all of the viruses listed in Table 1. The reader is referred to the appropriate chapters for discussion of the remaining viruses.

Herpes Simplex Virus

The herpes viruses, members of the *Herpesviridae* family, share the property of causing a persistent infection in the host. Herpes simplex virus (HSV), the most extensively studied of the herpes viruses, persists in its human host by establishing a latent infection in sensory neurons. The two types of herpes simplex virus, type 1 (HSV-1) and type 2

TABLE 4. *Viral strategies for evading the immune system*

Escape mechanism	Example[a]
1. Restricted gene expression; virus remains latent in the cell with minimal to no expression of viral proteins	HSV and VZV in latently infected neurons, EBV in B cells, HIV in resting T cells
2. Infection of sites not readily accessible to the immune system	HSV, VZV, measles, and rubella persistence in neurons/CNS CMV, polyomaviruses BK and JC in the kidney EBV and CMV in the salivary gland Papillomaviruses in the epidermis
3. Antigenic variation; virus rapidly evolves and mutates antigenic sites that are critical for recognition by antibody and T cells	Antibody escape variants in lentiviruses CTL escape variants in HIV, EBV, and HBV T cell receptor antagonism by HIV and HBV variants
4. Suppression of cell surface molecules required for T cell recognition	Suppression of MHC class I molecules by adenoviruses, CMV, HSV, and HIV Decreased expression of cell-adhesion molecules LFA-3 and ICAM-1 by EBV Suppression of MHC class II molecules by CMV, HIV, and measles
5. Viral defense molecules that interfere with the function of antiviral cytokines	Adenovirus proteins (E3-14.7K, E3-10.4K/14.5K, and E1B-19K) protect infected cells from lysis by TNF Adenovirus VA RNA, EBV EBER RNA, and HIV TAR RNA inhibit function of interferon-alpha/beta EBV protein BCRFI (a homologue of IL-10) blocks synthesis of cytokines such as IL-2 and interferon-γ
6. Immunological tolerance	Clonal deletion/anergy of virus-specific CTL in HBV carriers

[a] Only examples of viruses known to persist in humans are cited.

(HSV-2), generally infect the human host at different sites. HSV-1 undergoes a primary productive infection in the oral mucosa or oral cavity (gingivostomatitis). The virus enters sensory nerve endings and establishes a latent infection in sensory neurons of the trigeminal ganglion. At later times the virus reactivates from latent infection to cause recurrent lesions commonly known as cold sores or fever blisters. HSV-1 can also infect the cornea (herpes keratitis) as a result of primary infection or reactivation from latent infection. The host immune response to recurrent infections of the cornea can lead to progressive scarring of the cornea, clouding, and eventual blindness. Some lines of evidence suggest the possibility that HSV-1 may also establish a latent infection in corneal keratinocytes (177). Lastly, HSV-1 can travel along neuronal pathways into the central nervous system during primary infection or reactivation, resulting in a very serious encephalitis.

HSV-2 is acquired generally as a sexually transmitted disease. HSV-2 infects the genital mucosa, spreads into nerves, and is transported to sacral ganglia where it establishes a latent infection. Reactivation of this virus causes the recurrent lesions associated with genital herpes. In addition to genital infection, HSV-2 can spread systemically to cause meningitis in a limited number of cases. More serious, however, is the intrauterine or peripartum transmission of HSV-2 from a productively infected mother to her child, which can result in encephalitis and/or disseminated herpes infection in the newborn. Mortality of infected newborns is high despite the availability of antivirals to limit infection, and survivors frequently experience lifelong sequelae.

Despite the varied manifestations of HSV-induced disease, the stages of pathogenesis of HSV infection are relatively constant (Fig. 2). The process begins with productive infection of epithelial cells at the site of inoculation. Infection of these cells involves transcription and expression of immediate early (IE), or α, genes; expression of early (E), or β, gene products; viral DNA replication; late (L), or γ, gene expression; and assembly of progeny viral particles. Following release from these cells, the virus spreads to surrounding cells and eventually binds to and enters axon termini of sensory neurons. The virus or its nucleocapsid is translocated by retrograde axonal transport to the neuronal cell body, where it can undergo one of two infection pathways. It can establish a productive infection leading to release of progeny virus and possibly further spread through the nervous system. Alternatively, it can establish a latent infection during which very limited viral gene expression occurs and no infectious virus can be detected in the tissue. In experimental systems, latent infection is defined as the absence of infectious virus when homogenized ganglion tissue is assayed directly on sensitive cell cultures but the presence of infectious virus when intact ganglion tissue or neurons are cocultivated with permissive cells. In contrast to productive infection, during which HSV expresses more than 70 viral gene products, latent infection involves expression of only the latency-associated transcripts (LATs), as detected by *in situ* hybridization (Fig. 3). At later times, upon appropriate stimulation of the latently infected neuron, virus can reactivate, bring about a limited productive infection, and spread down the axon to the site innervated by the sensory neuron. There virus is released to establish a productive infection in epithelial tissue, resulting in a recurrent lesion at a site approximating that of the primary infection.

Establishment of Latent Infection

In the infected sensory neuron, at least in experimental animal systems, HSV can cause a productive infection lead-

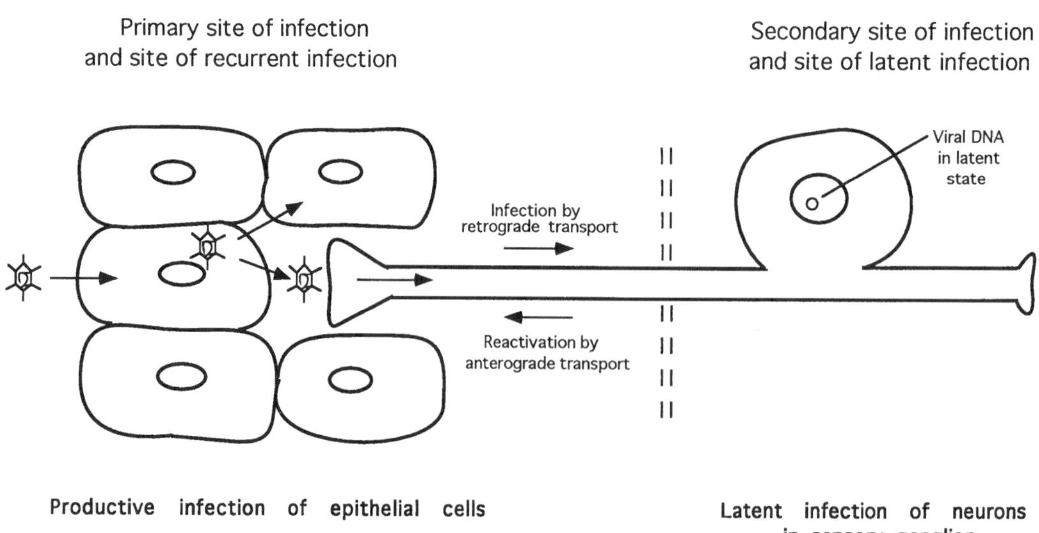

FIG. 2. Diagram of the stages of infection of the host by HSV.

ing to formation of progeny virus. Alternatively, the virus can establish a latent infection where viral gene expression is limited and the neuron survives with the viral genome maintained stably in the neuron in a quiescent form. Although the events in an infected neuron that lead to establishment of latent infection have not been defined, the limited viral gene expression during latent infection (75,315) led to the idea that restriction of viral IE gene transcription in sensory neurons contributes to or is the sole determinant of the nonpermissive state of these cells. Various hypotheses to explain the basis for the limited transcription of IE genes have been proposed: (i) Cellular transcription factors needed for expression of these viral genes are absent or present at levels too low to support productive infection in these neurons (reviewed in 269). (ii) Cellular inhibitors of IE gene transcription are present in these neurons (161). (iii) The HSV VP16 virion protein, which normally up-regulates transcription of IE genes, cannot function in these neurons to promote IE gene transcription (269). However, a recent study indicates that the lack of VP16 in the neuronal nucleus cannot be the sole factor leading to the nonpermissive state of the neuron and the resultant latent infection (289). (iv) It has also been suggested that limited IE gene expression may occur in neurons but that transactivation of later gene expression cannot take place (115). (v) Another hypothesis is that viral gene products such as ICP4, ICP27, or ICP8 may negatively regulate IE transcription in neurons (165,238,322). Obviously, one or more of these mechanisms could contribute to the nonpermissive infection of sensory neurons.

A modified regulatory pathway for HSV gene expression in neurons as compared to epithelial cells has also been proposed (169). In this model, IE gene transcription is limited upon infection of neuronal cells, but if enough IE and E viral gene products are expressed to permit viral DNA replication, IE and other viral gene expression is amplified to levels sufficient for productive infection to occur. Thus, if IE and E gene expression and DNA replication cannot occur, the virus establishes a latent infection. This proposed regulatory pathway in a sense provides a mechanism for the virus to sense the permissiveness of the host cell and establish a latent infection if the host cell is not permissive for IE and E gene transcription and viral DNA replication.

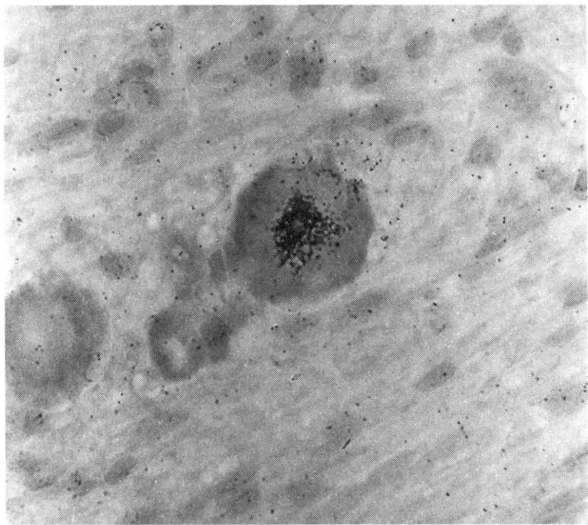

FIG. 3. Detection of the HSV LAT RNA in latently infected murine trigeminal ganglion neurons by *in situ* hybridization. Mice were infected with HSV-1 following corneal scarification. At 30 days postinfection, the mice were sacrificed, and trigeminal ganglia were frozen and sectioned. The sections were hybridized with a ^{3}H-labeled DNA probe detecting the LAT RNA as described (61). (Micrograph provided by Magdalena Kosz-Vnenchak.)

Viral genetic studies have attempted to address two issues regarding establishment of latent infection: (i) the stage in infection at which the latent versus productive infection pathways diverge, and (ii) the role of viral gene products in effecting this choice of infection pathways. With regard to the stage of infection at which the two infection pathways diverge, it is known that mutant viruses (61,158,168,336) or wild-type viruses, at least in some neuronal cells (198,292,303), can establish latent infections without complete replication or even substantial viral gene expression in neurons. This has been interpreted to mean that the latent infection pathway deviates very early from the productive infection pathway (61,158,198,336). However, the *in situ* hybridization techniques used in many of these studies were not sensitive enough to distinguish between the total absence and low levels of viral gene expression during establishment of latent infection. If low levels of viral gene expression do occur, the productive infection would be aborted at a later stage after limited viral gene expression. With regard to the second issue, the role of viral gene products in establishment of latency, all viral mutants tested to date are capable of establishing latent infection, and thus there is no evidence that expression of any viral gene product is required for latency (61,86,158,181–183,210,290,328). Some viral mutants, including LAT-negative mutant viruses, show reduced efficiency of establishment of latent infection (279). Although most workers have interpreted this as being caused by decreased spread of the mutant virus to the neurons, one group has postulated a specific role for LAT in promoting establishment of latent infection (279). Because no viral gene product appears to be absolutely essential for establishment of latent infection, the possibility that the neuron controls the establishment of latency has been raised (158,198). Nevertheless, it is conceivable that viral gene products may participate in or influence a choice of infection pathways in the neuronal host cell. The other implication of these results is that productive infection is not a prerequisite for establishment of latent infection.

Maintenance of Latent Infection

As described earlier, during the latent infection no infectious virus can be detected within sensory ganglia, but viral DNA can be demonstrated, as detected by Southern blot hybridization (266) and by polymerase chain reaction (PCR) amplification of viral DNA sequences within ganglion tissue (158). Southern blot hybridization studies have determined that the viral DNA is present in an "endless" form in which the viral DNA termini are not detected (211,266–267). Detailed studies have permitted the conclusion that the viral DNA is probably in a circular episomal form (211) associated with nucleosomes (77). Because neurons are nondividing, postmitotic cells, the viral DNA would not need to be replicated during the maintenance phase of the latent infection, and viral gene products required for replication would not need to be expressed. In fact, as described earlier, the only viral gene products that have been detected within latently infected tissue are the LAT transcripts (67,75,254,268,314,315), and no protein product in latently infected tissue has been associated with these viral transcripts. As discussed in the preceding section, the lack of viral protein expression would make the latent infection invisible to the host immune system. Furthermore, sensory neurons express few or no MHC molecules; thus, T cells cannot readily target infected neurons even if viral proteins were expressed.

The major form of the LAT transcript is a 2-kb RNA originally detected by *in situ* hybridization with viral probes. It accumulates within the nuclei of latently infected neurons (Fig. 3). The minor LAT transcript is a large, approximately 8.5-kb RNA whose coding sequences extend upstream and downstream from those encoding the 2-kb major LAT RNA species (80). The minor LAT is believed to be the primary transcript of this transcriptional unit, which is spliced to give the stable 2-kb species (92).

Genetic studies examining the role of the LAT transcripts in the maintenance of latent infection have demonstrated that the LAT transcripts are not essential for maintenance of latency because LAT-negative viruses can still persist in a latent state (126,182,309,333). Thus, maintenance of latent infection may involve only host cell factors that downregulate viral replication, and HSV gene products may not play an active role.

Reactivation from Latent Infection

The mechanism of HSV reactivation in response to neuronal stimuli has not been defined, but this process must in some way lead to expression of viral gene products needed for productive infection. One model proposes that an appropriate stimulus leading to expression of ICP0, an IE nonspecific transactivator of viral gene expression, may activate expression of other IE gene products such as ICP4 and ICP27 (181). These gene products would then in turn activate E and L viral gene expression. However, ICP0 is not absolutely required for reactivation because ICP0-negative mutants can reactivate, albeit less efficiently (181). Two other models suggest that amplification of HSV DNA in the latently infected neuron may trigger expression of viral lytic phase genes (169,269), although one model postulates that amplification of the viral DNA is accomplished with cellular enzymes (269,291), whereas the other postulates that viral DNA synthesis utilizes viral gene products (169).

With regard to the viral gene products needed for reactivation, ICP0 may play a role in promoting reactivation because ICP0-negative mutants reactivate less frequently than wild-type viruses, even though the amount of viral DNA present in the ganglia is approximately the same for mutant and wild-type viruses (181). Similarly, LAT-negative mutant viruses show reduced efficiency or delayed kinetics of reactivation (126,182,309,333), leading to the hypothesis that LAT promotes reactivation, even though it is not essential. Ultimately, reactivation involves replication of the virus, and thus all the gene products essential for productive infection must be required also for the reactivation event.

The stimuli for reactivation in the human host include immunosuppression, hormonal changes, stress, neurectomy, nerve damage, and ultraviolet light exposure. These stimuli are likely to cause changes in the physiologic status of the neuron, and it would be predicted that this could lead to activation of neuronal signaling pathways, causing activation of transcription factors or protein kinases. These factors in turn could activate expression of viral or cellular gene products, leading to an up-regulation of expression of viral IE gene products. In experimental animals, reactivation has been achieved by explant of the latently infected tissue into culture and cocultivation with permissive cells. In addition, *in vivo* reactivation has been induced by physical trauma to the animal, high temperature, uv irradiation, iontophoresis of adrenaline into the eye, or neurectomy. These treatments may cause damage to, or physiologic changes in, the neuron that would similarly be predicted to lead to changes in levels of protein kinases or transcription factors within the neuron. HSV seems to have evolved reactivation mechanisms that become operative when the neuronal host cell is perturbed so that the virus can get out of an injured or dying neuron and spread to a new host cell or to a new host organism.

Varicella-Zoster Virus

Varicella-zoster virus (VZV), a second member of the *Herpesviridae* family, is the causative agent of the acute disease called chickenpox (varicella) and the recrudescent disease called shingles (herpes zoster). Because VZV remains highly cell-associated during its replication in cultured cells, it has been difficult until recently to obtain high-titer stocks of cell-free virus. This has hampered studies of the replicative mechanisms of VZV. In addition, there is no good animal model system for pathogenesis studies. Thus, nearly all of our understanding about the pathogenesis of VZV has been limited to the information derived from clinical samples and studies.

VZV enters an individual by infection of mucosal epithelial cells in the upper respiratory tract, oropharynx, or conjunctiva. After primary replication in the epithelium, the virus is disseminated by the bloodstream to the reticuloendothelial system where viral replication leads to secondary viremia. Infection of capillary endothelial cells allows spread of the virus to epithelial cells of the epidermis, where focal cutaneous lesions are formed, causing the pocks characteristic of varicella. Host immunity limits the acute disease, but during spread through the epidermal epithelium, the virus also infects sensory nerve endings and is transported to sensory ganglia. Latent infection is established in the ganglia, in neurons (109,142,196), and/or in satellite cells (66,320). At a later time, reactivation may occur as a consequence of immunosuppression, nerve damage, or other stimuli. If the existing immune response cannot control the reactivating virus, extensive viral replication can occur in the ganglia. Virus also spreads to the periphery by axonal transport through neurons innervating a specific dermatome(s). There the virus productively infects the epithelium to cause the lesion characteristic of shingles.

Little is known about the mechanisms of establishment and maintenance of latent infection by VZV. It is believed that latent infection is established in a host cell nonpermissive for viral gene transcription. There have been reports of expression of viral transcripts and proteins from a variety of genes involved in VZV productive infection during what was presumed to be latent infection (66,142,329), but it is difficult to rule out that these viral gene products were not expressed in cells undergoing limited reactivation. A recent detailed study has found VZV transcripts only from genes 29 and 62 in latently infected human trigeminal ganglia, suggesting that selected transcription of VZV IE and E genes occurs during latent infection (209). Similarly, little is known about the mechanisms of reactivation.

Like HSV, which also undergoes a latent infection in sensory ganglia, VZV evades the immune response to establish latency in an immunologically privileged site and downregulates gene expression there. However, the clinical picture of VZV latency and reactivation shows many differences from that of HSV-1: (a) Most individuals undergo one episode of zoster, whereas HSV recurrences are numerous. (b) The probability of zoster increases with age, whereas the frequency of HSV recurrences decreases with time. (c) Neuropathy and cutaneous spread of VZV during zoster can be more extensive than during HSV recurrence. Some have postulated that the capacity of VZV to establish latent infection in satellite cells instead of, or in addition to, neurons could contribute to these differences (66), but the explanations for these differences will come only with a full understanding of VZV and HSV latent infection.

Epstein-Barr Virus

Epstein-Barr virus (EBV), a third member of the *Herpesviridae* family, is the causative agent in infectious mononucleosis and is a cofactor in the induction of the neoplastic diseases of Burkitt's lymphoma, nasopharyngeal carcinoma, posttransplant lymphoproliferative disease, non-Hodgkin's lymphoma, and oral hairy leukoplakia. Limited information exists about the replicative mechanisms of EBV because the virus does not efficiently undergo a productive infection in any cell type *in vitro*. However, EBV does efficiently establish a latent infection in B lymphocytes *in vitro*. Because no animal system faithfully reproduces the many other EBV disease manifestations seen in humans, much of our knowledge about EBV pathogenesis has been derived from clinical studies and studies on latently infected B cells.

Individuals become infected with EBV by the oral route, possibly by contact with saliva containing infectious virus. EBV can undergo a limited productive infection in oropharyngeal epithelial cells (299,300), which can continue as a chronic infection in some individuals (107). By an unknown mechanism, virus is transmitted to B lymphocytes, where a nonproductive, latent infection is usually established. For entry into B cells, the virus utilizes the CD21 cell surface protein as a receptor (99,102,230). Upon entry into the cell, the viral DNA is uncoated and circularizes to form the covalently closed circular genome that persists as an episome in the latently infected cell (3,233).

Unlike infection of neurons by HSV, EBV establishes latency in dividing cells and therefore must retain the potential for replication of its genome during latency. As many as 11 latency gene products are expressed, the number of viral gene products differing somewhat in different latently infected cells. The detection of Epstein-Barr nuclear antigens (EBNAs) by immunofluorescence is illustrated in Fig. 4. The Epstein-Barr virus nuclear antigen 1 (EBNA-1) protein is always expressed in latently infected cells because it plays an essential role in viral DNA maintenance in the latently infected B cell. EBNA-1 binds specifically to EBV DNA sequences (257) that serve as the origin for plasmid replication, oriP (356), and promotes the replication of the viral episome by host cell DNA polymerase during S phase of the host cell cycle (3,4,123). Other EBV latent gene products, such as EBNA-2, stimulate growth of the host B cells, in part by inducing expression of B-cell activation molecules (345). This state of continuously stimulated growth is often described as immortalization.

A small proportion of latently infected B cells may become permissive for viral replication, in which case viral IE proteins are induced, followed by expression of E proteins, viral DNA replication, L gene expression, and progeny virus assembly, very much as described earlier for HSV. Reactivation of latent virus from B lymphocytes in the human host is thought to allow the virus to reinfect epithelial surfaces, including the oropharynx, which would allow viral shedding and transmission. Induction of the lytic cycle in latently infected cells in culture can be achieved by addition of phorbol esters or halogenated nucleosides.

The mechanisms employed by EBV to evade the immune response may differ in latent versus chronic infection by EBV, and the virus has been shown to have a broad array of evasion strategies. Viral gene expression is restricted during latent infection of B cells, but because some viral proteins are expressed, the virus must employ additional strategies to escape immune detection. The T-cell adhesion molecules LFA-3 and ICAM-1 are down-regulated in EBV-transformed lymphoblasts (117), representing a possible evasion strategy for cells either latently or chronically infected with EBV. In chronic infection of nasopharyngeal epithelium by EBV, the virus infects a site that is immunologically privileged because of its relative inaccessibility to T cells. In addition, CTL escape variants have been demonstrated (71). Although involvement of many of these mechanisms in EBV persistence *in vivo* has yet to be demonstrated, EBV is a fine example of a virus that apparently has evolved multiple strategies to maintain a persistent infection.

EBV infection of B cells is a classical latent infection with persistence of the viral genome as an episomal plasmid whose replication is tied to the host cell cycle and promoted by the EBNA-1 protein. In addition, it is now clear that the virus can chronically infect oropharyngeal epithelial cells, although little is known about the mechanisms underlying this chronic infection. Thus, within the same individual, EBV may persist by two mechanisms: latent and chronic infection.

Human Papilloma Virus

Human papilloma viruses (HPV), the causative agents of human warts, establish a nonproductive infection of basal layer cells in the skin, but replicate productively when basal layer cells or their daughter cells differentiate into keratinized cells of the stratum spinosum and granular layers

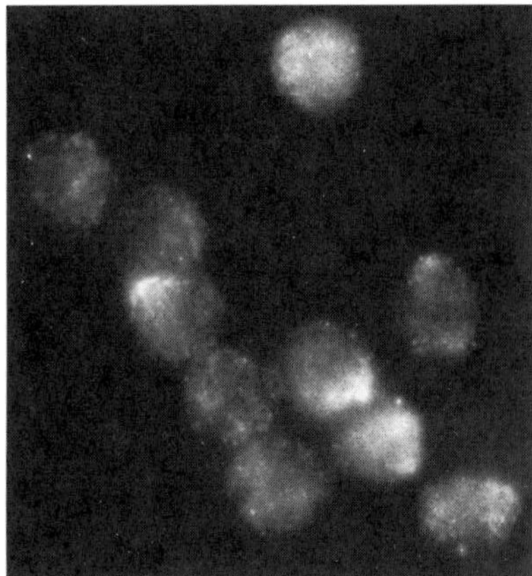

FIG. 4. Detection of Epstein-Barr nuclear antigens (EBNAs) in latently infected cells by immunofluorescence. Cells from an Epstein-Barr virus-positive Burkitt lymphoma line expressing EBNA complex were stained with human serum from a normal individual persistently infected with EBV, followed by anticomplement immunofluorescence. (Micrograph provided by E Kieff, J Minton, and F Wang.)

of the epidermis (Fig. 5). Papilloma viruses do not replicate in cultured cells except in very specialized systems, so much of the information available on their replication cycle and pathogenesis has been pieced together from studies of nonpermissive infections of cultured cells by bovine papilloma virus and, more recently, human papilloma viruses, and from histologic analyses of viral gene expression in clinical samples of wart tissue. Infection of basal layer skin cells by HPV is initiated when the virus is introduced into the lower layers of the epidermis as a consequence of trauma to the epithelium. Early viral transcription can occur in these cells (318), and the viral DNA is retained in these cells (283), probably replicated to low copy numbers as an extrachromosomal plasmid. However, the undifferentiated basal layer cells are not permissive for replication of viral DNA for packaging into progeny particles (a process called vegetative viral DNA replication) or for late gene transcription. The early viral gene products stimulate cell growth in the basal layer, leading to formation of a wart. Wart tissue consists of thickened layers of normal skin tissue (Fig. 5) with an intact basement membrane. As the basal layer cells divide and differentiate into keratinocytes, they become permissive for vegetative viral DNA replication, allowing late viral gene expression and formation of infectious viral particles. Thus, amplified levels of viral DNA and late transcripts and proteins can be found only in the most superficial layers of wart tissue (see Fig. 1 in Chapter 29; 318,319). According to our formal definition of latency given earlier, HPV infection of basal cells could be considered to constitute a form of latent infection during primary infection itself in that the cells contain the viral genome, but no viral proteins or infectious virus are produced. These cells are nonpermissive because vegetative viral DNA replication cannot take place. Differentiation of the basal cells leads to induction of vegetative viral DNA replication, most likely because either a missing host factor becomes available or an inhibitor is lost.

In addition, there is evidence for a true latent infection by HPV. HPV can remain quiescent without stimulating cellular differentiation and clinically evident wart production. This is observed when primary infection does not immediately lead to a wart or when a wart regresses, but wart formation is induced at later times by immunosuppression or other stimuli. Little is known about the mechanisms involved in establishment and maintenance of this latent infection, but basal cells may harbor the latent viral genomes because upon reactivation a wart is formed within differentiated epithelial tissue as in a primary infection (95,308). Therefore, the signal for productive infection by HPV from the latent infection as well as the primary infection appears to be differentiation of the host cell. It will be of interest to determine whether the viral DNA persists as a replicating plasmid as described earlier. The limited viral gene expression during latency probably contributes to the inability of the host immune response to detect and clear HPV-infected cells. In addition, productive infection occurs only in the epidermis, placing potentially immunogenic proteins and viral particles at a site physically separated by a basement membrane from immune effectors.

At a low frequency, after a period of 20–25 years, a wart or latently infected cell can undergo an oncogenic and in some cases malignant conversion (308). In carcinomas associated with HPV-6 and 11 infections, usually at low risk for malignant progression, the viral genome is extrachromosomal and transcriptionally active. In contrast, in neoplasia associated with the more frequently malignant HPV-16 and HPV-18 infections, the viral genome is integrated into host cell chromosomes (83), often in a way that disrupts the HPV E2 ORF (18,288). This inactivates the repressor function of E2, allowing expression of the E6 and E7 oncogene products (339). The HPV-16 and HPV-18 E6 proteins inactivate the host p53 tumor suppressor gene product by forming a ternary complex with p53, activating the ubiquitination of p53 and thereby effecting its subsequent degradation (140,141,280,281). The HPV-16 and HPV-18 E7 proteins displace pRB from a complex with the E2F transcription factor, giving free E2F, which can activate transcription of genes involved in progression of the cell cycle into S phase (84). Thus, the E6 and E7 proteins induce cell transformation by inactivating two of the major cell proteins that regulate cell growth, thereby inducing a multistep pathway to malignancy. Although this pathway is not a mechanism for persistence of viral genetic information, it illustrates how persistence of HPV in a host may ultimately lead to severe disease.

Adenovirus

Humans are susceptible to infection with nearly 50 serotypes of adenoviruses, DNA viruses of the *Adenoviridae* family. Adenoviruses enter a host through the respiratory or gastrointestinal tracts or the conjunctiva, most likely by interacting with a receptor on epithelial cells (193). Acute adenovirus infections are usually restricted to epithelial cells of respiratory and gastrointestinal epithelium and are readily resolved, but severe respiratory infections result in adenovirus pneumonitis, which can be fatal. During acute infection of cultured cells, adenovirus profoundly inhibits host cell macromolecular synthesis (111,132,250). Systemic infection can occur also in other organs, but knowledge of the pathogenesis of these infections is limited by the dearth of suitable animal models.

Adenovirus can persist in a host by establishing a latent infection in nonepithelial cells and tissues. In fact, adenovirus was initially isolated from adenoid tissue (272) that contained no detectable infectious virus prior to cocultivation with susceptible cells (1). More convincing evidence of adenovirus latency in human cells was obtained from tonsillar tissue in which adenovirus DNA was found, but even repeated passage failed to demonstrate infectious virus

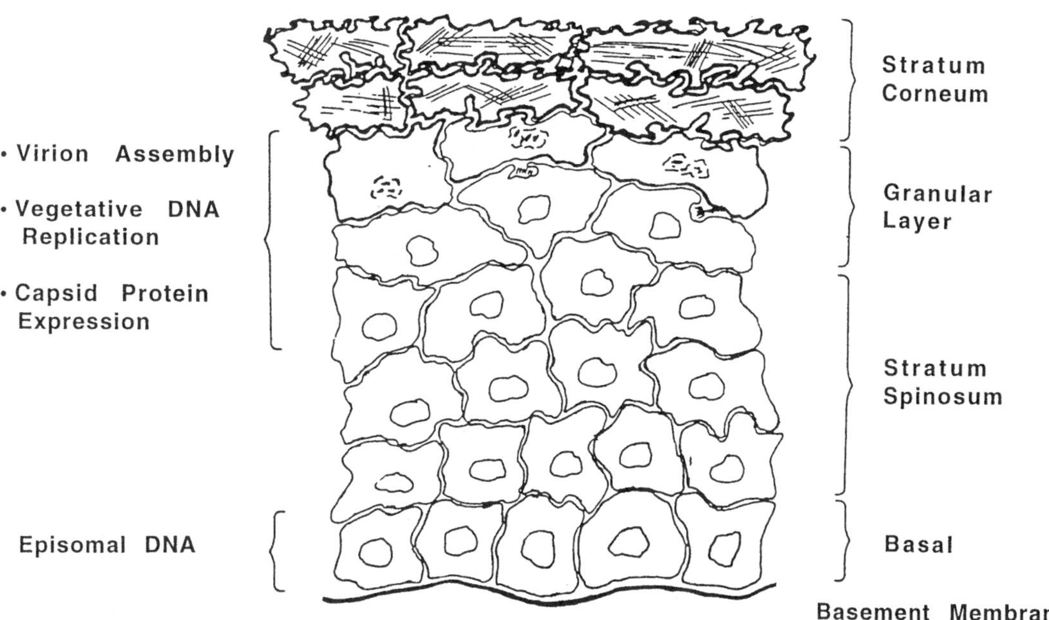

FIG. 5. Diagram of the stages of infection of the host by papilloma virus.(Provided by Michael Ceniceros and Peter Howley.)

(231). Adenovirus DNA has also been detected in peripheral blood lymphocytes from healthy adults, suggesting that circulating lymphocytes may be another source of persistent adenovirus (1,138). Latent adenovirus does not reactivate spontaneously, but it can be induced to reactivate by explant of latently infected tissue to permissive cell cultures or by transplantation of latently infected organs. The precise mechanisms by which adenovirus establishes and maintains a latent infection in organs and lymphoid cells, and whether its DNA is integrated into the host genome or remains episomal, have not yet been elucidated.

Adenoviruses also have the capacity to persist in a host by establishing a chronic infection. In some cases, shedding of virus from epithelial surfaces can be quite prolonged despite the induction of host immune responses to the virus (91,100). Epidemiologic studies have provided evidence that chronic shedding of adenovirus can occur in healthy human hosts for at least 2 years after infection (100,127). During chronic infection by adenovirus, replication is thought to be "smoldering," with low levels of progeny virus formation and limited spread to neighboring susceptible cells. Adenovirus may escape the host immune system to establish and maintain chronic infection with the aid of products of the adenovirus E3 transcription region. These viral gene products may prevent host clearance of virus by inhibiting TNF-mediated cytolysis and by abrogating CTL recognition through binding of the E3 19K protein to MHC class I molecules (reviewed in ref. 353). Much remains to be learned about adenovirus persistence and immune evasion *in vivo*, particularly in light of the proposed use of adenovirus vectors for gene therapy.

Adeno-Associated Virus

Adeno-associated virus (AAV) is a member of the *Dependovirus* genus of the *Parvoviridae* family, which can persist in a latent form in the human host and in cultured cells. Although there is limited information about the mechanisms by which parvoviruses persist in the human host, the mechanism by which AAV establishes a persistent infection in cultured cells, integration of the viral DNA into host chromosomes, is novel among the DNA viruses and warrants discussion here. AAV particles contain a single-stranded DNA genome encapsidated in a protein coat. AAV replication requires coinfection with a helper virus such as adenovirus or HSV. Because AAV is commonly found in a latent form within human embryonic kidney cells (134), latent infection *in vivo* is believed to be common. Nevertheless, AAV is not associated with any known disease.

All knowledge about the mechanisms of latent infection by AAV has come from studies of AAV infection of cells in culture. Upon infection of a host cell in the absence of helper virus, limited AAV gene expression occurs. Viral DNA replication has not been detected after viral entry into the host cell, but the viral DNA sequences become integrated as double-stranded DNA in the host cell DNA in the form of tandem repeats (55,124). Therefore, at least limited DNA replication must occur. The viral DNA is integrated specifically within one region of chromosome 19, although the precise site of integration may range over several hundred base pairs (170–172,277). The AAV rep protein appears to promote integration, either through down-regulation of viral gene expression and DNA replication or by binding to the preintegration site, but it is not absolutely essential for integration. Thus, AAV can persist within cultured cells and probably its human host by integration of its genomic DNA into the host chromosome. Upon superinfection of latently infected cells with a helper virus (134) or possibly upon other environmental changes, rescue of the latent genome occurs with production of infectious AAV leading to possible transmission to a new host.

Hepatitis B Virus

Perhaps the best example of a human virus causing an acute infection followed by a chronic infection is hepatitis B virus (HBV). HBV is a hepadna virus containing a partially double-stranded DNA genome. The genome is converted into a double-stranded form upon entry into the cell and is transcribed into RNA molecules, which serve as mRNA and genomic replicative intermediates. The replicative intermediate is then reverse-transcribed into the partially double-stranded form that is assembled into progeny virions. The productive life cycle is maintained during both the acute and chronic phases of infection. The mechanisms by which the virus might regulate replication during infection are not well understood and remain an area for further study.

HBV infects humans by transfusion of contaminated blood materials, by use of contaminated needles, or by perinatal or transplacental transmission of the virus from mother to child. Acute replication of the virus in hepatocytes and the immunopathology associated with the antiviral immune response lead to acute hepatitis. Subsequently, 2% to 10% of adults undergoing an acute HBV infection and more than 80% to 90% of acutely infected infants become chronic carriers. During the chronic infection, infectious virus is continuously produced by hepatocytes and possibly by other cells, resulting in viremia. Chronic infection can lead to transmission to other individuals and in some cases to development of hepatocellular carcinoma.

HBV-specific T- and B-cell responses during acute and chronic infection have been extensively studied (57). Individuals who clear HBV infection and those who become chronic carriers provide an interesting contrast in the immune responses they make against HBV.

The antibody response to HBV envelope proteins (HBsAg, PreS1, and PreS2) is thought to play a critical role in viral clearance and in preventing reinfection

(23,37,57,90,144,164,239). High levels of these anti-envelope antibodies are present in individuals who have resolved an acute HBV infection but are usually undetectable in the serum of chronic HBV carriers. In contrast, antibodies to HBV nucleocapsid antigens (HBcAg and HBeAg) and nonstructural proteins are readily detected in patients with chronic HBV infection (53,57,94, 214,311,312,350,360). These findings documenting the selective absence of anti-envelope antibody in carriers have led to the concept that neutralizing antibodies directed to HBV envelope proteins play an important role in viral clearance. However, a recent study has shown that anti-envelope antibodies are, in fact, present in HBV carriers, but they are often missed by conventional techniques because these antibodies are already complexed with viral surface antigens that are present in vast excess in the sera of HBV carriers (200). This is a potentially important finding, and additional quantitative studies should be done to address this issue in greater detail.

Several studies have examined both $CD4^+$ and $CD8^+$ T-cell responses during acute and chronic HBV infections (20,52,57,96–98,153,229,247). The general consensus is that in individuals who resolve the acute hepatitis and eventually clear the infection, the T-cell response to HBV is potent and directed to multiple epitopes. However, in individuals who become persistently infected, the T-cell response is relatively weak and oligoclonal (i.e., restricted to a few epitopes). Taken together, these results suggest that both the quality and the magnitude of the T-cell response to HBV are important determinants of viral clearance or persistence. However, the factors that determine why certain individuals generate an effective response and control the infection, whereas others make weaker responses and fail to eliminate the virus, are not well understood. Recent studies have documented the presence of CTL escape variants as well as HBV variants that interfere with CTL function (TCR antagonists) in HBV carriers and it is possible that such variants play a role in HBV persistence *in vivo* (26–27).

Measles Virus

Measles virus (MV) is a member of the *Paramyxoviridae* family of negative-strand RNA viruses, which replicate in the cytoplasm of infected cells. Infection of humans results in the acute disease called measles and, less commonly, measles pneumonia. Although measles is typically an acute disease, a high incidence of CNS involvement occurs, which can lead to development of acute postinfectious encephalitis or, rarely, subacute sclerosing panencephalitis (SSPE). SSPE is a progressive, dementing disease that develops years after apparent resolution of acute MV infection and appears to be a consequence of persistent MV infection of the CNS (330,347).

Humans constitute the only natural host for MV, so much of our information regarding its pathogenesis derives from studies of patient isolates. Many laboratory animals, including monkeys and rodents, can also be infected with MV and have been used to study both acute and persistent infection. From studies in these various systems it has been determined that MV initiates acute infection in the epithelium of the respiratory tract, oro- or nasopharynx, or conjunctiva (159,264). After replication in the mucosa, virus is transported to draining lymph nodes, possibly within macrophages. Virus undergoes secondary replication in the lymph nodes and then enters the bloodstream in leukocytes (118,143,150,246). This primary viremia disseminates virus to the reticuloendothelial system, especially of lymphoid tissue, where another round of virus replication causes lymphoid hyperplasia. Multinucleate giant cells pathognomonic of MV infection also form through fusion of infected cells with neighboring uninfected cells (14,47,159,296). Secondary viremia disseminates virus to multiple tissue sites where a generalized infection of vascular endothelium precedes epithelial infection of the gastrointestinal tract, respiratory tract, and conjunctiva and the appearance of the characteristic measles rash. Virus probably gains access to the CNS during the secondary viremic phase by infection of capillaries in the meninges, pia mater, and choroid plexus (217). In a normal individual, an immune response is induced that effectively controls acute infection.

Some MV-infected cells may escape immune clearance, however, to become persistently infected. Studies in cell culture indicate that addition of anti-MV monoclonal antibody to infected lymphocytes induces capping and subsequent shedding of the specific MV antigen (151), accompanied by selective decreases in other MV antigens (103,104). Modulation of viral antigens on the surface of infected cells has been postulated to subvert normal viral clearance, allowing MV to establish persistence in the host. MV has also been shown to suppress expression of MHC class I molecules (256), and interferon-γ-mediated up-regulation of MHC class II molecules (184). Although these phenomena have been demonstrated only *in vitro*, generalized decreases in T-cell responsiveness have been observed in MV-infected individuals, suggesting that these or other immune evasion strategies may be operative *in vivo*. Nonetheless, the establishment and maintenance of chronic infection has a virologic component because cells in culture can be persistently infected in the absence of immune effectors (234,347,358). These infections with MV differ from the classic definition of a chronic infection, however, because virus replicates slowly in the cultures with little or no production of extracellular virus (273). Restricted expression of the viral genome may occur through reduced synthesis of one or more structural components (120,298) or selection of defective viruses (17,151). These persistently infected cultures resemble infection in the brain of SSPE patients, in which viral nucleocapsids are retained intracellularly but the amount of various membrane antigens is reduced (17,189). This restricted viral gene expression, and infection of the privileged neuronal site, presumably plays a role in development of persistent infection in some MV-infected individuals.

The association of measles with SSPE was made first by observations of paramyxovirus nucleocapsids in neurons and glial cells and high titers of anti-MV antibody in the cerebrospinal fluid of SSPE patients. Infectious virus has been detected occasionally in SSPE tissue by cocultivation of infected brain tissue with cells permissive for MV, but more often no infectious virus is found. Mutation of genes encoding MV proteins as a mechanism of persistence was suggested by the observations that many of the viruses isolated from brain tissue of patients with SSPE show alterations in matrix (M) protein synthesis or expression (17,33,45,48,348). The M proteins of some SSPE strains are functionally distinct in that they remain cytosolic and cannot bind viral nucleocapsids at the plasma membrane or promote budding (128,129). M protein coding sequences from SSPE strains show evidence of clustered U to C substitutions in the positive strand sequences (postulated to result from a cellular RNA-modifying activity that alters regions of dsRNA) compared to sequences of acute MV isolates. These biased hypermutation events, originally identified in a case of measles inclusion body encephalitis (49), are thought to play a role in evolution of SSPE viruses from acute measles strains (354–355). Conserved hypermutation events in the coding sequences of cryptic M proteins from several SSPE virus isolates have been associated with loss of conformation-specific epitopes present in acute measles isolates (355). Consistent with these observations, M protein often cannot be detected immunohistochemically in brain tissue of SSPE patients (120–121), and many SSPE patients mount strong antibody responses to all MV proteins except the M protein (122,349). Presumably, these SSPE viruses fail to bud from infected cells due to defects in M protein, but the viral genome is amplified and defective virus is disseminated by direct cell-to-cell spread in the face of intact immunity.

M protein expression (17,235) and antibody to M protein (17,78,235–237) have been detected in other cases of SSPE, however, and in these cases defects in one or more genes encoding envelope proteins suggest a second class of genetic alterations in SSPE viruses. Some SSPE strains encode truncated or elongated fusion (F) proteins or F proteins with nonconservative amino acid substitutions in the cytoplasmic domain (51,282). Genetic variability in the gene encoding the hemagglutinin and antigenic variants have also been observed (30,46,296,297,331). Lastly, multiple strains resistant to the effects of interferon have been identified in acute MV infections and in SSPE patients (44).

The role of various viral mutations in establishment or maintenance of persistent MV infection and development of the rare cases of SSPE remains to be determined. Restriction of MV gene expression is observed in neurons and glia and may precede the accumulation of mutations. Recent evidence indicates that high levels of the antiviral protein MxA are rapidly induced upon infection of human brain cells, accompanied by marked restriction of MV RNA synthesis (284). This transcriptional attenuation is cell type specific and may contribute to establishment of MV persistence in the CNS.

One hypothesis to explain the genesis of persistent infection and ultimately of SSPE (50) proposes that host defense mechanisms promote the establishment of persistence through modulation of MV transcription and protein expression, preventing immunologic clearance of MV-infected cells in the process. During subsequent rounds of replication, the MV genome may accumulate mutations in multiple viral genes that would impair assembly of virions, in this case maintaining chronic infection without continual shedding of progeny virus. Selection of genetically altered viruses that are defective for progeny formation may eventually lead to evolution of a strain possessing a more pathogenic phenotype, shifting the infection from one that is undetectable to one that causes a progressive, degenerative and eventually fatal disease of the CNS.

Human Immunodeficiency Virus

Human immunodeficiency virus (HIV), the causative agent of acquired immunodeficiency syndrome (AIDS), is a human lentivirus in the *Retroviridae* family. Infection can occur by sexual transmission through breaks in the mucosal surface; by direct injection, as in puncture of the skin with contaminated needles; or by transfusion of contaminated blood products. Although the principal feature marking development of AIDS in HIV-infected individuals is the decline in numbers of $CD4^+$ T cells and the ensuing impairment of immune responses, current models of HIV pathogenesis describe a complex series of stages of infection that precede the state of immunodeficiency.

Acute Infection

Following entry of HIV into the host, the virus may replicate initially at the site of entry in Langerhans cells or other cells of phagocytic lineage and spread to local lymph nodes. Although the sites and mechanisms of primary replication and spread are currently unknown, the fact that initial isolates during acute infection are predominantly macrophage-tropic supports this conjecture (271,361). The blood and lymph then disseminate HIV widely through the body, where it becomes sequestered in macrophages or trapped by follicular dendritic cells in the spleen and other lymphoid organs (15,21,28,101,255,304,327). In these microenvironments, HIV may be concentrated on the surface of follicular dendritic cells from which it could be transmitted to $CD4^+$ lymphocytes as the lymphocytes migrate through the germinal center to the perifollicular mantle and paracortical areas of the lymph node and the spleen where most of the HIV-infected lymphocytes have been found (119).

Dissemination of virus corresponds to a burst of viremia during the acute infection that may be accompanied by acute clinical illness (58,68; Fig. 6). This phase lasts up to 2–4 weeks postinfection. The primary viremia then shows a rapid decline, usually between 3 and 6 weeks after infection. This decline correlates with appearance of a vigorous, HIV-specific CTL response that may play a crucial role in controlling virus replication (32,173,243,274). In contrast to the CTL response, neutralizing antibody appears much later (at or after 8 weeks). These observations strongly suggest that HIV-specific CTL are involved in eliminating productively infected cells.

Persistent Infection/Asymptomatic Phase

Despite an impressive control of the acute infection, the host immune response does not completely clear the virus. One crucial consequence of acute infection is the establishment of latency in monocytes and $CD4^+$ T cells in the lymph node and other lymphoid organs. Virus often persists for years before any obvious symptoms of immune deficiency are apparent, and infected individuals enter a period that has been called clinical latency. The very low level of productively infected lymphocytes circulating in the blood during clinical latency had led to hypotheses of a truly quiescent latent infection of limited numbers of cells infected by HIV during this asymptomatic phase of the disease (16, 93 and 130). Surprisingly, then, recent use of *in situ* polymerase chain reaction (PCR) to detect HIV DNA has shown that large numbers of cells within the lymph nodes contain viral DNA in asymptomatic individuals (89,244). An example of the large number of $CD4^+$ T helper lymphocytes containing HIV proviruses is shown in Fig. 7.

Infection of $CD4^+$ T cells by HIV involves reverse transcription and integration of the viral DNA in a provirus form. Apparently, if the T cell is incapable of transcribing the provirus, HIV may establish a latent infection. On the other hand, if the cell is able to transcribe viral mRNA from the provirus, a productive infection results, with spread of infection among $CD4^+$ T cells. Antigen-specific activation of $CD4^+$ T cells may be a key control switch that permits transcription of the viral RNA from the provirus (116) and conversion from latent to chronic infection. At early to late stages of infection, Embretson and colleagues (89) observed many provirus-containing cells in lymphoid tissue but few cells containing viral RNA. In contrast, Pantaleo and colleagues (244) observed many cells containing viral DNA and RNA, a situation more closely resembling a chronic infection. At this point, it is clear that there are many HIV-infected cells in lymphoid tissues during the asymptomatic phase, but the relative frequency of latently versus chronically infected cells is yet to be determined.

During the clinical latency period, infected individuals continue to exhibit vigorous T- and B-cell responses against HIV. They possess antiviral antibody, including neutralizing antibody as well as virus-specific CTL directed against several HIV proteins (130,133,174,263,265,340,341,351). Fresh PBLs obtained from "healthy" HIV carriers can directly mediate virus-specific CTL activity without any *in vitro* stimulation. This observation suggests an ongoing T-

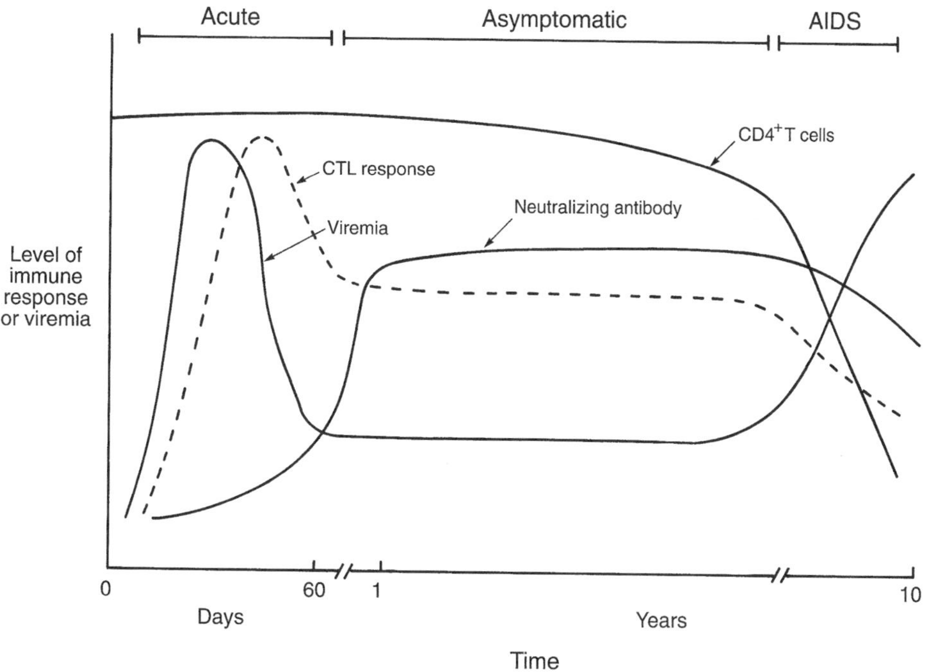

FIG. 6. Temporal changes in virus level, anti-HIV immune responses and total $CD4^+$ T cell counts during various stages of HIV infection. (adapted from ref. 174.)

cell response against the virus, which in turn implies chronic, low-level HIV replication during clinical latency (119). Why, then, is HIV not eliminated? The various mechanisms that viruses use to evade host immune responses were discussed in an earlier section (see Table 4), and many of these strategies apply to HIV. Perhaps the most relevant is that a large number of cells remain latently infected, thereby making it difficult for the immune system to deplete the reservoir of HIV-infected cells. Periodic activation of latently infected cells is thought to maintain this reservoir by resulting in productive replication and spread of virus to uninfected and perhaps newly generated CD4+ T cells. However, while the immune system remains healthy and CD4+ T-cell counts are high, the overall virus load is kept in check.

Progression to AIDS

Eventually, however, CD4+ T-cell numbers decline, possibly through attrition due to productive infection of mature CD4+ T cells or their progenitors (285). Loss of CD4+ T cells may also occur by immunopathologic mechanisms (93). Disease progression results. A recent report indicates that the level of HIV mRNA expression in peripheral blood cells predicts disease progression in the infected individual in that those who show high levels of mRNA within circulating T cells appear likely to become symptomatic (275). Thus, individuals infected with HIV may undergo simultaneously a latent infection and a chronic infection, with the relative proportion of cells exhibiting each type of persistent infection determining the stage of disease progression. The mechanism(s) by which the latent infection is reactivated or by which the chronic infection of lymph node cells is converted to an infection of circulating lymphocytes, as seen in symptomatic individuals, remains unknown.

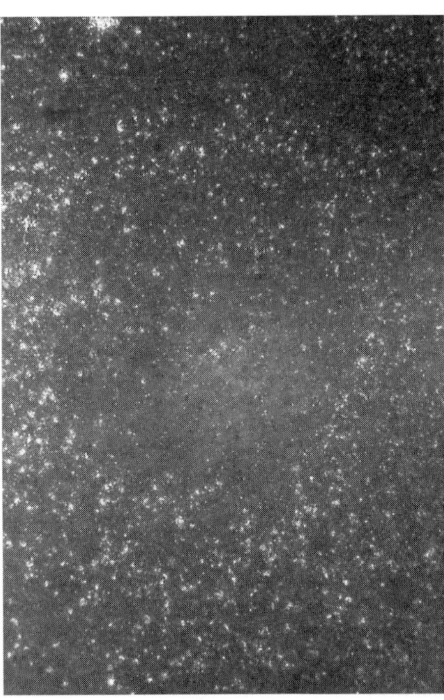

FIG. 7. Detection of HIV proviruses by *in situ* PCR. A section of a lymph node from an HIV-positive individual (89) was reacted first with monoclonal antibodies to mark CD4+ T cells. After subsequent amplification *in situ* and hybridization with an HIV-specific probe, the section was coated with emulsion, exposed, and developed. The photograph shows an autoradiograph photographed with epipolarizing illumination. The silver grains demonstrating hybridization appear white. The large number of CD4+ cells with HIV DNA in their nuclei are visible as discrete white foci. The central circular area with reduced signal corresponds to the germinal center, and most of the positive cells are present in the perifollicular mantle and cortical areas surrounding the germinal center. (Micrograph and description of methods generously provided by A. Haase.)

Profound immunosuppression and susceptibility to opportunistic infections are late events in the pathogenesis of AIDS (16,93,130). During the later stages of AIDS, the number of productively infected cells increases concomitant with an increase in virus load, whereas the number of CD4+ T cells declines dramatically and HIV-specific CTL responses are lost. HIV no longer evades the immune response, instead overwhelming the defenses it has weakened. Two events may contribute to overt disease progression. First, with impaired immune effectiveness the opportunistic infections that arise may increase the frequency of T-cell activation that promotes HIV transcription and productive infection. Second, viral variants emerge from the growing virus pool. These variants possess augmented replicative, and in some cases cytopathic, potential in CD4+ T cells as determined from sequential isolates from patients over time (62). These events set up vicious cycles that contribute to an accelerated decline in most AIDS patients and eventually to their demise.

SUMMARY AND PERSPECTIVES

Viruses can persist in host organisms by latent infection in which no infectious virus is produced or by a chronic infection in which infectious virus is continually shed. There are three basic requirements for persistent infection by a virus.

First, the virus must be able to infect cells without being cytopathic. Viruses that are not cytolytic can readily establish a chronic infection. Continual shedding of virus creates a carrier state in which the virus is readily transmissible to a new host. Viruses that normally are cytolytic undergo an alternative type of infection in some cells in order to establish a persistent infection. Bacteriophages undergo a lysogenic infection under certain conditions because virus-encoded gene products play an active role in latent infection. In the case of bacteriophage lambda, a combination of viral proteins, including the cI repressor molecule, controls the type of infection established. Animal DNA viruses utilize different mechanisms for establishment of a latent infection from those used by bacteriophage lambda for lysogenic infection. The animal DNA viruses undergo productive infection in permissive cell types and latent infection in nonpermissive cell types, not because of a viral repressor, but because of restrictions placed on viral replication by the host cell. Once latency is established, changes in the physiology on the nonpermissive host cell result in intracellular signals that indicate to the virus the need to reactivate and spread to new host cells. Thus it appears that animal viruses play a passive role in the establishment of persistent infections, but it is likely that these viruses have evolved to exploit nonpermissive cells within the host as a place to survive long term.

Second, there must be a mechanism for maintenance of the viral genome in the persistently infected cell. True latent infections are more common with DNA viruses, probably because the host cell carries out processes that can retain the viral DNA genome in a quiescent form, either integrated or as an extrachromosomal plasmid. In contrast, there are few, if any, mechanisms for long-term propagation or maintenance of RNA molecules in a host cell. Therefore, RNA viruses can persist only by a low-grade, chronic infection of some form or by conversion of the genome to DNA as carried out by the reverse transcriptase of retroviruses.

Third, the virus must evade the host immune response. This may occur by lack of expression of viral proteins in the infected cell, by infection of an immunologically privileged site, or by escape from or suppression of the host immune response.

Evolution of Viral Persistence to a Symbiotic Relationship

Because persistent viruses have evolved mechanisms to maintain themselves in the host organism, it is interesting to ask whether a true symbiotic relationship ever evolves between the virus and host or whether the virus always remains in the role of a parasite. A role for viruses in exchange of genetic material has been postulated as a force in evolution (302,323), but there is no evidence that this occurs to a significant extent with animal viruses. Also, it is conceivable that viruses affect host immunity in a way that is advantageous to the host. Many viruses have been reported to induce a polyclonal B-cell response that increases the level of IgG2a antibody specific for the inducing virus and other antigens (64,65,232). The increase in IgG2a is predominantly nonspecific; thus, activation of this arm of the immune response could provide some preexisting defense against other infectious agents. For example, periodic reactivation of HSV from latent infection could increase the frequency of B cells responsive to other viruses or bacteria and so would benefit the host in the case of a subsequent infection. Similarly, virus infection has been shown to increase the frequency of memory T cells responsive to heterologous viruses (294), possibly by exposure of the host to cross-reactive epitopes. It will be of interest to determine if persistent viral infection and recurrent infection affect the susceptibility of an individual to heterologous infections.

Therapeutic Treatment of Persistent Viral Infections

The considerable disease load caused by viruses that persist in humans (Table 1) necessitates the consideration of approaches to target viruses that have already established persistent infection in the host. It is not coincidental that only a few of the vaccines currently in use or in trials are directed against the viruses listed in Table 1, because persistent virus infections have been notoriously difficult to prevent with vaccines. Vaccines are available only for HBV and measles viruses, and these are used prophylactically and not therapeutically. Similarly, there are few antiviral drug approaches that target persistent viral infections. Therefore, one of the frontiers of virology is the design of approaches to prevent and treat persistent viral infections.

There has been considerable recent interest in the use of therapeutic vaccines, especially in the treatment of genital herpes infections. Studies in the guinea pig vaginal infection model first showed the feasibility of therapeutic vaccination to decrease recurrent infection (307). A recent clinical study has shown the ability of therapeutic immunization with a glycoprotein subunit vaccine to decrease the frequency of recurrent genital infections (321). These studies have shown the potential feasibility of therapeutic vaccination, but the mechanisms of the therapeutic effects remain to be defined. It would seem unlikely that the immune system is affecting the latent genome in the neuron; therefore, the augmented immune response may alter viral replication during reactivation in the neuron, prevent spread from the neuron, or reduce recurrent infection in the epithelium. The mechanisms underlying therapeutic immunization are exciting problems for the future in viral immunology.

Our goals in this chapter were to summarize the current knowledge about persistent viral infections and to attempt to outline the many fundamental questions remaining about viral persistence. Increased knowledge of the mechanisms of persistent infection should stimulate the design of preventive and therapeutic strategies needed for elimination of these virus infections.

REFERENCES

1. Abken H, Butzler C, Willecke K. Adenovirus type 5 persisting in human lymphocytes is unlikely to be involved in immortalization of lymphoid cells by fusion with cytoplasts or by transfection with DNA of mouse L cells. *Anticancer Res* 1987;7:553–558.
2. Ackermann WW, Kurtz H. Observations concerning a persisting in-

fection of HeLa cells with poliomyelitis virus. *J Exp Med* 1955;102:555–565.
3. Adams A, Lindahl T. Epstein-Barr virus genomes with properties of circular DNA molecules in carrier cells. *Proc Natl Acad Sci USA* 1975;72:1477–1481.
4. Adams A. Replication of latent Epstein-Barr virus genomes in Raji cells. *J Virol* 1987;61:1743–1746.
5. Ahmed R. Persistent viral infection. In: *Encyclopedia of Virology*.London: Academic Press; 1994.
6. Ahmed R, Chakrabarty PR, Fields BN. Genetic variation during lytic virus infection: high passage stocks of wild type reovirus contain temperature-sensitive mutants. *J Virol* 1980;34:285–287.
7. Ahmed R, Canning WM, Kauffman RS, Sharpe AH, Hallum JV, Fields BN. Role of the host cell in persistent viral infection: coevolution of L cells and reovirus during persistent infection. *Cell* 1981;25:325–332.
8. Ahmed R, Fields BN. Role of the S4 gene in the establishment of persistent reovirus infection in L cells. *Cell* 1982;28:605–612.
9. Ahmed R, Jamieson BD, Porter DD. Immune therapy of a persistent and disseminated viral infection. *J Virol* 1987;61:3920.
10. Ahmed R, Butler LD, Bhatti L. T4[+] T helper cell function *in vivo*: differential requirement for induction of anti-viral cytotoxic T cell and antibody responses. *J Virol* 1988;62:2102–2106.
11. Ahmed R, Salmi A, Butler LD, Chiller JM, Oldstone MBA. Selection of genetic variants of lymphocytic choriomeningitis virus in spleens of persistently infected mice: role in suppression of cytotoxic T lymphocyte response and viral persistence. *J Exp Med* 1984;60:521–540.
12. Andersson M, Paabo S, Nilsson T, Peterson PA. Impaired intracellular transport of class I MHC antigens as a possible means for adenoviruses to evade immune surveillance. *Cell* 1985;43:215–222.
13. Ando K, Guidotti LG, Cerny A, et al. CTL access to tissue antigen is restricted *in vivo*. *J Immunol* 1994;153:482–488.
14. Archibald RWR, Weller RO, Meadow SR. Measles pneumonia and the nature of the inclusion bearing giant cells: a light-and electron-microscope study. *J Pathol* 1971;103:27–34.
15. Armstrong JA, Horne R. Follicular dendritic cells and virus-like particles in AIDS-related lymphadenopathy. *Lancet* 1984;2:370–372.
16. Auger I, Thomas P, De Gruttola V, et al. Incubation periods for paediatric AIDS patients. *Nature* 1988;336:575–577.
17. Baczko K, Liebert UG, Billeter M, Cattaneo R, Budka H, ter Meulen V. Expression of defective measles virus genes in brain tissues of patients with subacute sclerosing panencephalitis. *J Virol* 1986;59: 472–478.
18. Baker CC, Phelps WC, Lindgren V, Braun MJ, Gonda MA, Howley PM. Structural and transcriptional analysis of human papillomavirus type 16 sequences in cervical carcinoma cell lines. *J Virol* 1987: 61:962–971.
19. Barker CF, Billingham RE. Immunologically privileged sites. *Adv Immunol* 1977;25:1.
20. Barnaba V, Franco A, Alberti A, Balsano C, Benvenuto R, Balsano F. Recognition of hepatitis B envelope proteins by liver-infiltrating T lymphocytes in chronic HBV infection. *J Immunol* 1989;143:2650–2655.
21. Baroni CD, Pezzella F, Mirolo M, Ruco LP, Rossi GB. Immunohistochemical demonstration of p24 HTLV III major core protein in different cell types within lymph nodes from patients with lymphadenopathy syndrome (LAS). *Histopathology* 1986;10:5–13.
22. Battegay M, Moskophidis D, Rahemtulla A, Hengartner H, Mak TW, Zinkernagel RM. Enhanced establishment of a virus carrier state in adult CD4[+] T-cell-deficient mice. *J Virol* 1994;68;4700–4704.
23. Beasley RP, Hwang LY, Stevens CE, Lin CC, Hsieh FJ, Wang KY, Sun TS, Szmuness W. Efficacy of hepatitis B immune globulin for prevention of perinatal transmission of the hepatitis B virus carrier state: final report of a randomized double-blind, placebo-controlled trial. *Hepatology* 1983;3:135–141.
24. Berke G. The binding and lysis of target cells by cytotoxic lymphocytes: molecular and cellular targets. *Annu Rev Immunol* 1994; 12:735–773.
25. Bernards R, Schrier PI, Houweling A, et al. Tumorigenicity of cells transformed by adenovirus type 12 by evasion of T-cell immunity. *Nature* 1983;305:776–779.
26. Bertoletti A, Sette A, Chisari FV, Penna A, Levrero M, De Carli M, Fiaccadori F, Ferrari C. Natural variants of cytotoxic epitopes are T cell receptor antagonists for antiviral cytotoxic T cells. *Nature* 1994;369:407.
27. Bertoletti A, Costanzo A, Chisari FV, Levrero M, Artini M, Sette A, Penna A, Giuberti T, Fiaccudori F, Ferrari C. Cytotoxic T lymphocyte response to a wild type hepatitis B virus epitope in patients chronically infected by variant viruses carrying substitutions within the epitope. *J Exp Med* 1994;180:933–943.
28. Biberfeld P, Chayt KJ, Marselle LM, Biberfeld G, Gallo RC, Harper ME. HTLV-III expression in infected lymph nodes and relevance to pathogenesis of lymphadenopathy. *Am J Pathol* 1986;125:436–442.
29. Biron CA. Cytokines in the generation of immune responses to and resolution of virus infection. *Curr Opin Immunol* 1994;6:530–538.
30. Birrer MJ, Bloom BR, Udem S. Characterization of measles polypeptides by monoclonal antibodies. *Virology* 1981;198:381–390.
31. Bjorkman PJ, Saper MA, Samraoui B, Bennett WS, Strominger JL, Wiley DC. Structure of the human class I histocompatibility antigen, HLA-A2. *Nature* 1987;329:506–512.
32. Borrow P, Lewicki H, Hahn BH, Shaw GM, and Oldstone MBA. Virus-specific CD8[+] cytotoxic T-lymphocyte activity associated with control of viremia in primary human immunodeficiency virus type 1 infection. *J Virol* 1994;68:6103–6110.
33. Brown HR, Goller NL, Thormar H, et al. Measles virus matrix protein gene expression in a subacute sclerosing panencephalitis patient brain and virus isolate demonstrated by cDNA hybridization and immunocytochemistry. *Acta Neuropathol* 1987;75:123–130.
34. Brown JH, Jardetzky TS, Gorga JC, Stern LJ, Urban RG, Strominger JL, Wiley DC. The three-dimensional structure of the human class II histocompatibility antigen HLA-DR1. *Nature* 1993;364:33–39.
35. Buchmeier MJ, Welsh RM, Dutko FJ, Oldstone MBA. The virology and immunobiology of lymphocytic choriomeningitis virus infection. *Adv Immunol* 1980;30:275–331.
36. Buchmeier NA, Cooper NR. Suppression of monocyte functions by human cytomegalovirus. *Immunology* 1989;66:278–283.
37. Budkowska A, Dubreuil P, Capel F, Pillot J Hepatitis B virus pre-S gene-encoded antigenic specificity and anti-pre-S antibody: relationship between anti-pre-S response and recovery. *Hepatology* 1986;6:360–368.
38. Burgert H-G, Kvist S. An adenovirus type 2 glycoprotein blocks cell surface expression of human histocompatibility class I antigens. *Cell* 1985;41:987–997.
39. Burgert H-G, Maryanski JL, Kvist S. E3/19K protein of adenovirus type 2 inhibits lysis of cytolytic T lymphocytes by blocking cell-surface expression of histocompatibility class I antigens. *Proc Natl Acad Sci USA* 1987;84:1356–1360.
40. Burns DP, Collignon C, Desrosiers RC. 1993. Simian immunodeficiency virus mutants resistant to serum neutralization arise during persistent infection of rhesus monkeys. *J Virol* 67:4104–4113.
41. Burns DP, Desrosiers RC. 1991. Selection of genetic variants of simian immunodeficiency virus in persistently infected rhesus monkeys. *J Virol* 65:1843–1856.
42. Butler JC, Peters CJ. Hantaviruses and hantavirus pulmonary syndrome. *Clin Infect Dis* 1994;19:387–394.
43. Byrne JA, Oldstone MBA. Biology of cloned cytotoxic T lymphocytes specific for lymphocytic choriomeningitis virus: clearance of virus *in vivo*. *J Virol* 1984;51:682–686.
44. Carrigan DR, Knox KK. Identification of interferon-resistant subpopulations in several strains of measles virus: positive selection by growth of the virus in brain tissue. *J Virol* 64:1606–1615.
45. Carter MJ, Willcocks MM, ter Meulen V. Defective translation of measles virus matrix protein in a subacute sclerosing panencephalitis cell line. *Nature* 1983;305:153–155.
46. Carter MJ, Willcocks MM, Loffler S, ter Meulen V. Relationships between monoclonal antibody-binding sites on the measles hemagglutinin. *J Gen Virol* 1982;63:113–120.
47. Cascardo MR, Karzon DT. Measles virus giant cell inducing factor (fusion factor). *Virology* 1965;26:311–325.
48. Cattaneo R, Rebmann G, Schmid A, Baczko K, ter Meulen V, Billeter MA. Altered transcription of a defective measles virus genome derived from a diseased human brain. *EMBO J* 1987;6:681–688.
49. Cattaneo R, Schmid A, Eschle D, Baczko K, ter Meulen V, Billeter MA. Biased hypermutation and other genetic changes in defective measles viruses in human brain infections. *Cell* 1988;55:255–265.
50. Cattaneo R, Schmid A, Billeter MA, Sheppard RD, Udem SA. Multiple viral mutations rather than host factors cause defective measles virus gene expression in a subacute sclerosing panencephalitis cell line. *J Virol* 1988;62:1388–1397.
51. Cattaneo R, Schmid A, Spielhofer P, et al. Mutated and hypermutated genes of persistent measles viruses which caused lethal human brain diseases. *Virology* 1989;173:415–425.

52. Celis E, Ou D, Otvos L. Recognition of hepatitis B surface antigen by human T lymphocytes: proliferative and cytotoxic responses to a major antigenic determinant defined by synthetic peptides. *J Immunol* 1988;140:1808–1815.
53. Chang LJ, Dienstag J, Ganem D, Varmus H. Detection of antibodies against hepatitis B polymerase antigen in hepatitis ZB virus-infected patients. *Hepatology* 1989;10:332–337.
54. Chen AX, Shen L, Miller MD, Ghim SH, Hughes AL, Letvin NL. Cytotoxic T lymphocytes do not appear to select for mutations in an immunodominant epitope of simian immunodeficiency virus gag. *J Immunol* 1992;149:4060.
55. Cheung AK-M, Hoggan MD, Hauswirth WW, Berns KI. Integration of the adeno-associated virus genome into cellular DNA in latently infected human Detroit 6 cells. *J Virol* 1980;33:739.
56. Chinami M, Nakamura E. Kakisako S, Xu B, Shingu M. Poliovirus resistant cells derived from HeLa cells. *Kurume Med J* 1986;33:125–129.
57. Chisari FV. Hepatitis B virus biology and pathogenesis. In: Friedmann T, ed. *Molecular genetic medicine*. San Diego: Academic Press; 1992:67–104.
58. Clark SJ, Saag MS, Decker WD, et al. High titers of cytopathic virus in plasma of patients with symptomatic primary HIV-1 infection. *N Engl J Med* 1991;324:954–960.
59. Clements JE, Narayan O. Immune selection of virus variants. In: Notkins AL, Oldstone MBA, eds. *Concepts in viral pathogenesis*. New York: Springer-Verlag; 1984:152–157.
60. Clements JE, D'Antonio N, Narayan O. Genomic changes associated with antigenic variation of visna virus. II. Common nucleotide sequence changes detected in variants from independent isolations. *J Mol Biol* 1982;158:415–434.
61. Coen DM, Kosz-Vnenchak M, Jacobson JG, et al. Thymidine kinase-negative herpes simplex virus mutants establish latency in mouse trigeminal ganglia but do not reactivate. *Proc Natl Acad Sci USA* 1989;86:4736–4740.
62. Conner RI, Ho DD. Transmission and pathogenesis of human immunodeficiency virus type 1. *AIDS Res Hum Retroviruses* 1994;10:321–323.
63. Couillin IB, Culmann-Penciolelli B, Gomard E, Levy J-P, Guillet JG, Saragosti S. Impaired CTL recognition due to genetic variations in the main immunogenic region of the HIV-1 nef protein. *J Exp Med* 1994;180:1129.
64. Coutelier J-P, van der Logt JTM, Heessen FWA, Warnier G, Snick JV. IgG2a restriction of murine antibodies elicited by viral infections. *J Exp Med* 1987;165:64.
65. Coutelier J-P, van der Logt JT, Heessen FWA, Vink A, Van Snick J Virally induced modulation of murine IgG antibody subclasses. *J Exp Med* 1988; 168:2373.
66. Croen KD, Ostrove JM, Dragovic LJ, Straus SE. Patterns of gene expression and sites of latency in human nerve ganglia are different for varicella-zoster and herpes simplex viruses. *Proc Natl Acad Sci USA* 1988;85:9773–9777.
67. Croen KD, Ostrove JM, Dragovic LJ, Smialek JE, Straus SE. Latent herpes simplex virus in human trigeminal ganglia: detection of an immediate-early gene "antisense" transcript by *in situ* hybridization. *N Engl J Med* 1987;317:1427–1432.
68. Daar ES, Moudgil T, Meyer RD, Ho DD. Transient high levels of viremia in patients with primary immunodeficiency virus type 1 infection. *N Engl J Med* 1991;324:961–964.
69. Dai LC, West K, Littaua R, Takahashi K, Ennis FA. Mutation of the human immunodeficiency virus type 1 at amino acid 585 on gp41 results in loss of killing by CD8+ A24–restricted cytotoxic T lymphocytes. *J Virol* 1992;66:3151.
70. Davis MM, Bjorkman PJ T-cell antigen receptor genes and T-cell recognition. *Nature* 1988;334:395–402.
71. De Campos-Lima P-O, Levitsky V, Brooks J, Lee SP, Hu LF, Rickinson AB, Masucci MG. T cell responses and virus evolution: loss of HLA A11–restricted CTL epitopes in Epstein-Barr virus isolates from highly A11–positive populations by selective mutation at anchor residues. *J Exp Med* 1994;179:1297.
72. De la Torre JC, Davila M, Sabrino F, Ortin J, Domingo E. Establishment of cell lines persistently infected with foot-and-mouth disease virus. *Virology* 1985;145:24–35.
73. De la Torre JC, Martinez-Salas E, Diez J, et al. Coevolution of cells and viruses in a persistent infection of foot-and-mouth disease virus in cell culture. *J Virol* 1988;62:2050–2058.
74. De la Torre JC, Martinez-Salas E, Diez J, Domingo E. Extensive cell heterogeneity during persistent infection with foot-and-mouth disease virus. *J Virol* 1989;63:59–63.
75. Deatly A, Spivack JG, Lavi E, and Fraser NW. RNA from an immediate early region of the HSV-1 genome is present in the trigeminal ganglia of latently infected mice. *Proc Natl Acad Sci USA* 1987;84:3204–3208.
76. Dermody TS, Nibert ML, Wetzel JD, Tong X, Fields BN. Cells and viruses with mutations affecting viral entry are selected during persistent infections of L cells with mammalian reoviruses. *J Virol* 1993;67:2055–2063.
77. Deshmane SL, Fraser JW. During latency, herpes simplex virus type 1 DNA is associated with nucleosomes in a chromatin structure. *J Virol* 1989;63:943–947.
78. Dhib-Jalbut S, McFarland HF, Mingiolo ES, Sever JL, McFarlin DE. Humoral and cellular immune responses to matrix protein of measles virus in subacute sclerosing panencephalitis. *J Virol* 1988;62:2483–2489.
79. Dietzschold B, Kao M, Zheng YM, Chen ZY, Maul G, Fu ZF, Rupprecht CE, Koprowski H. Delineation of putative mechanisms involved in antibody-mediated clearance of rabies virus from the central nervous system. *Proc Natl Acad Sci USA* 1992;89:7252–7256.
80. Dobson AT, Sedarati F, Devi-Rao G, Flanagan WM, Farrell MJ, Stevens JG, Wagner EK, Feldman LT. Identification of the latency-associated transcript promoter by expression of rabbit beta-globin mRNA in mouse sensory nerve ganglia latently infected with a recombinant herpes simplex virus. *J Virol* 1989;63:3844–3851.
81. Doherty PC, Allan W, Eichelberger M, Carding SR. Roles of αβ and γδ T cell subsets in viral immunity. *Annu Rev Immunol* 1992;10:123–151.
82. Dukto FJ, Oldstone MBA. Cytomegalovirus causes a latent infection in undifferentiated cells and is activated by induction of cell differentiation. *J Exp Med* 1981;154:1636–1651.
83. Durst M, Kleinheinz A, Hotz M, Gissmann L. The physical state of human papillomavirus type 16 DNA in benign and malignant genital tumors. *J Gen Virol* 1985:66:1515–1522.
84. Dyson N, Guida P, Munger K, Harlow E. Homologous sequences in adenovirus E1A and human papillomavirus E7 proteins mediate interaction with the same set of cellular proteins. *J Virol* 1992;66:6893–6902.
85. Eager KB, Pfizenmaier K, Ricciardi RP. Modulation of major histocompatibility complex (MHC) class I genes in adenovirus 12 transformed cells: interferon-γ increases class I expression by a mechanism that circumvents E1A induced-repression and tumor necrosis factor enhances the effect of interferon-γ. *Oncogene* 1989;4:39–44.
86. Efstathious S, Kemp S, Darby G, and Minson AC. The role of herpes simplex virus type 1 thymidine kinase in pathogenesis. *J Gen Virol* 1989;70:869–879.
87. Einhorn L, Ost A. CMV infection of human blood cells. *J Infect Dis* 1984;149:207–214.
88. Eisenlohr LC, Yewdell JW, Bennink JR. Flanking sequences influence the presentation of an endogenously synthesized peptide to cytotoxic T lymphocytes. *J Exp Med* 1992;175:481.
89. Embretson J, Supancic M, Ribas JL, et al. Massive covert infection of helper T lymphocytes and macrophages by HIV during the incubation period of AIDS. *Nature* 1993;362:359–362.
90. Emini EA, Larson V, Eichberg J, et al. Protective effect of a synthetic peptide comprising the complete pre-S2 region of the hepatitis B virus surface protein. *J Med Virol* 1989;28:7–12.
91. Evans AS. Latent adenovirus infections of the human respiratory tract. *Am J Hyg* 1958;67:256–266.
92. Farrell MJ, Dobson AT, Feldman LT. Herpes simplex virus latency associated transcript is a stable intron. *Proc Natl Acad Sci USA* 1991;88:790–794.
93. Fauci A. Immunopathogenesis of HIV infection. *AIDS* 1993;6:655–662.
94. Feitelson MA, Millman I, Duncan GD, Blumberg BS. Presence of antibodies to the polymerase gene products (s) of hepatitis B and woodchuck hepatitis virus in natural and experimental infections. *J Med Virol* 1988;24:121–126.
95. Ferenczy A, Mitao M, Nagai N, Silverstein SJ, Crum CP. Latent papillomavirus and recurring genital warts. *N Engl J Med* 1985;313:784–788.
96. Ferrari C, Bertoletti A, Penna A, et al. Identification of immunodominant T cell epitopes of the hepatitis B virus nucleocapsid antigen. *J Clin Invest* 1991;88:214–222.
97. Ferrari C, Penna A, Bertoletti A, et al. Cellular immune response to

hepatitis B virus–encoded antigens in acute and chronic hepatitis B virus infection. *J Immunol* 1990;145:3442–3449.

98. Ferrari C, Penna A, Giuberti T, Tong MJ, Ribera E, Fiaccadori F, Chisari FV. Intrahepatic, nucleocapsid antigen-specific T cells in chronic active hepatitis B. *J Immunol* 1987;139:2050–2058.
99. Fingeroth JD, Weis JJ, Tedder TF, Strominger JL, Biro PA, Fearon DT. Epstein-Barr virus receptor of human B lymphocytes is the C3d receptor CR2. *Proc Natl Acad Sci USA* 1984;81:4510–4516.
100. Fox JP, Brandt CD, Wassermann FE, et al. The Virus Watch Program: a continuing surveillance of viral infections in metropolitan New York families. VI. Observations of adenovirus infections: virus excretion patterns, antibody response, efficiency of surveillance, patterns of infection and relation to illness. *Am J Epidemiol* 1969;89:25–50.
101. Fox CH, Tenner-Racz K, Racz P, Firpo A, Pizzo PA, Fauci AS. Lymphoid germinal centers are reservoirs of human immunodeficiency virus type 1 RNA. *J Infect Dis* 1991;164:1051–1057.
102. Frade R, Barel M, Ehlin-Henriksson B, Klein G. gp140, the C3d receptor of human B lymphocytes, is also the Epstein-Barr virus receptor. *Proc Natl Acad Sci USA* 1985;82:1490–1493.
103. Fujinami RS, Oldstone MBA. Antiviral antibody reacting on the plasma membrane alters measles virus expression inside the cell. *Nature* 1979;279:935–940.
104. Fujinami RS, Oldstone MAB. Alterations in expression of measles virus polypeptides by antibody: molecular events in antibody-induced antigenic modulation. *J Immunol* 1980;125:78–85.
105. Fujinami RS, Oldstone MBA. Antibody initiates virus persistence: immune modulation and measles virus infections. In: Nokins AL, Oldstone MBA, eds. *Concepts in viral pathogenesis*. New York: Springer-Verlag; 1984:187–193.
106. Gartner S, Markovits P, Markovitz DM, Kaplan MH, Gallo RC, Popovic M. The role of mononuclear phagocytes in HTLV-III/LAV infection. *Science* 1986;233:215–219.
107. Gerber P, Lucas S, Nonoyama M, Perlin E, Goldstein LI. Oral excretion of Epstein-Barr viruses by healthy subjects and patients with infectious mononucleosis. *Lancet* 1972;2:988–989.
108. Germain RN. MHC-dependent antigen processing and peptide presentation: providing ligands for T lymphocyte activation. *Cell* 1994;76:287–299.
109. Gilden DH, Vafai A, Shtram Y, Becker Y, Devlin M, Wellish M. Varicella-zoster virus DNA in human sensory ganglia. *Nature* 1983;306:478–480.
110. Gilden DH, Cole GA, Nathanson N. Immunopathogenesis of acute central nervous system disease produced by lymphocytic choriomeningitis virus. II. Adoptive immunization of virus carriers. *J Exp Med* 1972;135:874–889.
111. Ginsberg HS, Bello LF, Levine AJ Control of biosynthesis of host macromolecules in cells infected with adenovirus. In: Colter JS, Paranchych W, eds. *The molecular biology of viruses.* New York: Academic Press; 1967:547–572.
112. Gonczol E, Andres PW, Plotkin SA. Cytomegalovirus replicates in differentiated but not in undifferentiated human embryonal carcinoma cells. *Science* 1984;224:159–161.
113. Goodglick L, Braun J Revenge of the microbes: superantigens of the T and B cell lineage. *Am J Pathol* 1994;144:623–636.
114. Gooding LR. Virus proteins that counteract host immune defenses. *Cell* 1992;71:5–7.
115. Green MT, Courtney RJ, and Dunkel EC. Detection of an immediate early herpes simplex virus type 1 polypeptide in trigeminal ganglia from latently infected animals. *Infect Immunol* 1981;34:987–992.
116. Greene WC. The molecular biology of human immunodeficiency virus type 1 infection. *N Engl J Med* 1991;324:308–317.
117. Gregory CD, Murray RJ, Edwards CF, Rickinson AB. Down regulation of cell adhesion molecules LFA-3 and ICAM-1 in Epstein-Barr virus-positive Burkitt's lymphoma underlies tumor cell escape from virus-specific T cell surveillance. *J Exp Med* 1988;167:1811–1824.
118. Gresser J, Chany C. Isolation of measles virus from the washed leucocytic fraction of blood. *Proc Soc Exp Biol Med* 1963;113:695–698.
119. Haase AT. The role of active and covert infections in lentivirus pathogenesis. *Ann NY Acad Sci* 1994;724:75–86.
120. Haase AT, Gantz D, Eble B, et al. Natural history of restricted synthesis and expression of measles virus genes in subacute sclerosing panencephalitis. *Proc Natl Acad Sci* 1985;82:3020–3024.
121. Hall WW, Choppin WP. Measles virus proteins in the brain tissue of patients with subacute sclerosing panencephalitis: absence of the M protein. *N Engl J Med* 1981;394:1152–1155.
122. Hall WW, Lamb RA, Choppin PW. Measles and subacute sclerosing panencephalitis virus protein: lack of antibodies to the M protein in patients with subacute sclerosing panencephalitis. *Proc Natl Acad Sci USA* 1979;76:2047–2051.
123. Hampar B, Tanaka A, Nonoyama M, Derge JG. Replication of the resident repressed Epstein-Barr virus genome during the early S phase (S-1 period) of nonproducer Raji cells. *Proc Natl Acad Sci USA* 1974;71:631–635.
124. Handa H, Shiroki K, Shimojo H. Establishment and characterization of KB cell lines latently infected with adeno-associated virus type 1. *Virology* 1977;82:84.
125. Hermiston TW, Tripp RA, Sparer T, Gooding LR, Wold WSM. Deletion mutation analysis of the adenovirus type 2 E3-gp 19K protein: identification of sequences within the endoplasmic reticulum luminal domain that are required for class I antigen binding and protection from adenovirus-specific cytotoxic T lymphocytes. *J Virol* 1993;67:5289–5298.
126. Hill JM, Sedarati F, Javier RT, Wagner EK, Stevens JG. Herpes simplex virus latent phase transcription facilitates *in vivo* reactivation. *Virology* 1990;174:117–125.
127. Hillis WO, Cooper MR, Bang FB. Adenovirus infection in West Bengal. I. Persistence of viruses in infants and young children. *Indian J Med Res* 1973;61:980–988.
128. Hirano A, Ayata M. Wang AH, Wong TC. Functional analysis of matrix proteins expressed from cloned genes of measles virus variants that cause subacute sclerosing panencephalitis reveals a common defect in nucleocapsid binding. *J Virol* 1993;67:1848–1853.
129. Hirano A, Wang AH, Gombart AF, Wong TC. The matrix proteins of neurovirulent subacute sclerosing panencephalitis virus and its acute measles virus progenitor are functionally different. *Proc Natl Acad Sci USA* 1992;89:8745–8749.
130. Ho DD, Pomerantz RJ, Kaplan JC. Pathogenesis of infection with human immunodeficiency virus. *N Engl J Med* 1987;317:278–286.
131. Ho DD, Rota TR, Hirsch MS. Infection of monocyte/macrophages by human T lymphotropic virus type III. *J Clin Invest* 1986;77:1712–1725.
132. Hodge LD, Scharff MD. Effect of adenovirus on host cell DNA synthesis in synchronized cells. *Virology* 1969;37:554–564.
133. Hoffenback A, Langlade-Demoyen P, Dadaglio G, et al. Unusually high frequencies of HIV-specific cytotoxic T lymphocytes in humans. *J Immunol* 1989;142:452–462.
134. Hoggan, MD, Thomas GF, Thomas FB, Johnson FB. Continuous carriage of adenovirus associated virus genome in cell cultures in the absence of helper adenoviruses. In: *Proceedings of the Fourth Lepetit Colloquium*, Cocoyac, Mexico. Amsterdam: North-Holland; 1972:243.
135. Holland, JJ, Spindler K, Horodyski F, Graham E, Nichol S, Vendepol S. Rapid evolution of RNA genomes. *Science* 1982;215:1577–1585.
136. Holland JJ, Kennedy SIT, Semmler BL, Jones CL, Roux L, Grabau EA. Defective interfering RNA viruses and the host–cell response. In: Fraenkel-Conrat H, Wagner R, eds. *Comprehensive virology*, Vol 16. New York: Plenum Press; 1980:137–192.
137. Holland JJ, de la Torre, JC, Steinhauer, DA. RNA virus populations as quasispecies. *Curr Top Microbiol Immunol* 1992;176:1–20.
138. Horvath J, Palkonyay L, Weber J Group C adenovirus DNA sequences in human lymphoid cells. *J Virol* 1986;59:189–192.
139. Howard CR. Arenaviruses. In: *Perspectives in medical virology*. Vol 2. Amsterdam: Elsevier; 1986.
140. Huibregtse JM, Scheffner M, Howley PM. A cellular protein mediates association of p53 with the E6 oncoprotein of human papillomavirus types 16 or 18. *EMBO J* 1991;10:4129–4135.
141. Huibregtse JM, Scheffner M, Howley PM. Cloning and expression of the cDNA for E6–AP, a protein that mediates the interaction of the human papillomavirus E6 oncoprotein with p53. *Mol Cell Biol* 1993;13:775–784.
142. Hyman RW, Ecker JR, Tenser RB. Varicella-zoster virus RNA in human trigeminal ganglia. *Lancet* 1983;2:814–816.
143. Hyypia T, Korkiamaki P, Vainionbaa R. Replication of measles virus in human lymphocytes. *J Exp Med* 1986;161:1261–1271.
144. Itoh Y, Takai E, Ohnuma H, et al. A synthetic peptide vaccine involving the product of the pre-S(2) region of hepatitis B virus DNA: protective efficacy in chimpanzees. *Proc Natl Acad Sci USA* 1986;83:9174–9178.

145. Jameson SC, Carbone FR, Bevan MJ Clone-specific T cell receptor antagonists of major histocompatibility complex class-I-restricted cytotoxic T cells. *J Exp Med* 1993;177:1541.
146. Jamieson BD, Butler LD, Ahmed R. Effective clearance of a persistent viral infection requires cooperation between virus-specific Lyt2+ T cells and nonspecific bone marrow-derived cells. *J Virol* 1987;61:3930–3937.
147. Jamieson BD, Somasundaram T, Ahmed R. Abrogation of tolerance to a chronic viral infection. *J Immunol* 1991;147:3521.
148. Jennings SR, Rice PL, Kloszewski ED, Anderson RW, Thompson KL, Tevethia SS. Effect of herpes simplex virus types 1 and 2 on surface expression of class I major histocompatibility antigens on infected cells. *J Virol* 1985;56:757–766.
149. Joly E, Mucke L, Oldstone MBA. Viral persistence in neurons explained by lack of major histocompatibility class I expression. *Science* 1991;253:1283.
150. Joseph BS, Lampert PW, Oldstone MBA. Replication and persistence of measles virus in defined subpopulations of human leukocytes. *J Virol* 1975;16:1638–1649.
151. Joseph BS, Oldstone MBA. Antibody-induced redistribution of measles virus antigens on the cell surface. *J Immunol* 1974;113:1205–1209.
152. Joseph J, Knobler RL, Lublin FD, Hart MN. Mouse hepatitis virus (MHV-4, JHM) blocks γ-interferon-induced major histocompatibility complex class II antigen expression on murine cerebral endothelial cells. *J Neuroimmunol* 1991;33:181–190.
153. Jung MC, Spengler U, Schraut W, et al. Hepatitis B virus antigen-specific T-cell activation in patients with acute and chronic hepatitis B. *J Hepatol* 1991;13:310–317.
154. Kägi D, Vignaux F, Ledermann B, et al. Fas and perforin pathways as major mechanisms of T cell–mediated cytotoxicity. *Science* 1994;265:528–530.
155. Kägi D, Ledermann B, Bürki K, Seiler P, Odermatt B, Olsen KJ, Podack ER, Zinkernagel RM, Hengartner H. Cytotoxicity mediated by T cells and natural killer cells is greatly impaired in perforin-deficient mice. *Nature* 1994;369:31–37.
156. Kapasi K, Rice GPA. Cytomegalovirus infection of peripheral blood mononuclear cells: effects of IL-1 and IL-2 production and responsiveness. *J Virol* 1988;62:3606–3607.
157. Kaplan G, Levy A, Racaniello VR. Isolation and characterization of HeLa cell lines blocked at different steps in the poliovirus life cycle. *J Virol* 1989;63:43–51.
158. Katz JP, Bodin ET, Coen DM. Quantitative polymerase chain reaction analysis of herpes simplex virus DNA in ganglia of mice infected with replication-incompetent mutants. *J Virol* 64:4288–4295.
159. Katz SL, Enders JF. Measles virus. In: Horsfall FL Jr, Tamm I, eds. *Viral and rickettsial infections of man,* 4th ed. Philadelphia: JB Lippincott.
160. Kauffman RS, Ahmed R, Fields BN. Selection of a mutant S1 gene during reovirus persistent infection of L cells: role in maintenance of the persistent state. *Virology* 1983;131:79–87.
161. Kemp LM, Dent CL, Latchman DS. Octamer motif mediates transcriptional repression of HSV immediate-early genes and octamer-containing cellular promoters in neuronal cells. *Neuron* 1990;4:215–227.
162. Klatzmann D, Barre-Sinoussi F, Nugeyre MT, et al. Selective tropism of lymphadenopathy associated virus (LAV) for helper-inducer T lymphocytes. *Science* 1984;225:59–62.
163. Klenerman P, Rowland-Jones S, McAdams S, et al. Cytotoxic T cell activity antagonized by naturally occurring HIV-1 gag variants. *Nature* 1994;369:403.
164. Klinkert MQ, Theilmann L, Pfaff E, Schaller H. Pre-S1 antigens and antibodies early in the course of acute hepatitis B virus infection. *J Virol* 1986;58:522–525.
165. Knipe DM. The role of viral and cellular nuclear proteins in herpes simplex virus replication. *Adv Virus Res* 1989;37:85–123.
166. Kono Y, Kobayashi K, Fukunaga Y. Serological comparison among various strains of equine infectious anemia virus. *Arch Gesamte Virusforsch* 1971;34:202–208.
167. Kono Y, Kobayashi K, Fukunaga Y. Antigenic drift of equine infectious anemia virus in chronically infected horses. *Arch Gesamte Virusforsch* 1973;41:1–10.
168. Kosz-Vnenchak M, Coen DM, Knipe DM. Restricted expression of herpes simplex virus lytic genes during establishment of latent infection by thymidine kinase–negative mutant viruses. *J Virol* 1990; 64:5396–5402.
169. Kosz-Vnenchak M, Jacobson J, Coen DM, Knipe DM. Evidence for a novel regulatory pathway for herpes simplex virus gene expression in trigeminal ganglion neurons. *J Virol* 1993;67:5383–5393.
170. Kotin RM, Berns KI. Organization of adeno-associated virus DNA in latently infected Detroit 6 cells. *Virology* 1989;170:460.
171. Kotin RM, Menninger JC, Ward DC, Berns KI. Mapping and direct visualization of a region-specific viral DNA integration site on chromosome 19q 13-qter. *Genomics* 1991;10:831.
172. Kotin RM, Siniscalco M, Samulski RJ, et al. Site-specific integration by adeno-associated virus. *Proc Natl Acad Sci USA* 1990;87:2211.
173. Koup RA, Safrit JT, Cao Y, Andrews CA, McLeod G, Borkowsky W, Farthing C, Ho DD. Temporal association of cellular immune responses with the initial control of viremia in primary human immunodeficiency virus type 1 syndrome. *J Virol* 1994;68:4650–4655.
174. Koup RA, Ho DD. Shutting down HIV. *Nature* 1994;370:416.
175. Koup RA. Virus escape from CTL recognition. *J Exp Med* 1994;180:779.
176. Koyanagi Y, O'Brien WA, Zhao JQ, Golde DW, Gasson JC, Chen ISY. Cytokines alter production of HIV-1 from primary mononuclear phagocytes. *Science* 1988;241:1673–1675.
177. Laycock KA, Lee SE, Stulting RD, et al. HSV-1 transcription is not detectable in quiescent human stromal keratitis by *in situ* hybridization. *Invest Opthalmol Vis Sci* 1993;34:285–292.
178. Lehmann-Grube F, Lohler J, Utermohlen O, Gegin C. Antiviral immune responses of lymphocytic choriomeningitis virus-infected mice lacking CD8+ T lymphocytes because of disruption of the β_2-microglobulin gene. *J Virol* 1993;67:332–339.
179. Lehmann-Grube F, Peralta LM, Bruns M, Lohler J. Persistent infection of mice with the lymphocytic choriomeningitis virus. In: Fraenkel-Conrat H, Wagner R, eds. *Comprehensive virology*. New York: Plenum Press; 1983:43–103.
180. Lehmann-Grube F. Portraits of viruses: arenaviruses. *Intervirology* 1984;22:121–145.
181. Leib DA, Coen DM, Bogard CL, Hicks KA, Yager DR, Knipe DM, Tyler KL, Schaffer PA. Immediate-early regulatory gene mutants define different stages in the establishment and reactivation of herpes simplex virus latency. *J Virol* 1989;63:759–768.
182. Leib DA, Kosz-Vnenchak M, Jacobson JG, et al. A deletion mutant of the latency-associated transcript of herpes simplex virus type 1 reactivates from the latent state with reduced frequency. *J Virol* 1989;63:2893–2900.
183. Leist TP, Sandri-Goldin RM, Stevens JA. Latent infections in spinal ganglia with thymidine kinase-deficient herpes simplex virus. *J Virol* 1989;63:4976–4978.
184. Leopardi R, Ilonen J, Mattila L, Salmi AA. Effect of measles virus infection on MHC class II expression and antigen presentation in human monocytes. *Cell Immunol* 1993;147:388–396.
185. Levine B, Griffin DE. Persistence of viral RNA in mouse brains after recovery from acute alphavirus encephalitis. *J Virol* 1992;66:6429–6435.
186. Levine B, Griffin DE. Molecular analysis of neurovirulent strains of Sindbis virus that evolve during persistent infection of scid mice. *J Virol* 1993;67:6872–6875.
187. Levine B, Hardwick M, Griffin DE. Persistence of alphaviruses in vertebrate hosts. *Trends microbiol* 1994;2:25–28.
188. Levine B, Hardwick JM, Trapp BD, Crawford TO, Bollinger RC, Griffin DE. Antibody-mediated clearance of alphavirus infection from neurons. *Science* 1991;254:856–860.
189. Liebert UG, Baczko K, Budka H, ter Meulen V. Restricted expression of measles virus proteins in brains from cases of subacute sclerosing panencephalitis. *J Gen Virol* 1986;67:2435–2444.
190. Lill NL, Tevethia MJ, Hendrickson WG, Tevethia SS. Cytotoxic T lymphocytes (CTL) against a transforming gene product select for transformed cells with point mutations within sequences encoding CTL recognition epitopes. *J Exp Med* 1992;176:449.
191. Littman DR. The structure of the CD4 and CD8 genes. *Annu Rev Immunol* 1987;5:561–584.
192. Lodmell DL, Esposito JJ, Ewalt LC. Rabies virus antinucleoprotein antibody protects against rabies virus challenge *in vivo* and inhibits rabies virus replication *in vitro*. *J Virol* 1993;67:6080–6086.
193. Londberg-Holm K, Philipson L. Early events of virus-cell interactions in an adenovirus system. *J Virol* 1969;4:323–338.
194. Lutley R, Petursson G, Palsson PA, Georgsson G, Klein J, Nathanson N. Antigenic drift in visna: virus variation during long-term infection of Icelandic sheep. *J Gen Virol* 1983;64:1433–1440.

195. Madden DR, Gorga JC, Strominger JL, Wiley DC. The three-dimensional structure of HLA-B27 at 2.1 resolution suggests a general mechanism for tight peptide binding to MHC. *Cell* 1992;70:1035–1048.
196. Mahalingam R, Wellish M, Wolf W, et al. Latent varicella-zoster viral DNA in human trigeminal and thoracic ganglia. *N Engl J Med* 1990;323:627–631.
197. Mahy BWJ Strategies of virus persistence. *Br Med Bull* 1985;41:50–55.
198. Margolis TP, Sedarati F, Dobson AT, Feldman LT, Stevens JG. Pathways of viral gene expression during acute neuronal infection with HSV-1. *Virology* 1992;189:150–160.
199. Marrack P, Kappler J Subversion of the immune system by pathogens. *Cell* 1994;76:323–332.
200. Maruyama T, McLachlan A, Iino S, et al. The serology of chronic hepatitis B infection revisited. *J Clin Invest* 1993;91:2586–2595.
201. Masucci MG, Torsteinsdottir S, Colombani J, Brautbar C, Klein E, Klein G. Down-regulation of class I HLA antigens and of the Epstein-Barr virus encoded latent membrane protein in Burkitt lymphoma lines. *Proc Natl Acad Sci USA* 1987;84:4567–4571.
202. Matloubian M, Concepcion RJ, Ahmed R. CD4⁺ T cells are required to sustain CD8⁺ cytotoxic T-cell responses during chronic viral infection. *J Virol* 1994; 68:8056–8063.
203. Matloubian M, Kolhekar SR, Somasundaram T, Ahmed R. Molecular determinants of macrophage-tropism and viral persistence: importance of single amino acid changes in the polymerase and glycoprotein of lymphocytic choriomeningitis virus. *J Virol* 1993;67:7340–7349.
204. Matzinger P. Tolerance, danger and the extended family. *Annu Rev Immunol* 1994;12:991–1045.
205. Maudsley DJ, Bateman WJ, Morris AG. Reduced stimulation of helper T cells by Ki-ras transformed cells. *Immunology* 1991;72:277–281.
206. Maudsley DJ, Morris AG. Regulation of IFN-gamma-induced host cell MHC antigen expression by Kirsten MSV and MLV. II. Effects on class II antigen expression. *Immunology* 1989;67:26–31.
207. Maudsley DJ, Morris AG. Kirsten murine sarcoma virus abolishes interferon γ-induced class II but not class I major histocompatibility antigen expression in a murine fibroblast line. *J Exp Med* 1988;167: 706–711.
208. Mazanec MB, Kaetzel CS, Lamm ME, Fletcher D, Nedrud JG. Intracellular neutralization of virus by immunoglobulin A antibodies. *Proc Natl Acad Sci USA* 1992;89:6901–6905.
209. Meier JL, Holman RP, Croen KD, Smialek JE, Straus SE. Varicella-zoster virus transcription in human trigeminal ganglia. *Virology* 1993;193:193–200.
210. Meignier B, Longnecker R, Roizman B. *in vivo* behavior of genetically engineered herpes simplex viruses R7017 and R7020: construction and evaluation in rodents. *J Infect Dis* 1988;158:602–614.
211. Mellerick DM, Fraser NW. Physical state of the latent herpes simplex virus genome in a mouse model system: evidence suggesting an episomal state. *Virology* 1987;158:265–275.
212. Meuer SC, Acuto O, Hergend T, Schlossman SF, Reinherz EL. The human T-cell receptor. *Annu Rev Immunol* 1984;2:23–50.
213. Meyerhans A, Dadaglio G, Vartanian J-P, et al. *In vivo* persistence of a HIV-1-encoded HLA-B27-restricted cytotoxic T lymphocyte epitope despite specific *in vitro* reactivity. *Eur J Immunol* 1991;21:2637.
214. Milich DR, McLachlan A. The nucleocapsid of hepatitis B virus is both a T-cell-independent and a T-cell-dependent antigen. *Science* 1986;234:1398–1401.
215. Mims CA. Parasite survival strategies and persistent infections. In: *Medical microbiology*. England: Mosby Europe; 1993.
216. Minton EJ, Tysee C, Sinclair JH, Sissons JCP. Human cytomegalovirus infection of the monocyte/macrophage lineage in bone marrow. *J Virol* 1994;68:4017–4021.
217. Moench TR, Griffin DE, Obriecht CR, Vaisberg AJ, Johnson RT. Acute measles in patients with and without neurological involvement: distribution of measles virus antigen and RNA. *J Infect Dis* 1988;158:433–442.
218. Montelaro CR, Parekh B, Orrego A, Issel CJ Antigenic variation during persistent infection by equine infectious anemia virus, a retrovirus. *J Biol Chem* 1984;259:10539–10544.
219. Moore KW, O'Garra A, de Waal Malefyt R, Vieira P, Mosmann TR. Interleukin-10. *Annu Rev Immunol* 1993;11:165–190.
220. Moskophidis D, Lechner F, Pircher H, Zinkernagel RM. Virus persistence in acutely infected immunocompetent mice by exhaustion of antiviral cytotoxic effector T cells. *Nature* 1993;362:758–761.
221. Moskophidis D, Cobbold SP, Waldmann H, Lehmann-Grube F. Mechanism of recovery from acute virus infection: treatment of lymphocytic choriomeningitis virus-infected mice with monoclonal antibodies reveals that Lyt2⁺T lymphocytes mediate clearance of virus and regulate the antiviral antibody response. *J Virol* 1987;61:1867–1874.
222. Moskophidis D, Lechner F, Pircher H, Zinkernagel RM. Viral persistence in acutely infected immunocompetent mice by exhaustion of antiviral cytotoxic effector cells. *Nature* 1993;362:758–761.
223. Mosmann TR, Coffman RL. TH1 and TH2 cells: different patterns of lymphokine secretion lead to different functional properties. *Annu Rev Immunol* 1989;7:145–173.
224. Muller D, Koller BH, Whitton JL, LaPan KE, Brigman KK, Frelinger JA. LCMV-specific, class II restricted cytotoxic T cells in β_2-microglobulin-deficient mice. *Science* 1992;255:1576–1578.
225. Narayan O, Zink MC, Huso D, et al. Lentiviruses of animals are biological models of the human immunodeficiency viruses. *Microbiol Pathogen* 1988;5:149–157.
226. Narayan O, Clements JE. Biology and pathogenesis of lentiviruses of ruminant animals. In: Wong-staal F, Gallo RC, eds. *Retrovirus biology: an emerging role in human disease.* New York: Marcel Dekker; 1988.
227. Narayan O, Clements J, Griffin DE, Wolinsky JS. Neutralizing antibody spectrum determines the antigenic profiles of emerging mutants of visna virus. *Infect Immun* 1981;32:1045–1050.
228. Narayan O, Griffin DE, Chase J Antigenic drift of visna virus in persistently infected sheep. *Science* 1977;197:376–378.
229. Nayersina R, Fowler P, Guilhot S, et al. HLA A2 restricted cytotoxic T lymphocyte responses to multiple hepatitis B surface antigen epitopes during hepatitis B virus infection. *J Immunol* 1993;150:4659–4671.
230. Nemerow G, Wolfert R, McNaughton M, Cooper N. Identification and characterization of the Epstein-Barr virus receptor on human B lymphocytes and its relationship to the C3d complement receptor CR2. *J Virol* 1985;55:347–351.
231. Neumann R, Genersch E, Eggers HJ Detection of adenovirus nucleic acid sequences in human tonsils in the absence of infectious virus. *Virus Res* 1987;7:93–97.
232. Nguyen L, Knipe DM, Finberg RW. Mechanism of virus-induced lg subclass shifts. *J Immunol* 1994;152:478–484.
233. Nonoyama M, Pagano JS. Separation of Epstein-Barr virus DNA from large chromosomal DNA in non-virus-producing cells. *Nature* 1972;333:41–45.
234. Norrby E, Chen S-N, Tagoshi T, Sheshberadaran H, Johnson CP. Five measles virus antigens demonstrated by use of mouse hybridoma antibodies in productively infected tissue culture cells. *Arch Virol* 1982;71:1–11.
235. Norrby E, Kristensson K, Brzosko WJ, Kapsenberg JG. Measles virus matrix proteins detected by immune fluorescence with monoclonal antibodies in the brain of patients with subacute sclerosing panencephalitis. *J Virol* 1985;56:337–340.
236. Norrby E, Orvell C, Vandvik B, Cherry DJ Antibodies against measles virus polypeptides in different disease conditions. *Infect Immun* 1981;34:718–724.
237. Ohara Y, Tashiro M, Takase S, Homma M. Detection of antibody to M protein of measles virus in patients with subacute sclerosing panencephalitis: a comparative study on immunoprecipitation. *Microbiol Immunol* 1985;29:709–723.
238. O'Hare P, Hayward GS. Three *trans*-acting regulatory proteins of herpes simplex virus modulate immediate-early gene expression in a pathway involving positive and negative feedback regulation. *J Virol* 1985;56:723–733.
239. Okamoto H, Usuda S, Imai M, et al. Antibody to the receptor for polymerized human serum albumin in acute and persistent infection with hepatitis B virus. *Hepatology* 1986;6:354–359.
240. Oldstone MBA, Blount P, Southern P. Cytoimmunotherapy for persistent virus infection reveals a unique clearance pattern from the central nervous system. *Nature* 1986;321:239–243.
241. Pääbo S, Nilsson T, Peterson PA. Adenoviruses of subgenera B, C, D, and E modulate cell-surface expression of major histocompatibility complex class antigens. *Proc Natl Acad Sci USA* 1986;83:9665–9669.
242. Palese P, Young JF. Variation of influenza A, B, and C viruses. *Science* 1982;215:1468–1474.
243. Pantaleo G, Demarest JF, Soudeyns H, et al. Major expansion of CD8⁺ T cells with a predominant Vβ usage during the primary immune response to HIV. *Nature* 1994;370:463–467.
244. Pantaleo G, Graziosi C, Demarest JF, et al. HIV infection is active and progressive in lymphoid tissue during the clinically latent stage of disease. *Nature* 1993;362:355–358.

245. Parker DC. T cell-dependent B-cell activation. *Annu Rev Immunol* 1993;11:331–360.
246. Peebles TC. Distribution of virus in blood components during the viremia of measles. *Arch Gesamte Virusforsch* 1967;22:43–47.
247. Penna A, Chisari FV, Bertoletti A, et al. Cytotoxic T lymphocytes recognize an HLA-A2-restricted epitope within the hepatitis B virus nucleocapsid antigen. *J Exp Med* 1991;174:1565–1570.
248. Petit AJC, Terpstra FG, Miedema F. Human immunodeficiency virus infection down-regulates HLA class II expression and induces differentiation in promonocytic U937 cells. *J Clin Invest* 1987;79:1883–1889.
249. Phillips RE, Rowland-Jones S, Nixon DF, et al. Human immunodeficiency virus genetic variation that can escape cytotoxic T cell recognition. *Nature* 1991;354:453.
250. Pina M, Green M. Biochemical studies on adenovirus multiplication. XIV. Macromolecule and enzyme synthesis in cells replicating oncogenic and nononcogenic human adenoviruses. *Virology* 1969;38: 573–586.
251. Pircher H, Moskophidis D, Rohrer U, Burki K, Hengartner H, Zinkernagel RM. Viral escape by selection of cytotoxic T cell-resistant virus variants *in vivo*. *Nature* 1990;346:629.
252. Pircher H, Burki K, Lang R, Hengartner H, Zinkernagel RM. Tolerance induction in double specific T-cell receptor transgenic mice varies with antigen. *Nature* 1989;342:559–561.
253. Porter DD. Persistent infections. In: Baron S, ed. *Medical microbiology*, 2nd ed. Menlo Park, CA: Addison-Wesley; 1985:784–790.
254. Puga A, Notkins AL. Continued expression of a poly(A)$^+$ transcript of herpes simplex virus type 1 in trigeminal ganglia of latently infected mice. *J Virol* 1987;61:1700–1703.
255. Racz P, Tenner-Racz K, Kahl C, Feller AC, Kern P, Dietrich M. Spectrum of morphologic changes of lymph nodes from patients with AIDS or AIDS-related complexes. *Prog Allergy* 1986;37:81–181.
256. Rager-Zisman B, Ju G, Rajan RV, Bloom BR. Decreased expression of H-2 antigens following acute measles virus infection. *Cell Immunol* 1981;59:319–329.
257. Rawlins DR, Milman G, Hayward SD, Hayward GS. Sequence specific DNA binding of the Epstein-Barr virus nuclear antigen (EBNA-1) to clustered sites in the plasmid maintenance region. *Cell* 1985;42:859–868.
258. Rawls WE, Chan MA, Gee SR. Mechanisms of persistence in arenavirus infections: a brief review. *Can J Microbiol* 1981;27:568–574.
259. Redfield RR, Markham PD, Salahuddin SZ, et al. Genetic variation in HTLV-III/LAV over time in patients with AIDS or at risk for AIDS. *Science* 1986;232:1548–1553.
260. Reznikoff C, Tegtmeyer P, Dohan C, Ender JF. Isolation of AGMK cells partially resistant to SV40: identification of the resistant step. *Proc Soc Exp Biol Med* 1972;141:740–746.
261. Rice GPA, Schrier RD, Oldstone MBA. CMV infects human lymphocytes and monocytes: virus expression is restricted to immediate-early gene products. *Proc Natl Acad Sci USA* 1984;81:6134–6138.
262. Rinaldo CR, Jr. Modulation of major histocompatibility complex antigen expression by viral infection. *Am J Pathol* 1994;144:637–650.
263. Riviere Y, Tanneau-Salvadori F, Regnault A, et al. HIV-specific cytotoxic responses of seropositive individuals: distinct types of effector cells mediate killing of targets expressing gag and env proteins. *J Virol* 1989;63:2270–2277.
264. Robbins FC. Measles: clinical features: pathogenesis, pathology, and complications. *Am J Dis Child* 1962;103:266–273.
265. Robert-Guroff M, Brown M, Gallo RC. HTLV-III neutralizing antibodies in patients with AIDS and AIDS-related complex. *Nature* 1985;316:72–74.
266. Rock DL, Fraser NW. Detection of HSV-1 genome in central nervous system of latently infected mice. *Nature* 1983;302:523–525.
267. Rock DL, Fraser NW. Latent herpes simplex virus type 1 DNA contains two copies of the virion DNA joint region. *J Virol* 1985;55:849–852.
268. Rock DL, Nesburn AB, Ghiasi H, et al. Detection of latency-related viral RNAs in trigeminal ganglia of rabbits latently infected with herpes simplex virus type 1. *J Virol* 1987;61:3820–3826.
269. Roizman B, Sears AE. An inquiry into the mechanism of herpes simplex virus latency. *Annu Rev Microbiol* 1987;41:543–571.
270. Ron D, Tal J. Coevolution of cells and virus as a mechanism for the persistence of lymphotropic minute virus of mice in L cells. *J Virol* 1985;55:424–430.
271. Roos MTL, Lange JMA, DeGoede REY, et al. Viral phenotype and immune response in primary human immunodeficiency virus type 1 infection. *J Infect Dis* 1992;165:427–432.
272. Rowe WP, Huebner RJ, Gilmore LK, Parrott RH, Ward TG. Isolation of a cytopathogenic agent from human adenoids undergoing spontaneous degeneration in tissue culture. *Proc Soc Exp Biol Med* 1953;84:570–573.
273. Rustigian R. Persistent infection of cells in culture by measles virus. II. Effect of measles antibody on persistently infected HeLa sublines and recovery of a HeLa clonal line persistently infected with incomplete virus. *J Bacteriol* 1966;92:1805–1811.
274. Safrit JT, Andrews CA, Zhu T, Ho DD, Koup RA. Characterization of human immunodeficiency virus type 1–specific cytotoxic T lymphocyte clones isolated during acute seroconversion: recognition of autologous virus sequences within a conserved immunodominant epitope. *J Exp Med* 1994;179:463–472.
275. Saksola K, Stevens C, Rubinstein P, Baltimore D. Human immunodeficiency virus type 1 messenger RNA in peripheral blood cells predicts disease progression independently of the numbers of CD4$^+$ lymphocytes. *Proc Natl Acad Sci USA* 1994;91:1104–1108.
276. Salinovich O, Payne LS, Montelaro RC, Hussain KA, Issel CJ, Schnorr KL. Rapid emergence of novel antigenic and genetic variants of equine infectious anemia virus during persistent infection. *J Virol* 1986; 57:71–80.
277. Samulski RJ, Zhu X, Xiao X, et al. Targeted integration of adeno-associated virus (AAV) into human chromosome 19. *EMBO J* 1991;10:3941.
278. Sattentau QJ, Weiss RA. The CD4 antigen: physiologic ligand and HIV receptor. *Cell* 1988;52:631–633.
279. Sawtell NM, Thompson RL. Herpes simplex virus type 1 latency-associated transcription unit promotes anatomical site-dependent establishment and reactivation from latency. *J Virol* 1992;66:2157–2169.
280. Scheffner M, Huibregtse JM, Vierstra RD, Howley PM. The HPV-16 E6 and E6–AP complex functions as a ubiquitin–protein ligase in the ubiquitination of p53. *Cell* 1993;75:495–505.
281. Scheffner M, Werness BA, Huibregtse JM, Levine AJ, Howley PM. The E6 oncoprotein encoded by human papillomavirus types 16 and 18 promotes the degradation of p53. *Cell* 1990;63:1129–1136.
282. Schmid A, Spielhofer P, Cattaneo R, Baczko K, ter Meulen V, Billeter MA. Subacute sclerosing panencephalitis is typically characterized by alterations in the fusion protein cytoplasmic domain of the persisting measles virus. *Virology* 1992;188:910–915.
283. Schneider A, Oltersdorf T, Schneider V, Gissmann L. Distribution of human papillomavirus 16 genomes in cervical neoplasia by molecular *in situ* hybridization of tissue sections. *Int J Cancer* 1987;39:717–721.
284. Schneider-Schaulies S, Schneider-Schaulies J, Schuster A, Bayer M, Pavlovic J, ter Meulen V. Cell type-specific MxA-mediated inhibition of measles virus transcription in human brain cells. *J Virol* 1994;68:6910–6917.
285. Schnittman SM, Denning SM, Greenhouse JJ, et al. Evidence for susceptibility of intrathymice T-cell precursors and their progeny carrying T-cell antigen receptor phenotypes TCR$\alpha\beta^+$ and TCR$\gamma\delta^+$to human immunodeficiency virus infection: a mechanism for CD4$^+$(T4) lymphocyte depletion. *Proc Natl Acad Sci USA* 1990;87:7727–7731.
286. Scholz M, Hamann A, Blaheta RA, Auth MKH, Encke A, Markus BH. Cytomegalovirus- and interferon-related effects on human endothelial cells: cytomegalovirus infection reduces upregulation of HLA class II antigen expression after treatment with interferon γ. *Hum Immunol* 1992;35:230–238.
287. Schrier PI, Bernards R, Vaessen RTMJ, Houweling A, van der Eb AJ. Expression of class I major histocompatibility antigens switched off by highly oncogenic adenovirus 12 in transformed rat cells. *Nature* 1983;305:771–775.
288. Schwarz E, Freese UK, Gissmann L, et al. Structure and transcription of human papillomavirus sequences in cervical carcinoma cells. *Nature* 1985;314:111–114.
289. Sears AE, Hukkanen V, Labow MA, Levine AJ, Roizman B. Expression of the herpes simplex virus 1 α transinducing factor (VP16) does not induce reactivation of latent virus or prevent the establishment of latency in mice. *J Virol* 1991;65:2929–2935.
290. Sears AE, Meignier B, Roizman B. Establishment of latency in mice by herpes simplex virus 1 recombinants that carry insertions affecting regulation of the thymidine kinase gene. *J Virol* 1985;55: 410–416.
291. Sears AE, Roizman B. Amplification by host factors of a sequence contained within the herpes simplex virus 1 genome. *Proc Natl Acad Sci USA* 1990;87:9441–9445.

292. Sedarati F, Margolis TP, Stevens JG. Latent infection can be established with drastically restricted transcription and replication of the HSV-1 genome. *Virology* 1993;192:687–691.
293. Sedmak DD, Guglielmo AM, Knight DA, Birmingham DJ, Huang EH, Waldman WJ. Cytomegalovirus inhibits major histocompatibility class II expression on infected endothelial cells. *Am J Pathol* 1994;144:683.
294. Selin LK, Nahill Sr, Welsh RM. Cross-reactivities in memory cytotoxic T lymphocyte recognition of heterologous viruses. *J Exp Med* 1994;179:1933–1943.
295. Shemesh J, Rotem-Yehudar R, Ehrlich R. Transcription and posttranscriptional regulation of class I major histocompatibility complex genes following transformation with human adenoviruses. *J Virol* 1991;65:5544–5548.
296. Sherman FE, Ruckle G. *in vivo* and *in vitro* cellular changes specific for measles. *Arch Pathol* 1958;65:587–599.
297. Sheshberadaran H, Chen S-H, Norrby E. Monoclonal antibodies against five structural components of measles virus. I. Characterization of antigenic determinants on nine strains of measles virus. *Virology* 128:341–353.
298. Sheshberadaran H, Norrby E, Rammohan KW. Monoclonal antibodies against five structural components of measles virus. II. Characterization of five cell lines persistently infected with measles virus. *Arch Virol* 1985;83:251–268.
299. Sixbey JW, Vesterinen EH, Nedrud JG, Raab-Traub N, Walton LA, Pagano JS. Replication of Epstein-Barr virus in human epithelial cells infected *in vitro*. *Nature* 1983;306:480–483.
300. Sixbey JW, Nedrud JG, Raab-Traub N, Hanes RA, Pagano JS. Epstein-Barr virus replication in oropharyngeal epithelial cells. *N Engl J Med* 1984;310:1225–1230.
301. Sloan-Lancaster J, Evavold BD, Allen PM. Induction of T-cell anergy by altered T-cell-receptor ligands on live antigen presenting cells. *Nature* 1993;363:156.
302. Sonea S. Bacterial viruses, prophages, and plasmids, reconsidered. *Ann NY Acad Sci* 1987;503:251–260.
303. Speck PG, Simmons A. Divergent molecular pathways of productive and latent infection with a virulent strain of herpes simplex virus type 1. *J Virol* 1991;65:4001–4005.
304. Spiegel H, Herbst H, Niedobitek G, Foss HD, Stein H. Follicular dendritic cells are a major reservoir for human immunodeficiency virus type 1 in lymphoid tissues facilitating infection of CD4+ T-helper cells. *Am J Pathol* 1992;140:15–22.
305. Spriggs MK. Cytokine and cytokine receptor genes captured by viruses. *Curr Opin Immunol* 1994;6:526–529.
306. Springer TA. Adhesion receptors of the immune system. *Nature* 1990;346:425–433.
307. Stanberry LR, Burke R, Myers MG. Herpes simplex virus glycoprotein treatment of recurrent genital herpes. *J Infect Dis* 1988;157:156–163.
308. Steinberg B, Topp W, Schneider PS, Abramson A. Laryngeal papillomavirus infection during clinical remission. *N Engl J Med* 1983;308:1261–1264.
309. Steiner I, Spivack JG, Deshmane SL, Ace CI, Preston CM, Fraser NW. A herpes simplex virus type 1 mutant containing a nontransinducing Vmw65 protein establishes latent infection *in vivo* in the absence of viral replication and reactivates efficiently from trigeminal ganglia. *J Virol* 1990;64:1630–1638.
310. Steinhauer D, Holland JJ Rapid evolution of RNA viruses. *Annu Rev Microbiol* 1987;41:409–433.
311. Stemler M, Hess J, Braun R, Will H, Schroder CH. Serological evidence for expression of the polymerase gene of human hepatitis B virus *in vivo*. *J Gen Virol* 1988;69:689–693.
312. Stemler M, Weimer T, Tu Z-X, et al. Mapping of B-cell epitopes of the human hepatitis B virus X protein. *J Virol* 1990;64:2802–2809.
313. Stephan W, Prince AM, Brotman B. Modulation of hepatitis B infection by intravenous application of an immunoglobulin preparation that contains antibodies to hepatitis B e and core antigens but not to hepatitis B surface antigen. *J Virol* 1984;51:420–424.
314. Stevens JG, Haarr L, Porter DP, Cook ML, Wagner EK. Prominence of the herpes simplex virus latency-associated transcript in trigeminal ganglia from seropositive humans. *J Infect Dis* 1988;158:117–123.
315. Stevens JG, Wagner EK, Devi-Rao GB, Cook ML, Feldman LT. RNA complementary to a herpesvirus alpha gene mRNA is prominent in latently infected neurons. *Science* 1987;253:1056–1059.
316. Stevens JG. Human herpesviruses: a consideration of the latent state. *Microbiol Rev* 1989;53:318–332.
317. Stevens JG. Overview of herpesvirus latency. *Semin Virol* 1994;5:191–196.
318. Stoler MH, Broker TR. *In situ* hybridization detection of human papilloma virus DNA and messenger RNA in genital condylomas and a cervical carcinoma. *Hum Pathol* 1986;17:1250–1258.
319. Stoler MH, Wolinsky SM, Whitbeck A, Broker TR, Chow LT. Differentiation-linked human papillomavirus types 6 and 11 transcription in genital condylomata revealed by *in situ* hybridization with message-specific probes. *Virology* 1989;172:331–340.
320. Straus SE. Clinical and biological differences between recurrent herpes simplex virus and varicella-zoster virus infections. *JAMA* 1989;262:3455–3458.
321. Straus SE, Corey L, Burke RL, Savarese B, Barnum G, Krause PR, et al. Placebo-controlled trial of vaccination with recombinant glycoprotein D of herpes simplex virus type 2 for immunotherapy of genital herpes. *Lancet* 1994. 343:1460–1463.
322. Su L, Knipe DM. Herpes simplex virus α protein ICP27 can inhibit or augment viral gene transactivation. *Virology* 1989;170:496–504.
323. Syvanen M. The evolutionary implications of mobile genetic elements. *Annu Rev Genet* 1984;18:271–293.
324. Takemoto KK, Habel K. Virus–cell relationship in a carrier culture of HeLa cells and coxsackie A9 virus. *Virology* 1959;7:28–44.
325. Tanaka K, Isselbacher KJ, Khoury G, Jay G. Reversal of oncogenesis by the expression of a major histocompatibility complex class I gene. *Science* 1985;228:26–30.
326. Taylor-Wiedman J, Sissons P, Sinclair J. Induction of endogenous human cytomegalovirus gene expression after differentiation of monocytes from healthy carriers. *J Virol* 1994;68:1597–1604.
327. Tenner-Racz K, Racz P, Bofill M, et al. HTLV-III/LAV viral antigens in lymph nodes of homosexual men with persistent generalized lymphadenopathy and AIDS. *Am J Pathol* 1986;123:9–15.
328. Tenser RB, Hay KA, Edris WA. Latency-associated transcript but not reactivatable virus is present in sensory ganglion neurons after inoculation of thymidine kinase–negative mutants of herpes simplex virus type 1. *J Virol* 1989;63:2161–2165.
329. Tenser RB, Hyman RW. Latent herpesvirus infections of neurons in guinea pigs and humans. *Yale J Biol Med* 1987;60:159–167.
330. Ter Meulen V, Stephenson JR, Kreth HW. Subacute sclerosing panencephalitis. *Compr Virol* 1983;18:105–185.
331. Ter Meulen V, Loffler S, Carter MJ, Stephenson JR. Antigenic characterization of measles and SSPE virus hemagglutinin by monoclonal antibodies. *J Gen Virol* 1981;57:357–364.
332. Thormar H, Barshatsky MR, Kozlowski PB. The emergence of antigenic variants is a rare event in long term visna virus infection *in vivo*. *J Gen Virol* 1983;64:1427–1432.
333. Trousdale M, Steiner I, Spivack JG, et al. *in vivo* and *in vitro* reactivation impairment of a herpes simplex virus type 1 latency-associated transcript variant in a rabbit eye model. *J Virol* 1991;65:6989–6993.
334. Tsomides TH, Aldovini A, Johnson RP, Walker BD, Young RA, Eisen HN. Naturally processed viral peptides recognized by cytotoxic T lymphocytes on cells chronically infected by human immunodeficiency virus type 1. *J Exp Med* 1994;180:1283–1293.
335. Upton C, Mossman K, McFadden G. Encoding of a homolog of the IFN-γ receptor by myxoma virus. *Science* 1992;258:1369–1372.
336. Valyi-Nagi T, Deshmane SL, Spivack JG, et al. Investigation of herpes simplex virus type 1 (HSV-1) gene expression and DNA synthesis during the establishment of latent infection by an HSV-1 mutant, in1814, that does not replicate in mouse trigeminal ganglia. *J Gen Virol* 1991;72:641–649.
337. Vogt M, Dulbecco R. Properties of HeLa cell culture with increased resistance to poliomyelitis virus. *Virology* 1958;5:425–434.
338. Volkert M. Studies on immunologic tolerance to LCM virus. 2. Treatment of virus carrier mice by adoptive immunization. *Acta Pathol Microbiol Scand* 1968;57:465–487.
339. von Knebel-Doeberitz M, Oltersdorf T, Schwarz E, Gissmann L. Correlation of modified human papillomavirus early gene expression with altered growth properties in C4–1 cervical carcinoma cells. *Cancer Res* 1988;48:3780–3785.
340. Walker BD, Flexner C, Paradis TJ, et al. HIV-1 reverse transcriptase is a target for cytotoxic T lymphocytes in infected individuals. *Science* 1988;240:64–66.
341. Walker CM, Moody DJ, Stites DP, Levy JA. CD8+ lymphocytes can control HIV infection *in vitro* by suppressing virus replication. *Science* 1986;124:1563–1566.

342. Walker DL. Persistent viral infection in cell cultures. In: Sanders M, Lennette EH, eds. *Medical and applied virology* 1968:99–110.
343. Walker DL. The viral carrier state in animal cell cultures. *Prog Med Virol* 1964;6:111–148.
344. Walsh CM, Matloubian M, Liu C-C, et al. Immune function in mice lacking the perforin gene. *Proc Natl Acad Sci* 1994; 91:10854–10858.
345. Wang F, Gregory CD, Rowe M, et al. Epstein-Barr virus nuclear antigen 2 specifically induces expression of the B-cell activation antigen CD23. *Proc Natl Acad Sci USA* 1987;83:3452–3457.
346. Webster RG, Laver WG, Air GM, Schild GC. Molecular mechanisms of variation in influenza viruses. *Nature* 1982;296:115–121.
347. Wechsler S, Meissner HC. Measles and SSPE viruses: similarities and differences. *Prog Med Virol* 1982;28:65–95.
348. Wechsler SL, Fields BN. Differences between the intracellular polypeptides of measles and subacute sclerosing panencephalitis virus. *Nature* 1978;272:458–460.
349. Wechsler SL, Weiner HL, Fields BN. Immune responses in subacute sclerosing panencephalitis: reduced antibody response to the matrix protein of measles virus. *J Immunol* 1979;123:884–889.
350. Weimer T, Weimer K, Tu Z-X, Jung C, Pape GR, Will H. Immunogenicity of human hepatitis B virus P–gene derived proteins. *J Immunol* 1989;143:3750–3759.
351. Weiss RA, Clapman PR, Cheingsong-Popou R, et al. Neutralization of human T lymphotropic virus type III by sera of AIDS and AIDS-risk patients. *Nature* 1985;316:69–72.
352. Wilson JH, De Pamphilis M, Berg P. Simian virus 40–permissive cell interactions: selection and characterization of spontaneously arising monkey cells that are resistant to simian virus 40 infection. *J Virol* 1976;20:391–399.
353. Wold WSM, Gooding LR. Region E3 of adenovirus: a cassette of genes involved in host immunosurveillance and virus–cell interactions. *Virology* 1991;184:1–8.
354. Wong TC, Ayata M, Hirano A, Yoshikawa Y, Tsuruoka H, Yamanouchi K. Generalized and localized biased hypermutation affecting the matrix gene of a measles virus strain that causes subacute sclerosing panencephalitis. *J Virol* 1989;63:5464–5468.
355. Wong TC, Ayata M, Ueda S, Hirano A. Role of biased hypermutation in evolution of subacute sclerosing panencephalitis virus from progenitor acute measles virus. *J Virol* 1991;65:2191–2199.
356. Yates J, Warren N, Reisman D, Sugden B. A *cis*-acting element from the Epstein-Barr viral genome that permits stable replication of recombinant plasmids in latently infected cells. *Proc Natl Acad Sci USA* 1984;81:3806.
357. York IA, Roop C, Andrews DW, Riddell SR, Graham FL, Johnson DC. A cytosolic herpes simplex virus protein inhibits antigen presentation to $CD8^+$ T lymphocytes. *Cell* 1994;77:525–535.
358. Young KKY, Heineki BE, Wechsler SL. M protein instability and lack of H protein processing associated with nonproductive persistent infection of HeLa cells by measles virus. *Virology* 1985; 143:536–545.
359. Youngner JS, Preble OT. Viral persistence: evolution of viral populations. In: Fraenkel-Conrat H, Wagner R, eds. *Comprehensive virology*, Vol 16. New York: Plenum Press; 1980:73–135.
360. Yuki N, Hayashi N, Kasahara A, et al. Detection of antibodies against the polymerase gene product in hepatitis B virus infection. *Hepatology* 1990;12:193–198.
361. Zhu T, Mo H, Wang N, Nam DS, Cao Y, Koup RA, Ho DD. Genotypic and phenotypic characterization of HIV-1 in patients with primary infection. *Science* 1993;261:1179–1181.
362. Zinkernagel RM, Doherty PC. MHC-restricted cytotoxic T cells: studies on the biological role of polymorphic major transplantation antigens determining T-cell restriction-specificity, function, and responsiveness. *Adv Immunol* 1979;27:51–177.
363. Zinkernagel RM, Welsh RM. H-2 compatibility requirement for virus-specific T cell–mediated effector functions *in vivo*. I. Specificity of T cells conferring antiviral protection against lymphocytic choriomeningitis virus is associated with H-2K and H-2D. *J Immunol* 1976;117: 1495–1502.

Fundamental Virology, Third Edition
edited by B.N. Fields, D.M. Knipe, P.M. Howley, et al.
Lippincott - Raven Publishers, Philadelphia © 1996

CHAPTER 8

Virus–Host Cell Interactions

David M. Knipe

By definition, viruses are unable to replicate on their own but must enter a host cell and use the host-cell macromolecular machinery and energy supplies to replicate. Thus, in many ways the study of viral replication is a study of the interactions of viruses with their host cells. The study of virus–host cell interactions has impacted on several areas of biology and biomedical science. First, investigation of the mechanisms by which viruses replicate in, alter, or take over their host cells has often led to insight into basic cell biological and molecular biological mechanisms. Viruses often target critical regulatory events in the host cell; thus, knowledge of the processes subverted by viruses has often highlighted cellular regulatory mechanisms. Second, during their replication within cells, viruses exploit host-cell molecules and processes at the expense of the host cell, which may result in cell injury or death. These injurious effects of viral replication in cells are one of the basic causes of viral disease. Therefore, precise knowledge of the pathogenic mechanisms by which viruses replicate in specific tissues, spread, and cause disease must come, in part, from studies of the intracellular replication of the virus. Over the past 40 years, increasing understanding of the mechanisms of viral replication has emerged from biochemical and cell biological studies of virus replication in cultured cells. Investigators of viral pathogenesis have recently attempted to define the molecular events occurring in different cell types during the series of stages that de-

D. M. Knipe: Department of Microbiology and Molecular Genetics, Harvard Medical School, Boston, Massachusetts 02115.

fine viral pathogenesis within a host organism. Third, the definition of antiviral strategies requires knowledge of the replicative mechanisms of viruses and the identification of replicative steps that involve virus-specific processes and not host-cell processes.

This chapter will focus on the interactions of viruses with an individual host cell. Chapter 6 discusses the events that lead to (a) spread of a virus from one cell to another within a host organism, (b) induction of disease, and (c) spread within the environment. Virus infection of a cell can lead to any of several possible outcomes. First, a nonproductive infection can occur in which viral replication is blocked, and the host cell may or may not survive. After the nonproductive infection, the viral genome may be lost from the cell. Alternatively, the viral genetic information may become integrated as DNA in the cellular genome or may persist as episomal DNA in these surviving cells. If the growth properties of the cell are altered to make it oncogenic, this would constitute an oncogenic transformation event (see Chapter 9). The virus may become dormant with little viral gene expression, and a latent infection results (see Chapter 7). Second, a productive viral infection may result in which the host cell dies and lyses. Third, the cell may survive and continue to produce virus at a low level, resulting in a chronic infection (see Chapter 7). Which of these possible scenarios is realized is determined by the nature of the interactions between the virus and the host-cell constituents. For example, a nonproductive infection may result if a host-cell component necessary for viral replication is not present. One of the main goals of this chapter is to describe the types of interactions between virus-encoded macromolecules and the host cell, which may define the ultimate outcome of a virus infection. This chapter will examine (a) the molecular and cell biological events that allow viral replication, (b) the ways in which viruses modify their host cells to promote their own replication, and (c) the kinds of mechanisms that may have evolved in cells to prevent virus infection. The types of experimental approaches used to obtain evidence for specific virus–host interactions will also be discussed.

The study of virus–cell interactions really started with the growth of viruses in cultured cells (73). Although infection of host organisms had given some indication of cell death resulting from viral infection, there was little clear evidence of other effects of viruses on the host cell before the infection of cultured cells and the identification of cytopathic effect (CPE) of viruses on cells (72). The elucidation of viral replication strategies in the 1950s and 1960s provided the broad outlines of virus replication. More recently, better probes for nucleic acids and proteins have allowed more precise descriptions of the molecular events of viral replication in a host cell. The techniques of molecular genetics and cell biology have begun to define the specific host-cell molecules and cellular compartments with which virus-encoded molecules interact. This provides one of the second themes of this chapter. In addition to defining the events of viral replication, molecular and cell biological studies of virus replication have used viruses as probes of the eukaryotic host cell. Viruses often subvert host-cell metabolism in such a subtle way that understanding viral replication can provide knowledge of critical metabolic events of the host cell. Thus, viruses often mimic their host cell (or have evolved from the host cell; see Chapter 5) in such a way that viruses frequently provide a prototype mechanism for a specific molecular process. In many cases, the initial evidence for a specific molecular event has come from the study of viruses and their intracellular replication processes.

In addition to classifying virus–host-cell interactions in terms of the final outcome of the infection, virus–cell interactions can also be described in molecular terms with regard to individual replication events. For example, the effects of viruses on the host cell can be mediated by addition or substitution of a virus-specific macromolecule to a cellular complex or structure. Alternatively, the virus may mediate a covalent or noncovalent modification of a host-cell molecule. Virus infection may cause a disassembly or rearrangement of a host-cell complex or structure, or virus infection may lead to the assembly of a new infected cell-specific complex or structure in the infected cell. After defining these molecular interactions, the challenge for us as virologists is to relate this molecular information back to the biology of the virus in the host cell and ultimately the host organism.

CYTOPATHIC EFFECTS OF VIRUS INFECTION

One of the classic ways of detecting virus replication in cells was the observation of changes in cell structure, or CPEs, resulting from virus infection (see Fig. 2 in Chapter 14 in *Fields Virology,* 3rd ed.). Some of the most com-

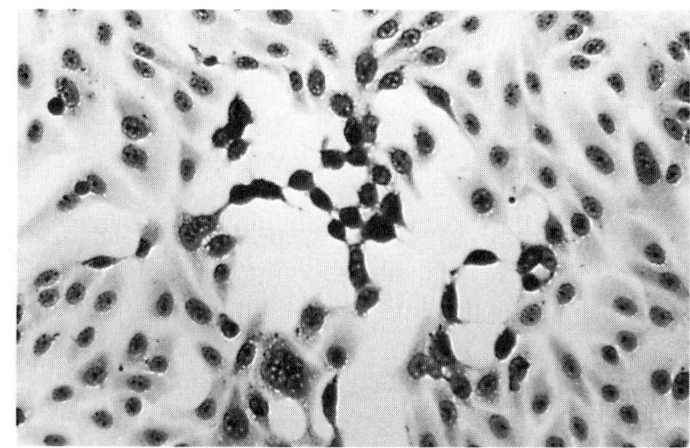

FIG. 1. Cytopathic effect (CPE) due to virus infection. The center portion of the figure shows monkey (Vero) cells rounding up and detaching from the substrate after infection with herpes simplex virus 1 (HSV-1). A normal monolayer of cells is visible around the focus of CPE. The cells were fixed with methanol and stained with Giemsa stain. Micrograph courtesy of M. Kosz-Vnenchak.

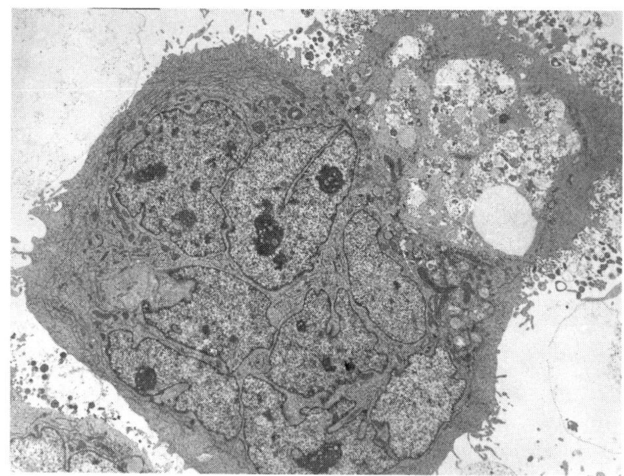

FIG. 2. A syncytium formed by HIV infection of T cells. Electron micrograph courtesy of J. Sodroski.

mon effects of viral infection are morphological changes such as (a) cell rounding and detachment from the substrate (Fig. 1), (b) cell lysis, (c) syncytium formation (Fig. 2), and (d) inclusion body formation (Fig. 3). The occurrence of cell morphological changes resulting from CPE has even led to classification schemes for viruses. Enders (72) proposed classifying viruses into the following groups: (a) those causing cellular degeneration; (b) those causing formation of inclusion bodies and cell degeneration; and (c) those causing formation of multinucleated cells or syncytial masses and degeneration, with or without inclusion bodies. However, as described in Chapter 2, other classification schemes based on virion and genome structure and modes of replication have provided much better ways of classifying viruses.

The CPE of viruses on cells has also been called *cell injury*. These terms tend to emphasize the pathology of the

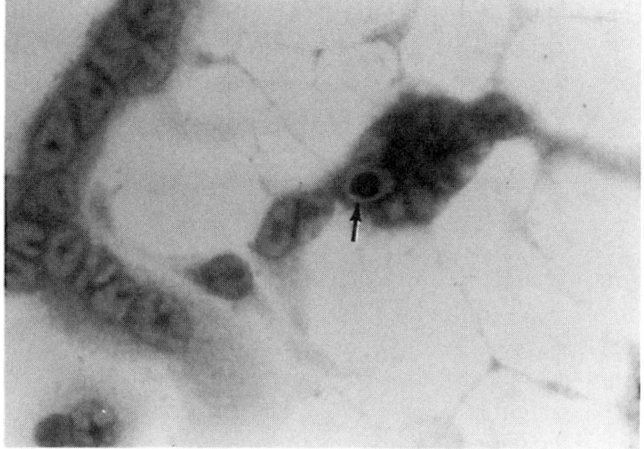

FIG. 3. Inclusion body formation in infected cells. The *arrow* indicates an intranuclear eosinophilic inclusion in a cell infected with HSV-1. The inclusion is surrounded by a clear halo. The cells were stained with hematoxylin and eosin. Micrograph courtesy of M. Kosz-Vnenchak.

host cell; however, from a virologist's point of view, we will see that many of the host-cell alterations by virus infection can now be explained as changes in the host cell that permit necessary steps in viral replication. Thus, many of the CPEs or cell injuries are secondary effects of the virus doing what it needs to do to replicate and are not simply toxic effects of viral gene products on the host cell. There are some viral gene products that cause toxic effects to the host with no apparent purpose, but as we learn more about the precise mechanisms of virus-cell interactions, many of these effects will likely be explained as aspects of essential steps in viral replication. For example, an adenovirus virion component was shown more than 30 years ago to cause monolayer cells to detach from the culture substrate (75,233), and this was later shown to be the penton base protein (27, 300, 319). This remained as an unexplained toxic effect until it was observed that the penton base protein binds to α_v integrins during adenovirus internalization (331). Therefore, penton base protein causes cell detachment by binding through RGD sequences to cell-surface integrins, thereby displacing vitronectin from the integrins. The interaction of vitronectin and integrins is involved in cell attachment and spreading on the culture substrate.

A number of reviews and monographs have examined the various aspects of cytopathology due to virus infection (84,151,274). The reader is referred to these sources for detailed discussion and references on these issues. Determining the primary cause of death of a cell resulting from viral infection can be a complex and difficult issue because of the numerous events occurring within the infected cell. It is also apparent now that there are at least two general pathways for cell death: necrosis and apoptosis (programmed cell death). Necrosis involves the death of a cell due to to physical damage or toxic agents. During necrosis, there is disruption of the mitochondria, swelling of the cell, disruption of organized structure, and lysis. Apoptosis involves the death of a cell in situations under the control of physiological stimuli during development and hormonal signalling in response to DNA damage. During apoptosis, there is DNA condensation and fragmentation, cell shrinkage, and membrane blebbing. Cell death due to viral infection could be due to either mechamism in that viruses could have toxic effects on the host cell, activate programmed cell death pathways, or both.

VIRUS INTERACTIONS WITH CELL UPTAKE MECHANISMS

Viruses must enter the host cell to replicate. Therefore, they must cross the cell plasma membrane to gain access to the cellular synthetic machinery in the cytoplasm and, in some cases, the nucleus of the host cell. Virus entry into the host cell has been divided into two events: (a) binding to cell-surface receptors and (b) penetration of the plasma membrane or entry. These two events will be discussed separately below. Virus particles serve to protect the viral genome during spread from cell to cell or from one host to the next. The lipid envelope and the protein capsid act to protect the genome from nucleolytic attack or chemical reaction in the extracellular environment. As the virus enters a new host cell, these components must be removed to allow the viral genome to be replicated. Thus, viruses have evolved to have a stable, protective particle that can be readily disassembled in a new host cell. For the enveloped virus, cell entry by fusion serves two functions: a means to cross the plasma membrane barrier and a means to remove its outer coat, the envelope.

Binding to Cell-Surface Receptors

The first event in viral infection of the host cell is binding of the virus to the cell surface. The cell-surface molecule with which the virus first specifically interacts or binds is called the virus receptor. This binding of the virus particle to the receptor involves numerous noncovalent interactions between the virion surface and the receptor, the sum of which leads to a high affinity, specific interaction between the virus particle and a cell-surface molecule. Presumably, viruses have evolved to use receptors that are present on cells that have special features desired by the virus, either a permissive environment for its replication or a unique environment for a persistent infection.

It has been difficult to define the physiological receptor for viruses for several reasons: (a) a virus may bind specifically or nonspecifically to a number of surface molecules; (b) a virus is a large ligand that can interact with a large surface area on the cell, thereby giving numerous cellular molecules that may cofractionate with a virus–receptor complex; or (c) a virus may have alternate receptors on individual or different cells. Nevertheless, specific receptors have been defined for several viruses. A complete list of putative viral receptors has been provided in Table 3 of Chapter 6. Several approaches have been used to attempt to identify virus receptors: (a) use of specific chemical compounds to compete for and block virus binding and infection of cells; (b) use of monoclonal antibodies specific for cell-surface proteins to block virus binding; (c) enzymatic treatment of the cell surface to remove receptor activity; (d) purification of virus–receptor complexes; (e) use of antiidiotypic antibodies to purify receptors; (f) gene transfer of receptor activity to receptor-negative cells, as well as cloning of transferred gene sequences; and (g) correlation of receptor activity with expression of a specific molecule on the cell surface.

Viruses utilize a wide variety of cell-surface molecules as their receptors in that they can use protein molecules (see below), carbohydrates (97), or glycolipids (184) as their cellular receptors. Some receptors are specific molecules such as the CD4 protein molecule, which serves as the receptor for human immunodeficiency virus (HIV) on T-lymphocytes (61,142). Other receptor molecules are widely distributed molecular moieties such as sialic acid, which serves as the receptor for influenza (97), or heparan sulfate, which serves as the initial cell receptor for herpes simplex virus (HSV) (339). Because a molecule such as CD4 is restricted to certain types of cells, receptor activity is restricted to a specific tissue. Other examples of tissue-specific molecules that could serve as receptors are (a) the C3d complement receptor on B cells, which is the receptor for Epstein-Barr virus (82,83,207), and (b) the acetylcholine receptor, which may act as a receptor for rabies virus (169). Thus, as their receptors, viruses can use cell-surface molecules that normally serve the host cells as receptors for other molecules. In this way, viruses are using the normal host-cell pathways for internalization of molecules or extracellular signals, and the tissue distribution of the viral receptor may define the tissue-tropism of viral infection.

Viral receptors can be species-specific also. For example, the poliovirus receptor is found only on primate cells and not on other mammalian cells (193). The block to poliovirus replication in murine cells is only at the surface because viral RNA introduced into murine cells is infectious (118). Using sensitivity to poliovirus infection as a screen, a human gene encoding a poliovirus receptor was transferred into murine cells (197). By identifying the human DNA sequences that correlated with receptor expression, the gene for a poliovirus receptor was isolated from these cells (198). The predicted amino acid sequence for the encoded protein indicates that the receptor is an integral membrane protein with characteristics of members of the immunoglobulin superfamily of proteins. Interestingly, the receptor for another picornavirus, human rhinovirus, is intercellular adhesion molecule-1 (ICAM-1) (103,297), which is also a member of the immunoglobulin superfamily of proteins.

Further study of known viral receptors has led to the hypothesis that other cofactors may form a part of the receptor complex or contribute to virus binding and entry. For example, the tissue distribution of the poliovirus receptor did not match the tissue tropism of viral replication (86,198). Recent studies have indicated that a 100-kd cellular glycoprotein expressed in tissues permissive for poliovirus infection (282) may play a role in receptor function and tissue tropism (283). Similarly, expression of the human CD4 protein in mouse cells does not always allow entry of HIV into these cells (181). One group has reported that the CD26 T-cell activation antigen could serve as a cofactor with CD4 for entry of HIV into cells (33,34), but others have not been able to confirm a role for CD26 (3,31,35,230). Therefore, the cofactor(s) for HIV entry remain to be identified.

Interaction of a virus with a cellular receptor represents the first interaction between virus and host cell. Obviously, if the virus cannot bind to the host cell, infection cannot be initiated, and minimal effects on the cell are likely to result. Even if the virus cannot enter the cell, virus binding to the cell surface may exert an effect on the host cell. For example, it has been hypothesized that binding of Epstein-Barr virus to the surface of a B-lymphocyte can ini-

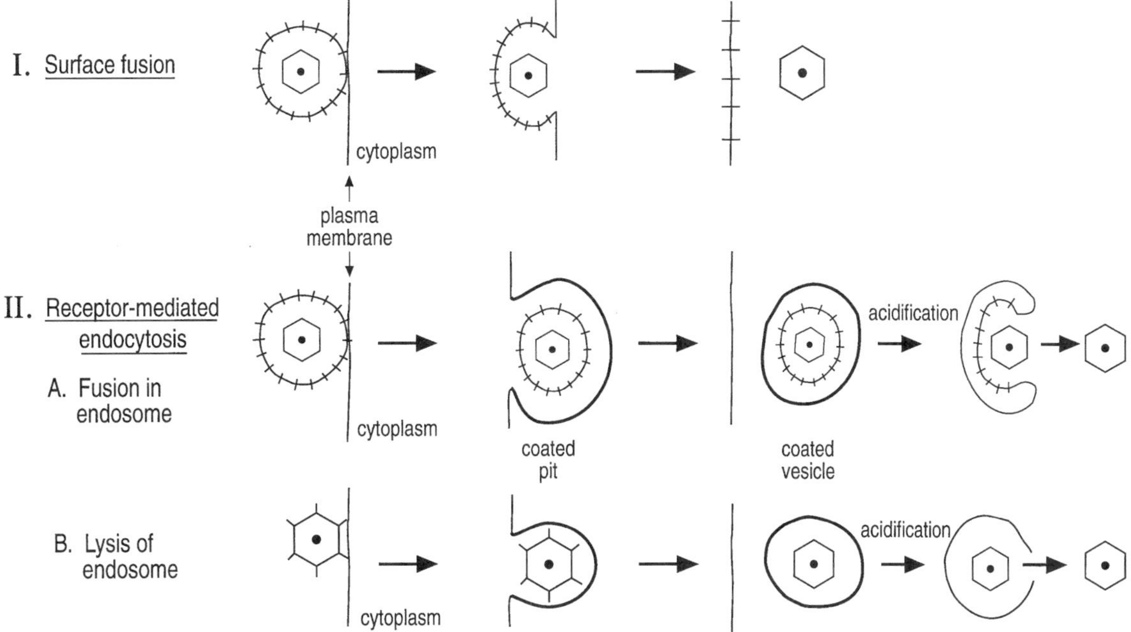

FIG. 4. Pathways for viral entry of the host cell.

tiate the activation of the B cell (96,126). Thus, the virus particle could act as a ligand to initiate the transfer of a signal to the interior of the host cell.

Entry into the Host Cell

After binding to its surface receptor, a virus must cross the plasma membrane to replicate. Two general pathways have been defined for virus entry: (a) surface fusion between the viral lipid envelope and the cell plasma membrane and (b) receptor-mediated endocytosis (Fig. 4). Enveloped viruses can gain entry to the cytoplasm by fusion of their lipid envelope with the plasma membrane or endosomal membrane, whereas nonenveloped viruses must use alternative strategies to cross the membrane.

Surface Fusion

Some enveloped viruses, notably paramyxoviruses, retroviruses, and herpesviruses, are capable of fusion with the cell plasma membrane at the cell surface. Binding of the virus to the cell-surface receptor leads to fusion between the virion lipid envelope and the cell plasma membrane. This releases the viral nucleocapsid into the cytoplasm of the host cell, effecting penetration of the host-cell plasma membrane. The surface fusion of these enveloped viruses is promoted by virion surface proteins. A series of events is believed to lead to fusion of the paramyxovirus envelope with the cell plasma membrane (162). For several of the paramyxoviruses—Sendai, Newcastle disease, human parainfluenza 2 and 3, and mumps viruses—both the hemagglutinin-neuraminidase (HN) and fusion (F) proteins are needed for efficient membrane fusion. The first step involves the binding of HN to its receptor, which may cause HN to undergo a conformational change. This triggers a conformational change in F, which exposes the fusion peptide and allows it to embed in and perturb the cell membrane. This ultimately leads to the fusion of the cell and viral lipid bilayers.

The F protein of paramyxoviruses is initially synthesized as a precursor molecule, which is cleaved by a host-cell protease to give a biologically active, disulfide-linked F_2-F_1 heterodimer. The NH_2-terminus of F_1 contains the hydrophobic fusion peptide (123), but this must not be exposed to the aqueous environment until the virion is in proximity to the appropriate cell membrane. Otherwise, the virions will undergo spurious hydrophobic interactions or aggregation, thereby reducing infectivity. Thus, the binding of HN to the virus receptor transmits a signal to F_1, activating it for fusion. The mechanism of this signal transduction poses an exciting problem for structural virology.

Endocytosis

Other viruses, notably the enveloped Semliki Forest and influenza viruses and the nonenveloped adeno and reoviruses, enter cells by an endocytotic pathway that exploits a normal cellular pathway for uptake of materials bound to cell-surface receptors (58, 285). After binding of the viruses to the cell surface, a clathrin-coated invagination of the plasma membrane, called a coated pit, is formed. An endocytotic vesicle then pinches off with the virion particle inside. Acidification of the interior of the vesicle, called an endosome, is then promoted by a proton pump in the membrane. As the pH drops, a virion surface protein(s) undergoes a conformational change, causing exposure of a hydrophobic portion of a virion surface protein. This hydrophobic region promotes fusion between the virion lipid envelope and the vesicle membrane, releasing the nucleocapsid into the cytoplasm. At this point, the viral genetic material has entered the host cell. Thus, for enveloped viruses, this mechanism is similar to surface fusion, except that fusion of the viral envelope with the cell membrane occurs within the host cell. The main differences appear to be trigger mechanisms activating fusion and thus the cellular sites, at which fusion occurs.

Several experimental approaches have been used to distinguish the surface fusion and endocytotic entry pathways (77,185,329). First, electron microscopy (EM) of newly infected cells may demonstrate a majority of virus particles being internalized in endosomes or undergoing fusion at the cell surface. The low specific infectivity (or high particle/plaque-forming unit ratio) for most animal viruses leaves open the possibility that virus particles may enter, or attempt to enter, the cell by nonproductive pathways. Thus, although this can be a very useful approach, EM studies must be interpreted with caution. Second, weak basic compounds, such as ammonium chloride, accumulate in acidic compartments of the cell, such as endosomes, and raise their internal pH. Because the pH increase is greatest within these organelles, the compounds inhibit the entry and infectivity of viruses through endocytosis more readily than that of viruses entering by surface fusion. Third, entry of viruses by endocytosis leads to internalization of viral surface proteins, whereas surface fusion leaves the virion envelope surface proteins on the cell surface as part of the plasma membrane. Thus, viral envelope proteins remain external and susceptible to protease digestion or reaction with antibodies after virus entry by surface fusion. Through experimental approaches such as these, these two pathways of virus entry can be distinguished.

The mechanism for activation of fusion during endocytosis has been most extensively studied for the influenza hemagglutinin (HA) molecule. Like the paramyxovirus F proteins, HA is synthesized as a precursor molecule that is inactive for fusion. Proteolytic cleavage in human influenza HA molecules occurs at monobasic cleavage sites by enzymes secreted from epithelial cells (143). The cleavage generates an HA_1-HA_2 heterodimer with the NH_2-terminus of HA_2 containing the hydrophobic fusion peptide. The NH_2-terminus of HA_2 is buried in the HA trimer at the subunit interfaces (334). For HA, the trigger for fusion activation is acidification of the endosome, which causes a conformational change in HA and exposure of the HA_2 fusion peptide (62). It has been proposed that the fusion peptide is moved 100Å toward the cell membrane by formation of an extended coiled-coil structure in HA_2 (37), and the structure of the low pH-induced form of influenza HA has confirmed this conformation change (333).

Nonenveloped viruses can also enter cells by receptor-mediated endocytosis, but entry cannot be achieved by membrane fusion. For adenovirus, it has been hypothesized that low pH activates a virion surface protein(s) to lyse the endosomal membrane, releasing the virion into the cytoplasm (229). Recent studies indicate that this lysis is due to hexon or penton base proteins of the adenovirus particle (101).

Reovirus, another unenveloped virus, is thought to enter the host cell by endocytosis under at least some conditions. The form of reovirus thought to cross the cell membrane is the *i*nfectious *s*ub*v*iral *p*article, or ISVP, a subvirion particle generated by proteolytic digestion of the virion or a conformationally altered form of the ISVP. These particles are believed to be generated in infected cells by digestion of virions by host-cell proteases in a low pH intracellular compartment (303). ISVP particles can bind to membranes and induce channels in lipid bilayers (313). Thus, proteolytic digestion of reo virions in endocytic vacuoles appears to activate a viral protein(s) to lyse or permeabilize the vacuolar membrane to allow access to the cytoplasm.

Separate cellular functions may be needed for internalization versus release of viral nucleocapsids. As described above, HIV requires the CD4 molecule as a receptor. When the gene encoding CD4 was transferred into human and murine non–T cells, HIV could bind to the cells and successfully infect some of the cells (181). However, some of the cell clones derived from murine cells could bind virus and internalize it, but none of the early events of infection ensued. Thus, there appear to be unique cellular products required after internalization of HIV for events such as release or uncoating of the virus.

Once within the host cell, the viral genome must be transported to the correct intracellular site for transcription or replication within either the cytoplasm or nucleus. Many observations have led to a hypothesized role for the cellular cytoskeleton in transport of the genome to the nucleus or to the correct place in the cytoplasm (see below), but few precise mechanisms for transport are known.

Uncoating

As described above, when viruses infect a host cell, they must efficiently disassemble the particle that protects the genome during the extracellular phase. Uncoating consists of the steps of disassembly involving removal of the envelope and removal of the protein shell or capsid from the viral genome. Uncoating events can occur at many stages in viral replication or sites in the cell, ranging from the cell surface to the nuclear envelope. Initiation of reovirus uncoating may even start before interaction with cells in that proteases in the gut can convert reovirions to infectious subviral particles (22). Picornaviruses undergo an initial uncoating event at the cell surface, a virion conformational shift upon binding to their receptor and loss of the VP4 capsid protein (56). The resulting particle is hydrophobic and likely interacts with membranes. It is believed that the viral RNA is released through the membrane via this particle (100), but the exact process and cellular site remain to be defined.

Uncoating events can occur in endosomal vesicles. We have already seen how proteases in endocytic vacuoles participate in uncoating of reoviruses. The low pH in the endosomes can also contribute to uncoating. The M2 protein of influenza virus forms a proton channel across the viral envelope through which protons can enter the virion and cause dissociation of the capsid (186,238,305,323). The antiviral drug amantadine can block the M2 pore and therefore block capsid acidification, uncoating, and viral infection in susceptible influenza strains.

Uncoating of some other viruses occurs in the cytoplasm. The capsid proteins of alpha viruses, Semliki Forest virus, and Sindbis virus are transferred from the nucleocapsid to the host-cell 28*S* ribosomal RNA (rRNA), thereby freeing the genomic RNA (286). DNA viruses such as adenovirus (59,100) and HSV (9) undergo uncoating at the nuclear pore, possibly because the viral capsids are too large to move through the pores.

VIRUS INTERACTIONS WITH THE CELLULAR TRANSCRIPTION APPARATUS

Viral messenger RNA (mRNA) is needed during replication of viruses to encode viral proteins for replication of the genomic nucleic acid and virion assembly. If the genomic RNA can be used as an mRNA directly (positive-strand RNA virus), then synthesis of viral RNA need not precede initial rounds of translation. However, if the genomic RNA is complementary to mRNA (negative-strand RNA viruses) or in the form of DNA, *de novo* synthesis of viral mRNA must occur. If the virion contains an RNA polymerase, then synthesis of viral mRNA may depend on cellular factors, albeit to a limited extent. However, if the virus uses cellular polymerases to synthesize mRNA, specific mechanisms may have evolved to promote transcription of viral DNA or to shift the host-cell RNA polymerase to the transcription of viral DNA. Viruses may encode a new transcription factor (23). The DNA viruses have several goals during infection of the host cell (Table 1). First, they must ensure efficient transcription of their viral DNA genome immediately upon entry into the cell, often known as immediate-early transcription. This has been accomplished by packaging a new polymerase in the virion (poxviruses), packaging a transcriptional adaptor molecule in the virion (HSV), or having a strong transcriptional enhancer upstream from the promoter for the immediate-early genes (e.g., adenovirus, papovaviruses, and cytomegalovirus). Second, they must ensure efficient transcription of the later viral genes, called early, delayed-early, and late genes. The strategy used by DNA viruses to achieve this last aim is to encode a transactivator protein as an immediate-early or early viral gene product that activates later viral gene expression. The following sections describe how viruses inhibit host-cell transcription and ensure efficient immediate-early transcription and later viral gene transcription.

Inhibition of Cellular Transcription

Infection with many viruses leads to an inhibition of transcription of cellular protein-coding genes by host RNA polymerase II. For RNA viruses, which do not use host-cell RNA polymerases for their replication, the presumed advantage conferred by this activity would be to provide higher pools of ribonucleoside triphosphate pools for viral RNA synthesis. For DNA viruses, inhibition of host transcription might allow the host-cell RNA polymerase II to transcribe the viral genome in addition to decreasing competition for triphosphate precursors and transcription factors.

Little is known about how DNA viruses cause an inhibition of host-cell transcription, except for possible competition for RNA polymerase II and cell transcription factors. A possible mechanism for inhibition of host transcription has been formulated for cells infected with the RNA virus vesicular stomatitis virus (VSV). VSV infection causes rapid inhibition of host RNA synthesis, and this inhibition requires transcription of the viral genome (190,325). Ultraviolet (UV)-inactivation studies indicated that transcription of a small virally encoded RNA called positive-strand leader RNA may be sufficient for inhibition of host transcription (104). The leader RNA accumulates in the cell nucleus early in infection (158) and associates with the cellular La protein (157) that binds

TABLE 1. *Mechanisms by which DNA viruses satisfy critical needs for replication*

Virus	Efficient transcription immediately after infection	High-level later expression	DNA replication proteins
Papova	Early gene enhancer to recruit cell transcription factors	Large T-antigen transactivator using cell transcription factors	Induce S phase in host cell
Adeno	Pre-early gene enhancer	E1A transactivator protein	Induce S phase in host cell
Herpes simplex	Virion VP16 transactivator protein	ICP4 transactivator protein	Encode DNA replication proteins, nucleotide metabolism enzymes
Pox	Virion RNA polymerase	Early, late viral transcription factors	Encode entire replication apparatus

transiently to nascent RNA transcripts (259). An oligodeoxynucleotide with a sequence identical to part of the leader RNA inhibited *in vitro* transcription of the adenovirus major late promoter and the VA RNA genes (106). A 65-kd HeLa cell protein binds to the oligonucleotide and can reverse the transcriptional block (106). Thus, this viral nucleic acid may bind a host-cell factor and prevent its binding to cellular promoters, thereby inhibiting host-cell transcription.

However, these experiments do not prove that the leader can inhibit transcription *in vivo*. Other workers have reported that viruses expressing very different amounts of identical leader RNAs can shut off host RNA synthesis equally well (69). Thus, there is no apparent quantitative relationship between leader RNA and shut-off of host transcription. These studies do not prove or disprove a role for leader RNA in inhibition of host transcription, but they have raised the idea that viral gene products other than protein products can exert effects on host-cell metabolism.

Mechanisms to Ensure High-Level Transcription upon Viral Entry

DNA viruses have several general mechanisms to promote efficient transcription of the viral DNA immediately after entry, and most involve subversion of the host transcriptional apparatus.

Packaging a Virally Encoded RNA Polymerase in the Virion

Poxviruses contain a double-stranded DNA genome but replicate in the cytoplasm. These viruses have evolved their own transcriptional and DNA replication machinery, allowing these processes to occur in the cytoplasm, almost completely independent of the host-cell nucleus. The poxvirus particle contains a virally encoded RNA polymerase (136) capable of transcribing about one half of the viral genome (25,228). The presence of the RNA polymerase in the virion ensures efficient transcription of the viral genome as soon as the virus enters the cell.

Packaging a Transcriptional Activator in the Virion

The best known example of a virion transactivator is the HSV VP16 protein. HSV VP16 (also called α TIF or Vmw65), is assembled into the virion and, when introduced into the cell, becomes a part of a transcriptional activatory complex that specifically stimulates viral immediate-early (α) gene expression (36,243). VP16 binds to host-cell proteins (192,244); specifically, it is thought to bind to at least one host factor, the C1 protein complex (156), and this complex binds to the transcription factor Oct-1 bound to DNA (92). One portion of VP16 binds to C1 and Oct-1, whereas another activates transcription (314) by binding to basal transcription factors. Thus, HSV provides for adequate transcription of its immediate-early genes by bringing into the cell in its virion a protein that binds to a cellular, sequence-specific DNA-binding protein, causing increased transcription from immediate-early viral gene promoters that contain this specific sequence.

Use of an Enhancer to Ensure High-Level Immediate Transcription

An alternative way that DNA viruses promote efficient transcription of viral genes immediately after infection is to have an enhancer sequence as part of the upstream sequences of these genes. Enhancers are blocks of DNA sequences containing multiple binding sites for transcription factors acting in synergy to promote high-level transcription from a nearby gene. They act in a position- and orientation-independent manner. Enhancers were, in fact, first identified in viral genomes (8,13). The early transcriptional units of SV40 virus (8,13), polyoma virus (318), human papilloma virus (55), the adenovirus E1A transcriptional unit (110), and the human cytomegalovirus ie1/ie2 transcriptional unit (26,310) all contain enhancers. These enhancer sequences provide for utilization of multiple cellular transcription factors in the transcription of the earliest genes of these viruses immediately after infection.

Mechanisms for Activation of Later Viral Genes

Many of the DNA viruses encode transactivators in their earliest gene products. The SV40 large T-antigen stimulates late transcription both by promoting viral DNA replication and by binding to host-cell transcription factors (107). The adenovirus E1A, the HSV ICP4, and the papilloma virus E2 proteins all serve to stimulate later viral gene transcription by altering the activity of host-cell RNA polymerase II.

The mechanism of transactivation has been most extensively studied with the adenovirus E1A gene function (20,132). The E1A gene encodes two mRNAs, 13S and 12S, by differential splicing, and much of the transactivation activity is due to the 13*S* gene product. The E1A proteins increase the expression of pol II–transcribed genes (102,127,210) and pol III–transcribed genes (17,91,117). The promoter requirements for E1A transactivation coincide with basal-level promoter elements (19,208). The E1A proteins do not bind to DNA specifically or efficiently, and there is now considerable evidence that E1A stimulates transcription through various effects on cellular transcription factors (165, 287, 338). The multiple specific effects of E1A are described in detail in chapter 67, but the general mechanisms are relevant here. First, E1A proteins directly interact with cell transcription factors. The 13*S* E1A protein can bind to the TATA-binding protein (TBP) (119,166), which could relieve repression by p53 or DR1 or serve as an adapter bridging TBP and other activators. Second, E1A can bind to complexes of host-cell proteins, releasing active transcription factors. For example, the E1A proteins bind to pRB and free an active form of the host E2F transcription factor (7). Third, E1A may lead to changes in the phosphorylation of host-cell transcription factors. E1A leads to an increase in activity of the TF III-C factor for pol III (117), and this has been hypothesized to be due to changes in phosphorylation of TF III-C (116). In general, E1A increases the amount of transcription of viral genes by increasing the rate of initiation. E1A also serves to activate expression of certain cellular genes, an important step for adenovirus replication as discussed below.

VIRUS INTERACTIONS WITH RNA PROCESSING PATHWAYS

RNA splicing and transport to the cytoplasm are cellular pathways often used by viruses to mature their mRNA from nucleus to cytoplasm. In fact, the first evidence for RNA splicing came when the adenovirus late mRNAs were mapped on the viral DNA by R-loop hybridization (18,47). These studies showed that these mRNAs are encoded by noncontiguous regions of the genome. Splicing of the viral mRNA precursors is accomplished by cellular enzymes recognizing splice donor and acceptor sequences in the viral RNA. However, some viruses use the cellular splicing mechanisms but regulate the extent to which the full-length transcript is spliced. For example, influenza and retroviruses have transcripts that are infrequently spliced (see Chapters 21 and 26). In cells infected with influenza virus, the viral NS1 and M1 RNAs are spliced to yield the NS2 and M2 RNAs, respectively, at a frequency of 10% (163,164). Although splicing of the NS1 RNA is inefficient, formation of the spliceosome complex involving the snRNPs U1, U2, U4, U5, and U6 is efficient (2). Thus, the block seems to occur after formation of the spliceosome complex. The block may be mediated by the structure of the RNA itself or by a virally encoded protein. This appears to be a situation in which virus infection regulates the extent of splicing of one of the viral RNAs, thereby regulating the levels of one of its own gene products.

Adenovirus inhibits maturation of cellular mRNA at a different stage. In adenovirus-infected cells, cell transcripts are synthesized and processed but do not accumulate in the cytoplasm (11). The adenoviral E1B-55K and E4-34K proteins are required for inhibition of transport of cellular mRNA as well as for promotion of cytoplasmic accumulation of viral mRNA (5,108,237). The E1B and E4 proteins have been observed to localize within and around the nuclear inclusions where viral transcription is believed to occur (221). Thus, these proteins may redistribute a cellular factor involved in mRNA transport. This system may provide important insights into the regulation and specificity of mRNA transport.

A similar observation has been made in cells infected with HSV, in which transport and processing of host-cell RNA is inhibited (322). In HSV-infected cells, small nuclear ribonucleoprotein particles become aggregated (187), and this may be related to impaired transport or splicing. The HSV IE protein ICP27 or IE63 has been reported to be necessary and sufficient for this effect (234). Only four of the more than 70 HSV genes contain introns. Thus, inhibition of host-cell splicing should contribute to host shut-off, but the mechanisms for specific retention of viral RNA splicing, if any exist, are unknown.

The regulation of RNA maturation may also provide a mechanism for temporal regulation of viral gene expression. HIV encodes several regulatory gene products from spliced mRNAs. One of these regulatory gene products, the *rev* gene product, stimulates the cytoplasmic accumulation of unspliced viral mRNAs that encode the viral structural proteins (78,144,182,269,291,307). The *rev* gene product may promote the export of newly synthesized viral transcripts to the cytoplasm so that the splicing pathway is avoided (183). Thus, there may be cellular pathways that regulate (a) the assembly of splicing complexes or (b) the maturational pathway into which a newly made transcript enters.

Influenza virus intervenes in the host-cell mRNA maturation pathway in another novel way. Influenza mRNA transcription from the genomic RNA segments occurs in the host-cell nucleus (113). Host-cell nascent transcripts are cleaved by a virus-encoded endonuclease, and the 5′ end of the host transcript is used as a primer for synthesis of viral mRNA from the viral genome (28,240). Thus, influenza virus transcriptional complexes subvert the host mRNA maturation pathway to obtain primer molecules for the viral transcriptional process.

VIRUS INTERACTIONS WITH THE TRANSLATIONAL APPARATUS

Once viral mRNA is available in the cytoplasm, it is translated by the host translational system to yield viral proteins. Many of the viral mRNAs are capped and contain a single major initiation site near the 5′ end. Thus, translation of these mRNAs is similar to that of host mRNA. In fact, much of the original evidence for the model for eukaryote ribosome scanning of an mRNA for an initiation codon came from the identification and comparison of ribosome-binding sites on viral mRNAs (151, 152). However, during lytic infection many viruses must subvert the host-cell translational apparatus to make the large amounts of structural proteins needed for progeny virus production. Initiation of protein synthesis is highly regulated in the eukaryotic cell, and viruses affect several steps in this process (Fig. 5). This section examines the mechanisms by which viruses inhibit host-cell translation or otherwise alter the translational apparatus of the host cell. More details can be obtained in recent reviews (71).

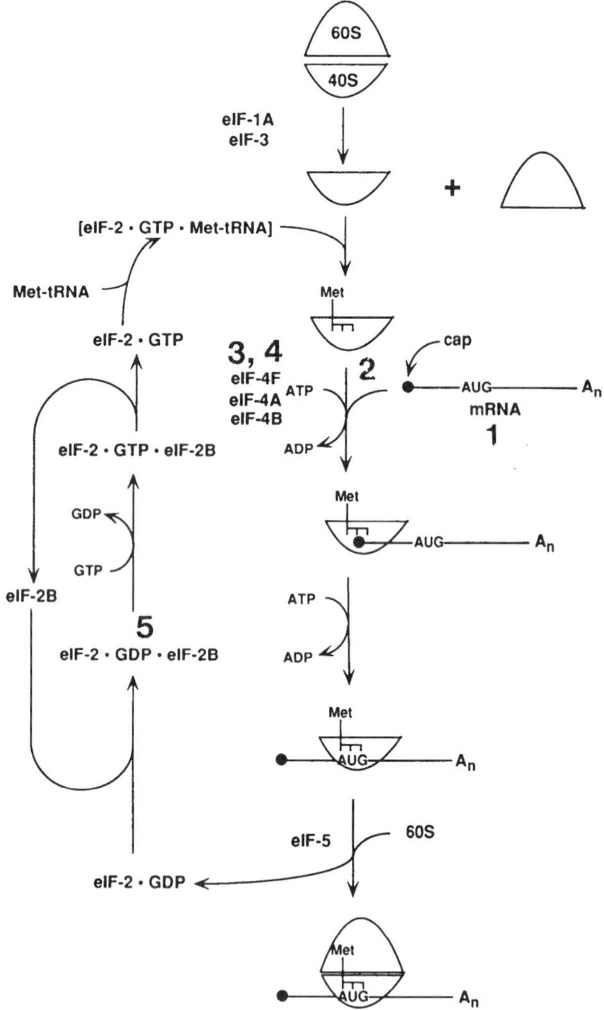

FIG. 5. Steps in initiation of translation and examples of their modification by viruses. Examples: 1, HSV *vhs* function causes degradation of mRNA; 2, VSV mRNA competes with host mRNA for binding to 40*S* ribosomal subunits; 3, poliovirus 2A protein activates a latent host protease that cleaves p220 (γ subunit) of eIF-4F; 4, adenovirus causes dephosphorylation of the 25K CBP, the α subunit of eIF-4F (also called eIF-4E); 5, adenovirus VAI RNA blocks the dsRNA-activated inhibitor that phosphorylates the α subunit of eIF-2 and freezes it in the eIF-2• GDP• eIF-2B complex. Modified from ref. 112 with permission.

Inhibition of Host Translation

After infection of cells by many viruses, the translation of host-cell mRNA is inhibited. Inhibition of translation of host-cell mRNA would provide the viral mRNA with increased availability of ribosomal subunits, translation factors, transfer RNAs, and amino acid precursors for protein synthesis. The extent to which shut-off of host translation is essential for efficient virus replication remains to be determined. There are viable mutant viruses that are impaired in their ability to shut off host translation (21,250,296). The *v*irion *h*ost *s*hut-off (*vhs*)-1 mutant of HSV is somewhat impaired for growth as compared with wild-type virus (160,250), but the relationship between limited host shut-off and poor growth of this mutant virus remains to be defined.

Inhibition of translation of host-cell mRNA can occur by many mechanisms, and it is conceivable that individual viruses could use more than one of the following mechanisms.

Degradation of Host mRNA

After infection of cells by HSV, influenza, or poxvirus, inhibition of host-cell translation occurs, and a decrease in the amount of intact host mRNA is observed (79,128,212,255). Thus, degradation of the host mRNA due to virus infection is one potential mechanism to provide free ribosomes and translational factors to preferentially translate the viral mRNAs (Fig. 5, Example 1). In some types of cells infected with HSV, *de novo* viral protein synthesis is needed for degradation of host mRNA (212,213), but in other cell types, a virion component can induce host polysome disaggregation and mRNA degradation (272). The *vhs* mutants of HSV produce virions that are unable to inhibit host protein synthesis and degrade host mRNA (250,302). The *vhs*-1 mutation has been mapped within the UL41 open-reading frame of the HSV-1 genome (160) which encodes a virion protein. Extracts from HSV-infected cells contain an activity that destabilizes mRNA in added polysomes (155,292). The mechanism by which the UL41 protein exerts its effect is not yet known. Because so little is known about enzymes degrading cell mRNA, it will be of considerable interest to determine whether UL41 acts as a nuclease itself or activates a cellular nuclease. The *vhs* gene product also destabilizes immediate-early and early viral mRNA (159, 222,223). Thus, there is no apparent discrimination between host and viral mRNA in this process. In addition to providing a means for inhibiting host translation, this viral function could promote the shut-off of immediate-early and early gene expression in the HSV lytic cycle and transition to late protein synthesis (222).

Competition for the Host Translational Apparatus

Some viruses may not use specific effects on the host-cell translational apparatus to allow efficient synthesis of viral proteins. For example, it has been reported that VSV protein synthesis occurs preferentially in infected cells because (a) large amounts of viral mRNA compete for limiting ribosomes (176,276) or (b) the viral mRNA has higher affinity for ribosomes than cell mRNA (216) (Fig. 5, Example 2). This has been a controversial area because others have reported that (a) a specific viral gene function is responsible for host translational inhibition (296), and (b) impairment of host translational factors eIF2 (42) or eIF3/4B (309) occurs in VSV-infected cells. There may be different mechanisms operating in different cells, or alternatively, more than one of the postulated mechanisms may be operating in one or more types of host cells. For example, a slight impairment of the host translational system would help switch translation from host mRNA to the more abundant or more efficiently initiating viral mRNA (175). Thus, more than one of the postulated mechanisms could be operating here, with the relative importance of each effect differing in various cellular situations.

Changing the Specificity of the Host Translational Apparatus

After poliovirus infection, translation of host mRNA is inhibited, but translation of viral mRNA occurs. Extracts from virus-infected cells can translate poliovirus RNA but not cellular or VSV mRNA (262). Thus, the specificity of the translational apparatus appears to change in poliovirus-infected cells so that viral RNA is preferentially translated. Further analysis of the poliovirus-infected cell extracts has shown that a host translational component, cap-binding protein (CBP) complex, or eIF-4F (Fig. 5, Example 3) is inactivated in poliovirus-infected cells due to cleavage of one of the constituent proteins, the p220 protein or the γ subunit of eIF-4F (74). The CBP complex is needed for efficient initiation of translation of capped RNA (280). Poliovirus virion RNA is linked to a protein at its 5′ end (114,214), but the protein is removed in the cytoplasm so that polysomal RNA has pUp at its 5′ end (4). Thus, poliovirus infection inactivates the eIF-4F so that translation of RNA not bearing capped 5′ ends can occur more efficiently. It has been shown that poliovirus 2A protease is required for cleavage of p220 (21), but it seems that this enzyme does not directly cleave p220 (174). Instead, it is thought to activate a latent cellular protease (341).

Initiation of translation can occur at an internal initiator codon on an mRNA molecule in poliovirus-infected cells, possibly by internal binding of the ribosomal subunits (232). This suggests that sequences within the untranslated 5′ region of poliovirus RNA can direct ribosome binding and initiation at internal sites within the RNA. Although this was thought to be a peculiarity of poliovirus RNA in infected cells, there is now evidence for internal initiation on cellular mRNAs (180,218). Thus, cellular translational mechanisms involve two potential initiation processes: cap recognition or internal binding of ribosomes (195). Accordingly, there are two mechanistic components involved in poliovirus conversion of the specificity of the translational machinery: (a) inactivation of the CBP complex so that initiation involving capped mRNA is decreased and (b) a sequence within the poliovirus RNA that promotes use of normal cellular, internal initiation mechanisms. This effectively results in a switch in translational specificity favoring translation of viral mRNA.

The adenovirus late mRNAs have a tripartite leader sequence that promotes their translation at late times after infection (177). This sequence confers a reduced requirement for eIF-4F, but this effect is thought to be due to a relaxed 5′ proximal secondary structure (345). eIF-4F activity is reduced in adenovirus-infected cells, possibly due to reduced phosphorylation of the α subunit (also known as eIF-4E or cap-binding protein) (124). Adenovirus may shift translation to its late mRNA by reducing the phosphorylation and activity of eIF-4F (Fig. 5, Example 4), allowing preferential translation of its late mRNAs with tripartite leader sequences (346).

Host Response to Virus Infection and Viral Defenses

Among the host responses to viral infection is the synthesis of interferon (see Chapter 11). Interferon is secreted from the infected cell, binds to a second cell, and initiates a series of events, at least two of which have an impact on the translation of viral mRNA. First, interferon activates an enzyme called the 2′,5′-oligoadenylate (2-5A) synthetase (140). The synthesis of 2′,5′-oligoadenylate by this enzyme activates a ribonuclease that degrades mRNA and rRNA (288). This effectively blocks viral mRNA translation. The second event is the induction of a kinase, the double-stranded RNA (dsRNA)-activated kinase, that phosphorylates the α subunit of the translation factor eIF-2 (Fig. 5, Example 5) (200). This inactivates the translation factor and inhibits protein synthesis. The dsRNA-activated protein kinase may also be activated by dsRNA produced in virus-infected cells, either in replicative or transcriptive intermediates of RNA viruses or from symmetrical transcription of DNA virus genomes.

Some viruses, such as VSV and influenza, are very sensitive to interferon. Others, such as poxvirus and adenovirus, are relatively resistant to interferon. Vaccinia virus expresses a factor that inhibits the dsRNA-activated kinase (225,257,327). Also, in vaccinia-infected cells, 2′,5′-oligoadenylate is produced, but the ribonuclease L is not activated (224,257). Thus, vaccinia virus seems to take active measures to prevent interferon-mediated inhibition of translation.

Adenovirus ensures efficient translation of late mRNA by encoding a small RNA known as VAI RNA (252). The VAI RNA is specifically required for late viral translation (275,308). The lack of VAI leads to an activation of the eIF-2α kinase and decreased eIF-2α activity (253,273,284). VAI RNA binds to the kinase (137,196) and prevents dsRNA activation (141,220). Thus, VAI would block interferon effects on the host cell (141) or activation of the kinase by dsRNA molecules produced by symmetrical transcription of the viral genome.

Translational Frameshifting

In addition to the normal ribosomal protein synthesis mechanisms, some viruses exploit a potentially inherent ability of the host-cell ribosomes to shift from one reading frame to another during protein synthesis. During translation of the Rous sarcoma virus *gag* protein, ~5% of the ribosomes shift reading frames to synthesize a *gag-pol* fusion protein (130). Specific sequences in the RNA are required for the frameshifting (129). Therefore, specific viral RNA sequences cause ribosomal slippage so that small amounts of reverse transcriptase can be synthesized. This could be a mechanism used by viruses to express a limited amount of a specific protein.

Suppression of Translational Termination

Expression of limited amounts of a viral protein can also be achieved by suppression of a nonsense codon and synthesis of a polyprotein. For certain retroviruses, this is the mechanism used to express the *pol* gene (148,235). Like frameshifting, this ribosomal effect requires *cis*-acting sequences on the viral mRNA (227). Again, viral nucleic acids direct cell proteins to perform specific processes that allow viral gene expression in limited amounts.

VIRUS INTERACTIONS WITH THE CELL DNA REPLICATION APPARATUS

Inhibition of Host-Cell DNA Replication

Both RNA and DNA viruses cause the inhibition of host-cell DNA synthesis. There are several possible reasons for viral inhibition of cell DNA synthesis: (a) to provide precursors for viral DNA synthesis, (b) to provide host-cell structures and/or replication proteins for viral DNA synthesis, or (c) a secondary effect of inhibiting cellular protein synthesis. The possible mechanisms by which virus infection might inhibit cellular DNA synthesis are discussed individually.

A Secondary Effect of Inhibiting Cell Protein Synthesis

There appears to be a small pool of an essential cell DNA replication protein(s) because the rate of DNA chain growth decreases within minutes after inhibition of protein synthesis (239,301). It has been proposed that some viruses, such as the herpesviruses, inhibit cellular DNA synthesis as a consequence of inhibiting cellular protein synthesis (134). HSV and adenovirus DNA synthesis do not require the limiting cellular factor, or the factor is not limiting in infected cells because their DNA synthesis continues independently of whether protein synthesis is ongoing or not (120,260).

Displacement of Cellular DNA from Its Normal Site of Replication

Herpesvirus infection has been variously reported to displace cellular DNA from the nuclear membrane (211) or to cause the displacement of cellular chromatin to the periphery of the nucleus (63,277). In either case, the cell DNA could be displaced from its normal site for replication because cell DNA synthesis has been reported to occur on the "nuclear cage" (191) or the nuclear matrix (16). The exposed sites on the nuclear matrix may provide a structural framework for viral DNA replication and late transcription (145).

Recruitment of Cell DNA Replication Proteins to Viral Structures

Recent studies have shown that HSV infection leads to a redistribution of the host-cell DNA replication apparatus (64). This type of event could serve the dual function of providing cellular factors for viral DNA replication and inhibiting cell DNA synthesis, thereby reserving deoxynucleoside triphosphates for viral DNA synthesis.

Degradation of Cellular DNA

Infection by vaccinia virus leads rapidly to a marked inhibition of cell DNA synthesis. A virion-associated DNAse enters the host-cell nucleus and acts on single-strand DNA (ssDNA) (241,242). This inhibition is mediated by the action of a virion component on the host cell.

Mechanisms by Which DNA Viruses Ensure the Availability of a DNA Replication Apparatus

DNA viruses need a replication apparatus for amplification of their genomic nucleic acid (Table 1). Although the host-cell enzymes may be used for DNA replication, they may not always be available for use by the virus. Thus, several strategies have evolved to ensure the availability of DNA replication machinery.

Viruses Can Encode a New Replication Apparatus

The poxviruses replicate entirely in the cytoplasm. Indeed, DNA synthesis can occur in enucleated cells. Therefore, they must encode all (or nearly all) of the proteins needed for replication of their DNA in the cytoplasm. Similarly, retrovirions contain an enzyme capable of copying the genomic RNA into DNA. This step is not possible in cells without the virion enzyme because a cellular reverse transcriptase is not available to copy the viral RNA. For these viruses, DNA synthesis is usually not restricted in different cells because the viral enzymes are viral-encoded.

The herpesviruses encode several proteins that form a major part of the replication complex. For example, HSV encodes seven viral proteins required for and directly involved in viral DNA replication (44,337). In addition, HSV encodes several other enzymes involved in providing deoxynucleoside triphosphate precursors for viral DNA synthesis: thymidine kinase, ribonucleotide reductase, and dUTPase. HSV replicates in the resting neuron and has apparently evolved to encode these DNA replication proteins because they are expressed at low levels in resting neurons. Some cellular proteins may be involved in HSV DNA replication. Although the identity and role of specific cell proteins in HSV DNA replication have not been defined, the cellular DNA replication apparatus is redistributed after HSV infection so as to colocalize with viral proteins in the cell nucleus (64). HSV DNA replication proteins also bind to the nuclear matrix (246) so cellular proteins may be required to anchor DNA replication complexes and thereby promote optimal HSV DNA synthesis. However, HSV blocks cells from progressing into S phase (64) so any cellular proteins required for HSV DNA synthesis may be present in non–S phase cells.

Viruses Can Induce Cellular DNA Synthesis

Some viruses, especially the DNA tumor viruses, rely on the host-cell DNA replication machinery, and to ensure its availability, they induce S phase in their host cells if they are not cycling or in G_0 phase (Table 1).

The papovaviruses encode one protein, large T-antigen, that essentially inserts itself into the host-cell replication complex and directs it to replicate viral DNA. The SV40 large T-antigen (a) binds to the SV40 DNA sequences that serve as the origin of DNA replication (311), (b) interacts with the cellular α-DNA polymerase (289), and (c) acts as a helicase (294). Through these and possibly other functions, T-antigen promotes replication of the SV40 chromosome. T-antigen could be viewed as an origin-binding protein that substitutes for an analogous cellular protein. The specific interaction of T-antigen with host proteins is apparent in other ways. SV40 DNA replication can occur in extracts prepared from only certain cell types (171), and the interaction between T-antigen and the α polymerase-primase complex seems to define or play a role in defining this specificity (202). Thus, the permissivity of a cell for viral growth can be defined by the ability of a viral DNA replication protein to interact with the cellular DNA replication apparatus.

However, the levels of several critical components of the cellular DNA replication apparatus are limiting except during S phase. Thus, the host DNA replication machinery would not be available at all stages of the cell cycle. Polyomaviruses have long been known to induce S phase in resting (G_0) cells (68), and the mechanisms for this effect may now be understood. T-antigen forms complexes with pRb and related proteins (76), releasing transcription factors of the E2F family that are important for expression of cellular DNA replication and cell cycle proteins such as c-fos, c-myc, DNA pol α, dihydrofolate reductase, thymidine kinase, thymidylate synthetase, and cdc-2 (210). Thus, activation of E2F contributes to the preparation of the host cell for viral DNA replication.

Adenovirus encodes a DNA polymerase, a terminal protein, and a ssDNA-binding protein for viral DNA replication (see Chapter 30), but the remainder of the replication proteins and nucleotide metabolism enzymes are host-cell proteins. Thus, adenovirus also induces its host cell to enter S phase. Like large T-antigen, the E1A proteins bind to pRb (330), releasing E2F and activating transcription of cell cycle genes. It is also becoming clear that E1A disrupts a number of protein complexes that are involved in regulation of the cell cycle. The adenovirus E1B-55K protein binds to the cellular p53 protein (270), inactivating the G1 cell cycle block imposed by p53. This also contributes to activation of the host-cell S phase and the optimal conditions for viral DNA synthesis.

Maintenance of Viral DNA Within the Host Cell

There are two types of mechanisms by which viral DNA is stably maintained within the host cell. First, the viral DNA can be integrated into the cellular chromosome and propagated as part of the cellular DNA. For example, retrovirus DNA is integrated into the cellular genome after its synthesis by the reverse transcriptase (see Chapter 26). Integration is promoted by the viral integrase function, but host functions can modulate the process. For example, the mouse *Fv-1* gene mediates a postpenetration block to murine leukemia virus (MuLV) DNA integration (109,172). Thus, this host gene can define the host range of MuLV in mouse cells.

Second, viral DNA can be maintained as an extrachromosomal circular molecule in the infected cell. For example, Epstein-Barr virus DNA is maintained in latently infected lymphocytes as an episomal molecule (1) and requires a specific sequence, oriP, for replication and propagation of the genome in growing cells (342). Similarly, the bovine papillomavirus genome contains sequences (called *plasmid maintenance sequences*) needed to maintain and replicate the DNA as an extrachromosomal element (122,179). Replication of these viral DNA molecules requires *trans*-acting factors encoded by the viral genome and the host-cell DNA replication apparatus (see Chapters 29 and 33). In addition, these extrachromosomal elements appear to be subject to the normal copy number control existing in the normal cell because there is a constant number of episomal copies of these viral DNA molecules per cell as the cells divide. In contrast, HSV DNA can persist during latent infection of neurons in a nonreplicating form because the host cell is not dividing.

VIRUS INTERACTIONS WITH CELLULAR PROTEIN MATURATION PATHWAYS

Maturation of viral proteins in infected cells involves mostly host-cell metabolic pathways, including localization mechanisms, folding proteins, and enzymes that modify the primary translation product. Because viral proteins so often exploit cellular pathways, their study has provided some very basic information about protein maturation pathways in the eukaryotic cell. However, exceptions exist where virally encoded proteins themselves can affect protein maturation or modify the cellular maturation pathways. In the cases where cellular mechanisms are altered, this may be a source of cytopathology.

Utilization of Host-Cell Pathways

Protein Targeting Mechanisms

Eukaryotic cell proteins often contain specific signals that target the protein to a compartment or organelle within the cell. Individual viral proteins can contain all of the necessary signals for intracellular localization because expression of individual viral proteins in cells either by transfection or from a vector system can lead to correct intracellular targeting of the viral protein (94,213,246, 261,263, 312). In other cases, more than one viral gene product is needed for efficient, correct localization (41,135,254,340).

Viral glycoproteins, especially the influenza HA and VSV glycoprotein (G), have been used extensively as prototypes for the study of biogenesis of plasma membrane proteins. These proteins use the signal receptor particle, endoplasmic reticulum enzymes, Golgi apparatus enzymes, and transport mechanisms to realize their proper structure and cell-surface location. Part of the evidence for the concept that transmembrane proteins are synthesized on membrane-bound polyribosomes and translocated cotranslationally across the endoplasmic reticulum membrane while peripheral membrane proteins are synthesized on free polyribosomes came from the study of virus membrane proteins (146,201). Detailed studies of *in vitro* insertion of viral membrane proteins into the endoplasmic reticulum membrane came from studies of the VSV G protein (265). Detailed genetic study of the VSV G protein has provided evidence for the loop model for signal sequence insertion into the endoplasmic reticulum (281). Similarly, viral nuclear proteins use cellular pathways to enter the nucleus. In fact, a nuclear localization signal was first identified in the SV40 large T-antigen (133).

The need for proper protein folding and assembly for correct intracellular localization has become apparent from studies of viral proteins (67). For example, correct folding and trimerization are needed for the influenza HA (52,53,66,93,154) to localize from the endoplasmic reticulum to the Golgi apparatus. Folding and assembly of the proteins have been monitored by (a) reactivity of the proteins with conformation-specific monoclonal antibodies, (b) protease sensitivity, and (c) sedimentation of protein complexes on velocity sedimentation gradients. Localization can be monitored by cell fractionation and immunocytochemistry. These studies also showed that mutant viral proteins or improperly folded wild-type proteins coprecipitate in immunoprecipitates with a cellular protein induced under stress conditions (93,278). This cellular protein, called BiP or grp78, is a member of the hsp70 family of proteins. It has been proposed that members of this family of proteins recognize incompletely or improperly folded proteins, catalyze their unfolding, and help them to attempt to fold into the correct conformation (231). It also has been proposed that BiP might prevent abnormally folded proteins from being exposed to the immune system, thereby increasing the repertoire of epitopes recognized as "self" (93). The viral glycoproteins provide systems to study further the role of these cellular "chaperone" proteins in the folding of normal proteins and in the metabolism of abnormal proteins.

Protein Modification

Many of the posttranslational modifications of proteins that occur in cells (i.e., cleavage, glycosylation, phosphorylation, acylation, or sulfation) are performed by cellular enzymes. Most of these enzymes are ubiquitous, and thus their presence is not usually limiting for viral replication in different cell types. One example of an exception to this is the presence of tissue-specific proteases that cleave specific virion surface glycoproteins, allowing the viral particles to become infectious (271).

Some protein modifications in infected cells are the result of the direct action of virally encoded gene products. Certain viral proteins, such as the v-*src* protein and related oncogene proteins (50,170), have intrinsic protein kinase activities. For the oncogene proteins, the major target proteins of the kinase activity are cellular proteins (125). Also, some viral proteins can act as proteases to make specific protein cleavages not readily made by cellular enzymes (153), such as the picornaviral and retrovirus proteases.

Cellular protein modification activities can also be modified by viral infection. The polyoma virus *hr-t* gene function is required for transformation of nonpermissive cells (8) and for proper assembly of polyoma virions (88). In cells infected with *hr-t* mutants altered in the middle T-antigen gene, phosphorylation of the polyoma major capsid protein is reduced. The polyoma middle T-antigen is known to bind to the cellular c-*src* protein and stimulate its tyrosine kinase activity (54). This kinase may phosphorylate another protein kinase, with the ultimate effect being an increased phosphorylation of VP-1. Thus, this alteration in the cellular enzyme activity may be necessary for efficient assembly of infectious virus particles.

EFFECTS OF VIRUSES ON CELL STRUCTURE

Effects of Viruses on the Cell Membrane

Viruses can alter the membranes of their host cell in at least two ways: (a) by promoting membrane fusion with neighboring cells and (b) by altering the permeability of the cell plasma membrane. Both of these effects may be exerted through the insertion of virally encoded proteins within the membrane.

Promotion of Cell Fusion

Some enveloped virions have cell-surface proteins that facilitate fusion of the virion envelope with the cell-surface membrane. This property can confer on the virion the ability to promote fusion between adjacent cells. For example, in sufficient amounts, Sendai virus can bind to and cause fusion between two neighboring cells, leading to a polykaryon. In addition, viral glycoproteins expressed within the infected cell can migrate to the cell surface and promote fusion with neighboring cells. This latter phenomenon has been referred to as fusion from within to distinguish it from fusion of adjacent cells caused by input virions or fusion from without (29).

The induction of cell fusion may be a form of cytopathology that is a side product of the membrane fusion activity of viral surface proteins that allows entry of the virus at the cell surface or endosome (see above). Fusion between neighboring cells may also promote cell-to-cell spread of virus.

Altering Plasma Membrane Permeability

Infection by viruses can cause an increase in permeability of the host-cell plasma membrane to ions, allowing, for example, an influx and increase of intracellular sodium ions. Because the translation of some viral mRNAs is more resistant to high sodium ion concentration than is the translation of cell mRNA (40), it has been hypothesized that increased membrane permeability may favor translation of viral mRNAs. Indeed, increased osmolarity of the culture medium inhibits host-cell translation (268,326) and allows preferential translation of viral RNAs (51,217). For Sindbis virus, the increase in permeability seems to correlate temporally with shut-off of host protein synthesis (89), but membrane permeability changes occur later than host shut-off in cells infected with picornaviruses (70,161,205). Furthermore, changes in ion flux do not account for host shut-off by certain other viruses (80,99).

Increased membrane permeability to antibiotics and toxins in infected cells has also been reported (51,81). This has been suggested as a possible antiviral strategy to kill infected cells (39).

The cause(s) of the permeability changes in infected cells has not been defined. Insertion of viral proteins into the membrane was hypothesized to alter the permeability of the plasma membrane (39). RNA virus mutants unable to transcribe viral mRNA failed to alter the membrane permeability (139). Therefore, viral gene products expressed in the infected cell are necessary for the membrane changes, but no specific genetic defects have defined individual gene products required for altering the cell membrane permeability.

One aspect of the mechanism of the change in permeability was studied by Garry et al. (90). They hypothesized that Sindbis virus might inhibit the sodium pump in the plasma membrane, which maintains the ionic balance within the cell. When they added ouabain (an inhibitor of the sodium pump) to uninfected cells, protein synthesis was inhibited. However, when ouabain was added to Sindbis virus–infected cells, viral protein synthesis was not affected. In addition, the Na^+ and K^+ concentrations did not change in infected cells upon ouabain treatment. These results indicate that Sindbis virus infection has an effect on cells similar to ouabain treatment. Therefore, for some viruses, increased membrane permeability may be a form of CPE that allows preferential biosynthesis of viral gene products.

Interactions Between Viruses and the Cytoskeleton

The cell cytoskeleton plays several roles in (a) the structure of the cell and (b) the transport and movement of organelles. Therefore, it should not be surprising that there are numerous documented associations between viral macromolecules and the cellular cytoskeleton in the infected cell, as well as observed effects of virus infection on the cytoskeleton.

Depolymerization of Cytoskeleton Filaments

Infection by many viruses leads to a disruption of one or more cytoskeletal fiber systems. For example, infection of cells by several viruses, including VSV (115,199,267), vaccinia virus (115,199), simian virus 40 (98,266), canine distemper virus (121), frog virus 3 (203), and HSV (10,111,335) cause a decrease in actin-containing microfilaments. Many of these studies have used immunofluorescence as an assay for microfilaments (Fig. 6). Some studies have used DNAse I inhibition as a quantitative assay for globular actin as a measure of actin filament depolymerization (e.g., 6). The mechanism by which this disassembly occurs has not been defined, but expression of HSV immediate-early and early proteins seems to be necessary for microfilament depolymerization (111). Many of these viruses, including HSV, canine distemper virus, and frog virus 3, also cause a depolymerization of microtubules in infected cells. In contrast, infection of cells by reovirus causes disruption of vimentin-containing intermediate fil-

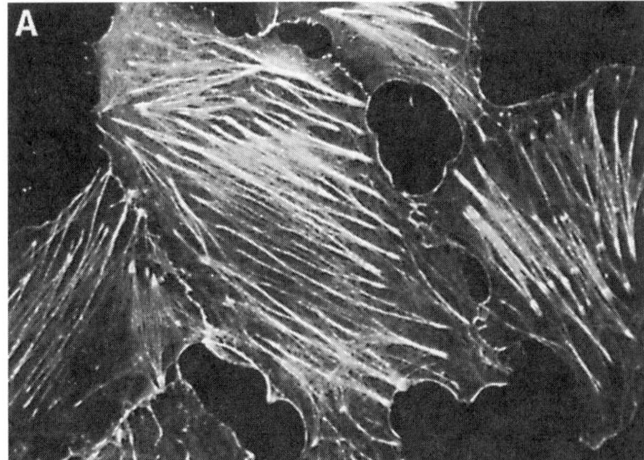

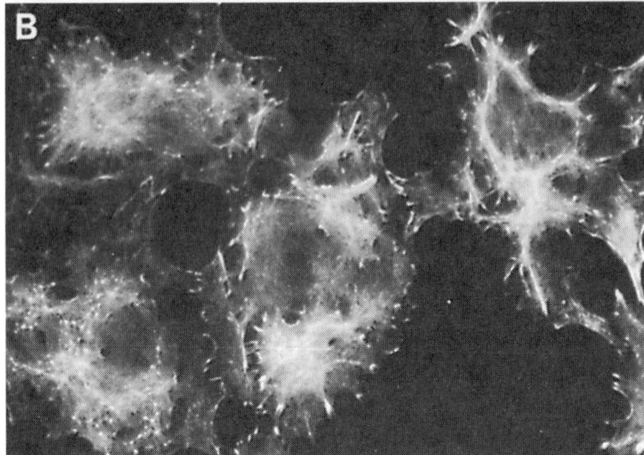

FIG. 6. Disruption of the cytoskeleton by virus infection. Uninfected (**A**) and HSV-infected (**B**) monkey cells were fixed, permeabilized, and stained with fluorescein-conjugated phalloidin. Phalloidin reacts with filamentous actin (f-actin) and thus shows the distribution of microfilament bundles. Micrographs courtesy of S. Rice.

aments but spares the microfilaments and microtubules (279). Although the cause of the cytoskeleton changes is not clear, it is clearly a potential cause of the structural changes that occur in infected cells, such as cell rounding, because the major cytoskeletal fibers play a role in maintaining cell morphology. The cytoskeletal changes may not be primary effects of viral infection, but instead other effects of virus infection, such as inhibition of macromolecular synthesis, may lead to cytoskeletal changes.

Incorporation of Cytoskeletal Components into Infected Cell Structures

As described below, new structures called *factories* or *inclusions* are assembled in the nucleus or cytoplasm for synthesis of viral nucleic acids and assembly of virions. In reovirus-infected cells, there is evidence that a specific cytoskeleton component is incorporated into the cytoplasmic inclusion bodies. EM studies showed that the inclusions contain several types of filaments (57,60). These include microtubules and 50- to 80-Å "kinky" filaments. Dales et al. (60) postulated that the 50- to 80-Å filaments represented cellular filaments reorganized into the cytoplasmic factory. Sharpe et al. (279) showed that anti-vimentin antibody stained filaments in inclusions and that these may therefore be the cellular filaments reorganized within the inclusions. Thus, viral infection may exploit cellular structural elements to build replication factories.

However, interpretation of some immunofluorescence data has been complicated. For example, anti-actin monoclonal antibodies stain intranuclear replication compartments in cells infected with HSV (256). However, the antiactin antibodies cross-react with the HSV immediate-early ICP4 protein molecule on Western blots. Although this could be an example of molecular mimicry (87) and could have some functional significance, the cross-reaction could be fortuitous. Most importantly for this discussion, the immunological cross-reaction of anti-actin monoclonal antibodies with ICP4 makes it difficult to localize actin by immunofluorescence after HSV infection.

Interactions of Viral Molecules and the Cytoskeleton

The cellular substructure—the cytoplasmic cytoskeleton, nuclear matrix, and membranes—provides a physical site and possibly some functional elements for metabolic activities of the cell. For example, the cytoskeleton provides (a) a substrate for polyribosomes, (b) structural integrity for the cell, (c) a structural framework for organelle movement, and (d) part of the system(s) for cellular movement. The nuclear matrix provides a substrate for chromatin and for transcription, as well as DNA replication complexes. Therefore, it is not surprising that many viral macromolecules appear to be associated with the cytoskeleton. From studies involving gentle detergent extraction procedures, some of the viral molecules shown to fractionate preferentially with the cytoskeleton and nuclear matrix are: the SV40 newly uncoated genomes, nascent RNA molecules, and RNA in transit from cytoplasm to nucleus (14,324); adenovirus DNA (344) and HSV proteins in transit from cytoplasm to nucleus (15,247); and VSV proteins during nucleocapsid assembly (45). Thus, it is likely that part of the subcellular compartmentalization of viral processes relies on the structural organization of the host cell. In addition, the virus provides for part of the compartmentalization by specifying the assembly of infected cell-specific structures, such as the replication factories described below.

Certain viral components have been shown to undergo specific associations with the cytoskeleton. For example, EM has shown adenovirus to be associated with microtubules in infected cells, an association believed to reflect nuclear transport of the parental virus (59). Adenovirus virions also can bind to microtubules *in vitro* (178). Similarly, reovirus particles are associated with microtubules in infected cells (57), and reovirus particles bind to microtubules *in vitro* (6). Also, viral proteins and viral factories codistribute with microtubules (5). Most importantly, colchicine does not lower virus yields (293) but does block axonal spread of reovirus (317). Thus, the cytoskeleton can play a role in virus transport in specialized cells. The cytoskeleton may also play a role in virus assembly, because disruption of portions of the cytoskeleton can block budding of some enveloped viruses (226,295). In general, the cytoskeleton is thought to play a role in providing a structure for viral replication or movement of macromolecules, but the precise molecular mechanisms remain to be defined.

Effects of Viruses on mRNA Association with the Cytoskeletal Framework

Using a gentle detergent extraction procedure to isolate the cytosketal elements, Lenk et al. (168) showed that polyribosomes were preferentially associated with the cytoskeleton fraction of the cell. Cervera et al. (43) showed that VSV mRNAs were associated with the cytoskeleton fraction while being translated. These observations led to the hypothesis that cytoskeleton association was necessary for translation of mRNA. As discussed above, viruses cause a disruption of portions of the cytoskeleton. Lenk and Penman (167) showed that poliovirus disrupted the cytoskeleton of HeLa cells and caused a release of host mRNAs from the cytoskeleton. It was also reported that adenovirus infection of human KB cells caused a dissociation of host mRNA from the cytoskeleton (320). The correlation between shut-off of host translation and dissociation of mRNA suggested that dissociation of host mRNA might be a general mechanism for virus inhibition of host translation. However, other reports have indicated that VSV (24) and adenovirus and influenza virus (138) inhibit host translation but do not cause a dissociation of host mRNA from the cytoskeleton. Thus, dissociation of the host mRNA from the cytoskeleton may not be required for viral inhibition of host translation. Poliovirus dissociation of host mRNA from the cytoskeleton has been confirmed (138). However, the events occurring as primary and secondary effects of poliovirus remain to be defined. The effects of poliovirus infection on the translational initiation apparatus were reviewed above. Poliovirus may cause a more drastic rearrangement of the cytoskeleton than other viruses (138,167), which leads to effects on the CBP complex (347), or, alternatively, the alteration of the CBP complex may displace host mRNA from the cytoskeleton. The effects on CBP complex and the cytoskeleton may even be unrelated effects of infection. If the two effects are related, further study of poliovirus-infected cells could give critical insight into the role of the cytoskeleton in protein synthesis.

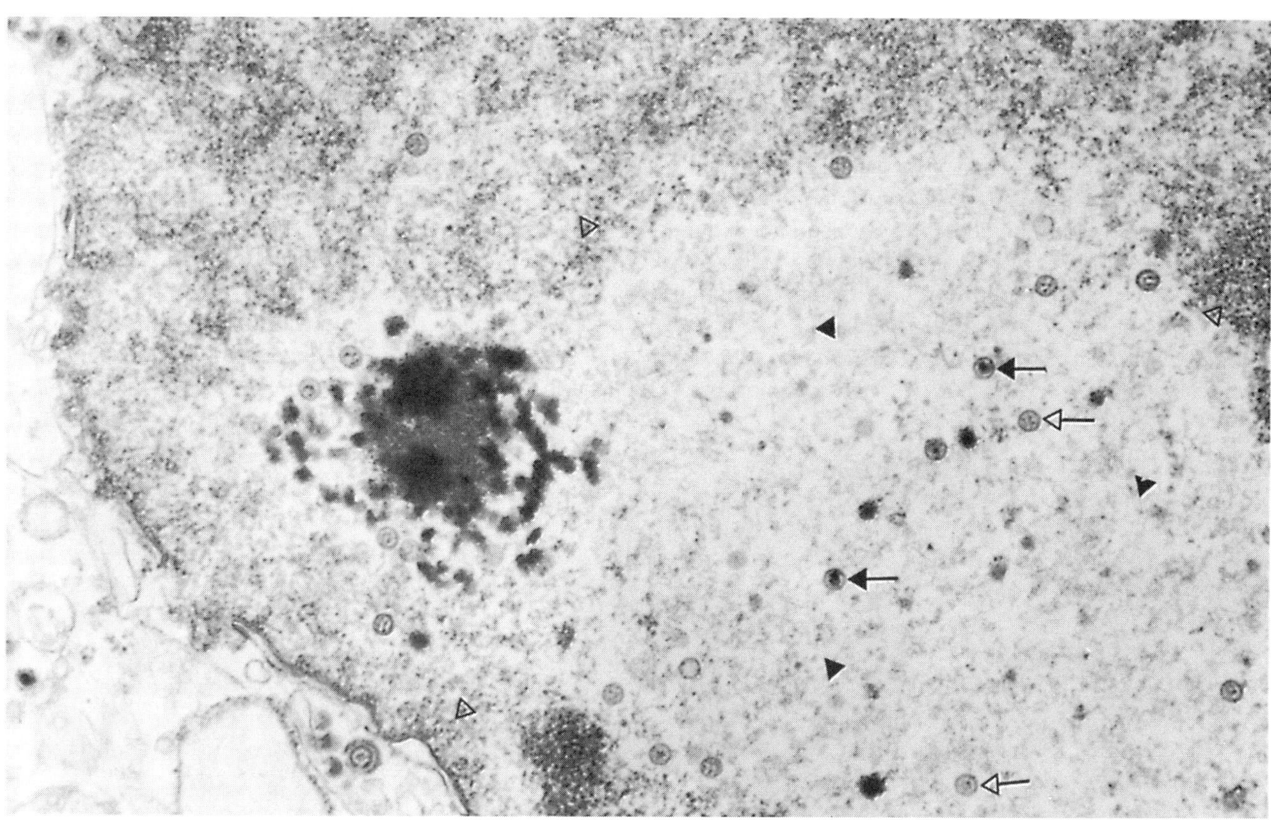

FIG. 7. EM visualization of nuclear inclusion areas. The *filled triangles* denote the electron-translucent nuclear inclusion area in a human HEp-2 cell infected with HSV-1.The *filled arrows* indicate full capsids, and the *unfilled arrows* indicate empty capsids. The *unfilled triangles* indicate host chromatin compressed to the periphery of the nucleus, apparently by the nuclear inclusion. Micrograph courtesy of D. Furlong and B. Roizman.

Assembly of Factories for Nucleic Acid Replication and Virion Assembly

Viral replication complexes, transcriptional complexes, replicative and assembly intermediates, and nucleocapsids and virions often accumulate in specific locations within the host cell and form structures called factories or inclusion bodies. The location of these structures obviously reflects the site of replication of a given virus. They may form in the nucleus, as with the inclusion bodies or factories in cells infected with herpesviruses (Fig. 7) or adenoviruses, or in the cytoplasm, as with the factories in reovirus- (Fig. 8) or poxvirus-infected cells or the Negri bodies in rabies virus–infected cells. Inclusion bodies have been useful in diagnostic virology because areas of altered staining can be detected at specific locations in the cell with specific staining properties, either basophilic or acidophilic, characteristic of individual groups of viruses. Crystalline arrays of capsids or nucleocapsids may accumulate in these inclusions (Fig. 8). The assembly of these new structures in the cell may alter or displace host-cell components and lead to one form of CPE.

Much as other viral proteins have provided probes for important cell biological questions, HSV DNA replication proteins and their localization to intranuclear inclusions have provided a situation in which to study intranuclear localization mechanisms. Initial studies characterized the nuclear inclusions of herpesvirus-infected cells and their molecular components. EM of herpesvirus-infected cells showed electron-translucent intranuclear inclusions surrounded by marginated and compacted cell chromatin (63,277) (Fig. 7). Light microscopic observation of nuclear inclusions shows an hourglass appearance of the inclusions at early times and an eosinophilic staining at later times (204,290) (Fig. 3). Immunofluorescence experiments using antibodies specific for HSV DNA replication proteins have shown that viral DNA replication proteins accumulate in intranuclear foci by 3 hr postinfection and that these foci enlarge into globular nuclear structures called *replication compartments* (64,246) (Fig. 9). The replication compartments are likely to be equivalent to (a) the translucent nuclear inclusions seen by EM (Fig. 7) and (b) the early nuclear inclusions seen by light microscopy.

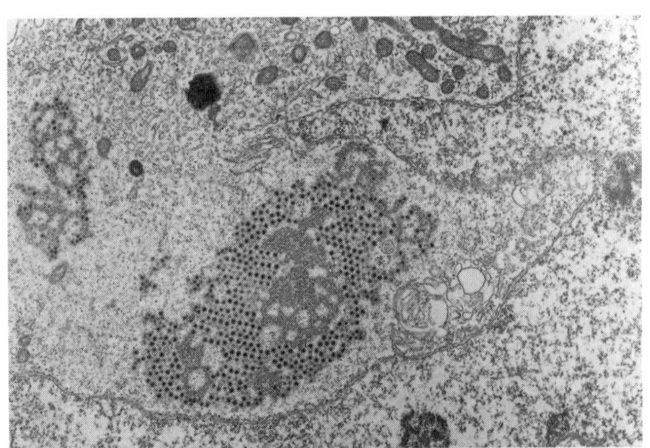

FIG. 8. EM visualization of cytoplasmic "factories" in a cell infected with reovirus. Micrograph courtesy of B. Fields.

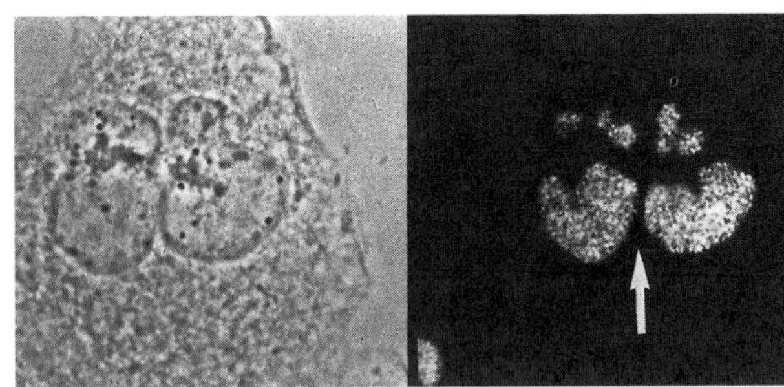

FIG. 9. Immunofluorescence detection of HSV replication compartments in binucleate cells. Micrograph courtesy of A. de Bruyn Kops.

These replication compartments are located in the interior of the nucleus as shown by EM (63,277) (Fig. 7) and by confocal microscopy (65). Bromodeoxyuridine pulse labeling followed by immunofluorescence detection of bromodeoxyuridine-substituted DNA has shown that viral DNA synthesis occurs in the replication compartments (64). This technique allows the determination of the cellular location of DNA synthesis. Similarly, *in situ* hybridization with a viral DNA probe has shown that progeny HSV DNA accumulates in replication compartments (Fig. 10). The initial sites of formation of these compartments likely takes place at specific locations in the nucleus defined by the availability of host-cell proteins or structures (65). The accumulation of progeny DNA, the probable template for late gene transcription and the viral transcriptional *trans*-activator protein ICP4 (147) in replication compartments suggests that late gene transcription occurs in the replication compartments. Thus, late gene transcription may be compartmentalized in infected cell-specific nuclear structures. Empty capsids appear to be assembled around dense bodies or in the inclusions within the infected cell nucleus, whereas encapsidation of viral DNA appears to occur within the inclusion body itself (Fig. 7). Thus, these intranuclear compartments in HSV-infected cells are sites of viral DNA replication, late gene transcription, and nucleocapsid assembly.

Studies of the nuclear inclusions in cells infected with HSV have raised several important cell biological questions related to the assembly of nuclear complexes and structures and compartmentalization of nuclear metabolic processes: (a) What signals target viral and cellular proteins to these compartments? Intranuclear localization mechanisms have not been studied extensively. Both viral and host proteins localize to these compartments. The HSV replication proteins, including the ICP8 DNA-binding protein, DNA polymerase, polymerase accessory protein, and helicase-primase complex (32,64,95,219,246,248) localize to compartments. These proteins undergo multiple in-

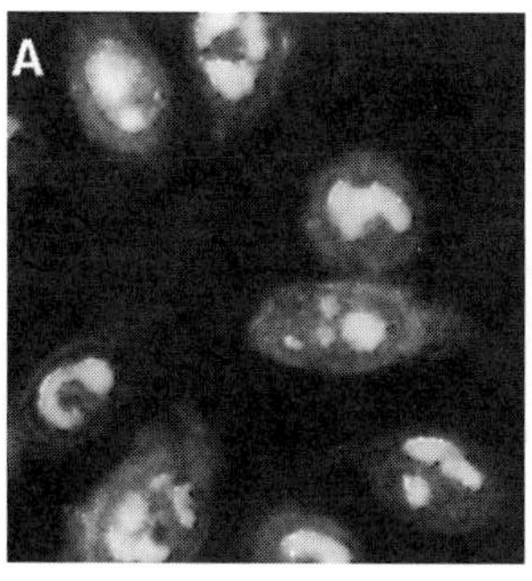

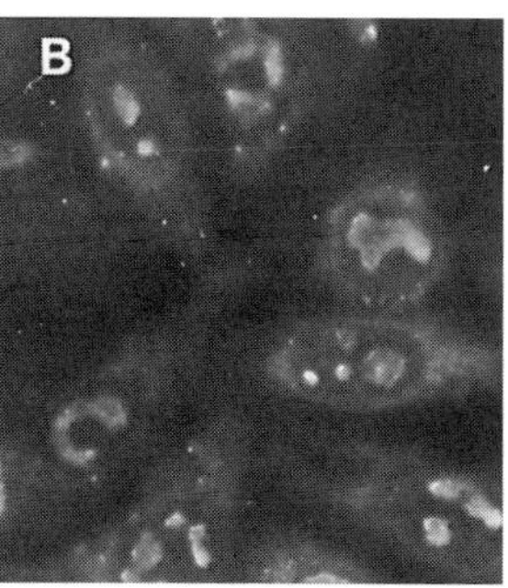

FIG. 10. *In situ* hybridization detection of the location of viral nucleic acids in infected cells. Monkey cells infected with HSV-1 were fixed in formaldehyde, permeabilized in acetone, and incubated with biotin-labeled HSV DNA. The cultures were heated to denature the cellular DNA and the probe, and the culture and solution were allowed to cool. The cells were then reacted with mouse anti-ICP4 monoclonal antibody followed by rhodamine-conjugated goat antimouse immunoglobulin antibody and fluorescein-conjugated avidin. **A:** Rhodamine fluorescence showing the location of the HSV trans-activator protein ICP4. **B:** Fluorescein fluorescence showing the location of HSV DNA in the infected cells. Micrographs courtesy of S. Rice.

teractions with each other; thus their targeting could involve binding to other proteins in the replication complex, binding to cellular proteins, or other targeting mechanisms. Cellular proteins, including pRb, p53, and DNA pol α (332), and proteins encoded by other viruses, such as SV40 large T-antigen (343), also localize to replication compartments. Genetic analysis of the SV40 large T-antigen indicates that sequences sufficient for localization to replication compartments are located in or near the T-antigen nuclear localization signal (NLS) (343). This raises the interesting possibility that an NLS may serve to bring a protein into the nucleus and to specific sites in the nucleus. (b) Is there a defined assembly pathway for viral protein complexes in these compartments? Experiments examining protein localization in cells infected with mutants defective for each of the HSV DNA replication proteins have shown that there is a functional order for assembly of HSV DNA replication proteins into nuclear structures (32,173). The helicase-primase complex of UL8, UL42, and U52 proteins, the UL9 origin-binding protein and the ICP8 ssDNA-binding proteins are the minimal HSV proteins needed to assemble into nuclear structures. The polymerase-UL42 complex then joins the structure. (c) Is there a receptor on the nuclear framework that anchors viral protein complexes? No cellular molecule has been identified as a receptor for the HSV replication proteins, and nuclear receptors remain as an important problem in targeting of nuclear proteins. One of the few situations where saturable binding of a viral protein to the nuclear matrix has been demonstrated is with the adenovirus preterminal protein (85). The identification of a nuclear molecule or receptor that anchors specific proteins in the cell nucleus remains as an important confirmation of the concept of the nuclear matrix, a nuclear structure believed to consist of nuclear lamins under the nuclear envelope and internal fibrous elements. (d) What defines the intranuclear location at which compartments are formed? In binucleate cells infected with HSV, replication compartments often show nearly identical patterns of compartments along a two-fold axis of symmetry between many of the sister nuclei (Fig. 9) (65). Therefore, the location of HSV replication compartments is defined by preexisting nuclear architecture, probably the internal nuclear matrix or receptors thereon.

Nuclear viral inclusions in adenovirus-infected cells (251,304,321) are also sites of viral DNA synthesis and accumulation of progeny DNA (30,189,321) and probably late viral transcription (188,336). Their structure has perhaps an added level of complexity over the HSV nuclear inclusions in that the E1B 55-kd/E4 34-kd complex of proteins is located in and around the inclusions and is believed to recruit critical components of the host-cell RNA transport process to the sites of late viral RNA synthesis (221).

RELEASE OF PROGENY VIRUS

The mechanism of release of progeny virus from the infected cell depends on the structure of the virus. Enveloped viruses exit from the infected cell either by budding through the plasma membrane (Fig. 11) or by fusion of secretory vesicles containing virus particles with the plasma membrane (298). Thus, nucleocapsids can bud through the plasma membrane (orthomyxoviruses, paramyxoviruses, rhabdoviruses, and retroviruses), directly producing extracellular virions, or through internal membranes such as the endoplasmic reticulum (ER) (rotaviruses), ER and/or Golgi apparatus (coronaviruses and bunyaviruses), or inner nuclear membrane (herpesviruses). The factors that determine the

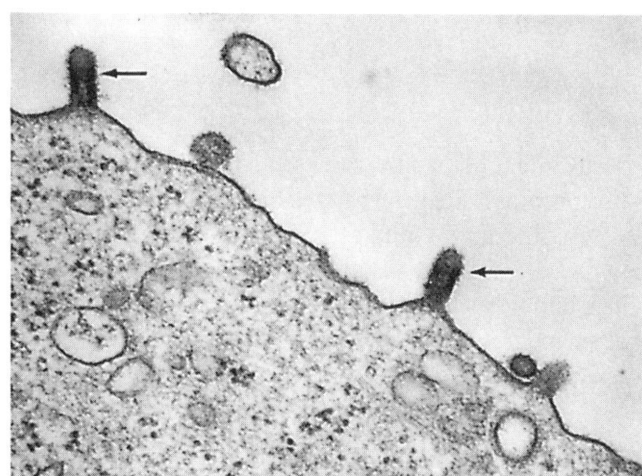

FIG. 11. Enveloped virus particles budding from the infected cell surface. The arrows indicate VSV particles budding from the surface plasma membrane of infected Chinese hamster ovary cells, as visualized by EM.

site of budding of a virus are not well understood, but the site of localization of the surface glycoproteins must be one important factor.

Polarized epithelial cells have differentiated apical and basal surface plasma membranes. Thus, viruses budding through the plasma membrane or vesicles containing virus particles can traffic specifically through either membrane. For example, orthomyxoviruses and paramyxoviruses bud at the apical surface, whereas VSV and retroviruses bud at the basal surface. Viral glycoproteins have an intrinsic ability to localize to specific surfaces of polarized cells (e.g., 131,245,263,299), and this may be an important factor in deciding the site of virus budding in polarized epithelial cells. Sorting of proteins destined for the different surfaces appears to occur within the Golgi apparatus (158). Studies of the targeting of chimeric glycoproteins expressed from recombinant clones have, in the words of a recent review, "met with mixed success" (315). Some studies indicated that the ectodomain contained the targeting signal (194,264), whereas other studies found the ectodomain was not sufficient (245,298). Recent results suggest that a glycosyl-phosphatidylinositol (GPI) anchor may target proteins to the apical surface, whereas a ß-turn structure in the cytoplasmic domain may target a protein to the basolateral surface (315). Thus, these targeting signals appear to be complex.

It has been generally assumed that unenveloped viruses are released by lysis of the cells. However, evidence exists for release of some unenveloped viruses without cell lysis (215), and one report indicates polarized release of SV40 from epithelial cells (48). Similarly, poliovirus is released almost exclusively from the apical surface of polarized human intestinal epithelial cells (316). Thus, there may be cellular mechanisms used by viruses for the active release of unenveloped viruses that are not the result of lysis of the host cell.

Once the progeny viruses have been released, they can initiate infection in new cells, and a whole new round of virus replication and interaction with a host cell can begin.

APOPTOSIS

Virus infection can trigger events that lead to apoptosis, or programmed death, which is characterized by cellular DNA fragmentation and cytoplasmic changes including membrane blebbing. Cellular situations hypothesized to induce apoptosis include DNA damage, high levels of p53 protein, and possibly uncontrolled DNA synthesis. Although little is known about the mechanisms of induction of apoptosis, study of the mechanisms by which viruses induce apoptosis may lead to a more precise understanding of this process. Furthermore, for the few viruses where this effect has been examined, they have evolved mechanisms to block the apoptotic pathway.

As described above, adenoviruses induce p53 expression, but they do not induce apoptosis unless the E1B genes are mutated (236,306,328). It is believed that the E1A gene products lead to increased p53 levels and apoptosis but the E1B gene products inhibit apoptosis (249).

Other viruses are known to encode gene products that block apoptosis. The Epstein-Barr virus latency membrane protein-1 (LMP-1) induces expression of the cellular *bcl-2* protein, which is known to block apoptosis (265). Baculovirus (49) and African swine fever virus (206) encode proteins that inhibit apoptosis. HSV encodes a gene product that has been reported to inhibit a total shutoff of protein synthesis characteristic of programmed cell death in neuronal cells (46). However, it is not clear that this is an apoptotic effect because there is no evidence yet for any of the usual characteristics of apoptosis, DNA fragmentation and cytoplasmic changes.

SUMMARY

The major points of this chapter can be summarized as follows:

1. The ability of a virus to replicate in a host cell can be determined by the availability of specific host macromolecules in the host cell. These molecules may be external, such as a receptor, or internal, such as a transcription or replication factor.
2. Some of the cytopathic effects of virus infection on a host cell are due to specific alterations in host-cell metabolism or structure that allow viral replication events. These cytopathic effects are not simply toxic side effects of virus infection. We presume that viruses have evolved to manipulate the environment of the host cell for their own optimal replication.
3. Interactions of viruses with host cells may involve subtle changes in the host cell, and understanding of the nature of the interaction between viral gene products and the host-cell molecules often provides insight into the metabolic processes and critical regulatory events of the host cell.

As stated above, our challenge for the future is to relate this information about the interactions of viruses with their host cells to the biology of viruses in their host organisms.

REFERENCES

1. Adams A, Lindahl T. Epstein-Barr virus genomes with properties of circular DNA molecules in carrier cells. *Proc Natl Acad Sci USA* 1975;72:1477–1481.
2. Agris CH, Nemeroff ME, Krug RM. A block in mammalian splicing occurring after formation of large complexes containing U1, U2, U4, U5 and U6 small nuclear ribonucleo proteins. *Mol Cell Biol* 1989;9:259–267.
3. Alizon M, Dragie T. CD26 and HIV fusion. *Science* 1994;264:1161–1162.
4. Ambros V, Petterson RF, Baltimore D. An enzymatic activity in uninfected cells that cleaves the linkage between poliovirion RNA and the 5′ terminal protein. *Cell* 1978;15:1439–1446.
5. Babiss LE, Ginsberg HS, Darnell JE. Adenovirus E1B proteins are re-

quired for accumulation of late viral mRNA and for effects on cellular mRNA translation and transport. *Mol Cell Biol* 1985;5:2552–2558.
6. Babiss LE, Luftig RB, Weatherbee JA, Weihing RR, Ray UR, Fields BN. Reovirus serotypes 1 and 3 differ in their *in vitro* association with microtubules. *J Virol* 1979;30:863–874.
7. Bagchi S, Raychaudhuri P, Nevins JR. Adenovirus E1A protein can dissociate heteromeric complexes involving the E2F transcription factor: a novel mechanism for E1A trans-activation. *Cell* 1990;62:659–669.
8. Banerji J, Rusconi L, Schaffner W. Expression of a ß-globin gene is enhanced by remote SV40 DNA sequences. *Cell* 1981;27:299–308.
9. Batterson W, Furlong D, Roizman B. Molecular genetics of herpes simplex virus. VII. Further characterization of a ts mutant defective in release of viral DNA and in other stages of viral reproductive cycle. *J Virol* 1983;45:397–407.
10. Bedows E, Rao KMK, Welsh MJ. Fate of microfilaments in Vero cells infected with measles virus and herpes simplex virus type 1. *Mol Cell Biol* 1983;3:712–719.
11. Beltz G, Flint SJ. Inhibition of HeLa cell protein synthesis during adenovirus infection: restriction of cellular messenger RNA sequences to the nucleus. *J Mol Biol* 1979;131:353–373.
12. Benjamin TL. Host range mutants of polyoma virus. *Proc Natl Acad Sci USA* 1970;67:394–399.
13. Benoist C, Chambon P. *in vivo* sequence requirements of the SV40 early promoter region. *Nature* 1981;290:304–310.
14. Ben-Ze'ev A, Abulafia R, Aloni Y. SV40 virions and viral RNA metabolism are associated with cell substructures. *EMBO J* 1982; 1:1225–1231.
15. Ben-Ze'ev A, Abulafia R, Bratosin S. Herpes simplex virus assembly and protein transport are associated with the cytoskeletal framework and the nuclear matrix in infected BSC-1 cells. *Virology* 1983;129:501–507.
16. Berezney R, Coffey DS. Nuclear protein matrix: association with newly synthesized DNA. *Science* 1975;189:291–293.
17. Berger SL, Folk WR. Differential activation of RNA polymerase III-transcribed genes by the polyoma virus enhancer and adenovirus E1A gene products. *Nucl Acids Res* 1985;13:1413–1428.
18. Berget SM, Moore C, Sharp PA. Spliced segments at the 5′ terminus of adenovirus 2 late mRNA. *Proc Natl Acad Sci USA* 1977; 74:3171–3175.
19. Berk AJ. Adenovirus promoters and E1a transactivation. *Annu Rev Genet* 1986;20:45–79.
20. Berk AJ, Lee F, Harrison T, Williams J, Sharp PA. Pre-early adenovirus 5 gene product regulates synthesis of early viral messenger RNAs. *Cell* 1979;17:935–944.
21. Bernstein HD, Sonenberg N, Baltimore D. Poliovirus mutant that does not selectively inhibit host protein synthesis. *Mol Cell Biol* 1985;5:2913–2923.
22. Bodkin DK, Nibert ML, Fields BN. Proteolytic digestion of reovirus in the intestinal lumens of neonatal mice. *J Virol* 1989;63:4676–4681.
23. Bohmann D, Bos TJ, Admon A, Nishimura T, Vogt PK, Tjian R. Human proto-oncogene c-*jun* encodes a DNA binding protein with structural and functional properties of transcription factor AP-1. *Science* 1987;238:1386–1392.
24. Bonneau AM, Darveau A, Sonenberg N. Effect of viral infection on host protein synthesis and mRNA association with the cytoplasmic cytoskeletal structure. *J Cell Biol* 1985;100:1209–1218.
25. Boone RF, Moss B. Sequence complexity and relative abundance of vaccinia virus mRNA's synthesized *in vivo* and *in vitro*. *J Virol* 1978;26:554–569.
26. Boshart M, Weber F, Jahn G, Dorsch-Hasler K, Fleckenstein B, Schaffner W. A very strong enhancer is located upstream of an immediate-early gene of human cytomegalovirus. *Cell* 1985;41:521–530.
27. Boudin M-L, Moncany M, D'Halluin J-C, Boulanger PA. Isolation and characterization of adenovirus type 2 vertex capsomer (penton base). *Virology* 1979;92:125–138.
28. Bouloy M, Plotch SJ, Krug RM. Globin mRNAs are primers for the transcription of influenza viral RNA *in vitro*. *Proc Natl Acad Sci USA* 1978;75:4886–4890.
29. Bratt MA, Gallaher WR. Preliminary analysis of the requirements for fusion from within and fusion from without by Newcastle disease virus. *Proc Natl Acad Sci USA* 1969;64:536–543.
30. Brigat DJ, Myerson D, Leary JJ, et al. Detection of viral genomes in cultured cells and paraffin-embedded tissue sections using biotin-labeled hybridization probes. *Virology* 1983;126:32–50.
31. Broder CC, Nussbaum O, Gutheil WG, Bachovchin WW, Berger EA. CD26 antigen and HIV fusion. *Science* 1994;264:1156–1159
32. Bush M, Yager D, Gao M, Weisshart K, Marcy A, Coen D, Knipe DM. Correct intranuclear localization of herpes simplex virus DNA polymerase requires the viral ICP8 DNA-binding protein. *J Virol* 1991;65:1082–1089.
33. Callebaut C, Jacotet E, Krust B, Hovanessian AG. CD26 antigen and HIV fusion. *Science* 1994;264:1162–1165.
34. Callebaut C, Krust B, Jacotot E, Hovanessian AG. T cell activation antigen, CD26, as a cofactor for entry of HIV in CD4ÿ2D cells. *Science* 1993;262:2045–2050.
35. Camerini D, Planelles V, Chen ISY. CD26 antigen and HIV fusion. *Science* 1994;264:1160–1161.
36. Campbell MEM, Palfreyman JW, Preston DM. Identification of herpes simplex virus DNA sequences which encode a transacting polypeptide responsible for stimulation of immediate early transcription. *J Mol Biol* 1984;180:1–19.
37. Carr CM, Kim PS. A spring-loaded mechanism for the conformation of influenza hemagglutinin. *Cell* 1993;73:823–832.
38. Carrasco L. The inhibition of cell functions after viral infection. A proposed general mechanism. *FEBS Lett* 1977;76:11–15.
39. Carrasco L. Membrane leakiness after viral infection and a new approach to the development of antiviral agents. *Nature* 1978; 272:694–699.
40. Carrasco L, Smith AE. Sodium ions and the shut-off of host cell protein synthesis by picornaviruses. *Nature* 1976;264:807–809.
41. Carswell S, Alwine JC. Simian virus 40 agnoprotien facilitates perinuclear–nuclear localization of VP1, the major capsid protein. *J Virol* 1986;60:1055–1061.
42. Centrella M, Lucas-Lenard J. Regulation of protein synthesis in vesicular stomatitis virus-infected mouse L-929 cells by decreased protein synthesis initiation factor 2 activity. *J Virol* 1982;41:781–791.
43. Cervera M, Dreyfuss G, Penman S. Messenger RNA is translated when associated with the cytoskeletal framework in normal and VSV-infected HeLa cells. *Cell* 1981;23:113–120.
44. Challberg MD. A method for identifying the viral genes required for herpesvirus DNA replication. *Proc Natl Acad Sci USA* 1986;83:9094–9098.
45. Chatterjee P, Cervera M, Penman S. Formation of vesicular stomatitis virus nucleocapsid cytoskeleton framework-bound N protein: possible model for structure assembly. *Mol Cell Biol* 1984;14:2231–2234.
46. Chou J, Roizman B. The γ:1 34.5 gene of herpes simplex virus 1 precludes neuroblastoma cells from triggering total shutoff of protein synthesis characteristic of programmed cell death in neuronal cells. *Proc Natl Acad Sci USA* 1992;89:3266–3270.
47. Chow LT, Gelinas RE, Broker TR, Roberts RJ. An amazing sequence arrangement at the 5′ ends of adenovirus 2 messenger RNA. *Cell* 1977;12:1–8.
48. Clayson ET, Brando LVJ, Compans RW. Release of simian virus 40 virions from epithelial cells is polarized and occurs without cell lysis. *J Virol* 1989;63:2278–2288.
49. Clem R, Feckheimer M, Miller L. Prevention of apoptosis by a baculovirus gene during infection of insect cells. *Science* 1991; 254:1388–1390.
50. Collett MS, Erikson RL. Protein kinase activity associated with the avian sarcoma virus src gene product. *Proc Natl Acad Sci USA* 1978;75:2021–2024.
51. Contreras A, Carrasco L. Selective inhibition of protein synthesis in virus-infected cells. *J Virol* 1979;29:114–122.
52. Copeland CS, Doms RW, Bolzau EM, Webster RG, Helenius A. Assembly of influenza virus hemagglutinin trimers and its role in intracellular transport. *J Cell Biol* 1986;103:1179–1191.
53. Copeland CS, Zimmer K-P, Wagner KR, Healey GA, Mellman I, Helenius A. Folding, trimerization and transport are sequential events in the biogenesis of influenza virus hemagglutinin. *Cell* 1988;53:197–209.
54. Courtneidge SA, Smith AE. Polyoma virus transforming protein associates with the product of the *c-src* cellular gene. *Nature* 1983;303:435–438.
55. Cripe TP, Haugen TH, Turk JP, et al. Transcriptional regulation of the human papilloma virus-16 E6-E7 promoter by a keratinocyte-dependent enhancer and by viral E2 *trans*-activator and repressor gene products: implications for cervical carcinogenesis. *EMBO J* 1987; 6:3745–3753.
56. Crowell RL, Philipson L. Specific alteration of coxsackavirus B3 eluted from HeLa cells. *J Virol* 1971;8:509–515.
57. Dales S. Association between the spindle apparatus and reovirus. *Proc Natl Acad Sci USA* 1963;50:268–275.

58. Dales S. Early events in cell-animal virus interactions. *Bact Rev* 1973;37:103–135.
59. Dales S, Chardonnet Y. Early events in the interaction of adenovirus with HeLa cells. IV. Association with microtubules and the nuclear pore complex during vectorial movement in the inoculum. *Virology* 1973;56:465–483.
60. Dales S, Gomatos PJ, Hsu KC. The uptake and development of reovirus in strain L cells followed with labelled viral ribonucleic acid and ferritin–antibody conjugates. *Virology* 1965;26:193–211.
61. Dalgleish AG, Beverley PCL, Clapham PR, Crawford DH, Greaves MF, Weiss RA. The CD4 (T4) antigen is an essential component of the receptor for the AIDS retrovirus. *Nature* 1984;312:763–767.
62. Daniels RS, Douglas AR, Skehel JJ, Waterfield MD, Wilson IA, Wiley DC. Studies of the influenza hemagglutinin in the pH 5 conformation. In: Laver WG, ed. *The origin of pandemic influenza viruses*. New York: Elsevier/North Holland; 1984:1–7.
63. Darlington RW, James C. Biological and morphological aspects of the growth of equine abortion virus. *J Virol* 1966;92:250–257.
64. De Bruyn Kops A, Knipe DM. Formation of DNA replication structures in herpes virus-infected cells requires a viral DNA-binding protein. *Cell* 1988;55:857–868.
65. De Bruyn Kops A, Knipe DM. Preexisting nuclear architecture defines the intranuclear location of herpesvirus DNA replication structures. *J Virol* 1994;68:3512–3516.
66. Doms RW, Keller DS, Helenius A, Balch WE. Role for adenosine triphosphate in regulating the assembly and transport of vesicular stomatitis virus G protein trimers. *J Cell Biol* 1987;105:1957–1959.
67. Doms RW, Lamb RA, Rose JK, Helenius A. Folding and assembly of viral membrane proteins. *Virology* 1993;193:545–562.
68. Dulbecco R, Harturell LN, Vogt M. Induction of cellular DNA synthesis by polyoma virus. *Proc Natl Acad Sci USA* 1965;53:403–408.
69. Dunigan DD, Baird S, Lucas-Lenard J. Lack of correlation between the accumulation of plus-strand leader RNA and the inhibition of protein and RNA synthesis in vesicular stomatitis infected mouse L cells. *Virology* 1986;150:231–246.
70. Egberts E, Hackett P, Traub P. Alteration of the intracellular energetic and ionic conditions by mengovirus infection of Ehrlich ascites tumor cells and its influence on protein synthesis in the midphase of infection. *J Virol* 1977;22:591–597.
71. Ehrenfeld E, ed. Translational regulation. *Semin Virology* 1993;4:1–268.
72. Enders JF. Cytopathology of virus infections. *Ann Rev Microbiol* 1954;8:473–502.
73. Enders JF, Weller TH, Robbins FC. Cultivation of the Lansing strain of poliomyelitis virus in cultures of various human embryonic tissues. *Science* 1949;109:85–87.
74. Etchison D, Milburn SC, Edery I, Sonenberg N, Hershey JWB. Inhibition of HeLa cell protein synthesis following poliovirus infection correlates with the proteolysis of a 220,000 dalton polypeptide associated with eukaryotic initiation factor 3 and a cap binding protein complex. *J Biol Chem* 1982;257:14806–14810.
75. Everett SF, Ginsberg HS. A toxinlike material separable from type 5 adenovirus particles. *Virology* 1958;6:770–771.
76. Ewen ME, Ludlow JW, Marsilio E, et al. An N-terminal transformation-governing sequence of SV40 large T antigen contributes to the binding of both p110Rb and a second cellular protein, p120. *Cell* 1989;58:257–267.
77. Fan DP, Sefton BM. The entry into host cells of Sindbis virus, vesicular stomatitis virus and Sendai virus. *Cell* 1978;15:985–992.
78. Feinberg MB, Jarrett RF, Aldovini A, Gallo RC, Wong-Staal F. HTLV-II expression and production involve complex regulation at the levels of splicing and translation of viral RNA. *Cell* 1986;46:807–817.
79. Fenwick ML, McMenamin MM. Early virion-associated suppression of cellular protein synthesis by herpes simplex virus is accompanied by inactivation of mRNA. *J Gen Virol* 1984;65:1225–1228.
80. Fenwick ML, Walker MJ. Suppression of the synthesis of cellular macromolecules by HSV. *J Gen Virol* 1978;41:37–51.
81. Fernandez-Puentes C, Carrasco L. Viral infection permeabilizes mammalian cells to protein toxins. *Cell* 1980;20:769–775.
82. Fingeroth JD, Weis JJ, Tedder TF, Strominger JL, Biro PA, Fearon DT. Epstein-Barr virus receptor of human B lymphocytes is the C3d receptor CR2. *Proc Natl Acad Sci USA* 1984;81:4510–4516.
83. Frade R, Barel M, Ehlin-Henriksson B, Klein G. gp140, the C3d receptor of human B lymphocytes is also the Epstein-Barr virus receptor. *Proc Natl Acad Sci USA* 1985;82:1490–1493.
84. Fraenkel-Conrat H, Wagner RR. Viral cytopathology. *Comprehensive virology*. Vol 19. New York: Plenum; 1984.
85. Fredman JN, Engler JA. Adenovirus precursor to terminal protein interacts with the nuclear matrix *in vivo* and *in vitro*. *J Virol* 1993;67:3384–3395.
86. Freistadt M, Kaplan G, Racaniello V. Heterogeneous expression of poliovirus receptor-related proteins in human cells and tissues. *Mol Cell Biol* 1990;10:5700–5706.
87. Fujinami RS, Oldstone MBA. Amino acid homology between the encephalitogenic site of myelin basic protein and virus: mechanism for autoimmunity. *Science* 1985;230:1043–1045.
88. Garcea R, Benjamin T. Host range transforming gene of polyoma virus plays a role in virus assembly. *Proc Natl Acad Sci USA* 1983;80:3613–3617.
89. Garry RF, Bishop JM, Parker S, Westbrook K, Lewis G, Waite M. Na^+ and K^+ concentrations and the regulation of protein synthesis in Sindbis virus-infected chick cells. *Virology* 1979;96:108–120.
90. Garry RF, Westbrook K, Waite MRF. Differential effects of ouabain on host- and Sindbis virus-specified protein synthesis. *Virology* 1979;99:179–182.
91. Gaynor RB, Feldman LT, Berk AJ. Transcription of class III genes activated by viral immediate early proteins. *Science* 1985;230:447–450.
92. Gerster T, Roeder RG. A herpesvirus *trans*-activating protein interacts with transcription factor OTF-1 and other cellular proteins. *Proc Natl Acad Sci USA* 1988;85:6347–6351.
93. Gething MJ, McCammon K, Sambrook J. Expression of wild-type and mutant forms of influenza hemagglutinin: the role of folding in intracellular transport. *Cell* 1986;46:939–950.
94. Gething MJ, Sambrook J. Cell surface expression of influenza virus hemagglutinin from a cloned DNA copy of the RNA gene. *Nature* 1981;293:620–625.
95. Goodrich LD, Schaffer PA, Dorsky DI, Crumpacker CS, Parris DS. Localization of the herpes simplex virus type 1 65-kilodalton DNA-binding protein and DNA polymerase in the presence and absence of viral DNA synthesis. *J Virol* 1990;64:5738–5749.
96. Gordon J, Walker L, Guy G, Rowe M, Rickinson A. Control of human B-lymphocyte replication. II. Transforming Epstein-Barr virus exploits three distinct viral signals to undermine three separate control points in B-cell growth. *Immunology* 1986;58:591–595.
97. Gottschalk A. Chemistry of virus receptors. In: Burnet FM, Stanley WM, eds. *The viruses*. Vol III. New York: Academic; 1959;51–61.
98. Graessman A, Graessman M, Tjian R, Topp WC. Simian virus 40 small-t protein is required for loss of actin cable networks in rat cells. *J Virol* 1980;33:1182–1191.
99. Gray MA, Micklem KJ, Pasternak CA. Protein synthesis in cells infected with Semliki Forest virus is not controlled by intracellular cation changes. *Eur J Biochem* 1983;135:299–302.
100. Greber UF, Singh I, Helenius A. Mechanisms of virus uncoating. *Trends Microbiol* 1994;53:52–56.
101. Greber UF, Willetts M, Webster P, Helenius A. Stepwise dismantling of adenovirus 2 during entry into cells. *Cell* 1993;75:477–486.
102. Green MR, Treisman R, Maniatis T. Transcriptional activation of cloned human B-globin genes by viral immediate early genes. *Cell* 1983;35:137–148.
103. Greve JM, Davis G, Meyer AM, et al. The major rhinovirus receptor is ICAM-1. *Cell* 1989;56:839–847.
104. Grinnell BW, Wagner RR. Comparative inhibition of cellular transcription by vesicular stomatitis virus serotypes New Jersey and Indiana: role of each viral leader RNA. *J Virol* 1983;48:88–101.
105. Grinnell BW, Wagner RR. Inhibition of DNA-dependent transcription by the leader RNA of vesicular stomatitis virus: role of specific nucleotide sequences and cell protein binding. *Mol Cell Biol* 1985;5:2502–2513.
106. Grinnell BW, Wagner RR. Nucleotide sequence and secondary structure of VSV leader RNA and homologous DNA involved in inhibition of DNA-dependent transcription. *Cell* 1984;36:533–543.
107. Gruda MC, Zabolotny JM, Xiao JH, Davidson I, Alurine JC. Transcriptional activation by simian virus 40 large T antigen: interactions with multiple components of the transcription complex. *Mol Cell Biol* 1993;13:961–969.
108. Halbert DN, Cutt JR, Shenk T. Adenovirus early region 4 encodes functions required for efficient DNA replication, late gene expression, and host cell shut-off. *J Virol* 1985;56:250–257.
109. Hartley JW, Rowe WP, Huebner RJ. Host range restrictions of murine

leukemia viruses in mouse embryo cell cultures. *J Virol* 1970; 5:221–225.
110. Hearing P, Shenk T. The adenovirus type 5 E1A transcriptional control region contains a duplicated enhancer element. *Cell* 1983;33:695–703.
111. Heeg U, Haase W, Brauer D, Falke D. Microtubules and microfilaments in HSV-infected cells. *Arch Virol* 1981;70:233–246.
112. Hershey JWB. Introduction to translational initiation factors and their regulation by phosphorylation. *Semin Virol* 1993;4:201–208.
113. Herz C, Stavnezer E, Krug RM, Gurney T. Influenza virus, an RNA virus, synthesizes its messenger RNA in the nucleus of infected cells. *Cell* 1981;26:391–400.
114. Hewlett MJ, Rose JK, Baltimore D. 5′ Terminal structure of poliovirus polyribosomal RNA is pUp. *Proc Natl Acad Sci USA* 1976;73:327–330.
115. Hiller G, Jungwirth C, Weber K. Fluorescence microscopical analysis of the life cycle of vaccinia virus in chick embryo fibroblasts. *Exp Cell Res* 1981;132:81–87.
116. Hoeffler WK, Kovelman R, Roeder RG. Activation of transcription factor IIIC by the adenovirus E1A protein. *Cell* 1988;53:907–920.
117. Hoeffler WK, Roeder RG. Enhancement of RNA polymerase III transcription by the E1A product of adenovirus. *Cell* 1985;41:955–963.
118. Holland JJ, McLaren LC, Syverton JT. The mammalian cell-virus relationship. IV. Infection of naturally insusceptible cells with enterovirus nucleic acid. *J Exp Med* 1959;110:65–80.
119. Horikoshi N, Maguire K, Kralli N, Maldonado E, Reinberg D, Weinmann R. Direct interaction between adenovirus E1A protein and the TATA box binding transcription factor IID. *Proc Natl Acad Sci USA* 1991;88:5124–5128.
120. Horwitz MJ, Brayton C, Baum SG. Synthesis of type 2 adenovirus DNA in the presence of cycloheximide. *J Virol* 1973;11:544–551.
121. Howard JM, Eckert BS, Bourguignon LYW. Comparison of cytoskeletal organization in canine distemper virus-infected and uninfected cells. *J Gen Virol* 1983;64:2379–2385.
122. Howley PM. The molecular biology of papilloma virus transformation. *Am J Pathol* 1983;113:413–421.
123. Hsu MC, Scheid A, Choppin PW. Activation of the Sendai virus fusion protein (F) involves a conformational change with exposure of a new hydrophobic region. *J Biol Chem* 1981;256:3557–3563.
124. Huang J, Schneider RJ. Adenovirus inhibition of cellular protein synthesis involves inactivation of cap binding protein. *Cell* 1991;65:271–280.
125. Hunter T. Proteins phosphorylated by RSV transforming function. *Cell* 1980;22:647–648.
126. Hutt-Fletcher LM. Synergistic activation of cells by Epstein-Barr virus and B-cell growth factor. *J Virol* 1987;61:774–781.
127. Imperiale MJ, Feldman LT, Nevins JR. Activation of gene expression by adenovirus and herpesvirus regulatory genes acting in *trans* and by a *cis*-acting adenovirus enhancer element. *Cell* 1983;35:127–136.
128. Inglis SC. Inhibition of host protein synthesis and degradation of cellular mRNAs during infection by influenza and herpes simplex virus. *Mol Cell Biol* 1982;2:1644–1648.
129. Jacks T, Madhani HD, Masiarz FR, Varmus HE. Signals for ribosomal frameshifting in the Rous sarcoma *gag-pol* region. *Cell* 1988;55:447–458.
130. Jacks T, Varmus HE. Expression of the Rous sarcoma virus by ribosomal frameshifting. *Science* 1985;230:1237–1242.
131. Jones LV, Compans RW, Davis AR, Bos TJ, Nayak DP. Surface expression of influenza virus neuraminidase, an amino-terminally anchored viral membrane glycoprotein, in polarized epithelial cells. *Mol Cell Biol* 1985;5:2181–2189.
132. Jones N, Shenk T. An adenovirus type 5 early gene function regulates expression of other early viral genes. *Proc Natl Acad Sci USA* 1979;76:3665–3669.
133. Kalderon D, Roberts BL, Richardson WD, Smith AE. A short amino acid sequence able to specify nuclear location. *Cell* 1984;39:499–509.
134. Kaplan AS. A brief review of the biochemistry of herpesvirus–host cell interaction. *Cancer Res* 1973;33:1393–1398.
135. Kasamatsu H, Nehorayan A. VP1 affects intracellular localization of VP3 polypeptide during simian virus 40 infection. *Proc Natl Acad Sci USA* 1979;76:2808–2812.
136. Kates JR, McAuslan B. Messenger RNA synthesis by a "coated" viral genome. *Proc Natl Acad Sci USA* 1967;57:314–320.
137. Katze M, DeCorato D, Safer B, Galabru J, Hovanessian A. Adenovirus VAI RNA complexes with the 68,000 Mr protein kinase to regulate its autophosphorylation and activity. *EMBO J* 1987;6:689–697.
138. Katze MG, Lara J, Wambach M. Nontranslated cellular mRNAs are associated with the cytoskeletal framework in influenza virus or adenovirus infected cells. *Virology* 1989;169:312–322.
139. Keranen S, Kaarianen L. Proteins synthesized by Semliki Forest virus and its 16 temperature-sensitive mutants. *J Virol* 1975;16:388–396.
140. Kerr IM, Brown RE. pppA2′p5′A2′p5′A: an inhibitor of protein synthesis synthesized with an enzyme fraction from interferon-treated cells. *Proc Natl Acad Sci USA* 1978;75:256–260.
141. Kitajewski J, Schneider RJ, Safer B, et al. Adenovirus VA1 RNA antagonizes the antiviral action of interferon by preventing activation of the interferon-induced eIF-2 kinase. *Cell* 1986;45:195–200.
142. Klatzmann D, Champagne E, Chamaret S, et al. T-lymphocyte T4 molecule behaves as the receptor for human retrovirus LAV. *Nature* 1984;312:767–768.
143. Klenk HD, Garten W. Host cell proteases controlling virus pathogenicity. *Trends Microbiol* 1994;2:39–43.
144. Knight DM, Flomerheit FA, Ghrayeb J. Expression of the art/trs protein of HIV and study of its role in viral envelope synthesis. *Science* 1987;236:837–840.
145. Knipe DM. The role of viral and cellular nuclear proteins in herpes simplex virus replication. *Adv Virus Res* 1989;37:85–123.
146. Knipe DM, Baltimore D, Lodish HF. Separate pathways of maturation of the major structural proteins of vesicular stomatitis virus. *J Virol* 1977;21:1128–1139.
147. Knipe DM, Senechek D, Rice SA, Smith J. Stages in the nuclear association of the herpes simplex virus transcriptional activator protein, ICP4. *J Virol* 1987;61:276–284.
148. Kopchich JJ, Jamjoon GA, Watson KF, Arlinghaus RB. Biosynthesis of reverse transcriptase from a Rauscher murine leukemia virus by synthesis and cleavage of a *gag-pol* read through viral precursor polyprotein. *Proc Natl Acad Sci USA* 1978;75:2016–2020.
149. Kovesdi I, Reichel R, Nevins JR. E1A transcription induction: enhanced binding of a factor to upstream promoter sequences. *Science* 1986;231:719–722.
150. Kovesdi I, Reichel R, Nevins JR. Identification of a cellular transcription factor involved in E1A *trans*-activation. *Cell* 1986;45:219–228.
151. Kozak M. How do eucaryotic ribosomes select initiation regions in messenger RNA? *Cell* 1978;15:1109–1123.
152. Kozak M. Regulation of protein synthesis in virus-infected animal cells. *Adv Virus Res* 1986;31:229–292.
153. Krausslich H-G, Wimmer E. Viral proteinases. *Annu Rev Biochem* 1988;57:701–754.
154. Kreis TE, Lodish HF. Oligomerization is essential for transport of vesicular stomatitis viral glycoprotein to the cell surface. *Cell* 1986;46:929–937.
155. Krikorian CR, Read GS. *in vitro* mRNA degradation system to study the virion host shutoff function of herpes simplex virus. *J Virol* 1991;65:112–122.
156. Kristie TM, Sharp PA. Purification of the cellular C1 factor required for the stable recognition of the oct-1 homeodomain by the herpes simplex virus α trans-induction factor (VP16). *J Biol Chem* 1993;268:6525–6534.
157. Kurilla MG, Keene JD. The leader RNA of vesicular stomatitis virus is bound by a cellular protein reactive with anti-La Lupus antibodies. *Cell* 1983;34:837–845.
158. Kurilla MG, Piwnica-Worms H, Keene JD. Rapid and transient localization of the leader RNA of vesicular stomatitis virus in the nuclei of infected cells. *Proc Natl Acad Sci USA* 1982;79:5240–5244.
159. Kwong AD, Frenkel N. Herpes simplex virus contains a function(s) that destabilizes both host and viral mRNAs. *Proc Natl Acad Sci USA* 1987;84:1926–1930.
160. Kwong AD, Kruper JA, Frenkel N. Herpes simplex virus virion host shutoff function. *J Virol* 1988;62:912–921.
161. Lacal JC, Carrasco L. Relationship between membrane integrity and the inhibition of host translation in virus-infected mammalian cells. *Eur J Biochem* 1982;127:359–366.
162. Lamb RA. Paramyxovirus fusion: a hypothesis for changes. *Virology* 1993;197:1–11.
163. Lamb RA, Choppin PW, Chanock RM, Lai C-J. Mapping of the two overlapping genes for polypeptides NS1 and NS2 on RNA segment 8 of influenza virus. *Proc Natl Acad Sci USA* 1980;77:1857–1861.
164. Lamb RA, Lai C-J, Choppin PW. Sequences of mRNAs derived from genomic RNA segment 7 of influenza virus: colinear and interrupted mRNAs code for overlapping proteins. *Proc Natl Acad Sci USA* 1981;78:4170–4174.

165. Lee KAW, Hai TY, Siva Raman L, et al. A cellular protein, activating transcription factor, activates transcription of multiple E1A-inducible adenovirus early promoters. *Proc Natl Acad Sci USA* 1987;84:8355–8359.
166. Lee WS, Kao CC, Bryant GO, Liu X, Berk AJ. Adenovirus E1A activation domain binds the basic repeat in the TATA box transcription factor. *Cell* 1991;67:365–376.
167. Lenk R, Penman S. The cytoskeletal framework and poliovirus metabolism. *Cell* 1979;16:289–301.
168. Lenk R, Ransom L, Kaufmann V, Penman S. A cytoskeletal structure with associated polyribosomes obtained from HeLa cells. *Cell* 1977;10:67–78.
169. Lentz TL, Burrage TG, Smith AL, Crick J, Tignor GH. Is the acetylcholine receptor a rabies virus receptor? *Science* 1982;215:182–184.
170. Levinson AD, Oppermann H, Levintow L, Varmus HE, Bishop JM. Evidence that the transforming gene of avian sarcoma virus encodes a protein kinase associated with a phosphoprotein. *Cell* 1978;15: 561–572.
171. Li JJ, Kelly TJ. Simian virus 40 DNA replication *in vitro*. *Proc Natl Acad Sci USA* 1984;81:6973–6977.
172. Lilly F, Pincus T. Genetic control of murine viral leukemogenesis. *Adv Cancer Res* 1973;17:231–277.
173. Liptak L, Weller SK, Knipe DM. Unpublished results.
174. Lloyd RE, Toyoda H, Etchison D, Wimmer E, Ehrenfeld E. Cleavage of the cap binding protein complex polypeptide p220 is not effected by the second poliovirus protease 2A. *Virology* 1986;150:299–303.
175. Lodish HF. Translational control of protein synthesis. *Annu Rev Biochem* 1976;45:39–72.
176. Lodish HF, Porter M. Vesicular stomatitis virus mRNA and inhibition of translation of cellular mRNA. Is there a P function in vesicular stomatitis virus? *J Virol* 1981;38:504–517.
177. Logan J, Shenk T. Adenovirus tripartite leader sequence enhances translation of mRNAs late after infection. *Proc Natl Acad Sci USA* 1984;81:3655–3659.
178. Luftig RB, Weihing RR. Adenovirus binds to rat brain microtubules *in vitro*. *J Virol* 1975;16:696–706.
179. Lusky M, Botchan MR. Characterization of the bovine papilloma virus plasmid maintenance sequences. *Cell* 1984;36:391–401.
180. Macejak DJ, Sarnow P. Internal initiation of translation mediated by the 5′ leader of a cellular mRNA. *Nature* 1991;353:90–94.
181. Maddon PJ, Dalgleish AG, McDougal JS, Clapham PR, Weiss RA, Axel R. The T4 gene encodes the AIDS virus receptor and is expressed in the immune system and the brain. *Cell* 1986;47:333–348.
182. Malim MH, Hauber J, Fenwick R, Cullen BR. Immunodeficiency virus *rev trans*-activator modulates the expression of the viral regulatory gene. *Nature* 1988;335:181–183.
183. Malim MH, Hauber J, Le S-Y, Maizel JV, Cullen BR. The HIV *rev trans*-activator acts through a structured target sequence to activate nuclear export of unspliced viral mRNA. *Nature* 1989;338:254–257.
184. Markwell MAK, Svennerholm L, Paulson JC. Specific gangliosides function as host cell receptors for Sendai virus. *Proc Natl Acad Sci USA* 1981;78:5406–5410.
185. Marsh M, Helenius A. Adsorptive endocytosis of Semliki Forest virus. *J Mol Biol* 1980;142:439–454.
186. Martin K, Helenius A. Nuclear transport of influenza virus ribonucleoproteins: the viral matrix protein (M1) promote export and inhibits import. *Cell* 1991;67:117–130.
187. Martin TE, Barghusen SC, Leser GP, Spear PG. Redistribution of nuclear ribonucleoprotein antigens during herpes simplex virus infection. *J Cell Biol* 1987;105:2069–2082.
188. Martinez-Palomo A. Ultrastructural study of the replication of human adenovirus type 12 in cultured cells. *Pathol Microbiol* 1968;31:147–164.
189. Martinez-Palomo A, Granboulan N. Electron microscopy of adenovirus 12 replication. II. High-resolution autoradiography of infected KB cells labeled with tritiated thymidine. *J Virol* 1967;1:1010–1018.
190. Marvaldi J, Sekellick M, Marcus P, Lucas-Lenard J. Inhibition of mouse L cell protein synthesis by ultraviolet-irradiated vesicular stomatitis virus requires viral transcription. *Virology* 1978;84:127–133.
191. McCready SJ, Godwin J, Mason DW, Brazell IA, Cook PR. DNA is replicated at the nuclear cage. *J Cell Sci* 1980;46:365–386.
192. McKnight JLC, Kristie TM, Roizman B. Binding of the virion protein mediating gene induction in herpes simplex virus 1-infected cells to its *cis* site requires cellular proteins. *Proc Natl Acad Sci USA* 1987;84:7061–7065.
193. McLaren LC, Holland JJ, Syverton JT. The mammalian cell–virus relationship. I. Attachment of poliovirus to cultivated cells of primate and nonprimate origin. *J Exp Med* 1959;109:475–485.
194. McQueen NL, Nayak DP, Stephens EB, Compans RW. Polarized expression of a chimeric protein in which the transmembrane and cytoplasmic domains have been replaced by those of vesicular stomatitis virus G protein. *Proc Natl Acad Sci USA* 1986;83:9318–9322.
195. Meerovitch K, Sonenberg N. Internal initiation of picornavirus RNA translation. *Semin Virol* 1993;4:217–227.
196. Mellits KH, Mathews MB. Effects of mutations in stem and loop regions on the structure of adenovirus VA RNA. *EMBO J* 1988;7:2849–2859.
197. Mendelsohn C, Johnson B, Kionetti BA, Nobis P, Wimmer E, Racaniello VR. Transformation of a human poliovirus receptor gene into mouse cells. *Proc Natl Acad Sci USA* 1986;83:7845–7849.
198. Mendelsohn CL, Wimmer E, Racaniello VR. Cellular receptor for poliovirus: molecular cloning, nucleotide sequence, and expression of a new member of the immunoglobulin super family. *Cell* 1989;56:855–865.
199. Meyer RK, Burger MM, Tschannen R, Schafer R. Actin filament bundles in vaccinia virus infected fibroblasts. *Arch Virol* 1981;67:11–18.
200. Miyamoto NG, Samuel CE. Mechanism of interferon action: interferon-mediated inhibition of reovirus mRNA translation in the absence of detectable mRNA degradation but in the presence of protein phosphorylation. *Virology* 1980;107:461–475.
201. Morrison TG, Lodish HF. Site of synthesis of membrane and nonmembrane proteins of vesicular stomatitis virus. *J Biol Chem* 1975;250:6955–6962.
202. Murakami Y, Wobbe CR, Weissbach L, Dean FB, Hurwitz J. Role of DNA polymerase and DNA primase in simian virus 40 DNA replication *in vitro*. *Proc Natl Acad Sci USA* 1986;83:2869–2873.
203. Murti KG, Goorha R. Interaction of frog virus-3 with the cytoskeleton. I. Altered organization of microtubules, intermediate filaments, and microfilaments. *J Cell Biol* 1983;96:1248–1257.
204. Naib ZM, Clepper AS, Elliott SR. Exfoliative cytology as an aid in diagnosis of ophthalmic lesions. *Acta Cytol* 1967;11:295–303.
205. Nair CN, Stowers JW, Singfield B. Guanidine-sensitive Na^+ accumulation by poliovirus-infected HeLa cells. *J Virol* 1979;31:184–189.
206. Neilan J, Lu Z, Alfonso C, Kutish G, Sussman M, Rock D. An african swine fever virus gene with similarity to the proto-oncogene bcl-2 and the Epstein-Barr virus gene BHRF1. *J Virol* 1993;67:4391–4394.
207. Nemerow GR, Wolfert R, McNaughton ME, Cooper NR. Identification and characterization of the Epstein-Barr virus receptor on human B lymphocytes and its relationship to the C3d complement receptor (CR2). *J Virol* 1985;55:347–351.
208. Nevins JR. Control of cellular and viral transcription during adenovirus infection. *CRC Crit Rev Biochem* 1986;19:307–322.
209. Nevins JR. E2F: a link between the Rb tumor suppressor protein and viral encoproteins. *Science* 1992;258:424-429.
210. Nevins JR. Induction of the synthesis of a 70,000 dalton mammalian heat shock protein by the adenovirus Ela gene product. *Cell* 1982;29:913–919.
211. Newton AA. The involvement of nuclear membrane in the synthesis of herpes-type viruses. In: Biggs PM, de The G, Payne LN, eds. *Oncogenesis and herpes viruses I*. No. 24. Lyon: International Agency for Research on Cancer Scientific Publications; 1972:489.
212. Nishioka Y, Silverstein S. Degradation of cellular mRNA during infection by HSV. *Proc Natl Acad Sci USA* 1977;74:2370–2374.
213. Nishioka Y, Silverstein S. Requirement of protein synthesis for the degradation of host mRNA in Friend erythroleukemia cells infected with HSV-1. *J Virol* 1978;27:619–627.
214. Nomoto A, Lee YF, Wimmer E. The 5′ end of poliovirus mRNA is not capped with $m^7G(5')pppG(5')$-Np. *Proc Natl Acad Sci USA* 1976;73:375–380.
215. Norkin LC, Ouellette J. Cell killing by simian virus 40: variation in the pattern of lysosomal enzyme release, cellular enzyme release, and cell death during productive infection of normal and simian virus 40-transformed simian cell lines. *J Virol* 1976;18:48–57.
216. Nuss DL, Koch G. Differential inhibition of vesicular stomatitis polypeptide synthesis by hypertonic initiation block. *J Virol* 1976;17:283–286.
217. Nuss DL, Oppermann H, Koch G. Selective blockage of initiation of host protein synthesis in RNA virus-infected cells. *Proc Natl Acad Sci USA* 1975;72:1258–1262.
218. Oh SK, Scott MP, Sarnow P. Homeotic gene *Antennapedia* mRNA

contains 5′ noncoding sequences that confer translational initiation by internal ribosome binding. *Genes Dev* 1992;6:1643–1653.
219. Olivo PD, Nelson NJ, Challberg MD. Herpes simplex virus type 1 gene products required for DNA replication: identification and overexpression. *J Virol* 1989;63:196–204.
220. O'Malley R, Duncan R, Hershey J, Mathews M. Modification of protein synthesis initiation factors and the shut-off of host protein synthesis in adenovirus infected cells. *Virology* 1989;168:112–118.
221. Ornelles D, Shenk T. Localization of the adenovirus early region 1B 55-kilodalton protein during lytic infection: association with nuclear viral inclusions requires the early region 4 34-kilodalton protein. *J Virol* 1991;65:424–439.
222. Oroskar AA, Read GS. Control of mRNA stability by the virion host shutoff function of herpes simplex virus. *J Virol* 1989;63:1897–1906.
223. Oroskar AA, Read GS. A mutant of herpes simplex virus exhibits increased stability of immediate early (alpha) mRNAs. *J Virol* 1987;61:604–606.
224. Paez E, Estaban M. Nature and mode of action of vaccinia virus products that block activation of the interferon-mediated ppp(A2′p)$_n$A-synthetase. *Virology* 1984;134:29–39.
225. Paez E, Estaban M. Resistance of vaccinia virus to interferon is related to an interference phenomenon between the virus and the interferon system. *Virology* 1984;134:12–28.
226. Panem S. Cell cycle-dependent inhibition of Kirsten murine sarcoma–leukemia virus release by cytochalasin B. *Virology* 1977;76:146–151.
227. Panganiban AT. Retroviral *gag* gene amber codon suppression is caused by an intrinsic *cis*-acting component of the viral mRNA. *J Virol* 1988;62:3574–3580.
228. Paoletti E, Grady LJ. Transcriptional complexity of vaccinia virus *in vivo* and *in vitro*. *J Virol* 1977;23:608–615.
229. Pastan I, Seth P, Fitzgerald D, Willingham M. Adenovirus entry into cells: some new observations on an old problem. In: Notkins A, Olstone MBA, eds. *Concepts in viral pathogenesis*. New York: Springer-Verlag; 1987:141–146.
230. Patience C, McKnight A, Clapham PR, Boyd MT, Weiss RA. CD26 and HIV fusion. *Science* 1994;264:1159–1160.
231. Pelham H. Heat shock proteins: coming in from the cold. *Nature* 1988;332:776–777.
232. Pelletier J, Sonenberg N. Internal initiation of translation of eukaryotic mRNA directed by a sequence derived from poliovirus RNA. *Nature* 1988;334:320–325.
233. Pereira HG. A protein factor responsible for the early cytopathic effect of adenoviruses. *Virology* 1958;6:601–611.
234. Phelan A, Carmo-Fonseca M, McLaughlin J, Lamond AI, Clements JB. A herpes simplex virus type 1 immediate-early gene product, IE63, regulates small nuclear ribonucleoprotein distribution. *Proc Natl Acad Sci USA* 1993;90:9056–9060.
235. Phillipson L, Anderson P, Olshevsky U, Weinberg R, Baltimore D, Gesteland R. Translation of murine leukemia and sarcoma virus RNAs in nuclease-treated reticulocyte extracts: enhancement of *gag-pol* polypeptide with yeast suppression tRNA. *Cell* 1978;13:189–199.
236. Pilder S, Logan J, ShenkT. Deletion of the gene encoding the adenovirus type 5 E1B-21K polypeptide leads to degradation of viral and host cell DNA. *J Virol* 1984;52:664–671.
237. Pilder S, Moore M, Logan J, Shenk T. The adenovirus E1B-55K transforming polypeptide modulates transport or cytoplasmic stabilization of viral and host cell mRNAs. *Mol Cell Biol* 1986;6:470–476.
238. Pinto LH, Holsinger LJ, Lamb RA. Influenza virus M_2 protein has ion channel activity. *Cell* 1992;69:517–528.
239. Planck SR, Mueller GC. DNA chain growth and organization of replicating units in HeLa cells. *Biochemistry* 1977;16:1808–1813.
240. Plotch SJ, Bouloy M, Ulmanen I, Krug RM. A unique cap (m^{7}GpppXm)-dependent influenza virus endonuclease cleaves capped RNAs to generate the primers that initiate viral RNA transcription. *Cell* 1981;23:847–858.
241. Pogo BGT, Dales S. Biogenesis of poxvirus. Further evidence for inhibition of host and virus DNA synthesis by a component of the invading inoculum particle. *Virology* 1974;58:377–386.
242. Pogo BGT, Dales S. Biogenesis of poxvirus. Inactivation of host DNA polymerase by a component of the invading inoculum partide. *Proc Natl Acad Sci USA* 1973;70:1726–1729.
243. Post LE, Mackem S, Roizman B. Regulation of alpha genes of herpes simplex virus: expression of chimeric genes produced by fusion of thymidine kinase with alpha gene promoters. *Cell* 1981; 24:555–566.
244. Preston DM, Frame MC, Campbell MEM. A complex formed between cell components and an HSV structural polypeptide binds to a viral immediate early gene regulatory DNA sequence. *Cell* 1988;52:425–434.
245. Puddington L, Woodgett C, Rose JK. Replacement of the cytoplasmic domain alters sorting of a viral glycoprotein in polarized cells. *Proc Natl Acad Sci USA* 1987;84:2756–2760.
246. Quinlan MP, Chen LB, Knipe DM. The intranuclear location of a herpes simplex virus DNA-binding protein is determined by the status of viral DNA replication. *Cell* 1984;35:857–868.
247. Quinlan MP, Knipe DM.. Nuclear localization of herpes viral proteins: potential role for the cellular framework. *Mol Cell Biol* 1983;3:315–324.
248. Randall RE, Dinwoodie N. Intranuclear localization of herpes simplex virus immediate early proteins: evidence that ICP4 is associated with progeny virus DNA. *J Gen Virol* 1986;67:2163–2177.
249. Rao L, Debbas M, Sabbatini P, Hockenbery D, Korsmeyer S, White E. The adenovirus E1A proteins induce apoptosis, which is inhibited by the E1B 19-kDa and Bcl-2 proteins. *Proc Natl Acad Sci USA* 1992;89:7742–7746.
250. Read GS, Frenkel N. Herpes simplex virus mutants defective in the virion-associated shutoff of host polypeptide synthesis and exhibiting abnormal synthesis of alpha (immediate early) polypeptides. *J Virol* 1983;46:498–512.
251. Reich N. Sarnow P, Duprey E, Levine AJ. Monoclonal antibodies which recognize native and denatured forms of the adenovirus DNA-binding protein. *Virology* 1983;128:480–484.
252. Reich PR, Rose J, Forget B, Weissman SM. RNA of low molecular weight in KB cells infected with adenovirus type 2. *J Mol Biol* 1966;17:428–439.
253. Reichel PA, Merrick WC, Siekierka J, Mathews MB. Regulation of a protein synthesis initiation factor by adenovirus VA-RNA$_I$. *Nature* 1985;313:196–200.
254. Resnick J, Shenk T. Simian virus 40 agnoprotein facilitates normal nuclear location of the major capsid polypeptide and cell-to-cell spread of the virus. *J Virol* 1986;60:1098–1106.
255. Rice AP, Roberts BE. Vaccinia virus induces cellular mRNA degradation. *J Virol* 1983;47:529–539.
256. Rice S, Knipe DM. Unpublished results.
257. Rice AP, Roberts WK, Kerr I. Interferon-mediated, double-stranded RNA-dependent protein kinase is inhibited in extracts from vaccinia virus-infected cells. *J Virol* 1984;50:229–236.
258. Rindler MJ, Ivanov IE, Plesken H, Rodriquez-Boulan EJ, Sabatini DD. Viral glycoproteins destined for apical or basolateral membrane domains traverse the same Golgi apparatus during their intracellular transport in Madin-Darby canine kidney cells. *J Cell Biol* 1984;98:1304–1319.
259. Rinke J, Steitz JA. Precursor molecules of human ES ribosomal RNA and tRNAs are bound by a cellular protein reactive with anti-La Lupus antibodies. *Cell* 1982;29:149–159.
260. Roizman B, Roane PHJ. The multiplication of HSV. II. The relation between protein synthesis and the duplication of viral DNA in infected HEp-2 cells. *Virology* 1964;22:262–269.
261. Rose JK, Bergmann JE. Expression from cloned cDNA of cell surface and secreted forms of the glycoprotein of vesicular stomatitis virus in eukaryotic cells. *Cell* 1982;30:753–762.
262. Rose JK, Trachsel H, Leong D, Baltimore D. Inhibition of translation by poliovirus: inactivation of a specific initiation factor. *Proc Natl Acad Sci USA* 1978;75:2732—2736.
263. Roth MG, Compans RW, Giusk L, et al. Influenza virus hemagglutinin expression is polarized in cells infected with recombinants SV40 viruses carrying cloned hemagglutinin DNA. *Cell* 1983;33:435–443.
264. Roth MG, Gundersen D, Patil N, Rodriquez-Boulan E. The large external domain is sufficient for the correct sorting of secreted or chimeric influenza virus hemagglutinins in polarized monkey kidney cells. *J Cell Biol* 1987;104:769–782.
265. Rothman JE, Lodish HF. Synchronized trans-membrane insertion and glycosylation of a nascent membrane protein. *Nature* 1977:269:755–778.
266. Rubin H, Figge J, Bladon MT, et al. Role of small t antigen in the acute transforming activity of SV40. *Cell* 1982;30:409–480.
267. Rutter G, Mannweiler K. Alterations of actin-containing structures in BHK-21 cells infected with Newcastle disease virus and vesicular stomatitis virus. *J Gen Virol* 1977;37:233–242.
268. Saborio JL, Pong S-S, Koch G. Selective and reversible inhibition of protein synthesis in mammalian cells. *J Mol Biol* 1974;85:195–211.

269. Sadaie MR, Benter T, Wong-Staal F. Site-directed mutagenesis of two *trans*-regulatory genes (tat-III, tis) of HIV-1. *Science* 1988;239:910–914.
270. Sarnow P, Ho Y-S, Williams J, Levine A. Adenovirus E1b-58kd tumor antigen and SV40 large tumor antigen are physically associated with the same 54kd cellular protein in transformed cells. *Cell* 1982; 28:387–394.
271. Scheid A, Choppin PW. Protease activation mutants of Sendai virus. Activation of biological properties by specific proteases. *Virology* 1976;69:265–277.
272. Schek N, Bachenheimer SL. Degradation of cellular mRNAs induced by a virion-associated factor during herpes simplex virus infection of Vero cells. *J Virol* 1985;55:601–610.
273. Schneider RJ, Safer B, Munemitsu SM, Samuel CE, Shenk T. Adenovirus VAI RNA prevents phosphorylation of the eukaryotic initiation factor 2 subunit subsequent to infection. *Proc Natl Acad Sci USA* 1985;82:4321–4325.
274. Schneider RJ, Shenk T. Impact of virus infection on host cell protein synthesis. *Annu Rev Biochem* 1987;56:317–332.
275. Schneider RJ, Weinberger C, Shenk T. Adenovirus VAI RNA facilitates the initiation of translation in virus-infected cells. *Cell* 1984;37:291–298.
276. Schnitzlein WM, O'Banion MK, Poirot MK, Reichmann ME. Effect of intracellular vesicular stomatitis virus mRNA concentration on the inhibition of host cell protein synthesis. *J Virol* 1983;45:206–214.
277. Schwartz J, Roizman B. Similarities and differences in the development of laboratory strains and freshly isolated strains of herpes simplex virus in HEp-2 cells: electron microscopy. *J Virol* 1969;4:879–889.
278. Sharma S, Rodgers L, Brandsma J, Gething MJ, Sambrook J. SV40 T antigen and the exocytic pathway. *EMBO J* 1985;4:1479–1489.
279. Sharpe AH, Chen LB, Fields BN. The interaction of mammalian reoviruses with the cytoskeleton of monkey kidney CV-1 cells. *Virology* 1982;120:399–411.
280. Shatkin AJ. Capping of eukaryotic mRNAs. *Cell* 1976;9:645–650.
281. Shaw AS, Rottier PJM, Rose JK. Evidence for the loop model of signal-sequence insertion into the endoplasmic reticulum. *Proc Natl Acad Sci USA* 1988;85:7592–7596.
282. Shepley M, Sherry B, Weiner H. Monoclonal antibody identification of a 100 kDa membrane protein in HeLa cells and human spinal cord involved in poliovirus attachment. *Proc Natl Acad Sci USA* 1988;85:7743–7747.
283. Shepley MP, Racaniello VR. A monoclonal antibody that blocks poliovirus attachment recognizes the lymphocyte homing receptor CD44. *J Virol* 1994;68:1301–1308.
284. Siekerka J, Mariano TM, Reichel PA, Mathews MB. Translational control by adenovirus: lack of virus-associated RNAI during adenovirus infection results in phosphorylation of initiation factor eIF-2 and inhibition of protein synthesis. *Proc Natl Acad Sci USA* 1985;82:1959–1963.
285. Simons K, Garoff H, Helenius A. How an animal virus gets into and out of cells. *Sci Am* 1982;246:58–66.
286. Singh I, Helenius A. Role of ribosomes in semliki forest virus nucleocapsid uncoating. *J Virol* 1992;66:7049–7058.
287. SivaRaman L, Subramanian S, Thimmapaya B. Identification of a factor in HeLa cells specific for an upstream transcriptional control sequence of an E1A-inducible adenovirus promoter and its relative abundance in infected and uninfected cells. *Proc Natl Acad Sci USA* 1986;83:5914–5918.
288. Slattery E, Gosh N, Samanta H, Lengyel P. Interferon, double-stranded RNA and RNA degradation: activation of an endonuclease by (2′–5′)An. *Proc Natl Acad Sci USA* 1979;76:4778–4782.
289. Smale ST, Tjian R. T-antigen–DNA polymerase complex implicated in simian virus 40 DNA replication. *Mol Cell Biol* 1986;6:4077–4087.
290. Smith RD, Sutherland K. The cytopathology of virus infections. In: Spector S, Lancz GJ, eds. *Clinical virology manual.* New York: Elsevier; 1986;53–69.
291. Sodroski J, Goh WC, Rosen C, Dayton A, Terwilliger E, Haseltine W. A second post-transcriptional trans-activator gene required for HTLV-III replication. *Nature* 1986;321:412–417.
292. Sorenson CM, Hart PA, Ross J. Analysis of herpes simplex virus-induced mRNA destablilizing activity using an *in vitro* mRNA decay system. *Nucl Acids Res* 1991;19:4459–4465.
293. Spendlove RS, Lennette EH, Chin JN, Knight CO. Effect of antimitotic agents on intracellular reovirus antigen. *Cancer Res* 1964;24:1826–1833.
294. Stahl H, Droge P, Knippers R. DNA helicase activity of SV40 large tumor antigen. *EMBO J* 1986;5:1939–1944.
295. Stallcup KC, Raine CS, Fields BN. Cytochalasin B inhibits the maturation of measles virus. *Virology* 1983;124:59–74.
296. Stanners CP, Franceour AM, Lam T. Analysis of VSV mutant with attenuated cyto-pathogenicity in viral function, P, for inhibition of protein synthesis. *Cell* 1977;11:273–281.
297. Staunton DE, Merluzzi VJ, Rothlein R, Barton R, Marlin SD, Springer TA. A cell adhesion molecule, ICAM-1, is the major surface receptor for rhinoviruses. *Cell* 1989;56:849–853.
298. Stephens EB, Compans RW. Assembly of animal viruses at cellular membranes. *Annu Rev Microbiol* 1988;42:489–516.
299. Stephens EB, Compans RW, Earl P, Moss B. Surface expression of viral glycoproteins in polarized epithelial cells using vaccinia virus vectors. *EMBO J* 1986;5:237–245.
300. Stewart PL, Fuller SD, Burnett RM. Difference imaging of adenovirus: bridging the resolution gap between X-ray crystallography and electron microscopy. *EMBO J* 1993;12:2589–2599.
301. Stimac E, Housman D, Huberman JA. Effects of inhibition of protein synthesis on DNA replication in cultured mammalian cells. *J Mol Biol* 1977;115:485–511.
302. Strom T, Frenkel N. Effects of herpes simplex virus on mRNA stability. *J Virol* 1987;61:2198–2207.
303. Sturzenbecker LJ, Nibert M, Furlong D, Fields BN. Intracellular digestion of reovirus particles requires a low pH and is an essential step in the viral infectious cycle. *J Virol* 1987;61:2351.
304. Sugawara K, Gilead Z, Wold WSM, Green M. Immunofluorescence study of the adenovirus type 2 single-stranded DNA binding protein in infected and transformed cells. *J Virol* 1977;22:527–539.
305. Sugrue RJ, Hay AJ. Structural characteristics of the M2 protein of influenza A viruses: evidence that it forms a tetrameric channel. *Virology* 1991;180:617–624.
306. Takemori N, Cladaras C, Bhat B, Conley A, Wold W. cyt gene of adenovirus 2 and 5 is an oncogene for transforming function in early region E1B and encodes the E1B 19,000-molecular-weight polypeptide. *J Virol* 1984;52:793–805.
307. Terwilliger E, Burghoff R, Sia R, Sodroski J, Haseltine W, Rosen C. The art gene product of human immunodeficiency virus is required for replication. *J Virol* 1988;62:655–658.
308. Thimmapaya B, Weinberger C, Schneider RJ, Shenk T. Adenovirus VAI RNA is required for efficient translation of viral mRNAs at late times after infection. *Cell* 1982;31:543–551.
309. Thomas JR, Wagner RR. Inhibition of translation in lysates of mouse L cells infected with vesicular stomatitis virus: presence of a defective ribosome-associated factor. *Biochemistry* 1983; 22:1540–1546.
310. Thomsen DR, Stenberg RM, Goins WF, Stinski MF. Promoter-regulatory region of the major immediate early gene of human cytomegalovirus. *Proc Natl Acad Sci USA* 1984;81:659–663.
311. Tjian R. The binding site of SV40 DNA for a T antigen-related protein. *Cell* 1978;13:165–179.
312. Tjian R, Fey G, Graessmann A. Biological activity of purified simian virus 40 T antigen proteins. *Proc Natl Acad Sci USA* 1978; 75:1279–1283.
313. Tosteson MT, Nibert ML, Fields BN. Ion channels induced in lipid bilayers by subvirion particles of the nonenveloped mammalian reoviruses. *Proc Natl Acad Sci USA* 1993;10549–10552.
314. Triezenberg SJ, Kingsbury RC, McKnight SL. Functional dissection of VP16, the transactivator of herpes simplex virus immediate early gene expression. *Genes Dev* 1988;2:718–729.
315. Tucker SP, Compans RW. Virus infection of polarized epithelial cells. *Adv Virus Res* 1993;42:187–247.
316. Tucker SP, Thornton CL, Wimmer E, Compans RW. Vectorial release of poliovirus from polarized human intestinal epithelial cells. *J Virol* 1993;67:4274–4284.
317. Tyler K, McPhee D, Fields B. Distinct pathways of viral spread in the host determined by reovirus S1 gene segment. *Science* 1986; 233:770–774
318. Tyndall C, LaMantia G, Thacker CM, Favaloro J, Kamen R. A region of the polyoma virus genome between the replication origin and late protein sequences is required in *cis* for both early gene expression and viral DNA replication. *Nuc Acids Res* 1981;9:6231–6250.
319. Valentine RC, Pereira HG. Antigens and structure of the adenovirus. *J Mol Biol* 1965;13:13–20.
320. Van Venrooji WJ, Sillekins PTG, Ekelen CAG, Rienders RJ. On the association of mRNA with the cytoskeleton in uninfected and adenovirus-infected human KB cells. *Exp Cell Res* 1981;135:79–91.

321. Voelkerding K, Klessig DF. Identification of two nuclear subclasses of the adenovirus type5-encoded DNA-binding protein. *J Virol* 1986;60:353–362.
322. Wagner EK, Roizman B. RNA synthesis in cells infected with herpes simplex virus. I. The patterns of RNA synthesis in productively infected cells. *J Virol* 1969;4:36–46.
323. Wang C, Takeuchi K, Pinto LH and Lamb RA. Ion channel activity of influenza A virus M_2protein: characterization of the amantadine block. *J Virol* 1993;67:5585–5594.
324. Watson JB, Gralla JD. Simian virus 40 associates with nuclear substructures at early times of infection. *J Virol* 1987;61:748–754.
325. Weck P, Wagner R. Transcription of vesicular stomatitis virus is required to shut off cellular RNA synthesis. *J Virol* 1979; 30:410–413.
326. Wengler G, Wengler G. Medium hypertonicity and polyribosome structure in HeLa cells. The influence of hypertonicity of the growth medium on polyribosomes. *Eur J Biochem* 1972;27:162–173.
327. Whitaker-Dowling PA, Youngner J. Vaccinia rescue of VSV from interferon-induced resistance: reversal of translational block and inhibition of protein kinase activity. *Virology* 1983;131:128–136.
328. White E, Grodzicker T, Stillman B. Mutations in the gene encoding the adenovirus early region 1B 19,000 molecular weight tumor antigen cause the degradation of chromosomal DNA. *J Virol* 1984; 42:410–419.
329. White J, Kartenbeck J, Helenius A. Fusion of Semliki forest virus with the plasma membrane can be induced by low pH. *J Cell Biol* 1980;87:264–272.
330. Whyte P, Buchkovich K, Horowitz J, Friend S, Raybuck M, Weinberg R, Harlow E. Association between an oncogene and an anti-oncogene: The adenovirus E1A proteins bind to the retinoblastoma gene product. *Nature* 1988;334:124–129.
331. Wickham TJ, Mathias P, Cheresh DA, Nemerow GW. Integrins $\alpha_v\beta_3$ and $\alpha_v\beta_5$promote adenovirus internalization but not virus attachment. *Cell* 1993;73:309–319.
332. Wilcock D, Lane D. Localization of p53 retinoblastoma and host replication proteins at sites of viral replication at sites of viral replication in herpes-infected cells. *Nature* 1991;349:429–431.
333. Wiley D. Unpublished results.
334. Wilson IA, Skehel JJ, Wiley DC. Structure of the hemagglutinin membrane glycoprotein of influenza virus at 3A resolution. Nature 1981;289:366–375.
335. Winkler M, Dawson GJ, Elizan TS, Beil S. Distribution of actin and myosin in rat neuronal cell line infected with herpes simplex virus. *Arch Virol* 1982;72:95–103.
336. Wolgemuth DJ, Hsu M-T. Visualization of nascent RNA transcripts and simultaneous transcription and replication in viral nucleoprotein complexes from adenovirus 2-infected HeLa cells. *J Mol Biol* 1981;147:247–268.
337. Wu CA, Nelson NJ, McGeoch DJ, Challberg MD. Identification of herpes simplex virus type 1 genes required for origin-dependent DNA synthesis. *J Virol* 1988;62:435–443.
338. Wu L, Rosser DSE, Schmidt MD, Berk A. A TATA box implicated in E1A transcriptional activation of a simple adenovirus 2 promoter. *Nature* 1987;326:512–515.
339. WuDunn D, Spear PG. Initial interaction of herpes simplex virus with cells is binding to heparan sulfate. *J Virol* 1989;63:52–58.
340. Wychowski C, Benichou D, Girard M. The intranuclear location of simian virus 40 polypeptides VP2 and VP3 depends on a specific amino acid sequence. *J Virol* 1987;61:3862–3869.
341. Wyckoff EE. Inhibition of host protein synthesis in poliovirus-infected cells. *Semin Virol* 1993;4:209–215.
342. Yates J, Warren N, Reisman D, Sugden B. A *cis*-acting element from the Epstein-Barr viral genome that permits stable replication of recombinant plasmids in latently infected cells. *Proc Natl Acad Sci USA* 1984;81:3806–3810.
343. Yeh K, Knipe DM. Unpublished results.
344. Younghusband HB, Maundrell K. Adenovirus DNA is associated with the nuclear matrix of infected cells. *J Virol* 1982;43:705–713.
345. Zhang Y, Dolph PJ, Schneider RJ. Secondary structure analysis of adenovirus tripartite leader. *J Biol Chem* 1989;264:10679–10684.
346. Zhang Y, Schneider RJ. Adenovirus inhibition of cellular protein synthesis and the specific translation of late viral mRNAs. *Semin Virol* 1993:4:229–236.
347. Zumbe A, Staehli C, Trachsel H. Association of M_r = 50,000 cap binding protein with the cytoskeleton in baby hamster kidney cells. *Proc Natl Acad Sci USA* 1982;79:2927–2931.

Fundamental Virology, Third Edition
edited by B.N. Fields, D.M. Knipe, P.M. Howley, et al.
Lippincott - Raven Publishers, Philadelphia © 1996

CHAPTER 9

Cell Transformation by Viruses

Joseph R. Nevins and Peter K. Vogt

GENERAL FEATURES OF CELL TRANSFORMATION

The science of virology includes studies directed at the mechanisms by which pathogenic viruses cause disease as well as the utilization of viruses as simple model systems to explore complex cellular events. In addition, viruses have served as probes for unraveling biochemical pathways by the deregulating or disrupting action of specific viral functions leading to oncogenic transformation of the cell.

Viruses that induce tumors occur in several taxonomic groups. All RNA tumor viruses belong to the retrovirus family. The DNA tumor viruses come from seven different groups encompassing naked and enveloped viruses (Table 1). Regardless of taxonomic provenance, the genetic strategy of virus replication is intimately linked to oncogenic potential. The replication of oncogenic retroviruses is not cytocidal, and therefore oncogenic transformation is compatible at the cellular level with the production of infectious progeny virus. However, that production is not a

J. R. Nevins: Howard Hughes Medical Institute, Department of Genetics, Duke University Medical Center, Durham, North Carolina 27710.

P. K. Vogt: Department of Molecular and Experimental Medicine, The Scripps Research Institute, La Jolla, California 92037.

TABLE 1. *Tumor viruses and associated cancers*

Taxonomic grouping	Examples	Tumor types
I. RNA viruses		
Retroviridae		
Mammalian B type	Mouse mammary tumor virus	Mammary carcinoma T-cell lymphoma
Mammalian C type	Murine leukemia viruses Gross leukemia virus Moloney leukemia virus Graffi leukemia virus Friend leukemia virus Moloney sarcoma virus Kirsten sarcoma virus Harvey sarcoma virus Feline leukemia viruses Gardner–Amstein feline sarcoma virus McDonough feline sarcoma virus Simian sarcoma virus	Leukemia, lymphoma, sarcoma, various other malignancies and pathologic conditions
Avian C type	Avian leukosis and sarcoma viruses Rous sarcoma virus Rous-associated viruses (RAV) Avian leukosis viruses Avian myeloblastosis virus Avian erythroblastosis virus Mill-Hill 2 virus Myelocytoma virus MC29	Sarcoma, B-cell lymphoma, myeloid and erythroid leukemia, various carcinomas and other tumors
HTLV-BLV	Human T-lymphotropic virus Bovine leukemia virus	T-cell leukemia[a] B-cell lymphoma
II. DNA viruses		
Adenoviridae	All types	Various solid tumors
Hepadnaviridae	Hepatitis B	Hepatocellular carcinoma[a]
Herpesviridae	EBV	Burkitts' lymphoma (African),[a] nasopharyngeal carcinomas[a]
Papovaviridae		
Polyomaviruses	SV40, polyoma	Various solid tumors
Papillomaviruses	HPV, Shope papillomavirus	Papillomas, carcinomas[a]
Poxviridae	Shope fibroma	Myxomas, fibromas

[a]Human tumors

prerequisite of oncogenesis (383,384). With DNA tumor viruses, the synthesis of infectious progeny is linked to cell death; hence oncogenic transformation can occur only if the viral life cycle is aborted (360).

Some viruses can act as carcinogens in the natural setting; others reveal their oncogenic potential only in experimental systems. The potency of viruses as inducers of tumors varies widely. The most effective retroviruses can initiate tumorous growth in the animal within a matter of days and in virtually all infected individuals. Most other viruses, however, require a much longer latent period, and only a small fraction of the infected hosts comes down with the virus-induced malignancy. All human tumors associated with virus infections belong in this category. The fact that tumor formation is not an inevitable consequence of virus infection reflects the multistep nature of oncogenesis in which each step constitutes an independent and irreversible genetic change that incrementally contributes to the deregulation of cell growth. Viral infection represents one of these steps; only if the others occur in the same cell does a cancer develop. In cell culture, viruses can transform cells to a fully oncogenic phenotype. Such cells display all the altered properties of the tumor cell and are capable of initiating a cancer in a susceptible host animal. In some instances, however, the transformation is only partial, discernible in culture by changes in morphology and growth property but not capable of inducing a tumor *in vivo*.

Virus-induced transformation in cell culture has played a major role in the conversion of modern cancer research into a genetic science. Much of what we know today about molecular mechanisms of oncogenesis has its origin in the study of tumor viruses. These studies have led to an almost universal consensus among cancer researchers on the oncogenic phenotype as resulting from discrete changes in key cellular control genes. Detailed mechanisms of virus-induced oncogenic transformation diverge widely, but all have important characteristics in common: (i) A single infectious virus particle is sufficient for transformation. There is no requirement for multiple infections of the same cell, nor one for cooperation between viral genomes in the same cell. Virus-induced transformation is a "single-hit" process. (ii) All or part of the viral genome persists in the transformed cell. However, there is often no production of infectious progeny virus. (iii) In virtually all cases of virus-induced transformation at least part of the viral genome is expressed in the transformed cell. (iv) Transformation results from corruption of normal cellular growth signals. (v) Reversion of the transformed cellular phenotype can be achieved by specific interference with the function of viral effector molecules (e.g., by transdominant negative mutants of oncogenes).

Genetic studies with tumor viruses have defined the transformed cellular phenotype. Viral mutants that are temperature sensitive in their transforming potential can be used to compare cellular properties at the permissive temperature, when the cell is transformed, with the nonpermissive temperature, when the cell is not transformed but all viral functions not required for the maintenance of transformation are still active. Expression vectors carrying single viral oncogenes have likewise been instrumental in determining the effects of these genes on cellular properties. The alteration of cellular parameters accompanying transformation are multiple. They can be divided into changes in cell growth and changes in cell morphology. The latter affect the cytoskeleton, cell surface, and extracellular matrix (Table 2). Both kinds of change are interrelated, but not all the changes are seen in every type of transformed cell.

Alterations in cell growth affect three parameters: saturation density, growth factor requirements, and anchorage dependence.

TABLE 2. *The oncogenic cellular phenotype*

Cellular property	Oncogenic change or component
Growth	Increase in saturation density Reduced growth factor requirement Increased nutrient transport Anchorage independence of growth
Cytoskeleton	Loss of "stress fibers" Redistribution of microfilaments
Cell surface	Increased agglutinability by lectins Increased lateral mobility of transmembrane proteins Increased production of surface proteases Decreased adhesion to solid substrates
Extracellular matrix	Decreased levels of fibronectins

When normal cells replicate to form a confluent layer, covering the available solid substrate surface, they stop dividing, even if ample nutrients are supplied. This saturation density of a culture is significantly higher when the same cells are transformed, reflecting an enhanced ability of the transformed cells to pile up and to grow in multiple layers (14). If a monolayer of normal cells is infected and transformed by a few individual virus particles, these changes in cellular growth behavior (as well as morphologic alterations discussed later) lead to the outgrowth of microtumors referred to as transformed cell foci (356) (Fig. 1).

Normal cells are also highly dependent for their growth on the presence of growth factors in the culture medium. In conventional cell culture these factors are supplied by the serum component of the medium, but for increasing numbers of cell types the specific growth factors and hormones needed for active replication have been determined (238). A special case of this independence from externally supplied growth factors is autocrine growth in which a growth factor is produced that acts as a mitogen for the producing cells (80). Viral transformation often abrogates or reduces these growth factor requirements.

Normal fibroblasts and epithelial cells must have a solid glass or plastic surface on which to spread before they can grow and divide. Transformation leads to a loss of this anchorage dependence. Transformed cells are able to form colonies if suspended in a semisolid gel nutrient medium. This anchorage independence allows for the selection of a few transformed cells occurring within a population of mostly normal cells (226). The latter will survive in the nutrient gel for a few days but will not grow and form colonies. (In contrast to fibroblasts and epithelial cells, normal hematopoietic cells can often form colonies in semisolid medium, provided the appropriate growth factors are supplied.)

As the name implies, transformation also leads to changes in cell morphology. They affect the cytoskeleton, cell surface, and extracellular matrix (360,383,384). Cytoskeletal changes are detectable by light microscopic inspection and show up as a loss of "stress fibers." At the electron microscopic level they are seen as a rearrangement and redistribution of microfilaments. Several proteins are involved in this reorganization. The major structural components are actin, myosin, and tropomyosin; near the plasma membranes the major structural components are a-actinin, vinculin, and talin (48,349). For instance, actin cables prominent in normal fibroblasts disappear upon transformation; the actin becomes diffusely distributed in the cytoplasm. One exciting recent advance in this area is the discovery of focal adhesion kinase (FAK) and its connection to the src oncogene product (64,160).

Changes in cell surface glycoproteins make transformed cells agglutinable by lectins, such as wheat germ agglutinin or conconavaline A (282). Viral transformation also stimulates the production of surface-bound and secreted

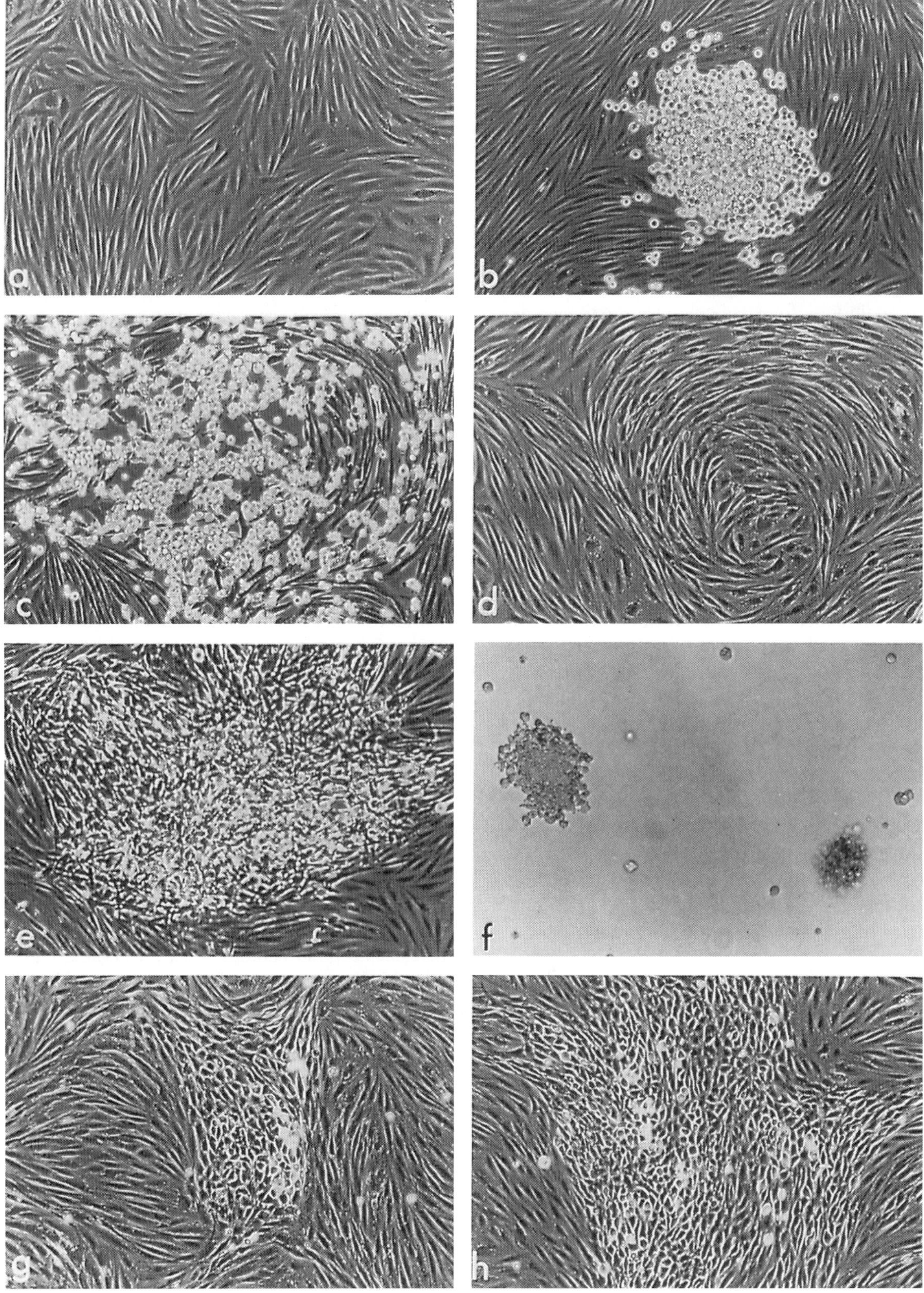

FIG. 1. Transformation of a rat fibroblast cell line by different viral oncogenes. **A**: Normal F-III cells; **B**: cells transformed by Rous sarcoma virus; **C**: cells transformed by Harvey murine sarcoma virus; **D**: cells transformed by Abelson leukemia virus; **E**: cells transformed by mouse polyomavirus; **F**: polyoma-transformed cells in soft agar; **G**: cells transformed by SV40; **H**: cells transformed by simian adenovirus-7. (Photographs courtesy of C. Ware and C. Riney.)

proteases (56,58,259). These have a mitogenic effect on the cell, and in the animal they facilitate tumor invasion and metastasis.

Finally, cells synthesize macromolecules that form an extracellular matrix important in the attachment, spreading, and movement of the cell. Major components of this matrix are fibronectin, laminin, and collagen. Matrix molecules bind to specific cell surface receptors called integrins, heterodimeric transmembrane proteins that form a connection between the extracellular matrix and the cytoskeleton (173). Transformed cells show decreased levels of fibronectin due to the action of extracellular proteases, reduced fibronectin synthesis, and reduced expression of fibronectin receptors (57,172,173,258).

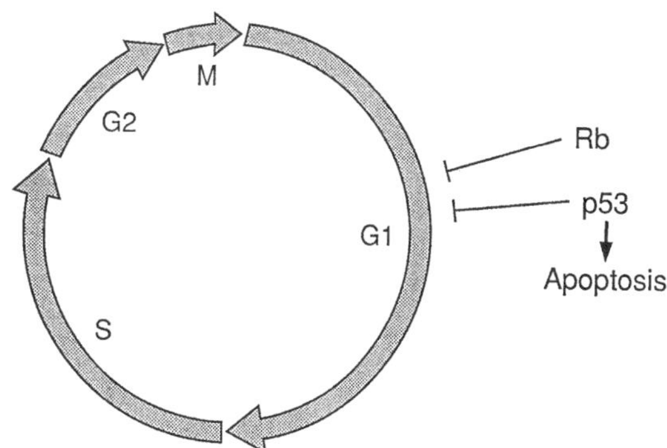

FIG. 2. Cell cycle targets of viral oncoproteins. Cell cycle progression is in G1 by the action of the Rb and p53 tumor suppressor proteins. p53 is also involved in the activation of a pathway of programmed cell death (apoptosis) initiated by a variety of stimuli, including viral infection and DNA damage. Various DNA tumor virus oncoproteins act to disrupt the control by Rb and p53.

VIRUS-INDUCED TRANSFORMATION: MODELS FOR CELL CYCLE CONTROL AND SIGNAL TRANSDUCTION PATHWAYS

Much of our present understanding of the cell cycle and cell growth control events derives from an elegant combination of genetics and biochemistry that the budding and fission yeast have provided (149). The ability to identify critical regulatory activities through mutation, to identify the genes that specify these activities, and to understand the biochemical mechanisms of action of these gene products has been enormously useful. Although the study of the cell cycle in mammalian cells clearly suffers from the lack of genetic analysis, studies of the oncogenic action of the proteins encoded by the tumor viruses, particularly the DNA tumor viruses, have provided important insights into the mechanisms controlling mammalian cell growth. Such studies have helped to elucidate roles for the retinoblastoma and p53 tumor suppressor proteins as regulators of mammalian cell growth (Fig. 2). Each of these proteins exerts a block to progression through G1. The p53 gene appears to play an additional role in triggering programmed cell death (apoptosis) in response to various signals. In each case, DNA tumor virus oncoproteins act to inhibit the action of these two key tumor suppressor proteins, thereby driving an otherwise quiescent cell to enter S phase.

In many respects the viral-mediated disruption of a cell growth control pathway is analogous to the genetic analysis afforded by yeast. For instance, the inactivation of p53 function through the action of the adenovirus E1B protein is, in principle, equivalent to the isolation of a p53 mutant cell; in essence, the viral protein is a mutagen. Moreover, the ability to understand this "mutation" in the context of a viral infectious cycle provides an additional physiologic context, namely, the evolution of a viral replication strategy, in which to view the mutation.

Likewise, the analysis of the oncogenes recovered in the RNA tumor viruses has been particularly important for the elucidation of signal transduction pathways linking the cell surface with the regulation of the genetic apparatus in the nucleus. Perhaps most striking is the fact that the oncogenes recovered in the RNA tumor viruses encode proteins involved in virtually every step in the pathway of signal transduction (Fig. 3). Indeed, the study of these viral oncogenes has identified many of the participants in these complex pathways. Unlike the oncogenes of the DNA tumor viruses, the RNA tumor virus oncogenes are not essential viral genes; rather, these are cellular genes that have been acquired during viral replication and selected based on transforming function. As such, the study of these genes cannot be placed in the context of a strategy of viral replication. Indeed, the presence of these genes renders the virus defective for replication. Nevertheless, the real utility of the RNA tumor viruses is the facility with which cellular genes encoding proteins that participate in cell growth control are trapped within the viral genome, altered in the process, and converted to a form that can radically change cell growth properties. The power of this approach is enormous, as evidenced by the shear number of cellular genes so identified.

Without question, the study of these tumor viruses has been of immense value in elucidating events of cell growth control, providing reagents for further dissection of pathways, and providing a perspective in which to view the action of various regulatory events. The contributions of the DNA tumor viruses and the RNA tumor viruses are distinct, each group providing unique advantages and approaches to the elucidation of mechanisms of oncogenesis. Like the combination of yeast genetics and biochemical analysis that has propelled the analysis of the cell cycle, the combination of these two viral oncogenic systems has provided a wealth of information and afforded an enormously valuable approach to the study of cell growth control and differentiation and the disruption of this control that is evident during oncogenic transformation.

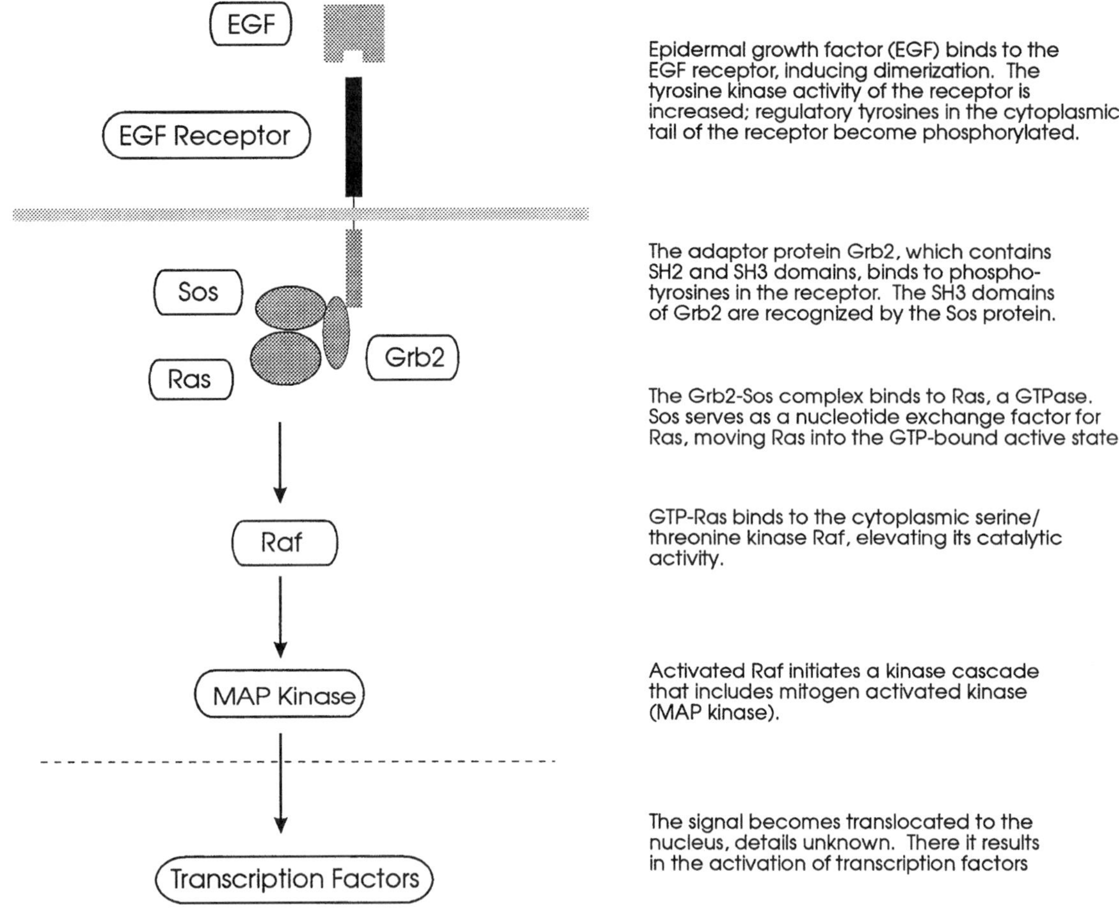

EGF
EGF Receptor
Epidermal growth factor (EGF) binds to the EGF receptor, inducing dimerization. The tyrosine kinase activity of the receptor is increased; regulatory tyrosines in the cytoplasmic tail of the receptor become phosphorylated.
Sos
Ras
Grb2
The adaptor protein Grb2, which contains SH2 and SH3 domains, binds to phosphotyrosines in the receptor. The SH3 domains of Grb2 are recognized by the Sos protein.
The Grb2-Sos complex binds to Ras, a GTPase. Sos serves as a nucleotide exchange factor for Ras, moving Ras into the GTP-bound active state.
Raf
GTP-Ras binds to the cytoplasmic serine/threonine kinase Raf, elevating its catalytic activity.
MAP Kinase
Activated Raf initiates a kinase cascade that includes mitogen activated kinase (MAP kinase).
Transcription Factors
The signal becomes translocated to the nucleus, details unknown. There it results in the activation of transcription factors

FIG. 3. A multistep signal transduction pathway. Mutated forms of the EGF receptor, Ras, Raf, and of various transcription factors have been found to function as retroviral oncoproteins.

MECHANISMS OF CELL TRANSFORMATION BY THE RNA TUMOR VIRUSES

Since early in this century, retroviruses have been known to cause cancer in animals. Lymphoid tumors in chickens were among the first diseases to be recognized as having a viral etiology; shortly thereafter, virus-induced sarcoma in fowl was discovered (104,302). But these tumors were regarded as somewhat of an oddity, having doubtful relevance for our understanding of cancer in higher animals, let alone cancer in humans. Only when mammary tumors, leukemias, and sarcomas in mice were traced to virus infections did the idea of virus-induced tumors gain some measure of acceptance and respectability as a worth-

TABLE 3. *Retroviruses: mechanisms of oncogenicity*

Virus category	Tumor latency period	Efficiency of tumor formation	Oncogenic effector	State of viral genome	Ability to transform cultured cells
Transducing retroviruses	Short (days)	High (can reach 100% of animals)	Cell-derived oncogene carried in viral genome	Viral-cellular chimera, replication defective	Yes
cis-Activating	Intermediate (weeks, months)	High to intermediate	Cellular oncogene activated *in situ* by provirus	Intact, replication competent	No
trans-Activating retroviruses	Long (months, years)	Very low (>5%)	Virus-coded regulatory protein controlling transcription	Intact, replication competent	No

while and relevant subject matter for cancer research (29,119,136,139,147,192,242).

The single most important development that opened the field of tumor virology to quantitative experimentation came from advances in cell culture. They made it possible to induce oncogenic transformation of individual cells inoculated with cell-free virus. These *in vitro* transformation assays, first established for Rous sarcoma virus (227,356), provided the basis for cellular and molecular studies that set RNA tumor viruses apart as a unique group of microbes (213,363,383,384) (see Chapter 51 in *Fields Virology,* 3rd ed.). The salient features of RNA tumor virus life cycle are (i) reverse transcription of the single-strand RNA viral genome into double-stranded DNA, (ii) integration of this DNA into the host chromosome, and (iii) expression of the integrated provirus under the control of viral transcriptional regulator sequences. Reverse transcription has given this viral group the name *retroviruses*. Integration into the host genome, together with an absence of cytocidal action, is the basis for the genetic permanence of retroviral infection. Retroviral genomes can become established in the germ line and are then transmitted as a set of Mendelian markers. Of particular importance for oncogenesis are two additional features of the retroviral growth cycle that are direct consequences of integration into the host genome: (i) the ability to acquire and transduce cellular genetic material and (ii) the insertional activation (and occasional inactivation) of cellular genes by the integrated provirus. Today retroviruses are recognized as significant natural carcinogens in several animal species: fowl, mice, cats, cattle, and monkeys. A rare but aggressive form of human leukemia has been linked to infection with a retrovirus, human T-lymphotropic virus-I (HTLV-I) (123,241,270,321,382). The analysis of retrovirus-induced transformation in cell culture and in animals has led to the discovery of a set of cellular genes, called *oncogenes*, that function as effectors of viral carcinogenesis and play key roles in the control of cell growth and differentiation (for reviews, see refs. 25–27, 43, 289, 363–365, 377–379). Retroviruses, through their interaction with oncogenes, have provided independent and often exclusive access to these important regulatory elements of the cell.

Retroviruses induce tumors and transform cells by three distinct mechanisms (Table 3). Most retroviruses effect oncogenesis through the action of oncogenes. This majority comprises two well-defined groups: (i) retroviruses that carry an oncogene within their genomes, called *transducing retroviruses*, and (ii) retroviruses that lack cellular information but transform by integrating in the vicinity of a cellular oncogene. We will refer to this group as the *cis*-activating retroviruses. Almost all the transducing retroviruses have lost some viral coding information in exchange for cellular sequences. Consequently, they are defective with respect to the production of progeny and depend on a closely related helper virus for reproduction. They are also highly efficient carcinogens that transform cells in culture and cause tumors within short latent periods, often within days after infection. Schematic representations of selected transducing retroviral genomes are compiled in Figs. 4 and 5. All the *cis*-activating retroviruses retain the full complement of viral genes and multiply efficiently in solitary infection. Characteristically, they induce tumors more slowly, within weeks or months. In cell culture, these *cis*-activating retroviruses fail to cause oncogenic transformation. Although the establishment of infection by transducing and by *cis*-activating retroviruses is dependent on viral coding sequences, such viral coding information does not play an essential and direct role in the processes of onco-

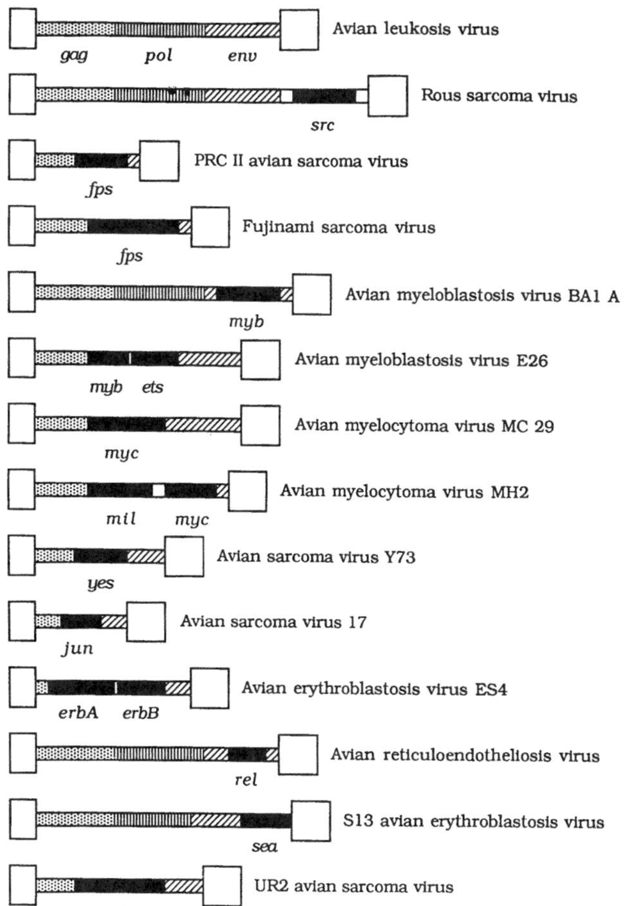

FIG. 4. Genomes of avian transducing retroviruses. Avian leukosis virus [e.g., Rous-associated virus (RAV)] is a prototypical retrovirus. Its genome contains three major coding regions termed *gag* (*stippled*), *pol* (*vertical shading*), *env* (*diagonal shading*) and regulatory sequences that constitute the long terminal repeat (LTR) (*boxed*) of the provirus. *Gag* encodes internal structural virion proteins, *pol* encodes enzymes involved in reverse transcription and integration of the viral genome, and *env* encodes the virion surface glycoproteins. In Rous sarcoma virus the oncogene *src* is added to the complete viral genome. In all other transducing retroviruses, some of the viral coding information is replaced by cell-derived oncogene sequences (marked black in the figure). Consequently, such transducing viruses are defective in replication.

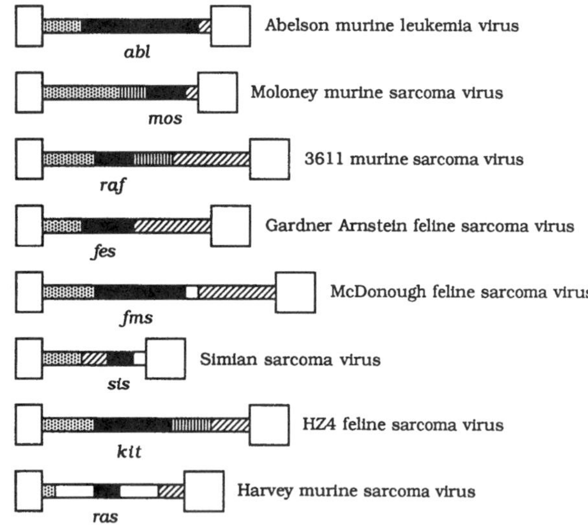

FIG. 5. Genomes of mammalian transducing retroviruses. Cell-derived oncogenes are marked in black; residual viral-coding sequences are *gag* (*stippled*), *pol* (*vertical shading*), and *env* (*diagonal shading*). For details, see Fig. 4.

genesis induced by these viruses. Transducing retroviruses often have none of their basic coding domains intact, and in *cis*-activating retroviruses these also become expendable once the provirus integrates near a cellular oncogene.

In the third, still hypothetical mechanism of retroviral oncogenesis, viral coding sequences not only are required for virus replication but also appear directly involved in oncogenesis. These sequences code for nonstructural regulatory proteins that function primarily to enhance transcription from the viral LTR but that may also interfere with the transcriptional control of specific cellular genes and thus induce tumors.

Oncogenes as Components of Cellular Regulatory Systems That Control Growth and Differentiation

Oncogenes found in retroviral genomes are derived from the cell. Most cellular proto-oncogenes or c-oncogenes are conserved over long evolutionary distances. For instance, the *ras* oncogene and parts of the *jun* oncogene are found in eukaryotes ranging from yeast to man. This durability of proto-oncogene sequences suggests that they are indispensable for a broad spectrum of life forms, fulfilling some fundamental function that permits little change. All oncogenes work through protein products, referred to as *oncoproteins*. There are no cell-derived oncogenes that consist only of noncoding, regulatory sequences. Oncoproteins must, by definition, harbor the potential for inducing oncogenic cellular transformation.

The carcinogenic action of oncogenes is usually a corruption of their normal physiologic function. To understand oncogene-mediated viral carcinogenesis, it is therefore important to discover the normal functions of cellular oncogenes. There are three principal approaches toward this goal. First, with the nucleotide and amino acid sequence in hand, one can search for a revealing homology to a known cellular gene. Second, one can study the expression of the proto-oncogene in the hope that a characteristic developmental or tissue-specific pattern emerges that might provide a clue as to function. Third, it is possible to look for specific biochemical properties of the oncoprotein, such as an enzymatic activity; binding of a hormone, growth factor, or low-molecular-weight ligand; or sequence-specific affinity for nucleic acid.

These approaches have been applied with varying degrees of success. Landmark discoveries have identified several oncogenes with known cellular genes and have determined the normal functions of these genes (4,35,94,96,125,201,254,336,372,376,381). For a significant fraction of oncogenes, however, information on physiologic roles remains general and tentative. But all data on oncogene function place these genes in the category of cellular growth regulators (43,67,289,364,365,377). They are components of a signaling network that receives input from outside the cell and propagates the signal to the nucleus where the information is converted into patterns of gene activity.

A typical growth signal may arrive at the cell surface in the form of a polypeptide growth factor and bind to its specific receptor (Fig. 3). The receptor, often an integral membrane protein with tyrosine-specific kinase activity, becomes activated by binding to the ligand (i.e., it dimerizes and its kinase activity rises, resulting in autophosphorylation of specific cytoplasmic residues of the receptor molecule). The signal is further propagated by sequential and specific protein/protein interactions involving numerous participants and ending up in the nucleus. The ultimate recipients of incoming signals are believed to be transcription factors that are in charge of specific sets of genes. Of special importance in the transduction of cellular signals are two protein domains that were first recognized in the Src oncoprotein, namely SH2 (Src homology region 2) and SH3 (Src homology region 3). These domains occur on several protein components of signal transduction chains; they mediate specific binding of interacting proteins. The SH2 domains bind to phosphotyrosine-containing sequences. The binding specificity is determined by the SH2 sequences in one molecule and the amino acid context of the phosphotyrosine in the other. SH3 domains have an affinity for proline-rich sequences. Proteins carrying SH2 or SH3 domains often also have an enzymatic activity, such as the Src protein tyrosine kinase. However, some of these proteins are devoid of any catalytic activity; they function as linker/adapter molecules and as regulators in binding to and controlling the activity of key enzymes (32,237,246,317). Table 4 contains a list of the principal retroviral oncogenes, arranged according to function. Figure 6 is a schematic summary of this information. Starting

TABLE 4. *Functional groups of transduced retroviral oncogenes*

Oncogene	Retrovirus	Viral oncoprotein	Function of cellular homolog
Growth factors			
sis	Simian sarcoma virus (SSV)	p28$^{env\text{-}sis}$	PDGF
Tyrosine kinase growth-factor receptors			
*erb*B	Avian erythroblastosis virus (AEV)-ES4,[b] AEV-R,[b] AEV-H	gp65^{erbB}	EGF receptor
fms	McDonough feline sarcoma virus (FeSV)	gp180$^{gag\text{-}fms}$	CSF-1 receptor
sea	S13 avian erythroblastosis virus	gp160$^{env\text{-}sea}$	Receptor; ligand unknown
kit	Hardy-Zuckerman-4 FeSV	gp80$^{gag\text{-}kit}$	Hematopoietic receptor; product of the mouse W locus
ros	UR2 avian sarcoma virus (ASV)	p68$^{gag\text{-}ros}$	Receptor, ligand unknown
mpl	Mouse myeloproliferative leukemia virus	p31$^{env\text{-}mpl}$	Member of the hematopoietin receptor family
eyk	Avian retrovirus RPL30	gp37^{eyk}	Receptor, ligand unknown
Hormone receptors			
*erb*A	AEV-ES4,[b] AEV-R[b]	p75$^{gag\text{-}erbA}$	Thyroid hormone receptor
G proteins			
H-*ras*	Harvey murine sarcoma virus (MSV)	p21^{ras}	GTPase
K-*ras*	Kirsten MSV	p21^{ras}	GTPase
Adaptor protein			
crk	CT10, ASV-1	p47$^{gag\text{-}crk}$	Signal transduction
Nonreceptor tyrosine kinases			
src	Rous sarcoma virus (RSV)	pp60^{src}	Signal transduction
abl	Abelson murine leukemia virus (MuLV)	p460$^{gag\text{-}abl}$	Signal transduction
fps[c]	Fujinami ASV	p130$^{gag\text{-}fps}$	Signal transducfion
	PRC 11 ASV	p105$^{gag\text{-}fps}$	
fes[c]	Snyder-Theilen FeSV	p85$^{gag\text{-}fes}$	Signal transduction
	Gardner-Amstein FeSV	p110$^{gag\text{-}fes}$	
fgr	Gardner-Rasheed FeSV	p70$^{gag\text{-}actin\text{-}fgr}$	Signal transduction
yes	Y73 ASV	p90$^{gag\text{-}yes}$	Signal transduction
	Esh ASV	p80$^{gag\text{-}yes}$	
Serine-threonine kinases			
mos	Moloney MSV	p37$^{env\text{-}mos}$	Required for germ-cell maturation
raf[d]	3611-MSV	p75$^{gag\text{-}raf}$	Signal transduction
mil[d]	MH2 avian myelocytoma virus[b]	p100$^{gag\text{-}mil}$	Signal transduction
Nuclear proteins			
jun	ASV17	p65$^{gag\text{-}jun}$	Transcription factor (AP-1 complex)
fos	Finkel-Biskis-Jenkins MSV	p55^{fos}	Transcription factor (AP-1 complex)
myc	MC29 avian myelocytoma virus	p100$^{gag\text{-}myc}$	Transcription factor
	CM II avian myelocytoma virus	p90$^{gag\text{-}myc}$	
	OK10 avian leukemia virus	p200$^{gag\text{-}pol\text{-}myc}$	
	MH2 avian myelocytoma virus[b]	p59$^{gag\text{-}myc}$	
myb	Avian myeloblastosis virus (AMV) BAI/A, AMV-E26[b]	p45^{myb} p135$^{gag\text{-}myb\text{-}ets}$	Transcription factor
ets	AMV-E26[b]	p135$^{gag\text{-}myb\text{-}ets}$	Transcription factor
rel	Avian reticuloendotheliosis virus T	p64^{rel}	Transcription factor
maf	Avian retrovirus AS42	p100$^{gag\text{-}maf}$	Transcription factor
ski	SKV ASV	p110$^{gag\text{-}ski\text{-}pol}$	Transcription factor
qin	Avian retrovirus ASV31	p90$^{gag\text{-}qin}$	Transcription factor of the forkhead/HNF-3 family

[a]These cumbersome but still widely used designations of viral oncoproteins contain basic structural data. p stands for protein, gp stands for glycoprotein, and pp stands for phosphoprotein; the latter is not applied consistently but, instead, is used mainly in conjunction with the *src* product. The numbers stand for estimated molecular weight in kilodaltons, and the superscript lists the genes from which the coding information is derived in 5' to 3' direction. In this chapter, we simply use the abbreviation of the oncogene; however, we use roman letters, with the first letter capitalized, to indicate the oncoprotein. Relevant contributions of viral genes to the oncoprotein are mentioned in the text.

[b]Transducing retrovirus with two oncogenes.

[c]*fps* and *fes* are the same oncogene derived from the avian and feline genomes, respectively.

[d]*raf* and *mil* are the same oncogene derived from the murine and avian genomes, respectively.

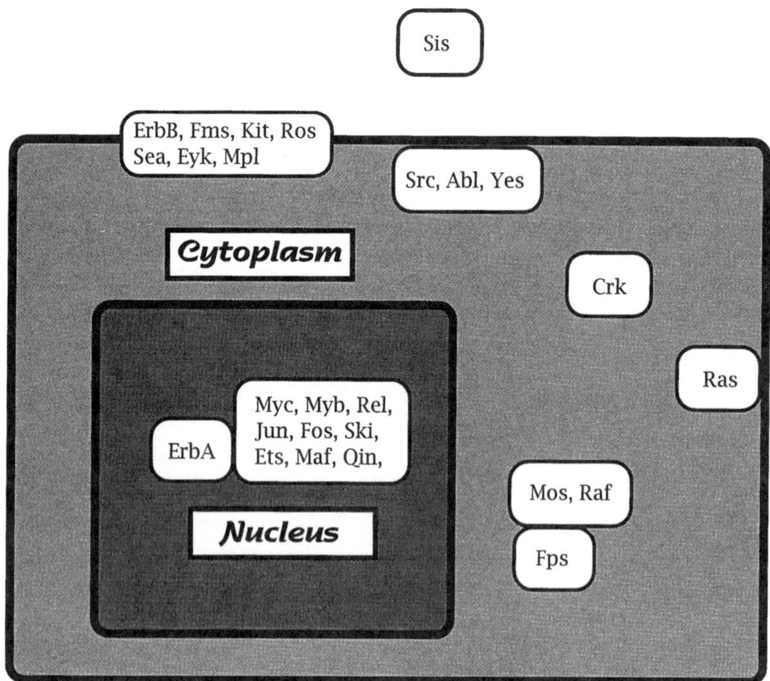

FIG. 6. Oncoproteins function as components of mitogenic signaling chains. The figure groups the products of important retroviral oncogenes into seven categories: growth factors (Sis); integral plasma proteins with the structure of growth factor receptors that are also tyrosine-specific protein kinases (ErbB, Fms, Kit, Ros, Sea, Eyk, Mpl); membrane-bound nonreceptor tyrosine kinases (Src, Abl, Yes); linker–adapter proteins (Crk); GTPases (Ras); cytosolic serine-threonine (Mos, Raf) and tyrosine protein kinase (Fps); transcription factors (Jun, Fos, Rel, Myb, Myc, Ets, Ski, Qin, Maf); and hormone receptors (ErbA).

at the cell periphery, we find that the oncogene *sis* is derived from the gene for a growth factor, PDGF (94,376). The *int*-2 oncogene, which has not been found in a retrovirus genome but which can be *cis*-activated by mouse mammary tumor virus, codes for a protein related to fibroblast growth factor (91). *Erb*B, *fms*, *kit*, and *mpl* code for altered growth factor receptors with known ligands (96,125,254,336,345); several other oncogenes produce proteins that have the general structure of a polypeptide growth factor receptor, but specific ligands remain to be identified. The products of the *ras* genes belong to a family of small GTPases that occupy central positions in signal transduction (34,38,130,236,338). The precise functions of *src*, *abl*, and other genes coding for nonreceptor tyrosine kinases are not clear; they may be active in the amplification stage of growth signals (47,69,71,324). A number of oncogenes code for cytoplasmic kinases that are either tyrosine (fps) or serine/threonine-specific (*raf*, *mos*). A recently discovered oncogene, *crk*, is of interest because it lacks enzymatic activity but contains both SH2 and SH3 domains. Similar to the Grb-2 protein, it belongs to a class of linker/adapter molecules that connect various catalytic steps of signal transduction (234). All the nuclear oncogenes control transcription, and some may play other roles (e.g., in DNA synthesis) as well. Notable are the Jun and Fos proteins, major components of the AP-1 transcription factor; Rel, a member of the NF-kappa B family; and Myc, which is of interest because of its involvement in human tumors (4,23,35,75,126,327). Other oncogenes are related to the control of development. Examples are *wnt*-1 and *qin*. *Wnt*-1 is frequently activated in MMTV-induced tumors. It shows a striking tissue-specific distribution during mammalian embryogenesis and is homologous to the *Drosophila* gene *wingless*, which participates in the control of segmentation polarity in *Drosophila* embryos (293). *Qin* is a transcription factor with homology to the *Drosophila* forkhead gene, which controls fore- and hindgut development in the early embryo (214).

It is self-evident that qualitative or quantitative alterations of these growth regulatory genes carry the risk of uncontrolled growth.

The Transforming Potential of Oncogenes: Mechanisms of Activation

Proto-oncogenes expressed in the appropriate cell type under normal cellular control are not oncogenic. What are the conditions that activate the latent oncogenic potential of these genes? A comparison between activated oncogenes and their cellular counterparts often reveals structural and functional changes. The definition of relevant structural changes is straightforward, whereas that of the important functional changes is more difficult. Transduced oncogenes are usually truncated at one or both ends (for reviews, see refs. 67, 289, 377). The remaining coding sequences carry point mutations and sometimes deletions. They are often fused with viral sequences that are themselves truncated. Commonly, the viral sequences constitute the 5′ end of the chimeric transcription unit, providing splice acceptor site and initiation codon. Occasionally, the oncogene sequences may also be fused to other, unrelated cellular sequences as, for instance, in viral *src* (353). If a proto-oncogene inserted into a retroviral vector and expressed at levels comparable to the viral gene fails to transform cells, then specific changes in the viral version must be responsible for eliciting the transforming potential. An example for an oncogene that is activated by structural and functional changes is *src*. The cellular *src* gene, overexpressed in a retrovirus vector, does not induce oncogenic transformation (174,263,328). As outlined later, specific changes in c-*src* are necessary to reveal its oncogenic trait. The activating structural changes of cellular oncogenes can be roughly defined by studying the properties of reciprocal recombinants between the viral and the cellular gene. Greater resolution is achieved by site-directed mutagenesis and nucleotide sequencing. This approach has defined oncogenically relevant structural changes for a number of oncogenes (e.g., *src*, *ras*, and *fms*) (67,289,377; see also later). These changes are always found in the oncogene proper; the viral sequences that are sometimes fused to it may facilitate oncogenicity by increasing the stability of the oncoprotein, determining its cellular location or enhancing efficiency of translation, but they do not participate directly in the transformation process. The structural changes cause functional changes. The role of these functional changes in oncogenesis remains the central open question for understanding many oncogenes. Where such functional changes have been studied, they have turned out to be quantitative (e.g., the increase or decrease of a catalytic activity). Qualitative changes have not been documented, but it is conceivable that they exist (e.g., changes in the substrate specificity of an oncoprotein).

There are proto-oncogenes that, if sufficiently overexpressed, can transform cells. The *mos* gene is an example of this category (395). For such oncogenes, structural changes of the transduced or insertionally activated oncogene are not essential to bring forth the oncogenic potential, although they may contribute to it. Increased dosage is then the main factor in oncogenesis.

Cellular Transformation by Retroviral Transduction of an Oncogene

Oncogenes in retroviral genomes were found about 20 years ago by a combination of genetic and biochemical studies on Rous sarcoma virus (RSV). The seminal discovery was the isolation of two kinds of RSV mutants: mutants that are temperature-sensitive for transformation and mutants that have nonconditionally lost transforming capacity (229,231,361,369). The temperature-sensitive transformation mutants fail to induce or maintain oncogenic transformation at the nonpermissive temperature while remaining fully oncogenic at the permissive temperature. Their replication is not temperature-sensitive. These temperature-sensitive mutants provide unequivocal evidence for the dependence of oncogenesis on viral genetic information. The properties of these mutants also show that viral information must be expressed continuously to maintain the transformed phenotype. The nonconditionally transformation-defective RSV represents a deletion mutant that has lost about 20% of the genome. This loss does not affect the replicative properties of the virus; the deletion mutant is competent to produce mutant progeny. The deleted sequences, although essential for transformation, are therefore dispensable for virus production. This nonessentialness of transforming information for the viral growth cycle presaged the cellular origin of retroviral oncogenes. In temperature-sensitive and transformation-defective mutants, the same gene is affected: The two types of mutants cannot complement each other or form wild-type virus by recombination (19). The gene of RSV initially defined by these mutants is now known as the *src* gene. The deletion of *src* sequences in transformation-defective RSV also allowed the production of the first nucleic acid probe specific for *src*. A cDNA transcript of wild-type virus was generated and then was hybridized to the genome of the transformation-defective deletion mutant lacking the *src* gene. The nonhybridizing single-stranded cDNA sequences of this reaction were separated from the double-stranded hybrids. These single-stranded sequences represented the viral oncogenic information. The cDNA *src* probe was found to hybridize to cellular DNA, providing conclusive evidence that *src* is not a viral gene but a cellular one (348).

The basic characteristics discovered with the *src* gene apply to transduced retroviral oncogenes in general (25–27,67,289,365,377). All are derived from the cell genome, and all are nonessential for virus replication. The virus serves as a vehicle of transduction, but it also exercises transcriptional control over the transduced oncogene. The cellular origin of retroviral oncogenes is of far-reaching significance because the transforming potential of a cellular oncogene can become activated independently of any viral intervention. Point mutations, gene amplification, and transcriptional up-regulation can uncover the malignant potential of an oncogene. Activated oncogenes play important roles in nonviral as well as viral carcinogenesis; they may be the common denominator of all kinds of cancer.

In several of the transduced oncogenes, temperature-sensitive and nonconditional mutants have been isolated that convincingly show that the gene is required for the initiation and maintenance of the transformed cellular phenotype (21,22,188,229,265,339,361,397). Transduced oncogenes can also be excised from their native viral genomes and inserted into an expression vector (166,167). They then confer oncogenicity to this new vector. In the transducing retrovirus, the cell-derived oncogene is the only coding information that is needed for oncogenicity. In fact, most or all retroviral coding information may be either nonfunctional or deleted without affecting the oncogenicity. Transduced retroviral oncogenes are the most efficient carcinogens known. They cause transformation within a record time; moreover, in appropriate target cells, transformation is an inevitable consequence of infection and of expression of the oncogene. With some retroviruses (e.g., RSV) it is possible to transform all cells of a culture synchronously with one inoculum in the first round of infection. Because transformation occurs with high frequency, tumors induced by transducing retroviruses are usually polyclonal in origin. Unlike most cancers that result from a multistep transformation process, the tumors caused by transducing retroviruses probably result from single-step carcinogenesis.

Some retroviruses are vectors for two oncogenes; for instance, avian erythroblastosis virus strain R carries the *erb*A and *erb*B genes, and MH2 avian carcinoma virus transduces *mil* and *myc* (28). In these cases, one of the two oncogenes is sufficient to induce transformation (*erb*B or *myc* in the preceding examples). The second enhances oncogenicity of the virus and conveys greater nutritional and growth-factor independence upon the tumor cells.

How are cellular oncogenes acquired by the retroviral genome? Since incorporation of an oncogene into a retroviral genome is a rare event, it can only be studied after the fact. A plausible model starts with the integration of a retrovirus upstream near a cellular oncogene. The provirus has a deletion that includes its 3′ LTR and is integrated in the same transcriptional orientation as the adjacent oncogene (65,133,351). Transcription from the 5′ LTR will extend through the oncogene, creating a chimeric RNA that contains the packaging signal located near the 5′-terminus of the viral genome. Complete introns are removed from this transcript, and the processed RNA can be exported in virus particles. The model also envisages infection of the same cell by a second virus that has not suffered a deletion and replicates complete progeny RNA as well as all structural proteins. Since retroviruses are diploid, such a cell will produce heterozygous progeny virus containing a complete viral genome and the defective chimeric RNA. In the next round of infection during reverse transcription of such a heterozygous particle, nonhomologous recombination between the two virion RNAs may take place, possibly as a result of template switching by the polymerase. This model proposes two formal steps of recombination. The first step involves (i) the integration of the defective provirus near the cellular oncogene, (ii) the formation of the 5′ junction of cellular with viral sequences, and (iii) the generation of the packageable chimeric RNA. This recombination event occurs at the DNA level. The second recombination event takes place during reverse transcription and involves the formation of the 3′ junction of cellular with viral information. The resulting recombinant is a provirus with two LTRs. Its 5′ coding information is derived from the chimeric RNA; 3′ sequences come from the nondefective retrovirus. Recent evidence supports this model (110,111). If chicken neuroretina cells are infected with the avian leukemia virus RAV-1, their growth is stimulated and the culture releases a virus that transduces the *raf* oncogene. The identifiable steps in this acquisition of a cellular gene are in accord with the preceding model. However, other possible models, suggested by exceptional provirus structures, have not been ruled out. RNA-splicing of viral and cellular sequences would be an alternative mechanism for recombination (175,232). A third possibility is that viral/cellular recombinants derive from packaged read-through transcripts of a nondefective retroviral genome and an oncogene located downstream (155). It is also possible that transducing retroviruses can be generated by more than one mechanism.

The defectiveness of transducing retroviruses is caused by the replacement of viral coding sequences through cellular oncogene sequences. Only in some strains of RSV is the *src* oncogene added to a complete viral genome (Fig. 4). The defective viruses go through a partial life cycle that includes reverse transcription, integration, RNA synthesis, and translation. The expression of the defective provirus leads to oncogenic transformation; however, since one or several of the essential virion proteins cannot be synthesized, no infectious progeny is made. Only if the same cell is infected by a replication-competent retrovirus that supplies the missing protein(s) in *trans* is infectious transforming virus released. This transforming progeny virus remains, however, genetically defective and helper dependent. The structures of transducing retroviral genomes are summarized in Figs. 4 and 5.

It is interesting that cellular oncogenes have been found only in one subfamily of retroviruses, formerly called the oncoviruses. They have not been seen in the genomes of the lenti- and spumaviruses, although the steps of replication that are thought to favor acquisition of cellular sequences are the same for all retroviruses. A possible explanation is the greater complexity of the lenti- and spumaviral genomes. Oncoviruses efficiently replicate without virus-coded regulatory proteins, whereas the replication of lenti- and spumaviruses depends on nonvirion proteins that are generated from the provirus by way of multiply spliced messages. Generation of these messages and their *cis*-acting targets would be destroyed by a cellular genetic insert, essentially inactivating the virus. Oncovirus genomes, on the other hand, can accept cellular inserts without such debilitating effects.

In the following sections we will discuss selected examples of transduced viral oncogenes representing components of cellular signaling chains; we will proceed from cell periphery to cell nucleus. For a more comprehensive account of oncogenes, the reader is referred to extensive reviews of the subject (26–28,36,67,289,364,365,367,377).

Sis: Growth Factor and Autocrine Transformation

Sis is the oncogene of simian sarcoma virus (SSV) (for reviews, see refs. 296, 386, 392); it is an insert that was acquired by the viral genome from the genome of the woolly monkey, a New World monkey. A homologous oncogenic determinant derived from the genome of a cat was discovered in the Parodi Irgens feline sarcoma virus. The amino acid sequence of the SSV Sis protein shows about 88% identity with the sequence of the B-chain of the human platelet-derived growth factor (PDGF). The discovery of this homology provided the first direct and exciting link between oncogenes and cellular growth signals. The immediate cellular precursor of the *sis* gene is most likely the PDGF-B gene of the woolly monkey genome, with the deviation from the human sequence being a species difference.

Sequence analysis of the SSV genome shows that v-*sis* is expressed as a fusion product containing, at its N-terminus, sequences derived from the retroviral envelope (Env) protein, including the Env signal sequence that effects the translocation of the Env/Sis protein into the vesicles of the endoplasmic reticulum. The initial v-*sis* product is a 28-kd glycoprotein that is rapidly dimerized to a gp56 dimer and then proteolytically processed to become a p24 homodimer, similar in structure to the homodimer of the PDGF-B chain. This Sis protein is firmly associated with cellular membrane fractions, but only about 1% is secreted by the SSV-transformed cell. Primary and processed v-*sis* products show PDGF-like activity: They react with anti-PDGF antibodies, bind to PDGF receptors, induce internalization of the receptor/ligand complex, induce phosphorylation of the receptor on a tyrosine residue, and stimulate the mitogenic response of the cell. This biologic activity of the v-*sis* product suggests an autocrine-transforming mechanism for SSV-infected cells. Additional support for this suggestion comes from the following observations: (i) SSV transforms only those cells that express a PDGF receptor. Transformation appears to be dependent on the interaction of the v-*sis* product with the PDGF receptor. (ii) The nascent v-*sis* product must be translocated into the endoplasmic reticulum in order to be transforming. Deletion of the Env signal sequence from the Env/Sis fusion protein abolishes transforming activity. Since the v-Sis protein is functionally equivalent to PDGF, the cellular PDGF gene can also function as an oncogene by autocrine mechanisms similar to v-Sis: The PDGF B-chain expressed under the control of a strong retroviral promoter induces oncogenic transformation in cells expressing the PDGF receptor.

The transforming signal of Sis could be generated by binding to receptors at the cell surface, but the possibility has also been considered that intracytoplasmic receptor binding plays a role in transformation. The PDGF receptor is processed along the same cellular compartments as the v-Sis protein, and it is conceivable that the two interact before reaching the cell surface. SSV-transformed cells, exposed to antibody against Sis, sometimes assume near normal phenotype. This observation was interpreted to suggest that the relevant interaction between growth factor and receptor, leading to transformation, is a cell surface event. However, if a signal sequence (KDEL) that effects retention in the endoplasmic reticulum is added to the PDGF B chain, the transforming potential for PDGF receptor-expressing cells is not abolished. Therefore, PDGF may interact with the receptor intracellularly and then become translocated to the cell surface as a complex.

ErbB and fms: Altered Receptors and Constitutive Mitotic Signals

The oncogene *erb*B has been studied in two avian retroviruses that cause erythroblastosis and fibrosarcoma in chickens–avian erythroblastosis virus strain H and strain ES4 (for reviews, see refs. 151, 230, 366). AEV-H contains *erb*B as its only oncogene. AEV strain ES4 carries, in addition to *erb*B, the *erb*A oncogene derived from a thyroid hormone receptor gene (381). However, *erb*B alone is both necessary and sufficient to induce the pathogenic effects associated with AEV infections, erythroblastosis and fibrosarcoma.

The nucleotide sequence of *erb*B shows close homology to that of the epidermal growth factor receptor (EGFR) gene (94). The *erb*B of AEV strains H and ES4 is derived from the chicken EGFR gene. The EGFR contains four major structural and functional domains: (i) an amino-terminal extracellular ligand-binding domain, (ii) a transmembrane segment, (iii) a kinase domain, and (iv) a carboxyl-terminal regulatory segment with at least five tyrosine phosphorylation sites, namely, P1 to P5. In *erb*B most of the growth factor–binding domain is deleted. Furthermore, carboxyl-terminal deletions have removed P1 from the *erb*B of AEV-H and have removed P1 as well as P2 from *erb*B of AEV ES4.

The mechanism of *erb*B-induced oncogenesis probably includes elements of the same mitotic signal that is initiated by the binding of epidermal growth factor (EGF) to its receptor (Figs. 2 and 7). In this signal, ligand binding appears to induce an allosteric change and oligomerization of the receptor (315–317). The kinase activity of EGFR is elevated, and the protein becomes autophosphorylated on tyrosine residues at its carboxyl terminal regulatory sites. EGFR substrates of potential importance in mitotic signaling include the SHC proteins (containing an SH2 domain but no recognizable catalytic domain), GAP (GTPase activating protein), PI3-kinase (phosphatylinositol 3-kinase), and phospholipase C. At least one of the phosphotyrosines in the cytoplasmic tail of EGFR is recognized and bound by the SH2 domain of a small adapter protein called Grb2 that also binds with its SH3 domains to the nucleotide releasing factor Sos. Sos, in turn, interacts with the product of the proto-oncogene *ras*, leading to an exchange of the *ras*-bound GDP for GTP, thereby activating the Ras protein. From Ras the signal is propagated through a kinase cascade involving the product of the proto-oncogene *raf* and the MAP kinase (mitogen-activated protein kinase), among other kinases. Translocation of the signal into the nucleus and activation of the ultimate recipient of the signal pathway, presumably a transcription factor, are still poorly understood. The tyrosine-specific kinase of EGFR is essential for the mitogenic effect of EGF. Binding of EGF to the receptor is followed by internalization of the receptor/ligand complex. This process removes avail-

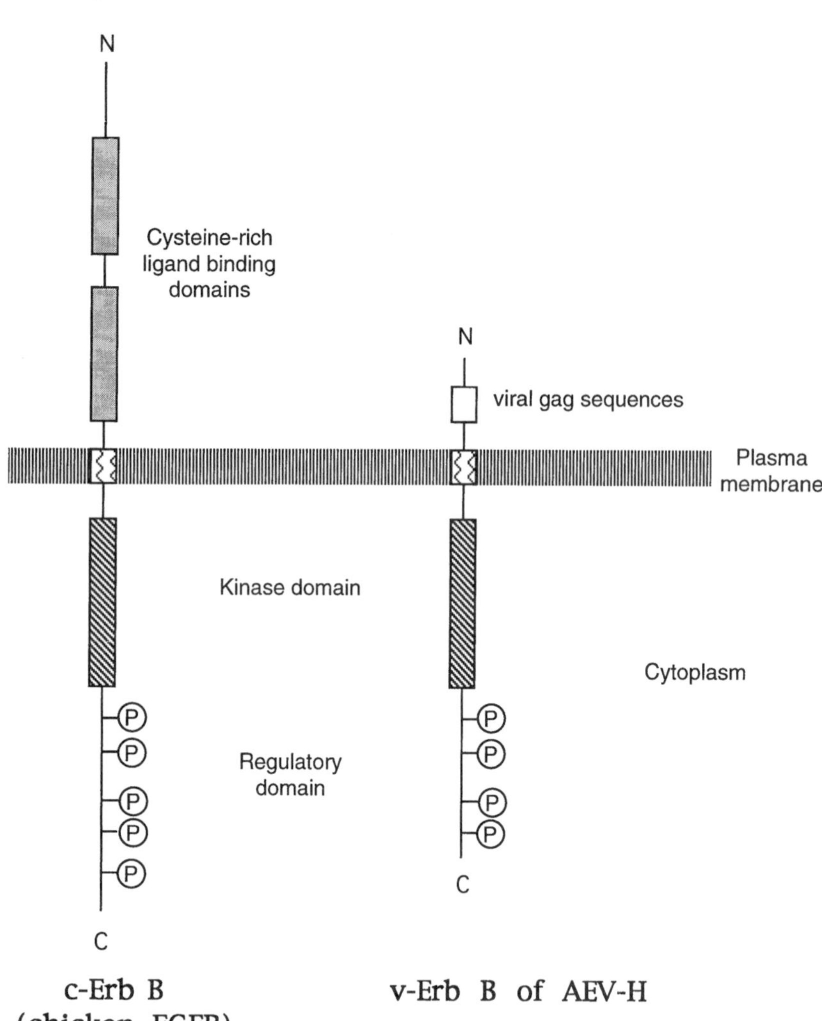

FIG. 7. Structural comparison of the cellular Erb B protein (the epidermal growth factor receptor, EGFR) with the transduced retroviral version, the v-Erb B protein of avian erythroblastosis virus—H (AEV-H). The oncogenic v-Erb B lacks a regulatory phosphorylation site in the carboxyl terminal of the protein and has lost the extracellular ligand binding domain. It functions as a constitutively activated EGFR.

able EGFR from the cell surface and results in a down-regulation of EGF responsiveness. Reduced EGF responsiveness can also be caused by an activated protein kinase C, probably through phosphorylation of a threonine residue near the EGFR kinase domain. The transforming activity of *erb*B has been analyzed with the help of several conditional and nonconditional mutants of AEV (135,253). The main conclusions drawn from these studies are that the ErbB protein needs to be located at the plasma membrane in order to be transforming and that truncation of the carboxyl-terminal regulatory domain of the receptor plays a role in bringing out the transforming potential of ErbB. The ErbB protein is glycosylated. Intracellular forms are gp65 and gp68 proteins; the mature cell surface molecule is a gp74. The *erb*B coding information lacks a conventional signal sequence; targeting to the plasma membrane appears to be a function of the transmembrane domain. Temperature-sensitive mutants of AEV ES4 that prevent membrane localization at the nonpermissive temperature are also temperature sensitive for the transformed cellular phenotype. Deletions from the carboxyl-terminal end of the c-ErbB protein have shown that loss of the P1 phosphorylation site is important for both erythroblast and fibroblast transformation. With increasing size these carboxyl-terminal deletions abolish erythroblast transformation while still permitting fibroblast transformation. A somewhat different picture emerges from the study of avian erythroblastosis viruses that have recently acquired *erb*B (239,279,280) (see section entitled "*cis*-Activation of Oncogenes"). These newer *erb*B-transducing agents carry a version of c-*erb*B that lacks the ligand-binding domain of EFGR, but some have an intact carboxyl terminus and yet can induce erythroblastosis in a highly susceptible chicken line. They do, however, fail to transform fibroblasts. A separate analysis of fibroblast transformation using mutants of AEV ES4 has shown that an intact kinase domain is essential for this biological effect. These observations are compatible with the proposal that ErbB transforms cells as a constitutively activated EGFR. Membrane localization and kinase activity are essential.

Ligand-independent activation of receptor tyrosine kinase also plays an important role in the oncogenicity of v-

fms (for reviews, see refs. 290, 299, 300, 334, 335). *Fms* is the oncogene of the McDonough strain of feline sarcoma virus (SM FeSV). It has also been found as the oncogenic insert in the Hardy Zuckerman 5 (HZ5) feline sarcoma virus. *Fms* is derived from the gene of the receptor for the macrophage colony stimulating factor CSF-1 (Fig. 8) (336). In SM FeSV it replaces most of the viral *gag* and practically all of the *pol* gene while the *env* gene of the virus has remained intact. *Fms* is fused in frame to the partial *gag* sequences. The crossover point in the *fms* gene is located in the 5′ untranslated region of c-*fms* messenger RNA and preserves the signal sequence of c-*fms*. The 3′ recombination point in *fms* is located 120 nucleotides upstream of the c-*fms* termination codon. SM-FeSV and HZ5-FeSV induce fibrosarcomas in cats but no hematopoietic malignancies, notwithstanding the normal function of CSF-1 as a hematopoietic growth factor.

The CFS-1 receptor has a structure similar to that of other receptors of peptide growth factors, such as the PDGF receptor, the Kit protein, and Flt (20,156,337,401) (Fig. 8). The extracellular domain consists of five immunoglobulin-like repeats; the three amino terminally located of these repeats form the ligand binding region. The fourth immunoglobulin-like domain functions in receptor dimerization; the point mutations contributing to the oncogenicity of v-*fms* are in or near this domain. Important cytoplasmic regions of Fms are the tyrosine kinase domain and the C-terminal tail that contains targets for negative regulation. Activation of Fms by ligand or by mutation results in autophosphorylation of several tyrosines in the cytoplasmic domain of the receptor. These phosphotyrosines are then bound by the SH2 regions of effector molecules, including PI-3, kinase, members of the Src family of proteins, and Grb2. From Grb2 propagation of the signal continues via SOS-1, Ras and Raf, and the Map kinase cascade. An alternative pathway to the nucleus may involve the interferon-stimulated gene factor 3, but molecular details of this signal chain are still sketchy.

The v-*fms* genes of SM-FeSV and of HZ5-FeSV differ from c-*fms* by several point mutations leading to amino

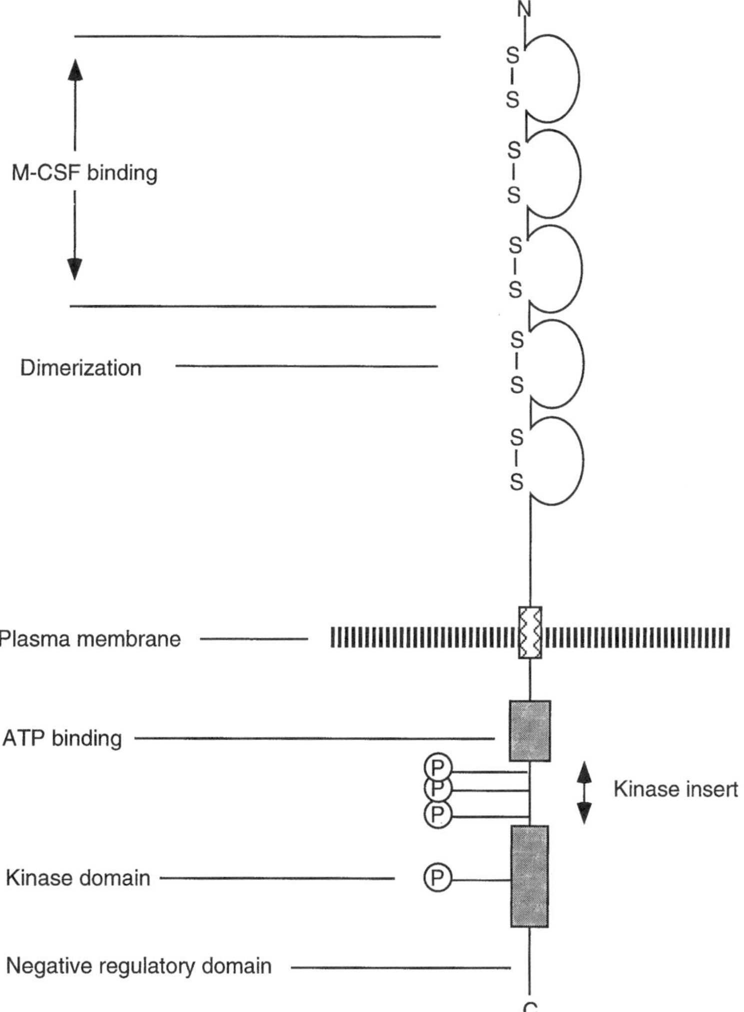

FIG. 8. The cellular homolog of the *fms* oncogene functions as a receptor for the macrophage colony stimulating factor. It has the basic structure of a receptor tyrosine kinase. The extracellular domains are responsible mainly for ligand binding and dimerization. Intracellular domains are catalytic and regulatory. Oncogenicity is activated by mutations in both extracellular and intracellular domains.

acid substitutions and by a deletion that removes 50 carboxyl terminal amino acids from the protein. The deleted residues are replaced in SM- and HZ5-FeSV by 14 and 11 unrelated amino acids, respectively. The oncogenicity of the v-*fms* genes can be attributed to two-point mutations in the extracellular domain and to the carboxy-terminal deletion. The deletion removes phosphorylation sites that negatively regulate the kinase activity of the receptor, and the point mutations in the extracellular domain lead to a ligand-independent activation of the receptor.

Src and *abl*: Membrane-Bound Nonreceptor Tyrosine Kinases

Src was the first retroviral oncogene discovered and defined (for reviews, see refs. 132, 324). It was also the first whose origin from the cellular genome was recognized and the first for which a protein product was identified (47,348). The discovery of the *src* gene product was soon followed by observations that showed that this protein is a tyrosine-specific protein kinase, providing suggestive evidence for a regulatory function of *src* and raising hopes for an immediate understanding of its oncogenic action (66,171,212). Throughout this period of rapidly expanding work on oncogenes, *src* has served as an important and pace-setting model. Despite this paradigmatic role and much additional data on structural and functional properties of cellular and viral *src*, the mechanism by which *src* transforms cells is only now emerging. *Src* codes for a 60-kd phosphoprotein that is bound to cytoplasmic membranes. (This protein has been termed $pp60^{src}$; we will refer to it as Src protein, to remain consistent with the terminology applied to other oncoproteins.) The cellular Src protein contains several well-defined functional domains: an amino-terminal myristilation domain, followed by a region that diverges among related nonreceptor tyrosine kinases, an SH3 and an SH2 domain, a catalytic domain, and a C-terminal regulatory region. The viral Src protein shows the same general structure but has suffered a deletion in the carboxy terminal regulatory domain. Since the c-*src* gene expressed at levels comparable to transduced v-*src* does not transform cells (174,263,328), structural differences between the cellular and viral version of the Src protein must activate oncogenic potential. The oncogenicity of *Src* requires binding to the plasma membrane and is correlated with increased kinase activity. The kinase activity of *Src* is controlled positively by phosphorylation of tyrosine 416 in the catalytic region and negatively by phosphorylation of tyrosine 527 in the C-terminal regulatory region. Tyrosine 527 is the target of Csk (C-terminal *Src* kinase), which appears to be a specific regulator of *Src* family kinases. The phosphotyrosine 527 is most likely bound intramolecularly by the SH2 domain of the Src protein (Fig. 9). This folding of the Src protein is believed to affect the catalytic region and to turn off kinase activity. Significantly, viral Src

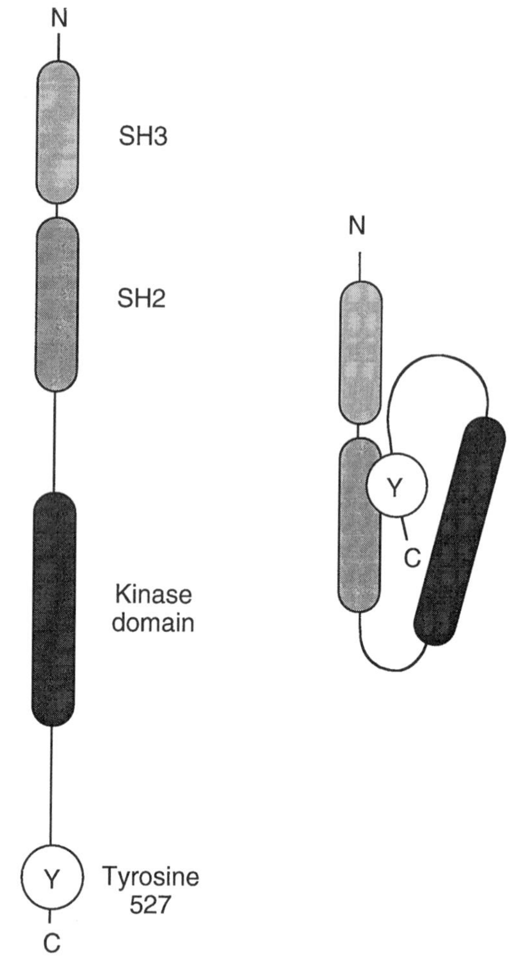

FIG. 9. Conformational changes regulate the activity of the Src tyrosine kinase. Phosphorylation of tyrosine 527 downregulates the Src kinase. This phosphotyrosine appears to be bound intramolecularly by the SH2 domain, inducing an allosteric change that shuts off kinase activity. Activation of the kinase could occur by dephosphorylation of tyrosine 527, but that in turn may be the result of an allosteric change induced by signal proteins that interact with SH2 or SH3.

lacks tyrosine 527. It is constitutively active as a kinase. Activation of Src in the presence of tyrosine 527 may involve dephosphorylation of this residue by a specific phosphatase. However, access to tyrosine 527 may depend on an allosteric change of the molecule that includes dissociation of phosphotyrosine 527 from the SH2 domain. Consequently, such allosteric interactions could target the SH2 and SH3 domains of Src that connect to other members of the signal transmission chains. Accordingly, oncogenic activation of Src can result from mutations in SH3, SH2, the kinase domain, and the C-terminal tail (68,264,355), suggesting that these may all be important in the allosteric transitions that affect Src kinase activity.

Over the years numerous proteins have been identified that can serve as downstream targets and substrates of Src. Most of these are probably not relevant in oncogenic transformation. Exceptions appear to be an SH2-containing protein (SHC) and focal adhesion kinase (FAK) (64,160,237,285,301). SHC is an adapter protein that is phosphorylated by Src and related kinases and then binds to Grb-2; the Src/Grb-2 complex then interacts with the SOS, making a nucleotide exchange factor that activates Ras. FAK is bound by the SH2 domain of Src, and this interaction appears to keep FAK in the activated state. The resulting overphosphorylation of cytoskeletal proteins may be responsible for the poor adhesion and altered shape of Src-transformed cells.

The normal functions of Src are not known.The cellular *src* gene is highly expressed in some cell types such as blood platelets, neurons, and osteoclasts. *Src*$^-$ mice develop osteopetrosis, the result of dysfunctional osteoclasts, but show no other major abnormalities in development and growth, possibly because the Src defect can be compensated by related kinases (344).

The product of the oncogene *abl* shares with the Src protein two important properties: (i) attachment to the cytoplasmic face of the plasma membrane via a myristic acid residue and (ii) tyrosine-specific protein kinase activity (for reviews, see refs. 288, 294, 393). *Abl* has been first identified as the oncogene of the Abelson murine leukemia virus (A-MuLV). More recently, it has also been found as a cell-derived oncogenic insert in the Hardy-Zuckerman HZ2 FeSV. In A-MuLV (with which most of the work on *abl* has been carried out), the *abl* gene is merged to partial *gag* sequences and is expressed as a Gag/Abl fusion protein. The *abl* insert in A-MuLV is truncated at its 5′-terminus as compared to the cellular gene, but the remaining *abl* coding sequences are unaltered and include the c-*abl* termination codon. The v-Abl protein contains three domains: (i) the amino-terminal Gag domain, (ii) an SH2 and SH3 domain mediating interactions with other signal transduction proteins, and (iii) a catalytic domain that is homologous to the corresponding domain of other tyrosine kinases. Mutational analysis of the *abl* gene shows that for oncogenic transformation to occur, two properties are essential: (i) membrane association through myristilation and (ii) elevated tyrosine kinase activity. In the v-Abl protein, the Gag domain provides the myristilation signal and therefore becomes indispensable for transformation—in contrast to the Gag domain of most other oncoproteins, which can be deleted without affecting oncogenicity. The Gag sequences of v-Abl also increase the half-life of the protein in hematopoietic cells. The c-Abl protein localizes predominantly in the nucleus; it is phosphorylated in a cell cycle-dependent manner (190,191). Recent observations suggest that it may function as a negative regulator of cell growth; transdominant negative mutants of c-*abl* disrupt the control of the cell cycle and enhance cellular susceptibility to oncogenic transformation (312). The SH2 and SH3 domains of Abl have complex regulatory functions. The SH3 domain appears to be the target of a negative control mechanism; deletion of this domain activates the oncogenic potential of Abl. The SH2 domain of Abl may have multiple functions—controlling the kinase activity of Abl by intramolecular binding to phosphotyrosine residues and interacting with signal transduction proteins apparently in a phosphotyrosine-independent manner. Replacement of the Abl SH2 domain with a heterologous SH2 domain may also activate the oncogenic potential of this nonreceptor kinase (63,246,247). A-MuLV is an effective transforming agent for early B cells. Other hematopoietic cells can be transformed, but special conditions and additional selective pressure are required to reveal these less frequent transforming events. A-MuLV can also transform fibroblasts. The HZ2 FeSV induces only sarcomas and does not transform hematopoietic cells either *in vitro* or *in vivo*. An activated *abl* oncogene is also found in human chronic myelogenous leukemia. The leukemic cells carry a chromosomal translocation that gives rise to the marker Philadelphia chromosome and joins *abl* sequences of chromosome 9 to *bcr* (breakpoint cluster region) sequences of chromosome 22 (153,341). *Bcr* and *abl* sequences are found in a single, fused transcript in leukemic cells. Although the junction point between *bcr* and *abl* varies in different patients, the regularity with which this translocation occurs in chronic myelogenous leukemia strongly suggests that the Bcr/Abl protein conveys a growth advantage to the leukemic cells.

***Ras*: Growth Regulatory GTPase**

Mammalian cells contain three *ras* genes: H-*ras*, K-*ras*, and N-*ras* (for reviews, see refs. 34, 38, 40–42, 44, 205, 211, and 236). All three are composed of four coding exons that contain information for a protein of about 190 amino acids or 21 kd, plus an upstream noncoding exon. The amino-terminal halves of these cellular p21 proteins are virtually identical. About two-thirds of the carboxyl-terminal halves are also closely related, and only the utmost 25 carboxyl-terminal amino acids show divergence. Two of the mammalian *ras* genes—namely, H-*ras* and K-*ras*—have been found transduced by retroviruses; N-*ras* has not been detected in a naturally occurring virus. There are more than half a dozen independently isolated retroviruses that transduce a *ras* gene; the oldest and best studied are Harvey and Kirsten murine sarcoma viruses (H-MuSV and K-MuSV). Their *ras* inserts are derived from the rat genome. They also carry sequences of an endogenous rat retrovirus VL30.

Ras proteins are GTPases that act as nodal points in cellular growth control, receiving inputs from several signal transmission chains and propagating the signal via kinase cascades toward the nucleus. Ras proteins belong to a superfamily of low-molecular-weight GTPases that also encompasses the Rho proteins, important in the control of the cytoskeleton, and the Rab proteins, which function in membrane protein trafficking. Other related proteins with GTPase activity and homology to Ras are the heterotrimeric

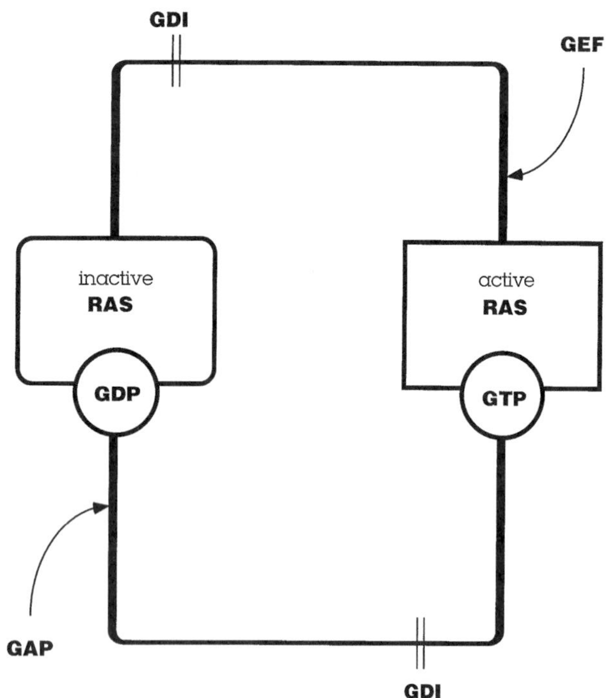

FIG. 10. The Ras GTPase functions as a binary switch. Bound to GTP it is active; bound to GDP it is inactive. Ras becomes activated by nucleotide exchange factors (GEF); it becomes inactivated by GTPase activating proteins (GAP). An additional layer of regulation is introduced by guanine nucleotide dissociation inhibitors (GDI) whose primary effect appears to be an interference with GEF.

G proteins involved in signal transduction and the elongation and initiation factors of the protein-synthesizing machinery.

Ras proteins need to be anchored in cellular membranes for proper functioning. They are posttranslationally modified by covalent attachment of prenyl groups to cysteines near their C-terminus, providing a lipophilic tail. Ras proteins can be considered as binary switches that are either on (bound to GTP) or off (bound to GDP) (Fig. 10). They are regulated by at least three types of proteins; the guanine nucleotide exchange factors (GEF), the GTPase-activating proteins (GAP), and guanine nucleotide dissociation inhibitors (GDI) (34). GEF proteins activate Ras through exchange of Ras-bound GDP for GTP. An example of a GEF is the SOS protein, which is instrumental in transmitting growth signals from tyrosine kinases to Ras. Gap proteins have an attenuating effect on Ras. They stimulate the conversion of active GTP Ras to the inactive GDP Ras. They are represented by P120 GAP and by NF1 (neurofibromin). The GDI proteins are the least understood regulators of Ras. They bind to GDP Ras, thereby interfering with the action of GEF activators and possibly also with GAP-like attenuators.

Oncogenicity of Ras is greatly elevated by specific point mutations, notably in codons 12, 13, and 61. These mutations inhibit the binding of GAP proteins to activated Ras and hence prevent the stimulation of Ras GTPase activity. The mutant Ras proteins remain bound to GTP and are constitutively active, contributing to oncogenic transformation. Retroviral Ras proteins carry these activating mutations. Additionally, the transforming potential of Ras can be elicited by changes in upstream nontranslated regions; the endogenous VL30 sequences can also have an enhancing effect on transformation.

Activated Ras with a mutation in one of the critical codons is also found in human cancers, notably in carcinomas of the bladder, colon, and lung. A significant minority of the primary tumors in these sites contains the mutated oncogene. Normal tissue surrounding the tumor is free of the mutation. The possibility that activated Ras plays an important role in human tumor development is also supported by animal studies. Virtually all tumors induced by certain chemical carcinogens contain an activated *ras* gene (13,350). Examples are mammary carcinomas in the rat induced by nitrosomethylurea and skin carcinomas in the mouse caused by dimethylbenzanthrazene.

Mos and *raf*: Cytoplasmic Serine/Threonine Kinases

The *mos* oncogene was originally found in Moloney sarcoma virus as a cell-derived insert into the retroviral *env* gene (for reviews, see refs. 36, 326, and 405). Moloney sarcoma virus induces rhabdomyosarcomas in mice and transforms cultured murine fibroblasts. The viral Mos protein is an Env/Mos fusion product; it has a molecular weight of 37 kd, and its first five codons are derived from the 5′ end of the retroviral *env* gene. The v-Mos protein is localized in the soluble portion of the cytoplasm. Several strains of the Moloney murine sarcoma virus exist and differ by point mutations and deletions. The HT1 strain, which is probably closest to the original isolate, carries a *mos* insert that is identical to c-*mos* if amino acid sequences are compared. Another *mos*-containing retrovirus, myeloproliferative sarcoma virus, has a frameshift mutation and appears to initiate *mos* translation from an internal site downstream of this mutation.

The cellular *mos* gene can transform cells in culture with virtually the same efficiency as the viral gene, suggesting that the transforming potential of this gene can be activated by deregulated expression of the cellular product (395). Mos shows homology to cyclic-AMP–dependent protein kinase and to the Src family of tyrosine kinases. The Mos protein is a serine/threonine kinase; the cellular *mos* is an intronless gene that is expressed at very low levels (less than one copy of *mos* RNA per cell) in most tissues. A notable exception are germ cells, where *mos* is highly expressed (134,275,306). Recent investigations show that the cellular Mos protein is a regulator of meiotic maturation. It is required for meiosis, and as a component of the cytostatic factor (CSF), it is essential for the normal arrest of meiosis at metaphase II. Mos also stabilizes maturation-promoting factor (MPF p34[cdc2]) in meiosis. It further associates with and phosphorylates tubulin and can activate MAP kinase. One hypothesis regarding *mos*-induced oncogenic transformation suggests that *mos* imposes a mitotic program onto the interphase cell and thus contributes to reduced adhesiveness, cell rounding, and cytoskeletal reorganization, all part of the transformed phenotype (405).

The *raf* oncogene has been found to be the sole oncogenic determinant in murine sarcoma virus 3611 but has also been isolated as a second oncogene, together with *myc*, from the genome of the MH2 avian carcinoma virus (for reviews see refs. 152 and 283). This avian homolog is still sometimes referred to as *mil*. In the mammalian genome, other *raf*-related genes have been identified, among them A-*raf* and B-*raf*. Cellular *raf* sequences inserted into a retroviral vector induce oncogenic transformation. These sequences are truncated at the 5′-end, a modification that is probably instrumental in activating the oncogenic potential. In cells infected with the 3611 murine sarcoma virus, *raf* is expressed as a 79-kd Gag/Raf fusion protein. This protein is myristilated in its Gag domain and is phosphorylated on serines and threonines. Since the murine Gag protein also occurs in a glycosylated form, there is a corresponding Gag/Raf glycoprotein in cells transformed by virus 3611. In contrast to the mammalian *raf* genes, the avian homolog *mil* does not induce transformation on its own. The *mil* insert of the MH2 avian carcinoma virus is also expressed as a fusion product with Gag. The 100-kd Gag/Mil and the 70-kd Gag/Raf proteins are both cytoplasmic serine/threonine kinases, and they show some homology to the *src* family of tyrosine kinases. Raf is emerging as a second important nodal point besides Ras for the propagation of growth signals. Its position in the signal transmission chain appears to be more proximal than that of Ras but upstream of the MAP kinase cascade. Activated Ras binds to the amino terminal regulatory domain of Raf, and there is a suggested association of Raf with MAP kinase kinase.

Jun, *myc*, and *erb*A: Oncogenes Coding for Transcriptional Regulators

Oncogenes that code for nuclear proteins are somewhat imprecisely referred to as *nuclear oncogenes*. In this category belong *myb*, *myc*, *fos*, *rel*, *ski*, *jun*, *erb*A, *maf*, and *qin* (5,76,77,79,100,106,157,261,273,287,347,366,371). These all code for transcriptional regulators that interact with the genome of the cell. In this section we first survey several nuclear oncogenes by giving brief summary statements and then present a more detailed account of three representative examples.

We begin these summaries with *myb*, which is the oncogene of avian myeloblastosis virus AMV BAI-1 and of the avian leukemia virus E26; it binds to a specific DNA sequence, and the viral oncogenic version of the gene may transform by interference with normal transcriptional control (23,92,224,333). The cellular *myb* gene is preferentially expressed in hematopoietic tissues, controls the expression of myeloid-specific genes, and is required for normal fetal hematopoiesis (245,278,287,322). The *rel* oncogene of avian reticular endotheliosis virus belongs to the NF-kappa B transcription factor family. The viral version of this gene corrupts normal transcriptional regulation by NF-kappa B (218,292). The *ets* oncogene is the second oncogene of E26 avian myeloblastosis virus. It is a member of an important transcription factor family (177,261,375) and can be activated by insertional mutagenesis as well as by transduction (15). The *fos* oncogene is the oncogene of the FBJ and FBR murine osteosarcoma viruses. It is a component of the cellular AP-1 transcription factor complex, and like other transcription factor oncoproteins it transforms by altering normal transcriptional controls (77,115). The product of the *maf* oncogene is structurally and functionally related to the AP-1 family of transcription factors. *Maf* was originally found in avian retrovirus AS42; it can interact with AP-1 proteins (187). The oncogene of avian sarcoma virus 31, *qin*, codes for a transcription factor of the fork-head/HNF-3 family. Fork-head proteins are important regulators of embryonal development (214).

The more detailed examples of nuclear oncogenes will deal with *jun, myc,* and *erb*A. The oncogene *jun* is the cell-derived insert in the genome of avian sarcoma virus 17 (5,77,371). In cells transformed by this virus, *jun* is expressed as a 65-kd Gag/Jun fusion protein concentrated in the nucleus. The cellular Jun protein is, like Fos, a component of the AP-1 transcription factor complex (4,35). It belongs to a multigene family together with the related genes *jun* B and *jun* D (161,305). Areas of homology among these genes roughly correspond to functional domains: (i) a carboxyl-terminal leucine zipper dimerization domain, (ii) an adjacent basic region that makes contact with DNA, and (iii) an amino-terminal domain required for transcriptional activation (77,206,257). Jun proteins can form homodimers, but under physiologic conditions are usually found in heterodimers with Fos family proteins, with proteins of the ATF transcription factor family, and with other leucine zipper proteins. In the heterodimers, the leucine zipper of Jun joins with that of the partner protein in a coiled coil structure (62,127,141,164,199,284,310,311,320,362). The DNA contact domain extends from the leucine zipper, forming a fork that interacts with the major groove for the DNA double helix. The principal functions of Jun show a hierarchical dependence: Dimerization is a prerequisite for DNA binding and transactivation, and DNA-binding is required for transactivation. All three domains, for dimerization, DNA binding, and transactivation, are needed in oncogenic transformation.

Transformation by Jun reflects a disturbance of normal transcriptional controls (370), but the molecular details of this process are still unknown. Viral Jun differs from cellular Jun by a 27-amino-acid deletion in the transactivation domain, by two amino acid substitutions affecting important regulatory sites of the molecule, and by the loss of the 3′ untranslated region, which appears to be responsible for the short half-life of Jun messenger RNA (Fig. 11). All these mutations contribute to the oncogenic potential of Jun. *Jun* acts as a single oncogene in transforming avian cells, and no auxiliary genetic changes appear to be required. In mammalian cells, *jun* transforms only in cooperation with a second oncogene such as activated *ras* or another growth stimulatory factor. Transactivation by the Jun protein is regulated through phosphorylation. Phosphorylation of amino-terminal serines and threonines increases transactivation potential, and phosphorylation of a serine and a DNA-binding domain inhibits transactivation. Transformation depends on the presence of the transactivation domain, but it is not correlated with a general increase in AP-1-dependent transcription that is controlled by AP-1 consensus sequences. The transforming potential of Jun reflects more subtle changes that may involve altered preference for heterodimer partners, for DNA consensus sequences, and hence for target genes. Although numerous target genes responsive to AP-1 are known, the target genes important in the oncogenic action of Jun have not been identified. The possibility that the oncogenic change in Jun reflects a loss of function rather than a gain of function has also not been ruled out.

The first nuclear oncogene discovered was *myc*; it is the oncogenic determinant in avian retroviruses MC29, CMII, MH2, and OK10 (106). Besides c-*myc*, the human genome contains several other *myc*-related genes; examples are N-*myc* and L-*myc* (these two have not been found as oncogenes in retroviruses). *Myc* plays an important yet poorly understood role in some human cancers. Thus, in all Burkitt lymphomas, *myc* is translocated into the vicinity of an immunoglobulin gene and, unlike the nontranslocated allele, is actively expressed (142). Our understanding of the Myc protein has made a great leap forward with the discovery of a small Myc dimerization partner, the Max protein (31,223,272) (Fig. 12). Max associates with Myc via a helix/loop/helix and a leucine zipper motif. Recent genetic experiments show that dimerization with Max is required not only for Myc DNA binding (consensus sequence CACGTG), but also for oncogenic transformation and for Myc-induced apoptosis (2,3). The transformation-specific targets of Myc remain to be identified.

An entirely different regulator of transcription is the product of the oncogene *erb*A (for reviews, see refs. 79, 128, 135, 273, and 366). *Erb*A occurs as the second oncogene in avian erythroblastosis virus (AEV) strains ES4 and R, which also carry the *erb*B oncogene. *Erb*B alone is sufficient for the induction of erythroblastosis and sarcoma. In cultured erythroblasts, expression of the v-*erb*B oncogene induces a high rate of self-renewal without completely blocking a low incidence of spontaneous differentiation to erythrocytes. V-*erb*A does not induce proliferation in erythroblasts, but it blocks spontaneous differentiation and allows cell growth in a wide range of Na- concentrations and pH. Erythroblasts transformed by *erb*B and *erb*A together are nutritionally less fastidious than erythroblasts transformed by *erb*B alone. Through its inhibitory effect on differentiation and by inducing nutritional vigor, *erb*A augments the oncogenicity of *erb*B. *Erb*A is the homolog of the thyroid hormone receptor (308,380,381,410) (Fig. 13). It belongs to a protein family that also includes receptors for glucocorticoid hormones, retinoic acid, and estrogen. These receptor proteins are related in structure and mechanism of action. They contain a ligand-binding domain, a transactivation domain, and a DNA-binding domain with two zinc fingers. They bind to similar but not identical palindromic DNA consensus sequences. Initially, the homology of v-*erb*A to the thyroid hormone receptor led to the suggestion that the oncogenic effect of the *erb*A is due to its interference with the thyroid receptor function. Indeed, the v-ErbA protein can act as a repressor of thyroid hormone-dependent transcription, functioning as a transdominant negative mutant. However, recent investigations have shown that the oncogenic potential of v-*erb*A is primarily due to its inhibitory effect on the retinoic

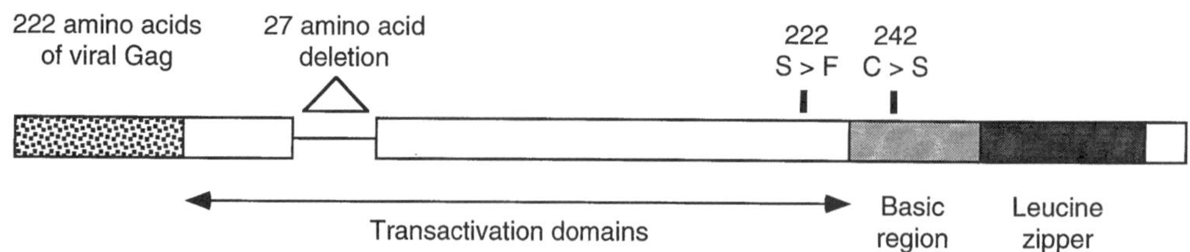

FIG. 11. The Jun protein of avian sarcoma virus 17 differs from its progenitor in chicken cells in the following ways: It is fused at its amino terminal to partial Gag sequences derived from an avian retrovirus. It has a 27-amino-acid deletion affecting the transactivation domains. Two amino acid substitutions interfere with normal posttranslational regulation, and an absence of 3′ untranslated sequences in the message contributes to the greater stability of the v-Jun mRNA. With the exception of the Gag sequences, all changes seen in viral Jun contribute to its enhanced transforming activity as compared to cellular Jun.

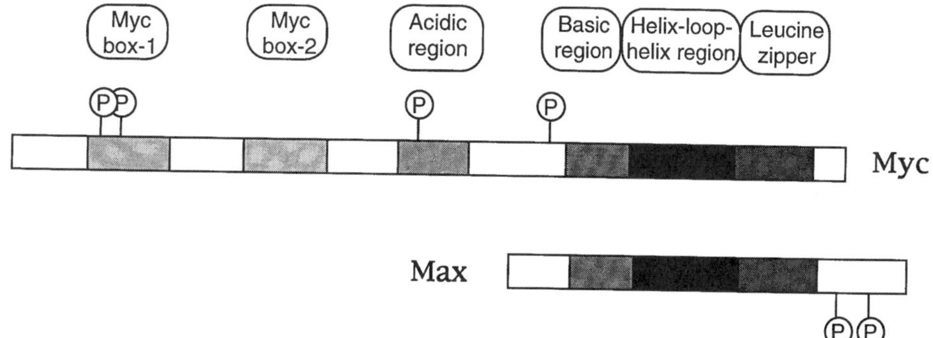

FIG. 12. Comparison of the Myc protein and its dimerization partner Max. Myc and Max associate in the leucine zipper/helix/loop/helix region. The heterodimer binds to the sequence CACGTG. Association with Max is necessary for transcriptional regulation and for transformation by Myc. Myc box-1 and box-2, and the acidic region represent sequences conserved in various myc genes.

acid receptor rather than to the thyroid hormone receptor (319,330). The transformation-related target genes that are dependent on retinoic acid (and possibly thyroid hormone) and affected by the v-*erb*A still remain to be determined.

Cis-Activation of Oncogenes

Many retroviruses that do not carry an oncogene in their genomes are nevertheless capable of inducing tumors in animals (for reviews, see refs. 202, 203, and 256). The kinds of tumor seen with these viruses are similar to the ones caused by transducing retroviruses. They include sarcomas, various forms of leukemia, and carcinomas. However, they uniformly differ from tumors induced by transduction in the latent period of tumor formation.

Nondefective retroviruses lacking an inserted oncogene induce tumors only after a long latent period of several weeks to several months. None of these viruses induces oncogenic transformation in cell culture. The tumors caused by these nontransducing retroviruses have several common properties that are important in understanding their origin. They all contain integrated provirus. The proviral sequences are found in the same chromosomal site in all cells of a given tumor; hence, each tumor is monoclonal, having originated from a single transformed cell. Although infection is initiated with a nondefective virus that undergoes multiple rounds of replication, the proviruses seen in tumors are usually defective, containing only part of the viral genome. The part that is always present and conserved is at least one LTR, and a portion or all of the viral coding sequences may be missing. This observation suggests that

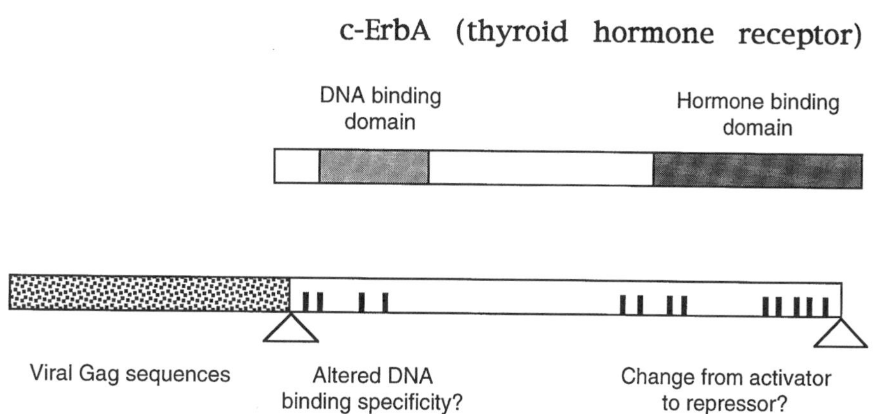

FIG. 13. Structural comparison between the cellular ErbA protein (the thyroid hormone receptor) and viral ErbA. The viral protein differs from its cellular counterpart by the presence of Gag sequences at the amino terminus, by 13-amino-acid substitutions in the coding sequences of ErbA, and by deletions at both termini of the ErbA coding region. These mutations convert ErbA from a cellular transcriptional activator to a viral transcriptional repressor. They also appear to induce changes in DNA binding specificity.

viral coding sequences are not required for the maintenance of the transformed state. Most important in the genesis of these tumors is the fact that viral sequences are integrated at preferred sites of the cellular genomes. In several tumor systems, these sites are located in the immediate vicinity of a cellular oncogene that has also been found in transducing retroviruses.

As a result of retroviral integration nearby, the transcription of the cellular oncogene is elevated in the tumor cells. We refer to this as *insertional activation* or *cis*-activation of an oncogene. Two principal mechanisms of *cis*-activation have been identified: promoter insertion and enhancer insertion. In promoter insertion, a chimeric mRNA is generated that combines R and U5 sequences of the viral LTR with sequences of the cellular oncogene. If the transcripts originate in the 5′ LTR of the provirus, they may also contain partial viral coding sequences. However, transcription starts in the 3′ LTR are more common; in these cases the provirus has usually suffered extensive deletions that extend into, and may include, the 5′ LTR. In fact, the activation of the 3′ LTR promoter may depend on the inactivation of the 5′ promoter. The preferred integration sites of promoter insertions may lie within the cellular oncogene and either truncate the coding sequences or effectively remove noncoding domains that contain negative regulatory elements. For enhancer insertion the provirus need not be integrated in the transcriptional orientation of the cellular oncogene, and its position may also be downstream of the cellular gene. In enhancer insertion the oncogene transcripts do not contain viral sequences. Rather than attempting an exhaustive review, we have again selected representative examples to discuss this mechanism of viral oncogenesis by *cis*-activation.

Avian Leukosis Virus: *cis*-Activation of Cellular-*myc* or Cellular-*erb*B

The Rous-associated viruses types 1 and 2 (RAV-1 and RAV-2) are replication-competent avian retroviruses that do not carry a cell-derived oncogene. In young chickens, these avian leukosis viruses induce B-cell tumors originating in the bursa of Fabricius. The tumors arise after a prolonged latent period, they are monoclonal with respect to proviral integration site, and they overexpress c-*myc*. Most contain a defective provirus that has suffered a 5′ deletion. The provirus is integrated in the vicinity of the cellular *myc* oncogene (152,266) (Fig. 14). The exact integration site varies from tumor to tumor. It is commonly located between exon 1 (which is noncoding) and exon 2 of c-*myc* or within exon 1. The *cis*-activation of c-*myc* then occurs by promoter insertion, generating a chimeric viral/cellular fusion transcript that starts in the 3′ viral LTR. In some tumors the provirus is found downstream of *myc*, and it may also be integrated in the opposite transcriptional orientation. These situations suggest that viral enhancer activity elevates *myc* transcription; the *myc* transcripts do not contain viral sequences. Although provirus insertion often leads to a truncation of the *myc* gene, the deleted sequences are noncoding. Their removal or inactivation seems to contribute to the elevated expression of c-*myc*. The Myc protein in avian B-cell lymphomas has been found to contain point mutations, but there is no evidence that these are necessary for transforming activity. The part played by *myc* in avian lymphomagenesis requires only overexpression of the normal cellular gene. Insertional activation of c-*myc* is also seen in murine and feline leukemias induced by nontransducing, replication-competent retroviruses.

In certain lines of chicken, notably those with the B5 MHC haplotype, RAV-1 induces erythroblastosis instead of lymphoma (120,239,279,280). These tumors also appear after a long latent period of almost 3 months, and they are monoclonal. Unlike the lymphomas, however, they contain an intact, nondefective provirus, integrated near the cellular *erb*B gene. *Erb*B expression is elevated in the tumor cells. The provirus integration sites in *erb*B are tightly clustered in a region that corresponds to the carboxyl terminus of the ligand-binding domain of the EGF receptor. Primary read-through transcripts are synthesized containing the

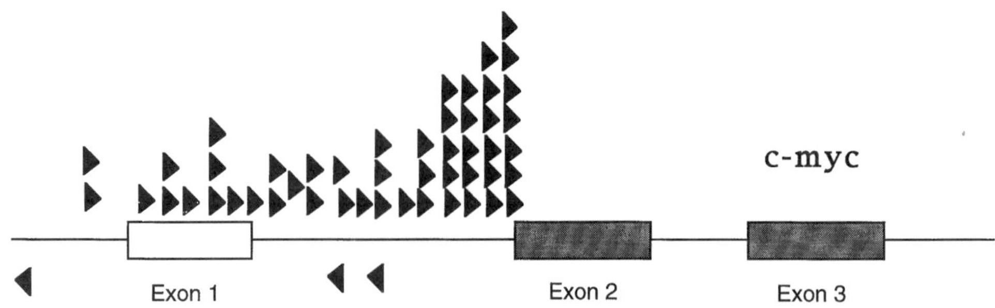

FIG. 14. Integration sites (marked by black triangles) in avian B-cell lymphoma. The figure illustrates the clustering of proviral integration in the noncoding exon 1 of the cellular myc gene and in intron 1. Most of the proviruses are integrated in the orientation of myc transcription (triangles pointing to the right), but some are found in opposite orientation.

entire proviral sequences and the truncated c-*erb*B sequences. RNA processing then creates an mRNA of uniform size with R, U5, partial *gag*, *env*, and *erb*B sequences. This overexpressed *erb*B message codes for an EGF receptor that has deleted the ligand-binding domain at the same site as transduced viral *erb*B. Its mechanism of transformation is probably similar to that of v-*erb*B: a constitutive mitogenic signal originating from the truncated EGF receptor. However, unlike v-*erb*B, the *cis*-activated gene contains the complete 3′-terminus, and the resulting protein retains the regulatory tyrosine phosphorylation sites of the EGF receptor. The absence of some of these sites in the *erb*B insert of avian erythroblastosis virus is therefore not the only factor determining increased tyrosine kinase and oncogenic activity. In RAV-1-induced erythroblastosis the activated *erb*B gene is also incorporated into the viral genome with comparatively high frequency. This results in the generation of new *erb*B-transducing viruses that now induce polyclonal erythroblastosis within a short latent period.

Multiple Targets of Mouse Mammary Tumor Virus

Another retrovirus that does not carry an oncogene in its genome and induces tumors by insertional activation of a cellular oncogene is mouse mammary tumor virus (MMTV). MMTV has several preferred integration sites in tumor target cells. The ones most frequently used are close to three cellular genes: *int*-1, *int*-2, and *int*-3 (for reviews, see refs. 203, 255, 256, and 267). These genes are located on different chromosomes; furthermore, despite the identical names, there is no sequence relationship between the three. In the mammary tumors the MMTV provirus enhances transcription of the adjacent cellular *int* locus, and this increased expression probably plays a key role in tumor development. Although none of the *int* sequences has been found as transduced oncogenes in naturally occurring retroviruses, there are good reasons to believe that these cellular loci represent oncogenes.

Int-1, now referred to as *wnt*-1 (an abbreviation derived from the name of the homologous segment polarity gene *wingless* of the *Drosophila* and *int*), belongs to a gene family that fulfills important functions in pattern formation during embryonal development from *Drosophila* to mammals (203,293). *Wnt*-1 codes for a glycoprotein that is secreted but remains closely associated with the cell of origin. It is probably a component of a short-range developmental signal. Ectopic expression of *wnt*-1 in mammary cell lines induces at least partial oncogenic transformation (45).

Int-2 codes for a protein that belongs to the fibroblast growth factor family (91). It also appears to be a secreted protein and may act as a growth factor. Another member of this family, *hst*, is occasionally activated in mammary tumors. *Hst* has also been isolated from human tumors (Kaposi's sarcoma and stomach cancer) as a transforming oncogene (85,352). The *int*-3 oncogene also codes for a protein with presumed developmental function and belongs to the *notch* family of developmental regulators (295).

Numerous loci have also been found insertionally activated in feline and murine retroviral leukemias. Significantly, those activated genes whose normal functions are known are all related to control of cell growth (252).

Transformation by Molecular Mimicry and Insertional Mutagenesis: Friend Leukemia Virus

The Friend leukemia virus occupies a unique position among retroviruses with respect to mechanism of oncogenic transformation. Key factors in this process are an altered viral envelope protein and preferred sites of provirus integration. The latter include *cis*-activation, justifying assignment of Friend virus in this category. However, Friend virus also inactivates the cellular p53 tumor suppressor gene, and some of this inactivation occurs by insertional mutagenesis. Friend leukemia virus causes erythroid malignancies that result from a two-step transformation (for reviews, see refs. 15, 78, 179, 182, 303, and 394). Virus stocks contain two agents: a replication-competent retrovirus and a replication-defective virus called *spleen-focus-forming virus* (SFFV), with the former acting as a helper virus for the latter.

Infection of mice with SFFV and its helper leads first to a polyclonal proliferation of infected erythroblasts, causing pronounced splenomegaly. The replicating erythroblasts are neither immortal nor transplantable in syngeneic animals. The pool of dividing cells is maintained and increased through recruitment of new target erythroblasts by replicating virus. The second step of transformation in Friend disease occurs only in one or a few cells in an infected animal. It results in immortality and tumorigenicity and is due to a discrete genetic change. A molecular analysis of this two-step leukemogenesis has revealed complex interactions between SFFV, helper virus, and target cells and tissues. Helper-free stocks of SFFV have been produced in packaging cell lines. These cell lines contain helper virus information necessary to produce infectious SFFV, but cannot synthesize infectious helper virus. Helper-free SFFV infects cells but does not produce infectious progeny. Its pathogenicity appears to be restricted to the first stage of Friend disease. Mutant analysis of the SFFV genome indicates that this initial, polyclonal proliferation of erythroblasts is caused by the expression of the altered Env protein of SFFV, gp55. Gp55 is the only gene product of SFFV; it acts as a mitogen for erythroblasts, probably by binding to and activating the erythropoietin receptor (78,215). This is a unique situation with viral coding sequences producing an altered virion protein that stimulates division of a specific target cell, possibly by molecular mimicry.

The second step in Friend virus–induced leukemia is due to a rare event that leads to the development of a monoclonal tumor consisting of transplantable, immortal, transformed erythroblasts. Second-stage Friend virus–induced malignancies have two important properties in common: (i) The provirus is integrated in either one of two preferred loci, referred to as *fli*-1 and *spi*-1. The cellular *fli*-1 and *spi*-1 genes code for members of the Ets family of transcription factors; the Fli-1 and Spi-1 proteins are functionally distinct, and their activation is associated with different variants of Friend virus disease. (ii) Many Friend virus–induced leukemias also show inactivation of p53. Some of these inactivations are the result of proviral integration in one of the p53 alleles, with consequent loss of wild-type function and reduction to an inactivated homozygous state (15,179). The activation of Ets-related transcription factors and the inactivation of p53 represent rare genetic changes probably contingent upon continuous helper virus–dependent replication that increases the probability for critical integration events.

Factors Affecting *cis*-Activation of Oncogenes

The active replication of a nontransducing competent retrovirus in a susceptible host and the presumably accidental insertional mutation or *cis*-activation of a cellular oncogene are not the only factors important in tumorigenesis by nontransducing retroviruses.

A systematic genetic analysis of the nontransducing retrovirus genomes for oncogenic potential has revealed a surprising multiplicity of important sequences and has demonstrated the complexity of their interactions. Among both avian and murine retroviruses, there are strains of high and low leukemogenicity. Recombinants between these strains, as well as mutants of these strains, show that the LTR is a major determinant (but not the sole one) of oncogenicity (46,87–89,297,298). The coding information for *gag*, *pol*, and *env* can be involved to varying degrees as well. In some murine leukemia viruses (e.g., Moloney leukemia virus), viral replication during the latent period of the disease gives rise to recombinants with endogenous retroviral information in the *env* gene (65,147). These recombinant “MCF” viruses multiply more actively in hematopoietic target tissue than does parental virus, and this increased replication is an important factor in leukemogenicity. The multiplicity of contributing factors suggests that every step of virus replication—from efficient entry into specific target cells to active transcription of the provirus and maturation of highly infectious progeny—is important in increasing the probability of *cis*-activation.

Several questions on the mechanism of *cis*-activation remain to be answered. One concerns the nature of the long latent period. The mere fact that insertional activation of the cellular oncogene is a rare event does not account for the fixed minimal duration of the latent period, which suggests the existence of some sort of threshold condition that has to be met. The monoclonality of the tumors indicates that the transformation to the oncogenic phenotype is rare. It is compatible with the idea that two or more independent events are necessary to convey tumorigenicity to the infected cell. The possibility of more than one necessary event is also suggested by the fact that retroviruses that cause cancer through *cis*-activation of a cellular oncogene have not been found to induce transformation in cell culture. If a single *cis*-activating event were sufficient, it should be detectable *in vitro*. Another puzzling aspect in *cis*-activation is the tissue tropism of oncogenesis. We do not understand why RAV activates *myc* in some chicken lines and in others affects *erb*B or why some mouse strains develop B-cell and others T-cell or myeloid lymphomas in response to infection with the same murine leukemia virus.

Oncogenicity Induced by *trans*-Activating Viral Proteins

The mechanism of viral oncogenesis by transduction of a cellular oncogene is supported by compelling genetic evidence. The *cis*-activation of an endogenous oncogene by a neighboring provirus is linked to viral oncogenesis by more circumstantial evidence. The third mechanism by which retroviruses may induce tumors is still largely hypothetical. According to this model, a nonstructural virus-coded protein acts as a transcriptional regulator to alter the expression of one or several cellular genes important in the control of cell growth; this in turn leads to genetic changes that ultimately result in the oncogenic phenotype.

Some observations on human T-cell lymphotropic virus (HTLV-I) and on human adult T-cell leukemia (ATL) associated with HTLV-I infection are in accord with this hypothetical mechanism of oncogenesis by transcriptional transactivation (for reviews, see refs. 123, 250, and 321). The fundamental features of HTLV-I in ATL are as follows: (i) All cases of ATL harbor the HTLV-I provirus. (ii) The provirus is found in the same chromosomal site in all cells of a given case of ATL; hence, ATL is a monoclonal disease. (iii) Integration of HTLV-I does not occur in preferred sites of the host genome. (iv) HTLV-I does not carry a cell-derived oncogene within its genome. (v) The HTLV-I provirus is not expressed in ATL cells. (vi) If ATL cells are placed in culture, viral gene expression is turned on.

The presence of HTLV-I provirus in all cases of ATL indicates an etiologic association between HTLV-I and ATL. The monoclonality of the tumors tells us that oncogenic transformation is a rare, and by no means inevitable, consequence of infection. It also suggests the necessity for a second and possibly third contributing event. Less than 1.0% of all individuals infected with HTLV-I later develop ATL; the latent period can be in excess of 20 years. The absence of a preferred integration site for HTLV-I in ATL shows that there is no *cis*-activation of a cellular oncogene;

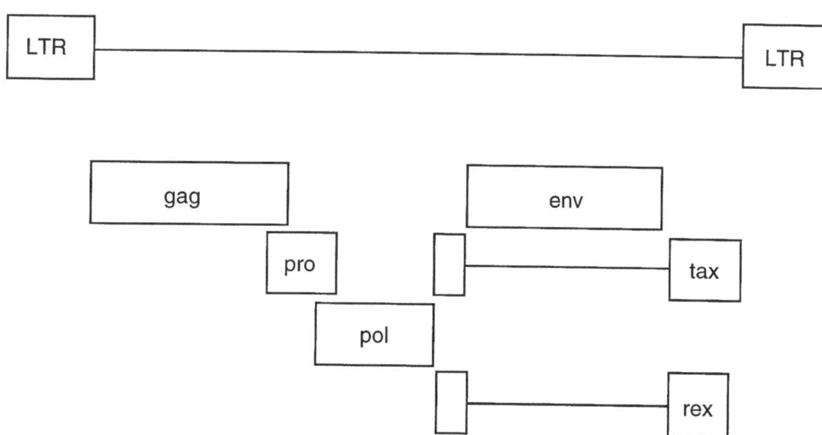

FIG. 15. Coding organization of human T lymphotropic virus 1 (HTLV-1). gag and env code for the major structural virion proteins; pro codes for a protease and pol codes for the reverse transcriptase. Two non-structural proteins are Tax and Rex. Tax is suspected of playing a role in HTLV-induced leukemogenesis.

and since there is no transduced oncogene in the HTLV-I genome, ATL must be caused by a unique mechanism of viral transformation. The lack of viral expression distinguishes ATL from most other retroviral tumors and implies that although a viral gene product may be required for the initiation of transformation, it is not required for the maintenance of the transformed phenotype. The absence of viral gene expression in the fully developed leukemia also supports the probable necessity for additional events that make the growth of the leukemic cell independent of continuous viral intervention.

The HTLV-1 genome contains coding information for several nonstructural proteins (Fig. 15). The best studied of these is the product of the *tax* gene, p40 (54). This protein functions as a transcriptional activator for the HTLV-I genome, probably in conjunction with cellular transcription factors. The target sequences are located in the LTR of the virus. Could *tax* be affecting the transcription of cellular genes? This possibility has received support from work with transgenic mice carrying transgene *tax* under the control of the HTLV-I LTR. A group of these animals expressed p40 in muscle tissue. These mice developed multiple soft-tissue sarcomas (249). The Tax protein has also been shown to transform fibroblasts in culture together with Ras (271). Similar results have also been obtained with transgenic mice that carry the *tat* gene of human immunodeficiency virus HIV. These animals develop Kaposi's sarcoma (368). Could *tax*, *tat*, and other regulatory genes of retroviruses function as oncogenes? Since *tax* is a viral gene without a cellular homolog and is required for efficient replication of HTLV, the situation is formally similar to that of DNA virus–induced transformation, where the oncogenic effectors are also essential viral genes. The data on ATL and on transgenic mice are compatible with the proposal that *tax* affects the expression of some key cellular regulatory genes. In the case of ATL, important targets appear to be the transcription factors CREB/ATF and NF-kappa B, which become stably up-regulated. Inhibition of NF-kappa B in leukemic cells with antisense constructs leads to tumor regression in the nude mouse model (197,342).

Another possible mechanism by which *tax* and similar retroviral regulatory genes could cause cancer is by inducing genetic instability in the host cell rather than by interfering with transcriptional controls. There is some evidence in favor of this suggestion (307).

MECHANISMS OF CELL TRANSFORMATION BY THE DNA TUMOR VIRUSES

The DNA Tumor Viruses

Each major DNA virus family includes viruses that have the capacity to elicit cellular transformation (see Table 1). This represents an extremely diverse group of viruses with very different structures, genome organizations, and strategies of replication. In some cases the DNA viruses have been shown to be responsible for inducing tumors in their natural hosts, including the induction of human cancers, whereas in other cases they are only known to induce transformation of cells in culture and form tumors in experimental animal systems. For instance, a variety of data strongly implicates various types of the human papillomaviruses in the development of cervical carcinoma (412). EBV has long been associated with the development of Burkitt's lymphoma as well as nasopharyngeal carcinoma, and epidemiologic data provide compelling evidence for a role of hepatitis B virus in the development of hepatocellular carcinoma (413). In contrast, although adenoviruses can form tumors in rodents and the viral genes E1A and E1B can transform human cells in culture, there is no evidence that adenoviruses are responsible for any human cancers (225). Similarly, SV40 and polyomavirus, which are clearly capable of forming tumors in experimental settings; have not been shown to be oncogenic in a natural setting. Nevertheless, it is these small DNA tumor viruses, including the adenoviruses, polyomaviruses, and papillomaviruses, that have been studied in greatest depth and that have provided much of our current understanding of viral-mediated oncogenic transformation. This is principally due to the utility of these viruses as model systems for the study of gene expression and DNA replication, to name but two examples. Thus, understanding of the oncogenic process has been aided by the concurrent elucidation of molecular mechanisms of action of the tumor virus oncoproteins.

Transformation by the DNA Tumor Viruses: The Relationship to Lytic Viral Growth

In most cases the oncogenic properties of the retroviruses (RNA tumor viruses) are unrelated to the requirements for productive viral growth. Indeed, in the vast majority of cases the oncogenic retroviruses are defective for replication because they lose essential viral sequences during formation. In sharp contrast, the oncogenic properties of the DNA tumor viruses are intimately associated with their ability to carry out a productive infection. Whereas the oncogenes of the retroviruses are cellular genes that have been acquired as a consequence of the integration of the viral genome into the host cell chromosome, and then selected because of their oncogenic activity, the DNA tumor virus oncogenes are essential viral genes that bear little or no relationship to cellular counterparts. Thus, infection of a cell that is permissive for viral replication will generally result in a productive infection, liberating increased numbers of viral particles and leading to eventual cell death. A transformation event that had been initiated would never be scored. It is only when the viral infection takes place under nonpermissive circumstances, in which the viral replication process is aborted, that a transformation event can be observed.

The frequency of cellular transformation following infection of a nonpermissive cell by adenovirus, SV40, or polyomavirus is very low, usually less than 10^{-5}. In large part, this inefficiency reflects the lack of a specific mechanism for integration of the DNA tumor virus genome into the host cell chromosome. Thus, the transformation process is extremely inefficient because of the low probability that the viral genes essential for transformation will be integrated intact and in a state that will allow transcription at appropriate levels. The lack of a system for analyzing a productive papillomavirus infection precludes such determinations for this virus group.

The oncogenic events mediated by the DNA tumor virus oncoproteins reflect the ability of these viruses to stimulate a quiescent, nongrowing cell to enter the cell cycle. For instance, a normal target for infection by adenovirus is a terminally differentiated epithelial cell that lines the upper respiratory tract. Since this cell is not in the cell cycle, the essential substrates for the replication of both viral and cellular DNA are in short supply and thus limiting. In particular, the levels of deoxyribonucleotides are tightly regulated during the cell cycle, and only rise to non-rate-limiting levels during S phase (357). The ability of the virus to stimulate such a cell to enter S phase, creating an environment for DNA replication, is thus critical for the efficient replication of the virus. This virus-mediated S-phase induction is dependent on the viral genes that also elicit transformation in other contexts. Thus, if the infection does not proceed to completion, either because of a nonpermissive cell type or as a consequence of a viral mutation that blocks viral growth, the disruption of cell growth control that was intended to prepare the cell for the viral infection may then lead to the development of a transformed cell.

The DNA Tumor Virus Oncogenes

A viral infection can be divided into an early and late phase as defined by the timing of viral gene expression relative to viral DNA replication. In general, the early genes encode proteins that prepare the infected cell to replicate viral DNA, whereas the late gene products include the structural components of the virion. For the DNA tumor viruses, it is these early gene products, or at least a subset of the early products, that are responsible for oncogenic transformation. What follows is a general description of the adenovirus, polyomavirus, and papillomavirus genes that encode the transforming activities. For a more detailed description of these viral genes and gene products, refer to the appropriate sections in the chapters on the specific viruses.

Adenovirus

The identification of the adenovirus genes responsible for oncogenic transformation was accomplished by the mapping of viral sequences that were retained in an integrated state in transformed cells (93,122) as well as by mapping the viral transcripts present in the transformed cell (331). Such analyses indicated that a full range of viral DNA sequences could be recovered from transformed cells but that it was the left end of the viral chromosome that was similar in every case. Transfection assays provided a direct demonstration that these viral sequences could indeed mediate transformation (137,138). Subsequent studies identified two distinct transcription units within this region that encode stable mRNAs found in the cytoplasm of transformed cells and during the early phase of a productive infection (18). The two transcription units, termed E1A and

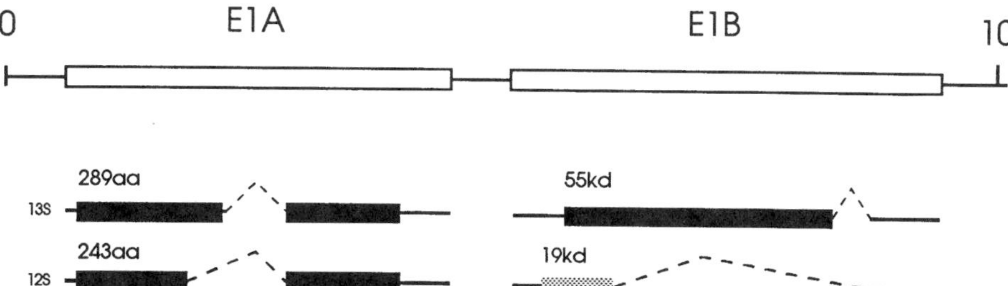

FIG. 16. The adenovirus-transforming genes. The left end of the adenovirus chromosome (zero to ten map units) is depicted that includes the E1A and E1B transcription units. The major mRNA products of each region are depicted as well as the coding sequences contained within each (black boxes). The E1A mRNAs are commonly referred to by their sedimentation coefficients (13S and 12S) and the protein products by the amino acid residues. The E1B products are generally identified by their molecular weights.

E1B, each encode two major messenger RNAs through alternative splicing of the two primary transcripts (see Fig. 16).

The major products of the E1A gene include proteins of 289 and 243 amino acids, encoded by the so-called $E1A_{13S}$ and $E1A_{12S}$ mRNAs, respectively, that are identical in sequence except for the additional 43 amino acids found in the middle of the $E1A_{13S}$ product. The $E1A_{13S}$ product is critical for the activation of viral transcription during a lytic infection (17,181), dependent on the CR3 sequences unique to this protein (291). This activity is not important, however, for transformation. Rather, the ability of distinct domains of the protein to interact with a series of cellular proteins, including the retinoblastoma gene product (Rb), does coincide with the transforming activity. The two major E1B gene products, a 55-kd protein of 495 amino acids and a 19-kd protein of 175 amino acids, are encoded by alternatively spliced mRNAs. The two E1B proteins do not share a sequence, and they appear to perform distinct although related functions. The 55-kd E1B protein, in concert with a 34-kd product of the early region 4 gene, functions to facilitate transport of viral mRNA from the nucleus to the cytoplasm (7,209,269). In addition, the 55-kd protein also interacts with at least one cellular protein, the p53 tumor suppressor (309), and it is this last event that correlates with the transforming function (406).

Polyomavirus

Although the polyomavirus group includes a large number of viruses that infect a variety of cell types and host species, the two best-studied viruses of this general group include the mouse polyomavirus and the monkey SV40 virus. These two viruses are very closely related in genome structure and DNA sequence, but they differ with respect to the organization and function of the early genes that carry the transforming activity of the virus (Fig. 17). Both viruses encode a multifunctional protein termed large T-antigen that is directly involved in the initiation of viral DNA replication through specific binding to the origin of replication (86,180). Although the origin-binding capacity, as well as associated ATPase (129,359) and helicase (81,346) activities, are essential for DNA replication and a productive infection, these activities are not important for transformation (6,131,184,228,274). Rather, the ability of the viral protein to bind to a variety of cellular proteins is critical for transformation. In this regard, the function of large T-antigen of SV40 and polyomavirus overlap but are also distinct. Specifically, although both bind to Rb, only SV40 large T-antigen binds to p53.

In addition, both viruses encode a low-molecular-weight protein termed small t-antigen that contains sequences in common with the amino terminus of large T-antigen in addition to unique sequences. Two activities have been associated with the small t-antigen: the ability to bind to the PP2A cellular protein phosphatase (260,373), inhibiting activity for some substrates (403), and the capacity to activate transcription dependent on polymerase II and polymerase III (220). At least one target for this transcription activation is the E2F transcription factor (221), which is also a target for activation by adenovirus E1A, SV40 large T-antigen, and the HPV E7 protein (251). Although small t-antigen does not appear to be essential in a lytic infection of tissue culture cells, it does contribute to transformation efficiency in conjunction with large T-antigen (24,39).

Unlike SV40, polyomavirus encodes a third distinct early protein, termed middle T-antigen, that is responsible for the principal transforming activity of the virus. Middle T-antigen also shares sequence with the N terminus of large T-antigen and small T-antigen but then, as a result of the use of an alternative splice acceptor, the reading frame is changed from that used to encode the large T-antigen (Fig. 17). Unlike the other polyomavirus and adenovirus transforming proteins, which are localized in the nucleus, the polyoma middle T-antigen is an integral membrane protein primarily associated with the plasma membrane (176,325). As we shall see later, middle T-antigen binds to and activates src family tyrosine kinases as well as the phosphatidylinositol-3 kinase, PI-3 (248).

SV40

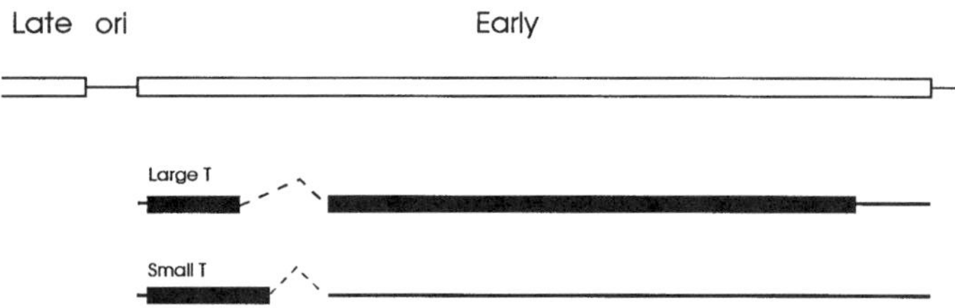

Polyoma

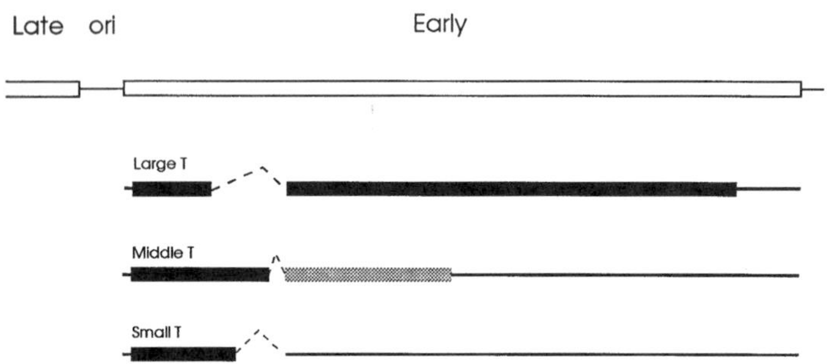

FIG. 17. The transforming genes of SV40 and polyomavirus. The early regions of SV40 and polyomavirus are depicted together with the major mRNA products. The region between the early and late coding sequences referred to as "ori" contains the origin of DNA replication as well as transcriptional regulatory sequences. Coding sequences for large-, middle-, and small T-antigen are shown by the boxes. The splice acceptor for polyoma large T and middle T are distinct, resulting in a change in the reading frame.

Papillomavirus

Although the papillomaviruses are grouped within the general category of the papovaviruses, they are quite distinct from the polyomaviruses in structure, gene organization, and nucleotide sequence. The majority of initial work directed at papillomavirus-mediated transformation focused on the bovine papillomaviruses, particularly BPV-1. Recently, however, considerable attention has been paid to the human papillomaviruses, particularly because of their association with human cervical carcinoma as well as the realization of the mechanistic similarities with the adenovirus and SV40 oncogene products.

The papillomavirus genetic information is read from only one strand, with distinct transcripts being produced by alternative transcription start sites, alternative splicing, and alternative polyadenylation (refer to Chapter 29). The genes responsible for oncogenic transformation have been identified through transfection assays using *in vitro* cell culture systems (98,222) as well as through determination of viral sequences that are retained and expressed in tumor cells (10,318,323,343). These studies have shown that the E5 gene of BPV encodes the major transforming activity of this virus. The E6 and the E7 gene products also contribute to transformation, but E5 is clearly the major activity. In contrast, E6 and E7 gene products of the human papillomaviruses (Fig. 18) are primarily responsible for eliciting transformation and are always found to be expressed in tumor cells. In addition, the integrated papillomavirus DNA found in human tumor cells usually is disrupted in the E2 gene, the product of which negatively regulates transcription of the E6, E7 promoter (10).

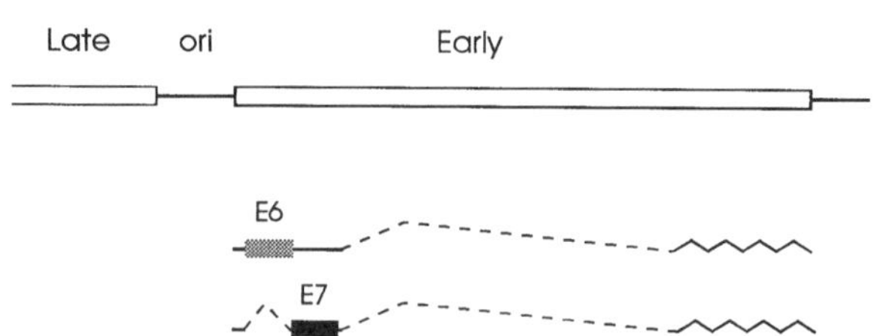

FIG. 18. The papillomavirus-transforming genes. The early region of the human papillomavirus genome is depicted together with the structure of the major mRNAs that encode the E6 and E7 products. Both RNAs derive from a common promoter that starts transcription at nucleotide 97. The wavy lines representing 3′ sequence in the mRNAs reflects the use of splice acceptors found in cellular sequences at the site of integration of the viral DNA.

Common Cellular Targets and Strategies for the DNA Tumor Virus Oncoproteins

Although the adenoviruses, polyomaviruses, and papillomaviruses are quite distinct and it is very likely that they do not share a common evolutionary relationship, the actions of the transforming proteins of the three groups of viruses appear to be remarkably similar. Much of the current understanding of the actions of the DNA tumor virus proteins has derived from the analysis the of physical associations of these viral proteins with a variety of cellular polypeptides. Initial studies, directed at the function of SV40 large T-antigen, led to the identification of a 53-kd cellular protein that was found in T-antigen immunoprecipitates (207,217). Further work demonstrated a specific interaction between T-antigen and this cellular protein, which became known as p53, and subsequent studies have shown that most of the small DNA tumor viruses encode a protein that interacts with p53 (see Table 2).

E1A Binding Proteins

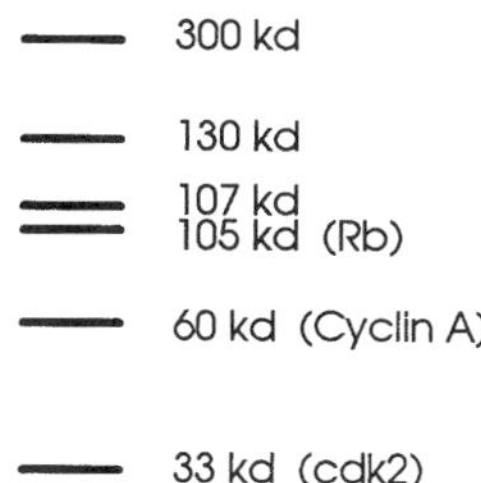

E1A Functional Domains

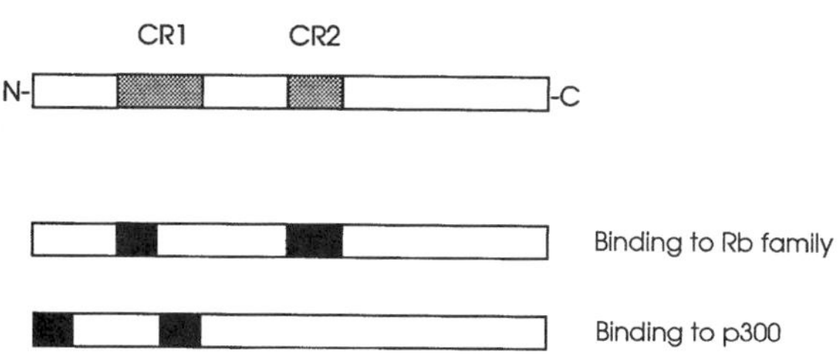

FIG. 19. Binding of E1A to cellular proteins. Schematic representation of the cellular proteins that are recovered in co-immunoprecipitation assays with the adenovirus E1A protein (145,404). Shown above is a schematic of the E1A gene indicating the positions of the CR1 and CR2 domains as well as the regions involved in binding to the various cellular proteins. The Rb family includes the p105 Rb protein as well as the p130 and p107 proteins. CR1 sequences appear to be involved in binding to both the p300 protein and the Rb family.

The one clear exception among the small DNA tumor viruses is polyomavirus, which shows no evidence of encoding a protein that binds to p53. Most important, the ability of the adenovirus 55-kd E1B protein (406), the human papillomavirus E6 product (385), and SV40 T-antigen to associate with p53 coincides with the transforming activity of the viral proteins.

Similar studies identified a large number of cellular polypeptides associated with the adenovirus E1A product (145,404). Analysis of E1A sequences essential for binding to these cellular proteins revealed two domains of interaction (Fig. 19). One, involving the N-terminal sequence and a portion of the CR1 domain, bound to a 300-kd protein, whereas the other, including the CR1 and CR2 domains, bound to a series of proteins ranging in molecular weight from 130 kd to 33 kd. Two subsequent findings on these E1A interactions were of enormous significance, both with respect to developing an understanding of the action of the DNA tumor virus oncoproteins as well as normal mechanisms of cell growth control. First, one of the E1A-associated proteins, a polypeptide of 105 kd, was identified as the product of the retinoblastoma susceptibility gene (391). This simple finding had immediate and far-reaching implications, since this was the first suggestion that the action of a DNA tumor virus–transforming protein could be viewed as inactivating the function of a cellular growth–suppressing protein.

Second, it soon became evident that the binding of the adenovirus E1A protein to the Rb tumor suppressor protein was an activity shared with the other DNA tumor viruses (Table 5). SV40 and polyomavirus large T-antigen, as well as the E7 product of the human papillomaviruses, were all found to bind to Rb (83,99). Moreover, the ability of each viral protein to bind to Rb was found to depend on viral sequences that were also important for oncogenic activity. Comparison of sequences of the viral proteins revealed a short region of homology that included the sequence important for binding to Rb (112) (Fig. 20). Although the homology between the viral proteins is limited to these sequences and is not consistent with an evolutionary relationship among the viruses, it does suggest that the viruses have each acquired a common activity.

Because the ability of the various DNA tumor virus proteins to interact with Rb and p53 coincides with the trans-

TABLE 5. *Cellular targets of the DNA tumor virus oncoproteins*

Virus	Gene product	Cellular target
Adenovirus	E1A	Rb
	E1B	p53
SV40	T antigen	Rb, p53
Papillomavirus	E7	Rb
	E6	p53

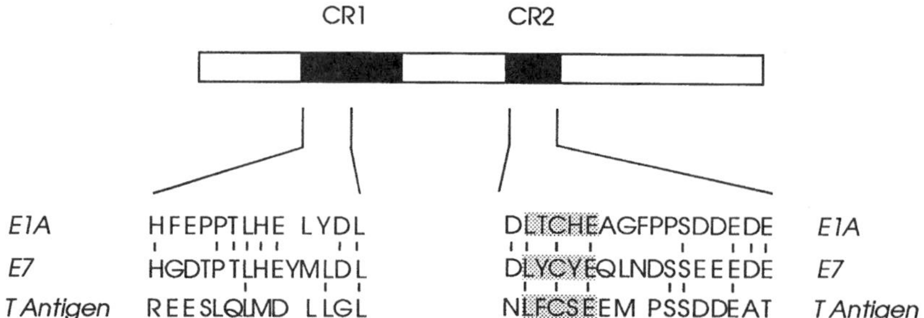

FIG. 20. Homology in the viral sequences involved in binding to the Rb family of proteins. The regions in E1A that exhibit homology with sequences in SV40 T-antigen and HPV E7 are depicted (112). The L-X-C-X-E motif found in the CR2 region of E1A, which is also shared with various cellular proteins (84), including the D-type cyclins (233,399), is indicated by the hatched box.

forming activity of the viral proteins, and the loss of Rb and p53 gene function is associated with tumors development, it is implied that the viral proteins achieve a common result by binding to the cellular regulatory proteins. This point is further emphasized by the relationship between HPV gene expression and the state of the Rb and p53 genes in cervical carcinoma cell lines (Table 6). The vast majority of cervical carcinomas are associated with the so-called high-risk types of HPV, primarily HPV 16 and HPV 18 (412). An analysis of cell lines derived from cervical carcinomas revealed the presence of normal wild-type Rb and p53 genes in those cells that were HPV positive and that expressed the E6 and E7 viral gene products, whereas in two cell lines that were HPV negative, mutations were found in the Rb and p53 genes (313). It is apparent from these results that loss of Rb and p53 function is a common event in the development of these human tumors, either as a result of mutation or through the action of the viral proteins.

Common Mechanisms of Action of the DNA Tumor Virus Oncoproteins

The common association of the DNA tumor virus proteins with two cellular growth regulatory proteins implies that these otherwise diverse viruses share a common need. Although the inactivation of Rb and p53 is most often viewed in the context of oncogenesis, the targeting of these proteins by DNA virus regulatory proteins implies a role in facilitating a productive infection. Given the common properties of the DNA tumor virus oncoproteins in binding to key cellular growth regulatory proteins, we will discuss the common aspects of their function rather than discuss each individual viral protein in isolation.

TABLE 6. *Relationship of HPV E6 and E7 expression and the state of the RB1 and p53 genes in human cervical carcinoma cell lines*[a]

Cell line	HPV DNA/RNA	RB1 gene	p53 gene
HeLa	Yes	WT	WT
C4-II	Yes	WT	WT
SiHa	Yes	WT	WT
CaSki	Yes	WT	WT
ME180	Yes	WT	WT
C-33A	No	Mutant	Mutant
HT-3	No	Mutant	Mutant

[a]Data from Scheffner et al., ref. 313.

Inactivation of Retinoblastoma Protein Function

As described earlier, adenovirus E1A, SV40 large T-antigen, and human papillomavirus E7 physically interact with the retinoblastoma protein. Various studies have now shown that these interactions involve viral protein sequences that are also critical for transforming activity, and thus it is believed that this association is part of the transforming function. Nevertheless, until recently the normal function of the Rb protein and the significance of this interaction with viral proteins was unclear. Two separate lines of investigation converged to provide insights into this interaction.

Studies directed at the action of the E1A proteins in viral transcription activation during a lytic infection led to the identification of a cellular transcription factor, termed E2F, that was important for transcription of the adenovirus E2 gene (200). Subsequent studies revealed that this cellular transcription factor was normally complexed to other cellular proteins in most cell types, that these interactions prevented the activation of E2 transcription, and that the E1A protein possessed the capacity to disrupt these complexes, releasing E2F that could be utilized for E2 transcription (8). The ability of E1A to carry out this dissociation was found to be dependent on viral sequences that were known to be important for oncogenic activity, including binding to the Rb protein (286). This finding suggested the relationship depicted in Fig. 21 in which the interaction of E1A with a cellular protein such as Rb, as previously described by immunoprecipitation assays, could be viewed as the consequence of E2F complex disruption. A variety of ex-

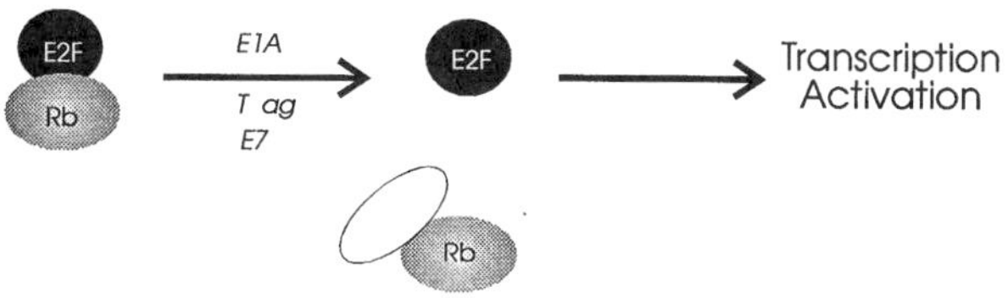

FIG. 21. Inactivation of Rb-mediated control of E2F activity. Depicted is the process of E1A- (and T-antigen- and E7-) mediated release of E2F from the complex with Rb. This action leaves the viral protein firmly associated with Rb and activates the transcriptional capacity of E2F.

periments have now shown that the Rb protein is a component of the E2F complexes, along with the majority of the other proteins previously identified as E1A-binding proteins (9,11,12,50,53,90,340).

At the same time, other studies demonstrated that cellular protein complexes involving the Rb protein could bind to DNA, and the elucidation of the sequence specificity of these interactions identified the E2F recognition sequence (61). Thus, studies with two distinct approaches came to the conclusion that the E2F transcription factor was physically associated with the Rb protein and that the E1A protein could disrupt this complex, leaving E1A bound to Rb. Additional assays demonstrated that the other DNA tumor virus proteins known to bind to the Rb protein, SV40 T-antigen and HPV E7, also possessed the ability to disrupt the E2F/Rb complex (52). Thus, it would appear that the common ability of the DNA tumor virus oncoproteins E1A, T-antigen, and E7 to bind to the Rb protein is a reflection of their ability to disrupt the E2F/Rb complex.

Two additional E1A-associated proteins identified by co-immunoprecipitation, the so-called p107 and p130 polypeptides, are now known to be related to Rb in sequence and function (108,143,216,235). Each member of this family of Rb-related proteins has been shown to bind to the E2F transcription factor, dependent on amino acid sequences that are common to the proteins. These are also the sequences to which the viral transforming proteins bind and that have been termed the Rb pocket.

The binding of Rb or related proteins to E2F inhibits the transcriptional activation capacity of the E2F factor, without affecting DNA-binding capacity (158,159,408). This finding is consistent with analyses of E2F1, a cDNA clone that encodes one member of the E2F family (154,183,329), in that the E2F1 amino acid sequences recognized by Rb are also sequences essential for transcription activation (74,154,183,329). Thus, the action of the viral oncoproteins in releasing E2F from the complex results in the activation of E2F transcription function.

The activation of E2F has direct benefit for adenovirus transcription during a productive infection, since E2F is utilized for transcription of the viral E2 gene (277). Indeed, a product of the viral E4 gene has evolved to maximize the use of E2F for E2 transcription by promoting the formation of a stable promoter complex (144,165,248). Nevertheless, there are no E2F binding sites in either the SV40 genome or the papillomavirus genome, and thus the activation of E2F by T-antigen and the E7 protein cannot directly benefit these viruses. The identification of a group of cellular genes that appear to be targets for the action of E2F has provided a potential explanation for the strategy of these DNA tumor virus early proteins in facilitating a productive infection as well as achieving oncogenic transformation through the activation of E2F.

Numerous experiments have shown that the infection of quiescent cells by adenovirus, SV40, and polyoma results in the activation of S phase in these infected cells, and several studies have documented the induction of cellular DNA synthesis as well as activities associated with entry into S phase that typically encode enzymes involved in both deoxynucleotide biosynthesis and DNA synthesis directly (97,116,117,148,150,186,193-196,208,332,402). For instance, dihydrofolate recductase, thymidine kinase, ribonucleotide reductase and DNA polymerase a are all induced when quiescent cells are infected with polyoma virus. This activation probably reflects the need of the virus to create an environment appropriate for viral DNA synthesis. For instance, the levels of deoxynucleotides are very low in quiescent cells, and they only rise as a consequence of the induction of ribonucleotide reductase activity when such cells are stimulated to grow and enter S phase (30,105). Thus, when a DNA tumor virus infects a quiescent, nongrowing cell, which is the normal target for infection by these viruses, the environment for DNA replication, whether it be viral or cellular, is not suitable. By activating genes encoding the enzymes that create the appropriate environment, the virus can efficiently replicate viral DNA.

The study of E2F function has provided a connection between the action of the viral oncoproteins in activating E2F and the ability of these viral proteins to drive quiescent cells into S phase. Specifically, studies of E2F indicate that it is the S-phase genes, which are induced when cells enter S phase and which have been known to be induced when quiescent cells are infected by the DNA tumor viruses, that appear to be targets for the action of the E2F transcription factor.

Although equivalent studies have not yet been performed with p130, it is clear from the initial analyses of Rb and p107 function that their ability to suppress cell growth directly correlates with their ability to interact with E2F and regulate the transcriptional activity of E2F (276,277). Introduction of an Rb gene or a p107 gene into Rb (-/-) cells results in the suppression of cell growth and an arrest of cells in the G1 phase of the cell cycle, and mutations of Rb or p107 that fail to bring about this growth arrest also fail to interact with E2F and to suppress the transcriptional activational capacity of E2F. Moreover, the introduction of the E2F1 cDNA into cells along with Rb overrides the ability of Rb to suppress cell growth (411). The ability of E2F1 to override p107 suppression of cell growth is only partial, possibly reflecting the fact that the normal target for p107 may not be the E2F1 product but rather another member of the E2F family. In short, these data all point to the conclusion that the control of E2F activity by the Rb family members is an important aspect of the growth-suppressing activity of these cellular regulatory proteins.

Finally, E2F is probably not the only Rb target with respect to the control of cell growth. Additional candidates include the transcription factors c-Myc (304), Elf-1 (374), and myoD (140). In addition, other experiments have suggested that a physical interaction of Rb with D-type cyclins may control the activity of these proteins in facilitating G1 progression (95).

Inactivation of p53 Function

Following the discovery of p53 as a protein in association with SV40 T-antigen (207,217), various analyses suggested that p53 functioned as an oncoprotein (103,178,262). Thus, early speculation centered on a role for the association of the viral protein with p53 in stimulating or somehow augmenting the action of this oncoprotein. A radical change in thinking occurred when it was realized that the p53 gene used for the initial studies of transformation was in fact a mutated version of the normal p53 gene and that the transforming activity of the previously assayed p53 gene was dependent on this mutation (102,114). Moreover, assays of the normal wild-type version of p53 revealed that it possessed characteristics of a tumor suppressor molecule (113). Indeed, it is now clear that the loss of p53 function, as a result of mutation or deletion of the gene, is coincident with the development of a large number of human cancers (210). In fact, these studies suggest that loss of p53 function may be an event involved in the majority of human tumors.

Once again, each of the DNA tumor viruses encodes a protein that has the ability to interact with p53 and, as a consequence of this interaction, inactivate p53 function. In contrast to the targeting of Rb, however, the viral protein sequences involved in these interactions, as well as the end result of the interaction, appear to be distinct among the different viruses. That is, an examination of amino acid sequences known to be involved in the targeting of p53 in adenovirus E1B, SV40 T-antigen, and human papillomavirus E6 show no evidence for sequence similarity. Moreover, it would appear that distinct domains of the p53 protein are recognized by the viral proteins. Finally, it is also apparent that the targeting of p53 represents a point of divergence in the polyomavirus group in that the primate polyomavirus SV40 has evolved a protein to target the p53 molecule, whereas the mouse polyomavirus lacks this function and shows no evidence of perturbing the function of p53.

Although the actual mechanism of E1B, T-antigen, and E6 action does not appear to be conserved, the ultimate consequence of the interaction of these viral proteins with p53 is the same—a loss of p53 function. For instance, whereas T-antigen interaction with p53 appears to stabilize the p53 protein, presumably in an inactive state, the interaction of the HPV E6 protein with p53 triggers a ubiquitin-mediated degradation of the p53 protein (314). The process of E6-mediated degradation of p53 has been studied in some detail since these events can be assayed in cell-free extracts. These studies have led to the identification of an additional cellular protein, termed E6-AP (E6 associated protein), that functions in concert with E6 to target p53 for destruction (168–170).

What is the normal function of p53 that is disrupted by these viral oncoproteins? Most likely, it is the ability of p53 to function as a transcription factor that is the activity targeted by the viral proteins. The p53 protein has been shown to bind to DNA in a sequence-specific manner (189), and as a consequence of this binding, it activates transcription if the binding site is part of a promoter element (121,409). Although in most cases this demonstration has utilized test promoters rather than actual *in vivo* targets, recent experiments have identified a cellular target for p53 that links p53 function to cell growth arrest. A gene termed WAF1 was identified as a target for p53 transcriptional activation (101). The product of the WAF1 gene is a 21-kd protein previously identified as an inhibitor of G1 cyclin/kinase activity (146,400). Thus, the p53-mediated suppression of cell growth may be due in part to the inhibition of G1 cyclin-dependent kinases, the activity of which is critical for G1 progression and cell growth. Interestingly, this finding also ties together the p53 growth arrest pathway and the Rb growth arrest pathway, since the control of Rb function through phosphorylation is probably mediated by the same G1 cyclin–dependent kinases.

Finally, the inactivation of p53 function by the viral oncoproteins may mimic the normal activity of the product of a cellular gene termed mdm2 (243). The mdm2 product has been shown to interact with p53 in a manner similar to that of the adenovirus E1B 55-kd protein (406), blocking sequences involved in transcriptional activation (55). Importantly, mdm2 has been associated with oncogenesis as a consequence of the amplification of the gene, leading to overproduction of the product and presumably inactivation of p53 function (49,109). Thus, mdm2 may play a role in regulating the action of p53, possibly during normal growth stimulation (396).

OTHER DNA TUMOR VIRUS TRANSFORMING ACTIVITIES

Although the inactivation of Rb and p53 function is an activity that is common to the small DNA tumor viruses, at least two other virus-specific activities have evolved that do not appear to be shared among all the DNA tumor viruses, but that do clearly contribute to the transforming capacity of the viruses.

Activation of Cellular Tyrosine Kinase Activity

Using an approach similar to that taken with E1A and large T-antigen, analysis of the polyomavirus middle T-antigen revealed the association of this viral protein with a particular cellular target. This cellular protein possessed kinase activity that later was identified as the product of the c-src proto-oncogene (72,73). The interaction of middle T with the c-src product results in an activation of the tyrosine kinase activity of the src protein (37,70) as a consequence of the prevention of a phosphorylation-mediated inhibition of the src kinase (51,70). Subsequent experiments have revealed that middle T can complex with other members of the src kinase family, including the yes and fyn kinases (59,163,198,204). In addition, formation of the middle T-antigen/src kinase complex leads to a recruitment of the phosphatidyl inositol-3 (PI-3) kinase (185,390), resulting in the formation of a ternary complex involving middle T, src, and the PI-3 kinase. The formation of this complex is dependent on the phosphorylation of middle T at tyrosine 315 by the src kinase (354).

One role for the middle-T–mediated activation of src kinase activity for a lytic viral infection would appear to be the promotion of phosphorylation of the late viral capsid protein VP1, an essential modification for viral morphogenesis (124).

The interaction of middle T with src family kinases and the PI-3 kinase appears to be a unique feature of polyomavirus, since an equivalent interaction is not found with proteins encoded by SV40, adenovirus, or papillomavirus. Moreover, this association and concomitant activation of src kinase activity appears to be sufficient for transformation by polyomavirus. Given the fact that this activity does mediate transformation, together with the observation that the other small DNA tumor viruses all appear to show an ability to drive quiescent cells into a cell cycle, one suspects that the action of middle T in activating the src kinase pathway achieves a similar result—namely, the creation of a proliferating cell environment that is conducive to viral replication. Presumably, this involves activation of downstream signal transduction pathways, including the Raf kinase (244). Nevertheless, the details of a src-dependent pathway for cellular proliferation, whether middle T initiated or otherwise, remain largely obscure.

A number of previous experiments have identified polyomavirus mutants with a host range and transformation (hr-t) defective phenotype (16) that have mutations in the small- and middle-T-specific sequences. These mutants are unable to induce a permissive state for virus replication in certain cell types, presumably because of the requirement for middle-T function in altering the growth properties of the target cell. Although in most instances there is a tight link between the two phenotypes, middle-T mutants have been isolated that continue to grow but fail to transform (124).

The analysis of middle T-antigen–mediated transformation and tumor induction *in vivo* has shown that the presence of the src kinase is not essential for this activity, since infection of src-negative mice with a middle-T–expressing retrovirus led to the formation of tumors with a frequency nearly the same as that for src-positive mice (358). Nevertheless, although the association of middle T with src may not be absolutely critical, it is certainly possible that other src family kinases compensate for the loss of src, since middle T is found in association with *yes* and *fyn*. These results do raise the question, however, of the nature of the underlying specificity in the interaction of middle T with one of the kinases.

Finally, although the action of middle T-antigen is clearly sufficient to transform cells, one suspects that the ability of polyoma large T-antigen to bind to and inactivate the Rb protein may well contribute in some circumstances. For instance, although polyomavirus mutants that are defective in large-T function with respect to Rb binding are still capable of transformation and induction of tumor formation, they are defective in immortalization assays (118). Thus, it may well be true that it is the combination of large-T function in inactivating Rb together with the action of middle T in activating the src kinase pathway that is necessary to facilitate fully the creation of a suitable environment for polyomavirus replication.

Inhibition of Cellular Apoptosis

Although the combined action of adenovirus E1A and E1B leads to the oncogenic transformation of susceptible cells in culture, E1A expression alone most often leads to cell death as a result of apoptosis (281,389). Expression of the adenovirus E1B gene suppresses this E1A-induced apoptosis response, allowing the cells to become stably transformed. This phenomenon was first observed in cells infected by adenovirus E1B mutants in which the infection led to the degradation of both viral and cellular DNA (268,388), a hallmark of the apoptosis response.

Further studies have shown that the induction of apoptosis in an adenovirus infection not only is dependent on E1A expression but also involves an induction of the p53 tumor suppressor protein. In particular, there is no such apoptotic response in cells lacking a functional p53 gene, but the response can be restored upon transfection of the wild-type p53 gene (82). Although the mechanism by which E1A induces apoptosis is unknown, the fact that the c-myc gene product also induces apoptosis (107,398) suggests the possibility that deregulated growth control or abnormal growth stimuli may be recognized in a p53-dependent manner (407).

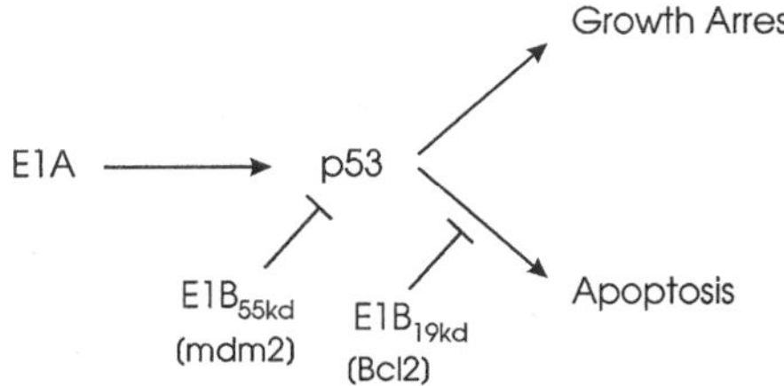

FIG. 22. Inhibition of p53-mediated growth arrest and apoptosis. Depicted is the relationship of E1A- and E1B- mediated stimulation of cell growth involving the inactivation of p53-mediated growth arrest and apoptosis. Shown in brackets are cellular gene products that appear to play similar roles in the control of p53 directly (mdm2) or p53-mediated apoptosis (bcl2). (Adapted from White, ref. 387.)

The action of E1B to prevent the E1A-induced apoptosis appears to be redundant, involving both of the E1B gene products (Fig. 22). The adenovirus E1B 55-kd gene product blocks p53 function directly, whereas the 19-kd product appears to inhibit a p53-dependent downstream event that triggers the apoptosis response. In many respects, the action of the E1B 19-kd protein is analogous to that of bcl-2, the product of a gene originally identified as a B-cell oncogene (162). Subsequent studies have shown that the Bcl2 product blocks the normal process of programmed cell death, leading to an expansion of B-cell populations and eventual tumor formation. Indeed, the Bcl2 gene can substitute for E1B in transformation assays in conjunction with E1A, and like E1B, bcl-2 can block a p53-dependent apoptosis response (60). Finally, the Bcl2 product and the E1B 19-kd protein both appear to localize to the nuclear membrane and share limited sequence homology.

Evolution of Common Strategies of the DNA Tumor Viruses

Because the small DNA tumor viruses represent a diverse group of viruses with little or no evolutionary relationship, it is striking that they share common targets and strategies for replication and transformation. Although the inactivation of Rb and p53 by the DNA tumor virus oncoproteins is generally considered in the context of oncogenic transformation, these events must be important for the normal process of a productive infection by these viruses, since it is the ability to replicate that provides the driving force for their evolution. Clearly, these evolutionarily distinct viruses share a common need and have developed a common strategy, not to transform cells but to replicate.

As discussed previously, these otherwise diverse viruses do have a common need to induce a quiescent, nondividing cell to enter S phase so as to create an environment that is favorable for viral DNA replication. The inactivation of Rb function through the action of E1A, T-antigen, or E7 would appear to facilitate this process by activating the E2F transcription factor and possibly the D-type G1 cyclins. At least for E2F activation, this would lead to an induction of various genes that create the environment for DNA replication.

The viral-mediated inactivation of p53 function also appears to facilitate entry into S phase, since expression of p53 can result in G1 arrest—particularly in response to DNA-damaging events (219), or to initiate a pathway of programmed cell death in response to various proliferative signals (387). As discussed in previous sections, the ability of E1A to stimulate S phase also coincides with the induction of apoptosis. The mechanism for this response remains to be determined but it does involve the p53 tumor suppressor protein. Indeed, there is an induction of p53 expression in cells expressing E1A. The E1B 55-kd protein, SV40 T-antigen, and HPV E6 all block the action of p53 and thus block the apoptosis pathway. Thus, one might view these actions as "allowing" the E1A-mediated process of S-phase induction to continue.

Given the common activities exhibited by adenovirus, SV40, and HPV, it is perhaps equally striking to find a distinct activity that is unique to polyomavirus, the middle-T-antigen–mediated induction of tyrosine kinase activity. Presumably, polyomavirus has evolved a mechanism to accomplish the same result, the creation of a favorable environment for viral replication, without the need to eliminate the Rb and p53 suppression events. Nevertheless, it is also true that polyoma large T-antigen does target Rb and, given the pairwise relationship between Rb inactivation and p53 inactivation seen with the other viral oncoproteins, one wonders if the ultimate action of middle T might lead to the same end result.

Finally, if the small DNA tumor viruses have evolved a common mechanism to drive a quiescent cell into S phase to allow for efficient viral DNA replication, one might also anticipate that other DNA viruses that must replicate DNA at a high level, such as the herpesviruses or the poxviruses, would also find this to be a beneficial event. Yet there is no compelling evidence that any of the herpesviruses or poxviruses encode proteins that inactivate Rb or p53. It is striking, however, that many of the genes whose products create the S-phase environment, and that are suspected to be activated as a consequence of the release of E2F from inhibition by Rb, are also found within the genomes of the large viruses of the herpesvirus and poxvirus family (1). Although the entire complement of genes is not clearly found within every virus of these groups, it is nevertheless true that each virus contains a ribonucleotide reductase, the rate-limiting enzyme in deoxynucleotide biosynthesis (357). Thus, a common strategy of the DNA viruses, whether oncogenic or not, may be to induce activities that create the appropriate environment for viral DNA replication to take place in an efficient manner. If this involves the disruption of normal cell growth control events to force a quiescent cell into S phase, then transformation can result if the infection does not go to completion.

CONCLUSIONS

The study of virus-induced oncogenic transformation has defined paradigmatic changes that differentiate cancer cells from their normal progenitors. These changes affect the components of signal transduction pathways and circuits that control the cell cycle. RNA tumor viruses mutate the sequence information of growth regulatory genes. These mutations are causative events in oncogenesis. DNA tumor viruses, on the other hand, disrupt various growth regulatory events as a consequence of the interaction of viral proteins with cellular regulatory proteins, temporarily inactivating the function of the cellular protein. The genes encoding these cellular proteins remain unaltered.

We have seen from the study of the RNA and DNA tumor viruses that the induction of oncogenic transformation results from the targeting of two distinct classes of cellular genes and their products: oncogenes and tumor suppressor genes. Oncogenes encode components of cellular signal transduction, and their activation leads to a constitutive growth signal. Most often, it is this process that is seen in the retroviruses. By contrast, tumor suppressor genes encode negative regulators of cell growth, particularly the cell cycle, and their inactivation removes such attenuating controls. The oncogenic activities of the DNA tumor viruses most often target the tumor suppressors.

The activation of cellular oncogenes can occur by all manner of mutagenic events, including in addition to viral intervention, point mutation, chromosomal translocation, and amplification. DNA transfection techniques have identified numerous genes that can confer oncogenic properties on recipient cells. Some of these genes are closely related to the oncogenes that have also been identified in retroviruses. Many others, however, have no known association with viruses. However, the normal functions of all these genes are related to the control of cell growth, providing strong evidence for their oncogenic potential as altered control elements.

The genetic inactivation of tumor suppressor genes clearly plays a dominant role in the heritable forms of human cancer. Although the action of the tumor virus proteins in achieving a phenotypic inactivation of these gene products is clearly not relevant to the understanding of the genetic basis of human cancer, somatic mutations of tumor suppressor genes are increasingly recognized as critical factors in sporadic cancers. Virus-induced and genetic inactivation of tumor suppressors have the same net effect, as is most dramatically seen in cervical carcinomas, namely, the loss of tumor suppressor activity; thus, the study of the events following inactivation by viral proteins has been and will continue to be an important approach for the study of human cancer.

Finally, despite strong evidence for the genetic etiology of cancer, it is likely that epigenetic factors also play important but much less well-defined and understood roles. In certain circumstances such factors can exert a dominant effect, thereby converting tumor cells into normal cells. A dramatic demonstration of such epigenetic reversion is seen with murine embryonal carcinoma cells that, when injected into a normal mouse blastocyst, become subject to developmental control and contribute to the development of normal mouse tissues (240). Because molecular biology is biased in favor of genetic analysis and interpretation, it is especially important to keep a watchful eye for such epigenetic phenomena. Although the influence of these events on the development of a tumor is often difficult to measure, it is nevertheless true that these epigenetic factors will also influence the development of virally induced tumors. Thus, one might expect that use of tumor viruses will play a role in the understanding of these more subtle events.

REFERENCES

1. Albrecht JC, Nicholas J, Biller D, Cameron KR, Biesinger B, Newman C, Wittmann S, Craxton MA, Coleman H, Fleckenstein B, et al. Primary structure of the herpesvirus saimiri genome. *J Virol* 1992; 66:5047–5058.
2. Amati B, Brooks MW, Levy N, Littlewood TD, Evan GI, Land H. Oncogenic activity of the c-Myc protein requires dimerization with Max. *Cell* 1993;72:233–245.
3. Amati B, Littlewood TD, Evan GI, Land H. The c-Myc protein induces cell cycle progression and apoptosis through dimerization with Max. *EMBO J* 1993;12:5083–5087.
4. Angel P, Allegretto EA, Okino ST. Oncogene jun encodes a sequence-specific trans-activator similar to AP-1. *Nature* 1988;332:166–171.
5. Angel P, Karin M. The role of jun, fos and the AP-1 complex in cell proliferation and transformation. *Biochim Biophys Acta* 1991; 1072:129–157.
6. Auborn K, Guo M, Prives C. Helicase, DNA binding, and immunological properties of replication defective simian virus 40 mutant T antigens. *J Virol* 1989;63:912–918.
7. Babiss LE, Ginsberg HS, Darnell JE, Jr. Adenovirus E1B proteins are required for accumulation of late viral mRNA and for effects on cellular mRNA translation and transport. *Mol Cell Biol* 1985;5:2522–2558.
8. Bagchi S, Raychaudhuri P, Nevins JR. Adenovirus E1A proteins can dissociate heteromeric complexes involving the E2F transcription factor: a novel mechanism for E1A trans-activation. *Cell* 1990; 62:659–669.
9. Bagchi S, Weinmann R, Raychaudhuri P. The retinoblastoma protein copurifies with E2F-I, an E1A-regulated inhibitor of the transcription factor E2F. *Cell* 1991;65:1063–1072.
10. Baker CC, Phelps WC, Lindgren V, Braun MJ, Gonda MA, Howley PM. Structural and transcriptional analysis of human papillomavirus type 16 sequences in cervical carcinoma cell lines. *J Virol* 1987;61:962–971.
11. Bandara LR, Adamczewski JP, Hunt T, La Thangue NB. Cyclin A and the retinoblastoma gene product complex with a common transcription factor. *Nature* 1991;352:249–251.
12. Bandara LR, La Thangue NB. Adenovirus E1a prevents the retinoblastoma gene product from complexing with a cellular transcription factor. *Nature* 1991;351:494–497.
13. Barbacid M. Involvement of *ras* oncogenes in the initiation of carcinogen-induced tumors. In: Aaronson SA, Bishop JM, Sugimura T, Terada M, Toyoshima K, Vogt PK, eds. *Oncogenes and cancer*. Japan Scientific Society; 1987:43–53.
14. Baserga R. *The biology of cell reproduction*. Cambridge, MA: Harvard University Press; 1985.
15. Ben-David Y, Bernstein A. Friend virus-induced erythroleukemia and the multistage nature of cancer. *Cell* 1991;66:831–834.
16. Benjamin TL. Host range mutants of polyoma virus. *Proc Natl Acad Sci USA* 1970;67:394–399.
17. Berk AJ, Lee F, Harrison T, Williams J, Sharp PA. Pre-early adenovirus 5 gene product regulates synthesis of early viral messenger RNAs. *Cell* 1979;17:935–944.
18. Berk AJ, Sharp PA. Sizing and mapping of early adenovirus mRNAs

by gel electrophoresis of S1 endonuclease digested hybrids. *Cell* 1977;12:721–732.
19. Bernstein A, MacCormick R, Martin GS. Transformation-defective mutants of avian sarcoma viruses: the genetic relationship between conditional and nonconditional mutants. *Virology* 1976;170:206–209.
20. Besmer P. The kit ligand encoded at the murine Steel locus: a pleiotropic growth and differentiation factor. *Curr Opin Cell Biol* 1991;3:939–946.
21. Beug H, Graf T. Transformation parameters of chicken embryo fibroblasts infected with the ts34 mutant of avian erythroblastosis virus. *Virology* 1980;100:348–356.
22. Beug H, Leutz A, Kahn P, Graf T. Ts mutants of E26 leukemia virus allow transformed myeloblasts, but not erythroblasts or fibroblasts, to differentiate at the nonpermissive temperature. *Cell* 1984; 39:579–588.
23. Biedenkapp H, Borgmeyer U, Sippel AE, Klempnauer H-H. Viral myb oncogene encodes a sequence-specific DNA-binding activity. *Nature* 1988;335:835–837.
24. Bikel I, Montano X, Agha ME, Brown M, McCormack M, Boltax J, Livingston DM. SV40 small t antigen enhances the transformation activity of limiting concentrations of SV40 large T antigen. *Cell* 1987;48:321–330.
25. Bishop JM. Cellular oncogenes and retroviruses. *Annu Rev Biochem* 1983;52:301–354.
26. Bishop JM. Viral oncogenes. *Cell* 1985;42:23–38.
27. Bishop JM. The molecular genetics of cancer. *Science* 1987; 235:305–311.
28. Bister K. Multiple cell-derived sequences in single retroviral genomes. *Adv Viral Oncol* 1986;6:45–70.
29. Bittner JJ. Some possible effects of nursing on the mammary gland tumor incidence in mice. *Science* 1936;84:162.
30. Bjorklund S, Skog S, Tribukait B, Thelander L. S-phase-specific expression of mammalian ribonucleotide reductase R1 and R2 subunit mRNAs. *Biochemistry* 1990;29:5452–5458.
31. Blackwood EM, Kretzner L, Eisenman RN. Myc and Max function as a nucleoprotein complex. *Curr Opin Genet Dev* 1992;2:227–235.
32. Blenis J. Signal transduction via the MAP kinases: proceed at your own RSK. *Proc Natl Acad Sci USA* 1993;90:5889–5892.
33. Boeuf H, Jansen Durr P, Kedinger C. Ela-mediated transactivation of the adenovirus EIIa early promoter is restricted in undifferentiated F9 cells. *Oncogene* 1990;5:691–699.
34. Boguski MS, McCormick F. Proteins regulating Ras and its relatives. *Nature* 1993;336:643–654.
35. Bohmann D, Bos TJ, Admon A, Nishimura T, Vogt PL, Tjian R. Human proto-oncogene c-jun encodes a DNA binding protein with structural and functional properties of transcription factor AP-1. *Science* 1987;238:1386–1392.
36. Bold RJ, Hannink M, Donoghue DJ. Functions of the *mos* oncogene. *Cancer Surv* 1986;5:243–256.
37. Bolen SB, Thiele CJ, Israel MA, Yonemoto W, Lipsich LA, Brugge JS. Enhancement of cellular src gene product associated tyrosine kinase activity following polyoma virus infection and transformation. *Cell* 1984;38:767–777.
38. Bollag G, McCormick F. Regulators and effectors of *ras* proteins. *Annu Rev Cell Biol* 1991;7:601–632.
39. Bouck N, Bealer N, Shenk T, Berg P, di Mayorca G. New region of simian virus 40 genome required for efficient viral transformation. *Proc Natl Acad Sci USA* 1978;75:2473–2477.
40. Bourne HR, Sanders DA, McCormick F. The GTPase superfamily: a conserved switch for diverse cell functions. *Nature* 1990;348:125–132.
41. Bourne HR, Sanders DA, McCormick F. The GTPase superfamily: conserved structure and molecular mechanism. *Nature* 1991; 349:117–127.
42. Bourne HR, Sullivan KA. Mammalian G proteins: models for ras proteins in transmembrane signalling? *Cancer Surv* 1986;5:257–274.
43. Bradshaw RA, Prentis S, eds. *Oncogenes and growth factors*. New York: Elsevier; 1987.
44. Broek D. Eukaryotic RAS proteins and yeast proteins with which they interact. *Curr Top Microbiol Immunol* 1989;147:155–169.
45. Brown AMC, Wildin RS, Prendergast TJ, Varmus HE. A retrovirus vector expressing the putative mammary oncogene *int*-1 causes partial transformation of a mammary epithelial cell line. *Cell* 1986; 46:1001–1009.
46. Brown DW, Blais BP, Robinson HL. Long terminal repeat (LTR) sequences, *env*, and a region near the 5′-LTR influence the pathogenic potential of recombinants between Rous-associated virus type 0 and 1. *J Virol* 1988;62:3431–3437.
47. Brugge JS, Erikson RL. Identification of a transformation-specific antigen induced by an avian sarcoma virus. *Nature* 1977;269:346–348.
48. Burridge K. Substrate adhesions in normal and transformed fibroblasts: organization and regulation of cytoskeletal, membrane and extracellular matrix components at focal contacts. *Cancer Rev* 1986;4:18–78.
49. Cahilly-Snyder L, Yang-Feng T, Francke U, George DL. Molecular analysis and chromosomal mapping of amplified genes isolated from a transformed mouse 3T3 cell line. *Somatic Cell Mol Genet* 1987; 13:235–244.
50. Cao L, Faha B, Dembski M, Tsai LH, Harlow E, Dyson N. Independent binding of the retinoblastoma protein and p107 to the transcription factor E2F. *Nature* 1992;355:176–179.
51. Cartwright CA, Kaplan PL, Cooper JA, Hunter T, Eckhart W. Altered sites of tyrosine phosphorylation in pp60c-src associated with polyoma virus middle tumor antigen. *Mol Cell Biol* 1986;6:1562–1570.
52. Chellappan S, Kraus VB, Kroger B, Munger K, Howley PM, Phelps WC, Nevins JR. Adenovirus E1A, simian virus 40 tumor antigen, and human papillomavirus E7 protein share the capacity to disrupt the interaction between transcription factor E2F and the retinoblastoma gene product. *Proc Natl Acad Sci USA* 1992;89:4549–4553.
53. Chellappan SP, Hiebert S, Mudryj M, Horowitz JM, Nevins JR. The E2F transcription factor is a cellular target for the RB protein. *Cell* 1991;65:1053–1061.
54. Chen ISY, Wachsman W, Rosenblatt JD, Cann AJ. The role of the *x* gene in HTLV associated malignancy. *Cancer Surv* 1986;5:329–341.
55. Chen J, Marechal V, Levine AJ. Mapping of the p53 and mdm-2 interaction domains. *Mol Cell Biol* 1993;13:4107–4114.
56. Chen J-M, Chen W-T. Fibronectin-degrading proteases from the membranes of transformed cells. *Cell* 1987;48:193–203.
57. Chen LB, Gallimore PH, McDougall JK. Correlation between tumor induction and the large external transformation sensitive protein on the cell surface. *Proc Natl Acad Sci USA* 1976;73:3570–3574.
58. Chen W-T, Olden K, Bernard BA, Chu FF. Expression of transformation-associated protease(s) that degrade fibronectin at cell contact sites.*J Cell Biol* 1984;98:1546–1555.
59. Cheng SH, Harvey R, Espino PC, Semba K, Yamamoto T, Toyoshima K, Smith AE. Peptide antibodies to the human c-fyn gene product demonstrate pp59c-fyn is capable of complex formation with the middle-T antigen of polyomavirus. *EMBO J* 1988;7:3845–3855.
60. Chiou S-K, Rao L, White E. Bcl-2 blocks p53-dependent apoptosis. *Mol Cell Biol* 1994;14:2556–2563.
61. Chittenden T, Livingston DM, Kaelin WG, Jr. The T/E1A-binding domain of the retinoblastoma product can interact selectively with a sequence-specific DNA-binding protein. *Cell* 1991;65:1073–1082.
62. Chiu R, Boyle WJ, Meek J, Smeal T, Hunter T, Karin M. The c-*Fos* protein interacts with c-*Jun*/AP-1 to stimulate transcription of AP-1 responsive genes. *Cell* 1988;54:541–552.
63. Cicchetti P, Mayer BJ, Thiel G, Baltimore D. Identification of a protein that binds to the SH3 region of Abil and is similar to Bcr and GAP-rho. *Science* 1992;257:803–806.
64. Cobb BS, Schaller MD, Leu T-H, Parsons JT. Stable association of pp60*src* and pp59*fyn* with the focal adhesion-associated protein tyrosine kinase, p125*FAK*. *Mol Cell Biol* 1994;14:147–155.
65. Coffin JM. Genetic diversity and evolution of retroviruses. *Curr Top Microbiol Immunol* 1992;176:143–164.
66. Collett MS, Erikson RL. Protein kinase activity associated with the avian sarcoma virus *src* gene product. *Proc Natl Acad Sci USA* 1978;75:2021–2024.
67. Cooper GM. *Oncogenes*. 2nd ed. Boston: Jones and Bartlett; 1995.
68. Cooper JA, Howell B. The when and how of Src regulation. *Cell* 1993;73:1051–1054.
69. Courtneidge SA. Activation of the pp60$^{\text{c-<src>}}$ kinase by middle T antigen binding or by dephosphorylation. *EMBO J* 1985;4:1471–1477.
70. Courtneidge SA. Further characterization of the complex containing middle T antigen and pp60. *Curr Top Microbiol Immun.* 1989; 144:121–128.
71. Courtneidge SA, Heber A. An 81 kd protein complexed with middle T antigen and pp60$^{\text{c-<src>}}$ a possible phosphatidylinositol kinase. *Cell* 1987;50:1031–1037.
72. Courtneidge SA, Smith AE. Polyoma virus transforming protein associates with the product of the c-src cellular gene. *Nature* 1983;303:435–439.

73. Courtneidge SA, Kypta RM, Ulug ET. Interactions between the middle T antigen of polyomavirus and host cell proteins. *Cold Spring Harbor Symp Quant Biol* 1988;53:153–160.
74. Cress WD, Johnson DG, Nevins JR. A genetic analysis of the E2F1 gene distinguishes regulation by Rb, p107, and adenovirus E4. *Mol Cell Biol* 1993;13:6314–6325.
75. Croce CM. Molecular biology of lymphomas. *Semin Oncol* 1993; 20:31–46.
76. Curran T. The *fos* oncogene. In: Reddy EP, Skalka AM, Curran T, eds. *The oncogene handbook*. New York: Elsevier; 1988:307–325.
77. Curran T, Vogt PK. Dangerous liaisons: Fos and Jun—oncogenic transcription factors. In: SL McKnight, KR Yamamoto, eds. *Transcriptional regulation*; Cold Spring Harbor, NY: Cold Spring Harbor Laboratory; 1991:797–831.
78. D'Andrea AD. The interaction of the erythropoietin receptor and gp55. *Cancer Surv* 1992;15:19–36.
79. Damm K. c-erbA: protooncogene or growth suppressor gene? *Adv Cancer Res* 1992;59:89–113.
80. De Larco JE, Todaro GF. Growth factors from murine sarcoma virus–transformed cells. *Proc Natl Acad Sci USA* 1978;75:4001–4005.
81. Dean FB, Bullock P, Murakami Y, Wobbe CR, Weissbach L, Hurwitz J. Simian virus 40 (SV40) DNA replication: SV40 large T antigen unwinds DNA containing the SV40 origin of replication. *Proc Natl Acad Sci USA* 1987;84:16–20.
82. Debbas M, White E. Wild-type p53 mediates apoptosis by E1A, which is inhibited by E1B. *Genes Dev* 1993;7:546–554.
83. DeCaprio JA, Ludlow JW, Figge J, Shew JY, Huang CM, Lee WH, Marsilio E, Paucha E, Livingston DM. SV40 large tumor antigen forms a specific complex with the product of the retinoblastoma susceptibility gene. *Cell* 1988;54:275–283.
84. Defeo-Jones D, Huang PS, Jones RE, Haskell KM, Vuocolo GA, Hanobik MG, Huber HE, Oliff A. Cloning of cDNAs for cellular proteins that bind to the retinoblastoma gene product. *Nature* 1991; 352:251–254.
85. Delli-Bovi P, Curatola AM, Kern FG, Greco A, Ittmann M, Basilico C. An oncogene isolated by transfection of Kaposi's sarcoma DNA encodes a growth factor that is a member of the FGF family. *Cell* 1987; 50:729–737.
86. DeLucia AL, Lewton BA, Tjian R, Tegtmeyer P. Topography of simian virus 40 A protein–DNA complexes: arrangement of pentanucleotide interaction sites at the origin of replication. *J Virol* 1983;46:143–150.
87. DesGroseillers L, Jolicoeur P. Mapping the viral sequences conferring leukemogenicity and disease specificity in Moloney and amphotropic murine leukemia viruses. *J Virol* 1984;52:448–456.
88. DesGroseillers L, Jolicoeur P. The tandem direct repeats within the long terminal repeat of murine leukemia viruses are the primary determinant of their leukemogenic potential. *J Virol* 1984; 52:945–952.
89. DesGroseillers L, Rassart E, Jolicoeur P. Thymotropism of murine leukemia virus is conferred by its long terminal repeat. *Proc Natl Acad Sci USA* 1983;80:4203–4207.
90. Devoto SH, Mudryj M, Pines J, Hunter T, Nevins JR. A cyclin A–protein kinase complex possesses sequence-specific DNA binding activity: p33cdk2 is a component of the E2F–cyclin A complex. *Cell* 1992;68:167–176.
91. Dickson C, Peters G. Potential oncogene product related to growth factors. *Nature* 1987;326:833
92. Dini PW, Lipsick JS. Oncogenic truncation of the first repeat of c-Myb decreases DNA binding *in vitro* and *in vivo*. *Mol Cell Biol* 1993;13:7334–7348.
93. Doerfler W. The fate of the DNA of adenovirus type 12 in baby hamster kidney cells. *Proc Natl Acad Sci USA* 1968;60:636
94. Doolittle RF, Hunkapiller MW, Hood LE. Simian sarcoma virus *onc* gene, v-*sis*, is derived from the gene (or genes) encoding a platelet-derived growth factor. *Science* 19;221:275–277.
95. Dowdy SF, Hinds PW, Louie K, Reed SI, Arnold A, Weinberg RA. Physical interaction of the retinoblastoma protein with human D cyclins. *Cell* 1993;73:499–511.
96. Downward J, Yarden Y, Mayes E. Close similarity of epidermal growth factor receptor and v-*erbB* oncogene protein sequences. *Nature* 1984;307:521–527.
97. Dulbecco R, Hartwell LH, Vogt M. Induction of cellular DNA synthesis by polyoma virus. *Proc Natl Acad Sci USA* 1965;53:403
98. Dvoretzky I, Shober R, Chattopadhyay SK, Lowy DR. A quantitative *in vitro* focus assay for bovine papillomavirus. *Virology* 1980; 103:369–375.
99. Dyson N, Howley PM, Munger K, Harlow E. The human papilloma virus-16 E7 oncoprotein is able to bind to the retinoblastoma gene product. *Science* 1989;243:934–937.
100. Eisenman RN. Nuclear oncogenes. *The oncogenes*. Cold Spring Harbor, NY: Cold Spring Harbor Laboratory; 1989:175–221.
101. el Deiry WS, Tokino T, Velculescu VE, Levy DB, Parsons R, Trent JM, Lin D, Mercer WE, Kinzler KW, Vogelstein B. WAF1, a potential mediator of p53 tumor suppression. *Cell* 1993;75:817–825.
102. Eliyahu D, Goldfinger N, Pinhasi-Kimhi O, Shaulsky G, Shurnik Y, Arai N, Rotter V, Oren M. Meth A fibrosarcoma cells express two transforming mutant p53 species. *Oncogene* 1988;3:313–321.
103. Eliyahu D, Raz A, Gruss P, Givol D, Oren M. Participation of p53 cellular tumour antigen in transformation of normal embryonic cells. *Nature* 1984;312:646–649.
104. Ellermann V, Bang O. Experimentelle Leukamie bei Huhnern. *Zentralb Bakteriol* 1908;46:595–609.
105. Engstrom Y, Eriksson S, Jildevik I, Skog S, Thelander L, Tribukait B. Cell cycle–dependent expression of mammalian ribonucleotide reductase: differential regulation of the two subunits. *J Biol Chem* 1985;260:9114–9116.
106. Ersman MD, Astrin SM. The *myc* oncogene. In: Reddy EP, Skalka AM, Curran T, eds. *The oncogene handbook*. New York: Elsevier; 1988:341–379.
107. Evan GL, Wyllie AH, Gilbert CS, Littlewood TD, Land H, Brooks M, Waters CM, Penn LZ, Hancock DC. Induction of apoptosis in fibroblasts by c-myc protein. *Cell* 1992;69:119–128.
108. Ewen ME, Xing YG, Lawrence JB, Livingston DM. Molecular cloning, chromosomal mapping, and expression of the cDNA for p107, a retinoblastoma gene product-related protein. *Cell* 1991;66:1155–1164.
109. Fakharzadeh SS, Trusko SP, George DL. Tumorigenic potential associated with enhanced expression of a gene that is amplified in a mouse tumor cell line. *EMBO J* 1991;10:1565–1569.
110. Felder M-P, Eychene A, Barnier JV, Calogeraki I, Calothy G, Marx M. Common mechanism of retrovirus activation and transduction of c-*mil* and c-R*mil* in chicken neuroretina cells infected with Rous-associated virus type 1. *J Virol* 1991;65:3633–3640.
111. Felder M-P, Laugier D, Eychene A, Calothy G, Marx M. Occurrence of alternatively spliced leader-delta*onc*Poly(A)transcripts in chicken neuroretina cells infected with Rous-associated virus type 1: implication in transduction of the c-*mil*/c-*raf* and c-R*mil*/B-*raf* oncogenes. *J Virol* 1993;67:6853–6856.
112. Figge J, Webster T, Smith TF, Paucha E. Prediction of similar transforming regions in simian virus 40 large T, adenovirus E1A, and myc oncoproteins. *J Virol* 1988;62:1814–1818.
113. Finlay CA, Hinds PW, Levine AJ. The p53 proto-oncogene can act as a suppressor of transformation. *Cell* 1989;57:1083–1093.
114. Finlay CA, Hinds PW, Tan TH, Eliyahu D, Oren M, Levine AJ. Activating mutations for transformation by p53 produce a gene product that forms an hsc70–p53 complex with an altered half-life. *Mol Cell Biol* 1988;8:531–539.
115. Forrest D, Curran T. Crossed signals: oncogenic transcription factors. *Curr Opin Genet Dev* 1992;2:19–27.
116. Frearson PM, Kit S, Dubbs DR. Deoxythymidylate synthetase and deoxythymidine kinase activities of virus-infected animal cells. *Cancer Res* 1965;25:737
117. Frearson PM, Kit S, Dubbs DR. Induction of dehydrofolate reductase activity by SV40 and polyoma virus. *Cancer Res* 1966;26:1653
118. Freund R, Bronson RT, Benjamin TL. Separation of immortalization from tumor induction with polyoma large T mutants that fail to bind to retinoblastoma gene product. *Oncogene* 1992;7:1979–1987.
119. Friend C. Cell-free transmission in adult Swiss mice of a disease having the character of a leukemia. *J Exp Med* 1957;105:307–319.
120. Fung Y-KT, Lewis WG, Crittenden LB, Kung H-J. Activation of the cellular oncogene c-*erbB* by LTR insertion: molecular basis for induction of erythroblastosis by avian erythroblastosis virus. *Cell* 1983;33:357–368.
121. Funk WD, Pak DT, Karas RH, Wright WE, Shay JW. A transcriptionally active DNA-binding site for human p53 protein complexes. *Mol Cell Biol* 1992;12:2866–2871.
122. Gallimore PH, Sharp PA, Sambrook J. Viral DNA in transformed cells. II. A study of the sequences of adenovirus 2 DNA in 9 lines of trans-

formed rat cells using specific fragments of the viral genome. *J Mol Biol* 1974;89:49
123. Gallo RC. The first human retrovirus. *Sci Am* 1986;255:88–98.
124. Garcea RL, Talmage DA, Harmatz A, Freund R, Benjamin TL. Separation of host range from transformation functions of the hr-t gene of polyomavirus. *Virology* 1989;168:312–319.
125. Geissler EN, Ryan MA, Housman DE. The dominant-white spotting (W) locus of the mouse encodes the c-*kit* proto-oncogene. *Cell* 1988;55:185–192.
126. Gelinas C, Temin HM. The v-*rel* oncogene encodes a cell-specific transcriptional activator of certain promoters. *Oncogene* 1988;3:349–355.
127. Gentz R, Rauscher FJI, Abate C, Curran T. Parallel association of *Fos* and *Jun* leucine zippers juxtaposes DNA binding domains. *Science* 1989;243:1695–1699.
128. Ghysdael J, Beug H. The leukaemia oncogene v-erbA: a dominant negative version of ligand dependent transcription factors that regulates red cell differentiation. *Cancer Surv* 19;14:169–180.
129. Giacherio D, Hager LP. A poly(dT)-stimulated ATPase activity associated with simian virus 40 large T antigen. *J Biol Chem* 1979; 254:8113–8116.
130. Gilman AG. G proteins: transducers of receptor-generated signals. *Annu Rev Biochem* 1987;56:615–649.
131. Gluzman Y, Davison T, Oren M, Winocour E. Properties of permissive monkey cells transformed by UV-irradiated simian virus 40. *J Virol* 1977;22:256–266.
132. Golden A, Brugge JS. The *src* oncogene. In: Reddy EP, Skalka AM, Curran T, eds. *The oncogene handbook*. New York: Elsevier; 1988:149–173.
133. Goldfarb MP, Weinberg RA. Generation of novel biologically active Harvey sarcoma viruses via apparent illegitimate recombination. *J Virol* 1981;38:136–150.
134. Goldman DS, Kiessling AA, Millette CF, Cooper GM. Expression of c-*mos* RNA in germ cells of male and female mice. *Proc Natl Acad Sci USA* 1987;84:4509–4513.
135. Graf T, Beug H. Role of the v-*erb*A and v-*erb*B oncogenes of avian erythroblastosis virus in erythroid cell transformation. *Cell* 1983;34:7–9.
136. Graffi A, Bielka H, Fey F, Scharsach F, Weiss R. Gehyauauftes Auftreten von Leukyauamien nach Injektion von Sarkom-Filtraten. *Wien Klin Wochenschr* 1955;105:61–64.
137. Graham FL, Abrahams PJ, Mulder C. Studies on *in vitro* transformation by DNA and DNA fragments of human adenoviruses and SV40. *Cold Spring Harbor Symp Quant Biol* 1974;39:637–650.
138. Graham FL, Abrahams PS, Mulder C, Heijneker HL, Warnaar SO, deVries FAJ, Fiers W, van der Eb AJ. Studies on *in vitro* transformation by DNA and DNA fragments of human adenoviruses and simian virus 40. *Cold Spring Harbor Symp Quant Biol* 1975;39:637
139. Gross L. Development and serial cell-free passage of a highly potent strain of mouse leukemia virus. *Proc Soc Exp Biol Med* 1957; 94:767–771.
140. Gu W, Schneider JW, Condorelli G, Kaushal S, Mahdavi V, Nadal Ginard B. Interaction of myogenic factors and the retinoblastoma protein mediates muscle cell commitment and differentiation. *Cell* 1993; 72:309–324.
141. Hai T, Curran T. Cross-family dimerization of transcription factors Fos/Jun and ATF/CREB alters DNA binding specificity. *Proc Natl Acad Sci USA* 1991;88:3720–3724.
142. Haluska FG, Tsujimoto Y, Croce CM. Mechanisms of chromosome translocation in B- and T-cell neoplasia. *Trends Genet* 1987;3:11–15.
143. Hannon GJ, Demetrick D, Beach D. Isolation of the Rb-related p130 through its interaction with cdk2 and cyclins. *Genes Dev* 1993; 7:2378–2391.
144. Hardy S, Engel DA, Shenk T. An adenovirus early region 4 gene product is required for induction of the infection-specific form of cellular E2F activity. *Genes Dev* 1989;3:1062–1074.
145. Harlow E, Whyte P, Franza BR, Jr., Schley C. Association of adenovirus early-region 1A proteins with cellular polypeptides. *Mol Cell Biol* 1986;6:1579–1589.
146. Harper JW, Adami GR, Wei N, Keyomarsi K, Elledge SJ. The p21 cdk-interacting protein Cip1 is a potent inhibitor of G1 cyclin-dependent kinases. *Cell* 1993;75:805–816.
147. Hartley JW, Wolford NK, Old LJ, Rowe WP. A new class of murine leukemia virus associated with development of spontaneous lymphomas. *Proc Natl Acad Sci USA* 1977;74:789–792.
148. Hartwell L, Vogt M, Dulbecco R. Induction of cellular DNA synthesis by polyoma: II. Increase in the rate of enzyme synthesis after infection with polyoma virus in mouse embryo kidney cells. *Virology* 1965;27:262
149. Hartwell LH, Weinert TA. Checkpoints: controls that ensure the order of cell cycle events. *Science* 1989;246:629–633.
150. Hatanaka M, Dulbecco R. Induction of DNA synthesis by SV40. *Proc Natl Acad Sci USA* 1966;56:736–740.
151. Hayman MJ. *erb*-B: growth factor receptor turned oncogene. In: Bradshaw RA, Prentis S, eds. *Oncogenes and growth factors*. New York: Elsevier; 1989:81–89.
152. Hayward WS, Neel BG, Astrin SM. Activation of a cellular *onc* gene by promoter insertion in ALV-induced lymphoid leukosis. *Nature* 1981;290:475–480.
153. Heisterkamp N, Stam K, Groffen J, de Klein A, Grosveld G. Structural organization of the *bcr* gene and its role in the Ph1 translocation. *Nature* 1985;315:758–761.
154. Helin K, Lees JA, Vidal M, Dyson N, Harlow E, Fattaey A. A cDNA encoding a pRB-binding protein with properties of the transcription factor E2F. *Cell* 1992;70:337–350.
155. Herman SA, Coffin JM. Efficient packaging of read-through RNa in ALV: implications for oncogene transduction. *Science* 1987; 236:845–848.
156. Herren B, Rooney B, Weyer KA, Iberg N, Schmid G, Pech M. Dimerization of extracellular domains of platelet-derived growth factor receptors: a revised model of receptor–ligand interaction. *J Biol Chem* 1993;268:15088–15095.
157. Herrlich P, Ponta H. "Nuclear" oncogenes convert extracellular stimuli into changes in the genetic program. *Trends Genet* 1989;5:112–115.
158. Hiebert SW. Regions of the retinoblastoma gene product required for its interaction with the E2F transcription factor are necessary for E2 promoter repression and pRb-mediated growth suppression. *Mol Cell Biol* 1993;13:3384–3391.
159. Hiebert SW, Chellappan SP, Horowitz JM, Nevins JR. The interaction of RB with E2F coincides with an inhibition of the transcriptional activity of E2F. *Genes Dev* 1992;6:177–185.
160. Hildebrand JD, Schaller MD, Parsons JT. Identification of sequences required for the efficient localization of the focal adhesion kinase. *J Cell Biol* 1993;123:993–1005.
161. Hiral SI, Ryseck R-P, Mechta F, Bravo R, Yaniv M. Characterization of *jun D*: a new member of the *jun* protooncogene family. *EMBO J* 1989;8:1433–1439.
162. Hockenbery D, Nunez G, Milliman C, Schreiber RD, Korsmeyer S. Bcl-2 is an inner mitochondrial membrane protein that blocks programmed cell death. *Nature* 1990;348:334–336.
163. Horak ID, Kawakami T, Gregory F, Robbins KC, Bolen JB. Association of p60fyn with middle tumor antigen in murine polyomavirus—transformed rat cells. *J Virol* 1989;63:2343–2347.
164. Hsu W, Kerppola TK, Chen P-L, Curran T, Chen-Kiang S. Fos and Jun repress transcription activation by NF-IL6 through association at the basic zipper region. *Mol Cell Biol* 1994;14:268–276.
165. Huang MM, Hearing P. The adenovirus early region 4 open reading frame 6/7 protein regulates the DNA binding activity of the cellular transcription factor, E2F, through a direct complex. *Genes Dev* 1989;3:1699–1710.
166. Hughes S, Greenhouse JJ, Petropoulos CJ, Sutrave P. Adapter plasmids simplify the insertion of foreign DNA into helper-independent retroviral vectors. *J Virol* 1987;61:3004–3012.
167. Hughes S, Kosik E. Mutagenesis of the region between env and src of the SR-A strain of Rous sarcoma virus for the purpose of constructing helper-independent vectors. *Virology* 1984;136:89–99.
168. Huibregtse JM, Scheffner M, Howley PM. A cellular protein mediates association of p53 with the E6 oncoprotein of human papillomavirus types 16 or 18. *EMBO J* 1991;10:4129–4135.
169. Huibregtse JM, Scheffner M, Howley PM. Localization of the E6-AP regions that direct human papillomavirus E6 binding, association with p53, and ubiquitination of associated proteins. *Mol Cell Biol* 1993; 13:4918–4927.
170. Huibregtse JM, Scheffner M, Howley PM. Cloning and expression of the cDNA for E6-AP, a protein that mediates the interaction of the human papillomavirus E6 oncoprotein with p53. *Mol Cell Biol* 1993;13:775–784.
171. Hunter T, Sefton BM. Transforming gene product of Rous sarcoma virus phosphorylates tyrosine. *Proc Natl Acad Sci USA* 1980; 77:1311–1315.

172. Hynes RO. Role of surface alterations in cell transformation: the importance of proteases and surface proteins. *Cell* 1974;1:147–156.
173. Hynes RO. Integrins: versatility, modulation, and signalling in cell adhesion. *Cell* 1992;69:11–25.
174. Iba H, Takeya T, Cross F, Hanafusa T, Hanafusa H. Rous sarcoma virus variants which carry the cellular *src* gene instead of the viral *src* gene cannot transform chicken embryo fibroblasts. *Proc Natl Acad Sci USA* 1984;81:4424–4428.
175. Ikawa S, Hagino-Yamagishi K, Kawai S, Yamamoto T, Toyoshima K. Activation of the cellular *src* gene by transducing retrovirus. *Mol Cell Biol* 1986;6:2420–2428.
176. Ito Y, Brocklehurst JR, Dulbecco R. Virus-specific proteins in the plasma membrane of cells lytically infected or transformed by polyoma virus. *Proc Natl Acad Sci USA* 1977;74:4666–4670.
177. Janknecht R, Nordheim A. Gene regulation by Ets proteins. *Biochim Biophys Acta* 1993;1155:346–356.
178. Jenkins JR, Rudge K, Currie GA. Cellular immortalization by a cDNA clone encoding the transformation associated phosphoprotein p53. *Nature* 1984;312:651–654.
179. Johnson P, Benchimol S. Friend virus induced murine erythroleukaemia: the p53 locus. *Cancer Surv* 1992;12:137–151.
180. Jones KA, Tjian R. Essential contact residues within SV40 large T antigen binding sites I and II identified by alkylation-interference. *Cell* 1984;36:155–162.
181. Jones N, Shenk T. An adenovirus type 5 early gene function regulates expression of other early viral genes. *Proc Natl Acad Sci USA* 1979;76:3665–3669.
182. Kabat D. Molecular biology of Friend viral erythroleukemia. *Curr Top Microbiol Immunol* 1989;148:1–42.
183. Kaelin WG, Krek W, Sellers WR, DeCaprio JA, Ajchenbaum F, Fuchs CS, Chittenden T, Li Y, Farnham PJ, Blanar MA, et al. Expression cloning of a cDNA encoding a retinoblastoma-binding protein with E2F-like properties. *Cell* 1992;70:351–364.
184. Kalderon D, Smith AE. *In vitro* mutagenesis of a putative DNA binding domain of SV40 large-T. *Virology* 1984;139:109–137.
185. Kaplan DR, Whitman M, Schaffhausen B. Phosphatidylinositol metabolism and polyoma-mediated transformation. *Proc Natl Acad Sci USA* 1986;83:3624–3628.
186. Kara J, Weil R. Specific activation of the DNA synthesizing apparatus in contact inhibited cells by polyoma virus. *Proc Natl Acad Sci USA* 1967;57:63
187. Kataoka K, Nishizawa M, Kawai S. Structure-function analysis of the maf oncogene product, a member of the b-Zip protein family. *J Virol* 1993;67:2133–2141.
188. Kawai S, Hanafusa H. The effects of reciprocal changes in the temperature on the transformed state of cells infected with a Rous sarcoma virus mutant. *Virology* 1971;46:470–479.
189. Kern SE, Kinzler KW, Bruskin A, Jarosz D, Friedman P, Prives C, Vogelstein B. Identification of p53 as a sequence specific DNA binding protein. *Science* 1991;252:1707–1711.
190. Kipreos ET, Wang JYJ. Differential phosphorylation of c-Abl in cell cycle determined by *cdc2* kinase and phosphatase activity. *Science* 1990;248:217–220.
191. Kipreos ET, Wang JYJ. Cell cycle–regulated binding of c-Abl tyrosine kinase to DNA. *Science* 1992;256:382–385.
192. Kirsten WH, Mayer LA. Morphologic responses to a murine erythroblastosis virus. *J Natl Cancer Inst* 1967;39:311–355.
193. Kit S, De Torres RA, Dubbs DR, Salvi ML. Induction of cellular deoxyribonucleic acid synthesis by simian virus 40. *J Virol* 1967;1:738
194. Kit S, Dubbs DR, Frearson PM. Enzymes of nucleic acid metabolism in cells infected with polyoma virus. *Cancer Res* 1966;26:638
195. Kit S, Dubbs DR, Frearson PM, Melnick JL. Enzyme induction in SV40-infected green monkey kidney cultures. *Virology* 1966;29:69
196. Kit S, Piekarski LJ, Dubbs DR. DNA polymerase induced by simian virus 40. *J Gen Virol* 1967;1:163
197. Kitajima I, Shinohara T, Bilakovics J, Brown DA, Xu X, Nerenberg M. Ablation of transplanted HTLV-I tax-transformed tumors in mice by antisense inhibition of MF-kappa B. *Science* 1992; 258:1792–1795.
198. Kornbluth S, Sudol M, Hanafusa H. Association of the polyomavirus middle-T antigen with c-yes protein. *Nature* 1987;325:171–173.
199. Kouzarides T, Ziff E. The role of the leucine zipper in the *fos–jun* interaction. *Nature* 1988;336:646–651.
200. Kovesdi I, Reichel R, Nevins JR. Identification of a cellular transcription factor involved in E1A trans-activation. *Cell* 1986; 45:219–228.
201. Krust A, Green S, Argos P. The chicken oestrogen receptor sequence: homology with v-*erbA* and the human oestrogen and glucocorticoid receptors. *EMBO J* 1986;5:891–897.
202. Kung H-J, Maihle NJ. Molecular basis of oncogenesis by nonacute avian retroviruses. In: De Boer GF, ed. *Avian leukosis*. Boston: Martinus Nijhoff; 1986:77–99.
203. Kung HJ, Vogt P, eds. Retroviral insertion and oncogene activation. *Curr Top Microbiol Immunol* 1991;171.
204. Kypta RM, Hemming A, Courtneidge SA. Identification and characterization of p59fyn (a src-like protein tyrosine kinase) in normal and polyoma virus transformed cells. *EMBO J* 1988;7:3837–3844.
205. Lacal JC, Tronick SR. The *ras* oncogene. In: Reddy EP, Skalka AM, Curran T, eds. *The oncogene handbook*. New York: Elsevier; 1988:257–304.
206. Landschulz WH, Johnson PF, McKnight SL. The leucine zipper: a hypothetical structure common to a new class of DNA binding proteins. *Science* 1988;240:1759–1764.
207. Lane DP, Crawford LV. T antigen is bound to a host protein in SV40-transformed cells. *Nature* 1979;278:261–263.
208. Ledinko N. Enhanced deoxyribonucleic acid polymerase activity in human embryonic kidney cultures infected with adenovirus 2 and 12. *J Virol* 1968;2:89–98.
209. Leppard KN, Shenk T. The adenovirus E1B 55 kd protein influences mRNA transport via an intranuclear effect on RNA metabolism. *EMBO J* 1989;8:2329–2336.
210. Levine AJ. The tumor suppressor genes. *Annu Rev Biochem* 1993; 62:623–651.
211. Levinson AD. Normal and activated *ras* oncogenes and their encoded products. In: Bradshaw RA, Prentis S, eds. *Oncogenes and growth factors*. New York: Elsevier; 1987:74–83.
212. Levinson AD, Oppermann H, Levintow L, Varmus HE, Bishop JM. Evidence that the transforming gene of avian sarcoma virus encodes a protein kinase associated with a phosphoprotein. *Cell* 1978;15:561–572.
213. Levy JA. *The Retroviridae*. Vol 1. New York: Plenum; 1992.
214. Li J, Vogt P. The retroviral oncogene *qin* belongs to the transcription factor family that includes the homeotic gene fork head. *Proc Natl Acad Sci USA* 1993;90:4490–4494.
215. Li J-P, D'Andrea A, Lodish HF, Baltimore D. Activation of cell growth by binding of Friend spleen focus-forming virus gp55 glycoprotein to the erythropoietin receptor. *Nature* 1990;343:762–764.
216. Li Y, Graham C, Lacy S, Duncan AMV, Whyte P. The adenovirus E1A-associated 130-kd protein is encoded by a member of the retinoblastoma gene family and physically interacts with cyclins A and E. *Genes Dev* 1993;7:2366–2377.
217. Linzer DI, Levine AJ. Characterization of a 54K dalton cellular SV40 tumor antigen present in SV40-transformed cells and uninfected embryonal carcinoma cells. *Cell* 1979;17:43–52.
218. Liou HC, Baltimore D. Regulation of the NF-Kappa B/rel transcription factor and I kappa B inhibitor system. *Curr Opin Cell Biol* 1993;5:477–487.
219. Livingstone LR, White A, Sprouse J, Livanos E, Jacks T, Tlsty TD. Altered cell cycle arrest and gene amplification potential accompany loss of wild-type p53. *Cell* 1992;70:923–935.
220. Loeken M, Bikel I, Livingston DM, Brady J. Trans-activation of RNA polymerase II and III promoters by SV40 small t antigen. *Cell* 1988;55:1171–1177.
221. Loeken MR. Simian virus 40 small t antigen *trans* activates the adenovirus E2A promoter by using mechanisms distinct from those used by adenovirus E1A. *J Virol* 1992;66:2551–2555.
222. Lowy DR, Dvoretzky I, Shober R, Law MF, Engel L, Howley PM. *In vitro* tumorigenic transformation by a defined sub-genomic fragment of bovine papilloma virus DNA. *Nature* 1980;287:72–74.
223. Luscher B, Eisenman RN. New light on Myc and Myb. Part I. Myc. *Genes Dev* 1990;4:2025–2035.
224. Luscher B, Eisenman RN. New light on Myc and Myb. Part II. Myb. *Genes Dev* 1990;4:2235–2241.
225. Mackey JK, Rigden PM, Green M. Do highly oncogenic group A human adenoviruses cause human cancer? Analysis of human tumors for adenovirus 12 transforming DNA sequences. *Proc Natl Acad Sci USA* 1976;73:4657–4661.
226. Macpherson I, Montagnier L. Agar suspension culture for the selective assay of cells transformed by polyoma virus. *Virology* 1964;23:291–294.

227. Manaker RA, Groupe V. Discrete foci of altered chicken embryo cells associated with Rous sarcoma virus in tissue culture. *Virology* 1956;2:838–840.
228. Manos MM, Gluzman Y. Genetic and biochemical analysis of transformation-competent, replication-defective simian virus 40 large T antigen mutants. *J Virol* 1985;53:120–127.
229. Martin GS. Rous sarcoma virus: a function required for the maintenance of the transformed state. *Nature* 1970;227:1021–1023.
230. Martin GS. The *erbB* gene and the EGF receptor. *Nature* 1970; 227:1021–1023.
231. Martin GS, Duesberg PH. The *a* subunit in the RNA of transforming avian tumor viruses. I. Occurrence in different virus strains. II. Spontaneous loss resulting in nontransforming variants. *Virology* 1972; 47:494–497.
232. Martin P, Henry C, Ferre F. Characterization of a *myc*-containing retrovirus generated by propagation of an MH2 viral subgenomic RNA. *J Virol* 1986;57:1191–1194.
233. Matsushime H, Roussel MF, Ashmun RA, Sherr CJ. Colony-stimulating factor 1 regulates novel cyclins during the G1 phase of the cell cycle. *Cell* 1991;65:701–713.
234. Mayer BJ, Hamaguchi M, Hanafusa H. A novel viral oncogene with structural similarity to phospholipase C. *Nature* 1988;332:272–278.
235. Mayol X, Grana X, Baldi A, Sang N, Hu Q, Giordano A. Cloning of a new member of the retinoblastoma gene family (pRb2) which binds to the E1A transforming domain. *Oncogene* 1993;8:2561–2566.
236. McCormick F. *Ras* GTPase activating protein: signal transmitter and signal terminator. *Cell* 1989;56:5–8.
237. McGlade J, Cheng A, Pelicci G, Pelicci PG, Pawson T. Shc proteins are phosphorylated and regulated by the v-Src and v-Fps protein-tyrosine kinases. *Proc Natl Acad Sci USA* 1992;89:8869–8873.
238. McKeehan WL, Barnes D, Reid L, Stanbridge E, Murakami H, Sato GH. Frontiers in mammalian cell culture. *In Vitro Cell Dev Biol* 1990;26:9–23.
239. Miles DB, Robinson HL. High frequency of transduction of c-*erbB* in avian leukosis virus–induced erythroblastosis. *J Virol* 1985;54:295–303.
240. Mintz B, Illmensee K. Normal genetically mosaic mice produced from malignant teratocarcinoma cells. *Proc Natl Acad Sci USA* 1975;72:3585–3589.
241. Miyoshi I, Kubonishi I, Yoshimoto S. Type C virus particles in a cord blood T-cell line derived from cocultivating normal human cord leukocytes and human leukaemic T cells. *Nature* 1981;294:770–771.
242. Moloney JB. A virus-induced rhabdomyosarcoma of mice. *Natl Cancer Inst Monogr* 1966;22:139–142.
243. Momand J, Zambetti GP, Olson DC, George D, Levine AJ. The mdm-2 oncogene product forms a complex with the p53 protein and inhibits p53-mediated transactivation. *Cell* 1992;69:1237–1245.
244. Morrison DK, Kaplan DR, Rapp U, Roberts TM. Signal transduction from membrane to cytoplasm: growth factors and membrane bound oncogene products increase c-raf phosphorylation and associated protein kinase activity. *Proc Natl Acad Sci USA* 1988;85:8855–8859.
245. Mucenski M, McLain K, Kier AB, Swerdlow SH, Schreiner CM, Miller TA, Pietryga DW, Scott WJJ, Potter SS. A functional c-*myb* gene is required for normal murine fetal hepatic hematopoiesis. *Cell* 1991;65:677–689.
246. Muller AJ, Pendergast AM, Havlik MH, Puil L, Pawson T, Witte ON. A limited set of SH2 domains binds BCR through a high-affinity phosphotyrosine-independent interaction. *Mol Cell Biol* 1992; 12:5087–5093.
247. Muller AJ, Pendergast AM, Parmar K, Havlik MH, Rosenberg N, Witte OH. En bloc substitution for the Src homology region 2 domain activates the transforming potential of the c-Abl protein tyrosine kinase. *Proc Natl Acad Sci USA* 1993;90:3457–3461.
248. Neill SD, Hemstrom C, Virtanen A, Nevins JR. An adenovirus E4 gene product trans-activates E2 transcription and stimulates stable E2F binding through a direct association with E2F. *Proc Natl Acad Sci USA* 1990;87:2008–2012.
249. Nerenberg M, Hinrichs SH, Reynolds RK, Khoury G, Jay G. The *tat* gene of human T-lymphotropic virus type 1 induces mesenchymal tumors in transgenic mice. *Science* 1987;237:1324–1329.
250. Nerenberg MI. Biological and molecular aspects of HTLV-1-associated diseases. In: Roos RP, ed. *Molecular neurovirology*. Clifton, NJ: Humana Press; 1992:225–247.
251. Nevins JR. E2F: a link between the Rb tumor suppressor protein and viral oncoproteins. *Science* 1992;258:424–429.
252. Nevins JR, Imperiale MJ, Kao HT, Strickland S, Feldman LT. Detection of an adenovirus E1A-like activity in mammalian cells. *Curr Top Microbiol Immunol* 1984;113:15–19.
253. Ng M, Privalsky ML. Structural domains of the avian erythroblastosis virus *erbB* protein required for fibroblast transformation: dissection by in-frame insertional mutagenesis. *J Virol* 1986;58:542–553.
254. Nocka K, Majumder S, Chabot B, et al. Expression of c-*kit* gene products in known cellular targets of *W* mutations in normal and *W* mutant mice—evidence for an impaired c-*kit* kinase in mutant mice. *Genes Dev* 1989;3:816–826.
255. Nusse R. The *int* genes in mammary tumorigenesis and in normal development. *Trends Genet* 1988;4:291–295.
256. Nusse R. The Wnt gene family in tumorigenesis and in normal development. *J Steroid Biochem Mol Biol* 1992;43:9–12.
257. O'Shea EK, Rutkowski R, Kim PS. Evidence that the leucine zipper is a coiled coil. *Science* 1989;243:538–542.
258. Olden K, Yamada KM. Mechanism of the decrease in the major cell surface protein of chick embryo fibroblasts after transformation. *Cell* 1977;11:957–969.
259. Ossowski L, Quigley JP, Kellerman GM, Reich E. Fibrinolysis associated with oncogenic transformation. *J Exp Med* 1973;138:1056–1064.
260. Pallas DC, Shahrik LK, Martin BL, Jaspers S, Miller TB, Brautigan DL, Roberts TM. Polyoma small and middle T antigens and SV40 small t antigen form stable complexes with protein phosphatase 2A. *Cell* 1990;60:167–176.
261. Papas TS, Blair DG, Fisher RJ. The *ets* oncogene. In: Reddy EP, Skalka AM, Curran T, eds. *The oncogene handbook*. New York: Elsevier; 1988:467–485.
262. Parada LF, Land H, Weinberg RA, Wolf D, Rotter W. Cooperation between gene encoding p53 tumour antigen and ras in cellular transformation. *Nature* 1984;312:649–651.
263. Parker R, Varmus HE, Bishop JM. Expression of v-*src* and chicken c-*src* in rat cells demonstrates qualitative differences between pp60$^{v-<src>}$ and pp60$^{v-<src>}$. *Cell* 1984;37:131–139.
264. Parsons JT, Weber MJ. Genetics of *src*: structure and functional organization of a protein tyrosine kinase. *Curr Top Microbiol Immunol* 1989;147:79–127.
265. Pawson T, Guyden J, King T-H, Radke K, Gilmore T, Martin GS. A strain of Fujinami sarcoma virus which is temperature-sensitive in protein phosphorylation and cellular transformation. *Cell* 1980;22:767–775.
266. Payne GS, Bishop JM, Varmus HE. Multiple arrangements of viral DNA and an activated host oncogene in bursal lymphomas. *Nature* 1982;259:209–214.
267. Peters G. The *int* oncogenes. In: Reddy EP, Skalka AM, Curran T, eds. *The oncogene handbook*. New York: Elsevier; 1988:487–494.
268. Pilder S, Logan J, Shenk T. Deletion of the gene encoding the adenovirus 5 early region 1b 21,000-molecular-weight polypeptide leads to degradation of viral and host cell DNA. *J Virol* 1984;52:664–671.
269. Pilder S, Moore M, Logan J, Shenk T. The adenovirus E1B-55K transforming polypeptide modulates transport or cytoplasmic stabilization of viral and host cell mRNAs. *Mol Cell Biol* 1986;6:470–476.
270. Poiesz BJ, Ruscetti FW, Gazdar AF, Bunn PA, Minna JD, Gallo RC. Detection and isolation of type C retrovirus particles from fresh and cultured lymphocytes of a patient with cutaneous T-cell lymphoma. *Proc Natl Acad Sci USA* 1980;77:7415–7419.
271. Pozzatti R, Vogel J, Jay G. The human T-lymphotropic virus type I tax gene can cooperate with the ras oncogene to induce neoplastic transformation of cells. *Mol Cell Biol* 1990;10:413–417.
272. Prendergast GC, Ziff EB. A new bind for Myc. *Trends Genet* 1992;8:91–96.
273. Privalsky ML. v-Erb A, nuclear hormone receptors, and oncogenesis. *Biochim Biophys Acta* 1992;1114:51–62.
274. Prives C, Covey L, Scheller A, Gluzman A. DNA-binding properties of simian virus 40 T-antigen mutants defective in viral DNA replication. *Mol Cell Biol* 1983;3:1958–1966.
275. Propst F, Van de Woude GF. Expression of c-*mos* proto-oncogene transcripts in mouse tissues. *Nature* 1985;315:516–518.
276. Qian Y, Luckey C, Horton L, Esser M, Templeton DJ. Biological function of the retinoblastoma protein requires distinct domains for hyperphosphorylation and transcription factor binding. *Mol Cell Biol* 1992;12:5363–5372.
277. Qin XQ, Chittenden T, Livingston DM, Kaelin WG, Jr. Identification of a growth suppression domain within the retinoblastoma gene product. *Genes Dev* 1992;6:953–964.

278. Queva C, Ness SA, Grasser FA, Graf T, Vandenbunder B, Stehelin D. Expression patterns of c-myb and of v-myb induced myeloid-1 (min-1) gene during the development of the chick embryo. *Development* 1992;114:125–133.
279. Raines MA, Haihle NJ, Moscovici C, Moscovici MG, Kung H-J. Molecular characterization of three *erbB* transducing viruses generated during avian leukosis virus-induced erythroleukemia: extensive internal deletion near the kinase domain activates the fibrosarcoma- and hemangioma-inducing potentials of *erbB*. *J Virol* 1988;62:2444–2452.
280. Raines MA, Maihle NJ, Moscovici C, Crittenden L, King H-J. Mechanism of c-*erbB* transduction: newly released transducing viruses retain Poly(A) tracts of *erbB* transcripts and encode C-terminally intact *erbB* proteins. *J Virol* 1988;62:2437–2443.
281. Rao L, Debbas M, Sabbatini P, Hockenbery D, Korsmeyer S, White E. The adenovirus E1A proteins induce apoptosis, which is inhibited by the E1B 19-kDa and Bcl-2 proteins [erratum appears in Proc Natl Acad Sci USA 1992;89(20):9974]. *Proc Natl Acad Sci USA* 1992;89:7742–7746.
282. Rapin AMC, Burger MM. Tumor cell surfaces: general alterations detected by agglutinins. *Adv Cancer Res* 1974;20:1–91.
283. Rapp UR, Cleveland JL, Bonner TI, Storm SM. The *raf* oncogene. In: Reddy EP, Skalka AM, Curran T, eds. *The oncogene handbook.* New York: Elsevier; 1988:213–253.
284. Rauscher FJI, Cohen DR, Curran T. *Fos*-associated protein p39 is the product of the *jun* proto-oncogene. *Science* 1988;240:1010–1016.
285. Ravichandran KS, Lee KK, Songyang Z, Cantley LC, Burn P, Burakoff SJ. Interaction of Shc with the zeta chain of the T cell receptor upon T cell activation. *Science* 1993;262:902–905.
286. Raychaudhuri P, Bagchi S, Devoto SH, Kraus VB, Moran E, Nevins JR. Domains of the adenovirus E1A protein required for oncogenic activity are also required for dissociation of E2F transcription factor complexes. *Genes Dev* 1991;5:1200–1211.
287. Reddy EP. The *myb* oncogene. *The oncogene handbook* 1988;Elsevier:327–340.
288. Reddy EP. The Abelson leukemia viral oncogene. *Chronic Myelogenous Leukemia* New York: Dekker; 1991:153–165.
289. Reddy EP, Skalka AM, Curran T. The oncogene handbook. *The oncogene handbook.*New York: Elsevier; 1988.
290. Rettenmier CW, Sherr CJ. The *fms* oncogene. *The oncogene handbook.*New York: Elsevier; 1988:73–99.
291. Ricciardi RP, Jones RL, Cepko CL, Sharp PA, Roberts BE. Expression of early adenovirus genes requires a viral encoded acidic polypeptide. *Proc Natl Acad Sci USA* 1981;78:6121–6125.
292. Rice NR, Gilden RV. The *rel* oncogene. *The oncogene handbook.* New York: Elsevier; 1988:495–508.
293. Rijsewijk F, Schuermann M, Wagenaar E, Parren P, Weigel D, Nusse R. The *Drosophila* homolog of the mouse mammary oncogene *int*-1 is identical to the segment polarity gene *wingless*. *Cell* 1987;50:649–657.
294. Risser R, Holland GD. Structures and activities of activated *abl* oncogenes. *Curr Top Microbiol Immunol* 1989;147:129–153.
295. Robbins J, Blondel BJ, Gallahan D, Callahan R. Mouse mammary tumor gene *int*-3: a member of the notch gene family transforms mammary epithelial cells. *J Virol* 1992;66:2594–2599.
296. Robbins KC, Aaronson SA. The *sis* oncogene. *The oncogene handbook* New York: Elsevier; 1988:427–452.
297. Robinson HL, Blais BM, Tsichlis PN, Coffin JM. At least two regions of the viral genome determine the oncogenic potential of avian leukosis viruses. *Proc Natl Acad Sci USA* 1982;79:1225–1229.
298. Robinson HL, Jensen L, Coffin JM. Sequences outside of the long terminal repeat determine the lymphomogenic potential of Rous-associated virus type 1. *J Virol* 1985;55:752–759.
299. Rohrschneider LR. The macrophage-colony stimulating factor (M-CSF) receptor. *Guidebook to cytokines and their receptors.* New York: Oxford University Press; 1994.
300. Rohrschneider LR, Woolford J. Structural and functional comparison of viral and cellular *fms*. *Semin Virol* 1991;2:385–395.
301. Rosakis-Adcock M, McGlade J, Mbamalu G, Pelicci G, Daly R, Li W, Batzer A, Thomas S, Brugge J, Pelicci PG. Association of the Shc and Grb2/Sem5 SH2-containing proteins is implicated in activation of the Ras pathway by tyrosine kinases. *Nature* 1992;360:689–692.
302. Rous P. A transmissible avian neoplasm: sarcoma of the common fowl. *J Exp Med* 1910;12:696–705.
303. Ruscetti S, Wolff L. Spleen focus-forming virus: relationship of an altered envelope gene to the development of a rapid erythroleukemia. *Curr Top Microbiol Immunol* 1984;112:21–44.
304. Rustgi AK, Dyson N, Bernards R. Amino-terminal domains of c-myc and N-myc proteins mediate binding to the retinoblastoma gene product. *Nature* 1991;352:541–544.
305. Ryder K, Lau LF, Nathans DA. A gene activated by growth factors is related to the oncogene v-*jun*. *Proc Natl Acad Sci USA* 1988; 85:1487–1491.
306. Sagata N, Oskarsson M, Copeland T, Brumbaugh J, Vande Woude GF. Function of c-*mos* proto-oncogene product in meiotic maturation in Xenopus oocytes. *Nature* 1988;335:519–525.
307. Saggioro D, Majone F, Forino M, Turchetto L, Chieco-Bianchi L. Studies on the mechanisms of HTLV-I leukemogenesis. *Leukemia* 1992;6:648–668.
308. Sap J, Munoz A, Damm K. The c-*erb*-A protein is a high-affinity receptor for thyroid hormone. *Nature* 1986;324:635–640.
309. Sarnow P, Ho YS, Williams J, Levine AJ. Adenovirus E1b-58kd tumor antigen and SV40 large tumor antigen are physically associated with the same 54-kd cellular protein in transformed cells. *Cell* 1982;28:387–394.
310. Sassone-Corsi P, Lamph WW, Kamps M, Verma IM. *fos*-associated cellular p39 is related to nuclear transcription factor AP-1. *Cell* 1988;54:553–560.
311. Sassone-Corsi P, Ransone LJ, Lamph WW, Verma IM. Direct interaction between *fos* and *jun* nuclear oncoproteins: role of the "leucine zipper" domain. *Nature* 1988;336:692–695.
312. Sawyers CL, McLaughlin J, Goga A, Havlik M, Witte O. The nuclear tyrosine kinase C-ABL negatively regulates cell growth. *Cell* 1994;in press.
313. Scheffner M, Munger K, Byrne JC, Howley PM. The state of the p53 and retinoblastoma genes in human cervical carcinoma cell lines. *Proc Natl Acad Sci USA* 1991;88:5523–5527.
314. Scheffner M, Werness BA, Huibregtse JM, Levine AJ, Howley PM. The E6 oncoprotein encoded by human papillomavirus types 16 and 18 promotes the degradation of p53. *Cell* 1990;63:1129–1136.
315. Schlessinger J. Allosteric regulation of the epidermal growth factor receptor kinase. *J Cell Biol* 1986;103:2067–2072.
316. Schlessinger J. Signal transduction by allosteric receptor oligomerization. *Trends Biochem Sci* 1988;13:443–447.
317. Schlessinger J. How receptor tyrosine kinases activate Ras. *Trends Biochem Sci* 1993;18:273–275.
318. Schnieder-Gadicke A, Schwarz E. Different human cervical carcinoma cell lines show similar transcription patterns of human papillomavirus type 18 early genes. *EMBO J* 1986;5:2285–2292.
319. Schroeder C, Gibson L, Beug H. The v-erbA oncogene requires cooperation with tyrosine kinase to arrest erythroid differentiation induced by ligand-activated endogenous c-erbA and retinoic acid receptor. *Oncogene* 1992;7:203–216.
320. Schuermann M, Neuberg M, Hunter JB. The leucine repeat motif in *Fos* protein mediates complex formation with *Jun*/AP-1 and is required for transformation. *Cell* 1988;56:507–516.
321. Schupbach J. *Human retrovirology facts and concepts*. Heidelberg: Springer-Verlag; 1989.
322. Schwab M. The *myc*-box oncogenes. *The oncogene handbook.* New York: Elsevier; 1988:381–388.
323. Schwarz E, Freese UK, Gissman L. Structure and transcription of human papillomavirus sequences in cervical carcinoma cells. *Nature* 1985;314:111–114.
324. Sefton BM, Hunter T. From c-*src* to v-*src*, or the case of the missing C terminus. *Cancer Surv* 1986;5:159–171.
325. Segawa K, Ito Y. Differential subcellular localization of *in vivo* phosphorylated and non-phosphorylated middle-sized tumor antigen of polyoma virus and its relationship to middle-sized tumor antigen phosphorylating activity *in vitro*. *Proc Natl Acad Sci USA* 1982;79:6812–6816.
326. Seth A, Vande Woude GF. The *mos* oncogene. *The oncogene handbook.* New York: Elsevier; 1988:195–211.
327. Setoyama C, Frunzio R, Liau G, Mudryj M, de Crombrugghe B. Transcriptional activation encoded by the v-*fos* gene. *Proc Natl Acad Sci USA* 1986;83:3213–3217.
328. Shalloway D, Coussens PM, Yaciuk P. Overexpression of the c-*src* protein does not induce transformation of NIH 3T3 cells. *Proc Natl Acad Sci USA* 1984;81:7071–7075.
329. Shan B, Zhu X, Chen PL, Durfee T, Yang Y, Sharp D, Lee WH. Molecular cloning of cellular genes encoding retinoblastoma-associated proteins: identification of a gene with properties of the transcription factor E2F. *Mol Cell Biol* 1992;12:5620–5631.

330. Sharif M, Privalsky ML. v-ErbA oncogene function in neoplasia correlates with its ability to repress retinoic acid receptor action. *Cell* 1991;166:885–893.
331. Sharp PA, Pettersson U, Sambrook J. Viral DNA in transformed cells. I. A study of the sequences of adenovirus 2 DNA in a line of transformed rat cells using specific fragments of the viral genome. *J Mol Biol* 1974;86:709
332. Sheinin R. Studies on the thymidine kinase activity of mouse embryo cells infected with polyoma virus. *Virology* 1966;28:47
333. Shen-Ong GL. The myb oncogene. *Biochem Biophys Acta* 1990; 1032:39–52.
334. Sherr CJ. The *fms* oncogene. *Biochim Biophys Acta* 1988;948:225–243.
335. Sherr CJ, Rettenmier CW. The *fms* gene and the CSF-1 receptor. *Cancer Surv* 1986;5:225–231.
336. Sherr CJ, Rettenmier CW, Sacca R, Roussel MF, Look AT, Stanley ER. The c-*fms* proto-oncogene product is related to the receptor for the mononuclear phagocyte growth factor, CSF-1. *Cell* 1985;41:665–676.
337. Shibuya M, Yamaguchi S, Yamane A, Ikeda T, Tojo A, Matsushime H, Sato M. Nucleotide sequence and expression of a novel human receptor-type tyrosine kinase gene (flt) closely related to the fms family. *Oncogene* 1990;5:519–524.
338. Shih TY, Papageorge AG, Stokes PE, Weeks MO, Scolnick EM. Guanine nucleotide-binding and autophosphorylating activities associated with the p21 *src* protein of Harvey murine sarcoma virus. *Nature* 1980;287:686–691.
339. Shih TY, Weeks MO, Young HA, Scolnick E. p21 of Kirsten murine sarcoma virus is thermolabile in a viral mutant temperature sensitive for the maintenance of transformation. *J Virol* 1979;31:546–556.
340. Shirodkar S, Ewen M, DeCaprio JA, Morgan J, Livingston DM, Chittenden T. The transcription factor E2F interacts with the retinoblastoma product and a p107-cyclin A complex in a cell cycle-regulated manner. *Cell* 1992;68:157–166.
341. Shtivelman E, Lifshitz B, Gale RP, Canaani E. Fused transcript of *abl* and *bcr* genes in chronic myeloid leukaemia. *Nature* 1985;315:550–554.
342. Smith MR, Green WC. Type I human T cell leukemia virus tax protein transforms rat fibroblasts through the cyclic adenosine monophosphate response element binding protein/activating transcription factor pathway. *J Clin Invest* 1991;88:1038–1042.
343. Smotkin D, Wettstein FO. Transcription of human papillomavirus type 16 early genes in a cervical cancer and a cancer-derived cell line and identification of the E7 protein. *Proc Natl Acad Sci USA* 1986;83:4680–4684.
344. Soriano P, Montgomery C, Geske R, Bradley A. Targeted disruption of the c-*src* proto-oncogene leads to osteopetrosis in mice. *Cell* 1991;64:693–702.
345. Souyri M, Vigon I, Penciolelli JF, Heard JM, Tambourin P, Wendling F. A putative truncated cytokine receptor gene transduced by the myeloproliferative leukemia virus immortalizes hematopoietic progenitors. *Cell* 1990;63:1137–1147.
346. Stahl H, Droge P, Knippers R. DNA helicase activity of SV40 large tumor antigen. *EMBO J* 1986;5:1939–1944.
347. Stavnezer E. The *ski* oncogenes. *The oncogene handbook.* New York: Elsevier; 1988:393–401.
348. Stehelin D, Varmus HE, Bishop JM, Vogt PK. DNA related to the transforming gene(s) of avian sarcoma viruses is present in normal avian DNA. *Nature* 1976;260:170–173.
349. Stossel TP, Chaponnier C, Ezzell RM. Nonmuscle actin-binding proteins. *Annu Rev Cell Biol* 1985;1:353–402.
350. Sukumar S. *Ras* oncogenes in chemical carcinogenesis. *Curr Top Microbiol Immunol* 1989;148:93–114.
351. Swanstrom R, Parker RC, Varmus HE, Bishop JM. Transduction of a cellular oncogene: the genesis of Rous sarcoma virus. *Proc Natl Acad Sci USA* 1983;80:2519–2523.
352. Taira M, Yoshida T, Miyagawa K, Sakamoto H, Terada M, Sugimura T. cDNA sequence of human transforming gene hst and identification of the coding sequence required for transforming activity. *Proc Natl Acad Sci USA* 1987;84:2980–2984.
353. Takeya T, Hanafusa H. Nucleotide sequence of c-*src*. *Cell* 1983; 32:881–890.
354. Talmage DA, Freund R, Young aT, Dahl J, Dawe CJ, Benjamin TL. Phosphorylation of middle T by pp60 c-src: A switch for binding of phosphatidylinositol 3-kinase and optimal tumorigenesis. *Cell* 1989;59:55–65.
355. Taylor SJ, Shalloway D. The cell cycle and c-Srf. *Curr Opin Gen Develop* 1993;3:26–34.
356. Temin HM, Rubin H. Characteristics of an assay for Rous sarcoma virus and Rous sarcoma cells in tissue culture. *Virology* 1958; 6:669–688.
357. Thelander L, Reichard P. Reduction of ribonucleotides. *Ann Rev Biochem* 1979;48:133–158.
358. Thomas JE, Aguzzi A, Soriano P, Wagner EF, Brugge JS. Induction of tumor formation and cell transformation by polyoma middle T antigen in the absence of Src. *Oncogene* 1993;8:2521–2529.
359. Tjian R, Robbins A. Enzymatic activities associated with a purified simian virus 40 T antigen-related protein. *Proc Natl Acad Sci USA* 1979;76:610–614.
360. Tooze J. *DNA tumor viruses: molecular biology of tumor viruses*, 2nd ed. Cold Spring Harbor, NY: Cold Spring Harbor Laboratory; 1981.
361. Toyoshima K, Vogt PK. Temperature sensitive mutants of an avian sarcoma virus. *Virology* 1969;39:930–931.
362. Turner R, Tjian R. Leucine repeats and an adjacent DNA binding domain mediate the formation of functional c*Fos*–c*Jun* heterodimers. *Science* 1989;243:1689–1694.
363. Varmus H. Retroviruses. *Science* 1988;240:1427–1435.
364. Varmus H, Bishop JM. Biochemical mechanisms of oncogene activity: proteins encoded by oncogenes (introduction). *Cancer Surv* 1986;5:153–158.
365. Varmus HE. The molecular genetics of cellular oncogenes. *Annu Rev Gen* 1984;18:553–612.
366. Vennstrom B, Damm K. The *erbA* and *erbB* oncogenes. In: Reddy EP, Skalka AM, Curran T, eds. *The oncogene handbook.* New York: Elsevier; 1988:25–37.
367. Verma IM. Proto-oncogene *fos*: a multifaceted gene. *Oncogenes and growth factors*. New York: Elsevier; 1987:67–73.
368. Vogel J, Hinrichs SH, Reynolds RK, Luciw PA, Jay G. The HIV *tat* gene induces dermal lesions resembling Kaposi's sarcoma in transgenic mice. *Nature* 1988;335:606–611.
369. Vogt PK. Spontaneous segregation of nontransforming viruses from cloned sarcoma viruses. *Virology* 1971;46:939–946.
370. Vogt PK. Oncogenic transformation by jun. In: P. Angel, P. Herrlich, eds. The *Fos* and *Jun families of transcription factors.* CRC Press 1994:203–220.
371. Vogt PK, Bos TJ. *Jun*: oncogene and transcription factor. *Adv Cancer Res* 1990;55:1–35.
372. Vogt PK, Bos TJ, Doolittle RF. Homology between the DNA-binding domain of the GCN4 regulatory protein of yeast and the carboxy-terminal region of a protein coded for by the oncogene jun. *Proc Natl Acad Sci USA* 1987;84:3316–3319.
373. Walter G, Ruediger R, Slaughter C, Mumby M. Association of protein phosphatase 2A with polyoma virus medium T antigen. *Proc Natl Acad Sci USA* 1990;87:2521–2525.
374. Wang CY, Petryniak B, Thompson CB, Kaelin WG, Leiden JM. Regulation of the Ets-related transcription factor Elf-1 by binding to the retinoblastoma protein. *Science* 1993;260:1330–1335.
374a. Wang HG, Rikitabe Y, Carter MC, Yaciuk P, Abraham SE, Zevler B, Moran E. Identification of specific adenovirus E1A N-terminal residues critical to the binding of cellular proteins and to the control of cell growth. *J Virol* 1993;67:476–488.
375. Wasylyk B, Hahn SL, Giovane A. The Ets family of transcription factors. *Euro J Biochem* 1993;211:7–18.
376. Waterfield MD, Scrace GT, Whittle N. Platelet-derived growth factor is structurally related to the putative transforming protein p28^{sis} of simian sarcoma virus. *Nature* 1983;304:35–39.
377. Weinberg R, Wigler M, eds. *Oncogenes and the molecular origins of cancer.* Cold Spring Harbor, NY: Cold Spring Harbor Laboratory; 1989.
378. Weinberg RA. Oncogenes of spontaneous and chemically induced tumors. *Adv Cancer Res* 1982;36:149–164.
379. Weinberg RA. The action of oncogenes in the cytoplasm and nucleus. *Science* 1985;230:770–776.
380. Weinberger C, Hollenberg SM, Rosenfeld MG, Evans RM. Domain structure of human glucocorticoid receptor and its relationship to the v-*erbA* oncogene product. *Nature* 1985;318:670–672.
381. Weinberger C, Thompson CC, Ong ES, Lebo R, Gruol DJ, Evans RM. The c-*erb*-A gene encodes a thyroid hormone receptor. *Nature* 1986;324:641–646.
382. Weiss R. Human T-cell retroviruses. In: Weiss R, Teich N, Varmus HE, Coffin J, eds. *RNA tumor viruses: molecular biology of tumor viruses*. Cold Spring Harbor, NY: Cold Spring Harbor Laboratory; 1985:405–485.
383. Weiss R, Teich N, Varmus HE, Coffin J, eds. In: *RNA tumor viruses:*

molecular biology of tumor viruses. Cold Spring Harbor, NY: Cold Spring Harbor Laboratory; 1982.

384. Weiss R, Teich N, Varmus HE, Coffin J, eds. In: *RNA tumor viruses: molecular biology of tumor viruses*. Supplements and Appendixes. Cold Spring Harbor, NY: Cold Spring Harbor Laboratory; 1985.
385. Werness BA, Levine AJ, Howley PM. Association of human papillomavirus types 16 and 18 E6 proteins with p53. *Science* 1990;248:76–79.
386. Westermark B, Heldin C. Platelet-derived growth factor in autocrine transformation. *Cancer Res* 1991;51:5087–5092.
387. White E. Death-defying acts: a meeting review on apoptosis. *Genes Dev* 1993;7:2277–2284.
388. White E, Grodzicker T, Stillman BW. Mutations in the gene encoding the adenovirus early region 1B 19,000-molecular-weight tumor antigen cause the degradation of chromosomal DNA. *J Virol* 1984;52:410
389. White E, Sabbatini P, Debbas M, Wold WS, Kusher DI, Gooding LR. The 19-kilodalton adenovirus E1B transforming protein inhibits programmed cell death and prevents cytolysis by tumor necrosis factor alpha. *Mol Cell Biol* 1992;12:2570–2580.
390. Whitman M, Kaplan DR, Schaffhausen B, Cantley L, Roberts TM. Association of phosphatidylinositol kinase activity and polyoma middle T competent for transformation. *Nature* 1985;315:239–242.
391. Whyte P, Buchkovich KJ, Horowitz JM, Friend SH, Raybuck M, Weinberg RA, Harlow E. Association between an oncogene and an anti-oncogene: the adenovirus E1A proteins bind to the retinoblastoma gene product. *Nature* 1988;334:124–129.
392. Williams LT. The *sis* gene and PDGF. *Cancer Surv* 1986;5:231–241.
393. Witte ON. Functions of the *abl* oncogene. *Cancer Surv* 1986;5:183–197.
394. Wolff L, Chung S-W, Ruscetti S. Molecular basis for the pathogenicity of the Friend spleen focus-forming virus. *Modern trends in virology*. New York: Springer-Verlag; 1988:123–133.
395. Wood TG, McGeady ML, Baroudy BM, Blair DG, Vande Woude GF. Mousec-*mos* oncogene activation is prevented by upstream sequences. *Proc Natl Acad Sci USA* 1984;81:7817–7821.
396. Wu X, Bayle JH, Olson D, Levine AJ. The p53-mdm-2 autoregulatory feedback loop. *Genes Dev* 1993;7:1126–1132.
397. Wyke JA. The selective isolation of temperature-sensitive mutants of Rous sarcoma virus. *Virology* 1973;52:587–590.
398. Wyllie AH, Rose KA, Morris RC, Steel CM, Foster E, Spandidos DA. Rodent fibroblast tumours expressing human myc and ras genes: growth, metastasis, and endogenous oncogene expression. *Br J Cancer* 1987;56:251–259.
399. Xiong Y, Connolly T, Futcher B, Beach D. Human D type cyclin. *Cell* 1991;65:691–699.
400. Xiong Y, Hannon GJ, Zhang H, Casso D, Kobayashi R, Beach D. p21 is a universal inhibitor of cyclin kinases. *Nature* 1993;366:701–704.
401. Yamaguchi TP, Dumont DJ, Conlon RA, Breitman ML, Rossant J. elk-1, an flt-related receptor tyrosine kinase is an early marker for endothelial cell precursors. *Development* 1993;118:489–498.
402. Yamashita T, Shimojo H. Induction of cellular DNA synthesis by adenovirus 12 in human embryo kidney cells. *Virology* 1969; 38:351–355.
403. Yang S, Lickteig RL, Estes R, Rundell K, Walter G, Mumby MC. Control of protein phosphatase 2A by simian virus 40 small-t antigen. *Mol Cell Biol* 1991;11:1988–1995.
404. Yee S, Branton PE. Detection of cellular proteins associated with human adenovirus type 5 early 1A polypeptides. *Virology* 1985;147:142–153.
405. Yew N, Strobel M, Vande Woude GF. Mos and the cell cycle: the molecular basis of the transformed phenotype. *Curr Opin Genet Dev* 1993;3:19–25.
406. Yew PR, Berk AJ. Inhibition of p53 transactivation required for transformation by adenovirus early 1B protein. *Nature* 1992;357:82–85.
407. Yonish-Rouach E, Grunwald D, Wilder S, Kimchi A, May E, Lawrence J-J, May P, Oren M. p53-mediated cell death: relationship to cell cycle control. *Mol Cell Biol* 1993;13:1415–1423.
408. Zamanian M, La Thangue NB. Transcriptional repression by the Rb-related protein p107. *Mol Biol Cell* 1993;4:389–396.
409. Zambetti GP, Bargonetti J, Walker K, Prives C, Levin AJ. Wild-type p53 mediates positive regulation of gene expression through a specific DNA sequence element. *Genes Dev* 1992;6:1143–1152.
410. Zenke M, Kahn P, Disela C. v-*erb* A specifically suppresses transcription of the avian erythrocyte anion transporter (band 3) gene. *Cell* 1988;52:107–119.
411. Zhu L, van den Heuvel S, Helin K, Fattaey A, Ewen M, Livingston D, Dyson N, Harlow E. Inhibition of cell proliferation by p107, a relative of the retinoblastoma protein. *Genes Dev* 1993;7:1111–1125.
412. zur Hausen H. Human papillomaviruses in the pathogenesis of anogenital cancer. *Virology* 1991;184:9–13.
413. zur Hausen H. Viruses in human cancers. *Science* 1991;254:1167–1173.

Fundamental Virology, Third Edition
edited by B.N. Fields, D.M. Knipe, P.M. Howley, et al.
Lippincott - Raven Publishers, Philadelphia © 1996

CHAPTER 10

Immune Response to Viruses

J. Lindsay Whitton and Michael B. A. Oldstone

Virus infection is a major cause of human morbidity and mortality. Although many viruses doubtless remain to be discovered, most of the agents known to cause widespread and severe human disease were identified and (at least partially) characterized many years ago. Vaccination has reduced the incidence of the more severe infections (polio, mumps, rubella, measles), at least in "developed" countries, and has of course eradicated a major scourge, smallpox. Nevertheless, and increasingly since the appearance of human immunodeficiency virus (HIV), viruses still exact a heavy toll in human suffering. Understanding the antiviral immune response is important for several reasons. It may help us evaluate clinical problems (at the most basic level, our understanding of the kinetics of the antibody response has been the mainstay of viral diagnostic procedures for many years); it may help in designing new or im-

J. L. Whitton and M. B. A. Oldstone: Department of Neuropharmacology, Division of Virology, Scripps Reseach Institute, La Jolla, California 92037.

proved antiviral vaccines; and it provides insight into the immune system, which may yield benefits in fields unrelated to virus infection.

The relationship between virus infection and immune response leading to immunity and/or tissue injury was noted over 195 years ago by Jenner (45), and associations of virus-induced immunosuppression and of immune complexes after antimicrobial therapy were recorded over 85 years ago by von Pirquet (133). Since then, multiple observations clearly indicate that infection of a host by a virus offers many opportunities for development of immune-mediated injury. Such immunopathologic responses against the virus and/or virus-infected cells reflect the normal mechanisms by which the host recognizes an infectious agent as foreign, and strives to remove it and the cells it infects. The replicating virus provides a supply of antigens and, in most if not all instances, elicits a host immune response. With acute viral infections, there is a race between viral replication and host immunity; the outcome is either termination of disease or death of the host. In chronic virus infections the time scale is lengthened, and both a continuous host immune response against the virus and ongoing virus replication occur. When virions or viral antigens are present in the serum or other fluids, interaction of viral antigen(s) and antiviral antibody(s) may result in the formation of viral antigen–antibody complexes with subsequent manifestations of immune complex disease. In other instances, antiviral antibody and/or T cells sensitized to the virus may react with and injure virus-infected cells. Indeed many, if not most, of the clinical signs and symptoms manifested during viral infection are a consequence of the host's immune response programmed to clear the virus and to remove infected cells.

OVERVIEW OF THE PRIMARY IMMUNE RESPONSE TO VIRUS INFECTION

Virus infections are countered by host immune responses. Several non–antigen-specific responses (e.g. interferons, natural killer cells, muco-ciliary responses, etc.) can influence the outcome, as can antigen-specific immune responses; it is this facet of host immunity that is our major focus in this chapter. Antigen-specific responses fall into two camps: antibody responses and T-cell responses. Antibody and T-cell responses rely on lymphocytes, the progenitors of which are hematopoietic stem cells. There are two broad types of lymphocytes: T (thymus-derived)-lymphocytes (which pass through the thymus during maturation) and B-lymphocytes (which give rise to antibody-producing plasma cells). The thymus is a retrosternal organ, highly active in youth but regressing in later years, in which lymphocytes are educated to fulfill their biological function. Lymphocytes entering the thymus undergo two rounds of selection: positive selection for those lymphocytes able to recognize the appropriate molecules on the surface of cells and negative selection to eradicate those T cells that may recognize self cells too well, with possible autoimmune consequences. Only ~1–2% of all lymphocytes entering the thymus eventually emerge as mature T cells; the remainder undergo deletion during the selection processes. It is from this 1–2% of cells that all virus-specific T cells are induced.

Both of the antigen-specific arms of the immune response play important roles in combatting microbial infection; there is built-in plasticity to the overall response, which dictates that the relative contributions of each arm varies with the nature of the infecting microbe. In countering viral infections, antibodies act to decrease antigen load and to diminish viral infectivity, thus diminishing the number of infected cells and thereby facilitating the eradication of this cell pool, a job that falls to T cells. Furthermore, antibodies are important in prophylaxis and treatment; passively-transferred antibodies can diminish the risk of infection, and in certain cases are of use in treating established infection. The major role of T cells is the recognition and eradication of infected cells, ideally soon after infection, before virus maturation has occurred. Thus, antibodies and T cells act in concert: antibodies reduce the load of infectious units in fluid phase, thereby decreasing the number of infected cells that T cells have to deal with, and T cells kill cells soon after infection, ensuring that the amount of virus released is minimized, thus easing the load on antibodies.

Primary virus infection (when the host is first exposed to the wild virus or to a vaccine containing viral antigens) in many cases induces readily detectable antiviral antibody and cytotoxic T-lymphocyte (CTL) responses. The T-cell responses peak early (7–10 days) and decrease (often following the resolving virus infection) usually within 2–3 weeks postinfection. Antibody responses usually peak later than CTL; antibodies are often barely detectable in the acute (symptomatic) stage of infection, but increase in number over a period of 2–4 weeks to a readily detectable level that often lingers for weeks or months, depending on the host and virus. In the case of some viruses (e.g., yellow fever and measles), detectable levels can linger for a lifetime. After subsequent infection (i.e., reexposure to the wild virus or initial exposure to wild virus after vaccination), the classical "anamnestic" or "memory" antibody response is seen, with antibody levels rapidly climbing much higher than before and usually being maintained much longer. Cytotoxic T-cell responses, too, are brisk, due to the presence of primed memory T cells. Therefore, the presence of a memory response is the hallmark of the antigen-specific immune response. The molecular requirements for memory remain somewhat controversial. Some workers argue that specific antigen must persist in order to continually restimulate memory T cells (86) and B cells; however, recent persuasive evidence indicates that memory cells can persist in the absence of specific antigen (1), perhaps being sporadically stimulated in a nonspecific manner by local cytokine release during other, antigenically unrelated, immune reactions.

In this chapter we present information about how antibodies and T cells act in controlling virus infections and evaluate their relative roles. We integrate *in vivo* findings with recent advances in our understanding of the underlying molecular events and show the evolutionary relationships between the invading microbes and the immune system that seeks to defeat them. We then discuss the detrimental consequences of antiviral immune responses, namely the immunopathological consequences that may arise from them.

THE STRUCTURE OF THE IMMUNE SYSTEM REFLECTS THE DIFFERENT TYPES OF MICROBES THAT CHALLENGE IT

The vertebrate immune system fights a ceaseless battle against infectious agents, and both of these opposing forces have been shaped by the constant conflict. Immunological pressures bear upon mutable microbes, selecting those best able to escape the unwelcome attentions of the immune system. Microbial evolution is thus driven, at least in part, by the efforts of the immune response to eradicate infection. These enforced microbial changes are mirrored in the vertebrate hosts: the immune system is molded by the nature of the microbes that it faces. Thus, hosts whose immune systems are less able to counter a specific infectious agent may die; those who live (and their progeny) are better able to resist similar organisms. Hence, the vertebrate gene pool is altered by exposure to specific microbes.

This battle for evolutionary supremacy is, in a sense, a no-win situation for the virus. If the virus is too readily eliminated by host immunity, and is in consequence unable to maintain itself in the host population, it may disappear. Conversely, if a virus is too virulent and cannot be controlled by host immunity, it may result in eradication of the host (e.g., in the 16th and 17th centuries, entire New World tribes were killed by infections introduced by European settlers, probably reflecting an inbred gene pool in these native tribes). Theoretically, eradication of the host may lead to disappearance of the virus. At both extremes, then, the virus may be eliminated. Thus, from the virus' viewpoint, it is best to be virulent and yet flexible enough to survive in the reservoir population, but to temper its pathogenicity to allow host survival. Frequently, the initial virulence of the virus is diminished by mutation to a less virulent form, and this, in concert with the selection of more resistant hosts, results in a fluctuating equilibrium, in which both host and microbe coexist. Many examples of both microbial and host changes have been documented. For example, influenza virus is the quintessence of antigenic variation in the face of host immunity, and those influenza viruses best able to evade host responses have fueled outbreaks of epidemic or pandemic proportions. The complementary outcome, in which the microbes select for survival those hosts best able to counter their pathogenic effects, is best demonstrated when the offspring have enhanced survival prospects. Such improved resistance in the progeny may, of course, be in part conferred by passive (maternally derived) immunity rather than by genetic fitness; but there are clear examples of the latter. For instance, wild Australian rabbits that survived the first onslaught by myxoma virus (a poxvirus deliberately administered in an attempt to control the exponential increase in the rabbit population) bred hardy offspring more resistant to the agent (26). Natives of the New World were devastated by smallpox and measles, introduced by the European invaders, and those who remained were better able to survive infection (73); and the appearance in dolphins (6,129) and seals (98,132) of a morbillivirus related to canine distemper virus caused high initial morbidity, which now is likely to be decreasing. It has been suggested that a contemporary example of such an evolving host-parasite relationship may be that between HIV and its human host. Several host genes responsible for determining the ability of the individual host to respond to an infectious agent have been characterized in experimental model systems. Many host gene systems are involved in this process, and we shall discuss several of them as we proceed.

In summary, the immune system has been shaped at least in part by the nature of the challenges it has faced. Above are cited some specific examples of such evolutionary constraints—but can we discern a more general global effect whereby different classes of organisms (e.g., viruses compared with bacteria) have caused the immune system to evolve in a particular way? Below we shall argue that this is indeed the case; that the two different facets of the antigen-specific immune response (antibodies and T cells) have evolved to control infection depending on the nature of the infecting organism (intracellular or extracellular); and that the deciding factor in determining the nature of the immune response is which of the two presentation pathways the microbial antigens best can enter.

THE ROLES OF ANTIBODIES AND T CELLS IN THE CONTROL OF MICROBIAL INFECTIONS: POINTERS FROM EXPERIMENTS OF NATURE

The role of antibodies in the control of many virus infections is clear. Indeed, the single major indicator of virus infection has been, historically, development of elevated levels of specific antibody. Thus, a patient presenting with an acute illness has blood drawn (acute phase serum sample), and at a later date, often 2–4 weeks later, after clinical recovery, a second sample is gathered (convalescent phase serum sample). For various candidate viruses, the levels of specific antibodies in the two samples are compared, and a rise of more than fourfold is usually considered diagnostic of recent infection. Similarly, the role of antiviral antibodies in vaccination has long been unquestioned. The excellent correlation between vaccine immunogenicity (as judged by induction of high levels of specific antibody) and vaccine efficacy (as judged by the vaccine's protective effect) has bolstered the belief that the antibodies are protective. However, although antibodies play a role in primary and secondary immunity, data suggest that their functions may be nonessential for many viruses. "Experiments of nature" in humans strongly suggest that antiviral antibody responses can be dispensed with at little cost to the host's ability to control several primary or subsequent virus infections. For example, children born with genetically determined deficiencies in antibody production, with no detectable immunoglobulins (Igs), do not show an increased susceptibility to most viral diseases, with the exception of rare enteroviral meningitides, caused most often by echovirus type 9 or 11 (72,77). Both the incidence of viral disease and disease severity remain indistinguishable in these antibody-deficient individuals from those of normal children. These agammaglobulinemic children show a marked increase in susceptibility to bacterial, rather than viral, infection (35,36). Similarly, the complement cascade is important as a final effector mechanism in some antibody-mediated responses, and children with genetic defects in components of this cascade are less able to cope with bacterial, particularly meningococcal, challenge, but their resistance to virus infection and disease appears normal (71,97). That T cells are critical in eradicating human viral infections is further suggested by several observations. Although it is difficult to find syndromes complementary to the agammaglobulinemias (i.e., in which T-cell functions are reduced but antibodies are unaffected), human disease again provides a clue. In humans with impaired T-cell responses—as observed in patients with Di George's syndrome (congenital thymic aplasia), acquired immunodeficiency syndrome (AIDS), or leukemia, or in recipients of immunosuppressive therapy—the frequency and severity of virus infections are markedly increased (35). In many of these cases, however, there is also some impairment of antibody response. In animals, models have been developed with clean mutations in various aspects of antibody or T-cell function; these are discussed later in this chapter.

Measles infection provides an excellent example of the role of T cells in humans. In normal children the disease is typified by the characteristic rash, and complete recovery is usual. In contrast, in T cell–deficient children the disease is often fatal (37,82,118). The rash itself is T cell mediated and does not develop in severely immunosuppressed children; indeed, the presence of a rash in an immunosuppressed child (e.g., in a child with leukemia who contracts measles) is considered a positive prognostic indicator (55). In agammaglobulinemic children the rash develops normally, and the course of the disease is unaltered by the absence of antibody responses. Furthermore, agammaglobulinemic children are subsequently immune to measles; thus, antibodies in this instance are required neither for control of the primary infection nor for resistance to secondary disease (36). Although antibodies may be dispensable components in the immune response to several viruses, this does not suggest that antibodies are functionless. Indeed the evidence to the contrary is overwhelming. Passive antibody therapy can protect against or modify the course of several human virus infections. For example, infusion of antibodies specific for the Junin arenavirus is beneficial in Argentinian hemorrhagic fever (24,68), whereas postexposure rabies prophylaxis relies on vaccination and concurrent administration of virus-specific Igs (64). Furthermore, in experimental models antibodies can lower viral titers and modulate disease. For instance, recovery from ocular herpesvirus infection is hastened by administration of anti-HSV antibody (62). However, the absence of these antibody-mediated functions is well tolerated by the human host in responding to many virus infections.

Thus, although both antibodies and T cells play important roles in the control of virus infection, disruption of the former has less effect on the host's ability to control virus infections but dramatically reduces the host's capacity to counter bacterial challenges. What is the molecular explanation for this phenomenon? To understand this question it is necessary to review how antibodies and T cells see virus antigens on the virus particle or at the surface of an infected cell and to describe in some detail how these virus antigens get there. We approach this first issue in the form of an overview, then describe in greater detail the individual genes and gene families involved.

THE ROLES OF ANTIBODIES AND T CELLS IN THE CONTROL OF MICROBIAL INFECTIONS: THE IMPORTANCE OF HOW ANTIGENS ARE RECOGNIZED

Antigen Recognition by Antibodies

Antibodies recognize (usually) whole protein antigen, through regions of hypervariable sequence. The antibody–antigen union has recently been subjected to crystallographic analyses (2,111) that indicated that the union is more "hand-in-glove" (in which components can, to some extent, alter their conformation to accommodate one another) than "lock and key" (in which both elements are fixed, each unable to modulate to the other). Antibodies recognize antigens as intact proteins, often in the fluid phase. By this means antibodies can neutralize viruses by agglutination (thereby reducing the number of infectious units), can coat viral particles, preventing them from infecting cells and, with the participation of complement, can directly lyse virions (93). In addition, antiviral antibodies can recognize viral antigens attached to cell (or microbial) membranes. Thus, antibodies can recognize free bacteria and viruses, as well as viral proteins (most often glycoproteins), on the surface of infected cells. From this vantage point, it is easy to imagine how antibodies can play a major role in the control of these organisms, by inactivating them as free extracellular entities. However, what happens when a microbe can evade antibody recognition by locating itself within a host cell? Viruses are obligate intracellular parasites; their replication is intracellular, often relying on subversion of host machinery. How can antibodies exert an antiviral effect at this stage? Certainly, recognition of viral proteins on the surface of infected cells is one mechanism. However, many viruses delay the display of cell surface proteins until late in the infective cycle. How then can a host detect an infected cell early in the infectious process, thus maximizing its immunological advantage? Here antibodies are less effective, being limited by their recognition requirements, whereas T cells play a more critical role. T cells, as will be detailed below, can recognize almost any viral protein, including those made immediately upon infection despite their not being cell surface proteins per se. When this early recognition capacity is allied to the ability to destroy such infected cells, T cells become a formidable component of the antiviral armamentarium.

Antigen Recognition by T Cells

T cells can be subdivided into several functional groups: cytotoxic (which kill target cells), suppressor (which suppress immune responses and may do so by killing host cells driving the immune response), and helper (which provide help in the form of cytokines, for example to B cells in driving antibody synthesis). There is a correlation between T-cell surface markers (CD4/CD8) and function, but this correlation is not as strong as that between surface markers and target cell major histocompatibility complex (MHC) molecules, which is described below. The majority of $CD8^+$ cells are cytotoxic (although helper phenotype $CD8^+$ cells have been described), and most $CD4^+$ cells are helper cells (although $CD4^+$ CTLs can be identified).

T cells recognize antigens via a cell surface heterodimer, the T-cell receptor (TCR). This molecule is structurally reminiscent of the Fab portion of an antibody molecule, but one critical aspect in which the nature of T-cell recognition differs from that of antibody recognition is that although antibodies recognize antigen in isolation, T cells react to antigen only when it is presented in association with host glycoprotein encoded in the MHC. This phenomenon, named MHC restriction, was discovered by analyses of host responses to virus infection (143,144). There are two classes of MHC molecule (class I and class II), and there is a tight relationship between the type of MHC/peptide complex recognized by a T cell and the surface marker (CD4 or CD8) borne by the T cell. MHC class I molecules are the "classical" molecules associated with graft rejection (the phenomenon that gave the MHC its name); they are expressed on most somatic cells [neurons being a rare exception (47)], and they interact with T cells bearing the CD8 surface marker. In contrast, MHC class II molecules, previously termed Ia or Ir (immune response gene) molecules, have a much more restricted expression, being found only on specialized antigen presenting cells (e.g., macrophages, B-lymphocytes, dendritic cells), and they interact with T cells carrying the CD4 surface marker (Table 1). The CD4/II and CD8/I relationships are determined by direct interactions between the respective molecules: CD4 binds to a specific conserved region of MHC class II (59,131), and a similar interaction occurs between CD8 and class I (117).

So, evolution has produced two classes of T cell, and the major feature distinguishing them is which of the two MHC/peptide complexes each can recognize. It is clear

TABLE 1. *Recognition requirements and functions of $CD4^+$ and $CD8^+$ T cells*

	$CD8^+$ T cells	$CD4^+$ T cells
MHC molecule that presents the antigen to T cells	Class I MHC	Class II MHC
Cell types that bear the MHC molecule and thus can present antigen to the T cells	Almost all nucleated cells, with probable exception of neurons	Specialized antigen-presenting cells (e.g., macrophages, B cells)
Type of antigen recognized	Usually endogenous antigen, made within the presenting cell	Usually exogenous antigen, taken into presenting cell by endocytosis
Such an antigen will be made during infection by:	Intracellular organisms (e.g. viruses), a few bacteria (e.g., *Listeria*, mycobacteria), and other parasites	Viruses, bacteria, and other parasites; infection results in microbial proteins released into circulation
Function of the T cells	Usually cytotoxic, killing target cells. Other possible mechanisms of antiviral effect. $CD8^+$ helper cells recently described	Usually provide "help" in the form of cytokines to B cells, thereby helping in antibody production; $CD4^+$ CTL increasingly recognized

how T cells distinguish between class I and class II (using the CD4/8 proteins); but what is the importance of their being able to make this distinction? Presumably there is some vital and fundamental difference between the class I and class II complexes, and the ability of the immune response to detect and respond to this class I/class II difference is of prime importance.

At the target cell surface, class I and class II molecules are similar in overall structure. The molecules are heterodimers, with four domains: two Ig-like domains at the cell membrane and two distal domains. The class I heterodimer comprises the class I heavy (H) chain (three domains) closely complexed with a non–MHC-encoded protein, ß2-microglobulin (ß2M). Expression of the class I H chain on the cell surface requires ß2M. The class II heterodimer consists of two similar chains, α and ß, each with two domains and encoded in the MHC. For both class I and class II, the two membrane-distal domains together form a structure graphically described as a "Venus fly trap," the "groove" of which binds the antigenic peptide in a sequence-specific manner and displays it on the cell surface for the perusal of T cells (8,9). The MHC molecule is hence said to "present" the antigenic peptide. MHC molecules (both I and II) are polymorphic, and many alleles exist in the population. Much of the polymorphism lies at or near this groove; as a result, different MHC molecules bind different peptides. Because MHC differs among different individuals, different people will mount T-cell responses to different parts (epitopes) of a given antigen (e.g., a virus). Thus, there are many common features between class I and class II complexes: both have four domains, both have a groove that presents antigenic peptide, and both are polymorphic around this groove. What is the evolutionary benefit of producing two such apparently similar molecular complexes? Presumably it relates to their interactions with the specific T-cell subsets. What, then, is the essential difference between the class I/peptide complex and the class II/peptide complex that makes it important that T cells can distinguish between them?

An early suggested explanation was that the two classes of MHC would present fundamentally different peptide sequences to T cells. By this argument, the existence of the two MHC classes allowed more of a microorganism's sequences to be presented to T cells. However, although the sizes of the peptides bound by the two MHC classes differ slightly (perhaps because of the way the peptide is processed before reaching the MHC molecule; see below), there is little evidence to sustain the contention that there are differences between "generic class I" and "generic class II" peptide sequences. Indeed, a single short region of a virus protein can be presented by both class I and class II (122). A second proposal is more promising. Perhaps the critical difference between the two complexes lies not in the complex itself but instead in how the MHC/peptide complex reaches the cell surface. The two classes of MHC reflect two different antigen-processing systems (described below), one of which (class I) is optimized to present intracellular antigen and the other (class II) to present extracellular antigen. Thus, when T cells distinguish between a class I complex and a class II complex, they are really discriminating between intracellular antigen and soluble (extracellular) antigen.

ANTIGEN PROCESSING: THE CLASS I AND CLASS II MHC PATHWAYS

There are two broad pathways of antigen presentation: (a) the class I pathway, in which endogenous antigen (usually made within the target cell) is processed, eventually to appear as a peptide fragment in the class I groove on the cell surface, and (b) the class II pathway, in which exogenous antigen, usually soluble protein applied to the outside of the presenting cell, is internalized and eventually also finds itself presented on the cell surface in an MHC/peptide complex. The key here is the location of the antigen; endogenous antigens are presented by class I MHC, exogenous antigens by class II MHC. As with most generalizations, this division is not absolute; there are examples of de novo synthesized antigen that enters the class II pathway and of protocols to introduce exogenous antigens into the class I pathway. By and large, however, the above generalization holds. Figure 1 summarizes the two pathways. A cell is shown, bisected by a dotted line; this hypothetical cell expresses both classes of MHC (it could therefore be, for example, an activated macrophage). Below and to the right of the dotted line the class I antigen processing/presentation pathway is outlined, and above and to the left of the line the class II pathway is shown.

Antigen Presentation by Class I MHC

The thick line (*1*) in Fig. 1 represents a virus protein, newly synthesized in the cell cytoplasm. This protein is degraded (*2*), perhaps by a cellular organelle named the proteasome [at least one gene of which is MHC encoded (34)], to generate shorter peptides. Potentially, at this step the sequences that flank a potential epitope might be critical in determining the susceptibility to proteasome cleavage. The fragments are next (*3*) passed into the endoplasmic reticulum (*E.R.*) by a transporter mechanism [the TAP proteins, which appear to belong to the ABC transporter group and are encoded within the MHC (54,105,127)]. Here are shown at least two peptides (● and ○), which have been transported into the E.R (*4*). One of these (○) is selected (presumably on the basis of binding affinity) by the class I MHC molecule as assembly of the class I H chain, the peptide, and ß2M occurs (*4*) (the precise organization and control of this remains obscure) and (*5*) the complex passes into the Golgi and *trans*-Golgi complexes, finally (*6*) reaching the cell surface. Note that there is another route whereby peptides can reach class I molecules, in a TAP-independent

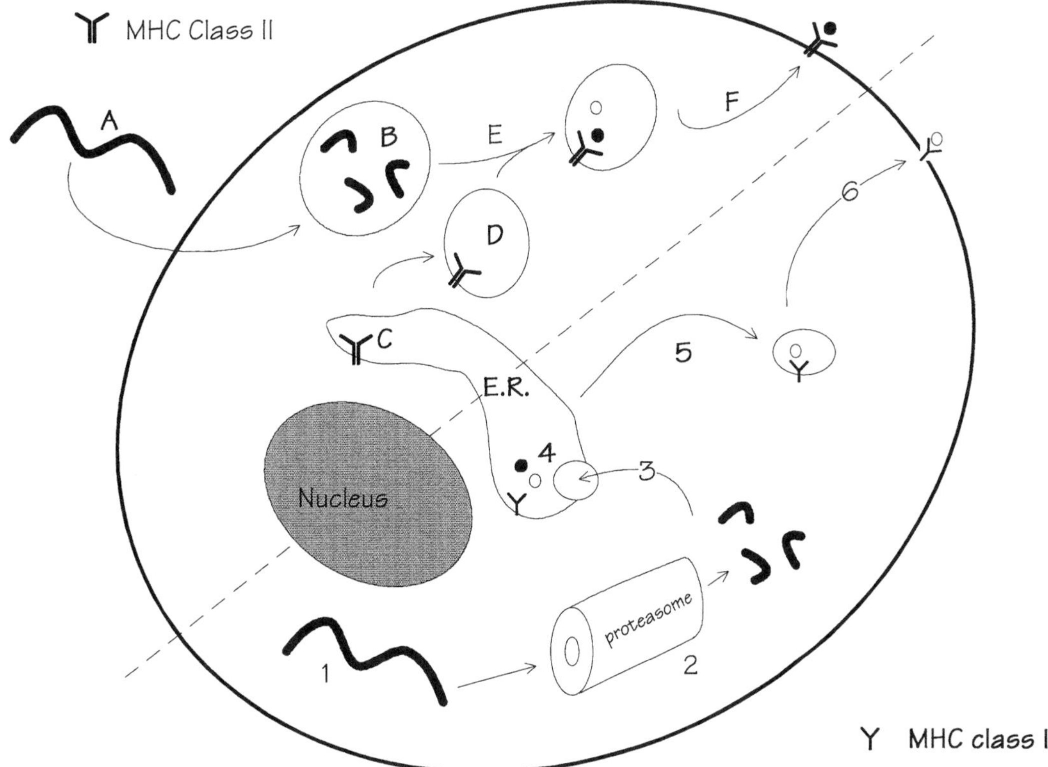

FIG. 1. Schematic outline of antigen processing. Both class I and class II pathways are represented. Details are provided in the accompanying text.

manner. This is the entry of peptides from proteins with signal sequences, translated on ribosomes bound to the rough endoplasmic reticulum (39). Of interest in this light is that one of the CTL epitopes found on lymphocytic choriomeningitis virus (LCMV) (137) is located within a putative signal sequence in the viral glycoprotein.

Antigen Presentation by Class II MHC

The thick line (*A*) in Fig. 1 represents a protein, but this time outside the cell. The protein is taken up by endocytosis and eventually reaches an acidic compartment where degradation occurs, generating fragments of protein (*B*). Meanwhile, the class II molecule is synthesized within the endoplasmic reticulum (*C*) and is transported out through the Golgi/*trans*-Golgi (*D*). The class II vacuole fuses with the endosome containing the fragmented internalized protein (*E*), and a peptide is bound in the groove, again, presumably selected on the basis of affinity. The mature complex is passed onto the cell membrane (*F*). In passing, note that the class II molecule in the endoplasmic reticulum (class II, *C*) does not bind peptides introduced by the transporters (class I, *4*). The reason for this is that the class II molecule in the endoplasmic reticulum is attached to another protein, the invariant chain (not shown), which prevents peptide binding. The invariant chain is lost as the class II molecule undergoes transport, so when fusion occurs with the endosome (class II, *E*), the class II molecule can bind to the peptide fragments therein (123).

BIOLOGICAL IMPORTANCE OF THE DIFFERENCE BETWEEN THE TWO ANTIGEN PRESENTATION PATHWAYS

T-cell recognition distinguishes between intracellular and extracellular sources of antigen; $CD8^+$ T cells, usually CTL, recognize peptides from endogenously synthesized proteins, and $CD4^+$ T cells, usually helpers, detect peptides derived from exogenously applied proteins. Why is it important for the different T-cell subsets to respond differently depending on the source of antigen?

Organisms that replicate intracellularly (viruses, as well as some bacteria and other parasites) will feed antigens into the class I pathway, as well as into the class II pathway (when antigens are shed, e.g., during cell death). Conversely, extracellular organisms (most bacteria) will be limited to introducing antigen mainly into the class II pathway. As shown in Fig. 2, bacteria (extracellular) will induce mainly $CD4^+$ and antibody responses, but minimal $CD8^+$ CTL; whereas the response to viruses will include $CD8^+$ T cells in addi-

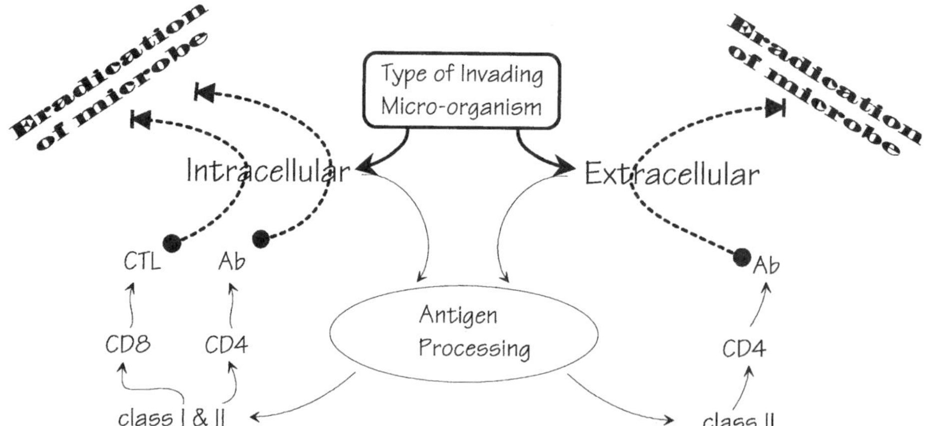

FIG. 2. Diagrammatic representation of the circular relationship between a microbe, the antigen presentation pathway that best presents its antigens, and the consequent nature of the immune response. As described in the text, the immune response induced is that best fitted to the eradication of the microbe.

tion to CD4⁺ T cells and antibodies. These considerations allow the experiments of nature to be viewed in a more molecular light. The absence of antibodies in patients with agammaglobulinemia will markedly reduce the host's ability to control bacterial infection, which relies primarily on antibodies, but will have much less effect on antiviral responses because the $CD8^+$ T-cell response, effective against intracellular organisms, remains intact. Furthermore, if CTLs have evolved in part to control infections by intracellular organisms, then the removal of the T-cell arm might be expected to most severely disable the host's capacity to combat such infections; this is the case. However, there are no clinical syndromes in which the class I presentation pathway alone is defective, and so it is difficult to infer how well antibody and CD4 responses could cope with virus infection in the absence of $CD8^+$ CTL. This question has been addressed experimentally in two transgenic mouse knockout models. First, mice lacking ß2M express little or no class I MHC on their cell surfaces, and in consequence do not develop $CD8^+$ T cells, because their maturation requires thymic expression of class I. Such mice cannot mount $CD8^+$ CTL responses, but develop readily detectable $CD4^+$ CTL (42,80). The outcome of virus infection in these $CD8^-CD4^+$ mice varies with the virus used, but in general there is delayed clearance of several viruses, including LCMV (19,80), Sendai virus (42), and influenza virus (22). These studies support the contentions (a) that $CD8^+$ CTLs play an important role in controlling virus infection, but their ablation reduces, but does not eliminate, the host's ability to clear virus and (b) that the second pathway of $CD4^+$ T cells and antibodies can play an important and effective role in control of viral infection. The second model system involves CD8 knockout mice. These animals can present antigen, but the T cells lack CD8, an important factor in recognition of class I MHC. In these animals, too, virus clearance is delayed (32). Thus, ablation of the $CD8^+$ CTL response has a marked effect on the host's ability to clear virus infection. The immunological responses that remain are able to control infections, but only partially.

Thus, there is a circular evolutionary relationship (Fig. 2) between the invading microbe, the MHC pathways that present its antigens, and the T cells that respond. Evolution appears to have honed the interactions to ensure that, upon infection, the immune response that predominates is that which is most appropriate to deal with the specific type of organism being faced. The fulcrum upon which this decision balances is the MHC antigen presentation pathway. Most bacterial (extracellular) infections will present antigen through the class II pathway and in consequence will induce a predominantly $CD4^+$ T-cell and antibody response; antibodies can recognize intact antigen, and use various effector mechanisms (e.g., opsonization, agglutination, and complement fixation) to destroy the bacteria. In contrast, during bacterial infection $CD8^+$ CTLs, which recognize antigen only if associated with MHC class I, are little induced and of limited value. Conversely, viral (intracellular) infections are susceptible to both antibodies and $CD8^+$ CTLs, both of which are induced, and play important roles in recovery from infection.

Certain aspects of the $CD8^+$ CTL response appear particularly well-suited to combat viral infections. First, the MHC allows the cell to present a fragment from a viral protein made at a very early stage of infection. Thus, an infected cell can signal its status before the virus has had a chance to replicate, and lysis of the cell by CTLs (at this stage beneficial to the host) as potential factories of virus progeny are eliminated. In contrast to the ability of CTLs to recognize early transcribed (usually nonstructural) antigens, antibodies usually recognize viral cell surface proteins, often made late in the viral life cycle during virus maturation. Second, $CD8^+$ CTLs are more likely to be of benefit than their $CD4^+$ counterparts because almost all cells, excepting neurons (47), express class I molecules. Conversely, most cells are class II negative, and virus in these cells would be "invisible" to $CD4^+$ T cells. These evolutionary lessons have implications for the design of antiviral vaccines.

INDIVIDUAL HOST GENES INVOLVED IN IMMUNE RESPONSES

Current immunological mechanisms have evolved in response to the variety of pathogenic challenges imposed on their predecessors; implicit in this assumption is the existence of primordial genes from which present day genes have evolved. Molecular analyses of the host genes described in the following section show that many of these genes are evolutionarily related, containing domains of similar protein folding. This has led to the inclusion of many of the genes in an Ig superfamily. The primitive gene(s) from which this superfamily has descended remains unidentified but presumably resides in a less complex organism, perhaps subserving an intercellular recognition function.

This Ig superfamily contains at least four families of proteins involved in the generation of an effective immune response. These are (a) the class I and class II glycoproteins of the MHC; (b) the TCR proteins, present in the membranes of T cells and responsible for the recognition of foreign antigen in the context of the above-mentioned MHC class I and class II glycoproteins; (c) proteins present in the TCR complex and necessary for T-cell function but that are not antigen specific; and (d) Ig molecules, the mediators of specific humoral immunity. Other genes, not identified as members of the Ig superfamily but important to the immune response, also are discussed below. These genes are the ABC transporter proteins, which lie in the MHC and are responsible for shuttling peptides from the cytoplasm into the endoplasmic reticulum for subsequent binding to and presentation by class I MHC molecules, and genes involved in antigen processing and cleavage, whereby intact cellular or microbial proteins are degraded in the cytoplasm before transfer to the endoplasmic reticulum by the transporter proteins.

MHC Class I and Class II Glycoproteins

The MHC of mice and humans, although not identical, bear strong similarities both in the overall organization of genes within the complexes and in the structure and sequences of individual genes. The murine MHC locus (on chromosome 17) comprises two regions (H-2 and Tla), whereas the human MHC locus (chromosome 6) is termed human leukocyte antigen (HLA). The genes within the locus are divided into three classes on the basis of structure and function. Even with the great strides made using molecular techniques to identify the genes, the large size of the MHC locus (some $1.6–2 \times 10^6$ base pairs) has so far precluded its complete cloning and sequencing. Consequently, data on gene numbers and relative positions remain incomplete. The classical members of the MHC gene family are the class I and class II glycoproteins, and these are described in detail below. However, additional genes lie within the MHC complex; although structurally unrelated to the class I and class II MHC gene, some of these genes encode immunological functions (i.e., complement proteins, tumor necrosis factor, transporter proteins). The linkage of these genes to the classical gene may be incidental or may reflect evolutionarily selected cosegregation.

MHC Class I Genes

Many texts refer to the MHC genes as a homogeneous group, but in fact the group supports at least one further subdivision, into classes A and B. Class IA proteins, by far the most studied class I molecules (and those referred to whenever the generic term class I is used) are the transplantation antigens, which led to the discovery of the MHC. These are the proteins that control MHC restriction, a phenomenon central to the specificity of cellular immune responses not only to transplanted tissues but to virus-infected cells and tumor cells. Class IB proteins are less well characterized; they are structurally similar to the class IA molecules, but many are soluble rather than membrane bound, and their functions remain obscure. The numbers of both class IA and class IB genes vary between different individuals. On average, for both mouse and human, the haploid gene number is 2–6 for class IA and perhaps 20–30 for class IB.

Three class IA loci have been defined; K, D, and L in mice and A, B, and C in humans. The precise number of genes at each locus varies (e.g., some mouse strains lack an L gene locus; some strains have four D genes, whereas others have a single D gene), which is consistent with the occurrence of gene duplication/deletion events. Sequence homology occurs among these genes both within and between species, suggesting evolution from a common ancestor. The interspecies sequence similarities extend to the functional level because a human HLA gene expressed in a transgenic mouse appears to function as well as the endogenous murine proteins. All share a genomic layout of eight exons that undergo splicing to generate a mature messenger RNA encoding the class I molecule, a glycoprotein of ~350 amino acids. This molecule has five domains: three extracellular ($\alpha 1$, $\alpha 2$, and $\alpha 3$), one transmembrane, and one cytoplasmic. A functional molecule is anchored in the cell membrane in association with ß2M, a 12-kd protein encoded outside the MHC. The crystallographic structure of the human MHC class I molecule has been solved to show the $\alpha 3$ domain and the ß2M molecule, both of which are evolutionarily related to the constant region of

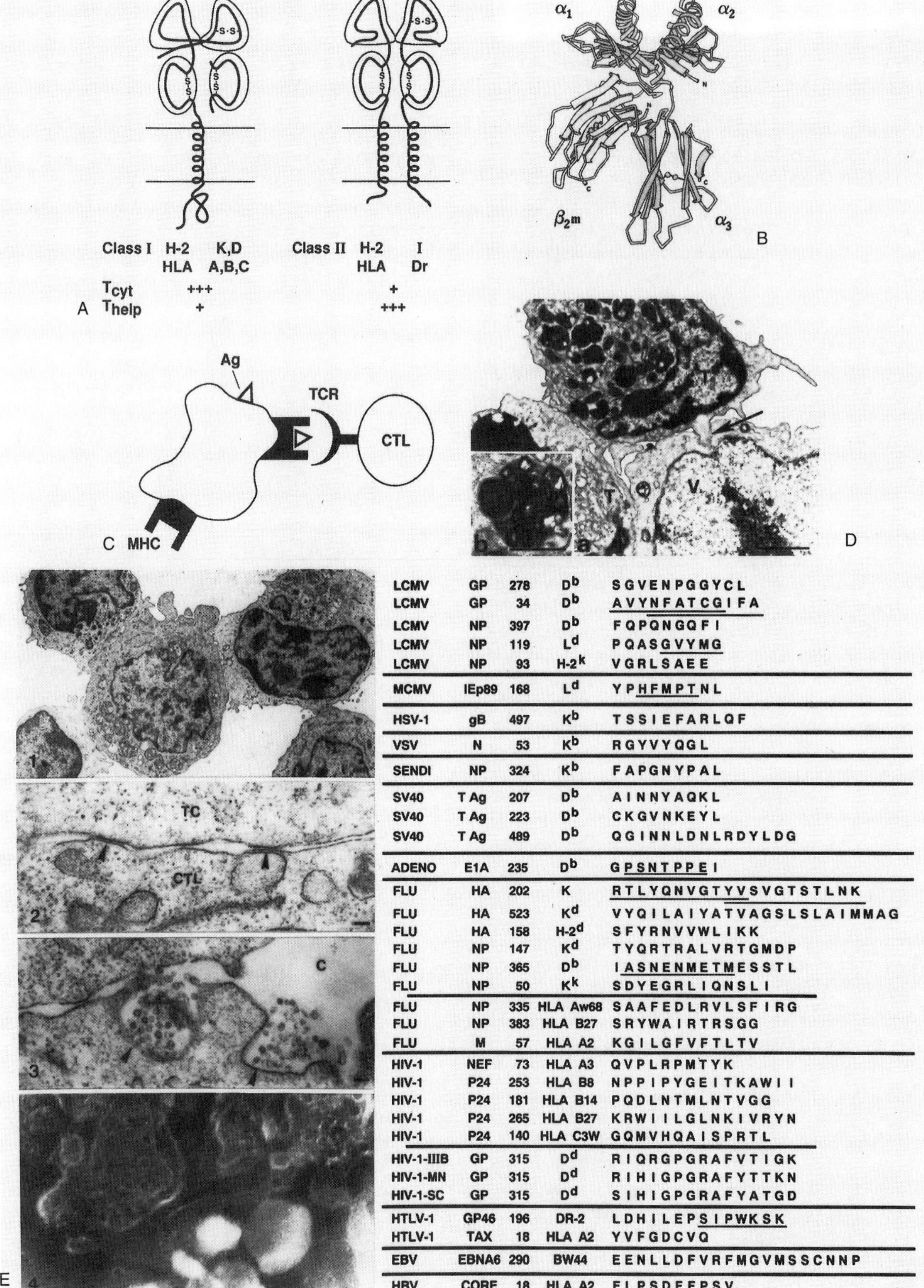

LCMV	GP	278	D^b	SGVENPGGYCL
LCMV	GP	34	D^b	AVYNFATCGIFA
LCMV	NP	397	D^b	FQPQNGQFI
LCMV	NP	119	L^d	PQASGVYMG
LCMV	NP	93	$H\text{-}2^k$	VGRLSAEE
MCMV	IEp89	168	L^d	YPHFMPTNL
HSV-1	gB	497	K^b	TSSIEFARLQF
VSV	N	53	K^b	RGYVYQGL
SENDI	NP	324	K^b	FAPGNYPAL
SV40	T Ag	207	D^b	AINNYAQKL
SV40	T Ag	223	D^b	CKGVNKEYL
SV40	T Ag	489	D^b	QGINNLDNLRDYLDG
ADENO	E1A	235	D^b	GPSNTPPEI
FLU	HA	202	K	RTLYQNVGTYVSVGTSTLNK
FLU	HA	523	K^d	VYQILAIYATVAGSLSLAIMMAG
FLU	HA	158	$H\text{-}2^d$	SFYRNVVWLIKK
FLU	NP	147	K^d	TYQRTRALVRTGMDP
FLU	NP	365	D^b	IASNENMETMESSTL
FLU	NP	50	K^k	SDYEGRLIQNSLI
FLU	NP	335	HLA Aw68	SAAFEDLRVLSFIRG
FLU	NP	383	HLA B27	SRYWAIRTRSGG
FLU	M	57	HLA A2	KGILGFVFTLTV
HIV-1	NEF	73	HLA A3	QVPLRPMTYK
HIV-1	P24	253	HLA B8	NPPIPYGEITKAWII
HIV-1	P24	181	HLA B14	PQDLNTMLNTVGG
HIV-1	P24	265	HLA B27	KRWIILGLNKIVRYN
HIV-1	P24	140	HLA C3W	GQMVHQAISPRTL
HIV-1-IIIB	GP	315	D^d	RIQRGPGRAFVTIGK
HIV-1-MN	GP	315	D^d	RIHIGPGRAFYTTKN
HIV-1-SC	GP	315	D^d	SIHIGPGRAFYATGD
HTLV-1	GP46	196	DR-2	LDHILEPSIPWKSK
HTLV-1	TAX	18	HLA A2	YVFGDCVQ
EBV	EBNA6	290	BW44	EENLLDFVRFMGVMSSCNNP
HBV	CORE	18	HLA A2	FLPSDFFPSV

Ig, in intimate contact with the cell membrane, whereas the α1 and α2 domains interact at the top of the molecule to form a groove. For many years it was thought that the MHC class I–restricted CTLs recognized MHC associated with a foreign antigen anchored in the cell membrane. However, studies over the past 10 years have shown that foreign antigens are seen by CTL not as native proteins but instead as short peptides, most often nonameric, presented to the immune system bound in the groove formed by the association of the MHC α1 and α2 domains. It is the resulting MHC/peptide antigen that is seen by the CTL receptor. Fig. 3 shows MHC class I and class II molecules and lists a number of viral peptides that have been mapped, along with the MHC alleles that present them. Class I molecules are found in the membrane of almost all nucleated cells, which are therefore presumably able to present antigen; neurons are an exception to this rule.

The peptides bound by MHC can originate from almost any protein made within the presenting cell: from microbial infection, from expression of new sequences and tumors, or from transplanted tissues. It has been estimated that a cell can present up to 1,000 different peptides (44) and that CTL induction requires a threshold presentation of at least 100 MHC/peptide complexes per cell (14). Viral nucleoproteins and polymerases as well as viral glycoproteins, can provide sequences to be presented to the cellular immune system. This provides a marked advantage to the host by allowing the entire coding region of the virus genome to be exposed to immune surveillance. Furthermore, the system allows presentation of virus antigens expressed early in infection, which often are non–membrane proteins; thus, infected cells may be recognized and lysed by immune effector cells before they have time to produce infectious progeny. Analyses of MHC structure/function relationships provide a molecular explanation for the variability of CTL responses in different mouse strains and in different individuals in a random-bred (e.g., human) populations. The MHC class I proteins are highly polymorphic, i.e., there are many functionally distinct alleles at each locus. Sequencing studies of many different alleles show that the variable residues cluster around sites close to the antigen-presenting groove. Thus, the functional polymorphism is caused by alterations affecting the nature of antigen binding and presentation by individual MHC molecules; different MHC molecules will bind and present peptides of different sequence. Hence, one variable in the immune response is the ability of the host class I molecule to bind microbial sequences and present them to the T-cell repertoire. If a host, when challenged with a particular virus, lacks the class I sequences necessary to bind and present viral peptides, then that host will be unable to mount class I–restricted CTL responses, and therefore will probably be more susceptible to disease caused by this particular agent. In contrast, a host whose genotype encodes MHC molecules capable of binding and presenting a peptide from this particular virus is more likely to be able to mount a CTL response that will limit viral replication, dissemination, and virus-induced disease. Given the risk of failure to present, one might ask why we have not retained more than the six or so MHC alleles in each host; if we had, for example, 20 alleles, the risk of our being unable to present epitopes from any microbe would be greatly reduced. The reason is straightforward. Some 98% of lymphocytes entering the thymus fail to complete the passage to mature T cells, and a major cause of this attrition is negative selection in which T cells with excessive affinity for self molecules (i.e., for MHC molecules associated with self peptides) are deleted. Thus, the more MHC alleles we carry, the more self peptides we can present; and as more self peptides are presented, so we clonally delete more T cells in the thymus. With our present complement of MHC molecules, we already delete some 98% of all of our T-cell

FIG. 3. Schematic and electron microscopic photographs displaying how a CTL reacts with a virus-infected cell. Shown at upper left **(A)** are the MHC class I and II structures and a list of functions in terms of cytotoxic (T cyt) or helper (T help) T lymphocyte activity. The upper **(B)** right panel shows a sketch of the three-dimensional x-ray crystallographic structure of MHC class I molecule HLA-A2 [see Bjorkman et al. (9) for details]. The viral peptide folds in the groove between the α1 and α2 arms of the MHC molecule. The β2M (L chain of MHC class I) is transcribed on a different chromosome than the H chain of MHC class I, which comprises the α1 and α2 binding arms and the transmembrane α3 region. Also shown is **(C)** the T-lymphocyte receptor (TCR) recognizing the viral antigen (ag, peptide)-MHC complex. **Panel D** is an electron micrograph showing the interaction of a cloned LCMV-specific CTL pushing a cytoplasmic finger (arrow) into the cytoplasm of an LCMV-infected cell (V). Note the close cell-cell contact that is essential for CTL-mediated lysis. After contact, a variety of proteolytic enzymes (granzymes) are released by the CTL from granules. **E** comprises four electron micrographs of CTL interactions with target cells **(Panels 1 and 2)**, the release of cytotoxic granules **(3)**, and demonstration of perforin-mediated lesions in the membrane of the virus-infected target cell **(E, panel 4)**. (Photomicrographs 1–3 courtesy of Dr. P. Peters, Utrecht; photomicrograph 4 courtesy of Steve Patterson, London.) **Table F** at lower right indicates a number of viral peptide sequences known to serve as epitopes for MHC class I CTL recognition. The optimal peptide can vary from eight to 11 amino acids. The longer peptides (>11 amino acids) shown have not yet been mapped to their optimal size. At lower affinities, CTL can recognize peptides with sequences as short as five amino acids (i.e., LCMV NP GVYMG).

specificities; increasing our MHC representation would most probably reduce our T-cell repertoire below effective levels. We are, therefore, in a state of equilibrium: our T-cell repertoire is balanced against our ability to present antigen for the perusal of T cells. We must present just the correct number of antigens; too many (too many MHC alleles) will decimate our population of mature T cells, whereas too few (too few MHC alleles) increases the risk that we will fall victim to failure to present.

Early attempts to predict viral (or other) sequences that may bind class I MHC molecules were not consistent and were thus unsuccessful, probably because they attempted to predict the viral epitopes without taking into consideration the specific polymorphism (i.e., allelism) of the MHC molecules. Recently, two factors have allowed more consistently correct predictive motifs to emerge for specific MHC alleles. First, the increasing database of CTL epitopes for the individual alleles has allowed comparison of the peptide sequences presented by several different MHC alleles, allowing allele-specific consensus sequences to emerge. Second, work by the Rammensee and Nathenson groups has shown that peptides can be eluted from MHC molecules and their sequences inferred or directly determined (44,112,113,128). By a combination of these two approaches, specific motifs have been identified for several murine and human MHC class I alleles. These motifs are summarized in Table 2. Most, but not all, epitopes are nonameric, with two highly conserved anchor residues critical for binding of the peptide to the MHC molecule. Thus, for example, the majority of peptides bound by the K^d molecule are nine amino acids long with a tyrosine at position 2 and usually a leucine or isoleucine at position 9.

TABLE 2. *Peptide motifs for binding to a variety of human and mouse MHC class I alleles*

		Peptide residue number								
Species	Allele	1	2	3	4	5	6	7	8	9[a]
Mouse	Db					N				M,I,L
Mouse	Kb					F,Y				L
Mouse	Kd		Y							L,I
Mouse	Kk		E,D							I
Mouse	Ld		P							L
Human	A2		L							V,L
Human	B35		P							Y
Human	B53		P							F,W
Human	B8			K		K				I
Human	B27		R							K,R
Human	B7	A	P	R						L,V
Human	A68		T,V							R
Human	A3.1		I,L	F						K,Y

[a]Residues shown at position 9 are the most probable C-terminal residues; these will often be at position 9 because most class I peptides are nonameric, but in some nonnonameric epitopes, this residue may not be no. 9. For example, the LCMV GP2 epitope presented by Db is SGVENPGGYCL. With the "N" at position 5, the presumed anchor L is at position 11. The data in this table are abstracted from the compilation recently published by Elliot et al. (23).

The MHC/peptide complex has been studied at high resolution. The initial crystallographic solution of MHC structure showed an electron-dense mass located in the groove. It was hypothesized at the time that this reflected a population of different peptides located in the MHC groove, unresolvable by crystallography because a different signal resulted from each peptide, producing an average signal showing an undefined mass (8,9). However, these results have recently been enhanced by crystallographic resolution of specific MHC molecules to which individual peptides have been bound. Several such structures have been solved, allowing the detailed analysis of the interaction between MHC class I molecular grooves and the peptides found therein (28,70,139). The MHC groove (Fig. 4) appears to contain several binding pockets, the most critical of which is that which binds the C-terminal residue of the peptide. The MHC residues that are critical in forming this binding pocket have been identified, and their nature allows fairly confident prediction of which C-terminal residues will be favorable for a particular MHC pocket. Because the polymorphism of the amino acids forming this MHC pocket is fairly restricted, so too are the C-terminal amino acids of the binding peptides. Thus, the great majority of peptides have as their C-terminal residue one of seven amino acids (I, L, V, R, K, Y, or F). Another pocket, located in the middle of the groove, appears important in determining allele-specific binding, i.e., in conferring fine specificity for specific peptide sequences (70); and the MHC molecule can slightly alter its conformation upon peptide binding, although the significance of this is uncertain (28).

However, despite the great leaps in our knowledge and understanding, several conundrums remain. Although for a particular MHC allele the majority of peptides presented by that allele conform to the predicted motif, not all peptides in a viral sequence that conform to the predicted motif appear able to induce T-cell responses via presentation by the appropriate MHC molecule. For example, a search for viral sequences fitting the D^b motif (N^5, MIL^9) may show multiple instances of peptides within the virus that fit the motif; nevertheless, biological studies that map the CTL responses to the virus on the H2b background show that only a small subgroup of this selection appears able to induce high levels of antiviral CTL activity. Although hypotheses are available to explain this apparent discrepancy, the final explanation has not been determined. Therefore we are not yet in a position to confidently predict which viral sequences will induce good CTL responses. Although computer analysis will identify a group of potential epitopes, and although probably the real epitopes will lie within this identified group, we are as yet unable to distinguish within this group which sequences will induce CTL responses and which are excluded from this activity.

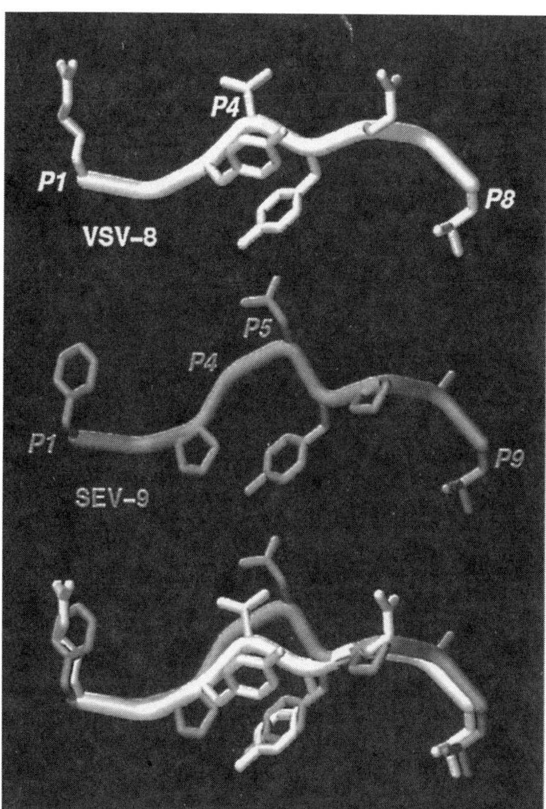

FIG. 4. The conformation of various viral peptide sequences situated in the α1-α2 groove display a similar motif. Peptides of vesicular stomatitis virus nucleoprotein amino acids 52-59: GRYVYQGL (VSV-8) and Sendai virus nucleoprotein 321-332: FAPGNYPAG (SEV-9), which are restricted by K^b MHC class I molecule, are shown singularly or combined. From ref. 28, with permission.

The "nonclassical" class I MHC molecules are less well studied and their functions are poorly understood. Several genes that have been sequenced show marked similarity to the classical class I proteins, and some are expressed as soluble proteins. They differ from classical antigens in their tissue distribution, being much more restricted and often detectable only on the surface of certain hematopoietic precursors. They also appear to exhibit less polymorphism than the classical molecules, suggesting that their roles may differ. In one case, a nonclassical molecule H-2M3 is able to bind and present a peptide from the bacterium *Listeria monocytogenes* (99). It has been suggested that these molecules may present N-formylated peptides, characteristic of prokaryotes (60). However, viral proteins too may be presented by these nonclassical molecules (76).

MHC Class II Genes

The existence in the mouse MHC of genes that regulate the humoral responses to antigen was clearly demonstrated more than 30 years ago. They were initially termed immune response (Ir) genes, subsequently termed Ia genes, and both terms have now been replaced by the term class II MHC glycoproteins. Like their class I counterparts, the overall structure and function of class II genes appear similar in both humans and mice. Mature class II molecules are, like class I proteins, membrane-bound heterodimers. However, although the class I structure results from the association of one MHC molecule with a non-MHC protein, ß2M, the class II structure results from the association of two MHC-encoded molecules, an α chain and a ß chain (Fig. 3). Six α and seven ß genes have been identified in the HLA-D region, divided into four clusters termed DR, DQ, DP, and DZ. Murine equivalents are present in the H-2 I region: two α and five ß genes divided into two clusters termed A and E. The genes encoding the α and ß chains are similar, although some variation between exon numbers is seen, even within a chain type. The resultant α or ß proteins have two extracellular domains: a transmembrane region and a cytoplasmic domain. The top (membrane-distal) domain is named α1 (ß1), whereas the lower one, adjacent to the membrane, is termed α2 (ß2). The α2 and ß2 domains share significant homology with the class I α3 domain and with ß2M, indicating once again common ancestry and possible functional relatedness. Indeed, there are many similarities in the overall topology of the two cell surface MHC classes. Both classes have four extracellular domains; both the membrane proximal domains are similar to Ig constant region, and the two distal domains interact in antigen presentation. The location of disulfide bridges is similar. The crystallographic structure of the class II molecule has been solved and appears similar to that of the class I heterodimer (10). In the case of class II molecules, the peptides, which derive from a different processing pathway (see above), are longer, usually between 15 and 20 amino acids. Unlike the class I peptides, in which the C-terminal residue is firmly anchored, class II-bound peptides "hang out" at both ends of the groove. Some motifs have begun to emerge for peptide binding to specific class II dimers, but this work lags behind the class I studies (115).

The polymorphism of class II molecules varies from gene to gene; for example, DQ α1 is highly polymorphic, DQ α2 much less so. However, the low polymorphism at certain loci is offset by the ability of class II genes to reassort. Thus, in theory the haploid genome encoding six α and seven ß genes might make 42 different class II molecules. In practice, reassortment is not entirely unrestricted but does occur. Such a mechanism cannot, of course, operate for the class I molecules because the heterodimer includes one monomorphic molecule (ß2M). Tissue distribution of class II molecules is markedly more restricted than that of class I. They are most abundant on B- and T-lymphocytes and cells of the macrophage monocyte lineage. The class II/peptide complex is recognized by T cells bearing the CD4 surface marker; these cells are most frequently helper T cells rather than cytotoxic T cells.

Genes Involved in Cleavage of Endogenous Antigen

The proteasome is a multiprotein cytoplasmic complex associated with the ubiquitin pathway, long known to be involved in breakdown of host proteins; its known functions made it a promising candidate as the source of antigenic peptides. This viewpoint received circumstantial support when several of the complex components were found to be encoded within the MHC complex. However, it has been difficult to clearly demonstrate the importance of these proteins, and initial studies tended to be conflicting (3,33,75). However, recent work suggests that the proteasome can cleave proteins to generate peptides with a variety of termini and that the MHC-encoded components may play a role in determining which termini are produced (21).

The TAP Transporter Genes

Two TAP (*t*ransporter associated with *a*ntigen *p*resentation) transporter genes are encoded in the MHC, called TAPs 1 and 2; one copy of each combines to form a functional heterodimeric transporter. The role of these proteins in antigen presentation has been clearly demonstrated (126). Cell lines defective in transport are unable to present endogenously synthesized antigen in a normal fashion, but when the defective TAP is restored, antigen presentation proceeds (4,108). *in vivo* studies confirm these observations: TAP knockout mice exhibit the expected phenotypes, having little or no cell surface MHC and no mature $CD8^+$ T cells (130). There is some selectivity in the peptides transported by these molecules, as demonstrated in the rat; the class I modifier (cim) locus contains (at least) two genes (named mtp-1 and mtp-2), the latter of which has (at least) two alleles that result in presentation of different peptides by identical MHC molecules. The cim locus encodes a transporter, and the two alleles permit the passage of different peptides, thus providing the MHC class I molecules in the endoplasmic reticulum with different peptide populations from which to select (65,66,106,107). Direct evidence has been obtained showing that different transporter proteins display preferences for specific peptide sequences (84). One cim allelic product could transport peptides terminating in H, K, or R, whereas the other could not (38). Allelism has recently been identified in humans by sequencing the human TAP equivalents (105), although functional distinctions between the alleles have not yet been demonstrated. Thus, although the major contributor to epitope selection probably remains the class I molecule, the transporter proteins may play some role in limiting the range of peptides to which the class I molecules are exposed. However, this role is likely to be minor, for several reasons. First, many cell lines have been transfected with nonhomologous MHC alleles, but these MHC molecules continue to present the same CTL epitopes as they did in their own background, suggesting that there is no major difference in the peptide pools generated in the differing backgrounds. Second, transporter heterodimers in which one component is murine and the other human appear to function the same as either of the natural heterodimers (141).

TCR Genes

TCR genes together comprise the second family of the Ig superfamily. Antigen specificity of T-lymphocytes is bestowed by a heterodimeric surface-bound glycoprotein, the TCR. Four distinct genomic regions encode different TCR chains, which have been named α, ß, γ, and δ. The great majority of circulating T cells, regardless of their helper/cytotoxic phenotype or MHC restrictions, bear αß receptors; γδ receptor-bearing T cells have been identified only recently and are less well characterized. The TCR genes are, like the Ig genes, products of somatic rearrangements that appose noncontiguous germline gene segments. The mature gene usually results from joining of V, D, J, and C regions; each chain type draws from its own pool of germline segments. The germline copy number of each of these regions varies among the chains but appears sufficient, when both combinatorial rearrangement and other V-region diversification mechanisms are invoked, to provide a TCR repertoire at least as large as that of antibodies. The mature TCR chains are glycoproteins of 260–300 residues in length, comprising a short cytoplasmic tail, a transmembrane region, and two extracellular domains—one constant and one variable. The antigen recognition site, which recognizes the complex of antigenic peptide and MHC, thus consists of the Vα and Vß regions. In overall structure the TCR molecule is similar to the distal (Fab) portion of an antibody molecule. The TCR differs from antibodies in several aspects. First, at the Ig H chain locus there are multiple nonallelic constant-region genes to which the V regions can be apposed. The effector function of an antibody is imposed on it by the constant region used in its generation. In contrast, the copy number of TCR constant-region genes is very low, and the same C regions are represented in both CTLs and T helper (Th) cells. Thus, the TCR C region does not dictate the T-cell's effector function. It had been suggested that the TCR might confer MHC restriction (i.e., class I or class II) on the T cell. However, the same Vα and Vß gene pools are used in both class I and class II restricted T cells, indicating that the TCR does not fundamentally distinguish between the two classes of MHC molecule. Rather, the TCR recognizes the antigen/MHC complex, and other T-cell components determine the MHC class specificity (CD4/class II and CD8/class I) and the effector function (helper/cytotoxic) of the cell.

How restricted at the molecular level are T-cell responses to virus infection? This has been studied at the level of *in vitro* specificity of T-cell lines and clones, but much more rigorous analysis can be achieved by studying the TCR sequences of individual clones. This allows us to ask if responses to immunological stimuli are monoclonal or polyclonal, which is interesting not only at an academic level, but also at a practical one. If responses are monoclonal (i.e., if all TCR are identical) then it may be possible to specifically ablate cells bearing that TCR, thus potentially diminishing immunopathological effects related to the antiviral immune response. In contrast, if such responses are polyclonal, such intervention is likely to be difficult, and unrewarding. The VDJ gene usage in TCR of cells responding to a variety of stimuli have been determined. In general, response to certain model antigens, such as cytochrome c, have shown quite restriction in VDJ usages, but responses to virus infection have shown less restricted usage (41,140). Thus, it appears unlikely that TCR-based ablation will be a valuable therapeutic approach in control of virus-induced pathology.

T-Cell Accessory Molecules

At the T-cell surface, both αß and γδ receptors are intimately associated with several other proteins, collectively referred to as the CD3 complex. The CD3 complex comprises at least four transmembrane proteins that are thought to relay a signal across the T-cell membrane when the TCR is appropriately stimulated. Also on the T-cell surface, although not stably linked to the CD3/TCR complex, are the CD4 and CD8 molecules, which play critical roles in T-cell function. These nonpolymorphic glycoproteins, again members of the Ig superfamily, are differentially expressed during ontogeny, but mature circulating T cells generally express only one of the two molecules. There is a very strong correlation between the CD4/CD8 characteristic of a T cell and the MHC class that presents the antigen to that T cell. Thus, almost all $CD4^+$ T cells are class II MHC restricted, whereas $CD8^+$ cells are class I restricted. There is also a correlation, although less absolute than the above, between the T cell's CD4/CD8 characteristic and its effector function; most $CD4^+$ (class II–restricted) T cells are of the helper phenotype, whereas most $CD8^+$ (class I–restricted) T cells are of the cytotoxic phenotype. The CD4/CD8 molecules serve to recognize specific regions of the MHC glycoproteins and presumably have a role in activating the transmembrane signal relayed by the CD3 complex. However, the outcome of signal transduction (help or cytotoxicity) is independent of the nature of the CD4/CD8 molecule or TCR; thus, there are many examples of $CD4^+$, MHC class II–restricted cytotoxic T cells. Interestingly, in animal model systems these cells appear to modulate the outcome of virus infection, despite their presumed inability to recognize the great majority of infected cells (which will be class II negative) (19,42,80).

Immunoglobulin Genes

The humoral response to virus infection plays an important role not only in controlling clearance of primary infection but also in preventing the establishment of widespread infection upon secondary exposure to the agent. All antibodies have a similar overall monomeric structure, comprising two pairs of identical chains joined by disulfide bridges. The central H chains usually have four domains: three constant and one N-terminal variable. The light (L) chains contain one constant domain and one variable domain. Each L and H variable region combines to form an antigen binding site: an antibody monomer is thus divalent in terms of antigen recognition. The L proteins are segmentally encoded in the germline, as variable (V), joining (J), and constant (C) regions, and are linked by somatic recombination. The H proteins are similarly arrayed, although they also have diversity (D) regions. Variability at the antigen binding site is brought about by a variety of mechanisms during somatic recombination (as for TCR). The rearranged VDJ region is then apposed to one of the constant regions; the nature of the constant region determines the class (and therefore the function) of antibody that results. There are five general classes of antibody, differing in function, time of appearance, and distribution in the body. Structurally, the classes differ in the constant region of their H chains. The classes are IgM (Vμ: H chain), IgG (Vγ), IgD (Vδ), IgE (Vε), and IgA (Vα). First exposure to a virus induces a rapid IgM response, which is usually independent of Th cells, and a slightly delayed IgG response, usually Th dependent and thus requiring antigen to be presented via MHC class II molecules. In contrast, secondary exposure results in a similar IgM response but a much more rapid and massive IgG response. In many instances, the precise specificity of the secondary IgG antibodies is identical to that of the initial IgM molecules: that is, the antigen binding sites have been retained, but the remainder of the molecule (which determines its effector functions) has been altered. This class switching occurs when, for example, an H chain VDJ region is moved from the Cμ: gene to the C-γ region; the new H chain has the same V region but a different C region. Two of these chains, in association with two unchanged L chains, generate an antibody identical to the original in specificity but different in effector function; it is now an IgG rather than an IgM. Initially, all B cells express membrane-bound IgM, but on exposure to specific antigen they develop into plasma cells, usually undergoing class switching. The five classes of antibody subserve distinct effector functions, but only three are thought to play a role in opposing virus infection. IgM molecules form pentameric masses, each with 10 antigen binding sites. These molecules are effective mediators of virus aggregation; they also activate both macrophages and the complement system. IgG, the main serum antibody, is monomeric, activates macrophages and complement, and can cross the placental barrier. IgA can exist as monomer, dimer, or trimer; it is transmitted to luminal surfaces (in saliva, tears, and intestinal and other secretions) to provide an initial defense against infection. IgE may play a role in immunity to parasite infection, but its major recognized task is the initiation of many allergic reactions, acting through histamine release from mast cells. The IgE-mediated release of histamine and other factors probably contributes to some of the respiratory manifestations of virus infection. The function of IgD remains obscure.

CELL-MEDIATED EFFECTOR MECHANISMS

We have outlined our current understanding of immune responsiveness at the molecular level. In the following section we discuss immune responsiveness at the biological level, describing the biological roles of several different cell types and the effector mechanisms through which they act.

Virus-Specific CTLs

These cells have been most extensively characterized in the mouse, but clearly equivalent findings occur in humans. First, the majority of CTLs are restricted by class I MHC proteins, although in humans a few have been identified that are class II restricted. The importance of CTLs in the control and eradication of virus has been documented for several experimental infections. In the mouse, CTLs are critical in combating infection with LCMV (137). In this case (57,58) and in murine cytomegalovirus (MCMV) infection (50), it is possible to fully protect a naive animal from virus challenge by immunization with a recombinant vaccinia virus (VV) expressing a single viral internal (i.e., nonmembrane) protein; the protection is mediated solely by CTLs. Indeed recombinant vaccines containing minigenes encoding isolated CTL epitopes as short as 11 residues can confer protection against normally lethal doses of challenge virus, and different epitopes can be strung together to protect on several MHC backgrounds (138). No LCMV-specific antibody responses are induced by these vaccines; the protective effects are mediated by cellular immune responses. In humans, it is clear that CTL responses can be generated against similar proteins. For example, the major CTL response to human cytomegalovirus (HCMV) is to a protein expressed immediately upon infection; a similar situation exists for varicella-zoster virus (VZV) and herpes simplex virus (HSV). Epstein-Barr virus (EBV) infects and transforms human B-lymphocytes, and the control of this cell population appears to be managed in large part by class I MHC-restricted CTLs. Indeed, some immunosuppressed individuals may develop EBV^+ lymphomata (110). Influenza virus infection induces CTLs directed against most viral components, but the major group is specific for the virus nucleoprotein. These CTLs, unlike antiinfluenza antibodies, are cross-reactive; that is, they lyse HLA-matched target cells infected with a serologically distinct strain of influenza virus. However their presence fails to confer long-term immunity to influenza virus infection. These results can be interpreted to challenge the hypothesis that the presence of CTLs confers protection against virus-induced disease. However, CTLs cannot protect against infection (indeed, CTL recognition requires infection); instead, CTLs limit virus production and dissemination and protect against consequent disease. Thus, the preexisting antiinfluenza CTLs, while failing to prevent infection and disease, may diminish the ensuing morbidity and mortality. CTLs have also been found against measles virus, mumps, respiratory syncytial virus (RSV), and other agents. In patients with AIDS-related complex and AIDS, CTLs are detectable against gag, pol, and env proteins. The role played by these cells is unclear, although CTLs to env correlate with clearance of initial viral load, and their absence heralds a return to high viral titers, and AIDS. Although it is difficult to determine their relative import in control of human infections, CTLs obviously constitute a frequent response. The molecules involved in CTL recognition of virus-infected cells are described above, and the biological consequences are shown in Fig. 3. Antigen-specific recognition mediated by the TCR initiates a series of events including signal transduction across the cell membrane and usually results in lysis of the infected cell. Several mechanisms have been proposed to explain the *in vivo* antiviral effects of CTLs. CTLs release perforin, a pore-forming protein. CTLs contain granules, which align with the target cell upon recognition, and the contents of these granules are released in a calcium-dependent manner onto the target cell membrane. These granules include a protein called perforin (103), which undergoes assembly into *trans*-membrane pores and thus "punches holes" in the cytoplasmic membrane of the target cell. Perforin shares immunologic cross-reactivity with the C9 component of complement, and recent cloning of the murine perforin gene has allowed identification of a short stretch of amino acid homology between the two proteins. The membrane lesions caused by perforin are similar to those induced by the complement C9 complex. Thus, CTLs and complement-mediated lysis seem to share one common mechanism of action. The importance of perforin in CTL activity *in vivo* has recently been demonstrated. Perforin knockout transgenic mice are much less effective in controlling infection by some (though not all) viruses (R.M. Zinkernagel, personal communication). CTLs can induce apoptosis in target cells. Apoptosis, or programmed cell death, is a well-recognized phenomenon responsible for several developmental processes, including the clonal deletion of T cells in the thymus. Its most characteristic features are nuclear blebbing and disintegration, resulting in a nucleosome stepladder of fragmented DNA. Recently it has been shown that CTLs may induce this process in virus-infected target cells (135,145). Both of these mechanisms are target cell

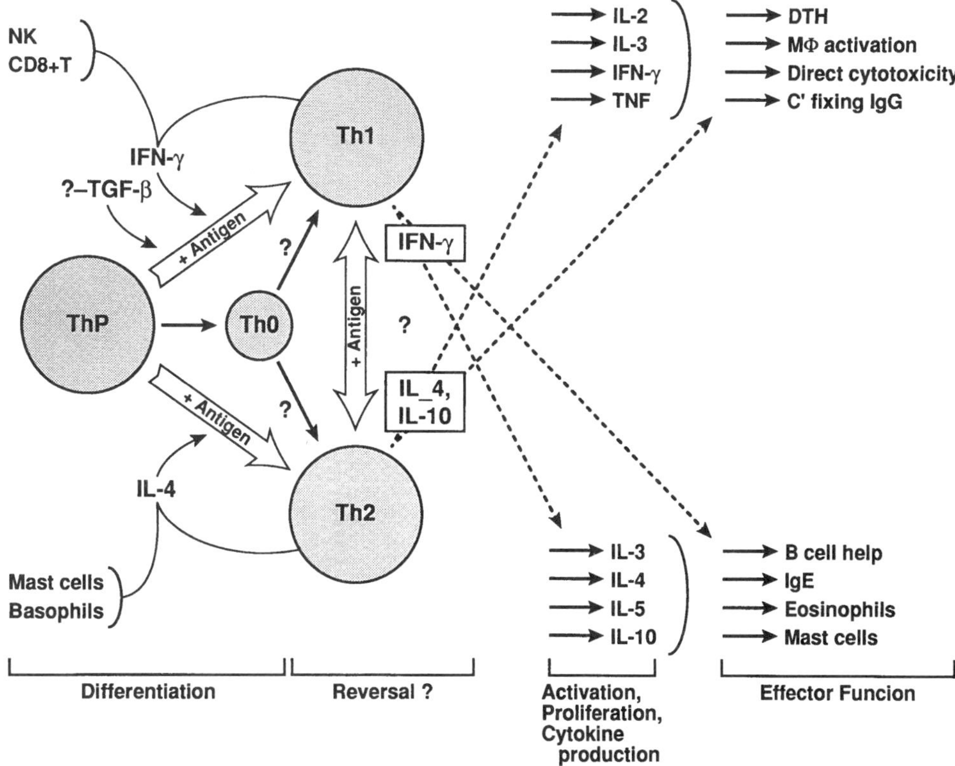

FIG. 5. Proposed generation of CD4+ T-lymphocyte subsets (Th1 and Th2) and the various cytokines made by them. Solid arrows signify stimulatory interactions; dashed arrows signify inhibitory interactions. Diagram courtesy of Robert Coffman, Palo Alto, California.

specific; nearby "bystander" cells are not harmed by the release of these toxins. CTLs release antiviral cytokines. Many $CD8^+$ T cells release high levels of cytokines, for example interferon-γ (IFNγ) (Fig. 5). It has been cogently argued that a major role of the TCR/MHC/peptide interaction is simply to hold $CD8^+$ T cells in the immediate proximity of virus-infected cells, thus focusing cytokines on the infected cell (109,114). The roles of interferons, including IFN-γ, in control of virus infection are clearly established, although the mechanism of antiviral effect remains to be determined. However, mice lacking the IFNγ receptor have increased susceptibility to several infections, despite apparently normal CTL and Th responses (43).

Natural Killer Cells

Natural killer (NK) cells exist as large granular lymphocytes in human peripheral blood and mediate cell lysis in a manner that does not show conventional immunological specificity. They may be characterized by cell surface markers such asialo GM1, and usually carry Fc receptors. They are not antigen specific and therefore do not exhibit immunological memory. They have a broad specificity and appear to play an important role in the control of certain virus infections.

A role for NK cells in natural immunity to viral infections has been established by clinical studies in humans and experimental manipulations of mice. In general, NK cell activity peaks earlier than that of T cells (~2–3 days) and diminishes rapidly thereafter. Correlations have been made between decreased NK cell activity as a consequence of disease or drug therapy and susceptibility to infections with HCMV and HSV-1. Examples of individuals with complete NK cell defects are rare, perhaps attesting to the importance of NK cells, but one adolescent with a complete and selective deficiency in NK cells presented with unusually severe infections with VZV, HCMV, and HSV-1 (7). Although the VZV and HCMV infections were life-threatening, each infection resolved completely, probably because T-cell and B-cell immunity was normal in this individual. Definitive studies on the role of NK cells in viral infections have been performed in the mouse model. Depletion of NK cells *in vivo* by administration of anti–NK cell antibodies leads to enhanced synthesis of some viruses, such as MCMV and VV, but not of other viruses, such as LCMV (134). NK cell–mediated resistance to MCMV is particularly profound. An age-dependent resistance to MCMV occurs in mice at about the third week of age and temporally correlates with the ontogeny of the developing NK-cell response. Furthermore, adoptive transfers of NK cells into suckling mice protect them from severe MCMV infection. Lymphokine-activated killer (LAK) cells, which are NK cells highly activated by growth in culture in the presence of interleukin-2 (IL-2) and which have been used in a variety of clinical and animal model tumor therapy protocols, provide marked resistance of suckling mice to MCMV. The presence of elevated NK-cell activity for a prolonged time period during persistent infections may select for NK cell–resistant viruses. Establishment of a persistent infection in severe combined immunodeficient (SCID) mice with an NK cell–sensitive variant of Pichinde virus resulted in the conversion of the virus into an NK cell–resistant genotype(s). The related arenavirus, LCMV, normally causes a persistent infection in mice in nature and is also NK cell resistant.

The structure on the target-cell membrane recognized by NK cells is unknown, as is the receptor on the NK cell itself. However, the presence of class I MHC often inhibits NK-mediated lysis (13). Several of the cell lines exquisitely sensitive to NK lysis express no class I proteins, and treatment of NK-sensitive cells with IFNγ, which induces class I expression, protects the cells. This has led to the "missing self" hypothesis, which argues that NK cells specialize in eradicating low MHC cells, which are for the most part abnormal (67). However the situation is not clear-cut, and it appears that only certain MHC molecules can inhibit recognition by certain NK cells, i.e., there is some form of MHC restriction, albeit different from that involved in T-cell recognition. Thus, the precise nature of the antigen-binding groove and/or the bound peptide on a target cell's surface appears to be critical to determining NK susceptibility, although the mechanism remains ill understood (53). Interestingly, the interferons also enhance the cytotoxic activity of NK cells, allowing the following speculative scenario. Virus infection induces interferon, which activates NK cells to elevate the overall level of NK cytotoxicity. This cytotoxic activity is subsequently obviated by interferon-induced expression of class I MHC on the target cells, rendering them less sensitive to NK-mediated lysis. One might anticipate that the increased levels of MHC expression would make the cells potential targets for CTLs. Interestingly, NK-cell activity usually peaks soon (e.g., 3 days) after virus infection and declines as the CTL activity appears (around 7 days); therefore, perhaps the NK cells represent a rapid, nonspecific, and self-limiting response to virus infection, which is superseded by specific CTL response. Many questions remain unresolved about this cell type. The mechanism of cytotoxicity probably involves release of perforin, which is found in abundance in the NK-cell granules.

Antibody-Dependent Cell-Mediated Cytotoxicity

Antibody-dependent cell-mediated cytotoxicity (ADCC) may be performed either by CTLs or by NK cells. In the latter case, the antibody, usually IgG, is specific for a structure on the target-cell membrane. The antibody "coats" the target cell, and the NK cell then attaches via its Fc receptors. The close approximation of killer and target triggers release of toxin and target cell death. ADCC can be brought about in a different way if the target cell bears Fc receptor. Incubation of uninfected cells with antibody to either the CD3 complex or the TCR idiotype itself causes a cytotoxic lymphocyte to attach to the target cell and lyse it. Note that in either case the specificity is conferred on the reaction by antibody; in one instance the antibody is target cell specific, whereas in the other it is effector cell specific. The contribution of ADCC *in vivo* is not known.

MACROPHAGES

Mononuclear phagocytes (blood monocytes, tissue macrophages, and dendritic cells) comprise a principal cellular element in the clearance and inactivation of most viral pathogens. Interestingly, these same cells also represent a major target cell and infectious reservoir for many viruses that persist, i.e., lentiviruses, cytomegaloviruses. The outcome of macrophage-virus interactions is the result of a complex series of reactions that rely on changes in macrophage differentiation in both the steady state and during immune responses. Such changes dramatically affect the ability of macrophages to act as scavenger cell, immune effector cell, or susceptible target cell for virus replication.

Mononuclear phagocytes have a vigorous phagocytic ability enabling them to eliminate virus from the circulation after a blood-borne infection. Their scavenger function constitutes an early line of defense for reducing the virus load either initially or in concert with the specific immune response. Macrophages can also restrict virus infection or replication by acting as a nonspecific effector cell for ADCC. Monocytes and macrophages can also initiate a series of immune reactions through the cytokine network (see below) that assists or injures the host. Macrophages may be activated for enhanced antiviral activity by cytokines produced by virus-specific T cells. Furthermore, these cells, especially dendritic cells, act as antigen-presenting cells taking up, processing, and delivering viral antigens to T cells in regional lymph nodes (116).

CYTOKINES

Cytokines, or for the purpose of this chapter, lymphokines and monokines, are proteins made by cells of the immune system, usually during cell activation. As soluble factors, they act as messengers on such cells and promote activities focused on regulation of the immune response or elimination of viruses and other pathogens. Figure 5 shows the cellular interactions of cytokines. Their activities and structure are reviewed elsewhere in this volume. Immune responses to antigens are often distinguished by a relatively distinct pattern of cytokines produced by subsets of $CD4^+$ Th cells. This in turn drives the immune system toward a predominant cell-mediated or humoral immune response (79). The subsets have been described in humans and mice. Th1 cells secrete IL-2, IFNγ, and lymphotoxin. They promote CTL-type and other delayed type hypersensitivity reactions frequently required for clearance of intracellular organisms. Th2 cells produce IL-4, IL-5, IL-6, IL-10, and IL-13. They promote allergic type responses and stimulate the differentiation of B-lymphocytes, mast cells, and eosinophils. The role of Th1 and Th2 cells and their cytokines in viral infections is just now being revealed. Viruses themselves can use cytokines, or cytokine homologs, to their own advantage. Certain viruses contain genes encoding cytokine homologs, the expression of which potentially enables the viruses to direct immune responses to best suit their life-styles of infection, replication pathogenesis, and survival. Examples include EBV/IL-10, poxviruses/tumor necrosis factor, and poxviruses/IL-1 receptor. Studies to assess the role these homologies play *in vivo* in affecting the immune system are currently underway.

HUMORAL EFFECTOR MECHANISMS

Generation of B-Cell Responses and Antibody

The B-cell receptor for antigen is Ig, which is initially expressed as a membrane protein. The structure and genetics of Ig are well defined and are described above. During their development, B cells initially express IgM on their surface, followed by expression of different H chain classes IgG(γ), IgA(α), or IgD(δ). IgM and IgD may be expressed in all B cells, but once a cell has expressed an H chain class other than IgM, it remains committed to secrete that class of Ig. When a B cell encounters antigen, it usually differentiates into mature plasma cells, secreting one class of Ig as specific antibody, and memory B cells also are generated. This differentiation of B cells is controlled by specific B-cell growth and differentiation factors produced by T cells, analogous to the factors involved in the differentiation of T cells themselves. As a general rule, the B-cell responses to viral and nonviral antigens are similar, with the caveat that viruses themselves are capable of infecting B cells and regulatory T cells and altering their function(s). Antibodies of all classes are usually generated in response to virus infection.

How Do the Various Antibody Classes Exert Their Antiviral Effects?

Recent animal experiments studying the effects of antiviral monoclonal antibodies on a wide variety of DNA and RNA infections have led to a new and important conceptual distinction between *in vitro* neutralization and *in vivo* protection. First, monoclonal antibodies that do not neutralize *in vitro* may nevertheless afford protection upon *in vivo* transfer. Second and conversely, antibodies that have high neutralizing activity *in vitro* may not provide protection when adoptively transferred *in vivo*. Nevertheless, *in vitro* studies have provided much information regarding the probable mechanisms underlying the antiviral efficacy of each antibody class. Note that intact antibodies are not a prerequisite for antiviral effectiveness; Fab fragments specific for RSV F glycoprotein, when instilled into the lungs of infected mice, were therapeutically effective (16). Such approaches hold promise, particularly in the light of recent advances in technologies that allow the rapid production of antibodies of any desired specificity (5,51).

IgA

During natural infection most viruses gain entry via mucosal surfaces. Hence, it is not surprising that mucosal immunity plays an important role in control of viral infections (87). Development of mucosal immunity and secretory IgA are important in preventing mucosally restricted viral respiratory and enteric infections, and virally specific circulating antibody plays an important role in limiting or aborting spread of virus to systemic sites by viremia. Thus, one of the issues in providing immunity by vaccination is the generation of effective and efficient mucosal immunity. For successful induction of a secretory IgA response in the gut, virus must come into contact with a Peyer's patch, the gut-associated lymphoid tissue. Once a local response has been generated, IgA-producing B memory cells recirculate to other areas in the gut. This has been demonstrated in humans by using oral polio vaccine. The protective effects of IgA are usually attributable to antibody directly preventing virus binding to the mucosa; in contrast to IgG, IgA does not recruit effector systems such as phagocytic cells or complement. The approaches and concepts are discussed in detail elsewhere in this volume.

IgG

There are several IgG subclasses in humans and mice. For example, human IgG falls into four subclasses—IgG1, IgG2, IgG3, and IgG4—which normally constitute 70%, 20%, 8%, and 2% of the total IgG, respectively. However, IgG subclass-specific antiviral antibodies are generated in different amounts in acute or persistent infections. For example, IgG1 is the dominant IgG isotype found in HIV-infected individuals during the quiescent stage of disease but declines in those with progressive illness. IgG1 is the major complement-binding and opsonizing antibody in humans (120), and in patients recovering from acute hepatitis B virus infection, IgG1 and IgG3 dominates with a switch (to non–complement-binding antibodies) from IgG3 to IgG4 during chronic infection. IgG1 in humans corresponds to mouse IgG2a (major complement-binding and opsonizing antibody). In experimentally manipulated mouse infection with LCMV, IgG2a antibody is the predominant isotype raised in acute infection with subsequent viral clearance (>90% of IgG isotype). However during persistent

LCMV infection there is a switch to >90% of IgG1, a non–complement-binding antibody (124). This switch may be related to activation of different Th subsets and the cytokines they produce: in the mouse, IgG2a is associated with the Th1 subset, IFNγ, IL-2, and lymphotoxin production, whereas IgG1 is associated with low levels of IFNγ and the Th2 subset. Complexing of viruses with IgG antibody will also facilitate their phagocytosis by mononuclear cells and polymorphonuclear leukocytes via Fc receptors on these phagocytic cells. However, this latter process may actually enhance the infectivity of some viruses (104).

Sindbis virus infection of mice provides an interesting animal model demonstrating the antiviral effect of IgG. Sindbis virus, an alphavirus, can establish persistent infection of mouse neurons. Because neurons express little or no class I MHC (47,48), the virus is effectively hidden from $CD8^+$ T cells, but infusion of antiviral antibodies, or even of antiviral Fab 2 fragments, clears virus infection almost entirely (63). The mechanism is unclear, but may involve modulation of viral replication, as demonstrated for measles virus in tissue culture (96).

IgM

This pentameric, decavalent molecule is produced early after virus infection, is usually independent of T-cell help, and acts as the initial antibody-mediated antiviral response. Later in infection, and upon secondary exposure, most IgM-producing cells switch classes to IgG but retain antigen specificity. There is good *in vitro* evidence that IgM is effective in neutralizing the infectivity of virus in the fluid phase. This may result from the antibody preventing virus attachment to specific cellular receptors and from the aggregation of virus particles, thereby reducing the number of infectious particles.

IgE

IgE antibody is present at low concentration in serum; its principal biological activity appears to be mediated by binding to mast cells via specific receptors. When antigen reacts with the cell-bound IgE, inflammatory mediators are released from the mast cell. This IgE-dependent mechanism accounts for several of the clinical features of "allergy," including wheezing with respiratory infections and pharmacologic relief with antihistamines. There has been little study on the antiviral IgE responses and the molecular mechanism(s) for its role in clinical complications.

Role of Complement

The human complement system comprises over 20 plasma proteins. Their sequence of interaction is now well understood at a mechanistic level (11,81). The classical pathway of complement is activated when the C1 macromolecule binds to IgG or IgM, which is either aggregated or complexed with antigen. Activation of C1r and C1s ensues after binding of the C1q subunit. Activated C1 sequentially cleaves C4 and C2 with resultant assembly of C4b2a, the C3 cleaving enzyme of the classical pathway. The activity of C4b2a is restrained by the lability of C2b and to a lesser extent by specific regulatory proteins that inactivate C4b (these are the C4 binding protein and factor I).

The major difference involved in the initiation of the alternative, or amplification, pathway of complement activation is that it does not depend on antibody. The internal thioester linkage of C3 undergoes slow spontaneous hydrolysis in plasma, producing a C3b-like molecule. Although most of these C3b-like molecules are rapidly inactivated by factor H and factor I, a proportion interact with factors B and D to form a C3 cleaving enzyme. Some of the C3b generated by this C3 convertase binds to surfaces. One property shared by activators of the alternative pathway, which are mostly particulate, is that when C3b is deposited on their surface from the fluid phase, it is relatively protected from inactivation by factors H and I. This protected C3b can then bind factor B, which is cleaved by factor D (a low molecular weight serine protease in plasma) to give C3bBb, the C3 cleaving enzyme of the alternative pathway. Its active site is on the Bb fragment and decays rapidly, but the half-life of the enzyme is prolonged by properdin, which binds to C3b in the bimolecular complex and retards this decay. An important point about the alternative pathway is that C3b, however generated (whether by the classical pathway, spontaneous hydrolysis, or cleavage of C3 by other proteolytic enzymes), serves to initiate the amplification of C3 cleavage. Thus, the presence of an activator surface unleashes a powerful biological positive-feedback loop. This emphasizes the importance of the regulatory protein factors H and I, without which trivial amounts of C3b would result in unrestrained activation of the pathway, even in the absence of an activator surface. After cleavage of C3, additional C3b molecules may be incorporated into either C3bBb or C4b2a, giving C3bnBb or C4b2a3b; these complexes have C5 cleaving activity. After C5 cleavage, sequential nonenzymatic binding of C6–9 to C5b results in assembly of the C5b–9 complex, which inserts into the lipid bilayer of cell membranes, producing the familiar complement membrane lesion so similar to the perforin-induced lesion caused by CTL and NK cells.

In addition to the proteins of the classical and alternative pathways and the C5b–9 membrane attack complex, the complement system also includes cell surface molecules that specifically bind activated or cleaved complement molecules in a receptor-mediated manner. These include receptors for C3b (CR1, CD35), iC3b (CR3, mac-1, CD11b, CD18), and C3dg (CR2, CD21). The complement system is regulated by a number of plasma proteins (factors H and I, C1 inhibitor, C4 binding protein, S protein) and cell membrane proteins (CD46 or membrane cofactor protein [MCF] and CD55 or decay accelerating factor). These regulatory molecules restrict inappropriate complement activation by sequestering activation fragments and/or facilitating their inactivation.

Complement enhances the neutralization of viruses by antibody. This is achieved by at least four different mechanisms. First, complement can blanket the sensitized (antibody-coated) virion, thus rendering it incapable of binding to its receptor. Second, complement can aggregate or clump sensitized virions, thereby reducing the number of infectious units. Third, by producing incorporation of C3b into virus/antibody complexes, such sensitized virions can bind to C3b receptors on phagocytic cells, leading to their ingestion and degradation by these cells. Fourth, antibody and complement can directly lyse viruses possessing a lipid envelope. Complement likely plays an important role early during infection when both limiting amounts and low avidity antibody are made. (For a comprehensive review of the area, the reader should consult reference 40.) Complement alone can inactivate certain viruses directly in the absence of antibody. Retroviruses from several species are lysed by human complement, independent of antibody (136). This occurs because a viral protein alone (rather than an Ab/Ag complex) can act as a receptor for C1q. The alternative pathway of complement can be activated on the surface of a variety of microorganisms, including bacteria and yeasts. Thus, an additional principal function of the complement pathway is as a nonspecific mechanism of host defense, capable of opsonizing or lysing a variety of microorganisms before the generation of specific immune responses. Finally, at least two of the complement proteins are used as receptors by viruses. These include the C3d receptor (CR2 or CD21) used by EBV (27,49,85), and the CD46 protein or MCF (20,69,83) used by measles virus.

Lysis of Virus-Infected Cells by Antibody and Complement

Antibody and complement can kill virus-infected cells *in vitro*. A variety of human cell lines (epithelioid, neural, and lymphoid) infected with a wide variety of human DNA and RNA viruses and expressing viral antigens on their surfaces are lysed by human serum when it contains both specific antibody to the virus and functional complement. Here, lysis appears primarily dependent on an intact alternative pathway of complement (15,119). A detailed analysis of this system using measles virus–infected cells as a model showed that the infected cell activates the alternative pathway in the absence of antibody, but that antibody is needed for lysis to occur. Despite the activation of the alternative pathway being independent of antibody, IgG enhances C3 deposition and thus facilitates lysis. The involvement of multiple cell types and viruses in similar events indicates the generality of the phenomenon. Although it is difficult to quantitate this mechanism *in vivo*, it is likely that cellular injury can be produced by antibody and complement *in vivo*.

Idiotypes and the Network Theory

The antigen-combining site of each Ig molecule itself carries antigenic determinants known as idiotypic determinants, or idiotypes. These idiotypes can be recognized by antigen-combining sites on other antibodies, and these latter "antiidiotypic" antibodies often represent the "internal image" of the antigen to which the antibody bearing the idiotype was directed. They can thus compete with antigen for the combining site. The "network theory," formulated by Jerne (46), stated this concept and extended it to view the whole immune system as a network of interacting idiotypes and antiidiotypes. Simply put, all antibodies are antiidiotypes, which also react with specific external antigens. When external antigen enters the system and produces expansion of specific clones of cells being particular idiotypes, the system is perturbed, and expansion of antiidiotypic, anti-antiidiotypic antibodies, and so on ensues. These were postulated to play a regulatory role in modulating the immune response. The same principles would apply to TCRs, which also bear idiotypes. These ideas have stimulated much current research. Antiidiotypic antibodies have been demonstrated in some human autoimmune diseases and infections. Does the network theory have any specific relevance for virus infections? Apart from their regulatory role in the immune response, antiidiotypic antibodies that have the "image" of antigen may compete with that antigen for specific cellular receptors (other than antibody). For example, monoclonal antiidiotypic antibodies raised against antibody to reovirus glycoprotein will prevent the adsorption of the reovirus to its specific receptor on the cell surface, presumably because the antiidiotype can act as a ligand for the receptor. Similarly, antiidiotype antibodies have been used as vaccines to raise antiviral immune responses, thus avoiding inoculation with the microorganism.

VIRUS STRATEGIES TO EVADE ELIMINATION BY ANTIBODIES AND T CELLS

Viruses are relatively unstable outside a competent host, and therefore for a virus to remain in circulation (i.e., to persist in the host population) it must be able to remain in one host or move efficiently among different hosts. One factor bearing on both is the virus' ability to evade host immunity as efficiently as possible. The evasion of host immunity is advantageous to the virus at several levels. For example, if there is widespread immunity in a host population, a new virus able to evade that preexisting immunity will have an increased pool of susceptible targets.

In other instances, evasion of immunity increases viral pathogenicity in the individual host, e.g., the virus may grow to higher titers in the host, perhaps increasing its chance of transmission to a naive recipient. In still other cases the virus may be able to persist in the individual host (virus persistence is the subject of Chapter 7 in this textbook); this requires that the virus be relatively noncytopathic and that it avoids detection and elimination by the immune system. Several microbes take advantage of immune privileged sites, which seem relatively inaccessible to immune responses (e.g., the anterior chamber of the eye, neurons), but we are concerned here with more specific events. Viruses have developed an impressive selection of evasive maneuvers, which are detailed below and listed in Table 3.

Strategies to Induce Immunosuppression

Viruses have developed several ways of suppressing the host immune response.

Infect T or B Cells and Abrogate Their Function

It is of interest that in the vast majority of persistent infections, the viruses infect cells of the immune system (Table 4). Conceptually and experimentally, many of these virus infections have been shown to initiate a suppression against the specific immune compartment required by the host to rid itself of that infecting microbial agent. Thus, by this means the virus is able to directly play a role in establishing its state of persistence.

Infect the Thymus and Induce Tolerance

If viruses can infect the thymus, then they may cause clonal deletion of maturing virus-specific T cells; i.e. the virus is viewed as self. Theoretically, neither antibody nor T-cell responses will be induced; there is no generalized immune suppression, and defects will be specific to the infecting virus. Recent studies indicate that despite replication in the thymus, negative selection may be incomplete with T cells escaping to the peripheral lymphoid organs and antiviral antibodies being generated. This occurs experimentally in LCMV infection and may similarly occur in HBV carriers. In both instances specific antiviral CTL are not generated, although antiviral antibodies are (88,95,100). Note that clearance of the virus, for example by adoptive transfer of virus-specific syngeneic T cells, restores to the host the capacity to make virus-specific CTL responses. The transferred CTLs remove thymic cells expressing viral peptide/MHC complexes, and new T cells mature in the thymus, which is now devoid of viral antigen (56). Consequently, the host antiviral CTL response is restored because those new T cells bearing TCR with specificity for the virus/MHC complex will not be deleted.

Destroy Specialized Antigen-Presenting Cells

Certain viruses can infect specialized antigen-presenting cells (e.g., macrophages), either destroying them directly or inducing an immunopathological antiviral attack that destroys them (74,142). The result will be a generalized immunosuppression against most immunogens, af-

TABLE 3. *Examples of how viruses can evade immune eradication: by residing in specialized cells, by producing immunomodulatory proteins, or by evading immune effector mechanisms*

Infection of specialized cell types		
Cell type employed	**Viruses**	**Reason**
Neurons	Various DNA and RNA viruses	Low/absent class I MHC; low/absent peptide transporters
Production of immunomodulatory proteins		
Viral protein	**Virus**	**Mechanism**
tat	HIV-1	↓ MHC class I transcription
UL18	HCMV	Homolog of class I H chain, associates with β2-microglobulin
E3 19K	Adenovirus	↓ MHC transport from ER
?	EBV	↓ LFA-3 and ICAM-1
Evasion of antiviral effector mechanisms		
Effector mechanism	**Virus**	**Method of evasion**
Antibody to HA	Influenza virus	Antigenic drift and shift, resulting in escape variants
Antibody to HA	Measles	↓ Surface expression of viral proteins, ↓ transcription of measles virus
CTL	LCMV	↓ Antigen-presenting cells, leading to ↓ anti-LCMV effector CTL

TABLE 4. *Viruses that infect lymphocytes and monocytes*

Virus	Host	Infected cells
Double-stranded DNA viruses		
Hepatitis B virus	Human, monkey	PBMC, T and B lymphocytes
Woodchuck hepatitis virus	Woodchuck	PBMC, bone marrow
Papovavirus	Human, monkey	PBMC
Group C adenoviruses	Human	T-, B-, and null lymphocytes
Herpes simplex virus	Human	T-lymphocytes
Epstein-Barr virus	Human	B-lymphocytes
Cytomegalovirus	Human	Lymphocytes, monocytes
Pox virus	Rabbit	Spleen cells
Single-stranded DNA viruses		
Porcine parvovirus	Pig	Spleen cells
Minute virus of mice	Mouse	Lymphocytes
Positive-strand RNA viruses		
Poliovirus	Human	Lymphocytes, monocytes
Rubella	Human	T- and B-lymphocytes
Negative-strand RNA viruses		
Measles	Human, monkey	T- and B- lymphocytes, monocytes
Mumps	Human	T- and B- lymphocytes
Respiratory syncytial virus	Human	Lymphocytes, monocytes
Vesicular stomatitis virus	Human, mouse	T-lymphocytes
Influenza A	Human	Lymphocytes, monocytes
Parainfluenza	Human	Lymphocytes, monocytes
Ambisense RNA viruses		
Lymphocytic choriomeningitis virus	Mouse	T- and B- lymphocytes, monocytes
Junin virus	Human	Peripheral blood mononuclear cells
Retroviruses		
Murine leukemia virus	Mouse	B-lymphocytes
Feline leukemia virus	Cat	T- and B-lymphocytes, monocytes
HTLV I, II	Human	T-, B-, and null lymphocytes
HTLV III	Human	T and B lymphocytes, monocytes
siv	Monkey	PBMC
Endogenous C-type virus	Mouse	Spleen cells
Bovine leukemia virus	Cow	PBMC

fecting both antibodies (78) and T cells. It has been suggested that this mechanism underlies the late immune failure that characterizes AIDS.

Virus Strategies to Evade Antibody Responses

Such strategies are best exemplified by influenza virus. This virus contains a segmented RNA genome, encased in a viral envelope studded with influenza proteins [hemagglutinin (HA) and neuraminidase (N)] critical for binding to, and entry into, target cells. These proteins, which are so important for the viral life cycle, are obvious targets for the host antibody response, and the host takes full advantage of the opportunity by producing neutralizing antibodies targeted to both proteins. The virus responds in two ways.

Mutation of Viral Gene Sequences

Here, the high mutation rate of RNA polymerases takes effect, generating mutant HA and N sequences, and thus providing the virus with the opportunity to evolve. This evolution is directed by two requirements: the mutant proteins must retain their function, i.e., the virus must be viable, but they should be able to escape detection by the preexisting antibodies. The resultant virus thus often has fairly subtle immunological changes, but these are sufficient to give it a temporary edge in the battle. The virus may have an expanded pool of susceptible hosts because it should be able to infect individuals carrying the original set of antibodies. This phenomenon has been termed antigenic drift and plays a role in the epidemics of influenza, which occur sporadically. The process of viral escape through mutation of its target proteins is in no way restricted to influenza virus. Examples have been found in almost all viruses in which they have been sought, and indeed the need to escape antibodies doubtless has driven the diversification of many viruses into different strains or serotypes that we recognize today. When we categorize viruses by the antibodies that do or do not recognize them, we are simply using the results of this ongoing interplay.

Wholesale Replacement of Viral Proteins

The second approach taken by influenza virus exploits the segmented nature of its genome, and thus is far less widespread among different virus families, few of which share this segmental structure. The influenza virus genome has eight RNA segments, and during coinfection of a single cell by two viruses (1 and 2), the segments from both infecting viruses can mix (reassort). The virus that emerges can therefore carry mostly the segments of virus 1, but with its HA-encoding segment replaced by that of virus 2, for example. This new virus will be identical to virus 1, except that its HA will be virus 2. The biological consequences of this event will be dictated in part by the degree of difference between HA1 and HA2. It appears likely that such antigenic shift is the cause of severe influenza epidemics and pandemics. The molecular shift appears to take place in nonhuman hosts, most probably avian, where there is coinfection of a human virus (which will infect birds only rarely) and an avian strain. If one of the resultant reassortants carries the avian HA but is able to reinfect humans, the potential exists that this virus will be immunologically unknown to, and highly pathogenic in, humans.

Down-Regulate Viral Protein Expression

A profound strategy used by viruses in evading antibodies is their intracellular localization, which hides most viral proteins from antibody detection. Of course, even then antibodies can detect viral cell surface proteins, and can lyse the infected cells, or may in some way modulate the ongoing infection (29). Viruses respond to this pressure by reducing the level of expression of glycoproteins, sometimes down-regulating them manifold in comparison with other, nonsurface, viral products (94). The tightest restriction of all is that practiced by herpes simplex virus, which establishes latency in neurons and appears to synthesize no proteins until the virus is reactivated. As has been stressed throughout this chapter, the detection of intracellular viral proteins is the major remit of T cells, in concert with the MHC. Thus, the major problem facing intracellular virus is the evasion of this aspect of the host's detection systems. This is discussed below.

Virus Strategies to Evade T-Cell Responses

Viruses have sought to evade T-cell recognition at several levels.

Infect Cells Lacking Class I MHC

Neurons express little or no class I MHC (47,48) and are a popular site for viral persistence. Herpesviruses establish latent infection therein, and several other viruses (measles, LCMV, alphaviruses) can infect these cells.

Inhibit MHC Function

Adenoviruses encode a 19K protein which binds to class I MHC and prevents its transport to the cell surface (61). In this way an adenovirus-infected cell is refractory to T-cell recognition. Analogously, MCMV appears to block transport of peptide/MHC complexes into the medial Golgi (18).

Down-Regulate Viral Protein Expression

T-cell recognition requires only around 100 MHC/peptide complexes to be displayed on the target cell, and this can be achieved with low levels of protein expression. Thus, it may be difficult for viruses to down-regulate expression beneath detectable levels; indeed, perhaps this is the driving force behind the development of the above-mentioned methods for interfering with antigen presentation. Some viruses, such as herpesviruses, may be able to shut down protein synthesis, but this luxury is available to only a few virus families.

Mutation of Viral Protein T-Cell Epitopes

In a manner analogous to evasion of antibody responses, viruses could mutate those sequences presented to the immune system as T-cell epitopes. Such mutation could act at any one of several stages in the process of antigen presentation (Fig. 1). It could prevent proteasome cleavage, TAP transport to the endoplasmic reticulum, binding to MHC, or recognition by T cells. The process of viral mutation toward CTL escape has been less widely studied than the antibody equivalent. In TCR-transgenic animals, almost all mature T cells express the transgene and thus are of monoclonal specificity. If transgenic animals express a TCR specific for a viral epitope and such animals are infected with the virus, the selective pressure is so strong that infection yields output virus in which the target epitope is mutated to a form not recognized by the transgenic TCR (102). Thus, in this highly skewed situation, a virus can evolve into a T-cell escape variant. The biological importance of this phenomenon has not been demonstrated, however. Because there are usually several CTL epitopes on any one virus, mutation of one of these might not confer upon the microbe much selective advantage because the other T-cell specificities would remain active. Nevertheless, it remains possible that any advantage, even if slight, would be biologically significant. Recent studies using HIV have shown that viruses can mutate *in vivo*, in a natural infection, into CTL escape variants (101); however, the biological significance of this has not been proven. Finally, circumstantial evidence of CTL escape variants comes from studies of a specific CTL epitope in EBV, presented by the HLA-A11 allele. EBV strains circulating in a population in which the HLA-A11 allele is infrequent carry the epitope, whereas virus circulating in an A11-rich population has a sequence change in the epitope that prevents binding to the A11 molecule (17).

IMMUNOPATHOLOGY OF VIRAL INFECTIONS: VIRUS-INDUCED IMMUNE COMPLEX DISEASE

One of the most common immunopathologic manifestations associated with viral infection (acute or persistent) is virus-induced immune complex formation. In many virus infections, antiviral antibody interacts with virus in the fluid phase, or with viral antigens on cells forming complexes that are shed into the fluid phase, resulting in the formation of virus/antibody (V-Ab) immune complexes. Such V-Ab complexes occur in most infections. When the infection is chronic, the continuous trapping of the complexes in renal glomeruli, arteries, and choroid plexus leads to arteritis, glomerulonephritis, and choroiditis (Fig. 6A, B, and C, respectively). V-Ab immune complex disease is the common pathogenic mechanism of animal nephritides, vasculitis associated with HBV infection, and glomerulonephritis seen in several chronic human infections, all presenting an immunopathologic picture similar to that seen in chronic human nephritis of unknown cause.

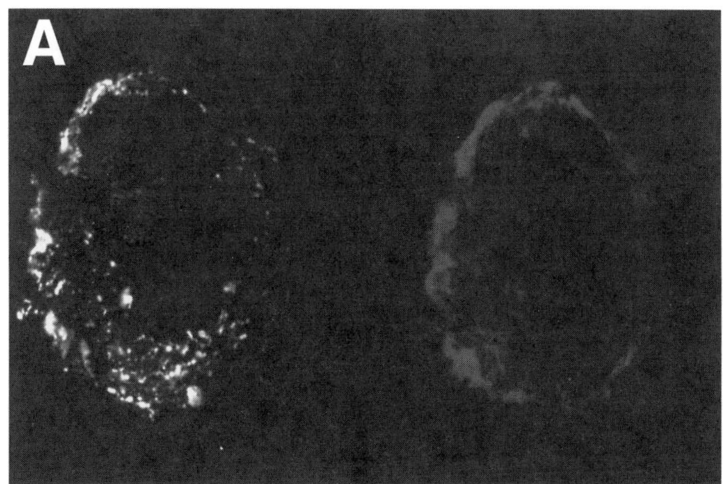

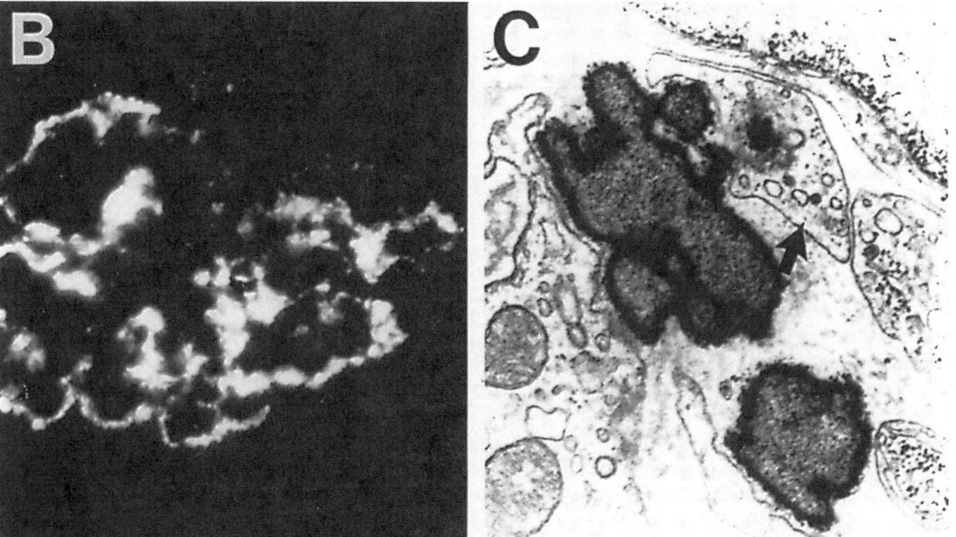

FIG. 6. Examples of virus-antiviral antibody immune complex–mediated tissue injury. **A:** Arteritis due to deposits of antibody to hepatitis B virus (left) and hepatitis B viral antigen(s) (right) in a medium sized artery of a patient with chronic hepatitis B virus infection. Detection is by immunofluorescence. **B:** Deposit of EBV antigen–anti-EBV antibody immune complexes in the renal glomerulus in a patient with chronic EBV infection leading to glomerulonephritis. Detection is by immunofluorescence. **C:** Immune complex of LCMV antigen and anti-LCMV antibody (dark material) being ingested by a macrophage (arrow) in the choroid plexus of the brain during persistent LCMV infection. Detection is by electron microscopy. From ref. 88, with permission.

Presence of immune complex disease is best documented by identification of viral antigen, host Ig, and complement in a granular pattern along the basement membrane of glomeruli or capillaries or in intima or media of arteries (Fig. 6). Identification and quantitation of specific antiviral antibodies are accomplished by elution of Ig by low ionic, high ionic, or low pH buffers to disassociate the V-Ab complex, recovery, and quantitation of the eluted Ig and determining the specific antiviral activity of the total Ig eluted (usually >75% compared with <1% specific viral Ig in circulation). Such immune complex depositions are constant companions of persistent viral infection of humans and animals (92,93).

Virus-Induced Autoimmune Disease

Viruses have been implicated in autoimmunity by three findings. First, autoimmune responses are made *de novo* or those already present are enhanced concomitant with infection by a wide variety of human DNA and RNA viruses (90,91). This point is strengthened by the second finding that in experimental animal models both acute and persistent virus infections can induce, accelerate, or enhance autoimmune disease in high-responder mice (125). Third, using an investigative approach that focuses on one potential mechanism whereby microbes cause autoimmunity, molecular mimicry, a number of etiologic agents have been identified as potential causes of autoimmune disease (89).

How can viruses induce autoimmune responses? Certain viruses [or their protein(s)] have a mitogenic effect on unique lymphocyte subsets and hence act as polyclonal activators. However, because agents such as mycoplasma can also activate lymphocytes and may contaminate viral stocks, stringent evidence must be presented that the activation is due to the virus and not mycoplasma contamination. Viruses can also infect lymphocytes and macrophages and directly or through their proteins cause release of lymphokines and monokines. These molecules can modulate immune responses in a variety of ways (Fig. 4), including as growth or differentiation factors or by regulating MHC class I and/or class II expression on cells. For example, epithelioid cells in the thyroid or ß cells in the islets of Langerhans in the pancreas usually express limited or negligible amounts of MHC molecules and can be upregulated and positioned to present self antigens. Finally, microbial agents share determinants with host self proteins. In this instance, an immune response mounted by the host against a specific determinant of the infecting agent may cross-react with the mimic (shared) host sequence, leading to autoimmunity and, in some cases, tissue injury and disease. Although many viral proteins share epitopes with host-cell proteins as determined by computer sequence analysis or by analysis of monoclonal antibodies raised against a large panel of DNA and RNA viruses in which nearly 4% of over 700 tested showed cross-reaction with host determinants expressed on uninfected tissues (121), the evidence that mimicry could cause disease came from experiments in which a viral protein sharing sequence conservation with the encephalitogenic site of myelin basic protein was able upon inoculation to cause the autoimmune disease allergic encephalomyelitis (30). These observations suggest that human diseases such as those of viral encephalopathies occurring after measles, mumps, vaccinia, or herpes zoster viral infection, in which recovery of a virus is unusual or rare, may have a similar pathogenic mechanism. In these instances, autoimmunity would occur only when the microbial and host determinants are similar enough to cross-react, yet different enough to break immunologic tolerance. The induction and breaking of tolerance at both the B- and T-cell levels have been established in heterologous serum protein models, and the same kinetics would likely govern the establishment and the break of tolerance to microbial agents cross-reacting with host proteins. Such cross-reactions could occur either through antibody- or cell-mediated immune responses.

Vaccine Considerations

At a time when new vaccines are being sought to counter new challenges (HIV, and even tumor vaccines) and because new methods of subunit vaccination are being suggested as replacements for older vaccines that have performed extremely well (e.g., live attenuated viruses), it is critical that we carefully determine the role of each aspect of the immune response in combating primary virus infection, and in preventing and/or limiting subsequent infection in an immune host. This can now be viewed in the light of the foregoing discussion. Historically, two approaches have been taken to developing antiviral vaccines. First, live vaccines have been developed, usually by attenuating the pathogenic virus by tissue-culture passage in a variety of cell lines. Attenuation was then tested in animal models. When pathogenicity was sufficiently reduced, the attenuated virus was considered a candidate. Secondly, the pathogenic virus could be inactivated (e.g., by formalin treatment) and the killed vaccine tested for immunogenicity. A major difference is evident between these two groups: the antigens of live vaccines will be presented by both class I and class II MHC, whereas those in killed vaccines will be presented mainly by the class II pathway. In Table 5 are listed the major vaccines in current use and their nature [live or dead (we have avoided the adjective killed because the hepatitis B vaccine is a recombinant protein, and thus killing of the hepatitis virus itself was not required for this vaccine)]. Included also are some of the immunological findings associated with each type of vaccine. Each vaccine has costs and benefits. Live vaccines give longer-lasting immune responses and against most or all viral antigens (partly because the vaccine replicates to give a large

TABLE 5. *Summary of currently available antiviral vaccines*

	Live	Dead
	Polio (Sabin) Measles Mumps Rubella Yellow Fever Vaccinia Varicella	Influenza Rabies Hepatitis B Polio (Salk)
Antibody induction	+++	+++
CD4 T cell	+++	+++
CD8 T cell	+++	–
All viral antigens?	Usually	Often not
Longevity of immunity	Months/years	Months
Cross-reactivity among viral strains	+++	+
Disease potentiation	? High-dose measles	RSV, measles
Risk of viral disease	+	–

antigenic dose), and they induce a balanced response that includes CD8 responses as well as CD4/antibody. Perhaps this latter point (induction of $CD8^+$ cells) may be in part responsible for the astonishing efficacy of the live vaccines listed above. Live vaccines carry the risk of vaccine-associated viral disease, often caused by reversion to pathogenicity of the vaccine strain during its replication in the vaccinee. For example, several cases of adult polio occur annually in developed countries and are almost invariably vaccine-associated; that is, the victims have been in recent close contact with vaccinees. Molecular analyses of the viruses shed in the feces of recently vaccinated infants demonstrated that the vaccine virus had rapidly reverted toward neurovirulence (12,25). In contrast, killed vaccines should carry no real risk of revertant-associated disease, although if inactivation is inadequate, as in an early batch of the Salk polio vaccine, then vaccinees may receive virulent virus. However, dead vaccines do not induce good $CD8^+$ T-cell responses; the immunity is short-lasting and often not detectable against all of the viral antigens (in part because inactivation may selectively modify or destroy the immunogenicity of the various virus proteins). Furthermore, there is a risk of disease potentiation. Vaccinees who are subsequently exposed to the wild virus may get much more severe disease than unvaccinated individuals. This has been well documented for killed measles (31) and RSV vaccines (52), which were withdrawn in large part for this reason. Such disease potentiation has rarely been shown for live vaccines, although recent experience with high-dose measles vaccine in Africa has indicated that occasional unpredictable problems might arise, thus counseling caution in use of new vaccines, and continual monitoring of vaccinee status.

The "ideal" vaccine therefore would combine the best features of both, while omitting the problems. It should deliver antigen into both class I and class II pathways to induce a balanced, long-lasting, and cross-reactive immunity without significant vaccine-associated risks.

REFERENCES

1. Ahmed R. Immunological memory against viruses. *Semin Immunol* 1992;4:105–109.
2. Arevalo JH, Taussig MJ, Wilson IA. Molecular basis of crossreactivity and the limits of antibody-antigen complementarity. *Nature* 1993;365:859–863.
3. Arnold D, Driscoll J, Androlewicz M, Hughes E, Cresswell P, T Spies. Proteasome subunits encoded in the MHC are not generally required for the processing of peptides bound by MHC class I molecules. *Nature* 1992;360:171–174.
4. Attaya M, Jameson S, Martinez CK, et al. Ham-2 corrects the class I antigen-processing defect in RMA-S cells. *Nature* 1992;355:647–649.
5. Barbas CF, Kang AS, Lerner RA, Benkovic SJ. Assembly of combinatorial antibody libraries on phage surfaces: the gene III site. *Proc Natl Acad Sci USA* 1991;88:7978–7982.
6. Barrett T, Visser IK, Mamaev L, Goatley L, van Bressem MF, Osterhaus AD. Dolphin and porpoise morbilliviruses are genetically distinct from phocine distemper virus. *Virology* 1993;193:1010–1012.
7. Biron CA, Byron KS, Sullivan JL. Severe herpesvirus infections in an adolescent without natural killer cells [see comments]. *N Engl J Med* 1989;320:1731–1735.
8. Bjorkman PJ, Saper MA, Samraoui B, Bennett WS, Strominger WL, Wiley DC. The foreign antigen binding site and T cell recognition regions of class I histocompatibility antigens. *Nature* 1987;329:512–518.
9. Bjorkman PJ, Saper MA, Samraoui B, Bennett WS, Strominger JL, Wiley DC. Structure of the human class I histocompatibility antigen HLA-A2. *Nature* 1987;329:506–512.
10. Brown JH, Jardetzky TS, Gorga JC, et al. Three-dimensional structure of the human class II histocompatibility antigen HLA-DR1. *Nature* 1993;364:33–39.
11. Campbell RD, Law SK, Reid KB, Sim RB. Structure, organization, and regulation of the complement genes. *Annu Rev Immunol* 1988;6:161–195.
12. Cann AJ, Stanway G, Hughes PJ, et al. Reversion to neurovirulence of the live-attenuated Sabin type 3 oral poliovirus vaccine. *Nucl Acids Res* 1984;12:7787–7792.
13. Carbone E, Racioppi L, La Cava A, et al. NK and LAK susceptibility varies inversely with target cell MHC class I antigen expression in a rat epithelial tumour system. *Scand J Immunol* 1991;33:185–194.
14. Christinck ER, Luscher MA, Barber BH, Williams DB. Peptide binding to class I MHC on living cells and quantitation of complexes required for CTL lysis. *Nature* 1991;352:67–70.
15. Cooper NR, Oldstone MBA. Virus infected cells, IgG and the alternative complement pathway. *Immunol Today* 1983;4:107–108.
16. Crowe JE, Murphy BR, Chanock RM, Williamson RA, Barbas CF III, Burton DR. Recombinant human RSV monoclonal antibody Fab is effective therapeutically when introduced directly into the lungs of respiratory syncitial virus-infected mice. *Proc Natl Acad Sci USA* 1994;91:1386–1390.
17. de Campos-Lima P, Gavioli R, Zhang Q, et al. HLA-A11 epitope loss isolates of Epstein-Barr virus from a highly A11+ population. *Science* 1993;260:98–100.
18. del Val M, Hengel H, Hacker H, et al. Cytomegalovirus prevents antigen presentation by blocking the transport of peptide-loaded major histocompatibility complex class I molecules into the medial-Golgi compartment. *J Exp Med* 1992;176:729–738.
19. Doherty PC, Hou S, Southern PJ. Lymphocytic choriomeningitis virus induces a chronic wasting disease in mice lacking class I MHC glycoproteins. *J Neuroimmunol* 1993;46:11–17.
20. Dorig RE, Marcil A, Chopra A, Richardson CD. The human CD46 molecule is a receptor for measles virus (Edmonston strain). *Cell* 1993;75:295–305.
21. Driscoll J, Brown MG, Finley D, Monaco JJ. MHC-linked LMP gene products specifically alter peptidase activities of the proteasome. *Nature* 1993;365:262–264.
22. Eichelberger M, Allan W, Zijlstra M, Jaenisch R, Doherty PC. Clearance of influenza virus respiratory infection in mice lacking class I major histocompatibility complex-restricted CD8+ T cells. *J Exp Med* 1991;174:875–880.

23. Elliot T, Smith M, Driscoll P, McMichael AJ. Peptide selection by class I molecules of the major histocompatibility complex. *Curr Biol* 1993;3:854–866.
24. Enria DA, Fernandez NJ, Briggiler AM, Levis SC, Maiztegui JI. Importance of dose of neutralizing antibodies in treatment of Argentine hemorrhagic fever with immune plasma. *Lancet* 1984;2:255–256.
25. Evans DM, Dunn G, Minor PD, et al. Increased neurovirulence associated with a single nucleotide change in a noncoding region of the Sabin type 3 poliovaccine genome. *Nature* 1985;314:548–550.
26. Fenner F. Myxomatosis in Australian wild rabbits—evolutionary changes in an infectious disease. *Harvey Lect* 1957;25–55.
27. Fingeroth JD, Weis JJ, Tedder TF, Strominger JL, Biro PA, Fearon DT. Epstein-Barr virus receptor of human B lymphocytes is the C3d receptor CR2. *Proc Natl Acad Sci USA* 1984;81:4510–4514.
28. Fremont DH, Matsumura M, Stura EA, Peterson PA, Wilson IA. Crystal structures of two viral peptides in complex with murine MHC class I H-2Kb. *Science* 1992;257:919–927.
29. Fujinami RS, Oldstone MB. Antiviral antibody reacting on the plasma membrane alters measles virus expression inside the cell. *Nature* 1979;279:529–530.
30. Fujinami RS, Oldstone MBA. Amino acid homology between the encephalitogenic site of myelin basic protein and virus: mechanism for autoimmunity. *Science* 1985;230:1043–1045.
31. Fulginiti VA, Eller JJ, Downie AW, Kempe CH. Atypical measles in children previously immunized with inactivated measles virus vaccine. *JAMA* 1967;202:1075–1080.
32. Fung-Leung WP, Kundig TM, Zinkernagel RM, Mak TW. Immune response against lymphocytic choriomeningitis virus infection in mice without CD8 expression. *J Exp Med* 1991;174:1425–1429.
33. Gaczynska M, Rock KL, Goldberg AL. Gamma-Interferon and expression of MHC genes regulate peptide hydrolysis by proteasomes. *Nature* 1993;365:264–267.
34. Glynne R, Powis SH, Beck S, Kelly A, Kerr LA, Trowsdale J. A proteasome-related gene between the two ABC transporter loci in the class II region of the human MHC. *Nature* 1991;353:357–360.
35. Good RA. Experiments of nature in the development of modern immunology. *Immunol Today* 1991;12:283–286.
36. Good RA, Zak SJ. Disturbance in gamma-globulin synthesis as "experiments of nature." *Pediatrics* 1956;18:109–149.
37. Gray MM, Hann IM, Glass S, Eden OB, Jones PM, Stevens RF. Mortality and morbidity caused by measles in children with malignant disease attending four major treatment centres: a retrospective review. *Br Med J* 1987;295:19–22.
38. Heemels M, Schumacher TNM, Wonigeit K, Ploegh HL. Peptide translocation by variants of the transporter associated with antigen processing. *Science* 1993;262:2059–2063.
39. Henderson RA, Michel H, Sakaguchi K, et al. HLA2.1-associated peptides from a mutant cell line: a second pathway of antigen presentation. *Science* 1992;255:1264–1266.
40. Hirsch RL. The complement system: its importance in the host response to viral infection. *Microbiol Rev* 1982;46:71–85.
41. Horwitz M, Yanagi Y, Oldstone MBA. T cell receptors from virus-specific CTL recognizing a single immunodominant nine amino acid viral epitope show marked diversity. *J Virol* 1994;68:352–375.
42. Hou S, Doherty PC, Zijlstra M, Jaenisch R, Katz JM. Delayed clearance of Sendai virus in mice lacking class I MHC-restricted CD8-T cells. *J Immunol* 1992;149:1319–1325.
43. Huang S, Hendriks W, Althage A, et al. Immune response in mice that lack the interferon-gamma receptor. *Science* 1993; 259:1742–1745.
44. Hunt DF, Henderson RA, Shabanowitz J, et al. Characterization of peptides bound to the class I MHC molecule HLA2.1 by mass spectrometry. *Science* 1992;255:1261–1263.
45. Jenner E. An inquiry into the causes and effects of the variolae vaccine, a disease discovered in some western counties of England, particularly Gloucestershire, and known by the name of cowpox. London: Cassell, 1896.
46. Jerne NK. The generative grammar of the immune system. *Science* 1985;229:1057–1059.
47. Joly E, Mucke L, Oldstone MBA. Viral persistence in neurons explained by lack of major histocompatibility class I expression. *Science* 1991;253:1283–1285.
48. Joly E, Oldstone MBA. Neuronal cells are deficient in loading peptides onto MHC class I molecules. *Neuron* 1992;8:1185–1190.
49. Jondal M, Klein G, Oldstone MB, Bokish V, Yefenof E. Surface markers on human B and T lymphocytes. VIII. Association between complement and Epstein-Barr virus receptors on human lymphoid cells. *Scand J Immunol* 1976;5:401–410.
50. Jonjic S, del Val M, Keil GM, Reddehase MJ, Koszinowski UH. A nonstructural viral protein expressed by a recombinant vaccinia virus protects against lethal cytomegalovirus infection. *J Virol* 1988; 62:1653–1658.
51. Kang AS, Barbas CF, Janda KD, Benkovic SJ, Lerner RA. Linkage of recognition and replication functions by assembling combinatorial antibody Fab libraries along phage surfaces. *Proc Natl Acad Sci USA* 1991;88:4363–4366.
52. Kapikian AZ, Mitchell RH, Chanock RM, Shvedoff RA, Stewart CE. An epidemiologic study of altered clinical reactivity to respiratory syncitial (RS) virus infection in children previously vaccinated with an inactivated RS vaccine. *Am J Epidemiol* 1969;89:404–421.
53. Karre K. Natural killer cells and the MHC class I pathway of peptide presentation. *Semin Immunol* 1993;5:127–145.
54. Kelly A, Powis SH, Kerr LA, et al. Assembly and function of the two ABC transporter proteins encoded in the human major histocompatibility complex. *Nature* 1992;355:641–644.
55. Kernahan J, McQuillin J, Craft AW. Measles in children who have malignant disease. *Br Med J* 1987;295:15–18.
56. King CC, Jamieson BD, Reddy K, Bali N, Concepcion RJ, Ahmed R. Viral infection of the thymus. *J Virol* 1992;66:3155–3160.
57. Klavinskis LS, Oldstone MBA, Whitton JL. Designing vaccines to induce cytotoxic T lymphocytes: protection from lethal viral infection. In: Brown F, Chanock R, Ginsberg H, Lerner R, eds. *Vaccines 89. Modern approaches to new vaccines including prevention of AIDS.* Cold Spring Harbor, NY: Cold Spring Harbor Laboratory; 1989:485–489.
58. Klavinskis LS, Whitton JL, Oldstone MBA. Molecularly engineered vaccine which expresses an immunodominant T-cell epitope induces cytotoxic T lymphocytes that confer protection from lethal virus infection. *J Virol* 1989;63:4311–4316.
59. Konig R, Huang LY, Germain RN. MHC class II interaction with CD4 mediated by a region analogous to the MHC class I binding site for CD8. *Nature* 1992;356:796–798.
60. Kurlander RJ, Shawar SM, Brown ML, Rich RR. Specialized role for a murine class I-b MHC molecule in prokaryotic host defenses. *Science* 1992;257:678–679.
61. Kvist S, Ostberg L, Curman B, Persson H, Philipson L, Peterson PA. Molecular association of a virus protein and transplantation antigens on tumor cell lines. *Scand J Immunol* 1978;8:162
62. Lausch RN, Staats H, Metcalf JF, Oakes JE. Effective antibody therapy in herpes simplex virus ocular infection. Characterization of recipient immune response. *Intervirology* 1990;31:159–165.
63. Levine B, Hardwick JM, Trapp BD, Crawford TO, Bollinger RC, Griffin DE. Antibody-mediated clearance of alphavirus infection from neurons. *Science* 1991;254:856–860.
64. Lin FT, Chen SB, Wang YZ, Sun CZ, Zeng FZ, Wang GF. Use of serum and vaccine in combination for prophylaxis following exposure to rabies. *Rev Infect Dis* 1988;10(suppl 4):766–770.
65. Livingstone AM, Powis SJ, Diamond AG, Butcher GW, Howard JC. A trans-acting major histocompatibility complex-linked gene whose alleles determine gain and loss changes in the antigenic structure of a classical class I molecule. *J Exp Med* 1989;170:777–795.
66. Livingstone AM, Powis SJ, Gunther E, Cramer DV, Howard JC, Butcher GW. Cim: an MHC class II-linked allelism affecting the antigenicity of a classical class I molecule for T lymphocytes. *Immunogenetics* 1991;34:157–163.
67. Ljunggren HG, Karre K. In search of the "missing self": MHC molecules and NK cell recognition. *Immunol Today* 1990; 11:237–244.
68. Maiztegui JI, Fernandez NJ, de Damilano AJ. Efficacy of immune plasma in treatment of Argentine haemorrhagic fever and association between treatment and a late neurological syndrome. *Lancet* 1979;2:1216–1217.
69. Manchester M, Liszewski MK, Atkinson JP, Oldstone MBA. Multiple forms of CD46 (membrane cofactor protein) serve as receptors for measles virus. *Proc Natl Acad Sci USA* (in press)
70. Matsumura M, Fremont DH, Peterson PA, Wilson IA. Emerging principles for the recognition of peptide antigens by MHC class I molecules. *Science* 1992;257:927–934.

71. McBride SJ, McCluskey DR, Jackson PT. Selective C7 complement deficiency causing recurrent meningococcal infection. *J Infect* 1991;22:273–276.
72. McKinney REJ, Katz SL, Wilfert SM. Chronic enteroviral meningoencephalitis in agammaglobulinemic patients. *Rev Infect Dis* 1987;9:334–356.
73. McNeill WH. Plagues and peoples: a natural history of infectious diseases. Garden City, NY: Anchor Press, Doubleday; 1976.
74. Merino E, Osuna J, Bolivar F, Soberon X. A general, PCR-based, method for single or combinatorial oligonuceotide-directed mutagenesis on pUC/M13 vectors. *Biotechniques* 1992;12:508–510.
75. Michalek MT, Grant EP, Gramm C, Goldberg AL, Rock KL. A role for the ubiquitin-dependent proteolytic pathway in MHC class I restricted antigen presentation. *Nature* 1993;363:552–554.
76. Milligan GN, Flaherty L, Braciale VL, Braciale TJ. Nonconventional (TL-encoded) major histocompatibility complex molecules present processed viral antigen to cytotoxic T lymphocytes. *J Exp Med* 1991;174:133–138.
77. Misbah SA, Spickett GP, Ryba PC, et al. Chronic enteroviral meningoencephalitis in agammaglobulinemia: case report and literature review. *J Clin Immunol* 1992;12:266–270.
78. Moskophidis D, Pircher H, Ciernik I, Odermatt B, Hengartner H, Zinkernagel RM. Suppression of virus-specific antibody production by CD8- class I-restricted antiviral cytotoxic T cells in vivo. *J Virol* 1992;66:3661–3668.
79. Mosmann TR, Coffman RL. Heterogeneity of cytokine secretion patterns and functions of helper T cells. *Adv Immunol* 1989;46:111–147.
80. Muller D, Koller BH, Whitton JL, LaPan K, Brigman KK, Frelinger JA. LCMV-specific, class II-restricted cytotoxic T cells in b2-microglobulin-deficient mice. *Science* 1992;255:1576–1578.
81. Muller-Eberhard HJ. The membrane attack complex of complement. *Annu Rev Immunol* 1986;4:503–528.
82. Nahmias AJ, Griffith D, Salsbury C, Yoshida K. Thymic aplasia with lymphopenia, plasma cells, and normal immunoglobulins. Relation to measles virus infection. *JAMA* 1967;201:729–734.
83. Naniche D, Varior-Krishnan G, Cervoni F, et al. Human membrane cofactor protein (CD46) acts as a cellular receptor for measles virus. *J Virol* 1993;67:6025–6032.
84. Neefjes JJ, Momburg F, Hammerling GJ. Selective and ATP-dependent translocation of peptides by the MHC-encoded transporter. *Science* 1993;261:769–771.
85. Nemerow GR, Wolfert R, McNaughton ME, Cooper NR. Identification and characterization of the Epstein-Barr virus receptor on human B lymphocytes and its relationship to the C3d complement receptor (CR2). *J Virol* 1985;55:347–351.
86. Oehen S, Waldner H, Kundig TM, Hengartner H, Zinkernagel RM. Antivirally protective cytotoxic T cell memory to lymphocytic choriomeningitis virus is governed by persisting antigen. *J Exp Med* 1992;176:1273–1281.
87. Ogra PL, Garofalo R. Secretory antibody response to viral vaccines. *Prog Med Virol* 1990;37:156–189.
88. Oldstone MBA. Arenaviruses: biology and immunotherapy. *Curr Top Microbiol Immunol* 1987;134.
89. Oldstone MBA. Molecular mimicry and autoimmune disease. *Cell* 1987;50:819–820.
90. Oldstone MBA. Molecular mimicry as a mechanism for the cause and a probe uncovering etiologic agent(s) of autoimmune disease. *Curr Top Microbiol Immunol* 1989;145:127–135.
91. Oldstone MBA. Overview: infectious agents as etiologic triggers of autoimmune disease. *Curr Top Microbiol Immunol* 1989;145:1–3.
92. Oldstone MBA. Virus induced immune complex formation and disease: definition, regulation and importance. In: Notkins AL, Oldstone MBA, eds. *Concepts in viral pathogenesis.* New York: Springer Verlag; 1984:201–209.
93. Oldstone MBA. Virus neutralization and virus-induced immune complex disease. Virus-antibody union resulting in immunoprotection or immunologic injury—two sides of the same coin. *Prog Med Virol* 1975;19:84–119.
94. Oldstone MBA, Buchmeier MJ. Restricted expression of viral glyco protein in cells of persistently infected mice. *Nature* 1982; 350:360–362.
95. Oldstone MBA, Dixon FJ. Lymphocytic choriomeningitis: production of antibody by "tolerant" infected mice. *Science* 1967; 158:1193–1195.
96. Oldstone MBA, Fujinami RS, Lampert PW. Membrane and cytoplasmic changes in virus infected cells induced by interactions of anti viral antibody with surface viral antigen. *Prog Med Virol* 1980; 26:45–93.
97. Orren A, Potter PC, Cooper RC, du Toit E. Deficiency of the sixth component of complement and susceptibility to Neisseria meningitidis infections: studies in 10 families and five isolated cases. *Immunology* 1987;62:249–253.
98. Osterhaus AD, Vedder EJ. Identification of virus causing recent seal deaths. *Nature* 1988;335:20
99. Pamer EG, Wang CR, Flaherty L, Lindahl KF, Bevan MJ. H-2M3 presents a listeria monocytogenes peptide to cytotoxic T lymphocytes. *Cell* 1992;70:215–223.
100. Penna A, Chisari FV, Bertoletti A, et al. Cytotoxic T lymphocytes recognize an HLA-A2-restricted epitope within the hepatitis B virus nucleocapsid antigen. *J Exp Med* 1991;174:1565–1570.
101. Phillips RE, Rowland-Jones S, Nixon DF, et al. Human immunodeficiency virus genetic variation that can escape cytotoxic T cell recognition. *Nature* 1991;354:453–459.
102. Pircher H, Moskophidis D, Rohrer U, Burki K, Hengartner H, Zinkernagel RM. Viral escape by selection of cytotoxic T cell-resistant virus variants in-vivo. *Nature* 1990;346:629–633.
103. Podack ER, Lowrey DM, Lichtenheld M, Hameed A. Function of granule perforin and esterases in T cell-mediated reactions. Components required for delivery of molecules to target cells. *Ann N Y Acad Sci* 1988;532:292–302.
104. Porterfield J, Cardosa MJ. Host range and tissue tropism: antibody-dependent mechanism. In: Notkins AL, Oldstone MBA, eds. *Concepts in viral pathogenesis.* New York: Springer Verlag; 1984:117–122.
105. Powis SH, Mockridge I, Kelly A, et al. Polymorphism in a second ABC transporter gene located within the class II region of the human major histocompatibility complex. *Proc Natl Acad Sci USA* 1992; 89:1463–1467.
106. Powis SJ, Deverson EV, Coadwell WJ, et al. Effect of polymorphism of an MHC-linked transporter on the peptides assembled in a class I molecule. *Nature* 1992;357:211–215.
107. Powis SJ, Howard JC, Butcher GW. The major histocompatibility complex class II-linked cim locus controls the kinetics of intracellular transport of a classical class I molecule. *J Exp Med* 1991; 173:913–921.
108. Powis SJ, Townsend AR, Deverson EV, Bastin J, Butcher GW, Howard JC. Restoration of antigen presentation to the mutant cell line RMA-S by an MHC-linked transporter. *Nature* 1991;354:528–531.
109. Ramsay AJ, Ruby J, Ramshaw IA. A case for cytokines as effector molecules in the resolution of virus infection. *Immunol Today* 1993;14:155–157.
110. Rickinson AB, Murray RJ, Brooks J, Griffin H, Moss DJ, Masucci MG. T cell recognition of Epstein-Barr virus associated lymphomas. *Cancer Surv* 1992;13:53–80.
111. Rini JM, Schulze-Gahmen U, Wilson IA. Structural evidence for induced fit as a mechanism for antibody-antigen recognition. *Science* 1992;255:959–965.
112. Rotzschke O, Falk K. Naturally-occurring peptide antigens derived from the MHC class-I-restricted processing pathway. *Immunol Today* 1991;12:447–455.
113. Rotzschke O, Falk K, Deres K, et al. Isolation and analysis of naturally processed viral peptides as recognized by cytotoxic T cells. *Nature* 1990;348:252–254.
114. Ruby J, Ramshaw IA. The antiviral activity of immune CD8+ T cells is dependent on interferon-gamma. *Lymphokine Cytokine Res* 1991;10:353–358.
115. Rudensky AY, Preston-Hurlburt P, Al-Ramadi BK, Rothbard J, Janeway CA. Truncation variants of peptides isolated from MHC class II molecules suggest sequence motifs. *Nature* 1992;359:429–431.
116. Russell SW, Gordon S. Macrophage biology and activation. *Curr Top Microbiol Immunol* 1992;181.
117. Salter RD, Benjamin RJ, Wesley PK, et al. A binding site for the T cell co-receptor CD8 on the a3 domain of HLA-A2. *Nature* 1990;345:41–46.
118. Siegel MM, Walter TK, Ablin AR. Measles pneumonia in childhood leukemia. *Pediatrics* 1977;60:38–40.
119. Sissons JG, Oldstone MBA. Antibody-mediated destruction of virus-infected cells. *Adv Immunol* 1980;29:209–260.
120. Spiegelberg HL. The role of interleukin-4 in IgE and IgG subclass formation. *Semin Immunopathol* 1990;12:365–383.

121. Srinivasappa J, Saegusa J, Prabhakar BS, et al. Molecular mimicry: frequency of reactivity of monoclonal antiviral antibodies with normal tissues. *J Virol* 1986;57:397–401.
122. Takahashi H, Germain RN, Moss B, Berzofsky JA. An immunodominant class I-restricted cytotoxic T lymphocyte determinant of human immunodeficiency virus type 1 induces CD4 class II-restricted help for itself. *J Exp Med* 1990;171:571–576.
123. Teyton L, O'Sullivan D, Dickson PW, et al. Invariant chain distinguishes between the exogenous and endogenous antigen presentation pathways. *Nature* 1990;348:39–44.
124. Tishon A, Salmi A, Ahmed R, Oldstone MBA. Role of viral strains and host genes in determining levels of immune complexes in a model system: implications for HIV infection. *AIDS Res Hum Retrovir* 1991;7:963–969.
125. Tonietti G, Oldstone MBA, Dixon FJ. The effect of induced chronic viral infections on the immunologic diseases of New Zealand mice. *J Exp Med* 1970;132:89–109.
126. Townsend A, Trowsdale J. The transporters associated with antigen presentation. *Semin Cell Biol* 1993;4:53–61.
127. Trowsdale J, Hanson I, Mockridge I, Beck S, Townsend A, Kelly A. Sequences encoded in the class II region of the MHC related to the abc superfamily of transporters. *Nature* 1990;348:741–744.
128. Van Bleek GM, Nathenson SG. Isolation of an endogenously processed immunodominant viral peptide from the class I H-2 Kb molecule. *Nature* 1990;348:213–216.
129. van Bressem MF, Visser IK, De Swart RL, et al. Dolphin morbillivirus infection in different parts of the Mediterranean Sea. *Arch Virol* 1993;129:235–242.
130. Van Kaer L, Ashton-Rickardt PG, Ploegh HL, Tonegawa S. TAP1 mutant mice are deficient in antigen presentation, surface class I molecules, and CD4-8- T cells. *Cell* 1992;71:1205–1214.
131. Vignali DA, Moreno J, Schiller D, Hammerling GJ. Species-specific binding of CD4 to the beta 2 domain of major histocompatibility complex class II molecules. *J Exp Med* 1992;175:925–932.
132. Visser IK, van Bressem MF, van de Bildt MW, et al. Prevalence of morbilliviruses among pinniped and cetacean species. *Rev Sci Tech* 1993;12:197–202.
133. von Pirquet C. Das ver halten der kutanen tuberkulin-reaktion wahren der masern. *Dtsch Med Wochenschr* 1908;37:1297.
134. Welsh RM. Regulation of virus infections by natural killer cells a review. *Nat Immun Cell Growth Regul* 1986;5:169–199.
135. Welsh RM, Nishioka WK, Antia R, Dundon PL. Mechanism of killing by virus-induced cytotoxic T lymphocytes elicited in vivo. *J Virol* 1990;64:3726–3733.
136. Welsh RMJ, Cooper NR, Jensen FC, Oldstone MBA. Human serum lyses RNA tumour viruses. *Nature* 1975;257:612–614.
137. Whitton JL. Lymphocytic choriomeningitis virus CTL. *Semin Virol* 1990;1:257–262.
138. Whitton JL, Sheng N, Oldstone MBA, McKee TA. A "string-of-beads" vaccine, comprising linked minigenes, confers protection from lethal-dose virus challenge. *J Virol* 1993;67:348–352.
139. Wilson IA, Fremont DH. Structural analysis of MHC class I molecules with bound peptide antigens. *Semin Immunol* 1993;5:75–80.
140. Yanagi Y, Tishon A, Lewicki H, Cubitt BA, Oldstone MBA. Diversity of T-cell receptors in virus-specific cytotoxic T lymphocytes recognizing three distinct viral epitopes restricted by a single major histocompatibility complex molecule. *J Virol* 1992;66:2527–2531.
141. Yewdell JW, Esquivel F, Arnold D, Spies D, Eisenlohr LC, Bennink JR. Presentation of numerous viral peptides to mouse major histocompatibility complex (MHC) class I-restricted T lymphocytes is mediated by the human MHC-encoded transporter or by a hybrid mouse-human transporter. *J Exp Med* 1993;177:1785–1790.
142. Zinkernagel RM. Virus-induced acquired immune suppression by cytotoxic T cell-mediated immunopathology. *Vet Microbiol* 1992;33:13–18.
143. Zinkernagel RM, Doherty PC. Immunological surveillance against altered self components by sensitised T lymphocytes in lymphocytic choriomeningitis. *Nature* 1974;251:547–548.
144. Zinkernagel RM, Doherty PC. Restriction of in vitro T cell-mediated cytotoxicity in lymphocytic choriomeningitis within a syngeneic or semiallogeneic system. *Nature* 1974;248:701–702.
145. Zychlinsky A, Zheng LM, Liu CC, Young JD. Cytolytic lymphocytes induce both apoptosis and necrosis in target cells. *J Immunol* 1991;146:393–400.

Fundamental Virology, Third Edition
edited by B.N. Fields, D.M. Knipe, P.M. Howley, et al.
Lippincott - Raven Publishers, Philadelphia © 1996

CHAPTER 11

Interferons and Other Cytokines

Jan Vilček and Ganes C. Sen

GENERAL FEATURES OF CYTOKINES

The cytokine family includes a large (and still growing) number of proteinaceous intercellular signaling molecules that mediate a variety of actions within and without the immune system. Because no brief definition can encompass all features of the members of this very diverse family, cytokines are best defined by a list of their characteristic properties (Table 1). Many of the major features of cytokines are shared by polypeptide hormones and growth factors. The resemblance is more than superficial because many common elements are being recognized in receptor structures and intracellular signaling pathways among the cytokines, polypetide hormones, and growth factors (11,127, 195,197). One difference between cytokines and classical hormones is that hormones are produced only by highly differentiated specialized cells (e.g., insulin is produced exclusively by β cells of the pancreas), whereas many cytokines are produced by less specialized cells and very often several unrelated cell types can produce the same cytokine. The dividing line between cytokines and growth factors is more tenuous because many cytokines do in fact act as growth factors (253). One difference is that "classical" growth factors (e.g., platelet-derived growth factor or epidermal growth factor) tend to be produced constitutively, unlike cytokines whose synthesis is tightly regulated.

The nomenclature of cytokines, like the field of cytokine research itself, reflects its evolution from many distinct sources (253). Some names are based on the original biological action associated with a particular cytokine. (For example, interferons *interfere* with virus replication.) More recently, many cytokines have been given the designation "interleukin" followed by a number, with the numbers increasing sequentially in chronological order. The first two interleukins (IL-1 and IL-2) were so named in 1979 (1), and the most recent one as of this writing (IL-13) in 1993 (149). There is no universally accepted system for group-

J. Vilček: Department of Microbiology, New York University Medical Center, New York, New York 10016.
G. C. Sen: Department of Molecular Biology, Research Institute, The Cleveland Clinic Foundation, Cleveland, Ohio 44195.

TABLE 1. *Characteristic features of cytokines*

Most cytokines are simple polypeptides or glycoproteins with a molecular weight ≤30 kD (but many cytokines form homodimers or homotrimers). One cytokine (IL-12) is a heterodimer.
Constitutive production of cytokines is usually low or absent; production is regulated by various inducing stimuli at the level of transcription or translation.
Cytokine production is transient and the action radius is usually short (typical action is autocrine or paracrine, not endocrine). However, some cytokines are found in the circulation during systemic infections.
Cytokines produce their actions by binding to specific high affinity cell surface receptors (Kd in the range of 10^{-9}–10^{-12} M).
Most cytokine actions can be attributed to an altered pattern of gene expression in the target cells. Phenotypically, cytokine actions lead to an increase (or decrease) in the rate of cell proliferation, change in cell differentiation state, and/or a change in the expression of some differentiated functions.
Although the range of actions displayed by individual cytokines can be broad and diverse, at least some actions of each cytokine are targeted at hematopoietic cells.
Cytokines often act by increasing (or decreasing) the production of other cytokines (cytokine cascade) or by transmodulating receptors for other cytokines.
Structurally dissimilar cytokines may have a similar spectrum of actions (redundancy).
A single cytokine often has multiple target cells and multiple actions (pleiotropism).
Because cells and tissues in the body are rarely, if ever, exposed to a single cytokine at a time, many cytokine actions reflect the result of synergistic or antagonistic interactions among several cytokines.

ing cytokines, and because many cytokines have more than one major function, a perfect system may not be possible (Table 2).

The aim of this chapter is to review the field of cytokines in the context of general virology. Rather than reviewing all major cytokine genes and proteins (an impossible task, given the limited length of the chapter), we shall concentrate on those cytokines that are known to be important in virus infections. The cytokines to be covered in greatest detail are the interferons (IFNs), the antiviral cytokines par excellence. IFN is in fact the oldest known cytokine. It was discovered in the course of studies on virus interference, as a product of influenza virus-infected chick embryo cells capable of inducing resistance to infection with homologous or heterologous viruses (103). In due course, virus-induced IFNs were demonstrated in all major classes of vertebrate species, except in amphibia (40). A functionally related protein (now called IFN-γ) was first described as an IFN-like virus-inhibitory protein produced by mitogen-activated human T-lymphocytes (261). Although both major classes, IFN-α/β and IFN-γ, mediate many diverse actions, their important roles in the defense against virus infections have been corroborated by recent studies in mice with targeted disruptions of the IFN-α/β and IFN-γ receptor genes (156).

The other cytokines to be described here much more briefly are those that have been shown to play some role in virus infections. Such cytokines include members of the tumor necrosis factor (TNF) and IL-1 families. These cytokines not only can be induced in the process of virus infection, but also affect the replication of some viruses both by direct actions on virus-infected cells and through their immunomodulatory effects. Because they represent components of the host defense to virus infection, some viruses have developed the capacity to suppress the actions of IFNs and of IL-1 or TNF. Recent analyses of the molecular strategies developed by viruses as a means to counteract cytokines have yielded interesting new information about pathogenetic mechanisms, to be reviewed in a separate section of this chapter.

CYTOKINE GENES AND PROTEINS

Type I Interferons

Members of the IFN-α/β or type I IFN superfamily represent the prototypical IFN molecules. They are further subdivided into four subfamilies, termed IFN-α, IFN-β, IFN-ω, and IFN-τ (Table 3). All genes and proteins comprising this family are related to each other structurally. The genes form a cluster (in the human species located on the short arm of chromosome 9), and it is widely believed that all type I IFN genes evolved from a single ancestral gene (42,259). IFN-β (earlier known as fibroblast IFN), the most divergent member of this family, shows about 25% to 30% homology with the numerous members of the IFN-α subfamily (earlier known as leukocyte IFN) at the amino acid level and about 45% homology in the coding sequence at the DNA level (225). There is also significant homology in the 5'-flanking regulatory regions of the IFN-α and IFN-β genes, reflecting the fact that they are often coordinately regulated (40). It has been estimated that the split between the IFN-α and IFN-β genes occurred between 200 and 400 million years ago, i.e., probably before the emergence of vertebrates (259). The split between the more closely related IFN-α and IFN-ω subfamily genes and the closely interrelated numerous IFN-α subspecies is thought to have occurred at least 85 million years ago (42).

Four conserved cysteines (positions 1, 29, 98, and 138) found in all human IFN-α proteins are thought to form two intramolecular disulfide bonds (Cys1-Cys98 and Cys29-Cys138) (260). Because the first four amino acid residues can be removed without significantly reducing biological activity, the integrity of the Cys1-Cys98 bond is not essential for IFN function (259). Cysteine residues in similar positions are preserved in IFN-α proteins of other animal species. Human IFN-α proteins lack potential *N*-glycosylation sites, and most members of the IFN-α sub-

TABLE 2. *A partial list of important cytokines, grouped according to their predominant function*

Group[a]	Name	Abbreviation	Subgroups or subtypes
Antiviral cytokines	Type I Interferon	IFN-α/β	IFN-α, IFN-β, IFN-ω, IFN-τ
	Type II Interferon	IFN-γ	
Inflammatory cytokines	Tumor necrosis factor	TNF	TNF-α Lymphotoxin (LT-α or TNF-β) Lymphotoxin-β (LT-β) Fas ligand
	Interleukin 1	IL-1	IL-1α, IL-1β, IL-1 receptor antagonist (IL-1ra)
	Interleukin 6	IL-6	
Chemotactic cytokines	Chemokines (interleukin 8 family)	IL-8	Many structurally related proteins (α and β subgroups)
Regulators of lymphocyte functions	Interleukin 2	IL-2	
	Interleukin 4	IL-4	
	Interleukin 5	IL-5	
	Interleukin 7	IL-7	
	Interleukin 9	IL-9	
	Interleukin 10	IL-10	
	Interleukin 12	IL-12	
	Interleukin 13	IL-13	
	Transforming growth factor β	TGF-β	TGF-β1, 2, 3
Hematopoietic colony-stimulating factors (CSFs)	Interleukin 3	IL-3	
	Macrophage CSF	M-CSF or CSF-1	
	Granulocyte-macrophage CSF	GM-CSF	
	Granulocyte CSF	G-CSF	
	Interleukin 11	IL-11	
	c-kit ligand	KL	

[a] Because almost all cytokines have multiple actions, the division into these groups is arbitrary. Most of the cytokines included in the table could be listed in more than one group.

families in their native state are not *N*-glycosylated (170). However, murine IFN-α proteins contain an *N*-glycosylation site. Three cysteine residues are present in human IFN-β, of which Cys41 and Cys141 form a disulfide bridge and Cys17 is free. Unlike the human IFN-α proteins, IFN-β has a potential *N*-glycosylation site at position 80, and the mature protein is known to be glycosylated (116). Although crystals of recombinant human IFN-α protein have been generated, results of x-ray crystallographic analysis have not yet been reported. The tertiary structure of IFN-β has been analyzed by nuclear magnetic resonance (NMR) spectrometry and x-ray crystallography (238,239). A model of the IFN-β molecule comprising five α-helices has been proposed. Both IFN-α (25) and IFN-β (240) appear to be active in monomeric form.

It is interesting that all animal species examined have large IFN-α subfamilies, but most have only one IFN-β gene. An exception are the ungulates, with at least five interrelated IFN-β genes identified in the cow, pig, horse, and blackbug (134,265). A distinct IFN-α/β subfamily termed trophoblast IFN (IFN-τ) has been identified in cattle and sheep (101). This IFN is produced in very large amounts in the epithelium of the early preimplantation embryo, and it has been implicated as a factor responsible for the preservation of the corpus luteum, needed for the suppression of the ovarian cycle and therefore essential for successful completion of pregnancy. Structurally, IFN-τ is most closely related to IFN-ω (223). Comparison of various IFN-α and IFN-β sequences among different animal species indicates that IFN-α genes and proteins are more highly conserved during evolution than are IFN-β genes and proteins, perhaps suggesting stronger constraints on the IFN-α proteins than on the IFN-β proteins (259).

Why evolution has favored the emergence of so many related type I IFN genes and proteins is not completely understood. All type I IFNs compete for binding to the same receptor, and in most experimental systems examined they generally exert similar biological activities, including multiple inhibitory effects on virus multiplication (42,171). However, some quantitative and even qualitative differences in biological actions have been seen with different type I IFN proteins. When the antiviral actions of various type I IFNs were examined in cells of heterologous species (e.g., human IFNs in rat or bovine cells), it was found that some types of cells were responsive to IFN-β but not to IFN-α, or vice versa (44,52). Differences also have been seen in the ability of individual recombinant IFN-α subspecies to inhibit the growth of tumor cell lines (61) or to produce activation of natural killer cells (162). These observations have not been explained. It is possible that there exist variants of the multicomponent type I IFN receptor (see below) which can selectively recognize different IFN-α/β proteins, even though the two known components of the receptor appear to be essential for responsiveness to all type I IFNs (156,160,242). Although IFN-τ appears to function as a factor that promotes embryo implantation in some animal species, it too binds to the type I IFN receptor and mediates antiviral activity (85,188). It is possible that IFN-τ differs from other members of the IFN-α/β family, mainly in the site, time, and magnitude of expression rather than in biological activity inherent in its molecular structure. Disruption of a subunit of the IFN-α/β receptor in mice did not affect embryonic development, suggesting that at least in the murine species type I IFN proteins are not essential for embryogenesis (156).

TABLE 3. *Classification and major features of the interferons*

	Type I IFN (IFN-α/β)	Type II IFN (IFN-γ)
Subfamilies	IFN-α (at least 14 potentially functional genes in humans) IFN-β IFN-ω IFN-τ	—
Structural genes	Chromosome 9 (human) Chromosome 4 (mouse) No introns	Chromosome 12 (human) Chromosome 10 (mouse) Three introns
Proteins[a]	IFN-α 165-166 amino acids (a.a.) IFN-β: 166 a.a. IFN-ω: 172 a.a. IFN-τ: 172 a.a.	146 a.a. (forms dimer)
Receptors	Genes for two chains located on chromosome 21 (human) and chromosome 16 (mouse) Additional components?	Gene for α chain on chromosome 6 (human) and chromosome 10 (mouse) Gene for β chain on chromosome 21 (human) and chromosome 16 (mouse) Additional components?
Major functions	Antiviral actions Regulation of cell growth and differentiation Induction of MHC class I antigens Embryo implantation in uterus (IFN-τ)	Macrophage activation Induction of MHC class I and II antigens Antiviral actions

[a]Polypeptide lengths refer to mature proteins, as predicted from cDNA sequences, after removal of cleavable signal peptide sequences. Some natural IFN proteins are known to undergo C-terminal processing so that shorter forms may be generated. Many IFN proteins are *N*- and *O*-glycosylated.

Although all type I IFNs are similar in their biological activities, they differ from one another in the regulation of their expression. In the human species, IFN-β is the predominant species, produced by various nonhematopoietic cells after virus infection or stimulation with double-stranded RNA (92). In contrast, cells of hematopoietic origin tend to produce more readily IFN-α and IFN-ω, together with variable amounts of IFN-β, after different forms of stimulation (42). Relative levels of induction of individual IFN-α subspecies also vary, depending on both the cell type and inducing stimulus. Thus, it appears that in the intact organism the relative biological significance of the individual members of the IFN-α/β superfamily is determined primarily by the site and magnitude of their production rather than by any unique functional properties intrinsic to their structural characteristics.

Type II Interferon

There is no obvious structural homology between type I and type II IFNs. The only reason why the protein now called IFN-γ or type II IFN was proposed to be named "interferonlike" (261) was because of its functional similarity to the type I IFNs, as exemplified by its ability to induce a characteristic antiviral state. Although type I and type II IFNs bind to distinct cell surface receptors, the signal transduction pathways activated by the IFN-α/β and IFN-γ receptors partly overlap (Fig. 1). This recently discovered overlap in intracellular signaling explains why many cellular genes are activated by both IFN-α/β and IFN-γ (205). It also explains why these structurally unrelated cytokines share so many biological functions (Table 4).

In contrast to the multiple type I IFN genes and proteins, a single IFN-γ gene has been found in all mammalian animal species examined. Also, unlike the intronless type I IFN genes, the IFN-γ gene contains three introns (Table

TABLE 4. *Major biological actions of interferons*

Activity	Observed with IFN-α/β	Observed with IFN-γ
Induction of antiviral state	+	+
Inhibition of cell growth	+	+
Induction of class I MHC antigens	+	+
Induction of class II MHC antigens	±	+
Activation of monocytes/macrophages	+	+
Activation of natural killer cells	+	+
Activation of cytotoxic T cells	+	+
Modulation of Ig synthesis in B cells	+	+
Induction of F_c receptors in monocytes	−	+
Inhibition of the growth of nonviral intracellular pathogens	−	+
Pyrogenic action	+	+

+, positive effect; ±, weak or variable effect; –, negative. Adapted from ref. 252.

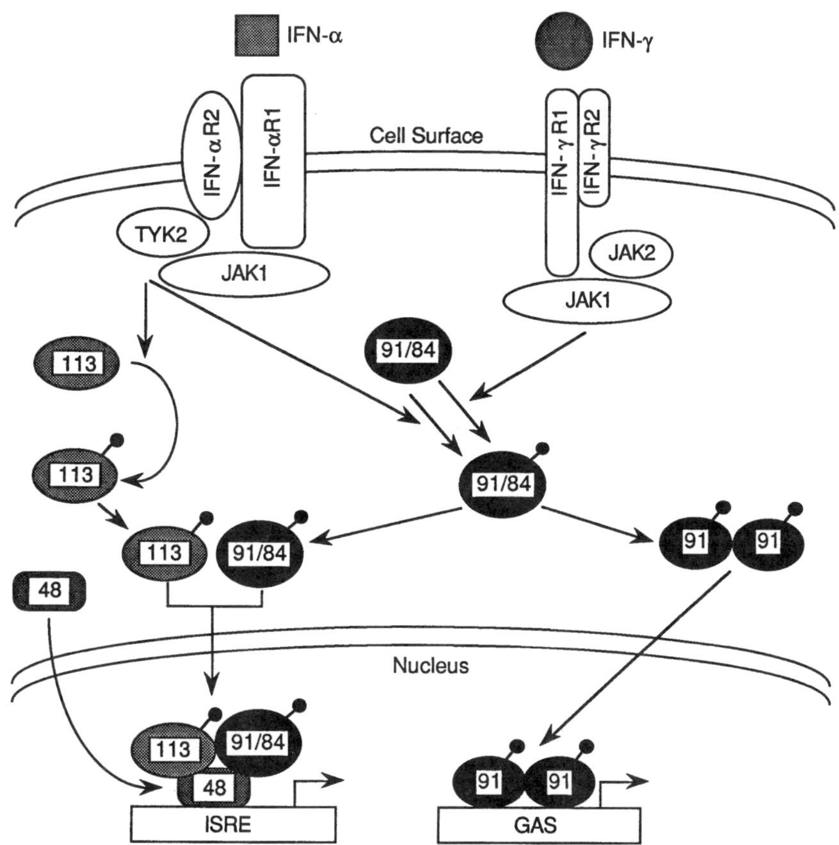

FIG. 1. Partially overlapping signal transduction pathways used for gene induction by IFN-α and IFN-γ. IFN-α and IFN-γ bind to different receptors. Two subunits of the IFN-α receptor and two subunits of the IFN-γ receptor have been identified so far. (IFN-β and other type I IFNs compete for binding to the same receptor as IFN-α.) Binding of IFN-α to its receptor leads to tyrosine phosphorylation of two essential tyrosine kinases, Tyk2 and JAK1, which is followed by tyrosine phosphorylation of the STAT proteins, p113, p91, and p84 (these three proteins are also termed ISGF-3α proteins). The phosphorylated STAT proteins and p48 (also termed ISGF-3γ) translocate to the nucleus to form the ISGF-3 complex that binds to the *cis*-element, ISRE, present in most IFN-α– and IFN-β–inducible genes. Binding of IFN-γ to its receptor leads to tyrosine phosphorylation of JAK1 and another tyrosine kinase, JAK2, resulting in the phosphorylation of STAT p91. Dimeric phosphorylated p91 translocates to the nucleus and binds to the GAS element present in many IFN-γ–inducible genes.

3). The gene encodes a 143– and 134–amino acid long basic mature protein in the human and murine species, respectively, with only ~40% homology between the two species (74). (As a result of this relative lack of conservation, IFN-γ tends to be strictly species specific in its actions.) Both human and murine IFN-γ proteins are *N*-glycosylated at two sites (111,186). Mature IFN-γ protein contains no cysteines, although two cysteine residues are present in the 23–amino acid long cleavable signal peptide sequence. Recombinant human and rabbit IFN-γ have been crystallized and their atomic structures have been determined by x-ray crystallographic analysis (198,233). In agreement with the prediction made on the basis of studies with natural IFN-γ (269), x-ray crystallographic analysis indicates that the active IFN-γ molecule is a homodimer. The sites binding to the receptor are likely to contain segments from the N and C termini of the two subunits (233).

Other Cytokines

Tumor Necrosis Factor

TNF was originally identified as a mediator of bacterial lipopolysaccharide (LPS)-induced hemorrhagic necrosis of tumors in animals (21). When TNF was purified and its gene cloned many years later, it was found that it is ~30% homologous at the amino acid level to lymphotoxin, a product of activated T cells (73,168). Lymphotoxin (now variously termed LT-α or TNF-β by different investigators) is the only known member of this family with characteristics of a secretory protein, as judged from the presence of a hydrophobic cleavable signal peptide sequence (158). TNF-α is synthesized as a type II membrane protein, having an extracellular, transmembrane, and intracellular domain. The 17-kD released form of TNF-α is derived from the integral transmembrane precursor by proteolytic cleavage at a membrane-proximal extracellular site (125). Numerous other members of the TNF family have now been identified (including LT-β and Fas ligand; see Table 2), all of which are type II transmembrane proteins (15,212). However, unlike TNF-α, no released forms of the other members of this family have been identified, suggesting that the extracellular portions of these transmembrane proteins function as signaling molecules during direct cell-cell interactions. Whether the intracellular domains of the TNF-α protein and related members of this family play a role in signaling is not known.

TNF-α and the released form of lymphotoxin (LT-α) bind to the same ubiquitously expressed cell surface receptors (to be reviewed below), and the spectrum of their varied biological activities is virtually indistinguishable (254). However, whereas TNF-α is produced by monocytes, macrophages, and a variety of other types of cells, LT-α is produced only by lymphocytes. In addition, LT-α, but apparently not TNF-α, forms a heterotrimeric complex with another member of this family (LT-β) on the surface of LT-β–expressing cells (19). The LT-α/LT-β heterotrimer, Fas ligand, and some other recently identified members of this cytokine family do not bind to TNF receptors, but recognize separate specific cell surface receptors (15,212).

Interleukin-1

This family consists of three gene products, two of which are agonists closely related in structure and function (IL-1α and IL-1β), whereas the third binds to the same IL-1 receptors but acts as a receptor antagonist (IL-1ra). Both IL-1α and IL-1β are initially synthesized as 31-kD precursors from which the 17-kD mature proteins are derived by proteolytic cleavage (49). Both forms lack a cleavable signal peptide sequence, and much of the newly synthesized IL-1 (especially IL-1α) remains cell associated, mostly in the cytoplasm (192). Intracellular cleavage of the precursor forms appears to be a regulated process, and proteases that cleave either IL-1α or IL-1β have been described (118,123). The IL-1β precursor completely lacks biological activity, and an IL-1β–converting enzyme that is coinduced with the precursor has been identified (16). How the IL-1 proteins are secreted from the cell is still not completely understood. IL-1α and IL-1β are about 25% homologous to each other in the human, murine, rabbit, and bovine species, and a high degree of amino acid conservation (60% to 80%) exists in evolution for both IL-1 forms (49). Human IL-1ra shows about 26% homology to IL-1β and 19% to IL-1α (54).

IL-1α and IL-1β bind to the same cell surface receptors, ubiquitously expressed on most types of cells (see below), and their biological actions are therefore indistinguishable (48). IL-1ra is unique among naturally occurring proteins in that it too binds the same receptors as the other members of the IL-1 family (although its affinity for one of the two IL-1 receptors is lower than that of IL-1α or IL-1β), but it fails to produce receptor activation and signaling (84). Instead, the binding of IL-1ra to the IL-1 receptor prevents IL-1α or IL-1β from interacting with the same receptors, leading to the suppression of IL-1 activity. It appears that the generation of IL-1ra in the intact organism represents a unique mechanism of negative regulation of IL-1 actions. Both IL-1α and IL-1β are produced by a variety of cells in response to stimuli generated during infections with bacteria, protozoa, or viruses. Despite a lack of structural relatedness, many similarities exist between the biological actions of IL-1 and TNF (48,128). Although much information about the signal transduction pathways activated by the binding of TNF and IL-1 to their respective receptors is still lacking (see below), the existence of so many similarities in their actions suggests common intracellular components.

Interleukin 10

Originally called "cytokine synthesis inhibitory factor" (59,152), IL-10 is best known for its ability to suppress the synthesis of IFN-γ, IL-2, and other cytokines in T-lymphocytes and to inhibit the production of several cytokines (including TNF-α, IL-1α, IL-1β, IL-6, and IL-8) in monocytes and macrophages. In addition, IL-10 is a potent B-cell growth factor (150). The human and murine IL-10 genes both encode 178–amino acid long polypeptides, including cleavable signal peptide sequences (250). Both human and murine IL-10 genes show a high degree of homology to an open reading frame in Epstein-Barr virus (EBV) termed BCRF1 (152,153). The viral BCRF1 protein shows biological activity similar to that of IL-10 (152). The ability of EBV to produce "viral IL-10" is representative of the many recently discovered mechanisms developed by viruses as a means to inhibit or emulate cytokine actions, to be described in a separate section of this chapter.

Interleukin 12

IL-12 is the only heterodimeric cytokine known, composed of two covalently linked chains (p35 and p40) encoded by separate genes (231,232). The p40 chain shows some homology to the α chain of the IL-6 receptor, raising the possibility that IL-12 evolved from a cytokine related to IL-6 that is covalently linked to its receptor.

IL-12, originally known as natural killer (NK) cell stimulatory factor, has pleiotropic effects on NK and T cells. Its major action is the stimulation of IFN-γ synthesis by NK cells and T-lymphocytes. In fact, recent evidence suggests that IL-12 is the major regulator of IFN-γ synthesis in the intact organism.

INDUCTION OF CYTOKINE SYNTHESIS

Type I Interferons

Cells do not synthesize IFNs unless they are activated by an IFN-inducing agent. IFN-α or IFN-β synthesis can be induced in almost all cell types, with the undifferentiated embryonic carcinoma cells being the most notable exception. Virus infection is the most common biological cause of IFN induction although other agents such as bacteria, mycoplasma and protozoa, as well as some of their constituents, can also induce IFN synthesis in certain types of cells, especially in mononuclear phagocytic cells. Bacterial LPS is a potent inducer of IFN-α/β in monocytes and macrophages. Several chemically defined low molecular weight compounds have been identified as IFN-α/β inducers in the intact animal (42). The mechanisms of IFN induction by nonviral inducers are not well understood. Viral induction of IFN synthesis is thought to be mediated by double-stranded RNA (dsRNA). dsRNA is not a usual constituent of an uninfected cell but it is produced during the replication of many RNA and DNA viruses either as an obligatory intermediate or as a side product. dsRNA introduced into a cell either by transfection or as the genome of a defective virus particle can strongly induce the synthesis of IFN-α/β. Consequently, many of the studies exploring the molecular mechanism of IFN synthesis have used exogenous dsRNA, such as poly(I)•poly(C), rather than virus infection as the IFN-inducing agent (42). However, not all forms of IFN induction by viruses can be attributed to dsRNA. In mononuclear leukocytes IFN production was shown to be triggered by viral envelope glycoproteins (42). Organs of normal individuals contain small amounts of constitutively expressed IFN messenger RNA (mRNA) and protein (230).

Synthesis of IFN-α/β is regulated at both transcriptional and posttranscriptional levels (140). The transcriptional induction involves both derepression of the genes by constitutive factors and activation by newly induced transcription factors. The posttranscriptional regulation is at the level of stabilization of IFN mRNAs, which have short half-lives due to the presence of destabilization sequences in their 3'-untranslated regions. Inhibitors of protein synthesis cause superinduction of IFN mRNA synthesis, presumably by blocking the synthesis of putative repressor proteins that are coinduced with IFNs. IFN induction is also enhanced by pretreatment of the cells with a low dose of IFN, a phenomenon called priming. Positive interactions between transcription factors activated by dsRNA and by IFN may play an important role in priming.

The *cis-* elements and the corresponding *trans*-acting factors that regulate transcription of the IFN-α and IFN-β genes have been studied extensively (129,205). These studies have shown that the underlying mechanism is complex and involves both positive and negative regulatory factors (Fig. 2). The regulatory region of the human IFN-β gene, for example, contains four positive regulatory domains PRDI to PRDIV and a negative regulatory element (70, 159). Some of these elements structurally overlap. The PRDI site has resemblance to the IFN-stimulated response element (ISRE) in IFN-responsive genes, and multimerized PRDI can respond not only to virus infection but also IFNs. Several protein factors that can bind to PRDI have been identified. These include IRF-1 and IRF-2, two structurally related host proteins that recognize the same DNA sequence but interact differently with other proteins involved in transcription (86,224). Consequently, IRF-1 stimulates transcription, whereas IRF-2 represses it. A third member of this family is ICSBP, an IFN-γ induced protein expressed primarily in cells of the immune system (50).

IRF-1 synthesis is induced by IFN-α, IFN-γ, TNF, IL-1, and IL-6 in many cell lines (67,90) and by prolactin in a lymphoma line (271). The transcription factor STAT (signal transducers and activators of transcription) p91 (see below) has been implicated in the induction of IRF-1 synthesis (174). The functional activity of the IRF-1 protein, on the other hand, appears to be regulated by a posttranslational process that is induced by virus infection but not by IFNs (256). The importance of IRF-1 in IFN-β induction depends on the system examined. Although the lack of IFN induction in undifferentiated embryonal carcinoma cells can be correlated with a deficiency of IRF-1 activation (88), IFN-β induction by virus infection of differentiated murine embryonic stem cells was not affected by the loss of both IRF-1 alleles (196). In cells of mice lacking the IRF-1 gene, IFN-β (and IFN-α) induction by dsRNA is decreased, but induction by virus is normal (143). Induction of nitric oxide synthase and of guanylate binding protein mRNAs by IFN-γ, however, is blocked in cells from IRF-1–less mice demonstrating the need for IRF-1 in these induction processes (107,114). There is strong experimental evidence to indicate that IRF-1 can have a major role in the regulation of growth of certain cell types. For example, the B cells of transgenic mice constitutively expressing IRF-1 do not survive (268). On the other hand, overexpression of IRF-2 in NIH 3T3 cells causes cellular transformation that is reversed by concomitant overexpression of IRF-1 (87). Thus, the two proteins antagonize each other in regulating the rate of cell proliferation. The possible role of IRF-1 as a tumor suppressor is also suggested by the frequent deletion of the genetic locus encompassing this gene in certain types of human leukemia (264).

Another protein that binds to the PRDI element is PRDI BF1 (112). This protein does not belong to the IRF-1 family; it has five zinc finger motifs and it is a potent repressor of the IFN-β gene. It is induced by virus infection. The PRDII site binds the NF-κB transcription factor, which is activated by dsRNA among many other agents (255). Induction of the IFN-β gene through the PRDII site requires another protein, HMGI (y). HMGI (y) also stimulates the binding of a virus-inducible complex containing ATF-2 to the PRDIV element (51). Thus, optimum transcription of

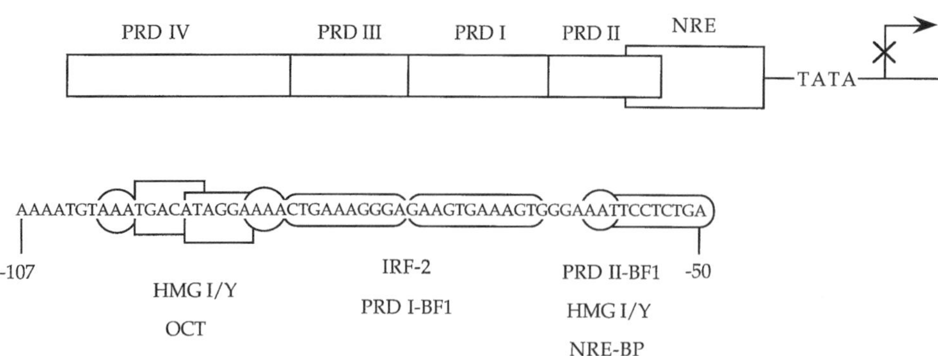

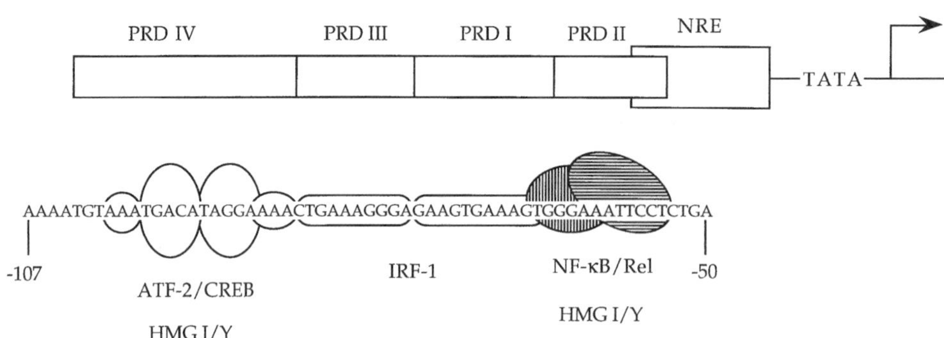

FIG. 2. A model for transcriptional induction of the human IFN-β gene. The *cis*-acting elements and the sequence of the human IFN-β promoter from –50 to –107 (+1 is the transcription start site) are shown. Transcription factors that may bind to different elements in this region (PRD I–IV, positive regulatory domains; NRE, negative regulatory element) are indicated below the sequence. (**A**) Uninduced cells. (**B**) Virus-induced cells. (Diagram courtesy of Dr. John Hiscott.)

the IFN-β gene requires the assembly of a multiprotein complex, many of whose components are activated by dsRNA or virus infection. Overlapping the PRDII element is an 11-bp negative element (NRE) that acts as a position-independent silencer of PRDII. The IFN-α genes are also induced by virus infection. However, these genes do not contain any NF-κB binding sites, although PRDI-like elements are present in them. As a result, the induction of IFN-α genes requires activation of PRDI binding proteins but not of NF-κB. Activation of additional factors that bind to the sequence GAAATG is also required for this process (138).

Type II Interferon

Synthesis of IFN-γ is largely restricted to T-lymphocytes and the large granular lymphocytes or NK cells (42). In general, agents that promote T-cell activation induce IFN-γ synthesis. In the body, it is stimulated by antigens to which the organism has been presensitized. Infection with bacteria and protozoa leads to a stimulation of IFN-γ production early in the infectious process, which is mediated by monocyte/macrophage-derived IL-12; IL-12 then acts on NK cells and T cells to promote IFN-γ synthesis (231,232). Stimulation of IFN-γ production during virus infection has been documented (26), but it is not clear if virus-induced IFN-γ production is IL-12 mediated or based on specific antigenic stimulation. Nonspecific T-cell activators such as phytohemagglutinin and concanavalin A also can serve as IFN-γ inducers. Many stimuli result in a coordinate enhancement of IFN-γ and IL-2 synthesis in T-lymphocytes. IFN-γ production is increased by IL-1, IL-2, some growth factors, estrogens, and IFN-γ itself (89), but downregulated by glucocorticoids, transforming growth factor β (TGF-β) and IL-10 (152,270). The induction of IFN-γ is regulated primarily at the transcription level. DNAse I hypersensitive sites are located in the 5'-flanking region and in the introns of the IFN-γ gene (28,270). A combination of positive and negative regulatory elements is thought to govern the lymphocyte-specific expression of the IFN-γ gene, but the details of this process have not yet been elucidated.

Induction of Other Cytokines by Viruses

Type I IFNs are not the only cytokines induced in the process of virus infection. Almost 30 years after the first demonstration that IFNs are virus-induced proteins, investigators have begun to analyze the induction of a variety of other cytokines by diverse viruses, both in cell culture and in the intact organism (Table 5). The mechanisms responsible for the induction of these cytokines by viruses have not been investigated in such detail as the induction of type I IFN (20). One extensively studied model of cytokine stimulation involves the roles of the Tax and Rex proteins of human T-lymphotropic virus type 1 (HTLV-1) in T-cell activation and immortalization mediated by induction of IL-2 and the IL-2 receptor (76). Studies on the mechanism of IFN-β induction (see above) showed that some viruses can produce NF-κB activation. The promoters of IL-6 (272), IL-8 (139), and some other cytokines contain NF-κB elements that are essential for induction by a variety of stimuli and are also likely to be involved in the activation of these genes by virus infection. What other virus-inducible transcription factors might be involved in the induction of cytokine expression by viruses is not known. It is interesting that poly(I)•poly(C) was shown to act as an inducing stimulus for the synthesis of several cytokines, including IL-6 (246,258), IL-8 (244), granulocyte-macrophage colony-stimulating factor (GM-CSF) (244), TNF-α, IL-1β (68), and monocyte chemotactic and activating factor (MCAF) (131). These results suggest that dsRNA, formed intracellularly in the process of virus infection, plays a role in the induction of these cytokines by viruses.

RECEPTORS AND SIGNAL TRANSDUCTION

Type I and Type II Interferons

Receptors

IFNs bind to specific receptors on the cell surface and elicit the signals necessary for transcriptional induction of the IFN-activatable genes (56,204). IFN-α and IFN-β compete for the same receptors, whereas IFN-γ signals through a different receptor. Binding of IFNs to receptors and hence their cellular actions are, in some instances, species specific; species specificity is more pronounced with IFN-γ than with type I IFNs. The type I IFN receptor to which IFN-α and IFN-β bind is present in low abundance on all major types of cells (2–5 x 10^3 binding sites/cell). Its affinity for the ligand is on the order of 10^{-10}M. The human type I receptor is encoded by a locus on chromosome 21. One gene in this locus has been shown to encode a glycopro-

TABLE 5. *Viruses activate cytokine networks*

Virus	Cell culture or animal host	Cytokine(s) Induced	References
Influenza (and many others)	Chick embryo cells (and many others)	IFN	103
Sendai	Human mononuclear leukocytes	TNF-α	3
VSV, EMC, adenovirus 2, herpes simplex	Human leukocytes	TNF-α, TNF-β	267
Influenza	Mice (bronchoalveolar washings)	TNF, IL-1	243
Influenza	Mouse splenocytes	IFN-γ, IL-2, IL-4, IL-5	72
NDV, Sendai, mengo, measles, rubella	Human fibroblasts	IL-6, GM-CSF, IFN-β	246
Measles	Human fibroblasts	IL-8	244
HIV-1	Human mononuclear cells	TNF-α, IL-1	145
HIV-1	Human monocytes	IL-6	157
HIV-1	Human myelomonoblastic cells	IL-1β, TNF-α	34
MuLV	Mouse organs	TNF-α, IL-1, IFN-γ	26
HTLV-1	Human T cells	IL-2	76
HTLV-1	Human T cells	TNF-β	235
NDV	Astrocytes	TNF-α, IL-6	137
Human CMV	Human promyelocytic cells	TNF-α, IL-1β, IL-8, M-CSF	53
Herpes simplex	Mouse macrophages	IL-1α, IL-1β	216
Human herpes virus 6, HSV-1, EBV	Human mononuclear cells	TNF-α, IL-1β, IL-6	71
LCM	Mouse brain	IL-6, IFN-γ, IL-4, IL-5, GM-CSF	154
Hepatitis B	Human fibrosarcoma cells	IL-8	139

tein of 554 residues, which, when expressed in mouse cells, confers responsiveness to certain subtypes of IFN-α (242). This cell surface protein has a single transmembrane domain and it represents the first member of the human IFN-α receptor family. By itself, however, it cannot make cells to respond to human IFN-β and several IFN-α subtypes. Additional proteins, also encoded by genes in chromosome 21, are probably necessary for eliciting the full response to type I IFNs. Several alternative or ancillary candidate proteins have been claimed to serve as IFN-α receptors. For example, type 2 complement receptor or a related protein has been reported to be the major IFN-α–binding protein in B lymphoma cells (41). Biochemical and immunological experiments strongly indicate that the physiological receptor is a multisubunit protein whose composition and response spectra may vary among different cell types (30). The cloning of another chain of the human IFN-α/β receptor, thought to represent the main binding component of this receptor, was recently reported (160). The latter receptor chain is a disulfide-linked dimer, consisting of two 51-kD subunits. Coprecipitation of the newly identified IFN-α/β receptor with the tyrosine kinase JAK1 (see below) provided the first evidence of a physical association of the IFN-α/β receptor with an intracellular kinase required for signaling (160).

The binding component of the IFN-γ receptor is also a cell surface protein with a single transmembrane domain (5,75). The extracellular domain of this protein dictates the species-specific interaction with its cognate ligand. Experiments with chimeric human–mouse receptors showed that there is an additional species-specific step. This involves interaction with another receptor-associated protein, provisionally called the accessory protein, which is required for the full cellular response to IFN-γ. One such factor, encoded by a gene on human chromosome 21, has recently been cloned (215). This transmembrane protein, along with the earlier identified human IFN-γ receptor, confers responsiveness to human IFN-γ. Mouse cells expressing these two proteins can bind human IFN-γ and respond to it by an increased expression of major histocompatibility complex (MHC) class I genes. However, they fail to block the replication of encephalomyocarditis (EMC) virus, suggesting that an additional human accessory factor is required for the antiviral effect. In contrast, in the murine system, one accessory factor (termed the β chain), in conjunction with its cognate receptor (the α chain), appears to be sufficient for both gene induction and antiviral activities (93). Binding of IFN-γ to its receptor leads to an enhanced phosphorylation of the receptor and its dimerization (78). The intracellular domain of the receptor is responsible for its signal transduction. Within this domain, specific residues have been identified whose mutation causes an inactivation of the receptor signaling process (32,55,77).

Although type I and type II IFN receptors are distinct, their components share structural similarities that are the basis for their grouping within an IFN receptor family. This family includes all identified chains of the human and murine IFN-α and IFN-γ receptors along with a secretory protein encoded by the myxoma virus, highly homologous to the binding chain of the IFN-γ receptor (12,46, 237,241). Another member of this family is a recently identified receptor for IL-10 (94). The IFN receptor family forms a distinct subgroup (class II) of the large cytokine receptor family (12).

Signal Transduction Pathways

Great advances have been made in understanding the mechanism of transcriptional signal transduction by IFNs (Fig. 1). Unlike the pathways used by many hormones, diffusible small molecular second messengers are not involved in this process. Moreover, IFN receptors lack intrinsic kinase activities. Instead, the key events here are ligand-induced activation of receptor-associated tyrosine kinases, resultant *tyr*-phosphorylation of specific cytoplasmic proteins and their translocation to the nucleus where they bind to cognate *cis*-acting sequences of the IFN-inducible genes and promote their transcription (37,167). Thus, the signal from the receptors on the cell surface to the gene in the nucleus is physically carried by proteins that also serve as transcriptional activators. This paradigm of signal transduction has now been shown to be operative for many other cytokines and growth factors.

The transcriptional induction of genes happens within minutes of cells coming in contact with IFN-α/β (185,203, 262). The induction is transient; it usually declines after a few hours. Inhibitors of protein synthesis do not block the induction process; instead, they prolong it. In contrast, for the induction of many, but not all, IFN-γ–inducible genes, ongoing protein synthesis is required. The kinetics of induction of genes by IFN-γ is also much slower. Most IFN-α/β–inducible genes analyzed so far contain a variation of the consensus sequence called ISRE (37). This *cis*-acting sequence is present in the 5'-flanking regions of the genes and it serves as the IFN-responsive enhancer of transcription of these genes. dsRNA, which is produced during virus infection, can also induce the transcription of many of these genes through ISRE. However, the signal transduction pathways used by IFN-α/β and dsRNA are quite distinct (229). In contrast to the genes induced by IFN-α/β, all IFN-γ–inducible genes do not contain one common *cis*-acting sequence. Several distinct *cis*-acting sequences mediate the IFN-γ response of different classes of genes (37,180). As a result, their induction characteristics are also quite different. For the IFN-α/β– and IFN-γ–inducible 9-27 gene, its ISRE sequence mediates the response to all IFNs (182). (However, not all ISRE-containing genes are induced by IFN-γ.) For another IFN-γ–inducible gene encoding a guanylate-binding protein (GBP), a distinct *cis*-acting sequence, IFN-γ activation site (GAS), has been identified.

In this gene, the GAS element physically overlaps an ISRE and both elements are needed for activation of the GBP gene by IFN-γ (37). The IFN-γ response element of the indoleamine dioxygenase gene also contains an ISRE, although it is not inducible by IFN-α/β (35). In another set of IFN-γ–inducible genes, such as MHC class II and the invariant chain genes, there are no ISRE and no GAS sequences. Instead, their IFN-γ inducibility is mediated by a region encompassing the Z (or H), X, and Y box elements, which are conserved in all MHC class II genes (234).

Some of the *trans*-acting factors that mediate the IFN-α/β and IFN-γ response have been identified. Biochemical analysis of the relevant *cis*-acting sequence binding proteins and genetic analyses of mutant cells, which fail to respond to IFNs due to mutations in genes encoding proteins involved in the signal transduction pathway, have led to the identification of many of these *trans*-acting factors (37). At least three ISRE-protein complexes are formed in the nuclear extract of many IFN-α–treated human cells. They can be separated by electrophoretic mobility shift assay and are called IFN-stimulated gene factors (ISGF)-1, -2, and -3 (222). The ISGF-1 complex is found in the extracts of untreated cells as well. ISGF-2 complex formation requires ongoing protein synthesis; this complex contains the earlier identified IRF-1 protein (86), which is also involved in the induction of the IFN-β gene (see above). The functional role of ISGF-2/IRF-1 is not completely clear. Although it was thought to be involved in downregulating ISRE-mediated transcription (175), it more likely functions as a positive transcription factor either in conjunction with ISGF-3 or independently from it (114). This view is supported by the demonstration that IRF-1 is structurally related to the p48 subunit of ISGF-3 (248). Because IFN-α/β, and especially IFN-γ, induce de novo synthesis of IRF-1 (114,174), stimulation of gene expression via IRF-1 is likely to be indirect and, therefore, not as rapid as ISGF-3–mediated transcriptional activation (see below). However, as already mentioned, recent evidence indicates that IRF-1 is essential for IFN-γ–induced expression of the inducible nitric oxide synthase (iNOS) and GBP genes, which in the iNOS gene is mediated by ISRE-like sites (107,114).

ISGF-3, the ISRE-binding *trans*-acting factor that mediates the transcriptional signal of IFN-α, is composed of proteins called STAT (208). The STAT complex interacts with the IFN-α/β receptor and links it with the ISRE in the IFN-αctivatable genes (Fig. 1). It consists of three proteins, p113, p91 (or p84), and p48, all of which are present in a latent form in the cell cytoplasm. They get activated upon IFN-α treatment of cells and translocate to the nucleus to stimulate transcription. The activation process involves phoshorylation of specific tyrosine residues of the p113, p91, and p84 proteins (ISGF-3α proteins) (201,207). These three proteins are related in structure and all contain the *src*-homology 2 (SH2) domain (65). p91 and p84 are the products of alternatively spliced mRNAs of the same gene, and p113 is encoded by a related gene (200). Upon their phosphorylation, p113 and p91/p84 form a complex with p48 (termed ISGF-3γ because the synthesis of p48 is inducible by IFN-γ). After translocation to the nucleus, the ISGF-3 complex binds the ISRE sequence, thereby activating gene expression. At least two tyrosine kinases, Tyk2 and JAK1, are essential for this signal transduction process (155,249). Both of these tyrosine kinases become phosphorylated on tyrosine residues upon IFN-α treatment of cells. They in turn cause phosphorylation of the p113, p91, and p84 STAT proteins. The exact sequential order of these phosphorylation events remains to be determined. Other as yet unidentified kinases may link Tyk2 and JAK1 to the receptor on one hand and the STAT proteins on the other. However, a direct association between JAK1 and a newly identified chain of the IFN-α/β receptor was recently demonstrated (160). IFN-β is thought to activate the same signal transduction pathway as IFN-α.

The IFN-α–induced STAT pathway partially overlaps with at least one other STAT pathway used by IFN-γ for activating genes containing the GAS element. The GAS-binding factor GAF contains *tyr*-phosphorylated p91. IFN-γ stimulates phosphorylation, dimerization, and nuclear translocation of p91 (77,206). IFN-γ–mediated p91 phosphorylation does not require the presence of p113, p84, p48, or Tyk2. However, it does require JAK1 and another tyrosine kinase, JAK2, both of which are phosphorylated in response to IFN-γ (209,257). Thus, one tyrosine kinase, JAK1 and a STAT, p91, are required for signal transduction by both IFN-α/β and IFN-γ, which signal from distinct receptors. Another level of complexity and an opportunity for cross-talk between various cytokines and growth factors was shown by the finding that p91 or a closely related protein is phosphorylated in response to a variety of agents such as epidermal growth factor, erythropoietin, IL-6, growth hormone, PDGF, CSF-1, and IL-10 (37,66,197). In addition, JAK2 is associated with the receptors of erythropoietin, growth hormone, and prolactin and is activated upon binding of the respective ligands. Much work remains to be done before it is fully understood how the specificity of response to different growth factors and cytokines is maintained and how a given cytokine can synergize or antagonize the action of another member of this complex network.

Tumor Necrosis Factor

The secreted forms of TNF-α and TNF-β (the latter is also known as lymphotoxin or LT-α) bind to the same cell surface receptors. Two distinct TNF receptors have been identified: the p55 receptor (or TNF-RI) and the p75 receptor (or TNF-RII) (227). The sequences of the extracellular domains of these receptors are similar to each other and to a large family of proteins that includes the low-affinity receptor for nerve growth factor, a receptor called Fas that signals apoptosis in T cells and, notably, two poxvirus gene products, T2 and A53R, which encode secreted proteins that can bind and inactivate TNF (15,212). No significant homology exists between the intracellular domains of the two TNF receptors, nor are they related to other known protein sequences (212,227). One exception is a weak homology between intracellular domains of TNF-RI and Fas, a receptor that signals programmed cell death (226). The absence of sequence homology in their intracellular domains suggests that TNF-RI and TNF-RII generate different signals. Indeed, the presence on the cell surface of either TNF-RI or TNF-RII alone is sufficient for high-affinity binding of TNF-α or TNF-β. Moreover, each of these receptors can elicit biological responses without the contribution of the other receptor. However, the types of biological responses generated by TNF binding to RI and RII are quite different. RI signals the majority of the many diverse TNF actions, including cytotoxicity, increased cell proliferation, and antiviral activity, whereas TNF-RII appears to be important in some actions of TNF on cells of hematopoietic origin, e.g., in the stimulation of thymocyte proliferation (18,226,227). Because RII binds TNF with a high affinity, it has been postulated that an important function of RII was to deliver TNF to RI (227).

The expression of RI and RII receptors on different cells varies, but in most cell lines both receptors are present, albeit in proportions that differ from one cell to another (43). X-ray crystallographic analysis of the complex of a soluble TNF-RI with TNF-β showed that three receptors bound one TNF-β trimer, consistent with the evidence that crosslinking of three RI molecules is needed (and sufficient) for the initiation of the biological responses (9). An ~80–amino acid domain near the C terminus of the intracellular region of RI (the Fas homology region) was identified as being essential for the antiviral and cytolytic activities (226).

TNF is one of the most pleiotropic cytokines, with a variety of important biological actions ascribed to it (14,161, 194). TNF action results in the activation or de novo synthesis of several major transcription factors, including NF-κB and AP-1. A very large number of target genes is known to be induced, whereas some genes are suppressed by TNF in target cells (254). Although these and other TNF actions have been extensively studied, the intracellular signaling pathways that mediate TNF actions are not well understood. One of the early events seen in TNF-treated cells is an increased phosphorylation of a set of intracellular proteins, but the mechanisms involved have not been fully explained (83,178,247). MAP kinase activation has been demonstrated in TNF-treated cells, but the events occurring farther upstream are not yet known (247,251). Components of the STAT complex and the JAK family tyrosine kinases (see above), important in the actions of the IFNs and numerous other cytokines (208), have not been implicated in TNF actions. It has been proposed that the lipid ceramide, generated by sphingomyelin hydrolysis, acts as a second messenger that activates a ceramide-dependent protein kinase to transduce the cytokine signal (120). The basis for this hypothesis is that TNF can increase ceramide concentration in some cells, and some TNF actions are mimicked by the treatment of cells with synthetic ceramide. A ceramide-activated protein kinase has been partially characterized.

Interleukin 1

Two IL-1 receptors have been identified. The 80-kD IL-1R1 is expressed predominantly on fibroblasts, T cells, and endothelial cells. The 68-kD IL-1R2 is the predominant IL-1 receptor on B cells and monocytic cells (49,144,210). IL-1R1 mediates a variety of IL-1 actions, whereas IL-1R2, which contains only a 29–amino acid long intracytoplasmic domain, probably is not a signaling receptor. It has been proposed that the role of IL-1R2 may be to act as the "decoy" that traps IL-1 (31,210). IL-1α and IL-1β bind to both receptors with a similar affinity, whereas the antagonist IL-1ra (see above) binds with much higher affinity to IL-1R1 (8).

IL-1 shares with TNF a number of important pleiotropic actions, but, as is also true for TNF, the molecular signaling pathways activated by IL-1 are not well understood (49). Although activation of many kinases and the induction of numerous potential second messengers have been reported in a variety of cells after exposure to IL-1, it is not clear what couples the IL-1 receptors to intracellular signaling components farther downstream. Ceramide is a potential second messenger for IL-1 as well as TNF (120). Recently, the C-terminal region of the intracellular domain of IL-1R1 was shown to be required for the IL-1–induced activation of the transcription factor NF-κB (135).

ACTIONS OF CYTOKINES ON VIRUS REPLICATION

Interferon-Induced Antiviral Pathways

Although the antiviral actions of the IFNs have been investigated for a long time, the detailed mechanism of their effects against most viruses remains elusive (199,205,220). However, several general features of these actions have emerged. First, different antiviral mechanisms are mediated by different IFN-induced proteins. Second, for the functioning of some of these antiviral pathways, viral gene products, such as dsRNA, are obligatory components. Third, for inhibiting the replication of different families of viruses, different antiviral pathways may be responsible. Fourth, one or more steps in the virus life cycle, namely viral penetration and uncoating, transcription of viral mRNA, translation of viral proteins, replication of viral genome, and assembly and release of progeny virions, can be inhibited in an IFN-treated cell. Which of these processes will be the primary target depends primarily on the nature of the virus; the host cell type, however, can also play a role. For some viruses, studies with partially IFN-responsive cell lines, in which the full repertoire of IFN-inducible genes is not expressed, led to the identification of specific IFN-inducible proteins responsible for inhibiting virus replication. The two best examples are the IFN-inducible gene products responsible for inhibiting orthomyxoviruses and picornaviruses. The IFN-inducible Mx family of proteins is responsible for inhibiting influenza viruses (220), whereas the dsRNA-dependent 2'-5' oligoadenylate/RNAse L pathway is responsible for inhibiting members of the picornavirus family (91).

The steps of viral multiplication that are affected by IFNs have been identified for several families of viruses (Table 6) (199,205). As mentioned above, more than one step is usually affected, and the degree of inhibition of a given step varies among different host cell types. With vesicular stomatitis virus (VSV), both primary transcription of viral mRNAs and their translation are impaired in IFN-treated cells. With encephalomyocarditis virus (EMCV) and mengovirus, translation and steady-state level of viral mRNAs are reduced, whereas with reovirus, translation of viral protein is affected. Synthesis and translation of primary viral RNAs are also affected with influenza virus. Retroviruses are affected at two stages. Early steps of viral DNA synthesis preceding provirus integration are impaired in IFN-treated cells. In cells already carrying integrated provirus, IFN affects assembly and release of progeny retroviral particles by interfering with their proper budding. Among the DNA viruses, IFNs affect early steps of SV40 infection, viral RNA metabolism is affected in vaccinia virus–infected cells, and both transactivation of immediate early genes and release of mature virions are affected with herpes simplex virus (HSV).

Two IFN-induced enzymatic pathways lead to an inhibition of protein synthesis. Both pathways require dsRNA, a viral by-product, for their activation. One pathway causes degradation of mRNA through the action of the IFN-induced enzyme 2'-5' oligoadenylate [2-5(A)] synthetase and RNAse L, whereas the other causes a block in translation initiation due to the action of the IFN-induced protein kinase PKR (Fig. 3).

2-5(A) Synthetase/RNAse L Pathway

IFNs induce the synthesis of a family of enzymes, called 2-5(A) synthetases, which polymerize ATP into 2'-5'–linked oligoadenylates of various lengths (205,263). These enzymes are inactive in the absence of the only known cofactor, dsRNA. Their product, 2-5(A), activates a latent ribonuclease, RNAse L, present in all animal cells. Activated RNAse L can cleave single-stranded RNAs. Most cells also contain a phosphodiesterase that can hydrolyze 2-5(A). The 2-5(A) synthetase/RNAse L pathway is responsible for inhibiting the replication of picornaviruses in IFN-treated cells. However, this pathway is neither necessary nor sufficient for inhibiting the replication of other RNA viruses such as VSV. The most cogent evidence establishing the function of this pathway has come from molecular genetic studies. Constitutive expression of a transfected 2-5(A) synthetase complementary DNA clone was shown to be sufficient, without IFN treatment, for inhibiting picornaviruses (24). However, a high level of 2-5(A) synthetase is not effective if the cellular level of RNAse L is too low (126) or if the function of RNAse L is blocked by a mutant RNAse L that acts as a transdominant inhibitor (91). Thus, both enzymes of the pathway are required for antiviral action. The necessary activation of the IFN-induced 2-5(A) synthetases is thought to be performed by viral dsRNA produced as an intermediate of picornavirus replication. Such viral dsRNA is complexed with 2-5(A) synthetase in IFN-treated virus-infected cells, which was demonstrated by their coimmunoprecipitation (80). The immunoprecipitated complex is enzymatically active without any added exogenous dsRNA. In line with these observations, accumulation of 2-5(A) and the resultant degradation of RNA have been demonstrated in IFN-treated EMCV-infected cells. Thus, there is overwhelming evidence to conclude that the 2-5(A) synthetase/RNAse L pathway is the major, if not the only, pathway that is physiologically used by the IFN system to inhibit the replication of picornaviruses. The possible role of this pathway in inhibiting the replication of viruses of any other family remains to be evaluated.

The 2-5(A) synthetases fall into three size classes of proteins encoded by three sets of IFN-inducible genes (205). The smallest family members have molecular weights in the range of 40–46 kD. Some of these isozymes are encoded

TABLE 6. *Multiple steps of viral replication inhibited by interferons*

Affected step	Viruses	Responsible IFN-induced proteins
Early: penetration, uncoating	SV40, retroviruses	Unknown
Transcription	Influenza, VSV, HSV	Mx proteins (and others?)
Translation	Picornaviruses	2-5(A) synthetase/RNAse L
	Reoviruses, adenovirus, vaccinia, VSV, influenza	DsRNA-dependent protein kinase (PKR)
Late: Maturation, assembly, release	Retrovirus, VSV, HSV	Unknown

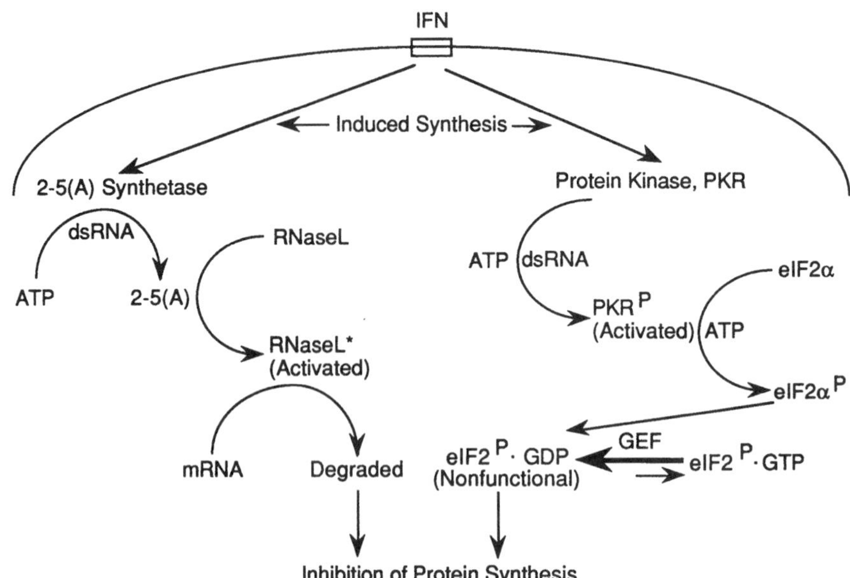

FIG. 3. Two independent pathways of translational inhibition by dsRNA in IFN-treated cells. IFN treatment of cells induces the synthesis of the 2-5(A) synthetase and dsRNA-dependent protein kinase (PKR). Both enzymes are activated by dsRNA produced by viral infection. Activated 2-5(A) synthetase polymerizes ATP into 2-5(A), which, in turn, activates the constitutively synthesized RNAse L. Activated RNAse L hydrolyzes mRNA, leading to inhibition of protein synthesis. PKR is activated by dsRNA-mediated autophosphorylation. Activated PKR phosphorylates the α subunit of the translation initiation factor eIF-2. Phosphorylated eIF-2 cannot be recycled from the inactive form (eIF-2•GDP) to the active form (eIF-2•GTP) by guanosine exchange factor (GEF). As a result, initation of protein synthesis is inhibited.

by different genes, whereas others are the products of alternatively spliced mRNAs encoded by the same gene (97). The different isozymes have different subcellular locations, thereby suggesting that they may have different cellular functions. The medium-sized human 2-5(A) synthetases are the products of alternatively spliced mRNAs of a single gene. The 69-kD isozyme of this class is myristoylated. Its two halves have strong sequence homology with each other and with the 40-kD isozyme, suggesting that its gene might have originated by gene duplication (141). The large family of 2-5(A) synthetases have a molecular weight of about 100 kD. They are activated by a lower concentrations of dsRNA than the other isozymes. It is conceivable that the small isozymes function as tetramers, the medium isozymes function as dimers, and the large isozymes as monomers (97). The 2-5(A)–dependent RNAse L has nine ankyrinlike repeats at its amino terminus (91,273). The 2-5(A) binding domain overlaps with some of these repeats. The protein also has a cysteine-rich domain and a protein kinase homology domain. The 2-5(A) synthetase/RNAse L system may be involved not only in cellular antiviral responses but also in the regulation of cellular growth (91).

dsRNA-Dependent Protein Kinase

The IFN-inducible dsRNA-dependent protein kinase, PKR, is also known as P68 kinase, P1, DAI, dsI, or eIF-2 kinase (96). Its involvement in the antiviral effects of IFNs is suggested by the existence of several inhibitors of this enzyme, which are either encoded or induced by various viruses (see below). PKR activation by dsRNA results in its autophoshorylation on several serine and threonine residues. The autophosphorylation is most probably intermolecular and occurs between two PKR molecules bound to the same dsRNA molecule. Consequently, high concentrations of dsRNA inhibit the autophosphorylation process. Polyanions such as heparin can also activate this enzyme. The only known physiological substrate of PKR is the α subunit of the translation initiation factor eIF-2. Phosphorylated eIF-2 cannot be recycled for repeated use in translation, thus causing an inhibition of cellular protein synthesis. Detectable levels of PKR are present in most cells even without IFN-treatment, and phosphorylation of PKR and eIF-2 have been observed in many virus-infected cells. Thus, PKR activation by viral dsRNA may be involved not only in the inhibition of viral protein synthesis in IFN-treated cells but also in the shut-off of host protein synthesis in virus-infected cells. The dsRNA-binding domain of PKR has been identified (57,147,165). It resides at the amino-terminal region. It contains structural motifs that are also present in other dsRNA-binding proteins (219). However, these motifs are not present in 2-5(A) synthetases, which belong to a different family of dsRNA-binding proteins (165). PKR may be important in the regulation of many other cellular functions. To evaluate the role of PKR in different cellular functions, investigators have frequently used a relatively specific inhibitor of this enzyme, 2-aminopurine. These studies have indicated that PKR may be involved in regulating the level of expression of newly transfected genes (106) and in the transduction of transcriptional signals elicited by many cytokines and dsRNA (228,274). It has also been implicated in growth arrest and adipocytic differentiation of mouse fibroblasts in culture (172) and growth suppression in yeast (27). Involvement of PKR in cellular growth regulation was most poignantly demonstrated by studying the effects of introduction of mutated PKR genes into NIH/3T3 cells (122,148). Mutant PKR proteins inhibit the functioning of the corresponding normal enzyme and cause unregulated cell proliferation. As a result, these cells undergo neoplastic transformation and form tumors in mice. These results suggest that PKR functions as a cellular antioncogenic suppressor gene. In accordance with this observation, overexpression of P58, a cellular inhibitor of PKR, also causes cellular transformation (10). The activity of PKR is blocked in cells infected with a variety of viruses (see below).

Mx Proteins

The Mx family of IFN-induced proteins mediates inhibition of orthomyxovirus replication (220). Some of these proteins are nuclear, whereas others are cytoplasmic. Murine and rat Mx1 proteins are nuclear, and both inhibit influenza virus replication. On the other hand, human MxA is localized in the cytoplasm. In addition to inhibiting the replication of influenza virus, it also inhibits the replication of VSV, a rhabdovirus (166). The antiviral effects of MxA against these two viruses can be dissociated by introducing a point mutation, which suggests that they are mediated by different mechanisms (275). In rat cells, nuclear Mx1 inhibits both viruses, whereas cytoplasmic Mx2 inhibits only VSV. The Mx proteins bind GTP and have an intrinsic GTPase activity that is necessary for their intracellular antiviral actions (95). MxA can bind to cytoskeletal proteins such as actin and tubulin, and it has a sequence homology with the yeast protein Vps1 that is implicated in vacuolar protein sorting (190). Their binding to viral transcriptases might be responsible for the antiviral actions of Mx proteins.

Other IFN-Induced Proteins

Although at the cellular level the antiviral effects of IFN are mediated primarily through the actions of intracellular antiviral proteins, additional distinct mechanisms operate in the intact organism. IFN helps to protect the organism against viral and other infections by augmenting its immunological defenses. One important mechanism is immune cell–modulated lysis of virus-infected cells. Virus-infected cells need to be recognized by cells of the immune system. Such recognition is mediated by cell surface molecules composed of MHC class I and class II proteins bound to unique peptides derived from viral proteins. Both type I and type II IFNs enhance MHC class I protein expression, whereas class II genes are activated much more readily by IFN-γ (42). The proteasomes and the permeases, which degrade and transport the viral peptides that are displayed in conjunction with the MHC protein, are also induced by IFNs in several cell types (113). IFNs therefore exert an important antiviral function by augmenting the expression of the machinery responsible for antigen processing, intracellular trafficking, and cell surface presentation of the viral peptides to T-lymphocytes. Another protein that can aid in the inhibition of virus replication in certain cell lines is the IFN-γ–inducible nitric oxide synthase enzyme because in mouse macrophages, IFN-γ–mediated inhibition of replication of ectromelia, vaccinia, and HSV-1 correlates with the production of nitric oxide (33, 110). A recent report has shown that replication of VSV is partially inhibited by the receptor of low-density lipoproteins, the extracellular release of which is augmented by IFNs (60).

Actions of TNF on Virus Replication

Under certain conditions, TNF was demonstrated to produce an IFN-like direct antiviral action in cultured cells. TNF-induced inhibition of replication was demonstrated with a variety of viruses, including EMC, VSV, adenovirus, and HSV (119,146,267). At least some of these actions involved IFN because the antiviral activity of TNF was eliminated or reduced in the presence of antibodies to type I IFNs (119,146,183). Several interrelated mechanisms have been shown to account for these antiviral actions. TNF can induce the expression of the 2-5(A) synthetase (146,267) and Mx (69) genes. In some cells, TNF can produce an increase in the synthesis of IFN-β (105,184). Finally, in many situations TNF can act synergistically with IFN-α/β or IFN-γ to activate intracellular antiviral pathways (193). It should be noted that in most experimental systems the capacity of TNF to produce a direct antiviral action is considerably weaker than that of the IFNs.

Another mechanism whereby TNF can limit virus replication is a selective lysis of virus-infected cells. This action, related to the long-recognized propensity of TNF to be cytolytic for cells in which RNA or protein synthesis is inhibited, was demonstrated in cells infected with VSV and many other viruses (4,193). Viruses may block the synthesis of cellular proteins needed to protect cells from the cytolytic action of TNF. Whether TNF-induced lysis of virus-infected cells is due to apoptosis or a more direct toxic action is largely unknown. Also unknown is whether TNF-induced lysis of virus-infected cells occurs in the intact organism and whether this mechanism contributes to the antiviral defenses. Still another mechanism whereby TNF can promote the lysis of virus-infected cells is through the induction of MHC class I and class II antigens on many types of cells (104,173).

In addition to the widely documented inhibition of the replication of various viruses by TNF, there is evidence that TNF can enhance the replication of some viruses. Most extensively studied is the activation of human immunodeficiency virus (HIV) expression induced by TNF (sometimes in concert with IL-6, GM-CSF, and other cytokines) in latently infected T cells or monocytic cells (62,177,189). The mechanism responsible for HIV activation was demonstrated to be the induction NF-κB family proteins that interact with recognition sequences in the HIV long terminal repeat (LTR) (163). These findings, coupled with the analysis of TNF production in HIV-infected patients, led to the widely accepted belief that TNF and related cytokines play an important role in the progression of HIV disease (189).

Actions of Other Cytokines on Virus Replication

Relatively few examples of direct antiviral actions have been reported with other cytokines. In view of the many similarities in the actions of TNF and IL-1, it is not surprising that at least under some conditions IL-1 apparently can suppress virus multiplication in cultured cells (245). Two cytokines, TGF-β (176) and IL-13 (151), were shown to inhibit HIV-1 replication in cultured monocytes. However, TGF-β appears to play a dual role in HIV infection because under some conditions it increased HIV-1 replication in U937 cells (176). The inhibitory effect of IL-13 was seen in cultures of tissue-derived macrophages, but not peripheral blood lymphocytes (151). The mechanisms of these inhibitory actions on HIV-1 replication have not been pinpointed. Protective actions have been seen with recombinant IL-1, IL-2, and M-CSF against HSV and Sendai virus infections in mice (13,100). It also has been demonstrated that IL-2 encoded by a recombinant vaccinia virus protected nude mice from the lethal effect of vaccinia virus by an indirect mechanism that involved the stimulation of IFN-γ and perhaps of TNF-α production at the sites of virus replication (179).

ROLES OF CYTOKINES IN VIRUS INFECTIONS IN THE INTACT ORGANISM

Interferons

Direct evidence for the role of endogenously produced IFN-α/β and IFN-g in virus infections in the intact organism has been derived from three types of experimental models: analysis of mouse strains that differ in their susceptibility to virus infection, investigation of the effect on the course of virus infections of antibodies that neutralize IFN, and studies in mice with targeted disruption of genes coding for IFN or IFN receptors. Genetic analysis of the A2G mouse strain, naturally resistant to influenza virus infection, has led to the recognition that resistance is determined by a single gene encoding the Mx protein (221). The Mx protein (see above) is not constitutively expressed, but its expression is induced at the transcription level by IFN-α or IFN-β. In $Mx^{+/+}$ or $Mx^{+/-}$ mice, IFN produced as a result of influenza virus infection induces sufficient amounts of Mx protein to confer resistance. However, most inbred mouse strains lack the Mx gene and succumb to influenza virus infection. In addition, $Mx^{-/-}$ mice cannot be protected against influenza virus infection by exogenously administered IFN-α/β (42).

Analysis of the effects of antibodies to IFNs on the course of experimental virus infections has generated a wealth of information. Most experiments were conducted with immunoglobulin isolated from the sera of sheep immunized with murine IFN-α/β; these antibody preparations neutralize all subspecies of IFN-α and IFN-β generated as a result of virus infections in mice. The effects of these antibodies on the morbidity, mortality, and extent of virus replication in the organs were examined in mice injected with a large variety of different viruses (79). The following infections in naturally susceptible mice were aggravated by the injection of antibody to IFN-α/β: Semliki Forest virus, EMC, HSV-1 and -2, Moloney sarcoma, Friend or Rauscher leukemia virus, and polyoma virus. These antibodies were also effective in exacerbating the course of infections to which various strains of mice were naturally resistant, including mouse hepatitis virus-3, influenza, Sindbis, EMC, mouse cytomegalovirus (CMV), and HSV-1 (79). These studies have clearly established the role of endogenous IFN-α/β production in the host resistance to many types of virus infection. Although most of these studies could not address the mechanism whereby endogenous IFN acts to protect the host, it is apparent that IFN produced early in the process of a primary infection is most important for determining the course of disease (42,79). Whether the major beneficial effects are due to a direct antiviral action or more indirect immunomodulatory effects is still not known. Administration of monoclonal antibodies to murine IFN-γ also helped to establish the role of endogenous IFN-γ in virus clearance and recovery from infection in mice infected with lymphocytic choriomeningitis (115,132) or vaccinia (109) viruses.

Mice with a targeted disruption of one chain of the IFN-α/β receptor have recently been generated (156). Cells of these mice are unresponsive to all IFN-α subtypes examined and to IFN-β, suggesting that the animals indeed lack a functional IFN-α/β system. These mice show a greatly increased susceptibility to infection with several viruses, especially to VSV, Semliki Forest, and vaccinia virus infection, confirming and supplementing the information generated earlier with the aid of neutralizing antibodies. Mice with a targeted disruption of the IFN-γ receptor gene

(98) or of the structural gene for IFN-γ (36) also were recently generated and used for analysis of susceptibility to some virus infections. Mice that lack the IFN-γ receptor were shown to have a defect in resistance to vaccinia virus manifested during an early stage of infection, whereas no significant change in their susceptibility to VSV and Semliki Forest virus infections was seen (98,156). The results suggested that in vivo IFN-α/β and IFN-γ exert their antiviral activities through different, nonredundant pathways and that the two IFN systems are functionally complementary in early defenses to viruses. No significant difference in the survival rate or the generation of a cytotoxic T-cell response was seen between mice that lack the structural gene for IFN-γ and wild-type mice after inoculation with influenza virus (72). Undoubtedly, additional studies will be conducted in these gene knockout mice that should help to better define the roles of type I and type II IFNs in virus infections and pinpoint their sites and mechanisms of action.

It should be noted that the generation of large amounts of IFN as a consequence of virus infection can be deleterious for the host, as has been extensively documented in mice infected with lymphocytic choriomeningitis (LCM) virus. In experiments performed almost 20 years ago, the administration of antibody against murine IFN-α/β to mice infected after birth with LCM virus increased virus multiplication up to 100-fold but suppressed disease symptoms and decreased mortality (187). Antibodies to IFN-α/β also decreased the mortality of adult mice injected intracerebrally with LCM virus (79). It has been pointed out that many of the general flulike symptoms of acute virus infections resemble the side effects seen in patients treated with recombinant IFN preparations (fever, myalgia, fatigue, leukopenia, etc.), suggesting that IFN induced in the course of virus infection, perhaps in conjunction with other endogenous cytokines, is responsible for some of the disease symptoms. One case in point is HIV infection; the serum of many symptomatic HIV-positive patients contains detectable IFN-α, the presence of which appears to correlate with disease progression and the severity of disease (45,124). Another example is Argentine hemorrhagic fever caused by Junin virus, in which the presence of high levels of circulating IFN-α (up to 64,000 U/ml) correlates with severe clinical symptoms and a poor prognosis (42,136).

Other Cytokines

Information about the role of other cytokines produced during virus infections on the extent of virus replication and course of disease is only starting to become available. Increasing use of the technique of targeted gene disruption in the study of cytokine functions is making it possible to analyze the role of a variety of cytokines in mouse models of virus infections. Mice with a disruption of the IL-6 gene were shown to have an increased susceptibility to infection with virulent vaccinia virus that correlated with a reduction in cytotoxic T-cell activity against vaccinia-infected cells and increased titers of virus in the lung and ovaries (121). In addition, upon infection with VSV, IL-6$^{-/-}$ mice produced less neutralizing immunoglobulin G than did wild-type mice.

STRATEGIES DEVELOPED BY VIRUSES TO COUNTERACT CYTOKINE ACTIONS

Antagonism with Interferon-Mediated Actions

Viruses are both inducers of IFN synthesis and the principal target of its actions. A virus that is a potent inducer of IFN and at the same time is highly sensitive to its inhibitory action would not be a successful pathogen. Thus, it is not surprising that evolution has favored the development of viral mechanisms that counteract the inhibitory effects of IFN on virus replication. The most simple mechanism available to some highly virulent viruses is a rapid inhibition of cellular RNA and protein synthesis that interferes with the ability of cells to produce IFN or to respond to it. In recent years it has become apparent that many viruses have developed much more elaborate means to antagonize IFN actions (133,199,205,217).

A survey of the known mechanisms of inhibition of IFN action by viruses (Table 7) indicates that a variety of viruses have developed a wide array of means to achieve a similar final result–the ability to escape the antiviral actions activated by the IFNs. It has been known since the early days of IFN research that many DNA viruses are resistant to IFN action. Several DNA viruses are now known to encode proteins that can inhibit the major signal transduction pathways activated as a result of IFN action. The most thoroughly studied example of such a mechanism is the ability of adenovirus E1a protein to inhibit IFN-α/β–or IFN-γ–induced signaling (205). The inhibitory function has been mapped to the conserved region 1 of E1a protein, and it is unrelated to the ability of E1a to bind several known cellular proteins, including Rb and p60 cyclin (82). The exact target of E1a protein, responsible for the inhibition of IFN signaling, has not been identified. In cells permanently transfected with the E1a protein, IFN-α failed to inhibit virus replication or to transcriptionally activate several IFN-inducible genes, and formation of the ISGF-3 transcription factor complex was also blocked (106). The response to IFN-γ or dsRNA, as measured by their ability to activate gene expression, was also blocked. A similar global inhibition of signaling by IFN-α, IFN-γ, or dsRNA was seen in cells transfected with the terminal protein of hepatitis B virus (HBV) (63). EBNA-2 protein of EBV also inhibits IFN signaling by a mechanism that appears to be different from adenovirus E1a or the terminal protein of HBV (108). In cells expressing the immortaliz-

TABLE 7. *Molecular mechanisms whereby viruses inhibit interferon actions*

Virus	Virus product	Specific target	Mechanism of action	References
Adenovirus	E1a protein	Signal transduction	Blocks signaling	2,82,106
	VA RNA	PKR	Blocks activation	142
Epstein-Barr	EBNA-2 protein	Signal transduction	Blocks signaling	108
	EBER RNA	PKR	Blocks activation	29
Vaccinia	E3L protein	PKR	Binds dsRNA	23
	K3L protein	PKR	Alternative substrate	38
Myxoma	M-T7 protein	IFN-γ	Neutralizes IFN-γ	237
Hepatitis B	Terminal protein	Signal transduction	Blocks signaling	63
Herpes simplex	2-5(A) analogs	RNAse L	Blocks activation	22
HIV-1	Tat protein	PKR	Degrades PKR?	191
	TAR RNA	PKR	Blocks activation	81
Influenza	Unknown	PKR	Activates cellular inhibitor	130
Poliovirus	Unknown	PKR	Activates cellular inhibitor	17
Reovirus	Sigma 3 protein	PKR	Binds dsRNA	102

ing EBNA-2 protein, the induction of several IFN-α–stimulated genes was decreased or abolished, but ISGF-3 formation was not inhibited. In addition, although the antiproliferative effects of IFN-α were inhibited in EBNA-2–expressing cells, its antiviral effects were not affected (7).

Many viruses have developed mechanisms to specifically counteract IFN actions responsible for the intracellular inhibition of virus replication (Table 7). The most common target of virus actions is the dsRNA-dependent protein kinase (PKR) whose activation results in the phosphorylation of initiation factor eIF-2 and resulting inhibition of protein synthesis (Fig. 3). The targeting of PKR by so many different viruses through several unrelated mechanisms provides evidence for the importance of PKR in the inhibition of virus replication by the IFNs (199,205). Indeed, adenovirus mutants that lack VA1 RNA and therefore cannot inhibit PKR activation are more strongly inhibited by IFN than are wild-type viruses (96). It is noteworthy that viral RNAs can both activate and inhibit PKR, and the balance between these opposite actions may determine the outcome of the process of virus infection even in the absence of IFN treatment (199).

A unique inhibitory mechanism specific for IFN-γ has been acquired by myxoma virus (237). A major secreted protein encoded by this poxvirus (M-T7 protein) has significant sequence similarity to the extracellular domain of the binding component of the IFN-γ receptor. Moreover, M-T7 protein specifically binds IFN-γ and neutralizes its activity. Together with the finding that IFN-γ is important in the control of vaccinia virus infection in the mouse (98,179), these data suggest a general role for IFN-γ in host resistance to poxvirus infection. By acquiring portions of cytokine receptor genes from animal cells and adapting them for their own use, poxviruses have developed the ability to neutralize several important cytokines (see below).

Antagonism with Other Cytokines

IFNs are not the only cytokines targeted by viral products. Viruses have developed elaborate forms of molecular mimicry whose only purpose appears to be to circumvent cytokine-mediated functions in the host defenses (Table 8). Dubbed as the ability of viruses to launch their own "star wars," most of these mechanism are based on the acquisition of genes for cytokines or cytokine receptors from host cells, allowing viruses to direct the synthesis of proteins that either neutralize cytokines or inhibit their synthesis. EBV DNA contains an open reading frame coding for the BCRF-1 protein, highly homologous to the murine and human IL-10 protein (152,153). The BCRF-1 protein was originally identified as a protein expressed late in the lytic phase of EBV replication (99). The protein encoded by the viral IL-10 gene is biologically active and similar to its human homologue in the spectrum of activities (152). Viral IL-10 is thought to aid the pathogenicity of EBV in two ways: by suppressing the production of immunostimulatory cytokines and by promoting B-cell growth and transformation. It has been shown that human B-lymphocytes infected with EBV actually express viral IL-10 also during early stages of infection (150). Moreover, antisense oligonucleotides specific for viral IL-10 inhibited B-cell transformation, indicating that this gene function is essential for the transforming capacity of EBV.

The presence and significance of the M-T7 gene in myxoma virus, coding for a soluble IFN-γ receptor (237), was already discussed in the preceding section. Two other poxvirus genes, T2 and B15R, encode soluble cytokine receptors specific for TNF and IL-1, respectively (Table 8). Myxoma virus in which the T2 genes were insertionally inactivated produced less severe infections than did wild-type virus in rabbits (217,236). The significance in pathogenesis of the B15R vaccinia virus protein that can neu-

TABLE 8. *Proteins encoded by viruses that inhibit cytokine synthesis or neutralize cytokine activity*

Virus	Product	Mechanism of action	References
Epstein-Barr	BCRF1 (viral IL-10)	Inhibits the synthesis of IFN-γ, IL-2, IL-1, TNF, and other cytokines; promotes B cell growth and transformation	150,152
Myxoma	M-T7 (soluble IFN-γ receptor)	Neutralization of IFN-γ	237
Shope fibroma, myxoma	T2 (soluble TNF receptor)	Neutralization of TNF	211,236
Vaccinia, cowpox	B15R (soluble IL-1 receptor)	Neutralization of IL-1	6,214,218
Cowpox	crmA (proteinase inhibitor)	Inhibits proteolytic activation of IL-1β	181

tralize IL-1 activity is less firmly established. A mutant vaccinia virus with a deletion of the B15R gene was less lethal for mice after intracerebral inoculation in mice (218), whereas after intranasal inoculation the virulence of a similar deletion mutant was actually increased and the animals died more rapidly than after inoculation with wild-type virus (6). It has been pointed out that not all vaccinia virus strains express the IL-1 receptorlike gene and that those strains that were known to cause a higher frequency of postvaccination complications tend not to express this protein (213).

A unique mechanism of inhibition of IL-1β generation was identified in cowpox virus (181). This virus encodes a proteinase inhibitor, now designated cytokine response modifier (crmA), that specifically blocks the IL-1β converting enzyme required to generate the biologically active form of IL-1β by proteolytic cleavage from a larger precursor form. The crmA gene of cowpox virus was earlier shown to mediate a suppression of the inflammatory response to virus inoculation and to influence the appearance of characteristic pocks produced by the virus on chorioallantoic membranes of chick embryos (164). Another anticytokine mechanism mediated by several adenovirus-encoded proteins is the protection of virus-infected cells from lysis by TNF (266).

THERAPEUTIC APPLICATIONS IN VIRUS INFECTIONS

As briefly reviewed in an earlier section of this chapter, a large body of evidence indicates that IFNs play an important role in natural resistance to virus infections. From the early days of IFN research it was believed that if IFN could promote recovery from virus infections then it should be possible to use it as an antiviral agent in humans (79). The therapeutic promise of IFN in the treatment of virus infections, reinforced in later years by optimism about the worth of IFN in the treatment of malignancies, proved to be a powerful impetus behind efforts to purify IFNs, to clone their genes, and to produce them by recombinant DNA techniques. By the early 1980s methods had been developed to produce sufficient quantities of several types of IFNs for clinical trials. These efforts have resulted in many undeniable successes, but they also have shown the limitations of the therapeutic use of IFNs in virus infections. One limitation is due to the timing of IFN administration. Experimental work in animal models has demonstrated that the administration of IFN can protect animals from a variety of virus infections (58). However, in most animal models IFN was most effective when administered before inoculation or during the early stages of infection, i.e., before the appearance of clinical signs of disease and before the peak of virus multiplication in target organs. Another limitation stems from the side effects caused by the administration of IFN in quantities needed for a systemic effect, as already mentioned earlier in this chapter. As a result, it has not been possible to use IFNs for the prophylaxis or therapy of many common virus infections.

One positive and unexpected achievement is the successful therapeutic application of type I IFN preparations in chronic active hepatitis caused by hepatitic C or hepatitis B viruses (39,47,169). Long-term administration of recombinant IFN-α (the two preparations licensed in the United States each contain a single IFN-α subtype) can produce virologic remission and normalization of serum aminotransferases in a significant portion of the treated patients. In other countries, other types of recombinant or "natural" (i.e., derived from human cells by induction with virus or dsRNA) IFN-α and IFN-β preparations are also being used for the treatment of chronic hepatitis infections. The mechanism whereby IFN therapy brings about an improvement in the clinical condition and, in some cases, a complete cure of chronic viral hepatitis is not completely understood. Immunomodulatory actions, rather than a more direct antiviral effect, may be responsible. Another virus-caused disease for which treatment with IFN-α has proved successful are genital warts, condylomata accuminata, caused by infection with papilloma viruses (64). The mechanism of action has not been fully explained.

IFNs are also used for the treatment of diseases not known to be caused by viruses, especially some neoplasias. In the United States recombinant IFN-α has been licensed for use in hairy cell leukemia and AIDS-related Kaposi's sarcoma. Recombinant IFN-β is being used for the management of relapsing-remitting forms of multiple sclerosis (117). IFN-γ is effective in the treatment of chronic granulomatous disease, a rare genetic disorder characterized by deficient phagocyte oxidative metabolism (202); IFN-γ reduces the incidence of severe bacterial infections in these patients.

Clinical trials are continuing with the IFNs, and other cytokines are being evaluated in a variety of conditions in experimental animals and in patients. It would not be surprising if new information useful for the treatment of viral infections emerged from these studies. The clinical evaluation of IL-12 is awaited with great interest. In view of its ability to increase IFN-γ production and to promote cellular immune responses (231,232), IL-12 is being considered for clinical trials in HIV-positive individuals.

REFERENCES

1. Aarden LA, Brunner TK, Cerottini J-C, et al. Revised nomenclature for antigen-nonspecific T cell proliferation and helper factors. *J Immunol* 1979;123:2928–2929.
2. Ackrill AM, Foster GR, Laxton CD, Flavell DM, Stark GR, Kerr IM. Inhibition of the cellular response to interferons by products of the adenovirus type 5 E1A oncogene. *Nucl Acids Res* 1991;19:4387–4393.
3. Aderka D, Holtmann H, Toker L, Hahn T, Wallach D. Tumor necrosis factor induction by Sendai virus. *J Immunol* 1986;136:2938–2942.
4. Aderka D, Novick D, Hahn T, Fischer DG, Wallach D. Increase of vulnerability to lymphotoxin in cells infected by vesicular stomatitis virus and its further augmentation by interferon. *Cell Immunol* 1985;92: 218–225.
5. Aguet M, Dembic Z, Merlin G. Molecular cloning and expression of the human inteferon-g receptor. *Cell* 1988;55:273–280.
6. Alcami A, Smith GL. A soluble receptor for interleukin-1 β encoded by vaccinia virus: a novel mechanism of virus modulation of the host response to infection. *Cell* 1992:71;153–167.
7. Aman P, von Gabain A. An Epstein-Barr virus immortalization associated gene segment interferes specifically with the IFN-induced antiproliferative response in human B-lymphoid cell lines. *EMBO J* 1990; 9:147–152.
8. Arend WP. Interleukin-1 receptor antagonist. *Adv Immunol* 1993;54: 167–227.
9. Banner DW, D'Arcy A, Janes W, et al. Crystal structure of the soluble human 55 kd TNF receptor-human TNF β complex: implications for TNF receptor activation. *Cell* 1993;73:431–445.
10. Barber GN, Thompson S, Lee TG, et al. The 58 kDa inhibitor of the interferon induced DS RNA activated protein kinase is a TPR protein with oncogenic properties. *Proc Natl Acad Sci USA* 1994;91:4278–4282.
11. Bazan JF. A novel family of growth factor receptors: a common binding domain in the growth hormone, prolactin, erythropoietin and IL-6 receptors, and the p75 IL-2 receptor β-chain. *Biochem Biophys Res Commun* 1989;164:788–795.
12. Bazan JF. Structural design and molecular evolution of a cytokine receptor superfamily. *Proc Natl Acad Sci USA* 1990;87:6934–6938.
13. Berkowitz C, Becker Y. Recombinant interleukin-1α, interleukin-2 and M-CSF-1 enhance the survival of newborn C57BL/6 mice inoculated intraperitoneally with a lethal dose of herpes simplex virus-1. *Arch Virol* 1992;124:83–93.
14. Beutler B, Cerami A. The biology of cachectin/TNF a primary mediator of the host response. *Ann Rev Immunol* 1989;7:625–655.
15. Beutler B, van Huffel C. Unraveling function in the TNF-ligand and receptor families. *Science* 1994;264:667–668.
16. Black RA, Kronheim SR, Sleath PR. Activation of IL-1β by a co-induced protease. *FEBS Lett* 1989;247:386–390.
17. Black TL, Safer B, Hovanessian A, Katze MG. The cellular 68,000-Mr protein kinase is highly autophosphorylated and activated yet significantly degraded during poliovirus infection: implications for translational regulation. *J Virol* 1989;63:2244–2251.
18. Brakebusch C, Nophar Y, Kemper O, Engelmann H, Wallach D. Cytoplasmic truncation of the p55 tumour necrosis factor (TNF) receptor abolishes signalling, but not induced shedding of the receptor. *EMBO J* 1992;11:943–950.
19. Browning JL, Ngam-ek A, Lawton P, et al. Lymphotoxin β, a novel member of the TNF family that forms a heteromeric complex with lymphotoxin on the cell surface. *Cell* 1993;72:847–856.
20. Campbell IL. Cytokines in viral diseases. *Curr Opin Immunol* 1991; 3:486–491.
21. Carswell EA, Old LJ, Kassel RL, Green S, Fiore N, Williamson B. An endotoxin-induced serum factor that causes necrosis of tumors. *Proc Natl Acad Sci USA* 1975;72:3666–3670.
22. Cayley PJ, Davies JA, McCullagh KG, Kerr IM. Activation of the ppp(A2'p)nA system in interferon-treated, herpes simplex virus-infected cells and evidence for novel inhibitors of the ppp(A2'p)nA-dependent RNase. *Euro J Biochem* 1984;143:165–174.
23. Chang H-W, Watson JC, Jacobs BL. The E3L gene of vaccinia virus encodes an inhibitor of the interferon-induced, double-stranded RNA-dependent protein kinase. *Proc Natl Acad Sci USA* 1992:89:4825–4829.
24. Chebath J, Benech P, Revel M, Vigneron M. Constitutive expression of (2'-5') oligo A synthetase confers resistance to picornavirus infection. *Nature* 1987;330:587–588.
25. Chelbi-Alix MK, Thang MN. Cloned human interferons alpha: differential affinities for polyinosinic acid and relationship between molecular structure and species specificity. *Biochem Biophys Res Commun* 1987;145:426–435.
26. Cheung SC, Chattopadhyay SK, Morse HC III, Pitha PM. Expression of defective virus and cytokine genes in murine AIDS. *J Virol* 1991; 65:823–828.
27. Chong KL, Feng L, Schappert K, et al. Human p68 kinase exhibits growth suppression in yeast and homology to the translational regulator GCN2. *EMBO J* 1992;11:1553–1562.
28. Chrivia JC, Wedrychowicz T, Young HA, Hardy KJ. A model of human cytokine regulation based on transfection of γ interferon gene fragments directly into isolated peripheral blood T lymphocytes. *J Exp Med* 1990;172:661–664.
29. Clarke PA, Schwemmle M, Schickinger J, Hilse K, Clemens MJ. Binding of Epstein-Barr virus small RNA EBER-1 to the double-stranded RNA-activated protein kinase DAI. *Nucl Acids Res* 1991;19:243–248.
30. Colamonici OR, Domanski P. Identification of a novel subunit of the type I interferon receptor localized to human chromosome 21. *J Biol Chem* 1993;268:10895–10899.
31. Colotta F, Re F, Muzio M, et al. Interleukin-1 type II receptor: a decoy target for interleukin-1 that is regulated by interleukin-4. *Science* 1993; 261:472–475.
32. Cook JR, Jung V, Schwartz B, Wang P, Pestka S. Structural analysis of the human IFN-γ receptor: a small segment of the intracellular domain is specifically required for class I major histocompatibility complex antigen induction and antiviral activity. *Proc Natl Acad Sci USA* 1992;89:11317–11321.
33. Croen KD. Evidence for antiviral effect of nitric oxide. Inhibition of herpes simplex virus type 1 replication. *J Clin Inv* 1993;91:2446–2452.
34. D'Addario M, Wainberg MA, Hiscott J. Activation of cytokine genes in HIV-1 infected myelomonoblastic cells by phorbol ester and tumor necrosis factor. *J Immunol* 1992;148:1222–1229.
35. Dai W, Gupta SL. Regulation of indoleamine 2,3-dioxygenase gene expression in human fibroblasts by interferon-γ. Upstream control region discriminates between interferon-γ and interferon-α. *J Biol Chem* 1990;265:19871–19877.
36. Dalton DK, Pitts-Meek S, Keshav S, Figari IS, Bradley A, Stewart TA. Multiple defects of immue cell function in mice with disrupted interferon-γ genes. *Science* 1993;259:1739–1742.
37. Darnell Jr JE, Kerr IM, Stark GR. Jak-STAT pathways and transcriptional activation in response to IFNs and other extracellular signaling proteins. *Science* 1994;264:1415–1421.
38. Davies MV, Chang HW, Jacobs BL, Kaufman RJ. The E3L and K3L vaccinia virus gene products stimulate translation through inhibition of the double-stranded RNA-dependent protein kinase by different mechanisms. *J Virol* 1993;67:1688–1692.
39. Davis GL, Balart LA, Schiff ER, et al. Treatment of chronic hepatitis C with recombinant interferon alpha: A multicenter randomized controlled trial. *N Engl J Med* 1989;321:1501–1505.
40. Degrave W, Derynck R, Tavernier J, Haegeman G, Fiers W. Nucleotide sequence of the chromosomal gene for human fibroblast (β1) interfer-

on and of the flanking regions. *Gene* 1981;14:137–143.
41. Delcayre A, Salas F, Mathur S, Kovats K, Lotz M, Lernhardt W. Epstein Barr virus/complement C3d receptor is an interferon α receptor. *EMBO J* 1991;10:919–926.
42. De Maeyer E, De Maeyer-Guignard J. *Interferons and other regulatory cytokines.* New York: John Wiley & Sons; 1988.
43. Dembic Z, Loetscher H, Gubler U, et al. Two human TNF receptors have similar extracellular, but distinct intracellular, domain sequences. *Cytokine* 1990;2:231–237.
44. Desmyter J, Stewart WE II. Molecular modification of interferon: attainment of human interferon in a conformation active on cat cells but inactive on human cells. *Virology* 1976;70:451–458.
45. DeStefano E, Friedman RM, Friedman-Kien AE, et al. Acid-labile human leukocyte interferon in homosexual men with Kaposi's sarcoma and lymphadenopathy. *J Infect Dis* 1982;146:451–455.
46. de Vos AM, Ultsch M, Kossiakoff AA. Human growth hormone and extracellular domain of its receptor: crystal structure of the complex. *Science* 1992;255:306–312.
47. Di Bisceglie AM, Martin P, Kassianides C, et al. Recombinant interferon α therapy for chronic hepatitis C: a randomized double blind placebo controlled trial. *N Engl J Med* 1989;321:1506–1510.
48. Dinarello CA, Wolff SM. The role of interleukin-1 in disease. *N Engl J Med* 1993;328:106–113.
49. Dower SK, Sims JE, Cerretti DP, Bird TA. The interleukin-1 system: receptors, ligands and signals. In: Kishimoto T, ed. *Interleukins: molecular biology and immunology.* Vol 51. Basel, Switzerland: Karger; 1992:33–64.
50. Driggers P, Ennist D, Gleason S, et al. An interferon γ-regulated protein that binds the interferon-inducible enhancer element of major histocompatibility complex class 1 genes. *Proc Natl Acad Sci USA* 1991; 87:3743–3747.
51. Du W, Thanos D, Maniatis T. Mechanisms of transcriptional synergism between distinct virus-inducible enhancer elements. *Cell* 1993; 74:887–898.
52. Duc-Goiran P, Galliot B, Chany C. Studies on virus-induced interferons produced by the human amniotic membrane and white blood cells. *Arch Ges Virusforsch* 1971;34:232–243.
53. Dudding L, Haskill S, Clark BD, Auron PE, Sporn S, Huang H-S. Cytomegalovirus infection stimulates expression of monocyte-associated mediator genes. *J Immunol* 1989;143:3343–3352.
54. Eisenberg SP, Evans RJ, Arend WP, et al. Primary structure and functional expression from complemenatary DNA of a human interleukin-1 receptor antagonist. *Nature* 1990;343:341–345.
55. Farrar MA, Campbell JD, Schreiber RD. Identification of a functionally important sequence in the C terminus of the inteferon-γ receptor. *Proc Natl Acad Sci USA* 1992;89:11706–11710.
56. Farrar MA, Schreiber RD. The molecular cell biology of interferon-γ and its receptor. *Ann Rev Immunol* 1993;11:571–611.
57. Feng GS, Chong K, Kumar A, Williams BR. Identification of double-stranded RNA-binding domains in the interferon-induced double-stranded RNA-activated p68 kinase. *Proc Natl Acad Sci USA* 1992;89: 5447–5451.
58. Finter NB, Oldham RK. *Interferon: In vivo and clinical studies.* Vol. 4. Amsterdam: Elsevier, 1985.
59. Fiorentino DF, Bond MW, Mosmann TR. Two types of mouse T helper cell. IV. Th2 clones secrete a factor that inhibits cytokine production by Th1 clones. *J Exp Med* 1989;170:2081–2095.
60. Fischer DG, Tal N, Novick D, Barak S, Rubinstein M. An antiviral soluble form of the LDL receptor induced by interferon. *Science* 1993; 262:250–253.
61. Fish EN, Banerjee K, Stebbing N. Human leukocyte interferon subtypes have different antiproliferative and antiviral activities on human cells. *Biochem Biophys Res Commun* 1983;112:537–545.
62. Folks TM, Clouse KA, Justement J, Rabson A, Duh E, Kehrl JH, Fauci AS. Tumor necrosis factor alpha induces expression of human immunodeficiency virus in a chronically infected T-cell clone. *Proc Natl Acad Sci USA* 1989;86:2365–2368.
63. Foster GR, Ackrill AM, Goldin RD, Kerr IM, Thomas HC, Stark GR. Expression of the terminal protein region of hepatitis B virus inhibits cellular responses to interferons α and γ and double-stranded RNA. *Proc Natl Acad Sci USA* 1991;88:2888–2892.
64. Friedman-Kien AE, Eron LJ, Conant MA, et al. Natural interferon alfa for treatment of condylomata acuminata. *JAMA* 1988;259:533–538.
65. Fu XY. A transcription factor wtih SH2 and SH3 domains is directly activated by an interferon α-induced cytoplasmic protein tyrosine kinase(s). *Cell* 1992;70:323–335.
66. Fu XY, Zhang JJ. Transcription factor p91 interacts with the epidermal growth factor receptor and mediates activation of the c-*fos* gene promoter. *Cell* 1993;74:1135–1145.
67. Fujita T, Reis LFL, Watanabe N, Kimura Y, Taniguchi T, Vilček J. Induction of the transcription factor IRF-1 and interferon-β mRNAs by cytokines and activators of second messenger pathways. *Proc Natl Acad Sci USA* 1989;86:9936–9940.
68. Gendelman HE, Friedman RM, Joe S, et al. A selective defect of interferon α production in human immunodeficiency virus-infected monocytes. *J Exp Med* 1990;172:1433–1442.
69. Goetschy JF, Zeller H, Content J, Horisberger MA. Regulation of the interferon-inducible IFI-78K gene, the human equivalent of the murine Mx gene, by interferons, double-stranded RNA, certain cytokines, and viruses. *J Virol* 1989;63:2616–2622.
70. Goodbourn S, Maniatis T. Overlapping positive and negative regulatory domains of the human β-interferon gene. *Proc Natl Acad Sci USA* 1988;85:1447–1451.
71. Gosselin J, Flamand L, D'Addario M, et al. Modulatory effects of Epstein-Barr, herpes simplex, and human herpes-6 viral infections and coinfections on cytokine synthesis. A comparative study. *J Immunol* 1992;149:181–187.
72. Graham MB, Dalton DK, Giltinan D, Braciale VL, Stewart TA, Braciale TJ. Response to influenza infection in mice with a targeted disruption in the interferon γ gene. *J Exp Med* 1993;178:1725–1732.
73. Gray PW, Aggarwal BB, Benton CV, et al. Cloning and expression of cDNA for human lymphotoxin, a lymphokine with tumour necrosis activity. *Nature* 1984;312:721–724.
74. Gray PW, Goeddel DV. Cloning and expression of murine immune interferon cDNA. *Proc Natl Acad Sci USA* 1983;80:5842–5846.
75. Gray PW, Leong S, Fennie E, et al. Cloning and expression of the cDNA for the murine interferon γ receptor. *Proc Natl Acad Sci USA* 1989;86:8479–8501.
76. Greene WC, Bohnlein E, Ballard DW. HTLV-1 and normal T cell growth: transcriptional strategies and surprises. *Immunol Today* 1989; 10:272–278.
77. Greenlund AC, Farrar MA, Viviano BL, Schreiber RD. Ligand-induced IFNγ receptor tyrosine phosphorylation couples the receptor to its signal transduction system (p91). *EMBO J* 1994;13:1591–1600.
78. Greenlund AC, Schreiber RD, Goeddel DV, Pennica D. Interferon-γ induces receptor dimerization in solution and on cells. *J Biol Chem* 1993;268:18103–18110.
79. Gresser I. Role of interferon in resistance to viral infection in vivo. In: Vilček J, De Maeyer E, eds. *Interferon.* Vol 2: Interferons and the immune system. Amsterdam: Elsevier Science, 1984:221–247.
80. Gribaudo G, Lembo D, Cavallo G, Landolfo S, Lengyel P. Interferon action: binding of viral RNA to the 40-kDa 2'-5'-oligoadenylate synthetase in interferon-treated HeLa cells infected with encephalomyocarditis virus. *J Virol* 1991;65:1748–1757.
81. Gunnery S, Rice AP, Robertson HD, Mathews MB. Tat-responsive region RNA of human immunodeficiency virus 1 can prevent activation of the double-stranded-RNA-activated protein kinase. *Proc Natl Acad Sci USA* 1990;87:8687–8691.
82. Gutch MJ, Reich NC. Repression of the interferon signal transduction pathway by the adenovirus E1A oncogene. *Proc Natl Acad Sci USA* 1991;88:7913–7917.
83. Guy GR, Cairns J, Ng SB, Tan YH. Inactivation of a redox-sensitive protein phosphatase during the early events of tumor necrosis/interleukin-1 signal transduction. *J Biol Chem* 1993;268:2141–2148.
84. Hannum CH, Wilcox CJ, Arend WP, et al. Interleukin-1 receptor antagonist activity of a human IL-1 receptor inhibitor. *Nature* 1990; 343:336–340.
85. Hansen TR, Kazemi M, Keisler DH, Malathy PV, Imakawa K, Roberts RM. Complex binding of the embryonic interferon, ovine trophoblast protein-1, to endometrial receptors. *J Interferon Res* 1989;9:215–225.
86. Harada H, Fujita T, Miyamoto M, et al. Structurally similar, but functionally distinct factors, IRF-1 and IRF-2, bind to the same regulatory elements of IFN and IFN-inducible genes. *Cell* 1989;58:729–739.
87. Harada H, Kitagawa M, Tanaka N, et al. Anti-oncogenic and oncogenic potentials of interferon regulatory factors -1 and -2. *Science* 1993; 259:971–974.
88. Harada H, Willison K, Sakakibara J, Miyamoto M, Fujita T, Taniguchi T. Absence of the Type I IFN system in EC cells: transcriptional acti-

vator (IRF-1) and repressor (IRF-2) genes are developmentally regulated. *Cell* 1990;63:303–312.
89. Hardy KJ, Sawada T. Human γ interferon strongly upregulates its own gene expression in peripheral blood lymphocytes. *J Exp Med* 1989; 170:1021–1026.
90. Harroch S, Revel M, Chebath J. Induction by interleukin-6 of interferon regulatory factor 1 (IRF-1) gene expression through the palindromic interferon response element pIRE and cell type-dependent control of IRF-1 binding to DNA. *EMBO J* 1994;13:1942–1949.
91. Hassel BA, Zhou A, Sotomayar C, Maran A, Silverman RH. A dominant negative mutant of 2-5A-dependent RNase suppresses antiproliferative and antiviral effects to interferon. *EMBO J* 1993;12:3297–3304.
92. Havell EA, Hayes TG, Vilček J. Synthesis of two distinct interferons by human fibroblasts. *Virology* 1978;89:330–334.
93. Hemmi S, Bohni R, Stark G, DiMarco F, Aguet M. A novel member of the interferon receptor family complements functionality of the murine interferon γ receptor in human cells. *Cell* 1994;76:803–810.
94. Ho ASY, Liu Y, Khan TA, Hsu D-H, Bazan JF, Moore KW. A receptor for interleukin 10 is related to interferon receptors. *Proc Natl Acad Sci USA* 1993;90:11267–11271.
95. Horisberger MA. Interferon-induced human protein MxA is a GTPase which binds transiently to cellular protein. *J Virol* 1992;66: 4705–4709.
96. Hovanessian A. The double-stranded RNA activated protein kinase induced by interferon. *J Interferon Res* 1989;9:641–647.
97. Hovanessian A. RNA-activated enzymes: a specific protein kinase and 2'-5' oligoadenylate synthetases. *J Interferon Res* 1991;11:199–205.
98. Huang S, Hendriks W, Althage A, et al. Immune response in mice that lack the interferon-γ receptor. *Science* 1993;259:1742–1745.
99. Hudson GS, Bankier AT, Satchwell SC, Barrell BG. The short unique region of the B95-8 Epstein-Barr virus genome. *Virology* 1985;147: 81–98.
100. Iida J, Saiki I, Ishihara C, Azuma I. Protective activity of recombinant cytokines against Sendai virus and herpes simplex virus (HSV) infections in mice. *Vaccine* 1989;7:229–233.
101. Imakawa K, Anthony RV, Kazemi M, Marotti KR, Polites HG, Roberts RM. Interferon-like sequence of ovine trophoblast protein secreted by embryonic trophectoderm. *Nature* 1987;330:377–379.
102. Imani F, Jacobs BL. Inhibitory activity for the interferon-induced protein kinase is associated with the reovirus serotype 1 sigma 3 protein. *Proc Natl Acad Sci USA* 1988;85:7887–7891.
103. Isaacs A, Lindenmann J. Virus interference. 1. The interferon. *Proc R Soc Lond [Biol]* 1957;147:258–267.
104. Israël A, Le Bail O, Hatat D, et al. TNF stimulates expression of mouse MHC class I genes by inducing an NF kappa B-like enhancer binding activity which displaces constitutive factors. *EMBO J* 1989;8: 3793–3800.
105. Jacobsen H, Mestan J, Mittnacht S, Dieffenbach CW. Beta interferon subtype 1 induction by tumor necrosis factor. *Mol Cell Biol* 1989;9: 3037–3042.
106. Kalvakolanu D, Bandyopadhyay S, Tiwari RK, Sen GC. Enhancement of expression of exogenous genes by 2-aminopurine. *J Biol Chem* 1991; 266:873–879.
107. Kamijo R, Harada H, Matsuyama T, et al. Requirement for transcription factor IRF-1 in NO synthase induction in macrophages. *Science* 1994;263:1612–1615.
108. Kanda K, Decker T, Aman P, Wahlstrom M, von Gabain A, Kallin B. The EBNA2-related resistance towards a interferon (IFN-α) in Burkitt's lymphoma cells effects induction of IFN-induced genes but not the activation of transcription factor ISGF-3. *Mol Cell Biol* 1992;12:4930–4936.
109. Karupiah G, Blanden RV, Ramshaw IA. Interferon γ is involved in the recovery of athymic nude mice from recombinant vaccinia virus/interleukin 2 infection. *J Exp Med* 1990;172:1495–1503.
110. Karupiah G, Xie Q-W, Mark R, et al. Inhibition of viral replication by interferon-γ-induced nitric oxide synthetase. *Science* 1993;261: 1445–1448.
111. Kelker HC, Le J, Rubin BY, Yip YK, Nagler C, Vilček J. Three molecular weight forms of natural human interferon-γ revealed by immunoprecipitation with monoclonal antibody. *J Biol Chem* 1984; 259:4301–4304.
112. Keller A, Maniatis T. Identification and characterization of a novel repressor of β-interferon gene expression. *Genes Dev* 1991;5:868–879.
113. Kelly A, Powis S, Glynne R, Radley E, Beck S, Trowsdale J. Second proteasome-related gene in the human MHC class II region. *Nature* 1991;353:667–668.
114. Kimura T, Nakayama K, Penninger J, et al. Involvement of the IRF-1 transcription factor in antiviral responses to interferons. *Science* 1994; 264;1921–1924.
115. Klavinskis LS, Geckeler R, Oldstone MB. Cytotoxic T lymphocyte control of acute lymphocytic choriomeningitis virus infection: interferon gamma, but not tumour necrosis factor alpha, displays antiviral activity in vivo. *J Gen Virol* 1989;70:3317–3325.
116. Knight E. Interferon: purification and initial characterization from human diploid cells. *Proc Natl Acad Sci USA* 1976;73:520–523.
117. Knobler RL, Greenstein JI, Johnson KP, et al. Systemic recombinant human interferon-β treatment of relapsing-remitting multiple sclerosis: pilot study analysis and six-year follow-up. *J Interferon Res* 1993; 13:333–340.
118. Kobayashi Y, Yamamoto K, Saido T, Kawasaki H, Oppenheim JJ, Matsushima K. Identification of calcium-activated neutral protease as a processing enzyme of human interleukin-1 alpha. *Proc Natl Acad Sci USA* 1990;87:5548–5552.
119. Kohase M, Henriksen-DeStefano D, May LT, Vilček J, Sehgal PB. Induction of β_2-interferon by tumor necrosis factor: a homeostatic mechanism in the control of cell proliferation. *Cell* 1986;45:659–666.
120. Kolesnick R, Golde DW. The sphingomyelin pathway in tumor necrosis factor and interleukin-1 signaling. *Cell* 1994;325–328.
121. Kopf M, Baumann H, Freer G, et al. Impaired immune and acute-phase responses in interleukin-6-deficient mice. *Nature* 1994;368:339–342.
122. Koromilas AE, Roy S, Barber GN, Katze MG, Sonenberg N. Malignant transfection by a mutant of the IFN-inducible dsRNA-dependent protein. *Science* 1992;257:1685–1689.
123. Kostura MJ, Tocci MJ, Limjuco G, et al. Identification of a monocyte specific pre-interleukin-1β convertase activity. *Proc Natl Acad Sci USA* 1989;86:5227–5231.
124. Kramer A, Biggar RJ, Hampl H, Friedman RM, Fuchs D, Wachter H, Goedert JJ. Immunologic markers of progression to acquired immunodeficiency syndrome are time-dependent and illness-specific. *Am J Epidemiol* 1992;136:71–80.
125. Kriegler M, Perez C, DeFay K, Albert I, Lu SD. A novel form of TNF/cachectin is a cell surface cytotoxic transmembrane protein: ramifications for the complex physiology of TNF. *Cell* 1988;53:45–53.
126. Kumar R, Choubey D, Lengyel P, Sen GC. Studies on the role of the 2'-5'-oligoadenylate synthetase-RNase L pathway in β interferon-mediated inhibition of encephalomyocarditis virus replication. *J Virol* 1988;62:3175–3181.
127. Larner AC, David M, Feldman GM, et al. Tyrosine phosphorylation of DNA-binding proteins by multiple cytokines. *Science* 1993;261: 1730–1733.
128. Le J, Vilček J. Tumor necrosis factor and interleukin 1: cytokines with multiple overlapping biological activities. *Lab Invest* 1987;56:234–248.
129. Leblanc J-F, Cohen L, Rodrigues M, Hiscott J. Synergism between distinct enhanson domains in viral induction of the human beta interferon gene. *Mol Cell Biol* 1990;10:3987–3993.
130. Lee TG, Tomita J, Hovanessian AG, Katze MG. Purification and partial characterization of a cellular inhibitor of the interferon-induced protein kinase of Mr 68,000 from influenza virus-infected cells. *Proc Natl Acad Sci USA* 1990;87:6208–6212.
131. Lee TH, Lee GW, Ziff EB, Vilček J. Isolation and characterization of eight tumor necrosis factor-induced gene sequences from human fibroblasts. *Mol Cell Biol* 1990;10:1982–1988.
132. Leist TP, Eppler M, Zinkernagel RM. Enhanced virus replication and inhibition of lymphocytic choriomeningitis virus disease in anti-gamma interferon-treated mice. *J Virol* 1989;63:2813–2819.
133. Lengyel P. Tumor-suppressor genes: news about the interferon connection. *Proc Natl Acad Sci USA* 1993;90:5893–5895.
134. Leung DW, Capon DJ, Goeddel DV. The structure and bacterial expression of three distinct bovine interferon-β genes. *Biotechnology* 1984;2:458–464.
135. Leung K, Betts JC, Xu L, Nabel GJ. The cytoplasmic domain of the interleukin-1 receptor is required for nuclear factor-kappa B signal transduction. *J Biol Chem* 1994;269:1579–1582.
136. Levis SC, Saavedra MC, Ceccoli C, et al. Endogenous interferon in Argentine hemorrhagic fever. *J Infect Dis* 1984;149:428–433.
137. Lieberman AP, Pitha PM, Shin HS, Shin ML. Production of tumor necrosis factor and other cytokines by astrocytes stimulated with lipopolysaccharide or a neurotropic virus. *Proc Natl Acad Sci USA* 1989;86:6348–6352.

138. MacDonald N, Kuhl D, Maguire D, et al. Different pathways mediate virus inducibility of the human IFN-α1 and IFN-β genes. *Cell* 1990; 60:767–779.
139. Mahe Y, Mukaida N, Kuno K, Akiyama M, Ikeda N, Matsushima K, Murakami S. Hepatitis B virus X protein transactivates human interleukin-8 gene through acting on nuclear factor κB and CCAAT/enhancer-binding protein-like cis-elements. *J Biol Chem* 1991;266: 13759–13763.
140. Maniatis T, Whittemore LA, Du W, et al. In: McKnight SL, Yamamoto K, eds. *Transcriptional regulation.* Part 2. Cold Spring Harbor, NY: Cold Spring Harbor Laboratory; 1992:1193–1220.
141. Marie I, Hovanessian A. The 69-kDa 2-5A synthetase is composed of homologous and adjacent functional domains. *J Biol Chem* 1992;267: 9933–9939.
142. Mathews MB, Shenk T. Adenovirus-associated RNA and translation control. *J Virol* 1991;65:5657–5662.
143. Matsuyama T, Kimura T, Kitagawa M, et al. Targeted disruption of IRF-1 or IRF-2 results in abnormal type 1 IFN gene induction and lymphocyte development. *Cell* 1993;75:83–97.
144. McMahan CJ, Slack JL, Mosley B, et al. A novel IL-1 receptor, cloned from B cells by mammalian expression, is expressed in many cell types. *EMBO J* 1991;10:2821–2832.
145. Merrill JE, Koyanagi Y, Isy C. Interleukin-1 and tumor necrosis factor-α can be induced from mononuclear phagocytes by human immunodeficiency virus type 1 binding to the CD4 receptor. *J Virol* 1989; 63:4404–4408.
146. Mestan J, Digel, W, Mittnacht S, et al. Antiviral effects of recombinant tumour necrosis factor in vitro. *Nature* 1986;323:816–819.
147. Meurs EF, Chong K, Galabru J, et al. Molecular cloning and characterization of the human double-stranded RNA-activated protein kinase induced by interferon. *Cell* 1990;62:379–390.
148. Meurs EF, Galabru J, Barber GN, Katze MG, Hovanessian AG. Tumor suppressor function of the interferon-induced-double stranded RNA-activated protein, kinase. *Proc Natl Acad Sci USA* 1992;90:232–236.
149. Minty A, Chalon P, Derocq JM, et al. Interleukin-13 is a new human lymphokine regulating inflammatory and immune responses. *Nature* 1993;362:248–250.
150. Miyazaki I, Cheung RK, Dosch H-M. Viral interleukin 10 is critical for the induction of B cell growth transformation by Epstein-Barr virus. *J Exp Med* 1993;178:439–447.
151. Montaner LJ, Doyle AG, Collin M, et al. Interleukin 13 inhibits human immunodeficiency virus type 1 production in primary blood-derived human macrophages in vitro. *J Exp Med* 1993;178:743–747.
152. Moore KW, O'Garra A, de Waal Malefyt R, Vieira P, Mosmann TR. Interleukin-10. *Annu Rev Immunol* 1993;11:165–190.
153. Moore KW, Vieira P, Fiorentino DF, Trounstine ML, Khan TA, Mosmann TR. Homology of cytokine synthesis inhibitory factor (IL-10) to the Epstein-Barr virus gene BCRFI. *Science* 1990;248:1230–1234.
154. Moskophidis D, Frei K, Lohler J, Fontana A, Zinkernagel RM. Production of random classes of immunoglobulins in brain tissue during persistent viral infection paralleled by secretion of interleukin-6 (IL-6) but not IL-4, IL-5, and gamma interferon. *J Virol* 1991;65: 1364–1369.
155. Muller M, Briscoe J, Laxton C, et al. The protein tyrosine kinase Jak1 complements defects in interferon-α/β and -γ signal transduction. *Nature* 1993;366:129–135.
156. Müller U, Steinhoff U, Reis LFL, et al. Functional role of type I and type II interferons in antiviral defense. *Science* 1994;264:1918–1921.
157. Nakajima K, Martinex-Masa O, Mirano T, et al. Induction of IL-6 (B-cell stimulatory factor-2/IFN-β2) production by HIV. *J Immunol* 1989; 142:531–536.
158. Nedwin G, Naylor S, Sakaguchi A, et al. Human lymphotoxin and tumor necrosis factor genes: structure, homology and chromosomal localization. *Nucl Acids Res* 1985;13:6361–6373.
159. Nourbakhsh M, Hoffmann K, Hauser H. Interferon-β promoters contain a DNA element that acts as a position-independent silencer on the NF-κB site. *EMBO J* 1993;12:451–459.
160. Novick D, Cohen B, Rubinstein M. The human interferon α/β receptor: characterization and molecular cloning. *Cell* 1994;77:391–400.
161. Old LJ. Tumor necrosis factor. *Sci Am* 1988;258:59–75.
162. Ortaldo JR, Herberman RB, Harvey C, et al. A species of human α interferon that lacks the ability to boost human natural killer activity. *Proc Natl Acad Sci USA* 1984;81:4926–4929.
163. Osborn L, Kunkel S, Nabel GJ. Tumor necrosis factor alpha and interleukin 1 stimulate the human immunodeficiency virus enhancer by activation of the nuclear factor kappa B. *Proc Natl Acad Sci USA* 1989; 86:2336–2340.
164. Palumbo GJ, Pickup DJ, Fredrickson TN, McIntyre LJ, Buller RM. Inhibition of an inflammatory response is mediated by a 38-kDa protein of cowpox virus. *Virology* 1989;171:262–273.
165. Patel R, Sen G. Identification of the double-stranded RNA binding domain of the human inteferon-inducible protein kinase. *J Biol Chem* 1992;267:7671–7676.
166. Pavlovic J, Zurcher T, Haller O, Staeheli P. Resistance to influenza and vesicular stomatitis virus conferred by human MxA protein. *J Virol* 1990;64:26–31.
167. Pellegrini S, Schindler C. Early events in signalling by interferons. *Trends Biochem Sci* 1993;18:338–342.
168. Pennica D, Nedwin GE, Hayflick JS, et al. Human tumour necrosis factor: precursor structure, expression and homology to lymphotoxin. *Nature* 1984;312:724–729.
169. Perrillo RP, Schiff ER, Davis GL, et al. A randomized controlled trial of interferon alpha 2b alone and after prednisone withdrawal for the treatment of chronic hepatitis B. *N Engl J Med* 1990;323:295–301.
170. Pestka S. The human interferons: from protein purification and sequence to cloning and expression in bacteria: before, between and beyond. *Arch Biochem Biophys* 1983;221:1–37.
171. Pestka S, Langer JA, Zoon KC, Samuel CE. Interferons and their actions. *Annu Rev Biochem* 1987;56:727–777.
172. Petryshyn R, Chen J-J, London I. Detection of activated double-stranded RNA-dependent protein kinase in 3T3-F442A cells. *Proc Natl Acad Sci USA* 1988;85:1427–1431.
173. Pfizenmaier K, Scheurich P, Schluter C, Kronke M. Tumor necrosis factor enhances HLA-A,B,C and HLA-DR gene expression in human tumor cells. *J Immunol* 1987;138:975–980.
174. Pine R, Canova A, Schindler C. Tyrosine phosphorylated p91 binds to a single element in the ISGF2/IRF-1 promoter to mediate induction by IFN α and IFN γ, and is likely to autoregulate the p91 gene. *EMBO J* 1994;13:158–167.
175. Pine R, Decker T, Kessler DS, Levy DE, Darnell JE Jr. Purification and cloning of interferon-stimulated gene factor 2 (ISGF2): ISGF2 (IRF-1) can bind to the promoters of both β interferon- and interferon-stimulated genes but is not a primary transcriptional activator of either. *Mol Cell Biol* 1990;10:2448–2457.
176. Poli G, Kinter AL, Justement JS, Bressler P, Kehrl JH, Fauci AS. Transforming growth factor β suppresses human immunodeficiency virus expression and replication in infected cells of the monocyte/macrophage lineage. *J Exp Med* 1991;173:589–597.
177. Popik W, Pitha PM. Role of tumor necrosis factor α in activation and replication of the tat-defective human immunodeficiency virus type 1. *J Virol* 1993;67:1094–1099.
178. Raines MA, Kolesnick RN, Golde DW. Sphingomyelinase and ceramide activate mitogen-activated protein kinase in myeloid HL-60 cells. *J Biol Chem* 1993;268:14572–14575.
179. Ramsay AJ, Ruby J, Ramshaw IA. A case for cytokines as effector molecules in the resolution of virus infection. *Immunol Today* 1993; 14:155–157.
180. Ransohoff RM, Sen GC. Interferon γ signaling: current status and evidence for cross-talk with other interferons and cytokines. In: Baron S, Coppenhaver DH, Dianzani F, eds. *Interferon: principles and medical applications.* Galveston, TX: Department of Microbiology, University of Texas Medical Branch; 1992:175–191.
181. Ray CA, Black RA, Kronheim SR, et al. Viral inhibition of inflammation: cowpox virus encodes an inhibitor of the interleukin-1β converting enzyme. *Cell* 1992;69:597–604.
182. Reid LE, Brasnett AH, Gilbert CS, et al. A single DNA response element can confer inducibility by both α- and γ-interferons. *Proc Natl Acad Sci USA* 1989;86:840–844.
183. Reis LFL, Le J, Hirano T, Kishimoto T, Vilček J. Antiviral action of tumor necrosis factor in human fibroblasts is not mediated by B cell stimulatory factor 2/IFN-β_2, and is inhibited by specific antibodies to IFN-β. *J Immunol* 1988;140:1566–1570.
184. Reis LFL, Lee TH, Vilček J. Tumor necrosis factor acts synergistically with autocrine interferon-β and increases interferon-β mRNA levels in human fibroblasts. *J Biol Chem* 1989;264:16351–16354.
185. Revel M, Chebath J. Interferon-activated genes. *Trends Biochem Sci* 1986;11:166–170.
186. Rinderknecht E, O'Connor BH, Rodriguez H. Natural human interfer-

on-γ. Complete amino acid sequence and determination of sites of glycosylation. *J Biol Chem* 1984;259:6790–6797.
187. Riviere Y, Gresser I, Guillon J-C, Tovey MG. Inhibition by anti-interferon serum of lymphocytic choriomeningitis virus disease in suckling mice. *Proc Natl Acad Sci USA* 1977;74:2135–2139.
188. Roberts RM, Imakawa K, Niwano Y, et al. Interferon production by the preimplantation sheep embryo. *J Interferon Res* 1989;9:175–187.
189. Rosenberg ZF, Fauci AS. Immunopathogenic mechanisms of HIV infection: cytokine induction of HIV expression. *Immunol Today* 1990; 11:176–180.
190. Rothman J, Raymond C, Gilbert T, O'Hara P, Stevens T. A putative GTP binding protein homologous to interferon-inducible Mx proteins performs an essential function in yeast protein sorting. *Cell* 1990; 61:1063–1074.
191. Roy S, Katze MG, Parkin NT, Edery I, Hovanessian AG, Sonenberg N. Control of the interferon-induced 68-kilodalton protein kinase by the HIV-1 tat gene product. *Science* 1990;247:1216–1219.
192. Rubartelli A, Cozzolino F, Talio M, Sitia R. A novel secretory pathway for interleukin-1 β, a protein lacking a signal sequence. *EMBO J* 1990;9:1503–1510.
193. Rubin BY. TNF and viruses: multiple interrelationships. In: Aggarwal BB, Vilček J, eds. *Tumor necrosis factors: structure, function, and mechanism of action.* New York: Marcel Dekker, 1992:331–340.
194. Ruddle NH. Tumor necrosis factor (TNF-α) and lymphotoxin (TNF-β). *Curr Opin Immunol* 1992;4:327–332.
195. Ruff-Jamison S, Chen K, Cohen S. Induction by EGF and interferon-γ of tyrosine phosphorylated DNA-binding proteins in mouse liver nuclei. *Science* 1993;261:1733–1736.
196. Ruffner H, Reis LFL, Naf D, Weissman C. Induction of type 1 interferon genes and interferon-inducible genes in embryonal stem cells devoid of interferon regulatory factor 1. *Proc Natl Acad Sci USA* 1993; 90:11503–11507.
197. Sadowski HB, Shuai K, Darnell JE, Gilman MZ. A common nuclear signal transduction pathway activated by growth factor and cytokine receptors. *Science* 1993;261:1739–1744.
198. Samudzi CT, Gribskov CL, Burton LE, Rubin JR. Crystallization and preliminary x-ray diffraction studies of recombinant rabbit interferon-γ. *Biochem Biophys Res Commun* 1991;178:634–640.
199. Samuel C. Antiviral actions of interferon: interferon-regulated cellular proteins and their surprisingly selective antiviral activities. *Virology* 1991;183:1–11.
200. Schindler C, Fu XY, Improta T, Aebersold R, Darnell JE Jr. Proteins of transcription factor ISGF-3: one gene encodes the 91- and 84-kDa ISGF-3 proteins that are activated by interferon α. *Proc Natl Acad Sci USA* 1992;89:7836–7839.
201. Schindler C, Shuai K, Prezioso VR, Darnell JE Jr. Interferon-dependent tyrosine phosphorylation of a latent cytoplasmic transcription factor. *Science* 1992;257:809–813.
202. Sechler JM, Malech HL, White CJ, Gallin JI. Recombinant human interferon-γ reconstitutes defective phagocyte function in patients with chronic granulomatous disease of childhood. *Proc Natl Acad Sci USA* 1988;85:4874–4878.
203. Sen GC. Transcriptional regulation of interferon-inducible genes. In: Cohen PP, Foulke J, eds. *Hormonal regulation of gene transcription.* Amsterdam: Elsevier, 1991:349–374.
204. Sen GC, Lengyel P. The interferon system: A bird's eye view of its biochemistry. *J Biol Chem* 1992;267:5017–5020.
205. Sen GC, Ransohoff RM. Interferon-induced antiviral actions and their regulation. *Adv Virus Res* 1992;42:57–102.
206. Shuai K, Horvath CM, Huang T, Gureshi SA, Cowburn D, Darnell JE Jr. Interferon activation of the transcription factor Stat91 involves dimerization through SH2-phosphotyrosyl peptide interactions. *Cell* 1994; 76:821–828.
207. Shuai K, Stark GR, Kerr IM, Darnell JE Jr. A single phosphotyrosine residue of Stat91 required for gene activation by interferon-γ polypeptide signalling to the nucleus through tyrosine phosphorylation of Jak and Stat protein. *Science* 1993;261:1744–1746.
208. Shuai K, Ziemiecki A, Wilks AF, et al. Polypeptide signalling to the nucleus through tyrosine phosphorylation of Jak and Stat protein. *Nature* 1993;366:580–583.
209. Silvennoinen O, Ihle JN, Schlessinger J, Levy DE. Interferon-induced nuclear signalling by Jak protein tyrosine kinases. *Nature* 1993;366:583–585.
210. Sims JE, Gayle MA, Slack JL, et al. Interleukin 1 signaling occurs exclusively via the type I receptor. *Proc Natl Acad Sci USA* 1993;90: 6155–6159.
211. Smith CA, Davis T, Anderson D, et al. A receptor for tumor necrosis factor defines an unusual family of cellular and viral proteins. *Science* 1990;248:1019–1023.
212. Smith CA, Farrah T, Goodwin RG. The TNF receptor superfamily of cellular and viral proteins: activation, costimulation, and death. *Cell* 1994;76:959–962.
213. Smith GL. Vaccinia virus glycoproteins and immune evasion. *J Gen Virol* 1993;74:1725–1740.
214. Smith GL, Chan YS. Two vaccinia virus proteins structurally related to the interleukin-1 receptor and the immunoglobulin superfamily. *J Gen Virol* 1991;72:511–518.
215. Soh J, Donnelly RJ, Kotenko S, et al. Identification and sequence of an accessory factor required for activation of the human interferon γ receptor. *Cell* 1994;76:793–802.
216. Sprecher E, Becker Y. Induction of interleukin-1 α and β gene transcription in mouse peritoneal exudate cells after intraperitoneal infection with herpes simplex virus-1. *Arch Virol* 1990;110:259–269.
217. Spriggs MK. Poxvirus-encoded soluble cytokine receptors. *Virus Res* 1994;33:1–10.
218. Spriggs MK, Hruby DE, Maliszewski CR, et al. Vaccinia and cowpox viruses encode a novel secreted interleukin-1-binding protein. *Cell* 1992;71:145–152.
219. St. Johnston D, Brown NH, Gall JG, Jantsch M. A conserved double-stranded RNA-binding domain. *Proc Natl Acad Sci USA* 1992;89: 10979–10983.
220. Staeheli P. Interferon-induced proteins and the antiviral state. *Adv Virus Res* 1990;38:147–200.
221. Staeheli P, Pravtcheva D, Lundin LG, et al. Interferon-regulated influenza virus resistance gene Mx is localized on mouse chromosome 16. *J Virol* 1986;58:967–969.
222. Stark GR, Kerr IM. Regulation of interferon-β gene: structure and function of *cis*-elements and transfactors. *J Interferon Res* 1992;12:147–151.
223. Stewart HJ, McCann SHE, Flint APF. Structure of an interferon-α2 gene expressed in the bovine conceptus early in gestation. *J Mol Endocrinol* 1990;4:275–282.
224. Taniguchi T. Regulation of interferon-β gene: structure and function of *cis*-elements and *trans*-factors. *J Interferon Res* 1989;9:633–640.
225. Taniguchi T, Mantei N, Schwarzstein M, Nagata S, Muramatsu M, Weissmann C. Human leukocyte and fibroblast interferons are structurally related. *Nature* 1980;285:547–549.
226. Tartaglia LA, Ayres TM, Wong GH, Goeddel DV. A novel domain within the 55 kd TNF receptor signals cell death. *Cell* 1993;74: 845–853.
227. Tartaglia LA, Goeddel DV. Two TNF receptors. *Immunol Today* 1992; 13:151–153.
228. Tiwari R, Kusari J, Kumar R, Sen G. Gene induction by interferons and double stranded RNA: selective inhibition by 2-aminopurine. *Mol Cell Biol* 1988;8:4289–4294.
229. Tiwari R, Kusari J, Sen G. Functional equivalents of interferon-mediated signals needed for induction of an mRNA can be generated by double-stranded RNA and growth factors. *EMBO J* 1987;6:3373–3378.
230. Tovey MG, Streuli M, Gresser I, et al. Interferon messenger RNA is produced constitutively in the organs of normal individuals. *Proc Natl Acad Sci USA* 1987;84:5038–5042.
231. Trinchieri G. Interleukin-12 and its role in the generation of T_H1 cells. *Immunol Today* 1993;14:335–337.
232. Trinchieri G, Wysocka M, D'Andrea A, et al. Natural killer cell stimulatory factor (NKSF) or interleukin-12 is a key regulator of immune response and inflammation. *Prog Growth Factor Res* 1992;4:355–368.
233. Trotta PP, Nagabhushan TL. Gamma interferon, protein structure and function. In: Baron S, Coppenhaver DH, Dianzani F, eds. *Interferon: principles and medical applications*. Galveston, TX: Department of Microbiology, University of Texas Medical Branch 1992:117–127.
234. Tsang SY, Nakanishi M, Peterlin BM. Mutational analysis of the DRA promoter: *cis*-acting sequences and *trans*-acting factors. *Mol Cell Biol* 1990;10:711–719.
235. Tschachler E, Robert-Guroff M, Gallo RC, Reitz MS Jr. Human T-lymphotropic virus I-infected T cells constitutively express lymphotoxin in vitro. *Blood* 1989;73:194–201.
236. Upton C, Macen JL, Schreiber M, McFadden G. Myxoma virus expresses a secreted protein with homology to the tumor necrosis factor receptor gene family that contributes to viral virulence. *Virology* 1991;

184:370–382.
237. Upton C, Mossman K, McFadden G. Encoding of a homolog of the IFN-γ receptor by myxoma virus. *Science* 1992;258:1369–1372.
238. Utsumi J, Shimizu H. Human interferon β, protein structure and function. In: Baron S, Coppenhaver DH, Dianzani F, eds. *Interferon: principles and medical applications*. Galveston, TX: Department of Microbiology, University of Texas Medical Branch 1992:107–116.
239. Utsumi J, Yamazaki S, Hosoi K, et al. Characterization of *E. coli*-derived recombinant human interferon-β as compared with fibroblast human interferon-β. *J Biochem* 1987;101:1199–1208.
240. Utsumi J, Yamazaki S, Kawaguchi K, Kimura S, Shimizu H. Stability of human interferon-β1: oligomeric human interferon-β1 is inactive but reactivated by monomerization. *Biochim Biophys Acta* 1989; 998:167–172.
241. Uzé G, Lutfalla G, Bandu MT, Proudhon D, Mogensen KE. Behavior of a cloned murine interferon α/β receptor expressed in homospecific or heterospecific background. *Proc Natl Acad Sci USA* 1992; 89:4774–4778.
242. Uzé G, Lutfalla G, Gresser I. Genetic transfer of a functional human interferon α receptor into mouse cells. Cloning and expression of its cDNA. *Cell* 1990;60:225–234.
243. Vacheron F, Rudent A, Perin S, Labarre C, Quero AM, Guenounou M. Production of interleukin 1 and tumour necrosis activities in bronchoalveolar washings following infection of mice by influenza virus. *J Gen Virol* 1990;71:477–479.
244. Van Damme J, Decock B, Conigs R, Lenaerts J-P, Opdenakker G, Billiau A. The chemotactic activity for granulocytes produced by virally infected fibroblasts is identical to monocyte-derived interleukin 8. *Eur J Immunol* 1989;19:1189–1194.
245. Van Damme J, De Ley M, Opdenakker G, Billiau A, De Somer P, Van Beeumen J. Homogeneous interferon-inducing 22K factor is related to endogenous pyrogen and interleukin-1. *Nature* 1985;314:266–268.
246. Van Damme J, Schaafsma MR, Fibbe WE, Falkenburg JHF, Opdenakker G, Billiau A. Simultaneous production of interleukin-6, interferon-β and colony-stimulating activity by fibroblasts after viral and bacterial infection. *Eur J Immunol* 1989;19:163–168.
247. Van Lint J, Agostinis P, Vandevoorde V, et al. Tumor necrosis factor stimulates multiple serine/threonine protein kinases in Swiss 3T3 and L929 cells. Implication of casein kinase-2 and extracellular signal-regulated kinases in the tumor necrosis factor signal transduction pathway. *J Biol Chem* 1992;267:25916–25921.
248. Veals SA, Schindler C, Leonard D, et al. Subunit of an α-interferon-responsive transcription factor is related to interferon regulatory factor and Myb families of DNA-binding proteins. *Mol Cell Biol* 1992; 12:3315–3324.
249. Velazquez L, Fellows M, Stark GR, Pellegrini S. A protein tyrosine kinase in the inteferon α/β signaling pathway. *Cell* 1992;70:313–322.
250. Vieira P, de Waal-Malefyt R, Dang MN, et al. Isolation and expression of human cytokine synthesis inhibitory factor cDNA clones: homology to Epstein-Barr virus open reading frame BCRFI. *Proc Natl Acad Sci USA* 1991;88:1172–1176.
251. Vietor I, Schwenger P, Li W, Schlessinger J, Vilček J. Tumor necrosis factor-induced activation and increased tyrosine phosphorylation of mitogen-activated protein (MAP) kinase in human fibroblasts. *J Biol Chem* 1993;268:18994–18999.
252. Vilček J. Interferons. In: Sporn MB, Roberts AB, eds. *Peptide growth factors and their receptors II, Handbook of experimental pharmacology*. Vol 95/II. Berlin: Springer-Verlag; 1990:3–38.
253. Vilček J, Le J. Immunology of cytokines: an introduction. In: Thomson AW, ed. *The cytokine handbook*. 2nd ed. London: Academic Press; 1994:1–19.
254. Vilček J, Lee TH. Tumor necrosis factor. New insights into the molecular mechanisms of its multiple actions. *J Biol Chem* 1991;266: 7313–7316.
255. Visvanathan KV, Goodbourn S. Double-stranded RNA activates binding of NF-κB to an inducible element in the human β-interferon promoter. *EMBO J* 1989;8:1129–1138.
256. Watanabe N, Sakakibara J, Hovanessian A, Taniguchi T, Fujita T. Activation of IFN-β element by IRF-1 requires a post translational event in addition to IRF-1 synthesis. *Nucl Acids Res* 1991;19:4421–4428.
257. Watling D, Guschin D, Muller M, et al. Complementation by the protein tyrosine kinase Jak2 of a mutant cell line defective in the inteferon-γ signal transduction pathway. *Nature* 1993;366:166–170.
258. Weissenbach J, Chernajovsky Y, Zeevi M, et al. Two interferon mRNAs in human fibroblasts: in vitro translation and *Escherichia coli* cloning studies. *Proc Natl Acad Sci USA* 1980;77:7152–7156.
259. Weissmann C, Weber H. The interferon genes. *Prog Nucl Acid Res Mol Biol* 1986;33:251–300.
260. Wetzel R. Assignment of the disulphide bonds of leukocyte interferon. *Nature* 1981;289:606–607.
261. Wheelock EF. Interferon-like virus-inhibitor induced in human leukocytes by phytohemagglutinin. *Science* 1965;149:310–311.
262. Williams BR. Transcriptional regulation of interferon-stimulated genes. *Eur J Biochem* 1991;200:1–11.
263. Williams BR, Silverman R. *The 2-5A system: clinical and molecular aspects of the interferon regulated pathway*. New York: Liss; 1985.
264. Willman CL, Sever CE, Pallavicini MG, et al. Deletion of IRF-1, mapping to chromosome 5q31.1, in human leukemia and preleukemic myelodysplasia. *Science* 1993;259:968–971.
265. Wilson V, Jeffreys AJ, Barrie PA, et al. A comparison of vertebrate interferon gene families detected by hybridization with human interferon DNA. *J Mol Biol* 1983;166:457–475.
266. Wold WS, Gooding LR. Region E3 of adenovirus: a cassette of genes involved in host immunosurveillance and virus-cell interactions. *Virology* 1991;184:1–8.
267. Wong GHW, Goeddel DV. Tumour necrosis factors a and b inhibit virus replication and synergize with interferons. *Nature* 1986;323: 819–822.
268. Yamada G, Ogawa M, Akagi K, et al. Specific depletion of the B-cell population induced by aberrant expression of human interferon regulatory factor 1 gene in transgenic mice. *Proc Natl Acad Sci USA* 1991; 88:532–536.
269. Yip YK, Barrowclough BS, Urban C, Vilček J. Molecular weight of human γ interferon is similar to that of other human interferons. *Science* 1982;215:411–413.
270. Young HA, Hardy KJ. Interferon-gamma: producer cells, activation stimuli, and molecular genetic regulation. *Pharmacol Ther* 1990; 45:137–151.
271. Yu-Lee LY, Hrachovy J, Stevens A, Schwarz L. Interferon-regulatory factor 1 is an immediate early gene under transcriptional regulation by prolactin in Nb2 cells. *Mol Cell Biol* 1991;10:3087–3094.
272. Zhang Y, Lin J-X, Vilček J. Interleukin-6 induction by tumor necrosis factor and interleukin-1 in human fibroblasts involves activation of a nuclear factor binding to a κB-like sequence. *Mol Cell Biol* 1990;10: 3818–3823.
273. Zhou A, Hassel BA, Silverman RH. Expression cloning of 2-5A-dependent RNase: a uniquely regulated mediator of interferon action. *Cell* 1993;72:753–765.
274. Zinn K, Keller A, Wittemore L-A, Maniatis T. 2-aminopurine selectively inhibits the induction of β-interferon, c-*fos*, and c-*myc* gene expression. *Science* 1988;240:210–213.
275. Zurcher T, Pavlovic J, Staeheli P. Mechanism of human MxA protein action: variants with changed antiviral properties. *EMBO J* 1992;11: 1657–1661.

Fundamental Virology, Third Edition
edited by B.N. Fields, D.M. Knipe, P.M. Howley, et al.
Lippincott - Raven Publishers, Philadelphia © 1996

CHAPTER 12

Plant Viruses

John G. Shaw

HISTORY

Descriptions of the symptoms and consequences of virus infections in plants have been recorded over a period of many centuries. Perhaps the most spectacular of these concerned the striking patterns of floral coloring in virus-infected tulips that were featured in the works of some of the famous seventeenth century Dutch painters—a disease that had the rather unusual effect of commanding the payment of astonishingly high prices for infected bulbs. However, most early accounts of plant diseases that would appear to have been caused by viruses described situations where the effects were considerably less beneficial to those who experienced them. It was not until the final years of the nineteenth century that the nature of such diseases and of a variety of unexplained maladies of humans and other animals began to be revealed. During this period, some

J. G. Shaw: Department of Plant Pathology, College of Agriculture, University of Kentucky, Lexington, Kentucky 40546.

experiments with diseased plants were being conducted, and they played the central role the founding of the science of virology.

During the nineteenth century, the investigations of Pasteur, Lister, Koch, and others with diseases of humans and other animals, and of Prévost, DeBary and others with diseased plants, had demonstrated the infectious nature of bacteria and fungi. Then, late in the century, some experiments with extracts of diseased tobacco plants led to the realization that such disorders did not involve these recognized types of pathogenic microbes but were caused by the agents we now know as viruses. In 1886, Adolf Mayer (89), working in Wageningen, described experiments in which tobacco plants that he had injected with crude extracts of plants with mosaic symptoms developed the disease. He had thus demonstrated that the disease was caused by an infectious agent, but his statements concerning the nature of the agent were to be proven incorrect. A few years later, Dmitrii Ivanowski (68) reported in St. Petersburg that extracts of mosaic-diseased tobacco plants were still infectious after passage through filtering devices that retained bacteria. However, he attributed the cause of the disease to a toxin or to a bacterium that could pass through the filter, even though some of his observations argued against the latter suggestion.

It was the report of his experiments in Delft by Martinus Beijerinck (6) that resolved the uncertainty and established the existence of a new class of pathogens. After conducting a series of experiments, some of which were similar to those of Ivanowski, Beijerinck concluded that the mosaic disease of tobacco "is not caused by microbes, but by a contagium vivum fluidum." With commendable foresight, he suggested that a number of other diseases of unknown cause would be shown to be of similar etiology. Today, less than a century later, it seems quite remarkable that a series of experiments involving such seemingly simple materials and facilities could have opened the way for the profound developments that have followed.

Tobacco mosaic virus (TMV), the cause of the mosaic disease that had been studied by Mayer, Ivanowski, and Beijerinck, was to continue to play a leading role in fundamental virus research. Efforts over the next several decades to isolate and characterize viruses led to Wendell Stanley's report (118) of the purification of TMV in a crystalline state and his initial conclusion that the virus was a globular protein. F.C. Bawden and N.W. Pirie (5) showed that similar preparations of TMV had the characteristics of nucleoproteins and, at about the same time, paracrystals of the virus were analyzed by X-ray diffraction (8). Tobacco mosaic virus was the first virus to be examined by electron microscopy (75). The intrinsic infectivity of viral RNA isolated from virus particles was first demonstrated with TMV (48). The virus was also the first with which infectious particles, indistinguishable from native virus particles, were reassembled *in vitro* from purified preparations of viral RNA and coat protein molecules (44).

Other plant viruses have played prominent roles in the growth of our knowledge of the fundamental characteristics of viruses. Some of the concepts underlying the structure of viruses with icosahedral symmetry were the result of early characterizations of a few plant viruses such as turnip yellow mosaic virus and tomato bushy stunt virus. Later, it was with two plant viruses that the structure of icosahedral viruses was first described at a level of resolution of a few Ångstroms.

Some of the smallest pathogenic agents were first discovered in plants. The satellite tobacco necrosis virus is a spherical particle of only 18 nm diameter whose RNA encodes only its own coat protein and which requires a helper virus for replication (73). The circular, independently replicating RNA molecules known as viroids are the smallest pathogens to have been characterized (36). A variety of types of satellite RNAs are associated with some plant viruses and are dependent on them for replication.

THE IMPACT OF PLANT VIRUSES

Crop plants in most parts of the world are subject to significant losses as a result of infections by viruses and viroids. There is at present no effective system for the periodic assessment of total, worldwide losses as a result of these types of diseases, but estimates of $60 billion per year have been mentioned. In addition, the adverse effects of virus diseases are exhibited in less direct ways. There are many cases where the persistent threat of an epidemic has made it impossible to raise a crop that is otherwise well suited to a particular region. Greater losses to nonviral pathogens are often sustained by plants whose vigor has been reduced by prior, perhaps chronic, infection with a virus. Extra costs are also incurred by the need for measures to prevent or minimize the effects of viruses, such as the application of pesticides for the control of agents that transmit viruses and the use of materials that have been certified to be virus-free for planting vegetatively propagated crops.

Crop plants of every type can be stricken by virus diseases. With long-lived plants, the effects can be particularly serious because, in addition to immediately sustained losses, considerable periods of time may elapse before replacements reach the age where they become productive. Two examples may serve to demonstrate the magnitude of losses that can occur in such crops. It has been estimated that some 50 million citrus trees have been lost or rendered unproductive as a result of a virus disease known as tristeza. The devastating swollen shoot disease has caused the destruction of some 200 million cocoa trees in parts of western Africa. Virus diseases in herbaceous plants may also be remarkably destructive. Rice plantings in southeast Asia and in South America sustain very large losses due to viruses. Barley yellow dwarf virus is one of the most serious pathogens of grain crops in many parts of the world.

Ornamental, vegetable, and field crops are susceptible to potyviruses wherever they are grown. In addition to cases where a virus or combination of viruses may almost completely destroy a crop in a single growing season, there are viruses that cause chronic but less debilitating infections in perennial plants.

As with all plant diseases, the consequences of those caused by viruses depend to a considerable extent on economic and social factors. In some parts of the world, the occurrence of some virus diseases may have limited economic consequences because of the availability of alternate commodities or sources. In others, and in particular those where human suffering may already be a sadly unrelenting condition, losses of plants to virus diseases may have catastrophic consequences.

It should be mentioned that the efforts of humans to promote the development, expansion, and modernization of agricultural and other economic enterprises can have profound effects on the epidemiology and ecology of plant viruses. The adverse consequences of some of these activities and the risks they continue to pose to crop production have been described (15).

SOME GENERAL FEATURES OF PLANT VIRUSES

As a result of the distinctive morphological, anatomical, and ecological features of higher plants and animals, there are some notable differences in the mechanisms used by viruses to establish infections in members of the two kingdoms. Some of the general characteristics of plant viruses that are briefly described in this section are the consequence of these differences.

The Transmission of Plant Viruses

Though the direct passage of virus from plant to plant is responsible for some diseases, most of the plant viruses depend upon mobile agents for their dissemination in natural settings. Invertebrate animals are the most important of these agents (64). Aphids, which can establish themselves in very large numbers on the leaves and shoots of plants, are particularly effective in this role. Viruses may be acquired and transmitted by aphids in a noncirculative manner, in which the association of virus and insect is quite brief, or by a circulative mode in which the virus is translocated to various organs and may multiply and persist for extended periods in the aphid. Other invertebrates that can introduce viruses into the above-ground parts of plants include leafhoppers, mealybugs, whiteflies, thrips, and mites.

Soil-inhabiting organisms may serve as agents for the transmission of viruses from plant to plant. Some viruses may be introduced into plants by nematodes feeding on young root tissues. Others are transported via the zoospores or resting spores of root-infecting fungi.

Some viruses are passed directly to progeny plants through infected or contaminated pollen or seeds. The use of virus-infected tubers, bulbs, and cuttings for the vegetative propagation of plants is another way in which viruses are spread. A few viruses can be transmitted as a result of contact with contaminated implements or by direct contact between neighboring plants.

The Initiation of Plant Virus Infections

Among the features of plants required for their survival is the presence on their surfaces of layers of protective materials consisting of waxes, cutin, and pectin. The underlying cells contain thick cellulosic walls. These substances form barriers to virus particles that must be breached if infections are to be established. Wounds made by the aforementioned agents of transmission during their efforts to obtain nourishment are the usual means whereby this is accomplished.

Infections can also be established by a process known as mechanical (or manual) inoculation. This procedure, which is widely used with many plant viruses for experimental purposes, usually involves gently rubbing the surface of a leaf with an implement (spatula, cloth pad, finger, etc.) that has been moistened with a suspension of virus particles. The presence of a fine-mesh abrasive substance during the rubbing process aids in creating the necessary wounds in the leaf surface. Some virus-host combinations result in the production of local lesions in mechanically inoculated leaves, and this response has formed the basis of a classic infectivity assay method for plant viruses that is analogous to the plaque assays of other types of viruses (60).

With most plant viruses, there is no evidence of the presence of the type of specific interactions between viral proteins and receptors on the surfaces of cells that are characteristic of many other virus-host systems. The process of inoculation, whether natural or manual, is usually presumed to result in the presence of infectious virus particles in the cytoplasm of cells that have survived the wounding event. The viral genome is then uncoated and the activities involved in the production of progeny virus particles commence. Some plant viruses, in particular those that can be replicated in the agents of their transmission, may employ mechanisms that are more typical of the early events in some animal virus infections.

The Movement of Virus in Infected Plants

Infection of a plant with a virus is generally of little consequence if it does not result in a systemic infection, i.e., if the virus does not spread from the initially infected cell (often in a leaf) to neighboring cells and then throughout most of the growing plant. Random passage of virus from cell to cell is prevented by the cell walls. However, plants contain channels (plasmodesmata) in the cell walls that provide an intercellular, symplastic continuum, and it is through these that virus particles or viral nucleic acids are transported by a process mediated by a virus-encoded protein (28). Cell-to-cell movement of virus in this manner is a relatively slow process, and in plants that resist systemic infection by a particular virus, it may not occur or may be limited to a small cluster of cells.

Movement to distant parts of the plant is a more rapid process and occurs through the companion cells and sieve elements of the phloem. The viral coat protein is thought to play a role in this process (63). The final distribution of virus in a systemically-infected plant is usually not uniform. In leaves displaying mosaic symptoms, there may be a much lower amount of virus in the dark green than in the light green areas. Some viruses are limited in their distribution to the phloem tissue in infected plants.

The Effects of Viruses on Plants

Perhaps the most enduring puzzle in plant virology has been the manner in which viruses cause diseases in plants. Many viral genes appear to be multifunctional, and changes in various parts of the genome of a particular virus may affect the type of symptom produced in an infected plant. Viral genes whose only role is to induce disease symptoms have not been reported. Disease responses may be triggered by specific interactions between host proteins (or other molecules) and viral genes or gene products that are quite incidental to the activities required for virus multiplication. A continuing problem in trying to identify these initial events is that of separating them from the array of secondary metabolic and physiological effects that usually follow.

The altered vigor and appearance of virus-infected plants is often the result of changes in such fundamentally important activities as photosynthesis, respiration, nutrient availability, and hormonal regulation of growth. The response of the plant to a particular virus can also be markedly influenced by environmental factors that affect the plant's physical condition. The symptoms produced by the plant may include mosaics, yellowing, mottling or other patterns of discoloration, stunting, wilting, necrosis, or other forms of aberrant growth and development. Most of these effects reduce the yields and commercial value of crops, and some of them result in the destruction of entire plantings.

The Control of Plant Virus Diseases

Virus diseases of plants cannot be prevented by the application of chemicals as can diseases caused by many fungal pathogens. Once a plant has become infected, the virus usually becomes distributed throughout the plant and remains viable for the life of the plant. The plant does not possess an immune system with which it can effect a cure, and there is no other practical or effective measure for ridding it of the virus.

Various cultural practices such as the elimination of agents that transmit viruses, the eradication of infected plants that may serve as sources of viruses, and the selection of suitable times and locations for planting are used in attempt to control virus diseases. The use of virus-free seeds is necessary for the prevention of some virus diseases of annually planted plants. In the case of some vegetatively propagated crop plants, chronic infections have made it necessary to produce clones of virus-free material for propagation and to establish regulations for the maintenance, certification, and distribution of such stocks.

The most effective methods for control of plant virus diseases have been the development and use of varieties of a given crop species that are resistant to a particular virus, i.e., the plants do not react adversely to inoculation by a particular virus. Resistance may be the outcome of a virus-host interaction in which the virus is localized to the inoculated cells and a few neighboring cells (the so-called hypersensitive response), with the result that the overall vigor and appearance of the plant is virtually unaffected. Another form of resistance is one in which a virus is unable to undergo replication in a particular variety of an otherwise susceptible plant species. Several reviews of the genetic basis of resistance have been published (46). Two major problems encountered with this approach to disease control are the scarcity of natural resistance genes and the tendency of resistance to be overcome by mutant viruses.

The phenomenon of cross-protection, which for decades has been a method for determining relationships between plant viruses, has been used to control a few specific virus diseases. Plants systemically infected with a strain of a virus that causes mild symptoms may be protected from infection with another strain of the same virus. Thus, in a few cases, viruses that cause severe diseases have been controlled by prior, deliberate infection of the plants they threaten with a related but benign virus.

During the past decade, there has been an intensive effort to apply recombinant DNA and plant transformation techniques for the development of novel control methods that, in some respects, are similar to the natural cross-protection phenomenon (65). Of particular importance to the technological achievements has been the transformation system involving *Agrobacterium tumefaciens* (22), a soil-inhabiting bacterium that infects and causes the formation of tumors in some plant species. The tumors are induced as a result of the integration into the plant's nuclear genome of part of a plasmid (Ti) that is borne by the bacterium. Ti plasmid constructs, modified to prevent tumor induction and to contain exogenous genes, transcription promoters and terminators, and selection markers, have been developed to enable the efficient transfer of specific DNA sequences into plant cells. Improved techniques for the regeneration of plants from the transformed cells have also been devised.

In 1986, it was reported that plants transformed with the coat protein gene of TMV were protected when inoculated with the virus (104). Later, part of the coding region of a replication-associated protein of the same virus was used to produce resistant transgenic plants (49). These demonstrations have generated extensive and ongoing efforts to produce plants that exhibit "pathogen-derived resistance" to virus diseases. Plants have been transformed with various viral structural and nonstructural genes (and fragments and mutants thereof), as well as satellite and defective-interfering RNA sequences, ribozymes, antisense RNA sequences, and antiviral antibody genes (42,128). There is a high level of expectation that application of some of these technologies will result in effective levels of resistance to viruses in plants grown in natural conditions, though such issues as the durability of resistance and the potential for undesirable ecological consequences arising from the presence of such sequences in plant genomes remain to be fully explored. In addition to these questions, the molecular mechanisms underlying the protective effect have not yet been determined, but the most interesting suggestion to date involves the induction, by the abnormal level or distribution of foreign RNA in the cell, of an antiviral state in which the RNA is recognized by some cytoplasmic factor (perhaps small cRNAs) that targets it for elimination by nucleolytic degradation (82).

Uses of Plant Viruses in Gene Transfer and Biotechnology Applications

In recent years, the development of new methods for the genetic modification of plants has promoted intensive efforts to produce plants with higher yields, enhanced nutritional value, greater resistance to pests and adverse environmental conditions, and other beneficial qualities. There is also considerable interest in the commercial development of plants as sources of pharmaceuticals and other high-value substances. The introduction of foreign genes into plants is a fundamental requirement of such efforts, and the plant viruses, especially the DNA viruses, have long been considered as possible vectors for these purposes. However, there are a number of technical problems with this approach, and other gene transfer systems, in particular the modified *Agrobacterium* Ti plasmid, have proven to be of more immediate and effective application. Nevertheless, there is considerable effort underway to use viral replicons, in particular those of some geminiviruses (116), for the development of transient gene expression systems with which to analyze foreign gene uptake and the regulation of gene expression in plants.

The characteristics of some RNA plant viruses are also being exploited to provide novel expression systems. An exciting example is the recently described introduction of epitopes derived from human immunodeficiency virus (HIV) or human rhinovirus 14 into one of the coat proteins of a virus that infects cowpea plants (103). The virus can easily be purified in large quantity from infected leaves and, in the case of the HIV chimera, has been shown to elicit anti-HIV antibodies in mice. Thus, this system offers an efficient epitope presentation system that may have important application to medical immunology.

The most significant contribution of a plant virus to the development of systems for constitutive and transient expression of foreign genes in plants has been that of the 35S promoter sequence of cauliflower mosaic virus (97). This strong, highly effective promoter has been included in the construction of a number of the *Agrobacterium* Ti plasmid vectors used for plant transformation. Another method of achieving enhanced levels of expression of foreign genes in plants has been the incorporation of the leader sequences of some RNA viruses into the 5' untranslated region of the transcription unit in gene transfer vectors (39).

THE PLANT VIRUSES

Over 1,000 plant viruses have been reported, and most have been placed in one of over 40 genera that are recognized by the International Committee on Taxonomy of Viruses (ICTV). The name by which a plant virus is known is usually a combination of the host plant in which the disease it causes was originally described and the type of symptom produced in that plant. There are single- and double-stranded DNA viruses that infect plants, but the great majority of plant viruses have RNA genomes, and most of these are nonenveloped, positive-sense ssRNA viruses (Fig. 1). None of the known plant viruses approaches the size and structural or genetic complexity of some of the largest viruses that infect animals.

Representatives of many of the genera of plant viruses are briefly described in this chapter. Some of them have been the subjects of intensive investigation of such features as the physical structure of the virions and the composition and organization of their genomes. With others, much less is known of these features, but they are of considerable interest because some aspects of their etiology or epidemiology are particularly fascinating or because they are responsible for diseases of serious consequence. Viruses that are not included should not be regarded as being of limited interest or importance. It will not be feasible to cite many of the important original publications that have contributed to our knowledge of the various viruses. Instead, reference will be made to pertinent review articles and, for a comprehensive treatment of plant virology, the reader may consult the excellent text and reference book by R.E.F. Matthews (88).

There are aspects of the subject about which little is known for any of the plant viruses, and treatment of some topics in the sections that follow will therefore be quite brief. The interactions that permit certain types of agents to transmit viruses to plants or that initiate the earliest events

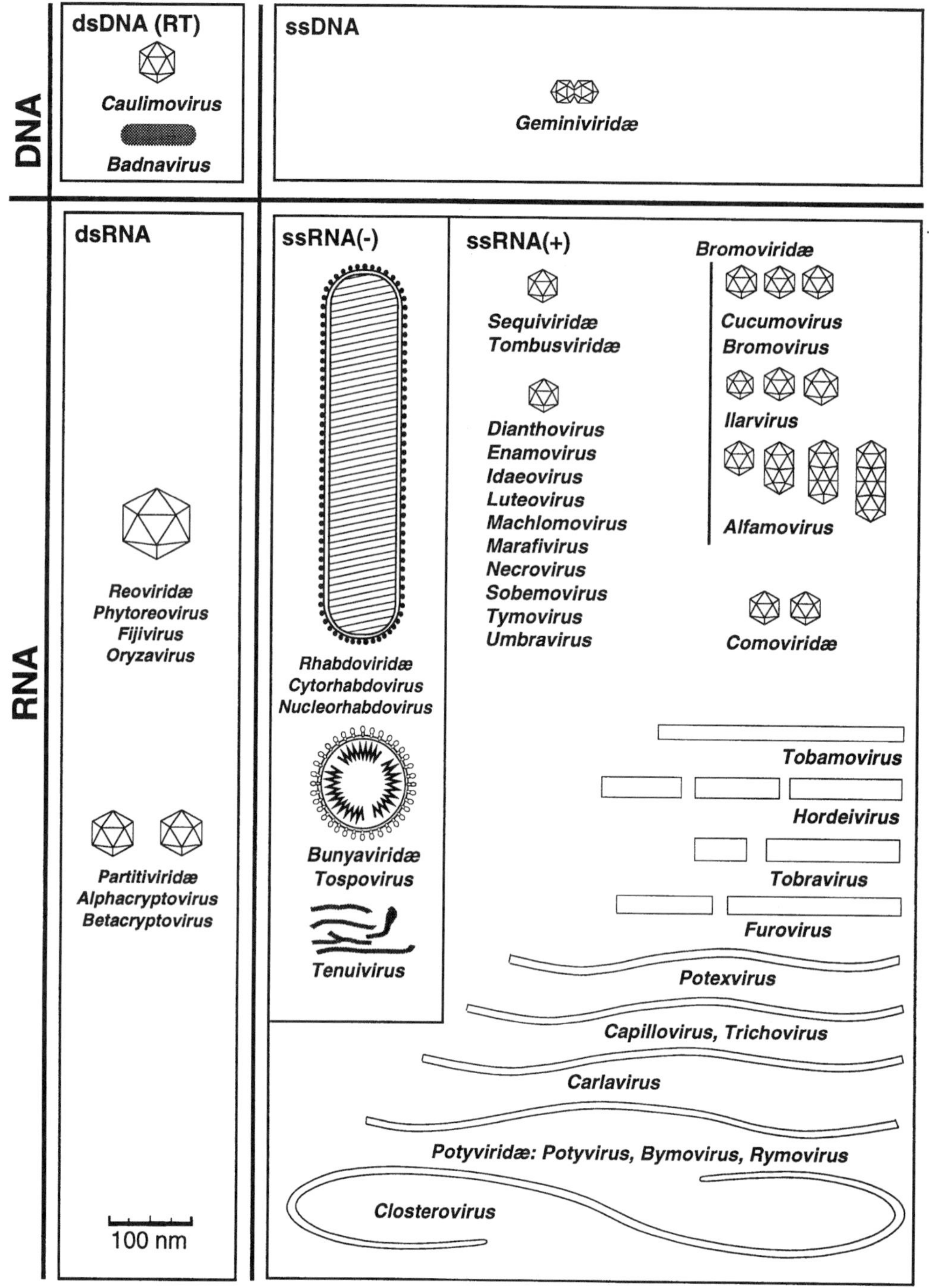

FIG. 1. Families and genera of plant viruses. (From Sixth Report of the International Committee on Taxonomy of Viruses, Springer-Verlag, with permission; courtesy of Dr. C.M. Fauquet.)

in the infection process are difficult to study and remain poorly understood. Detailed information concerning the replication of viral nucleic acids and the intercellular movement of virus in infected plants is just beginning to accumulate. The mechanisms by which viruses induce disease responses in plants have not yet been determined. Nevertheless, it seems certain that the increasingly powerful techniques of molecular and cellular biology should provide us with explanations of these phenomena within a relatively short period of time. On a broader and more serious scale, however, those who study viral ecology and epidemiology face problems of daunting complexity in trying to understand how viruses, vectors, plants, soils, weather, and the various activities of humans interact to produce epidemics.

dsDNA VIRUSES

Cauliflower Mosaic Virus

Classification

The dozen or more members of the genus Caulimovirus have rather narrow host ranges. The type member of the genus is cauliflower mosaic virus (CaMV), and, as noted above, this virus has played a very important role in the development of transformation and gene expression systems in plants. Like the hepadnaviruses (see Chapter 35), CaMV is referred to as a pararetrovirus because its genome is synthesized by reverse transcription though the viral DNA is not integrated into host chromosomes.

Virion Structure

Cauliflower mosaic virus consists of spherical particles of approximately 54 nm diameter, the components of which surround an empty core (27). The outer layer contains 420 protein subunits arranged with T=7 icosahedral symmetry. The DNA is associated with internally disposed domains of the coat protein. Other proteins may be present in the virion.

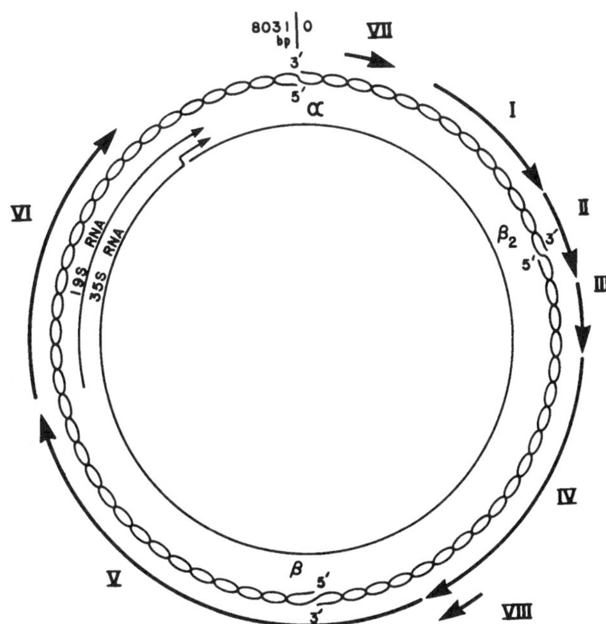

FIG. 2. Circular dsDNA genome of cauliflower mosaic virus (*interwoven lines*) showing three discontinuities (α, β_1, β_2); ORFs (*heavy arrows*) and the 19S and 35S RNA transcripts (*inner lines*) are represented. (Modified from Sheperd, ref. 114, with permission.)

Genome Structure And Organization

The genome of CaMV consists of circular dsDNA of ≈8 kbp that is organized into six closely-spaced or overlapping, major open reading frames (ORFs) (I–VI), one or two additional ORFs (VII, VIII), and a large intergenic region (Fig. 2). The DNA contains a single gap in the negative (transcribed) strand and two gaps (with overlapping residues) in the positive strand. A number of reviews of the caulimoviruses are available (30,114).

Open reading frame I encodes a 41- to 45-kD protein that is involved in cell-to-cell movement of virus in infected plants. The 18- to 19-kD protein product of ORF II mediates the transmission of virus particles from plant to plant by aphids, though the mechanism by which this activity occurs is not known. The 15-kD product of ORF III possesses nonspecific, double-stranded DNA binding activity, and an 11-kD processed fragment of the protein may be present in the surface of the virion. Open reading frame IV encodes a 57-kD polyprotein precursor of the viral coat protein. This is believed to be cleaved during or after encapsidation, perhaps by a protease encoded by ORF V, to yield the 37- to 42-kD subunits of mature virions that are phosphorylated by a virion-associated protein kinase. Open reading frame V also encodes the reverse transcriptase of CaMV. Unlike the gag-pol polyprotein in retroviruses, the 80-kD ORF V protein is not fused with the ORF IV protein but is the precursor of the 58-kD transcriptase and a 22-kD protease. The major constituent of the inclusion bodies in CaMV-infected cells is the 62-kD product of ORF VI. This regulatory protein is involved in the development of symptoms in infected plants and functions in posttranscriptional transactivation of the major RNA transcript.

Stages Of Infection

Cauliflower mosaic virus particles are transmitted from plant to plant by aphids in a nonpersistent manner, i.e., the insect is able to transmit virus for only a short period after it has been acquired. Infections can also be established by manual inoculation. There are very few details of the earliest events in the establishment of CaMV infections. Uncoated viral DNA is transported to the cell nucleus and, after removal of the overlapping nucleotides and closing of the gaps, it becomes associated with histones in a minichromosome configuration (Fig. 3). Two polyadenylated RNAs are produced by transcription of one strand of the viral DNA. These are the 35S RNA, a greater-than-genome-length transcript that is the template for DNA synthesis and the mRNA for most of the viral proteins, and the 19S RNA, which contains the ORF VI coding region. Other transcripts have been reported, but their importance has not been established.

The 35S promoter of CaMV has been well characterized (97) and has played a central role in the development and use of transient and constitutive transformation sys-

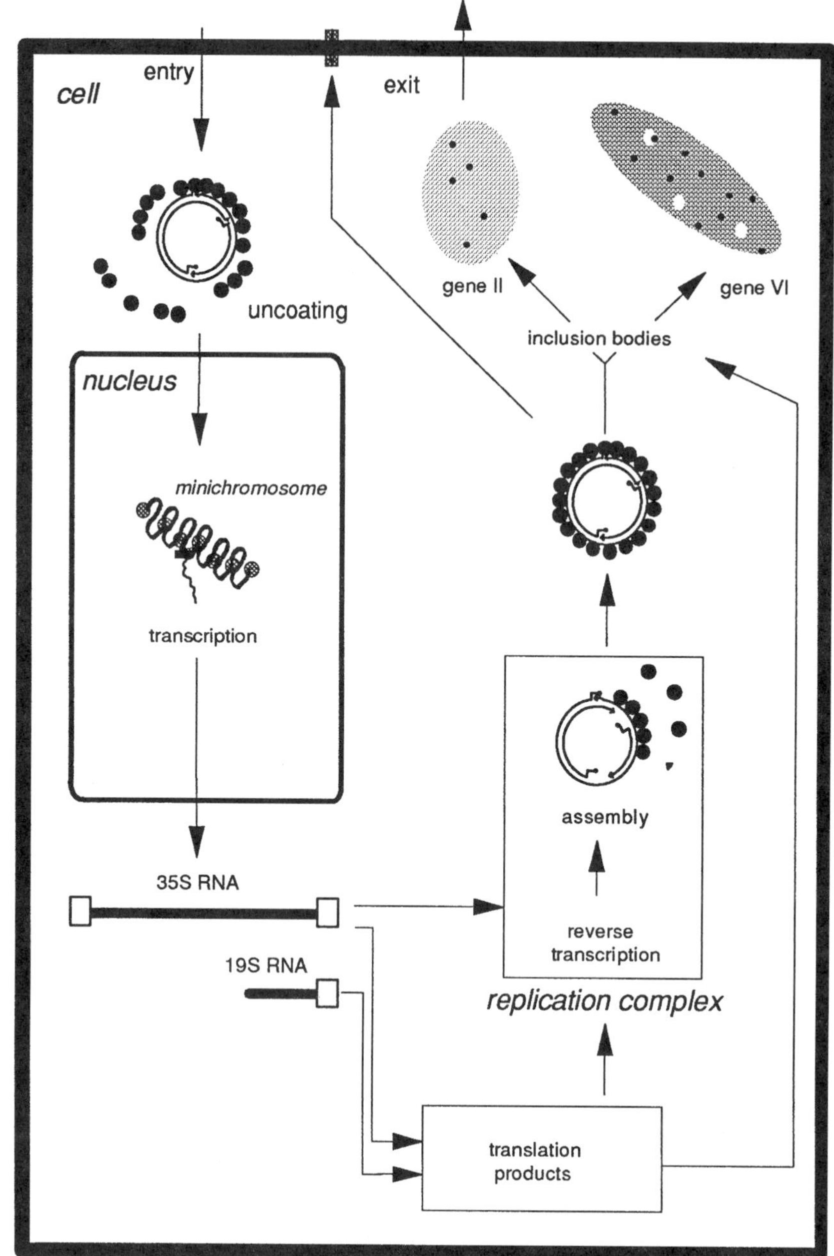

FIG. 3. Diagram of stages in CaMV infections. (Courtesy of Dr. S.N. Covey.)

tems in plants. This promoter consists of a TATA box 31 residues upstream from the transcription initiation site and an enhancer and several other sequence elements located in an upstream region of over 300 bp. Various domains and subdomains in the 35S promoter have been shown to act in various combinations to produce differential effects on the constitutive expression of genes in different parts of plants and at different stages of plant development (7). The 19S promoter has been less well characterized.

The two polyadenylated RNAs are transported to the cytoplasm, where they serve as mRNAs. The ORF VI protein is produced by translation of the 19S RNA. The polycistronic 35S RNA serves as the messenger RNA for the other proteins, and the manner in which these are expressed has for some time posed a perplexing and unresolved question. The 5' end of the 35S RNA consists of a 600-nt leader containing several very small ORFs, followed by ORF VII. Two mechanisms have been proposed to regulate the translation of the 35S RNA (59). *Cis*-acting elements in the leader sequence permit ribosomes to be shunted from the 5' end to a downstream position and to effect translation of the internal genes of the 35S RNA.

This requires a *trans*-acting mechanism involving the ORF VI protein (14,51).

The synthesis of progeny viral DNA is thought to occur in the cytoplasm in viral inclusion bodies composed of the ORF VI protein. A host methionyl tRNA serves as a primer to initiate reverse transcription of the RNA. A 180-nt direct repeat at each end of the RNA permits strand-switching at the 5' end of the template and the completion of synthesis of the negative-strand DNA. Removal of the used template, probably by the viral RNase H, leaves remnants of the RNA that may serve as primers for synthesis of plus strand DNA.

Reverse transcription of 35S RNA is thought to occur in a previrion complex with coat protein and, therefore, to be closely associated with the assembly of progeny virions. A zinc-finger-like motif in the coat protein may be involved in this interaction, but little is known about the mechanisms that are involved in assembly of caulimoviruses.

Not much is known about the manner in which viral activities in CaMV-infected cells are temporally regulated. The ratios of the amounts of supercoiled (minichromosome) DNA to transcript RNA are significantly higher at later than at earlier stages of infection, which may indicate a feedback mechanism, possibly regulated by the ORF VI protein, that ensures encapsidation rather than nuclear transport of progeny viral DNA (31).

There is not much information available concerning mechanisms involved in the interaction between CaMV and a susceptible plant that results in the onset of disease symptoms. Some of the efforts to gain an understanding of the pathogenesis of CaMV and of the problems encountered in such investigations have been reviewed (30).

Badnaviruses

A second genus with dsDNA genomes is the badnaviruses. Some members, such as cacao swollen shoot virus and rice tungro bacilliform virus (RTBV), cause devastating losses in important crop plants. The viruses have narrow host ranges and most are transmitted by mealybugs. Rice tungro bacilliform virus is transmitted by leafhoppers and is one of the two viruses that together cause the rice tungro disease. The badnaviruses, like CaMV, are pararetroviruses, but they consist of bacilliform rather than isometric particles. The movement protein, coat protein, protease, and reverse transcriptase may be encoded by a single ORF and arise from proteolytic processing.

ssDNA VIRUSES

Geminiviruses

Classification

Numerous very serious diseases of plants in tropical and subtropical regions are caused by geminiviruses, a family that contains approximately 50 members. These have been classified in three subgroups on the basis of host plant, insect vector, and genome structure. Geminiviruses in subgroup I (type member, maize streak virus) infect monocotyledonous plants; those in subgroups II (beet curly top virus) and III (tomato golden mosaic virus) infect dicotyledonous plants. Subgroup I and II geminiviruses have monopartite genomes and are transmitted by leafhoppers. Those in subgroup III have bipartite genomes and are transmitted by whiteflies. This classification scheme may be temporary, since new geminiviruses continue to be discovered and some of them have characteristics of more than one of the subgroups. In addition to their importance as destructive pathogens of important crops, the geminiviruses are of considerable interest as potential vectors for the expression of genes in plants (116). Recent reviews of the molecular biology of the geminiviruses are available (79,115).

Virion Structure

There have been no detailed structural analyses of geminivirus particles, but they are thought to consist of two joined, incomplete T=1 icosahedra containing 110 protein subunits (of approximately 28 kD) arranged in 22 capsomeres (54). The dimensions of each geminate particle are approximately 20×30 nm. The physical disposition of the DNA within the particle is not known. Particles of subgroup I viruses contain small DNA molecules that are thought to be primers for second-strand synthesis of DNA.

Genome Structure And Organization

Geminiviral genomes consist of one or two small (2.7–3.0 kb), covalently closed, circular ssDNA molecules (Fig. 4). All geminiviral DNAs are characterized by a bidirectional arrangement of conserved ORFs in which coding regions are positioned in both the virion (V) and complementary (C) senses of the DNA (92,121). The DNA components of the viruses with bipartite genomes are referred to as *A* and *B*. DNA A of African cassava mosaic virus (ACMV, a subgroup III virus) contains six ORFs, two in the virion and four in the opposite sense. DNA B has two ORFs, one in each sense. DNA A of another subgroup III virus, tomato golden mosaic virus, contains fewer ORFs. Each component contains an intergenic region of variable size (e.g., ≈300 b in DNA A and ≈600 b in DNA B of ACMV). With any given geminivirus, a sequence of ≈200 b (the common region) in the intergenic region of component A is the same as that in component B but, except for a highly conserved sequence of some 30 b, is different from the common region in other bipartite geminiviral genomes. The 30-b sequence has the potential to assume a stem-loop structure and contains an absolutely conserved nonanucleotide motif (TAATATTAC) in the loop sequence.

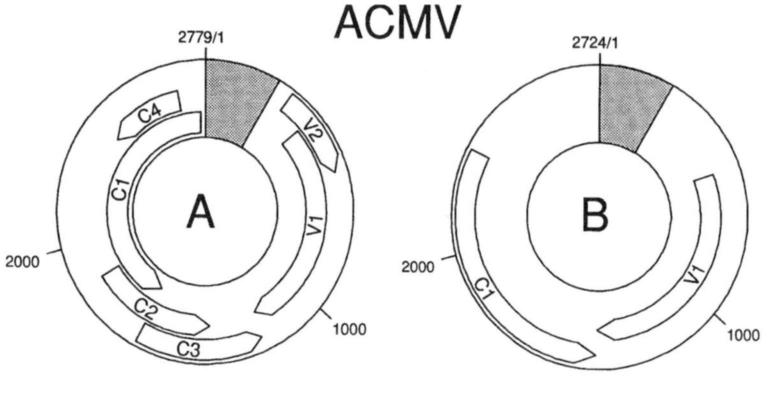

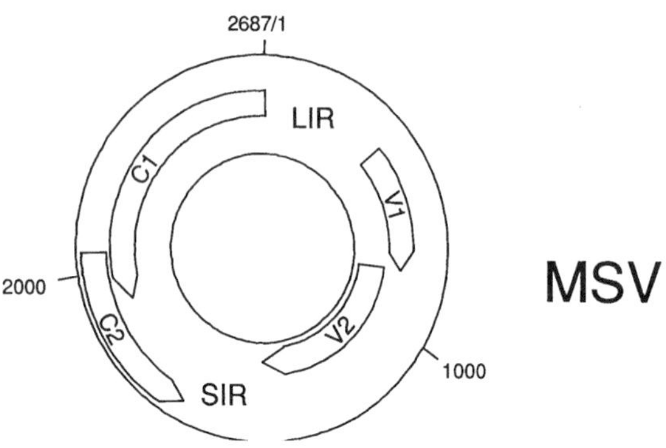

FIG. 4. Genome maps of two geminiviruses, African cassava mosaic virus (*ACMV*) and maize streak virus (*MSV*) showing locations of coding regions in the virion sense (*V1,V2*) and complementary sense (*C1,C2,C3,C4*). *Shaded portion of ACMV DNAs,* intergenic sequences; *LIR* and *SIR,* the large and small intergenic regions in the MSV genome. (Courtesy of Dr. J. Stanley.)

There are four ORFs in the genomes of the subgroup I monopartite geminiviruses, two (V1 and V2) in the positive strand and two (C1 and C2) in the complementary strand. The beet curly top virus (subgroup II) genome is actually similar to component A of the bipartite geminiviruses. These DNAs also contain a large intergenic region in which occurs the same strictly conserved nonanucleotide motif. In addition, subgroup I viruses contain a small intergenic region (SIR) in which is located the putative primer for second strand synthesis.

Functions have not yet been attributed to some of the ORFs in geminiviral genomes. The AV1 and V2 ORFs encode the capsid proteins of the bi- and monopartite genome geminiviruses, respectively, which, in addition to encapsidation of the viral DNA, appear to be determinants of insect vector specificity. The AC1 and AC3 proteins are reported to be involved in the replication of viral DNA (41,55). The DNA B gene products of the bipartite genome viruses appear to be involved in movement of virus in infected plants. In the monopartite genomes, C1 and C2 are expressed as a fusion protein that is required for replication.

Stages Of Infection

Geminiviruses are transmitted from plant to plant by leafhoppers or whiteflies. The viruses are able to persist in their insect vectors for many days or even for the life of the insect. However, they are not replicated in their vectors. Many of these viruses cannot be manually transmitted, and some are restricted to phloem-associated cells in the vascular system of the infected plant. In some cases, *Agrobacterium* has been a useful way to introduce geminiviral DNAs into plants that cannot be mechanically inoculated.

Replication and transcription of geminiviruses occur in the nuclei of infected cells, and it is likely that both are highly regulated processes. Transcription occurs from both strands of the genomic DNAs, and transcripts isolated from tissues infected with a number of geminiviruses have been mapped. Members of subgroup I may use a splicing mechanism to produce the fusion protein that is involved in replication. With both the monopartite and bipartite genome geminiviruses, complicated patterns of transcription have been observed and have yet to be fully resolved. Replication occurs via a rolling circle mechanism in the nuclei of infected cells. The origin of replication has been mapped to common region sequences that include the conserved stem-loop structure (43,79).

Investigations of parts of geminiviral genomes that are involved in disease development are underway (115), e.g., symptom determinants in one of the bipartite viruses have been identified in DNA B.

Defective interfering DNAs have been detected in geminivirus-infected plants (117).

dsRNA VIRUSES

Plant Reoviruses

The family Reoviridae, which is described in detail in Chapter 24 in this volume and Chapters 51 and 53 in *Fields Virology,* 3rd ed., contains three genera that infect plants. Particles of those in the genus Phytoreovirus [type member, clover wound tumor virus (WTV)] do not have spikes and have genomes containing 12 segments of dsRNA. Members of the genus Fijivirus (sugarcane Fiji disease virus) have ten-segmented genomes and spiked cores. Particles of the rice ragged stunt virus (genus Oryzavirus) are more similar to those of the Cypoviruses than to the other plant reoviruses; in other respects they resemble the Fijiviruses. The molecular biology of these viruses has been reviewed (96).

Though most of the plant reoviruses are limited in host range to members of the plant family Graminae, they cause some serious diseases in such important crops as rice, sugarcane, and maize. The viruses are transmitted by leafhoppers and planthoppers. It was with WTV that some of the classic experiments demonstrating the multiplication of plant viruses in the bodies of their insect vectors were performed (9).

Cryptoviruses

Two genera in the family Partitiviridae infect plants (11). The type members are white clover cryptic viruses I and II. These viruses cause no or very mild symptoms and accumulate to only very low levels in infected plants. Transmission vectors have not been identified.

RNA VIRUSES WITH NEGATIVE-SENSE OR AMBISENSE GENOMES

Members of the Rhabdoviridae and Bunyaviridae that infect plants have many of the general features of related viruses that infect animals. Detailed descriptions of these features are provided in Chapters 19 and 22 in this volume and Chapters 38 and 48 in *Fields Virology,* 3rd ed. and will therefore not be included here.

Rhabdoviridae

Rhabdoviruses infect a wide variety of plant species, though each has a restricted host range. Most of these viruses are transmitted by, and are able to be replicated within, leafhoppers, planthoppers, or aphids. Two genera have been established. Members of the genus Cytorhabdovirus (type member, lettuce necrotic yellows virus) mature in association with membranes of the endoplasmic reticulum and accumulate in the cytoplasm in viroplasms; those of the genus Nucleorhabdovirus (potato yellow dwarf virus) multiply in the nuclei and accumulate in the perinuclear spaces in infected cells. A review of these viruses has been presented (69).

Most plant rhabdoviruses consist of bacilliform rather than bullet-shaped particles. The characteristics of the particles, their negative-strand genomes, and the viral proteins are similar to those of the animal rhabdoviruses. The molecular biology of sonchus yellow net virus has probably been more extensively investigated than that of the other plant rhabdoviruses.

Tomato Spotted Wilt Virus

Tomato spotted wilt virus (TSWV) is the type member of the genus Tospovirus in the family Bunyaviridae (47). The virus infects an extremely broad range of plants, reportedly over 500 species in over 50 plant families, and can cause very serious losses of floral, fruit, and vegetable crops in many parts of the world. A wide range of types of symptoms is produced in infected plants. The Tospovirus genus contains a half-dozen or so other members or possible members.

Tomato spotted wilt virus particles are spherical, enveloped particles of approximately 100 nm diameter. The genome consists of a negative sense segment (L RNA) and two ambisense segments (the S and M RNAs). The L RNA encodes the putative polymerase; the other genome segments encode various structural and nonstructural proteins.

Tomato spotted wilt virus is transmitted by several species of thrips and is replicated in the insect vector. In infected plant cells, several types of cytopathic structures and aggregates of viral proteins can be observed and mature virus particles are produced by budding of the nucleocapsids through membranes of the endoplasmic reticulum.

Tenuiviruses

Classification

Rice stripe virus (RSV) is the type member of the genus Tenuivirus, of which there are about seven members. In several respects, these viruses are similar to the phleboviruses and uukuviruses of the family Bunyaviridae. Tenuiviruses infect graminaceous plants and cause serious diseases in rice and maize crops. They are transmitted from plant to plant by delphacid leafhoppers. Virus multiplies in the insects and is retained during moults and passed to the progeny.

Particle Structure

The segments of Tenuivirus genomes are separately packaged in narrow, filamentous, sometimes circular particles that have not yet been well characterized (67). The particles may be enveloped, and they probably contain an RNA-dependent RNA polymerase.

Genome Structure and Organization

The genome of maize stripe virus (MStV) consists of five ssRNA molecules, and that of RSV has been shown to have four but may also have a fifth component. Nucleotide sequences of RNAs 2–4 of RSV and 3–4 of MStV reveal these RNAs to have ambisense coding regions; RNA 5 of MStV is of negative polarity. The ORF in the complementary sense of RNA 3 encodes a 35-kD structural protein and that in the virion sense of RNA 4 encodes a major nonstructural protein of ≈20 kD that forms large inclusion bodies in the cytoplasm of infected cells. The functions of the other ORFs and the sequence of RNA 1 have not yet been reported.

The terminal regions of the viral RNAs contain complementary sequences that, in the virus particles, are thought to form panhandle-like structures. Heterogeneous nonviral leader sequences at the 5'-termini suggest that a cap snatching mechanism is involved in the initiation of synthesis of viral mRNAs (61).

POSITIVE-SENSE ssRNA VIRUSES

The majority by far of the viruses that infect plants are nonenveloped and have positive-sense ssRNA genomes. Most of these viruses consist of rod-shaped, filamentous or spherical particles (Fig. 5). In recent years, they have been placed in three supergroups as a result of computer analyses of amino acid sequences in some of the nonstructural proteins encoded by their genomes (77). The picornavirus-like members of supergroup I include the poty- , nepo- and comoviruses of plants, while the tombusviruses and some luteoviruses fit in supergroup II. Many other RNA plant viruses can be placed in the large alphavirus-like supergroup III. The following descriptions will include members of each supergroup but will be organized on the basis of the number of distinct molecules in their genomes.

MONOPARTITE GENOMES

The entire genome of these viruses consists of one molecule of ssRNA that is encapsidated in elongated or isometric particles. Representatives of both types of particle morphology will be presented here; information concerning others can be found elsewhere (88).

Tobacco Mosaic Virus

Classification

Tobacco mosaic virus (TMV) is the type member of the tobamovirus genus, which contains approximately 15 additional viruses. Tobamoviruses are found in most parts of the world and in some crop situations (e.g., greenhouse-grown tomatoes) can cause very significant losses.

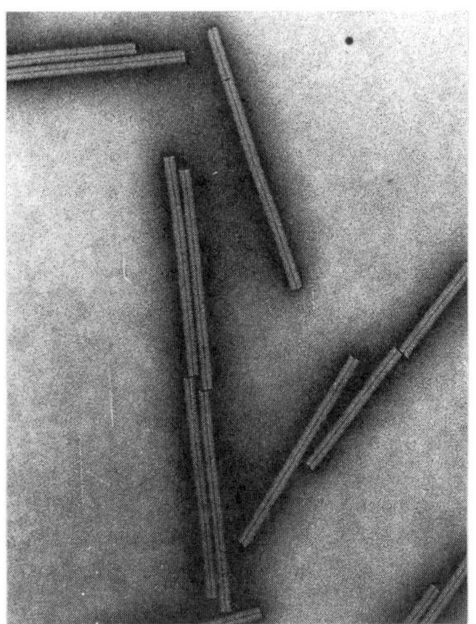

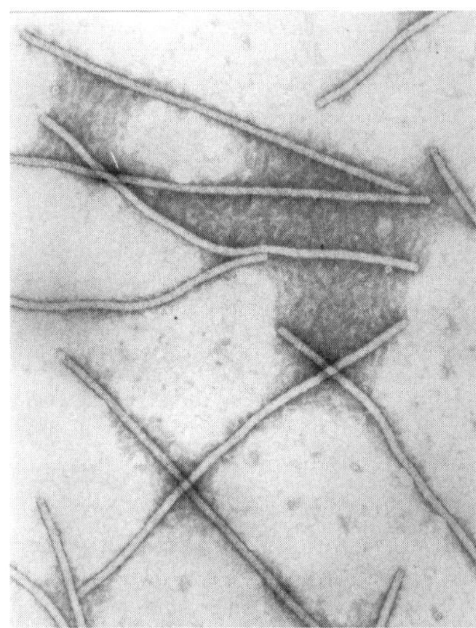

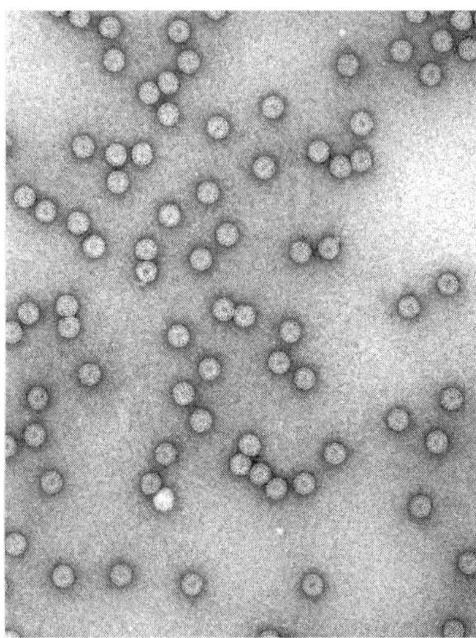

FIG. 5. Electron micrographs of plant viruses with rod-shaped, filamentous or icosahedral particles. **Left:** Tobacco mosaic virus (courtesy of Drs. J. Finch and P.J.G. Butler); some particles appear as end-to-end aggregates; the central hole and helical packing of subunits are visible. **Center:** Bean yellow mosaic virus (courtesy of Dr. S.A. Ghabrial). **Right:** Cucumber mosaic virus (from Palukaitis et al., ref. 98, with permission; courtesy of Dr. J.W. Randles). Particles in each panel are shown at approximately ×85,000 magnification.

Virion Structure

Many of the pioneering studies of the structure and chemical composition of viruses involved TMV. The virion is a rigid rod of approximately 18 nm diameter and 300 nm length with a central hole of 4 nm diameter and is extraordinarily stable (Figs. 5,6). The structures of the disk-shaped aggregate of the coat protein (10) and of the virion (95) have been determined by x-ray diffraction analysis. In the virion, the 2,130 coat protein subunits are arranged in a right-handed helix with $16^{1}/_{3}$ subunits per turn. The RNA is located between successive turns of the helix with the phosphates 4 nm from the axis of the helix.

Genome Structure and Organization

The genome of TMV (Fig. 7) is a monopartite, positive-sense, 6.4-kb ssRNA that contains 4 ORFs. The 5'-terminus is capped and the 3' end bears a tRNA-like configuration that is capable of accepting histidine. The 3' end also contains 5 pseudoknots, 2 of which are in the tRNA-like structure.

The first ORF encodes a 126-kD protein, but readthrough of the UAG termination codon generates a 183-kD protein. These two proteins are involved in replication of viral RNA, with the 126-kD protein containing consensus methyltransferase and helicase motifs and the readthrough part of the 183-kD protein containing amino acid sequences typical of viral RNA-encoded polymerases.

The last five codons of the 183-kD reading frame overlap the 30-kD reading frame, and the latter terminates two nucleotides prior to the initiation codon of the ORF that encodes the 17.5-kD coat protein. The 30-kD and coat proteins are translated from two 3'-coterminal subgenomic RNAs. A review of the organization and expression of the tobamoviral genome is available (33).

Stages of Infection

Entry and Uncoating

Some aspects of the interactions of TMV and other plant viruses with their hosts and of the difficulties encountered in trying to define these interactions have been reviewed (130). A diagram representing the "life cycle" of TMV is presented in Fig. 8; many of the features are shared by other positive-strand ssRNA viruses.

There do not appear to be insect vectors of TMV, but the virus is readily transmitted to plants during contact with infected plants or virus-contaminated implements. There has been some disagreement as to whether the small wounds produced in plant tissues during such contact, necessary for infection to be established, result in the placement of virus particles directly in the cytoplasm of epidermal cells (125). It has been proposed that uncoating begins by removal of a few coat protein subunits from the end of the virion bearing the 5'-terminus of the RNA and that ribosomes then drive a cotranslational disassembly process in which protein subunits are very rapidly dislodged while the 126-kD and 183-kD ORFs are being translated (113,127). It is not known how regions near the 3' end of the viral RNA become uncoated.

Translation, Transcription and Assembly

Synthesis of (+)- and (–)-sense viral RNA and subgenomic RNAs occurs in the cytoplasm of infected cells, probably in membrane-associated viroplasms. The 126-kD and 183-kD proteins are involved in these processes, but an *in vitro* replicase system that requires the addition of a specific template for activity has not yet been isolated and characterized. Synthesis of the 1.5-kb and 0.7-kb subgenomic RNAs, from which the 30-kD movement protein and the coat protein, respectively, are translated, is initiated at specific internal sites in full-length negative-sense RNA molecules. A third subgenomic RNA of ≈3 kb has also been detected in infected cells.

A detailed model for the *in vitro* assembly of virus particles was proposed many years ago (21,80), and it seems reasonable to suppose that this mechanism also operates in infected cells. Assembly begins by the association of an aggregate of coat protein molecules with a specific origin-of-assembly sequence (131) that includes nts 5444–5518 in the viral RNA (Fig. 9). The protein molecules have been assumed to be present as two-layer, "20 S" disks, each containing 34 protein molecules, though other types of aggregates have also been proposed. Part of the origin-of-assembly sequence is inserted—via the central hole of the disk—between two layers of protein molecules, and the disk then switches to a protohelical, "lockwasher-like" configuration. Disks are then successively and rapidly added to the growing helical rod as segments of the RNA bearing the 5'-terminus are drawn through the central cavity to become inserted between newly-formed turns of the helix (122). Encapsidation of the 3'-terminal part of the RNA occurs more slowly and probably involves smaller aggregates of the coat protein. The entire process of *in vitro* assembly of a virion takes approximately 6 minutes. It has recently been suggested that progeny TMV particles begin to accumulate some 40 minutes after inoculation of tobacco cells (129).

Massive amounts of TMV particles (10^6–10^7 per cell) are produced and may accumulate in large crystalline plates. Several grams of virus may be obtained from one kilogram (fresh-weight) of mosaic-diseased tobacco leaves.

Movement of Virus in Infected Plants

As noted above,the genomes of many plant viruses encode a protein that is required for the movement of virions or viral nucleic acids from cell to cell in the infected plant.

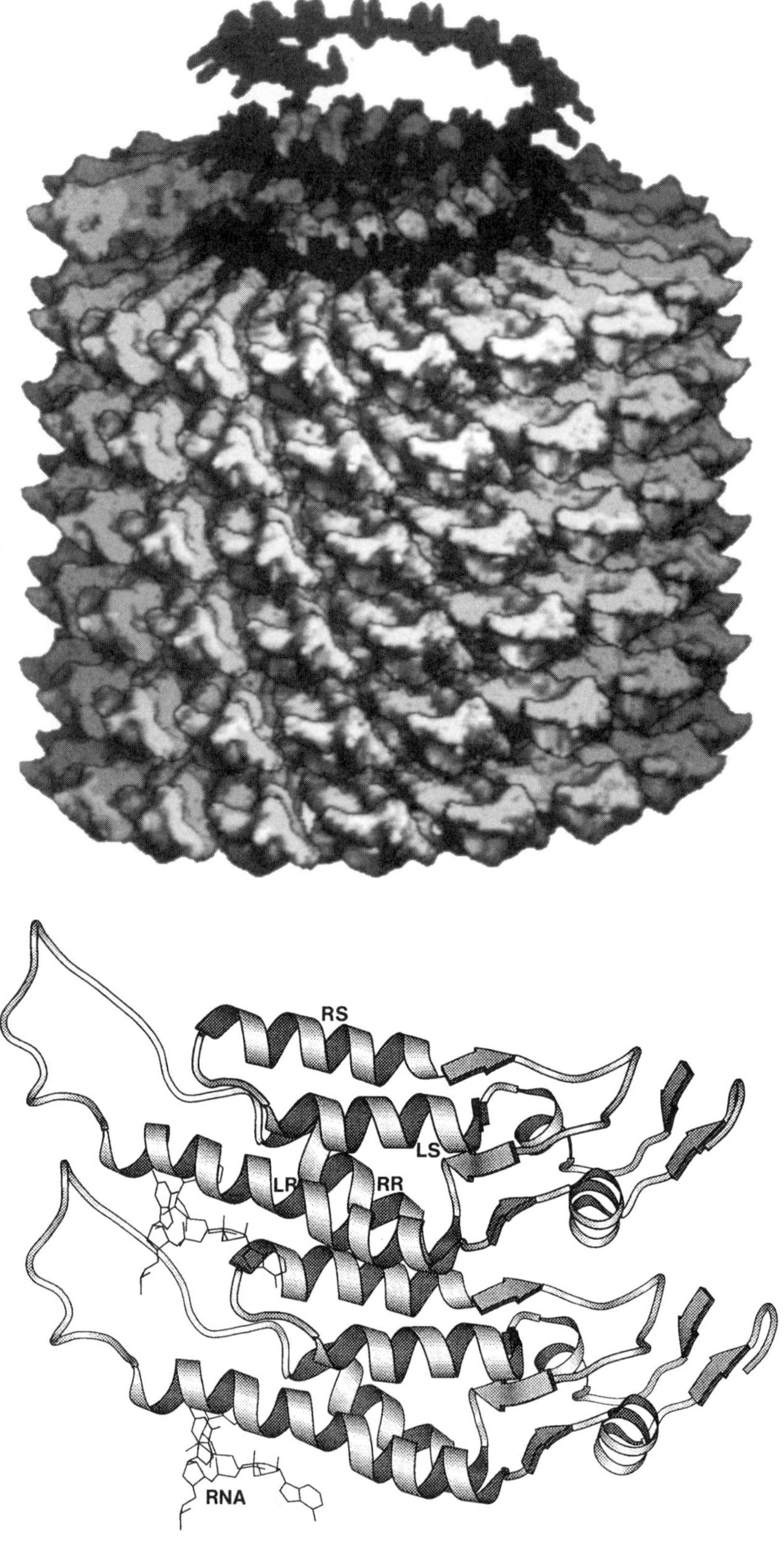

FIG. 6. Upper panel: Representation of part of TMV particle showing helical arrangement of coat protein subunits (*light gray*) and RNA (*dark gray*; shown extending beyond helix for clarity). (From Namba et al., ref. 94, with permission.) **Lower panel:** Ribbon diagram of two TMV coat protein subunits in adjoining turns of the helix. Four regions of α-helix (*LS, RS, LR* and *RR*) in the proteins and the positions of nucleotides (*RNA*) in the virion are shown. Central axis of particle is to the left; exterior of particle is to the right. (From Wang and Stubbs, ref. 126, with permission.)

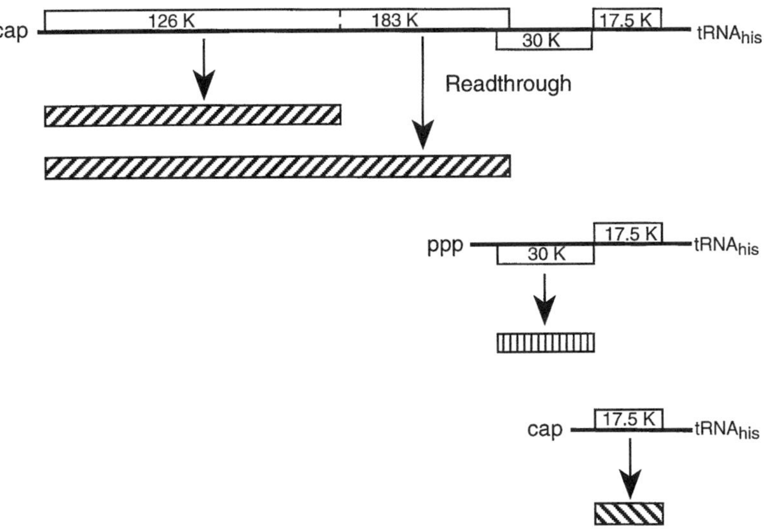

FIG. 7. Genome organization and expression of TMV. In this and similar figures, genomic and subgenomic RNAs are shown (*thick horizontal lines*), as are ORFs (*open rectangles*) and proteins (*shaded or open rectangles*). Replication-associated and coat proteins are shaded (*with diagonal lines / and \, respectively*) as are movement proteins (*shaded with vertical lines*).

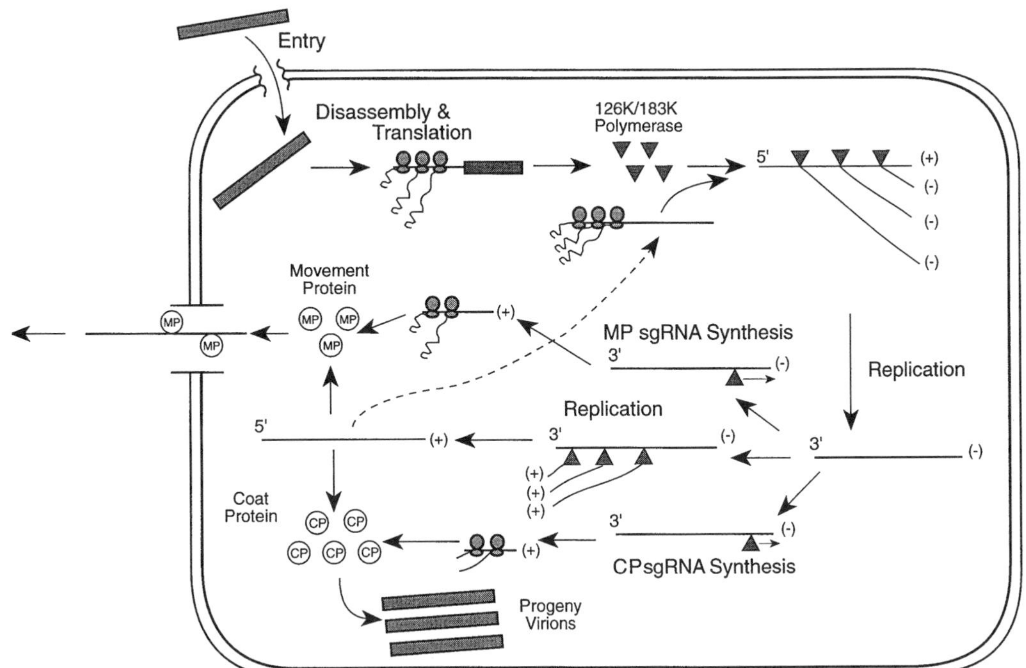

FIG. 8. Diagram of stages in TMV infections. All the events shown are presumed to occur in the cytoplasm of infected cells.

The 30-kD protein of TMV (81) is the best characterized of these movement proteins. This protein is thought to modify the plasmodesmata so as to permit TMV RNA molecules, configured in an extended, unfolded form by association with the 30-kD protein, to be transported to neighboring cells (Fig. 10). Rapid movement of virus to distant parts of the plant occurs through the companion cells and sieve elements of the phloem. The viral coat protein is involved in this process (32).

Tobacco varieties into which a single, dominant resistance gene (the N gene) has been introduced, from a species related to tobacco, produce localized necrotic lesions rather than systemic mosaic symptoms when inoculated with TMV (Fig. 11). In addition to the restriction of virus to the site of initial infection, this type of "hypersensitive response" may activate nonspecific defense mechanisms in the plant at some distance from the site of initial infection. These responses involve the synthesis of a number of host-

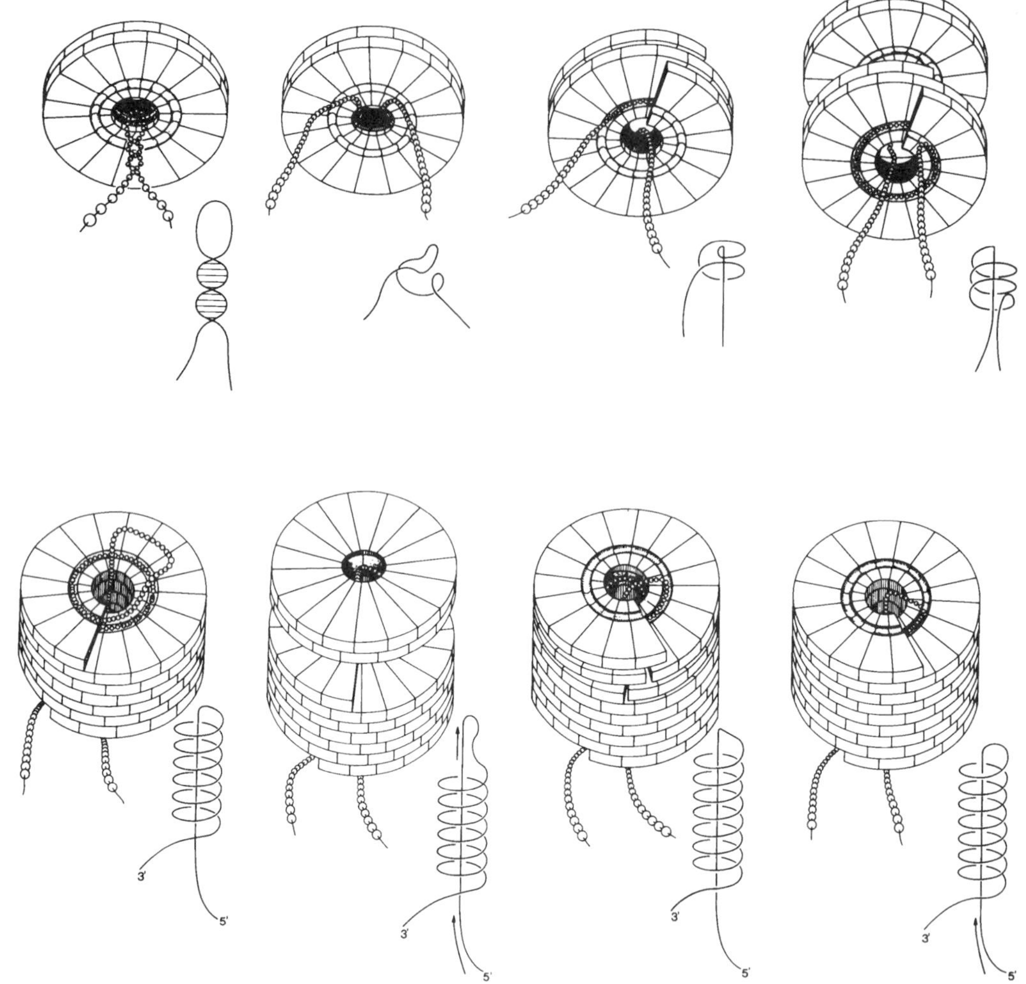

FIG. 9. Proposed mechanism of assembly of TMV. **Upper Row:** Initial interaction of origin-of-assembly sequence in viral RNA with 20 S disk and switch to helical configuration. **Lower Row**: Elongation of helical rod by addition of disks and insertion of RNA via the central hole. (From Butler, ref. 20, with permission.)

encoded proteins, among which are the so-called pathogenesis-related proteins (12).

Potato Virus X

Classification

Potato virus X (PVX) is the type member of the genus Potexvirus, which contains about 40 recognized or possible members (106).

Virion Structure

The virions are flexuous rod-shaped particles of ≈500 nm length and ≈13 nm diameter. Each particle consists of a single molecule of RNA and ≈1,400 coat protein subunits of ≈25 kD arranged in a helical structure with ≈9 subunits per turn of the helix.

Genome Structure and Organization

The genome of PVX is monopartite, positive-sense ssRNA of ≈6.4 kb that contains m^7GpppG at the 5′-terminus and is polyadenylated at the 3′ end (Fig. 12) (62). The largest of the five ORFs encodes the putative polymerase (166 kD). A "triple gene block" of partially overlapping ORFs encodes nonstructural proteins of 25 kD, 12 kD, and 8 kD. These proteins are thought to be involved in cell-to-cell movement of virus, though the 25-kD protein contains amino acid sequence motifs more similar to those in helicases than in

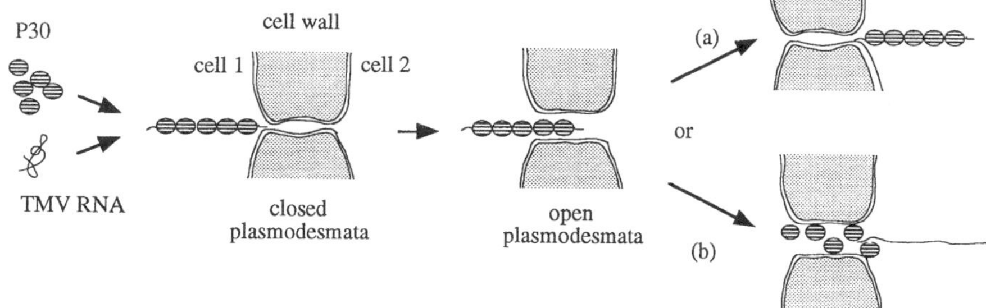

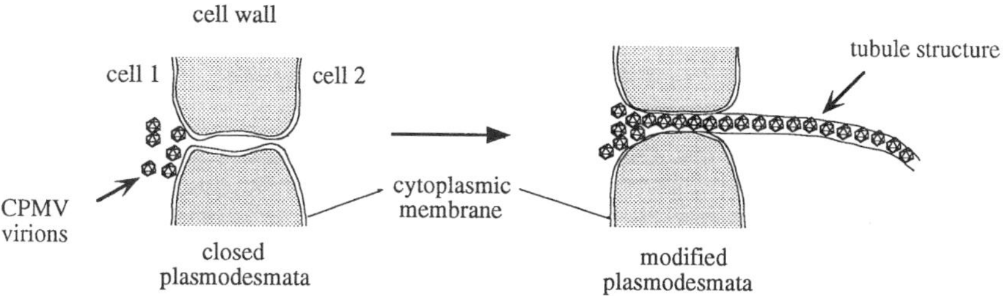

FIG. 10. Proposed mechanisms of cell-to-cell movement of virus through plasmodesmata in infected plant tissues. **Upper Diagrams:** Movement of TMV RNA mediated by virus-encoded 30K protein. **Lower Diagrams:** Movement of cowpea mosaic virus (*CPMV*) particles through tubular structures. (From McLean et al., ref. 90, with permission.)

movement proteins. A 2.1-kb subgenomic RNA is probably involved in the translation of the triple gene block proteins. The coat protein gene is near the 3′ terminus of the genome and is expressed via a 0.9-kb subgenomic RNA.

Stages of Infection

Little information is available concerning the early events in PVX infections. Plants usually become infected because of the use of infected stock for vegetative propagation or by mechanical or contact inoculation. Large inclusions are produced in the cytoplasm of infected cells. Assembly of potexvirus particles *in vitro* may begin by interaction between a double-layered, disc-like aggregate of $\approx$17 coat protein subunits and a sequence near the 5′ end of the viral RNA. Potato virus X particles are quite stable and are produced in large amounts in infected cells.

Some of the classic experiments involving the characterization of mixtures of viruses in infected plants and the resistance of plants to viruses were conducted with PVX some 60 years ago. Because of the availability of potato lines with varying degrees of resistance to PVX and of isolates of the virus that produce distinctive reactions in those lines, much of the recent research on PVX has been concerned with viral and plant factors that are involved in resistance mechanisms (50).

Potyviridae

Classification

The family Potyviridae is the largest of all the plant virus families. Well over 10% of the plant viruses that have been identified are members or probable members of this fam-

FIG. 11. Systemic mosaic symptoms (*patterns of light and dark green areas in leaf at left*) and localized necrotic lesions (*right*) in TMV-infected tobacco leaves. The leaf on the right is from a plant that bears the N gene (see text).

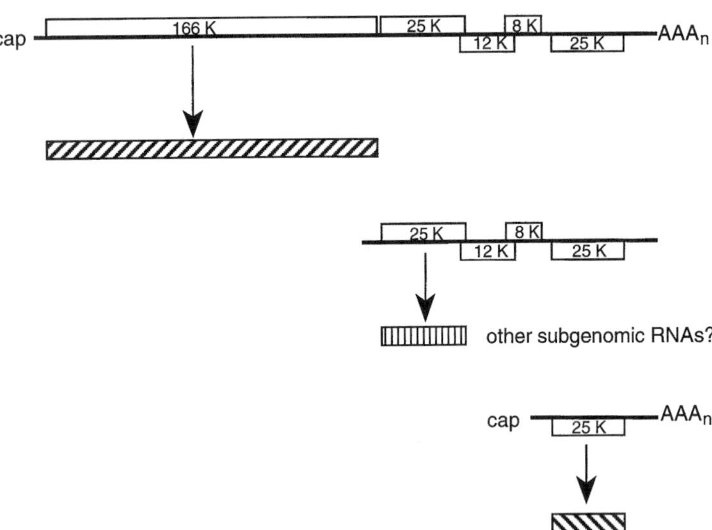

FIG. 12. Genome organization and expression of potato virus X.

ily, though this statistic may reflect a well-developed propensity for taxonomic subdivision among potyvirologists. Three genera have been designated on the basis of the agent of transmission of virus from plant to plant: Potyvirus (aphids), Bymovirus (fungi) and Rymovirus (mites). The type species of the potyviruses is potato virus Y, but the molecular biology of other viruses in the genus [plum pox virus, tobacco etch virus (TEV), and tobacco vein mottling virus (TVMV)] has been more extensively characterized. These are monopartite RNA viruses; the bymoviruses have bipartite genomes.

Collectively, the potyviruses are considered the most destructive of the plant viruses. Most of them have narrow host ranges, but most of the world's important vegetable and field-crop growing areas sustain serious losses to these viruses. In some cases, mixed infections involving more than one potyvirus or a potyvirus and another virus may produce much more serious diseases than are caused by single infections.

Virion Structure

Virions of potyviruses are flexuous, filamentous particles of 11–12 nm diameter and 680 nm or more in length (see Fig. 5). Little detail of the structure of these viruses has been reported.

Genome Structure and Organization

The potyviral genome is positive-sense ssRNA of ≈9–10 kb that is polyadenylated and has a protein (VPg) covalently linked to the 5′-terminus (Fig. 13). The RNA contains a single, long ORF encoding a polyprotein of ≈340 kD that is processed by three virus-encoded proteinases (P1, HCpro, and NIa) in co- and posttranslational cleavage reactions. The CI protein possesses RNA helicase activity, and the NIb protein is the putative RNA-dependent RNA polymerase. The 6K protein may regulate nuclear transport of the NIa protein of some potyviruses. The ≈30-kD coat protein encapsidates the viral RNA and is involved in the spread of virus from plant to plant by aphids. Some of the other potyviral proteins also have more than one function. A review of the potyviral genome has been published (111).

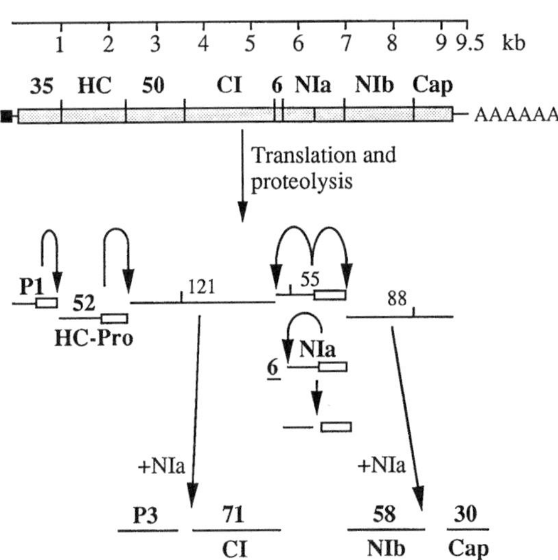

FIG. 13. Genome organization and expression of tobacco etch virus. The names and sizes (in kD) of the proteins (*vertical lines*) that are produced by processing of the polyprotein (*shaded rectangle*) are shown. The proteolytic domains of the NIa, HCpro, and P1 proteinases are shown (*open rectangles*), as are the cleavages reactions for which they are responsible (*curved arrows*). (From Restrepo-Hartwig and Carrington, ref. 109, with permission.)

Stages of Infection

Viruses in the genus Potyvirus can be transmitted manually but in natural settings are moved from plant to plant by aphids in a nonpersistent manner. The requirement for the HCpro protein for aphid transmission has been extensively investigated by activity assays and genetic analyses (3,120). In addition, an amino acid sequence motif near the N-terminus of the viral coat protein is a determinant of aphid transmissibility of virus particles (4). Some potyviruses are transmitted to new generations of plants through infected seeds.

Very little is known about the earliest stages of infections caused by potyviruses. However, there have been elegant genetic and biochemical studies of the processing of the TEV polyprotein and Fig. 13 shows the cleavages performed by the three virus-encoded proteinases. The serine-like NIa proteinase, which cleaves itself from the polyprotein and performs several other processing reactions, has been extensively characterized (24). The P1-HCpro and HCpro-P3 cleavages are carried out by the other two proteinases (23,124).

Potyvirus-infected cells contain one or more types of structures that have been extensively examined by electron microscopists. The CI protein aggregates in the cytoplasm of potyvirus-infected cells to produce intricate and distinctive structures (termed cylindrical inclusions) that, in different planes, may appear as pinwheels (Fig. 14), laminated aggregates, or scrolls. Two of the other nonstructural proteins, NIa and NIb, produce crystalline inclusions in the nuclei of cells infected with some potyviruses. In addition to being a proteinase, the NIa protein also has a VPg domain. A protein that mediates movement of virus from cell to cell has not yet been identified.

Closteroviruses

Another distinctive group of rod-shaped viruses are the ≈30 members of the genus Closterovirus, some of which cause very serious diseases (29). Infection of citrus trees by citrus tristeza virus (CTV) continues to produce particularly severe and long-term losses; beet yellows virus (BYV) causes large losses in sugarbeet crops. The ≈2,000 × 12 nm flexuous, filamentous particles of CTV are the longest of all the known plant viruses. The CTV genome is the largest (≈20 kb) of any of the ssRNA plant viruses but has not yet been fully characterized. The 15.5-kb genome of BYV has been shown to contain nine ORFs. A putative 348-kD fusion protein is thought to contain proteinase and replication-associated activities; other ORFs encode the coat protein and a heat shock protein-like product. The genomes of both viruses contain an ORF that could encode a protein very similar to the coat protein encoded by an adjacent ORF. Citrus tristeza virus and BYV are transmitted by aphids in a semi-persistent manner, i.e., the virus may be transmitted for a day or more after it has been acquired by the insect but does not multiply in the insect. Other closteroviruses have mealybug or whitefly vectors.

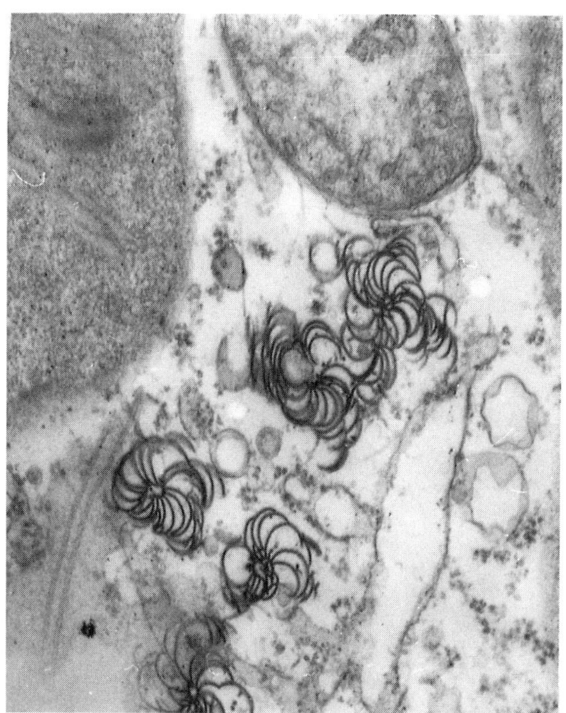

FIG. 14. Cylindrical inclusions (in the pinwheel configuration) produced in the cytoplasm of a tobacco cell infected with tobacco vein mottling virus. Parts of two chloroplasts (*top left and top right*) and a mitochondrion (*top center*) are present. (Courtesy of Dr. E. Ammar and Ms. U. Jälfors.)

Turnip Yellow Mosaic Virus

Classification

Experiments with turnip yellow mosaic virus (TYMV) provided one of the earliest indications of the infectious nature of viral RNA (84). Turnip yellow mosaic virus is the type member of the tymoviruses. There are about 20 other members, and they have rather narrow host ranges.

Virion Structure

Some of the pioneering studies of the structure of isometric viruses were conducted with TYMV (87). The particles are T=3 icosahedral structures of 29 nm diameter and the 180 20-kD protein subunits are arranged in distinct pentameric and hexameric clusters. Purified preparations contain a significant proportion of empty capsids.

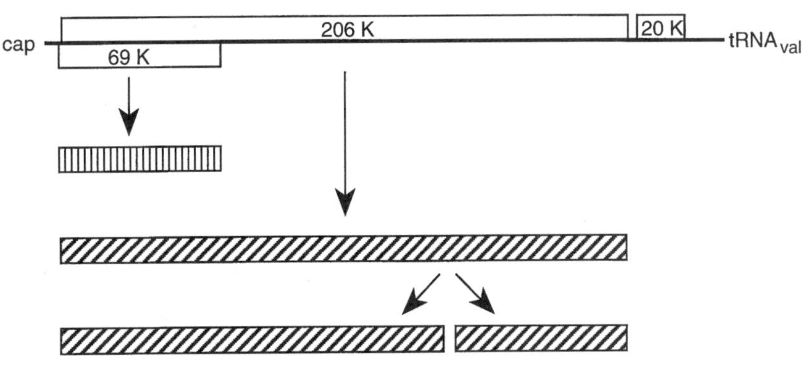

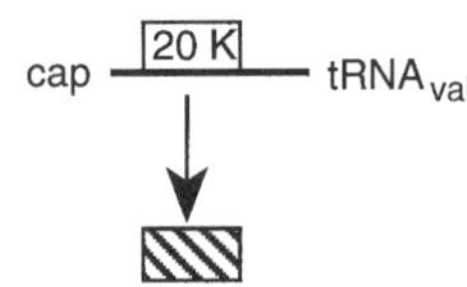

FIG. 15. Genome organization and expression of turnip yellow mosaic virus.

Genome Structure and Organization

The genome is capped, monopartite, positive-sense, 6.3-kb ssRNA (Fig. 15). Turnip yellow mosaic virus was the first virus shown to have an amino acid (valine) accepting, tRNA-like structure at the 3′ end of its RNA (101). There are three ORFs; the 206-kD protein encoded by the largest is proteolytically processed to yield proteins of 150-kD and 70-kD proteins which are involved in replication of viral RNA. An ORF overlapping this large ORF encodes a 69-kD protein that is probably involved in movement of virus in infected plants. The coat protein is encoded by a third ORF and is expressed from a subgenomic RNA.

Stages of Infection

Turnip yellow mosaic virus can be introduced into plants by mechanical inoculation or by beetles. Uncoating with this virus may occur by the loss of a few protein subunits from the capsid and the escape of the RNA through the resulting hole. Synthesis of viral RNA probably takes place in small vesicles that appear near the surfaces of the chloroplasts in infected cells. Assembly of progeny virus particles is thought to occur near the necks of the vesicles.

Detailed studies of the relationships between leaf ontogeny, the development of mosaic symptoms, and the distribution of virus in light and dark green areas in virus-infected leaves have been conducted with TYMV (88).

Tomato Bushy Stunt Virus

Classification

Tomato bushy stunt virus (TBSV) is the type member of the tombusvirus genus which consists of a dozen members. To a large extent, this virus has long been considered more as a model for fundamental studies than as a serious pathogen of plants. A review of the tombusviruses is available (85).

Virion Structure

Tomato bushy stunt virus was the first icosahedral virus for which the structure was determined to high resolution by x-ray crystallography (see Chapter 3). The virus particles are approximately 30 nm in diameter and contain 180 coat protein molecules in a T=3 icosahedral structure. There are distinct domains in the coat protein. The S domain forms the basic icosahedral shell and the P domain forms half of a surface protrusion. Another domain connects these two. The N-terminal part of the protein constitutes the R domain which is connected to the S domain. The S domain is folded into eight-stranded antiparallel β-sheets with jelly-roll topology. The viral RNA is situated internally and may interact with R domains of the protein subunits.

Genome Structure and Organization

The small, compact genome of TBSV is a monopartite, 4.8-nt, ssRNA that contains five ORFs (Fig. 16). The RNA is capped and nonpolyadenylated. Readthrough of the stop codon of the 5′-terminal ORF results in translation of the second ORF and produces the putative polymerase. The third ORF encodes the coat protein which is translated from a subgenomic RNA. The two 3′-terminal ORFs are overlapping and their products, one or both of which may be involved in cell-to-cell movement of virus, are probably translated from a second subgenomic RNA.

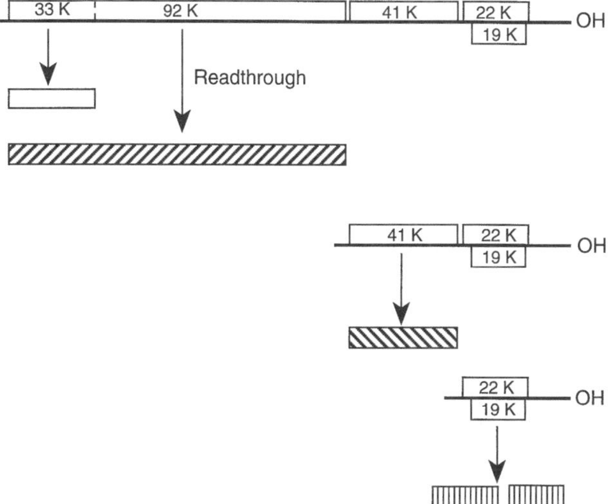

FIG. 16. Genome organization and expression of tomato bushy stunt virus.

Stages of Infection

Infections are readily established by manual inoculation procedures. The virions of TBSV and several other isometric viruses swell when divalent cations are removed and exterior carboxyl groups are deprotonated. This process may be involved in the uncoating of the viral RNA at the onset of infection of a plant cell (57). A model for the assembly of virions of similar structure, those of turnip crinkle virus, has been proposed (53).

Tomato bushy stunt virus was the first plant virus to be shown to produce DI RNAs in infected cells (58). These DI RNAs contain sequences from the 5′ untranslated region, the second (polymerase) ORF, and two regions near the 3′ end of the viral genome.

Luteoviruses

Classification

The genus Luteovirus is so named for the yellowish color of plants infected by the member viruses (Fig. 17). Barley yellow dwarf virus (BYDV), potato leafroll virus (PLRV) and beet western yellows virus (BWYV) are particularly serious pathogens. Barley yellow dwarf virus affects the production of grain crops in all parts of the world. The luteoviruses can be placed in two groups by the presence of distinctive ORFs in their genomes and by amino acid sequence differences in their putative polymerases. A number of reviews of the luteoviruses are available (86,112).

Virion Structure

Luteoviruses consist of small (25–26 nm diameter), isometric particles, and their capsids probably have 180 coat protein molecules (of about 24 kD) arranged with T=3 icosahedral symmetry.

Genome Structure and Organization

The monopartite, 5.5- to 6-kb ssRNA genomes of luteoviruses are nonpolyadenylated and have a 5′-terminal VPg (Fig. 18). These viruses employ a variety of strategies, including ribosomal frameshifting, internal initiation of translation, readthrough of in-frame termination codons, and generation of subgenomic RNAs for the expression of internal ORFs. Barley yellow dwarf virus (PAV strain) RNA contains six ORFs, the first two of which are overlapping in different reading frames, and are translated from the genomic RNA. The second ORF encodes the putative polymerase (60 kD) which is expressed by ribosome frameshifting as a fusion protein of 99 kD (16). The other four ORFs are probably translated from two subgenomic RNAs. The larger of these is involved in the synthesis of the coat protein, a smaller overlapping protein, and a fusion protein generated by readthrough of the coat protein termination codon (38). The 6.7-kD protein is probably translated from a smaller subgenomic RNA.

FIG. 17. Yellowing symptoms (*light areas*) produced in flag leaves of wheat plants infected by barley yellow dwarf virus (*BYDV*). (Courtesy of Dr. S.M. Gray.)

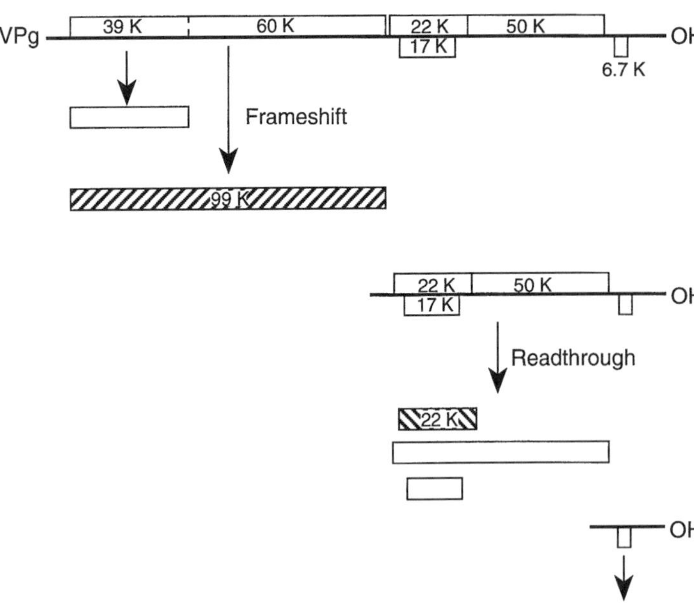

FIG. 18. Genome organization and expression of BYDV (PAV).

The genomes of PLRV and BWYV also consist of six ORFs but contain an extra 5′-terminal ORF and lack the 3′-terminal 6.7 ORF. The second and third ORFs contain helicase and polymerase motifs, and a fusion protein is produced by a translational frameshift in their overlapping regions (105,123). The three other ORFs are translated from a subgenomic RNA and probably utilize the same internal initiation and stop codon suppression mechanisms as described above.

Stages of Infection

Most luteoviruses have narrow host ranges. They are transmitted by aphids in which the virus does not multiply but may persist for prolonged periods, even for the life of the insect. Some luteoviruses exhibit a high degree of insect vector specificity, and some serve as helpers for aphid transmission from mixed infections. Since aphids are the only means of spread of luteoviruses in the field, they play the central role in the epidemiology of these viruses (66).

There is very little information concerning the early stages of infection or the replication of virus in luteovirus-infected cells. Virus particles are present in low amounts in infected plants. They are generally restricted to phloem cells in which a variety of cytopathological effects can be observed. The collapse of the phloem tissue accounts for the primary symptoms in diseased plants.

Virions of the ST9 strain of BWYV contain a novel type of RNA of ≈2.8 kb that, unlike a satellite RNA, is capable of replication in the absence of the viral RNA (99). The presence of this associated RNA results in a more serious disease and the accumulation of greatly increased amounts of virus in infected plants. Another member of the same group as BWYV contains a satellite RNA.

The etiology of some plant virus diseases is extraordinarily complex, and an interesting example involves another luteovirus. Groundnut rosette is a devastating disease of a vitally important crop in parts of Africa. Infected tissue contains the 4.6-kb, single-stranded genomic RNA of groundnut rosette virus (GRV) and subgenomic and satellite RNAs. These RNAs are thought to be encapsidated by the coat protein of groundnut rosette assistor virus (GRAV), a luteovirus. Aphid transmission of GRV is dependent on the presence of GRAV, and this dependence is mediated by the satellite RNA. In turn, the satellite RNA depends upon GRV for replication and on GRAV for transmission by aphids (93). Different types of symptoms may be produced by different isolates of the satellite RNA; satellite RNA-free GRV does not produce disease symptoms.

Sequiviridae

This is a newly recognized family of plant viruses whose monopartite RNA genomes encode single ORFs (108). Polyprotein processing presumably produces the individual proteins whose coding regions within the large ORF are arranged in a manner similar to those of the picornaviruses (see Chapter 16). One of these viruses is rice tungro spherical virus which, with a Badnavirus mentioned above, causes the serious tungro disease of rice.

BIPARTITE ssRNA VIRUSES

The first virus with which it was demonstrated that parts of the genome are separately encapsidated was tobacco rattle virus (TRV) (83). Like the monopartite RNA viruses of plants, those with bipartite genomes may consist of rod-shaped or isometric particles.

Tobacco Rattle Virus

Classification

The genus Tobravirus contains three members of which TRV is the type member (52). More than 100 species of plants are susceptible to this virus.

Virion Structure

The genome segments of TRV are contained in separate rigid, rod-shaped particles, the lengths of which vary among different isolates of the virus. RNA 1-containing particles are 185–197 nm in length, while those with RNA 2 are 50–115 nm in length. Both types are ≈25 nm in diameter and have a central, axial hole of ≈5 nm diameter. Each turn of the helix contains 25⅓ coat protein molecules, and the pitch of the helix is ≈2.5 nm. The RNA, which has a m^7GpppG cap at its 5′ end, is probably located between successive turns of coat protein subunits at a radius of ≈8 nm, with 4 nucleotides associated with each subunit.

Genome Structure and Organization

It was the discovery that the long TRV particles could infect plants in the absence of the short particles, but that progeny virus particles were not produced in such circumstances, that first demonstrated the existence of multipartite viral genomes (83). The 6.8-kb RNA 1 contains four ORFs (Fig. 19). A readthrough protein of 194 kD is N-coterminal with the 134-kD product of the first ORF and is probably the polymerase. The 29-kD protein is the movement protein, but the function of the 16-kD product of the 3′-proximal ORF is not known. These two proteins are translated from subgenomic RNAs.

RNA 2 encodes the 25-kD coat protein. This is the only gene in RNA 2 of one strain of TRV but, in RNA 2 of other strains of the virus, there may be an additional ORF of unknown function and/or a sequence identical to the 3′-terminal ORF of RNA 1. This variable organization of RNA 2 accounts for the wide range of lengths of the shorter particles of the virus. Subgenomic forms of RNA 2 have been detected.

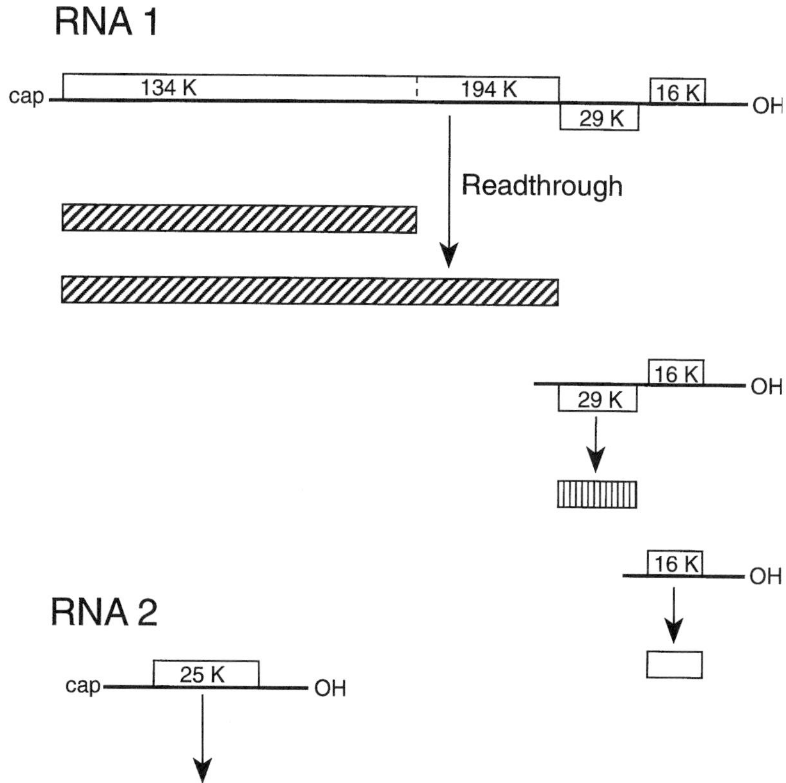

FIG. 19. Genome organization and expression of tobacco rattle virus.

Stages of Infection

Tobacco rattle virus is transmitted in nature by soil-inhabiting trichodorid nematodes in which the virus can be retained for prolonged periods but apparently does not multiply. Susceptible plants can also be readily inoculated by manual application of virus.

Very little is known about the mechanisms involved in uncoating and replication of viral RNA in infected cells. A wide variety of symptoms can be produced in the aboveground and underground parts of infected plants.

Cowpea Mosaic Virus

Classification

Cowpea mosaic virus (CPMV) is the type member of the genus Comovirus in the family Comoviridae. The genus contains ≈15 members, each of which has a narrow natural host range. Some of these viruses are capable of producing serious though localized outbreaks of disease in various parts of the world, particularly in leguminous plants.

Virion Structure

Three types of particle of ≈28 nm diameter are present in preparations of CPMV—the capsids contain one molecule of either RNA 1 or RNA 2 or are empty. The capsids are made up of 60 copies each of a large (42 kD) and a small (24 kD) coat protein. The large coat protein molecules contain two regions of eight-stranded, anti-parallel β-barrel structure and the small molecules contain one, and the 180 β-barrel domains are arranged in a pseudo T=3 pattern in the capsid (Fig. 20). Another comovirus, bean pod mottle virus, was the first icosahedral virus with which details of the interaction of coat proteins and the encapsidated viral RNA were revealed by high-resolution crystallographic analysis (25).

Genome Structure

The genome of CPMV (Fig. 21) consists of two molecules of positive-sense ssRNA of 5.9 kb (RNA 1) and 3.5 kb (RNA 2), each containing a small VPg (of ≈4 kD) at the 5′ end and a poly(A) sequence at the 3′ end. Each RNA contains a single long ORF encoding a polyprotein that is proteolytically processed to yield the various products found in infected cells. The structure and expression of the genome of CPMV has been reviewed (40).

Stages of Infection

Hosts of CPMV can be easily inoculated by manual application of virus. In nature, chrysomelid beetles serve as vectors and the virus can be transmitted through seeds to new generations of plants. There is almost no information concerning the uncoating of CPMV or other early stages in infection.

A proteinase encoded by RNA 1 is responsible for processing the proteins encoded by both genomic RNAs. Cleavage of the 200-kD polyprotein gives rise to the 24-kD proteinase, a 32-kD protein that regulates processing, the VPg, and the putative replicase (110 kD). Various pathways of processing of this polyprotein have been reported (35,100) and are shown in Fig. 21. RNA 2 encodes the 48-kD/58-kD proteins that are involved in cell-to-cell movement of virus and the 60-kD precursor of the two coat proteins.

Structures consisting of dense material and vesicular membranes appear in the cytoplasm of CPMV-infected cells. Viral RNA and proteins are associated with these structures, and they are probably the intracellular site of replication. Large amounts of progeny virus are produced in infected cells. Intact virions are transported from cell to cell in tubules that connect neighboring cells via the plasmodesmata (see Fig. 10); the 48-kD/58-kD proteins encoded by RNA 2 play a role in tubule formation (74).

Tobacco Ringspot Virus

The genus Nepovirus (family Comoviridae) contains over 30 members of which tobacco ringspot virus (TRSV) is the type species. Some of these viruses cause diseases of considerable economic consequence; grapevine fanleaf virus is one of the most serious. The virions are isometric particles of ≈28 nm diameter, but their structure remains to be firmly established. Nepoviruses share many of the general characteristics of the comoviruses, but the organization of their genomes has not been as well described. Unlike the comoviruses, they have wide host ranges and are transmitted in natural settings by soil-inhabiting nematodes of the family Longidoridae. Passage of virus from generation to generation through seeds is also a common form of transmission of nepoviruses. A characteristic symptom in many nepovirus-infected plants is the development of necrotic rings in the leaves.

Some plant virus particles contain satellite RNAs which, because they depend on the virus for replication but are not required by the virus for replication, may, in a sense, be considered molecular parasites of viruses (45). Satellite RNAs vary in size and shape (circular or linear) and may or may not encode a protein. In some cases, the presence of a satellite RNA leads to attenuation of the symptoms normally produced by the virus. There is therefore much interest in the possibility of exploiting these characteristics for the development of new strategies of virus disease control.

One of the more intensively investigated satellite RNAs is that of TRSV, a linear molecule of 359 nucleotides that does not have an ORF (17). Replication begins by circu-

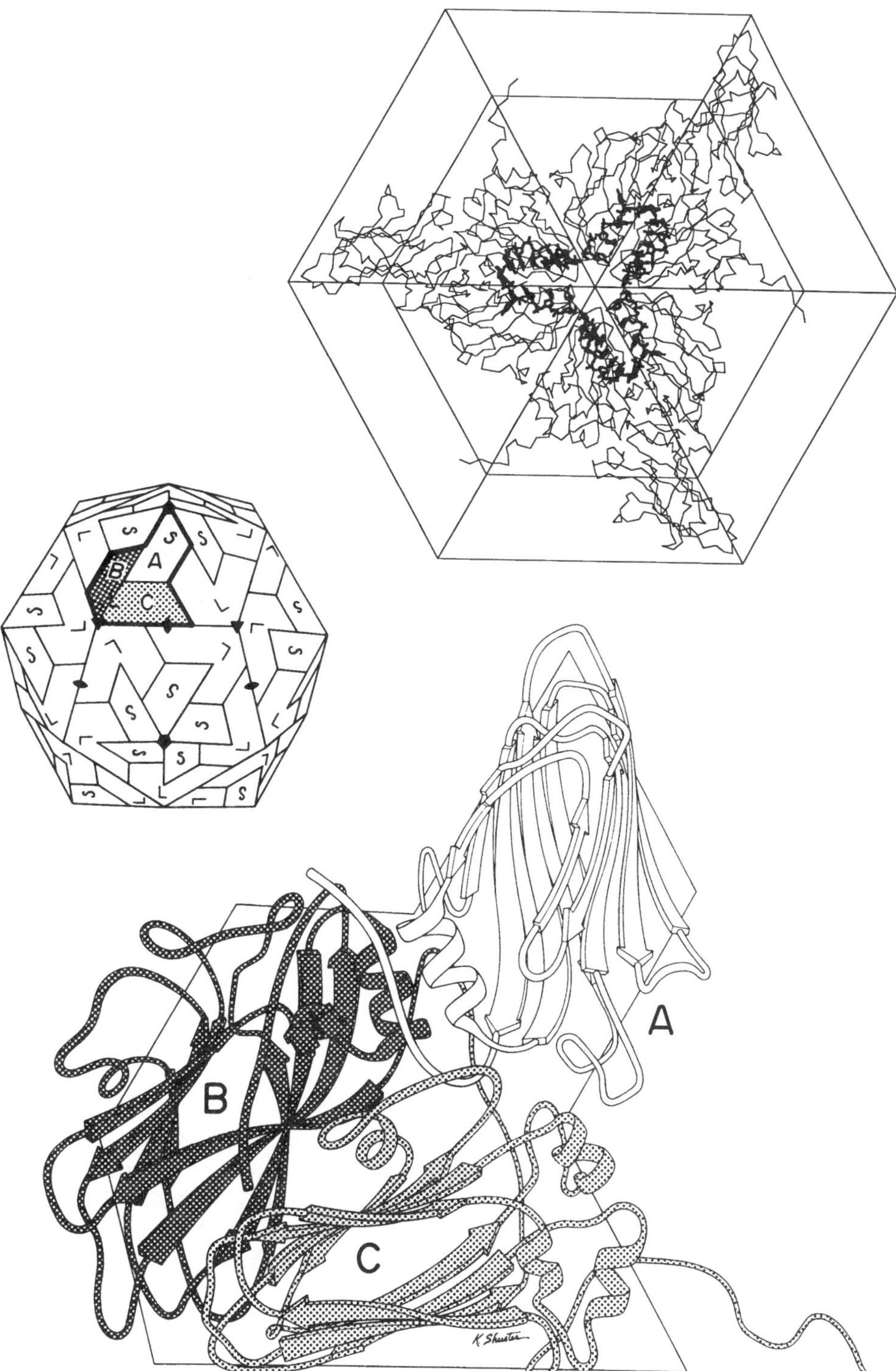

FIG. 20. Structure of bean pod mottle virus particle and its components. **Upper Diagram:** Interaction of viral RNA (*heavy lines*) with three large coat protein subunits at a threefold particle axis. **Center Diagram:** Positions of large (*L*) and small (*S*) coat protein subunits in icosahedral particle. **Lower Diagram:** Ribbon diagram showing the arrangement of three ß-barrel domains of coat proteins in the virion. The *A* domain is in the *S* subunit and the *B* and *C* domains are in the *L* subunit. (From Chen et al., ref. 26, with permission.)

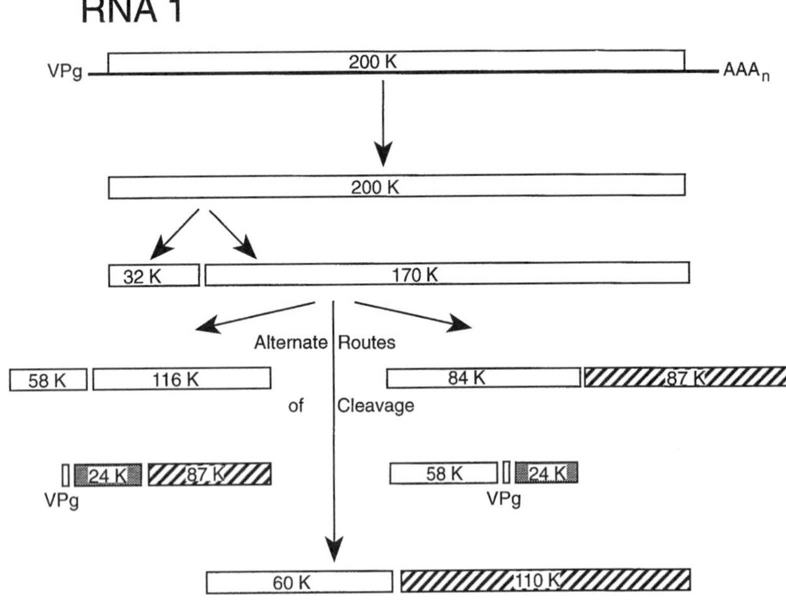

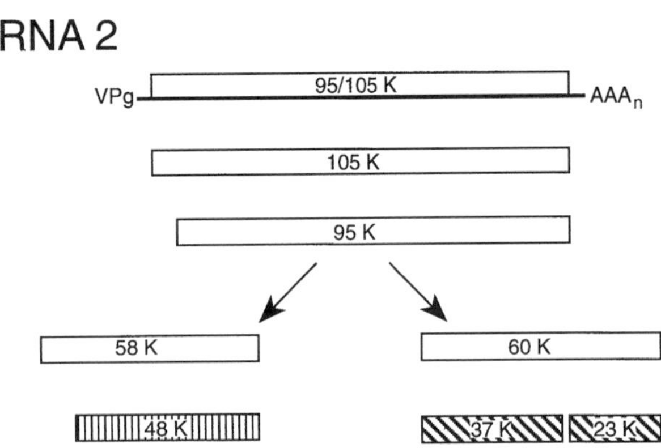

FIG. 21. Genome organization and expression of cowpea mosaic virus. The protein shown as a stippled area in this and other RNA genome maps is a virus-encoded proteinase.

larization of the RNA, followed by the production of "minus"-strand multimers via a rolling circle mechanism. These are cleaved and circularized, and transcribed into "plus" strand multimers which are also cleaved. The cleavage reactions are autolytic and, in the case of those involving the plus strands, are probably carried out by a hammerhead domain.

Larger satellite RNAs are associated with other nepoviruses, and these encode a protein that may have a role in the replication of the satellite RNA.

Pea Enation Mosaic Virus

Pea enation mosaic virus (PEMV), the only member of the genus Enamovirus, has some rather peculiar characteristics. The two components of the genome are encapsidated in separate particles. The organization and nucleotide sequence of the coding regions on RNA 1 are similar to those of the monopartite luteoviruses PLRV and BWYV, described above. RNA 2 contains several ORFs and, unlike its counterpart in most bipartite RNA viruses, is capable of independent replication. While it remains to be established with certainty that PEMV does not involve a helper-dependent virus situation, it may represent a symbiotic association of the genomes of different viruses (34). Like the luteoviruses, the virus is transmitted by aphids in which it persists but is not replicated. Unlike those of most RNA plant viruses, PEMV particles are found in the nuclei of infected cells.

Red Clover Necrotic Mosaic Virus

Another bipartite genome virus, red clover necrotic mosaic virus (genus Dianthovirus), has some characteristics in common with members of the monopartite Tombusviridae family but expresses its movement protein from the monocistronic RNA 2.

TRIPARTITE ssRNA VIRUSES

Before recombinant DNA procedures were developed and applied to the investigation of viral gene functions, the split genomes of some of the plant viruses were particularly useful in attempts to locate genome components responsible for certain viral activities. Studies with the viruses containing three-piece genomes continue to lead the way in these endeavors.

Brome Mosaic Virus

Classification

Brome mosaic virus (BMV) is the type member of the genus Bromovirus which contains 5 other members. The virus infects some graminaceous plants but is not considered a serious pathogen.

Virion Structure

The icosahedral (T=3) virus particles are ≈26 nm in diameter and contain 180 protein subunits. The three segments of the BMV genome are encapsidated in separate particles. Changes in pH or divalent cation content of suspensions of virions produce reversible changes in particle diameter and capsid conformation.

Genome Structure and Organization

It was with BMV that the first *in vitro* system for generating infectious transcripts of DNA copies of viral RNA was developed (2). Functional analysis of the BMV genome has been more extensive than that of the genome of any other plant virus. The genome is contained in three segments (Fig. 22): RNA 1 (3.2 kb), RNA 2 (2.8 kb), and RNA 3 (2.1 kb), each of which is capped and contains a

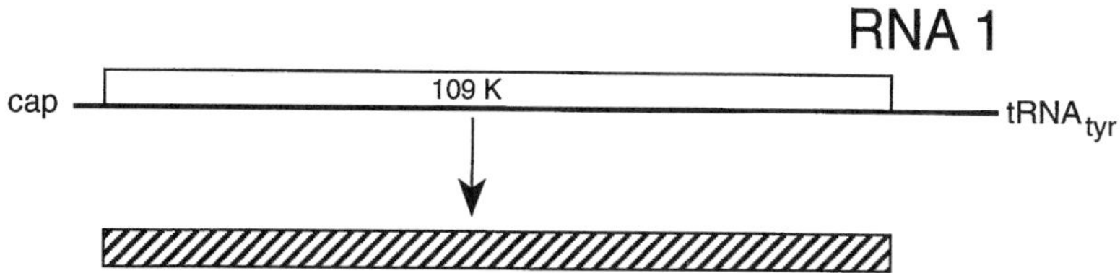

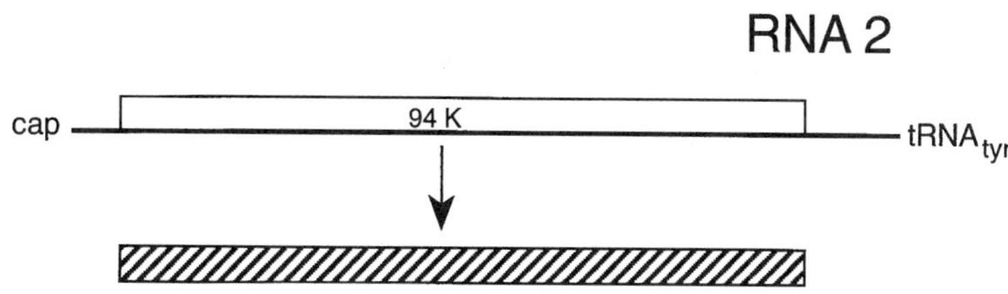

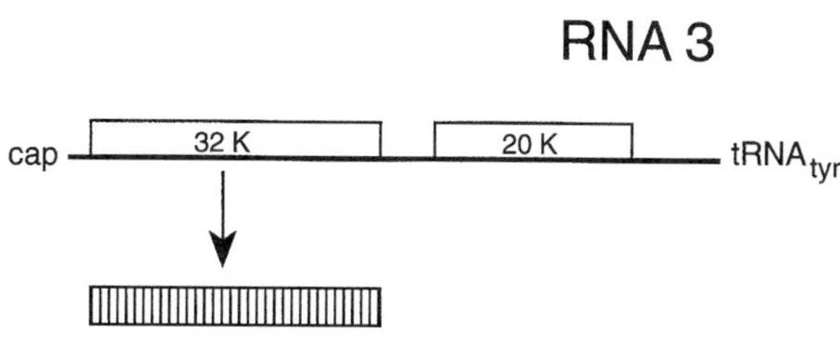

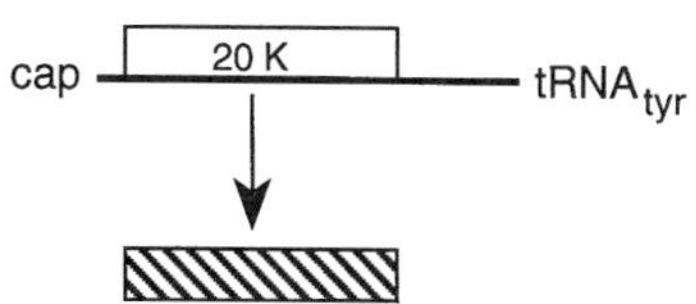

FIG. 22. Genome organization and expression of brome mosaic virus. The subgenomic RNA that encodes the coat protein (*bottom*) is often referred to as *RNA 4.*

tyrosine-accepting tRNA-like structure with a pseudoknot at the 3′ end.

RNA 1 encodes a single protein (109 kD) that contains amino acid sequence motifs similar to those in methyltransferases and helicases in the N- and C-terminal regions, respectively. The 94-kD product of the single ORF of RNA 2 contains a polymerase-like domain. These two proteins are involved in a coordinated manner in RNA-dependent replication of the viral RNA. RNA 3 has two ORFs, one of which plays a role in host plant specificity and encodes a 32-kD protein that mediates cell-to-cell movement of virus. The other ORF encodes the 20-kD coat protein which is expressed in infected cells from subgenomic RNA 4.

Brome mosaic virus has been the subject of extensive analyses of *cis*-acting sequences that are involved in synthesis of viral RNA (1). The 3′-proximal 200 nt of each genome component contain the polymerase promoter for negative-strand synthesis and nucleotidyl transferase and aminoacyl-tRNA synthetase recognition sites (78). Sequence motifs that are essential for the synthesis of positive strands from negative-strand templates have been identified in the 5′ noncoding region of the BMV RNAs (102). For synthesis of RNA 3 and RNA 4, elements in the region between the two ORFs are also required.

The manner in which subgenomic viral RNAs are produced was first demonstrated with BMV (91). Synthesis of RNA 4 is initiated in the intergenic region of the negative strand of RNA 3, and positive-sense copies of the 3′ end of this genome segment are produced. Promoter and other sequence elements that are involved in this process have been identified.

Stages of Infection

Brome mosaic virus is readily transmitted from plant to plant by manual inoculation and, in natural situations, it may be spread by beetles.

A mechanism for the uncoating of particles of cowpea chlorotic mottle virus (CCMV), another bromovirus, has been proposed (125). Initial destabilization of the particle in the cytoplasm would release a few capsid protein molecules and expose the 5′ end of the encapsidated RNA. Ribosomes would then commence translation and the RNA be pulled from the capsid which subsequently collapses.

A complex containing the proteins encoded by RNA 1 and RNA 2 and a host protein, a subunit of eIF-3, is involved in the replication of BMV RNA (107). The recent demonstration of the ability of the BMV RNAs to be expressed and replicated in yeast cells provides a powerful system for the study of host factors involved in the replication of eukaryotic viral RNA (71).

Recombination in RNA viruses of plants was first demonstrated with BMV (19), and a number of other viruses have since been shown to undergo recombination.

The coat proteins of bromoviruses have been used in many investigations of the assembly of virus particle-like structures *in vitro*. Elements in the BMV coat protein and RNAs that may be important in the initiation of virion assembly are beginning to be identified, but the underlying mechanism of the process remains unknown.

Progeny virus particles accumulate in the cytoplasm of infected cells. Both the movement and coat proteins encoded by RNA 3 are necessary for the development of systemic infections in inoculated plants.

Cucumber Mosaic Virus

In many respects, the cucumoviruses are very similar to the bromoviruses. The structure the virions (see Fig. 5) and the organization of the genomes are almost identical. However, in several important respects, the two genera are quite distinct.

Cucumber mosaic virus (CMV), the type member, infects perhaps the largest number of plant species of any virus and is found in every area of cultivation in the world. There are many isolates of this virus, and the genus contains two other members. In natural settings, CMV is effectively spread from plant to plant by aphids, though it does not persist within the insect. Transmission through seeds is another avenue of transmission, and plants can be readily infected by manual inoculation. A review of the virus is available (98).

The first purification of a functional eukaryotic viral RNA-dependent RNA polymerase that satisfies the definition of a replicase was accomplished with CMV (56). The replicase is template-dependent, highly template-specific, and is capable of synthesizing negative and positive strands from exogenously supplied positive-sense RNA.

Cucumber mosaic virus particles may contain a small satellite RNA that is dependent upon the virus for its replication and that may have profound effects on pathogenicity of the virus. Over 30 of these satellite RNAs of CMV have been sequenced—they vary between 332 and 380 b in size, and their nucleotide sequences are not similar to that of the viral RNA. The effects of the presence of a satellite RNA depend on the particular strain of the virus, the isolate of the satellite RNA, and the host plant cultivar. Often there is a reduction in replication of the virus and the severity of symptoms. However, situations such as a devastating systemic necrosis in virus-infected tomato crops in parts of Europe can also result from the presence of a satellite RNA. Specific sequences in viral and satellite RNAs that affect replication and pathogenicity have been identified, but the mechanisms involved in the interaction remain to be explained.

Alfalfa Mosaic Virus

Classification

Alfalfa mosaic virus (AlMV, genus Alfamovirus) is a member of the Bromoviridae. The virus has a world-wide distribution and infects a large number of plant species, particularly legumes. It is a serious pathogen of several important crop plants. Viruses in another genus in this family, the Ilarviruses, have a number of the characteristics of AlMV.

Virion Structure

Preparations of AlMV contain 4–5 major nucleoprotein particles and a number of minor components (72). The genome and subgenomic RNAs are packaged in bacilliform particles of dimensions 56 × 18 nm, 43 × 18 nm, and 35 × 18 nm that contain RNA 1, RNA 2, and RNA 3, respectively. RNA 4 is encapsidated in bacilliform (30 × 18 nm) and ellipsoidal particles. A detailed understanding of the arrangements of the 24-kD coat protein subunits in the capsids has not yet been achieved. RNA-protein interactions appear to be important in the stability of the particles.

Genome Structure and Organization

The genome of AlMV is very similar to that of BMV, though the RNA molecules do not contain a 3′-terminal tRNA-like structure.

Stages in Infection

Alfalfa mosaic virus is seed-transmitted and is readily spread from plant to plant by aphids. Infections in plants can be easily established by manual inoculation.

One of the most interesting features of AlMV is the phenomenon of coat protein "activation" (13). The genomic RNAs alone will not establish infections but require the presence of either some coat protein molecules or the coat protein subgenomic RNA 4. The coat protein may be essential for the recognition of template RNA by the viral polymerase or for the release of progeny viral RNA from replication complexes during the earliest stages of infection.

Barley Stripe Mosaic Virus

Classification

The genus Hordeivirus contains four members, of which BSMV is the type member. Because it is readily transmitted from one generation to the next through the seeds produced by infected plants, BSMV has at times been responsible for significant losses in yields of barley crops in many parts of the world. There are numerous strains of the virus, and they cause different types of symptoms.

Virion Structure

Barley stripe mosaic virus particles are rigid helical rods ≈20 nm in diameter and of length 110–150 nm, depending upon the strain of the virus and the genome segment that is encapsidated. There is a central hole of diameter ≈3.4 nm in the particle, and the RNA is believed to lie at a radius of ≈5.5 nm from the central axis. The pitch of the helix is ≈2.6 nm.

Genome Structure and Organization

Each of the three components (known as RNAs α, β, and γ) of the BSMV is capped and bears a tRNA-like structure at the 3′ end that can be aminoacylated by tyrosine (Fig. 23). There is a poly(A) tract upstream from this structure. The single ORF of RNA α (3.8 kb) encodes a protein with putative capping and helicase domains. RNA β (3.3 kb) has four ORFs, the 22-kD coat protein gene, and a triple gene block whose products appear to be involved in movement of virus in infected plants. RNA γ has two ORFs, the first of which encodes the putative polymerase. The second encodes a protein that may have a *trans*-activating regulatory role in expression of RNA β proteins and is probably expressed via a subgenomic RNA. In some strains of the virus, RNA γ contains a block of repeated nucleotide sequence near the 5′ end. The genome of BSMV is described in a review (70).

Stages of Infection

Infections in plants can be established by manual inoculation. In field situations, virus-containing seed and plant-to-plant contact account for outbreaks of disease. There is little information about molecular mechanisms involved in the establishment of infections by BSMV.

Beet Necrotic Yellow Vein Virus

Classification

A very important virus with a quadripartite genome is beet necrotic yellow vein virus (BNYVV). This is a member of the genus Furovirus, which contains about ten members. Some of these are widely distributed in temperate or tropical regions; BNYVV causes a particularly serious disease, known as rhizomania, in sugar beets (18).

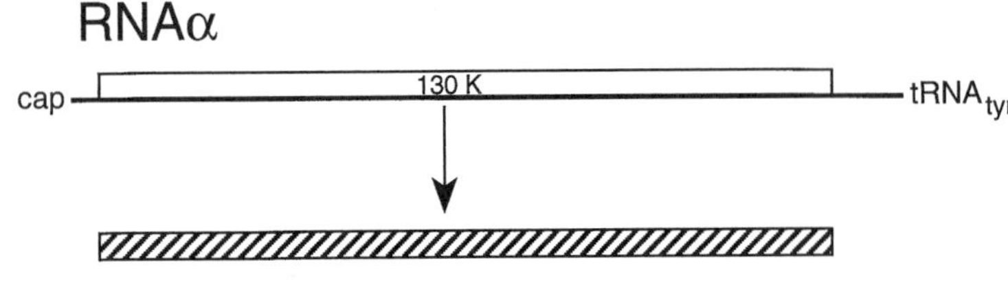

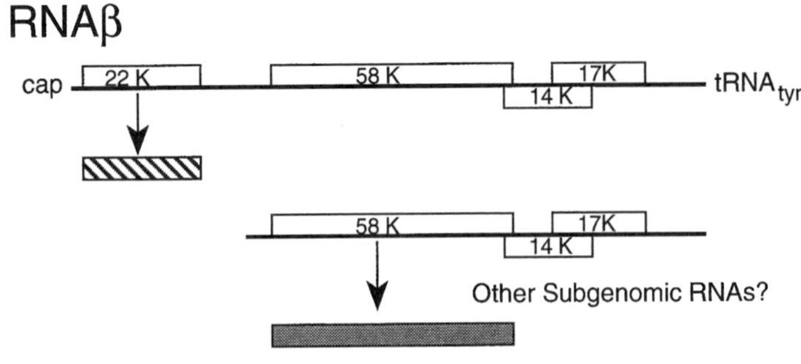

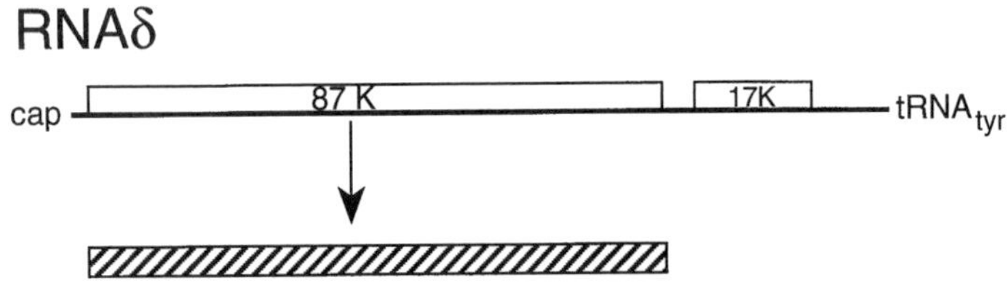

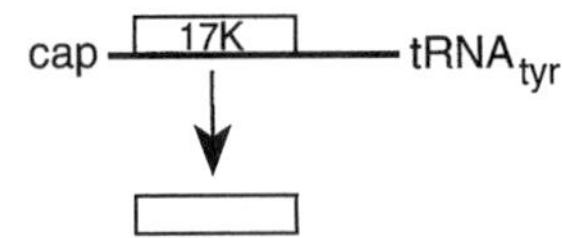

FIG. 23. Genome organization and expression of barley stripe mosaic virus.

Virion Structure

The virus particles are rigid rods and somewhat similar in appearance to TMV particles. They are ≈20 nm in diameter and 60–390 nm long, depending on the genome segment(s) they contain. The 21-kD coat protein subunits are arranged in a helical manner with 12¼ molecules per turn.

Genome Structure and Organization

The genome of BNYVV consists of four capped, polyadenylated RNA components (Fig. 24), though some isolates have a fifth segment (110). The single ORF of 6.8-kb RNA 1 encodes a 238-kD protein that is involved in replication of viral RNA. RNA 2 (4.7 kb) contains six ORFs. The first encodes the coat protein and, via readthrough of the UAG termination codon, part of a 75-kD fusion protein that seems to be involved in virion assembly and transmission. Three ORFs form a triple gene block encoding 42-kD, 13-kD, and 15-kD proteins that are probably involved in cell-to-cell movement; the 42-kD protein contains a helicase-like motif. Subgenomic RNAs that probably serve as mRNAs for these proteins are produced. The function of the 14-kD protein encoded by the 3′-proximal ORF has not been firmly established.

RNA 3 (1.9 kb) and RNA 4 (1.5 kb) encode proteins that are involved in transmission of the virus by its fungal vector and in the spread and symptom expression of virus in the roots of infected plants. A subgenomic form of RNA 3 is produced. In some cases, RNAs 3 and 4 may disappear when the virus is manually passaged in the leaves of a series of plants.

Stages in Infection

Beet necrotic yellow vein virus is transmitted from plant to plant in nature in zoospores of the plasmodiophoromycete fungus *Polymyxa betae*, an obligate, intercellular parasite of the roots of beets and some related plant

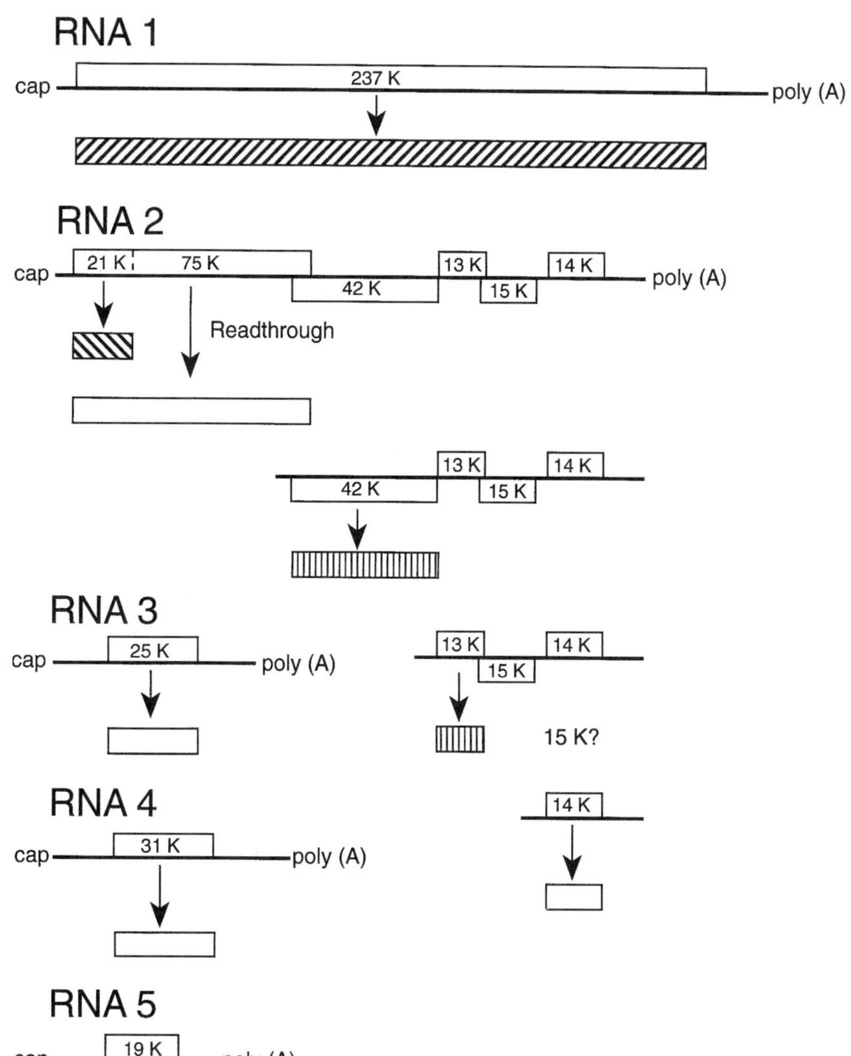

FIG. 24. Genome organization and expression of beet necrotic yellow vein virus.

species. The virus can also be transmitted by manual inoculation. Little is known of the early stages of infection, but the cytoplasm of infected cells may eventually contain crystalline arrays of virus particles.

VIROIDS

Classification

The viroids are small, unencapsidated ssRNA molecules and are the smallest known pathogenic agents of plants (119). Some 20–25 viroids have been identified and, on the basis of nucleotide sequence similarities, can be placed in several groups. Many viroids are serious pathogens of important crop plants. The type member is potato spindle tuber viroid (PSTVd), the cause of a disease which, until about 25 years ago, had been thought to be of viral etiology (36).

Structure

The single-stranded, covalently-closed circular RNAs of viroids range from 246 to 375 nt in size. The RNA contains highly base-paired regions separated by unpaired loops and thus forms small rod-shaped structures. Viroids do not have ORFs and do not serve as mRNAs. The RNA of PSTVd contains five structural domains (Fig. 25). The conserved central domain (C) consists of about 95 nucleotides and contains a short, inverted repeat sequence that flanks a strictly conserved helical region. Other regions include the pathogenicity (P) or virulence modulating (VM) domain, the variable (V) domain and two terminal domains (T1 and T2). Most of the viroids fall within two groups

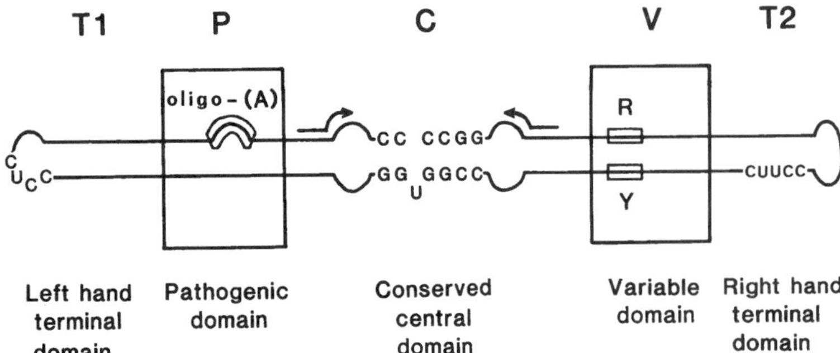

FIG. 25. Domains in potato spindle tuber viroid. The abbreviations and names of the domains are given (*above and below the map*). Shown are conserved inverted repeat sequences (*arrows*) and short helical regions (*R,Y*). (From Keese and Symons, ref. 76, with permission.

which can be distinguished on the basis of sequence similarities in the central domain.

Stages of Infection

Vegetative propagation of plants (e.g., tubers, cuttings) is a common means of transmission of viroids from crop to crop. Some viroids are transmitted through seeds. Most can also be spread by manual inoculation and by the use of contaminated implements.

Little is known about the early stages of viroid infections. Replication probably occurs by variations of a rolling circle mechanism. Linear, concatameric negative-strand RNA is synthesized on the circular viroid RNA (positive-strand) template. With most viroids, the negative-strand RNA is probably not cleaved but is copied to produce positive-strand RNA, which is cleaved to yield linear monomers. These, in turn, undergo ligation to produce circular monomeric progeny RNA molecules. Another viroid, avocado sunblotch viroid, contains a hammerhead structure and is capable of site-specific self-cleavage of both positive- and negative-strand RNAs. Replication is dependent on host enzymes and may involve RNA polymerase II. Potato spindle tuber viroid accumulates in nucleoli in infected cells.

In spite of the small size of viroids, the manner in which they cause disease is unknown (37). Sequences in the P domains appear to be determinants of symptom severity in some plants. The types of symptoms produced in viroid-infected plants are similar to those caused by some viruses. Many of them appear not to induce any symptoms. On the other hand, one of them produces a slowly developing but lethal disease of coconut palm trees.

REFERENCES

1. Ahlquist P. Bromovirus RNA replication and transcription. *Curr Opin Genet Dev* 1992;2:71–76.
2. Ahlquist P, French R, Janda M, Loesch-Fries LS. Multicomponent RNA plant virus infection derived from cloned viral cDNA. *Proc Natl Acad Sci USA* 1984;81:7066–7070.
3. Atreya CD, Pirone TP. Mutational analysis of the helper component-proteinase gene of a potyvirus: Effects of amino acid substitutions, deletions, and gene replacement on virulence and aphid transmissibility. *Proc Natl Acad Sci USA* 1993;90:11919–11923.
4. Atreya PL, Atreya CD, Pirone TP. Amino acid substitutions in the coat protein result in loss of insect transmissibility of a plant virus. *Proc Natl Acad Sci USA* 1991;88:7887–7891.
5. Bawden FC, Pirie NW. The isolation and some properties of liquid crystalline substances from solanaceous plants infected with three strains of tobacco mosaic virus. *Proc Roy Soc London*, Ser. B 1937; 123:274–320.
6. Beijerinck MW. Uber ein contagium vivum fluidum als Ursache der Flechenkrankheit der Tabaksblatter. *Verhandelingen der Koninklyke akademie van Wettenschappen te Amsterdam*. 1898;65:3–21.
7. Benfey PN, Chua N-H. The cauliflower mosaic virus 35S promoter: combinatorial regulation of transcription in plants. *Science* 1990; 250:959–966.
8. Bernal JD, Fankuchen I. Structure types of protein 'crystals' from virus-infected plants. *Nature* 1937;139:923–924.
9. Black LM, Brakke MK. Multiplication of wound tumor virus in an insect vector. *Phytopathology* 1952;42:269–273.
10. Bloomer AC, Champness JN, Bricogne G, Staden R, Klug A. Protein disk of tobacco mosaic virus at 2.8 A resolution showing the interactions within and between subunits. *Nature* 1978;276:362–368.
11. Boccardo G, Lisa V, Lusoni E, Milne RG. Cryptic plant viruses. *Adv Virus Res* 1987;32:171–214.
12. Bol JF, Linthorst HJM, Cornelissen BJC. Plant pathogenesis-related proteins induced by virus infection. *Ann Rev Phytopathol* 1990; 28:113–138.
13. Bol JF, Van Vloten-Doting L, Jaspars EMJ. A functional equivalence of top component a RNA and coat protein in the initiation of infection by alfalfa mosaic virus. *Virology* 1971;46:73–85.
14. Bonneville JM, Sanfaçon H, Fütterer J, Hohn T. Posttranscriptional *trans*-activation in cauliflower mosaic virus. *Cell* 1989;59:1135–1143.
15. Bos L. New plant virus problems in developing countries: A corollary of agricultural modernization. *Adv Virus Res* 1992;41:349–407.
16. Brault V, Miller WA. Translational frameshifting mediated by a viral sequence in plant cells. *Proc Natl Acad Sci USA* 1992;89:2262–2266.
17. Bruening G, Passmore BK, Van Tol H, Buzayan JM, Feldstein PA. Replication of a plant virus satellite RNA: evidence favors transcription of circular templates of both polarities. *Mol Plant-Microbe Interact* 1991;4:219–225.
18. Brunt AA, Richards KE. Biology and molecular biology of furoviruses. *Adv Virus Res* 1989;36:1–32.
19. Bujarski JJ, Kaesberg P. Genetic recombination between RNA components of a multipartite plant virus. *Nature* 1986;321:528–531.
20. Butler PJG. The current picture of the structure and assembly of tobacco mosaic virus. *J Gen Virol* 19984;65:253–280.
21. Butler PJG, Finch JG, Zimmern D. Configuration of tobacco mosaic virus RNA during virus assembly. *Nature* 1977;265:217–219.
22. Caplan A, Herrera-Estrella L, Inze D, Van Haute E, Van Montagu M, Schell J, Zambryski P. Introduction of genetic material into plant cells. *Science* 1983;222:815–821.
23. Carrington JC, Cary SM, Parks TD, Dougherty WG. A second proteinase encoded by a plant potyvirus genome. *EMBO J* 1989;8:365–370.
24. Carrington JC, Dougherty WG. Small nuclear inclusion protein encoded by a plant potyvirus genome is a protease. *J Virol* 1987;61:2540–2548.

25. Chen Z, Stauffacher C, Li Y, Schmidt T, Bomu W, Kamer G, Shanks M, Lomonossoff G, Johnson JE. Protein-RNA interactions in an icosahedral virus at 3.0 Å resolution. *Science* 1989;245:154–159.
26. Chen Z, Stauffacher CV, Johnson JE. Capsid structure and RNA packaging in comoviruses. *Semin Virol* 1990;1:453–466.
27. Cheng RH, Olson NH, Baker TS. Cauliflower mosaic virus: a 420 subunit (T=7), multilayer structure. *Virology* 1992;186:655–668.
28. Citovsky V, Zambryski P. How do plant virus nucleic acids move through intercellular connections. *BioEssays* 1991;13:373–379.
29. Coffin RS, Coutts RHA. The closteroviruses, capilloviruses and other similar viruses: A short review. *J Gen Virol* 1993;74:1475–1483.
30. Covey SN. Pathogenesis of a plant pararetrovirus: CaMV. *Semin Virol* 1991;2:151–159.
31. Covey SN, Turner DS. Comparison of viral nucleic acid intermediates at early and late stages of cauliflower mosaic virus infection suggests a feedback regulatory mechanism. *J Gen Virol* 1991;72:2603–2606.
32. Dawson WO. Tobamovirus-plant interactions. *Virology* 1992; 186:359–367.
33. Dawson WO, Lehto KM. Regulation of tobamovirus gene expression. *Adv Virus Res* 1990;38:307–342.
34. Demler SA, Rucker DG, De Zoeten GA. The chimeric nature of the genome of pea enation mosaic virus: The independent replication of RNA 2. *J Gen Virol* 1993;74:1–14.
35. Dessens JT, Lomonossoff GP. Sequence upstream of the 24K protease enhances cleavage of the cowpea mosaic virus B RNA-encoded polyprotein at the junction between the 24K and 87K proteins. *Virology* 1992;189:225–232.
36. Diener TO. Potato spindle tuber "virus." IV. A replicating, low molecular weight RNA. *Virology* 1971;45:411–428.
37. Diener TO, Owens RA, Hammond RW. Viroids: The smallest and simplest agents of infectious disease. How do they make plants sick? *Intervirology* 1993;35:186–195.
38. Dinesh-Kumar SP, Brault V, Miller WA. Precise mapping and *in vitro* translation of a trifunctional subgenomic RNA of barley yellow dwarf virus. *Virology* 1992;187:711–722.
39. Dowson Day MJ, Ashurst JL, Mathias SF, Watts JW, Wilson TMA, Dixon RA. Plant viral leaders influence expression of a reporter gene in tobacco. *Plant Mol Biol* 1993;23:97–109.
40. Eggen R, Van Kammen A. RNA replication in comoviruses. In: Ahlquist P, Holland J, Domingo E, eds. *RNA genetics*, vol 1. Boca Raton, FL: CRC Press, Inc., 1988;49–69.
41. Elmer JS, Brand L, Sunter G, Gardiner WE, Bisaro DM, Rogers SG. Genetic analysis of the tomato golden mosaic virus II. The product of the AL1 coding sequence is required for replication. *Nucleic Acids Res* 1988;16:7043–7060.
42. Fitchen JH, Beachy RN. Genetically engineered protection against viruses in transgenic plants. *Ann Rev Microbiol* 1993;47:739–763.
43. Fontes EPB, Luckow VA, Hanley-Bowdoin L. A geminivirus replication protein is a sequence-specific DNA binding protein. *Plant Cell* 1992;4:597–608.
44. Fraenkel-Conrat H, Williams RC. Reconstitution of active tobacco mosaic virus from its inactive protein and nucleic acid components. *Proc Natl Acad Sci USA* 1955;41:690–698.
45. Francki RIB. Plant virus satellites. *Ann Rev Microbiol* 1985;39:151–174.
46. Fraser RSS. The genetics of resistance to plant viruses. *Ann Rev Phytopathol* 1990;28:179–200.
47. German TL, Ullman DE, Moyer JW. Tospoviruses: diagnosis, molecular biology, phylogeny, and vector relationships. *Ann Rev Phytopathol* 1992;30:315–348.
48. Gierer A, Schramm G. Infectivity of ribonucleic acid from tobacco mosaic virus. *Nature* 1956;177:702–703.
49. Golemboski DB, Lomonossoff GP, Zaitlin M. Plants transformed with a tobacco mosaic virus nonstructural gene sequence are resistant to the virus. *Proc Natl Acad Sci USA* 1990;87:6311–6315.
50. Goulden MG, Köhm BA, Santa Cruz S, Kavanagh TA, Baulcombe DC. A feature of the coat protein of potato virus X affects both induced virus resistance in potato and viral fitness. *Virology* 1993;197:293–302.
51. Gowda S, Wu FC, Scholthof HB, Shepherd RJ. Gene VI of figwort mosaic virus (caulimovirus group) functions in posttranscriptional expression of genes on the full-length RNA transcript. *Proc Natl Acad Sci USA* 1989;86:9203–9207.
52. Harrison BD, Robinson DJ. Tobraviruses, p. 339–369. In: Van Regenmortel MHV, Fraenkel-Conrat H, eds. *The plant viruses*, vol 2. New York: Plenum Press, 1986.
53. Harrison SC. Common features in the structures of some icosahedral viruses: a partly historical overview. *Semin Virol* 1990;1:387–403.
54. Hatta T, Francki RIB. The fine structure of chloris striate mosaic virus. *Virology* 1979;92:428–435.
55. Hayes RJ, Buck KW. Replication of tomato golden mosaic virus DNA B in transgenic plants expressing open reading frames (ORFs) of DNA A: requirement of ORF AL2 for production of single-stranded DNA. *Nucleic Acids Res* 1989;17:10213–10222.
56. Hayes RJ, Buck KW. Complete replication of a eukaryotic virus RNA in vitro by a purified RNA-dependent RNA polymerase. *Cell* 1990; 63:363–368.
57. Heaton LA, Morris TJ. Structural implications for spherical plant virus disassembly in vivo. *Semin Virol* 1992;3:433–439.
58. Hillman BI, Carrington JC, Morris TJ. A defective interfering RNA that contains a mosaic of a plant virus genome. *Cell* 1987; 51:427–433.
59. Hohn T, Fütterer J. Transcriptional and translational control of gene expression in cauliflower mosaic virus. *Curr Opin Genet Dev* 1992;2:90–96.
60. Holmes FO. Local lesions in tobacco mosaic. *Bot Gaz* 1929;87:39–55.
61. Huiet L, Feldstein PA, Tsai JH, Falk BW. The maize stripe virus major noncapsid protein messenger RNA transcripts contain heterogeneous leader sequences at their 5' termini. *Virology* 1993;197:808–812.
62. Huisman MJ, Linthorst HJM, Bol JF, Cornelissen BJC. The complete nucleotide sequence of potato virus X and its homologies at the amino acid level with various plus-stranded RNA viruses. *J Gen Virol* 1988;69:1789–1798.
63. Hull R. The movement of viruses in plants. *Ann Rev Phytopath* 1989;27:213–240.
64. Hull R. Molecular biology of plant virus-vector interactions. In: Harris KF, ed. *Advances in disease vector research*, vol 10. New York: Springer-Verlag, 1994;361–386.
65. Hull R, Davies JW. Approaches to nonconventional control of plant virus diseases. *Crit Rev Plant Sci* 1992;11:17–33.
66. Irwin ME, Thresh JM. Epidemiology of barley yellow dwarf: A study in ecological complexity. *Ann Rev Phytopathol* 1990;28:393–424.
67. Ishikawa K, Omura T, Hibino H. Morphological characteristics of rice stripe virus. *J Gen Virol* 1989;70:3465–3468.
68. Ivanowski D. Uber die Mosaikkrankheit der Tabakspflanze. St. Petersb. *Acad Imp Sci Bul* 1892;35:67–70.
69. Jackson AO, Francki RIB, Zuidema D. Biology, structure and replication of plant rhabdoviruses. In: Wagner RR, ed. *The rhabdoviruses*. New York: Plenum, 1987;427–508.
70. Jackson AO, Hunter BG, Gustafson GD. Hordeivirus relationships and genome organization. *Ann Rev Phytopathol* 1989;27:95–121.
71. Janda M, Ahlquist P. RNA-dependent replication, transcription, and persistence of brome mosaic virus RNA replicons in S. cerevisiae. *Cell* 1993;72:961–970.
72. Jaspars EMJ. Interaction of alfalfa mosaic virus nucleic acid and protein. In: Davies JW, ed. *Molecular Plant Virology*, vol. I. Boca Raton, FL: CRC Press, Inc., 1985;155–221.
73. Kassanis B. Properties and behaviour of a virus depending for its multiplication on another. *J Gen Microbiol* 1962;27:477–488.
74. Kasteel D, Wellink J, Verver J, Van Lent J, Goldbach R, Van Kammen A. The involvement of cowpea mosaic virus M RNA-encoded proteins in tubule formation. *J Gen Virol* 1993;74:1721–1724.
75. Kausche GA, Pfankuch E, Ruska A. Die Sichtbormachung von pflanzlichem Virus im Ubermikroskop. *Naturwissenschaften* 1939; 27:292–299.
76. Keese P, Symons RH. The structure of viroids and virusoids. In: Semancik JS, ed. *Viroids and viroid-like pathogens*. Boca Raton, FL: CRC Press, Inc., 1987;1–47.
77. Koonin EV, Dolja VV. Evolution and taxonomy of positive-strand RNA viruses: Implications of comparative analysis of amino acid sequences. *Crit Rev Biochem Mol Biol* 1993;28:375–430.
78. Lahser FC, Marsh LE, Hall TC. Contributions of the Brome mosaic virus RNA-3 3'-nontranslated region to replication and translation. *J Virol* 1993;67:3295–3303.
79. Lazarowitz SG. Geminiviruses: genome structure and gene function. *Crit Rev Plant Sci* 1992;11:327–349.
80. Lebeurier G, Nicolaieff A, Richards KE. Inside-out model for self-assembly of tobacco mosaic virus. *Proc Natl Acad Sci USA* 1977; 74:149–153.
81. Leonard DA, Zaitlin M. A temperature-sensitive strain of tobacco mo-

saic virus defective in cell-to-cell movement generates an altered viral-encoded protein. *Virology* 1982;117:416–424.
82. Lindbo JA, Silva-Rosales L, Proebsting WM, Dougherty WG. Induction of a highly specific antiviral state in transgenic plants: Implications for regulation of gene expression and virus resistance. *Plant Cell* 1993;5:1749–1759.
83. Lister RM. Possible relationship between virus specific products of tobacco rattle infections. *Virology* 1966;28:350–353.
84. Markham R, Smith KM. Studies on the virus of turnip yellow mosaic. *Parasitology* 1949;39:330–342.
85. Martelli GP, Gallitelli D, Russo M. Tombusviruses. In: Koenig R, ed. *The plant viruses*, vol 3. New York: Plenum, 1988;13–72.
86. Martin RR, Keese PK, Young MJ, Waterhouse PM, Gerlach WL. Evolution and molecular biology of luteoviruses. *Ann Rev Phytopathol* 1990;28:341–363.
87. Matthews REF. Portraits of viruses: Turnip yellow mosaic virus. *Intervirology* 1981;15:121–144.
88. Matthews REF. *Plant Virology*. Academic Press, Inc., New York.1991
89. Mayer A. Uber die Mosaikkrankheit des Tabaks. *Die Landwirtschaftlichen Versuchs-Stationen.* 1886;32:451–467.
90. McLean BG, Waigmann E, Citovsky V, Zambryski P. Cell-to-cell movement of plant viruses. *Trends Microbiol* 1993;1:105–109.
91. Miller WA, Dreher TW, Hall TC. Synthesis of brome mosaic virus subgenomic RNA in vitro by internal initiation on (-)-sense genomic RNA. *Nature* 1985;313:68–70.
92. Morris-Krsinich BAM, Mullineaux PM, Donson J, Boulton MI, Markham PG, Short MN, Davies JW. Bidirectional transcription of maize streak virus DNA and identification of the coat protein gene. *Nucleic Acids Res* 1985;13:7237–7256.
93. Murant AF. Dependence of groundnut rosette virus on its satellite RNA as well as on groundnut rosette assistor luteovirus for transmission by *Aphis craccivora*. *J Gen Virol* 1990;71:2163–2166.
94. Namba K, Caspar DLD, Stubbs GJ. Computer graphics representation of levels of organization in TMV struture. *Science* 1985;227:773–776.
95. Namba K, Pattanayek R, Stubbs G. Visualization of protein-nucleic acid interactions in a virus. Refined structure of intact tobacco mosaic virus at 2·9 Å resolution by x-ray fiber diffraction. *J Mol Biol* 1989;208:307–325.
96. Nuss DL, Dall DJ. Structural and functional properties of plant reovirus genomes. *Adv Virus Res* 1990;38:249–306.
97. Odell JT, Nagy F, Chua N-M. Identification of DNA sequences required for activity of the CaMV 35S promoter. *Nature* 1985;313:810.
98. Palukaitis P, Roossinck MJ, Dietzgen RG, Francki RIB. Cucumber mosaic virus. *Adv Virus Res* 1992;41:281–339.
99. Passmore BK, Sanger M, Chin L-S, Falk BW, and G Bruening G. Beet western yellows virus-associated RNA: An independently replicating RNA that stimulates virus accumulation. *Proc Natl Acad Sci USA* 1993;90:10168–10172.
100. Peters SA, Voorhorst WGB, Wellink J, Van Kammen A. Processing of VPg-containing polyproteins encoded by the B-RNA from cowpea mosaic virus. *Virology* 1992;191:90–97.
101. Pinck M, Yot P, Chapeville F, Duranton H. Enzymatic binding of valine to the 3'-end of TYMV RNA. *Nature* 1970;226:954–956.
102. Pogue GP, Hall TC. The requirement for a 5' stem-loop structure in brome mosaic virus replication supports a new model for viral positive-strand RNA initiation. *J Virol* 1992;66:674–684.
103. Porta C, Spall VE, Loveland J, Johnson JE, Barker PJ, Lomonossoff GP. Development of cowpea mosaic virus as a high-yielding system for the presentation of foreign peptides. *Virology* 1994;202:949–955.
104. Powell Abel PA, Nelson RS, De B, Hoffmann N, Rogers SG, Fraley RT, Beachy RN. Delay of disease development in transgenic plants that express the tobacco mosaic virus coat protein gene. *Science* 1986;232:738–743.
105. Prüfer D, Tacke E, Schmitz J, Kull B, Kaufmann A, Rohde W. Ribosomal frameshifting in plants: A novel signal directs the -1 frameshift in the synthesis of the putative viral replicase of potato leafroll luteovirus. *EMBO J* 1992;11:1111–1117.
106. Purcifull DE, Edwardson JR. Potexviruses. In: Kurstak E, ed. *Handbook of plant virus infections and comparative diagnosis.* Amsterdam: Elsevier, 1981;627–693.
107. Quadt R, Kao CC, Browning KS, Hershberger RP, Ahlquist P. Characterization of a host protein associated with brome mosaic virus RNA-dependent RNA polymerase. *Proc Natl Acad Sci USA* 1993;90:1498–1502.
108. Reavy B, Mayo MA, Turnbull-Ross AD, Murant AF. Parsnip yellow fleck and rice tungro spherical viruses resemble picornaviruses and represent two genera in a proposed new plant picornavirus family (*Sequiviridae*). *Arch Virol* 1993;131:441–446.
109. Restrepo-Hartwig MA, Carrington JC. Regulation of nuclear transport of a plant potyvirus protein by autoproteolysis. *J Virol* 1992; 66:5662–5666.
110. Richards KE, Tamada T. Mapping functions on the multipartite genome of beet necrotic yellow vein virus. *Ann Rev Phytopathol* 1992;30:291–313.
111. Riechmann JL, Laín S, García JA. Highlights and prospects of potyvirus molecular biology. *J Gen Virol* 1992;73:1–16.
112. Rochow WF, Duffus JE. Luteoviruses and yellows diseases. In: Kurstak E, ed. *Handbook of plant virus infections and comparative diagnosis.* Amsterdam: Elsevier, 1981;147–170.
113. Shaw JG, Plaskitt KA, Wilson TMA. Evidence that tobacco mosaic virus particles disassemble cotranslationally in vivo. *Virology* 1986;148:326–336.
114. Shepherd RJ. Biochemistry of DNA Plant Viruses. In: Marcus A, ed. *The biochemistry of plants,* vol 15. New York: Academic Press, Inc., 1989;563–616.
115. Stanley J. The molecular determinants of geminivirus pathogenesis. *Semin Virol* 1991;2:139–149.
116. Stanley J. Geminiviruses: Plant viral vectors. *Curr Opin Genet Dev* 1993;3:91–96.
117. Stanley J, Frischmuth T, Ellwood S. Defective viral DNA ameliorates symptoms of geminivirus infection in transgenic plants. *Proc Natl Acad Sci USA* 1990;87:6291–6295.
118. Stanley WM. Isolation of a crystalline protein possessing the properties of tobacco-mosaic virus. *Science* 1935;81:644–645.
119. Symons RH. The intriguing viroids and virusoids: what is their information content and how did they evolve? *Mol Plant-Microbe Interact* 1991;4:111–121.
120. Thornbury DW, Hellmann GM, Rhoads RE, Pirone TP. Purification and characterization of potyvirus helper component. *Virology* 1985;144:260–267.
121. Townsend R, Stanley J, Curson SJ, Short MN. Major polyadenylated transcripts of cassava latent virus and location of the gene encoding coat protein. *EMBO J* 1985;4:33–38.
122. Turner DR, McGuigan CJ, Butler PJG. Assembly of hybrid RNAs with tobacco mosaic virus coat protein. Evidence for incorporation of disks in 5'-elongation along the major RNA tail. *J Mol Biol* 1989; 209:407–422.
123. Veidt I, Bouzoubaa SE, Leiser R-M, Ziegler-Graff V, Guilley H, Richards K, Jonard G. Synthesis of full-length transcripts of beet western yellows virus RNA: Messenger properties and biological activity in protoplasts. *Virology* 1992;186:192–200.
124. Verchot J, Koonin EV, Carrington JC. The 35-kDa protein from the N-terminus of the potyviral polyprotein functions as a third virus-encoded proteinase. *Virology* 1991;185:527–535.
125. Verduin BJM. Early interactions between viruses and plants. *Semin Virol* 1992;3:423–431.
126. Wang H, Stubbs G. Structure determination of cucumber green mottle mosaic virus by X-ray fiber diffraction. Significance for the evolution of tobamoviruses. *J Mol Biol* 1994;239:371–384.
127. Wilson TMA. Cotranslational disassembly of tobacco mosaic virus in vitro. *Virology* 1984;137:255–265.
128. Wilson TMA. Strategies to protect crop plants against viruses: Pathogen-derived resistance blossoms. *Proc Natl Acad Sci USA* 1993; 90:3134–3141.
129. Wu X, Xu Z, Shaw JG. Uncoating of tobacco mosaic virus RNA in protoplasts. *Virology* 1994;200:256–262.
130. Zaitlin M, Hull R. Plant virus-host interactions. *Ann Rev Plant Physiol* 1987;38:291–315.
131. Zimmern D. The nucleotide sequence at the origin for assembly on tobacco mosaic virus RNA. *Cell* 1977;11:463–482.

Fundamental Virology, Third Edition
edited by B.N. Fields, D.M. Knipe, P.M. Howley, et al.
Lippincott - Raven Publishers, Philadelphia © 1996

CHAPTER 13

Insect Viruses

Lois K. Miller

HISTORY OF INSECT VIRUSES

Literature and scientific reports of the sixteenth and seventeenth centuries contain sufficiently detailed descriptions of the "wilting" disease of silkworms to indicate that insect viruses have been of interest and concern for a long time. In the early nineteenth century, crystalline polyhedral bodies were found to be associated with the wilting disease of larvae and, by the late nineteenth century, these polyhedra were established as causal agents of the disease (9). By 1920, wilting disease was attributed to a filterable virus (9) and, in the 1940s, Bergold (10) made the crucial discovery of rod-shaped virions, characteristic of baculoviruses, embedded within the crystalline protein matrix of the polyhedral bodies.

In 1949, a broad stage for development of insect pathogens as biological pesticides was established; insect viruses, especially baculoviruses, were considered important candidates for development (160). The 1960s marked the discoveries of many new types of insect viruses, including iridoviruses, nodaviruses, polydnavirus-like particles, and entomopoxviruses. The 1960s also marked progress in establishing insect cell lines (69) and defining a distinct biological role for the occluded forms (e.g., polyhedra) of insect viruses (82). In 1975, the first insect virus was registered by the U.S. Environmental Protection Agency as a pesticide (92). By 1977, a catalog of viral diseases of insect and mite viruses (113) listed over 640 insect or mite species belonging to 10 different orders, in which one or more of 21 different types of viral disease had been reported.

The mid-1970s marked the beginning of the era of molecular biological research on insect viruses. This research, in turn, triggered additional interest in and application of insect viruses such as their use as gene expression vectors (see section discussing current impact of insect viruses, below). Basic molecular research on insect viruses has also contributed many insights into molecular, cellular, and organismal biology.

L.K. Miller: Departments of Genetics and Entomology, The University of Georgia, Athens, Georgia 30602.

CURRENT IMPACT OF INSECT VIRUSES

Insects comprise over 80% of existing animal species and are critically important to the ecosystem. A few insect species are pests, some are of biomedical concern, and others are of agricultural or silvicultural concern. Insect populations are held in natural balance by many different factors, including predation, parasitism, and microbial infection. The importance of each factor varies among species. Viral diseases are found in beneficial insects (e.g., bees, parasitoids, and silkworms) as well as pest species such as gypsy moths, corn earworms, and mosquitoes. Diseases of beneficial insects can lead to economic and/or ecological problems, while diseases in pest populations can alleviate medical, economic, and/or social problems.

The extent to which insect viruses regulate insect population levels varies with regard to both the virus and the insect species. Some viruses have little or no overt effects on insect population size but can influence the ``health'' of the population; infections by some tetraviruses and cypoviruses, for example, cause a gut disease similar to diarrhea. Such viruses are thought to be a stress factor on insect populations. However, insect viruses can also have very striking effects on population levels, causing widespread epizootics and morbidity in dense insect populations. Because most insect populations are naturally held in check by predation or parasitism, high population densities of specific insect species are usually observed under unnatural situations such as accidental introduction of a species into a new geographical area or intensive chemical pesticide application.

Natural baculovirus epizootics have been observed frequently in agricultural and forest settings but usually occur only when insect populations are very high. One of the most dramatic and documented instances of natural regulation of insect populations by a baculovirus was described for infestations of European spruce sawfly following their accidental introduction into North America. When the sawfly baculovirus arrived in North America in the 1930s, sawfly populations declined in a dramatic fashion and, in following years, depression of populations was attributed to the presence of this virus (2). Relying on naturally occurring epizootics for insect pest control, however, is not economically feasible in most agricultural settings because, at population levels high enough to initiate epizootics, pest damage to the crop exceeds economically and/or aesthetically acceptable thresholds.

Replacement of broad-range chemical pesticides with biological control agents is a future goal of a number of environmentally conscious scientists, and insect viruses provide one of the most promising avenues for pest-specific control. In Brazil, over a million hectares of land are treated annually with a baculovirus to control the soybean looper; in the South Pacific, a baculovirus is used on many islands to control coconut beetles; and in Europe, a baculovirus is used to control the apple maggot. Several baculoviruses are now registered in the U.S. as pesticides, but their implementation has been pursued primarily by government agencies; for example, the U.S. Forest Service has developed and employs a baculovirus for the control of tussock moth larvae in the Pacific Northwest. However, industrial interest in commercial development of insect virus pesticides is increasing in the U.S. and Europe as environmental concerns mount and as insects continue to gain resistance to chemical pesticides. Another factor contributing to increased industrial interest is the ability to genetically engineer the viruses to be more effective as pesticides (114, 129, 161, 178). Field tests of genetically improved baculovirus pesticides have been conducted in both the U.S. and U.K., and further commercial development is expected.

Baculoviruses are also currently used extensively as vectors for foreign gene expression; this application stemmed from basic molecular studies on baculoviruses and is a particularly striking example of utility that insect viruses can have in other areas of biology and biotechnology. Baculovirus vectors can stably accommodate large amounts of foreign DNA and are especially noted for their ability to produce very high levels of biologically active foreign eukaryotic proteins (96, 130). The posttranslational modification and tertiary protein folding which occur in insect cells are similar enough to mammalian cells to provide proteins of research, vaccine, diagnostic, or pharmaceutical utility. By 1993, over 500 different eukaryotic genes had been successfully expressed using baculovirus vectors. Expression in insect cells has also provided a useful environment to test the biological activity of the expressed proteins, being different enough from mammalian cells to sometimes provide an exceptional advantage for functional analysis (119).

In addition to baculovirus expression vectors, the unique properties of other insect viruses are also being exploited for expression vector work. It may be possible to use the RNA polymerase of nodaviruses to amplify RNA and achieve high-level gene expression in this manner (3). The ability of some insect viruses (e.g., LaCrosse virus) to replicate prolifically in insects without obvious impact could provide a very useful tool in exploring the function of insect genes and might be exploited for insect control and gene expression vector work in the future. Entomopoxviruses may also prove to be useful as expression vectors with broad host range capabilities.

Last but not by any means least, the study of insect viruses provides unique insights into fundamental biological processes. For example, the 7-methyl guanosine cap structure of eukaryotic mRNAs was first discovered during the study of cypoviruses. The study of polydnaviruses has uncovered a novel and fascinating mutualistic relationship between these viruses, which are vertically transmitted, and their host insect species (see polydnavirus section following). Insects and their viruses provide a unique opportunity to study host defenses to virus infection in the absence of an antibody immune response. Insect viruses are also proving to be particularly valuable in unraveling apoptosis (see the baculovirus virus-cell interaction section following) and virus structural biology (see nodavirus and tetravirus sections following).

The highly interactive nature of insects with other taxa suggests that associations of virus-infected insects with other organisms may have played a central, pivotal role in the evolution of many plant and animal viruses. An understanding of the evolution of many virus families (e.g., Rhabdoviridae, Reoviridae, Poxviridae) will probably depend on knowledge of their counterparts in insects. In addition, the evolution of some virus families may have been distinctively shaped by their co-evolution with insect-specific virus families. Some entomopoxvirus and orthomyxovirus genes, for example, appear to be related to baculovirus genes, suggesting a possible evolutionary link between these virus families.

Thus, insect viruses impact a wide variety of both basic and applied areas of biology. Research on the molecular biology, biotechnology, pathology, epidemiology, and environmental aspects of insect viruses is likely to accelerate and reveal additional biomedical and environmental applications in the future.

CLASSIFICATION OF INSECT VIRUSES

A list of viruses infecting invertebrates is presented in Table 1. Members of all virus genera listed have been found in insects, although the coltiviruses are found predominantly in ticks. The ascoviruses were discovered in the 1980s and remain to be classified taxonomically by the ICTV. Further classification of several virus families is described in individual chapter subsections.

This chapter describes only those families of viruses which are found solely or primarily in insects: Baculoviridae, Polydnaviridae, Ascoviruses, Tetraviridae, and Nodaviridae. Viruses which use insects as temporary hosts in their transmission between vertebrate animals or plants, e.g., arboviruses, are not discussed. Also excluded are those virus families which include extensively characterized vertebrate viruses covered in separate chapters (e.g., poxviridae, picornaviridae, etc.). However, readers should be aware that insect-specific members of these families often have novel properties which may not be discussed in those chapters. For example, the entomopoxviruses and cypoviruses have occluded forms which are probably similar in function to those of baculoviruses discussed in this chapter.

DESCRIPTION OF INSECT VIRUS FAMILIES

Baculoviridae and Nudiviruses

Classification of Baculoviruses and Nudiviruses

The family Baculoviridae is taxonomically characterized by a large, double-stranded, circular DNA genome which is packaged in a rod-shaped capsid and further enveloped by a unit membrane. The name is derived from the Latin word *baculum*, meaning "rod" or "stick." Until recently, the family was subdivided into two subfamilies, eubaculovirinae and nudibaculovirinae, which were distinguished by the presence or absence of an occluded form. However, the nudibaculovirinae were recently removed from the family, despite significant similarities with eubaculoviruses, and are currently unclassified until more molecular information is available. In this chapter, the terms

TABLE 1. *Families of viruses infecting invertebrates*

Characteristics	Virus family	Genus	Representative member
dsDNA, enveloped	Baculoviridae	Nucleopolyhedrovirus (NPV)	*Autographa californica* NPV (AcNPV)
		Granulovirus (GV)	Trichoplusia ni GV (TnGV)
	Nudiviruses[a]	Non-occluded virus	Hz-1, Oryctes rhinoceros virus
	Polydnaviridae	Ichnovirus	Campoletis sonorensis virus (CsV)
		Bracovirus	Cotesia melanoscela virus (CmV)
	Poxviridae	Entomopoxvirus A	Melontha melontha entomopoxvirus
		Entomopoxvirus B	Amsacta moorei entomopoxvirus
		Entomopoxvirus C	Chironomus luridus entomopoxvirus
	Ascoviruses[a]		Spodoptera ascovirus (SAV)
dsDNA, nonenveloped	Iridoviridae	Iridovirus	Chilo iridescent virus
		Chloriridovirus	Mosquito iridescent virus
ssDNA, nonenveloped	Parvoviridae	Densovirus	Galleria densovirus
dsRNA, nonenveloped	Reoviridae	Orbivirus	Bluetongue virus
		Coltivirus	Colorado tick fever virus
		Cypovirus	Cytoplasmic polyhedrosis viruses (CPVs)
	Birnaviridae	Birna virus	Drosphila X virus
ssRNA, enveloped	Togaviridae	Alphavirus	Sindbis virus
	Flaviviridae	Flavivirus	Yellow fever virus
	Rhabdoviridae	Vesiculovirus	Carajas virus
		Lyssavirus	Humpty Doo, Sigma virus
	Bunyaviridae	Bunyavirus	Anopheles A group
		Phlebovirus	Sandfly fever virus
		Nairovirus	Crimean-Congo hemorrhagic fever virus
ssRNA, nonenveloped	Picornaviridae	Unassigned	Cricket paralysis virus
	Tetraviridae	Unassigned	Nudarelia β virus
	Nodaviridae	Nodavirus	Nodamura virus, Flock House Virus

[a]Nudiviruses and ascoviruses are not currently taxonomically classified.

baculovirus and *nudivirus* are used to distinguish these types of viruses.

Baculoviruses have an occluded form during some stage of their life cycle. In the occluded form, the virions (enveloped nucleocapsids) are embedded in a crystalline protein matrix. Baculoviruses include two genera, nucleopolhedrovirus and granulovirus, commonly known as *nuclear polyhedrosis viruses* (NPVs) and *granulosis viruses* (GVs). Each occlusion body of a GV contains only a single virion, and, collectively, GV occlusion bodies have a "granular" appearance in the light microscope. In contrast, NPVs have numerous (e.g., >20) virions per occlusion body so that the occlusion bodies are large (up to 5 microns in diameter and easily visible under the light microscope) with polyhedral morphology (Fig. 1A and B). There are two types of NPVs: multiple nucleocapsid polyhedrosis viruses (MNPVs) and single nucleocapsid polyhedrosis viruses (SNPVs). The MNPVs have multiple (e.g., 2 or more) nucleocapsids embedded together (in a single envelope) in the occlusion bodies (Fig. 1B), whereas in SNPVs, nucleocapsids are individually enveloped and embedded in the occlusion body matrix (Fig. 1A). Occlusion bodies of NPVs are produced in the nucleus of infected cells. Occlusion bodies of GVs generally form in the cell after loss of the nuclear membrane. Baculoviruses usually produce a second type of virion known as the *budded virus* (BV) or extracellular virus (ECV) (Fig. 1C) (see virion structure below).

Nudiviruses lack an occluded form. Taxonomically, the difference between baculoviruses and nudiviruses is thought to extend beyond the simple presence or absence of occlusion-specific genes, though this has yet to be demonstrated at the molecular level. At least one nudivirus, Hz-1, efficiently establishes persistent infections in cell culture using a specific molecular mechanism to maintain persistence (23). Nudivirus nucleocapsids are rod-shaped, though some are reported to have a tail-like projection and are individually enveloped in a unit membrane. The envelope is acquired in the nucleus and, in some cases, a second unit membrane is acquired by budding from the plasma membrane of the infected cell.

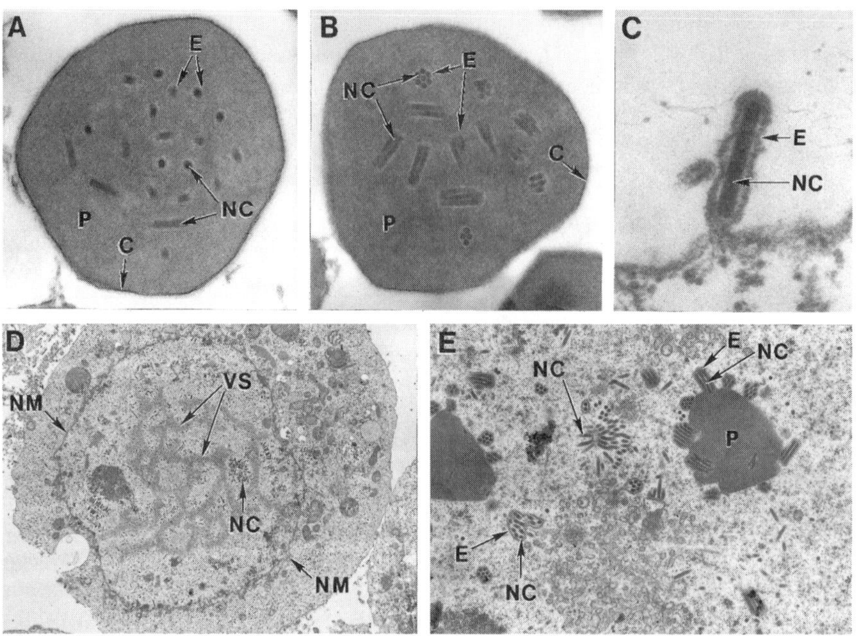

FIG. 1. Different morphological forms of nuclear polyhedrosis viruses and their development within an insect cell. **A:** Cross-section of a polyhedral occlusion body of an SNPV (ca. 2 microns in diameter): a single nucleocapsid (*NC*) within a unit membrane envelope (*E*) is embedded in a matrix of polyhedrin protein (*P*). The external surface is covered by a thin carbohydrate-rich calyx (*C*). **B:** Cross-section of a polyhedral occlusion body of an MNPV (ca. 3 microns in diameter) with a calyx (*C*) surface; virions, containing 2 or more nucleocapsids (*NC*) within a unit membrane envelope (*E*), are embedded within a polyhedrin matrix (*P*). **C:** A nucleocapsid in the process of budding from the plasma membrane of the infected cell. Budded virus of both MNPV and SNPVs usually contain only a single rod-shaped nucleocapsid (NC), approximately 30 nm in diameter and 250 nm in length—the same dimension as the nucleocapsids embedded in the occlusion bodies shown (*Panel B*). **D:** An infected cell during the late phase of MNPV replication. The nucleus, surrounded by the nuclear membrane (*NM*), contains the virogenic stroma (*VS*) with associated nucleocapsids. **E:** A transitional period between the late and very late phases of an MNPV infection. Nucleocapsids (*NC*) associate with intranuclear membrane envelopes (*E*) and enveloped nucleocapsids (virions) are then embedded in the polyhedral occlusion matrix (*P*). (Electron micrographs courtesy of Dr. Malcolm J. Fraser, University of Notre Dame.)

Baculoviruses are often named according to the host from which they were isolated, e.g., *Autographa californica* nuclear polyhedrosis virus (AcMNPV) was isolated from the alfalfa looper *Autographa californica*. This nomenclature, however, can be misleading. Baculoviruses do, however, have distinctive and relatively narrow host ranges, usually being limited to a few closely related species within a single genus or within a family.

Baculovirus Virion Structure

As noted above, a baculovirus virion is composed of a rod-shaped nucleocapsid surrounded by a membrane envelope [see (44) for a review of baculovirus virion structure]. Baculovirus nucleocapsids are composed of a rod-shaped (bacilliform) proteinaceous sheath known as the *capsid* and a nucleoprotein core. Capsids of baculoviruses average 250–300 nm in length and 30–60 nm in diameter. The capsid consists of ringlike subunits stacked with a 4.5-mm spacing to form the longitudinal axis (7). The ends of the capsids appear to be rings of diminishing diameter. The distinctive morphology of the apical end suggests a possible role in DNA packaging, budding, nuclear entry, and/or uncoating (55,181). The capsids of nudiviruses tend to be wider and longer. The major capsid protein of baculoviruses is approximately 35–40 kD (145,168,175), but additional minor proteins are also associated with the capsid (181). The nucleoprotein core within the capsid consists of a single molecule of circular, double-stranded DNA associated with protein. The major protein found in cores is a small (6.9K), basic protamine-like protein, rich in arginine (95,154,180,191).

Budded viruses of baculoviruses are composed of a nucleocapsid and a membrane envelope derived from budding of the nucleocapsid through the plasma membrane of the infected cell. In BVs, the space between the envelope and the nucleocapsid of the virion appears to be relatively unstructured and is probably primarily cytoplasmic in nature (Fig. 1C). Budded viruses are involved in secondary infection within the insect (see stages of replication, below).

The virions embedded within occlusions bodies obtain an envelope within the nucleus. The intranuclear membranes used for envelopment are often described in baculovirus literature as *de novo membranes* because electron microscopy reveals no obvious disruptions in the nuclear membrane during infection by some baculoviruses. This observation, however, does not exclude the possibility that the intranuclear membranes are a modified form of the nuclear membrane. In virions which are occluded, a 6- to 12-nm region of low to moderate electron density is located between the envelope and the nucleocapsid(s) and is referred to as the *intermediate layer* or *tegument* [see (44) for a review of baculovirus virion structure]. Occluded viruses (OVs) are involved in horizontal transmission of the virus among insect larvae and are responsible for the primary infection of the host.

Differences in the origins of BV and OV membranes result in differences in membrane proteins. For example, the major N-glycosylated protein of BV, gp64, is probably not present in OVs (154,183), whereas gp41, an abundant O-glycosylated protein of the OV virion tegument, is not found in BV (188). Differences in membrane proteins also influence the mechanism of viral entry into cells (discussed further below).

The matrix of occlusion bodies is composed primarily of a single polypeptide which forms a crystalline lattice around the virion(s). The matrix proteins of NPVs and GVs, known as *polyhedrin* and *granulin*, respectively, are closely related. Mature occlusion bodies also have an additional covering, known as the *calyx* or *polyhedron envelope*, on the surface of the occlusion body (Fig. 1A,B). The calyx contains protein and carbohydrate.

Baculovirus Genome Structure

The genome of baculoviruses consists of a large (80–230 kbp) double-stranded DNA molecule which is circular and supercoiled. The A+T content of baculovirus DNA varies widely among the members (40–70%). Extensive sequencing and transcriptional analysis of several NPV genomes is providing a broad picture of gene organization among this baculovirus genus. The most extensively studied baculovirus is AcMNPV, followed by OpMNPV and BmSNPV, isolated from the tussock moth *Orgyia pseudotsugata* and the silkworm *Bombyx mori*, respectively. The complete sequence of the AcMNPV genome is now available (1a) and a transcriptional map is maintained (98).

The circular 131-kbp DNA genome of AcMNPV is composed almost entirely of unique DNA sequences, though several small repeated sequences known as *homologous regions* (hrs), are interspersed within the DNA genome (1a,30,73). These hr sequences have been assigned roles as enhancers for early gene transcription (73,77) and as origins of DNA replication (99,144). The hrs form a complex directly or indirectly with a multifunctional viral regulatory protein, IE-1 (101) and with insect proteins (72). The basic unit of each hr of AcMNPV (and BmSNPV) is a conserved 60-bp motif which includes a highly conserved 28-bp imperfect inverted repeat. The 60-bp motifs are tandemly repeated from 2 to 8 times. (73,112). In the AcMNPV genome, the hr regions account for 3–4% of the total genomic DNA. Oryctes virus, a nudivirus, has a similar number of reiterated sequences interspersed in its genome (31).

Baculovirus open reading frames (ORFs) are closely spaced on either strand of the DNA. The information content is compacted in a highly efficient manner (see ref. 98 and refs. therein). Although there are several examples of overlapping ORFs, most ORFs are separated by 2–200 bps of DNA rich in A+T constituting promoter and termination regions. Of the 60 or more AcMNPV genes studied, only one gene, encoding the transcriptional regulator *ie-0/ie-1,* is known to have an intervening sequence. Frequently the translational termination codon, usually UAA, overlaps with the primary polyadenylation signal AAUAAA. Some promoters are located within neighboring ORFs. Frequently transcripts of one gene initiate within or extend into or through neighboring ORFs. Extension through neighboring ORFs is especially true of late or very late RNAs. There are numerous cases in which genes are transcribed as bicistronic or multicistronic RNAs, though the downstream ORFs of such RNAs may not be translated. There are also several examples of overlapping antisense 5′ ends and other examples of overlapping antisense 3′ ends for RNAs transcribed from opposite strands. Thus, a baculovirus genome of 133 kbp can easily encode 150 genes of average size. Sufficient information concerning nudivirus genomes is not yet available to provide an overview of gene organization.

Genes do not appear to be clustered with regard to function in NPV genomes. One exception is the partial clustering of several genes which have been assigned roles in early gene regulation, *ie-1, ie-2* (or *ie-n*), and *pe-38*. These genes are located within a region spanning about 5% of the genome (95–100 map units), though late genes are also present in this region. Genes with related function, such as those encoding virion structural proteins or those encoding late gene expression factors, are distributed in what appears to be a random fashion in the genome. Early, late, and very late genes are interspersed throughout the genome.

There is evidence, however, to indicate that the interspersion of different temporal classes of transcripts may have functional significance in some cases. Activation of the very late promoter of *polh*, the gene encoding the major protein (polyhedrin) of the occlusion body matrix of NPVs, down-regulates the levels of late transcripts initiated downstream of and overlapping in an antisense direction to *polh* (134). However, no evidence of promoter occlusion was observed for another set of overlapping RNAs which were all transcribed in the same direction (71).

Stages of Baculovirus Replication in Cell Culture

General Features

The steps occurring during the infection of the BV form of baculoviruses in cell culture are outlined in Fig. 2A and described in more detail in this section. The timing of each step is dependent on the virus and the cell line studied. The stated times postinfection (p.i.) are for a synchronous AcMNPV infection at a multiplicity of infection (MOI) of 20 pfu per cell in SF-21 cells, a cell line derived from a nocturnal moth, *Spodoptera frugiperda*.

Budding virus infection in cell culture is thought to represent general secondary infection in the insect host, and several aspects of its infection process contrasts with that of the occluded form of the virus which is responsible for primary infection of the host insect. Occluded virus infection is outlined in Fig. 2B and is described, along with other features of the organismal infection process (see further discussion below). Nudivirus replication may differ and has been reviewed (19).

Adsorption, Penetration, and Uncoating

Budded baculoviruses enter cells by what appears to be a typical receptor-mediated endocytosis mechanism (24,184). This pathway requires gp64, the major glycoprotein of BVs, which appears to be responsible for both receptor interaction (Fig 2A, step 1) and fusion within the endosomal membrane (18) (Fig. 2A, step 3). Although BVs bind at the cell surface, a cellular receptor has not yet been identified. A second, less efficient, membrane fusion pathway may also be used by BV (185). Nucleocapsids released into the cytoplasm migrate to the nucleus, apparently in association with actin polymerization induced by the virus particles (24). Nucleocapsids of AcMNPV interact end-on with the nuclear pore, enter the nucleus, and uncoat (70). The nucleoprotein core of the nucleocapsid is released from the capsid sheath (180), but whether viral DNA remains associated with the 6.9K protamine-like core protein is not clear. A virus-borne protein kinase may be involved in release (190), and the viral DNA may temporarily adopt a nucleosomal structure during the early phase (192). However, these important early steps in viral infection are not well characterized.

The Early Phase

During the early phase, the molecular environment of the cell is altered in preparation for the replication and expression of viral DNA. In the first 6 hours of infection, AcMNPV produces proteins required for timely viral DNA replication [e.g., genes encoding DNA polymerase (*dnapol*), a DNA helicase-like protein (p143), and a proliferating cell nuclear antigen homologue (*etl*)]. During this early phase, AcMNPV also expresses proteins involved in transcriptional regulation of early genes (e.g., *ie-1, ie-2, pe-38*) and factors involved in late gene expression (e.g., *lefs*, see below). Readers may also wish to consult prior reviews on baculovirus gene regulation (15,58,118,130).

In addition to setting up the molecular apparatus necessary to achieve a highly coordinated cascade of viral replication and gene expression, the virus is also involved in modifying many other aspects of the intracellular and extracellular environments during the early stage of infection. The cytoskeletal structure of infected cells changes significantly; the cells round and the nucleus becomes hypertrophied. In anticipation of triggering cellular apoptosis (see virus-cell interactions below), AcMNPV expresses the p35 gene to block this suicide response to infection (26). Expression of the early gene egt is an example of virus control of the extracellular environment. The egt gene encodes an enzyme which allows the virus to block the molting of its insect host by inactivating the hormones which trigger insect ecdysis (128) (see below). By blocking molting of the host organism, virus production is maximized (129).

Early baculovirus genes are transcribed by a preexisting RNA polymerase, presumably host RNA polymerase II. Naked baculovirus DNA is infectious in cell culture, indicating that host RNA polymerase(s) and factors are sufficient to initiate the infection cycle. Nuclear run-on assays indicate that early gene transcription is sensitive to alpha-amanitin (91). Early gene promoters are recognized *in vitro* by extracts derived from uninfected cells (87) and are also efficiently transcribed during infection in the presence of the protein synthesis inhibitor cycloheximide (151). However, most early promoters require the presence of the multifunctional regulatory protein IE-1 or its spliced gene product IE-0 to be efficiently transcribed in transient expression assays (76,150). Historically, the term *delayed early* has been used to describe early gene promoters which are strongly transactivated by IE-1 or IE-0 in transient assays, in contrast to "immediate early" genes (*ie-0, ie-1,* and *ie-n*) which are less dependent. A strict requirement of delayed early promoters for prior IE-1 or IE-0 production, however, may be an artifact of transient assays; IE-1, which is produced most abundantly late in infection, is present in budded virions (171). This could account for the apparent lack of dependence of delayed early promoters on prior viral protein synthesis in virus-infected cells. Nevertheless, *ie-1* is an important and essential gene (150). IE-1 is

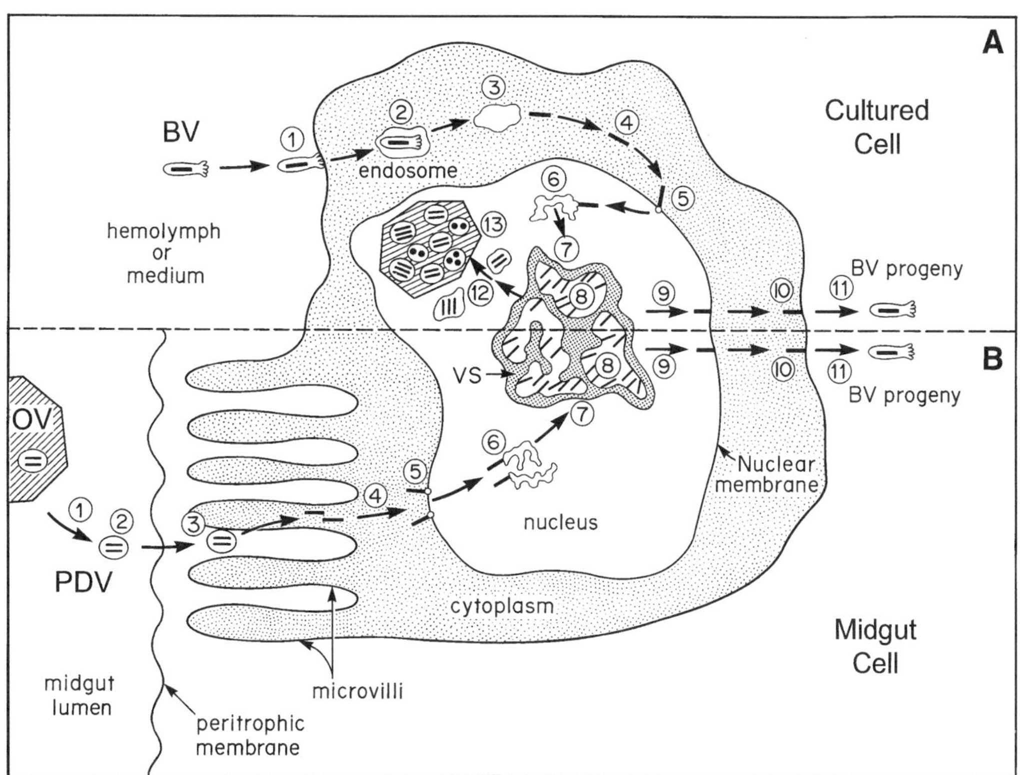

FIG. 2. The replication cycle of a baculovirus, specifically a MNPV. **A:** The process of BV infection of a cultured cell. A BV attaches to receptors on the cell surface (*1*) and enters by endocytosis (*2*). As the endosome acidifies, virus and endosomal envelopes fuse (*3*), releasing nucleocapsids into the cytoplasm. Nucleocapsids move to the nucleus (*4*) where they interact end-on with a nuclear pore (*5*). Upon entering the nucleus, the core is released (*6*) and the viral DNA is transcribed (*7*), replicated and packaged into nucleocapsids (*8*) in association with the virogenic stroma (*VS*). During the late stage of infection, nucleocapsids leave the nucleus (*9*) and travel to the plasma membrane where they bud from the membrane (*10*) to produce BV (*11*). During the very late phase, nucleocapsids are enveloped within the nucleus (*12*). Intranuclear virions are then embedded in a polyhedrin matrix to form occlusion bodies (*13*). **B:** The process of infection of a midgut epithelial cell: the primary site of infection in the insect. The polyhedrin matrix of an ingested occluded virus (*OV*) is dissolved in the midgut lumen, releasing (*1*) polyhedra-derived virions (*PDV*) which cross the peritrophic membrane (*2*) separating the lumen from the epithelial cells. The membranes of the PDVs fuse with the membranes of the microvilli of the cell, releasing nucleocapsids into the cytoplasm. The remaining events appear to be similar to those in cultured cells (*Panel A*) except that little or no occlusion body formation occurs in columnar cells of the midgut epithelium.

able to transactivate early promoters in the absence of hr sequences but also interacts directly or indirectly with hr sequences (101,102).

The patterns of transcription of different early genes are varied: some early gene transcripts appear early (e.g., 1 hr p.i.) and decline quickly (e.g., by 6 hr p.i.), whereas other early gene transcripts accumulate with time. Thus early transcription must be extensively and intricately controlled. In addition to control by IE-1/IE-0, some early genes are additionally controlled by the products of *ie-n* (also known as *ie-2*) and *pe-38*. Both IE-N and PE-38 contain a zinc-finger-like motif, $CX_2C..CXHX_{2-3}CX_2C...CX_2C$, also called a *C3HC4* or *ring-finger motif*, which is also found in at least two other AcMNPV gene products, CG30 and IAP, and a variety of other viral and eukaryotic genes thought or known to have DNA binding and/or gene regulatory function (29). Thus, AcMNPV has a family of proteins based on the C3HC4 motif. The function of these genes remain to be defined; at least one gene, *cg30*, appears to be nonessential for virus replication (141).

Most, if not all, early baculovirus promoters resemble host promoters. Many have TATA boxes 25–30 bp upstream of the transcriptional start point which sometimes, but not always, is located within the sequence CAGT. The TATA box, like mammalian TATA boxes, influences transcriptional initiation and is subject to activation by upstream elements (13,37,74,97). The CAGT sequence is similar to a conserved sequence found at the start site of many insect or arthropod transcripts and is important for efficient transcription initiation (12). Three of the early promoters which have been analyzed in detail, the promoters of *p35, gp64,* and *p39* (also known as *pp31*), also have a late promoter embedded within the early promoter, so they are transcribed both early and late in infection (13,37,74).

The Late Phase

The late phase, which begins approximately 6 hr p.i. and extends through 24 hr p.i., involves viral DNA replication, late gene expression, and the production of the budded form of the virus. Proteins which must be synthesized in abundance at this time include BV structural proteins (e.g., p6.9K, VP39, and gp64). Other late proteins include the minor virion structural proteins and a variety of other proteins such as: (a) the 39K protein (aka pp31), which may be a component of the virogenic stroma (see Fig. 1D); (b) a ubiquitin homologue (75); (c) a superoxide dismutase homologue (177); and (d) a polypeptide with structural similarity to omega-conotoxins (calcium ion channel regulators) (41).

Late baculovirus gene transcription appears to involve a promoter switching mechanism; a novel, virus-induced, alpha-amanitin resistant RNA polymerase activity is responsible for transcription from invariant TAAG initiation sites [reviewed in (15, 130)]. The basic observations suggesting a novel viral polymerase activity are: (a) alpha-amanitin-resistant transcription of late genes in nuclear run-on assays (90); (b) a new alpha-amanitin-resistant polymerase activity peak upon chromatographic analysis of infected cell lysates (63,196); and (c) infected cell extracts supporting *in vitro* late gene transcription which is resistant to tagetitoxin, an inhibitor of RNA polymerase III. The novel polymerase may be an entirely new one encoded by the virus, a modified host RNA polymerase, or some combination of these possibilities.

With very few (if any) exceptions, late transcripts initiate from a TAAG sequence motif which is the predominant element for late as well as very late promoter activity (123,136). Mutations within TAAG completely inactivate promoter activity; greater than 1000-fold reduced expression is observed when the TAAG of the vp39 proximal promoter is mutagenized. Fewer than 18 bp surrounding this initiation site is sufficient to direct late gene expression; it may be that TAAGs, which are underrepresented in the AcMNPV genome, can act as basal late promoters (123). The context of the TAAG, however, affects the level of transcription (123,136).

Some late genes have multiple TAAGs upstream of the ORF, each TAAG serving as an initiation site. The vp39 gene, for example, is transcribed from three TAAG transcriptional start sites at –57, –105, and –321 relative to the ORF (175). Successive deletion of DNA from the 5′ end progressively eliminates each start site, but the proximal site continues to function until the TAAG is deleted. The vp39 gene is also an interesting example of the complexity of overlapping late and early transcripts. The three TAAG-initiated *vp39* RNAs are bicistronic; they continue through the downstream gene, *cg30,* which is also transcribed as an early monocistronic RNA from a promoter located near the C-terminus of the *vp39* ORF (174). Another early promoter is located near the N-terminus of *vp39* ORF and initiates transcription of a late expression factor, *lef*-4, in the opposite direction of transcription (138). The overlapping of late *vp39* RNAs at both ends with early transcripts suggests the possibility of intricate transcriptional or posttranscriptional regulation by the physical arrangement of transcripts.

In the late phase of infection, baculovirus DNA adopts a unique nucleoprotein structure while host DNA remains in a typical nucleosomal structure (192). DNase I digestion of the viral DNA-specific nucleoprotein complexes yields discrete fragments of 90 and 120 bps, suggesting a periodicity in the binding of the protein(s) to the DNA. Whether this nucleoprotein structure is only the prelude to viral DNA packaging or whether it is a specific structure adapted by viral DNA for replication and transcription during the late phase remains unknown.

Late transcription appears to be coupled to DNA replication. Aphidicolin, an inhibitor of both viral and host DNA polymerases (120,179), blocks viral DNA replication and late viral gene transcription (151). Mutations in DNA replication-related genes also indicate a connection between

DNA replication and late gene expression. A ts mutation in *p143*, a homologue of DNA helicases (109), blocks both DNA replication and late gene expression at the nonpermissive temperature (67). A null mutation in *etl*, an early viral gene homologous to proliferating cell nuclear antigen, delays both DNA replication and late gene expression (32,127). The mechanism by which late gene expression is coupled to DNA replication remains to be determined.

A number of genes required for late but not early gene expression have been identified. Eighteen late expression factor (*lef*) genes were identified by a transient expression assay through the use of an overlapping set of AcMNPV genomic clones able to *trans*-activate a reporter gene under the control of a late promoter (108,110,124,137–139,142,143). Along with several additional genes known to be involved in DNA replication (e.g., *dnapol* and *p143*) or late gene expression, *p47* (21), these lef genes are able to support both DNA replication and late gene expression in transient expression assays (110a). Thus, approximately 20% of the genome of AcMNPV appears to be involved in late gene expression and/or DNA replication. However, the role of each of the *lef*s remains to be defined; at least one *lef, lef-8,* has a conserved sequence motif of RNA polymerases (142).

The hr sequences of AcMNPV have been implicated as origins of DNA replication (99,107,144) based on a demethylation assay for plasmid DNAs transiently introduced into baculovirus-infected cells. It remains to be demonstrated that hr sequences are used as viral DNA replication origins during virus infection. The presence of hr sequences is not required for expression from late reporter genes in transient expression assays (150); other sequences have also been assigned roles as origins of viral DNA replication (97a).

The Very Late or Occlusion Phase

Occluded viruses are produced in the final, "very late" stage of infection which begins approximately 18–24 hr p.i. and extends through 76 hr p.i until cell lysis. This occlusion phase is marked by a dramatic increase in the transcription of *polh*, the gene encoding polyhedrin (the major protein of the occlusion body matrix). The temporally regulated hyperexpression of *polh*, which is nonessential for BV production, made baculoviruses ideal vectors for foreign gene expression (111,116,146,159). By replacing *polh* with a foreign gene of interest, it is possible to obtain very high levels of foreign gene expression very late in infection, following BV production.

This expression system is now used in a variety of ways (96,119,130) and has been extended to the use of other promoters including the p10 promoter, which is also hyperexpressed very late in infection. P10 is nonessential for virus replication including virus occlusion and has been implicated in microtubule assembly (25). Null mutations in this gene result in the lack of nuclear lysis at the end of infection (132,189). In a normal infection, the nuclear membrane as well as the plasma membrane ruptures. The value of p10-associated nuclear disruption to the infection process is not clear but may have significance at the tissue or organismal levels.

Very late genes are also transcribed by a novel alpha-amanitin resistant, viral-induced polymerase from TAAG-based promoters (see the late phase above). Linker scan analysis of the *polh* promoter demonstrated the absolute dependence of very late transcription on the TAAG motif and also revealed the importance of downstream sequences in the rate of initiation from this promoter (136). The 50 bps specifying the leader of *polh* RNAs all appear to be required for the maximum "burst" of transcription observed in the very late phase. Mutations in the downstream region which decrease RNA levels do not decrease polh RNA stability, nor do they appear to decrease translation, but they do decrease the rate of transcriptional initiation (136). Mutations upstream of the TAAG tend to increase rather than decrease transcription but do not alter the temporal regulation of transcription. Thus, the *polh* promoter consists primarily, if not exclusively, of 50 nucleotides at and downstream of the TAAG initiator site. The structure is reminiscent of yeast mitochondrial promoters and T-odd bacteriophage promoters. The *polh* promoter contrasts with the late *vp39* promoter in that 50 bps downstream of the TAAG are required for the very late "burst" of expression, whereas only a few bps flanking the TAAG of the *vp39* promoter provide a significant level of "recognition" during the late phase (123). The *p10* promoter appears to be activated approximately 3 hr before the *polh* promoter (153). Late antisense transcription through *polh* may affect its temporal expression subtly (134).

The genes required for expression of the late *vp39* promoter in transient expression assays, e.g., the *lef* genes, are also required for expression from the *polh* promoter in these assays (108,124,137–143), but additional gene products and/or some specific intranuclear environment are probably necessary for the regulation of very late gene expression (114a). The presence of specific "burst" sequences in the polh promoter suggests the existence of a specific trans-activator for very late gene transcription (123,136). Hybrid late and very late promoters have been constructed by placing late (vp39) TAAG sequences in tandem with the *polh* promoter (123,176); each TAAG sequence retains its fundamental temporal regulation in this setting.

Virion Assembly and Occlusion

Nucleocapsids assemble during the late and very late phases in conjunction with the virogenic stroma, a virus-specific nucleoprotein structure near the center of the nucleus (see Fig. 1D). Viral capsids appear to assemble at the edge of the stromal matte (81,197). Electron micrographs of partially filled nucleocapsids with DNA strands connecting the core with the virogenic stroma suggest that the virogenic stroma may be the active site for viral DNA replication, condensation, and packaging (55,197). Electron microscopy, coupled with differential enzyme digestions, has implicated RNA as a possible structural component of the virogenic stroma (197).

In the late phase of infection, nucleocapsids leave the nucleus, possibly by budding or by exit through nuclear pores. Nucleocapsids within envelopes in the cytoplasm can be observed by electron microscopy but are naked by the time they reach the plasma membrane. By interacting end-on with gp64-rich sections of the plasma membrane, the nucleocapsids bud from the surface of the cell becoming extracellular virions (see Fig 1C). The rate of release of BV of AcMNPV from cells increases exponentially between 10 and 20 hr p.i. and continues at a reduced linear pace from 24 through 36 hr p.i. (105); other baculoviruses may exhibit different kinetics of BV release.

Occlusion occurs in the nucleus of AcNPV-infected cells from approximately 20 through 76 hr p.i. Occlusion involves a number of steps including the appearance of segments of phospholipid membrane inside the nucleus. Nucleocapsids align along the membrane segments to acquire an envelope which is a prerequisite for their incorporation into an occlusion body (see Fig. 1E). A series of mutant baculoviruses known as *few polyhedra* (FP) *mutants*, appear to be defective in membrane envelopment, and the nucleocapsids fail to be enveloped and packaged within the occlusion bodies [reviewed in (117)]. The FP mutants produce a higher titer of budded virus, suggesting that envelopment in the nucleus may be a factor in redirecting nucleocapsids from BV formation into preoccluded virion (POV) production.

Before cell lysis, occlusion bodies accumulate in the nucleus of the infected cell (Fig. 3). Maturation of the occluded form involves the deposition of an outer covering known as the *calyx* (or polyhedron envelope) over the surface of the occlusion body. The calyx is composed of carbohydrate and protein, and a gene involved in calyx formation has been identified (66). Fibrillar structures, composed of P10, can also be observed in both the nucleus and the cytoplasm (see Fig. 3).

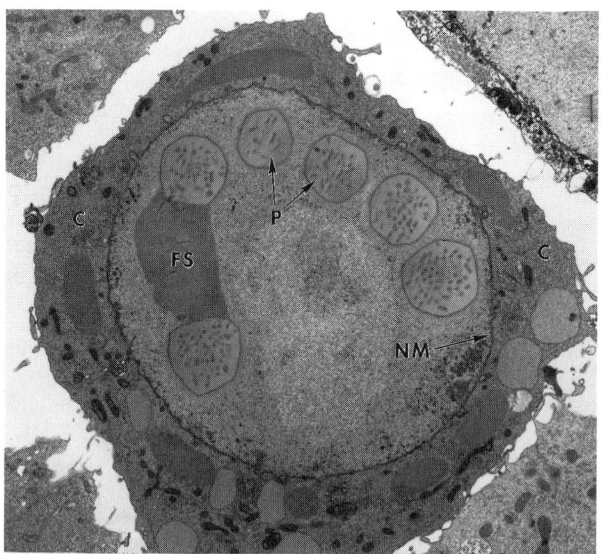

FIG. 3. The very late stage of AcMNPV infection. Multiple polyhedral occlusion bodies (*P*) are formed in enlarged nucleus which is still surrounded by a nuclear membrane (*NM*). Fibrillar structures (*FS*) are often present in both the nucleus and the cytoplasm (*C*). (Electron micrograph courtesy of Drs. Peter Faulkner and Gregory Williams, Queens University.)

Virus-Cell Interactions

Baculoviruses have a dramatic effect on the host cell, altering both cytoskeletal and nuclear structure dramatically and, during the late phase, decreasing the levels of host RNAs (135) and shutting off host protein synthesis (22). While the viral DNA adopts a very unique nucleoprotein structure, host chromatin remains largely intact and in a nucleosomal structure (192).

The ability of cells to undergo apoptosis, programmed cell death, during baculovirus infection leads to abortive virus replication and has a dramatic affect on the outcome of infection in the insect host (26,27). At least two different types of baculovirus genes, p35 and some members of the "inhibitor of apoptosis" (iap) gene family, are able to block cellular apoptosis in SF-21 cells (26,33). In the absence of a viral gene which can block apoptosis, infected cells undergo a rapid apoptotic response involving extensive blebbing on the cell surface and resulting in the release of membrane-bound vesicles, referred to as *apoptotic bodies* (26). During this process, the nucleus undergoes fragmentation and degradation of host chromatin into oligonucleosomal-length fragments while, in the cytoplasm, there is a general shutdown of host and viral protein synthesis (27,86).

Widespread apoptosis accompanies the infection of SF-21 cells with an AcMNPV mutant defective in *p35*, a gene responsible for blocking apoptosis in these cells (26), and results in the shutdown of protein and RNA synthesis (27,86). The titer of progeny BV is reduced by 100-fold or more (27,86), and occluded virus production is eliminated by this cellular apoptotic process which initiates during the late phase of virus infection (27,34).

In larvae of *S. frugiperda* (the parent organism of the SF-21 cell line), BV of p35 mutants have a 1,000-fold higher LD_{50} than wild-type virus, indicating that the ability to undergo apoptosis in response to viral infection can be an important antiviral response (27). Cellular suicide may be a particularly important defense system in multicellular organisms that lack an antibody immune response (e.g., invertebrates). The absence of p35, however, has little or no effect on AcMNPV replication in a cell line (TN-368) or in larvae of another permissive species, *Trichoplusia ni* (27). The nature of the gene(s) which a virus carries to block apoptosis can thus influence the host range properties of the viruses.

The AcMNPV p35 gene is also able to block SF-21 cellular apoptosis induced by actinomycin D in the absence of other viral genes (28). Thus, p35 is independently able to block apoptosis in insect cells. Furthermore, the p35 gene is able to block apoptosis in mammalian neuronal cells (148), indicating that p35 acts at an evolutionarily conserved point in the apoptotic pathway.

The iap genes derived from *Cydia pomonella* GV and *Orgyia pseudotsugata* NPV, *Cp-iap* and *Op-iap,* respectively, are also able to block apoptosis induced by actinomycin D in the absence of other viral genes (28). Although AcMNPV carries its own iap homologue, it is unable to block apoptosis induced by *p35* mutants or by actinomycin D in SF-21 cells. The AcMNPV iap gene was identified only by its sequence homology to the Cp-iap and Op-iap genes, and all three homologues share a C3HC4 (ring) finger motif at their C-termini and two tandem cys/his motifs, BIR (baculovirus *iap* repeat) motifs, in the N-terminal portion of the polypeptides (11). Domain swapping studies show that the sequences within both the N-terminal BIR and the C3HC4 finger are important for antiapoptotic function (28); these observations suggest that *iaps* have gene regulatory roles.

SF-21 cells appear to be "pretriggered" for apoptosis since treatment with actinomycin D or other inhibitors of RNA synthesis (e.g., alpha-amanitin) are able to induce rapid and widespread apoptosis even in the presence of the protein synthesis inhibitor, cycloheximide. Cessation of host RNA synthesis may be the signal that triggers apoptosis during AcMNPV infection (28).

Some nudiviruses are able to establish persistent infections in their host cells. During persistent infection with Hz-1, only a single viral RNA is transcribed. Persistently infected cells undergo viral replication at a low level in what appears to be a sporadic fashion (19). Viruses produced from these infections cause lytic infections in cells with no prior exposure to the virus. Persistently infected cells are at least partially protected from infection by the homologous virus. Persistently infected cell cultures undergo extensive, but not complete, apoptosis following homologous virus infection, and the apoptotic response may be correlated with protection (106).

Baculovirus infection also affects intracellular calcium ion distribution and cytoplasmic calcium concentrations (169). Calcium is sequestered into intracellular vesicles, probably the endoplasmic reticulum. While the basic mechanism responsible for calcium ion sequestration is not defined, a viral homologue of omega-conotoxins, *ctl*, has been implicated in modulating calcium ion concentrations (169).

Genomic Changes Upon Serial Passage

Extensive changes in the DNA genome occur when baculoviruses are serially propagated in cell culture (100,104). The effects differ depending whether the passages are performed at high or low MOI. Low MOI serial passage of AcMNPV results in the selection of viable viruses having deletions and/or insertions of DNA (56,104,115; reviewed in 117). The insertions are usually found in a single locus, affecting a 25K protein (5,56,104), and giving rise to a phenotype called the *FP phenotype*. Presumably, the disruption of 25K provides a strong selective advantage for growth of BV at the expense of occluded virus production; it is not cell culture conditions per se which cause insertion or selection, because FP mutants also arise upon serial passage in insects if the passage is performed by serial injections of BV into insect hemolymph (reviewed in 57,117). At least some of the FP mutants produce a higher titer of BV than wild-type virus and thus have a selective advantage for growth in serial passages involving only the budded form (147). Occlusion bodies of FP mutants are defective in primary infections.

Insertions are usually host-transposable elements, having the characteristic duplication of the target site at the point of insertion (6 and references therein). Several elements which transposed into the fp25 locus have been sequenced but have not been fully classified with regard to their relationship to other known transposable elements. Transposable element insertions into other regions of AcMNPV have also been observed and characterized (57,104,115,131). One of these insertions was an entire 7.2-kbp retrotransposon (59,115). It was proposed that baculoviruses play an important role in the horizontal movement of transposable elements among invertebrates (115); horizontal transmission of transposable elements appears to occur frequently within arthropods.

Deletions in the virus genome arise during both low and high MOI serial passages. During low MOI passages, there is a selection for viable mutants with deletions in the egt gene (see below). During high MOI passages, defective interfering virus particles are produced with a selection for retaining and amplifying DNAs carrying hr sequences and possibly other origins of DNA replication (100).

Disease Progression in the Insect

Insects are horizontally infected by baculoviruses primarily by eating food contaminated with virus, and the occluded form of the virus is specifically adapted for horizontal infection. The crystalline protein matrix of the occlusion body provides virions with protection from environmental stress/degradation during the interim period between (outside of) hosts. This protein matrix, however, is also designed to release the virions at the exact point in the alimentary canal of the insect, the midgut, where susceptible epithelial cells are exposed and virus infection can proceed. Most of the surface of the insect is covered by a waxy chitinous cuticle which is impervious to virus entry; but the insect midgut, where most food absorption occurs, is protected by another structure known as the *peritrophic membrane* containing protein and chitin. Some baculoviruses, especially GVs, contain proteins within the occlusion bodies which appear to be able to alter the structure of the peritrophic membrane and facilitate virus infection (83 and references therein). However, the major feature of the midgut which the virus utilizes in its infection strategy is the very high pH (>10) of the midgut lumen. The crystalline protein matrix of occlusion bodis is dissolved under these high-pH conditions, releasing the virions within the midgut lumen.

The primary site of infection is the midgut epithelium. Released virions bind to the brush border membrane of columnar cells (70) and appear to enter by direct membrane fusion with the plasma membrane of the columnar epithelial cells (Fig. 2B, steps 3 and 4). The remainder of the infection process is thought to be generally similar to that observed for BV infections in cell culture, with a few exceptions. In AcNPV infections, there is a clear polarity in the movement of gp64, which is directed primarily to the basal membrane (94). This directionality likely dictates the specific budding of nucleocapsids from the basal laminar side of the cell, allowing access to tracheal epithelial cells that cross the basement membrane bordering the hemocoel. Some NPVs produce only a few occluded viruses in midgut cells. Other baculoviruses primarily infect the midgut, and extensive occluded virus production is observed.

Secondary infection is achieved by viruses produced in midgut cells. Hemocytes (blood cells) and epithelial cells lining the tracheal network, which supplies oxygen, seem to be cells responsible for initiating the secondary infection process (94). In infections of a highly susceptible host by a highly infectious virus such as AcMNPV, secondary infection can eventually affect virtually all the internal tissues of the organism. In the final phases of organismal infection, the insect is literally converted into a milky white liquid consisting mostly of virus particles. This effect was described as the *wilting disease* in the nineteenth and early twentieth century and is now often referred to as *melting*. In a typical AcMNPV infection, over 1 billion occlusion bodies are produced per larva, constituting over 10% of the dry weight of the insect. Upon rupture of the cuticular exoskeleton, the OV are released into the environment for horizontal transfer to other insects. Thus, BVs and OVs differ not only morphologically and biochemically but differ functionally as well (70).

Baculoviruses are able to regulate their insect hosts at the organismal level as well as the cellular and molecular levels. The most dramatic demonstration of this is the ability of baculoviruses to control the molting of their insect hosts by controlling the activity of the steroid hormones which trigger the molting process, ecdysone and its derivatives (128). The AcMNPV egt gene encodes an ecdysteroid UDP-glucose/galactose transferase which is secreted from infected cells and transfers either glucose or galactose from a UDP-sugar (glucose or galactose) substrate to the 22-OH position of ecdysteroids, thereby inactivating the hormones and preventing insect molting (126,128). If this gene is deleted from the virus, the insects attempt to molt and die sooner, producing fewer virus progeny (129).

It is also relevant to note that not all baculoviruses cause a lethal infection under all circumstances. There are a number of reports regarding the presence of presumably persistent or latent baculoviruses in seeming healthy insects and the vertical transmission of these viruses to progeny. The most thoroughly described virus-host system is that of an NPV of *Mammestra brassicae* that persists in a laboratory insect colony without causing substantial mortality under these conditions (89).

Polydnaviridae

Classification and Structure of Polydnaviruses

The family Polydnaviridae is taxonomically characterized by segmented double-stranded DNA genomes. The genomes consist of multiple supercoiled, double-stranded DNA molecules of differing sizes ranging from approximately 2 kbp to 28 kbp (52,53,103,162). The name *polydnavirus* derives from *poly*disperse *DNA*s. Depending on the virus, there are approximately 15–30 different DNA molecules comprising the "collective" genome of polydnaviruses with an estimated genomic complexity of approximately 75–200 kbp (52,103). Polydnaviruses appear to be vertically transmitted within wasps, apparently as an integral part of their chromosomes (i.e., as proviruses). Amplification of viral DNA and virus particle formation, however, occurs only in the nuclei of calyx epithelial cells in the oviduct of females (see replication cycle below).

Polydnaviruses have been characterized only from endoparasitic hymenopteran insects, i.e., wasps which parasitize other insects. Polydnaviruses have been isolated from two families of wasps, Ichneumonidae and Braconidae (165). Virus-like particles have been observed in other families of wasps and other orders of parasitoids (e.g., diptera) but lack sufficient characterization to be classified as polydnaviruses (152,155). Characterized polydnaviruses are divided into two genera: the ichnoviruses and the bracoviruses. The ichnoviruses, found in ichneumonid wasps, are morphologically quite different from the bracoviruses of braconid wasps. (Not all species of braconids and ichneumonid wasps, however, have polydnaviruses.)

The nucleocapsids of bracoviruses are similar in structure to baculoviruses with rod-shaped (cylindrical) nucleocapsids of uniform diameter (ca. 35–40 nm) but variable (30–200 nm) length (see Fig. 4C). Bracoviruses have a single membrane envelope, derived within the nucleus during replication, and, like baculoviruses, some bracoviruses (e.g., *Chelonus nigriceps* virus) have a single nucleocapsid per envelope; whereas other bracoviruses (e.g., the type species from *Cotesia melanoscela*, CmV) have several nucleocapsids enveloped within a single envelope.

In contrast, the morphology of ichnoviruses is more variable and complex. Ichnovirus nucleocapsids are uniform in size: 80–90 nm in diameter and approximately 300 nm in length (see Fig. 4A). The shape of the nucleocapsids is

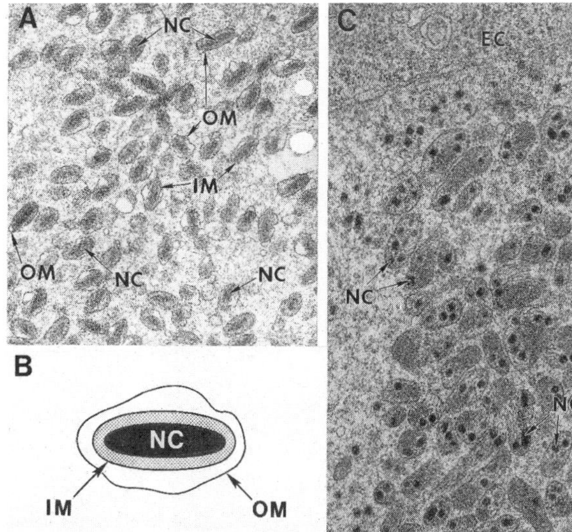

FIG. 4. Morphological features of polydnaviruses. **A:** An ichnovirus showing the lenticular nucleocapsid (*NC*) of uniform size surrounded by and inner membrane (*im*) and a second, outer membrane (*om*). **B:** A schematic of ichnovirus structure. **C:** Bracoviruses in the calyx fluid (*CF*) of an adult female parasitoid wasp. Nucleocapsids (*NC*) are observed end-on and in lateral view. In this bracovirus, multiple NC are usually found within a unit membrane envelope (*E*). An adjacent calyx epithelial cell (*EC*) is also observed in this cross-section. (Electron micrographs courtesy of Dr. Donald B. Stoltz, Dalhousie University. Modified from Stoltz, ref. 164, with permission.)

usually described as lenticular or quasi-cylindrical. The nucleocapsids of ichnoviruses are surrounded by two unit membranes (Fig. 4A,B), one derived within the nucleus and the other derived by budding from the plasma membrane of calyx epithelial cells.

Polydnaviruses, like baculoviruses, obtain an envelope from intranuclear membranes elaborated within the nucleus during replication. Approximately 20 different-sized polypeptides are associated with polydnavirus particles. The minimum genetic information needed to encode these proteins is estimated to be 30 kbp, assuming that the proteins are coded on separate, nonoverlapping genes (103). However, it is not yet known whether the DNAs which are packaged into virions encode the structural proteins of the virions since there is no evidence that the virions are able to replicate independently.

Life Cycle of Polydnaviruses

Before describing the genome organization of polydnaviruses, it is important to appreciate the mode of transmission and the complexity of the life cycle of these viruses (Fig. 5). The viruses are vertically transmitted, apparently as proviruses integrated into the genomes of the parasitic wasps (52,162). Parasitoids lay their eggs in insects of other species, often in one of the larval stages of the insect. The wasp eggs hatch, and young parasite larvae develop within the parasitized host using host nutrients as their food source until development is complete. The larval host usually dies if successfully parasitized by the wasp. However, when the wasp lays an egg, the larva would normally attempt to mount an immune response to the egg. This response, mediated at least in part by hemocytes (cells of the open circulatory system of the insect), results in "encapsulation" of the egg, the formation of a multilayered structure that effectively isolates the egg and prevents its survival. The encapsulation response is one of the primary immune defense systems of insects against a variety of "foreign" intrusions. Wasps, however, have counter-defenses against the host immune response and, in the case of many ichneumonid and braconid wasps, defense against egg encapsulation is dependent on the action of polydnaviruses which are deposited in the caterpillar by the wasp along with the egg (40,164,182).

The formation of polydnavirus particles occurs exclusively in the nuclei of calyx cells of the oviducts of female wasps. Polydnavirus replication is under wasp hormonal control (187) and is first detected at a specific stage in pupal development (125). In mature females, the virus particles are released from the calyx cells into the oviduct lumen, either by budding (ichnoviruses) or by cell lysis (bracoviruses); virus production and/or cell lysis does not appear to adversely affect the wasp. The fluid in the lumen (i.e., the calyx fluid) contains massive numbers of polydnavirus particles which are deposited with the egg along with secretions from accessory glands (e.g., venom glands) during oviposition. Replication of ichnoviruses, which bud from the calyx cells, does not seem to negatively influence the structure or function of the calyx cells. Bracoviruses, however, appear to be released from the calyx cells upon cell lysis.

Polydnavirus particles participate in the abrogation of the host immune defense system (40). If virus particles are artificially removed from the calyx fluid and/or are inactivated with a nucleic acid cross-linking agent such as psoralen, the injected egg is encapsulated rapidly. Addition of purified active polydnaviruses to washed eggs prior to injection protects the egg. Suppression of encapsulation by a polydnaviruses appears to involve alterations in populations of those hemocytes that participate in capsule formation, i.e., granulocytes and plasmatocytes (35,162,167).

In the case of some braconid parasitoids, the venom which is injected along with the egg and virus particles also plays a role in preventing encapsulation and other physiological effects (170). However, venom alone does not effectively prevent encapsulation. The venom appears to function in conjunction with the polydnaviruses, possibly by enhancing virus infectivity directly or indirectly (163).

Polydnaviruses may modulate other aspects of the physiology of the caterpillar during parsitization. Degeneration of the prothoracic gland has been observed following injection of purified particles of the *Campoletis sonorensis* polydnavirus (CsV or CsPDV) (38). Since prothoracic glands provide the developmental regulatory hormone

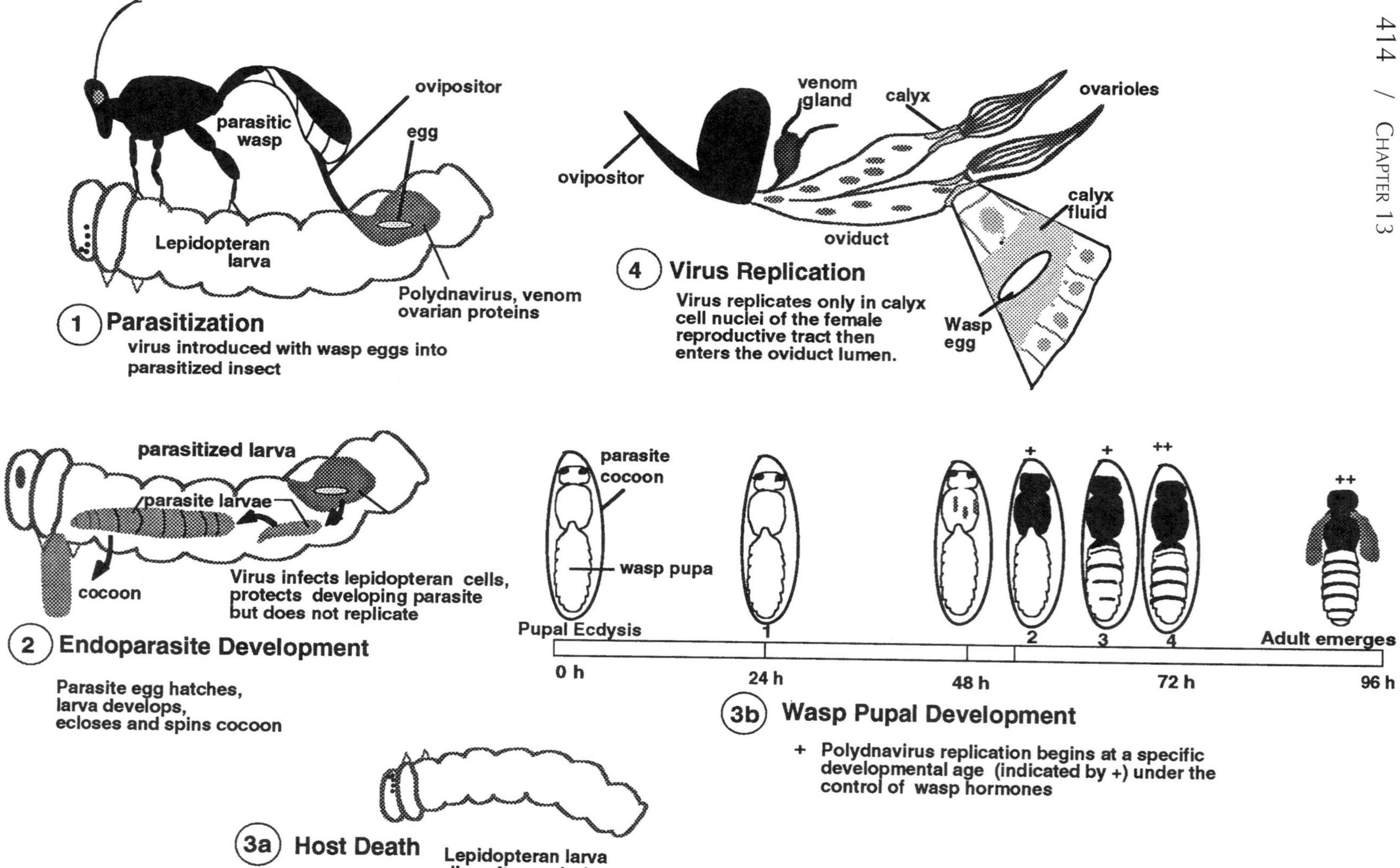

FIG. 5. Polydnavirus life cycle. Virus particles are injected into a lepidopteran larva along with a wasp egg during oviposition, the first step in parasitization (*1*) of the larva by the wasp. The wasp carries and vertically transmits the polydnavirus in proviral form. The egg develops within the parasitized larva (*2*) with the protective aid of the polydnavirus. The host lepidopteran larvae dies (*3a*) during or after parasite emergence. As the wasp develops (*3b*), the amplification of polydnavirus DNA in calyx cells of the female oviduct is first observed at a specific stage in wasp pupal development (ca. 48 hr after pupal ecdysis). The amplified polydnavirus DNA is packaged into viral particles which are released from epithelial cells in the calyx of adult female ovaries (*4*). The wasp egg is bathed in calyx fluid which contains polydnavirus particles. The fluid and wasp egg are injected, along with venom gland secretions, upon oviposition (*1*). (Courtesy of Dr. Bruce A. Webb, Rutgers University.)

ecdysone, the effects of gland deterioration would alter host physiology and development significantly. In braconids, defining additional polydnavirus functions during parasitization is complicated by the presence of products of the developing wasp egg/larva and by teratocytes, cells which are released from the egg upon hatching and which develop and actively secrete products during parasitzation. Nevertheless, multiple possible roles of the viruses in controlling the physiology of the caterpillar have been ascribed, either as indirect effects on the normal function of the infected cells or as a direct result of the production and secretion of virus-specific products from infected cells (reviewed in 8,52,162).

Polydnavirus DNA replication has not been detected in parasitized host larvae but does appear to persist and be transcribed throughout the course of parasitization (52,53,103,166). At least 12 different polyadenylated RNAs of CsV have been identified in parasitized or CsV-injected larvae (14). Some of these transcripts are spliced (16) using standard splice junction signals. CsV transcripts are detected within 2 hr, and transcripts are detected throughout development until parasitoid emergence and larval death (approximately 9 days postoviposition). Most transcripts have similar steady-state levels from approximately 12 hr to 6 days. At least some of these transcripts specify polypeptides; using the baculovirus expression system, two polydnavirus cDNAs produced secreted proteins of 23 or 25 kD (17).

Persistence and transcription of *Microplitis demolitor* polydnavirus (MdPDV) DNA has also been studied, and a similar picture has emerged (166). This bracovirus DNA persists for at least 6 days in the parasitized host but is not amplified. Transcription, including at least 6 different size classes of MdPDV RNA, is observed as early as 4 hours postparasitism, and transcripts persist for at least 6 days postparasitism. More viral DNA and RNA was detected in hemocytes than in other tissues examined.

Polydnaviruses cause no apparent pathology in the wasp harboring the virus. It is difficult to determine what effect, if any, the presence of polydnaviruses have on adult wasps, because the virus particles are found in every female of the species that are known to carry a polydnavirus. Although virus particles are confined to the female reproductive tract, viral DNA is present in all examined tissues of both adult males and females (54). The predominant form of the viral DNA found in males appears to be in the integrated "proviral" form, but a minor component of the "replicated" form of viral DNA is also found in males (see genome organization below). Transcripts from some genomic segments have been detected in wasp ovaries, and there is evidence that some of the transcripts produced in wasps differ from those produced in parasitized hosts (17,173).

In summary, the relationship between the virus and the wasp can be regarded as mutualistic: the relationship between the wasp and virus is sustained at the chromosomal level by integration in the wasp germline, is regulated by the endocrine system of the wasp in somatic tissues, and is maintained by selective pressure for wasp egg survival in the parasitized host larva.

Genome Organization of Polydnaviruses

The aggregate size of a polydnavirus genome is difficult to estimate because of the presence of multiple topological forms of the DNAs (e.g., superhelical, relaxed circular, etc.) in viral DNA preparations, nonequimolar concentrations of the genome segments, and the presence of restriction fragment length polymorphisms (RFLPs). The genome of CsV is estimated to be comprised of 28–30 different DNAs ranging from 5.5 to 21 kbp with a collective size of 200–250 kb (52,53,103). Another ichnovirus, *Hyposoter exiguae* virus (HeV), appears to have a considerably smaller genome of only approximately 75 kb (53,103).

Some polydnavirus genome segments share homology. For example, segments H and O^1 of CsV have multiple homologues of a 540-bp region found as a single copy in segment B (172). The nine characterized members of the 540-bp sequence family share only approximately 60–70% sequence homology. Theories abound as to the potential role of such homologous regions in polydnavirus replication or recombination. Unfortunately, very little sequence information is available for polydnavirus DNA. Thus, a clear picture for polydnavirus genome structure has not yet emerged. However, several examples of gene families in polydnavirus genomes have been described (16,36,194).

One CsV gene family encodes polypeptides with sequence similarity to omega-conotoxins of marine snail venoms (36). The three characterized genes in this family contain two introns at equivalent positions within the genes. Curiously, comparison of the sequences reveals significantly more identity in the introns than the exons. It was suggested that there may be a relationship between polydnavirus-encoded polypeptides and venom peptides and that this might account for the lack of dependence of ichnoviruses on venom gland secretions, in contrast to bracoviruses. It has also been reported that *C. sonorensis* venom gland proteins and CsV envelope proteins share some common epitopes (186). However, much remains to be established with regard to the nature and function of these gene products.

It is not known how many viral DNAs are packaged within each nucleocapsid or whether each nucleocapsid contains all segments, a specific subset of segments, or a random distribution of segments. In the case of CsV, the particles appear to be of uniform size and of uniform density. Compared to cylindrical viruses of similar size (such as baculoviruses), the nucleocapsids of CsV are estimated to be large enough to contain approximately 450 kb of DNA, or approximately twice the collective size of the DNA genome. Thus, random packaging of the segments could be an effective packaging strategy for this virus, especially if the collective genomic complexity is significantly less than the collective size. In contrast, nucleocapsids of bracoviruses vary in length, and there is some correlation between bracovirus nucleocapsid length and the frequency (molar ratio) of DNA segments with a given contour length (103). Thus, it has been suggested that bracoviruses may individually package different DNA segments.

Polydnavirus DNA appears to be inherited in a Mendelian fashion (162), but the organization of the DNA segments within the wasp chromosome and the means by which these segments are amplified and excised (or excised and amplified) is not well defined. In a study of segments B and Q of CsV, a different restriction pattern was found for B and Q DNA in male and nonoviduct female tissues than was found for DNA packaged in virus particles (51). Similar observations were made for the ichnovirus of Hyposoter fugitivus (195). Such "off-size" fragments provide the major molecular evidence supporting distinctive integrated (i.e., proviral) forms of these DNAs. Additional studies indicate that CsV DNA segments are integrated individually in the wasp chromosome(s); each segment appears to be integrated at a different locus. In the best-studied case, the integrated form of segment B is flanked by short (59 bp) imperfect direct repeats (53). Only a single copy of the terminal repeat is found in packaged polydnavirus DNA circles (51), suggesting the possibility that recombination within the direct repeats results in a circular DNA segment. Circular segments may be the ``replicative form" of polydnavirus DNAs in ovarian tissue, though direct replication of linear integrated DNA (e.g., "onion skin" replication) is also possible.

Ascoviridae

Classification and Structure of Ascoviruses

The ascoviruses have not yet been formally classified taxonomically but, based on several unique properties, they probably belong in a distinct virus family. The name *ascovirus* derives from the Greek word *askós*, meaning "sac," and refers to the large membrane-bound vesicles in which virus assembly is thought to proceed (see stages of replication below). Ascoviruses were first discovered in 1977 and, to date, have been found only in larvae of six different species of lepidopteran noctuids, a family of moths which includes the species *T. ni* and *S. frugiperda* (20,45,79). While some of the isolates appear to be only variants of the same virus, others appear to be distinct by DNA hybridization analysis (48). Since the gross pathology of ascovirus disease is not highly distinctive, it is likely that the disease has gone undetected in other insect taxa.

The virions of ascoviruses are typically allantoid (kidney-shaped, see Fig. 6A), though the ascovirus isolate from *S. frugiperda* has almost rod-shaped virions (48). The virions are approximately 400 nm long and 130 nm in diameter and appear to have two unit-membranes, an outer envelope and an inner, electron-translucent membrane layer surrounding a DNA-containing core (Fig. 6B). Both inner and outer membranes appear to have distinctive surface structure (see Fig. 6B), presumably composed of protein subunits that confer the reticular appearance. The virions are composed of at least 12 different polypeptides ranging in size from 11 to 200 kD (48).

Ascoviruses have a large, double-stranded, DNA genome ranging in size from approximately 140 to 180 kb (43,48). The genome appears to be linear in electron micrographs, but the DNA in these EM preparations may not be full-length. Size estimates are from restriction fragment analyses (48). No ascovirus genes have been described.

Stages of Ascovirus Replication

The stages of ascovirus replication are defined primarily by cytological observations in fat body tissues of infected insects (45). The earliest symptom of virus infection is nuclear hypertrophy followed by cellular hypertrophy, the diameter reaching 5–10 times that of uninfected cells

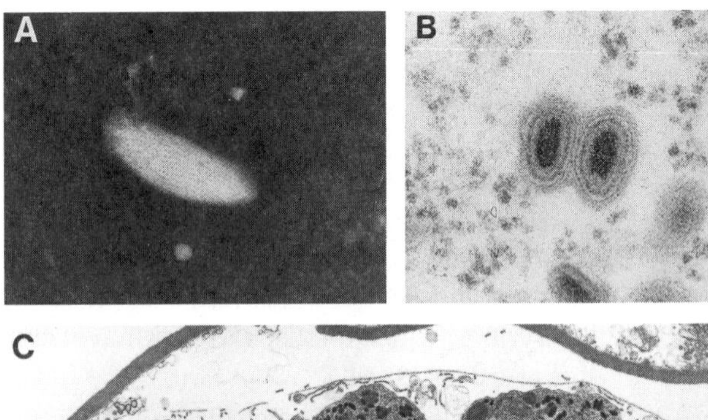

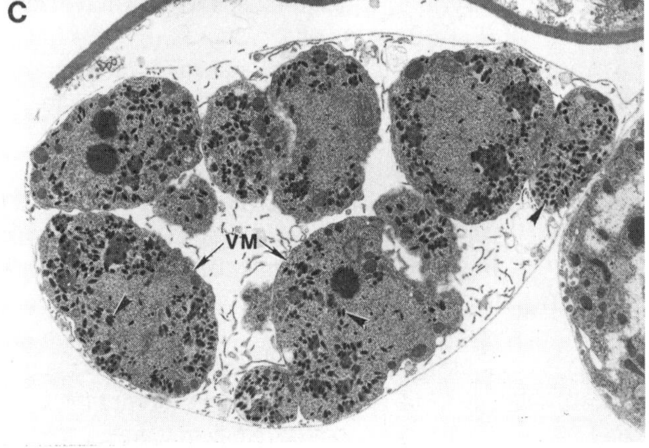

FIG. 6. Ascoviruses and sac-like vesicles containing ascoviruses. **A:** A kidney-shaped virion of the ascovirus of *Trichoplusia ni*; note the reticular appearance of the surface. **B:** Cross-section of two adjacent virions of the ascovirus of *Scotogramma*; note the multi-layered envelope structure surrounding the nucleocapsid. **C:** Vesicles, remnants of infected cells, containing *Autographa precationis* ascovirus particles (*arrowheads*) are surrounded by vesicular membrane envelopes (*VM*). Several vesicles of variable size are collectively bordered by what appears to be a basement membrane. (Courtesy of B.A. Federici, University of California-Riverside, and J.J. Hamm, USDA-ARS, Tifton, GA; modified from Federici et al., refs. 42,47, with permission.)

(47). The description of events following hypertrophy (45) are similar, if not identical, to those of cellular apoptosis, programmed cell death. Briefly, the nucleus fragments and the infected cells are divided into approximately 20–30 membrane-bound vesicles through the extensive invagination of the plasma membrane. The vesicles, once formed, dissociate from the rest of the cell and from each other (Fig. 6C). Virions may continue to assemble in these sac-like vesicles. As the vesicles accumulate, the basement membrane surrounding the fat body tissue is disrupted, releasing the vesicles into the hemolymph. As the disease progresses, the hemocoel becomes filled with vesicles (ca. 10^8 vesicles per mL of hemolymph). The ascoviruses may thus use a very unique viral replication strategy in which cellular apoptosis becomes an integral part of the viral replication mechanism (29).

Ascovirus Transmission and Disease Progression

In the laboratory, ascoviruses can be transmitted from insect to insect by piercing an infected insect with a fine pin and then piercing an uninfected insect with the vesicle-/virus-contaminated pin. An injection dose of 10 vesicles is sufficient to infect over 90% of the injected insects (45). In nature, ascoviruses may be transmitted by female endoparasitic wasps through contamination of their ovipositor during egg laying; the prevalence of ascovirus disease is correlated with a high rate of parasitism (45), and transmission by parasitoids can be observed under laboratory conditions (78). It may also be possible that piercing/sucking arthropods are able to transmit this disease among insects. The host range of the ascovirus isolated from *T. ni* extends to a variety of lepidopteran noctuids (79).

Different ascoviruses exhibit different tissue tropisms. The isolate from *T. ni* has a rather broad tissue tropism, infecting trachael matrix, epidermis, fat body, and connective tissue; whereas the *S. frugiperda* isolate infects only fat body. As the hemolymph fills with vesicles, it becomes milky (47,68).

During the early stages of the disease, there is little gross pathology associated with infected larvae. Although some larvae exhibit a milky coloration throughout the abdomen in the later stages of disease, the most notable feature of the disease is the lack of development in the insect following infection (46,68). Larvae may survive 2–6 weeks following infection. but they feed less and eventually die.

Nodaviridae

Classification of Nodaviruses

Nodaviridae are positive-strand RNA viruses characterized by a bipartite single-stranded RNA genome (i.e., two RNA segments) packaged in a single virion approximately 30 nm in diameter and having an icosahedral capsid with T=3 quasi-symmetry (65,85,93). The family name *Nodaviridae* originated from the first isolated member of the family, Nodamura virus (156), named after a Japanese village, Nodamura, which is now a city called *Nodashi*. Nodaviridae are morphologically similar to plant viruses with T=3 symmetry and share many taxonomic features with Tetraviridae (see below), which differ in isocohedral shell symmetry.

Nodamura virus is the type species of the genus Nodavirus—though the black beetle virus (BBV) and the Flock House virus (FHV) (*Flock House* is the name of an agricultural research station in New Zealand)—are more thoroughly characterized. All members of the family were originally isolated from insects—except Striped Jack nervous necrosis virus (SJNNV), which was isolated from a species of fish (122). Some members of the Nodaviridae multiply in vertebrates, and FHV can form virions in plants if RNA is used to transfect plant tissue. Sites of isolation of nodaviruses have been limited, to date, to Australasia.

Virion Structure

The structure of BBV and FHV capsids have been determined at the 3Å resolution level by x-ray diffraction analysis (49,88). Nodavirus capsids are assembled from 180 protein subunits referred to as *protomers*. During viral morphogenesis, a provirion structure is first assembled from 180 precursor proteins, alpha proteins, consisting of a single polypeptide of approximately 400 amino acids. Like picornaviruses, the provirions of nodaviruses mature into an infectious particle by autocatalytic cleavage of the precursor alpha proteins (158). Cleavage of FHV alpha protein near the carboxyl terminus is required for infectivity; two smaller proteins, beta and gamma, of 363 and 44 residues are generated. Maturation of the virions does not always involve cleavage of all 180 precursors, and uncleaved precursors may comprise between 10% and 50% of the capsid protein, depending on the virus as well as propagation and purification conditions.

Protomers are grouped as trimers into 60 triangular, asymmetric subunits forming pyramid-like structures (93) (Fig. 7), in which the trimers formed by three protomers (designated *A, B,* and *C* for easier visualizaton) form 1 of the 60 pyramidal triangles. Each asymmetric pyramid unit binds to 3 adjacent units and collectively, the 60 pyramids form the shell. Adjacent subunit binding is through contacts between adjacent coplanar C protomers. The RNA, which is likely in duplex form due to secondary structure, binds in a groove under the C-C protomer interface (49). The RNA and a helical protein domain of the protomer, which derives in part from the gamma peptide and extends inside the capsid, stabilizes the flat surface at the C-C interface. The N-terminus of the protomer is "disordered" and invisible in the electron density maps but is also likely to interact with the RNA internally.

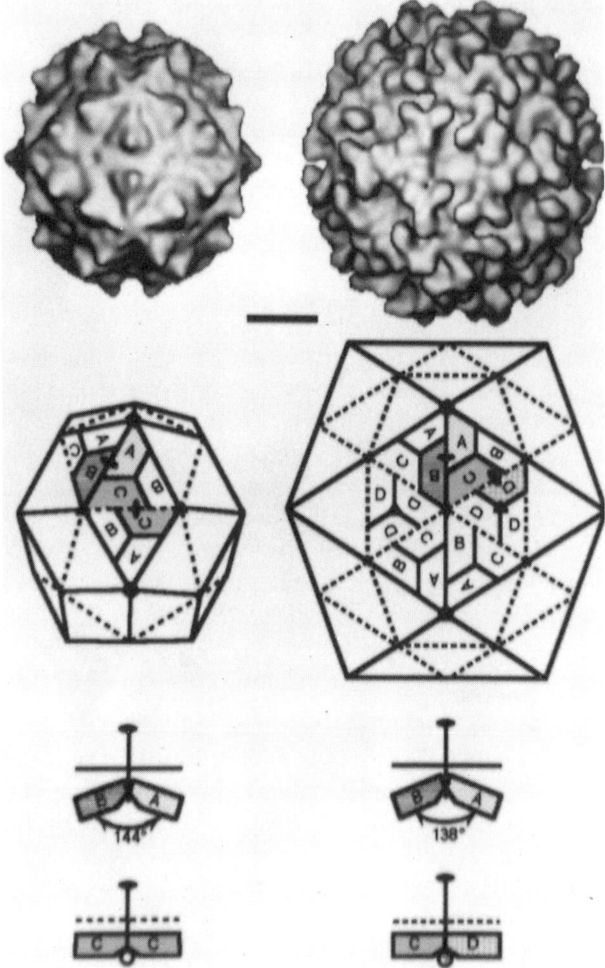

FIG. 7. Comparison of nodavirus and tetravirus structure. **Top:** Image reconstructions of a nodavirus capsid (*left*) and a tetravirus capsid (*right*). Notice the difference in capsid sizes and surface topology. **Bottom:** Diagrammatic illustration of the assembly of protamers in adjacent faces of the capsids of a nodavirus (*left*) and tetravirus (*right*). The protamers (*A–D*) are identical but are labeled to highlight their interaction with adjacent protamers. The interaction of adjacent C protamers on the surface of nodaviruses is planar while protamers *A* and *B* are at a 144° angle. Similarly, the adjacent C and D protamers of tetraviruses are planar while protamers *A* and *B* are at a 138° angle. (Courtesy of Dr. John E. Johnson, Purdue University.)

The nodaviruses, along with the tetraviruses (below) and picornaviruses, offer a powerful model to understand virus morphogenesis and a novel "structural" approach to understanding virus evolution.

Genome Organization and Replication

The genome consists of two single-stranded RNAs which are capped but not polyadenylated (34). The 3′ end of the RNAs is blocked, possibly by a protein. The larger of the two RNAs, RNA 1, is approximately 3.1 kb while the smaller RNA, RNA 2, is approximately 1.4 kb. Both RNAs are required for infectivity and are apparently packaged together in a single virion.

RNA 1 and RNA 2 are message-sense RNAs (85 and references cited therein). RNA 1 carries the information for at least two proteins; protein A, approximately 112 kD; and protein B, approximately 12 kD (Fig. 8). Protein A, which is probably an RNA polymerase or polymerase subunit, is derived by direct translation of RNA 1 and is essential for viral RNA replication. The message for protein B is a subgenomic RNA, RNA 3, which is transcribed from the 3′-terminal region of RNA 1 during infection but is not packaged into virions. RNA 3, approximately 0.4 kb, is capped and contains a single ORF specifying protein B. Protein B may play a role in the synthesis of positive (+) strand RNA. RNA 2 is directly translated into the coat precursor protein, alpha, which is approximately 43 kD. The negative strand of RNA 2 also contains a long ORF of approximately 270 codons.

Stages of Replication

The molecular biology of nodavirus replication has been extensively studied in the cases of BBV and the closely re-

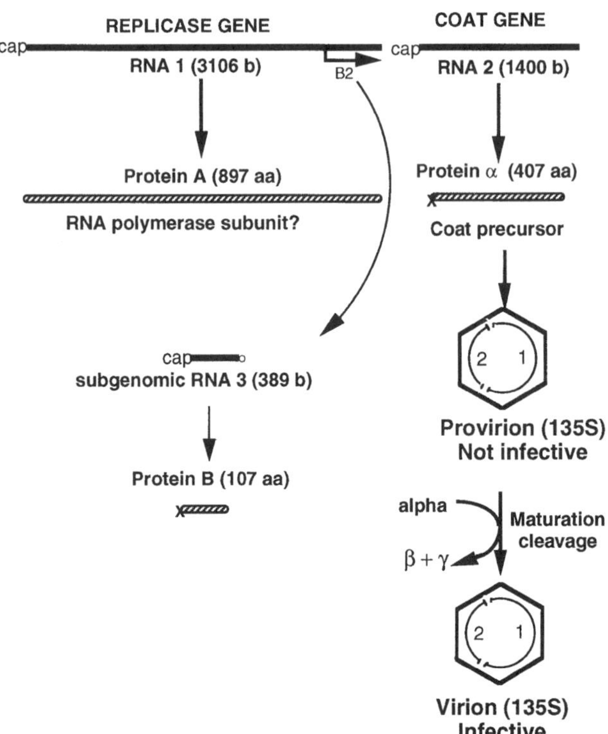

FIG. 8. Nodavirus replication strategy. The two RNA segments of the Flock House Virus genome (*top line*) are 3,106 and 1,400 nucleotides in length and encode the replicase (protein A) and coat precursor protein (protein alpha), respectively. RNA1 also specifies a subgenomic RNA of 389 nucleotides which encodes protein B. Provirions assemble from coat precursor proteins and subsequent cleavage allows maturation of the virion. (Courtesy of Dr. Roland R. Rueckert, University of Wisconsin-Madison.)

lated virus, FHV. Both of these viruses have been adapted to multiply prolifically in cell lines derived from *Drosophila melanogaster* (62). Unless passaged at low multiplicity of infection, defective interfering particles are generated readily. Persistently infected cultured cells, which resist superinfection, are easily established.

The virus replicates in the cytoplasm. In the case of BBV, virus-specific RNA is detected as early as 2 hr p.i. and, by 8 hr, viral RNA accounts for over 50% of the total RNA synthesized in the cell (60,61). Maximum BBV RNA synthesis occurs around 36 hr p.i. and declines by 60 hr p.i. Replication involves the formation of double-stranded RNA intermediates. RNA synthesis occurs in the presence of actinomycin D and an RNA-dependent RNA polymerase activity is found in infected cells; protein A copurifies with this activity. The RNA-dependent RNA polymerase activity from FHV-infected cells can direct the synthesis of negative-strand RNA *in vitro* and, in the presence of glycerophospholipids, this activity also efficiently produces new, positive single-strand RNAs (193).

Transfection of individual RNAs into cells has been used effectively to study the regulation of nodavirus RNA synthesis (64, reviewed in 85). Cells transfected with RNA 1 synthesize RNA 1 and RNA 3 but not RNA 2. Replication of RNA 2 requires cotransfection with RNA 1. The presence of RNA 2 is required for virion production, and RNA 2 strongly inhibits RNA 3 positive-strand synthesis (199). The inhibitory effect of RNA 2 is observed in the presence of cycloheximide and may involve competition for viral replication. The termini of RNA 2 are important *cis*-acting determinants of replication; 5′-terminal extensions are particularly inhibitory. An internal region between nucleotides 538 and 616 were also found to be important *cis*-acting elements for RNA 2 replication (4), while a 32-base region (bases 186–217) is required for packaging into virions (198). The self-replicative properties of RNA 1 may be employed for the amplification of heterologous RNA (3).

Nodavirus Genetics

The demonstration that transcripts derived from cloned cDNAs of FHV genomic RNAs are infectious in Drosophila cells (34) has added a powerful dimension to the genetic analysis of nodaviruses by recombinant DNA technology. Coupled with a good cell line which supports plaque formation and substantial virus replication, FHV is proving to be an excellent experimental system for studying nodaviruses. Expression of FHV genes using baculovirus expression vectors has recently been achieved, resulting in the production of viruslike particles. Thus, mutations which affect FHV viability can be studied via the baculovirus expression system (157). For example, a mutant which is defective in capsid protein cleavage forms provirions in the baculovirus expression system, and these have been crystallized and structurally analyzed (50)

Tetraviridae

Classification of Tetraviruses

Tetraviruses, formerly known as the *Nudaurelia beta virus (NβV) group*, contain single-stranded, positive-sense RNA in an icosahedral capsid, approximately 40 nm in diameter and having T=4 shell symmetry (84,121,149). The T=4 symmetry provides the basis for the family name; *tetra* is derived from the Greek word *tettares*, meaning "four." To date, all tetraviruses have been isolated from species of Lepidoptera (moths and butterflies). Nine viruses, either serologically related or having a capsid protein of similar size (>62 kD), have been assigned to the family, and approximately ten other known viruses are likely members of the family. The RNA genome is either a single (approximately 5.5 kb) molecule or is bipartite, containing two RNAs of 2.5 and 5.5 kb. The NβV has a single genomic RNA, whereas Nudarelia omega virus (NωV) and *Heliothis armigera* stunt virus (HaSV) have bipartite genomes. This difference may lead to further taxononmic subdivision of the tetraviridae into the "beta" and "omega" groups in the future (84).

Virion Structure

Image reconstruction of electron micrographs distinguished the basic structural features of the T=4 lattice symmetry (133). In addition, crystals of NβV and NωV which diffract x-rays to 2.7Å resolution have been produced, and an electron density map of NωV at 8Å resolution has been computed (93). The capsids are composed of 240 protein subunits (protomers). The protomers are composed of either a single, large polypeptide of 71–78 kD or a pair of cleavage products, the larger product being approximately 64–70 kD and the smaller product being 7–8 kD. Twelve protomers comprise each triangular face of the capsid and are packed so that each face is nearly planar.

A comparison of protomer arrangements to produce the different capsid symmetries of the nodavirus (T=3) and tetravirus (T=4) capsids is diagrammed in Fig. 7. Each protomer is indicated as an A, B, C, or D to highlight the arrangement of the protomers in each of the two different structures, but each of the protomers within a structure is identical in amino acid composition. There are 60 additional protomers (illustrated as protomer D) in tetravirus capsids, compared to the 180 total protomers of the nodavirus (T=3) capsids. Tetraviruses are larger (ca. 40 nm) than nodaviruses (ca. 30 nm).

The simple difference in protomer number and arrangement provides a dramatic difference in capsid morphology. It will be fascinating to trace the evolutionary relationships of the tetraviruses, nodaviruses, and picornaviruses.

Genome Structure

As noted above, tetravirus genomes are either monopartite, consisting of a single 5.5-kb RNA; or bipartite, consisting of RNA1, 5.5 kb; and RNA2, 2.5 kb. It is not yet known whether the two RNAs of the bipartite genome are packaged together in the same particle. Neither of the RNAs are polyadenylated and the 3′-termini of both RNAs are unblocked (80). The gene organization of monopartite tetravirus genomes is currently being studied.

The entire nucleotide sequence of the NωV 2.5-kb RNA has been determined (1) and was found to contain one long ORF of 644 codons within a total sequence of 2,448 nucleotides. This ORF corresponds to the coat protein protomer. The smaller cleavage product of the protomer is likely generated by proteolytic cleavage near the C-terminus of the protein. A short 5′-proximal ORF of 30 codons terminates one nucleotide before the initiating methionine codon of the coat protein ORF. Nucleotide sequence data has also been obtained recently for the HaSV 2.5-kb RNA. An additional ORF, capable of encoding a 17-kD protein, was found at the 5′ end of this RNA. In addition, the 3′ end of this RNA, as well as the NωV 2.5-kb RNA, can adopt a tRNA-like structure (T. Hanzlik, *personal communication*). Like the nodaviruses, it is likely that RNA1 of the bipartite genome encodes an RNA polymerase.

Stages of Replication

Because of the inability to replicate tetraviruses in cell culture to date, the stages of replication of the tetraviruses remain to be determined, but it seems likely that the basic replication strategies of these viruses will be similar to picornaviruses and nodaviruses. Research on genome expression has been primarily limited to *in vitro* translation (39,80) using rabbit reticulocyte and/or wheat germ extracts. These two cell-free translation systems show differences in the nature of the gene products and have not yet provided a clear indication of tetravirus replication strategy. Virus replication occurs in the cytoplasm of infected cells, and crystalline arrays of virus particles accumulate within cytoplasmic vesicles.

Pathology, Transmission, and Ecology

Insect larvae can acquire a tetravirus infection by ingesting food contaminated with virus (85). In the laboratory, infection of larvae can also be achieved by injecting virus particles into the hemolymph. Although infection of larvae is prevalent, NβV has also been isolated from adults and pupae. Replication of many tetraviruses appears to be largely restricted to the foregut and midgut epithelial cells. The infection can cause diarrhea-like symptoms, and virus particles are excreted in the frass (feces).

Insects can exhibit a wide range of pathogenic effects including rapid death, an extended wasting disease involving weight loss, and/or a delay in pupation. Most tetravirus infections involve larval death, usually accompanied by discoloration and flaccidity of the body, sometimes with complete liquifaction of the internal organs, reminiscent of the "melting" phenomenon of baculoviruses. (One concern of *in vivo* insect research is the possibility of activating persistent or dormant viruses.)

Tetravirus infections occur sporadically in the wild so that ecological aspects of virus reservoirs and spread are poorly understood. Tetraviruses have been observed to replicate in the same cells as baculoviruses and to be co-occluded in occlusion bodies of baculoviruses. Although tetraviruses have been isolated only from insects, antibodies cross-reacting with some tetraviruses have been detected in domestic as well as wild animals with no associated illness.

Tetraviruses serve as natural control agents of insects. Their importance in the natural control of certain insect pests has been noted. The *Darna trima* and *Gonometa* viruses have been applied in oil palm plantations as pest control agents (121). Because antibodies to tetraviruses were reported in humans, pursuit of these viruses as pesticides has been limited (121).

REFERENCES

1. Agrawal DK, Johnson JE. Sequence and analysis of the capsid protein of Nudaurelia capensis omega virus, an insect virus with T=4 icosahedral symmetry. *Virology* 1992;190:806–814.

1a. Ayers MD, Howard SC, Kuzio J, Lopez-Ferber M, Possee, RD. The complete sequence of *Autographa californica* nuclear polyhedrosis virus. *Virology* 1994;202:586–605.

2. Balch RE, Bird FT. A disease of the European spruce sawfly, *Gilpinia hercyniae* (Htg.) and its place in natural control. *Sci Agric* 1944; 25:65–73.
3. Ball LA. Cellular expression of a functional nodavirus RNA replicon from vaccinia virus vectors. *J Virol* 1992;66:2335–2345.
4. Ball LA, Li Y. Cis-acting requirements for the replication of Flock House Virus RNA2. *J Virol* 1993;67:3544–3551.
5. Beames B, Summers MD. Location and nucleotide sequence of the 25k protein missing from baculovirus few polyhedra FP mutants. *Virology* 1989;168:344–353.
6. Beames B, Summers MD. Sequence comparison of cellular and viral copies of host cell DNA insertions found in *Autogrupha californica* nuclear polyhedrosis virus. *Virology* 1990;174:354–363.
7. Beaton CD, Fishie BK. Comparative ultrastructural studies of insect granulosis and nuclear polyhedrosis viruses. *J Gen Virol* 1976; 31:151–164.
8. Beckage NE. Endocrine interactions between endoparasitic insects and their hosts. *Ann Rev Entomol* 1985;30:371–413.
9. Benz GA. Introduction: Historical perspectives. In: Granados RR, Federici BA, eds. *The biology of baculoviruses*. Boca Raton, FL: CRC Press, 1986:1–35.
10. Bergold GH. Die Isolierung des polyeder-virus und die natur der polyeder. *Z. Naturforsch Teil B* 1947;2b:122–143.
11. Birnbaum MJ, Clem RJ, Miller LK. An apoptosis-inhibiting gene from a nuclear polyhedrosis virus encoding a polypeptide with cys/his sequence motifs. *J Virol* 1994; in press.
12. Blissard GW, Kogan, PH, Wei, R, Rohrmann, GF. A synthetic early promoter from a baculovirus: roles of the TATA box and conserved start site CAGT sequence in basal levels of transcription. *Virology* 190:783–793.
13. Blissard G, Rohrmann G. Baculovirus gp64 gene expression: analysis of sequences modulating early transcription and transactivation by IE1. *J Virol* 1991;65:5820–5827.
14. Blissard G, Vinson S, Summers M. Identification, mapping, and in vitro translation of *Campoletis sonorensis* virus mRNAs from parasitized *Heliothis virescens* larvae. *J Virol* 1986;57(1):318–327.
15. Blissard GW, Rohrmann GF. Baculovirus diversity and molecular biology. In: Mittler,TE, ed. *Ann Rev Entomol* 1990:127–155.

16. Blissard GW, Smith OP, Summers MD. Two related viral genes are located on a single superhelical DNA segment of the multipartite *Campoletis sonorensis* virus genome. *Virology* 1987;160:120–134.
17. Blissard GW, Theilmann DA, Summers MD. Segment W of *Campoletis sonorensis* virus: expression, gene products, and organization. *Virology* 1989;169:78–89.
18. Blissard GW, Wenz JR. Baculovirus gp64 envelope glycoprotein is sufficient to mediate pH-dependent membrane fusion. *J Virol* 1992;66:6829–6835.
19. Burand JP. Molecular biology of the HZ-1 and Oryctes nonoccluded baculoviruses. In: Kurstak E, ed. *Viruses of Invertebrates*. New York: Marcel Dekker, 1991:111–126.
20. Carner GR, Hudson JS. Histopathology of virus-like particles in Heliothis spp. *J Invertebr Pathol* 1983;41:238–249.
21. Carstens EB, Lu AL, Chan HLB. Sequence, transcriptional mapping and overexpression of p47, a baculovirus gene regulating late gene expression. *J Virol* 1993;67:2513–2520.
22. Carstens EB, Tjia ST, Doerfler W. Infection of *Spodoptera frugiperda* cells with *Autographa californica* nuclear polyhedrosis virus. I. Synthesis of intracellular proteins after virus infection. *Virology* 1979; 99:386–398.
23. Chao Y, Wood HA, Chang C, Lee HJ, Shen W, Lee H. Differential expression of Hz-1 baculovirus genes during productive and persistent viral infections. *J Virol* 1992;66(3):1442–1448.
24. Charlton CA, Volkman LE. Penetration of *Autographa californica* nuclear polyhedrosis virus nucleocapsids into IPLB Sf21 cells induces actin cable formation. *Virology* 1993;197:245–254.
25. Cheley S, Kosik KS, Pakevich P, Bakalis S, Bayley H. Phosphorylated baculovirus p10 is a heat-stable microtubule-associated protein associated with process formation in Sf9 cells. *J Cell Sci* 1992; 102:739–752.
26. Clem RJ, Fechheimer M, Miller LK. Prevention of apoptosis by a baculovirus gene during infection of insect cells. *Science* 1991; 254:1388–1390.
27. Clem RJ, Miller LK. Apoptosis reduces both the in vitro replication and the in vivo infectivity of a baculovirus. *J Virol* 1993;67:3730–3738.
28. Clem RJ, Miller LK. Control of programmed cell death by the baculovirus genes p35 and inhibitor of apoptosis. *Molec Cell Biol* 1994;14:5212–5222.
29. Clem RJ, Miller LK. Induction and inhibition of apoptosis by insect viruses. In: Tomei LD, Cope FO, eds. *Apoptosis II: the molecular basis of apoptosis in disease*. Cold Spring Harbor, NY: Cold Spring Harbor Laboratory Press, 1994:89–110.
30. Cochran MA, Faulkner P. Location of homologous DNA sequences interspersed at five regions in the baculovirus *Autographa californica* nuclear polyhedrosis virus genome. *J Virol* 1983;45:961–970.
31. Crawford AM, Ashbridge K, Sheehan C, Faulkner P. A physical map of the Oryctes baculovirus genome. *J Gen Virol* 1985;66:2649–2657.
32. Crawford AM, Miller LK. Characterization of an early gene accelerating expression of late genes of the baculovirus *Autographa californica* nuclear polyhedrosis virus. *J Virol* 1988;62:2773–2781.
33. Crook NE, Clem RJ, Miller LK. An apoptosis-inhibiting baculovirus gene with a zinc finger motif. *J Virol* 1993;67:2168–2174.
34. Dasgupta R, Ghosh A, Dasmahapatra B, Guarino LA, Kaesberg P. Primary and secondary structure of black beetle virus RNA2, the genomic messenger for BBV coat protein precursor. *Nucl Acids Res* 1984;12:7215–7223.
35. Davies, DH, Strand, MR and Vinson, SB. Changes in differntial hemocyte count and in vitro behavior of plasmatocytes from host *Heliothis virescens* caused by *Campoletis sonorensis* polydnavirus. *J Insect Physiol* 33; 143–154.
36. Dib-Hajj SD, Webb BA, Summers MD. Structure and evolutionary implications of a "cysteine-rich" *Campoletis sonorensis* polydnavirus gene family. *Proc Natl Acad Sci USA* 1993;90:3765–3769.
37. Dickson JA, Friesen PD. Identification of upstream promoter elements mediating early transcription from the 35,000-molecular weight protein gene of *Autographa californica* nuclear polyhedrosis virus. *J Virol* 1991;65:4006–4016.
38. Dover BA, Davies DH, Vinson SB. Degeneration of last instar Heliothis virescens prothoracic glands by *Campoletis sonorensis* polydnavirus. *J Invertebr Pathol* 1988;51:80–91.
39. du Plessis DH, Mokhosi G, Hendry DA. Cell-free translation and identification of the replicative form of Nudaurelia beta virus RNA. *J Gen Virol* 1991;72:267–273.
40. Edson KM, Vinson SB, Stoltz DB, Summers MD. Virus in a parasitoid wasp suppression of the cellular immune response in the parasitoids host. *Science* (Wash D C) 1981;211:582–583.
41. Eldridge R, Li Y, Miller LK. Characterization of a baculovirus gene encoding a small conotoxinlike polypeptide. *J Virol* 1992;66:6563–6571.
42. Federici BA. A new type of insect pathogen in larvae of the clover cutworm *Scotogramma trifolii*. *J Invertebr Pathol* 1982;40:41–54.
43. Federici BA. Enveloped double-stranded DNA insect virus with novel structure and cytopathology. *Proc Natl Acad Sci USA*. 1983; 80:7664–7668.
44. Federici BA. Ultrastructure of baculoviruses. In: Granados RR, Federici BA, eds. *The biology of baculoviruses*. Boca Raton, FL: CRC Press, 1986:61–88.
45. Federici BA. Viral pathobiology in relation to insect control. In: Beckage NE, Thompson SN, Federici BA, eds. *Parasites and pathogens of insects*. San Diego, CA: Academic Press, Inc., 1993:81–101.
46. Federici BA, Govindarajan R. Comparative histopathology of three ascovirus isolates in larval noctuids. *J Invertebr Pathol* 1990;56:300–311.
47. Federici BA, Hamm JJ, Styer EL. Ascoviridae. In: Adams JR, Bonami JR, eds. *Atlas of invertebrate viruses*. Boca Raton, FL: CRC Press, 1991:339–349.
48. Federici BA, Vlak JM, Hamm JJ. Comparative study of virion structure, protein composition and genomic DNA of three ascovirus isolates. *J Gen Virol* 1990;71:1661–1668.
49. Fisher AJ, Johnson JE. Ordered duplex RNA controls capsid architecture in an icosahedral animal virus. *Nature* 1993;361:176–179.
50. Fisher AJ, McKinney BR, Scheemann A, Rueckert RR, Johnson JR. Crystallization of viruslike particles assembled from Flock House Virus coat protein expressed in a baculovirus system. *J Virol* 1993; 67:2950–2953.
51. Fleming JGW. Polydnavirus DNA is integrated in the DNA of its parasitoid wasp host. *Proc Natl Acad Sci USA* 1991;88:9770–9774.
52. Fleming JGW. Polydnaviruses: mutualists and pathogens. *Ann Rev Entomol* 1992;37:401–425.
53. Fleming JGW, Krell PJ. Polydnavirus genome organization. In: Beckage NE, Thompson SE, Federici BA, eds. *Parasites and pathogens of insects*. San Diego, CA: Academic Press, 1993:189–225.
54. Fleming JGW, Summers MD. *Campoletis sonorensis* endoparasitic wasps contain forms of *C. sonorensis* virus DNA suggestive of integrated and extrachromosomal polydnavirus DNAs. *J Virol* 1986; 57:552–562.
55. Fraser MJ. Ultrastructural observations of virion maturation in *Autographa californica* nuclear polyhedrosis virus infected *Spodoptera frugiperda* cell cultures. *J Ultrastruct Mol Struct Res* 1986;95(1-3):189–195.
56. Fraser MJ, Smith GE, Summers MD. Acquisition of host cell DNA sequences by baculoviruses: relationship between host DNA insertions and FP mutants of *Autographa californica* nuclear polyhedrosis virus and *Galleria mellonella* nuclear polyhedrosis viruses. *J Virol* 1983;47:287–300.
57. Friesen PD. Invertebrate transposable elements in the baculovirus genome: characterization and significance. In: Beckage NE, Thompson SN, Federici BA, eds. *Parasites and pathogens of insects*. San Diego: Academic Press, 1993.
58. Friesen PD, Miller LK. The regulation of baculovirus gene expression. In: Doerfler W, Boehm P ed. *The Molecular Biology of Baculoviruses*. Berlin: Springer-Verlag, 1986:31–50.
59. Friesen PD, Nissen MS. Gene organization and transcription of TED, a lepidopteran retrotransposon integrated within the baculovirus genome. *Mol Cell Biol* 1990;10:3067–3077.
60. Friesen PD, Rueckert RR. Synthesis of black beetle virus proteins in cultured Drosophila cells: Differential expression of RNAs 1 and 2. *J Virol* 1981;37:876–886.
61. Friesen PD, Rueckert RR. Early and late functions in a bipartite RNA virus: Evidence for translational control by competition between viral mRNAs. *J Virol* 1984;49:116–124.
62. Friesen PD, Scotti P, J. L, Rueckert RR. Propagation in Drosophila line 1 cells and an infection-resistant subline carrying endogenous black beetle virus-related particles. *J Virol* 1980;35:741–747.
63. Fuchs LY, Woods MS, Weaver RF. Viral transcription during *Autographa californica* nuclear polyhedrosis virus infection: a novel RNA polymerase induced in infected *Spodoptera frugiperda* cells. *J Virol* 1983;43:641–646.
64. Gallagher TM, Friesen PD, Rueckert RR. Autonomous replication and expression of RNA 1 from black beetle virus. *J Virol* 1983;46:481–489.
65. Garzon S, Charpentier G. Nodaviridae. In: Adams JR, Bonami JR, eds.

Atlas of invertebrate viruses. Boca Raton, FL: CRC Press, 1992: 351–371.
66. Gombart AF, Pearson MN, Rohrmann GF, Beaudreau GS. A baculovirus polyhedral envelope-associated protein genetic location nucleotide sequence and immunocytochemical characterization. *Virology* 1989;169:182–193.
67. Gordon JD, Carstens EB. Phenotypic characterization and physical mapping of a temperature sensitive mutant of *Autographa californica* nuclear polyhedrosis virus defective in DNA synthesis. *Virology* 1984;138:69–81.
68. Govindarajan R, Federici BA. Ascovirus infectivity and effects of infection on the growth and development of noctuid larvae. *J Invertebr Pathol* 1990;56:291–299.
69. Grace TDC. Establishment of four strains of cells from insect tissues grown *in vitro*. *Nature* 1962;195:788–789.
70. Granados RR, Williams KA. In vivo infection and replication of baculoviruses. In: Granados RR, Federici BA, eds. *The biology of baculoviruses*, vol. I. Biological Properties and Molecular Biology. Boca Raton, FL: CRC Press, 1986:89–108.
71. Gross CH, Rohrmann GF. Analysis of the role of 5' promoter elements and 3' flanking sequences on the expression of a baculovirus polyhedron envelope protein gene. *Virology* 1993;192:273–281.
72. Guarino LA, Dong W. Expression of an enhancer-binding protein in insect cells transfected with the *Autographa californica* nuclear polyhedrosis virus ie1 gene. *J Virol* 1991;65:3676–3680.
73. Guarino LA, Gonzalez MA, Summers MD. Complete sequence and enhancer function of the homologous DNA regions of Autographa californica nuclear polyhedrosis virus. *J Virol* 1986;60:224–229.
74. Guarino LA, Smith M. Regulation of delayed-early gene transcritpion by dual TATA boxes. *J Virol* 1992;66:3733–3739.
75. Guarino LA, Smith MW. Nucleotide sequence and characterization of the 39K gene region of *Autographa californica* nuclear polyhedrosis virus. *Virology* 1990;179:1–8.
76. Guarino LA, Summers MD. Functional mapping of a trans-activating gene required for expression of a baculovirus delayed-early gene. *J Virol* 1986;57:563–571.
77. Guarino LA, Summers MD. Interspersed homologous DNA of *Autographa californica* nuclear polyhedrosis virus enhances delayed-early gene expression. *J Virol* 1986;60:215–223.
78. Hamm JJ, Nordlund DA, Marti OG. Effects of a nonoccluded virus of *Spodoptera frugiperda* (Lepidotpera: Noctuidae) on the development of a parasitoid, *Cotesia marginiventris* (Hymenoptera: Braconidae). *Experimen Entomol* 1985;14:258–261.
79. Hamm JJ, Pair SD, Marti OG Jr. Incidence and host range of a new ascovirus isolated from fall armyworm, *Spodoptera frugiperda* (Lepidoptera: Noctuidae). *Flor Entomol* 1986;69:525–531.
80. Hanzlik TN, Dorrian SJ, Gordon KHJ, Christian PD. A novel small RNA virus isolated from the cotton bollworm, *Helicoverpa armigera*. *J Gen Virol* 1993;74:1805–1810.
81. Harrap K. The stucture of nuclear polyhedrosis viruses. III.Virus assembly. *Virology* 1972;50:133–139.
82. Harrap K, Robertson L. A possible infection pathway in the development of a nuclear polyhedrosis virus. *J Gen Virol* 1968;3:221–225.
83. Hashimoto Y, Corsaro BG, Granados RR. Location and nucleotide sequence of the gene encoding the viral enhancing factor of the *Trichoplusia ni* granulosis virus. *J Gen Virol* 1991;72:2645–2651.
84. Hendry D, Agrawal D. Small RNA viruses of insects: Tetraviridae. In: Webster RG, Granoff A, eds. *Encyclopedia of virology*. London: Academic Press, 1994.
85. Hendry DA. Nodaviridae of invertebrates. In: Kustak E, ed. *Viruses of invertebrates*. New York, NY: Marcel Dekker Inc., 1991:227–276.
86. Herschberger PA, Dickson JA, Friesen PD. Site-specific mutagenesis of the 35-kilodalton protein gene encoded by *Autographa californica* nuclear polyhedrosis virus: cell line-specific effects on virus replication. *J Virol* 1992;66:5525–5533.
87. Hoopes RR, Rohrmann GF. In vitro transcription of baculovirus immediate early genes—accurate messenger RNA initiation by nuclear extracts from both insect and human cells. *Proc Natl Acad Sci USA* 1991;88:4513–4517.
88. Hosur MV, Schmidt T, Tucker RC, et al. Structure of an insect virus at 3.0 A resolution. Proteins: *Struct Funct Genet* 1987;2:167–176.
89. Hughes DS, Possee RD, King LA. Activation and detection of a latent baculovirus resembling *Mamestra brassicae* nuclear polyhedrosis virus in *M. brassicae* insects. *Virology* 1993;194:608–615.
90. Huh NE, Weaver RF. Categorizing some early and late transcripts directed by the *Autographa californica* nuclear polyhedrosis virus. *J Gen Virol* 1990;71:2195–2200.
91. Huh NE, Weaver RF. Identifying the RNA polymerases that synthesize specific transcripts of the *Autographa californica* nuclear polyhedrosis virus. *J Gen Virol* 1990;71(1):195–202.
92. Ignoffo CM. Living microbial insecticides. In: Norris JR, Richmond MH, eds. *Essays in applied microbiology*. New York: John Wiley & Sons, 1981:2–31.
93. Johnson JE, Munshi S, Liljas L, et al. Comparative studies of T=3 and T=4 icosahedral RNA insect viruses. *Arch Virol* 1994;9:567–577.
94. Keddie BA, Aponte GW, Volkman LE. The pathway of infection of *Autographa californica* nuclear polyhedrosis virus in an insect host. *Science* (Washington D C) 1989;243:1728–1730.
95. Kelly DC, Brown DA, Ayres MD, Allen CJ, Walker IO. Properties of the major nucleocapsid protein of *Heliothis zea* singly enveloped nuclear polyhedrosis virus. *J Gen Virol* 1983;64:399–408.
96. King LA, Possee RD. *The baculovirus expression system: a laboratory guide*. London: Chapman & Hall, 1992:229.
97. Kogan PH, Blissard GW. A baculovirus gp64 early promoter is activated by host transcription factor binding to CACGTG and GATA elements. *J Virol* 1994;68:813–822.
97a. Kool M, Goldbach RW, Vlak JM. A putative non-hr origin of DNA replication in the Hind III-K fragment of *Autographa californica* multiple nuclear capsid nuclear polyhedrosis virus. *J Gen Virol* 1994; 75:3345–3352.
98. Kool M, Vlak JM. The structural and functional organization of the *Autographa californica* nuclear polyhedrosis virus genome. *Arch Virol* 1993;130:1–16.
99. Kool M, Voeten JTM, Goldbach RW, Tramper J, Vlak JM. Identification of seven putative origins of Autographa californica nuclear polyhedrosis virus DNA replication. *J Gen Virol* 1993;74:2661–2668.
100. Kool M, Voncken JW, Vanlier FLJ, Tramper J, Vlak JM. Detection and analysis of *Autographa californica* nuclear polyhedrosis virus mutants with defective interfering properties. *Virology* 1991;183:739–746.
101. Kovacs GR, Choi J, Guarino LA, Summers MD. Functional dissection of the *Autographa californica* nuclear polyhedrosis virus immediate-early 1 transcriptional regulatory protein. *J Virol* 1992;66:7429–7437.
102. Kovacs GR, Guarino LA, Summers MD. Novel regulatory properties of the IE1 and IE0 transactivators encoded by the baculovirus *Autographa californica* multicapsid nuclear polyhedrosis virus. *J Virol* 1991;65:5281–5288.
103. Krell PJ. The polydnaviruses: multipartite DNA viruses from parasitic Hymenoptera. In: Kurstak E, ed. *Viruses of invertebrates*. New York: Marcel Dekker, 1991:141–178.
104. Kumar S, Miller LK. Effects of serial passage of *Autographa californica* nuclear polyhedrosis virus in cell culture. *Virus Res* 1987; 7:335–350.
105. Lee HH, Miller LK. Isolation, complementation, and initial characterization of temperature sensitive mutants of the baculovirus *Autographa californica* nuclear polyhedrosis virus. *J Virol* 1979;31:240–252.
106. Lee JC, Chen HH, Wei HL, Chao YC. Super-infection induced apoptosis and its correlation with the reduction of viral progeny in cells persistently infected with Hz-1 baculovirus. *J Virol* 1993;67:6989–6994.
107. Leisy DL, Rohrmann GF. Characterization of the replication of plasmids containing hr sequences in baculovirus-infected *Spodoptera frugiperda* cells. *Virology* 1993;196:722–730.
108. Li Y, Passarelli AL, Miller LK. Identification, sequence, and transcriptional mapping of lef-3, a baculovirus gene involved in late and very late gene expression. *J Virol* 1993;67:5260–5268.
109. Lu A, Carstens EB. Nucleotide sequence of a gene essential for viral DNA replication in the baculovirus *Autographa californica* nuclear polyhedrosis virus. *Virology* 1991;181:336–347.
110. Lu A, Miller LK. Identification of three late expression factor genes within the 33.8- to 43.4-map-unit region of *Autographa californica* nuclear polyhedrosis virus. *J Virol* 1994;68:6710–6718.
110a. Lu A, Miller LK. The roles of eighteen baculovirus late expression factor genes in transcription and DNA replication. *J Virol* 1995;69:975–982.
111. Maeda S, Kawai T, Obinata M, et al. Production of human α-interferon in silkworm using a baculovirus vector. *Nature* 1985;315(6020): 592–594.
112. Majima K, Kobara R, Maeda S. Divergence and evolution of homologous regions of *Bombyx mori* nuclear polyhedrosis virus. *J Virol* 1993;67:7513–7521.

113. Martignoni ME, Iwai PJ. A catalog of viral diseases. *Pacific northwest forest and range experiment station*. Forest Service. USDA, 1977.
114. McCutchen BF, Choudary PV, Crenshaw R, et al. Development of a recombinant baculovirus expressing an insect-selective neurotoxin: potential for pest control. *Bio/Technology* 1991;9:848–852.
114a. McLachlin JR, Miller LK. Identification of vef-1, a baculovirus gene involved in very late gene expression. *J Virol* 1994;68:7746–7756.
115. Miller DW, Miller LK. A virus mutant with an insertion of a copia-like transposable element. *Nature* 1982;299:562–564.
116. Miller LK. A virus vector for genetic engineering in invertebrates. In: Panapoulous NJ, ed. *Genetic engineering in the plant sciences*. New York: Praeger, 1981:203–224.
117. Miller LK. The genetics of baculoviruses. In: Granados RR, Federici BA, eds. *The biology of baculoviruses*. Boca Raton, FL: CRC Press, 1986:217–238.
118. Miller LK. Baculoviruses as gene expression vectors. *Ann Rev Microbiol* 1988;42:177–199.
119. Miller LK. Baculoviruses: high level expression in insect cells. *Curr Opin Genet Develop* 1993;3:97–101.
120. Miller LK, Jewell JE, Browne D. Baculovirus induction of a DNA polymerase. *J Virol* 1981;40:305–308.
121. Moore NF. The Nudaurelia beta family of insect viruses. In: Kurstak E ed. *Viruses of Invertebrates*. New York: Marcel Dekker, Inc., 1991:277–285.
122. Mori KI, Nakai T, Muroga K, Arimoto M, Musiake K, Firusawa I. Properties of a new virus belonging to Nodaviridae found in larval Striped Jack (Pseudocaranx dentiex) with nervous necrosis. *Virology* 1992;187:368–371.
123. Morris TD, Miller LK. Mutational analysis of a baculovirus major late promoter. *Gene* 1994;140:147–153.
124. Morris TD, Todd JW, Fisher B, Miller LK. Identification of lef-7: a baculovirus gene affecting late gene expression. *Virology* 1994; 200:360–369.
125. Norton WN, Vinson SB. Correlating the initiation of virus replication with a specific pupal developmental phase of an ichneumonid parasitoid. *Cell Tissue Res* 1983;231:387–389.
126. O'Reilly DR, Brown MR, Miller LK. Alteration of ecdysteroid metabolism due to baculovirus infection of the fall armyworm *Spodoptera frugiperda:* host ecdysteroids are conjugated with galactose. *Insect Biochem Molec Biol* 1992;22:313–320.
127. O'Reilly DR, Crawford AM, Miller LK. Viral proliferating cell nuclear antigen. *Nature* 1989;337:606.
128. O'Reilly DR, Miller LK. A baculovirus blocks insect molting by producing ecdysteroid UDP-glucosyl transferase. *Science* 1989;245:1110–1112.
129. O'Reilly DR, Miller LK. Improvement of a baculovirus pesticide by deletion of the egt gene. *Bio/Technology* 1992;9:1086–1089.
130. O'Reilly DR, Miller LK, Luckow VA. *Baculovirus expression vectors: a laboratory manual*. New York: W.H. Freeman and Company, 1992:347.
131. O'Reilly DR, Passarelli AL, Goldman IF, Miller LK. Characterization of the DA26 gene in a hypervariable region of the *Autographa californica* nuclear polyhedrosis virus genome. *J Gen Virol* 1990;71:1029–1037.
132. Oers MM, Flipsen JTM, Reusken CBEM, Sliwinsky EL, Goldbach RW, Vlak JM. Functional domains of the p10 protein of *Autographa californica* nuclear polyhedrosis virus. *J Gen Virol* 1993;74:563–574.
133. Ohlson NH, Baker TS, Johnson JE, Hendry DA. The three dimensional structure of frozen hydrated Nudaurelia capensis beta virus, a T=4 insect virus. *J Struct Biol* 1990;105:11–122.
134. Ooi B, Miller LK. Transcription of the baculovirus polyhedrin gene reduces the levels of an antisense transcript initiated downstream. *J Virol* 1990;64:3126–3129.
135. Ooi BG, Miller LK. Regulation of host RNA levels during baculovirus infection. *Virology* 1988;166:515–523.
136. Ooi BG, Rankin C, Miller LK. Downstream sequences augment transcription from the essential initiation site of a baculovirus polyhedrin gene. *J Molec Biol* 1989;210:721–736.
137. Passarelli AL, Miller LK. Identification and characterization of lef-1, a baculovirus gene involved in late and very late gene expression. *J Virol* 1993;67:3481–3488.
138. Passarelli AL, Miller LK. Identification of genes encoding late expression factors located between 56.0 and 65.4 map units of the *Autographa californica* nuclear polyhderosis virus genome. *Virology* 1993;197:704–714.
139. Passarelli AL, Miller LK. Three baculovirus genes involved in late and very late gene expression: ie-1, ie-n, and lef-2. *J Virol* 1993;67:2149–2158.
140. Passarelli AL, Miller LK. Identification and transcriptional regulation of the baculovirus lef-6 gene. *J Virol* 1994;68:4458–4467.
141. Passarelli AL, Miller LK. In vivo and in vitro analyses of recombinant baculoviruses lacking a functional cg30 gene. *J Virol* 1994;68:1186–1190.
142. Passarelli AL, Todd JW, Miller LK. A baculovirus gene involved in late gene expression encodes a 102kDa polypeptide with a conserved motif of RNA polymerases. *J Virol* 1994;68:4673–4678.
143. Passarelli AL, Todd JW, Miller LK. Eighteen baculovirus genes, including lef-11, P35, 39K and P47, support late gene expression. *J Virol* 1994;69:968–974.
144. Pearson M, Bjornson R, Pearson G, Rohrmann G. The Autographa californica baculovirus genome: evidence for multiple replication origins. *Science* 1992;257:1382–1384.
145. Pearson MN, Russell RLQ, Rohrmann GF, Beaudreau GS. P39, A major baculovirus structural protein: immunocytochemical characterization and genetic location. *Virology* 1988;167(2):407–413.
146. Pennock GD, Shoemaker C, Miller LK. Strong and regulated expression of *Escherichia coli* β-galactosidase in insect cells with a baculovirus vector. *Mol Cell Biol* 1984;4(3):399–406.
147. Potter KN, Jaques RP, Faulkner P. Modification of *Trichoplusia ni* nuclear polyhedrosis virus passaged in vivo. *Intervirology* 1978;9:76–85.
148. Rabizadeh S, LaCount DJ, Friesen PD, Bredesen DE. Expression of the baculovirus p35 gene inhibits mammalian neural cell death. *J Neurochem* 1993;61:2318–2321.
149. Reinganum C. Tetraviridae. In: Adams JR, Bonami JR, eds. *Atlas of invertebrate viruses*. Boca Raton: CRC Press, Inc., 1991:388–392.
150. Ribeiro BM, Hutchinson K, Miller LK. A mutant baculovirus with a temperature-sensitive IE-1 transregulatory protein. *J Virol* 1994;68:1075–1084.
151. Rice WC, Miller LK. Baculovirus transcription in the presence of inhibitors and in nonpermissive *Drosophila* cells. *Virus Res* 1986;6:155–172.
152. Rizki RM, Rizki TM. Parasitoid virus-like particles destroy Drosophila cellular immunity. *Proc Natl Acad Sci USA* 1987;87:8388–8392.
153. Roelvink PW, van Meer MMM, de Kort CAD, Possee RD, Hammock BD, Vlak JM. Dissimilar expression of *Autographa californica* multiple nucleocapsid nuclear polyhedrosis virus polyhedrin and p10 genes. *J Gen Virol* 1992;73:1481–1489.
154. Rohrmann GF. Baculovirus structural proteins. *J Gen Virol* 1992;73:749–761.
155. Rotheram S. The surface of the egg of a parasitic insect. II. The ultrastructure of the particulate coat on the egg of Nemeritis. *Proc R Soc Lond* B 1973;183:195–204.
156. Scherer WF, Hurlbut HS. Nodamura virus from Japan: A new and unusual arbovirus resistant to diethyl ether and chloroform. *Am J Epidemiol* 1967;86:271–285.
157. Schneemann A, Dasgupta R, Johnson JE, Rueckert RR. Use of recombinant baculoviruses in synthesis of morphologically distinct virus-like particles of Flock House Virus, a nodavirus. *J Virol* 1993; 67:2756–2763.
158. Schneemann A, Zhong WD, Gallagher TM, Rueckert RR. Maturation cleavage required for infectivity of a nodavirus. *J Virol* 1992; 66:6728–6734.
159. Smith GE, Summers MD, Fraser MJ. Production of human β-interferon in insect cells infected with a baculovirus expression vector. *Mol Cell Biol* 1983;3:2156–2165.
160. Steinhaus EA. *Principles of insect pathology*. New York: McGraw-Hill, 1947.
161. Stewart LMD, Hirst M, Ferber ML, Merryweather AT, Cayley PJ, Possee RD. Construction of an improved baculovirus insecticide containing an insect-specific toxin gene. *Nature* 1991;352:85–88.
162. Stoltz DB. The polydnavirus life cycle. In: Beckage NE, Thompson SN, Federici BA, eds. *Parasites and pathogens of insects*. San Diego, CA: Academic Press, 1993:167–187.
163. Stoltz DB, Guzo D, Belland ER, Lucarotti CJ, MacKinnon EA. Venom promotes uncoating in vitro and persistence in vivo of DNA from a braconid polydnavirus. *J Virol* 1988;69:903–907.
164. Stoltz DB, Vinson SB. Viruses and parasitism in insects. *Adv Virus Res* 1979;24:125–171.
165. Stoltz DB, Whitfield JB. Viruses and virus-like entities in the parasitic Hymenoptera. *J Hym Res* 1992;1:125–139.
166. Strand MR, McKenzie DI, Grassl V, Dover BA, Aiken JM. Persistence

and expression of *Microplitis demolitor* polydnavirus in *Pseudoplusia includens*. *J Gen Virol* 1992;73:1627–1635.
167. Strand MR, Noda T. Alterations in the hemocytes of *Pseudoplusia includens* after parasitisim by *Microplitis demolitor*. *J Insect Physiol* 1991;37:839–850.
168. Summers MD, Smith GE. Baculovirus structural polypeptides. *Virology* 1978;84:390–402.
169. Sundset R, Keith C, Ribeiro B, Miller LK. Sequestration and regulation of intracellular calcium during baculovirus infection. 1994;submitted.
170. Tanaka T. Effect of the venom of the endoparasitoid, *Apanteles kariyai,* on the cellular defense reaction of the host, *Pseudaletia separata* Walker. *J Insect Physiol* 1987;33:413–420.
171. Theilmann DA, Stewart S. Analysis of the *Orgyia pseudotsugata* multicapsid nuclear polyhedrosis virus trans-activators IE-1 and IE-2 using monoclonal antibodies. *J Gen Virol* 1993;74:1819–1826.
172. Theilmann DA, Summers MD. Physical analysis of the *Campoletis sonorensis* virus multipartite genome and identification of a family of tandemly repeated elements. *J Virol* 1987;61:2589–2598.
173. Theilmann DA, Summers MD. Identification and comparison of *Campoletis sonorensis* virus transcripts expressed from four genomic segments in the insect hosts *Campoletis sonorensis* and *Heliothis virescens*. *J Virol* 1988;167:329–341.
174. Thiem SM, Miller LK. A baculovirus gene with a novel transcription pattern encodes a polypeptide with a zinc finger and a leucine zipper. *J Virol* 1989 ;63:4489–4497.
175. Thiem SM, Miller LK. Identification, sequence, and transcriptional mapping of the major capsid protein gene of the baculovirus *Autographa californica* nuclear polyhedrosis virus. *J Virol* 1989;63:2008–2018.
176. Thiem SM, Miller LK. Differential gene expression mediated by late, very late, and hybrid baculovirus promoters. *Gene* 1990;91:87–94.
177. Tomalski MD, Eldridge R, Miller LK. A baculovirus homolog of a Cu/Zn superoxide dismutase gene. *Virology* 1991;184:149–161.
178. Tomalski MD, Miller LK. Insect paralysis by baculovirus-mediated expression of a mite neurotoxin gene. *Nature* 1991;352:82–85.
179. Tomalski MD, Wu J, Miller LK. The location, sequence, transcription and regulation of a baculovirus DNA polymerase gene. *Virology* 1988;167:591–600.
180. Tweeten KA, Bulla LA Jr., Consigli RA. Characterization of an extremely basic protein derived from granulosis virus nucleocapsids. *J Virol* 1980;33:866–876.
181. Vialard JE, Richardson CD. The 1,629-nucleotide open reading frame located downstream of the *Autographa californica* nuclear polyhedrosis virus polyhedrin gene encodes a nucleocapsid-associated phosphoprotein. *J Virol* 1993;67:5859–5866.
182. Vinson SB, Iwantsch GF. Host regulation by insect parasitoids. *Q Rev Biol* 1980;55:143–165.
183. Volkman LE. The 64K envelope protein of budded *Autographa californica* nuclear polyhedrosis virus. In: Doerfler W, Bohm P, eds. *Current topics in microbiology and immunology: the molecular biology of baculoviruses*. New York: Springer-Verlag, 1986:103–118.
184. Volkman LE, Goldsmith PA. Mechanism of neutralization of budded *Autographa californica* nuclear polyhedrosis virus by a monoclonal antibody inhibition of entry by adsorptive endocytosis. *Virology* 1985;143:185–195.
185. Volkman LE, Goldsmith PA, Hess RT. Alternate pathway of entry of budded *Autographa californica* nuclear polyhedrosis virus fusion at the plasma membrane. *Virology* 1986;148:288–297.
186. Webb BA, Summers MD. Venom and viral expression products of the endoparasitic wasp *Campoletis sonorensis* share epitopes and related sequences. *Proc Natl Acd Sci USA* 1990;87:4961–4965.
187. Webb BA, Summers MD. Stimulation of polydnavirus replication by 20-hydroxyecdysone. *Experentia* 1992;48:1018–1022.
188. Whitford M, Faulkner P. Nucleotide sequence and transcriptional analysis of a gene encoding gp41, a structural glycoprotein of the baculovirus *Autographa californica* nuclear polyhedrosis virus. *J Virol* 1992; 66:4763–4768.
189. Williams GV, Rohel DZ, Kuzio J, Faulkner P. A cytopathological investigation of *Autographa californica* nuclear polyhedrosis virus p10 gene function using insertion-deletion mutants. *J Gen Virol* 1989; 70:187–202.
190. Wilson ME, Consigli RA. Functions of a protein kinase activity associated with purified capsids of the granulosis virus infecting *Plodia interpunctella*. *Virology* 1985;143:526–535.
191. Wilson ME, Mainprize TH, Friesen PD, Miller LK. Location transcription and sequence of a baculovirus gene encoding a small arginine-rich polypeptide. *J Virol* 1987;61:661–666.
192. Wilson ME, Miller LK. Changes in the nucleoprotein complexes of a baculovirus DNA during infection. *Virology* 1986;151:315–328.
193. Wu SX, Ahlquist P, Kaesberg P. Active complete in vitro replication of nodavirus RNA requires glycerophospholipid. *Proc Natl Acad Sci USA* 1992;89:11136–11140.
194. Xu D, Stoltz D. Polydnavirus genome segment families in the ichneumonid parasitoid *Hyposoter fugitivus*. *J Virol* 1993;67:1340–1349.
195. Xu D, Stoltz DB. Evidence for a chromosomal location of polydnavirus DNA in the ichneumonid parasitoid *Hyposoter fugitivus*. *J Virol* 1991; 65:6693–6704.
196. Yang CL, Stetler DA, Weaver RF. Structural comparison of the *Autographa californica* nuclear polyhedrosis virus-induced RNA polymerase and the three nuclear RNA polymerases from the host, *Spodoptera frugiperda*. *Virus Res* 1991;20:251–264.
197. Young JC, MacKinnon EA, Faulkner P. The architecture of the virogenic stroma in isolated nuclei of *Spodoptera frugiperda* cells in vitro infected by *Autographa californica* nuclear polyhedrosis virus. *J Struct Biol* 1993;110:141–153.
198. Zhong WD, Dasgupta R, Rueckert RR. Evidence that the packaging signal for nodaviral RNA2 is a bulged stem loop. *Proc Natl Acad Sci USA* 1992;89:11146–11150.
199. Zhong WD, Rueckert RR. Flock House Virus—down-regulation of subgenomic RNA3 synthesis does not involve coat protein and is targeted to synthesis of its positive strand. *J Virol* 1993;67:2716–2722.

Fundamental Virology, Third Edition
edited by B.N. Fields, D.M. Knipe, P.M. Howley, et al.
Lippincott - Raven Publishers, Philadelphia © 1996

CHAPTER 14

Viruses of Yeasts, Fungi and Parasitic Microorganisms

Reed B. Wickner

HISTORY

In 1948, a disease of cultivated mushrooms was observed in Pennsylvania at the La France Brothers mushroom farm (219,280). This "La France" disease became widespread, causing poor yields of misshapen mushrooms, and its study led to the first description, in 1962, of fungal viruses (110).

R.B. Wickner: Section on Genetics of Simple Eukaryotes, Laboratory of Biochemical Pharmacology, National Institute of Diabetes, Digestive and Kidney Diseases, National Institutes of Health, Bethesda, Maryland 20892.

Interferon Inducers

In the search for antiviral agents, statolon from culture filtrates of the penicillin-producing filamentous fungus *Penicillium stoloniferum* (196) and helenine from *Penicillium funiculosum* (218) were found to induce resistance to viral infection in mice. With the discovery of interferon (133; see also Chapter 11), it became clear that the effects of statolon and helenine were mediated by their induction of interferon (32,147). The inducing agents were then shown to be double-stranded RNA (dsRNA) from virus particles (148,155). Extensive use was made of dsRNA in the production of interferon until recombinant DNA methods became available. dsRNA, probably as the replicative form of RNA viruses, is both a natural

TABLE 1. *dsRNA, ssRNA, and dsDNA viruses of simple eukaryotes*[a]

Virus	Host species	Genome size (kbp)	Features (references)
dsRNA viruses			
Totiviridae			One segment, Gag-Pol, icosahedral particles
L-A	*S. cerevisiae*	4.6	Type species
M_1, M_2, M_3, M_{28},...	*S. cerevisiae*	1.6–1.8	Satellites of L-A; encode killer toxins
L-BC	*S. cerevisiae*	4.6	222
Hv190S	*Helminthosporium victoriae*	4.5	Coat protein phosphorylation (117)
P1-H, P4-H, P6-H	*U. maydis*	2.6–6.1	Associated killer phenomenon (149)
AfV-S, AfV-F	*A. foetidus*	3.0, 3.1	(34)
YlV	*Yarrowia lipolytica*	4.6	Can replicate in *Saccharomyces* (120,230)
LRV	*L. braziliensis*	5.3	Upstream ORF(s)
GLV	*G. lamblia*	6.1	Extracellular Infection; transformation
GgV-87-1-H	*Gaeumannomyces graminis*		
	Zygosaccharomyces bailii	4.0, 2.9, 1.9	Associated killer phenomenon (198)
Partitiviridae			Two segments, separate particles
PcV	*Penicillium crysogenum*		33
PsV-F, PsV-S	*Penicillium stoloniferum*	2.5, 1.4	33
AbV-4	*Agaricus bisporus*		La France disease of mushrooms (280)
RsV	*Rhizoctonia solani*	2.2, 2.0	(96)
Hypoviridae			
L = HAV = CHV1	*Cryphonectria parasitica*	12.7	Hypovirulence-associated virus, potyvirus-like, chestnut blight
Reoviridae			
C18, 9B21	*Cryphonectria parasitica*	11 segments	B. Hillman (personal communication)
Unclassified			
NB631 RNA	*Cryphonectria parasitica*	2.7	Mitochondrial, resembles 20S and 23S RNA, hypovirulence; B. Hillman (personal communication)
TVV	*Trichomonas vaginalis*	5.4, 5.5, 5.6	Host phenotypic variation
ssRNA replicons			
20S RNA (= W dsRNA)	*S. cerevisae*	2.9	Induced by N-starvation, high temperature
23S RNA (= T dsRNA)	*S. cerevisae*	2.5	Induced by N-starvation, high temperature
dsDNA viruses			
	Amoeba histolytica		Lytic, icosahedral, filamentous (66)
PBCV-1 and many others	*Chlorella*	333	Encode restriction enzymes, methylases, glycosidases (246)

[a]This incomplete list emphasizes the most heavily studied viruses and viruslike genomes of simple eukaryotes. More complete lists of viruses, particularly mycoviruses, may be found in other reports (34,151,160,162).

inducer of interferon and a trigger of the action of the interferon-induced protein kinase system and the 2′-5′ oligoA synthase. dsRNA-containing virus particles were soon discovered in many other species of fungi (Table 1) (23,34,151,160,162).

Viruses Affecting Virulence of the Chestnut Blight Fungus

The American chestnut tree was a major component of Eastern forests in the United States until the accidental introduction, in 1905, of the pathogenic fungus *Cryphonectria parasitica* along with an oriental variety of the chestnut tree (190). Over the following decades, the fungus virtually eliminated the upper parts of the chestnut trees in the Eastern United States. However, the root systems are not killed by the fungus and shoots continue to emerge, only to have their growth limited by an attack of the fungus.

Cryphonectria parasitica had a similar effect in Europe, but in the 1950s and 1960s, Italian and French workers noted the emergence of fungal strains with markedly reduced virulence to the trees. These hypovirulent strains (Fig. 1) could transmit their hypovirulence by hyphal anastomosis to virulent strains (119), and this became the basis of a successful biological control of chestnut blight in Europe. The basis for the hypovirulence was found to be dsRNAs transmitted by cytoplasmic exchange when hyphae fuse (7,64). These dsRNAs replicate in intracellular vesicles (124), and their structure is reminiscent of the plant (+) single-stranded RNA (ssRNA) potyviruses (190). The details now emerging of their specific suppression of pathogenicity of the fungus toward the chestnut tree is providing unique insights into virus-host interactions. Most recently, infectious comple-

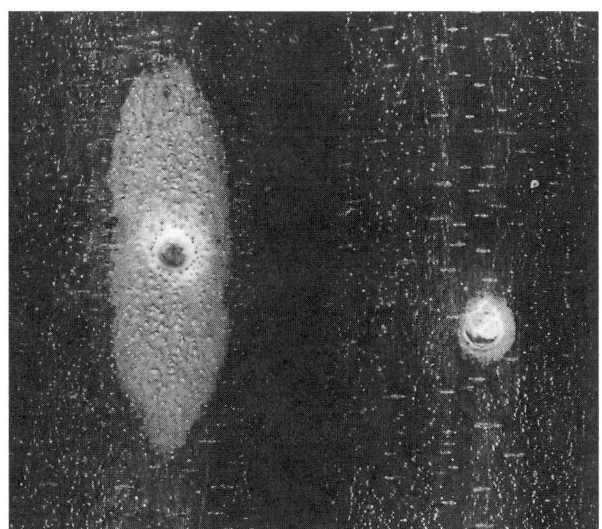

FIG. 1. Virulent (uninfected, left) and hypovirulent (virus-infected, right) *Cryphonectria parasitica* (chestnut blight fungus) innoculated into chestnut trees. Infection of the fungus by any of several hypoviruses attenuates its virulence toward the trees. Photo courtesy of Dr. Donald Nuss.

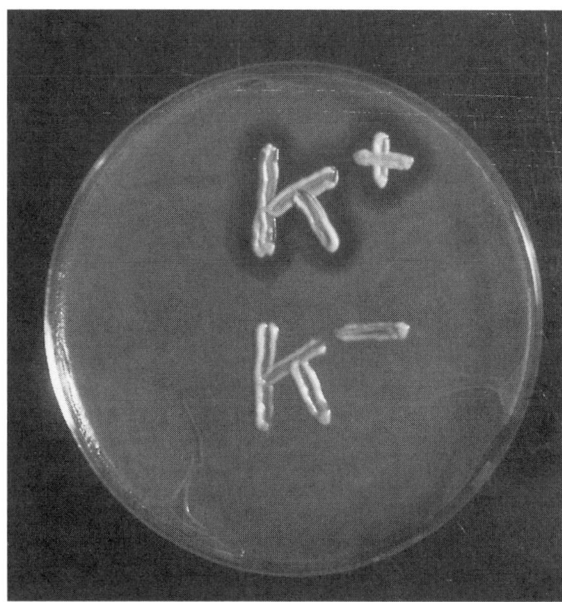

FIG. 2. The killer phenomenon of *S. cerevisiae*. A sensitive strain of yeast was spread as a lawn and streaks of a killer strain (above) or a nonkiller strain (below) were applied. After 2 days' incubation, the lawn of the sensitive strain has not grown in a zone around the killer strain. The secreted protein killer toxin and immunity to the toxin are encoded by M_1, a satellite dsRNA of the L-A dsRNA virus.

mentary DNA (cDNA) clones have been applied in attempts to control chestnut blight (47).

Yeast Killer Systems

The killer phenomenon was discovered in 1963 by Makower and Bevan (168). Some strains of *Saccharomyces cerevisiae* secrete a protein toxin that is lethal to other strains, but to which the secreting strains themselves are immune (Fig. 2) (275). The non-Mendelian inheritance of the ability to produce this toxin (221) led to the discovery that a dsRNA species (M_1 dsRNA) in intracellular viral particles determines the killer trait (14,35,247). The helper virus of M_1 is called L-A, and it initially obscured a less abundant dsRNA virus family, L-BC, which was eventually uncovered using host mutants that lose L-A (222). As will be developed below, studies of the *Saccharomyces* dsRNA viruses have made important contributions to both virology and to protein processing and secretion. *in vitro* template-dependent RNA replication, transcription, and packaging systems were developed. Detailed dissection of the viral sites and encoded proteins is continuing. In addition, analysis of host genes promoting viral replication and expression and host genes repressing viral propagation has yielded information not yet available for any other eukaryotic RNA virus system. The L-A virus seems to be typical of a now-emerging large class of single-segment viruses, the Totiviridae, found in fungi, and parasitic microorganisms (Table 1).

Several other dsRNA-based killer systems have been described in *Saccharomyces* (M_2, M_3, M_{28}, etc. (174,187,192), and many have been found in other yeasts (193,198), but only a few of the latter are virus based. One other virus-based killer system that has been extensively studied is that of the corn pathogen, *Ustilago maydis* (149,150,197,274).

Viruses of Parasitic Protozoa

Several different types of infectious virus particles have been described in *Entamoeba histolytica* (66). A cytoplasmic 80-nm icosahedral virus and a nuclear filamentous virus were described, but their nucleic acids have not yet been well defined.

More recently, dsRNA viruses have been found in *Trichomonas vaginalis* and *Giardia lamblia,* as well as in *Eimeria* and *Leishmania* species (252). The *Giardia* virus is infectious and can be transfected. The presence of the *Trichomonas* virus is associated with phenotypic variation of a cellular surface antigen.

Retroviruses (Retroelements, Retrotransposons) in Simple Eukaryotes

The Ty1 element of *S. cerevisiae* (20,111,210) was among the first retrotransposons recognized. Ty1 was first detected as a sequence present at about 35 copies per genome, but not always at the same locations in all strains (38). Ty1 is composed of a 5-kb unique sequence bounded by short direct repeats, called δ sequences. On cultiva-

TABLE 2. *Retroelements of simple eukaryotes*

Retrovirus		LTRs	ε (unique)	Group[a]	Reference
Ty1, Ty2	*S. cerevisiae*	δ, 334–338 bp	5.2 kb	*copia*	20
Ty3	*S. cerevisiae*	σ, 340 bp	4.7	*gypsy*	125
Ty4	*S. cerevisiae*	τ, 371 bp	5.6	*copia*	136,228,229
Ty5	*S. cerevisiae*	245 bp		*copia*	248
Tf1, Tf2	*Schizosaccharomyces pombe*	349–358 bp	4.4	*gypsy*	163
DIRS-1	*Dictyostelium discoideum*	ITRs	4.2		39
DRE	*Dictyostelium discoideum*	Complex TRs			169
Tp1	*Physarum polycephalum*	277 bp	8.3	*copia*	207
CfT-1	*Cladosporium fulvum*	427 bp	6.1	*gypsy*	175
CRE1	*Crithidia fasciculata*	–		LINE	108
SLACS, CZAR	*Trypanosoma brucei* *Trypanosoma cruzi*	–	6.7	LINE	
TOC1	*Chlamydomonas reinhardtii*	217, 237 bp	4.6		63
Tad	*Neurospora*	–	7.0	LINE	144,145

[a]Based on amino acid sequence homology and gene order, retrotransposons may be divided into those similar to the *copia* element or the *gypsy* element of *Drosophila* (74).
LINE-like elements are retroposons, lacking LTRs.

tion of a strain, new bands appeared on a Southern blot hybridizing with this sequence, suggesting transposition (38). Transposition of Ty1 next to any of several genes (encoding cytochrome c, arginase, alcohol dehydrogenase, etc.) abolished their normal control and placed them under control of the mating type, as was later to be found for the transcription of Ty1 itself (79,84).

The elegant experiment that demonstrated the already suspected retrotransposition of Ty1 used a GAL1-promoted Ty1 containing an artificial intron (19). The transposed copy was found to have precisely lost the intron, showing that it must have transposed via an RNA intermediate. Induction of transposition of this GAL1-Ty1 was accompanied by the accumulation of 60-nm virus particles containing reverse transcriptase (112). The structure of Ty1 (51) showed that it had substantial homology with retroviruses, including the presence of overlapping reading frames and long terminal repeats (LTRs), but that it had no env gene.

Retrotransposons are now known in many different simple eukaryotes, including *Physarum, Dictyostelium,* and *Schizosaccharomyces.* Furthermore, there are now five known distinct retrotransposons in *S. cerevisiae* (Table 2) (111).

The retrotransposons resemble mammalian retroviruses in most aspects except for their lack of env (and an envelope) and their lack of extracellular infectivity, although they may be transmitted by cell-cell fusion, as is common for fungal dsRNA viruses. Retroposons are further distinguished by their lack of LTRs and include LINE (long interspersed) elements in mammals and elements in *Neurospora* (144), *Crithidia* (108), and *Trypanosoma* (127). Most recently, retrotransposition of reverse transcriptase–encoding mitochondrial introns has been shown (142,181,214).

IMPACT

dsRNA Viruses

In addition to their early role in interferon research and in the biological control of chestnut blight, dsRNA viruses of simple eukaryotes have had a substantial scientific impact, often in unexpected areas. The discovery of the *KEX2* gene of *S. cerevisiae* by its requirement for secretion of killer toxin and of α mating pheromone (161) led to its definition as a protease specific for pairs of basic amino acid residues (138). This is just the specificity known for mammalian prohormone processing and has led to the discovery of *KEX2*-homologous proteases of mammals that perform these functions (225).

The M dsRNA–encoded killer toxins act on many different yeasts and fungi, including pathogenic species. They bind to cell wall components as the first step in their action (36). Because fungal cell walls are important targets for drug development, the killer toxins are useful tools in efforts to find antifungal antibiotics. Antiidiotypic antibodies that mimic killer toxin action have been produced, and efforts are underway to use their induction as a treatment of infections with susceptible fungi (195).

The ease with which fungi in general, and *Saccharomyces* in particular, are subject to genetic and molecular manipulation, has made these systems particularly useful in the study of the virus-host interaction (262). For example, the *SKI* broad-spectrum host antiviral system specifically inhibits translation of viral messenger RNA (mRNA) without shutting down all translation (269). An RNA transfection method (208) has been used to introduce Brome mosaic virus into *S. cerevisiae* specifically to take advantage of these features (135). The Brome mosaic virus segment 3 can replicate stably in yeast when supplied with the products of virus segments 1 and 2, and its interactions with host components can be studied more easily than in plants. The development of template-dependent *in vitro* systems for L-A dsRNA virus replication, transcription, and packaging (103,104) is as yet unique among dsRNA viruses of eukaryotes, and the information gleaned from such studies may be useful in studies of other systems. Studies of the genes affecting –1 ribosomal frameshifting by the L-A dsRNA virus of *S. cerevisiae* have suggested that drugs affecting this process might be useful as antiretrovirals (70).

Retroelements

Demonstration of the retrotransposition of Ty1 in *S. cerevisiae* led to the identification of this type of retroviral element in all groups of organisms, and the realization that some previously identified repeat sequences, such as the LINE elements of mammals, were retroelements. In fact, human LINE elements are now studied in part through the use of Ty1 (170). Tad is a LINE-like element of *Neurospora* that actively transposes between nuclei, making it a promising object for study (144,145).

One practical use proposed for Ty1 is in vaccine production and in developing antiviral strategies. TYA1, the Gag analog of Ty1, can form viral particles without Pol, and fusing TYA1 with antigenic proteins results in the production in yeast of particles carrying multiple copies of the antigen exposed on the surface (1). These particles are excellent immunogens. Fusing the highly active RNAse barnase to the TYA1 (Gag) gene makes cells nearly completely immune to Ty1 transposition, presumably because the barnase-fused Gag protein is incorporated into Ty1 viral particles and destroys the viral RNA (184). This strategy could be applied to blocking propagation of other viruses.

Studies of the several mechanisms of +1 ribosomal frameshifting used by all of the Ty elements to form their Gag-Pol fusion proteins has shed important light on translational mechanisms (13,94). The finding that the lariat debranching enzyme, which has a role in degrading spliced-out intron RNA, is also necessary for retrotransposition shows the usefulness of the yeast system in exploring the host components involved in the retroviral life cycle (43).

The insertion of retroelements in the control regions of host genes has profound effects on the expression of these genes and is often the basis of tumorigenesis by retroviruses. This area has been extensively studied using Ty elements, and much has been learned of its molecular basis (see below). The inserting Ty element separates the normal control sequences from the gene's promoter and introduces its own control sequences in close proximity. In some cases the host gene shows a similar pattern of control to that of the Ty element itself, whereas in others the gene is inactivated. The host transcriptional factors controlling the expression of Ty and of the nearby host genes also have been dissected in detail. In view of the profound effects that retroelements exert on genes adjacent to their site of insertion, the specificity of insertion is also of considerable importance. It has been found that Ty1 targets the control region of genes, probably by recognizing some feature of chromatin structure. Ty3 inserts at specific sites adjacent to the RNA polymerase III–expressed transfer RNA (tRNA), 5S recombinant RNA (rRNA), and U6 snurp genes. It probably recognizes RNA polymerase III transcription factors bound to the host DNA (see below). Evidence for a host antiretrotransposition system also has been developing (61), apparently acting by blocking the processing of Ty proteins.

CLASSIFICATION

General Properties of Viruses of Simple Eukaryotes

With the exception of the *G. lamblia* dsRNA virus and the DNA viruses of *Chlorella* and *Entamoeba,* all of the viruses and retroelements discussed in this chapter have forsaken the extracellular route of transmission, being transmitted either vertically or by cytoplasmic mixing, such as occurs in cell-cell mating or in hyphal anastomosis of fungi. They resemble, in this regard, the plant Cryptoviruses and intracisternal A type particles (retroelements) and LINEs (retroposons) of mammals. Viruses of *Gaeumannomyces graminis* (224) and *S. cerevisiae* (78) have been introduced into cells using polyethylene glycol or lithium acetate treatment (230). However, there is no evidence for a natural extracellular phase for these viruses. It has been argued that the high frequency of mating and hyphal fusion of fungi in nature makes extracellular transmission dispensable for these viruses (262), and it is difficult to find strains of *S. cerevisiae* that lack either L-A or L-BC dsRNA viruses, the 20S RNA replicon, or any of the Ty retrotransposons. Nor is the direct cell-to-cell route of infection completely neglected by mammalian viruses, those of the herpes group and human immunodeficiency virus being prime examples.

However, this mode of transmission means that these viruses must balance the need to spread and propagate against consideration for the viability of their host. Furthermore, the fact that most of these elements are widespread means that the hosts have learned to protect themselves. This is clearly seen in the Ty elements, which target sites that can tolerate an insertion. The Ty elements, particularly Ty3, regulate their transcription to largely limit transposition to the opportunity to infect the (potentially unoccupied) genome of a mating partner. The copy numbers of several of the RNA replicons (L-A dsRNA, L-BC dsRNA, and 20S ssRNA) are partially repressed by the host Ski proteins preventing cytopathology but allowing spread by cell fusion.

It is striking that most of the viruses and retroelements discussed use some form of ribosomal frameshifting to make a Gag-Pol fusion protein. This is not surprising for the retroelements, but was not expected for the dsRNA genomes.

This review will emphasize the L-A dsRNA virus and the Ty1 retroelement of *S. cerevisiae* because these are the most studied elements in this group. However, we also hope to make clear the lessons already learned from the study of other viruses and their further exciting potential. Many of the fungal hosts of these viral elements have well-developed genetics, which facilitates their study, particularly the interaction of virus and host.

Virus Families

The classification of RNA viruses of simple eukaryotes (Table 1) is in a rather early stage because relatively few have been studied sufficiently to define their essential characteristics. Nonetheless, several patterns are emerging.

There is a growing group of viruses, the Totiviridae, whose genome consists of a single segment of dsRNA encapsidated in icosahedral particles, encoding a major coat protein and an RNA-dependent RNA polymerase. The RNA polymerase is expressed using ribosomal frameshifting or another translational trick, as discussed below. The L-A dsRNA virus of *S. cerevisiae* is the most studied member of this group.

The Partitiviridae have a bipartite genome with the two segments separately encapsidated in particles containing the proteins encoded by both segments. This group includes viruses of the filamentous fungi *Penicillium* and *Aspergillus,* the mushroom *Agaricus,* as well as many other fungi. It also includes a large group of plant cryptoviruses (18,186), whose genome is dsRNA and whose biology is much like that of the systems discussed in detail here. However, few of the viruses tentatively classified in this group have actually been shown to have the properties thought to be general for the group.

A third group, comprised of RNA replicons is typified by that called simply 20S RNA. 20S RNA and 23S RNA of *S. cerevisiae* and NB631 of *Cryphonectria* are not encapsidated, and 20S RNA appears to be essentially naked in the soluble fraction of cell extracts.

The retroelements, including retrotransposons, retroposons, and retrotransposing introns, are another widespread group with varying similarity to retroviruses of mammalian cells (Table 2). The Ty1 element of *S. cerevisiae* is the most studied member of this group, but there are many unique aspects of other elements that have made their study of obvious importance.

dsRNA Viruses: *Sacchoromyces cerevisiae* (L-A and Its Satellites)

The L-A dsRNA virus of baker's (brewer's) yeast—*S. cerevisiae*—has a single 4.6-kbp segment (34,116,261, 262). It is the type species of the Totiviridae, so named because their single segment is, except for functions supplied by the host, sufficient for viral propagation. A number of satellite dsRNAs, called M_1, M_2, M_3, M_{28}, and so forth, encode secreted protein toxins whose action produces the killer phenotypes that first led to the discovery of yeast viruses. The satellite dsRNAs are packaged separately from the L-A helper virus (22). Internal deletion mutants of M_1 dsRNA, called S dsRNAs (97), and one internal deletion mutant of L-A, called X and retaining only 530 bp of L-A's ends (89), have been used extensively in studying viral functions.

Virion Structure

The L-A virion is an isometric 39-nm particle (35) with about 120 copies per particle of the 76-kd major coat protein (Gag) (88) and an estimated one to two copies per particle of a minor 180-kd protein (Gag-Pol) (102,222). The virion structure, from cryoelectron microscopy and image reconstruction, is icosahedral with T = 1 (Fig. 3) (9). The asymmetric unit is a dimer of Gag proteins. Thus, there are two different environments for Gag proteins. This result and the measured 1.9% efficiency of frameshifting (69) suggest that there are two Gag-Pol fusion proteins per particle, but direct measurements are not yet sufficiently accurate to confirm this. Each L-A particle contains a single molecule of L-A dsRNA, but particles with M satellite dsRNA may have either one or two dsRNA molecules per particle, and X particles may have up to eight (see Replication Cycles, below) (88,89,99).

Genome Structure

The L-A (+) strand has two overlapping open reading frames (ORFs; Fig. 4A) (132): Gag encoding the 76-kd major coat protein and Pol encoding the 100-kd Pol domain of the Gag-Pol fusion protein (68,102,130,132). The Pol ORF includes the consensus amino acid sequence patterns typical of viral RNA-dependent RNA polymerases. Three *in vitro* ssRNA binding domains also have been localized to Pol (102,200,201), the central one cryptic unless an adjacent inhibitory region is deleted (Fig. 4A). Pol residues 67–213 are necessary for packaging of the viral (+) strands (100,201). Neither 5′ cap structure nor 3′ polyA has been found on either strand of genomic L-A dsRNA or M_1 dsRNA (31), but there is, at each 3′ end, an uncoded base that can be either A or G (25,235). The functional significance and the mechanism of generation of this extra base extending from the 3′ ends are unknown.

The M dsRNAs each encode, in their 5′ portion, a preprotoxin protein (21,67,167,212) (Fig. 4B). The M_1 (+) strand has an internal encoded polyA region whose length shows frequent clonal variation in length, presumably due to transcriptase stuttering (123), and a substantial 3′ region that encodes no protein but contains essential *cis* sites (87,98,115,217). As will be detailed below, sites on the L-A and M_1 (+) strands necessary for packaging and (–) strand synthesis have been determined (Fig. 2), and some limits have been set on the possible transcription signals.

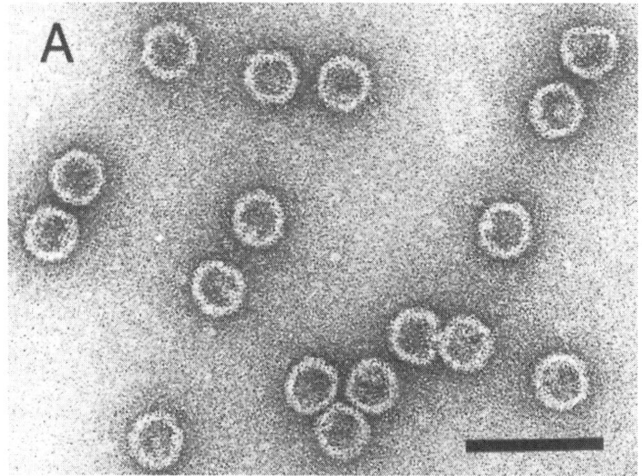

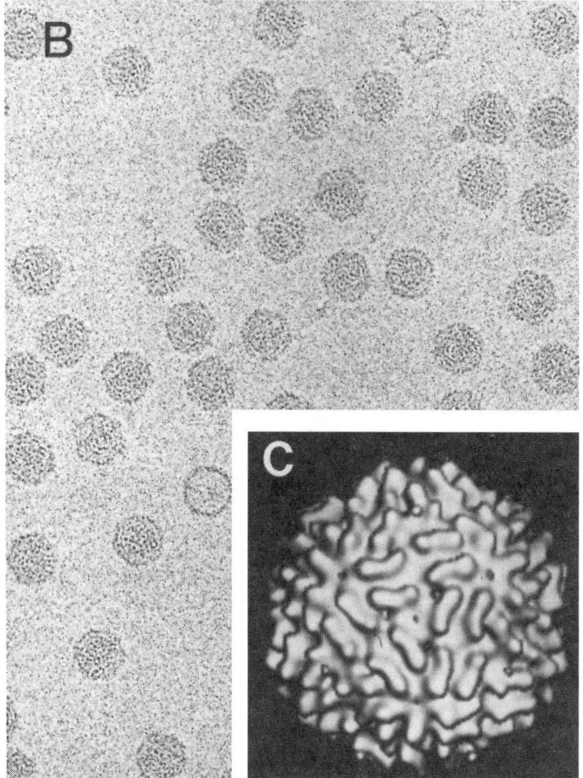

FIG. 3. Electron microscopy of L-A virus. **A:** Negatively stained L-A capsids. **B:** Unstained, frozen-hydrated specimen of L-A virions. **C:** This cryoelectron micrograph was used to reconstruct the three-dimensional structure of L-A. The surface-shaded representation, viewed along a three-fold axis of icosahedral symmetry, shows a T = 1 structure built up by pentons containing 10 subunits arranged in two concentric rings of five subunits each. The diameter of the virions is 39 nm (C). The bar (A and B) represents 100 nm (courtesy of Cheng et al., ref. 9).

Stages in the Replication Cycle

The L-A replication cycle and the closely related cycle of its satellites are shown in Fig. 5 (88,99). L-A dsRNA-containing viral particles synthesize (+) ssRNA in a conservative reaction (99,129), and these new (+) strands are then extruded from the particle (88,257). There they serve as mRNA for the production of the Gag and the Gag-Pol fusion proteins. These proteins (and possibly one or more host proteins) then assemble with a viral (+) strand to form new particles. These newly assembled viral particles perform the synthesis of (–) strands on the (+) strand template to form dsRNA and complete the cycle (99).

In addition to their appropriating viral proteins from L-A (22), the replication cycle of the satellite dsRNAs, M_1, S (deletion mutants of M_1), and X (a 530-bp deletion mutant of L-A) is similar to that of L-A itself, except that they replicate more than once within the viral particle until it is full (Fig. 5). Only a single (+) ssRNA is packaged per particle (98), so L-A and M are separately encapsidated. Because the particle capacity is determined by the structure of Gag to be sufficient to accommodate one full-length L-A dsRNA molecule and the M and X dsRNAs are much smaller, new M or X (+) strand transcripts are often retained within the viral particles where they are converted to a second (third, etc.) dsRNA. Then all new (+) strands are extruded from the particle (88,89). This is called the headful replication mechanism (in contrast to the headful packaging mechanism of many DNA bacteriophages) (88,89). This implies that extrusion of the transcripts is a mechanical consequence of the head being full, rather than an active process. Both (+) and (–) strands are synthesized within the viral particles, but at different points in the cycle, so the replication is said to be conservative, intraviral, and asynchronous, with headful replication.

Transcription Reaction [(+) Strand Synthesis]

As in the Reoviridae, the transcription reaction for L-A is conservative (99,270), resulting in the overall process of viral replication being conservative (189,213). However, in *Aspergillus foetidus* slow virus and *Penicillium stoloniferum* slow virus (33), dsRNA transcription is semiconservative. The difference between a conservative and semiconservative reaction concerns whether or not there is re-pairing of the template (–) strand with the parental (+) strand that was (in either case) displaced during the synthesis.

A template-dependent *in vitro* transcription reaction method has been developed for the L-A dsRNA virus (104). When L-A dsRNA-containing particles are treated with very low ionic strength conditions, they rupture, releasing the dsRNA. These opened empty particles can be reisolated free of RNA and perform a dsRNA template-dependent reaction that is, like the *in vivo* reaction, conservative (104). This reaction is highly template specific, using only L-A, M, and X dsRNAs (all physiological templates), but not L-BC, ϕ6, or rotavirus dsRNAs. Because X dsRNA retains only 25 bp of the end of L-A from which the transcription reaction starts (89), the signal recognized by the transcriptase is likely within this region, perhaps within the terminal 6 bp that are in common among L-A, M_1, and M_2 (122,235,236). The template-dependent transcription reaction requires very high concentrations of polyethylene

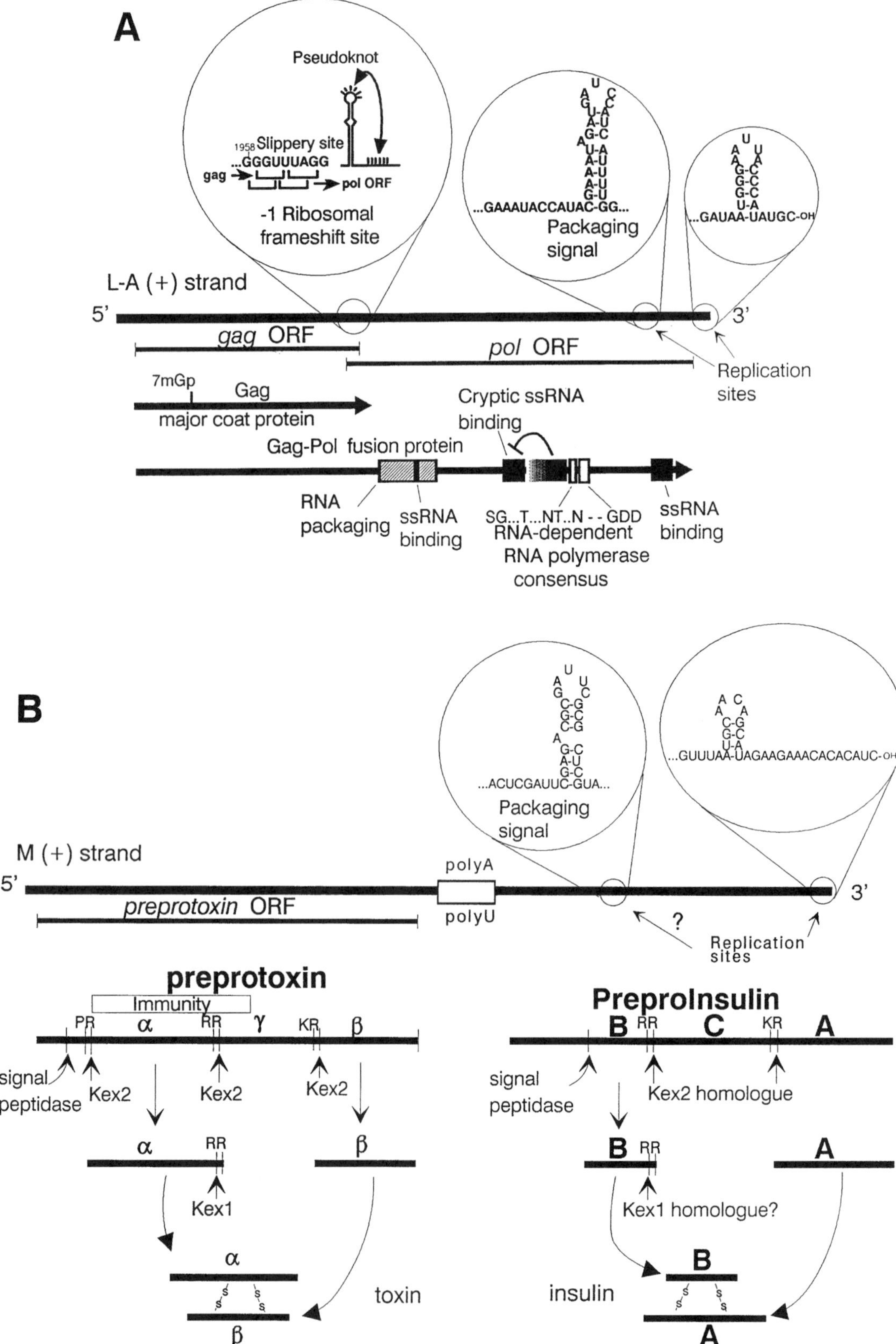

FIG. 4. A: Sites and encoded proteins of the L-A (+) strand. The sites responsible for ribosomal frameshifting, (+) ssRNA packaging and replication are shown. Functional domains in the Pol region are also indicated. "7mGp" indicates the cap-binding site at His154 of Gag. The sequences present in X dsRNA are shown as a hatched box at the 3′ and 5′ ends of L-A. **B:** Coding and *cis* sites of M_1 (+) strand. The analogous and homologous processing of the K_1 preprotoxin encoded by M_1 is compared with that of preproinsulin.

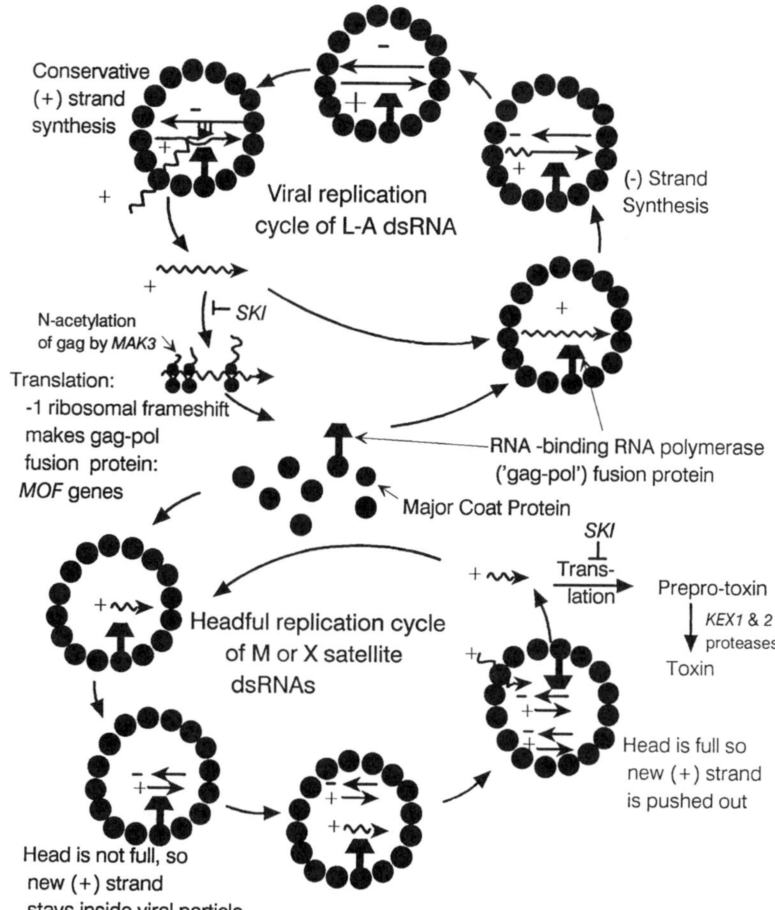

FIG. 5. Replication cycles of L-A virus and its satellite dsRNAs, M and X.

glycol (20%), suggesting that the transcriptase has a low affinity for the dsRNA template (104). Because the dsRNA is normally formed within and stays inside the viral particles, the RNA polymerase sees a high effective concentration and need not have a particularly high affinity.

Translation

The translation apparatus is a prime battleground of the fight between an RNA virus and its host. For example, interferon acts primarily by blocking viral translation. This can also be seen in picornaviruses, where cleavage of a cap-binding component inactivates host translation (see Chapter 16), or in influenza virus cap-stealing and antiinterferon measures (see Chapter 46 in *Fields Virology,* 3rd ed.). Studies of the replication mechanisms of the L-A virus and its satellites have likewise led to the suggestion that translation of viral proteins is the critical event determining the balance between virus and host.

Ribosomal Frameshifting. The Gag and Pol ORFs of L-A overlap by 130 nucleotides, and the Pol protein is expressed only as a fusion protein with the major coat protein, Gag (Fig. 4A) (102,132). The mechanism of formation of this Gag-Pol fusion protein is a –1 ribosomal frameshift (69), similar to those described in Rous sarcoma virus, many mammalian retroviruses, coronaviruses, and several plant viruses (8,128,134).

There are two components that determine the frequency of –1 ribosomal frameshifts (Fig. 6). The "slippery site" is a sequence of the form X XXY YYZ, where the Gag reading frame is shown. This sequence allows the tRNAs reading XXY and YYZ to each move back one base on the mRNA and still have their nonwobble bases correctly paired (134). X can be any base, but Y can only be A or U, probably because the frequency with which unpairing of the tRNAs from the 0 frame codons (XXY and YYZ) is also important and because A site pairing is stronger than P site pairing (27,69). Z can be any base but G, suggesting that specific tRNAs are more able to frameshift than others (42,128). The second component determining frameshift efficiency is the presence of a strong secondary structure, usually an RNA pseudoknot, just downstream of the slippery site, to slow ribosomal movement at this point (26,134). The location of this secondary structure is particularly critical (28), and a pseudoknot is far more effective in promoting frameshifting than is a simple stem loop of the same overall melting energy. The pseudoknot, more

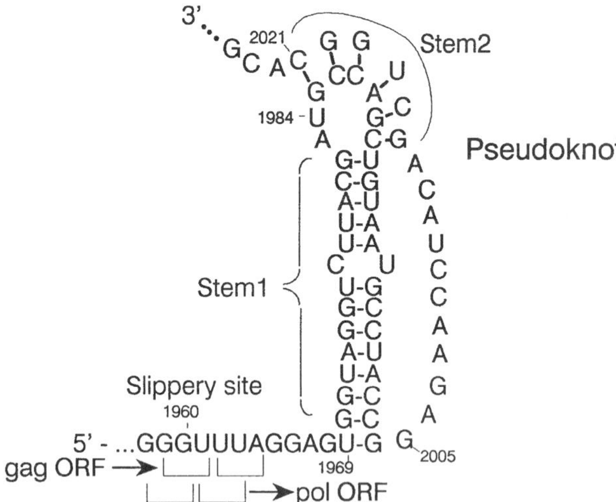

FIG. 6. The L-A site determining –1 ribosomal frameshifting to form the Gag-Pol fusion protein. The pseudoknot makes the ribosome pause over the slippery site, where the tRNAs can unpair from the mRNA and repair in the –1 frame with correct base pairing of the nonwobble bases.

than the simple stem loop, should resist melting of the first few bases of the stem, precisely positioning the ribosome with the slippery site in the A and P sites.

The efficiency of –1 frameshifting, and thus the ratio of Gag-Pol fusion protein to Gag protein produced, is critical for viral propagation (70), as is the efficiency of +1 frameshifting for Ty1 retrotransposition frequency (see below). A twofold change away from L-A's normal 1.9% efficiency results in failure to propagate the M1 satellite dsRNA, but L-A propagation is less sensitive. The Gag-Pol to Gag ratio is likely to be important for viral assembly (Fig. 7). Excess Gag-Pol (high frameshift efficiency) could result in starting too many particles and winding up with too little Gag to complete any of them. Excess Gag might result in particles closing before the packaging domain of Pol has had a chance to find a viral (+) strand (70). Recently, host genes controlling –1 frameshifting efficiency have been described (70,71). These genes, called *mof* for *m*aintenance *o*f *f*rame, are expected to encode ribosomal components or translation factors. Several *mof* mutations result in temperature-dependent frameshifting and are temperature sensitive for cell growth, with classical cell-cycle arrest phenotypes at the nonpermissive temperature. The differential effects of specific host mutations on frameshifting at specific slippery sites suggest that drugs similarly affecting frameshifting might target specific viruses (71).

Why do most retroviruses, L-A and other totiviruses, coronaviruses, and many other (+) strand RNA viruses use ribosomal frameshifting or readthrough of termination codons, but not splicing or RNA editing, to make their Gag-Pol or other fusion proteins? Retroviruses, (+) ssRNA viruses, and dsRNA viruses each use their (+) strands as (a) mRNA for viral proteins, (b) the species packaged to make new particles, and (c) the template for virus replication. Covalent modification of these (+) strands for translation purposes (e.g., splicing or editing) would create mutant virus unless the modification process removed essential signals for packaging or replication. Indeed, (+) ssRNA and dsRNA viruses are not known to splice or edit their (+) strands, and retroviruses, in splicing viral (+) strands to make env or other proteins, remove the packaging site (Psi), preventing propagation of the altered RNA. Ribosomal frameshifting (–1 or +1) and termination codon readthrough do not alter the (+) strands and so do not create mutants (132).

Gag's Decapitation of Cellular mRNAs Is Necessary for M Expression. In a search for yeast proteins capable of binding the 5′ cap structure of cellular mRNA, it was found that the L-A and L-BC Gag proteins were each capable of covalently binding this structure (16). L-A Gag covalently attaches 7mGMP from the cap to His154 in a reaction that requires only Mg^{2+} (17). Modification of His154 on an L-A cDNA expression clone destroys Gag's ability to covalently bind cap but does not prevent the clones ability to propagate the M_1 satellite virus (17). However, expression of the killer toxin encoded by M_1 is impaired by the His154 mutation. There is now evidence that Gag removes caps from cellular mRNAs *in vivo*. These decapitated mRNAs serve as decoys to distract a cellular 5′ →3′ exoribonuclease specific for uncapped RNAs from attacking the viral uncapped mRNAs (D. Masison et al., unpublished observation).

SKI *Antiviral System Selectively Controls Translation of Viral mRNA.* The sole essential function of the six *SKI* genes of S. cerevisiae is the repression of viral copy number, particularly control of M (199,203,223,238,269). The *SKI* proteins repress three unrelated viral systems: L-A and its satellites, L-BC (a dsRNA virus), and the ssRNA repli-

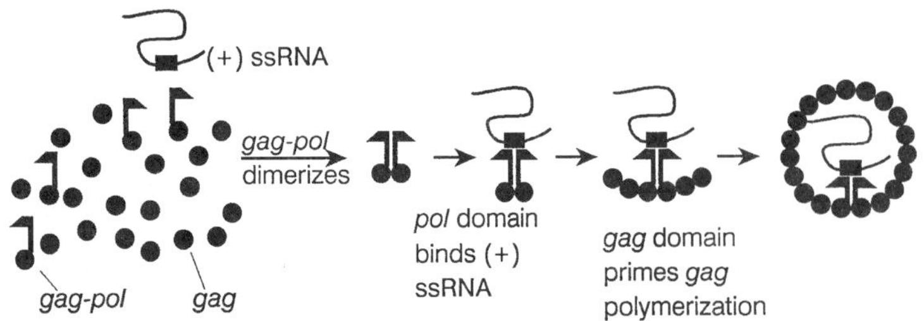

FIG. 7. Packaging model for the L-A dsRNA virus (70,102).

con called 20S RNA (11,171,238). Detailed studies of the *SKI2* gene suggested that the system acts by limiting translation of viral mRNA (269). Because all of these replicons are cytoplasmic and none were known to have either 5′ caps or 3′ polyA structures, it was speculated that the *SKI* system recognized the absence of one or both of these structures, typical of eukaryotic cellular mRNAs (269). That *SKI2* also represses translation of RNA polymerase I transcripts (probably lacking cap and/or polyA) supports this idea (269).

Recently, using an RNA transient expression assay (208), it has been shown that *SKI2* inhibits the translation of mRNA specifically if it lacks a 3′ polyA structure (D. Masison et al., unpublished observation).

The *SKI2, SKI3,* and *SKI8* genes have been characterized. Ski2p has helicase motifs and a glycine-arginine rich domain typical of nucleolar proteins (172,199,223,269). Ski3p is a nuclear protein with an amino acid repeat pattern (TPR repeats) of unknown function, and Ski8p has an amino acid repeat pattern first identified in β-transducin and also of unknown function.

Translation Proteins and Viral Propagation. Inhibitors of protein synthesis that act on ribosomal proteins produce curing of M dsRNA (but not L-A) (95,159), and two chromosomal genes needed for propagation of M, but not of L-A, have proven to encode ribosomal proteins. *MAK8/TCM1* encodes ribosomal protein L3 (266) and *MAK7/RPL4A* encodes L4 (Y. Ohtake and R. Wickner, unpublished observations). Mutants in several GCD genes, now known to encode translation factors, produce the Mak$^-$ phenotype (126). The mechanism of these effects is not yet clear, but it does not appear to involve frameshift efficiency.

Posttranslational Modification

KEX1 *and* KEX2 *Proteases and Discovery of Prohormone Proteases.* The *kex1* and *kex2* mutants were first isolated based on their inability to produce the "killer" toxin encoded by the M_1 satellite dsRNA (*KEX* = killer expression) (265). The *kex2* mutants were noted to have a defect in mating that was specific to cells of the α mating type, and *kex2/kex2* homozygous diploids were defective for sporulation (161). The α-specific mating defect was partially explained by the finding that the cells failed to secrete the α pheromone (161), a peptide that prepares cells of the opposite (a) mating type for mating by arresting them in the G1 phase of the cell cycle. The failure to secrete killer toxin and α pheromone was explained by the finding that *KEX2* encodes a protease that cleaves C-terminal to pairs of basic amino acid residues (138) and that *KEX1* encodes a carboxypeptidase that can remove the pair of basic amino acids (55,72). These were the cleavages needed to convert the toxin and α pheromone proproteins to their mature forms (Fig. 4B).

The importance of these discoveries lies in the fact that the specificities of the Kex proteases are just those needed to process preproinsulin (Fig. 4B), preproopiomelanocortin, and other mammalian prohormones, but the enzymes responsible for these maturation cleavages had been elusive. Using homology with Kex2p, several candidate genes and enzymes were identified, and evidence that they are involved in these prohormone processing steps is accumulating (225).

There is also evidence that the Kex2p-homologous enzymes are involved in the proteolytic processing of some mammalian viral proteins. Mutant CHO cells resistant to Sindbis virus and Newcastle disease virus were made sensitive by expression of the Kex2p-homologue, mouse furin, or by expression of Kex2p itself (179). This confirms earlier *in vitro* evidence for the role of furin in NDV protein processing (209).

N-*Acetylation of Gag by Mak3p Is Necessary for Assembly.* *MAK3* is one of the three chromosomal genes known to be necessary for the propagation of the L-A dsRNA virus. Its sequence shows homology with *N*-acetyltransferases, particularly with the rimI protein of *Escherichia coli*, which acetylates the *N*-terminal of ribosomal protein S18 (233). In fact, the L-A Gag protein is normally blocked (presumably acetylated) and is unblocked in a *mak3* mutant host, resulting in failure of viral assembly (234). *Mak3p* recognizes the *N*-terminal 4–amino acid residues of Gag (232), and this signal can be transferred to β-galactosidase, producing *MAK3*-dependent N-terminal acetylation. Like L-A, the major coat proteins of Rous sarcoma virus, tobacco mosaic virus, turnip yellow mosaic virus, alfalfa mosaic virus, and potato X virus are *N*-terminally acetylated, but the enzymes responsible have not been identified, nor can the role of acetylation for these viruses be examined without altering the primary protein sequence.

Viral Assembly

The headful replication mechanism implies that the coat protein determines the structure of the head, not the genome. This is typical of isometric viruses, and has now been confirmed by the finding that expression of the Gag protein alone produces empty particles that are morphologically indistinguishable from normal L-A virions (100). The L-A particles have a T = 1 icosahedral structure with an asymmetric unit consisting of a dimer of Gag (9). Each particle has only one or two Gag-Pol fusion proteins (perhaps as a dimer). The requirement for the Mak3p-catalyzed *N*-acetylation of Gag for assembly has been discussed above. The Mak10 protein may also affect assembly, but its precise role is as yet unclear (157).

The existence of a Gag-Pol fusion protein, the ssRNA binding activity of its Pol domain, and the fact that (+) ssRNA is the species encapsidated to form new viral particles suggested a model of assembly and packaging that has been supported by subsequent findings (Fig. 7) (102).

The Pol domain of the Gag-Pol fusion protein recognizes and binds to a packaging site on the viral (+) strands. Then (or concomitantly) the Gag domain of the fusion protein associates with the free Gag protein. This leads to encapsidation of a single (+) strand per particle if there is only one Gag-Pol fusion protein (or one dimer) per particle.

The isolation of full-length cDNA clones of L-A (132), which are capable of maintaining the satellite RNAs M_1 or X in the absence of the L-A virus (264), have made possible experiments to examine the functions of the L-A–encoded proteins. *in vivo* experiments using such a clone have shown that Pol residues 67–213 of the Gag-Pol fusion protein are, as predicted by the model, necessary for packaging of (+) strands but not for assembly of morphologically normal viral particles (100,201). One of the three single-stranded RNA binding domains of Pol is located within this region and is necessary for the packaging (201) (Fig. 4A).

The packaging site recognized is an internal stem-loop sequence with an A residue bulging from the 5′ side of the stem (Fig. 4) (98). This site, located about 400 nt from the 3′ end of the L-A (+) strand, was first identified as necessary for binding of X (+) ssRNA to the opened empty particles (87). Binding requires the stem structure, but not the sequence of the stem. In contrast, the sequence of the loop is important. The protruding A residue must be present and must be an A (98). A similar site similarly located on the M_1 (+) strand was found (98) by examining the predicted folded structure of the M_1 sequence (115), and another was found by study of sequences involved in exclusion of M_1 (217). Either the L-A (X) or M_1 stem loops are sufficient for binding, but addition of 10 bp from the 5′ side of either one improves the binding substantially (98,217). Both the L-A and M_1 sites contain direct repeat sequences whose significance has not yet been determined. A similar site has now also been found in a similar location in the M_{28} satellite dsRNA (M.J. Schmitt, personal communication).

The L-A and M_1 sites can each serve as a portable packaging signal *in vivo*, directing packaging of heterologous transcripts by L-A virus or by proteins produced from the L-A cDNA clone (98,100). The heterologous transcripts were packaged alone in viral particles, confirming the prediction of the headful replication model that a single (+) strand is initially packaged per particle.

The ratio of Gag-Pol fusion protein to Gag can be altered *in vivo* by adjusting the efficiency of ribosomal frameshifting, and, as discussed above, this ratio is critical for M_1 satellite dsRNA propagation. Surprisingly, substantial increases in this ratio appear to be compatible with unaltered propagation of L-A itself (71). Although M_1 or X particle assembly can fail in a Gag-poor cell, an L-A (+) strand could be its own supply of Gag that will not be exhausted until completion of the particle makes the L-A (+) strand no longer available to the ribosomes. This form of *cis*-packaging could explain the differential effect of frameshift efficiency on M_1 and L-A propagation.

Replication [(–) Strand Synthesis]

Newly assembled viral particles contain an L-A (+) strand and are capable of converting it to the dsRNA form when supplied with nucleoside triphosphates (99). The particles formed in this reaction have all the properties of mature L-A particles and can perform the transcription reaction (101).

In order to study the detailed mechanism of the replication reaction, it was necessary to establish a template-dependent *in vitro* reaction. The opened empty particles—when supplied with viral (+) strands, Mg^{2+}, NTPs, and a low concentration of polyethylene glycol—perform (–) strand synthesis to form dsRNA (103). Only L-A, M, or X (+) strand templates are active in this reaction, supporting the notion that it accurately reflects the *in vivo* reaction.

A maximally active template required both sequences at the 3′ end of the L-A (or X) plus strand and internal sequences (the internal replication enhancer) overlapping the packaging signal (87). The 3′ end of L-A (+) strands has a stem loop whose structure is necessary for template activity. Although the sequence of the loop and that of the 3′ terminal 4 nt are important, that of the stem is not. Despite the requirements for these structures and sequences in the context of L-A, the 3′ terminal 33 nucleotides of M_1's (+) strand can substitute for the L-A 3′ end, although there is little or no similarity between the two sequences (Fig. 4) (87). The internal replication enhancer and the 3′ end site must be bound together for optimal template activity in the replication reaction (105). However, they need not be covalently attached and can simply be hydrogen bonded. This suggests that the RNA polymerase binds first to the internal site and is thus brought close to the 3′ end site where polymerization is to begin.

Because the RNA-dependent RNA polymerase consensus domains defined by Kamer and Argos (140) have since been found in almost all (+) ssRNA and dsRNA viruses examined, the detailed mutagenesis of the most highly conserved of these regions was performed using the L-A cDNA clone (202). This has defined the extent of the domains necessary for the propagation of the M_1 satellite dsRNA. However, the precise function of these domains in the polymerization reactions remains unknown. Interestingly, homologous regions from reovirus or Sindbis virus RNA polymerases could partially substitute for that of L-A (202).

Host Genes Affecting Virus Propagation

The host *SKI* genes, which repress the L-A, L-BC, and 20S RNA replicons by partially blocking translation of cap+ poly(A)- viral mRNA, were discussed above, as were the *MAK3*-encoded *N*-acetyltransferase and the *MAK10* gene needed for both L-A and M propagation.

Mitochondrial Proteins (NUC1, POR1) that Repress L-A

Mutation of either *por1,* encoding the major mitochondrial outer membrane porin, or *nuc1,* encoding the major mitochondrial nuclease, result in derepression of L-A copy number (68,166). It has been suggested (68) that the elevated amounts of L-A virus may facilitate growth of porin-deficient cells. Neither mitochondrial function nor the mitochondrial genome is necessary for L-A or M propagation or expression. However, the absence of the mitochondrial genome largely suppresses a *mak10* mutation (259).

MAK Genes Needed for M, but not L-A

Mutations of nearly 30 chromosomal genes (called *MAK* for *ma*intenance of *k*iller) can result in the loss of M_1 dsRNA from a wild-type host. These genes include *MAK1/TOP1,* encoding DNA topoisomerase I (237); *MAK8/TCM1* and *MAK7/RPL4A,* encoding ribosomal proteins L3 and L4 (see above); *MAK11,* encoding an essential membrane-associated protein with the repeat patterns first identified in β-transducin (131); *SPE2,* encoding adenosylmethionine decarboxylase [spermidine biosynthesis (243)]; and *MAK16,* encoding a nuclear protein with a cell cycle control function (260). Except for mutations in *mak16, mak3, mak10,* and *pet18,* these mutations are all suppressed by *ski* mutations (239). Many are also suppressed by the L-A-HNB natural variant (244).

Viruses Reducing Virulence of *Cryphonectria parasitica,* the Chestnut Blight Fungus

Cryphonectria parasitica infects chestnut trees and kills the upper part of the tree. Strains of *C. parasitica* that have largely lost their pathogenicity were found first in Italy and later in Michigan (Fig. 1). These hypovirulent strains also show decreased asexual spore formation, decreased production of laccase (a phenol oxidase possibly involved in pathogenesis), and reduction in pigment formation (190).

The cytoplasmically inherited factor that reduces the pathogenicity of the fungus, and produces these hypovirulence-associated traits was identified as any of several dsRNA replicons found in the hypovirulent strains and not in the virulent strains (7,64). Curing the dsRNAs by growth in the presence of cycloheximide was accompanied by the return of virulence (106). Hybridization studies have shown that there are a variety of apparently unrelated dsRNA genomes, any one of which is capable of producing hypovirulence in the host, although the associated traits can vary (81,153).

Genome Structure

The L-dsRNA of hypovirulent *C. parasitica* strain EP713 is a 12,712-bp molecule whose (+) strand has two long ORFs, ORFA and ORFB (Fig. 8) (215). Each encodes a papainlike cysteine protease in its *N*-terminal portion that self-cleaves the primary translation products at least once, as shown in Fig. 8 (49,216). The presence of related papainlike proteases and helicase domains, as well as RNA-dependent RNA polymerase motifs, has led to the suggestion that L-dsRNA of *C. parasitica* is related to the potyviruses, a group of (+) ssRNA viruses of plants (152).

ORFA and ORFB overlap by a single nucleotide, with the UAA termination codon of the first overlapping with the AUG codon of the second (215). The mechanism by which ORFB is expressed has not yet been defined, but this structure implies that the mechanism will not involve ribosomal frameshifting or termination codon readthrough. Rather, reinitiation is the most likely mechanism. The translation of ORFA also poses an interesting problem because the 5′ noncoding region contains six short ORFs (215). Whether ribosomes initiate internally, as in the case of picornaviruses (see Chapter 16), or use these tiny ORFs for regulatory purposes, as in the GCN4 gene of *S. cerevisiae* (182), is not yet clear.

Virus Replication in Intracellular Vesicles

Unlike other mycoviruses, most of the *C. parasitica* dsRNAs described to date are not associated with virus parti-

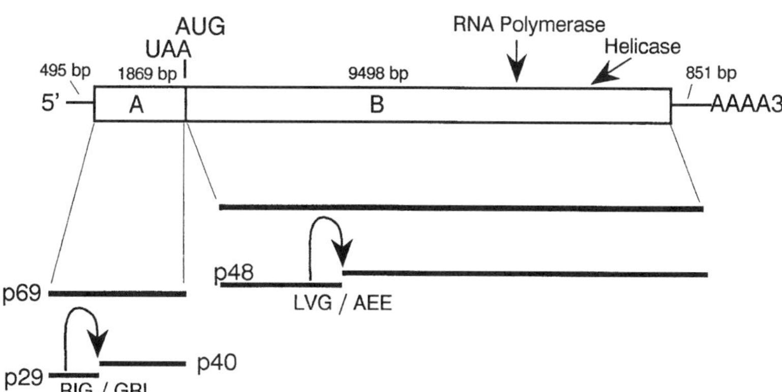

FIG. 8. Coding information and protein processing of the *Cryphonectria parasitica* virus, L. The sites of action of the p29 and p48 proteases are indicated. This virus is now designated CHV1-713, the type member of the Hypoviridae. Adapted from Nuss (190), with permission.

cles. Rather, they are found in intracellular vesicles (73,124). These vesicles have an RNA polymerase activity that produces both ssRNA and dsRNA (90). Most of the label is incorporated into viral (+) strands (90). This is reminiscent of the membrane association of *in vivo* RNA synthesis of many (+) strand RNA viruses, supporting the notion that this virus is not closely related to other fungal viruses.

Virus Induction of Hypovirulence

The L-dsRNA has very specific effects on certain genes, one of which is the laccase gene encoding a phenol oxidase believed to be involved in pathogenesis. Transcription of the laccase gene is repressed by L-dsRNA infection (46,204), an effect that can be produced by expression of just the p29 protease encoded by part of ORFA (48). Although p29 is sufficient to alter fungal phenotypes, it is neither necessary nor sufficient to cause hypovirulence (60). Deletion of p29 from the virus results in decreased induction of the hypovirulence-associated traits but no decrease in hypovirulence itself. Moreover, it is not the protease activity of p29 that produces hypovirulence (60). The effect of L-dsRNA on laccase appears to be transmitted by an influence on the inositol triphosphate–calcium signal transduction system of the fungus, possibly explaining the multiple phenotypic effects of hypovirulence (156). Cryparin, a hydrophobic cell surface protein, is also reduced in hypovirulent strains, an effect mediated at transcription (282). Another gene, *Vir2,* was isolated based on its decreased transcription in a hypovirulent strain. Although no effect was seen on virulence itself, deletion of this gene partially mimics some of the hypovirulent-associated traits with decreased asexual sporulation and fruiting body formation and impaired sexual crossing ability (281). The hypovirulence-associated traits are thus due to specific effects of viral gene products rather than to the presence of a replicating dsRNA.

Infectious cDNA Clones and Biological Control of Chestnut Blight

Introduction of complete cDNA clones of L-dsRNA under control of the *C. parasitica* glyceraldehyde-3-phosphate dehydrogenase promoter resulted in both a complete hypovirulence phenotype of the fungus and the launching of the RNA replicon in a form transmissible to other strains (47). In addition to its usefulness in studying the mechanism of the effects of L-dsRNA on the cell, this infectious cDNA clone method is an important advance in attempts to control chestnut blight. Although artificial innoculation of trees with hypovirulent fungal strains in Europe resulted in spread of the L-dsRNAs to virulent strains and control of the blight, this approach did not succeed in the United States. L-dsRNA spreads by fusion of the growing fungal cells with other cells, a process called hyphal anastomosis, but not by sexual crosses (5). Hyphal anastomosis requires that strains have identical alleles at several different loci, determining compatibility (6). The number of compatibility groups is apparently much greater in the United States, limiting the spread of the hypovirulence dsRNA.

In contrast, infectious cDNA incorporated into the *C. parasitica* genome, while generating dsRNA replicons, will also naturally spread to other mating and vegetative compatibility groups through sexual transmission (47,190). It is thus expected to be more effective in the biological control of chestnut blight than the natural L virus.

A Reovirus of Cryphonectria

Hillman et al. (82) have isolated two hypovirulent strains of *C. parasitica* in West Virginia carrying reoviruslike elements. Strains C18 and 9B21 each have 11 dsRNA segments ranging in size from about 1 to about 3 kb and associated with 60-nm virus particles, unlike the vesicles in which the potyviruslike Hypoviridae are found. These 11 segments are present in equimolar amounts, and transmission studies show that either all or none of the segments are transmitted, suggesting that they are parts of a single viral genome. Blotting experiments suggest that the C18 and 9B21 genomes are not closely related, but both apparently cause hypovirulence. This system promises to add new dimensions to both the study of hypovirulence and to the Reoviridae.

Mitochondrial NB631 dsRNA

Most of the hypovirulence-inducing potyvirus-like dsRNAs are 10 to 13 kbp and are located in the cytoplasm. However, in *C. parasitica* strain NB631, a small 2.7-kbp dsRNA has been found that is localized in the mitochondria (B. Hillman, personal communication). This dsRNA has a single ORF with multiple UGA codons (read as tryptophan in mitochondria) but no UAA or UAG terminators. The dsRNA also has been shown by fractionation experiments to be associated with the mitochondria. This NB631 dsRNA is associated with hypovirulence, as for the larger, more common Hypoviridae.

The sequence of NB631 dsRNA (B. Hillman, personal communication) predicts a single ORF (assuming UGA = Trp) encoding an RNA-dependent RNA polymeraselike protein with striking similarity to those of the 20S and 23S ssRNA replicons of *S. cerevisiae* (see below).

Leishmania dsRNA Viruses

Leishmania is a flagellated protozoan that causes cutaneous, visceral, or mucosal infections in humans. *Leish-*

mania virus (LRV1-1) was first discovered as an RNA species associated with cytoplasmic 32-nm virus particles (231). A similar virus (LRV1-4) was found in another strain and shown to be associated with an RNA polymerase activity (267). A survey of 71 *Leishmania* isolates showed virus present in 12, all of which were *Leishmania braziliensis* or *Leishmania guyanensis* originating in the Amazon basin (121). These species cause cutaneous and mucocutaneous forms of leishmaniasis, but not all such isolates carry the virus, nor is there evidence that the virus affects pathogenicity. All of the LRV isolates were related as judged by hybridization experiments.

The replication cycle of LRV appears to be similar to that of the L-A virus of *S. cerevisiae* (256). The structure of LRV likewise closely resembles L-A (227). The two major ORFs, ORF2 and ORF3, overlap by 71 nt, with ORF3 in the +1 frame relative to ORF2 (227) (Fig. 9). ORF2 encodes the major coat protein (37), and ORF3 has the consensus patterns typical of viral RNA-dependent RNA polymerases (140). This suggests that a Gag-Pol fusion protein might be made (227), and an *in vitro* translation product of the predicted size (37) is a candidate for this protein. The sequence of the overlap region suggests a possible Ty1-like +1 shift site or a pseudoknot-promoted hop (227). The similarity of the LRV1 Pol amino acid sequence with that of L-A of *S. cerevisiae* is striking and is far beyond that due to their simply sharing the RNA polymerase consensus domains (227).

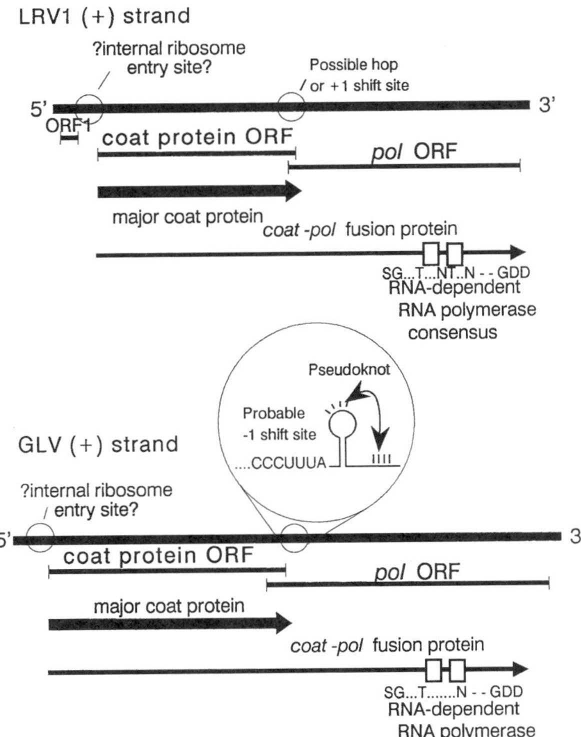

FIG. 9. Genome structure and expression of *Leishmania* virus [LRV1 (227)] and *Giardia* virus [GLV (253)]. Several similarities exist among these viruses and compared with the *Saccharomyces* virus (L-A), which has led to their classification together as totiviruses.

A small ORF of 228 nt (ORF1) is present upstream of ORF2 in LRV1-1 (227), and two such small upstream ORFs are found in the closely related LRV1-4 (211). The absence of the spliced leader sequence common to all *Leishmania* cellular mRNAs suggests that this upstream region may be untranslated.

Giardia Lamblia Virus

The *Giardia lamblia* virus (GLV) is an infectious Totivirus, with a single-segment 6.1 kb double-stranded RNA genome (250). Its host, *G. lamblia,* is a primitive eukaryote with two nuclei but no mitochondria, Golgi, or endoplasmic reticulum. This protozoan is a flagellated intestinal parasite of mammals with a wide distribution.

Unlike other known viruses of fungi or parasites discussed here, GLV is found in the culture supernatant and readily infects *G. lamblia* cells. Infected cells contain 36 nm isometric particles composed of a major 100-kd protein and a minor 190-kd protein (178). Purified viral (+) ssRNA can be reintroduced into a virus-free strain by electroporation, and a new infection can be initiated in this way (107). Because the entire GLV genome has been cloned and sequenced, it should be possible to manipulate this system easily.

The structure of the GLV genome (Fig. 9) (253) has several features in common with other Totiviridae and with certain animal viruses. There are two major ORFs. The 5′ ORF encodes the 100-kd major coat protein. The 3′ ORF has the typical RNA-dependent RNA polymerase motifs and is expressed only as a fusion protein with the 5′ ORF (253). Thus, expression in GLV resembles that of retroviruses, the *Saccharomyces* L-A virus, and the *Leishmania* dsRNA virus LRV-1 in forming a Gag-Pol fusion protein. The structure of the 210-nt overlap region of the Gag and Pol ORFs strongly suggests that the mechanism of fusion protein formation is by the same −1 ribosomal frameshift as used by many animal retroviruses and the yeast L-A dsRNA virus. A slippery site C CCU UUA (Gag frame shown) is followed closely by a predicted pseudoknot, as has been shown to be essential in these systems and in coronaviruses for programmed frameshifting.

The 5′ noncoding region of the GLV genome has seven AUG codons, each followed by a short reading frame (253). This feature appears to be common among the *Cryphonectria* L hypovirulence virus, LRV1 and GLV.

The possibility of using an RNA virus, combined with cDNA clones and an RNA transfection system to develop an expression system for this interesting organism has been one of the prime motivations for this work (252).

Trichomonas Virus and Host Phenotypic Variation

Many strains of *Trichomonas vaginalis* carry 33-nm viral particles composed of an 85-kd major coat protein and containing 5.5 kb dsRNA (251). This 5.5-kb dsRNA in fact consists of three distinct dsRNA species that do not cross-hybridize (143). Remarkably, the presence of the virus is correlated with a phenotypic variation of the host cells (249). Expression of a major surface antigen switches on and off in cells carrying the virus, but is always off in virus-free strains. Cells often lose the virus on serial passage, and loss of the virus is always accompanied by loss of phenotypic variation (249).

Retroviruses (Retroelements): Classification

Retroviruses, Retrotransposons, Retroposons, and Retrointrons

Retroelements (Table 3) all share their use of reverse transcriptase in their propagation. The retrotransposons of fungi and parasitic microorganisms resemble mammalian retroviruses in all essentials except for their lacking an env (envelope) gene and probably for this reason are restricted to propagation without leaving the intracellular environment. Nonetheless, the frequency with which these cells mate in nature is so high that most of these elements are widely distributed in their respective species. The retroposons are one step further removed in that they lack the LTR structure, and so their transposition process must differ in some details from those of the other groups. These elements resemble the mammalian LINE elements but have the advantage of readily detectable transposition and facile host genetics for their study. For example, Tad, of *Neurospora,* can retrotranspose between nuclei of a heterokaryon (144,145).

A group of reverse transcriptase–encoding introns in mitochondrial DNA has also been identified (142). These are introns aI1 and aI2 of the COX1 (cytochrome oxidase subunit I) gene. These introns are capable of retrotransposition into both their normal location and into heterologous locations, both *in vitro* (142) and *in vivo* (181). These retroelements lack LTRs and have only a single ORF. Their properties suggest that introns may have begun as parasitic elements.

The retrotransposons are also divided based on amino acid sequence homologies and gene order in the Pol domain into the *copia*- like and the *gypsy*-like elements. *Copia* and *gypsy* are retroelements of *Drosophila.* Because Ty3 is more similar to *gypsy* than to Ty1 (125,276), it is likely that Ty3 and Ty1 entered the yeast genomes at different times and thus that horizontal transfer of these elements does occur.

TABLE 3. *Groups of retroelements*

Element	env	LTRs	RT	Examples
Retrovirus	+	+	+	RSV, HIV,
Retrotransposon	–	+	+	Ty1-5
Retroposon	–	–	+	LINEs, Tad
Retrointron	–	–	+	Intron aI1 of COX1

Structure of Tys and Other Retroelements

The Ty elements of *S. cerevisiae* each have LTRs of 245–371 bp separated by a unique region (called ϵ) of 4.7 to 5.6 kb (51,125,229,248,255) (Fig. 10). The major Ty RNA transcript begins within the 5′ LTR (at base 241 from the 5′ end in Ty1) and ends at base 289 of the 3′ LTR, 45 bp from the 3′ end of the element (80) (Fig. 11). This provides the basis for the conventional division of retroviral LTRs into the U3 region (present only at the 3′ end of the Ty RNA, but located at the 5′ end of the LTR), the R region (repeated at both ends of Ty RNA), and the U5 region (present only at the 5′ end of Ty RNA, but at the 3′ end of the LTR).

Ty1–4 each have two overlapping ORFs: TYA, corresponding to Gag, and TYB, homologous to Pol of mammalian retroviruses. Like mammalian retroviruses and *gypsy,* Ty3 has the gene order protease (PR)–reverse transcriptase (RT)–RNAse H (RH)–integrase (IN) in the Pol ORF (125), whereas Ty1 and Ty2 have the *copia* order PR–IN–RT–RH (51,255).

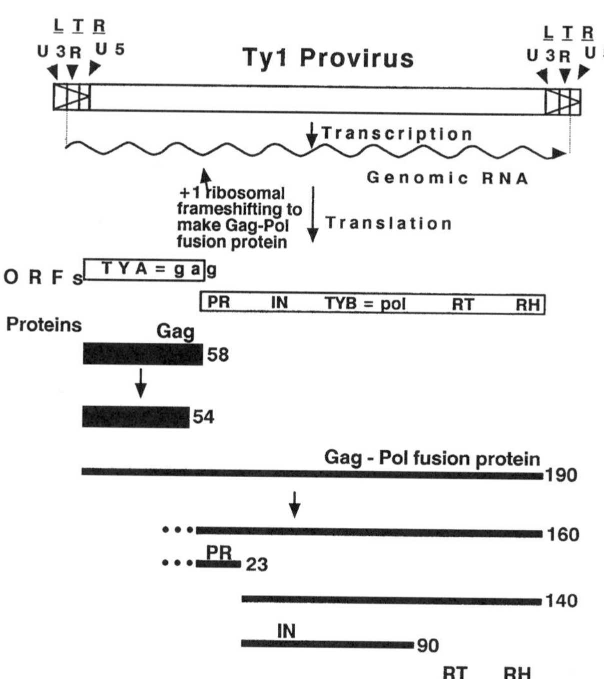

FIG. 10. Genome structure and expression of Ty1. Adapted from Garfinkel et al. (113), with permission.

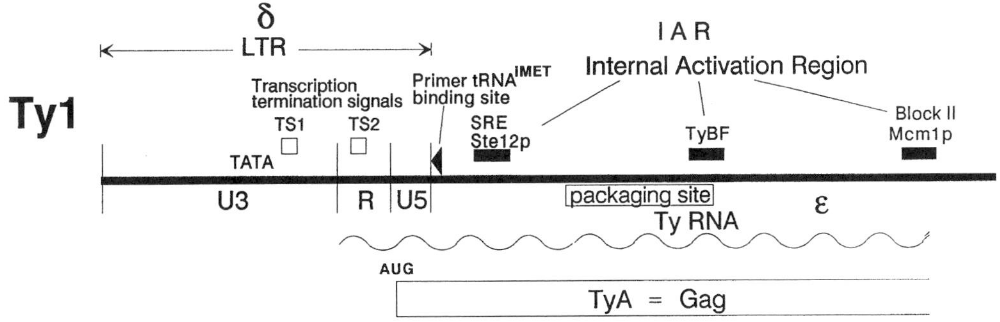

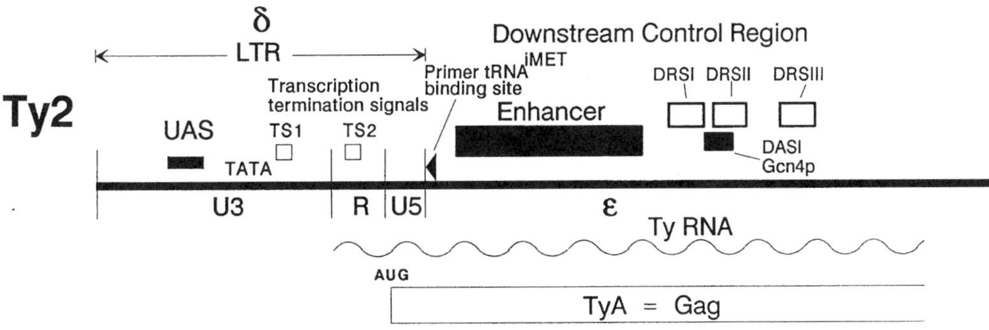

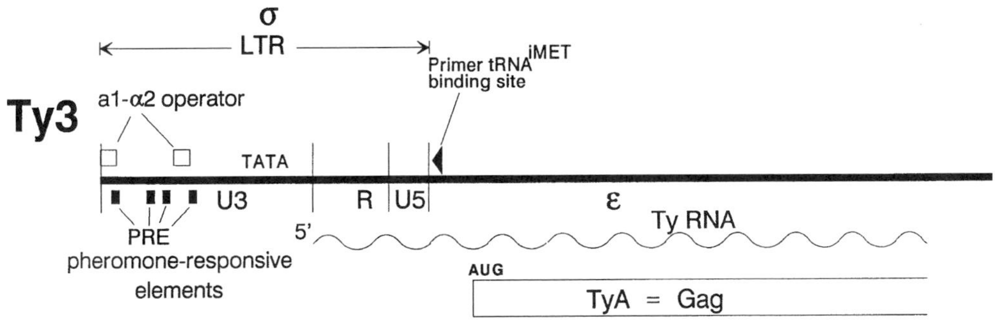

FIG. 11. Structure and functional regions of the Ty1, Ty2, and Ty3 long terminal repeats (LTRs). Although the LTRs of Ty1 and Ty2 are similar, the control sequences within ε, the unique regions, are different. SRE, Ste-responsive element; TyBF, Ty binding factor site; Block II, Mcm1p binding site. Ty3's control sequences are apparently all within the LTR, its primer binding site is two bases from the LTR, and its Gag protein initiation codon is within the unique region, not in the LTRs. DRS, downstream repression site; DAS, downstream activation site.

Replication Cycle of S. cerevisiae *Ty Elements*

The Ty replication cycle resembles that of mammalian retroviruses except that it begins and ends with the integrated form of the genome (Fig. 12). Ty transcripts made with RNA polymerase II are translated in the cytoplasm to make the Gag (TYA) and Gag-Pol fusion (TYA-TYB) proteins. These proteins assemble, packaging the Ty RNA to make particles that are analogous (and perhaps homologous) to core particles of retroviruses. The reverse transcriptase and RNAse H make a dsDNA copy of the genome, and integrase inserts this into the genome, producing a short repeat of the chromosomal integration site.

Reverse Transcription

The reverse transcriptase and RNAse H–homologous domains are present in the Ty1 particles as a 60-kd protein that is produced by proteolytic processing of the 190-kd Gag-Pol fusion protein (Fig. 10) (113). Isolated Ty1 particles have reverse transcriptase that is active on either the endogenous Ty1 RNA or externally added templates

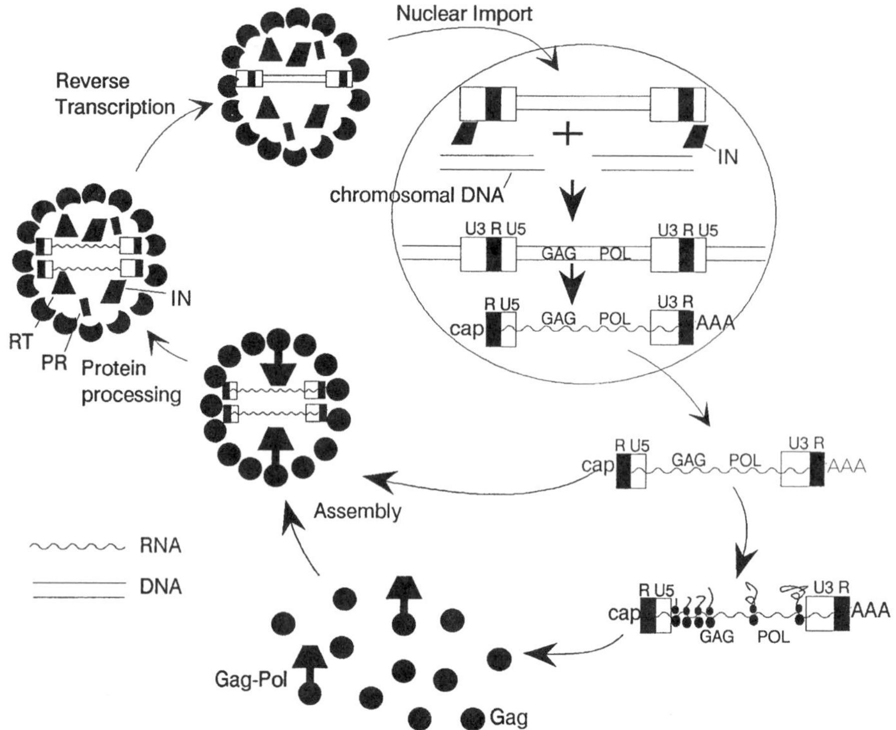

FIG. 12. The Ty replication cycle. It is likely that all retrotransposons follow this cycle.

(112,176). The Ty particles are also open to externally added enzymes, indicating that they are not impermeable shells, but have pores. This porous structure may help explain how pseudogene formation can occur (65).

The mechanism of reverse transcription of Ty RNA is believed to be the same as that for mammalian retroviruses (Fig. 13) (118), and several aspects have now been examined in detail. The primer for reverse transcription is $tRNA^{Met}_i$, with 10 nucleotides of the 3′ acceptor stem complementary to a site on the Ty1 (+) strand just 3′ to the 5′ LTR called the (–) primer binding site (–PBS) (44). This tRNA is specifically packaged in viral particles (44). Interestingly, although the primer function of the initiator $tRNA^{Met}$ depends on its complementarity with the –PBS, its packaging is independent of this complementarity (44). This indicates that the $tRNA^{Met}_i$ is recognized by some other component of the Ty virus particle and that it is probably not the acceptor stem of the tRNA that is recognized.

As predicted from the model of Gilboa et al. (118), information can be transferred from the 5′ LTR to the 3′ LTR and the reverse (19,183) (Fig. 13).

Integration (Fig. 12)

Ty3 shows a tight specificity for target sites. It integrates 16 or 17 bp upstream of the 5′ ends of tRNA coding regions (29,40,41). The DRE element of *Dictyostelium discoideum* and the spliced leader–specific elements SLACS, CZAR, and CRE1 of Trypanosomids have a similar integration target specificity (4,108,169). Other genes transcribed by RNA polymerase III, such as 5S ribosomal RNA and U6 small nuclear RNA, are targets for Ty3, with the same location specificity relative to the start site of transcription (41). However, no common consensus sequence could be deduced for the insertion sites, suggesting that the integration apparatus was recognizing some aspect of chromatin structure or the RNA polymerase III transcription apparatus itself at these sites. Substitution of four purines with pyrimidines at the site of transcription initiation without moving the promoter did not change the site of Ty3 insertion. Eliminating transcription by destroying the promoter (box B inside the tRNA gene) eliminated target activity. However, leaving the original transcription initiation site and Ty3 insertion site intact and moving box A of the promoter to a new site resulted in moving the site of transcription initiation and changing the Ty3 insertion site to a comparable position (41). This indicates that the integration apparatus recognizes the transcription apparatus but not the transcription initiation site or the promoter itself.

Ty1 is capable of insertion at many different sites, producing a 5-bp duplication of the target DNA (91,109). Although Ty1 does not recognize special DNA sequences, there is specificity when insertion sites are examined on a larger scale. Insertions tend to be in the 5′ part of the *URA3, LYS2,* and *CAN1* genes (75,158,185) with control regions targeted far more often than the ORFs.

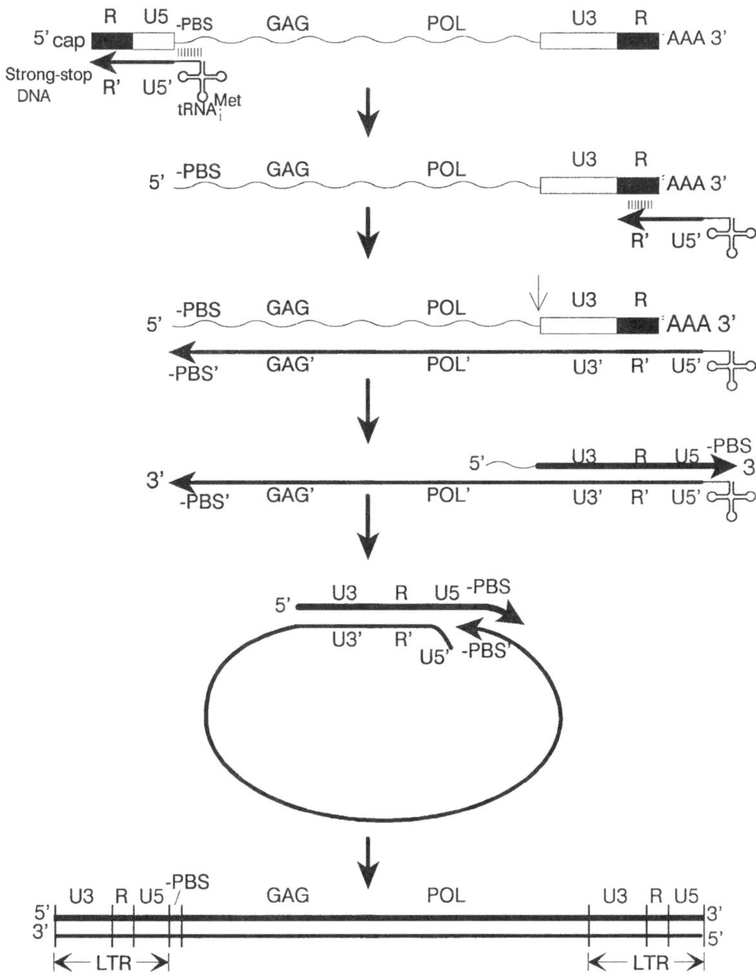

FIG. 13. Model of reverse transcription as applied to Ty1 Adapted from Coffin (53) with permission.

Strikingly, mutation of the cellular *rad6* gene, encoding a ubiquitin-conjugating enzyme, increases the frequency of Ty transposition at either *URA3* or *CAN1* (194). This effect was not due to altered Ty transcript levels and was seen even when retrotransposition from a GAL1-promoted Ty1 was studied. The *rad6* mutation is known to alter chromatin structure due to a failure to ubiquitinate histones, suggesting that its effect on Ty1 transposition is on the nature of the target. Indeed, Ty1's target-site specificity for gene control regions is apparently eliminated by the *rad6* mutation (158).

On an even larger scale, Ty1 is seen to target regions near tRNA genes or previously existing LTRs (137,191). The factors determining this specificity have not yet been defined, but they are presumed to be some feature of chromatin structure as in the case of the *rad6* mutants. If these are regions where a lethal hit is less likely, then Ty's integration apparatus has certainly been selected to detect such regions (59).

An *in vitro* integration system similar to that developed for Moloney murine leukemia virus (30) has been established for Ty1 using Ty1 viral particles produced from an element carrying a copy of the *E. coli supF* gene and, as target, λ DNA from a multiple amber mutant suppressible by *supF* (76,77). Using this system, it has been shown that linear dsDNA substrates carrying the terminal 12 bp at each end of the LTR are sufficient for the integration reaction to occur at normal efficiency (77). There are also no nucleotide requirements. The substrate DNA must have 3′ hydroxyls, suggesting that the reaction involves covalent attachment of these 3′ ends to the target DNA. Unlike mammalian retroviruses, Ty1's IN does not remove two terminal bases in the process of integration (77), probably because in Ty1 the –PBS is immediately adjacent to the U5 part of the LTR. Ty3, however, resembles mammalian retroviruses in this regard (S. Sandmeyer, personal communication).

In fact, the purified IN protein is capable of performing an integration model reaction without other components of the viral particle (180). This suggests that it may be sufficient for IN and the reverse transcript to enter the nucleus for integration.

Expression

Control of Transcription. Ty1 transcription is controlled by MAT, the yeast mating type locus. MATa/MATα diploid cells have 20-fold lower Ty1 transcript levels than do MATa or MATα haploid cells or MATa/MATa or MATα/MATα diploids (79). Why does Ty1 choose to be so regulated? In nature, the haploid phase (either the MATa or MATα sex) is transient and soon is followed by mating to form MATa/MATα diploid cells. Could Ty1 be preparing to transpose in anticipation of encountering a new genome potentially free of Ty1? This would make Ty1 infection a sort of sexually transmitted disease. Arguing against this idea is the posttranscriptional blockage of Ty1 transposition by treatment of cells with mating pheromones, a step that immediately precedes mating itself (279).

DNA damage, induced by ultraviolet irradiation or 4-nitroquinoline-1-oxide, also induces Ty1 transposition by inducing transcription of the element (24). Is this effect adaptive for Ty1 as a first step in finding a new home, or is it a consequence of a failure of the host anti-Ty system?

Retroelements all must cajole the cellular RNA polymerase II into transcribing their proviral form to make viral RNA. However, they are constrained to place the controlling sequences inside the limits of the element. Although most mammalian retroviruses have enhancers in their LTRs upstream of the start site of transcription, Ty1 and Ty2 elements have major transcriptional control sequences downstream of both the transcript start site and the translation start site, most of them inside the unique region (ϵ) (Fig. 11). Ty1 has two major downstream sites responsible for its haploid-specific transcription: the sterile-responsive element (SRE) binds Ste12p while another site (block II or PRTF) binds Mcm1p (54,83) (Fig. 11). Ty2 has one upstream activation site (165) and at least two downstream activation regions, one of which (DAS1) responds to Gcn4p (92,242). Ty2 also has several downstream repression sites (DRS) (93).

Ty3 transcription is, like that of Ty1, much higher in haploid cells than in diploids. Moreover, its transcription and transposition are derepressed by exposure of cells to the mating pheromones (52,245). In fact, transposition is induced in mating cells, and Ty3 transposing from the genome of one mating partner to that of the other occurs at high rates (146). Ty3 transcription is induced before mating and repressed after mating by mating type control, but the particles synthesized before mating are sufficient to give a burst of transposition from one genome into the other (146). Here the interpretation of preparation to transpose into the potentially Ty3-free genome of the mating partner seems clear.

In *Saccharomyces* cells, the MATa1-MATα2 protein complex (encoded by the mating type locus) represses haploid-specific cellular genes. The regulation of Ty3 transcription involves two a1-α2 consensus recognition sites, determining mating type repression, and three PRE sequences, determining *p*heromone *r*esponsive *e*xpression, all located within the 5′ LTR of the Ty3 element (15,245). An upstream repression sequence (URS) was located at the junction of the 5′ LTR and the adjacent tRNA gene (15). Thus, in contrast to Ty1 and Ty2, most of whose control regions are located within the unique region downstream of the transcription and translation start sites, Ty3's transcription control regions are largely upstream within the U3 region of the LTR (15).

Effect of Ty Insertion on Cellular Genes. Insertion of Ty elements into the control regions of cellular genes can activate, inactivate, or alter the control of the target gene (20,84,271). The Ty insertions often move the normal regulatory sequences 5 kb away from the target gene, eliminating the normal regulation. Insertions of Ty whose 5′ end is close to the 5′ end of the target gene impose the Ty transcriptional control on the target gene. This produces a divergent transcription of the Ty element and the target gene, but both are under mating type control because of the effect of the Ty control region.

Chromosomal Genes Regulating Ty Transcription. Mutant cellular genes that have come under control of Ty1 have been used extensively to investigate the cellular genes affecting the transcription of Ty1 itself (50,272,273). Second site mutations (suppressors) that restore the normal expression of the target genes have defined a large group of genes, called *SPT,* that include the TATA binding factor TFIID, the genes encoding histones, and many general transcription factors with effects on many genes (e.g., *SNF2, SNF5, SNF6, GAL11,* and *SIN1*). In addition, the control of the target genes by mating type and mating pheromones have led to studies of effects of the mating type and pheromone control pathway genes on Ty itself. Thus for example, *STE12* controls Ty1 transcription, as do the genes upstream of *STE12* in the mating type control kinase cascade (84,85).

+1 Ribosomal Frameshifting. Like mammalian retroviruses, Ty elements direct the synthesis of a Gag protein and a Gag-Pol fusion protein (Fig. 10). For reasons that are not yet clear, each of the *Saccharomyces* Ty elements use +1 ribosomal frameshifting, whereas animal retroviruses all use −1 frameshifting (or readthrough of a terminator) to make Gag-Pol (8,128). As shown by studies of the L-A dsRNA virus, yeast can perform −1 ribosomal frameshifts by the same simultaneous slippage mechanism used by retroviruses.

The mechanism of the +1 frameshifts in Ty1 or Ty2 (Fig. 14) and Ty3 each involve the combination of starvation for a rare tRNA and an unusual tRNA able to perform the frameshift (13,94). In the case of Ty1, the slippery site on the mRNA is CUU-AGG-C. The ribosomes are slowed by the AGG codon at the A ribosomal site. This codon is recognized by a $tRNA^{Arg}$ that is present in low abundance in yeast, so it is called a "hungry codon." While the A site is waiting for the AGG codon to be occupied, the $tRNA^{Leu}$ located at the P site has as its anticodon UAG, and is capable of pairing with either the 0 frame CUU codon or the +1 frame codon UUA. When this $tRNA^{Leu}$ slips into the +1 frame, the GGC codon can be easily recognized by an abundant $tRNA^{Gly}$ species, and the ribosomes then continue in the +1 *pol* reading frame to make Gag-Pol fusion protein (13).

The efficiency of the Ty1 +1 ribosomal frameshift depends on the scarcity of the $tRNA^{Arg}$ recognizing the AGG codon.

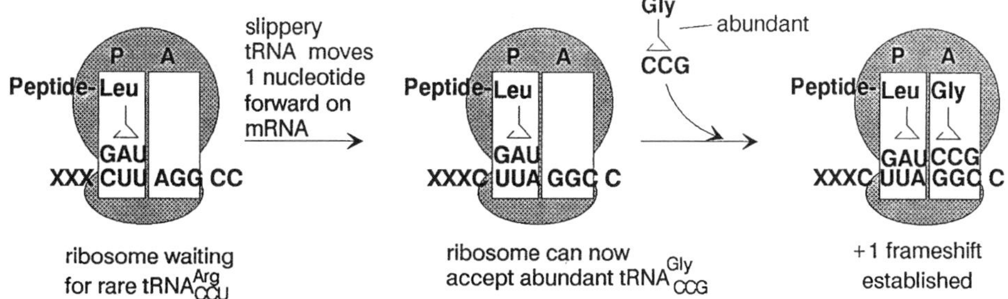

FIG. 14. Mechanism postulated for +1 ribosomal frameshifting of Ty1. Adapted from from Belcourt and Farabaugh (13), with permission.

Thus, artificially oversupplying this tRNA lowers frameshift efficiency (13) and also lowers transposition frequency (277). Likewise, deletion of the gene for this $tRNA^{Arg}$ increases the efficiency of frameshifting and lowers the frequency of transposition (141). Starvation of cells for spermidine elevates the efficiency of Ty1 +1 ribosomal frameshifting and results in decreased transposition efficiency (10). Like the similar experiments performed with the L-A dsRNA virus of *S. cerevisiae,* these results suggest that drugs affecting frameshifting efficiency might be useful as antivirals.

The Ty3 frameshift site is GCG-AGU-U, and, like that of Ty1, is based on a hungry codon in the ribosomal A site, namely AGU (94). AGU is recognized by a low-abundance $tRNA^{ser}$, and so the ribosome pauses at this point. But the $tRNA^{Ala}{}_{CGC}$ that decodes the GCG codon in the P site cannot slip +1 and re-pair. The mechanism is not yet clear, but appears to be different from other known programmed shifts.

Proteolytic Processing and Phosphorylation. Ty1 Gag is expressed as a primary translation product of about 58 kd, most of which is processed by the Ty1 protease to form the 54-kd major particle protein (2). Gag is phosphorylated (177), and this phosphorylation is increased concomitant with the inhibition of transposition that occurs when cells are treated with mating pheromone (279). After Pol is synthesized as a 190-kd Gag-Pol fusion protein, it is processed through several intermediates to form a 23-kd protease, the 90-kd integrase, and the 60-kd reverse transcriptase RNAse H (113) (Fig. 10).

Packaging: RNA Sites and Protein Requirements

The Ty1 RNA site determining packaging has been localized to within a 381 nt region between nt 239 and 620 (278) (Fig. 11). The RNA structure recognized and the parts of TYA or TYA-TYB proteins that recognize this region have not yet been determined.

Debranching Enzyme and Ty1

Mutation of *DBR1,* encoding the enzyme that debranches the lariat structure produced by intron excision by cleaving the 2′-5′ linkage at the branch point, results in a nine-fold decrease in Ty1 transposition efficiency (43). However, the explanation of this unexpected finding remains unclear.

Processing of TYB (Pol) Limits Transposition Efficiency

Although normal cells have about 35 chromosomal copies of Ty1, most of which are probably transposition competent (62), transposition is a relatively rare event. Most studies of the mechanism of transposition have used a high copy plasmid with a GAL1-promoted Ty1 carrying a marker (such as *HIS3*) included in the Ty1 to facilitate detection of transposition (19). When Ty1 transposition is induced with such a plasmid, although Ty1 RNA is only increased a few fold over that derived from the normal chromosomal Ty1 copies, the frequency of transposition increases about 100-fold (61). Detailed studies of the point at which transposition is blocked in uninduced cells indicate that processing of viral proteins, particularly the PR and IN proteins encoded by TYB, is limiting (61). This block might represent the effect of a cellular antiviral system, analogous to the *SKI* system that limits dsRNA and ssRNA replicon propagation (see above), or inhibition by defective Ty proteins having a dominant negative effect. Either such block might be overwhelmed by overproduction of Ty proteins from the plasmid (61).

Schizosaccharomyces pombe *Retroelements*

The Tf1 and Tf2 elements of *S. pombe* are unusual in that a single ORF encodes both Gag and Pol (163). From the primary 140-kd translation product, the viral protease cleaves the proteins that form the viral particles (164). In view of the strict requirement for the ratio of Gag to Gag-Pol in Ty1 (141,277) and M_1 (70), the assembly process in Tf1 and Tf2 must be significantly different.

Summary of Retroelements

The mating type and pheromone control of Ty transcription are clearly adapted to maximize transposition activity at the time when a potentially unoccupied genome becomes available (mating) and minimize potential damage to the host, whose health is indispensable for survival. There is also a posttranslational mechanism, probably cellular, that limits transposition.

Among the remaining interesting questions about Ty elements are: Why do all Tys use +1 frameshifting but mammalian retroviruses use –1 frameshifting or termination readthrough to make their Gag-Pol fusion proteins? How does RNA polymerase II know to stop in the 3′ LTR, but keep going in the identical sequence in the 5′ LTR? What does Dbr1, the debranching enzyme, do for Ty1?

ssRNA Replicons

20S RNA

In 1971, a stable species of RNA, intermediate in size between 18S and 25S rRNAs, dubbed 20S RNA, was found to appear specifically in cells exposed to the condition used to induce meiosis and sporulation, namely supplying acetate as the carbon source in the absence of a nitrogen source (139). The ability to produce 20S RNA was then found to be inherited as a non-Mendelian genetic element, distinct from other known elements (114). It was not connected with meiosis, except that the same culture conditions are used to induce both (114). 20S RNA was finally proven to be an independent RNA replicon whose copy number is inducible in acetate (171). Its sequence shows that its 2,500 nucleotides encode a single 95-kd protein with some similarities to the RNA-dependent RNA polymerases of RNA phages and RNA viruses (173,205). W dsRNA, a minor species inducible by growth at high temperature (258), proved to be the replicative form of 20S RNA (173,205). Electron micrographs of purified 20S RNA showed about 50% circular molecules (171), but biochemical experiments indicate that the RNA itself is not circular (206). The explanation of these findings remains unclear.

20S RNA is an apparently naked cytoplasmic replicon (268). The 95-kd RNA polymerase–like molecule has been detected in cells (W.R. Widner and R.B. Wickner, unpublished observations) and may be associated with the genome, but there is no associated protein of sufficient abundance to be considered a coat protein.

Replication intermediates of 20S RNA include W dsRNA, which is a unit length molecule, and larger, apparently double-length species (173). Single-stranded molecules of approximately dimer length have also been detected on denaturing gels (173).

23S RNA

T dsRNA was discovered as a minor species of dsRNA easily detected in cells lacking L-A and L-BC, and shown to be, like W dsRNA, an independent replicon inducible by growth of cells at high temperature (258). T is the replicative form of 23S ssRNA (86). 23S RNA has substantial homology with 20S RNA and likewise appears to encode an RNA-dependent RNA polymerase (86). All strains known to have T dsRNA also have W dsRNA, but many strains with W lack T (258). Whether T depends on W is not clear.

Control of 20S and 23S RNA Replication

20S RNA and 23S RNA copy number are controlled by media conditions, requiring acetate as the carbon source and the absence of a nitrogen source for their 10,000-fold induction (86,139,171,258). Both are also induced by growing cells at 37°C, and at least 20S is repressed by the *SKI* system (171,258). The mechanism of these effects is not yet clear.

Similarity to **Cryphonectria parasitica** *Mitochondrial Replicon*

Hillman et al. (personal communication) have described a mitochondrial dsRNA species in *C. parasitica* strain NB631 with substantial similarity to 20S and 23S RNAs (see above). The degree of similarity is such that these elements must be close relatives. Nevertheless, 20S RNA and 23S RNA are primarily ssRNA replicons with small amounts of dsRNA replicative form found in cells, whereas NB631 has been identified as a dsRNA element.

DNA Viruses: *Chlorella* Viruses

Chlorella is a unicellular eukaryotic alga with a rigid cell wall and a single chloroplast (246). Most *Chlorella* organisms are free living, but several species, called collectively zoochlorellae, live as endosymbionts (intracellular symbionts) of *Hydra* or *Paramecium.*

Attempts to isolate zoochlorellae free of their *Hydra* or *Paramecium* host often induces multiplication of a virus that grows and kills the zoochlorellae. These viruses are called HVCV (for *Hydra viridis Chlorella* virus) or PBCV (for *Paramecium bursaria Chlorella* virus) (246). Hundreds of *Chlorella* viruses have been isolated directly from fresh water and are found throughout the world. They are grown on cultured zoochlorellae and form lytic plaques on agar plates.

Chlorella viruses are large (150–230 nm), polyhedral particles containing 5% to 10% lipid. The dsDNA genome of PBCV-1, the best studied *Chlorella* virus, is 333 kb, one of the largest known of any virus. PBCV-1 DNA is linear with (a) terminal inverted repeats of 2.2 kb, (b) terminal hairpin structures, and (c) shorter direct repeats within the inverted terminal repeats (226). All three characteristics are in common with the poxviruses, vaccinia virus, and African swine fever virus. Some other features are shared with the iridoviruses.

Chlorella virus DNA is heavily methylated, with 5-methylcytosine accounting for as much as 47% of C residues and 6-methyladenine for up to 37% of A residues. Surprisingly, the *Chlorella* viruses encode their own methylases. They also have been found to encode a variety of restriction endonucleases, similar in properties and, in many cases, in specificity to bacterial type II restriction endonucleases. The variety of such methylases and restriction endonucleases has only begun to be explored, but it is clear that a wide variety of specificities will be found (188). The function of these enzymes is completely unclear, and there is evidence against their being required for either degradation of host DNA or exclusion of other coinfecting viruses (188).

It is not surprising to find glycosylated viral proteins and the 54-kd major capsid protein of PBCV-1 has a carbohydrate component of about 5 kd. What is unexpected is that the virus apparently encodes products that determine the glycosylation of Vp54 (254). Antivirus monoclonal antibodies directed against the carbohydrate moiety of Vp54 were used to select resistant mutants. The resulting virus mutants had altered glycoprotein moieties but unaltered Vp54 amino acid sequence, indicating that the virus either encodes its own glycosyl transferases or encodes products that control the host enzymes.

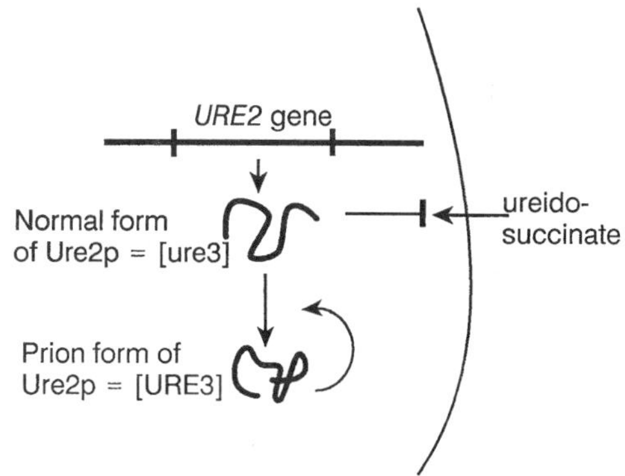

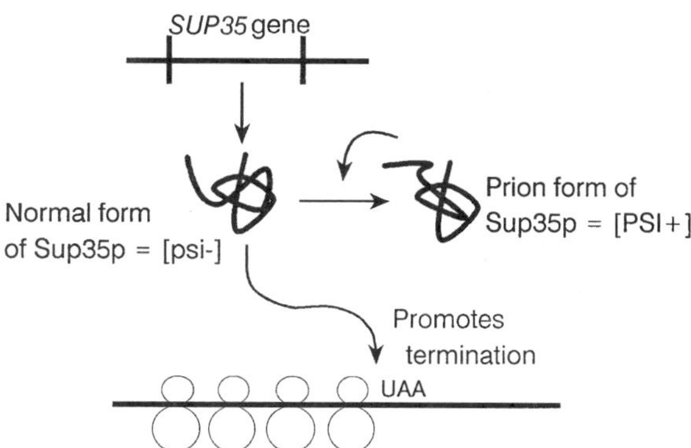

FIG. 15. Prion models of [URE3] (above) and [PSI] (below) as a modified form of the Ure2 and Sup35 proteins, respectively. It is proposed that [URE3] is the alteration of Ure2p such that it has lost its normal function (including blocking uptake of ureidosuccinate) and has acquired the ability to convert the normal Ure2p into this altered form. Similarly, the normal Sup35p promotes normal termination, whereas the prion [PSI] form does not, but can convert the normal form to the prion form. Adapted from Wickner (263), with permission.

Prions of *Saccharomyces cerevisiae*

[URE3] as a Prion Form of Ure2p

The spongiform encephalopathies are infectious neurological disorders of mammals that include scrapie of sheep, human Creutzfeld-Jakob disease and kuru, bovine spongiform encephalopathy, and transmissible mink encephalopathy. The resistance of the purified infectious agent to ultraviolet irradiation and the absence of a unique nucleic acid species in such preparations led to the idea of an infectious protein, or prion. Substantial genetic support for this notion has now appeared (reviewed in Chapters 90 and 91 in *Fields Virology,* 3rd ed.).

In 1971, Lacroute described a non-Mendelian genetic element of *S. cerevisiae,* called [URE3], that allowed cells to take up ureidosuccinate in media containing ammonium (154). Ureidosuccinate is normally produced by aspartate transcarbamylase in the uracil biosynthesis pathway, and its uptake is repressed by ammonium. Chromosomal *ure2* mutants have the same effects as carrying [URE3] but in addition fail to propagate the [URE3] element (3). Ure2p is now known to act posttranslationally on the transcriptional activator, Gln3p. Gln3p positively regulates many genes in nitrogen metabolism, including genes necessary for ureidosuccinate uptake (56):

$$\text{Ure2p} \dashv \text{Gln3p} \rightarrow \text{N enzyme genes}$$

Three lines of genetic evidence indicate that [URE3] is actually an altered form of Ure2p that has lost its activity in repressing nitrogen metabolic enzymes through Gln3p but has acquired the ability to convert the normal Ure2p to this altered form (263) (Fig. 15). [URE3] is curable by growth of cells on guanidine HCl, but the curing is reversible in that [URE3] derivatives may be again selected from the cured strain (263). The requirement of Ure2p for propagation of [URE3] is easily explained by the prion model (263) (Fig. 15) but difficult to explain if [URE3] is a typical DNA or RNA replicon. Finally, overproduction of Ure2p results in a 100-fold increase in the frequency with which cells become [URE3], a result predicted by the prion model (263). However, the nature of the alteration of Ure2p in [URE3] strains is unknown, and a direct demonstration of transmission of the [URE3] trait by isolated altered Ure2p has not yet been accomplished.

[PSI] as a Prion Form of Sup35p

Cox described a non-Mendelian genetic element that increases the efficiency of nonsense suppression and frameshift suppression by classical tRNA suppressors (57,58). Like [URE3], [PSI] is reversibly curable, in this case by high osmotic strength media (220,241). Propagation of [PSI] requires Sup35p, recessive mutants of which have an omnipotent suppressor phenotype similar to that of strains carrying [PSI] (57). Finally, overexpression of Sup35p results in a 100-fold increase in the frequency with which [psi–] strains acquire [PSI] (45). This logical parallel with [URE3] has led to the suggestion that [PSI] is also a prion form of Sup35p (263). Tuite et al. (240) described an inhibitor of *in vitro* translational suppression present only in [psi–] strains. The properties of this inhibitor are those expected of Sup35p, further supporting the prion hypothesis for [PSI] (263).

These prionlike phenomena in yeast suggest that the concept of infectious proteins is not restricted to the mammalian PrP gene, and may provide good experimental material for detailed studies of the mechanisms of prion generation, transmission, and action.

REFERENCES

1. Adams SE, Dawson KM, Gull K, Kingsman SM, Kingsman AJ. The expression of hybrid HIV:Ty virus-like particles in yeast. *Nature* 1987;329:68–70.
2. Adams SE, Mellor J, Gull K, et al. The functions and relationships of Ty-VLP proteins in yeast reflect those of mammalian retroviral proteins. *Cell* 1987;49:111–119.
3. Aigle M, Lacroute F. Genetical aspects of [URE3], a non-Mendelian, cytoplasmically inherited mutation in yeast. *Mol Gen Genet* 1975;136:327–335.
4. Aksoy S, Williams S, Chang S, Richards FF. SLACS retrotransposon from *Trypanosoma brucei gambiense* is similar to mammalian LINEs. *Nucleic Acids Res* 1990;18:785–792.
5. Anagnostakis SL. Biological control of chestnut blight. *Science* 1982; 215:466–471.
6. Anagnostakis SL. Genetic analysis of *Endothia parasitica:* linkage data for four single genes and three vegetative compatibility types. *Genetics* 1982;102:25–28.
7. Anagnostakis SL, Day PR. Hypovirulence conversion in *Endothia parasitica. Phytopathology* 1979;69:1226–1229.
8. Atkins JF, Weiss RB, Thompson S, Gesteland RF. Towards a genetic dissection of the basis of triplet decoding, and its natural subversion: programmed reading frame shifts and hops. *Annu Rev Genet* 1991; 25:201–228.
9. Cheng RH, Caston JR, Wang G-J, Gu F, Smith TJ, Baker TS, Bozarth RF, Trus BL, Cheng N, Wickner RB, Steven AC. Fungal virus capsids are cytoplasmic compartments for the replication of double-stranded RNA, formed as icosahedral shells of asymetric Gag dimers. *J Mol Biol* 1994;244:255–258.
10. Balasundaram D, Dinman JD, Wickner RB, Tabor CW, Tabor H. Spermidine deficiency increases +1 ribosomal frameshifting efficiency and inhibits Ty1 retrotransposition in *Saccharomyces cerevisiae. Proc Natl Acad Sci USA* 1994;91:172–176.
11. Ball SG, Tirtiaux C, Wickner RB. Genetic control of L-A and L-BC dsRNA copy number in killer systems of Saccharomyces cerevisiae. *Genetics* 1984;107:199–217.
12. Barbone FP, Williams TL, Leibowitz MJ. Yeast killer virus transcription initiation *in vitro. Virology* 1992;187:333–337.
13. Belcourt MF, Farabaugh PJ. Ribosomal frameshifting in the yeast retrotransposon Ty: tRNAs induce slippage on a 7 nucleotide minimal site. *Cell* 1990;62:339–352.
14. Bevan EA, Herring AJ, Mitchell DJ. Preliminary characterization of two species of dsRNA in yeast and their relationship to the "killer" character. *Nature* 1973;245:81–86.
15. Bilanchone VW, Claypool JA, Kinsey PT, Sandmeyer SB. Positive and negative regulatory elements control expression of the yeast retrotranposon Ty3. *Genetics* 1993;134:685–700.
16. Blanc A, Goyer C, Sonenberg N. The coat protein of the yeast double-stranded RNA virus L-A attaches covalently to the cap structure of eukaryotic mRNA. *Mol Cell Biol* 1992;12:3390–3398.
17. Blanc A, Ribas JC, Wickner RB, Sonenberg N. His154 is involved in the linkage of the *Saccharomyces cerevisiae* L-A double-stranded RNA virus gag protein to the cap structure of mRNAs and is essential for M_1 satellite virus expression. *Mol Cell Biol* 1994;14:2664–2674.
18. Boccardo G, Milne RG, Luisoni E, Lisa V, Accotto GP. Three seed-borne cryptic viruses containing double-stranded RNA isolated from white clover. *Virology* 1985;147:29–40.
19. Boeke JD, Garfinkel DJ, Styles CA, Fink GR. Ty elements transpose through an RNA intermediate. *Cell* 1985;40:491–500.
20. Boeke JD, Sandmeyer SB. Yeast transposable elements. In: Broach JR, Pringle JR, Jones EW, eds. *The molecular and cellular biology of the yeast Saccharomyces: genome dynamics, protein synthesis and energetics.* Cold Spring Harbor, NY: Cold Spring Harbor Laboratory Press; 1991:193–261.
21. Bostian KA, Elliott Q, Bussey H, Burn V, Smith A, Tipper DJ. Sequence of the preprotoxin dsRNA gene of type I killer yeast: multiple processing events produce a two-component toxin. *Cell* 1984;36:741–751.
22. Bostian KA, Sturgeon JA, Tipper DJ. Encapsidation of yeast killer double-stranded ribonucleic acids: dependence of M on L. *J Bacteriol* 1980; 143:463–470.
23. Bozarth RF. The physico-chemical properties of mycoviruses. In: Lemke PA, ed. *Viruses and plasmids in fungi.* New York: Marcel Dekker; 1979:43–91.
24. Bradshaw VA, McEntee K. DNA damage activates transcription and transposition of yeast Ty retrotransposons. *Mol Gen Genet* 1989; 218:465–474.
25. Brennan VE, Field L, Cizdziel P, Bruenn JA. Sequences at the 3′ ends of yeast viral dsRNAs: proposed transcriptase and replicase initiation sites. *Nucleic Acids Res* 1981;9:4007–4021.
26. Brierley I, Dingard P, Inglis SC. Characterization of an efficient coronavirus ribosomal frameshifting signal: requirement for an RNA pseudoknot. *Cell* 1989;57:537–547.
27. Brierley I, Jenner AJ, Inglis SC. Mutational analysis of the "slippery-sequence" component of a coronavirus ribosomal frameshifting signal. *J Mol Biol* 1992;227:463–479.
28. Brierley I, Rolley NJ, Jenner AJ, Inglis SC. Mutational analysis of the RNA pseudoknot component of a coronavirus ribosomal frameshifting signal. *J Mol Biol* 1991;220:889–902.
29. Brodeur GM, Sandmeyer SB, Olson MV. Consistent association between sigma elements and tRNA genes. *Proc Natl Acad Sci USA* 1983;80:3292–3296.
30. Brown PO, Bowerman B, Varmus HE, Bishop JM. Correct integration of retroviral DNA *in vitro. Cell* 1987;49:347–356.
31. Bruenn J, Keitz B. The 5′ ends of yeast killer factor RNAs are pppGp. *Nucleic Acids Res* 1976;3:2427–2436.
32. Buck KW. Fungal viruses, double-stranded RNA and interferon induction. In: Lemke PA, ed. *Viruses and plasmids in fungi.* New York: Marcel Dekker; 1979:1–42.
33. Buck KW. Replication of double-stranded RNA mycoviruses. In: Lemke PA, ed. *Viruses and plasmids in fungi.* New York: Marcel Dekker; 1979:93–160.
34. Buck KW. *Fungal virology.* Boca Raton, FL: CRC Press; 1986.
35. Buck KW, Lhoas P, Street BK. Virus particles in yeast. *Biochem Soc Trans* 1973;1:1141–1142.
36. Bussey H. Proteases and the processing of precursors to secreted proteins in yeast. *Yeast* 1988;4:17–26.

37. Cadd TL, Keenan MC, Patterson JL. Detection of *Leishmania* RNA virus 1 proteins. *J Virol* 1993;67:5647–5650.
38. Cameron JR, Loh EY, Davis RW. Evidence for transposition of dispersed repetitive DNA families in yeast. *Cell* 1979;16:739–751.
39. Cappello J, Handelsman K, Lodish HF. Sequence of *Dictyostelium* DIRS-1: an apparent retrotransposon with inverted terminal repeats and an internal circle junction sequence. *Cell* 1985;43:105–115.
40. Chalker DL, Sandmeyer SB. Transfer RNA genes are genomic targets for *de novo* transposition of the yeast retrotransposon Ty3. *Genetics* 1990;126:837–850.
41. Chalker DL, Sandmeyer SB. Ty3 integrates within the region of RNA polymerase II transcription initiation. *Genes Dev* 1992;6:117–128.
42. Chamorro M, Parkin N, Varmus HE. An RNA pseudoknot and an optimal heptameric shift site are required for highly efficient ribosomal frameshifting on a retroviral messenger RNA. *Proc Natl Acad Sci USA* 1992;89:713–717.
43. Chapman KB, Boeke JD. Isolation and characterization of the gene encoding yeast debranching enzyme. *Cell* 1991;65:483–492.
44. Chapman KB, Bystrom AS, Boeke JD. Initiator methionine tRNA is essential for Ty1 transposition in yeast. *Proc Natl Acad Sci USA* 1992;89:3236–3240.
45. Chernoff YO, Derkach IL, Inge-Vechtomov SG. Multicopy SUP35 gene induces de-novo appearance of psi-like factors in the yeast *Saccharomyces cerevisiae. Curr Genet* 1993;24:268–270.
46. Choi GH, Larson TG, Nuss DL. Molecular analysis of the Laccase gene from the chestnut blight fungus and selective suppression of its expression in an isogenic hypovirulent strain. *Mol Plant Microb Interact* 1992;5:119–128.
47. Choi GH, Nuss DL. Hypovirulence of chestnut blight fungus conferred by an infectious viral cDNA. *Science* 1992;257:800–803.
48. Choi GH, Nuss DL. A viral gene confers hypovirulence—associated traits to the chestnut blight fungus. *EMBO J* 1992;11:473–477.
49. Choi GH, Shapira R, Nuss DL. Cotranslational autoproteolysis involved in gene expression from a double-stranded RNA genetic element associated with hypovirulence of the chestnut blight fungus. *Proc Natl Acad Sci USA* 1991;88:1167–1171.
50. Ciriacy M, Freidel K, Lohning C. Characterization of trans-acting mutations affecting Ty and Ty-mediated transcription in *Saccharomyces cerevisiae. Curr Genet* 1991;20:441–448.
51. Clare J, Farabaugh PJ. Nucleotide sequence of a yeast Ty element: evidence for an unusual mechanism of gene expression. *Proc Natl Acad Sci USA* 1985;82:2828–2833.
52. Clark DJ, Bilanchone VW, Haywood LJ, Dildine SL, Sandmeyer SB. A yeast sigma composite element, Ty3, has properties of a retrotransposon. *J Biol Chem* 1988;263:1413–1423.
53. Coffin JM. Retroviridae and their replication. In: Fields BN, Knipe DM, Channock RM, eds. *Fields virology.* New York: Raven; 1990:1437–1500.
54. Company M, Errede B. Identification of a Ty1 regulatory sequence responsive to STE7 and STE12. *Mol Cell Biol* 1988;8:2545–2554.
55. Cooper A, Bussey H. Characterization of the yeast KEX1 gene product: a carboxypeptidase involved in processing secreted precursor proteins. *Mol Cell Biol* 1989;9:2706–2714.
56. Courchesne WE, Magasanik B. Regulation of nitrogen assimilation in *Saccharomyces cerevisiae:* roles of the URE2 and GLN3 genes. *J Bacteriol* 1988;170:708–713.
57. Cox BS. Psi phenomena in yeast. In: Hall MN, Linder P, ed. *The early days of yeast genetics.* Cold Spring Harbor, NY: Cold Spring Harbor Laboratory; 1993:219–239.
58. Cox BS, Tuite MF, McLaughlin CS. The psi factor of yeast: a problem in inheritance. *Yeast* 1988;4:159–179.
59. Craigie R. Hotspots and warmspots: integration specificity of retroelements. *Trends Genet* 1992;8:185–187.
60. Craven MG, Pawlyk DM, Choi GH, Nuss DL. Papain-like protease p29 as a symptom determinant encoded by a hypovirulence-associated virus of the chestnut blight fungus. *J Virol* 1993;67:6513–6521.
61. Curcio MJ, Garfinkel DJ. Posttranslational control of Ty1 retrotransposition occurs at the level of protein processing. *Mol Cell Biol* 1992;12:2813–2825.
62. Curcio MJ, Sanders NJ, Garfinkel DJ. Transcriptional competence and transcription of endogenous Ty elements in *Saccharomyces cerevisiae:* implications for regulation of transposition. *Mol Cell Biol* 1988; 8:3571–3581.
63. Day A, Rochaix JD. A transposon with an unusual LTR arrangement from *Chlamydomonas reinhardtii* contains an internal tandem array of 76 bp repeats. *Nucleic Acids Res* 1991;19:1259–1266.
64. Day PR, Dodds JA, Elliston JE, Jaynes RA, Anagnostakis SL. Double-stranded RNA in Endothia parasitica. *Phytopathology* 1977; 67:1393–1396.
65. Derr LK, Strathern JN, Garfinkel DJ. RNA-mediated recombination in S. cerevisiae. *Cell* 1991;67:355–364.
66. Diamond LS, Mattern CFT. Protozoal viruses. *Adv Virus Res* 1976; 20:87–112.
67. Dignard D, Whiteway M, Germain D, Tessier D, Thomas DY. Expression in yeast of a cDNA copy of the K2 killer toxin gene. *Mol Gen Genet* 1991;227:127–136.
68. Dihanich M, Van Tuinen E, Lambris JD, Marshallsay B. Accumulation of viruslike particles in a yeast mutant lacking a mitochondrial pore protein. *Mol Cell Biol* 1989;9:1100–1108.
69. Dinman JD, Icho T, Wickner RB. A –1 ribosomal frameshift in a double-stranded RNA virus of yeast forms a gag-pol fusion protein. *Proc Natl Acad Sci USA* 1991;88:174–178.
70. Dinman JD, Wickner RB. Ribosomal frameshifting efficiency and gag/gag-pol ratio are critical for yeast M_1 double-stranded RNA virus propagation. *J Virol* 1992;66:3669–3676.
71. Dinman JD, Wickner RB. Translational maintenance of frame: mutants of *Saccharomyces cerevisiae* with altered –1 ribosomal frameshifting efficiencies. *Genetics* 1994;136:75–86.
72. Dmochowska A, Dignard D, Henning D, Thomas DY, Bussey H. Yeast *KEX*1 gene encodes a putative protease with a carboxypeptidase B–like function involved in killer toxin and alpha factor precursor processing. *Cell* 1987;50:573–584.
73. Dodds JA. Association of type 1 viral-like dsRNA with club-shaped particles in hypovirulent strains of *Endothia parasitica. Virology* 1980;107:1–12.
74. Doolittle RF, Feng DF, Johnson MS, McClure MA. Origins and evolutionary relationships of retroviruses. *Q Rev Biol* 1989;64:1–30.
75. Eibel H, Philippsen P. Preferential integration of yeast transposable element Ty into a promoter region. *Nature* 1984;307:386–388.
76. Eichinger DJ, Boeke JD. The DNA intermediate in yeast Ty1 element transposition copurifies with virus-like particles: cell-free Ty1 transposition. *Cell* 1988;54:955–966.
77. Eichinger DJ, Boeke JD. A specific terminal structure is required for Ty1 transposition. *Genes Dev* 1990;4:324–330.
78. El-Sherbeini M, Bostian KA. Viruses in fungi: infection of yeast with the K1 and K2 killer virus. *Proc Natl Acad Sci USA* 1987;84:4293–4297.
79. Elder RT, John TPS, Stinchcomb DT, Davis RW. Studies on the transposable element Ty1 of yeast. I. RNA homologous to Ty1. *Cold Spring Harb Symp Quant Biol* 1981;45:581–591.
80. Elder RT, Loh EY, Davis RW. RNA from the yeast transposable element Ty1 has both ends in the direct repeats, a structure similar to retrovirus RNA. *Proc Natl Acad Sci USA* 1983;80:2432–2436.
81. Elliston JE. Characterization of dsRNA-free and dsRNA-containing strains of *Endothia parasitica* in relation to hypovirulence. *Phytopathology* 1985;75:151–158.
82. Enebak SA, Hillman BI, MacDonald WL. Examination of a hypovirulent isolate of *Cryphonectria parasitica* associated with multi-segmented dsRNA. *Mole Plant Microbe Interact* (in press)
83. Errede B. MCM1 binds to a transcriptional control element in Ty1. *Mol Cell Biol* 1993;13:57–62.
84. Errede B, Cardillo TS, Sherman F, Dubois E, Deschamps J, Wiame JM. Mating signals control expression of mutations resulting from insertion of a transposable repetitive element adjacent to diverse yeast genes. *Cell* 1980;22:427–436.
85. Errede B, Cardillo TS, Wever G, Sherman F. Studies on transposable elements in yeast. I. ROAM mutations causing increased expression of yeast genes: their activation by signals directed toward conjugation functions and their formation by insertion of Ty1 repetitive elements. *Cold Spring Harbor Symp Quant Biol* 1981;45:593–602.
86. Esteban LM, Rodriguez CN, Esteban R. T double-stranded RNA (dsRNA) sequence reveals that T and W dsRNAs form a new RNA family in Saccharomyces cerevisiae. Identification of 23 S RNA as the single-stranded form of T dsRNA. *J Biol Chem* 1992;267:10874–10881.
87. Esteban R, Fujimura T, Wickner RB. Internal and terminal cis-acting sites are necessary for *in vitro* replication of the L-A double-stranded RNA virus of yeast. *EMBO J* 1989;8:947–954.
88. Esteban R, Wickner RB. Three different M1 RNA-containing virus-

like particle types in Saccharomyces cerevisiae: *in vitro* M1 double-stranded RNA synthesis. *Mol Cell Biol* 1986;6:1552–1561.

89. Esteban R, Wickner RB. A deletion mutant of L-A double-stranded RNA replicates like M_1 double-stranded RNA. *J Virol* 1988;62:1278–1285.
90. Fahima T, Kazmierczak P, Hansen DR, Pfeifer P, Alfen NKV. Membrane-associated replication of an unencapsidated double-stranded RNA of the fungus, *Cryphonectria parasitica. Virology* 1993;195:81–89.
91. Farabaugh PJ, Fink GR. Insertion of the eukaryotic transposable element Ty1 creates a 5 base pair duplication. *Nature* 1980;286:352–356.
92. Farabaugh PJ, Liao XB, Belcourt M, Zhao H, Kapakos J, Clare J. Enhancer and silencerlike sites within the transcribed portion of a Ty2 tranposable element of Saccharomyces cerevisiae. *Mol Cell Biol* 1989;9:4824–4834.
93. Farabaugh PJ, Vimaladithan A, Turkel S, Johnson R, Zhao H. Three downstream sites repress transcription of a Ty2 retrotransposon in *Saccharomyces cerevisiae. Mol Cell Biol* 1993;13:2081–2090.
94. Farabaugh PJ, Zhao H, Vimaladithan A. A novel programmed frameshift expresses the POL3 gene of retrotranposon Ty3 of yeast: frameshifting without tRNA slippage. *Cell* 1993;74:93–103.
95. Fink GR, Styles CA. Curing of a killer factor in *Saccharomyces cerevisiae. Proc Natl Acad Sci USA* 1972;69:2846–2849.
96. Finkler A, Ben-Zvi BS, Koltin Y, Barash I. Transcription and *in vitro* translation of the dsRNA virus isolated from *Rhizoctonia solani. Virus Genes* 1988;1:205–219.
97. Fried HM, Fink GR. Electron microscopic heteroduplex analysis of "killer" double-stranded RNA species from yeast. *Proc Natl Acad Sci USA* 1978;75:4224–4228.
98. Fujimura T, Esteban R, Esteban LM, Wickner RB. Portable encapsidation signal of the L-A double-stranded RNA virus of S. cerevisiae.*Cell* 1990;62:819–828.
99. Fujimura T, Esteban R, Wickner RB. *in vitro* L-A double-stranded RNA synthesis in virus-like particles from Saccharomyces cerevisiae. *Proc Natl Acad Sci USA* 1986;83:4433–4437.
100. Fujimura T, Ribas JC, Makhov AM, Wickner RB. Pol of gag-pol fusion protein required for encapsidation of viral RNA of yeast L-A virus. *Nature* 1992;359:746–749.
101. Fujimura T, Wickner RB. L-A double-stranded RNA viruslike particle replication cycle in Saccharomyces cerevisiae: particle maturation *in vitro* and effects of mak10 and pet18 mutations. *Mol Cell Biol* 1987;7:420–426.
102. Fujimura T, Wickner RB. Gene overlap results in a viral protein having an RNA binding domain and a major coat protein domain. *Cell* 1988;55:663–671.
103. Fujimura T, Wickner RB. Replicase of L-A virus-like particles of Saccharomyces cerevisiae. *in vitro* conversion of exogenous L-A and M1 single-stranded RNAs to double-stranded form. *J Biol Chem* 1988;263:454–460.
104. Fujimura T, Wickner RB. Reconstitution of template-dependent *in vitro* transcriptase activity of a yeast double-stranded RNA virus. *J Biol Chem* 1989;264:10872–10877.
105. Fujimura T, Wickner RB. Interaction of two cis sites with the RNA replicase of the yeast L-A virus. *J Biol Chem* 1992;267:2708–2713.
106. Fulbright DW. Effect of eliminating dsRNA in hypovirulent *Endothia parasitica. Phytopathology* 1984;74:722–724.
107. Furfine ES, Wang CC. Transfection of the *Giardia lamblia* double-stranded RNA virus into *Giardia lamblia* by electroporation of a single-stranded RNA copy of the viral genome. *Mol Cell Biol* 1990;10:3659–3663.
108. Gabriel A, Yen TJ, Schwartz DC, et al. A rapidly rearranging retrotransposon within the miniexon gene locus of *Crithidia fasciculata. Mol Cell Biol* 1990;10:615–624.
109. Gafner J, Philippsen P. The yeast transposon Ty1 generates duplications of target DNA on insertion. *Nature* 1980;286:414–418.
110. Gandy DG, Hollings M. Die-back of mushrooms: a disease assocaited with a virus. *Rep Glasshouse Crops Res Inst* 1962;1961:103–108.
111. Garfinkel DJ. Retroelements in microorganisms. In: Levy JA, ed. *The retroviridae.* New York: Plenum; 1992:107–158.
112. Garfinkel DJ, Boeke JD, Fink GR. Ty element transposition: reverse transcriptase and virus-like particles. *Cell* 1985;42:507–517.
113. Garfinkel DJ, Hedge A-M, Youngren SD, Copeland TD. Proteolytic processing of pol-TYB proteins from the yeast retrotransposon Ty1. *J Virol* 1991;65:4573–4581.
114. Garvik B, Haber JE. New cytoplasmic genetic element that controls 20S RNA synthesis during sporulation in yeast. *J Bacteriol* 1978;134:261–269.
115. Georgopoulos DE, Hannig EM, Leibowitz MJ. Sequence of the M1-2 region of killer virus double-stranded RNA. *Basic Life Sci* 1986;40:203–213.
116. Ghabrial SA. New developments in fungal virology. *Adv Virus Res* 1994;43:303–388.
117. Ghabrial SA, Havens WM. The *Helminthosporium victoriae* 190S mycovirus has two forms distinguishable by capsid protein composition and phosphorylation state. *Virology* 1992;188:657–665.
118. Gilboa E, Mitra SW, Goff S, Baltimore D. A detailed model of reverse transcription and tests of crucial aspects. *Cell* 1979;18:93–100.
119. Grente J. Les formes hypovirulentes d'*Endothia parasitica* et les espoirs de lutte contre le contre le chancre du chataignier. *C R Acad Agric France* 1965;51:1033–1037.
120. Groves DP, Clare JJ, Oliver SG. Isolation and characterization of a double stranded RNA virus like particle from the yeast Yarrowia lipolytica. *Curr Genet* 1983;7:185–190.
121. Guilbride L, Myler PJ, Stuart K. Distribution and sequence divergence of LRV1 viruses among different *Leishmania* species. *Mol Biochem Parisitol* 1992;54:101–104.
122. Hannig EM, Leibowitz MJ. Structure and expression of the M2 genomic segment of a type 2 killer virus of yeast. *Nucleic Acids Res* 1985;13:4379–4400.
123. Hannig EM, Williams TL, Leibowitz MJ. The internal polyadenylate tract of yeast killer virus M1 double-stranded RNA is variable in length. *Virology* 1986;152:149–158.
124. Hansen DR, Alfen NKV, Gillies K, Powell WA. Naked dsRNA association with hypovirulence in *Endothia parasitica* is packaged in fungal vesicles. *J Gen Virol* 1985;66:2605–2614.
125. Hansen LJ, Chalker DL, Sandmeyer SB. Ty3, a yeast retrotransposon associated with tRNA genes, has homology to animal retroviruses. *Mol Cell Biol* 1988;8:5245–5256.
126. Harashima S, Hinnebusch AG. Multiple GCD genes required for repression of GCN4, a transcriptional activator of amino acid biosynthetic genes in *Saccharomyces cerevisiae. Mol Cell Biol* 1986; 6:3990–3998.
127. Hasan G, Turner MJ, Cordingley J. Complete nucleotide sequence of an unusual mobile element from *Trypanosoma brucei. Cell* 1984; 37:333–341.
128. Hatfield D, Levin JG, Rein A, Oroszlan S. Translational suppression in retroviral gene expression. *Adv Virol Res* 1992;41:193–239.
129. Herring AJ, Bevan EA. Yeast virus-like particles possess a capsid-associated single-stranded RNA polymerase. *Nature* 1977;268:464–466.
130. Hopper JE, Bostian KA, Rowe LB, Tipper DJ. Translation of the L-species dsRNA genome of the killer-associated virus-like particles of *Saccharomyces cerevisiae. J Biol Chem* 1977;252:9010–9017.
131. Icho T, Wickner RB. The MAK11 protein is essential for cell growth and replication of M double-stranded RNA and is apparently a membrane-associated protein. *J Biol Chem* 1988;263:1467–1475.
132. Icho T, Wickner RB. The double-stranded RNA genome of yeast virus L-A encodes its own putative RNA polymerase by fusing two open reading frames. *J Biol Chem* 1989;264:6716–6723.
133. Isaacs A, Lindenmann J. Virus interference: I. The interferon. *Proc R Soc Lond [B]* 1957;147:258–263.
134. Jacks T, Madhani HD, Masiarz FR, Varmus HE. Signals for ribosomal frameshifting in the Rous sarcoma virus gag-pol region. *Cell* 1988;55:447–458.
135. Janda M, Ahlquist P. RNA-dependent replication, transcription and persistence of Brome Mosaic virus RNA replicons in S. cerevisiae. *Cell* 1993;72:961–970.
136. Janetzky B, Lehle L. Ty4, a new retrotransposon from *Saccharomyces cerevisiae,* flanked by tau-elements. *J Biol Chem* 1992;267:19798–19805.
137. Ji H, Moore DP, Blomberg MA, et al. Hotspots for unselected Ty1 transposition events on yeast chromosome III are near tRNA genes and LTR sequences. *Cell* 1993;73:1007–1018.
138. Julius D, Brake A, Blair L, Kunisawa R, Thorner J. Isolation of the putative structural gene for the lysine–arginine–cleaving endopeptidase required for the processing of yeast prepro-alpha factor. *Cell* 1984;36:309–318.
139. Kadowaki K, Halvorson HO. Appearance of a new species of ribonucleic acid dirung sporulation in *Saccharomyces cerevisiae. J Bacteriol* 1971;105:826–830.
140. Kamer G, Argos P. Primary structural comparison of RNA-dependent

polymerase from plant, animal and bacterial viruses. *Nucleic Acids Res* 1984;12:7269–7282.

141. Kawakami K, Pande S, Faiola B, et al. A rare tRNA-Arg(CCU) that regulates Ty1 element ribosomal frameshifting is essential for Ty1 retrotransposition in *Saccharomyces cerevisiae. Genetics* 1993; 135:309–320.
142. Kennell JC, Moran JV, Perlman PS, Butow RA, Lambowitz AM. Reverse transcriptase activity associated with maturase-encoding group II introns in yeast mitochondria. *Cell* 1993;73:133–146.
143. Khoshnan A, Alderete JF. Multiple double-stranded RNA segments are associated with virus particles infecting *Trichomonas vaginalis. J Virol* 1993;67:6950–6955.
144. Kinsey JA. Tad, a LINE-like transposable element of *Neurospora,* can transposse between nuclei in heterokaryons. *Genetics* 1990;126:317–323.
145. Kinsey JA. Transnuclear retrotransposition of the Tad element of Neurospora. *Proc Natl Acad Sci USA* 1993;90:9384–9387.
146. Kinsey PT, Sandmeyer SB. Ty3 transposes in mating populations of yeast. *Genetics* (in press)
147. Kleinschmidt WJ, Cline JC, Murphy EB. Interferon production induced by statolon. *Proc Natl Acad Sci USA* 1964;52:741–744.
148. Kleinschmidt WJ, Ellis LF. Statolon as an inducer of interferon. In: Wolstenholme GEW, O'Conner M, ed. *Interferon.* London: Churchill; 1967:39–46.
149. Koltin Y. The killer system of *Ustilago maydis:* secreted polypeptides encoded by viruses. In: Koltin Y, Leibowitz MJ, eds. *Viruses of fungi and simple eukaryotes.* New York: Marcel Dekker; 1988:209–242.
150. Koltin Y, Day PR. Inheritance of killer phenotypes and double-stranded RNA in *Ustilago maydis. Proc Natl Acad Sci USA* 1976; 73:594–598.
151. Koltin Y, Leibowitz MJ, eds. *Viruses of fungi and simple eukaryotes.* New York: Marcel Dekker; 1988.
152. Koonin EV, Choi GH, Nuss DL, Shapira R, Carrington JC. Evidence for common ancestry of a chestnut blight hypovirulence-associated double-stranded RNA and a group of positive-strand RNA plant viruses. *Proc Natl Acad Sci USA* 1991;88:10647–10651.
153. L'Hostis B, Hiremath ST, Rhoads RE, Ghabrial SA. Lack of sequence homology between double-stranded RNA from European and American hypovirulent strains of Endothia parasitica. *J Gen Virol* 1985;66:351–355.
154. Lacroute F. Non-Mendelian mutation allowing ureidosuccinic acid uptake in yeast. *J Bacteriol* 1971;106:519–522.
155. Lampson GP, Tytell AA, Field AK, Nemes MM, Hilleman MR. Inducers of interferon and host resistance. I. Double-stranded RNA from extracts of *Penicillium funiculosum. Proc Natl Acad Sci USA* 1967;58:782–789.
156. Larson TG, Choi GH, Nuss DL. Regulatory pathways governing modulation of fungal gene expression by a virulence-attenuating mycovirus. *EMBO J* 1992;11:4539–4548.
157. Lee Y, Wickner RB. *MAK10,* a glucose-repressible gene necessary for replication of a dsRNA virus of *Saccharomyces cerevisiae,* has T cell receptor a-subunit motifs. *Genetics* 1992;132:87–96.
158. Leibman SW, Newman G. A ubiquitin-conjugating enzyme, RAD6, affects the distribution of Ty1 retrotransposon integration positions. *Genetics* 1993;133:499–508.
159. Leibowitz MJ. Role of protein synthesis in the replication of the killer virus of yeast. *Curr Genet* 1982;5:161–163.
160. Leibowitz MJ, Koltin Y, Rubio V. *Viruses of simple eukaryotes: molecular genetics and applications to biotechnology and medicine.* Newark, DE: University of Delaware Press; 1994.
161. Leibowitz MJ, Wickner RB. A chromosomal gene required for killer plasmid expression, mating, and spore maturation in Saccharomyces cerevisiae. *Proc Natl Acad Sci USA* 1976;73:2061–2065.
162. Lemke PA, ed. *Viruses and plasmids in fungi.* New York: Marcel Dekker; 1979.
163. Levin HL, Boeke JD. Demonstration of retrotransposition of the Tf1 element in fission yeast. *EMBO J* 1992;11:1145–1153.
164. Levin HL, Weaver DC, Boeke JD. Novel gene expression mechanism in a fission yeast retroelement: Tf1 proteins are derived from a single primary translation product. *EMBO J* 1993;12:4885–4895.
165. Liao XB, Clare JJ, Farabaugh PJ. The upstream activation site of a Ty2 element of yeast is necessary but not sufficient to promote maximal transcription of the element. *Proc Natl Acad Sci USA* 1987; 84:8520–8524.
166. Liu Y, Dieckmann CL. Overproduction of yeast virus–like particle coat protein genome in strains deficient in a mitochondrial nuclease. *Mol Cell Biol* 1989;9:3323–3331.
167. Lolle S, Skipper N, Bussey H, Thomas DY. The expression of cDNA clones of yeast M_1 double-stranded RNA in yeast confers both killer and immunity pheonotypes. *EMBO J* 1984;3:1283–1387.
168. Makower M, Bevan EA. The inheritance of a killer character in yeast (*Saccharomyces cerevisiae*). *Proc Int Congr Genet* 1963;11:202.
169. Marschalek R, Hofmann J, Schumann G, Gosseringer R, Dingermann T. Structure of DRE, a retrotransposable element which integrates with position speicificity upstream of *Dictyostelium discoideum* tRNA genes. *Mol Cell Biol* 1992;12:229–239.
169a. Masison DC, Blanc A, Ribas JC, Carroll K, Sonenberg N, Wickner RB. Decoying the cap⁻ mRNA degradation system by a dsRNA virus and poly(A)⁻ mRNA surveillance by a yeast antiviral system. *Mol Cell Biol* 1995;May.
170. Mathias SL, Scott AF, Kazazian HH, Boeke JD, Gabriel A. Reverse transcriptase encoded by a human transposable element. *Science* 1991;254:1808–1810.
171. Matsumoto Y, Fishel R, Wickner RB. Circular single-stranded RNA replicon in *Saccharomyces cerevisiae. Proc Natl Acad Sci USA* 1990;87:7628–7632.
172. Matsumoto Y, Sarkar G, Sommer SS, Wickner RB. A yeast antiviral protein, *SKI8,* shares a repeated amino acid sequence pattern with beta-subunits of G proteins and several other proteins. *Yeast* 1993;8:43–51.
173. Matsumoto Y, Wickner RB. Yeast 20 S RNA replicon. Replication intermediates and encoded putative RNA polymerase. *J Biol Chem* 1991;266:12779–12783.
174. Maule AP, Thomas PD. Strains of yeast lethal to brewery yeasts. *J Inst Brew (London)* 1973;79:137–141.ÆFN1Au: Cite in text.
175. McHale MT, Roberts IN, Noble SM, et al. CfT-1: an LTR-retrotransposon in *Cladosporium fulvum,* a fungal pathogen of tomato. *Mol Gen Genet* 1992;233:337–347.
176. Mellor J, Malim MH, Gull K, et al. Reverse transcriptase activity and Ty RNA are associated with virus-like particles in yeast. *Nature* 1985;318:583–586.
177. Mellor M, Fulton AM, Dobson MJ, Wilson SM, Kingsman SM,Kingsman AJ. A retrovirus-like strategy for the expression of a fusion protein encoded by yeast transposon Ty1. *Nature* 1985;313:243–246.
178. Miller RL, Wang AL, Wang CC. Purification and characterization of the *Giardia lamblia* double-stranded RNA virus. *Mol Biochem Parasitol* 1988;28:189–196.
179. Moehring JM, Inocencio NM, Robertson BJ, Moehring TJ. Expression of mouse furin in a Chinese hamster cell resistant to Pseudomonas exotoxin A and viruses complements the genetic lesion. *J Biol Chem* 1993;268:2590–2594.
180. Moore SP, Garfinkel DJ. Expression and partial purification of enzymatically active recombinant Ty1 integrase in *Saccharomyces cerevisiae. Proc Natl Acad Sci* 1994;91:1843–1847.
181. Mueller MW, Allmaier M, Eskes R, Schweyen RJ. Transposition of group II intron aI1 in yeast and invasion of mitochondrial genes at new locations. *Nature* 1993;366:174–176.
182. Mueller PP, Hinnebusch AG. Multiple upstream AUG codons mediate translational control of GCN4. *Cell* 1986;45:201–207.
183. Muller F, Laufer W, Pott U, Ciriacy M. Characterization of products of TY1-mediated reverse transcription in *Saccharomyces cerevisiae. Mol Gen Genet* 1991;226:145–153.
184. Natsoulis G, Boeke JD. New antiviral strategy using capsid-nulcease fusion proteins. *Nature* 1991;352:632–635.
185. Natsoulis G, Thomas W, Roghmann MC, Winston F, Boeke JD. Ty1 tranposition in *Saccharomyces cerevisiae* is nonrandom. *Genetics* 1989;123:269–279.
186. Natsuaki T, Natsuaki KT, Okuda S, et al. Relationships between the cryptic and temperate viruses of alfalfa, beet and white clover. *Intervirology* 1986;25:69–75.
187. Naumov GI, Naumova TI. Comparative genetics of yeast. XIII. Comparative study of killer strains of *Saccharomyces* from different collections. *Genetika* 1973;9:140–145.
188. Nelson M, Zhang Y, Van Etten JL. DNA methyltransferases and DNA site-specific endonucleases encoded by chlorella viruses. In: Jost JP, Saluz HP, eds. *DNA methylation: molecular biology and biological significance.* Basel, Switzerland: Birkhauser Verlag; 1993:186–211.
189. Newman AM, Elliot SG, McLaughlin CS, Sutherland PA, Warner RC. Replication of double-stranded RNA of the virus-like particles of *Saccharomyces cerevisiae. J Virol* 1981;38:263–271.
190. Nuss DL. Biological control of chestnut blight: an example of virus-

mediated attenuation of fungal pathogenesis. *Microbiol Rev* 1992;56:561–576.

190a. Ohtake Y, Wickner RB. Yeast virus propagation depends critically on free 60S ribosomal subunit concentration. *Mol Cell Biol* 1995; May.

191. Oliver SG, van der Aart QJ, Agostoni-Carbone ML. The complete DNA sequence of yeast chromosome III. *Nature* 1992;357:38–46.

192. Pfeiffer P, Radler F. Comparison of the killer toxin of several yeasts and the purification of a toxin of type K2. *Arch Microbiol* 1984;137:357–361.

193. Philliskirk G, Young TW. The occurence of killer character in yeasts of various genera. *Antonie van Leeuwenhoik J Microbiol Serol* 1975;41:147–151.

194. Picologlou S, Brown N, Leibman SW. Mutations in RAD6, a yeast gene encoding a ubiquitin-conjugating enzyme, stimulate retrotransposition. *Mol Cell Biol* 1990;10:1017–1022.

195. Polonelli L, Conti S, Gerloni M, Magliani W, Chezzi C, Morace G. Interfaces of the yeast killer phenomenon. *Crit Rev Microbiol* 1991;18:47–87.

196. Powell HM, Culbertson CG, McGuire JM, Hoehn MM, Baker LA. A filtrate with chemoprophylactic and chemotherapeutic action against MM and Semliki Forest viruses in mice. *Antibiot Chemother* 1952;2:432–437.

197. Puhalla JE. Compatibility reactions on solid medium and interstrain inhibition in *Ustilago maydis. Genetics* 1968;60:461–474.

198. Radler F, Herzberger S, Schonig I, Schwarz P. Investigation of a killer strain of *Zygosaccharomyces bailii. J Gen Microbiol* 1993;139:495–500.

199. Rhee SK, Icho T, Wickner RB. Structure and nuclear localization signal of the SKI3 antiviral protein of *Saccharomyces cerevisiae. Yeast* 1989;5:149–158.

200. Ribas JC, Fujimura T, Wickner RB. A cryptic RNA–binding domain in the Pol region of the L-A dsRNA virus Gag-Pol fusion protein. *J Virol* 1994;68:6014–6020.

201. Ribas JC, Fujimura T, Wickner RB. Essential RNA binding and packaging domains of the Gag-Pol fusion protein of the L-A double-stranded RNA virus of *Saccharomyces cerevisiae. J Biol Chem* 1994; 269:28420–28428.

202. Ribas JC, Wickner RB. RNA-dependent RNA polymerase consensus sequence of the L-A double-stranded RNA virus: definition of essential domains. *Proc Natl Acad Sci USA* 1992;89:2185–2189.

203. Ridley SP, Sommer SS, Wickner RB. Superkiller mutations in Saccharomyces cerevisiae suppress exclusion of M_2 double-stranded RNA by L-A-HN and confer cold sensitivity in the presence of M and L-A-HN. *Mol Cell Biol* 1984;4:761–770.

204. Rigling D, Alfen NKV. Regulation of laccase biosynthesis in the plant pathogenic fungus *Cryphonectria parasitica* by double-stranded RNA. *J Bacteriol* 1991;173:8000–8003.

205. Rodriguez CN, Esteban LM, Esteban R. Molecular cloning and characterization of W double-stranded RNA, a linear molecule present in *Saccharomyces cerevisiae.* Identification of its single-stranded RNA form as 20 S RNA. *J Biol Chem* 1991;266:12772–12778.

206. Rodriguez-Cousino N, Esteban R. Both yeast W double-stranded RNA and its single-stranded form 20S RNA are linear. *Nucleic Acids Res* 1992;20:2761–2766.

207. Rothnie HM, McCurrach KJ, Glover LA, Hardman N. Retrotransposon-like nature of Tp1 elements: implications for the organization of highly repetitive, hypermethylated DNA in the genome of *Physarum polycephalum. Nucleic Acids Res.* 1991;19:279–286.

208. Russell PJ, Hambidge SJ, Kirkegaard K. Direct introduction and transient expression of capped and non-capped RNA in *Saccharomyces cerevisiae. Nucleic Acids Res* 1991;19:4949–4953.

209. Sakaguchi T, Matsuda Y, Kiyokage R, et al. Identification of endoprotease activity in the trans Golgi membranes of rat liver cells that specifically processes *in vitro* the fusion glycoprotein precursor of virulent Newcastle disease virus. *Virology* 1991;184:504–512.

210. Sandmeyer SB. Yeast retrotransposons. *Curr Opin Genet Dev* 1992;2:705–711.

211. Scheffter S, Widmer G, Patterson JL. Complete sequence of *Leishmania* RNA virus 1-4 and identification of conserved sequences. *Virology* 1994;199:479–483.

212. Schmitt MJ, Tipper DJ. Genetic analysis of maintenance and expression of L and M double-stranded RNAs from yeast killer virus K28. *Yeast* 1992;8:373–384.

213. Sclafani RA, Fangman WL. Conservative replication of double-stranded RNA in *Saccharomyces cerevisiae. Mol Cell Biol* 1984;4:1618–1626.

214. Sellem CH, Lecellier G, Belcour L. Transposition of a group II intron. *Nature* 1993;366:176–178.

215. Shapira R, Choi GH, Nuss DL. Virus-like genetic organization and expression strategy for a double-stranded RNA genetic element assocaited with biological control of chestnut blight. *EMBO J* 1991;10:731–739.

216. Shapira R, Nuss DL. Gene expression by a hypovirulence-associated virus of the chestnut blight fungus involves two papain-like protease activities. *J Biol Chem* 1991;266:19419–19425.

217. Shen Y, Bruenn JA. RNA structural requirements for RNA binding, replication, and packaging in the yeast double-stranded RNA virus. *Virology* 1993;195:481–491.

218. Shope RE. An antiviral substance from *Penicillium funiculosum.* I. Effect upon infection in mice with swine influenza virus and Columbia SK encephalomyelitis viruses. *J Exp Med* 1953;97:601–625.

219. Sinden JW, Hauser E. Report on two new mushroom diseases. *Mushroom Sci* 1950;1:96–100.

220. Singh AC, Helms C, Sherman F. Mutation of the non-Mendelian suppressor [PSI] in yeast by hypertonic media. *Proc Natl Acad Sci USA* 1979;76:1952–1956.

221. Somers JM, Bevan EA. The inheritance of the killer character of yeast. *Genet Res* 1968;13:71–83.

222. Sommer SS, Wickner RB. Yeast L dsRNA consists of at least three distinct RNAs; evidence that the non-Mendelian genes [HOK], [NEX] and [EXL] are on one of these dsRNAs. *Cell* 1982;31:429–441.

223. Sommer SS, Wickner RB. Gene disruption indicates that the only essential function of the SKI8 chromosomal gene is to protect Saccharomyces cerevisiae from viral cytopathology. *Virology* 1987;157:252–156.

224. Stanway CA, Buck KW. Infection of protoplasts of the wheat take-all fungus, *Gaeumannomyces graminis* var. *tritici,* with double-stranded RNA viruses. *J Gen Virol* 1984;65:2061–2066.

225. Steiner DF, Smeekens SP, Ohagi S, Chan SJ. The new enzymology of precursor processing endoproteases. *J Biol Chem* 1992;267:23435–23438.

226. Strasser P, Zhang Y, Rohozinski J, Van Etten JL. The termini of the Chlorella virus PBCV-1 genome are identical 2.2 kbp inverted repeats. *Virology* 1991;180:763–769.

227. Stuart kd, Weeks R, Guilbride L, Myler PJ. Molecular organization of *Leishmania* RNA virus 1. *Proc Natl Acad Sci USA* 1992;89:8596–8600.

228. Stucka R, Lochmuller H, Feldmann H. Ty4, a novel low-copy number element in *Saccharomyces cerevisiae*: one copy is located in a cluster of Ty elements and tRNA genes. *Nucleic Acids Res* 1989;17:4993–5001.

229. Stucka R, Schwarzlose C, Lochmuller H, Hacker U, Feldmann H. Molecular analysis of the yeast Ty4 element: homology with Ty1, copia and plant retrotransposons. *Gene* 1992;122:119–128.

230. Sturley SL, El-Sherbeini M, Kho S-H, LeVitre JL, Bostian KA. Acquisition and expression of the killer character in yeast. In: Koltin Y, Leibowitz MJ, eds. *Viruses of fungi and simple eukaryotes.* New York: Marcel Dekker; 1988:179–208.

231. Tarr PI, Aline RF, Smiley BL, Scholler J, Keithly J, Stuart K. LR1: a candidate RNA virus of *Leishmania. Proc Natl Acad Sci USA* 1988;85:9572–9575.

232. Tercero JC, Dinman JD, Wickner RB. Yeast *MAK3* N-acetyltransferase recognizes the N-terminal four amino acids of the major coat protein (*gag*) of the L-A double-stranded RNA virus. *J Bacteriol* 1993;175:3192–3194.

233. Tercero JC, Riles LE, Wickner RB. Localized mutagenesis and evidence for post-transcriptional regulation of *MAK3*, a putative N-acetyltransferase required for dsRNA virus propagation in *Saccharomyces cerevisiae. J Biol Chem* 1992;267:20270–20276.

234. Tercero JC, Wickner RB. *MAK3* encodes an N-acetyltransferase whose modification of the L-A *gag* N-terminus is necessary for virus particle assembly. *J Biol Chem* 1992;267:20277–20281.

235. Thiele DJ, Hannig EM, Leibowitz MJ. Multiple L double-stranded RNA species of *Saccharomyces cerevisiae:* evidence for separate encapsidation. *Mol Cell Biol* 1984;4:92–100.

236. Thiele DJ, Leibowitz MJ. Structural and functional analysis of separated strands of killer double-stranded RNA of yeast. *Nucleic Acids Res* 1982;10:6903–6918.

237. Thrash C, Voelkel K, DiNardo S, Sternglanz R. Identification of *Saccharomyces cerevisiae* mutants deficient in DNA topoisomerase I. *J Biol Chem* 1984;259:1375–1379.

238. Toh-e A, Guerry P, Wickner RB. Chromosomal superkiller mutants of Saccharomyces cerevisiae. *J Bacteriol* 1978;136:1002–1007.

239. Toh-e A, Wickner RB. "Superkiller" mutations suppress chromoso-

mal mutations affecting double-stranded RNA killer plasmid replication in *Saccharomyces cerevisiae. Proc Natl Acad Sci USA* 1980;77:527–530.

240. Tuite MF, Cox BS, McLaughlin CS. A ribosome-associated inhibitor of *in vitro* nonsense suppression in [psiØ] strains of yeast. *FEBS Lett* 1987;225:205–208.
241. Tuite MF, Mundy CR, Cox BS. Reagents that cause a high frequency of genetic change from [*psi*+] to [*psi*Ø] in *Saccharomyces cerevisiae. Genetics* 1981;98:691–711.
242. Turkel S, Farabaugh PJ. Interspersion of an unusual GCN4 activation site with a complex transcriptional repression site in Ty2 elements of *Saccharomyces cerevisiae. Mol Cell Biol* 1993;13:2091–2103.
243. Tyagi AK, Wickner RB, Tabor CW, Tabor H. Specificity of polyamine requirements for the replication and maintenance of different double-stranded RNA plasmids in Saccharomyces cerevisiae.*Proc Natl Acad Sci USA* 1984;81:1149–1153.
244. Uemura H, Wickner RB. Suppression of chromosomal mutations affecting M1 virus replication in Saccharomyces cerevisiae by a variant of a viral RNA segment (L-A) that encodes coat protein. *Mol Cell Biol* 1988;8:938–944.
245. Van Arsdell SW, Stetler GL, Thorner J. The yeast repeated element sigma contains a hormone-inducible promoter. *Mol Cell Biol* 1987;7:749–759.
246. Van Etten JL, Lane LC, Meints RH. Viruses and viruslike particles of eukaryotic algai. *Microbiol Rev* 1991;55:586–620.
247. Vodkin MH, Katterman F, Fink GR. Yeast killer mutants with altered double-stranded ribonucleic acid. *J Bacteriol* 1974;117:681–686.
248. Voytas DF, Boeke JD. Yeast retrotransposon revealed. *Nature* 1992;358:717.
249. Wang A, Wang CC, Alderete JF. *Trichomonas vaginalis* phenotypic variation occurs only among trichomonads infected with the double-stranded RNA virus. *J Exp Med* 1987;166:142–150.
250. Wang AL, Wang CC. Discovery of a specific double-stranded RNA virus in *Giardia lamblia. Mol Biochem Parasitol* 1986;21:269–276.
251. Wang AL, Wang CC. The double-stranded RNA in *Trichomonas vaginalis* may originate from virus-like particles. *Proc Natl Acad Sci USA* 1986;83:7956–7960.
252. Wang AL, Wang CC. Viruses of the protozoa. *Annu Rev Microbiol* 1991;45:251–263.
253. Wang AL, Yang H-M, Shen KA, Wang CC. Giardiavirus double-stranded RNA genome encodes a capsid polypeptide and a gag-pol-like fusion protein by a translational frameshift. *Proc Natl Acad Sci USA* 1993;90:8595–8589.
254. Wang I-N, Li Y, Que Q, et al. Evidence for virus-encoded glycosylation specificity. *Proc Natl Acad Sci USA* 1993;90:3840–3844.
255. Warmington JR, Waring RB, Newlon CS, Inge KJ, Oliver SG. Nucleotide sequence characterization of Ty1-17, a class II transposon from yeast. *Nucleic Acids Res* 1985;13:6679–6693.
256. Weeks R, Aline RF, Myler PJ, Stuart K. LRV1 viral particles in *Leishmania guyanensis* contain double-stranded or single-stranded RNA. *J Virol* 1992;66:1389–1393.
257. Welsh JD, Leibowitz MJ, Wickner RB. Virion DNA-independent RNA polymerase from Saccharomyces cerevisiae. *Nucleic Acids Res* 1980;8:2349–2363.
258. Wesolowski M, Wickner RB. Two new double-stranded RNA molecules showing non-Mendelian inheritance and heat inducibility in *Saccharomyces cerevisiae. Mol Cell Biol* 1984;4:181–187.
259. Wickner RB. Deletion of mitochondrial DNA bypassing a chromosomal gene needed for maintenance of the killer plasmid of yeast. *Genetics* 1977;87:441–452.
260. Wickner RB. Host function of MAK16: G1 arrest by a mak16 mutant of *Saccharomyces cerevisiae. Proc Natl Acad Sci USA* 1988;85:6007–6011.
261. Wickner RB. Yeast RNA virology: the killer systems. In: Broach JR, Pringle JR, Jones EW, eds. *The molecular and cellular biology of the yeast Saccharomyces: genome dynamics, protein synthesis, and energetics.* Plainview, NY: Cold Spring Harbor Laboratory; 1991:263–296.
262. Wickner RB. Double-stranded and single-stranded RNA viruses of *Saccharomyces cerevisiae. Annu Rev Microbiol* 1992;46:347–375.
263. Wickner RB. Evidence for a prion analog in *S. cerevisiae:* the [URE3] non-Mendelian genetic element as an altered URE2 protein. *Science* 1994;264:566–569.
264. Wickner RB, Icho T, Fujimura T, Widner WR. Expression of yeast L-A double-stranded RNA virus proteins produces derepressed replication: a ski phenocopy. *J Virol* 1991;65:155–161.
265. Wickner RB, Leibowitz MJ. Two chromosomal genes required for killing expression in killer strains of Saccharomyces cerevisiae. *Genetics* 1976;82:429–442.
266. Wickner RB, Ridley SP, Fried HM, Ball SG. Ribosomal protein L3 is involved in replication or maintenance of the killer double-stranded RNA genome of *Saccharomyces cerevisiae. Proc Natl Acad Sci USA* 1982;79:4706–4708.
267. Widmer G, Comeau AM, Furlong DB, Wirth DF, Patterson JL. Characterization of a RNA virus from the parasite *Leishmania. Proc Natl Acad Sci USA* 1989;86:5979–5982.
268. Widner WR, Matsumoto Y, Wickner RB. Is 20S RNA naked? *Mol Cell Biol* 1991;11:2905–2908.
269. Widner WR, Wickner RB. Evidence that the *SKI* antiviral system of *Saccharomyces cerevisiae* acts by blocking expression of viral mRNA. *Mol Cell Biol* 1993;13:4331–4341.
270. Williams TL, Leibowitz MJ. Conservative mechanism of the *in vitro* transcription of killer virus of yeast. *Virology* 1987;158:231–234.
271. Williamson VM, Cox D, Young ET, Russell DW, Smith M. Characterization of transposable element-associated mutations that alter yeast alcohol dehydrogenase II expression. *Mol Cell Biol* 1983;3:20–31.
272. Winston F. Analysis of SPT genes: a genetic approach toward analysis of TFIID, histones, and other transcription factors of yeast. In: McKnight SL, Yamamoto KR, eds. *Transcription regulation.* Cold Spring Harbor, NY: Cold Spring Harbor Laboratory; 1992:1271–1293.
273. Winston F, Chaleff DT, Valent B, Fink GR. Mutations affecting Ty-mediated expression of the *HIS4* gene in *Saccharomyces cerevisiae. Genetics* 1984;107:179–197.
274. Wood HA, Bozarth RF. Heterokaryon transfer of virus-like particles associated with a cytoplasmically inherited determinant in *Ustilago maydis. Phytopathology* 1973;63:1019–1021.
275. Woods DR, Bevan EA. Studies on the nature of the killer factor produced by *Saccharomyces cerevisiae. J Gen Microbiol* 1968;51:115–126.
276. Xiong Y, Eickbush TH. Origin and evolution of retroelements based upon their reverse transcriptase sequences. *EMBO J* 1990;9:3353–3362.
277. Xu H, Boeke JD. Host genes that influence transposition in yeast: the abundance of a rare tRNA regulates Ty1 transposition frequency. *Proc Natl Acad Sci USA* 1990;87:8360–8364.
278. Xu H, Boeke JD. Localization of sequences required in *cis* for yeast Ty1 element transposition near the long terminal repeats: analysis of mini-Ty elemnets. *Mol Cell Biol* 1990;10:2695–2702.
279. Xu H, Boeke JD. Inhibition of Ty1 transposition by mating pheromones in *Saccharomyces cerevisiae. Mol Cell Biol* 1991;11:2736–2743.
280. Zaayen Av. Mushroom viruses. In: Lemke PA, ed. *Viruses and plasmids in fungi.* New York: Marcel Dekker; 1979:239–324.
281. Zhang L, Churchill ACL, Kazmierczak P, Kim D-H, Van Alfen NK. Hypovirulence-associated traits induced by a mycovirus of *Cryphonectria parasitica* are mimicked by targeted inactivation of a host gene. *Mol Cell Biol* 1993;13:7782–7792.
282. Zhang L, Villalon D, Sun Y, Kazmierczak P, Van Alfen NK. Virus-associated down-regulation of the gene encoding cryparin, an abundant cell-surface protein from the chestnut blight fungus, *Cryphonectria parasitica. Gene* 1994;139:59–64.

Fundamental Virology, Third Edition
edited by B.N. Fields, D.M. Knipe, P.M. Howley, et al.
Lippincott - Raven Publishers, Philadelphia © 1996

CHAPTER 15

Bacteriophages

Allan M. Campbell

HISTORY

Bacterial viruses (generally called bacteriophages or simply phages) were discovered independently by Twort (66) and d'Herelle (33) as filtrable, transmissible agents of bacterial lysis. Although their relevance to the study of animal and plant viruses was not universally accepted, phages proved so tractable to experimentation that information about their nature was soon forthcoming. For example, the ability of phage to form isolated plaques on bacterial lawns was noted; and d'Herelle correctly attributed the linearity of plaque assays to the self-replicating nature of the phage particles (as expected for a subcellular pathogen) rather than to the cooperative effect of many particles (as expected for a typical poison.)

The early hope that phage might be useful in combating pathogenic bacteria (either therapeutically or in decontamination of the human environment, such as in purifying water supplies) did not materialize, although occasional therapeutic applications are made. They are most effective where susceptible bacteria are concentrated in lesions or abscesses, where large amounts of phage can be injected. Starting in the 1920s, a more successful and widespread use of phage was in the typing of bacterial strains of medical interest by testing their sensitivity to a battery of phages.

A new era in phage research was initiated in the early 1940s by Max Delbrück, Salvador Luria, and their disciples. Their efforts laid the groundwork for the use of phage in the fundamental experiments of molecular biology; but the work itself is notable for its deemphasis of biochemistry and its focus on basic biological information relevant to the mechanism of self-replication. The techniques used were simple, mainly microbiological, and some of the questions investigated had already received indicative answers in the previous two decades. What was new was an emphasis on sharp, incisive answers, quantitative where appropriate, and the development of a few standardized systems suitable for obtaining such answers. Conditions were determined for the rapid, reproducible attachment of phage to bacterial cells, allowing synchronous infection of all the cells in a culture with the resulting "one-step" growth. The effect of varying the ratio of phage to bacteria (multiplic-

A.M. Campbell: Department of Biological Sciences, Stanford University, Stanford, CA 94305

ity of infection) was analyzed as a problem in random statistics, to which the Poisson distribution could properly be applied. It was also shown that synchronously infected cells pass through an eclipse phase, where infectious phage is unrecoverable from infected cells, and that infectious particles appear at a later time (2). The ability of one phage to exclude another and of coinfecting phage to interact productively either genetically (by recombination) or phenotypically (by complementation or phenotypic mixing) or both (as in the production of viable phage from cells multiply infected with radiation-damaged particles) was rigorously documented. Quantitative analysis at a comparable level was introduced into animal virology with the development of plaque assays on tissue monolayers (21). Thus, the foundations of modern animal virology (independent of its molecular aspects) go back to the phage work of the 1940s.

Studies of phage genetics started in the 1940s with the construction of recombinational maps (36) and continued in the 1950s with Benzer's analysis of fine structure genetics of the T4*r*II genes (8) and with the discovery and use of conditionally lethal mutations in λ and T4, which allowed dissection of the developmental programs of phages (12,23). The *r*II work depended conceptually on the prior discovery of phenotypic mixing (53), which showed that when mutants of the same phage coinfect a cell, phage gene products are drawn from a common pool, and the conditional lethal work depended on the operational protocol laid out in the *r*II studies.

Serious phage biochemistry was initiated in the 1930s, when Schlesinger determined that phage particles were about 50% DNA and 50% protein (60). It did not enter into the mainstream of phage work until the 1950s, when Hershey and Chase showed by radiolabeling that the phosphorus (DNA) entered the cell during T2 infection but that most of the sulfur (protein) did not (35). The experiment followed on the demonstrations of the eclipse phase and electron microscopy, showing that empty phage heads remain attached to the cell envelope and that the DNA of virions could be separated from the protein shell by osmotic shock (1). Concurrently, the enzymology of T4 infection was developed in detail, clearly indicating that the phage was introducing instructions for proteins present neither in the uninfected cell nor in the virion (17).

It was reported in the 1920s that many bacterial strains harbored phage permanently and secreted it into the medium, so that every culture contained phage. Such strains are called lysogenic. The Luria-Delbrück school initially chose to ignore lysogeny, in part because of the possibility that the phenomenon might be explainable simply by the adherence of extracellular phage particles to the bacterial surface and their occasional detachment from it. Increased rigor was introduced by Lwoff et al. around 1950 in a series of experiments showing first, that when individual cells were allowed to divide and daughter cells separated into different drops of liquid by micromanipulation, both daughters retained the ability to produce phage; second, that an occasional cell lysed and liberated a burst of phage (so that it was unnecessary to postulate secretion by living cells); and third, that certain treatments (such as ultraviolet light or other agents that induce the SOS response) induce mass lysis and phage liberation by all cells of the culture (49). As with infected cells in the eclipse phase, no infectious phages were found when lysogenic cells were artificially disrupted.

Phages capable of producing lysogenic complexes are called temperate, and the noninfectious form in which phage genetic information is transmitted is called prophage. Of the temperate phages, coliphage λ has been studied most intensively, partly because it grows in the K-12 strain of *Escherichia coli*, where conjugational and transductional mapping were possible by the 1950s. It was soon shown that prophage has a specific location on the bacterial map, and later that its genome is linearly inserted into the continuity of the bacterial chromosome (13,43).

From 1950 onward, the virtues of a simple identifiable genome with rigorous genetics that could be simultaneously introduced into (or induced to replicate in) all the cells of a culture made phage the favored experimental material for many of the classical experiments of molecular biology, such as the demonstration of DNA breakage and joining during genetic recombination and the proof of the triplet code. In the 1960s, λ, in particular, become the favored object for study of the genetic hierarchy controlling a simple developmental program. Although the initial investigations on conditionally lethal mutations were pursued in far more depth with T4 than with λ, λ had certain inherent advantages: the initiation of the entire program of the lytic cycle from two promoters directly controlled by repressor, the uniformity of the DNA molecules extracted from virions, and the specific probes for different parts of the genome provided by specialized transducing phages and deletions or substitutions (whose locations could be seen via heteroduplex mapping) made the effects of genetic manipulation on molecular events readily interpretable; and the existence of alternative pathways toward lysis or lysogeny added to the interest.

Since 1970, phage work merges with the rest of biology because the invention of artificial cloning allows the application to many systems of the techniques early available for phage. The background data on phage systems, plus their facile manipulation, continues to allow their study at a high level of sophistication.

VIRULENT PHAGES

Like viruses in general, phages can be divided into those with RNA genomes (mostly small and single stranded), those with small DNA genomes (generally less than 10 kb, mostly single stranded), and those with medium to large DNA genomes (30–200 kb). The last group includes most of the temperate phages, as well as the first virulent phages to be studied intensively.

Large DNA Phages

The classical T coliphages selected for study by the Luria-Delbrück school fall into four groups: T1 (about 50 kb); T2, T4, and T6 (about 170 kb); T3 and T7 (40 kb); and T5 (about 130 kb). The members of a group have similar virion morphology and can produce viable recombinants in mixed infection.

Phage T4

Virion

T4, the object of many classical experiments, is an appropriate prototype. More than 200 T4 genes have been identified through mutational studies. The virion contains 43 phage-encoded proteins: 16 are located in the head that encapsidates the DNA and 27 form the tail, through which DNA passes into the cell during infection. The head is an elongated T = 13 icosahedron with an extra equatorial row of capsomeres. The tail consists of a hollow core (surrounded by a contractile sheath) that terminates in a baseplate to which are attached six fibers. The whole apparatus functions as a syringe for injection of phage DNA into the interior of the cell. The tips of the tail fibers make initial contact with the lipopolysaccharide surface receptors on the bacterium. Once the phage is anchored by tail fiber attachment, random motion brings the baseplate in contact with the cell surface. Either this contact or tail fiber attachment itself triggers conformational changes in the tail: the center of the baseplate opens like a shutter, allowing the tip of the core to pass through, and the sheath contracts, driving the core through the cell envelope, after which DNA is released into the cell. The double-stranded DNA is distinctive in containing hydroxymethylcytosine (heavily glucosylated) rather than cytosine. It is cyclically permuted and terminally repetitious; i.e., the sequences of various molecules can be represented as ABCDEFGHAB, DEFGHABCDE, GHABCDEFGH, etc. This is most simply demonstrated by allowing the DNA to separate into single strands and then reannealing with complementary strands from other virions, generating circular duplexes with single-stranded tails.

DNA Transactions

Infection triggers a massive degradation of cytosine-containing host DNA with subsequent enzymatic conversion of deoxycytosine monophosphate to deoxyhydroxymethylcytosine monophosphate. The phage encodes a battery of enzymes that perform these and other conversions of deoxyribonucleotides that are then used in phage DNA replication. Bidirectional replication is initiated at several potential replication origins along the DNA, and elongation is performed by a complex of phage-encoded enzymes, leading and lagging strand synthesis proceeding coordinately from a single replication complex. Because there is no mechanism for priming synthesis at the extreme 5′ end of the lagging strand, this first round of replication produces linear molecules with protruding 3′ ends.

Soon after infection, these early replication origins cease to function (because replication there requires priming by RNA made by host RNA polymerase, whose promoter recognition specificity is altered). Late replication initiates at recombination intermediates formed by invasion of 3′ ends into homologous double-stranded DNA. Because of the terminal redundancy, such homologous sequences are available, even in single infection. As replication proceeds, repeated invasions of this kind generate a complex network of intracellular viral DNA, which includes end-to-end concatemers (the preferred packaging substrate for T4). In packaging, empty heads are first assembled, then filled with DNA. Cutting is coordinated with packaging, so that headful lengths are cut to size after the head is filled. The packaging length exceeds the genome length, hence the terminal redundancy. The position of molecular ends along the concatemer is close to random.

This picture of the DNA transactions has genetic consequences; in fact, much of it was inferred from genetic studies that preceded the biochemistry and molecular biology. First, the recombination rate in T4 infection is very high, producing a linkage map thousands of centimorgans in length. It was early noted that this corresponded to about one recombination per replication cycle. Second, the linkage map is circular, as expected if virion DNA has random endpoints. Third, when cells are infected with phages of two different genotypes, some of the progeny particles can carry different information within the terminal duplication; for example, in a mixed infection between T4 and T4*h*, some progeny plaques contain phage of both *h* and h^+ genotypes. Another mechanism that can also generate such mixed plaques is packaging of heteroduplex DNA recombination intermediates. Both types of heterozygous progeny are in fact demonstrable (63).

Finally, there are the genetic implications of headful packaging. If the length of the terminal overlap represents the difference between packaging length (determined by head size) and genome length, then that length should change in a predictable manner if either packaging length or genome length is deliberately altered. The first test was to alter genome length with a nonlethal deletion. The prediction that this would increase the length of the terminal overlap (and therefore the fraction of particles heterozygous for any locus in the genome) was fulfilled (65). Later experiments with other phages that use headful packaging, such as P22, showed that increasing the genome length by insertion of extra DNA has the reverse effect of decreasing the terminal overlap.

The converse experiment, where genome length remains constant but packaging length changes, was accomplished through examination of virions with abnormal morphology. Preparations of T4 contain a small fraction of virions whose heads are isometric rather than elongated, separable from the majority type by ultracentrifugation. Certain treatments (such as phage development in the presence of amino acid analogs) induce formation of giant particles with the same diameter as normal virions but additional rows of capsomeres, approaching helical tubes in shape. Missense mutations in the major capsid protein can increase the proportion of both isometric and giant particles (19). Both types of abnormal particle inject their DNA into infected cells. The packaging capacity of isometric particles is about 67% that of normal virions; accordingly, single infection is unproductive, but multiple infection allows a normal cycle through recombination between DNA molecules of less than genome length with random endpoints. Giant particles have longer than normal DNA (up to several genome lengths) and a correspondingly high degree of heterozygosity when coming from mixed infection. They are also highly resistant to ultraviolet light because of complementation and recombination between damaged genomes.

Regulation

The genes whose products are needed for phage DNA synthesis and host DNA breakdown, including those mediating nucleotide metabolism and the seven proteins that make up the replication complex, are all expressed immediately after infection, being transcribed from promoters recognizable by *E. coli* RNA polymerase with σ70. With time, early synthesis is shut off and other genes that code for virion components and lysis genes are activated. The shutoff of early genes is effected both at the transcriptional and the translational level. Translational control is sometimes autogenous, as for the DNA polymerase gp43 and the single-stranded binding protein gp32; many other early genes are controlled by a dedicated translational repressor, product of the *reg*A gene, itself an early gene subject to autogenous control (3,11,71).

Transcriptional control operates on three different types of T4 promoters (28). The early promoters have sequences approximating the consensus for *E. coli* σ70 (with a small but apparently real difference in consensus). Middle promoters require *E. coli* polymerase with σ70, in addition to a T4-specified positive regulator, *motA*. Their hallmark is a sequence TGCTT at around –32. Late promoters follow –12 consensus TATAAATA. They are transcribed by *E. coli* holoenzyme, with a phage-coded σ factor (gp55) replacing σ70. In normal infection, late promoter activation requires concurrent DNA replication. *In vitro*, activation can be achieved by nicks or gaps on the nontranscribed strand near the promoter; treatments that can create damage *in vivo* can also induce some replication-independent transcription. Thus, the timing of late transcription is prescribed by the requirement for phage-specific factors (including gp55) and also by the modification of the template through introduction of interruptions during DNA synthesis.

In vivo, late transcription requires the integrity of all the DNA replication genes. *In vitro*, activation of late transcription by nicks or gaps requires three replication proteins gp45 and gp42/66, which are associated with the phage-coded polymerase gp43 during replication. An RNA polymerase-associated protein, gp33, is also required. An attractive model is that gp45 and gp42/66 recognize the DNA interruption and gp33 receives the enhancement signal thus created. Unlike other enhancers, transcription cannot be stimulated by bringing the DNA interruption spatially close to the promoter through catenation of DNA molecules; enhancement is only seen when all signals are on the same DNA double helix (34).

Some other ancillary changes that occur during infection may reinforce the temporal sequence but are not essential for it. For example, the α subunit of RNA polymerase becomes adenosine diphosphate—ribosylated a few minutes after infection by the action of either of two phage gene products: gpalt, which is activated during virion assembly and injected along with the DNA, and gpmod. ADP-ribosylation modifies polymerase specificity, but both successful infection and shutoff of host promoters occur in its absence (in *alt*/*mod* double mutants.) Shutoff of host transcription is also effected by a T4-encoded transcriptional terminator (gpalc) which recognizes local substitution of HMC for C near the terminator site; thus the chemical marking of phage DNA by HMC is used not only in the specific degradation of host DNA but also in reducing transcription before degradation (39). And a T4-encoded inhibitor of σ70 could explain how gp55 supplants σ70 at late times, but again this protein is nonessential for late gene function (55). T4 also has some noteworthy features with no known regulatory role, such as genes that are discontinuous because of intron splicing or ribosomal skipping (6,70).

Assembly

Once late genes are expressed, the stage is set for assembling virions. The T4 virion was the first biological structure of comparable complexity whose pathways of assembly were worked out, from a combination of physical and genetic knowledge. The major technical factors facilitating the study were, first, the availability of conditionally lethal mutations in genes for individual virion proteins; second, the fact that the final steps of the pathway were readily executed *in vitro* so that reactions in the pathway could be distinguished from irrelevant interactions or side reactions that do not contribute to the yield of functional plaque-forming particles; and third, the facility of operations by then standard in phage work, such as assay for complexes stable to dilution.

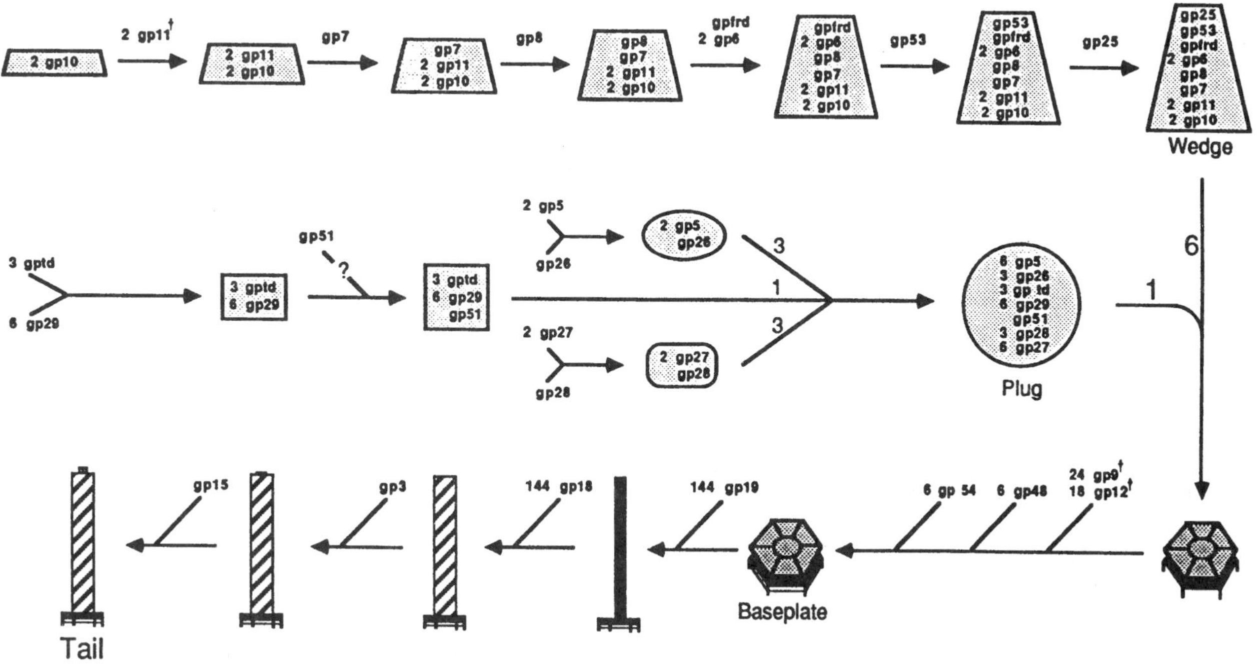

FIG. 1. Assembly pathway of T4 tails. Proteins must be added in the indicated order except that gp11, gp12, and gp9 (indicated by †) can be added at later times as well. From Casjens and Hendrix (14), with permission.

The result of the studies suggested several general principles whose rationalization as effective strategies and extension to other phages has been subsequently assessed and updated (26). One of these is the repeated use of subassembly pathways rather than a single linear pathway. Thus, heads, tails, tail fibers, and soluble protein (gp63) catalyzing tail fiber addition are all made separately. Heads and tails combine first to form complexes that are both visible in the electron microscope and stable to dilution, and finally tail fibers are added to the complex. The tail consists of a baseplate (to which the fibers attach) composed of a central hub surrounded by six wedges. Hub and wedges are made separately, then combined to form the baseplate (Fig. 1). Likewise, in tail fiber assembly, the distal and proximal segments are assembled separately, then combined immediately before addition to the head/tail complex. The most obvious advantage of subassembly pathways is that mistakes in assembly can be rejected for each pathway separately rather than ruining entire virions.

A second general rule is that linear elements are built from the end in (toward their junction with the rest of the virion). In tail assembly, baseplates are made first, then the core and sheath are added to them. Likewise, wedges are formed by adding different protein units successively from one of the outermost units, finishing with the one that interacts with the hub. This strategy ensures that the intermediates of the subassembly pathway cannot compete with the final product for assembly into the virion. The order of assembly of the wedges themselves may be prescribed by allosteric changes on addition to the complex or by the recognition of binding sites that contain elements of more than one of the proteins already present. In fiber addition to head-tail complexes, fibers do not add efficiently to base plates until heads and tails are joined; apparently because the bend in the fibers contacts the head tail junction and interacts with it. A protein (wac) in the collar serves this function *in vitro*, although gpwac is not essential for plaque formation *in vivo*. In free virions, fibers are frequently wrapped around the tail sheath, becoming extended only under specific conditions.

The joining of fibers to head/tail complexes attracted attention early because it proceeds by catalysis of a noncovalent interaction. The catalyst (gp63) also functions as an RNA ligase that apparently acts during (but is not essential for) DNA synthesis. Protein catalysis of a one-step noncovalent interaction was novel at the time, and the discovery was too far in advance of later work on chaperonins to have wide impact.

Tail assembly raised the question of length determination. The core and sheath are both formed by polymerization of single protein species. Core assembly is initiated through nucleation on the baseplate, followed by addition of subunits to build the structure. How is the number of subunits counted, so as to ensure the observed uniformity among virions? The answer (better documented for λ than for T4) is by use of a linear measuring stick protein that is embedded within the core and eliminated when assembly is complete (40).

The T4 tail fiber is a bent linear structure with several specific protein monomers or dimers arranged end to end, terminating in a dimer of gp37. This protein at the tip of the tail fiber is so arranged that its amino acid sequence (and hence its genetic map) is colinear with the fiber axis. This was shown in T2-T4 hybrids recombinant within gene 37, where specific antibodies decorate that part of the structure derived from the phage against which the antibodies are directed (5). The specificity for host cell receptors and neutralizing antibodies lies in the distal portion, near the tip, which is also the location of mutations for altered host range.

The pathway of capsid formation in T4 (as for isometric capsids of any virus) is still understood in less depth than tail synthesis, but several established features are noteworthy. Some of these are better understood for other phages (λ, P22) and inferred for T4. The capsid is synthesized first, and it is filled with DNA later. This fits with the determination of DNA content by head size, but is more directly established by the association of capsids assembled early with DNA made late in infection (47). Second, capsids are not built directly into their final form; rather, capsid proteins associate with an internal scaffolding protein, which fills the interior of the prohead and is later extruded through holes in the protein shell, which expands to its final shape before DNA enters. Both scaffolding protein (gp24) and the major head protein (gp23) have a shape-determining role [inferred from the aberrations in shape caused by certain missense mutations (11)]. Probably both proteins are added to the structure concurrently. How these proteins actually determine the elongated icosahedral shape is unknown, and the intermediate states are no better understood than for any other closed lattice. After the prohead expands and assumes a more obviously icosahedral shape, two decoration proteins, gphoc and gpsoc, are added at regular positions in the lattice; neither protein is essential for formation of functional virions. Third, assembly requires nucleation from one vertex, in which minor capsid proteins, noncapsid proteins, and host chaperonins (groEL) all participate. This unique vertex becomes the portal vertex through which DNA enters the virion and to which the tail later attaches. DNA packaging requires two phage proteins (gp16 and gp17) that probably act at the portal vertex but are not present in the final virion. gp17 has been identified as the endonuclease that cuts the DNA from the concatemer during packaging (26). Another packaging gene, gp49, is needed for processing the DNA substrate. Only linear DNA is packaged, but intracellular T4 DNA becomes highly branched during replication and recombination; gp49 cleaves Holliday junctions and other branched structures, creating packageable lengths of linear DNA (9). Fourth, processing of gp23 takes place concurrently with assembly. Figure 2 presents the general scheme for virion assembly in double-stranded DNA phages.

Lysis

The final stage of the T4 cycle is cellular lysis, releasing virions into the medium. Lysis requires two gene products: a lysozyme (gpe), which attacks the bonds joining *N*-acetylglucosamines in the rigid murein layer, and a holin (gpt), which creates holes in the inner membrane, allowing the lysozyme to reach its substrate. Other genes affect the timing of lysis. The status of gpt as a holin is not fully established and rests largely on analogy to λ gpS (73). In *e* t^+ mutants, the cell dies and the membrane potential collapses at the normal time of lysis, whereas in e^+ *t* mutants, the cell continues to produce lysozyme and intracellular phage long after the normal time of lysis.

It is obviously desirable that the time of lysis be coordinated with the rest of the lytic cycle, but no regulatory system serving this function has been identified. The timing of lysis appears to be determined largely by the accumulation of gpt, formed coordinately with other late proteins because of transcriptional control. One genetic system that regulates time of lysis under certain conditions is the *r*II locus. T4*r* mutants were discovered early by their ability to form larger than normal plaques, and the *r*II mutants have the related property of inability to plate on strains expressing the *rex* genes of λ [which provided the appropriate selective conditions for Benzer's classical studies on fine structure genetics (7)]. However, their effect on lysis and plaque size is attributable to the absence, in *r*II mutants, of lysis inhibition. When a cell infected with T4*r*+ is superinfected before lysis by another T4 particle, lysis is delayed and intracellular phage development continues, up to a period of several hours beyond the normal latent period of 24 min. The rIIA and rIIB membrane proteins must mediate this response and must likewise destroy membrane damage inflicted by the λ *rex* system in response to superinfection, but the mechanism in unknown.

Phage T7

Each group of T phages has its unique properties. A brief account of phage T7 can indicate some of the major differences from the T4 paradigm. The phage T7 virion contains 39,936 bp of linear double-stranded DNA (completely sequenced). It is terminally redundant (like T4) but not cyclically permuted. The length of the terminal overlap is 160 bp (22). Immediately after infection, host RNA polymerase initiates transcription from a promoter near the left end of the molecule, whose products include an antirestriction factor, a protein kinase that phosphorylates and inactivates *E. coli* RNA polymerase, a new T7-specific RNA polymerase, and a DNA ligase. After a few minutes, T7 polymerase replaces host polymerase and all transcription takes place on the 81% of the genome toward the right end. The first genes transcribed by T7 polymerase

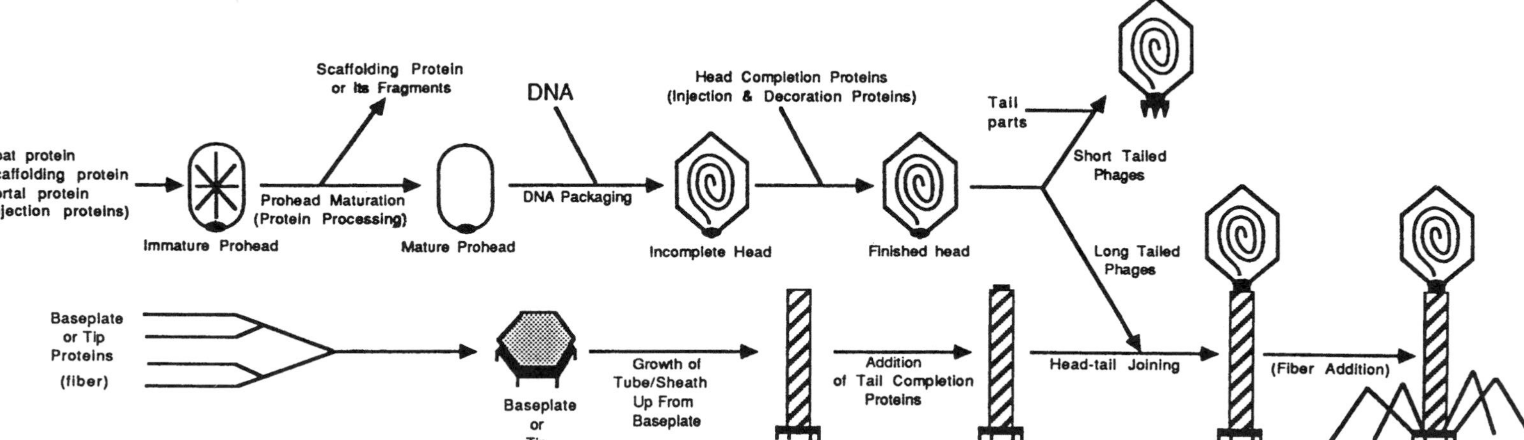

FIG. 2. Generalized assembly pathway for dsDNA phages with tailed virions, such as T4 and λ. From Casjens and Hendrix (14), with permission.

(nearer to the left end) include a DNA polymerase and recombination genes. Later, after replication ensues, the late genes encoding virion components and at least one lysis gene are expressed. Unlike T4, late gene expression does not require replication, and the host polymerase is replaced rather than reprogrammed. Even without inactivation of host polymerase, the T7 RNA polymerase competes so effectively for nucleoside triphosphates that it redirects almost all transcription to T7 promoters, a feature that has been exploited in the design of T7-based expression vectors for DNA cloning.

The first cycle of DNA replication proceeds bidirectionally from a unique origin of the linear monomer and proceeds toward the ends. As with T4, concatemers are probably generated through strand invasion of terminal segments into other molecules within the 160-bp terminal repeat. Replication then generates concatemers in which the unique 39,616 viral sequences alternate with single copies of the 160-bp repeat. If complete virion DNA molecules with repeats at both ends were cut and packaged from such concatemers, half of the genomes would lack terminal repeats. Watson (69) suggested that T4 might solve the problem by cutting the DNA in opposite strands at the two ends of the repeat, creating 160-base protruding 5′ ends, followed by DNA synthesis to produce flush ends. The actual solution is apparently more complicated and wasteful. Replication of one strand initiated from a nick near the 160-bp segment proceeds leftward through the 160-bp segment to a reverse repeat beyond the segment. The displaced strand is then nicked at the end of the reverse repeat, which folds back and initiates synthesis rightward through the 160-bp segment. Thus, two copies of the 160-bp segment are present, which can be trimmed to size and packaged (16). Packaging proceeds from right to left processively along the concatemer (64). Because T7 packages the DNA between two specific sites, the amount of DNA per head is not fixed by head size (as it is in T4). From study of T7 variants with deletions or duplications, the packaging limits of T7 are about 85% to 103% of normal genome length (F.W. Studier, personal communication).

Small DNA Phages

A class of phages package small covalently closed circles of (+) strand genomic DNA in their virions. There are two groups: isometric and filamentous. Although the two groups are not detectably related in sequence and differ in modes of attachment, packaging, and egress, their replication mechanisms are similar.

Isometric Phages

The prototype is ϕ X 174. The virion (25–35 nm) has an icosahedral T = 1 capsid, composed of three polypeptide species: gpF (60 copies), gpG (60 copies), and gpH (12 copies). gpG and gpH are considered spike proteins and can be physically removed without disrupting the integrity of the capsid. The spike proteins are necessary for attachment to cells, and gpH enters the infected cell with the DNA and participates in phage development. Such participation is not essential for all productive infection, because transfection by pure (+) strand DNA from virions leads to some phage production.

Once within the cell, the DNA becomes coated with single-stranded binding protein (Ssb), then begins to replicate. The first cycle [parental single strand → double-stranded replication form (RF)] is mediated entirely by host enzymes. Transcription by host DNA-dependent RNA polymerase at a unique site near a DNA hairpin provides a primer, which is extended by the host replication complex and then digested away. At completion of the cycle, the new DNA strand is ligated to give a covalently closed circle (RF). Transcription from RF by host RNA polymerase leads to production of phage proteins, among them a dimeric protein gpA, which works only in *cis* (i.e., an *A*- mutant cannot be complemented in *trans,* perhaps because the protein binds rapidly and irreversibly to DNA near the site of transcription). Rolling circle replication is initiated when gpA binds to RF at a specific origin site and nicks its parental (+) strand, which is then extended by action of a host-specified DNA helicase (Rep) followed by the host elongation complex. The displaced 5′ end is not free, but remains bound to the growing point through the gpA dimer. The displaced strand is coated with Ssb and remains single stranded (no lagging strand synthesis). When one genome length of DNA has been synthesized and the growing point comes to the origin, the displaced DNA is cut from the newly synthesized DNA and its ends are ligated, giving a single-stranded circle and a double-stranded (RF) circle, both of which can recycle through further rounds of replication. Thus, a fraction of the (+) strands in infected cells are linear and from monomer to dimer length, whereas both + and - strands are present as monomer circles.

With time, virion proteins accumulate and form empty proheads, and concurrently a phage-specified DNA-binding protein (gpC) is made. On some of the + strands, gpC competes with Ssb for binding to the Rep-gpA complex, and these interact with proheads to become encapsidated into virions. Thus, the DNA is directed toward packaging before it encounters the prohead (4). The replicating pool then ceases to expand as + strands are packaged. The cell cycle ends with lysis, mediated by the phage gpE protein. The mode of gpE action is uncertain, but does not involve either a lysozyme or a holin (73).

Filamentous DNA Phages

The virions of filamentous phages, of which phage fd is the prototype, differ from those of most helical viruses in that the capsid surrounds a DNA molecule that is not linear but circular and therefore is doubled back on itself, although unpaired. The replication cycle of the filamentous DNA phages is similar to that of the isometric phages. Infection is initiated by adsorption of the end of a virion to the terminus of an F-coded pilus, used during bacterial conjugation. The details of DNA entry are obscure. As with ϕ X 174, ssDNA formed at late times is coated with a phage-specified DNA-binding protein (gp5) that directs the DNA to packaging rather than to recycling. Packaging is concomitant with egress. The major coat protein (gp8) is embedded in the cell membrane. As phage DNA enters the membrane, gp5 is exchanged for a helical capsid of gp8. Subunits of gp8 are proteolytically processed as they are added to phage DNA, and virions are extruded through the cell envelope without loss of cell viability. Except for a few enveloped phages, these are the only phages whose egress does not require lysis.

The filamentous phages (especially M13) are popular as cloning vectors because of their high yields, the absence of a packaging limit, and the single-strandedness of the product. Cloning is generally accomplished through cutting and ligation of RF DNA, followed by transfection.

RNA Phages

The common RNA phages have isometric T = 1 virions each containing a single linear single-stranded RNA molecule about 4 kb in length, encoding three to four genes and capable of assuming extensive secondary structure. They enter bacterial cells by attachment to the sides of F pili (rather than the ends, as with the filamentous DNA phages). Like the filamentous phages, they are only able to infect bacteria that harbor the F plasmid. The four genes of phage MS2 encode the coat protein (C), an RNA-dependent RNA polymerase (P), a lysin (L), and a fourth protein (A for *a*ssembly) present in one copy per virion. The genes are arranged in the order 5′ A-C-(L)-P3′, where the L gene overlaps the 3′ end of C and the 5′ end of P.

In these phages, there is no known distinction between replication and transcription. All RNA copies are full length, and gene expression is controlled at the translational level. Nevertheless, they have developed a program of temporal control exquisitely tailored to their needs. The *A* gene at the 5′ end is available for translation only in nascent RNA that has not yet folded into its most stable secondary configuration, a reasonable strategy for obtaining one copy per virion. The *C* gene, on the other hand, is available for translation at all times. The *P* gene is under a more complex control. Translation of the *C* gene is needed to open the secondary structure that otherwise sequesters the *P* ribosome binding site; but coat protein itself, binding to the RNA upstream of the *P* gene, inhibits polymerase translation. The first control delays the onset of phage replication whenever conditions for protein synthesis are so poor that translation initiation of both genes is unlikely within a brief window of time; the second means that, as with T2 or T7, synthesis at late times is diverted to virion components (in this case, coat protein) with minimal competition from polymerase synthesis. The *L* gene lacks a ribosome binding site of its own. Lysin synthesis results from reattachment of ribosomes that initiated at the ribosome binding site for *C*, then dissociated due to high frequency frameshifting and reattached at the AUG for *L* (41). By tying the accumulation of lysin to the rate of coat protein synthesis, this arrangement should help coordinate the time of lysis with the rest of phage development.

In broad outline, the mechanism of RNA synthesis is similar to that of most linear single-stranded RNA viruses. RNA phage replication enjoys the distinction of being the first system in which *in vitro* replication of a nucleic acid was shown to initiate at a specific origin (32). In conjunction with three host proteins (elongation factors Tu and Ts and ribosomal protein S1, all part of the machinery for host protein synthesis), the phage polymerase initiates replication at the 3′ end of a (+) strand. Generally, several subsequent initiations take place at the same end before the first (-) strand is completed. When completed (-) strands dissociate from their (+) strand templates, they are used in a similar manner as templates for (+) strand synthesis.

Because of their small size and rapid mutation rate, RNA phage genomes have been used in studies of *in vitro* evolution, where enzyme is supplied and those RNA molecules structurally capable of the most rapid rate of multiplication are selected (37). The most consistent result was selection for much smaller molecules, which always include the terminal sequences recognized by the polymerase. The results are instructive with respect to the defective interfering particles that accumulate *in vivo* during the high multiplicity passage of many viruses. In absence of added template, the replicase can also create short sequences *de novo* and copy them (10).

TEMPERATE PHAGES

Temperate phages are mostly large (>20 kb) double-stranded DNA viruses. Three prototypes can be distinguished based on their modes of establishment of lysogeny. The first, represented by coliphage λ, insert their DNA into the chromosome at one or a few preferred sites. The second, represented by phage Mu-1, insert anywhere in the host chromosome by use of a phage-coded transposase. In the third, represented by phage P1, prophage DNA is not inserted into the chromosome but instead maintained as a plasmid. Some other phages, such as satellite phage P4, can be maintained either as inserted prophages or as plasmids. The inserted state can be stably maintained only if the phage genes for autonomous replication are repressed, whereas in the plasmid state some viral functions are expressed without killing the cell.

Phage λ

λ development has become a paradigm for gene control in developing systems. An infected cell has one of two options: either to lyse and produce viral progeny or to survive as a lysogen. Once a cell is committed to one of these pathways, there are several safeguards to stabilize the decision.

The λ virion is isometric with a T = 7 configuration of capsomeres and a long tail terminating in a single fiber. The virion DNA is a 48,502-bp linear molecule with complementary 12-base 5′ overhangs. Once injected into the cell, the DNA ends pair and are ligated by host ligase to generate a covalently closed circle.

Commitment

After infection, host RNA polymerase initiates transcription at three promoters: *p*L, *p*R, and *p*R′. The *p*L and *p*R transcripts each contain one gene (*N* and *cro*, respectively). The *p*R′ transcript contains no genes. The Cro protein binds to the operators controlling *p*L and *p*R and represses transcription from them; however, because *cro* itself is transcribed from *p*R, the Cro concentration never reaches a level where repression is complete. The gpN protein is an antiterminator, which allows transcription at *p*L and *p*R to read through ordinary transcription signals. The seminal early observations leading to this conclusion were, first, that the λ gene *exo*, which is leftward and downstream of *N*, requires for expression both a functional gpN (suppliable in *trans*) and a derepressed *p*L (in *cis*) and second, that terminators of ordinary bacterial operons such as *trp* are read through when placed downstream of *p*L in the presence of gpN (27,48).

Antitermination by gpN is effected by binding to specific RNA sequences (*nut* sites) that occur early in the transcripts from *p*L and *p*R. The *nutR* RNA includes a binding site (*boxA*) for host protein NusA and a stem-loop sequence (*boxB*) specific for λ's gpN, where NusA is the product of one of the *nus* genes, isolated by screening for mutants with an "*N* undersupply" phenotype, so that even λN^+ behaves like an N^- mutant on the *nus* mutant strains. Various λ-related phages have distinct *N* specificities determined by the sequence in the *boxB* loop. The gp*N*-NusA complex bound to *nut* RNA interacts with the RNA polymerase that has passed the *nut* site and is transcribing downstream from it and modifies transcription so that normal transcription signals are no longer recognized. The complex is stabilized by interaction with other host factors: NusB, NusG, and ribosomal protein S10 (50).

The result of antitermination is that both leftward and rightward transcription proceed beyond the primary terminators (*t*L1 and *t*R1, respectively.) The leftward transcript includes several genes whose functions are not essential to either lytic or lysogenic development, although some of them play ancillary roles in DNA replication (*exo*, β) or lysogenization (*cIII*). The rightward transcript includes *cII* (whose product promotes lysogenization), replication genes *O* and *P*, and gene *Q*, a positive regulator of late transcription.

The gpcII protein is a positive effector of transcription initiating at a site within *c*II and proceeding leftward beyond *p*R through genes *c*I and *rexAB*. The *c*I product (repressor) binds to the same operator sites as Cro and represses transcription from *p*L and *p*R, thus stopping all transcription characteristic of the lytic cycle.

If gpcII were expressed at a high rate in all infected cells, they should all become lysogenic. In fact, only some cells do. Although all cells in a culture are under the same ambient conditions, some become committed to lysis and others to lysogeny. The reason for that is still not very well understood, but a great deal is known about the players in the decision and some of the factors influencing the outcome. The gpcII protein works effectively only at high concentration (probably because the active form is a multimer); and it is metabolically unstable so that a high concentration can only be achieved by a high transcription rate (unlike Cro, which accumulates with time). The instability is due to *in vitro* proteolysis; when the host protease most responsible for gpcII degradation (HflA) is eliminated by mutation, the rate of lysogenization by λ (and especially by *c*III mutants) increases. The gpcIII protein acts by inhibiting gpcII degradation.

So the commitment to lysogeny is effected by achieving a sufficient gpcII concentration before it is too late. When is it too late? What clock is ticking? The clock is the steady accumulation of Cro, which causes a progressive decrease in the transcription rate from *p*R and therefore in the rate of gpcII synthesis. Once that drops below a critical level, the cell is irrevocably committed to the lytic cycle. The fraction of cells that commits to lysogeny increases with the multiplicity of infection; in fact, under some conditions, the lysogenization frequency of singly infected cells is essentially zero, as might happen if a single copy of the *c*II gene could not work fast enough to build up a high concentration of product.

In those cells where repression is established, further events are needed to ensure stable lysogeny. First, repressor synthesis must continue in the lysogen so that the lytic cycle functions are shut off permanently. Once repression is established, *p*L transcription stops and gpcII disappears. Repressor does not disappear as well, because *cI* transcription now commences from the maintenance promoter *p*M immediately to the left of *o*R.

The *o*R and *o*L operators are both tripartite, with three binding sites separated by short spacers. Both repressor and Cro recognize these binding sites, but with different relative affinities. Cro binds most tightly to *o*R3 (which represses leftward transcription from *p*M) and less tightly to *o*R2 and *o*R1 (repressing rightward transcription from *p*R). Repres-

TABLE 1. *Gene regulation by repressor (R) and Cro (C) at λ oR*

Leftward transciption from *p*M (cI⟵)	*o*R1	*o*R2	*o*R3	Rightward transcription from *p*R (cro⟶)
Low	—	—	—	On
On	—	R	R	Off
Off	R	R	R	Off
Off	C	—	—	On
Off	C	C	—	Off
Off	C	—	C	Off
Off	C	C	C	Off

Adapted from Ptashne (57) with permission.

sor, on the other hand, binds tightly and cooperatively to *o*R2 and *o*R1, which both represses transcription from *p*R and stimulates transcription from *p*M. At high repressor concentration (unsustainable in the steady state) repressor also binds to *o*R3 and represses its own synthesis (Table 1). Thus, once a high concentration of repressor is achieved through transcription from *p*E, repression is self sustaining (57). The inhibition by Cro of spontaneous initiation at *p*M reinforces the decision to follow the lytic pathway.

Commitment to lysogeny requires not only repression of lytic cycle genes but also insertion of phage DNA into the host chromosome. Insertion is mediated by the phage-coded integrase (Fig. 3). The structural gene for integrase, *int*, is transcribed from two promoters: the major leftward promoter *p*L and a *c*II-activated promoter *p*I. The latter is the important promoter for lysogenization. Its control by gpcII assures that insertion occurs in those cells where repression is established.

Integrase promotes reciprocal recombination between sites on phage and host DNA, which inserts the phage DNA into the continuity of the host chromosome. The recombination occurs within a 15-bp sequence that is identical in phage and host. Both *in vivo* and *in vitro*, the minimal length of specific sequence required in the *attB* partner is 21 bp and that in the *attP* partner is about 240 bp. A linear

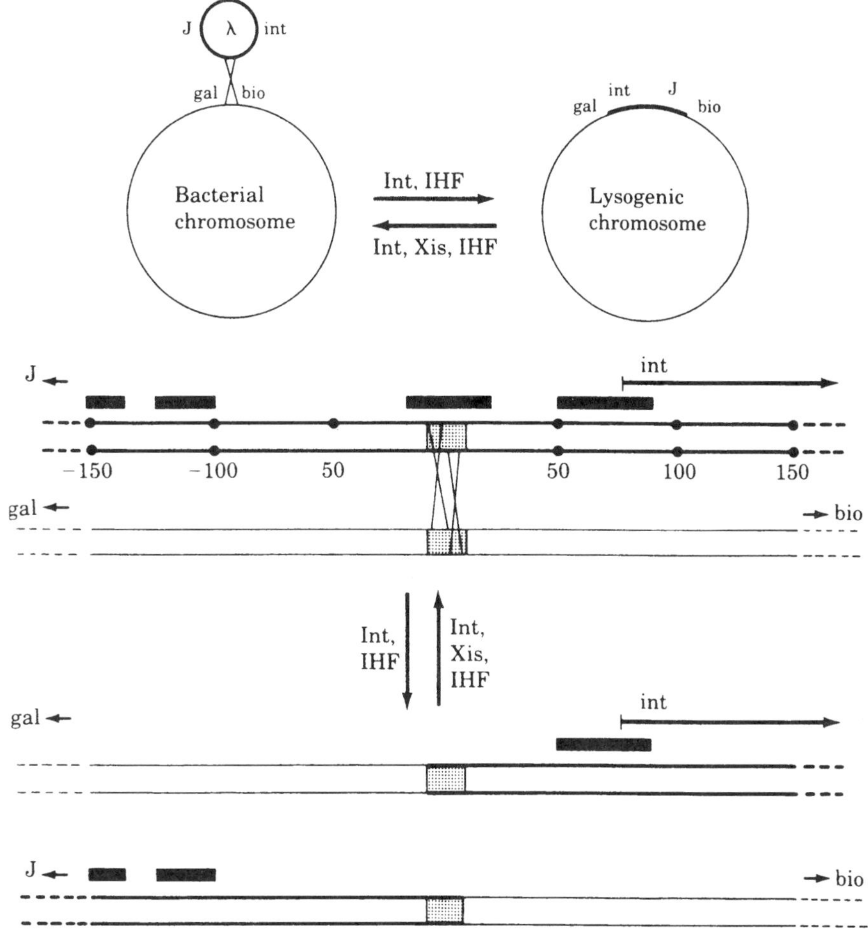

FIG. 3. Insertion of λ DNA into the bacterial chromosome. Top: Overall process. Bottom: Detail of action at the crossover point. The 15 bp identity between phage and host DNA is stippled. The heavy bars overlying the phage are Int-binding sites. The 3′ end of the *int* gene overlap the right site. (Reproduced from 12a with permission.)

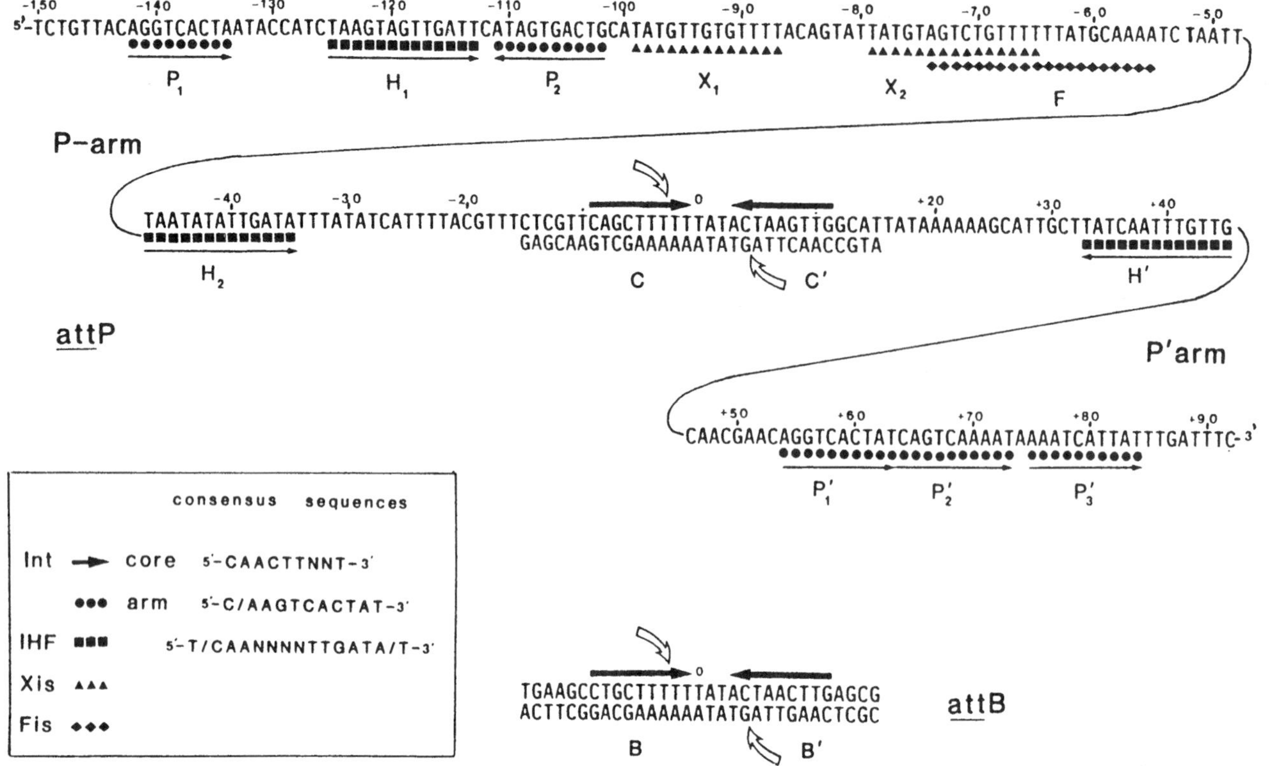

FIG. 4. Sequence of λ *attP* and λ*attB*, showing binding sites for proteins required for insertion and/or excision. P_1,P_2, P_1', p_2',P_3': Arm binding sites for Int. C,C′,B,B′: Core binding sites for Int. H_1,H_2,H′: IHF binding sites. X_1,X_2: Xis binding sites Fi: Fis binding site. Curved arrows indicate positions of strand exchange in core. (Reproduced from 42a with permission.)

attB molecule works *in vitro*, but *attP* must be supercoiled for optimal activity.

Integrase has two DNA binding sites. The strong binding sites (in the *N*-terminal part of the molecule) recognizes arm sites, which are located on *attP* at position −130, −80, +60, +70, and +80 (where 0 is the center of the 15-bp identity) (Fig. 4). Integrase molecules bound to these sites on DNA, bent appropriately by the host-coded DNA bending protein integration host factor (IHF), are so positioned that they can bind (in their *C*-terminal domains) to core sites, which are symmetrically disposed about the crossover point and approach the consensus 5′CAACTTNNT3′ (Fig. 5). Core sites are present in both *attP* and *attB*. The actual crossover then takes place by an exchange on the left between the "top" strands of the two double helices to create a Holliday junction, then branch migration 7 bp to the right, and finally resolution by exchange on the right between the bottom two strands. The left-right orientation here is determined by the positions of the arm sites on the *attP* partner, not by the core sequences themselves (42). Host and phage sites are aligned by protein-protein and DNA-protein interaction, and DNA-DNA recognition is restricted to the 7-bp overlap segment where branch migration occurs.

Commitment to lysogeny is further reinforced by a third *c*II-activated promoter pAQ, located within gene *Q* reading leftward. This antisense transcription apparently minimizes *Q*-promoted late gene transcription characteristic of the lytic cycle.

The lysis/lysogeny decision can be shifted by various external treatments, such as intracellular cyclic adenosine monophosphate concentration, ion concentration, and multiplicity of infection. Changing these variables can shift the proportion of infected cells that are channeled toward lysogeny rather than lysis. The most noteworthy (and still least understood) feature is how the system is poised so that a significant fraction of the cells almost always goes in each direction, over a wide range of ambient conditions; as though a large degree of indeterminacy were built into the system. One possible contributor to the indeterminacy may be the molecular fluctuations early in infection, such as whether the first molecule of gpN (which promotes lysogenization by increasing *cII* and *cIII* transcription) is made before or after the first molecule of Cro.

Lysogeny

Once established, the lysogenic state is quite stable. Lysogenic cells divide to give lysogenic progeny. Lysogeny can break down either through spontaneous switching to the lytic cycle (which happens about once every 10^4 cell divisions in

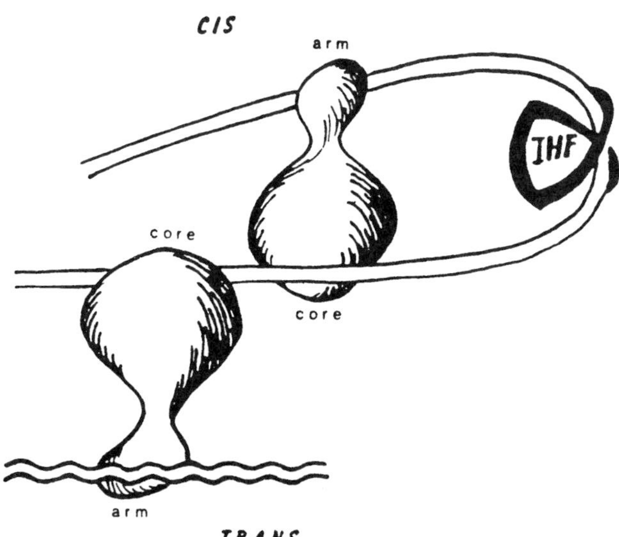

FIG 5. DNA looping by Int. The arm and core sites binding an Int monomer may either lie on the same partner (cis) or opposite partners (trans). DNA bending by IHF brings the cis sites into favorable juxtaposition. The exact configuration of the integration complex is unknown. (Reproduced from 42a with permission.)

an actively growing culture) or by spontaneous loss of prophage with cell survival (which happens less than once every 10^6 cell divisions). As long as repressor is present, *c*I transcription from *p*M continues, and more repressor is made. If repressor concentration rises so high as to saturate *o*R3, then repressor synthesis ceases until it returns to a lower level. Likewise, once the prophage is stably inserted into the chromosome, it remains so; there is a low rate of integrase synthesis from the *p*I promoter (not activated by gpcII) but essentially no excisionase (a phage-coded protein required for excision, though not for insertion).

Some ancillary factors reinforce this stability. One of these is the need for antitermination by gpN to initiate lytic development. This requirement puts a double lock on the system, because any transcripts made during a momentary unblocking of the operator sites will not get anywhere until the gpN that has been made by the first transcript can extend other transcripts. Because gpN is unstable and accidents should be rare, this will seldom happen. Stabilization of lysogeny is one reason that has been suggested for the use of antitermination by λ. Poising the lysis/lysogeny decision to depend on molecular fluctuations may be another. Explanations for antitermination have been sought because it seems simpler a priori to eliminate the terminators rather than to maintain and override them.

Escape from Lysogeny

About once every 10^4 cell divisions, stability breaks down. Detailed knowledge of what happens in those rare cells is unavailable, but the rate is strongly depressed in *recA*$^-$ hosts. This fact suggests that the rare spontaneous transitions to lytic development have the same primary basis as the mass induction of this transition by exposure to ultraviolet light or other DNA-damaging agents. In that case, the primary event is the activation of the proteolytic activity of RecA protein by products of DNA degradation, such as single-stranded DNA. RecA is not a conventional protease, but it accelerates the specific cleavage of the host LexA protein, which can autocatalyze the same cleavage under extreme conditions. LexA is a repressor for many DNA repair functions, and RecA-promoted LexA cleavage is the initial step in the SOS response that turns these genes on. Like LexA, λ repressor is sensitive to RecA-enhanced autoproteolysis (46).

Turning on the SOS system is one means of inducing the switch to lysis in almost all cells of a lysogenic culture. Another method makes use of λ *c*Its mutants, which form a thermolabile repressor. With an appropriate mutant, lysogeny is quite stable at 30°C, but a shift to 43°C rapidly inactivates repressor in all cells. After derepression, transcription initiates from *p*L and *p*R. One early consequence of *p*L transcription is excision of the prophage from the chromosome. Whereas insertion requires only one phage-coded protein, integrase, excision requires a second protein, excisionase, as well. Excisionase is encoded by a gene *xis*, upstream from *int* and overlapping it by 20 bp. The *c*II-dependent *p*I promoter used during lysogenization lies at the 5′ end of *xis*, and the start site is within *xis* so that the *p*I transcript makes only integrase. From the *p*L transcript, *int* and *xis* are expressed coordinately.

The mechanism of excisionase (Xis) action is not fully understood. Excisionase binds to two specific sites in the left arm of *attR*, bends the DNA, and interacts with integrase; one of the two excisionase molecules can be replaced by a host DNA-bending protein, Fis. Conventional enzymes are pure catalysts, and the nature of the catalysts cannot determine the direction of the reaction. Superficially, excisionase appears to reverse the integrase reaction; however, in excision as in insertion, branch migration proceeds from left to right; therefore, excision is not a true reversal of insertion. What remains unsolved is why integrase alone fails to promote the true reversal.

The *p*L transcript is subject to posttranscriptional control. In infected cells, either before commitment or after commitment to lysis, *int* and *xis* are cotranscribed; however, little integrase is made. This is because of exonucleolytic (3′ → 5′) degradation of the transcript after cleavage by host enzyme RNAse III at a site (*sib*) downstream of *att*. This process (retroregulation) prevents untimely or wasteful synthesis of integrase. Retroregulation does not affect *p*L transcripts made from a derepressed prophage because the *sib* site is at the other end of the prophage. It also does not affect *p*I transcripts, which terminate within *sib* and therefore are resistant to RNAse III. Because the *p*L transcript is antiterminated, it reads through *sib* and is degraded; this is another function for antitermination in the λ life cycle. Despite the elegance and apparent utility of the retroregulation mechanism, deletion or mutation of the *sib* site has no observed effect on λ development as normally studied in the laboratory.

The potentiality of controlling insertion and excision differentially provides a rationale for the use of arm sites by λ integrase. Some related site-specific recombinases recognize only core sites, but in those cases there is no recognition of directionality.

When lytic development is induced in a lysogen, a small fraction (about 10^{-5}) of the particles produced result from abnormal excision that follows from breakage and joining of DNA at sites other than the *att* sites. If the excised DNA includes the *cos* sites and is within the packaging limits of the λ virion, it can be packaged into virions and used to infect other cells. Thus, a lysate produced by induction can transfer host genes that lie close to the λ insertion site (such as *gal* or *bio*) into recipient cells. This specialized transduction is a form of natural cloning. The transduced bacterial DNA segment need not replace its homolog in the recipient; instead the cell becomes lysogenic for the specialized transducing phage and is diploid for the transduced segment (Figs. 6 and 7).

The genesis of specialized transducing phages may be relevant to the acquisition of oncogenes by retroviruses. Although the mechanisms may differ substantially, specialized transduction has provided a useful model because the entire process can be followed in the laboratory. With retroviruses, until very recently the critical events have generally occurred in nature from inferred progenitors.

Lytic Development

Lytic development can be studied either in infected cells or in derepressed lysogens. Because of the high degree of

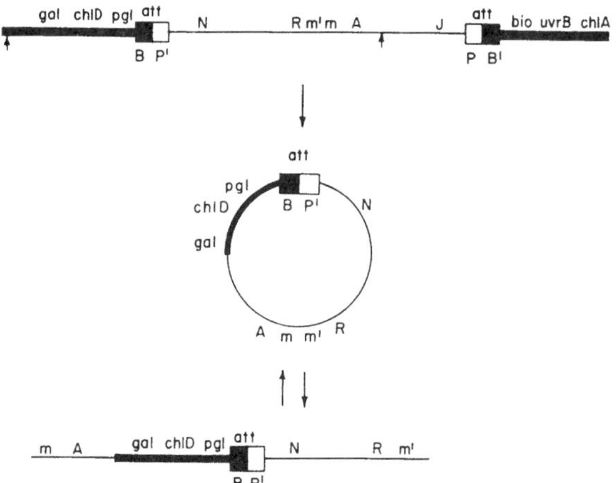

FIG. 6. Genesis of λ *gal* specialized transducing phage. Arrows indicate positions of breaking and joining of heterologous DNA in a particular isolate. m and m′ indicate the ends of mature λ DNA (equivalent to *cos*). (Reproduced from 12a with permission.)

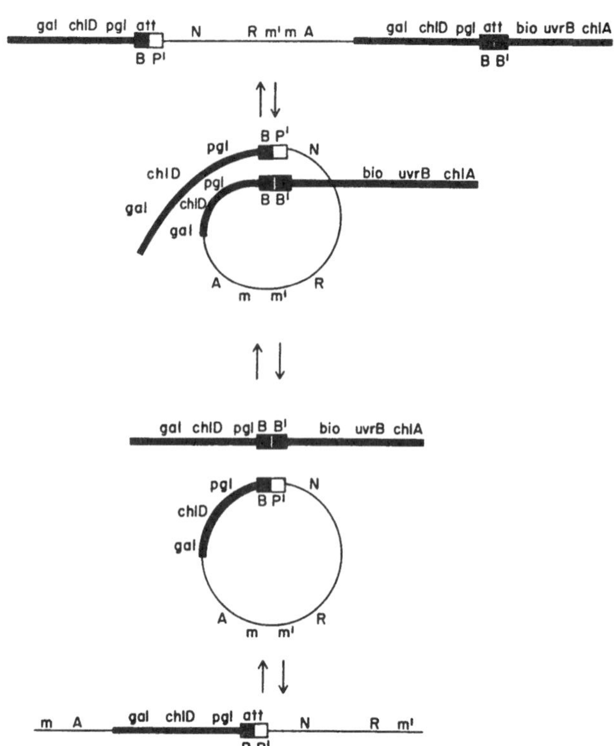

FIG. 7. Lysogenization by λ gal (bottom line) to produce a heterogenote with two copies of *gal* separated by λ DNA. (Reproduced from 12a with permission.)

synchrony obtainable, the method of choice has usually been thermal induction of lysogens carrying *cI ts* prophages. The genetic program and its readout are well understood. At early times, most transcription is from *pL* and *pR*. The only protein of the *pL* transcript that is essential for lytic development is gpN. The *p*R transcript contains three important genes: *O* and *P* (replication) and *Q* (late gene activation). As time proceeds, gpQ accumulates and causes antitermination of the short transcript starting at *p*R′ (downstream from *Q*). The mechanism of antitermination differs from that performed by gpN in that the recognition elements for antitermination are immediately downstream from the promoter and do not cause termination when further displaced from it (29). *Q* antitermination also requires a pause site close to the start site, whereas the role of pausing in *N*-antitermination is less certain.

At any rate, as gpQ accumulates, transcription of genes for lysis and virion formation becomes increasingly rapid. Concurrently, the rate of transcription from *p*L and *p*R decreases as Cro accumulates, producing an orderly transition from early gene expression to late gene expression.

Bidirectional DNA replication initiates from an origin within gene *O*. From studies of hybrid phages with different initiation specificities, it is clear that gpO recognizes the origin and also recognizes gpP, which in turn interacts with the host DnaB helicase. Disassembly of the initiation complex to leave an elongation complex requires three host heat shock proteins (DnaJ, DnaK, GrpE) not needed for host DNA synthesis. Elongation proceeds, as in the host, by the combined action of DnaB helicase, DnaG primase, and Pol III holoenzyme, the polymerase. Theta-form replication can reproduce monomer circles, but a switch to rolling circle replication occurs in some molecules. λ rolling circle replication (which predominates at late times) differs from that of ϕX in two major respects. First, the tails are double stranded rather than single stranded because lagging strand synthesis takes place on the displaced strand. Second, tails can grow to multigenomic lengths rather than stopping when a single genome has been spun off.

Multigenomic tails are the primary substrates for λ packaging. Empty proheads are first assembled, composed mainly of the major capsid protein gpE. Later, DNA enters the prohead, increases in size and becomes more conspicuously icosahedral, and the smaller decoration protein gpD is added to the shell, with gpD hexamers interspersed among gpE trimers in a T = 7 arrangement.

Efficient DNA packaging requires the presence of specific sequences (called *cos* for *co*hesive *s*ite") at both ends of the packaged genome; hence, multigenomic tails or multimeric circles are good packaging substrates and monomer circles are poor ones. Cutting at *cos* is effected by a heterodimeric protein, terminase, whose subunits are encoded by λ genes Nul and *A*. DNA molecules are packaged from left to right so a DNA monomer can be cut from the concatemer at its left end early in packaging and at its right end after the DNA has entered the virion. The *cos* site (about 200 bp in length) has three components (from right to left): a binding site (*cos B*), recognized by gpNul before the DNA is cut at either end; a nicking site (*cos N*), at which DNA is cut at sites 12 bp apart in the two strands to generate the packaged single strand overhangs of virion DNA; and a termination site (*cos Q*), required for the final nicking at the right end (18). Packaging is processive. After one genome is packaged, the next genome to the right can be packaged with no requirement for recognition of *cos B* on the second or subsequent genomes. This was demonstrated in the packaging of adjacent prophages with different *cosB* specificities determined by λ-related phages (24).

Because λ packages the DNA between two *cos* sites, the amount of DNA per head can be changed by deletions or insertions (as in T7). The packaging limits are from about 79% to 110% of normal λ length. Internal *cos* sites artificially placed within λ at less than or near the lower limit can be packaged uncut, as though scanning for *cos* sites becomes more efficient as the head fills.

Table 2 indicates the packaging limits for some phages described in this chapter. Those with headful packaging are listed with the same upper and lower limits. In fact, there is some random variation around the mean value. Alternative forms with different head sizes, such as the isometric and giant particles of T4 or the λ virions formed by D^- mutants are not considered.

Genome Organization

The functional clustering of genes on the λ map is notable. Not only do genes acting in a single pathway (head formation, tail formation, replication) lie together on the map, but genes for DNA binding lie close to their target sites, products of adjacent genes in a cluster frequently interact directly, and even within genes determinants are so ordered as to lie close to the elements with which they interact. For example, the *cosB* site is at the right end of *cos*, next to the Nul gene whose product recognizes *cosB*; Nul in turn lies next to *A*, whose product forms the terminase heterodimer with gpNul. In both head and tail clusters, the same tendency is observed, the tail fiber gene *J* being most distal from the head cluster. The *att* site is adjacent to the *int* gene, and cI and *cro* are close to their target sites. The N-terminus of gpO recognizes the *ori* site that lies within the 5′ end of *O*, whereas the C-terminus interacts with the product of the adjacent *P* gene. The clustering of major functional groups could increase regulatory efficiency, but the finer orderings may be more related to the disruptive effect of recombination on co-evolved functions.

Phage Mu-1 as a Model Transposon

Mu-1 first attracted attention as a phage capable of inducing bacterial mutations at a high rate. The mutations

TABLE 2. *DNA cutting and packaging*

Phage	Packaging substrate	First cut	Second cut	Packaging limits (% of genome)
T4	Concatemer	Random	Headful	102
P22	Concatemer	*pac* site	Headful	109
λ	Concatemer	*cos* site	*cos* site	79–110
T7	Concatemer	Terminal overlap	Terminal repeat	85–103
ϕX174	Dimer ss of rolling circle	Replication origin	Replication origin	Limited
M13	Dimer ss of rolling circle	Replication origin	Replication origin	Unlimited
Mu-1	Inserted monomer	Near left end	Headful	105
P1	Concatemer	*pac* site	Headful	110

proved to be insertions of Mu-1 into random sites on the host chromosome, disrupting genes or operons into which it inserted.

Like λ, Mu-1 can follow two alternative developmental pathways, leading on the one hand to a transcriptionally quiescent prophage and on the other to a lytic cycle with replication, packaging, and lysis. In λ, the DNA transactions in these two pathways are distinct. Insertion plays no role in λ replication, which proceeds normally if either the *int* gene or the *att* sites are deleted. With Mu-1, on the other hand, replication and insertion are minor variations of a common pathway, transposition. In Mu-1 replication, transposition happens many times during each cycle of infection, which has made it a system of choice for studying the biochemistry of transposition (compared with most bacterial transposons, which transpose naturally at rates such as 10^{-6} per cell division).

Mu-1 DNA is inserted into host DNA at all stages of its life cycle, even in the virion. When linear DNA from the virion is injected into a cell and the Mu-specific transposase is made, the first step in replication is transposition of Mu-1 DNA into the chromosome. Among bacterial transposons, the transposition mechanism may either be replicative (generating transposon copies in both the donor and the target site) or conservative (excising the transposon from the donor and inserting it into the target site). At least in cells destined to become lysogenic, the initial transposition is conservative rather than replicative. In cells destined to lyse, Mu-1 undergoes repeated rounds of replicative transposition.

The initial steps of replicative and conservative transposition are the same. Target sites are cut to generate protruding 5-base 5′ ends. The two donor strands are then cut at the 3′ ends of the transposon and ligated to the protruding 5′ ends of the target DNA. Replication then initiates from the recessed 3′ ends of the target DNA. In conservative transposition, the nicked donor DNA adjacent to the transposon is cut in the other strand and digested away; replication fills the gap, and ligation incorporates the parental transposon bodily into the target site (Fig. 8). In replicative transposition, the donor strands remain attached to the 5′ transposon ends, and replication proceeds through the transposon with ligation to the free 5′ donor ends at the opposite end of the transposon. This produces two semiconserved transposon copies, each joined to donor DNA at one end and target DNA at the other (51). Replicative transposition from a chromosomal donor produces two inserted copies accompanied by chromosomal rearrangement (inversion or excision); if replicative transposition happened from a linear infecting molecule (unaccompanied by subsequent recombination), it would linearize the chromosome.

In Mu-1 packaging, the left end of the inserted DNA is recognized, and cutting takes place within the flanking DNA, 50 to 150 bp to the left of the left end. Packaging is then by headfuls, and the headful length is longer than the genome length. Therefore 500 to 3,000 bp of DNA on the right end of the virion molecule is host DNA. So virion DNA molecules have random segments of host DNA at their termini. When virion DNA is separated into single strands and reannealed, the host sequences are visibly unpaired when viewed in the electron microscope.

An internal segment of Mu-1 DNA about 3 kb in length is invertible through the action of a phage-coded integrase, so reannealed molecules are sometime unpaired in this segment as well. The gene for the tail fiber protein spans one boundary of the invertible segments so that two types of tail fibers with different host specificities can be formed, depending on the orientation. Thus, if Mu grown on *E. coli* K-12 encounters an alternative host such as *Citrobacter*, the rare particle that attaches to *Citrobacter* can invade and replicate. The rate of inversion is about 10^{-5} per generation. Certain other phages, such as P1, have related invertible segments.

In lysogens, inserted Mu-1 can occasionally transpose without induction of a full lytic cycle. Thus, inserted Mu-1 behaves in all respects as a bacterial transposon.

Phage P1 as a Model Plasmid

The P1 virion contains about 100 kb of linear double-stranded DNA in an isometric T = 16 capsid. Like T4, the DNA is cyclically permuted and terminally redundant. Unlike T4 (but like *Salmonella* phage P22, a relative of λ), processive packaging does not initiate at random but instead at specific *pac* sites. After infection, reciprocal recombination within the terminal overlap creates a circular DNA molecule of about 90 kb. Some of this recombination is mediated by a site-specific recombinase (Cre) whose target site (*lox*) is close to the *pac* site.

The cell can now enter either a lytic cycle or a lysogenic cycle. Unlike λ or Mu-1, the lysogenic alternative entails replication of the circular P1 DNA as a plasmid and therefore requires expression of phage-coded replication genes. In lysogens, P1 is maintained at low copy number (about one per chromosome) and has an efficient partitioning mechanism so that both daughters of a dividing cell receive a P1 copy. The mechanism is similar to that used by other stringently controlled low copy number plasmids, such as the fertility factor F, and in some respects to that of the bacterial chromosome. The partitioning mechanism could potentially become ineffective if homologous recombination created a dimeric molecule, which could not pass into both daughter cells; this problem is greatly diminished by the site-specific *lox*-Cre recombination, which rapidly and reversibly equilibrates monomers with dimers. The elements for replication and partitioning (except *lox*-Cre) are localized in 1.5 kb of P1 DNA that can replicate as a plasmid when circularized; it includes a specific replication origin (*ori*R), its cognate replication initiator (*repA*) and a DNA sequence (*incA*) that controls replication by binding and sequestering *repA* (72).

In cells destined to lyse, P1 DNA replication (like λs) proceeds both in a theta form and rolling circle mode, the latter providing the packaging substrate. The replication origin used is distinct from the plasmid origin *oriR*. Rare mistakes lead to occasional packaging of random fragments of host DNA instead of phage DNA. Such fragments are responsible for the generalized transduction of bacterial genes. Generalized transduction at detectable frequencies is observed for most phages that use headful packaging but not by wild-type λ, probably because in λ both cutting (at both ends) and injection require the specific *cosN* sequence, whereas in P1 or P22 initial cutting at sequences resembling *pac* sites can initiate processive packaging. Mutations that relax the specificity and thereby increase the frequency of transduction have been isolated in both P22 and P1 (61,68).

DEFECTIVE PHAGES AND PHAGELIKE OBJECTS

The fact that Mu-1 can behave as a transposon and P1 as a plasmid underscores some of the commonality between phages and nonviral elements.

Transposable elements are common in bacteria, and most of them seem to use a biochemistry similar to that of Mu-1. Most bacterial transposons are smaller than Mu-1, going down to simple insertion sequences a few hundred base pairs long; and typical transposons carry genes affecting conspicuous bacterial traits such as antibiotic resistance. Conservative transposition is the usual mode for some elements, replicative transposition for others. Those using replicative transposition frequently resolve the resulting cointegrate structures through use of a site-specific recombinase that acts on a sequence internal to the element. Some transposons integrate and excise by a recombination mechanism similar in some respects to λ insertion (62).

Bacterial plasmids fall into two general groups: elements with small genomes, present in many copies per cell and replicating randomly with respect to the cell division cycle; and larger elements similar in size to phage P1, present in one or a few copies per cell and partitioned regularly at cell division. Conjugative plasmids resemble viruses as autonomous elements capable of infecting cells, but differ from them in the absence of an extracellular phase.

Paper scenarios in which phages have evolved into or from transposons or plasmids are easy to construct, and laboratory evolution from temperate phage to obligate plasmid or chromosomal element is readily achieved. The present sequence database does not provide a meaningful phylogeny placing various phages among their nonviral counterparts any more than the various phage groups can be clearly related to one another.

DNA sequences closely related to those of temperate phages are found in the genomes of many natural bacteria. For example, Southern hybridization of enteric bacteria with λ DNA usually turns up some λ-related sequences (58). Most of these are probably defective prophages: remnants of previous lysogenization that have since lost phage function through deletion or mutation. The K-12 strain of *E. coli* is naturally lysogenic for λ but also harbors four λ-related elements: DLP12, Rac, e14, and Qin. Such elements could in principle be of some value to the host or could even be precursors rather than descendants of complete phages; however, each defective prophage contains an array of genes from different functional clusters of the phage genome, frequently in the same order found in active phages, so that a role as host elements seems contrived. Defective lysogens arise from active lysogens in the laboratory by prophage mutations and frequently retain superinfection immunity, SOS-induced lysis, and production of structures resembling phage heads, tails, or complete virions.

Many natural strains liberate (or produce on induction) such phagelike or phage-related particles, which have not been shown to be infectious. Many bacteria also produce proteins (bacteriocins) that are toxic to other strains of the same bacterial species but not to the producing strain. Bacteriocins are sometimes encoded on plasmids and act through diverse mechanisms. Operationally, some phagelike objects qualify as bacteriocins.

For example, many natural strains of *Bacillus subtilis* and several other *Bacillus* species harbor a prophagelike

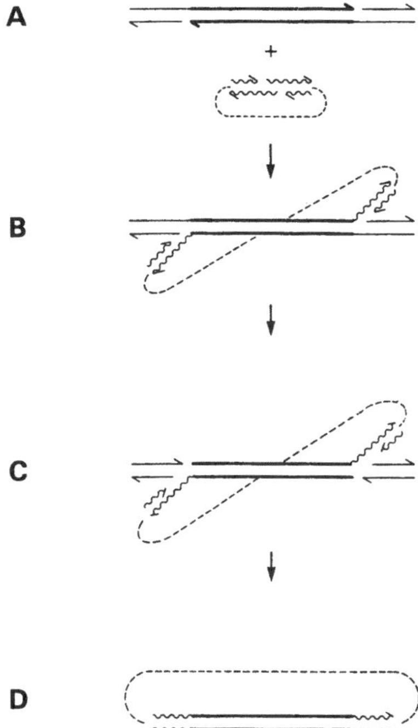

FIG 8. Conservative transposition by Mu-1. **A:** Linear donor molecule nicked at the 3′ ends of Mu-1 strands, plus circular target molecule cut with recessed 3′ ends. **B:** Protruding 5′ ends of target DNA are ligated to 3′ ends of Mu-1. **C:** Donor strands are nicked at 5′ Mu-1 ends (and probably degraded). **D:** DNA synthesis from 3′ target ends fills gaps, ending with ligation to 5′ Mu-1 ends. (Reproduced from 51 with permission.)

element Phage of *Bacillus subtilis,* (PBS), that has been localized on the chromosome. After SOS induction, the cells lyse and liberate many virionlike particles with tails and DNA-filled heads. This is not simply a phage whose sensitive host remains to be discovered because the large majority of particles contain random segments of host DNA with no significant preference for packaging prophage DNA. Like a bacteriocin, these particles kill other *Bacillus* cells. The producing strain is resistant, at least sometimes because the particles fail to be adsorbed. There are several varieties of PBS, each with a different killing specificity (74).

PBS has some similarity to mutants of generalized transducing phages that have lost (in this case, completely) their extreme preference for initiating packaging at phage-specified *pac* sites. Its wide distribution and the variety of killing specificities suggest that the bacteriocinlike activity confers a selective advantage to the host. Inasmuch as some bacteriocins are single protein molecules, a DNA-filled virion seems an especially clumsy and expensive kind of bacteriocin to have been so successful in evolution.

EVOLUTION AND NATURAL BIOLOGY OF PHAGES

For historical reasons, the coliphages have become workhorses of molecular biology and remain among the most thoroughly understood biological objects. Phages are found throughout the prokaryotic world in a variety comparable with that of eukaryotic viruses, including enveloped virions with double-stranded RNA genomes, (67), linear DNA genomes with protein covalently bound to their 5′ ends (59), and satellite viruses like P4 [which forms virions with the capsid protein of its helper virus P2, induces late gene function from P2 prophage, lysogenizes P2 lysogens using its own integrase but can establish a plasmid state in P2 nonlysogens, and represses its own genes by the use of antisense RNA rather than a protein repressor (44)]. Compared with our knowledge of phage development in infected cells, little is known about their natural ecology or population biology. Some aspects are worthy of note.

Abundance

In some natural environments phagelike particles are much more numerous than one might suspect from plaque assays on susceptible hosts. In estuaries, concentrations as high as 10^8 have been reported. If most of these are infectious (and they may very well not be), they potentially could play a major role in the natural cycling of the biosphere (30).

Host Defense Mechanisms

Bacteria have evolved mechanisms for phage resistance, and phages have evolved counterstrategies, presumably indicating in some measure the significant impact that phage infection can have. A host defense mechanism unique to prokaryotes is restriction/modification. Restriction enzymes damage DNA that has not undergone chemical modification at specific sites (most frequently methylation) by cleaving it either at those sites or nearby; nonspecific exonucleases then degrade the resulting linear fragments. Cells protect their own DNA from the restriction enzymes they make by modifying it. A phage that escapes restriction becomes modified and is fully infectious on a restricting host in the next cycle; so the strategy can only be effective for the host if the strains sensitive to a given phage have various restriction specificities. A strain with a rare restriction specificity will exclude almost all phage it encounters and will be favored over one with a common specificity; so eventually the rare specificity will become common. The recognition sites for restriction enzymes are generally symmetrical, even if the cleavage takes place at a variable distance from the recognition site; this allows the modification of either strand to prevent cleavage, which assures that, during semiconservative replication, a new strand that has not yet been modified is protected from destruction.

Phages have many strategies for avoiding restriction. The first level is a paucity of recognition sequence. Phages of *Bacillus*, where many known restriction enzymes recognize 4-base palindromes, have few such palindromes in their DNA; coliphages are deficient in 6-base palindromes, which are sites for some of the enzymes they naturally encounter (38). Various phages also have enzymes that destroy 5-adenosylmethionine, a cofactor for restriction (T3), expedite modification of phage DNA before restriction enzymes can cleave (λ), or block recognition sites by wholesale nonspecific DNA modification (T4).

Some restriction enzymes are encoded by chromosomal genes, but many are encoded by phages or plasmids. Many resistance mechanisms contributing to defense of the host may in fact be manifestations of competition between extrachromosomal elements preserving their cellular territories.

One element may exclude others by diverse mechanisms, and in many cases exclusion may be an incidental side effect of the phage life cycle. The repressors of temperate prophages (essential for the internal stability of lysogens) render the lysogens immune to superinfection by phage of the carried type, and this superinfection immunity probably selects for diversification of repressor specificities. Certain phages, such as the ϵ phages of *Salmonella*, encode enzymes that alter the polysaccharide surface receptors used in phage attachment, so that lysogens are not only immune but also cannot be penetrated by DNA of the same type. The alteration also takes place during the lytic cycle and reduces phage loss from attachment to fragments of cell envelope after lysis. The λ-related phage HKO22 has a variant of the λ*N* gene that competes with λN when λ infects an HKO22 lysogen and causes termination rather than antitermination of early λ transcripts (54). The λ Rex proteins trigger a collapse of membrane potential when a λ lysogen is infected with certain other phages (such as T4 *r*II mutants), preventing the production of viral progeny (56). These last two mechanisms do not protect the infected cell but may protect other cells in a colony by impeding viral replication.

Natural Recombination

Third, the study of natural isolates has provided some perspectives on the degree of variation and the frequency of recombination among phages in nature. The λ-related (lambdoid) phages are a good example.

A note on the way that phage workers have approached taxonomy may be useful because it differs substantially from the method used by many medical virologists. In medical virology, once a virus such as measles is identified as a disease agent, many isolates from different clinical sources are examined, and those sharing some acceptable degree of similarity to the type of virus are classified as measles virus. For epidemiology, some such approach is necessary. Phage workers are not epidemiologists, and their interest in natural variation is relatively recent. At least from the Luria-Delbrück era onward, investigators have been more interested in developing highly defined experimental systems. To that end, each isolate is given its own name. "λ" refers only to the prophage present in *E. coli* K-12 and its laboratory descendants; "K-12" refers to one bacterial isolate, not to other natural strains with similar properties. Workers with a different perspective might have classified all lambdoid phages as natural variants of λ, and the same can be said of other phage groups such as T2, T4, and T6.

The lambdoid phages have a common genetic map and the ability to generate viable hybrids when crossed in the laboratory. They have been isolated from various sources (usually from lysogenic bacteria) isolated over the past 70 years from Europe, Asia, and the United States). Genetic variation among the lambdoid phages is apparent both from functional specificity and DNA sequence. For example, phages with at least a dozen repressor specificities have been isolated; where the sequences are known, homologies between heterospecific repressor genes are barely discernible. Phages with different repressors have corresponding differences in their *cro* genes and operator sites.

If the λ sequence is compared with that of any other lambdoid phage (or, equivalently, if heteroduplexes of the two phages are constructed) the general result is that the sequences match closely in some parts of the genome, whereas homologies are weak or absent in others. This is not because some sequences are highly conserved, because another phage pair shows a different set of matching segments; instead, the whole pattern strongly indicates that it is generated by frequent natural recombination so that in any phage pair, some portions of their genomes are of recent common ancestry. When defective prophages are compared with known lambdoid phages, they likewise are closely related to different phages in different segments. For example, the defective prophage DLP12 has an *int* gene and a partially deleted *xis* gene related to phage P22, followed by segments homologous to λs *exo* and *P*/*ren* genes, followed by analogs (*qsr'*) of λ *Q S R* unrelated to any known λ phage, followed by *cos* DNA similar to λs (45). Thus, all the lambdoid phages, including the defective prophages, seem to be drawn from a common gene pool.

Among the lambdoid phages, some genes that may serve viral functions have been borrowed from outside the λ pool. Gene 12 of P22 occupies the same position in the genome as λ gene *P*, but gene 12 is a homolog of the bacterial *dnaB* gene, and λ gpP recruits the host DnaB helicase to participate in phage replication. Wildtype λ has a gene for side tail fibers (*stf*) that facilitate phage attachment. (In the λ commonly used in the laboratories, *stf* has been inactivated by a frameshift.) The *stf* gene is closely related to tail fiber genes of other groups of coliphages (31). Most lambdoid phages encode a true lysozyme that hydrolyzes glycosydic bonds; λ makes instead a transglycosylase that attacks the same bonds but has no detectable homology to lysozyme (73). The λ lysozyme may have been appropriated from some external source, as yet unidentified.

The natural function of genetic recombination is not obvious. In λ as in sexual eukaryotes, it is clear that recombination has a substantial evolutionary impact. But that does not tell us at what level natural selection acts to preserve recombination as a process or whether recombination is an accidental by-product of gene activities selected to function in repair or replication. One classical argument for the value of recombination is that an asexual line deteriorates through accumulation of deleterious mutations, which can be corrected by recombination with other lines that are wild type for the loci of those particular mutations (52). The first experimental demonstration of this Muller's ratchet effect came from phage ϕ6 (15); corroborative evidence from animal viruses has since appeared (20). Muller's ratchet may be central to the evolution of certain human viruses such as influenza, where the clonally selected epidemic strains are eventually replaced by reassortants with avian viruses (25).

In their natural ecology, phages exhibit many of the same features observed in eukaryotic viruses and will probably find increasing use as model systems.

REFERENCES

1. Anderson TF. The reactions of bacterial viruses with their host cells. *Botan Rev* 1949;15:464–505.
2. Anderson TF, Doermann AH. The intracellular growth of bacteriophages II. The growth of T3 studied by sonic disintegration and by T6-cyanide lysis of infected cells. *J Gen Physiol* 1952;35:657–667.
3. Andrake M, Guild N, Hsu T, Gold L, Tuerk C, Karam J. DNA polymerase of bacteriophage T4 is an autogenous translational repressor. *Proc Natl Acad Sci USA* 1988;85:7942–7946.

4. Aoyama A, Hamatake RK, Hayashi M. *in vitro* synthesis of bacteriophage ϕX174 by purified components. *Proc Natl Acad Sci USA* 1983;80:4195–4199.
5. Beckendorf SK. Structure of the distal half of the bacteriophage T4 tail fiber, *J Mol Biol* 1973;73:37–53.
6. Belfort M, Ehrenman K, Chandry PS. Genetic and molecular analysis of RNA splicing in *Escherichia coli*. *Methods Enzymol* 1990; 181:521–539.
7. Benzer S. The elementary units of heredity, In: McElroy WD, Glass B, eds. *The chemical basis of heredity*. Baltimore: Johns Hopkins Press; 1957:76–93.
8. Benzer S. On the topology of the genetic fine structure. *Proc Natl Acad Sci USA* 1959;45:1607–1620.
9. Bhattacharyya SP, Rao VB. A novel terminase activity associated with the DNA packaging protein gp17 of bacteriophage T4. *Virology* 1993;196:34–44.
10. Biebricher CK, Eigen M, McCaskill JS. Template-directed and template-free RNA synthesis by Qβ replicase. *J Mol Biol* 1993;231:174–179.
11. Black L, Showe M. Morphogenesis of the T4 head. In: Mathews C, Kutter E, Mosig G, Berget P, eds. *Bacteriophage T4* Washington, DC: ASM Publications; 1983:219–245.
12. Campbell A. Sensitive mutants of bacteriophage l. *Virology* 1961; 14:22–32.
12a. Campbell A. Genetic structure. In: Hershey AD, ed. *The bacteriophage lambda*. Cold Spring Harbor Laboratory, Cold Spring Harbor, NY; 1971;13–44.
12b. Campbell A. Transposons and their evolutionary significance . In: Nei M, Koehn RK, eds. *Evolution of genes and proteins*. Sunderland, MA: Sinauer.
13. Campbell AM. Episomes. *Adv Genet* 1962;11:101–146.
14. Casjens S, Hendrix R. Control mechanisms in dsDNA bacteriophage assembly. In: Calendar R, ed. *The bacteriophages*. New York: Plenum; 1988:15–91.
15. Chao L. Fitness of RNA virus decreased by Muller's ratchet. *Nature* 1990;348:454–455.
16. Chung Y-B, Nardone C, Hinkle DC. Bacteriophage T7 packaging. III. A "hairpin" end formed on T7 concatemers may be an intermediate in the primary reaction. *J Mol Biol* 1990;216:939–948.
17. Cohen SS. Virus-induced acquisition of metabolic function. *Fed Proc* 1961;20:641–649.
18. Cue D, Feiss M. A site required for termination of packaging of the phage λ chromosome. *Proc Natl Acad Sci USA* 1993;90:9290–9294.
19. Doermann AH, Eisenberg FA, Boehner L. Genetic control of capsid length in bacteriophage T4. I. Isolation and preliminary description of four new mutants. *J Virol* 1973;12;374–385.
20. Duarte EA, Clarke DK, Moya A, Elena SF, Domingo E, Holland J. Many-trillionfold amplification of single RNA virus particles fails to overcome the Muller's ratchet effect. *J Virol* 1993;67:3620–3623.
21. Dulbecco R. Production of plaques in monolayer tissue cultures by single particles of an animal virus. *Proc Natl Acad Sci USA* 1942; 38:747–752.
22. Dunn J, Studier W. Complete nucleotide sequence of bacteriophage T7 DNA and the location of T7 genetic elements. *J Mol Biol* 1983;166:477–535.
23. Epstein RH, Bolle A, Steinberg CM, et al. Physiological studies of conditional lethal mutants of bacteriophage T4D. *Cold Spring Harb Symp Q Biol* 1963;28:375–394.
24. Feiss M, Sippy J, Miller G. Processive action of terminase during sequential packaging of bacteriophage λ chromosomes. *J Mol Biol* 1985;186:759–771.
25. Fitch WM, Leiter JME, Li X, Palese P. Positive Darwinian evolution in human influenza A virus. *Proc Natl Acad Sci USA* 1991; 88:4270–4274.
26. Flemming M, Deumling B, Kemper B. Function of gene 49 of bacteriophage T4. III. Isolation of Holliday structures from very fast-sedimenting DNA. *Virology* 1993;196:910–913.
27. Franklin NC. Altered reading of genetic signals fused to the *N* operon of bacteriophage λ: genetic evidence for modification of polymerase by the protein product of the *N* gene. *J Mol Biol* 1974;89:33–48.
28. Geiduschek EP. Regulation of expression of the late genes of bacteriophage T4. *Annu Rev Genet* 1991;25:437–460.
29. Goliger JA, Roberts JW. Sequences required for antitermination by phage 82 Q protein. *J Mol Biol* 1989;210:461–471.
30. Goyal SM, Gerba CP, Bitton G, eds. *Phage ecology*. New York: John Wiley & Sons, 1987.
31. Haggard-Ljungquist E, Halling C, Calendar R. DNA sequences of the tail fiber genes of bacteriophage P2; evidence for horizontal transfer of tail fiber genes among unrelated bacteriophages. *J Bacteriol* 1992;174:1462–1977.
32. Haruna I, Spiegelman S. Specific template requirements of RNA replicases. *Proc Natl Acad Sci USA* 1965;54:579–587.
33. d'Herelle F. Sur un microbe invisible antagoniste des bacilles dysenteriques. *Compt Rend Acad Sci* 1917;165:373–375.
34. Herenden DR, Kassavetes GA, Geiduschek EP. A transcriptional enhancer whose function imposes a requirement that proteins track along DNA. *Science* 1992;256:1298–1303.
35. Hershey AD, Chase M. Independent functions of viral protein and nucleic acid in growth of bacteriophage. *J Gen Physiol* 1952;36:31–56.
36. Hershey AD, Rotman R. Linkage among genes controlling inhibition of lysis in a bacterial virus. *Proc Natl Acad Sci USA* 1948;34:89–96.
37. Kacian DL, Mills DR, Kramer FR, Spiegelman S. A replicating RNA molecule suitable for a detailed analysis of extracellular evolution and replication. *Proc Natl Acad Sci USA* 1972;69:3038–3042.
38. Karlin S, Burge C, Campbell AM. Statistical analysis of counts and distributions of restriction sites in DNA sequences. *Nucleic Acids Res* 1992;20:1363–1370.
39. Kashlev M, Nudler E, Goldfarb A, White T, Kutter E. Bacteriophage T4 Alc protein: a transcription termination factor sensing local modification of DNA. *Cell* 1993;75:147–154.
40. Katsura I, Hendrix R. Length determination in bacteriophage lambda tails. *Cell* 1984;39:691–698.
41. Kestelein RA, Remaut E, Fiers W, van Duin J. Lysis gene expression of RNA phage MS2 depends on a frameshift during translation of the overlapping coat protein gene. *Nature* 1982;285:35–41.
42. Kitts PA, Nash HA. Homology-dependent interactions in phage site-specific recombination. *Nature* 1987;329:346–348.
42a. Landy A. Dynamic, structural and regulatory aspects of λ site-specific recombination. *Amer Rev Biochem* 1989;58:913–949.
43. Lederberg EM, Lederberg J. Genetic studies of lysogenicity in *Escherichia coli*. *Genetics* 1953;38:51–64.
44. Lindquist BH, Dehó G, Calendar R. Mechanisms of genome propagation and helper exploitation by satellite phage P4. *Microbiol Rev* 193;57:683–702.
45. Lindsey DR, Mullin DA, Walker JR. Characterization of the cryptic lambdoid prophage DLP12 of *Escherichia coli* and overlap of the DLP12 integrase gene with the tRNA gene *argU*. *J Bacteriol* 1989;171: 6197–6205.
46. Little JW. Mechanism of specific LexA cleavage: autodigestion and the role of RecA coprotease. *Biochimie* 1991;73:411–422.
47. Luftig RB, Wood WB, Okinawa R. Bacteriophage T4 head morphogenesis. On the nature of gene 49-defective heads and their role as intermediates. *J Mol Bol* 1971;57:553–573.
48. Luzzati D. Regulation of λ exonuclease synthesis: role of the N gene product and l repressor. *J Mol Biol* 1970;49:515–519.
49. Lwoff A. Lysogeny. *Bacteriol Rev* 1953;17:269–337.
50. Mason SW, Li J, Greenblatt J. Host factor requirements for processive antitermination of transcription and suppression of pausing by the N protein of bacteriophage lambda. *J Biol Chem* 1992;267:19418–19426.
51. Mizuuchi K, Craigie R. Mechanism of bacteriophage Mu transposition. *Annu Rev Genet* 1986;20:385–430.
52. Muller HJ. The relation of recombination to mutational advance. *Mutation Res* 1964;1:2–9.
53. Novick A, Szilard L. Virus strains of identical phenotype but different genotype. *Science* 1951;113:34–35.
54. Oberto J, Weisberg RA, Gottesman ME. Structure and function of the *nun* gene and the immunity region of the lambdoid phage HK022. *J Mol Biol* 1989;207:675–693.
55. Orsine G, Brody EN. Phage T4 DNA codes for two distinct 10 kDa proteins which strongly bind to RNA polymerase. *Virology* 1988; 162:397–405.
56. Parma DH, Snyder M, Soboleosh S, Mowray M, Brody E, Gold L. The Rex system of bacteriophage λ: tolerance and altruistic cell death. *Genes Dev* 1992;6:497–510.
57. Ptashne M. *A genetic switch*. Cambridge, MA: Cell Press and Blackwell; 1992.
58. Riley M, Anilionis A. Conservation and variation of nucleotide sequences within related bacterial genomes: Enterobacteria. *J Bacteriol* 1980;143:366–376.
59. Salas M. Phages with protein attached to their DNA ends. In: Calendar R, ed. *The bacteriophages*. New York: Plenum; 1988:169–192.

60. Schlesinger M. The Feulgen reaction of the bacteriophage substance. *Nature* 1936;138:508–509.
61. Schmieger H, Backhaus H. The origin of DNA in transducing particles in P22-mutants with increased transduction-frequencies (HT-mutants). *Mol Gen Genet* 1973;120:181–190.
62. Scott JR. Sex and the single circle: conjugative transposons. *J Bacteriol* 1992;174:6005–6010.
63. Séchaud J, Streisinger G, Emrich J, et al. Chromosome structure in phage T4. II. Terminal redundancy and heterozygosis. *Proc Natl Acad Sci USA* 1965;54:1333–1339.
64. Son M, Watson RH, Sewer P. The direction and rate of phage T7 DNA packaging *in vitro*. *Virology* 1993;196:282–289.
65. Streisinger G, Emrich J, Stahl MM. Chromosome structure in phage T4. III. Terminal redundancy and length determination. *Proc Natl Acad Sci USA* 1967;57:292–295.
66. Twort FW. An investigation on the nature of ultramicroscopic viruses. *Lancet* 1915;189:1241–1243.
67. Van Etten, JL, Burbank DE, Cuppels PA, Lane LC, Vidaver AK. Semi-conservative replication of double-stranded RNA by a virion-associated RNA polymerase. *J Virol* 1980;33:769–783.
68. Wall JD, Harrison PD. Phage P1 mutants with altered transducing abilities for *Escherichia coli*. *Virology* 1974;59:532–544.
69. Watson JD. Origins of concatameric DNA. *Nature* 1972;239:197–201.
70. Weiss RB, Huang WM, Dunn DM. A nascent peptide is required for ribosomal bypass of the coding gap in bacteriophage T4 gene 60. *Cell* 1990;62:117–126.
71. Winter RB, Morrissey L, Gauss P, Gold L, Hsu T, Karam J. Bacteriophage T4 *regA* protein binds to mRNAs and prevents translation initiation. *Proc Natl Acad Sci USA* 1987;84:7822–7826.
72. Yarmolinsky MB, Sternberg N. Bacteriophage P1. In: Calendar R, ed: *The bacteriophages*. New York: Plenum; 1988:291–418.
73. Young R. Bacteriophage lysis: mechanism and regulation. *Microbiol Rev* 1992;56:430–481.
74. Zahler SA, Temperate bacteriophage of *Bacillus subtilis*. In: Calendar R, ed. *The bacteriophages*. New York: Plenum; 1988:559–592.

Fundamental Virology, Third Edition
edited by B.N. Fields, D.M. Knipe, P.M. Howley, et al.
Lippincott - Raven Publishers, Philadelphia © 1996

CHAPTER 16

Picornaviridae: The Viruses and Their Replication

Roland R. Rueckert

R. R. Rueckert: Institute for Molecular Virology, Graduate School, and Department of Biochemistry, College of Agricultural and Life Sciences, University of Wisconsin, Madison, Wisconsin 53706.

The *Picornaviridae*, among the smallest ribonucleic acid—containing viruses known, comprise one of the largest and most important families of human and agricultural pathogens. Poliovirus, human hepatitis A virus, and foot-and-mouth disease virus (FMDV) are all members of the picornavirus family. So, too, are the human rhinoviruses, the single most important etiologic agents of the common cold. The acronym picorna (*p*oliovirus, *i*nsensitivity to ether, *C*oxsackie, *o*rphan virus, *r*hinovirus, *rna*) is a useful mnemonic (272). The prefix pico designates a very small unit of measurement equivalent to 10^{-12}.

Because of the economic and medical importance of picornaviruses, it is not surprising that they have figured prominently in the development of modern virology. Foot and mouth disease, the most important single pathogen of livestock, was in fact the first animal virus to be recognized when Loeffler and Frosch in 1898 (235) discovered that the causative agent passes through Berkfeld filters and was therefore much smaller than other microorganisms then known to transmit disease. That poliomyelitis was also caused by a virus was announced about a decade later by Landsteiner and Popper (218), but it did not come to be called poliovirus until around 1955 (430).

Studies on the molecular biology of animal viruses were ushered in by the discovery that poliovirus could be propagated in cultured cells (122). Other crucial advances were the development of a plaque assay for infectivity (113) and the develpment of methods for purification and crystallization of poliovirus (381,407), thus opening the way for structural analysis via x-ray crystallography. Determination of the nutritional requirements of cultured cells (117) provided defined nutrient media so that it became practical to radiolabel viral RNA and protein in synchronously infected cultures (96) using precursors of known specific activity (97,98).

Discovery that isolated picornaviral RNA was infective (5,79) facilitated demonstration that susceptibility of cells is correlated with the presence of specific receptors (99,180).

Three additional discoveries, with important implications for molecular biology as a whole, were (a) demonstration of the first RNA- dependent RNA polymerase (25); (b) identification of a helical base-paired double-stranded replicative form of viral RNA (284); and (c) discovery that poliovirus synthesizes its gene products by proteolytically cleaving a large polyprotein (410). Recognition that proteolytic cleavage was a special way of expressing gene products within the limitations of the eukaryotic rule of translation provided the first insight into an important difference in the way prokaryotes and eukaryotes translate polycistronic messengers (192).

CLASSIFICATION

The quest for effective vaccines, particularly against FMDV (46) and poliovirus (41,67,274,311), led to the discovery of a host of other poliolike viruses, now numbering in excess of 217 serotypes (Table 1).

The family is currently divided into five genera: the *r*hinoviruses, the *e*nteroviruses, the *a*phthoviruses (FMDVs), the *c*ardioviruses, and the *h*epatoviruses (REACH). Comparison of genetic organization, cleavage patterns, and sequence relationships suggests that this classification requires further revision. Human rhinoviruses, for example, bear many sequence similarities to human enteroviruses and might well be classed as subgenera of a single group I (Fig. 1). The murine polioviruses (Theiler viruses) are closely related in sequence and polyprotein structure to encephalomyocarditis viruses and are now classified as cardioviruses (Fig. 1). Members of these five genera are not morphologically distinguishable by electron microscopy. However, they do exhibit distinctive surfaces at the higher resolution afforded by x-ray crystallography.

The enteroviruses, so-called because most inhabit the alimentary (enteric) tract, include not only the polioviruses (three serotypes), but also the coxsackieviruses (23 serotypes), the echoviruses (31 serotypes), human enteroviruses 68–71, human hepatitis virus A (1 serotype), and a number of nonhuman enteric viruses (273,274).

The human rhinoviruses, so called because of their special adaptation to the nasopharyngeal region, are the most important etiologic agents of the common cold in adults and children (105,156,201). Officially there are now 102 serotypes (158), including human rhinovirus 1B and Hanks, which was not assigned a number. Most of the picornaviruses isolated from the human respiratory system are acid labile, and this lability has become a defining characteristic of rhinoviruses (200,272,274,426).

Aphthoviruses (FMDVs) infect cloven-footed animals, especially cattle, goats, pigs, and sheep, and rarely infect even humans. Seven immunotypes have been identified: types A, C, O, Asia-1, and the South African Territory types SAT-1, SAT-2, and SAT-3. Within these seven types, at least 53 subtypes have been designated by complement-fixation tests. The serological characteristics of the subtypes are sufficiently different to cause difficulty in classification and immunization. The aphthoviruses are highly labile, being rapidly inactivated at pH <7.

The cardioviruses are represented by two serotypes. The encephalomyocarditis (EMC) viruses (encephalomyocarditis virus, Columbia SK virus, ME virus, MM virus, and mengovirus) represent one serotype. They are generally regarded as murine viruses, although their host range includes humans, monkeys, pigs, elephants, and squirrels, among others. The second serotype includes the Theiler's murine encephalomyelitis viruses (TMEV). They are divided into two groups, typified by strains called GDVII and TO. The GDVII viruses cause an acute poliolike disease in mice. The TO viruses are less virulent and cause a chronic demyelinating disease resembling multiple sclerosis and have thus become important models for study of this and other motor neuron diseases (233). Vilyuisk virus may be a divergent Theiler virus (336).

TABLE 1. *The family Picornaviridae*[a]

Genus	Number of serotypes	Members
Rhinovirus	102	Human rhinoviruses 1A–100, 1B, "Hanks"
	3	Bovine rhinoviruses 1,2,3
Enterovirus	3	Human polioviruses 1,2,3
	23	Human Coxsackieviruses A1–22,24 (A23-echovirus 9)[b]
	6	Human Coxsackieviruses B1-6[c] (swine vesicular disease virus is very similar to coxsackie B5 virus)
	30	Human echoviruses 1–7,9,11–27,29–34 Echo 8 is echo 1[b] Echo 10 is reovirus, type 1[b,d] Echo 28 is human rhinovirus 1A[b]
	4	Human enteroviruses 68–71
	1	Vilyuisk virus
	18	Simian enteroviruses 1–18
	2	Bovine enteroviruses 1,2
	8	Porcine enteroviruses 1–8
Aphthovirus	7	Foot-and-mouth disease virus 1–7 (serotypes A,C,O,SAT-1,2,3,Asia-1)
Cardiovirus	2	Encephalomyocarditis (EMC) virus[e], Theiler's murine encephalomyelitis (TME) virus (TO, GDVII)
Hepatovirus	1	Human hepatitis virus A[f]
Unassigned	3	Equine rhinoviruses 1,2, cricket paralysis virus, *Drosophila* C virus

[a]The so-called feline picornaviruses are now classified in the family *Caliciviridae*.
[b]Vacated numbers are now unused; Echovirus 22 is atypical in that it shows little sequence relationship to other picornaviruses (189).
[c]Coxsackieviruses are named after Coxsackie, New York, the town from which the initial isolates were made.
[d]Echo is an acronym for *e*nteric *c*ytopathic *h*uman *o*rphan (274).
[e]Also mengovirus, Maus Elberfeld (ME) virus, Columbia SK virus, MM virus.
[f]Formerly classified human enterovirus 72.

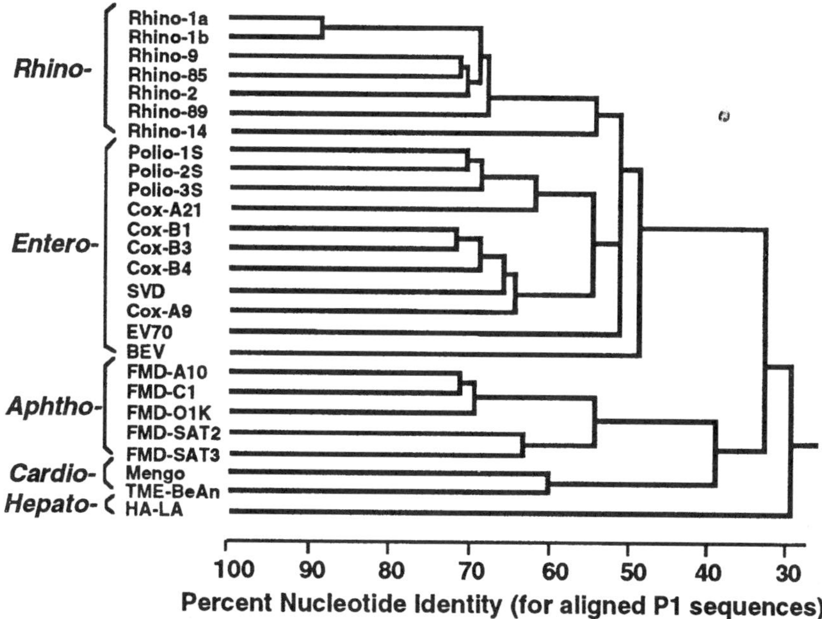

FIG. 1. Dendrogram constructed by comparing nucleotide alignments of coat protein sequences for 26 picornaviruses (courtesy Ann Palmenberg). Percent nucleotide identity is given by the vertical bar connecting virus pairs. For example, there is 88% identity between rhinovirus serotypes 1A and 1B but only 28% between hepatitis A and any other picornavirus genus. For rationale of computing such dendrograms, see Fitch (132); for use of dendrograms to establish epidemiologic relationships, see Rico-Hesse et al. (345).

pH Stability

The enteroviruses, cardioviruses, Theiler viruses, and hepatitis A viruses are acid stable, surviving pH 3 or even lower, whereas the rhinoviruses and aphthoviruses are labile at pH <6 (Table 2). This difference in pH stability probably represents adaptation to a way of life because the enteroviruses must pass through the acidic conditions of the stomach in order to gain access to their native habitat, the gut. The acid-labile rhinoviruses and aphthoviruses on the other hand inhabit the nasal and oropharyngeal regions with no apparent need for acid stability. Indeed it may be that the structure conferring acid lability to the virion conveys some still inapparent counteradvantage because the highly labile aphthoviruses are among the most contagious viruses known.

Buoyant Density of Picornaviruses

Picornaviruses differ widely in buoyant density (Table 2, column 3). The enteroviruses and cardioviruses band at a density of 1.34 g/ml, whereas the equine rhinoviruses and some of the aphthoviruses band at 1.45 g/ml. The latter density is characteristic of viruses with RNA contents on the order of 30% to 35% (387). Poliovirus is abnormally light because its cesium-impermeable shell prevents binding of these heavy atoms to the viral RNA (137). Cesium can be incorporated into poliovirions by growing infected cells in medium enriched in cesium chloride. In this way it is possible to relate the density of the virus to its cesium content and to show that incorporation of cesium is accompanied by elimination of potassium, the counterion of which normally neutralizes the phosphate residues in the poliovirus genome. When artificially substituted with a full complement of 7,500 cesium atoms, poliovirus has a predicted buoyant density of 1.45 g/ml (251). "Dense" particles of poliovirus of this density have been observed (364,435,441) but they are practically noninfectious. Such dense, poorly infective particles are also a common feature of other enteroviruses, including Coxsackieviruses, echoviruses, and hepatitis A virus.

The high buoyant density of the aphthoviruses and equine rhinoviruses (Table 2, column 3) indicates that cesium exchanges with almost all of the phosphate residues in the RNA. The intermediate buoyant density of the human rhinoviruses, near 1.40 g/ml, is due to incomplete substitution with cesium ion, about 5,000 atoms per virion. The remaining unexchanged sites are occupied by the polyamines spermine and spermidine, which are too large to diffuse out of the semipermeable protein envelope and cannot therefore be exchanged out of the RNA core (137). The reason for permeability of the aphthovirus shell is now evident from presence of prominent pores some 11 Å in diameter at its pentameric vertices (3), but there are no such prominent openings in HRV14.

THE RNA GENOME

Sensitivity to Ribonuclease

The picornaviral genome consists of a single strand of messenger-active RNA that can be extracted out of virions by shaking aqueous suspensions of virus with an equal volume of buffer-saturated phenol. When the resulting emulsion is separated, proteins partition to the phenol-rich phase while RNA remains in the aqueous phase. The specific infectivity of the naked RNA is about one millionth that of virions. That this infectivity is indeed due to free RNA and not to traces of surviving virions is shown by its extreme sensitivity to ribonuclease (<0.01 μg/ml). Hence, wearing of surgical gloves is commonly practiced when extracting infective RNA to avoid conta-

TABLE 2. *Physical properties of picornaviruses*[a]

Group	pH stability	Buoyant density (CsCl)		Sedimentation coefficient[b]
		Virions	Shells	
Rhinovirus (human)	Labile <6[c]	1.34	1.30	149S[c]
Enterovirus	Stable 3–9[d]	1.34	1.30	156S
Aphthovirus	Labile <7	1.43–1.45	1.30	142–146S
Cardiovirus	Stable 3–9[e]	1.34	1.30[f]	156S
Hepatitis A	Stable	1.34	1.30	156

[a]Newman (292).
[b]Relative to poliovirus (156S) in sucrose gradients (ionic strength 0.04–0.15, pH 7.3–7.6).
[c]Equilibrated against 0.1 mol/L potassium chloride. S value of human rhinovirus 14 is reversibly increased by up to 7% upon exposure to 2 mol/L CsCl (215).
[d]At low ionic strength some coxsackieviruses are thermolabile even at neutral pH (84).
[e]EMC-like viruses (EMC, ME, Columbia SK, and MM viruses), but not Theiler's viruses, are thermolabile in the pH region 4.5 to 6.5 if chloride or bromide ion are present at about 0.1 mol/L (367).
[f]EMC viral shells are stable only under carefully defined conditions (227).

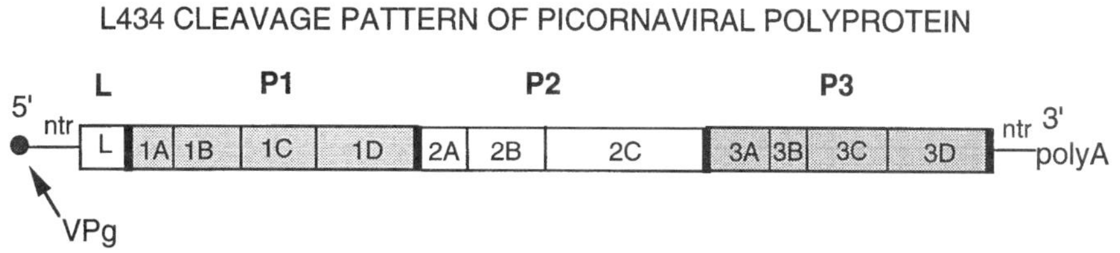

L434 CLEAVAGE PATTERN OF PICORNAVIRAL POLYPROTEIN

FIG. 2. Structure of picornaviral RNA and genetic organization of its polyprotein (open bar). The RNA is organized 5′-VPg-ntr-polyprotein-ntr-poly(A). ntr refers to nontranslated regions flanking the polyprotein. L434 is a mnemonic for recalling the polyprotein cleavage pattern; L specifies a leader protein found in cardioviruses, Theiler viruses, and aphthoviruses, but not in enteroviruses, human rhinoviruses, or human hepatitis virus A. P1, P2, and P3 refer to precursor proteins cleaved by virus-coded proteinases into four, three, and four end products, respectively.

mination by invisible flecks of nuclease-containing particles shed from the skin. With intact virions, by contrast, where the RNA resides inside a protective protein shell, the infectivity is completely resistant to ribonuclease even at millionfold higher concentrations. A single break in the RNA, whether free or inside the virus particle, is sufficient to destroy infectivity.

Structure of the RNA Genome

The genome consists of a single "plus" (messenger-active) strand that is polyadenylated at the 3′ terminus and carries a small protein (virion protein, genome; VPg) covalently attached to its 5′ end. The first picornaviral RNA to be completely sequenced and molecularly cloned into DNA was that of type 1 poliovirus (212,341). Since this landmark work, sequencing of many other picornaviral RNAs has shown a common organizational pattern (Fig. 2).

Sequence comparisons show significant variations in the size of picornaviral RNAs, which range in length from 7,209 to 8,450 bases (Table 3, column 2).

Aphthoviruses and encephalomyocarditis viruses carry a poly(C) tract (48) between VPg and the beginning of the protein coding region (Table 3). All of the RNAs encode a large polyprotein ranging in length from 2,178 amino acid residues for human rhinovirus 14 to 2,332 for aphthovirus. The length of the poly(A) tract also varies in size, being shortest in cardioviruses and longest in aphthoviruses (Table 3). The protein coding region is flanked on each end by nontranslated (ntr) regions (Fig. 2) whose sequences tend to be strongly conserved and carry signals initiating translation near the 5′ end and for initiation of RNA synthesis at the 3′ ends of the plus and minus strands, respectively (12). The small protein, VPg, is attached to the 5′ terminal -pUpUp of the RNA through a phosphodiester linkage to the phenolic (O^4) hydroxyl group of a tyrosine residue (438). The length of VPg (called 3B in L434 nomenclature; Fig. 2) varies only slightly in different viruses (Table 4, 3B column). The amino terminal tryptic fragment of polioviral VPg linked to the RNA is as follows:

O-pUpUp-[RNA]-poly(A)
|
NH_2-Gly-Ala-Tyr-Thr-Gly-Leu-Pro-Asn-Lys

TABLE 3. *Properties of some picornaviral RNAs*

Virus	No. of bases[a]	Poly(C) tract	Poly(A)[b]	Mol wt[c] in polyprotein	No. codons	Reference
Human rhinovirus 14	7,209	–	74	2.50×10^6	2,178	59
Poliovirus	7,433[d]	–	62	2.58×10^6	2,207	212
Aphthovirus	8,450	–	100	2.93×10^6	2,332	62
Encephalomyocarditis virus	7,840	–	35	2.70×10^6	2,290	306
Theiler's virus	8,098	–	?	2.81×10^6	2,303	319
Hepatitis A virus	7,478	–	?	2.60×10^6	2,227	73

[a]Excluding the variable-length poly(A) tract at the 3′ end of the RNA.
[b]Average number (4).
[c]Computed as the sodium salt of the VPg-containing RNA. Sodium and potassium, the major cations found in poliovirus, are required to neutralize each of the negatively charged phosphate residues in the tightly packed nucleic acid core. In human rhinovirus 14, about 2,000 of these metal cations are replaced by polyamines, principally spermidine and spermine (137).
[d]For the Mahoney strain of type 1 poliovirus [see also Racaniello and Baltimore (341)]. The Sabin vaccine strains (423) contain 7,441 (type 1), 7,440 (type 2), and 7,432 (type 3) bases, respectively.

TABLE 4. *Genome structure of representative picornaviruses*

			Number of amino acid residues[a]												
Virus	5′ ntr bases[b]	L	1A	1B	1C	1D	2A	2B	2C	3A	3B	3C	3D	3′ ntr bases	Reference
R															
HRV14	624	—	68	262	236	290	145	97	330	85	23	182	460	47	59,402
E															
Polio 1	740	—	68[c]	271	238	302	149	97	329	87	22	182	461	72	212,341
Cox B4	743	—	68	261	238	284	147	99	329	89	22	183	462	100	196
A															
FMDV	1,199	205	81	218	221	212	16	154	317	154	24[d]	214	470	87	62,347
C															
EMC	833	67	69	255	231	288	146	136	325	88	20	215	460	126	295
Theiler	1,064	76	71	267	232	276	142	127	326	88	20	217	461	125	319
H															
HepA	733	—	22	224	245	345	144	128	330	58	23	217	491	64	73

–, no leader protein is encoded.
[a]Deduced from nucleotide sequence.
[b]5′ nontranslated region; initiating AUG begins at next base, e.g., at base 741 in poliovirus.
[c]Not counting initiator methionine, which is removed before myristylation.
[d]Has 3 VPgs in tandem, each with 23 or 24 amino acids (136,210).

FMDV encodes three VPg genes in tandem (135), whereas the other picornaviruses encode one. Reports of two forms of VPg in the virion RNA of EMC and poliovirus may be an experimental artifact from desamidation of an asparagine residue in the aminoterminal tryptic fragment shown above (344). VPg appears to play an important role in initiation of picornaviral RNA synthesis (see section under Viral Synthesis and Assembly).

The 5′-nontranslated region of the RNA is unusually long compared to the homologous region of cellular messages; it ranges in size from 624 bases in HRV14 to nearly 1,200 in aphthovirus (Table 4).

Computer analysis of the 5′-nontranslated region of poliovirus RNA predicts a cloverleaflike configuration that is supported by site-directed mutagenesis and enzymatic analysis (43,293,393). Similar folds are seen in other picornaviruses; they fall into two distinct patterns (Fig. 3). These folds bind specific host cell proteins and play a key role in initiating synthesis of viral protein and RNA. Cells may differ in their spectrum of initiation factors, and this difference

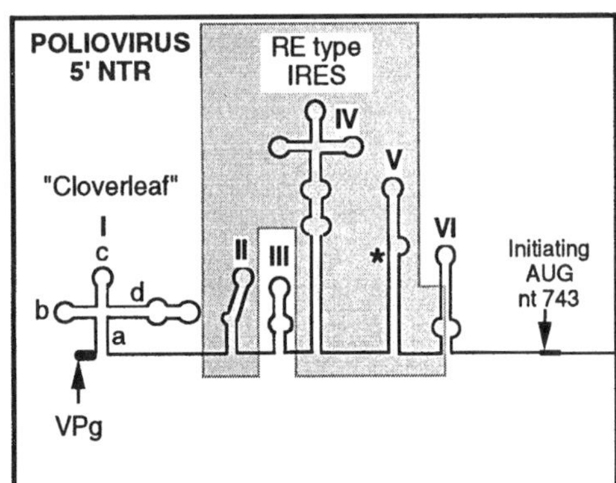

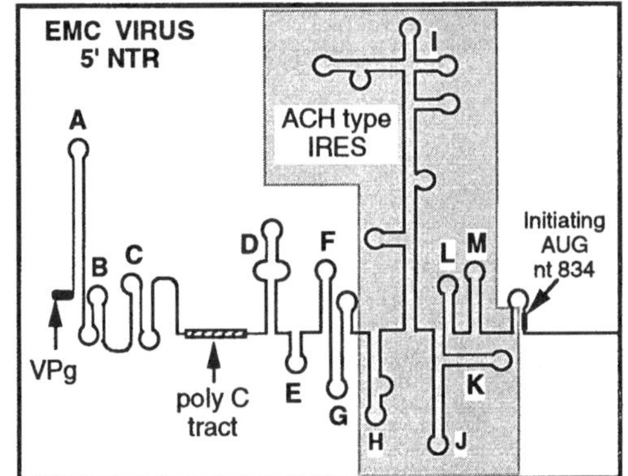

FIG. 3. Two folding patterns characteristic of the 5′-nontranslated regions of picornaviruses (439). The IRES (shaded) enables eukaryotic ribosomes to bind directly to the internal site without first scanning from the 5′ terminus. The RE-folds of rhinoviruses and enteroviruses resemble that of poliovirus (224,439) (top left), whereas the ACH-folds of aphthoviruses, cardioviruses, and hapatoviruses (110,223) resemble that of EMC virus. The IRES of EMC viruses is now widely used in high-level protein expression systems because it is one of the most active translational initiation sites known (120,121,205). A region about 100 bases upstream from the initiating AUG is complementary to 18S recombinant RNA and may play a role analogous to the Shine-Dalgarno sequence in translational initiation of prokaryotic mRNAs (223). The asterisk (left stem of poliovirus loop V) marks the approximate location of attenuating mutations in the Sabin poliovaccine strains (279). Shortening the long poly(C) regions markedly reduced the virulence of mengovirus (111) but not of aphthovirus.

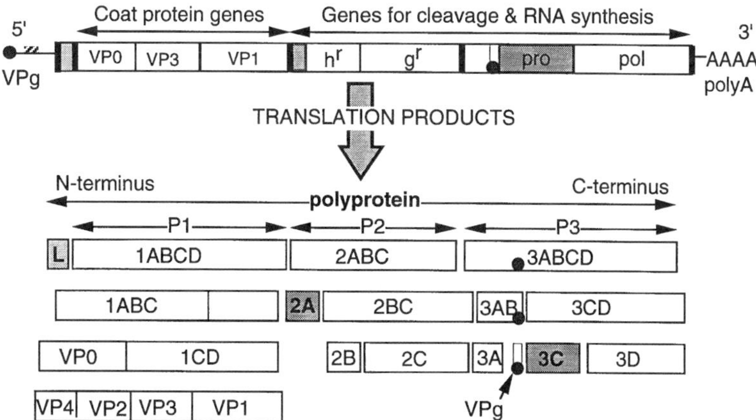

FIG. 4. Organization and expression of the picornaviral genome. The striped bar over the 5′-nontranslated region indicates the presence of a polycytidylic acid tract found in EMC-like viruses and aphthoviruses. Synthesis of the protein is from left (*N*-terminus) to right (*C*-terminus). Growth functions, i.e., proteins needed for RNA synthesis and proteinases required to cleave the polyprotein, are encoded downstream from the capsid protein. g^r represents the guanidine-resistance marker, a genetic locus affecting the action of a drug thought to block initiation of RNA synthesis. The 2B gene, h^r, carries a host range determinant (443) involved in RNA synthesis (378). Cleavage of the polyprotein is accomplished by three virus-coded proteinases: the M or maturation proteinase, the early 2A (424) or 2AB proteinase (inactive in aphthoviruses, which have a unique early proteinase, L, not found in other picornaviruses; see Fig. 5); and the 3C (307) (or 3CD) proteinase. The maturation cleavage (VP0 → VP4 + VP2) occurs only after the RNA has been packaged in the protein shell (see Fig. 15, step 10). The VP0 cleavage site lies buried inside the shell near the RNA; thus the M proteinase responsible for this cleavage is not yet precisely known (358). Proteins L and 2A perform early cleavages of the polyprotein as shown in Fig. 5. All other cleavages are performed by proteinase 3C or a precursor form, 3CD. The amino termini of coat proteins P1 and 1A are blocked by a myristyl group (68); HAV appears to be an exception (417). Proteinase 3C is turned over by proteolytic degradation (298).

may play a significant role in determining host range and neurovirulence (412). A single base change from C to U at position 472 of Sabin type 3 poliovaccine, for example, is sufficient to increase neurovirulence. Point mutations modify the translation efficiency of the RNA (412), yet large portions of the ntr can be deleted without destroying messenger activity in cell-free protein synthesizing extracts from rabbit reticulocytes. Conservation of specific sequences in the 5′ntr also provides a basis for developing new more rapid diagnostic tests for picornaviruses (452).

The poly(C) tracts found in cardioviruses and aphthoviruses are located in a noncoding region upstream from the initiation site for translation Fig. 4, top). In cardioviral RNA the tract lies about 150 bases from the 5′ end (70,317), whereas in aphthoviral RNA it resides about 400 bases from the end (161,363).

The size of the poly(C) tract is homogeneous within a given isolate of virus but varies in length in different virus isolates (80–250 for cardioviruses; 100–170 for aphthoviruses) (39). Thus, the length of the poly(C) tract is a useful tool for identifying virus strains. Natural isolates often contain longer poly(C) tracts than do laboratory strains; longer length may be more virulent in cardioviruses but not apparently in aphthoviruses (85,112,161).

The 3′ ntr is relatively short, ranging in length from 47 bases for HRV14 to 126 for EMC virus. The function is unknown but may be important at some stage of replication because an 8-base insertion in this region of poliovirus produces a temperature-sensitive phenotype (378).

The poly(A) tract located at the 3′ end of the RNA is heterogeneous in length. This heterogeneity is not eliminated by plaque-purifying virus stocks. However, the mean length of the heterogeneous tract is genetically determined, ranging from 35 residues in EMC virus to 100 in aphthoviruses (Table 3, column 3).

The function of poly(A) is not known, but RNA molecules with short poly(A) tracts have a lower specific infectivity (184,377,401). Minus strands from replicative intermediates contain poly(U) of similar length and heterogeneity at the 5′ end (444), indicating that the 3′ terminus is genetically encoded; the role of posttranscriptional addition (375) has not yet been critically evaluated.

The Protein-Coding Region

The RNA contains a single long reading frame encoding a long polypeptide chain, the polyprotein, which is normally cleaved during translation so that the full-length protein is not formed. These cleavages are performed by three virus-coded proteinases that ultimately generate 11 end

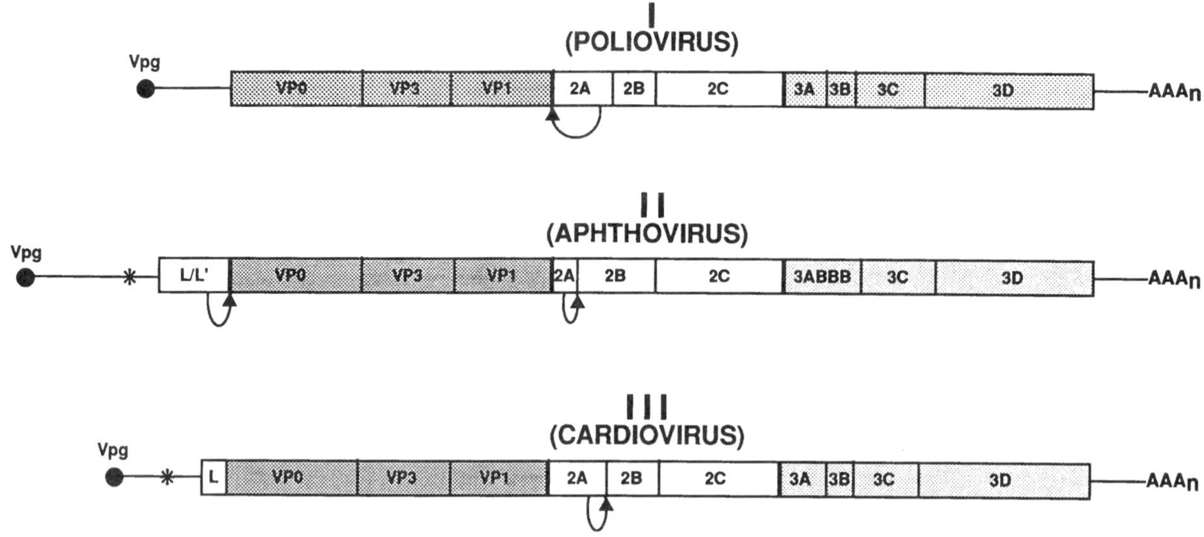

FIG. 5. Early cleavage patterns I, II, and III are characterized by differences in proteolytic activity of proteins 2A and L. All cleavages, except those shown by arrows, are performed by proteinase 3C or its precursors. Cleavage of VP0, performed by the maturation proteinase, is not shown. Cleavages due to the 2A, 2AB, or L/L′ proteinases (arrows) can cleave the nascent polyprotein even before proteinase 3C has been synthesized. In cleavage pattern I (poliovirus, Coxsackievirus, and human rhinovirus), the P1 capsid precursor is released from the nascent polyprotein by 2A cleaving at its amino terminus. In cleavage pattern II (FMDV) the 2A protein is short, containing only 16 or so amino acid residues, and is proteolytically inactive. In cleavage pattern III (cardio- and aphthoviruses), the 2A protein is proteolytically inactive except in conjunction with 2B sequences and is cleaved at its carboxyterminus (2A-2B site) to release L-P1-2A, and the P1-2A junction is cleaved by 3C (308,374). In aphthoviruses the L protein cleaves itself autocatalytically from the L-P1 site (234), whereas in cardioviruses the shorter L protein is proteolytically inactive and is removed from P1 by proteinase 3C (308). In hepatitis A, protein 3C is reported to cleave all sites including P1-2A and 2A-2B (384). Note that the polyprotein of FMDV contains three copies of 3B (VPg) in tandem; other picornaviruses encode only one VPg.

products, or 12 in the case of viruses encoding a leader (L) protein (Fig. 4). The P1 region encoding the coat protein is ultimately cleaved into four segments called 1A (VP4), 1B (VP2), 1C (VP3), and 1D (VP1) in the L434 nomenclature. Proteinase 3C, or its precursor 3CD, mediates most of the cleavages in the polyprotein (159,307,411); exceptions involving "early" cleavages of the nascent polyprotein and the "maturation" cleavage (VP0 → PV4 + VP2) are outlined in Figs. 4 and 5. Product 3B represents VPg, which is derived from a precursor 3AB thought to be involved in initiating RNA synthesis (303,389). Product 3D is an enzyme capable of elongating nascent RNA chains from an RNA template (134).

Protein 3CD of polioviruses (304,371), human rhinoviruses (266), and aphthoviruses (106,376) can be cleaved in two different ways (266,371) to yield either 3C + 3D or 3C′ + 3D′. The second pathway, also called the alternate cleavage pathway, appears to be absent in cardiovirus-infected cells (51,52). In poliovirus the alternate cleavage occurs at a tyr-gly site and is mediated in poliovirus by protein 2A (424), but the cleavage is not essential for infectivity (226). The main function of protein 2A in poliovirus is to cleave the capsid precursor protein P1 away from the nascent polypeptide (424) and translation slows markedly if P1 is not quickly removed from the nascent chain.

Early work suggested that the 2C gene is involved in RNA synthesis, whereas the 2B gene is a host range determinant. Thus, studies on FMDV (379) and poliovirus (10) mapped the guanidine resistance marker to the 2C region. The drug guanidine hydrochloride blocks synthesis of many picornaviruses at millimolar concentrations (58,93,346) and is thought to block initiation of RNA synthesis (58,416). Mutants of human rhinovirus 2, selected for extended host range by ability to grow in murine cells, appear to be selectively altered in the sequences found in protein 2B (443).

The 2C region is the most strongly conserved sequence in the polyprotein. This is particularly true of the central third of the protein (19), to which sensitivity or resistance of multiplication in the presence of guanidine has now been mapped (211,329-331). Studies on the effects of mutations introduced at known sites in complementary DNA clones suggest that proteins 2B (33) and 2C (231) are both involved in RNA synthesis and that protein 2A (34) is involved in shut-off of host protein synthesis.

THE VIRION

Physical Properties

Poliovirus is one of the most thoroughly characterized viruses. Some of its physical properties are listed in Table 5.

The virion is roughly spherical, with no lipid envelope. Hence, its infectivity is generally unaffected by shaking virus-containing fluids with organic solvents such as ether. Particle diameters reported from electron micrographs range from 24 to 30 nm. This wide range in size is due to flattening of particles or variable penetration of heavy-metal stains during the drying and staining procedures required in preparing samples for electron microscopy. Lyophilized preparations lose 99.99% or more of their initial infectivity, indicating that water is essential for the integrity of the nucleocapsid. Nondestructive methods, such as sedimentation-equilibrium (50) and small angle x-ray scattering and x-ray diffraction analysis (123,213), which measure the diameter of wet particles, indicate diameters in the range of 29.8 to 30.7 nm.

TABLE 5. *Physical properties of the poliovirion*[a]

Shape	Roughly spheroidal
Diameter (hydrated)	About 30.5 nm
Symmetry (x-ray)	5:3:2 (icosahedral)
Capsomers (EM) (32,42, or 60)	Indistinct
$S_{20,w}$	156S
$D_{20,w}$	1.40×10^{-7} cm²/sec
Partial specific volume ($\bar{v}$)	0.685 ml/g
Virion mass[b]	8.43×10^{6}
% RNA (as K-salt)[b]	31.6
% Protein	68.4
Virions/mg	7.07×10^{13}
Virions/OD_{260}-unit	9.4×10^{12}

[a]Data from Rueckert (368).
[b]Computed from complete nucleotide sequence.

The virion contains an RNA core that when unraveled and fully extended for measurement in the electron microscope has a length of about 2,500 nm. The tightly packed RNA resides in the central cavity of a thin protein shell (see arrow in Fig. 6A). This shell can sometimes be ob-

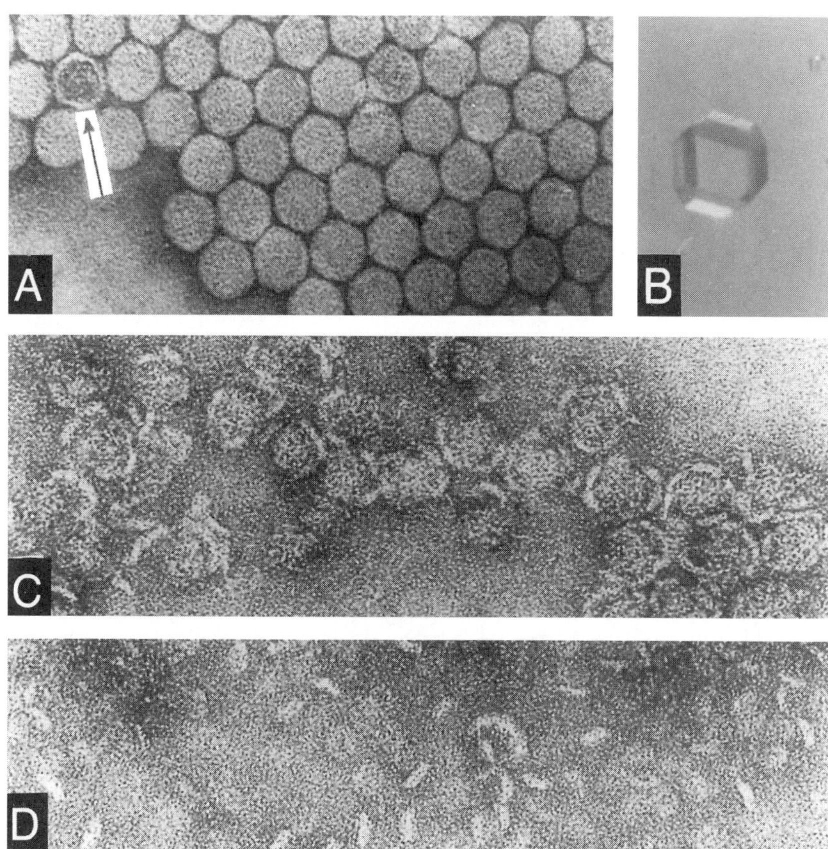

FIG. 6. Stages in disassembly of FMDV. **A:** Purified virions, aligned in a regular two-dimensional array, consist of a core of tightly coiled RNA surrounded by a protein shell about 5 mm in thickness. Arrow identifies a hollow, RNA-free protein shell called the natural top component. **B:** A well-formed crystal of poliovirus (courtesy F. Schaeffer). Tiny crystalline arrays of virus are also often found in the cytoplasm of infected cells. **C:** Disintegrating virions (pH 6, 20°C) showing peripheral protein shells partially stripped away from a still intact RNA core. **D:** More completely disintegrated virions (pH 5, 37°C) showing curved 12S platelets called skullcaps (429). The cores are now completely unraveled into RNA strands too thin to be visible by this type of staining (2% uranyl acetate) (photos courtesy of C. Vasquez).

served peeling away from the RNA core (Fig. 6C) before dissociating into thin skullcap-shaped subunits (Fig. 6D). This simple RNA core-protein shell design was first deduced from small-angle x-ray scattering studies with small plant viruses (9).

Dodecahedral Nature of the Protein Capsid

Picornaviruses typically contain four polypeptide chains (virion proteins): VP1, VP2, VP3, and VP4 (Fig. 7, bottom). There are many possible ways of constructing an icosahedral shell with four nonidentical proteins (367). Evidence that these chains are all elements of identical four-segmented subunits, now called protomers (Fig. 7, top), and that these protomers are organized into pentameric units (dodecahedral model of shell structure), was first provided by analysis of the acid dissociation products of ME virus, a cardiovirus (116,367,368,370). In the presence of 0.1 mol/L chloride or bromide ion at pH 6, the virion can be thermally dissociated into infectious RNA, VP4, and two kinds of pentamers (Fig. 8).

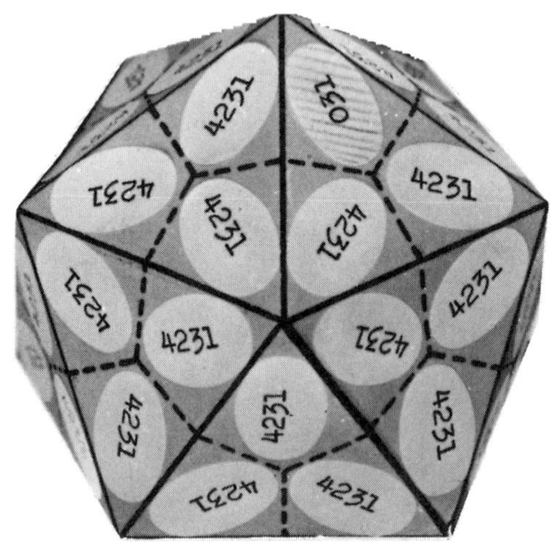

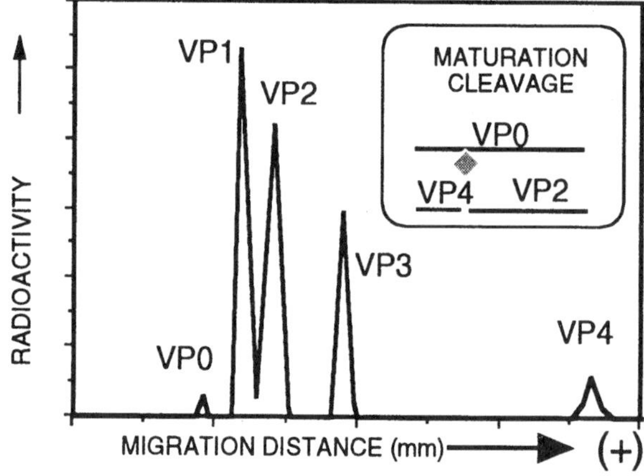

FIG. 7. Structural organization of the picornaviral capsid. **Top:** The shell contains 60 subunits called protomers. Of these subunits, 60 – n are identical mature protomers (VP1,2,3,4), and n are immature protomers (VP0,1,3) (see periphery at one o'clock position). VP0 represents an uncleaved precursor of chains VP2 and VP4. Picornavirions rarely if ever contain fewer than two immature subunits. **Bottom:** Electrophoretic profile on an SDS-polyacrylamide gel of protein from EMC virions radiolabeled with a mixture of [^{14}C]amino acids.

The 13.4S subunits (249) are curved plates about 4.8 nm thick and 16 nm wide (248). When treated with 2 mol/L urea they dissociate into 4.7S subunits (370). Measurements of chain stoichiometry in concert with molecular weight determinations on both subunits show that the 4.7S subunit is a monomer composed of one chain each of VP1, VP2, and VP3, whereas the 13.4S subunit is a pentamer of the 4.7S monomer (Table 6).

Thus, the protein shell of EMC virus behaves as an assembly of 12 pentamers held together by one kind of acid-sensitive bonding domain. The pentamers, in turn, are held together by a second type of urea-sensitive domain. In other words, the dissociation of the shell can be explained by the presence of just two kinds of bonding domains within each of the identical protein subunits. One domain holds pentamers together; the other binds monomers into pentamers. Assembly of the picornaviral shell can be understood as the operation of these domains in reverse order (Fig. 9).

The 12S subunit produced by acid dissociation of FMDV also contains equimolar amounts of VP1, VP2, and VP3 (414) and was originally thought to be a trimer [$(VP1,2,3)_3$]. Later, electron microscopy showed the 12S subunit to be roughly a half virion in diameter (Fig. 6C and D), more consistent with that of a pentamer (429). It is now apparent from the smaller size of the FMDV protomer (Table 7) that the difference in sedimentation rates between the aphthoviral 12S subunit and that of the EMC viral 13.4S sub-

TABLE 6. *Two protein particles derived from cardiovirus*[a]

	Molecular weight		
Subunit	Electrophoretic	Sedimentation equilibrium	Deduced structure
4.7S	88,000	86,000	(VP1,2,3)
13.4S	440,000	420,000	$(VP1,2,3)_5$

[a]Data from Dunker and Rueckert (115,116).

TABLE 7. *Calculated S values of some picornaviral pentamers*

Virus	Pentamer	Molecular weight	Predicted $S_{20,3}$[a]
FMDV	VP123	361,000	11.9
FMDV	VP013	403,000	12.9
EMC	VP013	466,000	14.2
Polio	VP013	486,000	14.6

[a]Computed from the relation $S1/S2 = (MW1/MW2)^{2/3}$ using the known MW and S value of the EMC viral 13.4S subunit.

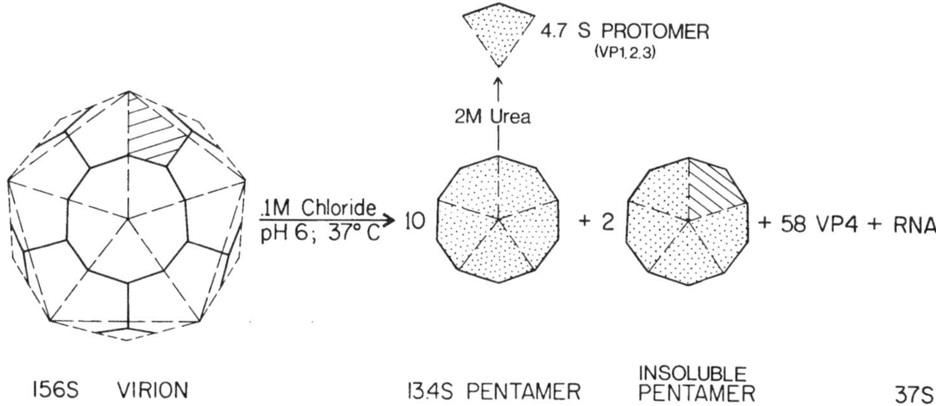

FIG. 8. Experimental basis for the dodecahedral model of picornavirus dissociation and assembly (116). When warmed to 37°C in 0.1 mol/L chloride ion at pH 6, cardiovirus (ME strain) dissociates to release infectious RNA, VP4, and a soluble 13.4S pentamer, the structure of which $(VP1,2,3)_5$ was established by electrophoresis and sedimentation equilibrium analysis (see Table 7). The 13.4S pentamer can be further dissociated with 2 mol/L urea into 4.7S protomers (VP1,2,3) represented by the stippled triangles. This stepwise dissociation suggests that pentamers are joined to form the shell by a chloride-sensitive bond, whereas the protomers are held in the pentamer by a second kind of urea-sensitive bond. About one sixth of the total protein, containing VP0:VP1:VP2:VP3 in the molar proportion 1:5:4:5, was recovered in an insoluble precipitate that has been attributed to a still hypothetical structure (insoluble pentamer) containing four (VP1,2,3) monomers (stippled triangles) and one (VP0,1,3) monomer (shaded triangle) (116).

unit is explained by the different masses of the two pentamers (Table 7).

Particle Mass of Picornavirions

The masses computed for 60-subunit viruses, using molecular weights computed from sequence data, show significant differences in masses of the particles (Table 8). The smaller masses of rhinoviruses and aphthoviruses conform to the smaller sedimentation coefficients reported for these viruses (Table 2).

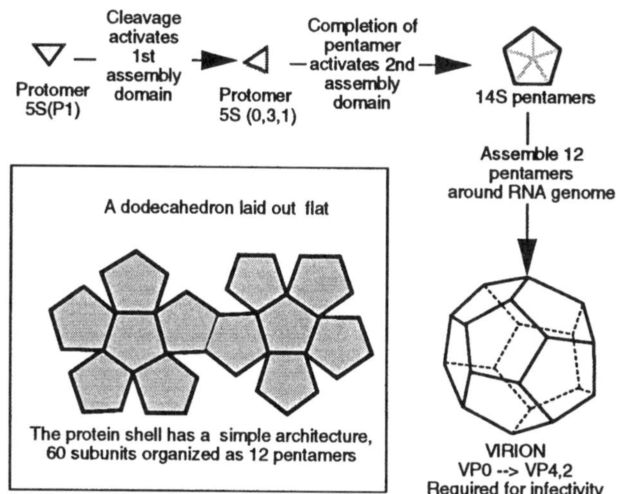

FIG. 9. Dodecahedral model of picornavirus structure and assembly (367,370).

TABLE 8. *Particle mass of some picornaviruses*[a]

Virus	Mass (Md)
HRV14	8.16
Polio	8.43
FMDV	7.77
Hepatitis A	7.85

[a]Computed from the data in Tables 3 and 4.

Ratio of Particles to Infective Units

The concentration of virus particles can be computed from spectrophotometric measurements (372) on highly purified preparations of virus. An optical density of 1.00 (1 cm light path) at a wave length of 260 nm corresponds to a concentration of 9.4×10^{12} particles per milliliter. Once the particle concentration is known, it is then a straightforward procedure to measure the number of particles required to produce a plaque-forming unit (PFU) in a standard assay. The value found from analysis of the plate shown in Figure 10: is 200 particles per PFU, corresponding to an infection efficiency of only 0.5%.

The plaque-forming efficiency of picornaviruses is typically low, in the range of 0.1% to 2%, even when every effort is taken to avoid defective or inactivated particles (198,215,268,322,386). This is probably not due to genetic errors in most of the RNA molecules because the infectivity of aphthoviral RNA, when microinjected into cells, is nearly one infectious unit per molecule (30). Chemical

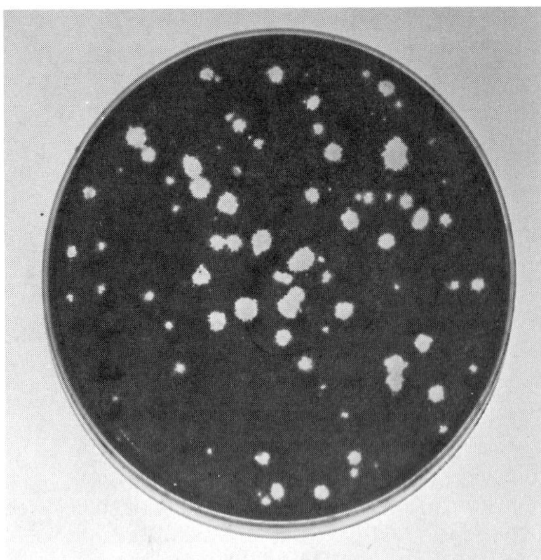

FIG. 10. Plaques of human rhinovirus 14 on a lawn of HeLa cells. About 15,000 virus particles in 0.2 ml of buffer were applied to a monolayer (4 million cells) attached to the bottom of the dish. After a suitable time period, unattached virus was rinsed away and the cells were overlaid with nutrient agar, which forms a gel. The gel prevents convection currents but allows diffusion of virus from each singly infected cell to its neighbors, thus producing growing islands (plaques) of virus-killed cells. When the plaques had grown to an appropriate size (48 hr at 34°C), the agar overlay was removed and the sheet of cells remaining in the dish was stained with an alcoholic solution of crystal violet. The unstained holes in the cell monolayer result from detachment of the virus-killed cells when the agar overlay was removed. Each of the 73 plaques is scored as 1 PFU. From this particular experiment we conclude that the number of virus particles needed to form 1 PFU was 15,000/73, or about 200. Differences in plaque size may be attributable to mutants in the parental virus population or to wide variation in the rate at which individual particles initiate infection (attach, penetrate, and uncoat their RNAs) (183). Which explanation is correct can often be determined by replating virus from agar plugs extracted over well-isolated plaques.

changes, such as oxidation of thiol or methionine residues, or thermal denaturation, loss of VP4, or fragmentation of the RNA in the core, are common sources of trouble that can be minimized by harvesting virus promptly after it reaches peak titer and by maintaining a neutral pH and low temperature during purification of the virus. Protective thiols, such as mercaptoethanol or dithiothreitol, in purification buffers also help minimize oxidation or substitution of reactive side chains (cysteine, methionine, tyrosine, and tryptophan). Echoviruses, for example, exhibit dramatic sensitivity to reagents that substitute sulfhydryl groups (324). However, accidents encountered along the infection pathway, such as abortive elution, are probably the single most important reason for the low efficiency of infection (198,390).

Atomic Structure of the Protomer

With general acceptance of the 60-protomer model of capsid structure, attention focused on the structural organization the four-segmented protomer. The feasibility of attacking this problem by the methods of x-ray diffraction analysis was signaled by success in solving the atomic structure of two spherical plant viruses, tomato bushy stunt virus in 1978 (163) and southern bean mosaic virus in 1980 (1). By 1985 the protein structure of human rhinovirus 14 was solved (358); that of type 1 poliovirus followed within a matter of weeks (173).

A significant innovation used in solving the atomic structure of the rhinovirus shell was successful application of the molecular replacement method in a large molecule (358). This powerful method, first proposed in 1962 as a way of solving the phase problem (357,359), exploits the speed of modern computers, reduces dependence on unpredictable success in finding satisfactory (isomorphous) heavy atom derivatives, and greatly facilitated the structural analysis of many other picornaviruses.

The Maturation Proteinase

Crystallographic analysis shows that the cleavage site of VP0 (yielding VP4 and VP2) is deeply buried within the particle and is therefore probably cleaved autolytically. An early hypothesis that the active site might be Ser-10 of VP2 in HRV14 (358) seems disproved by the finding that maturation cleavage is not blocked by replacing this serine with an alanine or cysteine (160,228).

Folding Motifs and the Packing Arrangement of VP1, VP2, and VP3

Crystallographic analysis confirmed the 60-protomer T = 1 model developed from biophysical studies. It also showed that VP1, VP2, and VP3 occupy distinct domains with pseudo T = 3 symmetry, thus confirming a conjecture earlier invoked to reconcile apparently contradictory findings of 32, 42, and 60 capsomers in electron micrographs of picornaviruses (367,370). The crystallographic work also supported earlier conclusions based on work with antibodies (101,243,428) and other surface-labeling reagents (237) that VP1, VP2, and VP3 are exposed at the virion surface, whereas VP4 lies buried (197,432) in close association with the RNA core (278,433).

The most prominent feature of the protomer, evident in HRV14 even before resolution reached the atomic level (366), was the presence of a prominent cleft or "canyon" on its surface (Fig. 11A). This canyon was later shown to be the acceptor site for the receptor used by the virus to infect susceptible host cells (300). Each of the three major picornaviral proteins (VP1, VP2, and VP3) had the same wedge-shaped folding pattern (Fig. 12 inset) as that found in the protein subunits of the two plant viruses mentioned above.

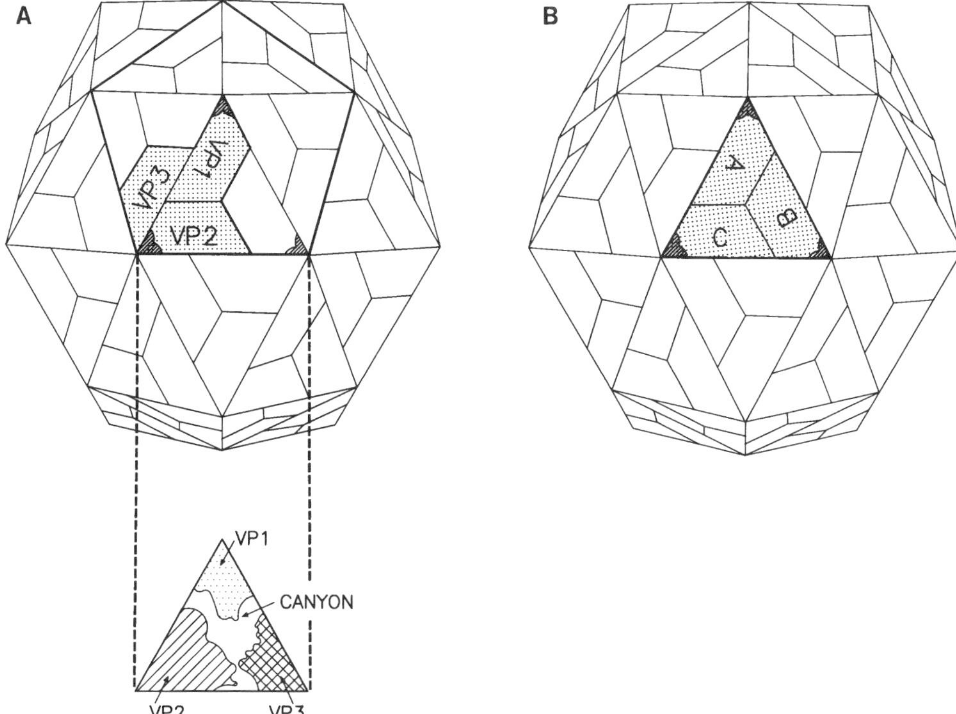

FIG. 11. Pseudoequivalent packing arrangement of VP1, VP2, and VP3 domains in the 60-subunit picornaviral shell and its relationship to the protein shell containing 180 identical subunits. **A:** Each of the three proteins has a common folding pattern (see inset, Fig. 12). Hatched arrow heads locate a "prow" (see Fig. 12A), which enables each protein to pack tightly at the five- or threefold symmetry axes. VP4 is buried deep inside that particle at the base of the protomer and is not an integral component of the framework making up the shell. The biological protomer (stippled) is not identical to the icosahedral asymmetric unit. As a result, the positions of VP2 and VP3 in the protomer are reversed relative to those in the crystallographic subunit (triangle). Note the similarity in packing arrangement of VP1, VP2, and VP3 of viruses with triangulation number T = 3. For a discussion of triangulation numbers see Caspar and Klug (63). **B:** Arrangement of protein subunits in the shell of T = 3 viruses containing 180 subunits. Each of the 60 triangular asymmetric units contains three identical polypeptide chains packed in non-identical quasiequivalent domains A, B, and C.

One of the most exciting discoveries emerging from the structure of HRV14 was the realization that the folding patterns of VP1, VP2, and VP3 were similar to each other (Fig. 12). Moreover, this folding pattern was similar to that observed in the protein subunits of several other RNA viruses.

The similar folding patterns in VP1, VP2, and VP3 suggested that the three proteins evolved by fusions of a common gene, a fact not apparent from amino acid sequence comparisons. Because protein folding patterns are conserved over longer time spans than amino acid sequences, these common folding motifs in capsid protein subunits has excited speculation about "ancient" evolutionary links between picornaviruses and other spherical viruses (360).

THE PROTOMER, OR PROTEIN SUBUNIT

Coat protein serves several key functions: (a) it protects the RNA genome from nucleases in the environment; (b) it recognizes specific cell-coded receptors in the plasma membrane and is therefore one important determinant of host range and tissue tropism (disease pathology); (c) it determines antigenicity; (d) it carries directions for selecting and packaging the viral genome and provides a proteinase involved in maturation of the virion (see section on Morphogenesis); and (e) it disgorges the RNA genome and delivers it through the cell membrane and into the cytosol of susceptible host cells. The atomic structure has provided significant insight into many of these functions.

Key Features of Picornavirus Architecture

X-ray crystallographic structures have now been determined for at least one member of each picornavirus genus except hepatitis A virus. All share similar features: a protein shell roughly 5 nm thick and 30 nm in diameter surrounding a single strand of messenger sense RNA (Fig.13). The RNA core, including VPg, although depicted in the

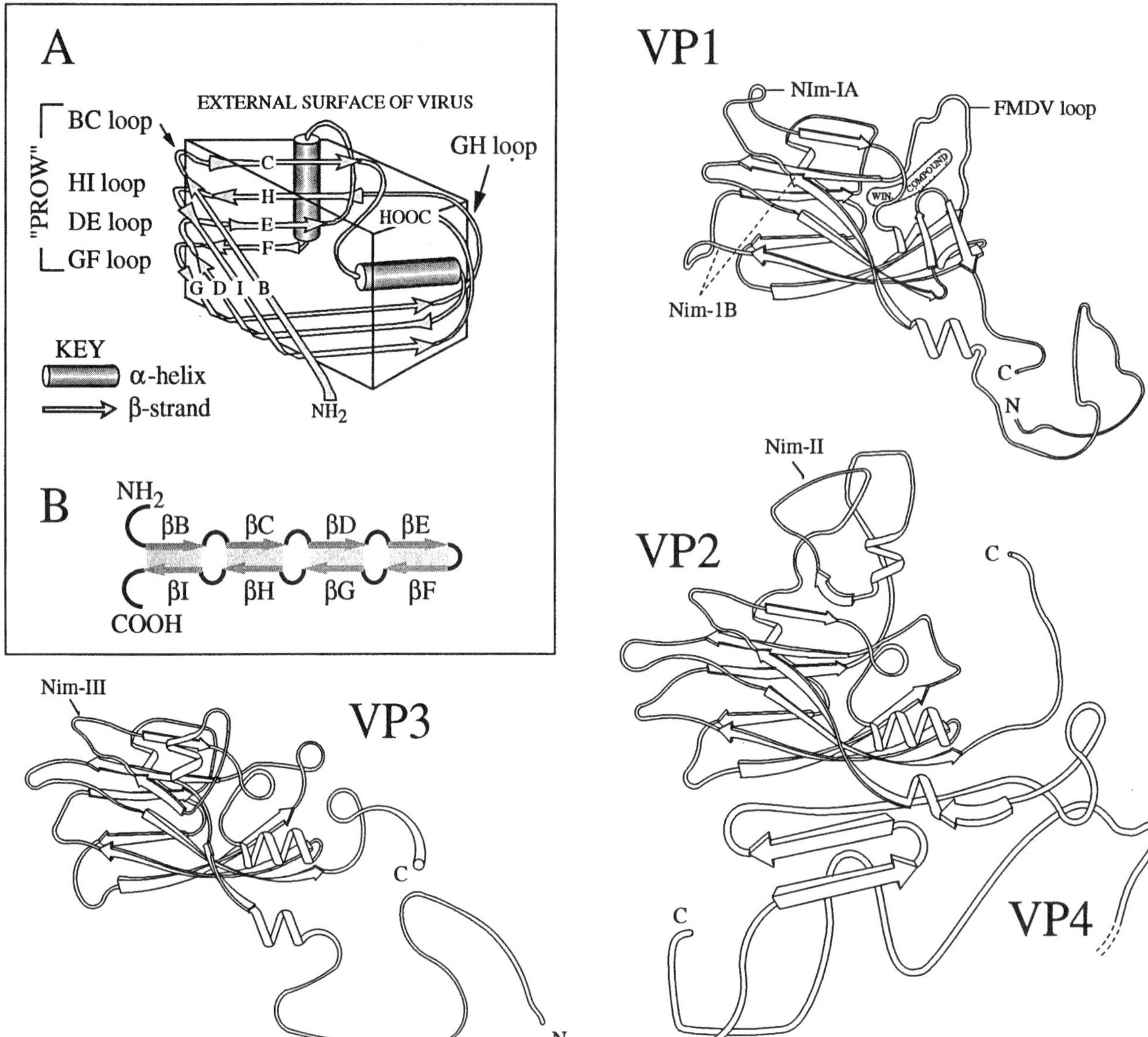

FIG. 12. "Swiss roll" fold found not only in VP1, VP2, and VP3 but also in most other RNA virus capsid proteins so far examined (162,360). This fold gives the protein subunit the wedgelike shape needed to fit compactly at threefold and fivefold vertices in small spherical (icosahedral) viruses. It has variously been called the eight-stranded antiparallel beta barrel, the Swiss roll, the jelly roll, the Greek key, or the RNA virus capsid domain (162). Pairing of the beta strands is most clearly evident in the topological representation of the Swiss roll (inset B). Inset A shows how the beta strands (arrows, top left) identify eight beta strands (B-H) connected by loops; the beta A strand found in tomato bushy stunt virus and southern bean mosaic virus is absent in picornaviral folds. In the beta strand, the polypeptide backbone is almost fully extended rather than tightly coiled as in an alpha helix. The extended state of the beta strand is stabilized by hydrogen bonds connecting amide and carbonyl groups in the backbone of adjacent strands running in opposite (antiparallel) directions. The walls of the wedgelike domain are formed by two sets of four beta strands (CHEF in back and BIDG in front) which are joined somewhat like the prow of a boat by loops (BC, HI, DE, FG) at the left edge. Note that the amino (NH_2) and carboxy (COOH) termini (black) of the chains are separated by nearly 50 Å (149, 173) and must therefore have moved after cleavage of the P1 precursor. The loops connecting the beta strands can accommodate extra segments of protein without disrupting the assembly framework, and it is variations in these loops and tails that give each subunit and each picornavirus its distinctive morphology and antigenicity. Thus, in many picornaviruses the neutralizing immunogenic sites lie in the BC loops of VP1 and VP3 but in the EF loop of VP2.

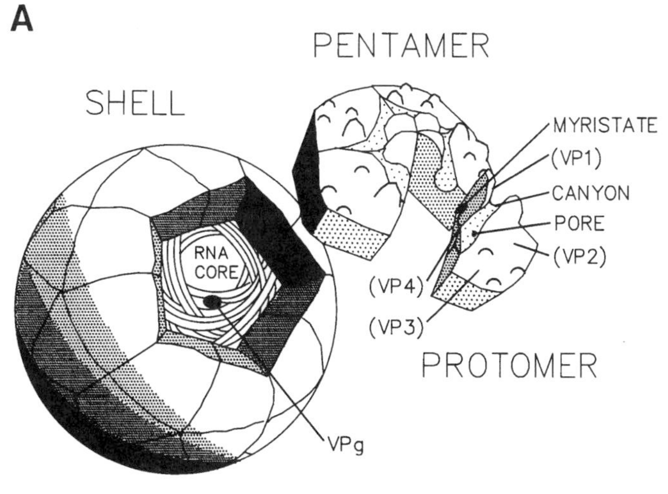

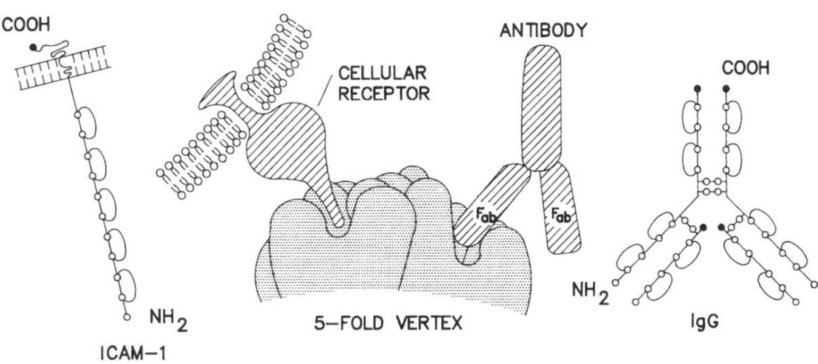

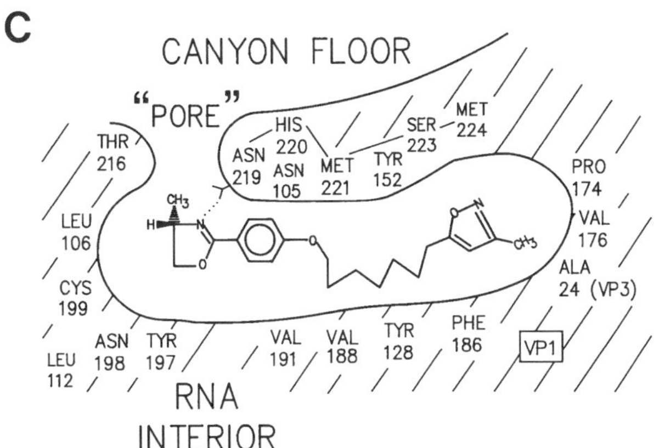

FIG. 13. Key features of human rhinovirus 14. **A:** Exploded diagram showing surface features of one of the pentameric units with a canyon (stippled) encircling a plateau at the pentamer center. Note the internal location of VP4 with its myristate residues clustered at the center of the pentamer. **B:** Binding of cellular receptor to the floor of the canyon. Note that the binding site of the ICAM-1 molecule, identified as major rhinovirus receptor, has a diameter roughly half that of an IgG antibody. **C:** Location of a drug binding site in VP1 of HRV14 and identity of amino acid residues lining the wall. The pentamer vertex lies to the right. The drug, WIN 52084, depicted in **C**, prevents attachment of HRV14. Crystallographic studies show the presence in most picornaviruses of still uncharacterized hydrophobic molecules (pocket factors) that are displaced from the pocket by WIN drugs. Each virus species seems to contain a different pocket factor (299). The function is not known but is thought to be involved in assembly and uncoating (see Fig. 19B).

diagram, is largely invisible to x-ray crystallographic analysis because, unlike the protein shell, the RNA has no symmetry and can occupy 60 different orientations in the crystal lattice. Therefore, its contribution to the crystallographic diffraction pattern is too smeared to be solved by current methods.

The protein shell is built up from 60 protomers arranged as 12 pentamers. Each protomer is composed of four proteins (VP1, VP2, VP3, and VP4). However, the surface topography of each type of virus is as varied and characteristic as faces in the human race, hence the many serotypes listed in Table 1. Human rhinoviruses and enteroviruses, for example, feature prominent canyons, like ragged moats encircling a plateau at the center of each pentameric unit in the shell; cardioviruses and aphthoviruses, on the other hand, lack such canyons.

Binding Sites for Cellular Receptors

That the prominent cleft in the protomer of HRV14 might be a receptor binding site was originally proposed (358,366) on the grounds that the canyon was too narrow (1.2–3.0 nm wide) to admit deep penetration of antibody molecules (Fig. 13B); this strategy presumably offers an advantage to a virus by permitting it to conserve the structure of its crucial cell attachment site in the face of antibody driven evolution. This idea was supported by evidence that binding affinity to isolated cellular receptors can be changed by site-directed alteration of amino acid residues in the canyon floor (78) and by demonstration that drugs that selectively alter the topography of the canyon floor also block attachment of HRV14 to HeLa cells (167) and to isolated receptors (320). That the receptor binding site is indeed located deep in the canyon has now been confirmed by direct visualization, using image reconstruction of a complex between human rhinovirus 16 and a soluble portion of intercellular adhesion molecule-1 (ICAM-1) (300).

Of the four proteins, VP1 exhibits the greatest sequence variability and VP4 the least. VP1 is also the dominant protein, playing key roles in surface topography and in several viral functions, including antigenicity, receptor attachment, and probably also viral uncoating. Many of these variations occur in the BC or GH loops (boxed inset, Fig. 12), which emerge at the surface of the virus. In the case of poliovirus and human rhinovirus, for example, the dominant neutralizing immunogenic sites reside in the BC loop of VP1; in aphthovirus, on the other hand, the major antigenic site and the receptor binding site reside in the GH loop.

Protein VP1 of aphthovirus has been a focus of attention ever since early work showed that it is not only a major antigenic site of FMDV, but also the locus of a receptor recognition site (65,437). When treated with trypsin, FMDV loses infectivity; this is accompanied by a single cleavage in VP1. The trypsin-modified particle is not only unable to attach to susceptible host cells, but loses its potency as a killed vaccine; that is, it is no longer able to stimulate production of neutralizing antibody. Studies of the isolated protein of FMDV, as well as synthetic peptides, demonstrated a major antigenic site in the region of amino acid residues 141 to 160 and a minor one at residues 200 to 213 (22,362,408). An amino acid sequence RGD at positions 145 to 147 and residues from positions 203 to 213 have been implicated in the receptor-binding site on VP1 (138). The RGD sequence has previously been implicated in cell-attachment activity of a host of other proteins, including fibronectin, vitronectin, discoidin I, fibrinogen, type I collagen, osteopontin, complement protein fragment C3bi, van Willebrand factor. and a phage receptor in *Escherichia coli* (138). These proteins, as well as some echoviruses and coxsackie A viruses, attach to a group of receptors called integrins, which have the ability to discriminate between different RGD-containing proteins.

Binding Site for Antivirals that Inhibit Attachment or Uncoating

A number of hydrophobic sausage-shaped compounds are reported to inhibit attachment or uncoating of picornaviruses (15,17,139,267,289,320,451); some have already progressed to the stage of clinical trials (451). Among the most extensively studied are the so-called WIN compounds, a third generation of neutralizing antivirals derived first from rhodanine (118) and then from arildone (267). Rhinoviruses have been classified into two groups, A and B (16), based on their similarity of inhibition by a panel of antiviral agents, but the structural basis of this classification is unclear. WIN-like drugs either block uncoating of poliovirus types 1, 2, and 3 and of human rhinovirus type 2 (139,289,449) or they prevent attachment to the cell receptor, as observed for HRV14 and HRV16 (167,299,320). WIN compounds insert into a hydrophobic pocket within the beta barrel of VP1 (397); this pocket lies just beneath the floor of the canyon (Fig. 13C). In HRV14, drug binding induces conformational changes in three regions of VP1 that comprise the roof of the drug-binding pocket and the canyon floor (23). Deformation is substantial, up to 4 Å, representing an uplifting of about one sixth the depth of the canyon. WIN 52084, the compound shown in Fig. 13C, binds with its isoxazole ring pointing toward the pentamer vertex. Many related WIN compounds bind in the reverse orientation, but the active compounds induce the same deformation in the canyon floor (23). However, in the case of HRV16, in which WIN drugs also prevent attachment, the drug produces no apparent deformation of the canyon floor. These observations suggest that proper seating of ICAM-1 into rhinoviruses requires downward deformation of the canyon floor (299).

Natural Pocket Factor

Crystallographic analysis of poliovirus showed the presence in the drug-binding pocket of types 1 and 3 poliovirus of a long sphingosinelike molecule (174), which has not yet been chemically characterized. Human rhinoviruses 1A and 16 also contain uncharacterized but shorter pocket molecules, whereas the pocket of HRV14 is empty, possibly lost during purification of the virus. Observation of these natural pocket factors raises the possibility that neutralizing antiviral molecules are actually analogs of viral pocket molecules involved in assembly or uncoating of the virus. It has been suggested that the empty pocket allows movement of polypeptide chains during the uncoating process (174,397). Lack of information on the chemical identity of any of the pocket factors has impeded progress, but results of studies on drug-dependent mutants of poliovirus suggest that one role of pocket factor is to stabilize the virion during transit from one host cell to the next (289).

Myristate Clusters at the Pentamer Base

Myristic acid, also called *n*-tetradecanoic acid, is covalently linked to amino terminal glycine residues on VP4, VP0, and P1 of most picornaviruses (68); hepatitis A virus appears to be an exception (417). Cellular enzymes are known to recognize the signal for acylation myristate-Gly-X-X-X-Ser/Thr (X = any amino acid). In entero- and rhinoviruses the *N*-terminal glycine requires removal of a methionine residue, whereas in the cardio- and aphthoviruses myristylation of the *N*-terminal glycine requires proteolytic removal of a leader peptide. Myristylation is required for assembly of pentamers (18,286) and may also play a role in early stages of infection, e.g., reorientation of the pentamer and release of VP4 (68).

Serotype

One of the dominant features of picornaviruses is the large number of serotypes, now over 217 and still counting (Table 1). A serotype is defined by the ability of a "monospecific" antiserum to neutralize viral infectivity (275). Monospecific refers to antiserum raised in an animal not previously exposed to a "related" virus. For example an isolate of poliovirus is identified as type 1 if its infectivity is reduced by some arbitrarily prescribed factor (e.g., 100-fold) after treatment with a reference antiserum raised against a prototype virus defined as type 1. A second serotype is defined by failure of a second genetically pure virus to be neutralized by the reference antiserum and by failure of antiserum raised against the second virus to neutralize the first. The concept of serotype has been particularly valuable in designing protective vaccines. Experience indicates that serum neutralization titer against reference strains is a fairly reliable indicator of protective immunity, but the atomic structure of picornavirus shells has so far provided no clear insight into factors that delineate one serotype from another or the reason why rhinoviruses (102 serotypes) are so variable relative to polioviruses (three serotypes). However, this variability does not appear to be determined by the number of neutralizing sites or by the depth or width of the canyon. Perhaps then serovariability is more related to the effect of habitat on interaction with the receptor or immune system than to virus structure per se.

Localization of Neutralizing Antigenic Sites by Mapping "Escape" Mutations

Identification of antigenic sites on the surface of the virion is of importance for studies on the mechanism of neutralization, as a foundation for understanding the molecular basis of serotyping, and as a starting point for developing subunit vaccines. Neutralizing monoclonal antibodies (NmAbs) have been used with considerable success to identify antigenic sites. Each monoclonal antibody is targeted to one specific site that is of necessity on the surface of the virion. Sequencing the RNA of mutants resistant to neutralization to a given NmAb (escape mutants) shows that mutation to resistance always involves replacement of a single amino acid side chain. When the structure of HRV14 was solved, it became clear for the first time that these mutations could be traced to single amino acid residues protruding radially from the surface of the virion (358).

It was found in the HRV14 study (366,392), and in a number of other picornaviruses since, that mutants resistant to different NmAbs targeted to a particular site select for amino acid substitutions at the same amino acid residue, or an immediately adjacent one, and that these adjacent escape residues always reside on the surface. Mutants selected with NmAb targeted to a second site are altered at the second site. In other words, antibodies selected escape mutants on the basis of surface topography, which was not immediately evident from coding location in the genome.

The collection of antibody-resistant mutants can also be used to segregate antibodies according to target sites. Thus, mutants resistant to one antibody are often resistant to neutralization by different antibodies directed against the same target site, but these same mutants are always neutralized by antibodies targeted to a different antigenic site. The patches of mutations found in this way correlate strongly with segregation of antibodies according to their ability to neutralize sets of resistance mutants; the only exceptions are rare double mutants with replacements in two sites (302).

Escape mutation mapping, despite its great utility, is unsuited to complete description of an antigenic surface. The number of sites identified by this method is commonly reported to be three or four (Fig. 14) (259,280,302,321,406). These patches represent the immunodominant or most antigenic regions of the virion surface and do not imply that none of the rest of the surface is antigenic. The footprint of a NmAb bound to the NIm-IA site of HRV14 shows a matching pattern of complementary charges between epitope and antibody (399).

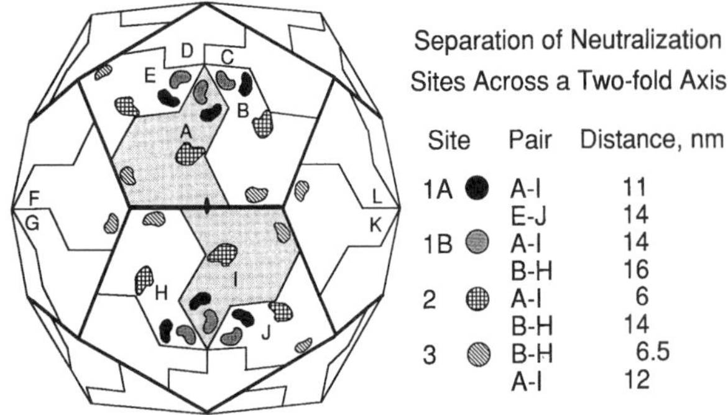

FIG. 14. Location of four neutralizing patches on each protomer of human rhinovirus 14. Proteins are as defined in Fig. 11A. The IgG molecule has twofold symmetry; hence, bivalent attachment is likeliest to occur between protomers paired by a twofold symmetry axis, for example shaded protomers A and I. Note that for any site there is more than one pairing distance. For example, the 1A sites are 11 nm apart in the A-I pair of protomers, but 14 nm apart in the E-J pair. Adapted from Mosser et al. (288), with permission.

Neutralization

Several mechanisms have been proposed for neutralization. Early studies by Dulbecco et al. (114) suggested a single-hit model for neutralization based on apparent first-order kinetics. According to this model, a single antibody binding at only one of a number of critical sites on the surface of the virus is sufficient to inactivate infectivity. Mandel (250) noted that poliovirus is able to exist in two reversible isoelectric forms: A (pI = 7) and B (pI = 4). Mandel proposed that antibody inactivated poliovirus by locking it in the A state. Attempts to confirm this hypothesis have been inconclusive, largely because of the complicating, and much ignored, effect of aggregation. Aggregation may well represent the major mechanism of neutralization (419), but only antibodies that do not aggregate can fruitfully be used as a tool for studying details such as attachment and uncoating.

The reversibility of the isoelectric A ↔ B transition suggests that picornaviruses "breathe"; that is, they periodically expose segments of protein that are normally buried in crystallized particles. That buried *N*-terminal portions of VP1 (residues 61–80, 182–201, and 222–264) are reversibly exposed to the surface was first suggested by the finding that antibodies raised against synthetic peptides to the buried regions were able to neutralize poliovirus (69,348). The breathing concept is reinforced by the finding that antibodies directed against VP4 can also neutralize (aggregate) poliovirus at 37°C but not at 23°C, and that neutralization can be reversed by dissociating the immune complexes (232).

Isoelectric changes are commonly associated with binding of neutralizing antibody, but results of quantitative studies have shown no regular relationship between the amount of antibody needed to shift isoelectric point and the amount needed to neutralize (45). Another complication is that neutralization is probably not an all-or-nothing phenomenon; some antibodies probably reduce the probability of success in infection (crippling mechanism) rather than completely inactivating the particle. There is one unconfirmed report in the literature that a single immunoglobulin G (IgG) molecule from rabbit antiserum is sufficient to neutralize poliovirus (434).

Evidence that a neutralizing IgG against human rhinovirus 14 can cross-link adjacent pentamers, the fundamental unit of capsid assembly and dissociation, has been documented by image reconstruction (398). It might be expected that this cross-linking interferes with uncoating; however, demonstration that this is so is complicated by the observation that this same antibody interferes with attachment; six IgG molecules are sufficient to reduce attachment by 50% (399).

OVERVIEW OF INFECTION CYCLE

Multiplication of picornaviruses occurs entirely in the cytoplasm. The initial event in infection is attachment of the virion to specific receptor units embedded in the plasma membrane (Fig. 15, step 1). The function of receptors is twofold: to position the virion to within striking distance of the membrane (step 1), then to trigger a conformational change in the virion (step 2), which involves loss of an internally located protein and delivery of the viral RNA genome across the membrane and into the cytosol (step 3), where translation can begin (step 4).

Translation is a crucial step because synthesis of new viral RNA cannot begin until the virus has successfully manufactured the virus-coded RNA-synthesizing machinery. By confiscating ribosomes and other protein-synthesizing machinery of the host cell, the incoming RNA strand directs synthesis of a polyprotein, which is then cleaved into segments while still in the process of synthesis. Translation of a viral message is not restricted to a single ribosome; indeed, polysomes carrying up to 40 ribosomes have been reported in poliovirus-infected cells. In poliovirus, the first fragment released from the nascent polyprotein is a coat precursor protein (P1); the next released is a midpiece precursor protein (P2); and the last segment released is P3. Each segment is released from the polyprotein by proteinases encoded in the polyprotein (Figs. 5 and 15).

Protein P3 can be further cleaved by two separate mechanisms (309,373). One mechanism is a monomolecular or *cis* cleavage, i.e., a concentration-independent self-cleavage that yields (a) a proteinase, 3C or 3CD; (b) a protein, 3AB, which is involved in initiating RNA synthesis; and (c) an RNA polymerase, 3D, which can elongate a primer RNA bound to a template RNA. The other mechanism is *trans* cleavage, i.e., bimolecular cleavage involving one P3

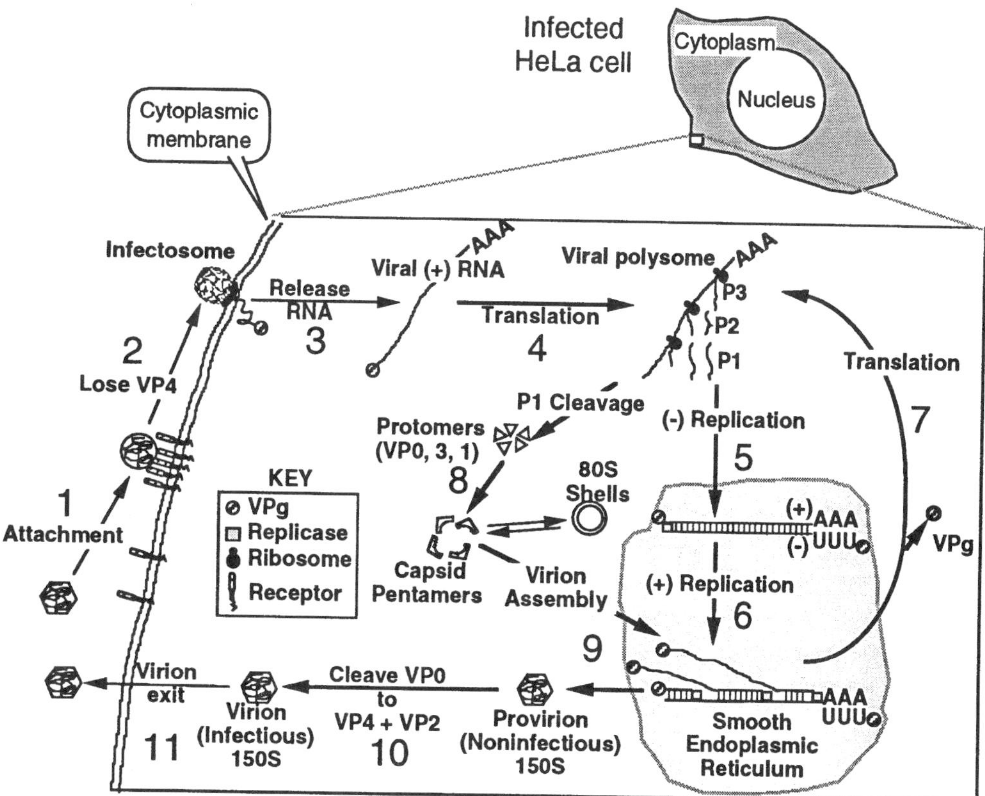

FIG. 15. Overview of the picornaviral infection cycle.

molecule and the 3C proteinase generated by cleavage of another P3 precursor molecule. *cis* cleavage of P3 presumably dominates the early stages of infection when the concentration of viral proteinases in the cytosol is low. Insights from the crystal structure (66,254) suggest that proteinase 3C cannot be active in the monomeric form. *cis* or intramolecular cleavage might therefore be accomplished by pairing of nascent P3 molecules on closely spaced ribosomes even before they have left the viral polysome.

The first step in synthesis of new viral RNA is to copy the incoming genomic RNA to form complementary minus-strand RNA (Fig. 15, step 5), which then serves as a template for synthesis of new plus strands (step 6). Synthesis of plus-stranded RNA, which occurs on smooth endoplasmic reticulum, is initiated so rapidly (20- to 50-fold that of minus strands) that it generates multistranded replicative intermediates consisting of one minus-stranded template and many plus-stranded copies. During the early steps of replication, newly synthesized plus-strand RNA molecules are recycled to form additional replication centers (step 7 → step 5 → step 6) until, with an ever-expanding pool of plus-stranded RNA, a greater and greater fraction of the plus-stranded RNA in the replication complex is packaged into virions.

Virion assembly (steps 8 and 9) is controlled by a number of events. One is that, before assembly can begin, coat precursor protein P1 must be cleaved to form immature protomers composed of three tightly aggregated proteins (VP0,3,1. Early in the infection cycle this cleavage is likely very slow because the concentrations of P1 and the necessary proteinase (3C or 3CD) are low. Later, with increasing proteinase activity, the rising concentration of immature (5S) protomers triggers assembly into pentamers (step 8), which then package the plus-stranded VPg-RNA to form provirions (step 9). Infected cells often contain empty 80S protein shells that are reversibly dissociable into 15S subunits. Whether pentamers condense around RNA or RNA is threaded into 80S shells is still a subject of debate.

Provirions are not infectious. Formation of infective 150-160S particles (step 10) requires a "maturation cleavage" in which most of the VP0 chains are cleaved to form the mature four-chain subunits (VP4,2,3,1) characteristic of picornavirions. Completed virus particles, which often form crystals in infected cells, are ultimately released by infection-mediated disintegration of the host cell (step 11).

The time required for a complete multiplication cycle, from infection to completion of virus assembly, generally ranges from 5 to 10 hr. The precise timing depends on variables such as pH, temperature, the virus, the host cell, the nutritional vigor of the cell, and the number of particles that infect the cell (27). Some viruses, such as hepatitis A, set up nonlytic infections that persist indefinitely (337). The precise characteristics must be established experimentally for each virus-cell system. This can often be done by constructing a single step growth experiment in which an entire population of cells is infected more or less simultaneously by inoculating cultured cells with enough virus to infect every cell (Fig. 16).

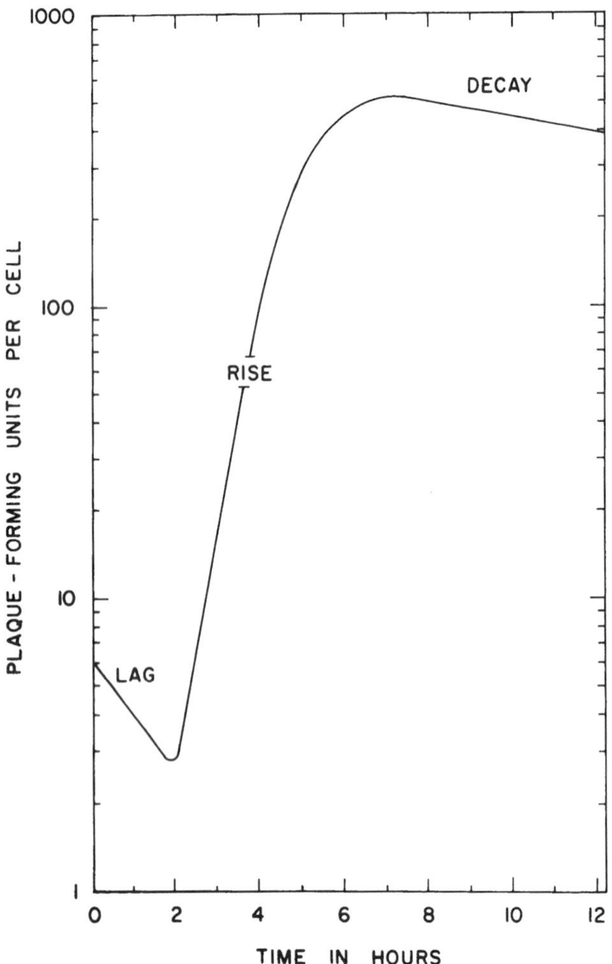

FIG. 16. One-step growth experiment for human rhinovirus 14 in a HeLa cell suspension culture. The amount of virus added (6 PFU/cell) was sufficient to infect every susceptible cell in the culture (4 × 10^6 cells/ml). After a suitable attachment period, the cells were sedimented to remove unattached virus and then resuspended in fresh nutrient medium, and the suspension was incubated at 35°C. At intervals thereafter, aliquots were removed and diluted tenfold into 1% SDS in distilled water at room temperature. This treatment lyses cells and releases both free and membrane-bound virions. Most picornaviruses survive this detergent treatment. The detergent lysates were diluted at least 1,000-fold in neutral isotonic saline containing 1% bovine serum albumin, before titering for infectivity. The large dilution eliminates toxic effects of the detergent on the cell monolayer used in the plaque assay. The albumin plays two protective roles: (a) it sequesters detergent, thus reducing its toxicity, and (b) it minimizes adsorptive loss of highly diluted virus onto tube, pipet, and dish surfaces.

During the initial stages of the cycle there is a lag period, also called the latent or eclipse period, in which infectivity is due to residual inoculum virus that fails to be uncoated. The decline or eclipse of infectivity during the lag period reflects particles engaged in the process of infection, i.e., releasing their RNA, synthesizing protein and RNA, but not yet assembling new virions. The length of the lag period typically lasts 2–4 hr; it can often be reduced by increasing the multiplicity of infection (27).

The lag period is followed by a rise period, which consists of an exponential phase during which the concentration of infectivity doubles at a constant rate followed by a declining rate of increase that finally reaches a maximum or plateau. Yields of 25,000 to 100,000 virus particles per cell are typical for picornaviruses growing under favorable conditions. Infectivity yields, in PFU per cell, are one to three orders of magnitude lower, depending on the particle to PFU ratio. Timing of virus release depends on the virus-cell system. For example, the strain of EMC virus used in the author's laboratory is released from HeLa cells very rapidly, whereas poliovirus types 1, 2, and 3 remain cell associated for hours after assembly has been completed.

The plateau is often followed by a "decay" period of declining infectivity, particularly with rhinoviruses and aphthoviruses, in which the RNA genome may fragment spontaneously inside the otherwise intact virion (100,143). In the case of aphthoviruses, this fragmentation is attributed to the presence of one or more 3D proteins packaged with each particle (291), but the reason why an RNA polymerase would cause fragmentation is unclear.

ATTACHMENT, ENTRY, AND UNCOATING

Measuring Attachment Rates

One way of measuring attachment rate is to mix cells and virus at a low multiplicity of infection and to determine, after various exposure times, the number of infected cells. This is done by mixing the exposed cells with other susceptible (indicator) cells, which furnish a "lawn" to detect infections arising from single infected cells (infectious center assay). If the cells are already attached to the bottom of a Petri dish, after the method described in Fig. 10, the rate of attachment to the monolayer can be determined by simply removing excess virus at intervals and scoring the subsequent appearance of plaques.

A second way of studying attachment, which does not require low multiplicity of infection, is to add virions to a suspension of cells, then remove aliquots at intervals, sediment the cell-attached virus by low-speed centrifugation, and measure the residual concentration of unattached infectivity, which remains in the supernatant fluid. The results of such an experiment are illustrated in Fig. 17.

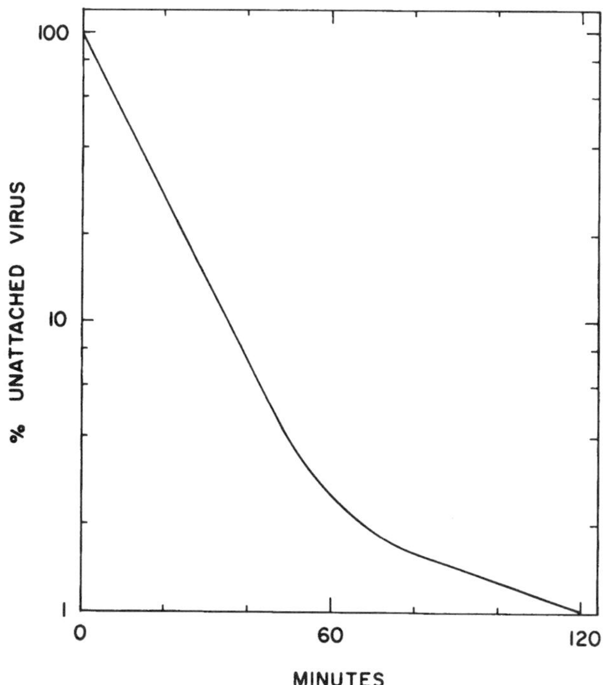

FIG. 17. Typical curve for attachment of virus to a suspension of susceptible HeLa cells. The linear portion of the curve can be used to compute a rate constant, K, for attachment. For example, attachment was 90% complete after 35 minutes; the cell concentration, C, was 4×10^6/ml. The constant, K, is calculated with the aid of Equation 3; thus,

$$K = 2.3 \log(V_o/V)/Ct$$
$$= 2.3 \log(100/10)/[(4 \times 10^6)(35)]$$
$$= 1.6 \times 10^{-8}\ cm^3min^{-1}$$

The fraction of unattached virus (dV/V) after a given increment of time, dt, is given by Equation 1:

$$dV/dt = -KCV \qquad (1)$$

where V is the concentration of unattached virus, C is the cell concentration, and K is the velocity constant for formation of virus-cell (VC) complexes (Equation 2):

$$V + C \xrightarrow{K} VC \qquad (2)$$

Integrating Equation 1 and taking the logarithm of both sides yields

$$\log(V/V_o) = (-KC/2.3)t \qquad (3)$$

where V_O is the initial concentration of unattached virus. Equation 3 is a straight line of slope (–KC/2.3) provided the ratio of virus particles to cells is not so high as to saturate the number of virus-binding sites, typically 10^3 to 10^5 per cell. In practice, virus preparations usually exhibit some nonhomogeneity with respect to attachment, typically following first-order kinetics until 90% to 99% of the virus is bound.

In the experiment illustrated in Fig. 17, virus attachment is 90% complete in about 35 min when the cell concentration is 4×10^6/ml. It should be apparent from inspection of Equation 1 that increasing the cell concentration tenfold speeds the rate of virus attachment by a similar factor. Thus, concentrating the cells to 4×10^7/ml reduces the time required to attach 90% of the virus from 35 to only 3.5 min. This direct proportionality between cell concentration and attachment rate justifies the common practice of using high cell concentrations to speed attachment. Similarly diluting such suspensions provides a rapid and convenient method of (virtually) stopping attachment.

Attachment rates, determined as described in Fig. 17, have been measured on a variety of viruses and host cells. They vary over a wide range. Coxsackievirus B3, for example, with a rate constant of 10^{-8}, attaches to HeLa cells 10,000-fold faster than does Coxsackievirus B2 ($K = 10^{-12}$) (91). All three serotypes of poliovirus, on the other hand, attach to HeLa cells at rather similar rates ($K = 2$ to 4×10^{-9}). Some viruses such as EMC attach to murine and HeLa cells so rapidly ($K = 10^{-7}$) that the rate is diffusion limited (260).

Electrostatic Nature of the Virus Attachment Step

With the aid of methods such as those described above, it is also possible to determine the effects of temperature, pH, and composition of the solvent on virus attachment. The results suggest that ionizable groups on the surface of the virus or cell play an important role in the attachment process. The attachment of group B Coxsackieviruses and echovirus 6, for example, change markedly with pH, each with a distinct pH profile (91). Attachment of the polioviruses, on the other hand, is almost independent of pH over the range 4.5 to 8.5.

Additional evidence that the initial attachment of virion to cell is governed by electrostatic events comes from a study of ionic composition on virus attachment. For example binding of poliovirus is greatly accelerated by monovalent ions (180). A number of picornaviruses require the divalent ions Ca^{2+} or Mg^{2+}. These include rhinoviruses (131,241), Coxsackieviruses A9 (265) and A13 (84), and FMDV (47). The chelating agent ethylene diamine tetraacetic acid inhibits attachment of viruses requiring divalent ions (239). Thus, electrostatic attraction appears to be important in fostering optimum attachment of virus to cell. Structural studies show the presence of a calciumlike ion at each fivefold rotation axis (pentamer center) in human rhinovirus 14 (145).

Determinants of Cell Tropism

Virus diseases are characterized by host specificity and a specific set of symptoms (syndrome). Poliovirus and human rhinovirus, for example, each cause specific diseases in humans, whereas aphthoviruses afflict cloven-footed animals. Symptoms can often be correlated with infection of specific tissues (cell tropism). In humans, for example, poliovirus infects cells of the nasopharynx and gut; it is these cells that the virus normally uses to propagate itself in nature. The paralytic form of poliovirus also has a high specificity for anterior horn cells of the spinal cord, which accounts for the special paralytic features of the disease. A similar kind of specificity is seen in the Coxsackieviruses, which, like poliovirus, infect nasopharyngeal cells but, unlike poliovirus, have a propensity for skeletal and heart muscles. Moreover, Coxsackieviruses are able to infect baby, but not adult, mice.

The mechanisms responsible for cell tropisms are only partially understood. There are numerous reports of "abortive" or "restricted" infection in which the viral RNA genome is successfully uncoated and initiates an infection only to be aborted at some critical step in the replication cycle (109,185,334,385,443). This implies requirement for specific cellular molecules in multiplication of the virus. Possible candidates are host factors involved in function of viral proteinase (40,159) or viral proteins involved in synthesis of its RNA genome (14,170,285).

However, one important determinant of cell tropism is the receptor surface–specific molecules necessary for viral invasion. The notion that viruses require receptors originally emerged from the observation that human kidney and amnion, which are not normally infected in polio victims, became highly susceptible to poliovirus when the cells were grown in culture (202). Subsequent work showed that this susceptibility is correlated with appearance on the cell surface of sites, called receptors, with the ability to bind and eclipse the virus (175,176,179,339,391). The ability of Coxsackievirus A to produce myositis only in baby mice is similarly correlated with specific receptors present in differentiating myoblasts but not in most other mouse embryo tissues (265). Ability of encephalomyocarditis virus to produce diabetes in mice has been traced to a variant virus (EMC-D) that is specific for the insulin-producing islet cells in the pancreas (445). Such work stimulated a great deal of interest in the concept that virus receptors are important determinants of disease, but early attempts to correlate receptor with the pattern of pathogenesis in paralytic disease (164,217) were frustrated by lack of proper tools.

Receptor Families

Discovery that isolated picornaviral RNA was infective (5,79) facilitated the demonstration that susceptibility of cells is correlated with the presence of specific receptors (99,180). The earliest clues that picornaviruses do not all use the same receptor were based on differences in their susceptibility to digestive enzymes (77). For example, poliovirus and rhinovirus receptors are destroyed by trypsin; however, receptors for echovirus, group B Coxsackievirus, and cardiovirus are not destroyed by trypsin. Receptors for the latter two viruses are inactivated by chymotrypsin. Neuraminidase inactivates receptors for cardiovirus, bovine enterovirus, equine rhinovirus, and human rhinovirus 87, indicating that they contain sialic acid.

However, the most useful insights into receptor classification emerged from development of two methods: (a) competition analysis (29,89,238) and (b) isolation of monoclonal antibodies that protect cells from infection by viruses.

The use of competition analysis is illustrated in Table 9. The three poliovirus serotypes compete for a common receptor but not for the receptor used by the group B Coxsackieviruses (92,339,340). It also provides evidence of separate receptors for Coxsackieviruses A21 and human rhinovirus 2 and 14. The competition method was later used in preliminary classification of rhinoviruses into two rhinovirus receptor groups: a minor group and a major group (2). Unrelated viruses also have been found to share the same receptor. For example, adenovirus 2 blocks attachment of Coxsackie B viruses and partially blocks human rhinovirus 2 (238).

Cytofluorimetric analysis, using biotinylated echovirus that can then be tagged with fluorescent avidin, has now been developed to displace the use of radioactive virus for studying competition binding (257).

TABLE 9. *Representative picornavirus receptor families*

	Interferes with attachment[a] of				
Virus used to saturate receptors	Polio 1–3	Cox A21	Cox B1–6	HRV2	HRV14
Polio 1–3	+	–	–	–	–
Cox A21	–	+	–	–	+
Cox B1-6	–	–	+–	–	–
HRV2[b]	–	–	–	+	–
HRV14[c]	–	+	–	–	+

[a]In a typical experiment one measures the attachment of a nonsaturating amount of radiolabeled virus in the presence of excess unlabeled virions. Excess type 1 poliovirus, for example, blocks attachment of type 1, 2, or 3 poliovirus but does not block coxsackieviruses A21 or B1-6 or human rhinoviruses 2 or 14.

[b]RV2 also inhibits attachment of human rhinoviruses 1A and 1B.

[c]RV14 also inhibits attachment of human rhinoviruses 3, 5, 15, 39, 41, and 51.

TABLE 10. *Picornavirus receptor groups*

Virus	No. of serotypes	Receptor	Reference
Rhinoviridae			
Human rhinovirus (major)[a]	91	ICAM-1	150,403,422
Human rhinovirus (minor)[b]	10	LDLR	172
Human rhinovirus 87	1	Sialoprotein[c]	427
Enteroviridae			
Poliovirus	3	PVR	276
Coxsackievirus A	(A13,A18,A21)	ICAM-1	349,422
Coxsackievirus A	(A9)	Integrin	256
Coxsackievirus B	6	?	90
Echovirus	1.8[d]	VLA-2	31
Echovirus	6,7,13,21,29,33	DAF	31
Aphthoviridae	7	Integrin?	3,253,321
Cardioviridae			
Encephalomyocarditis virus	1	VCAM-1	186
Theiler's virus	1	?	
Hepatoviridae	1	?	

VLA-2 is a member of the integrin family.

ICAM-1, intercellular adhesion molecule 1; LDLR, low-density lipoprotein receptor; PVR, poliovirus receptor; VCAM-1, vascular cell adhesion molecule; DAF, decay-accelerating factor.

[a]Includes all rhinovirus serotypes except HRV87 and the minor group HRVs. Monoclonal antibody directed against ICAM-1 protects cells from infection by viruses of major group HRVs but not against viruses using other receptors. By this criterion, some group A Coxsackieviruses (A13, A18, and A21) also use ICAM-1 as receptor.

[b]Includes rhinovirus serotypes 1A, 1B, 2, 29, 30, 31, 44, 47, 49, and 62.

[c]Attachment of serotype 87 is not blocked by other rhinovirus serotypes; treating HeLa cells with neuraminidase at pH 6 blocks plaque titer 500-fold.

[d]All echovirus serotypes except 22 and 23.

Monoclonal antibodies, and even specially prepared antisera (20,290), directed against cellular surface proteins have been exceptionally useful in identifying receptor groups. For example, monoclonal antibody directed against the poliovirus receptor protects cells from infection against all three serotypes of poliovirus, but not from infection against other picornaviruses (see footnotes, Table 10). Cells coated with antibody directed against ICAM-1 are protected from infection by some 90 serotypes of human rhinoviruses and some Coxsackie A viruses known to infect the respiratory tract. By and large, pathogenesis correlates with receptor group.

Nature of Receptors

The receptors for poliovirus (86,269,277,395), Coxsackie B virus (87), echovirus (87), and the major serogroup of rhinoviruses (150) have all been mapped to human chromosome 19. The receptors for poliovirus and human rhinovirus have been identified as members of the Ig superfamily, whereas the receptor for echoviruses has been identified as a member of the integrin family (32). Such molecules, commonly found in cellular plasma membranes, are typically involved in binding specific growth factors or cells containing special adhesive receptors called integrins.

A number of picornaviruses agglutinate erythrocytes, which are an abundant source of plasma membrane. Erythrocyte receptors have been partially characterized for Coxsackievirus B3 and echovirus 7 (323), as well as for EMC virus; the latter resembles glycophorin A (49), a major membrane-spanning sialoglycoprotein that is primarily responsible for the MN blood group antigen. However, because erythrocytes play no known role in viral pathogenesis, the significance of such hemagglutinin receptors in relation to virus-eclipsing receptors of susceptible host cells is not yet clear.

Use of protective monoclonal antibodies has been useful, not only for receptor classification but also as reagents for cloning and characterizing the responsible receptors. Several receptor genes have now been isolated by transfecting cells that lack receptor with human DNA and isolating from them cells that bound monoclonal antibody against the virus receptor (i.e., antibody that specifically protected susceptible cells from infection by poliovirus or rhinovirus). Some of these transfected cells were then susceptible to infection. Sequencing of the relevant genes showed the Ig nature of the molecules, predicted sizes of about 45 kD for the poliovirus receptor (276) and 95 kD for the major rhinovirus receptor (150,405). This led to the recognition that the major rhinovirus receptor (422) was identical to ICAM-1, previously characterized by its ability to bind to an integrin molecule (LFA-1) on lymphocytes (405). ICAM-1 was originally identified using a monoclonal antibody that inhibits adhesion of lymphocytes and induces expression of cytokines associated with the in-

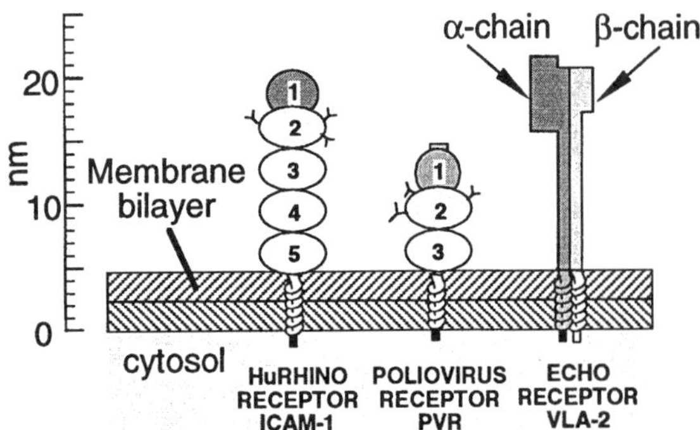

FIG. 18. Three classes of picornavirus receptors. The receptors for poliovirus and rhinovirus consist of single chains and are members of the immunoglobulin supergene family. ICAM-1 has five domains; PVR has only three. The mRNA for poliovirus receptor is spliced in such a way as to produce four different proteins; two of these isoforms are secreted because they lack the hydrophobic transmembrane domain (214). Virus attaches to domain 1 in both receptors, but domain 2 (N-glycosylated) must also be attached for high binding affinity. Both ICAM-1 and PVR are *N*-linked glycoproteins (76,214,276,405); cell-dependent variations in glycosylation thus provide a potential mechanism by which individual cell types might modulate affinity differences. Domain 1 of poliovirus receptor mediates infection even if fused to ICAM-1 (388). VLA-2, the echovirus receptor, is a member of the integrin family and consists of two polypeptide chains, called alpha and beta. Neither ICAM-1 (404) nor poliovirus (214,276) receptors require a cytoplasmic tail to mediate infection.

flammatory response. Expression of ICAM-1 is stimulated by factors such as gamma-interferon, interleukin-1, and tumor necrosis factor, raising the possibility that rhinoviruses exploit this response to facilitate the spread of infection. ICAM-1 also facilitates interactions between T cells and antigen-presenting cells (405).

Availability of receptor genes in cloned form provides new insights in understanding the mechanisms by which picornaviruses penetrate susceptible host cells and opens the way for the crystallographic analysis needed to visualize atomic interactions involved in virus-receptor complexes (Fig. 18).

Subparticles Produced by Soluble Receptors

Virions of poliovirus and human rhinovirus when mixed with purified "soluble" receptors produce reversible complexes that are stable at low temperatures (Fig. 19, top). Receptors are made soluble by deleting the hydrophobic transmembrane domain, thereby eliminating aggregation of the receptors in aqueous buffers. Virus-receptor complexes formed at low temperatures are reversibly dissociable. Saturated complexes are unable to attach to cellular receptor and therefore lack infectivity; upon dilution, however, virion infectivity gradually reappears as the virus-receptor complexes dissociate.

Upon elevating the temperature, virus-receptor complexes may undergo an eclipse transition to generate two kinds of noninfectious subparticles: (a) A particles, which are noninfectious, sediment more slowly than do virions and lack VP4, but still contain an intact RNA genome; and (b) empty capsids or "shells," which lack both VP4 and RNA, sediment roughly half as fast as virions. A particles and empty shells generally suffer a marked decline in receptor binding affinity. A particles are "sticky," i.e., have a greater tendency than virions to adsorb to surfaces.

The precise conditions required to trigger eclipse and the exact spectrum of subparticles produced vary with virus, temperature, and pH, and likely also with the number of receptor molecules per virion. With poliovirus, for example, stable complexes form at 4°C and eclipse occurs when the temperature is increased to 37°C; uncoating stops with the A particle (203). With human rhinovirus 14, the virion–ICAM-1 complex uncoats readily at 34°C to yield both A particles or empty shells depending on the pH (190). HRV16, on other hand, even when saturated with ICAM-1, forms virus-receptor complexes that are stable at 34°C, and HRV3 exhibits intermediate behavior (151,182).

Entry and Uncoating

Receptors must ultimately mediate passage of the viral genome through a lipid bilayer. Until recently, clues to events between attachment of virion to the cell and genome penetration depended on methods such as electron microscopy (94) and interpretation of changes in affinity of virus for the cell (91,166,236). Poliovirus, for example, attaches reversibly to cells at 4°C but upon increasing the temperature to 37°C becomes progressively more difficult to recover with reagents such as low pH sodium deoxycholate, sodium dodecyl sulfate (SDS), 8 mol/L urea or 6 mol/L lithium chloride when virus-cell complexes are incubated at 37°C (178,179). An early model (44) attributed this progressively temperature-dependent binding to receptor "recruiting," i.e., incremental envelopment of virus with more and more receptor units whose lateral mobility in the plasma membrane depends on the fluidity of the lipid bilayer (Fig. 19B, step 2). Attachment to a single receptor unit (step 1) accounts for loose reversible association of virus attached to cells at low temperatures (0°–10°C), and recruiting of receptors, especially at temperatures high enough to afford receptor mobility, accounts for progressively tighter association of virus with the cell (step 2); this process might simultaneously draw the membrane around the virion to begin invagination and endocytosis.

Eclipse (step 3) corresponds to an irreversible conformational transition characterized by formation of a membrane-bound intermediate that can be released from the cell with special detergents to yield noninfectious RNA-containing A particles. The decreased sedimentation coef-

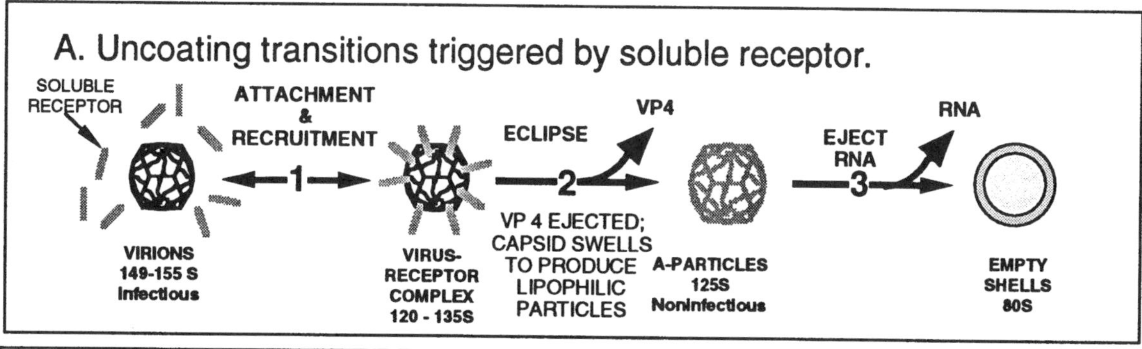

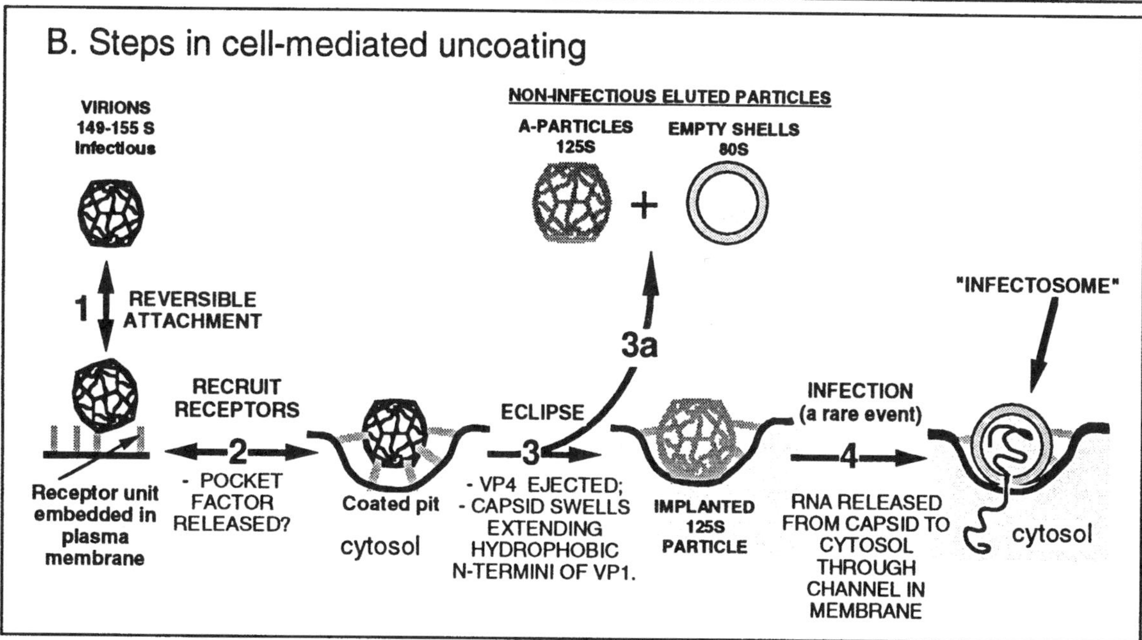

FIG. 19. Receptor-mediated transition and model for cell mediated uncoating. **A:** Soluble receptor lacks the hydrophobic transmembrane domain and cytoplasmic tail shown in Fig. 18. Affinity for soluble receptor varies with virus serotype; so do the temperature and pH required to trigger eclipse of virus-receptor complexes and ejection of RNA (151,203). Virus-receptor complexes are reversibly dissociable, i.e., regenerate infective virions when greatly diluted. **B:** Model for cell-mediated uncoating. Step 1: Loose reversible attachment of receptor-binding site on the virion's surface to one of the membrane-associated receptor units. Step 2: Attachment becomes tighter as mobile receptor units are recruited from the membrane surface; this process might simultaneously draw the membrane around the virion to begin invagination and endocytosis. (For explanation of pocket factor, see Fig. 13.) Step 3: Conformational alteration of the protein shell, as indicated by susceptibility of the particle to digestion with proteinases, and loss of VP4 to form sticky A particles. This pathway is subjected to many abortive events, with 50% to 90% of attached particles eluting or sloughing as noninfective subparticles (125 and 80S particles in the cases of polio-, coxsackie-, echo-, and human rhinoviruses or free RNA and 12-14S protein subunits in the case of cardio- and aphthoviruses, whose empty shells are less stable). Much of the uncoated RNA is destroyed by nucleases. A small fraction of the infective virions may fail to be uncoated and may be recovered in the progeny harvest. Step 4: Infection or entry, i.e., delivery of the RNA genome through the membrane and into the cytosol. The term infectosome applies to the transient intermediate that successfully transfers its genome into the cytosol and thereby initiates an infection. Provirions (see Fig. 15) lack infectivity because they cannot complete step 4 (228). Infectosomes have yet to be experimentally demonstrated, likely because 1% or fewer particles typically initiate a successful infection and because the infectosome is probably short lived.

ficient, 125S compared with 149S for rhinovirions, implies a nearly 20% expansion in the hydrodynamic radius of the A particle. These swollen A particles are sticky, i.e., they adhere to glass surfaces and liposomes apparently because of extrusion of hydrophobic *N*-termini of VP1 (141). A particles are also sensitive to digestion by proteinases (102,141,155,179). The final step of uncoating, infection (step 4), requires that the RNA genome be successfully transferred across the cell membrane and into the cytosol when synthesis of viral polymerase protein must occur before replication can begin. However, the high particle:PFU ratio, typically greater than 50 for most picornaviruses (Fig. 10), implies that only a tiny fraction of the attached particles ever succeed in reaching step 4.

Elution of Surface-Attached Virus

One major reason for the low infective efficiency of picornaviruses is elution or sloughing of noninfectious A particles or empty shells (step 3a). Some 50% to 90% of all surface-attached virions are typically lost to this abortive pathway (128,157,198). Abortive elution of cell-attached virions also occurs with aphthoviruses (29,47,64) and with the EMC cardioviruses (157,260); with these viruses, however, the eluted products are free RNA and pentameric subunits, perhaps because A particles of cardioviruses and aphthoviruses are too labile to survive. Some viruses, such as polio and coxsackie B3, undergo further modification, which may correspond to loss of VP2 (B particles) (102,155,240,262).

Receptor-Mediated Endocytosis

The role of receptor-mediated endocytosis (Fig. 20) in uncoating of picornaviruses is still controversial, in part because support comes mainly from studies with inhibitors. For example, uncoating of poliovirus (measured by ability of virus to inhibit incorporation of radioactive leucine into cultured cells) is inhibited by monensin, chloroquine and by N,N′-dicyclohexyl carbodiimide (DCCD), all reagents that arrest receptor-mediated endocytosis from coated pits (245,246). Monensin acts by exchanging protons in the vesicle for monovalent cations on the other side of the membrane; chloroquine and other weak bases buffer against acidification; and DCCD interferes with ATPases involved in pumping protons into

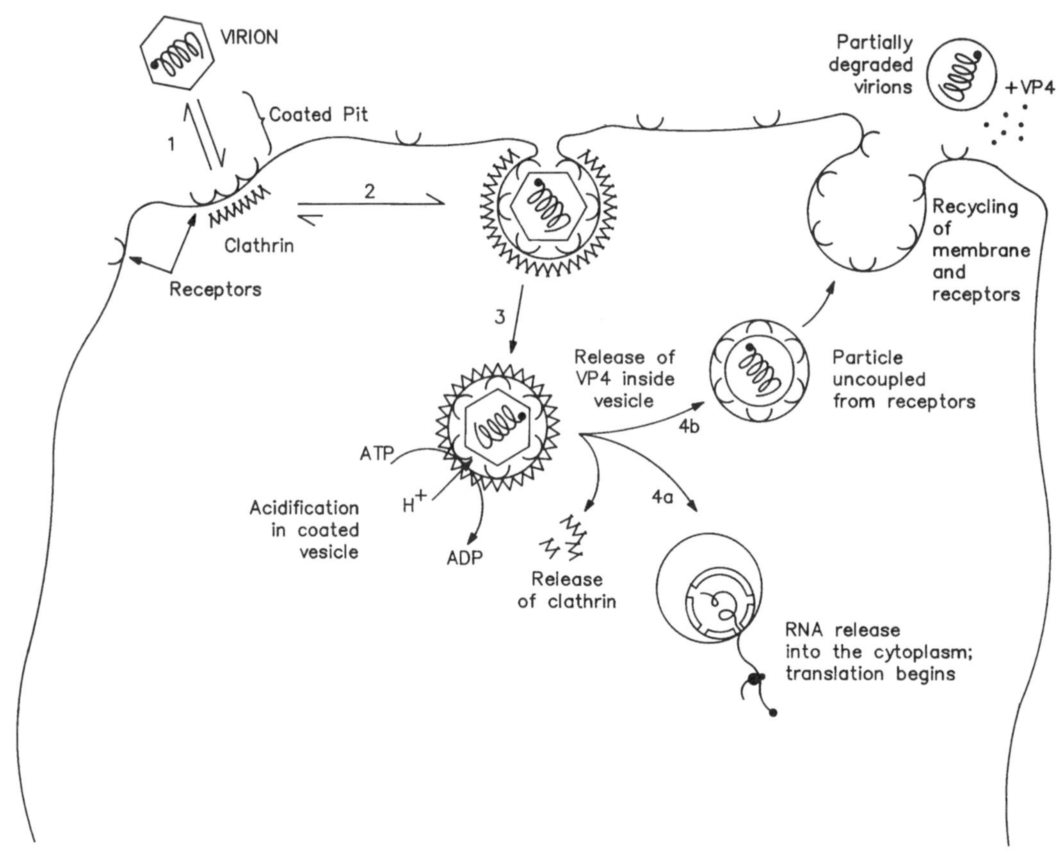

FIG. 20. Model for receptor-mediated endocytosis (245,246). The virus exploits cell surface receptors, glycoproteins that normally interact with specific nonviral molecules important to the economy of the cell surface. Electron microscopy (94,450) indicates that freshly attached virus is initially distributed more or less evenly on the cell surface. Clustering of receptors at clathrin-coated pits, representing a few percent of the cell surface, is followed by invagination and internalization (endocytosis) to form clathrin-coated vesicles. Acidification inside the coated vesicles, brought about by an energy-requiring ATPase-coupled proton pump, triggers release of VP4 and unfolding of hydrophobic patches previously buried inside the viral capsid. Fusion of the lipid bilayer with hydrophobic patches in the acid-unfolded capsid protein presumably triggers transfer of RNA from virion to cytosol where ribosomes can begin translating the genome. Fusion of uncoated vesicles with other kinds of intracellular lysosomelike vesicles may also be involved in the uncoating process. The role of endocytosis in picornavirus infection is still controversial.

endocytic vesicles. The inhibitory effect of these drugs was fully overcome by acidifying the nutrient medium to pH 5.5 (245,246,450). Results of studies with the detergent Triton X114 (245) suggest that poliovirus transfers reversibly from the aqueous phase to a hydrophobic phase when the pH is reduced below 5, suggesting that at this low pH normally hidden hydrophobic domains of the virus become exposed so they can interact with lipid in the membrane. The affinity of eluted A particles for liposomes is correlated with extrusion of amino terminal arms of VP1 (141).

However, the role of receptor-mediated endocytosis is clouded by secondary effects of inhibitors (216) and by findings that soluble receptors are able to mediate uncoating transitions at neutral pH for some picornaviruses but not for others (182,203). It may turn out that the site of uncoating—plasma membrane or endocytic vesicle—is determined by uncoating sensitivity of the virion; for example, if virus-receptor complexes can uncoat at neutral pH they may uncoat at the plasma membrane, but if some additional stimulus such as low pH is required, uncoating may be delayed until they are transported, by normal trafficking of receptors, into the endosome.

Why Is Maturation Cleavage Required for Infectivity?

Provirions are complete virus particles in which the maturation cleavage (VP0 → VP2 + VP4) has not yet taken place (Fig. 15). They have long been regarded as noninfective (154). Studies on maturation-defective rhinovirus 14 support this position (228) and provided an opportunity to determine the step at which infectivity was blocked. It was found that provirions of HRV14 attach to HeLa cells normally and undergo all of the measurable transitions (steps 1, 2, 3, and 3a in Fig. 19, panel B) at the same rate as mature infectious virions, yet they are not infective (228). This result suggests that maturation cleavage is required for the function of a still uncharacterized membrane-associated particle, the "infectosome," which transfers genomic viral RNA through the membrane. The role of maturation cleavage in conferring infectivity is suggested by the model described in Fig. 21B.

Is RNA Uncoated Through a Channel in the Pentamer Vertex?

Crystallographic evidence (68,132,299,358) shows that VP4 with its hydrophobic myristate residues is clustered

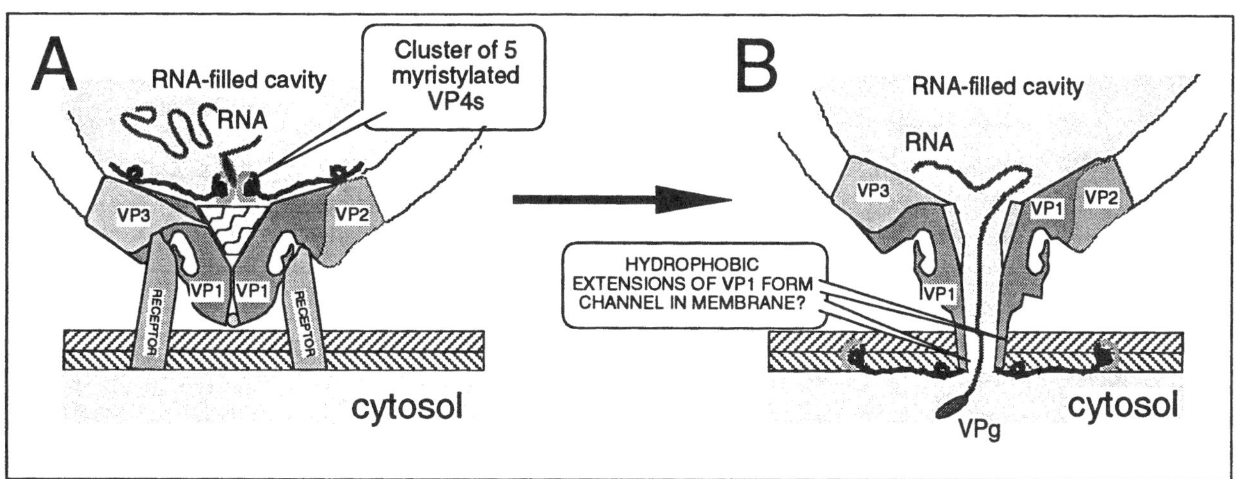

FIG. 21. Receptors mediate infection by positioning virions, then triggering them, to implant in the membrane bilayer. **A:** Multiple receptor units, recruited from the membrane, position the particle close enough to implant into the cell membrane. A case can be made that virions "breathe," i.e., periodically expose normally buried elements such as VP4 and VP1 (see section on Neutralization). This raises the possibility that eclipse is triggered by progressive stimulation of breathing as more and more receptors are recruited from the membrane. Breathing also may be stimulated by other factors such as increasing temperature and possibly by loss of pocket molecules (see Fig. 19). The effect of particle distance from the membrane on implantation success is not yet clear (e.g., note that PVR is shorter than ICAM-1 or VLA-2). The explanation for elution is also not clear but might be explained by implantation failure during the eclipse transition (see step 3, Fig. 19), followed by release from the cell surface because of decreased affinity of receptors for eclipsed particles. **B:** Hypothetical structure of a picornaviral infectosome showing how extrusion of hydrophobic capsid elements might form a channel or pore (infection tube), enabling the RNA genome to pass through the membrane and into the cytosol (141). This is homologous to the formation of fusion pores by the hemagglutinin of influenza virus (436). Lack of infectivity in the provirion would be explained if VP4 must clear the channel before RNA can exit. The role of the maturation cleavage may thus be to free VP4 from VP0 and thereby prevent channel clogging during the VP4 exit step (228,287).

flowerlike at the base of each pentameric subunit [see also Fig. 5A in Hogle et al. (173)]. Evidence that poliovirions "breathe" (369), that is, periodically expose normally buried segments of VP1 and VP4 on the virion surface, has already been discussed in the section on "Neutralization." Extrusion of these elements from the pentamer center to implant hydrophobic fusion tubes into the adjacent lipid bilayer then provides a plausible basis for establishing a fusion pore in the membrane. The absence of VP4 in A particles suggests that extrusion is a concerted process (i.e., all 60 VP4s are extruded simultaneously) and that the expansion of A particles actually corresponds to extrusion of 12 infection tubes; should RNA be extruded from only one of the 12 tubes, the orientation of the emerging RNA genome (i.e., pointing toward the membrane or away from it) might also be a factor in initiating a successful infection. High-resolution electron microscopy and x-ray analysis of crystallized A particles will likely shed light on some of these issues, but new approaches are also critically needed.

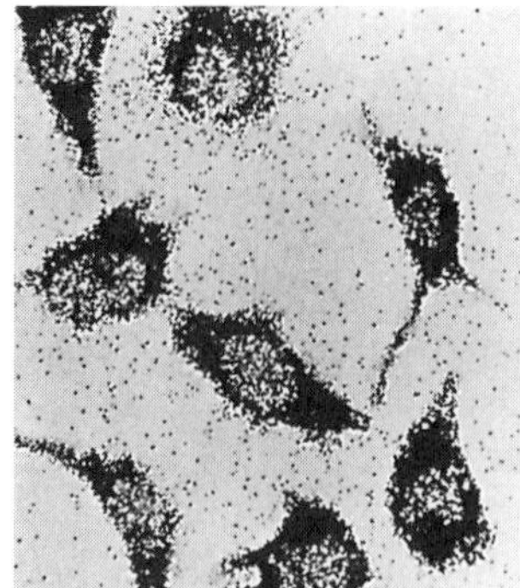
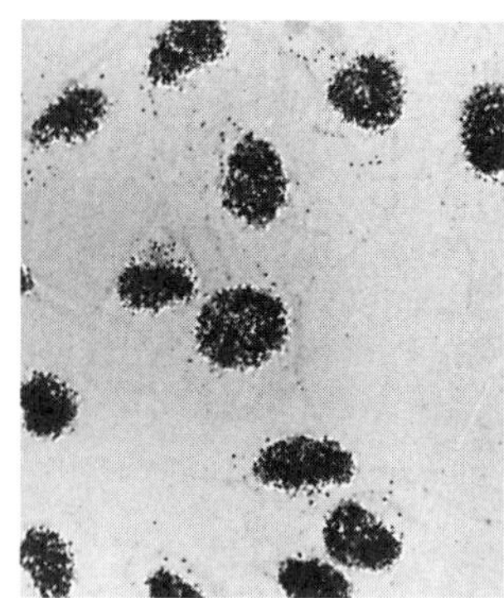

FIG. 22. Autoradiograph of mouse cells 5 hr after infection with ME virus, a cardiovirus. **Left:** Tritiated uridine was added to the culture 5 minutes before fixing the cells with acid alcohol; the incorporation period was chosen to be short relative to the 30 minutes or so required for transport of labeled RNA from nucleus to cytoplasm in normal cells. The distribution of black grains indicates that virus-infected cells synthesize all of their RNA in the cytoplasm. Note also that infection shuts off synthesis of nuclear RNA synthesis normally seen in uninfected cells (**right**). (Photo courtesy H. Hausen).

EFFECTS ON THE HOST CELL

Most of our knowledge of the biochemical effects of picornaviruses on cellular functions comes from studies on viruses that grow readily in cultured cells and produce dramatic, easily studied changes in macromolecular metabolism of the infected cells (140). Much is therefore known of the effects of cytocidal viruses, such as poliovirus, but little is known of viruses such as hepatitis A, which characteristically establish only persistent infections.

Contribution of the Host Cell

The susceptible cell contributes energy and precursors for synthesis of viral components and all of the machinery (ribosomes, transfer RNA, enzymes, etc.) required for synthesis of viral protein. It also provides receptors necessary for infection, membranes needed for virion assembly, and at least one factor required for synthesis of viral RNA. The viruses encode most of the proteins required for synthesis of the viral genome and evidently require no nuclear functions because they are able to multiply in the cytoplasm of enucleated cells (88). Enucleated cells are also susceptible to infection by single-stranded poliovirus RNA but not to the double-stranded RI (103); the latter is normally infective in nucleated cells (38,284,332).

Additional evidence of their independence from cellular nucleic acid synthesis emerges from studies on metabolic inhibitors. Inhibitors of DNA synthesis, such as 5-bromouracil and 5-fluorodeoxyuridine, do not generally inhibit synthesis of poliovirus. Neither does the drug actinomycin D, which inhibits synthesis of cellular RNA by intercalating itself between the bases of double-stranded DNA. This is not to say that such drugs never inhibit picornaviral multiplication. With certain exceptional strains of poliovirus, actinomycin inhibits the yield to less than 1% of normal (380). Sensitivity to the drug is greatest during the first half hour of infection (148) and is relieved by exposing cells to insulin 24 hr before infection (80).

Infection by picornaviruses causes dramatic changes in macromolecular metabolism (140). The rate of RNA synthesis declines shortly after infection. This is accompanied by synthesis of a vigorously expanding pool of viral RNA in the cytoplasm. The ability of picornaviruses to inhibit cellular RNA synthesis in the nucleus and establish viral RNA synthesis in the cytoplasm is illustrated in Fig. 22.

Picornaviruses inhibit cellular protein synthesis soon after infection. In the case of poliovirus-infected HeLa cells, the most intensively studied system, shut-off occurs quickly, within the half hour or so required for attachment, penetration, and uncoating of the RNA genome (Fig. 23). The declining phase of protein synthesis, which is studied by incorporation of radioactive amino acids into acid-insoluble material, reaches a minimum at about 2 hr. The following wave of incorporation, which peaks at about 3 hr, reflects vigorous synthesis of viral proteins. Staining infected cells with fluorescent antibody against coat protein shows virus protein in the cytoplasm, first in the perinuclear area, then throughout the cytoplasm and in blebs near the plasma membrane (165). Finally there is a decline of viral protein synthesis associated with gross leakage of intracellular components and cell death.

Cytopathic Effects

Most picornaviruses induce characteristic morphological changes in the infected cells. One of the earliest ef-

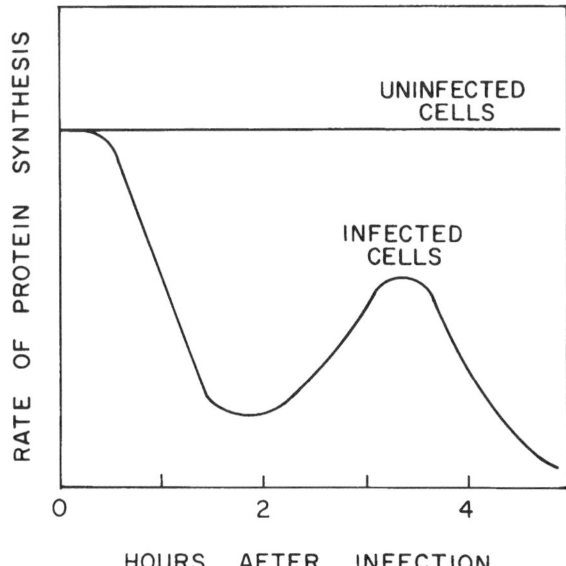

FIG. 23. Synthesis of protein in poliovirus-infected HeLa cells infected at high multiplicity. The rate of decline and the magnitude and timing of the new wave of protein synthesis vary significantly with virus, multiplicity of infection, and identity of the host cell. Use of the drug actinomycin D, which blocks synthesis of cellular RNA but not of picornaviral RNA (342), hastens shut-off of host cell protein synthesis. Adapted from Ehrenfeld (119), with permission.

fects, beginning within an hour of infection, is "margination of the chromatin" where the nuclear material loses its normally homogeneous microscopic texture and accumulates along the inside of the nuclear envelope (140). Alterations in histone composition have been reported in cells infected with cardioviruses (181,425) and with an aphthovirus (152).

About 2.5 to 3 hr after infection, membranous vesicles appear in the cytoplasm, beginning first in the vicinity of the nuclear envelope and spreading outward until the entire cytoplasm is involved (95). Spreading of vesicles is associated with changes in permeability of the plasma membrane, loss of ability to concentrate vital stains (83), stimulated incorporation of choline into cell membranes (6), and later, leakage of intracellular components, followed by shriveling of the entire cell. Crystals of virus are often found in the cytoplasm during late stages of infection.

It has been suggested that this so-called cytopathic effect (CPE), so characteristic of picornavirus-infected cells, is mediated in part by a redistribution of lysosomal enzymes, possibly phospholipase (343,440). Double-stranded RNA was once thought to be involved in CPE, but this hypothesis has fallen into disfavor (75).

Cellular RNA, protein, and DNA synthesis begin to decline within the first few hours of infection. However, inhibition of host macromolecular synthesis is not sufficient to account for the marked cytopathic effects of picornaviral infection, and indirect evidence indicates that full expression of CPE requires synthesis of viral protein (21). Attempts have been made to correlate CPE with distribution of viral proteins in host organelles, but the results have not been definitive (36).

Inhibition of Cellular RNA Synthesis

Synthesis of ribosomal and messenger RNA begins to decline soon after infection with many picornaviruses, including poliovirus, mengovirus, EMC virus, human rhinoviruses, and FMDV (242). The kinetics of decline appears to vary with the host cell. In murine L cells, for example, mengovirus inhibits synthesis of ribosomal and messenger RNA to less than 10% of normal within an hour after infection (140), whereas in HeLa cells infected with the same virus the decline is much slower (261).

The inhibition of nuclear RNA synthesis is paralleled by a decline in the ability of isolated nuclei or crude DNA-protein complexes from infected cells to make RNA. Measurements on alpha-amanitin–sensitive transcription indicate that polymerase II is most strongly affected; polymerase I and III are also inhibited but to a smaller extent. When freed from their nucleoprotein complexes, however, all three polymerases are active. Recent studies on poliovirus suggest that the viral proteinase 3C plays a role in modifying transcription factor complexes of all three polymerases; one mechanism is to cleave the TATA-binding protein subunit of transcription factor IID (71,72,365). Proteinase 3C has also been implicated in inhibition of cellular transcription by an aphthovirus; this is correlated with cleavage of histone H3 in the case of aphthovirus but not of poliovirus (127,418).

How Do Picornaviruses Inhibit Synthesis of Host Proteins Without at the Same Time Blocking Synthesis of Viral Proteins?

The mechanism by which cellular protein synthesis is shut off by picornaviruses has variously been attributed to accumulation of double-stranded RNA, toxic effects of viral coat protein, increase in cytoplasmic sodium and decrease in cytoplasmic potassium concentration associated with changes in the plasma membrane, inactivation of factors required for initiation of protein synthesis, and ability of viral RNA to outcompete host messenger RNA (mRNA) for critical components of the protein synthetic machinery (61,119).

The poliovirus-induced decline in synthesis of cellular proteins (Fig. 23) is accompanied by dissociation of ribosomes from host mRNA (316). This is not caused by degradation of host mRNA molecules because, when extracted from infected cells, they remain fully able to direct protein synthesis in a cell-free translation system (206).

One important key to the puzzle was the discovery (129,169,296) that poliovirus mRNA lacked the m⁷G cap group found at the 5′ terminus of most cellular mRNAs. A second key discovery (125,126) was that host protein shut-off in poliovirus-infected HeLa cells is due to inactivation of the cap-binding complex (CBC). Binding of CBC at the m⁷G cap is a prerequisite for translation of most cellular mRNAs by ribosomes (270). The CBC, which can be purified by affinity chromatography using its ability to bind to mRNA cap analogs, contains several proteins. The largest, p220, is cleaved into smaller, antigenically related peptides early after infection (126). The third key discovery (314) was that poliovirus RNA does not use the cap-dependent ribosomal scanning process employed in the translation of most eukaryotic mRNA; rather, 40S ribosomal units are able to attach directly to the highly folded region called the ribosomal landing pad or internal ribosomal entry site (IRES) (Fig. 24).

Several cellular proteins (called p52, p57, p70, and p100 on the basis of their apparent masses in kilodaltons on SDS-polyacrylamide gels) have been shown by cross-linking to bind to various portions of the 5′-nontranslated regions of poliovirus and EMC virus (439), but binding can be correlated with translation for only two of them (p52 and p57), and the roles of all four proteins in initiating translation are still ill-defined. One of the proteins, p52, has been identified as La (271), a protein that binds the 3′ terminus of nascent RNA polymerase III transcripts in uninfected cells. The other three (p57, p70, and p100) appear to form a complex. p57 has been identified as PTB (polypyrimidine tract binding protein), a nuclear protein involved in splice site selection or RNA metabolism (168). It is speculated that the different translational efficiencies of picornaviral RNA in different cells is based on cellular differences in the spectrum of such proteins, but this remains to be shown.

That protein 2A was implicated in host protein shut-off was suggested by the finding that a poliovirus mutant, containing a single amino acid insertion in protein 2A, failed to cleave p220 during infection and altered the host protein shut-off pattern (34). Attempts to demonstrate direct cleavage of p220 by purified 2A of poliovirus have been negative, but the result is now controversial because the corresponding purified 2A proteins of rhinovirus 2 and a Coxsackievirus A have been found active against p220 or peptides derived therefrom (400). Protein 2A also cleaves p220 in cells infected by other enteroviruses and rhinoviruses, including human rhinovirus 14 (124), Coxsackievirus B3, and enterovirus 70 (234). Aphthoviruses and cardioviruses (EMC and Theiler viruses) display significant differences in sequence and function of their 2A proteins. In the case of cardiovirus-infected cells, p220 is not cleaved (234,377), nor is the CBC inactivated; rather, host-protein synthesis is shut off gradually by the ability of the expanding pool of viral RNA to outcompete some, though not all, cellular mRNAs in initiating protein synthesis (195,242,301). In aphthovirus-infected cells, on the other hand, p220 is cleaved not by the 2A protein but by the L protein (104).

In summary, picornavirus-induced shut-off of host protein synthesis appears to be mediated by protein 2A in poliovirus, human rhinovirus 14, Coxsackievirus B3, and enterovirus 70. But shut-off is mediated by the leader protein L in aphthoviruses. In cardiovirus-infected cells p220 is not cleaved, and shut-off is accomplished by an exceptionally active IRES, which enables the viral RNA to outcompete host messages. This behavior has been correlated (104,234) with an 18–amino acid sequence present in the 2A protein of viruses with early cleavage pattern I (Fig. 5) and in the L protein of viruses with early cleavage pattern II but in neither protein of viruses with early cleavage pattern III. Cleavage of p220 may not be sufficient for the observed inhibition of host-protein synthesis (42).

A
Cap-binding complex (CBC), needed to initiate translation, includes p220 which is cleaved (indirectly) through the agency of viral protein 2A
p220
7MeG
cellular mRNA
An
B
Inititiation of translation at the internal ribosomal entry site (IRES) found in all picornaviruses bypasses need for CBC.

IRES
AUG
5′
U
picornavirus mRNA
An

FIG. 24. Differences in initiating translation of cellulor (**A**) and Picornaviral (**B**) mRNA. **A:** Translation of most cellular mRNAs cannot be initiated until a set of proteins (CBC) binds to the m7G cap. Poliovirus and human rhinovirus 14 (124) inactivate the CBC by cleaving p220, one of the CBC proteins; this cleavage is mediated by viral protein 2A. **B:** The IRES, found on all picornaviral mRNAs (see Fig. 3), enables ribosomes to bind downstream from the 5′ end, thus bypassing the cellular requirement for a CBC. Cardioviruses, which lack an active 2A protein, do not cleave p220; instead they have evolved an unusually efficient IRES that enables the virus to simply outcompete host messages for ribosomes.

Other Events

Infection halts cell division within an hour. Premitotic cells stop entering prophase within minutes, whereas those cells entering prophase at the time of infection proceed to division (420). DNA synthesis declines after 4 hr or so. These effects are probably all secondary consequences of the shut-off of protein and RNA synthesis (140).

SYNTHESIS OF VIRAL RNA AND PROTEIN

The First Step

The initial events in the picornaviral multiplication cycle are not readily observable for two reasons. First, the fraction of virus particles that actually succeed in establishing a productive infection is small, generally less than 1% of the input virus. Second, the minute amount of viral material synthesized is masked by vigorous macromolecular metabolism of the initially healthy host cells. The first task of the virus is to translate its genome and thereby supply the viral proteins necessary for replication of its RNA.

The viral proteins are synthesized by translating a single large coding region on the genome, and the protein products are then produced by cleavage of the nascent polyprotein (Fig. 1). Cleavage of the tyr-gly pairs that connect coat precursor P1 to P2-P3 and 3C′-3D′ in poliovirus is accomplished by proteinase 2A (424), but the latter cleavage is not essential for viability of the virus (226). The remaining cleavage sites in P2-P3 are at Gln-Gly pairs recognized by the virus-coded proteinase $3C^{pro}$ (212,304). To deal with the special problem of accomplishing these cleavages before the required proteinase has been freed, the picornaviral P3 molecule releases $3C^{pro}$ by intramolecular self-cleavage (309).

RNA Synthesis

The time required for a ribosome to translate the RNA genome of poliovirus from one end to the other is about 10 to 15 minutes (53,315,413). Because replication cannot begin until the polymerase gene has been completed, this probably represents the minimum time between the time of uncoating and the moment when synthesis of viral RNA can begin. Elongation of the RNA molecule is performed by protein $3D^{pol}$, some of which is found in the cytoplasm of infected cells tightly associated with its RNA template (133,244) and with cellular membranes (57). Most $3D^{pol}$ is found in soluble form. It is not likely engaged in RNA synthesis but may simply result from the excess produced by the polyprotein strategy of expression.

Studies on cell-free synthesis of minus strands from the plus template indicate that at least three proteins are required: virus protein $3C^{pol}$, a VPg donor, and one or more host factor proteins [a terminal uridylyltransferase (TUT) (14), and perhaps a protein kinase (285)]. Two models have been proposed for polioviral RNA synthesis (Fig. 25A). In the first model a VPg donor, such as protein 3AB, donates a primer such as VPg-pUpU, which is then elongated by protein 3D using the plus strand as template (174,407,449). In the second model a host enzyme adds poly(U) to the 3′ end; this forms a hairpin that can then self-prime; VPg addition and cleavage then releases the product strand (446). If VPg primes, then there is no role for TUT. The two models are not mutually exclusive in the sense that one may be used for initiation of plus strand synthesis and the other for initiation of minus strand synthesis. Genetic studies also implicate virus-coded proteins 2A, 2B, and 2C (an NTPase), as well as 3AB and 3CD in RNA synthesis (333,439). For example, binding of 3C (229) or 3CD (12,13) at the 5′ end of the plus strand RNA appears to be important for initiation of RNA synthesis.

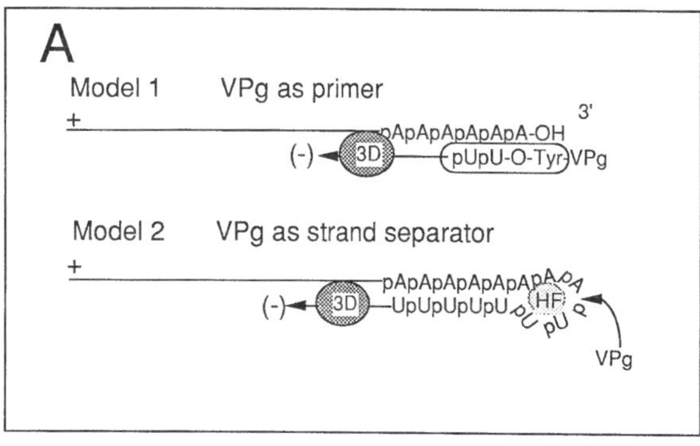

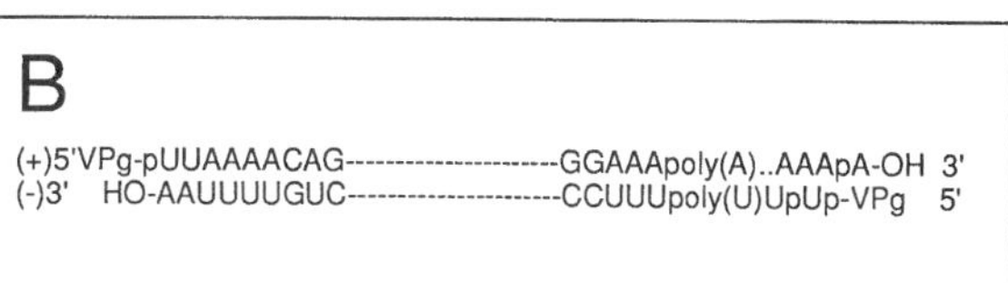

FIG. 25. A: Two models for polioviral RNA synthesis. In Model 1 the newly forming RNA strand is initiated by elongation of a pUpU-VPg primer derived from protein 3AB. In model 2, the template strand is elongated at the 3′ end by TUT, which adds US to the end of the poly(A) tract. The oligo(U) tract then folds back to hybridize with the poly(A) tract, permitting the nascent strand to be elongated by the viral 3D protein. Cleavage of the A-U bond is thought to occur by transesterification using tyrosyl group of VPg (446). The two models are not incompatible in the sense that one may be used for initiation of plus strand synthesis and the other for initiation of minus strand synthesis. The role of host factor (HF) is still unclear. **B:** Replicative form (RF) of poliovirus RNA illustrating key role of A and U stretches at the ends of the plus and minus strands. -pApA residues at or near the 3′ terminus of each strand provide a potential recognition site for a UpUp-VPg primer. Adenylation occurs by transcription of a poly(U) stretch at the 5′ end of minus strands. A typical signal for adenylation by end addition is lacking (108); however, the poly(A) tract is reported to be longer than the poly(U) tract by up to 100 bases (219).

Little is yet known of the requirements for the next step, synthesis of plus strands from minus templates, nor is it yet known if the proteins required for synthesis of plus strands are identical to those that make minus strands, but stem loops at the 5′ ends of both plus and minus strands may be involved (11,12). Replication takes place via a structure known as the replication intermediate (RI) which consists of a full-length template strand with some six to eight nascent daughter strands (24). The RI is partially single stranded (297,344), with stretches of base pairing between (Fig. 15, step 4). The time required for synthesis of each RNA molecule has been estimated at about 45 seconds (24). Double-stranded RNA molecules, called replicative form (RF), are also found in infected cells (Fig. 25); the role of the RF in replication is not clear, but some of the RF in infected cells is in fact hairpinned, i.e., contains a covalented attached plus and minus strand (439).

Role of Membranes in RNA Synthesis

Replicating forms of picornaviral RNA and newly made virions are associated with smooth membranes (37,55,57,221) that proliferate as viral RNA multiplies (35). The proliferation and vacuolization is blocked by 1 to 3 mmol/L guanidine, a drug that inhibits initiation of both plus and minus RNA strands of poliovirus (58,415); this block is accompanied by simultaneous cessation of RNA packaging, suggesting that packaging is linked to RNA synthesis.

Kinetics of RNA Synthesis

Once initiated, polioviral RNA synthesis proceeds exponentially, producing new templates and thus continually increasing the rate of synthesis (27,230). During this early stage of synthesis the rate doubles every 15 minutes until about 10% of the final yield has been produced. At this point the rate of synthesis becomes constant, accumulating linearly for an additional hour until the number of RNA molecules reaches about 4×10^5 per cell. Some 2% to 5% of the total viral RNA in infected cells consists of minus strands (26,194). The mechanism controlling this differential synthesis of plus and minus strands is not clear. A substantial fraction of the RNA synthesized during the exponential phase of synthesis is destined to become mRNA (27), whereas about 50% of that made during the linear phase is packaged into virions (24). The switch from exponential to linear synthetic rate may reflect siphoning of the RNA pool into virions.

Among all the forms of polioviral RNA in the infected cell, only that isolated from polysomes (mRNA) lacks VPg; instead, the mRNA terminates in pUpUpA (129,169,296). Viral mRNA is as infective as genomic RNA; thus, VPg is not essential for infectivity (134).

Uninfected HeLa cells contain an activity, called unlinking enzyme, that removes VPg from polioviral RNA by cleaving the phosphodiester bond between tyrosine and the 5′ terminal pUp (7,8). The normal role of the enzyme is likely related to other types of metabolic activities because host mRNAs carrying VPg-like proteins have not yet been discovered.

Removal of VPg is apparently not mandatory for messenger activity because the genomic VPg RNA is itself a competent messenger in cell-free extracts (146), and VPg RNA is sometimes found in polysomes of infected cells late in infection (376). Despite the in vitro findings, it seems likely that removal of VPg plays an important regulatory effect within the intact cell by favoring flow of viral RNA to ribosomes early during the infection cycle. The resulting polysomes then associate with membrane to form patches of rough endoplasmic reticulum, a site of viral protein synthesis (361).

Genetic Recombination

That recombination takes place in RNA viruses, including picornaviruses, was deduced from genetic studies conducted over two decades ago (171,225,335), but direct biochemical confirmation has been obtained only recently (209,353). The pioneering studies of Cooper (81) demonstrated that it was feasible to construct a genetic map based on recombination frequencies obtained from crosses between temperature-sensitive (ts) mutants of poliovirus with defects in coat protein and in RNA synthesis. It has been proposed that recombination occurs when viral RNA polymerase detaches from its original template (Fig. 26), then completes the nascent transcript after reassociating with a different template (82). Studies supporting this model (211) suggest that the strand switching leading to recombination

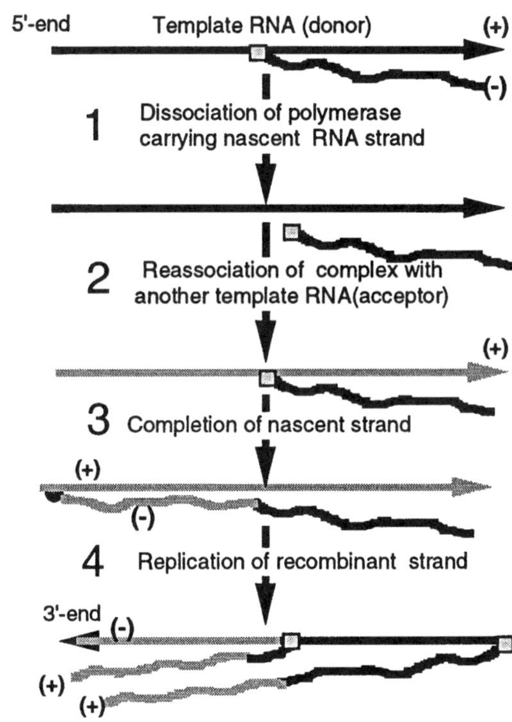

FIG. 26. Strand-switching model for genetic recombination during replication of poliovirus RNA. The polymerase carrying a nascent strand detaches occasionally from its template (step 1). For clarity, the nascent strand is depicted as diffusing to the second (acceptor) template. In practice, the acceptor template is thought to hybridize with the donor (352) or nascent (194) strand before polymerase transfers to the acceptor template. After dissociating from the donor template (step 2) and reassociating with the acceptor template (step 3), the nascent strand is completed to form the recombinant strand, which is then copied (step 4). Homologous recombination results when the polymerase complex switches to the same position on the acceptor template. Nonhomologous recombination results when polymerase switches to a different position on acceptor template (produces a deletion or an insertion).

occurs most frequently during synthesis of minus strands (step 5 in Fig. 15). Recombination occurs most frequently during synthesis of minus strands, likely because the concentration of positive RNA strands in infected cells exceeds that of negative strands by 20- to 50-fold; nascent polymerase complexes might then be expected to have a correspondingly higher probability of finding a plus RNA template than a minus RNA template (194).

The length of the picornaviral recombination map, about 1.2% for poliovirus (81) and 3% for aphthovirus (258), implies that recombination occurs with very high frequency. For parents of the same serotype, 10% to 20% of all viral RNA molecules produced in a single growth cycle are recombinant (208). The error frequency during RNA replication on the other hand is estimated to be on the order of 10^{-4} per base position (107,177) or roughly one substitution error per genome. High recombination frequency may be a mechanism for coping with the high error rate in synthesis of the RNA in the face of poor ability of picornaviral proteins to complement each other (208).

When crosses are made between serotypes, recombination frequency decreases rapidly. With poliovirus serotypes 1 and 3, for example, recombination frequencies were on the order of 10^{-5} to 10^{-4} (421). Recombination is not a mere laboratory curiosity but also occurs in nature. Some recombinants involve all three serotypes of oral poliovaccine virus (207), arising possibly when live oral poliovaccine is administered as a mixture of all three serotypes (trivalent vaccine), thus making possible the infection of a single intestinal cell by all three viruses. One study on a normal healthy child given trivalent vaccine documented the appearance of two different recombinants within several weeks, the time required for the infection to clear (281).

Translation and Cleavage of the Polyprotein

As noted in Fig. 23, the virus-induced decline in cellular protein synthesis is followed by an increase due to vigorous synthesis of viral proteins. This was first demonstrated by feeding radioactive amino acids to infected cells and analyzing the radioactive products by electrophoresis on SDS-polyacrylamide gels, a procedure that separates proteins according to size. In this way Summers et al. (410) discovered a complex spectrum of proteins like that shown in Fig. 27 and thereby set the stage for unraveling the synthesis and cleavage of the virus-encoded proteins.

Recognition that all of the proteins were derived from a single large polyprotein made it possible to order the genes by analyzing the specific activity of viral proteins synthesized in the presence of the drug pactamycin (53,409,413). The drug inhibits initiation of new protein chains while allowing nascent proteins on the viral polysome to be completed (run-off chains). Thus, when a radioactive amino acid is added to infected cells with pactamycin-blocked polysomes, proteins encoded at the 5′ end of the polysomal mRNA incorporate less radiolabel than do those at the 3′ end as synthesis is completed. The order of the completed proteins is then deduced from the gradient of specific activity, which increases in the direction of translation (5′ → 3′).

Precursor-product relationships were established by relating the patterns of proteins fragmented with trypsin or cyanogen bromide, and by analyzing the flow of radioactivity through proteins (52–54). VPg was first mapped by tryptic analysis to precursor $3AB^{VPg}$ (303), and its location on the right side of the precursor was then determined by radiochemical sequencing (212).

End group analysis showed that the aminoterminus of coat proteins P1 and 1A are blocked (212,453). Radiochemical sequencing of polioviral proteins, which became feasible once the genome sequence was known, confirmed the map constructed from earlier studies (371). In later work, they identified the precise location of the cleavage sites and showed that the termini of the products are not further trimmed (212,304).

The leader protein L, found in aphtho- and cardioviruses but not in rhino-, entero-, or hepatitis A viruses, was first detected during studies on the cell-free translation products of cardioviral RNA (60,220,396). Improvement of the rabbit reticulocyte system, pretreated with nuclease to remove endogenous message (313), greatly increased the efficiency of cell-free translation and opened the way for identification (147,307,312,394) of the virus coded proteinase and mapping it to $3C^{pro}$ (307). Until the nomenclature of picornaviral proteins was standardized, each picornavirus had different names. A key to the early nomenclature can be found elsewhere [see Table 12 in Rueckert (369)].

MORPHOGENESIS

Picornaviruses are attractive models for studying the mechanism of viral assembly because the 60-subunit shell is relatively simple, involving assembly of only 12 pentamers (370). Their morphogenesis can be described (Fig. 28) as a tandem three- (or four-) stage process in which monomers are converted to a pentamer (step 1). Pentamers are then assembled into a provirion consisting of the RNA genome within a 12-plated capsid (step 2). RNA is probably packaged at step 2a, although evidence interpreted as favoring step 2d is not ruled out. The final step during which mature virus is formed (step 3) involves cleavage of most, if not all, of the VP0 chains in the 60-subunit capsid.

Capsid assembly can be understood in terms of latent assembly domains encoded in the protomer (370). Orderly activation of these latent domains, including two intersubunit bonds, an RNA packaging domain, and maturation cleavage, may be controlled by configurational changes

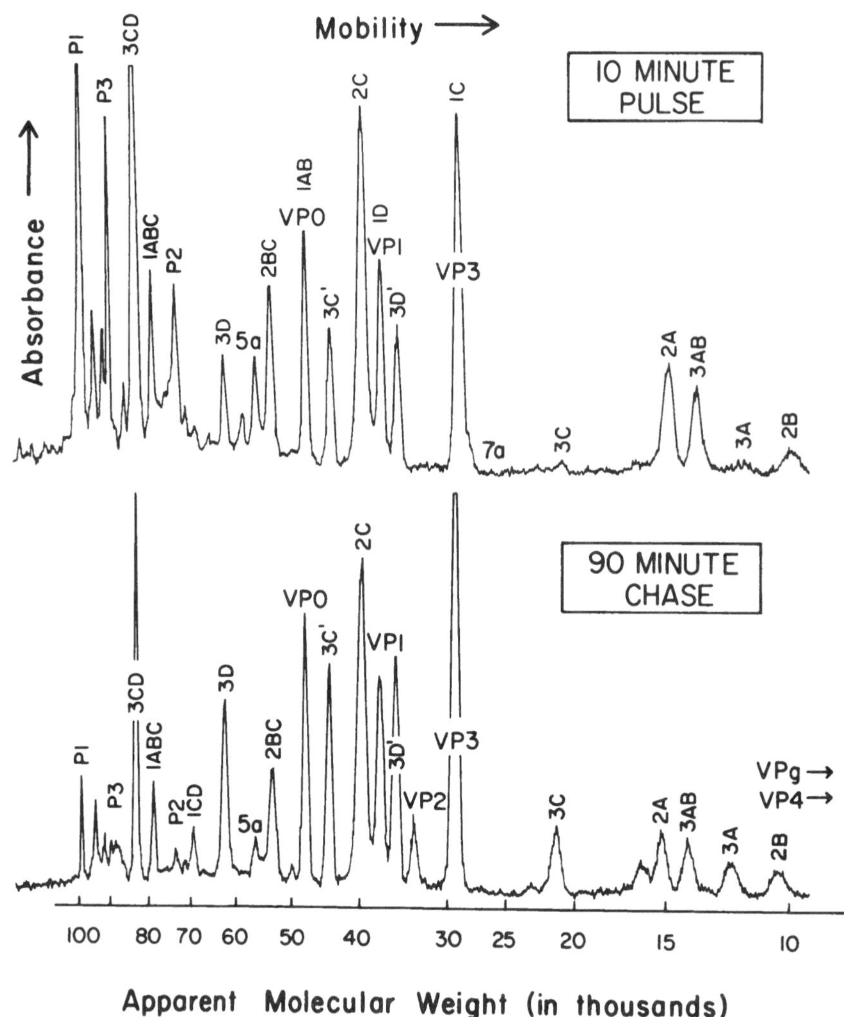

FIG. 27. Scan of an autoradiogram showing the spectrum of proteins synthesized in poliovirus-infected HeLa cells at 3.5 hr after infection. **Top:** Cells were exposed to radioactive [^{35}S]methionine for a 10-minute incorporation period. They were then sedimented and dissolved in detergent (SDS), and proteins in the lysate were separated by electrophoresis on polyacrylamide gels containing the same detergent. Migration was toward the positive electrode. The method separates proteins by size, with the largest proteins at the left and the smallest at the right. **Bottom:** The same procedure as described above was followed, except that cells were further incubated for 90 minutes in medium lacking radiolabel to allow additional time for cleavage (304,371).

after each intermediate step (368,370). Experimental support for the existence of configurational changes during assembly comes in part from serological studies and in part from structural studies. In what follows we trace the problem from its roots with the discovery of viral antigens to the current status summarized in Fig. 28.

Natural Empty Capsids

In 1957 Mayer et al. (255), using rate-zonal centrifugation, separated partially purified preparations of poliovirus into four bands designated A, B, C, and D. The first two were impurities. The fastest sedimenting band, D, contained infective virions, whereas band C contained virion-sized particles subsequently identified as RNA-free shells, now called natural empty capsids (NECs). Empty capsids band above virions when centrifuged in CsCl density gradients; hence, they are often called natural top components.

Each of the virus-related bands, C and D, exhibits independent precipitin lines in agar. Band D reacts preferentially with certain convalescent human sera, whereas band C (C particles) reacts better with human serum taken from patients in the acute phase of poliomyelitis (350,351). C particles from unheated cell extracts were ultimately found to be indistinguishable in serologic tests from an antigen produced by heating poliovirus (188,351).

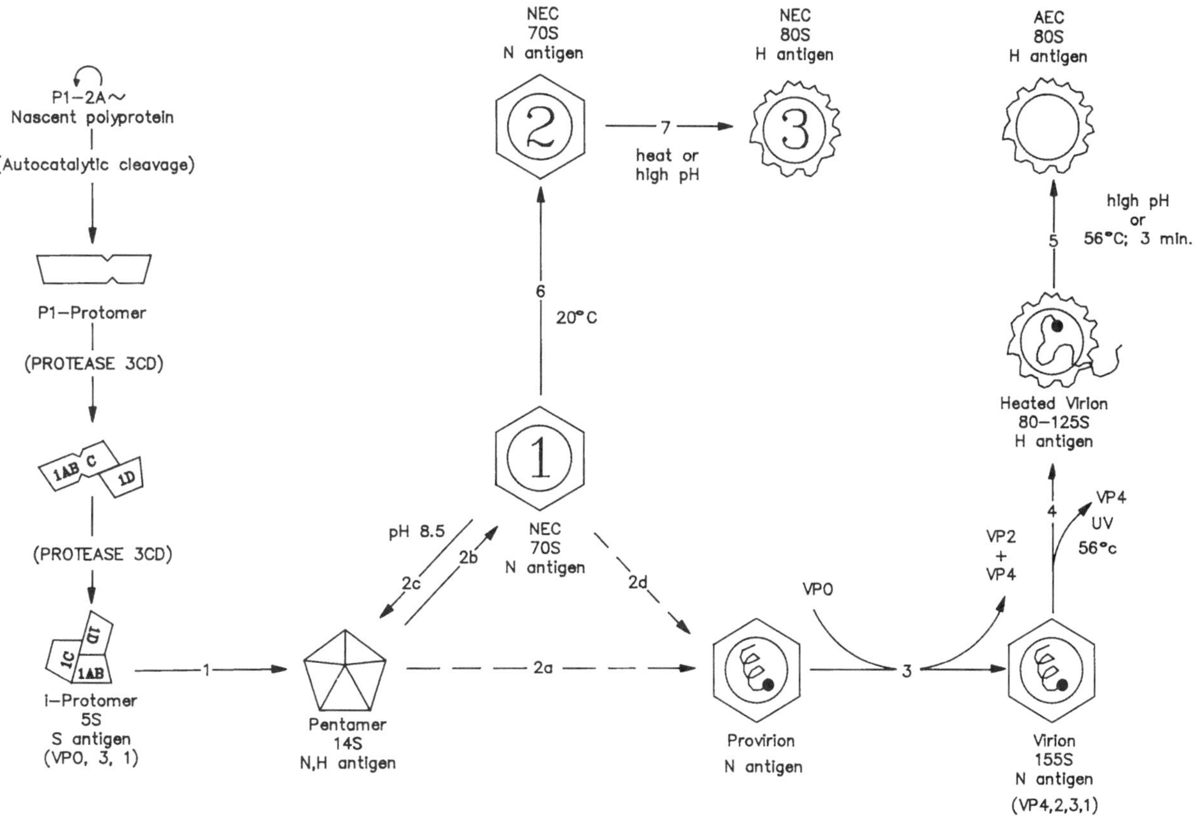

FIG. 28. Picornaviral morphogenesis. Main assembly pathway is shown horizontally at the bottom (see also Fig. 9). After release from the polyprotein precursor, P1 is cleaved twice. A 50-Å separation of the cleaved ends of VP1,2,3 in the virion (see Fig. 12) implies that the polypeptide backbone in the I protomer (VP0,1,3) undergoes a substantial rearrangement after P1 cleavage. Assembly does not occur unless both cleavages take place (310). In the case of poliovirus, protomer cleavage is performed much more efficiently by 3CD than by proteinase 3C (199,294,447). Twelve pentamers are then assembled into shells. Note changes in antigenicity (from S to N, H to N) as assembly proceeds from monomer to virion (horizontal pathway, bottom). Maturation cleavage of chain VP0 → VP4 + VP2 (step 3) gives rise to infective virions (N antigen). Dashed arrows indicate controversial steps.

The Antigenic N to H Transformation

When infective or "native" poliovirions (called N antigen or, occasionally, D antigen from the centrifugation studies described above) are heated in neutral saline for 2 or 3 minutes at 56°C, RNA is released from the particle and the protein shell is simultaneously converted to a new antigenicity, called H antigen (for *h*eated) (Fig. 28, step 4). Little or no immune cross-reaction takes place between the H and N components (351). Thus, antiserum raised against H antigen does not neutralize the infectivity of native virions nor indeed does it even bind to them. Finally, heated particles are no longer able to attach to cells. In other words when native virions are heated, all of its characteristic surface antigens vanish. Empty shells, produced by heating virions, also lack VP4 (247) and are called artificial empty capsids or artificial top components.

Conversion from N to H antigenicity can also be brought about by other means such as ultraviolet (UV) irradiation, high pH, mercurials, phenol, and desiccation (222). Loss of RNA is not required for conversion to H antigenicity. Thus, the virion can also be converted to H antigenicity without loss of its encapsidated RNA by shorter times of heating at 56°C or by treatment with UV light (204,222,351). The UV dose required for the N to H conversion is greatly in excess of that needed for inactivation of infectivity (204,351,435).

The conversion from N to H antigenicity is irreversible and occurs by a process called concerted transition; i.e., particles are either N antigenic or H antigenic. Mosaic virions containing a mixture of H and N determinants are rare or nonexistent (187,188). This implies that once the N to H transition begins, it is quickly and completely propagated throughout all other subunits in the particle. This all-or-not change of all the subunits is also sometimes called concerted reorientation.

S Antigen

Still another antigen, designated S antigen, is detected by immunization with a 2S protein from poliovirus degraded with 6 mol/L guanidine hydrochloride. This antibody, but not N- or H-specific antibody, reacts with a 5S antigen in extracts from infected cells (383). It also reacts with nascent viral protein associated with polysomes (382).

In vitro Assembly of NECs from 14S Subunits

Extracts from poliovirus-infected cells contain, in addition to 5S subunits and NECs, a 14S particle (325), first reported as 10S (431). A study of the polypeptide composition of these particles showed that the NEC, the 14S, and probably also the 5S component contained roughly similar amounts of VP0, VP1, and VP3 (328), but lacked two chains, VP2 and VP4, found in the mature virion; extracts from infected cells had the remarkable ability to promote assembly of the 14S particles into NECs. That this *in vitro* assembly-promoting activity was virus specific was shown by the lack of activity in extracts from uninfected cells.

Morphopoietic Factor

Phillips and coworkers subsequently showed that there are twokinds of assembly: self-assembly and extract-mediated assembly (326,338). Both display similar assembly kinetics, pH, salt, and temperature (37°C) optima. Self-assembly requires high concentrations of 14S particles, and the shells produced are H antigenic. Extract-mediated assembly occurs at much lower 14S concentrations, and the shells are N antigenic. The assembly-catalyzing activity, which is sensitive to trypsin but not to ribonuclease, was traced to rough membranes carrying 14S, NECs, and an unidentified 110S structure containing capsid proteins that could be released by solubilizing the membrane with deoxycholate (318). Extracts from cells infected with defective-interfering (DI) particles, a form of poliovirus whose only known defect is a deletion in its coat protein (74), lacked assembly-promoting activity (327). Recent studies indicate that morphopoietic factor associated with rough endoplasmic reticulum probably corresponds of 14S subunits with the correct (native) antigenicity (354).

5S particles synthesized in a cell-free translation system from rabbit reticulocytes are able to self-assemble into 14S subunits, albeit abnormally slowly: 5 to 18 hr (153,305) compared with a few minutes in infected cells. Little is yet known about the conditions required to promote their assembly.

Role of Smooth Membranes

Other experiments suggest that morphogenesis takes place in association with the viral replication complexes found on smooth, rather than rough, membranes. The crucial evidence is that virus particles associated with fractionated complexes from cells labeled for a short time with radioactive uridine reportedly exhibit a specific activity some three- to eightfold higher than virions in other fractions (56). The higher specific activity of these virions suggest that they are the ones most recently made and are therefore formed on the replicating complex. Stable empty shells can be recovered by treating these membrane-associated replication complexes with ribonuclease (442).

Formation of Virions and the Maturation Cleavage

Additional evidence for a link between RNA synthesis and virion formation comes from studies using 1 mmol/L guanidine hydrochloride, which blocks flow of newly made RNA from the replication complex into virions; this cessation of virion assembly is accompanied by accumulation of empty shells, a halt in cleavage of VP0, and a block in the flow of radioactive protein into virions (24). The effect is reversible; when the drug is removed, RNA synthesis resumes, label in empty shells flows rapidly and completely into virions, and VP0 is cleaved in parallel (191,193). This work established that the difference in polypeptide composition of the capsid and the capsid precursors was due to a maturation cleavage of VP0 $\rightarrow$ VP2 + VP4.

This picture of a final maturation cleavage of VP0 was supported by identification of an RNA-containing particle called provirion, originally reported to sediment at 125S (130) but later reported to cosediment with mature 155S virions (154). Provirions lack infectivity until the maturation cleavage takes place (228).

Structure of the 5S and 14S Subunits

The hypothesis (370) that 14S particles are pentamers of immature 5S protomers (VP0,3,1) was given strong support by measurements on sedimentation velocity and chain stoichiometry (263). The 14S particles from EMC virus-infected HeLa cells sediment 6% faster than the 13.4S (VP2,3,1)-containing, pentamers from mature virions, exactly as predicted by sedimentation theory (Table 7). Moreover, 14S subunits can be dissociated into smaller subunits with the 5S sedimentation coefficient and 1:1:1 proportion of VP0, VP3, and VP1 chains predicted for an immature protomer (263,264).

Mechanics of RNA Packaging

The most elusive problem in picornavirus synthesis is the mechanism by which the RNA becomes packaged into virions. The possibility of a rapid equilibration between a shell and a subunit was originally considered and rejected because NECs were known to be exceedingly stable, dissociable only by reagents that denatured them to polypeptide chains.

Faced with these experimental findings, Jacobson and Baltimore proposed the procapsid hypothesis, in which the final step in morphogenesis is association of the RNA with the preformed shell (191). They suggested two possible mechanisms by which this might happen: (a) that RNA is threaded through a pore in the empty shell (threading model); or (b) that the RNA wraps around the procapsid, fitting into appropriate channels and thereby triggering concerted reorientation of the subunits in such a way as to internalize the RNA and precipitate maturation cleavages. The latter hypothesis (transfiguration model) had the virtue of explaining the large surface difference between the H-antigenic NECs and the N-antigenic virions.

Three Kinds of NECs

The ensuing years have seen increasing evidence that the two fundamental properties of NECs that led Jacobson and Baltimore to the procapsid hypothesis, namely high stability and H antigenicity, are artifacts of the procedure used for extracting them from infected cells. Thus, the native antigenicity of NECs is probably N, not H (142,355,356). When extracted under the usual conditions, room temperature, neutral pH, and low ionic strength, NECs are H antigenic or a mixture of N and H antigenic. When extracted at pH 6 to 6.5 at low temperature (4ºC), however, they are purely N antigenic and sediment at 70S (NEC-2 in Fig. 28). When incubated at 37ºC these N-antigenic particles are irreversibly converted to an H-antigenic form that sediments at 80S (NEC-3).

Evidence for a dissociable form of NEC (labile NEC or NEC-1) was provided by La Colla et al. (252), who showed that when extracted at low temperature NEC dissociates easily into 14S subunits. These 14S subunits reassemble into NECs, even at very low concentration and without benefit of morphopoietic membranes. However, when labile NECs were warmed, even briefly at room temperature, they became nondissociable (stable NEC) NEC-2 or NEC-3, depending on the degree of mistreatment.

CONCLUSION

The mechanism of RNA packaging remains the central problem of picornaviral morphogenesis. With evidence that NECs are N antigenic, the case for the transfiguration hypothesis vanishes. Similarly, discovery of reversibly dissociable NEC reopens the possibility that these shells are merely storage reservoirs for 14S subunits.

At this writing, the most widely accepted assembly model is accretion of 14S pentamers around a condensed core of RNA (144). However, the procapsid hypothesis cannot yet be considered dead. Lacking firm evidence that NECs indeed contain a full complement of 60 subunits, it is possible that they are missing one or two pentamers (227), thus leaving a hole through which RNA might be threaded. Another proposal is that RNA is packaged walnutlike between a pair of 53S half-shells (227). All of these models raise a nontrivial question as to how the RNA is condensed into a ball tight enough to fit into the central cavity where the RNA concentration is on the order of 50% to 70%.

Available evidence, such as the rapid cessation of RNA packaging after addition of guanidine (24) and evidence for newly synthesized virions on replication complexes associated with smooth membrane (56,442) suggest a close tie between synthesis and packaging of RNA. Synthesis might then provide the driving energy necessary for inserting RNA into a procapsid or for condensing it into a compact core. Development of cell-free systems capable of net synthesis of infectious poliovirus from genomic RNA is an important step toward the goal of elucidating the final steps in packaging viral RNA (28,282,283).

REFERENCES

1. Abad-Zapatero C, Abel-Meguid SS, Johnson JE, et al. Structure of southern bean mosaic virus at 2.8 angstroms resolution. *Nature* 1980;286:33–39.
2. Abraham G, Colonno RJ. Many rhinovirus serotypes share the same cellular receptor. *J Virol* 1984;51:340–345.
3. Acharya R, Fry E, Stuart D, Fox G, Rowlands D, Brown F. The three-dimensional structure of foot-and-mouth disease virus at 2.9 A resolution. *Nature* 1989;337:709–715.
4. Ahlquist P, Kaesberg P. Determination of the length distribution of poly(A) at the 3′-terminus of the virion RNAs of EMC virus, poliovirus, rhinovirus, RAV-61 and CPMV and of mouse globin mRNA. *Nucleic Acids Res* 1979;7:1195–1204.
5. Alexander HE, Koch G, Mountain IM, Sprunt K, Van Damme O. Infectivity of ribonucleic acid of poliovirus on HeLa cell monolayers. *Virology* 1958;5:172–173.
6. Amako K, Dales S. Cytopathology of mengovirus infection. II. Proliferation of membranous cisternae. *Virology* 1967;32:201–215.
7. Ambrose V, Baltimore D. Purification and properties of a HeLa cell enzyme able to remove the 5′ terminal protein from polioviral RNA. *J Biol Chem* 1980;255:6739–6744.
8. Ambrose V, Pettersson RF, Baltimore D. An enzymatic activity in uninfected cells that cleaves the linkage between poliovirion RNA and the 5′ terminal protein. *Cell* 1978;5:1439–1446.
9. Anderegg JW. X-ray scattering from small RNA viruses in solution. In: Brumberger H, ed. *Small-angle x-ray scattering*. New York: Gordon and Breach; 1957:243–265.
10. Anderson-Sillman K, Bartal S, Tershak D. Guanidine-resistant poliovirus mutants produce modified 37-kilodalton proteins. *J Virol* 1984;50:922–928.
11. Andino R, Rieckhof GE, Achacoso PL, Baltimore D. Poliovirus RNA synthesis utilizes an RNP complex formed around the 5′-end of viral RNA. *EMBO J* 1993;12:3587–3598.
12. Andino R, Rieckhof GE, Baltimore D. A functional ribonucleoprotein complex forms around the 5′ end of poliovirus RNA. *Cell* 1990;63:369–380.
13. Andino R, Rieckhof GE, Trono D, Baltimore D. Substitutions in the protease (3Cpro) gene of poliovirus can suppress a mutation in the 5′ noncoding region. *J Virol* 1990;64:607–612.
14. Andrews NC, Baltimore D. Purification of a terminal uridylyltransferase that acts as host factor in the in vitro poliovirus replicase reaction. *Proc Natl Acad Sci USA* 1986;83:221–225.
15. Andries K, Dewindt B, De Brabander M, Stokbroekx R, Janssen PAJ. *In vitro* activity of R61837, a new antirhinovirus compound. *Arch Virol* 1988;101:155–167.
16. Andries K, Dewindt B, Snoeks J, et al. Two groups of rhinoviruses revealed by a panel of antiviral compounds present sequence divergence

and differential pathogenicity. *J Virol* 1990;64:1117–1123.

17. Andries K, Dewindt B, Snoeks J, et al. In vitro activity of pirodavir (R77975), a substituted phenoxypyridazinamine with broad-spectrum antipicornaviral activity. *Antimicrob Agents Chemother* 1992; 36:100–107.
18. Ansardi DC, Porter DC, Morrow CD. Myristoylation of poliovirus capsid precursor P1 is required for assembly of subviral particles. *J Virol* 1992;66:4556–4563.
19. Argos P, Kamer G, Nicklin MJH, Wimmer E. Similarity in gene organization and homology between proteins of animal picornaviruses and a plant comovirus suggest common ancestry of these virus families. *Nucleic Acids Res* 1984;12:7251–7267.
20. Axler DA, Crowell RL. Effect of anticellular serum on the attachment of enteroviruses to HeLa cells. *J Virol* 1968;2:813–821.
21. Bablanian R, Eggers HJ, Tamm I. Studies on the mechanism of poliovirus-induced cell damage. II. The relation between poliovirus growth and virus-induced morphological changes in cells. *Virology* 1965;26:114–121.
22. Bachrach HL, Morgan DO, McKercher PD, Moore DM, Robertson BH. Foot and mouth disease virus: immunogenicity and structure of fragments derived from capsid protein vp3 and of virus containing cleaved VP3. *Vet Microbiol* 1982;7:85–96.
23. Badger J, Minor I, Dremer MJ, Oliveira MA, Smith TJ, Griffith JP. Structural analysis of a series of antiviral agents complexed with human rhinovirus 14. *Proc Natl Acad Sci USA* 1988;85:3304–3308.
24. Baltimore D. The replication of picornaviruses. In: Levy HB, ed.*The biochemistry of viruses*. New York: Marcel Dekker; 1969:101–176.
25. Baltimore D, Franklin RM, Callender J. Mengovirus-induced inhibition of host ribonucleic acid and protein synthesis. *Biochim Biophys Acta* 1963;76:425–430.
26. Baltimore D, Girard M. An intermediate in synthesis of poliovirus RNA. *Proc Natl Acad Sci USA* 1966;56:741–748.
27. Baltimore D, Girard M, Darnell JE. Aspects of the synthesis of poliovirus RNA and the formation of virus particles. *Virology* 1966;29:179–189.
28. Barton DJ, Flanegan JB. Coupled translation and replication of poliovirus RNA in vitro: synthesis of functional 3D polymerse and infectious virus. *J Virol* 1993;67:822–831.
29. Baxt B, Bachrach HL. Early interactions of foot-and-mouth disease virus with cultured cells. *Virology* 1980;104:42–55.
30. Belsham G, Bostock CJ. Studies on the infectivity of foot-and-mouth disease virus RNA using microinjection. *J Gen Virol* 1988;69:265–274.
31. Bergelson JM, Chan M, Solomon KR, St. John NF, Lin H, Finberg RW. Decay-accelerating factor (CD55), a glycosylphosphatidylinositol-anchored complement regulatory protein, is a receptor for several echoviruses. *Proc Natl Acad Sci USA* 1994;91:6245–6248.
32. Bergelson JM, Shepley MP, Chan BM, Hemler ME, Finberg RW. Identification of the integrin VLA-2 as a receptor for echovirus 1. *Science* 1992;255:1718–1720.
33. Bernstein HD, Sarnow P, Baltimore D. Genetic complementation among poliovirus mutants derived from an infectious cDNA clone. *J Virol* 1986;60:1040–1049.
34. Bernstein HD, Sonenberg N, Baltimore D. Poliovirus mutant that does not selectively inhibit host cell protein synthesis. *Mol Cell Biol* 1985;5:2913–2923.
35. Bienz K, Egger D, Rasser Y, Bossart W. Kinetics and location of poliovirus macromolecular synthesis in correlation to virus-induced cytopathology. *Virology* 1980;100:390–399.
36. Bienz K, Egger D, Rasser Y, Bossart W. Intracellular distribution of poliovirus proteins and the induction of virus-specific cytoplasmic structures. *Virology* 1983;131:39–48.
37. Bienz K, Egger D, Troxler M, Pasamontes L. Structural organization of poliovirus RNA replication is mediated by viral proteins of the P2 genomic region. *J Virol* 1990;64:1156–1163.
38. Bishop JM, Koch G. Purification and characterization of poliovirus-induced infectious double-stranded RNA. *J Biol Chem* 1967; 242:1736–1743.
39. Black DN, Stephenson P, Rowlands DJ, Brown F. Sequence and location of the poly C tract in aphtho- and cardiovirus RNA. *Nucleic Acids Res* 1979;6:2381–2390.
40. Blair WS, Li X, Semler BL. The cellular cofactor facilitates efficient 3CD cleavage of the poliovirus P1 precursor. *J Virol* 1993; 67:2336–2343.
41. Bodian D, Horstmann DM. Polioviruses. In: Horsfall F, Tamm I, eds. *Viral and rickettsial infections of nan.* 4th ed. Philadelphia: JB Lippincott; 1965:430–473.
42. Bonneau A-M, Sonenberg N. Proteolysis of the p220 component of the cap-binding protein complex is not sufficient for complete inhibition of host cell protein synthesis after poliovirus infection. *J Virol* 1987;61:986–991.
43. Borman AM, Deliat FG, Kean KM. Sequences within the poliovirus internal ribosome entry segment control viral RNA synthesis. *EMBO J* 1994;13:3149–3157.
44. Boulanger P, Lonberg-Holm K. Components of non-enveloped viruses which recognize receptors. In: Lonberg-Holm K, Philipson L, eds. *Receptors and recognition.* Series B, Vol. 8: Virus receptors. Part 2: Animal viruses. New York: Chapman and Hall; 1981:21–46.
45. Brioen P, Rombaut B, Boeye A. Lack of quantitative correlation between the neutralization of poliovirus and the antibody-mediated pI shift of the virion. *J Gen Virol* 1985;66:609–613.
46. Brooksby JB. Foot-and-mouth disease virus. *Intervirology* 1982; 18:1–23.
47. Brown F, Cartwright B, Stewart DL. Further studies on the infection of pig-kidney cells by foot-and-mouth disease virus. *Biochim Biophys Acta* 1962;55:768–774.
48. Brown F, Newman JFE, Stott EJ, et al. Poly C in animal viral RNAs. *Nature* 1974;251:342–344.
49. Burness ATH. Glycophorin and sialylated components as virus receptors. In: Lonberg-Holm K, Philipson L, eds. *Receptors and recognition.* Series B, Vol. 8: Virus receptors. Part 2: Animal viruses. New York: Chapman and Hall; 1980:63–84.
50. Burness ATH, Clothier FW. Particle weight and other biophysical properties of encephalomyocarditis virus. *J Gen Virol* 1970;6:381–393.
51. Butterworth BE. A comparison of the virus-specific polypeptides of encephalomyocarditis virus, human rhinovirus-1A and poliovirus. *Virology* 1973;56:439–453.
52. Butterworth BE, Hall L, Stoltzfus CM, Rueckert RR. Virus-specific proteins synthesized in encephalomyocarditis virus-infected HeLa cells. *Proc Natl Acad Sci USA* 1971;68:3083–3087.
53. Butterworth BE, Rueckert RR. Gene order of encephalomyocarditis virus as determined by studies with pactamycin. *J Virol* 1972;9:823–828.
54. Butterworth BE, Rueckert RR. Kinetics of synthesis and cleavage of encephalomyocarditis virus-specific polypeptides. *Virology* 1972;50:535–549.
55. Butterworth BE, Shimshick EJ, Yin FH. Association of the polioviral RNA complex with phospholipid membranes. *J Virol* 1976;19:457–466.
56. Caliguiri LA, Mosser AG. Protein associated with the poliovirus RNA replication complex. *Virology* 1971;46:375–386.
57. Caliguiri LA, Tamm I. Characterization of poliovirus-specific structures associated with cytoplasmic membranes. *Virology* 1970;42:112–122.
58. Caliguiri LA, Tamm I. Guanidine and 2-(alpha-hydroxybenzyl)benzimidazole (HBB): selective inhibitors of picornavirus multiplication. In: Carter W, ed. *Selective inhibitors of viral functions*. Cleveland: CRC Press; 1973:257–294.
59. Callahan PL, Mizutani S, Colonno RJ. Molecular cloning and complete sequence determination of RNA genome of human rhinovirus type 14. *Proc Natl Acad Sci USA* 1985;82:732–736.
60. Campbell EA, Jackson RJ. Processing of the encephalomyocarditis virus capsid precursor protein studied in rabbit reticulocyte lysates incubated with *N*-formyl-^{35}S-methionine-tRNA. *J Virol* 1982; 45:439–441.
61. Carrasco L, Castrillo JL. Regulation of translation in picornavirus-infected cells. In: Pevear DC, Calenoff M, Rozhon E, Lipton HL, eds. *Mechanisms of viral toxicity in animal cells*. Boca Raton, FL: CRC Press; 1987:115–146.
62. Carroll AR, Rowlands DJ, Clarke BE. The complete nucleotide sequence of the RNA coding for the primary translation product of foot-and-mouth disease virus. *Nucleic Acids Res* 1984;12:2461–2472.
63. Caspar DLD, Klug A. Physical principles in the construction of regular viruses. *Cold Spring Harbor Symp Quant Biol* 1962;27:1–24.
64. Cavanagh D, Rowlands DJ, Brown F. Early events in the interaction between foot-and-mouth disease virus and primary pig kidney cells. *J Gen Virol* 1978;41:255–264.
65. Cavanagh D, Sangar DV, Rowlands DJ, Brown F. Immunogenic and cell attachment sites of FMDV: further evidence for their location in a single capsid polypeptide. *J Gen Virol* 1977;35:149–158.
66. Chemaia MM, Malcolm BA, Allaire M, James M. Hepatitis A virus 3C protease: some properties, crystallization and preliminary crystallographic characterization. *J Mol Biol* 1993;234:890–893.

67. Cherry JD. Enteroviruses: the forgotten viruses of the 80's. In: de la Maza LM, Peterson EM, eds. *Medical virology*. Vol. VII. New York: Elsevier Science; 1988:1–32.
68. Chow M, Newman JFE, Filman D, Hogle JM, Rowlands DJ, Brown F. Myristylation of picornavirus capsid protein VP4 and its structural significance. *Nature* 1987;327:482–486.
69. Chow M, Yabrov R, Bittle J, Hogle J, Baltimore D. Synthetic peptides from four separate regions of the poliovirus type 1 capsid protein VP1 induce neutralizing antibodies. *Proc Natl Acad Sci USA* 1985;82:910–914.
70. Chumakov KV, Agol VI. Poly(C) sequence is located near the 5′ end of the encephalomyocarditis virus RNA. *Biochem Biophys Res Commun* 1976;71:551–557.
71. Clark ME, Hammerle T, Wimmer E, Dasgupta A. Poliovirus proteinase 3C converts an active form of transcription factor IIIC to an inactive form: a mechanism for inhibition of host cell polymerase III transcription by poliovirus. *EMBO J* 1991;10:2941–2947.
72. Clark ME, Lieberman PM, Berk AJ, Dasgupta A. Direct cleavage of human TATA-binding protein by poliovirus protease 3C in vivo and in vitro. *Mol Cell Biol* 1993;13:1232–1237.
73. Cohen JI, Ticehurst JR, Purcell RH, Buckler-White A, Baroudy BM. Complete nucleotide sequence of wild-type hepatitis A virus: comparison with different strains of hepatitis A virus and other picornaviruses. *J Virol* 1987;61:50–59.
74. Cole CN. Defective interfering (DI) particles of poliovirus. *Prog Med Virol* 1975;20:180–207.
75. Collins FD, Roberts WK. Mechanism of mengovirus-induced cell injury in L-cells: use of inhibitors of protein synthesis to dissociate virus-specific events. *J Virol* 1972;10:969–978.
76. Colonno RJ. Cell surface receptors for picornaviruses. *BioEssays* 1986;5:270–274.
77. Colonno RJ, Callahan PL, Long WJ. Isolation of a monoclonal antibody that blocks attachment of the major group of human rhinoviruses. *J Virol* 1986;57:7–12.
78. Colonno RJ, Condra JH, Mizutani S, Callahan PL, Davies M-E, Murcko MA. Evidence for the direct involvement of the rhinovirus canyon in receptor binding. *Proc Natl Acad Sci USA* 1988;85:5449–5453.
79. Colter JS, Bird HH, Brown RA. Infectivity of ribonucleic acid from Ehrlich ascites tumor cells infected with mengo encephalitis. *Nature* 1957;179:859–860.
80. Cooper PD. The inhibition of poliovirus growth by actinomycin D and the prevention by pretreatment of the cells with serum or insulin. *Virology* 1966;28:663–678.
81. Cooper PD. Genetics of picornavirus. In: Fraenkel-Conrat H, Wagner R, eds. *Comprehensive virology*. Vol. 9. New York: Plenum; 1977:133–207.
82. Cooper PD, Steiner-Pryor S, Scotti PD, Delong D. On the nature of poliovirus genetic recombinants. *J Gen Virol* 1974;23:41–49.
83. Cordell-Stewart B, Taylor MW. Effect of double-stranded viral RNA on mammalian cells in culture. *Proc Natl Acad Sci USA* 1971; 68:1326–1330.
84. Cords CE, James CG, McLaren LC. Alteration of capsid proteins of coxsackievirus A13 by low ionic concentrations. *J Virol* 1975;15:244–252.
85. Costa Giomi MP, Bergmann IE, Scodeller EA, Auge de Mello P, Gomez I, La Torre JL. Heterogeneity of the polyribocytidylic acid tract in aphthovirus: biochemical and biological studies of viruses carrying polyribocytidylic acid tracts of different lengths. *J Virol* 1984;51:799–805.
86. Couillin P, Bouye A, Rebourcet R, van Cong N. Permissivity of mouse-man hybrid cell clones to three enteroviruses: Poliovirus 2, coxsackie B3 and echovirus. II. Role of human chromosome F. 19. *Pathol Biol* 1976;24:195–203.
87. Couillin P, Huyghe F, Grisard MC, et al. Echovirus 6, 11, 19; coxsackie B3 sensitivities and poliovirus I, II, III sensitivities are on chromosome 19 respectively on 19pter-q133 and 19q131-q133. *Cytogenet Cell Genet* 1987;46:579–586.
88. Crocker TT, Pfendt E, Spendlove R. Poliovirus growth in non-nucleate cytoplasm. *Science* 1964;145:401–403.
89. Crowell RL. Specific cell-surface alteration by enteroviruses as detected by viral-attachment interference. *J Bacteriol* 1966;91:198–204.
90. Crowell RL, Field AK, Schleif WA, et al. Monoclonal antibody that inhibits infection of HeLa and rhabdomyosarcoma cells by selected enteroviruses through receptor blockade. *J Virol* 1986;57:438–445.
91. Crowell RL, Landau BJ. Receptors in the initiation of picornavirus infections. In: Fraenkel-Conrat H, Wagner RR, eds. *Comprehensive virology*. Vol. 18. New York: Plenum; 1983:1–42.
92. Crowell RL, Syverton JT. Specific cell-surface alteration by enteroviruses as selected by viral-attachment interference. *J Exp Med* 1961;113:419–435.
93. Crowther D, Melnick JL. Studies of the inhibitory action of guanidine on poliovirus multiplication in cell culture. *Virology* 1961;15:65–74.
94. Dales S. An electron microscope study of the early association between two mammalian viruses and their hosts. *J Cell Biol* 1962;13:303–322.
95. Dales S, Eggers HJ, Tamm I, Palade GE. Electron microscopic study of the formation of poliovirus. *Virology* 1965;26:379–389.
96. Darnell JE. Adsorption and maturation of poliovirus in single and multiply infected HeLa cells. *J Exp Med* 1958;107:633–641.
97. Darnell JE, Levintow L. Poliovirus protein: source of amino acids and time course of synthesis. *J Biol Chem* 1960;235:74–77.
98. Darnell JE, Levintow L, Thoren MM, Hooper JL. The time course of synthesis of poliovirus RNA. *Virology* 1961;13:271–279.
99. Darnell JE, Sawyer TK. Variation in plaque-forming ability among parental and clonal strains of HeLa cells. *Virology* 1959;8:223–229.
100. Denoya CD, Scodeller EA, Vasquez C, La Torre JL. Foot-and-mouth disease virus. II. Endoribonuclease activity within purified virions. *Virology* 1978;89:67–74.
101. Dernick R, Heukeshoven J, Hilbrig M. Induction of neutralizing antibodies by all three structural poliovirus polypeptides. *Virology* 1983;130:243–245.
102. DeSena J, Mandel B. Studies on the *in vitro* uncoating of poliovirus. II. Characterization of the membrane-modified particle. *Virology* 1977;78:554–566.
103. Detjen BM, Lucas J, Wimmer E. Poliovirus single-stranded RNA and double-stranded RNA: differential infectivity in enucleate cells. *J Virol* 1978;27:582–594.
104. Devaney MA, Vakharia VN, Lloyd RE, Ehrenfeld E, Grubman MJ. Leader protein of foot-and-mouth disease virus is required for cleavage of the p220 component of the cap-binding protein complex. *J Virol* 1989;62:4407–4409.
105. Dick EC, Inhorn SL. Rhinoviruses. In: Feigin RD, Chesney JD, eds. *Textbook of pediatric infectious diseases*. 3rd ed. Philadelphia: WB Saunders; 1992:1507–1532.
106. Doel TR, Sangar DV, Rowlands DJ, Brown F. A reappraisal of the biochemical map of foot-and-mouth disease virus RNA. *J Gen Virol* 1978;41:395–404.
107. Domingo E, Holland JJ. High error rates, population equilibrium and evolution of RNA replication systems. In: Domingo E, Holland JJ, Ahlquist P, eds. *RNA genetics*. Vol. III. Variability of RNA genomes. Boca Raton, FL: CRC Press; 1988:3–36.
108. Dorsch-Haesler K, Yogo Y, Wimmer E. Evidence from in vitro RNA synthesis that poly(A) of the poliovirus genome is genetically coded. *J Virol* 1975;16:1512–1527.
109. Dubois M-F. Localization of the restrictive event of EMC virus replication in semi-permissive monkey and monkey-mouse hybrid cells. *J Gen Virol* 1977;34:115–125.
110. Duke GM, Hoffman MA, Palmenberg AC. Sequence and structural elements that contribute to efficient encephalomyocarditis viral RNA translation. *J Virol* 1992;66:1602–1609.
111. Duke GM, Osorio JE, Palmenberg AC. Attenuation of mengovirus through genetic engineering of the 5′ noncoding poly(C) tract. *Nature* 1990;343:474–476.
112. Duke GM, Palmenberg AC. Cloning and synthesis of infectious cardiovirus RNAs containing short, discrete poly(C) tracts. *J Virol* 1989;63:1822–1826.
113. Dulbecco R, Vogt M. Plaque formation and isolation of pure lines with poliomyelitis virus. *J Exp Med* 1954;99:167–182.
114. Dulbecco R, Vogt M, Strickland AGR. A study of the basic aspects of neutralization of two animal viruses, western equine encephalitis virus and poliomyelitis virus. *Virology* 1986;59:479–485.
115. Dunker AK, Rueckert RR. Observations on molecular weight determination on polyacrylamide gels. *J Biol Chem* 1969;224:5074–5080.
116. Dunker AK, Rueckert RR. Fragments generated by pH dissociation of ME virus and their relation to the structure of the virion. *J Mol Biol* 1971;58:217–235.
117. Eagle H. Nutritional needs of mammalian cells in tissue culture. *Science* 1955;122:501–505.
118. Eggers HJ. Selective inhibition of uncoating of echovirus 12 by rhodanine. A study on early virus-cell interactions. *Virology* 1977;78:241–252.

119. Ehrenfeld E. Picornavirus inhibition of host cell protein synthesis. In: Fraenkel-Conrat H, Wagner RR, eds. *Comprehensive Virology, Vol. 19*. New York: Plenum Press, 1984;177–221.
120. Elroy-Stein O, Fuerst TR, Moss B. Cap-independent translation of mRNA conferred by encephalomyocarditis virus 5′ sequence improves the performance of the vaccinia virus/bacteriophage T7 hybrid expression system. *Proc Natl Acad Sci USA* 1989;86:6126–6130.
121. Elroy-Stein O, Moss B. Cytoplasmic expression system based on constitutive synthesis of bacteriophage T7 RNA polymerase in mammalian cells. *Proc Natl Acad Sci USA* 1990;87:6743–6747.
122. Enders JF, Weller TH, Robbins FC. Cultivation of the Lansing strain of poliomyelitis virus in cultures of various human embryonic tissues. *Science* 1949;109:85–87.
123. Erickson JW, Frankenberger EA, Rossmann MG, Fout GS, Medappa KC, Rueckert RR. Crystallization of a common cold virus, human rhinovirus 14: isomorphism with poliovirus crystals. *Proc Natl Acad Sci USA* 1983;80:931–934.
124. Etchison D, Fout S. Human rhinovirus 14 infection of HeLa cells results in the proteolytic cleavage of the p220 cap-binding complex subunit and inactivates globin mRNA translation *in vitro*. *J Virol* 1985;54:634–638.
125. Etchison D, Hansen J, Ehrenfeld E, et al. Demonstration *in vitro* that eucaryotic initiation factor 3 is active but that a cap-binding protein complex is inactive in poliovirus-infected HeLa cells. *J Virol* 1984;51:832–837.
126. Etchison D, Milburn SC, Edery I, Sonenberg N, Hershey JWB. Inhibition of HeLa cell protein synthesis following poliovirus infection correlates with the proteolysis of a 220,000-dalton polypeptide associated with eucaryotic initiation factor 3 and a cap binding protein complex. *J Biol Chem* 1982;257:14806–14810.
127. Falk MM, Grigera PR, Bergmann IE, Zibert A, Multhaup G, Beck E. Foot-and-mouth disease virus protease 3C induces specific proteolytic cleavage of host cell histone H3. *J Virol* 1990;64:748–756.
128. Fenwick ML, Cooper PD. Early interactions between poliovirus and ERK cells. Some observations on the nature and significance of the rejected particles. *Virology* 1962;18:212–223.
129. Fernandez-Munoz R, Darnell J. Structural differences between the 5′-termini of viral and cellular mRNA in poliovirus infected cells: possible basis for the inhibition of host protein synthesis. *J Virol* 1976;18:719–726.
130. Fernandez-Tomas CB, Baltimore D. Morphogenesis of poliovirus. II. Demonstration of a new intermediate, the provirion. *J Virol* 1973;12:1122–1130.
131. Fiala M, Kenny GE. Effect of magnesium on replication of rhinovirus HGP. *J Virol* 1967;1:489–493.
132. Fitch W. On the problem of discovering the most parsimonious tree. *Am Naturalist* 1977;111:223–257.
133. Flanegan JB, Baltimore D. Poliovirus-specific primer-dependent RNA polymerase able to copy poly(A). *Proc Natl Acad Sci USA* 1977;74:2677–2680.
134. Flanegan JB, Pettersson RF, Ambrose V, Hewlett MJ, Baltimore D. Covalent linkage of a protein to a defined nucleotide sequence at the 5′ terminus of the virion and replicative intermediate RNAs of poliovirus. *Proc Natl Acad Sci USA* 1977;74:961–965.
135. Forss S, Schaller H. A tandem repeat gene in a picornavirus. *Nucleic Acids Res* 1982;10:6441–6450.
136. Forss S, Strebel K, Beck E, Schaller H. Nucleotide sequence and genome organization of foot-and-mouth disease virus. *Nucleic Acids Res* 1984;12:6587–6603.
137. Fout GS, Medappa KC, Mapoles JE, Rueckert RR. Radiochemical determination of polyamines in poliovirus and human rhinovirus 14. *J Biol Chem* 1984;259:3639–3643.
138. Fox G, Parry NR, Barnett PV, McGinn B, Rowlands DJ, Brown F. The cell attachment site on foot-and-mouth disease virus includes the amino acid sequence RGD (arginine-glycine-aspartic acid. *J Gen Virol* 1989;70:625–637.
139. Fox MP, Otto MJ, McKinlay MA.The prevention of rhinovirus and poliovirus uncoating by WIN 51711: A new antiviral drug. *Antimicrob Agents Chemother* 1986;30:110–116.
140. Franklin RM, Baltimore D. Patterns of macromolecular synthesis in normal and virus-infected mammalian cells. *Cold Spring Harbor Symp Quant Biol* 1962;27:175–194.
141. Fricks CE, Hogle JM. Cell-induced conformational change in poliovirus: externalization of the amino terminus of VP1 is responsible for liposome binding. *J Virol* 1990;64:1934–1945.
142. Gauntt C, Gilbert SF, Grieves J, Anderegg J, Rueckert RR. A neutralizing monoclonal antibody against poliovirus and its reaction with related antigens. *Virology* 1981;115:211–215.
143. Gauntt C, Griffith MM. Fragmentation of RNA in virus particles of rhinovirus type 14. *J Gen Virol* 1974;21:253–267.
144. Ghendon Y, Yakobson E, Mikhejva A. Study of some stages of poliovirus morphogenesis in MiO cells. *J Virol* 1972;10:261–266.
145. Giranda VL, Heinz BA, Oliveira MA, et al. Acid-induced structural changes in human rhinovirus 14: possible role in uncoating. *Proc Natl Acad Sci USA* 1992;89:10213–10217.
146. Golini F, Semler BL, Dorner AJ, Wimmer E. Protein-linked RNA of poliovirus is competent to form an initiation complex of translation *in vitro*. *Nature* 1980;287:600–603.
147. Gorbalenya AE, Svitkin YV, Kazachkov YA, Agol VI. Encephalomyocarditis virus-specific polypeptide p22 is involved in the processing of the viral precursor polypeptides. *FEBS Lett* 1979;108:1–5.
148. Grado C, Fisher S, Contreras G. The inhibition by actinomycin D of poliovirus multiplication in Hep 2 cells. *Virology* 1965;27:623–625.
149. Grant RA, Filman DJ, Fujinami RS, Icenogle JP, Hogle JM. Three-dimensional structure of Theiler virus. *Proc Natl Acad Sci USA* 1992;89:2061–2065.
150. Greve JM, Davis G, Meyer AM, et al. The major human rhinovirus receptor is ICAM-1. *Cell* 1989;56:839–847.
151. Greve JM, Forte CP, Marlor CW, et al. Mechanisms of receptor-mediated rhinovirus neutralization defined by two soluble forms of ICAM-1. *J Virol* 1991;65:6015–6023.
152. Grigera PR, Tisminetsky SG. Histone H3 modification in BHK cells infected with foot-and-mouth disease virus. *Virology* 1984;136:10–19.
153. Grubman M. *In vitro* morphogenesis of foot-and-mouth disease virus. *J Virol* 1984;49:760–765.
154. Guttman N, Baltimore D. Morphogenesis of poliovirus. IV. Existence of particles sedimenting at 150S and having the properties of provirion. *J Virol* 1977;23:363–367.
155. Guttman N, Baltimore D. A plasma membrane component able to bind and alter virions of poliovirus type 1: studies on cell-free alteration using a simplified assay. *Virology* 1977;82:25–36.
156. Gwaltney JM. Rhinoviruses. In: Evans AS, ed. *Viral infections of humans: epidemiology and control*. 3rd ed. New York: Plenum; 1989: 593–615.
157. Hall L, Rueckert R. Infection of mouse fibroblasts by cardioviruses. Premature uncoating and its prevention by elevated pH and magnesium chloride. *Virology* 1971;43:152–165.
158. Hamparian VV, Colonno RJ, Cooney MK, et al. A collaborative report: rhinovirusóextension of the numbering system from 89 to 100. *Virology* 1987;159:191–192.
159. Hanecak R, Semler BL, Anderson CW, Wimmer E. Proteolytic processing of poliovirus polypeptides: antibodies to a polypeptide P3-7c inhibit cleavage at glutamine-glycine pairs. *Proc Natl Acad Sci USA* 1982;79:3973–3977.
160. Harber JJ, Bradley J, Anderson CW, Wimmer E. Catalysis of poliovirus VP0 maturation cleavage is not mediated by serine 10 of VP2. *J Virol* 1991;65:326–334.
161. Harris TJR, Brown F. The location of the poly(C) tract in the RNA of foot-and-mouth disease virus. *J Gen Virol* 1976;33:493–501.
162. Harrison SC. Common features in the design of small RNA viruses. In: Notkins AL, Oldstone MBA, eds. *Concepts in viral pathogenesis*. New York: Springer; 1989:3–19.
163. Harrison SC, Olson A, Schutt CE, Winkler FK, Bricogne G. Tomato bushy stunt virus at 2.9 Angstroms resolution. *Nature* 1978; 276:368–373.
164. Harter DH, Choppin PW. Adsorption of attenuated and neurovirulent poliovirus strains to central nervous system tissues of primates. *J Immunol* 1965;95:730–736.
165. Hausen H. Cytologische Studien uber die Vermehrung des ME-Virus in L-Zellen. *Z Naturforsch* 1962;17b:158–160.
166. Haywood AM. Virus receptors: binding, adhesion strengthening, and changes in viral structure. *J Virol* 1994;68:1–5.
167. Heinz BA, Shepard DA, Rueckert RR. Escape mutant analysis of a drug-binding site can be used to map functions in the rhinovirus capsid. In: Laver G, Air G, eds. *Use of x-ray crystallography in design of antiviral agents*. New York: Academic; 1989:173–186.
168. Hellen CUT, Witherell GW, Schmid M, et al. pPTB and p57 are identical RNA binding proteins involved in nuclear splicing and cytoplas-

mic translation. *Proc Natl Acad Sci USA* 1992;90:7642–7646.
169. Hewlett MJ, Rose JK, Baltimore D. 5′ Terminal structure of poliovirus polyribosomal RNA is pUp. *Proc Natl Acad Sci USA* 1976; 73:327–330.
170. Hey TD, Richards OC, Ehrenfeld E. Host factor-induced template modification during synthesis of poliovirus RNA in vitro. *J Virol* 1987;61:802–811.
171. Hirst G. Genetic recombination with Newcastle disease virus, poliovirus and influenza. *Cold Spring Harbor Symp Quant Biol* 1962;27:303–309.
172. Hofer F, Gruenberger M, Howalski H, et al. Members of the low density lipoprotein receptor family mediate cell entry of a minor-group common cold virus. *Proc Natl Acad Sci USA* 1994;91:1839–1842.
173. Hogle JM, Chow M, Filman DJ. Three-dimensional structure of poliovirus at 2.9A resolution. *Science* 1985;229:1358–1365.
174. Hogle JM, Chow M, Frick CE, Minor PD, Filman DJ. The three-dimensional structure of poliovirus: its biological implications. In: Oxender DL, ed. *Protein structure, folding and design 2*. New York: Alan R Liss; 1987:505–519.
175. Holland J. Receptor affinities as major determinants of enterovirus tissue tropisms in humans. *Virology* 1961;15:312–326.
176. Holland J. Enterovirus entrance into specific host cells and subsequent alterations of cell protein and nucleic acid synthesis. *Bacteriol Rev* 1964;28:3–13.
177. Holland J, Spindler K, Horodyski F, Grabau E, Nichol S, Vande Pol S. Rapid evolution of RNA genomes. *Science* 1982;215:1577–1585.
178. Holland JJ. Irreversible eclipse of poliovirus by HeLa cells. *Virology* 1962;16:163–176.
179. Holland JJ, Hoyer BH. Early stages of enterovirus infection. *Cold Spring Harbor Symp Quant Biol* 1962;27:101–111.
180. Holland JJ, McLaren LC. The mammalian cell-virus relationship. II. Adsorption, reception and eclipse of poliovirus by HeLa cells. *J Exp Med* 1959;109:487–504.
181. Holoubeck V, Crocker TT. Histones and RNA metabolism in EAT cells infected with Maus-Elberfeld virus. *Virology* 1971;43:527–530.
182. Hoover-Litty H, Greve JM. Formation of rhinovirus-soluble ICAM-1 complexes and conformational changes in the virion. *J Virol* 1993;67:390–397.
183. Howes DW. The growth cycle of poliovirus in cultured cells. II. Maturation and release of virus in suspended cell populations. *Virology* 1959;9:96–109.
184. Hruby DE, Roberts WK. Variations in polyadenylic acid content and biological activity. *J Virol* 1957;19:325–330.
185. Hsiung GD, Melnick JL. Adsorption, multiplication and cytopathogenicity of enteroviruses in susceptible and resistant monkey kidney cells. *J Immunol* 1957;80:45–52.
186. Huber AA. VCAM-1 is a receptor for encephalomyocarditis virus on murine vascular endothelial cells. *J Virol* 1994;68:3453–3458.
187. Hummeler K, Anderson TF, Brown RA. Identification of poliovirus particles of different antigenicity by specific agglutination as seen in the electron microscope. *Virology* 1962;16:84–90.
188. Hummeler K, Hamparian VV. Studies on the complement fixing antigens of poliomyelitis. I. Demonstration of type and group specific antigens in native and heated viral preparations. *J Immunol* 1958; 81:499–505.
189. Hyypia T, Horsnell C, Maaronen M, et al. A distinct picornavirus group identified by sequence analysis. *Proc Natl Acad Sci USA* 1992;89:8847–8851.
190. Icenogle J, Hong S, Duke G, Gilbert S, Rueckert R, Anderegg J. Neutralization of poliovirus by a monoclonal antibody: kinetics and stoichiometry. *Virology* 1983;127:412–425.
191. Jacobson M, Baltimore D. Morphogenesis of poliovirus. I. Association of the viral RNA with the coat protein. *J Mol Biol* 1968;33:369–378.
192. Jacobson M, Baltimore D. Polypeptide cleavages in the formation of poliovirus proteins. *Proc Natl Acad Sci USA* 1968;61:77–84.
193. Jacobson MF, Asso J, Baltimore D. Further evidence on the formation of poliovirus proteins. *J Mol Biol* 1970;49:657–669.
194. Jarvis TC, Kirkegaard K. The polymerase in its labyrinth: mechanisms and implications of RNA recombination. *Trends Genet* 1991; 7:186–191.
195. Jen G, Thach RE. Inhibition of host translation in encephalomyocarditis virus-infected L cells: a novel mechanism. *J Virol* 1982;43:250–261.
196. Jenkins O, Booth JD, Minor PD, Almond JW. The complete nucleotide sequence of coxsackie B4 and its comparison to other members of the picornaviridae. *J Gen Virol* 1987;68:1835–1848.
197. Johnston MD, Martin SJ. Capsid and procapsid proteins of a bovine enterovirus. *J Gen Virol* 1971;11:71–79.
198. Joklik WK, Darnell JE. The adsorption and early fate of purified poliovirus in HeLa cells. *Virology* 1961;13:439–447.
199. Jore J, de Geus B, Jackson RJ, Pouwels PH, Engervalk BE. Poliovirus protein 3CD is the active protease for processing of the precursor protein P1 *in vitro*. *J Gen Virol* 1988;69:1627–1636.
200. Kapikian AZ. Rhinoviruses: a numbering system. *Nature* 1967; 213:761–762.
201. Kapikian AZ. Rhinoviruses. In: Lennette EH, Schmidt NJ, eds.*Diagnostic procedures for viral and rickettsial infections*. 4th ed. New York: American Public Health Association; 1969:603–640.
202. Kaplan AS. The susceptibility of monkey kidney cells to poliovirus *in vivo* and *in vitro*. *Virology* 1955;1:377–392.
203. Kaplan G, Freistadt MS, Racaniello VR. Neutralization of poliovirus by cell receptors expressed in insect cells. *J Virol* 1990;64:4697–4702.
204. Katagiri S, Hinuma Y, Ishida N. Biophysical properties of poliovirus particles irradiated with ultraviolet light. *Virology* 1967;32:337–343.
205. Kaufman RJ, Davies MV, Wasley LC, Michnick D. Improved vectors for stable expression of foreign genes in mammalian cells by use of the untranslated leader sequence from EMC virus. *Nucleic Acids Res* 1991;16:4485–4490.
206. Kaufmann Y, Goldstein E, Penman S. Poliovirus-induced inhibition of polypeptide initiation *in vitro* on native polyribosomes. *Proc Natl Acad Sci USA* 1976;73:1834–1838.
207. Kew O, Nottay BK. Evolution of the oral poliovaccine strains in humans occurs by both mutation and intramolecular recombination. In: Channock RM, Lerner RA, eds. *Modern approaches to vaccines*. Cold Spring Harbor, NY: Cold Spring Harbor Laboratory; 1984:357–362.
208. King AMQ. Genetic recombination in positive strand RNA viruses. In: Domingo E, Holland JJ, Ahlquist P, eds. *RNA genetics*. Vol. II. Retroviruses, viroids and RNA recombination. Boca Raton, FL: CRC Press; 1988:149–165.
209. King AMQ, McCahon D, Slade WR, Newman JWI. Recombination in RNA. *Cell* 1982;29:921–928.
210. King AMQ, Sangar DV, Harris TJR, Brown F. Heterogeneity of the genome-linked protein of foot-and-mouth disease virus. *J Virol* 1980;34:627–634.
211. Kirkegaard K, Baltimore D. The mechanism of RNA recombination in poliovirus. *Cell* 1986;47:433–443.
212. Kitamura N, Semler B, Rothberg PG, et al. Primary structure, gene organization and polypeptide expression of poliovirus RNA. *Nature* 1981;291:547–553.
213. Klug A, Caspar DLD. The structure of small viruses. *Adv Virus Res* 1960;7:225–325.
214. Koike S, Ise I, Nomoto A. Functional domains of the poliovirus receptor. *Proc Natl Acad Sci USA* 1991;88:4104–4108.
215. Korant BD, Lonberg-Holm K, Noble J, Stasny JT. Naturally occurring and artificially produced components of three rhinoviruses. *Virology* 1972;48:71–86.
216. Kronenberger P, Vrijsen R, Boeye A. Chloroquine induces empty capsid formation during poliovirus eclipse. *J Virol* 1991;65:7008–7011.
217. Kunin CM. Cellular susceptibility to enteroviruses. *Bacteriol Rev* 1964;28:382–390.
218. Landsteiner K, Popper E. Ubertragung der Poliomyelitis acuta auf Affen. *Z Immunitatsforsch Orig* 1909;2:377–390.
219. Larsen GR, Dorner AJ, Harris TJR, Wimmer E. The structure of poliovirus replicative form. *Nucleic Acids Res* 1980;8:1217–1229.
220. Lawrence C, Thach RE. Identification of a viral protein involved in post-translational maturation of the EMC virus capsid precursor. *J Virol* 1975;15:918–928.
221. Lazarus LH, Barzilai R. Association of foot-and-mouth disease virus replicase with RNA template and cytoplasmic membranes. *J Gen Virol* 1974;23:213–218.
222. Le Bouvier GL. The D to C change in poliovirus particles. *Br J Exp Pathol* 1959;40:605–620.
223. Le SY, Chen JH, Sonenberg N, Maizel JV. Conserved tertiary structural elements in the 5′ non-translated region of cardiovirus, aphthovirus and hepatitis A virus RNAs. *Nucleic Acids Res* 1993;21:2445–2451.
224. Le SY, Zuker M. Common structures of the 5′ non-coding RNA in enteroviruses and rhinoviruses: thermodynamical stability and statistical significance. *J Mol Biol* 1990;216:729–741.
225. Ledinko N. Genetic recombination with poliovirus type 1. Studies of crosses between a normal horse serum-resistant mutant and several

guanidine-resistant mutants of the same strain. *Virology* 1963;20:107–119.

226. Lee C-K, Wimmer E. Proteolytic processing of poliovirus polyproteins: elimination of 2A^{pro}-mediated, alternative cleavage of polypeptide 3CD by *in vitro* mutagenesis. *Virology* 1988;166:405–414.
227. Lee PWK, Colter JS. Further characterization of mengo subviral particles: a new hypothesis for picornavirus assembly. *Virology* 1979;97:266–274.
228. Lee W-M, Monroe SS, Rueckert RR. Role of maturation cleavage in infectivity of picornaviruses: activation of an infectosome. *J Virol* 1993;67:2110–2122.
229. Leong LEC, Walker PA, Porter AG. Human rhinovirus-14 protease 3C binds specifically to the 5′ noncoding region of the viral RNA. Evidence that 3C has different domains for the RNA binding and proteolytic activities. *J Biol Chem* 1993;268:25735–25739.
230. Levintow L. The reproduction of picornaviruses. In: Fraenkel-Conrat H, Wagner RR, eds. *Comprehensive virology*. Vol. 2. New York: Plenum; 1974:109–169.
231. Li J-P, Baltimore D. Isolation of poliovirus 2C mutants defective in viral RNA synthesis. *J Virol* 1988;62:4016–4021.
232. Li Q, Yafal AG, Lee YM, Hogle J, Chow M. Poliovirus neutralization by antibodies to internal epitopes of VP4 and VP1 results from reversible exposure of these sequences at physiological temperature. *J Virol* 1994;68:3964–3970.
233. Lipton HL, Rozhon EJ. Theiler's murine encephalomyelitis viruses. In: Bhatt P, ed. *Viral and mycoplasma infection of laboratory rodents*. New York: Academic; 1986:253–276.
234. Lloyd RE, Grubman MJ, Ehrenfeld E. Relationship of p220 cleavage during picornavirus infection to 2A proteinase sequencing. *J Virol* 1988;62:4216–4223.
235. Loeffler F, Frosch P. Report of the Commission for Research on Foot-and-Mouth Disease. In: Hahon N, ed. *Selected papers on virology*. Englewood Cliffs, NJ: Prentice-Hall; 1964:64–68.
236. Lonberg-Holm K. Attachment of animal virus to cells, an introduction. In: Lonberg-Holm K, Philipson L, eds. *Receptors and recognition*. Series B, Vol. 8: Virus receptors. Part 2, Animal viruses. London: Chapman and Hall; 1980:1–20.
237. Lonberg-Holm K, Butterworth BE. Investigation of the structure of polio- and human rhinovirus through the use of selective chemical reactivity. *Virology* 1976;71:207–216.
238. Lonberg-Holm K, Crowell RL, Philipson L. Unrelated animal viruses share receptors. *Nature* 1976;259:679–716.
239. Lonberg-Holm K, Korant BD. Early interaction of rhinoviruses with host cells. *J Virol* 1972;9:29–40.
240. Lonberg-Holm K, Philipson L. Early interaction between animal viruses and cells. *Monogr Virol* 1974;9:67–70.
241. Lonberg-Holm K, Whiteley NM. Physical and metabolic requirements for early interaction of poliovirus and human rhinovirus with HeLa cells. *J Virol* 1976;19:857–870.
242. Lucas-Lenard JM. Inhibition of cellular protein synthesis after virus infection. In: Perez-Bercoff R, ed. *The molecular biology of picornaviruses*. New York: Cambridge University Press; 1979:73–99.
243. Lund GA, Ziola BR, Salmi A, Scraba DG. Structure of the Mengo virion. V. Distribution of the capsid polypeptides with respect to the surface of the virus particle. *Virology* 1977;78:35–44.
244. Lundquist RE, Ehrenfeld E, Maizel JV. Isolation of a viral polypeptide associated with the poliovirus replication complex. *Proc Natl Acad Sci USA* 1974;71:4774–4777.
245. Madshus IH, Olsnes S, Sandvig K. Mechanism of entry into the cytosol of poliovirus type 1: requirement for low pH. *J Cell Biol* 1984;98:1194–1200.
246. Madshus IH, Olsnes S, Sandvig K. Requirements for entry of poliovirus RNA into cells at low pH. *EMBO J* 1984;3:1945–1950.
247. Maizel JV, Phillips BA, Summers DF. Composition of artificially produced and naturally occurring empty capsids of poliovirus type I. *Virology* 1967;32:692–699.
248. Mak TW, Colter JS, Scraba DG. Structure of the mengovirion. II. Physico-chemical and electron microscopic analysis of degraded virus. *Virology* 1974;57:543–553.
249. Mak TW, O'Callaghan DJ, Colter JS. Studies of the pH inactivation of three variants of mengo encephalomyelitis virus. *Virology* 1970;40:565–571.
250. Mandel B. Neutralization of poliovirus: a hypothesis to explain the mechanism and the one-hit character of the neutralization reaction. *Virology* 1976;69:500–510.
251. Mapoles JE, Anderegg JW, Rueckert RR. Properties of poliovirus propagated in medium containing cesium chloride. Implications for picornaviral structure. *Virology* 1978;90:103–111.
252. Marongiu ME, Pani A, Corrias MV, Sau M, La Colla P. Poliovirus morphogenesis. I. Identification of 80S dissociable particles and evidence for the artifactual production of procapsids. *J Virol* 1981; 39:341–347.
253. Mason PW, Rieder E, Baxt B. RGD sequence of foot-and-mouth disease virus is essential for infecting cells via the natural receptor but can be bypassed by an antibody-dependent enhancement pathway. *Proc Natl Acad Sci USA* 1994;91:
254. Matthews DA, Smith WW, Ferre RA, et al. Structure of human rhinovirus 3C protease reveals a trypsin-like polypeptide fold, RNA-binding site, and means for cleaving precursor polyprotein. *Cell* 1994;77:761–771.
255. Mayer MM, Rapp HJ, Roizman B, et al. The purification of poliomyelitis virus as studied by complement fixation. *J Immunol* 1957;78:435–455.
256. Mbida AD, Gaudin OG, Sabido O, Pozzetto B, Le Bihan JC. Monoclonal antibody specific for the cellular receptor of echoviruses. *Intervirology* 1992;33:17–22.
257. Mbida AD, Pozzetto B, Sabido O, et al. Competition binding studies with biotinylated echovirus 11 in cytofluorimentry analysis. *J Virol Methods* 1991;34:169–176.
258. McCahon D. The genetics of aphthovirus. *Arch Virol* 1981;69:1–23.
259. McCahon D, Crowther JR, Belsham GJ, et al. Evidence for at least four antigenic sites on type O foot-and-mouth disease virus involved in neutralization: identification by single and multiple site monoclonal antibody-resistant mutants. *J Gen Virol* 1989;70:625–637.
260. McClintock PR, Billups LC, Notkins AL. Receptors for encephalomyocarditis virus on murine and human cells. *Virology* 1980; 106:261–272.
261. McCormick W, Penman S. Inhibition of RNA synthesis in HeLa and L cells by mengovirus. *Virology* 1967;31:135–141.
262. McGeady ML, Crowell RL. Stabilization of "A" particles of coxsackievirus B3 by a HeLa cell plasma membrane extract. *J Virol* 1979; 32:790–795.
263. McGregor S, Hall L, Rueckert RR. Evidence for the existence of protomers in the assembly of encephalomyocarditis virus. *J Virol* 1975;15:1107–1120.
264. McGregor S, Rueckert RR. Picornaviral capsid assembly: similarity of rhinovirus and enterovirus precursor subunits. *J Virol* 1977;21:548–553.
265. McLaren LC, Holland JJ, Syverton JT. The mammalian cell-virus relationship. V. Susceptibility and resistance of cells *in vitro* to infection by coxsackie A9 virus. *J Exp Med* 1960;112:581–594.
266. McLean C, Matthews TJ, Rueckert RR. Evidence of ambiguous processing and selective degradation in the noncapsid proteins of rhinovirus 1A. *J Virol* 1976;19:903–914.
267. McSharry JJ, Caliguiri LA, Eggers HJ. Inhibition of uncoating of poliovirus by arildone, a new antiviral drug. *Virology* 1979;97:307–315.
268. Medappa KC, McLean C, Rueckert RR. On the structure of rhinovirus 1A. *Virology* 1971;44:259–270.
269. Medrano L, Green H. Picornavirus receptors and picornavirus multiplication in human-mouse hybrid cell lines. *Virology* 1973;54:515–524.
270. Meerovitch K, Sonenberg N. Internal initiation of picornavirus RNA translation. *Sem Virol* 1993;4:217–227.
271. Meerovitch K, Svitkin YV, Lee HS, et al. La autoantigen enhances and corrects aberrant translation of poliovirus RNA in reticulocyte lysate. *J Virol* 1993;67:3798–3807.
272. Melnick JL. Picornaviridae. *Intervirology* 1974;4:303–316.
273. Melnick JL. Enteroviruses. In: Evans AS, ed. *Viral infections of humans*. New York: Plenum Medical; 1976:163–208.
274. Melnick JL. Portraits of viruses: the picornaviruses. *Intervirology* 1983;20:61–100.
275. Melnick JL, Wenner HA. The enteroviruses. In: Lennette EH, Schmidt NJ, eds. *Diagnostic procedures for viral and rickettsial disease*. 4th ed. New York: American Public Health Association; 1969:529–602.
276. Mendelsohn CL, Wimmer E, Racaniello VR. Cellular receptor for poliovirus: Molecular cloning, nucleotide sequence and expression of a new member of the immunoglobulin superfamily. *Cell* 1989; 56:855–865.
277. Miller DA, Miller OJ, Dev VG, et al. Human chromosome 19 carries

a poliovirus receptor gene. *Cell* 1974;1:167–173.
278. Miller RL, Plagemann PGW. Effect of ultraviolet light on mengovirus: formation of uracil dimers, instability and degradation of capsid, and covalent linkage of protein to viral RNA. *Virology* 1974;13:729–739.
279. Minor PD. The molecular biology of poliovaccines. *J Gen Virol* 1992;73:3065–3077.
280. Minor PD, Ferguson M, Evans DMA, Almond JW, Icenogle JP. Antigenic structure of polioviruses of serotypes 1, 2 and 3. *J Gen Virol* 1986;67:1283–1291.
281. Minor PD, John A, Ferguson M, Icenogle JP. Antigenic and molecular evolution of the vaccine strain of type 3 poliovirus during the period of excretion by a primary vaccinee. *J Gen Virol* 1986;67:693–706.
282. Molla A, Paul AV, Wimmer E. Cell-free de novo synthesis of poliovirus. *Science* 1991;254:1647–1651.
283. Molla A, Paul AV, Wimmer E. Effects of temperature and lipophilic agents on poliovirus formation and RNA synthesis in a cell-free system. *J Virol* 1993;67:5932–5938.
284. Montagnier L, Sanders FK. Replicative form of encephalomyocarditis virus ribonucleic acid. *Nature* 1963;199:664–667.
285. Morrow CD, Gibbons GF, Dasgupta A. The host protein required for in vitro replication of poliovirus is a protein kinase that phosphorylates eukaryotic initiation factor-2. *Cell* 1985;40:913–921.
286. Moscufo N, Chow M. Myristate-protein interactions in poliovirus: interactions of VP4 threonine 28 contribute to the structural conformation of assembly intermediates and the stability of assembled virions. *J Virol* 1992;66:6849–6857.
287. Moscufo N, Yafal AG, Rogove A, Hogle J, Chow M. A mutation in VP4 defines a new step in the late stages of cell entry by poliovirus. *J Virol* 1993;67:5075–5078.
288. Mosser AG, Leippe DM, Rueckert RR. Neutralization of picornaviruses: support for the pentamer bridging hypothesis. In: Semler B, Ehrenfeld E, eds. *Molecular aspects of picornavirus infection and detection.* Washington, DC: ASM Publications; 1988:155–167.
289. Mosser AG, Rueckert RR. WIN 51711-dependent mutants of poliovirus type 3: evidence that virions decay after release from cells unless drug is present. *J Virol* 1993;67:1246–1254.
290. Much DH, Zajac I. Homology of surface receptors for poliovirus on mammalian cell lines. *J Gen Virol* 1973;21:385–390.
291. Newman JF, Piatti PG, Gorman BM, et al. Foot-and-mouth disease virus particles contain replicase protein 3D. *Proc Natl Acad Sci USA* 1994;91:733–737.
292. Newman JFE, Rowlands DJ, Brown F. A physicochemical subgrouping of the mammalian picornaviruses. *J Gen Virol* 1973;18:171–180.
293. Nicholson R, Pelletier J, Le SY, Sonenberg N. Structural and functional analysis of the ribosome landing pad of poliovirus type 2: in vivo translation studies. *J Virol* 1991;65:5886–5894.
294. Nicklin MH, Harris KS, Pallai PV, Wimmer E. Poliovirus proteinase 3C: large-scale expression, purification and specific cleavage activity on natural and synthetic substrates *in vitro*. *J Virol* 1988;62:4586–4593.
295. Ninomiya Y, Ohsawa C, Aoyama M, Umeda I, Suhara Y, Ishipsuka H. Antivirus agent, Ro 09-0410, binds to rhinovirus specifically and stabilizes the virus conformation. *Virology* 1984;34:269–276.
296. Nomoto A, Lee YF, Wimmer E. The 5′-end of poliovirus mRNA is not capped with $m^7G(5')pppG(5')Np$. *Proc Natl Acad Sci USA* 1976;73:375–380.
297. Oberg BF, Philipson L. Replicative structures of poliovirus RNA *in vivo*. *J Mol Biol* 1971;58:725–737.
298. Oberst MD, Gollan TJ, Gupta M, et al. The encephalomyocarditis virus 3C protease is rapidly degraded by an ATP-dependent proteolytic system in reticulocyte lysate. *Virology* 1993;193:28–40.
299. Oliveira MA, Zhao R, Lee W-M, et al. The structure of human rhinovirus 16. *Structure* 1993;1:51–68.
300. Olson NH, Kolatkar PR, Oliveira MA, et al. Structure of a human rhinovirus complexed with its receptor molecule. *Proc Natl Acad Sci USA* 1993;90:507–511.
301. Otto MJ, Lucas-Lenard J. The influence of the host cell on the inhibition of viral protein synthesis in cells doubly-infected with vesicular stomatitis virus and mengovirus. *J Gen Virol* 1980;50:293–307.
302. Page GS, Mosser AG, Hogle JM, Filman DJ, Rueckert RR, Chow M. Three-dimensional structure of poliovirus serotype 1 neutralizing determinants. *J Virol* 1988;62:1781–1794.
303. Pallansch MA, Kew OM, Palmenberg AC, Golini F, Wimmer E, Rueckert RR. Picornaviral VPg sequences are contained in the replicase precursor. *J Virol* 1980;35:414–419.
304. Pallansch MA, Kew OM, Semler BL, et al. Protein processing map of poliovirus. *J Virol* 1984;49:873–880.
305. Palmenberg AC. *In vitro* synthesis and assembly of picornaviral capsid intermediate structures. *J Virol* 1982;44:900–906.
306. Palmenberg AC, Kirby EM, Janda MR, et al. The nucleotide and deduced sequences of the encephalomyocarditis viral polyprotein coding region. *Nucleic Acids Res* 1984;12:2969–2985.
307. Palmenberg AC, Pallansch MA, Rueckert RR. Protease required for processing picornaviral coat protein resides in the viral replicase gene. *J Virol* 1979;32:770–778.
308. Palmenberg AC, Parks GD, Hall DJ, Ingraham RH, Seng TW, Pallai PV. Proteolytic processing of the cardioviral P2 region: primary 2A/2B cleavage in clone-derived precursors. *Virology* 1992;190:754–762.
309. Palmenberg AC, Rueckert RR. Evidence for intramolecular self-cleavage of picornaviral replicase precursors. *J Virol* 1982;41:244–249.
310. Parks GD, Palmenberg AC. Site-specific mutations at a picornavirus VP3/VP1 cleavage site disrupt *in vitro* processing and assembly of capsid precursors. *J Virol* 1987;61:3680–3687.
311. Paul J. *A history of poliomyelitis virus*. New Haven, CT: Yale University Press; 1971.
312. Pelham HR. Translation of encephalomyocarditis virus RNA *in vitro* yields an active proteolytic enzyme. *Eur J Biochem* 1978;85:457–462.
313. Pelham HR, Jackson RJ. An efficient mRNA-dependent translation system from reticulocyte lysates. *Eur J Biochem* 1976;67:247–256.
314. Pelletier J, Sonenberg N. Internal initiation of translation of eukaryotic mRNA directed by a sequence derived from poliovirus RNA. *Nature* 1988;334:320–325.
315. Penman S, Becker Y, Darnell JE. A cytoplasmic structure involved in the synthesis and assembly of poliovirus components. *J Mol Biol* 1964;8:541–555.
316. Penman S, Summers D. Effects on host cell metabolism following synchronous infection with poliovirus. *Virology* 1965;27:614–620.
317. Perez-Bercoff R, Gander M. The genomic RNA of mengovirus. I. Location of the poly(C) tract. *Virology* 1977;80:426–429.
318. Perlin M, Phillips BA. *In vitro* assembly of polioviruses. III. Assembly of 14S particles into empty capsids by poliovirus-infected HeLa cell membranes. *Virology* 1973;53:107–114.
319. Pevear DC, Calenoff M, Rozhon E, Lipton HL. Analysis of the complete nucleotide sequence of the picornavirus Theiler's murine encephalomyelitis virus indicates that it is closely related to cardioviruses. *J Virol* 1987;61:1507–1516.
320. Pevear DC, Fancher MJ, Felock PJ, et al. Conformational change in the floor of the human rhinovirus canyon blocks adsorption to HeLa cell receptors. *J Virol* 1989;63:2002–2007.
321. Pfaff E, Thiel H-J, Beck E, Strohmaier K, Schaller H. Analysis of neutralizing epitopes on foot-and-mouth disease virus. *J Virol* 1988;62:2033–2040.
322. Philipson L. The early interaction of animal viruses and cells. *Prog Med Virol* 1963;5:43–78.
323. Philipson L, Bengtsson S, Brishammer S, Svennerholm L, Zeterquist O. Purification and chemical analysis of the erythrocyte receptor for hemagglutinating enteroviruses. *Virology* 1964;22:580–590.
324. Philipson L, Choppin PW. Inactivation of enteroviruses by 2,3-dimercaptopropanol (BAL). *Virology* 1962;16:404–413.
325. Phillips BA. *In vitro* assembly of polioviruses. I. Kinetics of the assembly of empty capsids and the role of extracts from infected cells. *Virology* 1969;39:811–821.
326. Phillips BA. *In vitro* assembly of poliovirus. II. Evidence for the self-assembly of 14S particles into empty capsids. *Virology* 1969;44: 307–316.
327. Phillips BA, Lundquist RE, Maizel JV. Absence of subviral particles and assembly activity in HeLa cells infected with defective-interfering (DI) particles of poliovirus. *Virology* 1980;100:116–124.
328. Phillips BA, Summers DF, Maizel JV Jr. *In vitro* assembly of poliovirus-related particles. *Virology* 1968;35:216–226.
329. Pincus S, Diamond D, Emini E, Wimmer E. Guanidine-selected mutants of poliovirus: mapping of point mutations to polypeptide 2C. *J Virol* 1986;57:638–646.
330. Pincus S, Rohl H, Wimmer E. Guanidine-dependent mutants of poliovirus: identification of three classes with different growth requirements. *Virology* 1987;157:83–88.
331. Pincus S, Wimmer E. Production of guanidine-resistant and -dependent poliovirus mutants from cloned cDNA: mutations in polypeptide

2C are directly responsible for altered guanidine sensitivity. *J Virol* 1986;60:793–796.
332. Pons M. Infectious double-stranded poliovirus RNA. *Virology* 1964;24:467–473.
333. Porter AG. Picornavirus nonstructural proteins: emerging roles in virus replication and inhibition of host cell functions. *J Virol* 1993;67:6917–6921.
334. Prather SO, Taylor MW. Host-dependent restriction of mengovirus replication. III. Effect of host restriction on late viral RNA synthesis and viral maturation. *J Virol* 1975;15:872–881.
335. Pringle CR. Evidence of genetic recombination in foot-and-mouth disease virus. *Virology* 1965;25:48–54.
336. Pritchard AE, Strom T, Lipton HL. Nucleotide sequence identified Vilyuisk virus as a divergent Theiler's virus. *Virology* 1992;191:469–472.
337. Provost PJ, Hilleman MR. Propagation of human hepatitis A virus in cell culture *in vitro*. *Proc Soc Exp Biol Med* 1979;160:213–221.
338. Putnak JR, Phillips BA. Picornaviral structure and assembly. *Microbiol Rev* 1981;45:287–315.
339. Quersin-Thiry L. Interactions between cellular extracts and animal viruses. I. Kinetic studies and some notes on the specificity of the interaction. *Acta Virol (Praha)* 1961;5:141–152.
340. Quersin-Thiry L, Nihoul E. Interaction between cellular extracts and animal viruses. II. Evidence for the presence of different inactivators corresponding to different viruses. *Acta Virol (Praha)* 1961;5:283–293.
341. Racaniello VR, Baltimore D. Molecular cloning of poliovirus cDNA and determination of the complete nucleotide sequence of the viral genome. *Proc Natl Acad Sci USA* 1981;78:4887–4891.
342. Reich E, Franklin RM, Shatkin AJ, Tatum EL. Effect of actinomycin D on cellular nucleic acid synthesis and virus production. *Science* 1961;134:556–557.
343. Rice JM, Wolff DA. Phospholipase in the lysosomes of HEp-2 cells and its release during poliovirus infection. *Biochim Biophys Acta* 1975;381:17–21.
344. Richards OC, Morton K, Martin SC, Ehrenfeld E. Two forms of poliovirus VPg result from amino acid modification of a single viral protein. *Virology* 1984;136:453–456.
345. Rico-Hesse R, Pallansch MA, Nottay BK, Kew OM. Geographic distribution of wild poliovirus type 1 genotypes. *Virology* 1987;160:311–322.
346. Rightsel WA, Dice J, McAlpine R, et al. Antiviral effect of guanidine. *Science* 1961;134:558–575.
347. Robertson BH, Grubman MJ, Wenddell GN, et al. Nucleotide and amino acid sequence coding for polypeptides of foot-and-mouth disease virus type A12. *J Virol* 1985;54:651–660.
348. Roivainen M, Piirainen L, Rysa T, Narvanen A, Hovi T. An immunodominant N-terminal region of VP1 protein of poliovirion that is buried in crystal structure can be exposed in solution. *Virology* 1993;195:762–765.
349. Roivaninen M, Hyypia T, Piirainen L, Kalkkinen N, Stanway G, Hovi T. RGD-dependent entry of coxsackievirus A9 into host cells, and its bypass after cleavage of VP1 protein by intestinal proteases. *J Virol* 1991;65:4735–4740.
350. Roizman B, Hopken W, Mayer MM. Immunological studies of poliovirus. II. Kinetics of the formation of infectious and noninfectious type I poliovirus in three cell strains of human derivation. *J Immunol* 1957;80:386–395.
351. Roizman B, Mayer MM, Roane PR. Immunochemical studies of poliovirus. IV. Alteration of the immunologic specificity of purified poliomyelitis virus by heat and ultraviolet light. *J Immunol* 1959;82:19–25.
352. Romanova LI, Blinov VM, Tolskaya EA, et al. The primary structure of crossover regions of intertypic poliovirus recombinants: a model of recombination between RNA genomes. *Virology* 1986;155:202–213.
353. Romanova LI, Tolskaya EA, Kolesnikova MS, Agol VI. Biochemical evidence for intertypic recombination of polioviruses. *FEBS Lett* 1980;118:109–112.
354. Rombaut B, Foriers A, Boeye A. In vitro assembly of poliovirus 14S subunits: identification of the assembly promoting activity of infected cell extracts. *Virology* 1991;180:781–787.
355. Rombaut B, Vrijesen R, Boeye A. *In vitro* assembly of poliovirus empty capsids: antigenic consequences and immunological assay of the morphopoietic factor. *Virology* 1984;135:546–550.
356. Rombaut B, Vrijsen R, Brioen P, Boeye A. A pH-dependent antigenic conversion of empty capsids of poliovirus studied with the aid of monoclonal antibodies to N and H antigen. *Virology* 1982;122:215–218.
357. Rossmann MG. *The molecular replacement method*. New York: Gordon Breach Publishers; 1972.
358. Rossmann MG, Arnold E, Erickson JW, et al. The structure of a human common cold virus (Rhinovirus 14) and its functional relations to other picornaviruses. *Nature* 1985;317:145–153.
359. Rossmann MG, Blow DM. The detection of subunits within the crystallographic asymmetric unit. *Acta Crystallogr* 1962;15:24–31.
360. Rossmann MG, Johnson JE. Icosohedral RNA virus structure. *Ann Rev Biochem* 1989;58:533–573.
361. Roumiantzeff M, Summers DF, Maizel JV. *In vitro* protein synthetic activity of membrane-bound poliovirus polyribosomes. *Virology* 1971;44:249–258.
362. Rowlands DJ, Clarke BE, Carroll AR, et al. Chemical basis of antigenic variation in foot and mouth disease virus. *Nature* 1983;306:694–697.
363. Rowlands DJ, Harris TJR, Brown F. More precise location of the polycytidylic acid tract in foot and mouth disease virus RNA. *J Virol* 1978;26:335–343.
364. Rowlands DJ, Shirley MW, Sangar DV, Brown F. A high density component in several vertebrate enteroviruses. *J Gen Virol* 1975;29:223–234.
365. Rubenstein SJ, Hammerle T, Wimmer E, Dasgupta A. Infection of HeLa cells with poliovirus results in modification of a complex that binds to the rRNA promoter. *J Virol* 1992;66:3062–3068.
366. Rueckert R, Sherry B, Mosser A, Colonno R, Rossmann M. Location of four neutralization antigens on the three-dimensional surface of a common-cold picornavirus, human rhinovirus 14. In: Crowell RL, Lonberg-Holm K, eds. *Virus attachment and entry into cells*. Washington, DC: American Society for Microbiology; 1986:21–27.
367. Rueckert RR. Picornaviral architecture. In: Maramorosch K, Kurstak E, eds. *Comparative virology*. New York: Academic; 1971:255–306.
368. Rueckert RR. On the structure and morphogenesis of picornaviruses. In: Fraenkel-Conrat H, Wagner RR, eds. *Comprehensive virology*. Vol. 6. New York: Plenum; 1976:131–213.
369. Rueckert RR. Picornaviruses and their replication. In: Fields BN, Knipe DM, Chanock RM, et al., eds. *Virology*. 2nd ed. New York: Raven; 1990:507–548.
370. Rueckert RR, Dunker AK, Stoltzfus CM. The structure of Maus-Elberfeld virus: a model. *Proc Natl Acad Sci USA* 1969;62:912–919.
371. Rueckert RR, Matthews TJ, Kew OM, Pallansch M, McLean C, Omilianowski D. Synthesis and processing of picornaviral polyprotein. In: Perez-Bercoff R, ed. *The molecular biology of picornaviruses*. New York: Plenum; 1979:113–125.
372. Rueckert RR, Pallansch MA. Preparation and characterization of encephalomyocarditis virus. In: Pestka S, ed. *Methods in enzymology*. Vol. 78. New York: Academic; 1981:315–325.
373. Rueckert RR, Palmenberg AC, Pallansch MA. Evidence for a self-cleaving precursor of virus-coded protease, RNA-replicase and VPg. In: Koch G, Richter D, eds. *Biosynthesis, modification and processing of cellular and viral polyproteins*. New York: Academic; 1980:263–275.
374. Ryan MD, King AMQ, Thomas GP. Cleavage of foot-and-mouth disease virus polyprotein is mediated by residues located within a 19 amino acid sequence. *J Gen Virol* 1991;72:2727–2732.
375. Sachs A, Wahle E. Poly(A) tail metabolism and function of eucaryotes. *J Biol Chem* 1993;268:22955–22958.
376. Sangar DV. The replication of picornaviruses. *J Gen Virol* 1979;45:1–11.
377. Sarnow P. Role of the 3′ end sequences in infectivity of poliovirus transcripts made *in vitro*. *J Virol* 1989;63:467–470.
378. Sarnow P, Bernstein HD, Baltimore D. A poliovirus temperature-sensitive RNA synthesis mutant located in a noncoding region of the genome. *Proc Natl Acad Sci USA* 1986;83:571–575.
379. Saunders K, King AMQ. Guanidine-resistant mutants of aphthovirus induce the synthesis of an altered nonstructural polypeptide, P34. *J Virol* 1982;42:389–394.
380. Schaffer FL, Gordon M. Differential inhibitory effects of actinomycin D among strains of poliovirus. *J Bacteriol* 1966;91:2309–2316.
381. Schaffer FL, Schwerdt CE. Crystallization of purified MEF-1 poliomyelitis virus particles. *Proc Natl Acad Sci USA* 1955;41:1020–1023.
382. Scharff MD, Levintow L. Quantitative study of the formation of poliovirus antigens in infected HeLa cells. *Virology* 1963;19:491–500.
383. Scharff MD, Maizel JV, Levintow L. Physical and immunological properties of a soluble precursor of the poliovirus capsid. *Proc Natl Acad Sci USA* 1964;51:329–337.
384. Schultheiss T, Kusov YY, Gauss-Muller V. Proteinase 3C of hepatitis

A virus cleaves the HAV polyprotein P2-P3 at all sites including VP1/2A and 2A/2B. *Virology* 1994;198:275–281.

385. Schultz M, Crowell RL. Acquisition of susceptibility to coxsackievirus A2 by the rat L8 line during myogenic differentiation. *J Gen Virol* 1980;46:39–49.
386. Schwerdt CE, Fogh J. The ratio of physical particles per infectious unit observed for poliomyelitis virus. *Virology* 1957;4:590–593.
387. Sehgal OP, Jean J-H, Bhalla RB, Soong MM, Drause GF. Correlation between buoyant density and ribonucleic acid content in viruses. *Phytopathology* 1970;60:1778–1784.
388. Selinka H-C, Zibert A, Wimmer E. Poliovirus can enter and infect mammalian cells by way of an intercellular adhesion molecule 1 pathway. *Proc Natl Acad Sci USA* 1991;88:3598–3602.
389. Semler BL, Anderson CW, Kitamura N, Rothberg PG, Wishart WL, Wimmer E. Poliovirus replication proteins: RNA sequence encoding P3-1b and the sites of proteolytic processing. *Proc Natl Acad Sci USA* 1981;78:3464–3468.
390. Shepard DA, Heinz BA, Rueckert RR. WIN 52035-2 inhibits both attachment and eclipse of human rhinovirus 14. *J Virol* 1993;67:2245–2254.
391. Shepley MP, Sherry B, Weiner HL. Monoclonal antibody identification of a 100-kDa membrane protein in HeLa cells and human spinal cord involved in poliovirus attachment. *Proc Natl Acad Sci USA* 1988;85:7743–7747.
392. Sherry B, Mosser AG, Colonno RJ, Rueckert RR. Use of monoclonal antibodies to identify four neutralization immunogens on a common cold picornavirus, human rhinovirus 14. *J Virol* 1986;57:246–257.
393. Shi J-P, Fersht AR. Fidelity of DNA replication under conditions used for oligodeoxynucleotide-directed mutagenesis. *J Mol Biol* 1984;177:269–278.
394. Shih DS, Shih CT, Kew O, Pallansch M, Rueckert R, Kaesberg P. Cell free synthesis and processing of the proteins of poliovirus. *Proc Natl Acad Sci USA* 1978;75:5807–5811.
395. Siddique T, McKinney R, Hung W, et al. The poliovirus sensitivity gene is on chromosome 19q12-q13.2. *Genomics* 1988;3:156–160.
396. Smith AE. The initiation of protein synthesis directed by the RNA from encephalomyocarditis virus. *Eur J Biochem* 1973;33:301–313.
397. Smith TJ, Kremer MJ, Luo M, et al. The site of attachment in human rhinovirus 14 for antiviral agents that inhibit uncoating. *Science* 1986;233:1286–1293.
398. Smith TJ, Olson NH, Cheng RH, Chase ES, Baker TS. Structure of a human rhinovirus-bivalently bound antibody complex: implications for viral neutralization and antibody flexibility. *Proc Natl Acad Sci USA* 1993;90:7015–7018.
399. Smith TJ, Olson NH, Cheng RH, et al. Structure of human rhinovirus complexed with Fab fragments from a neutralizing antibody. *J Virol* 1993;67:1148–1158.
400. Sommergruber W, Ahorn H, Klump H, et al. 2A Proteinases of coxsackie- and rhinovirus cleave peptides derived from eIF-4gamma via a common recognition motif. *Virology* 1994;198:741–745.
401. Spector DH, Baltimore D. Requirement of 3′-terminal polyadenylic acid for the infectivity of poliovirus RN. *Proc Natl Acad Sci USA* 1974;71:2983–2987.
402. Stanway G, Hughes P, Mountford R, Minor P, Almond J. The complete nucleotide sequence of a common cold virus: human rhinovirus 14. *Nucleic Acids Res* 1984;12:7859–7875.
403. Staunton DE, Dustin ML, Springer TA. Functional cloning of ICAM-2, a cell adhesion ligand for LFA-1 homologous to ICAM-1. *Nature* 1989;339:61–64.
404. Staunton DE, Gaur A, Chan P-Y, Springer TA. Internalization of a major group human rhinovirus does not require cytoplasmic or transmembrane domains of ICAM-1. *J Immunol* 1992;148:3271–3274.
405. Staunton DE, Merluzzi VJ, Rothlein R, Barton R, Marlin SD, Springer TA. A cell adhesion molecule, ICAM-1, is the major surface receptor for rhinoviruses. *Cell* 1989;56:849–853.
406. Stave JW, Card JL, Morgan DO, Vikharia V. Neutralization sites of type 10 foot-and-mouth disease virus defined by monoclonal antibodies and neutralization-escape virus variants. *Virology* 1988;162:21–29.
407. Steere RL, Schaffer FL. The structure of crystals of purified Mahoney poliovirus. *Biochim Biophys Acta* 1958;28:241–246.
408. Strohmaier K, Franze R, Adam K-H. Location and characterization of the antigenic portion of the FMDV immunizing protein. *J Gen Virol* 1982;59:295–306.
409. Summers DF, Maizel JV. Determination of the gene sequence of poliovirus with pactamycin. *Proc Natl Acad Sci USA* 1971;68:2852–2856.
410. Summers DF, Maizel JV, Darnell JE. Evidence for virus-specific noncapsid proteins in poliovirus-infected cells. *Proc Natl Acad Sci USA* 1965;54:505–513.
411. Svitkin YV, Gorbalenya AE, Kazachkov YA, Agol VI. Encephalomyocarditis virus-specific polypeptide p22 possessing a proteolytic activity: preliminary mapping of the viral genome. *FEBS Lett* 1979;108:6–9.
412. Svitkin YV, Pestova TV, Maslova SV, Agol VI. Point mutations modify the response of poliovirus RNA to a translation initiation factor: a comparison of neurovirulent and attenuated strains. *Virology* 1988;166:394–404.
413. Taber R, Rekosh D, Baltimore D. Effect of pactamycin on synthesis of poliovirus proteins: a method for genetic mapping. *J Virol* 1971;8:395–410.
414. Talbot P, Brown F. A model for foot-and-mouth disease virus. *J Gen Virol* 1972;15:163–170.
415. Tamm I, Eggers HJ. Specific inhibition of replication of animal viruses. *Science* 1963;142:24–33.
416. Tershak DR. Inhibition of poliovirus polymerase by guanidine in vitro. *J Virol* 1982;41:313–318.
417. Tesar M, Jia XY, Summers DR, Ehrenfeld E. Analysis of a potential myristoylation site in hepatitis A virus capsid protein VP4. *Virology* 1993;194:616–626.
418. Tesar M, Marquardt O. Foot-and-mouth disease virus protease 3C inhibits cellular transcription and mediates cleavage of histone H3. *Virology* 1990;174:364–374.
419. Thomas AA, Vrijsen MR, Boeye A. Relationship between poliovirus neutralization and aggregation. *J Virol* 1986;59:479–485.
420. Tobey RA, Petersen DF, Anderson EC. Mengovirus replication. IV. Inhibition of Chinese hamster ovary cell division as a result of infection. *Virology* 1965;27:17–22.
421. Tolskaya EA, Romanova LI, Kolesnikova MS, Agol VI. Intertypic recombination in poliovirus: genetic and biochemical studies. *Virology* 1983;124:121–132.
422. Tomassini JE, Colonno RJ. Isolation of a receptor protein involved in attachment of human rhinoviruses. *J Virol* 1986;58:290–295.
423. Toyoda H, Kohara M, Kataoka Y, et al. Complete nucleotide sequences of all three poliovirus serotype genomes. Implication for genetic relationship, gene function and antigenic determinants. *J Mol Biol* 1984;174:561–585.
424. Toyoda H, Nicklin MJH, Murray MG, et al. A second virus-encoded proteinase involved in proteolytic processing of poliovirus polyprotein. *Cell* 1986;45:761–770.
425. Traub V, Traub P. Changes in the microheterogeneity of histone H1 after mengovirus infection of EAT cells. *Hoppe-Seyler's Z Physiol Chem* 1978;359:581–592.
426. Tyrrell DAJ, Chanock RM. Rhinoviruses: a description. *Science* 1963;141:152–153.
427. Uncapher CR, DeWitt CM, Colonno RJ. The major and minor group receptor families contain all but one human rhinovirus serotype. *Virology* 1991;180:814–817.
428. Van der Marel P, Hazendonk TG, Heneke MAC, van Wezel AL. Induction of neutralizing antibodies by poliovirus capsid polypeptides, VP1, VP2 and VP3. *Vaccine* 1983;1:17–22.
429. Vasquez D, Denoya CD, LaTorre JL, Palma EL. Structure of foot-and-mouth disease virus capsid. *Virology* 1979;97:195–200.
430. von Magnus H, Gear JHS, Paul JR. A recent definition of poliomyelitis viruses. *Virology* 1955;1:185–189.
431. Watanabe Y, Watanabe K, Hinuma Y. Synthesis of poliovirus-specific proteins in HeLa cells. *Biochim Biophys Acta* 1962;61:976–977.
432. Wetz K, Habermehl K-O. Topographical studies on poliovirus capsid proteins by chemical modification and cross-linking with bifunctional reagents. *J Gen Virol* 1979;44:525–534.
433. Wetz K, Habermehl K-O. Specific cross-linking of capsid proteins to virus RNA by ultraviolet irradiation of poliovirus. *J Gen Virol* 1982;59:397–401.
434. Wetz K, Willingmann P, Zeichhardt H, Habermehl KO. Neutralization of poliovirus by polyclonal antibodies requires binding of a single IgG molecule per virion. *Arch Virol* 1986;91:207–220.
435. Wetz K, Zeichardt H, Willingmann P, Habermehl K-O. Dense particles and slow sedimenting particles produced by ultraviolet irradiation of poliovirus. *J Gen Virol* 1983;64:1263–1275.
436. White JM. Integrins as virus receptors. *Curr Biol* 1993;3:596–599.

437. Wild TF, Burroughs JN, Brown F. Surface structure of foot-and-mouth disease virus. *J Gen Virol* 1969;4:313–320.
438. Wimmer E. Genome-linked proteins of viruses. *Cell* 1982;28:199–201.
439. Wimmer E, Hellen CUT, Cao X. Genetics of poliovirus. *Annu Rev Genet* 1993;27:353–436.
440. Wolff DA, Bubel HC. The disposition of lysosomal enzymes as related to specific viral cytopathic effects. *Virology* 1964;24:502–505.
441. Yamaguchi-Koll U, Wieger KJ, Drzeniek R. Isolation and characterization of "dense particles" from poliovirus-infected HeLa cells. *J Gen Virol* 1975;26:307–319.
442. Yin FH. Involvement of viral procapsid in the RNA synthesis and maturation of poliovirus. *Virology* 1977;82:299–307.
443. Yin FH, Lomax NB. Host range mutants of human rhinovirus in which nonstructural proteins are altered. *J Virol* 1983;48:410–418.
444. Yogo Y, Wimmer E. Sequence studies of poliovirus RNA. III. Polyuridylic acid and polyadenylic acid as components of the purified poliovirus replicative intermediate. *J Mol Biol* 1975;92:467–477.
445. Yoon JW, Onodera T, Notkins A. Virus-induced diabetes mellitus. XV. Beta cell damage and insulin-dependent hyperglycemia in mice infected with coxsackievirus B4. *J Exp Med* 1978;148:1068–1080.
446. Young DC, Dunn BM, Tobin GJ, Flanegan JB. Anti-VPg antibody precipitation of product RNA synthesized in vitro by the poliovirus polymerase and host factor is mediated by VPg on the poliovirion RNA template. *J Virol* 1986;58:715–723.
447. Ypma-Wong MF, Dewalt PG, Johnson VH, Lamb JG, Semler BL. Protein 3CD is the major poliovirus proteinase responsible for cleavage of the P1 capsid precursor. *Virology* 1988;166:265–270.
448. Zajac I, Crowell RL. Effect of enzymes on the interaction of enteroviruses with living HeLa cells. *J Bacteriol* 1965;89:1097–1100.
449. Zeichhardt H, Otto MJ, McKinlay MA, Willingmann P, Habermehl K-O. Inhibition of poliovirus uncoating by disoxaril. *Virology* 1987;160:281–285.
450. Zeichhardt H, Wetz K, Willingmann P, Habermehl KO. Entry of poliovirus type 1 and Mouse Elberfeld (ME) virus into HEp-2 cells: receptor-mediated endocytosis and endosomal or lysosomal uncoating. *J Gen Virol* 1985;66:483–492.
451. Zhang A, Nanni RG, Oren DA, Rozhon EJ, Arnold E. Three-dimensional structure-activity relationships for antiviral agents that interact with picornavirus capsids. *Sem Virol* 1992;3:453–472.
452. Zhang HY, Yousef GE, Bowles NE, Archard LC, Manna GF, Mowbray JF. Detection of enterovirus RNA in experimentally infected mice by molecular hybridization: specificity of subgenomic probes in quantitative slot blot and in situ hybridization. *J Med Virol* 1988;26:375–386.
453. Ziola BR, Scraba DG. Structure of the mengo virion. IV. Amino and carboxyl-terminal analyses of the major capsid polypeptides. *Virology* 1976;71:111–121.

Fundamental Virology, Third Edition
edited by B.N. Fields, D.M. Knipe, P.M. Howley, et al.
Lippincott - Raven Publishers, Philadelphia © 1996

CHAPTER 17

Togaviridae: The Viruses and Their Replication

Sondra Schlesinger and Milton J. Schlesinger

TOGAVIRUSES

The *Togaviridae* now consist of two genera—the alphaviruses and the rubiviruses. This family was originally much larger and included flaviviruses, pestiviruses, and other viruses that had not been well characterized. They were all grouped together on the basis of size, on their having a single-strand nonsegmented RNA genome that functions as a messenger RNA, and on the ability of many of the members to replicate in and be transmitted by mosquitos. More detailed knowledge of the genome structure and replication strategy of many of these viruses led to the establishment of the *Flaviviridae* family which now seems almost as diverse as the *Togaviridae* family once was (see Chapter 30 in *Fields Virology,* 3rd ed.).

The alphavirus genus is defined by a group of 27 different members which appear to be very similar in structure. There is not yet enough information to know how distinct their replication strategies will be; however, it is clear that they differ in their ability to cause disease (see Chapter 28 in *Fields Virology,* 3rd ed.). The type-specific member of the alphaviruses is Sindbis virus whose structure and replication have been studied in great detail. In addition, both Sindbis and Semliki Forest viruses have provided valuable models for examining the synthesis, posttranslational modifications, and localization of membrane glycoproteins. Most of the information presented in this chapter was obtained by studies with these two viruses, although other alphaviruses are beginning to contribute to our knowledge of the replication strategy and diversity of this genus.

Rubella virus, the sole member thus far of the rubivirus genus, is well known for its ability to cause disease in humans. Alphaviruses and rubella virus share many features which suggest that they evolved from a common ancestor. A more detailed scrutiny of homologies, however, makes it difficult to propose a straighforward evolutionary relatedness (see Fig. 3 and Chapter 5).

Virion Structure

Togaviruses are among the most simple enveloped animal viruses. Their genome consists of a single strand of RNA of positive (+) polarity encapsidated in an icosahedral protein shell composed of a single species of protein (the capsid protein) and enveloped by a lipid bilayer de-

S. Schlesinger and M. J. Schlesinger: Department of Molecular Microbiology and Immunology, Washington University School of Medicine, St. Louis, Missouri 63110-1093.

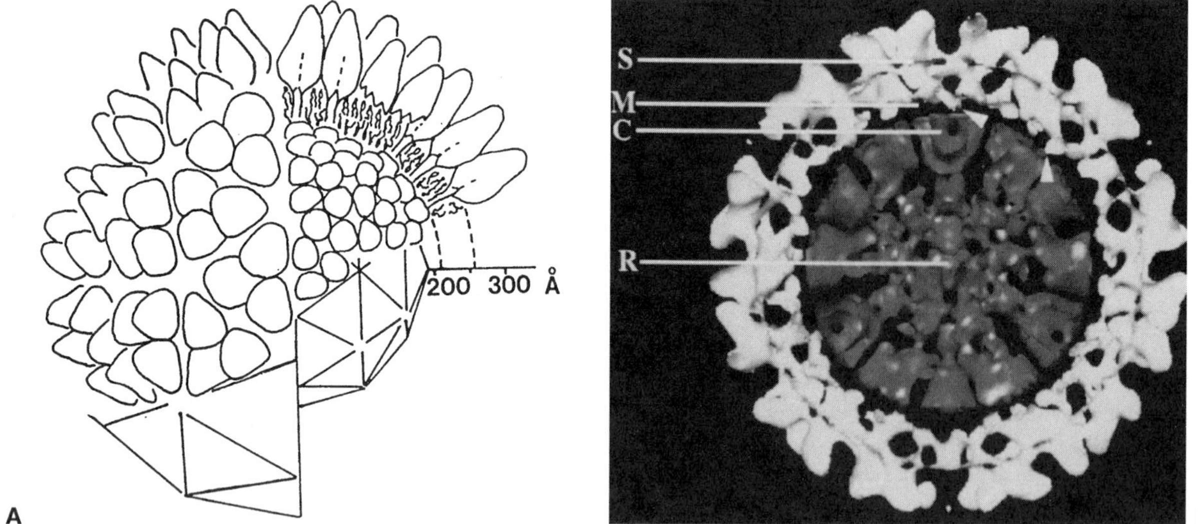

FIG. 1. Views of Sindbis virus particles. **A:** Cartoon showing the icosahedral nucleocapsid surrounded by a lipid bilayer containing heterodimer spikes (*pear shaped*) arranged as trimers on the virus surface in an icosahedral lattice (provided by S. Harrison, Harvard University, Cambridge, Massachusetts). **B:** Cross-section of a three-dimensional image reconstruction from cryoelectron micrographs of Sindbis virus. Shown is a 50-Å-thick structure close to the center of the particle and perpendicular to the strict threefold axis. *S,* spike proteins; *M,* lipid bilayer; *C,* nucleocapsid protein; *R,* genomic RNA. From Paredes et al. (117), with permission; provided by D. T. Brown, University of Texas, Austin.

rived from the host-cell plasma membrane (Fig. 1A). Projecting from the bilayer and embedded in it are the viral-encoded glycoproteins designated *E1* and *E2*. A three-dimensional structure of the Sindbis virion, to a resolution of 28 Å and based on data obtained by cryo-electron microscopy (Fig. 1B), shows a very tight arrangement of glycoprotein spikes distributed as 80 trimers on its outer surface, which are located at the local and strict threefold axis of a T=4 icosahedral lattice (117). Each trimer contains three E1-E2 heterodimers that appear intertwined but flare out in a triangular shape from the membrane. Based on studies with cross-linking reagents (4), the proximal portions of the trimers are postulated to have close E1-E1 interactions with these glycoproteins forming the inner cores of the trimers. Protein mass was found immediately underlying the membrane and is taken as supportive evidence for the hypothesis that a 33-amino acid cytoplasmic domain of the E2 glycoprotein functions as a site to which nucleocapsids bind.

The nucleocapsid of Sindbis virus consists of 240 copies of the capsid protein which contains 264 amino acids arranged in a T=4 icosahedral lattice (19,117,118). An atomic structure of the Sindbis virus capsid polypeptide based on x-ray diffraction to a resolution of 3 Å has been solved (Fig. 2) (19,165). Amino acids from positions 114 of the capsid sequence to the carboxy terminus at amino

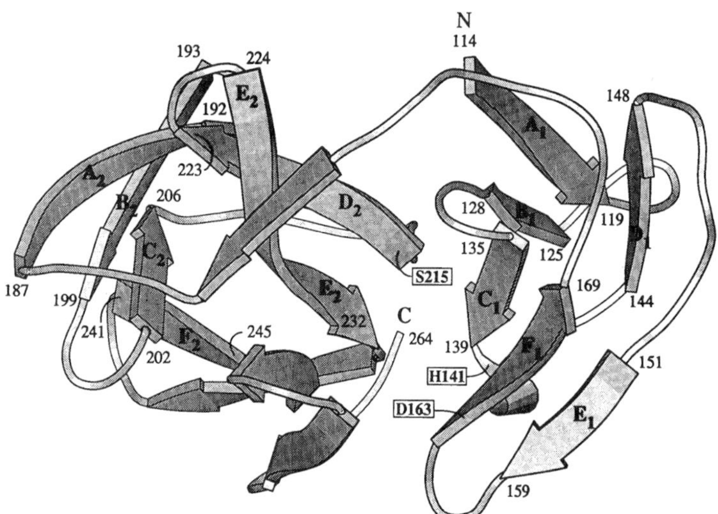

FIG. 2. Schematic drawing of the polypeptide chain of Sindbis virus capsid protein with secondary structure nomenclature, based on x-ray diffraction analysis. The catalytic triad of the active site serine-proteinase are in boxes. The amino terminal sequence to residue 114 is not structured. From Choi et al. (19), with permission; provided by M. G. Rossmann, Purdue University, West Lafayette, Indiana.

acid 264 were traced, but the amino terminal domain was largely unstructured. The monomer fold is like mammalian serine proteases of the chymotrypsin family and even more closely related to the bacterial serine proteinases such as α-lytic proteinase, a result consistent with previous observations that the Sindbis virus capsid has an autoprotease activity (3,45,93). The chain is folded into two Greek key β-barrel domains with the C-terminal tryptophan located between the domains and occupying the active proteinase site. The structure of the latter reveals the catalytic triad of serine, histidine, and aspartate—conserved among the serine proteinases—surrounding a hydrophobic pocket occupied by tryptophan, which is the substrate for cleavage of the capsid from the nascent polypeptide (19,165).

Genome Structure and Organization

The overall organization of the genomes of alphaviruses and rubella virus is shown in Fig. 3 where similarities and distinctions between them are noted. The genomes of alphaviruses are about 12 kb in length (158); the genome of rubella virus is significantly smaller, containing only 9,756 bases (25). The genomes are arranged in two modules: the 5′ two-thirds codes for the nonstructural proteins (nsPs) required for transcription and replication of the RNA and the 3′ one-third codes for the structural proteins. The 5′ terminus is capped with a 7-methylguanosine, and the 3′ terminus is polyadenylated. During replication a discrete subgenomic mRNA species, identical in sequence to the 3′ terminal one-third of the genomic RNA, is synthesized (see also Fig. 4). This RNA is also capped and polyadenylated, and serves as the mRNA for the synthesis of the viral structural proteins [reviewed in more detail in (158,159)].

Sequence comparisons of different alphaviruses indicate that four regions of the genome are highly conserved (112–114). These include (i) the 19 nucleotides at the 3′ terminus; (ii) 21 nucleotides that span the junction between the nonstructural and structural genes and include the start of the subgenomic 26S RNA (the junction region); (iii) 51 nucleotides near the 5′ terminus; and (iv) the 5′ terminus, though this region appears to be conserved more in potential secondary structure than in sequence. The latter three regions are also conserved in rubella virus (25), and there is increasing evidence that all four regions play an important role in the regulation of viral RNA synthesis. Sequence analyses also revealed similarities in the genome organization and in the sequences of the nonstructural proteins of alphaviruses and a number of plant viruses, some of which are quite distinct from alphaviruses in their overall structure (2,159,193). This not only suggested an evolutionary relatedness between the plant and alphaviruses but also led to the definition of a Sindbis-like superfamily (described in Chapter 6).It was this type of comparison which suggested that rubiviruses may be more distantly related to alphaviruses than are some of the plant viruses (25,32).

The cloned cDNAs of alphaviruses (22,64,79,128) and rubella virus (174) can be transcribed into infectious RNAs,making it possible to apply reverse genetics for analyzing replication and assembly of these viruses. A number of examples are described in this chapter. Important differences between the alphaviruses and rubella virus exist, and their replication strategies and assembly are described separately.

ALPHAVIRUSES

Attachment, Entry, and Uncoating

Alphaviruses have a wide host range and replicate in a variety of different species ranging from mammals to in-

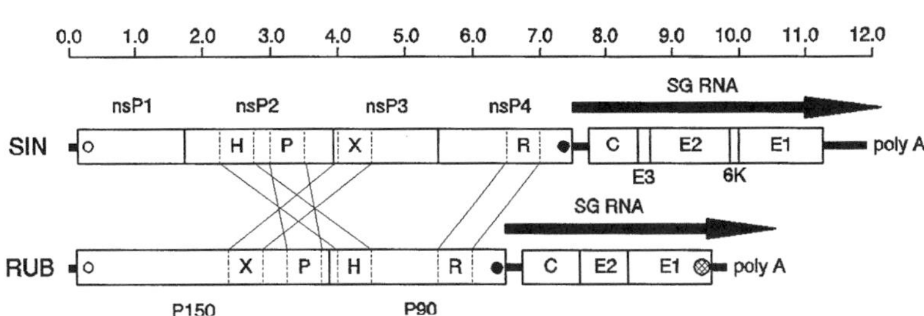

FIG. 3. Diagrams of the genomes of the alphavirus, Sindbis virus and the rubivirus, rubella virus. The 5′ two-thirds of these genomes code for the nonstructural proteins (nsPs) and the 3′ one-third code for the structural proteins, translated from subgenomic (SG) RNAs. Untranslated regions are designated (*black lines*), as are open reading frames (*open boxes*). Designations are for individual proteins which are processed from precursors. The scale at the top of the figure is in kilobases. The open circles near the 5′-termini and the closed circles at the start of the subgenomic RNAs indicate conserved sequences in the alpha- and rubella virus genomes. The location of the amino acid motifs for helicases (*H*), replicases (*R*) and cysteine proteases (*P*) are indicated. See text for more details. *X* denotes a small region of homology between the deduced amino acid sequences in the Sindbis virus nsP3 and the nsP of rubella virus. Adapted from Dominguez et al. (25) and Frey (32); provided by T. Frey, Georgia State University, Atlanta, Georgia.

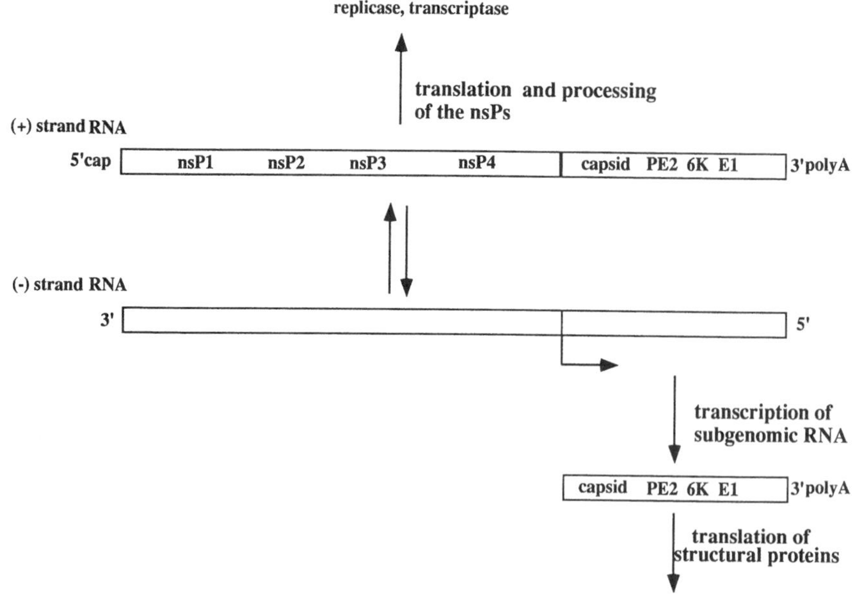

FIG. 4. Replication and transcription of alphavirus RNA. The first two-thirds of the (+) strand genomic RNA serves as the mRNA for translation of the nsPs which are eventually processed into 4 proteins (see text and Fig. 5). The (–) complementary strand is the template for the synthesis of both new genomic RNAs and the subgenomic 26S RNA which codes for the structural proteins.

sects as well as in many different cell types. Because attachment of a virion to a cell surface is one of the steps that can define host range, it seemed that these viruses must use a variety of different molecules for attachment or else utilize a ubiquitous surface molecule as a receptor. The two cell types—chicken embryo fibroblasts (CEF) and baby hamster kidney (BHK) cells—most frequently used to study the replication of Sindbis virus appear to have different receptors. The high affinity 67-kd laminin receptor is a receptor for Sindbis virus in BHK and other mammalian cells (176) but not in CEF where a 63-kd protein acts as the receptor (175). Mouse neuronal cells have other receptors: a 74-kd protein was found on neuroblastoma cells that are loosely adherent, and a 10-kd protein was identified on firmly adherent cells (168). These putative mouse receptor proteins were detected on brain cells of newborn mice, but half of the cells no longer expressed the receptors 4 days after birth. H2 and HLA histocompatibility antigens were reported to be receptors for Semliki Forest virus (51), but their presence on the cell surface is not essential for infectivity (111).

The viral glycoprotein E2 is most likely the protein that interacts with these receptors. Antibodies directed primarily against the E2 protein neutralize viral infectivity, and amino acid changes in the E2 protein can affect both the efficiency of binding of the virus to neural cells and virulence in mice (148,167). Convincing evidence for the importance of E2 in binding virus to the cell surface comes from recent experiments in which Sindbis virus mutants unable to bind to CEF were shown to be altered in the E2 protein (27). Ionic interactions between Sindbis virus and the host cell surface were also found to be important for attachment and initiation of infection (122,162,163).

Attachment of Sindbis virus to cells leads to the exposure on both glycoproteins of new transitional epitopes, which are indicative of conformational changes (30,96). These epitopes were detected also when purified virus was treated with heat, reducing agents, or low pH (95). Recent studies indicate that some of these changes may involve reduction of critical disulfide bonds in the glycoproteins, which may disrupt protein-protein associations in the envelope and be important for disassembly of the particle (1,4,5).

In most cells, alphaviruses gain entry to the cell via the endocytic pathway which normally functions for uptake of receptor-ligand complexes (90). Bound virus accumulates in coated pits which are endocytosed to form coated vesicles. These vesicles are subsequently uncoated to form acidified endosomes, providing conditions that trigger fusion of the virus membrane with the vesicle membrane. The mechanism of receptor-mediated endocytosis for alphavirus penetration into cultured cells is supported by many studies, which include (a) the inability of bound Sindbis virus to effect antibody-mediated, complement-dependent cell lysis, indicating a rapid loss of virus glycoprotein from the cell surface (29); (b) the ability of lysosomotropic amines which raise the pH of endocytic vesicles to inhibit initiation of virus replication—demonstrating that acidified vesicles are required for early virus replication events; (c) a requirement for a low pH (5 to 6) for fusogenic activity by Sindbis and Semliki Forest virus glycoproteins (170,187–189); and (d) the detection of virus in coated vesicles immediately following uptake from the cell surface, as revealed by electron micrographs of virus entry (189). Alternative pathways not involving acidified endosomes may exist in arthropod cells, acidified endocytic vesicles do not seem to be essential for infectivity (15), and a Chinese hamster ovary cell mutant defective in endosome acidification was able to replicate Sindbis virus (28).

Studies with Semliki Forest virus have demonstrated that the low pH environment of the endosome leads to a dissociation of the E1E2 heterodimer and a concomitant trimerization of the E1 subunits (172,173). These alterations suggest that the E1 trimer is the fusion-active form of the protein with a structure possibly analogous to that now attributed to the coiled-coil configuration present in the low-pH-induced structure of the influenza virus HA2 protein (18). Fusion of Semliki Forest virus also requires the presence of cholesterol in the host cell membrane (121).

Fusion leads to release of the nucleocapsid into the cytoplasm and must be followed by an uncoating event to permit the RNA to become accessible to ribosomes for initiation of translation. Several reports, based on *in vivo* and *in vitro* experiments, suggest that the binding of the nucleocapsid to ribosomes triggers the uncoating process (153,169,182,184,185). The capsid protein itself appears to have a ribosome-binding site between amino acids 94 and 106 of the capsid sequence (186), which is in the same region of the protein that interacts with viral RNA during encapsidation (37).

Transcription, Translation, and Replication of the Genome

The genomic RNA serves as the messenger RNA for translation of the viral nonstructural proteins and as a template for the synthesis of the complementary minus strand. The minus strand, in turn, provides the template for the synthesis of both new genomic RNA and the 26S subgenomic RNA [Fig. 4 and for a more detailed review see (158,159)]. Translation of the virion RNA is initiated at a single AUG near the 5′ terminus of the RNA and proceeds uninterrupted for two-thirds of the mRNA until encountering three termination codons located just downstream of the start of the sequences corresponding to the subgenomic RNA. The polyprotein that is synthesized is co- and posttranslationally cleaved to give rise to four distinct polypeptides designated nsP1, nsP2, nsP3 and nsP4, according to the order of the genes on the RNA. The protease activity responsible for these cleavages has been localized to the C-terminal domain of nsP2 (24,48,154). In the genome of Sindbis virus and several other alphaviruses there is an opal termination codon a few codons before the nsP4 gene, and nsP4 is produced by readthrough of the opal codon followed by proteolytic cleavage (156). Mutagenesis of the opal codon to one that encodes an amino acid or to the other two translation termination codons adversely affects viral replication at low multiplicities of infection (76). An opal codon at this position in the genome of alphaviruses is common but not universal, and no termination codon exists at this position in Semliki Forest or O'Nyong-nyong viruses (155). Despite this difference, the amount of nsP4 produced in cells infected with these viruses is no greater than in those in which the level of nsP4 is controlled by the extent of readthrough. A proteolytic activity has been reported for Semliki Forest virus nsP4 and could control levels of this polypeptide (164). An nsP4 protease has not been found in Sindbis virus-infected cells.

There are a number of activities that must be carried out by the nsPs in their role as replicative enzymes. The isolation, characterization and mapping of temperature-sensitive (ts) mutants defective in viral RNA synthesis at the nonpermissive temperature has provided one means of identifying the functions of each of the polypeptides (16,17,46,47,157). The RNA$^-$ ts mutants fell into four complementation groups which have now been correlated with specific nsPs. nsP1 has been implicated in the synthesis of minus-strand RNA and also appears to be important for methylation of the 5′ terminal cap structure on the genomic RNA (47,98,138,177). A methyl transferase activity was found for the nsP1 expressed in bacterial cells transformed with a plasmid carrying the Sindbis virus nsP1 gene (98). Members of two different RNA$^-$ complementation groups mapped to the nsP2 gene (47). Based on the phenotypes of the different mutants, nsP2 appears to be involved in the regulation of minus-strand RNA synthesis (135,177) and in the initiation of subgenomic RNA synthesis. As mentioned earlier, it also functions in the proteolytic processing of the nsPs (24,48,154). About half of the Semliki Forest virus nsP2 was found to be associated with the nuclear fraction in virus-infected BHK cells although its function there is not known (119). Under conditions in which nsP2 was expressed in the absence of other viral proteins there was specific enrichment in nucleoli (129). A pentapeptide in the carboxy-terminal region of nsP2 appears to be required for nuclear localization (129). The nsP3 is phosphorylated but has not yet been assigned a function (75,120). Recent studies indicate that mutants in this gene define an additional complementation group (66). Mutations in the nsP4 gene affect the synthesis of all viral RNAs and *in vitro* studies of Sindbis virus-infected cells point to nsP4 as the viral polymerase (8,46,134).

In addition to studies with ts mutants, functions for the nsPs have been deduced by gene-sequence homologies with known proteins [reviewed in (40,159,193) and see Chapter 5]. Three of the four nsPs contain domains that are conserved among the alphavirus-like super family. The nsP1 protein contains a methyltransferase domain, and the amino terminal region of nsP2 contains nucleotide-triphosphate binding motifs homologous to those found in bacterial helicases [(41) and see Fig 3]. nsP4 has a GDD motif,

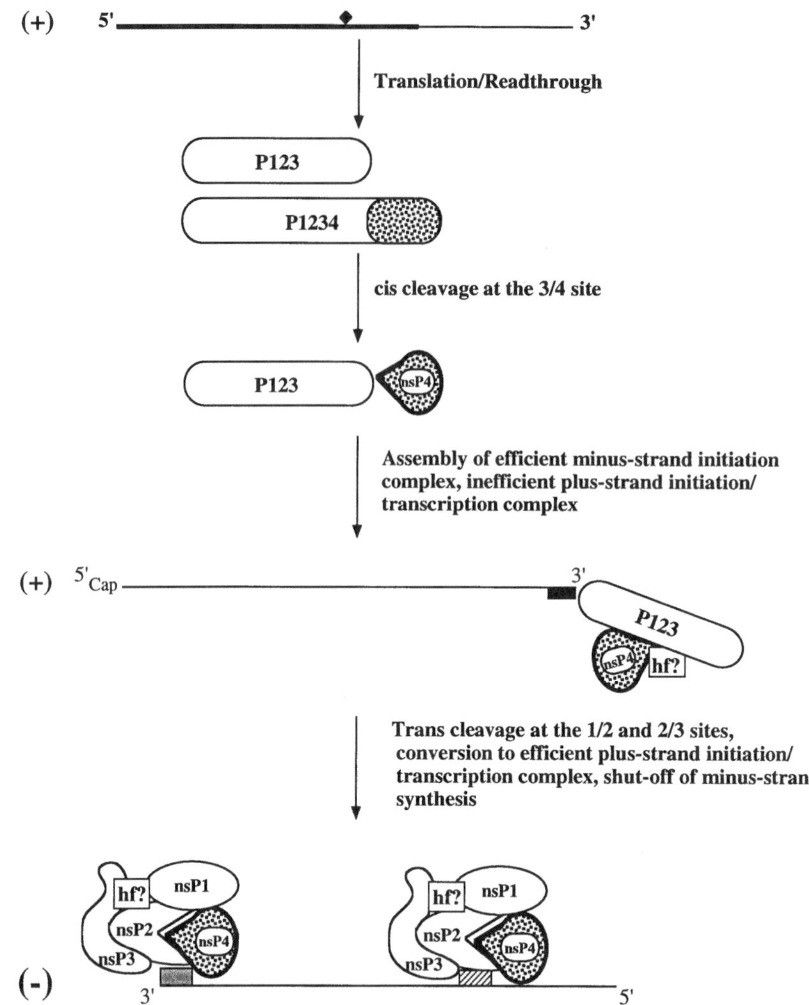

FIG. 5. A model for the temporal regulation of minus and plus-strand RNA synthesis. The protein complex that initiates (–) strand RNA synthesis is composed of P123 and nsP4. This complex is also able to synthesize (+) strands, but inefficiently. Efficient synthesis of (+) strand genomic and subgenomic RNAs occurs after trans-cleavage at the 1/2 and 2/3 sites of the polyprotein which also results in the shut-off of (–) strand synthesis. Host factors (hf) are most likely involved in these reactions. From Lemm et al. (70), with permission; provided by J. Lemm and C. M. Rice, Washington University, St. Louis, Missouri.

a triplet that is present in a number of viral RNA polymerases (58). The assignment of nsP4 as the viral polymerase is consistent with other data identifying this protein as the polymerase.

Replication and transcription of alphavirus RNAs occur on cellular membranes (7,34). Recent studies with Sindbis virus have defined the proteolytic processing scheme for the nsPs and have led to an elegant model for the regulation of viral RNA synthesis. Proteolytic processing has been analyzed by translation of mRNAs *in vitro* (23,48), by expression of nsP genes in a vaccinia virus transient expression system (68–70), and by expression in cultured cells of Sindbis virus genomic RNAs containing site-directed mutations in the nsP genes (151,152). Each of these has provided important information for the proposed scheme illustrated in Fig. 5. The nsPs are initially translated as two polyproteins, P123 and P1234, depending on whether the in-frame opal termination codon is suppressed. Cleavage of the latter at the 3/4 site occurs rapidly to form the complex P123 and nsP4 which is the complex postulated to initiate minus-strand synthesis. Evidence for this is based on the result that nsP4 plus a P123 which is cleavage defective, but not a cleavage-defective P1234, is able to synthesize both plus- and minus-strand RNAs. In addition, in a vaccinia virus expression system functional replication complexes were not formed when the nsPs were expressed as individual polypeptides or as different combinations of individual nsPs (68,69). Only the independent expression of P123 and nsP4 led to formation of a functional complex. The most convincing argument for a role for P123 in the synthesis of minus strands is based on studies in which the Asn_{614} of nsP2 was changed to Asp

(70,154). This mutation leads to more efficient processing of P123 *in vitro*, but RNAs containing this mutation do not give rise to viable Sindbis virus in transfection assays (154). With the vaccinia virus expression system, minus-strand RNA synthesis could not be detected when this mutation was present in P123, but it was restored when the mutation was inserted into a P123 protein that was also cleavage-defective (70). The latter result not only supported the essential role of the uncleaved protein in minus-strand RNA synthesis, but it also indicated that the uncleaved protein could synthesize plus strands, though much less efficiently than the cleaved products. In the absence of mutations affecting cleavage, the P123 precursor would be cleaved to nsP1, nsP2, and nsP3 which in conjunction with nsP4 form the complexes capable of carrying out plus-strand RNA synthesis.

This model explains some aspects of the regulation of alphavirus RNA synthesis, in particular, the low level of minus strands synthesized by the replication complexes, the continued requirement for protein synthesis for production of minus strands, and the cessation of minus-strand synthesis several hours postinfection (136,137). *In vitro* translation studies show that processing of P123 occurs in *trans* and, in infected cells, the accumulation of P123 should accelerate its cleavage leading to cessation of minus-strand synthesis. It is the fully processed complexes that carry out plus-strand RNA synthesis, which continues throughout the infection cycle. Not all of the earlier studies can be explained by this model, and there are mutations in both nsP4 and in nsP2 which permit continued synthesis of minus strands even after cleavage of the proteins has occurred (134,135). To explain these latter results it was proposed that proteolytic cleavage leads to conformational changes that influence template specificity of the replication complex, but mutational alterations may permit these changes despite a cleaved complex (70).

The four conserved sequences, noted earlier, in the alphavirus genome were presumed to be replicase-binding sites and to function as promoter elements. Strauss and his colleagues have carried out extensive mutagenesis on three of these regions in the Sindbis virus genome: the 3′ nontranslated region (63), the 51 nucleotides located between nucleotides 155 and 205 (107), and the 5′ nontranslated 44 nucleotides (106). They show an important role for these sequences as most of the mutations affect viral growth, frequently in a host cell-dependent manner with strikingly different effects in CEF and mosquito cells. These results support the contention that host proteins are involved in alphavirus RNA replication.

The fourth conserved region of the alphavirus genome surrounds the start of the subgenomic RNA. The function of this region as a promoter for subgenomic RNA synthesis was first shown by the newly acquired ability of defective RNAs carrying an insertion of this region to produce subgenomic RNAs in cells coinfected with Sindbis virus (73). The latter provided, by *trans* complementation, the nsPs required for transcription of the subgenomic RNA. The minimal promoter region contains 19 nucleotides upstream and 5 nucleotides downstream of the start site for transcription of the subgenomic RNA, which coincides with the conserved region (73). Promoter activity is enhanced, however, when additional sequences in this region are included in the RNA (73,126). Additional studies of this promoter were carried out by the construction of Sindbis virus derivatives with two subgenomic RNA promoters: the wild-type promoter which was used for expression of the structural protein genes and a second promoter which was placed upstream of the chloramphenicol acetyltransferase gene and was subjected to modifications (126). Minimal promoters from other alphaviruses, but not from rubella virus, could be utilized by the Sindbis virus nonstructural proteins (52).

A role for host cell proteins in alphavirus RNA synthesis was inferred from observations made in both vertebrate and invertebrate cells. Alphaviruses can replicate in vertebrate cells, but not in *Aedes albopictus* mosquito cells treated with dactinomycin or that have been enucleated [reviewed in (15)]. Extended treatment of vertebrate cells with either dactinomycin or a-amanitin, however, inhibited viral replication (6), suggesting that in both vertebrate and invertebrate cells there were host components required for the synthesis of viral RNA. Studies of Sindbis virus mutants provided further evidence for the involvement of host proteins in replication. Some mutations in nsP4 affectedviral replication to a greater extent in mosquito cells than in CEF (67)and, as noted above, some of the mutations in the conserved regions of the Sindbis virus genome have different effects in the two cell types. More definitive evidence has come from experiments using a gel-retardation assay which show that the 3′ end of the Sindbis virus minus-strand RNA binds three proteins of 42-, 44-, and 52-kd from extracts of CEF and two proteins of 50- and 52-kd from mosquito cell extracts (115,116). Four domains were identified in the first 250 nucleotides of the RNA. They all bind to the same mosquito cell proteins, three with high affinity and one with low affinity (115). These same proteins also interacted with the corresponding regions of RNA from Semliki Forest and Ross River viruses. A similar region from rubella virus RNA bound weakly to these proteins, which would be consistent with the inability of this virus to grow in mosquito cells.

Expression of Virus Structural Genes

Alphavirus structural proteins are translated from the 26S subgenomic RNA early (2 to 3 hours postinfection) in the replication cycle, probably as soon as the 26S mRNA is formed. Ribosomes bind to a single site near the 5′ end of this RNA and proceed uninterrupted for some 3,000 nucleotides to a termination site about 300 nucleotides from

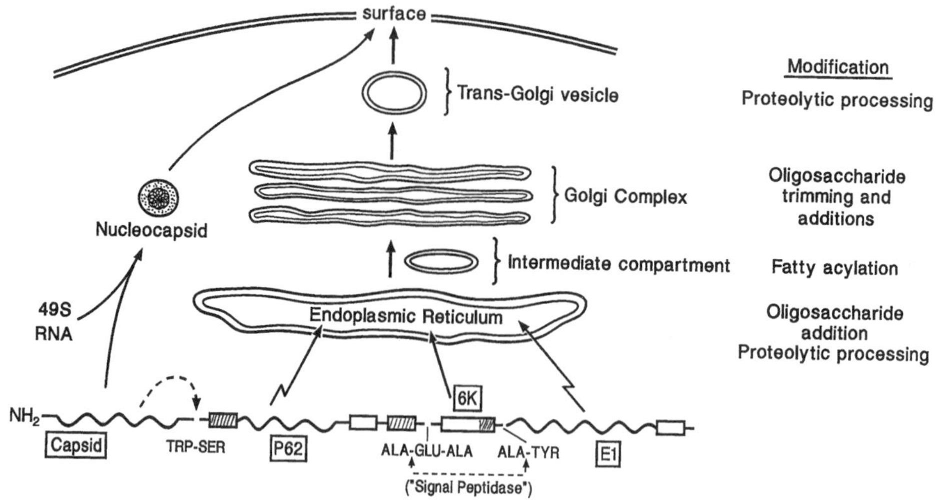

FIG. 6. Processing and modifications of the Sindbis virus structural proteins. Capsid is released co-translationally by autoproteolysis and assembles to encapsidate the 49S genomic RNA. The P62, 6K, and E1 proteins are modified during transport to the cell surface. Signal sequence (*shaded rectangle*); hydrophobic sequences (*open rectangle*).

the 3′ poly A terminus [reviewed in (158)]. The first sequences to be translated encode the viral capsid polypeptide, which folds to form a serine-protease catalytic site (see Fig. 2) that cleaves the capsid from the nascent polyprotein at a tryptophan-serine bond. Continued translation of the mRNA produces a sequence which binds to the host cell signal-recognition particle and facilitates translocation of the remaining polypeptide into the lumen of the endoplasmic reticulum (Fig. 6). As the protein emerges into this vesicle, it is modified by covalent attachment of oligosaccharides (150) and, later, by proteolytic cleavages carried out by host cell signalase (77). Asparagine 14 of the P62 is glycosylated. Signalase cleavage does not occur at that site, and the signal sequence is retained intact except for acetylation at the amino-terminal serine of the P62 (9,10). The final products of 26S mRNA translation are the capsid, two large (45 kd) type I transmembrane glycoproteins, noted P62 and E1, and a small (6 kd) membrane-embedded protein; the topology of these three membrane-bound proteins is shown in Fig. 7. Much of the data supporting this model derive from (a)assignments of domains with 22 hydrophobic amino acids in the sequences of the P62, 6K, and E1 as either signal- or stop-transfer sequences which span the membrane (Fig. 6); (b) expression of cDNA constructs of Semliki Forest virus 26S RNA under the control of an heterologous promoter or expression of the Semliki Forest virus RNA genome which show that the carboxy termini of either P62 or the 6-kd protein can function as a signal sequence for E1 (77,92); and (c) a change in the localization of the E2 carboxyl terminus to the inner leaflet of the bilayer late in its transport to the cell surface (81). The P62, E1, and 6K proteins appear to move as a cohort through the secretory vesicles from their site of synthesis on the endoplasmic reticulum to their ultimate lo-

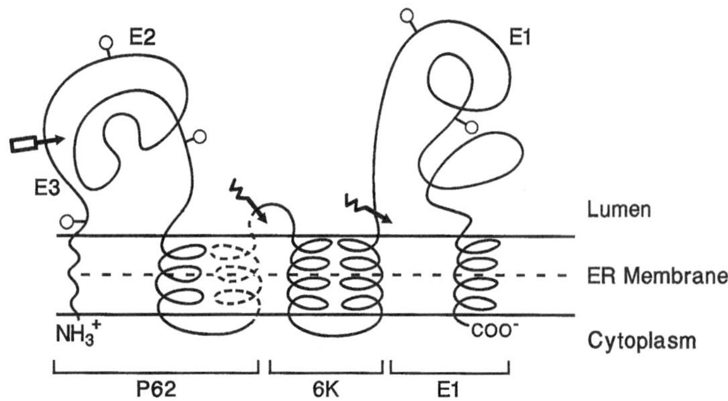

FIG. 7. Topology of the Sindbis virus P62 and E1 glycoproteins and the 6K protein in the endoplasmic reticulum membrane. Part (*dashed line*) of P62 is removed from its transmembrane configuration and binds to the inner face of the lipid bilayer during transport to the cell surface. Host cell, [symbol]; signalase activity, host cell furin-type protease, [symbol]. Refer to text for detail.

cation at the plasma membrane of the cell (Fig.6). This process takes about 30 min at physiological temperatures and leads to several posttranslational modifications. Among these are the trimming of and additions to the three oligosaccharides on Sindbis virus P62 and to the two on E1 (141). Cellular enzymes localized to various compartments of the secretory system carry out these modifications in a manner resembling that found for cellular proteins destined for the plasma membrane or secretion to the extracellular fluid (62). Thus, the precise composition of the oligosaccharides on the virus is dependent on the host cell in which the virus replicates (59). The carbohydrates play an important role in glycoprotein folding (39) but several chaperone activities in the endoplasmic reticulum are required to form the P62-E1 heterodimer and, ultimately, the trimers (26).

Palmitate acylation occurs at 3 sites on the Sindbis virus P62, four sites on the 6K protein, and one site on E1 during transport of the proteins after they have exited from the endoplasmic reticulum and prior to entrance into the Golgi complex [Fig. 6 and (12, 87,147)]. Cysteines oriented to the cytoplasmic face of the lipid bilayer are the sites of fatty acylation. The role of these fatty acids in glycoprotein function is unknown, but they probably affect interactions of the proteins with the lipid bilayer. Blocking their attachment by cerulenin (140) or by mutationally altering cysteine attachment sites (35,56) interferes with virus assembly and budding.

An important step in glycoprotein maturation is a proteolytic cleavage in P62 to form E2 and E3. This cleavage occurs at a sequence enriched in arginines and lysines at residue 65 of the P62 by a host cell furin-type enzyme localized to a trans-Golgi vesicle (100,178). Failure of the proteolytic cleavage does not interfere with assembly and budding of particles in vertebrate cells (124,131), but conversion of P62 to E2 is essential for particle formation in insect cells and in most cases is required for infectivity of the newly replicated particles in vertebrate cells (27,125, 131,132). The failure of an alphavirus containing P62-E1 spikes to replicate in a cell is due to the inability of the E1 to undergo the low pH-induced conformation change required for its fusogenic activity (82,83). Thus, unlike the E2-E1 complex, the P62-E1 heterodimer is insensitive to pH changes that lead to a tight E1 trimer conformation, which functions in membrane fusion (57,172,173). One variant of Sindbis virus, which has an uncleaved P62, contains additional mutations in the protein and is just as infectious as the wild-type virus particle (50).

Other modifications to these structural proteins have been reported. One is a transient change in topology of the carboxy-terminal 33 amino acids of the P62/E2 from a transmembrane-lumenal orientation (see Fig. 6) to one in which most of these amino acids are at the cytoplasmic face of the membrane (81). Changes in the E1 glycoprotein have been detected during maturation by monoclonal antibodies (146) and by differences in mobilities analyzed by polyacrylamide electrophoresis (11). These changes may be due to formation and rearrangement of disulfide bonds during the maturation process (102). In addition, the Sindbis virus glycoproteins appear to be transiently phosphorylated; a modification which may be important in the maturation process as inhibitors of phosphorylation inhibit production of infectious virus (80).

Assembly of Nucleocapsids and Enveloped Virions

For most alphaviruses, the assembly of nucleocapsids involves a selection of the genomic RNA for encapsidation. Although the subgenomic RNA is present in molar excess, it is not encapsidated in the infected cell. This discrimination is not observed *in vitro;* the Sindbis virus capsid protein can assemble into corelike particles in the presence of any RNA (183,185). Competition experiments, however, show a preference for encapsidation of the Sindbis virus genomic RNA(144). Evidence that selection *in vivo* is due to an encapsidation signal in the genomic RNA came from studies by Weiss et al. (180) who noted that defective interfering RNAs of Sindbis virus (see below) that contain nucleotides 746 to 1,225 of the genomic RNA, but not RNAs lacking these sequences, were packaged into virion particles. Furthermore, this same region of the genome, which lies within the nsP1 gene, binds to purified capsid protein with a specificity that is maintained by a fragment of only 132 nucleotides extending from nucleotide 945 to 1,076 of the Sindbis virus genome (179). The domain of the capsid protein required for the specific RNA binding consists of a 32 amino acid region extending from amino acids 76 to 107 (37). As mentioned earlier, this region of the capsid protein interacts also with ribosomal RNA and may be important in uncoating of the RNA (186).

The virions of the Aura alphavirus are an exception to the almost complete selection of genomic over subgenomic RNA for encapsidation. These virions package both RNAs, although the ratio of genomic to subgenomic RNA is three- to tenfold higher in virions than in infected cells (130). This result suggests that the Aura 26S RNA may contain an encapsidation signal. It is also possible that the Aura virus encapsidation signal present in the genomic RNA is not as efficient as that in other alphaviruses, resulting in less specificity in packaging as was found in the in vitro assembly experiments (183,185).

The final stages in the replication of alphaviruses occur when preformed nucleocapsids interact with the host cell plasma membrane at sites occupied by viral transmembrane glycoproteins. Immediately after this nucleation event, additional virus glycoproteins diffuse into the site, bind to the nucleocapsid, and bend the membrane around the contours of the capsid until the bilayers meet and fuse to release the enveloped particle. Electron micrographs graphically illustrate this process (Fig. 8), which is similar to the budding and release of other kinds of enveloped viruses as

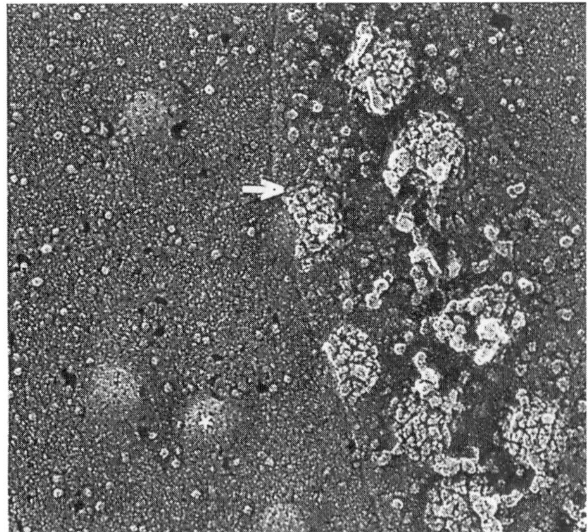

FIG. 8. Budding of Sindbis virions from infected chicken embryo fibroblasts. Cells were quick frozen and freeze fractured. The electron micrograph shows the fracture across the surface membrane of the cell (*white arrow*). The outer surface of the cell with clusters of virus glycoproteins at sites of budding (right) and the inner leaflet of the surface lipid bilayer with viral nucleocapsids (*left, white star*) bound to the inner cytoplasmic side of the membrane are depicted. Sample preparation for electron microscopy was carried out by J. Heuser, Washington University School of Medicine, St. Louis, Missouri.

well as cellular vesicles involved in intracellular trafficking. Budding of alphaviruses requires the presence of both nucleocapsids and the virus-encoded glycoproteins (192). The specific molecular interactions between nucleocapsids and spike glycoproteins have not been identified; however, a region within the hydrophobic 33 amino acid sequence at the carboxy terminus of E2 has been found to be important for assembly, based on studies with a chimeric virus containing genomic sequences from the Ross River and Sindbis viruses (85). No viable virions could be recovered when this chimera contained the Sindbis virus capsid sequences and all of the remaining genes from Ross River virus, but when 7 amino acids of the Sindbis virus E2 cytoplasmic sequence were substituted for Ross River virus sequences, significant levels of infectious particles were recovered. The implication of these results is that there is a complementarity between capsid surface and the E2 carboxyl terminal amino acids. Further support for an E2-capsid "match" are data showing that site-directed mutations at several sites within the E2 carboxyl terminus forming the cytoplasmic domain inhibit particle release at the stage of assembly and budding (36,56). In addition, a very hydrophobic 6-amino acid peptide corresponding to a conserved sequence within the E2 carboxy-terminal domain could selectively inhibit Sindbis and Semliki Forest virus assembly and budding (21). Similar kinds of peptides with sequences identical to those in the E2 carboxyl terminus of Semliki Forest virus were reported to bind capsids (94). Initial reports (171) that anti-idiotypic antibodies, which were raised against monoclonal antibodies that recognized E2 carboxy terminal sequences, reacted with nucleocapsids were subsequently shown to be flawed (160). Images from cryo-EM of Sindbis virus, referred to earlier, show protein mass at the inner side of the virus membrane—another result consistent with the model in which the cytoplasmic part of E2 acts as a docking site for the nucleocapsid.

In most infected vertebrate cells, the plasma membrane is the site of alphavirus assembly and budding. In insect cells, however, budding occurs intracellularly and virus-loaded vesicles are later disgorged to the extracellular fluid (99). In general, the lipid composition of alphaviruses closely resembles that of the membrane of the host cell (71). The ratio of cholesterol to phospholipid, however, is much higher in the membrane of alphaviruses than in the cellular membranes (127), and this may explain why Semliki Forest virus budding requires the presence of cholesterol in the host cell membrane (88). Virus membranes are also much more densely packed with protein and have a greater curvature than the corresponding cellular membranes; thus, their membranes are less fluid than cellular membranes (149).

Mutations in at least two alphavirus genes can affect the budding process both quantitatively and qualitatively. For example, the substitution of valine for alanine at position 344 in the Sindbis virus E2 glycoprotein produced aberrant particles with multiple nucleocapsids within a single membrane (44). Multicored particles form also during infection of cells with mutations in the Sindbis virus 6K protein (35). An important but nonessential role for this small hydrophobic membrane protein in virus assembly was shown in studies with a mutant Semliki Forest virus with a deletion of the 6K gene (79) and a Sindbis virus 6K insertion mutant, in which the abnormal 6K protein was degraded (139). In both mutants, the amount of virus secreted from BHK cells was greatly reduced relative to a comparable infection with the parental virus. The E1-E2 glycoprotein heterodimers that were made in these mutants accumulated at the plasma membrane, and particles released from these cells were indistinguishable from the wild-type virus with regard to morphology and ability to bind and initiate all the early steps in virus replication [(139); Liljeström, personal communication].

Defective Interfering Particles

Defective interfering (DI) particles have been isolated from a wide spectrum of different viruses including alphaviruses (142). DI RNAs of both Sindbis and Semliki Forest viruses, generated by high multiplicity passaging of virus in cultured cells, contain extensive deletions and rearrangements of the genome (145). They depend on *trans*-acting proteins from the intact virus genome, present in the same cell, to provide the components needed for replication and packaging into a virus particle. DI RNAs generated by high-multiplicity passaging of Sindbis or Semliki Forest viruses in cultured cells can play an important role in the establishment and maintenance of persistent infections (145). DI RNAs generated by recombinant DNA techniques have helped to define *cis*-acting promoter-type sequences in the viral genome (74,166), though some deletions at the 5′ terminus which were lethal for replication of DI RNAs were tolerated quite well in the genomic RNA, and others that seemed not to affect replication of a DI RNA had a significant effect when placed in the virion genomic RNA (106). Sindbis virus DI RNAs have been particularly useful in identifying the encapsidation signal in genome RNA (180) and the promoter for transcription of the subgenomic RNA (73) as both lie in coding regions of the genome where extensive modifications would destroy the viability of the virus.

Alphaviruses as Vectors for the Expression of Heterologous Proteins

Two types of RNA expression vectors have been derived from the infectious Sindbis and Semliki Forest virus genomes; both take advantage of the high level of gene expression obtained from the subgenomic RNA species produced in alphavirus-infected cells. In one vector the heterologous gene replaces the structural protein genes (191). This vector is self-replicating (a replicon) but requires complementation to be packaged and released from cells as virion particles. DI RNAs modified to become “defective helper” RNAs can provide the structural proteins for replicons which can then be packaged under conditions in which the helper itself is not packaged (14,78). Such particles are infectious but self-limiting as they produce nsPs as well as genomic and subgenomic RNAs but, in the absence of structural proteins, new particles will not be formed. Some of the defective helpers derived from Sindbis virus retain the region of the genome that includes the packaging signal. These defective RNAs can be copackaged with the replicon, creating a Sindbis virus with a bipartite genome (38). When replicons are packaged with this latter type of defective helper they can be amplified through multiple passages (14).

The other type of vector contains two subgenomic RNA promoters; one controls the synthesis of the subgenomic mRNA that codes for the viral structural proteins and the other controls the synthesis of the subgenomic RNA that codes for the heterologous protein (42,126). This vector is self-replicating and also produces infectious virus particles. Both types of vectors offer useful tools for production of proteins in a variety of eukaryotic cells and for studies of protein expression and modification (13,143).

A completely different approach to the use of alphaviruses as vectors has been to insert heterologous amino acid sequences into the viral structural proteins in such a way that the virus is still viable but the heterologous epitope is expressed and can function as an immunogen. This approach was used by London et al. (84) who made random insertions of an epitope from Rift Valley Fever virus into the structural protein genes of Sindbis virus. Viable viruses containing the insert were then selected. The data showed not only that insertions were tolerated but also that the chimeric Sindbis virus could be used as a vaccine to make mice resistant to infection by Rift Valley Fever virus.

Recombination

Recombination between nonsegmented viral RNA genomes appears to be a much more common occurrence than had originally been appreciated (61,65). Sequence analysis show that at least one alphavirus, western equine encephalitis virus, is a recombinant between eastern equine encephalitis virus and a Sindbis-like virus (43). Three types of RNA recombination have been defined (65). The first, homologous recombination, involves recombinational events in which the crossover is precise. The second is nonhomologous or illegitimate, as the RNAs have no obvious homology. This type of recombination can explain the presence of a modified cellular $tRNA^{Asp}$ at the 5′ terminus of naturally-occurring DI RNAs of Sindbis virus (101) as well as several other examples in which sequences from cellular genes have been identified in the genomes of RNA viruses (60,97). The third type of recombination is termed aberrant homologous recombination because the two RNAs undergoing recombination are homologous but the crossover is not precise. This type of recombination has also been shown to occur with Sindbis virus and may predominate as a consequence of the modular nature of the alphavirus genome. In several different crosses, the two parental RNAs that were able to undergo recombination were each defective in different modules. Analysis of the recombinant RNAs demonstrated that each parental RNA contributed its intact module and that the crossovers occurred within the defective modules (181). The recombinational events giving rise to infectious virion RNAs could create deletions, rearrangements, or insertions as long as they occurred outside of each functional module.

RUBIVIRUSES

One of the first characteristics that distinguished rubella virus from the alphaviruses was a limited host range: rubella virus is only found in humans and it does not grow in insect cells. Furthermore, rubella virus grows slowly and to low titers in cultured vertebrate cells. Infection is much less cytopathic than that observed with alphaviruses, and persistent infections are readily established [reviewed in (32)].

Transcription, Translation, and Replication of the Genome

The overall strategy of the replication of rubella virus is similar to that of the alphaviruses: the genomic RNA is first translated to produce the nonstructural proteins which are required for the synthesis of the complementary minus-strand RNA, the genomic RNA, and the subgenomic RNA. The subgenomic RNA is the mRNA for the viral structural proteins (see Fig. 3). Details about the nonstructural proteins and their functions are now emerging, based on sequence comparisons (25) and expression studies (89). This region of the genome contains a single open reading frame that can be translated to a 2,115 amino acid polypeptide. Motifs corresponding to a helicase and a methyl transferase domain were identified within the sequence (Fig. 3). The cDNA for this open reading frame, under the control of a heterologous promoter, has been expressed in transfected cells and gave rise to three rubella virus specific proteins with apparent molecular masses of 200-, 150- and 90-kd. Mutation of the cysteine at residue 1,151 led to the accumulation of the 200-kd protein at the expense of the smaller proteins, providing evidence that the larger protein was the precursor of the smaller and that the cysteine was essential for proteinase activity (89). Region-specific antibodies were used to locate the 150-kd protein at the amino terminus, and the 90-kd protein at the carboxyl terminus of the polyprotein (31).

As mentioned earlier, three regions of the rubella virus genome share nucleotide homology with the alphavirus genome. In the minus-strand rubella RNA, the 3′ terminal sequences and the region surrounding the junction between the nonstructural and structural protein genes are the putative promoters for synthesis of the plus-strand genome and the subgenomic RNA, respectively. In the rubella virus genome the region homologous to the alphavirus subgenomic RNA promoter is located 23 nucleotides upstream from the start of the subgenomic RNA (25,32) suggesting that the minimal promoter for transcription of the rubella virus subgenomic RNA may be larger than the one identified in the alphavirus genome .

Host cell proteins have been implicated in the replication of rubella virus RNA. Nakhashi and his colleagues (104) first showed by UV-light-induced covalent crosslinking and gel-retardation assays that a region at the 3′ terminus of the genomic plus-strand RNA bound three cellular proteins of molecular masses 61, 63, and 68 kd. Binding activity was inhibited by prior treatment of cell extracts with alkaline phosphatase. A similar analysis was carried out using two minus-strand RNA probes: the 3′ terminal stem-loop region and 46 nucleotides located 224 nucleotides from the 5′ terminus (103). The latter are homologous to the 51 nucleotides located near the 3′ terminus of the minus strand (the 5′ terminus of the plus strand) of the alphavirus genome. Three proteins of molecular masses 97, 79, and 56 kd bound to the 3′ terminal stem-loop of the minus strand; the 56-kd protein appears to be identical to the protein originally identified as binding to the 3′ terminus of the plus strand and designated as 61 kd. These proteins also bound to the 3′ terminus of the minus strands of both Sindbis and eastern equine encephalitis viruses. The two regions from the rubella virus minus strand RNA (the 46 conserved nucleotides and the 3′ terminus) competed for the same three proteins. In contrast, the 51 nucleotide conserved region from Sindbis virus only competed for the 56-kd protein.

Two host proteins, one 59 and the other 52 kd, that bind to the 5′ terminus of the plus-strand genomic RNA have also been identified (123). The protein-RNA complexes could be immunoprecipitated with a Ro-type human polyclonal serum obtained from patients with autoimmune disorders. Autoantibodies from these patients are directed against discrete classes of small ribonucleoprotein particles. Anti-Ro antibodies recognize small cytoplasmic ribonucleoproteins; the major protein in these particles is a polypeptide of 60 kd (190). The result with rubella virus RNA, taken together with a report that the human La autoantigen is involved in the correct translation of poliovirus RNA (91), led to the speculation that the Ro autoantigen may be involved in translation of rubella virus RNA (123). In this regard, the 56-kd protein that binds the 3′ terminal sequences of both the plus and minus strands of rubella virus RNA was reported to be identical to calreticulin, the human protein which is associated with the cytoplasmic ribonucleoprotein complexes that are components of the Ro/SS-A autoantigen complex (105).

Synthesis of Structural Proteins and Virus Assembly

Translation of the rubella subgenomic mRNA produces a polyprotein that is proteolytically processed by a pathway different from that of the alphaviruses (20,108–110). The rubella virus capsid protein lacks an autoprotease activity. Stretches of 23 and 20 hydrophobic amino acids precede the amino termini of the E2 and E1 glycoproteins and are postulated to act as signals for the translocation of the glycoproteins into the lumen of the endoplasmic reticulum. Cleavage between the capsid and E2 and between E2 and E1 are catalyzed by the host cell signalase. After cleavage, the E2 signal sequence remains attached to the capsid protein, which becomes anchored to the membrane (53,161). The latter interaction is probably important in targeting the assembly of the virus nucleocapsid to intracellular membranes. There is no small hydrophobic "linker" protein between the two glycoproteins, nor is there a precursor form of E2.

The two rubella virus glycoproteins resemble those of the alphaviruses in name, but their structure and synthesis are distinct [(20,108–110) and reviewed in (32)]. Transit of the rubella glycoproteins from their site of synthesis in the endoplasmic reticulum to the Golgi complex is slow with a $t_{1/2}$ of 60 to 90 minutes (54,55), two to three times longer than the transit time of the alphavirus glycoproteins (26). Both rubella virus glycoproteins contain N-linked oligosaccharides and E2 also contains O-linked glycans (86,133). These proteins are retained in the Golgi organelle and only a small fraction reach the plasma membrane. The accumulation of the viral glycoproteins in the Golgi complex may explain why virus assembly appears to occur mainly at this site.

PERSPECTIVES

Togaviruses and, in particular, alphaviruses, are among the best-characterized RNA enveloped viruses. This chapter has documented much of our knowledge of the structure and replication of these viruses, but many unanswered questions remain. Studies on the three-dimensional structure of the virus (49,117) and the viral proteins (19,165) have only just begun. Proteolytic cleavage of the nsP complex appears to play an essential role in the regulation of viral RNA synthesis (70,152), but the details have not yet been worked out. It will be important to learn if this type of regulation extends to the synthesis of rubella virus RNAs. Host cell proteins involved in these steps are now being identified.

Many alphaviruses can be highly cytopathic to vertebrate cells in culture, and infection of a variety of these cells by Sindbis virus results in apoptotic cell death (72). Recent studies with Sindbis virus replicons provide convincing evidence that early steps in alphavirus replication are responsible for the inhibition of host cell protein synthesis which accompanies infection of vertebrate cells (33). This inhibition occurs in the absence of the expression of the viral structural proteins and under conditions in which the synthesis of the subgenomic RNA is too low to be detected. The synthesis of the viral glycoproteins, however, does lead to a more rapid appearance of cytopathogenicity. An important future goal will be to correlate these observations in cultured cells with pathogenesis in humans and animals. Pathogenesis of alpha and rubella viruses is described in the following chapters.

REFERENCES

1. Abell BA, Brown DT. Sindbis virus membrane fusion is mediated by reduction of glycoprotein disulfide bridges at the cell surface. *J Virol* 1993;67:5496–5501.
2. Ahlquist P, Strauss EG, Rice CM, Strauss JH, Haseloff J, Zimmern D. Sindbis virus proteins nsP1 and nsP2 contain homology to nonstructural proteins from several RNA plant viruses. *J Virol* 1985;53: 536–542.
3. Aliperti G, Schlesinger MJ. Evidence for an autoprotease of Sindbis virus capsid protein. *Virology* 1978;90:366–369.
4. Anthony RP, Brown DT. Protein-protein interactions in an alphavirus membrane. *J Virol* 1991;65:1187–1194.
5. Anthony RP, Paredes AM, Brown DT. Disulfide bonds are essential for the stability of the Sindbis virus envelope. *Virology* 1992;190:330–336.
6. Baric RS, Carlin LJ, Johnston RE. Requirement for host transcription in the replication of Sindbis virus. *J Virol* 1983;45:200–205.
7. Barton DJ, Sawicki S, Sawicki DL. Solubilization and immunoprecipitation of alphavirus replication complexes. *J Virol* 1991;65:1496–1506.
8. Barton DJ, Sawicki SG, Sawicki DL. Demonstration in vitro of temperature-sensitive elongation of RNA in Sindbis virus mutant ts6. *J Virol* 1988;62:3597–602.
9. Bell JR, Rice CM, Hunkapiller MW, Strauss JH. The N-terminus of PE2 in Sindbis virus-infected cells. *Virology* 1982;119:255–267.
10. Bell JR, Strauss JH. In vivo NH_2-terminal acetylation of Sindbis virus proteins. *J Biol Chem* 1981;256:8006–8011.
11. Bonatti S, Cancedda FD. Posttranslational modifications of Sindbis virus glycoproteins: electrophoretic analysis of pulse-chase labeled infected cells. *J Virol* 1982;42:64–70.
12. Bonatti S, Migliaccio G, Simons K. Palmitylation of viral membrane glycoproteins takes place after exit from the endoplasmic reticulum. *J Biol Chem* 1989;264:12590–12595.
13. Bredenbeek P, Rice CM. Animal RNA virus expression systems. *Semin Virology* 1992;3:297–310.
14. Bredenbeek PJ, Frolov I, Rice CM, Schlesinger S. Sindbis virus expression vectors: packaging of RNA replicons by using defective helper RNAs. *J Virol* 1993;67:6439–6446.
15. Brown DT, Condreay LD. Replication of alphaviruses in mosquito cells In: Schlesinger S, Schlesinger MJ, eds. *The Togaviridae and Flaviviridae*. New York: Plenum Press, 1986;171–207.
16. Burge BW, Pfefferkorn ER. Complementation between temperature-sensitive mutants of Sindbis virus. *Virology* 1966;30:214–223.
17. Burge BW, Pfefferkorn ER. Isolation and characterization of conditional-lethal mutants of Sindbis virus. *Virology* 1966;30:204–213.
18. Carr CM, Kim PS. A spring-loaded mechanism for the conformational change of influenza hemagglutinin. *Cell* 1993;73:823–832.
19. Choi H-K, Tong L, Minor W, Dumas P, Boege U, Rossmann MG, Wengler G. Structure of Sindbis virus core protein reveals a chymotrypsin like serine protease and the organization of the virion. *Nature* 1991;354:37–43.
20. Clarke DM, Loo TW, Hui I, Chong P, S. G. Nucleotide sequence and *in vitro* expression of rubella virus 24S subgenomic messenger RNA encoding the structural proteins E1, E2 and C. *Nucleic Acids Res* 1987;15:3041–3057.
21. Collier NC, Adams SP, Weingarten H, Schlesinger MJ. Inhibition of enveloped RNA virus formation by peptides corresponding to glycoprotein sequences. *Antiviral Chemistry and Chemotherapy* 1992;3:31–36.
22. Davis NL, Willis LV, Smith JF, Johnston RE. In vitro synthesis of infectious Venezuelan equine encephalitis virus RNA from a cDNA clone: analysis of a viable deletion mutant. *Virology* 1989;171:189–204.
23. de Groot RJ, Hardy WR, Shirako Y, Strauss JH. Cleavage-site preferences of Sindbis virus polyproteins containing the non-structural proteinase. Evidence for temporal regulation of polyprotein processing *in vivo*. *EMBO J* 1990;9:2631–2638.
24. Ding M, Schlesinger MJ. Evidence that Sindbis virus nsP2 is an autoprotease which processes the virus nonstructural polyprotein. *Virology* 1989;280–284.
25. Dominguez G, Wang, C-Y, Frey TK. Sequence of the genome RNA of rubella virus: evidence for genetic rearrangement during Togavirus evolution. *Virology* 1990;177:225–238.
26. Doms RW, Lamb RA, Rose JK, Helenius A. Folding and assembly of viral membrane proteins. *Virology* 1993;193:545–562.
27. Dubuisson J, Rice CM. Sindbis virus attachment: isolation and characterization of mutants with impaired binding to vertebrate cells. *J Virol* 1993;67:3363–3374.
28. Edwards J, Brown DT. Sindbis virus infection of a Chinese hamster ovary cell mutant defective in the acidification of endosomes. *Virology* 1991;182:28–33.
29. Fan DP, Sefton BM. The entry into host cells of Sindbis virus, vesicular stomatitis virus and Sendai virus. *Cell* 1978;15:985–992.
30. Flynn DC, Meyer WJ, Mackenzie JMJ, Johnston RE. A conformational change in Sindbis virus glycoproteins E1 and E2 is detected at the plasma membrane as a consequence of early virus-cell interaction. *J Virol* 1990;64:3643–3653.
31. Forng R-Y, Frey TK. Identification of the rubella virus nonstructural proteins. *Virology* 1995;206:843–853.

32. Frey TK. Molecular biology of rubella virus. *Advances in Virus Research* 1994;44:69–160.
33. Frolov I, Schlesinger S. Comparison of the effects of Sindbis virus and Sindbis virus replicons on host cell protein synthesis and cytopathogenicity in BHK cells. *J Virol* 1994;68:1721–1727.
34. Froshauer S, Kartenbeck J, Helenius A. Alphavirus RNA replication is located on the cytoplasmic surface of endosomes and lysosomes. *J Cell Biol* 1988;107:2075–2086.
35. Gaedigk-Nitschko K, Ding M, Levy MA, Schlesinger MJ. Site-directed mutations in the Sindbis virus 6K protein reveal sites for fatty acylation and the underacylated protein affects virus release and virion structure. *Virology* 1990;175:282–291.
36. Gaedigk-Nitschko K, Schlesinger MJ. Site-directed mutations in Sindbis virus E2 glycoprotein's cytoplasmic domain and the 6K protein lead to similar defects in virus assembly and budding. *Virology* 1991;183:206–214.
37. Geigenmüller-Gnirke U, Nitschko H, Schlesinger S. Deletion analysis of the capsid protein of Sindbis virus: identification of the RNA binding region. *J Virol* 1993;67:1620–1626.
38. Geigenmüller-Gnirke U, Weiss B, Wright R, Schlesinger S. Complementation between Sindbis viral RNAs produces infectious particles with a bipartite genome. *Proc Natl Acad Sci USA* 1991;88:3253–3257.
39. Gibson R, Kornfeld S, Schlesinger S. A role for oligosaccharides in glycoprotein biosynthesis. *Trends Biochem Sci* 1980;5:290–293.
40. Goldbach R, Le Gall O, Wellink J. Alpha-like viruses in plants. *Seminars in Virology* 1991;2:19–25.
41. Gorbalenya AE, Koonin EV, A.P. D, Blinov VM. A novel superfamily of nucleoside triphosphate-binding motif containing proteins which are probably involved in duplex unwinding in DNA and RNA replication and recombination. *FEBS Lett* 1988;235:16–24.
42. Hahn CS, Hahn YS, Braciale TJ, Rice CM. Infectious Sindbis virus transient expression vectors for studying antigen processing and presentation. *Proc Natl Acad Sci USA* 1992;89:2679–2683.
43. Hahn CS, Lustig S, Strauss EG, Strauss JH. Western equine encephalitis virus is a recombinant virus. *Proc Natl Acad Sci USA* 1988;85:5997–6001.
44. Hahn CS, Rice CM, Strauss EG, Lenches EM, Strauss JH. Sindbis virus ts103 has a mutation in glycoprotein E2 that leads to defective assembly of virions. *J Virol* 1989;63:3459–3465.
45. Hahn CS, Strauss EG, Strauss JH. Sequence analysis of three Sindbis virus mutants temperature-sensitive in the capsid protein autoproteinase. *Proc Natl Acad Sci USA* 1985;82:4648–4652.
46. Hahn YS, Grakoui A, Rice CM, Strauss EG, Strauss JH. Mapping of RNA- temperature-sensitive mutants of Sindbis virus: complementation group F mutants have lesions in nsP4. *J Virol* 1989;63:1194–1202.
47. Hahn YS, Strauss EG, Strauss JH. Mapping of RNA- temperature-sensitive mutants of Sindbis virus:assignment of complementation groups A, B, and G to nonstructural proteins. *J Virol* 1989;63:3142–3150.
48. Hardy WR, Strauss JH. Processing the nonstructural polyproteins of Sindbis virus: nonstrutural proteinase is in the C-terminal half of nsP2 and functions both in cis and in trans. *J Virol* 1989;63:4653–4664.
49. Harrison SC, Strong RK, Schlesinger S, Schlesinger MJ. Crystallization of Sindbis virus and its nucleocapsid. *J Mol Biol* 1992;226:277–280.
50. Heidner HW, McKnight KL, Davis NL, Johnston RE. Lethality of PE2 incorporation into Sindbis virus can be suppressed by second-site mutations in E3 and E2: pleiotropic effects of PE2 incorporation on replication in cultured vertebrate cells, mosquito cells, and neonatal mice. *J Virol* 1994;68:2683–2692.
51. Helenius A, Morein B, Fries E, et al. Human (HLA-A and HLA-B) and murine (H-2K and H-2D) histocompatibility antigens are cell surface receptors for Semliki Forest virus. *Proc Natl Acad Sci USA* 1978;75:3846–3850.
52. Hertz JM, Huang HV. Utilization of heterologous alphavirus junction sequences as promoters by Sindbis virus. *J Virol* 1992;66:857–864.
53. Hobman TC, Gillam S. In vitro and in vivo expression of rubella virus E2 glycoprotein: the signal peptide is located in the C-terminal region of capsid protein. *Virology* 1989;173:241–250.
54. Hobman TC, Woodward L, Farquhar MG. The rubella virus E1 glycoprotein is arrested in a novel post-ER, pre-Golgi compartment. *J Cell Biol* 1992;118:795–811.
55. Hobman TC, Woodward L, Farquhar MG. The rubella virus E2 and E1 spike glycoproteins are targeted to the Golgi complex. *J Cell Biol* 1993;121:269–281.
56. Ivanova L, Schlesinger MJ. Site-directed mutations in the Sindbis virus E2 glycoprotein identify palmitoylation sites and affect virus budding. *J Virol* 1993;67:2546–2551.
57. Justman J, Klimjack MR, Kielian M. Role of spike protein conformational changes in fusion of Semliki Forest virus. *J Virol* 1993;67:7597–7607.
58. Kamer G, Argos P. Primary structural comparison of RNA-dependent polymerases from plant, animal and bacterial viruses. *Nucleic Acids Res* 1984;12:7269–7282.
59. Keegstra K, Sefton B, Burke D. Sindbis virus glycoproteins: effect of the host cell on the oligosaccharides. *J Virol* 1975;16:613–620.
60. Khatchikian D, Orlich M, Rott R. Increased viral pathogenicity after insertion of a 28S ribosomal RNA sequence into the haemagglutinin gene of an influenza virus. *Nature* 1989;340:156–157.
61. King AMQ. Genetic recombination in positive strand RNA viruses In: Domingo E, Holland JJ, Ahlquist P, eds. *RNA Genetics*. Boca Raton: CRC Press, Inc, 1988;150–185.
62. Kornfeld R, Kornfeld S. Assembly of asparagine-linked oligosaccharides. *Ann Rev Biochem* 1985;54:631–664.
63. Kuhn RJ, Hong Z, Strauss JH. Mutagenesis of the 3′ nontranslated region of Sindbis virus RNA. *J Virol* 1990;64:1465–1476.
64. Kuhn RJ, Niesters HGM, Hong Z, Strauss JH. Infectious RNA transcripts from Ross River virus cDNA clones and the construction and characterization of defined chimeras with Sindbis virus. *Virology* 1991;182:430–441.
65. Lai MMC. RNA recombination in animal and plant viruses. *Microbiol Rev* 1992;56:61–79.
66. LaStarza MW, Lemm JA, Rice CM. Genetic analysis of the nsP3 region of Sindbis virus: evidence for roles in minus-strand and subgenomic RNA synthesis. *J Virol* 1994;68:5781–5791.
67. Lemm JA, Durbin RK, Stollar V, Rice CM. Mutations which alter the level or structure of nsP4 can affect the efficiency of Sindbis virus replication in a host-dependent manner. *J Virol* 1990;64:3001–3011.
68. Lemm JA, Rice CM. Assembly of functional Sindbis virus RNA replication complexes: requirement for coexpression of P123 and P34. *J Virol* 1993;67:1905–1915.
69. Lemm JA, Rice CM. Roles of nonstructural polyproteins and cleavage products in regulating Sindbis virus RNA replication and transcription. *J Virol* 1993;67:1916–1926.
70. Lemm JA, Rumenapf T, Strauss EG, Strauss JH, Rice CM. Polypeptide requirements for assembly of functional Sindbis virus replication complexes: a model for the temporal regulation of minus and plus-strand RNA synthesis. *EMBO J* 1994;13:2925–2934.
71. Lenard J. Lipids of alphaviruses. In: Schlesinger RW, eds. *The Togaviruses*. New York: Academic Press, 1980;335–341.
72. Levine B, Huang Q, Isaacs JT, Reed JC, Griffin DE, Hardwick JM. Conversion of lytic to persistent alphavirus infection by the *bcl-2* cellular oncogene. *Nature* 1993;361:739–742.
73. Levis R, Schlesinger S, Huang HV. Promoter for Sindbis virus RNA-dependent subgenomic RNA transcription. *J Virol* 1990;64:1726–1733.
74. Levis R, Weiss BG, Tsiang M, Huang H, Schlesinger S. Deletion mapping of Sindbis virus DI RNAs derived from cDNAs defines the sequences essential for replication and packaging. *Cell* 1986;44:137–145.
75. Li G, La Starza MW, Hardy WR, Strauss JH, Rice CM. Phosphorylation of Sindbis virus nsP3 in vivo and in vitro. *Virology* 1990;179:416–427.
76. Li G, Rice CM. Mutagenesis of the in-frame opal termination codon preceding nsP4 of Sindbis virus: studies of translational readthrough and its effect on viral replication. *J Virol* 1989;63:1326–1337.
77. Liljeström P, Garoff H. Internally located cleavable signal sequences direct the formation of Semliki Forest virus membrane proteins from a polyprotein precursor. *J Virol* 1991;65:147–154.
78. Liljeström P, Garoff H. A new generation of animal cell expression vectors based on the Semliki Forest virus replicon. *Bio/Technology* 1991;9:1356–1361.
79. Liljeström P, Lusa S, Huylebroeck D, Garoff H. In vitro mutagenesis of a full-length cDNA clone of Semliki Forest virus: the small 6,000-molecular-weight membrane protein modulates virus release. *J Virol* 1991;65:4107–4113.
80. Liu N, Brown DT. Phosphorylation and dephosphorylation events play critical roles in Sindbis virus maturation. *Virology* 1993;196:703–711.
81. Liu N, Brown DT. Transient translocation of the cytoplasmic (endo) domain of a type I membrane glycoprotein into cellular membranes. *J Cell Biol* 1993;120:877–883.
82. Lobigs M, Garoff H. Fusion function of the Semliki Forest virus spike

is activated by proteolytic cleavage of the envelope glycoprotein precursor p62. *J Virol* 1990;64:1233–1240.
83. Lobigs M, Wahlberg JM, Garoff H. Spike protein oligomerization control of Semliki Forest virus fusion. *J Virol* 1990;64:5214–5218.
84. London SD, Schmaljohn AL, Dalrymple JM, Rice CM. Infectious enveloped RNA virus antigenic chimeras. *Proc Natl Acad Sci USA* 1991;89:207–211.
85. Lopez S, Yao J-S, Kuhn RJ, Strauss EG, Strauss JH. Nucleocapsid-glycoprotein interactions required for assembly of alphaviruses. *J Virol* 1994;68:1316–1323.
86. Lundström ML, Mauraccher CA, Tingle AJ, . Characterization of carbohydrates linked to rubella virus glycoprotein E2. *J Gen Virol* 1991;72:843–850.
87. Magee AI, Koyama AH, Malfer C, Wen D, Schlesinger MJ. Release of fatty acids from virus glycoproteins by hydroxylamine. *Biochim Biophys Acta* 1984;798:156–166.
88. Marquardt MT, Phalen T, Kielian M. Cholesterol is required in the exit pathway of Semliki Forest virus. *J Cell Biol* 1993;123:57–65.
89. Marr LD, Wang C-Y, Frey TK. Expression of the rubella virus nonstructural protein ORF and demonstration of proteolytic processing. *Virology* 1994;198:586–592.
90. Marsh M, Helenius A. Virus entry into animal cells. *Adv Virus Research* 1989;36:107–151.
91. Meerovitch K, Svitkin YV, Lee HS, Lejbkowicz F, Kenan DJ, Chan EKL, Agol VI, Keene JD, Sonenberg N. La autoantigen enhances and corrects aberrant translation of poliovirus RNA in reticulocyte lysate. *J Virol* 1993;67:3798–3807.
92. Melancon P, Garoff H. Reinitiation of translocation in the Semliki Forest virus structural polyprotein:identification of the signal for the E1 glycoprotein. *EMBO J* 1986;5:1551–1560.
93. Melancon P, Garoff H. Processing of the Semliki Forest virus structural polyprotein: role of the capsid protease. *J Virol* 1987; 61:1301–1309.
94. Metsikko K, Garoff H. Oligomers of the cytoplasmic domain of the p62/E2 membrane protein of Semliki Forest virus bind to the nucleocapsid in vitro. *J Virol* 1990;64:4678–4683.
95. Meyer WJ, Gidwitz S, Ayers VK, Schoepp RJ, Johnston RE. Conformational alteration of Sindbis virion glycoproteins induced by heat, reducing agents, or low pH. *J Virol* 1992;66:3504–3513.
96. Meyer WJ, Johnston RE. Structural rearrangement of infecting Sindbis virions at the cell surface: mapping of newly accesible epitopes. *J Virol* 1993;67:5117–5125.
97. Meyers G, Tautz N, Dubovi EJ, Thiel H-J. Viral cytopathogenicity correlated with integration of ubiquitin-coding sequences. *Virology* 1991;180:602–616.
98. Mi S, Durbin R, Huang HV, Rice CM, Stollar V. Association of the Sindbis virus RNA methyltransferase activity with the nonstructural protein nsP1. *Virology* 1989;1770:385–391.
99. Miller ML, Brown DT. Morphogenesis of Sindbis virus in three subclones of *Aedes albopictus* (mosquito) cells. *J Virol* 1992;66:4180–4190.
100. Moehring JM, Inocencio NM, Robertson BJ, Moehring TJ. Expression of mouse furin in a Chinese hamster cell resistant to *Pseudomonas* exotoxin A and viruses complements the genetic lesion. *J Biol Chem* 1993;268:2590–2594.
101. Monroe SS, Schlesinger S. RNAs from two independently isolated defective interfering particles of Sindbis virus contain a cellular tRNA sequence at their 5′ ends. *Proc Natl Acad Sci USA* 1983;80:3279–3283.
102. Mulvey M, Brown DT. Formation and rearrangement of disulfide bonds during maturation of the Sindbis virus E1 glycoprotein. *J Virol* 1994;68:805–812.
103. Nakhasi HL, Cao X-Q, Rouault TA, Liu T-Y. Specific binding of host cell proteins to the 3′ -terminal stem-loop structure of rubella virus negative-strand RNA. *J Virol* 1991;65:5961–5967.
104. Nakhasi HL, Rouault TA, Haile DJ, Liu T-Y, Klausner RD. Specific high-affinity binding of host cell proteins to the 3′ region of rubella virus RNA. *New Biol* 1990;2:255–264.
105. Nakhasi HL, Singh NK, Pogue GP, Cao X-Q, Rouault TR. Identification and characterization of host factor interactions with *cis*-acting elements of rubella virus RNA. In: Brinton MA, Rueckert R, eds. *Proceedings of the 3rd International Positive Strand RNA Virus Conference*. Washington, D.C.: American Society for Microbiology, 1994;in press.
106. Niesters HGM, Strauss JH. Defined mutations in the 5′ nontranslated sequence of Sindbis virus RNA. *J Virol* 1990;64:4162–4168.
107. Niesters HGM, Strauss JH. Mutagenesis of the conserved 51-nucleotide region of Sindbis virus. *J Virol* 1990;64:1639–1647.
108. Oker-Blom C. The gene order for rubella virus structural proteins is NH_2-C-E2-E1-COOH. *J Virol* 1984;51:354–358.
109. Oker-Blom C, Kalkkinen N, Kääriäinen L, Pettersson RF. Rubella virus contains one capsid protein and three envelope glycoproteins, E1, E2a, and E2b. *J Virol* 1983;46:964–973.
110. Oker-Blom C, Ulmanen I, Kääriäinen L, Petterson RF. Rubella virus 40S genome specifies a 24S subgenomic mRNA that codes for a precursor to structural proteins. *J Virol* 1984;49:403–408.
111. Oldstone MBA, Tishon A, Dutko FJ, Kennedy SIT, Holland JJ, Lampert PW. Does the major histocompatibility complex serve as a specific receptor for Semliki Forest virus. *J Virol* 1980;34:256–265.
112. Ou J-H, Rice CM, Dalgarno L, Strauss EG, Strauss JH. Sequence studies of several alphavirus genomic RNAs in the region containing the start of the subgenomic RNA. *Proc Natl Acad Sci USA* 1982; 79:5235–5239.
113. Ou J-H, Strauss EG, Strauss JH. Comparative studies of the 3′ -terminal sequences of several alphavirus RNAs. *Virology* 1981; 109:281–289.
114. Ou J-H, Strauss EG, Strauss JH. The 5′ -terminal sequences of the genomic RNAs of several alphaviruses. *J Mol Biol* 1983;168:1–15.
115. Pardigon N, Lenches E, Strauss JH. Multiple binding sites for cellular proteins in the 3′ end of Sindbis alphavirus minus-sense RNA. *J Virol* 1993;67:5003–5011.
116. Pardigon N, Strauss JH. Cellular proteins bind to the 3′ end of Sindbis virus minus-strand RNA. *J Virol* 1992;66:1007–1015.
117. Paredes AM, Brown DT, Rothnagel R, Chiu W, Schoepp RJ, Johnston RE, Prasad BVV. Three-dimensional structure of a membrane-containing virus. *Proc Natl Acad Sci USA* 1993;90:9095–9099.
118. Paredes AM, Simon MN, Brown DT. The mass of the Sindbis virus nucleocapsid suggests it has T=4 icosahedral symmetry. *Virology* 1992; 187:329–332.
119. Peränen J, Rikkonen M, Liljeström P, Kääriäinen L. Nuclear localization of the Semliki Forest virus-specific nonstructural protein nsP2. *J Virol* 1990;64:1888–1896.
120. Peränen J, Takkinen K, Kalkkinen N, Kääriäinen L. Semliki Forest virus-specific non-structural protein nsP3 is a phosphoprotein. *J Gen Virol* 1988;69:2165–2178.
121. Phalen T, Kielian M. Cholesterol is required for infection by Semliki Forest virus. *J Cell Biol* 1991;112:615–623.
122. Pierce JS, Strauss EG, Strauss JH. Effects of ionic strength on the binding of Sindbis virus to chick cells. *J Virol* 1975;13:1030–1036.
123. Pogue GP, Cao X-Q, Singh NK, Nakhasi HL. 5′ Sequences of rubella virus RNA stimulate translation of chimeric RNAs and specifically interact with two host-encoded proteins. *J Virol* 1993; 67:7106–7117.
124. Presley JF, Brown DT. The proteolytic cleavage of PE2 to envelope glycoprotein E2 is not strictly required for the maturation of Sindbis virus. *J Virol* 1989;63:1975–1980.
125. Presley JF, Polo JM, Johnston RE, Brown DT. Proteolytic processing of the Sindbis virus membrane protein precursor PE2 is nonessential for growth in vertebrate cells but is required for efficient growth in invertebrate cells. *J Virol* 1991;65:1905–1909.
126. Raju R, Huang HV. Analysis of Sindbis virus promoter recognition in vivo, using novel vectors with two subgenomic mRNA promoters. *J Virol* 1991;65:2531–2510.
127. Renkonen O, Kääriäinen L, Simons K, Gahmberg CG. The lipid composition of Semliki Forest virus and of plasma membranes of the host cells. *Virology* 1971;46:318–426.
128. Rice CM, Levis R, Strauss JH, Huang HV. Production of infectious RNA transcripts from Sindbis virus cDNA clones: mapping of lethal mutations, rescue of a temperature-sensitive marker, and in vitro mutagenesis to generate defined mutants. *J Virol* 1987;61:3809–19.
129. Rikkonen M, Peränen J, Kääriäinen L. Nuclear and nucleolar targeting signals of Semliki Forest virus nonstructural protein nsP2. *Virology* 1992;189:462–473.
130. Rümenapf T, Strauss EG, Strauss JH. Subgenomic mRNA of Aura alphavirus is packaged. *J Virol* 1994;68:56–62.
131. Russell DL, Dalrymple JM, Johnston RE. Sindbis virus mutations which coordinately affect glycoprotein processing, penetration and virulence in mice. *J Virol* 1989;63:1619–1629.
132. Salminen A, Wahlberg JM, Lobigs M, Liljeström P, Garoff H. Membrane fusion process of Semliki Forest virus II: cleavage-dependent reorganization of the spike protein complex controls virus entry. *J Cell Biol* 1992;116:349–357.
133. Sanchez A, Frey TK. Vaccinia-vectored expression of the rubella virus structural proteins and characterization of the E1 and E2 glycosidic

linkages. *Virology* 1991;183:636–646.
134. Sawicki DL, Barkhimer DB, Sawicki SG, Rice CM, Schlesinger S. Temperature sensitive shut-off of alphavirus minus strand RNA synthesis maps to a nonstructural protein nsP4. *Virology* 1990;174:43–52.
135. Sawicki DL, Sawicki SG. A second nonstructural protein functions in the regulation of alphavirus negative-strand RNA synthesis. *J Virol* 1993;67:3605–3610.
136. Sawicki DL, Sawicki SG, Keränen S, Kääriäinen L. Specific Sindbis virus coded functions for minus strand RNA synthesis. *J Virol* 1981;39:348–358.
137. Sawicki SG, Sawicki DL, Kääriäinen L, Keränen S. A Sindbis virus mutant temperature-sensitive in the regulation of minus-strand RNA synthesis. *Virology* 1981;115:161–172.
138. Scheidel LM, Stollar V. Mutations that confer resistance to mycophenolic acid and ribavirin on Sindbis virus map to the nonstructural protein nsP1. *Virology* 1991;181:490–499.
139. Schlesinger MJ, London SD, Ryan C. An in-frame insertion into the Sindbis virus 6K gene leads to defective proteolytic processing of the virus glycoproteins, a *trans*-dominant negative inhibition of normal virus formation, and interference in virus shut off of host-cell protein synthesis. *Virology* 1993;193:424–432.
140. Schlesinger MJ, Malfer C. Cerulenin blocks fatty acid acylation of glycoproteins and inhibits vesicular stomatitis and Sindbis virus particle formation. *J Biol Chem* 1982;257:9887–9890.
141. Schlesinger MJ, Schlesinger S. Domains of virus glycoproteins In: Maramorosch K, Murphy FA, Shatkin AJ, eds. *Advances in virus research.* Orlando: Academic Press, 1987;1–44.
142. Schlesinger S. The generation and amplification of defective interfering RNAs. In: Domingo E, Holland JJ, Ahlquist P, eds*RNA Genetics*. Boca Raton: CRC Press, 1988;167–185.
143. Schlesinger S. Alphaviruses—vectors for the expression of heterologous genes. *Trends Biotechnol* 1993;11:18–22.
144. Schlesinger S, Weiss B, Nitschko H. Sindbis virus RNAs bind the viral capsid protein specifically and are preferentially encapsidated. In: Brinton MA, Heinz FX, eds. *New aspects of positive-strand RNA viruses*. Washington, DC: American Society for Microbiology, 1990;237–244.
145. Schlesinger S, Weiss BG. Defective RNAs of Alphaviruses. In: Schlesinger S, Schlesinger MJ, eds. *The Togaviridae and Flaviviridae.* New York: Plenum Press, 1986;149–166.
146. Schmaljohn AL, Kokuban KM, Cole GA. Protective monoclonal antibodies define maturational and pH-dependent antigenic changes in Sindbis virus E1 glycoprotein. *Virology* 1983;130:144–154.
147. Schmidt MFG, Schlesinger MJ. Relation of fatty acid attachment to the translation and maturation of vesicular stomatitis and Sindbis virus membrane glycoproteins. *J Biol Chem* 1980;255:3334–3339.
148. Schoepp RJ, Johnston RE. Directed mutagenesis of a Sindbis virus pathogenesis site. *Virology* 1993;193:149–159.
149. Sefton B, Gaffney BJ. Effect of the viral proteins on the fluidity of the membrane lipids in Sindbis virus. *J Mol Biol* 1974;90:343–358.
150. Sefton BM. Immediate glycosylation of Sindbis virus membrane proteins. *Cell* 1977;10:659–668.
151. Shirako Y, Strauss JH. Cleavage between nsP1 and nsP2 initiates the processing pathway of Sindbis virus nonstructural polyprotein P123. *Virology* 1990;177:54–64.
152. Shirako Y, Strauss JH. Regulation of Sindbis virus RNA replication: uncleaved P123 and nsP4 function in minus strand RNA synthesis whereas cleaved products from P123 are required for efficient plus strand RNA synthesis. *J Virol* 1994;68:1874–1885.
153. Singh I, Helenius A. Role of ribosomes in Semliki Forest virus nucleocapsid uncoating. *J Virol* 1992;66:7049–7058.
154. Strauss EG, De Groot RJ, Levinson R, Strauss JH. Identification of the active site residues in the nsP2 proteinases of Sindbis virus. *Virology* 1992;191:932–940.
155. Strauss EG, Levinson R, Rice CM, Dalrymple J, Strauss JH. Nonstructural proteins nsP3 and nsP4 of Ross River and O'Nyong-nyong viruses: sequence and comparison with those of other alphaviruses. *Virology* 1988;164:265–274.
156. Strauss EG, Rice CM, Strauss JH. Sequence coding for the alphavirus nonstructural proteins is interrupted by an opal termination codon. *Proc Natl Acad Sci USA* 1983;80:5271–5275.
157. Strauss EG, Strauss JH. Mutants of alphaviruses: Genetics and physiology. In: Schlesinger RW, eds. *The Togaviruses*. New York: Academic Press, 1980;393–426.
158. Strauss EG, Strauss JH. Structure and Replication of the Alphavirus Genome. In: Schlesinger S, Schlesinger MJ, eds. *The Togaviridae and Flaviviridae*. New York: Plenum Press, 1986;35–82.
159. Strauss JH, Strauss EG. Evolution of RNA Viruses. *Ann Rev Microbiol* 1988;42:657–683.
160. Suomalainen M, Garoff H. Alphavirus spike-nucleocapsid interaction and network antibodies. *J Virol* 1992;66:5106–5109.
161. Suomalainen M, Garoff H, Baron MD. The E2 signal sequence of rubella virus remains part of the capsid protein and confers membrane association in vitro. *J Virol* 1990;64:5500–5509.
162. Symington J, Schlesinger MJ. Isolation of a Sindbis virus variant by passage on mouse plasmacytoma cells. *J Virol* 1975;15:1037–1041.
163. Symington J, Schlesinger MJ. Characterization of a Sindbis virus variant with altered host range. *Arch Virol* 1978;58:127–136.
164. Takkinen K, Peränen J, Keränen S, Söderlund H, Kääriäinen L. The Semliki Forest virus-specific nonstructural protein nsP4 is an autoproteinase. *Eur J Biochem* 1990;189:33–38.
165. Tong L, Wengler G, Rossmann MG. Refined structure of Sindbis virus core protein and comparison with other chymotrypsin-like serine proteinase structures. *J Mol Biol* 1993;230:228–247.
166. Tsiang M, Weiss BG, Schlesinger S. Effects of 5′ -terminal modifications on the biological activity of defective interfering RNAs of Sindbis virus. *J Virol* 1988;62:47–53.
167. Tucker PC, Griffin DE. Mechanism of altered Sindbis virus neurovirulence associated with a single-amino-acid change in the E2 glycoprotein. *J Virol* 1991;65:1551–1557.
168. Ubol S, Griffin D. Identification of a putative alphavirus receptor on mouse neural cells. *J Virol* 1991;65:6913–6921.
169. Ulmanen I, Söderlund H, Kääriäinen L. Semliki Forest virus capsid protein associates with the 60S ribosomal subunit in infected cells. *J Virol* 1976;20:203–210.
170. Vaananen P, Kääriäinen L. Fusion and haemolysis of erythrocytes caused by three togaviruses; Semliki Forest, Sindbis and rubella. *J Gen Virol* 1980;46:467–475.
171. Vaux DJT, Helenius A, Mellman I. Spike-nucleocapsid interactions in Semliki Forest virus reconstructed using network antibodies. *Nature* 1988;336:36–42.
172. Wahlberg JM, Bron R, Wilschut J, Garoff H. Membrane fusion of Semliki Forest virus involves homotrimers of the fusion protein. *J Virol* 1992;66:7309–7318.
173. Wahlberg JM, Garoff H. Membrane fusion process of Semliki Forest virus I: low pH-induced rearrangement in spike protein quaternary structure precedes virus penetration into cells. *J Cell Biol* 1992;116:339–348.
174. Wang C-Y, Dominguez G, Frey TK. Characterization of rubella virus genomic RNA and construction of an infectious cDNA clone. 1994;*submitted.*
175. Wang K-S, Kuhn RJ, Strauss EG, Ou S, Strauss JH. High-affinity laminin receptor is a receptor for Sindbis virus in mammalian cells. *J Virol* 1992;66:4992–5001.
176. Wang K-S, Schmaljohn AL, Kuhn RJ, Strauss JH. Antiidiotypic antibodies as probes for the Sindbis virus receptor. *Virology* 1991;181:694–702.
177. Wang Y-F, Sawicki SG, Sawicki DL. Sindbis nsP1 functions in negative-strand RNA synthesis. *J Virol* 1991;65:985–988.
178. Watson DG, Moehring JM, Moehring TJ. A mutant CHO-K1 strain with resistance to *Pseudomonas* exotoxin A and alphaviruses fails to cleave Sindbis virus glycoprotein PE2. *J Virol* 1991;65:2332–2339.
179. Weiss B, Geigenmüller-Gnirke U, Schlesinger S. Interactions between Sindbis virus RNAs and a 68 amino acid derivative of the viral capsid protein further defines the capsid binding site. *Nucleic Acids Res* 1994;22:780–786.
180. Weiss B, Nitschko H, Ghattas I, Wright R, Schlesinger S. Evidence for specificity in the encapsidation of Sindbis virus RNAs. *J Virol* 1989;63:5310–5318.
181. Weiss BG, Schlesinger S. Recombination between Sindbis virus RNAs. *J Virol* 1991;65:4017–4025.
182. Wengler G. The mode of assembly of alphavirus cores implies a mechanism for the disassembly of the cores in the early stages of infection. *Arch Virol* 1987;94:1–14.
183. Wengler G, Boege U, Wengler G, Bischoff H, Wahn K. The core protein of the alphavirus Sindbis virus assembles into core-like nucleoproteins with the viral genome RNA and with other single-stranded nucleic acids *in vitro*. *Virology* 1982;118:401–410.
184. Wengler G, Wengler G. Identification of a transfer of viral core protein to cellular ribosomes during the early stages of alphavirus infection. *Virology* 1984;134:435–442.

185. Wengler G, Wengler G, Boege U, Wahn K. Establishment and analysis of a system which allows assembly and disassembly of alphavirus core-like particles under physiological conditions *in vitro*. *Virology* 1984;132:401–410.
186. Wengler G, Würkner D, Wengler G. Identification of a sequence element in the alphavirus core protein which mediates interaction of cores with ribosomes and the disassembly of cores. *Virology* 1992; 191:880–888.
187. White J, Helenius A. pH-Dependent fusion between the Semliki Forest virus membrane and liposomes. *Proc Natl Acad Sci USA* 1980;77:3273–3277.
188. White J, Kartenbeck J, Helenius A. Fusion of Semliki Forest virus with the plasma membrane can be induced by low pH. *J Cell Biol* 1980;87:264–272.
189. White J, Matlin K, Helenius A. Cell fusion by Semliki Forest, influenza, vesicular stomatitis viruses. *J Cell Biol* 1981;89:674–679.
190. Wolin SL, Steitz JA. The Ro small cytoplasmic ribonucleoproteins: identification of the antigenic protein and its binding site on the Ro RNAs. *Proc Natl Acad Sci USA* 1984;81:1996–2000.
191. Xiong C, Levis R, Shen P, Schlesinger S, Rice CM, Huang HV. Sindbis virus: an efficient, broad host range vector for gene expression in animal cells. *Science* 1989;243:1188–1191.
192. Zhao H, Garoff H. Role of cell surface spikes in alphavirus budding. *J Virol* 1992;66:7089–7095.
193. Zimmern D. Evolution of RNA viruses. In: Domingo E, Holland JJ, Ahlquist P, eds. *RNA Genetics*. Boca Raton: CRC Press, 1988;211–240.

Fundamental Virology, Third Edition
edited by B.N. Fields, D.M. Knipe, P.M. Howley, et al.
Lippincott - Raven Publishers, Philadelphia © 1996

CHAPTER 18

Coronaviridae: The Viruses and Their Replication

Kathryn V. Holmes and Michael M. C. Lai

CLASSIFICATION

Coronaviruses, a genus in the family *Coronaviridae*, are large, enveloped, plus-strand RNA viruses that cause highly prevalent diseases in humans and domestic animals (Table 1). They have the largest genomes of all RNA viruses and replicate by a unique mechanism which results in a high frequency of recombination. Virions mature by budding at intracellular membranes, and infection may induce cell fusion.

Coronaviruses were initially recognized as a separate virus group by their distinctive virion morphology in negatively stained preparations (159,225,226) (Fig. 1). The viral envelopes are studded with long, petal-shaped spikes and the nucleocapsids are long, flexible helices. Characteristics now used to classify *Coronaviridae* also include their unique RNA replication strategy, genome organization and mRNA structure, nucleotide sequence homology, and properties of the structural proteins (34,38,112,200, 205,215). Recently, *toroviruses* were classified as a genus within the *Coronaviridae*, and *arteriviruses* were proposed as a new genus in this family because they are plus-strand RNA viruses, their replication strategies appear to be similar to that of coronaviruses, and they have some nucleotide sequence homology to coronaviruses (25,38,56,203). Toroviruses are discussed in Chapter 2, and this chapter is devoted to coronaviruses.

Three serologically distinct groups of coronaviruses have been identified (Table 1). Within each serogroup, viruses are identified by their natural hosts and by nucleotide sequence and serological relationships. Most coronaviruses naturally infect only one species or several closely related species. Virus replication *in vivo* can be disseminated, causing systemic infections, or restricted to a few cell types (often the epithelial cells of the respiratory or enteric tracts and macrophages), causing localized infections (Table 1). Numerous serologically and biologically different strains of avian infectious bronchitis virus (IBV) and mouse hepatitis virus (MHV) have been isolated from infected birds and mice (15,33). The extent of strain variation in human coronaviruses is unknown because few human coronaviruses have been isolated and characterized. Nucleotide sequences have been determined for the entire genomes of several coronaviruses, and partial sequences are available

K. V. Holmes: Department of Microbiology, Box B-175, University of Colorado Health Sciences Center, 4200 East 9th Avenue, Denver, Colorado 80262.

M. M. C. Lai: Howard Hughes Medical Institute, Department of Microbiology, University of Southern California School of Medicine, Los Angeles, California 90033-1054.

TABLE 1. *Coronaviruses, natural hosts and diseases*

Antigenic group	Virus	Host	Respiratory infection	Enteric infection	Hepatitis	Neurologic infection	Other[a]
I	HCV-229E	Human	X				
	TGEV	Pig	X	X			X
	CCV	Dog		X			
	FECV	Cat		X			
	FIPV	Cat	X	X	X	X	X
	RbCV	Rabbit					X
II	HCV-OC43	Human	X	?			
	MHV	Mouse	X	X	X	X	
	SDAV	Rat				X	
	HEV	Pig	X	X		X	
	BCV	Cow		X			
	RbEVC	Rabbit		X			
	TCV	Turkey	X	X			
III	IBV	Chicken	X		X		X

[a]Other diseases caused by coronaviruses include infectious peritonitis, immunological disorders, runting, nephritis, pancreatitis, parotitis, and adenitis.

HCV-229E, human respiratory coronavirus; TGEV, porcine transmissible gastroenteritis virus; CCV, canine coronavirus; FECV, feline enteric coronavirus; FIPV, feline infectious peritonitis virus; TCV, turkey coronavirus; HCV-OC43, human respiratory coronavirus; MHV, mouse hepatitis virus; SDAV, sialodacryoadenitis virus; HEV, porcine hemagglutinating encephalomyelitis virus; BCV, bovine coronavirus; RbCV, rabbit

for many others (21,82,83,95,96,119,120,161,162). At this time, there is no infectious cDNA clone for any coronavirus genome.

VIRION STRUCTURE

The structure of coronavirus virions is shown in Fig. 2. The genomic RNA of coronaviruses is 27 to 32 kb in size, the largest of all RNA virus genomes (21,24,112,120,205). The genome is a single-stranded, positive-sense RNA, which is capped and polyadenylated. Genomic RNA isolated from coronavirus virions is infectious when introduced into eukaryotic cells (136,188). The RNA genome is associated with the nucleocapsid phosphoprotein, N, (50 to 60 kd) to form a long, flexible, helical nucleocapsid (144,216). In thin sections, the nucleocapsids appear as tubular strands 9 to 11 nm in diameter. The nucleocapsid lies within a lipoprotein envelope formed by budding from intracellular membranes (65,80,170,223). The envelopes of all coronaviruses contain two viral glycoproteins, membrane protein (M, formerly called E1) and spike protein (S, formerly called E2) (34,213). A third envelope glycoprotein, hemagglutinin-esterase protein (HE), is found in the envelopes of many of the coronaviruses in antigenic group II, including human respiratory coronavirus (HCV-0C43), bovine coronavirus (BCV), porcine hemagglutinating encephalomyelitis virus (HEV), turkey coronavirus (TCV), and some strains of MHV (52,88,97,103,192, 246,248). The properties and functions of the coronavirus structural proteins (Table 2) are described below.

The M glycoprotein differs markedly from other viral glycoproteins. The deduced amino acid sequence of the M genes, structural analysis of the protein, and expression of cloned M cDNA suggest that only a short N-terminal domain is exposed on the exterior of the viral envelope; the glycoprotein penetrates the lipid bilayer three times via three hydrophobic, alpha-helical domains; and a large carboxy-terminal domain lies beneath the bilayer (3,37,80, 135,141,182). Glycosylation of the N-terminal domain is O-linked for MHV and N-linked for IBV and porcine transmissible gastroenteritis virus (TGEV) (63,90,118,134, 166,167). Antibody to the external domain of M neutralizes virus, but only in the presence of complement (41). M protein of some coronaviruses can induce alpha interferon (40). The M proteins of coronaviruses are targeted to the Golgi apparatus and do not appear on the plasma membrane (142,143,219). The M glycoprotein of MHV binds to isolated nucleocapsid *in vitro* (216). Like the nonglycosylated matrix glycoproteins of the negative-stranded viruses, coronavirus M probably binds the helical nucleocapsid to the viral envelope during virus budding (65). The accumulation of M in the Golgi apparatus and its absence from the plasma membrane may explain why coronaviruses bud intracellularly instead of at the plasma membrane (213,223).

The S glycoprotein is the structural protein of the large, petal-shaped spikes. The deduced amino acid sequences of many coronavirus S genes reveal a large number of potential N-linked glycosylation sites and 4 structural domains: a short carboxyterminal cytoplasmic domain, a transmembrane domain, and two large external domains called S1 and S2. The cytoplasmic domain is rich in cysteine residues, suggesting a complex tertiary structure that may play a role in assembly of the spikes or interaction

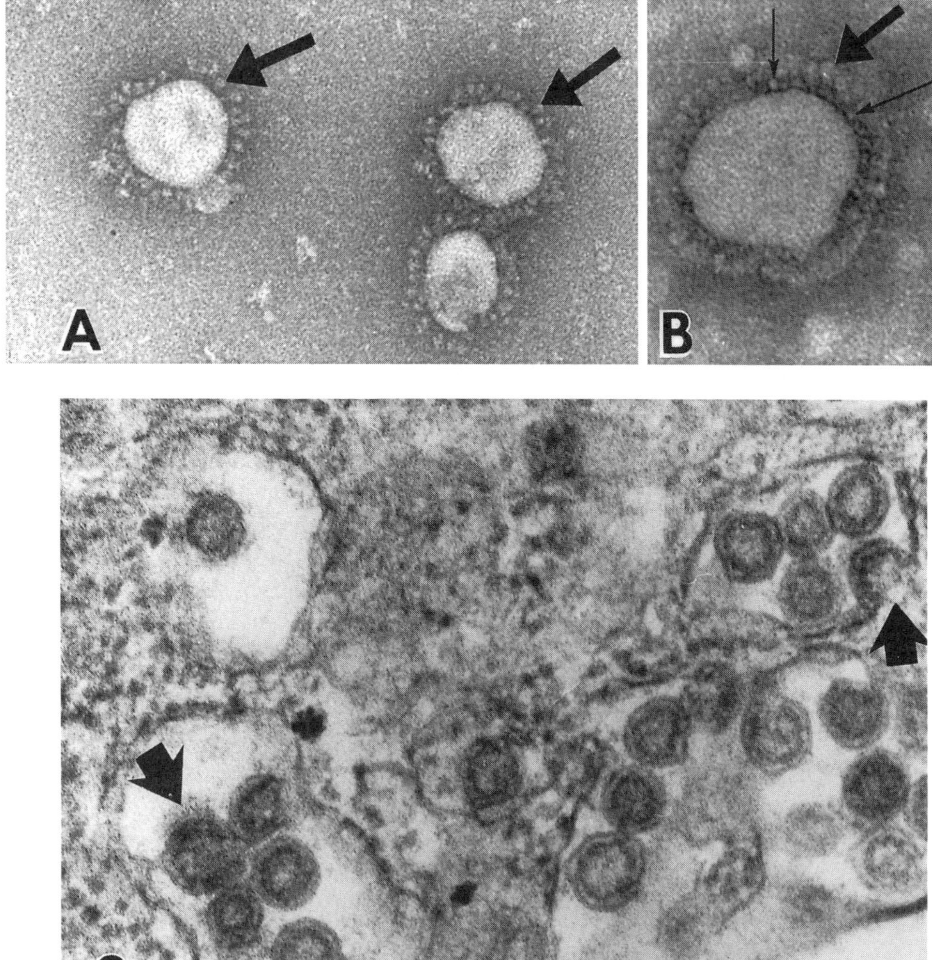

FIG. 1. Morphology of coronaviruses. **A:** Human respiratory coronavirus HCV-OC43 (×90,000) and **B:** Turkey enteric coronavirus (TCV) in negatively stained preparations. The large, petal-shaped spikes, composed of the S glycoprotein, seen on the envelopes of both viruses distinguish coronaviruses from other enveloped viruses. Some coronaviruses in group II also exhibit a fringe of shorter spikes composed of the hemagglutinin-esterase (HE) glycoprotein (small arrow). (×207,000) Courtesy of P. Tijssen. **C:** Human cell infected with human respiratory coronavirus HCV-229E. Spherical virions are seen budding at membranes of the RER and smooth-walled vesicles (arrows). The electron-dense helical nucleocapsid is visible as slender, flexible tubules within the virions. Virions released into the lumen of the intracellular vesicles apparently migrate to the plasma membrane and are released by exocytosis. ×60,000.

with other viral proteins. The transmembrane domain is attached to the S2 domain, which shows approximately 30% homology among coronaviruses. S2 is acylated and has two heptad repeat motifs that suggest a coiled-coil structure like that seen for other viral glycoproteins (50). For some coronaviruses, between S1 and S2 lies a cluster of basic amino acids that is a trypsin-cleavage motif. The amino-terminus of S2 that would be generated by cleavage is not a hydrophobic region like those adjoining the cleavage sites of the influenza HA or paramyxovirus F glycoproteins (241). There is considerable diversity in both the lengths and nucleotide sequences of the S1 glycoproteins of different coronaviruses and even of different strains of a single coronavirus (9,36,49,76,173). This diversity in S1 probably results from mutation and recombination between coronaviruses and strong positive selection *in vivo*. Several hypervariable regions in the S1 sequence where large deletions or insertions commonly occur may represent externally oriented loops that are not essential for the structure of the spikes (9,36,76,119,173,232). Changes in S1 have been associated with altered antigenicity and virulence (69,119).

The tertiary structure of S has not yet been determined. Each spike appears to be a trimer of S glycoproteins held together by noncovalent bonds (54). Trypsin treatment of MHV virions can cleave all of the S glycoprotein to S1 and

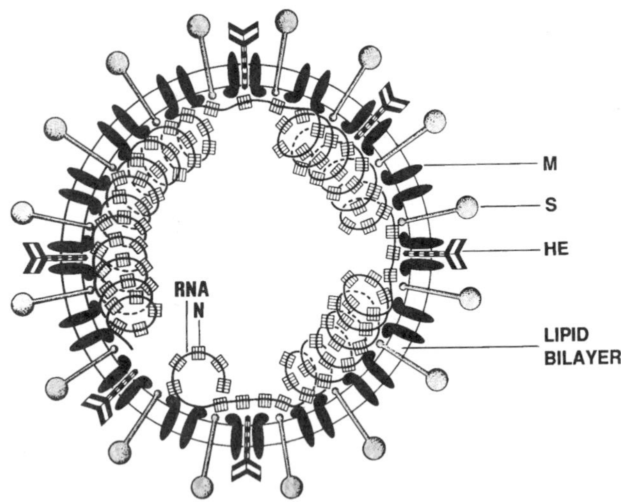

FIG. 2. Model of coronavirus structure. The viral nucleocapsid is a long, flexible helix composed of the 30-kb long, plus-strand genomic RNA and the phosphorylated nucleocapsid protein, N. The lipid bilayer in the envelope is derived by budding from intracellular membranes of the host cell. The large spikes are composed of the S glycoprotein. The membrane glycoprotein (M) traverses the lipid bilayer three times. Its cytoplasmic domain may interact with the nucleocapsid inside the virion. The HE glycoprotein is expressed only by some group II coronaviruses. Not shown in this diagram is a small, hydrophobic membrane protein (SM) of unknown function recently found in the virions of several coronaviruses.

S2 without disrupting the spikes (218,237). However, S1 can be released from IBV virions by treatment with urea (35) or from trypsin-treated MHV virions by treatment at pH 8.0 and 37°C (217), making the virions noninfectious.

S has many important biological functions (Table 2). Monoclonal antibodies to S can neutralize virus infectivity and inhibit membrane fusion. Some MAb epitopes on S detect linear epitopes, and others are conformation-dependent (49,55,139). Recombinant S glycoproteins of some coronaviruses can induce fusion of susceptible cell lines (51,220,253). For BCV or some strains of MHV, cleavage of the S glycoprotein may be required to stimulate cell-fusing activity and/or plaque-forming activity for certain host cells (47,206,217). However, S proteins of some MHV strains or mutants and of feline infectious peritonitis virus (FIPV) and TGEV which lack the protease motif can also induce cell fusion (51,117,206,220). Several other biological activities have been associated with the coronavirus S glycoprotein (Table 2). The presence of S on the plasma membrane of infected cells can make them susceptible to cell-mediated cytotoxicity (92,239), and S glycoprotein of MHV has an IgG Fc type III receptor domain that can bind rabbit immunoglobulin (168). The S glycoproteins of BCV and IBV bind to neuraminic acid-containing glycans and have hemagglutinating activity (189,190). Thus, S is a large, multifunctional protein that plays a complex and central role in the biology and pathogenesis of coronavirus infections.

The HE glycoprotein is found on virions of some group II coronaviruses as a 130- to 140-kd, disulfide-linked dimer of a 65- to 70-kd protein that forms short spikes (Figs. 1,2) (88,97,103,176,192,252). It is not yet known whether each short spike is composed of more than one 140-kd dimer. The gene encoding HE is found in the genomes of group II coronaviruses, but not group I or III coronaviruses. The coronavirus HE genes have amino acid sequence homology with the HE glycoprotein of influenza C virus and may have been derived by a recombination between an HE mRNA of influenza C and the genomic RNA of an ancestral coronavirus (138). Like HE of influenza C, coronavirus HE glycoproteins bind to 9-O-acetylated neuraminic acid residues on glycoproteins or glycolipids (85,191,192,231). Coronaviruses that express HE cause hemagglutination and hemadsorption, which can be used to titer virus or to identify infected tissue culture cells. HE also has acetylesterase activity, which cleaves acetyl groups from the substrate, potentially eluting adsorbed virions and destroying the HE-binding activity of the glycans on the cell membrane (192,230,248). The HE glycoprotein may permit initial adsorption of the virus to cell membranes, but subsequent interaction of the S glycoprotein with its glycoprotein re-

TABLE 2. *Properties and functions of coronavirus structural proteins*

Nucleocapsid protein, N
- Phosphoprotein
- Binds to viral RNA
- Forms nucleocapsid
- Elicits cell-mediated immunity
- May be involved in the regulation of viral RNA synthesis

Membrane glycoprotein, M
- May determine budding site on intracellular membranes
- Essential for envelope formation
- May interact with viral nucleocapsid
- May induce alpha interferon

Spike glycoprotein, S
- Binds to specific host cell receptor glycoprotein
- May induce fusion of viral envelope with cell membrane[a]
- Induces cell fusion[a,b]
- Binds immunoglobulin at Fc receptor site[c]
- Binds to 9-O-acetylated neuraminic acid[d]
- Induces neutralizing antibody
- Elicits cell-mediated immunity

Hemagglutinin-esterase glycoprotein, HE[e]
- Binds to 9-O-acetylated neuraminic acid on cell membranes
- Causes hemagglutination
- May cause hemadsorption
- Esterase cleaves acetyl groups from 9-O-acetyl neuraminic acid

[a]Fusion is activated by host cell-dependent proteolytic cleavage of some coronaviruses in groups II and III.
[b]Coronavirus fusion depends upon the virus strain and host cell type.
[c]Demonstrated for MHV-A59 (168).
[d]Demonstrated for BCV (190).
[e]Present only in some group II coronaviruses.

ceptor may be required for fusion of the viral envelope with cell membranes. HE is not required for infectivity *in vitro*, as some group II strains do not express it (138,205,246).

The nucleocapsid protein N, is a 50- to 60-kd phosphoprotein that binds viral genomic RNA to form the helical nucleocapsid. N has three structural domains, and RNA binding requires the middle domain (156). Possibly, N binds to the cytoplasmic domain of the M protein during virus budding. Interaction of nucleocapsid with M has been demonstrated *in vitro* (216), but the region of N responsible for this is not known. A temperature-sensitive point mutation causing thermolability of MHV virions lies within a highly conserved region of the N protein (157). N may play a role in replication of viral RNA, since antibody directed against N inhibits an *in vitro* RNA polymerase reaction (43). Interestingly, an antigenic peptide of the N protein is recognized on the surface of infected cells by T-cells (18,19,211,234).

A small (9- to 12-kd) hydrophobic protein (sM), encoded by a so-called nonstructural gene of several coronaviruses was recently found to be incorporated into the virus envelope, where its function is unknown (77,131).

Coronaviruses, like other plus-strand RNA viruses, do not contain any RNA-dependent RNA polymerase in their virions. Protein kinase activity is present in purified virions, but it is not yet clear whether it is a viral or cellular enzyme (201).

GENOME STRUCTURE AND ORGANIZATION

The genomes of several coronaviruses are illustrated in Fig. 3. Coronavirus genomic RNAs are capped and polyadenylated. They can function as mRNAs and are infectious (112). At the 5' end is the 60- to 80-nucleotide-long leader RNA which is followed by about 200 to 500 untranslated nucleotides. The first 60% of the length of the genome from the 5' end, roughly 20 kb, consists of two overlapping open reading frames (ORF 1a and ORF 1b) that encode the viral RNA-dependent RNA polymerase, proteases, and other as yet uncharacterized proteins (22,112,120, 171,205). At the overlapping region between ORF 1a and ORF1b are a 7-nucleotide slippery sequence and a pseudoknot that are essential for ribosomal frameshifting (26,84,204). The gene order of the polymerase and the three structural proteins shared by all coronaviruses is Pol-S-M-N. In addition, there are several open reading frames encoding nonstructural proteins and the HE protein, which differ markedly between coronaviruses in number, nucleotide sequence, and gene order (112,205,224,238) (Fig. 3). The functions of these gene products are unknown. Several of the nonstructural proteins are not essential for virus replication *in vitro* (193,249), but it is uncertain whether they play a role in virus pathogenesis *in vivo*. Possibly these genes were inserted into different sites in the coronavirus genomes by RNA recombination.

GROWTH OF CORONAVIRUSES

Most coronaviruses, e.g., MHV, can infect cells of only one host species or, at most, a few closely related species. In contrast, several coronaviruses, e.g., BCV, infect cells of many species *in vitro* and have a fairly broad host range *in vivo*. Primary isolation of most human respiratory coro-

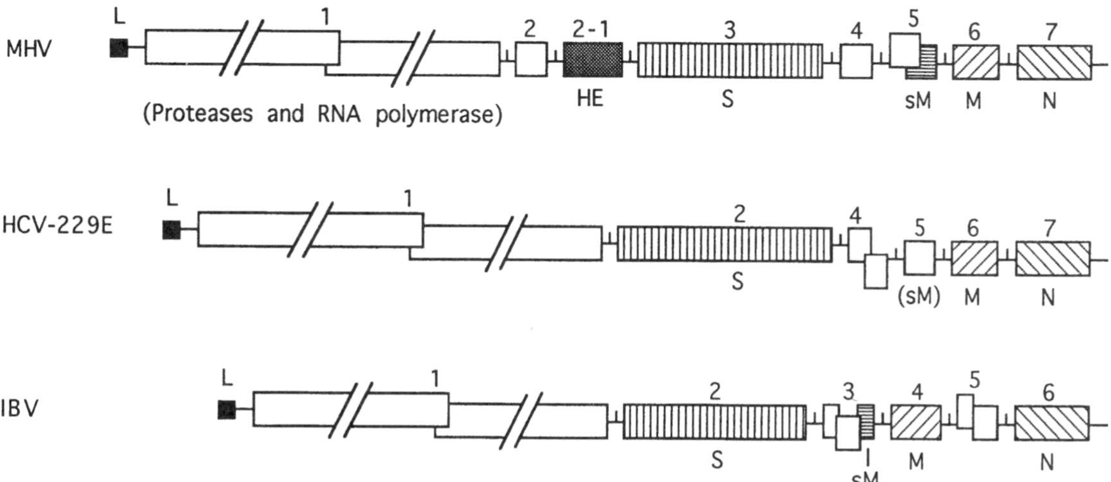

FIG. 3. Genomic organization of coronaviruses. The MHV genome is 31.2 kb, while the IBV genome is 27.6 kb. The sizes of the genes are drawn approximately to scale, except for gene 1. Open reading frames encoding structural proteins are shown (*filled boxes*). *Vertical lines* represent intergenic sequences; the areas between the two vertical lines represent one gene, within which open reading frames are translated from one mRNA by internal initiation or other mechanisms. For HCV-229E, the origin and nature of mRNA 3 is not clear.

naviruses requires human fetal tracheal organ cultures (see Chapter 35). In certain cell lines, some coronaviruses can be propagated and form plaques only in the presence of trypsin (47). Thus, appropriate host cells and growth conditions are particularly important for isolation and propagation of coronaviruses.

Coronavirus replication takes place in the cytoplasm of infected cells and can occur normally in enucleated cells (242). In tissue culture, coronaviruses have a latent period of about 5 to 7 hours. Infectious virus can be isolated from culture fluids and from cells disrupted by freezing and thawing. Infectivity of virions is fairly stable at pH 6.0, but rapidly inactivated at mildly alkaline pH (218). Coronaviruses can cause either cytocidal or persistent infections of cells *in vitro* and *in vivo*, depending on the virus strain and the host cell. In cytocidal coronavirus infections, cells may form multinucleated syncytia and/or lyse. For example, syncytia are observed in MHV-infected intestinal epithelial cells and in cultured fibroblasts (16). Trypsin-induced cleavage of the viral S protein is required for BCV-induced fusion of some cell types but not others (47,212). Occasionally, viral nucleocapsids accumulate to form inclusions in infected cells (65). Plaques are visualized by cell lysis, fusion, neutral red staining, or, for BCV, by hemadsorption. Persistent human respiratory coronavirus (HCV-229E) infection can be established in some human cell lines, producing virus for weeks without cytopathic effects or cell death (39). Following an acute lytic infection, many coronaviruses establish persistently infected carrier cultures with minimal cytopathic effects and resistance to superinfection by the wild-type virus (39,45,78,87). Small plaque variants with mutations in the S gene and mutants with attenuated transcription have been isolated from some persistently infected cultures (78, 87,221).

Coronaviruses and their replication *in vitro* and *in vivo* can be detected by sensitive immunoassays with many different monoclonal antibodies, nucleic acid hybridization, and polymerase chain reaction; as well as by immunofluorescence, hemadsorption, or plaque assay (33,42,163, 164,178,233). These new diagnostic tools are now being used to study the epidemiology and pathogenesis of human coronavirus infections.

CORONAVIRUS REPLICATION

The events in coronavirus replication are summarized in Fig. 4.

Attachment and Penetration

The first step in the replicative cycle is binding of virions to the plasma membranes of susceptible cells. Coronaviruses in group II that express the HE glycoprotein may bind to 9-O-acetylated neuraminic acid moieties on membrane macromolecules via HE or the S glycoproteins (190–192,231). The S protein then binds to its specific receptor glycoprotein (Table 3). Inhibition by diisopropylfluorophosphate of the esterase activity of HE on BCV virions markedly reduces viral infectivity, suggesting that binding of HE to 9-O-acetylated neuraminic acid residues can facilitate virus infection (230). Enzymatic removal of the 9-O-acetylated neuraminic acid from cell membranes or treatment with HE-specific monoclonal antibodies inhibits BCV infection (62,191). Most coronaviruses do not express HE, and their S glycoproteins bind directly to the specific receptor glycoproteins on the cell surface (Table 3).

The S glycoproteins of all strains of MHV recognize a common receptor glycoprotein, MHVR, a biliary glycoprotein (BGP) that is a member of the carcinoembryonic antigen family of glycoproteins in the immunoglobulin superfamily (67,91,243) (Table 3). Isoforms of the BGP proteins are homophilic cell adhesion proteins, bind calmodulin, have ecto-ATPase activity, and play a role in taurocholate transport (17,128,158,202). MHVR has four immunoglobulin-like domains, a transmembrane region, and a short intracytoplasmic domain. The viral S glycoprotein and a monoclonal anti-MHVR antibody that blocks virus attachment bind to the N-terminal Ig-like domain (68). Some posttranslational modification of the glycoprotein is required for recognition by the virus (174). Splice variants of MHVR and of another allele of MHVR called *mmCGM2* can also serve as MHV receptors when their cDNAs are expressed in hamster or monkey cells (66,250, 251). MHV binds to an epitope of the BGP proteins that is expressed only on murine cells. This specificity of virus binding may, at least in part, explain the species specificity of MHV infection (44).

Two coronaviruses in group I utilize receptors unrelated to MHVR. TGEV and HCV-229E use aminopeptidase N (APN), a zinc-binding protease, as their receptors (53,244) (Table 3). Some monoclonal antibodies against porcine or human APN block binding of TGEV or HCV-229E virions to their respective APN receptors. Expression of cDNAs encoding the human or porcine APN glycoprotein in cells from species that are normally resistant renders them susceptible to infection with HCV-229E or TGEV, respectively. The protease activity of APN is not re-

TABLE 3. *Coronavirus receptors*

Virus	Host	Receptor glycoprotein
MHV	Mouse	MHVR, and several additional carcino-embryonic-related glycoproteins in the immunoglobulin superfamily
TGEV	Pig	Porcine aminopeptidase N, a metallo-protease
HCV-229E	Human	Human aminopeptidase N, a metallo-protease
BCV	Cow	9-0-acetylated neuraminic acid

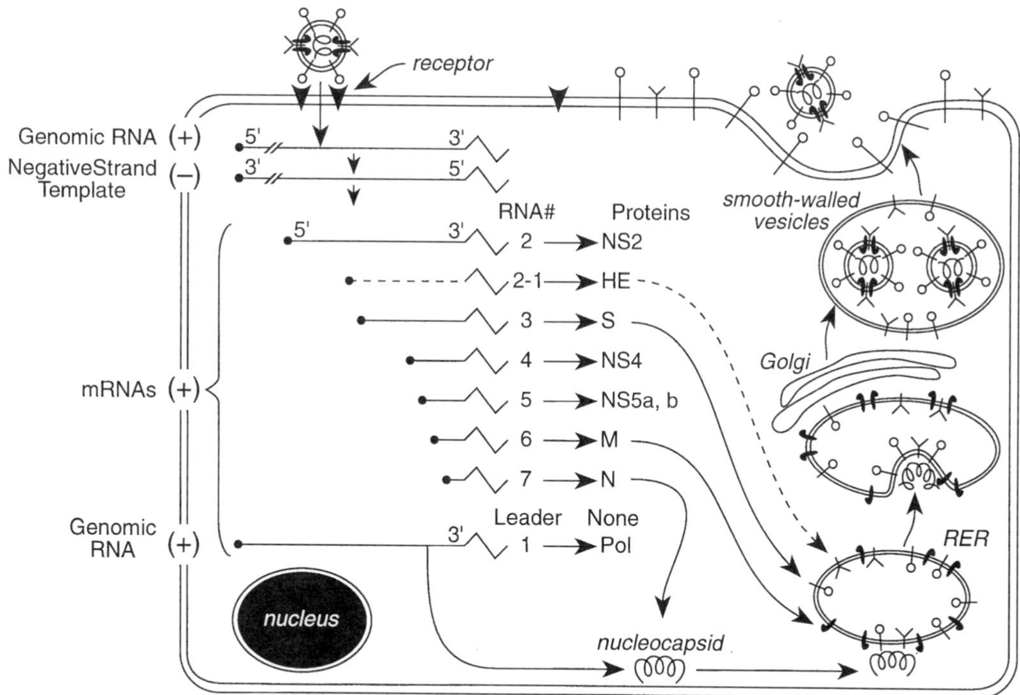

FIG. 4. Model of coronavirus replication based on MHV. Virions bind to the plasma membrane by interaction of S proteins in the large spikes with specific receptor glycoproteins (see Table 3). Binding of HE glycoprotein in the short spikes to 9-O-acetylated neuraminic acid residues on the plasma membrane may serve as a prereceptor interaction for some coronaviruses in group II. Penetration occurs by S protein-mediated fusion of the viral envelope with the plasma membrane or, for some strains, with endocytic membranes. The genomic RNA is translated to form a polyprotein (>800 kd) which is co- or posttranslationally processed to yield multiple proteins that serve as a virus-specific, RNA-dependent RNA polymerase and may play other as yet undetermined roles in the transcription and replication of viral RNAs. The viral RNA-dependent RNA polymerase uses the positive-strand genomic RNA as a template for full-length, negative-strand RNA which is replicated to form new positive-strand genomic RNA. Overlapping nested sets of 3′ coterminal subgenomic mRNAs and subgenomic negative-strand RNAs and leader RNA are made in infected cells, as shown in the alternative models in Fig. 5. The genomic RNA and mRNAs are capped and polyadenylated. At the 5′ end of each mRNA is a common leader sequence about 70 nucleotides long that is encoded by the 3′ end of the negative-strand template. The genome-length and subgenomic, negative-strand RNAs each have a sequence complementary to the leader sequence at their 3′ ends. With few exceptions, each of the polycistronic mRNAs is translated to yield only the polypeptide encoded at the 5′ end of the mRNA. For example, mRNA 6 encodes the M protein. The N protein and newly formed genomic RNA assemble in the cytoplasm to form helical nucleocapsids. The HE, S, and M glycoproteins are encoded by mRNAs 2-1, 3 and 6, respectively, and translated on membrane-bound polysomes. The S glycoprotein is cotranslationally glycosylated at asparagine residues, trimerized, and transported through the Golgi apparatus, where it is acylated, its oligosaccharides are trimmed, and it may be proteolytically cleaved to form two 90K subunits. Excess S protein that is not incorporated into virions is transported to the plasma membrane, where it may participate in cell-cell fusion. Some coronaviruses do not encode HE or express mRNA 2-1. The matrix glycoprotein M is transported to the Golgi apparatus, where it accumulates and may be glycosylated at serine or threonine residues, but M is not transported to the plasma membrane. Virions are formed in a budding compartment between the RER and the Golgi apparatus, but virions do not bud from the plasma membrane. Virions are apparently released by fusion of smooth-walled, virion-containing vesicles with the plasma membrane. Numerous virions may remain adsorbed to the plasma membranes of infected cells. Adapted from Sturman and Holmes (215), with permission.

quired for virus receptor activity, since protease inhibitors do not block infection (53,244), and point mutations that abolish enzymatic activity retain receptor function (Holmes K.J., Look A.R., and Ashmun R., in preparation).

After binding of S to its glycoprotein receptor, the next step in replication is viral entry into cells, which involves fusion of the viral envelope with either the plasma membrane or endosomal membranes (75,107). Murine cell lines that express virus receptors differ in susceptibility to MHV (4,108,245), so there may be additional host-dependent stages in the replicative cycle. Like paramyxoviruses and HIV, which fuse at the plasma membrane, MHV and presumably other coronaviruses exhibit a neutral or slightly alkaline pH optimum for cell fusion (125,218,237). However, in some studies coronavirus infectivity was found to be reduced by treatment with lysosomotropic drugs like viruses that enter via endosomes (160), while other studies did not show significant inhibition of coronavirus infectivity by these drugs (75,107). A mutant of MHV appears to enter via the endocytic pathway (75). A conformational change in the spike glycoprotein leading to increased hydrophobicity is induced at pH 8.0 and 37°C (218,237). This may be analogous to the pH-induced conformational changes associated with membrane fusion for other enveloped viruses (177,241).

Primary Translation

For all plus-strand RNA viruses, the first macromolecular synthesis event following entry of the viral genome into the cytoplasm is the translation of the viral genomic RNA to yield an RNA-dependent RNA polymerase, which is translated from gene 1. In all coronaviruses, gene 1 contains two overlapping ORFs (1a and 1b) which could potentially be translated into one protein of more than 700 kd by a ribosomal frame-shifting mechanism (24,27,29,82,120). The deduced amino acid sequence includes two papainlike protease domains, a domain similar to poliovirus 3C protease, and consensus sequences for polymerase and helicase (79,120). This primary gene product has not been detected. The primary translation product is co- or posttranslationally processed into multiple proteins by viral and cellular proteases. The first processed product, p28, is generated from the N-terminus of the ORF 1a protein by a virus-specific papainlike protease located downstream of the cleavage site (7,58). P28, p290, p50, p240, and several additional cleavage products have been detected in virus-infected cells and by *in vitro* translation of the ORF 1a portion of the viral genomic RNA (7,8,57,59,60). Inhibition of protein synthesis at any time after infection inhibits virus RNA synthesis, suggesting that polymerase is probably translated continuously throughout the replication cycle (175,184). The sequence of processing events and the functions of the gene 1 products are not yet fully understood.

Transcription of Viral RNA

The plus-strand viral genomic RNA is transcribed into minus-strand RNA, which is then used as the template for synthesis of plus-strand viral mRNAs and genomic RNA. The synthesis of minus-strand RNA peaks somewhat earlier and falls off more rapidly than the synthesis of plus-strand RNA (184). Infected cells contain 10- to 100-fold more plus-strand viral RNA than minus-strand RNA (175,184). Though initial studies suggested that all minus-strand RNA was of genome length (115), it is now clear that infected cells also contain subgenomic minus-strand RNAs that correspond in length and relative abundance to the viral mRNAs (185,195,196). All minus-strand RNA is found in double-stranded forms, and no free minus-strand RNA is detected (175,184,185). The mechanism of synthesis and the functions of the subgenomic- and genomic-length minus-strand RNAs are matters of controversy. However, it is clear that they serve as templates for subgenomic mRNA synthesis because pulse-labeled replicative intermediates contain both genomic-length and subgenomic-length templates (185). Several models of coronavirus RNA synthesis will be discussed below.

All coronavirus mRNAs and the plus-strand genomic RNA are capped and polyadenylated. Depending on the virus species, there are 5 to 7 subgenomic mRNAs which are numbered in order of decreasing size (112). Messenger RNAs discovered after the viral mRNAs were named are denoted by a hyphen and a second number (e.g., mRNA 2-1)(34). The mRNAs form a nested set with a common 3' end (Fig. 4). Although each of the mRNAs except the smallest one is polycistronic, i.e., contains two or more ORFs, only the ORF at the 5' end is translated, with minor exceptions noted below (124,199).

A characteristic feature of coronavirus subgenomic mRNAs is the presence at their 5' ends of a leader sequence 60 to 80 bases long that is identical to the 5' end of the genomic RNA. This leader sequence is not found elsewhere in the genome, but between each ORF on the genome there is a short intergenic sequence (IS, Fig. 5) that contains a 7- to 18-nucleotide sequence homologous with a sequence near the 3' end of the leader (198). The two most favored models to explain how the leader RNA is joined to each mRNA and how the subgenomic mRNAs are generated are illustrated in Fig. 5. The models must account for the observations that, late in the replication cycle, the UV target size of each mRNA is similar to its physical size (93,208,247) and that in cells simultaneously infected with two different MHV strains, approximately half of the mRNAs have leader from one strain and coding sequences from the other strain (153).

One model for coronavirus mRNA synthesis postulates leader-primed transcription (112, Fig. 5a). Transcription of the leader RNA would begin at the 3' end of the full-length, minus-strand template RNA and terminate with the dissociation of the leader from the template, either alone or with attached polymerase protein(s). The leader would then bind to intergenic sequences downstream on the minus-strand template and serve as the primer for mRNA transcription. This model was based on early studies that showed that infected cells contained only genome-length, negative-strand templates (115). Splicing of the genome was ruled out because the rate of inactivation by ultraviolet light of the synthesis of genomic and subgenomic mRNA was proportional to the length of the RNA (93,208). This suggested that the leader RNA was joined to the body of subgenomic mRNA during positive-strand synthesis. Unclear in this model are the cause of termination of the initial transcription of the

a) Leader-primed transcription during positive-strand synthesis

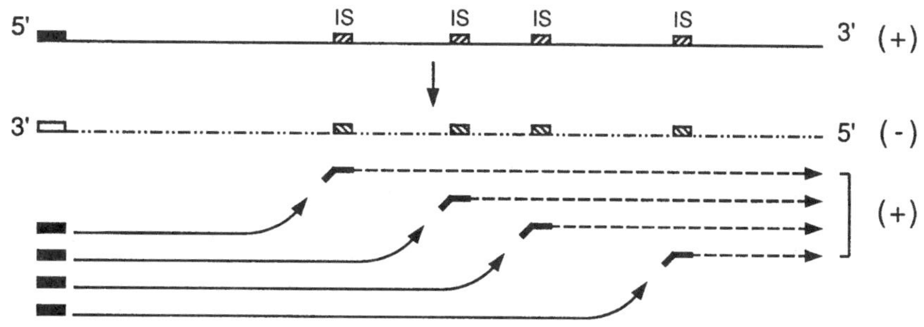

b) Discontinuous transcription during negative-strand synthesis

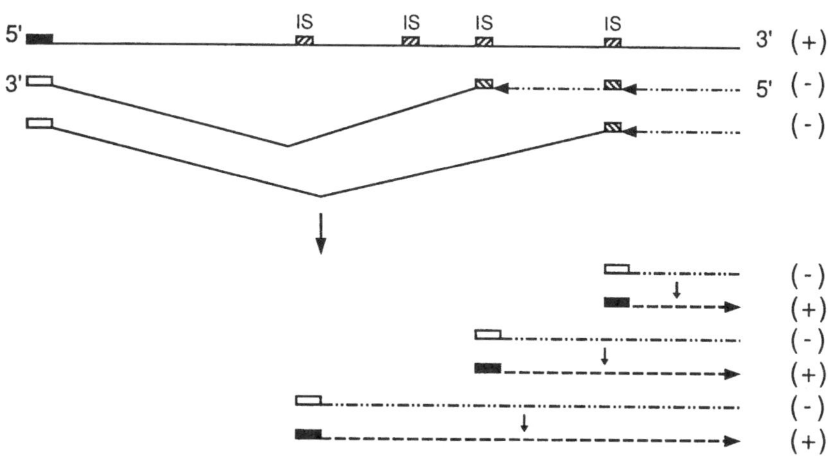

FIG. 5 Models for discontinuous transcription of subgenomic mRNAs. Generation of subgenomic (+) and (−) strand RNAs could be generated either by leader primed transcription during (+) strand synthesis (A) or by discontinuous transcription during (−) strand synthesis (B). *Filled box*, positive-strand leader; *open box*, antileader; *solid line*, genomic RNA; *dashed line*, subgenomic (+) strand RNAs; *dashed and dotted line*, (−) strand template RNAs. **A:** The genomic RNA is transcribed into a full-length, (−) strand RNA. The leader RNA is then transcribed from the 3′ end of (−) strand RNA and transcription terminates via an unknown mechanism somewhere downstream from the leader sequence. The leader RNA is dissociated, probably together with RNA polymerase, from the template and subsequently binds to intergenic sequence (IS) elements downstream. The bound leader RNA then serves as a primer for continuing mRNA synthesis until the polymerase reaches the 5′ end of the template. The leader RNA may be derived from a different RNA molecule *in trans*. In this mechanism IS elements serve as transcription initiation sites. **B:** The genomic RNA serves as a template for (−) strand RNA synthesis. Transcription stops at an IS element, and then jumps via an unknown mechanism to the leader at the 5′ end of the (+) strand template, generating subgenomic (−) strand RNAs that contain an antileader sequence at the 3′ end. Transcription termination is presumably caused by the interaction between the leader RNA and IS elements. Each subgenomic (−) strand RNA then serves as template for faithful, uninterrupted transcription of (+) strand subgenomic RNAs. In this mechanism, IS elements serve as transcription terminators.

leader and whether subsequent binding of the leader at intergenic sequences is mediated by RNA sequence alignment or recognition by the RNA polymerase. Several lines of evidence support this leader-primed transcription model.

1. MHV-infected cells contain free leader RNAs ranging in size from 60 to 90 nucleotides (14).
2. A temperature-sensitive mutant of MHV accumulates leader RNA, but not mRNAs, at the nonpermissive temperature (14,186).
3. Leaders transcribed from two different MHV strains in the same cell are randomly joined to mRNAs from either strain (153).
4. In an *in vitro* transcription system derived from MHV-infected cells, exogenously added leader RNA can be incorporated into mRNAs (6).

5. Engineering of an intergenic sequence into the middle of recombinant defective-interfering (DI) RNA, and expression of the DI in cells infected with a helper MHV lead to the transcription of an RNA containing the leader of the helper virus and the DI RNA downstream of the inserted intergenic sequence (147,254).

In the leader-primed transcription model, the intergenic sequences serve as promoters for mRNA transcription. The consensus intergenic sequence is conserved in most coronavirus species. Seven nucleotides within the intergenic sequence constitute the core promoter sequence (94,146), and additional upstream sequences from the leader RNA and 5' end sequence of genomic RNAs are also required for initiation of transcription (127). There is some variability in the intergenic sequences that precede different ORFs of a virus strain and between intergenic sequences preceding the same ORF in different MHV strains. Certain mRNAs, such as MHV mRNA 2-1, can only be transcribed when the leader RNA contains a compatible sequence (111,197). The leader sequence of some mRNAs differs slightly from the leader sequence of the corresponding genomic RNA, suggesting that the leader priming process may require some sequence processing at the leader-mRNA fusion sites (6,152). The leader-primed transcription model is compatible with generation of mRNAs from either full-length or subgenomic minus-strand templates, but the model does not account for the generation of the observed subgenomic minus-strand RNAs or subgenomic replicative intermediate RNAs (185,196).

An alternative model for coronavirus RNA synthesis postulates discontinuous transcription during minus-strand RNA synthesis (185) (Fig. 5b). During minus-strand RNA synthesis from the genomic template, the polymerase would pause at one of the intergenic sequences and then jump to the 3' end of the leader sequence near the 5' end of the genomic RNA template, generating subgenomic minus-strand RNA with an antisense leader sequence at its 3' end. These subgenomic minus-strand RNAs and a full-length minus strand could then serve as templates for uninterrupted transcription of plus-strand mRNAs and genomic RNA. In this model, the intergenic sequence serves as the transcriptional termination sequence and/or as a sequence that interacts with leader RNA to permit polymerase jumping during minus-strand RNA synthesis. This model is supported by several experimental observations.

1. Subgenomic minus-strand RNAs have a poly(U) sequence at the 5' end and an antisense leader sequence at the 3' end (86,195). Thus, they appear to be exact complementary copies of the viral mRNAs.
2. The relative abundance of different subgenomic minus-strand RNAs in infected cells is the same as the relative abundance of the corresponding viral mRNAs (196).
3. Subgenomic replicative intermediate (RI) RNAs are actively replicating in infected cells. When subgenomic RNAs were separated by size and then denatured, smaller RIs generated smaller mRNAs, while the larger RIs generated only the larger mRNAs. This suggests that each subgenomic mRNA may be transcribed from a corresponding subgenomic-sized, minus-strand template (185).

A third model is based upon the finding that subgenomic mRNAs are incorporated into BCV virions along with the full-length genomic RNA. This model proposes that, following virus penetration, the incoming plus-strand subgenomic mRNAs are used directly as templates for transcription of the subgenomic minus-strand RNAs, which then serve as templates for synthesis of additional subgenomic mRNAs (195,196). This straightforward model is attractive, but does not explain how the various RNA species originated, how leaders from different strains can be randomly incorporated into mRNAs, or what factors regulate the relative abundance of different mRNA species. Evidence against this model is that virion-associated subgenomic mRNAs have not been detected in all coronavirus species, and transfected mRNAs cannot be amplified (127).

Data available at this time do not unequivocally rule out either of the first two models of coronavirus transcription. Possibly elements from each of these models will be found to be correct. Comparative information about RNA transcription mechanisms in other enveloped, plus-strand RNA viruses related to coronaviruses such as toroviruses and arteriviruses may help elucidate coronavirus RNA transcription (25,56,203). The polymerase genes of these viruses are considerably smaller than the coronavirus polymerase gene, and there are no leader sequences on torovirus subgenomic mRNAs.

Coronavirus RNA can be transcribed *in vitro* in extracts of infected cells (6,43,121). The polymerase is associated with membrane fractions (23,61). One study suggests that there are two types of RNA polymerase complexes, one for genomic RNA synthesis and another for synthesis of subgenomic mRNAs (23). The polymerase complex may include several products of gene 1 and possibly also N protein, since anti-N antibody inhibits MHV RNA synthesis *in vitro* (43). Host components may also be involved in a polymerase complex.

Replication of Viral RNA

Unlike the discontinuous synthesis of mRNA, the production of full-length, plus-strand genomic coronavirus RNA requires uninterrupted synthesis using the full-length, minus-strand template. Therefore, the mechanisms of RNA replication and subgenomic mRNA transcription probably differ in some respects. The study of RNA replication has been aided by cloned DI RNA, which can replicate with the help of replication machinery provided by a wild-type MHV virus. Replication of the DI RNA was found to require sequences of approximately 400 nucleotides at both the 3' and 5' ends of the genomic RNA. An internal sequence of approximately 135 nucleotides may also be required for replication of some DI RNAs of MHV (105,129,157). During replication, the leader sequence of the DI RNA can be rapidly replaced by that of the helper virus (149), suggesting that RNA replication also involves a free leader RNA, similar to transcription. However, the *cis*-acting signal for replication is different from that for transcription (127). The precise mechanism of coronavirus RNA replication remains to be determined.

Translation of Viral Proteins

Though all but the smallest coronavirus mRNAs are polycistronic, in general, only the ORF at the 5' end of each mRNA is translated (Fig. 4). Recently, some exceptions to this generalization have been discovered. As described above, for all coronaviruses mRNA 1 contains 2 large ORFs that overlap by 43 to 76 nucleotides and is translated into one polyprotein by a ribosomal frameshifting mechanism (21,28,82,120,171). The polyprotein is apparently co- or posttranslationally processed into multiple proteins by viral and host proteases (7,8,59,60). Several other coronavirus mRNAs are bicistronic or tricistronic. For example, mRNA 3 of IBV contains three slightly overlapping ORFs that are all translated *in vitro* and *in vivo* (130). Translation of the third ORF is regulated by an upstream internal ribosomal entry site (IRES) sequence, which allows ribosomes to bypass upstream ORFs and translate ORF 3C by a cap-independent translation mechanism (132) similar to picornavirus translation. The highly hydrophobic 10- to 12-kd protein encoded by IBV ORF 3C may be a viral envelope protein called sM, similar to a TGEV protein encoded by monocistronic mRNA 4 (77,224) and to the second ORF of MHV mRNA 5 (123). MRNA 5 of IBV contains two ORFs that are both translated *in vitro* and *in vivo* (133), and an internal ORF within the N gene of BCV is translated *in vivo* (194).

The number of ORFs that encode nonstructural proteins, their order on the genome, and their mechanisms of translation differ among coronavirus species (Fig. 3). The functions of the nonstructural proteins are largely unknown. Several nonstructural proteins, e.g., MHV genes 2, 4, and 5a, are not essential for virus production *in vitro* (193,249).

Processing and Intracellular Transport of Viral Proteins

S Glycoprotein

The spike glycoprotein is cotranslationally inserted into the RER, cotranslationally glycosylated with N-linked glycans, reduced to form numerous disulfide bonds, and oligomerized into noncovalently linked homotrimers (54,169). Transportation of the protein to the Golgi complex is associated with trimming of high mannose oligosaccharides, addition of terminal sugars, and fatty acylation of the S2 domain (187,217). The S proteins of some group II and III coronaviruses can be cleaved near the center by trypsinlike proteases in the Golgi apparatus (possibly furin) or by extracellular enzymes in the respiratory or enteric tracts to form two peptides that remain associated with the oligomeric spike (1,63,72,140,179). Protease cleavage of the S proteins of some group II and III coronaviruses yields the 85- to 100-kd N-terminal S1 and the C-terminal S2 proteins, but the S proteins of group I coronaviruses and of some mutants of group II viruses lack the protease-sensitive sequence and are not cleaved (51,78,183,205,220). Some S protein accumulates in the Golgi, while a fraction of the S oligomers is slowly transported to the plasma membrane, where it plays a role in cell-cell fusion (80,229).

M Glycoprotein

The M glycoprotein is synthesized on membrane-bound polysomes in the rough endoplasmic reticulum (RER). Pulse labeling and *in vitro* translation studies suggest that M may be inserted into the ER membranes by an internal signal sequence (80,181). M protein is highly hydrophobic and spans the membrane 3 times (3,20,141,182). A small N-terminal domain of M located on the luminal side of the ER is cotranslationally glycosylated with N-linked glycans for IBV (207), but this domain of the M protein of MHV is posttranslationally glycosylated by the addition of O-linked glycans in the Golgi apparatus (90,98,134, 165,166). The mature M protein accumulates in the Golgi apparatus and, unlike S, is not transported to the plasma membrane. Analysis of recombinant M proteins suggests that the Golgi targeting of IBV M protein depends upon four uncharged, polar amino acids on one face of the first alpha-helical, membrane-spanning domain, but targeting of MHV M protein to the Golgi appears to be determined by different structural elements of M (80,135,141).

HE Glycoprotein

The 65- to 70-kd monomer of HE is cotranslationally glycosylated by the addition of N-linked glycans in the RER (88,248,252). The monomers form disulfide-linked dimers of 165 to 170-kd (88,103,252), but whether these homodimers form larger oligomers is not yet known. HE is transported to the Golgi, where the N-linked glycans are trimmed (248), and then to the plasma membrane, where it can cause hemadsorption (103).

N Phosphoprotein

Large amounts of the 50- to 60-kd N protein are translated on free polysomes from the most abundant mRNA species. N is rapidly phosphorylated in the cytosol (210), but the extent of phosphorylation and its functional significance are not yet known. *In vitro* assays show that N is an RNA-binding protein (12,180,209). N protein may be associated with membranes (2,210).

Nonstructural Proteins

The complex proteolytic processing of the gene 1 polyprotein is discussed above. Other nonstructural proteins appear to be synthesized on free polysomes. The nonstructural ns2 protein of BCV is phosphorylated (46).

Assembly and Release of Virions

The initial binding of N to viral RNA to form helical nucleocapsids may require a specific nucleotide sequence, but addition of subsequent N proteins must occur in a non-sequence-specific manner to permit encapsidation of the long viral RNA. It was initially believed that only the full-length, plus-strand genomic RNA was incorporated into virions, but encapsidation of subgenomic RNAs has been detected in virions of some coronaviruses, although genomic RNA is always present in great excess (195,196,255). It is not clear whether the packaging of subgenomic mRNAs is specific. N protein binds to RNA in both sequence- and non-sequence-specific manners (180,209) and has a preferred binding site within the leader RNA (209). This finding would explain how both genomic and leader-containing, subgenomic plus-strand RNAs and some DI RNAs could be encapsidated, but it would not account for the large excess of genome-length RNA found in virions. Packaging of genomic RNA requires a specific signal that consists of a stretch of 61 nucleotides in the 3' end of gene 1b, approximately 20 kb from the 5' end of the genome (71,227). This signal is found only in genomes and certain DI RNAs, but not in subgenomic mRNAs. The interactions between RNA and N protein to form nucleocapsids clearly require several steps whose details remain to be elucidated.

In budding virions and in cytoplasmic inclusions, nucleocapsids are seen by electronmicroscopy as long, flexible, tubular structures, but isolated nucleocapsids are uncoiled or very loosely coiled (144,216). Thus, the lateral binding between adjacent N protomers appears to be much stronger than the binding between N protomers on adjacent turns of the helical nucleocapsid. The molecular structure of coronavirus nucleocapsids has not yet been determined.

Virions are formed by binding of nucleocapsids to intracellular membranes that contain M protein. An *in vitro* assay demonstrated temperature-dependent binding of nucleocapsids to M protein (216), but the nature of the molecular interactions is unknown. Virus budding is first detected at specialized membranes called a *budding compartment,* located between the RER and the Golgi (64,222,223). S glycoprotein is not required for the formation or release of virions; MHV virions released from tunicamycin-treated cells contain normal amounts of nucleocapsid and M protein but no S protein, and are therefore noninfectious (90). Also, at the nonpermissive temperature, a temperature-sensitive mutant of MHV forms noninfectious virions that contain nucleocapsid and M protein but lack S (Ricard, *personal communication*). Host proteins are excluded from budding virions. Late in the infectious cycle, budding virions are also observed in the RER and at the outer leaflet of the nuclear envelope, but not at the plasma membrane. This is probably because the M protein, which is required for budding, is not found on the plasma membrane.

The large and small spikes formed by S and HE, respectively, are incorporated into virions at the time of budding. It is not yet clear whether the Golgi-associated post-translational processing of these glycoproteins occurs before or after their incorporation into virions. Since the budding compartment apparently precedes the Golgi, retrograde transport of S or HE would be required if the glycoproteins were completely processed before budding. Alternatively, processing of virion-associated glycoproteins may occur as virions pass through the Golgi. After budding, virions accumulate in large, smooth-walled vesicles which apparently fuse with the plasma membrane to release virus (80). Infectious virus is released efficiently from intact cells. Numerous virions are observed adsorbed to the plasma membranes of coronavirus-infected cells *in vitro* and *in vivo* (170).

Effects on Host Cells

Coronavirus-induced cytopathic effects vary with the virus strain and the host cell. Infection with certain coronaviruses, such as MHV, BCV, IBV or FIPV, induces cell fusion beginning about 6 hours postinfection (p.i.). Fusion is a function of the S glycoprotein and does not occur below pH 6.5 (125,218). These coronaviruses may also induce fusion in infected tissues *in vivo*. Host-dependent protease cleavage of the viral S glycoprotein can affect the extent of cell fusion induced by MHV and BCV (47,72). Other coronaviruses and mutants of the cell-fusing coronaviruses cause little or no cell fusion, but may induce rounding and lysis of cells by about 24 to 48 hours p.i. (75,78).

Persistent coronavirus infection occurs readily in cell culture. Generally, after an initial cytocidal infection, carrier cultures arise in which only a fraction of the cells produce infectious virus at any time, and virus production continues for months (78,89,116,137). Cells in persistently infected cultures are resistant to superinfection with wild-type virus, and small plaque mutants are rapidly selected (75,78,89). Alterations in the protease cleavage site of the S glycoprotein are found in such small plaque mutants, and such viruses may show delayed penetration and altered pathogenesis *in vivo* (78). It is not known whether DI viruses play a role in coronavirus persistence.

Although most coronavirus infections *in vivo* are acute and self-limited, persistent coronavirus infection *in vivo* can occur in immunocompromised hosts such as nude mice or newborns (15,16,235). Some MHV strains can cause persistent infection in the central nervous system (31). It is not clear how long coronaviruses can be shed from persistently infected individuals, providing a reservoir for virus. In general, immunity to local coronavirus infection appears to be fairly short-lived, and reinfection can occur with the same strain of virus (32).

CORONAVIRUS GENETICS

Virus Mutants

Like most RNA viruses, coronaviruses have a high frequency of mutation. Considering the length of the genome and the calculated error frequencies of RNA polymerases from other viruses, coronavirus genomic RNA probably accumulates several point mutations during each round of RNA replication. Thus, even a plaque-purified coronavirus stock would be a population of related quasi-species rather than homogeneous virus. Nucleotide sequence analysis of different isolates of a coronavirus reveals extensive sequence variability that may affect virus replication and pathogenesis (104,105,129,228).

Temperature-sensitive coronavirus mutants have been grouped into five to seven complementation groups (11,73,109,122,186). Four or five of these complementation groups are RNA-negative mutants which do not replicate RNA at the nonpermissive temperature. All of these mutations are mapped within gene 1, suggesting that at least four or five separate functions of gene 1 products are required for viral RNA synthesis. Several complementation groups of RNA-positive mutants replicate viral RNA at the nonpermissive temperature but show altered cytopathic effects and/or fail to produce infectious virions (11,122,214). The RNA-positive mutants have defects in the genes encoding the S or N structural proteins (11). *In vivo*, some ts mutants differ in virulence from wild-type virus. For example, while wild-type MHV-JHM causes acute encephalitis in mice, some ts mutants cause a subacute demyelinating disease (81,109).

Mutants of MHV selected from virus stocks by neutralizing monoclonal antibodies contain either point or deletion mutations in the S gene (48,76,173,232). Viruses with single residue changes in the cleavage site of the MHV S glycoprotein have been isolated from persistently infected cultures (78). In general, S mutants differ significantly from wild-type virus in cytopathic effects, virulence, and tissue tropism (48,69,70,78,236). In addition to ts mutants, field isolates of coronavirus and coronavirus stocks show a surprisingly high incidence of deletion mutations (5,9,36,76, 119). For example, many natural strains of MHV have deletions of up to 200 amino acids in a hypervariable region and show altered biological properties *in vivo* (9,76,173).

The most striking example of the biological importance of deletion mutations is the emergence of porcine respiratory coronavirus (PRCV) from TGEV. TGEV causes epizootic enteric infection of pigs. In the early 1980s, PRCV emerged in Europe as a new virus that causes widespread, devastating epizootics of respiratory disease in pigs (119). Nucleotide sequence comparison of the two viruses showed that PRCV was derived from TGEV and had a large deletion within the S1 glycoprotein (30,172,240). Similar PRCV strains arose independently in the U.S. as deletion mutants in the S gene of TGEV (5,119).

RNA Recombination

A unique feature of coronavirus genetics is the high frequency of RNA recombination (113). While the nonsegmented RNA genomes of most other viruses exhibit very low or undetectable recombination frequencies, the recombination frequency for the entire coronavirus genome has been calculated to be as high as 25% (11). RNA recombinants were first isolated during coinfection with several ts mutants of MHV at the nonpermissive temperature (114), and additional recombinants have been isolated using selectable genetic markers such as neutralization escape mutants and strains that differ in cell fusion activity (99,101,102,114,148). Recombination occurs both in tissue culture and in experimental and natural animal infections (100,233). RNA recombination in MHV has been detected almost everywhere in the viral genome, but serial passage usually selects for certain types of recombinants (10,113).

The high frequency of recombination in coronaviruses is probably the result of the unique mechanism of coronavirus RNA replication which involves discontinuous transcription and polymerase jumping (113) (Fig. 5). Possibly, polymerase associated with incomplete nascent RNAs dissociates at a random point from its template and switches to a homologous region on a different RNA template to complete RNA synthesis by a copy-choice mechanism like that of other RNA viruses (13,113).

RNA recombination is an important factor in the natural evolution of coronaviruses. For example, new strains of IBV can arise in poultry flocks by natural recombination between different field strains (33,110,233).

Recombination may have played a role in the evolution of different coronavirus species. For example, reversal of the order of the M gene and gene 5 on the IBV and MHV genomes (Fig. 3) could have arisen by homologous recombination at consensus intergenic regions between the two genes in a progenitor of one of the viruses. The variation in the number and order of the genes encoding nonstructural proteins of coronaviruses suggest that some of these genes may have been acquired from cellular mRNAs by nonhomologous recombination with coronavirus genomes. Recombination may also explain the acquisition of the HE gene by a progenitor of the group II coronaviruses. Possibly, nonhomologous recombination occurred between coronavirus genomic RNA and the influenza C virus mRNA encoding HE (138). This fortuitous insertional mutation provided the ancestral group II coronavirus with an additional virus attachment protein and an esterase that might facilitate virus penetration of mucous secretions.

RNA recombination has recently been used as a genetic tool to introduce desired sequences into coronavirus genomes. Coronavirus-infected cells are transfected with an RNA fragment or a DI RNA containing the desired sequence, and the desired recombinant is selected (106,126,157,228). This is an important tool for the genetic manipulation of coronaviruses, as no full-length, infectious cDNA clones of coronavirus genomic RNA are yet available.

Defective-Interfering RNA

When MHV is passaged at a high multiplicity of infection, new subgenomic-sized RNA species often emerged (154). Because these RNAs have extensive deletions compared with the genomic RNA and some interfere with the replication of genomic RNA, they are called *DI RNAs* (154). Three different types of DI RNAs have been detected in MHV-infected cells (145). The first type, typified by DIssE RNA, replicates efficiently but is not packaged efficiently into virions. However, small amounts of this DI RNA are nonspecifically incorporated into virions and can be passaged for a few generations (150). The second type of DI RNA, typified by DIssF, contains a packaging signal that enables it to be incorporated into virions efficiently (155,227). DIssE and DIssF both require a helper virus for RNA replication. A third type of subgenomic RNA, not a true DI, is typified by DIssA, which contains multiple small deletions as compared to the wild-type genome, is nearly the size of genomic RNA, encodes a functional polymerase, and can replicate independently of a helper virus (150).

Both DIssE and DIssF RNAs contain the 5' and 3' ends and one or several discontinuous segments of internal sequence from the MHV genome, consistent with the idea that the 5' and 3' ends contain essential replication signals (104,129). All of the DI RNAs also contain an ORF (151). It is not yet known whether the ORF requires translation to allow replication of the DI RNA.

During serial passage of viruses in cell culture, DI RNAs are generated by recombination or polymerase jumping, and evolve rapidly; i.e., old DI RNAs disappear and are replaced by new DI RNAs which usually differ in size (145). Loss of old DI RNAs may be due to inefficient incorporation into virions and/or to competition with other DI RNAs for replication factors. Recombination can occur between the original DI RNA and the RNA of the helper virus (74). DI RNAs with a larger ORF have a selective advantage over those with a shorter ORF, thus contributing to the evolution of DI RNA (105,228). DI RNAs have not been detected during natural coronavirus infections.

CONCLUSIONS

Progress on the molecular biology of coronaviruses has been rapid. The complete nucleotide sequence of the IBV, MHV, and HCV-229E genomes have been determined, and large portions of the genomes of other coronaviruses have been sequenced. Genes that encode structural and nonstructural coronavirus proteins have been expressed *in vitro* and *in vivo*, and nucleotide sequences are available for all coronavirus genes. Monoclonal antibodies and antibodies to synthetic peptides are being used to identify functional domains of viral polypeptides. *In vitro* assays for coronavirus RNA polymerase and a variety of induced, monoclonal antibody-selected and site-directed mutants of many coronaviruses are available for study. Unresolved questions about coronavirus replication are: How is the coronavirus polymerase synthesized and processed, and how does it function? How is the leader RNA added to mRNA? How is synthesis of mRNAs, genome, and minus-strand template regulated? What is the function of the subgenomic minus-strand RNAs? What are the structure-function relationships of coronavirus proteins? What viral and host factors determine the tissue tropism, pathogenicity, and species specificity of coronaviruses?

REFERENCES

1. Abraham S, Kienzle TE, Lapps W, Brian DA. Deduced sequence of the bovine coronavirus spike protein and identification of the internal proteolytic cleavage site. *Virology* 1990;176:296–301.
2. Anderson R, Wong F. Membrane and phospholipid binding by murine coronaviral nucleocapsid N protein. *Virology* 1993;194:224–232.
3. Armstrong J, Niemann H, Smeekens S, Rottier P, Warren G. Sequence and topology of a model intracellular membrane protein, E1 glycoprotein, from a coronavirus. *Nature* 1984;308:751–752.
4. Asanaka M, Lai MM. Cell fusion studies identified multiple cellular factors involved in mouse hepatitis virus entry. *Virology* 1993;197: 732–741.
5. Bae I, Jackwood DJ, Benfield DA, Saif LJ, Wesley RD, Hill H. Differentiation of transmissible gastroenteritis virus from porcine respiratory coronavirus and other antigenically related coronaviruses by using cDNA probes specific for the 5' region of the S glycoprotein gene. *J Clin Microbiol* 1991;29:215–218.
6. Baker SC, Lai MM. An in vitro system for the leader-primed transcription of coronavirus mRNAs. *EMBO J* 1990;9:4173–4179.
7. Baker SC, Shieh CK, Soe LH, Chang MF, Vannier DM, Lai MM. Identification of a domain required for autoproteolytic cleavage of murine coronavirus gene A polyprotein. *J Virol* 1989;63:3693–3699.
8. Baker SC, Yokomori K, Dong S, Carlisle R, Gorbalenya AE, Koonin EV, Lai MM, et al. Identification of the catalytic sites of a papain-like cysteine proteinase of murine coronavirus. *J Virol* 1993;67:6056–6063.
9. Banner LR, Keck JG, Lai MM. A clustering of RNA recombination sites adjacent to a hypervariable region of the peplomer gene of murine coronavirus. *Virology* 1990;175:548–555.
10. Banner LR, Lai MM. Random nature of coronavirus RNA recombination in the absence of selection pressure. *Virology* 1991;185:441–445.

11. Baric RS, Fu K, Schaad MC, Stohlman SA. Establishing a genetic recombination map for murine coronavirus strain A59 complementation groups. *Virology* 1990;177:646–656.
12. Baric RS, Nelson GW, Fleming JO, et al. Interactions between coronavirus nucleocapsid protein and viral RNAs: implications for viral transcription. *J Virol* 1988;62:4280–4287.
13. Baric RS, Shieh CK, Stohlman SA, Lai MM. Analysis of intracellular small RNAs of mouse hepatitis virus: evidence for discontinuous transcription. *Virology* 1987;156:342–354.
14. Baric RS, Stohlman SA, Razavi MK, Lai MM. Characterization of leader-related small RNAs in coronavirus-infected cells: further evidence for leader-primed mechanism of transcription. *Virus Res* 1985;3:19–33.
15. Barthold SW. Mouse Hepatitis Virus Biology and Epizootiology. In: Bhatt PN, Jacoby RO, Morse HC,III, New AE, eds. *Viral and mycoplasmal infections of laboratory rodents. Effects on biomedical research.* Orlando, FL: Academic Press, 1986:571–601.
16. Barthold SW, Smith AL, Lord PF, Bhatt PN, Jacoby RO, Main AJ. Epizootic coronaviral typhlocolitis in suckling mice. *Lab Anim Sci* 1982;32:376–383.
17. Blikstad I, Wikstr^m T, Aurivillius M, Obrink B. C-CAM (Cell-CAM 105) is a calmodulin binding protein. *FEBS Lett* 1992;302:26–30.
18. Boots AM, Kusters JG, van Noort JM, et al. Localization of a T-cell epitope within the nucleocapsid protein of avian coronavirus. *Immunology* 1991;74:8–13.
19. Boots AM, Van Lierop MJ, Kusters JG, Van Kooten PJ, van der Zeijst BA, Hensen EJ. MHC class II-restricted T-cell hybridomas recognizing the nucleocapsid protein of avian coronavirus IBV. *Immunology* 1991;72:10–14.
20. Boursnell ME, Brown TD, Binns MM. Sequence of the membrane protein gene from avian coronavirus IBV. *Virus Res* 1984;1:303–313.
21. Boursnell ME, Brown TD, Foulds IJ, Green PF, Tomley FM, Binns MM. Completion of the sequence of the genome of the coronavirus avian infectious bronchitis virus. *J Gen Virol* 1987;68:57–77.
22. Boursnell ME, Brown TD, Foulds IJ, Green PF, Tomley FM, Binns MM. The complete nucleotide sequence of avian infectious bronchitis virus: analysis of the polymerase-coding region. *Adv Exp Med Biol* 1987;218:15–29.
23. Brayton PR, Lai MM, Patton CD, Stohlman SA. Characterization of two RNA polymerase activities induced by mouse hepatitis virus. *J Virol* 1982;42:847–853.
24. Bredenbeek PJ, Pachuk CJ, Noten AF, et al. The primary structure and expression of the second open reading frame of the polymerase gene of the coronavirus MHV-A59; a highly conserved polymerase is expressed by an efficient ribosomal frameshifting mechanism. *Nucleic Acids Res* 1990;18:1825–1832.
25. Bredenbeek PJ, Snijder EJ, Noten FH, et al. The polymerase gene of corona- and toroviruses: evidence for an evolutionary relationship. *Adv Exp Med Biol* 1990;276:307–316.
26. Brierley I, Boursnell ME, Binns MM, et al. An efficient ribosomal frame-shifting signal in the polymerase-encoding region of the coronavirus IBV. *EMBO J* 1987;6:3779–3785.
27. Brierley I, Digard P, Inglis SC. Characterization of an efficient coronavirus ribosomal frameshifting signal: requirement for an RNA pseudoknot. *Cell* 1989;57:537–547.
28. Brierley I, Jenner AJ, Inglis SC. Mutational analysis of the slippery-sequence component of a coronavirus ribosomal frameshifting signal. *J Mol Biol* 1992;227:463–479.
29. Brierley I, Rolley NJ, Jenner AJ, Inglis SC. Mutational analysis of the RNA pseudoknot component of a coronavirus ribosomal frameshifting signal. *J Mol Biol* 1991;220:889–902.
30. Britton P, Mawditt KL, Page KW. The cloning and sequencing of the virion protein genes from a British isolate of porcine respiratory coronavirus: comparison with transmissible gastroenteritis virus genes. *Virus Res* 1991;21:181–198.
31. Buchmeier MJ, Dalziel RG, Koolen MJ. Coronavirus-induced CNS disease: a model for virus-induced demyelination. *J Neuroimmunol* 1988;20:111–116.
32. Callow KA, Parry HF, Sergeant M, Tyrrell DA. The time course of the immune response to experimental coronavirus infection of man. *Epidemiol Infect* 1990;105:435–446.
33. Cavanagh D. Recent advances in avian virology. *Br Vet J* 1992;148: 199–222.
34. Cavanagh D, Brian DA, Enjuanes L, et al. Recommendations of the Coronavirus Study Group for the nomenclature of the structural proteins, mRNAs, and genes of coronaviruses. *Virology* 1990; 176:306–307.
35. Cavanagh D, Davis PJ. Coronavirus IBV: removal of spike glycopolypeptide S1 by urea abolishes infectivity and haemagglutination but not attachment to cells. *J Gen Virol* 1986;67:1443–1448.
36. Cavanagh D, Davis PJ, Mockett AP. Amino acids within hypervariable region 1 of avian coronavirus IBV (Massachusetts serotype) spike glycoprotein are associated with neutralization epitopes. *Virus Res* 1988;11:141–150.
37. Cavanagh D, Davis PJ, Pappin DJ. Coronavirus IBV glycopolypeptides: locational studies using proteases and saponin, a membrane permeabilizer. *Virus Res* 1986;4:145–156.
38. Cavanagh D, Horzinek MC. Genus Torovirus assigned to the Coronaviridae [news]. *Arch Virol* 1993;128:395–396.
39. Chaloner-Larsson G, Johnson-Lussenburg CM. Establishment and maintenance of a persistent infection of L132 cells by human coronavirus strain 229E. *Arch Virol* 1981;69:117–129.
40. Charley B, Laude H. Induction of alpha interferon by transmissible gastroenteritis coronavirus: role of transmembrane glycoprotein E1. *J Virol* 1988;62:8–11.
41. Collins AR, Knobler RL, Powell H, Buchmeier MJ. Monoclonal antibodies to murine hepatitis virus-4 (strain JHM) define the viral glycoprotein responsible for attachment and cell-cell fusion. *Virology* 1982;119:358–371.
42. Collisson EW, Li JZ, Sneed LW, Peters ML, Wang L. Detection of avian infectious bronchitis viral infection using in situ hybridization and recombinant DNA. *Vet Microbiol* 1990;24:261–271.
43. Compton SR, Rogers DB, Holmes KV, Fertsch D, Remenick J, McGowan JJ. In vitro replication of mouse hepatitis virus strain A59. *J Virol* 1987;61:1814–1820.
44. Compton SR, Stephensen CB, Snyder SW, Weismiller DG, Holmes KV. Coronavirus species specificity: Murine coronavirus binds to a mouse-specific epitope on its carcinoembryonic antigen-related receptor glycoprotein. *J Virol* 1992;66:7420–7428.
45. Coulter-Mackie MB, Flintoff WF, Dales S. In vivo and in vitro models of demyelinating disease X. A Schwannoma-L-2 somatic cell hybrid persistently yielding high titres of mouse hepatitis virus strain JHM.*Virus Res* 1984;1:477–487.
46. Cox GJ, Parker MD, Babiuk LA. Bovine coronavirus nonstructural protein ns2 is a phosphoprotein. *Virology* 1991;185:509–512.
47. Cyr-Coats KS, Storz J. Bovine coronavirus-induced cytopathic expression and plaque formation: host cell and virus strain determine trypsin dependence. *Zentralbl Veterinarmed [B]* 1988;35:48–56.
48. Dalziel RG, Lampert PW, Talbot PJ, Buchmeier MJ. Site-specific alteration of murine hepatitis virus type 4 peplomer glycoprotein E2 results in reduced neurovirulence. *J Virol* 1986;59:463–471.
49. Daniel C, Anderson R, Buchmeier MJ, et al. Identification of an immunodominant linear neutralization domain on the S2 portion of the murine coronavirus spike glycoprotein and evidence that it forms part of complex tridimensional structure. *J Virol* 1993;67:1185–1194.
50. de Groot RJ, Luytjes W, Horzinek MC, van der Zeijst BAM, Spaan WJ, Lenstra JA. Evidence for a coiled-coil structure in the spike proteins of coronaviruses. *J Mol Biol* 1987;196:963–966.
51. de Groot RJ, Van Leen RW, Dalderup MJ, Vennema H, Horzinek MC, Spaan WJ. Stably expressed FIPV peplomer protein induces cell fusion and elicits neutralizing antibodies in mice. *Virology* 1989;171:493–502.
52. Dea S, Garzon S, Tijssen P. Identification and location of the structural glycoproteins of a tissue culture-adapted turkey enteric coronavirus. *Arch Virol* 1989;106:221–237.
53. Delmas B, Gelfi J, L'Haridon R, et al. Aminopeptidase N is a major receptor for the entero-pathogenic coronavirus TGEV. *Nature* 1992;357:417–420.
54. Delmas B, Laude H. Assembly of coronavirus spike protein into trimers and its role in epitope expression. *J Virol* 1990;64:5367–5375.
55. Delmas B, Rasschaert D, Godet M, Gelfi J, Laude H. Four major antigenic sites of the coronavirus transmissible gastroenteritis virus are located on the amino-terminal half of spike glycoprotein S. *J Gen Virol* 1990;71:1313–1323.
56. den Boon JA, Snijder EJ, Chirnside ED, de Vries AA, Horzinek MC, Spaan WJ. Equine arteritis virus is not a togavirus but belongs to the coronaviruslike superfamily. *J Virol* 1991;65:2910–2920.
57. Denison M, Perlman S. Identification of putative polymerase gene prod-

uct in cells infected with murine coronavirus A59. *Virology* 1987; 157:565–568.
58. Denison MR, Perlman S. Translation and processing of mouse hepatitis virus virion RNA in a cell-free system. *J Virol* 1986;60:12–18.
59. Denison MR, Zoltick PW, Hughes SA, et al. Intracellular processing of the N-terminal ORF 1a proteins of the coronavirus MHV-A59 requires multiple proteolytic events. *Virology* 1992;189:274–284.
60. Denison MR, Zoltick PW, Leibowitz JL, Pachuk CJ, Weiss SR. Identification of polypeptides encoded in open reading frame 1b of the putative polymerase gene of the murine coronavirus mouse hepatitis virus A59. *J Virol* 1991;65:3076–3082.
61. Dennis DE, Brian DA. RNA-dependent RNA polymerase activity in coronavirus-infected cells. *J Virol* 1982;42:153–164.
62. Deregt D, Gifford GA, Ijaz MK, et al. Monoclonal antibodies to bovine coronavirus glycoproteins E2 and E3: demonstration of in vivo virus-neutralizing activity. *J Gen Virol* 1989;70:993–998.
63. Deregt D, Sabara M, Babiuk LA. Structural proteins of bovine coronavirus and their intracellular processing. *J Gen Virol* 1987; 68: 2863–2877.
64. Dubois-Dalcq ME, Doller EW, Haspel MV, Holmes KV. Cell tropism and expression of mouse hepatitis viruses (MHV) in mouse spinal cord cultures. *Virology* 1982;119:317–331.
65. Dubois-Dalcq ME, Holmes KV, Rentier B. Assembly of RNA viruses. New York: Springer-Verlag, 1984;7.
66. Dveksler GS, Dieffenbach CW, Cardellichio CB, et al. Several members of the mouse carcinoembryonic antigen-related glycoprotein family are functional receptors for the coronavirus mouse hepatitis virus-A59. *J Virol* 1993;67:1–8.
67. Dveksler GS, Pensiero MN, Cardellichio CB, et al. Cloning of the mouse hepatitis virus (MHV) receptor: expression in human and hamster cell lines confers susceptibility to MHV. *J Virol* 1991;65: 6881–6891.
68. Dveksler GS, Pensiero MN, Dieffenbach CW, et al. Mouse hepatitis virus strain A59 and blocking antireceptor monoclonal antibody bind to the N-terminal domain of cellular receptor. *Proc Natl Acad Sci U S A* 1993;90:1716–1720.
69. Fazakerley JK, Parker SE, Bloom F, Buchmeier MJ. The V5A13.1 envelope glycoprotein deletion mutant of mouse hepatitis virus type-4 is neuroattenuated by its reduced rate of spread in the central nervous system. *Virology* 1992;187:178–188.
70. Fleming JO, Trousdale MD, el-Zaatari FA, Stohlman SA, Weiner LP. Pathogenicity of antigenic variants of murine coronavirus JHM selected with monoclonal antibodies. *J Virol* 1986;58:869–875.
71. Fosmire JA, Hwang K, Makino S. Identification and characterization of a coronavirus packaging signal. *J Virol* 1992;66:3522–3530.
72. Frana MF, Behnke JN, Sturman LS, Holmes KV. Proteolytic cleavage of the E2 glycoprotein of murine coronavirus: host-dependent differences in proteolytic cleavage and cell fusion. *J Virol* 1985;56:912–920.
73. Fu K, Baric RS. Evidence for variable rates of recombination in the MHV genome. *Virology* 1992;189:88–102.
74. Furuya T, Macnaughton TB, La Monica N, Lai MM. Natural evolution of coronavirus defective-interfering RNA involves RNA recombination. *Virology* 1993;194:408–413.
75. Gallagher TM, Escarmis C, Buchmeier MJ. Alteration of the pH dependence of coronavirus-induced cell fusion: effect of mutations in the spike glycoprotein. *J Virol* 1991;65:1916–1928.
76. Gallagher TM, Parker SE, Buchmeier MJ. Neutralization-resistant variants of a neurotropic coronavirus are generated by deletions within the amino-terminal half of the spike glycoprotein. *J Virol* 1990;64:731–741.
77. Godet M, L'Haridon R, Vautherot JF, Laude H. TGEV corona virus ORF4 encodes a membrane protein that is incorporated into virions. *Virology* 1992;188:666–675.
78. Gombold JL, Hingley ST, Weiss SR. Fusion-defective mutants of mouse hepatitis virus A59 contain a mutation in the spike protein cleavage signal. *J Virol* 1993;67:4504–4512.
79. Gorbalenya AE, Koonin EV, Donchenko AP, Blinov VM. Coronavirus genome: prediction of putative functional domains in the non-structural polyprotein by comparative amino acid sequence analysis. *Nucleic Acids Res* 1989;17:4847–4861.
80. Griffiths G, Rottier P. Cell biology of viruses that assemble along the biosynthetic pathway. *Semin Cell Biol* 1992;3:367–381.
81. Haspel MV, Lampert PW, Oldstone MB. Temperature-sensitive mutants of mouse hepatitis virus produce a high incidence of demyelination. *Proc Natl Acad Sci USA* 1978;75:4033–4036.
82. Herold J, Raabe T, Schelle-Prinz B, Siddell SG. Nucleotide sequence of the human coronavirus 229E RNA polymerase locus. *Virology* 1993;195:680–691.
83. Herold J, Raabe T, Siddell S. Molecular analysis of the human coronavirus (strain 229E) genome. *Arch Virol Suppl* 1993;7:63–74.
84. Herold J, Siddell SG. An "elaborated" pseudoknot is required for high frequency frameshifting during translation of HCV 229E polymerase mRNA. *Nucleic Acids Res* 1993;21:5838–5842.
85. Herrler G, Szepanski S, Schultze B. 9-O-acetylated sialic acid, a receptor determinant for influenza C virus and coronaviruses. *Behring Inst Mitt* 1991;177–184.
86. Hofmann MA, Brian DA. The 5′ end of coronavirus minus-strand RNAs contains a short poly(U) tract. *J Virol* 1991;65:6331–6333.
87. Hofmann MA, Senanayake SD, Brian DA. A translation-attenuating intraleader open reading frame is selected on coronavirus mRNAs during persistent infection. *Proc Natl Acad Sci U S A* 1993;90: 11733–11737.
88. Hogue BG, Kienzle TE, Brian DA. Synthesis and processing of the bovine enteric coronavirus haemagglutinin protein. *J Gen Virol* 1989;70:345–352.
89. Holmes KV, Behnke JN. Evolution of a coronavirus during persistent infection in vitro. *Adv Exp Med Biol* 1981;142:287–299.
90. Holmes KV, Doller EW, Sturman LS. Tunicamycin resistant glycosylation of coronavirus glycoprotein: demonstration of a novel type of viral glycoprotein. *Virology* 1981;115:334–344.
91. Holmes KV, Dveksler GS. Specificity of coronavirus-receptor interactions. In: Wimmer E, ed. *Virus Receptors*. Cold Spring Harbor, NY: Cold Spring Harbor Press, 1994:
92. Holmes KV, Welsh RM, Haspel MV. Natural cytotoxicity against mouse hepatitis virus-infected target cells. I. Correlation of cytotoxicity with virus binding to leukocytes. *J Immunol* 1986;136:1446–1453.
93. Jacobs L, Spaan WJ, Horzinek MC, van der Zeijst BAM. Synthesis of subgenomic mRNA's of mouse hepatitis virus is initiated independently: evidence from UV transcription mapping. *J Virol* 1981;39:401–406.
94. Joo M, Makino S. Mutagenic analysis of the coronavirus intergenic consensus sequence. *J Virol* 1992;66:6330–6337.
95. Jouvenne P, Mounir S, Stewart JN, Richardson CD, Talbot PJ. Sequence analysis of human coronavirus 229E mRNAs 4 and 5: evidence for polymorphism and homology with myelin basic protein. *Virus Res* 1992;22:125–141.
96. Jouvenne P, Richardson CD, Schreiber SS, Lai MM, Talbot PJ. Sequence analysis of the membrane protein gene of human coronavirus 229E. *Virology* 1990;174:608–612.
97. Künkel F, Herrler G. Structural and functional analysis of the surface protein of human coronavirus OC43. *Virology* 1993;195:195–202.
98. Kapke PA, Tung FY, Brian DA, Woods RD, Wesley R. Nucleotide sequence of the porcine transmissible gastroenteritis coronavirus matrix protein gene. *Adv Exp Med Biol* 1987;218:117–122.
99. Keck JG, Makino S, Soe LH, Fleming JO, Stohlman SA, Lai MMC. RNA recombination of coronavirus. *Adv Exp Med Biol* 1987;218: 99–107.
100. Keck JG, Matsushima GK, Makino S, et al. In vivo RNA-RNA recombination of coronavirus in mouse brain. *J Virol* 1988;62:1810–1813.
101. Keck JG, Soe LH, Makino S, Stohlman SA, Lai MM. RNA recombination of murine coronaviruses: recombination between fusion-positive mouse hepatitis virus A59 and fusion-negative mouse hepatitis virus 2. *J Virol* 1988;62:1989–1998.
102. Keck JG, Stohlman SA, Soe LH, Makino S, Lai MM. Multiple recombination sites at the 5'-end of murine coronavirus RNA. *Virology* 1987;156:331–341.
103. Kienzle TE, Abraham S, Hogue BG, Brian DA. Structure and orientation of expressed bovine coronavirus hemagglutinin-esterase protein. *J Virol* 1990;64:1834–1838.
104. Kim YN, Jeong YS, Makino S. Analysis of cis-acting sequences essential for coronavirus defective interfering RNA replication. *Virology* 1993;197:53–63.
105. Kim YN, Lai MM, Makino S. Generation and selection of coronavirus defective interfering RNA with large open reading frame by RNA recombination and possible editing. *Virology* 1993;194:244–253.
106. Koetzner CA, Parker MM, Ricard CS, Sturman LS, Masters PS. Repair and mutagenesis of the genome of a deletion mutant of the coronavirus mouse hepatitis virus by targeted RNA recombination. *J Virol* 1992;66:1841–1848.

107. Kooi C, Cervin M, Anderson R. Differentiation of acid-pH-dependent and -nondependent entry pathways for mouse hepatitis virus. *Virology* 1991;180:108–119.
108. Kooi C, Mizzen L, Alderson C, Daya M, Anderson R. Early events of importance in determining host cell permissiveness to mouse hepatitis virus infection. *J Gen Virol* 1988;69:1125–1135.
109. Koolen MJ, Osterhaus AD, Van Steenis G, Horzinek MC, van der Zeijst BAM. Temperature-sensitive mutants of mouse hepatitis virus strain A59: Isolation, characterization and neuropathogenic properties. *Virology* 1983;125:393–402.
110. Kusters JG, Niesters HG, Lenstra JA, Horzinek MC, van der Zeijst BA. Phylogeny of antigenic variants of avian coronavirus IBV. *Virology* 1989;169:217–221.
111. La Monica N, Yokomori K, Lai MM. Coronavirus mRNA synthesis: identification of novel transcription initiation signals which are differentially regulated by different leader sequences. *Virology* 1992;188:402–407.
112. Lai MM. Coronavirus: organization, replication and expression of genome. *Annu Rev Microbiol* 1990;44:303–333.
113. Lai MM. Genetic recombination in RNA viruses. *Curr Top Microbiol Immunol* 1992;176:21–32.
114. Lai MM, Baric RS, Makino S, et al. Recombination between nonsegmented RNA genomes of murine coronaviruses. *J Virol* 1985;56:449–456.
115. Lai MM, Patton CD, Stohlman SA. Replication of mouse hepatitis virus: negative-stranded RNA and replicative form RNA are of genome length. *J Virol* 1982;44:487–492.
116. Lamontagne LM, Dupuy JM. Persistent infection with mouse hepatitis virus 3 in mouse lymphoid cell lines. *Infect Immun* 1984;44:716–723.
117. Laude H, Rasschaert D, Delmas B, Godet M, Gelfi J, Charley B. Molecular biology of transmissible gastroenteritis virus. *Vet Microbiol* 1990;23:147–154.
118. Laude H, Rasschaert D, Huet JC. Sequence and N-terminal processing of the transmembrane protein E1 of the coronavirus transmissible gastroenteritis virus. *J Gen Virol* 1987;68:1687–1693.
119. Laude H, Van Reeth K, Pensaert M. Porcine respiratory coronavirus: molecular features and virus-host interactions. *Vet Res* 1993;24:125–150.
120. Lee HJ, Shieh CK, Gorbalenya AE, et al. The complete sequence (22 kilobases) of murine coronavirus gene 1 encoding the putative proteases and RNA polymerase. *Virology* 1991;180:567–582.
121. Leibowitz JL, DeVries JR. Synthesis of virus-specific RNA in permeabilized murine coronavirus-infected cells. *Virology* 1988;166:66–75.
122. Leibowitz JL, DeVries JR, Haspel MV. Genetic analysis of murine hepatitis virus strain JHM. *J Virol* 1982;42:1080–1087.
123. Leibowitz JL, Perlman S, Weinstock G, et al. Detection of a murine coronavirus nonstructural protein encoded in a downstream open reading frame. *Virology* 1988;164:156–164.
124. Leibowitz JL, Weiss SR, Paavola E, Bond CW. Cell-free translation of murine coronavirus RNA. *J Virol* 1982;43:905–913.
125. Li D, Cavanagh D. Coronavirus IBV-induced membrane fusion occurs at near-neutral pH. *Arch Virol* 1992;122:307–316.
126. Liao CL, Lai MM. RNA recombination in a coronavirus: recombination between viral genomic RNA and transfected RNA fragments. *J Virol* 1992;66:6117–6124.
127. Liao CL, Lai MM. Requirement of the 5'-end genomic sequence as an upstream cis-acting element for coronavirus subgenomic mRNA transcription. *J Virol* 1994;68:4727–4737.
128. Lin SH, Guidotti G. Cloning and expression of a cDNA coding for a rat liver plasma membrane ecto-ATPase. The primary structure of the ecto-ATPase is similar to that of the human biliary glycoprotein I. *J Biol Chem* 1989;264:14408–14414.
129. Lin YJ, Lai MM. Deletion mapping of a mouse hepatitis virus defective interfering RNA reveals the requirement of an internal and discontiguous sequence for replication. *J Virol* 1993;67:6110–6118.
130. Liu DX, Cavanagh D, Green P, Inglis SC. A polycistronic mRNA specified by the coronavirus infectious bronchitis virus. *Virology* 1991;184:531–544.
131. Liu DX, Inglis SC. Association of the infectious bronchitis virus 3c protein with the virion envelope. *Virology* 1991;185:911–917.
132. Liu DX, Inglis SC. Internal entry of ribosomes on a tricistronic mRNA encoded by infectious bronchitis virus [published erratum appears in J Virol 1992 Nov;66(11):6840]. *J Virol* 1992;66:6143–6154.
133. Liu DX, Inglis SC. Identification of two new polypeptides encoded by mRNA5 of the coronavirus infectious bronchitis virus. *Virology* 1992;186:342–347.
134. Locker JK, Griffiths G, Horzinek MC, Rottier PJ. O-glycosylation of the coronavirus M protein. Differential localization of sialyltransferases in N- and O-linked glycosylation. *J Biol Chem* 1992;267:14094–14101.
135. Locker JK, Rose JK, Horzinek MC, Rottier PJ. Membrane assembly of the triple-spanning coronavirus M protein. Individual transmembrane domains show preferred orientation. *J Biol Chem* 1992;267:21911–21918.
136. Lomniczi B. Biological properties of avian coronavirus RNA. *J Gen Virol* 1977;36:531–533.
137. Lucas A, Coulter M, Anderson R, Dales S, Flintoff W. In vivo and in vitro models of demyelinating diseases. II Persistence and host-regulated thermosensitivity in cells of neural derivation infected with mouse hepatitis and measles viruses.*Virology* 1978;88:325–337.
138. Luytjes W, Bredenbeek PJ, Noten AF, Horzinek MC, Spaan WJ. Sequence of mouse hepatitis virus A59 mRNA 2: Indications for RNA recombination between coronaviruses and influenza C virus. *Virology* 1988;166:415–422.
139. Luytjes W, Geerts D, Posthumus W, Meloen R, Spaan W. Amino acid sequence of a conserved neutralizing epitope of murine coronaviruses. *J Virol* 1989;63:1408–1412.
140. Luytjes W, Sturman LS, Bredenbeek PJ, et al. Primary structure of the glycoprotein E2 of coronavirus MHV-A59 and identification of the trypsin cleavage site. *Virology* 1987;161:479–487.
141. Machamer CE, Grim MG, Esquela A, et al. Retention of a cis Golgi protein requires polar residues on one face of a predicted alpha-helix in the transmembrane domain. *Mol Biol Cell* 1993;4:695–704.
142. Machamer CE, Mentone SA, Rose JK, Farquhar MG. The E1 glycoprotein of an avian coronavirus is targeted to the cis Golgi complex. *Proc Natl Acad Sci U S A* 1990;87:6944–6948.
143. Machamer CE, Rose JK. A specific transmembrane domain of a coronavirus E1 glycoprotein is required for its retention in the Golgi region. *J Cell Biol* 1987;105:1205–1214.
144. Macnaughton MR, Davies HA, Nermut MV. Ribonucleoprotein-like structures from coronavirus particles. *J Gen Virol* 1978;39:545–549.
145. Makino S, Fujioka N, Fujiwara K. Structure of the intracellular defective viral RNAs of defective interfering particles of mouse hepatitis virus. *J Virol* 1985;54:329–336.
146. Makino S, Joo M. Effect of intergenic consensus sequence flanking sequences on coronavirus transcription. *J Virol* 1993;67:3304–3311.
147. Makino S, Joo M, Makino JK. A system for study of coronavirus mRNA synthesis: a regulated, expressed subgenomic defective interfering RNA results from intergenic site insertion. *J Virol* 1991;65:6031–6041.
148. Makino S, Keck JG, Stohlman SA, Lai MM. High-frequency RNA recombination of murine coronaviruses. *J Virol* 1986;57:729–737.
149. Makino S, Lai MM. High-frequency leader sequence switching during coronavirus defective interfering RNA replication. *J Virol* 1989;63:5285–5292.
150. Makino S, Shieh CK, Keck JG, Lai MM. Defective-interfering particles of murine coronavirus: mechanism of synthesis of defective viral RNAs. *Virology* 1988;163:104–111.
151. Makino S, Shieh CK, Soe LH, Baker SC, Lai MM. Primary structure and translation of a defective interfering RNA of murine coronavirus. *Virology* 1988;166:550–560.
152. Makino S, Soe LH, Shieh CK, Lai MM. Discontinuous transcription generates heterogeneity at the leader fusion sites of coronavirus mRNAs. *J Virol* 1988;62:3870–3873.
153. Makino S, Stohlman SA, Lai MMC. Leader sequences of murine coronavirus mRNAs can be freely reassorted: evidence for the role of free leader RNA in transcription. *Proc Natl Acad Sci USA* 1986;83:4204–4208.
154. Makino S, Taguchi F, Fujiwara K. Defective interfering particles of mouse hepatitis virus. *Virology* 1984;133:9–17.
155. Makino S, Yokomori K, Lai MM. Analysis of efficiently packaged defective interfering RNAs of murine coronavirus: localization of a possible RNA-packaging signal. *J Virol* 1990;64:6045–6053.
156. Masters PS. Localization of an RNA-binding domain in the nucleocapsid protein of the coronavirus mouse hepatitis virus. *Arch Virol* 1992;125:141–160.
157. Masters PS, Koetzner CA, Kerr CA, Heo Y. Optimization of targeted RNA recombination and mapping of a novel nucleocapsid gene mutation in the coronavirus mouse hepatitis virus. *J Virol* 1994;68:328–337.
158. McCuaig K, Turbide C, Beauchemin N. mmCGM1a: A mouse carcinoembryonic antigen gene family member, generated by alternative

splicing, functions as an adhesion molecule. *Cell Growth Differ* 1992; 3:165–177.
159. McIntosh K. Coronaviruses. A comparative review. *Curr Top Microbiol Immunol* 1974;63:85–129.
160. Mizzen L, Hilton A, Cheley S, Anderson R. Attenuation of murine coronavirus infection by ammonium chloride. *Virology* 1985;142: 378–388.
161. Mounir S, Talbot PJ. Human coronavirus OC43 RNA 4 lacks two open reading frames located downstream of the S gene of bovine coronavirus. *Virology* 1993;192:355–360.
162. Mounir S, Talbot PJ. Molecular characterization of the S protein gene of human coronavirus OC43. *J Gen Virol* 1993;74:1981–1987.
163. Myint S, Harmsen D, Raabe T, Siddell SG. Characterization of a nucleic acid probe for the diagnosis of human coronavirus 229E infections. *J Med Virol* 1990;31:165–172.
164. Myint S, Siddell S, Tyrrell D. Detection of human coronavirus 229E in nasal washings using RNA: RNA hybridisation. *J Med Virol* 1989;29:70–73.
165. Niemann H, Boschek B, Evans D, Rosing M, Tamura T, Klenk H-D. Post-translational glycosylation of coronavirus glycoprotein E1: Inhibition by monensin. *EMBO J* 1982;1:1499–1504.
166. Niemann H, Geyer R, Klenk H-D, Linder D, Stirm S, Wirth M. The carbohydrates of mouse hepatitis virus (MHV) A59: structures of the O-glycosidically linked oligosaccharides of glycoprotein E1. *EMBO J* 1984;3:665–670.
167. Niemann H, Klenk H-D. Coronavirus glycoprotein E1, a new type of viral glycoprotein. *J Mol Biol* 1981;153:993–1010.
168. Oleszak EL, Perlman S, Leibowitz JL. MHV S peplomer protein expressed by a recombinant vaccinia virus vector exhibits IgG Fc-receptor activity. *Virology* 1992;186:122–132.
169. Opstelten DJ, de Groote P, Horzinek MC, Vennema H, Rottier PJ. Disulfide bonds in folding and transport of mouse hepatitis coronavirus glycoproteins. *J Virol* 1993;67:7394–7401.
170. Oshiro LS. Coronaviruses. In: Dalton AJ, Haguenau F, eds.*Ultrastructure of animal viruses and bacteriophages: an atlas.* New York: Academic Press, 1973: 331–343.
171. Pachuk CJ, Bredenbeek PJ, Zoltick PW, Spaan WJ, Weiss SR. Molecular cloning of the gene encoding the putative polymerase of mouse hepatitis coronavirus, strain A59. *Virology* 1989;171:141–148.
172. Page KW, Mawditt KL, Britton P. Sequence comparison of the 5' end of mRNA 3 from transmissible gastroenteritis virus and porcine respiratory coronavirus. *J Gen Virol* 1991;72:579–587.
173. Parker SE, Gallagher TM, Buchmeier MJ. Sequence analysis reveals extensive polymorphism and evidence of deletions within the E2 glycoprotein gene of several strains of murine hepatitis virus. *Virology* 1989;173:664–673.
174. Pensiero MN, Dveksler GS, Cardellichio CB, et al. Binding of the coronavirus mouse hepatitis virus A59 to its receptor expressed from a recombinant vaccinia virus depends on posttranslational processing of the receptor glycoprotein. *J Virol* 1992;66:4028–4039.
175. Perlman S, Ries D, Bolger E, Chang LJ, Stoltzfus CM. MHV nucleocapsid synthesis in the presence of cycloheximide and accumulation of negative strand MHV RNA. *Virus Res* 1986;6:261–272.
176. Pfleiderer M, Routledge E, Herrler G, Siddell SG. High level transient expression of the murine coronavirus haemagglutinin-esterase. *J Gen Virol* 1991;72(6):1309–1315.
177. Puri A, Booy FP, Doms RW, White JM, Blumenthal R. Conformational changes and fusion activity of influenza virus hemagglutinin of the H2 and H3 subtypes: effects of acid pretreatment. *J Virol* 1990;64: 3824–3832.
178. Raabe T, Schelle-Prinz B, Siddell SG. Nucleotide sequence of the gene encoding the spike glycoprotein of human coronavirus HCV 229E. *J Gen Virol* 1990;71:1065–1073.
179. Rasschaert D, Laude H. The predicted primary structure of the peplomer protein E2 of the porcine coronavirus transmissible gastroenteritis virus. *J Gen Virol* 1987;68:1883–1890.
180. Robbins SG, Frana MF, McGowan JJ, Boyle JF, Holmes KV. RNA-binding proteins of coronavirus MHV: detection of monomeric and multimeric N protein with an RNA overlay-protein blot assay. *Virology* 1986;150:402–410.
181. Rottier P, Armstrong J, Meyer DI. Signal recognition particle-dependent insertion of coronavirus E1, an intracellular membrane glycoprotein. *J Biol Chem* 1985;260:4648–4652.
182. Rottier PJ, Welling GW, Welling-Wester S, Niesters HG, Lenstra JA, van der Zeijst BAM. Predicted membrane topology of the coronavirus protein E1. *Biochemistry* 1986;25:1335–1339.
183. Sawicki SG. Characterization of a small plaque mutant of the A59 strain of mouse hepatitis virus defective in cell fusion. *Adv Exp Med Biol* 1987;218:169–174.
184. Sawicki SG, Sawicki DL. Coronavirus minus-strand RNA synthesis and effect of cycloheximide on coronavirus RNA synthesis. *J Virol* 1986;57:328–334.
185. Sawicki SG, Sawicki DL. Coronavirus transcription: subgenomic mouse hepatitis virus replicative intermediates function in RNA synthesis. *J Virol* 1990;64:1050–1056.
186. Schaad MC, Stohlman SA, Egbert J, et al. Genetics of mouse hepatitis virus transcription: identification of cistrons which may function in positive and negative strand RNA synthesis. *Virology* 1990;177: 634–645.
187. Schmidt MFG. Acylation of viral spike glycoproteins, a feature of enveloped RNA viruses. *Virology* 1982;116:327–338.
188. Schochetman G, Stevens RH, Simpson RW. Presence of infectious polyadenylated RNA in the coronavirus avain infectious bronchitis virus. *Virology* 1977;77:772–782.
189. Schultze B, Cavanagh D, Herrler G. Neuraminidase treatment of avian infectious bronchitis coronavirus reveals a hemagglutinating activity that is dependent on sialic acid-containing receptors on erythrocytes. *Virology* 1992;189:792–794.
190. Schultze B, Gross HJ, Brossmer R, Herrler G. The S protein of bovine coronavirus is a hemagglutinin recognizing 9-O-acetylated sialic acid as a receptor determinant. *J Virol* 1991;65:6232–6237.
191. Schultze B, Herrler G. Bovine coronavirus uses N-acetyl-9-0-acetylneuraminic acid as a receptor determinant to initiate the infection of cultured cells. *J Gen Virol* 1992;73:901–906.
192. Schultze B, Wahn K, Klenk HD, Herrler G. Isolated HE-protein from hemagglutinating encephalomyelitis virus and bovine coronavirus has receptor-destroying and receptor-binding activity. *Virology* 1991;180: 221–228.
193. Schwarz B, Routledge E, Siddell SG. Murine coronavirus nonstructural protein ns2 is not essential for virus replication in transformed cells. *J Virol* 1990;64:4784–4791.
194. Senanayake SD, Hofmann MA, Maki JL, Brian DA. The nucleocapsid protein gene of bovine coronavirus is bicistronic. *J Virol* 1992;66:5277–5283.
195. Sethna PB, Hofmann MA, Brian DA. Minus-strand copies of replicating coronavirus mRNAs contain antileaders. *J Virol* 1991;65:320–325.
196. Sethna PB, Hung SL, Brian DA. Coronavirus subgenomic minus-strand RNAs and the potential for mRNA replicons. *Proc Natl Acad Sci U S A* 1989;86:5626–5630.
197. Shieh CK, Lee HJ, Yokomori K, La Monica N, Makino S, Lai MM. Identification of a new transcriptional initiation site and the corresponding functional gene 2b in the murine coronavirus RNA genome. *J Virol* 1989;63:3729–3736.
198. Shieh CK, Soe LH, Makino S, Chang MF, Stohlman SA, Lai MM. The 5'-end sequence of the murine coronavirus genome: implications for multiple fusion sites in leader-primed transcription. *Virology* 1987;156: 321–330.
199. Siddell S. Coronavirus JHM: coding assignments of subgenomic mRNAs. *J Gen Virol* 1983;64:113–125.
200. Siddell SG, Anderson R, Cavanagh D, et al. Coronaviridae. *Intervirology* 1983;20:181–189.
201. Siddell SG, Barthel A, ter Meulen V. Coronavirus JHM: a virion-associated protein kinase. *J Gen Virol* 1981;52:235–243.
202. Sippel CJ, Suchy FJ, Ananthanarayanan M, Perlmutter DH. The rat liver ecto-ATPase is also a canalicular bile acid transport protein. *J Biol Chem* 1993;268:2083–2091.
203. Snijder EJ, den Boon JA, Horzinek MC, Spaan WJ. Comparison of the genome organization of toro- and coronaviruses: evidence for two non-homologous RNA recombination events during Berne virus evolution. *Virology* 1991;180:448–452.
204. Somogyi P, Jenner AJ, Brierley I, Inglis SC. Ribosomal pausing during translation of an RNA pseudoknot. *Mol Cell Biol* 1993;13: 6931–6940.
205. Spaan W, Cavanagh D, Horzinek MC. Coronaviruses: structure and genome expression. *J Gen Virol* 1988;69:2939–2952.
206. Stauber R, Pfleiderera M, Siddell S. Proteolytic cleavage of the murine coronavirus surface glycoprotein is not required for fusion activity. *J Gen Virol* 1993;74:183–191.

207. Stern DF, Sefton BM. Coronavirus proteins: structure and function of the oligosaccharides of the avian infectious bronchitis virus glycoproteins. *J Virol* 1982;44:804–812.
208. Stern DF, Sefton BM. Synthesis of coronavirus mRNAs: kinetics of inactivation of infectious bronchitis virus RNA synthesis by UV light. *J Virol* 1982;42:755–759.
209. Stohlman SA, Baric RS, Nelson GN, Soe LH, Welter LM, Deans RJ. Specific interaction between coronavirus leader RNA and nucleocapsid protein. *J Virol* 1988;62:4288–4295.
210. Stohlman SA, Fleming JO, Patton CD, Lai MM. Synthesis and subcellular localization of the murine coronavirus nucleocapsid protein. *Virology* 1983;130:527–532.
211. Stohlman SA, Kyuwa S, Polo JM, Brady D, Lai MM, Bergmann CC. Characterization of mouse hepatitis virus-specific cytotoxic T cells derived from the central nervous system of mice infected with the JHM strain. *J Virol* 1993;67:7050–7059.
212. Storz J, Zhang XM, Rott R. Comparison of hemagglutinating, receptor-destroying, and acetylesterase activities of avirulent and virulent bovine coronavirus strains. *Arch Virol* 1992;125:193–204.
213. Sturman L, Holmes KV. The novel glycoproteins of coronaviruses. *Trends Biochem Sci* 1985;10:17–20.
214. Sturman LS, Eastwood C, Frana MF, et al. Temperature-sensitive mutants of MHV-A59. *Adv Exp Med Biol* 1987;218:159–168.
215. Sturman LS, Holmes KV. The molecular biology of coronaviruses. *Adv Virus Res* 1983;28:35–112.
216. Sturman LS, Holmes KV, Behnke J. Isolation of coronavirus envelope glycoproteins and interaction with the viral nucleocapsid. *J Virol* 1980;33:449–462.
217. Sturman LS, Ricard CS, Holmes KV. Proteolytic cleavage of the E2 glycoprotein of murine coronavirus: Activation of cell-fusing activity of virions by trypsin and separation of two different 90K cleavage fragments. *J Virol* 1985;56:904–911.
218. Sturman LS, Ricard CS, Holmes KV. Conformational change of the coronavirus peplomer glycoprotein at pH 8.0 and 37 degrees C correlates with virus aggregation and virus-induced cell fusion. *J Virol* 1990;64:3042–3050.
219. Swift AM, Machamer CE. A Golgi retention signal in a membrane-spanning domain of coronavirus E1 protein. *J Cell Biol* 1991;115:19–30.
220. Taguchi F. Fusion formation by the uncleaved spike protein of murine coronavirus JHMV variant cl-2. *J Virol* 1993;67:1195–1202.
221. Taguchi F, Ikeda T, Makino S, Yoshikura H. A murine coronavirus MHV-S isolate from persistently infected cells has a leader and two consensus sequences between the M and N genes. *Virology* 1994;198: 355–359.
222. Tooze J, Tooze S, Warren G. Replication of coronavirus MHV-A59 in sac- cells: determination of the first site of budding of progeny virions. *Eur J Cell Biol* 1984;33:281–293.
223. Tooze J, Tooze SA. Infection of AtT20 murine pituitary tumour cells by mouse hepatitis virus strain A59: virus budding is restricted to the Golgi region. *Eur J Cell Biol* 1985;37:203–212.
224. Tung FY, Abraham S, Sethna M, et al. The 9-kda hydrophobic protein encoded at the 3' end of the porcine transmissible gastroenteritis coronavirus genome is membrane-associated. *Virology* 1992;186:676–683.
225. Tyrrell DA, Alexander DJ, Almeida JD, et al. Coronaviridae: 2nd report. *Intervirology* 1978;10:321–328.
226. Tyrrell DA, Almeida JD, Berry DM, et al. Coronaviruses. *Nature* 1968;220:650.
227. van der Most RG, Bredenbeek PJ, Spaan WJ. A domain at the 3' end of the polymerase gene is essential for encapsidation of coronavirus defective interfering RNAs. *J Virol* 1991;65:3219–3226.
228. van der Most RG, Heijnen L, Spaan WJ, de Groot RJ. Homologous RNA recombination allows efficient introduction of site-specific mutations into the genome of coronavirus MHV-A59 via synthetic co-replicating RNAs. *Nucleic Acids Res* 1992;20:3375–3381.
229. Vennema H, Heijnen L, Zijderveld A, Horzinek MC, Spaan WJ. Intracellular transport of recombinant coronavirus spike proteins: implications for virus assembly. *J Virol* 1990;64:339–346.
230. Vlasak R, Luytjes W, Leider J, Spaan W, Palese P. The E3 protein of bovine coronavirus is a receptor-destroying enzyme with acetylesterase activity. *J Virol* 1988;62:4686–4690.
231. Vlasak R, Luytjes W, Spaan W, Palese P. Human and bovine coronaviruses recognize sialic acid-containing receptors similar to those of influenza C viruses. *Proc Natl Acad Sci U S A* 1988;85:4526–4529.
232. Wang FI, Fleming JO, Lai MM. Sequence analysis of the spike protein gene of murine coronavirus variants: study of genetic sites affecting neuropathogenicity. *Virology* 1992;186:742–749.
233. Wang L, Junker D, Collisson EW. Evidence of natural recombination within the S1 gene of infectious bronchitis virus. *Virology* 1993;192: 710–716.
234. Wege H, Schliephake A, Korner H, Flory E. An immunodominant $CD4^+D$ T cell site on the nucleocapsid protein of murine coronavirus contributes to protection against encephalomyelitis. *J Gen Virol* 1993;74:1287–1294.
235. Wege H, Siddell S, ter Meulen V. The biology and pathogenesis of coronaviruses. *Curr Top Microbiol Immunol* 1982;99:165–200.
236. Wege H, Winter J, Meyermann R. The peplomer protein E2 of coronavirus JHM as a determinant of neurovirulence: definition of critical epitopes by variant analysis. *J Gen Virol* 1988;69:87–98.
237. Weismiller DG, Sturman LS, Buchmeier MJ, Fleming JO, Holmes KV. Monoclonal antibodies to the peplomer glycoprotein of coronavirus mouse hepatitis virus identify two subunits and detect a conformational change in the subunit released under mild alkaline conditions. *J Virol* 1990;64:3051–3055.
238. Weiss SR, Zoltick PW, Leibowitz JL. The ns 4 gene of mouse hepatitis virus (MHV), strain A 59 contains two ORFs and thus differs from ns 4 of the JHM and S strains. *Arch Virol* 1993;129:301–309.
239. Welsh RM, Haspel MV, Parker DC, Holmes KV. Natural cytotoxicity against mouse hepatitis virus-infected cells. II. A cytotoxic effector cell with a B lymphocyte phenotype. *J Immunol* 1986;136:1454–1460.
240. Wesley RD, Woods RD, Cheung AK. Genetic analysis of porcine respiratory coronavirus, an attenuated variant of transmissible gastroenteritis virus. *J Virol* 1991;65:3369–3373.
241. White JM. Viral and cellular membrane fusion proteins. *Annu Rev Physiol* 1990;52:675–697.
242. Wilhelmsen KC, Leibowitz JL, Bond CW, Robb JA. The replication of murine coronaviruses in enucleated cells. *Virology* 1981;110:225–230.
243. Williams RK, Jiang GS, Holmes KV. Receptor for mouse hepatitis virus is a member of the carcinoembryonic antigen family of glycoproteins. *Proc Natl Acad Sci U S A* 1991;88:5533–5536.
244. Yeager CL, Ashmun RA, Williams RK, et al. Human aminopeptidase N is a receptor for human coronavirus 229E. *Nature* 1992;357:420–422.
245. Yokomori K, Asanaka M, Stohlman SA, Lai MM. A spike protein-dependent cellular factor other than the viral receptor is required for mouse hepatitis virus entry. *Virology* 1993;196:45–56.
246. Yokomori K, Banner LR, Lai MM. Heterogeneity of gene expression of the hemagglutinin-esterase (HE) protein of murine coronaviruses. *Virology* 1991;183:647–657.
247. Yokomori K, Banner LR, Lai MM. Coronavirus mRNA transcription: UV light transcriptional mapping studies suggest an early requirement for a genomic-length template. *J Virol* 1992;66:4671–4678.
248. Yokomori K, La Monica N, Makino S, Shieh CK, Lai MM. Biosynthesis, structure, and biological activities of envelope protein gp65 of murine coronavirus. *Virology* 1989;173:683–691.
249. Yokomori K, Lai MM. Mouse hepatitis virus S RNA sequence reveals that nonstructural proteins ns4 and ns5a are not essential for murine coronavirus replication. *J Virol* 1991;65:5605–5608.
250. Yokomori K, Lai MM. The receptor for mouse hepatitis virus in the resistant mouse strain SJL is functional: implications for the requirement of a second factor for viral infection. *J Virol* 1992;66:6931–6938.
251. Yokomori K, Lai MMC. Mouse hepatitis virus utilizes two carcinoembryonic antigens as alternative receptors. *J Virol* 1992;66: 6194–6199.
252. Yoo D, Graham FL, Prevec L, et al. Synthesis and processing of the haemagglutinin-esterase glycoprotein of bovine coronavirus encoded in the E3 region of adenovirus. *J Gen Virol* 1992;73:2591–2600.
253. Yoo DW, Parker MD, Babiuk LA. The S2 subunit of the spike glycoprotein of bovine coronavirus mediates membrane fusion in insect cells. *Virology* 1991;180:395–399.
254. Zhang X, Liao CL, Lai MM. Coronavirus leader RNA regulates and initiates both in trans and in cis subgenomic mRNA transcription. *J Virol* 1994;in press.
255. Zhao X, Shaw K, Cavanagh D. Presence of subgenomic mRNAs in virions of coronavirus IBV. *Virology* 1993;196:172–178.

Fundamental Virology, Third Edition
edited by B.N. Fields, D.M. Knipe, P.M. Howley, et al.
Lippincott - Raven Publishers, Philadelphia © 1996

CHAPTER 19

Rhabdoviridae: The Viruses and Their Replication

Robert R. Wagner and John K. Rose

CLASSIFICATION

Viruses of the family *Rhabdoviridae* (in the order Mononegavirales) are perhaps more widely distributed in nature than those of any other virus family. Rhabdoviruses infect vertebrates and invertebrates, as well as many species of plants. The rhabdoviruses that cause rabies and economically important diseases of fish appear to have life cycles confined to vertebrate species. All other rhabdoviruses are thought to be transmitted to vertebrates and plants by infected arthropods, which may be the original hosts from which all rhabdoviruses evolved. Shope and Tesh (105) provided an excellent review of the ecology and classification of rhabdoviruses that infect vertebrates. Characteristically, all rhabdoviruses have a wide host range, although many have adapted to grow in specific hosts and at their particular ambient temperatures.

R. R. Wagner: Department of Microbiology, University of Virginia School of Medicine, Charlottesville, Virginia 22908.
J. K. Rose: Departments of Pathology and Cell Biology, Yale University School of Medicine, New Haven, Connecticut 06510.

In addition to numerous plant rhabdoviruses (44), more than 70 rhabdoviruses of vertebrates have been identified and classified thus far, and many more await identification. The viruses of the family *Rhabdoviridae* known to infect mammals including humans, have been classified into two genera: the *Vesiculovirus* genus stemming from vesicular stomatitis virus (VSV) and the *Lyssavirus* genus otherwise known as the rabies and rabieslike viruses. The well-characterized viruses of these two genera are listed in Table 1, which has been adapted from the review by Shope and Tesh (105). Not represented in this table is the recently characterized Borna disease virus of horses, the genome of which closely resembles that of rhabdoviruses. Unlike other vertebrate rhabdoviruses that replicate in the cytoplasm, the Borna disease virus apparently replicates in the nucleus of infected cells as do plant rhabdoviruses (20).

Among the animal rhabdoviruses, many of those belonging to the genus *Vesiculovirus* infect insects, and perhaps other arthropods, but it is uncertain whether they transmit infection to vertebrates; however, identical VSV–New Jersey viruses were recovered from black flies and diseased horses during the 1982 epizootic of vesicular stomatitis in

TABLE 1. *Members of two major rhabdoviridae genera*

Virus	Source of virus in nature
Vesiculovirus genus	
VSV–New Jersey	Mammals, mosquitoes, midges, blackflies, houseflies
VSV-Indiana	Mammals, mosquitoes, sandflies
VSV-Alagoas	Mammals, sandflies
Cocal	Mammals, mosquitoes, mites
Jurona	Mosquitoes
Carajas	Sandflies
Maraba	Sandflies
Piry	Mammals
Calchaqui	Mosquitoes
Yug Bogdanovac	Sandflies
Isfahan	Sandflies, ticks
Chandipura	Mammals, sandflies
Perinet	Mosquitoes, sandflies
Porton-S	Mosquitoes
Lyssavirus genus	
Rabies	Dogs, cats, wild carnivores, bats, cattle, humans
Lagos bat	Bats
Mokola	Shrews, humans, dogs, cats
Duvenhage	Humans, bats
Obodhiang	*Mansonia* mosquitoes
Kotonkan	*Culicoides* mosquitoes

Adapted from Shope and Tesh (105), with permission.

Colorado (100). VSV, which can be divided into two antigenically distinct species (serotypes) called VSV-Indiana and VSV–New Jersey, appear to infect insects and mammals (Table 1). They generally cause nonfatal disease of significant economic importance in cattle and swine. Various substrains of the Hazelhurst strain of VSV–New Jersey are apparently responsible for the widespread infections of cattle and swine in North America and northern South America. Rare vesicular stomatitis infections in humans have been observed and result in influenzalike symptoms. These infections have occurred in the laboratory or after exposure to infected animal carcasses. The nature, pathogenesis and immunology of rabies viruses are described in the following chapter.

VIRION AND GENOME STRUCTURE

Morphology

Figure 1 is a schematic diagram of the prototype rhabdovirus, vesicular stomatitis virus, showing the arrangement of the structural components. The gene order along the RNA genome is diagrammed below with symbols for the proteins. Figure 2 is an electron micrograph showing intact, bullet-shaped VSV particles as well as two partially disrupted particles in which the helical nucleocapsid can be seen. The particle size and shape, 180 nm long and 75 nm wide, is typical of all rhabdoviruses except certain plant virions, which are bacilliform in shape and almost twice the length (44). Table 2 gives the composition of VSV particles and some of their physical characteristics.

Exceedingly common in preparations of all rhabdoviruses, particularly in uncloned preparations propagated by undiluted passage, are truncated or defective-interfering (DI) particles of about the same width but only 20% to 50% the length of standard infectious virions. These DI particles have about the same complement of proteins and lipids but are not infectious because 50% to 80% of the genome is deleted (40).

Rhabdovirus virions are all composed of two major structural components: a nucleocapsid or ribonucleoprotein (RNP) core and an envelope in the form of a lipoprotein bilayer membrane closely surrounding the RNP core. The virion RNP core is tightly wound into 35 coils in the infectious form of VSV. Extending from the outer surface of the envelope is an array of spikelike projections.

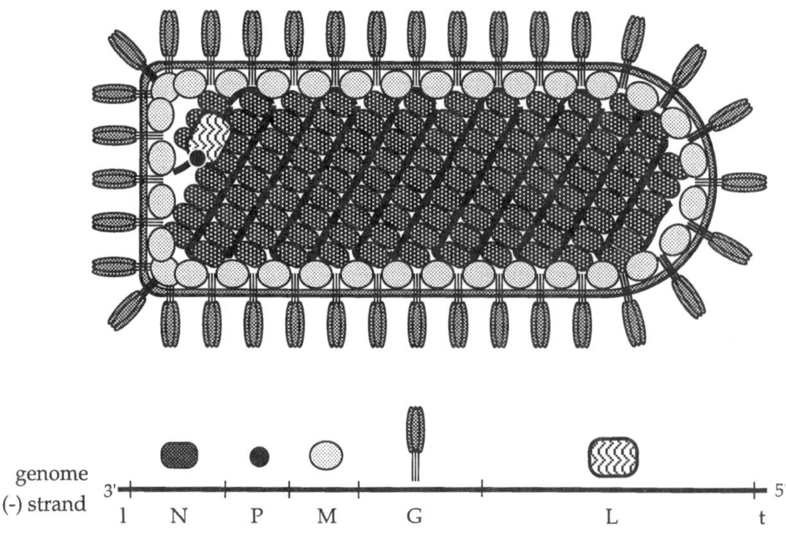

FIG. 1. Schematic representation of the morphology of VSV showing the two major structural components: the nucleocapsid (RNP) core, which contains single-stranded RNA tightly encased in the major N protein and the minor polymerase proteins L and P, and the bilayer membrane associated with two major proteins, the integral glycosylated (G) protein that traverses the bilayer and the peripheral matrix (M) protein, which adheres to the inner membrane surface and binds to the RNP core. Shown below is the gene order of the VSV genome with symbols representing the expressed proteins. Courtesy of R. Owens.

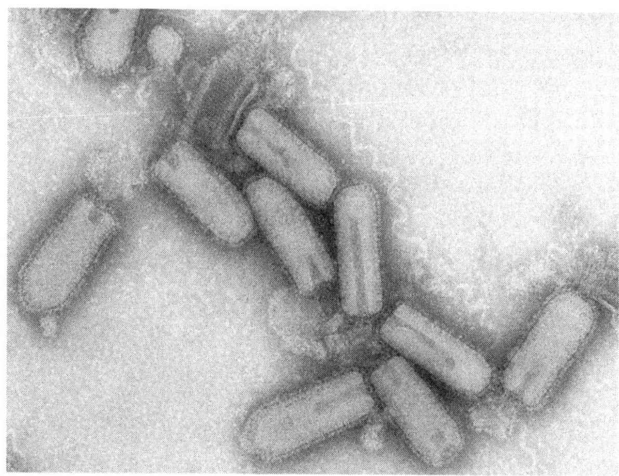

FIG. 2. Electron micrograph of VSV, negatively stained with phosphotungstic acid. Two virions are partially disrupted to show internal nucleocapsid. Surface of virions shows protruding glycoprotein spikes. Micrograph by M. Whitt.

Nucleocapsid

The infectious component of VSV and all rhabdoviruses is the RNP core. VSV serves as an excellent experimental system for studying viral replication and cytopathology because of the relative simplicity of its structure, genetics, and physiology. The VSV-Indiana genome is an unsegmented single strand of RNA containing 11,161 nucleotides (94). The genomic RNA is the negative sense strand and therefore requires its own endogenous RNA polymerase for transcription of plus sense messenger RNA (3). The RNA genome is tightly encased by the nucleocapsid (N) protein consisting of 422 amino acids (31) to form the RNP core. Also associated with the RNP core are two minor proteins, L (for large) and P (for phosphoprotein, originally designated NS); collectively the L and P proteins, in association with the core N protein, serve as the viral transcriptase (29). The L-protein gene represents 60% of the coding potential of the VSV genome (6,380 nucleotides), whereas the P gene (822 nucleotides) codes for a protein of 222 amino acids. The P protein is phosphorylated to varying degrees, the most highly phosphorylated forms of which apparently have the greatest potential for supporting transcription (28,47). Using dark-field scanning transmission electron microscopy, Thomas et al. (111) calculated the length of the VSV-Indiana nucleocapsid to be 3.3 to 3.7 mm, containing 3.7 megadaltons of RNA, and the numbers of protein molecules shown in Table 2. When it contains full-length genomic RNA and a full complement of the three RNP proteins, the VSV nucleocapsid alone is infectious at an efficiency of 10^{-5} to 10^{-6} that of the membrane-enclosed complete virion (114).

TABLE 2. *Composition and physical characteristics of vesicular stomatitis virus*

Proteins	Calculated molecular weight[a]	Molecules per virion
N	47,355	1,258
P (NS)	29,878	466
M	26,064	1,826
G	63,416	1,205
L	240,707	50

Adapted from Rose and Schubert (94) and Thomas et al. (111), with permission.

Length, 180 nm; width, 75 nm; $S_{20,w}$, 660 Svedbergs; density, 1.16 g/ml (in sucrose); RNA, 11,161 nucleotides.

[a]Calculated from the predicted amino acid sequences, including 6,000 daltons for the two N-linked glycans on G protein.

Genome Structure

The negative-strand rhabdovirus genome RNA contains five to six genes in the order 3' N-P-M-G-(X)-L 5' (94). X represents a nonstructural gene in the fish infectious hematopoietic necrosis virus (51), a nonstructural glycoprotein in the bovine ephemeral fever virus (117), and a pseudogene in the rabies virus (112). In the sonchus yellow net virus (SYNV), a plant virus, an extra gene exists between the P and M genes (39). The sigma virus of drosophila exhibits another variation in gene structure, having three genes between N and G as well as a 33-nucleotide overlap of the G gene with the preceding gene (110).

The VSV P gene has recently been shown to encode two additional basic proteins of 55 and 65 amino acids from a second reading frame (108). Translation of these overlapping proteins is apparently initiated at codons 68 and 41 nucleotides downstream of the P protein initiation codon. The proteins are present in the cytoplasm of infected cells but not in virions. Potential to encode these small highly basic proteins is highly conserved among the vesiculoviruses, suggesting an important but unknown function in viral replication.

The VSV genome has an extremely compact arrangement with only two nucleotides between each of the genes encoding the five messenger RNAs (mRNAs) (94). This simplicity of genetic structure is generally seen in other rhabdoviruses and paramyxoviruses as well, although some significant variations exist. In addition, the 5' and 3' noncoding regions in the mRNAs are generally quite short.

Virion Membrane Components

Two reviews provide an in-depth analysis of rhabdovirus membranes (80,82). The VSV membrane is composed of 50% lipid and 50% protein. The lipids are derived entirely from the host cell but are selected in somewhat different proportions than are those in the host cell plasma membrane; the principal lipid differences in the VSV membranes are the larger proportion of cholesterol and, among the phospholipids, much less phosphatidylcholine compared with sphingomyelin and larger amounts of amino phospholipids (82). This altered lipid composition contributes significantly to the greater viscosity of the VSV membrane compared with that of the host cell membrane from which it is derived. Cholesterol depletion results in a slight loss in VSV infectivity, probably related to a decrease in the membrane fusion competence (80,81).

The VSV membrane, and that of all rhabdoviruses, contains two proteins: an externally oriented, membrane-spanning glycoprotein (G) and a peripheral matrix (M) protein that ostensibly lines the inner surface of the virion membrane (82). The G protein is the major antigenic determinant responsible for type specificity; it also gives rise to and reacts with neutralizing antibody (114). The M protein appears to serve as a bridging molecule that attaches the nucleocapsid to the cell plasma membrane where the G protein is inserted; the M protein is quite basic (pI $\approx$ 9.1) and inhibits transcription by binding to the nucleocapsid (15,123).

Both the G and M proteins can be reconstituted into or onto membranes of lipid vesicles (80,82). The G protein has a long stretch of 20 hydrophobic amino acids near its carboxyl terminus and readily partitions into and traverses neutral lipid (phosphyatidycholine) vesicles, whereas the peripheral and positively charged M protein attaches superficially only to vesicles containing negatively charged phospholipids such as phosphotidylserine or phosphatidylglycerol. Both the G and M protein profoundly alter the dynamic properties of interacting lipid bilayers (80,82).

Virion Membrane: The Glycoprotein

All rhabdoviruses contain a single glycoprotein of similar size. The VSV membrane contains approximately 1,200 molecules of this single transmembrane G protein that form 400 trimeric spikes (19,22,49) arranged in a tightly packed coat on the virion membrane (Figs. 1 and 2). The VSV-Indiana G protein is synthesized as a precursor of 511 amino acids (93) from which an *N*-terminal signal sequence of 16 amino acids is cleaved after the protein is inserted into the endoplasmic reticulum (ER). The VSV G protein is a typical type I membrane glycoprotein with the majority of the amino acids exposed on the virion surface. There are two sites in the external domain where N-linked glycans are attached. A hydrophobic domain of 20 amino acids spans the membrane, and a 29–amino acid cytoplasmic domain extends into the virion. A single molecule of palmitate is esterified to a cysteine residue in the cytoplasmic domain of some rhabdovirus G proteins (92,98), but the function of the palmitate is unclear because a mutant G protein lacking palmitate has normal membrane fusion activity and assembles normally into virions (121).

The VSV G protein has served as an important model for many studies on protein folding, assembly, and transport to the cell surface. Immediately after synthesis and before the correct disulfide bonds are formed, G protein is found associated with a major protein in the ER called BiP (GRP78), which assists in protein folding (21,67). Correctly folded monomers are released from BiP within about 3 minutes after synthesis, and these then assemble into trimers in about 7 minutes before leaving the ER. Transport to the Golgi occurs within about 15 minutes, and protein appears on the cell surface after about 30 minutes.

Mutations in all three domains of the VSV G proteins that affect transport from the ER have been analyzed in detail. Mutations in the extracellular domain typically prevent correct folding of G monomers and induce aggregation with BiP protein before trimerization. When VSV G protein monomers have severe folding defects, for example, because of an absence of glycosylation sites (85), they are found in large, disulfide-bonded aggregates that remain permanently bound to BiP protein in the ER (67,68). Mutations in the transmembrane domain also can affect transport by inducing aggregation and preventing transport, whereas mutations in the cytoplasmic domain that reduce the transport rate do not appear to affect the folding of the extracellular domain. These results suggest that the cytoplasmic tail of the glycoprotein contains a signal that accelerates protein exit from the ER, perhaps by signaling concentration of the protein at sites of vesicle budding from the ER (23).

Recent studies have shown that the VSV G protein trimers are not tightly associated, but instead are in a constant and rapid equilibrium with monomeric subunits in the ER (127) and at the cell surface (126), or when examined in solution after solubilization with detergent (66). Once assembled into virions it is likely that a trimeric or higher order G protein structure is stabilized by lateral interactions among trimers on the virion surface and by interactions with the internal virion components.

VSV G protein undergoes a conformational change at pH below about 6.1, which stabilizes the trimer (22). This change presumably exposes a hydrophobic domain that can insert into a target membrane and mediate membrane fusion (30,81). A linear hydrophobic domain typical of fusion domains in other viral fusion proteins is not obvious in the VSV G protein sequence, although the N terminus is somewhat hydrophobic. Initial studies with small hydrophobic peptides corresponding to the N terminus of the mature VSV G protein showed that these could mediate hemolysis, and it was suggested that hemolysis might be equated with membrane fusion activity (96). However, a change in the peptide sequence that abolished hemolysis had no effect on fusion activity when introduced at the N terminus of the VSV G protein (124). These results indicate that the hemolysis assay using peptides is not always relevant to membrane fusion, but they do not rule out a direct role of the N terminus in the membrane fusion process.

More recent studies with mutants of the VSV G protein have implicated the involvement of other domains in the fusion process. A site for N-linked glycosylation introduced at residue 117 completely abolished the membrane fusion activity of G protein without blocking folding, transport to the cell surface, or incorporation into virions (122). The proximity of this mutation to a sequence of 19 uncharged residue (118 to 136) suggested that this domain might be involved in fusion. Analysis of linker insertion mutants that failed to induce cell fusion identified the same site as well as two others elsewhere in G protein (57). These results raise the possibility that several domains may be involved in fusion or that mutations in domains unrelated to fusion may be able to prevent the conformational change required for exposure of the fusion domain. Three-dimensional structures of both the neutral and low pH forms of the VSV G protein may be required for identification of the fusion domain(s).

Virion Membrane: The Matrix Protein

The location of the VSV matrix protein within the virion is controversial. Because studies showed that some M protein is associated with membranes in infected cells (48,71) and that it associates with membranes in vitro (128), it has generally been thought that M protein occupies the position shown in Fig. 1, acting as a bridge between the viral envelope and the nucleocapsid. Such a model is supported by the finding that some M protein expressed alone in cells associates with membranes having the density of plasma membranes (17). Reports that M protein can stabilize G protein trimers *in vitro* (64) and that membranes from cells expressing M protein bind nucleocapsids (16) also support the idea that M bridges the gap between the nucleocapsid and the G protein. Also, the amino terminus of M appears to be membrane associated because it can be cross-linked to a membrane-bound compound (24,56,113).

A recent electron microscopy study has challenged this view by suggesting that the matrix protein may reside inside the coiled nucleocapsid (6). This conclusion relies on immuno-gold labeling for identification of M protein in core material protruding from the blunt end of disrupted virions. Because these protrusions might contain M protein bound to partially uncoiled nucleocapsid, it is difficult to reach a definitive conclusion about M-protein localization from these or previous (77) electron microscopy studies alone. The two views of M localization might be partially reconciled if M bound to the membrane initially acted as a nucleation site for coiling of the nucleocapsid. The majority of M protein could then be inside the coiled nucleocapsid, with a small amount bound to the membrane at the ends of the nucleocapsid.

Expression of the M protein in the absence of other VSV proteins results in cell rounding (11), suggesting a role for M protein in cytopathogenesis during infection. This effect apparently results from disruption of the cytoskeleton, leading to cell rounding. The VSV M protein is phosphorylated on serine and threonine residues between amino acids 20 and 35, but the function of this modification, if any, is not known (45).

Virus Mutations

A critical point that must be considered when evaluating and comparing studies using rhabdoviruses is their extreme genetic variability. As an example, 16 point mutations were detected among complementary DNAs (cDNAs) prepared from viral mRNAs spanning the 6,380 nucleotides of the L gene (101). This high frequency of substitutions is probably largely attributable to the infidelity of the VSV transcriptase but also could be due to mistakes made by the reverse transcriptase. The high frequency with which temperature-sensitive mutants and revertants occur spontaneously (75) is undoubtedly a manifestation of the error frequency of the VSV polymerase. Steinhauer and Holland (109) described a method for direct quantitation and sequence analysis of base substitution levels at predetermined single nucleotides in clones of the VSV genome or its transcripts. In one of the most highly conserved sites, nucleotide substitution frequency averaged 1 to 4×10^{-4} substitutions per base incorporated at this single site. They concluded "if polymerase error frequency averaged as high at all other sites in the 11-kilobase VSV genome, then every member of a cloned VSV population would differ from other genomes in that clone at a number of nucleotide positions. The preservation of a consensus sequence in such variable RNA virus genomes then could only result from strong biological selection." An extension of this hypothesis is that many of these frequent base substitutions by the nonproofreading VSV polymerase could lead to lethal mutations, which may explain why fewer than one in ten morphologically indistinguishable progeny virions in a cloned VSV population is infectious. On the other hand, mutations in large populations of virus particles can also result in greater "fitness," as evidenced by increased pathogenicity (18).

Also, DI particles with large deletions of the genome are exceedingly common in VSV stocks. These DI particles are of two major types. In one type, half to three quarters of the 3' end of the VSV genome is deleted; such 5' DI particles are transcriptionally inert except for synthesis of a small 46-nucleotide leader RNA (27) and can replicate only in the presence of a helper infectious virus. Another type of DI particle that is far less common has a genome in which 50% of the 5' end is deleted; these 3' DI particles contain a template that can transcribe messengers of the four genes that have not been deleted. Both types of DI particles interfere with replication of standard infectious virus, but the 3' DI particle effectively interferes only with homotypic VSV. Convincing evidence has been provided (73) for a copy choice mechanism of replication for generation of DI particle RNA of the four types described by Lazzarini et al. (53). Details concerning the molecular nature and biological activities of DI particles are presented in a review by Holland (40).

Conditional lethal mutants of VSV, usually identified as temperature-sensitive (ts) mutants, can be induced by various mutagenic agents, but they also arise spontaneously at high frequency. As described in detail in a recent review (88), ts mutant growth in various cells is restricted at 40°C, but all functions including yield of viral progeny are relatively normal at the permissive temperature of 30°C. All VSV-Indiana *ts* mutants fall into five complementation groups, each of which has been mapped to one of five specific cistrons on the genome, and each mutation results in a phenotypically defective function in one of the five structural proteins. It seems likely that each mutant is the result of a single base substitution leading to replacement of a single amino acid. It is likely that the changed phenotype of each ts mutant is the result of a conformational change in the affected protein resulting from a single amino acid substitution. The function of each protein is altered at restrictive temperature, ranging from defective transcription in the case of lesions in the L-protein polymerase gene or block in assembly and budding of virions in the case of lesions in the G-protein and M-protein genes. In any case, the hallmark of the conditional ts phenotype is markedly reduced yield of progeny at restrictive temperature, a phenomenon that can be readily reversed by temperature downshift. These ts mutants served as enormously useful tools for probing and mapping the diverse functions of VSV and, more recently, other rhabdoviruses (88).

CYCLE OF INFECTION

When an infectious virion of the family *Rhabdoviridae* encounters a susceptible host cell, this typically results in a series of events that terminate in release of progeny virions and death of the cell. Although these events can proceed simultaneously, it is convenient to consider the process of infection depicted in Fig. 3 as a linear series in which each event depends on occurrence of the preceding event in the following order: adsorption, penetration, uncoating, transcription, translation, replication, assembly, and budding.

Adsorption

Rhabdoviral infection is initiated by attachment of virus to a receptor on the host cell surface. VSV adsorption is not energy-dependent because it occurs at 4°C but is quite inefficient and rather difficult to quantitate. VSV binding was found to be pH-dependent and fails to reach equilib-

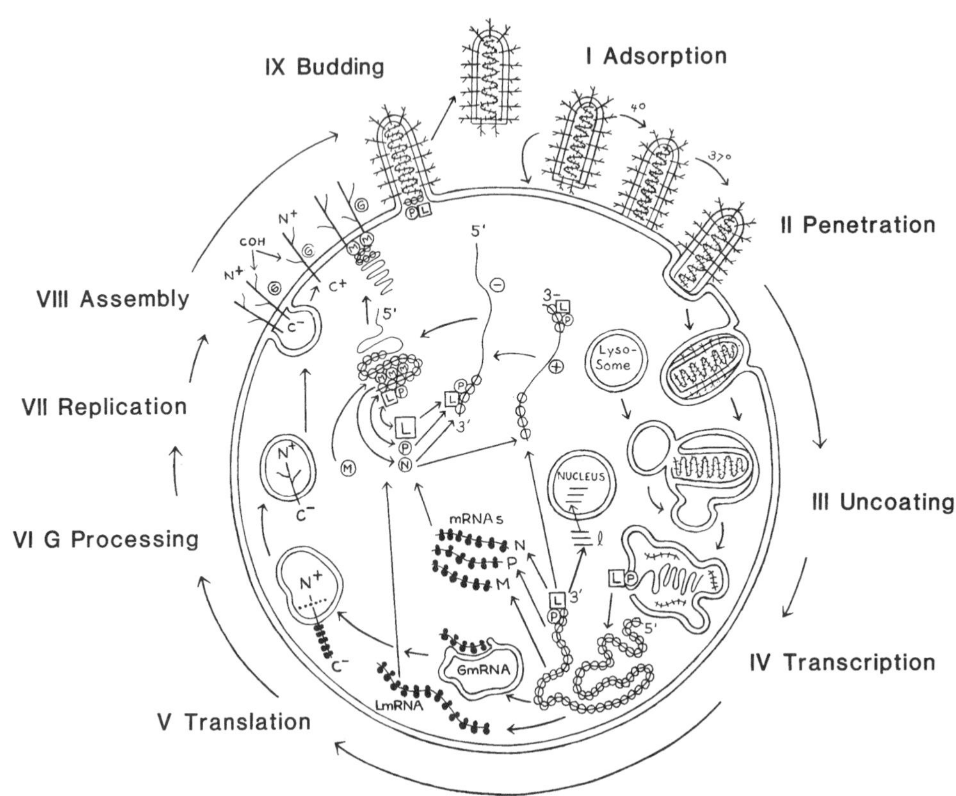

FIG. 3. Cycle of infection schematically depicting the series of events during VSV infection of a susceptible cell from the onset of adsorption to final budding and release of infectious progeny virions. From Wagner (114), with permission.

rium even at optimal pH of 6.5. At least 90% of the virus particles adsorbed on MDCK cells are removable by protease treatment and only a limited amount of the bound virus is internalized by warming the cells from 4° to 37°C (70). VSV binds to the cell surface through its glycoprotein (G) spike, and removal of this protein by protease reduces infectivity by at least 10^5-fold; partitioning intact G protein into spikeless virions can restore infectivity 100-fold (8). The oligosaccharide chains of the VSV G protein apparently do not contribute to adsorption because oligosaccharide-free virions released from cells grown in the presence of tunicamycin are fully infectious. Moreover, addition of specific antibody, sufficient to neutralize the infectivity of VSV, did not impede its binding to Vero cells at 4°C, nor did the antibody reduce internalization of VSV into cells warmed to 37°C (95).

Unlike receptors for myxoviruses and paramyxoviruses, the receptors for attachment of rhabdoviruses have been more difficult to identify. Kinetic analysis of VSV binding to Vero cells led to the postulate that two separate sites (saturable and unsaturable) existed for attachment of VSV (97). Vero cell membranes disrupted with the dialyzable, nonionic detergent octyl-β-D-glucopyranoside yielded an extract that specifically inhibits the saturable, high-affinity binding of ^{35}S-methionine–labeled VSV to Vero cells (95). The binding inhibitor was resistant to protease and neuraminidase, but was inactivated by phospholipase C, suggesting it was a phospholipid. Of all phospholipids tested, only phosphatidylserine totally inhibited the high-affinity binding by VSV to Vero cells and also inhibited VSV plaque formation by 80% to 90% but did not block herpesvirus plaque formation (97). It appears, therefore, that phosphatidylserine may be one of the VSV receptors, at least on Vero cells.

Penetration and Uncoating

Penetration of host cells by VSV closely follows adsorption (Fig. 3). An event that occurs soon after penetration is called uncoating or, more properly in the case of VSV, envelope removal. Whereas adsorption can occur efficiently at 4°C, entry of the virus into the cell (penetration) is an energy-dependent event and requires a physiological temperature. It is now generally accepted that VSV enters cells by endocytosis, and that subsequent reductions in the pH of the endocytic compartment eventually trigger the membrane fusion reaction catalyzed by the G protein. This fusion reaction releases the viral cores into the cytoplasm, allowing infection to proceed (70).

In an effort to define the reactive components in VSV and the target cell membrane that promote fusion and uncoating, Eidelman et al. (26) designed a system in which unilammelar PC vesicles, into which VSV glycoprotein had been inserted by the method of detergent dialysis, were tested for their capacity to fuse with other vesicles containing different phospholipids. Fusion was monitored by electron microscopy and fluorescence energy transfer. It was found that the G protein vesicles fused only with receptor vesicles containing acidic phospholipids (phosphatidylserine or phosphatidic acid); the fusion reaction was pH-dependent with a pK of about 4 and an apparent energy of activation for the fusion reaction of 16 ± 1 kcal/mol (26). Similar pH-dependent fusion was demonstrated when intact VSV virions, biologically labeled with fluorescent pyrene phospholipds, were allowed to interact with phosphatidylserine-containing vesicles (81).

Transcription

After viral entry and release of nucleocapsids into the cytoplasm, the negative-strand rhabdovirus RNA genome cannot encode protein until it has been transcribed into mRNAs by the RNA-directed RNA polymerase carried within the virion (3). This primary transcription process, which occurs in the absence of protein synthesis, has been studied in the most detail for VSV. Transcription begins at the exact 3' end of the genome, producing a 48-nucleotide leader RNA followed by sequential synthesis of the individual mRNAs encoding the N, P, M, G, and L proteins (Fig. 1). Sequential transcription was first shown through the effect of increasing doses of ultraviolet (UV) irradiation on synthesis of the individual transcripts (1,2) and was later demonstrated directly by nucleic acid hybridization of sequences synthesized *in vitro* (43). Transcription order follows the gene order determined later by direct sequencing.

The transcriptase pauses, and transcription is also attenuated 20% to 30% at each of the gene junctions (42), resulting in a gradient of mRNA production following the gene order N>P>M>G>L. The individual mRNAs are capped and polyadenylated, the latter process apparently occurring by repetitive copying of a U7 sequence present at the end of each gene. There are two basic models for the generation of individual transcripts after an obligatory start of transcription at the 3' end of the genome. In the first, polymerase would reach the end of a gene, add poly(A), and then continue to transcribe the next gene (4). RNA cleavage would be required to remove the intergenic dinucleotide, and partial termination would be required to explain attenuation. In the second model, polymerase would also initiate transcription at the 3' end of the genome and terminate transcription after adding poly(A), and then 70% to 80% of the polymerases would reinitiate transcription of the next gene or induce initiation by other polymerase molecules already bound at the promoter (27). We favor the second model because more evidence supports it and because the existence of cleavage or uncleaved precursors has not been demonstrated.

The template for VSV transcription is the genome RNA tightly complexed with the N protein in a ribonuclease-resistant form. Both the L and P proteins are required for polymerase activity (29). This transcription complex will synthesize, cap, methylate, and polyadenylate mRNAs produced *in vitro* (27). Specific enzymatic functions have not been assigned definitively to either L or P, but the polymerase function can be expressed by an L-gene clone (102) and almost certainly resides in L protein because its N-terminal half contains sequence motifs common to all eucaryotic polymerases (86). Also, sequence analyses of a temperature-sensitive mutant and its revertants have indicated that the L protein is involved in transcription, methylation, capping, and polyadenylation (41).

Promising systems for analyzing the roles of L and P proteins in transcription have been described recently (14,106). In these systems, L and P proteins produced by expression from cDNA clones are used to reconstitute transcription from nucleocapsids *in vitro,* and the effects of *in vitro* mutagenesis can be examined. Using the reconstituted system using bacterially produced P protein, which is not phosphorylated, Barik and Banerjee (7) showed an absolute requirement for phosphorylated P in transcription. Phosphorylation of P to a transcription-competent state appears to involve both a cellular kinase and a kinase activity associated with the L protein, possibly as an enzymatic function in the L protein itself (38).

Moyer et al. (76) have developed an *in vitro* system in which short, synthetic RNAs are reconstituted into nucleocapsids and then transcribed by the VSV polymerase complex *in vitro.* Using this system, they have shown that transcription absolutely requires the terminal sequence 3'-UGC and that optimal transcription requires the natural 3'-terminal 15 to 17 nucleotides as a promoter (107).

Another regulator of VSV transcription is the matrix (M) protein (15), which serves the dual function of virion assembly by binding to nucleocapsids and membranes. Temperature-sensitive (ts) mutants in the M gene (complementation group III) give rise to M proteins, which lose their capacity to down-regulate transcription; frequently appearing revertants arising by second mutations at distant sites in the M gene regain transcription-inhibitory activity to a certain extent (75). Monoclonal antibodies (MAbs) have been used for partial mapping of the antigenic determinants (epitopes) of the wild-type (wt) VSV M protein; one of these MAbs (directed to epitope 1) reverses M-protein inhibition of transcription and maps to N-terminal amino acids 19 to 43 (78). The M protein of mutant *ts*O23 completely loses both epitope 1 and its capacity to inhibit transcription based on a mutation leading to substitution of glutamic acid for glycine at amino acid position 23, as confirmed by experiments with expression vectors of wt and mutant genes (59,60) and with synthetic oligopeptides the amino acid sequences of which correspond to positions 19 to 31 of the wt and *ts*O23 mutant (104). However, the ts phenotype of *ts*O23, as predicted by Morita et al. (75), resides in a separate amino acid substitution (leucine to phenylalanine) at position 111 (60). These studies on the transcription-inhibition activity, antigenic properties, and *ts* phenotype of the VSV M protein highlight the significance of the conformation adopted by a protein in determining its biological activities.

Replication

Unlike transcription, VSV replication requires active, ongoing translation, particularly of viral N and P proteins. This was found initially because inhibition of protein synthesis by cycloheximde would shut down RNA replication but not transcription. Therefore, all *in vitro* VSV replication systems are based on coupled translation and replication, as well as assembly of nucleocapsids. The most efficient systems for *in vitro* replication are those originating from nucleocapsids isolated from VSV-infected cells (84), suggesting that a host cell factor(s) is required for efficient replication. However, the most critical component is the RNP template complexed with the L and P proteins, which clearly serve as the RNA polymerase for viral replication as well as for transcription.

The major complication in understanding VSV replication is that two distinct events must occur sequentially: first production of a complete positive RNA strand complementary to the entire parental template, followed by production of complete negative-strand (progeny) RNAs. It is now clear that continuous synthesis of N protein by a coupled translation system is required to make both positive and negative-strand viral RNAs, both of which are completely encased by N protein to make RNP complexes. The original concept that the N protein alone is sufficient for replication of VSV RNPs has been superseded by evidence that ongoing synthesis of both N and P proteins, complexed in a 1:1 molar ratio, is far more efficient for supporting viral replication (84).

During transcription, the VSV polymerase responds to signals telling it to generate the discrete leader and mRNAs. In contrast, during replication, the polymerase must ignore these signals and synthesize genome-length positive-strand RNA. This RNA is then replicated to form the genomic minus-strand RNA. An accepted model to explain the switch from transcription to replication states that newly synthesized N protein binds to nascent leader RNA and prevents recognition of termination signals. This model is supported by the finding that newly synthesized N protein selectively encapsidates the leader RNA (12). This model is attractive because immediately after virus entry, transcription would be favored over replication. At later times, when the concentration of N protein increases, replication would be favored. Two recent reviews cover aspects of transcription and replication in greater detail (5,119).

A promising system for studying replication of VSV has been described recently (83). In this system a cDNA encoding a VSV DI particle RNA is transcribed in cells to generate an RNA that is replicated in the presence of VSV N, P, and L proteins. The replicated RNA is also packaged into infectious DI particles when G and M proteins are expressed along with N, P, and L proteins. A significant conclusion from this study is that the exact 3' sequence is critical for replication.

Both rabies virus (98a) and VSV (52a) have recently been recovered from complete DNA representations of their genomes. These systems will provide critical tools in future studies of all aspects of rhabdovirus transcription, replication, and assembly.

Despite significant progress, there are still major questions remaining to be answered concerning rhabdoviral replication. The mechanism that allows the polymerase to switch from transcription to replication is still at the model stage, and there is no information on the mechanism that allows synthesis of sixfold more full-length minus strands than plus strands. These questions will likely be answered through futher study of *in vitro* and *in vivo* systems combined with analyses using *in vitro* mutagenesis.

Assembly and Budding

Newly synthesized N, P, and L proteins associate in the cytoplasm with newly replicated genomic RNA to form ribonucleoprotein cores. Such cores isolated from virions can be largely stripped of M protein. M protein will associate with these cores *in vitro* (77) to form a tightly coiled structure called a skeleton that resembles the coiled structure seen within VSV virions. Thus, it is likely that M protein also binds to and condenses RNP cores in the cytoplasm of infected cells. As described above, a recent immunoelectron microscopy study suggests that M protein may reside inside the coiled skeleton (6). In addition, a fraction of M protein synthesized in VSV infected cells or in the absence of other viral proteins associates with the plasma membrane (16). Because plasma membranes containing M protein will bind nucleocapsids *in vitro* (17,18) it is likely that M protein plays a dual role in assembly, both condensing RNP cores in the cytoplasm and binding cores to the membrane in preparation for virus budding. Vector-expressed M protein also causes evagination of cytoplasmic membrane and is released in the form of lipoprotein vesicles budding from the cell surface (58).

The VSV G protein also has a critical role in virus assembly. An early report that spikeless VSV particles lacking G protein could be produced at low levels in cells infected with a G protein ts mutant (*ts*O45) suggested that G protein was not absolutely required for the budding process (99). However, a role for G was resurrected when it was found that these particles, which were devoid of intact G protein, contained C-terminal fragments of G including the cytoplasmic domain (74). Studies using a system for phenotypic rescue of the *ts*O45 mutant, showed that the presence of the cytoplasmic domain of G protein was critical for incorporation of G protein into particles and thus for rescue of infectivity (120). In a related study it was found that the cytoplasmic domain of the G protein was sufficient to direct a foreign envelope protein into *ts*O45 VSV particles lacking G (79). CD4 or CD4 bearing the VSV G cytoplasmic domain was found to be incorporated with equal efficiency into wt VSV particles containing a normal complement of G protein (103). The lack of a requirement for the G tail in these experiments may reflect nonspecific copackaging of CD4 during normal budding of particles containing G, a situation that may be different than assembling proteins into *ts*O45 particles lacking G protein. The fact that the cytoplasmic domain of G can be both a necessary and sufficient assembly signal suggests that it interacts with an internal virion component, most likely the M or N proteins.

Although far from complete, the available data support a model of assembly in which condensed nucleocapsids bind to regions of the plasma membrane enriched in both M and G proteins. There may be interactions between M and G that promote coclustering at the plasma membrane (64). Once nucleocapsids bind to these regions, interactions between the cytoplasmic tail of G or between the membrane bound M and nucleocapsids presumably drive the envelopment of the particles and subsequent budding. Coexpression of M, G, and N proteins in the absence of VSV replication, results in evagination and budding of membrane vesicles containing M but not G or N proteins (58). These results suggest a crucial role for M protein in membrane interaction and directing the budding process. In the presence of correctly assembled nucleocapsids and G protein, M protein could direct a highly efficient assembly of particles containing G and nucleocapsids. A complete understanding of the roles of individual proteins and of host factors involved in budding will likely require the development of *in vitro* systems for viral assembly. In addition, development of a system in which infectious virus can be recovered from a DNA clone will enhance our ability to use site-directed mutagenesis to analyze the protein domains involved in assembly.

PATHOGENESIS

Despite close similarity in structure and function, viruses of the genus *Lyssavirus* generally cause relatively slow, progressive disease compared with those of the genus *Vesiculovirus*, which cause acute, often self-limiting infections. A subsequent chapter addresses the pathogenicity of rabies and rabieslike viruses. VSV is enzootic in cattle, particularly in the Caribbean, and occasionally causes widespread epizootics in the United States. Disease also occurs in horses and pigs, but rarely in humans. The molecular basis for the pathogenicity of VSV has been studied extensively and reviewed recently (115). Although cells vary somewhat in their response to VSV infection, all vertebrate cells that have been tested, and to some extent invertebrate cells, are susceptible. It is wise to remember that the times at which cytopathic effects (CPEs) occur in cell cultures depend on the multiplicity of infection. At low doses of infection, VSV CPE is not evident for 24 hr or longer. However, at high multiplicities, cell rounding occurs in several hours, progressing rapidly to membrane permeability to supravital dyes and terminating in cell detachment from adherent surfaces. It has been shown by use of a ts mutant that rounding of infected cells is caused by M protein, probably by disorganization of the cytoskeleton (11). An even earlier event, within the first hour or so, is inhibition of cellular macromolecular synthesis, resulting in the progressive shutdown of cellular RNA, DNA, and protein synthesis (118). These early effects on cellular macromolecular synthesis were once thought to be due to toxicity of input virion proteins at high multiplicity of UV-irradiated or inert DI particles of VSV. Later studies showed that the DI particles are often contaminated with infectious VSV and that certain viral functions can withstand relatively high doses of UV light (69). It is now generally accepted that newly synthesized viral products, and not the preexisting components of the infecting virion, are responsible for CPE and inhibition of cellular macromolecular synthesis caused by VSV (116).

Inhibition of Cellular RNA Synthesis

It seemed logical that one or more of these viral products (leader RNA, mRNAs, or proteins) is responsible for inhibiting RNA synthesis in cells infected with pathogenic VSV. The two principal methods used to study VSV inhibition of cellular RNA synthesis have been to test ts or DI mutants restricted in transcription and to inactivate by varying doses of UV irradiation the specific regions of the wt viral genome required for transcription of each viral gene. A complementation group I mutant, *ts*G114, restricted in transcriptional activities, failed to shut-off host cell RNA metabolism in MPC-11 cells incubated at the restrictive temperature of 39°C but did so at the permissive temperature of 31°C (118). Similar results in L cells infected with ts mutants from all five complementation groups were obtained by Wu and Lucas-Lenard (125), who also reported an effect on cellular RNA synthesis by the putative P protein mutant *ts*G22(II) but not those of other complementation groups. MPC-11 cells infected with transcription-defective purified DI particles derived from the 5' end of the VSV genome also exhibited no significant reduction in host-cell RNA synthesis even at a multiplicity of infection (MOI) equivalent to 10,000 particles per MPC-11 cell (116). Thus, even if a very short leader sequence is transcribed by the 5' DI particles, it apparently plays no role in the inhibition of cellular RNA synthesis. These and other experiments suggested that primary transcription of the VS viral genome is essential for early inhibition of host cell RNA synthesis.

The genetic evidence that transcription of VSV is required to shut off cellular RNA synthesis led two laboratories to reinvestigate the effects of UV-irradiated VS virus on cellular RNA synthesis (116,125). Heavily UV-irradiated virus retained much of its capacity to shut off cellular RNA synthesis. Using a value of 104 ergs/mm^2 for the 37% (1/e) UV survival dose for VSV infectivity and a VSV genome size of 12,000 nucleotides, the UV target size for survival of the capacity of VSV to inhibit cellular RNA synthesis was calculated as approximately 85 nucleotides (34), compared with the actual size of 48 nucleotides for the VSV leader RNA. These data suggested that the VSV plus-strand leader RNA had the capacity to inhibit cellular RNA synthesis.

The cellular targets for VSV inhibition of host cell RNA synthesis have been examined in some detail and, in general, such factors as nucleoside incorporation, nucleotide pools, cellular RNA transport and processing, and altered cellular RNA degradation rates have been ruled out. By comparing nuclei and isolated chromatin from uninfected and VSV-infected cells, evidence was obtained for inhibition of initiation of transcription for ribosomal, messenger, and transfer RNA in VSV-infected cells (116). These experiments suggested that RNA chain elongation was not affected by VSV infection but the number of polymerases able to initiate transcription is reduced in the VSV-infected cells (116). These studies were subsequently greatly extended by using a cell-free system in which HeLa-cell extracts supplied polymerases and cofactors for transcription of the adenovirus major late promoter (MLP) and the virus-associated (VA) genes excised from recombinant DNA clones (35,72). This series of experiments showed that the wt VSV leader RNA, but not the DI leader, inhibited the activities of both polII and polIII in initiation of synthesis of MLP mRNA and VA 5S transfer RNA, respectively. The major difference between the wt leader and the DI leader is a sequence 5'-AUUAUUAUCAUUA-3' from nucleotides 18 to 30 that are present in the wt leader but not in the DI leader. Synthetic oligodeoxynucleotides of the same sequence as the wt leader RNA also inhibited transcription of the MLP and VA genes, but base substitutions (e.g. T to G) resulted in loss of *in vitro* transcription-inhibition activity (35,36). It seems clear from these studies that the AU-rich region of the VSV wt leader (and DNA homologues) of two distinct VSV serotypes is responsible, at least in large measure, for inhibiting DNA-dependent transcription. The site of action of these leader RNAs is not known, but several studies appear to rule out direct interaction of wt leader with polymerases or with the DNA template. However, preliminary evidence has been presented for the presence of a HeLa cell cofactor, a 65-kd protein, that binds to wt leader or homologous oligodeoxynucleotides but not to DI leader RNA; this 65-kd protein also possesses the capacity to reverse transcription inhibition by the wt leader RNA (36).

The *wt* leader was found to be made in the cytoplasm of VSV-infected cells and then migrates to the nucleus (34,52). A temporal correlation between *in vivo* synthesis of wt VSV leader RNA and the degree of cellular RNA synthesis inhibition was also described (34) but Dunegan et al. (25) could not detect a correlation between the accumulation of plus-strand leader RNA and inhibition of protein and RNA synthesis in VSV-infected cells.

Newly synthesized M protein also has been proposed as an inhibitor of cellular transcription, particularly because it could be detected in the nucleus of VSV-infected cells but only under special detergent conditions (65). Later studies showed that vector-expressed VSV wt M protein alone was capable of inhibiting cell-directed expression of a cotransfected target gene encoding chloramphenicol acetyltranferase (CAT) (9). CAT transcription-inhibition was found to be a function of M-gene expression, but this effect on target transcription required up to 12 hr posttransfection compared with inhibition of cellular transcription within 1 hr after VSV infection (34,118). Subsequently, the same investigators showed that expression of the mutant M gene derived from VSV *ts*O82 failed to inhibit CAT-gene expression, whereas expression of the M gene in which residues 4 to 21 are deleted retained its capacity to inhibit expression of the target CAT gene. These data show that the assembly and transcription-inhibition functions are located on separate regions of the M protein (10). These data indicate a separate cellular transcription-inhibition activity for the M protein of VSV presumably functioning later in infection than the transcription-inhibition activity of the VSV wt leader RNA.

Inhibition of Cellular DNA and Protein Synthesis

The same VSV factor(s) that inhibits cellular RNA synthesis also appears to shut off cellular DNA synthesis as judged by similar dose response, kinetics of inhibition, lack of effect of 3' DI particles, and inactivation by equivalent UV doses (116). Implication of the VSV wt leader RNA as the inhibitor of cellular DNA synthesis is supported by its capability to block adenovirus DNA replication in a cell-free system (90,91). Those deoxynucleotide sequences homologous to wt leader that inhibit *in vitro* adenovirus transcription also inhibit adenovirus DNA replication, probably by binding to the preterminal protein (pTP)-dCMP complex that initiates DNA replication; there is striking homology of the wt VSV leader RNA to the 5'-terminal adenovirus sequences and nuclear factor I, which bind to the adenovirus replication-initiation site (90).

VSV infection also results in inhibition of cellular translation by mechanisms different from those that inhibit cellular transcription and replication as judged by dissimilar kinetics and UV-inactivation studies, which implicate at least the N gene of VSV (25). Moreover, neither wt leader RNA nor any of the mRNAs of VSV have any effect on *in vitro* translation, although double-stranded RNA in association with VSV nucleocapsids reduces translation of endogenous and globin mRNAs in an *in vitro* reticulocyte system (116). Translation-inhibition by VSV appears to occur at initiation, but it is not clear whether the target is eIF-3/eIF-4B or eIF-2 (116).

IMMUNITY

Vertebrates acquire immunity to rhabdoviruses in much the same manner as they do in other acute viral infections. Primary infection results in a humoral immune response developing within a week after first exposure. Cellular immune responses also occur after infection, but these are probably of greater importance in subacute infections, such as those with rabies virus. Antibodies generated during the humoral immune response are directed to two major antigens, the group-specific N protein and the type-specific G protein. The type-specific VSV G protein is the viral antigen that gives rise to and reacts with neutralizing antibody (114).

Monoclonal Antibodies

The use of hybridoma cells that secrete MAb has been extensively exploited by Flamand et al. to study the major antigenic determinant (G protein) of rabies viruses, including specific site mutations that alter antigenic determinants and virulence (87). More details of such studies will be discussed in the following chapter. MAbs raised against the G proteins of VSV-Indiana and VSV–New Jersey reacted with nine to 11 distinct epitopes of each serotype (13,54). Several of these MAbs cross-react with the G protein of the opposite serotype, reflecting 50% amino acid homology of the two G proteins (94). Keil and Wagner (46) have provided fairly definitive maps of the antigenic determinants of the two G proteins by testing the reactivity with epitope-specific MAbs of deletion mutants and chimeras of the two proteins expressed by cDNA recombinants. The epitopes of the VSV-Indiana G protein were found to be widely distributed, but almost all the epitopes of VSV–New Jersey G protein mapped to the central third (residues 193 to 289), including all four neutralization epitopes. Reduction of disulfide bonds in the VSV–New Jersey G protein resulted in complete loss of its capacity to bind all neutralizing MAbs and most non-neutralizing MAbs, emphasizing the critical role of secondary and probably tertiary structure in availability of antigenic determinants for antibody binding (13).

Glycosidase removal of carbohydrate chains does not affect antigenic specificity of the VSV–New Jersey G protein, but prevention of glycosylation of nascent G proteins synthesized in the presence of tunicamycin resulted in significant loss in MAb reactivity of conformational and neutralization epitopes clustered in the center of the VSV–New Jersey G protein (33). These findings are consistent with the evidence that N-linked glycosylation influences G protein folding and the formation of the correct disulfide bonds and hence the recognition of conformational epitopes by antibodies (68). Further evidence for the importance of disulfide bonds for the structural integrity of discontinuous neutralization epitopes was provided by site-directed mutations in the VSV–New Jersey G gene, in which 10 of the 12 cysteines were individually converted to serines. In addition to certain NH_2-terminal cysteines, mutations in cysteines at residues 235, 240, or 273 resulted in the disappearance of neutralization epitopes VII and VIII, which map to this region of the VSV–New Jersey G protein (32). These studies, coupled with cleavage at residue 110 by *Staphylococcus* V8 protease, show the presence of a loop structure in the G protein proximal to residue 193. The exact location of the antigenic determinants and functional domains of the VSV G proteins awaits determination of the three-dimensional structure by crystallographic x-ray diffraction.

The marked genetic variability of the G proteins of VSV has been noted in two series of experiments using epitope-specific neutralizing MAbs to select variants of VSV with altered antigenic specificity (62,113). In one series, mutants of VSV-Indiana were selected for their capacity to grow in the presence of monoclonal and/or polyclonal antibodies directed to the homotypic G protein; sequencing the RNAs of the variant G genes by primer extension showed a series of mutations, specific for each MAb, distributed rather widely along the coding sequence of the VSV-Indiana G gene (113). In the case of VSV–New Jersey, short exposure to neutralizing MAbs for four separate epitopes gave rise in each case to mutant virus clones, the G proteins of which had lost the antigenic determinant specific for the MAb that selected that neutralization-escape mutant (62). The G gene of each epitope-specific VSV–New Jersey mutant was sequenced by cDNA primer extension, showing a single base change leading to a single amino-acid substitution, specific for each MAb that gave rise to that mutant. These neutralizing MAb-selected mutations were clustered in the middle third of the 517–amino acid VSV–New Jersey G protein in the region mapped by Keil and Wagner (46). Mutations generated by an error-prone polymerase presumably generate conformational changes in G proteins that then allow escape from neutralization. It is surprising to learn therefore that enzootic vesicular stomatitis in Costa Rica is characterized by yearly reinfection of the same cattle with antigenically indistinguishable virus, even in the presence of high levels of neutralizing antibody (L. Rodriguez, personal communication). Moreover, only limited antigenic variations, resulting from mutations in epitope-specific map locations of the VSV–New Jersey G gene, were found in field isolates from diseased cattle during the widespread epizootic of vesicular stomatitis in 1982 to 1985 in the western United States (63).

Cellular Immunity

Cellular immune responses to viruses have been shown to be mediated by polymorphic major histocompatability antigens (129). Viral infection generates cytotoxic T-lymphocytes (CTLs) specific for both a viral antigen and a self component coded by the K and/or D loci of the H-2 histocompatability complex. VSV has provided a nice model system for such studies because the well-characterized G and N proteins appear to be the major antigens that mediate the CTL response (37,89). Of considerable interest is the finding that cells infected with either VSV-Indiana or VSV–New Jersey are both lysed by effector T cells generated by the heterotypic virus. This cross-reactivity of CTLs is quite different from that of the non–cross-reactivity of neutralizing antibody to the two serotypes, suggesting that the epitopes common to the VSV-Indiana and VSV–New Jersey G and/or N proteins are the targets for CTLs. Attempts to block CTL lysis of VSV-infected cells with anti-G sera have yielded conflicting results. VSV polyclonal antisera raised in mice did not block CTL activity, but hyperimmune VSV rabbit antisera did inhibit the cytotoxic effect by CTLs. LeFrancois and Lyles (55) were able to block the lysis of both VSV-Indiana and VSV–New Jersey cells by homotypic and heterotypic CTLs by cross-reactive MAbs to the two heterotypic G proteins. They obtained better blocking with a combination of MAbs, but the results clearly showed that MAb to VSV-Indiana G protein inhibits lysis by CTL of cells infected with either VSV serotype (55). It seems likely, therefore, that the cross-reactive non-neutralizing MAbs are the primary blockers of CTLs, suggesting that these shared G-protein epitopes are the primary targets of CTLs.

One of the likely effects of antibody to the G protein in VSV-infected cells is the redistribution of the G protein and H-2 antigens, particularly H-$2K_b$, which could markedly effect the action of CTLs. The nature of the H-2 haplotype in association with the VSV G protein, as shown with reconstituted vesicles, also determines the specificity of the CTL that are generated (61). Clearly, many diverse factors influence the action of CTLs in cellular immunity to VSV infection and to the relationship with histocompatability antigens. However, fewer protective CTL epitopes than T-helper epitopes are present on VSV-infected cells; protective CTL responsiveness against three VSV proteins were found to be H-2 linked and inducible in only half of 15 combinations tested in five H-2 haplotypes, whereas T-cell responses were found to be inducible for all three proteins in all five haplotypes (50).

REFERENCES

1. Abraham G, Banerjee AK. Sequential transcription of the genes of vesicular stomatitis virus. *Proc Natl Acad Sci USA* 1976;73:1504–1508.
2. Ball LA, White CN. Order of transcription of genes of vesicular stomatitis virus. *Proc Natl Acad Sci USA* 1976;73:442–446.
3. Baltimore D, Huang AS, Stampfer M. Ribonucleic acid synthesis of vesicular stomatitis virus. II. An RNA polymerase in the virion. *Proc Natl Acad Sci USA* 1970;66:572–576.
4. Banerjee, AK. Transcription and replication of rhabdoviruses. *Microbiol Rev* 1987;51:66–87.
5. Banerjee AK, Barik S. Gene expression of vesicular stomatitis virus genome RNA. *Virology* 1992;188:417–428.

6. Barge A, Gaudin Y, Coulon P, Ruigrok RWH. Vesicular stomatitis virus M protein may be inside the ribonucleocapsid coil. *J Virol* 1993;67:7246–7253.
7. Barik S, Banerjee AK. Cloning and expression of the vesicular stomatitis virus phosphoprotein gene in Escherichia coli: analysis of phosphorylation status versus transcriptional activity. *J Virol* 1991;65: 1719–1726.
8. Bishop DHL, Repik P, Obijeski JF, Moore NF, Wagner RR. Restitution of infectivity to spikeless vesicular stomatitis virus by solubilized virus components. *J Virol* 1975;16:75–84.
9. Black, BL, Lyles DS. Vesicular stomatitis virus matrix protein inhibits host cell-directed transcription of target genes *in vivo*. *J Virol* 1992;66:4058–4064.
10. Black BL, Rhodes RB, McKenzie M, Lyles DS. The role of vesicular stomatitis virus matrix protein in inhibition of host-directed gene expression is genetically separable from its function in virus assembly. *J Virol* 1993;67:4814–4821.
11. Blondel D, Harmison GG, Schubert M. Role of matrix protein in cytopathogenesis of vesicular stomatitis virus. *J Virol* 1990;64:1716–1725.
12. Blumberg BM, Giorgi C, Kolakofsky D. N protein of vesicular stomatitis virus selectively encapsidates leader RNA *in vitro*. *Cell* 1983;32:559–567.
13. Bricker BJ, Snyder RM, Fox JW, Volk WA, Wagner RR. Monoclonal antibodies to the glycoprotein of vesicular stomatitis virus (New Jersey serotype): a method for preliminary mapping of epitopes. *Virology* 1987;151:533–540.
14. Canter DM, Jackson RL, Perrault J. Faithful and efficient *in vitro* reconstitution of vesicular stomatitis virus transcription using plasmid-encoded L and P proteins. *Virology* 1993;194:518–529.
15. Carroll AR, Wagner RR. Role of the membrane (M) protein in endogenous inhibition of *in vitro* transcription of vesicular stomatitis virus. *J Virol* 1979;29:134–142.
16. Chong LD, Rose JK. Membrane association of functional vesicular stomatitis virus matrix protein *in vivo*. *J Virol* 1993;67:407–414.
17. Chong LD, Rose JK. Interaction of normal and mutant vesicular stomatitis virus matrix proteins with the plasma membrane and nucleocapsids. *J. Virol.* 1994;68:441–447.
18. Clarke DR, Duarte EA, Moyer A, Elena SF, Domingo E, Holland JJ. Genetic bottlenecks and population passages cause profound fitness differences in RNA viruses. *J Virol* 1993;67:222–228.
19. Crise B, Ruusala A, Zagouras P, Shaw A, Rose JK. Oligomerization of glycolipid-anchored and soluble forms of the vesicular stomatitis virus glycoprotein. *J Virol* 1989;63:5328–5333.
20. Cubitt B, Oldstone C, de la Torre, JC. Sequence and genome organization of Borna disease virus. *J Virol* 1994;68:1382–1396.
21. de Silva A, Braakman I, Helenius A. Posttranslational folding of vesicular stomatitis virus G protein in the ER: involvement of noncovalent and covalent complexes. *J Cell Biol* 1993;120:647–655.
22. Doms RW, Keller DS, Helenius A, Balch WE. Role for adenosine triphosphate in regulating the assembly and transport of vesicular stomatitis viryus G protein trimers. *J Cell Biol* 1987;105:1957–1969.
23. Doms RW, Ruusala A, Machamer C, Helenius J, Helenius A, Rose JK. Differential effects of mutations in three domains on folding, quarternary structure, and intracellular transport of vesicular stomatitis virus G protein. *J Cell Biol* 1988;107:89–99.
24. Dubovi EJ, Wagner RR. Spatial relationships of the proteins of vesicular stomatitis virus: induction of reversible oligomers by cleavable protein cross-linkers and oxidation. *J Virol* 1977;22:500–509.
25. Dunigan DD, Baird S, Lucas-Lenard J. Lack of correlation between the accumulation of plus-strand leader RNA and its inhibition of protein and RNA synthesis in vesicular stomatitis virus infected cells. *Virology* 1986;150:231–246.
26. Eidelman O, Schlegel R, Tralka T, Blumenthal R. pH-dependent fusion induced by vesicular stomatitis virus glycoprotein reconstituted into phospholipid vesicles. *J Biol Chem* 1984;259:4622–4628.
27. Emerson SU. Transcription of vesicular stomatitis virus. In: Wagner RR, ed. *The rhabdoviruses*. New York: Plenum; 1987:245–269.
28. Emerson SU, Schubert M. Location of the binding domains for the RNA polymerase L and the ribonucleocapsid template within different halves of the NS phosphoprotein of vesicular stomatitis virus. *Proc Natl Acad Sci USA* 1987;84:5655–5659.
29. Emerson SU, Yu Y-H. Both NS and L proteins are required for *in vitro* RNA synthesis by vesicular stomatitis virus. *J Virol* 1975;15:1348–1356.
30. Florkiewicz RZ, Rose JK. A cell line expressing the vesicular stomatitis virus glycoprotein fuses at low pH. *Science* 1984;225:721–723.
31. Gallione C, Greene J, Iverson L, Rose JK. Nucleotide sequences of the mRNA's encoding the vesicular stomatitis virus N and NS proteins. *J Virol* 1981;39:529–535.
32. Grigera PR, Keil W, Wagner RR. Disulfide-bonded discontinuous epitopes on the glycoprotein of vesicular stomatitis virus (New Jersey serotype). *J Virol* 1992;66:3749–3757.
33. Grigera PR, Mathieu ME, Wagner RR. Effect of glycosylation on the conformational epitopes of the glycoprotein of vesicular stomatitis virus (New Jersey serotype). *Virology* 1991;180:1–9
34. Grinnell BW, Wagner RR. Comparative inhibition of cellular transcription by vesicular stomatitis virus serotypes New Jersey and Indiana: role of each viral leader RNA. *J Virol* 1983;48:88–101.
35. Grinnell BW, Wagner RR. Nucleotide sequence and secondary structure of VSV leader RNA and homologous DNA involved in inhibition of DNA-dependent transcription. *Cell* 1984;36:533–543.
36. Grinnell BW, Wagner RR. Inhibition of DNA-dependent transcription by the leader RNA of vesicular stomatitis virus: role of specific nucleotide sequences and cell protein binding. *Mol Cell Biol* 1985;5: 2502–2513.
37. Hale VH, Witte ON, Baltimore D, Eisen HN. Vesicular stomatitis virus glycoprotein is necessary for H-2 restricted lysis of infected cells by cytoxic T lymphocytes. *Proc Natl Acad Sci USA* 1978;75:970–974.
38. Hammond DC, Haley BE, Lesnaw JA. Identification and characterization of serine/threonine protein kinase activity intrinsic to the L protein of vesicular stomatitis virus New Jersey. *J Gen Virol* 1992;73: 67–75.
39. Heaton LA, Hillman BI, Hunter BG, Zuidema D, Jackson AO. Physical map of the genome of sonchus yellow net virus, a plant virus with six genes and conserved gene junction sequences. *Proc Natl Acad Sci USA* 1989;86:8665–8668.20.
40. Holland JJ. Defective-interfering rhabdoviruses. In: Wagner RR, ed. *The rhabdoviruses*. New York: Plenum; 1987:297–360.
41. Hunt DM, Hutchinson KL. Amino acid changes in the L polymerase protein of vesicular stomatitis virus which confer aberrant polyadenylation and temperature sensitive phenotypes. *Virology* 1993;193: 786–793.
42. Iverson LE, Rose JK. Localized attenuation and discontinuous synthesis during vesicular stomatitis virus transcription. *Cell* 1981;23: 477–484.
43. Iverson LE, Rose JK. Sequential synthesis of 5'-proximal vesicular stomatitis virus mRNA sequences. *J Virol* 1982;44:356–365.
44. Jackson AO, Francki RIB, Zuidema D. Biology, structure, and replication of plant rhabdoviruses. In: Wagner RR, ed.*The rhabdoviruses*. New York: Plenum; 1987:427–508.
45. Kaptur PE, McCreedy GJ Jr, Lyles DS. Sites of *in vivo* phosphorylation of vesicular stomatitis virus matrix protein. *J Virol* 1992;66: 5384–5392.
46. Keil W, Wagner RR. Epitope mapping by deletion mutants and chimeras of two vesicular stomatitis virus glycoprotein genes expressed by a vaccinia virus vector. *Virology* 1989;170:392–407.
47. Kingsford L, Emerson SU. Transcriptional activity of different phosphorylated species of NS protein purified from vesicular stomatitis virions and cytoplasm of infected cells. *J Virol* 1980;33:1097–1105.
48. Knipe DM, Baltimore D, Lodish HF. Separate pathways of maturation of the major structural proteins of vesicular stomatitis virus. *J Virol* 1977;21:1128–1139.
49. Kreis TE, Lodish HF. Oligomerization is essential for transport of vesicular stomatitis virus glycoprotein to the cell surface. *Cell* 1986;46: 929–937.
50. Kundig TM, Cantelmus I, Bachmann M, Abraham D, Binder D, Hengartner R, Zinkernagel RM. Fewer protective cytoxic T-cell epitopes than T-helper cell epitopes on vesicular stomatitis virus. *J Virol* 1993;67:3680–3683.
51. Kurath G, Ahern KG, Pearson GD, Leong JC. Molecular cloning of the six mRNA species of infectious hematopoietic necrosis virus, a fish rhabdovirus, and gene order determined by R-loop mapping. *J Virol* 1985;53:469–476.
52. Kurilla MC, Piwnica-Worms H, Keene JD. Rapid and transient localization of the leader RNA of vesicular stomatitis virus in the nucleus of infected cells. *Proc Natl Acad Sci USA* 1982;79:5240–5244.

52a. Lawson N, Stillman EA, Whitt MA, Rose JK. Recombinant vesicular stomatitis viruses from DNA. *Proc Natl Acad Sci USA* 1995;92: 4477–4481.

53. Lazzarini RA, Keene JD, Schubert M. The origin of defective-interfering particles of the negative-strand RNA viruses. *Cell* 1982;26: 145–154.

54. LeFrancois L, Lyles DS. The interaction of antibody with the major surface glycoprotein of vesicular stomatitis virus. I. Analysis of neutralizing epitopes with monoclonal antibodies. *Virology* 1982;121: 157–167.
55. LeFrancois L, Lyles DS. Cytotoxic T lymphocytes reactive with vesicular stomatitis virus: analysis of specificity with monoclonal antibodies directed to the viral glycoprotein. *J Immunol* 1983;130:1408–1412.
56. Lenard J, Vanderoef R. Localization of the membrane-associated region of vesicular stomatitis virus M protein at the N terminus, using the hydrophobic, photoreactive probe ^{125}I-TID. *J Virol* 1990;64: 3486–3491.
57. Li Y, Drone C, Sat E, Ghosh HP. Mutational analysis of the vesicular stomatitis virus glycoprotein G for membrane fusion domains. *J Virol* 1993;67:4070–4077.
58. Li, Y, Luo L, Schubert M, Wagner RR, Kang CY. Viral liposomes released from insect cells infected with recombinant baculovirus expressing the matrix protein of vesicular stomatitis virus. *J Virol* 1993;67:4415–4420.
59. Li Y, Luo L, Snyder RM, Wagner RR. Site-specific mutations in vectors that express antigenic and temperature-sensitive phenotypes of the M gene of vesicular stomatitis virus. *J Virol* 1988;62:
60. Li Y, Luo, LZ, Wagner, RR. Transcription inhibition site on the M protein of vesicular stomatitis virus located by marker rescue of mutant *ts*O23 (III) with M-gene expression vectors. *J Virol* 1987;63:2841–2843.
61. Loh D, Ross AH, Hale AH, Baltimore D, Eisen HN. Synthetic phospholipid vesicles containing purified viral antigen and cell membrane proteins stimulate the development of cytotoxic T lymphocytes. *J Exp Med* 1979;150:1067–1074.
62. Luo L, Li Y, Snyder RM, Wagner RR. Point mutations in glycoprotein gene of vesicular stomatitis virus (New Jersey serotype) selected by resistance to neutralization by epitope-specific monoclonal antibodies. *Virology* 1988;163:341–348.
63. Luo L, Li Y, Snyder RM, Wagner RR. Spontaneous mutations leading to antigenic variations in the glycoproteins of vesicular stomatitis field isolates. *Virology* 1990;174:70–78.
64. Lyles DS, McKenzie M, Parce JW. Subunit interactions of vesicular stomatitis virus envelope glycoprotein stabilized by binding to viral matrix protein. *J Virol* 1992;66:349–358.
65. Lyles DS, Puddington L, McCreedy BS. Vesicular stomatitis virus M protein in the nuclei of infected cells. *J Virol* 1988;62:4387–4392.
66. Lyles DS, Varela VA, Parce JW. Dynamic nature of the quaternary structure of the vesicular stomatitis virus envelope glycoprotein. *Biochemistry* 1990;29:2442–2449.
67. Machamer CE, Doms RW, Bole DG, Helenius A, Rose JK. Heavy chain binding protein recognizes incompletely disulfide-bonded forms of vesicular stomatitis virus G protein. *J Biol Chem* 1990;265: 6879–6883.
68. Machamer CE, Rose JK. Vesicular stomatitis virus G proteins with altered glycosylation sites display temperature-sensitive intracellular transport and are subject to aberrant intermolecular disulfide bonding. *J Biol Chem* 1988;263:5955–5960.
69. Marcus PI, Sekellick MJ. Cell killing by viruses. II. Cell killing by vesicular stomatitis virus: a requirement for virion-derived transcription. *Virology* 1975;63:176–190.
70. Matlin KS, Reggio H, Helenius A, Simons K. Pathway of vesicular stomatitis virus entry leading to infection. *J Mol Biol* 1982;156:609–631.
71. McCreedy BJ Jr, Lyles DS. Distribution of M protein and nucleocapsid protein of vesicular stomatitis virus in infected cell plasma membranes. *Virus Res* 1989;14:189–205.
72. McGowan JJ, Emerson SU, Wagner RR. The plus-strand leader RNA of VSV inhibits DNA-dependent transcription of adenovirus and SV40 genes in a soluble whole cell extract. *Cell* 1982;28:325–333.
73. Meier E, Harrison G, Keene JD, Schubert M. Sites of copy choice replication involved in generation of vesicular stomatitis virus defective-interfering particle RNAs. *J Virol* 1984;51:515–521.
74. Metsikko K, Simons K. The budding mechanism of spikeless vesicular stomatitis virus particles. *EMBO J* 1986;5:1913–1920.
75. Morita K, Vanderoef R, Lenard J. Phenotypic revertants of temperature-sensitive M protein mutants of vesicular stomatitis virus: sequence analysis and functional characterization. *J Virol* 1987;61:256–263.
76. Moyer SA, Smallwood-Kentro S, Haddad A, Prevec L. Assembly and transcription of synthetic vesicular stomatitis virus nucleocapsids. *J Virol* 1991;65:2170–2178.
77. Newcomb WW, Tobin GJ, McGowan JJ, Brown JC. *In vitro* reassembly of vesicular stomatitis virus skeletons. *J Virol* 1982;41:1055–1062.
78. Ogden JR, Pal R, Wagner RR. Mapping regions of the matrix protein of vesicular stomatitis virus which bind to ribonucleocapsids, liposomes, and monoclonal antibodies. *J Virol* 1986;58:860–868.
79. Owens RJ, Rose JK. Cytoplasmic domain requirement for incorporation of a foreign envelope protein into vesicular stomatitis virus. *J Virol* 1993;67:360–365
80. Pal R, Barenholz Y, Wagner RR. Vesicular stomatitis virus membrane proteins and their interaction with lipid bilayers. *Biochim Biophys Acta* 1987;906:175–193.
81. Pal R, Barenholz Y, Wagner RR. Pyrene phospholipids as biological fluorescent probes for studying fusion of virus membranes with liposomes. *Biochemistry* 1988;27:30–36.
82. Pal R, Wagner RR. Rhabdovirus membrane and maturation. In: Wagner RR, ed. *The rhabdoviruses*. New York: Plenum; 1987:75–128.
83. Pattnaik AK, Ball LA, LeGrone AW, Wertz GW. Infectious defective interfering particles of VSV from transcripts of a cDNA clone. *Cell* 1992;69:1011–1020.
84. Peluso RW, Moyer SA. Viral proteins required for the in vitro replication of vesicular stomatitis virus defective-interfering particle genome RNA. *Virology* 1988;162:369–376.
85. Pitta AM, Rose JK, Machamer CE. A single amino acid substitution eliminates the stringent carbohydrate requirement for the intracellular transport of a viral glycoprotein. *J Virol* 1989;63:3801–3809.
86. Poch O, Sauvaget I, Delarue M, Tordo N. Identification of four conserved motifs among the RNA-dependent polymerase coding elements. *EMBO J* 1989;8:3867–3874
87. Prehaud C, Coulon P, Lafay F, Thiers C, Flamand A. Antigenic site II of the rabies virus glycoprotein: structure and role in viral virulence. *J Virol* 1988;62:1–7.
88. Pringle CR. Rhabdovirus genetics, In: Wagner RR, ed. *The rhabdoviruses*. New York: Plenum; 1987:167–243.
89. Puddington L, Bevan MJ, Rose JK, LeFrancois L. N protein is the predominant antigen recognized by vesicular stomatitis virus-specific cytotoxic T cells. *J Virol* 1986;60:708–717.
90. Remenick J, Kenny MK, McGowan JJ. Inhibition of adenovirus DNA replication by vesicular stomatitis virus leader RNA. *J Virol* 1988;62: 1286–1292.
91. Remenick J, McGowan JJ. A small RNA transcript of vesicular stomatitis virus inhibits the initiation of adenovirus replication *in vitro*. *J Virol* 1986;59:660–668.
92. Rose JK, Adams GA, Gallione CJ. The presence of cysteine in the cytoplasmic domain of the vesicular stomatitis virus glycoprotein is required for palmitate addition. *Proc Natl Acad Sci USA* 1984;81: 2050–2054.
93. Rose JK, Gallione CJ. Nucleotide sequences of the mRNAs encoding the vesicular stomatitis virus G and M proteins determined from cDNA clones containing the complete coding regions. *J Virol* 1981;39:519–528.
94. Rose JK, Schubert M. Rhabdovirus genomes and their products. In: Wagner RR, ed. *The rhabdoviruses*. New York: Plenum; 1987:129–166.
95. Schlegel R, Wade M. Neutralized vesicular stomatitis virus binds to host cells by a different "receptor." *Biochem Biophys Res Commun* 1983;114:774–778.
96. Schlegel R, Wade M. Biologically active peptides of the vesicular stomatitis virus glycoprotein. *J Virol* 1985;53:319–323.
97. Schlegel R, Willingham C, Pastan I. Saturable binding sites for vesicular stomatitis virus on the surface of Vero cells. *J Virol* 1982;43: 871–875.
98. Schmidt MF, Schlesinger MJ. Fatty acid binding to vesicular stomatitis virus glycoprotein: a new post-translational modification of the viral glycoprotein. *Cell* 1979;17:813–819.
98a. Schnell MJ, Mabatsion T, Conzelmann KK. Infectious rabies viruses from cloned cDNA. *EMBO J.* 1994;13:4195–4203.
99. Schnitzer TJ, Dickson C, Weiss RA. Morphological and biochemical characterization of viral particles produced by the *ts*O45 mutant of vesicular stomatitis virus at restrictive temperature. *J Virol* 1979;29: 185–195.
100. Schnitzlein WM, Reichmann ME. Characterization of New Jersey vesicular stomatitis virus isolates from horses and black flies during the 1982 outbreak in Colorado. *Virology* 1985;142:426–431.
101. Schubert M, Harmison G, Meier E. Primary structure of the vesicular stomatitis virus polymerase (L) gene: evidence for a high frequency of mutations. *J Virol* 1984;51:505–514.
102. Schubert M, Harmison G, Richardson CD, Meier E. Expression of a cDNA encoding of functional 241-kilodalton vesicular stomatitis virus RNA polymerase. *Proc Natl Acad Sci USA* 1985;82:7984–7988.

103. Schubert M, Joshi B, Blondel D, Harmison GG. Insertion of the human immunodeficiency virus CD4 receptor into the envelope of vesicular stomatitis virus particles. *J Virol* 1992;66:1579–1589.
104. Shipley JB, Pal R, Wagner RR. Antigenicity, function and conformation of synthetic oligopeptides corresponding to amino terminal sequences of wild-type and mutant matrix proteins of vesicular stomatitis virus. *J Virol* 1988;62:2569–2577.
105. Shope RE, Tesh RB. The ecology of rhabdoviruses that infect vertebrates. In: Wagner RR, ed.*The rhabdoviruses*. New York: Plenum; 1987:509–534.
106. Sleat DE, Banerjee AK. Transcriptional activity and mutational analysis of recombinant vesicular stomatitis virus RNA polymerase. *J Virol* 1993;67:1334–1339.
107. Smallwood S, Moyer SA. Promoter analysis of the vesicular stomatitis virus RNA polymerase. *Virology* 1993;192:254–263.
108. Spiropoulou CF, Nichol ST. A small highly basic protein is encoded in overlapping frame within the P gene of vesicular stomatitis virus. *J Virol* 1993;67:3103–3110.
109. Steinhauer DA, Holland JJ. Direct method for the quantitation of extreme polymerase error frequency at selected single base sites in viral RNA. *J Virol* 1986;57:219–228.
110. Teninges D, Bras F, Dezélée S. Genome organization of the sigma rhabdovirus: six genes and a gene overlap. *Virology* 1993;193:1018–1023.
111. Thomas D, Newcomb WW, Brown JC, et al. Mass and molecular composition of vesicular stomatitis virus: a scanning transmission electronmicroscopy analysis. *J Virol* 1985;54:598–607.
112. Tordo N, Poch O, Ermine A, Keith G, Rougeon F. Walking along the rabies genome: is the large G-L intergenic region a remnant gene? *Proc Natl Acad Sci USA* 1989;83:3914–3918.
113. Vandepol SB, LeFrancois L, Holland JJ. Sequence of the major antibody binding epitopes of the Indiana serotype of vesicular stomatitis virus. *Virology* 1986;148:312–325.
114. Wagner RR. Rhabdovirus biology and infection: an overview. In: Wagner RR, ed. *The rhabdoviruses*. New York: Plenum; 1987:9–74.
115. Wagner RR. Molecular basis of rhabdovirus pathogenicity. In: Notkins AL and Oldstone MBA, eds. *Concepts in viral pathogenesis*. New York: Springer-Verlag; 1989:268–274.
116. Wagner RR, Thomas JR, McGowan JJ. Rhabdovirus cytopathology: effects on cellular macromolecular synthesis. In: Fraenkel-Conrat H, Wagner RR, eds. *Comprehensive virology*. Vol. 19. New York: Plenum; 1984:223–295.
117. Walker P, Byrne KA, Riding GA, Cowley JA, Wang Y, McWilliam S. The genome of bovine ephemeral fever virus contains two related glycoprotein genes. *Virology* 1992;191:49–61.
118. Weck PK, Wagner RR. Transcription of vesicular stomatitis virus is required to shut off cellular RNA synthesis. *J Virol* 1979;30:410–413.
119. Wertz GW, Davis NL, Patton J. The role of proteins in vesicular stomatitis virus RNA replication. In: Wagner RR, ed. *The rhabdoviruses*. New York: Plenum; 1987:271–296.
120. Whitt M, Chong L, Rose JK. Glycoprotein cytoplasmic domain sequences required for rescue of the VSV glycoprotein mutant. *J Virol* 1989;63:3569–3578.
121. Whitt MA, Rose JK. Fatty acid acylation is not required for membrane fusion activity or glycoprotein assembly into VSV virions. *Virology* 1991;185:875–878.
122. Whitt MA, Zagouras P, Crise B, Rose JK. A fusion-defective mutant of the vesicular stomatitis virus glycoprotein. *J Virol* 1990;64: 4907–4913.
123. Wilson T, Lenard J. Interaction of wild-type and mutant M protein of VSV with nucleocapsids *in vitro*. *Biochemistry* 1981;20:1349–1354.
124. Woodgett C, Rose JK. Amino-terminal mutations of the vesicular stomatitis virus glycoprotein does not affect its cell fusion activity. *J Virol* 1986;59:486–489.
125. Wu F-S, Lucas-Lenard J. Inhibition of ribonucleic acid accumulation in mouse L cells infected with vesicular stomatitis virus requires viral ribonucleic acid transcription. *Biochemistry* 1980;19:804–810.
126. Zagouras P, Rose JK. Dynamic equilibrium between vesicular stomatitis virus glycoprotein monomers and trimers in the Golgi and at the cell surface. *J Virol* 1993;67:7533–7538.
127. Zagouras P, Ruusala A, Rose JK. Dissociation and reassociation of oligomeric viral glycoprotein subunits in the endoplasmic reticulum. *J Virol* 1991;65:1976–1984.
128. Zakowski JJ, Petri WA, Wagner RR. Role of matrix protein in assembling the membrane of vesicular stomatitis virus: reconstitution of matrix proteins with negatively charged phospholipid vesicles. *Biochemistry* 1981;23:3902–3907.
129. Zinkernagel RM. Cellular immune responses to viruses and the biological role of polymorphic major transplantation antigens. In: Fraenkel-Conrat H, Wagner RR, eds. *Comprehensive virology*. Vol. 15. New York: Plenum; 1979:171–204.

Fundamental Virology, Third Edition
edited by B.N. Fields, D.M. Knipe, P.M. Howley, et al.
Lippincott - Raven Publishers, Philadelphia © 1996

CHAPTER 20

Paramyxoviridae: The Viruses and Their Replication

Robert A. Lamb and Daniel Kolakofsky

The viruses of the family Paramyxoviridae are enveloped negative-stranded RNA viruses that have special relationships to two other families of negative-strand RNA viruses, namely the Orthomyxoviridae (for the biological properties of the envelope glycoproteins) and the Rhabdoviridae (for the similarity of organization of the nonsegmented genome and its expression). The genomic RNA of negative strand RNA viruses has to serve two functions: first as a template for synthesis of messenger RNAs (mRNAs) and second as a template for synthesis of the antigenome (+) strand. Negative strand RNA viruses encode and package their own RNA transcriptase, but mRNAs are only synthesized once the virus has been uncoated in the infected cell. Viral replication occurs after synthesis of the mRNAs and requires the continuous synthesis of viral proteins. The newly synthesized antigenome (+) strand serves as the template for further copies of the (–) strand genomic RNA.

R. A. Lamb: Howard Hughes Medical Institute, Department of Biochemistry, Molecular Biology and Cell Biology, Northwestern University, Evanston, Illinois 60208-3500.

D. Kolakofsky: Department of Genetics and Microbiology, University of Geneva School of Medicine, CH1211 Geneva, Switzerland.

CLASSIFICATION

The family Paramyxoviridae was reclassified in 1993 by the International Committee on the Taxomony of Viruses into two subfamilies: the Paramyxovirinae and the Pneumovirinae. The Paramyxovirinae contains three genera, Parainfluenzavirus, Rubulavirus and Morbillivirus. The sub-family Pneumovirinae contains the genus Pneumovirus. The new classification is based on morphological criteria, the organization of the genome, the biological activities of the proteins, and the sequence relationship of the encoded proteins now that most of the genome sequences have been obtained. The morphological distinguishing feature among enveloped viruses for the subfamily Paramyxovirinae is the size and shape of the nucleocapsids (diameter 18 nm, 1 μm in length, a pitch of 5.5 nm), which have a left-handed helical symmetry. The biological criteria are (a) antigenic cross-reactivity between members of a genus and (b) the presence (*Parainfluenzavirus* and *Rubulavirus*) or absence (*Morbillivirus*) of neuraminidase activity. In addition, the differing coding potentials of the P genes are considered and there is the presence of an extra gene (SH) in rubulaviruses. The pneumoviruses can be distinguished from Paramyxovirinae morphologically because they contain narrower nucleocapsids. In addition, pneumoviruses have major differences in the number of encoded proteins and an attachment protein that is very different from that of Paramyxovirinae. Examples of members of the three genera are shown in Table 1.

VIRION STRUCTURE

The Paramyxoviridae contain a lipid bilayer envelope that is derived from the plasma membrane of the host cell in which the virus is grown (44,144,145). Paramyxoviridae are generally spherical and 150 to 350 nm in diameter but can be pleiomorphic in shape; filamentous forms can be observed. Inserted into the envelope are glycoprotein spikes that extend approximately 8 to 12 nm from the surface of the membrane and can be readily visualized by electron microscopy (EM). Inside the viral membrane is the internal nucleocapsid core (sometimes called the ribonucleoprotein core), which contains the approximately 15,000-nucleotide single-stranded RNA genome (see Fig. 1 for schematic diagram of virion and Fig. 2 for EM photographs).

The helical nucleocapsid, rather than the free genome RNA, is the template for all RNA synthesis. For Sendai virus, each nucleocapsid is composed of approximately 2,600 nucleocapsid (NP), 300 P, and 50 L proteins (162). The NP and genome RNA together form a core structure, to which the P and L proteins are attached. This nucleocapsid core is remarkably stable, as it withstands the high salt and gravity forces of CsCl density gradient centrifugation and bands at 1.31 g/ml. It is also this core structure that is seen in the EM photographs; the P and L proteins have so far only been visualized with the aid of antibodies (219,220). Holonucleocapsids (NP:RNA plus P and L) have the capacity to transcribe mRNAs *in vitro,* presumably mimicking primary transcription *in vivo,* and they are the only subviral structures thought to retain infectivity.

When negatively stained preparations of paramyxoviral nucleocapsids are viewed via EM, the most tightly coiled forms resemble the *Tobamovirus* tobacco mosaic virus (TMV); i.e., a relatively rigid coiled rod 18 nm in diameter, with a central hollow core of 4 nm and a helical pitch of approximately 5 nm (54,55,86). Unlike TMV, however, in which the nucleocapsid disassembles so its (+) RNA genome can function (28), paramyxoviral nucleocapsids function without disassembling; as far as we know, these nucleocapsids never disassemble naturally. This remarkable property is undoubtedly associated with the finding that, unlike TMV, these nucleocapsids are not rigid. They uncoil in response to increased salt concentrations and reversibly recoil when the ionic strength of the solution is reduced (106,107). They also become more rigid on trypsin treatment (189).

More importantly, for the life-cycle of the virus, Sendai virus nucleocapsids exist in several distinct morphological states at normal salt concentration (75). The most prevalent form in negatively stained preparations is the most tightly coiled one, with a helical pitch of 5.3 nm. Two other forms, one with a slightly larger pitch of 6.8 nm and another with a much larger pitch of 37.5 nm, also have been noted. The fact that no structures of intermediate pitch have been found indicates that these are distinct states. The transition from the 5.3 nm to the 6.8 nm pitch is marked by a large increase in flexibility of the coil, and the highly supercoiled nucleocapsids in virions are probably in the 6.8-nm state. The most extended form, with a pitch of 37.5 nm, in which the coil is almost completely unwound, may reflect the form in which the template is being copied by the viral polymerase. Because the template is copied without dissociation of NP from the nucleocapsid core, the uncoiling of the nucleocapsid may be necessary for the polymerase to gain access to the RNA bases, especially if the RNA chain, as in TMV, is found near the inside surface of the ribbon (192). In either the 5.3-nm or 6.8-nm pitch states, it appears unlikely that the polymerase can easily maneuver (75,106,107). Individual nucleocapsids in the EM photographs may be seen when a transition from a tighter coil to the 37.5 nm state occurs. It is possible that these viral polymerases traverse their nucleocapsid template by uncoiling the helix in front and recoiling it once they have

TABLE 1. *Examples of members of the family* Paramyxoviridae

Family Paramyxoviridae
Subfamily Paramyxovirinae
Genus *Paramyxovirus*
Sendai virus (mouse parainfluenza virus type 1)
Human parainfluenza virus type 1 and type 3
Bovine parainfluenza virus type 3
Genus *Rubulavirus*
Simian virus 5 (Canine parainfluenza virus type 2)
Mumps virus
Newcastle disease virus (Avian paramyxovirus 1)
Human parainfluenza virus type 2, type 4a and 4b
Genus *Morbillivirus*
Measles virus
Dolphin morbillivirus
Canine distemper virus
Peste-des-petits-ruminants virus
Phocine distemper virus
Rinderpest virus
Subfamily Pneumovirinae
Genus *Pneumovirus*
Human respiratory syncytial virus
Bovine respiratory syncytial virus
Pneumonia virus of mice
Turkey rhinotracheitis virus

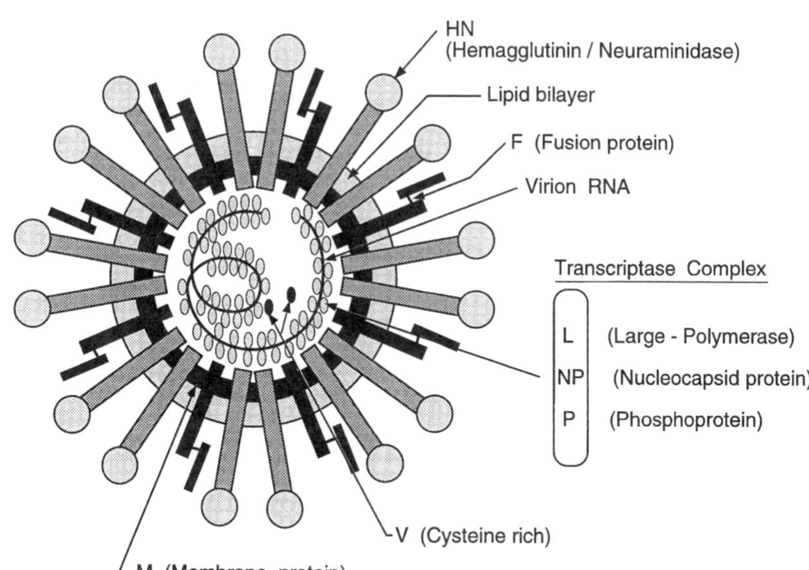

FIG. 1. Schematic diagram of a paramyxovirus (not drawn to scale). The lipid bilayer is shown as the gray concentric circle, and, underlying the lipid bilayer, the viral matrix protein is shown as a black concentric circle. Inserted through the viral membrane are the hemagglutinin-neuraminidase (HN) attachment glycoprotein and the (F) fusion glycoprotein. The HN protein is thought to have a stalk region and a globular head, and the F protein consists of two sulfide-linked chains F_1 and F_2. However, the shapes used are purely schematic, and no detailed structural information is available on these proteins. Inside the virus is the negative-strand virion RNA, which is encapsidated with the NP. The RNA is never found free of NP in the life cycle of the virus. Associated with the nucleocapsid are the L and P proteins, and together this complex has RNA-dependent RNA transcriptase activity. For the Rubulaviruses, the cysteine-rich protein V is found as an internal component of the virion, whereas for other members of the family, the V protein is only found in virus-infected cells. The nature of possible interactions between the cytoplasmic tails of the glycoprotein spikes and the matrix protein and the interactions between the matrix protein and the nucleocapsid have not been fully elucidated and no attempt has been made to illustrate them.

passed, much the same as cellular RNA polymerase would generate its template "bubble" in traversing double-stranded DNA.

As expected, the diameter of the nucleocapsid decreases as the pitch increases and the nucleocapsid lengthens, and for Sendai virus the diameter is 3.5 nm less for the 6.8-nm form than for the 5.3-nm pitch form. These latter values are similar to those of *Pneumovirus* nucleocapsids, which also have a pitch of 7 nm. As discussed above, these differences in nucleocapsid morphology are used to distinguish the different Paramyxoviridae, but they probably relate mainly to which form predominates in negatively stained preparations.

THE PARAMYXOVIRIDAE GENOMES AND THEIR ENCODED PROTEINS

The genome of the family Paramyxoviridae is nonsegmented single-stranded RNA of negative polarity. For many years the genome RNA was known as 50S for its sedimentation coefficient on sucrose gradients. Today, we know the complete nucleotide sequence for many members of the family Paramyxoviridae, including Sendai virus, human parainfluenza virus (hPIV)-3, SV5, mumps virus, measles virus, canine distemper virus (CDV), and respiratory syncytial virus (RS virus) (92). The genome is approximately 15,500 nucleotides in length and contains a 3′ extracistronic region of approximately 50 nucleotides known as the leader and a 5′ extracistronic region of approximately 50 nucleotides known as the trailer [or (–) leader]. These control regions, which are essential for transcription and replication, flank the six genes (seven for rubulaviruses and ten for pneumoviruses). (Note: by the convention used for paramyxoviruses, the term "gene" refers to the genome RNA sequence encoding a single mRNA, even if that mRNA contains more than one open reading frame and encodes more than one protein.) The coding capacity of the genome of Paramyxovirinae is extended by the use of overlapping reading frames in the P gene. The gene order of a representative member of each genus is shown in Fig. 3. At the beginning and end of each gene are conserved transcriptional control sequences that are copied into mRNA. Between the gene-end boundaries are intergenic regions (Fig. 4). These are precisely three nucleotides long for the parainfluenzaviruses and morbilliviruses but are variable in length for the rubulaviruses (1 to 47 nucleotides) and pneumoviruses (1 to 56 nucleotides) (Table 2).

The Nucleocapsid Protein (NP)

The NP serves several functions in virus replication, including encapsidation of the genome RNA into an

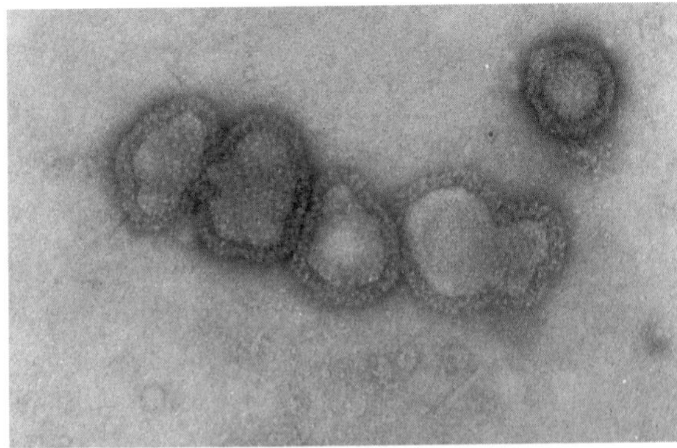

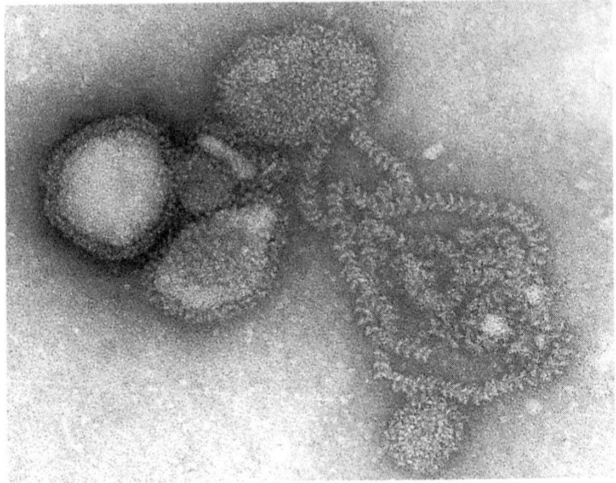

FIG. 2. Ultrastructure of SV5 virions shown by negative staining. In the top panel the glycoprotein spikes on intact 150- to 300-nm virus particles can be observed. In the lower panel a virus particle has ruptured on the EM grid and the relaxed extended helical nucleocapsid structure can be clearly seen (×104,000). Courtesy of G. Leser, Northwestern University, Evanston, IL.

RNAse-resistant nucleocapsid (the template for RNA synthesis), association with the P-L polymerase during transcription and replication, and most likely interacting with the M protein during virus assembly. The intracellular concentration of unassembled NP is also thought to be the major factor that controls the relative rates of transcription and replication from the genome templates by analogy to studies made on the *Rhabdovirus* vesicular stomatitis virus (VSV) (20,21).

The sequence of many NPs has now been predicted (9,47,78,87,132,134,135,152,169,178,194,203,232,239,240, 273,288). The NPs range from 489 to 553 amino acids (M_r 53,167 to 57,896) and, unexpectedly for a protein that interacts with RNA, has a net charge of –7 to –12 with the exception of mumps virus (+2). A comparison of these protein sequences, coupled with data obtained from protease digestions of nucleocapsids, suggest that NP contains two domains. The *N*-terminal 80% of the protein of approximately 500 residues is relatively well conserved among related viruses, whereas the *C*-terminal 20% is poorly conserved, although this domain always contains a highly charged and mostly negative region (59,203). This hypervariable *C*-terminus appears to be a tail extending from the

TABLE 2. *Examples of gene start, gene end, and intercistronic nucleotide sequences of members of the family* Paramyxovirinae

	Start	End	ICS
Sendai virus	UCCCANUUNC	$UNAUUCU_5$	GAA [GGG for HN-L]
hPIV3	UCCUNNUUUC	$UUNAU^{AU}_{UC}U_5$	GAA
SV5	$U^{U}_{C}CN^{U}_{G}NCUUG$	$A^{GUA}_{AAU}UCU_{4\text{-}7}$	Variable 1–22 nt
NDV	$UG^{C}_{G}CCAUC^{U}_{A}N$	$A^{A}_{U}UCU_6$	Variable 1–31 nt

nt, nucleotide.

surface of a globular *N*-terminal body, based on the observation that the *C*-terminal sequences of the SV5 and Sendai virus NP in nucleocapsids are hypersensitive to trypsin digestion, leaving a 48-kd *N*-terminal core (56,107,189). However, the overall structure of the nucleocapsid as seen in the EM photographs and the resistance of the genomic RNA within this structure to nuclease attack are mostly unchanged (141). These data indicate that the determinants for the helical nature of the nucleocapsid, as well as the RNA binding domains, must lie within the highly conserved *N*-terminal body. An extensive deletion mutant analysis indicates that the *N*-terminal region of Sendai virus NP is required for nucleocapsid assembly (24). The invariant sequence F-X_4-Y-X_4-S-Y-A-M-G (where X is any residue) is found near the middle of the NP in a region predicted to be hydrophobic, and it could be involved in NP:NP contacts during assembly (185), but the presence of several conserved aromatic residues also suggests that it may be involved in RNA binding. In any event, NP does not appear to be a classical RNA binding protein in that it does not contain any previously recognized RNA-binding motifs (73), nor does it interact with RNAs in Northwestern blots. The *C*-terminal tail, on the other hand, contains most of the protein's phosphorylation sites and antigenic sites (126,237). For measles virus, the *N*-terminal body is unstructured when initially synthesized and undergoes a conformational maturation before assembling into nucleocapsids (98).

The structure-function relationship of the Sendai virus NP was examined using a complementary DNA (cDNA)-encoded genome of a defective-interfering (DI) particle the intracellular replication of which was directed by NP, P, and L protein expressed from cotransfected cDNAs (59). The entire *C*-terminal tail of NP was found to be dispensable for simply making an encapsidated, complementary copy of the template, whereas deletions anywhere within the *N*-terminal body of the protein eliminated this activity. The *C*-terminal tail is therefore not essential for assembling the nascent chain and helping the polymerase to ignore the junctions. However, templates that were themselves assembled with tail-less NPs were unable to act as

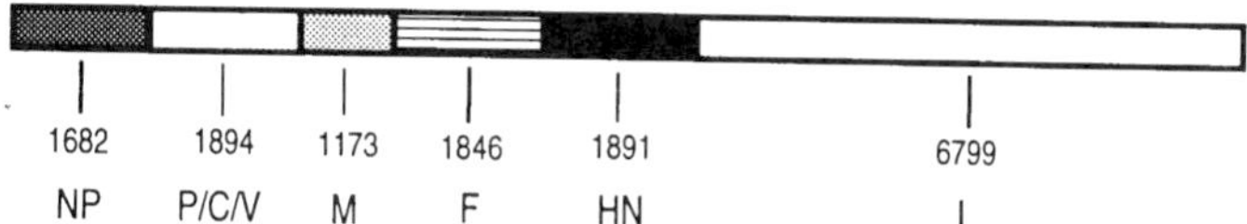

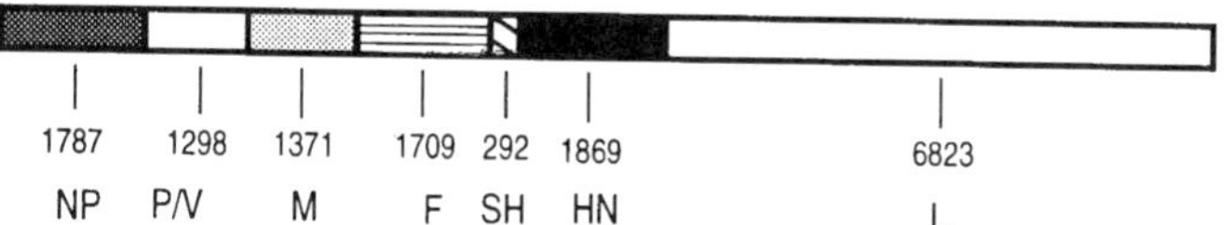

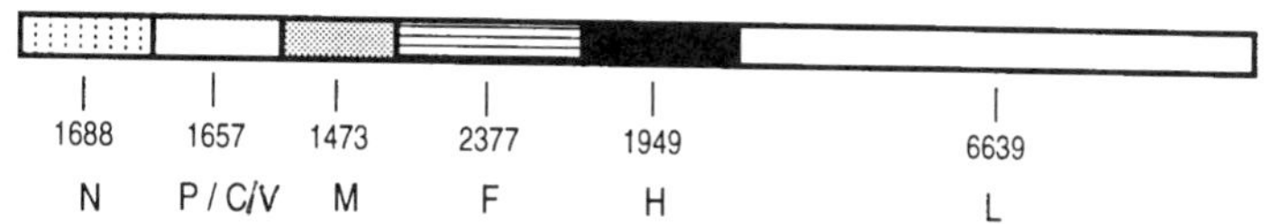

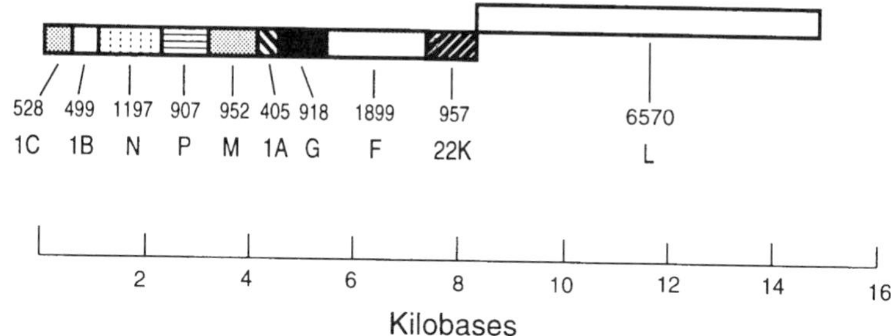

FIG. 3. Genetic map of a typical member of each genus of the family Paramyxoviridae. The gene size is drawn to scale. Gene boundaries are shown by vertical lines. The pneumovirus L gene transcription overlaps that of the 22K (M2) gene and is thus shown in a staggered format: this overlap configuration is seen in human and animal viruses, but not in other pneumoviruses.

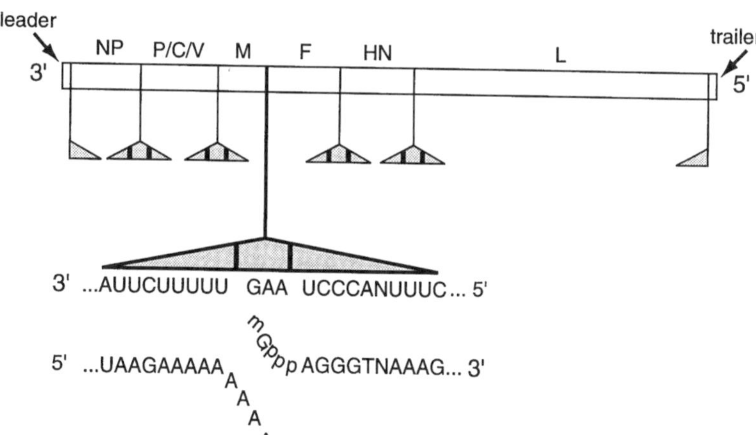

FIG. 4. Schematic diagram of the Sendai virus genome. The positions of the extragenic 3′-terminal leader region and 5′-terminal trailer region are shown. The conserved transcription regulatory sequences at the gene boundaries are indicated. These are sometimes known as E (end), I (intergenic), and S (start) sequences.

templates for new rounds of genome replication. The highly charged domain of the *C*-terminal tail appears to be essential for this function because deletions downstream of this domain have little or no effect. The precise template function of the NP *C*-terminal tail is unclear, but it may be required for the transition between helical states because nucleocapsids with tail-less NP appear more rigid in the EM photographs (189). Moreover, the tail appears to mediate P protein binding to nucleocapsids because binding of antibodies specific to the *C*-terminal tail leads to the release of the normally tightly bound P proteins, both for Sendai virus and hPIV-1 (237). Whether P binds directly to this *C*-terminal tail domain is unclear, but if so it could participate in the helical transitions thought necessary for template function.

The P Gene and Its Encoded Proteins

For the Paramyxovirinae the P gene represents an extraordinary example of a virus compacting as much genetic information as possible into a small gene. The P gene gives rise to a plethora of polypeptide products by means of using overlapping reading frames and by a remarkable process of transcription known as RNA editing or pseudotemplated addition of nucleotides, the consequence of which is a reading frame shift on translation. The mechanisms of expression of the protein products derived from the P gene are described in the forthcoming section on virus adsorpiton and entry. In contrast to the Paramyxovirinae, the Pneumovirinae have a P gene that encodes a single protein, P. Thus, this constitutes a major difference between the virus subfamilies.

The Phosphoprotein (P)

The P proteins were named for their highly phosphorylated nature. The sequence of many P proteins has now been predicted (11,15,16,36,79,88,96,177,180,207,239,254, 257,266,270). Within the virus family, they are quite variable in length. Those of the parainfluenzaviruses and morbilliviruses are 507 to 603 amino acids long, those of the rubulaviruses are 245 to 397 residues long, and those of the pneumoviruses are the smallest, 241 residues long. The protein P is a modular protein that plays a central role in all RNA synthesis. Together with the L protein it forms the viral polymerase (P-L), and together with unassembled NP (NP^0) forms a complex (P-NP^0), which is probably the active form in RNA encapsidation (105,122). In reconstitution experiments, only NP^0 that is coexpressed with P can act in genome replication, and this activity correlates with P-NP^0 complex formation (122). It also has been suggested that this complex prevents NP^0 from assembling RNA nonspecifically (59,176).

Based on sequence conservation, the Sendai virus P protein appears to be composed of *N*- and *C*-terminal domains, separated by a hypervariable region containing the site where the V open reading frame (ORF) is fused in place by mRNA editing (63,177) (Fig. 5). The P protein *C*-terminal domain is further subdivided into a domain responsible for the protein's trimerization, one that tightly binds the L protein, and finally one that binds NP^{NC} and probably also NP^0 (65,235,236). All of these domains appear to be essential because deletions anywhere within the *C*-terminal half of the protein renders P nonfunctional in genome replication (57).

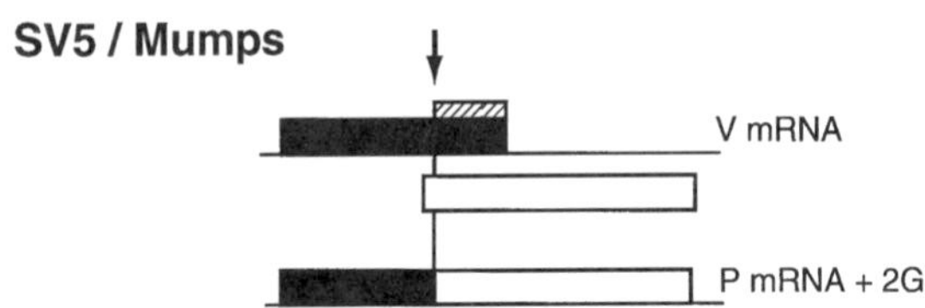

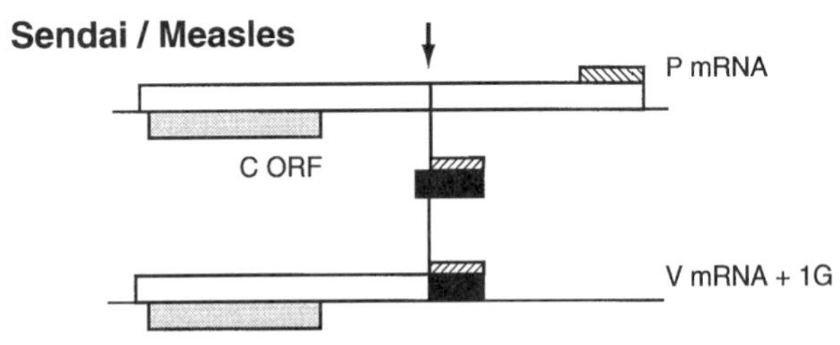

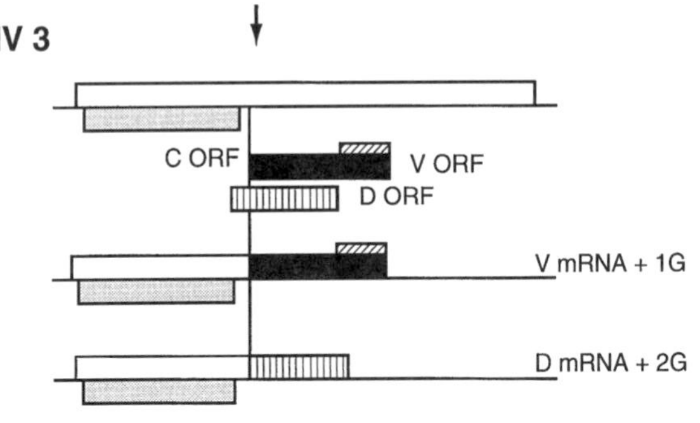

FIG. 5. Schematic diagram of the P gene of representative paramyxoviruses. The unedited version of the SV5 and mumps virus "P" mRNA encodes the V protein (black shading). The cysteine-rich domain is indicated by the overlaid rectangle with cross-hatching. The RNA editing site, at which two nontemplated nucleotides are added to the mRNA to yield the P protein mRNA, is indicated by an arrow. The P protein is a fusion of the *N*-terminal portion of the V reading frame with the second ORF (white rectangle). For Sendai virus and measles virus, the unedited mRNA encodes the P protein. RNA editing with the addition of one G nucleotide at the editing site produces an mRNA that encodes the V protein. The Sendai virus and measles virus P mRNA also encode the C protein ORF at the 5′ end of the mRNA independant of editing status (light stippling). bPIV-3 P mRNA encodes the P/C ORFs, and RNA editing can produce mRNAs that encode either the V protein or the D protein. In this case all three reading frames are used after the RNA editing site.

The P protein *N*-terminal domain is distinguished by containing most of the sites at which the protein is phosphorylated (70,277) and because it is present in two other viral proteins, namely, W/I and V. The P protein *N*-terminal half also contains stretches rich in acidic residues, and these stretches presumably account for the anomolously slow migration of the *Parainfluenzavirus* and *Morbillivirus* P, V, and W proteins in sodium dodecyl sulfate–polyacrylamide gel electrophoresis (SDS-PAGE) (58), a suggestion that was used previously to explain the anomolously slow migration of the yeast transcriptional activator GCN4 (121) and the VSV P protein (8). The immediate *N*-terminal region (residues 1 to 78) of the P protein also contains a region essential for RNA synthesis, but the manner by which this region acts is unclear. Interestingly, two separate regions near the *N*-terminus could provide this function. The functional redundancy in this region and its general acidic nature are reminiscent of acidic activation domains of cellular transcription factors, as well as the VSV P protein (8). The *N*-terminus also contains a unique region specifically required for the RNA encapsidation step of genome replication (65).

Most of the *N*-terminal half of Sendai virus P protein (residues 78 to 325) can be deleted without eliminating the activity of this protein in RNA synthesis and nascent chain encapsidation (65). This "dispensable" region contains most of the protein's phosphorylation sites. It is, therefore, possible that the P protein performs tasks in virus replication other than RNA synthesis. Alternatively, this region of the P protein would perform a more subtle task in RNA synthesis, one that cannot be seen with the present-day systems of study.

The C Protein

The C proteins are relatively small (180 to 204 residues), basic proteins expressed from the P gene of all members of *Paramyxovirus* and *Morbillivirus,* from an ORF that overlaps the *N*-terminal portion of the P gene, in the +1 frame. The first of these proteins to be described was found in cells infected with Sendai virus (84,158,160,162). Because it was originally only observed in virus-infected cells, the C protein was considered to be a nonstructural protein. However, more recent observations indicate that it is present in small amounts in both virions and in nucleocapsids isolated from cells and virions (163,218). The number of C-related proteins expressed from each of these viral P genes varies; there are four from Sendai virus (designated C, C, Y_1, and Y_2) (62) (Fig. 6), two or three from hPIV-1, but only a single C protein is made from the hPIV-3 and measles virus P genes. This diversity of C-related proteins is due to the use of a variable number of in-phase start codons, including ACG and GUG codons, to initiate this protein (23,60,103). On a molar basis, the C rather than the P proteins are the major translation products of the P gene, at least for some of these viruses, in part because C protein expression is independent of the editing status of the mRNA (see forthcoming section on primary transcription).

The Sendai virus C protein is not required for mRNA synthesis or for genome replication. Rather, the presence of this protein was found to inhibit mRNA synthesis *in vitro.* Genome replication *in vitro,* on the other hand, is only slightly affected by the C protein, and this selective inhibition of RNA synthesis may play a role in helping the transition from primary transcription to genome replication (64). Rubulaviruses do not express a C protein.

The V Protein and W, I, and D Proteins

The P gene mRNAs (except those of the pneumoviruses and hPIV-1) are cotranscriptionally edited (see forthcoming section on primary transcription), and this always occurs downstream of the C protein ORF. The consequence of this unusual transcription process is that upon translation, the *N*-terminal half of the P protein sequences can be joined not only to those of P protein *C*-terminal half, or instead can be fused to a very highly conserved, cysteine-rich domain expressed from the V protein ORF (in the –1 frame relative to that of P) (Fig. 5). This fusion protein is referred to as V. The cysteine-rich domain of the measles virus V protein can bind zinc (167), presumably to form a specific structure, but the function of this structure is unclear.

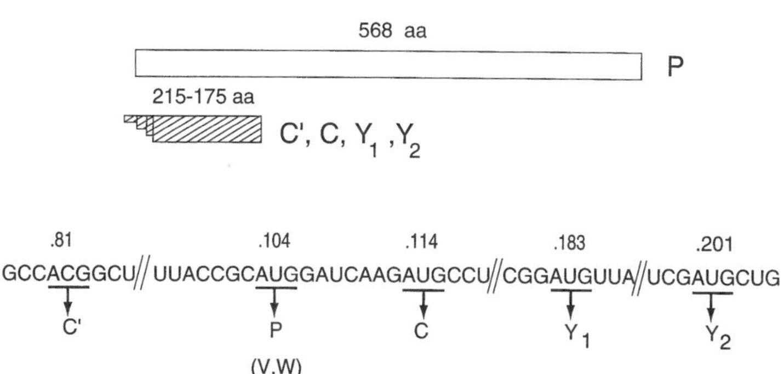

FIG. 6. Representation of the Sendai virus larger ORFs to illustrate the use of multiple initiation codons. The nucleotide sequences surrounding the initiation codons for C´, P, C, Y_1, and Y_2 are indicated. Note that the AUG codon at position 104 also starts the V and W proteins in the edited P gene mRNAs. Except for the first initiation codon (the ACG for C´), all are in a suboptimal context for initiation. Adapted from Lamb and Paterson (163), with permission.

For the *Parainfluenzavirus* and *Morbillivirus* P genes, which appear to be organized similarly, these V proteins would be missing the P protein trimerization site and the L and NP binding sites found in the P protein *C*-terminal half. This is consistent with the observation that these V proteins are not found in virions and are not associated with viral nucleocapsids intracellularly (58,97,281). The Sendai virus V protein was found to inhibit genome replication in a dose-dependent fashion (57). However, it had little or no effect on mRNA synthesis, and it appears to specifically interfere with the RNA encapsidation step of genome replication (65). Although the manner of this inhibition remains to be determined, it is independent of the cysteine-rich domain because the Sendai virus W protein has similar inhibitory properties.

Less is known about the function of the *Rubulavirus* V proteins, but in contrast to the parainfluenza V proteins, they appear to be present in virions (267,270). Moreover, the *N*-terminal half of the *Rubulavirus* P and V proteins is predicted to be basic (like that of the C proteins) rather than acidic in character. The *Rubulavirus* P proteins are presumably organized differently from their *Parainfluenzavirus* and *Morbillivirus* counterparts. Because the P gene encodes different protein modules that are shuffled via mRNA editing, it is also possible that the *Rubulavirus* P proteins contain the equivalent of the *Parainfluenzavirus* or *Morbillivirus* C protein sequence as their *N*-terminal half.

The Large (L) Protein

The L protein is the least abundant of the structural proteins (approximately 50 copies per virion), and its mRNA is the most 5′ promoter-distal in the transcriptional map and thus the last to be transcribed. Its low abundance, large size, and its localization to the transcriptionally active viral cores suggest that it might be the viral polymerase. The L genes of many members of the family, including SV5, Sendai virus, Newcastle Disease Virus (NDV), measles virus, CDV, hPIV-3, and RS virus have now been sequenced (18,90,116,140,184,203,254,255,259,289), and they are all of very similar length (approximately 2,200 amino acids), but there is little overall sequence homology outside of the subfamily. However, there are five short regions of high homology near the center of these proteins that are also conserved in the RNA-dependent RNA polymerases of other virus families (136,216). These regions are thought to represent structural features of a common, ancestral polymerase fold (69,216). To date, however, there are no reports of structure-function studies of the paramyxovirus L protein.

The P and L proteins form a complex, and both of these components are required for polymerase activity with NP:RNA templates (64,105,122,175,261). The precise composition of the viral polymerase is unclear, but transcriptionally active nucleocapsids contain five to ten P proteins/L protein, and this complex can make mRNA *in vitro* that is both capped at its 5′ end and contains a polyA tail at the 3′ end. Polyadenylation is thought to result from polymerase stuttering on a short stretch of U residues, but the capping step requires both guanylyl and methyl transferase activities. These latter activities are thought to be provided by the L protein, by analogy to VSV (see Chapter 37), and because similar reactions have not been described for cellular enzymes that operate in the cytoplasm. The Sendai virus L protein purified from virions also was found to phosphorylate the NP and P proteins, and Einberger et al. (76) have suggested that L is the kinase that has long been known to be associated with the viral core (156,230). However, at least the Sendai virus P and V proteins appear to be highly phosphorylated when expressed in transfected cells in the absence of L (67), suggesting that the majority of phosphates added to these proteins are the result of cellular kinases.

The Matrix (M) Protein

The paramyxovirus matrix (M) protein is the most abundant protein in the virion. The sequence of many members of the family have been predicted (14,22,42,77,89,112,138, 166,168,239,244,252,253). These proteins contain 341 to 375 residues (M_r approximately 38,500 to 41,500) and are basic proteins (net change at neutral pH of +14 to +17) and somewhat hydrophobic, although there are no domains of sufficient length to span a lipid bilayer, and this protein is made on free ribosomes. In electron micrographs of virions, an electron-dense layer is observed underlying the viral lipid bilayer, and this is thought to represent the location of this protein. Fractionation studies of virions indicate that the M protein is peripherally associated with membranes and is not an intrinsic membrane protein. Reconstitution studies of purified M protein and fractionation studies of infected cells indicate that the M protein can associate with membranes (85,158,191,241).

As a purified protein, the Sendai virus M protein can self-associate and form two-dimensional paracrystalline arrays (sheets and tubes) in low salt conditions (6,108,110, 111), and there is a paracrystalline array of identical periodicity at the inner surface of infected cells when examined by freeze-fracture techniques via EM (6). In addition, the M protein is also associated with nucleocapsids (263,287). The M protein probably contains amphipathic α-helices that insert themselves into the inner leaflet of the lipid bilayer to coat this surface and organize its contacts with the helical nucleocapsid (22). Evidence that the M protein of Sendai virus interacts specifically with membranes expressing individually the F and HN glycoproteins also has been obtained, which implies that there is an interaction of the F and HN cytoplasmic tails with the M protein (242). Thus, the M protein is considered to be the central organizer of viral morphogenesis, making interactions with the cytoplasmic tails of the integral membrane proteins, the lipid bilayer, and the nucleocapsids. The self-association of M and its contact with the nucleocapsid may be the driving force in forming a budding virus particle (211). The relative abundance of basic residues in the M protein may reflect their importance in ionic interactions with the acidic NPs.

Consistent with its central role in virus budding, M protein is often inactivated in persistent paramyxovirus infections in which budding fails to occur. For example, in subacute sclerosing panencephalitis (SSPE), an invariable fatal persistent measles virus infection of the brain, the M protein is either absent for a variety of reasons (37,38,40,41) or when present is not associated with budding structures *in vivo* and is unable to bind to viral nucleocapsids *in vitro* (117,118). Moreover, in model systems of persistent Sendai virus infection in culture in which the normally lytic infection is converted to a persistent one using DI particles, the change from a lytic to a persistent infection correlates mainly with M protein instability and an absence of budding structures (231).

The M protein of several paramyxoviruses is phosphorylated, and for Sendai virus a large proportion of the M protein is phosphorylated yet the M protein found in virions is not phosphorylated (159). However, a clear role for phosphorylation of M on budding has not been elucidated. For the *Rhabdovirus* VSV, there is clear evidence that the VSV M protein that is associated with the RNP structure inhibits viral transcription (33) (see Chapter 37). For Paramyxovirinae, it is presumed that association of RNPs with M protein shuts down RNA synthesis in preparation for virus assembly, but this has not been shown experimentally.

In addition to a matrix protein, the *Pneumovirus* RS virus encodes a second nonglycosylated membrane-associated protein, the M2 protein (formerly called 22K), which has no counterpart in the Paramyxovirinae. The predicted amino acid sequence of the M2 protein does not contain a region of sufficient length and hydrophobicity to indicate that it can span a membrane (53). The available evidence indicates that the M2 protein is an inner component of the viral envelope (130). However, the function of the M2 protein is not known.

Pneumovirus Nonstructural Proteins NS1 and NS2

The *Pneumovirus* RS virus contains two genes, NS1 and NS2, located adjacent to the leader RNA (3′ leader-NS1-NS2-N . . .) that encode proteins of 139 and 124 residues, respectively. There is no counterpart for these proteins in the Paramyxovirinae. NS1 and NS2 are abundantly expressed in RS virus–infected cells and seem not to be incorporated into virions (130). However, little is known about these proteins in the life cycle of the RS virus.

Envelope Glycoproteins

All Paramyxoviridae possess two integral membrane proteins, one of which is involved in cell attachment and the other in mediating pH-independent fusion of the viral envelope with cellular membranes. The assignment of specific biological activities to individual paramyxovirus glycoproteins was originally made on the basis of purification and reconstitution studies, mainly for the Sendai virus and SV5 proteins (245,247). For the parainfluenzaviruses and rubulaviruses, the attachment glycoprotein binds to cellular sialic acid–containing receptors, and these can be glycoproteins or glycolipids. The binding is of sufficiently high affinity that these viruses agglutinate mammalian and avian erythrocytes (hemagglutination). The attachment proteins of parainfluenzaviruses and rubulaviruses also have neuraminidase activity, and the proteins have been designated hemagglutinin-neuraminidase (HN). The *Morbillivirus* attachment protein (H) can cause agglutination of primate erythrocytes but lacks detectable neuraminidase or esterase activity. The restricted host range of measles virus for primate cells made it unlikely that sialic acid was the primary receptor for measles virus. Recently it has been shown that measles virus has a specific cellular receptor, the complement-binding protein known as either the membrane cofactor protein (MCP) or CD46 (72,193). The *Pneumovirus* RS virus does not cause detectable hemagglutination, and the cellular receptor for RS virus is unknown. However, the *Pneumovirus,* pneumonia virus of mice, can hemagglutinate mouse cells (46). After attachment of a Paramyxoviridae particle to the host-cell receptor, the viral envelope fuses with the host-cell plasma membrane, and the major viral protein involved in this process is the fusion (F) glycoprotein. A prominent feature of the cytopathic effect of paramyxoviruses is syncytium formation.

Attachment Protein

The *Parainfluenzavirus* and *Rubulavirus* surface glycoprotein HN is a multifunctional protein as well as the major antigenic determinant of the paramyxoviruses. The protein is responsible for the adsorption of the virus to sialic acid containing cellular molecules. In addition, HN mediates enzymatic cleavage of sialic acid from the surface of virions and the surface of infected cells. By analogy to the role of influenza virus neuraminidase, it seems likely that the role of the paramyxovirus neuraminidase activity is to prevent self-aggregation of viral particles during budding at the plasma membrane. It is interesting to note that these dual activities of HN can be modulated by halide ion concentration and pH (182). Whereas the halide ion concentration and pH of the extracellular environment is optimal for hemagglutination, *Paramyxovirus* neuraminidases have an acidic pH optima (pH 4.8 to 5.5), suggesting that neuraminidase acts in the acidic *trans* Golgi network to remove sialic acid from the HN carbohydrate chains and from the F protein carbohydrate chains. In addition, to the hemagglutinating and neuraminidase activities of HN, for many paramyxoviruses HN also has a fusion-promoting activity, i.e., coexpression of HN and F is required for cell-cell fusion to be observed.

Analysis of the nucleotide sequence of the molecularly cloned HN protein mRNA has been conducted for the vast majority of paramyxoviruses and indicates that the encoded proteins contain 565 to 582 residues (1,17,68,94,113, 139,154,183,264,265,272,282,284). For some strains of NDV, HN is synthesized as a biologically inactive precursor (HN_0), and 90 residues from the *C*-terminus are removed to activate the molecule (190,191). For a schematic diagram of important features of the HN amino acid sequence, see Fig. 7. The HN proteins are type II integral membrane proteins that span the membrane once. The HN protein contains a single hydrophobic domain, located close to the *N*-terminus, that acts as a combined signal/anchorage domain, targeting the nascent chain as it emerges from the ribosome to the membrane of the endoplasmic reticulum (ER) and during translocation of the polypeptide chain across the membrane, bringing about the stable anchoring of the protein in the lipid bilayer. This orientation of HN (Fig. 8) is analogous to that of the influenza virus neuraminidase. Interestingly, all of the enzymes that are resident in compartments of the exocytic pathway and use carbohydrates as substrates are type II integral membrane proteins, suggesting evolution of all carbohydrate modifying enzymes from a common progenitor. The HN glycoproteins contain four to six sites for the addition of *N*-linked carbohydrate chains. For the SV5 HN molecule, it has been shown that

A

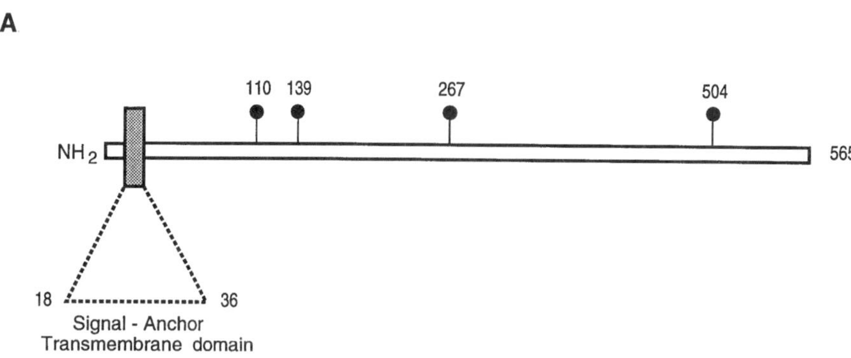

B

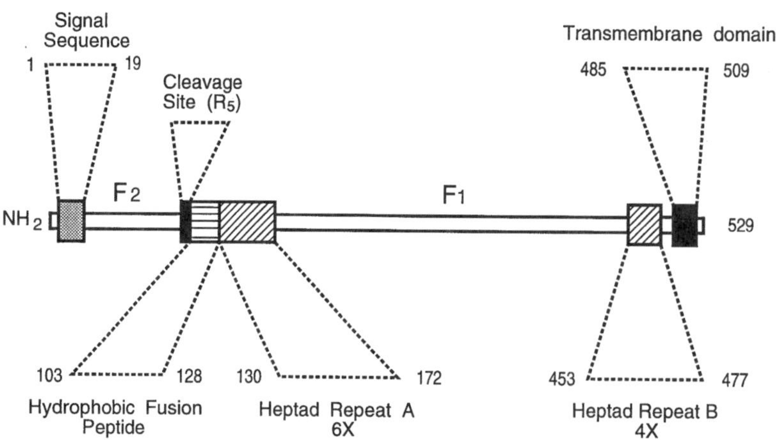

FIG. 7. Schematic diagram to show important domains and features of the paramyxovirus glycoprotein. **A:** Hemagglutinin-neuraminidase attachment protein [based on the predicted sequence of the SV5 HN gene (113)]. The signal anchor transmembrane domain and the sites used for addition of *N*-linked carbohydrate (195) are indicated. **B:** Fusion protein [based on the predicted sequence of the SV5 F gene (204)]. The position of the signal sequence, the transmembrane domain, the cleavage site, the hydrophobic fusion peptide, and the heptad repeats A and B are indicated.

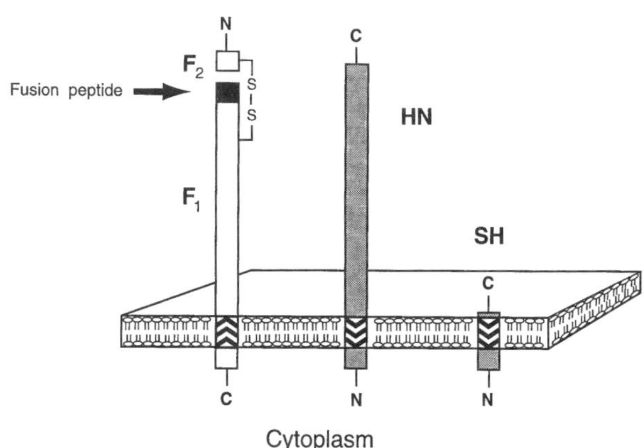

FIG. 8. Schematic diagram showing the orientation of paramyxovirus integral membrane proteins. The F and HN protein orientations are shown. In Paramyxovirnae, only SV5 has been shown to encode an SH protein, although mumps virus also has an SH ORF.

four sites are used (195). The paramyxovirus HN glycoproteins have conserved cysteine, proline, and glycine residues between related paramyxoviruses, strongly suggesting a similarity of protein structure. HN forms an oligomer consisting of disulfide-linked homodimers that form a noncovalently linked tetramer. The oligomeric form of HN has been examined in detail by using bifunctional cross-linking reagents and sucrose density gradient analysis (170,188,196,234,271). Sendai virus HN molecules that were proteolytically removed from the surface of virions (at residue 131) retain enzymatic activity and in the EM appeared as box-shaped molecules (10 × 10 nm) consisting of four subunits with fourfold symmetry (271). The HN type II membrane orientation, the presence of a protease accessible stalk that is not essential for enzymatic activity, and the tetrameric oligomeric form all parallel influenza virus neuraminidase. In addition, *Paramyxovirus* HN proteins contain a conserved sequence N-R-K-S-K-S that is similar to the known sialic acid binding site of influenza virus neuraminidase (187,274). However, an x-ray structure of *Paramyxovirus* HN has not yet been obtained.

It has been a matter of debate as to whether HN molecules contain combined or separate active sites for hemagglutinating and neuraminidase activities. Some evidence has been interpreted to indicate a single site that binds sialic acid tightly (hemagglutinating) but hydrolyses the molecule slowly (neuraminidase) (246), whereas studies with temperature-sensitive mutants and monoclonal antibody inhibition of activities have been interpreted to suggest that there are separate sites for the two activities (217,221). If the analogy to influenza virus neuraminidase structure is brought to bear, the influenza A virus N9 subtype neuraminidase also has hemagglutinating activity, and those activities are located at separate sites (164).

The *Morbillivirus* attachment protein (H) which is thought to interact with the measles virus cellular receptor molecule, CD46 (72,193), is also a type II integral membrane protein with some sequence homology and similarity to the HN glycoprotein of other members of the Paramyxovirinae, especially the paramyxoviruses (187). However, as discussed above, the *Morbillivirus* H glycoprotein lacks detectable neuraminidase activity. The measles virus receptor CD46 is a low-abundance cell surface protein and is therefore unlikely to cause virus aggregation during virus budding. Measles virus and morbilliviruses in general may not need a neuraminidase activity in the manner in which paramyxoviruses need a neuraminidase activity to free themselves from the cell surface.

The structure of the *Pneumovirus* attachment protein (G) has some overall structural analogy to the attachment protein of the Paramyxovirinae but also has some major differences. The RS virus G protein has neither hemagglutinating nor neuraminidase activity, and the RS virus cell surface receptor molecule is not known. The nucleotide sequence of the RS virus G gene predicts that the protein is of 298 amino acids (M_r, 32,587) and is a type II integral membrane protein with a single *N*-terminal hydrophobic signal/anchor domain (243,284). The G protein is found in virus-infected cells in both membrane-bound and proteolytically cleaved soluble forms. The distinguishing feature of the RS virus G protein is the extent of its carbohydrate modification. On SDS-PAGE, the protein migrates with an apparent M_r of approximately 84,000 to 90,000, and the dramatic increase in molecular weight of 8 to 12 kd over that predicted for the polypeptide chain is attributable to addition of *N*-linked carbohydrate (four potential addition sites) and 40 to 50 kd is attributable to the addition of *O*-linked glycosylation (77 potential acceptor serine or threonine residues; 30% of total residues) (46,285).

Fusion Protein

Paramyxovirus fusion proteins are synthesized as an inactive precursor (F_0) that is cleaved by a host-cell proteolytic enzyme to release the new *N*-terminus of F_1, thus forming the biologically active protein consisting of the disulfide-linked chains F_1 and F_2 (120,247). Analysis of the nucleotide sequence of the molecularly cloned F protein mRNAs indicates that the encoded proteins contain 540 to 580 residues (19,42,81,128,137,181,204,225,254, 258,265,272,279,283,284) (Fig. 7). Paramyxovirus fusion proteins are type I integral membrane proteins that span the membrane once and contain at their *N*-terminus a cleavable signal sequence that targets the nascent polypeptide chain synthesis to the membrane of the endoplasmic reticulum (ER). At their carboxy-termini, a hydrophobic stop-transfer domain (transmembrane-domain) anchors the protein in the membrane, leaving a short cytoplasmic tail (approximately 20 to 40 residues). Comparison of the amino acid sequences of paramyxovirus F proteins (187) does not show overall major regions of homology, but the similar placement of cysteine, glycine, and proline residues together with the overall hydrophobicity of the F proteins suggests a similar structure for all F proteins. The *Parainfluenzavirus* and *Rubulavirus* F_2 and F_1 subunits are glycosylated. There are a total of three to six potential sites for the addition of *N*-linked carbohydrate, but a determination of which sites are used has not been made. The *Morbillivirus* F_2 subunit contains three potential sites and the F_1 subunit none for *N*-linked glycosylation. For measles virus all three sites in the F_2 subunit are used (3).

Protein sequencing studies of the F protein and nucleotide sequencing studies of the F genes have indicated that the *N*-terminal 20 residues of F_1 (fusion peptide) are extensively hydrophobic. This region of the F protein is highly conserved among *Paramyxovirus* F proteins (up to 90% identity) (Fig. 9). The paramyxovirus fusion peptides are thought to intercalate into target membranes initiating the fusion process, and evidence for direct insertion into bilayers has been obtained using hydrophobic photoaffinity labeling probes (197). It also has been shown that the fusion peptides are sufficiently hydrophobic that they can act as a transmembrane anchor domain to convert a formerly soluble protein to a membrane-bound form (206). The invariant nature of many of the residues of the fusion peptide between paramyxovirus F proteins suggests a role more complex than preservation of hydrophobic residues required for a membrane-intercalating domain; neither signal sequences nor membrane anchorage domains show sequence conservation beyond their hydrophobic nature (280). It has been noted that if the fusion peptide is assumed to be an α-helix, then the invariant residues are located on one face of the helix (251). Interestingly, in a study on the conserved residues of the fusion peptide, when the glycine residues at position 3, 7, or 12 were changed to alanine and the altered F proteins were expressed by using a vector, a dramatic increase in syncytium formation was observed (123). Thus, the invariant amino acids in the fusion peptide may preserve a balance between high fusion activity and successful viral replication because high fusion activity is deleterious to cell viability.

It would be remarkable if the hydrophobic fusion peptides could be freely exposed to an aqueous environment without aggregation occurring, either before or after cleavage of F_0. Thus, the argument is compelling that F proteins undergo a conformational change to release the fusion peptide. *Paramyxovirus* F proteins contain two heptad repeats (Fig. 7). Heptad repeat A is adjacent to the fusion peptide (43), and heptad repeat B is adjacent to the transmembrane domain (26). Heptad repeats can form a triple-stranded coil formed of three α-helices (210). Thus, by analogy to the model for the conformational change in influenza virus HA (27,32), the *Paramyxovirus* F protein heptad repeat A

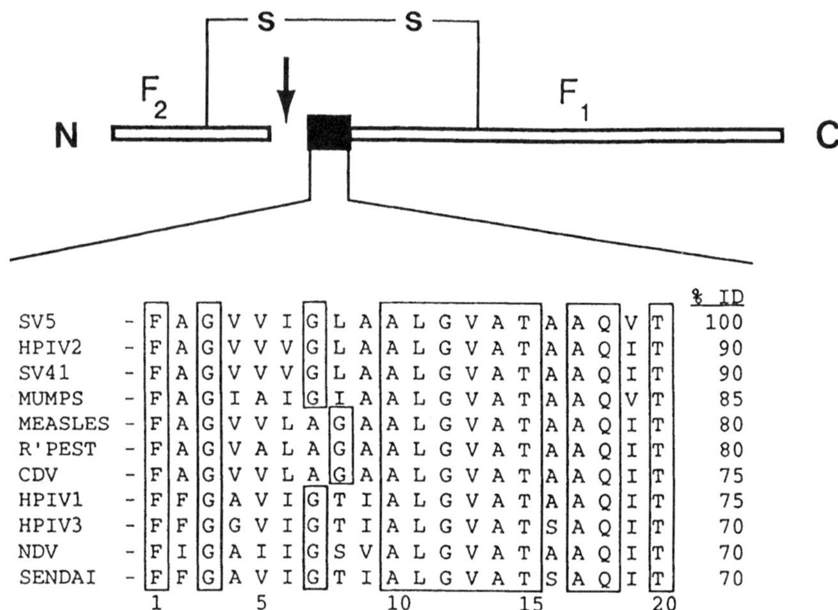

FIG. 9. Schematic diagram of the SV5 F protein fusion peptide domain and its amino acid sequence conservation between paramyxoviruses. The open box represents the disulfide-linked F protein subunits F_1 and F_2; the arrow represents the cleavage site. The shaded box represents the F_1 NH_2-terminal fusion peptide. The expanded section shows a comparison of the amino acid sequence of SV5 and ten other paramyxovirus fusion proteins, with the boxed region indicating amino acid identity. %ID, percentage identity to the SV5 sequence in the NH_2-terminal 20 amino acids. Adapted from Horvath and Lamb (123), with permission.

may change from one conformation in a presumptive metastable native form of F to a coil in a presumably more stable fusogenic form. It is presumed that the fusion peptide is relocated from a shielded position in the protein structure toward the target membrane. Heptad repeat B may also be involved in directly promoting fusion because mutations in this domain of measles virus F protein can abrogate fusion activity without changing the oligomeric form or level of surface expression of the protein (25).

In making predictions for the presumptive conformational change in the F protein leading to release of the fusion peptide, the oligomeric form of the F protein has to be considered. Earlier data obtained from cross-linking the Sendai virus F protein (249) were interpreted to indicate that the *Paramyxovirus* F protein is a tetramer. However, a more recent analysis using cross-linkers of SV5, NDV, and hPIV-3 F proteins indicate a major cross-linked oligomeric species with an M_r of approximately 195,000, a minor cross-linked species with an M_r of approximately 130,000, and uncross-linked monomers with an M_r of approximately 65,000. The largest F cross-linked species migrated on gels faster than did trimeric influenza virus hemagglutinin. These data are consistent with those that would be expected if paramyxovirus F proteins were homotrimers (234).

The *Paramyxovirus* F protein is the major viral glycoprotein involved in virus-cell and cell-cell fusion. When the SV5 and measles virus F cDNAs are expressed in mammalian cells, syncytium formation is observed (2,3,124,205, 208), although many more cells expressed the F protein than could be found in multinucleated cells (205). However, for many other paramyxoviruses, including Newcastle disease virus, hPIV-3, bPIV-3, mumps virus, and CDV (39,74,124,127,186,238,268,286), co-expression of F and HN (H for the morbilliviruses) is required for syncytium formation, and expression in the same cells of homotypic F and HN and not a heterotypic HN molecule is required (124,127,250). Thus, it has been suggested that a type-specific interaction occurs between the HN and F protein that is necessary for fusion to occur (127,250). An important question that remains to be answered is the mechanism of the trigger for the proposed conformational change in F (see forthcoming section on virus adsorption and entry).

Cleavage Activation

The precursor F_0 molecule is biologically inactive, and cleavage of F_0 to the disulfide-linked subunits F_1 and F_2 activates the protein, rendering the molecule fusion active and permitting viral infectivity. Thus, cleavage of F_0 is a key determinant for infectivity and pathogenicity. Proteolytic activation of F_0 involves the sequential action of two enzymes: (a) the host protease that cleaves at the carboxyl side of an arginine residue and (b) a carboxypeptidase. The Paramyxoviridae can be divided into two groups: those that have F proteins with multibasic residues at the cleavage site and those with F proteins that have a single basic residue at the cleavage site (Table 3). Cleavage of F proteins containing multibasic residues at the cleavage site occurs intracellularly during transport of the protein through the *trans* Golgi network.

Furin is a cellular protease localized to the *trans* Golgi network, its sequence specificity for cleavage is R-X-K/R-R (10,125), and the available evidence suggests that furin, a subtilisinlike endoprotease, is the (or one of the) protease(s) that cleaves F proteins intracellularly (143,200). Paramyxoviruses that have F proteins with single basic residues in the cleavage site (e.g., Sendai virus) are not usually cleaved when grown in tissue culture. Thus, only a single cycle of growth is obtained. However, the F_0 precursor that is expressed at the cell surface and incorporated into released virions can be cleavage-activated by the addition of exogenous protease (247), and cleavage leads to multiple rounds of replication. Purification of a protease from the allantoic fluid of embryonated chicken eggs has indicated that the endoprotease responsible for Sendai virus activation is homologous to the blood clotting factor X, a member of the prothrombin family (100,101). A protease with a similar substrate specificity is secreted from Clara cells of the bronchial epithelium in rats and mice, and this enzyme is probably responsible for activating paramyxoviruses in the respiratory tract. For Newcastle disease virus, the nature of the cleavage site correlates with virulence of the virus. Those strains with multibasic residues in the F_0 cleavage site are virulent strains for their hosts, whereas those strains with F_0 molecules that have single basic residues are avirulent (45,190).

Other Envelope Proteins

The rubulaviruses SV5 and mumps virus both contain a small gene located between F and HN designated SH. In SV5-infected cells, a monocistronic mRNA transcript has

TABLE 3. *Amino acid sequences upstream of the F protein cleavage site of some members of the family* Paramyxoviridae

Sendai virus	G - V - P - Q - S - R ↓
Human PIV1	D - N - P - Q - S - R ↓
Human PIV3	D - P - **R** - T - **K** - **R** ↓
SV5	T - R - **R** - R - **R** - **R** ↓
Mumps	S - R - **R** - H - **K** - **R** ↓
NDV (virulent strain)	G - R - **R** - Q - **R/K** - **R** ↓
NDV (avirulent strain)	G/E - G - K/R - Q - G/S - R ↓
Measles	S - R - **R** - H - **K** - **R** ↓
RS virus	K - K - **R** - K - **R** - **R** ↓

Consensus sequence for furin protease cleavage is **R** - X - **R/K** - **R** ↓

Data from Hosaka et al. (125).

been observed that is translated to yield a 44-residue protein, SH (114,115). SH is an integral membrane protein that is expressed at the plasma membrane and has the orientation of a type II integral membrane protein (115) (Fig. 8). The functional role of SH in the replicative cycle of SV5 is unknown. For mumps virus, although an mRNA transcript derived from the SH gene has been detected (80,82), attempts to detect a translated protein have as yet been unsuccessful, but the putative mumps virus SH protein does contain a hydrophobic domain sufficient to span a lipid bilayer.

The *Pneumovirus* RS virus encodes a small glycosylated integral membrane protein designated SH (or 1A) (50,52,199). The SH protein contains 64 amino acids and has a single internal hydrophobic domain and sites for the potential addition of carbohydrate at the *N*-terminal region and the *C*-terminal region. By using site-specific antibodies raised to synthetic peptides predicted by the SH gene sequence and immunofluorescent staining techniques, the available data suggest that SH is expressed at the plasma membrane of RS virus-infected cells and that the *N*-terminal domain is cytoplasmic and the *C*-terminal domain extracellular. This membrane orientation of SH as a type II integral membrane protein implies that the *C*-terminal glycosylation site is used. In RS virus–infected cells, four SH-related polypeptide species have been identified: M_r 4,800, M_r 7,500, M_r 13,000 to 15,000, and M_r 21,000 to 30,000. The M_r 4,800 species is thought to result from the initiation of protein synthesis at an internal AUG codon; the M_r 7,500 species is unglycosylated SH; the M_r 13,000 to 15,000 species is SH containing one high mannose *N*-linked carbohydrate chain; and the M_r 21,000 to 30,000 species is generated by the addition of polylactosaminoglycan to the *N*-linked carbohydrate chain (5,46,199). The role of the SH protein in the RS virus life cycle is not known.

STAGES OF REPLICATION

General Aspects

As far as is known, all aspects of the replication of Paramyxoviridae take place in the cytoplasm. An overview of the life cycle of the virus is shown schematically in Fig. 10, and a schematic diagram indicating the differences between transcription and replication is shown in Fig. 11. Unlike the situation for influenza viruses, Paramyxoviridae mRNA synthesis is insensitive to DNA-intercalating drugs such as actinomycin D (44), and the Paramyxoviridae can replicate in enucleated cells (213). In cell culture, single cycle growth curves are generally of 14 to 20 hr duration, but can be as short as 10 hr for virulent strains of NDV. The effect of viral replication on host macromolecular synthesis is quite variable, ranging from almost complete shutoff late in infection for NDV and Sendai virus to no obvious effect with SV5.

Virus Adsorption and Entry

For the parainfluenzaviruses and rubulaviruses it has long been accepted that molecules containing sialic acid (sialoglycoconjugates) serve as cell surface receptors. This is based on the fact that sialidase of *Vibrio cholerae* acted as a receptor-destroying enzyme and protected the host cell from infection (172). Sialic acid, the acyl derivative of neuraminic acid, is found on both glycoproteins and on lipids (sialoglycolipids or gangliosides). For Sendai virus it has been shown that gangliosides function as both the attachment factor and the receptor for the virus (171,173,174). As described above, the cellular receptor for the *Morbillivirus* measles virus is the cell surface protein CD46, and the cellular receptor for pneumoviruses is not known. On adsorption of the virus to the cellular receptor, the viral membrane fuses with the cellular plasma membrane at the neutral pH found at the cell surface, the consequence of which is the release into the cytoplasm of the helical nucleocapsids. As discussed above, the attachment protein of many Paramyxovirinae (HN or H) has fusion promoting activity. A model that would rationalize the involvement of HN (H) in the fusion process is one in which the hypothesized conformational change in F to release the fusion peptide is highly regulated. For those Paramyxovirinae that require HN (H) and F to be coexpressed to observe fusion, the first step would be the binding of HN (H) to its receptor. On binding ligand, the HN (H) protein would undergo its own conformational change, which in turn could trigger a conformational change in F to release the fusion peptide. In this way, F and HN operate as a coupled molecular scaffold to release and direct the fusion peptide to the target membrane (157). For viruses such as SV5, for which coexpression of HN only weakly influences cell-cell fusion, the presumptive F conformational change could either be hair-triggered by contact of F with a target membrane or is triggered after docking of F with an unrecognized receptor located on the target membrane (157).

In the virus particle, the M protein is thought to make numerous contacts with the nucleocapsid, and these M-NP contacts are thought to be responsible for inhibiting transcription during virus assembly. On fusion of the viral envelope with the cell plasma membrane and release of the nucleocapsid into the cytoplasm, a mechanism needs to exist to disrupt the M-NP contacts. With influenza A virus, the factor that alters the equilibrium between self-assembly and disassembly is thought to be the difference in pH between the acidic uncoating compartment (endosomes) and the assembly site (plasma membrane). The driving force for paramyxovirus uncoating is not known.

Primary Transcription

The early evidence indicating that the mRNA of paramyxoviruses was complementary to virion RNA, the

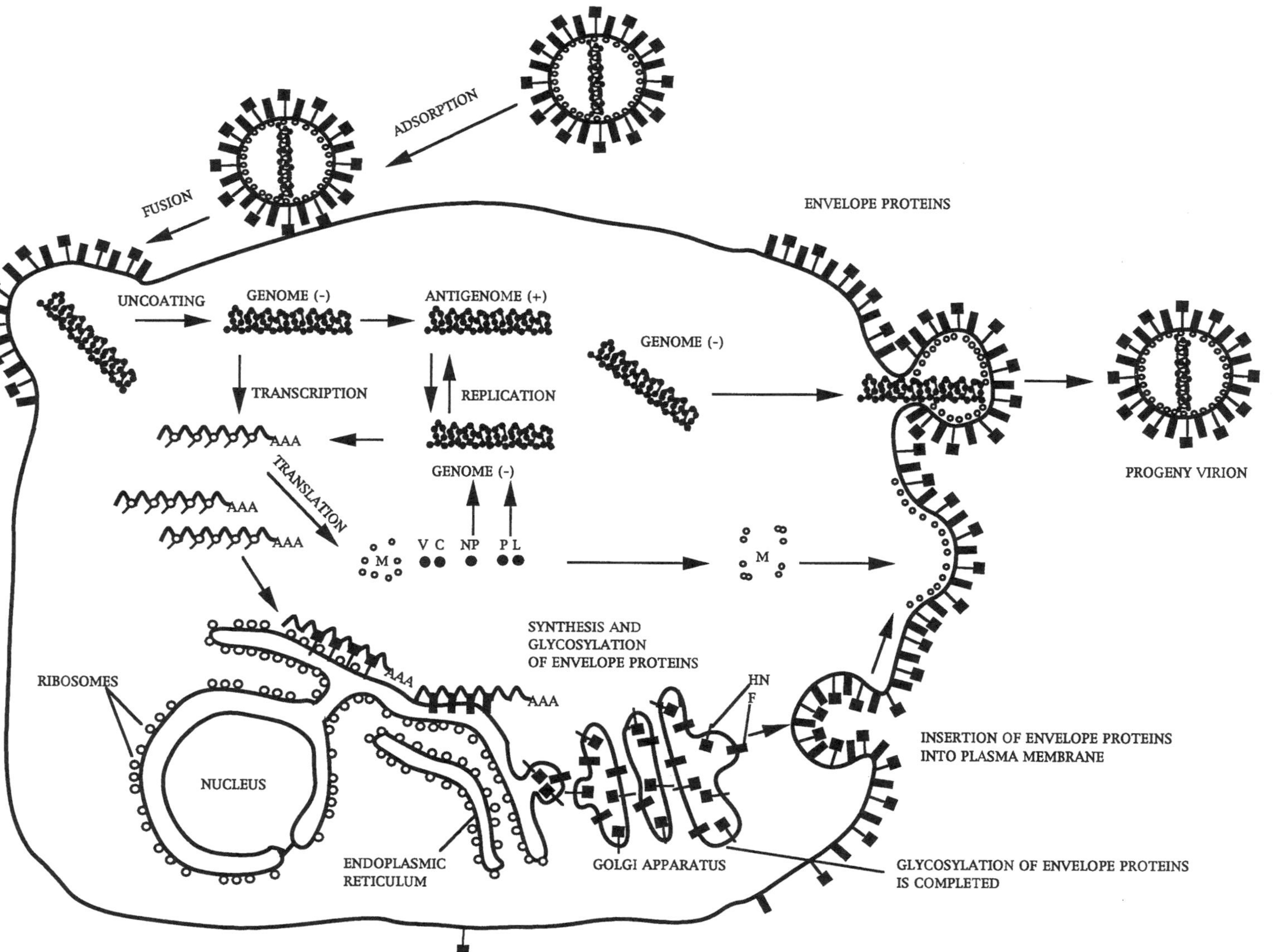

FIG. 10. Schematic represention of the life cycle of a paramyxovirus.

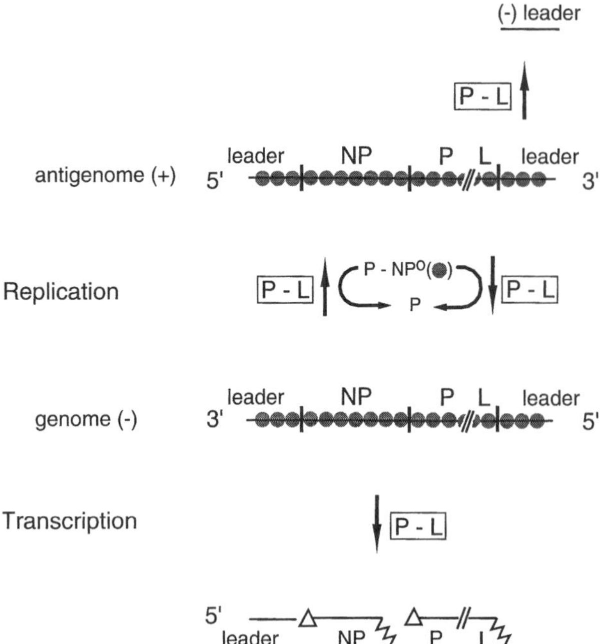

FIG. 11. Paramyxovirus RNA synthesis. Viral nucleocapsids, the templates for RNA synthesis, are shown as a linear array of NP subunits (circles), with short verticle lines indicating the gene junctions. The viral polymerase (P-L) transcribes the genome template, starting at its 3′ end, to generate the plus leader RNA and the successive capped (triangle) and polyadenylated (squiggly line) mRNAs, by stopping and restarting at each junction. Once these primary transcripts have generated sufficient viral proteins, unassembled NP (as a P-NP° complex) begins to assemble the nascent leader chain, and the coordinate assembly and synthesis of the RNA causes the polymerase to ignore the junctions, yielding the antigenome nucleocapsid. In this capacity, P acts as a chaperone to deliver NP° to the nascent RNA. The P-L polymerase can also initiate RNA synthesis at the 3′ end of the antigenome in the absence of sufficient P-NP°; however, only a minus leader RNA is made in this case. Note that genomic and antigenomic RNAs never appear as naked RNAs.

lack of infectivity of naked viral RNA, and the inability to detect RNA-dependent RNA polymerase activity in uninfected cells suggested that a transcriptase might exist in the virions of paramyxoviruses. The first negative strand virus virion-associated polymerase was identified in the Rhabdovirus VSV (7) and shortly thereafter was indentified in Newcastle disease virus (129) and Sendai virus (228,262). After fusion of the viral envelope with the host cell plasma membrane has occurred, the infecting nucleocapsids enter the cytoplasm carrying multiple copies of the P/L polymerase. Polymerases are thought to enter the (–) genome template only at its 3′ end, but the virion polymerases are scattered over the length of the template (219,220), perhaps frozen in place from when they were shut down for packaging into virions. RNA synthesis is expected to begin as soon as the genome encounters the ribonucleoside triphosphates in the cytoplasm, with the synthesis of the (+) leader RNA. The leader sequences are thought to play a critical role in the control of viral gene expression. The (+) leader sequence separates the 3′ end of the genome from the beginning of the first gene, whereas the (–) leader or trailer sequence separates the 3′ end of the antigenome from the end of the last gene. These are short sequences, approximately 50 nucleotides long, the first 12 nucleotides of which are identical for the viruses in each genera (and well conserved across the entire subfamily Paramyxovirinae); the remainder of the leader sequence is mostly A and U rich.

In contrast to cellular transcription, in which there are a vast number of genes and the critical choice is often whether to initiate at a given promoter, the promoters of the Paramyxoviridae (or polymerase entry sites) are always "on," and the viral polymerases are restricted to these templates. During primary transcription, the polymerase terminates at the end of the (+) leader region and reinitiates at the beginning of the first gene. Reinitiation appears to be a critical event because it defines a polymerase that transcribes the template to make mRNAs, as opposed to one that replicates the genome. Reinitiation leads to a transcript whose 5′ end is capped and whose internal sequences, at specific sites and at specific frequencies, can be cotranscriptionally edited by G insertions. The transcript is also polyadenylated at its 3′ end. When the mRNAs of influenza viruses, rhabdoviruses, and paramyxoviruses are aligned on their (–)genome templates, the beginning of the poly(A) tails in all cases line up with a stretch of four to seven U residues on the template. Because polyadenylation of all these viral mRNAs also occurs in polymerase reactions with purified virions, it was suggested that the tails were created by the reiterative copying of the oligo(U) stretch by the polymerase (96,102,104, 133,179,226,248). For the *Rhabdovirus* VSV, the polymerase spends as much time crossing the gene junctions as in crossing the much larger coding sequences (133), and virus mutants have been found in which abnormally long poly(A) tails are made (131). Both these findings have reinforced the view that poly(A) addition occurs by polymerase stuttering. More recently, the finding that P gene mRNAs are cotranscriptionally edited by G insertion has suggested that a more controlled form of reiterative copying (or pseudotemplated transcription) also occurs for this process (202,270,276).

The length of the poly(A) tail (approximately 200 residues) is thought to be limited by chain termination. To reinitiate mRNA synthesis at the next start sequence, the polymerase would then skip the intergenic region (which can be as short as a single base or as long as 56 nucleotides for the *Rubulavirus* and pneumoviruses, but is precisely three nucleotides for the *Paramyxovirus* and morbilliviruses). The polymerase would continue transcription in this stop-start fashion until it completed the L mRNA. Thus, except for the short intergenic regions and the trailing (–) leader region, the entire genome is transcribed into mRNA;

for Sendai virus, only 70 of the 15,384 nucleotides are not present in mRNA.

The frequency with which the polymerase restarts mRNA synthesis at each junction is high, but not perfect. Thus, there is always less mRNA made from downstream genes relative to their upstream neighbors, and there is a gradient of mRNA abundance according to the position of the gene relative to the 3′ end of the template. This gradient of mRNA abundance is not due to different half-lives of the mRNAs. For measles virus, this gradient of mRNA abundance is relatively smooth, indicating that the polymerase has an equal probability of not reinitiating at each junction, whereas for Sendai virus, there are larger disparities at the M-F and HN-L boundaries (119). This is then one way that each virus can fine tune the relative amounts of each gene product. For measles virus, moreover, infection of human brains and cell lines of neural origin, in which the virus grows more slowly and leads to persistent infections, gives much steeper gradients than do infections in cell lines (e.g., HeLa) in which the virus is lytic (37,38). There may then be cellular factors that can affect the rate of mRNA reinitiation at the junctions. However, whether these factors represent specific cellular proteins or nonspecific effects such as the cytoplasmic concentration of adenosine triphosphate is unknown.

Two other mechanisms of attenuation of gene expression at the transcriptional level have been noted, both of which expand the ways that these viruses can fine tune the relative amounts of their mRNAs. The first mechanism involves inefficient transcriptional termination and polyadenylation. Specifically, some viruses read through some junctions at a higher frequency than usual and produce bicistronic mRNAs. Only the most upstream of the ORFs is normally translated, and so the downstream ORFs on these mRNAs is not expressed. However, the polymerase is still free to reinitiate downstream, so in this way genes in the middle of the map can also be selectively downregulated. Measles virus, for example, uses this mechanism to downregulate M gene expression during long-term persistent infections of the human brain, known as SSPE (37,117).

The second method involves genes encoding overlapping mRNA. For the *Pneumovirus* RS virus, the start sequence of its L gene is situated upstream within the preceding M2 gene (51). This overlap places the end sequence of the M2 gene 56 to 68 nucleotides into the L gene (Fig. 3). Because of this unusual arrangement, most transcripts that initiate at the L start site terminate shortly thereafter at the M2 end site and are apparently by-products of unknown and limited function. Only a minority of transcripts read through the M2 end sequence and continue onto the true end of the L gene (51). For RS virus, transcriptional read-through then becomes the obligatory mechanism by which the L mRNA is made. Curiously, this superattenuation mechanism by transcriptional overlap of the M2 and L genes has not been described for the *Pneumovirus* pneumonia virus of mice.

An additional mechanism of attenuation of gene expression is known as biased hypermutation. In some cases of SSPE, the M gene is inactivated by a large number of mutations, most of which are U-to-C changes in the mRNA strand. Because the limits of these mutations were precisely the M gene transcriptional boundaries, and a cellular activity that converts adenosines to inosines in double-stranded RNA has recently been described (12), this has led to a model of how these hyperbiased mutations occur (13,35,161). In terms of gene expression, this is another way that measles virus can specifically downregulate the level of M protein during a persistent infection. Hyperbiased mutations are also known to occur in the RSV G gene during infection of cultured cells (233).

The Extraordinary Diversity of P Gene Expression

Except for the P gene, all the viral genes are monocistronic, using a single start codon and producing a single primary translation product. However, the P gene can use multiple start codons and produce proteins from all three ORFs.

Diversity Due to Translational Choice

Parainfluenzavirus and *Morbillivirus* P genes differ from those of rubulaviruses in that the former contain the C protein ORF, which overlaps the amino-terminal sequences of the P ORF in the +1 frame. The existence of mRNAs coding for two separate proteins led to one of the early group of examples that required modification of the scanning model for ribosomal initiation and introduced an element of choice about start site use, namely, that scanning ribosomes can occasionally bypass 5′ proximal AUGs in an unfavorable context (155) and start on downstream AUGs. However, the C proteins themselves were found as a pair of polypeptides (C and C′), and these were also subsequently shown to be the result of ribosomal choice. Both Sendai virus and hPIV-1 are known to express C′ from a 5′ proximal non-AUG start codon (ACG or GUG) (23,60, 66,103). For Sendai virus, moreover, two more C proteins were found (called Y_1 and Y_2) that were similarly shown to initiate on other downstream AUGs (62,63), and the initiation of a *C*-terminal fragment of the P protein (called X) was reported to initiate from an AUG more than 1,500 nucleotides from the 5′ end (61). Sendai virus, which has been the most intensively studied, may use as many as six independent start sites for protein synthesis: two for the P ORF and four for the C ORF. Whether the simple modified scanning hypothesis can accommodate all these different ribosomal choices remains unclear (60–62).

Diversity Due to Transcriptional Choice

The P genes have been considered unusual for some time. In every case multiple proteins were found to be expressed from this gene, and this was also true for the rubulaviruses, which do not contain the overlapping C ORF. For example, the mumps virus P gene also expresses the nonstructural proteins NS1 and NS2 (since renamed V and I for consistency with SV5) (109,207). However, until the correct sequence of the first P gene in this group was determined, it was unclear whether ribosomal choice alone produced these multiple proteins.

The first *Rubulavirus* P gene determined correctly was that of SV5 (270) (Fig. 5) and yielded the unexpected result that there was no single ORF large enough to code for the P protein. Rather, the gene contained two separate ORFs at each end, which briefly overlapped in the middle. mRNA transcribed from these cDNA *in vitro* could be translated into the shorter V protein encoded by the 5′ proximal ORF but not into the longer P protein. The solution to this dilemma became clear when it was found that there was a second group of mRNAs that were identical to the first except that they had two G residues inserted within a short run of Gs in a region where the two ORFs overlapped (arrowhead in Fig. 5). This fused the separate ORFs at each end into a single continuous unit, and mRNAs prepared from these latter clones translated into P (but not V) proteins *in vitro.*

A second unexpected finding emerged from this study. Whereas the P protein sequences are among the least conserved of the viral proteins between different paramyxoviruses, that region of V that is not shared with P (i.e., the *C*-terminal region) is cysteine rich and among the best conserved of all the viral sequences. This cysteine-rich V domain was in fact present in all the P genes of the paramyxoviruses and morbilliviruses previously sequenced but had not been recognized because it was relatively short (approximately 70 amino acids), in the middle of the genes, and generally lacked a start codon. Thomas et al. (209,270) predicted that these other viruses would also express V proteins by a similar mechanism, except that for the *Parainfluenza* and morbilliviruses 1G would need to be inserted to switch from the P to the V ORF.

This prediction was first verified for measles virus, when it was found that some P gene mRNAs generated from cDNA clones could be translated to yield the normal sized P protein, whereas other mRNAs could only be translated to synthesize proteins of smaller molecular weight, designated V (36). When the various cDNAs were sequenced, the only difference between them was that those that made the shorter V protein contained a single extra G within a short run of Gs, which was also absent in the genome (36). The generality of the RNA editing process of G insertions in P gene mRNAs was shown for many, but not all, members of the Paramyxoviridae family; e.g., Sendai virus (275), mumps virus (83,207), hPIV-2 (198,256), hPIV-4 (153), bPIV-3 (212), hPIV-3 (91), LPMV (16), and NDV (260).

The Mechanism of the G Insertions

The formation of the poly(A) tail on paramyxovirus mRNAs is thought to take place by the viral polymerase reiteratively copying the U_{4-7} stretch within the termination sequence at the end of each gene. Because the G insertions appear to take place only in the P gene mRNAs and not in the P genes (i.e., during transcription but not during genome replication), and because there was also some homology between the sequences that precedes the U_{4-7} stretch and those of the G insertion site, it was suggested that the G insertions arose by reiterative copying of the short C stretch on the genome template at the insertion site (36,209,270).

Support for the notion that G insertion was due to reiterative copying was obtained with Sendai virus (275,276). All paramyxovirions contain a template-bound polymerase, and purified virions can make mRNA *in vitro*; however, only for Sendai virus had the conditions for efficient mRNA synthesis been well characterized. When P gene mRNAs made with purified Sendai virions were examined and compared with those made *in vivo,* they were found to be “edited” very similarly. This suggested that the insertions were due to viral proteins. When the P gene mRNAs were expressed from recombinant vaccinia viruses in cells coinfected with Sendai virus, those made from the Sendai virus genome were modified as before, whereas those made from the vaccinia genome were not modified. The inability of the insertion activity to act in *trans* supports the idea that the insertions occur during mRNA synthesis, rather than on preformed mRNA. The G insertions also do not occur when the mRNA is made *in vivo* from plasmid DNA via bacteriophage T7 polymerase or from an RNA vector that uses the Sindbis virus polymerase (276). All these data are consistent with a cotranscriptional mechanism for G insertion. A “stuttering” model, in which these events are largely controlled by the genome sequence, has been proposed (150,212,276) based on the available data. The model proposes that the polymerase pauses at or near the insertion site because of an intrinsic property of the template, and then when the pause is sufficiently long, the slippage of the nascent mRNA occurs by one nucleotide (e.g., Sendai virus and measles virus) or two nucleotides (SV5 and mumps).

The mechanism of RNA editing has been further examined by using an artificial minigenome of Sendai virus that can be expressed, replicated, and packaged by wild-type Sendai virus (201). Park and Krystal (202) developed a means of examining the minimum size of the P gene sequences surrounding the editing site needed for the process to occur. The RNA minigenome consisted of the antisense chloramphenicol acetyltransferase (CAT) coding region flanked by 145 and 119 5′ and 3′-terminal sequences of the Sendai virus genome. An analogous strategy also has been developed for examining respiratory syncytial virus replication (48,49). A synthetic Sendai virus 42-base RNA molecule derived from the RNA editing site of the P/C/V gene was inserted in the middle of the CAT gene, and it was found to be sufficient to allow the Sendai virus polymerase to transcribe mRNAs with G-nucleotide insertions, albeit at a reduced efficiency. By progressively deleting from the 3′ end it was found that a 24-nucleotide sequence spanning the G-insertion site was sufficient for editing to occur (202).

A remarkable aspect of paramyxovirus cotranscriptional editing is that in every case the G insertion frequencies are matched to the ORF organizations of each gene (Fig. 5). Thus, for Sendai virus and the *Morbillivirus,* a 1G insertion is the predominant event when insertions occur; for the rubulaviruses, a 2G insertion is the predominant event; and for bPIV3, one to six Gs are added at equal frequencies, presumably to accommodate the two overlapping ORFs contained in this gene.

The various products of the different P genes are listed in Fig. 5 and Table 4. Except for the highly phosphorylated P protein, which is essential for RNA synthesis and is also the longest polypeptide made from this gene, none of the others are consistently expressed from this gene. In terms of mRNA editing and P gene expression, however, there a pattern emerges according to virus group.

For parainfluenzaviruses and morbilliviruses, the P ORF is found in one continuous stretch with the C ORF overlapping the *N*-terminal P sequences. The V ORF is found near the middle of the gene, and V is made from an mRNA with a 1G insertion. About half the mRNAs have insertions, and among these, the insertion of 1G predominates. However, mRNA with two or more Gs are also found. The 2G insertion generally closes the reading frame shortly thereafter and produces a protein representing the *N*-terminal half of P alone, referred to as W.

This pattern is invariant for the morbilliviruses, but there are some interesting exceptions with the parainfluenzaviruses. For bPIV-3 and hPIV-3, there are two ORFs overlapping that of P, the highly conserved cysteine-rich V ORF, and one called D. For both hPIV-3 and bPIV-3, the 2G-inserted mRNA expresses the D protein, which is a fusion protein that is analagous to V except that it involves an alternate *C*-terminal protein sequence. Because bPIV-3 also expresses a V protein from a 1G-inserted mRNA, this virus has the unusual distinction, along with HIV and polyoma virus, of expressing protein from the same nucleotide sequence in all three frames. In the hPIV-3 P gene, on the other hand, although the cysteine-rich V ORF has been conserved, there are four stop codons between the insertion site and the cysteine-rich ORF, and so the 1G-inserted mRNA codes for a W rather than a V protein. It is then unclear whether hPIV-3 expresses a V protein at all.

TABLE 4. *Expressible P gene ORFs*

Paramyxovirus	Host	Ribosomal choice P	C′	C	mRNA editing D	V
Sendai	Mouse	+	+	+	−	+
hPIV1	Human	+	+	+	−	−
bPIV3	Cattle	+	−	+	+	+
hPIV3	Human	+	−	+	+	(+)
Mumps	Human	+	−	−	−	+
Simian virus 5	Dog	+	−	−	−	+
hPIV2 and hPIV4	Human	+	−	−	−	+
NDV	Fowl	+	−	−	−	+
Measles	Human	+	−	+	−	+
Canine distemper	Dog	+	−	+	−	+
Rinderpest	Cattle	+	−	+	−	+
Phocine distemper	Seals	+	−	+	−	+

(+), not definitely proven

Like hPIV-3 and bPIV-3, Sendai virus (also referred to as murine PIV-1) and hPIV-1 are a very closely related pair of viruses. Nevertheless, hPIV-1 does not possess the consensus sequence for G insertions, and the V ORF is interrupted by no less than nine stop codons; therefore, hPIV-1 cannot express a V protein.

For rubulaviruses, there is no overlapping C ORF, and the P ORF is not found as a continuous stretch in the genome. Rather, the genome RNA sequence is equivalent to the complement of the 1G-inserted Sendai virus/measles virus mRNA, with the cysteine-rich ORF already fused in place. In the rubulaviruses, the downstream P ORF sequences are re-fused to the *N*-terminal portion by the insertion of two additional G residues to restore the intact P ORF. The insertion of a single G is a rare (or nonexistent) event in this virus group. However, the insertion of 4Gs has been found, and this, similar to the W situation above, leads to a third protein called I.

The only exception to this rule among the rubulaviruses is NDV, the only nonmammalian paramyxovirus. For NDV, the P gene, although relatively short, is organized and expressed similarly to the parainfluenzaviruses and morbilliviruses (260), except that it does not contain the overlapping C ORF.

The Function of the Various P Gene Products

Although it has so far proved impossible to generate infectious cDNA clones of paramyxoviruses, it has been possible to recreate certain aspects of the virus life cycle, both inside cells and *in vitro,* by using cDNAs that express DI RNAs and cDNAs that express the viral proteins. Using these systems, the NP, P, and L genes were all found to be required for genome replication as expected. Of the various Sendai virus P gene products, only the P protein itself was found to be essential for RNA synthesis. The C, V, and W proteins, in contrast, inhibited RNA synthesis when present and in a dose-dependent fashion (57). When these inhibitory effects were examined more closely, it was found that the C protein primarily inhibited mRNA synthesis and had little effect on genome replication. The V and W proteins, in contrast, primarily inhibited genome replication and had little or no effect on mRNA synthesis. It would appear that the Sendai virus P gene codes not only for an essential component of the polymerase complex but also for two groups of proteins that can modify its action, and in opposite directions. If these nonstructural proteins also interact with cellular components (and this could be the reason for the high conservation of the V amino acid sequence with the Paramyxovirinae when all the other ORFs of this gene are not highly conserved), they may be useful to the virus in sensing the metabolic state of the host cell and allow it to respond by either speeding up or slowing down virus replication according to what the cellular traffic can bear. If so, these other P gene products would fall in the category of "supplemental" genes, and this would also help explain why some paramyxoviruses, even those that are so closely related, do not express one or the other of the proteins.

Genome Replication

The Switch from Transcription to Replication

The (–) genome replicates via a full-length complementary copy, called the (+) antigenome, which like the (–) genome is found only in an assembled form. A schematic diagram showing the essential differences between transcription and replication is shown in Fig. 11. Compared with Rhabdoviridae, there are considerable amounts of antigenome nucleocapsids in paramyxovirus-infected cells (10% to 40% for Sendai virus). Antigenomes can also represent 5% to 20% of the genome-sized RNAs in virus particles because discrimination against antigenome nucleocapsids is poor during virus assembly (148,149,227). However, they contain no ORFs of any note, and no mRNAs are known to be transcribed from them. Their sole function is thought to be as an intermediate in genome replication.

After translation of the primary transcripts and accumulation of the viral proteins, antigenome synthesis begins. Here, ostensibly the same polymerase that until then was engaged in mRNA synthesis copies the same template but now ignores all the junctional stop signals (and editing sites) and synthesizes an exact, complementary copy. Transcription and replication were first distinguished by finding that when infected cells are treated with drugs that inhibit protein synthesis, mRNA synthesis continued normally, but genome synthesis was lost quickly (229). Because genome synthesis and encapsidation appear to occur concomitantly, this requirement for ongoing protein synthesis for genome replication has been associated with the continued supply of unassembled NP (NP^0) necessary for genome encapsidation. The coupling of genome assembly and synthesis is thought to somehow cause the polymerase to ignore the junctional and editing signals (151).

This coupling of genome assembly and synthesis also leads to a self-regulatory system for controlling the relative levels of transcription and replication. For VSV, the site for the initiation of antigenome encapsidation has been mapped to the first 14 nucleotides of the leader sequence; for paramyxoviruses, a similar situation appears to apply. In measles virus–infected cells, a rare polyadenylated transcript representing the leader and NP gene sequences fused together is found, but only as a nucleocapsid, whereas the NP mRNA is found only in a nonassembled form (34). Because the (+) leader sequences contain the encapsidation site, the leader must be separated from the body of the first mRNA (by termination and reinitiation at the first junction) to prevent the first mRNA from ending up in an assembled and untranslatable form. The proposed self-regulatory mechanism would then be quite straightforward; when NP^0 is limiting, polymerases would preferentially be engaged in mRNA synthesis, raising the intracellular levels of all the viral proteins, including NP^0. When NP^0 levels are sufficient, the polymerases would be switched to replication, thereby lowering the levels of NP^0 because each initiation of encapsidation would commit approximately 2,600 NP monomers to finish the assembled chain (147).

The rare measles virus leader-NP mRNA (34) described above was presumably made by a polymerase that began the chain as an antigenome but nevertheless polyadenylated/terminated the chain at the second junction. The assembly and synthesis steps can then sometimes become uncoupled, possibly because assembly of the nascent chain does not keep pace with polymerase movement as it approaches a junction. Reading through the first junction without continuing on to finish the antigenome chain is usually a rare event, but in the pol^R mutants of VSV (215) and the Z strain of Sendai virus (278), the polymerase reads through the leader-NP junction at a frequency of 20% or more without encapsidating the nascent RNA. However, unlike the rare measles virus transcript, these transcripts terminate heterogenously well before the second junction (278). The inability of these polymerases to reach the second junction was unexpected and suggests that if the polymerase has crossed the first junction without the nascent chain being concurrently assembled, it is in a relatively nonprocessive mode; hence, it terminates (or stalls) in a heterogenous fashion shortly thereafter. In this view, the polymerase would always initiate in this nonprocessive mode but would continue to the ends of the template as long as the nascent chain continued to be assembled. For mRNA synthesis, however, the polymerase would presumably be converted to a mode where its processivity was independent of concurrent assembly of the nascent chain, due to its reinitiation at the first junction. In the transcriptive mode, the polymerase would further cap the 5′ ends and stutter on the oligo (U) runs at the junctions to form the poly(A) tails of the mRNAs, as well as edit the P gene mRNAs. In this model, the leader/NP junction becomes a critical check-point because before the polymerase arrives at this site there is no way of distinguishing whether it is engaged in making mRNAs or replicating the genome.

Genome synthesis from antigenome templates is thought to take place in a similar fashion to that for antigenome synthesis in that the promoter at the 3′ end of the antigenome is also always "on," and (–) leader RNAs are made from this region independent of ongoing protein synthesis. However, in contrast to synthesis from (–) genome templates, there are no reinitiation sites on the antigenome template, and so termination of the (–) leader RNAs serves only to recycle the polymerase. Under conditions of sufficient intracellular concentrations of NP^0, encapsidation of the nascent (–) leader chain would begin before the polymerase has reached the L gene, and this would both prevent termination and lead to the synthesis on an encapsidated (–) genome.

The Rule of Six

Efficient replication of genome RNAs requires that their total length be a multiple of six (Fig. 12). This unexpected rule was found when it became possible to perform genome replication starting with cDNA copies of natural Sendai virus DI RNAs that expressed genome RNAs containing the exact 5′ and 3′ end (29,30). Using a variety of restriction sites, Calain and Roux (31) found that the addition (or deletion) of a few nucleotides at any site destroyed efficient replication, which could be restored by further addition or insertion, but only if this restored the total length to a multiple of six. Inefficient replication in the absence of the rule of six was not due to the lack of encapsidation of the cDNA-expressed DI genome, suggesting that it was the inability of the polymerase to initiate at the 3′ end of these nucleocapsids that was at fault.

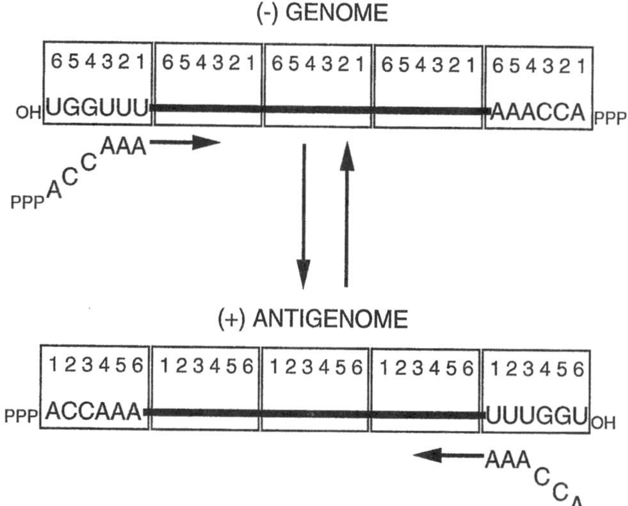

FIG. 12. A schematic representation of the rule of six for replicating the Sendai virus genome (31).

The rule of six is probably related to the fact that each NP monomer is associated with precisely six nucleotides (75). Nucleocapsid assembly presumably begins with the first nucleotide at the 5′ end and continues by assembling six nucleotides at a time until the 3′ end is reached. The efficiency of the promoter at the 3′ end then presumably depends on its position relative to the NP subunits, and this is determined by the total number of nucleotides in the genome chain. Remarkably, a similar rule does not appear to apply to the replication of rhabdovirus genomes, in which, in any event, each N monomer is associated with nine rather than six nucleotides (269).

Defective Interfering Genomes

DI genomes arise as a result of replicative errors, in which the polymerase, carrying the nascent genome or antigenome chain, leaves its template and either (a) rejoins the same template further downstream to finish the chain, thus creating an internal deletion, or (b) rejoins the nascent chain itself, and by finishing synthesis creates an inverted terminal repeat. Both these polymerase transfers are intramolecular events: intermolecular transfers that lead to true recombinants have never been reported for Paramyxoviridae. The latter DI genomes are called "copy backs," and of the two possible forms, only those that have duplicated the 3′ end of the antigenome (closed square, Fig. 13), and therefore do not express mRNAs, have been found (4,146,165,223,224).

DI genomes are thought to interfere with their nondefective helper virus because they somehow have a competitive advantage, which is independent of their size (222). The advantage for copy-backs appears obvious. Because there are five times as many genomes as antigenomes in an infected cells, and each is the template for the other, then synthesis of genomes from antigenomes occurs 25 times as frequently (on a molar basis) than the other way around. This could be because the antigenome promoter is stronger (either in itself or because it does not also have to act as a promoter for transcription) or because the 5′ end of the nascent genome chain competes better for NP^0. Copy-backs have the stronger promoter on genomes as well as antigenomes (Fig. 13). This explanation appears to apply for the *Rhabdovirus* VSV, for which only a single internal deletion DI has been described, and this DI quickly evolved to a copy-back (214).

For Sendai virus, however, internal deletion DIs are not rare, and they can coexist stably even in stocks with copy-backs, regardless of their size (4). Their ability to compete with helper virus as well as copy-back DIs has remained an enigma. Their ability to interfere with the helper virus, however, could be due to the altered proteins they express. A deleted P protein containing only the *C*-terminal half of P is expressed in a persistent Sendai virus infection containing both types of DI genomes, and this naturally de-

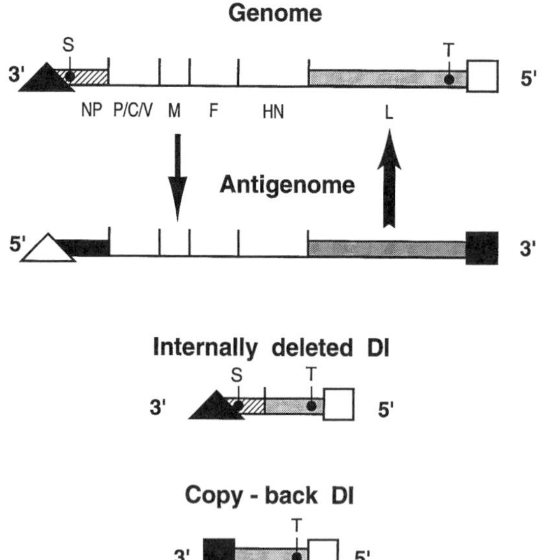

FIG. 13. Structures of Sendai virus internally deleted and copy-back defective-interfering RNAs. The 3′ end of the genome RNA is shown as a filled triangle and the 5′ terminus as the open square. The complementary antigenome termini are shown as an open triangle (5′) and filled square (3′). Internally deleted DI RNAs retain the 3′ and 5′ ends of the genome RNA and contain fragments of the NP and L gene retaining the NP transcription start signal (S) and the L gene termination and polyadenylation signal (T). Copy-back DI RNAs consist of 5′-terminal fragments of the genome RNA and have the 3′ end of the antigenome. The arrows (and their thickness) indicate the frequency of use of the templates for the generation of new RNA molecules. Adapted from Re (222), with permission.

rived protein was shown to be inhibitory for genome replication (93). Because persistent infections with DI genomes undergo rapid evolution, this internal deletion DI may have arisen after the copy-backs and is retained because the P fragment is useful in maintaining the persistent infection, by modulating the effects of the standard virus. Sendai virus DI particles are also implicated in the establishment of a persistent infection. When copy-backs are added to standard virus infections, nearly all the cells survive as a persistent infection, and this change from a normally lytic infection mainly correlates with the ablation of virus budding and the instability of the M protein (231).

Sendai virus DI genomes can be replicated by other helper parainfluenzaviruses, such as hPIV-1 and bPIV-3, but not by measles virus. However, this DI replication only occurs if the endogenous Sendai virus polymerase initiates the RNA chain (67). Presumably the heterologous virus cannot initiate RNA synthesis from the Sendai virus nucleocapsid but can assembly the nascent chain started by the Sendai virus polymerase. Once assembled by the heterologous NP, the heterologous polymerase can now amplify the chimeric DI genome. All parainfluenza viruses share the first 12 nucleotides, whereas parainfluenza and morbilliviruses share only the first nine. The inability of measles virus to amplify the Sendai virus DI genome may be due to its inability to assembly the nascent Sendai virus chain.

Virion Assembly and Release

Like all other events in the life cycle of replication of paramyxoviruses, the intracellular site of nucleocapsid assembly is the cytoplasm. The nucleocapsids are thought to be assembled in two steps: first, association of free NP subunits with the genome or template RNA to form the helical RNP structure, and second, the association of the P-L protein complex (142). By analogy to the mechanism of assembly of TMV nucleocapsid, which uses a defined enucleation point for the association of the first protein subunit with the RNA, and the observation that the paramyxovirus mRNAs are not encapsidated like the template positive-stranded antigenome intermediate, it has been assumed that the leader and trailer regions of the genome contain specific sequences for initiating encapsidation (20).

The assembly of the second part of the virus, the envelope, is at the cell surface. In polarized epithelial cells, the Paramyxovirinae bud only from the apical surface. The viral integral membrane proteins are synthesized in the ER and undergo a stepwise conformational maturation before transport through the secretory pathway. The mechanisms of viral glycoprotein folding, oligomeric assembly, and sorting have been extensively characterized in recent years, and an increasingly detailed view of how virus proteins acquire their functional structure has been obtained. Folding and conformational maturation of glycoproteins are not spontaneous events that occur in the cell. Rather, they are assisted by numerous folding enzymes and molecular chaperones. Viral integral membrane proteins, particularly the hemagglutinin and neuraminidase of influenza A virus, the G protein of vesicular stomatitis virus, and the HN protein of SV5 and Sendai virus have been used extensively as paradigms of cellular glycoproteins, and much of the current understanding of the process occurring in the exocytotic pathway has been learned from studying these proteins (71). When the nascent polypeptide chain of an integral membrane protein containing a signal or signal/anchor domain emerges from the protease inaccessible space in the translating ribosome, it engages with the cellular machinery for insertion into the membrane of the ER. On translocation of the peptide chain through the proteinaceous translocon (95,99), the proteins begin to fold cotranslationally. *N*-linked oligosaccharides are added to the nascent chain from dolichol phosphate, and disulfide bonds begin to form. Sequential folding from the beginning of the polypeptide chain to its end, with independently folded subdomains, probably helps account for the efficiency of folding *in vivo*. When the monomeric subunit is extensively folded, subunit-subunit recognition occurs and oligomer assembly follows. Only correctly folded and assembled proteins are generally transported out of the ER. In the Golgi apparatus, the carbohydrate chains may be modified from the high mannose to the complex form, and for those F proteins with multibasic cleavage sites, cleavage occurs in the *trans* Golgi network. Finally, the glycoproteins are transported to the plasma membrane.

The mechanism by which the virus particle is assembled at the plasma membrane is unknown. As discussed above, the viral M proteins are thought to play a major role in bringing the assembled ribonucleoprotein core to the appropriate patch on the plasma membrane to form a budding virion. The protein-protein interactions that are made to assemble a virion must be specific because cellular membrane proteins are largely excluded from the virions. It is presumed that the glycoprotein cytoplasmic tails make important contacts with the M protein, which, in turn, associates with the nucleocapsid. Those Paramyxovirinae that have neuraminidase activity contain glycoproteins that lack sialic modification of their carbohydrate chains, and it is thought that the HN neuraminidase activity serves the same purpose as the neuraminidase of influenza virus, to prevent self-binding.

REFERENCES

1. Alkhatib G, Briedis DJ. The predicted primary structure of the measles virus hemagglutinin. *Virology* 1986;150:479–490.
2. Alkhatib G, Richardson C, Shen SH. Intracellular processing, glycosylation, and cell-surface expression of the measles virus fusion protein (F) encoded by a recombinant adenovirus. *Virology* 1990;175:262–270.
3. Alkhatib G, Shen S-H, Briedis D, et al. Functional analysis of N-linked glycosylation mutants of the measles virus fusion protein synthesized by recombinant vaccinia virus vectors. *J Virol* 1994;68:1522–1531.
4. Amesse LS, Pridgen CL, Kingsbury DW. Sendai virus DI RNA species with conserved virus genome termini and extensive internal deletions. *Virology* 1982;118:17–27.
5. Anderson K, King AM, Lerch RA, Wertz GW. Polylactosaminoglycan modification of the respiratory syncytial virus small hydrophobic (SH) protein: a conserved feature among human and bovine respiratory syncytial viruses. *Virology* 1992;191:417–430.
6. Bachi T. Intramembrane structural differentiation in Sendai virus maturation. *Virology* 1980;106:41–49.
7. Baltimore D, Huang AS, Stampfer M. Ribonucleic acid synthesis of vesicular stomatitis virus 2. An RNA polymerase in the virion. *Proc Natl Acad Sci USA* 1970;66:572–576.
8. Banerjee AK, Barik S. Gene expression of vesicular stomatitis virus genome RNA. *Virology* 1992;188:417–428.
9. Barr J, Chambers P, Pringle CR, Easton AJ. Sequence of the major nucleocapsid protein gene of pneumonia virus of mice: sequence comparisons suggest structural homology between nucleocapsid proteins of pneumoviruses, paramyxoviruses, rhabdoviruses and filoviruses. *J Gen Virol* 1991;72:677–685.
10. Barr PJ. Mammalian subtilisins: The long-sought dibasic processing endoproteases. *Cell* 1991;66:1–3.
11. Barrett T, Shrimpton SB, Russell SE. Nucleotide sequence of the entire protein coding region of canine distemper virus polymerase-associated (P) protein mRNA. *Virus Res* 1985;3:367–372.
12. Bass BL, Weintraub H. An unwinding activity that covalently modifies double-stranded RNA substrate. *Cell* 1988;55:1089–1098.
13. Bass BL, Weintraub H, Cattaneo R, Billeter MA. Biased hypermutation of viral RNA genomes could be due to unwinding/modification of double-stranded RNA (Letter). *Cell* 1989;56:331.
14. Bellini WJ, Englund G, Richardson CD, Rozenblatt S, Lazzarini RA. Matrix genes of measles virus and canine distemper virus: cloning, nucleotide sequences, and deduced amino acid sequences. *J Virol* 1986;58:408–416.
15. Bellini WJ, Englund G, Rozenblatt S, Arnheiter H, Richardson CD. Measles virus P gene codes for two proteins. *J Virol* 1985;53:908–919.
16. Berg M, Hjertner B, Moreno LJ, Linne T. The P gene of the porcine paramyxovirus LPMV encodes three possible polypeptides P, V and C: the P protein mRNA is edited. *J Gen Virol* 1992;73:1195–1200.
17. Blumberg B, Giorgi C, Roux L, et al. Sequence determination of the Sendai virus HN gene and its comparison to the influenza virus glycoproteins. *Cell* 1985;41:269–278.
18. Blumberg BM, Crowley JC, Silverman JI, et al. Measles virus L protein evidences elements of ancestral RNA polymerase. *Virology* 1988;164:487–497.
19. Blumberg BM, Giorgi C, Rose K, Kolakofsky D. Sequence determination of the Sendai virus fusion protein gene. *J Gen Virol* 1985; 66:317–331.
20. Blumberg BM, Kolakofsky D. Intracellular vesicular stomatitis virus leader RNAs are found in nucleocapsid structures. *J Virol* 1981; 40:568–576.
21. Blumberg BM, Leppert M, Kolakofsky D. Interaction of VSV leader RNA and nucleocapsid protein may control VSV genome replication. *Cell* 1981;23:837–845.
22. Blumberg BM, Rose K, Simona MG, Roux L, Giorgi C, Kolakofsky D. Analysis of the Sendai virus M gene and protein. *J Virol* 1984; 52:656–663.
23. Boeck R, Curran J, Matsuoka Y, Compans R, Kolakofsky D. The parainfluenza virus type 1 P/C gene uses a very efficient GUG codon to start its C′ protein. *J Virol* 1992;66:1765–1768.
24. Buckland R, Wild F. Leucine zipper motif extends. *Nature* 1989;338:547.
25. Buckland R, Malvoisin E, Beauverger P, Wild F. A leucine zipper structure present in the measles virus fusion protein is not required for its tetramerization but is essential for fusion. *J Gen Virol* 1992; 73:1703–1707.
26. Buchholz CJ, Spehner D, Drillien R, Neubert WJ, Homann HE. The conserved N-terminal region of Sendai virus nucleocapsid protein NP is required for nucleocapsid assembly. *J Virol* 1993;67:5803–5812.
27. Bullough PA, Hughson FM, Skehel JJ, Wiley DC. Structure of influenza haemagglutinin at the pH of membrane fusion. *Nature* 1994;371:37–43.
28. Butler PJG. The current picture of the structure and assembly of tobacco mosaic virus. *J Gen Virol* 1984;65:253–279.
29. Calain P, Curran J, Kolakofsky D, Roux L. Molecular cloning of natural paramyxovirus copy-back defective interfering RNAs and their expression from DNA. *Virology* 1992;191:62–71.
30. Calain P, Roux L. Generation of measles virus defective interfering particles and their presence in a preparation of attenuated live-virus vaccine. *J Virol* 1988;62:2859–2866.
31. Calain P, Roux L. The rule of six, a basic feature for efficient replication of Sendai virus defective interfering RNA. *J Virol* 1993; 67:4822–4830.
32. Carr CM, Kim PS. A spring-loaded mechanism for the conformational change of influenza hemagglutinin. *Cell* 1993;73:823–832.
33. Carroll AR, Wagner RR. Role of the membrane (M) protein in endogenous inhibition of in vitro transcription by vesicular stomatitis virus. *J Virol* 1979;29:134–142.
34. Castaneda SJ, Wong TC. Leader sequence distinguishes between translatable and encapsidated measles virus RNAs. *J Virol* 1990;64:222–230.
35. Cattaneo R, Billeter MA. Mutations and A/I hypermutations in measles virus persistent infections. *Curr Top Microbiol Immunol* 1992;176:63–74.
36. Cattaneo R, Kaelin K, Baezko K, Billeter MA. Measles virus editing provides an additional cysteine-rich protein. *Cell* 1989;56:759–764.
37. Cattaneo R, Rebmann G, Baczko K, ter Meulen V, Billeter M. A. Altered ratios of measles virus transcripts in diseased human brains. *Virology* 1987;160:523–526.
38. Cattaneo R, Rebmann G, Schmid A, Baczko K, ter Meulan V, Billeter MA. Altered transcription of a defective measles virus genome derived from a diseased human brain. *EMBO J* 1987;6:681–688.
39. Cattaneo R, Rose JK. Cell fusion by the envelope glycoproteins of persistent measles viruses which cause lethal human brain disease. *J Virol* 1993;67:1493–1502.
40. Cattaneo R, Schmid A, Billeter MA, Sheppard RD, Udem SA. Multiple viral mutations rather than host factors cause defective measles virus gene expression in a subacute sclerosing panencephalitis cell line. *J Virol* 1988;62:1388–1397.
41. Cattaneo R, Schmid A, Eschle D, Baczko K, ter Meulen V, Billeter MA. Biased hypermutation and other genetic changes in defective measles

virus in human brain infections. *Cell* 1988;55:255–265.
42. Chambers P, Millar NS, Emmerson PT. Nucleotide sequence of the gene encoding the fusion glycoprotein of Newcastle disease virus. *J Gen Virol* 1986;67:2685–2694.
43. Chambers P, Pringle CR, Easton AJ. Heptad repeat sequences are located adjacent to hydrophobic regions in several types of virus fusion glycoproteins. *J Gen Virol* 1990;71:3075–3080.
44. Choppin PW, Compans RW. Reproduction of paramyxoviruses. In: Fraenkel-Conrat H, Wagner RR, eds. *Comprehensive virology*. New York: Plenum; 1975:95–178
45. Choppin PW, Scheid A. The role of viral glycoproteins in adsorption, penetration, and pathogenicity of viruses. *Rev Infect Dis* 1980;2:40–61.
46. Collins PL. The molecular biology of human respiratory syncytial virus (RSV) of the genus Pneumovirus. In: Kingsbury DW, ed. *The Paramyxoviruses*. New York: Plenum; 1991:103–162
47. Collins PL, Anderson K, Langer SJ, Wertz GW. Correct sequence for the major nucleocapsid protein mRNA of respiratory syncytial virus. *Virology* 1985;146:69–77.
48. Collins PL, Mink MA, Hill MG III, Camargo E, Grosfeld H, Stec DS. Rescue of a 7502-nucleotide (49.3% of full-length) synthetic analog of respiratory syncytial virus genomic RNA. *Virology* 1993;195:252–256.
49. Collins PL, Mink MA, Stec DS. Rescue of synthetic analogs of respiratory syncytial virus genomic RNA and effect of truncations and mutations on the expression of a foreign reporter gene. *Proc Natl Acad Sci USA* 1991;88:9663–9667.
50. Collins PL, Mottet G. Membrane orientation and oligomerization of the small hydrophobic protein of human respiratory syncytial virus. *J Gen Virol* 1993;74:1445–1450.
51. Collins PL, Olmsted RA, Spriggs MK, Johnson PR, Buckler-White AJ. Gene overlap and site-specific attenuation of transcription of the viral polymerase L gene of human respiratory syncytial virus. *Proc Natl Acad Sci USA* 1987;84:5134–5138.
52. Collins PL, Wertz GW. The 1A protein gene of human respiratory syncytial virus: nucleotide sequence of the mRNA and a related polycistronic transcript. *Virology* 1985;141:283–291.
53. Collins PL, Wertz GW. The envelope-associated 22K protein of human respiratory syncytial virus: nucleotide sequence of the mRNA and a related polytranscript. *J Virol* 1985;54:65–71.
54. Compans RW, Choppin PW. Isolation and properties of the helical nucleocapsid of the parainfluenza virus SV5. *Proc Natl Acad Sci USA* 1967;57:949–956.
55. Compans RW, Choppin PW. The length of the helical nucleocapsid of Newcastle disease virus. *Virology* 1967;33:344–346.
56. Compans RW, Mountcastle WE, Choppin PW. The sense of the helix of paramyxovirus nucleocapsids. *J Mol Biol* 1972;65:167–169.
57. Curran J, Boeck R, Kolakofsky D. The Sendai virus P gene expresses both an essential protein and an inhibitor of RNA synthesis by shuffling modules via mRNA editing. *EMBO J* 1991;10:3079–3085.
58. Curran J, de Melo M, Moyer S, Kolakofsky D. Characterization of the Sendai virus V protein with an anti-peptide antiserum. *Virology* 1991;184:108–116.
59. Curran J, Homann H, Buchholz C, Rochat S, Neubert W, Kolakofsky D. The hypervariable C-terminal tail of the Sendai paramyxovirus nucleocapsid protein is required for template function but not for RNA encapsidation. *J Virol* 1993;67:4358–4364.
60. Curran J, Kolakofsky D. Ribosomal initiation from an ACG codon in the Sendai virus P/C mRNA. *EMBO J* 1988;7:245–251.
61. Curran J, Kolakofsky D. Scanning independent ribosomal initiation of the Sendai virus X protein. *EMBO J* 1988;7:2869–2874.
62. Curran J, Kolakofsky D. Scanning independent ribosomal initiation of the Sendai virus Y proteins in vitro and in vivo. *EMBO J* 1989;8:521–526.
63. Curran J, Kolakofsky D. Sendai virus P gene produces multiple proteins from overlapping open reading frames. *Enzyme* 1990;44:244–249.
64. Curran J, Marq JB, Kolakofsky D. The Sendai virus nonstructural C proteins specifically inhibit viral mRNA synthesis. *Virology* 1992;189:647–656.
65. Curran J, Pelet T, Kolakofsky D. An acidic activation-like domain of the Sendai virus P protein is required for RNA synthesis and encapsidation. *Virology* 1994;202:875–884.
66. Curran JA, Kolakofsky D. Rescue of a Sendai virus DI genome by other parainfluenza viruses: implications for genome replication. *Virology* 1991;182:168–176.
67. Curran J, Richardson C, Kolakofsky D. Ribosomal initiation at alternate AUGs on the Sendai virus P/C mRNA. *J Virol* 1986;57:684–687.
68. Curran MD, Clarke DK, Rima BK. The nucleotide sequence of the gene encoding the attachment protein H of canine distemper virus. *J Gen Virol* 1991;72:443–447.
69. Delarue M, Poch O, Tordo N, Moras D, Argos P. An attempt to unify the structure of polymerases. *Protein Eng* 1990;3:461–467.
70. Deshpande KL, Portner A. Monoclonal antibodies to the P protein of Sendai virus define its structure and role in transcription. *Virology* 1985;140:125–134.
71. Doms RW, Lamb RA, Rose JK, Helenius A. Folding and assembly of viral membrane proteins. *Virology* 1993;193:545–562.
72. Dorig RE, Marcil A, Chopra A, Richardson CD. The human CD46 molecule is a receptor for measles virus (Edmonston strain). *Cell* 1993;75:295–305.
73. Dreyfuss G, Matunis MJ, Pinol-Roma S, Burd CG. hnRNP proteins and the biogenesis of mRNA. *Annu Rev Biochem* 1993;62:289–329.
74. Ebata SN, Coté MJ, Kang CY, Dimock K. The fusion and hemagglutinin-neuraminidase glycoproteins of human parainfluenza virus 3 are both required for fusion. *Virology* 1991;183:437–441.
75. Egelman EH, Wu SS, Amrein M, Portner A, Murti G. The Sendai virus nucleocapsid exists in at least four different helical states. *J Virol* 1989;63:2233–2243.
76. Einberger H, Mertz R, Hofschneider PH, Neubert WJ. Purification, renaturation and reconstituted protein kinase activity of the Sendai virus large (L) protein: L protein phosphorylates the NP and P proteins in vitro. *J Virol* 1990;64:4274–4280.
77. Elango N. Complete nucleotide sequence of the matrix protein mRNA of mumps virus. *Virology* 1989;168:426–428.
78. Elango N. The mumps virus nucleocapsid mRNA sequence and homology among the paramyxoviriade proteins. *Virus Res* 1989;12:77–86.
79. Elango N, Kovamees J, Norrby E. Sequence analysis of the mumps virus mRNA encoding the P protein. *Virology* 1989;169:62–67.
80. Elango N, Kovamees J, Varsanyi TM, Norrby E. mRNA sequence and deduced amino acid sequence of the mumps virus small hydrophobic protein gene. *J Virol* 1989;63:1413–1415.
81. Elango N, Satake M, Coligan JE, Norrby E, Camargo E, Venkatesan S. Respiratory syncytial virus fusion glycoprotein: nucleotide sequence of mRNA, identification of cleavage activation site and amino acid sequence of N-terminus of F1 subunit. *Nucleic Acids Res* 1985; 13:1559–1574.
82. Elliott GD, Afzal MA, Martin SJ, Rima BK. Nucleotide sequence of the matrix, fusion, and putative SH protein genes of mumps virus and their deduced amino acid sequences. *Virus Res* 1989;12:61–75.
83. Elliott GD, Yeo RP, Afzal MA, Simpson EJB, Curran JA, Rima BK. Strain-variable editing during transcription of the P gene of mumps virus may lead to the generation of non-structural proteins NS1 (v) and NS2. *J Gen Virol* 1990;71:1555–1560.
84. Etkind PR, Cross RK, Lamb RA, Merz DC, Choppin PW. In vitro synthesis of structural and nonstructural proteins of Sendai and SV5 viruses. *Virology* 1980;100:22–33.
85. Faaberg KS, Peeples ME. Association of soluble matrix protein of Newcastle disease virus with liposomes is independent of ionic conditions. *Virology* 1988;166:123–132.
86. Finch JT, Gibbs AJ. Observations on the structure of the nucleocapsids of some paramyxoviruses. *J Gen Virol* 1970;6:141–150.
87. Galinski MS, Mink MA, Lambert DM, Wechsler SL, Pons MM. Molecular cloning and sequence analysis of the human parainfluenza 3 virus RNA encoding the nucleocapsid protein. *Virology* 1986;149:139–151.
88. Galinski MS, Mink MA, Lambert DM, Wechsler SL, Pons MW. Molecular cloning and sequence analysis of the human parainfluenza 3 virus mRNA encoding the P and C proteins. *Virology* 1986;155:46–60.
89. Galinski MS, Mink MA, Lambert DM, Wechsler SL, Pons MW. Molecular cloning and sequence analysis of the human parainfluenza 3 virus gene encoding the matrix protein. *Virology* 1987;157:24–30.
90. Galinski MS, Mink MA, Pons MW. Molecular cloning and sequence analysis of the human parainfluenza 3 virus gene encoding the L protein. *Virology* 1988;165:499–510.
91. Galinski MS, Troy RM, Banerjee AK. RNA editing in the phosphoprotein gene of the human parainfluenza virus type 3. *Virology* 1992;186:543–550.
92. Galinski MS, Wechsler SL. The molecular biology of the Paramyxovirus genus. In: Kingsbury DW, ed. *The Paramyxoviruses*. New York: Plenum; 1991:41–82
93. Garcin D, De Melo M, Roux L, Kolakofsky D, Curran J. Presence of a

truncated form of the Sendai virus P protein in a long-term persistent infection: implications for the maintenance of the persistent state. *Virology* 1994;201:19–25.

94. Gerald C, Buckland R, Barker R, Freeman G, Wild TF. Measles virus haemagglutinin gene: cloning, complete nucleotide sequence analysis and expression in COS cells. *J Gen Virol* 1986;67:2695–2703.
95. Gilmore R. Protein translocation across the endoplasmic reticulum: a tunnel with toll booths at entry and exit. *Cell* 1993;75:589–592.
96. Giorgi C, Blumberg BM, Kolakofsky D. Sendai virus contains overlapping genes expressed from a single mRNA. *Cell* 1983;35:829–836.
97. Gombart AF, Hirano A, Wong TC. Expression and properties of the V protein in acute measles virus and subacute sclerosing panencephalitis virus strains. *Virus Res* 1992;25:63–78.
98. Gombart AF, Hirano A, Wong TC. Conformational maturation of measles virus nucleocapsid protein. *J Virol* 1993;67:4133–4141.
99. Gorlich D, Rapoport TA. Protein translocation into proteoliposomes reconstituted from purified components of the endoplasmic reticulum membrane. *Cell* 1993;75:615–630.
100. Gotoh B, Ogasawara T, Toyoda T, Inocencio NM, Hamaguchi M, Nagai Y. An endoprotease homologous to the blood clotting factor X as a determinant of viral tropism in chick embryo. *EMBO J* 1990;9:4189–4195.
101. Gotoh B, Yamauchi F, Ogasawara T, Nagai Y. Isolation of factor Xa from chick embryo as the amniotic endoprotease responsible for paramyxovirus activation. *FEBS Lett* 1992;296:274–278.
102. Gupta KC, Kingsbury DW. Conserved polyadenylation signals in two negative-strand RNA virus families. *Virology* 1982;120:518–523.
103. Gupta KC, Patwardhan S. ACG, the initiator codon for a Sendai virus protein. *J Biol Chem* 1988;263:8553–8556.
104. Gupta M, Kingsbury D. Complete sequences of the intergenic and mRNA start signals in the Sendai virus genome: homologies with the genome of vesicular stomatitis virus. *Nucleic Acids Res* 1984;12:3829–3841.
105. Hamaguchi M, Yoshida T, Nishikawa K, Naruse H, Nagai Y. Transcriptive complex of Newcastle disease virus 1. Both L and P proteins are required to constitute an active complex. *Virology* 1983;128:105–117.
106. Heggeness MH, Scheid A, Choppin PW. Conformation of the helical nucleocapsids of paramyxoviruses and vesicular stomatitis virus: reversible coiling and uncoiling induced by changes in salt concentration. *Proc Natl Acad Sci USA* 1980;77:2631–2635.
107. Heggeness MH, Scheid A, Choppin PW. The relationship of conformational changes in the Sendai virus nucleocapsid to proteolytic cleavage of the NP polypeptide. *Virology* 1981;114:555–562.
108. Heggeness MH, Smith PR, Choppin PW. In vitro assembly of the nonglycosylated membrane protein (M) of Sendai virus. *Proc Natl Acad Sci USA* 1982;79:6232–6236.
109. Herrler G, Compans RW. Synthesis of mumps virus polypeptides in infected vero cells. *Virology* 1982;119:430–438.
110. Hewitt JA. Studies on the subunit composition of the M-protein of Sendai virus. *FEBS Lett* 1977;81:395–398.
111. Hewitt JA, Nermut MV. A morphological study of the M-protein of Sendai virus. *J Gen Virol* 1973;34:127–136.
112. Hidaka Y, Kanda T, Iwasaki K, Nomoto A, Shioda T, Shibuta H. Nucleotide sequence of a Sendai virus genome region covering the entire M gene and the 3′ proximal 1013 nucleotides of the F gene. *Nucleic Acids Res* 1984;12:7965–7973.
113. Hiebert SW, Paterson RG, Lamb RA. Hemagglutinin-neuraminidase protein of the paramyxovirus simian virus 5: nucleotide sequence of the mRNA predicts an N-terminal membrane anchor. *J Virol* 1985;54:1–6.
114. Hiebert SW, Paterson RG, Lamb RA. Identification and predicted sequence of a previously unrecognized small hydrophobic protein, SH, of the paramyxovirus simian virus 5. *J Virol* 1985;55:744–751.
115. Hiebert SW, Richardson CD, Lamb RA. Cell surface expression and orientation in membranes of the 44-amino-acid SH protein of simian virus 5. *J Virol* 1988;62:2347–2357.
116. Higuchi Y, Miyahara Y, Kawano M, et al. Sequence analysis of the large (L) protein of simian virus 5. *J Gen Virol* 1992;73:1005–1010.
117. Hirano A, Ayata M, Wang AH, Wong TC. Functional analysis of matrix proteins expressed from cloned genes of measles virus variants that cause subacute sclerosing panencephalitis reveals a common defect in nucleocapsid binding. *J Virol* 1993;67:1848–183.
118. Hirano A, Wang AH, Gombart AF, Wong TC. The matrix proteins of neurovirulent subacute sclerosing panencephalitis virus and its acute measles virus progenitor are functionally different. *Proc Natl Acad Sci USA* 1992;89:8745–8749.
119. Homann HE, Hofschneider PH, Neubert WJ. Sendai virus expression in lytically and persistently infected cells. *Virology* 1990;177:131–140.
120. Homma M, Tamagawa S. Restoration of the fusion activity of L cell-borne Sendai virus by trypsin. *J Gen Virol* 1973;19:423–426.
121. Hope IA, Struhl K. Functional dissection of a eukaryotic transcriptional activator protein, GCN4 of yeast. *Cell* 1986;46:885–894.
122. Horikami SM, Curran J, Kolakofsky D, Moyer SA. Complexes of Sendai virus NP-P and P-L proteins are required for defective interfering particle genome replication in vitro. *J Virol* 1992;66:4901–4908.
123. Horvath CM, Lamb RA. Studies on the fusion peptide of a paramyxovirus fusion glycoprotein: roles of conserved residues in cell fusion. *J Virol* 1992;66:2443–2455.
124. Horvath CM, Paterson RG, Shaughnessy MA, Wood R, Lamb RA. Biological activity of paramyxovirus fusion proteins: factors influencing formation of syncytia. *J Virol* 1992;66:4564–4569.
125. Hosaka M, Nagahama M, Kim W-S, et al. Arg-X-Lys/Arg-Arg motif as a signal for precursor cleavage catalyzed by furin within the constitutive secretory pathway. *J Biol Chem* 1991;266:12127–12130.
126. Hsu C-H, Kingsbury DW. Topography of phosphate residues in Sendai virus proteins. *Virology* 1982;120:225–234.
127. Hu X, Ray R, Compans RW. Functional interactions between the fusion protein and hemagglutinin-neuraminidase of human parainfluenza viruses. *J Virol* 1992;66:1528–1534.
128. Hu XL, Compans RW, Matsuoka Y, Ray R. Molecular cloning and sequence analysis of the fusion glycoprotein gene of human parainfluenza virus type 2. *Virology* 1990;179:915–920.
129. Huang AS, Baltimore D, Bratt M A. Ribonucleic acid polymerase in virions of Newcastle disease virus: comparison with the vesicular stomatitis virus polymerase. *J Virol* 1971;7:389–394.
130. Huang YT, Collins PL, Wertz GW. Characterization of the 10 proteins of human respiratory syncytial virus: identification of a fourth envelope-associated protein. *Virus Res* 1985;2:157–173.
131. Hunt DM, Smith EG, Buckley DW. Aberrant polyadenylation by a vesicular stomatitis virus mutant is due to an altered L protein. *J Virol* 1984;52:515–521.
132. Ishida N, Taira H, Omata T, et al. Sequence of 2617 nucleotides from the 3′ end of Newcastle disease virus genome RNA and the predicted amino acid sequence of viral NP protein. *Nucleic Acids Res* 1986;14:6551–6564.
133. Iverson LB, Rose JK. Localized attenuation and discontinuous synthesis during vesicular stomatitis virus transcription. *Cell* 1981;23:477–484.
134. Jambou RC, Elango N, Venkatesan S, Collins PL. Complete sequence of the major nucleocapsid protein gene of human parainfluenza type 3 virus: comparison with other negative strand viruses. *J Gen Virol* 1986;67:2543–2548.
135. Kamata H, Tsukiyama K, Sugiyama M, Kamata Y, Yoshikawa Y, Yamanouchi K. Nucleotide sequence of cDNA to the rinderpest virus mRNA encoding the nucleocapsid protein. *Virus Genes* 1991;5:5–15.
136. Kamer G, Argos P. Primary structural comparison of RNA-dependent polymerases from plant, animal, and bacterial viruses. *Nucleic Acids Res* 1984;12:7169–7282.
137. Kawano M, Bando H, Ohgimoto S, et al. Sequence of the fusion protein gene of human parainfluenza type 2 virus and its 3′ intergenic region: lack of small hydrophobic (SH) gene. *Virology* 1990;178:289–292.
138. Kawano M, Bando H, Ohgimoto S, et al. Complete nucleotide sequence of the matrix gene of human parainfluenza type 2 virus and expression of the M protein in bacteria. *Virology* 1990;179:857–861.
139. Kawano M, Bando H, Yuasa T, et al. Sequence determination of the hemagglutinin-neuraminidase (HN) gene of human parainfluenza type 2 virus and the construction of a phylogenetic tree for HN proteins of all the paramyxoviruses that are infectious to humans. *Virology* 1990;174:308–313.
140. Kawano M, Okamoto K, Bando H, et al. Characterizations of the human parainfluenza type 2 virus gene encoding the L protein and the intergenic sequences. *Nucleic Acids Res* 1991;19:2739–2746.
141. Kingsbury DW, Darlington RW. Isolation and properties of Newcastle disease virus nucleocapsid. *J Virol* 1968;2:248–255.
142. Kingsbury DW, Hsu CH, Murti KG. Intracellular metabolism of Sendai virus nucleocapsids. *Virology* 1978;91:86–94.
143. Klenk H-D, Garten W. Host cell proteases controlling virus pathogenicity. *Trends Microbiol* 1994;2:39–43.
144. Klenk HD, Choppin PW. Lipids of plasma membranes of monkey and hamster kidney cells and of parainfluenza virions grown in these cells. *Virology* 1969;38:255–268.
145. Klenk HD, Choppin PW. Plasma membrane lipids and parainfluenza

virus assembly. *Virology* 1970;40:939–947.
146. Kolakofsky D. Isolation and characterization of Sendai virus DI-RNAs. *Cell* 1976;8:547–55.
147. Kolakofsky D, Blumberg BM. A model for the control of non-segmented negative strand viruses genome replication. In: *Virus Persistence Symposium 33.* Cambridge, MA: Society for General Microbiology Ltd, Cambridge University Press; 1982.
148. Kolakofsky D, Boy de la Tour E, Bruschi A. Self-annealing of Sendai virus RNA. *J Virol* 1974;14:33–9.
149. Kolakofsky D, Bruschi A. Antigenomes in Sendai virions and Sendai virus-infected cells. *Virology* 1975;66:185–91.
150. Kolakofsky D, Curran J, Pellet T, Jacques J-P. Paramyxovirus P gene mRNA editing. In: Benne R, ed. *RNA editing.* London, England: Ellis Horwood; 1993:105–123.
151. Kolakofsky D, Vidal S, Curran J. Paramyxovirus RNA synthesis and P gene expression. In: Kingsbury DW, ed. *The Paramyxoviruses.* New York: Plenum; 1991:215–233
152. Kondo K, Bando H, Kawano M, et al. Sequencing analyses and comparison of parainfluenza virus type 4A and 4B NP protein genes. *Virology* 1990;174:1–8.
153. Kondo K, Bando H, Tsuradome M, Kawano M, Nishio M, Ito Y. Sequence analysis of the phosphorprotein (P) genes of human parainfluenza type 4A and 4B viruses and RNA editing at transcript of the P genes: the number of G residues added is imprecise. *Virology* 1990;178:321–326.
154. Kovamees J, Norrby E, Elango N. Complete nucleotide sequence of the hemagglutinin-neuraminidase (HN) mRNA of mumps virus and comparison of paramyxovirus HN proteins. *Virus Res* 1989;12:87–96.
155. Kozak M. Bifunctional messenger RNAs in eukaryotes. *Cell* 1986;47:481–483.
156. Lamb RA. The phosphorylation of Sendai virus proteins by a virus particle-associated protein kinase. *J Gen Virol* 1975;26:249–263.
157. Lamb RA. Paramyxovirus fusion: a hypothesis for changes. *Virology* 1993;197:1–11.
158. Lamb RA, Choppin PW. The synthesis of Sendai virus polypeptides in infected cells.II. Intracellular distribution of polypeptides. *Virology* 1977;81:371–81.
159. Lamb RA, Choppin PW. The synthesis of Sendai virus polypeptides in infected cells. III. Phosphorylation of polypeptides. *Virology* 1977;81:382–397.
160. Lamb RA, Choppin PW. Determination by peptide mapping of the unique polypeptides in Sendai virions and infected cells. *Virology* 1978;84:469–478.
161. Lamb RA, Dreyfuss G. RNA structure. Unwinding with a vengeance. *Nature* 1989;337:19–20.
162. Lamb RA, Mahy BW, Choppin PW. The synthesis of Sendai virus polypeptides in infected cells. *Virology* 1976;69:116–131.
163. Lamb RA, Paterson RG. The nonstructural proteins of paramyxoviruses. In: Kingsbury DW, ed. *The Paramyxoviruses.* New York: Plenum; 1991:181–214
164. Laver WG, Colman PM, Webster RG, Hinshaw VS, Air GM. Influenza virus neuraminidase with hemagglutinin activity. *Virology* 1984;137:314–323.
165. Leppert M, Kort L, Kolakofsky D. Further characterization of Sendai virus DI-RNAs: a model for their generation. *Cell* 1977;12:539–552.
166. Limo M, Yilma T. Molecular cloning of the rinderpest virus matrix gene: comparative sequence analysis with other paramyxoviruses. *Virology* 1990;175:323–327.
167. Liston P, Briedis DJ. Measles virus V protein binds zinc. *Virology* 1994;198:399–404.
168. Luk D, Masters PS, Sanchez A, Banerjee AK. Complete nucleotide sequence of the matrix protein mRNA and three intergenic junctions of human parainfluenza virus type 3. *Virology* 1987;156:189–192.
169. Lyn D, Gill DS, Scroggs RA, Portner A. The nucleoproteins of human parainfluenza virus type 1 and Sendai virus share amino acid sequences and antigenic and structural determinants. *J Gen Virol* 1991;72:983–987.
170. Markwell MA, Fox CF. Protein-protein interactions within paramyxoviruses identified by native disulfide bonding or reversible chemical cross-linking. *J Virol* 1980;33:152–166.
171. Markwell MA, Portner A, Schwartz AL. An alternative route of infection for viruses: entry by means of the asialoglycoprotein receptor of a Sendai virus mutant lacking its attachment protein. *Proc Natl Acad Sci USA* 1985;82:978–982.
172. Markwell MAK. New frontiers opened by the exloration of host cell receptors. In: Kingsbury DW, ed. *The Paramyxoviruses.* New York: Plenum; 1991:407–425
173. Markwell MAK, Fredman P, Svennerholm L. Specific gangliosides are receptors for Sendai virus. *Adv Exp Biol* 1984;174:369–379.
174. Markwell MAK, Moss J, Hom BE, Fishman PH, Svennerholm L. Expression of gangliosides as receptors at the cell surface controls infection of NCTC 2071 cells by Sendai virus. *Virology* 1986;155:356–364.
175. Marx PA, Portner A, Kingsbury DW. Sendai virion transcriptase complex: polypeptide composition and inhibition by virion envelope proteins. *J Virol* 1974;13:107–112.
176. Masters PS, Banerjee AK. Complex formation with vesicular stomatitis virus phosphoprotein NS prevents binding of nucleocapsid protein N to nonspecific RNA. *J Virol* 1988;62:2658–2664.
177. Matsuoka Y, Curran J, Pelet T, et al. The P gene of human parainfluenza virus type 1 encodes P and C protiens but not a cysteine-rich V protein. *J Virol* 1991;65:3406–3410.
178. Matsuoka Y, Ray R. Sequence analysis and expression of the human parainfluenza type 1 virus nucleoprotein gene. *Virology* 1991; 181:403–407.
179. McGeoch DJ. Structure of the gene N: gene NS intercistronic junction in the genome of vesicular stomatitis virus. *Cell* 1979;17:673–681.
180. McGinnes L, McQuain C, Morrison T. The P protein and the nonstructural 38K and 29K proteins of Newcastle disease virus are derived from the same open reading frame. *Virology* 1988;164:256–264.
181. McGinnes LW, Morrison TG. Nucleotide sequence of the gene encoding the Newcastle disease virus fusion protein and comparisons of paramyxovirus fusion protein sequences. *Virus Res* 1986;5:343–356.
182. Merz DC, Prehm P, Scheid A, Choppin PW. Inhibition of the neuraminidase of paramyxoviruses by halide ions: a possible means of modulating the two activities of the HN protein. *Virology* 1981;112:296–305.
183. Millar NS, Chambers P, Emmerson PT. Nucleotide sequence analysis of the haemagglutinin-neuraminidase gene of Newcastle disease virus. *J Gen Virol* 1986;67:1917–1927.
184. Morgan EM, Rakestraw KM. Sequence of the Sendai virus L gene: open reading frames upstream of the main coding region suggest that the gene may be polycistronic. *Virology* 1986;154:31–40.
185. Morgan EM, Re GG, Kingsbury DW. Complete sequence of the Sendai virus NP gene from a cloned insert. *Virology* 1984;135:279–287.
186. Morrison T, McQuain C, McGinnes L. Complementation between avirulent Newcastle disease virus and a fusion protein gene expressed from a retrovirus vector: requirements for membrane fusion. *J Virol* 1991; 65:813–822.
187. Morrison T, Portner A. Structure, function, and intracellular processing of the glycoproteins of Paramyxoviridae. In: Kingsbury DW, ed. *The Paramyxoviruses.* New York: Plenum; 1991:347–382
188. Morrison TG, McQuain C, O'Connell KF, McGinnes LW. Mature, cell-associated HN protein of Newcastle disease virus exists in two forms differentiated by posttranslational modifications. *Virus Res* 1990;15:113–133.
189. Mountcastle WE, Compans RW, Lackland H, Choppin PW. Proteolytic cleavage of subunits of the nucleocapsid of the paramyxovirus simian virus 5. *J Virol* 1974;14:1253–1261.
190. Nagai Y, Klenk HD. Activation of precursors to both glycoproteins of Newcastle disease virus by proteolytic cleavage. *Virology* 1977; 77:125–134.
191. Nagai Y, Ogura H, Klenk HD. Studies on the assembly of the envelope of Newcastle disease virus. *Virology* 1976;69:523–538.
192. Namba K, Pattanayek R, Stubbs G. Visualization of protein-nucleic acid interactions in a virus: refined structure of intact Tobacco Mosaic virus at 2.9 A resolution by x-ray fiber diffraction. *J Mol Biol* 1989;208:307–325.
193. Naniche D, Varior-Krishnan G, Cervoni F, et al. Human membrane cofactor protein (CD46) acts as a cellular receptor for measles virus. *J Virol* 1993;67:6025–6032.
194. Neubert WJ, Eckerskorn C, Homann HE. Sendai virus NP gene codes for a 524 amino acid NP protein. *Virus Genes* 1991;1:25–32.
195. Ng DTW, Hiebert SW, Lamb RA. Different roles of individual N-linked oligosaccharide chains in folding, assembly, and transport of the simian virus 5 hemagglutinin-neuraminidase. *Mol Cell Biol* 1990; 10:1989–2001.
196. Ng DTW, Randall RE, Lamb RA. Intracellular maturation and transport of the SV5 type II glycoprotein hemagglutinin-neuraminidase: specific and transient association with GRP78-BiP in the endoplasmic reticulum and extensive internalization from the cell surface. *J Cell Biol*

1989;109;3273–3289.

197. Novick SL, Hoekstra D. Membrane penetration of Sendai virus glycoproteins during the early stage of fusion with liposomes as determined by hydrophobic affinity labeling. *Proc Natl Acad Sci USA* 1988;85:7433–7437.
198. Ohgimoto S, Bando H, Kawano M, et al. Sequence analysis of P gene of human parainfluenza type 2 virus: P and cysteine-rich proteins are translated by two mRNAs that differ by two nontemplated G residues. *Virology* 1990;177:116–123.
199. Olmsted RA, Collins PL. The 1A protein of respiratory syncytial virus is an integral membrane protein present as multiple, structurally distinct species. *J Virol* 1989;63:2019–2029.
200. Ortmann D, Ohuchi M, Angliker H, Shaw E, Garten W, Klenk H-D. Proteolytic cleavage of wild type and mutants of the F protein of human parainfluenza virus type 3 by two subtilisin-like endoproteases, furin and Kex2. *J Virol* 1994;68:2772–2776.
201. Park KH, Huang T, Correia FF, Krystal M. Rescue of a foreign gene by Sendai virus. *Proc Natl Acad Sci USA* 1991;88:1–5.
202. Park KH, Krystal M. In vivo model for pseudo-templated transcription in Sendai virus. *J Virol* 1992;66:7033–7039.
203. Parks GD, Ward CD, Lamb RA. Molecular cloning of the NP and L genes of simian virus 5: identification of highly conserved domains in paramyxovirus NP and L proteins. *Virus Res* 1992;22:259–279.
204. Paterson RG, Harris TJR, Lamb RA. Fusion protein of the paramyxovirus simian virus 5: nucleotide sequence of mRNA predicts a highly hydrophobic glycoprotein. *Proc Natl Acad Sci USA* 1984;81:6706–6710.
205. Paterson RG, Hiebert SW, Lamb RA. Expression at the cell surface of biologically active fusion and hemagglutinin/neuraminidase proteins of the paramyxovirus simian virus 5 from cloned cDNA. *Proc Natl Acad Sci USA* 1985;82:7520–7524.
206. Paterson RG, Lamb RA. Ability of the hydrophobic fusion-related external domain of a paramyxovirus F protein to act as a membrane anchor. *Cell* 1987;48:441–452.
207. Paterson RG, Lamb RA. RNA editing by G-nucleotide insertion in mumps virus P-gene mRNA transcripts. *J Virol* 1990;64:4137–4145.
208. Paterson RG, Shaughnessy MA, Lamb RA. Analysis of the relationship between cleavability of a paramyxovirus fusion protein and length of the connecting peptide. *J Virol* 1989;63:1293–1301.
209. Paterson RG, Thomas SM, Lamb RA. Specific nontemplated nucleotide addition to a simian virus 5 mRNA: prediction of a common mechanism by which unrecognized hybrid P-cysteine-rich proteins are encoded by paramyxovirus "P" genes. In: Kolakofsky D, Mahy BWJ, eds.*Genetics and pathogenicity of negative strand viruses.* London: Elsevier; 1989:232–245
210. Pauling L, Corey RB. Compound helical configurations of polypeptide chains: structure of protein of the α-keratin type. *Nature* 1953;171:59–61.
211. Peeples ME. Paramyxovirus M proteins: pulling it all together and taking it on the road. In: Kingsbury DW, ed. *The Paramyxoviruses.* New York: Plenum; 1991:427–456
212. Pelet T, Curran J, Kolakofsky D. The P gene of bovine parainfluenza virus 3 expresses all three reading frames from a single mRNA editing site. *EMBO J* 1991;10:443–448.
213. Pennington TH, Pringle CR. Negative strand viruses in enucleate cells. In: Mahy BWJ, Barry RD, eds. *Negative strand viruses and the host cell.* New York: Academic; 1978:457–464
214. Perrault J. Origin and replication of defective interfering particles. *Curr Top Microbiol Immunol* 1981;93:151–207.
215. Perrault J, Clinton GM, McClure MA. RNP template of vesicular stomatitis virus regulates transcription and replication functions. *Cell* 1983;35:175–185.
216. Poch O, Blumberg BM, Bougueleret L, Tordo N. Sequence comparison of five polymerases (L proteins) of unsegmented negative-strand RNA viruses: theoretical assignment of functional domains. *J Gen Virol* 1990;71:1153–1162.
217. Portner A. The HN glycoprotein of Sendai virus: analysis of site(s) involved in hemagglutinating and neuraminidase activities. *Virology* 1981;115:375–384.
218. Portner A, Gupta KC, Seyer JM, Beachey EH, Kingsbury DW. Localization and characterization of Sendai virus nonstructural C and C′ proteins by antibodies against synthetic peptides. *Virus Res* 1986;6:109–121.
219. Portner A, Murti KG. Localization of P, NP, and M proteins on Sendai virus nucleocapsid using immunogold labeling. *Virology* 1986; 150:469–478.
220. Portner A, Murti KG, Morgan EM, Kingsbury DW. Antibodies against Sendai virus L protein: distribution of the protein in nucleocapsids revealed by immunoelectron microscopy. *Virology* 1988;163:236–239.
221. Portner A, Scroggs RA, Marx PS, Kingsbury DW. A temperature-sensitive mutant of Sendai virus with an altered hemagglutinin-neuraminidase polypeptide: consequences for virus assembly and cytopathology. *Virology* 1975;67:179–187.
222. Re GD. Deletion mutants of paramyxoviruses In: Kingsbury DW, ed. *The Paramyxoviruses.* New York: Plenum; 1991:275–298
223. Re GG, Gupta KC, Kingsbury DW. Genomic and copy-back 3′ termini in Sendai virus defective interfering RNA species. *J Virol* 1983; 45:659–664.
224. Re GG, Gupta KC, Kingsbury DW. Sequence of the 5′ end of the Sendai virus genome and its variable representation in complementary form at the 3′ ends of copy-back defective interfering RNA species: identification of the L gene terminus. *Virology* 1983;130:390–396.
225. Richardson C, Hull D, Greer P, et al. The nucleotide sequence of the mRNA encoding the fusion protein of measles virus (Edmonston strain): a comparison of fusion proteins from several different paramyxoviruses. *Virology* 1986;155:508–523.
226. Robertson JW, Schubert JS, Lazzarini RA. Polyadenylation sites for influenza virus mRNA. *J Virol* 1981;38:157–163.
227. Robinson WS. Self-annealing of subgroup 2 myxovirus RNAs. *Nature* 1970;225:944–945.
228. Robinson WS. Ribonucleic acid polymerase activity in Sendai virions and nucleocapsid. *J Virol* 1971;8:81–86.
229. Robinson WS. Sendai virus RNA synthesis and nucleocapsid formation in the presence of cycloheximide. *Virology* 1971;44:494–502.
230. Roux L, Kolakofsky D. Protein kinase associated with Sendai virions. *J Virol* 1974;13:545–547.
231. Roux L, Waldvogel FA. Instability of the viral M protein in BHK-21 cells persistently infected with Sendai virus. *Cell* 1982;28:293–302.
232. Rozenblatt S, Eizenberg O, Ben-Levy R, Lavie V, Bellini WJ. Sequence homology within the morbilliviruses. *J Virol* 1985;53:684–690.
233. Rueda P, Delgado T, Portela A, Melero JA, Garcia-Barreno B. Premature stop codons in the G glycoportein of human respiratory suncytial viruses resistant to neutralization by monoclonal antibodies. *J Virol* 1991;65:3374–3378.
234. Russell R, Paterson RG, Lamb RA. Studies with cross-linking reagents on the oligomeric form of the paramyxovirus fusion protein. *Virology* 1994;199:160–168.
235. Ryan KW, Morgan EM, Portner A. Two noncontiguous regions of Sendai virus P protein combine to form a single nucleocapsid binding domain. *Virology* 1991;180:126–134.
236. Ryan KW, Portner A. Separate domains of Sendai virus P protein are required for binding to viral nucleocapsids. *Virology* 1990;174:515–521.
237. Ryan KW, Portner A, Murti KG. Antibodies to paramyxovirus nucleoproteins define regions important for immunogenicity and nucleocapsid assembly. *Virology* 1993;193:376–384.
238. Sakai Y, Shibuta H. Syncytium formation by recombinant vaccinia viruses carrying bovine parainfluenza 3 virus envelope protein genes. *J Virol* 1989;63:3661–3668.
239. Sakai Y, Suzu S, Shioda T, Shibuta H. Nucleotide sequence of the bovine parainfluenza 3 virus genome: its 3′ end and the genes of NP, P, C and M proteins. *Nucleic Acids Res* 1987;15:2927–2944.
240. Sanchez A, Banerjee AK, Furuichi Y, Richardson MA. Conserved structures among the nucleocapsid proteins of the paramyxoviriade: complete nucleotide sequence of the human parainfluenza virus type 3 NP mRNA. *Virology* 1986;152:171–180.
241. Sanderson CM, McQueen NL, Nayak DP. Sendai virus assembly: M protein binds to viral glycoproteins in transit through the secretory pathway. *J Virol* 1993;67:651–663.
242. Sanderson CM, Wu H-H, Nayak DP. Sendai virus M protein binds independently to either the F or the HN glycoprotein in vivo. *J Virol* 1993;68:69–76.
243. Satake M, Coligan JE, Elango N, Norrby E, Venkatesan S. Respiratory syncytial virus envelope glycoprotein (G) has a novel structure. *Nucleic Acids Res* 1985;13:7795–7812.
244. Satake M, Venkatesan S. Nucleotide sequence of the gene encoding respiratory syncytial virus matrix protein. *J Virol* 1984;50:92–99.
245. Scheid A, Caliguiri LA, Compans RW, Choppin PW. Isolation of paramyxovirus glycoproteins. Association of both hemagglutinating and neuraminidase activities with the larger SV5 glycoprotein. *Virology* 1972;50:640–652.
246. Scheid A, Choppin PW. The hemagglutinating and neuraminidase pro-

tein of a paramyxovirus: interaction with neuraminic acid in affinity chromatography. *Virology* 1974;62:125–133.
247. Scheid A, Choppin PW. Identification of biological activities of paramyxovirus glycoproteins. Activation of cell fusion, hemolysis, and infectivity of proteolytic cleavage of an inactive precursor protein of Sendai virus. *Virology* 1974;57:475–490.
248. Schubert M, Keene JD, Herman RC, Lazzarini RA. Site on the vesicular stomatitis virus genome specifying polyadenylation and the end of the L gene mRNA. *J Virol* 1980;34:550–559.
249. Sechoy O, Phillipot JR, Bienvenue A. F protein-F protein interaction within the Sendai virus identified by native bonding or chemical cross-linking. *J Biol Chem* 1987;262:11519–11523.
250. Sergel T, McGinnes LW, Peeples ME, Morrison TG. The attachment function of the Newcastle disease virus hemagglutinin-neuraminidase protein can be separated from fusion promotion by mutation. *Virology* 1993;193:717–726.
251. Server AC, Smith JA, Waxham MN, Wolinsky JS, Goodman HM. Purification and amino-terminal protein sequence analysis of the mumps virus fusion protein. *Virology* 1985;144:373–383.
252. Sharma B, Norrby E, Blixenkrone MM, Kovamees J. The nucleotide and deduced amino acid sequence of the M gene of phocid distemper virus (PDV). The most conserved protein of morbilliviruses shows a uniquely close relationship between PDV and canine distemper virus. *Virus Res* 1992;23:13–25.
253. Sheshberadaran H, Lamb RA. Sequence characterization of the membrane protein gene of paramyxovirus simian virus 5. *Virology* 1990;176:234–243.
254. Shioda T, Iwasaki K, Shibuta H. Determination of the complete nucleotide sequence of the Sendai virus genome RNA and the predicted amino acid sequences of the F, HN, and L proteins. *Nucleic Acids Res* 1986;14:1545–1563.
255. Sidhu MS, Menonna JP, Cook SD, Dowling PD, Udem SA. Canine distemper virus L gene: sequence and comparison with related viruses. *Virology* 1993;193:50–65.
256. Southern JA, Precious B, Randall RE. Two non-templated nucleotide additions are required to generate the P mRNA of parainfluenza virus type 2 since the RNA genome encodes protein V. *Virology* 1990;177:388–390.
257. Spriggs MK, Collins PL. Sequence analysis of the P and C protein genes of human parainfluenza virus type 3: patterns of amino acid sequence homology among paramyxovirus proteins. *J Gen Virol* 1986;12:2705–2719.
258. Spriggs MK, Olmsted RA, Venkatesan S, Coligan JE, Collins PL. Fusion glycoprotein of human parainfluenza virus type 3: nucleotide sequence of the gene, direct identification of the cleavage-activation site, and comparison with other paramyxoviruses. *Virology* 1986; 152:241–251.
259. Stec DS, Hill MG, Collins PL. Sequence analysis of the polymerase L gene of human respiratory syncytial virus and predicted phylogeny of nonsegmented negative-strand viruses. *Virology* 1991;183:273–287.
260. Steward M, Vipond IB, Millar NS, Emmerson PT. RNA editing in Newcastle disease virus. *J Gen Virol* 1993;74:2539–2547.
261. Stone HO, Kingsbury DW, Darlington RW. Sendai virus-induced transcriptase from infected cells: polypeptides in the transcriptive complex. *J Virol* 1972;10:1037–1043.
262. Stone HO, Portner A, Kingsbury DW. Ribonucleic acid transcriptases in Sendai virions and infected cells. *J Virol* 1971;8:174–180.
263. Stricker R, Mottet G, Roux L. The Sendai virus matrix protein appears to be recruited in the cytoplasm by the viral nucleocapsid to function in viral assembly and budding. *J Gen Virol* 1994;75:1031–1042.
264. Sundqvist A, Berg M, Moreno LJ, Linne T. The haemagglutinin-neuraminidase glycoprotein of the porcine paramyxovirus LPMV: comparison with other paramyxoviruses revealed the closest relationship to simian virus 5 and mumps virus. *Arch Virol* 1992;122:331–340.
265. Suzu S, Sakai Y, Shioda T, Shibuta H. Nucleotide sequence of the bovine parainfluenza 3 virus genome:the genes of the F and HN glycoproteins. *Nucleic Acids Res* 1987;15:2945–2958.
266. Takeuchi K, Hishiyama M, Yamada A, Sugiura A. Molecular cloning and sequence analysis of the mumps virus gene encoding the P protein: mumps virus P gene is monocistronic. *J Gen Virol* 1988; 69:2043–2049.
267. Takeuchi K, Tanabayashi K, Hishiyama M, Yamada YK, Yamada A, Sugiura A. Detection and characterization of mumps virus V protein. *Virology* 1990;178:247–253.
268. Taylor J, Pincus S, Tartaglia J, et al. Vaccinia virus recombinants expressing either the measles virus fusion or hemagglutinin glycoprotein protect dogs against canine distemper virus challenge. *J Virol* 1991;65:4263–4274.
269. Thomas D, Newcomb WW, Brown JC, et al. Mass and molecular composition of vesicular stomatitis virus: a scanning transmission electron microscopy analysis. *J Virol* 1985;54:598–607.
270. Thomas SM, Lamb RA, Paterson RG. Two mRNAs that differ by two nontemplated nucleotides encode the amino coterminal proteins P and V of the paramyxovirus SV5. *Cell* 1988;54:891–902.
271. Thompson SD, Laver WG, Murti KG, Portner A. Isolation of a biologically active soluble form of the hemagglutinin-neuraminidase protein of Sendai virus. *J Virol* 1988;62:4653–4660.
272. Tsukiyama K, Sugiyama M, Yoshikawa Y, Yamanouchi K. Molecular cloning and sequence analysis of the rinderpest virus mRNA encoding the hemagglutinin protein. *Virology* 1987;160:48–54.
273. Tsurudome M, Naohiro O, Higuchi Y, et al. Molecular relationships between human parainfluenza virus type 2 and simian viruses 41 and 5: determination of nucleoprotein gene sequences of simian viruses 41 and 5. *J Gen Virol* 1991;72:2289–2292.
274. Varghese JN, Laver WG, Colman PM. Structure of the influenza virus glycoprotein antigen neuraminidase at 2.9 Å resolution. *Nature* 1983;303:35–40.
275. Vidal S, Curran J, Kolakofsky D. Editing of the Sendai virus P/C mRNA by G insertion occurs during mRNA synthesis via a virus-encoded activity. *J Virol* 1990;64:239–246.
276. Vidal S, Curran J, Kolakofsky D. A stuttering model for paramyxovirus P mRNA editing. *EMBO J* 1990;9:2017–2022.
277. Vidal S, Curran J, Orvell C, Kolakofsky D. Mapping of monoclonal antibodies to the Sendai virus P protein and the location of its phosphates. *J Virol* 1988;62:2200–2203.
278. Vidal S, Kolakofsky D. Modified model for the switch from Sendai virus transcription to replication. *J Virol* 1989;63:1951–1958.
279. Visser IK, van der Heijden RW, van de Bildt MW, Kenter MJ, Orvell C, Osterhaus AD. Fusion protein gene nucleotide sequence similarities, shared antigenic sites and phylogenetic analysis suggest that phocid distemper virus type 2 and canine distemper virus belong to the same virus entity. *J Gen Virol* 1993;74:1989–1994.
280. von Heijne G. On the hydrophobic nature of signal sequences. *Eur J Biochem* 1981;116:419–422.
281. Wardrop EA, Briedis DJ. Characterization of V protein in measles virus-infected cells. *J Virol* 1991;65:3421–3428.
282. Waxham MN, Aronowski J, Server AC, Wolinsky JS, Smith JA, Goodman HM. Sequence determination of the mumps virus HN gene. *Virology* 1988;164:318–325.
283. Waxham MN, Server AC, Goodman HM, Wolinsky JS. Cloning and sequencing of the mumps virus fusion protein gene. *Virology* 1987;159:381–388.
284. Wertz GW, Collins PL, Huang Y, Gruber C, Levine S, Ball LA. Nucleotide sequence of the G protein gene of human respiratory syncytial virus reveals an unusual type of viral membrane protein. *Proc Natl Acad Sci USA* 1985;82:4075–4079.
285. Wertz GW, Krieger M, Ball LA. Structure and cell surface maturation of the attachment glycoprotein of human respiratory syncytial virus in a cell line deficient in O glycosylation. *J Virol* 1989;63:4767–4776.
286. Wild TF, Malvoisin E, Buckland R. Measles virus: both the haemagglutinin and fusion glycoproteins are required for fusion. *J Gen Virol* 1991;72:439–442.
287. Yoshida T, Nakayama Y, Nagura H, et al. Inhibition of the assembly of Newcastle disease virus by monensin. *Virus Res* 1986;4:179–195.
288. Yuasa T, Bando H, Kawana M, et al. Sequence analyses of the 3′ genome end and NP gene of human parainfluenza type 2 virus: sequence variation of the gene-starting signal and the conserved 3′ end. *Virology* 1990;179:777–784.
289. Yusoff K, Millar NS, Chambers P, Emmerson PT. Nucleotide sequence analysis of the L gene of Newcastle disease virus: homologies with Sendai and vesicular stomatitis viruses. *Nucleic Acids Res* 1987;15:3961–3976.

Fundamental Virology, Third Edition
edited by B.N. Fields, D.M. Knipe, P.M. Howley, et al.
Lippincott - Raven Publishers, Philadelphia © 1996

CHAPTER 21

Orthomyxoviridae: The Viruses and Their Replication

Robert A. Lamb and Robert M. Krug

The *Orthomyxoviridae* are enveloped viruses with a segmented, single-stranded RNA (ssRNA) genome that has been termed *negative-stranded* because the viral messenger RNAs are transcribed from the viral RNA segments; by convention mRNA is plus-stranded (13). *Orthomyxoviridae* have special relationships to another family of negative-strand RNA viruses, namely, the *Paramyxoviridae*, with regard to the biological properties of the envelope glycoproteins (see Table 1). The genomic RNA of negative-strand RNA viruses has to serve two functions: first as a template for synthesis of mRNAs and second as a template for synthesis of the antigenome (+) strand. Negative-strand RNA viruses encode and package their own RNA-dependent RNA transcriptase, but mRNAs are synthesized only after the virus has been uncoated in the infected cell. Viral replication occurs after synthesis of the mRNAs and requires synthesis of viral proteins. The newly synthesized antigenome (+) strand (for influenza virus often termed *template RNA* or *cRNA*) serves as the template for further copies of the (–) strand genomic RNA. Among the RNA viruses, influenza virus is very special in that all of its RNA synthesis—transcription and replication—takes place in the nucleus of the infected cell. The nucleus provides the environment for the synthesis of influenza virus mRNAs

R.A. Lamb: Howard Hughes Medical Institute and Department of Biochemistry, Molecular Biology, and Cell Biology, Northwestern University, Evanston, Illinois 60208-3500.
R.M. Krug: Department of Molecular Biology and Biochemistry, Rutgers University, Piscataway, New Jersey 08855-1179.

TABLE 1. *Comparison of Orthomyxoviridae and Paramyxoviridae*

Similarities

Single-stranded RNA genome of negative sense
Helical nucleocapsid
Virion contains RNA-dependent RNA polymerase
Lipid-containing virion envelope
Sialic acid used as a receptor (except *Morbilliviruses*)
Two surface glycoprotein species (except influenza C virus)
Hemagglutination (most paramyxoviruses)
Neuraminidase activity (influenza A and B virus) and many paramyxoviruses (but not *Morbilliviruses*); (note, influenza C viruses have an esterase activity);
Proteolytically activated glycoprotein that mediates virus-cell fusion
Respiratory pathogens, primarily

Differences

Property	Ortho-myxovirus	Para-myxovirus
Genome organization	Segmented	Nonsegmented
RNase sensitivity of RNA in nucleocapsid	Sensitive	Insensitive
Cellular location of RNA synthesis	Nucleus	Cytoplasm
Inhibition by actinomycin D	Sensitive	Insensitive
Primer requirement for transcription	Yes	No
Endogenous capping activity catalyzed by RNA ranscriptase complex	No	Yes
Rate of genetic reassortment/recombination	High	Not detected
Site of fusion of virus with cell	Endosomal	Plasma membrane

in an unusual process as initiation requires $m^7GpppXm$-containing capped primers that are generated from a subset of host cell RNAs by an influenza virus-encoded cap-dependent endonuclease. In addition to stealing caps, influenza virus mRNAs make use of another aspect of host cell nuclear function, namely, the splicing machinery. Influenza virus mRNA transcripts provide the only known example of splicing of RNA that is not transcribed from DNA by RNA polymerase II.

In recent years a remarkable catalog of mechanisms has been identified by which eukaryotic cells have expanded their genome coding capacity beyond the linear array of nucleotide sequences usually thought to encode a protein. Such coding strategies create diversity by increasing the number of proteins encoded; in addition, they provide a means by which the expression of these proteins can be regulated. Eukaryotic viruses have usually provided the prototypic example of molecular mechanisms that serve to maximize coding potentials. Indeed, the compactness of the viral genome may be an important factor contributing to the success of a virus as a cellular parasite. The influenza viruses provide some remarkable examples of genome diversity: spliced mRNAs and overlapping reading frames, bicistronic mRNAs and overlapping reading frames, and coupled translation of tandem cistrons. The variety of mechanisms used for the synthesis of proteins by influenza virus provides a paradigm of successful exploitation of a genome.

CLASSIFICATION

The family *Orthomyxoviridae* (from the Greek *orthos*, "standard, correct" and *myxo*, "mucus") contains two genera: influenza A and B viruses; and influenza C virus. This name was given to influenza viruses in the 1960s [and approved by the International Committee on Taxonomy of Viruses (ICTV)], Chapter 2) because of their ability to bind to mucus (hemagglutination) and to distinguish them from another family of enveloped negative-strand RNA viruses (*Paramyxoviridae*). Other family members, D, are tick-borne viruses that are structurally and genetically similar to influenza A, B, and C viruses.

The influenza A, B, and C viruses can be distinguished on the basis of antigenic differences between their nucleocapsid (NP) and matrix (M) proteins. Influenza A viruses are further divided into subtypes based on the antigenic nature of their hemagglutinin (HA) and neuraminidase (NA) glycoproteins. Other important characteristics that distinguish influenza A, B, and C viruses are:

1. Influenza A viruses naturally infect humans and several other mammalian species, including swine and horses, and a wide variety of avian species. Influenza B virus appears to naturally infect only humans. Influenza C virus has been isolated mainly from humans but also from swine in China.
2. The surface glycoproteins of influenza A viruses exhibit much greater amino acid sequence variability than their counterparts in the influenza B viruses. Influenza C virus has only a single multifunctional glycoprotein.
3. There are morphological features that distinguish influenza A and B viruses from influenza C viruses.
4. Though the strategy by which influenza A, B, and C viruses encode their proteins are similar overall, each virus type has distinct mechanisms for encoding proteins distinct to the virus type.
5. Influenza A and B viruses each contain eight distinct RNA segments, whereas influenza C viruses contain seven RNA segments.

VIRION STRUCTURE

The *Orthomyxoviridae* are composed of 0.8% to 1% RNA, 70% protein, 20% lipid and 5% to 8% carbohydrate [reviewed in (66)]. The lipid envelope of influenza viruses is derived from the plasma membrane of the host cell in which the virus is grown [reviewed in (66)]. Influenza A and B virions are morphologically indistinguishable, whereas influenza C virions can be distinguished from the other genus as the glycoprotein spike is organized into orderly hexagonal arrays (66,97,130,388). Influenza A or B virions grown in eggs or in tissue culture cells have a fairly regular appearance in the electron microscope (EM) with particles of 80 to 120 nm in diameter. In contrast, virus strains isolated from man or animals and propagated by single passage in culture exhibit greater heterogeneity and pleomorphism (146,149). The morphological characteristics of influenza A viruses are a genetic trait (179). Although the gene(s) that specify spherical morphology have

TABLE 2. *Influenza A virus genome RNA segments and coding assignments*

Segment	Length[a] (nucleotides)	mRNA length (nucleotides)[b]	Encoded polypeptide	Nascent polypeptide length[c](aa)	Mol. wt. predicted	Approx. no. molecules/ virion	Remarks
1	2341	2320	PB2	759	85,700	30–60	[7]Me-Gppp Nm (cap) recognition of host-cell RNA; component of RNA transcriptase complex.
2	2341	2320	PB1	757	86,500	30–60	Catalyzes nucleotide addition; component of RNA transcriptase and replication complex.
3	2233	2211	PA	716	84,200	30–60	Component of RNA transcriptase and replicase complex; function unknown.
4	1778	1757	HA	566	61,468[d]	500	Major surface glycoprotein; sialic acid binding; proteolytic cleavage activation; fusion activity at acid pH; major antigenic determinant; trimer; x-ray structure known.
5	1565	1540	NP	498	56,101	1000	Monomer binds to RNA to form coiled ribonucleoprotein; involved in switch from mRNA to template RNA synthesis; and in virion RNA synthesis.
6	1413	1392	NA	454	50,087[d]	100	Surface glycoprotein; neuraminidase activity; tetramer; antigenic determinant; x-ray structure known.
7	1027	1005	M_1	252	27,801	3000	Major component of virion, probably underlies lipid bilayer; no known enzymatic activity.
		315	M_2	97	11,010	20-60	Integral membrane protein; ion channel activity blocked by antiviral drug amantadine.
		276	?	? (9)	—	—	Spliced mRNA sequence predicts that 9 amino acid peptide could be made; no evidence for its synthesis.
8	890	868	NS_1	230	26,815	—	High abundance, nonstructural protein found predominantly in nucleus: Inhibits nuclear export of poly(A)-containing mRNAs and pre-mRNA splicing.
		395	NS_2	121	14,216	130–200	Minor component of virions; cytoplasmic, and nuclear location; function unknown.

[a] For A/PR/8/34 strain.
[b] Deduced from RNA sequence, excluding poly(A) tract.
[c] Determined by nucleotide sequence analysis and protein sequencing.
[d] Contribution of carbohydrate makes observed molecular weight larger.

not been identified, the trait segregates independently of the HA and NA proteins (183).

The most striking feature of the influenza A virion is a layer of about 500 spikes radiating outward (10 to 14 nm) from the lipid envelope. These spikes are of two types: rod-shaped spikes of HA and mushroom-shaped spikes of NA (221). The ratio of HA to NA varies but is usually 4–5 to 1. The third integral membrane protein of influenza A virus, the M_2 ion channel protein which biochemical evidence indicates is present only in a few copies in virions (422), has been observed by EM only indirectly using immunoelectron microscopy (153,159) (see Fig. 2). The viral matrix protein (M_1) is thought to underlie the lipid bilayer, and it is the most abundant virion protein (Table 2).There is no evidence for the existence of a stable lipid-free core structure other than the ribonucleoprotein (RNP).

Inside the virus, observable by thin-sectioning of virus or by disrupting particles, are the RNP structures which can be separated into different size classes (86,302,314) and contain eight different segments of ssRNA (see Fig. 1 for schematic diagram of virion and Fig. 2 for EM photographs). The RNPs have the appearance of flexible rods

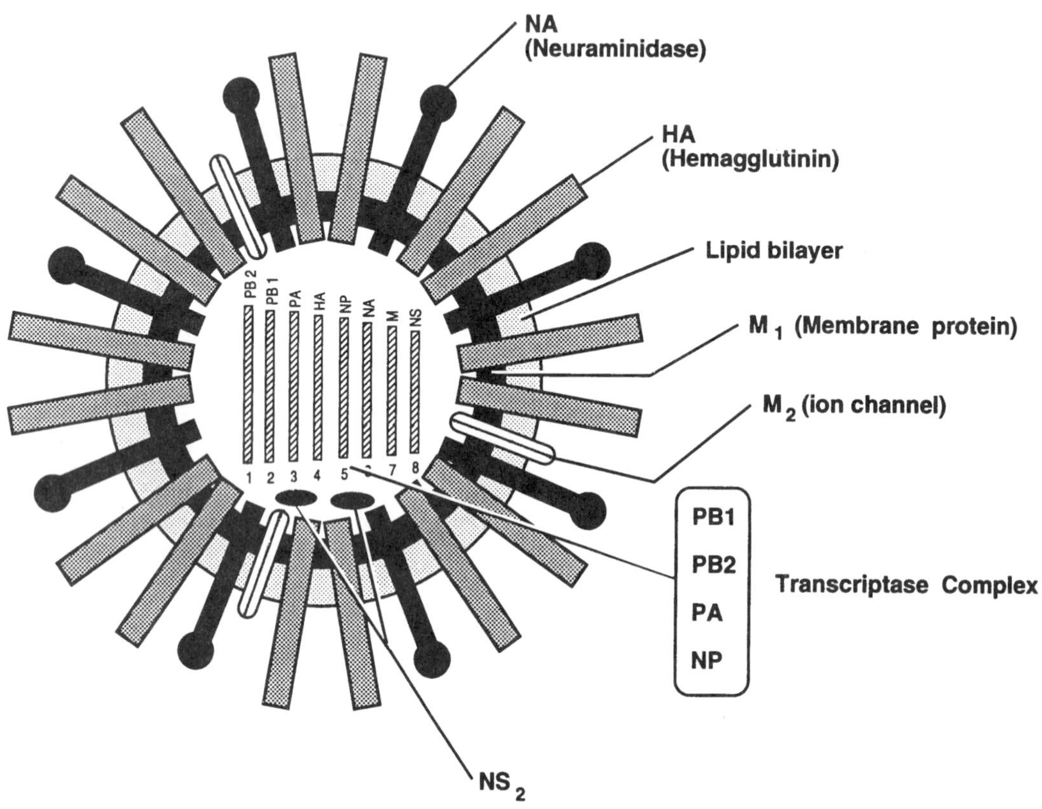

FIG. 1. Structure of the influenza A virus particle. Three types of integral membrane protein—hemagglutinin (*HA*), neuraminidase (*NA*), and small amounts of the M_2 ion channel protein—are inserted through the lipid bilayer of the viral membrane. The virion matrix protein M_1 is thought to underlie the lipid bilayer but also to interact with the helical ribonucleoproteins (*RNPs*). Within the envelope are eight segments of single-stranded genome RNA (ranging from 2,341 to 890 nucleotides) contained in the form of an RNP. Associated with the RNPs are small amounts of the transcriptase complex, consisting of the proteins PP1, PB2, and PA. The coding assignment of the eight RNA segments are also illustrated. RNA segments 7 and 8 each code for more than one protein (M_1 and M_2, and NS_1 and NS_2, respectively). NS_1 is found only in infected cells and is not thought to be a structural component of the virus, but small amounts of NS_2 are present in purified virions. Influenza B virus does not encode an M_2 integral membrane protein, but the NB glycoprotein encoded by RNA segment 6 (which also encodes NA) is of similar structure to M_2. Influenza C virus contains seven RNA segments, lacking a RNA segment for NA. However, the single-glycoprotein HEF encoded by influenza C virus is multifunctional and has hemagglutinating, neuraminate-O-acetylesterase activity and fusion activity. An equivalent of M_2 or NB has yet to be identified in influenza C viruses.

(304); depending on the salt concentration during sample preparation or method of staining, the flexibility of the RNPs appears to vary (67,124,304). The RNP strands often exhibit loops on one end and a periodicity of alternating major and minor grooves, suggesting that the structure is formed by a strand that is folded back on itself and then coiled on itself to form a type of twin-stranded helix. Larger helical structures, interpreted to be formed by the orderly aggregation of the smaller segments, have occasionally been seen in partially disrupted virions (7,11,149,260). These structures might represent ordered complexes of the eight separate RNPs.

The RNPs consist of four protein species and RNA. NP is the predominant protein subunit of the nucleocapsid and each NP subunit interacts with approximately 20 nucleotides. Associated with the RNPs are the three P (polymerase) proteins (PB1, PB2, and PA), which are present only at 30 to 60 copies per virion (158,206). They have been visualized on the RNP by immunoelectron microscopy and were found to be localized to one end of the RNP (262). The RNA in the RNP complex remains sensitive to digestion by RNAase, and the RNA can be displaced by polyvinylsulfate (104,185,304). This suggests that the structure of the influenza virus RNP is very different from the RNAase-insensitive *Paramyxoviridae* RNP (see Chapter 20). The RNP complex is the RNA-dependent transcriptase complex, and purified RNP cores carry out the cap-binding, endonuclease, RNA synthesis and polyadenylation reactions described below (22–24,65,297–300).

The NS_2 protein, long thought to be a nonstructural protein, has been shown recently to be a structural component of the virion present at 130 to 200 molecules per virion and forming an association with the M_1 protein (317,416).

INFLUENZA VIRUS GENOME ORGANIZATION AND ENCODED PROTEINS

Influenza A and B viruses each contain eight segments of ssRNA, and influenza C viruses contain seven segments of ssRNA (influenza C viruses lack a neuraminidase gene). The gene assignment for influenza A virus is as follows: RNA segment 1 codes for PB2, 2 for PB1, 3 for PA, 4 for HA, 5 for NP, 6 for NA, 7 for M_1 and M_2, and 8 for NS_1 and NS_2 (see Table 2). The complete nucleotide sequence of influenza virus A/PR/8/34 and many RNA segments of other subtypes were obtained by 1982 [reviewed in (205, 211)], and today the complete sequences of many influenza A and B viruses and one influenza C virus have been obtained. The PR/8/34 strain of influenza A virus contains

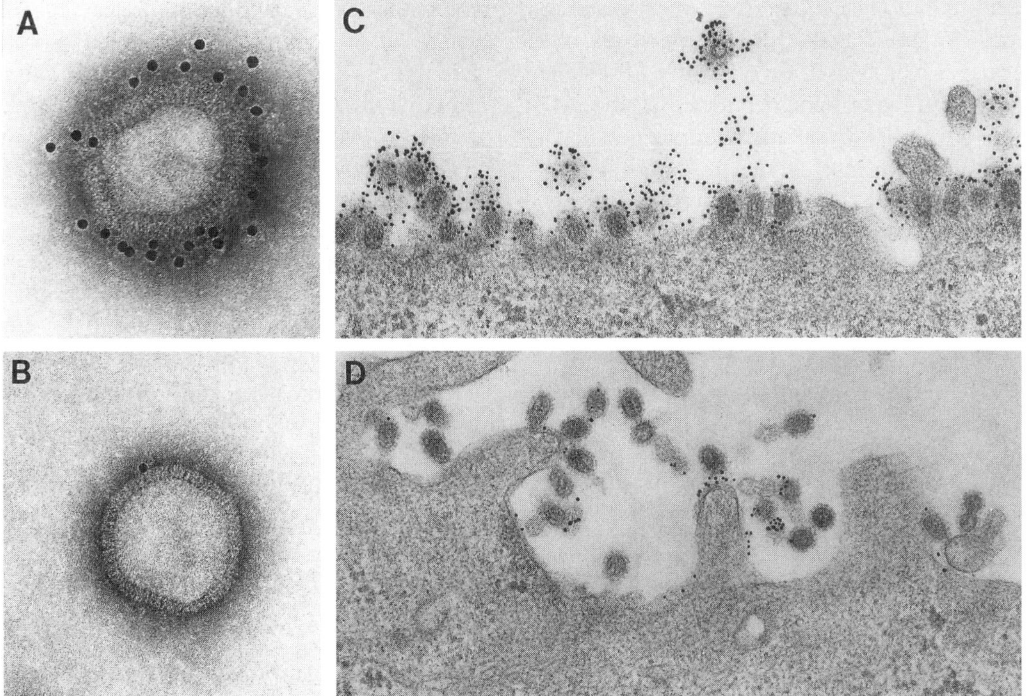

FIG. 2. Electron micrographs of purified influenza virus virions and virions budding from the surface of MDCK cells. **A:** Influenza A/Udorn/72 virus negatively stained with HA decorated with 10 nm gold (×159,250). **B:** Influenza A/Udorn/72 virus negatively stained with M_2 decorated with 10 nm gold (×159,250). **C:** Thin section of an influenza A/Udorn/72 virus-infected MDCK cell with HA decorated with 10 nm gold (H&E, ×40,600). **D:** Thin section of an influenza A/Udorn/72 virus-infected MDCK cell with M_2 decorated with 10 nm gold (×40,600). Courtesy of George Leser, Northwestern University, Evanston, IL.

13,588 nucleotides. In this chapter no attempt has been made to cite the original publication of the several hundred influenza virus nucleotide sequences obtained and deposited in the EMBL/GenBank database. Although the nucleotide sequence of the RNA segments provides the definitive proof of the coding assignment of the genes, these assignments were made earlier by ingenious experiments using genetic methods and hybrid-arrest of translation strategies [reviewed in (211)].

The first 12 nucleotides at the 3′ end and the first 13 nucleotides at the 5′ end of each vRNA segment are conserved in all eight RNA segments (see Fig. 15). Recent studies using RNAs of altered sequence and reconstitution of the transcriptase activity indicate a promoter role for these sequences in transcription (see below).

The Polymerase Proteins and their Genes

The three largest RNA segments encode the PB1, PB2, and PA proteins (apparent M_r ~96,000, 87,000, and 85,000, respectively) (158,206). Because of anomalous migration of influenza virus RNA segments on gels, the proteins encoded by RNA segments 1 and 3 from different subtypes are different. Therefore, the proteins are named after their behavior on isoelectric focusing gels: two P proteins were found to be basic (PB1 and PB2) and one acidic (PA) (144, 372,411). RNA segments 1 and 2 are both 2,341 nucleotides in length and code for proteins of 759 amino acids (PB2) and 757 amino acids (PB1), respectively, and PB1 and PB2 are basic proteins at pH 6.5 with a net charge of +28. As the sizes of the two polypeptides PB1 and PB2 are very similar, the ability to separate them on polyacrylamide gel electrophoresis must be due to factors such as differential binding of SDS. RNA segment 3 is 2,233 nucleotides in length and codes for a protein of 716 amino acids that has a charge of −13.5 at pH 6.5. Of the three polymerase proteins, PB1 has been implicated in catalytic activity including nucleotide polymerization and chain elongation (35). All viral RNA-dependent RNA polymerases examined to date contain four conserved motifs (301), and a mutational analysis of these conserved motifs in PB1 indicates that invariant residues in each of these motifs are critical in PB1 function (25). The three P proteins form a complex in the cytoplasm and nucleus of cells that is largely resistant to disruption by normal immunoprecipitation buffers and the complex sediments on sucrose gradients at 11 to 22S (76). After synthesis in the cytoplasm, the P proteins are transported to the nucleus, possibly as a complex. However, expression of the individual P proteins from cDNA has shown that each P protein migrates to the nucleus and thus contains a karyophilic signal (4,163,354). A fuller description of the known functions of the PB2, PB1, and PA proteins in RNA synthesis is presented below.

Influenza B and C viruses also encode three proteins which show extensive sequence identity and homology to those of influenza A virus, and it is presumed these proteins have roles and properties similar to those of influenza A virus.

The Nucleocapsid Protein and its Gene

The NP is the major structural protein that interacts with the RNA segments to form the RNP. Nucleocapsid protein is also one of the type-specific antigens that distinguishes between the influenza A, B, and C viruses. The NP protein is also the major target of cross-reactive cytotoxic T lymphocytes generated against all influenza virus subtypes in mice and man [reviewed in (418)]. The NP protein is encoded by RNA segment 5 which for influenza A virus is 1,565 nucleotides in length. The NP protein contains 498 amino acids with a predicted M_r of 56,101; the protein is rich in arginine residues and has a net positive charge of +14 at pH 6.5 (410). However, unlike what might be expected for a protein that interacts with the acidic phosphate residues of RNA, there are no clusters of basic residues which suggest that probably many regions of the NP molecule interact with RNA. The NP is phosphorylated (293, 306,308), but it is not clear what percentage of NP molecules are phosphorylated or whether phosphorylation is essential for function. After synthesis in the cytoplasm, NP molecules are transported to the nucleus. Nuclear targeting of NP is an intrinsic property of the protein (235,332), and a molecular analysis has indicated that NP residues 327 to 345 are sufficient and necessary to target NP to the nucleus (74). Interestingly, the NP karyophilic sequence has little resemblance to the prototype basic residue sequence identified in SV40 T-antigen (165,166).

The mechanism of assembly of the nucleocapsid is poorly understood. Both virion RNA (– strand) and template (+ strand) RNAs are found associated with NP molecules, whereas the viral mRNAs (+ strand) are not encapsidated (121,302); therefore, there must exist a mechanism that prevents NP from association with mRNAs. Although NP protein synthesized from cDNA in bacteria associates with many RNA species (and DNA), this interaction lacks specificity (184). Influenza virus must form RNPs with all eight RNA segments, and if there is a common nucleation site it may lie in the conserved 3′ and 5′-terminal nucleotides which are found on vRNA and template RNA strands. Because template (+ strand) RNAs are encapsidated with NP, whereas mRNAs are not, it seems possible that the putative nucleation site includes the 13 common 5′ vRNA nucleotides whose complement is lacking in mRNAs (see below).

The influenza B virus NP protein contains 560 amino acids and is 47% homologous to influenza A virus NP, and the influenza C virus NP protein contains 565 amino acids and has several regions of homology with the NP of influenza A and B viruses [reviewed in (205)].

The Hemagglutinin Protein and its Gene

The hemagglutinin was originally named because of the ability of the virus to agglutinate erythrocytes (136,251) by attachment to specific sialic glycoprotein receptors. Today we know HA has three major roles during the influenza virus replicative cycle:

1. HA binds to a sialic acid-containing receptor on the cell surface, bringing about the attachment of a virus particle to the cell.
2. Hemagglutinin is responsible for penetration of the virus into the cell cytoplasm by mediating the fusion of the membrane of the endocytosed virus particle with the endosomal membrane, the consequence of which is that viral nucleocapsids are released into the cytoplasm.
3. Hemagglutinin is the major antigen of the virus against which neutralizing antibodies are produced and influenza virus epidemics are associated with changes in its antigenic structure.

Hemagglutinin is encoded by RNA segment 4, and the polypeptide is synthesized in the endoplasmic reticulum (ER) as a single polypeptide HA_0 (M_r ~76,000). An N-terminal signal peptide targets the nascent chain to the ER membrane and is cleaved by signal peptidase, which makes HA a prototype type I integral membrane protein (Fig. 3). The native protein for the H3 subtype consists of an ectodomain of 512 residues, a carboxyl-terminal proximal transmembrane domain of 27 residues, and a cytoplasmic tail of 10 residues (377). Hemagglutinin is co(post)-translationally modified by the addition of up to seven oligosaccharide chains added to the ectodomain, and three palmitate residues are added by a thioether linkage to the three C-terminal proximal cysteine residues (264,265,336). The HA spike glycoprotein is a homotrimer (401,403,408) of noncovalently linked monomers. An x-ray structural determination of both the HA trimer (395, 396,404,408) and the trimer bound to a receptor analogue, sialyl lactose, has been obtained (394), as well as a proteolytic fragment of the low pH conformation (46). Depending on the virus strain, host-cell type, and growth conditions, HA is uncleaved or cleaved into two disulfide-linked chains HA_1 (M_r ~47,000) and HA_2 (M_r ~29,000). Cleavage is required for the virus to be infectious and is thus a critical determinant in pathogenicity and in the spread of infection [reviewed in (205)]. The newly liberated N-terminus of HA_2 (fusion peptide) is hydrophobic, is highly conserved in HAs of different influenza virus strains, and has been implicated as an essential participant in HA fusion activity.

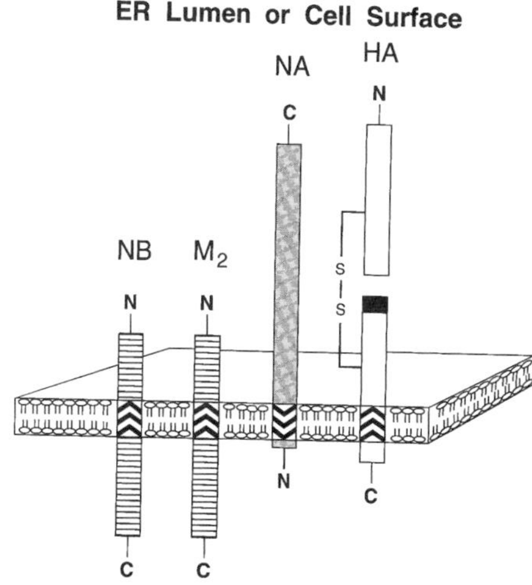

FIG. 3. Schematic representation of the orientation of the influenza virus integral membrane proteins HA, NA, M_2, and NB. See text for details of the individual proteins.

RNA Segment 4: Gene Structure and HA Amino Acid Sequence

The first gene of influenza virus to have its nucleotide sequence determined was HA (305). Since then, the nucleotide sequences of RNA segment 4 of all 14 known HA antigenic subtypes and many variants within a subtype have been determined [reviewed in (205,277)]). RNA segment 4 ranges from 1,742 to 1,778 nucleotides and encodes a polypeptide of 562 to 566 residues. The HA_1 chain is from 319 to 326 residues, and HA_2 is from 221 to 222 residues. Depending on the subtype, the number of residues lost on proteolytic cleavage between HA_1 and HA_2 ranges from one to six residues. For the A/Aichi/68 H3 subtype, for which the x-ray crystallographic structure was obtained, the cleaved signal sequence contains 16 residues, native HA_1 contains 328 residues, and HA_2 contains 221 residues (377). A single arginine residue is lost on proteolytic cleavage of HA_0, suggesting that two enzymes, a trypsin-like enzyme and an exopeptidase of the carboxypeptidase B type, are involved in activation of HA (82,101). The deduced amino acid sequence of the influenza B/Lee/40 HA has 24% homology to HA_1 and 39% homology to HA_2 of A/PR/8/34, suggesting a close evolutionary relationship between influenza A and B virus HAs (202,203).

Three-Dimensional Structure of HA

Although intact HA "spikes" can be isolated from purified influenza virions and infected cells by detergent solubilization of membranes, when the detergent is removed, the hydrophobic transmembrane domains aggregate and HA "rosettes" form (221). However, bromelain treatment of virus released an antigenically and structurally intact trimeric ectodomain of HA (BHA) which is water-soluble and contains all of HA_1 and the first 175 of the 221 residues of HA_2 (36,350,404). The BHA can be crystallized, and from x-ray diffraction data the three-dimensional structure of A/Aichi/68 H3 HA was determined (404,408) (Fig. 4).

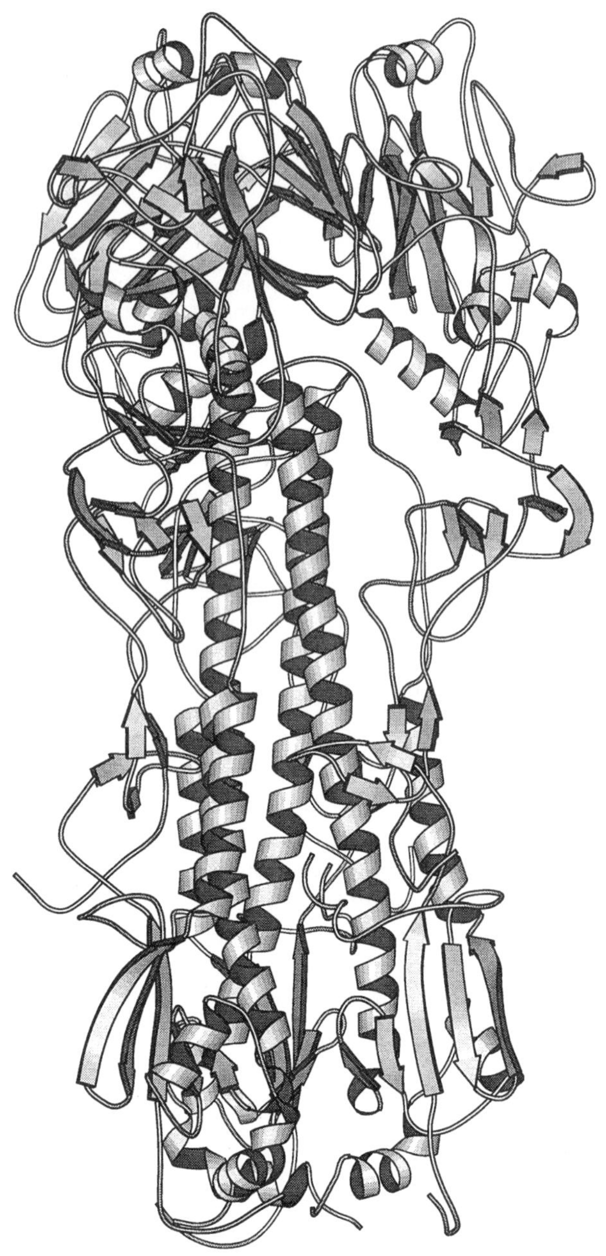

FIG. 4. Schematic diagram of the three-dimensional structure of the HA trimer as determined from x-ray diffraction data. Regions of β-strands (*flat, twisted arrows*) and regions of α-helix (*helices*) are represented. Drawn using Molscript (191). Data from Weis et al. (395). Courtesy of F. Hughson and D. Wiley, Howard Hughes Medical Institute, Harvard University, Cambridge, Massachusetts.

The HA trimer extends 135Å from the membrane and is composed of two predominant regions: (a) a long, fibrous stem extending 76Å from the membrane, which forms a triple-stranded coiled-coil of α-helices and is derived from HA_2 residues; and (b) the globular head of the molecule which is derived from HA_1 residues and consists predominantly of antiparallel β-sheet (Fig. 4). The hydrophobic "fusion peptides" are buried in the subunit interfaces in the stem of the trimers about 35Å from the viral membrane and approximately 100Å from the distal head of the molecule. As the N-terminus of HA_2 is separated from the C-terminus of HA_1 by 21Å, despite only one residue being removed by proteolytic cleavage, it indicates that a conformational change accompanies the cleavage of HA_0.

The HA receptor-binding site is a pocket located on each subunit at the distal end of the molecule. The pocket is inaccessible to antibody, and the residues forming the pocket (tyr 98, trp 153, his 183, glu 190, leu 194) are largely conserved between subtypes (408) [reviewed in (398,402)]. Confirmation that the pocket is the receptor-binding site was obtained by isolating HAs with single amino acid substitutions at residue 226 that conveyed an altered specificity for sialic acid linked to galactose by either α2,6- or α2,3-linkages (323,394). Because the receptor-binding specificities differ among humans, birds, and horses, the number of types of HA molecules that can transfer successfully from one species to another may be restricted.

The three-dimensional structure of HA indicates that the six carbohydrate chains on HA_1 and one chain on HA_2 lie on the lateral surface of HA, except for one which appears to stabilize the oligomeric contacts between globular units at the top of the structure (408). Although there is no known required function for carbohydrate on the native HA molecule, addition of carbohydrate is needed for the correct folding of HA in the ER [reviewed in (81)].

Five antigenic sites have been mapped on the three-dimensional structure of HA by determining the location of amino acid changes in the HAs of natural influenza virus isolates and in HAs of antigenic variants selected by growth in the presence of monoclonal antibodies (404). The regions of antigenic variation cover much of the surface of the globular head of HA, including residues around the antibody-inaccessible receptor-binding pocket. Monoclonal antibodies to each of the five antigenic sites neutralize the infectivity of the virus.

The Low pH Fusion-Activation HA Conformational Change

Exposure of HA to low pH (335,399) triggers a progression of irreversible conformational changes which ultimately activate the fusion activity of the virus. In influenza virus-infected cells, it is exposure of HA to acid pH found in endosomal compartments that is the physiologically relevant trigger to release the hydrophobic fusion peptides, although heat can also induce this change (330) [reviewed in (402)]. Thus, it is now thought that native (neutral) HA is a metastable protein, and the low pH form is a more stable and irreversible form.

The HA conformational change has been studied extensively [reviewed in (348)]. For example, it has been found that as discrete trypsin cleavage sites become exposed (80,347), the fusion peptide becomes sensitive to thermolysin digestion (72) and the single disulfide bond which links the HA_1 and HA_2 subunits becomes exposed (108). In addition, new antigenic epitopes are revealed with the concomitant loss of others (70,389,400). Nonetheless, the HA molecule remains trimeric and the globular heads retain their sialic acid binding sites and most of their antigenic epitopes. Partial dissociation of the most distal region of the trimer must occur because cross-linking of the globular head of the trimer, by rational engineering of cysteine residues into the head of the molecule such that intersubunit disulfide bonds could form, prevented the low pH conformational change (103,176). A large number of contacts between the monomeric HA subunits must change, and a considerable number of HA residues have been identified which, when protonated, would destabilize intersubunit contacts (71).

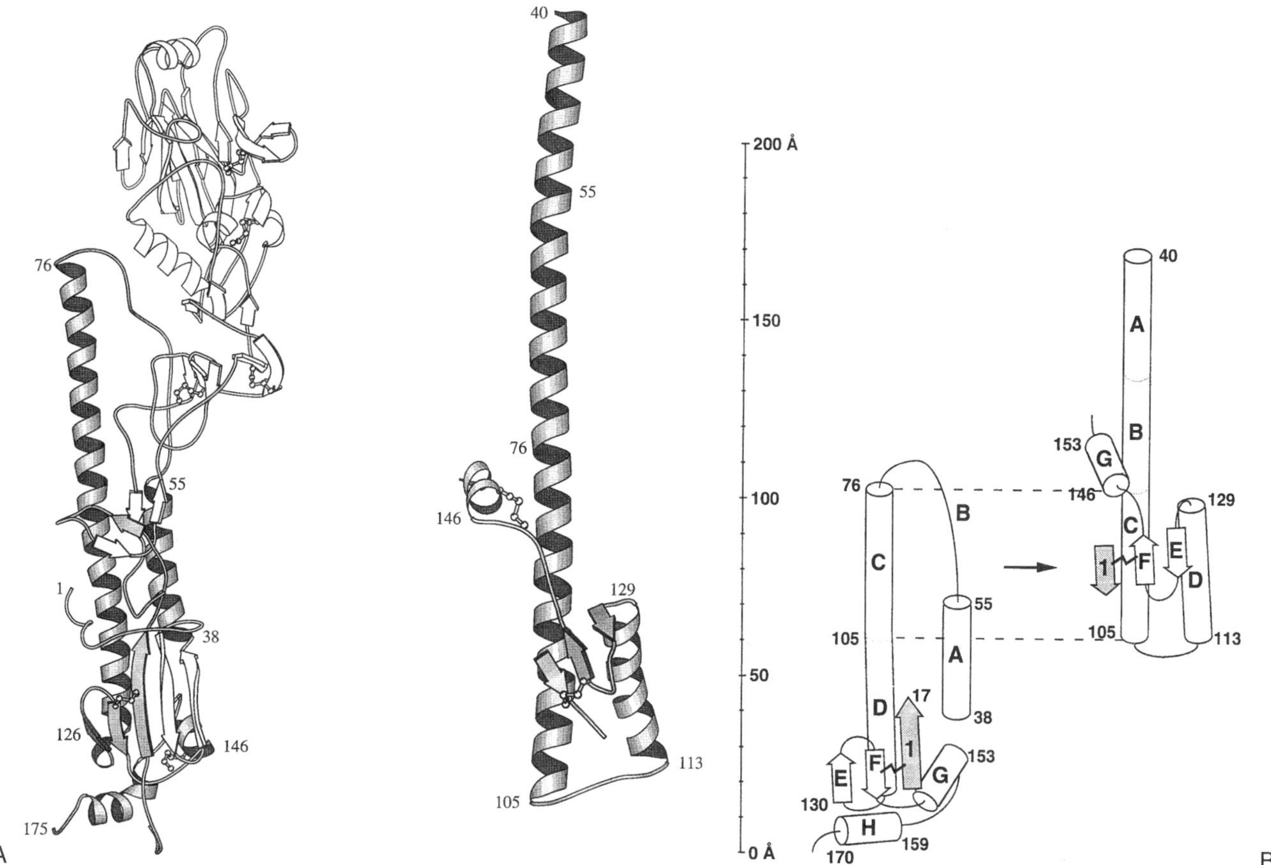

FIG. 5. The three-dimensional structure of the influenza virus HA at the pH of membrane fusion. **A: Left:** Schematic diagram of the HA monomer at neutral pH. Regions of β-chain (*flat, twisted arrows*) and regions of α-helix (*helices*) are represented. Regions (most of HA_1 plus the first 37 amino acids of HA_2) which were proteolytically removed in generating the low pH fragment (shown at right) are without shading, while the remaining regions (most of HA_2 plus the first 27 residues of HA_1) are shaded. **Right:** Schematic diagram of a proteolytic fragment of HA ($TBHA_2$) in a low pH induced conformation. Only a monomer is shown. The long α-helix, extending 100Å from near the proteolytic cleavage site at HA_2 residue 38 to HA_2 residue 105, participates in forming a triple-stranded coiled-coil in the trimer. In addition, the membrane-proximal portion of HA_2 swings up to pack against this long coiled coil; residues HA_2 106-112, helical in the neutral pH structure form an extended loop. Drawn using Molscript (191). Data from Bullough, et al. (46). Courtesy of F. Hughson and D. Wiley, Howard Hughes Medical Institute, Harvard University, Cambridge, Massachusetts. **B:** The structure of a $TBHA_2$ monomer (*right*) in schematic form is compared to the corresponding region of a BHA (bromelain released neutral pH) monomer (*left*). Consecutive regions of the HA_2 chain are labeled A-H to aid in following the rearrangement of the low pH induced conformation. H is apparently disordered in $TBHA_2$. Region 1 is the remaining disulfide-bonded fragment of HA_1 remaining in $TBHA_2$. From Bullough et al. (46), with permission.

Recently, the crystal structure of a proteolytic fragment of the low pH conformation of HA has been determined (46) and, in the large part, the structure fits well with a model predicted for the low pH form (51). In the pH neutral form HA_2 residues 76 to 126 form a long α-helix, residues 56 to 75 form an extended loop, and residues 38 to 55 form a short α-helix (408). In comparison, in the low pH HA conformation, HA_2 residues 40 to 105 form a triple-stranded α-helical coiled-coil extending 100Å. Residues 106 to 112, which in the neutral structure are α-helical, uncoil to form an extended loop, and the membrane proximal region of HA_2 swings up to pack against the long coiled-coil (46). Thus, the fusion peptide, which is in an interior position in the neutral HA structure, is relocated 100Å toward the target membrane (Fig. 5). The mechanism by which two bilayers of two opposing membranes are fused together is not known, but some important observations have been made:

1. The conformational change in HA has to occur at the right time and in the right place as premature release of the HA fusion peptide causes oligomers to aggregate (331) [reviewed in (348)].
2. The fusion peptide can intercalate into lipid bilayers [reviewed in (43)].
3. Several trimers are required to form a competent fusion pore (89).
4. Hemagglutinin requires its transmembrane domain for complete fusion to occur. A genetically engineered HA anchored to membranes with a glycosyl phosphatidylinositol anchor promotes hemifusion but not complete fusion (177).

Cleavage Activation of HA and Its Correlation with Pathogenicity

As discussed above, cleavage of the HA_0 precursor to HA_1 and HA_2 is a prerequisite for the conformational change to the low pH form and is thus a prerequisite for virus infectivity (187,223). Those HAs that contain the sequence R-X-K/R-R in the connecting peptide are cleaved intracellularly by furin, a protease resident in the *trans* Golgi network (360) [reviewed in (186)]. Those HAs with a single arginine residue in the connecting peptide (including all human strains) are not cleaved when grown in tissue culture cells, but these HAs can be cleaved by addition of exogenous trypsin. When grown in embryonated eggs, influenza viruses with HAs that have a single arginine residue at the cleavage site are cleaved, probably by a factor Xa-like protease by analogy to studies done on paramyxoviruses (280) (see Chapter 20). It was proposed that the cleavage site would correlate with the virulence of the virus and virulent strains would contain the furin recognition motif, whereas the avirulent strains would contain only a single arginine residue (27,175,276); there is experimental support for this notion derived from work done with avian influenza viruses [reviewed in (186)].

Influenza C Virus Glycoprotein, HEF

Influenza C virus contains only a single glycoprotein hemagglutinin, esterase, fusion (HEF) which has receptor-binding activity for 9-O-acetyl-N-acetylneuraminic acid, a fusion activity, and a receptor-destroying activity which is a neuraminate-O-acetyl esterase (and not a neuraminidase) (127–129,131–133,322,378). The HEF is a polypeptide of 655 residues which is cleaved by a trypsin-like protease (with a concomitant gain of low pH fusion activity and infectivity) to two disulfide-linked subunits (126,282). Although HEF does not have obvious amino acid homology with either influenza A or B virus HA, the observation that some of the cysteine residues are at homologous positions suggests the three-dimensional structures may be related (266).

The Neuraminidase and Its Gene

The NA integral membrane protein is the second subtype-specific glycoprotein of the influenza virion. It is important both for its biological activity in removing sialic acid from glycoproteins and as a major antigenic determinant that undergoes antigenic variation. Neuraminidase (acylneuraminyl hydrolase, EC 3.2.1.18) catalyzes the cleavage of the α-ketosidic linkage between a terminal sialic acid and an adjacent D-galactose or D-galactosamine (105). The role of NA in the influenza virus life cycle is still unclear. Influenza viruses containing a *ts* mutation in NA aggregate at the nonpermissive temperature at the surface of infected cells, suggesting that one function of NA is the removal of sialic acid from HA, NA, and the cell surface (287). The NA may also permit transport of the virus through the mucin layer present in the respiratory tract, enabling the virus to find its way to the target epithelial cells. Some neuraminidases (N1 and N9 subtypes) have a receptor-binding site (causing hemagglutination) as well as a neuraminidase active site, although the receptor specificity and function of this second activity is not clear (118,220). The NA is a homotetramer (M_r ~220,000) that can be released from the virion membrane with protease (374). In the EM, NA appears as a mushroom-shaped structure containing a stalk and head, with the molecules forming aggregated rosettes via an interaction at the ends of the stalk (221). Neuraminidase molecules lacking the hydrophobic domain and part of the stalk can be readily obtained by pronase digestion and the isolated head molecules (100×100×60Å) retain their enzymatic and antigenic properties; these molecules have been crystallized and their three-dimensional structure obtained [reviewed in (60,205)] (Fig. 6).

RNA Segment 6: Gene Structure and NA Amino Acid Sequence

The nucleotide sequence of all nine NA subtypes has been obtained [reviewed in (60,205)]. For the A/PR/8/34 subtype, RNA segment 6 is 1,413 nucleotides in length and encodes a polypeptide of 453 residues. The NA polypeptide contains only one hydrophobic domain sufficient to span a lipid bilayer which is located near its N-terminus (residues 7 to 35). This domain acts as a combined uncleaved signal/anchor domain, both targeting NA to the membrane of the ER and bringing about its stable attachment in the membrane. Thus, NA is orientated with its N-terminus in the cytoplasm and is a prototype class II integral membrane protein. There are five sites for the potential addition of N-linked carbohydrate.

Comparison of NA sequences among the nine subtypes of influenza A virus and influenza B virus reveals varying

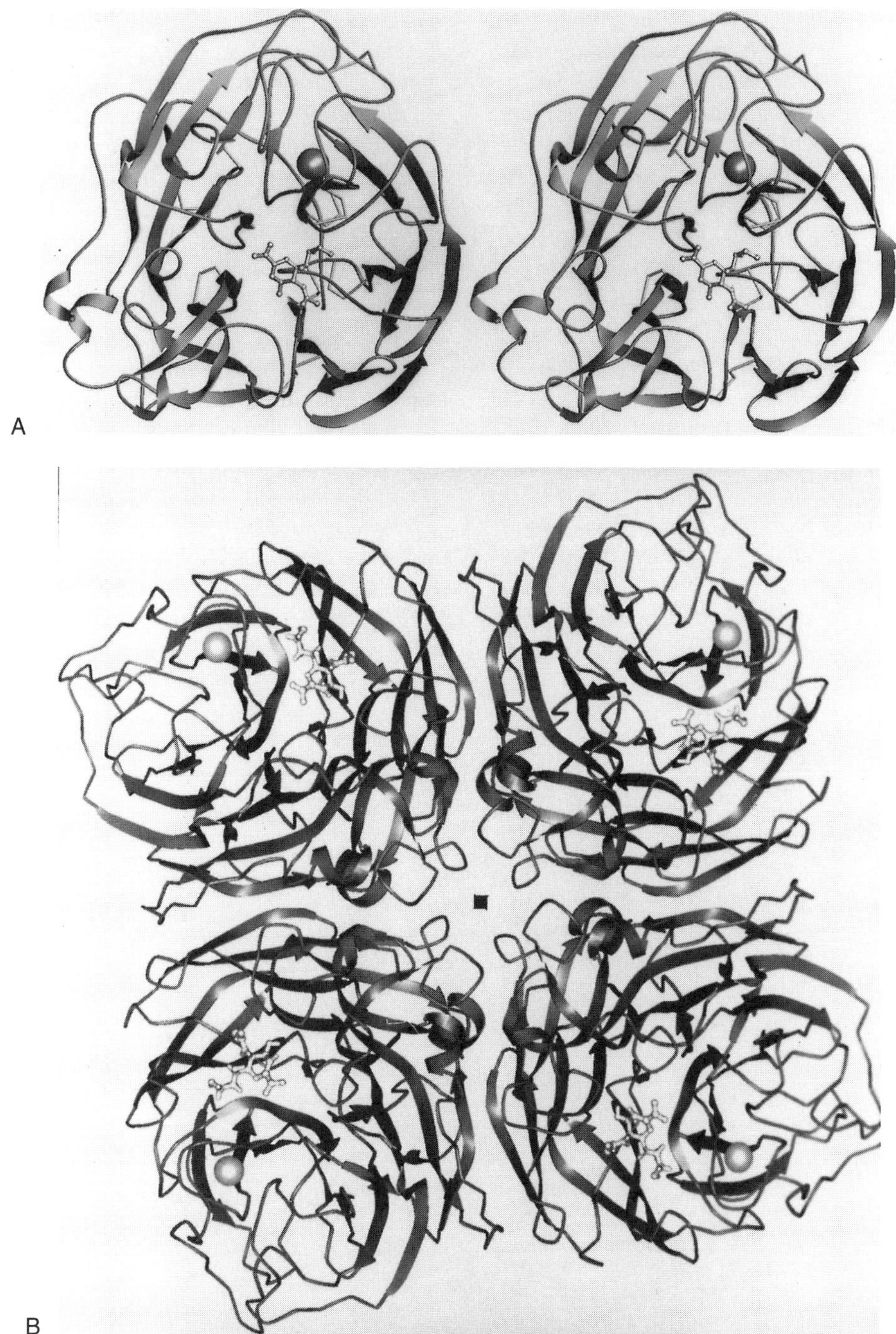

FIG. 6. Chain trace of the influenza neuraminidase monomer (subtype N9). **A:** The stereo ribbon drawing shows an inhibitor, 2-deoxy 2,3-dehydro-N-acetyl neuraminic acid, bound in the active site. The sphere is a Ca^{2+} atom. The B sheets (*arrows*) and ribbons show the two short helical segments, one being the C-terminal residues. The darkest shading corresponds to the most rigid segments; note that many of these line the active site crate. The fourfold axis is at *lower left.* **B:** Schematic ribbon drawing of the NA homotetramer to show the alignment of the four monomers. The diagrams were drawn using the program Ribbons (52). Data from Bossart-Whitaker et al. (28). Courtesy of Gillian Air, University of Alabama, Birmingham, Alabama.

degrees of homology for the cytoplasmic tail, the transmembrane domain, the stalk, the head overall, and the catalytic active site. The cytoplasmic tail N-Met-Asn-Pro-Asn-Gly-Lys is conserved among influenza A subtypes (perhaps for a role in viral assembly), but not between type A and B viruses. The transmembrane domains of the A virus subtypes and B viruses share only the common property of hydrophobicity and not sequence homology. The stalks show a great deal of variability, even between influenza A virus subtypes, both in amino acid sequence and in length (62 to 82 residues). The head domain shows extensive homology (42% to 57%) between influenza A virus subtypes and some homology (29%) between the type A and B viruses. The positions of the cysteine residues are conserved, and the expected similarity of protein fold was confirmed in a comparison of the three-dimensional structure of the type A and B NAs (28,49). Most significantly, within the head region, 11 conserved residues bind N-acetyl neuraminic acid. These residues are located discontinuously on the primary sequence but are known to comprise the active site in the three-dimensional structure; they are conserved among both influenza A and B viruses (28,49,375).

Influenza B virus RNA segment 6, in addition to containing the open reading frame (ORF) which encodes NA (466 residues), contains a second ORF of 100 residues (NB) that overlaps the NA reading frame by 292 nucleotides (342) (Fig. 7). The influenza B virus NA polypeptide is synthesized using the second AUG initiation codon from the 5′ end of the mRNA. The NB ORF begins with the first AUG codon from the 5′ end of the mRNA, and it is separated from the second AUG codon by four nucleotides (342). The NB ORF is translated in influenza B virus-infected cells to yield the NB glycoprotein; this protein is discussed further below.

Three-Dimensional Structure of NA

The pronase-released neuraminidases of A/Tokyo/3/67, A/RI/5⁻/57, A/Tern/Australia/75, and B/Beijing/87 have been crystallized and from x-ray diffraction data, the three-dimensional structure has been determined to 2.9Å (28,49, 61,62,370,374). The box-shaped head (100Å×100Å×60Å) has circular, four-fold symmetry stabilized in part by calcium ions believed to be bound on the four-fold axis of symmetry among a cluster of eight acidic groups. Each monomer of NA folds into six topographically identical, four-stranded, antiparallel β-sheets arranged in the manner of the blades of a propeller (374). The amino acids involved in the active site of NA were identified based on locating the substrate sialic acid in the molecule and obtaining an electron difference map (62). The substrate-binding site was identified to be a large pocket on the surface of each subunit rimmed by charged residues and consisting of nine acidic and six basic residues that are conserved between influenza A and B viruses. Mutagenesis of these conserved residues indicate that changes in these sites result in a loss of enzymatic activity (230).

In a manner analogous to that done for HA, antigenic sites were mapped on NA by analysis of changes in naturally existing strains and in variants selected with monoclonal antibodies. Four antigenic sites were identified in

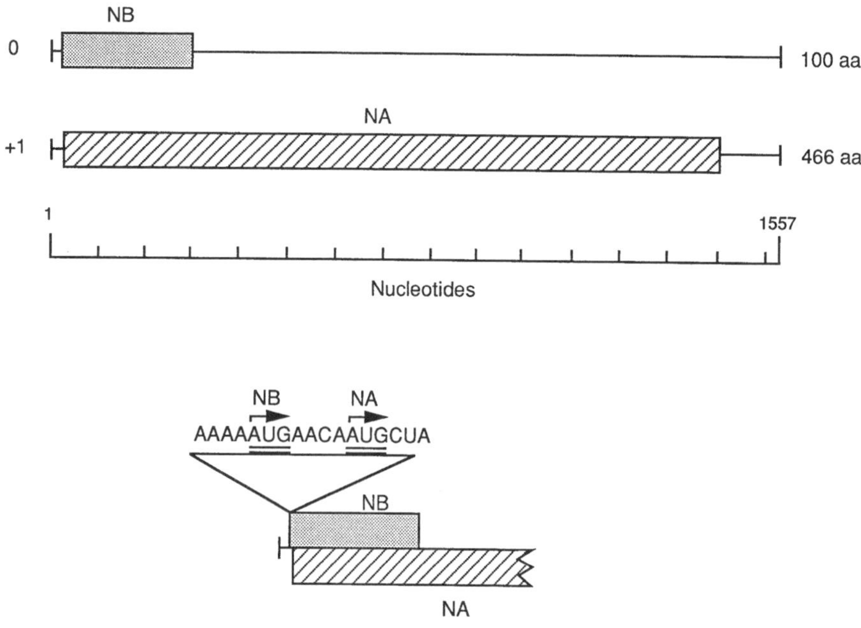

FIG. 7. Schematic representation of ORFs in influenza B virus RNA segment 6. The NB and NA overlapping reading frames are shown. Lower section nucleotide sequences surrounding the AUG codons used to initiate the synthesis of NB and NA glycoproteins. From Shaw et al. (340,342).

influenza A NA, each consisting of multiple epitopes (390) and these cluster into the distal surface loops which connect the various strands of β-sheet. It should be noted that antibodies to NA are not neutralizing but they inhibit plaque size enlargement when incorporated into the overlay medium (182).

The three-dimensional structure of a complex between influenza A virus N9 neuraminidase and the Fab fragment of a monoclonal antibody has also been obtained. The site recognized by the antibody involves 19 NA residues in four loops covering 600Å on the rim of the active site. Although it was first thought that both the catalytic site of NA and the antibody undergo a small conformational change on binding (61) the refined x-ray structure indicates that this is not the case (371).

The Influenza B Virus NB Glycoprotein

The NB ORF on influenza B virus RNA segment 6 encodes the NB glycoprotein (Fig. 7). NB is an integral membrane protein that is abundantly expressed at the plasma membrane of influenza B virus-infected cells (340), and genetic and biochemical data has indicated that NB is oriented in membranes with an 18-residue N-terminal ectodomain, a 22-residue transmembrane domain, and a 60-residue cytoplasmic tail (405). NB lacks a cleavable signal sequence; it instead contains a single hydrophobic domain that is thought to both target NB to membranes and anchor NB in a stable manner. The N-terminal ectodomain of NB contains two N-linked carbohydrate chains attached to residues 3 and 7 (405). In influenza B virus-infected cells, NB is detected within 2 minutes of synthesis as an M_r ~11,000 unglycosylated species; with time, it acquires one carbohydrate chain (NB M_r ~15,000) and by 10 minutes has two carbohydrate chains in the high mannose form (NB M_r ~18,000). These carbohydrate chains are further modified by the addition of a number of repeating units of galactose β1→4-N-acetylglucosamine β1→3 (Gal β1→4-GlcNAcβ1→3) attached to a $(mannose)_3$ $(GlcNAc)_2$ core oligosaccharide, known as *polylactosaminoglycan modification* (405,406). Originally, NB was thought to be lacking from virions but recent analysis of virions, after endo-β-galactosidase digestion to remove polylactosaminoglycan, suggests that NB is a component of virions present at about 10–50 copies per virion (37). The function of NB in the influenza virus life cycle is not proven. The influenza B virus NB glycoprotein has the same uncommon membrane orientation [type III integral membrane protein (380)] as the influenza A virus M_2 protein encoded by RNA segment 7 (see below and Fig. 3). The influenza A virus M_2 protein has ion channel activity, but influenza B virus lacks a direct M_2 counterpart. As the replication strategies of the influenza type A and B viruses are similar, there has been considerable speculation that the influenza B virus NB protein may have ion channel activity.

The available evidence indicates that the mRNA derived from influenza B virus RNA segment 6 is bicistronic. The initiation of protein synthesis for eukaryotic mRNAs usually occurs at the 5′ proximal AUG codon. Most examples of translational initiation can be accommodated by the modified scanning hypothesis of eukaryotic translation (191), in which 40S ribosomal subunits enter the 5′ end of the mRNA and "scan" the nucleotide sequence, examining each AUG codon for its potential use as an initiation codon. From examination of the NB/NA mRNA sequence in conjunction with the rules of the modified scanning hypothesis, it would be predicted that only NB and not NA would be synthesized in infected cells. However, NB and NA accumulate in a 0.6:1 ratio, suggesting that approximately 60% of ribosome preinitiation complexes scanning the NB/NA mRNA do not initiate protein synthesis at the first AUG codon but continue scanning to the second downstream AUG codon. The importance of nucleotide sequences surrounding the two AUG codons was investigated by altering many of the residues flanking the initiation codons. Most of the changes had very little effect on the ratio of NB:NA that accumulated. However, deletion of 40 of the 46 influenza virus-specific 5′-untranslated region nucleotides decreased NA synthesis tenfold, suggesting that an unrecognized feature of the entire region, such as a specific secondary structure, also has a major role in the initiation of protein synthesis at the second AUG codon (407).

RNA Segment 7 of Influenza Virus

RNA segment 7 of influenza A virus encodes two known proteins, the matrix protein M_1 which lies inside the lipid envelope and constitutes the most abundant polypeptide in the virion, and the M_2 protein which is a minor compo-

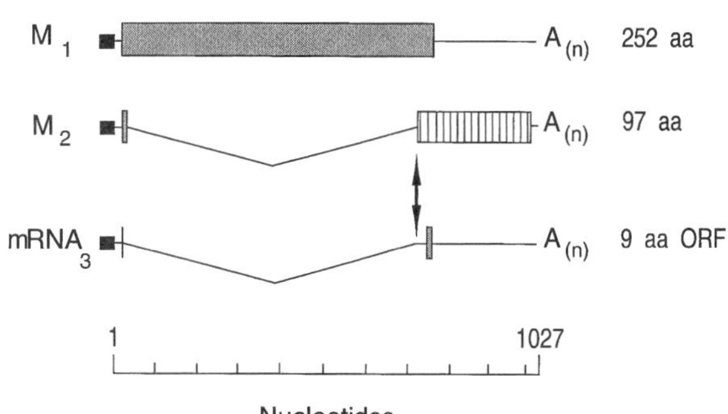

FIG. 8. Model for the arrangement of the influenza A virus M_1, M_2 and $mRNA_3$ and their coding regions. *Thin lines* at the 5′ and 3′ termini of the mRNAs represent untranslated regions. *Shaded or hatched areas* represent coding regions in 0 or +1 reading frames, respectively. The introns in the mRNAs are shown (*V-shaped lines*). *Rectangles* at 5′-ends of the mRNAs represent heterogeneous nucleotides derived from cellular RNAs that are covalently linked to viral sequences. No evidence has yet been obtained that $mRNA_3$ is translated *in vivo*. Adapted from Lamb et al. (218).

nent of virions and has ion channel activity. Three mRNA transcripts have been identified that are derived from influenza A virus RNA segment 7: a colinear transcript encoding M_1 protein; a spliced mRNA encoding the M_2 protein; and an alternatively spliced mRNA ($mRNA_3$) which has the potential to encode a 9 amino acid peptide, but it has not been recognized. Influenza B virus RNA segment 7 encodes two polypeptides using tandem cistrons, the matrix protein M_1, and the 109-residue BM_2 protein of unknown function. The equivalent RNA segment in influenza C virus (RNA segment 6) encodes the matrix protein which the available evidence indicates is translated from a spliced mRNA.

RNA Segment 7 Gene Structure and Encoded Proteins

Influenza A Virus

RNA segment 7 of influenza A virus encodes two known proteins, M_1 and M_2. The gene is 1,027 nucleotides in chain length and has one large ORF of 237 residues which encodes the M_1 protein (6,215,409) (Fig. 8). A colinear transcript mRNA encodes the M_1 protein, whereas the M_2 protein is encoded by a spliced mRNA (157,218). The M_2 mRNA contains a 51-nucleotide virus-specific leader sequence, a 689-nucleotide intron, and a 271-nucleotide [excluding poly(A) tail] body region. The leader sequence of the M_2 mRNA body region encodes 88 residues in the +1 reading frame and overlaps the M_1 protein by 14 residues. A second alternatively spliced mRNA ($mRNA_3$) has also been identified and it has a 5′ leader sequence of 11 virus-specific nucleotides, and shares the same 3′ splice site as the M_2 mRNA. No evidence has been obtained to indicate that M $mRNA_3$ is translated but if it is it would yield a 9-residue peptide identical to the C-terminus of the M_1 protein (218).

The M_1 protein contains 252 residues (M_r=27,801) and requires 0.5 M KCl to be solubilized (219), although it is not an integral membrane protein. The M_1 protein is a type-specific antigen of influenza virus and comparison of its predicted amino acid sequence among influenza A virus subtypes indicates it is highly conserved [reviewed in (207)].

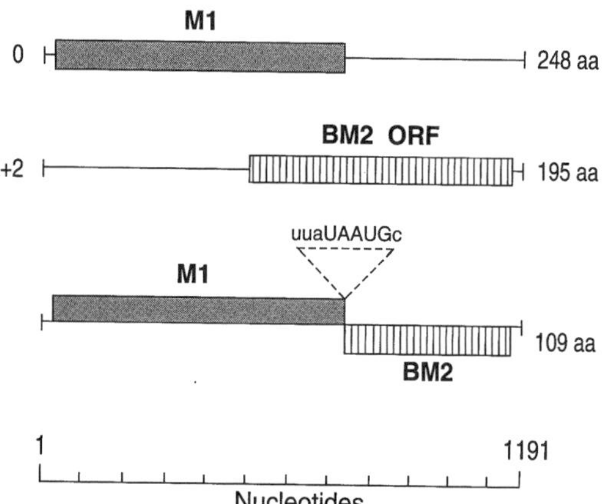

FIG. 9. Schematic representation of the ORFs in influenza B virus RNA segment 7. **Top two lines:** the ORF encoding the M_1 protein contains 248 residues, and the BM2 ORF consists of 195 residues (41). **Third line:** the extent of the BM2 ORF used to translate the BM2 protein found in influenza B virus infected cells. The pentanucleotide at which M_1 translation stops and M_2 translation starts is shown in capital letters. From Horvath et al. (148), with permission.

Influenza B Virus

RNA segment 7 of influenza B virus encodes two known proteins, M_1 and BM_2. The gene is 1,191 nucleotides in length and contains two ORFs (Fig. 9). The first ORF in the 0 reading frame begins at the AUG codon at nucleotides 25 to 27 and continues to a termination codon at nucleotides 769 to 771. This ORF encodes the 248 amino acid M_1 protein, 63 residues of which are identical with those of the influenza A virus M_1 protein (41). A second ORF in the +2 frame, overlapping the M_1 protein ORF by 86 residues, has a coding capacity of 195 residues and is designated BM_2 (41,135,148). A polypeptide, BM_2, derived from the BM_2 ORF was identified in cells infected with influenza B virus by using an antisera generated to a β-galactosidase-BM_2 ORF fusion protein (148). The BM_2 protein appears to be a soluble and cytoplasmically located protein of unknown function. In an attempt to understand the mechanism by which the BM_2 protein is synthesized, a mutational analysis of the cloned RNA segment 7 was performed. The data indicate that the BM_2 protein initiation codon overlaps with the termination codon of the M_1 protein in a translational stop-start pentanucleotide UAAUG, and that expression of the BM_2 protein requires termination of M_1 synthesis adjacent to the 5′ end of the BM_2 coding region; thus, termination of translation and the reinitiation event are tightly coupled (148) (see Fig. 9). Reinitiation of translation at downstream AUG codons in eukaryotes is uncommon but has been found to occur with artificially constructed bicistronic mRNAs (191). However, in prokaryotes, coupled tandem cistrons with the termination codon of one gene overlapping the initiation codon for a downstream gene is a common situation for coordinating regulated bacterial genes (e.g., *trp* operon) [reviewed in (278)].

Influenza C Virus

RNA segment 6 of influenza C virus contains 1,180 nucleotides and contains a single ORF of 374 residues that could encode a polypeptide of M_r ~41,700 (415) (Fig. 10). However, the M_1 protein of influenza C virus has an M_r ~28,000 on polyacrylamide gels (64,362). Hybrid-selec-

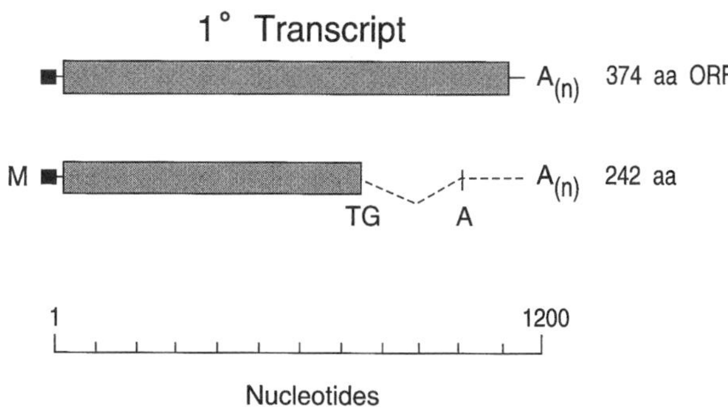

FIG. 10. Schematic representation of the influenza C virus mRNAs derived from RNA segment 6. The spliced mRNA is used to translate the matrix protein. *Thin lines* at the 5′ and 3′ termini of the mRNAs represent untranslated regions. *Stippled areas* represent the reading frame. The intron in the mRNA is shown (*V-shaped line*). *Rectangles* at the 5′ end of the mRNAs represent the presumptive host-cell nucleotides used as a transcription primer. Data from Yamashita et al. (415).

tion translation experiments of mRNAs isolated from influenza C virus-infected cells yielded a M_1 protein of M_r ~28,000, and analysis of the structure of the selected mRNA indicates that the RNA is spliced such that a translational termination codon is introduced after 242 residues of the ORF (see Fig. 10) (415). To date, a protein product translated from the unspliced transcript mRNA has not been identified.

The Matrix Protein

It is now virtually dogma that the matrix protein underlies the viral lipid envelope and provides rigidity to the membrane. In addition, it is widely believed that the M_1 protein interacts with the cytoplasmic tails of the HA, NA, and M_2 proteins and interacts with the RNP structures. However, experimental evidence for these very plausible interactions has been difficult to obtain. Solubility properties of purified M_1 protein (soluble in chloroform/methanol (112) and 0.5 M KCl (219) are consistent with the protein being a peripheral membrane protein. An interaction between M_1 and lipid has been shown *in vivo* using light-activated cross-linking (113) and *in vitro* using purified M_1 protein and liposomes (44). Contacts between the M_1 protein and the RNPs are suggested by finding that purified RNPs often contain M_1 protein (314), and by using immuno-EM it has been found that the RNPs are heavily labeled with M_1 antibody (261). Purified M_1 protein inhibits transcription *in vitro* (417,427), and if M_1 protein is not dissociated from the RNPs *in vivo* (in the presence of amantadine—see section below on the M_2 ion channel function), the RNPs fail to be transported to the nucleus (247). In addition, evidence has been presented that transport of the M_1 protein into the nucleus is required for exit of the newly assembled RNPs from the nucleus (247) (see below). An interaction between the M_1 protein and RNA has been demonstrated by using filter-binding assays and blotting procedures (383,417). The M_1 protein contains a zinc-binding motif (cys-cys-his-his type—residues 148 to 162 of influenza virus A/PR/8/34 M_1 protein), and this domain which is conserved in influenza A and B virus M_1 proteins can bind zinc (90). However, the zinc content of M_1 does not influence binding to RNA and thus the zinc-domain may be involved in protein-protein interactions. An interaction between M_1 and the M_2 protein has been suggested based on the observation that virus growth restriction by anti-M_2 antibodies can be overcome by mutations in M_1 (423) (see below). Finally, an interaction has been proposed between M_1 and the NS_2 protein found in purified virions (416).

The M_2 Protein and Its Ion Channel Activity

As discussed above, the M_2 integral membrane protein (97 amino acids) is encoded by a spliced mRNA derived from genome RNA segment 7 (210,218). The M_2 protein is abundantly expressed at the plasma membrane of virus-infected cells but is greatly underrepresented in virions, as only a few (on average 20 to 60) molecules are incorporated into virus particles (219,422,424). In polarized cells the M_2 protein is expressed at the apical cell surface (153), the surface at which influenza virus particles bud. The M_2 protein spans the membrane once, and by using domain-specific antibodies and specific proteolysis, it has been shown that M_2 protein is oriented such that it has 24 N-terminal extracellular residues, a 19-residue transmembrane (TM) domain, and a 54-residue cytoplasmic tail (219). The presence of an N-terminal extracellular domain in the absence of a cleavable signal sequence indicates that the M_2 protein is a model type III integral membrane protein [nomenclature of von Heijne and Gavel (380)] that is dependent on the signal recognition particle for cotranslational insertion into the ER membrane (154) (Fig. 3). The M_2 protein cytoplasmic domain is posttranslationally modified by phosphorylation on serine (residue 64) (141) and palmitylation on cysteine (residue 50) (141,364,376).

The native form of the M_2 protein is minimally a homotetramer consisting of either a pair of disulfide-linked dimers or disulfide-linked tetramers (139,288,365) (Fig. 11).

An indication as to one function of the M_2 protein in the influenza virus life cycle came from studies with a monoclonal antibody (14C2) specific for the N-terminal domain of M_2. When this antibody was included in an agarose overlay of a standard plaque assay titration, it was found to restrict the size of plaque growth of a variety of influenza A virus strains (422). Variant viruses resistant to the antibody were isolated, and they were found to have compensating changes in the cytoplasmic tail of M_2 as well as in the N-

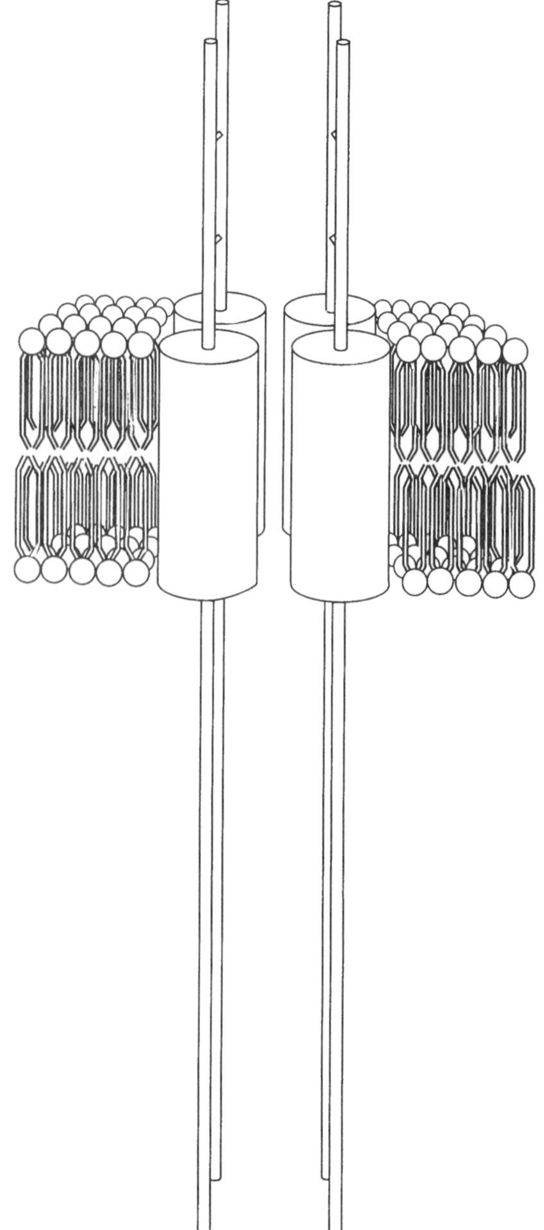

FIG. 11. Schematic diagram of the influenza virus M_2 protein ion channel in a membrane. The influenza virus M_2 protein is a disulfide-linked homotetramer with each chain consisting of 97 amino acid residues with 24 residues exposed extracellularly, a 19-residue transmembrane domain, and a 54-residue cytoplasmic tail. The disufide bonds can form either between the same subunit partner as shown, or once the first bond is made the second disulfide can link to another partner to form the fully disulfide-linked tetramer. The M_2 protein has a pH-activated ion channel activity that conducts principally hydronium ions. The channel is specifically blocked by the antiviral drug amantadine hydrochloride. The pore of the channel has been shown to be the transmembrane domain of the M_2 protein. The ion channel activity is essential for the uncoating of influenza virus in endosomal compartments in the infected cell. Data from Lamb et al. (219).

terminal domain of the M_1 protein (423), suggesting that the antibody may be interfering with critical M_1/M_2 interactions during virus assembly and budding.

Amantadine (1-aminoadamantane hydrochloride) displays a specific antiinfluenza A virus action (75) and has been used in the prophylaxis and treatment of influenza A virus infections [reviewed in (285)]. The M_2 protein TM domain was deduced to be the target of the antiinfluenza virus drug amantadine, as mutants which arose that were resistant to amantadine contained amino acid changes in the M_2 protein TM domain (123). From studies on the effect of amantadine on virus replication (see section concerning virus adsorption and entry and section on virion assembly), it was proposed that the M_2 protein functions as an ion channel that permits ions to enter the virion during uncoating and also acts as an ion channel that modulates the pH of intracellular compartments (365) [reviewed in (119)].

Direct evidence that the M_2 protein has ion channel activity was provided by expressing the M_2 protein in oocytes of *Xenopus laevis* and in mammalian cells, then measuring membrane currents. It was found that the M_2 ion channel activity is blocked by amantadine and that the channel activity is regulated by changes in pH, with the channel being activated at the lowered pH found in endosomes and the *trans* Golgi network (TGN) (140,295,384,385). Altered M_2 proteins containing changes in the TM domain—which, when found in virus lead to resistance to amantadine—when expressed in oocytes, exhibited ion channel activities that were not affected by the drug (295). The ion selectivity of the M_2 ion channel activity indicated the M_2 channel is permeable to Na^+ ions, although the major monovalent cation conductance is the hydronium ion (295). Specific changes in the M_2 protein TM domain alter the kinetics and ion selectivity of the channel, providing strong evidence that the M_2 TM domain constitutes the pore of the channel (295). This notion is supported by the finding that when a peptide corresponding to the M_2 protein TM domain was incorporated into planar membranes, a proton translocation susceptible to block by amantadine could be detected (87). Reconstitution of purified M_2 protein into planar bilayers has revealed a pH-activated and amantadine-sensitive channel activity, which formally demonstrates that the observed channel activity is intrinsic to the M_2 protein (369).

RNA Segment 8 of Influenza Virus: Unspliced and Spliced mRNAs Encode the NS_1 and NS_2 Proteins

RNA segment 8 of influenza A and B viruses and the smallest RNA segment of influenza C virus (RNA 7) encode two proteins, NS_1 and NS_2. NS_1 is encoded by a colinear mRNA transcript, whereas NS_2 is encoded by a spliced mRNA. The finding of a spliced NS_2 mRNA was the first evidence for splicing with an RNA virus (one that does not use a DNA intermediate in its replication).

RNA Segment 8 Gene Structure and Encoded Proteins

RNA segment 8 of influenza A virus is 890 nucleotides in length and contains a large ORF encoding the NS_1 protein (M_r ~26,000) ranging from 202 to 237 residues, depending on the virus. In addition to the NS_1 protein, genetic and biochemical evidence has shown that RNA segment 8 encodes a second protein, NS_2 (M_r ~14,000) and that NS_2 is translated from a small mRNA (~350 nucleotides) (155,156,208,209,212,213). Nucleotide sequencing studies indicated that the NS_1 mRNA is unspliced, directly encoding the NS_1 protein, whereas the NS_2 mRNA contains a 473-nucleotide intron. NS_1 and NS_2 share a 56-nucleotide leader sequence which contains the AUG codon used for initiation of protein synthesis such that NS_1 and NS_2 share 9 N-terminal amino acids before the intron, then translation of the body of the NS_2 mRNA continues in the +1 ORF which overlaps the NS_1 frame by 70 residues (214). The arrangement of the NS_1 and NS_2 mRNAs and their ORFs is shown in Fig. 12.

RNA segment 8 of influenza B virus is 1,096 nucleotides in length and contains a large ORF of 281 residues encoding NS_1 (281 residues M_r=32,026). An analogous arrangement of unspliced and spliced mRNA transcripts is found to encode NS_1 and NS_2 as described above for influenza A virus. The NS_2 mRNA has a 5′ leader sequence of 75 nucleotides which are shared with the NS_1 mRNA and contains the AUG codon used for initiation of protein synthesis such that 10 N-terminal residues are shared between NS_1 and NS_2. The 350-nucleotide body region of the NS_2 mRNA encodes in the +1 reading frame 112 residues which overlaps the NS_1 ORF by 52 residues (39,40,279).

RNA segment 7 of influenza C virus contains 934 nucleotides. The NS_1 protein contains 286 residues (267). The NS_2 protein (122 residues) is encoded by a spliced mRNA: 62 N-terminal residues are shared by NS_1 and NS_2 and after the splice junction, translation continued in the +1 ORF for 59 residues. The second ORF of influenza virus RNA segment 7 is completely overlapped by that of the NS_1 protein (268).

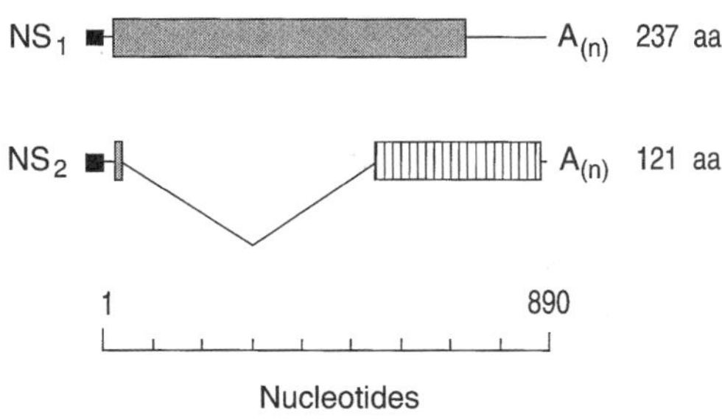

FIG. 12. Model for the arrangement of the influenza A virus NS_1 and NS_2 mRNAs and their coding regions. *Thin lines* at the 5′ and 3′ termini of the mRNAs represent untranslated regions. *Shaded or hatched areas* represent coding regions in 0 or +1 reading frames, respectively. The intron in the mRNA is shown (*V-shaped line*). *Rectangles* at 5′ ends of the mRNAs represent heterogeneous nucleotides derived from cellular RNAs that are covalently linked to viral sequences. Adapted from Lamb et al. (212).

The NS_1 Protein

The NS_1 protein (M_r ~26,000) is abundant in influenza virus-infected cells but it has not been detected in virions, hence the designation *NS* for *nonstructural* (198,224,344). NS_1 is a phosphoprotein (306–308), and the protein is found in infected cells associated with polysomes and also in the nucleus and nucleolus (63,198,201,224). With certain influenza A virus strains late in infection, NS_1 forms electron-dense paracrystalline inclusion bodies (257,341) which contain a mixture of cellular RNA species (419). The nuclear localization signals within NS_1 have been mapped, and it has been found that the protein contains two separate signals (residues 34 to 38 and within residues 203 to 237) (111).

In influenza virus A/Turkey/Oregon/71 the NS_1 protein is of only 124 residues (279), whereas in other subtypes the protein is of 237 residues. Where the NS_1 and NS_2 ORFs overlap (in NS_1 proteins of 237 residues) the amino acid sequences indicate that NS_2 is conserved at the expense of NS_1 [reviewed in (205)]. These observations support the original suggestion (412) that NS_1 and NS_2 may have been colinear on the vRNA but not overlapping and that read-through of a termination codon at the C-terminus of NS_1 allowed NS_1 protein to become longer.

Recently, some of the functions of the NS_1 protein in the influenza virus life cycle have been determined, and these are described below.

The NS_2 Protein

The NS_2 protein (M_r ~11,000) until recently was thought to be nonstructural but is now thought to exist in virions (130 to 200 molecules on average) and to form an association with the M_1 protein (317,416). The subcellular localization of NS_2 has been indicated to be nuclear (110) and cytoplasmic (354), and the role of the protein in the replicative cycle is unknown.

STAGES OF REPLICATION OF INFLUENZA VIRUSES

Virus Adsorption, Entry, and Uncoating

Influenza viruses bind to sialic acid residues present on cell surface glycoproteins or glycolipids via the receptor-binding site in the distal tip of the HA molecules (see above). Different influenza viruses have different specificities for sialic acid linked to galactose by α2,6- or α2,3-linkages, and this is dependent on specific residues in the HA receptor-binding pocket (394). Although the interaction of HA with sialic acid is of fairly low affinity, a high avidity of the virus for cell surfaces is probably achieved by multiple low-affinity interactions.

Influenza viruses enter cells by a process of engulfment, historically called *viropexis* (95), which today is called *receptor-mediated endocytosis*, the high-capacity endocytic activity by which components of the medium are internalized in clathrin-coated membrane-bound vesicles formed by invagination of specialized coated-pit domains of the plasma membrane. Following internalization, the clathrin coat is removed and vesicles fuse with endosomes (a series of organelles of increasingly acidic pH), beginning with mildly acidic primary endosomes and progressing to late endosomes. The acidification of endosomes is brought about by H^+-ATPases. The uncoating of influenza virions in endosomes is dependent on the acidic pH of this compartment, and agents such as acidotropic weak bases (ammonium chloride, chloroquine) and carboxylic ionophores (monesin and nigericin) raise the pH of endosomes and block influenza virus uncoating (249). However, these acidotropic weak bases do not have a specific antiinfluenza virus effect: raising the intralumenal pH of endosomes blocks the uncoating of most viruses that enter cells by the endocytic pathway [reviewed in (246)]. For the influenza virus RNPs to penetrate into the cytosol, they have to cross the membrane of the virion and that of endosomes. This is accomplished by HA-mediated fusion of the viral membrane with the cellular membrane (see above) (Figs. 13, 14). The precise time and location of penetration depends on the pH dependence of the transition of HA to its low-pH form. For human strains such as A/Japan/305/57, the transition to the HA low-pH form occurs at pH 5.3 which is found in late endosomes and occurs with a $t_{1/2}$ ~20 to 35 min after virus entry (359).

Studies with the antiviral drug amantadine have lead to the notion that the M_2 ion channel activity is essential for the uncoating process. For all influenza A virus strains, the amantadine block to viral replication occurs at an early stage in the virus life cycle between the steps of virus penetration and uncoating (45,168,349). Recent findings suggest that in the presence of the drug, the influenza virus M_1 protein fails to dissociate from the RNPs, and transport of the RNPs to the nucleus is blocked (247). It is general-

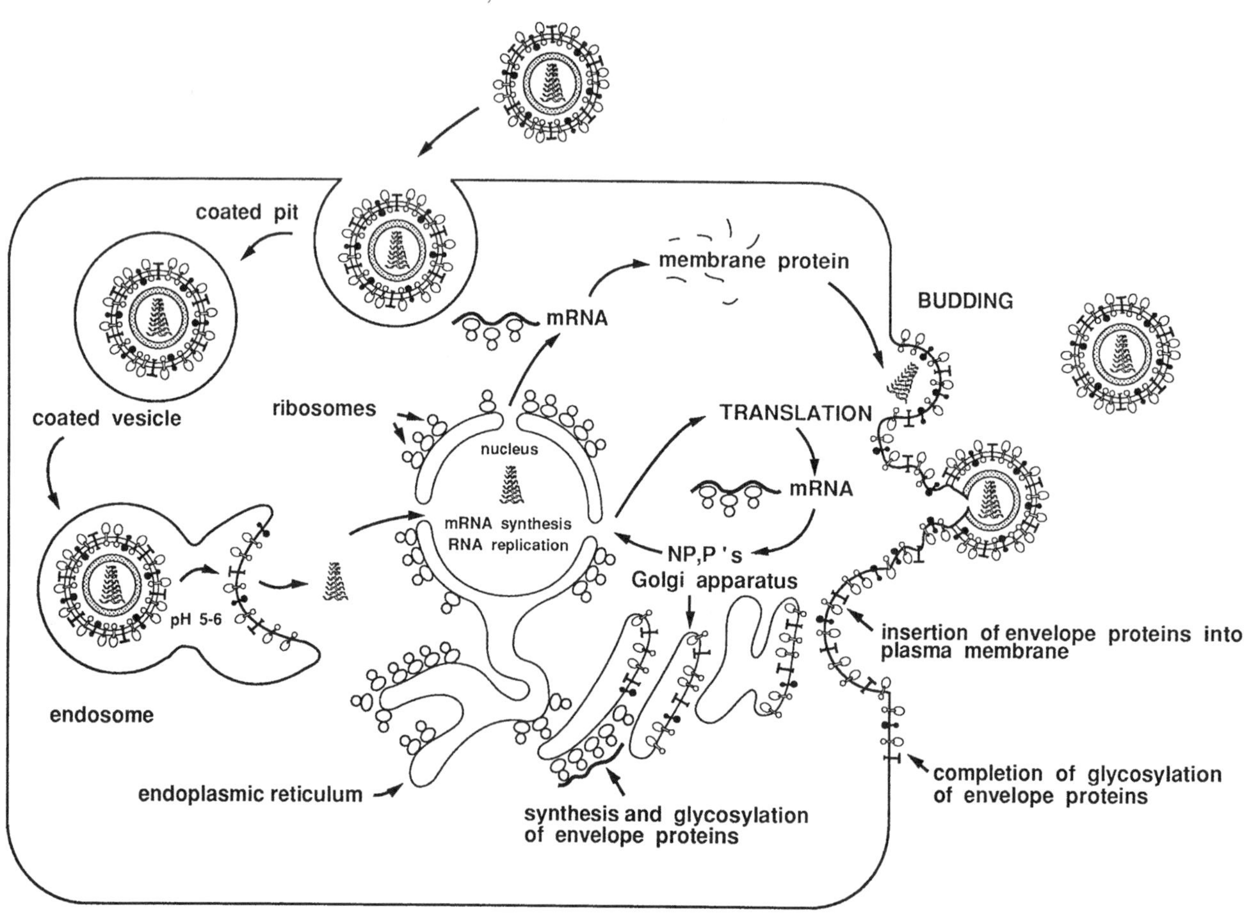

FIG. 13. Schematic diagram of the life cycle of influenza virus. See text for details of the model.

ly believed that once a virion particle has been endocytosed, the low pH activated ion channel activity of the virion-associated M_2 protein (140,295,385) permits the flow of ions from the endosome to the virion interior to disrupt protein-protein interactions and free the RNPs from the M_1 protein (Fig. 14) [reviewed in (119,125,245,346)]. Support for this idea comes from detergent solubilization studies that demonstrate that the interaction between M_1 and the RNPs can be disrupted by mildly acidic pH (426). An attractive feature of this hypothesis for virus uncoating is that the low pH of endosomes where uncoating occurs, in contrast to the neutral pH at the plasma membrane where assembly occurs, would push the equilibrium of the disassembly-assembly process in favor of uncoating. Another consequence of the M_2 protein ion channel activity in virions during the uncoating process may be to prepare HA for the fusion process. Recently, it has been found that fusion of influenza virus with liposomes was slowed in the presence of amantadine, and it was suggested that intraviral low pH facilitates influenza virus fusion, possibly by weakening interactions of the cytoplasmic tail of HA with the M_1 protein and/or RNPs (42,397).

Transcription of the input RNPs into mRNA occurs in the nucleus (see below). Microinjection and immunoelectron microscopy experiments indicate that the RNPs enter the nucleus through the nuclear pore as an intact RNP. Transport is not dependent on intact microfilaments, microtubules or an intermediate filament network. The M_1 protein which has dissociated from the RNP, or newly synthesized M_1 protein, enters the nucleus by passive diffusion as it is below the 70-kD cutoff for energy-dependent selective nuclear uptake mechanisms. An examination of the fate of the proteins of the infecting virion indicate that whereas HA is degraded in lysosomes ($t_{1/2}$ ~120 min), over 30% to 35% of M_1 and NP persists for many hours (248).

Overview of Transcription and RNA Replication

The vast majority of the research on influenza virus gene expression and RNA replication has been carried out with influenza A viruses. As already indicated above, influenza A viruses contain eight genome segments. Each genome segment or virion RNA (vRNA) contains 12 and 13 con-

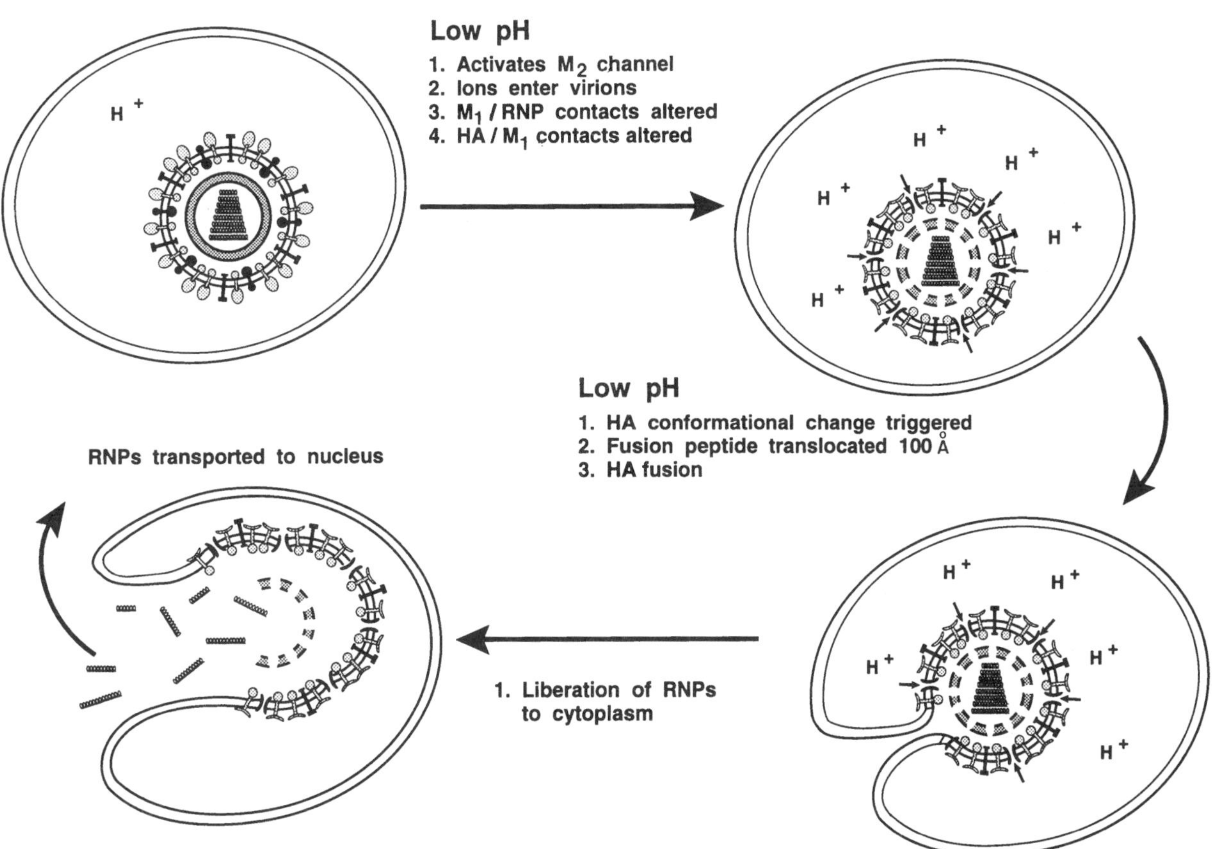

FIG. 14. Schematic diagram of the proposed role of the M_2 ion channel activity in virus entry. The M_2 ion channel activity is thought to facilitate the flow of ions from the lumen of the endosome into the virion interior, bringing about dissociation of protein-protein interactions between the HA cytoplasmic tail and M_1, M_1 and lipid and/or RNPs and M_1 from the RNPs.

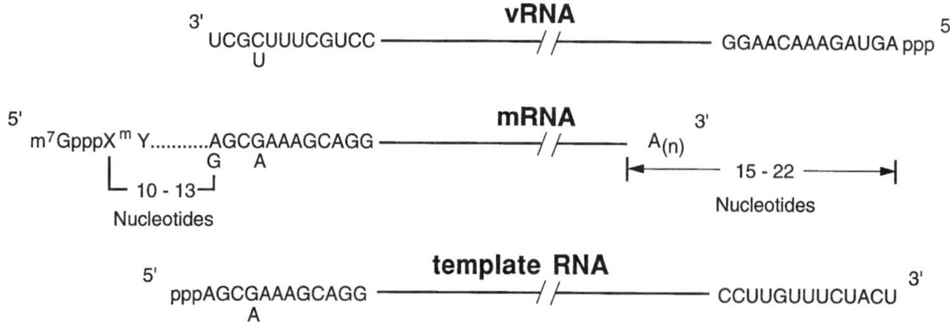

FIG. 15. Schematic diagram to illustrate the differences between influenza virus virion RNA (vRNA) segments, mRNAs and full-length cRNA or template RNA. The conserved 12 nucleotides at the 3′ end and 13 nucleotides at the 5′ end of each influenza A virus vRNA segment are indicated. The mRNAs contain a m^7GpppX^m cap structure and on average 10 to 13 nucleotides derived from a subset of host cell RNAs (see Fig. 16 and text). Polyadenylation of the mRNAs occurs at a site 15 to 22 nucleotides before the 5′ end of the vRNA segment. The template RNA contains at its 5′ terminus pppA and it is a complete copy of the vRNA segment. Redrawn from Lamb and Choppin (211).

served nucleotides at its 3′ and 5′ end, respectively (Fig. 15). In infected cells the vRNAs are both transcribed into messenger RNAs (mRNAs) and replicated. In the early 1960s it was found that influenza virus replication was inhibited by actinomycin D (17), an inhibitor of DNA-dependent RNA transcription (315,316), and a role of the nucleus in the replication of this RNA virus was proposed (17,328). These findings were extended when it was found that α-amanatin, a specific inhibitor of DNA-dependent RNA polymerase II (96), inhibited the transcription of influenza virus *in vivo* but not *in vitro* (207,241,355). This led to a search for the role of cellular RNA transcription in the *in vivo* transcription of influenza virus mRNAs. The first data suggesting that the transcription process was unusual was the finding that the dinucleotide ApG, which is complementary to the first two nucleotides at the 3′ end of each vRNA segment (3′-UpCpG-..........-5′) greatly stimulated transcription, and the ApG was found to be incorporated directly into the 5′ end of the newly synthesized transcript (253). This was followed by the observation, when using a coupled transcription/translation system in rabbit reticulocyte lysates, that globin mRNA greatly stimulated transcription and that the globin mRNA m^7GpppG cap structure was transferred to the influenza virus mRNA transcript (297). Today, we know that the distinctive feature of influenza viral mRNA synthesis is that it is primed by 5′ capped ($m^7GpppNm$-containing) fragments derived from newly synthesized host-cell RNA polymerase II transcripts (30,31,134,193,196,297,298). The mRNA chains are elongated until a stretch of uridine residues is reached 15 to 22 nucleotides before the 5′ ends of the vRNAs, where transcription terminates and polyadenylate [poly(A)] is added to the mRNAs (120,320). For replication to occur, an alternative type of transcription is required that results in the production of full-length copies of the vRNAs. The full-length transcripts, or template RNAs, are initiated without a primer and are not terminated at the poly(A) site used during mRNA synthesis (120–122). The second step in replication is the copying of the template RNAs into vRNAs. This synthesis also occurs without a primer, since the vRNAs contain 5′ triphosphorylated ends (421). The three types of virus-specific RNAs—mRNAs, template RNAs, and vRNAs—are all synthesized in the nucleus (338,339,353). In the nucleus, the viral mRNAs undergo at least some of the same processing steps as cellular RNA precursors. Internal adenosine residues of influenza virus mRNAs are methylated (199,271), and two of the viral mRNAs act as precursors and are processed by splicing to form smaller mRNAs (see below).

Viral mRNA Synthesis

Viral mRNA synthesis in the nucleus requires initiation by host-cell primers, specifically capped ($m^7GpppNm$-containing) RNA fragments derived from host-cell RNA polymerase II transcripts (31,193,297,298). It is thought that the need for a continuous supply of newly synthesized host-cell primers explains the requirement for a functional cellular RNA polymerase II and the inhibition of viral mRNA synthesis by actinomycin D and α-amanatin. Host-cell primers are generated in the nucleus by a viral cap-dependent endonuclease that cleaves the capped cellular RNAs 10 to 13 nucleotides from their 5′ ends, preferentially at a purine residue (298). Priming does not require hydrogen bonding between the capped primer fragments and the 3′ ends of the vRNA templates (193,194,197). Rather, priming requires the presence of a 5′-methylated cap structure (30,297,298). Transcription is initiated by the incorporation of a G residue onto the 3′ end of the resulting fragments, directed by the penultimate C residue of the vRNAs (298). Viral mRNA chains are then elongated until a stretch of 5 to 7 uridine (U) residues is reached 15 to 22 nucleotides before the 5′ ends of the vRNAs. Termination occurs apparently as a result of stuttering or reiterative copying of the stretch of U residues, thereby adding a poly(A) tail to the 3′ ends of the viral mRNAs (238,320). The stretch of U residues in the vRNAs is adjacent to a potential panhandle structure formed between the 5′ and 3′ ends of the vRNAs: the conserved 3′ and 5′ sequences of the vRNAs, along with an additional two to three segment-specific bases at the 3′ and 5′ ends, show inverted complementarity (142,150). One model for polyadenylation is that the panhandle provides a physical barrier against transcription, thereby causing the stuttering. However, recent data has shown that the polymerase complex binds specifically and with high affinity to the 5′ end of the viral RNA and not the 3′ end (98,368). If the polymerase remains bound to the 5′ end during transcription, then the polymerase itself may provide the physical barrier needed for polyadenylation to occur. A model to illustrate the hypothesis is shown in Fig. 16.

A great deal of the knowledge about the mechanism of viral mRNA synthesis has been obtained from *in vitro* studies of the polymerase complex isolated from virions by detergent disruption of virions and isolation of the RNP cores. Viral mRNA synthesis is catalyzed by viral nucleocapsids (158,298) which consist of the individual vRNAs associated with four viral proteins: the NP protein and the three P proteins (PB1, PB2, and PA) (158,372). The P proteins are responsible for viral mRNA synthesis, and some of their roles have been determined by analyses of the *in vitro* reaction catalyzed by virion nucleocapsids. From data obtained by using ultraviolet (UV)-light-induced cross-linking, it was shown that the three P proteins exist in the form of a complex (35). The PB2 protein in this complex recognizes and binds to the cap of the primer RNA (26,35, 372). Cap recognition by PB2 was verified by analysis of the *in vitro* transcription pattern of the nucleocapsids derived from PB2 temperature-sensitive (ts) mutants (373). The UV-cross-linking experiments also showed that the PB1 protein, which is initially found at the first residue (a G residue) added onto the primer, moves as part of the P-protein complex to the 3′ ends of the growing viral mRNA chains, indicating that it most likely catalyzes each nucleotide addition (35). On the basis of the relative positions of PB1 and PB2 on the nascent chains, it was concluded that the P-protein complex most likely has the PB1 protein at its leading edge and the PB2 protein at its trailing edge (35). No specific role for PA in viral mRNA synthesis has been found.

In the last 5 years the characterization of the RNA signals required for chain initiation has become possible once it was found that the viral transcriptase activity could be reconstituted using polymerase proteins and recombinant RNA (transcribed from DNA). One source of polymerase proteins has been protein fractions isolated from intact virus (142,143,291,294). By using these preparations it was found that the 3′ end of the vRNA template was sufficient for transcriptase activity. However, a different result was obtained using an alternative source of the polymerase proteins: polymerase complexes formed in cells infected by recombinant vaccinia viruses encoding the three P proteins (115) or micrococcal nuclease-treated RNPs (98,337). In these latter cases, both the 3′ and 5′ ends of the vRNA were found to be required for initiation of transcription suggesting a role for both viral RNA ends in transcription (98,115). The contrasting results obtained with these systems probably reflects the source of enzyme used. Polymerase obtained from virions may still contain the 5′ end of the vRNA, which it binds with high affinity. Polymerase formed by recombinant gene expression has never encountered virion RNA. This recombinant polymerase has also been used to provide insights into the assembly and activation of functional enzyme. A polymerase complex formed by coexpression in cells of PB2, PB1, and PA (by using recombinant vaccinia viruses that express the polymerase proteins) is inactive in all *in vitro* assays designed to investigate the various functions of the enzyme. Addition of RNA containing the 5′ end of vRNA stimulates the cap binding activity of the polymerase, but not the endonuclease function (59). Subsequent addition of RNA containing the 3′ end sequence from vRNA stimulates the endonuclease activity, leading to a functional transcriptase complex. It is likely that the transcriptase remains bound to the 5′ end of vRNA as the 3′ end of vRNA is transcribed. The current model for assembly of the transcriptase complex and transcription is shown in Fig. 16.

Replication of Virion RNA

Replication of virion RNA occurs in two steps: (a) the synthesis of template RNAs, the full-length copies of the vRNAs; and (b) the copying of template RNAs into vRNAs. To switch from the synthesis of viral mRNAs to that of template RNAs, it is necessary to change from capped RNA-primed initiation to unprimed initiation and to antiterminate at the poly(A) site, 15 to 22 nucleotides from the 5′ ends of the vRNA templates, used during viral mRNA synthesis. Because the switch from mRNA to template RNA synthesis requires protein synthesis *in vivo* (15, 122), one or more newly synthesized virus-specific proteins can be presumed to be needed for either unprimed synthesis or antitermination, or both. It has not been determined which, if any, protein is required for unprimed synthesis, whereas it has been established that NP proteins not associated with nucleocapsids are required for antitermination. This was shown by adding supernatant fractions (lacking nucleocapsids but containing soluble NP subunits) isolated from infected cells to transcribing infected-cell nucleocapsids and finding that transcription was antiter-

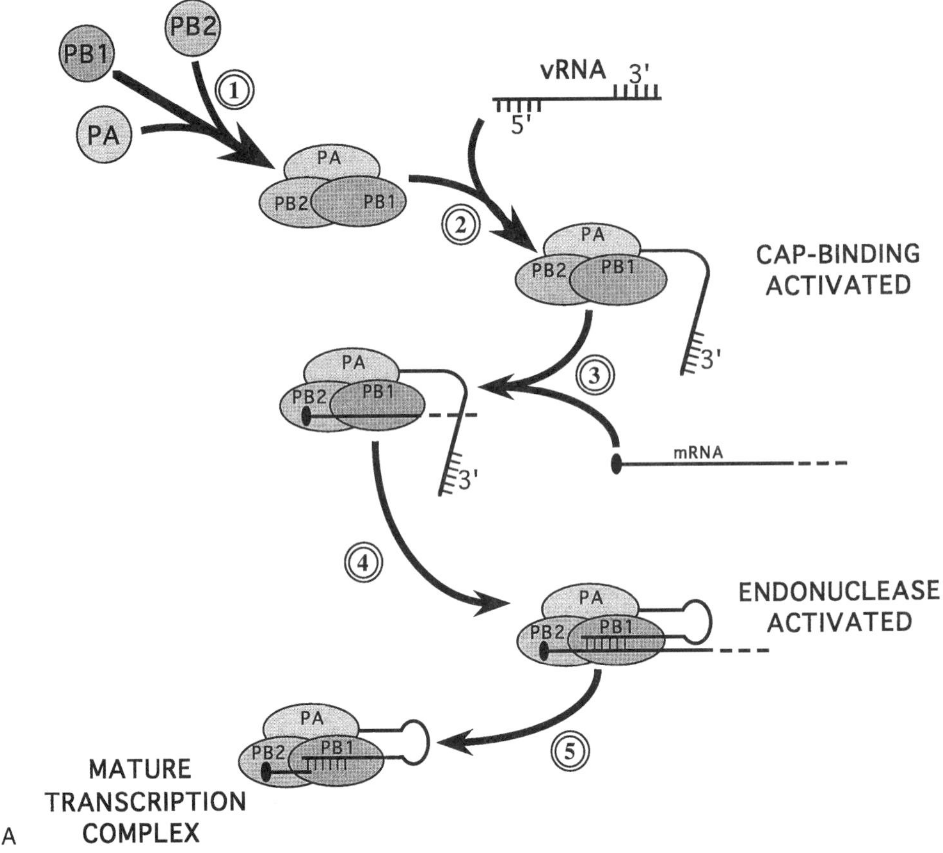

FIG. 16. Model for assembly of the polymerase complex and subsequent viral transcription. **A:** The three polymerase proteins assemble initially into an inactive complex. *1.* Upon binding to the 5′ end sequence of vRNA, mRNA cap-binding activity associated with the PB2 subunit is acquired. *2.* Interaction with the 3′ end of vRNA activates the endonuclease activity of the polymerase complex. This can occur prior to or subsequent to binding to the capped mRNA primer. Cleavage of the capped mRNA then occurs, creating a mature transcription complex. Data from Ciana et al. (59). **B:** Initiation usually occurs at the penultimate base and transcription then proceeds as the 3′ ends of vRNAs thread through the polymerase complex. The 5′ end sequences of vRNA remains tightly bound to the polymerase throughout transcription. The cap structure initially remains bound to the PB2 subunit, but becomes disengaged soon after elongation ensues. Polyadenylation occurs upon encountering the oligo U sequence adjacent to the 5′ end through a slippage mechanism, as the polymerase bound to the 5′ end does not allow read-through. Courtesy of L. Tiley and M. Krystal, Bristol-Meyers Squibb Pharmaceutical Research Institute, Wallingford, CT.

minated (19), and by finding that the antitermination activity was depleted using an antiserum directed against the NP protein. In addition, using a virus containing a *ts* mutation in the NP protein, template RNA but not mRNA synthesis was *ts* both *in vivo* and *in vitro* (339). The most likely explanation for these results is that the nonnucleocapsid NP protein molecules bind to the common 5′ ends of the nascent transcripts. The ensuing addition of NP molecules to the transcripts would permit read through when the termination site is reached, probably by forcing the panhandle to open.

The copying of template RNAs to form vRNAs also requires nonnucleocapsid-bound NP protein molecules (339). No vRNA-sense RNAs of discrete sizes are made in the absence of these soluble NP molecules, indicating that elongation of vRNA chains most likely ceases at any time at which NP is not available. It is possible that the replicase for influenza virus is a different enzyme complex from the transcriptase. The replicase would also catalyze the synthesis of template RNAs. In contrast, mRNAs would be synthesized by the transcriptase, which is independent of nonnucleocapsid NP molecules. Consequently, at least two types of P-protein complexes can be postulated to exist in infected cells. One type, which is also present in virion nucleocapsids, uses capped primer fragments to initiate mRNA synthesis and therefore requires the participation of the PB2 protein, but apparently not the PA protein (as indicated above). The second type of complex would initiate the synthesis of either template RNA or vRNA chains without a primer, would require the presence of nonnucleocapsid NP molecules, and is hypothesized to involve the action of the PA but not the PB2 protein.

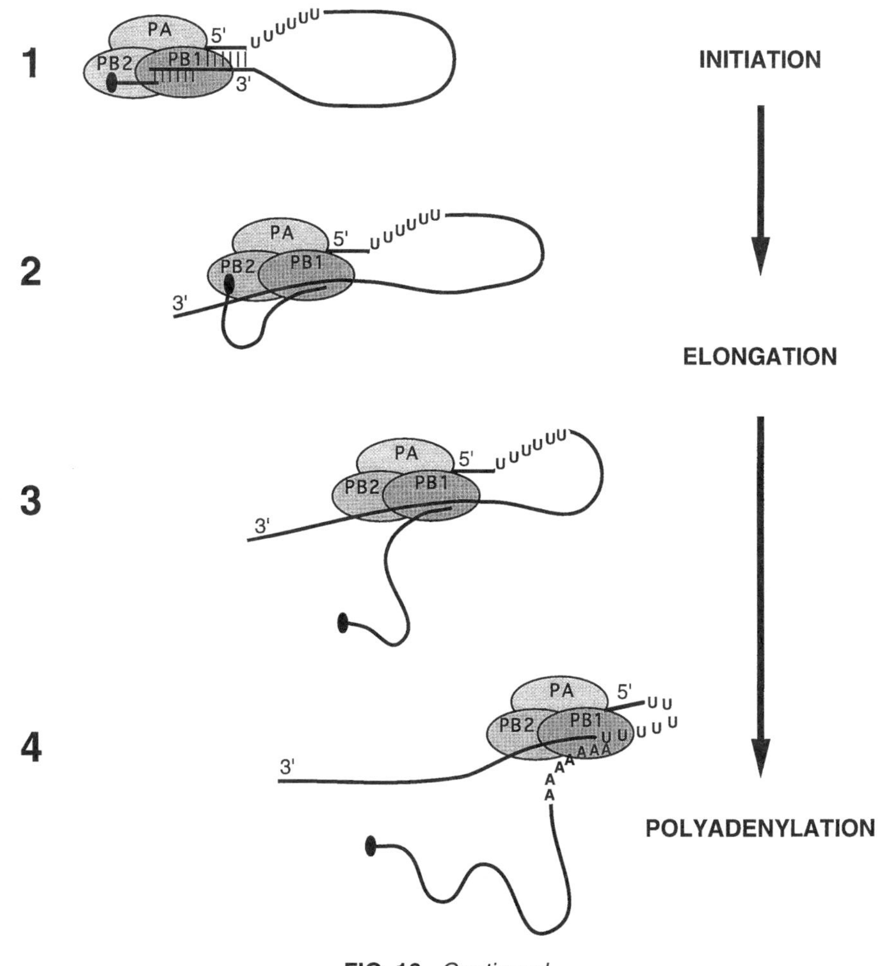

FIG. 16. *Continued.*

As described below, it has been found possible to replicate an artificial gene construction which contains the 5′ and 3′ terminal nucleotide sequences of influenza virus flanking a reporter gene [chloroamphenicol acetyltransferase (CAT)], when the synthetic RNA was mixed with a solubilzed virus polymerase complex and transfected into cells (239,291). To test which influenza virus proteins are required for genome replication, the CAT RNP was transfected into cells which were previously infected with vaccinia recombinant viruses that expressed various influenza virus proteins. It was found that the minimum subset of influenza virus proteins needed for specific replication of the synthetic CAT RNP is PB1, PB2, PA, and NP (152). Thus, although it had long been speculated upon that NS_1 and NS_2 might be involved in RNA replication, these data effectively eliminated this possibility.

Regulation of Viral Gene Expression In Infected Cells

Influenza virus infection can be divided into an early and late phase of gene expression. However, unlike DNA viruses where specific patterns of transcription occur before and after DNA replication, the phases of influenza virus gene expression are quantitative changes of transcription of individual RNA segments as oppposed to qualitative changes (206,345). During the early phase, the synthesis of specific vRNAs, viral mRNAs, and viral proteins are coupled (121,338,353). The first event detected after primary transcription is the synthesis of template RNAs, presumably copied off the parental vRNAs. Approximately equimolar amounts of each of the template RNAs are made. The peak rate of template RNA synthesis occurs early and then sharply declines. Specific template RNAs are then selectively transcribed into vRNAs. Specifically, the NS and NP vRNAs are preferentially synthesized early, whereas the synthesis of M vRNA is delayed. The rate of synthesis of a particular vRNA correlates with, and therefore most likely determines, the rate of synthesis of the corresponding mRNA and of its encoded protein. The NS_1 and NP mRNAs and proteins are preferentially synthesized at early times, whereas the synthesis of the M_1 mRNA and protein are delayed. Thus, the control of viral protein synthesis during the early phase is predominantly a direct consequence of the regulation of vRNA synthesis, i.e., the selective copying of specific template RNAs into vRNAs.

During the late phase, the relationships between the syntheses of vRNAs, viral mRNAs, and viral proteins change dramatically (339). The synthesis of all the viral mRNAs reaches its peak rate at the beginning of the late phase, and the rate of synthesis of all the mRNAs then decreases precipitously. In contrast, the rate of synthesis of all the vRNAs remain at or near maximum during the second phase. Thus, vRNA and viral mRNA synthesis are not coupled during the second phase. In addition, viral mRNA and protein synthesis are not coupled, as the synthesis of all the viral proteins continues at maximum levels during the second phase. Previously synthesized viral mRNAs undoubtedly direct viral protein synthesis during the second phase (see below). The pattern of protein synthesis during the second phase differs from that in the first phase. During the second phase two structural proteins, M_1 and HA, which are poorly synthesized in the first phase, are synthesized at high rates along with the other structural proteins.

As already discussed, the synthesis of all three types of virus-specific RNAs occurs in the nucleus. The template RNAs, which are synthesized only at early times, remain in the nucleus to direct vRNA synthesis throughout infection (339). In contrast, the vRNAs are efficiently transported to the cytoplasm, particularly at later times (see below). The export of viral mRNAs from the nucleus will be discussed below.

A significant part of the replication control system established by influenza virus in infected cells is directed at the preferential synthesis of the NP and NS_1 proteins early and at delaying the synthesis of the M_1 protein. The NP protein is synthesized early presumably because it is needed for template RNA and vRNA synthesis. The NS_1 protein is needed for another early function (see below). It is likely that the synthesis of the M_1 protein is delayed because this protein stops the transcription of vRNA into viral mRNA (427) and mediates the transport of vRNA (in the form of nucleocapsids) from the nucleus to the cytoplasm (247).

Role of the NS_1 Protein in Viral and Cellular Gene Expression

Two functions of the influenza virus NS_1 protein in infected cells have been identified recently: (a) NS_1 regulates the nuclear export of mRNA and (b) NS_1 inhibits pre-mRNA splicing. Like the Rev (and Rev-like) proteins of the lentiviruses [which includes human immunodeficiency virus 1 (HIV-1)] (114,242,243), the influenza virus NS_1 protein regulates the nuclear export of mRNA (10,99). However, whereas the HIV-1 Rev protein facilitates the nuclear export of unspliced viral pre-mRNAs, the NS_1 protein inhibits the nuclear export of spliced viral and cellular mRNAs. It was shown by using gel shift assays that the NS_1 protein binds specifically to poly(A) sequences (310). This binding specificity of NS_1 protein for RNA was also observed *in vivo* (310). The NS_1 protein inhibited the nuclear export of every poly(A)-containing mRNA that was tested. In contrast, the NS_1 protein failed to inhibit the nuclear export of a mRNA whose 3′ end was generated by cleavage without subsequent addition of poly(A). Addition of poly(A) to this mRNA enabled the NS_1 protein to inhibit the nuclear export of the mRNA.

Two functional domains required for regulating nuclear export of mRNA have been identified in the NS_1 protein of influenza A/Udorn/72 virus through an extensive mutational analysis (309). A domain near the N-terminal end of NS_1 (residues 19 through 38) was shown to be the RNA-binding domain as assayed by the elctrophoretic migrational shift of oligonucleotides when they bind protein (gel shift). This domain is well conserved among some influenza A virus strains (94% to 100% identity), is partially conserved among other influenza A virus strains (71% to 89% identity), and is only weakly conserved among a few influenza A virus strains (57% identity) (48). The RNA-binding domain sequence of the influenza virus A/Udorn/72 NS_1 protein is not homologous with the RNA binding domains of cellular poly(A)-binding proteins. The poly(A)-binding domains in cellular proteins consist of one or more RNA recognition motifs about 80 nucleotides in length, each of which contains both an 8-amino-acid and a 6-amino-acid conserved sequence (called *RNP-1* and *RNP-2*, respectively) [reviewed in (85)] and such a sequence does not exist in the NS_1 protein. In addition, the NS_1 RNA binding domain is not arginine-rich, as in some RNA-binding domains (225,242), does not even have a high net positive change (of the 20 amino acids, 6 are basic and 4 are acidic) and does not have homology with other known RNA-binding proteins.

The second NS_1 functional domain, which is located in the C-terminal 50% of the molecule, can be presumed to be the effector domain that interacts with host nuclear proteins to carry out the nuclear RNA export function, by analogy with the effector domain of the Rev proteins of HIV and other *Lentiviridae* which facilitate rather than inhibit nuclear RNA export (242). The NS_1 protein has a 10-amino-acid sequence that is similar to the consensus sequence in the effector domains of *Lentivirus* Rev proteins, including two crucial leucines at positions 7 and 9 of this sequence (242,309) (Fig. 17). However, the effector domains of the NS_1 and Rev (HIV-1) proteins differed in several significant ways: (a) unlike the HIV-1 Rev protein the influenza virus NS_1 protein effector domain mutants were found to be negative-recessive rather than negative-dominant; (b) the influenza virus NS_1 protein effector domain is about three times larger than the effector domain of the HIV-1 Rev protein; and (c) unlike the HIV-1 Rev protein, the influenza virus NS_1 protein effector domain mutants exhibited a changed intracellular/intranuclear distribution, compared with the wild-type protein (309). These differences strongly suggest that the effector domains of the NS_1 and

HIV-1 REV

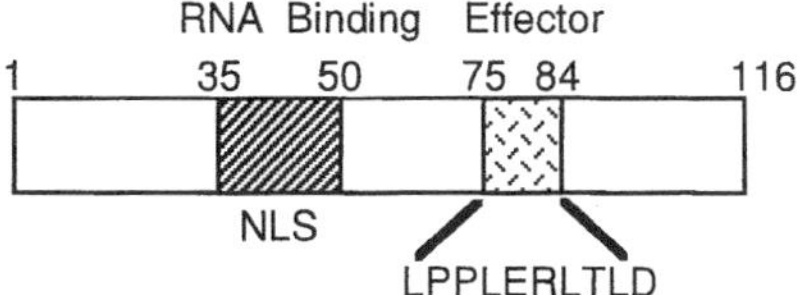

Influenza A Virus NS1

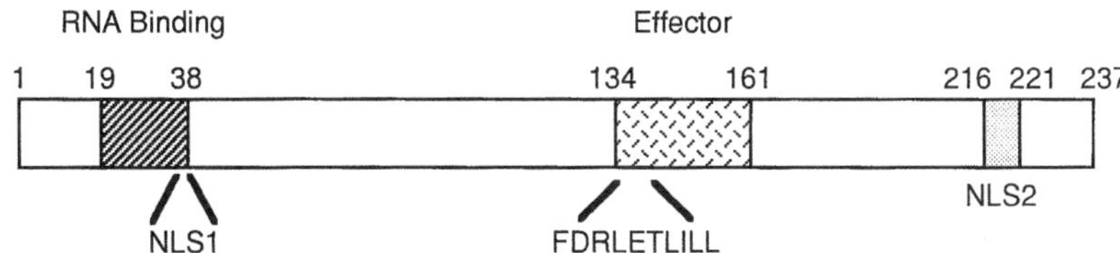

FIG. 17. Comparison of the functional domains of the influenza A virus NS_1 protein with those found in the HIV-1 Rev protein. The approximate locations of RNA-binding and effector domains are indicated. The 10-amino-acid effector domain sequence is shown with the two critical leucine residues (*underlined*). Redrawn from Qian et al. (309).

Rev proteins interact with different nuclear protein targets, and this would explain the opposite effect of these two proteins on nuclear mRNA export. Nonetheless, the data obtained to date indicate that two types of proteins that regulate the nuclear export of mRNA, *Lentivirus* Rev (and Rev-like) proteins and the influenza virus NS_1 protein, have the common property of containing two functional domains, an RNA-binding domain and an effector domain. One type (the Rev proteins) facilitates nuclear export, whereas the other type (the NS_1 protein) inhibits nuclear export. This is comparable to the situation with transcriptional activators and repressors, both of which usually contain two major domains: a DNA-binding domain and an effector domain which either activate or repress transcription (83). Analogous to transcriptional activators and repressors, the effector domain of the Rev proteins and of the NS_1 protein can function with a heterologous RNA-binding domain.

The other known function of the NS_1 protein is to inhibit pre-mRNA splicing both *in vivo* (99,236) and *in vitro* (236). The mode by which the NS_1 protein inhibits splicing has not been reported previously. Other proteins that regulate pre-mRNA splicing apparently operate by inhibiting or enhancing the formation of spliceosomes at specific splice site(s) [reviewed in (109)]. In contrast, the NS_1 protein (at moderate levels) allows the pre-mRNA to associate with the U1, U2, U4, U5, and U6 small nuclear ribonucleoproteins (snRNPs) to form spliceosomes, but then the NS_1 protein remains associated with these spliceosomes to inhibit the catalytic steps of splicing normally carried out by spliceosomal components (236). The RNA-binding domain of the NS_1 protein is required for the inhibition of splicing and for the interaction of the protein with spliceosomes (236). Whereas the NS_1 protein binds to a poly(A) sequence when it inhibits the nuclear export of mRNAs (310), as discussed above, when the NS_1 protein inhibits splicing it most likely interacts with a different RNA sequence. No poly(A) sequences were present in the *in vitro* splicing reactions, and the addition of a cellular poly(A)-binding protein had no effect on *in vitro* splicing (236). Furthermore, experiments in which the NS_1 protein selected its specific binding site from a random collection of oligoribonucleotides led to the demonstration that the NS_1 protein specifically binds to a sequence (nucleotides 83 to 101) at the 3′ end of U6 snRNA (311). This region of U6 snRNA includes the nucleotides (86 to 95) that form a helix (called *helix II*) with the 5′ end of U2 snRNA during splicing (73,413). The NS_1 protein blocks the formation of this U6-U2 helix (311), thereby indicating at least one mechanism whereby the NS_1 protein inhibits splicing. At higher levels of the NS_1 protein, the interaction of U6 snRNA with U4 snRNA is also blocked, leading to an inhibition of spliceosome formation (311). Because immunoprecipation of the NS_1 protein from extracts of influenza virus-infected cells precipitated the U6 snRNA but not other spliceosomal U snRNAs (236), it is likely that the NS_1 protein also inhibits pre-mRNA splicing in infected cells.

A model that rationalizes the advantage for influenza virus to have an NS_1 protein that both inhibits pre-mRNA splicing and the nuclear export of all poly(A)-containing mRNAs in infected cells is illustrated schematically in Fig. 18. First, the NS_1 protein function might be needed at appropriate times of infection to make the cellular mRNAs accessible to cleavage by the viral polymerase for the generation of the capped primers which are needed for viral mRNA synthesis. At the beginning of infection (Fig. 18, panel I), cellular pre-mRNAs would be spliced and spliced mRNAs would freely exit from the nucleus. Utilizing

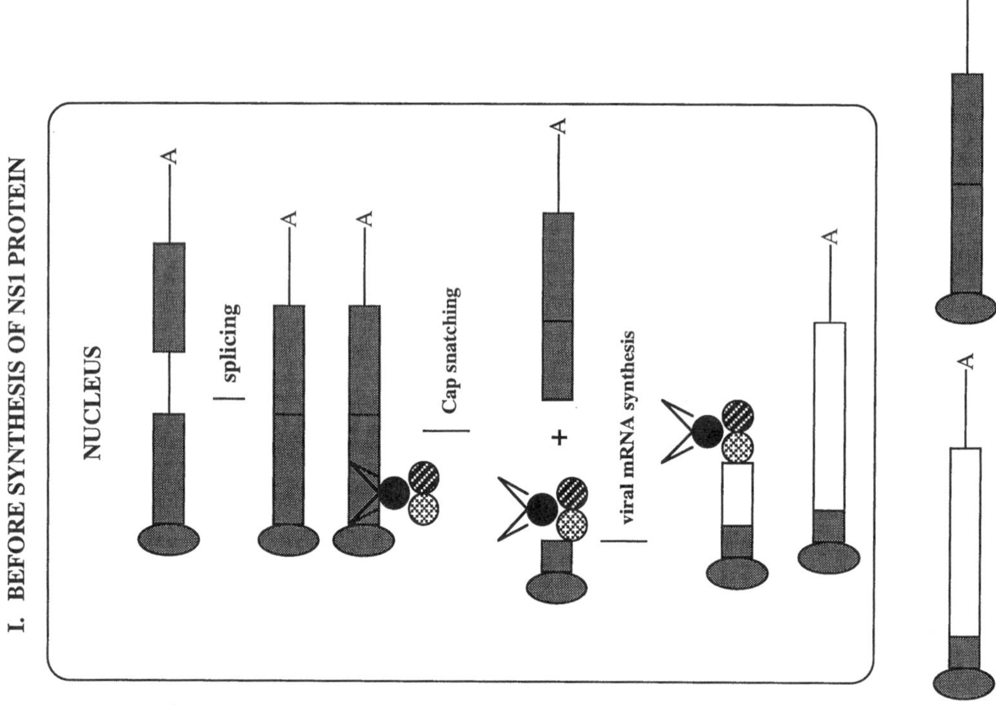
I. BEFORE SYNTHESIS OF NS1 PROTEIN
NUCLEUS
A
splicing
A
A
Cap snatching
A
+
viral mRNA synthesis
A

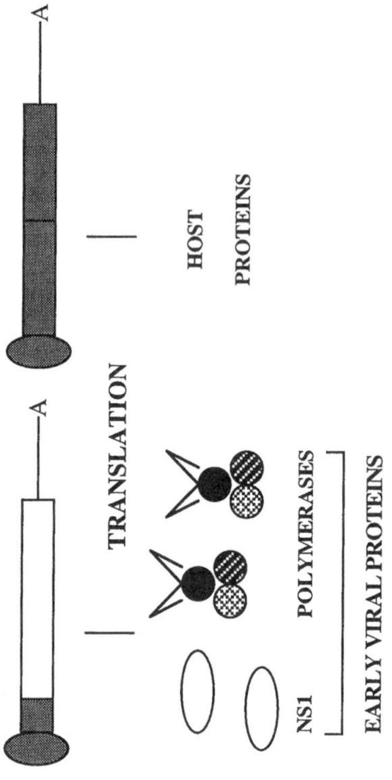
A
A
TRANSLATION
HOST
PROTEINS
NS1
POLYMERASES
EARLY VIRAL PROTEINS

II. LOW LEVEL OF NS1 PROTEIN
NUCLEUS
A
A
spliceosome
U6
A
spliceosome
U6
+
A
EXPORT
BLOCK

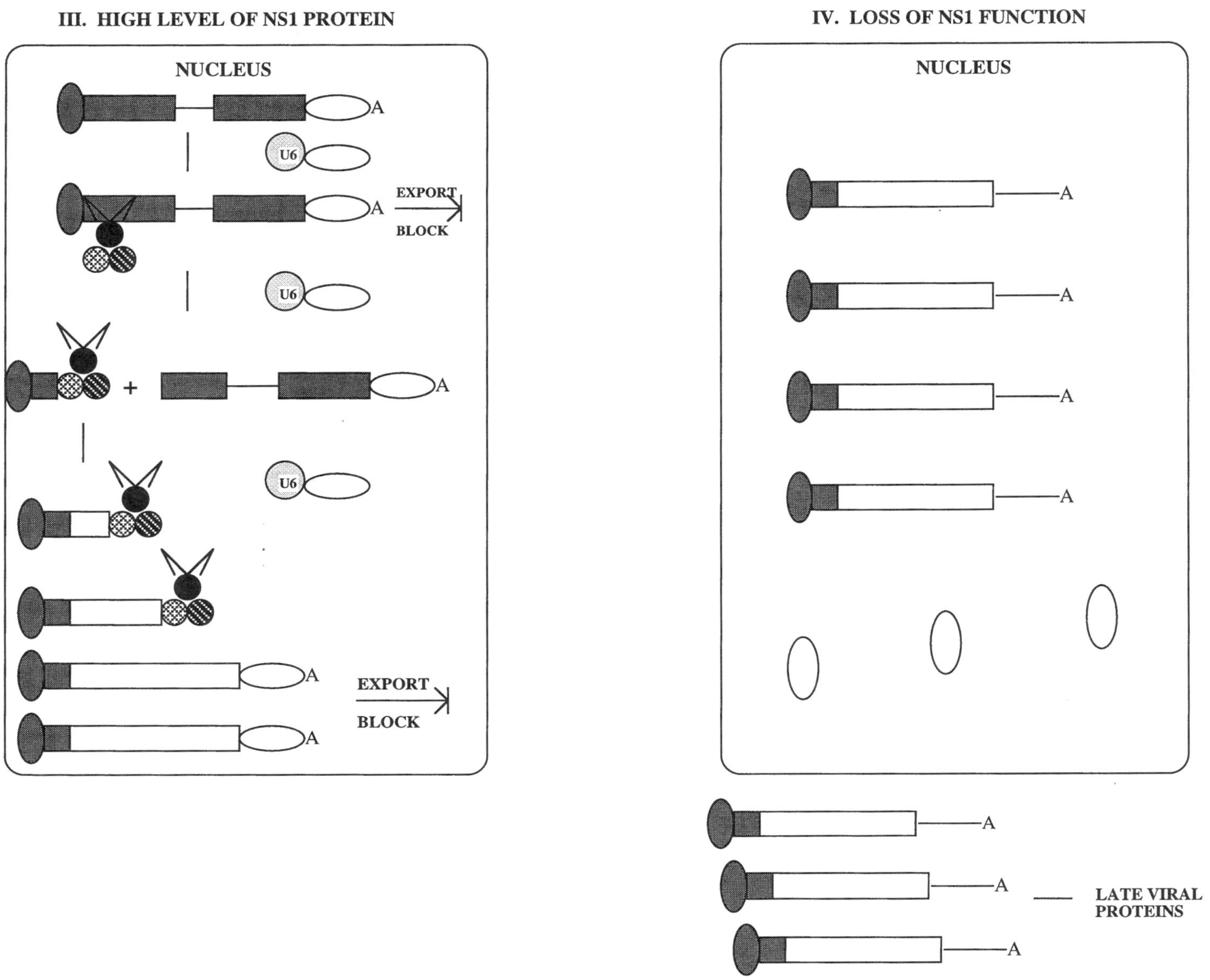

FIG. 18. Cartoon to illustrate the postulated functions of the NS_1 protein during infection. See text for a description of the model. Data from Lu et al. (236) and Qiu and Krug (310).

primers generated from cellular pre-mRNAs before they are spliced and/or spliced mRNAs before they exit from the nucleus, the viral polymerase would synthesize early mRNAs, particularly those encoding the NS_1 protein and the constituents of the polymerase-nucleocapsid complex. As discussed above, the NS_1 protein and the nucleocapsid protein are preferentially made at early times after infection. When enough NS_1 protein has accumulated in the nucleus, it would inhibit the splicing of cellular pre-mRNAs (Fig. 18, panel II). By allowing cellular pre-mRNAs to form spliceosomes but inhibiting most of the subsequent catalytic steps of splicing, the NS_1 protein would sequester cellular pre-mRNAs to discrete sites in the nucleus, where these pre-mRNAs would then be accessible to the viral cap-dependent endonuclease for the generation of the capped primers that are needed for viral mRNA synthesis. Viral mRNA synthesis would then occur with the required efficiency and cellular pre-mRNAs, which would have lost their protective 5′ capped ends, would be degraded in the nucleus. Adding credence to the hypothesis is the finding that the degradation of cellular mRNAs in the nuclei of infected cells occurs (see below). When higher levels of the NS_1 protein accumulate in the nucleus (Fig. 18, panel III), the formation of spliceosomes would be blocked, and the cellular pre-mRNAs would be sequestered in the nucleus by virtue of the NS_1 protein-medicated block in export of poly(A)-containing RNAs. These cellular pre-mRNAs would thus remain accessible to the viral cap-dependent endonuclease.

The inhibition of the nuclear export of mRNAs would serve a second purpose. Because little or no spliced cellular mRNAs would be generated in infected cells (236), the NS_1 protein-mediated inhibition of nuclear export of poly(A)-containing mRNAs (310) is most likely directed against viral mRNAs (Fig. 18, panels II and III), most of which are not generated by splicing (see below). As a consequence, the nuclei would become packed with viral mRNAs. In order for the viral mRNAs to exit from the nucleus, the NS_1 protein would have to lose its ability to bind to the poly(A) at the 3′ ends of the mRNAs (and presumably also to the U6 snRNA in spliceosomes), or the other functional domain of the NS_1 protein (309) might be inactivated (Fig. 18, panel IV). Such a loss of function could result from a posttranslational modification of the NS_1 protein. The NS_1 protein contains such modifications: it is a phospho-protein and may also contain another modification, because two-dimensional gel electrophoresis indicates the presence of two nonphosphorylated forms of the NS_1 protein (308). Support for this part of the hypothesis comes from the results of experiments using a protein kinase inhibitor (H7) or a methyltransferase inhibitor (379). Either inhibitor caused late viral mRNAs to accumulate in the nucleus, suggesting that blocking posttranscriptional modification of the NS_1 protein causes it to retain its function in inhibiting the nuclear export of mRNAs. Consequently, a second role for the NS_1 protein would be the regulation of the switch from early to late protein synthesis by retaining the late viral mRNAs in the nucleus until the appropriate time for their expression. The inactivation of the NS_1 protein would also serve to allow viral mRNA splicing to occur (see below).

Regulated Splicing of the Viral NS_1 and M_1 mRNAs

As discussed above, as for cellular polymerase II transcripts, influenza viral mRNAs are synthesized in the nucleus and contain 5′-terminal methylated cap structures ($m^7GpppNm$). In addition, influenza viral mRNAs undergo two nuclear processing steps that polymerase II transcripts undergo: methylation of internal A residues and splicing (195,214,216–218,271).

The role of the methylation of internal A residues in the metabolism of cellular polymerase II transcripts has not been established. Methylation occurs in the consensus sequences Gm^6AC and Am^6AC (50,78,393), and only a specific subset of the available GAC and AAC sequences in several polymerase II transcripts has been shown to be methylated (8,20,147,167). It has been postulated that internal m^6A residues might play a role in splicing (55,167, 361,425). However, the results obtained with influenza viral mRNAs have not provided support for this postulate. The entire population of influenza viral mRNA molecules contains an average of 3 m^6A residues per molecule, and these m^6A residues are found in the consensus AAC and GAC sequences (271). The distribution of these m^6A residues among the individual viral mRNAs is unexplained (271). The HA and NA mRNAs contain the most m^6A residues (8 and 7 residues per molecule, respectively), whereas the M_1 and NS_1 mRNAs, the two mRNAs that are spliced, as well as two of the P mRNAs, contain fewer m^6A residues (1 to 3 residues per molecule). Consequently, there is no obvious correlation between m^6A content and splicing.

As discussed above, two of the influenza A viral mRNAs are spliced. The NS_1 and M_1 mRNAs are spliced to form mRNAs coding for two other proteins, NS_2 and M_2, respectively (214,216–218). Except for a short region at their 5′ ends (corresponding to the 5′ exon), the spliced NS_2 and M_2 mRNAs are translated in the +1 reading frame relative to their unspliced precursor. The M_1 mRNA is also spliced to form another mRNA, $mRNA_3$, which has a coding potential for only 9 amino acids. The splice junctions are similar to those found in polymerase II transcripts, and most, though not all (see below), of these splice junctions were used when the NS_1 and M_1 genes were expressed using DNA vectors (216,217) (see later in this section). Consequently, it can be concluded that splicing of the NS_1 and M_1 mRNAs is catalyzed by host-cell nuclear enzymes.

As both the unspliced (NS_1 and M_1) and spliced (NS_2 and M_2) mRNAs code for proteins, the extent of splicing is regulated such that some of the unspliced precursor is preserved at the same time that a sufficient amount of the spliced product is reduced. In influenza virus-infected cells, this regulation results in a steady-state amount of the spliced mRNAs that is only about 10% of that of the unspliced mRNAs (212,218). This type of splicing regulation is not restricted to influenza virus. It also occurs with retroviruses, both "simple" retroviruses (such as avian and murine leukemia viruses) and "complex" retroviruses (*Lentiviridae*, which include HIV-1) (169,243). In these systems, not only is the extent of splicing regulated, but also the unspliced pre-mRNA as well as the spliced mRNAs are exported from the nucleus. Usually pre-mRNAs are retained in the nucleus by virtue of the binding of splicing factors to the 5′ splice site, 3′ splice site, and/or branchpoint, thereby committing the pre-mRNA to spliceosome formation (229). This constitutes the "spliceosome retention" hypothesis.

Most of the studies concerning influenza virus splicing have focused on the splicing and nuclear export of NS_1 mRNA. It was shown that the rate of splicing of NS_1 mRNA encoded in an adenovirus recombinant is not significantly different from the splicing rate of NS_1 mRNA in influenza virus-infected cells (4 hours postinfection) (9). The most likely conclusion is that the rate of splicing of NS_1 mRNA in influenza virus-infected cells is controlled solely by *cis*-acting sequences in NS_1 mRNA itself. Candidates for such *cis*-acting sequences have been identified in *in vitro* splicing experiments. Though NS_1 mRNA contains 5′ and 3′ splice sites that closely fit the splicing consensus sequence, the NS_1 mRNA was not detectably spliced *in vitro* using nuclear extracts from uninfected (or infected) HeLa cells (2). Only extremely small amounts of a splicing intermediate, intron lariat-3′ exon, were detected. Nonetheless, NS_1 mRNA, like other splicing precursors, formed ATP-dependent 55S complexes containing the snRNAs characteristic of spliceosomes (2). These 55S complexes did not catalyze splicing. When certain intron and 3′ exon sequences were removed, the block in splicing was alleviated, and the 5′ and 3′ splice sites and branchpoint of NS_1 mRNA functioned efficiently (275). It can be proposed that these intron and 3′ exon sequences would also operate *in vivo*, where they would decrease the rate of splicing rather than block splicing completely.

It is likely that the extent of splicing of NS_1 mRNA is determined by competition between the splicing rate and the rate of nuclear export of NS_1 mRNA. As discussed above, the available evidence indicates that the rate of splicing is intrinsic to the NS_1 mRNA sequence. In contrast, the rate of nuclear export of NS_1 mRNA varies depending on whether the NS_1 mRNA is expressed via a DNA vector or via influenza virus machinery in infected cells. When NS_1 mRNA is expressed with an adenovirus vector, its transport is totally blocked because NS_1 mRNA is committed to the splicing pathway (9). Inactivation of the 3′ splice site of NS_1 mRNA by mutation caused NS_1 mRNA to be transported to the cytoplasm. This block in transport of wild-type NS_1 mRNA determines the extent of splicing. Even though *cis*-acting sequences probably suppress the rate of NS_1 mRNA splicing (275), NS_1 mRNA resides in the nucleus for relatively long periods of time and is extensively spliced (9). In contrast, in influenza virus-infected cells, transport of NS_1 mRNA is efficient, and the rate of splicing largely, if not totally, determines the extent of splicing. It should be emphasized that this nuclear export of NS_1 mRNA in infected cells is most likely occurring at times of infection after the inactivation of the NS_1 protein-mediated block in the transport of poly(A)-containing mRNAs.

To address the question of whether transport of unspliced NS_1 mRNA from the nucleus to the cytoplasm requires the action of a virus-specific protein, NS_1 mRNA was expressed using a transient transfection DNA vector. In these transfected cells about 20% to 30% of unspliced NS_1 mRNA is transported to the cytoplasm (10), indicating that in this uninfected cell system some unspliced NS_1 mRNA escaped from nuclear spliceosomes and was transported. This transport did not require the expression of any viral proteins. However, because this transport was much less efficient than that occurring in influenza virus-infected cells, this leaves open the possibility that in infected cells other factors or conditions operate to increase the efficiency of the nuclear export of unspliced NS_1 mRNA.

The splicing of NS_1 mRNA is not inhibited by the NS_1 protein, in contrast to all other pre-mRNAs examined (236). This resistance is due to sequences in NS_1 mRNA because in the same cells in which NS_1 mRNA splicing was not affected by the NS_1 protein, the splicing of a globin pre-mRNA was effectively inhibited. In addition, this result indicates that the viral NS_1 and NS_2 mRNAs do not affect the ability of the NS_1 protein to inhibit the splicing of other pre-mRNAs in the same cells. Therefore, it does not seem likely that the NS_1 and NS_2 mRNAs titrate out the NS_1 protein. It is not known how these NS_1 mRNA sequences confer resistance to the NS_1 protein which inhibits catalytic step(s) of splicing that are presumably common to all other pre-mRNAs.

As already indicated, the extent of splicing of the M_1 mRNA of influenza A virus is also regulated. The splicing of M_1 mRNA is complicated by the use of alternative 5′ splice sites (218). When the 5′ splice site at position 11 (the first virus-coded nucleotide at the 5′ end of the mRNA is assigned position number 1) is used, $mRNA_3$ is generated. This RNA has only a 9-amino acid ORF; it is unknown if this RNA is a functional mRNA and it is possible that the RNA has some other function during infection. The $mRNA_3$ 5′ splice site, CAG || *GU*AGAU, fits the consensus sequence (C/AAG || GUA/GAGU) closely. By contrast, the 5′ splice site for the M_2 mRNA (at position 51), AAC || *GU*AUGU, does not fit the consensus as well in that there is a C rather than a G at the 3′ end of the 5′ exon. The M_2 mRNA encodes the essential M_2 ion channel protein (see above). The common 3′ splice site, $G_7G(A)_4(U)_3$ GC*AG* || G, deviates from the consensus ($U/C_nNC/UAG$ || G) in that it does not contain an immediately adjacent polyprimidine tract. When the M_1 gene was expressed in uninfected cells using a transient transfection DNA vector, only the strong mRNA$_3$ 5′ splice site and not the weaker M$_2$ 5′ splice site was used (10,216). This raises the question of how the M$_2$ mRNA 5′ splice site as well as the mRNAs 5′ splice site is used in influenza virus-infected cells. Unlike the splicing of NS$_1$ mRNA, the splicing of M$_1$ mRNA is inhibited by the NS$_1$ protein (236). This occurs at early times of infection and would not be expected to occur at times late in infection when the NS$_1$ protein is postulated to lose its function (see above).

Splicing also occurs with influenza B and C viruses. As with influenza A virus, the NS$_1$ mRNA of influenza B virus is spliced to form a smaller NS$_2$ mRNA (39). In contrast, the M$_1$ mRNA of influenza B virus is not spliced. Both the ORF and the intron of the NS$_1$ mRNA of influenza virus B/Lee/40 are larger than that of the NS$_1$ mRNA of influenza virus A/Udorn/72; the sizes of the NS$_2$ mRNAs of the two virus strains are similar. The 3′ splice site, GAG || *GU*GGGU, of the NS$_1$ mRNA of influenza B virus fits the consensus sequence closely, whereas the 3′ splice site, GAUCGGAC*AG* || U, deviates from the consensus sequence in two ways: (a) absence of a pyrimidine tract immediately upstream of the 3′-terminal AG of the intron; and (b) the presence of a U rather than a G at the 5′ end of the 3′ exon. The extent of splicing of the influenza B virus NS$_1$ mRNA is regulated similar to that of the influenza A virus NS$_1$ mRNA; the steady-state level of the spliced influenza B virus NS$_2$ mRNA is only about 5% to 10% of that of the unspliced influenza B virus NS$_1$ mRNA (39).

With influenza C virus, it has been shown that a spliced mRNA encodes the M$_1$ protein (415). The 5′ splice site (AUG || *GU*UAGU) deviates from the consensus most notably by the presence of a U rather than an A at the 3′ penultimate position of the 5′ exon, and the 3′ splice site (GCUCGGC*AG* || A) deviates substantially from the splicing consensus sequence. Influenza C virus differs from both influenza A and B viruses in that its M$_1$ protein is encoded not in an unspliced mRNA that is colinear with the genomic RNA but rather in an interrupted mRNA that arises by splicing of the colinear mRNA (see above). In addition, the M gene of influenza C virus does not contain a second reading frame. Finally, the extent of production of the interrupted M$_1$ mRNA of influenza C virus is apparently not regulated like the splicing of the NS$_1$ and M$_1$ mRNAs of influenza A virus and of the NS$_1$ mRNA of influenza B virus. In cells infected by influenza C virus, only a small amount of a colinear M transcript was detected; this transcript might actually represent M template RNA (415).

Suppression of Host Gene Expression Leading to the Selective Translation of Influenza Virus mRNAs

During influenza virus infection, a dramatic switch from cellular to viral protein synthesis occurs (206,224,344,345). This switch results from both nuclear and cytoplasmic processes. In the nucleus, newly synthesized cellular polymerase II transcripts are degraded. It is likely that this degradation is initiated by the cleavage of the 5′ ends of polymerase II transcripts by the viral cap-dependent endonuclease. The resulting decapped RNAs would likely be more susceptible to degradation by cellular nucleases, as it has been shown that the 5′ cap structure stabilizes RNAs against nucleolytic degradation both *in vivo* and in cell extracts (14). This degradation would be expected to be significantly enhanced by the action of the viral NS$_1$ protein in blocking the splicing of cellular pre-mRNAs (see above).

The degradation of newly synthesized cellular mRNAs in the nucleus is not a sufficient explanation to understand the shutoff of host-cell mRNA translation, because high levels of functional host-cell mRNAs remain in the cytoplasm (170). In contrast to cellular transcripts in the nucleus, a variety of cellular mRNAs in the cytoplasm were found to be stable, and they were not significantly reduced in amount until relatively late times of infection (170). Moreover, these cellular mRNAs isolated from infected cells were found to be efficiently translated *in vitro* in reticulocyte extracts, indicating that influenza virus did not significantly inactivate cytoplasmic host mRNAs. Consequently, the shutoff of host-cell protein synthesis, which is complete by about 3 hours postinfection in several cell lines, did not result from the degradation or modification of cytoplasmic host-cell mRNAs. The step at which the block in the translation of host-cell mRNAs occurred was determined by examining the polysome association of several representative cellular mRNAs in uninfected and infected cells (171). The results indicated that both the initiation and the elongation step in the translation of host-cell mRNAs are blocked. The molecular mechanisms by which influenza virus imposes this selective block against the initiation and elongation of cellular, but not viral, proteins have not been determined. Selectivity at initiation is most likely due at least in part to the fact that influenza viral mRNAs are very efficient initiators of translation. Convincing evidence for the efficient initiation of translation of influenza virus mRNAs came from experiments in which adenovirus-infected cells at late times of infection were superinfected with influenza virus: influenza viral mRNAs were selectively translated over late adenovirus mRNAs (170,172).

Maintenance of Efficient Translation in Infected Cells

The level of translation in cells can be regulated by the level of phosphorylation of the α-subunit of the initiation factor eIF-2. This initiation factor forms the ternary complex (eIF-2)· GTP· (met-tRNA$_i$) that binds to the initiating 40S ribosomal subunit before mRNA is bound (160). Phosphorylation of the α-subunit of eIF-2 prevents the recycling of eIF-2-GDP to form the functional form of eIF-2, eIF-2-GTP (189,289,333). Recycling has been shown to be catalyzed by the factor eIF-2B, which is trapped in an inactive complex with eIF-2-GDP when the α-subunit of eIF-2 is phosphorylated (189,289,333). Without this recycling of eIF-2 by eIF-2B, protein synthesis initiation is effectively blocked. During infection by several viruses, including influenza virus, a kinase is activated that phosphorylates eIF-2α (172,173). The kinase is identical to the double-stranded RNA-activated P68 kinase that is induced by interferon. Most likely this kinase is activated during infection by viral double-stranded RNAs. Unless the virus develops a mechanism for suppressing this kinase, all protein synthesis would be inhibited.

Unlike most other viruses, influenza virus does not encode a viral gene product that directly inactivates the P68 kinase. Rather, influenza virus activates a cellular protein of 58,000 daltons that in turn inhibits the P68 kinase (228). The experimental evidence indicates that this 58,000-dalton cellular protein interacts directly with P68 kinase, and that influenza virus activates the P68 inhibitory activity of the 58,000-dalton protein by dissociating it from its own natural inhibitor. Based on its amino acid sequence the 58,000-dalton cellular protein has been categorized as a member of the TPR (tetratricopeptide repeat) family of proteins (227).

Defective Interfering (DI) Particles

When influenza viruses are passaged at high multiplicity, DI particles are produced (382). These particles contain one or more DI RNAs derived from one or more of the P (polymerase) genomic RNAs (161,343). All the subgenomic DI RNAs are internally deleted and retain both the 5′ and 3′ termini of the progenitor genomic RNA segment (226,272). The mechanism by which these DI RNAs are generated has not been established, but the hypothesis that is most consistent with current data is the rollover/loop-out model (272–274). In this model, the viral polymerase complex remains associated with its template (either vRNA or template RNA) and rolls over to a new site on the template that has been brought into juxtaposition. The two sites on the template might be brought close together by the formation of transient secondary/tertiary structures that cause some regions of the template to loop out. Interestingly, there is no consensus sequence at the junctions where sequences of the progenitor genomic RNA are deleted (226,272).

It is not clear how DI RNAs exert a replicative advantage over standard genomic RNAs, thereby suppressing the replication of the latter. Although subgenomic RNAs can be produced from all genomic RNA segments, only those derived from the polymerase RNA segments appear to evolve into DI RNAs (161,272,273). Evidence has been obtained indicating that the DI RNAs derived from a given polymerase RNA segment interfere with the packaging of the progenitor polymerase RNA segment and are packaged in their place (5,269).

Finally, the role of DI virus particles in natural infections has not been established. However, some evidence has been presented that DI viruses may attenuate the pathogenic effects of standard virus infections [e.g., see (77, 100,244,312)].

Intracellular Transport of Viral Proteins, Viral Assembly, and Release

Nuclear-Cytoplasmic Transport

As discussed above, the infecting RNPs are transported into the nucleus for transcription and replication, and newly assembled RNPs are exported back from the nucleus to the cytoplasm. Thus, the switch between import and export is an important factor in viral replication. As discussed above, the RNP penetrates the cytosol after fusion of the input virion with the late endosomal membrane; the RNPs separate from the M_1 protein and are transported into the nucleus (see above). Under conditions in which the M_1 protein fails to dissociate from the RNP (blockade of the M_2 ion channel by amantadine) (45), the RNPs fail to enter the nucleus (247) and primary transcription does not occur (349). In influenza virus-infected cells, M_1 protein is localized to both the cytoplasm and nucleus (38). M_1 entry into the nucleus is thought to be by passive diffusion as it is insensitive to wheat germ agglutinin and to antibodies to the nuclear pore that block active import but do not block passive diffusion into the nucleus (248). In contrast, NP, PB1, PB2, and PA all contain karyophilic signals (see above) and the RNPs are assembled in the nucleoplasm. Transport of these newly assembled RNPs back to the cytoplasm requires association with the M_1 protein (247), and the process can be inhibited by microinjection of antibodies to the M_1 protein into the cytosol. Thus, the M_1 protein plays a critical role in modulating the nuclear-cytoplasmic transport of the RNPs.

Intracellular Transport of the Integral Membrane Proteins

The influenza virus integral membrane proteins HA, NA, M_2, and NB are synthesized on membrane-bound ribosomes and are translocated across the membrane of the ER in an signal recognition particle (SRP)-dependent manner (88,154). For HA the N-terminal signal sequence is cleaved in the ER by signal peptidase (250): NA, M_2, and NB do not contain cleavable signal sequences. N-linked carbohydrate chains are transferred to HA, NA, and NB (M_2 is not glycosylated) *en bloc* from a dolicyl lipid carrier, and trimming of terminal glucose residues from mannose-rich oligosaccharides occurs in the ER. The HA and NA have been used as prototypes from which much of what we know about the process of folding and oligomerization of an integral membrane protein has been learned [reviewed in (81)]. For HA a large body of data indicates that there is a step-wise conformational maturation of the protein with independent folding of specific domains in the HA monomer, followed by trimerization and the completion of folding to the pH-neutral form of HA (29,32–34,79,102) [reviewed in (326)]. Once correctly folded and assembled, proteins are transported out of the ER to the Golgi apparatus. There, oligosaccharide chains may be further processed to the complex form [reviewed in (190)]. For influenza virus A/Aichi/68 HA, five carbohydrate chains are in the complex form and two are in the high mannose form (386). HA and NA lack terminal sialic acid on their complex carbohydrate chains in influenza virus-infected cells, presumably due to the action of the viral neuraminidase (188). However, when HA is expressed from cDNA, the HA complex carbohydrate chains contain sialic acid (334). Palmitoylation of the HA three C-terminal proximal cysteine residues (two in the cytoplasmic tail and one in the presumed transmembrane domain) and palmitylation of the M_2 protein cytoplasmic tail residue 50 (141,364,376) is thought to occur in the cis Golgi complex. For those strains of HA with a cleavage site containing multi-basic residues, cleavage occurs in the *trans* Golgi network by a subtilisin-like endoprotease, furin (360). In polarized epithelial cells, influenza viruses assemble and bud at the apical surface of cells (321,325) and it has been shown that HA, NA, and M_2, when expressed alone from cDNAs in polarized cells, are transported to the apical surface (153, 164,324,327). Finally, the integral membrane proteins are expressed at the plasma membrane. Though HA appears to be diffusely distributed over the surface, M_2, NA, and the influenza C virus HEF glycoprotein appear to cluster in patches (68,69,130,153,164,219,422,424).

Role of M_2 Ion Channel Activity on Maturation of HA in the trans *Golgi Network*

In addition to the early effect of amantadine on influenza virus replication described above, for the avian influenza virus A/chicken/Germany/34 (H7N1) fowl plague virus (FPV) Rostock, the drug has a second "late" effect. These observations were of great importance for the development of the hypothesis that the M_2 protein has ion channel activity. The accumulated data from several studies indicate that addition of amantadine to FPV Rostock-infected cells late in infection brings about a premature conformational change in HA that occurs in the TGN during the transport of HA to the cell surface (57,58,106,107,363). By immunological and biochemical criteria this form of HA is indistinguishable from the low pH-induced form of HA (347) [reviewed in (402)] that results following exposure of native HA to low pH. Recently, it has been reported that it is possible to perform a reconstruction experiment in the absence of an influenza virus infection. When the FPV HA was expressed from cDNA by using a recombinant DNA vector, the HA undergoes its transition to the low-pH form unless a functional M_2 protein is coexpressed (281,366). Thus, the M_2 ion channel activity is thought to function in the TGN and associated transport vesicles to regulate intracompartmental pH and keep the pH above the threshold at which the FPV Rostock HA conformational change occurs (57,58,363). The consequence of the amantadine-induced low-pH HA conformational change is that it brings about the extrusion of the fusion peptide at the wrong time in the infectious cycle and in the wrong subcellular compartment. Under these conditions, viral budding is observed to be greatly restricted (329), possibly because the hydrophobic HA fusion peptides aggregate. However, it is clear that the avian influenza virus FPV Rostock, even for an influenza virus with an HA that is cleaved intracellularly, is an exceptional case in requiring a functional M_2 protein ion channel activity in the TGN as both influenza virus A/chicken/Germany/27 (H7N7) (FPV Weybridge) and A/equine/Cornell/74 (H7N7) have an HA that does not require M_2 function for transport through the TGN (107,367).

The Budding Process

Influenza viruses can be observed morphologically to bud from the plasma membrane of infected cells (12,56, 68,256,259) (see Fig. 2). Electron microscopic observations also have provided evidence for the order by which viral components associate at the plasma membrane as the glycoprotein spikes can be detected in the absence of RNPs (12,68). This has led to the suggestion that the precursor to the envelope of a budding virion is a patch of cell membrane containing viral envelope proteins [reviewed in (66)]. Host-cell membrane proteins appear to be excluded from virions and, even among virus-encoded proteins, there must be a positive selection or a rejection process; as in comparison to HA, the M_2 protein is greatly underrepresented in virions yet, at the plasma membrane, is expressed at about a fourfold lower level than HA (219,422). The nature of the protein-protein contacts necessary to form a virion are not known. As discussed above, the M_1 protein forms contacts with the RNPs and can associate with the lipid bilayer. It is commonly assumed that the cytoplasmic tails of the integral membrane proteins interact with the internal viral components. In this regard, it has been found that variants of influenza virus which can grow in the presence of an antibody to the M_2 protein extracellular domain contain mutations in the M_2 protein cytoplasmic tail or the M_1 protein (423), which suggests these proteins make important contacts. However, perhaps unexpectedly, it has been found that influenza viruses that contain a HA that lack a cytoplasmic tail assemble and replicate almost as efficiently as virions containing normal HA (162). The mechanism by which the bud pinches off from the plasma membrane is not known. To release fully formed virions from the cell surface, the action of the viral neuraminidase is thought to be needed as cells infected with mutants defective in NA produce virus particles containing neuraminic acid which form large aggregates, reducing the yield of released virus (287).

Packaging a Segmented Genome

For an influenza virus to be infectious, it requires one of each of the eight RNA segments. It is not known if this occurs randomly or whether there is a selective mechanism for packaging eight RNA segments. The evidence for a selective mechanism consists of: (a) early reports that the RNPs are physically linked (7); (b) the RNA segments are isolated from a population of virions in approximately equimolar abundance (252); (c) in the generation of defective-interfering RNAs, the progenitor RNA to the deleted RNA is preferentially lost in packaged virus (5,269). Evidence for a random packing is as follows: (a) aggregates of virus produced by addition of nucleohistone have an enhanced infectivity, suggesting that complementation occurred between two or more virus particles, each of which alone contains less than the eight RNA segments (138); (b) in cells infected with a low multiplicity of infection of influenza virus, some cells lack the expression of some antigens, suggesting that some virions lack a full set of viral RNA segments (247); and (c) an engineered influenza virus has been obtained under selection conditions which contains nine RNA segments (93). If the probability of packaging 8 discrete RNA segments of influenza virus into a virion on a random basis is calculated, only 12 segments need be randomly packaged to have 10% of virions containing 1 copy of each of the 8 segments (93,204).

GENETICS OF INFLUENZA VIRUSES

RNA Segment Reassortment

The emergence of new pandemic strains of influenza A virus usually results from the appearance of a new virus subtype containing a novel HA and/or NA that is immunologically distinct from those of the previous circulating strain. In the 20th century, there was the appearance of swine influenza in 1918 to 1919, with an HA related to those of swine viruses or H1 subtype viruses. Viruses of this subtype circulated until 1957, when viruses of the H2N2 subtype (Asian strains) were isolated. The H2 subtype HA has little or no cross-reactivity with the H1 HA. In addition to containing an H2 HA, the Asian strains had a new NA (N2). For 11 years, the H2N2 strains of influenza virus spread and changed until the next pandemic in 1968, with the introduction of a new H3 subtype (Hong Kong strains). These extensive changes in the HA and/or NA that lead to new subtypes of human influenza viruses are known as *antigenic shifts* [reviewed in (180,181,352,392)] (see also Chapter 46 in *Fields Virology*, 3rd ed.). These drastic antigenic changes come about from the reassortment of previously circulating human viruses and influenza viruses of animal origin. There is a great deal of evidence for the reassortment of RNA segments between human and animal viruses *in vivo* (391) and among human viruses in nature (420). It has been established that the A/Hong Kong/68 H3N2 virus contains the NA and all genes encoding internal proteins, except that PB1 from an Asian (H2N2) strain and the genes for HA and PB1 are thought to be derived from a strain that is very closely related to A/duck/Ukraine/63 and A/equine/2/Miami/63 (H3N8 viruses) (94,174,222,387). Historically, the high rate of genetic changes in mixed infections using laboratory influenza viruses, in comparison to the low rate of genetic exchanges observed with many other viruses, led to the proposal that influenza virus had a segmented genome (137). This was reinforced by finding multiplicity-dependent reactivation of UV-inactivated virus (16) and was eventually demonstrated biochemically (18,252,286,303,318,319).

Influenza type A, B, and C viruses can form reassortants *in vivo* between members of the homotypic virus type but not between types. In nature, reassortants have only been detected with influenza A viruses, presumably due to the lack of circulating subtypes of the influenza B and C viruses in animal and avian species.

Theoretically, it might be expected that *in vivo* the 8 RNA segments of influenza A virus could give rise to 256 possible reassortants. However, a random segregation does not occur, as at the protein level some proteins need their matched strain-specific partner. Nowhere is this more clearly demonstrated than with the requirement of the A/chicken/Germany/34 (H7N1) FPV Rostock HA (encoded by RNA segment 4) to cosegregate with its cognate M_2 protein (encoded by RNA segment 7) (106,107).

RNA Mutations

Minor changes in antigenic character of influenza viruses occur due to the accumulation of amino acid changes due to mutation: of principal importance are changes in HA and NA. This process is known as *antigenic drift*, and it occurs with influenza A, B, and C viruses [reviewed in (180,181,352,392)]. However, influenza A viruses show a high number of substitutions accumulating in a sequential pattern with time, whereas influenza C viruses isolated decades apart show much lesser degrees of variability [reviewed in (352)]. The mutation rate for the H3 HA (HA1 domain) has been estimated to be 6.7×10^{-3} substitutions per site per year, whereas the mutation rate for the cognate NS gene was estimated to be 1.9×10^{-3} substitutions per site per year. The observation that the nucleotide substitutions in HA were found predominantly in HA_1 suggests that many substitutions may be selected for by the immune pressure [see (352) and references therein]. Many studies have been performed to analyze the rate of mutation of RNA viruses in tissue-culture. Recent calculations that take into consideration many artifacts of earlier studies indicate the influenza virus mutation rate at 1.5×10^{-5} mutations per nucleotide per infectious cycle (290). However, although the influenza virus mutation rate is high compared to mammalian DNA genomes (10^{-8} to 10^{-11} per incorporated nucleotide), the mutation rate of influenza A virus is not markedly different from that of other RNA viruses [reviewed in (351,352)].

RNA Recombination

RNA recombination is the joining of two separate RNA chains by crossing-over or copy-choice mechanisms. Historically for influenza virus, *RNA recombination* has been used to mean *RNA segment reassortment*. Unlike the situation with the positive-stranded RNA viruses, the *Picornaviridae* and *Coronaviridae* (see Chapters 16 and 18), where RNA recombination occurs at quite high frequency, until recently had not been measured for influenza viruses. However, natural examples of RNA recombination have been found when HAs were identified which contain at the HA_1/HA_2 cleavage site either an insertion of 54 nucleotides derived from host 28S ribosomal RNA (178,284) or 60 nucleotides derived from the NP gene (283). Presumably, this is one of the few places in the influenza virus genome where such an insertion could be tolerated without an adverse consequence on the viability of the virus. In addition, using molecular genetic techniques (see below), RNA recombination within an RNA segment has been shown to occur (21).

Temperature Sensitive Mutants

To understand the functions of the proteins encoded by each RNA segment, many temperature-sensitive mutants have been isolated following chemical mutagenesis of virus [reviewed in (240)]. Although some of these mutants have been useful in understanding protein function (e.g., mutants of NA and M_1, see sections above), studies using other mutants, particularly those of NS_1 and NS_2, were not as rewarding. Some host-range mutants have been isolated, e.g., mapping to the NS_1 protein (47), and some cold-adapted mutants have been obtained for application to vaccine development (see Chapter 46 in *Fields Virology*, 3rd ed.).

Molecularly Engineered Genetics (Reverse Genetics)

From the beginnings of the era of recombinant DNA technology, it has been possible to genetically alter DNA viruses to study properties of the altered virus and to use DNA viruses as vectors for the expression of foreign genes [e.g., (258)]. For many RNA viruses which contain an infectious, positive-strand genome (e.g., poliovirus, Sindbis virus, and many plant viruses), it has been found possible to transcribe synthetic RNA genomes from cloned DNA and obtain infectious virus [e.g., (3,313,414)]. The negative-stranded RNA viruses require that the virion RNA is assembled with an active transcriptase complex for the genome to initiate replication, and until recently for influenza virus the inability to solubilize and reconstitute the transcriptase complex on naked RNA has prevented molecular engineering of the genome.

The finding that it was possible to isolate an active polymerase complex from influenza virus RNP cores by using CsCl-glycerol gradient centrifugation (143) made it possible to transcribe *in vitro* short synthetic RNA molecules containing the influenza virus 5′ and 3′ terminal nucleotide sequences and to analyze the promoter signals (291). These studies were then extended by finding it was possible to replicate *in vivo* an artificial gene construct containing the 5′ and 3′ terminal nucleotide sequences flanking a CAT reporter gene. The synthetic CAT RNA was mixed with the solubilized RNA polymerase complex and transfected into cells. RNP transfections were either preceded or followed by infection with helper influenza virus. The artificial CAT gene was replicated, packaged into virions, and could be passaged several times in cells (239). The technology was later improved by finding that the efficiency of isolating a virus containing a foreign gene is enhanced if the transcription of the syntheic RNA is performed in the presence of the influenza virus RNA polymerase complex (92). Another procedure for producing an active polymerase complex for transcribing RNA *in vitro* is to treat isolated RNPs with micrococcal nuclease and then to add exogenous synthetic RNA templates (337).

Thus, as described above, systems have been developed which permit the introduction of defined changes into the influenza virus genome. However, at the present time the systems are dependent on helper virus; in addition, selection of a new virus [termed *transfectant* (92)] requires a good selection scheme, e.g., antibody selection of the new virus. The system has been used to examine *cis*-acting signals in the promoter and *cis*-acting signals needed for polyadenylation (233,234,238,294). In addition, the system has been used to introduce mutations into the influenza virus genome to study protein function. A NA gene containing defined mutations has been rescued (91). An influenza A virus was generated that contains 5′ and 3′ noncoding regions of influenza B virus NS gene flanking influenza A virus NA: this virus was found to be considerably attenuated for disease in mice (263). Novel viruses containing NA with altered stalk lengths have been generated (54,237), and foreign epitopes have been inserted into NA and HA (53,231,232).

SPECIFIC INHIBITORS OF INFLUENZA VIRUS REPLICATION

Several approaches have been employed to identify compounds that specifically inhibit influenza virus replication.

The antiviral drug amantadine exemplifies a compound that was identified by the mass screening empirical approach. As discussed above, amantadine inhibits all natural isolates of influenza A virus, and its target site of action has been shown to be the M_2 ion channel protein transmembrane domain (see above). A considerable effort is currently underway to identify better drugs with a related mode of action to amantadine in the hope that the virus could not so readily generate resistant mutants.

Another approach, rational drug design, has recently been used to devise inhibitors of the NA. As discussed above, the sialic acid binding site on NA was identified as a large pocket on the surface of each of the four subunits (62). The pocket is lined by amino acids that are invariant in all influenza A and B viruses that have been characterized. The crystal structure indicated that there should be an empty cavity in this pocket when sialic acid is in the pocket, which led to the prediction that substitution of the 4-hydroxyl group of sialic acid with an amino or guanidinyl group would increase the binding interaction of the resulting compound (381). x-ray crystallography confirmed this prediction. Two compounds (4-amino or 4-guanidino 2-deoxy-2,3 didehydro-D-N-acetylneuraminic acid) (4-amino or 4-guanidino–Neu 5Ac2en) were synthesized and they were found to be extremely effective inhibitors of the NAs of both influenza A and B viruses, and these componds were orders of magnitude less active against an array of nonviral NAs. These compounds, particularly 4-guanidino–Neu 5Ac2en, inhibited influenza virus plaque formation and inhibited both virus replication and disease symptoms in ferrets (381). In 1994 testing of these compounds in humans was begun. One concern is the possible generation of virus mutants resistant to these drugs. As the cavity in the active site pocket does not appear to be needed for the natural substrate, sialic acid, the viral NA could mutate to eliminate this cavity, thereby making the NA resistant to the 4-amino and 4-guanidino–Neu 5Ac2en compounds.

One target which has received considerable attention is the viral transcriptase complex, because the process of capped RNA-primed transcription is catalyzed by viral and not host-cell enzymes. Currently various groups are screening a large number of chemicals for their activity against the viral transcriptase *in vitro*. Another approach for antiviral therapy stems from the action of the interferon (IFN)-induced Mx protein. In mouse cells the antiviral state induced by IFN α/β is controlled by the host Mx gene (116, 117,356). The mouse Mx gene product, the Mx1 protein, is a 72-kD protein that accumulates in the nucleus (84,145). The Mx1 protein has been shown to mediate specific resistance to influenza virus infection independent of other IFN-induced proteins (357). Mx proteins with related antiviral activities have been found in other species (1,254, 292). The Mx proteins possess GTPase activity and a tripartite GTP-binding consensus element in their N-terminal region (145,255,270). The GTP-binding domain of the murine Mx1 protein is required for its antiviral activity (296). It has been established that the murine Mx1 protein blocks influenza virus replication by inhibiting primary transcription, i.e., synthesis of viral mRNA by the inoculum transcriptase (200,292). Recent experiments have indicated that overexpression of the PB2 protein subunit of the polymerase can neutralize the inhibitory action of the Mx1 protein (151,358). This suggests that the Mx1 protein binds to the PB2 protein or that the Mx1 protein competes with the PB2 protein for a common substrate. One long-range goal could be the development of a small compound that mimics the mode of action of the Mx1 protein.

REFERENCES

1. Aebi M, Fah J, Hurt N, Samuel CE, Thomas D, Bazzigher L, Pavlovic J, Haller O, Staeheli P. cDNA structures and regulation of two interferon-induced human Mx proteins. *Mol Cell Biol* 1989;9:5062–5072.
2. Agris CH, Nemeroff ME, Krug RM. A block in mammalian splicing occurring after formation of large complexes containing U1, U2, U4, U5, and U6 small nuclear ribonucleoproteins. *Mol Cell Biol* 1989;9: 259–267.
3. Ahlquist P, French R, Janda M, Loesch-Fries, S. Multicomponent RNA plant virus infection derived from cloned viral cDNA. *Proc Natl Acad Sci* USA 1984;81:1066–1070.
4. Akkina RK, Chambers TM, Londo DR, Nayak DP. Intracellular localization of the viral polymerase proteins in cells infected with influenza virus and cells expressing PB_1 protein from cloned cDNA. *J Virol* 1987;61:2217–2224.
5. Akkina RK, Chambers TM, Nayak DP. Mechanisim of interference by defective-interfering particles of influenza virus: differential reduction of intracellular synthesis of specific polymerase proteins. *Virus Res* 1984;1:687–702.
6. Allen H, McCauley J, Waterfield M, Gething MJ. Influenza virus RNA segment 7 has the coding capacity for two polypeptides. *Virology* 1980; 107:548–551.
7. Almeida JD, Brand CMA. Morphological study of the internal component of influenza virus. *J Gen Virol* 1975;27:313–318.
8. Aloni Y, Dhar R, Khoury G. Methylation of nuclear simian virus 40 RNAs. *J Virol* 1979;32:52–60.
9. Alonso-Caplen FV. Krug RM. Regulation of the extent of splicing of influenza virus NS_1 mRNA: role of the rates of splicing and of the nucleocytoplasmic transport of NS_1 mRNA. *Mol Cell Biol* 1991;11: 1092–1098.
10. Alonso-Caplen FV, Nemeroff ME, Qiu Y, Krug RM. Nucleocytoplasmic transport: the influenza virus NS_1 protein regulates the transport of spliced NS_2 mRNA and its precursor NS_1 mRNA. *Genes Dev* 1992;6:255–267.
11. Apostolov K, Flewett TH. Internal structure of influenza virus. *Virology* 1965;26:506–508.
12. Bachi T, Gerhard W, Lindermann J, Muhlethaler, K. Morphogenesis of influenza A virus in Ehrlich ascites tumor cells as revealed by thin sectioning and freeze-etching. *J Virol* 1969;4:769–776.
13. Baltimore D. Expression of animal virus genomes. *Bacteriol Rev* 1971; 35:235–241.
14. Banerjee AK. 5′-terminal cap structure in eukaryotic messenger ribonucleic acids. Microbiol. Rev. 1980;44:175–205.
15. Barrett T, Wolstenholme AJ, Mahy BWJ. Transcription and replication of influenza virus RNA. *Virology* 1979;98:211–225.
16. Barry RD. The multiplication of influenza virus. 2. Multiplicity reactivation of ultraviolet irradiated virus. *Virology* 1961;14:398–405.
17. Barry RD, Ives DR, Cruickshank JG. Participation of deoxyribonucleic acid in the multiplication of influenza virus. *Nature* 1962;194: 1139–1140.
18. Bean WJ, Simpson RW. Transcriptase activity and genome composition of defective influenza virus. *J Virol* 1976;18:365–369.
19. Beaton AR. Krug RM. Transcription antitermination during influenza viral template RNA synthesis requires the nucleocapsid protein and the

absence of a 5′ capped end. *Proc Natl Acad Sci USA* 1986;83:6282–6286.

20. Beemon K, Keith J. Localization of N^6-methyladenosine in the Rous sarcome virus genome. *J Mol Biol* 1977;113:165–179.
21. Bergmann M, Garcia SA, Palese P. Transfection-mediated recombination of influenza A virus. *J Virol* 1992;66:7576–7580.
22. Bishop DHL, Obijeski JF, Simpson RW. Transcription of the influenza ribonucleic acid genome by a virion polymerase. 1. Optimal conditions for in vitro activity of the ribonucleic acid-dependent ribonucleic acid polymerase. *J Virol* 1971;8:66–73.
23. Bishop DHL, Obijeski JF, Simpson RW. Transcription of the influenza ribonucleic acid genome by a virion polymerase. 2. Nature of the in vitro polymerase product. *J Virol* 1971;8:74–80.
24. Bishop DHL, Roy P, Bean WJ, Simpson RW. Transcription of the influenza ribonucleic acid genome by a virion polymerase. III. Completeness of the transcription process. *J Virol* 1972;10:689–697.
25. Biswas SK, Nayak DP. Mutational analysis of the conserved motifs of influenza A virus polymerase basic protein 1. *J Virol* 1994;68:1819–1826.
26. Blaas D, Patzelt E, Kuechler E. Cap-recognizing protein of influenza virus. *Virology* 1982;116:339–348.
27. Bosch FX, Garten W, Klenk H-D, Rott R. Proteolytic cleavage of influenza virus hemagglutinins: primary structure of the connecting peptide between HA1 and HA2 determines proteolytic cleavability and pathogenicity of Avian influenza viruses. *Med Welt* 1981;32:1008–1009.
28. Bossart-Whitaker P, Carson M, Babu YS, Smith CD, Laver WG, Air GM. Three-dimensional structure of influenza A N9 neuraminidase and its complex with the inhibitor 2-deoxy 2, 3-dehydro-N-acetyl neuraminic acid. *J Mol Biol* 1993;232:1069–1083.
29. Boulay F, Doms RW, Webster RG, Helenius A. Post-translational oligomerization and cooperative acid activation of mixed influenza hemagglutinin trimers. *J Cell Biol* 1988;106:629–639.
30. Bouloy M, Morgan MA, Shatkin AJ, Krug RM. Cap and internal nucleotides of reovirus mRNA primers are incorporated into influenza viral complementary RNA during transcription in vitro. *J Virol* 1979; 32:895–904.
31. Bouloy M, Plotch SJ, Krug RM. Globin mRNAs are primers for the transcription of influenza viral RNA in vitro. *Proc Natl Acad Sci USA* 1978;75:4886–4890.
32. Braakman I, Helenius J, Helenius A. Manipulating disulfide bond formation and protein folding in the endoplasmic reticulum. *EMBO J* 1992;11:1717–1722.
33. Braakman I, Helenius J, Helenius A. Role of ATP and disulphide bonds during protein folding in the endoplasmic reticulum. *Nature* 1992;356: 260–262.
34. Braakman I, Hoover LH, Wagner KR, Helenius A. Folding of influenza hemagglutinin in the endoplasmic reticulum. *J Cell Biol* 1991;114: 401–411.
35. Braam J, Ulmanen I, Krug RM. Molecular model of a eucaryotic transcription complex: functions and movements of influenza P proteins during capped RNA-primed transcription. *Cell* 1983;34:609–618.
36. Brand CM, Skehel JJ. Crystalline antigen from the influenza virus envelope. Nature New Biol. 1972;238:145–147.
37. Brassard DL. Lamb RA. Influenza B virus NB glycoprotein is a structural protein (*submitted*) 1995.
38. Briedis DJ, Conti G, Munn EA, Mahy BWJ. Migration of influenza virus-specific polypeptides from cytoplasm to nucleus of infected cells. *Virology* 1981;111:154–164.
39. Briedis DJ. Lamb RA. Influenza B virus genome: sequences and structural organization of RNA segment 8 and the mRNAs coding for the NS_1 and NS_2 proteins. *J Virol* 1982;42:186–193.
40. Briedis DJ, Lamb RA, Choppin PW. Influenza B virus RNA segment 8 codes for two nonstructural proteins. *Virology* 1981;112:417–425.
41. Briedis DJ, Lamb RA, Choppin PW. Sequence of RNA segment 7 of the influenza B virus genome: Partial amino acid homology between the membrane proteins (M_1) of influenza A and B viruses and conservation of a second open reading frame. *Virology* 1982;116:581–588.
42. Bron R, Kendal AP, Klenk H-D, Wilschut J. Role of the M_2 protein in influenza virus membrane fusion: Effects of amantadine and monensin on fusion kinetics. *Virology* 1993;195:808–811.
43. Brunner J. Tsurudome M. Fusion protein membrane interactions as studied by hydrophobic photolabeling. In: Bentz J, ed.*Viral fusion mechanisms*. Boca Raton, FL: CRC Press, 1993;104–123.
44. Bucher DJ, Kharitonenkov IG, Zakomirdin JA, Grigoriev VB, Klimenko SM, Davis JF. Incorporation of influenza virus M-protein into liposomes. *J Virol* 1980;36:586–590.
45. Bukrinskaya AG, Vorkunova NK, Kornilayeva GV, Narmanbetova RA, Vorkunova GK. Influenza virus uncoating in infected cells and effect of rimantadine. *J Gen Virol* 1982;60:49–59.
46. Bullough PA, Hughson FM, Skehel JJ, Wiley DC. Structure of influenza haemagglutinin at the pH of membrane fusion. *Nature* 1994;371:37–43.
47. Buonagurio DA, Krystal M, Palese P, DeBorde DC, Maassab HF. Analysis of an influenza A virus mutant with a deletion in the NS segment. *J Virol* 1984;49:418–425.
48. Buonagurio DA, Nakada S, Parvin JD, Krystal M, Palese P, Fitch WM. Evolution of human influenza A viruses over 50 years: rapid, uniform rate of change in NS gene. *Science* 1986;232:980–982.
49. Burmeister WP, Ruigrok RWH, Cusack S. The 2·2 Å resolution crystal structure of influenza B neuraminidase and its complex with sialic acid. *EMBO J* 1992;11:49–56.
50. Canaani D, Kahana C, Lavi CS, Groner Y. Identification and mapping of N^6-methyladenosine containing sequences in simian virus 40 RNA. *Nucleic Acid Res* 1979;6:2879–2899.
51. Carr CM, Kim PS. A spring-loaded mechanism for the conformational change of influenza hemagglutinin. *Cell* 1993;73:823–832.
52. Carson M. Bugg CE. Algorithm for ribbon models of proteins.*J Mol Graph* 1986;4:121–122.
53. Castrucci MR, Hou S, Doherty PC, Kawaoka Y. Protection against lethal lymphocytic choriomeningitis virus (LCMV) infection by immunization of mice with an influenza virus containing an LCMV epitope recognized by cytotoxic T lymphocytes. *J Virol* 1994;68:3486–3490.
54. Castrucci MR. Kawaoka Y. Biologic importance of neuraminidase stalk length in influenza A virus. *J Virol* 1993;67:759–764.
55. Chen-Kiang S, Nevins JR, Darnell JE. N-^{6}-methyladenosine in adenovirus type 2 nuclear RNA is conserved in the formation of messenger RNA. *J Mol Biol* 1973;135:733–752.
56. Choppin PW. Multiplication of two kinds of influenza A2 virus particles in monkey kidney cells. *Virology* 1963;21:342–352.
57. Ciampor F, Bayley PM, Nermut MV, Hirst EM, Sugrue RJ, Hay AJ. Evidence that the amantadine-induced, M_2-mediated conversion of influenza A virus hemagglutinin to the low pH conformation occurs in an acidic *trans* Golgi compartment. *Virology* 1992;188:14–24.
58. Ciampor F, Thompson CA, Grambas S, Hay AJ. Regulation of pH by the M_2 protein of influenza A viruses. *Virus Res* 1992;22:247–258.
59. Ciani C, Tiley L, Krystal M. Differential activation of the influenza virus polymerase via template template RNA binding (*submitted*) 1995.
60. Colman PM. Neuraminidase: Enzyme and Antigen. In: Krug RM, ed. *The influenza viruses*. New York: Plenum Press, 1989;175–218.
61. Colman PM, Laver WG, Varghese JN, Baker AT, Tulloch PA, Air GM, Webster RG. Three-dimensional structure of a complex of antibody with influenza virus neuraminidase. *Nature* 1987;326:358–363.
62. Colman PM, Varghese JN, Laver WG. Structure of the catalytic and antigenic sites in influenza virus neuraminidase. *Nature* 1983;303:41–44.
63. Compans, RW. Influenza virus proteins. II. Association with components of the cytoplasm. *Virology* 1973;51:56–70.
64. Compans RW, Bishop DHL, Meier-Ewert H. Structural components of influenza C virions. *J Virol* 1977;21:658–665.
65. Compans RW, Caliguiri LA. Isolation and properties of an RNA polymerase from influenza virus-infected cells. *J Virol* 1973;11:441–448.
66. Compans RW, Choppin PW. Reproduction of myxoviruses. In: Fraenkel-Conrat H, Wagner RR, eds. *Comprehensive Virology*, vol IV. New York: Plenum, 1975;179–252.
67. Compans RW, Content J, Duesberg PH. Structure of the ribonucleoprotein of influenza virus. *J Virol* 1972;10:795–800.
68. Compans RW, Dimmock NJ. An electron microscopic study of single-cycle infection of chick embryo fibroblasts by influenza virus. *Virology* 1969;39:499–515.
69. Compans RW, Dimmock NJ, Meier EH. Effect of antibody to neuraminidase on the maturation and hemagglutinating activity of an influenza A2 virus. *J Virol* 1969;4:528–534.
70. Daniels RS, Douglas AR, Skehel JJ, Wiley DC. Analyses of the antigenicity of influenza haemagglutinin at the pH optimum for virus-mediated membrane fusion. *J Gen Virol* 1983;64:1657–1662.
71. Daniels RS, Douglas AR, Skehel JJ, Wiley DC. Antigenic and amino acid sequence analyses of influenza viruses of the H1N1 subtype isolated between 1982 and 1984. *Bull WHO* 1985;63:273–277.
72. Daniels RS, Downie JC, Hay AJ, Knossow M, Skehel JJ, Wang ML,

Wiley DC. Fusion mutants of the influenza virus hemagglutinin glycoprotein. *Cell* 1985;40:431–439.
73. Datta B, Weiner AM. Genetic evidence for base pairing between U2 and U6 snRNA in mammalian mRNA splicing. *Nature* 1991;352: 821–824.
74. Davey J, Dimmock NJ, Colman A. Identification of the sequence responsible for the nuclear accumulation of the influenza virus nucleoprotein in xenopus oocytes. *Cell* 1985;40:667–675.
75. Davies WL, Grunert RR, Haff RF, McGahen JW, Neumayer EM, Paulshock M, Watts JC, Wood TR, Herrman EC, Hoffman CE. Antiviral activity of 1-Adamantanamine (Amantadine). *Science* 1964;144: 862–863.
76. Detjen BM, St. Angelo C, Katze MG, Krug RM. The three influenza virus polymerase (P) proteins not associated with viral nucleocapsids in the infected cell are in the form of a complex. *J Virol* 1987;61:16–22.
77. Dimmock NJ, Beck S, McLain L. Protection of mice from lethal influenza: evidence that defective interfering virus modulates immune response and not virus multiplication. *J Gen Virol* 1986;67:839–850.
78. Dimock K, Stoltzfus CM. Sequence specificity of internal methylation in B77 avian sarcoma virus RNA subunits. *Biochemistry* 1977;16: 471–478.
79. Doms RW, Helenius A. Quaternary structure of influenza virus hemagglutinin after acid treatment. *J Virol* 1986;60:833–839.
80. Doms RW, Helenius A, White J. Membrane fusion activity of the influenza virus hemagglutinin. The low pH-induced conformational change. *J Biol Chem* 1985;260:2973–2981.
81. Doms RW, Lamb RA, Rose JK, Helenius A. Folding and assembly of viral membrane proteins. *Virology* 1993;193:545–562.
82. Dopheide TA, Ward CW. The carboxyl-terminal sequence of the heavy chain of a Hong Kong influenza hemagglutinin. *Eur J Biochem* 1978; 85:393–398.
83. Drapkin R, Merino A, Reinberg D. Regulation of RNA polymerase II transcription. *Curr Opin Cell Biol* 1993;5:469–576.
84. Dreiding P, Staeheli P, Haller O. Interferon-induced protein Mx accumulates in nuclei of mouse cells expressing resistance to influenza viruses. *Virology* 1985;140:192–196.
85. Dreyfuss G, Matunis MJ, Pinol-Roma S, Burd CG. hnRNP proteins and the biogenesis of mRNA. *Annu Rev Biochem* 1993;62:289–329.
86. Duesberg P. Distinct subunits of the ribonucleoprotein of influenza virus. *J Mol Biol* 1969;42:485–499.
87. Duff KC, Ashley RH. The transmembrane domain of influenza A M_2 protein forms amantadine-sensitive proton channels in planar lipid bilayers. *Virology* 1992;190:485–489.
88. Elder KT, Bye JM, Skehel JJ, Waterfield MD, Smith AE. In vitro synthesis, glycosylation, and membrane insertion of influenza virus haemagglutinin. *Virology* 1979;95:343–350.
89. Ellens H, Bentz J, Mason D, Zhang F, White JM. Fusion of influenza hemagglutinin-expressing fibroblasts with glycophorin-bearing liposomes: role of hemagglutinin surface density. *Biochemistry* 1990;29: 9697–9707.
90. Elster C, Fourest E, Baudin F, Larsen K, Cusack S, Ruigrok RWH. A small percentage of influenza virus M_1 protein contains zinc but zinc does not influence in vitro M_1-RNA interaction. *J Gen Virol* 1994;75: 37–42.
91. Enami M, Lutyjes W, Krystal M, Palese P. Introduction of site-specific mutations into the genome of influenza virus. *Proc Natl Acad Sci USA* 1990;87:3802–3805.
92. Enami M, Palese P. High-efficiency formation of influenza virus transfectants. *J Virol* 1991;65:2711–2713.
93. Enami M, Sharma G, Benham C, Palese P. An influenza virus containing nine different RNA segments [published erratum appears in *Virology* 1992 Feb;186(2):798]. *Virology* 1991;185:291–298.
94. Fang R, Min Jou W, Huylebroeck D, Devos R, Fiers W. Complete structure of A/duck/Ukraine/63 influenza hemagglutinin gene: animal virus as progenitor of human H3 Hong Kong 1968 influenza hemagglutinin. *Cell* 1981;25:315–323.
95. Fazekas de St. Groth S. Viropexis: The mechanism of influenza virus infection. *Nature* 1948;162:294–296.
96. Fiume L, Wieland T. Amanatins: Chemistry and action. *FEBS Lett* 1970;8:1–5.
97. Flewett TH, Apostolov KA. Reticular structure in the wall of influenza C virus. *J Gen Virol* 1967;1:297–304.
98. Fodor E, Pritlove DC, Brownlee GG. The influenza virus panhandle is involved in the initiation of transcription. *J Virol* 1994;68:4092–4096.
99. Fortes P, Beloso A, Ortin J. Influenza virus NS_1 protein inhibits pre-mRNA splicing and blocks mRNA nucleocytoplasmic transport. *EMBO J* 1994;13:704–712.
100. Gamboa ET, Harter DH, Daffy PE, Hsu KC. Murine influenza virus encephalomyeitis. III. Effect of defective interfering particles. *Acta Neuropathol (Berl)* 1975;34:157–169.
101. Garten W, Bosch FX, Linder D, Rott R, Klenk H-D. Proteolytic activation of the influenza virus hemagglutinin: the structure of the cleavage site and the enzymes involved in cleavage. *Adv Exp Med Biol* 1981; 142:119–131.
102. Gething M-J, McCammon K, Sambrook J. Expression of wild-type and mutant forms of infuenza hemagglutinin: the role of folding in intracellular transport. *Cell* 1986;46:939–950.
103. Godley L, Pfeifer J, Steinhauer D, Ely B, Shaw G, Kaufmann R, Suchanek E, Pabo C, Skehel JJ, Wiley DC. Introduction of intersubunit disulfide bonds in the membrane-distal region of the influenza hemagglutinin abolishes membrane fusion activity. *Cell* 1992;68: 635–645.
104. Goldstein EA, Pons MW. The effect of polyvinyl-sulfate on the ribonucleoprotein of influenza virus. *Virology* 1970;41:382–384.
105. Gottschalk A. The specific enzyme of influenza virus and Vibrio cholerae. *Biochim Biophys Acta* 1957;23:645–646.
106. Grambas S, Bennett MS, Hay AJ. Influence of amantadine resistance mutations on the pH regulatory function of the M_2 protein of influenza A viruses. *Virology* 1992;191:541–549.
107. Grambas S, Hay AJ. Maturation of influenza A virus hemagglutinin—estimates of the pH encountered during transport and its regulation by the M_2 protein. *Virology* 1992;190:11–18.
108. Graves PN, Schulman JL, Young JF, Palese P. Preparation of influenza virus subviral particles lacking the HA1 subunit of hemagglutinin: unmasking of cross-reactive HA2 determinants. *Virology* 1983;126: 106–116.
109. Green MR. Pre-mRNA processing and mRNA nuclear export. *Curr Opin Cell Biol* 1989;1:519–525.
110. Greenspan D, Krystal M, Nakada S, Arnheiter H, Lyles DS, Palese P. Expression of influenza virus NS_2 nonstructural protein in bacteria and localization of NS_2 in infected eucaryotic cells. *J Virol* 1985;54:833–843.
111. Greenspan D, Palese P, Krystal M. Two nuclear location signals in the influenza virus NS_1 nonstructural protein. *J Virol* 1988;62:3020–3026.
112. Gregoriades A. The membrane protein of influenza virus: extraction from virus and infected cell with acidic chloroform-methanol. *Virology* 1973;54:369–383.
113. Gregoriades A, Frangione B. Insertion of influenza M protein into the viral lipid bilayer and localization of site of insertion. *J Virol* 1981;40: 323–328.
114. Hadzopoulou-Cladaras M, Felber BK, Cladaras C, Athanossopoulos A, Tse A, Pavlakis GN. The rev/trs/art protein of human immunodeficiency virus type 1 affects viral mRNA and protein expression via a *cis*-acting sequence in the env region. *J Virol* 1989;63:1265–1274.
115. Hagen M, Thomas DY, Chung J, Butcher A, Krystal M. Recombinant influenza virus polymerase: Requirement of both 5′ and 3′ viral ends for endonuclease activity. *J Virol* 1994;68:1509–1515.
116. Haller O. Inborn resistance of mice to orthomyxoviruses. *Curr Top Microbiol Immunol* 1981;92:25–52.
117. Haller O, Acklin M, Staeheli P. Influenza virus resistance of wild mice: Wild-type and mutant Mx alleles occur at comparable frequencies. *J Interferon Res* 1987;7:647–656.
118. Hausman J, Kretzschmar E, Ohuchi M, Garten W, Klenk HD. N1 neuraminidase of influenza virus A/FPV/Rostock/34 has hemagglutinin activity. (Abstract) *12th Annual Meeting American Society for Virology* 1993;34.
119. Hay AJ. The action of adamantanamines against influenza A viruses: inhibition of the M_2 ion channel protein. *Sem Virol* 1992;3:21–30.
120. Hay AJ, Abraham G, Skehel JJ, Smith JC, Fellner P. Influenza virus messenger RNAs are incomplete transcripts of the genome RNAs. *Nucleic Acid Res* 1977;4:4197–4209.
121. Hay AJ, Lomniczi B, Bellamy AR, Skehel JJ. Transcription of the influenza virus genome. *Virology* 1977;83:337–355.
122. Hay AJ, Skehel JJ, McCauley J. Characterization of influenza virus RNA complete transcripts. *Virology* 1982;116:517–522.
123. Hay AJ, Wolstenholme AJ, Skehel JJ, Smith MH. The molecular basis of the specific anti-influenza action of amantadine. *EMBO J* 1985;4: 3021–3024.
124. Heggeness MH, Smith PR, Ulmanen I, Krug RM, Choppin PW. Stud-

ies on the helical nucleocapsid of influenza virus. *Virology* 1982;118: 466–470.
125. Helenius A. Unpacking the incoming influenza virus. *Cell* 1992;69: 577–578.
126. Herrler G, Compans RW, Meier-Ewert H. A precursor glycoprotein in influenza C virus. *Virology* 1979;99:49–56.
127. Herrler G, Durkop I, Becht H, Klenk HD. The glycoprotein of influenza C virus is the haemagglutinin, esterase and fusion factor. *Virology* 1988; 165:428–437.
128. Herrler G, Klenk HD. Structure and function of the HEF glycoprotein of influenza C virus. *Arch Virol* 1991;120:289–296.
129. Herrler G, Multhaup G, Beyreuther K, Klenk HD. Serine 71 of the glycoprotein HEF is located at the active site of the acetylesterase of influenza C virus. *Virology* 1989;168:274–280.
130. Herrler G, Nagele A, Meier EH, Bhown AS, Compans RW. Isolation and structural analysis of influenza C virion glycoproteins. *Virology* 1981;113:439–451.
131. Herrler G, Reuter G, Rott R, Klenk HD, Schauer R. N-acetyl-9-O-acetylneuraminic acid, the receptor determinant for influenza C virus, is a differentiation marker on chicken erythrocytes. *Vopr Virusol* 1987; 32:300–303.
132. Herrler G, Rott R, Klenk HD, Muller HP, Shukla AK, Schauer R. The receptor-destroying enzyme of influenza C virus is neuraminate-O-acetyl esterase. *EMBO J* 1985;4:2711–2720.
133. Herrler G, Rott R, Klenk HD. Neuraminic acid is involved in the binding of influenza C virus to erythrocytes. *Virology* 1985;141:144–147.
134. Herz C, Stavnezer E, Krug RM, Gurney T Jr. Influenza virus, an RNA virus, synthesizes its messenger RNA in the nucleus of infected cells. *Cell* 1981;26:391–400.
135. Hiebert SW, Williams MA, Lamb RA. Nucleotide sequence of RNA segment 7 of influenza B/Singapore/222/79: maintenance of a second large open reading frame. *Virology* 1986;155:747–751.
136. Hirst GK. Agglutination of red cells by allantoic fluid of chick embryos infected with influenza virus. *Science* 1941;94:22–23.
137. Hirst GK. Genetic recombination with Newcastle disease virus, poliovirus, and influenza. *Cold Spring Harbor Symp Quant Biol* 1962; 27:303–308.
138. Hirst GK, Pons MW. Mechanism of influenza recombination. II. virus aggregation and its effect on plaque formation by so-called noninfective virus. *Virology* 1973;56:620–631.
139. Holsinger LJ, Lamb RA. Influenza virus M_2 integral membrane protein is a homotetramer stabilized by formation of disulfide bonds. *Virology* 1991;183:32–43.
140. Holsinger LJ, Nichani D, Pinto LH, Lamb RA. Influenza A virus M_2 ion channel protein: a structure-function analysis. *J Virol* 1994;68: 1551–1563.
141. Holsinger LJ, Shaughnessy MA, Pinto LH, Lamb RA. Analysis of the post-translational modification of the influenza virus M_2 ion channel protein. *J Virol* 1994;68:1551–1563.
142. Honda A, Udea K, Nagata K, Ishihama A. Identification of the RNA polymerase binding site on genome RNA of the influenza virus. *J Biochem (Tokyo)* 1987;102:1241–1249.
143. Honda A, Ueda K, Nagata K, Ishihama A. RNA polmerase of influenza virus: role of NP on RNA chain elongation. *J Biochem (Tokyo)* 1988; 104:1021–1026.
144. Horisberger MA. The large P proteins of influenza A viruses are composed of one acidic and two basic polypeptides. *Virology* 1980;107: 302–305.
145. Horisberger MA. Interferon-induced human protein MxA is a GTPase which binds transiently to cellular proteins. *J Virol* 1992;66:4705–4709.
146. Horne RW, Waterson AP, Wildy P, Farnham AE. The structure and composition of the myxoviruses. I. Electron microscope studies of the structure of myxovirus particles by negative staining techniques. *Virology* 1960;11:79–98.
147. Horowitz S, Horowitz A, Nilsen TW, Munns TW, Rottman FM. Mapping of N^6-methyladenosine residues in bovine prolactin mRNA. *Proc Natl Acad Sci USA* 1984;81:5667–5671.
148. Horvath CM, Williams MA, Lamb RA. Eukaryotic coupled translation of tandem cistrons: identification of the influenza B virus BM2 polypeptide. *EMBO J* 1990;9:2639–2647.
149. Hoyle L, Horne RW, Waterson AP. The structure and composition of the myxoviruses. II. Components released from the influenza virus particle by ether. *Virology* 1961;13:448–459.
150. Hsu MT, Parvin JD, Dupta S, Krystal M, Palese P. Genomic RNAs of influenza viruses are held in a circular conformation in virions and in infected cells by a terminal panhandle. *Proc Natl Acad Sci USA* 1987; 84:8140–8144.
151. Huang T, Pavloic J, Staeheli P, Krystal M. Overexpression of the influenza virus polymerase can titrate out inhibition by the murine Mx1 protein. *J Virol* 1992;66:4154–4160.
152. Huang TS, Palese P, Krystal M. Determination of influenza virus proteins required for genome replication. *J Virol* 1990;64:5669–5673.
153. Hughey PG, Compans RW, Zebedee SL, Lamb RA. Expression of the influenza A virus M_2 protein is restricted to apical surfaces of polarized epithelial cells. *J Virol* 1992;66:5542–5552.
154. Hull JD, Gilmore R, Lamb RA. Integration of a small integral membrane protein, M_2, of influenza virus into the endoplasmic reticulum: analysis of the internal signal-anchor domain of a protein with an ectoplasmic NH_2 terminus. *J Cell Biol* 1988;106:1489–1498.
155. Inglis SC, Almond JW. An influenza virus gene encoding two different proteins. *Phil Trans R Soc Lond* 1980;288:375–381.
156. Inglis SC, Barrett T, Brown CM, Almond JW. The smallest genome RNA segment of influenza virus contains two genes that may overlap. *Proc Natl Acad Sci USA* 1979;76:3790–3794.
157. Inglis SC, Brown CM. Spliced and unspliced RNAs encoded by virion RNA segment 7 of influenza virus. *Nucleic Acids Res* 1981;9: 2727–2740.
158. Inglis SC, Carroll AR, Lamb RA, Mahy BW. Polypeptides specified by the influenza virus genome. I. Evidence for eight distinct gene products specified by fowl plague virus. *Virology* 1976;74:489–503.
159. Jackson DC, Tang XL, Murti KG, Webster RG, Tregear GW, Bean WJ. Electron microscopic evidence for the association of M_2 protein with the influenza virion. *Arch Virol* 1991;118:199–207.
160. Jagus R, Anderson WF, Safer B. The regulation of initiation of mammalian protein synthesis. *Prog Nucleic Acid Res* 1981;25:127–185.
161. Jennings PA, Finch JT, Winter G, Robertson JS. Does the higher order structure of the influenza virus ribonucleoprotein guide sequence rearrangements in influenza viral RNA? *Cell* 1983;34:619–627.
162. Jin H, Leser G, Lamb RA. The influenza virus hemagglutinin cytoplasmic tail is not essential for virus assembly or infectivity. *EMBO J* 1994;13:5504–5514.
163. Jones IM, Reay PA, Philpott KL. Nuclear location of all three influenza polymerase proteins and a nuclear signal in polymerase PB2. *EMBO J* 1986;5:2371–2376.
164. Jones LV, Compans RW, Davis AR, Bos TJ, Nayak DP. Surface expression of influenza virus neuraminidase, an amino-terminally anchored viral membrane glycoprotein, in polarized epithelial cells. *Mol Cell Biol* 1985;5:2181–2189.
165. Kalderon D, Richardson WD, Markham AF, Smith AE. Sequence requirements for nuclear localization of SV40 large-T antigen. *Nature* 1984;311:33–38.
166. Kalderon D, Roberts BL, Richardson WD, Smith AE. A short amino acid sequence able to specify nuclear location. *Cell* 1984;39:499–509.
167. Kane SE, Beemon K. Precise localization of m^6A in Rous sarcoma virus RNA reveals clustering of methylation sites: implications for RNA processing. *Mol Cell Biol* 1985;5:2298–2306.
168. Kato N, Eggers HJ. Inhibition of uncoating of fowl plague virus by 1-adamantanamine hydrochloride. *Virology* 1969;37:632–641.
169. Katz RA, Skalka AM. Control of retroviral RNA splicing through maintenance of suboptimal processing signals. *Mol Cell Biol* 1990;10: 696–704.
170. Katze MG, Chen YT, Krug RM. Nuclear-cytoplasmic transport and VAI RNA-independent translation of influenza viral messenger RNAs in late adenovirus-infected cells. *Cell* 1984;37:483–490.
171. Katze MG, DeCorato D, Krug RM. Cellular mRNA translation is blocked at both initiation and elongation after infection by influenza virus or adenovirus. *J Virol* 1986;60:1027–1039.
172. Katze MG, Detjen BM, Safer B, Krug RM. Translational control by influenza virus: suppression of the kinase that phosphorylates the alpha subunit of initiation factor eIF-2 and selective translation of influenza viral mRNAs. *Mol Cell Biol* 1986;6:1741–1750.
173. Katze MG, Krug RM. Translational control in influenza virus-infected cells. *Enzyme* 1990;44:265–277.
174. Kawaoka Y, Krauss S, Webster RG. Avian-to-human transmission of the PB1 gene of influenza A viruses in the 1957 and 1968 pandemics. *J Virol* 1989;63:4603–4608.
175. Kawaoka Y, Webster RG. Sequence requirements for cleavage activation of influenza virus hemagglutinin expressed in mammalian cells. *Proc Natl Acad Sci USA* 1988;85:324–328.
176. Kemble GW, Bodian DL, Rose J, Wilson IA, White JM. Intermonomer

disulfide bonds impair the fusion activity of influenza virus hemagglutinin. *J Virol* 1992;66:4940–4950.

177. Kemble GW, Danieli T, White JM. Lipid-anchored influenza hemagglutinin promotes hemifusion, not complete fusion. *Cell* 1994;76:383–391.
178. Khatchikian D, Orlich M, Rott R. Increased viral pathogenicity after insertion of a 28S ribosomal RNA sequence into the haemagglutinin gene of an influenza virus. *Nature* 1989;340:156–157.
179. Kilbourne ED. Influenza virus genetics. *Prog Med Virol* 1963;5:79–126.
180. Kilbourne ED. Molecular epidemiology-influenza as archetype. *Harvey Lect* 1979;73:225–258.
181. Kilbourne ED. *Influenza*. New York: Plenum Medical Book Company, 1987.
182. Kilbourne ED, Laver WG, Schulman JL, Webster RG. Antiviral activity of antiserum specific for an influenza virus neuraminidase. *J Virol* 1968;2:281–288.
183. Kilbourne ED, Schulman JL, Schild GC, Schloer G, Swanson J, Bucher D. Correlated studies of a recombinant influenza-virus vaccine. I. Derivation and characterization of virus and vaccine. *J Infect Dis* 1971;124:449–462.
184. Kingsbury DW, Jones IM, Murti KG. Assembly of influenza ribonucleoprotein in vitro using recombinant nucleoprotein. *Virology* 1987;156:396–403.
185. Kingsbury DW, Webster RG. Some properties of influenza virus nucleocapsids. *J Virol* 1969;4:219–225.
186. Klenk HD, Garten W. Host cell proteases controlling virus pathogenicity. *Trends in microbiology* 1994;2:39–43.
187. Klenk HD, Rott R, Orlich M, Blodorn J. Activation of influenza A viruses by trypsin treatment. *Virology* 1975;68:426–439.
188. Klenk HD, Compans RW, Choppin PW. An electron microscopic study of the presence or absence of neuraminic acid in enveloped viruses. *Virology* 1970;42:1158–1162.
189. Konieczny A, Safer B. Purification of the eukaryotic initiation factor 2-eukaryotic initiation factor 2B complex and characterization of its guanine nucleotide exchange activity during protein synthesis. *J Biol Chem* 1983;256:3402–3408.
190. Kornfeld R, Kornfeld S. Assembly of asparagine-linked oligosaccharides. *Ann Rev Biochem* 1985;54:631–664.
191. Kozak M. The scanning model for translation: an update. *J Cell Biol* 1989;108:229–241.
192. Kraulis PJ. MOLSCRIPT: a program to produce both detailed and schematic plots of protein structures. *J Appl Crystallogr* 1991;24:946–950.
193. Krug RM. Priming of influenza viral RNA transcription by capped heterologous RNAs. *Curr Top Microbiol Immunol* 1981;93:125–149.
194. Krug RM. Transcription and replication of influenza viruses. In: Palese P, Kingsbury DW, eds. *Genetics of influenza viruses*.Vienna: Springer-Verlag, 1983;70–98.
195. Krug RM, Bouloy M, Plotch SJ. RNA primers and the role of host nuclear RNA polymerase II in influenza viral RNA transcription. *Philos Trans R Soc Lond Biol* 1980;288:359–370.
196. Krug RM, Broni B, Bouloy M. Are the 5′ ends of influenza viral mRNAs synthesized in vivo donated by host mRNAs? *Cell* 1979;18:329–334.
197. Krug RM, Broni BA, LaFiandra AJ, Morgan MA, Shatkin AJ. Priming and inhibitory activities of RNAs for the influenza viral transcriptase do not require base pairing with the virion template RNA. *Proc Natl Acad Sci USA* 1980;77:5874–5878.
198. Krug RM, Etkind PR. Cytoplasmic and nuclear virus-specific proteins in influenza virus-infected MDCK cells. *Virology* 1973;56:334–348.
199. Krug RM, Morgan MA, Shatkin AJ. Influenza viral mRNA contains internal N^6-methyladenosine and 5′-terminal 7-methylguanosine in cap structures. *J Virol* 1976;20:45–53.
200. Krug RM, Shaw M, Broni B, Shapiro G, Haller O. Inhibition of influenza viral mRNA synthesis in cells expressing the interferon-induced Mx gene product. *J Virol* 1985;56:201–206.
201. Krug RM, Soeiro R. Studies on the intranuclear localization of influenza virus-specific proteins. *Virology* 1975;64:378–387.
202. Krystal M, Elliott RM, Benz EW Jr, Young JF, Palese P. Evolution of influenza A and B viruses: conservation of structural features in the hemagglutinin genes. *Proc Natl Acad Sci USA* 1982;79:4800–4804.
203. Krystal M, Young JF, Palese P, Wilson IA, Skehel JJ, Wiley DC. Sequential mutations in hemagglutinins of influenza B virus isolates: definition of antigenic domains. *Proc Natl Acad Sci USA* 1983;80:4527–4531.
204. Lamb RA. The influenza virus RNA segments and their encoded proteins. In: Palese P, Kingsbury DW, eds. *Genetics of influenza viruses*. Vienna: Springer-Verlag, 1983;26–69.
205. Lamb RA. Genes and proteins of the influenza viruses. In: Krug RM, ed. *The influenza viruses*. New York: Plenum Press, 1989;1–87.
206. Lamb RA, Choppin PW. Synthesis of influenza virus proteins in infected cells: translation of viral polypeptides, including three P polypeptides, from RNA produced by primary transcription. *Virology* 1976;74:504–519.
207. Lamb RA, Choppin PW. Synthesis of influenza virus polypeptides in cells resistant to alpha-amanitin: evidence for the involvement of cellular RNA polymerase II in virus replication. *J Virol* 1977;23:816–819.
208. Lamb RA, Choppin PW. Segment 8 of the influenza virus genome is unique in coding for two polypeptides. *Proc Natl Acad Sci USA* 1979;76:4908–4912.
209. Lamb RA, Choppin PW. A ninth unique influenza virus-coded polypeptide. *Philos Trans R Soc Lond (Biol)* 1980;288:327–333.
210. Lamb RA, Choppin PW. Identification of a second protein (M_2) encoded by RNA segment 7 of influenza virus. *Virology* 1981;112:729–737.
211. Lamb RA, Choppin PW. The gene structure and replication of influenza virus. *Ann Rev Biochem* 1983;52:467–506.
212. Lamb RA, Choppin PW, Chanock RM, Lai CJ. Mapping of the two overlapping genes for polypeptides NS_1 and NS_2 on RNA segment 8 of influenza virus genome. *Proc Natl Acad Sci USA* 1980;77:1857–1861.
213. Lamb RA, Etkind PR, Choppin PW. Evidence for a ninth influenza viral polypeptide. *Virology* 1978;91:60–78.
214. Lamb RA, Lai CJ. Sequence of interrupted and uninterrupted mRNAs and cloned DNA coding for the two overlapping nonstructural proteins of influenza virus. *Cell* 1980;21:475–485.
215. Lamb RA, Lai CJ. Conservation of the influenza virus membrane protein (M_1) amino acid sequence and an open reading frame of RNA segment 7 encoding a second protein (M_2) in H1N1 and H3N2 strains. *Virology* 1981;112:746–751.
216. Lamb RA, Lai CJ. Spliced and unspliced messenger RNAs synthesized from cloned influenza virus M DNA in an SV40 vector: expression of the influenza virus membrane protein (M_1). *Virology* 1982;123:237–256.
217. Lamb RA, Lai CJ. Expression of unspliced NS_1 mRNA, spliced NS_2 mRNA, and a spliced chimera mRNA from cloned influenza virus NS DNA in an SV40 vector. *Virology* 1984;135:139–147.
218. Lamb RA, Lai CJ, Choppin PW. Sequences of mRNAs derived from genome RNA segment 7 of influenza virus: Colinear and interrupted mRNAs code for overlapping proteins. *Proc Natl Acad Sci USA* 1981;78:4170–4174.
219. Lamb RA, Zebedee SL, Richardson CD. Influenza virus M_2 protein is an integral membrane protein expressed on the infected-cell surface. *Cell* 1985;40:627–633.
220. Laver WG, Colman PM, Webster RG, Hinshaw VS, Air GM. Influenza virus neuraminidase with hemagglutinin activity. *Virology* 1984;137:314–323.
221. Laver WG, Valentine RC. Morphology of the isolated hemagglutinin and neuraminidase subunits of influenza virus. *Virology* 1969;38:105–119.
222. Laver WG, Webster RG. Studies on the origin of pandemic influenza. 3. Evidence implicating duck and equine influenza viruses as possible progenitors of the Hong Kong strain of human influenza. *Virology* 1973;51:383–391.
223. Lazarowitz SG, Choppin PW. Enhancement of the infectivity of influenza A and B viruses by proteolytic cleavage of the hemagglutinin polypeptide. *Virology* 1975;68:440–454.
224. Lazarowitz SG, Compans RW, Choppin PW. Influenza virus structural and nonstructural proteins in infected cells and their plasma membranes. *Virology* 1971;46:830–843.
225. Lazinski D, Grzadzielska E, Das A. Sequence-specific recognition of RNA hairpins by bacteriophage antiterminator requires a conserved arginine-rich motif. *Cell* 1989;59:207–218.
226. Lazzarini RA, Keene JD, Schubert M. The origin of defective interfering particles of the negative strand RNA viruses. *Cell* 1981;26:145–154.
227. Lee TG, Tang N, Thompson S, Miller J, Katze MG. The 58,000 dalton cellular inhibitor of the interferon-induced DS RNA activated protein kinase (PKR) is a member of the TPR family of proteins. *Mol Cell Biol* 1994;14:2331–2342.
228. Lee TG, Tomita J, Hovanessian AG, Katze G. Characterization and regulation of the 58,000-dalton cellular inhibitor of the interferon-in-

duced, dsRNA-activated protein kinase. *J Biol Chem* 1992;267:14238–14243.
229. Legrain P, Rosbash M. Some cis- and trans-acting mutants for splicing target pre-mRNA to the cytoplasm. *Cell* 1989;57:573–583.
230. Lentz MR, Webster RG, Air GM. Site-directed mutation of the active site of influenza neuraminidase and implications for the catalytic mechanism. *Biochemistry* 1987;26:5321–5358.
231. Li S, Polonis V, Isobe H, Zaghouani H, Guinea R, Moran T, Bona C, Palese P. Chimeric influenza virus induces neutralizing antibodies and cytotoxic T cells against human immunodeficiency virus type 1. *J Virol* 1993;67:6659–6666.
232. Li S, Schulman JL, Moran T, Bona C, Palese P. Influenza A virus transfectants with chimeric hemagglutinin containg epitopes from different subtypes. *J Virol* 1992;66:399–404.
233. Li X, Palese P. Mutational analysis of the promoter required for influenza virus virion RNA synthesis. *J Virol* 1992;66:4331–4338.
234. Li X. Palese, P. Characterization of the polyadenylation signal of influenza virus RNA. *J Virol* 1994;68:1245–1249.
235. Lin BC, Lai CJ. The influenza virus nucleoprotein synthesized from cloned DNA in a simian virus 40 vector is detected in the nucleus. *J Virol* 1983;45:434–438.
236. Lu Y, Qian XY, Krug RM. The influenza virus NS_1 protein: A novel inhibitor of pre-mRNA splicing. *Genes Dev* 1994;8:1817–1828.
237. Luo G, Chung J, Palese P. Alterations of the stalk of the influenza virus neuraminidase: deletions and insertions. *Virus Res* 1993;29:141–153.
238. Luo G, Luytjes W, Enami M, Palese P. The polyadenylation signal of influenza virus RNA involves a stretch of uridines followed by the RNA duplex of the panhandle structure. *J Virol* 1991;65:2861–2867.
239. Luytjes W, Krystal M, Enami M, Pavin JD, Palese P. Amplification, expression, and packaging of foreign gene by influenza virus. *Cell* 1989;59:1107–1113.
240. Mahy BWJ. Mutants of influenza virus In: Palese P, Kingsbury DW, eds. *Genetics of influenza virus*. New York: Springer-Verlag, 1983; 192–254.
241. Mahy BWJ, Hastie ND, Armstrong SJ. Inhibition of influenza virus replication by a-amantinin: Mode of action. *Proc Natl Acad Sci USA* 1972;69:1421–1424.
242. Malim MH, Bohnlein S, Hauber J, Cullen BR. Functional dissection of the HIV-1 Rev trans-activator—derivation of a trans-dominant repressor of Rev function. *Cell* 1989;58:205–214.
243. Malim MH, Hauber J, Le S-Y, Maizel JV, Cullen BR. The HIV-1 Rev trans-activator acts through a structural target sequence to activate nuclear export of unspliced viral mRNA. *Nature* 1989;338:254–257.
244. Manire GP. Studies on the toxicity for mice of incomplete influenza virus. *Acta Pathol Microbiol Scand* 1957;40:501–510.
245. Marsh M. Keeping the viral coat on. *Curr Biol* 1992;2:379–381.
246. Marsh M, Helenius A. Virus entry into animal cells. *Adv Virus Res* 1989;36:107–151.
247. Martin K, Helenius A. Nuclear transport of influenza virus ribonucleoproteins: the viral matrix protein (M_1) promotes export and inhibits import. *Cell* 1991;67:117–130.
248. Martin K, Helenius A. Transport of incoming influenza virus nucleocapsids into the nucleus. *J Virol* 1991;65:232–244.
249. Matlin KS, Reggio H, Helenius A, Simons K. The entry of enveloped viruses into an epithelial cell line. *Prog Clin Biol Res* 1982;91:599–611.
250. McCauley J, Bye J, Elder K, Gething MJ, Skehel JJ, Smith A, Waterfield MD. Influenza virus haemagglutinin signal sequence. *FEBS Lett* 1979;108:422–426.
251. McClelland L, Hare R. The adsorption of influenza virus by red cells and a new in vitro method of measuring antibodies for influenza virus. *Can J Public Health* 1941;32:530–538.
252. McGeoch D, Fellner P, Newton C. The influenza virus genome consists of eight distinct RNA species. *Proc Natl Acad Sci USA* 1976;73: 3045–3049.
253. McGeoch D, Kitron N. Influenza virion RNA-dependent RNA polymerase: stimulation by guanosine and related compounds. *J Virol* 1975; 15:686–695.
254. Meier E, Fah J, Grob MS, End R, Staeheli P, Haller O. A family of interferon-induced Mx-related mRNAs encodes cytoplasmic and nuclear proteins in rat cells. *J Virol* 1988;62:2386–2393.
255. Melen K, Ronni T, Broni B, Krug RM, von Bonsdorff CH, Julkunen I. Interferon-induced Mx proteins form oligomers and contain a putative leucine zipper. *J Biol Chem* 1992;267:25898–25907.
256. Morgan C, Hsu KC, Rifkind RA, Knox AW, Rose HM. The application of ferritin-conjugated antibody to electron microscopic studies of influenza virus in infected cells. *J Exp Med* 1961;114:825–832.
257. Morrongiello MP, Dales S. Characterization of cytoplasmic inclusions formed during influenza/WSN virus infection of chick embryo fibroblasts. *Intervirology* 1977;8:281–293.
258. Mulligan RC, Howard BH, Berg P. Synthesis of rabbit -globin in cultured monkey kidney cells following infection with a SV40-globin recombinant genome. *Nature* 1979;277:108–114.
259. Murphy JS, Bang FB. Observations with the electron microscope on cells of the chick chorioallantoic membrane infected with influenza virus. *J Exp Med* 1952;95:259–271.
260. Murti KG, Bean WJ, Webster RG. Helical ribonucleoproteins of influenza virus: An electron microscope analysis. *Virology* 1980;104: 224–229.
261. Murti KG, Brown PS, Bean WJ, Webster RG. Composition of the helical internal components of influenza virus as revealed by immunogold labeling/electron microscopy. *Virology* 1992;186:294–299.
262. Murti KG, Webster RG, Jones IM. Localization of RNA polymerases on influenza viral ribonucleoproteins by immunogold labeling. *Virology* 1988;164:562–566.
263. Muster T, Subbarao EK, Enami M, Murphy BR, Palese P. An influenza A virus containing influenza B virus 5′ and 3′ noncoding regions on the neuraminidase gene is attenuated in mice. *Proc Natl Acad Sci USA* 1991;88:5177–5181.
264. Naeve CW, Williams D. Fatty acids on the A/Japan/305/57 influenza virus hemagglutinin have a role in membrane fusion. *EMBO J* 1990; 9:3857–3866.
265. Naim HY, Roth MG. Basis for selective incorporation of glycoproteins into the influenza virus envelope. *J Virol* 1993;67:4831–4841.
266. Nakada S, Creager RS, Krystal M, Aaronson RP, Palese P. Influenza C virus hemagglutinin: Comparison with influenza A and B virus hemagglutinins. *J Virol* 1984;50:118–124.
267. Nakada S, Graves PN, Desselberger U, Creager RS, Krystal M, Palise P. Influenza C virus RNA 7 codes for a nonstructural protein. *J Virol* 1985;56:221–226.
268. Nakada S, Graves PN, Palese P. The influenza C virus NS gene: evidence for a spliced mRNA and a second NS gene product(NS_2 protein). *Virus Res* 1986;4:1–11.
269. Nakajima K, Ueda M, Sugiura A. Origin of small RNA in von Magnus particles of influenza virus. *J Virol* 1979;29:1142–1148.
270. Nakayama M, Nagata K, Kato A, Ishihama A. Interferon-inducible mouse Mx1 protein that confers resistance to influenza virus is GTPase. *J Biol Chem* 1991;266:21404–21408.
271. Narayan P, Ayers DF, Rottman FM, Maroney PA, Nilsen TW. Unequal distribution of N^6-methyladenosine in influenza virus mRNAs. *Mol Cell Biol* 1987;7:1572–1575.
272. Nayak DP, Chambers TM, Akkina RK. Defective-interfering (DI) RNAs of influenza viruses: origin, structure, expression, and interference. *Curr Top Microbiol Immunol* 1985;114:104–151.
273. Nayak DP, Chambers TM, Akkina RK. Structure of defective-interfering RNAs of influenza viruses and their role in interference. In: Krug RM, ed. *The influenza viruses*. New York: Plenum Press, 1989;269–317.
274. Nayak DP, Sivasubramanian N. The structure of the influenza defective interfering (DI) RNAs and their progenitor genes. In: Palese P, Kimgsbury DW, eds. *Genetics of influenza viruses*. Vienna: Springer-Verlag, 1983;255–279.
275. Nemeroff ME, Utans U, Kramer A, Krug RM. Identification of cis-acting intron and exon regions in influenza virus NS_1 mRNA that inhibit splicing and cause the formation of aberrantly sedimenting presplicing complexes. *Mol Cell Biol* 1992;12:962–970.
276. Nestorowicz A, Kawaoka Y, Bean WJ, Webster RG. Molecular analysis of the hemagglutinin genes of Australian H7N7 influenza viruses: Role of passerine birds in maintenance or transmission? *Virology* 1987; 160:411–418.
277. Nobusawa E, Aoyama T, Kato H, Suzuki Y, Tateno Y, Nakajima K.Comparison of complete amino acid sequences and receptor-binding properties among 13 serotypes of hemagglutinin of influenza A viruses. *Virology* 1991;182:475–485.
278. Normark S, Bergstrom S, Edlund T, Grundstrom T, Jaurin B, Lindberg FP, Olsson O. Overlapping genes. *Ann Rev Genet* 1983;17:499–525.
279. Norton GP, Tanaka T, Tobita K, Nakada S, Buonagurio DA, Greenspan D, Krystal M, Palese P. Infectious influenza A and B virus variants with long carboxyl terminal deletions in the NS_1 polypeptides. *Virology* 1987;156:204–213.

280. Ogasawara T, Botoh B, Suzuki H, Asaka J, Shimokata K, Rott R, Nagai Y. Expression of factor X and its significance for the determination of paramyxovirus tropism in the chick embryo. *EMBO J* 1992;11:467–472.
281. Ohuchi M, Cramer A, Vey M, Ohuchi R, Garten W, Klenk H-D. Rescue of vector-expressed fowl plague virus hemagglutinin in biologically active form by acidotropic agents and coexpressed M_2 protein.*J Virol* 1994;68:920–926.
282. Ohuchi M, Ohuchi R, Mifune K. Demonstration of hemolytic and fusion activities of influenza C virus. *J Virol* 1982;42:1076–1079.
283. Orlich M, Gottwald H, Rott R. Nonhomologous recombination between the HA gene and the NP gene of an influenza virus. *Virology* 1994; 204:462–465.
284. Orlich M, Khatchikian D, Teigler A, Rott R. Structural variation occuring in the hemagglutinin of influenza virus A/turkey/Oregon/71 during adaptation to different cell types. *Virology* 1990;176:531–538.
285. Oxford JS, Galbraith A. Antiviral activity of amantadine: A review of laboratory and clinical data. *Pharmacol Ther* 1980;11:181–262.
286. Palese P, Schulman JL. Differences in RNA patterns of influenza A viruses. *J Virol* 1976;17:876–884.
287. Palese P, Tobita K, Ueda M, Compans RW. Characterization of temperature sensitive influenza virus mutants defective in neuraminidase. *Virology* 1974;61:397–410.
288. Panayotov PP, Schlesinger RW. Oligomeric organization and strain-specific proteolytic modification of the virion M_2 protein of influenza A H1N1 viruses. *Virology* 1992;186:352–355.
289. Panniers R, Henshaw EC. A GDP/GTP exchange factor essential for eukaryotic initiation factor 2 cycling in Ehrlich ascites tumor cells and its regulation by eukaryotic initiation factor 2 phosphorylation. *J Biol Chem* 1983;258:7982–7934.
290. Parvin JD, Moscona A, Pan WT, Leider JM, Palese P. Measurement of the mutation rates of animal viruses: influenza A virus and poliovirus type 1. *J Virol* 1986;59:377–383.
291. Parvin JD, Palese P, Honda A, Ishihama A, Krystal M. Promoter analysis of influenza virus RNA polymerase. *J Virol* 1989;63:5142–5152.
292. Pavlovic J, Zurcher T, Haller O, Staeheli P. Resistance to influenza virus and vesicular stomatitis virus conferred by expression of human MxA protein. *J Virol* 1990;64:3370–3375.
293. Petri T, Dimmock NJ. Phosphorylation of influenza virus nucleoprotein in vivo. *J Gen Virol* 1981;57:185–190.
294. Piccone ME, Fernandez-Sesma A, Palese P. Mutational analysis of the influenza virus vRNA promoter. *Virus Res* 1993;28:99–112.
295. Pinto LH, Holsinger LJ, Lamb RA. Influenza virus M_2 protein has ion channel activity. *Cell* 1992;69:517–528.
296. Pitossi F, Blank A, Schroder A, Schwarz A, Hussi P, Schwemmle M, Pavloic J, Staeheli P. A functional GTP-binding motif is necessary for antiviral activity of Mx proteins. *J Virol* 1993;67:6726–6732.
297. Plotch SJ, Bouloy M, Krug RM. Transfer of 5′-terminal cap of globin mRNA to influenza viral complementary RNA during transcription in vitro. *Proc Natl Acad Sci USA* 1979;76:1618–1622.
298. Plotch SJ, Bouloy M, Ulmanen I, Krug RM. A unique cap(m^7Gpp-pXm)-dependent influenza virion endonuclease cleaves capped RNAs to generate the primers that initiate viral RNA transcription. *Cell* 1981; 23:847–858.
299. Plotch SJ, Krug RM. Influenza virion transcriptase: synthesis in vitro of large, polyadenylic acid-containing complementary RNA. *J Virol* 1977;21:24–34.
300. Plotch SJ, Krug RM. Segments of influenza virus complementary RNA synthesized in vitro. *J Virol* 1978;25:579–586.
301. Poch O, Sauvaget I, Delarue M, Tordo N. Identification of four conserved motifs among the RNA-dependent polymerase encoding elements. *EMBO J* 1989;8:3867–3874.
302. Pons MW. Isolation of influenza virus ribonucleoprotein from infected cells. Demonstration of the presence of negative-stranded RNA in viral RNP. *Virology* 1971;46:149–160.
303. Pons MW. A re-examination of influenza single-and double-stranded RNAs by gel electrophoresis. *Virology* 1976;69:789–792.
304. Pons MW, Schulze IT, Hirst GK, Hauser R. Isolation and characterization of the ribonucleoprotein of influenza virus. *Virology* 1969;39: 250–259.
305. Porter AG, Barber C, Carey NH, Hallewell RA, Threlfall G, Emtage JS. Complete nucleotide sequence of an influenza virus haemagglutinin gene from cloned DNA. *Nature* 1979;282:471–477.
306. Privalsky ML, Penhoet EE. Phosphorylated protein component present in influenza virions. *J Virol* 1977;24:401–405.
307. Privalsky ML, Penhoet EE. Influenza virus proteins: identity, synthesis, and modification analyzed by two-dimensional gel electrophoresis. *Proc Natl Acad Sci USA* 1978;75:3625–3629.
308. Privalsky ML, Penhoet EE. The structure and synthesis of influenza virus phosphoproteins. *J Biol Chem* 1981;256:5368–5376.
309. Qian X-Y, Alonso-Caplen F, Krug RM. Two functional domains of the influenza virus NS_1 protein are required for regulation of nuclear export of mRNA. *J Virol* 1994;68:2433–2441.
310. Qiu Y, Krug RM. The influenza virus NS_1 protein is a poly A-binding protein that inhibits the nuclear export of mRNAs containing poly A. *J Virol* 1994;68:2425–2432.
311. Qiu Y, Krug RM. The influenza virus NS_1 protein binds to a specific structure in U6 snRNA and inhibits U6-U2 and U 6-U4 interactions during splicing splicing (*submitted*). 1995.
312. Rabinowitz SG, Huprikar J. Influence of defective-interfering particles of the PR8 strain of influenza A virus on the pathogenesis of pulmonary infection in mice. *J Infect Dis* 1979;140:305–315.
313. Racaniello V, Baltimore D. Cloned poliovirus complementary DNA is infectious in mammalian cells. *Science* 1981;214:916–919.
314. Rees PJ, Dimmock NJ. Electrophoretic separation of influenza virus ribonucleoproteins. *J Gen Virol* 1981;53:125–132.
315. Reich E, Franklin RM, Shatkin AJ, Tatum EL. Effect of actinomycin D on cellular nucleic acid synthesis and virus production. *Science* 1961; 134:556–557.
316. Reich E, Goldberg IH, Rabinowitz M. Structure activity correlations of actinomycins and their derivatives. *Nature* 1962;196:743–748.
317. Richardson JC, Akkina RK. NS_2 protein of influenza virus is found in purified virus and phosphorylated in infected cells. *Arch Virol* 1991; 116:69–80.
318. Ritchey MB, Palese P, Kilbourne ED. RNAs of influenza A, B, and C viruses. *J Virol* 1976;18:738–744.
319. Ritchey MB, Palese P, Schulman JL. Mapping of the influenza virus genome. III. Identification of genes coding for nucleoprotein, membrane protein, and nonstructural protein. *J Virol* 1976;20:307–313.
320. Robertson JS, Schubert M, Lazzarini RA. Polyadenylation sites for influenza mRNA. *J Virol* 1981;38:157–163.
321. Rodriguez-Boulan E, Sabatini DD. Asymmetric budding of viruses in epithelial monolayers; a model system for study of epithelial polarity. *Proc Natl Acad Sci USA* 1978;75:5071–5075.
322. Rogers GN, Herrler G, Paulson JC, Klenk H-D. Influenza C virus uses 9-O-acetyl-N-acetylneuraminic acid as a high affinity receptor determinant for attachment to cells. *EMBO J* 1986;5:1359–1365.
323. Rogers GN, Paulson JC, Daniels RS, Skehel JJ, Wilson IA, Wiley DC. Single amino acid substitutions in influenza haemagglutinin change receptor binding specificity. *Nature* 1983;304:76–78.
324. Roth MG, Compans RW, Giusti L, Davis AR, Nayak DP, Gething M-J, Sambrook J. Influenza virus hemagglutinin expression is polarized in cells infected with recombinant SV40 viruses carrying cloned hemagglutinin DNA. *Cell* 1983;33:435–443.
325. Roth MG, Fitzpatrick JP, Compans RW. Polarity of influenza and vesicular stomatitis virus maturation MDCK cells: lack of a requirement for glycosylation of viral glycoproteins. *Proc Natl Acad Sci USA* 1979;76: 6430–6434.
326. Roth MG, Gething M-J, Sambrook J. Membrane insertion and intracellular transport of influenza virus glycoproteins. In: Krug RM, ed. *The influenza viruses*. New York: Plenum Press, 1989;219–267.
327. Roth MG, Gundersen D, Patil N, Rodriguez-Boulan E. The large external domain is sufficient for the correct sorting of secreted or chimeric influenza virus hemagglutinins in polarized monkey kidney cells. *J Cell Biol* 1987;104:769–782.
328. Rott R, Saber S, Scholtissek C. Effect on myxovirus of mitomycin C, actinomycin D and pretreatment of the host cell with ultraviolet light. *Nature* 1965;205:1187–1190.
329. Ruigrok RWH, Hirst EMA, Hay AJ. The specific inhibition of influenza A virus maturation by amantadine: an electron mircoscopic examination. *J Gen Virol* 1991;72:191–194.
330. Ruigrok RWH, Martin SR, Wharton SA, Skehel JJ, Bayley PM, Wiley DC. Conformational changes in the hemagglutinin of influenza virus which accompany heat-induced fusion of virus with liposomes. *Virology* 1986;155:484–497.
331. Ruigrok RWH, Wrigley NG, Calder LJ, Cusack S, Wharton SA, Brown EB, Skehel JJ. Electron microscopy of the low pH structure of influenza virus hemagglutinin. *EMBO J* 1986;5:41–49.
332. Ryan KW, Mackow ER, Chanock RM, Lai CJ. Functional expression

of influenza A viral nucleoprotein in cells transformed with cloned DNA. *Virology* 1986;154:144–154.
333. Safer B. 2B or not 2B: Regulation of the catalytic utilization of eIF-2. *Cell* 1983;33:7–8.
334. Sambrook J, Rodgers L, White J, Gething MJ. Lines of BPV-transformed murine cells that constitutively express influenza virus hemagglutinin. *EMBO J* 1985;4:91–103.
335. Sato SB, Kawasaki K, Ohnishi S-I. Haemolytic activity of influenza virus haemagglutinin glycoproteins activated in mildly acidic environments. *Proc Natl Acad Sci USA* 1983;80:3153–3157.
336. Schmidt MFG. Acylation of viral spike glycoproteins: a feature of enveloped RNA viruses. *Virology* 1982;116:327–338.
337. Seong BL, Brownlee GG. A new method for reconstituting influenza polymerase and RNA in vitro: a study of the promoter elements for cRNA and vRNA synthesis in vitro and viral rescue in vivo. *Virology* 1992;186:247–260.
338. Shapiro GI, Gurney T Jr, Krug RM. Influenza virus gene expression: control mechanisms at early and late times of infection and nuclear-cytoplasmic transport of virus-specific RNAs. *J Virol* 1987;61: 764–773.
339. Shapiro GI, Krug RM. Influenza virus RNA replication in vitro: synthesis of viral template RNAs and virion RNAs in the absence of an added primer. *J Virol* 1988;62:2285–2290.
340. Shaw MW, Choppin PW, Lamb RA. A previously unrecognized influenza B virus glycoprotein from a bicistronic mRNA that also encodes the viral neuraminidase. *Proc Natl Acad Sci USA* 1983;80:4879–4883.
341. Shaw MW, Compans RW. Isolation and characterization of cytoplasmic inclusions from influenza A virus-infected cells. *J Virol* 1978;25: 608–615.
342. Shaw MW, Lamb RA, Erickson BW, Briedis DJ, Choppin PW. Complete nucleotide sequence of the neuraminidase gene of influenza B virus. *Proc Natl Acad Sci USA* 1982;79:6817–6821.
343. Sivasubramanian N, Nayak DP. Defective interfering influenza RNAs of polymerase 3 gene contain single as well as multiple internal deletions. *Virology* 1983;124:232–237.
344. Skehel JJ. Polypeptide synthesis in influenza virus-infected cells. *Virology* 1972;49:23–36.
345. Skehel JJ. Early polypeptide synthesis in influenza virus-infected cells. *Virology* 1973;56:394–399.
346. Skehel JJ. Influenza virus. Amantadine blocks the channel [news]. *Nature* 1992;358:110–111.
347. Skehel JJ, Bayley PM, Brown EB, Martin SR, Waterfield MD, White JM, Wilson IA, Wiley DC. Changes in the conformation of influenza virus hemagglutinin at the pH optimum of virus-mediated membrane fusion. *Proc Natl Acad Sci USA* 1982;79:968–972.
348. Skehel JJ, Daniels RS, Hay AJ, Ruigrok RWH, Wharton SA, Wrigley NG, Weiss W, Wiley DC. Structure changes in influenza virus haemagglutinin at the pH of membrane fusion. *Biochem Soc Trans* 1986;14: 252–253.
349. Skehel JJ, Hay AJ, Armstrong JA. On the mechanism of inhibition of influenza virus replication by amantadine hydrochloride. *J Gen Virol* 1978;38:97–110.
350. Skehel JJ, Waterfield MD. Studies on the primary structure of the influenza virus hemagglutinin. *Proc Natl Acad Sci USA* 1975;72:93–97.
351. Smith DB, Inglis SC. The mutation rate and variability of eukaryotic viruses: an analytical review. *J Gen Virol* 1987;68:2729–2740.
352. Smith FL, Palese P. Variation in influenza virus genes: epidemiology, pathogenic, and evolutionary consequences. In: Krug RM, ed. *The influenza viruses*. New York: Plenum Press, 1989.
353. Smith GL, Hay AJ. Replication of the influenza virus genome. *Virology* 1982;118:96–108.
354. Smith GL, Levin JZ, Palese P, Moss B. Synthesis and cellular location of the ten influenza polypeptides individually expressed by recombinant vaccinia viruses [published erratum appears in *Virology* 1988 Mar;163(1):259]. *Virology* 1987;160:336–345.
355. Spooner LLR, Barry RD. Participation of DNA-dependent RNA polymerase II in replication of influenza viruses. *Nature* 1977;268:650–652.
356. Staeheli P, Haller O. Interferon-induced Mx protein: a mediator of cellular resistance to influenza virus. *Interferon* 1987;8:1–23.
357. Staeheli P, Haller O, Boll W, Lindenmann J, Weissmann C. Mx protein: Constitutive expression in 3T3 cells transformed with cloned Mx cDNA confers selective resistance to influenza virus. *Cell* 1986;44: 147–158.
358. Standen AM, Staeheli P, Pavlovic J. Function of the Mouse Mx1 protein is inhibited by overexpression of the PB2 protein of influenza virus. *Virology* 1993;197:642–651.
359. Stegmann T, Morselt HWM, Scholma J, Wilschut J. Fusion of influenza virus in an intracellular acidic compartment measured by fluorescence dequenching. *Biochim Biophys Acta* 1987;904:165–170.
360. Stieneke-Grober A, Vey M, Angliker H, Shaw E, Thomas G, Roberts C, Klenk H-D, Garten W. Influenza virus hemagglutinin with multibasic cleavage site is activated by furin, a subtilisin-like endoprotease. *J Gen Virol* 1992;11:2407–2414.
361. Stoltzfus CM, Dane RW. Accumulation of spliced avian retrovirus mRNA is inhibited in S-adenosylmethionine-depleted chicken embryo fibroblasts. *J Virol* 1982;42:918–931.
362. Sugawara K, Nakamura K, Homma M. Analyses of structural polypeptides of seven different isolates of influenza C virus. *J Gen Virol* 1983; 64:579–587.
363. Sugrue RJ, Bahadur G, Zambon MC, Hall SM, Douglas AR, Hay AJ. Specific structural alteration of the influenza haemagglutinin by amantadine. *EMBO J* 1990;9:3469–3476.
364. Sugrue RJ, Belshe RB, Hay AJ. Palmitoylation of the influenza A virus M_2 protein. *Virology* 1990;179:51–56.
365. Sugrue RJ, Hay AJ. Structural characteristics of the M_2 protein of the influenza A viruses: evidence that it forms a tetrameric channel. *Virology* 1991;180:617–624.
366. Takeuchi K, Lamb RA. Influenza virus M_2 protein ion channel activity stabilizes the native form of fowl plague virus hemagglutinin during intracellular transport. *J Virol* 1994;68:911–919.
367. Takeuchi K, Shaughnessy MA, Lamb RA. Influenza virus M_2 protein ion channel activity is not required to maintain the equine-1 hemagglutinin in its native form in infected cells. *Virology* 1994;202: 1007–1011.
368. Tiley LS, Hagen M, Matthews JT, Krystal M. Sequence-specific binding of the influenza virus RNA polymerase to sequences located at the 5 ends of the viral RNAs. *J Virol* 1994;68:5108–5116.
369. Tosteson MT, Pinto LH, Holsinger LJ, Lamb RA. Reconstitution of the influenza virus M_2 ion channel in lipid bilayers. *J Membr Biol* 1994; 142:117–126.
370. Tulip WR, Varghese JN, Baker AT, van Donkelaar A, Laver WG, Webster RG, Colman PM. Refined atomic structures of N9 subtype influenza virus neuraminidase and escape mutants. *J Mol Biol* 1991;221:487–497.
371. Tulip WR, Varghese JN, Laver WG, Webster RG, Colman PM. Refined crystal structure of the influenza virus N9 neuraminidase-NC41 Fab complex. *J Mol Biol* 1992;227:122–148.
372. Ulmanen I, Broni BA, Krug RM. Role of two of the influenza virus core P proteins in recognizing cap 1 structures (m^7GpppNm) on RNAs and in initiating viral RNA transcription. *Proc Natl Acad Sci USA* 1981; 78:7355–7359.
373. Ulmanen I, Broni BA, Krug RM. Influenza virus temperature-sensitive cap(m^7GpppNm)-dependent endonuclease. *J Virol* 1983;45:27–35.
374. Varghese JN, Laver WG, Colman PM. Structure of the influenza virus glycoprotein antigen neuraminidase at 2.9 Å resolution. *Nature* 1983; 303:35–40.
375. Varghese JN, Webster RG, Laver WG, Colman PM. Structure of an escape mutant of glycoprotein N2 neuraminidase of influenza virus A/Tokyo/3/67 at 3 Å. *J Mol Biol* 1988;200:201–203.
376. Veit M, Klenk H-D, Kendal A, Rott R. The M_2 protein of influenza A virus is acylated. *Virology* 1991;184:227–234.
377. Verhoeyen M, Fang R, Min Jou W, Devos R, Huylebroeck D, Saman E, Fiers W. Antigenic drift between the haemagglutinin of the Hong Kong influenza strains A/Aichi/2/68 and A/Victoria/3/75. *Nature* 1980; 286:771–776.
378. Vlasak R, Krystal M, Nacht M, Palese P. The influenza C virus glycoprotein (HE) exhibits receptor-binding (hemagglutinin) and receptor-destroying (esterase) activities. *Virology* 1987;160:419–425.
379. Vogle U, Kunerl M, Scholtissek C. Influenza A virus late mRNAs are specifically retained in the nucleus in the presence of a methyltransferase or a protein kinase inhibitor. *Virology* 1994;198:227–233.
380. von Heijne G, Gavel Y. Topogenic signals in integral membrane proteins. *Eur J Biochem* 1988;174:6711–6718.
381. von Itzstein M, Wu WY, Kok GB, Pegg MS, Dyason JC, Jin B, Van Phan T, Smythe ML, White HF, Oliver SW, Colman PM, Varghese JN, Ryan MD, Woods JM, Bethell RC, Hotham VJ, Cameron JM, Penn CR. Rational design of potent sialidase-based inhibitors of influenza virus replication. *Nature* 1993;363:418–423.

382. von Magnus P. Incomplete forms of influenza virus. *Adv Virus Res* 1954;2:59–78.
383. Wakefield L, Brownlee GG. RNA-binding properties of influenza A virus matrix protein M_1. *Nucleic Acid Res* 1989;17:8569–8580.
384. Wang C, Lamb RA, Pinto LH. Measurement of the influenza virus M_2 ion channel activity in mammalian cells. *Virology* 1994;205:133–140.
385. Wang C, Takeuchi K, Pinto LH, Lamb RA. Ion channel activity of influenza A virus M_2 protein: characterization of the amantadine block. *J Virol* 1993;67:5585–5594.
386. Ward CW, Dopheide TAA. The Hong Kong (H3) hemagglutinin. complete amino acid sequence and oligosaccharide distribution for the heavy chain of A/Memphis/102/72. In: Laver G, Air G, eds. *Structure and variation in influenza virus*. New York: Elsevier/North-Holland, 1980;27–38.
387. Ward CW, Webster RG, Inglis AS, Dopheide TA. Composition and sequence studies show that A/duck/Ukraine/1/63 haemagglutinin (Hav7) belongs to the Hong Kong (H3) subtype. *J Gen Virol* 1981;53: 163–168.
388. Waterson AP, Hurrell JMW, Jensen KE. The fine structure of influenza A, B, and C viruses. *Arch ges Virusforsch* 1963;12:487–495.
389. Webster RG, Brown LE, Jackson DC. Changes in the antigenicity of the hemagglutinin molecule of H3 influenza virus at acidic pH. *Virology* 1983;126:587–599.
390. Webster RG, Brown LE, Laver WG. Antigenic and biological characterization of influenza virus neuraminidase (N2) with monoclonal antibodies. *Virology* 1984;135:30–42.
391. Webster RG, Campbell CH, Granoff A. The "in vivo" production of "new" influenza A viruses. I. Genetic recombination between avian and mammalian influenza viruses. *Virology* 1971;44:317–328.
392. Webster RG, Laver WG, Air GM, Schild GC. Molecular mechanisms of variation in influenza viruses. *Nature* 1982;296:115–121.
393. Wei C-M, Moss B. Nucleotide sequence at the N^6-methyladenosine sites of HeLa cell messenger ribonucleic acid. *Biochemistry* 1977;16: 1672–1676.
394. Weis W, Brown JH, Cusack S, Paulson JC, Skehel JJ, Wiley DC. Structure of the influenza virus haemagglutinin complexed with its receptor, sialic acid. *Nature* 1988;333:426–431.
395. Weis WI, Brunger AT, Skehel JJ, Wiley DC. Refinement of the influenza virus hemagglutinin by simulated annealing. *J Mol Biol* 1990; 212:737–761.
396. Weis WI, Cusack SC, Brown JH, Daniels RS, Skehel JJ, Wiley DC. The structure of a membrane fusion mutant of the influenza virus haemagglutinin. *EMBO J* 1990;9:17–24.
397. Wharton SA, Belshe RB, Skehel JJ, Hay AJ. Role of virion M_2 protein in influenza virus uncoating: specific reduction in the rate of membrane fusion between virus and liposomes by amantadine. *J Gen Virol* 1994; 75:945–948.
398. Wharton SA, Weis W, Skehel JJ, Wiley DC. Structure, function, and antigenicity of the hemagglutinin of influenza virus. In: Krug RM, ed. *The influenza viruses*. New York: Plenum Press, 1989;153–173.
399. White J, Matlin K, Helenius A. Cell fusion by Semliki Forest, influenza, and vesicular stomatitis viruses. *J Cell Biol* 1981;89:674–679.
400. White JM, Wilson IA. Anti-peptide antibodies detect steps in a protein conformational change: low-pH activation of the influenza virus hemagglutinin. *J Cell Biol* 1987;105:2887–2896.
401. Wiley DC, Skehel JJ. Crystallization and x-ray diffraction studies on the haemagglutinin glycoprotein from the membrane of influenza virus [letter]. *J Mol Biol* 1977;112:343–347.
402. Wiley DC, Skehel JJ. The structure and function of the hemagglutinin membrane glycoprotein of influenza virus. *Ann Rev Biochem* 1987;56: 365–394.
403. Wiley DC, Skehel JJ, Waterfield MD. Evidence from studies with a cross-linking reagent that the hemagglutinin of influenza virus is a trimer. *Virology* 1977;79:446–448.
404. Wiley DC, Wilson IA, Skehel JJ. Structural identification of the antibody-binding sites of Hong Kong influenza haemagglutinin and their involvement in antigenic variation. *Nature* 1981;289:373–378.
405. Williams MA, Lamb RA. Determination of the orientation of an integral membrane protein and sites of glycosylation by oligonucleotide-directed mutagenesis: Influenza B virus NB glycoprotein lacks a cleavable signal sequence and has an extracellular NH_2-terminal region. *Mol Cell Biol* 1986;6:4317–4328.
406. Williams MA, Lamb RA. Polylactosaminoglycan modification of a small integral membrane glycoprotein, influenza B virus NB. *Mol Cell Biol* 1988;8:1186–1196.
407. Williams MA, Lamb RA. Effect of mutations and deletions in a bicistronic mRNA on the synthesis of influenza B virus NB and NA glycoproteins. *J Virol* 1989;63:28–35.
408. Wilson IA, Skehel JJ, Wiley DC. Structure of the haemagglutinin membrane glycoprotein of influenza virus at 3 Å resolution. *Nature* 1981; 289:366–373.
409. Winter G, Fields S. Cloning of influenza cDNA into M13: the sequence of the RNA segment encoding the A/PR/8/34 matrix protein. *Nucleic Acid Res* 1980;8:1965–1974.
410. Winter G, Fields S. The structure of the gene encoding the nucleoprotein of human influenza virus A/PR/8/34. *Virology* 1981;114:423–428.
411. Winter G, Fields S. Nucleotide sequence of human influenza A/PR/8/34 segment 2. *Nucleic Acid Res* 1982;10:2135–2143.
412. Winter G, Fields S, Gait MJ, Brownlee GG. The use of synthetic oligodeoxynucleotide primers in cloning and sequencing segment 8 of influenza virus (A/PR/8/34). *Nucleic Acid Res* 1981;9:237–245.
413. Wu JA, Manley JL. Base pairing between U2 and U6 snRNAs is necessary for splicing of a mammalian pre-mRNA. *Nature* 1991;352: 818–821.
414. Xiong C, Levis R, Shen P, Schlesinger S, Rice CM, Huang HV. Sindbis virus: an efficient broad range vector for gene expression in animal cells. *Science* 1989;243:1188–1191.
415. Yamashita M, Krystal M, Palese P. Evidence that the matrix protein of influenza C virus is coded for by a spliced mRNA. *J Virol* 1988;62: 3348–3355.
416. Yasuda J, Nakada S, Kato A, Toyoda T, Ishihama A. Molecular assembly of influenza virus: association of the NS_2 protein with virion matrix. *Virology* 1993;196:249–255.
417. Ye ZP, Baylor NW, Wagner RR. Transcription-inhibition and RNA-binding domains of influenza A virus matrix protein mapped with anti-idiotypic antibodies and synthetic peptides. *J Virol* 1989;63:3586–3594.
418. Yewdell JW, Hackett CJ. Specificity and function of T lymphocytes induced by influenza A viruses. In: Krug RM, ed. *The influenza viruses*. New York: Plenum Press, 1989;361–429.
419. Yoshida T, Shaw MW, Young JF, Compans RW. Characterization of the RNA associated with influenza A cytoplasmic inclusions and the interaction of NS_1 protein with RNA. *Virology* 1981;110:87–97.
420. Young JF, Palese P. Evolution of human influenza A viruses in nature: recombination contributes to genetic variation of H1N1 strains. *Proc Natl Acad Sci USA* 1979;76:6547–6551.
421. Young RJ, Content J. 5′-terminus of influenza virus RNA. *Nature New Biol* 1971;230:140–142.
422. Zebedee SL, Lamb RA. Influenza A virus M_2 protein: monoclonal antibody restriction of virus growth and detection of M_2 in virions. *J Virol* 1988;62:2762–2772.
423. Zebedee SL, Lamb RA. Growth restriction of influenza A virus by M_2 protein antibody is genetically linked to the M_1 protein. *Proc Natl Acad Sci USA* 1989a;86:1061–1065.
424. Zebedee SL, Richardson CD, Lamb RA. Characterization of the influenza virus M_2 integral membrane protein and expression at the infected-cell surface from cloned cDNA. *J Virol* 1985;56:502–511.
425. Zeitlin S, Efstratiadis A. In vivo splicing products of the rabbit-globin pre-mRNA. *Cell* 1984;39:589–602.
426. Zhirnov OP. Solubilization of matrix protein M_1/M from virions occurs at different pH for orthomyxo- and paramyxoviruses. *Virology* 1990;176:274–279.
427. Zvonarjev AY, Ghendon YZ. Influence of membrane (M) protein on influenza A virus virion transcriptase activity in vitro and its susceptibility to rimantadine. *J Virol* 1980;33:583–586.

Fundamental Virology, Third Edition
edited by B.N. Fields, D.M. Knipe, P.M. Howley, et al.
Lippincott - Raven Publishers, Philadelphia © 1996

CHAPTER 22

Bunyaviridae: The Viruses and Their Replication

Connie S. Schmaljohn

The family *Bunyaviridae* was established in 1975 to encompass a large group of arthropod-borne viruses sharing morphological, morphogenic, and antigenic properties (70,177,178). The family includes more than 300 serologically distinct members divided into five genera (22,107, 142). The *Bunyavirus, Hantavirus, Nairovirus,* and *Phlebovirus* genera comprise viruses that infect animals, and the *Tospovirus* genus contains viruses known to infect more than 400 species in 50 plant families (113,170). The type species for each respective genus is Bunyamwera (BUN), Hantaan (HTN), Congo-Crimean hemorrhagic fever (CCHF), sandfly fever Sicilian (SFS), and tomato spotted wilt (TSW) viruses (142). Until recently, Uukuniemi (UUK) virus and related viruses were classified in a separate genus in the family, the *Uukuvirus* genus (34); however, because of their strong molecular similarities to phleboviruses, uukuviruses have been reclassified to be members of the *Phlebovirus* genus (34,142).

Most viruses in the family have been isolated from or are transmitted by arthropods, primarily mosquitoes, ticks, sand flies, or thrips (138,142,144). Hantaviruses are exceptions; these viruses are primarily rodent-borne and have no known arthropod vector but instead are transmitted by means of aerosolized rodent excreta (123). Some members of the family Bunyaviridae are associated with severe or fatal human infections: Rift Valley fever (RVF), HTN, Crimean-Congo hemorrhagic fever (CCHF), and La Crosse (LAC) viruses; but many others are not known to infect humans. (See Chapter 48 in *Fields Virology*, 3rd ed. for details on the ecology, epidemiology, and medical significance of viruses in this family.) Serological cross-reactivity has not been found among viruses in different genera of the family *Bunyaviridae*. However, these viruses do share several common structural, genetic, replicative, and morphogenic properties, which are discussed below.

C. S. Schmaljohn: Virology Division, United States Army Medical Research Institute of Infectious Diseases, Fort Detrick, Frederick, Maryland, 21702-5011.

VIRION MORPHOLOGY AND STRUCTURE

Morphology

Morphological properties vary among viruses in each of the five genera; however, virions generally are spherical, 80 to 120 nm in diameter, and display surface glycoprotein projections of 5 to 10 nm that are embedded in a lipid bilayered envelope approximately 5 nm thick. Representatives of each genus display unique external features.

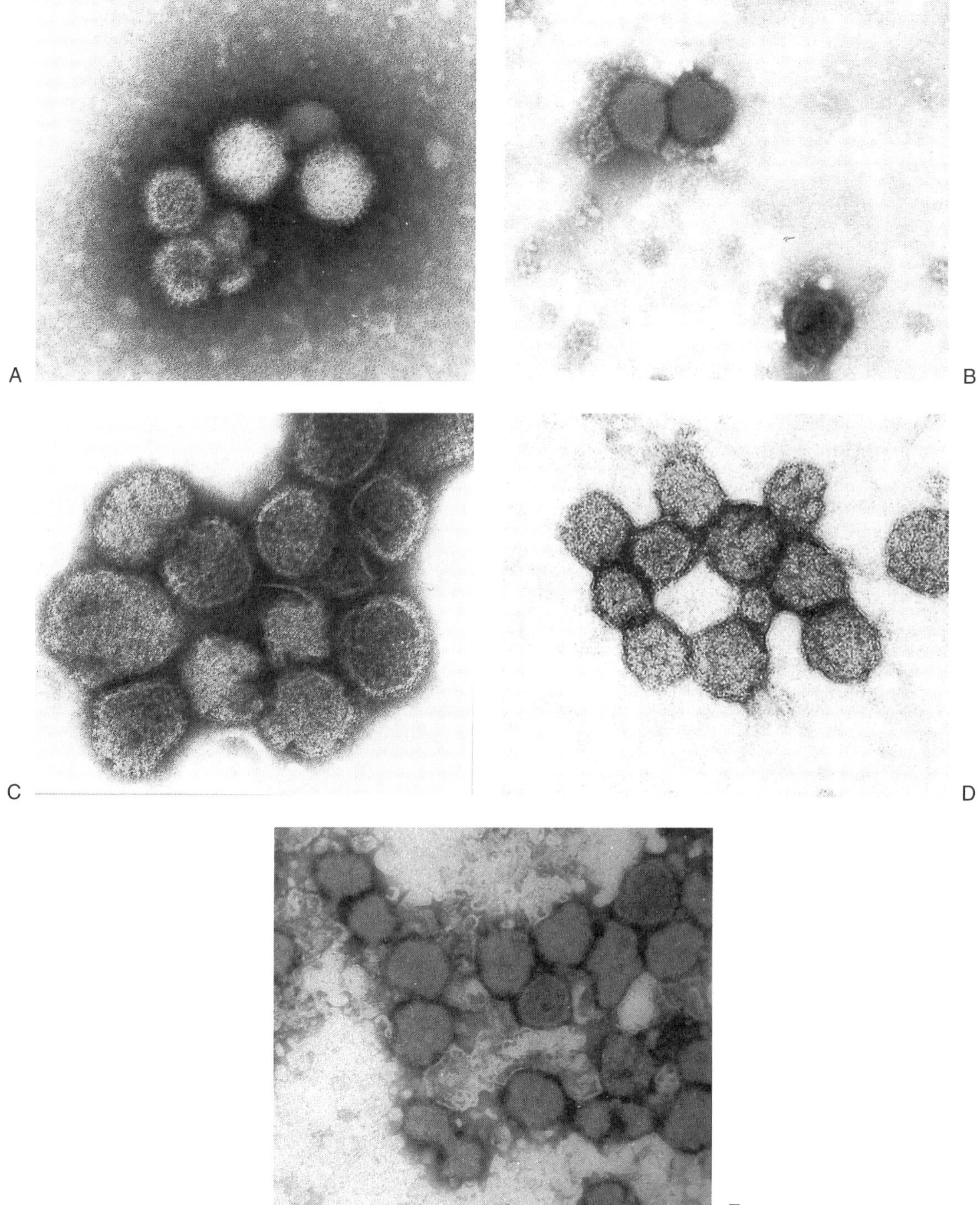

FIG. 1. Ultrastructure of viruses representing each genus in the family *Bunyaviridae*. Electron micrographs are of virion particles either fixed with glutaraldehyde (**A–D**) or unfixed (**E**) and negatively stained. All particles are 80 to 120 nm in diameter. **A:** Glycoprotein spikes are clearly defined on the surface of Rift Valley fever virus (*Phlebovirus* genus). **B:** Indistinct peripheral fringe with occasional small morphological subunits appear on Crimean-Congo hemorrhagic fever virus (*Nairovirus* genus). **C:** An organized, gridlike surface structure is evident on Hantaan virus (*Hantavirus* genus). **D:** La Crosse virus (*Bunyavirus* genus) exhibits knoblike surface units. **E:** Tomato spotted wilt virus (*Tospovirus* genus) displays an indeterminate surface morphology and a membrane fringe. Courtesy of K. Kuhl and J.E. White, US Army Medical Research Institute of Infectious Diseases, Frederick, MD (A and C); J.F. Smith and T. Drier, US Army Medical Research Institute of Infectious Diseases, Frederick, MD (B); M.L. Martin, Centers for Disease Control, Atlanta, GA (D); and S.T. Adkins, T-J. Choi, K.E. Richmond, and T.L. German, University of Wisconsin, Madison, WI (E).

The surface structure of the phlebovirus UUK is defined by clustered glycoproteins that form hollow, cylindrical morphological units. Negative staining after glutaraldehyde fixation, as well as freeze-etching, were used to demonstrate that the surface units of UUK virus are penton-hexon clusters arranged in a T = 12, P = 3 icosahedral surface lattice with hexon-hexon distances estimated at 12.5 to 16 nm for stained viral particles and 17 nm for freeze-etched samples (230). The subunits of other phleboviruses, such as Punta Toro (PT) (212), SFS, and RVF (134) (Fig. 1A), appear similar to those described for UUK virus. The surface structures of the nairoviruses, CCHF (Fig. 1B) (54,134) and Qalyub (42), clearly differ from those of the phleboviruses and sometimes exhibit less distinct, small morphological surface subunits. Hantaviruses display an unusual gridlike pattern on their surfaces, and elongated particles (110 to 210 nm long) are often observed (Fig. 1C) (95,96,134,139,239). Bunyaviruses exhibit knoblike morphological units with no distinct pattern when observed after glutaraldehyde fixation and negative staining (Fig. 1D) (134), but a lipid bilayer and well-defined surface spikes can be observed in vitrified-hydrated preparations of LAC virus (219). After unfixed particles are stained with 1% methylamine tungstate, the tospovirus TSW displays a fringe of surface projections typical of the family (Fig. 1E).

Structural Components

The internal organization of virion particles has been inferred from biochemical and morphological studies. An overall chemical composition of 2% RNA, 58% protein, 33% lipid, and 7% carbohydrate was estimated for UUK virion particles (149). All viruses in the family have three single-stranded RNA genome segments designated as large (L), medium (M), and small (S). Consensus 3′ terminal nucleotide sequences have been found on the L, M, and S genome segments of viruses within each genus, but differ from those of viruses in other genera (Table 1). Complementary sequences at the 5′ termini of each segment have been postulated to allow formation of stable, base-paired, panhandle structures (Fig. 2) (45,49,51,61,113,158,190, 205,207) and probably are the basis of the noncovalently closed circular RNA structures observed by electron microscopy (91). The complementary ends also may play a role in replication, possibly by serving as a transcriptase recognition structure.

The RNA segments, together with numerous copies of a nucleocapsid protein (N), form individual L, M, and S nucleocapsids that appear to be helical in structure (230, 231). The nucleocapsids can be released by nonionic detergent disruption of virion particles and often, like the naked RNA, may appear as circular structures in electron micrographs, suggesting that the complementary RNAs can base pair even when complexed with protein in an estimated ratio of 4% RNA to 96% protein (148,173,182). This assumption is supported by the ability to cross-link the ends of nucleocapsid-enclosed RNAs by treating them with photoreactive, nucleic acid cross-linking agents, such as psoralens (182). Equal numbers of nucleocapsids may not always be packaged in mature virions, as evidenced by varying, nonequimolar ratios of L, M, and S RNA molecules extracted from purified virions (25). Talmon et al.

TABLE 1. *Terminal nucleotide sequences of the L, M, and S genome segments of representative members of the family Bunyaviridae*

Genus	Virus	Gene segment	3′ terminus[a]	5′ terminus[b]	GenBank accession no.
Bunyavirus	Bunyamwera	S	3′ UCAUCAUAUGAGGUG	5′ AGUAGUGUGCUCCAC	D00353
		M	3′ UCAUCAUAUGAUGGC	5′ AGUAGUGUGCUACCG	M11852
		L	3′ UCAUCAUAUGAGGAU	5′ AGUAGUGUGCUCCUA	X14383
Hantavirus	Hantaan	S	3′ AUCAUCAUCUGAGGG	5′ UAGUAGUAGUGCUCCC	M14626
		M	3′ AUCAUCAUCUGAGGC	5′ UAGUAGUAGACACCG	M14627
		L	3′ AUCAUCAUCUGAGGG	5′ UAGUAGUAGACUCCC	X55901
Nairovirus	Dugbe	S	3′ AGAGUUUCUGUUUGC	5′ UCUCAAAGAUAAAUCG	M25150
		M	3′ AGAGUUUCUGUAUGG	5′ UCUCAAAGAUAGCGUG	M94133
		L	3′ AGAGUUUCUUUA	5′ UCUCAAAGACATCAA	U15018
Phlebovirus	Rift Valley fever	S	3′ UGUGUUUCGAGGGAUC	5′ ACACAAAGACCCCCT	X53771
		M	3′ UGUGUUUCUGCCACGU	5′ ACACAAAGACCGGUG	M11157
		L	3′ UGUGUUUCUGGCGGGU	5′ ACACAAAGGCGCCCA	X56464
Tospovirus	Tomato spotted wilt	S	3′ UCUCGUUAGCACAGUU	5′ AGAGCAAUUGUGUCAG	D00645
		M	3′ UCUCGUUAGUCACGUU	5′ AGAGCAAUCAGUGCAU	S48091
		L	3′ UCUCGUUAGUCCAUUU	5′ AGAGCAAUCAGGUACA	D10066

[a]3′ terminal sequences were obtained by direct sequencing of virion RNAs or by analysis of cDNA clones. Sequences identical on all three genome segments are underlined.

[b]5′ terminal sequences were determined by sequence analysis of cDNA.

Data are from the following sources: Bunyamwera virus (58,59,124); Hantaan virus (196,201,205,207); Dugbe virus (131, 234, A.C. Marriott and P.A. Nuttall, personal communication); Rift Valley fever virus (45,80,141); Tomato spotted wilt virus (49–51,113).

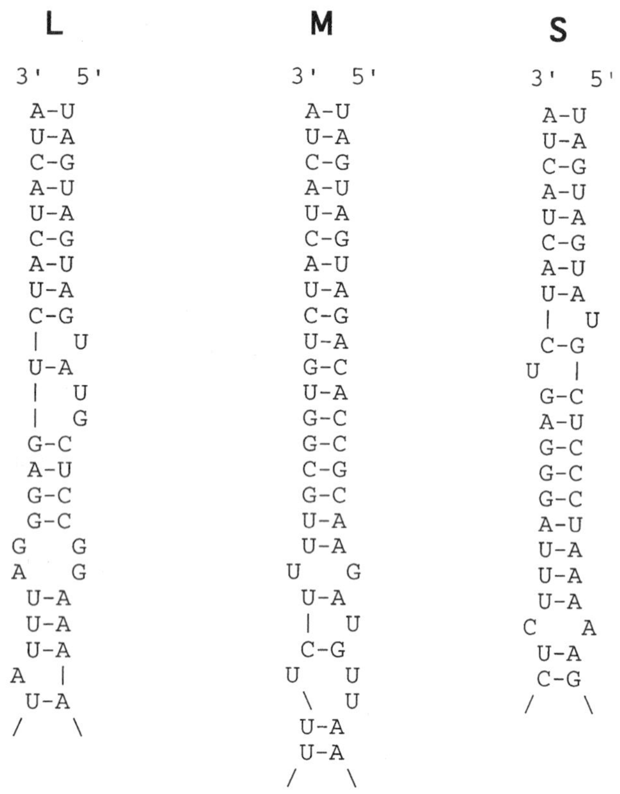

```
    L
3'   5'
 A-U
 U-A
 C-G
 A-U
 U-A
 C-G
 A-U
 U-A
 C-G
 |  U
 U-A
 |  U
 |  G
 G-C
 A-U
 G-C
 G-C
G   G
A   G
 U-A
 U-A
 U-A
A  |
 U-A
/   \
```

```
    M
3'   5'
 A-U
 U-A
 C-G
 A-U
 U-A
 C-G
 A-U
 U-A
 C-G
 U-A
 G-C
 U-A
 G-C
 G-C
 C-G
 G-C
 U-A
 U-A
U   G
 U-A
 |  U
 C-G
U   U
 \  U
 U-A
 U-A
/   \
```

```
    S
3'   5'
 A-U
 U-A
 C-G
 A-U
 U-A
 C-G
 A-U
 U-A
 |  U
 C-G
U  |
 G-C
 A-U
 G-C
 G-C
 G-C
 A-U
 U-A
 U-A
 U-A
C   A
 U-A
 C-G
/   \
```

FIG. 2. Possible base-pairing of the 3′ and 5′ terminal nucleotide sequences of the L, M, and S segments of Hantaan virus (*Hantavirus* genus). The calculated free energies of the S and M structures respectively are –23.6 and –28.2 kcal/mol.

(219) speculated that the size differences of virion particles observed by electron microscopy may directly relate to the number of nucleocapsids incorporated into individual virions. Interestingly, the phlebovirus, UUK, was found to package virus-complementary RNA (cRNA) as well as virus RNA (vRNA) of the S segment (but not the M segment) in a ratio of about 1:10 (209). Similarly, both the M and S segment cRNAs of TSW virus, a tospovirus, were found to be packaged (114), suggesting that encapsidation of vRNAs might relate to the ambisense coding strategies of these RNAs. For the bunyavirus LAC, S segment cRNA was detected in virions synthesized in insect cells, but not in mammalian cells (182). The significance of this finding, if any, is not known.

A virion-associated polymerase, which is necessary to copy the negative-sense viral genome into messenger-sense RNA(s), has been described for the bunyaviruses [Lumbo, Germiston (GER), and LAC (28,77,157)]; the phlebovirus [UUK (184)]; and the hantavirus [HTN (199)]. Expression of the L protein of BUN virus yielded a functional polymerase protein, thus formally verifying the long-held assumption that the *Bunyaviridae* L protein is the viral transcriptase (102). Nothing is known about the precise location of the L protein within virions.

In contrast to members of four other negative-sense RNA virus families (*Orthomyxoviridae, Paramyxoviridae, Rhabdoviridae,* and *Filoviridae*), viruses in the family *Bunyaviridae* have no internal matrix (M) protein (see Chapters 20 and 21, and Chapter 39 in *Fields Virology*, 3rd ed.). No enzymatic activity (other than transcriptase activity and associated endonucleolytic cleavage activity of the L protein described below) has been found associated with any of the viral proteins.

The envelope proteins of several members of the family *Bunyaviridae* will hemagglutinate goose erythrocytes at a characteristic (generally low) pH optimum (12,18,82). The carbohydrate moieties on the envelope proteins of all viruses in the family for which data are available have been found to be *N*-linked and mostly of the high-mannose rather than complex type, and no evidence for the presence of *O*-linked oligosaccharides has been obtained (119,120,129,167,203).

CODING STRATEGIES OF VIRAL GENES

Coding strategies of the complete genomes of representative members of all genera are known. Both similarities and marked differences in the strategies used to generate viral proteins have been described. Although members of the family *Bunyaviridae* are commonly thought of as negative-strand viruses, this designation is not accurate for all genera. Viruses in each genus do encode their structural proteins (i.e, N, G1, G2 and L in the cRNA), but phleboviruses also encode a nonstructural protein in the vRNA of their S segments. Tospoviruses encode nonstructural proteins in the vRNA of both their M and S segments. Thus, these viruses use an ambisense coding strategy to generate nonstructural proteins. As discussed below, the significance of nonstructural proteins for the replication of viruses in the family Bunyaviridae is still a matter of conjecture, and viruses in the *Hantavirus* and *Nairovirus* genera apparently do not require such proteins. A summary of available coding strategy information is illustrated in Figs. 3–5.

S Segment Strategies

Bunyaviruses

The first coding strategy elucidated for any member of the family was that of the S segment of the bunyavirus snowshoe hare (SSH). Coding of the 26.5-kd N protein and also a 7.4-kd nonstructural protein (NS_S) was localized to the viral S genome segment by *in vitro* translation of hybrid-selected messenger RNAs (mRNAs) (37). These results were confirmed by analysis of polypeptides obtained from cell cultures infected with SSH and LAC reassortant viruses (i.e., viruses with various complements of L, M, or S from either SSH or LAC) (71,72,76). Similar results with less closely related viruses in the *Bunyavirus* genus identified the encoding of N in the S genome segment as

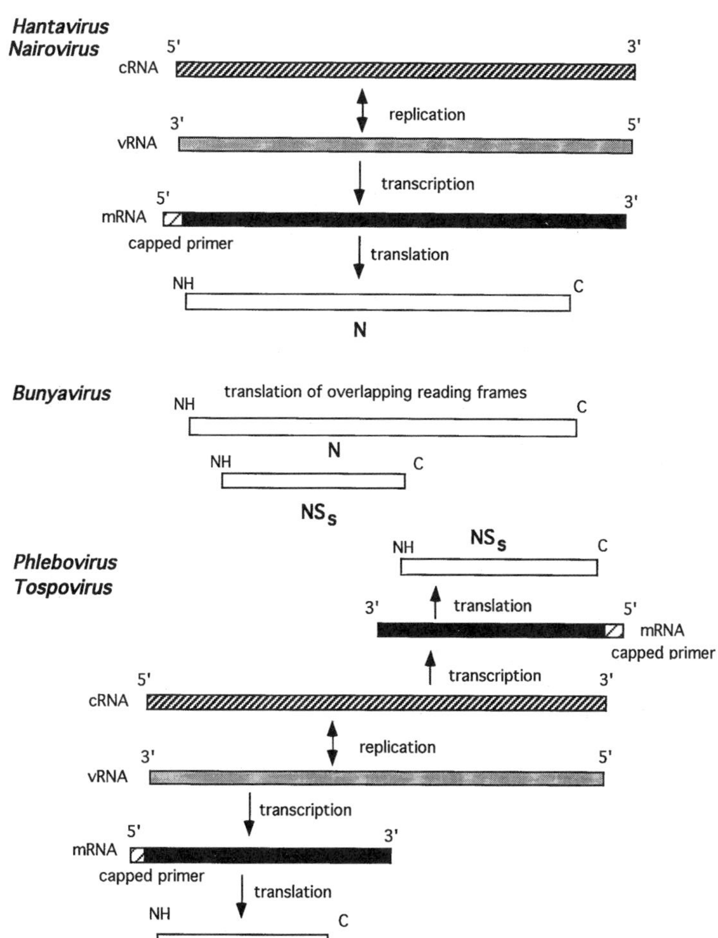

FIG. 3. Coding strategies of the S genome segments of viruses in the family *Bunyaviridae*. Three different coding strategies have been described. Viruses in the *Phlebovirus* and *Tospovirus* genera use an ambisense strategy to encode their nucleocapsid protein (N) in the viral complementary-sense RNA (cRNA) and a nonstructural (NS$_S$) protein in viral-sense RNA (vRNA). Viruses in the *Bunyavirus* genus encode N and NS$_S$ in ORFs of the cRNA. A single mRNA, truncated at the 3′ termini as compared with virion RNA, is believed to code for both proteins. Viruses in the *Hantavirus* and *Nairovirus* genera use a simple negative-sense strategy to encode N. No evidence for NS$_S$ proteins has been obtained. mRNAs are truncated at the 3′ terminus. Host-derived transcriptional primers are found on the 5′ termini of all mRNAs. Diagram is not to scale.

a generic property (30,57). Cloning, sequence analyses, and hybridization studies with cDNA representing the S genome segments of SSH, LAC, Aino, and GER and several other viruses showed that the N and NS$_S$ polypeptides are produced from open reading frames (ORFs) in the cRNA (3,4,23,33,55,78). Bunyaviruses, therefore, use the same nucleotide sequences, but in different reading frames, to encode two polypeptides. Only one S segment mRNA species has been detected for the bunyaviruses SSH, LAC, and GER (30,67,157). Thus, a single mRNA appears to be used to translate both N and NS$_S$ (Fig. 3).

Phleboviruses and Tospoviruses

The coding strategies of the S segments of phleboviruses and tospoviruses are even more unusual. Like the bunyaviruses, phleboviruses and tospoviruses encode both N and NS$_S$ in their S genome segments, but their much larger S segments (i.e., 1.7 to 1.9 kb for phleboviruses and 2.9 kb for tospoviruses versus 0.98 kb for bunyaviruses) have sufficient coding potential for both N and NS$_S$, without accessing overlapping ORFs. Studies with phleboviruses showed both N and NS$_S$ proteins in lysates of cells infected with the UUK, RVF, and Karimabad viruses (212, 215, 223). *In vitro* translation studies and Northern blot analyses with viral RNAs demonstrated that these polypeptides are generated from separate, subgenomic messages (152,209,223). Two subgenomic messages were also identified in lysates of cells infected with the related phleboviruses PT (99) and SFS (130); however, corresponding NS$_S$ proteins were not identified. Cloning, sequence analysis, and cell-free translation studies revealed that the 5′ half of the S segment of PT virus has an ORF encoding N (97,150). A second ORF capable of encoding a 29- to 37-kd polypeptide was located at the 5′ end of vRNA. This polypeptide was expressed in a baculovirus system, and antisera prepared to it could be used to precipitate small amounts of a similarly sized protein from PT-infected lysates (150). It was postulated that this polypeptide is analogous to the NS$_S$ protein identified in RVF virus–infected cells. Analysis of cDNA clones representing the S segments of the phleboviruses SFS (130), RVF, Toscana (TOS) (80), and UUK (209) indicate that ambisense coding is a general property of this genus. Tomato spotted wilt virus also was found to have ambisense coding potential in the S

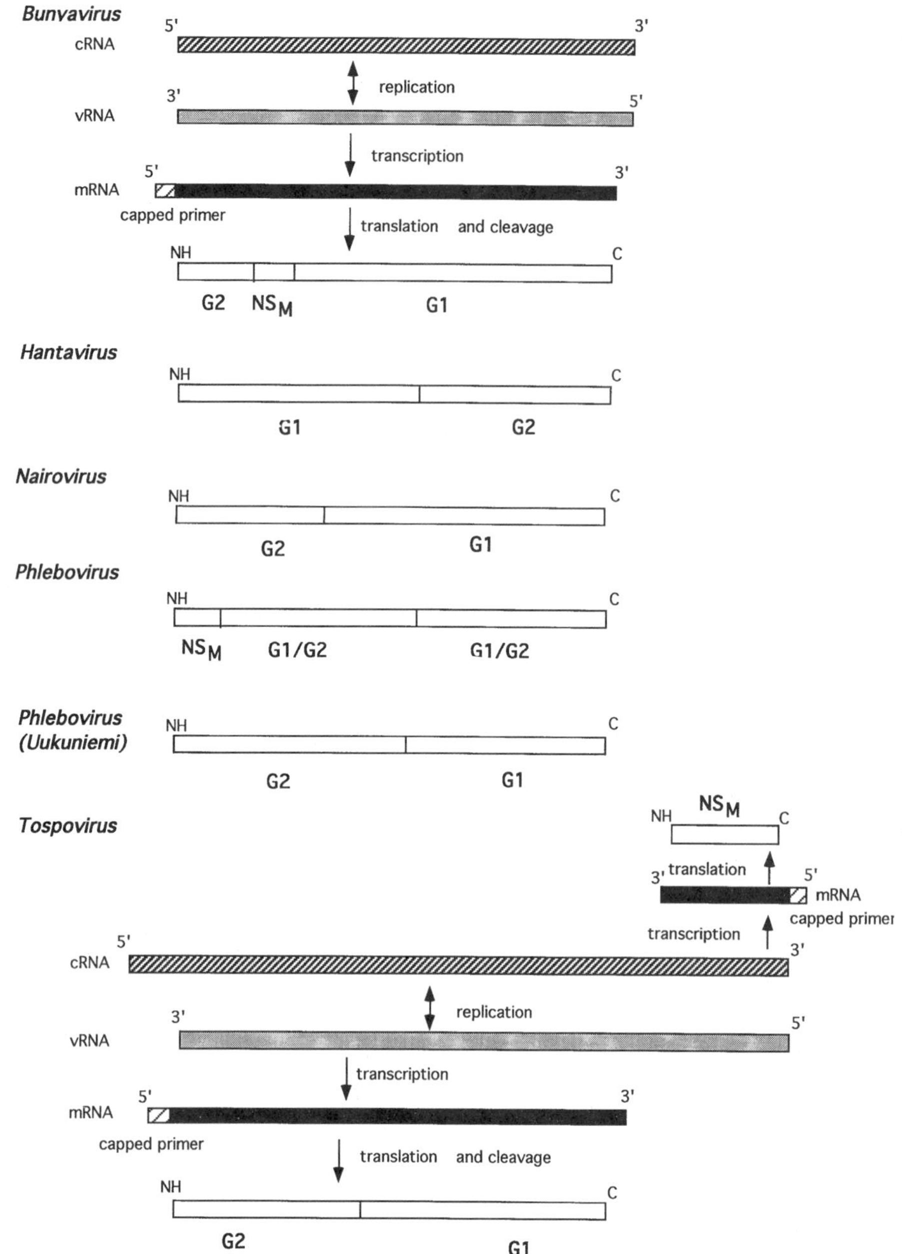

FIG. 4. Coding strategies of the M genome segments of viruses in the family *Bunyaviridae*. Viruses in all genera except the *Tospovirus* genus encode their envelope glycoproteins (G1 and G2) in a continuous ORF in the cRNA, which are processed by cotranslational cleavage. Viruses in the *Phlebovirus* and *Bunyavirus* genera also encode nonstructural (NS_M) polypeptides, but those in the *Hantavirus* and *Nairovirus* genera probably do not; although the G1 protein of nairoviruses may be processed from a precursor. mRNAs are truncated at their 3′ termini and have host-derived transcriptional primers attached to their 5′ termini. Tospoviruses use an ambisense coding strategy to encode G1 and G2 and the cRNA and NS_M in the vRNA. Host-derived transcriptional primers are found on the 5′ termini of both subgenomic mRNAs. Diagram is not to scale.

genome segment (51). However, the tospovirus NS_S protein is approximately twice the size of the N protein; i.e, 52.4 and 28.8 kd, respectively. Subgenomic mRNAs for N and NS_S were detected by Northern blot analysis of TSW virus–infected plant extracts and by cell-free translation of RNA transcripts generated from cDNA representing the S segment of TSW virus (51). The function(s) or significance of the NS_S protein has not been clearly defined for either phleboviruses or tospoviruses. NS_S of TSW virus was found to be distributed throughout the cytoplasm of infected plants

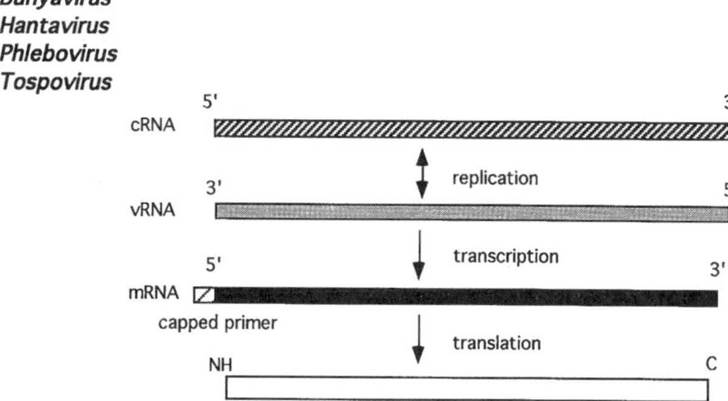

FIG. 5. Coding strategies of the L genome segments of viruses in the family *Bunyaviridae.* Viruses in all genera encode a single protein (L) in the cRNA. The mRNAs have host-derived transcriptional primers and are nearly the same length as vRNA. Diagram is not to scale.

or, for some isolates, associated with fibrillar structures of unknown function (115). For RVF virus, the NS_S protein was reported to be phosphorylated and to accumulate in the nuclei of infected cells (215). Although this is not known to be a general finding for phleboviruses, few of them have been examined with immune sera specific for the NS_S protein. UUK virus NS_S, expressed by recombinant baculoviruses or vaccinia viruses, is not phosphorylated but associates with the 40S ribosomal subunit of the host cells (210). Because NS_S appears relatively late in infection and appears to not be included in UUK virions, it is unlikely to have a role in primary transcription, unless it can be generated in small amounts directly by translation of the vRNA (210). Also, there is no evidence that this protein is absolutely required for replication. Preliminary studies with a deletion mutant of RVF virus, missing 600 nucleotides of the coding information (deleted in-frame) for NS_S, is still replication competent in Vero and mosquito cells but is defective in human lung diploid, MRC-5, cells (141a,210). If these results can be confirmed, the role of NS_S in replication may prove to be related to host-range properties of the virus. It is interesting to speculate that the extremely large NS_S protein of TSW virus provides this virus with the means to replicate in the wide variety of plants that tospoviruses are known to infect.

Hantaviruses and Nairoviruses

Unlike bunyaviruses, phleboviruses, and tospoviruses, members of the *Hantavirus* and *Nairovirus* genera apparently do not encode NS_S polypeptides. Hantaviruses and nairoviruses have much larger nucleocapsid proteins than do viruses in other genera—approximately 48 to 54 kd—yet originate from S RNA segments similar in size to those of phleboviruses (approximately 1.8 kb) (36,132,202,205, 234). The mRNA encoding HTN N is nearly equivalent in size to genomic S RNA, and no subgenomic messages were detected in infected cells (204,205). RNA transcripts of cDNA, which corresponded to Hantaan S cRNA, were able to program a cell-free translation system to yield N, indicating that the single, continuous ORF in the HTN S segment cRNA encoded N. Numerous hantavirus isolates have now been cloned and sequenced, and a comparison of these genes indicates that there are no additional conserved ORFs among hantaviruses, either in vRNA or cRNA, that could encode an NS_S polypeptide (7). Similarly, sequence analysis of cDNA representing the S segments of Dugbe (DUG), Hazara, and CCHF viruses (132) showed no significant ORFs in addition to that encoding N. Only a single mRNA, nearly equivalent in size to genomic S, could be detected in DUG virus–infected cells (234). Until the role of NS_S proteins in the replication of at least one virus in the family *Bunyaviridae* can be determined, it is not possible to comment on the significance of their absense in hantaviruses and nairoviruses.

M Segment Strategies

Bunyaviruses

Like the S segments, the M segments of viruses in the family *Bunyaviridae* use a variety of coding strategies to generate proteins (Fig. 4). Nucleotide sequence analyses of cDNA representing the M genome segments of the bunyaviruses BUN, SSH, GER, and LAC showed a continuous ORF in the cRNA (66,69,86,124,151). The gene order within the 4,527-nucleotide M segment of SSH, determined by amino- and carboxy-terminal sequencing of G1 and G2, was 5′ G2-G1 3′ (66,69). A large intergenic region, with a coding potential of 19 kd of protein, was discovered between the SSH virus G2 and G1 coding sequences, which is more than sufficient to encode the 11-kd nonstructural M (NS_M) polypeptide previously identified in SSH virus–infected cell lysates (69,72). Antisera to synthetic peptides representing amino acids predicted from the intergenic coding region were able to immune precipitate 10- and 11-kd polypeptides from SSH virus–infected cell lysates, thus supporting the conclusion that NS_M proteins are encoded in that region. Antiserum raised to synthetic peptide representing the NS_M region of BUN virus was found to immune precipitate a 16-kd protein from BUN virus–infected cells (145).

Phleboviruses

M segment coding properties of phleboviruses also include the generation of G1, G2, and NS_M polypeptides in a single ORF in cRNA. The 3,885-nucleotide RVF virus M segment has a potential coding capacity of 133 kd in the major ORF of cRNA (45). Amino-terminal sequence analysis of purified G1 and G2 showed a gene order of 5′ G2-G1 3′ with respect to cRNA and indicated that a coding potential of approximately 17 kd existed between the first ATG translation initiation codon and the amino terminus of G2. This preglycoprotein, or NS_M region, has five in-frame ATG codons that precede G2, the first four of which can be used to efficiently initiate synthesis of G2 and G1. *In vitro* translation of an RNA transcript of cDNA containing all five of the initiation codons produced a primary translation product of 133 kd that could be cleaved to yield the two viral glycoproteins in the presence of microsomal membranes. The cleavage was demonstrated to occur cotranslationally, evidenced by the absence of cleavage if the microsomal membranes were added after synthesis of the precursor polypeptide (217). In addition to G1 and G2, this transcript yielded 78- and 14-kd polypeptides, corresponding in size to polypeptides previously observed in lysates of cells infected with RVF virus (45,105,217). Results obtained by vaccinia virus or baculovirus expression of cDNAs that had been altered to remove one or more of the initiation codons and immune precipitation of expressed products by antipeptide antibodies demonstrated that the 78-kd protein initiates at the first ATG and the 14-kd protein at the second ATG (105,206,218). Pulse-chase experiments showed no precursor/product relationship between the 78- and 14-kd proteins; thus, it appears that use of the first and second ATGs, respectively, is what dictates generation of these proteins (44). The reason why the 78-kd protein is not processed at the NS_M-G2 junction has not been fully ascertained. Interestingly, however, a potential *N*-linked glycosylation site located in the NS_M region and shared by the 78- and 14-kd polypeptides was found to be used only in the 78-kd product. Removing of the glycosylation signal by site-directed mutagenesis appeared to have no effect on production of the 14- and 78-kd proteins, suggesting that glycosylation alone is not what prevents cleavage in the 78-kd protein at the NS_M-G2 junction (44,105). The significance of these NS_M polypeptides is not known; however, the apparent routine use of two in-frame initiation codons to generate proteins clearly discernible in infected cells suggests some functional role in replication.

Complete sequence information for the M genome segments of two other phleboviruses, PT and UUK, also have been reported (98,190). The M segment of PT virus is larger than that of RVF virus (4,330 nucleotides) and has a larger (30-kd) potential preglycoprotein coding region, although no NS_M proteins analogous to those of RVF virus have been detected in PT virus–infected cells. The gene order is reversed (i.e., 5′ G1-G2 3′), as compared with that of RVF virus; however, this has no biological significance because the G1 and G2 designations refer only to relative mobilities on polyacrylamide gels and do not correlate with functional properties of the proteins. This is clearly illustrated by computer-predicted homologies indicating that RVF virus G1 is related to PT virus G2 and that RVF virus G2 is related to PT virus G1 (98). No homology was detected between predicted NS_M products. Like RVF virus, PT virus G1 and G2 are efficiently expressed by vaccinia virus recombinants in the absence of NS_M coding sequences (137). Thirteen in-frame initiation codons are located in the preglycoprotein coding sequences of the PT virus M segment, but neither *in vivo* nor *in vitro* data are yet available concerning their usage. Three potential glycosylation sites reside in the preglycoprotein coding information of PT virus (98). It will be interesting to determine whether a selective use of glycosylation sites similar to that observed in RVF virus is used in PT virus NS_M gene product(s).

The UUK virus M genome segment consists of 3,231 nucleotides and, like those of RVF and PT viruses, encodes the G1 and G2 proteins (both 55 kd) in a single ORF of the cRNA (190). The envelope proteins also are processed by cotranslational cleavage of a polyprotein precursor (190, 223). However, the M segment coding strategy of UUK virus appears to differ from those of other phleboviruses in that there is no preglycoprotein coding region, and the amino terminus of G1 is located 17 amino acids downstream of the first (and only) initiation codon. Thus, UUK virus G1 appears similar to the RVF virus 78-kd protein. Although dot matrix comparison of the predicted G1 and G2 amino acid sequences of UUK, RVF, and PT viruses showed an obvious but low degree of homology among polypeptides encoded in the 5′ portion of the cRNAs of each virus and also among those encoded in the 3′ portion of the cRNAs, no homology with the NS_M coding regions of either PT or RVF viruses could be identified (190). Because few or none of the polypeptides encoded in the preglycoprotein region are detected in virion preparations of RVF or PT viruses (although a small amount of 78-kd polypeptide is found in RVF virion preparations), these polypeptides must be cleaved proteolytically before mature virus particles are extruded (45,98). Until a function can be assigned to these NS_M polypeptides, it is impossible to determine whether UUK virus replicates in the absence of such a function or accomplishes whatever function is required without removing a portion of the amino terminus of G1.

Carboxy-terminal sequence information has not been obtained for phlebovirus envelope proteins; consequently, the exact point of cleavage between G1 and G2 has not been identified. However, little or no extra coding sequences reside between G1 and G2. An intergenic region of less than 27 amino acids was demonstrated between G2 and G1 coding information of RVF virus by immune precipitation of viral proteins with antisera to synthetic peptides representing amino acid sequences at –16 to –27 upstream of the amino terminus of G1 (105). For UUK virus, a possible intergenic region of 81 amino acids was found between the predicted transmembrane, carboxy-terminal anchor of G1 and the putative signal sequence of G2 (190).

Hantaviruses

Like UUK virus, the hantavirus HTN encodes G1 and G2 in a continuous ORF in the cRNA of the M genome segment and does not appear to encode NS_M polypeptide(s). The M genome segment of HTN virus consists of 3,616 nucleotides and has a short, 17–amino acid leader sequence between the first initiation codon and the amino terminus of G1 (207). The carboxyl terminus of HTN virus G1 was localized to within 34 amino acids of the amino terminus of G2 by immune precipitation of viral proteins with antisera prepared to synthetic peptides representing M segment sequences. It was not determined if the 34 amino acids represented an intergenic region; however, at least 19 of these amino acids have characteristics that suggest they constitute a signal sequence for G2 (207). Immune precipitation with antipeptide antibodies also was used to determine that the carboxyl terminus of G2 extends to the end of the coding information of M segment cRNA. No other significant ORFs were detected in either the vRNA or cRNA of the M segment of HTN virus. Consequently, with the possible exception of a very short intergenic region, no coding information for an NS_M polypeptide exists. The complete M segments of a number of hantaviruses have now been cloned and sequenced, confirming the coding strategy described above for HTN virus as a generic property (8,11,79,153,197,225,242–244).

Nairoviruses

The M segment of the nairovirus DUG consists of 4,888 nucleotides and has coding information extending from nucleotides 48 to 4703 of the cRNA. A gene order of 5′ G2-G1 3′ was determined by expressing portions of the genes with recombinant baculoviruses and by analyzing the amino-terminal sequence of authentic G1 (131). The amino terminus of G1 does not follow a predicted signal sequence, but rather resides about 50 residues downstream of the nearest hydrophobic region. This observation was postulated to indicate that mature G1 arises by proteolytic processing of a precursor protein (131). Pulse-chase studies in DUG virus–infected cells and immune precipitation with G1-specific monoclonal antibodies showed the presence of 85- and 110-kd polypeptides in addition to G1. These proteins correspond in size to those previously reported for two other nairoviruses, Qalyub and Clo Mor (41,42,236). The 85-kd protein diminished over time as levels of G1 increased, suggesting that it is a precursor of the mature 73-kd G1. The amount of the 110-kd protein did not appear to change over time; thus, this protein probably has no precursor significance and may be a dimer of G1 or heterodimer of G1 and G2. Because it was difficult to detect the 35-kd G2 protein in infected cells, it was not possible to determine if it had a precursor product relationship with any larger protein (131). However, the data to this point suggest that the M segment of nairoviruses encodes only G1 and G2, although a G1 precursor may be proteolytically processed to yield mature G1.

Tospoviruses

Molecular cloning and sequence analyses of the M segments of two tospoviruses, TSW and impatiens necrotic spot (INS), showed sizes of 4,821 and 4,972 nucleotides, respectively (113,122). Like the S segments, the M segments of these viruses display nonoverlapping ORFs in both the vRNA and cRNA. The G1 and G2 proteins of INS are localized to the cRNA and an NS_M protein to the vRNA by cell-free translation of RNA transcripts corresponding to the two ORFs. A gene order of 5′ G2-G1 3′ with respect to cRNA was identified (122). As reported for other viruses in this family, a single ORF in cRNA encodes the two envelope glycoproteins of tospoviruses. For INS and TSW viruses, a 125- to 128-kd polyprotein precursor is processed to yield 78-kd G1 and 58-kd G2 proteins (113,122). Separate, subgenomic messages have been identified for the G1-G2 precursor and for the 34-kd NS_M protein (113,122). Tospoviruses, therefore, are the first viruses in the family *Bunyaviridae* that have two ambisense gene segments.

L Segment Strategies

All L segments of viruses in the family Bunyaviridae studied to date display conventional negative-sense coding strategies (Fig. 2). The L protein of bunyaviruses was first conclusively mapped to the L genome segment by using reassortant viruses generated by coinfecting cultured cells with LAC and Tahyna viruses, which have discernibly different L proteins when analyzed by polyacrylamide gel electrophoresis (65). Molecular cloning and sequence studies with the L genome segments of viruses in each genus demonstrated that the L protein is encoded in a single, large ORF in the cRNA (2,9,49,58,60,141,196,213, A.C. Marriott and P.A. Nuttall, personal communication). For each L segment described thus far, there are fewer than 200 nucleotides of total noncoding information (49,58,60,141,196). No evidence for additional coding regions in either the cRNA or the vRNA has been discovered. The L segments described for bunyaviruses, hantaviruses, and phleboviruses are all similar, consisting of approximately 6,400 to 6,700 nucleotides and encoding a polypeptide of about 240 to 260 kd. In contrast, the L segment of the nairovirus, DUG, is more than 12,000 nucleotides long and encodes a 460 kd polypeptide (A.C. Marriott and P.A. Nuttall, personal communication). The plant-infecting tospovirus TSW is approximately 8,900 nucleotides, with a coding potential of about 331 kd. An L protein of this predicted size was identified in purified TSW virions, indicating that the protein is not processed (224). The function of the L protein as the viral polymerase was confirmed by using L protein expressed from vaccinia virus recombinants to transcribe BUN virus ribonucleocapsid templates (102). An endonucleolytic activity also was demonstrated with expressed BUN virus L protein, thus providing evidence that the L protein is also responsible for generating the capped primers needed for transcription (103).

STAGES OF REPLICATION

The principal stages of the replication process for viruses in the family Bunyaviridae are illustrated in Fig. 6 and are summarized as follows:

1. Attachment, mediated by an interaction of viral proteins and host receptors.
2. Entry and uncoating, probably by endocytosis of virions and fusion of viral membranes with endosomal membranes.
3. Primary transcription; that is, the synthesis of virus-complementary mRNA species from genome templates, host cell–derived primers, and the virion-associated polymerase.
4. Translation of primary L and S segment mRNAs by free ribosomes, translation of M segment mRNAs by mem-

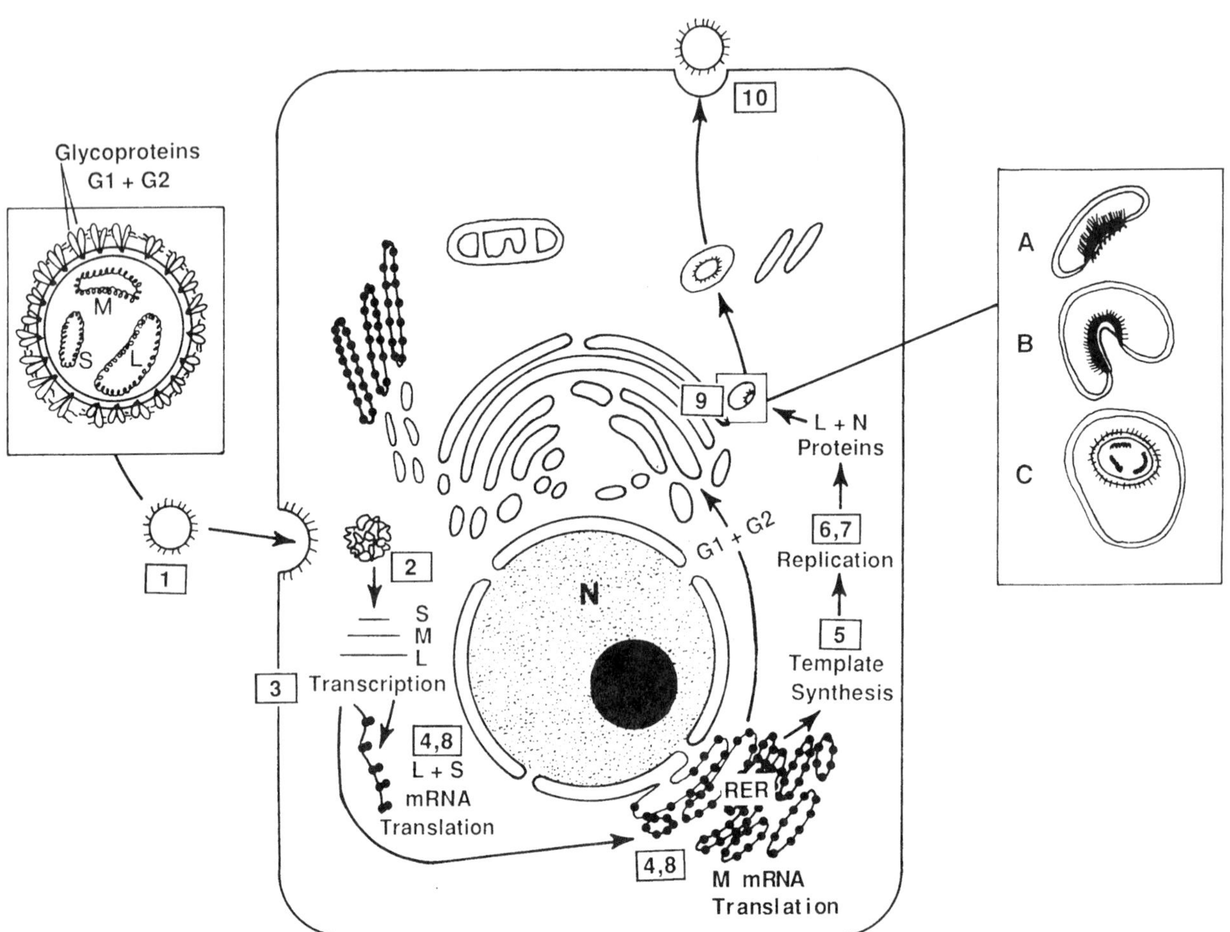

FIG. 6. Summary of probable replication processes for viruses in the family *Bunyaviridae*. Virion particles (inset on left) contain large (L), medium (M), and small (S) viral RNAs complexed with nucleocapsid proteins. The bilaminar, lipid envelope has integral virus-specified glycoproteins (G1 and G2), which interact across the membrane with the ribonucleoprotein structures. The precise location of the virion-associated polymerase is not known. Numbered events corresponding to those listed in the text include (1) attachment; (2) entry and uncoating; (3) primary transcription to yield viral mRNAs; (4) translation of L and S segment mRNA on free ribosomes [translation M segment mRNA on membrane-bound ribosomes and primary glycosylation of the gene products (G1 and G2)]; (5) synthesis of anti-genome templates; (6) genome replication; (7) secondary transcription; (8) further translation; (9) terminal glycosylation of G1 and G2 and assembly of viral particles by budding into Golgi vesicles; and (10) transport of cytoplasmic vesicles to the cell surface, and fusion and release of mature virions. The stages of budding into smooth membrane vesicles (inset on right) are as follows: (A) ribonucleoprotein structures accumulate on the cytoplasmic face of membranes that have G1 and G2 embedded into them and are exposed on the luminal side; (B) involution of membranes occurs; and (C) completion of budding takes place to yield a morphologically mature virion within a cytoplasmic vacuole (152). Abbreviations for cellular substructures: N, nucleus; RER, rough endoplasmic reticulum.

brane-bound ribosomes, and primary glycosylation of nascent envelope proteins.
5. Synthesis and encapsidation of cRNA to serve as templates for genomic RNA or, in some cases, for subgenomic mRNA.
6. Genome replication.
7. Secondary transcription; that is, the amplified synthesis of the mRNA species and ambisense transcription.
8. Continued translation and RNA replication.
9. Morphogenesis, including accumulation of G1 and G2 in the Golgi, terminal glycosylation, and acquisition of modified host membranes, generally by budding into the Golgi cisternae.
10. Fusion of cytoplasmic vesicles with the plasma membrane and release of mature virions.

Attachment and Entry

The early events in the infection process of members of the family Bunyaviridae are not well defined. Like other enveloped viruses, one or both of the integral viral envelope proteins mediate attachment to host cell receptors. This was first demonstrated by proteolytic enzyme treatment of purified LAC, which resulted in "spikeless" viral particles, and a five-log reduction in infectivity as compared with that of nondigested virions (148). The nature of cell receptors involved in attachment has not been identified thus far for any member of the family. However, the viral proteins involved in attachment have been examined indirectly by using polyclonal and monoclonal antibodies to block infection or hemagglutinating activity. For example, neutralization of LAC can be obtained with antibodies directed against G1, indicating a relationship of that protein to infectivity (82,87,109,110). Although neutralization may not necessarily be related to attachment, for LAC virus a relationship was proposed because it was determined that treatment of LAC virus with bromelain or pronase, which degraded portions of G1 but left G2 uncleaved, rendered the virus completely noninfectious (109). Attachment was postulated to be different for insect and vertebrate cells in that the LAC G1 protein bound to cultured mammalian (Vero) and mosquito (MAT) cells, but not to intact mosquito midguts. In contrast, G2 bound to the MAT cells and midguts, but not to Vero cells (127). Affinity-purified G1 competitively inhibited the binding of radiolabeled G1 to the Vero cells. To competitively inhibit binding of G2 to MAT cells, both G1 and G2 were required. Thus, it was proposed that the G1 protein might be the viral attachment protein for vertebrates, whereas G2 is used for arthropod infections (127). Confirmation of this as a general feature of viruses in the *Bunyavirus* genus is needed. For viruses in other genera, there is no evidence that one or the other of the glycoproteins is the attachment protein. The finding that neutralizing and hemagglutination inhibiting (HI) sites are present on both the G1 and G2 proteins of RVF virus (108), PT virus (175), and HTN virus, which is not known to infect arthropods (12,48), suggests that both proteins may be involved in attachment, either directly or due to conformational requirements.

Fusion of infected cells at acidic pH has been reported for viruses in the family *Bunyaviridae* (13,83–85) as well as for numerous other enveloped viruses (68,93,125,133, 135). The pH-dependent fusion is generally believed to relate to early events in the infection process, particularly the translocation of RNA and proteins into cell cytoplasm. Electron microscopy of the infection process of RVF virus showed that viral particles appeared to enter cells in phagocytic vacuoles (62). This observation is consistent with a mode of entry similar to that first described for alphaviruses in which the virus is endocytosed in coated vesicles (133). These endosomes subsequently become acidified (221), triggering a fusion of viral membranes and endosomal membranes, which releases the nucleocapsid into the cell cytoplasm. Direct evidence for this process with viruses in the *Bunyaviridae* has not yet been obtained. Furthermore, whether one or both of the envelope proteins are necessary for fusion has not been defined for most members of the family; however, selecting a viral mutant with a defective fusion function suggested that the G2 protein of LAC mediates such fusion (84).

Transcription

After uncoating of viral genomes, primary transcription of negative-sense vRNA to complementary mRNA is believed to occur by interaction of the virion-associated polymerase and the three viral RNA templates (28,184). The mechanism by which viruses in the *Bunyaviridae* family initiate transcription of their mRNA is less well defined but displays some marked similarities to that used by influenza viruses (see Chapter 21). Influenza viruses initiate transcription by using an endonuclease associated with the polymerase complex to cleave host cell mRNAs, primarily at purine residues, 10 to 13 nucleotides from the capped 5′ end (118). The presence of a methylated 5′ cap structure on the host mRNA is an absolute requirement for cleavage to occur (176). This capped oligonucleotide is used by the viral transcriptase to initiate mRNA synthesis. During elongation, the polymerase complex remains associated with the cap structure while the first 11 to 15 nucleotides are synthesized (31). Like those of influenza virus, mRNAs of at least one virus in each genus of the family Bunyaviridae possess 5′ terminal extensions of approximately 10 to 18 nucleotides that are heterogeneous in sequence and are not templated from vRNA (21,24,29,43,52,67,99,103,104,116,157,211). Direct evidence for the presence of capped structures on the termini of the extensions was obtained by using anti-cap antibodies to immuno-select mRNAs (89,228). These studies substantiated earlier studies that showed stimulation of LAC virus transcription by the presence of cap analogs such as m^7GpppAm or naturally capped RNAs (156). These same studies also showed an endonuclease associated with LAC virions, which was thought to be involved in the acquisition of the host-derived oligonucleotides. Although heterogeneous in sequence, the 5′-terminal extensions described for various viruses in the family Bunyaviridae display preferences for specific mono-, di-, or trinucleotides at the –1 to –3 positions with respect to the 5′ terminus of their mRNAs. For example, almost all cDNA clones of the S mRNA of the bunyavirus GER virus displayed U or G at the –1 position (29). Similarly, the bunyavirus BUN preferred UG, but SSH virus had A most commonly as the 3′-terminal nucleotide of the S mRNA extensions (21,67,103). For phleboviruses, C was most often found at the –1 position of the M mRNA of RVF virus and also at the –1 position of the N and NS_S mRNAs of UUK virus (43,211). For the hantavirus HTN, the preferred –1 nucleotide was G (52) and for the nairovirus DUG virus, C (104). The reason for such preferences could imply that a restricted or specific subset of host mRNAs are used for "cap-snatching" because of a need for limited base pairing with the viral genome, but this has not been demonstrated. Alternatively, similarities between the 3′-most nucleotides of the stolen host sequences and the 5′-terminal viral sequences might result in a backward slippage of the viral polymerase on the template after the first two or three nucleotides are transcribed, resulting in a partial reiteration of the 5′-terminal sequence (104). This mechanism would explain findings with GER where there was an apparent insertion of U or GU between the primer and viral sequences (228). It might also explain results obtained with BUN virus mRNAs synthesized by vaccinia virus–expressed L where insertions of GU or AGU were found (103). In addition to such insertions, for bunyavirus and phlebovirus mRNAs, but not for the nairovirus mRNAs studied, the +1 nucleotide in the mRNA transcript was sometimes missing (21,29,103, 104,211), and for hantavirus RNAs, the first three nucleotides were often absent (52). Whether this indicates primary transcription initiated at nucleotides other than the terminal one or perhaps are artifacts of the cloning procedures used remains to be determined.

Distinct differences as well as similarities between primary transcription of influenza virus and viruses in the family *Bunyaviridae* are known. A nuclear step is obligate for transcription of influenza genomes due to the requirement for primers cleaved from precursors of cellular RNA. Both heterogeneous nuclear RNA and mRNA transcribed by RNA polymerase II in the cell nucleus are possible candidates for these primers (100). The nuclear step is probably also prerequisite for splicing the viral mRNAs. Unlike influenza virus, viruses in the family *Bunyaviridae* do not have spliced mRNAs and appear to replicate solely in the cytoplasm. Goldman et al. (81) reported that the California encephalitis bunyavirus could produce progeny in enucleated cells. Although LAC virus did not replicate in enucleated cells (163), other evidence suggests that replication occurs exclusively in the cytoplasm. Pulse-labeling infected cells at various times after infection and examination of both cytoplasmic and nuclear fractions for labeled mRNA showed LAC virus S genome transcription only in the cytoplasm (191). Moreover, unlike influenza virus, LAC virus is resistant to the effects of actinomycin D, a drug that inhibits DNA-dependent, RNA polymerases, such as host-cell polymerase II (149). Therefore, it is believed that bunyaviruses acquire the primers needed for transcription from a stable pool of host cell messages rather than from newly synthesized nuclear transcripts and consequently have no requirement for ongoing host RNA synthesis as do influenza viruses.

Inhibitors of host cell protein synthesis, such as cycloheximide and puromycin, have been found to have no effect on primary transcription (i.e., copying of viral genomes to cRNA by the virion-associated polymerase) in other negative-strand RNA virus families such as the *Orthomyxoviridae, Rhabdoviridae,* and *Paramyxoviridae*. There are conflicting data concerning the requirement for ongoing host protein synthesis and primary transcription of viruses in the *Bunyavirus* genus. Both BUN and Akabane viruses were sensitive to cycloheximide treatment (1). Similarly, either no or greatly reduced amounts of S segment mRNA could be detected in LAC virus–infected cell cultures by hybridization with probes complementary to the 3′ end of mRNA if cells were treated with cycloheximide or puromycin (157,179,180). Also with LAC virus, the virion-associated polymerase produced incomplete transcripts *in vitro* unless rabbit reticulocyte lysates were added to provide a coupled transcription-translation system. In the coupled system, drugs that inhibit protein synthesis also inhibited full-length mRNA synthesis and resulted in the reappearance of the incomplete transcripts (19,180). This led to a hypothesis that translation of the nascent bunyavirus S segment mRNA is required to prevent premature termination of primary transcription products and that the nascent chain may be interacting with its template to cause premature termination. Such a translation requirement is postulated to be similar to that observed in certain bacterial systems, whereby ribosomes prevent RNA-RNA interaction and thus premature termination (180). In discord with these findings were demonstrations of transcription in cell-free systems with virion preparations of HTN, UUK Lumbo, LAC, and GER viruses (28,52,77,156, 184,199). For GER virus, although S segment mRNA was inhibited in cell culture by either anisomycin or cycloheximide, full-length S transcripts could be obtained in an *in vitro* transcription system without added translational capabilities (77,78). However, further studies showed that these full-length transcripts were obtained only under increased temperature and that premature termination products were obtained under more usual transcription conditions. Adding translating ribosomes from reticulocyte lysate improved transcription significantly and restored the ability for full-length transcription to occur (228). The mechanism of the inhibition of transcription by proteinase inhibitors was more clearly defined for GER virus in that adding the translation inhibitor cycloheximide, puromycin, or anisomycin, but not edeine, inhibited transcription. Because edeine inhibits translation by preventing the 40S and 60S ribosomal subunits from complexing but still allows the 40S subunit to scan the transcript, the need for ongoing protein synthesis for GER virus is postulated to be at the level of ribosome scanning and to be independent of the need for actual translation (228). These data are consistent with an earlier hypothesis that scanning ribosomes prevent premature termination by preventing base-pairing between the template RNA and the transcript, which would cause the polymerase to halt and terminate prematurely. This hypothesis is based on the finding that if inosine was substituted for guanosine in the *in vitro* transcription reaction, thus destabilizing base pairing, there was less premature termination (19). Results conflicting with these findings were obtained with the bunyavirus SSH in that strand-specific cDNA probes were able to detect full-length S segment mRNA (but not vRNA) readily in the presence of puromycin or cycloheximide (67). These results suggest that SSH primary transcription does not depend on either ongoing host protein synthesis or ribosomal scanning. A similar finding was obtained with LAC virus transcription in mosquito cells in that translation inhibitors had no effect on transcription, suggesting that host-cell factors may influence the ability of the viral polymerase to function in the presence of such inhibitors (183). At present, it is unclear whether the variations reported in transcriptional requirements for viruses in this genus are due to differences in their transcriptional properties or to differences in the sensitivities of methods used.

Differences between vRNA and mRNA are found not only at their 5′ termini but also at their 3′ termini. For gene segments with simple negative-sense coding, mRNAs are truncated at the 3′ termini by approximately 100 nucleotides as compared with cRNA (29,30,37,43,67,157,159,174,181). Potential transcription termination sites have been proposed for the S segments of LAC (157) and SSH viruses (66,67) at or near the genomic sequence, 3′-G/CUUUUU. A U-rich proposed transcription termination site for the M and S segments of GER virus were also found (3′AUGUUUUUGUU and 3′GGGGUUUGUU, respectively) (29). Similar GU-rich regions were found to be present in the 3′ noncoding region of the S segments of seven other bunyaviruses (55). Despite the presence of these motifs, which are simnilar to other negative-strand RNA viral transcription termination-polyadenylation signals (88,187), the mRNAs of viruses in the family *Bunyaviridae* are probably not polyadenylated, as evidenced by their low affinity for oligo dT columns (159,174,223). Transcription termination sites without the homopolymeric U_5 or U_6 tract have been proposed for the S and M segment mRNAs of SSH virus (5′-GGUGGGGGGUGGGG and 5′-GGUGGGGGGUGGGG, respectively) (66) and the M segments of the phleboviruses RVF and PT viruses (5′-UGGGGUGGUGGGGU and 5′-GGUGAGAGUGUAGAAAG, respectively) (43). If these purine-rich sequences are involved in transcription termination, the mechanism by which they are recognized will require further study. Interestingly, the polymerase protein of BUN, expressed with a recombinant vaccinia virus, was able to transcribe virion-derived nucleocapsids; some of the S transcripts terminated at the authentic transcription termination site, but most did not (103). These results suggest that an additional factor(s), either viral or cellular, may be required for consistent transcription termination. One suggested transcription termination mechanism involves an interaction between the transcriptase and a newly synthesized viral protein, such as NS_S (103,228). This could not be a familial characteristic, however, because hantaviruses and nairoviruses do not encode NS_S proteins.

Ambisense Transcription

For phleboviruses and tospoviruses, transcription of the S segment is complicated by the ambisense coding strategy. As described above, both the N and NS_S subgenomic mRNAs have the same sort of heterogeneous, nonviral, 5′-terminal extensions as other mRNAs (99,116,211). Hybridization studies with the phleboviruses PT and UUK confirmed earlier predictions based on sequence analysis that the NS_S message was the same polarity as vRNA and that the N message was complementary to vRNA (97–99,209). Similarly, for tospoviruses, both the subgenomic NS_S and NS_M messages are the same polarity as vRNA (51,116,122). In the presence of cycloheximide, PT virus N mRNA was still detected, but NS_S mRNA was not. These results are consistent with a model in which replication of full-length, encapsidated cRNA must occur before synthesis of the NS_S mRNA. This suggests that the NS_S protein is not involved in the early stages of replication (i.e., primary transcription).

Transcription termination for the ambisense genomes may involve RNA secondary structure. The transcription termination sites for both the N and NS_S mRNAs of PT virus were mapped by hybridizing a series of synthetic oligonucleotides corresponding to vRNA or cRNA to the messages. The results indicated that the 3′ termini of both mRNAs are within 40 nucleotides of one another (64). Computer analysis of the intergenic region and the sequences encoding the 3′ termini of the messages showed a long, inverted complementary sequence that could potentially form a hairpin structure. Similar stable hairpin structures are predicted to occur in the intergenic regions of the S and M segments of tospoviruses (51,116,122). For the S segment of the phlebovirus UUK, although there is a noncoding region of 70 nucleotides between the N and NS_S genes, hybridization studies demonstrated that the 3′ ends of the subgenomic messages overlap one another by about 100 nucleotides. Thus, the 3′ end of the NS_S mRNA extends into the coding region of N and the 3′ end of the N mRNA terminates just before the coding sequences for N. A short palindromic sequence in the intergenic region (including the 3′ ends of each mRNA) was predicted to allow formation of an A/U-rich hairpin structure (211). Similar (but shorter), energetically favored structures have been identified in intergenic regions of the S segments of the arenaviruses Pichinde and lymphocytic choriomeningitis, which also use an ambisense coding strategy (188). Such structures also have been observed in other eukaryotic systems and are postulated to be involved in transcription termination (20). For the family *Bunyaviridae*, such structures could affect transcription termination only if they can form while the genome is complexed with N. To date there is no direct evidence that this occurs. In addition to secondary structure, it is possible that particular sequences or motifs also may play a role in transcription termination of ambisense mRNAs. Comparison of the S segment sequences of the phleboviruses RVF, TOS, PT, UUK, and SFS showed the presence of G-rich sequences and similar sequence motifs in the intergenic regions of RVF, TOS, and SFS viruses, but not for PT and UUK viruses (80). Currently, there is no conclusive evidence that either of these mechanisms (secondary structure or gene sequence) is required for transcription termination with viruses in this family.

Genome Replication

The L protein of BUN, expressed by a recombinant vaccinia virus, is able to synthesize both vRNA and cRNA, suggesting that this protein is involved both in primary transcription and genome replication (103). In negative-strand viruses, the change from primary transcription to replication requires a switch from mRNA synthesis to synthesis of full-length cRNA templates and then vRNA. For viruses in the family *Bunyaviridae*, the polymerase protein, either acting alone or in concert with undefined viral or cellular factors, must first function as a cap-dependent

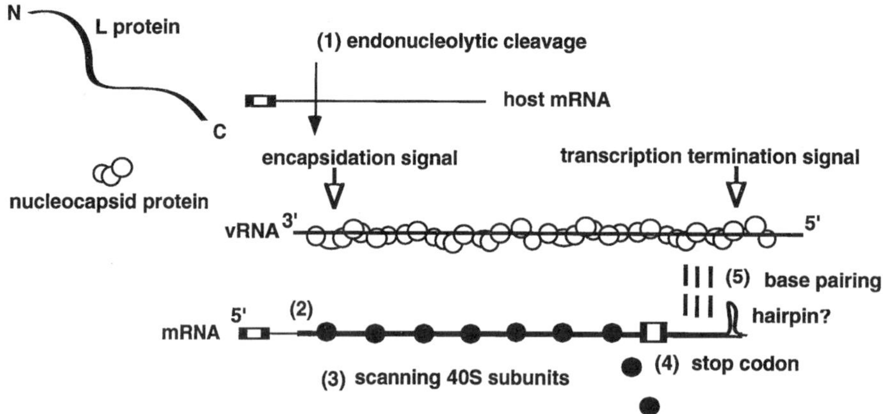

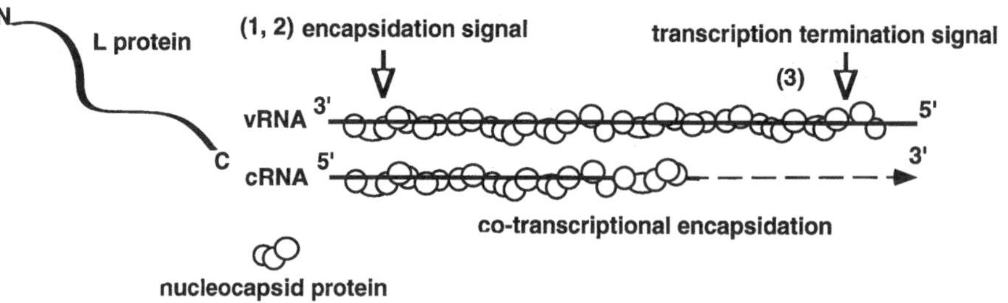

FIG. 7. A model for transcription versus replication for viruses in the family *Bunyaviridae*. **A:** Primary transcription. The L protein associated with infectious virions cleaves host cell mRNAs 12 to 18 nucleotides downstream of the 5′ cap structure (1) and uses those capped oligonucleotides to prime transcription on the encapsidated vRNA templates (2). Nascent mRNAs are scanned by 40S ribosomal subunits, until (3) they reach a tranlational stop codon (4), at which point they fall off the template. Transcription is terminated due to either base pairing beyond the translational stop, or possible energetically favored (hairpin) secondary structure of the RNA (5). The encapsidation signal is not recognized on mRNAs either due to the presence of added host sequences or the scanning ribosomes. **B:** Replication. The switch from transcription to replication may depend on the presence of newly synthesized nucleocapsid protein. The nascent L protein copies encapsidated vRNA to yield cRNA (1). The encapsidation signal is recognized, and cRNA is cotranscriptionally complexed with nucleocapsid protein (2). The transcription termination signal is not recognized, possibly because of the presence of nucleocapsid protein (3).

endonuclease to generate a primer for transcription of a nonencapsidated transcript of subgenomic length. At some point, the polymerase must switch to a process of independently initiating transcription at the precise 3′ end of the template and producing an encapsidated, full-length transcript. The processes involved in making that switch from primary transcription to genome replication have not been defined completely for any member of the family Bunyaviridae. Presumably, some viral or host factor is required to signal a suppression of the transcription termination signal responsible for generation of truncated mRNA and also to prevent the addition of the capped and methylated structures to the 5′ termini of the cRNAs. There is no question that genome replication and subsequent secondary transcription are prevented by translational inhibitors such as cycloheximide. These results indicate that continuous protein synthesis is required for replication of the genome. Although it is not known which proteins are required, they are likely to be of viral origin.

For the rhabdovirus vesicular stomatitis virus (VSV) and the paramyxovirus Sendai virus, the switch to antigenome synthesis appears to be controlled by the N protein (14,15, 26,160,161,229,238). Encapsidation by N seems to serve as an antitermination signal, thus allowing full-length genome synthesis. It has been suggested that the VSV NS protein is also involved and acts to control the availability of N (92). A similar mechanism appears plausible for the viruses in the family *Bunyaviridae*, whereby N would function to regulate replication and would be tempered by the presence of nonstructural proteins where they exist. The factors dictating that vRNA and cRNA should complex with N to form nucleocapsid structures, while mRNAs should not, are not known, but it has been suggested that the added capped host cell sequences on the 5′ ends of viral messages may somehow prevent encapsidation (180). However, this is evidently not an absolute preclusion to encapsidation in that capped, encapsidated S segment mRNAs were isolated by immuno-selection with anti-cap antibody in mosquito cells infected with LAC virus. This finding was suggested to represent a means of controlling N protein synthesis in insect cells (90). Similar findings have not yet been reported in mammalian cells. A model for genome transcription and replication based on information presented above and modified slightly from that proposed by Simons (208) is presented in Fig. 7.

Morphogenesis

Synthesis and Processing of Viral Proteins

Viral polypeptides are synthesized shortly after infection, suggesting that mRNAs are transcribed and translated rapidly. Time-course studies of the synthesis of viral polypeptides for the phlebovirus RVF virus showed that, at a high multiplicity of infection (moi), radiolabeled N and NS_S proteins can both be detected as early as 2 hr after infection and the envelope glycoproteins shortly thereafter (152). Consequently, if the ambisense strategy described above is correct (i.e., transcription of both cRNA and vRNA is required before synthesis of NS_S mRNA), then these events must occur very quickly. Similar kinetics of protein synthesis were observed in cell cultures infected with the bunyaviruses BUN and Trivittatus (164) and LAC; protein synthesis was found to reach a maximum at 3 to 5 hr after infection (129). The phlebovirus UUK (223) and the hantavirus HTN (200) exhibited slower rates of protein synthesis in that viral proteins could not be detected until approximately 6 hr after infection. The nairovirus DUG also demonstrated slower kinetics of synthesis; however, only low moi studies were performed (36).

As described above, the N and NS_S proteins encoded by the vRNA and cRNA of the S segments of bunyaviruses, phleboviruses, and tospoviruses and the NS_M protein encoded in the M segment cRNA of tospoviruses do not originate from precursor polypeptides; therefore, processing of proteins is not required. Hantaviruses and nairoviruses encode only one product in their S segments. Posttranslational modifications (e.g., phosphorylation or amidation) in general have not been defined for S segment products except for the NS_S of RVF virus, which appears to be phosphorylated (152,215).

In contrast, the cRNA M segment gene products are both processed and modified. Both G1 and G2 (and NS_M in the *Bunyavirus* and *Phlebovirus* genera) are translated from a single M segment mRNA species as a polyprotein precursor and are cotranslationally cleaved. The deduced amino acid sequences of the polyprotein precursors of numerous viruses in the family have been determined (see Coding Strategies). Except for DUG virus, all predicted polyprotein precursors have hydrophobic regions preceding both G1 and G2 that are consistent with signal sequences, variable numbers of potential transmembrane regions, and a hydrophobic sequence at the carboxy-terminus, indicating a membrane anchor region. Thus, these proteins are typical class 1 membrane proteins, i.e., with the amino terminus exposed on the surface of the virion and the carboxy terminus anchored in the membrane. Trypsin digestion studies with the bunyavirus SSH confirmed that the amino terminus of G1 (the second encoded protein) was on the virion surface (69). DUG virus differs in that there is no signal sequence immediately preceding G1 (the second polypeptide encoded in the ORF). This and results of pulse-chase studies suggest that G1 may be processed from a precursor polypeptide (131).

Although cotranslational cleavage of a precursor is undoubtedly the primary means of biogenesis of G1 and G2 (119,217,223), some studies suggest that independent initiation of translation also may be possible. This was demonstrated with RVF virus by deleting initiation codons preceding the amino terminus of G2 (the first encoded protein). Under these conditions, no G2 was detected in cell-free systems; however, G1 was still synthesized, indicating that an in-frame ATG codon somewhere upstream of the amino terminus of G1 functioned as a translation initiator (217). Similarly, the second protein in the M segment ORF (G2) of the hantavirus HTN could be expressed independently (165,198); however, the presence of an independent translation initiation codon was not a requirement for G2 biogenesis (106). Whether such independent initiation of translation of the envelope proteins is a means of regulating relative amounts of proteins or has any significance at all is not yet known.

A common property of all M segment gene products predicted from cDNA sequences studied so far is their high cysteine content (4% to 7%). Within individual genera, another common property is the conservation of the position of the cysteine residues (7,86,98,113,122,124,131,190). These findings suggest that extensive disulfide-bridge formation may occur and that the positions may be crucial for determining correct polypeptide folding. The secondary structure of the proteins also can play a role in immunogenicity, in that neutralizing or protective epitopes are often made up of nonlinear amino acids (16,233).

Glycosylation

All of the envelope proteins examined to date possess *N*-linked oligosaccharides. The gene sequences that define a potential *N*-linked glycosylation site are those that encode Asn-X-Ser/Thr, where X is not Pro (94). Five such sites were identified in the glycoprotein coding sequences of the phleboviruses RVF (45) and PT (98) and eight in UUK viruses (190). The bunyaviruses SSH (69), LAC (86), and BUN (124) had four sites; the hantaviruses HTN, SEO, PUU, PH, and THAI had five to seven potential sites (7, 242); and the tospoviruses TSW and INS had eight and nine potential sites, respectively, in the G1 and G2 proteins (113,122). The number of sites actually used in mature virion proteins has not been defined for all viruses; however, G2 of the phlebovirus RVF is glycosylated at its single available site and at least three of four possible G1 sites (105). Analysis of the predicted amino acid sequences of the envelope proteins of SSH (66,69) and examination of glycosylated tryptic oligopeptides (232) suggest that at least one of two sites on G1 and all three potential glycosylation sites on G2 are used. However, because one of the G2 sites is carboxy-terminal with respect to the only potential transmembrane regions, it is likely that only two sites are actually used. The single glycosylation site available in the G2 protein of HTN virus is used, and this site is conserved among numerous other hantaviruses (7,203). Potential *N*-linked sites also have been identified in the NS_M proteins of phleboviruses and tospoviruses and, at least for RVF virus, are used to some extent (44,105). The NS_M of BUN appears not to be glycosylated (124).

Two broad classes of asparagine-linked oligosaccharides, complex or high-mannose (simple), generally are found on mature glycoproteins (126). Often both types are attached to the same polypeptide chain. As described for the hemagglutinin protein of influenza virus, as well as for other glycoproteins (112,117), for oligosaccharides to evolve from the high-mannose type to the complex type, they are normally transported through the Golgi, where mannose residues are trimmed and terminal residues added. Examining the oligosaccharides attached to the G1 and G2 proteins of UUK virus in infected cells showed that G2 has mostly high-mannose glycans, whereas G1 contains both complex and a novel intermediate-type oligosaccharide (167). Similar results were obtained with Inkoo and LAC viruses (129,167), whereas the glycoproteins of HTN virus are mostly of the high-mannose type (10,194,203).

The type and amount of oligosaccharides attached to viral proteins correlate to some extent with the mode of viral maturation. For example, shortly after primary glycosylation of nascent proteins at the endoplasmic reticulum (ER), oligosaccharides are susceptible to cleavage by endoglycosidase H (endo-H), an enzyme that cleaves only high-mannose residues. Later, after removal of glucose residues at the rough ER, migration of the glycoproteins to the smooth ER and Golgi, trimming of residues, and attachment of peripheral sugars, the oligosaccharides are no longer susceptible to endo-H cleavage. This acquired resistance to endo-H therefore generally indicates that the proteins have been processed through the Golgi. The time required to convert between endo-H susceptibility and resistance correlates with the time needed for protein transport from the ER to the Golgi (see Transport). The endo-H susceptibility of the high-mannose and intermediate-type glycans found on the envelope proteins of viruses in the family *Bunyaviridae* suggests that they are incompletely processed through the Golgi complex. It was postulated initially that vacuolization of the Golgi—which accompanies *Bunyaviridae* infection of cells—might prevent further processing of viral proteins by that organelle (119,120). This appears not to be the case, because both viral and nonviral glycoproteins are still efficiently transported through the Golgi, despite vacuolization in response to infection with UUK virus (73–75). Thus, the reasons for incomplete processing of the G1 and G2 carbohydrates are currently not known.

Transport

One of the earliest notable features found to distinguish members of the family *Bunyaviridae* from all other negative-strand RNA viruses was that the viral particles are formed intracellularly by a budding process at smooth-surface vesicles in the Golgi area (Fig. 6) (25,128,143,231). The Golgi complex actually consists of several subcompartments, including the *cis*-, *medial*, and *trans*-Golgi (140). By using the fungal antibacterial reagent Brefeldin A, Sandoz Ltd., Basel, Switzerland which inhibits transport of proteins out of the ER and redistributes the Golgi component to the ER, the glycoproteins of PT virus were found to be localized in the *cis*/medial Golgi membranes (38). Similar redistribution of the G1 and G2 proteins of BUN virus from the Golgi to the ER after Brefeldin A treatment occurred in cells infected with vaccinia virus recombinants (145). Because Brefeldin A did not change the apparent glycosylation pattern of the G1 and G2 proteins of HTN virus as analyzed by polyacrylamide gel electrophoresis, it was postulated that the viral proteins were not processed through the Golgi at all (10); however, this was not confirmed by studies involving visualization of subcellular organelles (194).

The reason(s) for virus maturation in the Golgi complex as opposed to the more usual mode of viral morphogenesis, i.e., budding at the plasma membrane, is not known. Because the ionophore monensin, which exchanges protons for sodium ions, inhibits budding of UUK virus, it was postulated that the pH or ionic conditions present in the Golgi complex might be critical for viral budding (121,172). It is also not known which signals are responsible for the accumulation of viral structural proteins in the Golgi complex; however, one possibility is a structural or molecular resemblance between G1 and/or G2 and Golgi-specific proteins, such as the glycosyltransferases that are normally retained in the Golgi (73). Another possibility is that viral structural proteins lack the signal required for transport from the Golgi to the plasma membrane (120). Although no specific amino acid sequence, such as the C-terminal peptide KDEL identified on several soluble cellular proteins as an ER retention signal (162), has been observed on G1 or G2 proteins, some other specific retention signal may yet be discovered. Whatever the reason, it is clear that the envelope proteins possess signals necessary to localize to the Golgi without additional requirements for other viral proteins or nucleic acid. However, in some cases, only one of the two proteins possesses whatever signal is required, and the other is transported to the Golgi, only if both proteins are expressed and dimerized. Dimerization of the G1 and G2 proteins in the ER, shortly after their synthesis, has been reported for the phleboviruses UUK (189) and PT (38,136,137) and the hantavirus HTN (194). For PT virus, kinetic studies indicate that heterodimerization occurs between newly synthesized G1 and G2 within 3 min after protein synthesis and that the dimers are linked by disulfide bonds. The dimeric G1/G2 proteins were observed both during transport and after accumulation in the Golgi complex (38). The NS_M proteins of RVF and PT viruses are not needed for transport in that G1 and G2 are targeted to the Golgi when expressed from an M segment from which the NS_M coding region is deleted (137,235). When the entire M segment was expressed, the 14- and 78-kd NS_M proteins of RVF virus colocalized to the Golgi with the G1 and G2 proteins (235). Similar findings with a 16-kd protein of BUN virus were reported (145), but in neither case was it determined if the NS_M proteins served any role in morphogenesis.

By using recombinant vaccinia viruses to independently express the G1 and G2 proteins of PT virus, Chen et al. (40) discovered that in the absence of G1, G2 was transported out of the ER and expressed on the cell surface. Removing the carboxy-terminal anchor sequence of G1 resulted in secretion of this protein. When both proteins were expressed, however, both G1 and G2 were found in the Golgi. These studies suggest that the Golgi transport and retention signal is located on the G1 protein of PT virus (40). Similar results were obtained with the phlebovirus UUK in that expression of G1 and G2 independently in COS cells resulted in translocation of only one of the two proteins, in this case G1, to the Golgi, whereas G2 remained in the ER, suggesting that the signal for transport resides on G1 and that G2 is transported only by association with G1 (189). These studies confirmed earlier findings with temperature-sensitive mutants of UUK virus, in which G1 and G2 (but not N) localized to the Golgi, despite the absence of virus maturation. Those same studies demonstrated that the transport of viral envelope proteins from their site of synthesis on the rough ER through the Golgi occurred at an estimated two to three times slower rate than that of most viral membrane glycoproteins destined to be transported to the plasma membrane (73). For example, endo-H resistance is achieved at 45 and 90 to 150 minutes for G1 and G2, respectively, of UUK virus (119), in contrast to 15 to 20 minutes for the hemaglutinin (HA) protein of influenza virus or the G protein of VSV (46,53,214). PT virus G1 and G2 have similar kinetics to those of UUK virus (39,40,137). The finding that UUK virus G1 and G2 have different transport kinetics (i.e., G1 is incorporated into virions 20 min faster than G2) suggests that the dimers may arise from different precursor proteins, possibly because faster folding G1 cannot dimerize with slower folding G2 until G2 has reached its correct conformation (241). In this same study, the G1 and G2 proteins of UUK virus exited the ER quickly but did not enter the Golgi for 15 to 20 minutes. The investigators interpreted these findings as indicating that the G1 and G2 proteins may dimerize in an intermediate compartment between the ER and Golgi (241). In contrast to results with UUK and PT viruses, independent expression of G1 and G2 HTN virus by vaccinia virus recombinants resulted in the inability of either protein to exit the ER, but coexpression resulted in efficient transport of both proteins to the Golgi (165,166,194). Endo-H resistance could not be used as a measure of transport time because neither protein achieved resistance (10,194).

The transport of tospovirus proteins within plant cells and insect cells has not been studied extensively. These viruses replicate in insects (222,240) as well as in plants. Thus, in addition to the transport mechanisms described for other arthropod-borne viruses in this family, tospoviruses must also be able to penetrate the plant cell wall. The mechanism for this is not known; however, repeated mechanical transmission of TSW virus among plants readily results in point mutations in M and the accumulation of defective, nonenveloped particles. It was suggested that the G1 and G2 proteins (and envelope) are not required for replication in plants but only in their insect vector (185,227).

Assembly and Release

Electron microscopy of the phleboviruses (212) UUK (120), PT and Karimabad (212), and RVF (6) and the nairovirus DUG (27) showed maturation mostly, but not exclusively, in perinuclear regions associated with smooth membranes, presumably Golgi membranes (Fig. 8A). The sequence of events leading to assembly of viral particles differs for viruses in the family Bunyaviridae as compared with those of other negative-strand RNA virus families, such as *Rhabdoviridae, Orthomyxoviridae, Paramyxoviridae,* and probably *Filoviridae*. The M protein of these viruses is responsible for bridging the gap between the integral viral envelope proteins and their nucleocapsids and acts as the nucleating step to signal assembly of virions at the cell surface (see Chapters 19, 20, and 21, and Chapter 39 in *Fields Virology*, 3rd ed.). Unlike these viruses, members of the family *Bunyaviridae* do not have an M protein. Instead, the early events of assembly include an interaction between viral ribonucleoprotein (RNP) structures, which accumulate on the cytoplasmic side of vesicular membranes, and viral envelope proteins, which are displayed as transmembrane proteins, principally on the luminal side (212) (inset, Fig. 6). The viral RNP and spike structures have been observed only on the portion of the Golgi vesicle membrane directly involved in the budding process and not on adjacent areas of the same membrane. RNPs could not be found under membranes with no spikes, suggesting that some sort of transmembranal recognition between the viral glycoproteins and the N protein is prerequisite to budding (152). Candidate transmembrane regions are predicted from hydropathic characteristics of derived amino acid sequences representing the envelope proteins of all members of the family *Bunyaviridae* examined to date. Direct examination of the phlebovirus Karimabad, by enzymatically digesting exposed proteins embedded in intracellular membranes, demonstrated that approximately 12% of G1 and/or G2 is exposed on the cytoplasmic face of membranes in infected cells and is accessible to digestion. A large protease-resistant fragment was identified, which is

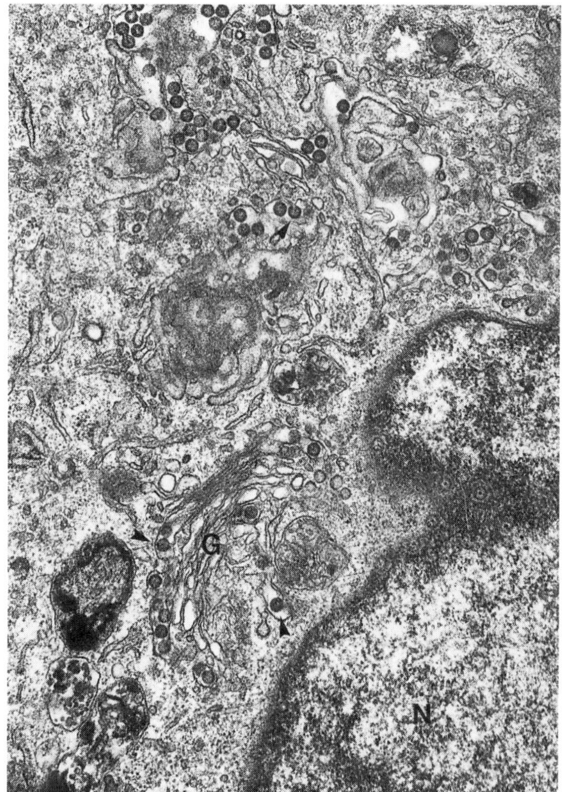

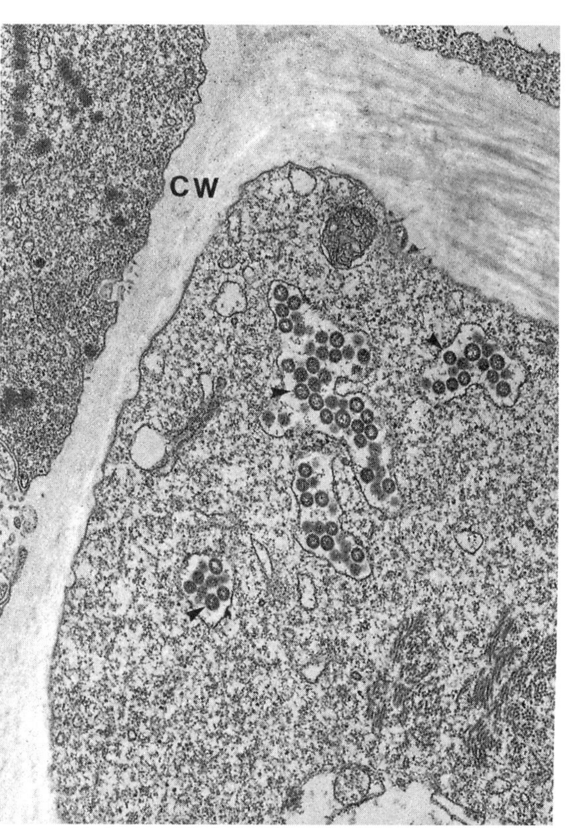

FIG. 8. A: Electron micrograph of a thin section of a Rift Valley fever virus–infected rat hepatocyte showing budding at internal smooth membranes including Golgi cisternae. The field contains several particles in the process of budding (arrowheads) as well as mature virions. G, Golgi; N, nucleus. Bar represents 600 nm. Reprinted from Anderson and Smith (6), with permission. **B:** Electron micrograph of a thin section of plant cells infected with tomato spotted wilt virus showing accumulated virions (arrowheads) in the cisternae of distended endoplasmic reticulum. CW, cell wall. Courtesy of J. Moyer, North Carolina State University, Raleigh, NC.

presumably sequestered in the membrane in a manner that renders it safe from enzymatic digestion (212). These enzyme-resistant fragments may therefore represent transmembrane regions of proteins, which could provide the interaction between RNPs and the cellular membranes required for envelopment. For UUK virus, it has been suggested that the cytoplasmic tail of G1, which is at least 70 residues long, is a logical candidate for interaction with the nucleocapsids (172,190).

After the particles bud into the Golgi cisternae, it is believed that they are released in individual small vesicles in a manner analogous to secretory granules of other cell types (32,193,212). The release of virus from infected cells presumably occurs when the cytoplasmic, virus-containing vesicles fuse with the cellular plasma membrane, that is, normal exocytosis. PT virions are released from a polarized epithelial cell line only at the basolateral surface (39). It was postulated that such polarized release of Sendai virus, and possibly viruses in the family *Bunyaviridae*, may facilitate the spread of virus during natural infection to produce systemic disease (39,220).

Although morphogenesis in association with the Golgi appears to be a *Bunyaviridae* familial trait, it is not an absolute requirement for virion production. Studies on RVF virus infection of primary rat hepatocytes demonstrated that mature virions can bud from the plasma membrane as well as into Golgi cisternae of the same cells (6). Because RVF virus is a hepatotropic virus *in vivo* and can cause liver necrosis and death in animals, these results raise questions of possible differences in morphogenesis in target versus nontarget cell types that might affect pathogenesis and/or immune defense mechanisms.

Although less well studied, morphogenesis of tospoviruses, both in plant cells and in plants, appears to be similar to that observed with other viruses in the family in that plant cells infected with TSW virus display numerous clusters of virions within dilated cisternae of the ER (111) (Fig. 8B). In the insect vector thrips, viruses accumulate in the salivary glands, and it is suggested that the presence of viral particles in secretory vesicles in this organ may indicate a direct involvement of the Golgi in viral budding (240). The steps involved in this budding have not yet been well-defined for tospoviruses.

EFFECTS OF VIRAL REPLICATION ON HOST CELLS

The cytopathic effects observed in cultured cells infected with members of the family *Bunyaviridae* vary widely, depending both on the virus and the type of host cell studied. Viruses in all genera except the *Hantavirus* genus are capable of alternately replicating in vertebrates (or plants for tospoviruses) and arthropods and generally are cytolytic for their vertebrate/plant hosts but cause little or no cytopathogenicity in their invertebrate hosts (17,101,128,240). Some viruses display a narrow host range, especially for arthropod vectors. Although the reason for this has not been defined completely, studies on LAC variant and revertant viruses suggests that the specificity is related to G1, probably at the level of viral attachment to susceptible cells (216). In natural infections of mammals, viruses often target to a particular organ or cell type. For example, bunyaviruses such as LAC virus appear to be neurotropic (154), the phlebovirus RVF is primarily hepatotropic (5,6,168,169), and the hantavirus HTN persists in rodent lungs (123). It will be interesting to determine whether this targeting is due solely to host-cell receptors or to other factors such as differences in effects on host-cell metabolism in target cell types versus the unnatural situation in cultured vertebrate cell lines.

Effects on Host Cell Metabolism

In vertebrate cells, bunyaviruses and phleboviruses reduce host cell protein synthesis, which decreases further as the infection progresses. A decline in BUN-infected BSC-1 host cell protein synthesis was observed at 5 hr postinfection and by 7 hr was almost completely abolished (164). Similar results were obtained in LAC virus–infected BHK cells (129). RVF virus–infected Vero cells displayed reduced host protein synthesis that gradually became more pronounced from 4 to 20 hr after infection (152).

No such reduction in host protein synthesis, even late in infection, has been observed in mammalian cells infected with the phlebovirus UUK (171,223) or the nairovirus DUG (36), both of which are transmitted by ticks rather than mosquitoes. However, host protein synthesis was somewhat inhibited in a *Xenopus laevis* (frog) cell line infected with the tick-borne nairovirus Clo Mor or the phlebovirus St. Abb's head (237). Hantaviruses not only do not detectably reduce host macromolecular synthesis (56,200) but routinely establish persistent, noncytolytic infections in susceptible mammalian host cells, a finding consistent with their nonpathogenic persistence in natural rodent hosts (123). Hantaan virus and other hantaviruses can produce plaques in certain cell types, however, suggesting transient, incomplete cytopathic effect.

The arthropod-borne members of the family, like most other arboviruses, cause little detectable cytopathology in mosquito cell cultures, and viral persistence is readily established (35,62,146,147,192). Unlike cultured vertebrate cells, mosquito cells infected with the bunyavirus Marituba displayed no reduction in host macromolecular synthesis; thus, viral infection apparently does not drastically interfere with normal cellular processes (35). One suggested reason for this is that in arthropod cells excess viral proteins do not accumulate in the cells but rather are more efficiently processed into mature virions (146). In agreement with this, only N proteins of BUN and LAC viruses could readily be detected in persistently infected mosquito cultures, although that protein accumulated (62,192). Alternatively, it was suggested that the viral transcriptase may be less active in arthropod cells than in mammalian cells and that the endonuclease activity of the polymerase (which is used to acquire transcriptional primers) is detrimental to host cell messages. A reduced level of activity of the viral transcriptase would therefore cause less damage to host cell messages and consequently to protein synthesis (192).

Another possible explanation for persistence, both in insect and mammalian cells, is the generation of defective or interfering viruses. Temperature-sensitive and plaque morphology mutants were recovered from persistent infections of mosquito cells infected with the bunyavirus BUN (62,146). The carrier cultures were resistant to superinfection with BUN virus, and at least some of the viruses recovered from the persistent infections effectively inhibited infection of normal mosquito cells by normal virus (62). In these studies, evidence for deleted RNAs was not obtained; thus, a persistence mechanism involving classical defective interfering (DI) particles could not be postulated (62,192). In a later study (195), deleted L segment BUN RNAs were found to be produced in persistently infected mosquito cells, but these RNAs were not encapsidated. The exact nature of the interference remains to be determined. Conventional DI particles, which displayed deletions only in L, were also described for BUN. The defect in L was a single internal deletion, amounting to 72% to 77% of the L RNA segment (155). Persistence of the phlebovirus TOS in cultured mammalian cells (Vero cells) also was reported, but the defect in the virus was not determined (226). Defective L RNA segments and resultant DI particles were also reported for the bunyavirus GER (47) and for the tospovirus TSW (185,186). The TSW virus deletions were generated by repeated high multiplicity passage and were found to be similar to those of BUN virus in that single internal deletions amounting to 60% to 80% of the L segment were found (186). Interestingly, the L deletions identified both in the TSW and BUN DI particles were in-frame, thus allowing translation of truncated L polypeptides (155,186). The significance of this finding is not known, but conservation of the terminal sequences and the observed encapsidation of these shortened L segments suggest that they may maintain all signals necessary for replication, transcription, and translation; thus, they may be able to interfere at several steps in the replication process.

CONCLUSION

The family *Bunyaviridae* consists of a large and widely diverse group of viruses and is divided into five genera based on molecular properties and serological relationships among its members. Representatives of each genus display common morphological, biochemical, and genetic attributes, yet have unique replicative properties and often extremely different biological characteristics. All viruses have three-segmented, single-stranded RNA genomes and, for those studied, encode their N, G1/G2, and L proteins in the cRNA of their S, M, and L genome segments. To generate the nonstructural proteins found in some but not all genera, three different expression strategies for S segments and two for M are known. Viruses in this family generally display an unusual Golgi-associated morphogenesis and usually acquire their envelopes by budding into intracytoplasmic vacuoles. The replication strategies of members of the family *Bunyaviridae* have only recently been elucidated, but already the possibilities for manipulating viral genomes, by means of infectious clones and reverse genetics appear to be not-too-distant prospects. With this will come an even greater understanding of these fascinating viruses.

REFERENCES

1. Abraham G, Pattnaik AK. Early RNA synthesis in Bunyamwera virus-infected cells. *J Gen Virol* 1983;64:1277–1290.
2. Accardi L, Gro MC, Di Bonito P, Giorgi C. Toscana virus genomic L segment: molecular cloning, coding strategy and amino acid sequence in comparison with other negative strand RNA viruses. *Virus Res* 1993; 27:119–131.
3. Akashi H, Bishop DHL. Comparison of the sequences and coding of La Crosse and snowshoe hare bunyavirus S RNA species. *J Virol* 1983; 45:1155–1158.
4. Akashi H, Gay M, Ihara T, Bishop DHL. Localized conserved regions of the S RNA gened products of bunyaviruses are revealed by sequence analysis of the Simbu serogroup Aino viruses. *Virus Res* 1984;1:51–63.
5. Anderson GW, Slone TW, Peters CJ. Pathogenesis of Rift Valley fever virus (RVFV) in inbred rats. *Microbial Pathogen* 1987;2:283–293.
6. Anderson GW, Smith JF. Immunoelectron microscopy of Rift Valley morphogenesis in primary rat hepatocytes. *Virology* 1987;161:91–100.
7. Antic D, Kang CY, Spik K, Schmaljohn C, Vapalahti O, Vaheri A. Comparison of the deduced gene products of the L, M and S genome segments of hantaviruses. *Virus Res* 1992;24:35–46.
8. Antic D, Lim BU, Kang CY. Molecular characterization of the M genomic segment of the Seoul 80-39 virus; nucleotide and amino acid sequence comparisons with other hantaviruses reveal the evolutionary pathway. *Virus Res* 1991;19:47–58.
9. Antic D, Lim BU, Kang CY. Nucleotide sequence and coding capacity of the large (L) genomic RNA segment of Seoul 80-39 virus, a member of the hantavirus genus. *Virus Res* 1991;19:59–65.
10. Antic D, Wright KE, Kang CY. Maturation of Hantaan virus glycoproteins G1 and G2. *Virology* 1992;189:324–328.
11. Arikawa J, Lapenotiere HF, Iacono Connors L, Wang ML, Schmaljohn CS. Coding properties of the S and the M genome segments of Sapporo rat virus: comparison to other causative agents of hemorrhagic fever with renal syndrome. *Virology* 1990;176:114–125.
12. Arikawa J, Schmaljohn AL, Schmaljohn CS, Dalrymple JM. Characterization of Hantaan virus envelope glycoprotein antigenic determinants by monoclonal antibodies. *J Gen Virol* 1989;70:615–624.
13. Arikawa J, Takashima I, Hashimoto N. Cell fusion by haemorrhagic fever with renal syndrome (HFRS) viruses and its application for titration of virus infectivity and neutralizing antibody. *Arch Virol* 1985;86: 303–313.
14. Arnheiter H, Davis NL, Wertz G, Schubert M, Lazzarini RA. Role of the nucleocapsid proteins in regulating vesicular stomatitis virus RNA synthesis. *Cell* 1985;41:259–267.
15. Banerjee A. Transcription and replication of rhabdoviruses. *Microbiol Rev* 1987;51:66–87.
16. Battles JK, Dalrymple JM. Genetic variation among geographic isolates of Rift Valley fever virus. *Am J Trop Med Hyg* 1988;39:617–631.
17. Beaty BJ, Calisher CH. Bunyaviridae natural history. *Curr Top Microbiol Immunol* 1991;169:27–78.
18. Beaty BJ, Shope RE, Clarke DH. Salt-dependent hemagglutination with Bunyaviridae antigens. *J Clin Microbiol* 1977;5:548–550.

19. Bellocq C, Raju R, Patterson J, Kovlakofsky D. Translational requirement of La Crosse virus s-mrna synthesis: in vitro studies. *J Virol* 1987; 61:87–95.
20. Birchmeier C, Folk W, Birnstiel ML. The terminal RNA stem-loop structure and 80 bp of spacer DNA are required for the formation of 3′ termini of sea urchin H2A mRNA. *Cell* 1983;35:433–440.
21. Bishop D, Gay ME, Matsuoko Y. Nonviral heterogeneous sequences are present at the 5′ ends of one species of snowshoe hare bunyavirus S complementary RNA. *Nucleic Acids Res* 1983;11:6409–6419.
22. Bishop DH, Calisher CH, Casals J, et al. Bunyaviridae. *Intervirology* 1980;14:125–143.
23. Bishop DH, Gould KG, Akashi H, Clerx-van Haastern, CM. The complete sequence and coding content of snowshoe hare bunyavirus small (S) viral RNA species. *Nucleic Acids Res* 1982;10:3703–3713.
24. Bishop DHL, Rud E, Belloncik S, et al. Coding analyses of bunyavirus RNA species. In: Compans R, Bishop D, ed. *Segmented negative strand viruses, arenavirus, bunyaviruses and othomyxoviruses*. Orlando, FL: Academic; 1984:3–11.
25. Bishop DHL, Shope RE. Bunyaviridae. *Compr Virol* 1979;14:1–156.
26. Blumberg BM, Giorgi C, Kolakofsky D. N protein of vesicular stomatitis virus selectively encapsidates leader RNA *in vitro*. *Cell* 1983;32: 559–567.
27. Booth TF, Gould EA, Nuttall PA. Structure and morphogenesis of Dugbe virus (Bunyaviridae, Nairovirus) studied by immunogold electron microscopy of ultrathin cryosections. *Virus Res* 1991;21:199–212.
28. Bouloy M, Hannuoun C. I. RNA-dependent RNA polymerase associated with virions. *Virology* 1976;69:258–264.
29. Bouloy M, Pardigon N, Vialat P, Gerbaud S, Girard M. Characterization of the 5′ and 3′ ends of viral messenger RNAs isolated from BHK21 cells infected with Germiston virus (Bunyavirus). *Virology* 1990;175:50–58.
30. Bouloy M, Vialat M, Girard M, Pardigon N. A transcript from the S segment of the Germiston bunyavirus is uncapped and codes for the nucleoprotein and a nonstructural protein. *J Virol* 1984;49:717–723.
31. Braam J, Ulmanen I, Krug RM. Molecular model of a eucaryotic transcription complex: functions and movements of influenza P proteins during capped RNA-primed transcription. *Cell* 1983;34:609–618.
32. Broadwell RD, Oliver C. Golgi apparatus, GERL, and secretory granule formation within neurons of the hypothalamo-neurohypophysial system of control and hyperosmotically stressed mice. *J Cell Biol* 1981; 90:474–484.
33. Cabradilla CD, Holloway BP, Obijeski JF. Molecular cloning and sequencing of the La Crosse virus S RNA. *Virology* 1983;128:463–468.
34. Calisher C. Classification and nomenclature of viruses. Fifth Report of the International Committeee on Taxonomy of Viruses. *Arch Virol* 1991;(suppl 2):273–283.
35. Carvalho MGC, Frugulhetti IC, Rebello MA. Marituba (Bunyaviridae) virus replication in cultured *Aedes albopictus* cells and in L-A9 cells. *Arch Virol* 1986;90:325–335.
36. Cash P. Polypeptide synthesis of Dugbe virus, a member of the *Nairovirus* genus of the *Bunyaviridae*. *J Gen Virol* 1985;66:141–148.
37. Cash P, Vezza AC, Gentsch JR, Bishop D. Genome complexities of the three mRNA species of snowshoe hare bunyavirus and *in vitro* translation of S mRNA to viral N polypeptide. *J Virol* 1979;31:685–694.
38. Chen SY, Compans RW. Oligomerization, transport, and Golgi retention of Punta Toro virus glycoproteins. *J Virol* 1991;65:5902–5909.
39. Chen SY, Matsuoka Y, Compans RW. Assembly and polarized release of Punta Toro virus and effects of Brefeldin A. *J Virol* 1991;65: 1427–1439.
40. Chen SY, Matsuoka Y, Compans RW. Golgi complex localization of the Punta Toro virus G2 protein requires its association with the G1 protein. *Virology* 1991;183:351–365.
41. Clerx J, Casals J, Bishop D. Structural characteristics of nairoviruses (genus *Nairovirus, Bunyaviridae*). *J Gen Virol* 1981;55:165–178.
42. Clerx JP, Bishop DH. Qalyub virus, a member of the newly proposed *Nairovirus* genus (*Bunyavividae*). *Virology* 1981;108:361–372.
43. Collett MS. Messenger RNA of the M segment RNA of Rift Valley fever virus. *Virology* 1986;151:151–156.
44. Collett MS, Kakach L, Suzich JA, Wasmoen TL. Gene products and expression strategy of the M segment of the phlebovirus Rift Valley fever virus. In: Mahy B, Kolaksofsky D, eds. *Genetics and pathogenicity of negative strand viruses*. New York: Elsevier Biomedical; 1989: 49–57.
45. Collett MS, Purchio AF, Keegan K, et al. Complete nucleotide sequence of the M RNA segment of Rift Valley fever virus. *Virology* 1985;144: 228–245.
46. Copeland CS, Zimmer KP, Wagner KR, Healy GA, Melman I, Helenius A. Folding, trimerization and transport are sequential events in the biogenesis of influenza virus hemagglutinin. *Cell* 1988;53:197–209.
47. Cunningham C, Szilagyi JF. Viral RNAs synthesized in cells infected with Germiston bunyavirus. *Virology* 1987;157:431–439.
48. Dantas JR, Okuno Y, Asada H, et al. Characterization of glycoproteins of virus causing hemorrhagic fever with renal syndrome (HFRS) using monoclonal antibodies. *Virology* 1986;151:379–384.
49. de Haan P, Kormelink R, de Oliveira RR, van Poelwijk F, Peters D, Goldbach R. Tomato spotted wilt virus L RNA encodes a putative RNA polymerase. *J Gen Virol* 1991;72:2207–2216.
50. de Haan P, Wagemakers L, Peters D, Goldbach R. Molecular cloning and terminal sequence determination of the S and M RNAs of tomato spotted wilt virus. *J Gen Virol* 1989;70:3469–3473.
51. de Haan P, Wagemakers L, Peters D, Goldbach R. The S RNA segment of tomato spotted wilt virus has an ambisense character. *J Gen Virol* 1990;71:1001–1007.
52. Dobbs M, Kang CY. Hantaan virus mRNAs contain non-viral 5′ end sequences and lack poly(A) at the 3′ end. *Abstracts, IXth International Congress of Virology* 1993;P44–16:280.
53. Doms RW, Keller DS, Helenius A, Balch WE. Role for adenosine triphosphate in regulating the assembly and transport of vesicular stomatitis virus G protein trimers. *J Cell Biol* 1987;105:1957–1969.
54. Donets MA, Chumakov MP, Korolev MB, Rubin SG. Physicochemical characteristics, morphology and morphogenesis of virions of the causative agent of Crimean hemorrhagic fever. *Intervirology* 1977;8: 294–308.
55. Dunn EF, Pritlove DC, Elliott RM. The S RNA genome segments of Batai, Cache Valley, Guaroa, Kairi, Lumbo, Main Drain and Northway bunyaviruses: sequence determination and analysis. *Gen Virol* 1994;75:597–608.
56. Elliott LH, Kiley MP, McCormick JB. Hantaan virus: identification of virion proteins. *J Gen Virol* 1984;65:1285–1293.
57. Elliott RM. Identification of nonstructural proteins encoded by viruses of the Bunyamwera serogroup (family Bunyaviridae). *Virology* 1985; 143:119–126.
58. Elliott RM. Nucleotide sequence analysis of the large (L) genomic RNA segment of Bunyamwera virus, the prototype of the family Bunyaviridae. *Virology* 1989;173:426–436.
59. Elliott RM. Nucleotide sequence analysis of the small (S) RNA segment of Bunyamwera virus, the prototype of the family Bunyaviridae. *J Gen Virol* 1989;70:1281–1285.
60. Elliott RM, Dunn E, Simons JF, Pettersson F. Nucleotide sequence and coding strategy of the Uukuniemi virus L RNA segment. *J Gen Virol* 1992;73:1745–1752.
61. Elliott RM, Schmaljohn CS, Collett MS. Bunyaviridae genome structure and gene expression. *Curr Top Microbiol Immunol* 1991;169: 91–141.
62. Elliott RM, Wilkie ML. Persistent infection of *Aedes albopictus* C6/36 cells by Bunyamwera virus. *Virology* 1986;150:21–32.
63. Ellis DS, Shirodaria PV, Fleming E, Simpson DIH. Morphology and development of Rift Valley fever virus in Vero cell cultures. *J Med Virol* 1988;24:161–174.
64. Emery VC. Characterization of Punta Toro S mRNA species and identification of an inverted complementary sequence in the intergenic region of Punta Toro phlebovirus ambisense S RNA that is involved in mRNA transcription termination. *Virology* 1987;156:1–11.
65. Endres MJ, Jacoby DR, Janssen RS, Gonzales-Scarano F, Nathanson N. The large viral RNA segment of California serogroup bunyaviruses encodes the large viral protein. *J Gen Virol* 1989;70:223–228.
66. Eshita Y, Bishop DHL. The complete sequence of the M RNA of snowshoe hare Bunyavirus reveals the presence of internal hydrophobic domains in the viral glycoprotein. *Virology* 1984;137:227–240.
67. Eshita Y, Ericson B, Romanowski V, Bishop DHL. Analyses of the mRNA transcription processes of snowshoe hare bunyavirus S and M RNA species. *J Virol* 1985;55:681–689.
68. Fan DP, Sefton BM. The entry into host cells of Sindbis virus, vesicular stomatitis virus and Sendai virus. *Cell* 1978;15:985–992.
69. Fazakerly JK, Gonzalez-Scarano F, Strickler J, Dietzschold B, Karush F, Nathanson N. Organization of the middle RNA segment of snowshoe hare bunyavirus. *Virology* 1988;167:422–432.
70. Fenner R. The classification and nomenclature of viruses. *Intervirolo-*

gy 1975;6:1–12.
71. Fuller F, Bhown AS, Bishop DHL. Bunyavirus nucleoprotein, N, and a non-structural protein, NSS are coded by overlapping reading frames in the S RNA. *J Gen Virol* 1983;64:1705–1714.
72. Fuller F, Bishop DHL. Identification of virus-coded nonstructural polypeptides in bunyavirus-infected cells. *J Virol* 1982;41:643–648.
73. Gahmberg N, Kuismanen E, Keranen S, Petterson RF. Uukuniemi virus glycoproteins accumulate in and cause morphological changes of the Golgi complex in the absence of virus maturation. *J Virol* 1986;57: 899–906.
74. Gahmberg N, Peltonen L. Efficient export of secretory proteins through a vacuolized Golgi complex. *Cell Biol Int Rep* 1987;11:547–555.
75. Gahmberg N, Petterson RF, Kaariainen L. Efficient transport of Semliki Forest virus glycoproteins through a Golgi complex morphologically altered by Uukuniemi virus glycoproteins. *EMBO J* 1986;5: 3111–3118.
76. Gentsch JR, Bishop DHL. Small viral RNA segment of bunyaviruses codes for viral nucleocapsid protein. *J Virol* 1978;28:417–419.
77. Gerbaud S, Pardigon N, Vialat P, Bouloy M. The S segment of Germiston bunyavirus genome: coding strategy and transcription. In: Mahy B, Kolakofsky D, ed. *The biology of negative strand viruses*. Amsterdam: Elsevier Science; 1987:191–198.
78. Gerbaud S, Vialat P, Pardigon N, Wychowski C, Girard M, Bouloy M. The S segment of the Germiston virus RNA genome can code for three proteins. *Virus Res* 1987;8:1–13.
79. Giebel LB, Stohwasser R, Zöller L, Bautz EKF, Darai G. Determination of coding capacity of the M genome segment of nephropathia epidemica virus strain Hällnäs B1 by molecular cloning and nucleotide sequence analysis. *Virology* 1989;172:498–505.
80. Giorgi C, Accardi L, Nicoletti L, et al. Sequences and coding strategies of the S RNAs of Toscana and Rift Valley fever viruses compared to those of Punta Toro, Sicilian Sandfly fever, and Uukuniemi viruses. *Virology* 1991;180:738–753.
81. Goldman N, Presser I, Sreevalson T. California encephalitis virus: some biological and biochemical properties. *Virology* 1977;76:352–364.
82. Gonzalez Scarano F, Shope RE, Calisher CE, Nathanson N. Characterization of monoclonal antibodies against the G1 and N proteins of LaCrosse and Tahyna, two California serogroup bunyaviruses. *Virology* 1982;120:42–53.
83. Gonzalez-Scarano F. La Crosse virus G1 glycoprotein undergoes a conformational change at the pH of fusion. *Virology* 1985;140:209–216.
84. Gonzalez-Scarano F, Janssen RS, Najjar JA, Pobjecky N, Nathanson N. An avirulent G1 glycoprotein variant of La Crosse bunyavirus with defective fusion function. *J Virol* 1985;54:757–763.
85. Gonzalez-Scarano F, Pobjecky N, Nathanson N. LaCrosse bunyavirus can mediate pH-dependent fusion from without. *Virology* 1984;132: 222–225.
86. Grady LJ, Sanders ML, Campbell WP. The sequence of the M RNA of an isolate of La Crosse virus. *J Gen Virol* 1987;68:3057–3071.
87. Grady LJ, Srihongse S, Grayson MA, Deibel R. Monoclonal antibodies against La Crosse virus. *J Gen Virol* 1983;64:1699–1704.
88. Gupta KC, Kingsbury DW. Conserved polyadenylation signals in two negative-strand RNA virus families. *Virology* 1982;120:518–523.
89. Hacker D. Anti-mRNAs in La Crosse bunyavirus-infected cells. *J Virol* 1990;64:5051–5057.
90. Hacker D, Raju R, Kolakofsky D. La Crosse virus nucleocapsid protein controls its own synthesis in mosquito cells by encapsidating its mRNA. *J Virol* 1989;63:5166–5174.
91. Hewlett MJ, Petterson RF, Baltimore D. Circular forms of Uukuniemi virion RNA: an electron microscopic study. *J Virol* 1977;21:1085–1093.
92. Howard M, Davis N, Patton J, Wertz G. Roles of vesicular stomatitis virus (VSV) N and NS proteins in viral RNA replication. In: Mahy B, Kolakofsky D, ed. *The biology of negative strand viruses*. Amsterdam: Elsevier Science; 1987:134–149.
93. Huang RTC, Rott R, Klenk HD. Influenza viruses cause hemolysis and fusion of cells. *Virology* 1981;110:243–247.
94. Hubbard S, Ivatt R. Synthesis and processing of asparagine-linked oligosaccharides. *Annu Rev Biochem* 1981;50:55–83.
95. Hung T, Chou Z, Zhao T, Xia S, Hang C. Morphology and morphogenesis of viruses of hemorrhagic fever with renal syndrome (HFRS). *Intervirology* 1985;23:97–108.
96. Hung T, Xia EM, Song G, et al. Viruses of classical and mild forms of hemorrhagic fever with renal syndrome isolated in China have similar bunyavirus-like morphology. *Lancet* 1983;1:589–591.
97. Ihara T, Akashi H, Bishop D. Novel coding strategy (ambisense genomic RNA) revealed by sequence analysis of Punta Toro phlebovirus S RNA. *Virology* 1984;136:293–306.
98. Ihara T, Dalrymple JM, Bishop DHL. Complete sequences of the glycoprotein and M RNA of Punta Toro phlebovirus compared to those of Rift Valley fever virus. *Virology* 1985;144:246–259.
99. Ihara T, Matsuura Y, Bishop DHL. Analysis of the mRNA transcription processes of Punta Toro phlebovirus (Bunyaviridae). *Virology* 1985;147:317–325.
100. Ishihama A, Nagata K. Viral RNA polymerases. *CRC Crit Rev Biochem* 1988;23:27–76.
101. James WS, Millican D. Host-adaptive antigenic variation in bunyaviruses. *J Gen Virol* 1986;67:2803–2806.
102. Jin H, Elliott RM. Expression of functional Bunyamwera virus L protein by recombinant vaccinia viruses. *J Virol* 1991;65:4182–4189.
103. Jin H, Elliott RM. Characterization of Bunyamwera virus S RNA that is transcribed and replicated by the L protein expressed from recombinant vaccinia virus. *J Virol* 1993;67:1396–1404.
104. Jin H, Elliott RM. Non-viral sequences at the 5′ ends of Dugbe nairovirus S mRNAs. *J Gen Virol* 1993;74:2293–2297.
105. Kakach LT, Suzich JA, Collett MS. Rift Valley fever virus M segment: phlebovirus expression strategy and protein glycosylation. *Virology* 1989;170:505–510.
106. Kamrud KI, Schmaljohn CS. Expression strategy of the M genome segment of Hantaan virus. *Virus Res* 1994;31:109–122.
107. Karabatsos N. Supplement to International Catalogue of Arboviruses including certain other viruses of vertebrates. *Am J Trop Med Hyg* 1978; 27:372–373.
108. Keegan K, Collett MS. Use of bacterial expression cloning to define the amino acid sequencins of antigenic determinants on the G2 glycoprotein of Rift Valley fever Virus. *J Virol* 1986;58:263–270.
109. Kingford L, Hill DW. The effects of proteolytic enzymes on structure and function of La Crosse G1 and G2 glycoproteins. In: Bishop D, Compans R, eds. *The replication of negative strand viruses*. New York: Elsevier; 1981:111–116.
110. Kingsford L, Ishizawa LD, Hill DW. Biological activities of monoclonal antibodies reactive with antigenic sites mapped on the G1 glycoprotein of La Crosse virus. *Virology* 1983;90:443–455.
111. Kitajima EW, De Avila AC, Resende R de O, Goldbach RW, Peters D. Comparative cytological and immunogold labelling studies on different isolates of tomato spotted wilt virus. *J Submicrosc Cytol Pathol* 1992;24:1–4.
112. Klenk HD, Rott R. Cotranslational and posttranslational processing of viral glycoproteins. *Curr Top Microbiol Immunol* 1980;90:19–48.
113. Kormelink R, de Haan P, Meurs C, Peters D, Goldbach R. The nucleotide sequence of the M RNA segment of tomato spotted wilt virus, a bunyavirus with two ambisense RNA segments. *J Gen Virol* 1992; 73:2795–2804.
114. Kormelink R, de Haan P, Peters D, Goldbach R. Viral RNA synthesis in tomato spotted wilt virus-infected *Nicotiana rustica* plants. *J Gen Virol* 1992;73:687–693.
115. Kormelink R, Kitajima EW, De Haan P, Zuidema D, Peters D, Goldbach R. The nonstructural protein (NSs) encoded by the ambisense S RNA segment of tomato spotted wilt virus is associated with fibrous structures in infected plant cells. *Virology* 1991;181:459–468.
116. Kormelink R, van Poelwijk F, Peters D, Goldbach R. Non-viral heterogeneous sequences at the 5′ ends of tomato spotted wilt virus mRNAs. *J Gen Virol* 1992;73:2125–2128.
117. Kornfeld R, Kornfeld S. Assembly of asparagine-linked oligosaccharides. *Annu Rev Biochem* 1985;54:631–664.
118. Krug RM. Priming of influenza virus RNA transcription by capped heterologous RNAs. *Curr Top Microbiol Immunol* 1981;93:125–149.
119. Kuismanen E. Posttranslational processing of Uukuniemi virus glycoproteins G1 and G2. *J Virol* 1984;51:806–812.
120. Kuismanen E, Hedman K, Saraste J, Pettersson RF. Uukuniemi virus maturation: accumulation of virus particles and viral antigens in the Golgi complex. *Mol Cell Biol* 1982;2:1444–1458.
121. Kuismanen E, Saraste J, Pettersson RF. Effect of monensin on the assembly of Uukuniemi virus in the Golgi complex. *J Virol* 1985;55: 813–822.
122. Law MD, Speck J, Moyer JW. The M RNA of impatiens necrotic spot tospovirus (*Bunyaviridae*) has an ambisense genomic organization. *Virology* 1992;188:732–741.
123. Lee HW, Lee PW, Baek LJ, Song CK, Seong IW. Intraspecific trans-

mission of Hantaan virus, etiologic agent of Korean hemorrhagic fever, in the rodent Apodemus agrarius. *Am J Trop Med Hyg* 1981;30: 1106–1112.
124. Lees JF, Pringle CR, Elliot RM. Nucleotide sequence of the Bunyamwera virus M RNA segment: conservation of structural features in the bunyavirus glycoprotein gene product. *Virology* 1986;148:1–14.
125. Lenard J, Miller DK. Uncoating of enveloped viruses. *Cell* 1982;28:5–6.
126. Lennarz WJ, ed. *The biochemistry of glycoproteins and proteoglycans*. New York: Plenum; 1980.
127. Ludwig GV, Israel BA, Christensen BM, Yuill TM, Schultz KT. Role of La Crosse virus glycoproteins in attachment of virus to host cells. *Virology* 1991;181:564–571.
128. Lyons MJ, Heyduk J. Aspects of the developmental morphology of California encephalitis virus in cultured vertebrate and arthropod cells and in mouse brain. *Virology* 1973;54:37–52.
129. Madoff DH, Lenard J. A membrane glycoprotein that accumulates intracellularly: cellular processing of the large glycoprotein of LaCrosse virus. *Cell* 1982;28:821–829.
130. Marriott A, Ward V, Nuttall P. The S RNA segment of sandfly fever Sicilian virus: evidence for an ambisense genome. *Virology* 1989;169: 341–345.
131. Marriott AC, el Ghorr AA, Nuttall PA. Dugbe nairovirus M RNA: nucleotide sequence and coding strategy. *Virology* 1992;190:606–615.
132. Marriott AC, Nuttall PA. Comparison of the S RNA segments and nucleoprotein sequences of Crimean-Congo hemorrhagic fever, Hazara, and Dugbe viruses. *Virology* 1992;189:795–799.
133. Marsh M, Helenius A. Adsorptive endocytosis of Semliki Forest virus. *J Mol Biol* 1980;142:439–454.
134. Martin ML, Regnery HL, Sasso DR, McCormick JB, Palmer E. Distinction between *Bunyaviridae* genera by surface structure and comparison with Hantaan virus using negative stain electron microscopy. *Arch Virol* 1985;86:17–28.
135. Matlin KS, Reggio H, Helenius A, Simons K. Pathway of vesicular stomatitis virus entry leading to infection. *J Mol Biol* 1982;156:609–631.
136. Matsuoka Y, Chen SY, Compans RW. Bunyavirus protein transport and assembly. *Curr Top Microbiol Immunol* 1991;169:161–179.
137. Matsuoka Y, Ihara T, Bishop DH, Compans RW. Intracellular accumulation of Punta Toro virus glycoproteins expressed from cloned cDNA. *Virology* 1988;167:251–260.
138. McClintock J. Mosquito-virus relationships of American encephalitides. *Annu Rev Entomol* 1978;23:17–37.
139. McCormick JB, Palmer EL, Sasso DR, Kiley MP. Morphological identification of the agent of Korean haemorrhagic fever (Hantaan virus) as a member of the Bunyaviridae. *Lancet* 1982;1:765–767.
140. Mellman I, Simons K. The Golgi complex: *In vitro veritas*? *Cell* 1992; 68:829–840.
141. Muller R, Argentini C, Bouloy M, Prehaud C, Bishop DH. Completion of the genome sequence of Rift Valley fever phlebovirus indicates that the L RNA is negative sense or ambisense and codes for a putative transcriptase-replicase. *Nucleic Acids Res* 1991;19:5433.
141a. Muller R, Saluzzo J-F, Lopez N, Dreier T, Turell M, Smith J, Bouloy M. Characterization of clone 13, a naturally attenuated avirulent isolate of Rift Valley Fever Virus, which is altered in the S segment. *Am J Trop Med Hyg* (in press).
142. Murphy FA, Fauquet CM, Bishop DHL, et al. Virus taxonomy. Sixth report of the International Committee on Taxonomy of Viruses. Springer-Verlag;1995.
143. Murphy FA, Harrison AK, Whitfield SG. Morphologic and morphogenetic similarities of Bunyamwera serological supergroup viruses and several other arthropod-borne viruses. *Intervirology* 1973;1:297–316.
144. Murphy FA, Whitfield SG, Sudia WD. Interactions of vector with vertebrate pathogenic viruses. In: Maramorosch K, Shope R, eds. *Invertebrate immunity*. Orlando, FL: Academic; 1975:25–37.
145. Nakitare GW, Elliott RM. Expression of the Bunyamwera virus M genome segment and intracellular localization of NSm. *Virology* 1993; 195:511–520.
146. Newton SE, Short NJ, Dalgarno L. Bunyamwera virus replication in cultured *Aedes albopictus* (mosquito) cells: establishment of a persistent viral infection. *J Virol* 1981;38:1015–1024.
147. Nicoletti L, Verani P. Growth of *Phlebovirus* Toscana in a mosquito (*Aedes pseudoscutellaris*) cell line (AP-61): establishment of a persistent infection. *Arch Virol* 1985;85:35–45.
148. Obijeski JF, Bishop DHL, Palmer EL, Murphy FA. Segmented genome and nucleocapsid of LaCrosse Virus. *J Virol* 1976;20:664–675.
149. Obijeski JF, Murphy FA. Bunyaviridae: recent biochemical developments. *J Gen Virol* 1977;37:1–14.
150. Overton HA, Ihara T, Bishop DH. Identification of the N and NSS proteins coded by the ambisense S RNA of Punta Toro phlebovirus using monospecific antisera raised to baculovirus expressed N and NS_S proteins. *Virology* 1987;157:338–350.
151. Pardigon N, Vialat P, Gerbaud S, Girard M, Bouloy M. Nucleotide sequence of the M segment of Germiston virus: comparison of the M gene product of several bunyaviruses. *Virus Res* 1988;11:73–85.
152. Parker MD, Smith JF, Dalrymple JM. Rift Valley fever virus intracellular RNA: a functional analysis. In: Compans R, Bishop D, eds. *Segmented negative strand viruses*. Orlando, FL: Academic; 1984:21–28.
153. Parrington MA, Lee PW, Kang CY. Molecular characterization of the Prospect Hill virus M RNA segment: a comparison with the M RNA segments of other hantaviruses. *J Gen Virol* 1991;72:1845–1854.
154. Parsonson I, McPhee DA. Bunyavirus pathogenesis. *Adv Virus Res* 1985;30:279–316.
155. Patel AH, Elliott RM. Characterization of Bunyamwera virus defective interfering particles. *J Gen Virol* 1992;73:389–396.
156. Patterson JL, Holloway B, Kolakofsky D. La Crosse virions contain a primer-stimulated RNA polymerase and a methylated cap-dependent endonuclease. *J Virol* 1984;52:215–222.
157. Patterson JL, Kolakofsky D. Characterization of La Crosse virus small-genome segment transcripts. *J Virol* 1984;49:680–685.
158. Patterson JL, Kolakofsky D, Holloway BP, Obijeski JF. Isolation of the ends of LaCrosse virus small RNA as a double-stranded structure. *J Virol* 1983;45:882–884.
159. Pattnaik AK, Abraham G. Identification of four complementary RNA species in Akabane virus-infected cells. *J Virol* 1983;47:452–462.
160. Patton JT, Davis NL, Wertz GW. N protein alone satisfies the requirement for protein synthesis during RNA replication of vesicular stomatitis virus. *J Virol* 1984;49:303–309.
161. Patton JT, Davis NL, Wertz GW. Role of vesicular stomatitis virus proteins in RNA replication. In: Bishop D, Compans R, eds. *Nonsegmented negative strand viruses. Paramyxoviruses and rhabdoviruses*. Orlando, FL: Academic; 1984:147–152.
162. Pelham HRB. Control of protein exit from the endoplasmic reticulum. *Annu Rev Cell Biol* 1989;5:1–23.
163. Pennington TH, Pringle CR. Negative strand viruses in enucleate cells. In: Mahy B, Barry R, eds. *Negative strand viruses and the host cell*. Orlando, FL: Academic; 1978:457–464.
164. Pennington TH, Pringle CR, McCrae MA. Bunyamwera virus-induced polypeptide synthesis. *J Virol* 1977;24:397–400.
165. Pensiero MN, Hay J. The Hantaan virus M-segment glycoproteins G and G2 can be expressed independently. *J Virol* 1992;66:1907–1914.
166. Pensiero MN, Jennings GB, Schmaljohn CS, Hay J. Expression of the Hantaan virus M genome segment by using a vaccinia virus recombinant. *J Virol* 1988;62:696–702.
167. Pesonen M, Kuismanen E, Pettersson RF. Monosaccharide sequence of protein-bound glycans of Uukuniemi virus. *J Virol* 1982;41:390–400.
168. Peters CJ, Jones D, Trotter R, et al. Experimental Rift Valley fever in rhesus macaques. *Arch Virol* 1988;99:31–44.
169. Peters CJ, Liu CT, Anderson GW Jr, Morrill JC, Jahrling PB. Pathogenesis of viral hemorrhagic fevers: Rift Valley fever and Lassa fever contrasted. *Rev Infect Dis* 1989;11:743–S749.
170. Peters D, de Avila AC, Kitajima EW, Resende R, de Haan P, Goldbach R. An overview of tomato spotted wilt virus. In: Hsu H-T, Lawson RH, eds. *Virus-thrips-plant interactions of TSWV. Proceedings of USDA Workshop*. Beltsville, MD: National Technology Information Service; 1991:1–14.
171. Pettersson RF. Effect of Uukuniemi virus infection on host cell macromolecule systhesis. *Med Biol* 1974;52:90–97.
172. Pettersson RF. Protein localization and virus assembly at intracellular membranes. *Curr Top Microbiol Immunol* 1991;170:67–106.
173. Pettersson RF, Bonsdorf CH. Ribonucleoproteins of Uukuniemi virus are circular. *J Virol* 1975;15:386–392.
174. Pettersson RF, Kuismanen E, Rauonnholm R, Ulmanen I. mRNAs of Uukuniemi virus, a bunyavirus. In: Becker Y, ed. *Viral messenger RNA transcription, processing, splicing, and molecular structure*. Boston: Nijhoff; 1985:283–300.
175. Pifat DY, Osterling MC, Smith JF. Antigenic analysis of Punta Toro virus and identification of protective determinants with monoclonal antibodies. *Virology* 1988;167:442–450.
176. Plotch SJ, Bouloy M, Ulmanen I, Krug RM. A unique cap (m^7Gpp-

pXm)-dependent influenza virion endonuclease cleaves capped RNAs to generate the primers that initiate viral RNA transcription. *Cell* 1981; 23:847–858.
177. Porterfield JS, Casals J, Chumakov MP, et al. Bunyaviruses and Bunyaviridae. *Intervirology* 1973/1974;2:270–272.
178. Porterfield JS, Casals J, Chumakov MP, et al. Bunyaviruses and Bunyaviridae. *Intervirology* 1975/76;6:13–24.
179. Raju R, Kolakofsky D. Inhibitors of protein synthesis inhibit both LaCrosse virus S-mRNA and S genome syntheses in vivo. *Virus Res* 1986;5:1–9.
180. Raju R, Kolakofsky D. Translational requirement of La Crosse virus S-mRNA synthesis. *J Virol* 1987;63:122–128.
181. Raju R, Kolakofsky D. Unusual transcripts in La Crosse virus-infected cells and the site for nucleocapsid assembly. *J Virol* 1987;61: 667–672.
182. Raju R, Kolakofsky D. The ends of La Crosse virus genome and antigenome RNAs within nucleocapsids are base paired. *J Virol* 1989; 63:122–128.
183. Raju R, Raju L, Kolakofsky D. The translational requirement for complete La Crosse virus mRNA synthesis is cell-type dependent. *J Virol* 1989;63:5159–5165.
184. Ranki M, Pettersson RF. Uukuneimi virus contains an RNA polymerase. *J Virol* 1975;16:1420–1425.
185. Resende R de O, de Haan P, de Avila AC, et al. Generation of envelope and defective interfering RNA mutants of tomato spotted wilt virus by mechanical passage. *J Gen Virol* 1991;72:2375–2383.
186. Resende R de O, de Haan P, van de VE, de Avila AC, Goldbach R, Peters D. Defective interfering L RNA segments of tomato spotted wilt virus retain both virus genome termini and have extensive internal deletions. *J Gen Virol* 1992;73:2509–2516.
187. Robertson JS, Schubert M, Lazzarini RA. Polyadenylation sites for influenza virus mRNA. *J Virol* 1981;38:157–163.
188. Romanowski V, Bishop DHL. Conserved sequences and coding of two strains of lymphocylic choriomeningitis virus (WE and ARM) and Pichinde arenavirus. *Virus Res* 1985;2:35–51.
189. Rönnholm R. Localization to the Golgi complex of Uukuniemi virus glycoproteins G1 and G2 expressed from cloned cDNAs. *J Virol* 1992; 66:4525–4531.
190. Rönnholm R, Pettersson RF. Complete nucleotide sequence of the M RNA segment of Uukuniemi virus encoding the membrane glycoproteins G1 and G2. *Virology* 1987;160:191–202.
191. Rossier C, Patterson J, Kolakofsky D. La Crosse virus small genome mRNA is made in the cytoplasm. *J Virol* 1986;58:647–650.
192. Rossier C, Raju R, Kolakofsky D. LaCrosse virus gene expression in mammalian and mosquito cells. *Virology* 1988;165:539–548.
193. Rothman JE. The Golgi apparatus: two organelles in tandem. *Science* 1981;213:1212–1218.
194. Ruusala A, Persson R, Schmaljohn CS, Pettersson RF. Coexpression of the membrane glycoproteins G1 and G2 of Hantaan virus is required for targeting to the Golgi complex. *Virology* 1992;186:53–64.
195. Scallan MF, Elliott RM. Defective RNAs in mosquito cells persistently infected with Bunyamwera virus. *J Gen Virol* 1992;73:53–60.
196. Schmaljohn CS. Nucleotide sequence of the L genome segment of Hantaan virus. *Nucleic Acids Res* 1990;18:6728.
197. Schmaljohn CS, Arikawa J, Hasty SE, et al. Conservation of antigenic properties and sequences encoding the envelope proteins of prototype Hantaan virus and two virus isolates from Korean haemorrhagic fever patients. *J Gen Virol* 1988;69:1949–1955.
198. Schmaljohn CS, Chu YK, Schmaljohn AL, Dalrymple JM. Antigenic subunits of Hantaan virus expressed by baculovirus and vaccinia virus recombinants. *J Virol* 1990;64:3162–3170.
199. Schmaljohn CS, Dalrymple JM. Analysis of Hantaan virus RNA: evidence for a new genus of Bunyaviridae. *Virology* 1983;131:482–491.
200. Schmaljohn CS, Dalrymple JM. Biochemical characterization of Hantaan virus. In: Compans RW, Bishop DHL, eds. *Segmented negative strand viruses*. New York: Elsevier; 1984:117–124.
201. Schmaljohn CS, Hasty SE, Dalrymple JM, et al. Antigenic and genetic properties of viruses linked to hemorrhagic fever with renal syndrome. *Science* 1985;227:1041–1044.
202. Schmaljohn CS, Hasty SE, Harrison SA, Dalrymple JM. Characterization of Hantaan virions, the prototype virus of hemorrhagic fever with renal syndrome. *J Infect Dis* 1983;148:1005–1012.
203. Schmaljohn CS, Hasty SE, Rasmussen L, Dalrymple JM. Hantaan virus replication: effects of monensin, tunicamycin and endoglycosidases on the structural glycoproteins. *J Gen Virol* 1986;67:707–717.
204. Schmaljohn CS, Jennings GB, Dalrymple JM. Identification of Hantaan virus messenger RNA species. In: Mahy B, Kolakofsky D, eds. *The biology of negative strand viruses*. Amsterdam: Elsevier; 1987: 116–121.
205. Schmaljohn CS, Jennings GB, Hay J, Dalrymple JM. Coding strategy of the S genome segment of Hantaan virus. *Virology* 1986;155:633–643.
206. Schmaljohn CS, Parker MD, Ennis WH, et al. Baculovirus expression of the M genome segment of Rift Valley fever virus and examination of antigenic and immunogenic properties of the expressed proteins. *Virology* 1989;170:184–192.
207. Schmaljohn CS, Schmaljohn AL, Dalrymple JM. Hantaan virus M RNA: coding strategy, nucleotide sequence, and gene order. *Virology* 1987;157:31–39.
208. Simons JF. Exploring the molecular biology of Uukuniemi virus. Studies on the S segment, its mRNAs and protein products and the L segment. Ludwig Institute for Cancer Research, Stockholm Branch and Department of Physiological Chemistry, Karolinska Institute, Stockholm, Sweden, 1992
209. Simons JF, Hellman U, Pettersson RF. Uukuniemi virus S RNA segment: ambisense coding strategy, packaging of complementary strands into virions, and homology to members of the genus Phlebovirus. *J Virol* 1990;64:247–255.
210. Simons JF, Persson R, Pettersson RF. Association of the nonstructural protein NSs of Uukuniemi virus with the 40S ribosomal subunit. *J Virol* 1992;66:4233–4241.
211. Simons JF, Pettersson RF. Host-derived 5′ ends and overlapping complementary 3′ ends of the two mRNAs transcribed from the ambisense S segment of Uukuniemi virus. *J Virol* 1991;65:4741–4748.
212. Smith JF, Pifat DY. Morphogenesis of sandfly viruses (Bunyaviridae family). *Virology* 1982;121:61–81.
213. Stohwasser R, Raab K, Darai G, Bautz EKF. Primary structure of the large (L) RNA segment of nephropathia epidemica virus strain Hällnas B1 coding for the viral RNA polymerase. *Virology* 1991;183: 386–391.
214. Strous GJAM, Lodish HF. Intracelluair transport of secretory and membrane proteins in hepatoma cells infected with vesicular stomatitis virus. *Cell* 1980;22:709–717.
215. Struthers JK, Swanepoel R. Identification of a major non-structural protein in the nuclei of Rift Valley fever virus-infected cells. *J Gen Virol* 1982;60:381–384.
216. Sundin DR, Beaty BJ, Nathanson N, Gonzalez-Scarano F. A G1 glycoprotein epitope of La Crosse virus: a determinant of infection of *Aedes triseriatus*. *Science* 1987;235:591–593.
217. Suzich JA, Collett MS. Rift valley fever virus M segment: cell-free transcription and translation of virus-complementary RNA. *Virology* 1988;164:478–486.
218. Suzich JA, Kakach LT, Collett MS. Expression strategy of a phlebovirus: biogenesis of proteins from the Rift Valley fever virus M segment. *J Virol* 1990;64:1549–1555.
219. Talmon Y, Prasad BV, Clerx JP, Wang GJ, Chiu W, Hewlett MJ. Electron microscopy of vitrified-hydrated La Crosse virus. *J Virol* 1987; 61:2319–2321.
220. Tashiro M, Yamakawa M, Tobita T, Seto JT, Klenk H-D, Rott R. Altered budding site of a pantropic mutant of Sendai virus, F1-R, in polarized epithelial cells. *J Virol* 1990;64:4672–4677.
221. Tycko B, Maxfield FR. Rapid acidification of endocytic vesicles containing alpha-2-macroglobulin. *Cell* 1982;28:643–651.
222. Ullman DE, German TL, Sherwood JL, Westcot DM, Cantone FA. Tospovirus replication in insect vector cells: immunocytochemical evidence that the nonstructural protein encoded by the S RNA of tomato spotted wilt tospovirus is present in thrips vector cells. *Phytopathology* 1993;83:456–463.
223. Ulmanen I, Seppala P, Pettersson RF. In vitro translation of Uukuniemi virus-specific RNAs: identification of a nonstructural protein and a precursor to the membrane glycoproteins. *J Virol* 1981;37:72–79.
224. van Poelwijk F, Boye K, Oosterling R, Peters D, Goldbach R. Detection of the L protein of tomato spotted wilt virus. *Virology* 1993;197: 468–470.
225. Vapalahti O, Kallio-Kokko H, Salonen EM, Brummer-Korvenkontio M, Vaheri A. Cloning and sequencing of Puumala virus Sotkamo strain S and M segments: evidence for strain variation in hantaviruses and expression of the nucleocapsid protein. *J Gen Virol* 1992;73:829–838
226. Verani P, Nicoletti L, Marchi A. Establishment and maintenance of

persistent infection by the Phlebovirus Toscana in Vero cells. *J Gen Virol* 1984;65:367–375.
227. Verkleij FN, Peters D. Characterization of a defective form of tomato spotted wilt virus. *J Gen Virol* 1983;64:677–686.
228. Vialat P, Bouloy M. Germiston virus transcriptase requires active 40S ribosomal subunits and utilizes capped cellular RNAs. *J Virol* 1992;66: 685–693.
229. Vidal S, Kolakofsky D. Modified model for the switch from Sendai virus transcription to replication. *J Virol* 1989;63:1951–1958.
230. von Bonsdorff C-H, Pettersson R. Surface structure of Uukuniemi virus. *J Virol* 1975;95:1–7.
231. von Bonsdorff C-H, Saikku P, Oker-Blom N. Electron microscopy study on the development of Uukuniemi virus. *Acta Virol* 1970;14:109–114.
232. Vorndam AV, Trent DW. Oligosaccharides of the California encephalitis viruses. *Virology* 1979;95:1–7.
233. Wang MW, Pennock DG, Spik KW, Schmaljohn CS. Epitope mapping studies with neutralizing and non-neutralizing monoclonal antibodies to the G1 and G2 envelope glycoproteins of Hantaan virus. *Virology* 1993;197:757–766.
234. Ward VK, Marriott AC, el Ghorr AA, Nuttall PA. Coding strategy of the S RNA segment of Dugbe virus (*Nairovirus;* Bunyaviridae). *Virology* 1990;175:518–524.
235. Wasmoen TL, Kakach LT, Collett MS. Rift Valley fever virus M segment: cellular localization of M segment-encoded proteins. *Virology* 1988;166:275–280.
236. Watret GE, Elliott RM. The proteins and RNAs specified by Clo Mor virus, a Scottish nairovirus. *J Gen Virol* 1985;66:2513–2516.
237. Watret GE, Pringle CR, Elliot RM. Synthesis of bunyavirus-specific proteins in a continuous cell line (ETC-2) derived from *Xenopus laevis*. *J Gen Virol* 1985;66:473–482.
238. Wertz GW. Replication of vesicular stomatitis virus defective interfering particle RNA in vitro: transition from synthesis of defective interfering leader RNA to synthesis of full-length defective interfering RNA. *J Virol* 1983;46:513–522.
239. White JD, Shirey FG, French GR, Huggins JW, Brand OM, Lee H-W. Hantaan virus, etiological agent of Korean haemorrhagic fever, has Bunyaviridae-like morphology. *Lancet* 1982;1:768–771.
240. Wijkamp I, van Lent J, Kormelink R, Goldbach R, Peters D. Multiplication of tomato spotted wilt virus in its insect vector, *Frankliniella occidentalis*. *J Gen Virol* 1993;74:341–349.
241. Wikstrom L, Persson R, Pettersson RF. Intracellular transport of the G1 and G2 membrane glycoproteins of Uukuniemi virus. In: Kolakosky D, Mahy B, eds. *Genetics and pathogenicity of negative strand viruses*. Amsterdam: Elsevier; 1989:33–41.
242. Xiao S-Y, LeDuc JW, Chu YK, Schmaljohn CS. Phylogenetic analyses of virus isolates in the genus *Hantavirus*, family Bunyaviridae. *Virology* 1994;198:205–217.
243. Xiao S-Y, Liang M, Schmaljohn CS. Molecular and antigenic characterization of HV114, a hantavirus isolated from a patient with hemorrhagic fever with renal syndrome in China. *J Gen Virol* 1993;74: 1657–1659.
244. Xiao S-Y, Spik KW, Li D, Schmaljohn CS. Nucleotide and deduced amino acid sequences of the M and S genome segments of two Puumala virus isolates from Russia. *Virus Res* 1993;30:97–103.

Fundamental Virology, Third Edition
edited by B.N. Fields, D.M. Knipe, P.M. Howley, et al.
Lippincott - Raven Publishers, Philadelphia © 1996

CHAPTER 23

Arenaviridae: The Viruses and Their Replication

Peter J. Southern

Members of the arenaviridae have been isolated from a diverse group of mammalian hosts in defined geographic locations (Table 1). These viruses merit significant attention both as manipulable experimental models to study acute and persistent infections and as clinically important human pathogens. The prototype virus, lymphocytic choriomeningitis virus (LCMV), has been studied for more than 60 years in infected mice (5,143,173,174) and significant new biological information continues to be amassed from the application of contemporary experimental procedures. Additional studies of both the biological properties and molecular aspects of virus infection for several other arenaviruses have successfully identified general features common to the arenaviridae and begun to define unique properties of the individual viruses (Table 1). Considerable research activity is now concentrated on defining how regulation of viral gene expression from the segmented ambisense RNA genome relates to the establishment and maintenance of persistent infection, defining more completely the nature of host immune responses to virus infection and identifying the molecular mechanisms of virus-induced disease. Several landmark publications have appeared over the years (15,29,41,75,79,94,95,118,119,133,154), culminating in a recent comprehensive volume (156), and the reader is referred to these texts for additional detail and more extensive literature citations than can be accommodated in this overview.

P.J. Southern: Department of Microbiology, University of Minnesota, Minneapolis, Minnesota 55455.

VIRION STRUCTURE

The virions are approximately spherical, enveloped particles that range in diameter from 50 to 300 nm. More uniform particles (90–110 nm in diameter) have been observed by cryoelectron microscopy, suggesting that some size variability may be attributable to procedures used for virion purification and/or microscopy sample preparation (32). The surface of the virion is smooth with T-shaped spikes, composed of viral glycoproteins, extending 7–10 nm from the envelope (32). The virions are relatively unstable and can be rapidly inactivated by either UV or gamma irradiation, by heating to 56°C or by exposure to pH outside the range 5.5–8.5. This pH sensitivity has recently been explained in terms of irreversible conformational changes in the viral glycoproteins that result in release of the hydrophilic head domain of the spike and disruption of binding to the cell surface (47). Infectivity can also be destroyed by disruption of the envelope with solvents or detergents. Virions are normally purified from infected tissue culture cell supernatants by polyethylene glycol precipitation and

TABLE 1. *The Arenaviridae*

Virus	Original reference(s)	Known Distribution	Known potential for human disease	Genomic sequence information reference(s)
Old World				
LCM	5	Virtually worldwide	Very mild to severe	153,159–161
Lassa	30,55	West Africa	Severe, often fatal	10,40
Mopeia	192	Southern Africa		189
Mobala	71	Central African Republic		
New World				
Junin	132	Argentina	Severe	68
Machupo	85	Bolivia	Severe	72
Guanarito	155	Venezuela	Severe	
Tacaribe[a]	48	(Trinidad)		59,82,83
Amapari	135	Brazil		
Flexal	136	Brazil		
Pichinde	172	Colombia		11
Latino	86	Bolivia, Brazil		
Parana	180	Paraguay		
Tamiami	35	Florida, USA		

[a] Tacaribe was isolated from bats and mosquitoes between March 1956 and December 1958 in Trinidad, but no subsequent isolations have been reported.

density gradient centrifugation in sucrose or amidotrizoate compounds (24). Buoyant densities are approximately 1.17–1.18 g cm in sucrose, 1.19–1.20 g/cm^3 in cesium chloride, and 1.14 g/cm^3 in amidotrizoate reagents.

The viral genomic RNAs are present within virions as helical nucleocapsid structures in circular configurations ranging between 400 and 1,300 nm in length (193). The interior of the virion often contains electron-dense granules that have been identified as host ribosomes (52). This characteristic accounts for the term *arena*, which was taken from *arenosus*, the Latin for "sandy." Further evidence in support of encapsidation of ribosomes is provided by the analysis of virion RNA preparations that frequently include abundant 28S and 18S species (36,49,133,167), shown by RNA fingerprinting to be identical to host ribosomal RNAs. A heterogeneous low-molecular-weight RNA component, 4–7S, is also present in total virion RNA preparations that consists of a mixture of host- and viral-derived species. Ribosomes do not appear to be necessary for infectivity because virions harvested at early times in acute infection display normal infectivity and significantly diminished ribosome content. The strongest evidence against the involvement of virion-associated ribosomes in the infectious process was obtained from the finding of normal Pichinde infectivity at the nonpermissive temperature after propagation in host cells with a *ts* lesion in a ribosomal protein (98).

GENOME ORGANIZATION

The arenavirus genome consists of two single-strand RNA molecules, designated L and S, that contain essentially nonoverlapping sequence information. There are minor differences in the lengths of the genomic RNA segments for the individual viruses (L approximately 7,200 bases and S approximately 3,400 bases), but the general organization of the viral genomes, based on current sequence information, is well preserved across the virus family. A short stretch of conserved sequence (17/19 nucleotides are identical) is present at the precise 3′-terminus of the L and S RNAs (7,8), and the inverted complement of this sequence is positioned at the 5′-termini of the genomic RNAs. Thus, through complementarity at the termini of the genomic RNAs, it is possible to form intra- and intermolecular complexes (159) that are consistent with the sizes of nucleocapsids observed in the electron microscope (193). The conserved sequence element is retained in all arenavirus genomes examined to date. A minor modification has been reported in the 5′ S sequence for Junin (68), but the arrangement of 3′/5′ complementarity is still preserved. It has been suggested that this conserved terminal sequence may represent part or all of a binding site for the viral RNA-dependent RNA polymerase. Models that have been proposed to account for the initiation of replication suggest that intramolecular duplexes may be formed at the termini of individual L or S RNA molecules and/or that intermolecular duplexes are generated by annealing between the termini of L and S RNA molecules (157).

The L and S genomic RNAs are not present in equimolar amounts, for under any condition of infection, S is always more abundant. The S segment encodes the major structural components of the virion—the internal nucleoprotein, NP and the two external glycoproteins, GP-1 and GP-2 (11,28,59,167). The L segment encodes the viral RNA-dependent RNA polymerase, L and a potential structural and/or regulatory protein, Z (82,83,159,160,163). Thus, the coding capacity is limited to four defined open reading frames that yield five mature proteins; the mature glycoproteins, GP-1 and GP-2, are derived by posttranslational cleavage of the primary glycoprotein translation product, GP-C (27,28). Reports have indicated some uncertainty over the exact nature of the mature glycoproteins for Tacaribe and Tamiami (17,66) because only one size class of glycoprotein has been detected by polyacrylamide gel electrophoresis. Recent work has established that the Tacaribe glycoprotein precursor is cleaved into two comigrating products that are homologous to GP-1 and GP-2 of LCMV (J.W. Burns and M.J. Buchmeier, manuscript submitted). The size of the Tacaribe glycoprotein coding region is consistent with this explanation. However, the dibasic amino acid cleavage site identified in other arenavirus glycoprotein precursors (28,68) is missing in Tacaribe, and it appears that cleavage of the precursor occurs at a single basic residue. Additional open reading frames have been identified by computer analysis of the genomic RNA sequences, most notably the X region at the 5′-end of the L RNA (159), but there is no current evidence to suggest either the presence of appropriate subgenomic mRNAs or the actual synthesis of these predicted proteins in LCMV-infected cells.

Sequence information derived from cDNA clones of the genomic S RNA segments provided the first clues that arenaviruses did not have a conventional negative strand coding arrangement (11,59,153,167). The NP coding region is transcribed into a genomic complementary mRNA, whereas the GP-C coding region is transcribed into a genomic sense mRNA (Fig. 1). The term *ambisense* has been coined to describe this situation in which, in one region, the S RNA is negative sense and in a second, nonoverlapping region the S RNA is pseudo-positive sense. The qualifier *pseudo* is added because, although the GP-C coding region is contained immediately within the genomic S RNA, there is no evidence that this RNA is translated to produce GP-C. Subsequent analysis of L-derived cDNA clones revealed that the L segment also has an ambisense coding arrangement with a small coding region (Z) located in the genomic sense at the 5′-end of the genomic L RNA (83,159). The ambisense coding arrangement provides a potential mechanism for temporal regulation of

Genomic Segments
L RNA 5′ Z L 3′
S RNA 5′ GP NP 3′

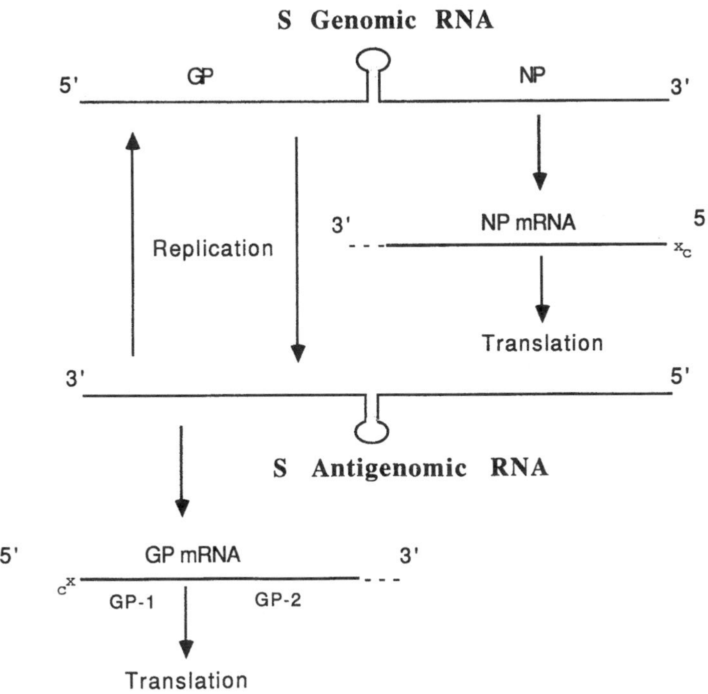

FIG.1. Schematic representations of the coding arrangements for the genomic L and S RNA segments. The stem-loop structures depict the intergenic noncoding regions. The lower part of the figure summarizes the transcription and replication events for the S genomic RNA, and this general scheme also applies to the genomic L RNA segment. As a consequence of the ambisense coding arrangement, GP mRNA (and Z mRNA) can be transcribed only after the initiation of genomic RNA replication. The S-derived subgenomic mRNAs contain a short stretch of nontemplated nucleotides and a cap (designated "x" and "c," respectively, in the diagram) at the 5′ ends and terminate at the 3′ ends at multiple sites within the intergenic region. Modified from Meyer and Southern (107).

gene expression because, while NP and L mRNAs can be transcribed from the incoming genomic RNA segments, GP-C and Z mRNAs are only transcribed from RNA antigenomic templates that also function as intermediates in the replication of genomic RNAs. The situation in reality is probably more complex because Z mRNA is a major component of the 4–7S RNA found within virions, suggesting that Z mRNA or protein may be essential at an early stage in the virus life cycle (159). In addition, L and S antigenomic RNAs can be incorporated into virions (108), so that all the known viral proteins could theoretically be synthesized *de novo* in infected cells prior to the initiation of RNA replication. Treatment of infected cells with protein synthesis inhibitors in the early stages of infection allows continued synthesis of NP mRNA but prevents RNA replication and transcription of GP-C mRNA (59). In populations of infected cells, NP mRNA and NP are found to accumulate earlier than GP-C mRNA and the mature glycoproteins. Furthermore, temporal analysis of the early stages of Tacaribe infection clearly shows the accumulation of newly synthesized NP prior to the onset of RNA replication (59), and although this sequential arrangement would be predicted for the other viruses, the finding for LCMV was that NP accumulation and replication became detectable simultaneously (60), perhaps reflecting different kinetics for LCMV and Tacaribe infections (59,99).

In the ambisense coding arrangement, the protein coding regions do not overlap and the intergenic, noncoding, regions in both the L and S segments have the potential to form relatively stable stem-loop structures. Translation termination codons are located on the proximal sides of the stems. For the S RNA, transcription termination occurs at multiple sites in the stem (58,107), suggesting that polymerase molecules may be displaced from the template RNAs by recognition of a structure rather than specific termination sites. There is a single stem-loop in the S RNA for Pichinde (11), LCMV (153,161), and Lassa (10,40), but the S sequence for Tacaribe (58,59), Junin (68), and Mopeia (189) can be folded to produce two distinct stem-loops that are positioned just beyond the translation termination codons for NP and GP-C. S1 nuclease mapping with mRNAs extracted from Tacaribe-infected cells has established that transcription termination occurs at the base of the hairpin on the distal sides. The intergenic region of the L segment also has the potential to form a stem-loop structure, but a stretch of polycytidylic acid, with a propensity to undergo deletion either during cDNA synthesis and/or cloning and propagation in bacteria, has created difficulties in sequencing this region for LCMV (159). cDNA clones spanning the LCMV L intergenic region were consistently about 30 nucleotides shorter than expected from direct RNA sequencing, but even the RNA sequence indicated heterogeneity within populations of RNA molecules (157). The L intergenic region for Tacaribe can be folded into a single hairpin structure and the 3′-termini of the L and Z mRNA map to the base of the distal sides of this hairpin. As a consequence of this arrangement, the 3′-termini of the mRNAs overlap by approximately 24–34 nucleotides (58). These mapping studies with Tacaribe assume additional importance by providing the first direct evidence for a subgenomic L mRNA (58).

The major structural proteins, nucleoprotein (NP) and the glycoprotein precursor (GP-C), are both encoded by the S RNA segment (11,28,144). NP is associated with genomic RNAs both within infected cells and virions in the form of ribonucleoprotein complexes (RNP). It is probable that the intracellular content of NP is influential in regulating the relative levels of transcription and replication within the infected cell, but precise molecular details for this regulation are not currently available. There are some indications for posttranslational modification to NP, but the unmodified form represents the single most abundant species. A phosphorylated derivative of NP has been detected at late times in acute infection and at greater abundance in extracts of persistently infected cells (23). This suggests that phosphorylation of NP may be involved with the attenuation of viral gene expression that marks the progression from acute to persistent infection (21,23). It is presently unclear whether this phosphorylation of NP can be linked to the activity of a kinase with specificity for serine and threonine residues that has been detected in purified LCM virions (80). A second modification to NP occurs through the generation of reproducible breakdown products that can be detected both in infected cells and in virions. A potential function for these NP fragments remains obscure, but it is conceivable, given the limited coding capacity of the arenavirus genome, that fragments of NP serve a role as minor structural proteins (43). A 28-kd degradation fragment of NP has been shown to accumulate in the nucleus of Pichinde-infected cells (194).

GP-C, the glycoprotein precursor (498 amino acids for LCMV, observed molecular weight 70,000–75,000), is cleaved by a trypsinlike protease to release the mature glycoproteins: GP-1 (40,000–46,000 molecular weight) and GP-2 (35,000 molecular weight). GP-1 is derived from the amino terminal of the precursor, and a putative cleavage signal is located between residues 262 and 263 (-Arg-Arg-) of the precursor (28). The cleavage is dependent upon prior glycosylation of the precursor and occurs in either the Golgi or post-Golgi compartments (190). A 58-residue peptide with characteristics resembling a signal sequence is cleaved from the amino terminus of GP-1 and, after further proteolytic processing, a fragment derived from this putative signal sequence (defined minimally by a synthetic peptide covering amino acid residues 34–40) actually constitutes one of the major GP epitopes recognized by LCMV-specific CTL on the H2b background (34,92). The predicted amino acid sequence of LCMV GP-C indicates five or six potential N-linked glycosylation sites within GP-1, depending upon the virus strain, and two sites within GP-2. Because GP-1 migrates as a heterogeneous species on SDS-polyacrylamide gels, it appears that not all potential glycosylation sites within GP-1 are utilized all the time (191). Neutralizing antibodies predominantly recognize conformational epitopes within GP-1, and this recognition is dependent upon glycosylation and retention of disulfide bonds (131,190). The external spike glycoprotein structure of LCMV is organized in the envelope such that GP-2 is an integral membrane protein and GP-1 is a peripheral membrane protein. GP-1 is located at the top of the spike, away from the envelope, and is held in place by ionic interactions with GP-2. Both GP-1 and GP-2 are present as homotetramers, and GP-2 is oriented such that the amino terminus provides a tether for GP-1 and the carboxy terminus is retained inside the envelope, possibly providing an anchor for ribonucleoprotein complexes (RNP) within the virion (31,32). Cross-linking studies have demonstrated complex formation between GP-2 and NP (31).

There is quite extensive conservation of size and amino acid sequence for the individual viral proteins when compared across the arenaviridae (38,152,165). NP and GP-2 are more highly conserved than GP-1, and cross-reacting antibodies have been described (25,181). Immunologic relatedness has also been established in rodent cross-protection studies, where, for example, infection with LCMV or Tacaribe can confer protection against ordinarily lethal challenges with Lassa or Junin. Predicted amino acid sequences have recently been used to establish phylogenetic trees for the relationships between the arenavirus structural proteins(38).

CELL SURFACE RECEPTORS AND VIRAL TROPISM

LCMV is capable of infecting and replicating efficiently in a wide variety of cell types from many different host species. This conclusion is emphatically supported by the recent description of a fatal hepatitis in captive marmosets and tamarins that has been linked to LCMV infection (109,141). In contrast, there are cell types—lymphocytes (20) and terminally differentiated neurons (45)—in which LCMV replication is severely restricted. Multiple cell types support LCMV replication in the newborn mouse (53,149, 166), and infectious virus can be recovered, at some level, from most major organs. These observations suggest that the cell surface component(s) required for virus adsorption and penetration must be widely distributed, highly conserved molecules. Some progress has been made with the demonstration of LCMV binding to a high-molecular-weight glycoprotein(s) in the VOPBA (virus overlay protein blot assay) procedure using protein extracts from susceptible rodent fibroblastic cell lines (18), but neither the normal function of this cellular protein nor information on the distribution across different species/cell types has yet been reported. Rodent lymphoid cell lines that are known to be refractory to LCMV infection did not bind LCMV and therefore appear to lack this particular cell surface component (18). The spike glycoprotein complex in virions is responsible for the initial interactions with the cell surface, and some antibodies directed against GP-1 can neutralize virus infectivity or even prevent virus binding to cells (32,131). Virus entry is inhibited by lysosomotropic agents, suggesting that virions are taken up by vesicles and subsequently released into the cytoplasm in a pH-dependent fusion step (19,47,70). This conclusion has been confirmed by immunoelectronmicroscopy (19). Other studies have recently shown that the spike glycoprotein complex undergoes irreversible changes at acid pH that result in the release of GP-1 from virions, loss of conformational epitopes on GP-1, and exposure of previously concealed GP-2 epitopes (47).

TRANSCRIPTION AND REPLICATION

Transcription and replication are confined to the cytoplasm of virus-infected cells (Fig. 2), although the information on whether a nuclear component may facilitate arenavirus replication is still contradictory. The nucleus may be required to provide capped cellular mRNAs for priming of arenavirus transcripts and/or the nuclear membrane may provide structural support for replication and transcription. There is now clear evidence for both Tacaribe and LCMV that the 5′-ends of the S-derived subgenomic mRNAs extend beyond the end of the genomic RNA template (63,107,140). The extensions are variable in length (1–7 nucleotides) and terminate with 5′-cap structures. There is no information yet to account for the derivation of these nontemplated nucleotides and the cap. "Cap-stealing," as displayed by influenza and bunyaviruses, can be neither confirmed nor excluded by the information currently available. The nucleoside analog ribavirin does inhibit LCMV replication (67), but because of remaining uncertainty over the mechanism of action, it is presently unclear whether capping and/or extension of newly synthesized RNA chains are disrupted by ribavirin. The virus life cycle can be completed in cells treated with actinomycin D (an inhibitor of RNA polymerase II), but yields of infectious progeny virions are reduced relative to untreated cells (100,142). Likewise, the timing of enucleation has a major effect on virus replication (12). It may be that disruption of host transcription with actinomycin D or enucleation leaves sufficient residue of preexisting cellular mRNA to prime some arenavirus transcription, but it is still far from clear how the caps and 5′-nontemplated extensions on arenavirus mRNAs are synthesized. The subgenomic mRNAs are not polyadenylated (163,167), but stabilization of the 3′-termini could be achieved via the formation of a terminal hairpin when transcription termination occurs on the distal side of the intergenic stem-loop structure (58,59,107).

There have been several attempts to identify the protein and enzymatic constituents of the viral RNA-dependent

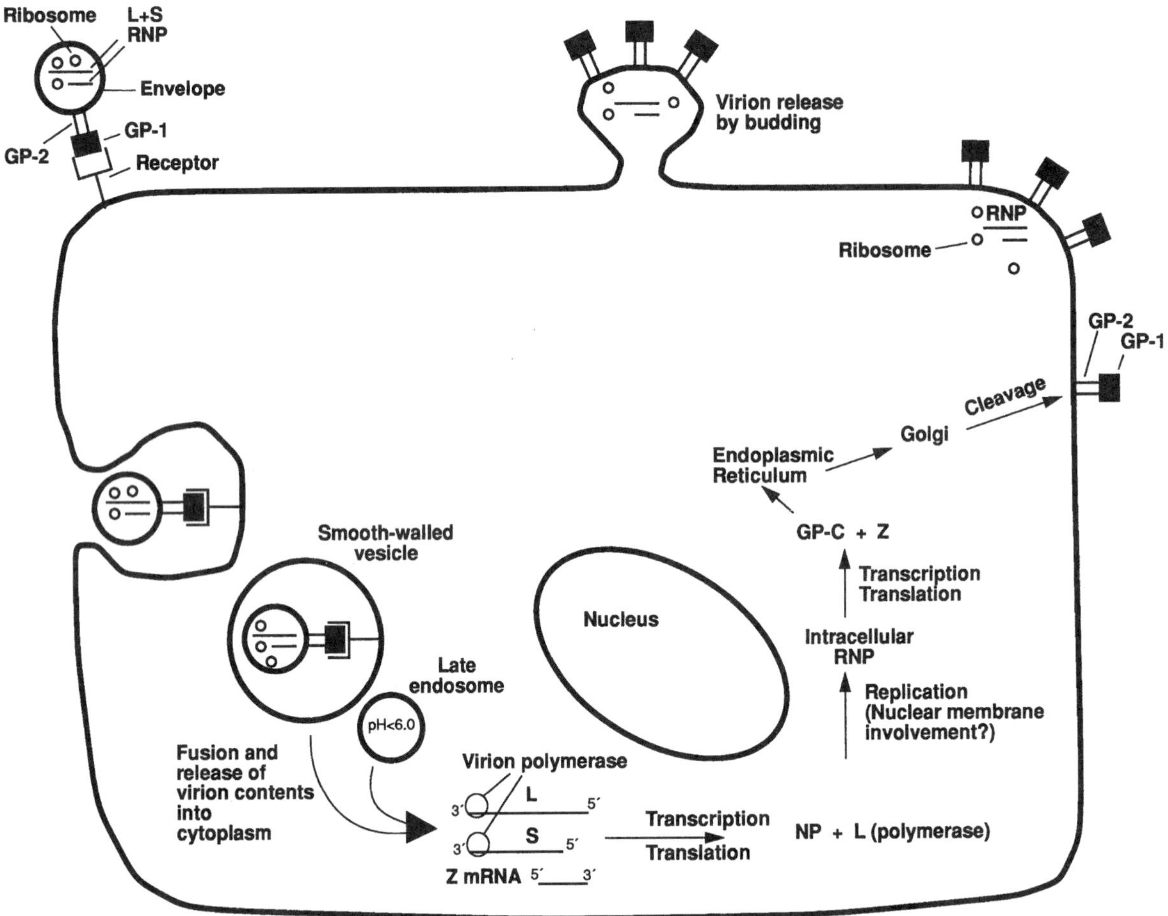

FIG. 2. Outline of the replication cycle for arenaviruses. The genome organization and profiles of viral transcription and replication are presented in Fig. 1.

RNA polymerase. By analogy with other negative-strand viruses, it has been assumed that the L protein (2,211 amino acids for LCMV, predicted molecular weight 254,529 and observed molecular weight about 200,000) constitutes part or all of the viral polymerase. Inspection of the predicted amino acid sequence for the L protein has identified conserved residues that are found in other RNA-dependent RNA polymerases (58,160). L protein is detectable in virions at low levels, as would be expected for a virion-associated polymerase, although reproducible detection of L was difficult until antipeptide antibodies became available (163). The L protein for Tacaribe contains 2,210 amino acids and shares 40% amino acid identity with LCMV (82).

An enzymatic activity was detected *in vitro* with extracts of purified Pichinde virions (97), but equivalent conditions failed to disclose any similar activity for LCM virions. One difficulty with the early work lay in distinguishing between the activities of the *bona fide* viral polymerase and poly-U and poly-A polymerases that were known to be associated with ribosomes. In a subsequent study, it was possible to demonstrate LCMV polymerase activity *in vitro* using extracts of acutely infected cells as a source of both polymerase and RNP templates (61). The LCMV system as described probably only supports the elongation of preinitiated chains, whereas, more recently, an *in vitro* system for Tacaribe has been developed in which there is clear evidence for chain initiation *in vitro* (64). Based on their accumulated experimental data, Garcin and Kolakofsky have proposed a novel scheme to explain the mechanism of initiation for Tacaribe RNA replication (64). A cornerstone of this replication initiation model is provided by a proposed explanation for a single, nontemplated nucleotide that has been detected at the precise 5′-terminus of arenavirus genomic S RNA segments (140). The model suggests that after initiation at the terminus of the template and synthesis of two nucleotides, the newly synthesized product slips backward on the template to create the appearance of a "nontemplated" nucleotide. Alternatively, a preexisting dinucleotide or a newly synthesized dinucleotide, derived from either a cellular polymerase or the

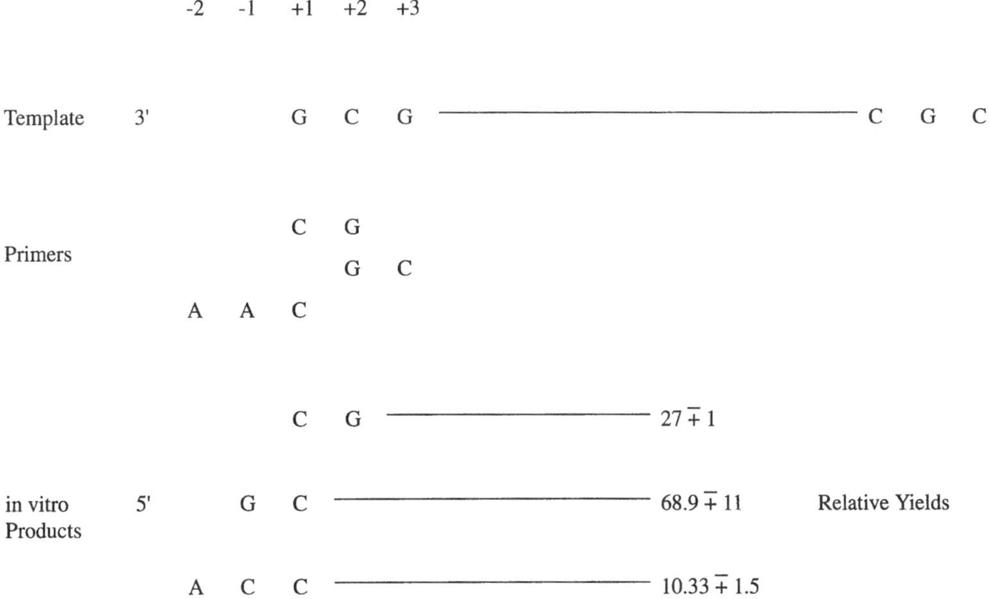

FIG. 3. A model to account for the initiation of arenavirus RNA replication. The template RNA contains a single nontemplated G (bold type) at the 5′-end, and this model provides a mechanism to generate a single nontemplated G at the 5′-end of a newly initiated chain. The annealing positions of exogenous oligonucleotide primers are indicated at the 3′-end of the RNA template. The 5′-termini of products synthesized in primer-dependent *in vitro* polymerase reactions are shown, together with relative yields of the predominant products. Note that the dinucleotide GC would anneal at positions +2/+3 and yet this primer directs the synthesis of a product where GC, at the 5′-end, now maps to positions −1/+1. These findings suggest that after annealing to the template, the GC primer slips backward two residues on the template prior to the initiation of RNA synthesis. In this manner, newly synthesized genomic and antigenomic RNAs would acquire a single nontemplated G at the 5′-ends. Summarized from Garcin and Kolakofsky (64).

viral polymerase, could anneal to the template as part of an initiation complex and then slip backward (Fig. 3). Analysis of products from the Tacaribe *in vitro* polymerase assay has shown that exogenous primers, complementary to varying positions at the terminus of the RNA template, can stimulate the synthesis of products with the predictable presence or absence of 5′-extensions (64, Fig. 3).

The intracellular concentration of L protein has been found to increase dramatically over the time course of acute infection (163). However, results derived from the LCMV *in vitro* polymerase assay suggest that an inverse correlation exists between the amount of L protein and the level of detectable polymerase activity (61). This finding could explain the observed down-regulation of viral transcription and replication that occurs at late times in acute infections (167) and suggests a potential regulatory role for the (level of) L protein. These results with LCMV are in general agreement with conclusions from a VSV complementation system in which transfected cell lines, expressing L (polymerase) gene cDNA clones, were used to propagate mutant viruses with defects in the L gene. The efficiency of VSV complementation was found to be inversely related to the level of L protein expressed, again suggesting that high levels of polymerase may actually be inhibitory to viral replication (106).

The presence of nontemplated nucleotides at the 5′-end of arenavirus subgenomic mRNAs is likely to be a general property for all the viruses. Similarly, there does not appear to be precise termination at the 3′-ends of the subgenomic mRNAs. Termination occurs instead at multiple sites within the intergenic hairpin region. There is no current explanation to account for the differential termination of transcription and read-through synthesis for replicating molecules in the intergenic region. The mRNAs appear to be synthesized at late times in acute infection (167), as assessed from populations of infected cells, so a simple switch from the hairpin being present to a situation of the hairpin being disrupted (e.g., by coating with NP) is not sufficient. It has been suggested that polymerase complexes might differ for replication versus transcription and that this could differentiate between termination and continued synthesis in the hairpin region. There is some evidence to suggest that Z may be involved with transcriptional regulation, but precise molecular details are still lacking (65,157). There are numerous similarities between arenavirus transcription and influenza virus transcription where the 3′-ends of the mRNAs are internal to the ends of the templates and the mRNAs are polyadenylated, whereas the replication intermediates represent exact full-length copies of the template RNAs. The presence or absence of the 5′-cap may determine for influenza virus whether the polymerase stops copying prior to the end of the template and then adds a poly A tail.

ASSEMBLY AND RELEASE

Only limited information is available on the processes of virus assembly and release (42). Virions have been observed budding from the plasma membrane of acutely infected cells in regions that display thickening of the membrane and aggregation of ribosome-like structures adjacent to budding virions (103), but essentially nothing is known about the assembly of viral nucleocapsids and how the interaction with the glycoproteins occurs. Cross-linking studies have indicated that the cytoplasmic tail of GP-2 may interact with NP (31) and that NP can also be cross-linked to Z (158). The incorporation of host ribosomes into virions, variability in the ratios of L:S genomic RNAs and evidence for diploid virions all suggest that encapsidation may not be precise. The arenaviruses are somewhat unusual because, unlike most other enveloped viruses, there is no evidence for a "matrixlike" protein that would be involved in organizing the virion components prior to assembly. The Z protein is present within virions and therefore could be involved in virion assembly (157,158). One study of persistently infected cells reported budding of virions into intracellular vesicles and the retention of infectivity within infected cells, but it is not clear how widely these conclusions may apply to other persistent infections (176,177).

INTERFERING PARTICLES AND DEFECTIVE INTERFERING RNAs

LCMV infection of a wide variety of cultured cell lines is not normally cytolytic, and there is a ready progression from acute infection, characterized by active virus replication and transcription and release of progeny virions, to persistent or chronic infection in which virus gene expression is significantly down-regulated (22,67,96,121,169). Cultures of persistently infected cells should perhaps be considered as in a dynamic state because most cells contain cytoplasmic viral antigen (NP) but lack expression of intact viral glycoproteins (GP-1 and GP-2) on the cell surface (123). Some cells may support a complete virus life cycle and therefore release infectious virions, and some cells may be free of viral nucleic acid and therefore fully susceptible to reinfection (76). Virion particles are continuously released into the supernatant medium, but these particles may lack infectivity and may actually be capable of interfering with the infectivity of standard infectious virus preparations (81,99,139). Although the biological properties of interfering particles have been well characterized (44,50,69,84,137,138,168,183,184), there has been no biochemical explanation of how the particles may actually be defective. The protein constituents of interfering particles are closely related to the proteins of standard virus preparations (182), although an alteration in the phosphorylation state of NP has been observed in persistently infected cells (23). Several independent studies have observed novel viral RNA species in persistently infected cells, but there is no clear correlation between the presence of deleted RNAs and the persistent infection. Unlike the classical systems for defective interfering RNAs in which a single deleted species vastly predominates over full-length genomic RNAs (13), the single most abundant RNA species in LCMV persistence is frequently an apparently full-length genomic RNA (56,57), although deleted RNAs have also been observed in numerous independent studies (51,56,57, 108,176,177). The relative molar ratio of genomic L:S segments is shifted further toward S in persistent infection (56), suggesting that particles may be formed that lack any L RNA. Persistence *in vivo* is associated with the accumulation of a collection of heterogeneously sized viral RNAs (57) rather than with the discrete and limited number of deleted RNAs that have been observed *in vitro,* but the situation *in vivo* may simply reflect multiple independent foci of infection each associated with the formation of one or a few deleted viral RNA species. Sequence analysis of individual RNA molecules, achieved via RT-PCR and cloning of the amplified products, has disclosed unexpected heterogeneity at the termini of genomic and antigenomic LCMV RNAs, and the possibility remains that such terminally altered molecules may represent a novel class of defective or defective interfering RNAs (108). Such deleted species would not be easily distinguishable from standard genomic RNAs by gel electrophoresis. There is, however, indirect evidence from the finding of the same types of terminal deletions in both genomic and antigenomic RNAs that molecules with terminal alterations may be capable of replicating and therefore potentially interfering with the replication of standard LCMV RNAs. The finding that the subgenomic mRNAs extend up to and actually beyond the ends of the corresponding genomic template RNAs (107) suggests that terminally deleted molecules may not function as efficient templates for transcription. Thus, the terminally deleted species would appear to satisfy the two basic properties of DI RNAs: proficiency for replication combined with an inability to support transcription and/or translation of mRNAs to yield standard viral proteins.

VIRUS EFFECTS ON THE HOST CELL

Virus propagation *in vitro* is normally performed in fibroblasts (BHK, L and Vero cells are common hosts). *In vivo*, macrophages are frequently infected, and at least for LCMV, this is directly related to the establishment of persistent infections (88,104). *In situ* hybridization studies with mice persistently infected with LCMV have documented widespread infection of multiple cell types (53), and in experimentally infected newborn mice, there is an extensive accumulation of viral nucleic acid sequence that is retained throughout the life span of the animal (166). In congenitally infected mice, the developing embryos are infected and accumulation of LCMV sequence has already occurred by birth (166).

Although infection with LCMV is not normally cytolytic, other combinations of arenaviruses and host cells (e.g., Tacaribe and Vero cells) can result in lytic infections (81). Host cell DNA, RNA (predominantly ribosomal RNA), and protein synthesis have been found to be disrupted in Tacaribe-infected cells (99). Plaque assays, even for LCMV, can be performed with Vero cell monolayers and an agarose overlay (24). Presumably, the combined stress of virus infection plus growth under agarose causes sufficient alteration in the growth properties of the infected cells to render the plaques visible, even prior to staining with neutral red. Access to a reliable plaque assay has been a critical component in understanding specific details of arenavirus biology as it permits both quantitation of infectious particles and, via plaque purification, the fractionation of virus preparations into individual and phenotypically distinct components.

There are now several examples in which infection by LCMV, either *in vitro* or *in vivo*, results in subtle alterations in host macromolecular synthesis (46,89,90,120,125,126, 128,171,175). In particular, specialized or "luxury" functions of differentiated cells may be compromised by virus infection. In the most extensively studied example, there is clear evidence that LCMV replication in the cytoplasm of infected cells can disrupt transcription of the growth hormone gene in the nucleus, but complete molecular explanations are still lacking (46,90,122,175).

VIRUS EFFECTS ON THE MAMMALIAN HOST

Detailed studies on the biology of LCMV infections of laboratory mice have provided considerable insight into the general areas of cellular immunology and molecular pathogenesis. Highly significant findings include the description of immune-mediated pathology associated with virus infection (26,75,154), the realization that the activity of cytotoxic T lymphocytes (CTL) is influenced by host MHC genes (MHC restriction, 195), demonstration of the importance of CTL for virus clearance *in vivo* (34,62,87,113), and the potential for a direct link between host MHC genes and disease susceptibility (196). More recent work has attempted to explain transient generalized alterations in the immune status of adult immunocompetent mice that arise during the clearance of peripheral LCMV infection (4,33, 114) and to provide mechanistic details for direct establishment of persistent infection in adult mice (112) and the slow appearance of neutralizing antibody in acutely infected mice (14). In general, because of the detailed information available relating to the infection of normal mice, LCMV has become a popular tool to examine the immune responses in genetically altered, "knockout," mice (62,87,113).

REASSORTANT VIRUSES

The description of several different strains of LCMV in terms of unique biological properties and nucleotide sequences (49,122,171) and the fact of the segmented arenavirus genome suggested that simple genetic mapping could establish correlations between biological properties and genomic segments (101,144–146,148,151,179). The basic approach has involved dual infection with two distinct parental LCMV strains and screening progeny virions, normally by nucleic acid hybridization procedures, for reassortant viruses possessing one genomic segment from each of the parental viruses. Thus, in a dual infection with LCMV Armstrong (Arm) and Pasteur (Past) strains, two distinct reassortants could be anticipated: L(Arm), S(Past) and L(Past),S(Arm). This approach assumes that all possible combinations will generate infectious virus and that there will not be strong selection pressure for compensatory mutations when the viral polymerase is forced to transcribe and replicate a heterologous genomic RNA segment. There is also the implicit assumption that the formation of hybrid strands by recombination, or more strictly, template switching during replication, does not occur. No evidence for this type of genetic exchange has been found, although the derivation and possible biological function of "longer than unit length" (LUS) RNAs have not yet been explained (74,162). Experiments with reassortants have produced clear-cut results indicating that these theoretical concerns were not warranted. Studies with reassortant viruses have also been extremely informative in analyzing the genotype and phenotype of variant viruses (see later).

MUTANT AND VARIANT ARENAVIRUSES

Several groups have attempted direct mutagenesis of arenavirus genomic RNA segments and the characterization of emergent mutant viruses (74,178). This task is complicated by the likelihood of multiple copies of genomic S RNA per virion (151) and a selective growth advantage for wild-type over mutant genomes. The generation and analysis of *ts* mutations of Junin and Pichinde stand out as significant, ongoing studies (37,74,162). Temperature-sensitive mutant viruses have been recovered during the normal course of persistent Junin infection in cell culture, and information is emerging on the relationship between different mutations in the L gene and polymerase function in Pichinde-infected tissue culture cells. In contrast, naturally arising variant viruses have been studied extensively in several different laboratories over a number of years. The earliest studies documented that mouse-passaged LCMV consisted of a mixture of at least two distinct virus types as assessed by plaque morphology with BHK-21/13S cells in agarose suspension (77,78). A clear plaque variant induced cytopathic effects in the indicator cells and was associated with a lethal choriomeningitis following intracerebral infection of adult mice. Conversely, a turbid plaque variant induced minimal cytopathology in the BHK-21/13S indicator cells and did not result in death following intracerebral challenge of adult mice. Clear plaque variants were most frequently recovered from the brain, whereas turbid plaque variants were isolated from the liver, in the original study. Spontaneously arising variants (Aggressive and Docile variants of LCMV UBC) have also been characterized that differ with respect to pathogenic potential by virtue of differences in the levels of interferon that are induced in the acutely infected mice (134). The aggressive variant induces high levels of interferon, and adult mice infected intracerebrally die with classic LCM-immune-mediated choriomeningitis. The docile variant replicates to substantially higher titers and induces only modest changes in interferon levels. Most of the adult mice infected intracerebrally survive the virus challenge (84,134). These studies have been extended to assess the relationship between LCMV and α/β interferon and γ interferon in acute infection and the probability with which different strains of LCMV may establish persistent infections in adult immunocompetent mice (111).

An extensive series of related experiments has been performed to characterize variant viruses recovered from spleen, macrophages, and blood of LCMV persistently infected mice (1–3,20,88,104). When a low-passage inoculum of a triple-plaque-purified stock (LCMV Armstrong) was used to infect newborn Balb/c pups and establish lifelong persistent infection, variant viruses could be readily isolated at 6 weeks postinfection (3). It is likely, given the highly reproducible character of the variants that have been recovered, that the *in vivo* growth conditions allowed preferential amplification of preexisting mutant genomic RNA segments. Infection of newborn Balb/c pups with the Armstrong strain of LCMV yields a predominance of clear plaques from the brain and more turbid plaques from the spleen. As originally noted by Hotchin (77,78), clear plaques are associated with lethal choriomeningitis, and turbid plaques are associated with long-term persistent infection when injected intracerebrally into adult mice. Variant viruses, as defined by antigenic changes, have also been documented in Junin-infected cricetids, and it has been suggested that selection pressure from antiviral antibody may be contributing to the emergence of variant viruses (43).

The LCMV variants recovered from persistently infected mice have been characterized as lymphotropic, macrophage-tropic, or amphotropic (indicating the property of equally efficient growth in either lymphocytes or macrophages), but the cell-type specificity is relative rather than absolute and all isolates replicate with equal efficiency in BHK cells (1,88). By far the most extensively studied variant is LCMV clone13 (1,3,20,104,161), and it has been shown to be representative of an entire class of variant viruses with highly conserved biological properties and conserved point mutations in the genomic RNA segments. The precise mechanism whereby clone13 infection of adult mice induces lifelong persistent infection has not yet been fully defined, but it is clear that persistence is linked to the absence of a vigorous CTL response (3,20). In immunocompetent mice, CTLs are responsible for clearance of infectious virus following peripheral infection (34) and trigger the immune-mediated choriomeningitis following intracerebral infection. Recent work has suggested that clone13 infection may induce CTLs transiently, but for unknown reasons, possibly linked in part to a high titer inoculum and to target cell specificity for the virus, this CTL response fails to achieve complete development and clone13 is able to persist (112). CTL recognition of clone13-infected target cells has been shown to occur normally, placing the defect in clone13 infection at the CTL induction/expansion stage. Sequence comparisons of genomic RNA segments have revealed only five differences between Arm and clone13 out of a total of 10,600 residues (1,104, 161). Only two of these nucleotide differences result in amino acid changes: residue 260 of GP-C is changed from Phe (Arm) to Leu (clone13) and residue 1079 in L is changed from Lys (Arm) to Glu (clone13). LCMV strains other than Arm also have Leu at residue 260 in GP-C yet these strains are fully proficient in inducing effective CTL responses. Furthermore, clone13 revertants, which have regained the ability to induce CTLs and therefore cause a lethal choriomeningitis after intracerebral injection into adult mice, retain Leu at residue 260 of GP-C. There is now strong evidence implicating a contribution for the change in L to the overall clone13 phenotype (104) with the possibility that changes to the viral polymerase could exert a major effect on the efficiency of viral replication in different cell types and thereby determine pathogenic potential directly. A similar relationship between the viral polymerase and pathogenesis was suggested by reassortant studies between LCMV Arm and WE strains where it was found that the lethal infection of adult guinea pigs with WE was duplicated by the L(WE),S(Arm) reassortant (147). From this same Arm/WE reassortant experiment, it was also noted that mixing of genomic segments can result in novel pathogenic viruses because, unlike both parental viruses, the L(WE),S(Arm) reassortant was found to cause an interferon-associated liver necrosis and death in neonatally infected Balb/c mice (147).

EXPRESSION OF ARENAVIRUS SEQUENCES FROM cDNA CLONES

Vaccinia Virus Vectors

In considering options for expression of arenavirus cDNAs in eukaryotic cells, the fact of the cytoplasmic arenavirus life cycle suggested that vaccinia, another virus with a cytoplasmic life cycle, might be an effective vector system (102,130). The vaccinia vectors would provide a mechanism to prevent exposure of arenavirus transcripts to potential destabilization in the nucleus. Accordingly, complete and partial cDNA clones for NP and GP-C coding regions have been incorporated into standard vaccinia virus vectors. Significant new information has been derived either from infecting target cells *in vitro* with recombinant vaccinia virus vectors to produce target cells for CTL assays (185–188) or from attempting vaccination *in vivo* with recombinant vaccinia virus vectors followed by virus challenge with the corresponding intact virus (6, 9,39,54,73,92,93,110). Whitton and colleagues expressed progressive carboxyl-terminal truncations of LCMV cDNA genes in recombinant vaccinia virus vectors to map linear epitopes and then defined short synthetic peptides that are recognized by CTL within NP and GP-C of LCMV (129, 185). Expression of a single CTL epitope from a 59-amino-acid LCMV GP-C mini-gene was subsequently shown to confer protection against virus challenge *in vivo* (92). Recombinant vaccinia viruses, expressing either Lassa virus NP or GP-C, have been shown to protect guinea pigs from an ordinarily lethal Lassa challenge (9,39,110). However, in rhesus monkeys, while the vaccinia/Lassa GP-C recombinant conferred protection against Lassa challenge (54), there was no protection, and possibly an exacerbation of disease, when animals treated with the vaccinia/Lassa NP recombinant were challenged (6). These seemingly paradoxical results of effective protection or disease enhancement, depending upon the viral epitopes presented by the vaccinia recombinant, have also been duplicated in the LCMV system (115).

Alternative Vector Systems

The SV40 early region promoter has been used to drive expression of a Junin cDNA clone for NP either in transient transfections or after stable integration into chromosomal locations (150) in cells cotransfected with a dominant selectable marker (164). The accumulation of NP in cytoplasmic (basophilic) granules was readily detectable by immunofluorescence in these cells, and this may suggest a link between the presence of NP and the cytopathic effects observed in some arenavirus infections (150). It has also been possible to express the cDNA clone for GP-C from LCMV (WE strain) in a baculovirus vector (105) but the potential of this system for production of large amounts of individual viral proteins has yet to be fully exploited. Similarly, an intriguing report of a coupled *in vitro* transcription/translation system for Tacaribe (16) has not been explored further.

Transgenic Mouse Studies

There are numerous examples of viral genes that have been incorporated into eukaryotic expression cassettes in the attempt to derive transgenic mice expressing viral proteins in a regulated and physiologically relevant manner. Many of these systems lack the opportunity of challenging the transgenic mice with the homologous virus to determine whether and how the outcome of infection may be altered in the presence of preexisting transgene expression. In two independent studies, cDNA genes for either LCMV Armstrong or WE strains have been linked to the rat insulin promoter to obtain expression of LCMV transgenes in pancreatic β cells (116,117,124). Infection of these transgenic mice by peripheral inoculation with LCMV resulted in induction of anti-LCMV CTL responses and β cells were destroyed by cellular infiltrates in the absence of virus replication in pancreatic β cells. These observations suggest that peripheral immunologic tolerance can be disrupted by LCMV challenge. Intracerebral infection of adult transgenic mice frequently results in lethal choriomeningitis (J. A. Zeller and PJS, unpublished), and it is unclear whether this T-cell-mediated immunopathology also represents disruption of the nonresponsive state previously established in the transgenic mice and/or whether epitopes encoded on the genomic L RNA segment may also be recognized by T cells. Although recognition of CTL epitopes encoded by the genomic L segment does not occur at high efficiency in the context of a normal virus infection, this response may be sufficient to provoke immunopathology and a lethal outcome when the transgenic mice are challenged by intracerebral infection. Studies with CTL clones have found that immunopathology and death can be triggered by as few as 100–1,000 cells when these cells are directly injected into the CNS (91). Similarly, studies involving transfer of CTLs into transgenic mice (127) may prove valuable in dissecting the CTL response to viral epitopes and may provide further insight into the nature of viral gene expression in the transgenic mice.

CONCLUSION

The convergence of molecular, immunologic and pathologic investigations during the past decade has facilitated enormous progress in understanding key features of the arenavirus life cycle. There is solid justification to anticipate that this same rate of progress can and will be maintained in the next decade with the goals of defining a complete set of molecular parameters for a virus/host interaction in a model system, developing effective vaccine strategies, and identifying interventive therapies that will reduce the severity of acute arenavirus infections in patients.

REFERENCES

1. Ahmed R, Hahn CS, Somasundaram T, Villarete L, Matloubian M, Strauss JH. Molecular basis of organ-specific selection of viral variants during chronic infection. *J Virol* 1991;65:4242–4247.
2. Ahmed R, Oldstone MBA. Organ specific selection of viral variants during chronic infection. *J Exp Med* 1988;167:1719–1724.
3. Ahmed R, Salmi A, Butler LD, Chiller J, Oldstone MBA. Selection of genetic variants of lymphocytic choriomeningitis virus in spleens of persistently infected mice: role in suppression of cytotoxic T lymphocyte response and viral persistence. *J Exp Med* 1984;60:521–540.

4. Althage A, Odermatt B, Moskophidis D, Kundig T, Hoffman-Rohrer, Hengartner H, Zinkernagel R. Immunosuppression by lymphocytic choriomeningitis virus infection: competent effector T and B cells but impaired antigen presentation. *Eur J Immunol* 1992;22:1803–1812.
5. Armstrong C, Lillie RD. Experimental lymphocytic choriomeningitis of monkeys and mice produced by a virus encountered in studies of the 1933 St. Louis encephalitis epidemic. *Public Health Rep* (Washington) 1934;49:1019–1027.
6. Auperin DD. Construction and evaluation of recombinant virus vaccines for Lassa fever. In: Salvato MS, ed. *The Arenaviridae*. New York: Plenum; 1993:259–280.
7. Auperin DD, Compans RW, Bishop DHL. Nucleotide sequence conservation at the 3′ termini of the virion RNA species of New World and Old World arenaviruses. *Virology* 1982;121:200–203.
8. Auperin D, Dimock K, Cash P, Rawls WE, Leung W-C, Bishop DHL. Analyses of the genome of prototype Pichinde arenavirus and a virulent derivative of Pichinde Munchique: evidence for sequence conservation at the 3′ termini of their viral RNA species. *Virology* 1982;116: 363–367.
9. Auperin DD, Esposito JJ, Lange JV, Bauer SP, Knight J, Sasso DR, McCormick JB. Construction of a recombinant vaccinia virus expressing the Lassa virus glycoprotein and protection of guinea pigs from a lethal Lassa virus infection. *Virus Res* 1988;9:233–248.
10. Auperin DD, McCormick JB. Nucleotide sequence of the Lassa virus (Josiah strain) S genome RNA and amino acid sequence comparison of the N and GPC proteins to other arenaviruses. *Virology* 1988;168: 421–425.
11. Auperin DD, Romanowski V, Galinski M, Bishop DHL. Sequencing studies of Pichinde arenavirus S RNA indicate a novel coding strategy, an ambisense viral S RNA. *J Virol* 1984;52:897–904.
12. Banerjee SN, Buchmeier M, Rawls WE. Requirement of a cell nucleus for the replication of an arenavirus. *Intervirology* 1976;6:190–196.
13. Barrett ADT, Dimmock NJ. Defective interfering viruses and infections of animals. *Curr Top Microbiol Immunol* 1986;128:55–84.
14. Battegay M, Moskophidis D, Waldner H, Brundler M-A, Fung-Leung W-P, Mak TW, Hengartner H, Zinkernagel R. Impairment and delay of neutralizing antiviral antibody responses by virus-specific T cells. *J Immunol* 1993;151:5408–5415.
15. Bishop DHL. Ambisense RNA genomes of arenaviruses and phleboviruses. *Adv Virus Res* 1986;31:1–51.
16. Boersma DP, Compans RW. Synthesis of Tacaribe virus polypeptides in an in vitro coupled transcription and translation system. *Virus Res* 1985;2:261–271.
17. Boersma DP, Saleh F, Nakamura K, Compans RW. Structure and glycosylation of Tacaribe viral glycoproteins. *Virology* 1982;123:452–456.
18. Borrow P, Oldstone MBA. Characterization of lymphocytic choriomeningitis virus-binding receptor protein(s): a candidate cellular receptor for the virus. *J Virol* 1992;66:7270–7281.
19. Borrow P, Oldstone MBA. Mechanism of lymphocytic choriomeningitis virus entry into cells. *Virology* 1994;198:1–9.
20. Borrow P, Tishon A, Oldstone MBA. Infection of lymphocytes by a virus that aborts cytotoxic T lymphocyte activity and establishes persistent infection. *J Exp Med* 1991;174:203–212.
21. Bruns M, Gessner A, Lother H, Lehmann-Grube F. Host cell-dependent homologous interference in lymphocytic choriomeningitis virus infection. *Virology* 1988;166:133–139.
22. Bruns M, Kratzberg T, Zeller W, Lehmann-Grube F. Mode of replication of lymphocytic choriomeningitis virus in persistently infected cultivated mouse L cells. *Virology* 1990;177:615–624.
23. Bruns M, Zeller W, Rohdewohld H, Lehmann-Grube F. Lymphocytic choriomeningitis virus. IX. Properties of the nucleocapsid. *Virology* 1986;151:77–85.
24. Buchmeier MJ, Elder JH, Oldstone MBA. Protein structure of lymphocytic choriomeningitis virus: identification of virus structure and cell associated polypeptides. *Virology* 1978;89:133–145.
25. Buchmeier MJ, Lewicki H, Tomori O, Johnson KM. Monoclonal antibodies to lymphocytic choriomeningitis virus react with pathogenic arenaviruses. *Nature* 1980;280:486–487.
26. Buchmeier MJ, Oldstone MBA. Virus-induced immune complex disease: identification of specific viral antigens and antibodies deposited in complexes during chronic lymphocytic choriomeningitis virus infection. *J Immunol* 1978;120:1297–1304.
27. Buchmeier MJ, Oldstone MBA. Protein structure of lymphocytic choriomeningitis virus: evidence for a cell-associated precursor of the virion glycopeptides. *Virology* 1979;99:111–120.
28. Buchmeier MJ, Southern PJ, Parekh BS, Wooddell MK, Oldstone MBA. Site-specific antibodies define a cleavage site conserved among arenavirus GP-C glycoproteins. *J Virol* 1987;61:982–985.
29. Buchmeier MJ, Welsh RM, Dutko FJ, Oldstone MBA. The virology and immunobiology of lymphocytic choriomeningitis virus infection. *Adv Immunol* 1980;30:275–331.
30. Buckley SM, Casals J. Lassa fever, a new virus disease of man from West Africa. III. Isolation and characterization of the virus. *Am J Trop Hyg Med* 1970;19:680–691.
31. Burns JW, Buchmeier MJ. Protein–protein interactions in lymphocytic choriomeningitis virus. *Virology* 1991;183:620–629.
32. Burns JW, Buchmeier MJ. Glycoproteins of arenaviruses. In: Salvato MS, ed. *The Arenaviridae*. New York: Plenum; 1993:17–35.
33. Butz EA, Hostager BS, Southern PJ. Macrophages in mice acutely infected with lymphocytic choriomeningitis virus are primed for nitric oxide synthesis. *Microb Pathog* 1994;16:283–295.
34. Byrne J, Oldstone MBA. Biology of cloned cytotoxic T lymphocytes specific for lymphocytic choriomeningitis virus: clearance of virus in vivo. *J Virol* 1984;51:682–686.
35. Calisher CH, Tzianabos T, Lord RD, Coleman PH. Tamiami virus, a new member of the Tacaribe group. *Am J Trop Med Hyg* 1970;19: 520–526.
36. Carter MF, Biswal N, Rawls WE. Characterization of the nucleic acid of Pichinde virus. *J Virol* 1973;11:61–68.
37. Ceriatti FS, Damonte EB, Mersich SE, Coto CE. Partial characterization of two temperature sensitive mutants of Junin virus. *Microbiologica* 1986;9:343–351.
38. Clegg JCS. Molecular phylogeny of the arenaviruses and guide to published sequence data. In: Salvato MS, ed. *The Arenaviridae*.New York: Plenum; 1993:175–187.
39. Clegg JCS, Lloyd G. Vaccinia recombinant expressing Lassa virus internal nucleocapsid protein protects guinea pigs against Lassa fever. *Lancet* 1987;8552:186–187.
40. Clegg JCS, Wilson SM, Oram JD. Nucleotide sequence of the S RNA of Lassa virus (Nigerian strain) and comparative analysis of arenavirus gene products. *Virus Res* 1990;18:151–164.
41. Cole GA, Nathanson N. Lymphocytic choriomeningitis virus pathogenesis. *Prog Med Virol* 1974;18:94–110.
42. Compans RW. Arenavirus ultrastructure and morphogenesis. In: Salvato MS, ed. *The Arenaviridae*. New York: Plenum; 1993:3–16.
43. Coto CE, Damonte EB, Alche LE, Scolaro L. Genetic variation in Junin virus. In: Salvato MS, ed. *The Arenaviridae*. New York: Plenum; 1993: 85–101.
44. D'Aiutolo AC, Coto CE. Vero cells persistently infected with Tacaribe virus: role of interfering particles in the establishment of the infection. *Virus Res* 1986;6:235–244.
45. De la Torre JC, Rall G, Oldstone C, Sanna PP, Borrow P, Oldstone MBA. Replication of lymphocytic choriomeningitis virus is restricted in terminally differentiated neurons. *J Virol* 1993;67:7350–7359.
46. De la Torre JC, Oldstone MBA. Selective disruption of growth hormone transcription machinery by viral infection. *Proc Natl Acad Sci USA* 1992;89:9939–9943.
47. Di Simone C, Zandonatti MA, Buchmeier MJ. Acidic pH triggers LCMV membrane fusion activity and conformational change in the glycoprotein spike. *Virology* 1994;198:455–465.
48. Downs WG, Anderson CR, Spence L, Aitken THG, Greenhall AH. Tacaribe virus, a new agent isolated from Artibeus bats and mosquitoes in Trinidad, West Indies. *Am J Trop Med Hyg* 1963;12:640–646.
49. Dutko FJ, Oldstone MBA. Genomic and biological variation among commonly used lymphocytic choriomeningitis virus strains. *J Gen Virol* 1983;64:1689–1698.
50. Dutko FJ, Pfau CJ. Arenavirus defective interfering particles mask the cell-killing potential of standard virus. *J Gen Virol* 1978;38:195–208.
51. Dutko FJ, Wright EA, Pfau CJ. The RNAs of defective interfering Pichinde virus. *J Gen Virol* 1976;31:417–427.
52. Farber FE, Rawls WE. Isolation of ribosome-like structures from Pichinde virus. *J Gen Virol* 1975;26:21–31.
53. Fazakerley JK, Southern P, Bloom F, Buchmeier MJ. High resolution in situ hybridization to determine the cellular distribution of lymphocytic choriomeningitis virus RNA in the tissues of persistently infected mice: relevance to arenavirus disease and mechanisms of viral persistence. *J Gen Virol* 1991;72:1611–1625.
54. Fisher-Hoch SP, McCormick JB, Auperin DD, Brown BG, Castor M, Perez G, Ruo S, Conanty A, Brammer L, Bauer S. Protection of rhesus monkeys from fatal Lassa fever by vaccination with a recombinant

vaccinia virus containing the Lassa virus glycoprotein gene. *Proc Natl Acad Sci USA* 1989;86:317–321.
55. Frame JE, Baldwin JM, Gocke DJ, Troup JM. Lassa fever, a new disease of man from West Africa. I. Clinical description and pathological findings. *Am J Trop Med Hyg* 1970;19:670–676.
56. Francis SJ, Southern PJ. Deleted viral RNAs and lymphocytic choriomeningitis virus persistence in vitro. *J Gen Virol* 1988;69:1893–1902.
57. Francis SJ, Southern PJ. Molecular analysis of viral RNAs in mice persistently infected with lymphocytic choriomeningitis virus. *J Virol* 1988;62:1251–1257.
58. Franze-Fernandez M-T, Iapalucci S, Lopez N, Rossi C. Subgenomic RNAs of Tacaribe virus. In: Salvato MS, ed. *The Arenaviridae*. New York: Plenum; 1993:113–132.
59. Franze-Fernandez M-T, Zetina C, Iapalucci S, Lucero MA, Bouissou C, Lopez R, Rey O, Deheli M, Cohen GN, Zakin MM. Molecular structure and early events in the replication of Tacaribe arenavirus S RNA. *Virus Res* 1987;7:309–324.
60. Fuller-Pace FV, Southern PJ. Temporal analysis of transcription and replication during acute infection with lymphocytic choriomeningitis virus. *Virology* 1988;162:260–263.
61. Fuller-Pace FV, Southern PJ. Detection of virus-specific RNA-dependent RNA polymerase activity in extracts from cells infected with lymphocytic choriomeningitis virus: in vitro synthesis of full-length viral RNA species. *J Virol* 1989;63:1938–1944.
62. Fung-Leung W-P, Kundig TM, Zinkernagel RM, Mak TW. Immune response against lymphocytic choriomeningitis virus infection in mice without CD8 expression. *J Exp Med* 1991;174:1425–1429.
63. Garcin D, Kolakofsky D. A novel mechanism for the initiation of Tacaribe arenavirus genome replication. *J Virol* 1990;64: 6196–6203.
64. Garcin D, Kolakofsky D. Tacaribe arenavirus RNA synthesis in vitro is primer dependent and suggests an unusual model for the initiation of genome replication. *J Virol* 1992;66:1370–1376.
65. Garcin D, Rochat S, Kolakofsky D. The Tacaribe arenavirus small zinc finger protein is required for both mRNA synthesis and genome replication. *J Virol* 1993;67:807–812.
66. Gard GP, Vezza AC, Bishop DHL, Compans RW. Structural proteins of Tacaribe and Tamiami virions. *Virology* 1977;83:84–95.
67. Gessner A, Lother H. Homologous interference of lymphocytic choriomeningitis virus involves a ribavirin-susceptible block in virus replication. *J Virol* 1989;63:1827–1832.
68. Ghiringhelli PD, Rivera-Pomar RV, Lozano ME, Grau O, Romanowski V. Molecular organization of Junin virus S RNA: complete nucleotide sequence, relationship with other members of the Arenaviridae and unusual secondary structures. *J Gen Virol* 1991;72:2129–2141.
69. Gimenez HB, Compans RW. Defective interfering Tacaribe virus and persistently infected cells. *Virology* 1980;107:229–239.
70. Glushakova SE, Lukashevich IS. Early events in arenavirus replication are sensitive to lysosomotropic compounds. *Arch Virol* 1989;104: 157–161.
71. Gonzalez JP, McCormick JB, Saluzzo JF, Herve JP, Georges AJ, Johnson KM. An arenavirus from wild-caught rodents (*Praomys* species) in the Central African Republic. *Intervirology* 1983;19:105–112.
72. Griffiths CM, Wilson SM, Clegg JCS. Sequence of the nucleocapsid gene of Machupo virus: close relationship with another South American pathogenic arenavirus, Junin. *Arch Virol* 1992;124:371–377.
73. Hany M, Oehen S, Schultz M, Hengartner H, Mackett M, Bishop DHL, Zinkernagel RM. Anti-viral protection and prevention of lymphocytic choriomeningitis or of the local footpad swelling reaction in mice by immunisation with vaccinia recombinant virus expressing LCMV-WE nucleoprotein or glycoprotein. *Eur J Immunol* 1989;19:417–424.
74. Harnish DG, Polyak SJ, Rawls WE. Arenavirus replication: molecular dissection of the role of protein and RNA. In: Salvato MS, ed. *The Arenaviridae*. New York: Plenum; 1993:157–174.
75. Hotchin J. The biology of lymphocytic choriomeningitis infection: virus-induced immune disease. Cold Spring Harbor Symp Quant Biol 1962;27:479–499.
76. Hotchin J. Transient virus infection; spontaneous recovery mechanisms of lymphocytic choriomeningitis virus-infected cells. *Nature* (New Biology) 1973;241:270–272.
77. Hotchin J, Kinch W, Benson L. Lytic and turbid plaque-type variants of lymphocytic choriomeningitis virus as a cause of neurological disease or persistent infection. *Infect Immun* 1971;4:281–286.
78. Hotchin J, Kinch W, Benson L, Sikora E. Role of substrains in persistent lymphocytic choriomeningitis virus infection. *Bull WHO* 1975;52: 457–463.
79. Howard CR. *Arenaviruses: perspectives in medical virology*. Vol 2. Amsterdam: Elsevier; 1986.
80. Howard CR, Buchmeier MJ. A protein kinase activity in lymphocytic choriomeningitis virus and identification of the phosphorylated product using a monoclonal antibody. *Virology* 1983;126:538–547.
81. Iapalucci S, Chernavsky A, Rossi C, Burgin MJ, Franze-Fernandez M-T. Tacaribe virus gene expression in cytopathic and non-cytopathic infections. *Virology* 1994;200:613–622.
82. Iapalucci S, Lopez R, Rey O, Lopez N, Franze-Fernandez M-T, Cohen GN, Lucero M, Ochoa A, Zakin MM. Tacaribe virus L gene encodes a protein of 2210 amino acid residues. *Virology* 1989;170:40–47.
83. Iapalucci S, Lopez N, Rey O, Zakin MM, Cohen GN, Franze-Fernandez M-T. The 5′ region of Tacaribe virus L RNA encodes a protein with a potential metal binding domain. *Virology* 1989;173:357–361.
84. Jacobson S, Pfau CJ. Virus pathogenesis and resistance to defective interfering particles. *Nature* 1980;283:311–313.
85. Johnson KM, Kuns ML, Mackenzie RB, Webb PA, Yunker CE. Isolation of Machupo virus from wild rodent *Calomys calosus*. *Am J Trop Med Hyg* 1966;15:103–106.
86. Johnson KM, Webb PA, Justines G. Biology of Tacaribe complex viruses. In: Lehmann-Grube F, ed. *Lymphocytic choriomeningitis virus and other arenaviruses*. Berlin: Springer-Verlag; 1973:241–258.
87. Kagi D, Ledermann B, Burki K, Seiler P, Odermatt B, Olsen KJ, Podack ER, Zinkernagel RM, Hengartner H. Cytotoxicity mediated by T cells and natural killer cells is greatly impaired in perforin-deficient mice. *Nature* 1994;369:31–37.
88. King CC, de Fries R, Kolhekar SR, Ahmed R. In vivo selection of lymphocyte-tropic and macrophage-tropic variants of lymphocytic choriomeningitis virus during persistent infection. *J Virol* 1990;64:5611–5616.
89. Klavinskis LS, Oldstone MBA. Lymphocytic choriomeningitis virus can persistently infect thyroid epithelial cells and perturb thyroid hormone production. *J Gen Virol* 1987;68:1867–1873.
90. Klavinskis LS, Oldstone MBA. Lymphocytic choriomeningitis virus selectively alters differentiated but not housekeeping functions: block in expression of growth hormone gene is at the level of transcriptional initiation. *Virology* 1989;168:232–235.
91. Klavinskis LS, Tishon A, Oldstone MBA. Efficiency and effectiveness of cloned virus specific cytotoxic T lymphocytes in vivo. *J Immunol* 1989;143:2013–2016.
92. Klavinskis LS, Whitton JL, Joly E, Oldstone MBA. Vaccination and protection from a lethal viral infection: identification, incorporation and use of a cytotoxic T lymphocyte glycoprotein epitope. *Virology* 1990;178:393–400.
93. Klavinskis LS, Whitton JL, Oldstone MBA. Molecularly engineered vaccine which expresses an immunodominant T-cell epitope induces cytotoxic T lymphocytes that confer protection from lethal virus infection. *J Virol* 1989;63:4311–4316.
94. Lehmann-Grube F. Portraits of viruses: arenaviruses. *Intervirology* 1984;22:121–145.
95. Lehmann-Grube F, Martinez-Peralta L, Bruns M, Lohler J. Persistent infection of mice with lymphocytic choriomeningitis virus. *Comp Virol* 1983;18:43–103.
96. Lehmann-Grube F, Slenczka W, Tees R. A persistent and inapparent infection of L cells with the virus of lymphocytic choriomeningitis. *J Gen Virol* 1969;5:63–81.
97. Leung W-C, Leung MFKL, Rawls WE. Distinctive RNA transcriptase, polyadenylic acid polymerase and polyuridylic acid polymerase activities associated with Pichinde virus. *J Virol* 1979;30:98–107.
98. Leung W-C, Rawls WE. Virion-associated ribosomes are not required for the replication of Pichinde virus. *Virology* 1977;81:174–176.
99. Lopez R, Franze-Fernandez M-T. Effect of Tacaribe virus infection on host cell protein and nucleic acid synthesis. *J Gen Virol* 1985;66:1753–1761.
100. Lopez R, Grau O, Franze-Fernandez M-T. Effect of actinomycin D on arenavirus growth and estimation of the generation time for a virus particle. *Virus Res* 1986;5:123–220.
101. Lukashevich IS. Generation of reassortants between African arenaviruses. *Virology* 1992;188:600–605.
102. Mackett M, Smith GL, Moss B. Vaccinia virus: a selectable eukaryotic cloning and expression vector. *Proc Natl Acad Sci USA* 1982;79: 7415–7419.
103. Mannweiler K, Lehmann-Grube F. Electron microscopy of LCM virus-infected L cells. In: Lehmann-Grube F, ed. *Lymphocytic choriomeningitis virus and other arenaviruses*. Berlin: Springer-Verlag; 1973:37–48.

104. Matloubian M, Kohlhekar SR, Somasundaram T, Ahmed R. Molecular determinants of macrophage tropism and virus persistence: importance of single amino acid changes in the polymerase and glycoprotein of lymphocytic choriomeningitis virus. *J Virol* 1993;67:7340–7349.
105. Matsuura Y, Possee RD, Bishop DHL. Expression of the S-coded genes of lymphocytic choriomeningitis arenavirus using a baculovirus vector. *J Gen Virol* 1986;67:1515–1529.
106. Meier E, Hermison GG, Schubert M. Homotypic and heterotypic exclusion of vesicular stomatitis virus replication by high levels of polymerase protein L. *J Virol* 1987;61:3133–3142.
107. Meyer BJ, Southern PJ. Concurrent sequence analysis of 5′ and 3′ RNA termini by intramolecular circularization reveals 5′ nontemplated bases and 3′ terminal heterogeneity for lymphocytic choriomeningitis virus mRNAs. *J Virol* 1993;67:2621–2627.
108. Meyer BJ, Southern PJ. Sequence heterogeneity in the termini of lymphocytic choriomeningitis virus genomic and antigenomic RNAs. *J Virol* 1994;68:7659–7664.
109. Montali RJ, Scanga CA, Pernikoff D, Wessner DR, Ward R Holmes KV. A common-source outbreak of calltrichid hepatitis in captive marmosets and tamarins. *J Infect Dis* 1993;167:946–950.
110. Morrison HG, Bauer S, Lange JV, Esposito JJ, McCormick JB, Auperin DD. Protection of guinea pigs from Lassa fever by vaccinia virus recombinants expressing the nucleoprotein or the envelope glycoproteins of Lassa virus. *Virology* 1989;171:179–188.
111. Moskophidis D, Battegay M, Bruendler M-A, Laine E, Gresser I, Zinkernagel RM. Resistance of lymphocytic choriomeningitis virus to alpha/beta interferon and to gamma interferon. *J Virol* 1994;68:1951–1955.
112. Moskophidis D, Lechner F, Pircher H, Zinkernagel RM. Viral persistence in acutely infected immunocompetent mice by exhaustion of antiviral cytotoxic effector T cells. *Nature* 1993;362:758–761.
113. Muller D, Koller BH, Whitton JL, LaPan KE, Brigman KK, Frelinger JA. LCMV-specific, class-II restricted cytotoxic T cells in β2-microglobulin-deficient mice. *Science* 1992;255:1576–1578.
114. Odermatt B, Eppler M, Leist TP, Hengartner H, Zinkernagel RM. Virus-triggered acquired immunodeficiency by cytotoxic T-cell-dependent destruction of antigen-presenting cells and lymph follicle structure. *Proc Natl Acad Sci USA* 1991;88:8252–8256.
115. Oehen S, Hengartner H, Zinkernagel RM. Vaccination for disease. *Science* 1991;251:195–198.
116. Ohashi PS, Oehen S, Aichele P, Pircher H, Odermatt B, Herrera P, Higuchi Y, Buerki K, Hengartner H, Zinkernagel RM. Induction of diabetes is influenced by the infectious virus and local expression of MHC class I and tumor necrosis factor-α. *J Immunol* 1993;150:5185–5194.
117. Ohashi PS, Oehen S, Buerki K, Pircher H, Ohashi CT, Odermatt B, Malissen B, Zinkernagel RM, Hengartner H. Ablation of "tolerance" and induction of diabetes by virus infection in viral antigen transgenic mice. *Cell* 1991;65:305–317.
118. Oldstone MBA, ed. Arenaviruses: genes, proteins and expression. *Curr Top Microbiol Immunol* 1987;133.
119. Oldstone MBA, ed. Arenaviruses: biology and immunotherapy. *Curr Top Microbiol Immunol* 1987;134.
120. Oldstone MBA. Prevention of type I diabetes in nonobese diabetic mice by virus infection. *Science* 1988;239:500–502.
121. Oldstone MBA. Molecular anatomy of virus persistence. *J Virol* 1991;65:6381–6386.
122. Oldstone MBA, Ahmed R, Buchmeier MJ, Blount P, Tishon A. Perturbation of differentiated functions during viral infection in vivo. I. Relationship of lymphocytic choriomeningitis virus and host strains to growth hormone deficiency. *Virology* 1985;142:158–174.
123. Oldstone MBA, Buchmeier MJ. Restricted expression of viral glycoprotein in cells of persistently infected mice. *Nature* 1982;300:360–362.
124. Oldstone MBA, Nerenberg M, Southern P, Price J, Lewicki H. Virus infection triggers insulin dependent diabetes mellitus in a transgenic model: role of anti-self (virus) immune response. *Cell* 1991;65:319–331.
125. Oldstone MBA, Rodriguez M, Daughaday WH, Lampert PW. Viral perturbation of endocrine function: disordered cell function leads to disturbed homeostasis and disease. *Nature* 1984;307:278–281.
126. Oldstone MBA, Sinha YN, Blount P, Tishon A, Rodriguez M, von Wedel R, Lampert PW. Virus-induced alterations in homeostasis: alterations in differentiated functions of infected cells in vivo. *Science* 1982;218:1125–1127.
127. Oldstone MBA, Southern PJ. Trafficking of activated cytotoxic T lymphocytes into the central nervous system: use of a transgenic model. *J Neuroimmunol* 1993;46:25–32.
128. Oldstone MBA, Southern P, Rodriguez M, Lampert P. Virus persists in β cells of the islets of Langerhans and is associated with chemical manifestations of diabetes. *Science* 1984;224:1440–1443.
129. Oldstone MBA, Whitton JL, Lewicki H, Tishon A. Fine dissection of a nine amino acid glycoprotein epitope, a major determinant recognized by lymphocytic choriomeningitis virus-specific class I-restricted H-2Db cytotoxic T lymphocytes. *J Exp Med* 1988;168:559–570.
130. Panicali D, Paoletti E. Construction of poxviruses as cloning vectors: insertion of the thymidine kinase gene from herpes simplex virus into the DNA of an infectious vaccinia virus. *Proc Natl Acad Sci USA* 1982;79:4927–4931.
131. Parekh BS, Buchmeier MJ. Proteins of lymphocytic choriomeningitis virus: antigenic topography of the viral glycoproteins. *Virology* 1986;153:168–178.
132. Parodi AJ, Greenway DJ, Rugiero HR, Rivero E, Frigerio MJ, Mettler NE, Garzon F, Boxaca M, de Guerrero LB, Nota NR. Sobre el etiologia del brote epidemico en Junin. *Diagn Med* 1958;30:2300–2302.
133. Pedersen IR. Structural components and replication of arenaviruses. *Adv Virus Res* 1979;24:277–330.
134. Pfau CJ, Gresser I, Hunt KD. Lethal role of interferon in lymphocytic choriomeningitis virus–induced encephalitis. *J Gen Virol* 1983;64:1827–1830.
135. Pinheiro FP, Shope RE, de Andrade AHP, Bensabath G, Cacios GV, Casals J. Amapari, a new virus of the Tacaribe group from rodents and mites of Amapa territory, Brazil. *Proc Soc Exp Biol Med* 1966;122:531–535.
136. Pinheiro FP, Woodall JP, Travassos Da Rosa APA, Travassos Da Rosa JF. Studies of arenaviruses in Brazil. *Medicina* (Buenos Aires) 1977;37:175–181.
137. Popescu M, Lehmann-Grube F. Diversity of lymphocytic choriomeningitis virus: variation due to replication in the mouse. *J Gen Virol* 1976;30:113–122.
138. Popescu M, Lehmann-Grube F. Defective interfering particles in mice infected with lymphocytic choriomeningitis virus. *Virology* 1977;77:78–83.
139. Popescu M, Schaefer H, Lehmann-Grube F. Homologous interference of lymphocytic choriomeningitis virus: detection and measurement of interference focus forming units. *J Virol* 1976;20:1–8.
140. Raju R, Raju L, Hacker D, Garcin D, Compans RW, Kolakofsky D. Nontemplated bases at the 5′ ends of Tacaribe virus mRNAs. *Virology* 1990;174:53–59.
141. Ramsay EC, Montali RJ, Worley M, Stephensen CB, Holmes KV. Callitrichid hepatitis: epizootiology of a fatal hepatitis in zoo tamarins and marmosets. *J Zool Wildl Med* 1989;20:178–183.
142. Rawls WE, Banerjee SN, McMillan CA, Buchmeier MJ. Inhibition of Pichinde virus replication by actinomycin D. *J Gen Virol* 1976;33:421–434.
143. Rivers TM, Scott TFM. Meningitis in man caused by a filterable virus. *Science* 1935;81:439–440.
144. Riviere Y, Ahmed R, Southern PJ, Buchmeier MJ, Dutko FJ, Oldstone MBA. The S RNA segment of lymphocytic choriomeningitis virus codes for the nucleoprotein and glycoproteins 1 and 2. *J Virol* 1985;53:966–968.
145. Riviere Y, Ahmed R, Southern PJ, Buchmeier MJ, Oldstone MBA. Genetic mapping of lymphocytic choriomeningitis virus pathogenicity: virulence in guinea pigs is associated with the L RNA segment. *J Virol* 1985;55:704–709.
146. Riviere Y, Ahmed R, Southern P, Oldstone MBA. Perturbation of differentiated functions during viral infection in vivo. Viral reassortants map growth hormone defect to the S RNA of the lymphocytic choriomeningitis virus genome. *Virology* 1985;142:175–182.
147. Riviere Y, Oldstone MBA. Genetic reassortants of lymphocytic choriomeningitis virus: unexpected disease and mechanism of pathogenesis. *J Virol* 1986;59:363–368.
148. Riviere Y, Southern PJ, Ahmed R, Oldstone MBA. Biology of cloned cytotoxic T lymphocytes specific for lymphocytic choriomeningitis virus. V. Recognition is restricted to gene products encoded by the viral S RNA segment. *J Immunol* 1986;136:304–307.
149. Rodriguez M, Buchmeier MJ, Oldstone MBA, Lampert PW. Ultrastructural localization of viral antigens in the CNS of mice persistently infected with lymphocytic choriomeningitis virus (LCMV). *Am J Pathol* 1983;110:95–100.

150. Romanowski V. Genetic organization of Junin virus, the etiological agent of Argentine hemorrhagic fever. In: Salvato MS, ed. *The Arenaviridae*. New York: Plenum; 1993:51–83.
151. Romanowski V, Bishop DHL. The formation of arenaviruses that are genetically diploid. *Virology* 1983;126:87–95.
152. Romanowski V, Bishop DHL. Conserved sequences and coding of two strains of lymphocytic choriomeningitis virus (WE and Arm) and Pichinde arenavirus. *Virus Res* 1985;2:35–51.
153. Romanowski V, Matsuura Y, Bishop DHL. Complete sequence of the S RNA of lymphocytic choriomeningitis virus (WE strain) compared to that of Pichinde arenavirus. *Virus Res* 1985;3:101–114.
154. Rowe WP. Studies on pathogenesis and immunity in lymphocytic choriomeningitis infection of the mouse. *Res Rep Naval Med Res Inst* 1954; 12:167–220.
155. Salas R, de Manzione N, Tesh RB, Rico-Hesse R, Shope RE, Betancourt A, Godoy O, Bruzual R, Pacheco ME, Ramos B, Taibo ME, Tamayo JG, Jaimes E, Vasquez C, Araoz F, Querales J. Venezuelan haemorrhagic fever. *Lancet* 1991;338:1033–1036.
156. Salvato MS, ed. *The Arenaviridae*. New York: Plenum; 1993.
157. Salvato MS. Molecular biology of the prototype arenavirus, lymphocytic choriomeningitis virus. In: Salvato MS, ed. *The Arenaviridae*. New York: Plenum; 1993:133–156.
158. Salvato MS, Schweighofer KJ, Burns J, Shimomaye EM. Biochemical and immunological evidence that the 11-kDa zinc-binding protein of lymphocytic choriomeningitis virus is a structural component of the virus. *Virus Res* 1992;22:185–198.
159. Salvato MS, Shimomaye EM. The completed sequence of lymphocytic choriomeningitis virus reveals a unique RNA structure and a gene for a zinc finger protein. *Virology* 1989;173:1–10.
160. Salvato M, Shimomaye EM, Oldstone MBA. The primary structure of the lymphocytic choriomeningitis virus L gene encodes a putative RNA polymerase. *Virology* 1989;169:377–384.
161. Salvato M, Shimomaye E, Southern P, Oldstone MBA. Virus–lymphocyte interactions: IV. Molecular characterization of LCMV Armstrong (CTL+) small genomic segment and that of its variant, clone 13 (CTL–). *Virology* 1988;164:517–522.
162. Shivaprakash M, Harnish D, Rawls W. Characterization of temperature-sensitive mutants of Pichinde virus. *J Virol* 1988;62:4037–4043.
163. Singh MK, Fuller-Pace FV, Buchmeier MJ, Southern PJ. Analysis of the genomic L RNA segment of lymphocytic choriomeningitis virus. *Virology* 1987;161:448–456.
164. Southern PJ, Berg P. Transformation of mammalian cells to antibiotic resistance with a bacterial gene under control of the SV40 early region promoter. *J Mol Appl Genet* 1982;1:327–341.
165. Southern PJ, Bishop DHL. Sequence comparisons among arenaviruses. *Curr Top Microbiol Immunol* 1987;133:19–39.
166. Southern P, Blount P, Oldstone MBA. Analysis of persistent virus infections by in situ hybridization to whole-mouse sections. *Nature* 1984; 312:555–558.
167. Southern PJ, Singh MK, Riviere Y, Jacoby DR, Buchmeier MJ, Oldstone MBA. Molecular characterization of the genomic S RNA segment from lymphocytic choriomeningitis virus. *Virology* 1987;157: 145–155.
168. Staneck LD, Pfau CJ. Interfering particles from a culture persistently infected with Parana virus. *J Gen Virol* 1974;22:437–440.
169. Stanwick TL, Kirk BE. Analysis of baby hamster kidney cells persistently infected with lymphocytic choriomeningitis virus. *J Gen Virol* 1976;32:361–367.
170. Swanepoel R, Leman PA, Shepherd AJ, Shepherd SP, Kiley MP, McCormick JB. Identification of Ippy as a Lassa-fever-related virus. *Lancet* 1985;1:639.
171. Tishon A, Oldstone MBA. Persistent virus infection associated with chemical manifestations of diabetes. II. Role of viral strain, environmental insult and host genetics. *Am J Pathol* 1987;126:61–72.
172. Trapido H, Sanmartin C. Pichinde virus: a new virus of the Tacaribe group from Columbia. *Am J Trop Med Hyg* 1971;20:631–641.
173. Traub E. A filterable virus recovered from white mice. *Science* 1935 81;298–299.
174. Traub E. Persistence of lymphocytic choriomeningitis virus in immune animals and its relation to immunity. *J Exp Med* 1936;63:533–546.
175. Valsamakis A, Riviere Y, Oldstone MBA. Perturbation of differentiated functions *in vivo* during persistent viral infection. III. Decreased growth hormone mRNA. *Virology* 1987;156:214–220.
176. van der Zeijst BAM, Bleumink N, Crawford LV, Swyryd EA, Stark GR. Viral proteins and RNAs in BHK cells persistently infected by lymphocytic choriomeningitis virus. *J Virol* 1983;48:262–270.
177. van der Zeijst BAM, Noyes BE, Mirault M-E, Parker B, Osterhaus ADME, Swyryd EA, Bleumink N, Horzinek MC, Stark GR. Persistent infection of some standard cell lines by lymphocytic choriomeningitis virus: transmission of infection by an intracellular agent. *J Virol* 1983: 48:249–261.
178. Vezza AC, Bishop DHL. Recombination between temperature sensitive mutants of the arenavirus Pichinde. *J Virol* 1977;24:712–715.
179. Vezza AC, Cash P, Jahrling P, Eddy G, Bishop DHL. Arenavirus recombination: the formation of recombinants between prototype Pichinde and Pichinde Munchique viruses and evidence that arenavirus S RNA codes for N polypeptide. *Virology* 1980;106:250–260.
180. Webb PA, Johnson KM, Hibbs JB, Kuns ML. Parana, a new Tacaribe complex virus from Paraguay. *Arch Gesamte Virusforsch* 1970;32: 379–388.
181. Weber EL, Buchmeier MJ. Fine mapping of a peptide sequence containing an antigenic site conserved among arenaviruses. *Virology* 1988; 164:30–38.
182. Welsh RM, Buchmeier MJ. Protein analysis of defective interfering lymphocytic choriomeningitis virus and persistently infected cells. *Virology* 1979;96:503–515.
183. Welsh RM, Lampert PW, Oldstone MBA. Prevention of virus-induced cerebellar disease by defective interfering lymphocytic choriomeningitis virus. *J Infect Dis* 1977;136:391–399.
184. Welsh RM, Oldstone MBA. Inhibition of immunologic injury of cultured cells infected with lymphocytic choriomeningitis virus: role of defective interfering virus in regulating viral antigen expression. *J Exp Med* 1977;145:1449–1468.
185. Whitton JL, Gebhard JR, Lewicki H, Tishon A, Oldstone MBA. Molecular definition of a major cytotoxic T-lymphocyte epitope in the glycoprotein of lymphocytic choriomeningitis virus. *J Virol* 1988;62: 687–695.
186. Whitton JL, Oldstone MBA. Class I MHC can present an endogenous peptide to cytotoxic T lymphocytes. *J Exp Med* 1989;170:1033–1038.
187. Whitton JL, Southern PJ, Oldstone MBA. Analyses of the cytotoxic T lymphocyte responses to glycoprotein and nucleoprotein components of lymphocytic choriomeningitis virus. *Virology* 1988;162:321–327.
188. Whitton JL, Tishon A, Lewicki H, Gebhard J, Cook T, Salvato M, Joly E, Oldstone MBA. Molecular analysis of a five-amino-acid cytotoxic T-lymphocyte (CTL) epitope: an immunodominant region which induces nonreciprocal CTL cross-reactivity. *J Virol* 1989;63:4303–4310.
189. Wilson SM, Clegg JCS. Sequence analysis of the S RNA of the African arenavirus Mopeia: an unusual secondary structure feature in the intergenic region. *Virology* 1991;180:543–552.
190. Wright KE, Salvato MS, Buchmeier MJ. Neutralizing epitopes of lymphocytic choriomeningitis virus are conformational and require both glycosylation and disulfide bonds for expression. *Virology* 1989;171: 417–426.
191. Wright KE, Spiro RC, Burns JW, Buchmeier MJ. Posttranslational processing of the glycoproteins of lymphocytic choriomeningitis virus. *Virology* 1990;177:175–183.
192. Wulff H, McIntosh BM, Hammer DB, Johnson KM. Isolation of an arenavirus closely related to Lassa virus from *Mastomys natalensis* in south east Africa. *Bull WHO* 1977;55:441–444.
193. Young PR, Howard CR. Fine structure of Pichinde virus nucleocapsids. *J Gen Virol* 1983;64:833–842.
194. Young PR, Chanas AC, Lee SR, Gould EA, Howard CR. Localization of an arenavirus protein in the nuclei of infected cells. *J Gen Virol* 1987; 68:2465–2470.
195. Zinkernagel RM, Doherty PC. Restriction of *in vitro* T cell–mediated cytotoxicity in lymphocytic choriomeningitis virus. *Nature* 1974;251: 547–548.
196. Zinkernagel RM, Pfau CJ, Hengartner H, Althage A. Susceptibility to murine lymphocytic choriomeningitis maps to class I MHC genes: a model for MHC/disease associations. *Nature* 1985;316:814–817.

Fundamental Virology, Third Edition
edited by B.N. Fields, D.M. Knipe, P.M. Howley, et al.
Lippincott - Raven Publishers, Philadelphia © 1996

CHAPTER 24

Reoviruses and Their Replication

Max L. Nibert, Leslie A. Schiff, and Bernard N. Fields

CLASSIFICATION

Orthoreovirus is one of nine recognized genera in the family *Reoviridae*. Genomes in this family are composed of 10, 11, or 12 segments of double-stranded (ds) RNA. Their infectious particles have characteristic sizes (70 to 85 nm), no lipid envelope, and proteins arranged in two or three concentric icosahedral capsids. A distinguishing feature of their replication cycles is synthesis of viral mRNAs by enzymes packaged within the icosahedral particles. Despite these similarities, viruses in the different genera exhibit significant genetic, biochemical, and biological differences.

Prototype viruses of the genus *Orthoreovirus*, the mammalian reoviruses, are the focus of this chapter. Reference is also made to the avian reoviruses. Viral isolates have been placed in this genus (rather than another genus in the family) because of similar traits, including a genome com-

M.L. Nibert: Institute for Molecular Virology, University of Wisconsin, Madison, Wisconsin 53706.
L.A. Schiff: Department of Microbiology, University of Minnesota Medical School, Minneapolis, Minnesota 55455.
B.N. Fields: Departments of Microbiology and Molecular Genetics and Medicine, The Shipley Institute of Medicine and Harvard Medical School, Boston, Massachusetts 02115.

posed of ten segments of dsRNA, size and morphology of particles by electron microscopy, number and arrangement of structural proteins within particles, and serologic reactivities. Sequencing studies should allow these viruses to be classified more precisely in the future.

Mammalian Reoviruses

Mammalian reoviruses are ubiquitous agents that infect a variety of mammalian species (322,386). There is little evidence for restrictions to the range of mammalian species in which a given isolate can replicate. The name *reovirus* includes an acronym for *r*espiratory and *e*nteric *o*rphan (328), reflecting that these viruses can be isolated from human respiratory and enteric tracts but are not associated with serious human disease. Mammalian reoviruses can be found in a variety of water sources (244,386), consistent with their predominantly enteric route of infection. Literature concerning the mammalian reoviruses has been notably reviewed in the past (131,186–189,284,307,337, 349,352,354,364,441). Chapter 53 in *Fields Virology*, 3rd ed. contains a current discussion of these viruses as agents of disease and tools for studies of viral pathogenesis. Some important properties of mammalian reoviruses are listed in Table 1.

TABLE 1. *Properties of the mammalian reoviruses*[a]

Genome
Double-stranded RNA
10 gene segments in three size classes (L, M, S)
Total size 23,500 base pairs
Gene segments encode either one or two proteins each.
Gene segments are transcribed into full-length mRNAs.
Plus strands of gene segments have 5′ caps.
Nontranslated regions at segment termini are short.
Gene segments can undergo reassortment between virus strains.
Particles
Spherical, with icosahedral (5:3:2) symmetry
Nonenveloped
Total diameter 85 nm (excluding σ1 fibers)
Two concentric protein capsids: outer capsid subunits in T=13 lattice, arrangement of inner capsid subunits unknown
8 structural proteins: 4 proteins in outer capsid (λ2, μ1 [mostly as cleavage fragments μ1N and μ1C], σ1, and σ3) and 4 proteins in inner capsid (λ1, λ3, μ2, and σ2)
Subviral particles (ISVPs and cores) can be generated from fully intact particles (virions) by controlled proteolysis.
Cell-attachment protein σ1 can extend from the virion and ISVP surface as a long fiber.
Protein λ2 forms pentamers that protrude from the core surface.
Replication
Fully cytoplasmic
Sialic acid can serve as cell surface receptor for recognition by cell-attachment protein σ1.
Proteolytic processing of outer capsid proteins σ3 and μ1/μ1C is essential to infection and can occur either extracellularly or in endo/lysosomes.
Uncoating of parent particles is incomplete: genomic dsRNA does not exit particles to enter the cytoplasm.
Transcription and capping of viral mRNAs occur within particles and are mediated by particle-associated enzymes.
Segment assortment and packaging involves mRNAs.
Minus-strand synthesis occurs within assembling particles.
Mature virions are inefficiently released from infected cells by lysis.

[a] See text for references.

Mammalian reoviruses share a common group antigen (328,386); however, three serotypes are identified by neutralization and hemagglutination-inhibition tests (321, 328,386). An isolate from a healthy child is the prototype for reovirus type 1 (strain Lang, abbreviated T1L); an isolate from a child with diarrhea is the prototype for reovirus type 2 (strain Jones, T2J); and isolates from a child with diarrhea (strain Dearing, T3D) and a child with an upper respiratory illness (strain Abney, T3A) are prototypes for reovirus type 3 (310,323,328). Although these prototype isolates have been studied most extensively, other isolates have been studied increasingly in recent years (173). Henceforth, we will commonly refer to the mammalian reoviruses simply as *reoviruses*.

Avian Reoviruses

Avian reoviruses are less well characterized than their mammalian counterparts [for review see (317,324)]. They can be isolated from poultry, in which they cause several diseases including arthritis. They share a common group antigen but are divided into five or more serotypes by neutralization tests (195,424). Some, but not all, isolates can be adapted for growth in mammalian cells (279,376,424) or will grow in mammalian cells under certain conditions (233,234). The particle forms and proteins of mammalian (see "Particles" and "Proteins" sections, below) and avian reoviruses have several features and properties in common (279,280,343,375,424). Unlike mammalian reoviruses, however, most avian reoviruses promote the formation of multinucleated syncytia within infected cultures and lack the capacity to agglutinate red blood cells (194,279).

Other Reoviruses

Nelson Bay virus, a reovirus isolated from an Australian flying fox, exhibits characteristics intermediate between mammalian and avian reoviruses. It contains the group-specific antigen of mammalian reoviruses and replicates in mammalian cells. However, like avian reoviruses, it induces formation of multinucleated syncytia within infected cultures (142,143,431).

dsRNA GENOME

Properties of the reovirus genome have been reviewed (188,352). An observation that reovirus-infected cells stain orthochromatically with acridine orange led to the discovery that the genomic RNA is double-stranded (162). Physicochemical properties of the virion RNA are all consistent with its double-stranded nature: (a) a high, sharp melting profile; (b) a buoyant density in Cs_2SO_4 of 1.61 g/cm^3, lower than expected for single-stranded (ss) RNA; (c) relative resistance to RNase I; and (d) susceptibility to RNase III, which is specific for dsRNA (39,161,176,351). In addition, the base composition of the RNA indicates a purine:pyrimidine ratio of 1, suggesting that it contains two complementary strands (39,161). X-ray diffraction studies indicate that the reovirus dsRNA forms a right-handed double helix with ~10 base pairs per turn, a 30-Å pitch (3 Å translation per base pair), and nucleotides oriented at a 75°-to-80° angle to the long axis (16,17).

Segmented Character

The genomic dsRNA of reoviruses occurs in discrete linear segments (Fig. 1). Evidence for dsRNA segments in three size classes was obtained by electron microscopy, sedimentation, and chromatography (36,115,159,231,400, 412). Polyacrylamide gel electrophoresis yielded the first evidence for ten discrete dsRNA segments (358,413). Other evidence that the segments are not artifactually generated fragments of a longer molecule includes observations (a) that neither denatured segments from the different size classes nor mRNAs transcribed from them cross-hybridize (39,412–414); (b) that there are 20 unblocked 3′ hydroxyl termini within each reovirus particle (255); and (c) that every segment possesses an identical blocking group (cap) at the 5′ terminus of its protein-coding (plus) strand (79,138,256; see "Replicaton Cycle" section, below). Whether the 5′ terminus of each genomic minus strand is blocked by an unknown structure or is an unblocked diphosphate remains unknown (26,79,256). Reports of long RNA molecules released from reovirus particles by mild disruption (115,159,163,193) may reflect that the discrete dsRNA segments are linked noncovalently within particles, but proof for this hypothesis is lacking. The dsRNA segments completely resist digestion by the single-strand-specific S1 nuclease, indicating that the plus and minus strands are colinear and complementary (274). The plus and minus strands of denatured segments can be separated by agarose gel electrophoresis in 7M urea (372).

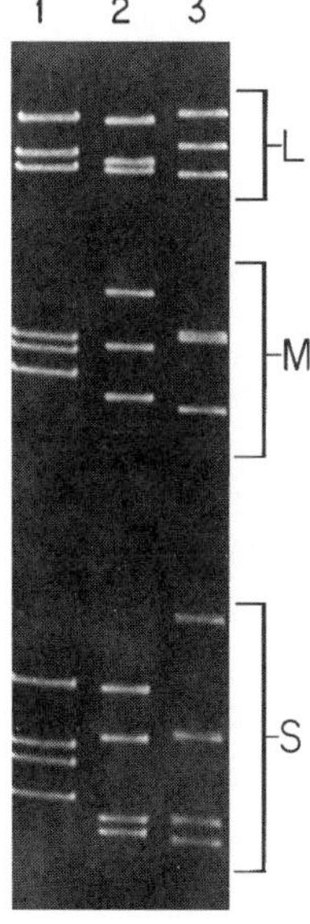

FIG. 1. Electropherotyping gel for reovirus gene segments. The ten dsRNA segments from strains T2J (*lane 1*), T1L (*lane 2*), and T3D (*lane 3*) were separated by electrophoresis on a 10% SDS-polyacrylamide gel. They cluster as three large (L), three medium (M), and four small (S) segments. Differences in mobility between analogous segments from these strains can be exploited to define the electropherotypes of reassortant viruses for use in genetic studies.

Three large (L1, L2, L3), three medium (M1, M2, M3), and four small (S1, S2, S3, S4) segments can be distinguished on polyacrylamide gels (Fig. 1), and purified virions contain equimolar quantities of these ten species (358). Homologous segments from different isolates, including prototypes of the three serotypes, often exhibit differences in their electrophoretic mobilities (303). The fact that reoviruses have segmented genomes has important biological and experimental consequences due to the exchange of gene segments between isolates within a genus and the generation of reassortant progeny viruses (see "Genetics" section and other sections in this chapter and Chapter 53 in *Fields Virology*, 3rd ed.). Use of electrophoretic migration as a genetic marker has facilitated the assignment of genes in reassortant viruses to one of the two parent viruses (272, 308,350).

TABLE 2. *References for reovirus gene sequences*

Gene	T1L	T2J	T3D	Other isolates	Mutants
L1	427	427	427	—	—
L2	—	—	343	—	—
L3	—	—	27	—	—
M1	442	—	424	—	441
M2	393,426	426	184,393, 426	—	420
M3	—	—	424	—	—
S1	70,112, 268,281	70,112, 281	30,70, 275	100	32
S2	151,103	103	71,428, 103	75	87,428
S3	150,425	425	313,425	—	425
S4	19,342	342	154	—	95

Sequence Analysis and Protein-Coding Strategies

Sequence information for the reovirus genes was first obtained by chemical and enzymatic methods that identified the 5′-terminal residues of mRNAs and the 3′-terminal residues of both strands of genomic RNA (reviewed in 188). All of the genes from prototype strain T3D (encompassing 23,549 base pairs in total) (429), have now been sequenced (Table 2) either from cDNAs (69,71,178) or directly from genomic RNA (32). Many of the genes from prototype strains T1L and T2J have also been sequenced, as have several genes from other isolates (Table 2).

Proteins encoded by the ten gene segments can be resolved by electrophoresis on different types of SDS-polyacrylamide gels (90,226,303,373,448). Names for the 11 distinct primary translation products are indicated in Fig. 2. Eight of these proteins are "structural," i.e., present within mature reovirus particles, whereas three (μNS, σNS, σ1s) are "nonstructural" and mediate functions during the intracellular steps in reovirus replication. Protein-coding assignments were made by *in vitro* translation of mRNAs (56,248) and by genetic analysis with reassortant viruses (273). Each gene segment is transcribed into a full-length mRNA, which includes one long, nearly full-length open reading frame (ORF). In each case, this long ORF encodes a predicted protein that corresponds in size to the larger protein assigned to that segment (Fig. 2). Translation usually initiates from an AUG at the beginning of the long ORF, but a recent study suggests that translation of μ2 from the S1 mRNA within cells initiates from the second in-frame AUG (319). The M1 gene segment is unique in encoding a second protein product, σ1s, in a separate but overlapping reading frame from the larger product, σ1 (120,183,336). The initiation codon for σ1s translation is the first out-of-frame AUG within the ORF for σ1 (30, 70,181,275). The M3 gene segment may also encode a second protein, μNSc, but in this case translation is proposed to start from the second AUG within the same ORF as the larger product, μNS (426). Additional small ORFs are

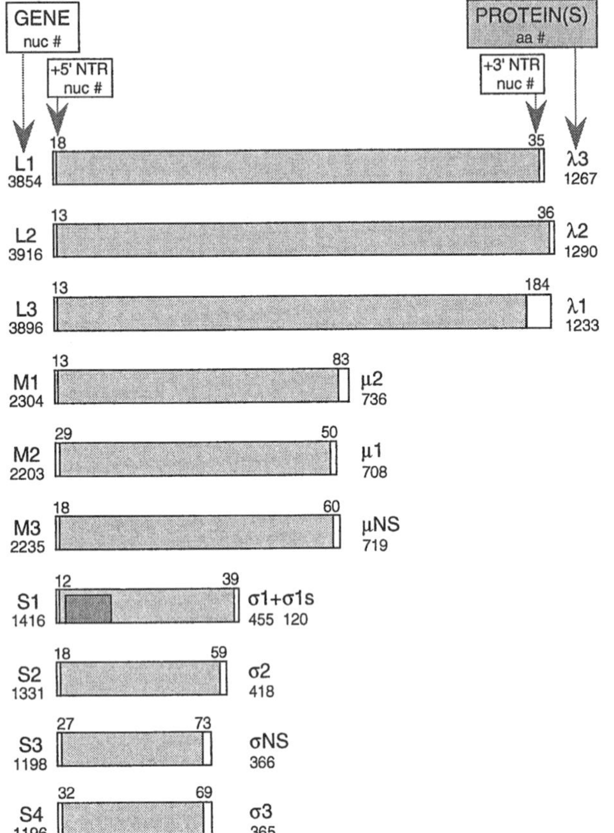

FIG. 2. Coding strategies of the ten reovirus gene segments. Segments are drawn approximately to scale and are oriented so that left-to-right corresponds to 5′-to-3′ for the protein-coding plus strands. The segment names and lengths in nucleotides (nuc no.) are listed *(left)*. The portion of each segment encompassed by protein-coding sequences is shaded in gray. Short nontranslated regions (NTR) at the ends of each segment remain *unshaded*, and the lengths of these regions in nucleotides (nuc no.) are designated above. Names of encoded proteins and their lengths in amino acids (aa no.) are listed *(right)*. The S1 segment encodes two proteins as shown: the σ1s protein *(darker gray)* initiates at a second initiator codon in a different reading frame from σ1 *(lighter gray)*. A second protein (μNSc) has also been proposed to arise from the M3 segment but is not shown, pending confirmation by other investigators (see text). Uncertainties in lengths of the λ1 and μ2 proteins and the 5′ or 3′ nontranslated region of the corresponding L3 and M1 gene segments are also not indicated, but are discussed in the text.

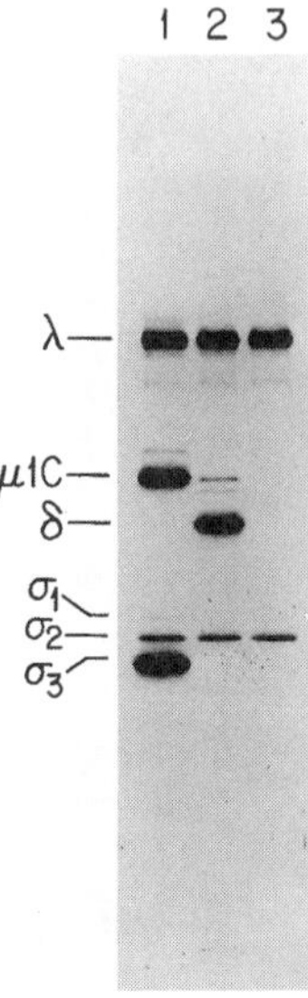

FIG. 3. Gel of proteins in different reovirus particle forms. Proteins from virions *(lane 1)*, ISVPs *(lane 2)*, and cores *(lane 3)* of strain T3D were separated on a 10% SDS-polyacrylamide gel. The δ cleavage product of μ1C is evident in ISVPs, but the ϕ fragment was not resolved on this gel. The minor bands immediately above μ1C in virions and δ in ISVPs represent the μ1 and μ1δ proteins, respectively. The σ1 protein is poorly visualized in this reproduction.

found in many reovirus genes, but proof for their translation in infected cells is lacking. There is no evidence that proteins are translated from the minus strands of reovirus gene segments.

Nontranslated Sequences

Sequencing studies (Table 2) revealed that nontranslated regions at the 5′ and 3′ termini of reovirus genes represent a small part of the genome (Fig. 2). The lengths of the 5′ nontranslated regions range between 12 (T3D S1 gene) and 32 bases (S4 genes); however, a recent report suggests that M1 contains a 5′ nontranslated region of 160 bases due to initiation of μ2 translation from the second AUG in the long ORF of that gene (319). The 3′ nontranslated regions range between 32 (L1 genes) and 181 bases (L3 gene); however, a recent report suggests that addition of a single base near the 3′ end of L3 would extend the λ1 ORF and give a 3′ nontranslated region of only 35 nucleotides in that gene, indicating the largest 3′ nontranslated region to be the 80 bases in M1 (103). Between homologous gene segments from different strains, lengths of the nontranslated sequences may not be entirely conserved (281). Though a minimal portion of the genome, the nontranslated regions are likely to include sequences for mRNA packaging, recognition by the viral RNA polymerase for initiating plus- and minus-strand synthesis, and determining translational efficiency (see "Replication Cycle" section, below). The plus strands of all gene segments analyzed to date contain the 5′ tetranucleotide 5′-GCUA and the 3′ pentanucleotide UCAUC-3′ (13).

PARTICLES

Virions and Subviral Particles

The mature reovirus particle containing a full complement of the 8 structural proteins (Fig. 3) is called the *virion*. Virions can be converted into two distinct types of subviral particles: *infectious* (355) or *intermediate* (46), *subviral particles* (ISVPs) and *cores* (357,373). These subviral particles differ in protein composition (Fig. 3) and conformation as well as in other physicochemical and biological properties (Table 3; see below). The generation of ISVPs and cores is routinely accomplished by treating virions with purified proteases *in vitro* (see below). ISVPs and cores, or particles related to these forms, also occur naturally in the course of infection and play specific roles in both the early and later stages of reovirus replication (see "Replication Cycle" section, below). Some authors have referred simply to *subviral particles* (SVPs), in which case they are usually describing ISVP-like particles (18,74,363,366).

TABLE 3. *Characteristics of reovirus particles*[a]

Feature or property	Virions	ISVPs	Cores
Buoyant density in CsCl (g/cm^3)	1.36	1.38	1.43
Sedimentation value (S)	730	630	470
Molecular weight (MDa)			
From diffusion coefficient	130	ND	52
Estimated from components	127	103	51
Particles per ml at 1 OD_{260} ($\times 10^{12}$)	2.1	2.7	4.2
Diameter (nm)			
Negative-stain electron microscopy	73	64	51
Cryoelectron microscopy	85	80	60
Low-angle x-ray diffraction	83	ND	60
Outer capsid proteins:			
σ3	Present	Absent (degraded)	Absent (degraded)
σ1	Present	Present as conformer (extended)	Absent
μ1/μ1C	Present	Present as μ1δ/δ + ϕ cleavage fragments	Absent (degraded)
λ2	Present	Present	Present as conformer (open fivefold channel)
Inner capsid proteins	Present	Present	Present
10 dsRNA gene segments	Present	Present	Present
Oligonucleotides	Present	Variable	Absent
Infectious (after attachment to cell surface)	Yes	Yes	No
Interactions with membrane bilayers	No	Yes	No
Transcription-related activities:			
Initiation and capping	Yes	Yes	Yes
Elongation	No	No	Yes

[a] See text for references and further explanations.
ND, not determined.

The "potentially infectious" and "infectious" reovirus particles described by other authors (2,380) probably correspond to virions and ISVPs, respectively.

Appearance in Electron Micrographs

A representative study in which all three reovirus particle types were compared by negative-stain electron microscopy indicated diameters of 73 nm for virions, 64 nm for ISVPs, and 51 nm for cores (46). Diameters obtained from cryoelectron micrographs, in which shrinkage artifacts common to negative staining (170) are minimal, were 85, 80, and 60 nm for virions, ISVPs, and cores (excluding the core spikes), respectively (110,253). The latter values concur with those obtained by low-angle x-ray diffraction (169).

Virions and subviral particles have distinct morphologies, as seen by negative-stain electron microscopy (10,46, 94,135,198,230,245,252,295,373,380,401,423). Virions appear roughly spheroidal, with smooth perimeters, except that when viewed in certain orientations, their perimeters are marked by flattened areas. Few details of subunit arrangement are apparent in negatively stained virions. ISVPs appear even more spheroidal than virions; furthermore, regions of their perimeters often have a saw-toothed appearance, suggesting structural subunits. Long fibers are sometimes seen extending from the surfaces of virions and ISVPs, as discussed below. Negatively stained cores are more distinctive in that they have prominent spikes protruding from their surfaces at axes of fivefold symmetry.

Concentric Protein Capsids

Observations from negative-stain electron micrographs first defined the organization of reovirus particles as in Fig. 4: two concentric protein shells (inner capsid and outer capsid) surrounding the dsRNA genome (10,230,245,295, 401). The description of two separate protein layers within reovirus particles is firmly supported by data from dynamic light scattering (170), low-angle x-ray diffraction (169), and cryoelectron microscopy and image reconstruction (110,253). These studies also show the outer capsid to be substantially thicker than the inner capsid and to have a higher water content.

The reovirus inner and outer capsids exhibit icosahedral (5:3:2) symmetry (245,401). Earlier studies using negative-stain electron microscopy to describe symmetry relationships within the capsids (10,198,230,245,252,295,401) were largely superceded by ones using cryoelectron microscopy and image analysis (Fig. 5) (110,253). The latter indicate that the outer capsid has subunits arranged in a skewed, T=13 lattice. The lattice is formed primarily by 600 mole-

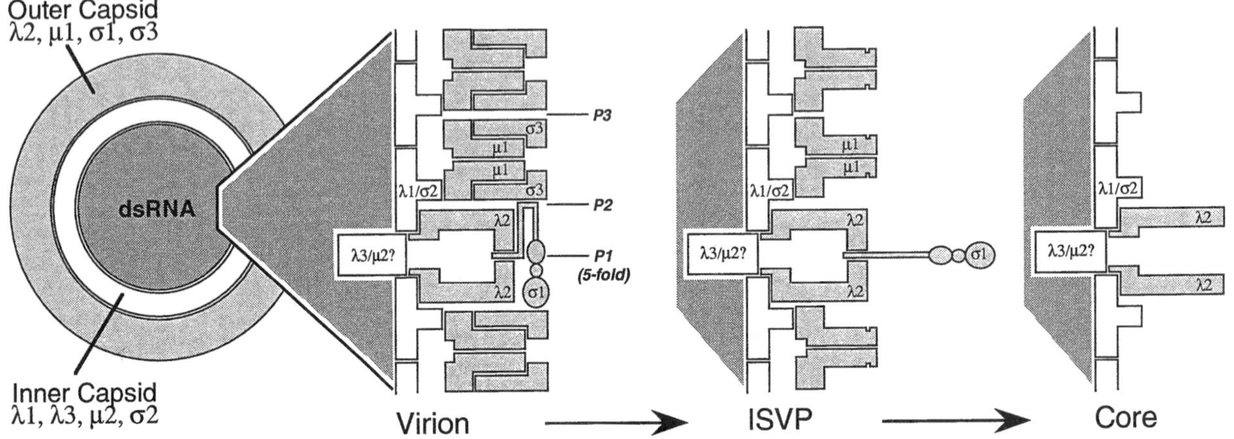

FIG. 4. Reovirus structural elements. The two concentric capsids and centrally located dsRNA genome in reovirus virions are drawn in cross-section *(left)*. Proteins in the outer capsid *(shading)* and inner capsid *(no shading)* are designated. A wedge of the virion capsids, encompassing an axis of fivefold symmetry *(labeled)*, is drawn in greater detail and shows approximate locations and interactions between reovirus proteins *(labeled)* as currently understood. Positions of open or potential channels through the outer capsid at axes of fivefold (P1) or sixfold (P2, P3) symmetry are indicated (nomenclature from Metcalf et al. [253]; also see Fig. 5). Relative positions of the λ1 and σ2 proteins in the inner capsid are unknown. The position and interaction indicated for the λ3 and μ2 proteins are hypothetical. The position of the nonextended σ1 protein in virions is also hypothetical. Comparable wedges from a reovirus ISVP and core are shown *(right)* and reveal the loss and change in conformation of outer capsid proteins in these particle forms. The ISVP is distinguished by absence of the σ3 protein, by an endoproteolytically cleaved μ1 protein *(notching)*, and by a conformationally altered *(extended)* σ1 protein. The core is distinguished by additional absence of the μ1 and σ1 proteins and by a conformationally altered λ2 protein. The conformational change in λ2 and loss of σ1 has opened a channel through the λ2 "spike" that protrudes from the core surface. From Nibert and Fields (283), with permission.

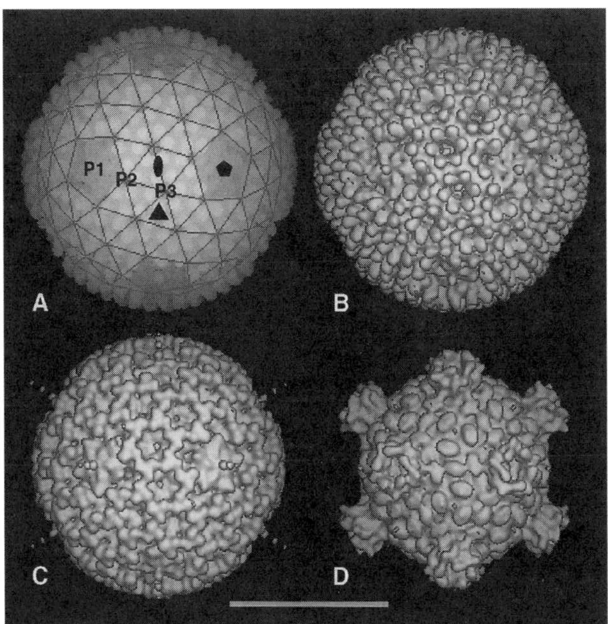

FIG. 5. Surface projections of reovirus particles obtained by image reconstruction from cryoelectron micrographs. Features contributed by individual reovirus proteins are described in the text. **A:** Model for T=13 *(laevo)* surface lattice of virions. Representative two-, three-, and fivefold axes are indicated (*oval, triangle,* and *pentagon*, respectively). Representative P1, P2, and P3 lattice positions (nomenclature from Metcalf et al. [253]), at which solvent channels are centered in different viral particles, are also indicated. **B:** Virion. **C:** ISVP. **D:** Core. Bar = 50 nm. From Dryden et al. (110), with permission.

cules of protein μ1, which fail to account for 180 of the expected 780 (13×60) subunits in a classical T=13 structure. The remaining positions are occupied instead by 60 molecules of protein λ2, which substitute for μ1 around each fivefold axis such that one molecule of λ2 occupies three subunit positions. This lattice exhibits a very similar structure in virions and ISVPs. Previous studies with negative-stain electron microscopy of both partially degraded virions (252) and intact ISVPs (198) suggested that the T=13 lattice of the outer capsid exhibits a left-handed (*laevo*) skew. The nature of the lattice in which subunits are arranged within the inner capsid remains difficult to define.

Components of Viral Particles

Outer Capsid

As noted above, proteins λ2 (60 copies) and μ1 (600 copies) together form the T=13 lattice of the outer capsid (Fig. 4). The remaining two outer capsid proteins, σ1 [36 to 48 copies (see below)] and σ3 (600 copies), occupy more external positions than λ2 and μ1 and may be said to decorate the primary lattice. Components of the outer capsid are considered in detail below.

μ1 and σ3

The μ1 protein (see "Proteins" section for information about its different fragments) is organized in trimeric complexes within virions and ISVPs (110,198,252). The trimers are most apparent at higher axial radii, where adjacent trimers make contact across local twofold axes (110) (Fig. 5). At lower radii, other intertrimer contacts are seen, giving μ1 a structure that interlocks across the outer capsid. Because the μ1 network is substituted around each fivefold axis by λ2, not all μ1 subunits are in identical positions. The copies of μ1 adjacent to λ2 contact that protein directly. Some μ1 subunits contact nodules projecting from the surface of the inner capsid (see below), and these contacts may contribute to binding the inner and outer capsids together.

The mass of μ1 is arranged around large solvent channels that pass axially through the outer capsid (110,253). Two different types of channels are present due to substitution in the μ1 lattice by λ2. The two channel types are designated *P2* and *P3* and range in diameter from 2.5 to 11 nm at different axial radii in ISVPs. Each P2 channel is partially blocked by a λ2 subunit. The diameters of both P2 and P3 channels are decreased at higher radii in virions by the presence of σ3 (see below). The purpose of these channels is unknown.

An observation that the μ1 and σ3 proteins bind to each other in solution (212,396) is reflected in intimate contacts between these proteins within the virion outer capsid (110). Difference maps between images of virions and ISVPs permit densities to be attributed specifically to σ3 and suggest that each σ3 subunit is elongated in an axial direction. The σ3 subunits project 3.4 nm above the outer extent of μ1 and appear as "fingers" on the virion surface (110,253) (Fig. 5). Each copy of σ3 contacts as many as three adjacent μ1 subunits and may in this way serve to stabilize the outer capsid structure. Six and 4 subunits of σ3 are placed around the perimeters of the P3 and P2 channels, respectively, and contact each other side-to-side to form complete and partial hexamers. Because of the presence of λ2, not all σ3 subunits are in identical positions. One of the σ3 copies adjacent to each λ2 subunit contacts that protein directly.

The arrangement of μ1 in virions and ISVPs exhibits only minor differences, despite the fact that μ1 has undergone an additional endoproteolytic cleavage in ISVPs (see "Proteins" section). Thus, the major difference between these two particle types is the loss of 600 σ3 subunits without a major perturbation of underlying structures in the outer and inner capsids.

λ2 Core Spikes

Spikes protruding from the surface of cores were attributed to λ2 by showing loss of both with exposure to high pH (301,423). Spikes eluted in this fashion remain pentameric (301); however, λ2 expressed in isolation fails to assemble into pentamers, suggesting that other core proteins are required for λ2 oligomerization (235). Some antibodies directed against λ2 can react with virions, providing evidence that λ2 is exposed on the virion surface (171,213,404). Findings from cryoelectron microscopy and image analysis have confirmed the pentameric nature of λ2 in viral particles and have shown that a significant portion of λ2 is exposed on the surface of virions, ISVPs, and cores (110,253) (Fig. 5).

Previously, λ2 was considered to be a constituent of the inner capsid because of its presence in cores and its enzymatic activities in forming 5′ caps on reovirus mRNAs (see "Replication Cycle" section). Observations that particles resembling inner capsids (a) remain after λ2 is removed from cores by high pH (423) and (b) are formed de novo when λ1 and σ2 are co-expressed without λ2 (435) suggest that λ2 is not integral to the structure of the inner capsid. In addition, λ2 is found in empty outer capsids formed by one reovirus temperature-sensitive (ts) mutant (243). Findings from cryoelectron microscopy and image analysis demonstrate that most of λ2 resides within the outer capsid and suggest that λ2 is best considered an outer capsid protein that remains bound to the inner capsid in cores (110). Work with expressed proteins has shown that λ2 can associate independently with inner capsid proteins λ1 and λ3 (387).

Each pentameric spike of λ2 in cores encloses an axial channel 8 nm in diameter (110). This channel may provide a pathway for exit of mRNAs as they are nascently transcribed by cores (see "Replication Cycle" section). The arrangement of λ2 in cores is significantly different from that in virions and ISVPs: while the fivefold channel is open in cores, it is closed off in virions and SVPs by a morphologic domain of λ2 that reaches toward the fivefold axis to contact μ1 (110). The conformational change in λ2 inferred to have taken place in cores may be relevant to the mechanism of transcriptase activation (see Replication Cycle" section).

σ1 Fibers

In negatively stained images of some virions and ISVPs, σ1 is seen as a long fiber that projects out 40 nm from the particle surface (135). The spacing of these fibers suggests location at fivefold axes, as does work with antibodies indicating that σ1 and λ2 are in close proximity within virions (213). The fibers have a head-and-tail morphology, with the head distal to the particle-bound end. The σ1 protein isolated from particles by heating (134,135) or after vaccinia expression (23) has a very similar morphology. The appearance of σ1 as an extended fiber was hypothesized after sequence analysis revealed an amino-terminal region of heptad repeats of hydrophobic residues, suggesting the capacity of σ1 to form an α-helical coiled coil (30,135).

Image processing of negatively stained, isolated σ1 fibers revealed regions of distinct morphology within the tail, which are proposed to be formed by different supersecondary structures (see "Proteins" section) (134,281). Current evidence indicates that regions important for cell attachment, a primary function of σ1, are located within both head-and-tail regions (see "Proteins" section). Other data suggest that sequences near the amino terminus of σ1 form a virus-attachment region that anchors the protein to viral particles (see "Proteins" section). This region is likely to be located near the tip of the σ1 tail (23,134,135) and to attach to virions and ISVPs at a binding site created by 5 subunits of λ2 (see above).

The order of the σ1 oligomer is a matter of debate. Some data suggest dimers (30,438) or pentamers (253) of σ1, but more convincing observations favor trimers or tetramers. Evidence for σ1 trimers comes from chromatography and sedimentation of oligomers (390) and from gel analysis of hetero-oligomers containing both full-length and truncated versions of σ1 (220,390). Evidence for σ1 tetramers comes from cross-linking (31) and from image-processed electron micrographs (134). Either trimers or tetramers of σ1 would present a symmetry mismatch for association with the pentameric complexes of λ2.

Accumulating evidence suggests that σ1 can assume two distinct conformations in virions and ISVPs (Fig. 4). Original observations from negative-stain electron microscopy were that σ1 can extend as a long fiber from the surface of virions. The frequency of this finding varied between strains and between preparations of virions (135), suggesting that σ1 conformation is sensitive to genetic and environmental parameters. Data also suggest that extended σ1 fibers are seen more frequently in ISVPs than virions of many strains (135,284). Reconstructed images from cryoelectron micrographs support this notion (110) (Fig. 5). It remains unknown whether the nonextended form of σ1 is completely folded at the particle surface or partially buried within the outer capsid and what molecular rearrangements are involved in its transition to the extended form. The importance of different σ1 conformations may relate to its role as cell attachment protein (see "Proteins" and "Replication Cycle" sections. The σ1 protein is absent from cores, possibly due to elution after a conformational change in λ2 (see above).

Inner Capsid

Proteins λ1 (120 copies) and σ2 (copy number estimated at 120 to 180) are major proteins in the core and form the primary icosahedral lattice of the inner capsid (423) (Fig. 4). Particles resembling inner capsids are obtained when λ1 and σ2 are co-expressed (435). Image reconstructions from cryoelectron micrographs show Y-shaped structures around each fivefold axis in viral particles, which may represent portions of λ1 (110). Such localization is consistent with an observation that λ1 becomes subject to greater surface iodination after λ2 is removed from cores by high pH (423). Previous studies involving surface iodination disagree as to whether σ2 is exposed on the core surface (240,423), but it has been generally accepted that σ2 is located more internally than λ1. Nodules protruding from the inner capsid may nevertheless be formed from both λ1 and σ2 (110) (Fig. 5). Regions of both λ1 and σ2 are expected to contact the genome since both can bind dsRNA *in vitro* (339,340). There is no direct evidence for oligomers of σ2 within the inner capsid, but an amphipathic α-helix near the carboxy terminus of σ2 (103) may play a role in σ2:σ2 interactions (87). The minor proteins λ3 and μ2 (copy number estimated at 12 each) have yet to be localized within the inner capsid but may exist in complex near each fivefold axis (see "Proteins" section). Vaccinia-expressed λ3 protein can bind to proteins λ1 and λ2 independently (387).

The inner capsid shows little change between particle forms (110), suggesting that the conformational flexibility common among outer capsid proteins is not shared by inner capsid proteins. The inner capsid is not penetrated by solvent channels, except perhaps at the fivefold axes, and thus may serve as a barrier between the genomic dsRNA and external factors.

Centrally Condensed dsRNA

Studies with low-angle x-ray diffraction and cryoelectron microscopy both indicate that the genomic dsRNA is found in reovirus particles within a central sphere, 49 nm in diameter, and is well ordered such that adjacent helices are locally parallel and separated by distances of 25 to 27 Å (110,169). Regions of local disorder are expected to permit the RNA to pack into the available space. Given the measured interhelix distances and the size of the ten segments, calculations indicate that the central cavity should provide just enough room to accommodate the total genome. Thus, large quantities of protein are probably not bound to the packaged dsRNA, and random assortment of more than ten segments per particle is an unlikely explanation for the high particle-to-pfu ratio of reoviruses. This arrangement of the genomic dsRNA may have relevance to the mechanism of reovirus transcription (see "Replication Cycle" section).

Oligonucleotides

About 25% of the RNA in purified virions is in the form of small, single-stranded oligonucleotides (37,356), which can be grouped into two major classes (34). Quantities can vary widely with conditions of virus growth, including temperature (208), as well as between virions obtained from the same purification (125).

About 70% of the oligonucleotides (2,000 copies per virion) terminate with a 5′GC(U) (A) and are 2 to 9 residues long (35,38,286). The conserved 5′ sequences suggest that these are the products of abortive transcription (see "Replication Cycle" section). They are often termed *initiator* oligonucleotides, and those packaged into virions are generated in the final stages of morphogenesis (see "Replication Cycle" section). The 5′ termini of most are triphosphorylated, but di- and monophosphorylated termini are also found. About 10% are guanylylated (5′GpppGC), but none are methylated (64).

Oligoadenylates constitute a second major class of oligonucleotides, which are 2 to 20 residues long and present in about 850 copies per virion (34,286,388,389). The oligoadenylates are variably tri-, di-, or monophosphorylated at their 5′ ends. Like the initiator oligonucleotides, they appear to result from an alternative activity of the viral transcriptase and are generated in the final stages of morphogenesis (see "Replication Cycle" section).

While virions contain both classes of oligonucleotides, cores lack them (67,185,355,373). Whether they are present in ISVPs is controversial (74,185,355). Their exact locations within virions remain unknown (169,253). Functions of the oligonucleotides are also unknown, but there is evidence that they are not required for infectivity (67).

Molecular Weights

Molecular weights for virions and cores of strain T3D were determined from diffusion coefficients to be 129.5 and 52.3 megadaltons (MDa), respectively (125). The total weight of the genomic RNA is calculated to be 15.1 MDa from sequences of the ten T3D genes (110). Oligonucleotides are estimated to contribute another 5 MDa to the weight of virions. Thus, proteins are expected to contribute ~109 and 37 MDa (~85% and 70% of totals) to the molecular weights of virions and cores, respectively (125). These weights correspond well with those obtained by calculation from sequence-predicted masses of the proteins and estimates for their copy numbers in particles (Table 3). Similar estimates can be made for ISVPs (Table 3). Despite the preceding data, some uncertainties remain as to the molecular weights of reovirus proteins as they occur in particles (Fig. 2, Table 4). It is assumed that the proteins incorporated into virions are full-length, given the lack of evidence for cleavage (except for μ1 and some copies of λ2) and good correlations between molecular weights calculated from sequences and those estimated from gels (303,373).

TABLE 4. *Reovirus proteins*[a]

Encoding segment	Protein	Mass (kd)	Copy number per virion	Location in virions	Presence in particle forms	Function or property
L1	λ3	142	12[b]	Inner capsid	VIC	RNA-dependent RNA polymerase
L2	λ2	145	60	Outer capsid, core spike	VIC	Guanylyltransferase, methyltransferase?
L3	λ1	137(143)[b]	120	Inner capsid	VIC	Binds dsRNA, zinc metalloprotein
M1	μ2	83(78)[b]	12[b]	Inner capsid	VIC	Unknown
M2	μ1	76	600	Outer capsid	VI	N-myristoylated, cleaved into fragments, role in penetration, role in transcriptase activation
M3	μNS	80	0	Nonstructural	—	Binds ssRNA, associates with cytoskeleton, role in assortment, role in secondary transcription
S1	σ1	49	36–48[b]	Outer capsid	VI	Cell-attachment protein, hemagglutinin, primary serotype determinant
	σ1s	14	0	Nonstructural	—	Unknown
S2	σ2	47	120–180[b]	Inner capsid	VIC	Binds dsRNA
S3	σNS	41	0	Nonstructural	—	Binds ssRNA, role in assortment
S4	σ3	41	600	Outer capsid	V	Sensitive to protease degradation, binds dsRNA, zinc metalloprotein, effects on translation

[a] See text for references.
[b] Uncertainties in these values are discussed in the text.
V, virion; I, ISVP; C, core.

Preparation of Viral Particles

Mouse L cells grown in suspension are routinely used for plaque assays (135,188,325) and for preparing purified virions. Virions are released from infected cells by sonication and dissociated from cellular components by treatment with deoxycholate and extraction with freon (39,108, 135,188,351,373). Final purification of virions was traditionally achieved by consecutive centrifugations through sucrose and CsCl gradients (371); however, the sucrose gradient is now frequently omitted. Values for sedimentation and buoyant density of reovirus particles are shown in Table 3 (46,125,135,373). Banded virions are dialyzed into buffer (commonly 150 mM NaCl, 10 mM $MgCl_2$, 10 mM Tris, pH 7.5) for storage at 4°C, where they can remain stable and infectious for extended periods. An average yield from this procedure is 1×10^{13} virions (~1.8 mg) per 1×10^{8} cells initially infected. The equivalence 1 OD_{260} = 185 mg protein/ml = 2.1×10^{12} virions/ml is routinely used for estimating the concentration of purified virions (373).

Chymotrypsin or trypsin is commonly used to convert virions into ISVPs or cores *in vitro* (46,86,135,169,185, 282,355,357,373,380,381,392). At fixed conditions (buffer, enzyme concentration, temperature, and time of treatment), the concentration of virions included in the mixture determines whether ISVPs or cores are obtained as products (46, 49,52,55,135,185). The particular conditions yielding one or the other particle form can vary between strains (86,107, 185). ISVPs or cores generated by treatment are either used directly or purified by centrifugation through CsCl gradients (46,86,135,229,282,373). Purified ISVPs are commonly stored in the same buffer as virions, but a different buffer (1 M NaCl, 100 mM $MgCl_2$, 20 mM HEPES, pH 8.0) was found to maintain cores in a nonaggregated state (86). Comparisons of protein quantities in virions and subviral particles suggest the following equivalences: 1 OD_{260} = 2.7×10^{12} ISVPs/ml (K.M. Coombs, personal communication) and 1 OD_{260} = 4.2×10^{12} cores/ml (86).

When preparations of purified virions are tested for infectivity by plaque assay in L cells, a particle-to-pfu ratio between 100:1 and 1000:1 is typically obtained (15,119, 188,392). The basis for this high ratio is unknown but could reflect either that defective virions are present within the preparation or that productive infection involves a complex process that not all potentially infectious particles can complete. In a few studies, a ratio approaching 1:1 between particles and infectious units has been reported; however, these studies included special circumstances such as protease activation of viral particles and measurement of infectious units by a fluorescent focus assay (380). In general, the efficiency of infection is enhanced in preparations of ISVPs and reduced by several orders of magnitude in preparations of cores (see "Replication Cycle" section).

Cell-lysate stocks containing nonpurified viral particles are used as a source of infectious reovirus in many experiments. Such stocks contain a mixture of viral particles which were released from cells during infection or by artificial lysis (e.g., freeze-thawing). Most infectious particles within these stocks behave like virions, and not ISVPs (45,102,392).

Top Component Particles

"Empty" virions, which contain markedly reduced quantities of dsRNA gene segments and oligonucleotides, account for a large proportion of the particles purified from reovirus-infected cells (314,373). Their buoyant density is ~1.29 g/cm^3, lower than full virions and consistent with their lack of RNA (125,135,210,373); thus, they migrate above full virions in CsCl gradients and are called *top component* particles. Both their protein content and morphology are very similar to those of full virions, except that negative stains can enter their central cavity (135,373,423). Protease treatments of empty virions can be used to generate empty ISVPs (229) and empty cores (373,423). Empty particles have been used in studies that address the structure and function of reovirus proteins (167,229,347). The derivation of empty virions is unknown; however, they are unlikely to be precursors of full virions during morphogenesis (188). Methods for increasing the relative yields of empty virions from infected cells have been described (210).

Particle Stability

Reovirus virions are remarkably stable, as evidenced by their retention of infectivity over long periods of storage (see above). Like other nonenveloped viruses, they show relative resistance to treatments with organic solvents and detergents (162,328). Virions can also resist exposure to extreme ionic strengths, pH values between 2 and 9, and temperatures approaching 55°C (10,108,109,162,314,386, 408). Exposure to ultraviolet radiation reduces infectivity (172,246,262,359), as does exposure to visible light in the presence of photoreactive dyes (246,407). Reoviruses are stable in aerosols, especially at high humidity (2). In one set of studies, the mechanisms by which different physicochemical agents destroy the infectivity of virions was studied by biochemical and genetic techniques (108,109). The results indicate that proteins in the outer capsid determine differences between reovirus strains in sensitivity to these agents and suggest that each agent primarily affects one or more outer capsid proteins to accomplish inactivation. Older literature about the stability of reovirus particles has been reviewed (352). The effects of proteases on reovirus particles are described elsewhere in this chapter.

The stabilities of ISVPs and virions have not been systematically compared; however, in some cases ISVPs are clearly less stable (52,381). The σ3 protein is proposed to give virions greater stability than ISVPs, which may be needed for virions to survive efficiently when moving between host animals in nature (284). Other data indicate that cores can resist disruption by a variety of denaturants and proteases except at extreme conditions (188,373).

PROTEINS

Structural Proteins

Proteins incorporated into mature virions (structural proteins) can have several particle-related functions, some of which are critical to the initiation of infection. These same proteins can also play nonstructural roles within the infected cell. General characteristics and functions of these proteins are discussed below. Structural details are discussed further in "Particles" section.

λ1

The λ1 protein is a major component of the inner capsid, where it interacts with proteins σ2 and λ3 from the inner capsid and λ2 from the outer capsid (387,423,435). It may also interact with outer capsid protein μ1 (see below). The deduced amino acid sequence of λ1 includes a putative nucleotide-binding motif near its amino terminus, suggesting that it may have an enzymatic activity relating to transcription (27) (see "Replication Cycle" section). The λ1 sequence also contains a zinc-finger motif near its amino terminus (27). Zn^{2+} has been detected in the reovirus core, and one or more of the λ proteins binds Zn^{2+} in a blotting assay (339,340). The identification of a zinc-finger sequence motif in λ1, but not λ2 and λ3, suggests that λ1 is the Zn^{2+}-binding protein in the core. The role of Zn^{2+} in λ1 function is unknown; however, λ1 has also been shown to bind dsRNA, suggesting that the zinc finger may be involved in that activity (339).

λ2

Pentamers of λ2 form the spikes projecting from the fivefold axes of cores (302,423). The λ2 protein is now considered to be a component of the outer capsid (110,171, 243). There is evidence that λ2 interacts with σ1, σ3, and μ1 in the outer capsid and λ1 and λ3 in the inner capsid (110,171,213,251,387,423). Temperature-sensitive (ts) mutants whose lesions map to the L2 gene encoding λ2 (272) form corelike particles, but not whole virions, at nonpermissive temperature, suggesting a role for λ2 in assembly of the outer capsid (133,264,307). A λ2 cleavage product that lacks a 15-kD fragment represents 10% of the λ2 molecules in virions (212), but the significance of this fragment is unknown.

Biochemical studies showed that λ2 is the reovirus guanylyltransferase (see "Replication Cycle" section) (80,353). The sequence of λ2 includes two consensus elements for GTP-binding and a region of similarity to the catalytic subunit of the vaccinia virus capping enzyme (345). Reovirus cores incubated with [α-^{32}P]GTP form covalent λ2-GMP complexes, consistent with the activity of λ2 as guanylyltransferase (80,353). One study localized the GMP linkage to lysine 226 in the amino-terminal half of λ2 (126), despite the fact that the consensus elements for GTP-binding are found in its carboxy-terminal half. A GTP-dependent pyrophosphate exchange activity in cores (405) probably represents the forward and reverse reactions of the guanylyltransferase. Vaccinia-expressed λ2, which does not assemble into pentamers but instead remains monomeric, mediates the full range of enzymatic activites expected of the guanylyltransferase (235).

The λ2 protein may also act as the methyltransferase that modifies guanine bases at the 5′ end of reovirus mRNAs in that it is the only viral protein labeled by 8-azido-*S*-adenosyl [^{35}S]methionine (345). This conclusion is supported by limited sequence similarity found between other viral and cellular methyltransferases and a central region of λ2 (201); however, studies with expressed, monomeric λ2 have failed to demonstrate this activity (235).

Reassortant studies have mapped several strain-dependent differences to the L2 gene segment. These phenotypes include the capacity of particular strains (a) to generate defective interfering stocks during serial passage at high multiplicity of infection (MOI) (59); (b) to survive and replicate efficiently in mouse enteric tissue (44) and to undergo efficient transmission between newborn mouse littermates (197); and (c) to form crystals of core particles (86). The molecular basis for the role of λ2 in each of these phenotypes is unknown.

λ3

The λ3 protein is a minor component of the inner capsid and has yet to be definitively localized within cores. It has been shown to interact with proteins λ1 and λ2 (387) and suggested to interact with μ2 (242,284). A ts mutant whose lesion maps to the L1 gene encoding λ3 (308) is severely affected in RNA synthesis (89,180) and forms empty particles at nonpermissive temperature (133). An early reassortant study indicated that the L1 gene determines the pH optimum of core transcription activity, suggesting that λ3 may be the viral transcriptase (107). The deduced amino acid sequence of λ3 includes the GDD sequence motif common to all viral RNA polymerases (61,199,267). Furthermore, vaccinia-expression of λ3 has revealed a protein with poly(C)-dependent poly(G) polymerase activity (387). These findings suggest that λ3 contains the catalytic site of the reovirus RNA-dependent RNA polymerase. Association of λ3 with other viral or cellular proteins may be required to impart complete activity and specificity for reovirus templates. As polymerase, λ3 would be expected to function during both transcription and minus-strand synthesis and may require different cofactors in each case. The L1 gene encoding λ3 was found to play a primary role in determining viral yields in L cells and an important role (in addition to M1) in determining viral yields in cultured mouse heart cells (242).

μ1

The μ1 protein is a major component of the outer capsid, where it interacts with proteins λ2 and σ3 (110,184, 212,251,373). Genetic evidence suggests that it also interacts with inner capsid protein λ1 (251). The μ1 protein is modified by addition of myristic acid to its extreme amino terminus (285) (Fig. 6). The presence of a myristoyl group on μ1 suggested a hypothesis that this protein functions in the penetration of cell membranes (see "Replication Cycle" section).

In the virion outer capsid, most μ1 exists in the form of a 72-kD fragment, μ1C, that is generated from μ1 by proteolytic cleavage (90,212,373,447). Amino-terminal sequencing localized the cleavage between residues 42 and 43 in μ1 (184,299), indicating that μ1C is a carboxy-terminal fragment of μ1 (Fig. 6). The cleavage of μ1 to μ1C should also yield a 4.2-kD amino-terminal fragment (184, 373,395,428). Early studies suggested that a virion-associated peptide of ~8 kD ("component viii") represented this species (373). A more recent study identified the myristoylated 4.2-kD amino-terminal fragment and named it *μ1N* (285). This cleavage of μ1 is highly sensitive to the residues at the μ1N:μ1C cleavage junction (396).

The μ1 and μ1C proteins both undergo a distinct proteolytic cleavage near their shared carboxy terminus after exposure to proteases *in vitro* (46,184,185,282,355), within endocytic compartments of cultured cells early in infection (54,74,363,392), and within the intestinal lumen of newborn mice after peroral inoculation (29,45). This cleavage, which occurs between residues 581 and 582 with chymotrypsin and between residues 584 and 585 with trypsin,

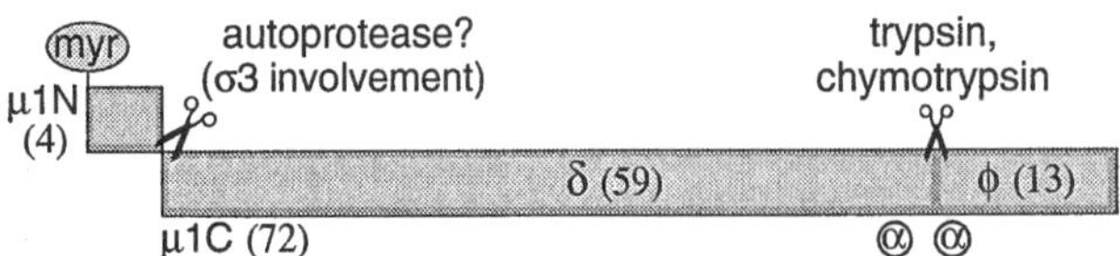

FIG. 6. Outer capsid protein μ1. Most (95%) of the μ1 protein (76 kD) in virions undergoes an assembly-related cleavage (*scissors*) into small amino-terminal and large carboxy-terminal fragments designated μ1N (4 kD, as indicated) and μ1C (72 kD), respectively. An autoprotease activity appears to be involved in this cleavage, as does the σ3 protein. The μ1N fragment (like uncleaved μ1) is modified by an amide-linked myristoyl group *(myr)*, probably at its extreme amino terminus. When ISVPs are generated by the action of exogenous proteases, such as trypsin or chymotrypsin, on virion proteins, the μ1C fragment undergoes additional cleavage (*scissors*) nearer its carboxy terminus. The products are large amino-terminal and small carboxy-terminal fragments designated δ (59 kD) and ϕ (13 kD), respectively. The δ:ϕ cleavage junction is flanked by predicted long amphipathic α-helices (α), which are proposed to play a role in membrane interaction and penetration by reoviruses. The small amount of uncleaved μ1 protein in virions is also cleaved during generation of ISVPs to produce fragments μ1δ [μ1N+δ] and ϕ.

generates an acidic amino-terminal fragment named *μ1δ* when derived from μ1 and named δ when derived from μ1C as well as a smaller, highly basic, carboxy-terminal fragment named ϕ. Both the μ1δ/δ and ϕ fragments remain bound to ISVPs in stoichiometric amounts (282) (Fig. 6). The carboxy-terminal cleavage of μ1/μ1C is proposed to be important for the penetration of cell membranes (see "Replication Cycle" section).

In the cytoplasm of infected cells, 90% of free M2-encoded protein is μ1, whereas 95% of the protein complexed with σ3 has been cleaved to form μ1C (212). This finding suggests that cleavage of μ1 may be linked to formation of a complex with σ3. This hypothesis is further supported by the observation that μ1C is detected in eukaryotic cells expressing both the M2 and S4 genes while intact μ1 is detected in cells expressing only the M2 gene (396). Furthermore, μ1 expressed from an M2 gene that lacks the signal for myristoylation neither forms complexes with σ3 nor undergoes cleavage to μ1C (396).

Several studies suggest that a fraction of μ1 and its fragments undergo other modifications. It was reported that 2% to 4% of μ1C molecules in virions are O-glycosylated on serine or threonine residues (205). Evidence was also presented for the existence of one or more phosphoserine residues per molecule of μ1C in virions (206). It has been suggested that reovirus particles contain an oligo(A) polymerase (see "Replication Cycle" section) and that some μ1C molecules become polyadenylated and/or ADP-ribosylated by this enzyme (65,66). The μ1 protein is reported to exist in the virion outer capsid as a disulfide-bonded dimer (175,184,373).

Despite marked sequence conservation between the M2 genes of different strains (428), genetic analyses have associated M2 with a number of biologic functions. A strain difference in condition dependence for conversion of virions to cores and activation of the viral transcriptase was mapped to M2 (107). Biochemical studies support an association between loss or rearrangement of μ1 and transcriptase activation (see "Replication Cycle" section). Strain differences in sensitivity of virions to inactivation with ethanol or phenol were mapped to M2 (108). The isolation and genetic characterization of ethanol-resistant mutants identified a region of μ1 that may be important for stability of the outer capsid (422). The reduced neurovirulence of strain T3 clone 8 after inoculation into newborn mice was also found to be determined by its M2 gene (174).

μ2

The μ2 protein is a minor component of the inner capsid. It has not been localized within the core but is suggested to interact with the other minor core protein, λ3 (242,284). The function of μ2 is unknown. It was reported that ts mutants with a lesion in the M1 gene encoding μ2 fail to synthesize dsRNA (426), suggesting a role for μ1 early in assembly or in RNA synthesis (360). Reassortant analyses have suggested that μ2 can determine strain differences in plaque size and extent of cytopathic effect in L cells (260), capacity to cause myocarditis in newborn mice (360), and extent of viral replication in cultured mouse heart cells (242). It was hypothesized that the role of μ2 as regulator of viral cytopathology may be related to its low level of translation *in vivo* (319).

σ1

Though only a minor component of the outer capsid, σ1 serves as the reovirus cell-attachment protein (23,213,241, 298,416,417,419,438). It is also the hemagglutinin (51, 241,420,438) and contains the primary epitopes against which the serotype-specific, humoral immune response is directed (62,384,404,418,438). The σ1 protein interacts with λ2 and perhaps σ3 in the outer capsid (110,171,213, 251,404). It occurs in virions and ISVPs in oligomeric form and can assume an extended conformation in which it appears as a long fiber with head-and-tail morphology projecting from the particle surface (see "Particles" section).

Sequence analysis led to a model in which residues in the amino-terminal third of σ1 assume an α-helical coiled-coil structure (30,135,218). Processed images of isolated σ1 fibers later revealed significant substructure within the tail region (134). Related work identified repeats of hydrophobic residues within a central region of σ1 sequences and suggested that different morphologic regions in the tail result from sequences assuming distinct supersecondary structures: alternating regions of α-helical coiled-coil and a β-sheet motif as shown in Fig. 7 (100,134,281). The head appears to be formed by sequences in the carboxy-terminal third of σ1 and to have a more complex globular structure (30,113,220,281,398). Protease cleavage of the ma-

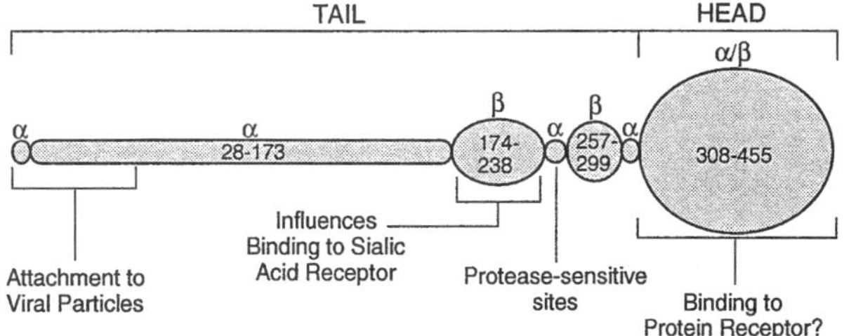

FIG. 7. Cell-attachment protein σ1. Oligomers of the reovirus σ1 protein can be isolated from viral particles as extended fibers characterized by tail and head domains as shown. Distinct regions within the tail are proposed to be formed by sequences that assume α-helical coiled-coil and cross-β sheet supersecondary structures as labeled [sequences assigned to the larger tail regions and head in σ1 proteins from type 3 (T3) reovirus strains are indicated by position number]. Properties or activities associated with different regions of the type 3 σ1 proteins are indicated.

ture T3D σ1 oligomer divides it into amino- and carboxy-terminal oligomeric fragments, which are physically separated and remain stably folded in the absence of detergent (113,390,439). Biophysical studies support the idea of an extended σ1 structure (390).

Several studies have implicated the σ1 head in receptor-binding activity. Sequencing of S1 genes from neutralization-resistant mutants identified two residues in the carboxy-terminal third of σ1 which determine neutralization by the monoclonal antibody G5 (30,32). Evidence that these mutants exhibit altered tropisms within the mouse brain (382,383) contributed to a hypothesis that this region of σ1 mediates receptor binding. Sequence similarities between T3D σ1 and an antiidiotypic monoclonal antibody directed against G5 identified other residues in the carboxy-terminal third of σ1 that may serve as receptor contacts (60,432,433). In other work, a carboxy-terminal protease fragment of σ1 (113,390,439) (see above), as well as deletion mutants that retain a large carboxy-terminal portion of σ1 (276), were found to retain L-cell binding capacity. Mutagenesis of conserved residues within a carboxy-terminal portion of σ1 suggest that noncontiguous amino acids contribute to the structure and function of the receptor-binding domain (398).

Hemagglutination by T3 reovirus strains involves binding of σ1 to sialic acid residues on the erythrocyte surface (15,148,296), most of which are associated with the protein glycophorin A (76,297). Monoclonal antibodies identify distinct epitopes on σ1 involved in neutralization and hemagglutination inhibition (62,384). Recently, analysis of deduced σ1 sequences from T3 strains which differ in capacity to agglutinate erythrocytes and bind glycophorin suggested that, although the carboxy-terminal head of σ1 is implicated in L-cell binding, a region in the σ1 tail is important for these properties (101) (Fig. 7). Findings with protease fragments (439) and deletion mutants of σ1 (276) provide additional evidence that L-cell and glycophorin binding are separable functions of σ1 that may be mediated by different regions of the protein.

Sequences near the amino terminus of σ1 form a distinct "virus-attachment region" that anchors this protein to viral particles (Fig. 7). A small region of hydrophobic residues at the extreme amino terminus was first predicted to fulfill this function through a coupling of sequence analysis and electron microscopy (23,134,135,281). Subsequent work with expressed, carboxy-terminally truncated σ1 proteins demonstrated that residues between positions 3 and 34 (of strain T3D) are absolutely required for virion anchoring (219,232).

A number of studies suggest that σ1 folds into its native conformation via a multistep process. Monomers of σ1 do not bind to cells and exhibit extreme protease sensitivity, indicating the importance of oligomerization for σ1 structure and function (217). An amino-terminal region of σ1, which includes a portion of the predicted long α-helical coiled-coil, possesses a capacity to undergo oligomerization independent of other sequences (24,113,218,220,398); in contrast, sequences from the carboxy-terminal half of σ1 depend on prior oligomerization by amino-terminal sequences to direct their folding (217,220). Additional maturation events in both amino- and carboxy terminal regions are required to generate the final σ1 structure (220). Cellular chaperone proteins may be involved in one or more of these steps in σ1 folding (220).

Many of the phenotypes mapped to the S1 gene by reassortant studies may relate to the role of σ1 as cell-attachment protein. Properties relating to this attachment function, many of which are important determinants of pathogenesis, are discussed below (see "Replication Cycle" and "Effects on Host Cells" sections; see also Chapter 53) in *Fields Virology,* 3rd ed.

σ2

The σ2 protein interacts with λ1 to form the inner capsid (243,423,435). It has weak affinity for reovirus dsRNA (103,339). Sequence analyses suggest that the protein has a two-domain structure and includes an amphipathic α-helix with similarity to the β subunit of *E. coli* DNA-dependent RNA polymerase (103). A ts mutant whose lesion maps to the S2 gene encoding σ2 (308) shows a defect in RNA synthesis and forms empty outer capsids at nonpermissive temperature, suggesting that σ2 plays a role in morphogenesis (89,133,180,243); however the σ2 protein from this mutant is not defective in its dsRNA binding activity (103). The ts mutant S2 gene has three nucleotide changes compared to that from wt T3D, each of which results in an amino acid change in σ2 (430). Selection of ts+ revertant viruses and subsequent sequence determinations indicated that reversion with this ts mutant is exclusively intragenic and involves true reversion at residue 383 (87).

σ3

The σ3 protein, together with μ1 (μ1N and μ1C), forms the bulk of the virion outer capsid (373). Available evidence is that σ3 interacts with proteins λ2, μ1, and perhaps σ1 in virions (110,184,212,213,251,404). Genetic studies implicate σ3 in determining the sensitivity of virions to inactivation with heat and SDS (109). The σ3 protein may play a role in granting virions increased stability for survival in environments outside cells (284) but must be removed from virions by proteolysis for infection to proceed (see "Replication Cycle" section). A ts mutant whose lesion maps to the S4 gene encoding σ3 accumulates core-like particles at nonpermissive temperature, suggesting that σ3 plays an important role in assembling the outer capsid (95,264). The S4 gene sequence of this mutant predicts a number of amino acid differences between the ts and wt σ3 proteins (95). The σ3 protein may also associate with reovirus mRNAs during the earliest stages of morphogenesis (14) (see "Replication Cycle" section).

Biochemical studies showed that σ3 binds dsRNA in a sequence-independent manner (175), and a dsRNA-binding activity was localized to a carboxy-terminal fragment of σ3 (254,339). This carboxy-terminal fragment contains two copies of a basic amino acid motif which may contribute to dsRNA binding (254). The σ3 protein occurs as a zinc metalloprotein in virions; furthermore, σ3 from virions binds Zn^{2+} in a blotting assay (339,340). A zinc finger motif is found in an amino-terminal portion of σ3, and Zn^{2+}-binding activity was localized to an amino-terminal fragment that includes these sequences (339). The function of Zn^{2+} in σ3 is unknown but may contribute to its folding or capacity to bind μ1. At present, the Zn^{2+} and dsRNA binding activities of σ3 are thought to be separate.

Genetic studies suggest that mutations in the S4 gene play a role in the establishment of persistent reovirus infection (7) and that strain differences in S4 influence the inhibition of cellular RNA and protein synthesis (348). Other studies suggest roles for σ3 in the regulation of viral transcription (18) and translation. The σ3 protein was found to be enriched in the initiation-factor fraction of eukaryotic cells expressing a cloned S1 gene (214) and was suggested to stimulate the translation of late viral mRNAs within infected cells (215,216). Several studies found that expression of σ3 can stimulate expression of cotransfected reporter genes (155,239,344,396). It was hypothesized that the effect of σ3 on translation relates to its capacity to bind dsRNA and thereby sequester the activator of an interferon-induced protein kinase which phosphorylates eIF2α and inhibits translation initiation (179,225) (see "Effects on Host Cells" section). Coexpression of the M2 and S4 genes abrogates the stimulatory effect of S4, suggesting that association with μ1 may interfere with the dsRNA binding activity of σ3 (396). One study has argued that the carboxy-terminal third of σ3 is responsible for stimulation of reporter gene activity; however, the instability of σ3 proteins expressed from some transfected genes makes this conclusion tentative (239).

Nonstructural Proteins

Examination of proteins in reovirus-infected cells identified two virally encoded nonstructural proteins (448): μNS (previously called μ0 or μ4) and σNS (previously called σ2A). These proteins are synthesized in large amounts suggesting that, although they are not components of mature virions, they may play structural roles in the assembly of virions. Nucleotide sequencing (30,70,268,275) plus *in vitro* translation of viral mRNAs and comparison with proteins synthesized in reovirus-infected L cells (72,120,183,336) led to the identification of an additional nonstructural protein, σ1s (189), encoded by a second ORF in the S1 gene (203). This protein has also been called p14 (72,122,123), σlbNS (181,183,268), σs (298,336), and σ1NS (40), and is present in infected cells in relatively small amounts.

σNS

The σNS protein is rich in cysteines and in sequences predicted to form α-helices (150,315,427). It shows strong affinity for ssRNA, including reovirus mRNAs (157,175, 316,385). Recent coprecipitation studies confirmed that σNS, as well as μNS and σ3, associate with viral mRNAs soon after transcription (14). The σNS protein binds to reovirus mRNAs in a manner which may be unique for members of each mRNA size class (385); thus, the ssRNA-binding activity of σNS may be involved in condensing reovirus mRNAs into assortment complexes during morphogenesis (see "Replication Cycle" section). A ts mutant whose lesion maps to the S3 gene encoding σNS (308) is severely affected in RNA synthesis at nonpermissive temperature (89,180). Particles containing σNS were reported to have poly(C)-dependent RNA polymerase activity (158), but this activity is probably attributable to λ3 that may also have been present in those particles (316,387).

μNS

Two closely sized proteins, μNS and μNSc, are made in similar amounts within infected cells (212,426). Although first thought to be a cleavage product of μNS, recent evidence suggests that μNSc is formed by an alternative initiation of translation at the second in-frame AUG in the M3 mRNA (426) (see "dsRNA Genome" section). Since translation of each is proposed to proceed from the same long ORF, μNS would differ from μNSc only in having an extra 5 kD of sequence at its amino terminus. Functional differences between μNS and μNSc are unknown. The deduced amino acid sequence of μNS suggests a high α-helical content and similarity with myosins (426). The μNS protein associates with the cytoskeleton, to which it may anchor viral structures (261). In addition, reovirus mRNAs bind to μNS shortly after synthesis, even before they associate with other viral proteins (14); thus, the μNS protein may be involved in forming assortment complexes during morphogenesis (see "Replication Cycle" section). Since newly assembled, secondary transcriptase particles appear to differ from replicase particles and mature virions in having μNS attached to them (265), μNS may also regulate assembly of the outer capsid or play a role in secondary transcription (see "Replication Cycle" section). One genetic study found that the M3 gene encoding μNS determines which gene segments undergo deletion during serial passage of viruses at high MOI (59).

σ1s

Sequencing studies indicate that σ1s is a basic protein (30,70,268,275) and that an amino-terminal cluster of basic residues is conserved between σ1s proteins which are otherwise highly variable (70,100,112,281). A role for this basic protein in binding to nucleic acid has been proposed (183). There is dispute as to whether σ1s accumulates in both cytoplasm and nucleus or in cytoplasm only (41,68,72). It may be responsible for some of the biologic functions that have been mapped to the S1 gene and attributed to the σ1 protein, such as inhibition of cellular DNA synthesis (see "Effects on Host Cells" section) and injury to specific cells within infected animals (see Chapter 53 in *Fields Virology*, 3rd ed.). When expressed alone or in combination with σ1 in mammalian cells, σ1s does not change the kinetics of cellular DNA replication; however, its overexpression in reovirus-infected cells does potentiate the cytopathic effects of infection as well as the viral inhibition of cellular DNA synthesis (123).

GENETICS

Reoviruses have been exploited as a genetic system, due largely to their capacity to undergo reassortment of gene segments during coinfections in cell culture. Older literature relating to reovirus genetics has been comprehensively reviewed (307).

Mutants

Temperature-Sensitive Mutants

Reovirus ts mutants are defined by reduced capacities to replicate at 39 versus 31°C. Ts mutants from seven of the ten genetic groups of strain T3D were first isolated after chemical mutagenesis with nitrous acid (groups B, D, and E), nitrosoguanidine (groups C, F, and G), or proflavin (group A) (132,177). Mutants from the other three groups were isolated from T3D stocks subjected to serial passage at high MOI (group H) or from persistently infected cultures (groups I and J) (4,5,9). Ts mutations were definitively assigned to discrete gene segments (Table 5) by reassortant analyses (272,302,307,308). Sequencing studies have identified nucleotide and amino acid changes in mutants from groups C (87,430), E (427), and G (95). Ts mutants in groups A to G have been analyzed extensively with respect to their morphologic, biochemical, and biological properties (Table 5).

Intra- and Extragenic Suppressor Mutants

Reversion of ts mutants to the ts+ phenotype might occur through a change in the same gene, either at the site of the original mutation (true reversion) or at a second site (*intragenic suppression*). Alternatively, mutation in a second gene might suppress the defect produced by the original mutation (*extragenic suppression*). A recent study found that one ts mutant (tsC[447]) undergoes true reversion exclusively (87); however, the majority of revertants of reovirus *ts* mutants result from extragenic suppression (304,306,309). A possible mechanism for extragenic suppression involves mutations in gene products that interact directly. Interactions between reovirus protein pairs μ1:σ3, μ1:λ1, and λ2:σ1 were inferred in this manner (251) and borne out by other experimental data (see "Proteins" section). Another possible mechanism for extragenic suppression involves a mutant transcriptase, which introduces reversion mutations during transcription or minus-strand synthesis (307).

Deletion Mutants and Defective Stocks

Some reovirus strains accumulate deletion mutants when subjected to serial passage at high MOI (6,59,290,292, 342). Segments L1 and L3 are most frequently deleted, but deletions in L2 and M1 are also seen. A series of deleted

TABLE 5. *Genetic map and properties of reovirus temperature-sensitive mutants*[a]

ts Group	Prototype isolate	Mutant gene segment	Mutant protein	% dsRNA synthesis[b]	% ssRNA synthesis[b]	% Protein synthesis[b]	Particle morphology[b]
ts*A*	*ts*A(201)	M2	μ1	100	100	100	Virions
ts*B*	*ts*B(352)	L2	λ2	25–50	25–50	25	Corelike
ts*C*	*ts*C(447)	S2	σ2	0.1	5	5–10	Empty
ts*D*	*ts*D(357)	L1	λ3	0.1	5	10–20	Empty
ts*E*	*ts*E(320)	S3	σNS	1	5	5–10	None
ts*F*	*ts*F(556)	(M3)[b]	(μNS)[b]	50–100	50–100	50–100	Virions
ts*G*	*ts*G(453)	S4	σ3	15–25	20	25	Corelike
ts*H*[c]	*ts*H(26/8)	M1	μ2	0	ND	ND	ND
ts*I*[c]	*ts*I(138)	L3	λ1	ND	ND	ND	ND
ts*J*[c]	*ts*J(128)	S1	σ1/σ1s	ND	ND	ND	ND

[a]See text for references.
[b]At nonpermissive temperature. Values expressed as percentage of synthesis seen with wild-type virus.
[c]More recently isolated and less completely characterized.
ND, not determined.
From Ramig and Fields (305) with modifications. Courtesy of Marcel Dekker, Inc.

M1 genes were analyzed and found to lack internal regions of sequence (443). Sequences of the smallest deleted M1 genes showed that minimal sizes for retained 5′- and 3′-terminal regions of the plus strand were 132 to 135 and 182 to 185 bases, respectively. One genetic study compared strains T1L and T3D, which differ in their capacity to accumulate deletion mutants (59). This study showed that the L2 gene determines whether or not viruses accumulate deletions and that the M3 gene determines in which segments the deletions occur. The L2 and M3 gene products λ2 and μNS may play a role in synthesis or packaging of reovirus RNA such that they influence the generation or amplification of deletion mutants.

Some reovirus stocks that have been serially passaged at high MOI can be classified as defective interfering (DI) stocks (4,6–8,59,102,290,293,342). The traits of defective stocks include (a) the presence of deletion mutants; (b) the capacity to interfere with growth of wt virus (see below); and (c) the capacity to establish persistent infections. The role of deletion mutants in the latter properties is unclear since defective stocks contain a variety of mutants, including a high frequency of *ts*G (S4 gene) mutants (4,6). Given that the nondeleted gene segments in defective stocks are normal with respect to transcription and reassortment, it was suggested that interference and facilitation of persistent infections may be linked to specific mutations in full-length segments rather than properties of the deleted segments (6,59) (see "Persistent Infections" section).

Neutralization-Resistant Mutants

Neutralizing monoclonal antibodies directed against the σ1 protein were used to select neutralization-resistant mutants of strain T3D, and changes in the mutants were localized to two residues in the carboxy-terminal third of σ1 (32,196). The mutants were found to be altered in their pathogenic behaviors within newborn mice (382,383), possibly because their mutations fall in or near the cell-attachment region of σ1. These mutants have also contributed to the model for σ1 structure (30,135,281).

Inactivation-Resistant Mutants

Physicochemical agents (e.g., heat, phenol, SDS) can give large reductions in reovirus infectivity, and strain differences in sensitivities to these agents can be associated with specific genes (108,109). A strain difference in sensitivity to ethanol was mapped to the M2 gene (108), and ethanol-resistant mutants were isolated in which relevant mutations were localized to M2 (422). These mutants exhibit a slowed growth rate and decreased sensitivity to generation of cores by chymotrypsin; thus, mutants selected with physicochemical agents may be used to study the structure and function of reovirus particles, as well as to define the mechanisms of inactivation by these agents.

Other Types of Mutants

Other types of reovirus mutants derived by experimental manipulation have been described. These include small-plaque mutants (8), cold-sensitive mutants (9), other mutants associated with the maintenance of persistent infections in L cells (see "Persistent Infections" section), and mutants with increased virulence obtained after passage in newborn mice (361).

Infectious RNA and Mutants Obtained by Genetic Engineering

A recent report noted that reovirus mRNAs can be infectious after transfection into cells (320). Although this report requires confirmation, its findings promise to facilitate molecular analysis of reovirus gene function by permitting site-directed mutations in cloned genes to be recovered into infectious particles for studies in cultured cells or host animals. To date, the genetic engineering of reovirus genes has been restricted to studies in which mutant gene products are analyzed for activity *in vitro* or in transfected cells.

Natural Variation Between Strains

Most progress in associating traits with individual reovirus genes has come from analysis of naturally occurring strains that exhibit phenotypic differences, or polymorphisms. The classic approach has been to identify a polymorphism between strains, define the electropherotypes of reassortants obtained from coinfections with the strains, and analyze the reassortants for the polymorphic character in an effort to associate it with a specific gene. Early studies applied this approach to identify the S1 gene as the primary determinant of a difference between strains in tropism and virulence for newborn mice after intracranial inoculation (417,419). Numerous other studies have succeeded in mapping strain differences to single reovirus genes. In some instances, polymorphisms have mapped to more than one gene at a time, and the rank sum test of Wilcoxon was used to evaluate the statistical significance of the segregations (44,360).

The availability of reovirus gene sequences has permitted a new type of study, in which the sequences of individual genes and proteins from naturally occurring strains are compared to link sequence differences with a polymorphic phenotype. In this manner a small region of σ1 sequence was implicated in determining the capacity of some type 3 σ1 proteins to bind sialylated proteins on the erythrocyte surface (101). Shared structural elements within proteins from different strains have also been suggested by this approach (103,281,427).

Interactions Between Viruses

Reassortment

Reassortment of genes between reovirus strains occurs not only in cultured cells but also in mice (421). In addition, sequencing studies indicate that reassortment between strains in nature is a major determinant of reovirus evolution (see below). Reassortment was first demonstrated in two-factor crosses (130,132). L cells were coinfected with two ts mutants at the permissive temperature, and reassortants were detected by assaying for progeny at the nonpermissive temperature. If a cross between two ts mutants did not produce ts+ reassortants, the mutants were considered to have lesions in the same gene segment and were placed in the same genetic group. Conversely, if a cross between two ts mutants resulted in ts+ reassortants, the mutants were considered to have lesions in different gene segments. In this way, ts mutants were separated into ten different genetic groups, one per gene segment (89,132, 306). In experiments involving crosses between a ts mutant and wt viruses whose gene segments were electrophoretically distinguishable, each ts group was assigned to a particular gene segment (272,302,307,308). If reassortment was a totally random process, 25% of the progeny from a mixed infection with two ts mutants would be reassortants; however, the highest percentage of reassortants seen in such crosses was 20%, and other crosses produced as low as 3% reassortants (130). Unknown factors thus contribute to reovirus reassortment being a nonrandom process.

The discovery of electrophoretic polymorphisms for protein μ1/μ1C in certain ts mutants provided a new marker (μ+ versus μ– to indicate wild-type versus faster mobility) that could be used in genetic crosses to evaluate reassortment (91). Three-factor crosses were performed using this mobility marker in combination with the ts phenotype (92). A hypothetical three-factor cross would take the form: *ts*P*ts*Q+μ–×tsP+*ts*Qμ+. The segregation of the mobility marker would then be evaluated in *ts*P+*ts*Q+ reassortants. Such studies confirmed that reovirus gene segments behave as discrete elements during reassortment.

Complementation and Interference

Complementation refers to interactions of viral gene products in mixedly infected cells that result in enhanced yield of one or both parental viruses without a change in their genotypes. *Interference* refers to an ability of certain viruses to decrease the yield of other viruses upon co- or superinfection. Although reassortants were readily identified after coinfections with reovirus ts mutants, enhanced yields of progeny at the nonpermissive temperature were often not observed (130,132). This lack of complementation was explained by the phenomenon of interference. Ts mutants from groups A, B, C, and G were found to interfere with the growth of wt virus, whereas those from groups D and E were not (73). Genetic crosses established that interference is a dominant trait. The degree of interference with the growth of wt virus is substantial (50% or more) and increases if the MOI of the interfering ts mutant is increased (73). Significant complementation occurs only if both parental ts mutants are noninterfering (73,180). Interference with growth of wt virus is also a property of defective high-passage stocks, which may reflect the frequent presence of interfering group G ts mutants (4,6).

The molecular basis of interference with reoviruses remains undefined; however, one model proposes that ts mutants from interfering groups encode mutant proteins that become incorporated into progeny virions and reduce their infectivity (6). Recent work identified dominant-negative mutations in σ1 that interfere with the capacity of wt σ1 to assemble into functional oligomers (220): mutant proteins from interfering ts mutants might act similarly to interfere with proper assembly and function of reovirus protein complexes.

Other Types of Interactions

There is no evidence for *recombination* between either homologous or heterologous reovirus genes. Reoviruses can undergo *multiplicity reactivation* after inactivation with genome-damaging agents like ultraviolet radiation (246), probably reflecting the capacity of viruses to reassort mRNAs produced from nondamaged gene segments. *Phenotypic mixing* in the progeny produced from coinfections with two reovirus strains remains to be formally documented; however, σ1 proteins expressed from transfected plasmids were shown to be incorporated into viral particles after the cells were superinfected with reovirus (219,232).

Evolution of Reoviruses

While useful for genetic studies, differences in electrophoretic mobility between genes and proteins of different strains (173,303) are minimally useful for classifying reoviruses. Tryptic peptide mapping of reovirus proteins has not been pursued sufficiently for its usefulness for classification to be evaluated (146). Nucleotide and deduced amino acid sequences provide the most reliable tools for studies of reovirus evolution (200).

Sequences are available from prototype strains of all three serotypes (T1L, T2J, T3D) for six of the ten gene segments (Table 2). Except for S1, homologous genes from the prototypes are identically sized, optimally aligned without gaps, and similar in sequence [75% to 96% and 86% to 99% identity in nucleotide and deduced amino acid sequences, respectively, in pairwise comparisons (103,344, 429)]. For every gene except S1 examined from all three prototypes, the T2J sequence is most distinct, which led to a suggestion that T2J diverged from a common progenitor strain prior to T1L and T3D (429).

The S1 genes of reoviruses exhibit greater variability than other genes. The variability includes differences in size, the need for gaps to align deduced σ1 protein sequences, and reduced sequence similarity (only 26% to 49% identity in σ1 sequences in pairwise comparisons) (112,281). Among the prototypes, the T3D S1 sequence is most divergent. Examination of S1 genes from a set of 11 type 3 strains revealed no differences in size or need for gaps to optimize σ1 alignments, and sequence diversity similar to that between prototypes for other genes was seen [79% and 86% identity in nucleotide and deduced σ1 protein sequences, respectively, in pairwise comparisons (100)]. These findings suggest that three versions of the S1 gene, representing the three serotypes, arose from progenitor strains at different times in the past and have diverged at comparable rates to other genes since those times.

Reassortment of gene segments between strains is an important force in reovirus evolution. Variable rates of change in third codon bases seen for different genes from T1L and T3D suggest that genes in these two strains came together by reassortment after having had distinct evolutionary histories (429). A study of S2 genes from 11 type 1 and type 3 strains revealed that diversity in these genes is independent of serotype (75). In addition, topological differences were noted in phylogenetic trees for S1 and S2 sequences from 7 type 3 strains (75,100), indicating that these genes have had distinct evolutionary histories and came together in the current strains after reassortment in nature. These studies also demonstrated that diversity in S1 and S2 is independent of the host of origin. Despite evidence for evolution at the level of individual segments, one study has argued for the existence of coevolving sets of reovirus outer capsid proteins (404).

REPLICATION CYCLE

A diagram of the reovirus replication cycle is shown in Fig. 8. Subsequent sections of this chapter discuss details of the depicted events, but a brief description follows here. The first step in reovirus infection is attachment of the virion (or ISVP) to receptor molecules on the cell surface. Following *attachment*, virion particles are taken up from the cell surface by receptor-mediated *endocytosis* and delivered into vacuoles resembling endosomes and lysosomes. Within these vacuoles, reovirus outer capsid proteins undergo specific proteolytic cleavages, which are acid-dependent and represent an essential step in the infection process. In some situations, ISVPs are generated by *extracellular proteolysis*, and the entry of such particles is acid-independent and may not require endocytosis. The mechanism by which reoviruses interact with and penetrate the vacuolar or plasma membrane barrier is unknown but appears dependent upon preceding *proteolysis*, as only the ISVP form exhibits *membrane interaction* in available assays. Coincident with or following *penetration*, a derivative of the ISVP (perhaps a core particle as shown in Fig. 8) becomes transcriptionally active and begins synthesis of the ten capped viral mRNAs. These products of *primary transcription* are used for the translation of viral proteins by the cellular *protein synthesis* machinery. In addition, the early transcripts associate with newly made viral proteins to form progeny "*RNA assortment complexes*." Within subsequently derived "replicase particles," each of the ten packaged plus strands serves as template for *minus-strand synthesis*. Viral mRNAs are also transcribed by progeny subviral particles (*secondary transcription*), which have a unique protein composition. These "late" transcripts are reported to be uncapped and serve as the primary templates for viral protein synthesis later in infection. The final steps of virion *capsid assembly* are not well defined, but may involve the addition of preformed complexes of outer capsid proteins to corelike particles. Mature virions undergo an inefficient process of *release* from infected cells following cell lysis.

Early Steps in Replication: Entry into Cells

Attachment to Cell Surface Receptors

A number of studies have investigated the attachment of reoviruses to erythrocytes and other cells, some of which are capable of undergoing productive infection. Most studies have utilized cells *in vitro* (see below), but some have investigated the binding of reoviruses to different cells within host animals (see Chapter 53 in *Fields Virology*, 3rd ed.). These studies generally indicate that defined "receptor" moieties, exposed on the cell surface, are responsible for binding reovirus particles via specific interactions with the viral hemagglutinin and cell-attachment protein σ1 (see "Proteins" section). Attachment to surface receptors is considered a required first step in the productive infection of cells. Interactions of cell surface proteins with other reovirus proteins, contributing to attachment or subsequent steps in infection, remain possible but undefined.

Hemagglutination

Hemagglutination has been used as a convenient assay to study the attachment capabilities of reovirus particles. Isolates from all three serotypes agglutinate human type O erythrocytes, but only T3 isolates agglutinate bovine red cells (116). This difference between strains T1L and T3D was mapped to their S1 genes, implicating the encoded structural protein σ1 as the reovirus hemagglutinin (420). Hemagglutination can also be mediated by purified σ1, and biochemical and molecular genetic studies have suggested that a central region of sequences in T3 σ1 proteins is involved in this activity (see "Proteins" section).

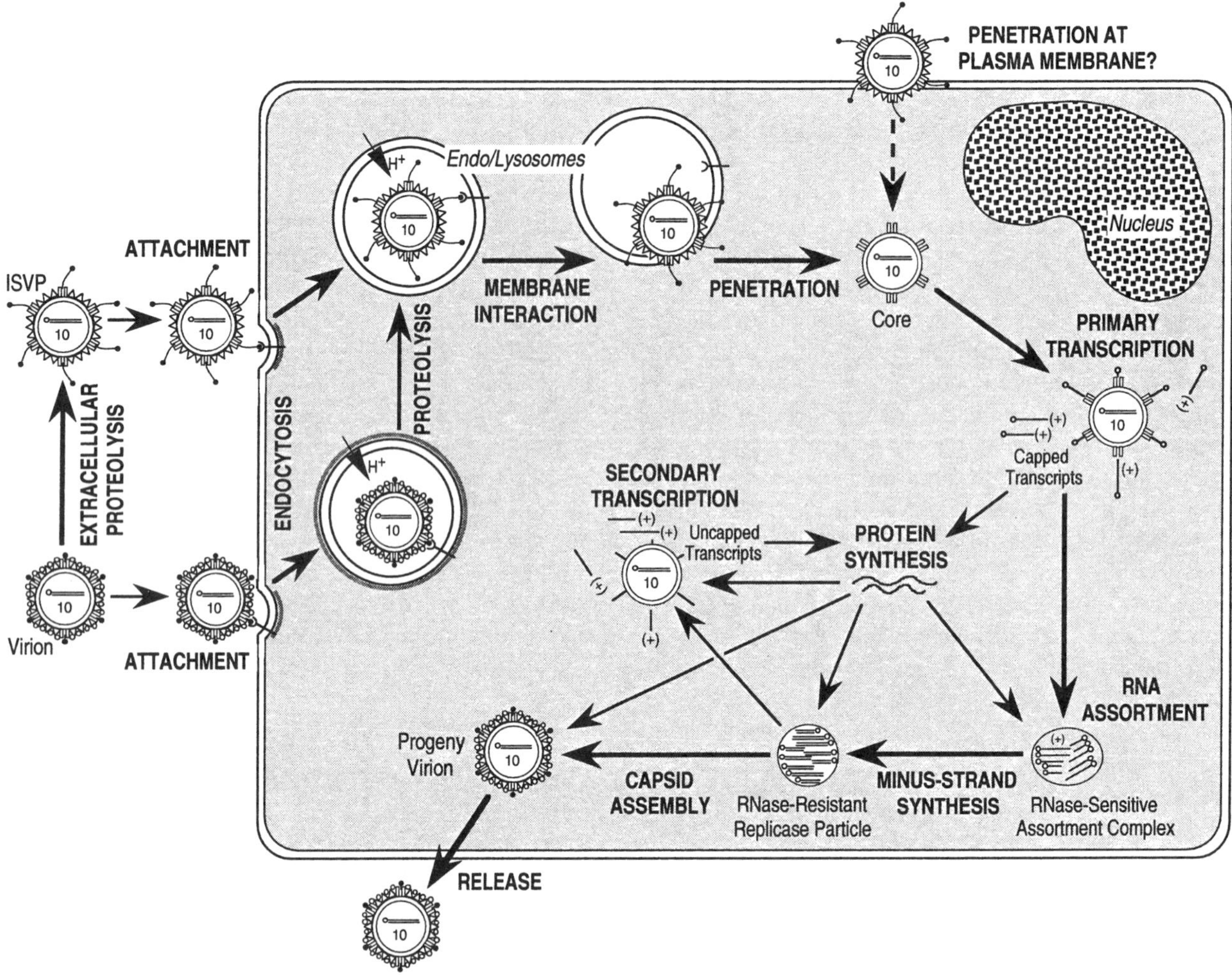

FIG. 8. Reovirus replication cycle. See text for detailed explanation. The primary steps in replication are labeled (*bold capital letters*). Virions, ISVPs, and cores, as well as other assembly-related particle forms are indicated. Capped plus-stranded RNAs—as part of the ten dsRNA genome segments, as cytoplasmic transcripts, and as assortment elements—are indicated (5′-terminal *open circles* represent caps). Uptake from the cell surface is shown to involve receptor-mediated endocytosis, with viral particles sequentially associated with clathrin-coated pits, clathrin-coated vesicles, endosomes, and lysosomes (the last three compartments are acidic, as shown). The roles played by most individual reovirus proteins are not indicated here.

A decrease in hemagglutination is commonly noted after erythrocytes are treated with periodate or proteases, suggesting that one or more glycoproteins on the red cell surface serves as the receptor for virus binding prior to agglutination (15,352). There is strong evidence, including the sensitivity of hemagglutination to treatment of red cells with neuraminidase and to treatment of virions with sialoglycoproteins, that sialic (N-acetylneuraminic) acid residues are essential components of this receptor in the case of T3 reoviruses (15,148,297). Glycophorin A is the primary sialoglycoprotein found on human erythrocytes and is capable, in purified form, of binding to the $\sigma1$ proteins of T3D and other T3 strains (101,276,297,404,439); thus, glycophorin A was specifically proposed to represent the erythrocyte receptor for (T3) reoviruses (297).

Hemagglutination by T1 and T2 reoviruses has been studied less thoroughly than that by T3 reoviruses but has some different characteristics (116). Unlike T3D and some other T3 strains, reovirus T1L is deficient in binding to sialoglycoproteins, including glycophorin A, as determined by its S1 gene (101,294); thus, the nature of the erythrocyte receptor for T1 and T2 reoviruses remains undefined.

Attachment to Other Cells

Studies of reovirus attachment have been performed with a variety of cultured cells, utilizing different virus strains and techniques of analysis. Attachment to L cells by strain T3D appears to be mediated by one or more populations of saturable receptors. Estimates for the number of attachment sites per cell range between 1×10^4 and 5×10^5, and the receptors exhibit a dissociation constant for virus binding of 0.25 to 13 nM (11,15,119,147,338). Maximal attachment is generally achieved after 30 to 90 minutes, regardless of temperature (15,119,147). Studies in other cell types or with strain T1L have provided similar estimates for the copy numbers and dissociation constants of receptors (11,119,147,237,338,402). Competition experiments have disputed whether T3D and T1L bind to the same (78,213) or different (11,119,415) receptors on L cells; a recent study argues for one receptor being shared by these two strains plus a second being restricted to T3D (11). There is general agreement that attachment is mediated primarily or exclusively by the σ1 protein (see "Particles" and "Proteins" sections and below).

The nature of the L-cell receptor(s) for reovirus T3D is well studied but remains incompletely defined [for review see (337)]. There is generally consistent evidence for involvement of a sialoglycoprotein based on the decreased virus-binding capacity of L cells treated with proteases, periodate, neuraminidase, or particular lectins (15,78,147, 149,326). The capacity of T3D virions or σ1 protein molecules to bind L cells is similarly decreased in the presence of sialoglycoproteins (149,294,297,433), gangliosides (15), or free sugars that contain α-linked sialic acid residues (296). The latter observations have been used to argue that the L-cell receptor, like the erythrocyte receptor described above, includes sialic acid. One study suggested that the α-anomer of sialic acid is the minimal receptor determinant required for reovirus (T3D) recognition (296). Correlation between the capacity of certain T3 strains to bind glycophorin and their capacity to bind and productively infect murine erythroleukemia cells (327) is consistent with a model in which binding to sialic acid is critically important for infection of at least some cells.

Other data suggest that one or more specific proteins, perhaps sialylated glycoproteins, can serve as cell surface receptors for T3D. Initial data in this regard was acquired using an antiidiotype approach [for review see (131,434) and Chapter 53 in *Fields Virology*, 3rd ed.]. Antibodies were raised against the anti-σ1 monoclonal antibody G5, which blocks the binding of T3D to cells and thereby neutralizes its infectivity. These preparations of G5-specific antibodies, as well as similarly derived monoclonal antibodies, exhibited behaviors predicted for antiidiotypic antibodies: they were capable of blocking the binding of T3D to cells, presumably by mimicking within their own structures a portion of the attachment site on σ1 and thus binding to σ1-specific receptors on the cell surface (60,278,293,432,433). A monoclonal antiidiotypic antibody was subsequently used to identify a 67-kD membrane glycoprotein, having notable structural similarity to the β-adrenergic receptor, as a probable receptor for reovirus T3D (81,82). Subsequent studies have provided mixed support for this hypothesis (76,77,78,105,338), and it currently seems unlikely that this protein represents the sole or even primary receptor for reoviruses on many cells. Other reports describe receptors for reovirus T3D that include the EGF receptor (394), a 54-kD protein on endothelial cells recognized by both T3D and T1L (402), and multiple sialoglycoproteins on L and human erythroleukemia cells (76,78). Reports describing specific receptors for reovirus T1L are less frequent but include a 47-kD protein from mouse intestinal epithelial cells (415) and undefined moieties on Caco-2 cells (11) and mouse ependymal cells (278,293).

Whether sialic acid or another protein element(s) alone is sufficient to mediate reovirus attachment to cells resulting in productive infection remains in question. A two-step mechanism involving initial binding to sialic acid and subsequent interaction with a specific protein receptor has been suggested (296). Available data suggest that differences in receptors are likely to exist between cell types and between reovirus serotypes or strains. It may also be important to distinguish attachment which leads to productive infection from that which is nonproductive; the capacity of attachment to lead to virus uptake from the cell surface by endocytosis may be an important distinguishing feature in this case (76,326) (see below). The importance of multivalent attachment via two or more of the σ1 oligomers present in viral particles remains undefined. Differences in the attachment process for virions and ISVPs are also unknown (see below).

Endocytic Uptake and Proteolytic Activation

Thin-section electron micrographs of cells soon after infection reveal reovirus particles associated with clathrin-coated pits or vesicles at or near the plasma membrane, suggesting that uptake from the cell surface can occur by receptor-mediated endocytosis (51,54,326,392) (Fig. 8). Particles are subsequently observed within vacuoles resembling endosomes and lysosomes (51,54,153,326,365, 392). Treatment with colchicine and nocodazole does not inhibit endocytosis, but inhibits the movement of these virus-containing vacuoles toward perinuclear regions, suggesting that vacuole translocation occurs along microtubules (153). Only some of the proteins to which reovirus particles bind on the cell surface may be capable of signaling endocytic uptake (76,326).

Experiments with radiolabeled virions provide evidence for proteolysis of viral proteins within hydrolytic compartments of the endocytic pathway (late endosomes or lysosomes), starting 20 to 30 minutes after infection at 37°C (365,366,392). The proteolytic cleavages are notable in that they involve only certain outer capsid components. Specifically, degradation of protein σ3 and conversion of protein μ1C to its stable cleavage product δ are observed, generating particles very similar to ISVPs that result from *in vitro* proteolysis (54,74,363,392). The reproducibility of these limited cleavages suggested that the generation of ISVP-like particles by proteolysis within lysosomes may be a required step in reovirus infection.

More direct evidence for proteolytic activation as a required step in infection arose from an observation that infection is inhibited by the weak base ammonium chloride, which raises the pH within endosomes and lysosomes (63). Later experiments revealed that an early step in infections initiated by virions, but not ISVPs, is inhibited by this agent (45,102,236,392). In addition, the extent of inhibition was found to correlate with the extent to which cleavages of proteins σ3 and μ1C in infecting virions are blocked. It was suggested that, by raising the pH within endosomes and lysosomes, ammonium chloride inhibits the activity of acidic proteases within those compartments and blocks the generation of ISVPs. ISVPs may resist the block to infection by ammonium chloride because they have undergone proteolysis *in vitro* and thus no longer contain an acid-dependent step in their replication cycle (392). Proteolysis appears to activate reovirus particles for the next steps in infection, membrane interaction and penetration (see below).

In certain settings, proteolytic activation of virions occurs outside cells, before attachment and endocytosis (Fig. 8). When newborn mice are perorally inoculated with virions, proteolysis to generate ISVPs occurs within the lumen of the small intestine and is probably mediated by pancreatic serine proteases (29,45). Extracellularly generated ISVPs are apparently required to initiate infection of mouse tissues by that route (29), but the basis for this requirement is unknown. ISVPs generated *in vitro* may have the capacity to bypass endocytic uptake and to penetrate into the cytoplasm directly through the plasma membrane (see below).

Interaction with and Penetration of Cell Membranes

As with other nonenveloped viruses, the mechanism by which reoviruses cross the membrane barrier during entry into cells (Fig. 8) is poorly defined. Two assays have demonstrated a direct interaction between reovirus particles and lipid bilayers. One involves the capacity of particles to effect the release of ^{51}Cr from preloaded L cells (51,229). More recent studies revealed the capacity of particles to induce conductance through artificial planar bilayers (397). Membrane interaction in both these assays was restricted to ISVPs; virions and cores did not exhibit the activity. Moreover, the membrane interactions did not require acid pH. These findings are consistent with a model in which ISVPs are required intermediates in the process of reovirus entry into cells and in which the acid-dependent step in reovirus entry (the one inhibited by ammonium chloride) is proteolytic activation rather than membrane interaction (229,284,392,397).

Available data implicate the outer capsid protein μ1, which covers most of the surface of ISVPs, in mediating direct interactions with membranes. A strain difference in the capacity of ISVPs to mediate ^{51}Cr release from L cells was mapped to the M2 gene, which encodes μ1 (229). Features of the μ1 protein noted to be consistent with a role in membrane interaction include (a) the modification of it and its amino-terminal fragment μ1N with a myristoyl group (285) and (b) the presence of a predicted pair of long, amphipathic α-helices flanking the cleavage junction between the μ1δ/δ and ϕ fragments of μ1/μ1C that are found in ISVPs (282) (see "Proteins" section). A conformational change in μ1 is expected to precede reovirus-membrane interaction according to one model (229,282,284). The relative importance to membrane interaction of σ3 degradation and μ1/μ1C cleavage into fragments μ1δ/δ and ϕ remains unclear; however, evidence for an intimate interaction between these proteins obtained by cryoelectron microscopy and image analysis (110) suggest that σ3 might interfere with an activity attributable to μ1.

Direct interaction of viral components with a cell membrane is expected to be one step in a process that gives reovirus particles access to the cytoplasm, where there are substrates for transcribing and capping the viral mRNAs and ribosomes for translating them (Fig. 8). As nonenveloped viruses, reoviruses cannot use fusion as a general mechanism for this process, and little is known about the steps that follow membrane interaction and permit reovirus components to penetrate into the cytoplasm. Suggested possibilities are (a) that membrane interaction results in local disruption of the bilayer such that virus enters the cytoplasm directly through a large breach and (b) that a more precise, porelike structure is formed by viral proteins within the membrane, through which viral components enter the cytoplasm (51,54,282,284,397) (also see "Transcription" section).

A current model (Fig. 8) suggests that when ISVP-like particles are generated from virions by proteolytic activation within late endosomes or lysosomes, membrane interaction and penetration are most likely to occur within and from those vacuoles (51,54,284,392). When ISVPs are generated *in vitro*, however, they may be capable of penetrating directly through the plasma membrane, prior to endocytosis (51,54,229) (Fig. 8), although other work suggests that ISVPs also undergo endocytic uptake and sequestration into vacuoles prior to penetration (392). Whether penetration through plasma membrane by extracellularly generated ISVPs occurs routinely during reovirus infections remains to be determined.

Transcription

Transcription by purified reovirus particles was first described in 1968 (47,357) [for review see (354)]. Notable features include the stability and productivity of the transcriptase and its encasement within icosahedral particles (25,47,107,190,221,357,368). A common system for *in vitro* transcription includes reovirus cores, nucleoside triphosphates, and an ATP-regenerating system. The products of transcription are ssRNAs, which are released from the transcribing particle (25,28,156).

Reovirus transcription is conservative and asymmetric (25,221,368). Ten distinct transcripts have been identified, which by hybridization and sequencing (98,204,224,247) appear to represent full-length copies of the ten genomic plus strands. Except for repetitive initiation (see below), there is no evidence that any of the gene segments give rise to less-than-full-length transcripts as part of the viral replication strategy.

If appropriate substrates are provided, transcripts produced by reovirus particles *in vitro* can undergo capping and methylation at their 5′ ends, indicating that several transcription-related enzymes are present within the particles (see below). Capping imparts increased stability to reovirus mRNAs (136) and increases the efficiency with which many mRNAs are translated within eukaryotic cells (354). Reovirus transcripts are not polyadenylated.

Discrete Enzyme Activities in Transcription and Capping

Each of the steps in transcription and capping (139) (Fig. 9) is mediated by an enzyme present in virions and probably encoded by a viral gene. The 5′ cap of reovirus transcripts includes a 7-N-methyl guanosine linked by three phosphate groups to the 5′-terminal guanosine present in all reovirus mRNAs (127,137). The templated 5′-terminal guanosine is 2′-O-methylated (127,137). The next residue

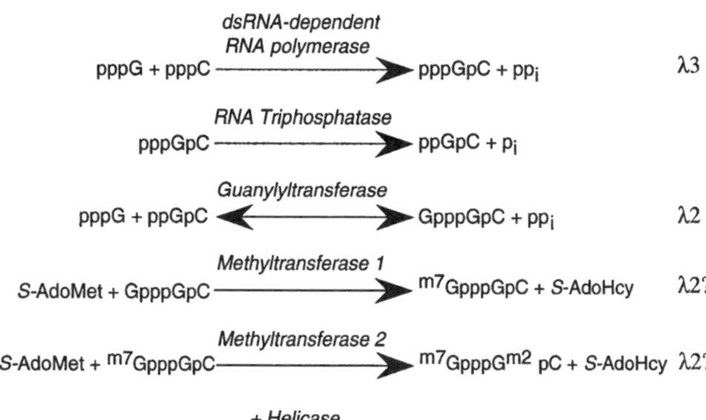

FIG. 9. Enzymatic activities in transcription and capping by reoviruses. Polymerization of nucleoside triphosphates (shown here for the first nucleotide pair) as specified by a genomic dsRNA template is mediated by the dsRNA-dependent RNA polymerase. The RNA triphosphatase generates a nucleosidyl diphosphate at the 5′ end of the nascent mRNA and sets the stage for formation of the dimethylated 5′ cap by sequential actions of the guanylyltransferase and methyltransferase(s). The role of a helicase in reovirus transcription is proposed but not characterized. Individual reovirus proteins proven or suspected to mediate each activity are indicated (*right*). *S*-AdoMet, *S*-adenosyl methionine; *S*-AdoHcy, *S*-adenosyl homocysteine.

in all reovirus mRNAs is cytidine, which can also undergo 2′-O-methylation, but by a cytoplasmic methyltransferase. Reovirus capping activities can be uncoupled from transcription and readily function when exogenous substrates, including preformed oligonucleotides and reovirus mRNAs in some cases, are incubated with viral particles (140,263,277).

Polymerase

Early work failed to associate polymerase activity with any individual reovirus proteins after disruption of cores (188,190,423). The transcriptase (dsRNA-dependent RNA polymerase) was later suggested to be formed by portions of both core proteins λ1 and λ2 due to their apparent labeling by pyridoxal phosphate during transcription (262, 263). Recent work indicates that the minor core protein λ3 is more likely to be the catalytic subunit of the transcriptase (61,107,199,267,387). A limited spectrum of activity [poly(C)-dependent poly(G) polymerase] has been reported for λ3 expressed in isolation, which probably indicates that other core proteins are needed to form the complete transcriptase (387). The λ3 protein may also be a primary component of the ssRNA-dependent RNA polymerase that operates during minus-strand synthesis (see "Later Steps in Replication" section). Sequences in reovirus RNAs that are recognized by the polymerase (e.g., initiation sites) have yet to be defined.

Helicase

The double-stranded nature of the reovirus genome may necessitate an RNA helicase activity for transcription to proceed. This activity might be inherent to the polymerase but also might be localized to another core protein (387). One report suggested that the antiviral drug ribavirin inhibits reovirus replication by interfering with helicase activity (311).

RNA Triphosphatase

An RNA triphosphatase (polynucleotide phosphohydrolase) should initiate formation of the 5′ cap on reovirus mRNAs by releasing inorganic phosphate from the original 5′ triphosphate of the transcript and generating a 5′ diphosphate (Fig. 9). A nucleoside triphosphatase has been found in reovirus virions and cores (48,191) and may represent this enzyme. The activity has yet to be ascribed to a particular core protein (235).

Guanylyltransferase

Protein λ2 has been defined as the reovirus guanylyltransferase (see "Proteins" section). Its enzymatic mechanism includes formation of a covalent λ2-GMP intermediate at lysine 226, using GTP as substrate (80,126,353). The GMP is subsequently transferred to the 5′ diphosphate of reovirus mRNAs (Fig. 9). The guanylyltransferase is promiscuous in its specificity for GMP-acceptors in that PP_i, GDP, GTP, and both ppG- and ppA-terminated oligonucleotides work as substrates (80,139). A GTP-dependent, pyrophosphate exchange activity associated with cores (405) represents guanylylation plus its reverse reaction. After methylation, transcript termini become resistant to pyrophosphate exchange (139); thus, reversibility of guanylylation may be required to assure that proper GpppG caps are formed on transcripts to act as substrates for methylation (see below).

Methyltransferases

Direct evidence is lacking for the protein(s) that act as methyltransferases to modify the 5′-terminal guanosines of reovirus mRNAs (Fig. 9), but a role for λ2 is proposed (see "Proteins" section). S-adenosyl methionine can serve as methyl group donor for these reactions (128,139). The 5′-terminal structure of reovirus mRNAs is specifically recognized by the methyltransferases since GpppG and GpppGC can act as methyl acceptors but GppG, GppppG, and GpppGA cannot (354).

Other Transcription-Related Activities

A nucleotide pyrophosphatase activity is found in both reovirus particles and lysates of infected cells and might represent a viral enzyme needed to generate the monophosphorylated 5′ ends of uncapped late mRNAs (369,370,405, 442) (see "Later Steps in Replication" section). The synthesis of initiator oligonucleotides and oligoadenylates represents alternative activities of the viral transcriptase (see below and "Later Steps in Replication" section). Putative nucleotide-binding sequences in λ1 (27) suggest that it may have one of the transcription-related activities not yet assigned to a reovirus protein. The dsRNA-binding activities of λ1 and σ2 (see "Proteins" section) may be important for the positioning or movement of dsRNA templates during transcription.

Structural Organization and Mechanism

The sequential action of enzymes that synthesize reovirus mRNAs implies a precise structural organization within the core. Some studies suggest that capping of nascent mRNAs occurs when they are only 2 to 15 nucleotides long (139,263), in which case the transcriptase and capping enzymes must be close together. Since preformed oligonucleotides and reovirus mRNAs can be capped by cores (140,263,277), the capping enzymes must be accessible to external solvent; however, antibodies have not been found to inhibit any transcription-related activity (354).

Transcription-related enzymes may be located in complexes near the fivefold axes of reovirus particles. Electron micrographs of transcribing cores suggest that mRNAs are released through the λ2 spikes that protrude upward around the fivefold axes of cores (28,156,423). This idea is supported by reconstructed images from cryoelectron microscopy showing a channel through the λ2 pentamer that is open in cores (110,253). In recent work λ3, the putative catalytic subunit of the transcriptase (see above), was shown to bind to λ2 (387). The low copy number in which λ3 is found in cores (12 copies) has also been used to argue that it is located near the fivefold axes.

Each of the ten reovirus gene segments may represent an independent transcriptional unit. Both *in vitro* and in cells later in infection, transcripts from the ten segments are produced at rates which are approximately proportional to the reciprocals of their lengths (25,187,368,413,447). Given that transcript elongation is the rate-limiting step in transcription (see below), this finding suggests a model for transcription in which each of the ten genes is transcribed independently and not as part of a linked array of genes (187,188,354). A feature of this model is that it involves at least ten transcriptases within the reovirus core (368). Electron micrographs of transcribing cores with multiple mRNAs being simultaneously extruded around the particle perimeter support this idea (28,156). Both moving transcriptase and moving template models for reovirus transcription have been proposed (188,354,437). Experiments in which transcription was inhibited by crosslinking genomic dsRNA are consistent with both these models (277,354); however, the moving template model is generally accepted. According to this model, a set of transcription-related enzymes is fixed near the base of each λ2 spike within the core. It is proposed that the entire length of each dsRNA gene segment moves past the fixed transcriptase and that the nascent mRNA is directed past the capping enzymes and out the λ2 spike. The liquid crystalline state of the packaged genomic dsRNA (110, 169) may permit it the freedom of movement required by this model.

Transcriptase Activation

Early studies showed that the virion-associated transcriptase can be activated by heat (47) or by treatment with proteases like chymotrypsin (357). For protease treatments yielding cores as end-products, transcriptase activation was correlated with loss of protein μ1 from particles (50,107, 185). In addition, a strain difference in the treatment conditions needed to generate cores and activate transcription was mapped to the M2 gene, which encodes μ1 (107). A number of parameters (e.g., the concentration and nature of monovalent cations, divalent cations, proteases, and viral particles) were found to affect whether transcriptase activation occurs during protease treatments, in general correlating with whether cores were produced (46,49,50,55, 107,185).

While some studies suggest that ISVPs are transcriptionally active (18,74,355,363), others indicate that the transcriptase is latent in ISVPs but can be activated under certain conditions without added proteolysis (49,50,53, 121,335). One study correlated the switch-on of transcriptase activity in ISVPs with a change in the μ1δ fragment of μ1 (121). From these studies, a model was proposed in which virions undergo transcriptase activation via two steps (50,54,121,284). The first step requires proteolysis and produces ISVPs. The second step is protease-independent, produces transcriptionally active particles, and may be linked to penetration during entry into cells (see "Early Steps in Replication" section).

Although virions cannot make full-length transcripts, they readily synthesize short oligonucleotides when given appropriate substrates (437). This important finding suggests that the transcriptase is constitutively active in virions but capable of only limited elongation. Capping enzymes are also active in virions (64,191,437); thus, transcriptase activation must primarily involve relief of a block to elongation. The nature of this block is unknown but may relate to movement of the dsRNA templates (188, 354). Changes in the outer capsid are essential for activation (see above), implying that interactions between outer and inner capsid proteins may produce the elongation block. One study suggested that structural changes in the core occur at the onset of transcription (300), and it may be that these changes are required for elongation but inhibited by the outer capsid proteins in virions and ISVPs. A conformational change in λ2 noted between ISVPs and cores in reconstructed images from cryoelectron micrographs (110) (see "Particles" section) may be important for transcriptase activation in that it opens a fivefold channel through which the substrates or products of transcription may be required to pass.

Initiator oligonucleotides are the major transcription products produced *in vitro* by both virions and cores, even though cores can at low frequency elongate these into full-length transcripts (436,440). This and other findings indicate that elongation is the rate-limiting step in reovirus transcription (190,436,437,440). Two phases of transcriptase activity are thus suggested: (a) efficient, repetitive initiation yielding large amounts of initiator oligonucleotides and (b) inefficient elongation yielding smaller amounts of full-length transcripts (437).

Primary Transcription Within Cells

Two types of transcription by reoviruses are considered to occur within infected cells, based on the derivation of the transcribing particle. Primary transcription is mediated by subviral particles derived from input particles (54, 74,363,365,366) (see below and "Early Steps in Replication" section). It results in production of capped transcripts that serve both as mRNAs for translation and as templates for minus-strand synthesis within progeny particles. Primary transcripts are detected by 2 hr after infection, reach a maximum at 6 to 8 hr, and decrease to undetectable levels by 12 hr (186,365,447). The primary transcriptase particle may be inactivated when newly synthesized outer capsid proteins are reassembled onto it (18). Secondary transcription is mediated by newly assembled particles and is the greater source of transcripts during infection (180,441). It is reported to involve production of uncapped mRNAs (see "Later Steps in Replication" section).

Primary transcription may be subject to additional regulation within cells, though some authors have disputed these findings (186,364,447). Transcripts from four "pre-early" genes (L1, M3, S3, and S4) were reported to be made preferentially during the first several hours of infection (291,377,413); moreover, transcription from the six other genes was suggested to require new protein synthesis by being blocked with cycloheximide (211,355,413). One hypothesis is that a cellular repressor prevents transcription of the six later genes and that a protein encoded by a pre-early gene interferes with its activity (211,291,441).

The nature of the primary transcriptase particle remains in question; however, because of their transcriptional activity *in vitro*, cores are assigned this role in one model (Fig. 8). In one study, transcriptionally active particles derived from input particles were isolated from infected cells and shown to have the characteristics of cores, perhaps indicating that additional uncoating occurs in association with or subsequent to penetration (54). Because proteinase K was used to separate the particles from cell debris in this study, degradation of other viral components may have occurred before the particles were characterized (441). Other studies suggest that the primary transcriptase particle is more closely related to ISVPs (18,74,355,363,366) (see above).

The intracellular site of primary transcription also remains undetermined. Thin-section electron micrographs have been reported to show small numbers of corelike particles free within the cytoplasm of newly infected cells (44,54) and to suggest that the primary transcriptase particle is introduced fully into the cytoplasm (Fig. 8) (see "Early Steps in Replication" section). In fact, most reovirus particles in such micrographs are located within endocytic vacuoles (54,365,392). It is not clear whether these particles have not yet succeeded in initiating infection or whether they represent primary transcriptase particles. Thus, another model was proposed in which the primary transcriptase particle remains within a vacuole but inserts one or more of its λ2 pentamers through the membrane so that mRNAs can be extruded into the cytoplasm (392).

Later Steps in Replication: Generation of Progeny Virions

Translation of Viral mRNAs

Shortly after the onset of infection, there is a gradual increase in synthesis of reovirus proteins and a decrease in synthesis of cellular proteins (160,348). Newly synthesized viral proteins are detected by 2 hr postinfection; by 10 hr most of the proteins synthesized in the infected cell are viral in origin (447). The capacity of viral mRNA to associate with ribosomes may be a host-range determinant, since certain avian reoviruses can enter mammalian cells and transcribe their early mRNAs, yet fail to initiate protein synthesis (42,234). The mechanism by which viral protein synthesis comes to predominate in infected cells is not well understood (see "Effects on Host Cells" section).

In addition to regulation of translation of viral versus cellular mRNAs, the translation of individual reovirus transcripts is regulated. The different gene segments are transcribed at variable rates, inversely proportional to segment size, and the transcripts have similar stabilities (see "Transcription" section); thus, differences in the amounts of each viral mRNA provide one source of variability in the amounts of each protein that are synthesized (211,447). A second source of variability is in the translation frequency of the different mRNAs. Reovirus mRNAs differ in translation frequency by as much as 100-fold (187). Such variability has been found in both infected cells and *in vitro* translation systems (20,141,164,222,249,250,447,448). The observation that the M1 mRNA is translated much less efficiently *in vivo* than *in vitro* has suggested that its translation may be influenced by negative regulatory factors in cells (319). Nucleotide sequences of reovirus mRNAs at the –3 and +4 positions flanking the initator AUG contribute to differing translation efficiencies *in vitro* (202); however, other nucleotide positions may influence the efficiency of translation of these mRNAs *in vivo* (41,270, 318). For the S1 mRNA, from which two unrelated proteins are translated, it was proposed that translation rates are regulated at the level of polypeptide elongation (41, 106,122). Reovirus mRNAs also may differ in capacity to compete for components of the translation machinery (57, 58,234,312,406). Several studies suggest that levels of translation may be affected by the influence of different viral mRNAs on negative regulators of translation, such as the dsRNA-activated protein kinase (41,43,331).

Assortment and Packaging of the Ten Viral mRNAs

The mechanism ensuring that ten unique dsRNA segments are assembled into each newly formed virion remains unknown. Pieces of evidence that the reovirus genes are assorted selectively are that a particle-to-pfu ratio as low as 1 has been reported (380) and that the central cavity of reovirus particles is not large enough to accommodate many more than the ten unique segments (110,169). It is generally accepted that assortment of the ten segments occurs at the level of the reovirus mRNAs and that minus-strand synthesis to generate dsRNA occurs later, within a nascent viral particle (409,410,445,446) (see below and Fig. 8). A recent report suggested that assortment and minus-strand synthesis may be more intimately linked than previously interpreted (14).

Assortment is likely to involve signals that identify an RNA as viral and not cellular as well as signals that specify each of the ten segments as unique. The relevance of either RNA:protein or RNA:RNA interactions to this process is unclear. All ten reovirus gene segments contain short conserved regions at the 5′ and 3′ ends of their plus and minus strands, but it is not known if these sequences play any role in packaging (see "dsRNA Genome" section). Sequences near the termini of some reovirus ssRNAs may interact in a panhandle-like structure which may be important for packaging (75,223,224,443). Analysis of mutants containing internal deletions of their M1 genes identified sequences within 140 to 190 nucleotides of the 5′ and 3′ termini that contain the minimum required signals for packaging (443). The plus strand of each genomic segment appears to contain a 5′ cap identical to that on mRNAs made by primary transcriptase particles (138) (see "Transcription" section), so the cap might serve as a packaging signal as well. Whether uncapped late mRNAs may also undergo assortment and packaging and give rise to genomic dsRNA remains controversial (see below).

One early study hypothesized that σNS plays a role in mRNA assortment and packaging because of its capacity to bind ssRNA (175). Protein assemblies, containing primarily σNS and apparently capable of binding to selected regions of reovirus mRNAs, were subsequently isolated from infected cells, consistent with the idea that this protein plays a significant role in the earliest stages of particle assembly (157,158,385). In a more recent study, protein-specific antisera were used to precipitate protein:RNA complexes from infected cells in an attempt to identify components of the assorting particle (14). This work suggested that σNS, as well as μNS and σ3, associate with individual reovirus mRNAs to form small nucleoprotein complexes soon after transcription, but that μNS or σ3 (not σNS) binds to the mRNAs first.

Minus-Strand Synthesis: The Replicase Reaction

Each packaged plus strand serves as a template for synthesis of a minus strand (1,329,330,341,445). Minus-strand synthesis proceeds from an initiation point at the 3′ end of each template, and once the minus strand is synthesized, it remains within the nascent progeny particle. The viral replicase catalyzes a single round of minus-strand synthesis. Initiation of minus-strand synthesis, but not elongation, depends upon continuing protein synthesis (413).

Although preparations of replicase particles are often heterogeneous and include particles with transcriptase activity (265,329,411,446), it is generally accepted that these two activities are associated with discrete types of particles (265). Particles with replicase activity have a buoyant density similar to virions, are morphologically indistinct, and contain the viral core proteins with reduced amounts of λ2 and other outer capsid proteins relative to virions

(265,446). The replicase activity associated with nascent particles is susceptible to RNase, but once minus-strand synthesis is complete, the particles become RNase-resistant (1). Recent work suggests that λ3 acts as the ssRNA-dependent dsRNA polymerase of reoviruses (387), but the roles of other proteins in minus-strand synthesis remain undefined.

Secondary Transcription

The appearance of "late" viral mRNAs represents transcription by progeny subviral particles (*secondary transcriptase particles*) (265,370,442,446). These particles are probably derived from those with replicase activity, but are distinguishable by sedimentation behavior, by the presence of a full complement of dsRNA, by the presence of variable amounts of protein μNS (μ0), and by reduced amounts of λ2 and the other outer capsid proteins (265,370,449). They appear to have latent capping enzymes and produce uncapped mRNAs (370,371,441,442). These uncapped viral mRNAs are proposed to be translated preferentially at later times in infection (see "Effects on Host Cells" section). Late viral transcripts begin to appear 4 to 6 hr after infection, reach maximum amounts (perhaps greatly exceeding the quantity of primary transcripts) at 12 hr, and subsequently decrease (207,371,413,414,441). Whether they serve only for translation into proteins or whether they may also be packaged into progeny particles and serve as templates for minus-strand synthesis remains controversial (130,266,341,441).

Assembly of the Viral Capsids

A great deal remains to be understood about the protein and nucleoprotein complexes involved in assembling the inner and outer capsids in reovirus particles (Fig. 8). The earliest reovirus particles may be ones that contain reovirus mRNAs primarily in complex with σ3 and the nonstructural proteins μNS and σNS (14,157). The next particle along the assembly pathway is described to lack nonstructural proteins, to include both inner and outer capsid proteins, and to be RNase-sensitive (1,265). These "replicase particles" contain a ssRNA-dependent dsRNA polymerase activity and appear to be engaged in minus-strand synthesis (265). With the completion of minus-strand synthesis, the replicase particles become RNase-resistant, probably reflecting the occurrence of structural rearrangements within the particle (1,441). Additional copies of the outer capsid proteins may be added to form complete virion particles (265,449). A final step in morphogenesis is synthesis of the virion-associated oligonucleotides (see "Particles" section), which may occur as the viral transcriptase undergoes inhibition during outer capsid assembly (18,38, 285,362,449) (see "Transcription" section). An alternative assembly pathway may involve the addition of nonstructural protein μNS to RNase-resistant replicase particles, which may interfere with the completion of outer capsid assembly and yield immature particles with secondary transcriptase activity (265). It is not known if particles that engage in secondary transcription can go on to form mature virions.

Many of the reovirus structural proteins exhibit some degree of self-assembly in the absence of viral RNA or a full complement of viral proteins. For example, cells coexpressing the major inner capsid proteins λ1 and σ2 form particles that are morphologically similar to core shells (435). When λ2 and λ3 are coexpressed with λ1 and σ2, they are incorporated into particles as well, the former creating spikes that project from the inner capsid surface as expected (435) (see "Particles" section). Other findings suggest that λ1 and σ2 may self-assemble into oligomers (dimers or trimers) before or after interacting in larger assemblies (87,110). The λ2 protein, however, does not form pentamers in the absence of other proteins (235). The exact order in which these proteins come together in forming nascent infectious particles and the roles of RNA and the minor core protein μ2 in this process are unknown.

Assembly of the outer capsid is even less well characterized. The major outer capsid proteins σ3 and μ1 bind together in solution to form hetero-oligomers (110,175, 212,396,447). Whether the σ3-associated cleavage of μ1 into fragments μ1N and μ1C (212,285,396) is required for assembly is unknown. The fact that a mutant with a ts lesion in S4 (which encodes σ3) results in the accumulation of corelike particles as assembly intermediates (133,264) may suggest that preformed complexes of μ1 and σ3 are required to bind to nascent particles. Similar observations with a ts mutant in L2 (which encodes λ2) may suggest a role for λ2 in initiating the binding of these complexes to particles (133,264,307). A significant body of work has addressed the self-assembly of σ1 oligomers (see "Proteins" section). Sequences near the amino terminus of σ1 are likely to mediate its interactions with λ2 within virions (23,110,134,135,219,232,281).

Release from Cells

Lysis of infected cells results in an inefficient release of mature virions from the cell debris; however, the steps leading to cell lysis are unknown. There is little evidence for another mechanism (such as membrane budding or vesicular transport) by which reovirus particles might be released in the absence of cytolysis.

EFFECTS ON HOST CELLS

Viral Factories and Involvement of the Cytoskeleton

Reovirus-infected cells develop typical cytoplasmic inclusions which are not bound by cellular membranes (12,

162,313) (Fig. 10). Inclusions first appear as phase-dense granules scattered in the cytoplasm, but these move toward the nucleus and increase in size as infection progresses (133,313,346,379). Viral factories contain dsRNA (162, 367), viral polypeptides (94,133,313,379), and both complete and incomplete particles (21,94,133,313), but they do not contain ribosomes (346). Particles are often found in crystalline arrays (Fig. 10). The characteristics and temporal development of viral factories can vary with virus strain and growth conditions (21,133).

Immunocytochemical studies reveal progressive disruption and reorganization of the vimentin (intermediate) filament network in reovirus-infected cells (346). Vimentin filaments are specifically incorporated into viral factories and thus may play a direct role in replication (93,94,346, 349). The nonstructural protein μNS may associate with vimentin filaments to anchor structures involved in viral assembly to the network (261). Disruption or reorganization of intermediate filaments may cause changes in cell shape that typify the cytopathic effect of reovirus infection and may contribute directly to cell injury (346,349).

Immunocytochemical studies reveal little change in the distribution of microtubules and microfilaments within reovirus-infected cells (346). Microtubules extend directly through viral factories (21,94,133,346,379), and there is evidence to suggest that proteins σ1 (21) and μNS (261) may both associate with microtubules. Experiments with the microtubule-disrupting agent colchicine nonetheless suggest that intact microtubules are not required for productive infection in cultured cells (21,93,378). Other studies implicate microtubules in the transport of reovirus particles (a) within endocytic vacuoles early in infection (153) and (b) during the neural spread of reovirus type 3 by fast axonal transport within infected mice (399).

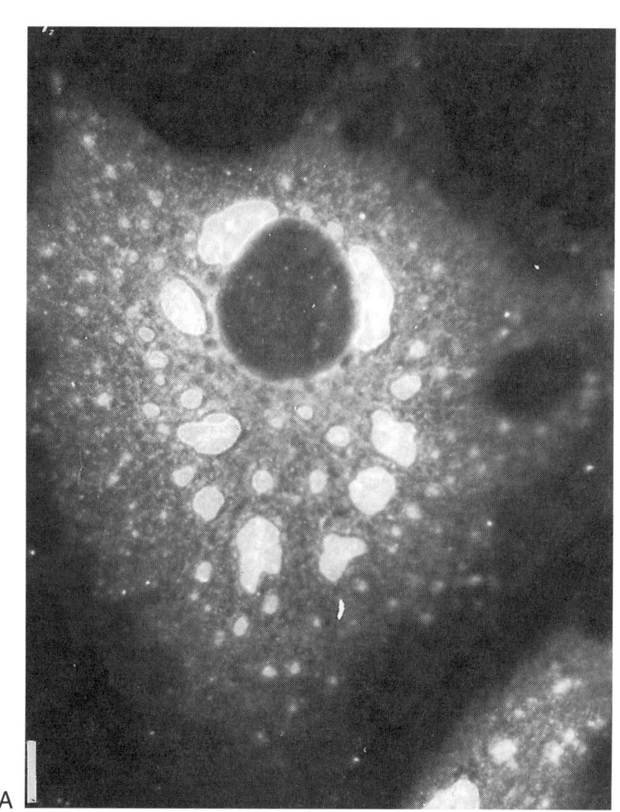

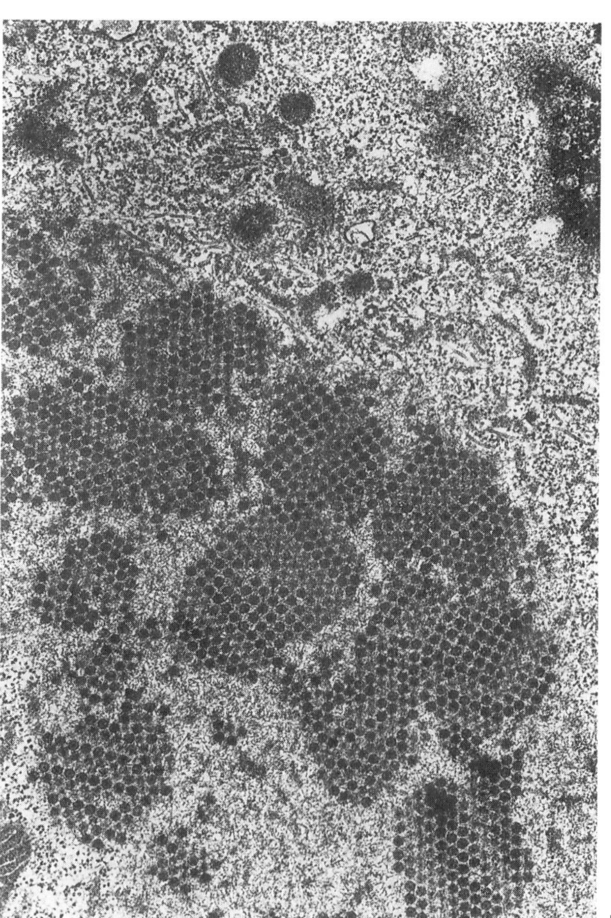

FIG. 10. The reovirus-infected cell. **A:** Viral antigens in CV-1 cells at 48 hr after infection with reovirus type 3 Dearing (T3D). Cells were stained with rabbit antireovirus serum and fluorescein-conjugated goat antirabbit serum. Cytoplasmic inclusions (viral factories) are represented by numerous white globular areas within the cytoplasm and range widely in size. Larger inclusions are routinely found in perinuclear regions as seen here. Bar = 20 mm. **B:** Electron micrograph of an inclusion from a T3D-infected CV-1 cell. The inclusion includes viral particles in crystalline arrays. (H&E stain,×4200.) From Sharpe et al. (346), with permission.

Inhibition of Cellular DNA Synthesis

One consequence of reovirus infection can be the inhibition of cellular DNA synthesis (111,117,160,349). Inhibition of DNA synthesis is detected 8 to 12 hr after infection of L cells with type 3 strains (117,160,347), occurs more rapidly as the MOI is increased (359), and specifically affects progression into S phase so that cells accumulate in G1 (88,118,166,168). Inhibition of L-cell DNA synthesis also occurs with ultraviolet light (UV)-inactivated virions, but not with cores or empty virions (167, 210,347,359). The genetic basis for a strain difference in inhibition by both live and UV-inactivated virions of T1L and T3D was mapped to the S1 gene (347), which encodes both the cell-attachment protein σ1 and the nonstructural protein σ1s. A study involving treatment of neuroblastoma-derived cell lines with T3D virions or antiidiotypic antireceptor antibodies suggested that inhibition of DNA synthesis occurs through a receptor-linked signaling pathway (144). This suggestion is consistent with findings that neither σ1 nor σ1s expressed alone or in combination within mammalian cells results in the inhibition of cellular DNA synthesis (123).

Effects on Cellular RNA and Protein Synthesis

Reovirus infection can also result in the inhibition of cellular RNA and/or protein synthesis (160,207,227,349, 447). Viral replication is required for this effect; moreover, inhibition is MOI-dependent, suggesting that increasing amounts of viral products have increasing effects (348). The extent of inhibition of cellular protein synthesis correlates with rate of viral multiplication, but not with final viral yield (96,269). In one case, a strain difference in the inhibition of cellular RNA and protein synthesis was mapped to the S4 gene, which encodes the σ3 protein (348). The mechanism of inhibition remains unknown (see below), but the demonstration of a dsRNA-binding activity in σ3 led to the suggestion that σ3 mediates this effect by binding to cellular RNA (348).

Several studies indicate that σ3 can downregulate the interferon-induced, dsRNA-activated protein kinase, probably by sequestering its dsRNA activator (179,225). A role for free σ3, versus σ3 complexed with μ1/μ1C, in this activity has been specifically proposed (155,179,225,396). Activation of the dsRNA-activated protein kinase results in phosphorylation of eIF-2α and inhibition of translation initiation. By interfering with this enzyme, σ3 might have a stimulatory effect on translation in some cases, and several studies have shown that transfection of the T3D S4 gene can stimulate translation from a reporter gene (155, 239,344,396). One hypothesis to explain how σ3 can stimulate translation in transfected cells yet play a role in inhibition of cellular protein synthesis during infection is that viral protein synthesis within infected cells is localized and that σ3 downregulates the dsRNA-activated protein kinase within those local regions only, thus sparing viral but not cellular protein synthesis.

Late reovirus transcripts may lack a 5′ cap (370,442), and some studies suggest that reovirus infection induces a modification of the cell translation machinery such that uncapped mRNAs are preferentially translated over capped (cellular) mRNAs (369,371). In one study, extracts from reovirus-infected cells were shown not to support translation of capped mRNAs (374). It has been speculated that a cellular protein required for capped mRNA synthesis may be inactivated in these infected cells (441); however, one recent study found no evidence for cleavage of the eIF-4F subunit p220 in reovirus-infected cells (114). Other studies imply that infection does not result in modification of the existing translation machinery but instead induces a factor that stimulates translation of uncapped mRNAs (216); moreover, protein σ3 can apparently function as such a factor *in vitro* (214,215). Reports in conflict with the preceding ones indicate that reovirus-infected cells can translate capped and uncapped mRNAs with equal efficiency (104, 271), suggesting a hypothesis that the preferential translation of reovirus mRNAs in infected cells is mediated by mRNA competition for a limiting translation factor (57, 58,234,312,406). The reasons for discrepancies between studies are unclear but may reflect the complexity of mechanisms that contribute to regulation of translation in reovirus-infected cells.

Some cell lines support reovirus infection, yet are resistant to inhibition of cellular protein synthesis (97,104, 259,312). In some of these cases, persistent infections are readily established (97,259). The resistance of these cells to cytolysis may involve a higher constitutive level of the dsRNA-activated protein kinase (97).

Induction of Interferon and Other Cytokines

Reoviruses can induce infected cells to express interferon (172,209,228,349,370). With ts mutants, interferon induction correlates with yields of infectious virus, but not with amounts of viral dsRNA, ssRNA, or protein made in infected cells (209). Interferon induction can also vary between wt strains (209,349). Cores and empty virions do not induce interferon, consistent with their poor infectivity (172,209).

It is generally accepted that the genomic dsRNA of reoviruses can serve as a potent inducer of interferon during infection (172,209,228,238,370). The reovirus replication cycle, throughout which the genome remains encased within a protein capsid (see “Replication Cycle” section), may have evolved in part to prevent the genomic dsRNA from inducing a florid interferon response (284). Regions of dsRNA formed within particular reovirus mRNAs (43,75, 443) might also play a role in interferon induction.

Treatment with UV has been used to study interferon induction by reoviruses (172,209). Interferon induction by viral particles is far more resistant to UV inactivation than is infectivity. UV-treated reoviruses induce interferon more rapidly than occurs during productive infection, suggesting that a different mechanism is involved. A range of UV doses was found to increase the potency of strain T3D as an interferon inducer; in addition, UV treatment turned one ts mutant (*ts*C[447]) into an extremely potent inducer, even at nonpermissive temperature. It was proposed that UV treatment destabilizes the inner capsid so that genomic dsRNA from input particles escapes into the cytoplasm (172).

Interferons may inhibit reovirus replication by a variety of mechanisms. In general, interferon α/β appears to block the translation of early reovirus mRNAs (425). Interferon-induced expression of the dsRNA-dependent protein kinase that phosphorylates eIF-2α and ribosomal protein P1 may explain the inhibition of viral protein synthesis in some reovirus-infected cells (99,165,287,332); however, interferon-δ is a poor inducer of this kinase and decreases the yield of progeny virions without having a major effect on viral protein synthesis (152,333,334). In other studies, cells treated with interferon α/β were found to contain an 2′,5′-oligo(A)-dependent endoribonuclease (RNase L) that cleaves reovirus mRNAs and may facilitate the antiviral state (22,257,258,288); however, whether an endoribonuclease contributes to inhibiting the translation of reovirus mRNAs within cells in which the dsRNA-activated protein kinase has been proposed to function remains controversial (257,258,289). One study has questioned whether either of these antiviral effector enzymes have significant effects on reovirus infection (129). Reovirus strains can differ in the extent to which their replication is inhibited by interferon (182).

A recent report showed that reovirus infection of peritoneal macrophages induces secretion of another cytokine, tumor necrosis factor-alpha (TNF-α) (124). Since UV-inactivated reovirus retained this capacity, stimulation of TNF-α expression might occur through receptor-linked signaling pathways.

Effects on Other Functions in Differentiated Cells

Some studies have identified reovirus effects on differentiated cells. In MDCK (polarized epithelial) cells, reoviruses produced persistent infections characterized by failure to form tight junctions, reduced EGF receptor expression, and increased expression of the adhesion molecule VCAM (259). In 3T3 (primary fibroblast) cells, reoviruses also produced persistent infections, characterized in this case by decreased expression of EGF receptors and increased expression of insulin receptors (403). Thus, reovirus persistence in a variety of differentiated cells can significantly alter cellular functions, including some involved in cellular growth control. An involvement of the EGF receptor-linked signal transduction pathway in the efficiency of reovirus infection in differentiated cells was recently proposed (391).

Other studies investigated how interactions between the reovirus cell-attachment protein and its cell surface receptor(s) affect functions in differentiated cells. In optic nerve glial cultures, ligand binding to the reovirus type 3 receptor stimulated expression of galactocerebroside and myelin basic protein and downregulated other cell surface markers, accelerating differentiation into oligodendrocytes (84,85). Ligand binding to the type 3 receptor in rats also caused changes in oligodendrocyte function, resulting in altered myelin expression (83). In another study, interaction of reovirus with thyroid follicular epithelial cells caused induction of class II MHC molecules (145). Since this occurred even in the absence of viral replication, reovirus attachment to cell surface receptors may be sufficient to affect the expression of cellular proteins, probably through receptor-linked signaling pathways.

PERSISTENT INFECTIONS

Although commonly characterized by the death of cells in which they replicate (lytic infections), reoviruses can produce nonlytic, persistent infections in a variety of cultured cells (3,5,7–9,33,59,97,102,111,259,393,403). Reoviruses are not reported to cause persistent infections of host mammals within the natural world. Mechanisms that distinguish persistent infections are poorly defined but must include a decrease in the cellular injury that accompanies lytic infections. Persistent infections of L cells by reoviruses are best described as *carrier cultures* (3,7,102) and appear to involve two primary phases: *establishment* and *maintenance* (Fig. 11).

Establishment

Viral characteristics influence the establishment of persistent reovirus infections. Persistent infection of L cells can be facilitated with defective stocks obtained by serially passaging viruses at high MOI (7,8,59,102) (Fig. 11). Defective stocks contain a variety of mutants (see "Genetics" section), including some that favor the establishment of persistence. Current evidence is that mutant S4 genes from defective stocks have a primary effect in this regard (7). Because establishment of persistent infection requires that cellular injury be attenuated, it was proposed (7) that mutations in S4 might alter its role in inhibiting cellular RNA and protein synthesis (348) (see "Effects on Host Cells" section) or in interfering with the growth of lytic, wt virus (4,6) (see "Genetics" section). Other work

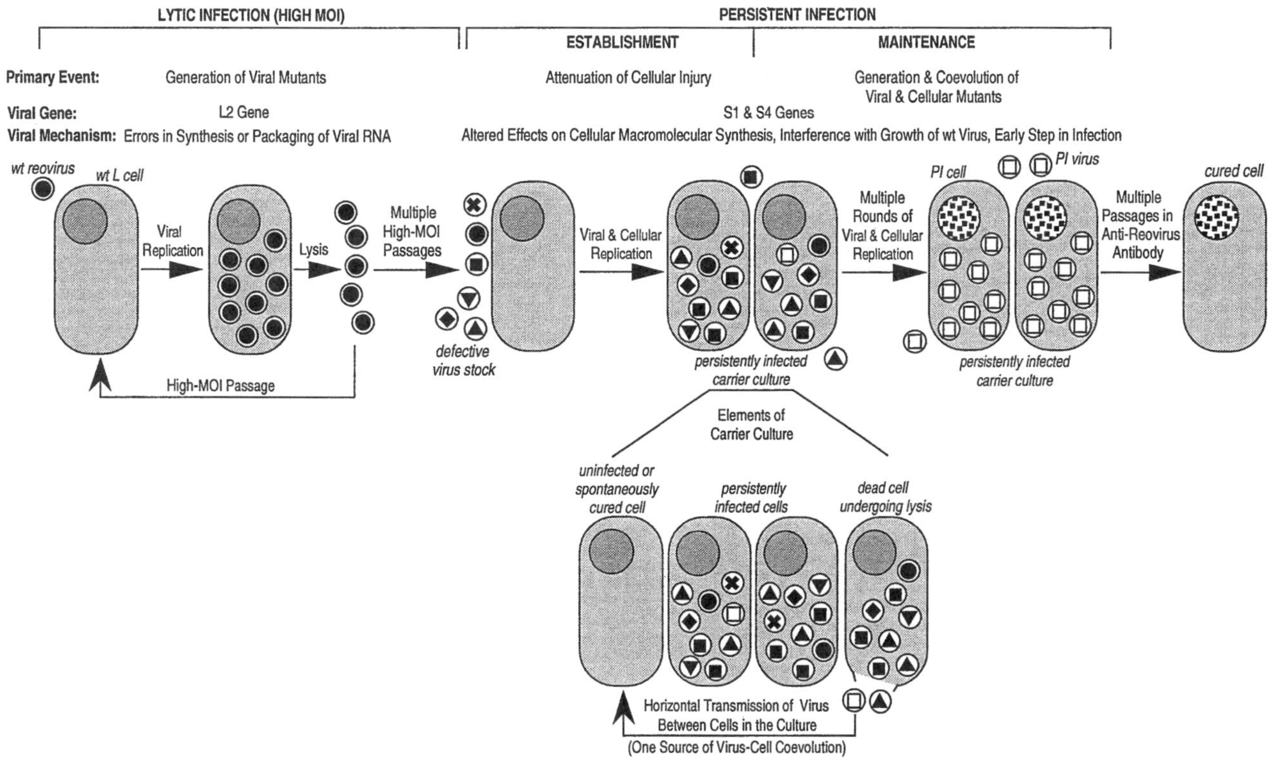

Fig. 11. A model for persistent infection by reoviruses. See text for detailed explanations of serial passage at high multiplicities of infection (MOI) and the establishment and maintenance of persistent reovirus infections in L cells. wt reoviruses are indicated (*solid circles*). Mutant viruses arising or undergoing selection during serial passage at high MOI or during the maintenance of persistent infection are indicated by different symbols. Mutant cells undergoing selection during the maintenance of persistent infection are indicated (*stippled nuclei*). Viral genes indicated (*top*) have been associated with different steps in the process by genetic analyses; primary events and viral mechanisms are proposed or indicated by experimental findings.

indicates that reovirus strains serially passaged at high MOI can differ in their capacities to establish persistent infections in L cells and that these differences are determined by their L2 genes (59). A current model links L2 to the generation or amplification of different types of mutants within defective stocks, including those that favor establishment of persistent infection (Fig. 11).

Cellular characteristics also influence the establishment of persistent reovirus infections. wt reoviruses can establish persistence in several, more differentiated cell lines (33,97,111,259, 393,403). The basis for an apparent relationship between cell differentiation state and establishment of persistent infection is unknown, but a role for the resistance of some cells to reovirus inhibition of cellular macromolecular synthesis was proposed (97,111). wt reoviruses can also establish persistent infections in L cells (which routinely undergo lytic infection) when the cells are treated with ammonium chloride during the first few days of infection (63). Ammonium chloride is a weak base that raises intralysosomal pH and inhibits virus entry into cells (see "Replication Cycle" section), but how such treatment facilitates the establishment of persistence is unknown. The finding that most cells treated with ammonium chloride survive infection and enter the persistent state disfavors the notion that either viral or cellular mutants are involved in establishment in this case, but rather suggests that the functional state of wt cells can influence whether a persistent infection is established.

Maintenance

A variety of viral mutants accumulate in cultures during persistent reovirus infections. Some of these mutants may play no role in the maintenance of persistence but reflect that reoviruses undergo extensive genetic changes during long-term culture (5,7–9). On the other hand, cell lines cured of persistent reovirus infection by passage with neutralizing antibodies (Fig. 11) are found to support reduced growth of wt reoviruses relative to viruses isolated from persistently infected cultures (PI viruses) (3,102,192). This finding argues that both viruses and cells undergo mutation during the maintenance of persistent infection (Fig. 11); furthermore, the nature of these changes suggests a process of virus-cell coevolution that may be essential to maintenance (3,102,192). In a genetic study involving one cloned PI virus, the mutation permitting increased replication in cured cells was mapped to the S1 gene (192), suggesting that mutations in S1 are especially important for the maintenance of reoviruses within persistently infected cultures.

In a recent study (102), cured cells were found to support reduced growth of wt virus when virions, but not when ISVPs, were used for infection; thus, one mutation in cured cells relates to an early step in infection that is unique to virions (e.g., proteolytic processing to generate an ISVP-like particle) (see "Replication Cycle" section). In addition, PI viruses were found to have increased resistance to ammonium chloride, which affects an early step in infection (392) (see "Replication Cycle" section). The involvement of an early step for both cellular and viral mutants in these studies suggests that coevolution at a step in virus entry into cells occurs during maintenance of persistent reovirus infection. These findings further suggest that reovirus persistence in L cells involves a carrier culture in which continuing horizontal transmission of virus between cells is required for maintenance (102) (Fig. 11).

REFERENCES

1. Acs G, Klett H, Schonberg M, Christman J, Levin DH, Silverstein JC. Mechanism of reovirus double-stranded RNA synthesis in vivo and in vitro. *J Virol* 1971;8:684–689.
2. Adams DJ, Spendlove JC, Spendlove RS, Barnett BB. Aerosol stability of infectious and potentially infectious reovirus particles. *Appl Environ Microbiol* 1982;44:903–908.
3. Ahmed R, Canning WM, Kauffman RS, Sharpe AH, Hallum JV, Fields BN. Role of the host cell in persistent viral infection: coevolution of L cells and reovirus during persistent infection. *Cell* 1981;25:325–332.
4. Ahmed R, Chakraborty PR, Fields BN. Genetic variation during lytic virus infection: high passage stocks of wild-type reovirus contain temperature-sensitive mutants. *J Virol*1980;34:285–287.
5. Ahmed R, Chakraborty PR, Graham AF, Ramig RF, Fields BN. Genetic variation during persistent reovirus infection: presence of extragenically suppressed temperature-sensitive lesions in wild-type virus isolated from persistently infected cells. *J Virol* 1980;34:383–389.
6. Ahmed R, Fields BN. Reassortment of genome segments between reovirus defective interfering particles and infectious virus: construction of temperature-sensitive and attentuated viruses by rescue of mutations from DI particles. *Virology* 1981;111:351–363.
7. Ahmed R, Fields BN. Role of the S4 gene in the establishment of persistent reovirus infection in L cells. *Cell* 1982;28:605–612.
8. Ahmed R, Graham AF. Persistent infections in L cells with temperature-sensitive mutants of reovirus. *J Virol* 1977;23:250–262.
9. Ahmed R, Kauffman RS, Fields BN. Genetic variation during persistent reovirus infection: isolation of cold-sensitive and temperature-sensitive mutants from persistently infected L cells. *Virology* 1983;131: 71–78.
10. Amano Y, Katagari S, Ishida N, Watanabe Y. Spontaneous degradation of reovirus capsid into subunits. *J Virol* 1971;8:805–808.
11. Ambler L, Mackay M. Reovirus 1 and 3 bind and internalize at the apical surface of intestinal epithelial cells. *Virology* 1991;184:162–169.
12. Anderson N, Doane FW. An electron microscope study of reovirus type 2 in L cells. *J Pathol Bacteriol* 1966;92:433–439
13. Antczak JB, Chmelo R, Pickup DJ, Joklik WK. Sequences at both termini of the ten genes of reovirus serotype 3 (strain Dearing). *Virology* 1982;121:307–319.
14. Antczak JB, Joklik WK. Reovirus genome segment assortment into progeny genomes studied by the use of monoclonal antibodies directed against reovirus proteins. *Virology* 1992;187:760–776.
15. Armstrong GD, Paul RW, Lee PW. Studies on reovirus receptors of L cells: virus binding characteristics and comparison with reovirus receptors of erythrocytes. *Virology* 1984;138:37–48.
16. Arnott S, Hutchinson F, Spencer M, Wilkins MHF. X-ray diffraction studies of double helical ribonucleic acid. *Nature* 1966;211:227–232.
17. Arnott S, Wilkins MHF, Fuller W, Langridge R. Molecular and crystal structures of double-helical RNA. *J Mol Biol* 1967;27:525–532.
18. Astell C, Silverstein SC, Levin DH, Acs G. Regulation of the reovirus RNA transcriptase by a viral capsomere protein. *Virology* 1972;48: 648–654.
19. Atwater JA, Munemitsu SM, Samuel CE. Biosynthesis of reovirus-specified polypeptides. Molecular complementary DNA cloning and nucleotide sequence of the reovirus serotype 1 Lang strain S4 messenger RNA which encodes the major capsid surface polypeptide $\sigma3$. *Biochem Biophys Res Commun* 1986;136:183–192.
20. Atwater JA, Munemitsu SM, Samuel CE. Biosynthesis of reovirus-specified polypeptides. Efficiency of expression of cDNAs of the reovirus S1 and S4 genes in transfected animal cells differs at the level of translation. *Virology* 1987;159:350–357.
21. Babiss LE, Luftig RB, Weatherbee JA, Weihing RR, Ray UR, Fields BN. Reovirus serotypes 1 and 3 differ in their in vitro association with microtubules. *J Virol* 1979;30:863–874.
22. Baglioni C, DeBenedetti A, Williams GJ. Cleavage of nascent reovirus mRNA by localized activation of the 2′-5′-oligoadenylate-dependent endoribonuclease. *J Virol* 1984;52:865–871.
23. Banerjea AC, Brechling KA, Ray CA, Erikson H, Pickup DJ, Joklik WK. High-level synthesis of biologically active reovirus protein $\sigma1$ in a mammalian expression vector system. *Virology* 1988;167:601–612.
24. Banerjea AC, Joklik WK. Reovirus protein $\sigma1$ translated in vitro, as well as truncated derivatives of it that lack up to two-thirds of its carboxyl-terminal portion, exists as two major tetrameric molecular species that differ in electrophoretic mobility. *Virology* 1990;179:460–462.
25. Banerjee AK, Shatkin AJ. Transcription in vitro by reovirus-associated ribonucleic acid-dependent polymerase. *J Virol* 1970;6:1–11.
26. Banerjee AK, Shatkin AJ. Guanosine-5′-diphosphate at the 5′ termini of reovirus RNA: evidence for a segmented genome within the virion. *J Mol Biol* 1971;61:643–653.
27. Bartlett JA, Joklik WK. The sequence of the reovirus serotype 3 L3 genome segment which encodes the major core protein $\lambda1$. *Virology* 1988;167:31–37.
28. Bartlett NM, Gillies SC, Bullivant S, Bellamy AR. Electron microscope study of reovirus reaction cores. *J Virol* 1974;14:315–326.
29. Bass DM, Bodkin D, Dambrauskas R, Trier JS, Fields BN, Wolf JL. Intraluminal proteolytic activation plays an important role in replication of type 1 reovirus in the intestines of neonatal mice. *J Virol* 1990; 64:1830–1833.
30. Bassel-Duby R, Jayasuriya A, Chatterjee D, Sonenberg N, Maizel JV Jr, Fields BN. Sequence of the reovirus haemagglutinin predicts a coiled-coil structure. *Nature* 1985;315:421–423.
31. Bassel-Duby R, Nibert ML, Homcy CJ, Fields BN, Sawutz DG. Evidence that the sigma 1 protein of reovirus serotype 3 is a multimer. *J Virol* 1987;61:1834–1841.
32. Bassel-Duby R, Spriggs DR, Tyler KL, Fields BN. Identification of attenuating mutations on the reovirus type 3 S1 double-stranded RNA segment with a rapid sequencing technique. *J Virol* 1986;60:64–67.
33. Bell TM, Ross MGR. Persistent latent infection of human embryonic cells with reovirus type 3. *Nature* 1966;212:412–414.
34. Bellamy AR, Hole LV. Single-stranded oligonucleotides from reovirus type 3. *Virology* 1970;40:808–819.
35. Bellamy AR, Hole LV, Baguley BC. Isolation of the trinucleotide pppG-pCpU from reovirus. *Virology* 1970;42:415–420.
36. Bellamy AR, Joklik WK. Studies on reovirus RNA. II. Characterization of reovirus messenger RNA and of the genome RNA segments from which it is transcribed. *J Mol Biol* 1967;29:19–26.
37. Bellamy AR, Joklik WK. Studies on the A-rich RNA of reovirus. *Proc Natl Acad Sci USA* 1967;58:1389–1395.
38. Bellamy AR, Nichols JL, Joklik WK. Nucleotide sequences of reovirus oligonucleotides: evidence for abortive RNA synthesis during virus maturation. *Nature* 1972;238:49–51.
39. Bellamy AR, Shapiro L, August JT, Joklik WK. Studies on reovirus RNA. I. Characterization of reovirus genome RNA. *J Mol Biol* 1967; 29:1–17.

40. Belli BA, Samuel CE. Biosynthesis of reovirus-specified polypeptides: Expression of reovirus S1-encoded σ1NS protein in transfected and infected cells as measured with serotype specific polyclonal antibody. *Virology* 1991;185:698–709.
41. Belli BA, Samuel CE. Biosynthesis of reovirus-specified polypeptides. Identification of regions of the bicistronic reovirus-S1 messenger-RNA that affect the efficiency of translation in animal cells. *Virology* 1993; 193:16–27.
42. Benavente J, Shatkin AJ. Avian reovirus mRNAs are nonfunctional in infected mouse cells: translational basis for virus host-range restriction. *Proc Natl Acad Sci USA* 1988;85:4257–4261.
43. Bischoff JR, Samuel CE. Mechanism of interferon action: Activation of the human P1/eIF-2-a protein kinase by individual reovirus s-class messenger RNA: S1 messenger RNA is a potent activator relative to S4 messenger RNA. *Virology* 1989;172:106–115.
44. Bodkin DK, Fields BN. Growth of reovirus in intestinal tissue: role of the L2 and S1 genes. *J Virol* 1989;63:1188–1193.
45. Bodkin DK, Nibert ML, Fields BN. Proteolytic digestion of reovirus in the intestinal lumens of neonatal mice. *J Virol* 1989;63:4676–4681.
46. Borsa J, Copps TP, Sargent MD, Long DG, Chapman JD. New intermediate subviral particles in the in vitro uncoating of reovirus virions by chymotrypsin. *J Virol* 1973;11:552–564.
47. Borsa J, Graham AF. Reovirus RNA polymerase activity in purified virions. *Biochem Biophys Res Commun* 1968;33:895–901.
48. Borsa J, Grover J, Chapman JD. Presence of nucleoside triphosphate phosphohydrolase activity in purified virions of reovirus. *J Virol* 1970; 6:295–302.
49. Borsa J, Long DG, Copps TP, Sargent MD, Chapman JD. Reovirus transcriptase activation in vitro: further studies on the facilitation phenomenon. *Intervirology* 1974;3:15–35.
50. Borsa J, Long DG, Sargent MD, Copps TP, Chapman JD. Reovirus transcriptase activation in vitro: involvement of an endogenous uncoating activity in the second stage of the process. *Intervirology* 1974;4:171–188.
51. Borsa J, Morash BD, Sargent MD, Copps TP, Lievaart PA, Szekely JG. Two modes of entry of reovirus particles into L cells. *J Gen Virol* 1979;45:161–170.
52. Borsa J, Sargent MD, Copps TP, Long DG, Chapman JD. Specific monovalent cation effects on modification of reovirus infectivity by chymotrypsin digestion in vitro. *J Virol* 1973;11:1017–1019.
53. Borsa J, Sargent MD, Ewing DD, Einspenner M. Perturbation of the switch-on of transcriptase activity in intermediate subviral particles from reovirus. *J Cell Physiol* 1982;112:10–18.
54. Borsa J, Sargent MD, Lievaart PA, Copps TP. Reovirus: evidence for a second step in the intracellular uncoating and transcriptase activation process. *Virology* 1981;111:191–200.
55. Borsa J, Sargent MD, Long DG, Chapman JD. Extraordinary effects of specific monovalent cations on activation of reovirus transcriptase by chymotrypsin in vitro. *J Virol* 1973;11:207–217.
56. Both GW, Lavi S, Shatkin AJ. Synthesis of all the gene products of the reovirus genome in vivo and in vitro. *Cell* 1975;4:173–180.
57. Brendler T, Godefroy-Colburn T, Carlill RD, Thach RE. The role of mRNA competition in regulating translation. II. Development of a quantitative in vitro assay. *J Biol Chem* 1981;256: 11747–11754.
58. Brendler T, Godefroy-Colburn T, Yu S, Thach RE. The role of mRNA competition in regulating translation. III. Comparison of in vitro and in vivo results. *J Biol Chem* 1981;256: 11755–11761.
59. Brown EG, Nibert ML, Fields BN. The L2 gene of reovirus serotype 3 controls the capacity to interfere, accumulate deletions and establish persistent infection. In: Compans RW, Bishop DHL, eds. *Double-stranded RNA viruses*. New York: Elsevier, 1983;275–287.
60. Bruck C, Co MS, Slaoui M, Gaulton GN, Smith T, Fields BN, Mullins JI, Greene MI. Nucleic acid sequence of an internal image-bearing monoclonal anti-idiotype and its comparison to the sequence of the external antigen. *Proc Natl Acad Sci USA* 1986;83:6578–6582.
61. Bruenn JA. Relationships among the positive strand and double-stranded RNA viruses as viewed through their RNA-dependent RNA polymerases. *Nucleic Acids Res* 1991;19:217–226.
62. Burstin SJ, Spriggs DR, Fields BN. Evidence for functional domains on the reovirus type 3 hemagglutinin. *Virology* 1982;117:146–155.
63. Canning WM, Fields BN. Ammonium chloride prevents lytic growth of reovirus and helps to establish persistent infection in mouse L cells. *Science* 1983;219:987–988.
64. Carter CA. Methylation of reovirus oligonucleotides in vivo and in vitro. *Virology* 1977; 80:249–259.
65. Carter CA. Polyadenylation of protein in reovirus. *Proc Natl Acad Sci USA* 1979;76:3087–3091.
66. Carter C, Lin B, Metley M. Polyadenylation of reovirus proteins. *J Biol Chem* 1980;255: 6479–6485.
67. Carter CA, Stolzfus CM, Banerjee AK, Shatkin AJ. Origin of reovirus oligo(A). *J Virol* 1974;13:1331–1337.
68. Cashdollar LW, Blair P, Van Dyne S. Identification of the σ1s protein in reovirus serotype 2-infected cells with antibody prepared against a bacterial fusion protein. *Virology* 1989;168:183–186.
69. Cashdollar LW, Chmelo R, Esparza J, Hudson GR, Joklik WK. Molecular cloning of the complete genome of reovirus serotype 3. *Virology* 1984;133:191–196.
70. Cashdollar LW, Chmelo RA, Wiener JR, Joklik WK. Sequences of the S1 genes of the three serotypes of reovirus. *Proc Natl Acad Sci USA* 1985;82:24–28.
71. Cashdollar LW, Esparza J, Hudson GR, Chmelo R, Lee PWK, Joklik WK. Cloning the double-stranded RNA genes of reovirus: sequence of the cloned S2 gene. *Proc Natl Acad Sci USA* 1982;79:7644–7648.
72. Ceruzzi M, Shatkin AJ. Expression of reovirus p14 in bacteria and identification in the cytoplasm of infected mouse L cells. *Virology* 1986; 153:35–45.
73. Chakraborty PR, Ahmed R, Fields BN. Genetics of reovirus: the relationship of interference to complementation and reassortment of temperature-sensitive mutants at non-permissive temperatures. *Virology* 1979;94:119–127.
74. Chang C-T, Zweerink HJ. Fate of parental reovirus in infected cell. *Virology* 1971;46:544–555.
75. Chappell JD,Goral MI, Rodgers SE, DePamphilis CW, Dermody TS. Sequence diversity within the reovirus S2 gene: reovirus genes reassort in nature, and their termini are predicted to form a panhandle motif. *J Virol* 1994; 68:750–756.
76. Choi AHC. Internalization of virus binding proteins during entry of reovirus into K562 erythroleukemia cells. *Virology* 1994;200:301–306.
77. Choi AHC, Lee PWK. Does the beta-adrenergic receptor function as a reovirus receptor? *Virology* 1988;163:191–197.
78. Choi AHC, Paul RW, Lee PWK. Reovirus binds to multiple plasma membrane proteins of mouse L fibroblasts. *Virology* 1990;178:316–320.
79. Chow N-L, Shatkin AJ. Blocked and unblocked 5′ termini in reovirus genome RNA. *J Virol* 1975;15:1057–1064.
80. Cleveland DR, Zarbl H, Millward S. Reovirus guanylyltransferase is L2 gene product. *J Virol* 1986;60:307–311.
81. Co MS, Gaulton GN, Fields BN, Greene MI. Isolation and biochemical characterization of the mammalian reovirus type 3 cell-surface receptor. *Proc Natl Acad Sci USA* 1985;82:1494–1498.
82. Co MS, Gaulton GN, Tominaga A, Homcy CJ, Fields BN, Greene MI. Structural similarities between the mammalian beta-adrenergic and reovirus type 3 receptors. *Proc Natl Acad Sci USA* 1985;82:5315–5318.
83. Cohen JA, Sergott RC, Williams WV, Hill SJ, Brown MJ, Greene MI. In vivo modulation of oligodendrocyte function by an anti-receptor antibody. *Pathobiology* 1992;27:847–847.
84. Cohen JA, Williams WV, Geller HM, Greene MI. Anti-reovirus receptor antibody accelerates expression of the optic nerve oligodendrocyte developmental program. *Proc Natl Acad Sci USA* 1991;88: 1266–1270.
85. Cohen JA, Williams WV, Weiner DB, Geller HM, Greene MI. Ligand binding to the cell surface receptor for reovirus type 3 stimulates galactocerebroside expression by developing oligodendrocytes. *Proc Natl Acad Sci USA* 1990;87:4922–4926.
86. Coombs KM, Fields BN, Harrison SC. Crystallization of the reovirus type 3 Dearing core. Crystal packing is determined by the λ2 protein. *J Mol Biol* 1990;215:1–5.
87. Coombs KM, Mak S-C, Petrycky-Cox LD. Studies of the major reovirus core protein σ2: reversion of the assembly-defective mutant tsC447 is an intragenic process and involves back mutation of asp-383 to asn. *J Virol* 1994;68:177–186.
88. Cox DC, Shaw JE. Inhibition of initiation of cellular DNA synthesis after reovirus infection. *J Virol* 1974;13:760–761.
89. Cross RK, Fields, BN. Temperature-sensitive mutants of reovirus type 3: studies on the synthesis of viral RNA. *J Virol* 1972;50:799–809.
90. Cross RK, Fields BN. Reovirus-specific polypeptides: analysis using discontinuous gel electrophoresis. *J Virol* 1976;19:162–173.
91. Cross RK, Fields BN. Temperature-sensitive mutants of reovirus type 3: evidence for aberrant μ1 and μ2 polypeptide species. *J Virol* 1976; 19:174–179.

92. Cross RK, Fields BN. Use of aberrant polypeptide as a marker in three-factor crosses: further evidence for independent reassortment as the mechanism of recombination between temperature-sensitive mutants of reovirus type 3. *Virology* 1976;74:345–362.
93. Dales S. Association between the spindle apparatus and reovirus. *Proc Natl Acad Sci USA* 1963;50:268–275.
94. Dales S, Gomatos P, Hsu KC. The uptake and development of reovirus in strain L cells followed with labelled viral ribonucleic acid and ferritin-antibody conjugates. *Virology* 1965;25:193–211.
95. Danis C, Garzon S, Lemay G. Further characterization of the ts453 mutant of mammalian orthoreovirus serotype 3 and nucleotide sequence of the mutated S1 gene. *Virology* 1992;190:494–498.
96. Danis C, Lemay G. Protein synthesis in different cell lines infected with orthoreovirus serotype 3: Inhibition of host-cell protein synthesis correlates with accelerated viral multiplication and cell killing. *Biochem Cell Biol* 1993;71:81–85.
97. Danis C, Mabrouk T, Garzon S, Lemay G. Establishment of persistent reovirus infection in SC1 cells: absence of protein synthesis inhibition and increased level of double-stranded RNA-activated protein kinase. *Virus Res* 1993;27:253–265.
98. Darzynkiewicz E, Shatkin AJ. Assignment of reovirus mRNA ribosome binding sites to virion genome segments by nucleotide sequence analyses. *Nucleic Acids Res* 1980;8:337–350.
99. DeBenedetti A, Williams GJ, Baglioni C. Inhibition of binding to initiation complexes of nascent reovirus mRNA by double-stranded RNA-dependent protein kinase. *J Virol* 1985;54:408–413.
100. Dermody TS, Nibert ML, Bassel-Duby R, Fields BN. Sequence diversity in S1 genes and S1 translation products of 11 serotype 3 reovirus strains. *J Virol* 1990;64:4842–4850.
101. Dermody TS, Nibert ML, Bassel-Duby R, Fields BN. A σ1 region important for hemagglutination by serotype 3 reovirus strains. *J Virol* 1990;64:5173–5176.
102. Dermody TS, Nibert ML, Wetzel JD, Tong X, Fields BN. Cells and viruses with mutations affecting viral entry are selected during persistent infections of L cells with mammalian reoviruses. *J Virol* 1993;76: 2055–2063.
103. Dermody TS, Schiff LA, Nibert ML, Coombs KM, Fields BN. The S1 gene nucleotide sequences of prototype strains of the three reovirus serotypes: characterization of reovirus core protein σ2. *J Virol* 1991; 65:5721–5731.
104. Detjen BM, Walden WE, Thach RE. Translational specificity in reovirus-infected mouse fibroblasts. *J Biol Chem* 1982;257:9855–9860.
105. Donta ST, Shanley JD. Reovirus type 3 binds to antagonist domains of the β-adrenergic receptor. *J Virol* 1990;64:639–641.
106. Doohan JP, Samuel CE. Biosynthesis of reovirus-specified polypeptides. Ribosome pausing during the translation of reovirus-S1 messenger-RNA. *Virology* 1992;186:409–425.
107. Drayna D, Fields BN. Activation and characterization of the reovirus transcriptase: genetic analysis. *J Virol* 1982;41:110–118.
108. Drayna D, Fields BN. Genetic studies on the mechanism of chemical and physical inactivation of reovirus. *J Gen Virol* 1982;63:149–160.
109. Drayna D, Fields BN. Biochemical studies on the mechanism of chemical and physical inactivation of reovirus. *J Gen Virol* 1982;63:161–170.
110. Dryden KA, Wang G, Yeager M, Nibert ML, Coombs KM, Furlong DB, Fields BN, Baker TS. Early steps in reovirus infection are associated with dramatic changes in supramolecular structure and protein conformation: analysis of virions and subviral particles by cryoelectron microscopy and image reconstruction. *J Cell Biol* 1993;122:1023–1041.
111. Duncan MR, Stanish SM, Cox DC. Differential sensitivity of normal and transformed human cells to reovirus infection. *J Virol* 1978;28: 444–449.
112. Duncan R, Horne D, Cashdollar LW, Joklik WK, Lee PWK. Identification of conserved domains in the cell attachment proteins of the three serotypes of reovirus. *Virology* 1990;174:399–409.
113. Duncan R, Horne D, Strong JE, Leone G, Pon RT, Yeung MC, Lee PWK. Conformational and functional analysis of the carboxyl-terminal globular head of the reovirus cell attachment protein. *Virology* 1991; 182:810–819.
114. Duncan RF. Protein synthesis initiation factor modifications during viral infections: Implications for translational control. *Electrophoresis* 1990;11:219–227.
115. Dunnebacke TH, Kleinschmidt AK. Ribonucleic acid from reovirus as seen in protein monolayers by electron microscopy. *Z Naturforsch* 1967;22b:159–164.
116. Eggers HJ, Gomatos PJ, Tamm I. Agglutination of bovine erythrocytes: a general characteristic of reovirus type 3. *Proc Soc Exp Biol Med* 1962; 110:879–882.
117. Ensminger WD, Tamm I. Cellular DNA and protein synthesis in reovirus-infected cells. *Virology* 1969;39:357–359.
118. Ensminger WD, Tamm I. The step in cellular DNA synthesis blocked by reovirus infection. *Virology* 1969;39:935–938.
119. Epstein RL, Powers ML, Rogart RB, Weiner HL. Binding of iodine-125 labelled reovirus to cell surface receptors. *Virology* 1984;133: 46–55.
120. Ernst G, Shatkin AJ. Reovirus hemagglutinin mRNA codes for two polypeptides in overlapping reading frames. *Proc Natl Acad Sci USA* 1985;82:48–52.
121. Ewing DD, Sargent MD, Borsa J. Switch-on of transcriptase function in reovirus: analysis of polypeptide changes using 2-D gels. *Virology* 1985;144:442–456.
122. Fajardo JE, Shatkin AJ. Effects of elongation on the translation of a reovirus bicistronic mRNA. *Enzyme* 1990;44:235–243.
123. Fajardo JE, Shatkin AJ. Expression of the two reovirus S1 gene products in transfected mammalian cells. *Virology* 1990;178:223–231.
124. Farone AL, O'Brien PC, Cox DC. Tumor necrosis factor-α induction by reovirus serotype 3. *J Leukocyte Biol* 1993;53:133–137.
125. Farrell JA, Harvey JD, Bellamy AR. Biophysical studies of reovirus type 3. I. The molecular weight of reovirus and reovirus cores. *Virology* 1974;62:145–153.
126. Fausnaugh J, Shatkin AJ. Active site localization in a viral mRNA capping enzyme. *J Biol Chem* 1990;265:7669–7672.
127. Faust M, Hastings KEM, Millward S. M^7G5'-ppp5′ G^mpCpUp at the 5′ terminus of reovirus messenger RNA. *Nucleic Acids Res* 1975;2: 1329–1343.
128. Faust M, Millward S. In vitro methylation of nascent reovirus mRNA by a virion-associated methyl transferase. *Nucleic Acids Res* 1974;1: 1739–1752.
129. Feduchi E, Esteban M, Carrasco L. Reovirus type 3 synthesizes proteins in interferon-treated HeLa cells without reversing the antiviral state. *Virology* 1988;164:420–426.
130. Fields BN. Temperature-sensitive mutants of reovirus type 3: features of genetic recombination. *Virology* 1971;46:142–148.
131. Fields BN, Greene MI. Genetic and molecular mechanisms of viral pathogenesis: implications for prevention and treatment. *Nature* 1982; 300:19–23.
132. Fields BN, Joklik WK. Isolation and preliminary genetic and biochemical characterization of temperature-sensitive mutants of reovirus. *Virology* 1969;37:335–342.
133. Fields BN, Raine CS, Baum SG. Temperature-sensitive mutants of reovirus type 3: defects in viral maturation as studied by immunofluorescence and electron microscopy. *Virology* 1971;43:569–578.
134. Fraser RDB, Furlong DB, Trus BL, Nibert ML, Fields BN, Steven AC. Molecular structure of the cell attachment protein of reovirus: correlation of computer-processed electron micrographs with sequence-based predictions. *J Virol* 1990;64:2990–3000.
135. Furlong DB, Nibert ML, Fields BN. Sigma 1 protein of mammalian reoviruses extends from the surfaces of viral particles. *J Virol* 1988; 62:246–256.
136. Furuichi Y, LaFiandra A, Shatkin AJ. 5′-Terminal structure and mRNA stability. *Nature* 1977;266:235–239.
137. Furuichi Y, Morgan M, Muthukrishnan S, Shatkin AJ. Reovirus messenger RNA contains a methylated blocked 5′-terminal structure $M^7G(5')ppp(5')G^mpCp$-. *Proc Natl Acad Sci USA* 1975;72:362–366.
138. Furuichi Y, Muthukrishnan S, Shatkin AJ. 5′-Terminal $M^7G(5')ppp(5')G^mp$ in vivo: identification in reovirus genome RNA. *Proc Natl Acad Sci USA* 1975;72:742–745.
139. Furuichi Y, Muthukrishnan S, Tomasz J, Shatkin AJ. Mechanism of formation of reovirus mRNA 5′-terminal blocked and methylated sequence M^7GpppG^mpC. *J Biol Chem* 1976;251: 5043–5053.
140. Furuichi Y, Shatkin AJ. 5′-Termini of reovirus mRNA: ability of viral cores to form caps post-transcriptionally. *Virology* 1977;77:566–578.
141. Gaillard RK, Joklik WK. The relative translation efficiencies of reovirus messenger RNAs. *Virology* 1985;147:336–348.
142. Gard G, Compans RW. Structure and cytopathic effects of Nelson Bay virus. *J Virol* 1970;6:100–106.
143. Gard G, Marshall ID. Nelson Bay virus: a novel reovirus. *Arch Virol* 1973;43:34–42.
144. Gaulton GN, Greene MI. Inhibition of cellular DNA synthesis by re-

ovirus occurs through a receptor-linked signaling pathway that is mimicked by antiidiotypic, antireceptor antibody. *J Exp Med* 1989;169: 197–211.

145. Gaulton GN, Stein ME, Safko B, Stadecker MJ. Direct induction of Ia antigen on murine thyroid-derived epithelial cells by reovirus. *J Immunol* 1989;142:3821–3825.
146. Gentsch JR, Fields BN. Genetic diversity in natural populations of mammalian reoviruses: tryptic peptide analysis of outer capsid polypeptides of murine, bovine, and human type 1 and 3 reovirus strains. *J Virol* 1984;49:641–651.
147. Gentsch JR, Hatfield JW. Saturable attachment sites for type 3 mammalian reovirus on murine L cells and human HeLa cells. *Virus Res* 1984;1:401–414.
148. Gentsch JR, Pacitti AF. Differential interaction of reovirus type 3 with sialylated receptor components on animal cells. *Virology* 1987;161: 245–248.
149. Gentsch JR, Pacitti AF. Effect of neuraminidase treatment of cells and effect of soluble glycoproteins on type 3 reovirus attachment to murine L cells. *J Virol* 1985;56:356–364.
150. George CX, Atwater JA, Samuel CE. Biosynthesis of reovirus-specified polypeptides. Molecular cDNA cloning and nucleotide sequence of the reovirus serotype 1 Lang strain S3 messenger RNA which encodes the nonstructural RNA-binding protein σNS. *Biochem Biophys Res Commun* 1986;139:845–851.
151. George CX, Crowe A, Munemitsu SM, Atwater JA, Samuel CE. Biosynthesis of reovirus-specified polypeptides. Molecular cDNA cloning and nucleotide sequence of the reovirus serotype 1 Lang strain S2 mRNA which encodes the virion core polypeptide σ2. *Biochem Biophys Res Commun* 1987;147:1153–1161.
152. George CX, Samuel CE. Mechanism of interferon action: expression of reovirus S3 gene in transfected COS cells and subsequent inhibition at the level of protein synthesis by type I but not type II interferon. *Virology* 1988;166:573–582.
153. Georgi A, Mottola-Hartshorn C, Warner A, Fields B, Chen LB. Detection of individual fluorescently labeled reovirions in living cells. *Proc Natl Acad Sci USA* 1990;87:6579–6583.
154. Giantini M, Seliger LS, Furuichi Y, Shatkin AJ. Reovirus type 3 genome segment S4: nucleotide sequence of the gene encoding a major virion surface protein. *J Virol* 1984;52:984–987.
155. Giantini M, Shatkin AJ. Stimulation of chloramphenicol acetyltransferase mRNA translation by reovirus capsid polypeptide σ3 in cotransfected COS cells. *J Virol* 1989;63:2415–2421.
156. Gillies S, Bullivant S, Bellamy AR. Viral RNA polymerases: electron microscopy of reovirus reaction cores. *Science* 1971;174:694–696.
157. Gomatos PJ, Prakash O, Stamatos NM. Small reovirus particles composed solely of sigma NS with specificity for binding different nucleic acids. *J Virol* 1981;39:115–124.
158. Gomatos PJ, Stamatos NM, Sarkar NH. Small reovirus-specific particle with polycytidylate-dependent RNA polymerase activity. *J Virol* 1980;36:556–565.
159. Gomatos PJ, Stoeckenius W. Electron microscope studies on reovirus RNA. *Proc Natl Acad Sci USA* 1964;52:1449–1455.
160. Gomatos PJ, Tamm I. Macromolecular synthesis in reovirus-infected L cells. Biochim Biophys Acta 1963;72:651–653.
161. Gomatos PJ, Tamm I. The secondary structure of reovirus RNA. *Proc Natl Acad Sci USA* 1963;49:707–714.
162. Gomatos PJ, Tamm I, Dales S, Franklin RM. Reovirus type 3: physical characteristics and interactions with L cells. *Virology* 1962;17: 441–454.
163. Granboulan N, Niveleau A. Etude au microscope electronique du RNA de reovirus. *J Microsc* 1967;6:23–30.
164. Graziadei WD III, Roy D, Konigsberg W, Lengyel P. Translation of reovirus RNA synthesized in vitro into reovirus proteins in a mouse L cell extract. *Arch Biochem Biophys* 1973; 158:266–275.
165. Gupta SL, Holmes SL, Mehra LL. Interferon action against reovirus: activation of interferon induced protein kinase in mouse L929 cells upon reovirus infection. *Virology* 1982; 120:495–499.
166. Hand R, Ensminger WD, Tamm I. Cellular DNA replication in infections with cytocidal RNA viruses. *Virology* 1971;44:527–536.
167. Hand R, Tamm I. Reovirus: effect of noninfective viral components on cellular deoxyribonucleic acid synthesis. *J Virol* 1973;11:223–231.
168. Hand R, Tamm I. Initiation of DNA synthesis in mammalian cells and its inhibition by reovirus infection. *J Mol Biol* 1974;82:175–183.
169. Harvey JD, Bellamy AR, Earnshaw WD, Schutt C. Biophysical studies of reovirus type 3. IV. Low-angle X-ray diffraction studies. *Virology* 1981;112:240–249.
170. Harvey JD, Farrell JA, Bellamy AR. Biophysical studies of reovirus type 3. II. Properties of the hydrated particle. *Virology* 1974;62:154–160.
171. Hayes EC, Lee PWK, Miller SE, Joklik WK. The interaction of a series of hybridoma IgGs with reovirus particles: demonstration that the core protein λ2 is exposed on the particle surface. *Virology* 1981;108: 147–155.
172. Henderson DR, Joklik WK. The mechanism of interferon induction by UV-irradiated reovirus. *Virology* 1978;91:389–406.
173. Hrdy DB, Rosen L, Fields BN. Polymorphism of the migration of double-stranded RNA segments of reovirus isolates from humans, cattle, and mice. *J Virol* 1979;31:104–111.
174. Hrdy DB, Rubin DN, Fields BN. Molecular basis of reovirus neurovirulence: role of the M2 gene in avirulence. *Proc Natl Acad Sci USA* 1982;79:1298–1302.
175. Huismans H, Joklik WK. Reovirus-coded polypeptides in infected cells: isolation of two native monomeric polypeptides with high affinity for single-stranded and double-stranded RNA, respectively. *Virology* 1976; 70:411–424.
176. Iglewski WK, Franklin RM. Purification and properties of reovirus ribonucleic acid. *J Virol* 1967;1:302–307.
177. Ikegami N, Gomatos PJ. Temperature-sensitive conditional lethal mutants of reovirus 3. *Virology* 1968;36:447–458.
178. Imai M, Richardson MA, Ikegami N, Shatkin AJ, Furuichi Y. Molecular cloning of double-stranded RNA virus genomes. *Proc Natl Acad Sci USA* 1983;80:373–377.
179. Imani F, Jacobs BL. Inhibitory activity for the interferon-induced protein kinase is associated with the reovirus serotype 1 σ3 protein. *Proc Natl Acad Sci USA* 1988;83:7887–7891.
180. Ito Y, Joklik WK. Temperature-sensitive mutants of reovirus. I. Patterns of gene expression by mutants of groups C, D, and E. *Virology* 1972;50:189–201.
181. Jacobs BL, Atwater JA, Munemitsu SM, Samuel CE. Biosynthesis of reovirus-specified polypeptides. The S1 mRNA synthesized in vivo is structurally and functionally indistinguishable from in vitro-synthesized S1 mRNA and encodes two polypeptides, σ1a and σ1bNS. *Virology* 1985;147:9–18.
182. Jacobs BL, Ferguson RE. The Lang strain of reovirus serotype 1 and the Dearing strain of reovirus serotype 3 differ in their sensitivities to beta interferon. *J Virol* 1991;65:5102–5104.
183. Jacobs BL, Samuel CE. Biosynthesis of reovirus-specified polypeptides: the reovirus S1 mRNA encodes two primary translation products. *Virology* 1985;143:63–74.
184. Jayasuriya AK, Nibert ML, Fields BN. Complete nucleotide sequence of the M2 gene segment of reovirus type 3 Dearing and analysis of its protein product μ1. *Virology* 1988;163:591–602.
185. Joklik WK. Studies on the effect of chymotrypsin on reovirions. *Virology* 1972;49:700–801.
186. Joklik WK. Reproduction of reoviridae. In: Fraenkel-Conrat H, Wagner RR, eds. *Comprehensive Virology*, vol 2. New York: Plenum Press, 1974;231–334.
187. Joklik WK. Structure and function of the reovirus genome. *Microbiol Rev* 1981;45:483–501.
188. Joklik WK. The reovirus particle. In: Joklik WK, ed. *The Reoviridae*. New York: Plenum Press, 1983;9–78.
189. Joklik WK. Recent progress in reovirus research. *Annu Rev Genet* 1985; 19:537–575.
190. Kapuler AM. An extraordinary temperature dependence of the reovirus transcriptase. *Biochemistry* 1970;9:4453–4457.
191. Kapuler AM, Mendelsohn N, Klett H, Acs G. Four base-specified nucleoside 5′-triphosphatases in the subviral core of reovirus. *Nature* 1970;225:1209–1213.
192. Kauffman RS, Ahmed R, Fields BN. Selection of a mutant S1 gene during reovirus persistent infection of L cells: role in maintenance of the persistent state. *Virology* 1983;131:79–87.
193. Kavenoff R, Talcove D, Mudd JA. Genome-sized RNA from reovirus particles. *Proc Natl Acad Sci USA* 1975;72:4317–4321.
194. Kawamura H, Shimizu F, Maeda M, Tsubahara H. Avian reovirus: its properties and serological classification. *Natl Inst Anim Health Q (Tokyo)* 1965;5:115–124.
195. Kawamura H, Tsubahara H. Common antigenicity of avian reoviruses. *Natl Inst Anim Health Q (Tokyo)* 1966;6:187–193.
196. Kaye KM, Spriggs DR, Bassel-Duby R, Fields BN. Genetic basis for

altered pathogenesis of an immune-selected antigenic variant of reovirus type 3 Dearing. *J Virol* 1986;59:90–97.
197. Keroack M, Fields BN. Viral shedding and transmission between hosts determined by reovirus L2 gene. *Science* 1986;232:1635–1638.
198. Khaustov VI, Korolev MB, Reingold VN. The structure of the capsid inner layer of reoviruses. Brief report. *Arch Virol* 1987;93:163–167.
199. Koonin EV, Gorbalenya EE, Chumakov KM. Tentative identification of RNA-dependent RNA polymerases of dsRNA viruses and their relationship to positive strand RNA viral polymerases. *FEBS Lett* 1989; 252:42–46.
200. Koonin EV. Evolution of double-stranded RNA viruses: a case for polyphyletic origin from different groups of positive-stranded RNA viruses. *Sem Virol* 1992;3:327–340.
201. Koonin EV. Computer-assisted identification of a putative methyltransferase domain in ns5 protein of flaviviruses and λ2 protein of reovirus. *J Gen Virol* 1993;74:733–740.
202. Kozak M. Possible role of flanking nucleotides in recognition of the AUG initiator codon by eukaryotic ribosomes. *Nucleic Acids Res* 1981;9:5233–5252.
203. Kozak M. Analysis of ribosome binding sites from the S1 message of reovirus. Initiation at the first and second AUG codons. *J Mol Biol* 1982;156:807–820.
204. Kozak M. Sequences of ribosome-binding sites from the large size class of reovirus mRNA. *J Virol* 1982;42:467–473.
205. Krystal G, Perrault J, Graham AF. Evidence for a glycoprotein in reovirus. *Virology* 1976;72:308–321.
206. Krystal G, Winn P, Millward S, Sakuma S. Evidence for phosphoproteins in reovirus. *Virology* 1975;64:505–512.
207. Kudo H, Graham AF. Selective inhibition of reovirus induced RNA in L cells. *Biochem Biophys Res Commun* 1966;24:150–155.
208. Lai KC, Bellamy AR. Factors affecting the amount of oligonucleotides in reovirus particles. *Virology* 1971;45:821–823.
209. Lai M-HT, Joklik WK. The induction of interferon by temperature-sensitive mutants of reovirus, UV-irradiated reovirus, and subviral reovirus particles. *Virology* 1973;51:191–204.
210. Lai M-HT, Werenne JJ, Joklik WK. The preparation of reovirus top component and its effect on host DNA and protein synthesis. *Virology* 1973;54:237–244.
211. Lau RY, Van Alstyne D, Berckmans R, Graham AF. Synthesis of reovirus-specific polypeptides in cells pretreated with cycloheximide. *J Virol* 1975;16:470–478.
212. Lee PWK, Hayes EC, Joklik WK. Characterization of anti-reovirus immunoglobulins secreted by cloned hybridoma cell lines. *Virology* 1981; 108:134–146.
213. Lee PWK, Hayes EC, Joklik WK. Protein σ1 is the reovirus cell attachment protein. *Virology* 1981;108:156–163.
214. Lemay G, Millward S. Expression of the cloned S1 gene of reovirus serotype 3 in transformed eukaryotic cells: enrichment of the viral protein in the crude initiation factor fraction. *Virus Res* 1986;6: 133–140.
215. Lemieux R, Lemay G, Millward S. The viral protein σ3 participates in translation of late viral messenger RNA in reovirus-infected L cells. *J Virol* 1987;61:2472–2479.
216. Lemieux R, Zarbl H, Millward S. Messenger RNA discrimination in extracts from uninfected and reovirus infected L cells. *J Virol* 1984; 51:215–222.
217. Leone G, Duncan R, Lee PWK. Trimerization of the reovirus cell attachment protein (σ1) induces conformational changes in σ1 necessary for its cell-binding function. *Virology* 1991;184:758–761.
218. Leone G, Duncan R, Mah DCW, Price A, Cashdollar LW, Lee PWK. The amino-terminal heptad repeat region of reovirus cell attachment protein σ1 is responsible for σ1 oligomer stability and possesses intrinsic oligomerization function. *Virology* 1991;182:336–345.
219. Leone G, Mah DCW, Lee PWK. The incorporation of reovirus cell attachment protein σ1 into virions requires the amino-terminal hydrophobic tail and the adjacent heptad repeat region. *Virology* 1991; 182:346–350.
220. Leone G, Maybaum L, Lee PWK. The reovirus cell attachment protein possesses two independently active trimerization domains: basis of dominant negative effects. *Cell* 1992;71;479–488.
221. Levin DH, Mendelsohn N, Schonberg M, Klett H, Silverstein S, Kapuler AM. Properties of RNA transcriptase in reovirus subviral particles. *Proc Natl Acad Sci USA* 1970;66:890–897.
222. Levin KH, Samuel CE. Biosynthesis of reovirus-specified polypeptides: purification and characterization of the small-sized class mRNAs of reovirus type 3. Coding assignments and translational efficiencies. *Virology* 1980;106:1–13.
223. Li JK-K, Keene JD, Scheible PP, Joklik WK. Nature of the 3′-terminal sequence of the plus and minus strands of the S1 gene of reovirus serotypes 1, 2, and 3. *Virology* 1980;105:41–51.
224. Li JK-K, Scheible PP, Keene JD, Joklik WK. The plus strand of reovirus gene S2 is identical with its in vitro transcript. *Virology* 1980; 105:282–286.
225. Lloyd, RM, Shatkin AJ. Translational stimulation by reovirus polypeptide σ3: substitution for VA1-RNA and inhibition of phosphorylation of the a-subunit of eukaryotic initiation factor-II. *J Virol* 1992;66: 6878–6884.
226. Loh PC, Shatkin AJ. Structural proteins of reovirus. *J Virol* 1968;2: 1353–1359.
227. Loh PC, Soergel M. Macromolecular synthesis in cells infected with reovirus type 2 and the effect of ARA-C. *Nature* 1967;214:622–623.
228. Long WF, Burke DC. Interferon production by double-stranded RNA: a comparison of interferon induction by reovirus RNA to that by a synthetic double-stranded polynucleotide. *J Gen Virol* 1971;12:1–11.
229. Lucia-Jandris P, Hooper JW, Fields BN. Reovirus M2 gene is associated with chromium release from mouse L cells. *J Virol* 1993;67:5339–5345.
230. Luftig RB, Kilham S, Hay A, Zweerink HJ, Joklik WK. An ultrastructure study of virions and cores of reovirus type 3. *Virology* 1972; 48:170–181.
231. Lyubchenko YL, Jacobs BL, Lindsay SM. Atomic force microscopy of reovirus dsRNA - a routine technique for length measurements. *Nucleic Acids Res* 1992;20:3983–3986.
232. Mah DC, Leone G, Jankowski JM, Lee PW. The N-terminal quarter of reovirus cell attachment protein σ1 possesses intrinsic virion-anchoring function. *Virology* 1990;179:95–103.
233. Mallo M, Martinez-Costas J, Benavente J. Avian reovirus S1133 can replicate in mouse L cells: effect of pH and cell attachment status on viral infection. *J Virol* 1991;65:5499–5505.
234. Mallo M, Martinez-Costas J, Benavente J. The stimulatory effect of actinomycin D on avian reovirus replication in L cells suggests that translational competition dictates the fate of the infection. *J Virol* 65: 5506–5512.
235. Mao ZX, Joklik WK. Isolation and enzymatic characterization of protein λ2, the reovirus guanylyltransferase. *Virology* 1991;185:377–386.
236. Maratos-Flier E, Goodman MJ, Murray AH, Kahn CR. Ammonium inhibits processing and cytotoxicity of reovirus a nonenveloped virus. *J Clin Invest* 1986;78:1003–1007.
237. Maratos-Flier E, Kahn CR, Spriggs DR, Fields BN. Specific plasma membrane receptors for reovirus on rat pituitary cells in culture. *J Clin Invest* 1983;72:617–625.
238. Marcus PI. Interferon induction by viruses: one molecule of dsRNA as the threshold for interferon induction. *Interferon* 1983;5:116–180.
239. Martin PE, McCrae MA. Analysis of the stimulation of reporter gene expression by the sigma 3 protein of reovirus in co-transfected cells. *J Gen Virol* 1993;74:1055–1062.
240. Martin SA, Pett D, Zweerink HJ. Studies on the topography of reovirus and bluetongue virus capsid proteins. *J Virol* 1973;12:194–198.
241. Masri SA, Nagata L, Mah DC, Lee PW. Functional expression in *Escherichia coli* of cloned reovirus S1 gene encoding the viral cell attachment protein σ1. *Virology* 1986;149;83–90.
242. Matoba Y, Sherry B, Fields BN, Smith TW. Identification of the viral genes responsible for growth of strains of reovirus in cultured mouse heart cells. *J Clin Invest* 1991;87:1628–1633.
243. Matsuhisa T, Joklik WK. Temperature-sensitive mutants of reovirus. V. Studies on the nature of the temperature-sensitive lesion of the group C mutant ts447. *Virology* 1974;60:380–389.
244. Matsuura K, Ishikura M, Nakayama T, et al. Ecological studies on reovirus pollution of rivers in Toyama Prefecture. II. Molecular epidemiological study of reoviruses isolated from river water. *Microbiol Immunol* 1993;37:305–310.
245. Mayor HD, Jamison RM, Jordan LE, Mitchell MV. Reoviruses. II. Structure and composition of the virion. *J Bacteriol* 1965;89:1548–1556.
246. McClain ME, Spendlove RS. Multiplicity reactivation of reovirus particles after exposure to ultraviolet light. *J Bacteriol* 1966;92:1422–1429.
247. McCrae MA. Terminal structure of reovirus RNAs. *J Gen Virol* 1981; 55:393–403.
248. McCrae MA, Joklik WK. The nature of the polypeptide encoded by each of the ten double-stranded RNA segments of reovirus type 3. *Virology* 1978;89:578–593.

249. McDowell MJ, Joklik WK. An in vitro protein synthesizing system from mouse L fibroblasts infected with reovirus. *Virology* 1971;45: 724–733.
250. McDowell MJ, Joklik WK, Villa-Komaroff L, Lodish HF. Translation of reovirus messenger RNAs synthesized in vitro into reovirus polypeptides by several mammalian cell-free extracts. *Proc Natl Acad Sci USA* 1972;69:2649–2653.
251. McPhillips TH, Ramig RF. Extragenic suppression of temperature sensitive phenotype in reovirus mapping suppressor mutations. *Virology* 1984;135:428–439.
252. Metcalf P. The symmetry of the reovirus outer shell. *J Ultrastruct Res* 1982;78:292–301.
253. Metcalf P, Cyrklaff M, Adrian M. The 3-dimensional structure of reovirus obtained by cryoelectron microscopy. *EMBO J* 1991;10:3129–3136.
254. Miller JE, Samuel CE. Proteolytic cleavage of the reovirus sigma 3 protein results in enhanced double-stranded RNA-binding activity. Identification of a repeated amino-acid motif within the C-terminal binding region. *J Virol* 1992;66:5347–5356.
255. Millward S, Graham AF. Structural studies on reovirus: discontinuities in the genome. *Proc Natl Acad Sci USA* 1970;65:422–426.
256. Miura K-I, Watanabe K, Sugiura M, Shatkin AJ. The 5′-terminal nucleotide sequences of the double-stranded RNA of human reovirus. *Proc Natl Acad Sci USA* 1974;71:3979–3983.
257. Miyamoto NG, Jacobs BL, Samuel CE. Mechanism of interferon action: effect of double-stranded RNA and the 5′-O-monophosphate form of 2′,5′-oligoadenylate on the inhibition of reovirus mRNA translation in vitro. *J Biol Chem* 1983;258:15232–15237.
258. Miyamoto NG, Samuel CE. Mechanism of interferon action: interferon-mediated inhibition of reovirus mRNA translation in the absence of detectable mRNA degradation but in the presence of protein phosphorylation. *Virology* 1980;107:461–475.
259. Montgomery LB, Kao CYY, Verdin E, Cahill C, Maratosflier E. Infection of a polarized epithelial cell line with wild-type reovirus leads to virus persistence and altered cellular function. *J Gen Virol* 1991;72: 2939–2946.
260. Moody MD, Joklik WK. The function of reovirus proteins during the reovirus multiplication cycle: analysis using monoreassortants. *Virology* 1989;173:437–446.
261. Mora M, Partin K, Bhatia M, Partin J, Carter C. Association of reovirus proteins with the structural matrix of infected cells. *Virology* 1987;159: 265–277.
262. Morgan EM, Kingsbury DW. Pyridoxal phosphate as a probe of reovirus transcriptase. *Biochemistry* 1980;19:484–489.
263. Morgan EM, Kingsbury DW. Reovirus enzymes that modify messenger RNA are inhibited by perturbation of the lambda proteins. *Virology* 1981;113:565–572.
264. Morgan EM, Zweerink HJ. Reovirus morphogenesis: core-like particles in cells infected at 39° with wild-type reovirus and temperature-sensitive mutants of groups B and G. *Virology* 1974;59:556–565.
265. Morgan EM, Zweerink HJ. Characterization of transcriptase and replicase particles isolated from reovirus infected cells. *Virology* 1975;68: 455–466.
266. Morgan EM, Zweerink HJ. Characterization of the double-stranded RNA in replicase particles in reovirus infected cells. *Virology* 1977; 77:421–423.
267. Morozov SY. A possible relationship of reovirus putative RNA-polymerases to polymerases of positive-strand RNA viruses. *Nucleic Acids Res* 1989;17:5394–5394.
268. Munemitsu SM, Atwater JA, Samuel CE. Biosynthesis of reovirus-specified polypeptides. Molecular cDNA cloning and nucleotide sequence of the reovirus serotype 1 Lang strain bicistronic S1 mRNA which encodes the minor capsid polypeptide σ1a and the nonstructural polypeptide σ1bns. *Biochem Biophys Res Commun* 1986;140: 508–514.
269. Munemitsu SM, Samuel CE. Biosynthesis of reovirus-specified polypeptides. Multiplication rate but not yield of reovirus serotypes 1 and 3 correlates with the level of virus-mediated inhibition of cellular protein synthesis. *Virology* 1984;136:133–143.
270. Munemitsu SM, Samuel CE. Biosynthesis of reovirus-specified polypeptides: effect of point mutation of the sequences flanking the 5′-proximal AUG initiator codons of the reovirus S1 and S4 genes on the efficiency of messenger RNA translation. *Virology* 1988;163:643–646.
271. Munoz A, Alonso MA, Carrasco L. The regulation of translation in reovirus-infected cells. *J Gen Virol* 1985;66:2161–2170.
272. Mustoe TA, Ramig RF, Sharpe AH, Fields BN. A genetic map of reovirus. III. Assignment of the double-stranded RNA mutant groups A, B, and G to genome segments. *Virology* 1978;85:545–556.
273. Mustoe TA, Ramig RF, Sharpe AH, Fields BN. Genetics of reovirus: identification of the dsRNA segments encoding the polypeptides of the m and s size classes. *Virology* 1978;89:594–604.
274. Muthukrishnan S, Shatkin AJ. Reovirus genome RNA segments: resistance to S1 nuclease. *Virology* 1975;64:96–105.
275. Nagata L, Masri SA, Mah DC, and PWK Lee. Molecular cloning and sequencing of the reovirus (serotype 3) S1 gene which encodes the viral cell attachment protein σ1. *Nucleic Acids Res* 1984;12:8699–8710.
276. Nagata L, Masri SA, Pon RT, Lee PWK. Analysis of functional domains on reovirus cell attachment protein σ1 using cloned S1 gene deletion mutants. *Virology* 1987;160:162–168.
277. Nakashima K, LaFiandra, Shatkin AJ. Differential dependence of reovirus-associated enzyme activities on genome RNA as determined by psoralen photosensitivity. *J Biol Chem* 1979; 254:8007–8014.
278. Nepom JT, Weiner HL, Dichter MA, et al. Identification of a hemagglutinin-specific idiotype associated with reovirus recognition shared by lymphoid and neuronal cells. *J Exp Med* 1982;155:155–167.
279. Ni Y, Ramig RF. Characterization of avian reovirus-induced fusion: the role of structural proteins. *Virology* 1993;194:705–714.
280. Ni Y, Ramig RF, Kemp MC. Identification of proteins encoded by avian reoviruses and evidence for post-translational modification. *Virology* 1993;193:466–469.
281. Nibert ML, Dermody TS, Fields BN. Structure of the reovirus cell-attachment protein: a model for the domain organization of σ1. *J Virol* 1990;64:2976–2989.
282. Nibert ML, Fields BN. A carboxy-terminal fragment of protein μ1/μ1C is present in infectious subvirion particles of mammalian reoviruses and is proposed to have a role in penetration. *J Virol* 1992;66:6408–6418.
283. Nibert ML, Fields BN. Early steps in reovirus infection of cells. In: Wimmer E, ed. *Cell receptor for animal viruses*. Cold Spring Harbor, NY: Cold Spring Harbor Laboratory Press, 1995;341–364.
284. Nibert ML, Furlong DB, Fields BN. Mechanisms of viral pathogenesis. Distinct forms of reoviruses and their roles during replication in cells and host. *J Clin Invest* 1991;88:727–734.
285. Nibert ML, Schiff LA, Fields BN. Mammalian reoviruses contain a myristoylated structural protein. *J Virol* 1991;65:1960–1967.
286. Nichols JL, Bellamy AR, Joklik WK. Identification of the nucleotide sequences of the oligonucleotides present in reovirions. *Virology* 1972; 49:562–572.
287. Nilsen TW, Maroney PA, Baglioni C. Inhibition of protein synthesis in reovirus-infected HeLa cells with elevated levels of interferon-induced protein kinase activity. *J Biol Chem* 1982; 257:14593–14596.
288. Nilsen TW, Maroney PA, Baglioni C. Synthesis of (2′-5′)oligoadenylate and activation of an endoribonuclease in interferon-treated HeLa cells infected with reovirus. *J Virol* 1982;42:1039–1045.
289. Nilsen TW, Maroney PA, Baglioni C. Maintenance of protein synthesis in spite of mRNA breakdown in interferon-treated HeLa cells infected with reovirus. *Mol Cell Biol* 1983;3:64–69.
290. Nonoyama M, Graham AF. Appearance of defective virions in clones of reovirus. *J Virol* 1970;6:693–694.
291. Nonoyama M, Millward S, Graham AF. Control of transcription of the reovirus genome. *Nucleic Acids Res* 1974;1:373–385.
292. Nonoyama M, Watanabe Y, Graham AF. Defective virions of reovirus. *J Virol* 1970;6: 226–236.
293. Noseworthy JH, Fields BN, Dichter MA, et al. Cell receptors for the mammalian reovirus. I. Syngeneic monoclonal anti-idiotypic antibody identifies a cell surface receptor for reovirus. *J Immunol* 1983;131: 2533–2538.
294. Pacitti AF, Gentsch JR. Inhibition of reovirus type 3 binding to host cells by sialylated glycoproteins is mediated through the viral attachment protein. *J Virol* 1987;61:1407–1415.
295. Palmer EL, Martin ML. The fine structure of the capsid of reovirus type 3. *Virology* 1977;76:109–113.
296. Paul RW, Choi AHC, Lee PWK. The α-anomeric form of sialic acid is the minimal receptor determinant recognized by reovirus. *Virology* 1989;172:382–385.
297. Paul RW, Lee PWK. Glycophorin is the reovirus receptor on human erythrocytes. *Virology* 1987;159:94–101.
298. Pelletier J, Nicholson R, Bassel-Duby R, Fields BN, Sonenberg N. Expression of reovirus type 3 Dearing σ1 and σs polypeptides in *Escherichia coli*. *J Gen Virol* 1987;68:135–145.

299. Pett DM, Vanaman TC, Joklik WK. Studies on the amino- and carboxyl-terminal amino acid sequences of reovirus capsid polypeptides. *Virology* 1973;52:174–186.
300. Powell KFH, Harvey JD, Bellamy AR. Reovirus RNA transcriptase: evidence for a conformational change during activation of the core particle. *Virology* 1984;137:1–8.
301. Ralph SJ, Harvey JD, Bellamy AR. Subunit structure of the reovirus spike. *J Virol* 1980;36:894–896.
302. Ramig RF, Ahmed R, Fields BN. A genetic map of reovirus: assignment of the newly defined mutant groups H, I, and J to genome segments. *Virology* 1983;125:299–313.
303. Ramig RF, Cross RK, Fields BN. Genome RNAs and polypeptides of reovirus serotypes 1, 2, and 3. *J Virol* 1977;22:726–733.
304. Ramig RF, Fields BN. Method for rapidly screening revertants of reovirus temperature sensitive mutants for extragenic suppression. *Virology* 1977;81:170–173.
305. Ramig RF, Fields BN. Reoviruses. In: Nayak DP, ed. *The molecular biology of animal viruses*, vol I. New York: Marcel Dekker 1977; 383–433.
306. Ramig RF, Fields BN. Revertants of temperature-sensitive mutants of reovirus: evidence for frequent extragenic suppression. *Virology* 1979; 92:155–167.
307. Ramig R, Fields BN. Genetics of reovirus. In: Joklik W, ed.*The Reoviridae*. New York: Plenum Press, 1983;197–228.
308. Ramig RF, Mustoe TA, Sharpe AH, Fields BN. A genetic map of reovirus. II. Assignment of the double-stranded RNA-negative mutant groups C, D, and E genome segments. *Virology* 1978;85:531–534.
309. Ramig RF, White RM, and Fields BN. Suppression of the temperature sensitive phenotype of a mutant of reovirus type 3. *Science* 1977;195: 406–407
310. Ramos-Alvarez M, Sabin AB Enteropathogenic viruses and bacteria. Role in summer diarrheal diseases of infancy and early childhood. *JAMA* 1958;167:147–156.
311. Rankin UT Jr, Eppes SB, Antczak JB, Joklik WK. Studies on the mechanism of the antiviral activity of ribavirin against reovirus. *Virology* 1989;168:147–158.
312. Ray BK, Brendler TG, Adya S, et al. Role of mRNA competition in regulating translation: further characterization of mRNA discriminatory initiation factors. *Proc Natl Acad Sci USA* 1983;80:663–667.
313. Rhim JS, Jordan LE, Mayor HD. Cytochemical, fluorescent-antibody and electron microscopic studies on the growth of reovirus (ECHO 10) in tissue culture. *Virology* 1962;17:342–355.
314. Rhim JS, Smith KO, Melnick JL. Complete and coreless forms of reovirus (ECHO 10): ratio of number of virus particles to infective units in the one-step growth cycle. *Virology* 1961;15:428–435.
315. Richardson MA, Furuichi Y. Nucleotide sequence of reovirus genome segment S3, encoding nonstructural protein σNS. *Nucleic Acids Res* 1983;11:6399–6408.
316. Richardson MA, Furuichi Y. Synthesis in *Escherichia coli* of the reovirus nonstructural protein σNS. *J Virol* 1985; 56:527–533.
317. Robertson MD, Wilcox GE. Avian reoviruses. U 1986;56:155–174.
318. Roner MR, Gaillard RK, Joklik WK. Control of reovirus messenger RNA translation efficiency by the regions upstream of initiation codons. *Virology* 1989;168:302–311.
319. Roner MR, Roner LA, Joklik WK. Translation of reovirus RNA species m1 can initiate at either of the first two in-frame initiation codons. *Proc Natl Acad Sci USA* 1993;90:8947–8951.
320. Roner MR, Sutphin LA, Joklik WK. Reovirus RNA is infectious. *Virology* 1990:179:845–852.
321. Rosen L. Serologic groupings of reovirus by hemagglutination-inhibition. *Am J Hyg* 1960;71:242–249.
322. Rosen L. Reoviruses in animals other than man. *Ann NY Acad Sci* 1962; 101:461–465.
323. Rosen L, Hovis JF, Mastrota FM, Bell JA, Huebner RJ. Observations on a newly recognized virus (Abney) of the reovirus family. *Am J Hyg* 1960;71:258–265.
324. Rosenberger JK, Olson NO. Reovirus infection. In: Calnek BW et al, eds. *Diseases of poultry*, 9th ed. Ames, Iowa: Iowa State Univ Press 1991;639–647.
325. Rubin DH, Fields BN. Molecular basis of reovirus virulence: role of the M2 gene. *J Exp Med* 1980;152:853–868.
326. Rubin DH, Weiner DB, Dworkin C, Greene MI, Maul GG, Williams WV. Receptor utilization by reovirus type 3: Distinct binding sites on thymoma and fibroblast cell lines result in differential compartmentalization of virions. *Microbial Pathogen* 1992;12:351–365.
327. Rubin DH, Wetzel JD, Williams WV, Cohen JA, Dworkin C, Dermody TS. Binding of type 3 reovirus by a domain of the σ1 protein important for hemagglutination leads to infection of murine erythroleukemia cells. *J Clin Invest* 1992;90:2536–2542.
328. Sabin AB. Reoviruses. *Science* 1959;130:1387–1389.
329. Sakuma S, Watanabe Y. Unilateral synthesis of reovirus double-stranded ribonucleic acid by a cell free replicase system. *J Virol* 1971;8: 190–196.
330. Sakuma S, Watanabe Y. Reovirus replicase-directed synthesis of double-stranded ribonucleic acid. *J Virol* 1972;10:628–638.
331. Samuel CE, Brody, MS. Biosynthesis of reovirus-specified polypeptides: 1-Aminopurine increases the efficiency of translation of reovirus s1 messenger RNA but not s4 messenger RNA in transfected cells. *Virology* 1990;176:106–113.
332. Samuel CE, Duncan R, Knutson GS, Hershey JWB. Mechanism of interferon action: increased phosphorylation of protein synthesis initiation factor eIF-2 alpha in interferon-treated reovirus-infected mouse L-929 fibroblasts in vitro and in vivo. *J Biol Chem* 1984;259:13451–13457.
333. Samuel CE, Knutson GS. Mechanism of interferon action: human leukocyte and immune interferons regulate the expression of different genes and induce different antiviral states in human amnion U cells. *Virology* 1983;130:474–484.
334. Samuel CE, Ulker N, Knutson GS, Zhang X, Masters PS. Molecular mechanisms of the antiviral action of human alpha and gamma interferons. In: Kirchner H, Schellekens H, eds. *Biology of the interferon system*. New York: Elsevier, 1985;131–140.
335. Sargent MD, Borsa J. Effects of calcium and magnesium on the switch-on of transcriptase function in reovirus in vitro. *Can J Biochem Cell Biol* 1984;62:162–169.
336. Sarkar G, Pelletier J, Bassel-Duby R, Jayasuriya A, Fields BN, Sonenberg N. Identification of a new polypeptide coded by reovirus gene S1. *J Virol* 1985;54:720–725.
337. Sauve GJ, Saragovi HU, Greene MI. Reovirus receptors. *Adv Virus Res* 1992;42:232–341.
338. Sawutz DG, Bassel-Duby R, Homcy CJ. High affinity binding of reovirus type 3 to cells that lack beta adrenergic receptor activity. *Life Sci* 1987;40:399–406.
339. Schiff LA, Nibert ML, Co MS, Brown EG, Fields BN. Distinct binding sites for zinc and double-stranded RNA in the reovirus outer capsid protein σ3. *Mol Cell Biol* 1988;8:273–283.
340. Schiff LA, Nibert ML, Fields BN. Characterization of a zinc blotting technique: evidence that a retroviral gag protein binds zinc. *Proc Natl Acad Sci USA* 1988;85:4195–4199.
341. Schonberg M, Silverstein SC, Levin DH, Acs G. Asynchronous synthesis of the complementary strands of the reovirus genome. *Proc Natl Acad Sci USA* 1971;68:505–508.
342. Schuerch AR, Matshuhisa I, Joklik WK. Temperature-sensitive mutants of reovirus. VI. Mutants ts447 and ts556 particles lack one or two L genome RNA segments. *Intervirology* 1974;3:36–46.
343. Schnitzer TJ, Ramos T, Gouvea V. Avian reovirus polypeptides: analysis of intracellular virus specified products, virions, top component, and cores. *J Virol* 1982;43:1006–1014.
344. Seliger LS, Giantini M, Shatkin AJ. Translational effects and sequence comparisons of the 3 serotypes of the reovirus S4 gene. *Virology* 1992; 187:202–210.
345. Seliger LS, Zheng K, Shatkin AJ. Complete nucleotide sequence of reovirus L2 gene and deduced amino acid sequence of viral mRNA guanylyltransferase. *J Biol Chem* 1987;262:16289–16293.
346. Sharpe AH, Chen LB, Fields BN. The interaction of mammalian reoviruses with the cytoskeleton of monkey kidney CV-1 cells. *Virology* 1982;120:399–411.
347. Sharpe AH, Fields BN. Reovirus inhibition of cellular DNA synthesis: role of the S1 gene. *J Virol* 1981;38:389–392.
348. Sharpe AH, Fields BN. Reovirus inhibition of cellular RNA and protein synthesis: role of the S4 gene. *Virology* 1982;122:381–391.
349. Sharpe AH, Fields BN. Pathogenesis of reovirus infection. In: Joklik WK, ed. *The Reoviridae*. New York: Plenum Press, 1983;229–285.
350. Sharpe AH, Ramig RF, Mustoe TA, Fields BN. A genetic map of reovirus. I. Correlation of genome RNAs between serotypes 1, 2, and 3. *Virology* 1978;84:63–74.
351. Shatkin AJ. Inactivity of purified reovirus RNA as a template for *E. coli* polymerases in vitro. *Proc Natl Acad Sci USA* 1965;54:1721–1728.
352. Shatkin AJ. Viruses containing double-stranded RNA. In: Fraenkel-

Conrat H, ed. *Molecular basis of virology*. New York: Reinhold Book Corp, 1968;351–392.
353. Shatkin AJ, Furuichi Y, LaFiandra AJ, Yamakawa M. Initiation of mRNA synthesis and 5′-terminal modification of reovirus transcripts. In: Compans RW, Bishop DHL, eds. *Double-stranded RNA viruses*. New York: Elsevier, 1983;43–54.
354. Shatkin AJ, Kozak M. Biochemical aspects of reovirus transcription and translation. In: Joklik WK, ed. *The Reoviridae*. New York: Plenum Press, 1983;79–106.
355. Shatkin AJ, LaFiandra AJ. Transcription by infectious subviral particles of reovirus. *J Virol* 1972;10:698–706.
356. Shatkin AJ, Sipe JD. Single-stranded adenine-rich RNA from purified reoviruses. *Proc Natl Acad Sci USA* 1968;59:246–253.
357. Shatkin AJ, Sipe JD. RNA polymerase activity in purified reoviruses. *Proc Natl Acad Sci USA* 1968;61:1462–1469.
358. Shatkin AJ, Sipe JD, Loh PC. Separation of 10 reovirus genome segments by polyacrylamide gel electrophoresis. *J Virol* 1968;2:986–991.
359. Shaw JE, Cox DC. Early inhibition of cellular DNA synthesis by high multiplicities of infections and UV-irradiated reovirus. *J Virol* 1973; 12:704–710.
360. Sherry B, Fields BN. The reovirus M1 gene, encoding a viral core protein, is associated with the myocarditic phenotype of a reovirus variant. *J Virol* 1989;63:4850–4856.
361. Sherry B, Schoen FJ, Wenske E, Fields BN. Derivation and characterization of an efficiently myocarditic reovirus variant. *J Virol* 1989; 63:4840–4849.
362. Silverstein SC, Astell C, Christman J, Klett H, Acs G. Synthesis of reovirus oligoadenylic acid in vivo and in vitro. *J Virol* 1974;13:740–752.
363. Silverstein SC, Astell C, Levin DH, Schonberg M, Acs G. The mechanisms of reovirus uncoating and gene activation in vivo. *Virology* 1972;47:797–806.
364. Silverstein SC, Christman JK, Acs G. The reovirus replication cycle. *Annu Rev Biochem* 1976;45:375–408.
365. Silverstein SC, Dales S. The penetration of reovirus RNA and initiation of its genetic function in L-strain fibroblasts. *J Cell Biol* 1968;36: 197–230.
366. Silverstein SC, Levin DH, Schonberg M, Acs G. The reovirus replicative cycle: conservation of parental RNA and protein. *Proc Natl Acad Sci USA* 1970;67:275–281.
367. Silverstein SC, Schur PH. Immunofluorescent localization of double-stranded RNA in reovirus-infected cells. *Virology* 1970;41:564–566.
368. Skehel JJ, Joklik WK. Studies on the in vitro transcription of reovirus RNA catalyzed by reovirus cores. *Virology* 1969;39:822–831.
369. Skup D, Millward S. Reovirus-induced modification of cap dependent translation in infected L cells. *Proc Natl Acad Sci USA* 1980;77: 152–156.
370. Skup D, Millward S. mRNA capping enzymes are masked in reovirus progeny subviral particles. *J Virol* 1980;34:490–496.
371. Skup D, Zarbl H, Millward S. Regulation of translation in L-cells infected with reovirus. *J Mol Biol* 1981;151:35–55.
372. Smith RE, Morgan MA, Furuichi Y. Separation of the plus and minus strands of cytoplasmic polyhedrosis virus and human reovirus double-stranded genome RNAs by gel electrophoresis. *Nucleic Acids Res* 1981; 9:5269–5286.
373. Smith RE, Zweerink HJ, Joklik WK. Polypeptide components of virions, top component and cores of reovirus 3. *Virology* 1969;39: 791–810.
374. Sonenberg N, Skup D, Trachsel H, Millward S. In vitro translation in reovirus and poliovirus-infected cell extracts: effects of anti-cap binding protein monoclonal antibody. *J Biol Chem* 1981;256:4138–4141.
375. Spandidos DA, Graham AF. Physical and chemical characterization of an avian reovirus. *J Virol* 1976;19:968–976.
376. Spandidos DA, Graham AF. Nonpermissive infection of L cells by an avian reovirus: restricted transcription of the viral genome. *J Virol* 1976;19:977–984.
377. Spandidos DA, Krystal G, Graham AF. Regulated transcription of the genomes of defective virions and temperature-sensitive mutants of reovirus. *J Virol* 1976;18:7–19.
378. Spendlove RS, Lennette EH, Chin JN, Knight CO. Effect of antimitotic agents on intracellular reovirus antigen. *Cancer Res* 1964;24: 1826–1833.
379. Spendlove RS, Lennette EH, Knight CO, Chin JH. Development of viral antigen and infectious virus on HeLa cells infected with reovirus. *J Immunol* 1963;90:548–553.
380. Spendlove RS, McClain ME, Lennette EH. Enhancement of reovirus infectivity by extracellular removal or alteration of the virus capsid by proteolytic enzymes. *J Gen Virol* 1970;8:83–93.
381. Spendlove RS, Schaffer FL. Enzymatic enhancement of infectivity of reovirus. *J Bacteriol* 1965;89:597–602.
382. Spriggs DR, Bronson RT, Fields BN. Hemagglutinin variants of reovirus type 3 have altered central nervous system tropism. *Science* 1983;220:505–507.
383. Spriggs DR, Fields BN. Attenuated reovirus type 3 strains generated by selection of hemagglutinin antigenic variants. *Nature* 1982;297:68–70.
384. Spriggs DR, Kaye K, Fields BN. Topological analysis of the reovirus type 3 hemagglutinin. *Virology* 1983;127:220–224.
385. Stamatos NM, Gomatos PJ. Binding to selected regions of reovirus mRNAs by a nonstructural reovirus protein. *Proc Natl Acad Sci USA* 1982;79:3457–3461.
386. Stanley NF. Reoviruses. *Br Med Bull* 1967;23:150–154.
387. Starnes MC, Joklik WK. Reovirus protein λ3 is a poly(C)-dependent poly(G) polymerase. *Virology* 1993;193:356–366.
388. Stoltzfus CM, Banerjee AK. Two oligonucleotide classes of single-stranded ribopolymers in reovirus A-rich RNA. *Arch Biochem Biophys* 1972;152:733–743.
389. Stoltzfus CM, Morgan M, Banerjee AK, Shatkin AJ. Poly(A) polymerase activity in reovirus. *J Virol* 1974;13:1338–1345.
390. Strong JE, Leone G, Duncan R, Sharma RK, Lee PW. Biochemical and biophysical characterization of the reovirus cell attachment protein σ1: evidence that it is a homotrimer. *Virology* 1991;184:23–32.
391. Strong JE, Tang D, Lee PWK. Evidence that the epidermal growth factor receptor on host cells confers reovirus infection efficiency. *Virology* 1993;197:405–411.
392. Sturzenbecker LJ, Nibert M, Furlong D, Fields BN. Intracellular digestion of reovirus particles requires a low pH and is an essential step in the viral infectious cycle. *J Virol* 1987;61: 2351–2361.
393. Taber R, Alexander V, Whitford W. Persistent reovirus infection of CHO cells resulting in virus resistance. *J Virol* 1976;17:513–525.
394. Tang D, Strong JE, Lee PWK. Recognition of the epidermal growth factor receptor by reovirus. *Virology* 1993;197:412–414.
395. Tarlow O, McCorquodale JG, McCrae MA. Molecular cloning and sequencing of the gene S1 encoding the major virion structural protein μ1—μ1C of serotypes 1 and 3 of mammalian reovirus. *Virology* 1988; 164:141–146.
396. Tillotson L, Shatkin AJ. Reovirus polypeptide σ3 and N-terminal myristoylation of polypeptide μ1 are required for site-specific cleavage to μ1C in transfected cells. *J Virol* 1992; 66:2180–2186.
397. Tosteson MT, Nibert ML, Fields BN. Ion channels induced in lipid bilayers by subvirion particles of the nonenveloped mammalian reoviruses. *Proc Natl Acad Sci USA* 1993; 90:10549–10552.
398. Turner DL, Duncan R, Lee PWK. Site-directed mutagenesis of the C-terminal portion of reovirus protein σ1. Evidence for a conformation-dependent receptor-binding domain. *Virology* 1992;186:219–227.
399. Tyler KL, McPhee DA, Fields BN. Distinct pathways of viral spread in the host determined by reovirus S1 gene segment. *Science* 1986;233: 770–774.
400. Vasquez C, Kleinschmidt AK. Electron microscopy of RNA strands released from individual reovirus particles. *J Mol Biol* 1968;34:137–147.
401. Vasquez C, Tournier P. The morphology of reovirus. *Virology* 1962; 17:503–510.
402. Verdin EM, King GL, Maratos-Flier E. Characterization of a common high affinity receptor for reovirus serotypes 1 and 3 on endothelial cells. *J Virol* 1989;63:1318–1325.
403. Verdin EM, Maratos-Flier E, Carpentier JL, Kahn CR. Persistent infection with a nontransforming RNA virus leads to impaired growth factor receptors and response. *J Cell Physiol* 1986;128:457–465.
404. Virgin HW IV, Mann MA, Fields BN, Tyler KL. Monoclonal antibodies to reovirus reveal structure/function relationships between capsid proteins and genetics of susceptibility to antibody action. *J Virol* 1991;65:5721–5731.
405. Wachsman JT, Levin DH, Acs G. Ribonucleoside triphosphate-dependent pyrophosphate exchange of reovirus cores. *J Virol* 1970;6: 563–565.
406. Walden WE, Godefroy-Colburn T, Thach RE. The role of mRNA competition in regulating translation. I. Demonstration of competition in vivo. *J Biol Chem* 1981;256:11739–11746.
407. Wallis C, Melnick JL. Irreversible photosensitization of viruses. *Virology* 1964;23:520–527.
408. Wallis C, Smith KO, Melnick JL. Reovirus activation by heating and inactivation by cooling in $MgCl_2$ solution. *Virology* 1964;22:608–619.

409. Ward RL, Banerjee AK, LaFiandra A, Shatkin AJ. Reovirus-specific ribonucleic acid from polysomes of infected L cells. *J Virol* 1972;9:61–69.
410. Ward RL, Shatkin AJ. Association of reovirus mRNA with viral proteins: a possible mechanism for linking the genome segments. *Arch Biochem Biophys* 1972;152:378–384.
411. Watanabe Y, Gauntt CJ, Graham AF. Reovirus-induced ribonucleic acid polymerase. *J Virol* 1968;2:869–877.
412. Watanabe Y, Graham AF. Structural units of reovirus ribonucleic acid and their possible functional significance. *J Virol* 1967;1:665–677.
413. Watanabe Y, Millward S, Graham AF. Regulation of transcription of the reovirus genome. *J Mol Biol* 1968;36:107–123.
414. Watanabe Y, Prevec L, Graham AF. Specificity in transcription of the reovirus genome. *Proc Natl Acad Sci USA* 1967;58:1040–1047.
415. Weiner DB, Girard K, Williams WV, McPhillips T, Rubin DH. Reovirus type 1 and type 3 differ in their binding to isolated intestinal epithelial cells. *Microbial Pathogen* 1988;5:29–40.
416. Weiner HL, Ault KA, Fields BN. Interaction of reovirus with cell surface receptors. I. Murine and human lymphocytes have a receptor for the hemagglutinin of reovirus type 3. *J Immunol* 1980;124:2143–2148.
417. Weiner HL, Drayna D, Averill DR Jr, Fields BN. Molecular basis of reovirus virulence: role of the S1 gene. *Proc Natl Acad Sci USA* 1977; 74:5744–5748.
418. Weiner HL, Fields BN. Neutralization of reovirus: the gene responsible for the neutralization antigen. *J Exp Med* 1977;146:1305–1310.
419. Weiner HL, Powers ML, Fields BN. Absolute linkage of virulence and central nervous system tropism of reoviruses to viral hemagglutinin. *J Infect Dis* 1980;141:609–616.
420. Weiner HL, Ramig RF, Mustoe TA, Fields BN. Identification of the gene coding for the hemagglutinin of reovirus. *Virology* 1978;86: 581–584.
421. Wenske EA, Chanock SJ, Krata L, Fields BN. Genetic reassortment of mammalian reoviruses in mice. *J Virol* 1985;56:613–616.
422. Wessner DR, Fields BN. Isolation and genetic characterization of ethanol-resistant reovirus mutants. *J Virol* 1993;67:2442–2447.
423. White CK, Zweerink HJ. Studies on the structure of reovirus cores: selective removal of polypeptide λ2. *Virology* 1976;70:171–180.
424. Wickramasinghe R, Meanger J, Enriquez CE, Wilcox GE. Avian reovirus proteins associated with neutralization of virus infectivity. *Virology* 1993;194:688–696.
425. Wiebe ME, Joklik WK. The mechanism of inhibition of reovirus replication by interferon. *Virology* 1975;66:229–240.
426. Wiener JR, Bartlett JA, Joklik WK. The sequences of reovirus serotype 3 genome segments M1 and M3 encoding the minor protein μ2 and the major nonstructural protein μNS, respectively. *Virology* 1989;169: 293–304.
427. Wiener JR, Joklik WK. Comparison of the reovirus type 1, 2, and 3 S3 genome segments encoding the nonstructural protein σNS. *Virology* 1987;161:332–339.
428. Wiener JR, Joklik WK. Evolution of reovirus genes: a comparison of serotype 1, 2, and 3 M2 genome segments which encode the major structural capsid protein μ1C. *Virology* 1988;163: 603–613.
429. Wiener JR, Joklik WK. The sequences of the reovirus serotype 1, 2, and 3 L1 genome segments and analysis of the mode of divergence of the reovirus serotypes. *Virology* 1989;169: 194–203.
430. Wiener JR, McLaughlin T, Joklik WK. The sequence of the S2 genome segment of reovirus serotype 3 and of that of the dsRNA-negative mutant ts447. *Virology* 1989;170:340–341.
431. Wilcox GE, Compans RW. Cell fusion induced by Nelson Bay Virus. *Virology* 1982;123: 312–322.
432. Williams WV, Guy HR, Rubin DH, et al. Sequences of the cell-attachment sites of reovirus type 3 and its anti-idiotypic/antireceptor antibody: modeling of their three-dimensional structures. *Proc Natl Acad Sci USA* 1988;85:6488–6492.
433. Williams WV, Kieber-Emmons T, Weiner DB, Rubin DH, Greene MI. Contact residues and predicted structure of the reovirus type 3 receptor interaction. *J Biol Chem* 1991;266:9241–9250.
434. Williams WV, Weiner DB, Greene MI. Development and use of antireceptor antibodies to study interaction of mammalian reovirus type 3 with its cell surface receptor. *Methods Enzymol* 1989;178:321–341.
435. Xu P, Miller SE, Joklik WK. Generation of reovirus core-like particles in cells infected with hybrid vaccinia viruses that express genome segments L1, L2, L3, and S2. *Virology* 1993;197:726–731.
436. Yamakawa M, Furuichi Y, Nakashima K, LaFiandra AJ, Shatkin AJ. Excess synthesis of viral mRNA 5′-terminal oligonucleotides by reovirus transcriptase. *J Biol Chem* 1981;256:6507–6514.
437. Yamakawa M, Furuichi Y, Shatkin AJ. Reovirus transcriptase and capping enzymes are active in intact virions. *Virology* 1982;118:157–168.
438. Yeung MC, Gill MJ, Alibhai SS, Shahrabadi MS, Lee PWK. Purification and characterization of the reovirus cell attachment protein σ1. *Virology* 1987;156:377–385.
439. Yeung MC, Lim D, Duncan R, Shahrabadi MS, Cashdollar LW, Lee PWK. The cell attachment proteins of type 1 and type 3 reovirus are differentially susceptible to trypsin and chymotrypsin. *Virology* 1989; 170:62–70.
440. Zarbl H, Hastings KEM, Millward S. Reovirus core particles synthesize capped oligonucleotides as a result of abortive transcription. *Arch Biochem Biophys* 1980;202:348–360.
441. Zarbl H, Millward S. The reovirus multiplication cycle. In: Joklik WK, ed. *The Reoviridae*. New York: Plenum Press, 1983;107–196.
442. Zarbl H, Skup S, Millward S. Reovirus progeny subviral particles synthesize uncapped mRNA. *J Virol* 1980;34:497–505.
443. Zou, S, Brown EG. Identification of sequence elements containing signals for replication and encapsidation of the reovirus M1 genome segment. *Virology* 1992;186:377–388.
444. Zou S, Brown EG. Nucleotide sequence comparison of the M1 genome segment of reovirus type 1 Lang and type 3 Dearing. *Virus Res* 1992; 22:159–164.
445. Zweerink HJ. Multiple forms of ss—dsRNA polymerase activity in reovirus-infected cells. *Nature* 1974;247:313–315.
446. Zweerink HJ, Ito Y, Matsuhisa T. Synthesis of reovirus double-stranded RNA within virion-like particles. *Virology* 1972;50:349–358.
447. Zweerink HJ, Joklik WK. Studies on the intracellular synthesis of reovirus-specified proteins. *Virology* 1970;41:501–518.
448. Zweerink HJ, McDowell MJ, Joklik WK. Essential and non-essential non-capsid reovirus proteins. *Virology* 1971;45:716–723.
449. Zweerink HJ, Morgan EM, Skyler JS. Reovirus morphogenesis: characterization of subviral particles in infected cells. *Virology* 1976;73: 442–453.

Fundamental Virology, Third Edition
edited by B.N. Fields, D.M. Knipe, P.M. Howley, et al.
Lippincott - Raven Publishers, Philadelphia © 1996

CHAPTER 25

Rotaviruses and Their Replication

Mary K. Estes

Rotaviruses were initially characterized as agents associated with gastroenteritis in animals. The history of the discovery of the rotaviruses exemplifies how studies of animal diseases can provide critical insights into human viral disease (see Chapter 55 in *Fields Virology*, 3rd ed.). Numerous epidemiologic studies have shown that the rotaviruses are the most significant cause of severe gastroenteritis in young children and animals (see Chapter 55 in *Fields Virology*, 3rd ed.). Rotaviruses cause diarrheal disease primarily in the young, but infection and disease in older children and adults are also common. Some outbreaks in adults have been caused by recently recognized non–group A rotaviruses. Rotaviruses also have been reported in association with syndromes other than diarrhea (e.g., exanthem subitum, otitis media, necrotizing enterocolitis, liver abscesses). However, at this time no convincing evidence proves that rotaviruses cause any of these syndromes.

The clinical significance of the rotaviruses stimulated basic research on these agents. Studies since 1973 on the biochemistry, biology, and molecular and antigenic properties of the rotaviruses have resulted in a fairly detailed understanding of virus structure and the lytic cycle of replication. Discovery that the rotaviruses code for two glycoproteins that are targeted to and retained exclusively in the endoplasmic reticulum (ER) led to studies of these glycoproteins as general models of intracellular polypeptide processing. These observations have brought the rotaviruses into prominence as biologic tools for understanding ER functions as well as virus-host interactions in the gastrointestinal tract. Studies on the structure of rotavirus and rotavirus-antibody complexes using electron cryomicroscopy and three-dimensional image processing techniques have highlighted the usefulness of these new methods to obtain significant biologic and structural information about large, complex viruses. Recognition that four of the five rotavirus nonstructural proteins bind nucleic acids has resulted in the realization that these proteins have unique properties, and understanding their properties is providing new fundamental information about RNA synthesis and expression. The clinical syndromes, epidemiology, techniques for viral cultivation, diagnosis of rotavirus infections, and vaccine status are discussed in Chapter 55 in *Fields Virology*, 3rd ed.

VIRUS CLASSIFICATION

The rotaviruses comprise a genus within the family *Reoviridae,* and rotaviruses share common morphologic and biochemical properties (Table 1). The most salient features are listed as follows:

M. K. Estes: Division of Molecular Virology, Baylor College of Medicine, Houston, Texas 77030.

TABLE 1. *General characteristics of rotaviruses*

Structure
- 65 to 75 nm icosahedral particles
- Triple-layered protein capsid
- Nonenveloped (resistant to lipid solvents)
- Capsid contains all enzymes for mRNA production

Genome
- 11 segments of dsRNA
- Purified RNA segments are not infectious
- Each RNA segment codes for at least one protein
- RNA segments from different viruses reassort at high frequency during dual infections of cells

Replication
- Cultivation facilitated by proteases
- Cytoplasmic replication
- Inclusion body formation
- Unique morphogenesis involves transient enveloped particles
- Virus normally released by cell lysis

1. Mature virus particles are approximately 75 nm (750 Å) in diameter and possess a triple-layered icosahedral protein capsid composed of an outer layer, an intermediate layer, and a inner core layer.
2. Sixty spikes extend 120 Å from the smooth surface of the outer shell.
3. Particles contain an RNA-dependent RNA polymerase and other enzymes capable of producing capped RNA transcripts.
4. The virus genome contains 11 segments of double-stranded RNA (dsRNA).
5. The viruses are capable of genetic reassortment.
6. Virus replication occurs in the cytoplasm of infected cells.
7. Virus cultivation *in vitro* is facilitated by treatment with proteolytic enzymes, which enhances infectivity by cleavage of an outer capsid spike polypeptide.
8. The viruses exhibit a unique morphogenic pathway (i.e., virus particles are formed by budding into the ER and enveloped particles are evident transiently at this stage of morphogenesis; mature particles are nonenveloped and these virions are liberated from infected cells by cell lysis).

Rotaviruses are classified serologically by a scheme that allows for the presence of multiple groups (serogroups) and for the existence of multiple serotypes within each group. A rotavirus group (or serogroup) includes viruses that share cross-reacting antigens detectable by a number of serologic tests such as immunofluorescence, enzyme linked immunosorbent assay (ELISA), and immune electron microscopy (IEM) (92,236). Rotaviruses comprise six distinct groups (A through F). Group A, B, and C rotaviruses are those currently found in both humans and animals, whereas viruses in groups D, E, and F have been found only in animals to date. Viruses within each group are capable of genetic reassortment, but reassortment does not occur among viruses in different groups (338). The group antigenic determinants, or common antigens, are found on most (if not all) of the structural proteins and probably on many of the nonstructural proteins as well. This has been documented by showing that monospecific antisera and some monoclonal antibodies (MAbs) specific for individual polypeptides cross-react with strains in addition to those to which they were made. However, cross-reactive epitopes on the inner capsid protein (VP6) are those usually detected in diagnostic ELISAs.

Group A rotaviruses have clearly been established as causing significant diarrheal disease in the young. Group B rotaviruses have been associated with annual epidemics of severe diarrhea, primarily in adults in China (56,219,297). Group C viruses have been sporadically reported in fecal specimens from children with diarrhea and in several outbreaks; their clinical significance remains unclear. Recently, rapid diagnostic tests (ELISAs) and MAbs to detect non–group A rotaviruses have been established, and these are beginning to facilitate determining the clinical importance of these viruses (39,219,220,339). Very few non–group A rotavirus strains (a single group C porcine, group C human virus, and group B porcine virus) have been successfully cultivated (182,230,277,332). This inability to grow most non–group A viruses has hampered obtaining information on these viruses, although gene-coding assignments and sequence data have been obtained. Unless noted otherwise, this chapter focuses on information about the group A rotaviruses. Reviews on the non–group A rotaviruses and comparisons between the proteins of the group A and non–group A viruses have been published recently (35,198,276).

Within each group, rotaviruses are classified into serotypes defined by reactivity of viruses in plaque reduction (or fluorescent foci reduction) neutralization assays using hyperimmune serum prepared in antibody-negative animals (135,324). Using such assays, 14 VP7 serotypes have been identified (see Tables 2 and 3 and Chapter 55 in *Fields Virology*, 3rd ed.), and strains of animal and human origin may fall within the same serotype (Tables 2 and 3). Neutralization assays can measure reactivity of antibody with the two outer capsid-neutralizing antigens (VP4 and VP7). However, in most cases the predominant reactivity measured is with the glycoprotein VP7. This may be because VP7 comprises a greater percentage of the virion outer capsid (Table 4), or alternatively with hyperimmunization, VP7 selectively induces highly specific antibodies. Identical classification of the same virus isolates using MAbs to VP7 unequivocally demonstrates that plaque reduction neutralization assays with hyperimmune serum primarily measure reactivities with VP7 (24,67,311).

In some cases, a rotavirus strain will not react clearly in reciprocal neutralization assays with hyperimmune antiserum. This is usually because the two viruses being compared possess distinct immunologic forms of VP4 (the second outer capsid protein), which is also a neutralization antigen (133). Because the genes that encode these separate

TABLE 2. *Rotavirus VP7 (G) serotype*

	Strains from indicated species of origin	
Serotype	Human	Animal
1	Wa, KU, K8, M37, D	Bo/T449
2	DS-1, S2, RV-5, 1076, KUN	
3	YO, P, M, MO, RV-3, AU-1, Ito, HCR3, RO1845, McN13	Si/SA11, Si/SA114fm, Si/RRV, Ca/K9, Ca/CU-1, Eq/H-2, Fe/FRV-1, Fe/FRV64, Fe/Cat2, Fe/Cat97, La/Ala, Mu/EDIM, Mu/EB, Mu/EHP, Po/CRW-8, Po/Ben-307, Po/MDR-13[a]
4	Hochi, Hosokawa, ST3, VA70	Po/Gottfried, Po/SB-2
5		Po/OSU, Po/TFR-41, Eq/H-1, Po/MDR-13[a]
6	PA151, PA169	Bo/NCDV, Bo/UK, Bo/C486, Bo/WC3, Bo/KN4, Bo/641
7		Ch/Ch2,Ty/Ty1, Bo/993/83, Pigeon/PO-13
8	69M, B37, HAL1166	Bo/678, Bo/A5
9	WI61, F45, AU32, Mc323, Mc345, 116E	
10	Mc35, I321	Bo/B223, Bo/A44, Bo/KK3, Bo/61A, Bo/V1005, Ov/Lp14
11		Po/YM
12	L26,L27	
13		Eq/L338
14		Eq/FI23

Species of origin of animal strains indicated by abbreviations: Si, simian; Mu, murine; Ca, canine; La, lapine; Eq, equine; Po, porcine; Bo, bovine; Ch, chicken (avian); Ty, turkey (avian); Ov, ovine; Fe, feline.

Data were compiled from the following sources: 2,27,57,76,92,110,110a, 132a,135,141,194,214a,215,282,286a,309a,310a,319a.

[a]Porcine MDR-13 shows reactivity as two G serotypes, G3 and G5 (138).

TABLE 3. *Rotavirus VP4 genotype [P] and serotype (P)*

		Strains from indicated species of origin	
Genotype[a]	Serotype	Human	Animal
1	6		Bo/C486, Si/SA114fm, Bo/NCDV, Bo/A5
2			Si/SA11
3	5	RO1845, HCR3	Si/RRV, Ca/K9, Ca/CU-1, Fe/FRV64, Fe/Cat97
4	1B	RV-5, DS-1, S2, L26	
5	7		Bo/UK, BoWC3, Bo/641, Bo/61A, Bo/678, Bo/V1005
6	2A	M37, 1076, RV-3, ST3, McN13	
	2B		Po/Gottfried
7	9		Po/OSU, Po/TFR-41, Po/SB-1A, Po/YM, Po/CRW-8, Po/Ben-307
8	1A	Wa, KU, P, YO, MO, VA70, D, Hochi, Hosokawa, WI61, F45	
9	3	K8, AU-1, PA151	Fe/FRV-1, Fe/Cat2
10	4	69M, 57M	
11	8	116E, I321	Bo/B223, Bo/A44, Bo/KK3
12			Eq/H2,.Eq/FI-14
13			Po/MDR-13[b]
14		Mc35, PA169, HAL1166	
15			Ov/Lp14
16	10		Mu/EB, Mu/EDIM
17			Bo/993/83, Pigeon/PO-13
18			Eq/L338
19			Po/4F
20			Mu/EHP

Species of origin of animal strains indicated by abbreviations: Si, simian; Mu, murine; Ca, canine; La, lapine; Eq, equine; Fe, feline; Po, porcine; Bo, bovine; Ch, chicken (avian); Ty, turkey (avian), Ov, (ovine).

Data were compiled from the following sources: 2,27,57,76,92,110,110a, 132a,135,141,194,214a,215,282,286a, 309a,310a,319a.

[a]P genotypes indicate predicted VP4 amino acid sequence identity of 89% and/or nucleic acid hybridization data, and these are still tentative because a biologic correlation between VP4 genotype and serotype is not yet clearly established (132a). Genotype numbers beyond P10 have been adapted from previously published lists (132a, 309a) and numbers beyond 18 have followed the date of acceptance of publication. It has been proposed that new genotype numbers be obtained from the WHO rotavirus reference laboratory (132). Not all viruses in each P genotype have been characterized into a P serotype by reciprocal or one-way neutralization assays and preliminary studies indicate that viruses currently in a single P genotype may eventually be classified into two P serotypes. For example, the viruses classified as P[3] appear to represent two distinct P serotypes but these have not yet been given formal numbers (214a) . Designation of each P genotype is to be done by including the number in square brackets after the term P. Thus, the human rotavirus Wa strain would be P1A[8], G1, and the equine rotavirus H2 strain would be P[12], G3 (132).

[b]Porcine MDR-13 shows reactivity as two G serotypes, G3 and G5 (138).

neutralization antigens can segregate (reassort) independently, it is not surprising that some virus isolates can possess heterologous neutralization (VP4, VP7) antigens (133). Classification of rotaviruses by a binary system (similar to that used for influenza viruses), in which distinct serotypes of VP4 and VP7 are named, has been proposed and accepted (122,135,267). However, a lack of readily available typing serum or MAbs to different VP4 types has hampered classification of VP4 (P) serotypes. Instead, properties of VP4 have been studied by sequence analysis, and current evidence suggests the existence of at least 19 different genotypes of VP4 (Tables 2 and 3). Genotypes of VP4 and VP7 are determined by sequence analysis, whereas serotypes are determined by reactivity with polyclonal or monoclonal antisera (see Chapter 55 in *Fields Virology*, 3rd ed.). For VP7, a correlation between genotype and serotype has been established. Such a correlation is less clear for VP4, although

TABLE 4. *Rotavirus proteins*

Genome segment	Protein product[a]	Nascent polypeptide (M_r)[b]	Mature protein modified	Location in virus particles	Number of molecules per virion[c]	Function[d]
1	VP1	125,005 (125K)	—	Inner core	ND	RNA polymerase?
2	VP2	102,431 (94K)	Cleaved	Inner core	120	RNA binding
3	VP3	98,120 (88K)	—	Inner core	ND	Guanylyltransferase
4	VP4 (VP5* + VP8*)	86,782 (88K)	Cleaved VP5* (529) VP8 (247)[e]	Outer capsid	120	Hemagglutinin, neutralization antigen, protease-enhanced infectivity, virulence, putative fusion region, cell attachment
5	NSP1 (NS53)	58,654 (53K)	—	Nonstructural		Slightly basic, zinc finger, RNA binding
6	VP6	44,816 (41K)	—	Inner capsid	780	Hydrophobic, trimer, subgroup antigen
7	NSP3 (NS34)	34,600 (34K)	—	Nonstructural		Slightly acidic, RNA binding
8	NSP2 (NS35)	36,700 (35K)	—	Nonstructural		Basic, RNA binding
9	VP7	37,368 (38K)	Cleaved signal sequence, high mannose glycosylation and trimming	Outer capsid	780	RER integral membrane glycoprotein neutralization antigen, bicistronic gene?, two hydrophobic NH_2-terminal regions
10	NSP4 (NS28)	20,290 (28K)	29K–28K, uncleaved signal sequence, high mannose glycosylation and trimming	Nonstructural		RER transmembrane glycoprotein, role in morphogenesis, three hydrophobic NH_2-terminal regions
11	NSP5 (NS26)	21,725 (26K)	28K, phosphorylated *O*-glycosylated	Nonstructural		Slightly basic, rich in serine and threonine, RNA binding

[a] The virion polypeptides are designated as proposed by Mason et al. (191) and modified by Liu et al. (172) and Maftion et al. (198). VP3 is the protein product of gene segment 3, and VP4 is the protein product of gene segment 4. Recently, the nonstructural proteins were renamed NSP1 to NSP5 to facilitate comparisons among these proteins of different virus strains (198), and this new nomenclature is used in this chapter. The parentheses in this table show the names of the nonstructural proteins as designated previously (NS followed by a number indicating their apparent molecular weight in thousands determined by electrophoresis in polyacrylamide gels containing sodium dodecyl sulfate).

[b] Molecular weights are for SA11 proteins and are calculated from the deduced amino acid sequences from the nucleotide sequence and from the longest potential ORF. Molecular weights in parentheses signify apparent molecular weights (in thousands) for the SA11 proteins from analyses on polyacrylamide gels containing SDS. See Table 5 for number of amino acids for each polypeptide.

[c] ND, not yet determined. Calculated from the mass density and structural studies of purified virions (253–256,337).

[d] See Tables 5 and 6 and text for references.

[e] There are two trypsin cleavage sites in SA11 VP4 at amino acids 241 and 247. The indicated mature products number of animo acids are those based on use of only the preferred second cleavage site (173,174).

sequence variation between amino acids 84 and 180 have been suggested to be useful to define P type-specific epitopes (168).

Serotype designation thus reflects the expression of neutralization epitopes on both VP4 and VP7. Increasingly, epitope expression on these outer capsid proteins has been found to be dependent on the specific combinations of VP4 and VP7 in this complex capsid (47,48). Although the molecular basis of these interactions is not yet well understood, new structural information clearly documents the presence of specific interactions between VP4, VP6, and VP7 within this complex capsid. The serology of the epitopes in proteins that interact in this complex capsid is clearly complicated and likely will not be fully understood until high-resolution structural data on the capsid are available.

VIRIONS

Structure

The morphologic appearance of rotavirus particles is distinctive, and three types of particles can be observed by electron microscope (EM) (Figs. 1 and 2). The complete particles (approximately 75 nm in diameter) resemble a wheel with short spokes and a well-defined smooth outer rim. The name rotavirus (from the Latin *rota,* meaning wheel) was suggested based on this morphology (101). The complete infectious particles historically have been called double-shelled particles, although new structural data show that they have three layers. Double-layered particles (previously called single-shelled particles) lacking the outer shell are often described as rough particles because their periphery shows projecting trimeric subunits of the inner capsid. Single-layered core particles are seen less frequently; they usually lack genomic RNA and are aggregated.

Early structural studies agreed that rotavirus particles possessed icosahedral symmetry, but the triangulation (T) number of the virus was reported to vary from T = 3 to T = 16 (86,159,190,269). The application of new techniques (that avoid double-sided images produced by negative staining) to study rotavirus structure resolved the early discrepancies. Roseto et al. (269) studied the structure of double-layered rotaviruses using a freeze-drying technique and reported the existence of 132 capsomeres arranged in a skew symmetry with T = 13. They also showed that the outer layer contains small holes that correspond one-to-one with holes in the inner capsid. This structure for the inner capsid was subsequently confirmed by Ludert et al. (178) using chemically disrupted particles and the same technique.

The three-dimensional structure of triple- and double-layered rotavirus particles (produced without protease treatment) was first determined at 40 Å resolution and more recently refined at 26 Å resolution using electron cryomicroscopy and image processing techniques (253,255, 285,337). These studies unequivocally establish a T = 13/(levo) icosahedral surface lattice for the two outer layers. A distinctive feature of the virus structure is the presence of 132 large channels spanning both shells and linking the outer surface with the inner core. One hundred twenty channels are along the 6-coordinated centers and 12 are along the 5-coordinated centers.

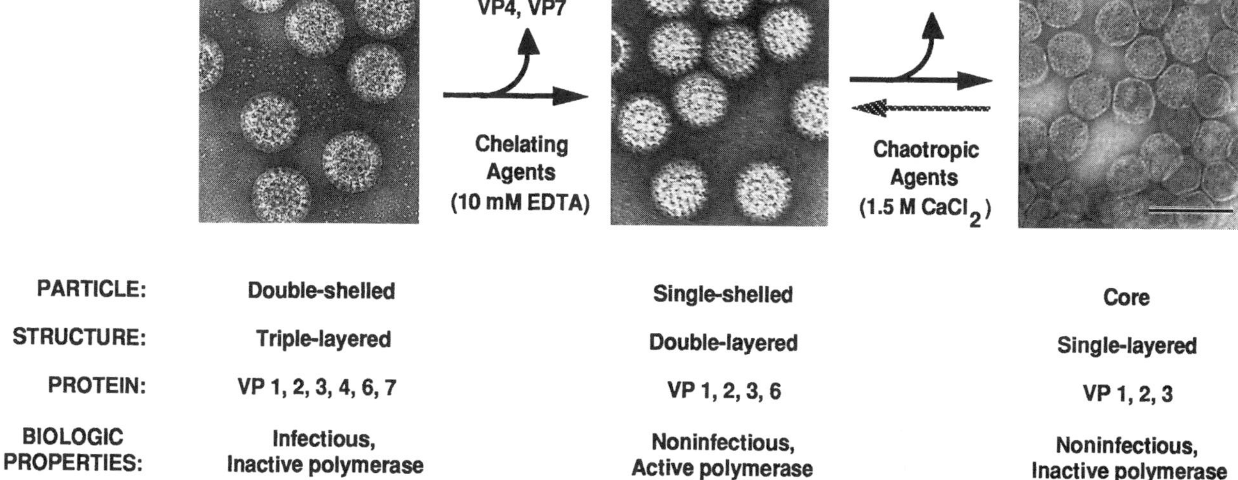

FIG. 1. Structural and biological properties of rotavirus particles. Electron micrographs show typical triple-layered, double-layered, and single-layered core particles seen after staining with 1% ammonium molybdate. The double-layered and core particles can be produced by sequential degradation of infectious triple-layered particles as shown. The proteins and biologic properties of the particles are detailed in the text. Bar = 100 nm. Modified from Estes and Tanaka (97), with permission.

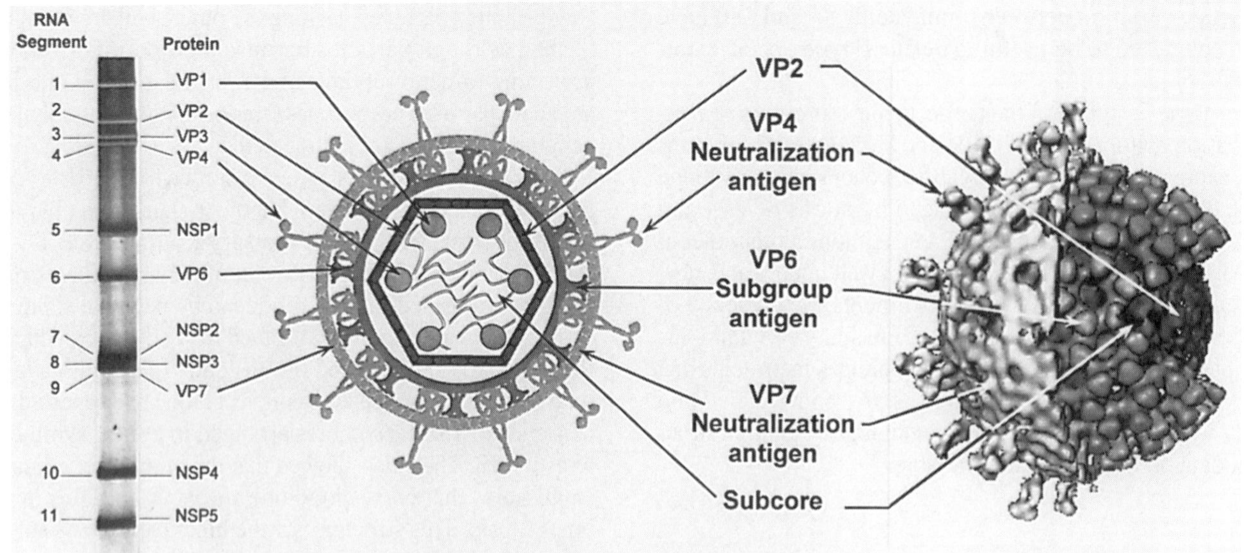

FIG. 2. Gene coding assignments and the three-dimensional structures of rotavirus particles. (Left) A polyacrylamide gel shows the 11 segments of dsRNA that comprise the rotavirus genome (A:Si:SA11 strain in this case) and the proteins encoded by each of these genes. (Center) A schematic of the complete rotavirus particle. The proteins in the different shells are indicated. (Right) The 28Å three-dimensional structure of a complete rotavirus particle, in which a portion of the outer shell mass and inner shell mass have been removed, showing the middle and inner shells. This structure was determined by image processing of electron micrographs of particles embedded in vitreous ice (253–255). Modified from Conner, Matson, and Estes (65), with permission, with an unpublished structure of Prasad and Shaw (kindly prepared by A. Shaw).

Three types of channels can be distinguished based on their position and size (Figs. 2 and 3). Type I channels run down the icosahedral fivefold axes, type II channels are those on the 6-coordinated positions surrounding the fivefold axes, and type III channels are those on the 6-coordinated positions around the icosahedral threefold axes. Type III channels are about 140 Å in depth and about 55 Å wide at the outer surface of the virus. On entering the particle, these channels constrict before widening to their maximum width, which is close to the surface of the inner shell. Similar features and dimensions are seen in the other two types of channels, except that type I channels have a narrower (approximately 40 Å) opening at the outer surface of the virus. The function of these channels is not yet known, but it is possible that they are involved in importing the metabolites required for RNA transcription and exporting the nascent RNA transcripts for subsequent viral replication processes.

These studies of non–protease-treated virus particles also showed structural features not seen previously (253). Sixty spikes approximately 120 Å in length with a knob at the distal end extend from the smooth surface of the outer shell (Figs. 2 and 3). These protein spikes are situated at an edge of the type II channels surrounding the fivefold icosahedral axes. These spikes are composed of dimers of the hemagglutinin (VP4), which was first demonstrated by showing that two Fab subunits of a MAb that recognizes VP4 bind on the sides near the tips of the spikes (254). Higher resolution reconstructions of native particles and of a reassortant lacking spikes or of spikeless particles (formed by removing VP4 by treatment at high pH) have confirmed that VP4 is the spike and have shown that the spike is multidomained with a radial length of approximately 200 Å with approximately 120 Å projecting from the surface of the virus (285,337). The head of the spike is bilobed and attached to a square-shaped body formed by two rods. The anchoring base displays a pseudo-sixfold symmetry, and this may represent a novel folding motif in which a single polypeptide of VP4 contributes similar but nonequivalent domains to form the arms of the hexameric base (285,337).

In the VP4 spike, no twofold axis of symmetry is strikingly evident. The location of the spike does not correspond to any local or strict twofold axis of the icosahedron; thus, the dimeric configuration is apparently asymmetric (Fig. 3). This result is consistent with the observation that the two Fab fragments of the VP4 MAb bind to the sides near the tip of the spike at different angles, indicating that the antigenic sites are not in structurally identical environments (254). It is unclear whether this asymmetry in the spike serves a specific function. It is possible that this asymmetry is required for stabilizing interactions between quasi-equivalent VP7 and VP6 molecules. On the outer surface, it appears that VP4 interacts with two molecules of VP7,

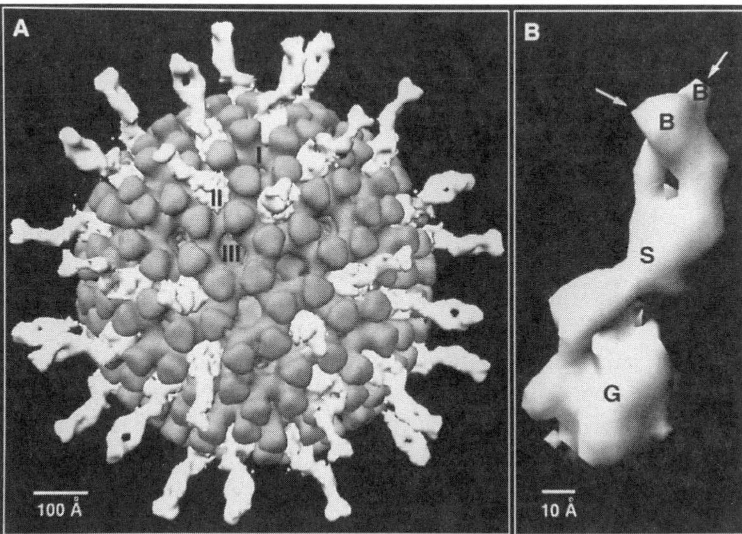

FIG. 3. Surface representation of double-layered particles interacting with the VP4 spikes. The left panel shows the spikes emerging from the type II channels in the VP6 shell. The globular interior portion of VP4 covers the channel. The right panel shows the structure of an isolated spike and illustrates that two molecules appear to participate in spike formation. The dimeric nature of the spike is most pronounced in the exterior portion, whereas the globular shape of the interior portion may allow the VP4 spike to make multiple contacts with the VP6 timers that line the type II channels. Modified from Shaw et al. (285), with permission (kindly provided by A. Shaw and B. V. V. Prasad).

but inside it appears that the large globular domain interacts with all six of the VP6 molecules surrounding the type II channel. However, there may be more specific interactions between the VP4 dimer and two of the six VP6 molecules.

The VP4 spike extends inward into the outer virion surface approximately 90 Å and interacts with both VP7 and the inner capsid protein VP6 (Figs. 2 and 3). The extensive VP4-VP7 and VP4-VP6 interactions imply that VP4 may participate in maintaining the precise geometric register between the inner and outer capsids as well as affect functional domains. The VP7 outer layer appears to consist of trimers, or 780 molecules, of this protein. VP7 interacts with the tips of the VP6 trimers.

The double-layered particle is approximately 705 Å in diameter, composed of 260 morphologic units or 780 molecules of VP6 arranged in trimers and positioned at all the local and strict threefold axes of the T = 13 icosahedral lattice. The structure of the VP6 shell is similar to that of the VP7 shell of bluetongue virus (257). The VP6 layer is approximately 100 Å thick and interacts with the inner VP2 shell. The VP2 shell is approximately 510 Å in diameter and is approximately 35 Å thick. This shell surrounds a subcore with a diameter of approximately 440 Å. Some features of the subcore appear similar to the reovirus core structure (270). Comparisons of high-resolution structural data of bluetongue virus, rotavirus, and reovirus cores should be useful to understand the process of transcription.

The existence of each shell and the ability of VP7 and VP4 to interact directly with VP6 has been demonstrated by the production of viruslike particles composed of VP2, VP2/6, VP2/4/6, VP2/6/7, and VP2,4,6,7 by expression of the respective proteins (68,167) or coassembly of purified proteins of each layer onto preformed VP2 particles (51, 52,341). These results show that the structural proteins have the intrinsic properties of self assembly and suggest high affinity interactions among the proteins.

Although VP4 appears to be a dimer, its elongated shape with multiple domains resembles the hemagglutinin of influenza virus, which is a trimer and has distinct rodlike and globular domains (see Chapter 45 in *Fields Virology*, 3rd ed.). The rodlike portion of the influenza hemagglutinin forms a triple-stranded coiled-coil of α helices. VP4 is predicted to contain extended α-helical regions packed as coiled-coils (177,198), and higher resolution analyses will be required to discern whether the rodlike domains of VP4 are folded as α helices. A similar coiled-coil region is probably also present in the reovirus σ1 hemagglutinin, although this protein is much longer and structurally distinct (75; see also Chapter 24).

Physicochemical Properties

The different rotavirus particle types possess distinct biophysical and biologic properties (Fig. 1). Infectivity depends on the presence of the outer capsid layer (75,83), and treatments with calcium-chelating agents (e.g., EDTA and EGTA) remove the outer capsid layer and result in a loss of infectivity (36,94). Calcium has been detected in triple-layered but not double-layered particles, and this may represent binding to potential calcium-binding sites present in the protein sequence of VP7 (11,73,284). Single-layered core particles can be produced by disruption of double-layered particles with chaotropic agents such as sodium thiocyanate or high concentrations of calcium chloride (4,63).

Triple- and double-layered particles can be separated by centrifugation in gradients of cesium chloride or sucrose, where they possess distinct densities and sedimentation values. Triple-layered particles have a density of 1.36 g/cm^3 in CsCl and sediment at 520 to 530 S in sucrose, whereas double-layered particles with a density of 1.38 g/cm^3 sediment at 380 to 400 S (305). Single-layered core particles have a density of 1.44 g/cm^3 in CsCl and a sedimentation coefficient of 280 S (23). The three types of particles also can be separated in agarose gels (108).

Rotavirus infectivity and particle integrity are generally resistant to fluorocarbon extraction and exposure to ether, chloroform, or deoxycholate (94,99,305,330), reflecting the absence of an envelope on mature particles. Chloroform treatment reduces infectivity slightly and destroys hemagglutinating activity (26,94). Sodium dodecyl sulfate (0.1%) inactivates infectivity, but exposure to nonionic detergents can enhance infectivity, presumably by disrupting aggregates (326).

Rotavirus infectivity is relatively stable to inactivation. Infectivity is stable within the pH range of 3 to 9 (94,186, 231,329,330), and virus samples from cattle and humans have retained infectivity for months at 4° or even 20°C, when stabilized by 1.5 mM $CaCl_2$ (288). The infectivity of bovine and simian rotaviruses is relatively stable, even at 45° to 50°C, but hemagglutinating activity is lost rapidly at 45°C. Infectivity and hemagglutinating activity are destroyed by repeated freezing and thawing (18,26,94), and the hemagglutinating spikes are removed by treatment at high pH (7).

Virus infectivity can be inactivated by disinfectants such as phenols, formalin, chlorine, and betapropiolactone (280, 322). Ethanol (95%), perhaps the most effective disinfectant, exerts its effect by removing the outer capsid (18,26, 34,166,306).

In general, studies of the structure and inactivation of rotaviruses have yielded similar results whether the viruses studied were of animal (bovine or simian) or human origin. However, human rotavirus strains are distinguishable by being more difficult to cultivate, and their hemagglutinin, if present, is more difficult to detect (157). In addition, human strains seem to lose their outer capsid more easily than do other strains. For example, purification of triple-layered human rotavirus particles is generally difficult in gradients of cesium chloride due to loss of the outer capsid; instead, metrizamide gradients may be required for purification of these viruses (310). Analysis of the determinants of density and stability by characterization of reassortants has shown that a reassortant may have an intermediate density compared with two parental viruses, and physical interactions among the structural proteins appear to be responsible for heterogeneity in particle density (50). In addition, the stability of infectivity during fluorocarbon extraction, CsCl gradient purification, and subsequent storage at 4°C can be dependent on the type of genome segment 4 (VP4) in particles (50).

Genome Structure

The viral genome of 11 segments of dsRNA is contained within the virus core capsid. Deproteinized rotavirus dsRNAs are not infectious, reflecting the fact that virus particles contain their own RNA-dependent RNA polymerase to transcribe the individual RNA segments into active messenger RNAs (mRNAs). Hydrodynamic studies of the flexibility or "stiffness" of isolated rotavirus RNA segments in solution indicate that packaging of these RNA segments into the rotavirus capsid requires intimate protein-RNA interactions (150). The proteins directly responsible for segment packaging remain unknown; the structural proteins present in core particles (VP1, VP2, and VP3) are obvious candidates, but nonstructural proteins also may play a role.

The nucleotide sequence of all of the 11 rotavirus RNA segments for several rotavirus strains is now known. Table 5 summarizes properties of each RNA segment of the prototype simian SA11 strain, which was the first genome completely sequenced. The sequences from different rotavirus strains have shown general features (Fig. 4) about the structure of each of the genome segments. Each RNA segment starts with a 5′ guanylate followed by a set of conserved sequences that are part of the 5′ noncoding sequences. An open reading frame (ORF) coding for the protein product and ending with the stop codon follows, and then another set of noncoding sequences that contain a subset of conserved terminal 3′ sequences and end with a 3′-terminal cytidine is found. The lengths of the 3′ and 5′ noncoding sequences vary for different genes, and no polyadenylation signal is found at the 3′ end of the genes. All of the sequenced genes possess at least one long ORF after the first initiation codon.

This usually is a strong initiation codon based on Kozak's rules (163,164). Although some of the genes possess additional in-phase (genes 7, 9, and 10) or out-of-phase (gene 11) ORFs, current evidence indicates all the genes are monocistronic, except possibly genes 9 and 11 (43,44,199).

The rotavirus gene sequences are A + T rich (58% to 67%), and this bias against CGN and NCC codons is shared with many eukaryotic and viral genes (247). The dsRNA segments are base paired end to end, and the plus-sense strand contains a 5′cap sequence m^7 $GpppG^{(m)}GPy$ (142, 202). Similar features of the RNA termini (capped structures and 5′ and 3′ conserved sequences) are found in the primary structures of the genome segments of other viruses (e.g., reovirus, cytoplasmic polyhedrosis virus, orbivirus) in the family *Reoviridae* (107,165,207) and in other virus families with segmented genomes (*Orthomyxoviridae, Arenaviridae,* and *Bunyaviridae*). The strong conservation of terminal sequences in genome segments suggests that they contain signals important for transcription, RNA transport, replication, or assembly of the viral genome segments.

Rotaviruses are the only known agents of mammals or birds that contain 11 segments of dsRNA. In most cases,

TABLE 5. *Nucleotide sequences of rotavirus genome segments*

Segment[a]	Base pairs	G + C (%)	Noncoding sequences[b] 5′	Noncoding sequences[b] 3′	First AUG is a favored sequence for initiation	Additional long ORF[c]	Amino acids[d]	References
1	3,302	34.6	18	17	Yes	No	1,088	211
2	2,690	32.9	16	28	Yes	No	881	211
3	2,591	28.9	49	35	Yes	No	835	170,211
4	2,362	34.7	9	22	Yes	No	776	222
5	1,581	33.9	32	73	Yes	No	491	211
6	1,356	38.6	23	139	Yes	No	397	95
7	1,104	33.5	25	131	No	Yes (2-i)	315 (312,306)	29
8	1,059	35.5	46	59	Yes	No	317	30
9	1,062	35.9	48	33	No	Yes (1-i)	326 (297)	9,31
10	751	40.2	41	182	Yes	Yes (2-i)	175	32
11	667	38.6	21	49	Yes	Yes (1-o)	198 (92)	210

[a]The numbering of the segment corresponds to the order of the prototype simian SA11 segments; data in other columns pertain only to SA11. The sequences of other viruses have been reviewed (69,198).

[b]The number of 5′ noncoding sequences was based on the position of the first AUG; the number of 3′ noncoding sequences does not include the termination codon.

[c]Indicates presence of additional long ORFs; values in parentheses indicate the number of ORFs and whether they are in phase (i) or out of phase (o) with the first ORF.

[d]The number of amino acids was derived from the longest potential reading frame. Values in parentheses signify number of amino acids in additional potential ORFs if first AUG is not favored for initiation.

the electrophoretic pattern of the genome of the group A viruses is composed of four high molecular weight dsRNA segments (1 to 4), five middle-sized segments (5 to 9), including a distinctive triplet of segments (7 to 9), and two smaller segments (10 and 11). When this basic pattern is not seen, the rotavirus being analyzed may be a group A avian virus, a non–group A virus, a group A virus that contains rearrangements within individual genome segments (Fig. 5), or a new unique group A virus. Analysis of genomic electropherotypes is a relatively easy, rapid, and popular technique for virus detection and is used for molecular epidemiology studies to monitor virus outbreaks and transmission. However, because distinct RNA patterns can arise by different mechanisms (shift, drift, rearrangements), these profiles cannot be used as the sole criterion for classification of a virus strain (45,90).

The electrophoretic migration rate of cognate (segments encoding the same protein) RNA segments in different virus strains often shows heterogeneity. In contrast, sequence data show that cognate genes from different strains usually contain the same number of nucleotides. Therefore, the heterogeneity in RNA segment mobility, observed among the cognate RNA segments of different virus strains, is attributable to sequence differences and secondary or tertiary structure that remain during electrophoresis of these segments.

Nucleic acid hybridization combined with Northern blots is a second assay used to classify viruses based on the relatedness of genome segments (97). This method has been found to be useful to classify genetically related viruses into different groups (218). This method also has been useful to characterize viruses possibly involved in cross-species transmission (216,217).

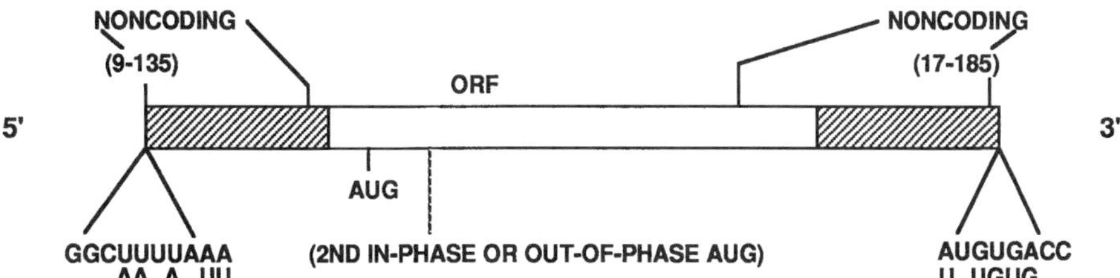

FIG. 4. Major features of rotavirus gene structure. Schematic shows the overall structure of rotavirus genes from the published sequences of genes 1 to 11 (see Table 4 for references). These genes lack a polyadenylation signal, are A + T rich, and contain conserved consensus sequences at their 5′ and 3′ ends.

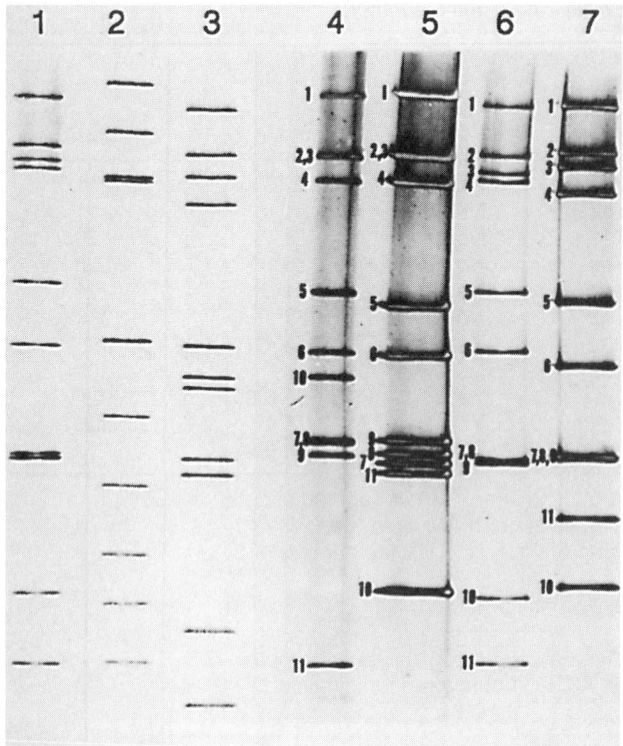

FIG. 5. Electropherogram of rotavirus RNA segments. The RNA segments were separated by electrophoresis in a 10% polyacrylamide gel and visualized by staining with silver nitrate. The RNA patterns of a group A rotavirus (SA11, lane 1), a group B rotavirus (adult diarrhea rotavirus isolate from China, lane 2) and a group C rotavirus (porcine pararotavirus, lane 3) are shown. The rearranged RNA patterns of three group A rabbit rotavirus strains (C11, lane 4; Ala, lane 5; and R2, lane 7) and their cognate segments compared with those of SA11 (lane 6) are also shown. The cognate genes were identified by hybridization with cDNAs for each SA11 RNA segment (307).

In viruses with genome rearrangements, typical RNA segments are missing or are decreased in concentration in an electrophoretic profile, and these are replaced by additional more slowly (or rarely more rapidly) migrating bands of dsRNA (Fig. 5). The slowly migrating bands represent concatemeric forms of dsRNA containing sequences specific for the missing RNA segments (139,195,237,244,307). The more rapidly migrating bands appear to represent deletions. Viruses with genome rearrangements of this type have been isolated most frequently from immunodeficient, chronically infected children (140,237), asymptomatically infected immunocompetent children (22,193), and animals [calves (244), pigs (21), or rabbits (307,314)]. They also have been obtained *in vitro* after serial passage of a tissue culture–adapted bovine rotavirus at high multiplicity of infection (139). Isolates with rearrangements in segments 5, 6, 8, 10, and 11 have been characterized, with the greatest number having rearrangements in segment 11. It is unknown if the rearrangements in segment 11 occur more frequently or if viruses with a rearranged segment 11 have some selective advantage (better growth or stability), so they are detected more easily.

Viruses containing rearranged genome segments are generally not defective, and the rearranged segments can reassort and replace normal RNA segments structurally and functionally (3,25,121). Biophysical characterization of such particles has shown that up to 1,800 additional base pairs can be packaged in particles without causing detectable changes in particle diameter or apparent sedimentation values. However, the density of particles containing rearranged genomes may be increased, and the increase in density is directly proportional to the number of additionally packaged base pairs (203). These results indicate that rotaviruses have considerable capacity to package additional genomic RNA, although the upper limit is unknown. Whereas a total of 11 RNA segments are invariantly packaged, there seems to be much less constraint on the length of individual RNA segments assembled into the maturing virus particle.

In most cases, the profiles of virus-coded proteins in cells infected with rotaviruses with rearranged genomes are similar to those seen in cells infected with standard rotavirus strains (3,25,195,205,244,281,307). This indicates that the rearrangement of the sequences in a segment apparently has left the normal reading frame intact and its expression unaltered. Sequence analyses of rearranged genome segments have confirmed this and have shown mechanisms by which the rearrangements arise.

In most cases, the rearrangements have resulted from a head-to-tail duplication that occurs immediately downstream from the normal ORF; hence, the rearranged segment retains its capacity to express its normal protein product. For one rearranged virus, the protein profile was distinct from that seen with standard virus. In this case, the gene rearrangement was in genome segment 5, and no protein of normal or larger size was detected in infected cells (139). Further characterization of this rearranged segment showed that due to the introduction of a point mutation in the ORF, a truncated protein lacking the C-terminal half of the protein is made. Because this rearranged virus is nondefective, the C-terminal half of this protein is nonessential for rotavirus replication, at least in cell culture (137). Genome rearrangements (concatemerization/deletion) are thought to be a third mechanism of evolution (in addition to antigenic shift and drift) of rotaviruses (316).

CODING ASSIGNMENTS AND ROTAVIRUS PROTEINS

The coding assignments and properties of the proteins encoded in each of the 11 genome segments are now fairly well established (Table 4 and Fig. 2). These assignments have been determined by *in vitro* translation using mRNA or denatured dsRNA (79,191,192,201,291) and by analyses of reassortant viruses (112,126,147,172,223,224,226, 320). The information on SA11 is complete, and SA11 has become the prototype virus. Comparative studies of other rotavirus strains have shown that the absolute order of migration of cognate genes may differ among virus strains. Therefore, identification of cognate genes must be based on hybridization with gene-specific probes (78,97), reassortant analysis (112,113), or biochemical or immunologic identification of the protein synthesized in a cell-free translation system programmed with mRNA specific to the gene. The ability to obtain sequence information directly from dsRNA or single-stranded RNA (ssRNA) (123,124) and the accumulating nucleic acid sequence databases (69,198) also make it possible to identify cognate genes based solely on sequence homology.

The rotavirus genome segments code for structural proteins found in virus particles and nonstructural proteins found in infected cells but not present in mature particles (Table 4, Figs. 2 and 6). The consensus is that the protein products (VP1-8) of six of the genome segments are structural proteins found in virus particles. The other proteins are nonstructural. Early studies often presented conflicting conclusions concerning the numbers and locations of the rotavirus proteins. Many of these conflicts were resolved, as reviewed elsewhere (20,96,131,198), when it was recognized that posttranslational modifications (glycosylation, trimming of carbohydrate residues, and proteolytic cleavages) occur after polypeptide synthesis. In addition, strain variations (such as the presence of more than one glycosylation site on VP7 in some bovine and human rotavirus strains) have been clearly shown (161, 279), and these variations provide explanations for other discrepancies.

The nomenclature of the viral proteins (as originally proposed for SA11 proteins) designates structural proteins as VP followed by a number, with VP1 being the largest molecular weight protein, and proteins generated by cleavage of a larger precursor being indicated by an asterisk [VP4 is cleaved to produce VP5* and VP8* (87,91)]. Initial studies failed to unequivocally identify a protein product from genome segment 3; hence, the protein product of genome segment 4 was called VP3 in early publications (192). When the protein product of genome segment 3 was confirmed to be a structural protein located in the inner core, the genome 3 product was designated VP3 and the genome 4 product was renamed VP4 (172). Initial studies also referred to the nonstructural proteins as NS followed by a number indicating the protein's molecular weight. This nomenclature has been replaced by NSP1 to NSP4 to facilitate comparisons among cognate nonstructural proteins of different molecular weights (Table 4) (198). The new nomenclature is used throughout this chapter, but it should be noted that much of the literature published before 1988 on "VP3" refers to the genome segment 4 product, and before, 1994, NS53, NS35 refers to NSP1, NSP2, and so forth (Table 4).

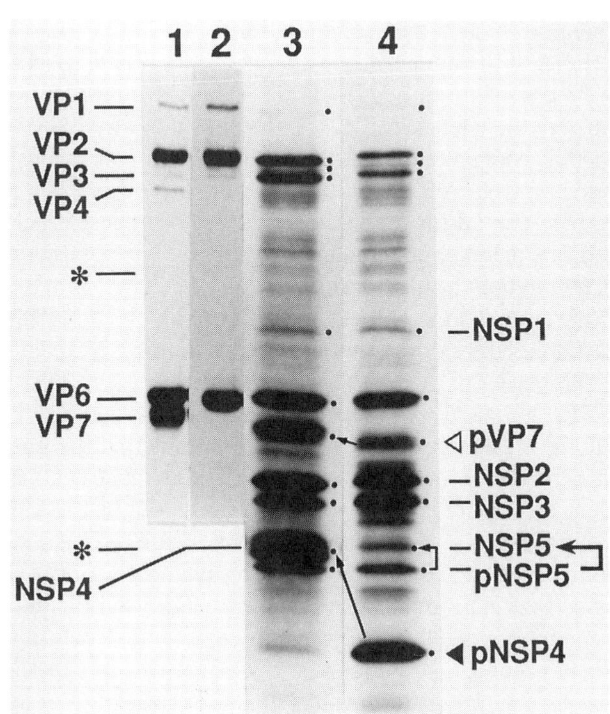

FIG. 6. Rotavirus polypeptides in purified virus particles and in infected cells. ^{35}S-methionine–labeled polypeptides in purified triple-layered virus (lane 1), double-layered virus (lane 2), or in monkey kidney cells infected with SA11 and labeled for 10 min 6 hr postinfection (lanes 3 and 4), were separated on 10% polyacrylamide gels and the polypeptides were visualized by fluorography. The glycoprotein precursors (p) to VP7 (◁) and to NSP4 (◀) are seen only when *N*-linked glycosylation is blocked by treatment of the cells with the inhibitor tunicamycin (lane 4). The structural polypeptides and mature NSP4 are labeled on the left, and the nonstructural polypeptides are labeled on the right. In lanes 3 and 4 the structural and nonstructural polypeptides and the precursor to VP7 are highlighted by dots. The two NSP5 bands (an unmodified and a modified form) are only seen clearly in cells treated with tunicamycin (lane 4). To resolve VP3 and VP4, the polypeptides in purified virus were separated on gels with a different cross-linking ratio than that used in the gels to analyze the infected-cell polypeptides (172). Therefore, in this composite the high molecular weight polypeptides in lanes 1 and 2 do not comigrate exactly with those in lanes 3 and 4. The triple-layered virus was grown in the absence of protease so only VP4 [and not its cleavage products VP5* (approximately 60 K) and VP8* (28 K)] are seen (lane 1). The positions at which VP5* and VP8* would be seen are noted by asterisks.

Except for their functions as structural components of virions, properties of the proteins of core particles (VP1, VP2, VP3) are poorly characterized. Some (if not all) of these proteins appear to function as the RNA-dependent RNA polymerase (transcriptase), VP1 is labeled by nucleotide analogs (321), and all VP1 sequences share the four common motifs conserved in the sequences of all RNA-dependent RNA polymerases (62,198) suggesting that this is part of the transcriptase present in particles. VP2 can bind RNA (33), and temperature-sensitive (ts) mutants mapping to genome segments 1 and 3 do not synthesize ssRNA at the nonpermissive temperature (Table 6) (49). VP3 covalently binds α-^{32}P-GTP, indicating that it is a guanylyltransferase (171,243). Recent studies have shown that the core proteins also function to replicate exogenously added mRNA (53). VP2, but not VP6, has been reported to be essential for replicase activity (187). These results indicate that the core proteins, or a subset of these proteins, function both as the transcriptase and replicase. Whether different domains of these proteins are involved in these different enzymatic activities remains to be determined.

Properties of most of the nonstructural proteins (NSP1, NSP2, NSP3, NSP5) are only beginning to be understood. However, the predictions or observations that the nonstructural proteins possess a basic charge (Table 4) (196), their demonstrated ability to bind RNA (33,136,151,197, 245), their presence in subviral particles with replicase activity (232), and the dsRNA negative phenotype of ts mutants that map to their genome segments (49,114) have been interpreted to suggest that these proteins function as part of replication complexes, as chaperones to transport RNAs or proteins to the sites of RNA replication or assembly, or to gather the genome segments for packaging. A better understanding of the functions of the nonstructural proteins in the replication cycle is being obtained as these proteins are expressed and studied using recombinant DNA methods.

Three of the rotavirus structural proteins (VP4, VP6, VP7) and one nonstructural protein (NSP4) have been studied extensively because of their unique biochemical, antigenic, and biologic roles in rotavirus replication and assembly (Fig. 7). Known functions of these proteins are described below. Antigenic properties and the role of these proteins in pathogenesis and immunity are considered in Chapter 55 in *Fields Virology*, 3rd ed..

VP4, the protein product of genome segment 4, is a nonglycosylated outer capsid protein (10,191,201). Knowledge of the diverse properties of VP4 resulted in an increased awareness of the importance of this protein in the biology of the rotaviruses. VP4 is the hemagglutinin in many virus strains (146) and probably the cell attachment protein (68,271). Proteolytic cleavage of VP4 into VP5* [molecular weight (M_r) approximately 60,000 (60K)] and VP8* (M_r approximately 28K) results in enhancement of viral infectivity (87,91). Cleavage of VP4 enhances penetration (but not binding) of virus into cells (59,105,148). Particles containing cleaved VP4 also possess lipophilic activity and may cause rapid release of chromium-51 (^{51}Cr) from cells and of fluorescent dyes from liposomes or intestinal vesicles (272). VP4 also is associated with restriction of growth of certain rotavirus strains in tissue culture cells (125) and with protease-enhanced plaque formation (146). Finally, VP4 has been implicated as a virulence determinant in mice and piglets (224). VP4 induces neutralizing antibodies (133,223), and antibodies directed to VP4 neutralize virus *in vitro* (39,66,127,133,155,310) and passively protect mice against heterologous rotavirus challenge *in vivo* (229). Further studies have shown that VP4 effectively induces protective immunity in animals (185,226) and is immunogenic in children and animals (64,286,304).

Features of VP4 with potential biologic relevance have been found by analyzing nucleotide and predicted amino acid sequence data (Fig. 7). Direct amino acid sequence analysis showed the sites of proteolytic cleavage of VP4 of SA11 (174). Both VP4 and VP8* have blocked NH_2-termini, and VP5* of SA11 is composed of two polypeptide species with slightly different amino acid sequences at their NH_2-termini. Comparison of these data with the nucleotide sequence of VP4 identified two trypsin cleavage sites (arginine 241 and arginine 247), with position 247 being the preferred cleavage site (174). The two trypsin cleavage sites are conserved in every rotavirus VP4 sequence analyzed (117,149,175,184,198,222,252).

TABLE 6. *SA11 genetic map*

Mutant group	Prototype mutant	% wild-type synthesis at 39°C[a]			Host shut-off	Mapped to genome segment (protein)
		ssRNA	dsRNA	Protein		
A	tsA(778)	100	75	100	+	4 (VP4)
B	tsB(339)	5	5	25	+	3 (VP3)
C	tsC(606)	5	5	25	+	1 (VP1)
D	tsD(975)	100	100	150	+	NA
E	tsE(400)	<5	<5	20	+	8 (NSP2)
F	tsF(2124)	100	<5	20	−	2 (VP2)
G	tsG(2130)	100	<5	20	−	6 (VP6)
H	tsH(2384)	100	100	100	−	NA
I	tsI(2403)	100	100	100	−	NA
J	tsJ(2131)	100	75	100	+	NA

Data from Gombold and Ramig (114) and Ramig and Petrie (263).
NA, no assignment.
[a]Determined in infected cells. Data compiled from the following sources: 49,187,263.

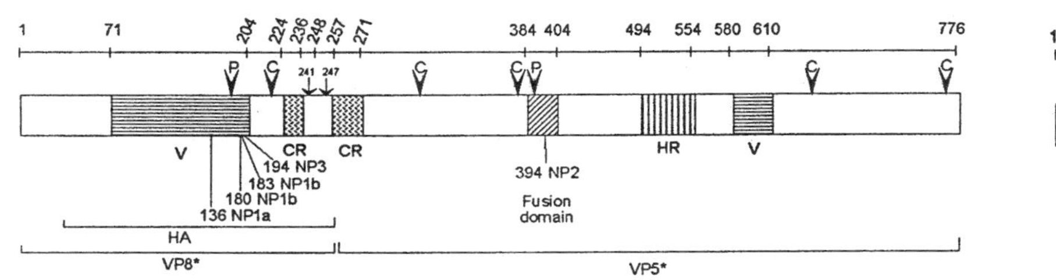

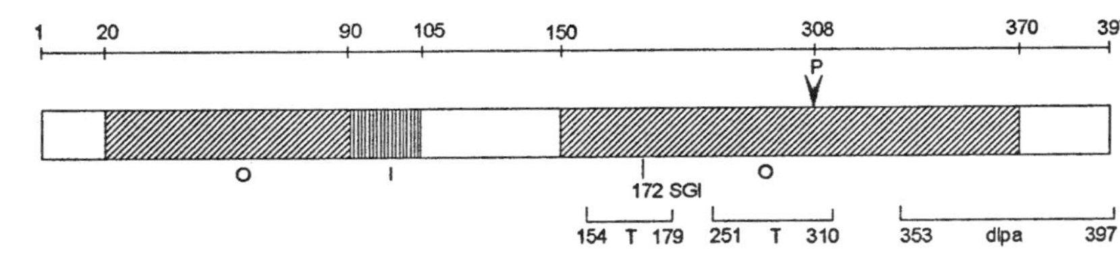

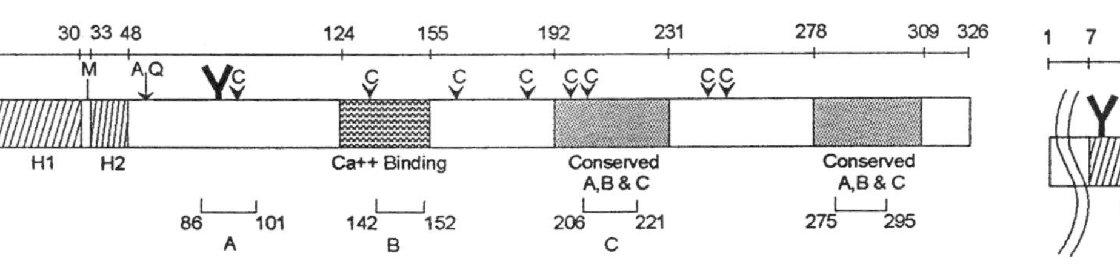

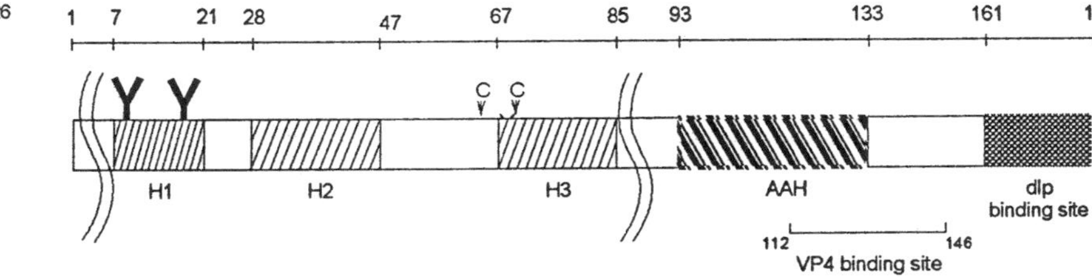

FIG. 7. Features of four rotavirus proteins, VP4, VP6, VP7, and NSP4. **Top left panel:** Schematic of structural and antigenic properties of VP4 based on the analyses of nucleotide sequences of different virus strains and escape mutants selected with neutralizing MAbs. The trypsin cleavage sites for SA11 (↓), predicted heptad repeat (HR), variable region (V), conserved region (CR), conserved cysteines (C) and prolines (P), the neutralization positive (NP) NP1a (amino acid 136), NP1b (amino acids 180 and 183), NP2 (amino acid 394), and NP3 (amino acid 194) MAb sites are indicated. The numbering shown here for VP4 is based on a protein of 776 amino acids as determined for SA114fM, rhesus rotavirus MMU18006, and bovine rotavirus C486. VP4 of human rotavirus strains are 775 amino acids, having lost an amino acid between residues 134 and 136. Based on sequence data (69,174; see Table 4 for additional references) and on analyses of neutralization escape mutants (184,309). **Top right panel:** Schematic of structural properties of VP6. The hydrophobic (O), hydrophilic (I), trimerization (T), double-layered particle assembly domain (dlpa), conserved proline (P) and SGI MAb 255 binding site are indicated. **Bottom left panel:** Schematic of structural and antigenic properties of VP7. The two hydrophobic domains H1 and H2 (amino acids 6 to 23 and amino acids 33 to 44); conserved sequence domains among group A, B, and C rotaviruses; putative Ca^{2+} binding site; conserved cysteines (C); signal cleavage site (↓); second in-frame methionine with a strong initiation codon (M); glycosylation site (Y); and neutralization epitopes A, B, and C are shown. The mature VP7 starts at amino acid Q 51. Based on sequence data (296; see Table 4 for additional references) and on analyses of neutralization escape mutants (80,183,308). **Bottom right panel:** Schematic of structural properties of NSP4 and its proposed topography in the membrane of the endoplasmic reticulum. The three hydrophobic domains (HI, H2, and H3), putative VP4 and double-layered particle (dlp) binding sites, predicted amphipathic alpha-helix (AAH), two glycosylation sites (Y), conserved cysteines (C); and amino-terminal domain located in the ER membrane are shown. Based on sequence data (see Table 4 for references) and biochemical analyses (43,85,144,242).

In the majority of animal rotavirus strains studied to date, VP4 contains 776 amino acids. In most human rotavirus strains, the deduced VP4 contains 775 amino acids instead of 776 amino acids, with the difference being that the human strains have one less amino acid between residues 134 and 136 (117,149). VP8* of the human strains is therefore one amino acid shorter. In several human rotavirus strains, a third potential trypsin cleavage site of either lysine or arginine is found just upstream of the 3′-proximal arginine site [amino acid 246 in human strains, amino acid 247 in animal strains) (118,232)]. This additional cleavage site in some human viruses may correlate with virus virulence (118), but to date there is no direct evidence to support this idea. It is also unknown whether cleavage at both sites (aa241 and 247) results in removal of the small intervening peptide and whether these cleavage sites shown for SA11 are really the same for other virus strains.

In recent years, the VP4s of animal (bovine B223) and human rotavirus strains containing 772 amino acids have been described (77,109,128). Comparison of the properties of the VP4s of distinct sizes is useful to map conserved domains (198,234). Unfortunately, the mechanism of cleavage activation of infectivity is not clear from either sequence or biochemical analyses. It is assumed that cleavage of VP4 activates an early step of replication that may be triggered by one or both of the terminal regions generated by cleavage or by a possible conformational change in the VP4 cleavage products. The new NH_2-terminus generated by cleavage in the rotavirus VP4 is not hydrophobic; however, it contains many polar amino acids (some of which may be charged at neutral pH) and cleaved VP4 possesses lipophilic activity (221,272). The mechanism of activation of rotavirus infectivity appears to be different from that postulated for other viruses, such as the orthomyxo- and paramyxoviruses, in which cleavage activation of the hemagglutinin or fusion protein generates a highly conserved apolar amino terminus, which is brought toward the membrane by a significant conformational change of the spike protein, and in which virus infectivity can be blocked by oligopeptides that mimic this sequence (see Chapters 20 and 21).

A region in VP4 (residues 384 to 401) has been identified that shares some homology with internal fusion sites in Semliki Forest and Sindbis virus proteins (184). This region in the alphavirus E1 protein has been implicated in cell fusion, leading to the hypothesis that these sequences might perform a similar function in VP4 (184). If this sequence is indeed a fusion domain, it remains unknown whether the fusion activity acts to mediate virus entry into cells or to mediate virus maturation during budding across the membrane of the ER, or both. Conservation of this putative fusion region among rotaviruses lends credence to the suggestion that it has an important function in the replication cycle.

The predicted amino acid sequence of VP4 indicates that the amino terminal 70% of the protein is rather hydrophobic (177,198). The predicted amino acid sequence of VP4 has a net negative charge at pH 7.0, and secondary structure predictions show that the amino terminus before the trypsin cleavage site contains numerous random coils and turns, suggesting a globular structure. Four cysteines (at amino acids 216, 318, 380, and 774) are conserved in the VP4 of all rotavirus strains sequenced (117,149,184) except those of the bovine rotavirus B223 (128). An additional cysteine is present at position 266 in the human M37 strain (117) and at amino acids 203 in SA114fM, bovine 486 and rhesus rotavirus. *In vitro* studies of the intrachain disulfide bonds in the VP5* and VP8* cleavage fragments of the rhesus and simian SA11 viruses have indicated that many animal rotaviruses contain two intrachain disulfide bonds: one in VP8* and one in VP5* (234). In contrast, human rotaviruses appear to contain only a single bond in the VP5* fragment (234). Sequences flanking the trypsin cleavage sites (amino acids 224 to 235 and 257 to 271) are relatively conserved in different virus strains and may be structurally important in keeping the trypsin cleavage site accessible.

The sequences of segment 4 from viruses in different serotypes have been compared in order to seek regions of divergence. Sequence variation is not evenly distributed across VP4 (Fig. 7). The greatest variation is seen between amino acids 71 and 204 in VP8*, and sequence analyses between amino acids 84 and 180 have been found to correlate with VP4 P types (168,169). Because VP8* contains the greatest sequence diversity, the binding sites of strain-specific neutralizing MAbs, and hemagglutination activity (100), whereas VP5* contains cross-reactive neutralizing MAb binding sites mapped to VP5* (185,309), it initially was tempting to speculate that VP8* is more externally exposed on the virus particle. A mechanism of immune selection could then explain the regions of sequence diversity in VP8*. Contrary to this hypothesis, a cross-reactive neutralizing MAb that induced mutations at amino acid 393 in VP5* was shown to bind to the sides near the tip of the spike (254). Thus, it appears that at least part of the VP8 domain may interact with VP6 and that VP5* is more external. This is consistent with the observation that a neutralizing MAb that destabilizes the outer capsid and appears to affect VP4-VP7 interactions induces substitutions in neutralization escape mutants at amino acids 180 and 183, which flank a conserved proline (342). High-resolution structural information will resolve these issues.

VP6 is the major structural component of virions, and it plays a key role in virion structure because of its interactions with both the outer capsid proteins VP4 and VP7 and the core protein VP2. VP6 spontaneously forms trimers (88) and is extremely stable; these characteristics and the presence of conserved epitopes among many virus strains may explain why this is the major antigen targeted in diagnostic assays. VP6 trimers can be disassociated and reassembled by changing pH (318). Removal of VP6 from double-layered particles results in a loss of polymerase activity, but it is unknown if VP6 plays a structural or functional role in this process (23,278). VP6 is a hydrophobic protein that is highly antigenic and immunogenic (88,119, 275), but it remains unclear whether VP6 plays a role in inducing protective immunity. Analyses of deletion mutants, virus variants, ts mutants, and chimeric proteins are beginning to dissect the domains of VP6 (Fig. 7). Thus, the trimerization domain first was reported to be between amino acids 105 and 326, and the domain necessary for the formation of double-layered particles is located between amino acids 281 and 397 (55). A refinement of this map of VP6 has indicated that amino acids 154 to 179 and 251 to 310 are needed for trimerization and amino acids 353 to 397 for double-layered particle formation (Fig. 7) (318). Residue 172 has been shown to contain a residue critical for binding of the subgrouping MAb 255/60 (176). Proline 308 has been implicated in trimer stability (287). The N-terminus is predicted to be an amphiphatic alpha helix and to be critical for virus assembly, possibly by functioning in transporting VP6 to viroplasmic inclusions (188). Studies with MAbs to VP6 have suggested that amino acids 58 to 62 are not accessible in viral particles and thus could be located at the interface with VP2; other domains may be accessible through the outer shell of the virus (317).

VP7 is an outer capsid glycoprotein that is the second most abundant protein species in the virion. VP7 was initially thought to be the only antigen containing neutralization determinants (125,147), and this stimulated numerous studies of its structure, biosynthesis, and function. Biochemical analyses rapidly determined that VP7 is a glycoprotein that contains only N-linked high mannose oligosaccharides, which are processed by trimming (10,84,144). $Man_8GlcNAc_2$ and $Man_6GlcNAc_2$ oligosaccharide residues are found on intracellular VP7, and $Man_6GlcNAc_2$ (and to a lesser extent $Man_5GlcNAc_2$) is found on mature virus particles (144,145). VP7 is cotranslationally glycosylated as it is inserted into the membrane of the ER, and insertion is directed by a cleavable signal sequence found at the amino terminus of the protein (10,31,85,144).

Comparisons of the deduced amino acid sequences of VP7 from a number of animal and human strains show interesting and potentially biologically relevant features about VP7 (Fig. 7). The nucleotide sequence predicts an ORF of 326 codons beginning with an initiation codon with a weak consensus sequence. A second, in-frame initiation codon with a strong consensus sequence lies 30 codons downstream. Each of these first two initiation codons precedes a region of hydrophobic amino acids (H1 and H2) that could act as the signal sequence to direct VP7 to the ER. A third in-frame initiation codon is also present downstream from the second hydrophobic domain in some strains. One glycosylation site, at amino acid 69, is used in the SA11 VP7. Other virus strains contain up to three potential glycosylation sites; only two glycosylation sites are apparently used, but the locations of the used sites are unknown (161,162,279).

Studies to determine the site of cleavage of the signal peptide in VP7 have identified glutamine 51 as the amino-terminal residue in at least some molecules of VP7 from purified virus (296). Gln 51 is conserved in all strains, and this amino-terminal residue is blocked by pyroglutamic acid (296). Studies of the expression and processing of VP7 from complementary cDNA constructs containing altered initiation codons indicate that the primary translation products produced from either initiation codon are cleaved at the same site and that either of the hydrophobic regions can direct transport of VP7 across the ER membrane (248, 296,333). However, initiation from the second codon is thought to be the major site used in cells. Therefore, mature VP7 lacks both hydrophobic domains and contains an amino-terminal Gln 51. Studies to determine the signals specifying retention of VP7 in the ER have shown that VP7 does not contain the sequence KDEL found to confer retention for some other ER proteins (212).

Recently, two regions (one spanning amino acids 51 to 61 and the second between amino acids 61 and 111) that mediate retention of VP7 in the ER have been identified (248,251,296,333). Further sequence analyses identified a consensus peptide LPXTG [STGAE], in which X indicates any other amino acid within the ER retention domain of VP7 (198). This sequence is of interest because it is found in bacterial surface proteins and is proposed as being responsible for a posttranslational modification necessary for proper anchoring of proteins to the bacterial plasma membrane. The LPITGS sequence is highly conserved and found in rotavirus VP7 between amino acids 57 and 63 (198), and mutation analyses have shown that the ITG residues are essential for ER retention (181). Although these residues are critical for retention, the method by which VP7 remains in the ER is unresolved. After its insertion into membranes, VP7 is resistant to digestion with proteolytic enzymes, suggesting that it is not a membrane-spanning protein (84,144). VP7 remains membrane associated after high salt treatment and release of microsomal contents at alkaline pH (144). These findings suggest that VP7 is an integral membrane protein with a luminal orientation. Further studies are required to determine why VP7 is retained in the ER and whether posttranslational modification or conformational properties are responsible for membrane association. Structural reconstructions of virion VP7 have shown that it is trimeric, and VP7 has been shown to form oligomers with other proteins (VP4 and NSP4) in infected cells (180). These oligomers, and protein rearrangements and interactions with calcium, appear to be important in the assembly of VP7 into the outer capsid (73, 249,250,283).

The normal site of VP7 synthesis and processing is exclusively in the ER, and rotaviruses are not transported to the Golgi apparatus. However, this pathway of protein targeting can be altered. Deletion of the two regions shown to mediate VP7 retention in the ER results in the processing of VP7 through the Golgi and into the extracellular medium. When transported through the Golgi, VP7 is modified to a form that contains complex carbohydrates (6,251). These results may be of practical significance for the production of soluble forms of VP7 for vaccines. Although these studies have only followed the fate of an individual protein, it also would be interesting to determine if infectious virus could be processed through such alternative pathways. Unfortunately, a method to alter a specific rotavirus gene and recover it in infectious virus is not yet available.

Two bands of VP7 are seen in polyacrylamide gel electropherograms of the proteins in virus particles. Cell-free translation of VP7 in wheat germ lysates and biochemical studies of VP7 in infected cells suggest that these bands result from the distinctive processing of VP7 initiated from each of the two initiation codons (44). This implies that genome segment 9 is bicistronic, but currently there is no direct proof of this hypothesis. The data indicating that mature VP7 lacks all amino acids proximal to Gln 51 raise the question of why so many conserved amino acids (18 of 35) are present before the second hydrophobic domain. In addition, none of the 11 amino acids in the second hydrophobic domain is highly conserved, raising the question of what causes this extensive sequence variation. VP7 is the dominant target for cytotoxic T-lymphocyte (CTL) activity in mice (103,225), and the second hydrophobic domain has been shown to be the CTL target within VP7 (103). Three of five other regions showing sequence variation in VP7 are major neutralization sites in virions (80; see also Chapter 55 in *Fields Virology*, 3rd ed.). It remains possible that the first initiation codon is used *in vivo* to produce low levels of a protein with a distinct function.

NSP4 is a nonstructural protein synthesized as a primary translation product with an apparent M_r of 20K. In the presence of microsomal membranes, NSP4 is cotranslationally glycosylated to an M_r 29K species that undergoes subsequent oligosaccharide processing to the mature M_r 28K protein that is a transmembrane protein of the ER (84, 144). The amino terminus of NSP4 contains three hydrophobic domains, and two *N*-linked high mannose glycosylation sites are found in the first hydrophobic domain (Fig. 7). The location of these glycosylation sites and analyses of the biosynthesis of this protein indicate that NSP4 contains a noncleavable signal sequence (32,85,144). Oligosaccharide processing of the $Man_9GlcNAc2$ carbohydrate added to NSP4 stops at $Man_8GlcNAc2$ with the mannose-9 species predominating (32,144).

Studies of the topography of NSP4 indicate that the amino terminus of the molecule is maintained in the membrane and that the carboxy terminus, which is hydrophilic, extends into the cytoplasm of infected cells (Fig. 7) (43). The cytoplasmic domain of NSP4 has been proposed to act as a receptor to bind subviral particles (double-layered particles) and mediate their budding into the lumen of the ER during viral morphogenesis (85,145,239). A receptor role for NSP4 is supported by the observation that double-layered particles bind to ER membranes containing only NSP4 (11,13,208). The binding domain on NSP4 for VP6 has been localized to the cytoplasmic domain of NSP4 between amino acids 161 and 175 (12,312,313).

Glycosylation of NSP4 or of a cellular protein is required for removal of the transient envelope from budding particles. This is known because if cells are infected with a rotavirus variant that possesses a nonglycosylated VP7 and treated with tunicamycin (which only prevents glycosylation of NSP4), all of the particles observed in the ER remain enveloped (85,131,238,301). Glycosylation of NSP4 is not required for its binding activity to double-layered particles or for oligomerization (13,313). NSP4 also has a binding site for VP4 and is cytotoxic (12,315). Heterooligomers of NSP4, VP4, and VP7 have been detected in enveloped particles (181,249), and calcium has been shown to be important for oligomerization of these proteins in the ER (250) and for proper folding of VP7 epitopes and outer capsid assembly (73,283). The precise mechanism of how (a) the envelope on particles is removed, (b) the heterooligomeric complexes function in particle budding through the ER, and (c) the outer capsid is assembled remains poorly understood. It also has been proposed that the formation of enveloped particles results from inefficient virus maturation, which leads to the formation of double-layered and not mature triple-layered particles (300). Rotavirus particle transport, maturation, and assembly remain an interesting model to understand the transport of protein complexes across the ER membrane as well as enveloped particle formation.

THE ROTAVIRUS REPLICATION CYCLE

Overview of the Replication Cycle

Figure 8 shows a schematic of the rotavirus replication cycle. Many of the details of this cycle are inferred from the closely related reovirus because studies specific to the rotaviruses are lacking. However, efforts to develop new assays to probe the specific steps in the replication cycle based on the expression and interaction of individual proteins and RNAs in *in vitro* replication systems are beginning to provide the currently missing information.

Rotavirus replication has been studied primarily in continuous cell cultures derived from monkey kidneys. In these cells, the replication cycle is fairly rapid, with maximum yields of virus being found after 10 to 12 hr at 37°C or 18 hr at 33°C when cells are infected with high multiplicities (10 to 20 plaque-forming units per cell) (58,200,259). Although the natural cell tropism for rotaviruses is the differentiated enterocyte in the small intestine, no direct studies to examine rotavirus replication in intestinal cell lines have been reported. This reflects in part the difficulties of

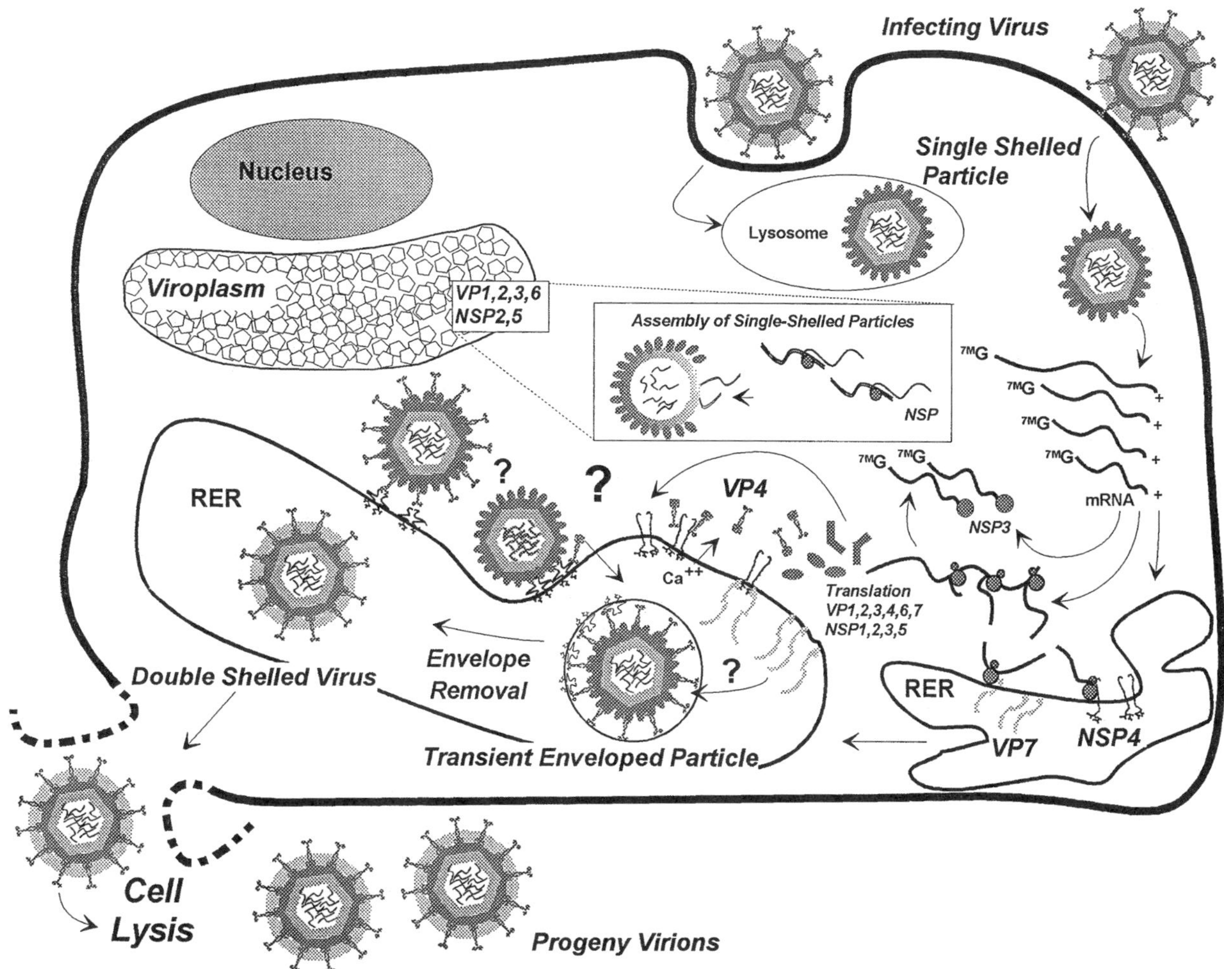

FIG. 8. Major features of the rotavirus replication cycle. For details, see section on the replication cycle.

culturing differentiated intestinal cells. However, EM studies of virus replication in intestinal cells indicate that the replication process outlined below (based on cell culture studies) is probably similar to that in intestinal cells.

The general features of rotavirus replication (based on studies in cultures of monkey kidney cells) are as follows:

1. Cultivation of most virus strains requires the addition of exogenous proteases to the culture medium. This assures activation of viral infectivity by cleaving the outer capsid protein VP4.
2. Replication is totally cytoplasmic.
3. Cells do not contain enzymes to replicate dsRNA, so the virus must supply the necessary enzymes.
4. Transcripts function both to produce proteins and as a template for production of a minus strand. Once the complementary minus strand is synthesized, it remains associated with the plus strand.
5. The dsRNA segments are formed within nascent subviral particles, and free dsRNA or free minus strand ssRNAs are never found in infected cells.
6. Subviral particles form in association with viroplasms, and these particles mature by budding through the membrane of the ER. In this process particles acquire their outer capsid proteins.
7. Cell lysis releases particles from infected cells. Details of the stages of rotavirus replication are summarized below.

Stages of the Replication Cycle

Adsorption, Penetration, and Uncoating

The initial stages of virus replication have been examined by biochemical and morphologic (electron microscopic) procedures. Only triple-layered particles containing VP4 attach to cells when monitored by EM (240) or by cell binding (68) or infectivity assays (36). Increasing direct evidence indicates that virus attachment occurs via VP4 (69,271), although early studies reported that VP7 (105,194,274) and NSP2 had cell binding activity (16).

Binding to cells does not require cleaved VP4 (59,105) or glycosylated VP7 (239). It remains possible that the entry steps require both VP4 and VP7.

The identity of the cellular receptor for rotaviruses is not known, but a study of the binding of radiolabeled SA11 to MA104 cells found approximately 13,000 receptor units per cell. Initial binding is sodium-dependent, pH insensitive between 5.5 and 8, and is dependent on sialic acid residues in the membrane; virus binds but is not internalized at 4°C (104,153,240,334) and probably involves more than one molecular species on the cell (294). A glycoprotein with an M_r of approximately 300 to 340K on murine enterocytes that binds to rhesus rotavirus and to VP4 has been reported, but whether this is a receptor remains unknown because this crude enterocyte extract was not purified or tested for its ability to directly compete for rotavirus binding to host enterocytes (17).

Bastardo and Holmes (18) were the first to show that neuraminidase treatment of red blood cells reduces virus binding, indicating a role of sialic acid in virus attachment. Sialic acid–containing compounds such as fetuin and mucin also inhibit virus binding to cells (153,340). These results add rotaviruses to an increasing number of viruses (such as reoviruses and influenza) that require sialic acid for binding to cells. However, these studies have not determined whether virus binds directly to sialic acid or whether sialic acid maintains the configuration of the binding site without directly interacting with the virus.

Increasing evidence indicates that the type of glycoconjugate (i.e., ganglioside, neutral glycolipid, glycoprotein) that actually functions as an *in vivo* receptor for a given rotavirus strain depends on the rotavirus serotype and/or type of host cell being studied. For example, the use of mouse-derived host cells with a heterologous simian rotavirus may not be the most advantageous combination to identify an *in vivo* relevant simian rotavirus receptor. Most animal rotaviruses apparently recognize sialic acid as an active receptor epitope (104), whereas infectivity of tissue culture cells by human rotaviruses is resistant to neuraminidase (104,340). This apparent dichotomy of binding activities of different rotavirus strains has been partially resolved by the demonstration that binding to sialic acid is not an essential step for animal rotavirus entry into cells (206); thus, sialic acid–resistant mutants were selected and found to retain full infectivity for cells, indicating that there are at least two binding sites on animal rotaviruses. Human rotavirus strains apparently initiate infection of cultured cells by a sialic acid–independent mechanism (104).

Studies of the rotavirus-receptor interaction of a homologous porcine rotavirus with porcine enterocytes have led to the characterization of a cell surface monoganglioside that may function as an *in vivo* relevant receptor for this rotavirus strain, and sialic acid is required for virus binding activity (268). Taken together, all the studies on rotavirus binding and entry suggest that unusual mechanisms may be involved. These include binding at two different sites, possibly with two different proteins, and direct membrane penetration of bound viruses rather than conventional receptor-mediated endocytosis. It is possible that rotavirus binding occurs initially by interaction with sialic acid–specific receptors followed by a separate interaction with a second, nonsialylated receptor. Conformational changes in viral binding proteins induced by initial binding events also may mediate some of the early events.

After binding occurs, virus is internalized. The enhancement of rotavirus infectivity by proteolysis is reportedly due to facilitating the penetration step (59). Internalization will not take place at 0° to 4°C, indicating that this step requires active cellular processes (153,240). All virus is internalized by 60 to 90 min after binding (153). The mechanism of internalization (penetration) into cells is controversial.

Both morphologic and biochemical approaches have been used to investigate the mode of entry of rotaviruses into cells. Early EM studies suggested that virus entry (SA11 strain) occurs by endocytosis (240,258) and that incoming particles are rapidly transported to lysosomes. Clear documentation of uptake of trypsin-treated virus particles into coated pits, coated vesicles, and secondary lysosomes by EM confirms that rotavirus particles (porcine OSU strain) enter cells by receptor-mediated endocytosis and suggests that uncoating might occur by the effect of lysosomal enzymes (179). Use of the calcium ionophore A23187 to increase the intracellular Ca^{2+} concentration during the early stages of replication was also found to block uncoating (179). These results support early hypotheses that low Ca^{2+} concentrations in the intracellular microenvironment may be responsible for uncoating. This idea was originally proposed because it was known that removal of the outer capsid of particles and activation of the endogenous polymerase could be accomplished by calcium chelation (63,131).

Studies of the uptake of a human rotavirus whose infectivity is reportedly absolutely dependent on trypsin cleavage suggest that the mode of rotavirus entry into cells differs depending on whether virus particles have been pretreated with trypsin (298). Infectious virus that was pretreated with trypsin was observed to enter cells by direct penetration of particles through the cell membrane into the cell cytoplasm. In contrast, non–trypsin-treated particles were taken up by phagocytosis, and such virions were sequestered into lysosomes 20 minutes after virus attachment to the cell membrane. Mere phagocytosis of particles into lysosomes was thought to be unrelated to rotavirus replication (298). Direct release of nucleic acid through spaces that form between the capsomeres and cell membrane pores after the attachment of trypsin-activated virus particles onto cells also has been proposed; this mechanism was considered to be analogous to injection of nucleic acid into bacteria by phages (299). The inability to detect virus shells lacking nucleic acid attached to cells makes this observation difficult to interpret.

The concept that trypsin-treated and non–trypsin-treated virus enter cells by different mechanisms is supported by a study of the kinetics of entry of rhesus rotavirus into MA104 cells (149). Trypsin-activated rhesus rotavirus was internalized with a half-time of 3 to 5 minutes, whereas nonactivated virus disappeared from the cell surface with a half-time of 30 to 50 minutes. Only trypsin-activated virus resulted in productive infection. Trypsin-treated virus was shown to mediate the rapid release of ^{51}Cr from infected cells, and this activity did not occur at 4°C. The specificity of the permeability alterations was examined by showing that double-layered particles did not mediate them and that neutralizing anti-VP4 MAbs (but not nonneutralizing antibodies to VP6 or VP7) inhibited the ^{51}Cr release. Unfortunately, permeability results with non–trypsin-treated virus were not reported. However, these data suggest that virus entry occurs by direct penetration of the cell membrane, which results in permeabilization of the cell membrane (148).

Other viruses that initiate infection by mechanisms involving receptor-mediated endocytosis often depend on the acidification of endosomes for partial uncoating or entry into the cell. The importance of acidification of endosomes for the initiation of infection of rotaviruses has been studied by several groups (106,148,152,179). In all cases, lysosomotropic agents (ammonium chloride, chloroquine, methylamine, amantadine) had little inhibitory effect on virus replication as measured by RNA synthesis, polypeptide synthesis, or virus yields. Thus, it seems clear that acidification of endosomes is not important for the entry of rotavirus into cells, unlike other viral systems, including reovirus (40,189). Energy inhibitors (sodium azide and dinitrophenol) have a minimal effect on rotavirus infection, and this has been taken to suggest that rotaviruses do not use endocytosis to enter cells (148).

It seems clear that the passage of rotaviruses from endocytic vesicles to the cytoplasm does not occur via a pH-dependent fusion mechanism, but this does not rule out the possibility that rotaviruses are still taken up by endocytosis. Direct demonstration of virus fusion with membranes or hemolysis is lacking, and, without this, the data suggesting direct penetration could have other interpretations. Protease cleavage of VP4 is important for rapid entry into cells and particles containing cleaved VP4 possess lipophilic activity and can affect release of fluorescent dyes from liposomes and isolated membrane vesicles. These observations are consistent with the hypothesis that virus enters cells after direct interactions with the plasma membrane (221,272). It is also possible that more than one mechanism, including endocytosis and direct passage, is operative for rotaviruses, as has been proposed for poliovirus and reoviruses (28,72). Further studies are needed to determine whether the common endocytosis-mediated entry pathway exists for rotaviruses as originally thought.

A single report that the protein product of genome segment 10 influences virus adsorption to cells further complicates the question of the mechanism of virus attachment to cells (273). This study of a series of reassortants suggests that the nonstructural protein NSP4 may be indirectly involved in adsorption by influencing the arrangement of outer-shell proteins assembled in infectious particles. This observation remains to be confirmed.

Transcription and Replication

The synthesis of viral transcripts is mediated by an endogenous viral RNA-dependent RNA polymerase (transcriptase) that has a number of enzymatic activities. The transcriptase is a component of the virion, and properties of this enzyme (or enzyme complex) have been inferred by studying the characteristics of products from *in vitro* transcription reactions (61,191,201,278). Rotavirus particles presumably contain the same enzymatic activities found in reoviruses (see Chapter 24), including transcriptase, nucleotide phosphohydrolase, guanylyltransferase, and two methylases. These activities are inferred because rotavirus transcripts made *in vitro* in the presence of S-adenosyl methionine possess a methylated 5′-terminal cap structure, m^{7}GpppGm (142), and transcription is inhibited by pyrophosphate (293). Particles also contain a poly(A) polymerase activity whose precise function remains unknown; it has been postulated to be responsible for the synthesis of oligo(A) molecules (116).

The virus-associated transcriptase is latent in triple-layered particles and can be activated *in vitro* by treatment with a chelating agent or by heat shock treatment (63,293). Such treatments result in removal of the outer capsid proteins with conversion of triple-layered particles to double-layered particles (63). In infected cells, triple-layered particles have been shown to be uncoated to double-layered particles, so it is thought that transcription in cells occurs from such particles (60). Consistent with this idea, cells are susceptible to infection by liposome-mediated transfection of double-layered particles, indicating that simple delivery of these particles into the cell cytoplasm permits transcription to proceed (15). Transcription is asymmetric, and all transcripts are full-length plus strands made off the minus dsRNA strand (202). The exact site of transcription within the cytoplasm has not been precisely localized, although it is thought to occur in viroplasms.

Activation of transcriptase activity is a process that is not well understood. “Activation” may be a misnomer because in reoviruses it has been suggested that this process does not actually modify the enzyme complex but instead releases the templates from structural constraints, allowing them to move past the transcriptase catalytic site (143). Rotavirus transcription requires a hydrolyzable form of adenosine triphosphate (ATP), and studies with analogs that inhibit transcription suggest that ATP is required in reactions other than polymerization (293). ATP may be used for initiation or elongation of RNA molecules, as has been described for vesicular stomatitis virus or vaccinia virus RNA polymerases.

The synthesis of plus- and minus-strand RNA has been studied in SA11-infected cells (295) and in a cell-free system using extracts from infected cells (232). These studies were facilitated by optimization of an electrophoretic system that allows separation of the plus and minus strands of rotavirus RNAs in acid urea agarose gels (235). In these gels, the complementary strands migrate at different rates. For SA11 the plus strand migrates faster than its complementary minus strand. Analysis of the kinetics of RNA synthesis in infected cells showed that plus and minus strand RNAs are detected initially at 3 hr postinfection (295), in agreement with other studies that looked at the time of incorporation of [^{3}H]-uridine into rotavirus RNA (200). After 3 hr, the level of transcription increased until 9 to 12 hr, at which time the levels of plus-strand RNAs were maximal. The ratio of plus- to minus-strand RNA synthesis changes during infection and the maximal level of minus-strand RNA synthesis occur several hours before the peak of plus-strand RNA synthesis.

The delay in obtaining maximal plus-strand RNA synthesis has been hypothesized to be due to a requirement for the accumulation of stoichiometric amounts of a protein (e.g., VP6) necessary for the assembly of transcriptase particles. Both newly synthesized and preexisting plus-strand RNA were found to act as templates for minus-strand RNA synthesis throughout infection (232), an unexpected result based on earlier studies with reoviruses (1). The observation that the level of RNA replication does not increase continually in conjunction with the increasing levels of plus-strand RNA suggests that RNA replication is regulated by factors other than simply the level of plus-strand RNAs in the infected cell.

Use of this cell-free system to study the replication of rotavirus RNA isolated from infected cells was hampered by a low efficiency of minus-strand synthesis (232). However, based on nuclease sensitivity in this system, approximately 20% of the RNA made *in vitro* is double stranded and 80% is single stranded. Although not examined directly, it is assumed that rotavirus RNA replication, like that of reovirus, takes place in a conservative fashion; that is, both parental dsRNA strands remain within partially uncoated particles. The synthesis of dsRNA *in vitro* has been determined to be an asymmetrical process in which a nuclease-sensitive plus-strand RNA acts as template for the synthesis of minus-strand RNA. After its synthesis, dsRNA remains associated with subviral particles, suggesting that free dsRNA is not found in cells. Characterization of the subviral particles (complexes separable by sedimentation through sucrose gradients and by equilibrium centrifugation in CsCl gradients) in which dsRNA synthesis occurs both in infected cells and in the cell-free system indicated that these replicase complexes consist of core proteins, VP1 and VP2, small amounts of the protein VP6, large amounts of the nonstructural protein NSP3, and lesser amounts of NSP1 and NSP2 (130). These types of results indicate that some of the NSPs (NSP2, NSP3, NSP5) might be involved in the RNA replication process based on (a) the presence of these NSPs in replication complexes isolated from infected cells (109,233), (b) the nucleic acid binding properties of some of the NSPs (37,137,151, 197,245,331), (c) localization of some NSPs to viroplasms (8,241), and (d) the RNA-negative phenotype of ts mutants mapped to the genes that encode these NSPs (49,115).

A model was hypothesized in which rotavirus double-layered particles were assembled by the sequential addition of VP2 and VP6 to a precore replication intermediate consisting of VP1, VP3, NSP2, NSP3, and NSP5 (108). Because of the inability to completely separate particles with transcriptase and replicase activities with these methods and problems of contamination of some fractions with proteins from neighboring fractions, it has remained unclear whether the components characterized in this system were necessary for RNA replication or simply present in these complexes for other reasons. A foreign reporter molecule also was engineered into a rotavirus genome segment and tested for expression in cells coinfected with a helper virus (120). This system identified *cis*-acting nucleotides at the 3′ end of the genome necessary for gene expression, but the lack of direct evidence of synthesis of minus (−) strands did not permit determination of whether these signals were necessary for RNA replication. Unfortunately, the efficiency of these systems was too low to assure clear evaluation of the roles of *cis*- and *trans*-acting sequences and proteins required for replication.

The role of individual proteins and specific protein complexes in RNA replication and viral morphogenesis will probably not be solved until they are studied *in vitro* with pure species of native rotavirus proteins and viral RNAs. Progress toward this goal has come from the unexpected discovery that a template-dependent replication system can be established using only viral core proteins and exogenously added purified mRNAs (53). Viruslike particles expressed and self-assembled in insect cells also possess replicase activity (53). The conversion of exogenous mRNA to dsRNA by subviral particles is an exciting result that will allow studies of (a) the specificity of viral proteins in recognition and replication of rotavirus mRNAs, and (b) the effect of adding exogenous synthetic RNAs containing specific mutations on replication. These results suggest that a cell-free system to support rotavirus RNA replication, transcription, and the assembly of subviral particles can be established.

This *in vitro* template-dependent replication system was surprising because it does not absolutely require the nonstructural proteins for replicase activity (53). However, the nonstructural proteins may play a role in increasing the efficiency of replication or play other roles in the replication cycle. This latter idea is supported by the recent demonstration that NSP1 and NSP3 show a diffuse cytoplasmic staining in cells and NSP3 fractionates with cellular cytoskeletal elements rather than concentrating in viroplasms (8,138,197). NSP3 also binds specifically to a consensus sequence on the 3′ end of each of the 11 mRNAs (245). NSP3 may function to stabilize the viral mRNAs that lack poly(A) tails or transport those mRNAs from their site of synthesis to the viroplasms. The NSPs may function to recruit or move other viral proteins from their sites of synthesis into the viroplasms, or they may facilitate movement of nascent viral particles from the viroplasms to the ER membrane. Clearly much remains to be learned about the NSPs and RNA binding, transport, replication, and assembly. It is anticipated that these studies will be of general interest because the rotaviruses contain a larger number of NSPs than some of the other viruses within the *Reoviridae*. In addition, studies of the RNA-binding properties of NSP3 indicate that this may represent a prototype of a new kind of sequence-specific RNA binding proteins that require only four nucleotides for RNA recognition (246).

Assembly

The distinctive feature of rotavirus morphogenesis is that subviral particles, which assemble in cytoplasmic viroplasms, bud through the membrane of the ER, and maturing particles are transiently enveloped. This is one of the more interesting aspects of rotavirus replication, differing from members of other genera in the Reoviridae family. The envelope acquired in this process appears to be lost as particles move toward the interior of the ER, and it is replaced by a thin layer of protein that ultimately comprises the outer capsid of mature virions.

The sites of synthesis or localization of the viral proteins have been examined by ultrastructural immunocytochemistry using polyclonal monospecific or monoclonal antibodies (241,265) and by studying the distribution of proteins by immunofluorescence or by subcellular fractionation (145,292). Taken together, the morphologic and biochemical data are consistent with rapidly assembling double-layered particles serving as an intermediate stage in the formation of triple-layered virions. Most of the rotavirus structural proteins and all of the nonstructural proteins are synthesized on the free ribosomes, although the nascent proteins on free ribosomes have not been analyzed. Instead, this conclusion has been drawn based on the absence of signal sequences that would indicate targeting to the ER and lack of protection from digestion in *in vitro* protease protection studies (85,144). In contrast, the glycoproteins VP7 and NSP4 are synthesized on ribosomes associated with the membrane of the ER and are cotranslationally inserted into the ER membrane due to signal sequences at their amino termini.

The sites and precise details of RNA replication remain unclear. However, electron-dense viroplasms are probably the sites of synthesis of the double-layered particles that contain RNA (242,265). This conclusion is based on the localization of several of the viral proteins (VP2, NSP2, NSP5) to viroplasms and of VP4 and VP6 to the space between the periphery of the viroplasm and the outside of the ER (242), and on the observation that particles emerging from these viroplasms often seem to directly bud into the ER that contains VP7 and NSP4.

Two pools of VP7 that can be separated biochemically by fluorocarbon extraction or by immunological methods using two classes of antibodies have been detected in the ER of SA11-infected cells (145). One pool is found only on intact particles and is detected only by a neutralizing MAb. The second pool of VP7 is unassembled, remains associated with the ER membrane, and is detected by a polyclonal antibody made to denatured VP7 (145). Distinction of these two forms of VP7 permitted a kinetic study of the assembly of VP7 and of other structural proteins into particles. The incorporation of the inner capsid proteins into double-layered particles was found to occur rapidly, whereas VP4 and VP7 appeared in mature triple-layered particles with a lag time of 10 to 15 minutes. Kinetic analyses of the processing of the oligosaccharides on the two pools of VP7 showed that the virus-associated VP7 oligosaccharides had a 15-minute lag compared with that of the membrane-associated form, suggesting that the latter is the precursor to virion VP7. This lag appears to represent the time required for virus budding and outer capsid assembly (145).

Rotavirus maturation reportedly is a calcium-dependent process, based on the observation that virus yields are decreased when produced in cells maintained in calcium-depleted medium (284). Viruses produced in the absence of calcium were found to be exclusively double-layered, and budding of virus particles into the ER was not observed (283). Among the viral proteins, reduced levels of VP7 were observed, and subsequent studies showed that such reduced levels were due to the preferential degradation, and not to the impaired synthesis, of VP7 (283). An interesting finding of these studies was that unglycosylated (but not mature) VP7 made in the presence of tunicamycin was relatively stable in a calcium-free environment. Although the role of calcium in morphogenesis was not completely shown in these studies, it is possible that calcium stabilizes or modulates folding or compartmentalization of the newly glycosylated VP7 for subsequent assembly into particles. VP7 expressed alone does not fold properly unless it is expressed with other rotavirus proteins, and calcium must be present in cells for correct epitope formation (73,74). Alternatively, calcium deprivation may destabilize the ER or ER proteins required for the stable association of glycosylated VP7 with the membrane. Outer capsid assembly also requires proper formation of disulfide bonds on VP7 (302).

Understanding viral morphogenesis has been facilitated by the expression of the rotavirus structural proteins individually or in combinations in insects cells using recombinant baculoviruses (68,167,264,341). This approach first showed that the single-layered VP2 particle shell self-

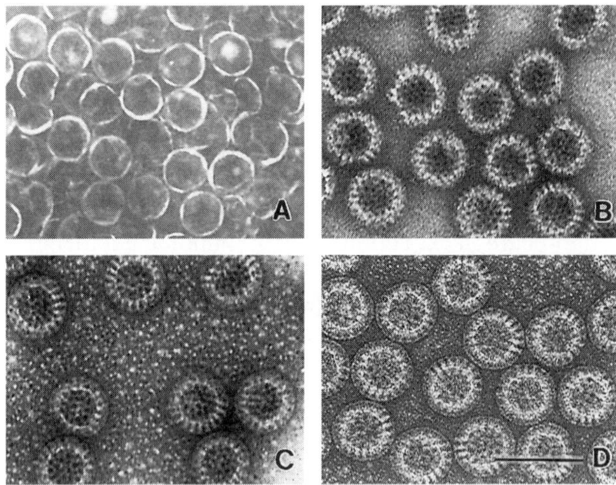

FIG. 9. Rotaviruslike particles self-assemble when the capsid proteins are coexpressed in eukaryotic cells. **A**, VP2; **B**, VP2/b; **C**, VP2/6/7; **D**, VP2/4/6/7 viruslike particles. Bar = 100 nm. Modified from Crawford et al. (68), with permission.

assembles when VP2 is expressed alone, and all of the other capsid proteins have been shown to self assemble into virus-like particles when coexpressed in the proper combinations (Fig. 9). Viruslike particles composed of VP2, VP1/2, VP1/2/3, VP2/3, VP2/6, VP2/6/7, VP2/4/6/7, and VP1/2/3/6 have been made and characterized (68,270,341). The outer and inner capsid proteins of different virus strains also have been shown to reassociate and be able to be transcapsidated onto other virus strains (51,52). These results demonstrate that the structural proteins contain the intrinsic information required to form particles and that coexpression of mutant proteins will be a feasible approach to analyze the domains responsible for the structural interactions between the proteins comprising the virus particles. These particles also are useful to probe the inner structure of particles by analyzing difference maps of particles with distinct protein compositions (270) and to analyze RNA transcription and replication (53,159a) and RNA packaging and assembly. These new developments offer hope that a reverse genetics system will be established soon for the rotaviruses.

Virus Release

EM studies have shown that the infectious cycle ends when progeny virus is released by host cell lysis (5,46,204). Extensive cytolysis during infection and drastic alterations in the permeability of the plasma membrane of infected cells resulting in the release of cellular and viral proteins has been demonstrated (213). Despite cell lysis, most double-layered and many triple-layered particles remain associated with the cellular debris, suggesting that these particles interact with structures within cells. Interactions with cell membranes and the cell cytoskeleton have been suggested (213). Whether the cytoskeleton provides a means of transport of viral proteins and particles to discrete sites in the cell for assembly or acts as a stabilizing element at the assembly site and in the newly budded virions or if particles are simply trapped by the cytoskeleton remains to be determined.

GENETICS (TYPES OF MUTANTS/VARIANTS)

Genetic studies have only begun to be exploited to understand rotavirus gene structure and function. Temperature-sensitive mutant collections have been isolated from several rotaviruses, including SA11, UK, and RRV (98, 126,259,261,262). The most extensive mutant collection is for SA11, in which 10 of the expected 11 reassortment groups have been identified and in which six of these have been assigned to genome segments (Table 6). The phenotypes of some of these mutants are consistent with functions inferred from biochemical data or recent sequence information (Table 4). For example, mutant groups B and C, which mark genome segments 3 and 1, respectively, are defective in the synthesis of ssRNA and dsRNA (49). Because these proteins are likely candidates to be part of the transcriptase particle, further biochemical and sequence analyses of these mutants or establishment of complementation systems with expressed gene products may permit dissection and localization of specific enzymatic domains on VP1 or VP3, or both.

Characterization of the genetic interactions between ts mutants has shown that recombination between pairs of ts mutants of SA11 occurs in the "all or none" fashion expected for recombination by a mechanism of reassortment of genome segments (259; see also Chapter 5). Genetic interactions of rotaviruses are influenced by factors similar to those that affect such interactions of other segmented viruses, including multiplicity of infection and the time of infection (260). Maximal, or near maximal, reassortment frequency is obtained at the earliest times when reassortment can be detected. This suggests that reassortment occurs early in the infectious cycle and that there is a single round of mating, an observation consistent with the particle association of progeny dsRNA. Reassortment is efficient at nonpermissive temperature, and mutants from all reassortment groups can interfere with the growth of wild-type virus at both permissive and nonpermissive temperatures. These results are important to consider when trying to produce reassortants.

Reassortment of genome segments among the progeny of *in vitro* mixed infections has been exploited to derive reassortant viruses with desired subsets of genome segments from each parent. Such reassortants have been used to assign ts mutants to specific genome segments and to correlate biologic properties with specific genome segments. For example, the genetic approach has been used to show that genome segment 4 is associated with the properties of hemagglutination, restriction of growth in tissue culture cells, protease-enhanced plaquing, reactivity with neutralizing antibody, and induction of diarrhea in a mouse model (115).

This approach will continue to be used to probe rotavirus gene function, so it is worthwhile to note the criteria needed to reach firm conclusions. The early studies to examine the genome segments responsible for inducing neutralizing antibody identified one member of the genome segment 7, 8, or 9 complex with this property. However, in these studies no reassortants containing genome segment 4 from a human virus were cultivated and characterized (125). Subsequent studies showed that genome segment 4 also contains neutralization determinants, highlighting the importance of analyzing reciprocal reassortants in such studies, and the usefulness of selection with specific neutralizing antibodies to obtain reassortants with desired combinations of VP4 and VP7 (158).

In addition, unexpected properties of parental viruses may influence the interpretation of studies of reassortant viruses. For example, two bovine viruses (UK and NCDV) have been found to share highly related VP7 antigens, but their VP4 antigens are distinct. Reassortant analyses between these two viruses will only allow examination of properties related to the distinct genome segment 4. Because the common properties of parental viruses may not be known in advance, it may be useful to confirm results from reassortment studies using different parental virus strains. The identity of naturally occurring intertypic rotavirus strains (133,134) and the direct demonstration that rotaviruses can undergo reassortment *in vivo* (113) highlight the fact that the genetic background of any rotavirus isolate cannot be assumed.

Naturally occurring rotavirus variants also have been found to be useful probes of gene function. Studies of a variant of the simian rotavirus SA11 that lacked a glycosylated VP7 showed that glycosylation was not essential for virus infectivity or hemagglutination (93,238) and that glycosylation of NSP4 was required for removal of the transient envelope from maturing particles (239). Further studies with neutralization escape mutants made from this variant have shown that glycosylation can affect rotavirus antigenicity (42). Studies with another variant of SA11 that has an altered genome segment 4 also have shown that the phenotype of certain reassortant constellations of genome segments depend on the parental strains used for analysis (47). This phenomenon has been confirmed with other reassortants containing specific combinations of VP4 and VP7, and this is beginning to explain some of the unexpected antigenic properties of some virus strains (48,76, 154,160,214,336,342).

EFFECTS ON HOST CELLS

Rotaviruses have a rather limited tissue tropism, in most cases only infecting the villus epithelium of the small intestine. In addition, the virus generally shows a substantial host range restriction, and symptomatic infection is usually restricted to the young of a species (see Chapter 55 in *Fields Virology*, 3rd ed.). It is thought that the diarrheal disease resulting from infection is a consequence of a cytocidal infection. In tissue culture, rotaviruses are normally cytocidal viruses that rapidly kill the permissive cells they infect. Adaption of rotaviruses to culture can be difficult and may require initial passage of virus in primary monkey kidney cells before adaptation to growth in continuous monkey kidney cell lines. In permissive cells, cytopathic effect is generally evident and cell death is preceded by the shut-off of host RNA, DNA, and protein synthesis (41,84,200). Cell death probably results from the function of a viral gene(s) on a specific target rather than from cumulative effects on host metabolism because certain ts mutants (ts groups F, G, H, and I) do not efficiently shut off host cell protein synthesis (Table 6) (263). NSP4 may be one protein that mediates cell death by causing intracellular calcium levels to increase (209,315).

Rotaviruses also have the ability to establish persistent infections in cell cultures (14,54,89) and in mice with severe combined immunodeficiency (SCID) syndrome (266). Persistently infected cells that release low levels of virus for multiple passages have been established in semipermissive (simian rotavirus in rabbit kidney cells) and in fully permissive (bovine rotavirus in bovine kidney) cells.

Characterization of these carrier cultures showed that a small percentage of cells were infected, and host-cell resistance to infection was partially overcome by the addition of trypsin to the media (54,89). In another case, a particular bovine rotavirus strain (2352) that failed to destroy infected BSC-1 cells was apparently important in inducing persistence. In these cells, a majority of the cells (greater than 80%) synthesized viral antigen during the S phase of the cell cycle, and the number of cells expressing antigen decreased as the cells moved into mitosis and cell division (14). Recognition that rotaviruses cause persistent infections in SCID mice suggests that the *in vitro* observations of carrier cultures may have clinical relevance, but that normally such infections are cleared by immune mechanisms. The prolonged excretion and occasional extraintestinal spread of rotaviruses in immunocompromised children indicates that cell-mediated immunity may be important in limiting infections (111,140,237,335). The cell-mediated immune response to rotavirus infection is just beginning to be investigated (70,71,102,103,225,227,228), but it is clear that rotaviruses, like the reoviruses and orbiviruses, have the ability to produce persistent infections. However, the mechanisms controlling these virus-cell interactions remain to be elucidated.

Rotavirus infection in polarized human intestinal epithelial (CaCO-2) cells has been studied to determine if infection occurs asymmetrically from either the apical or basolateral membranes. Rhesus rotavirus was found to infect polarized Madin-Darby canine kidney (MDCK) and

CaCO-2 monolayers in a symmetric manner as does influenza virus (303). MDCK cells supported only restricted replication of rotavirus, but in both cell types, infected cells showed transmembrane leaks or opening of tight junctions before the development of cytopathic effect and extensive virus release. Attachment and entry of virus into CaCO-2 cells did not cause any measurable transmembrane leak during the first hour of infection. These results suggest that infected epithelial cells could be sloughed from the intestine due to the loss of tight junctions early in the viral replication cycle and before the development of cytopathic effect.

Rotavirus replication also has been examined in polarized neuronal cells in which cytopathic effect again was not obvious. These infections are of interest because expression of VP7 is targeted to axons and dendrites by a pathway that does not involve the Golgi apparatus (327). In contrast, NSP4 was localized solely in the cell body. Rotavirus infection was found to stimulate alterations of specific cellular proteins. Thus, the distribution of microtubule-associated protein 2, which is normally restricted to nerve cell bodies and dendrites, was altered and found in axons of cultured dorsal root ganglia and spinal cord neurons. This study indicates that selective interactions between certain viral and neuronal cytoskeletal proteins can occur and that noncytolytic viral infection can cause alterations in the polarized sorting of neuronal proteins (328). These studies are of interest because they highlight the usefulness of monitoring rotavirus protein transport to understand new intracellular trafficking and signaling pathways.

INHIBITORS OF REPLICATION AND THEIR MECHANISM OF ACTION

Relatively few studies have evaluated inhibitors of rotavirus replication. Replication is inhibited by fetal bovine serum, most probably due to its antiprotease activities (94, 290) or to other components in serum such as lecithin whose mechanism of action is uncharacterized (325).

Evaluations of the effects of antiviral agents as inhibitors of rotavirus replication have yielded the following conclusions:

1. Nucleoside analogues are effective, probably due to inhibition of S-adenosylhomocysteine hydrolase, a key enzyme regulating methylation of viral mRNA (156).
2. Ribavirin and 3-deazaguanine prevent rotavirus polypeptide synthesis and may exert their antiviral effect by inhibiting the production or modification (capping) of viral-specific mRNA. Polymerase activity is not affected (289).
3. 9-(S)-(2,3 dihydroxypropyl) adenine is an effective inhibitor of replication with low toxicity *in vivo* but its mechanism of action is not known (289).
4. Clioquinal, an 8-hydroxyguinaline derivative, may reduce infection of mice if given at frequent intervals, possibly by removing the outer virus capsid (19).
5. Mucin, protease inhibitors and human milk reduce infectivity in cultured cells and/or animals (81,82,129, 323,340).

Additional inhibitors can be expected to be found as the structure and function of the unique rotavirus proteins are characterized and the molecular mechanisms controlling virus-cell interactions begin to be understood.

REFERENCES

1. Acs G, Klett H, Schonberg M, Christman J, Levin DH, Silverstein SC. Mechanism of reovirus double-stranded ribonucleic acid synthesis *in vivo* and *in vitro*. *J Virol* 1971;8:684–689.
2. Albert MJ, Unicomb LE, Bishop RF. Cultivation and characterization of human rotaviruses with "super short" RNA patterns. *J Clin Microbiol* 1987;25:183–185.
3. Allen AM, Desselberger U. Reassortment of human rotaviruses carrying rearranged genomes with bovine rotavirus. *J Gen Virol* 1985; 66:2703–2714.
4. Almeida JD, Bradburne AF, Wreghitt TG. The effect of sodium thiocyanate on virus structure. *J Med Virol* 1979;4:269–277.
5. Altenburg BC, Graham DY, Estes MK. Ultrastructural study of rotavirus replication in cultured cells. *J Gen Virol* 1980;46:75–85.
6. Andrew ME, Boyle DB, Coupar BE, Whitfeld PL, Both GW, Bellamy AR. Vaccinia virus recombinants expressing the SA11 rotavirus VP7 glycoprotein gene induce serotype-specific neutralizing antibodies. *J Virol* 1987;61:1054–1060.
7. Anthony ID, Bullivant S, Dayal S, Bellamy AR, Berriman JA. Rotavirus spike structure and polypeptide composition. *J Virol* 1991; 65:4334–4340.
8. Aponte C, Mattion NM, Estes MK, Charpilienne A, Cohen J. Expression of two bovine rotavirus non-structural proteins (NSP2 and NSP3) in the baculovirus system and production of monoclonal antibodies directed against these proteins. *Arch Virol* 1994;133:85–95.
9. Arias CF, López S, Bell JR, Strauss JH. Primary structure of the neutralization antigen of simian rotavirus SA11 as deduced from CDNA sequence. *J Virol* 1984;50:657–661.
10. Arias CF, López S, Espejo RT. Gene protein products of SA11 simian rotavirus genome. *J Virol* 1982;41:42–50.
11. Au K-S, Chan W-K, Estes MK. Rotavirus morphogenesis involves an endoplasmic reticulum transmembrane glycoprotein. In: Compans R, Helenius A, Oldston M, eds. *Cell biology of virus entry. Replication and pathogenesis.* New York: Alan R Liss; 1989:257–267.
12. Au K-S, Mattion NM, Estes MK. A subviral particle binding domain on the rotavirus nonstructural glycoprotein NS28. *Virology* 1993;194: 665–673.
13. Au KS, Chan WK, Burns JW, Estes MK. Receptor activity of rotavirus nonstructural glycoprotein NS28. *J Virol* 1989;63:4553–4562.
14. Babiuk LA, Misra V. Persistent infection of BSC-1 cells with rotavirus. In: Bricout F, Flewett TH, Pensaert M, Scherrer R, eds. *Viral enteritis in humans and animals.* Paris: INSERM; 1979:91–94.
15. Bass DM, Baylor MR, Chen C, Mackow EM, Bremont M, Greenberg HB. Liposome-mediated transfection of intact viral particles reveals that plasma membrane penetration determines permissivity of tissue culture cells to rotavirus. *J Clin Invest* 1992;90:2313–2320.
16. Bass DM, Mackow ER, Greenberg HB. NS35 and not VP7 is the soluble rotavirus protein which binds to target cells. *J Virol* 1990;64: 322–330.
17. Bass DM, Mackow ER, Greenberg HB. Identification and partial characterization of a rhesus rotavirus binding glycoprotein on murine enterocytes. *Virology* 1991;183:602–610.
18. Bastardo JW, Holmes IH. Attachment of SA-11 rotavirus to erythrocyte receptors. *Infect Immun* 1980;29:1134–1140.
19. Bednarz-Prashad AJ, John EI. Effect of clioquinol, an 8-hydroxyquinoline derivative, on rotavirus infection in mice. *J Infect Dis* 1983;148:613

20. Bellamy AR, Both GW. Molecular biology of rotaviruses. *Adv Virus Res* 1990;38:1–44.
21. Bellinzoni RC, Mattion NM, Burrone O, Gonzalez A, La Torre JL, Scodeller EA. Isolation of group A swine rotaviruses displaying atypical electropherotypes. *J Clin Microbiol* 1987;25:952–954.
22. Besselaar TG, Rosenblatt A, Kidd AH. Atypical rotavirus from South African neonates. Brief report. *Arch Virol* 1986;87:327–330.
23. Bican P, Cohen J, Charpilienne A, Scherrer R. Purification and characterization of bovine rotavirus cores. *J Virol* 1982;43:1113–1117.
24. Birch CJ, Heath RL, Gust ID. Use of serotype-specific monoclonal antibodies to study the epidemiology of rotavirus infection. *J Med Virol* 1988;24:45–53.
25. Biryahwaho B, Hundley F, Desselberger U. Bovine rotavirus with rearranged genome reassorts with human rotavirus. Brief report. *Arch Virol* 1987;96:257–264.
26. Bishai FR, Blaskovic P, Goodwin D. Physicochemical properties of Nebraska calf diarrhea virus hemagglutinin. *Can J Microbiol* 1978;24:1425–1430.
27. Blackhall J, Bellinzoni R, Mattion N, Estes MK, La Torre JL, Magnusson G. A bovine rotavirus serotype 1: serologic characterization of the virus and nucleotide sequence determination of the structural glycoprotein VP7 gene. *Virology* 1992;189:833–837.
28. Borsa J, Morash BD, Sargent MD, Copps TP, Lievaart PA, Szekely JG. Two modes of entry of reovirus particles into L cells. *J Gen Virol* 1979;45:161–170.
29. Both GW, Bellamy AR, Siegman LJ. Nucleotide sequence of the dsRNA genomic segment 7 of simian 11 rotavirus. *Nucleic Acids Res* 1984;12:1621–1626.
30. Both GW, Bellamy AR, Street JE, Siegman LJ. A general strategy for cloning double-stranded RNA: nucleotide sequence of the Simian-11 rotavirus gene 8. *Nucleic Acids Res* 1982;10:7075–7088.
31. Both GW, Mattick JS, Bellamy AR. Serotype-specific glycoprotein of simian 11 rotavirus: coding assignment and gene sequence. *Proc Natl Acad Sci USA* 1983;80:3091–3095.
32. Both GW, Siegman LJ, Bellamy AR, Atkinson PH. Coding assignment and nucleotide sequence of simian rotavirus SA11 gene segment 10: location of glycosylation sites suggests that the signal peptide is not cleaved. *J Virol* 1983;48:335–339.
33. Boyle JF, Holmes KV. RNA-binding proteins of bovine rotavirus. *J Virol* 1986;58:561–568.
34. Brade L, Schmidt WAK, Gattert I. Relative effectiveness of disinfectants against rotaviruses. *Zentralbl Bakteriol Hyg* 1981;174:151–159.
35. Bridger JC. Non-group A rotaviruses. In: Kapikian AZ, ed. *Viral infections of the gastrointestinal tract.* New York: Marcel Dekker; 1994:369–407.
36. Bridger JC, Woode GN. Characterization of two particle types of calf rotavirus. *J Gen Virol* 1976;31:245–250.
37. Brottier P, Nandi P, Bremont M, Cohen J. Bovine rotavirus segment 5 protein expressed in the baculovirus system interacts with zinc and RNA. *J Gen Virol* 1992;73:1931–1938.
38. Burns JW, Greenberg HB, Shaw RD, Estes MK. Functional and topographical analyses of epitopes on the hemagglutinin (VP4) of the simian rotavirus SA11. *J Virol* 1988;62:2164–2172.
39. Burns JW, Welch SK, Nakata S, Estes MK. Characterization of monoclonal antibodies to human group B rotavirus and their use in an antigen detection enzyme linked immunosorbent assay. *J Clin Microbiol* 1989;27:245–250.
40. Canning WM, Fields BN. Ammonium chloride prevents lytic growth of reovirus and helps to establish persistent infection in mouse L cells. *Science* 1983;210:987–988.
41. Carpio MM, Babiuk LA, Misra V, Blumenthal RM. Bovine rotavirus-cell interactions: effect of virus infection on cellular integrity and macromolecular synthesis. *Virology* 1981;114:86–97.
42. Caust J, Dyall-Smith ML, Lazdins I, Holmes IH. Glycosylation, an important modifier of rotavirus antigenicity. *Arch Virol* 1987;96:123–134.
43. Chan WK, Au KS, Estes MK. Topography of the simian rotavirus nonstructural glycoprotein (NS28) in the endoplasmic reticulum membrane. *Virology* 1988;164:435–442.
44. Chan WK, Penaranda ME, Crawford SE, Estes MK. Two glycoproteins are produced from the rotavirus neutralization gene. *Virology* 1986;151:243–252.
45. Chanock SJ, Wenske EA, Fields BN. Human rotaviruses and genome RNA. *J Infect Dis* 1983;148:49–50.
46. Chasey D. Different particle types in tissue culture and intestinal epithelium infected with rotavirus. *J Gen Virol* 1977;37:443–451.
47. Chen D, Burns JW, Estes MK, Ramig RF. Phenotypes of rotavirus reassortants depend upon the recipient genetic background. *Proc Natl Acad Sci USA* 1989;86:3743–3747.
48. Chen D, Estes MK, Ramig RF. Specific interactions between rotavirus outer capsid proteins VP4 and VP7 determine expression of a cross-reactive, neutralizing VP4-specific epitope. *J Virol* 1992;66:432–439.
49. Chen D, Gombold JL, Ramig RF. Intracellular RNA synthesis directed by temperature-sensitive mutants of simian rotavirus SA11. *Virology* 1990;178:143–151.
50. Chen D, Ramig RF. Determinants of rotavirus stability and density during CsCl purification. *Virology* 1992;186:228–237.
51. Chen D, Ramig RF. Rescue of infectivity by *in vitro* transcapsidation of rotavirus single-shelled particles. *Virology* 1993;192:422–429.
52. Chen D, Ramig RF. Rescue of infectivity by sequential *in vitro* transcapsidation of rotavirus core particles with inner capsid and outer capsid proteins. *Virology* 1993;194:743–751.
53. Chen D, Zeng CQY, Wentz MJ, Gorziglia M, Estes MK, Ramig RF. Template-dependent, *in vitro* replication of rotavirus RNA. *J Virol* 1994;68:7030–7039.
54. Chiarini A, Arista S, Giammanco A, Sinatra A. Rotavirus persistence in cell cultures: selection of resistant cells in the presence of foetal calf serum. *J Gen Virol* 1983;64:1101–1110.
55. Clapp LL, Patton JT. Rotavirus morphogenesis: domains in the major inner capsid protein essential for binding to single-shelled particles and for trimerization. *Virology* 1991;180:697–708.
56. Clark CS, Linnemann CC Jr, Gartside PS, Phair JP, Blacklow N, Zeiss CR. Serologic survey of rotavirus, Norwalk agent and Prototheca wickerhamii in wastewater workers. *Am J Public Health* 1985;75:83–85.
57. Clark HF, Hoshino Y, Bell LM, et al. Rotavirus isolate W161 representing a presumptive new human serotype. *J Clin Microbiol* 1987;25:1757–1762.
58. Clark SM, Barnett BB, Spendlove RS. Production of high-titer bovine rotavirus with trypsin. *J Clin Microbiol* 1979;9:413–417.
59. Clark SM, Roth JR, Clark ML, Barnett BB, Spendlove RS. Trypsin enhancement of rotavirus infectivity: mechanism of enhancement. *J Virol* 1981;39:816–822.
60. Clark SM, Spendlove RS, Barnett BB. Role of two particle types in bovine rotavirus morphogenesis. *J Virol* 1980;34:272–276.
61. Cohen J. Ribonucleic acid polymerase activity associated with purified calf rotavirus. *J Gen Virol* 1977;36:395–402.
62. Cohen J, Charpilienne A, Chilmonczyk S, Estes MK. Nucleotide sequence of bovine rotavirus gene 1 and expression of the gene product in baculovirus. *Virology* 1989;171:131–140.
63. Cohen J, Laporte J, Charpilienne A, Scherrer R. Activation of rotavirus RNA polymerase by calcium chelation. *Arch Virol* 1979;60:177–186.
64. Conner ME, Estes MK, Graham DY. Rabbit model of rotavirus infection. *J Virol* 1988;62:1625–1633.
65. Conner ME, Matson DO, Estes MK. Rotavirus vaccines and vaccination potential. In: Ramig RF, ed. *Rotaviruses.* Berlin: Springer-Verlag; 1994:286–337.
66. Coulson BS, Fowler KJ, Bishop RF, Cotton RG. Neutralizing monoclonal antibodies to human rotavirus and indications of antigenic drift among strains from neonates. *J Virol* 1985;54:14–20.
67. Coulson BS, Unicomb LE, Pitson GA, Bishop RF. Simple and specific enzyme immunoassay using monoclonal antibodies for serotyping human rotaviruses. *J Clin Microbiol* 1987;25:509–515.
68. Crawford SE, Labbe M, Cohen J, Burroughs MH, Zhou Y, Estes MK. Characterization of virus-like particles produced by the expression of rotavirus capsid proteins in insect cells. *J Virol* 1994;68:5945–5952.
69. Desselberger U, McCrae MA. In: Ramig RF, ed. *The Rotavirus genome. Rotaviruses.* Berlin: Springer-Verlag; 1994:31–66.
70. Dharakul T, Labbe M, Cohen J, et al. Immunization with baculovirus-expressed recombinant rotavirus proteins VP1, VP4, VP6, and VP7 induces CD8+ T lymphocytes that mediate clearance of chronic rotavirus infection in SCID mice. *J Virol* 1991;65:5928–5932.
71. Dharakul T, Roft L, Greenberg HB. Recovery from chronic rotavirus infection in mice with severe combined immunodeficiency: virus clearance mediated by adoptive transfer of immune CD8+ T lymphocytes. *J Virol* 1990;64:4375–4382.
72. Dimmock NJ. Initial stages in infection with animal viruses. *J Gen Virol* 1982;59:1–22.
73. Dormitzer PR, Greenberg HB. Calcium chelation induces a confor-

mational change in recombinant herpes simplex virus-1-expressed rotavirus VP7. *Virology* 1992;189:828–832.
74. Dormitzer PR, Ho DY, Mackow ER, Mocarski ES, Greenberg HB. Neutralizing epitopes on herpes simplex virus-l-expressed rotavirus VP7 are dependent on coexpression of other rotavirus proteins. *Virology* 1992;187:18–32.
75. Dryden KA, Wang G, Yeager M, et al. Early steps in reovirus infection are associated with dramatic changes in supramolecular structure and protein conformation: analysis of virions and subviral particles by cryoelectron microscopy and image reconstruction. *J Cell Biol* 1993; 122:1023–1041.
76. Dunn JD, Burns JW, Cross TI, et al. Comparisons of VP4 and VP7 of five murine rotavirus strains. *Virology* 1994;203:250–259.
77. Dunn SJ, Greenberg HB, Ward RL, et al. Serotypic and genotypic characterization of human serotype 10 rotaviruses from asymptomatic neonates. *J Clin Microbiol* 1993;31:165–169.
78. Dyall-Smith ML, Azad AA, Holmes IH. Gene mapping of rotavirus double-stranded RNA segments by northern blot hybridization: application to segments 7, 8, and 9. *J Virol* 1983;46:317–320.
79. Dyall-Smith ML, Holmes IH. Gene-coding assignments of rotavirus double-stranded RNA segments 10 and 11. *J Virol* 1981;38:1099–1103.
80. Dyall-Smith ML, Lazdins I, Tregear GW, Holmes IH. Location of the major antigenic sites involved in rotavirus serotype-specific neutralization. *Proc Natl Acad Sci USA* 1986;83:3465–3468.
81. Ebina T, Sato A, Umezu K, et al. Prevention of rotavirus infection by oral administration of cow colostrum containing antihuman rotavirus antibody. *Med Microbiol Immunol (Berl)* 1985;174:177–185.
82. Ebina T, Tsukada K. Protease inhibitors prevent the development of human rotavirus-induced diarrhea in suckling mice. *Microbiol Immunol* 1991;35:583–588.
83. Elias MM. Distribution and titres of rotavirus antibodies in different age groups. *J Hyg (Lond)* 1977;79:365–372.
84. Ericson BL, Graham DY, Mason BB, Estes MK. Identification, synthesis, and modifications of simian rotavirus SA11 polypeptides in infected cells. *J Virol* 1982;42:825–839.
85. Ericson BL, Graham DY, Mason BB, Hanssen HH, Estes MK. Two types of glycoprotein precursors are produced by the simian rotavirus SA11. *Virology* 1983;127:320–332.
86. Esparza J, Gil F. A study on the ultrastructure of human rotavirus. *Virology* 1978;91:141–150.
87. Espejo RT, López S, Arias C. Structural polypeptides of simian rotavirus SA11 and the effect of trypsin. *J Virol* 1981;37:156–160.
88. Estes MK, Crawford SE, Penaranda ME, et al. Synthesis and immunogenicity of the rotavirus major capsid antigen using a baculovirus expression system. *J Virol* 1987;61:1488–1494.
89. Estes MK, Graham DY. Establishment of rotavirus persistent infection in cell culture. Brief report. *Arch Virol* 1980;65:187–192.
90. Estes MK, Graham DY, Dimitrov DH. The molecular epidemiology of rotavirus gastroenteritis. *Prog Med Virol* 1984;29:1–22.
91. Estes MK, Graham DY, Mason BB. Proteolytic enhancement of rotavirus infectivity: molecular mechanisms. *J Virol* 1981;39:879–888.
92. Estes MK, Graham DY, Petrie BL. Antigenic structure of rotaviruses. In: vanRegenmortel MHV, Neurath AR, eds. *Immunochemistry of viruses.* 1985:389–405.
93. Estes MK, Graham DY, Ramig RF, Ericson BL. Heterogeneity in the structural glycoprotein (VP7) of simian rotavirus SA11. *Virology* 1982; 122:8–14.
94. Estes MK, Graham DY, Smith EM, Gerba CP. Rotavirus stability and inactivation. *J Gen Virol* 1979;43:403–409.
95. Estes MK, Mason BB, Crawford S, Cohen J. Cloning and nucleotide sequence of the simian rotavirus gene 6 that codes for the major inner capsid protein. *Nucleic Acids Res* 1984;12:1875–1887.
96. Estes MK, Palmer EL, Obijeski JF. Rotaviruses: a review. *Curr Top Microbiol Immunol* 1983;105:123–184.
97. Estes MK, Tanaka T. Nucleic acid probes for rotavirus detection and characterization. In: Tenover F, ed. *DNA probes for infectious diseases.* Boca Raton: CRC Press; 1989:79–100.
98. Faulkner-Valle GP, Clayton AV, McCrae MA. Molecular biology of rotaviruses. III. Isolation and characterization of temperature-sensitive mutants of bovine rotavirus. *J Virol* 1982;42:669–677.
99. Fernelius AL, Ritchie AE, Classick LG, Norman JO, Mebus CA. Cell culture adaptation and propagation of a reovirus-like agent of calf diarrhea from a field outbreak in Nebraska. *Arch Ges Virusforschung* 1972;37:114–130.
100. Fiore L, Greenberg HB, Mackow ER. The VP8 fragment of VP4 is the rhesus rotavirus hemagglutinin. *Virology* 1991;181:553–563.
101. Flewett TH, Bryden AS, Davies H, Woode GN, Bridger JC, Derrick JM. Relation between viruses from acute gastroenteritis of children and newborn calves. *Lancet* 1974;2:61–63.
102. Franco MA, Lefevre P, Willems P, Tosser G, Lintermanns P, Cohen J. Identification of cytotoxic T cell epitopes on the VP3 and VP6 rotavirus proteins. *J Gen Virol* 1994;75:1–8.
103. Franco MA, Prieto I, Labbe M, Poncet D, Borras-Cuesta F, Cohen J. An immunodominant cytotoxic T cell epitope on the VP7 rotavirus protein overlaps the H2 signal peptide. *J Gen Virol* 1994;74:2579–2586.
104. Fukudome K, Yoshie O, Konno T. Comparison of human, simian, and bovine rotaviruses for requirement of sialic acid in hemagglutination and cell adsorption. *Virology* 1989;172:196–205.
105. Fukuhara N, Yoshie O, Kitaoka S, Konno T. Role of VP3 in human rotavirus internalization after target cell attachment via VP7. *J Virol* 1988;62:2209–2218.
106. Fukuhara N, Yoshie O, Kitaoka S, Konno T, Ishida N. Evidence for endocytosis-independent infection by human rotavirus. *Arch Virol* 1987; 97:93–99.
107. Furuichi Y, Morgan MA, Muthukrishnan S, Shatkin AJ. Reovirus messenger RNA contains a methylated, blocked 5′-terminal structure; m7GpppGmC. *Proc Natl Acad Sci USA* 1975;72:362–366.
108. Gallegos CO, Patton JT. Characterization of rotavirus replication intermediates: a model for the assembly of single-shelled particles. *Virology* 1989;172:616–627.
109. Gentsch JR, Das BK, Jiang B, Bhan MK, Glass RI. Similarity of the VP4 protein of human rotavirus strain 116E to that of the bovine B223 strain. *Virology* 1993;194:424–430.
110. Gerna G, Sarasini A, Di Matteo A, Parea M, Orsolini P, Battaglia M. Identification of two subtypes of serotype 4 human rotavirus by using VP7-specific neutralizing monoclonal antibodies. *J Clin Microbiol* 1988;26:1388–1392.
110a. Gerna G, Sears J, Hoshino Y, Steele AD, Nakagomi O, Sarasimi A, Flores J. Identification of a new VP4 serotype of human rotaviruses. *Virology* 1994;200:66–71.
111. Gilger MA, Matson DO, Conner ME, Rosenblatt HM, Finegold MJ, Estes MK. Extraintestinal rotavirus infections in children with immunodeficiency. *J Pediatr* 1992;120:912–917.
112. Gombold JL, Estes MK, Ramig RF. Assignment of simian rotavirus SA11 temperature-sensitive mutant groups B and E to genome segments. *Virology* 1985;143:309–320.
113. Gombold JL, Ramig RF. Analysis of reassortment of genome segments in mice mixedly infected with rotaviruses SA11 and RRV. *J Virol* 1986; 57:110–116.
114. Gombold JL, Ramig RF. Assignment of simian rotavirus SA11 temperature sensitive mutant groups A, C, F, and G to genome segments. *Virology* 1987;161:463–473.
115. Gombold JL, Ramig RF. Genetics of the rotaviruses. In: Ramig RF, ed. *Rotaviruses.* Berlin: Springer-Verlag; 1994:129–177.
116. Gorziglia M, Esparza J. Poly (A) polymerase activity in human rotavirus. *J Gen Virol* 1981;53:357–362.
117. Gorziglia M, Green K, Nishikawa K, Taniguchi K, Jones R, Kapikian AZ, Chanock RM. Sequence of the fourth gene of human rotaviruses recovered from asymptomatic or symptomatic infections. *J Virol* 1988; 62:2978–2984.
118. Gorziglia M, Hoshino Y, Buckler-White A, et al. Conservation of amino acid sequence of VP8 and cleavage region of 84-kDa outer capsid protein among rotaviruses recovered from asymptomatic neonatal infection [erratum appears in *Proc Natl Acad Sci USA* 1987;84:2062]. *Proc Natl Acad Sci USA* 83:7039–7043.
119. Gorziglia M, Larrea C, Liprandi F, Esparza J. Biochemical evidence for the oligomeric (possibly trimeric) structure of the major inner capsid polypeptide (45K) of rotaviruses. *J Gen Virol* 1985;66:1889–1900.
120. Gorziglia MI, Collins PL. Intracellular amplification and expression of a synthetic analog of rotavirus genomic RNA bearing a foreign marker gene: mapping cis-acting nucleotides in the 3′-noncoding region. *Proc Natl Acad Sci USA* 1992;89:5784–5788.
121. Graham A, Kudesia G, Allen AM, Desselberger U. Reassortment of human rotavirus possessing genome rearrangements with bovine rotavirus: evidence for host cell selection. *J Gen Virol* 1987;68:115–122.
122. Graham DY, Estes MK. Proposed working serologic classification system for rotaviruses. *Ann Inst Pasteur* 1985;136:5–12.
123. Green KY, Midthun K, Gorziglia M, Hoshino Y, Kapikian AZ, Chanock

RM, Flores J. Comparison of the amino acid sequences of the major neutralization protein of four human rotavirus serotypes. *Virology* 1987; 161:153–159.
124. Green KY, Sears JF, Taniguchi K, et al. Prediction of human rotavirus serotype by nucleotide sequence analysis of the VP7 protein gene. *J Virol* 1988;62:1819–1823.
125. Greenberg HB, Flores J, Kalica AR, Wyatt RG, Jones R. Gene coding assignments for growth restriction, neutralization and subgroup specificities of the W and DS-1 strains of human rotavirus. *J Gen Virol* 1983; 64:313–320.
126. Greenberg HB, Kalica AR, Wyatt RG, Jones RW, Kapikian AZ, Chanock RM. Rescue of noncultivatable human rotavirus by gene reassortment during mixed infection with *ts* mutants of a cultivatable bovine rotavirus. *Proc Natl Acad Sci USA* 1981;78:420–424.
127. Greenberg HB, Vaidesuso J, van Wyke K, et al. Production and preliminary characterization of monoclonal antibodies directed at two surface proteins of rhesus rotavirus. *J Virol* 1983;47:267–275.
128. Hardy ME, Gorziglia M, Woode GN. Amino acid sequence analysis of bovine rotavirus B223 reveals a unique outer capsid protein VP4 and confirms a third bovine VP4 type. *Virology* 1992;191:291–300.
129. Hatta H, Tsuda K, Akachi S, Kim M, Yamamoto T, Ebina T. Oral passive immunization effect of anti-human rotavirus IgY and its behavior against proteolytic enzymes. *Biosci Biotechnol Biochem* 1993;57: 1077–1081.
130. Heimberger-Jones M, Patton JT. Characterization of subviral particles in cells infected with simian rotavirus SA11. *Virology* 1986;155: 655–665.
131. Holmes IH. Rotaviruses. In: Joklik WK, ed. *The Reoviridae.* New York: Plenum; 1983:359–423.
132. Holmes IH, Bishop RF, Coulson BS, et al. Nomenclature of rotavirus serotypes. *Arch Virol* (submitted for publication).
132a. Hoshino Y, Kapikian AZ. Rotavirus vaccine development for the prevention of severe diarrhea in infants and young children. *Trends in Mibcrobiol* 1994;2:242–249.
133. Hoshino Y, Sereno MM, Midthun K, Flores J, Kapikian AZ, Chanock RM. Independent segregation of two antigenic specificities (VP3 and VP7) involved in neutralization of rotavirus infectivity. *Proc Natl Acad Sci USA* 1985;82:8701–8704.
134. Hoshino Y, Serono MM, Midthun K, Flores J, Chanock RM, Kapikian AZ. Analysis by plaque reduction neutralization assay of intertypic rotaviruses suggests that gene reassortment occurs *in vivo*. *J Clin Microbiol* 1987;25:290–294.
135. Hoshino Y, Wyatt RG, Greenberg HB, Flores J, Kapikian AZ. Serotypic similarity and diversity of rotaviruses of mammalian and avian origin as studied by plaque-reduction neutralization. *J Infect Dis* 1984; 149:694–702.
136. Hua J, Chen X, Patton JT. Deletion mapping of the rotavirus metalloprotein NS53 (NSPI): the conserved cysteine-rich region is essential for virus-specific RNA binding. *J Virol* 1994;68:3990–4000.
137. Hua J, Patton JT. The carboxyl-half of the rotavirus nonstructural protein NS53 (NSPL) is not required for virus replication. *Virology* 1994; 198:567–576.
138. Huang J, Nagesha HS, Holmes IH. Comparative sequence analysis of VP4s from five Australian porcine rotaviruses: implication of an apparent new P type. *Virology* 1993;196:319–327.
139. Hundley F, Biryahwaho B, Gow M, Desselberger U. Genome rearrangements of bovine rotavirus after serial passage at high multiplicity of infection. *Virology* 1985;143:88–103.
140. Hundley F, McIntyre M, Clark B, Beards G, Wood D, Chrystie I, Desselberger U. Heterogeneity of genome rearrangements in rotaviruses isolated from a chronically infected immunodeficient child. *J Virol* 1987;61:3365–3372.
141. Ikegami N, Akatani K, Hosaka T, Ushijima H. Prevalence of a new serotype of group A human rotaviruses in Japan [Abstract 113]. Abstracts of the VII International Congress of Virology, Alberta, Canada, 1988.
142. Imai M, Akatani K, Ikegami N, Furuichi Y. Capped and conserved terminal structures in human rotavirus genome double-stranded RNA segments. *J Virol* 1983;47:125–136.
143. Joklik WK. Recent progress in reovirus research. *Ann Rev Genet* 1985; 19:537–575.
144. Kabcenell AK, Atkinson PH. Processing of the rough endoplasmic reticulum membrane glycoproteins of rotavirus SA11. *J Cell Biol* 1985; 101:1270–1280.
145. Kabcenell AK, Poruchynsky MS, Bellamy AR, Greenberg HB, Atkinson PH. Two forms of VP7 are involved in assembly of SA11 rotavirus in endoplasmic reticulum. *J Virol* 1988;62:2929–2941.
146. Kalica AR, Flores J, Greenberg HB. Identification of the rotaviral gene that codes for hemagglutination and protease-enhanced plaque formation. *Virology* 1983;125:194–205.
147. Kalica AR, Greenberg HB, Wyatt RG, et al. Genes of human (strain Wa) and bovine (strain UK) rotaviruses that code for neutralization and subgroup antigens. *Virology* 1981;112:385–390.
148. Kaijot KT, Shaw RD, Rubin DH, Greenberg HB. Infectious rotavirus enters cells by direct cell membrane penetration, not by endocytosis. *J Virol* 1988;62:1136–1144.
149. Kantharidis P, Dyall-Smith ML, Holmes IH. Marked sequence variation between segment 4 genes of human RV-5 and simian SA11 rotaviruses. *Arch Virol* 1987;93:111–121.
150. Kapahnke R, Rappold W, Desselberger U, Riesner D. The stiffness of dsRNA: hydrodynamic studies on fluorescence-labelled RNA segments of bovine rotavirus. *Nucleic Acids Res* 1986;14:3215–3228.
151. Kattoura MD, Clapp LL, Patton JT. The rotavirus nonstructural protein, NS35, possesses RNA-binding activity *in vitro* and *in vivo*. *Virology* 1992;191:698–708.
152. Keljo DJ, Kuhn M, Smith A. Acidification of endosomes is not important for the entry of rotavirus into the cell. *J Pediatr Gastroenterol Nutr* 1988;7:257–263.
153. Keljo DJ, Smith AK. Characterization of binding of simian rotavirus SA-11 to cultured epithelial cells. *J Pediatr Gastroenterol Nutr* 1988; 7:249–256.
154. Kirkwood C, Masendycz PJ, Coulson BS. Characteristics and location of cross-reactive and serotype-specific neutralization sites on VP7 of human G type 9 rotaviruses. *Virology* 1993;196:79–88.
155. Kitaoka S, Fukuhara N, Tazawa F, et al. Characterization of monoclonal antibodies against human rotavirus hemagglutinin. *J Med Virol* 1986;19:313–323.
156. Kitaoka S, Konno T, De Clercq E. Comparative efficacy of broad-spectrum antiviral agents as inhibitors of rotavirus replication *in vitro*. *Antiviral Res* 1986;6:57–65.
157. Kitaoka S, Suzuki H, Numazaki T, et al. Hemagglutination by human rotavirus strains. *J Med Virol* 1984;13:215–222.
158. Kobayashi N, Taniguchi K, Urasawa T, Urasawa S. Efficient production of antigenic mosaic reassortants of rotavirus with the aid of anti-VP4 and anti-VP7 neutralizing monoclonal antibodies. *J Virol Methods* 1993;44:25–34.
159. Kogasaka R, Akihara M, Horino K, Chiba S, Nakao T. A morphological study of human rotavirus. *Arch Virol* 1979;61:41–48.
159a. Kohli E, Pothier P, Tosser G, Cohen J, Sandino AM, Spencer E. *In vitro* reconstitution of rotavirus transcriptional activity using viral cores and recombinant baculovirus expressed VP6. *Arch Virol* 1993;133: 451–458.
160. Kool DA, Matsui SM, Greenberg HB, Holmes IH. Isolation and characterization of a novel reassortant between avian Ty-1 and simian RRV rotaviruses. *J Virol* 1992;66:6836–6839.
161. Kouvelos K, Petric M, Middleton PJ. Comparison of bovine, simian and human rotavirus structural glycoproteins. *J Gen Virol* 1984;65: 1211–1214.
162. Kouvelos K, Petric M, Middleton PJ. Oligosaccharide composition of calf rotavirus. *J Gen Virol* 1984;65:1159–1164.
163. Kozak M. Possible role of flanking nucleotides in recognition of the AUG initiator codon by eukaryotic ribosomes. *Nucleic Acids Res* 1981; 9:5233–5250.
164. Kozak M. Bifunctional messenger RNAs in eukaryotes. *Cell* 1986;47: 481–483.
165. Kuchino Y, Nishimura S, Smith RE, Furuichi Y. Homologous terminal sequences in the double-stranded RNA genome segments of cytoplasmic polyhedrosis virus of silkworm. *J Virol* 1982;44:538–543.
166. Kurtz JB, Lee TW, Parsons AJ. The action of alcohols on rotavirus, astrovirus and enterovirus. *J Hosp Infect* 1980;1:321–325.
167. Labbé M, Charpilienne A, Crawford SE, Estes MK, Cohen J. Expression of rotavirus VP2 produces empty corelike particles. *J Virol* 1991; 65:2946–2952.
168. Larratde G, Gorziglia M. Distribution of conserved and specific epitopes on the VP8 subunit of rotavirus VP4. *J Virol* 1992;66:7438–7443.
169. Larratde G, Li B, Kapikian AZ, Gorziglia M. Serotype-specific epitope(s) present on the VP8 subunit of rotavirus VP4 protein. *J Virol* 1991;65:3213–3218.
170. Liu M, Estes MK. Nucleotide sequence of the simian rotavirus SA11

genome segment 3. *Nucleic Acids Res* 1989;17:7991

171. Liu M, Mattion NM, Estes MK. Rotavirus VP3 expressed in insect cells possesses guanylyltransferase activity. *Virology* 1992;188:77–84.
172. Liu M, Offit PA, Estes MK. Identification of the simian rotavirus SA11 genome segment 3 product. *Virology* 1988;163:26–32.
173. López S, Arias CF. The nucleotide sequence of the 5′ and 3′ ends of rotavirus SA11 gene 4. *Nucleic Acids Res* 1987;15:4691.
174. López S, Arias CF, Bell JR, Strauss JH, Espejo RT. Primary structure of the cleavage site associated with trypsin enhancement of rotavirus SA11 infectivity. *Virology* 1985;144:11–19.
175. López S, Arias CF, Mendez E, Espejo RT. Conservation in rotaviruses of the protein region containing the two sites associated with trypsin enhancement of infectivity. *Virology* 1986;154:224–227.
176. López S, Greenberg HB, Arias CF. Mapping the subgroup epitopes of rotavirus protein VP6. *Virology* 1994;204:153–162.
177. López S, López I, Romero P, Méndez E, Soberón X, Arias CF. Rotavirus YM gene 4: analysis of its deduced amino acid sequence and prediction of the secondary structure of the VP4 protein. *J Virol* 1991; 65:3738–3745.
178. Ludert JE, Gil F, Liprandi F, Esparza J. The structure of the rotavirus inner capsid studied by electron microscopy of chemically disrupted particles. *J Gen Virol* 1986;67:1721–1725.
179. Ludert JE, Michelangeli F, Gil F, Liprandi F, Esparza J. Penetration and uncoating of rotaviruses in cultured cells. *Intervirology* 1987;27: 95–101.
180. Maass DR, Atkinson PH. Rotavirus proteins VP7, NS28, and VP4 form oligomeric structures. *J Virol* 1990;64:2632–2641.
181. Maass DR, Atkinson PH. Retention by the endoplasmic reticulum of rotavirus VP7 is controlled by three adjacent amino acid residues. *J Virol* 1994;68:366–378.
182. Mackow ER, Chou R, Fay ME, Dowling W, Chen GM. Gene segment 6 of group B rotavirus, ADRV, encodes a putative fusion protein [Abstract 237]. Abstracts of the American Society for Virology 13th Annual Meeting, Madison, Wisconsin, 1994.
183. Mackow ER, Shaw RD, Matsui SM, Vo PT, Benfield DA, Greenberg HB. Characterization of homotypic and heterotypic VP7 neutralization sites of rhesus rotavirus [erratum appears in *Virology* 1988;167:660]. *Virology* 1988;165:511–517.
184. Mackow ER, Shaw RD, Matsui SM, Vo PT, Dang MN, Greenberg HB. The rhesus rotavirus gene encoding protein VP3: location of amino acids involved in homologous and heterologous rotavirus neutralization and identification of a putative fusion region. *Proc Natl Acad Sci USA* 1988;85:645–649.
185. Mackow ER, Vo PT, Broome R, Bass D, Greenberg HB. Immunization with baculovirus-expressed VP4 protein passively protects against simian and murine rotavirus challenge. *J Virol* 1990;64:1698–1703.
186. Malherbe HH, Strickland-Cholmley M. Simian virus SA11 and the related 0 agent. *Arch Ges Virusforschung* 1967;22:235–245.
187. Mansell EA, Patton JT. Rotavirus RNA replication: VP2, but not VP6, is necessary for viral replicase activity. *J Virol* 1990;64:4988–4996.
188. Mansell EA, Ramig RF, Patton JT. Temperature-sensitive lesions in the capsid proteins of the rotavirus mutants tsF and tsG that affect virion assembly. *Virology* 1994;204:69–81.
189. Maratos-Flier E, Goodman MJ, Murray AH, Kahn CR. Ammonium inhibits processing and cytotoxicity of reovirus, a nonenveloped virus. *J Clin Invest* 1986;78:1003–1007.
190. Martin ML, Palmer EL, Middleton PJ. Ultrastructure of infantile gastroenteritis virus. *Virology* 1975;68:146–153.
191. Mason BB, Graham DY, Estes MK. *In vitro* transcription and translation of simian rotavirus SA11 gene products. *J Virol* 1980;33:1111–1121.
192. Mason BB, Graham DY, Estes MK. Biochemical mapping of the simian rotavirus SA11 genome. *J Virol* 1983;46:413–423.
193. Matsui SM, Mackow ER, Matsuno S, Paul PS, Greenberg HB. Sequence analysis of gene 11 equivalents from "short" and "super short" strains of rotavirus. *J Virol* 1990;64:120–124.
194. Matsuno S, Inouye S. Purification of an outer capsid glycoprotein of neonatal calf diarrhea virus and preparation of its antisera. *Infect Immun* 1983;39:155–158.
195. Mattion NM, Gonzalez SA, Burrone O, Bellinzoni R, La Torre JL, Scodeller EA. Rearrangement of genomic segment 11 in two swine rotavirus strains. *J Gen Virol* 1988;69:695–698.
196. Mattion NM, Bellinzoni RC, Blackhall JO, et al. Genome rearrangements in porcine rotaviruses: biochemical and biological comparisons between a supershort strain and its standard counterpart. *J Gen Virol* 1990;71:355–362.
197. Mattion NM, Cohen J, Aponte C, Estes MK. Characterization of an oligomerization domain and RNA-binding properties on rotavirus nonstructural protein NS34. *Virology* 1992;190:68–83.
198. Mattion NM, Cohen J, Estes MK. The Rotavirus proteins. In: Kapikian A, ed. *Viral infections of the gastrointestinal tract.* New York: Marcel Dekker, 1994:169–249.
199. Mattion NM, Mitchell DB, Both GW, Estes MK. Expression of rotavirus proteins encoded by alternative open reading frames of genome segment 11. *Virology* 1991;181:295–304.
200. McCrae MA, Faulkner-Valle GP. Molecular biology of rotaviruses. I. Characterization of basic growth parameters and pattern of macromolecular synthesis. *J Virol* 1981;39:490–496.
201. McCrae MA, McCorquodale JG. The molecular biology of rotaviruses. II. Identification of the protein-coding assignments of calf rotavirus genome RNA species. *Virology* 1982;117:435–443.
202. McCrae MA, McCorquodale JG. Molecular biology of rotaviruses. V. Terminal structure of viral RNA species. *Virology* 1983;126:204–212.
203. McIntyre M, Rosenbaum V, Rappold W, Desselberger M, Wood D, Desselberger U. Biophysical characterization of rotavirus particles containing rearranged genomes. *J Gen Virol* 1987;68:2961–2966.
204. McNulty MS, Curran WL, McFerran JB. The morphogenesis of a cytopathic bovine rotavirus in Madin-Darby bovine kidney cells. *J Gen Virol* 1976;33:503–508.
205. Mendez E, Arias CF, López S. Genomic rearrangements in human rotavirus strain Wa; analysis of rearranged RNA segment 7. *Arch Virol* 1992;125:331–338.
206. Mendez E, Arias CF, López S. Binding of sialic acids is not an essential step for the entry of animal rotaviruses to epithelial cells in culture. *J Virol* 1994;67:5253–5259.
207. Mertens PPC, Sanger DV. Analysis of the terminal sequence of the genome segments of four orbiviruses. In: Barger TL, Yochim MM, eds. *Bluetongue and related orbiviruses.* New York: Alan R Liss; 1985: 371–387.
208. Meyer JC, Bergmann CC, Bellamy AR. Interaction of rotavirus cores with the nonstructural glycoprotein NS28. *Virology* 1989;171:98–107.
209. Michelangeli F, Ruiz MC, Del Castillo JR, Ernesto Ludert J, Liprandi F. Effect of rotavirus infection on intracellular calcium homeostasis in cultured cells. *Virology* 1991;181:520–527.
210. Mitchell DB, Both GW. Simian rotavirus SA11 segment 11 contains overlapping reading frames. *Nucleic Acids Res* 1988;16:6244.
211. Mitchell DB, Both GW. Completion of the genomic sequence of the simian rotavirus SA11: nucleotide sequences of segments 1, 2, and 3. *Virology* 1990;177:324–331.
212. Munro S, Pelham HRB. A C-terminal signal prevents secretion of luminal ER proteins. *Cell* 1987;48:899–907.
213. Musalem C, Espejo RT. Release of progeny virus from cells infected with simian rotavirus SA11. *J Gen Virol* 1985;66:2715–2724.
214. Nagesha HS, Huang J, Hum CP, Holmes IH. A porcine rotavirus strain with dual VP7 serotype specificity. *Virology* 1990;175:319–322.

214a. Nakagomi O, Isegawa Y, Hoshino Y, Aboudy Y, Shif I, Silberstein I, Nakagomi T, Ueda S, Sears J, Flores J. A new serotype of the outer capsid protein VP4 shared by an unusual human rotavirus strain RO1845 and canine rotaviruses. *J Gen Virol* 1993;74:2771–2774.

215. Nakagomi O, Isegawa Y, Ueda S, et al. Nucleotide sequence comparison of the VP8* gene of rotaviruses possessing the AU-1 gene 4 allele. *J Gen Virol* 1993;74:1709–1713.
216. Nakagomi O, Nakagomi T. Genetic diversity and similarity among mammalian rotaviruses in relation to interspecies transmission of rotavirus. *Arch Virol* 1991;120:43–55.
217. Nakagomi O, Nakagomi T. Interspecies transmission of rotaviruses studied from the perspective of genogroup. *Microbiol Immunol* 1993; 37:337–348.
218. Nakagomi O, Nakagomi T, Akatani K, Ikegami N. Identification of rotavirus genogroups by RNA-RNA hybridization. *Mol Cell Probes* 1989; 3:251–261.
219. Nakata S, Estes MK, Graham DY, et al. Antigenic characterization and ELISA detection of adult diarrhea rotaviruses. *J Infect Dis* 1986;154: 448–455.
220. Nakata S, Estes MK, Graham DY, Wang SS, Gary GW, Melnick JL. Detection of antibody to group B adult diarrhea rotaviruses in humans. *J Clin Microbiol* 1987;25:812–818.
221. Nandi P, Charpilienne A, Cohen J. Interaction of rotavirus particles

with liposomes. *J Virol* 1992;66:3363–3367.
222. Nishikawa K, Taniguchi K, Torres A, et al. Comparative analysis of the VP3 gene of divergent strains of the rotaviruses simian SA11 and bovine Nebraska calf diarrhea virus. *J Virol* 1988;62:4022–4026.
223. Offit PA, Blavat G. Identification of the two rotavirus genes determining neutralization specificities. *J Virol* 1986;57:376–378.
224. Offit PA, Blavat G, Greenberg HB, Clark HF. Molecular basis of rotavirus virulence: role of gene segment 4. *J Virol* 1986;57:46–49.
225. Offit PA, Boyle DB, Both GW, et al. Outer capsid glycoprotein VP7 is recognized by crossreactive, rotavirus-specific, cytotoxic T lymphocytes. *Virology* 1991;184:563–568.
226. Offit PA, Clark HF, Blavat G, Greenberg HB. Reassortant rotaviruses containing structural proteins VP3 and VP7 from different parents induce antibodies protective against each parental serotype. *J Virol* 1986; 60:491–496.
227. Offit PA, Dudzik KI. Rotavirus-specific cytotoxic T lymphocytes cross-react with target cells infected with different rotavirus serotypes. *J Virol* 1988;62:127–131.
228. Offit PA, Hoffenberg EJ, Pia ES, Panackal PA, Hill NL. Rotavirus specific helper T cell responses in newborns, infants, children, and adults. *J Infect Dis* 1992;165:1107–1111.
229. Offit PA, Shaw RD, Greenberg HB. Passive protection against rotavirus induced diarrhea by monoclonal antibodies to surface proteins VP3 and VP7. *J Virol* 1986;58:700–703.
230. Oseto M, Yamashita Y, Hattori M, Mori M, Inouye H, Ishimaru Y, Matsuno S. Successful propagation of human group C rotavirus in a continuous cell line and characterization of the isolated virus. Abstracts of 27th Joint Viral Diseases Meeting of the US-Japan Cooperative Medical Sciences Program, 1993.
231. Palmer EL, Martin ML, Murphy FA. Morphology and stability of infantile gastroenteritis virus: comparison with reovirus and bluetongue virus. *J Gen Virol* 1977;35:403–414.
232. Patton JT. Synthesis of simian rotavirus SA11 double-stranded RNA in a cell-free system. *Virus Res* 1986;6:217–233.
233. Patton JT, Gallegos CO. Structure and protein composition of the rotavirus replicase particle. *Virology* 1988;166:358–365.
234. Patton JT, Hua J, Mansell EA. Location of intrachain disulfide bonds in the VP5* and VP8* trypsin cleavage fragments of the rhesus rotavirus spike protein VP4. *J Virol* 1993;67:4848–4855.
235. Patton JT, Stacy-Phipps S. Electrophoretic separation of the plus and minus strands of rotavirus SA11 double-stranded RNAS. *J Virol Methods* 1986;13:185–190.
236. Pedley S, Bridger JC, Brown JF, McCrae MA. Molecular characterization of rotaviruses with distinct group antigens. *J Gen Virol* 1983; 64:2093–2101.
237. Pedley S, Hundley F, Chrystie I, McCrae MA, Desselberger U. The genomes of rotaviruses isolated from chronically infected immunodeficient children. *J Gen Virol* 1984;65:1141–1150.
238. Petrie BL. Biologic activity of rotavirus particles lacking glycosylated proteins. In: Compans RW, Bishop DHL, eds. *Double-stranded RNA viruses.* New York: Elsevier; 1983:146–156.
239. Petrie BL, Estes MK, Graham DY. Effects of tunicamycin on rotavirus morphogenesis and infectivity. *J Virol* 1983;46:270–274.
240. Petrie BL, Graham DY, Estes MK. Identification of rotavirus particle types. *Intervirology* 1981;16:20–28.
241. Petrie BL, Graham DY, Hanssen H, Estes MK. Localization of rotavirus antigens in infected cells by ultrastructural immunocytochemistry. *J Gen Virol* 1982;63:457–467.
242. Petrie BL, Greenberg HB, Graham DY, Estes MK. Ultrastructural localization of rotavirus antigens using colloidal gold. *Virus Res* 1984; 1:133–152.
243. Pizarro JM, Pizarro JL, Fernández J, Sandino AM, Spencer E. Effect of nucleotide analogues on rotavirus transcription and replication. *Virology* 1991;184:768–772.
244. Pocock DH. Isolation and characterization of two group A rotaviruses with unusual genome profiles. *J Gen Virol* 1987;68:653–660.
245. Poncet D, Aponte C, Cohen J. Rotavirus protein NSP3 (NS34) is bound to the 3′ end consensus sequence of viral mRNAs in infected cells. *J Virol* 1993;67:3159–3165.
246. Poncet D, Laurent S, Cohen J. Four nucleotides are the minimal requirement for RNA recognition by rotavirus non-structural protein NSP3. *EMBO J* 1994;13:4165–4173.
247. Porter AG, Barber C, Carey NH, Hallewell RA, Threlfall G, Emtage JS. Complete nucleotide sequence of an influenza virus hemagglutinin gene from cloned DNA. *Nature* 1979;282:471–477.
248. Poruchynsky MS, Atkinson PH. Primary sequence domains required for the retention of rotavirus VP7 in the endoplasmic reticulum. *J Cell Biol* 1988;107:1697–1706.
249. Poruchynsky MS, Atkinson PH. Rotavirus protein rearrangements in purified membrane-enveloped intermediate particles. *J Virol* 1991;65: 4720–4727.
250. Poruchynsky MS, Maass DR, Atkinson PH. Calcium depletion blocks the maturation of rotavirus by altering the oligomerization of virus-encoded proteins in the ER. *J Cell Biol* 1991;114:651–661.
251. Poruchynsky MS, Tyndall C, Both GW, Sato F, Bellamy AR, Atkinson PH. Deletions into an NH2-terminal hydrophobic domain result in secretion of rotavirus VP7, a resident endoplasmic reticulum membrane glycoprotein. *J Cell Biol* 1985;101:2199–2209.
252. Potter AA, Cox G, Parker M, Babiuk LA. The complete nucleotide sequence of bovine rotavirus C486 gene 4 CDNA [erratum appears in *Nucleic Acids Res* 1987;12;15:8124]. *Nucleic Acids Res* 1987;15:4361
253. Prasad BV, Wang GJ, Clerx JP, Chiu W. Three-dimensional structure of rotavirus. *J Mol Biol* 1988;199:269–275.
254. Prasad BVV, Burns JW, Marietta E, Estes MK, Chiu W. Localization of VP4 neutralization sites in rotavirus by three-dimensional cryo-electron microscopy. *Nature* 1990;343:476–479.
255. Prasad BVV, Chiu W. Structure of rotaviruses. In: Ramig R, ed. *Rotaviruses.* Berlin: Springer-Verlag; 1994:9–29.
256. Prasad BVV, Estes MK. Molecular basis of rotavirus replication: structure-function correlations. In: Chiu W, Burnett RM, Garcea R, eds. *Structural biology of viruses.* New York: Oxford Press; 1995.
257. Prasad BVV, Yamaguchi S, Roy P. Three-dimensional structure of single shelled bluetongue virus. *J Virol* 1994;66:2135–2142.
258. Quan CM, Doane FW. Ultrastructural evidence for the cellular uptake of rotavirus by endocytosis. *Intervirology* 1983;20:223–231.
259. Ramig RF. Isolation and genetic characterization of temperature-sensitive mutants of simian rotavirus SA11. *Virology* 1982;120:93–105.
260. Ramig RF. Factors that affect genetic interaction during mixed infection with temperature-sensitive mutants of simian rotavirus SA11. *Virology* 1983;127:91–99.
261. Ramig RF. Genetic studies with simian rotavirus SA11. In: Compans RW, Bishop DHL, eds. *Double-stranded RNA viruses.* New York: Elsevier; 1983:321–327.
262. Ramig RF. Isolation and genetic characterization of temperature-sensitive mutants that define five additional recombination groups in simian rotavirus SA11. *Virology* 1983;130:464–473.
263. Ramig RF, Petrie BL. Characterization of temperature-sensitive mutants of simian rotavirus SA11: protein synthesis and morphogenesis. *J Virol* 1984;49:665–673.
264. Redmond MJ, Ijaz MK, Parker MD, Sabara MI, Dent D, Gibbons E, Babiuk LA. Assembly of recombinant rotavirus proteins into virus-like particles and assessment of vaccine potential. *Vaccine* 1993;11: 273–281.
265. Richardson SC, Mercer LE, Sonza S, Holmes IH. Intracellular localization of rotaviral proteins. *Arch Virol* 1986;88:251–264.
266. Riepenhoff-Taity M, Dharakul T, Kowalski E, Michalak S, Ogra PL. Persistent rotavirus infection in mice with severe combined immunodeficiency. *J Virol* 1987;61:3345–3348.
267. Rodger SM, Holmes IH. Comparison of the genomes of simian, bovine, and human rotaviruses by gel electrophoresis and detection of genomic variation among bovine isolates. *J Virol* 1979;30:839–846.
268. Rolsma MD, Gelberg HB, Kuhlenschmidt MS. Assay for evaluation of rotavirus-cell interactions: identification of an enterocyte ganglioside fraction that mediates group A porcine rotavirus recognition. *J Virol* 1994;68:258–268.
269. Roseto A, Escaig J, Delain G, Cohen J, Scherrer R. Structure of rotaviruses as studied by the freeze-drying technique. *Virology* 1979;98: 471–475.
270. Rothnagel R, Zeng Q, Estes MK, Chiu W, Prasad BVV. 3-D structural studies on baculovirus-expressed rotavirus sub-assemblies. Abstracts of 1994 Biophysical Society Annual Meeting, 1994.
271. Ruggeri FM, Greenberg HB. Antibodies to the trypsin cleavage peptide VP8* neutralize rotavirus by inhibiting binding of virions to target cells in culture. *J Virol* 1991;65:2211–2219.
272. Ruiz M, Alonso-Torre SR, Charpilienne A, et al. Rotavirus interaction with isolated membrane vesicles. *J Virol* 1994;68:4009–4016.
273. Sabara M, Babiuk LA. Identification of a bovine rotavirus gene and gene product influencing cellular attachment. *J Virol* 1984;51:489–496.

274. Sabara M, Gilchrist JE, Hudson GR, Babiuk LA. Preliminary characterization of an epitope involved in neutralization and cell attachment that is located on the major bovine rotavirus glycoprotein. *J Virol* 1985; 53:58–66.
275. Sabara M, Ready KF, Frenchick PJ, Babiuk LA. Biochemical evidence for the oligomeric arrangement of bovine rotavirus nucleocapsid protein and its possible significance in the immunogenicity of this protein. *J Gen Virol* 1987;68:123–133.
276. Saif LJ, Jiang B. Nongroup A rotaviruses of humans and animals. In: Ramig RF, ed. *Current topics in microbiology and immunology.* Berlin: Springer-Verlag; 1994:339–371.
277. Saif LJ, Terrett LA, Miller KL, Cross RF. Serial propagation of porcine group C rotavirus (pararotavirus) in a continuous cell line and characterization of the passaged virus. *J Clin Microbiol* 1988;26:1277–1282.
278. Sandino AM, Jashes M, Faundez G, Spencer E. Role of the inner protein capsid on *in vitro* human rotavirus transcription. *J Virol* 1986;60: 797–802.
279. Sato T, Suzuki H, Kitaoka S, Konno T, Ishida N. Patterns of polypeptide synthesis in human rotavirus infected cells. *Arch Virol* 1986;90: 29–40.
280. Sattar SA, Raphael RA, Lochnan H, Springthorpe VS. Rotavirus inactivation by chemical disinfectants and antiseptics used in hospitals. *Can J Microbiol* 1983;29:1464–1469.
281. Scott GE, Tarlow O, McCrae MA. Detailed structural analysis of a genome rearrangement in bovine rotavirus. *Virus Res* 1989;14:119–127.
282. Sereno MM, Gorziglia MI. The outer capsid protein VP4 of murine rotavirus strain Eb represents a tentative new P type. *Virology* 1994;199: 500–504.
283. Shahrabadi MS, Babiuk LA, Lee PW. Further analysis of the role of calcium in rotavirus morphogenesis. *Virology* 1987;158:103–111.
284. Shahrabadi MS, Lee PW. Bovine rotavirus maturation is a calcium-dependent process. *Virology* 1986;152:298–307.
285. Shaw AL, Rothnagel R, Chen D, Ramig RF, Chiu W, Venkataram Prasad BV. Three-dimensional visualization of the rotavirus hemagglutinin structure. *Cell* 1993;74:693–701.
286. Shaw RD, Fong KJ, Losonsky GA, et al. Epitope-specific immune responses to rotavirus vaccination. *Gastroenterology* 1987;93:941–950.
286a. Shen S, Burke B, Desselberger U. Nucleotide sequences of the VP4 and VP7 genes of a Chinese lamb rotavirus: evidence for a new P type in a G10 type virus. *Virology* 1993;197:497–500.
287. Shen S, Burke B, Desselberger U. Rearrangement of the VP6 gene of a group A rotavirus in combination with a point mutation affecting trimer stability. *J Virol* 1994;68:1682–1688.
288. Shirley JA, Beards GM, Thouless ME, Flewett TH. The influence of divalent cations on the stability of human rotavirus. *Arch Virol* 1981; 67:1–9.
289. Smee DF, Sidwell RW, Clark SM, Barnett BB, Spendlove RS. Inhibition of rotaviruses by selected antiviral substances: mechanisms of viral inhibition and *in vivo* activity. *Antimicrob Agents Chemother* 1982;21: 66–73.
290. Smith EM, Estes MK, Graham DY, Gerba CP. A plaque assay for the simian rotavirus SA11. *J Gen Virol* 1979;43:513–519.
291. Smith ML, Lazdins I, Holmes IH. Coding assignments of double-stranded RNA segments of SA11 rotavirus established by *in vitro* translation. *J Virol* 1980;33:976–981.
292. Soler C, Musalem C, Loroño M, Espejo RT. Association of viral particles and viral proteins with membranes in SA11-infected cells. *J Virol* 1982;44:983–992.
293. Spencer E, Arias ML. *In vitro* transcription catalyzed by heat-treated human rotavirus. *J Virol* 1981;40:1–10.
294. Srnka CA, Tiemeyer M, Gilbert JH, et al. Cell surface ligands for rotavirus: mouse intestinal glycolipids and synthetic carbohydrate analogs. *Virology* 1992;190:794–805.
295. Stacy-Phipps S, Patton JT. Synthesis of plus- and minus-strand RNA in rotavirus-infected cells. *J Virol* 1987;61:3479–3484.
296. Stirzaker SC, Whitteld PL, Christie DL, Bellamy AR, Both GW. Processing of rotavirus glycoprotein VP7: implications for the retention of the protein in the endoplasmic reticulum. *J Cell Biol* 1987;105:2897–2903.
297. Su CQ, Wu YL, Shen HK, Wang DB, Chen YH, Wu DM, He LN, Yang ZL. An outbreak of epidemic diarrhoea in adults caused by a new rotavirus in Anhui Province of China in the summer of 1983. *J Med Virol* 1986;19:167–173.
298. Suzuki H, Kitaoka S, Konno T, Sato T, Ishida N. Two modes of human rotavirus entry into MA 104 cells. *Arch Virol* 1985;85:25–34.
299. Suzuki H, Kitaoka S, Sato T, et al. Further investigation on the mode of entry of human rotavirus into cells. *Arch Virol* 1986;91:135–144.
300. Suzuki H, Konno T, Numazaki Y. Electron microscopic evidence for budding process-independent assembly of double-shelled rotavirus particles during passage through endoplasmic reticulum membranes. *J Gen Virol* 1993;74:2015–2018.
301. Suzuki H, Sato T, Konno T, Kitaoka S, Ebina T, Ishida N. Effect of tunicamycin on human rotavirus morphogenesis and infectivity. Brief report. *Arch Virol* 1984;81:363–369.
302. Svensson L, Dormitzer PR, von Bonsdorff C, Maunula L, Greenberg HB. Intracellular manipulation of disulfide bond formation in rotavirus proteins during assembly. *J Virol* 1994;68:5204–5215.
303. Svensson L, Finlay BB, Bass D, Von Bonsdorff C-H, Greenberg HB. Symmetric infection of rotavirus on polarized human intestinal epithelial (CaCO-2) cells. *J Virol* 1991;65:4190–4197.
304. Svensson L, Sheshberadaran H, Vene S, Norrby E, Grandien M, Wadell G. Serum antibody responses to individual viral polypeptides in human rotavirus infections. *J Gen Virol* 1987;68:643–651.
305. Tam JS, Szymanski MT, Middleton PJ, Petric M. Studies on the particles of infantile gastroenteritis virus (orbivirus group). *Intervirology* 1976;7:181–191.
306. Tan JA, Schnagi RD. Inactivation of a rotavirus by disinfectants. *Med J Aust* 1981;1:19–23.
307. Tanaka TN, Conner ME, Graham DY, Estes MK. Molecular characterization of three rabbit rotavirus strains. *Arch Virol* 1988;98:253–265.
308. Taniguchi K, Hoshino Y, Nishikawa K, et al. Cross-reactive and serotype-specific neutralization epitopes on VP7 of human rotavirus: nucleotide sequence analysis of antigenic mutants selected with monoclonal antibodies. *J Virol* 1988;62:1870–1874.
309. Taniguchi K, Maloy WL, Nishikawa K, et al. Identification of cross-reactive and serotype 2-specific neutralization epitopes on VP3 of human rotavirus. *J Virol* 1988;62:2421–2426.
309a. Taniguchi K, Urasawa S. Diversity in rotavirus genomes. *Virology* 1995;6:123–131.
310. Taniguchi K, Urasawa S, Urasawa T. Preparation and characterization of neutralizing monoclonal antibodies with different reactivity patterns to human rotaviruses. *J Gen Virol* 1985;66:1045–1053.
310a. Taniguchi K, Urasawa T, Urasawa S. Species specificity and interspecies relatedness in VP4 genotypes demonstrated by VP4 sequence analysis of equine, feline, and canine rotavirus strains. *Virology* 1994;200:390–400.
311. Taniguchi K, Urasawa T, Morita Y, Greenberg HB, Urasawa S. Direct serotyping of human rotavirus in stools by an enzyme-linked immunosorbent assay using serotype 1-, 2-, 3-, and 4-specific monoclonal antibodies to VP7. *J Infect Dis* 1987;155:1159–1166.
312. Taylor JA, Meyer JC, Legge MA, et al. Transient expression and mutational analysis of the rotavirus intracellular receptor: the C-terminal methionine residue is essential for ligand binding. *J Virol* 1992;66: 3566–3572.
313. Taylor JA, O'Brien JA, Lord VJ, Meyer JC, Bellamy AR. The RER-localized rotavirus intracellular receptor: a truncated purified soluble form is multivalent and binds virus particles. *Virology* 1993;194: 807–814.
314. Thouless ME, DiGiacomo RF, Neuman DS. Isolation of two lapine rotaviruses: characterization of their subgroup, serotype and RNA electropherotypes. *Arch Virol* 1986;89:161–170.
315. Tian P, Hu Y, Schilling WP, Lindsay DA, Eiden J, Estes MK. The nonstructural glycoprotein of rotavirus affects intracellular calcium levels. *J Virol* 1994;68:251–257.
316. Tian Y, Tarlow O, Ballard A, Desselberger U, McCrae MA. Genomic concatemerization/deletion in rotaviruses: a new mechanism for generating rapid genetic change of potential epidemiological importance. *J Virol* 1994;67:6625–6632.
317. Tosser G, Delaunay T, Kohli E, Grosclaude J, Pothier P, Cohen J. Topology of bovine rotavirus (RF strain) VP6 epitopes by real-time biospecific interaction analysis. *Virology* 1994;204:8–16.
318. Tosser G, Labbé M, Brémont M, Cohen J. Expression of the major capsid protein VP6 of group C rotavirus and synthesis of chimeric single-shelled particles by using recombinant baculoviruses. *J Virol* 1992;66: 5825–5831.
319. Tursi JM, Albert MJ, Bishop RF. Production and characterization of neutralizing monoclonal antibody to a human rotavirus strain with a "super-short" RNA pattern. *J Clin Microbiol* 1987;25:2426–2427.

319a. Urasawa T, Taniguchi K, Kobayashi N, Mise K, Hasegawa A, Yamazi Y, Urasawa S. Nucleotide sequence of VP4 and VP7 genes of a unique human rotavirus strain Mc35 with subgroup 1 and serotype 10 specificity. *Virology* 1993;195:766–771.
320. Urasawa S, Urasawa T, Taniguchi K. Genetic reassortment between two human rotaviruses having different serotype and subgroup specificities. *J Gen Virol* 1986;67:1551–1559.
321. Valenzuela S, Pizarro J, Sandino AM, et al. Photoaffinity labeling of rotavirus VP1 with 8-azido-ATP: identification of the viral RNA polymerase. *J Virol* 1991;65:3964–3967.
322. Vaughn JM, Chen YS, Thomas MZ. Inactivation of human and simian rotaviruses by chlorine. *Appl Environ Microbiol* 1986;51:391–394.
323. Vonderfecht SL, Miskuff RL, Wee SB, et al. Protease inhibitors suppress the *in vitro* and *in vivo* replication of rotavirus. *J Clin Invest* 1988;82:2011–2016.
324. W.H.O. Steering Committee of the Scientific Working Group on Viral Diarrheas. Nomenclature of human rotaviruses: designation of subgroups and serotypes. *Bull WH0* 1984;62:501–503.
325. Walsh DS, Kappes JC, Duell GA, Tsuchiya Y, Nutini LG. *In vitro* suppression of rotavirus replication using a purified fraction of bovine lecithin. *IRCS Med Sci* 1985;13:595–596.
326. Ward RL, Ashley CS. Comparative study on the mechanisms of rotavirus inactivation by sodium dodecyl sulfate and ethylenediaminetetraacetate. *Appl Environ Microbiol* 1980;39:1148–1153.
327. Weclewicz K, Kristensson K, Greenberg HB, Svensson L. The endoplasmic reticulum-associated VP7 of rotavirus is targeted to axons and dendrites in polarized neurons. *J Neurocytol* 1993;22:616–626.
328. Weciewicz K, Svensson L, Billger M, Holmberg K, Wallin M, Kristensson K. Microtubule-associated protein 2 appears in axons of cultured dorsal root ganglia and spinal cord neurons after rotavirus infection. *J Neurosci Res* 1994;36:173–182.
329. Weiss C, Clark HF. Rapid inactivation of rotaviruses by exposure to acid buffer or acidic gastric juice. *J Gen Virol* 1985;66:2725–2730.
330. Welch AB, Thompson TL. Physicochemical characterization of a neonatal calf diarrhea virus. *Can J Compr Med* 1973;37:295–301.
331. Welch SK, Crawford SE, Estes MK. Rotavirus SA11 genome segment 11 protein is a nonstructural phosphoprotein. *J Virol* 1989;63:3974–3982.
332. Welter MW, Welter CJ, Chambers DM, Svensson L. Adaptation and serial passage of porcine group C rotavirus in ST-cells, an established diploid swine testicular cell line. *Arch Virol* 1991;120:297–304.
333. Whitteld PL, Tyndall C, Stirzaker SC, Bellamy AR, Both GW. Location of sequences within rotavirus SA11 glycoprotein VP7 which direct it to the endoplasmic reticulum. *Mol Cell Biol* 1987;7: 2491–2497.
334. Willoughby RE, Yolken RH, Schnaar RL. Rotaviruses specifically bind to the neutral glycosphingolipid asialo-GM1. *J Virol* 1990;64:4830–4835.
335. Wood DJ, David TJ, Chrystie IL, Totterdell B. Chronic enteric virus infection in two T-cell immunodeficient children. *J Med Virol* 1988; 24:435–444.
336. Xu Z, Woode GN. Studies on the role of VP4 of G serotype 10 rotavirus (B223) in the induction of the heterologous immune response in calves. *Virology* 1993;196:294–297.
337. Yeager M, Berriman JA, Baker TS, Bellamy AR. Three-dimensional structure of the rotavirus haemagglutinin VP4 by cryo-electron microscopy and difference map analysis. *EMBO J* 1994;13:1011–1018.
338. Yolken R, Arango-Jaramillo S, Eiden J, Vonderfecht S. Lack of genomic reassortment following infection of infant rats with group A and group B rotaviruses. *J Infect Dis* 1988;158:1120–1123.
339. Yolken R, Wee SB, Eiden J, Kinney J, Vonderfecht S. Identification of a group-reactive epitope of group B rotaviruses recognized by monoclonal antibody and application to the development of a sensitive immunoassay for viral characterization. *J Clin Microbiol* 1988;26:1853–1858.
340. Yolken RH, Willoughby R, Wee SB, Miskuff R, Vonderfecht S. Sialic acid glycoproteins inhibit *in vitro* and *in vivo* replication of rotaviruses. *J Clin Invest* 1987;79:148–154.
341. Zeng Q, Labbé M, Cohen J, Prasad BVV, Chen D, Ramig RF, Estes MK. Characterization of rotavirus VP2 particles. *Virology* 1994;201:55–65.
342. Zhou Y, Burns JW, Morita Y, Tanaka T, Estes MK. Localization of rotavirus VP4 neutralization epitopes involved in antibody-induced conformational changes of virus structure. *J Virol* 1994;68:3955–3964.

Fundamental Virology, Third Edition
edited by B.N. Fields, D.M. Knipe, P.M. Howley, et al.
Lippincott - Raven Publishers, Philadelphia © 1996

CHAPTER 26

Retroviridae: The Viruses and Their Replication

John M. Coffin

INTRODUCTION AND GROUP DEFINITION

No group of infectious agents has received as much attention from scientists in recent years as retroviruses. The intense scrutiny given these viruses reflects not only their importance as human and animal pathogens, but also their remarkable value as experimental objects which in turn is a consequence of their intimate association with the host. The unique replication cycle of these viruses leads—directly or indirectly—to a variety of special biological features which include:

- The existence, within the constraints of a common virion structure, genetic organization, and replication cycle, of a large number of virus strains with very different lifestyles and pathogenic effects.
- A wide variety of interactions between virus and host, ranging from the completely benign infections displayed by endogenous viruses through moderate exogenously acquired infections to the generally fatal consequences of viruses such as HIV and the rapidly oncogenic viruses.
- The ability to acquire and alter the structure and function of host derived sequences and, as a consequence, present them to the researcher as oncogenes—the study of which has provided our most fundamental insights into molecular mechanisms of carcinogenesis.
- The ability to insert themselves into the germline of a host as endogenous proviruses and behave like transposable elements, with genetic consequences that have been important in vertebrate evolution.
- The ability to cause certain types of genetic damage—such as activation or inactivation of specific genes near the site of integration of the provirus.
- The ability to rapidly alter their genomes by mutation and recombination and thus to change in response to altered environmental conditions.

J. M. Coffin: Department of Molecular Biology and Microbiology, Tufts University School of Medicine, Boston, Massachusetts 02111.

- The ability to serve as vectors for foreign genes inserted in the laboratory and to carry and express these genes in a predictable way into a wide variety of cells and organisms.

This chapter will describe the general features of retrovirus biology, as illuminated by study of all virus groups used more or less interchangeably. Special features of the human retroviruses: the human T-cell lymphoma viruses (HTLVs) and human immunodeficiency virus (HIV), as well as the animal lentiviruses are covered in detail in other chapters and will be discussed only in passing here. Similarly, retroviral oncogenesis is thoroughly discussed in Chapter 9 and will be only touched on here.

Even with these limitations, it is impossible to cover every well-studied aspect of retrovirology in any significant depth. The two-volume *RNA Tumor Viruses* (789,790), comprising about 2,500 pages, has been a standard for over 10 years and is still worth consulting on some topics. A few more recent works, including the three volumes of *The Retroviridae* (439) and some others (252,716) are more current. A new edition of *RNA Tumor Viruses* (to be called *Retroviruses*) is presently in preparation and should be available soon. There are also numerous recent reviews on specialized aspects of the subject, many of which will be cited in the appropriate sections of this chapter. Due to the very large size of the retroviral literature, this chapter will make no attempt to be comprehensive in citations. In particular, references to papers prior to the appearance of the second volume of *RNA Tumor Viruses* in 1985 will be minimal, and the reader is encouraged to consult "the bible."

The *Retroviridae* comprise a large family of viruses, primarily of vertebrates, although they are also found in other animals (such as insects and mollusks). Both in the wild and in the laboratory, they are associated with many diseases, including rapid as well as long-latency malignancies, wasting diseases, neurological disorders, and immunodeficiencies. Retrovirus infection can also lead to lifelong viremia in the absence of any obvious ill effects. Despite the variety of host species and of interactions with the host, all retrovirus isolates are quite similar in virion structure, genome organization, and mode of replication.

The virion is enveloped, about 100 nm in diameter, and its surface is decorated by a single protein structure consisting of a multimer of two protein subunits, products of the *env* gene. The internal nucleocapsid, or core, is an ill-defined, roughly spherical to conical structure made up of the three or four products of the *gag* gene. Also included in the core are several proteins which have important catalytic roles during replication. These include a protease encoded in the *pro* gene and two products of the *pol* gene: the reverse transcriptase whose several enzymatic activities cooperate to convert the genetic information from single-stranded RNA to double-stranded DNA; and integrase, necessary for covalently joining virus to cell DNA to form the provirus.

The genome consists of two, usually identical, molecules of single-stranded RNA, ranging from about 7 to 10 kb in length, modified in ways reminiscent of cell mRNAs, including capping at the 5′ end and polyadenylation at the 3′ end. The order of the genes encoding structural proteins is invariably *gag-pro-pol-env.*

A number of other genes involved in regulation of virus expression are present in some virus groups. Viruses that contain such genes are called *complex*; viruses lacking them are referred to as *simple* retroviruses.

The replication cycle can be thought of as proceeding in two phases. The first phase includes entry of the virion core into the cytoplasm, synthesis of double-stranded DNA using the single-stranded genome as template, transfer of the core structure to the nucleus, and integration of the DNA into the host genome. These steps are mediated by proteins found within the virion and proceed in the absence of viral gene expression. The second phase includes synthesis and processing of viral genomes, mRNAs, and proteins using host cell systems including RNA polymerase, sometimes aided by the presence of specific viral gene products. Virion assembly proceeds by encapsidation of the genome by unprocessed precursors of the *gag*, *pro*, and *pol* genes, association of the nucleocapsids with the cell membrane, release of the virion by budding, and finally processing of the precursors to the finished products.

CLASSIFICATION

Retroviruses were traditionally divided into three subfamilies, based primarily on pathogenicity rather than on genome relationships (78,728) (see Chapter 2). When nucleotide sequence relationship and genome structure are used as criteria, seven groups of viruses comprising all well-analyzed isolates of birds and mammals can be recognized (Fig. 1). The International Committee on the Taxonomy of Viruses (ICTV) has recently redefined the taxonomy of the *Retroviridae* to recognize these groups as distinct genera (133) (Chapter 2). Recently defined viruses of fish (588) and insects (388) have not yet been included. Also not included are retrovirus-like DNA sequences found in the genomes of most eukaryotes with which no infectious virus has been associated. The older division into subfamilies is no longer used. The genera and some well-known strains are listed in Table 1 along with the commonly used abbreviations. The names assigned to genera are based on routine usage and should be considered provisional. Viruses can be further classified according to virion structure (types A to D and others; see later); utilization of particular cell receptors; lifestyle, whether endogenous (i.e., passed from parent to offspring as a provirus integrated into the germline) or exogenous; presence or absence of an oncogene; and other pathogenic properties. The genera include the following.

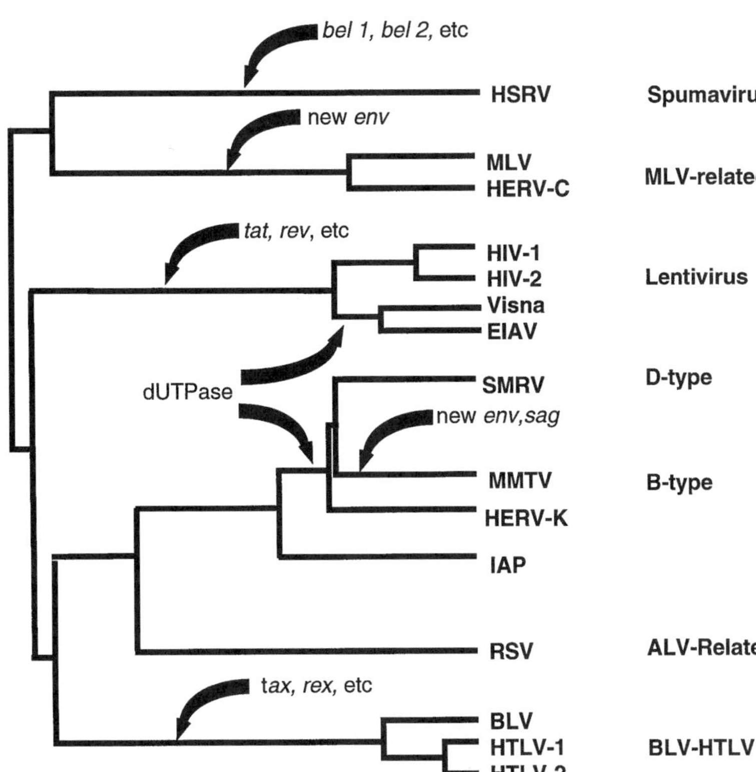

FIG. 1. Relationships of retrovirus groups. The relationships are based on amino acid sequence similarities in the RT protein of the groups shown (173,483). The insertion of sequence from another source is indicated (*arrows*). Note that the scale is approximate and not necessarily linear. HERV-C, an ancient endogenous provirus of humans (611); SMRV, squirrel monkey retrovirus; other abbreviations as in Table 1.

The Avian Leukosis-Sarcoma Virus (ALSV) Group

This genus comprises both exogenous and closely related endogenous viruses of birds (570). Viruses of this group have C-type virions and genomes which encode only virion structural genes (*gag, pol, env*), and are thus considered simple. Many isolates of exogenous virus are further modified by the presence of oncogenes, such as *src* in Rous sarcoma virus. They are further divided according to host range (i.e., receptor utilization) into ten subgroups, denoted A to J. The first four subgroups are characteristic of exogenous viruses from chickens; subgroup E belongs to endogenous viruses of chickens; and subgroups F and G are from endogenous viruses of pheasants.

The Mammalian C-Type Virus Group

These viruses include a large number of endogenous and exogenous viruses and are represented by isolates from many groups of mammals, including rodents (408), carnivores (291), and primates, as well as some exogenous viruses of birds (i.e., the reticuloendotheliosis viruses) and reptiles. The human genome is sprinkled with closely related defective endogenous proviruses, but no replicating human viruses of this group have been isolated. Like the avian viruses, which they superficially resemble, nondefective mammalian C-type viruses are simple retroviruses, and many oncogene-containing isolates are known. Viruses of mice and cats can be further classified by host range. The murine viruses are (unfortunately) designated by the species distribution of their receptors; *ecotropic* viruses replicate only in mouse cells; *xenotropic* viruses use receptors found on cells of most species except mice; and *polytropic* and *amphotropic* viruses use different receptors found in both mouse and nonmurine species. Exogenous murine viruses are ecotropic or amphotropic; endogenous viruses are found with eco, poly, or xenotropic host ranges (408). Feline leukemia viruses are classified by receptor use into subgroups A to C, with endogenous viruses belonging to subgroup B (291).

The B-Type Virus Group

This group includes as infectious agents only the mouse mammary tumor virus (MMTV), isolated as both endogenous and exogenous (but vertically transmitted) viruses. Related, but defective, sequences can be found in the genomes of many species, including rats, monkeys, and humans. These viruses have simple genomes, containing *gag, pol*, and *env* genes as well as an additional coding region, *sag*, which encodes a superantigen activity. No oncogene-containing variants have been described.

The D-Type Virus Group

This group of simple retroviruses includes both endogenous viruses and exogenous isolates from primates,

TABLE 1. *Retrovirus Genera*

Genus	Example Isolates	Comments
avian-leukosis-sarcoma	Rous sarcoma virus (RSV)	exogenous, oncogene-containing (*src*)
	avian myeloblastosis virus (AMV)	exogenous, oncogene-containing (*myb*)
	avian erythroblastosis virus (AEV)	exogenous, oncogene-containing (*erb-A* and *-B*)
	avian myelocytomatosis virus (MC) 29	exogenous, oncogene-containing (*myc*)
	Rous-associated virus (RAV) 1 to 50	exogenous, cause B-lymphoma, osteopetrosis, and other diseases
	RAV-0	endogenous, benign
mammalian C-type	Moloney murine leukemia virus (Mo-MLV)	exogenous, causes T-cell lymphoma
	Harvey murine sarcoma virus (Ha-MSV)	exogenous, oncogene-containing (H-*ras*)
	Abelson murine leukemia virus (A-MuLV)	exogenous, oncogene-containing (*abl*)
	AKR-MuLV	endogenous, benign
	Feline leukemia virus (FeLV)	exogenous, causes T-cell lymphoma, immunodeficiency, and other disease
	Simian sarcoma virus (SSV)s	exogenous, oncogene-containing (*sis*)
	numerous endogenous and exogenous viruses mostly in mammals	
	reticuloendotheliosis virus (REV)	exogenous viruses of birds
	Spleen necrosis virus (SNV)	
B-Type viruses	Mouse mammary tumor virus (MMTV)	endogenous and exogenous, mostly milk-borne, cause mostly mammary carcinoma, some T-lymphoma
D-Type viruses	Mason-Pfizer monkey virus (MPMV)	exogenous, unknown pathogenicity
	"SAIDS" viruses	immunodeficiencies in monkeys
HTLV-BLV group	Human T-cell leukemia (or lymphotropic) virus (HTLV)-1 and -2	causes T-cell lymphoma, neurological disorders
	Bovine leukemia virus (BLV)	causes B-cell lymphoma
Lentivirus	Human immunodeficiency virus (HIV)-1 and -2	cause of AIDS
	Simian immunodeficiency virus (SIV)	causes AIDS-like disease in certain monkeys; usually benign in natural host
	Feline immunodeficiency virus (FIV)	not closely related to HIV, SIV
	Bovine Immunodeficiency virus (BIV)	not closely related to HIV, SIV
	Visna/maedi virus	causes neurological and lung disease in sheep
	Caprine arthritis-encephalitis virus (CAEV)	
	Equine infectious anemia virus (EIAV)	
Spumavirus	Simian foamy virus (SFV)	exogenous, apparently benign
	Human foamy virus (HFV) or	exogenous, apparently benign
	human spumareetrovirus (HSRV)	human origin uncertain
	Feline syncytium-forming virus (FeSV)	exogenous, apparently benign

such as the Mason-Pfizer virus, which was isolated from a mammary carcinoma of a rhesus monkey but is of uncertain pathogenicity. More recent isolates include a virus associated with an immunodeficiency syndrome (SAIDS) in some captive monkey colonies, and a virus associated with ovine pulmonary adenocarcinoma (also called *Jaagsiekte*) (817). No oncogene-containing isolates have been described. The B and the D type viruses are relatively closely related to one another and have a common morphogenetic pathway (formation of intracytoplasmic A particles, see below) shared only with spumaviruses.

The HTLV-BLV Group

This group includes complex exogenous viruses associated with B-cell lymphoma in cattle and T-cell lymphoma as well as some neurological diseases in humans (see Chapter 59 in *Fields Virology*, 3rd ed.). No endogenous relatives or oncogene-containing viruses of this group are known. In addition to genes encoding virion proteins, these viruses contain at least two genes (*tax* and *rex*) encoding nonvirion proteins important for gene expression.

Lentiviruses

This genus includes complex exogenous viruses responsible for a variety of neurological and immunological diseases, but not directly implicated in any malignancies. The prototype members of this family were the "slow" viruses visna, equine infectious anemia virus, and caprine arthritis-encephalitis virus. More recent isolates include the related human and simian immunodeficiency viruses (HIV and SIV) and the more distantly related feline and bovine immunodeficiency viruses (FIV and BIV). Genomes of these viruses are characterized by a complex combination of genes in addition to *gag*, *pol*, and *env*.

Spumaviruses

The foamy viruses are the least well characterized of the retroviruses. They have been isolated as agents which cause vacuolation ("foaming") of cells in culture from a number of mammalian species, including monkeys, cattle, cats, and humans (226,458), although the true provenance of the "human" foamy virus is still debated. Persistent infection with these viruses is not associated with any known disease. Sequence and expression analysis of the simian and human foamy viruses reveals that they are complex viruses encoding transactivating and other proteins.

Evolutionary and Taxonomic Relationships

Analysis of the nucleotide sequence of a large number of retrovirus isolates has yielded phylogenetic trees similar to the one shown in Fig. 1 (173). Although it is impossible to affix a uniform time scale to such a tree, and attempts to do so (675) should not be taken seriously, there is no reason to believe it does not accurately reflect the relationships among the groups shown. These relationships provide a basis for the current taxonomy: between genera the relationship is barely visible, even in the most highly conserved regions of the genome; within a genus the relationship of genome structure and sequence is quite obvious. The two most similar genera—the B and D type viruses—do not differ in sequence from one another by much more than species within some of the other genera. Their separate placing is due to the acquisition of a distinct *env* gene in the B-type viruses as well as the presence in these viruses of *sag*—a gene encoding superantigen activity and not found in other genera.

Also shown on the phylogeny are the likely points at which major genome changes are inferred to have occurred. Such events include the appearance of transactivator and regulatory genes as well as major recombination events which led to acquisition or exchange of genes, including *sag* and a sequence encoding a dUTPase activity. Note in particular that there is no apparent clustering of complex or simple viruses: the complex lifestyle seems to have arisen more than once. This conclusion is consistent with the very different way in which this lifestyle is implemented among the three genera of complex viruses.

VIRION STRUCTURE

Types of Particles

Retroviruses are united by a common virion structure, shown in a highly schematic cartoon form in Fig. 2. Some of the earliest studies using the electron microscope to probe the virion and its synthesis revealed some differences in detail, which could be used to distinguish among the various groups and led to the classification into four morphological groups (types A to D), a classification still used to some extent today, although it is not current with the more recently described viruses.

A particles were first viewed as the immature intracellular forms of MMTV, and the term is still reserved for strictly intracellular structures, comprising fully formed immature cores of B-type viruses, D-type viruses, and spumaviruses. Virtually indistinguishable structures are

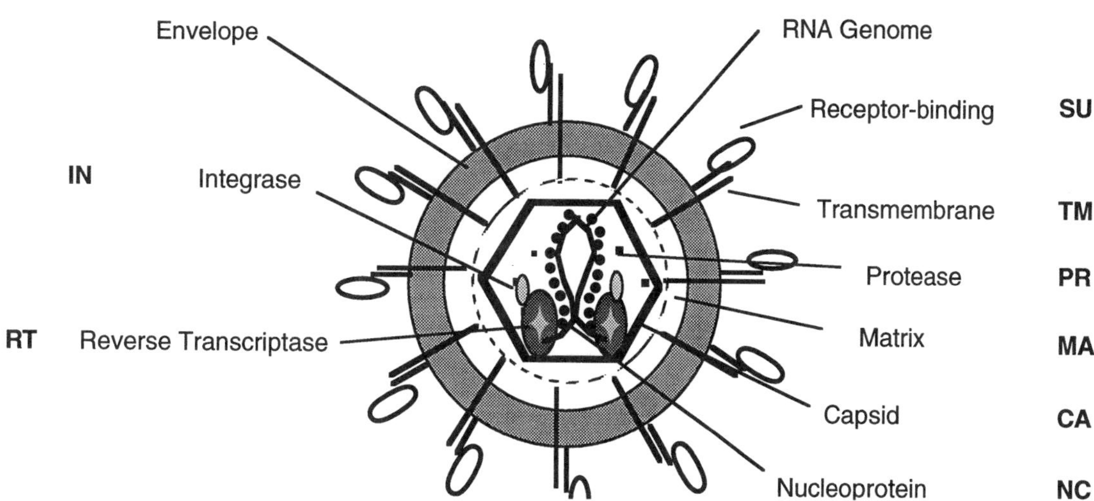

FIG. 2. The retrovirus virion. This highly schematic figure shows the relative locations of the various structures and proteins. It is not intended to be an accurate depiction of their organization.

also found in uninfected cell lines derived from a number of rodent species (mouse myelomas are a famous example). These are the product of a fairly high copy number endogenous provirus like element known as IAP (intracisternal A particle) (413,698). As the name implies, this form is usually seen within intracellular membranes, although apparently similar cytoplasmic forms can also be seen. Similar structures are sometimes seen in human tumors and cell lines and are the products of one or more defective endogenous proviruses (799). A particles are hollow, roughly spherical structures, about 60 to 90 nm in diameter with a double-walled appearance.

B particles are the enveloped, extracellular form of MMTV. Freshly budded forms resemble enveloped A particles, but they rapidly mature to a form with a tightly condensed, acentric nucleocapsid. MMTV virions are also characterized by prominent surface spikes, comprising the env protein.

C particles characterize most of the simple avian and mammalian viruses studied to date, despite the lack of close sequence relationship between the various groups. Typically, complete intracellular forms are not observed; rather, the first visible forms are crescent-shaped patches at the site of budding from the cell membrane. Budding and assembly seem to occur simultaneously; nearly complete immature forms are within nearly completed envelopes. As with B particles, immature, freshly budded forms display hollow nucleocapsids; with maturation comes condensation of the core into a central, electron-dense structure. This morphological change is a consequence of cleavage of the gag precursor protein (see below). C-type virions generally have barely visible surface projections, significantly less prominent than those of B-type viruses.

D particles resemble B-type viruses, having a complete intracellular nucleocapsid and in the eccentric location of the core in mature particles. They differ in that they have less prominent surface projections.

Other virion types not assigned with a letter are seen in the other virus groups. BLV and the HTLVs resemble C-type viruses in their mode of budding and the central nucleocapsid, but differ in the appearance of the envelope proteins. Lentiviruses, including HIV, also bud like C-type viruses, but the mature virion has a distinctive bar-shaped (or truncated cone) nucleocapsid. Spumaviruses display intracellular A-particle-like cores and virions with prominent surface projections.

Genome Organization

The retrovirus genome is unique among viruses in several aspects, including its physical organization, its mode of synthesis, and its functions in replication. It is the only diploid virus genome; the only one synthesized and processed by the cell mRNA handling machinery; the only one to be associated with a specific RNA whose function is to prime replication; and the only plus-stranded RNA genome that does not serve as mRNA early after infection.

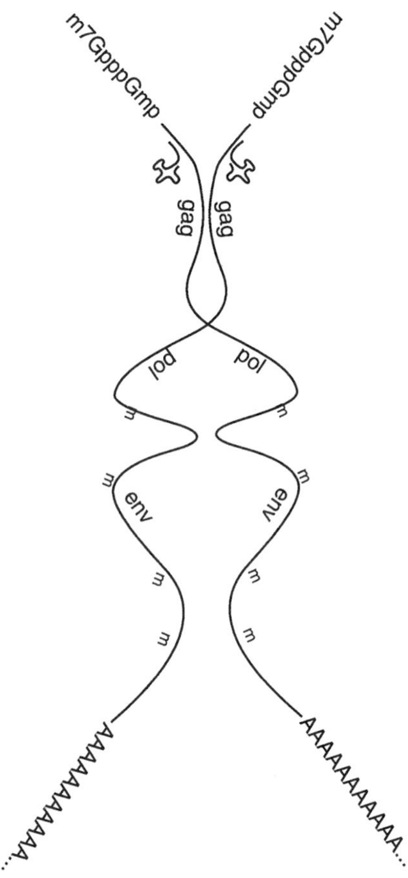

FIG. 3. Features of the retrovirus genome. The dimeric genome includes the following, from 5′ to 3′; the m[7]GPPP capping group; the primer tRNA, the coding regions, M_6A residues (m), and the 3′ poly(A) sequence.

Physical Structure

Figure 3 illustrates physical features common to all retrovirus genomes. These featyres include both modifications to the RNA imposed by the cellular machinery responsible for its synthesis and processing and noncovalent arrangements imposed on the genome during virion assembly:

A Capping Group

In all retrovirus genomes analyzed to date, this modification, applied to the 5′ end during synthesis by the cellular transcription machinery, takes the form of m7G5′ppp5′G_mp. This structure is presumably important for the translation of those molecules which serve as messages. Another role, if any, in virus replication is unknown.

Internal Methylation

Retroviral genomes, like cell mRNAs, are also modified posttranscriptionally by methylation on the 6 position of occasional A residues. Careful analysis of the sites of methylation on the RSV genome reveals that the methylation is somewhat sequence-specific (364), with the favored sequence of the form . . . RGm^{6}ACU . . . Not all such sites are modified, however. The suggestion that the 6-methylation might be important for the regulation of splicing (697) remains to be tested.

Polyadenylation

All retrovirus genomes contain a string of about 200 A residues at their 3′ end, a modification typical of most eukaryotic mRNAs. Again, the poly(A) sequence is added as a posttranscriptional modification to newly made transcripts by the cell's mRNA processing machinery. Most genomes contain a canonical signal of the form AAUAAA within about 20 bases of the site of poly(A) addition. A notable exception to this is the genome of viruses in the HTLV-BLV group, in which the nearest such signal is more than 250 bases upstream of the poly(A) site. In this case, the high degree of secondary structure of the intervening region brings the poly(A) signal very near to the point of cleavage (4).

As with normal cell mRNA, The AAUAAA sequence, perhaps in combination with less well characterized signals, specifies cleavage of a longer precursor RNA and, in a coupled reaction, subsequent poly(A) addition. It has not been directly determined whether poly(A) addition is a necessary feature of retrovirus genomes or simply reflects of their mode of synthesis, although indirect evidence implies the former (180). The *site* of polyadenylation seems relatively unimportant. Mutations which partly or completely inactivate the ALSV polyadenylation signal have only a small effect on genome packaging and replication, even though most or all of the genomes are elongated into cell sequences at their 3′ ends and polyadenylated (712).

Dimer Structure

If isolated from virions under nondenaturing conditions, retrovirus genomes have physical properties—sedimentation and electrophoretic mobility—consistent with a molecule about twice the size of the subunit size determined for denatured molecules. Careful determination of the sequence complexity of the genome revealed that the two molecules must have identical sequence, i.e., the genome must be diploid.

The structure of the complex is very poorly understood. The two genomes must be associated at numerous points along their length; however, titration of the complexes with denaturing agent preparatory to examination in the electron microscope reveals the most stable joining point to be near the 5′ end of each genome, in a region also containing signals important in packaging of genomic RNA, reverse transcription, and integration of viral DNA. Synthetic RNA fragments containing this region can form dimers *in vitro* (22,158,476,636). Complicated models predicting secondary structure have been presented (368,517); including participation of purine-rich sequences in four-stranded (quadriplex) RNA structures (709). It is not yet clear how seriously these should be taken.

The role of diploidy in the retrovirus lifestyle remains to be determined. While it has been proposed that viral DNA synthesis makes use of the two RNA molecules to synthesize one of DNA, there is no obvious necessity for this in our current model of reverse transcription (see below). One clear consequence of having two genomes per virion is the extraordinarily high rate of recombination seen in retroviruses, and it may well be that this recombination frequency permits the effective repair of physical damage to the genome (see below).

Association With Other RNA Molecules

Denaturation of native genomes releases a number of small RNA molecules—largely tRNAs and small ribosomal RNAs. The most important of these is a single molecule per genome of tRNA which is associated via base pairing of its 3′ terminal 18 nucleotides to a complementary sequence located 100 to 500 bases from the 5′ end of the genome. This tRNA molecule serves a key role in replication: it is the primer on which reverse transcriptase initiates DNA synthesis. The specific tRNA which is appropriated by the virus for this role varies from genus to genus. At present, five tRNAs are known which serve this function (Table 2). In general, the specific tRNA primer used is highly conserved within a retrovirus genus, but among different MLVs two different tRNAs—pro and gln—have been found to be used (140).

Other Virion Nucleic Acids

In addition to the genome and associated RNAs, other nucleic acid molecules can be found in retrovirus virions. These include small RNAs—a variety of tRNAs, 5S ribosomal RNAs, as well as others—and traces of mRNA, such as globin. Although they are not merely a random representation of cellular nucleic acids, there is presently no good reason to ascribe functional significance to these inclusions in the virion. It should be borne in mind, however, that such molecules might—as a rare event—be copied into DNA which could be subsequently integrated into the cell genome and thus give rise to processed pseudogenes or other genomic rearrangements (135,449).

TABLE 2. *Terminal regions of retrovirus genomes*

	Approximate Size (bases)					
Virus genus	U3	R	U5	Leader	3′ Untrans	Primer tRNA
ASLV	150–250	18–21	80	250	150	trp
Mam C-type	500	60	75	475	40	pro/gln*
B-type	1200	15	120	130	*	lys[3]
D-type	240[+]	15[+]	95[+]	145	100	lys[1,2]
HTLV/BLV	250–350	120–240	100–200	50–100	*	Pro
Lenti	450	100	60–80	150	*	lys[1,2]
Spuma	800	200	150	90	*	lys[1,2]

*The 3′ open reading frame overlaps U3.
From references 513, 590, 679, 678.

Also present in a fraction of virions of at least some viruses are molecules of DNA, which are products of partial reverse transcription (461,742,829). Apparently, virion reverse transcriptase can sometimes use deoxynucleotides included in the virion to begin synthesis of DNA. Whether these molecules can be extended into complete molecules after infection is not known.

Cis-Acting Regions

Roughly speaking, retrovirus genomes are arranged so that all (or almost all) noncoding sequences that contain important recognition signals for DNA or RNA synthesis and processing are located in terminal regions, with internal regions given over virtually entirely to protein coding functions. The general sequence organization of a typical retrovirus genome is displayed in Fig. 4. Reading from the 5′ end, the important noncoding features common to all retroviruses include:

R

All retrovirus genomes are terminally redundant, containing the identical sequence at the 5′ end, adjacent to the capping group and at the 3′ end immediately preceding the poly(A) tract. This sequence plays an important role during reverse transcription in permitting the transfer of nascent DNA from one end of the genome to the other. The R sequence varies considerably in length, from a low of 12 bases in MMTV to a high of 235 in BLV.

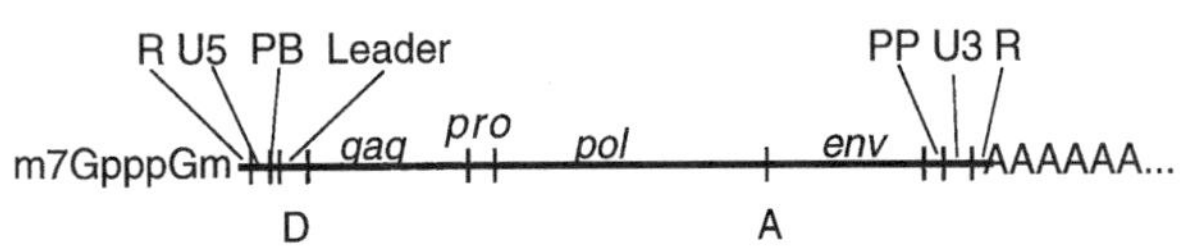

FIG. 4. Sequence features of retrovirus genomes.

U_5

This region containing unique information near the 5′ end of the genome is defined by its flanking sequences—R and the primer binding site. U_5 is the first region copied into DNA during reverse transcription, and becomes the 3′ end of the LTR. Mutational analysis implies multiple roles of this region. Some U_5 sequences are essential for initiation of reverse transcription (5,6,528). The 3′ end contains one of the *att* sites necessary for integration.

PB

The 18 nucleotides that form the primer binding site are invariably perfectly complementary to the 3′ terminal nucleotides of the tRNA primer. The length of the PB region is defined by the position of an uncopyable base in the tRNA—an m^1 A residue (251).

Leader

Like some other RNA viruses, retroviruses have unusually long sequences preceding the first translated region. The untranslated sequence between the primer binding site and the beginning of *gag* has at least two functional roles. Usually, it contains the donor site for the generation of all spliced subgenomic mRNAs encoded by the virus. In many viruses, there is only one such message known—that encoding the *env* proteins. Even in the highly complex sets of spliced mRNAs displayed by lentiviruses, all subgenomic species are spliced from this donor, with variation in the downstream site used and the presence of additional splices distinguishing the individual species.

The other important function encoded in the leader region is to specify incorporation of genome RNA into virions (450). The so-called packaging signal, called Ψ or *E*, has been roughly localized by analysis of spontaneous or introduced mutations (13,377,472). In at least one case—

Moloney MuLV—it seems to extend into *gag* (41), and insertion of a sequence encompassing this region is sufficient to confer the ability to be packaged into virions upon unrelated RNAs (2,18,177). Although it can be moved to other locations in the genome and still function (17,356), in most retroviruses, the location of the packaging signal downstream of the splice donor site ensures its absence from the viral mRNAs and consequently prevents incorporation of these into virions. In viruses of the ALSV group, by contrast, important signals for packaging seem to be largely upstream of the splice donor site near the beginning of *gag*. How mRNAs are excluded from virions is not well understood. The selection may involve negative as well as positive signals, since ALV RNAs generated by splicing from the viral donor to nonviral sequence can be packaged efficiently (17,713).

Internal Signals

In the majority of retroviruses, the sequence from the beginning of *gag* to the end of *env* is translated in its entirety; there is no internal nontranslated region. Indeed, there is some overlapping of reading frames. Internal *cis*-acting signals are primarily the splice acceptor sites used to form the various mRNAs. A few exceptions are notable: in lentiviruses and HTLV, complex internal regulatory sequences interact with systems specified by both virus and cell to regulate relative concentrations of mRNA for structural and regulatory proteins (see Chapter 27, and Chapter 59 in *Fields Virology*, 3rd ed.) (531). In some viruses, the DNA form of a sequence within *gag* contains enhancer activity when added to standard assays. In general, however, sequences required in cis for virus replication are not found interspersed with coding regions. This division allows replication of retrovirus genomes in which coding sequences are deleted or replaced by virtually any sequence (such as an oncogene), and is of central importance to development of retrovirus vectors.

The 3′ Untranslated Region

In most simple viruses, there is a short and rather variable untranslated sequence separating the end of *env* and the beginning of U_3. In the ALSV group, mutations in this region diminish incorporation of genomes into virions (680). A possible role for the 3′ nontranslated region of MPMV region in permitting transport of unspliced mRNA—a function requiring a virus-encoded protein in complex viruses—is suggested by recent experiments showing that it can provide this function to HIV even in the absence of *rev* function (69).

The Polypurine Tract

All retrovirus genomes contain a characteristic sequence—a run of at least nine A and G residues—immediately preceding the beginning of U_3. This sequence contains the initiation site for synthesis of the plus strand of viral DNA; it escapes digestion by RNase H during the synthesis of the first strand of viral DNA and can thus serve as primer for second-strand synthesis (251). In some viruses (including lentiviruses and spumaviruses), at least one additional functional polypurine run is found elsewhere in the genome (113), and its function is required for replication (see below).

U_3

The U_3 region is defined as the region between the initiation site of plus-strand DNA synthesis and the beginning of R. It forms the 5′ portion of the LTR, and, in the DNA form, contains a number of *cis*-acting signals necessary for virus replication. The sequence at the immediate 5′ end of U_3 is a signal, called *att*, recognized by the integration machinery, and is an approximate inverted copy of the matching signal in U_5. Since U_3 ends up at the 5′ end of the provirus, it also contains signals that are recognized by the cellular transcription machinery and responsible for much or all of the transcriptional control of virus expression. Near its 3′ end are other transcriptionally important signals—the canonical consensus sequences that mark most eukaryotic promoters. In some groups (those with short R regions; see Table 2) the consensus polyadenylation signal is also found in the 3′ portion of U_3.

R

At the 3′ end of the genome, between U_3 and the poly(A) sequence is the other copy of the R region, which sometimes contains the poly(A) addition signal.

Coding Regions

Figure 5 shows the distribution of protein coding sequences in the genomes of the various retrovirus groups. Considerable variation is apparent within the common framework. All viruses have the genes encoding virion proteins arranged in a common order, but each group has special features which distinguish it from the rest. The common features are discussed first. Special features which characterize the HTLV and lentivirus groups will be discussed only briefly. Complete details can be found in the appropriate chapters.

gag

The 5′-most gene of all retrovirus genomes is named in honor of the first recognition of the proteins encoded by it as **g**roup-specific **a**nti**g**ens. The *gag* gene is translated from the full-length RNA to yield a precursor polyprotein which

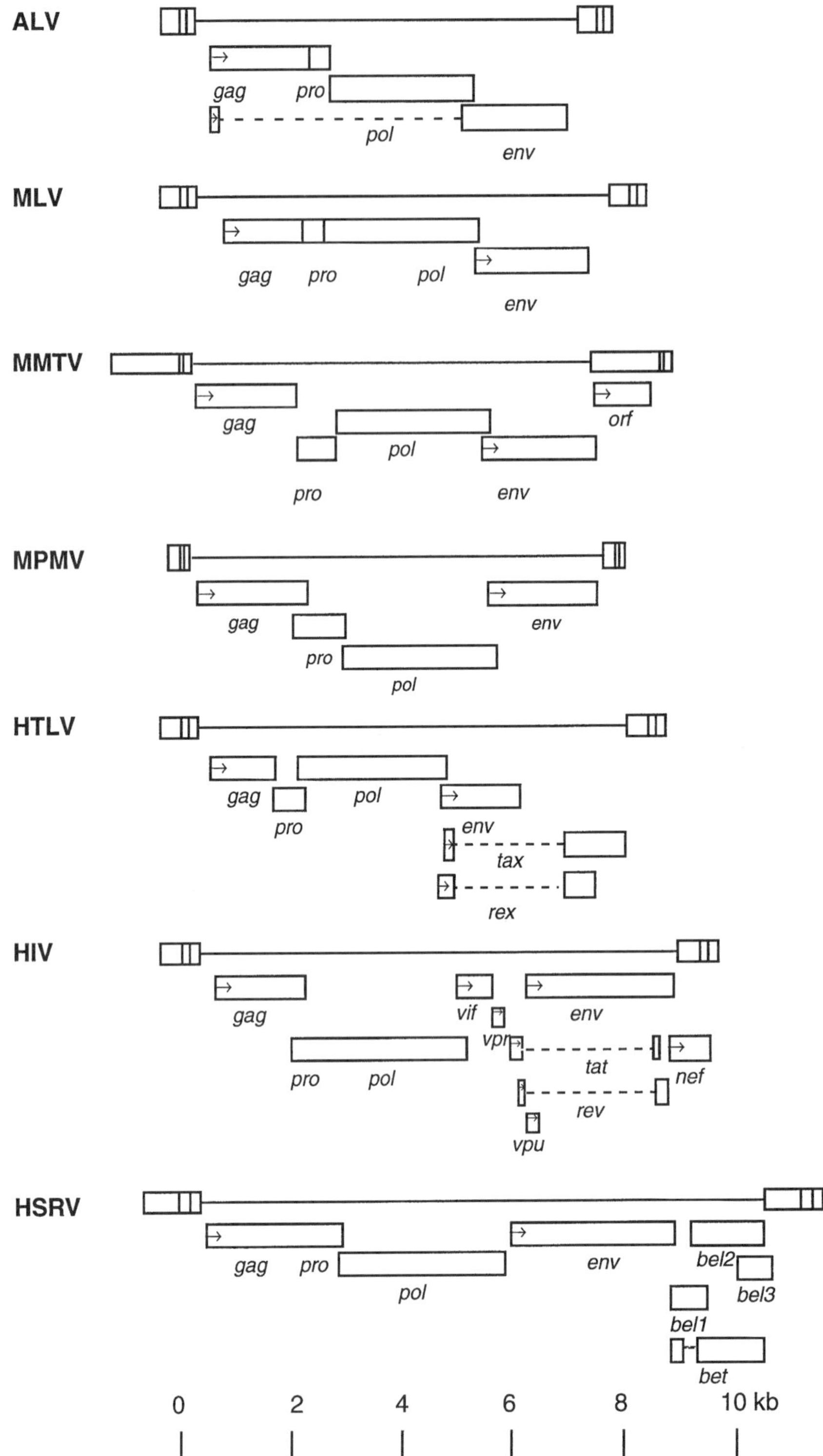

FIG. 5. Coding regions of retrovirus genomes. Open reading frames in the genes of representative members of each retrovirus group are shown relative to the proviral structure. Sites of translation initiation are shown *(small arrows)*. Note the different orientation of *pro* and *pol* relative to *gag*, and the overlap of *gag* and *env* in most viruses. Sequences from refs. 756,481,678,513. The human spumavirus (HSRV) reading frames are predicted from the sequence but have not been verified.

is subsequently cleaved to yield three to five capsid proteins. The three invariant proteins are (in their order of translation): the matrix (MA) protein; the capsid (CA) protein, and the nucleic acid-binding protein (NC). In the avian and mammalian C-type viruses, an additional cleavage product, found in virions but of unknown function, occupies the region between the MA and CA peptides. Other, quite small *gag* cleavage products are also found. The AUG initiation codon for *gag* is about 300 to 500 bases from the 5′ end and is rarely the first potential initiation codon. Rather, *gag* is often preceded by short open reading frames (ORFs) whose function remains uncertain. In some mammalian C-type viruses, occasional use of an additional initiation codon in frame with and upstream of *gag* leads to synthesis of a cell-surface variant of the *gag* protein, whose role in virus replication remains unclear.

pro

The *pro* region encodes the protease (PR) responsible for the cleavage of the *gag* and *pol* polyproteins and sometimes part of *env* as well. Strictly speaking, neither *pro* nor *pol* should be considered as distinct genes, since they are translated only as carboxy-terminal extensions of a fraction of the *gag* proteins, and all precursor proteins which contain them also contain the entire *gag* precursor. The *pro* reading frame is variously arranged with respect to the *gag* and *pol* regions that flank it and the various devices employed by different groups of viruses to express it are described in a later section.

pol

The *pol* gene encodes the two proteins containing the two activities needed by the virus early in infection—the reverse transcriptase (RT) and the integrase (IN) protein needed for integration of viral into cell DNA. In ALSV, there is an additional short, apparently nonfunctional cleavage product encoded by the 3′ end of *pol* (374). As with *pro*, translation of *pol* into protein is made possible only by an occasional slip of the translation machinery which causes the preceding termination codon to be bypassed and translation to continue in the *pol* reading frame.

In most retrovirus genera, the 3′ end of *pol* overlaps the beginning of *env* in a different reading frame, so there is no intervening untranslated sequence .

env

The *env* gene encodes the two envelope glycoproteins which are themselves cleaved from a larger precursor. The larger of the two, the surface (SU) protein, is responsible for recognition of cell-surface receptors. The smaller transmembrane (TM) protein anchors the complex to the virion envelope and contains domains responsible for fusion of viral and cellular membranes. Unlike *gag*, *pro*, and *pol*, *env* is translated from a spliced subgenomic RNA, and a characteristic splice acceptor sequence can always be identified in the appropriate location upstream of the initiation codon.

Other Genes

While the simple retroviruses such as the avian and mammalian C-type virus groups require only the *gag, pro, pol,* and *env* genes and their products to fill all virus coded roles necessary for replication, other virus genera encode additional proteins which play other roles in replication. Understanding of the functions and products of these genes is a major research area of retroviral research. Many of them encode proteins whose role is to actively modify the rate and pattern of gene expression of the provirus. A number of these proteins are transactivators which increase the extent of transcription of the provirus; a number of others serve a regulatory function by modifying the relative levels of the various mRNAs. In lentiviruses, still other proteins are also encoded by the so-called accessory genes.

B-Type Viruses

MMTV was the first retrovirus to reveal an additional gene, 3′ of *env* and extending well into U_3. For many years, this gene was known as *orf*, standing for "open reading frame" and reflecting the lack of attributable function. Recently, it has been discovered that it encodes a superantigen activity which serves to stimulate proliferation and infection of lymphocytes in newly infected mice. It is now known as *sag* in honor of this function.

HTLV and BLV

These viruses have a long region between *env* and U_3, which encodes two proteins in overlapping ORFs. This region was originally called *X* or *lor* (for "long open reading frame"). Both genes are translated from a doubly spliced mRNA, which contains the initiation codon and a few additional codons from *env*. These genes include:

tax. As suggested by the name, the protein encoded by *tax* is the HTLV transactivator derived from the X region. The presence of the 40,000 molecular weight (MW) protein encoded by it is essential for LTR-driven expression of these viruses.

rex. The name *rex* stands for "regulator from the X region." *rex* encodes a protein of about MW 27,000, which is essential for the expression of full-length and *env* mRNAs. In its absence, only mRNAs encoding *tax* and *rex* are found in infected cells.

Lentiviruses. HIV and its relatives hold the unchallenged retroviral record for the number of genes identified—at least nine are known to encode products, although clearly defined functions are still lacking for some. This group of viruses is unique in having a region between *pol* and *env,* which encodes at least part of several proteins. There is also considerable variability among lentiviruses in the nature, number, and pattern of these products (see Chapter 62 in *Fields Virology*, 3rd ed.). The following list gives the HIV pattern. Chapter 27 should be consulted for the complete picture.

tat. This gene is named for its transactivating function, which greatly stimulates expression from the HIV LTR, although by a mechanism quite different from that of *tax*. It is also quite different in size, and contains only about 90 amino acids. Despite its small size, the translated portion is divided over two exons, and it is translated from a multiply spliced mRNA.

rev. This small protein is a regulator of virus expression. It is expressed in a fashion like *tat* from an overlapping reading frame. Although very different in size and structure, it has a function similar to the HTLV rex protein in that the pattern of spliced mRNA is strongly affected by it and, indeed, under certain conditions, one protein can substitute for the other in virus replication.

nef. This ORF 3′ of *env* extending into U_3 encodes a small, myristylated protein which localizes to the inner face of the cell membrane. It was named for its apparent function as a "negative factor" for virus replication. It is now known to cause the removal of CD4, the HIV receptor, from the cell surface (42,246), but the biological importance of this function is still debated. An additional function is apparently to interact with the signaling pathway of T cells and induce an activated state, which in turn may promote more efficient virus gene expression (539). Some HIV and SIV isolates have mutations in this region which cause them to not encode functional protein. This has the apparent effect of giving these viruses a growth advantage in cell culture over their wild-type parents. Such viruses are severely compromised in their replication and pathogenesis *in vivo* (387).

vif. The product of this short ORF overlapping *pol* does not seem to be a virion protein, but encodes a protein necessary for production of infectious virions (683, 779).

vpu. This short ORF is located just upstream of *env* and apparently expressed as an alternative reading frame in the *env* mRNA. Its product is not absolutely essential for virus replication, but seems to serve interesting accessory functions, e.g., in degradation of the CD4 receptor shortly after synthesis, allowing efficient expression and processing of the *env* gene product (115) and in permitting efficient budding and release of virions (651).

vpr, vpx. The virus proteins encoded by reading frame r and x have been detected in cells infected with HIV-1 and HIV-2, respectively, but little is known about their function. vpr is found in virions and it has been proposed that it helps to transport the viral DNA from cytoplasm to nucleus in nondividing cells (300). It also seems to have effects on expression of certain cellular genes (438).

Spumaviruses. As with HTLV and BLV, the nucleotide sequence of human and simian spumavirus genomes reveals the presence of several overlapping ORFs between *env* and U_3 (227,494).

bel 1. In HFV, from "between env and LTR", also known as *taf* in SFV, this reading frame encodes a transcriptional transactivator analogous—although not obviously related—to the HLTLV *tax* gene product.

bel 2. Encodes a protein product not essential for replication in some cell cultures, but possibly in hemapoietic cells, and whose function is not known.

bel 3. This protein has been visualized, but its function is not known.

bet. A 62-kd fusion proton formed by splicing bel 1 to bel 2.

The Capsid

The structure of the nucleocapsid of retroviruses is not well understood. Our primary source of knowledge has been genetic analysis supplemented with some information gleaned from fractionation studies. When virions are carefully treated with mild detergent, two rather fragile structures are formed, depending on virus and conditions. The nucleoprotein contains the virus RNA, the NC protein, and a fraction of the reverse transcriptase activity. The core contains, in addition, the CA protein. Electron microscopy of whole or fractionated virions has not been particularly revealing of the fine structure—the particles are somewhat pleomorphic and do not reveal obvious overall symmetries although local icosahedrally symmetric regions can be seen (535,536). Immature virions containing unprocessed Gag proteins (see below) seem to have a much better defined structure (692), and these can probably be exploited to gain further useful information on such issues as the exact number and organization of core proteins.

The majority of the structural features observed are attributable to *gag* proteins. Indeed, almost normal-looking virions can be formed in the absence of *pol* and *env* proteins as well as of genome RNA. The major gag proteins are most readily studied as the cleavage products found in virions. It should be borne in mind, however, that they act in two different forms at different times. During virion assembly they are unified as a precursor protein; it is, however, the cleavage products that enter the cell to initiate the infection cycle. Thus, domains filling important roles in assembly may not be neatly distributed among the cleavage products (803). As a consequence of the mode of assembly, it can be expected that all *gag* proteins will be present in equal numbers in the virion. This expectation seems to be borne out (at least approximately) and it has been estimated (for ALSV) that there are about 2,000 to 4,000 copies of each cleavage product per virion (160).

TABLE 3. *Virion proteins of retroviruses*

Protein	ALSV	Mam C-type	B-Type	D-Type	HTLV/BLV	Lenti	Spuma
	(RSV)	(MLV)	(MMTV)	(MPMV)	(HTLV-I)	(HIV)	(HFV)
MA	p19	p15	p10	p10	p19/15	p17	
?	p10	p12	p21	p24	NP	NP	NP
CA	p27	p30	p27	p27	p24	p24	p33
NC	p12	p10	p14	p14	p12	p7	p15
DU	NP	NP	p30	-	p15[a]	NP	NP
PR	p15	p14	p13	-	p14	p14	p10
RT	p68	p80	-	-	-	p65/51	p80
IN	p32	p46	-	-	-	-	p40
SU	gp85	gp70	gp52	gp70	gp60	gp46	gp130
TM	gp37	p15E	gp36	gp22	gp30	gp21	gp48

[a] From FIV
The order is 5´ to 3´, from top to bottom.
NP, not present.
Data from refs. 430, 537, 538.

Conventionally, virion proteins are designated by a two-letter mnemonic suggesting their function (430). Traditionally, they were named according to their apparent molecular weights, by affixing a prefix (p, gp, pp, or Pr—for "protein", "glycoprotein", "phosphoprotein," or "precursor") to the MW in thousands, and they appear this way in much of the current literature. The correspondence between the two naming systems among the various groups is shown in Table 3. The organization of the *gag, pro,* and *pol* proteins along their respective precursors is shown in Fig. 6.

The MA Protein

The MA protein is the *gag* protein in closest association with the membrane. It is the only protein which can be chemically cross-linked to lipid in virions (577). [The idea that this protein is also a specific RNA-binding protein seems to have been generally discarded (688).] Consistent with its membrane association, the amino terminus of most MA proteins is modified by the addition of a fatty acid—invariably myristic acid—group, a modification characteristic of many proteins that lie on the internal face of the

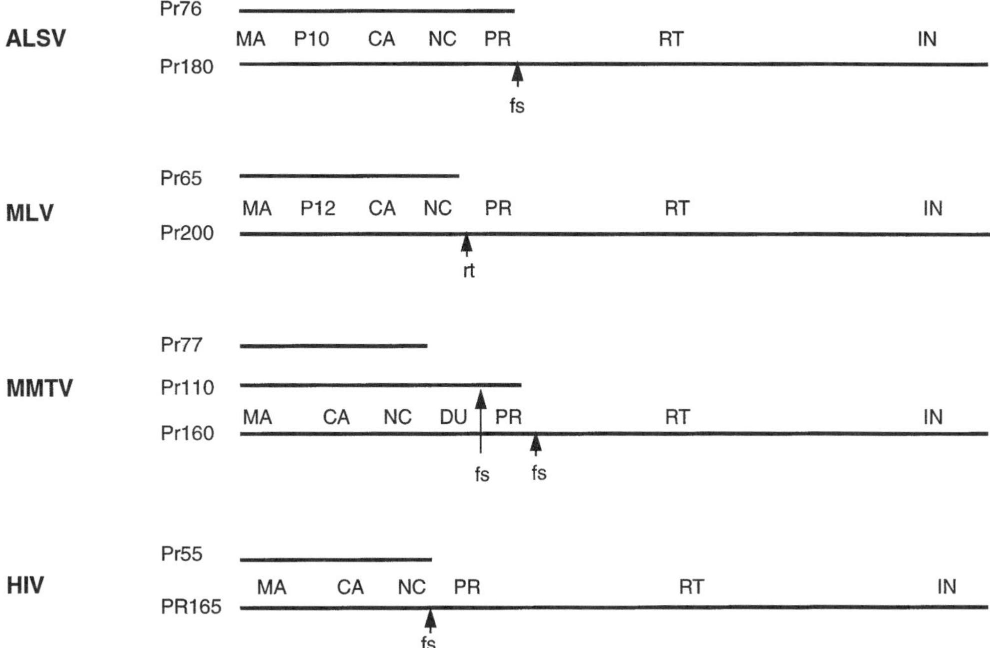

FIG. 6. Expression of the *gag, pro* and *pol* genes in various retroviruses. The disposition of the coding regions for the final cleavage products is shown on each of the precursor molecules. Note the use of read-through suppression (*rt*) of frameshift suppression (*fs*) to accomplish the synthesis of the nested products.

cell membrane. Interestingly, the ALSV MA protein has only an added acetyl group at its amino terminus and also lacks any significant stretch of hydrophobic amino acids that might provide for membrane attachment. In all viruses studied, the important determinants for the site of capsid assembly have been found to lie within the MA region (202,617,828).

The CA Protein

The CA protein is hydrophobic and forms the major internal structural feature of the virion—the core shell. Beyond this little is known about its structure or mechanism of action. Many mutations in CA eliminate virus assembly altogether (290). The sequence of this protein is not highly conserved among retrovirus groups, although a stretch of about 20 amino acids known as the major homology region can be seen in all retroviruses and seems essential for virion formation (703).

The NC Protein

This is a small basic protein found in the virion in association with the genome RNA. It is usually phosphorlated on serine residues, and it has been suggested that the phosphorylation state of this protein may be an important determinant of viral assembly or disassembly (234). When the structure of this protein is compared among the different groups (and among some retrotransposons as well), a conserved feature emerges, in addition to the overall basic nature of the protein. This is a sequence of the form Cys-x-x-Cys-x-x-x-x-His-x-x-x-x-Cys, present twice in NC proteins of most retrovirus groups but only once in most mammalian C-type viruses and not at all in spumaviruses. These "Cys-His boxes" resemble zinc finger domains well known in some DNA-binding proteins and binding of zinc *in vitro* to NC proteins has been observed (649,682). The issue of the presence of zinc in virions has been somewhat controversial, but seems to have been decided in the affirmative (46).

Other gag Proteins

Two retrovirus genera, the ALSV group and the mammalian C-type viruses, encode additional *gag* cleavage products, called p10 and p12, respectively. In both cases, this protein occupies the portion of the precursor between the MA and CA domains, and it is found in the corresponding location in virions, i.e., just outside the core shell. To date, no function has been assigned to these proteins.

Adjacent to p10 in ALSV is a small *gag* cleavage product known as P2, which seems to fill an important role in regulating the assembly process (101,576). An additional HIV cleavage product, known as P6, contains sequences required for efficient release of newly budded virions from the infected cell (267).

Virion Enzymes

The products of the *pro* and *pol* regions constitute the enzymatic activities found in retrovirus virions (376). Their mode of synthesis, as occasional extension products of *gag,* results in their presence in much smaller numbers—less than a hundred as compared to thousands of copies per virion.

RT

The primary product of *pol* is cleaved by the virion protease to yield the amino terminal RT peptide, which contains the activities necessary for DNA synthesis (RNA and DNA-directed DNA polymerase, ribonuclease H) and the carboxyterminal IN protein [for review, see (252)]. In the case of ALSV, the cleavage between the two occurs in only a fraction of molecules, and results in two proteins with reverse transcriptase activity. The 572 amino acid form is called α and the larger, uncleaved molecule is called ß. The ALSV RT is most commonly present as an αß heterodimer, although other monomeric and homodimeric forms are also present. The activities of each are not well sorted out. This pattern is not well conserved among the genera: MLV RT is an apparent monomer of fully cleaved RT when free in solution [but a dimer when active (731)]; HIV RT is a heterodimer of full-length RT (p66) and a cleavage product (p51) lacking the carboxyterminal RNase H domain. Despite the relationship of the two molecules, their conformation in the molecule is very different.

A fraction of the RT in virions is in association with the viral nucleoprotein structure where it seems to be bound specifically to the primer tRNA, at least in ALSV. This binding may be important in virion assembly; ALSV mutants lacking *pol* are also devoid of primer tRNA (580). In MLV, specific primer tRNA-reverse transcriptase binding has not been demonstrated.

The two major activities of reverse transcriptase are the DNA polymerase and ribonuclease H. Fortunately, the retrovirus enzyme (unlike most other virion polymerases) is readily separable from the virion, and several have been successfully produced in popular cloning systems in bacteria and insect cells. Both the ALSV and MLV enzymes are commercially available. The DNA polymerase activity displayed by RT uses RNA and DNA as templates more or less interchangeably, and like all DNA polymerases known, is unable to initiate DNA synthesis *de novo*, but instead requires a preexisting molecule to serve as primer. In the test tube, either RNA or DNA will do; in the natural case it is always RNA: the tRNA used for synthesis of the first (minus) strand; and the genome itself nicked within the polypurine tract for second (plus) strand synthesis. Reverse transcriptase activity is readily detected within disrupted virions. In the absence of added template, the "endogenous" activity uses the genome-primer combination to synthesize minus-strand DNA. Addition of an appropriate hetero- or homopolymeric template-primer pair to disrupted virus leads to a substantially greater synthesis of the complementary DNA. The use of an RNA homopolymer template and a DNA oligonucleotide primer provides some (although far from absolute) specificity relative to adventitious cellular DNA polymerases. Poly rC•oligo dG seems to be best in this regard. This fairly sensitive assay is still used routinely in the laboratory to detect retrovirus infections in cell culture, to quantitate the amount of virus produced, and to partially characterize the virus. A potentially very useful recent modification is to use polymerase chain reaction (PCR) to amplify products of an RT reaction and provide a very sensitive test for the possible presence of retroviruses (602). Many of the most recently discovered retroviruses were first seen by virtue of their reverse transcriptase activity (793). Unfortunately, there is also a long history of false positives detected in the same way and reported as possible human retroviruses (792).

All reverse transcriptases have an absolute requirement for divalent cation for polymerase activity *in vitro*. In all groups but one, the preferred ion is Mg^{2+} at about 10 mM. The exceptional group is the mammalian C-type viruses, whose RT prefers Mn^{2+} at about 3 mM, especially for use on homopolymer templates. This difference is occasionally useful for classification.

The RNase H activity inherent in all RT proteins plays the essential role early in replication of removing the RNA genome as DNA synthesis proceeds (see below). It selectively degrades the RNA from an RNA-DNA hybrid molecule. On model substrates it behaves as both exonuclease and endonuclease. On the one hand, it usually requires a base-paired RNA end; it will not cleave molecules with overhanging single-stranded regions, and it cleaves progressively from either a 5′ or 3′ end. On the other hand, the cleavage products are small oligonucleotides, and it makes specific endonucleolytic cuts at strategic points in the process of DNA synthesis. In the presence of Mn^{2+}, many or all RT-associated RNase H activities will also cleave certain double-stranded RNA substrates (316). The role of this activity, originally called *RNase D* [now *RNase* H^* (317)], *in vivo* is unresolved.

The structure of RT was originally inferred from extensive mutational analysis which revealed that the polymerase and ribonuclease H activities occupy separate, nonoverlapping domains, with the polymerase covering approximately the amino-terminal two-thirds of the molecule (591,724) (Fig. 6). Recently, reasonably high resolution structures have been obtained for the HIV RT, complexed with antibody and either complexed with an inhibitor (400,401) or bound to a model template-primer structure (342) (Fig. 7). These analyses reveal a number of important aspects of RT structure and function.

First, the two subunits, although completely overlapping in sequence, are very different in structure and function. Although the common portions have the same four do-

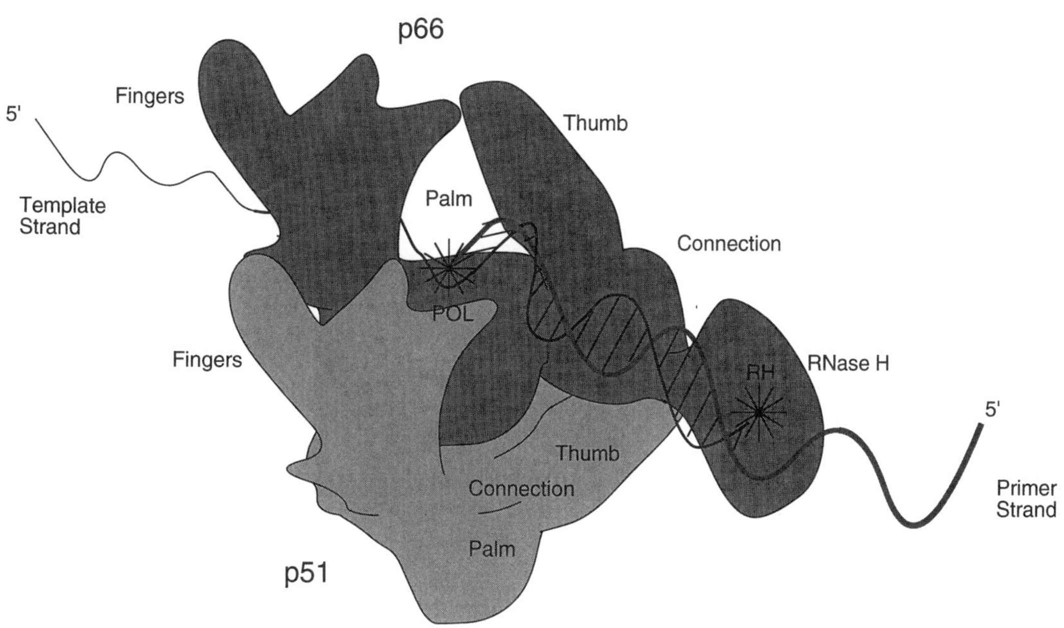

FIG. 7. The structure of RT. The cartoon (not perfectly to scale) is derived from the crystal structures of HIV reverse transcriptase determined by Jacobo-Molina et al., ref. 342, and Kohlstaedt et al., ref. 400.

mains, the domains are oriented quite differently, with the result that p51 is more compact and unable itself to bind nucleic acid, and all catalytic activity resides in the p66 subunit. The function of the smaller subunit thus remains obscure, and it will be instructive to learn the exact nature of the subunit interactions in the RTs of other retroviruses.

The p66 catalytic subunit is folded into five distinct subdomains. The amino terminal 3 of these embody the portion with reverse transcriptase activity, and their organization can be likened to a slightly curled right hand with "fingers," "palm," and "thumb" domains. Carboxyterminal to these are the "connection" domain (at the base of the hand) and the RNase H (RH) domain on the "wrist."

The template-primer lies along the palm and connection subdomains in the cleft between the thumb and fingers and appears to be held in place by contacts with the palm, thumb, and fingers. The active site for polymerization is in the middle of the palm, with the participation of three aspartic acid residues which may coordinate a metal ion involved in catalyzing nucleotide addition.

The double-stranded template-primer pair is DNA-DNA in the structure determined (342) but is also RNA-RNA or RNA-DNA at various times during reverse transcription (see below). It does not extend straight across the RT molecule, but rather takes a sharp (45°) bend a few bases from the active site and changes conformation from A to B form as a consequence. The duplex contacts the RH domain about 18 bases from the polymerization site, in good agreement with the size of the contact region, as determined from other experiments (235,262).

Finally, there is considerable structural similarity with the few other polymerases for which there is structural information, including the Klenow fragment of *E. coli* DNA polymerase 1 (813), a fragment of mammalian DNA polymerase ß (575), and the RNA polymerase of bacteriophage T7 (681). Although a primary sequence similarity is not readily discernible, the relationship suggests a common evolutionary origin for these enzymes. The nature of the shared ancestor remains obscure.

RT is one of the most highly conserved proteins encoded by the retroviral genome. Although the overall relationship from one virus group to the next is not close, there is enough amino acid sequence similarity between the various enzymes to detect a clear relationship, to infer the family tree shown in Fig. 1, and to discern certain key structural features. Some of these features are in common with the much more distantly related reverse transcriptases of retrotransposons of flies, yeast, and plants; hepadnaviruses (Chapter 35); and cauliflower mosaic virus, as well as even the recently discovered enzyme in certain bacteria species [for review, see (173,482,483,812)]. A noteworthy characteristic, for example, is the presence of the highly conserved Tyr-x-Asp-Asp sequence that is associated with the catalytic site.

IN

The IN protein is the smaller, carboxy terminal cleavage product of the *pol* region. Genetic analysis provides clear evidence for the role of IN in replication. Mutations that affect its structure block virus replication; in many viruses, such mutations have been shown to affect integration but not DNA synthesis (170,416,561,642). Some caution is required in interpreting these experiments, since IN mutations can also affect other *pol* as well as *gag* functions (665), probably due to effects on *gag-pol* precursor structure.

The two enzymatic functions—DNA cleavage and strand transfer—of this protein have recently been elucidated using pure IN purified from virions or prepared from expression systems [for reviews, see (80,271)]. IN recognizes and binds to double-stranded DNA corresponding to the viral *att* site at the end of the LTR (referred to here as substrate DNA). It catalyzes the removal of two bases from the 3′ end of the substrate DNA and the subsequent joining of the free 3′ OH to DNA (the target DNA) with little or no specificity for target DNA sequence. The latter reaction is accomplished by direct attack of the hydroxyl group of the substrate DNA on a phosphodiester bond of the target, leading to joining of the substrate to one strand of the target and creation of an adjacent break in the target DNA (196, 775).

Crystallization and structural determination of IN remain to be accomplished; however, some inferences as to its structure-function relationship can be drawn from *in vitro* mutagenesis (93,95,182,358,372,414,758,759). The aminoterminal region of IN contains a conserved region characterized by a cys-his "zinc finger" array resembling those found in well-characterized DNA-binding proteins. This region is thought to confer specific binding to the substrate LTR DNA. The central region contains a domain with highly conserved aspartic and glutamic acid residues [called D,D,35 E (414)], believed to contain the active site used for both reactions. The carboxy terminal domain is less well conserved and may participate in target DNA binding. A cartoon illustrating the basic reactions carried out by IN *in vitro* is shown in Fig. 8.

Reactions with purified components generally join only a single end of the LTR substrate to target DNA (the half reaction) (91,148,373). In the context of virus replication, however, the enzyme must accomplish simultaneous joining of both LTR ends about half a turn (4 to 6 bp) of DNA apart. This requirement implies that IN must function at least as a dimer during replication. Evidence based on complementation of mutations in different regions of the protein for activity *in vitro* implies that the half reaction is carried out with dimeric enzyme and therefore that the viral form is a tetramer (93,760).

All IN proteins exhibit nonspecific DNA binding activity; until recently, specific binding to substrate DNA had not been convincingly demonstrated. This problem probably reflects predominance of DNA binding to the "target" site

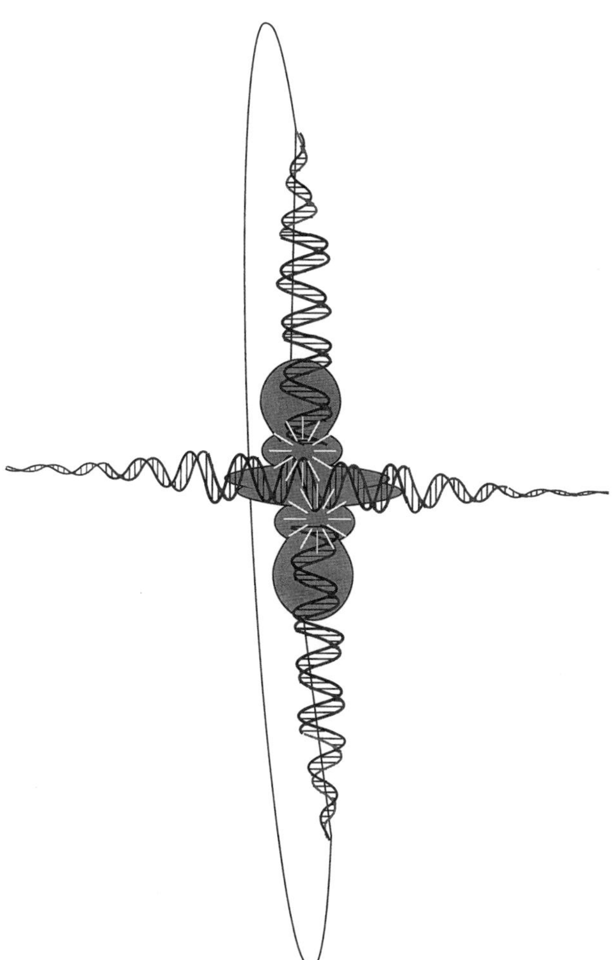

FIG. 8. Possible structure of IN. This imaginary structure shows the three inferred domains of IN (substrate DNA binding, catalytic, target DNA binding) interacting with the two ends of a viral DNA molecule (*vertical*) and a target DNA molecule (*horizontal*).

and the short life of bound substrate before integration. Temporary blocking of integration does allow indirect detection of specific substrate DNA-integrase complexes (193).

PR

The viral protease is responsible for the cleavages that separate the *gag* and *pol* proteins from one another. Although its gene (*pro*) is located in the same place—between *gag* and *pol*—in all retrovirus genomes, the variable orientation of the reading frames in this region cause it to be expressed quite differently in the various groups (Fig. 6). In ALSV, for example, it is expressed as part of *gag*, and therefore much more abundantly than in other viruses. In MLV and HIV, *pro* is in the *pol* frame; in MMTV as well as BLV and HTLV, it is in a frame distinct from both *gag* and *pol*.

PR shares sequence features and sensitivity to inhibitors with a large class of aspartic proteases, including a highly conserved sequence (Leu-Leu/Val-Asp-Thr-Gly-Ala-Asp-Lys) around the active site [for reviews, see (431,552, 674)]. The relationship to other aspartic proteases has been confirmed by development of a three-dimensional structural model for both the ALSV (Fig. 9) and the HIV enzymes based on x-ray crystallographic analysis (347,505,784,808). PR differs in structure from the other proteases in that the active form is a dimer. In all other aspartic proteases, the active form is a monomer with two copies of the active site consensus. The unique structure of the retroviral enzyme is believed to be important in keeping the enzyme inactive in the precursor until after assembly has taken place. The retroviral proteases are also characterized by a distinctive "flap" structure over the active site.

The amino acid sequences cleaved by PR are quite variable, but they have some common features, including a tendency toward hydrophobic residues on either side of the cleavage site with preference toward Tyr or Phe preceding the cut and Pro following it (571). A number of small synthetic peptides containing identified cleavage sites has been shown to provide suitable substrates for the cleavage activity (403). The structure of PR bound to substrate or inhibitors reveals that the important interactions are between the first three amino acids on either side of the cleavage site (called P1 to P3 and P1′ to P3′, respectively) and a series of pockets in the each subunit (referred to as subsites S1 to S3 and S1′ to S3′) (100,279). Because of the symmetry of the dimeric enzyme, the S1 and S1′ subsites are identical, as are the S2 and S2′ and S3 and S3′. The fit of substrate side chains into the subsites is a major determinant of cleavage efficiency (52,102). Although there is considerable variation in cleavage sites (especially as compared to some other viruses), cleavage is nevertheless highly specific. Indeed, many sizable foreign proteins are completely resistant to cleavage by PR,

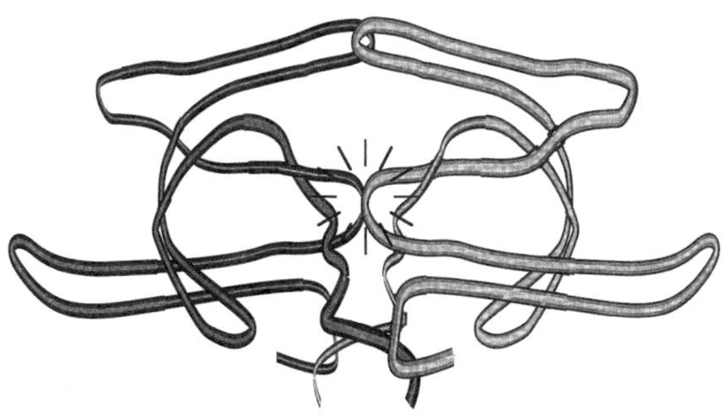

FIG. 9. The structure of PR. The structure shown is a somewhat simplified and slightly inaccurate rendition of an ALSV PR dimer (505,552). The polypeptide chain to be cleaved (*not shown*) would extend into and out of the page. Note that the active site (*starburst*) is formed by combining sequences of each subunit.

although others can be cleaved at least a few times (553,667).

DU

There are two retroviral lineages which have acquired an additional sequence, albeit in somewhat different places. This sequence—originally suggested to be "protease-like" on the basis of sequence homology (484)—is found as part of *gag* in B and D-type viruses and as part of *pol* (between RT and IN) in all lentiviruses except HIV and SIV. Recently, the protein encoded by this sequence has been found in abundance in virions of both types of viruses (402). Its enzymatic activity has been identified, not as a protease, but as a deoxyuridinetriphosphatase (192). Its distribution among the genera, its different location in the two groups that contain it, and the presence of a related enzyme in some herpesviruses (487), raise the possibility that it was acquired independently by ancestors of the two groups of viruses. The function of this virion enzyme remains rather unclear; mutants of FIV lacking this enzyme are still capable of replication, although their ability to replicate in macrophages relative to T cells is impaired (781) (J. Elder, personal communication). It seems probable that the function of DU is to prevent the incorporation of dUTP into nascent viral DNA. Such incorporation might be mutagenic; alternatively, DU removal by cell systems (whose function is to monitor the spontaneous deamination of C to U in cellular DNA) could lead to irreparable damage to the nascent viral DNA.

Other Enzymatic Activities

Over the years, sensitive assays have been used to detect a number of other functions, particularly those of potential relevance to DNA synthesis, such as DNA ligase, in retrovirus virions (166). These additional activities have not been associated with any known virus-coded protein, and it seems likely that they are either adventitiously included in virions or within a small fraction of cellular debris that inevitably copurifies with virions.

The Envelope

As with most enveloped viruses, the retroviral envelope is derived by budding from the cell membrane and comprises preexisting lipids and other membrane components modified by the insertion of viral proteins in place of normal cellular components. The composition of the membrane reflects that of the host, and there is no evidence for any virus-induced modifications to the lipids or other small molecules.

Envelope Proteins

All retroviruses have a similar complement of envelope proteins, displayed in cartoon form in Fig. 10. The complex comprises two different polypeptide chains held together by disulphide bonds as well as noncovalent interactions. Although this complex does not display any of the activities seen with other enveloped viruses (such as hemagglutinin, fusion activity, or neuraminidase), its size and organization closely resemble those of the much better studied HA protein of influenza virus (see Chapters 3 and 21). The three-dimensional structure of a retroviral *env* protein has yet to be determined. Until such time as this information is available, the influenza model will provide a useful framework on which to pin an overall concept of the *env* protein (332).

The organization of the *env* protein into higher order structures remains somewhat cloudy. There is good evidence, based on sedimentation analysis, that the ALSV protein exists as a trimer when released from the surface of the virion (190). This could be another point of similarity between the retroviral structure and that of influenza and other enveloped viruses. However, measurements of the structure of the HIV *env* proteins have given conflicting results, ranging from dimers to tetramers (108,188,653,788). Similarly, structures resembling both dimers and tetramers have been observed in MLVs, leading to the suggestion

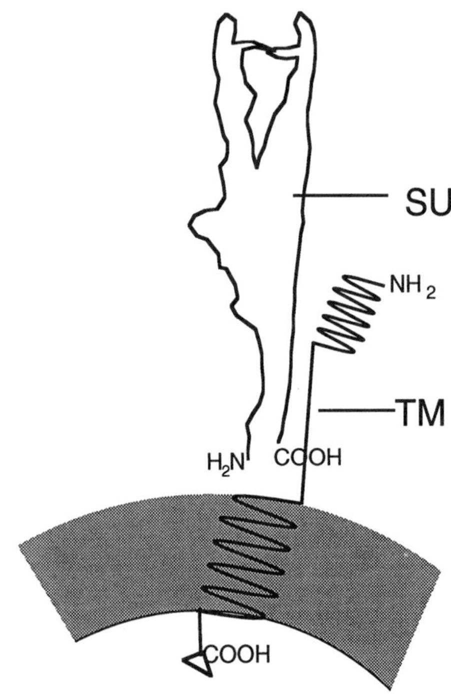

FIG. 10. The envelope protein. This cartoon indicates the inferred relationship between the TM and SU proteins and the virus envelope. The helical regions denote hydrophobic amino acid sequences involved in membrane attachment and fusion.

that the basic structure is a dimer which itself is capable of further oligomerization (362,752,815).

SU

The larger (SU) polypeptide contains the receptor-binding function and is also the primary (although not the only) antigen capable of eliciting synthesis of neutralizing antibodies in the infected host. This protein is invariably glycosylated, sometimes quite heavily. The ALSV SU protein is predicted to have 14 sites of the sequence Asn-X-Ser/Thr which serve as addition points for N-linked carbohydrate side chains. Careful analysis suggests that the actual number of side chains corresponds with the prediction (330), implying that all predicted sites are used. The contribution of carbohydrate to the mass of the protein is considerable; the ALSV protein alone has a predicted MW of around 38,000; its mobility in polyacrylamide gels is consistent with a value of about 85,000. The situation in lentiviruses is even more extreme: HIV SU proteins have over 30 such predicted sites for glycosylation (530). The distribution of glycosylation sites and other features on the ALSV and HIV *env* proteins is shown in Fig. 11. In addition to the N-linked carbohydrate, O-linked polysaccharide chains have also been found or suggested on *env* proteins in some virus groups, including HIV, MLV and FeLV (44,586), but little is known about them.

In some virus groups, it has been possible to localize regions of the SU protein important for receptor binding. In ALSV, this has been accomplished by taking advantage of the polymorphism of cellular receptors and the corresponding diversity of receptor-binding ability among closely related subgroups of viruses (see below). Genetic analysis of *env* proteins from six subgroups implicates the two separated regions in approximately the middle of the protein shown in Fig. 11. This analysis also implies that these regions work together somehow, perhaps by folding to form a single receptor-binding region (64,178,179). In MLV, similar sorts of analyses are not quite so clear-cut, but imply a region nearer the amino terminus of SU than ALSV as conferring receptor specificity (273,514,554). In HIV, analysis of deletion mutants and competitive antibody binding studies also implicate contribution from several different regions of the protein, particularly those which are highly conserved among isolates (511) (Fig. 11, and see Chapter 27).

TM

The smaller *env* protein is the transmembrane, or TM, protein. In most retrovirus genera, with the exception of the mammalian C-type viruses, the TM protein is also glycosylated, although usually much less extensively than its partner. In at least some viruses, this protein contains the domain(s) responsible for multimer (trimer or tetramer) formation (191). As its name implies, TM has three domains. The external region is attached to the SU protein and also contains at its amino terminus a hydrophobic region responsible for fusion with the cell membrane (232, 242). This region may be folded into a "coiled coil" structure which unfolds to interact with the membrane of the target cell analogous to the corresponding region in HA (107).

The membrane-spanning domain is made up of a string of 20 to 25 exclusively hydrophobic amino acids (except in HIV, where it is interrupted by a lysine residue). The cytoplasmic domain is usually quite short (about 20 to 30 amino acids), except in some lentiviruses such as HIV, where it is considerably longer. In MLV and some other viruses, the C-terminal region (about 16 amino acids in the case of MLV) is removed from TM during virion assembly; recent evidence implies that this modification is necessary to activate the fusion ability of the complex (622) (see below). In some lentiviruses such as SIV, deletion of a portion of this region does not inhibit replication in cell culture and may also enhance fusion activity (831).

In many retroviruses, association between the SU and TM proteins is rather tenuous, and SU is rapidly lost from virions, contributing to poor infectivity of virus preparations and instability to manipulations such as concentration by ultracentrifugation (90). In at least the case of ALSV, the TM protein is capable of forming trimers

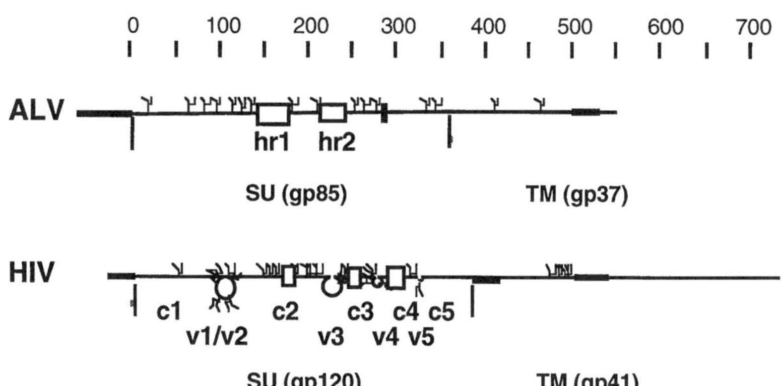

FIG. 11. Organization of env gene products. The scale is in amino acids. *Thick bars* depict hydrophobic regions. Note the (*from left*) signal peptide; the hydrophobic "fusion" sequence, and the membrane anchor region. Sites of cleavage by cellular enzymes are shown (*arrows*). *Branched structures* depict the predicted location of N-linked polysaccharide chains. *Loops* in the HIV structure show the location of sequences which exhibit considerable variability from one isolate to the next (see Chapter 27). *Boxes* represent regions implicated in receptor binding. Modified from Coffin, ref. 130.

when expressed in the absence of SU, indicating that some or all of the sequences mediating oligomerization reside in TM (191).

The obvious assumption that a specific interaction between the intracellular region of TM and a *gag* protein facilitates incorporation of the *env* protein onto the virion surface has not fared well in the face of ALSV mutants which encode protein lacking the cytoplasmic region (578). Such viruses not only have normal levels of virus production and infectivity; they also have a normal complement of *env* protein, leaving open the source of the specific interaction responsible for ensuring the association of envelope and capsid. In other groups, however, there is better evidence that a specific interaction involving the C terminus of TM is necessary for *env* protein incorporation into the virion envelope (822).

Other Proteins

Although the two *env*-encoded proteins are the only retrovirus gene products on the virion surface, other proteins can be present in large amount, brought there by mechanisms which remain unclear. For example, expression of the human CD4 glycoprotein in cells infected with ALSV leads to its efficient incorporation onto virions (820). Such inclusions, while accidental and nonspecific, can nevertheless be biologically relevant. The presence of large amounts of human major histocompatibility (MHC) proteins on SIV virions grown in human cells is an important contributor to their ability to stimulate an immune response in monkeys (20). Also, retrovirus virions can readily incorporate *env* proteins of other retroviruses (390,419,721), or even viruses of other families, including vesicular stomatitis virus (90,194) and influenza virus (172).

THE RETROVIRUS REPLICATION CYCLE

The replication cycle followed by all retroviruses is schematically outlined in Fig. 12 [for other reviews, see (466,716,767)]. In brief, it consists of the following steps:

1. Attachment of the virion to a specific cell surface receptor.
2. Penetration of the virion core into the cell.

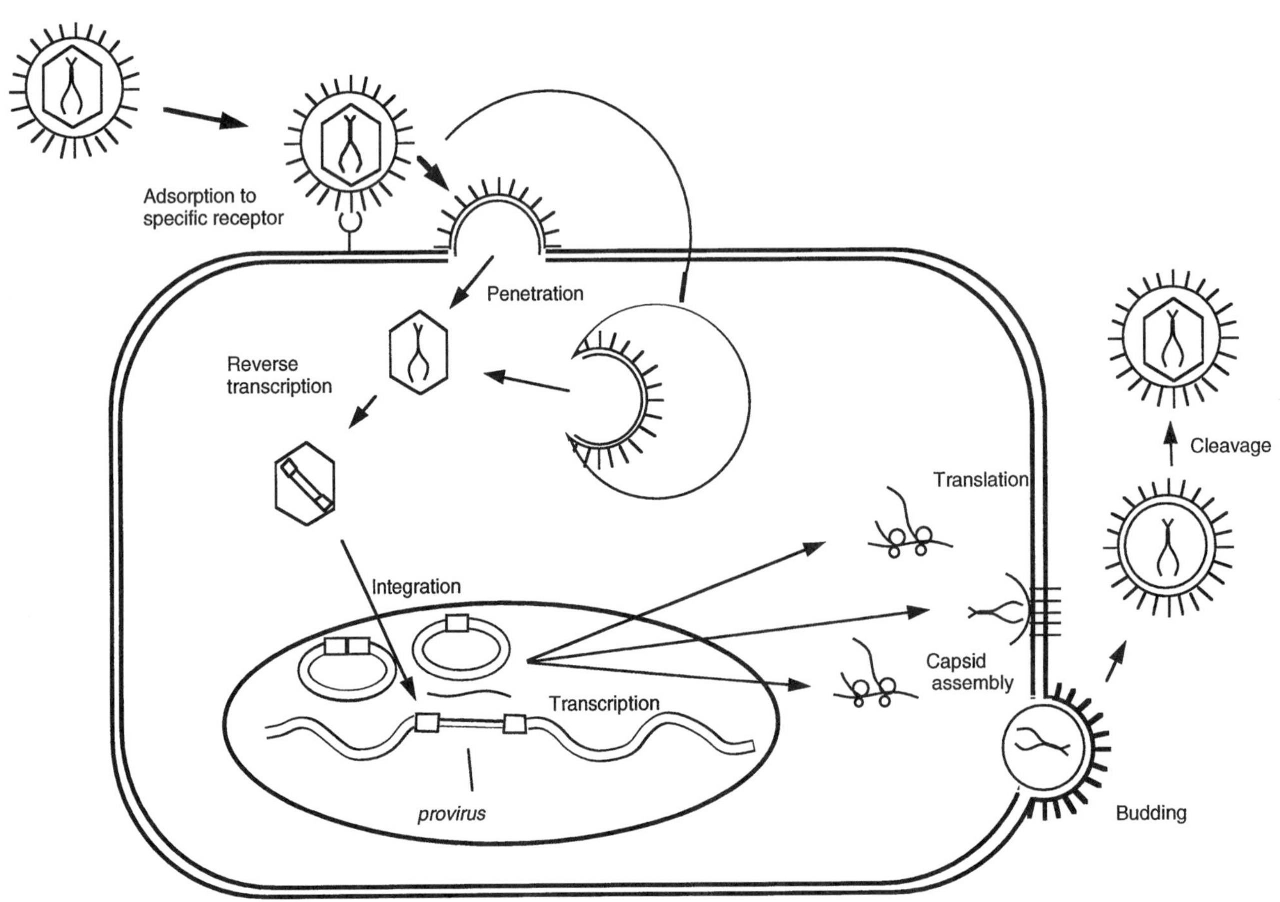

FIG. 12. Overview of retrovirus replication.

3. Reverse transcription within the core structure to copy the genome RNA into DNA.
4. Transit of the DNA, still associated with incoming virion proteins, to the nucleus.
5. Integration of the viral DNA into more or less random sites in cell DNA to form the provirus.
6. Synthesis of viral RNA by cellular RNA polymerase II using the integrated provirus as a template.
7. Processing of the transcripts to genome and mRNAs.
8. Synthesis of virion proteins.
9. Assembly and budding of virions.
10. Proteolytic processing of capsid proteins.

In the detailed discussion that follows, there are several key points that should be kept in mind. First, the cycle has two obvious phases, divided by the integration step. The first phase is carried out by incoming virion systems that remain in a structure derived from the virion core; the second phase is accomplished by cell machinery. Second, the provirus, once integrated, is stable. It is replicated and passed to progeny cells as part of the chromosomal complement of DNA. There is no evidence for any mechanism for removal of a provirus from the cell DNA; for its direct transposition to another site, or for its independent replication. Third, with many retroviruses, expression of the provirus does not require the help of any virus gene products. Finally, in most cases, the replication process proceeds without significant effect to the infected cell, which often continues dividing and is unaltered in any obvious way except that it is now continually producing new virions. A corollary of this relatively benign association is that the process of replication must be under some sort of control that prevents an infected cell from being repeatedly reinfected by the virions it produces, a series of events that would inevitably be fatal to the infected cell.

Attachment

Like all viruses, retroviruses have a specific requirement for interaction with a cell-surface receptor molecule for infection [for review, see (794)]. In all cases known (and suspected), this molecule is a protein which interacts specifically with the SU protein on the virion envelope. With the possible exception of HIV and relatives (see Chapter 27), each retrovirus has only a single receptor. If this molecule is missing (as when its expression is specific for certain cell types) or mutated in the binding region (as it may be in species other than the usual host, or even in some individuals of the natural host species), then the replication cycle cannot proceed and infectivity can be reduced by many orders of magnitude. That such a reduction is due to a block in an early event is shown by the use of cell fusion techniques to artificially fuse virus and cell membranes, partially bypassing the need for a specific receptor (336), and by the use of pseudotype virions of vesicular stomatitis virus with a retrovirus *env* protein substituting for the usual G protein. These display the same receptor requirements for infection as do the retrovirus *env* protein donors, ruling out any effects on events much later than the initial virus-cell interaction (791). The pseudotype system has been particularly useful for studying virus-receptor interactions, for example in identifying the HIV receptor (157).

Identification of Receptors

In recent years retrovirus receptor hunting has become a major area of activity among retrovirologists. Early attempts at identifying receptors by binding activity generally failed, and it is now recognized that it is insufficient to identify a protein as a receptor solely on the basis of binding studies; at a minimum, a number of distinct criteria must be met to confirm such an identification, including:

1. When expressed in ordinarily nonpermissive cells, it should confer sensitivity to infection.
2. The distribution of its expression among cell types should match sensitivity to infection.
3. At least some antibodies to the receptor should block infection.
4. Its chromosomal location should correspond to that of a gene conferring sensitivity to infection.

To date, six retrovirus receptors have been identified, with reasonable certainty, and a seventh provisionally (Fig. 13). The receptors for ecotropic MLV, subgroup A ALV, and GALV were discovered by use of an elaborate cloning strategy in which a few nonpermissive cells transfected with DNA of permissive cells become receptor-positive and can be selected by infection with virus containing a selectable drug resistance gene (10). The receptor gene can then be identified by its linkage to specific repeated DNA from the donor species, cloned, and analyzed in detail.

The most apparent characteristic of the set of known retrovirus receptors is the lack of a common theme. As expected, all are cell surface proteins, yet they differ widely in structure, tissue distribution, and function.

CD4, the receptor for HIV (157,396), is an immunoglobulin superfamily molecule containing a large exterior region of four Ig-like domains, a single transmembrane region, and an intracellular kinase domain (798). It plays an important role in signaling interaction of helper T cells with antigen-presenting cells. This function is irrelevant to the infection process, since a portion of the external domain can be attached to other cell surface molecules and still provide receptor activity (255,346).

In striking contrast to CD4, the receptors for ecotropic MLV (the eco receptor, also called Rec-1), amphotropic MLV (Ram-1), as well as GALV (Glvr-1) are similar proteins with minimal extra- and intracellular domains and extensive numbers (14 in the case of the eco receptor) of transmembrane domains (10,502,546,765). The resemblance of such proteins to membrane transport systems

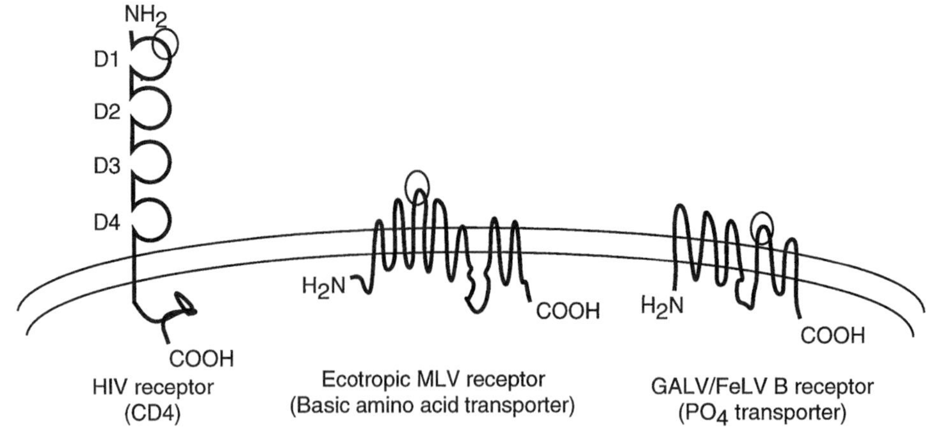

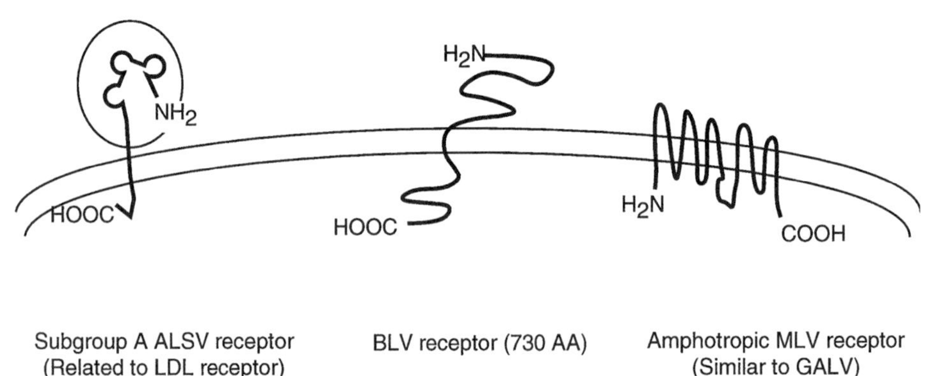

Subgroup A ALSV receptor (Related to LDL receptor) BLV receptor (730 AA) Amphotropic MLV receptor (Similar to GALV)

FIG. 13. Retrovirus receptors. The organization of the various known receptors for retroviruses as inferred from nucleotide sequence and other indirect analysis is shown relative to the plasma membrane (*thin lines*). The circles indicate the region of virus Env protein binding, to the best currently known resolution. Redrawn in part from ref. 794 and derived from refs. 27,33,502,765.

was affirmed by the demonstrations that the eco receptor functions as a transporter of basic amino acids (389,782). Similarly, Ram-1 and Glvr1 are related to a protein that serves as a phosphate transporter in *Neurospora* (352) and can act as phosphate transporters when expressed in oocytes and mammalian cells (379). Again, the natural function of these proteins seems unrelated to their role in virus infection. Interestingly, the GALV receptor serves also as receptor for two other viruses—FeLV (subgroup B) and simian sarcoma virus (719,722). All of these viruses are distinct species but are within the same genus and are quite closely related. It will be interesting to see if all viruses within a taxonomic group use receptors of similar structure.

The receptor for subgroup A ALSV (Tva) belongs to yet a different class of cell-surface proteins (33,819). It is quite a small molecule with a single transmembrane domain and a cysteine-rich extracellular region which resembles one domain of the much more complex, low-density lipoprotein receptor. The function of this protein is unknown. Again, the extracellular domain alone is sufficient to confer receptor activity when expressed with a glycophosphatidylinositol linkage to the cell surface. Given the polymorphism in receptor usage within this genus, identification of receptors for other subgroups of ALSV is eagerly awaited.

The receptor for BLV is another protein with apparently a single transmembrane domain and extensive extracellular sequence (27). It is unrelated to any of the other known retrovirus receptors, and its function remains to be identified.

A candidate receptor for FIV was identified as a cell surface protein which binds tightly to virions (315). This protein has a tissue distribution consistent with that of sensitivity to FIV infection, and a monoclonal antibody against it blocks FIV infection, but it has yet to be shown to confer infectibility when expressed in nonpermissive cells. It has been identified as the feline homologue of the CD9 protein (800). CD9 is a small (24 kD) protein found on the surface of T and B lymphocytes, platelets, and some other cell types (63,420,638). It differs from other identified retrovirus receptors by the presence of a predicted four transmembrane domains.

Virion-Receptor Interactions

The interaction between virion and receptor has been fruitfully studied by comparison of known receptors with homologues from other species which are not functional as receptors. In the case of the mouse eco receptor, for example, the human relative is very similar in sequence, but human cells are not infectible with ecotropic MLV. By making recombinants between the two proteins and using these to guide site-directed mutagenesis, the key region of the receptor for virus binding has been identified as a rather small sequence within the largest extracellular domain (9). Analogous studies have also identified a region in GLVR that confers specificity for GALV in human cells relative to mouse cells (353). Interestingly, these studies also show that different regions of the GLVR protein may be involved in interaction with the other viruses which use it as a receptor (719).

By far the best studied retrovirus-receptor interaction is that of HIV with the CD4 protein (103,510,512). This interaction will be considered in more detail in Chapter 27. In brief, the binding site on the SU protein (gp120), tentatively identified as a combination of regions shown in Fig. 11 (548,811), interacts with a fairly small region on the receptor identified by the use of site-directed mutagenesis and recombinants with the mouse CD4 protein (which is closely related to the human one but does not bind HIV) (125,589,620). The region of CD4 so identified is not directly involved in cell-cell recognition or any other known function of this protein. This binding per se seems insufficient to initiate the replication cycle, since mouse cells expressing human CD4 bind HIV but are not infected by it (346). This observation has given rise to the idea that there may be a second human-cell-specific factor necessary for HIV infection. Despite considerable effort and some apparently false leads (11,73,97,98,99,569), this factor remains to be identified.

Nonreceptor genetic control of early events in infection is not unique to HIV, although the specific mechanism may be. In both mice and chickens, for example, the presence of certain endogenous proviruses creates a block to infection by expression of env genes and induction of interference (77,335,447). In hamsters, receptors for a number of mammalian C-type viruses are present but masked by the production of a soluble factor which is also found in hamster serum (503,805). As a consequence, hamster cell lines, so useful for genetics and biotechnology, have proven almost useless for retrovirology.

Genetics and Distribution of Receptors

Retroviruses are highly unusual among animal viruses in the polymorphism of receptor utilization displayed by otherwise very closely related viruses. In the absence of direct isolation, receptors can be identified and studied in one of several ways (791,794).

Interference

Cells infected by retroviruses display a very strong resistance to superinfection by viruses that utilize the same receptor as the preinfecting virus and complete susceptibility to viruses that use a different receptor. This is easily assessed if the second virus encodes a dominant marker such as an oncogene. This phenomenon obviously arises from interaction of the *env* protein of the preinfecting virus with the receptor, but the level of this interaction is not known. Virion formation is not necessary for interference, indicating that the *env* protein and receptor can interact *in cis* in the same membrane. The interference is virtually absolute; titers of superinfecting viruses can be reduced by more than 10^7-fold, suggesting that more than simple competition may be involved. In the case of HIV infection, there can be a striking loss of receptor protein from the surface of chronically infected cells, suggesting that the interaction may occur during synthesis and/or processing of the two proteins, leading to failure of the receptor to be correctly processed. It is likely that interference in HIV infection is much more complex than with other, simple, retroviruses. At least two other genes probably contribute to the process. *vpu* interacts with newly synthesized CD4, causing its destruction if it is associated with the *env* protein in the endoplasmic reticulum (115,432), and *nef* expression leads to loss of CD4 from the cell surface (42,245). The involvement of three gene products to induce interference, where other viruses apparently use only one, may reflect the higher concentration of CD4 on the cell surface as compared to other known retrovirus receptors and the correspondingly increased difficulty of inducing effective interference.

Cloning of the ecotropic MLV receptor has allowed detailed analysis of the process of interference. Expression of the receptor and the env gene simultaneously completely prevents infection of cells, but has only a relatively modest effect on cell surface expression of the receptor and its transport activity (J. Cunningham, personal communication). Thus, the important effect in interference seems to be direct blocking of virus-cell interaction by continued association of receptor and cell surface *env* protein. This association does not greatly interfere with the normal function of the receptor protein, thus resolving—in this specific case—the general concern that downregulation or blocking of a receptor might exert pathogenic side effects by interfering with an essential normal cell function. The coexpression of env and receptor does lead to a dramatic decrease in the extent of glycosylation of the receptor (189), implying that interaction of env and receptor takes place soon after synthesis of both proteins.

A related phenomenon called early interference can be demonstrated by preincubating cells with virions, disrupted virions, or *env* protein. This effect requires large amounts of protein, is not as strong as superinfection resistance, and is clearly due to interaction of *env* protein with receptor on the cell surface.

Species Distribution

Use of different receptors can result in a distinctive host range that reflects the presence or absence of a receptor in animals of different species. Murine C-type viruses can be distinguished as follows. *Ecotropic* viruses, like Moloney MLV, use a receptor found predominantly on mouse cells, the one described above. *Xenotropic* viruses utilize a receptor found on many nonmouse species but seemingly specifically lacking in most, but not all, mice. *Amphotropic* viruses recognize receptors found on both mouse and nonmouse species, as do *polytropic* viruses. Assignment of the latter two to different groups is based on their failure to interfere with one another (608). The xenotropic and polytropic host ranges are found only in endogenous viruses of mouse viruses that have established proviruses in the mouse germline and are passed from parent to progeny as Mendelian genes (see below).

The assignment of subgroup by chance distribution of the receptor among different species, although firmly entrenched among specialists in this area, is one of a number of unfortunate aspects of retroviral nomenclature that has generated considerable confusion among those trying to understand the field.

Polymorphism Within a Species

In chickens, individual animals (or lines of animals) of varying sensitivity to infection with different ALSV isolates are commonly found. Based on this variation in susceptibility, as well as interference patterns, no less than nine distinct subgroups, denoted A through J have been distinguished (one subgroup, D, is a variant of another, B, and uses the same receptor but differs slightly in other properties). Subgroups A to D are found among exogenous viruses of chickens; E is unique to endogenous viruses of chickens; and F and G are found in endogenous viruses of pheasants. The ability of a cell to resist infection by a specific subgroup is denoted by a slash (or bar) following a letter standing for the species and followed by the subgroups to which the cell is resistant. Thus *C/A* ("C bar A") are chicken cells sensitive to all virus subgroups except A; *T/BD* ("T bar BD") are turkey cells resistant to B and D subgroups, and *C/0* ("C bar oh") cells are susceptible to all known subgroups.

Genetic analysis has revealed the existence of three unlinked autosomal loci responsible for these resistance patterns. These are known as *tv-a*, *tv-b*, and *tv-c*. In all cases, sensitivity is dominant to resistance, implying that the loci encode the receptors themselves. *tv-a* and *-c* have two alleles, encoding sensitivity and resistance. The cloned receptor for subgroup A ALV, as expected, is genetically linked to the *tv-a* locus (33) and presumably represents the s allele. Examination of the r allele will be quite rewarding, but it remains to be cloned. Molecular identification and examination of the *tv-b* locus which controls sensitivity to both B and E subgroup viruses will also be very rewarding, since this locus is genetically more complex. In chickens, three alleles are known, encoding resistance to both subgroups of virus, sensitivity to both, and resistance to subgroup E (the most common pattern), respectively. The fourth combination, sensitivity to subgroup E virus and resistance to subgroup B virus, is not seen in chickens but is the most common pattern in other birds such as turkeys and quail.

In mice, a similar polymorphism in sensitivity to MLV exists. Although most laboratory strains of mice cannot be infected by xenotropic MLV, some wild mouse strains do express a suitable receptor (406). The gene encoding this receptor is allelic to that for polytropic virus (331), analogous to the situation with *tv-b.*

Endogenous viruses of mice and birds—subgroup E ALV and xenotropic MLV—exhibit a property called xenotropism, the inability to infect the animal whose germline they inhabit. The polymorphism among receptors for retroviruses in a given species seems to reflect genetic changes in the species in response to selective pressure for resistance to infection with the virus some time in the past. It is important to bear in mind that the lack of receptors in a given species probably does not reflect lack of the protein, but rather sequence differences in the protein itself from one species to another. The use of such differences can be of great help in localizing the specific binding domains on a receptor molecule, as in the case of CD4, mentioned above.

Penetration and Uncoating

The mechanism by which retroviruses enter cells is one of the most poorly understood aspects of the virus life cycle. In the absence of a good structural model for the *env* glycoprotein and in the face of a very poor ratio of infectious to total virions (meaning that the vast majority of observable events are nonproductive ones), only indirect and inferential evidence as to the mechanism involved is available. After binding of the SU protein to its receptor, the virus envelope and the cell membrane fuse to release the virion core into the cytoplasm. The site of this fusion (directly at the cell surface or following endocytosis) varies among retroviruses and even among cells infected with the same virus (485). In the case of HIV, this fusion occurs at the cell surface immediately after binding. This conclusion is supported by the ability of cells expressing HIV *env* protein to fuse directly to uninfected, receptor-positive cells, a property which has provided convenient assays for infectivity and viral entry (72) (see Chapter 27). Furthermore, penetration of HIV occurs at a pH typical of the cell surface, not the relatively acid conditions that are present following endocytosis (690). Some other retroviruses do not cause cell fusion, and infection by some of these is sensitive to high pH (587). Like many enveloped viruses, these seem to be internalized via receptor-mediated endocytosis followed by fusion of viral envelope and endosomal membrane, possibly provoked by the lower pH typical of the endosomal contents.

Little is known about how the fusion itself occurs. It seems likely that this step is mediated by the region of hydrophobic amino acids at the amino terminus of the smaller TM protein. Site-directed mutagenesis of this region in HIV and other viruses has yielded results consistent with this idea (486,579). Mutations that prevent cleavage of the two proteins to generate the hydrophobic end of TM, as well as mutations which insert charged amino acids into it, preserve receptor binding but block infectivity and cell fusion ability of the virus. The important interaction of the fusion domain could either be with the membrane itself or with some component (such as a protein) within it. Modeling suggests the possibility that, like influenza, the fusion domain may form a pair of coiled coil structures whose interaction with each other in the virion prevents their interaction with the cell membrane. Structural rearrangements in the *env* protein following receptor binding would cause the region to switch to an extended coil structure which could insert directly into the cell membrane. There is good evidence for this process with influenza virus (see Chapter 21), and indirect evidence (based on inhibition of infection by certain peptides) in HIV (E. Hunter, personal communication). The presence of a specific "fusion receptor" for HIV has been invoked to explain both the lack of infectibility of mouse cells expressing human CD4 protein and the sequence similarity between the hydrophobic region in HIV and the F protein of some paramyxoviruses that infect human cells (242).

The process by which the fusion activity of the *env* protein is activated following receptor interaction is unclear. Unlike the case of influenza virus, it seems unlikely to be a rearrangement of the protein due to reduced pH; in the case of HIV, there is some evidence that cleavage of SU in the V3 region by a cell surface protease is required (510); the release of a fragment of SU might then allow the hydrophobic end of TM to interact with the cell membrane.

It is important to bear in mind that the role played by the receptor appears to be strictly a passive one: provision of a signal consisting of a few amino acids to be in the correct conformation recognized by SU. To date, no active role for the receptor has been identified. On the contrary, CD4 mutants, themselves incapable of endocytosis, nonetheless serve as receptors (36); and the receptors for MLV and GALV, although capable of importing small molecules, do not function in endocytosis.

The fate of the various capsid proteins upon fusion is not well known. Clearly, RT, IN, and NC proteins must remain with the genome, and there is good reason to believe that CA does as well (see below). One would expect MA to remain associated with the cell membrane; however, there is some evidence that at least some of the MA protein in incoming HIV virions stays with the capsid and plays a role in nuclear transport (see below and Chapter 27).

Synthesis of Viral DNA

Overview

After entry of the core into the cytoplasm, the process of reverse transcription of the RNA genome into double-stranded DNA occurs, using the structure and enzymatic activities that entered the cell in the virion (252,767). An outline of the process of synthesis of viral DNA is shown in Fig. 14. The mechanism by which this occurs is a complex but elegant solution to two problems that a virus replicating by this mode would inevitably face. First, the requirement of the system for RNA primers precludes precise, end-to-end copying of the genome into DNA, since there would be no way to copy the primer binding region. Second, since new viral genomes are to be made by cellular RNA polymerase II, some way must be provided to ensure that signals appropriate to direct the synthesis of RNA lie upstream of the initiation site of synthesis, outside of the region to be copied. Obviously, such signals cannot simultaneously be inside and outside of the viral genome. The resolution of this paradox evolved by retroviruses and related elements is to provide for the synthesis at each end of the DNA molecule of an extra copy of sequences present only once in the RNA genome. These extra sequences together form a structure, the long terminal repeat (LTR), which contains virtually all of the *cis*-acting sequences necessary for events that take place at the DNA level—integration and expression of the provirus.

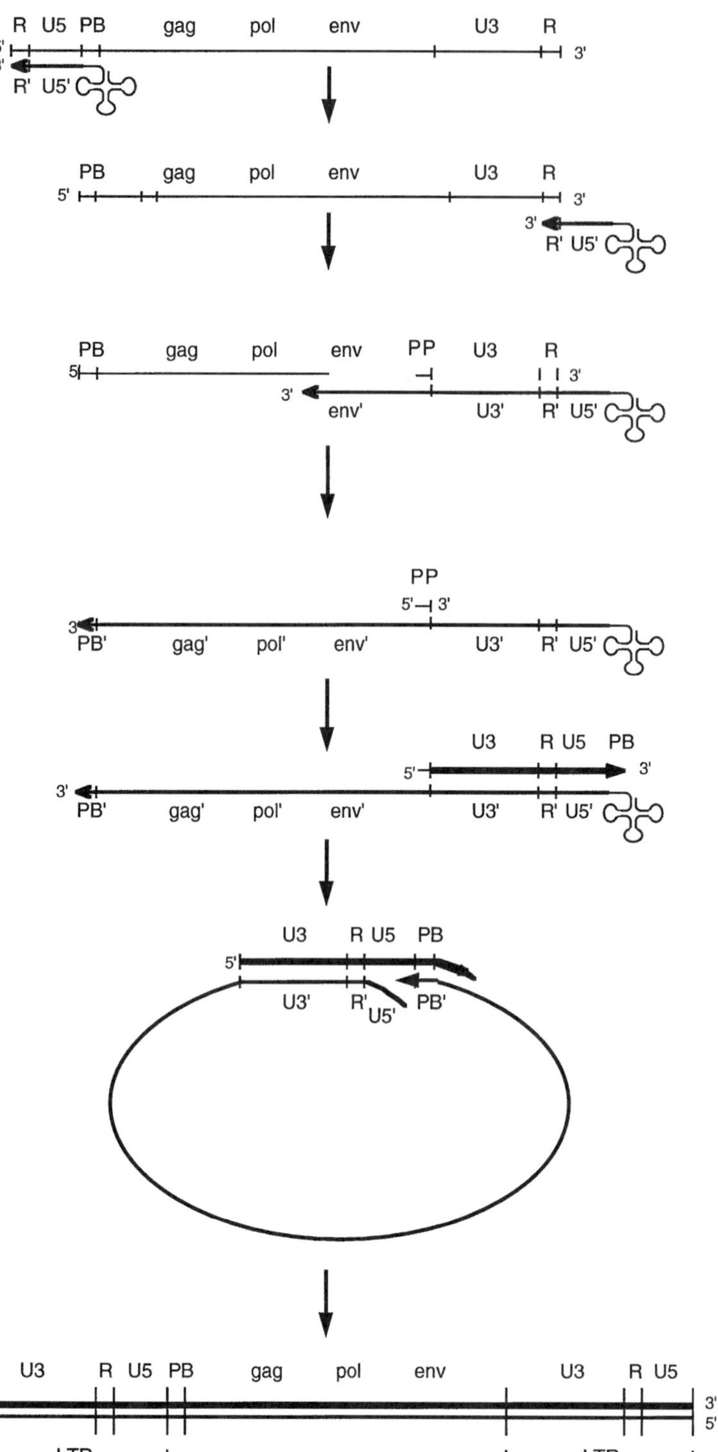

FIG. 14. Mechanism of viral DNA synthesis. *Thin lines* depict RNA; *medium lines*, minus-strand DNA; *thick lines* plus-strand DNA. Minus sense sequences are also indicated (′). Drawing courtesy of S. Herman.

Only one enzymatically active protein molecule—the reverse transcriptase—is necessary to accomplish this unusual metamorphosis, although completely correct synthesis also requires the structural milieu provided by the capsid proteins. The secret of viral DNA synthesis is in the "jumps"—the ability of the enzyme, on encountering a block to DNA synthesis, to transfer the growing DNA chain to a similar sequence elsewhere and then continue elongation. Two such transfers form the LTR by moving copies of the U_5 and U_3 regions flanking R to either end of the viral DNA. Specific sequences at the tips of these regions then provide signals for integration.

The In Vitro *Reaction*

The first inkling that retroviruses replicated through a DNA intermediate—a heretical idea when first proposed and for many years afterward (732)—was provided by experiments on the ability of certain inhibitors of DNA synthesis to block early steps in replication as well as of inhibitors of DNA-directed RNA synthesis to block late steps. This idea received dramatic confirmation with the demonstration that retrovirus virions contain an enzymatic activity which can copy the virion RNA into DNA (26,736). This activity can readily be seen when virions, rendered permeable with a nonionic detergent, are incubated with an appropriate buffer containing deoxynucleoside triphosphates, one of which is usually radioactively labeled. Synthesis of DNA complementary to at least a fraction of the genome then ensues. Under carefully controlled conditions, a full-length, double-stranded and biologically active copy of the genome, complete with LTRs, is synthesized, validating the use of the *in vitro* reaction to study the mechanism. A useful adjunct to this reaction is the addition of actinomycin D to block DNA-directed (plus strand) synthesis, but not RNA-directed (minus strand) synthesis. In this way, it has been possible to obtain a very detailed picture of the overall process.

Initiation of DNA Synthesis

As noted previously, all retrovirus genomes have a characteristic molecule of tRNA joined by base pairing to the PB region near the 5′ end of the genome, and reverse transcription is initiated using this molecule as primer (429). Primer tRNA specificity is highly conserved within virus genera (Table 2). However, in at least some viruses, including MLV, this specificity can be altered by changing the PB sequence (467). Thus, tRNAs other than the natural one can be functional, albeit with reduced activity.

Analyses using site-directed mutations and model structures have indicated that the initiation process may have more complex requirements than this simple picture would suggest, since mutations in U5 downstream of the initiation site can greatly affect the rate of viral DNA synthesis *in vitro* and in infected cells (126,127,528). In many retroviruses, the U5 regions can form base pairs with part of the TψC loop of the tRNA, and formation of this base-paired region, as well as another between U5 and leader sequences to make an overall rather complex secondary structure (5,6), seems necessary for initiation to occur. The exact function of this structure is uncertain.

Synthesis of Minus-Strand DNA

If the *in vitro* reaction products are separated by their size, a number of discrete species representing points at which reverse transcriptase pauses during elongation can be seen. The earliest products are complementary to the genome RNA and are still attached to the tRNA primer at their 5′ ends. The most prominent of such products is a molecule called *strong stop DNA*, which is a copy of the short region consisting of R and U_5 lying between the primer binding site and the 5′ end of the genome (137). Its presence as a major species *in vitro* implies that DNA synthesis initiates by elongation from the 3′ end of the primer tRNA until the 5′ end of the genome is reached. In cells newly infected with ALSV, by contrast, strong stop DNA is not a major species (427), suggesting that the pause at the 5′ end of the template is an artifact of *in vitro* conditions.

Once strong stop DNA is made, the newly made strand must transfer, presumably along with the reverse transcriptase, to the other end of the genome in order for synthesis to continue. Several features of the system make this transfer, or "jump," possible. First, the R sequence, present near each end of the genome, permits appropriate base pairing to direct continued synthesis to the correct position. Second, the RNase H activity of reverse transcriptase removes the newly copied RNA, leaving the DNA free to pair with the R sequence at the other end. Consistent with this requirement, mutations in *pol*, which specifically affect RNase H activity, strongly inhibit this transfer (110,725, 730). Third, some aspect of the capsid structure is necessary for this process. The synthesis of full-length minus-strand DNA by disrupted virions drops off dramatically at concentrations of detergent even slightly greater than optimal. Instead, the major product at greater than optimal detergent is strong stop DNA along with a significant fraction of incorrectly elongated molecules, as if the jump had been to the wrong part of the genome. It appears that the driving force for jumping is more likely to be affinity of reverse transcriptase for its template, and that the major responsibility for ensuring correct jumping lies in capsid structure, not in sequence homology. The complementarity between the R region at the 3′ end of the genome and its copy in the DNA serves to align synthesis properly but does not itself direct the jumping. Indeed, experiments designed to test the heritability of mutations in either copy of R suggest that transfer can occur prior to copying its full length (452,604). Strand transfer prior to the end of a template can also be seen in model studies *in vitro* (163,574). The improved fidelity of correct jumping of reverse transcriptase on RNA templates in the presence of NC protein (12) implies that the conformation of the template imposed by NC is at least one aspect of capsid structure which facilitates alignment of the growing strand and the new template.

The role of NC protein in this process may be still more complex. Based on inhibition of appearance of full-length HIV DNA after infection in the presence of protease inhibitors (532) and the sensitivity of NC to PR cleavage in nucleocapsids *in vitro* (625), it has been proposed that PR activity may play a role early in the infection cycle—possibly through modification (or even destruction) of NC.

The presence of two molecules of RNA in the virion raises the issue of the relative use of the two ends for transfer. Although some evidence suggested an obligatory jump to the strand other than the one on which initiation occurred (558), it is now clear that the same RNA strand can be used as template for the whole process (357). Thus, the diploid genome does not seem to be necessary for completion of reverse transcription. Rather, it is more likely that diploidy provides a means for repairing physical damage to the genome by allowing recombination during the reverse transcription process.

Once the jump has occurred, synthesis of the minus strand can proceed unchecked to the 5′ end of the template, which is now the 5′ end of PB, since R and U5 have been removed earlier by RNase H. Elongation of the growing DNA chain on the RNA template occurs simultaneously with degradation of the template by the RNase H domain of the same molecule, about 18 bases behind the point of synthesis (235,262). The low processivity of RT as compared to other DNA polymerases (326,367) implies that numerous enzyme molecules participate in the synthesis of one DNA molecule, and that there is frequent pausing along the way. It is also likely that the process is frequently interrupted as reverse transcriptase encounters breaks in the RNA template. In this case, jumps analogous to those occurring at the end of the genome may occur, leading to recombination and repair of genomic damage (see below).

The molecule that results from copying the genome is a complete, but slightly permuted, complement of the genome with the structure

3′—PB-*gag-pro-pol-env-PP-*U_3 *R*U_5—*(tRNA)*—5′

with the tRNA primer still attached at the 5′ end.

Synthesis of Plus-Strand DNA

To obtain a completed molecule, it is necessary for reverse transcriptase to copy the minus-strand DNA molecule just made. This process also requires a primer molecule to initiate synthesis. In this case, the 3′ end that serves this role does not preexist; it must be created. This is accomplished by the PP sequence at just 5′ of U3 on the RNA genome. This sequence resists digestion by RNase H, and therefore remains base-paired with the newly synthesized DNA where its 3′ end can prime plus-strand DNA synthesis. Generation of the plus-strand primer can be readily duplicated *in vitro* using highly purified reverse transcriptase and model substrates (327,606). All molecules are cleaved to yield an end at the same location, since close observation of the DNA molecules primed at this site both *in vitro* (222) and *in vivo* (240,427) reveals no detectable heterogeneity. The signal for this specific cleavage lies in the polypurine (PP) tract characteristic of all retroviruses and retrotransposons, as well as the RNase H (468), although the precise nature the signal remains to be worked out.

Cleavage adjacent to the PP tract is quite precise and occurs with every molecule of RNA copied. However, it is possible to detect initiation of plus strand synthesis at other points on the genome as well, both in vitro and in the infected cell. At least some of these are probably the result of the same kind of reaction; the sequence upstream of these sites resembles the PP region . In other cases, the preexisting breaks in the genome mentioned above could serve as priming sites after minus strand synthesis. The extent of use of such additional priming sites varies from one virus group to another; analysis of DNA made in infected cells shows a much greater proportion of fragmented plus strands in ALSV than in MLV infections (427). In the case of spuma and lentiviruses, there is an additional PP region in *pol* which provides a single additional site of initiation of plus strand DNA synthesis, leading to the formation of a discontiuity at this location in the finished molecule (53, 113, 492). Rather surprisingly, this structural feature is essential for continued progress through the HIV replication cycle; if the second PP sequence is removed, the virus is incapable of replicating (112). Restoration of the signal in a different location restores infectivity.

Failure to use the correct site for initiation of plus-strand synthesis can result in the generation of viral DNA forms containing either more or less than a full LTR. Such forms are frequently seen in cloned viral DNA molecules (547,664). Since the lack of a proper LTR tip will block integration, such molecules will be overrepresented among the unintegrated DNA molecules often used for cloning.

Following initiation, elongation of the plus strands is carried out by reverse transcriptase toward the 5′ end of the minus strand, which is soon reached. As with the minus strand, another jump must occur to permit complete synthesis. In this case the redundant sequence used is formed by the copy of the primer binding site at the 5′ end of the minus strand. As can be seen in Fig. 14, the plus strand is elongated by copying the minus strand DNA through the U3, R, and U5 regions, and into the primer itself. Copying the first 18 bases of the primer yields a direct copy of the primer binding site which can then form yet another template-primer pair with its complement at the 3′ end of the minus strand (251). The species of plus-strand RNA just prior to the jump is a prominent product of reactions *in vitro* and *in vivo*, and it is called plus strong stop DNA. The termination point at the 3′ end of plus strong stop DNA is set not by the end of the tRNA primer, but rather by the presence of a modified base in the tRNA molecule itself, an m^1A residue, which cannot be copied. Termination at this point ensures a perfect match of the 3′ end of plus strong stop DNA with the PB sequence at the end of the minus-strand DNA.

Completion of the full-length double-stranded DNA can now be accomplished by the synthesis of each strand to its logical end. All that is required is to clean up some details.

The primer RNAs must be removed from the 5′ ends of each strand. This reaction uses a special feature of the RNase H activity. It can be readily duplicated with purified reverse transcriptase and artificial substrates (110,549). Note that the genome RNA is completely degraded by the action of RNase H during the course of minus-strand synthesis. Retrovirus DNA synthesis can thus be considered as destructive replication, in which there is no net gain of genome-related molecules. Rather, there is a metamorphosis of the genome from two molecules of single-stranded RNA to one of double-stranded DNA. As noted above, a single molecule of RNA appears to be sufficient for all individual steps of DNA synthesis. However, in the usual case, an infected cell contains only a single provirus, and it is not known whether it is possible to derive two proviruses from a single infecting virion. The low frequency of plus-strand transfers from one minus strand DNA to another (357,558) may mean that formation of two complete minus strands from one virion is quite rare.

Sites and Structures for DNA Synthesis and Transport

The site of viral DNA synthesis also remains to be well clarified. Cell fractionation using detergent treatment implied that DNA synthesis takes place in a soluble cytoplasmic fraction.

A number of lines of evidence give strong support to the idea that early events in retrovirus replication do not take place in more or less free solution (as they are usually drawn), but rather within a structure derived from the viral capsid. First, proper completion of viral DNA synthesis *in vitro* seems to be critically dependent on the capsid structure. Second, there is a genetic locus in mice (called *Fv-1*) that encodes intracellular resistance to infection by some MLV strains (585). The block to infection seems to occur following viral DNA synthesis but prior to integration (354,598). Virus mutations which relieve the block map to the region encoding the CA protein (555,632), implying that this protein is still in association with the virus at the end of viral DNA synthesis and possibly later. Third, association between newly made viral DNA and the CA protein has been found for both MLV (65) and ALSV (D. Schenkein, personal communication), and is well established for the retrovirus-like yeast retrotransposon Ty 1 and other reverse transcribing systems (241). In both MLV and ALSV, CA protein derived from the viral capsid and unintegrated DNA cosediment in a structure much larger than the size expected for DNA alone. These structures are poorly understood. In addition to CA and DNA, it is reasonable to suppose (but not directly shown) that they also contain RT and IN proteins, and possibly NC as well. The structures are not very tight and do not protect the DNA from digestion with nucleases (65). In HIV, there is evidence that MA (or the MA domain of a larger polyprotein) remains in association with the DNA (85)

Following DNA synthesis, which takes place rather asynchronously within the first 4 to 8 hours after infection, the viral DNA becomes tightly associated with or within the nucleus. The mechanism of nuclear transport remains to be fully elucidated and may differ among different retroviruses and host cells. The apparent size of preintegration complexes isolated from infected cells seems to preclude passage through a nuclear pore (65). Indeed, in the case of MLV, complexes active for integration can be found in the cytoplasm many hours before their appearance in the nucleus and subsequent integration. Inhibitors that delay the onset of mitosis delay transport and integration until somewhat later, leading to the conclusion that breakdown of the nuclear membrane at mitosis is necessary to allow access of the viral DNA-protein complex to chromosomal DNA (630). This conclusion is consistent with the observed segregation of MLV proviruses into only one daughter cell at the first division after infection, a result which implies that integration is into postreplication DNA (285).

In the case of lentiviruses, integration of viral DNA does not seem to require that the infected cell enter mitosis (440,441,515), implying the presence of a different mechanism of nuclear transport. Somewhat controversial evidence from several laboratories suggests that nuclear localization signals in MA (83,778) and/or Vpr (300) direct the preintegration complex through the intact nuclear membrane by a process that requires ATP hydrolysis (84). The inferred role for MA (or, more precisely, an MA-containing polypeptide) in nuclear transport is rather surprising, since this protein would have been expected to remain in association with the cell membrane following virus entry.

Two additional viral DNA forms appear in the nucleus. These are covalently closed circles containing either one or two LTRs. The circles containing one LTR could be formed from linear molecules by homologous recombination across the LTRs, but it is more probable that they represent an aberration of the DNA synthetic process in which the 5′ end of the plus strand is not displaced from the minus strand at the final step (207,427) (see Fig. 14). The two-LTR circles include a variety of structures derived either by ligation of the ends of the linear molecule or integration of the viral DNA into itself (see below). Once thought to be intermediates in integration (560), it is now abundantly clear that they are side products and functional dead ends usually derived from aberrant reverse transcripts which would have been incapable of integration (184,605). The circles have been quite useful, since they are much easier to clone than the integrated or linear forms. They are also easy to detect by PCR and have been used as markers for nuclear transport of the viral DNA (83).

Integration

Integration is the process most unique to retrovirus replication (80,147,271,672,673,767). Other animal and plant viruses use reverse transcription as part of their replication cycle; no other has a regular mechanism for stably associating itself with the host DNA. Although there are occasional reports to the contrary (292), integration of viral DNA is most likely a key part of every replication cycle. The majority of the viral DNA molecules synthesized are integrated during every round of replication, and integration is probably necessary for proper expression of the provirus as well (221). These properties clearly distinguish retroviral integration from the sort of "integration" that is an occasional aberration of some DNA virus infections (Chapters 9, 28). Indeed, in eucaryotic systems, true integration is found only in retroviruses and the related retrotransposable elements. There is no known precedent, either in other viruses or in normal cellular function. In bacteria, the closest analogy is found in bacteriophage mu for which the biochemical mechanism is virtually identical (507), but with considerable difference in the overall process.

Structure of Integrated Proviruses

Nucleotide sequence analysis of a large number of integrated proviruses has revealed a collection of features common among all virus groups, including retrotransposons (Fig. 15).

First, the provirus is colinear with the viral DNA as synthesized by reverse transcriptase containing the genes in the order they are found in the genome, flanked by LTRs.

Second, both viral and cellular DNA have undergone characteristic changes as a result of the integration process. The viral DNA has been shortened, usually by the loss of two bases (usually TT) at each end. The cell DNA flanking the site of integration has not been grossly perturbed. A short sequence (4 to 6 base pairs) at the cell target immediately adjoining the viral DNA is always duplicated. The length of the duplication is characteristic of the virus group, not of the cell type, showing that specificity for the duplication (created by repair of staggered ends in the target DNA; see below) resides in the virus.

Third, the ends of the viral DNA are always 5′ TG . . . . . . CA 3′. The CA dinucleotide is found at the 3′ ends of virtually all retroviruses, retrotransposable elements, and even bacterial transposable elements. This striking conservation would seem to imply that these specific bases play some indispensable role in the process. It is thus surprising that at least one of them (the A) can be changed to a G without ill effect on MLV replication (141).

Finally, the ends of the LTR are usually characterized by an approximate inverted repeat spanning 2 to 10 bases, which provide a specific recognition signal for the integration machinery. In general, mutations introduced into these sequences have a substantial damping effect on the integration process and, as a consequence, on virus replication (141,559).

Specificity of the Integration Reaction

On the part of the virus, integration is a highly specific process, nearly always making the joint to cell DNA at the

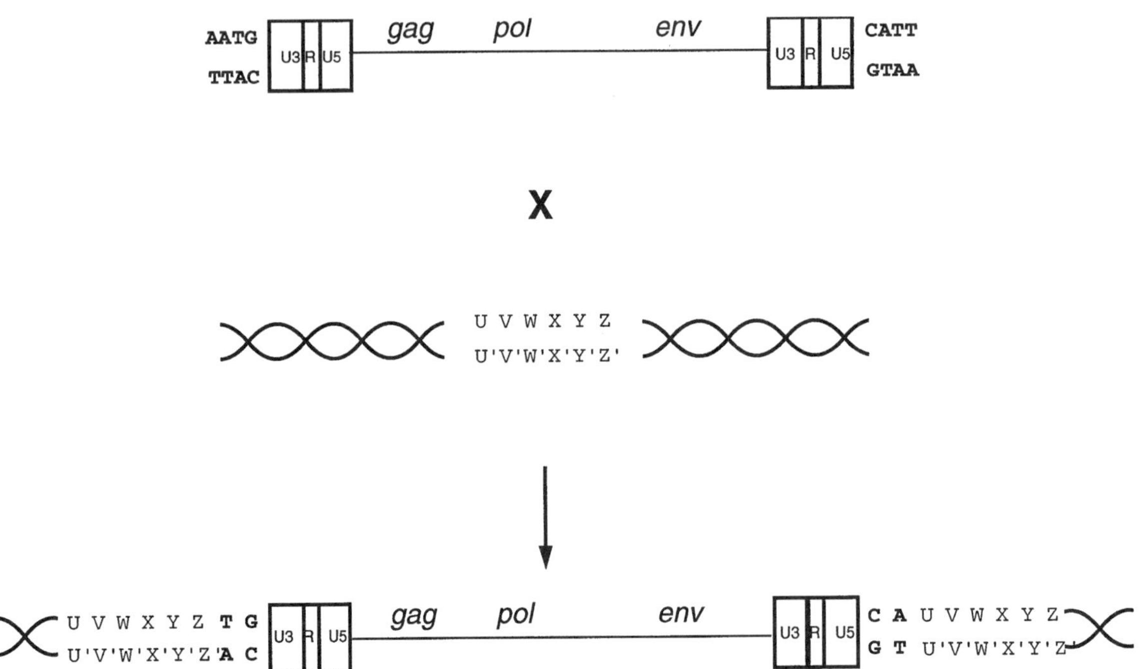

FIG. 15. Properties of retrovirus integration. **Top**: The juxtaposed ends of unintegrated ALSV DNA and target cell DNA. **Bottom:** The structure after integration. Note the loss of 2 base pairs from each end of the provirus and the reduplication of 6 (in this case) base pairs of cell DNA.

same place in the viral sequence. Analysis of the effects of mutations near the ends of the LTRs on integration *in vivo* and *in vitro* show that this specificity is provided both by recognition of a specific sequence containing the inverted repeat at each end and by proximity of this sequence to the ends of the viral DNA (141,774). This sequence, sometimes called *att,* probably forms the only signal required in *cis* by the viral integration machinery. Mutational analysis has narrowed the sequence required to as few as 6 bases from the ends of the LTR (92,527,660).

From the perspective of the cellular target, most approaches to the problem suggest that integration is a much more random process. Comparison of cellular sequences flanking a modest number of independent integration events (12 to 20 in most cases studied) reveal no common sequence which might serve as a cellular target. Similarly, analysis of the distribution of integration events into target DNA *in vitro* is not suggestive of any specificity for a particular site (81), although it may be possible to define loose "consensus" sequences by analyzing large numbers of integration joints (223). More incisive analyses using PCR to determine the pattern of integration into small defined regions of target DNA reveals that integration targets can be found in all regions examined, but that there is a decidedly nonrandom pattern of usage of specific sites within any region (395,597). This pattern differs from one virus to another (Y. Kitamura and J. M. Coffin, unpublished observations) and seems to reflect the interaction of the integrase system with local structural features. Incorporation of a DNA target into chromatin does not block integration but does alter its specificity in a striking way, such that integration targets appear with 10-base periodicity where nucleosomes are present (599). Similarly, integration is not inhibited by C-methylation of DNA, and this modification can even create strong target sites (395). Strong target sites for integration *in vitro* can also be created by introducing bends into the target DNA (596). The role of such features in the infected cell remains to be examined.

Another approach to the issue of integration specificity is to compare sites of integration selected for insertion into the same general region, e.g., in tumors induced by activation of proto-oncogenes. This type of analysis can be complicated by selection for integration in specific regions by effects on cell growth. In B-cell lymphomas induced by ALV inoculation of chickens (see below), the large majority of tumors have an ALV provirus inserted within the first intron of the c-*myc* protooncogene (259,628). Since proviruses have not been found in this region in any other experiment, there is no reason to believe that integration into it is especially favored. Their presence is a consequence of selection of cells transformed by provirus alteration of c-*myc* expression. Selection for proviruses integrated in specific regions can also be more subtle, as a consequence of reduced growth factor dependence of tissue culture cells (746), for example. For this reason, inferences regarding integration targeting are reliably drawn only from analysis of newly infected cells.

Several reports that suggest a tendency for integration to occur in regions of DNA that might tend to be transcriptionally active, e.g., regions characterized by relatively "open" chromatin structure. Proviruses cloned from MLV-infected cells were often found to be integrated in the vicinity of active genes or DNase-sensitive sites which often mark actively transcribed genes (631,648,772), and the *myc*-associated ALV-proviruses mentioned above showed a tendency to be located in proximity to one of five DNase-hypersensitive sites in c-*myc* (628). Another level of specificity has been reported for a fraction of integrations of ALSV into avian cell DNA (662), which suggested that some 20% of integrations are into one of about 1,000 specific sites.

A more recent analysis which makes use of PCR to detect and localize single integration events in large cultures of cells yields a rather different picture of the specificity issue (807). Targets for ALSV integration in newly infected cells were found to be distributed very much like targets *in vitro*. All tested regions of the genome contained integration sites, with very local hot spots. On average, the frequency of targets per region was like that expected on a purely random basis, suggesting the absence of any strong regional targeting.

Mechanism of Integration

Studies of the mechanism of integration have been made possible by the availability of powerful *in vitro* systems. The first of these consisted simply of extracts of cells made at a time (about 12 to 24 hours after infection) when viral DNA synthesis is complete but integration is still ongoing. If an appropriate target DNA is added to such an extract (either from nuclei or cytoplasm), then integration of the viral DNA mediated by preintegration complexes in the extract can be readily detected. Suitable detection methods include the use of a selectable marker and cloning after integration into a suitable vector (81,224), Southern blotting to reveal integration into a closed circular DNA target (206,240,426), and PCR using primers derived from the target DNA and the LTR (395,599). These reactions are quite efficient. In all cases, detailed analysis of the reaction products shows that the *in vitro* reaction is correct. There is a loss of the appropriate number of bases from the end of the viral DNA and a duplication of the correct number of bases of cellular sequence flanking the provirus.

Success with reactions from preintegration complexes led to the development of model reactions using IN protein purified from virions or produced by recombinant techniques and simple labeled double-stranded oligonucleotides as substrates (94,373,378). Although the purified systems are much less efficient than reactions using preintegration complexes, and their biochemical properties (such as divalent cation requirement) are somewhat different, they have permitted identification of IN as the source of all necessary catalytic activity, as well as in-depth analysis and understanding of the integration mechanism. In these reactions, incubation of suitable substrate oligonucleotides with IN, followed by analysis by electrophoresis in denaturing polyacrylamide gels, reveals two major classes of product: a molecule two bases shorter than the substrate, representing the product of the cleavage reaction which removes the 3′ dinucleotide and a heterogeneous collection of larger molecules resulting from the strand transfer reaction in which the 3′ end of the substrate oligonucleotide is integrated into an internal position on another (target) molecule. In the simplest system, substrate and target are the same; however, integration into other target molecules such as plasmid DNA can also be assayed. This provides the basis for some rapid, quantitative assays for integrase activity (149,524).

Using these approaches, the following information has been gleaned regarding the integration process.

First, concordant with the rationale, integration in extracts of infected cells is mediated by the preintegration complexes; the activity of complexes containing only linear DNA molecules implies the linear form of DNA as the important intermediate, a conclusion well substantiated by subsequent biochemical analysis. At least in the case of MLV and HIV, structures purified from soluble cellular components by sedimentation or chromatography are still active in integration, implying that no soluble cell factors are required (65,206). Indeed, active HIV complexes containing only DNA and IN can be isolated (208), indicating that other viral proteins, though often present, may also be dispensable.

Second, the reaction carried out by the MLV complexes is limited to integration into the added target. No side reactions, such as circle formation or autointegration, are seen (82,240). In the case of ALSV and HIV, however, the situation seems to be somewhat more complex. Circular products due to integration of the viral DNA into itself as well as single LTR circles are found among the reaction products in high yield (Fig. 12) (207,426,427), although their formation can be suppressed by appropriate reaction conditions (427). Presumably, there are specific mechanisms to block autointegration *in vivo*, although small amounts of one- and two-LTR circles, as well as the circular products arising from integration of viral DNA molecules into themselves, can be found in infected cells (666). Recent evidence suggests that a host cell factor, tightly bound to the MLV complex, may suppress autointegration in this virus (R. Craigie, personal communication). How this important aim is accomplished in other viruses remains unclear. In the case of ALSV, the DNA in cytoplasmic preintegration complexes is incomplete (427). Perhaps there is a block to some event required for completion of DNA synthesis (such as the plus-strand jump) that causes a delay until the complex is in the nucleus where integration targets are abundant.

Third, most of the 3′ ends of the viral DNA in the complexes have already been cleaved by IN to remove the terminal two bases, while the 5′ end of each strand is precisely at the site of its initiation, and therefore has been modified only by removal of the primer. The ends thus have the sequence

5′ AATG ....CA 3′
3′ AC.............GTAA 5′

Since the majority of unintegrated forms can have this structure some time prior to the integration reaction, its formation must not be tightly coupled kinetically to the rest of the integration reaction. However, cleavages at the ends of a molecule are coupled to one another; mutations in MLV at one *att* site block cleavage at both (529).

The preintegration complex reaction joins the previously formed 3′ end to a 5′ end of the target DNA (82,240). The 5′ ends of the viral DNA are unaltered and remain unjoined. This result strongly implicated a linear molecule as the integration intermediate rather than a two-LTR circular intermediate, as was previously thought. This conclusion was confirmed by the suitability of small oligonucleotides as substrates for integrase.

Fourth, mutational analysis implies that the cleavage and strand transfer reactions take place at the same active site (195,372,414,759), requiring the highly conserved D,D,35 E motif. While the coincidence of the two different reactions at the same site might at first seem contradictory, the underlying mechanism of the two reactions is, in fact, the same (196,759) (Fig. 16). In both cases, the reaction involves a direct attack on a phosphate group by a hydroxyl group, leading to exchange of an internucleotide bond. In the usual cleavage reaction, the OH donor is water; however, other molecules, such as glycerol or the 3′ OH of the same DNA (leading to a cyclic dinucleotide product) can also participate (196). In the strand transfer reaction, the OH donor is the 3′ end of the newly cleaved DNA, leading to a direct transesterification. Thus, the integration reaction is concerted, rather than involving separate cleavage and ligation reactions.

Fifth, there is no obvious requirement for a particular sequence or structure of the target DNA. Indeed, as noted above, both incorporation of DNA into chromatin (599) and extensive methylation of dC residues (395)—modifications associated with reduced transcriptional activity—can create strong target sequences for integration *in vitro*. Their *in vivo* effects remain to be tested.

Finally, the reaction proceeds efficiently in the absence of ATP or any other added energy-generating system (81,94). This independence is consistent with the coupled breakage-rejoining reaction. Also consistent is the observation that IN can catalyze the reverse reaction, referred to as *disintegration*. If provided with a DNA molecule resembling the integration intermediate, IN will

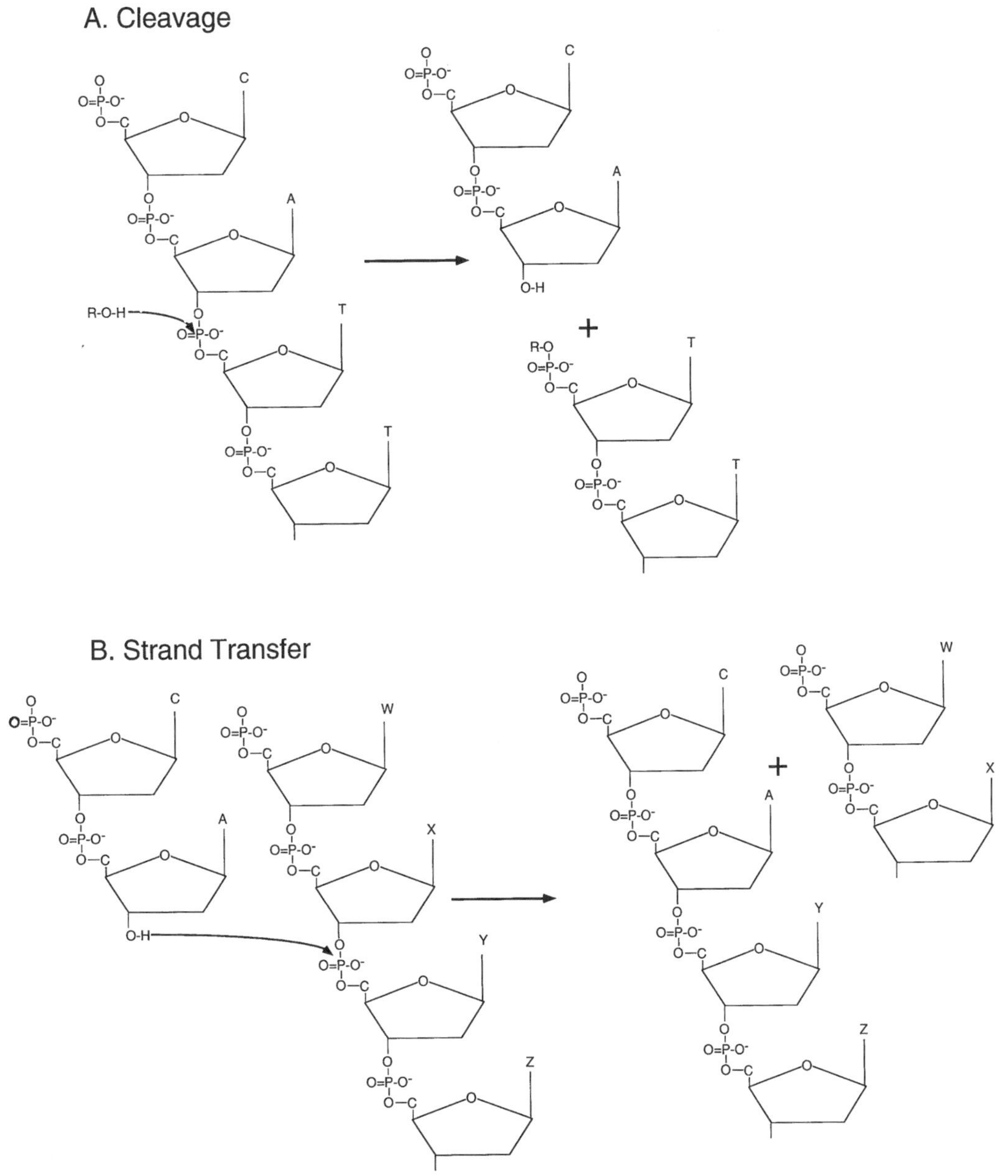

A. Cleavage

B. Strand Transfer

FIG. 16. IN-mediated cleavage and strand transfer. Note that the underlying mechanism (OH-mediated attack on an internucleotide phosphate bond) is the same for both reactions (196,775).

separate it into two molecules, resealing the adjacent nick in the process (120).

All of these considerations have led to formulation of the pathway shown in Fig. 17.

1. Following viral DNA synthesis, the core structure containing linear DNA, and the CA, IN, and possibly RT and NC proteins, enters the nucleus.
2. The 3′ terminal two bases at either end are removed by the cleavage reaction of IN, leaving a 3′ OH end. This reaction may occur before entry into the nucleus.
3. The strand transfer reaction simultaneously joins the two ends of the viral DNA to cellular DNA about half a turn of the helix apart, with the precise spacing determined by the geometry of the IN multimer.
4. A cellular DNA repair system fills in the resulting gap in the molecule, displacing the two mismatched bases at the 5′ end of the provirus and ligating the remaining ends. This gap repair of the initial staggered joints generates the characteristic duplication of cell DNA flanking the provirus.

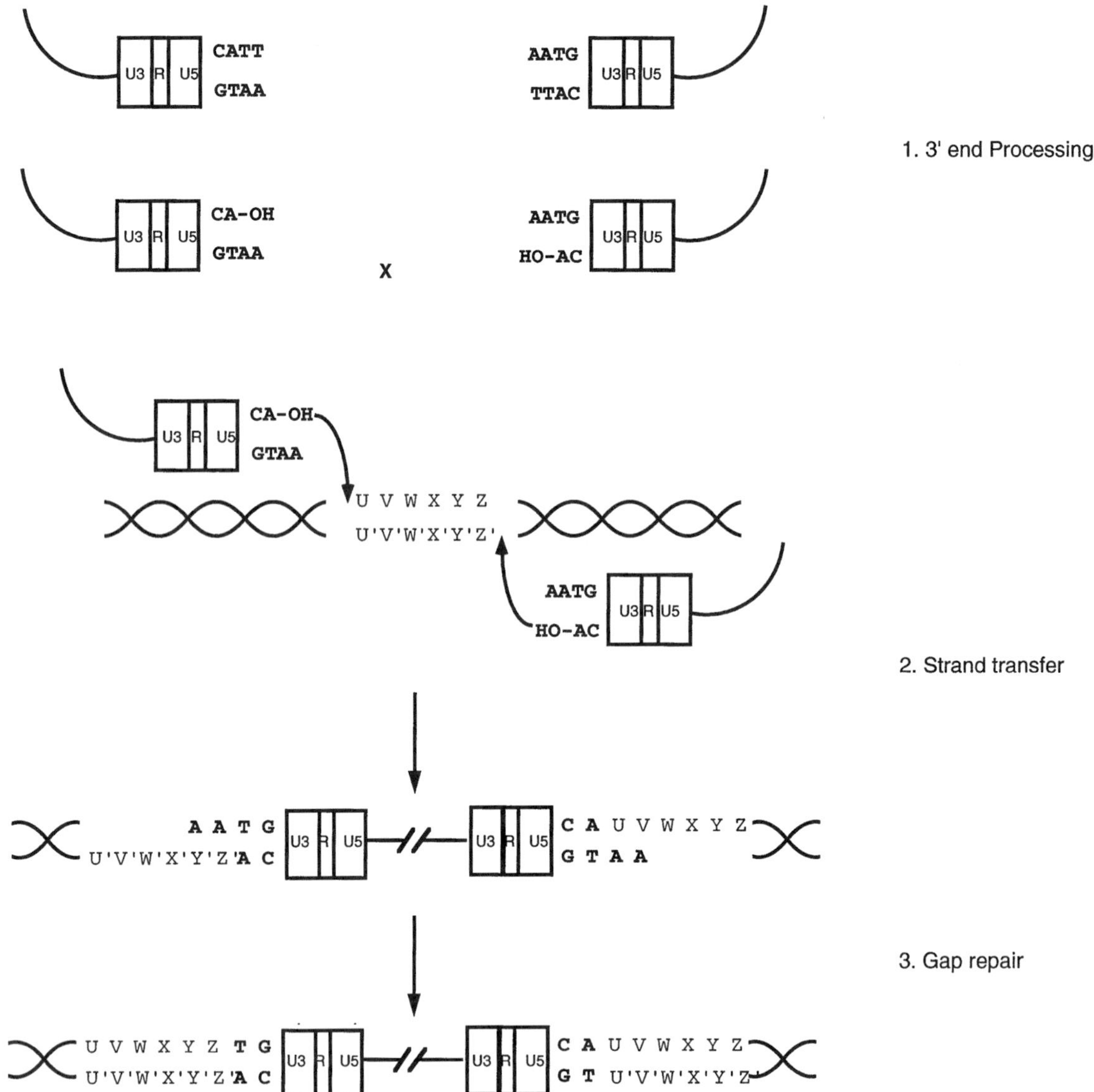

FIG. 17. The overall mechanism of integration. Note that the generation of the recessed 3′ end is temporally unlinked to the subsequent events, but that cleavage of the target and joining of the viral DNA occur as a concerted reaction. The final step may be accomplished by a cellular repair system.

Stability of the Provirus

Once integrated, the provirus, for almost all purposes, can be considered to be perfectly stable. There is no specific mechanism by which proviruses can be excised, moved from one location to another, or replicated independently of the chromosome in which they reside. When occasional loss of a provirus is seen, it appears to be due to random cellular processes, not to virus-encoded functions. Such processes can include deletion of a larger region of cell DNA including the provirus or, more commonly, recombination between the LTRs, leaving behind a single LTR. The best studied case is that of the endogenous provirus of mice whose insertion was responsible for the *dilute* coat color mutation of mice (349). Since recombination across the LTRs causes reversion of the mutation and large numbers of *dilute* mice are raised each year, it is relatively easy to screen for such revertants. When such a survey was done, a reversion rate of about $4–5 \times 10^{-6}$ per generation was estimated (144,657). Somewhat lower values (about 10^{-7})have been estimated for reversion of the d mutation in somatic tissue and for proviruses in tissue culture cells (768).

Expression of the Provirus

Once the provirus is integrated into cell DNA, the virion systems have accomplished their purpose and all further replication is via transcription of the provirus into RNA using cellular systems. This process can be quite efficient:

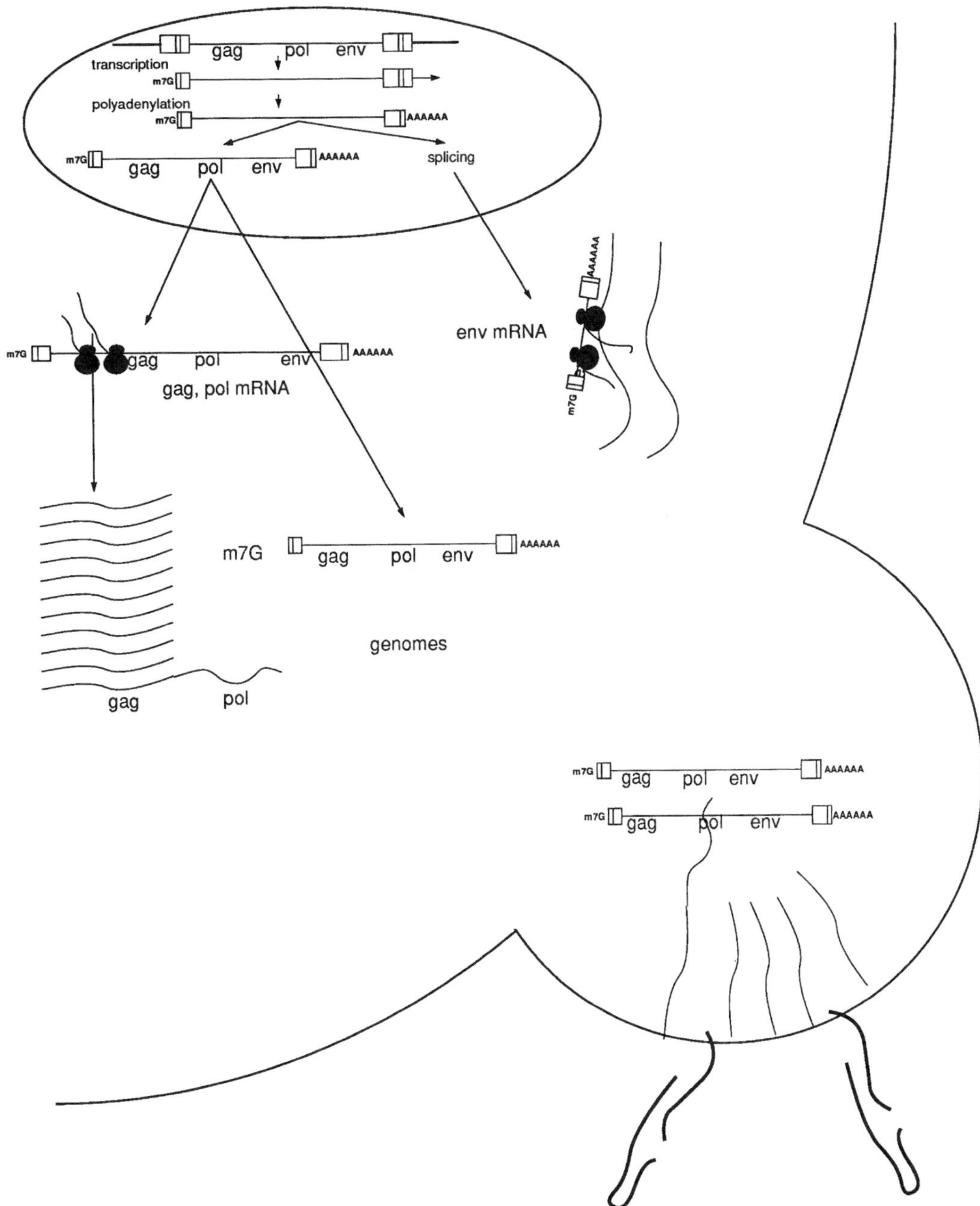

FIG. 18. The strategy of retroviral gene expression.

up to 10% of the mRNA in an infected cell can be derived from one or a few integrated proviruses (J. P. Stoye, personal communication). The overall strategy of expression of retroviruses is shown in Fig. 18. In most cases, the provirus is transcribed into a single RNA precursor, which is subsequently processed by polyadenylation at the 3′ end of R to yield a genome-length molecule and by splicing of a fraction of the transcripts to generate at least one subgenomic mRNA species. In at least one case (the spumaviruses), RNAs encoding regulatory proteins are derived from transcripts initiated at an internal site rather than the LTR site used for all genomes and mRNAs for virion proteins. After transport to the cytoplasm, a fraction of the full-length RNA is reserved for genomes; the remainder is used as mRNA for *gag*, *pro*, and *pol*. These genes are expressed as a nested set of precursor polyproteins. The *gag* and/or *gag-pro*, and *gag-pro-pol* precursors are found in various combinations in different viruses, in graded amounts such that the *pol* peptides are about 5% as abundant as *gag*. This relationship results from partial bypassing of the translational stop signal at the end of *gag*, and/or *gag-pro*. The spliced RNA is translated on membrane-bound polyribosomes to give the *env* precursor protein. Other spliced mRNAs, if present (as they are in the complex viruses), yield a variety of gene products most of which are not found in virions, but rather that act in varied and complex ways to modify machinery of the infected cell and directly or indirectly to regulate expression of the provirus.

Transcription

A major function of the LTR is to provide signals recognized by cellular transcription machinery for the efficient expression of the provirus (470). Transcription of the provirus is initiated at the site where the capping group is placed, by definition the U_3-R junction (or cap site), and proceeds through the 3′ LTR into flanking cell DNA, with the final 3′ end determined at the end of R by cleavage and poly(A) addition (138,280). All retrovirus genomes are synthesized by RNA polymerase II, the same enzyme responsible for synthesis of cell mRNA, and the LTRs of different retroviruses all contain recognition sequences clearly identifiable with corresponding sequences in normal cellular genes, although often in complex combinations. Most of these sequences lie upstream of the initiation site of transcription, and thus within U_3. The disposition within several different LTRs of sequences known or suspected to be important for transcription is shown in Fig. 19. Several different lines of evidence have pointed to the role of these sequences in permitting and regulating virus expression.

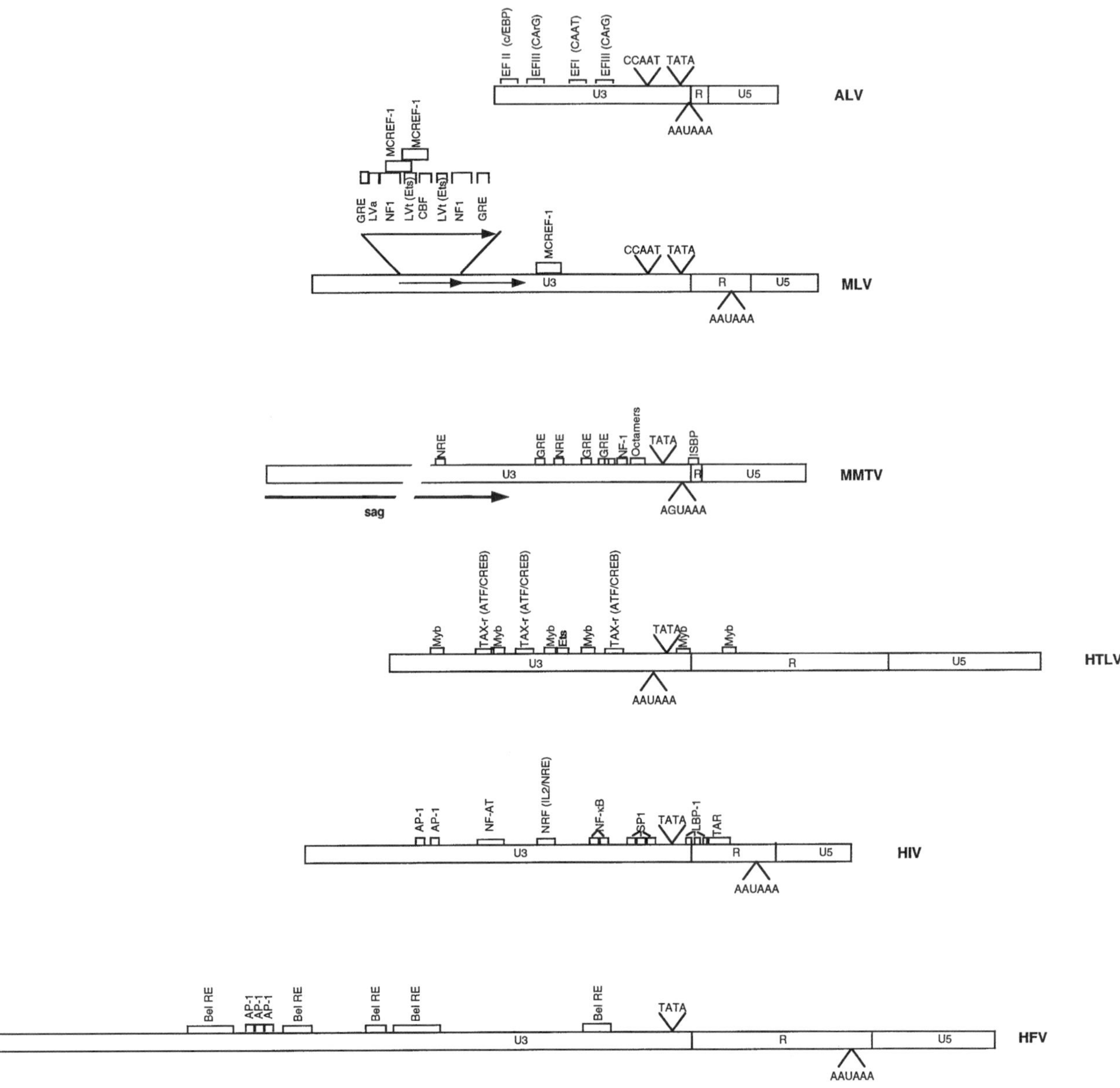

FIG. 19. Transcriptional signals in LTRs. The presence of specific sequence signals used for transcription and binding sites for known or suspected transcription factors is indicated for each LTR. See references (62,282,425,471,481,530,582,654,655,708,823) and Chapter 27, and Chapter 59 in *Fields Virology*, 3rd ed.

Within most groups of viruses, the greatest sequence divergence is often found within U_3. Among members of the ALSV group, for example, the U_3 region of endogenous viruses is quite different from that of their exogenous counterparts, despite the close similarity of the rest of the genome. This difference confers a 10- to 30-fold difference in replication rate (156,744) The reduced replication rate is most likely an essential feature of endogenous viruses since it reduces their ability to induce disease in the infected host (626) and thus permits their benign association with the germline.

A similar example comes from the murine leukemia virus group (204). Again, differences in sequence affecting pathogenesis, either the ability to induce tumors (71,668) or the specific target cell for infection and or transformation (289,445,696,809), are found to lie within U_3. In some cases, these differences may be quite small; only a few bases can determine the relative specificity. These differences lie within the region of U_3 that contains the enhancer elements and, as one might expect, seem to affect the interaction of LTR sequences with one or another cellular transcription factor. Introduced mutations such as the insertion of regulatory regions from other LTRs or even specific enhancer sequences from other groups of viruses can have similar effects. For example, most MLV strains do not replicate in cells of early developmental stages, such as stem cells *in vivo* (109,266), or undifferentiated embryonal (EC) cells in culture (142,220,729). However, when EC cells are induced to differentiate, they become permissive for MLV infection, implying the appearance during development of a different panel of transcription factors. This failure has been traced to the lack of appropriate transcription signals in the LTR as well as the presence of negative regulators (7,220,383,459,749,750,751). Variants can be isolated which provide a somewhat different U_3 sequence capable of supporting expression and replication in these cells (7, 278,594). One interesting such variant has a point mutation in PB which apparently creates an enhancer functional in these cells (32). [As expected from the mechanism of reverse transcription (Fig. 14), half of the progeny of cells infected with this virus revert to the wild-type PB and lose ability to be expressed in EC cells (45).] Also, insertion into the U_3 region of the enhancer region of a mutant strain of polyoma virus capable of growing in EC cells relieves this block to MLV expression (451). Interestingly, expression of MLV proviruses introduced into undifferentiated cells is not induced by differentiation (542,751), an effect related to the permanent silencing of endogenous proviruses (see below) and involving DNA methylation (248).

U_3 sequences can also provide temporal (or conditional) regulation as well as cell-type specific regulation. Expression of MMTV in infected cells is strongly dependent on the presence of exogenously supplied glucocorticoid hormone, a dependence which is important to the transmission mode of this virus—from mother to offspring via milk. In general, MMTV expression in infected mice is confined to lactating mammary glands and lymphoid cells, consistent with the hormonal dependence (although other factors must be involved as well). This specificity can be traced, by mutational analysis, to the portion of U3 just upstream of the cap site, and has presented a valuable model for analysis of the mechanism of hormonal regulation of gene expression in general. Other examples of this sort of regulation can be found in the human retroviruses and their relatives. HTLV-1 expression is strongly enhanced by the action of the *tax* protein, acting through a region of U_3 (Fig. 19 and Chapter 27) (and this regulatory ability can also be transferred into the MLV LTR (392, 393), and HIV expression appears to be positively regulated by the interaction with the NF-κB transcription factor, which is found in activated lymphocytes and has been postulated to effect activation of latent proviruses in vivo.

Standard assays for enhancer and promoter structure and function have been extensively used to dissect the specific sequences involved in the control of expression. These assays usually involve measuring expression of the LTR linked to a reporter gene whose product is easily assayed [such as the bacterial chloramphenicol acetyl transferase (CAT) or β-galactosidase] (543) after their introduction into appropriate cultured cells. It should be kept in mind when reviewing such experiments that they reflect a rather artificial situation in that the activity is being taken out of its context as an integrated provirus and is being introduced in much higher copy number than is the usual case. It has been repeatedly observed, for example, that expression of artificially introduced "proviral" DNA is much lower than that of the same DNA introduced as an integrated provirus by a standard infection (334,642). While the transfection technique has been quite successful, results should always be interpreted cautiously until they are verified in a more natural setting.

Taken together, the results of the genetic and biochemical analyses have permitted characterization of the LTR signals important for transcription. The following general conclusions have emerged.

1. With the exception of a few viral-coded transactivators (such as *tax*), viral transcription control elements are acted upon by the same factors as are used by the cell for expression of its own genes.
2. Consistent with this, the viral signals have sequences similar to the normal cell counterparts.
3. These sequences are usually short, as few as 8 bases, and active in either orientation.
4. The factor recognition signals are often found repeated two or more times per LTR. Such repeats sometimes include only single elements; at times extended regions of sequence containing multiple elements are duplicated. In some MLV strains, for example, a central region containing about 70 bases of U_3 is often duplicated in tandem. This duplication has important ef-

fects on the ability of the virus to induce disease in infected animals .

5. The signals are often—perhaps always—present in very complex patterns and combinations. Presumably, combinatorial subtleties in the arrangement are crucial to the activity and specificity of the LTR. The potential for turning this organization to our advantage to create viruses with tailored expression and replication patterns is obvious, but the rules by which this game might be played are still largely unknown.
6. Small sequence changes affecting the signals themselves can have dramatic effects upon expression. Much of the U_3 sequence, however, is not included in these signals and can be changed with only subtle or no changes in expression (205,453,573,607).
7. The signals themselves can have quite different effects in different cell types. Several of the MLV sequences that activate transcription in fibroblasts and other relatively highly differentiated cell types can have a negative effect on expression in stem cells as well as EC cells.

The specific signals relevant to transcription initiation and found within U_3 include the following.

TATA and CCAAT Boxes

Sequences with a consensus suggested by the names are found in most LTRs (and in most other virus and cellular genes) in relatively fixed location upstream of the cap site. Their presence and relative location are indispensable for virus expression, and they are often seen to be strongly conserved among otherwise highly divergent LTRs. They cannot be deleted, inverted, or moved elsewhere without greatly reducing virus expression. Families of cellular factors interacting with these elements in cellular genes (TF IID in the case of TATA and C/EBP with CCAAT sequences) have been identified. Where studied, these were also found to interact with the virus counterparts (272,640).

Enhancers

Enhancers are defined as control sequences (or groups of control sequences), usually found upstream of the cap site, which increase the frequency of initiation of transcripts at nearby promoters, but do not themselves specify or provide the sites for transcription. Within broad limits, they can be placed at different locations and orientations relative to the promoter elements they control. The most thoroughly dissected enhancer region is that found in the 70-bp repeat of the MLV LTR (Fig. 19) (256). No less than seven binding sites for six different transcription factors can be identified (471,684,708), and variation in the details of the precise sequence of this region and the relationship of the sites to one another is important in determining the level and cell specificity of viral expression and pathogenesis (288,462,463,516,572,685). For example, viruses with only one copy of the enhancer region seem to be less capable of inducing thymic lymphoma in mice than those in which the region is duplicated (311). Indeed, rearrangements involving this region can be observed in animals during the course of infection and tumor induction by viruses containing only a single element (see below).

Corresponding regions of other LTRs have similar roles, but differ considerably in the nature and arrangement of the factor-binding sequences. The HIV U_3 region, for example, can be divided into four subregions, in order of increasing distance from the cap site:

1. The TATA box.
2. Repeated binding sites for the ubiquitous cellular transcription factor SP1.
3. The site responsive to T-cell stimulation through the action of NF-κB.
4. A negative regulatory region.

Binding sites for many well characterized transcription factors can also be seen.

It is possible that, in the context of a complete provirus, enhancers and other transcriptional regulatory elements may reside in locations other than upstream of the cap site. For instance, there is some evidence that the downstream LTR may contribute enhancer activity to the upstream one, since structures with 2 LTRs are more active in transcriptional assays than those with only one (544). Also, a set of sequences with enhancer activity which bind the C/EBP transcription factor has been reported to lie within the *gag* region of ALSV (19,105). Mutagenesis of these gag enhancers shows that they contribute significantly to, but are not absolutely essential for, virus replication (104,641). Similarly, binding of transcription factors to internal regions of the HIV provirus can also be seen (761,762,770,771).

Regulation of Expression

The best studied example of regulated expression of a retrovirus is the induction of transcription from the MMTV provirus by glucocorticoid hormones. As a model, this system has provided the key insights necessary for understanding the mechanism of regulation of hormone responsive genes in general and has provided a promoter useful in genetic engineering, since its level of expression can be controlled from outside (647) and expression in whole animals can be limited to specific tissues. This system has been particularly useful in elucidating the role of chromatin structure in suppression and activation of expression (14,584). In the noninduced state, active glucocorticoid receptor is not present in the nucleus, and nucleosomes are aligned over the LTR so that the NF-1 site (Fig. 19) is blocked. In the presence of hormone, the receptor is released into the nucleus and binds to the GRE sites. This interaction causes repositioning of the nucleosomes and frees the NF-1 site to interact with its cognate factor. It is this interaction, not the direct binding of receptor, that stimulates the transcription complex to initiate RNA synthesis.

In addition to this regulation, other tissue-specific elements, which have been identified by introduction of mutant LTRs into specialized cell lines or transgenic mice (428,508,635) also contribute in an important way to the pattern of expression and replication of the virus in the whole animal (see below).

Transactivation

In simple retroviruses the regulation of expression is passive in that the only contribution of the virus is the signals in the LTR (and elsewhere) which are themselves recognized by cellular factors. By contrast, the complex viruses of at least three groups—lentiviruses, spumaviruses, and the HTLV-related viruses—take a more active role. With these viruses, products of the virus genome are required to achieve high levels of expression. This phenomenon resembles that seen with many DNA viruses in which expression of virus genes is dependent on the presence of one or more viral gene products in the cell. Although the effect is very similar in the two virus groups, the mechanisms are quite different.

The HTLV *tax* protein seems to act as an accessory transcription factor. The target for its action is a 21-base sequence (the TaxRE or TRE) repeated three times within the U_3 region (239,477,656,663) (see Chapter 59 in *Fields Virology*, 3rd ed.). This sequence does not directly bind *tax* protein, implying that the effect of *tax* is indirect. Indeed, the tax recognition motif is the same as that used by transcription factors of the cyclic AMP response (CREB) family, and current thinking is that tax binds to a CREB factor and either stimulates its transcriptional activity or provides an activator domain to complement the DNA binding domain provided by the CREB protein (464,478,540,711,827). A host of other interactions of tax and transcription factors has also been described (16,106,238,305,448,464). In addition to its action on the expression of the homologous LTR, *tax* can also cause the transcriptional activation of a number of cellular genes (at least in cotransfection experiments, see caveat above), including genes thought to be important in regulation of growth of the target T cell, such as T-cell growth factor, IL-2 and its receptor, the proto-oncogene c-fos, and others, including the HIV LTR (via NF-κB) (15,237,488,748,801, 826,830). This effect may be important to the ability of these viruses to stimulate multiplication of these cells in culture and *in vivo*. Induction of growth factors and related genes has been postulated to play an important role in HTLV-mediated leukemia, although direct evidence is so far not available.

A similar sort of transactivation is also found in spumaviruses (226,492), also directed by products encoded by the region between *env* and the LTR, named *bel-1* in the case of HFV and *taf* in SFV (226,382,494,495,612,613). Like HTLV, these proteins stimulate transcription initiation by interacting with multiple sites in U_3 (423,493). Also, they do not bind DNA directly but, like *tax,* must act through interaction with some other factor or factors. Despite these similarities, the spumavirus transactivators, although distantly related to one another (495), show no obvious relationship to *tax*. Furthermore, the various U_3 target sequences do not share any common motifs (425,493), suggesting that each of the transactivators may interact with multiple different DNA-binding factors.

The HIV *tat* protein is unrelated in size, sequence, and mechanism to the HTLV and spumavirus transactivators and has been the subject of intensive study (see Chapter 27). It is clearly not a usual sort of transcription factor. Its target (called TAR) is found in the R portion of the LTR, and is thus present in the transcript, where it can fold into a characteristic stem-loop structure with which it interacts, specifically at a small bulge in the stem (162,167,168,216, 658, 726). The novelty of this interaction has made its mechanism of action quite difficult to study, and several different possibilities have been proposed and are still debated (51, 155,269,348,465,659). These include bypassing of a transcriptional termination site, stimulation of transcription initiation by simultaneous interaction with nascent RNA and transcription factors, and combinations of mechanisms. Tat may also contribute in important ways to HIV pathogenesis, since it has been shown to affect expression, both positively and negatively, of a number of cellular genes (87,88, 121,122,318,361,366,718,797). It also has the unusual property of being released from infected cells and efficiently taken up by uninfected cells (198,228,474). These properties may combine to stimulate abnormal growth of cells in the vicinity of infected cells. This effect has been invoked as an explanation for the origin of Kaposi's sarcoma, an important complication of AIDS (29,197,776).

A special sort of transactivation can occur between completely unrelated viruses (737). Expression of the transactivating gene from a number of DNA viruses, including several herpesviruses (199,314,380,475) and hepatitis B virus (436,670,753,754) and others (219) have been shown to activate expression from the HIV LTR, apparently by acting through several different mechanisms. Transactivation between retroviruses of different genera can also be observed; both HTLV and spumaviruses can activate expression from the HIV LTR (58,381,423). The ubiquity of this phenomenon probably reflects the relatively small number of different transcriptional activation pathways available in the cell and the consequent use of the same ones by unrelated viruses. It is easy to imagine how these mechanisms might have important pathogenic consequences if there are significant numbers of cells harboring both viruses in doubly infected individuals, but a direct demonstration of this has yet to be provided.

3′ vs. 5′ LTR

The mode of synthesis of viral DNA enforces identity of sequence between the two LTRs, yet their functions in virus replication are quite distinct. While the use of transcription initiation signals to generate transcripts initiating in the 3′ LTR is an important method of viral pathogenesis (see below and Chapter 9), it is a side effect of no obvious benefit to the virus. Indeed, close examination of RNA in cells infected with ALSV has revealed no trace of such transcripts (303). What suppresses the 3′ LTR? It has been proposed, and supported with some experimental evidence, that the transit of the transcription machinery across the U_3 region may inhibit its use as a site of initiation (77). This "promoter occlusion" mechanism has solid precedent in some other systems, but insufficient evidence is available to form a firm conclusion. In at least one case, a signal adjacent to the 5′ LTR specifies its use for transcriptional initiation. The *gag* enhancer would be a reasonable candidate for such an activity, but its role has not been directly tested. Consistent with the idea of a 5′ signal is the observation of 5′ deletions, often not affecting the LTR, in proviruses which have activated expression of a proto-oncogene via transcription initiated in the 3′ LTR (57,259).

Position Effects

It is frequently observed, especially by retrovirovectorologists, that the level of expression of a provirus can differ by a factor of 10 or more from one clone of infected cells to another (8,32,213,221,308,343). This effect is usually attributed to the variable influence of surrounding DNA on provirus expression. In support of a role of position in expression are the observations that integrated proviruses are expressed much more efficiently than unintegrated viral DNA and that proviruses acquired by infection yield higher levels of transcripts than do similar DNA molecules introduced by transfection techniques (334). The clearest evidence for such effects, albeit in a special case, comes from analysis of MLV proviruses introduced into EC cells. As noted above, the MLV LTR does not provide a usable promoter function in these cells. However, if cells which express viral genes are selected, a small proportion of surviving cells is found, and independent infections use nearby integration sites (720,787), indicative of selection of rare cells in which integration had occurred in a region capable of directing expression of the adjacent LTR.

Other Sites of Transcriptional Initiation

Until recently, it had been thought that all normal virus transcripts shared the same start and end points, differing only in the pattern of splicing. It has, however, been known for some time that this rule was not absolute, since genes in retrovirus vectors could be efficiently expressed from internal promoters. More recently, evidence has emerged that at least some viruses include "nonstandard" transcription initiation sites as part of their expression strategy. The clearest case is that of the human spumavirus, which has been shown to direct the synthesis of transcripts initiated at a site in the 3′ end of *env*, just upstream of *bel-1* (456). These transcripts, of which there are several splicing variants, (455), can be translated to yield the *bel-1* and *bel-2* peptides. Their expression is dependent on provision of signals in the 3′ LTR, which in turn are transactivated by *bel-1* (454,456). It is likely that this mechanism is part of a strategy the viruses uses to achieve temporal regulation of gene expression, analogous to that of the lentiviruses (455), but how this works remains unclear.

Other, less well characterized examples of nonstandard retroviral transcripts have also recently appeared. No less than two additional transcripts have been proposed for the expression of the MMTV *sag* gene, one in the U_3 portion of the 5′ LTR, well upstream of the usual site (281), and one in the 5′ portion of *env* (501). So far, the *env* site has been seen only in infected and activated T cells, but it may also be important for *sag* expression in B cells as well (U. Beutner, B. Huber, F. Reuss, personal communication).

Processing of Viral Transcripts

Most genomic and mRNA species are derived from a primary transcript that begins at the cap site and extends through the provirus into flanking cell DNA. Responsibility for processing this molecule into finished genomes and mRNAs for the various proteins lies with cellular systems. Again, with many retroviruses, the virus provides only signals; HIV and HTLV and their relatives take a more active role in part of the process.

Polyadenylation

With all retroviruses the genome and most mRNAs have identical ends: the 5′ end as determined by the site of initiation of transcription and the 3′ end at the R-U_5 border set by the site of poly(A) addition. The polyadenylation reaction seems in all important respects identical to that used by the cell to process transcripts of its own genes (280,595). Although primary retroviral transcripts have not been studied, they most likely extend well beyond the eventual 3′ ends. Specific signals in the transcript, including the hexanucleotide AAUAAA and a GU-rich sequence downstream of the poly(A) site, specify coupled cleavage and poly(A) addition reactions, commonly following a CA sequence 12 to 20 bases downstream of the hexanucleotide. An interesting exception to this virtually universal pattern is found in the 276-base separation of the poly(A) signal and the poly(A) addition site in HTLV and BLV (Fig. 19). In this case, most of the R region separating the AAUAAA sequence from the poly(A) site folds into a highly ordered structure, juxtaposing the signal to the correct point (4,28). [This highly ordered structure also forms the recognition site for the rex regulatory protein (see below).]

An interesting problem with understanding retrovirus expression stems from the presence of identical poly(A) signals in both LTRs. As with transcriptional initiation, why is only the correct site used and not the other? With ALSVs and some other retroviruses, the answer is obvious: the AAUAAA poly(A) signal is in U_3 and therefore present only once in the transcripts. With most others, primary transcripts have two copies of this sequence: one in each R region. There are two likely solutions to this problem: either that some sequence upstream of R in the transcript (and therefore found only at its 3′ end) is required for efficient polyadenylation, or there is a minimum spacing requirement such that polyadenylation cannot occur too near the 5′ end (138). To date, experimental evidence on this issue is somewhat conflicting. On the one hand, deletions in the HIV U_3 region decrease polyadenylation efficiency (79,164,165). On the other, a short region containing the signal but lacking U_3 sequences efficiently directs polyadenylation of foreign genes, but not if it is at the 5′ end of the transcript (338,339,786). In HIV, this effect seems to be intensified by the binding of tat protein to the TAR sequence near the 5′ end (785). To complicate the issue still further, sequences far upstream of the poly (A) site can also influence the efficiency of its use, most likely by an indirect effect involving splicing efficiency (165, 504).

In the natural case, polyadenylation at the correct site is not always accomplished with perfection. In cells infected with ALSV, some 15% of transcripts contain complete LTR sequences, followed by up to 2 or more kb of additional nonviral sequence at their 3′ end (304). These RNA molecules, called readthrough transcripts, are the result of poly(A) addition to a long primary transcript at sites derived from host cell sequence flanking the provirus (714). Readthrough transcripts are likely to be important in the acquisition of oncogenes in some retrovirus groups (713) (see below).

Although they are most probably irrelevant to normal retrovirus replication, such molecules are biologically active. They can be incorporated into virions and serve as functional genomes (712). Indeed, although the poly (A) signal is absolutely conserved among retroviruses, it is not absolutely required. Its inactivation in ALSV causes all transcripts to be inappropriately polyadenylated, yet leads to only a small reduction in replication efficiency.

Splicing

The single primary, polyadenylated transcript serves to provide both genomes and most necessary mRNA species (Fig. 20). All retroviruses use full-length transcripts for genomes and as mRNA for the *gag* and *pol* genes, and all yield in addition at least one spliced mRNA—that encoding *env*. In the complex viruses, elaborate splicing patterns give rise to many more subgenomic species encoding not only *env* but also the various regulatory and accessory proteins (124,153,212,226,531,755). In all retroviruses, the spliced RNAs share 5′ and 3′ ends with the genome, and all subgenomic RNAs in a given virus have a splice starting at the common 5′ donor. This site in most virus groups is located in the leader region upstream of *gag,* and genes encoded by subgenomic species supply their own initiation codon. There are some exceptions to this. In HTLV

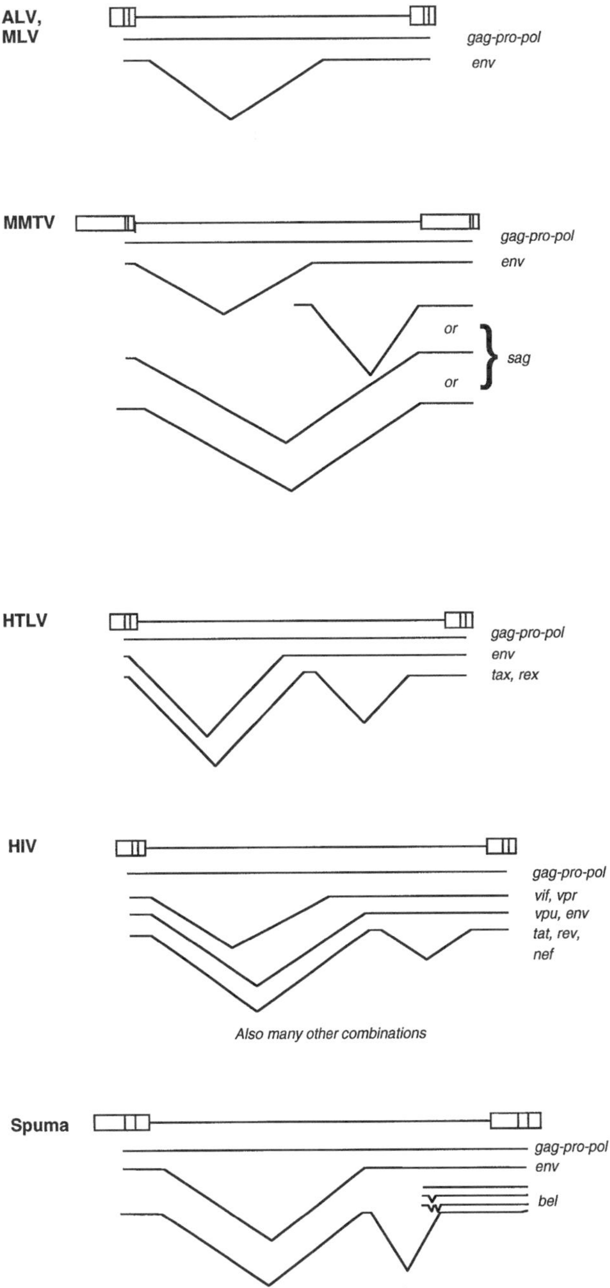

FIG. 20. Splicing patterns of some retrovirus mRNAs.

the donor is located in R about 130 bases from the 5′ end; in ALSV it is found within *gag* such that the complete leader and the first 6 codons of *gag* are appended to the 5′ end of smaller mRNAs.

The necessity for some sort of control of the splicing process is obvious: an appropriate balance between amounts of the various proteins and genomes has to be maintained for optimal replication rates. How this balancing is accomplished is only partly understood. In most cellular genes, splicing of transcripts is complete: RNAs with complete introns are not found in the cytoplasm. This feature involves not only the efficiency of the splicing process, but also an active block to the transport of incompletely spliced or unspliced RNA. Regulation of relative mRNA levels in retroviruses probably involves both regulation of splicing efficiency and provision of signals to permit export from the nucleus of the incompletely spliced RNAs. With most—perhaps all—retroviruses, regulation of splicing efficiency appears to be passive, and splicing is not directly affected by viral gene products themselves. It follows then that signals in the transcripts must specify the partial use of their splicing signals. Sequences of the gene expressed by a given mRNA itself are probably not important; experience with several types of retroviral vectors has shown that many eukaryotic and prokaryotic genes can be placed downstream of splice signals without greatly affecting splicing patterns. The sequence of the identifiable splicing signals themselves also does not seem to play a major role. Virus consensus sequences for recognition by the splicing machinery closely resemble their normal cell counterparts, and the acceptor sequences used by some oncogenes, such as *src*, have been incorporated into the genome essentially unaltered from the ancestral proto-oncogenes, yet now display the splicing pattern characteristic of the virus.

These considerations suggest that sequences affecting the extent of splicing should be sought within the portion of the RNA that serves as intron—i.e., within the *gag* and *pol* coding regions. Examination of these regions has indeed revealed effects of the expected sort—many different mutations introduced into them have strong effects on the relative levels of spliced and unspliced RNAs. The effects observed in this sort of experiment are very complex, and multiple regions of the genome, both well upstream and in the vicinity of the 3′ splice site, have been implicated. For example, insertion in the ALSV genome of a short sequence 12 bases upstream of the splice acceptor for *env* causes the acceptor to be used much more efficiently and leads to loss of infectivity of the resulting virus (371,375). A variety of single-base mutations in the general vicinity of the duplicated sequence, but not necessarily directly affecting either it or the consensus acceptor sequence, can restore the normal phenotype. Other mutations affecting splicing have also been found in this region, as well as in the *gag* region of the same virus (43,489,490). The upstream region identified by these mutations, called *NRS* (for "negative regulator of splicing") by one group (490), can reduce splicing efficiency if inserted into an intron of a cellular gene (489), apparently by interaction with components of the splicing machinery (258).

In contrast to its effect on the virus, nonviral transcripts containing the NRS are retained in the nucleus. This result implies the presence of another signal in the viral transcript necessary for transport of unspliced mRNA into the cytoplasm. With the complex viruses, this signal is apparently provided by interaction of the regulator (*rev* in lentiviruses, *rex* in HTLV and BLV) with its target in the RNA (see below and Chapter 27, and Chapter 59 in *Fields Virology*, 3rd ed.). The recent observation that a fragment of the MPMV genome from the untranslated region between *env* and the LTR can relieve the dependence of HIV expression on *rev* (69) implies the presence of such a signal in at least one simple virus, and it will most likely be found in the others as well.

In the case of at least the two groups of complex viruses, regulation of the relative levels of completely and incompletely spliced RNAs is an active process. Both these viruses encode gene products—*rex* and *rev*, respectively—which, although quite different in size, structure, and mode of expression, affect relative mRNA levels in a similar fashion (154,633), possibly through the same pathway. These proteins act to regulate the relative mRNA levels in a particular way. In the presence of *rex* or *rev*, the full complement of mRNAs is generated. In their absence only multiply spliced mRNAs accumulate in the cytoplasm; unspliced and singly-spliced *env* mRNAs are found at reduced levels and only in the nucleus. The inferred mechanism for this effect is that (as with cellular mRNAs) there is a block to transport of RNAs which retain one or more complete introns, and that interaction of the regulatory protein with the RNA relieves the block. Presumably, the *rev* and *rex* proteins also interact with cellular transport machinery. The HIV RRE sequence (for *rev*-responsive element) is a region of RNA within the *env* gene predicted to have a complex secondary structure. The functionally corresponding region in HTLV, known as *RxRE*, also is very highly structured, but in a completely different part of the genome—constituting most of the R region (741). Despite the difference in location of their response elements, *rev* and *rex* probably act by highly similar mechanisms. Indeed, *rex* can substitute for *rev* in HIV expression, at least to a point (621).

Proteins regulating relative amounts of spliced mRNAs could feed back into the overall replication cycle in important ways. Since the transactivating proteins tax and tat and the regulatory proteins rex and rev are expressed from multiply spliced RNAs, they would accumulate early after infection or induction of expression, in the absence of expression of *gag*, *pol* and *env*. Later, as the regulatory proteins increased in amount, the amount of *gag-pol* and *env* mRNA would increase rapidly, and relatively more virion proteins would be synthesized. Thus, a sort of temporal regulation of expression could be achieved, crudely analogous to that seen with many DNA viruses. This pattern of expression is observed after HIV infection of cells (397).

Translation

After transport to the cytoplasm, the spliced RNAs serve as mRNA for *env* as well as for regulatory genes or some oncogenes when present. The full-length RNA has two different fates. Some molecules become new genomes; others serve as message for the several proteins encoded in the *gag, pro*, and *pol* genes. How molecules are selected to enter either the mRNA or the genome pool is not known. Perhaps initial binding of the *gag* precursor is sufficient to repress translation and commit an RNA to becoming a genome. When the concentration of a *gag* protein is low, RNA would instead become associated with polysomes and would subsequently be unavailable for packaging. This question is intimately associated with the issue of packaging of genomes, and deserves further study.

gag-pro-pol

Compared to cellular mRNAs, retroviruses have unusually long leader regions preceding the initiation site for translation of *gag* (see Table 2). Unlike picornaviruses, which seem to use a special mechanism for initiating protein synthesis at internal translational initiation sites (IRES sites) located at some distance from the 5′ end of the genome (see Chapter 16), the translation of retrovirus mRNA apparently follows the standard "scanning" model in which ribosomal subunits bind initially to the capping group and move along the RNA until the *gag* AUG initiation codon is encountered (581). [However, picornavirus IRES sites can provide functional initiation sites if inserted into retrovirus vectors (3,250).] At least in the case of ALSV, ribosome scanning means traversing several small ORFs, each of which has an initiation codon in a less-than-optimal context but at least one of which is translated into a small peptide (284). The function of these reading frames is unclear, although their presence seems to be required for efficient replication (520,521). The ALSV leader seems to be unusually specific for the class of host cell. In mammalian cells, it is utilized very inefficiently (370). This type of specificity is not observed in other retrovirus groups; certain murine leukemia viruses can replicate efficiently in avian cells.

An unusual feature of mammalian C-type viruses is the presence of two *gag* gene products. In both murine and feline leukemia viruses, translation of a significant fraction of RNA molecules initiates at a GUG codon well upstream of and in the same frame as the standard *gag* initiation codon (756). The resulting protein thus contains all of *gag* sequence with additional amino acids at its 5′ end. The additional amino acids contain a signal peptide which directs membrane translocation and then is removed. The resulting uncleaved *gag* precursor is then processed like an extracellular protein. It is glycosylated and exported from the cell and ends up in the extracellular matrix (583). The function of the glycosylated *gag* protein is not entirely clear. Although its strong conservation implies a functional role, introduction into the MLV genome of mutations expected to prevent its synthesis has little effect on the replication of the virus in cell culture, but does diminish spread and pathogenicity of Friend MLV in mice (146).

In all retroviruses the *gag*, *pro*, and *pol* genes can be viewed as forming a single translational unit which is expressed to yield a set of nested primary translation products of the sort

gag

gag-pro

gag-pro-pol

such that only one-tenth to one-twentieth as much *pol* as *gag* protein is made. [Note that most viruses encode only one or the other of the first two products.] This strategy serves both to ensure a proper ratio of the various proteins to one another and to provide the *pro* and *pol* proteins in association with *gag* so that they can be incorporated into virions (see Fig. 6). There are two different mechanisms which retroviruses have evolved to accomplish this proportional synthesis, both of which take advantage of quirks in the eukaryotic translation machinery and involve provision of a translational termination signal at the *gag-pro* and/or *pro-pol* boundary that is occasionally bypassed by the translating ribosomes to incorporate the adjacent downstream protein into the same molecule (296,340,434).

The first such case to be recognized was that of MLV (Fig. 6). In this instance, the *gag* and the *pro-pol* reading frames are in the same reading frame but separated by a single UAG (amber) translational terminator. Examination of the amino terminal amino acid sequence of the mature PR protein revealed the presence of a glutamine residue, usually encoded by a CUG (or, less likely, CUA) codon, at this site (818). Thus, approximately one time in twenty, the translational machinery must misinsert a glutamine at the position of the terminator, allowing continued translation through *pro* and *pol*. This sort of nonsense suppression is a well-known mechanism in bacteria, but is not known to occur in normal eukaryotic gene expression. It has been observed in the expression of the nonstructural proteins of alphavirus genomes, where it apparently plays a similar role (704).

Nonsense suppression is not a general characteristic of UAG terminators or of glutamine tRNA. Other terminators (UAA and UAG) can be inserted within the same sequence and also give similar levels of suppression but insert different amino acids (217). Comparison of viral sequences and mutagenesis studies implicate two regions of the RNA: a conserved purine-rich sequence immediately downstream of the terminator and a 49-base region containing a stem-loop structure, followed by a sequence that can base-pair with the loop, forming a structure called a pseudoknot (215,218,313,804). It is likely that these sequences act together to slow or halt transit of the ribosome over the site of suppression, but the underlying mechanism is still unclear.

In all other retrovirus genera, the *gag* (and/or *gag-pro*) reading frame ends in a translational terminator and is also in a reading frame different from that of *pol* (Fig. 6). Thus, in the majority of cases, simple nonsense suppression cannot suffice for readthrough into the next coding region; a shift of reading frame must occur as well. In most retrovirus genera, the shift of frame must be in the -1 direction. In the case of spumaviruses (as well as the retrovirus-like Ty elements of yeast), however, a +1 frameshift is required (538). As with nonsense suppression, there is precedent for frameshifting in certain prokaryotic systems, but none had been known in eukaryotes, and the existence of a specific spliced mRNA was at one time considered the more probable mechanism. Clear demonstration that the mechanism of readthrough involved "frameshift suppression" was obtained initially for the junction between the *gag-pro* and *pol* frames of ALSV when RNA which was synthesized *in vitro* by purified prokaryotic enzymes was used as an mRNA *in vitro*. This RNA, which could not possibly have been spliced, could still serve as mRNA for the appropriate ratios of the two polyproteins when presented to eukaryotic ribosomes. Using this system, it was possible to define precisely the sequences necessary for the phenomenon and to determine the amino acid sequence at the site of frameshifting. In the ALSV case, the following sequences were found:

PR—...Leu Thr Asn Leu ***

UUGACAAAUUUAUAGGGAGGGCC . . .

RT—...Leu Thr Asn Leu

Ile Gly Arg Ala . . .

Here there is apparently an occasional slippage of the ribosome to the -1 frame so that the sequence around the terminator is read *AUA-G* . . . instead of the expected . . . *A-UAG*. This slippage requires the seven nucleotides shown in boldface as well as the presence of a region with a predicted stem-loop structure just beyond the frameshift site. As in the case of the mammalian C-type viruses, the stem-loop joins with a sequence downstream to form a pseudoknot, which is absolutely required for frameshifting to occur (203,533,563). These requirements seem to be general. In all viruses studied, a specific (but different) seven-nucleotide A-U rich sequence at the frameshift site along with a pseudoknot structure following it seem to be the essential features for shifting of translation to the -1 frame. Apparently the specific sequence permits (but does not force) the ribosome containing the charged tRNA in the A site to slip back one base when it encounters the obstacle formed by the secondary structure. The efficiency of this event is adjustable over a wide range so that similar ratios of *gag* and *pol* products always result, no matter the configuration. In ALSV, where only one such event is required, the probability of frameshifting is about 5%. In MMTV and HTLV-2, where *gag*, *pro*, and *pol* are in three different reading frames, requiring two frameshifts (Fig. 6), the first takes place about 25% of the time; the second about 10%, giving an overall ratio of gag-pro-pol precursor to gag precursor of about 2.5% (306,341,469).

The mechanism of the apparent +1 frameshift in spumaviruses remains to be worked out. In Ty1, a similar shift is induced by a much smaller site (37).

Again, this mechanism has not yet been seen in eukaryotic gene expression, but is used in at least one other virus group—the coronaviruses (70) (see Chapter 18).

env

Unlike the translation of *gag*, *pro*, and *pol*, which takes place on free polyribosomes using full-length RNA; the *env* protein product, like other cell-surface proteins, is synthesized on polyribosomes associated with the rough endoplasmic reticulum using a spliced subgenomic mRNA. To direct this process, all primary *env* protein products contain an unremarkable signal peptide at their amino termini. As with cellular proteins, this peptide directs the system to initiate transmembrane synthesis and is itself cleaved off after insertion into the rough ER by the normal cell machinery for this process. Following translation, all retroviral *env* precursor proteins remain anchored to the membrane by the hydrophobic membrane-spanning domain near the carboxy terminus; deletion of this sequence causes the protein to be exported and released from the cell in soluble form (578). The first reaction in the addition of N-linked carbohydrate side chains occurs during or shortly after translation. Further modification of the carbohydrate (trimming of the high mannose core, addition of other carbohydrates to the branches, addition of terminal sialic acid residues) takes place in the Golgi apparatus, as does cleavage of the precursor into the SU and TM peptides, which remain associated with one another (332). The cleavage uses a cellular enzyme system, presumably the same furin used to process the influenza HA protein, as well as other viral surface proteins (see Chapter 21). It always occurs at a characteristic amino acid sequence—following a string of at least three of four basic amino acid residues, usually of the form Arg/Lys-x-Lys-Arg (171). Failure of this cleavage to occur does not necessarily inhibit further processing or assembly, but does yield noninfectious virions.

Virion Assembly

Assembly of virions is a poorly understood aspect of retroviral replication (803). In the electron microscope, two

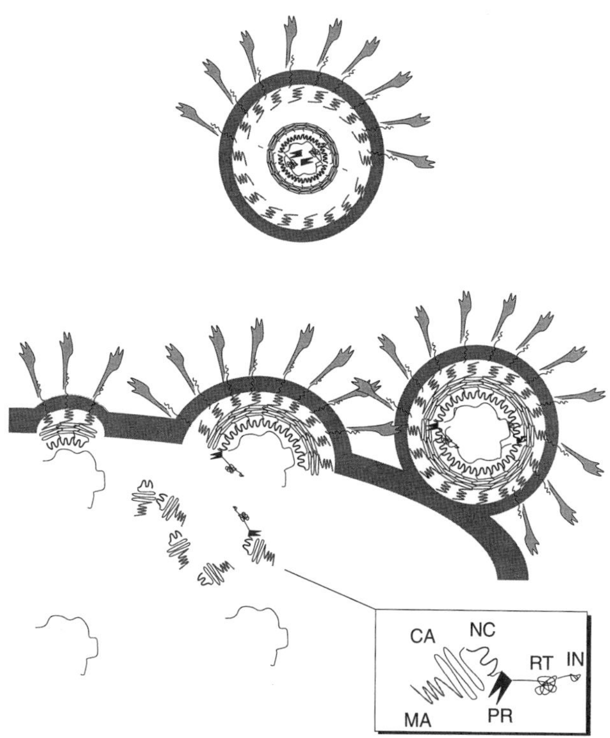

FIG. 21. A model for retrovirus assembly. The scheme would apply to most retroviruses (except B and D types). Note that cleavage of proteins and assembly of the genome do not take place until around the point of release of virus.

patterns can be discerned, differing in the site of assembly. With most virus groups, assembly of capsids and budding are simultaneous. The assembling particles are first visible as crescent-shaped patches on the inner face of the cell membrane, which then appear to extend until the ends meet to form a hollow sphere as the membrane wraps around to form the envelope. With B- and D-type viruses, capsid assembly takes place in the cytoplasm, leading to formation of an A particle, which only later associates with the membrane and buds out. In both cases, a structural rearrangement of the capsid to a more condensed form is visible during or shortly after release of the virion.

A diagram of how this process is currently thought to occur, based on a model first proposed over 15 years ago (60), is shown in Fig. 21. In outline, assembly involves interaction of the gag precursor protein with the genome, with itself, and with the cell membrane in such a way that the C-terminal (NC) end associates with the genome, the middle (CA) with the CA region of other precursor molecules, and the N-terminal (MA) end with the cell membrane. In the case of B- and D-type viruses, the membrane interaction takes place only after the others; in all other viruses, the three interactions are more or less simultaneous. Although it would seem obvious that the various domains would be responsible for the same interactions in the precursor as after cleavage, recent evidence suggests that regions important for assembly are not coincident with the cleavage products. The same interactions on the fraction of molecules extended through *pro* and/or *pol* serves to bring these domains into the virion in the appropriate location and amount. Following assembly, proteolytic processing mediated by the PR protein separates the domains into individual proteins, thus preparing the system for reverse transcription when it encounters the proper environment. The separately processed and exported *env* protein complex is incorporated into the virion envelope via some sort of interaction with one of the gag proteins, presumably MA.

This scheme has a few interesting features. First, the capsid proteins function in different forms at different times in the life cycle—in a polyprotein precursor during virion synthesis and as cleaved proteins during early events—penetration, reverse transcription, and integration.

Second, the process is unidirectional. It would probably be impossible to assemble a capsid from the cleaved components, since it is not assembled that way in nature. Also, the unprocessed form of the capsid is inactive for reverse transcription (150,693,694) (although the MLV gag-pro-pol precursor has some reverse transcriptase activity (214), preventing intracellular buildup of viral DNA, a potentially lethal event for the infected cell (see below).

Third, although all proteins and the genome are required for full infectivity, most are dispensable for assembly of recognizable virions. Indeed, virions are produced in the absence of genomes, the pol, env, and pro products, and the NC domain of gag. The only two domains required for the process are MA and CA. In fact, a small (180 of 700 amino acids) fragment from the amino terminus of the ALSV gag precursor containing the MA and some additional sequence, but none of the CA domains, is sufficient for release of particles, albeit devoid of genome and of aberrant density (795). Thus, there is likely to be a self-assembly domain in the N-terminal portion of the gag precursor outside of CA.

Role of Specific Regions

Genetic and physical analyses have been applied to determine the role of the various domains of the precursor proteins in the formation of retrovirus virions. In general, the analyses support the model presented above (and in Fig. 21) and illuminate some of the details of the process. Because of the nature of the assembly process, it is necessary to interpret negative effects on assembly in such experiments with great care. Mutations in all parts of the precursors are likely to affect their overall folding or stability or to create signals for inappropriate targeting. For example, effects on assembly caused by relatively small changes in protease or integrase (562,665) cannot be interpreted to mean that these proteins play important roles in the normal assembly process, since mutants which lack *pro* and *pol* altogether assemble normally.

MA

In most retrovirus groups, the amino terminal domain of gag is modified by the addition of a hydrophobic myristate chain. The expectation that this modification might be important for membrane association is well supported by experiment: mutations in MLV which eliminate its ability to be so modified also eliminate membrane association, capsid formation, and budding (609). Interestingly, such mutations also eliminate cleavage of the gag polyprotein (652), consistent with the idea that cleavage requires assembly. The specificity of the myristylation requirement is better demonstrated in D-type viruses where a myristylation-negative mutant does not inhibit capsid assembly, but only membrane association and budding (616). The gag protein of ALSV (and some lentiviruses) is an exception to the rule of myristylation, and its amino-terminal acetyl group can hardly be sufficiently hydrophobic to support membrane association, implying the existence of a specific (but unknown) receptor. Interestingly, a mutation which replaces the normal ALSV gag N terminus with a myristylation site does not affect assembly or infectivity (200).

In addition to the myristylation requirement, mutant analysis reveals sequences in the MA domain required for stability (perhaps proper folding) of the gag precursor (359,618) and for targeting of the precursor to the cytoplasmic face of the cell membrane (202), where a basic region in the MA domain apparently interacts with acidic phospholipids (828). In the case of D type viruses, a similar signal directs assembly at an intracellular site, and a single base change can redirect assembly to the plasma membrane, leading to a morphogenetic pathway indistiguishable from C-type viruses although the resulting virions are noninfectious (617).

At one time it was thought that the MA protein was also an RNA binding protein—indeed, that specificity for recognition of the genome RNA resided in it. More recent evidence suggests that the original observations might have failed to distinguish completely among the gag proteins (496), and current thinking assigns this role to NC (688).

The final role in assembly attributed to MA is interaction with the env protein complex to facilitate its association with the virion envelope. As noted below, it is not clear that this interaction occurs with all viruses. In the case of HIV, mutations in MA that do not affect virion formation and gag processing often reduce or eliminate the incorporation of env protein into virions (174,821). The specific interaction implied by these results is supported by the observation that virions containing chimeric HIV gag proteins with the MA region from visna virus do not incorporate HIV env proteins (174). The nature of this interaction remains to be elucidated.

CA

There is no reason to doubt that CA forms the core shell prominent in electron micrographs and necessary for formation of capsids. Despite its apparent importance, relatively little is known about the structure or mechanism of action of this protein in virion assembly. Mutations in CA can have a variety of phenotypes, ranging from failure to assemble to assembly of noninfectious but normal-appearing virions to no detectable effect (290,703). Altogether, these have not been as useful as mutations in other domains for probing the function of the region.

NC

The NC protein, as befits its RNA-binding properties (496,688), has been shown to be required for inclusion of RNA in virions (265). In both ALSV and MLV, mutations in the region encoding the NC domain, particularly in and around the cys-his boxes supposed to mark binding sites, lead to synthesis of virions lacking genome RNA, and in some cases any RNA at all (185,264,497,498,499). Although most binding assays characterize NC proteins as nonspecific RNA binding (and annealing) proteins of high affinity, they most likely confer the necessary specificity for packaging genome RNA as well. Some evidence for this idea comes from studies of chimeric gag proteins with the MLV NC region in an ALV backbone (186); however, this issue needs to be better addressed. One could imagine that a level of specificity might exist in the uncleaved precursor but then be lost from the cleaved form; alternatively, otherwise-modified variants, e.g., variably phosphorylated forms (234), may also be important in permitting both specific and nonspecific interactions.

Genetic and biochemical analyses of NC proteins reveal a second possible role in assembly: mediation of RNA-RNA interactions. Certain mutants defective in the Cys-His boxes of the protein package genome RNA but do not support its dimerization (236,498). Annealing studies show that addition of NC to mixtures of short complementary RNAs (like the primer tRNA and its binding site) hastens the rate of duplex formation to a considerable extent (592). This activity could be quite important in catalyzing the formation of short duplex structures, like that between the primer and its binding site or between the two subunits, which might not easily form by themselves.

Other gag Domains

In addition to the three major gag proteins, all gag precursors contain several other regions which are cleaved out and remain in the virion as peptides of various sizes. Since it has not been possible to assign locations or structural roles to these components, and since they are not consistently present, they remain unnamed. In ALSV, mutations in the 62 amino acid peptide (p10) between MA and CA seem not to affect gag assembly or processing, but lead to reduced virus yields, suggestive of an effect on a late stage of assembly (187). A still smaller peptide (p2) is cleaved from between MA and p10. The p2-p10 cleavage site is very inefficient, and it has been proposed that it may play a role in regulating the activity of PR and thus contribute to correct timing of proteolytic processing (101). Another small (9 amino acid) "spacer" peptide between CA and NC in ALSV, called p1, is not necessary for generation of the correct cleavage products or for budding but, in its absence, only noninfectious virions are made (151). Perhaps it serves to regulate the order of proteolytic and rearrangement events.

The most interesting of the additional gag peptides is the C-terminal region of the gag precursor of HIV, called p6. Mutations in this region do not block assembly or cleavage but have a striking effect on release, leaving fully formed (but immature-appearing) virions tethered to the cell surface (267). An analogous peptide is not seen in other genera, and it is not yet known whether the same activity is present in other viruses.

PR

The protease domain is not required for assembly or budding: mutants lacking activity synthesize virions which are noninfectious, contain only uncleaved precursor, and do not exhibit morphological maturation (369,399,692). These results imply that the PR protein is not active until a late stage of viral maturation and must not be freely present in the cytoplasm of the infected cell. Premature cleavage of gag by intracellular protease is deleterious to virion synthesis (365,411). Consistent with the requirement of assembly for protease activity is the lack of self-processing of the ALSV Gag-Pro precursor *in vitro*, or of the MLV precursor *in vivo* when assembly is blocked by mutations in the myristylation site (652). Thus, the PR domain in the precursor form must have little or no activity until itself cleaved out. At one time it was supposed that another cellular protease might be necessary to initiate the process; it now seems more likely that some special feature of virion assembly activates the enzyme, since precursors containing the ALSV or HIV protease undergo spontaneous cleavage when expressed in bacteria or yeast (404,409,519). Conceivably, the activation is a simple matter of attaining a sufficient critical mass. If the enzyme is very slightly active in precursor form, and then only in trans, no cleavage could occur until a very high local concentration of protein formed in the budding virion. Once the initial cleavage(s) occurred, active enzyme molecules would be released which would cleave out more PR proteins, allowing them to dimerize, and leading to a rapid chain reaction and virtually instantaneous processing of the entire capsid. In its simplest form, this model is unlikely, since the A particle precursors to B- and D-type viruses are fully assembled in the cytoplasm, yet remain completely unprocessed until they bud from the cell membrane. This stability cannot be simply a matter of time, since mutations in MA which block transport to the cell membrane also block processing (616). It is therefore likely that some other factor (a change in ionic environment, for example) that accompanies budding is also necessary to initiate the cleavage process.

pol

Mutants lacking pol proteins assemble, bud, and process their capsids normally but are, of course, noninfectious. Because at least some RT proteins exhibit specific binding of the "correct" tRNA primer and because mutants of ALSV defective in *pol* do not incorporate the tRNA into virions, it has been supposed that binding of RT to tRNA is an important mechanism for ensuring inclusion of the primer in virions. However, some viruses can use alternative tRNA primers if forced to do so by mutation of the PB site (467), implying that RT is not the only selective agent.

The proportion of pol containing precursors is also apparently critical. Introduction into the MLV genome of a point mutation altering the *gag* terminator to a glutamine codon, so that all molecules synthesized resemble the 5% of suppression products seen in the normal case, causes a failure of assembly to occur, perhaps because the core structure cannot accommodate a large number of RT molecules (214).

env

As with pol, the env proteins are not necessary for virion assembly or budding. In their absence, normal-appearing virions (except for the absence of surface projections) are made and released. Such virions are also infectious if introduced into cells by artificial means, implying that env proteins are required only for initial events (adsorption and entry) in infection.

As noted above, the signals that specify incorporation of env proteins into virions are a mystery at present. A priori, it seems reasonable to suppose that the usually short C-terminal cytoplasmic domain of the TM protein specifically interacts with the capsid. Surprisingly, however, deletion of this entire region from the ALSV TM protein has no detectable effect on the synthesis, processing, or incorporation into virions of the env proteins (578). Conceivably, there is no specific signal, and any glycoprotein not attached to something else can be incorporated into the envelope (E. Hunter, pers. comm.) That the process may not be tightly specific is implied by the ability of many retroviruses to incorporate and use env proteins from unrelated retroviruses, other viruses, or even normal cellular proteins into functional virions, albeit at variable efficiency (20, 418, 419). Clearly this issue deserves more study. In other viruses, including B- and D-type viruses and lentiviruses, there is more evidence (noted above) for a direct interaction between TM and MA, but the sites and signals for this interaction remain to be defined.

In some groups, including MLV and related viruses and the B-type viruses, a final modification of the *env* gene product is the removal of a small C-terminal fragment (once called the *R peptide*) from the end of TM. This cleavage occurs during budding and is catalyzed by PR (75). Since a mutation which truncates the MPMV TM protein at the same point causes cells expressing the mutant env protein to fuse into large syncytia (74), it is likely that the role of this small peptide is to suppress the ability of the env protein complex to bind or fuse to cell membranes until after budding. Taking the virus's point of view, it is easy to see how the formation of syncytia could be an undesirable side effect of replication. In the same regard, it is noteworthy that syncytium-forming variants of HIV-2 and the closely related SIV also have mutations which truncate the homologous region of TM (525,831).

The RNA

In all retroviruses studied to date, the principal determinant of packaging on the genome is in the packaging signal in the leader region between U_5 and *gag* (450). Deletions in this region do not affect expression, translation, or assembly, but do result in the production of virions devoid of genome RNA (13,297,377,773), often containing other RNA molecules, either from related endogenous proviruses or adventitious cellular sequences (243,385,646). The sequence so identified as essential for packaging has been termed ψ (472) or E (783). In no case has the minimum essential sequence been precisely defined. The region of the MLV genome extending from just downstream of the splice donor to immediately preceding the *gag* initiation codon can permit packaging of subgenomic MLV RNA if placed at other locations, or even of unrelated RNAs containing no other viral sequence. For full efficiency, however, a longer sequence, probably extending well into *gag*, is required (2). In the case of ALV, a similarly long signal can also confer packaging ability on foreign RNAs (18).

The results with the mammalian C-type viruses explain the selectivity of packaging for genome RNA, since the packaging signal is absent from the spliced mRNAs. More difficult to explain is the situation in ALSV. Since the ALSV splice donor is within *gag*, spliced mRNAs contain the entirety of the signal identified so far. Subgenomic ALSV RNAs are also discriminated against during packaging, but perhaps not as effectively, since there are a number of reported cases of proviruses derived from reverse transcription of mRNA molecules which must have been incorporated into virions (54). Since the minimum sequence sufficient for packaging has not been identified, it remains possible that unidentified essential sequences lie within *gag*. Such a sequence, if it exists, must lie very near the 5′ end of *gag*, since vector constructs which include only 50 codons of *gag* are capable of being packaged (543). Furthermore, small RNAs in which the 5′ end of the ALSV genome is spliced onto a foreign sequence can be efficiently packaged, while ALSV mRNAs containing the identical 5′ sequence are not (17,713). This result implies that some of the selectivity against mRNA packaging may be due to negative signals in the mRNAs.

Except for size, there seems to be little in the way of additional requirements for packaging of RNA. To date, a very large number of different sequences have been passaged in retrovirus vectors. No failure of an RNA sequence to be packaged has been reported, implying that the constraints for packaging of any sequence are few. The precise 3′ end of the RNA is also unimportant; genome RNA molecules extended into cell-derived sequence at their 3′ ends are efficiently incorporated into infectious virions (712,714).

The most severe limitation on packaged RNA is that imposed by size. No lower size limit has yet been reported: the smallest ALSV known to be packaged is less than 1 kb (713); the smallest MLV about 3 kb. In several studies, RNAs larger than about 10 kb were found not to be packaged into infectious virions (249). Although these experiments were not done systematically, they suggest a strong constraint. The level of this restriction—whether on physical assembly into virions, on virion structure or stability, or on ability of the genome to be reverse transcribed—remains to be determined. The observation that 3′ extended genomes as large as 11 kb can be seen in virions raises the possibility that the restriction might be at some level other than virion assembly (304).

SPECIAL FEATURES OF RETROVIRUS BIOLOGY

Host Cell Effects

The retrovirus replication cycle does not require that the infected cell be significantly harmed or otherwise affected by the virus-specified events, and many retroviruses have little or no effect on the infected cell, other than to render it permanently capable of producing virus at a low level. However, the unique association retroviruses share with the host cell leads to numerous special features of retrovirus biology and results in a remarkably wide range of effects on the host organism. These effects range from malignant or degenerative disease through benign viremia to short- or long-term genetic effects such as insertional activation or inactivation of cellular genes. Most of the effects discussed here are considered in more detail in specific chapters, and will be presented only briefly.

Transformation

Many retrovirus groups first came to light as agents isolated from naturally occurring tumors and that were capable of causing such tumors in appropriate hosts. It soon became apparent that these isolates could be divided into two distinct groups: one that causes tumors rapidly (1 to a few weeks) and causes readily visible transformation of appropriate target cells in culture; and one that induces malignant disease only slowly (with a typical latency of 6 months to 1 year) and has no visible effect on cells in culture. Generically, the first group of viruses are often referred to as transforming retroviruses; the second group as leukemia or leukosis viruses (ignoring the fact that many of them cause disease other than leukemia), or even nontransforming viruses. The molecular distinction between these groups is now very clear. The transforming viruses act on cells through the expression of a specific oncogene usually unrelated to viral sequence; the leukemia viruses lack such a sequence and transform cells by other means.

Viral Oncogenes

Alone among animal viruses, retroviruses have the ability to incorporate fragments of certain cellular genes (known as c-*onc* genes or proto-oncogenes) into their genome and alter their structure and expression in ways that give them the ability to directly transform a normal cell into a malignant one. Over two dozen different viral oncogenes (v-*onc* genes) have been found in retroviruses—some of them many times over—and study of these has been an important and rewarding aspect of fundamental cancer research over the last two decades. Detailed analysis of their structure and function as well as their relationship to their normal counterparts has provided insights into molecular mechanisms of oncogenesis, growth control, and signal transduction to a depth impossible to match in any other system. Discussion of these aspects is found elsewhere (Chapter 9); here are described a few general aspects of the association of these genes with the viruses that carry them and the cells they transform.

All retroviral oncogenes are very recently derived from the genome of the host—most likely within the animal from which the transforming virus was isolated. There is no evidence that transforming viruses are efficiently passed from one individual to another and good reason to believe that they are not. Retroviruses in general are transmitted horizontally very inefficiently, and transforming viruses kill the infected host very rapidly. The most productive sources of new viral oncogenes have been populations of animals in which large numbers are routinely screened for tumors and within which leukemia virus infection is common. In practice, these have been chickens raised for slaughter and pet cats presenting to veterinarians with tumors. Although the events which give rise to oncogene-containing viruses are probably very rare, the selection is powerful since small numbers (perhaps one) of events per animal are amplified into a visible effect. Nevertheless, oncogene-containing viruses cannot be considered to be natural infectious agents. Rather, they should be thought of as rare aberrations which, if not given a good home in the laboratory, would die with the animal in which they arose.

Viral oncogenes differ from their cellular antecedents in a number of important ways. First, they often contain only a portion of their corresponding proto-oncogenes, limited to a subset of the region transcribed into mRNA. Second, they are derived from processed versions of the transcripts, with all internal introns removed by splicing. In some cases (such as *src* in RSV), a small fragment of intron along with the 5′ splice acceptor site remains and provides an acceptor for processing of the subgenomic v-*src* mRNA. Third, they have been separated from the cellular controls on their expression, including both the normal cellular promoter as well as other controlling sequences such as those which confer instability on the corresponding mRNA. Their expression has come completely under the control of the viral LTR and thus can differ from that of the proto-oncogene in level, in lack of regulation (by growth factors, for example), and in cell type specificity. With some viral oncogenes, such as *myc* and *mos,* differences in regulation of expression seem to be sufficient to convert a proto-oncogene into an oncogene. Fourth, many oncogenes have suffered deletions or other rearrangements of sequence affecting the structure of the protein product. For example, v-*src* differs from c-*src* most prominently by the replacement of a short amino acid sequence at the very C-terminus of the protein. This difference includes loss of a tyrosine residue, the phosphorylation of which apparently regulates the tyrosine-specific protein kinase activity of the c-*src* protein. Similarly, v-*erb*-B differs from its cellular relative (now known to be the receptor for epidermal growth factor) by the deletion of a large amount of sequence, amounting to most of the extracellular region, including the ligand binding site. Changes such as these are presumed to have the effect of allowing a signal transducing function of the proteins to occur independently of the normal signals. Finally, viral oncogenes are often joined to viral genes in ways important for modifying their function. For example, v-*abl* is expressed as a fusion protein with a portion of the MLV *gag* protein as its amino terminus. This fusion provides to the *gag-abl* protein the *gag* myristylation site and directs the protein to a membrane site essential for its activity. The *fms* oncogene (derived from the receptor for the growth factor CSF-1) is also expressed as a fusion with the amino-terminal portion of the feline leukemia virus *gag* gene. In this case, the important feature is the signal peptide provided by the glycosylated form of the *gag* protein, which allows proper placement of the protein into the cell membrane.

With the exception of some strains of RSV, virtually all oncogene-containing retroviruses are defective for replication since the oncogene has replaced all or part of the essential protein coding region of the genome but retains all necessary *cis*-acting sequences. Nevertheless, such viruses are capable of replicating as mixtures with the corresponding nondefective nontransforming virus. In this context, such a virus is known as a helper virus. In a doubly infected cell containing proviruses of both the transforming and the helper virus, the proteins synthesized by the helper can provide all the functions necessary to encapsidate the defective virus genome and allow it to generate an integrated provirus in the next round of infection. This works only because retroviruses capable of acquiring oncogenes do not contain *cis*-acting signals for expression within the coding region, nor do they encode transactivating or

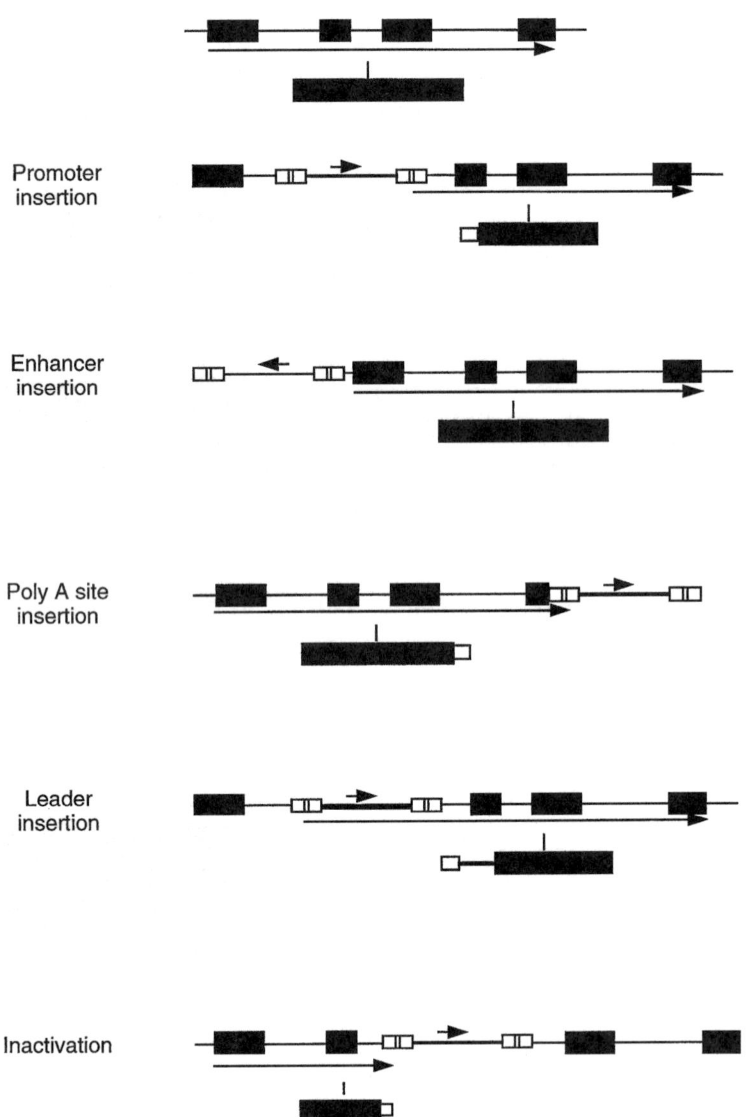

FIG. 22. Effects of retroviral integration. Exons of cellular genes are shown (*solid bar*), as is the primary transcript (*arrow)* and the spliced mRNA (*narrow bar*).

other regulatory proteins necessary for their expression. Cell lines containing only the transforming provirus can be obtained by cloning infected cells shortly after infection with a mixed population of virus and testing for the absence of virus production. Such cell lines are called nonproducer cells and have been quite useful in studying oncogene function in the past.

Insertional Activation

Quite commonly, nondefective retroviruses lacking oncogenes, although unable to transform cells in culture, are capable of inducing a variety of malignancies (204,746, 763). For example, avian leukosis virus, when inoculated into young chickens, frequently causes a B-cell lymphoma with a latency of 6 months or more. Close examination of the virus-cell relationship in such tumors reveals a striking result: virtually all tumors have a provirus inserted in a similar portion of the genome, specifically within the proto-oncogene c-*myc* (298). In the majority of tumors, the *myc*-associated provirus is similarly located—it almost always lies within an intron between the first (noncoding) and the second exon in the same transcriptional orientation (Fig. 22). The effect of this insertion is to bring the two coding exons of c-*myc* under the transcriptional control of the 3′ LTR. Transcripts encoding c-*myc* are found at higher levels than normal and are structurally different in that they contain the R-U_5 sequence derived from the LTR at their 5′ ends. Thus, synthesis of these transcripts must be initiated at the promoter sequence within the 3′ LTR. The altered regulation of the *myc* gene in these tumors mimics the case with v-*myc*, and presumably is a sufficient event to initiate the tumorigenic process in these animals, although other events seem to follow and be necessary for transformation to a fully malignant phenotype.

This mechanism of activation of c-*myc* by ALV is usually referred to as *promoter insertion*. The use of the 3′ LTR may require some additional sequence rearrangement in the provirus. Most proviruses in the tumors are altered by deletions near, but often not including, the upstream LTR (259,628). These deletions might be important to relieve a block to expression from the downstream LTR.

Insertional activation of a number of proto-oncogenes has been found to be an important feature of pathogenesis by nontransforming retroviruses (415) and has been quite valuable in identifying potential new oncogenes not seen in other ways. Several different mechanisms have been found to be involved (Fig. 22). Interestingly, the promoter insertion mechanism seen in ALV-induced lymphomas is relatively uncommon, perhaps reflecting the necessity of viral mutations as well as the insertion. Other known mechanisms of gene activation by provirus insertion include:

1. *Enhancer insertion* in which insertion of a provirus upstream of a natural or cryptic promoter of a proto-oncogene but in the opposite transcriptional orientation causes deregulated expression from that promoter.
2. *Leader insertion* where expression of a portion of a proto-oncogene is in the form of multiply spliced readthrough transcripts initiated within the 5′ LTR of a provirus inserted within an intron (261).
3. *Terminator insertion* in which the poly(A) addition signal of the 5′ LTR of a provirus inserted near the 3′ end of a gene results in a truncated mRNA lacking a signal promoting rapid turnover, thereby increasing its effective concentration.

It should be remembered that insertion of a provirus in a position to do this sort of damage must be quite rare on a per-cell basis, but considering the total number of infected cells in the target organ, such an insertion can be statistically certain in the organ. Also, in many cases there is good evidence that the initial insertion does not suffice for the full transformation of the target cell to a tumor cell, but that additional mutagenic events must take place, themselves often involving activation by proviral insertion. Two models for studying such progression have been particularly rewarding. In one, mice are made to carry a transgene for one of the activated proto-oncogene loci such as *pim-1* (764). Tumors arising after infection with MLV then carry proviruses inserted specifically at a cooperating locus (c-*myc* or N-*myc* in this case). In the other, thymomas induced in rats by MLV are selected for a "progressed" phenotype (such as growth factor independence) by growth in cell culture (747). The sites of integration of new proviruses in these cells often reveal loci involved in the phenotypic change (30,422).

Other Mechanisms

Envelope Genes and Oncogenesis

Until recently, the genes encoding virion proteins were not thought to play a direct role in the oncogenic process, but a number of observations have changed this view. The several strains of Friend murine leukemia virus induce a variety of diseases of red blood cell precursors, characterized by a very rapid but nonmalignant overgrowth of erythroblasts which often progresses to a malignant, transplantable, erythroleukemia (408). While these diseases can be induced, albeit with reduced efficiency, by nondefective virus alone, many independent strains contain a defective component which accelerates the process and gives rise to rapid focal proliferation of erythroblasts in spleens of infected mice as well as colonies of erythroid cells after infection of bone marrow cell cultures. This component, known as spleen focus-forming virus (SFFV), is replication-defective and encodes a characteristic product known as gp55, which is necessary and sufficient for the "transforming" properties of the virus (47,123,810). In contrast to transforming genes of all other known retroviruses, gp55 is not the product of a typical oncogene. Rather, it is a modified *env* gene product, derived by a series of events including acquisition of an altered sequence by recombination with the polytropic *env* gene of an endogenous provirus; a deletion extending from the middle of SU into the TM coding region, and a rearrangement extending the membrane-spanning domain.

The role of this protein has become clear with the observation that it binds to and stimulates the receptor for erythropoetin (epo) (444), and in so doing, can induce the growth of epo-dependent cell lines (669,832). Thus, the premalignant phase of the disease is most likely a consequence of stimulation of overgrowth of normal erythroblasts by the interaction between gp55 and the epo receptor. Progression of the erythroblastosis to malignant erythroleukemia occurs as a result of additional mutations, most commonly insertional mutagenesis by the helper virus, leading to clonal outgrowth of truly transformed cells. Both activation of proto-oncogenes (38,39) and inactivation of the recessive oncogene p53 (40,159,522) have been seen.

The ability of MLV *env* proteins to mimic hematopoietic growth factors is not limited to gp55. Rather it is a property of the polytropic *env* gene from which gp55 is partly derived, as reflected by the similar, albeit somewhat less efficient, pathogenesis of Friend MLV recombinants with intact *env* genes (called Friend mink cell focus-forming—Fr-MCF—virus). The other rearrangements characteristic of gp55 presumably serve to make the interaction more efficient, albeit at the expense of replication ability. Other MCF viruses (so called because some types can infect and cause cytopathic effects in mink cell lines) have been derived by recombination of endogenous polytropic *env* genes with exogenous or other endogenous MLVs (see below) and are the proximal agents of thymoma induction in mice. It has been proposed that the *env* gene products (specifically the SU protein) contribute to oncogenesis by stimulating the IL-2 receptor (443). Infection with MCF virus alone is sufficient to give IL-2 independence to primary rat thymomas in the model described above (743). Interestingly, further progression of these cell lines involves insertional activation of the IL-9 receptor, implicating IL-9 rather than IL-2 as the growth factor being mimicked (225).

The ability of these retroviruses to stimulate growth factor receptor pathways can hardly be by chance. Rather, it must contribute in an important way to the growth of the virus *in vivo*. Most likely, this stimulation gives a slight growth advantage to an infected cell over its uninfected siblings, and thus gives the virus an extra replication advantage by increasing the number of infected cells.

Retroviral Superantigens

MMTV has an unusual biology in that it is transmitted almost exclusively from mother to offspring via milk. In the adult animal, it therefore needs to be expressed only in the lactating mammary gland, and special features of the LTR that contribute to this expression were discussed in a previous section. In the newborn mouse, the ingested virus has a different problem—to make its way from the site of infection (the gut) to mammary tissue. Current thinking is that it does so by infection of B cells, probably in the Peyer's patches of the gut. Some fraction of these infected cells eventually find their way to the mammary gland and produce virus which infects the mammary epithelium. This process is likely to be quite inefficient, and MMTV has discovered a unique way to improve the efficiency—the superantigen encoded by the *sag* gene (1,134,345,557).

Superantigens were originally described as proteins secreted by Staphylococcus and certain other pathogenic bacteria that react with cells of the immune system in an unusual way. Normal antigens stimulate an immune response through presentation by MHC molecules on B cells to that very small fraction of T cells whose T-cell receptor (TCR) genes happen to be rearranged in a specific way. The stimulated T cells both divide and release cytokines which activate the presenting cell to divide as well. Superantigens, by contrast, bind simultaneously to MHC class II molecules and to *all* TCR molecules expressing a particular class of ß-subunit variable (V_β) chains (Fig. 23). Since there are about two dozen different V_β genes, some 5% of the T cells can be stimulated at one time—an enormous response by the usual standards. Thus, following infection of a newborn mouse with MMTV, there is a very large response, leading initially to expansion of T cells of a specific V_β class (or a few classes). As with other immune responses, this is later followed by a loss (or anergy) of reactive T cells.

At one time, it was thought that the importance of this effect was to stimulate the division of T cells and thus permit their productive infection (134), and T cells are known to be targets for infection and transformation by MMTV variants [containing specific deletions in the LTR (24,183, 814)]. However, close examination of the kinetics of virus and cell growth after experimental infection reveals that the major effect is a very large increase—primarily by cell division, but possibly also by infection of activated bystander cells—in the number of infected B cells (302). Infection of T cells seems to happen much later (48), if at all. Thus, it seems probable that the infection is carried by infected B cells homing to the mammary epithelium.

Although it induces a very strong T-cell activation *in vivo* and in mixed lymphocyte reactions *in vitro*, the *sag* gene product has proven quite difficult to study due to very low levels of expression on B cells. From its sequence and

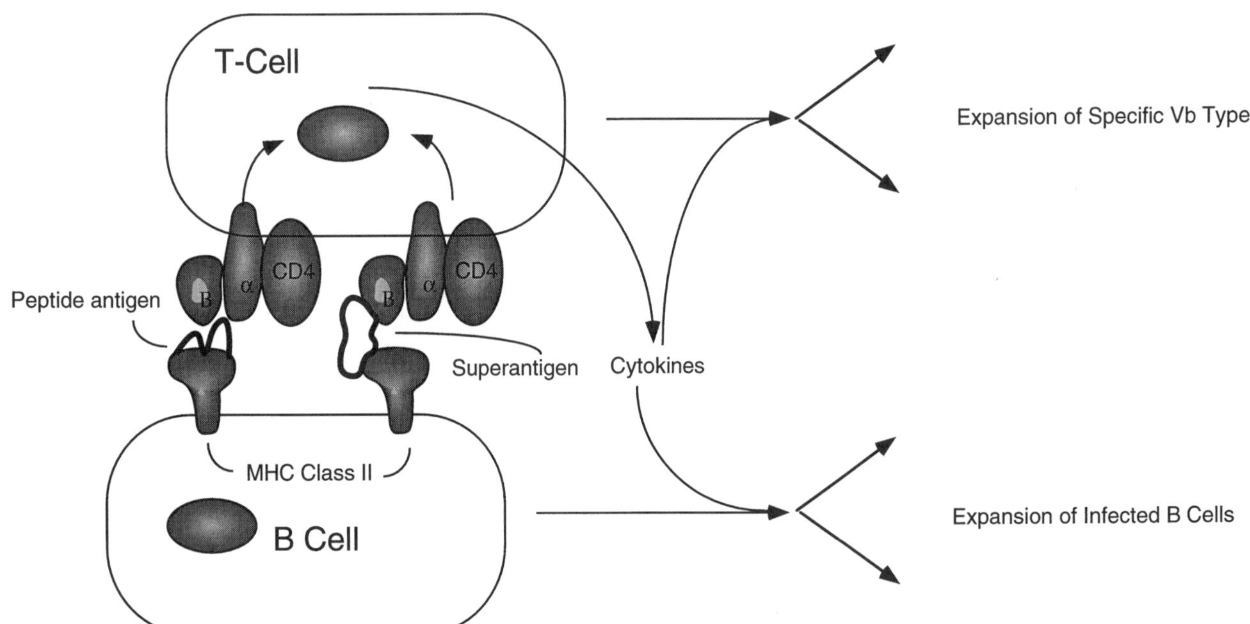

FIG. 23. Superantigen activation of B and T cells.

studies using expression vectors, it has been determined to be a class II membrane protein, oriented with the amino terminal end toward the cytoplasm (49,118). Although closely related to one another, *sag* proteins from different strains of exogenous MMTV and different endogenous proviruses react with different subsets of V_{β} genes. Based on sequence comparison and exchange of regions of the protein, the specificity for this interaction resides at the extreme C-terminus of the molecule (119,398,816). Quite possibly, a cleavage event releases the bulk of the extracellular region, and only a small fraction of this is able to associate with MHC molecules, the rest being lost (806).

The superantigen activity of MMTV is clearly a highly specialized adaptation to particular problems of its unique lifestyle. Nevertheless, it remains possible that a similar strategy might be used by other retroviruses for other reasons. Indeed, it has been suggested that a superantigen-like activity in HIV (421) and the MLV variant that causes the murine AIDS (MAIDS)-like syndrome (363), but these ideas have not received strong support. At present, the only other virus for which there is solid evidence for superantigen activity is rabies virus (417).

HTLV

As discussed in more detail in Chapter 59 in *Fields Virology*, 3rd ed., HTLV-1 and related viruses can immortalize lymphoid cells in culture, and infection of cells *in vivo* can lead to their malignant transformation, albeit rarely, and with a very long latent period (275,645,707,723). The mechanism of transformation by these viruses must be different from other retroviruses since no oncogene is present and consistent proviral integration sites of the type that characterize insertional activation of proto-oncogenes cannot be found. Because expression of the transactivator *tax* can also affect the expression of a number of other genes, in particular the T-cell growth factor IL-2 and its receptor as well (25,274,276, 277,637), it has been hypothesized that this sort of activation might lead to an autocrine mechanism of transformation in which a cell, by secreting its own growth factor, might stimulate itself to divide continuously (479). Effects consequent to binding of the transcription factor NF-κB or other transcription factors have also found some experimental support (237,238,394,541). Direct proof for such a mechanism is lacking, since expression of *tax* alone is insufficient to immortalize cells in culture. Expression of *tax* in transgenic mice does lead to a benign tumor, but the tumor is very different from the natural disease (534).

Osteopetrosis

This disease can be induced in chickens by certain strains of ALV (570). It involves the overgrowth of osteoblasts, particularly in the long bones of the leg, and leads to gross thickening of these bones. It is apparently not a clonal disease and is not a malignancy. The mechanism of its induction is far from clear. Close examination of infected tissue reveals unusually high levels of virus replication and large amounts of unintegrated virus DNA (21), characteristics more commonly associated with cytopathic effects by retroviruses (see below). Perhaps the extensive virus replication results in disruption of the complicated balance between bone growth and dissolution that characterize the normal tissue. Genetic exchange of fragments between osteopetrotic and lymphoma-inducing strains implicate a region at the beginning of *gag* as important in conferring the difference in one model (627, 629), and a region in *env* in another (355). A good mechanistic explanation is not yet forthcoming.

Cytopathic Effects

Although retrovirus infection can and often does proceed without noticeable detriment to the infected cell, there are a number of well-characterized virus-cell interactions that lead to its death. The most prominent cytopathic virus-cell interaction is found with HIV and other lentiviruses, discussed extensively in Chapters 61 and 62 in *Fields Virology*, 3rd ed.. In addition to these well-known examples, there are a number of models for degenerative disease caused by various oncoviruses closely related to the noncytopathic tumor viruses. These include wasting and immunodeficiency diseases inducible in cats, mice, and monkeys by variants of FeLV, MLV, and a D-type virus, respectively (23,526). Unfortunately, the temptation to name these conditions *FAIDS, MAIDS,* and *SAIDS* was too great to resist. Other variants of MLV have been associated with neurological disease, and some reticuloendotheliosis virus strains cause degenerative disease (717). At present, there is little insight available into the mechanism by which retrovirus infection leads to death of the host cell. A number of studies have been undertaken to compare the genomes of strains of varying pathology (by exchange of restriction fragments, for example) and thus hope to identify gene products important to the difference. In most cases, differences between pathogenic and nonpathogenic variants have been mapped to the *env* gene, in particular, the region encoding the SU protein (320,196, 421). In the MAIDS and FAIDS cases, an additional interesting feature of the pathogenic isolates is the repeated isolation of a variety of replication-defective variants of the virus, often carrying large deletions and also harboring some other determinant necessary for disease induction (23,556). Defective viruses also arise along with highly cytopathic ALV variants after rapid repeated passage in cell culture, but in this case do not seem necessary for the cytopathic effect (139,780).

There are several possible mechanisms by which a retrovirus infection might lead to cell destruction: immune system-mediated killing of infected cells; direct toxicity of a virus gene product; extensive replication of the virus leading to overwhelming of functions necessary for survival of the cell; and indirect effects of viral gene products on cell-cell interactions necessary for their function and survival. Experimental models exist for all of these mechanisms, and all have been put forward as explanations for HIV-associated pathology. In cell cultures, certain ALV strains and reticuloendotheliosis viruses exhibit strong cytopathic effects; and useful plaque assays, analogous to those used routinely for other types of viruses, have been developed. Results using the cell culture systems are concordant with the hypothesis that cell killing is due to overreplication of virus within the infected cell. Typically, cultures infected with a cytopathic virus have large numbers of copies of unintegrated DNA a few days after infection, indicating that they have been subjected to repeated cycles of virus replication (386). Survivors that outgrow in the culture typically contain only a small number of integrated proviruses. That the appearance of excessive amounts of DNA can be due to complete replication cycles is suggested by the observation that cytopathic ALV variants exhibit reduced resistance to superinfection (796). Also, detailed mapping studies indicate that the only significant difference between cytopathic and noncytopathic viruses is the ability of the former to use a specific receptor—that for subgroup B virus (178). The model that ensues from these considerations is that cytopathology can be a consequence of the breakdown of the mechanisms which usually act to prevent "replication" of viral DNA within a single infected cell—the combination of postbudding maturation to infection-competent virions and a very strong and rapidly developing level of resistance to superinfection. Variation in receptor used or properties of the *env* protein can apparently decrease the effectiveness of the resistance and permit repeated infection of the cell with the progeny of the provirus it carries, probably leading to an excessive burden of the viral DNA or some other product and consequent death of the cell. This type of direct cell killing is most likely the major mechanism operative in the induction of immunodeficiency in cats by FeLV (169,412,610,738).

In the case of the immunodeficiency in mice caused by the MAIDS virus, the mechanism seems to be quite different. Unlike the ALV and FeLV models, virus replication cannot be directly involved since the defective virus component alone, if introduced into mice as a pseudotype, can also give the same disease (324). MAIDS virus has a characteristic deletion in *gag* and *pol* (23), leading to synthesis of a 60-kD unprocessed fusion protein which is necessary and probably sufficient for the disease induction (114,323). Interestingly, this precise deletion is also found in an endogenous virus present in some strains of mice, suggesting a recombinational origin (N. Copeland and N. Jenkins, *personal communication*). Although MAIDS resembles AIDS in its deficiency of mature CD4 T cells, the important target for infection is a B cell, in which expression of the *gag* fusion protein induces a polyclonal stimulation, outgrowth, and eventually lymphoma (293, 324, 325). In some way, the affected B cells then lead to loss of T cell function (671). As noted above, superantigen activity has been suggested as an explanation for this effect (329,363) but seems quite unlikely, given its magnitude.

The mechanisms important in killing by HIV remain to be worked out. Certainly, very large amounts of unintegrated DNA are often found in infected cells before they die, and with the correct virus-cell combination, very high levels of virus protein have been reported (677), consistent with overreplication of the virus. The issue is still open to debate, however. The formation of syncytia due to interaction of the *env* protein on the surface of one cell with CD4 on its neighbor is an obvious part of the cytopathology of HIV in cell culture (446), but unlikely to be important *in vivo*. Differences in *env* gene sequence between cytopathic and noncytopathic variants are often seen (117,676), as are differences in other genes (117,161,643). Still more complex is the issue of the importance of these mechanisms to T-cell loss and pathogenesis *in vivo*, where immune-mediated cell killing, induction of programmed cell death (apoptosis), and secondary effects on the ability of the body to replace the lost cells are all likely to be important (103,201, 209–211,233,295,307,634) (see Chapter 61 in *Fields Virology*, 3rd ed.).

Insertional Inactivation

The insertion of a provirus into the cell genome is a mutagenic event. Obviously, a provirus inserted into a coding region of a gene will disrupt the function of that gene (although insertions within introns are sometimes tolerated). To date, no pathogenic process other than oncogenesis has been conclusively shown to involve this mechanism. This is not too surprising, since most infected cells are diploid and the probability of insertion into the two copies of the same gene at the same time is very small. Also, the occasional loss of gene function in an individual cell usually would go unnoticed unless the event could be amplified, such as by inactivation of a recessive oncogene, as with the inactivation of p53 in F-MLV-induced malignancy, discussed above.

Insertional inactivation of cellular genes has been observed in several cases of endogenous provirus integration (see below) and in several model systems in tissue culture. For example, MLV infection has been used to inactivate expression of *src* from a resident RSV provirus (768), of hypoxanthine phosphoribosyl transferase (HPRT) in mouse teratocarcinoma cells (whose inability to express the virus renders them repeatedly infectible) (391), and of ß-2 microglobulin on mouse B cells (229). In all cases, the insertion events were less frequent than expected, and their detection required both a very powerful selection and that the target gene be haploid. The use of retroviral insertion as an approach to identifying genes encoding a selectable function in cell culture has been repeatedly proposed, but to date no new genes have been identified using this sort of "tagging."

A variant of this approach is to select for integration of proviruses randomly within genes and then test for function of the gene containing the provirus. Infection of embryonic stem (ES) cells with MLV, followed by regeneration of mice from the infected cells and breeding the proviruses to homozygosity is in principle a way to identify and tag interesting genes. In practice, integration into genes is relatively infrequent, and this process would be very laborious. To improve the odds, a gene for a selectable marker lacking its own control sequences can be placed within U_3. This gene is expressed only when the provirus is integrated downstream of an active promoter. Thus, selection for cells expressing the marker gene will yield clones with a high frequency of insertional inactivation (61,111,607,777). This approach works well in model systems and may serve to reveal interesting new genes.

Genetics

Retroviruses display a wealth of genetic phenomena which contribute a remarkable level of plasticity to the genome. First, they have a high mutation rate and can, under the appropriate selective conditions, rapidly accumulate both point mutations and major rearrangements of sequence. Second, they display a frequency of genetic recombination unapproached by other systems. Finally, they have the unique ability to acquire foreign sequences such as oncogenes and express them as part of their genomes. All of these features together give this virus group the capacity to rapidly evolve into new niches (131,132,135,136, 531,735). This sort of evolution can be seen in many settings—in some strains of mice, there is a predictable evolution leading from several benign endogenous proviruses to one capable of causing a tumor and resulting in the premature death of almost every mouse in the strain (408,702). In HIV, it results in a high level of sequence and antigenic diversity in the *env* gene and its product (130,268,480,545), as well as in the appearance of an impressive variety of drug-resistant variants in virtually all AIDS patients treated with antiviral compounds (650) (see Chapter 61 in *Fields Virology*, 3rd ed.). In some ALV, MLV, and FeLV strains it leads to the reproducible isolation of a specific oncogene in animals infected with a nontransforming virus (500,603,695).

Sequence Variation

Before considering mechanisms of variation, it must be emphasized that the rate of genetic variation is a composite of three variables: the relevant mutation rate per replication cycle, the number of replication cycles per unit time, and the selective advantage or disadvantage possessed by the variant virus. Usually, the third variable is the most important in determining the rate of variation, and it is likely, given the complex combinations of functions encoded within the genome, that no mutation is selectively neutral (132). For this reason, it is a mistake to attempt to apply the principles of neutral theory (253) to retroviruses.

Point Mutations

Although much work (discussed below) has been done on both the error rate of reverse transcriptases *in vitro* and the extent of variation *in vivo*, only a relatively small number of studies have addressed the mutation rate in infected cells. The experimental problem is that, in order to eliminate effects of selection on mutant frequency, the accumulation of mutations in a single replication cycle must be measured. This has been accomplished by molecular cloning of proviruses derived by infection of cells with virus progeny of a single provirus in a clone of cells (181). If the provirus contains a gene such as the *lac* repressor whose inactivation can be easily detected in bacteria, then the frequency and nature of mutations can be readily determined (566,567). With this assay, the substitution rate for a single round of replication was estimated to be about $1–2 \times 10^{-5}$ per base for SNV (567), and about half that value for BLV (473). Other clonal, single-step assays have yielded values ranging from 2×10^{-5} (509) down to 2×10^{-6} (766). The difference probably reflects the fact that the latter study looked at mutation at a specific codon, while the former surveyed a substantial fraction of the genome. Unfortunately, similar results have not yet been reported for HIV, whose RT is sometimes reputed to have a greater error rate *in vitro* (624).

The point mutation rates are in the range of rates estimated for RNA viruses in general (312). Somewhat higher figures have been obtained for bacteriophage Qß (34), and vesicular stomatitis virus (691), while similar rates have been estimated for poliovirus and influenza virus (564). Thus, the popular misconception that retroviruses in general (and most likely HIV in particular) have extraordinarily high error rates is incorrect. The extraordinary genetic diversity of HIV populations *in vivo* is a consequence of its very rapid rate of replication, not a very high mutation rate (132).

The assays used to detect point mutation also reveal a variety of other mutational events, including deletions, duplications, and insertions of sequence, as well as frameshifts (deletion or insertion of one or two bases) and hypermutation (566). The latter event—also seen in HIV proviruses cloned from infected patients (244,769)—is characterized by a very large number of G to A changes, particularly within GA sequences. The occurrence of hypermutation in patches within occasional genomes suggests that it may be the product of a variant RT molecule. Genomes with hypermutant stretches are so extensively altered that they are unlikely to survive, and the contribution of hypermutation to functional genetic diversity is doubtful. As with point mutations whose frequency is strongly influenced by local secondary structure (568), the frequency of other types of mutation is strongly influenced by local features. For example, some 40% of progeny of a provirus containing run of 10 to 12 As had a different number of bases in the run (89).

It is common to refer all blame for mutation to reverse transcriptase. It should, however, be kept in mind that retroviruses are unique in their use of three different enzyme systems for their replication—reverse transcriptase, cell DNA polymerase, and RNA polymerase II. The potential contribution of each of these to mutations will be considered separately.

Reverse Transcriptase

Reverse transcriptase lacks an exonucleolytic (proofreading) activity and seems to have a corresponding high frequency of base misincorporation *in vitro* (35, 802). Error frequencies ranging over a factor of 100 from 3×10^{-3} to 3×10^{-5} misinsertions per copying have been reported for the enzyme from various viruses (263,506,624). Since the extremely high rates quoted are based on misincorporation with homopolymeric templates and probably do not closely resemble the natural case, a more reliable estimate would lie within a tenfold range centered on about one error in 10^4 bases incorporated, a value usually somewhat higher than the measured mutation rate; although, with one exception (766), a direct comparison has not been performed. The reverse transcriptase of HIV seems to be more error-prone when compared head-to-head with that of ALSV or MLV (593). Given that the error rate in *in vitro* reactions depends on conditions (like nucleotide and salt concentration) that are unknown at the *in vivo* site of synthesis, it is difficult to know how to interpret relative values such as these.

It is important to note that the rate of misincorporation is not likely to be uniform across the whole genome. Rather, it seems to be strongly dependent on the specific context, with some types of errors much more likely than others (619,623). The error rate *in vitro* also depends on the template, with RNA-directed synthesis having a much higher rate than DNA-directed (67,328). Also, HIV RT inserts mutations in runs of A residues at a very high frequency (351), consistent with the frequent frameshift mutations noted above. As with the infected cell results, these studies implicate sites of RT pausing as mutational hot spots. A particularly frequent type of mutation *in vitro* is the addition of extra, untemplated bases at the end of a nascent DNA molecule. On model templates, as many as one in three newly synthesized DNA molecules were found to be extended by an additional one to four extra nucleotides beyond the end of the template (565), and similar errors are seen on forced strand transfer *in vitro* (574). If this is also true *in vivo*, it would be an important source of mutation wherever an end was encountered or created. However, the lack of evidence for high frequencies of mutation at sites corresponding to known jumps or recombinational events (825) suggests that this type of mutation may be suppressed under natural conditions.

DNA Polymerase

Provirus replication occurs only as a regular part of chromosome replication and must have a similar error rate, one that is likely to be so low as to be negligible for the purposes of this discussion. For example, analysis of a 200-base region of five endogenous proviruses inserted independently into the mouse germline prior to inbreeding and therefore separated by at least 100 animal generations from one another revealed no difference in sequence—an error frequency of less than 10^{-5} per animal generation (699). Indeed, an endogenous provirus known to have been inserted into the primate germline more than 5 million years ago still displays considerable sequence similarity to exogenous viruses of mice (611,689).

RNA Polymerase

The error rate of RNA polymerase II is not known (and could be very difficult to measure), but is unlikely to be much different from that of other RNA-synthesizing enzymes (or of reverse transcriptase) i.e.. about 10^{-4}, plus or minus a factor of 3. If so, then it is worth noting that errors in genomic RNA synthesis may be as important as those in reverse transcription in generating sequence variation. Also noteworthy is that there would be no point for the virus to encode a highly accurate reverse transcriptase if half the replication cycle is carried out by relatively inaccurate *pol* II.

Rearrangements

Retrovirus genomes are subject to a high rate of intragenomic rearrangements—deletions, duplications, inversions, or combinations of these (132,135,735). Indeed, a survey of the literature on cloning of viral DNA from infected cells indicates that perhaps half of the DNAs so obtained have suffered some sort of rearrangement. In the majority of cases, the rearrangements are so severe as to render the genome inactive, but there are some important exceptions. For example, rearrangements in the LTR of murine leukemia viruses are important determinants of differential pathogenicity (311,702); genetic variation within the *env* gene of different HIV isolates seems to include reduplication of short sequences (530); and specific types of defective mutants, such as SFFV and the MAIDS and FAIDS viruses, can be important pathogens in their own right.

Rearranged genomes are largely a consequence of errors in reverse transcription. While rearrangements involving integrated viral DNA can sometimes be detected if the selection is strong enough [e.g., see (433)], such events seem to occur at a very low rate (a few in a million) and cannot account for the vast majority of such events. Lesions in the process of reverse transcription include mispriming, premature termination of one end or the other, incorrect end-to-end strand transfer, and foldback copying of a newly completed strand to form inversions (601). The most common type of rearrangement, however, is reduplication or deletion of sequence, most likely due to aberrant "jumping" of reverse transcriptase from one RNA template to another or within a template. In about half the cases examined in one study, there was a short additional sequence at the deletion point, implying the involvement of two crossover events (600). Together these two types of mutation were observed at a rate of about 1 event in 25 proviruses. Mutations of this sort can be considered to be the consequence of an incorrect recombination event (see next section). It is noteworthy that, while deletion and reduplication often involve the use of short stretches of identical sequence (550,566), they do not always do so, and there are instances of such rearrangements in the absence of detectable homology or other sequence features (713,780).

Recombination

One of the most remarkable features of retrovirus genetics is the high rate of recombination. There is probably no other biological system which displays the capacity for exchange of genetic information to the same extent as retroviruses; all retrovirus systems tested, including HIV, undergo recombination at high rates (128,131,310,319,734, 735). Although most commonly observed between infecting exogenous viruses, recombination can also readily be observed between exogenous and endogenous viruses and between virus and unrelated host cell information (as in oncogene capture). The frequency of recombination is so high that in a usual experiment (i.e., coinfecting a cell culture with a mixture of viruses differing in two selectable markers, such as transforming ability and host range, and then selecting a recombinant between them), markers as close as 1 kb are found to segregate independently, as if unlinked. More recent estimates of recombination rate have been made using single-step assays for restoration of a selectable marker gene containing different inactivating mutations. These vary from about 20% per genome per replication for SNV (320,322) down to a lower values (<1%) for MLV (705). The basis for the discrepancy is unclear. This free exchange implied by the higher figure suggests that a population of virus containing sequence variants and allowed to interact (by coinfection of the same cells) should be considered to be homogenized across the genome such that all possible combinations of variants are present at any one time. This has important implications for the variation of pathogens such as HIV and the potential of populations of viruses to generate new combinations of variants which might have very different properties.

It is clear that recombination occurs only following infection with heterozygous virions, i.e., those produced by cells coinfected with two parental viruses (322). If a cell is simply doubly infected, recombinants are not observed in that cell. From this requirement, it follows that recombination must be an early event, occurring prior to integration, presumably related to the process of viral DNA synthesis. Although, as discussed above, recombination can occur between completely unrelated sequences—often, but not always, at very short stretches of identical sequence—such nonhomologous recombination is much less frequent. In one study, recombination frequency was found to depend in a very sensitive way on the length of identical sequence (825), and was 1,000-fold less frequent across completely unrelated sequence than at highly related sites (824).

Two fundamentally different models for retrovirus recombination have been debated for some 15 years, and recent observations regarding the mechanism of reverse transcription suggest a third. These are illustrated in Fig. 24. The first (360) proposes that exchange of a newly made fragmentary plus strand copied from one genome onto a minus-strand copy of the other could lead to heteroduplex proviruses which would segregate wild-type and recombinant virus after integration and cell division. While concordant with electron microscope observation of the products of *in vitro* reverse transcriptase reactions, this scheme requires that both members of the diploid genome serve as templates for minus-strand DNA synthesis simultaneously, and it does not readily explain recombination involving large regions of nonhomology, such as around oncogenes. Also, the production of fragmentary plus-strand DNA molecules is a characteristic of ALSV not found in most other retroviruses.

At present, the weight of evidence favors some sort of strand transfer during minus-strand synthesis (321), and there are two possibilities. The first model (128) is based on the observation that retrovirus genomes can have considerable numbers of breaks and yet remain associated in an intact complex. Thus, when reverse transcriptase encounters such a break, it can behave exactly as it would at the end of a genome; i.e., switch to the homologous sequence on the other genome and continue synthesis. The product of this reaction would be a minus strand containing a mosaic of information from both genomes (depending on the number of preexisting breaks). This mechanism could not only promote high-frequency exchange of information, but would have the important additional benefit of repairing preexisting RNA breakage and relieving the virus of using extraordinary measures to shield its genome from the ravages of the extracellular environment (128,734). Since reverse transcriptase can make strand transfers in the absence of extended homology (indicating that affinity of enzyme for template may be a stronger driving force for jumping than base pairing), a similar mechanism can be readily in-

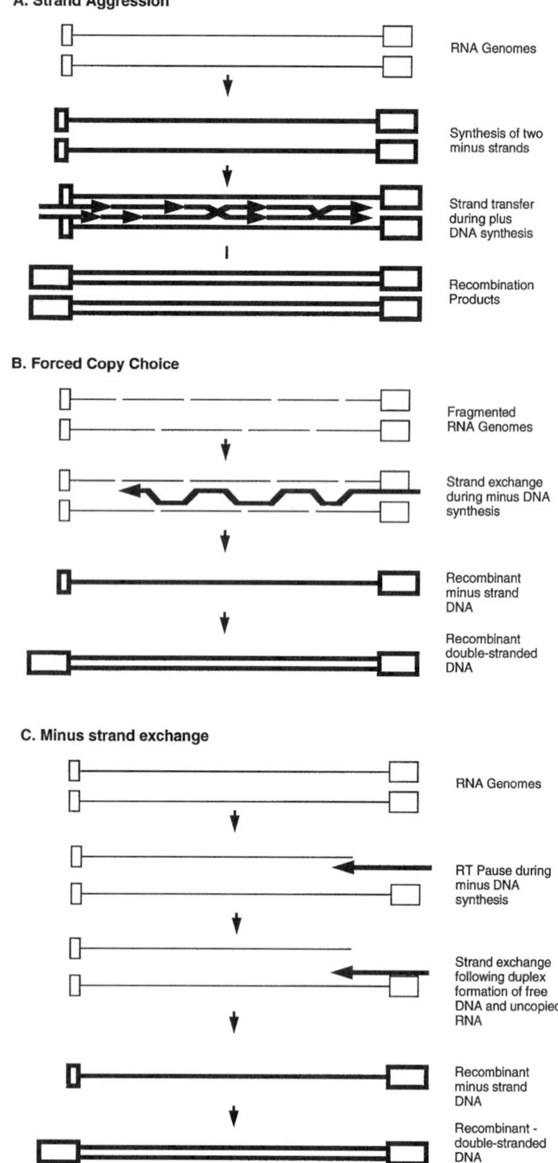

FIG. 24. Possible mechanisms of retrovirus recombination. In both cases, DNA synthesis beginning with a virus with a heterozygous genome leads to the formation of recombinants. In the first, two full-length minus strands are synthesized and recombination occurs by strand exchange during plus strand synthesis. In the second, "jumps" by reverse transcriptase lead to synthesis of a recombinant minus strand.

voked to explain other types of genomic rearrangements, as well as capture of cellular sequences.

A third possibility does not require preexisting breaks in the RNA, but is based on the observations that reverse transcriptase has a relatively low processivity and that the RNA template is degraded some 18 nucleotides from the point of DNA synthesis, leaving an extended single-stranded DNA molecule trailing behind. This molecule is then free to form a hybrid duplex with the other RNA molecule. If the RT molecule detaches from the template, then there is a good chance that the short (18-bp) hybrid with the original template will be displaced by branch migration of the second RNA, which will then become the template for continued synthesis. Although the weight of evidence is strongly in favor of recombination at the level of minus-strand synthesis, there is at present insufficient evidence to decide between the latter two models.

Oncogene Capture

Retroviruses are the only group of animal viruses which acquire cellular genes and convert them to oncogenes, so the mechanism of oncogene acquisition must be connected with special features of retroviruses. As with deletions and other rearrangements, the incorporation of oncogenes (or other nonviral sequences) into retrovirus genomes is a consequence of illegitimate recombination events, with the distinctions that at least two such events (one at each end of the oncogene) are required, and the participating sequences are derived from different genes. Over the years, a number of model systems have been developed to study the mechanism underlying this process, relying on the generation of selectable marker genes to mimic the process (254,260,287,706,824). Based on these studies, it was initially proposed that oncogene capture was a three-step process (Fig. 25):

1. The provirus integrates upstream from a proto-oncogene. As discussed above, integrations of this sort can also lead directly to oncogene "activation."
2. A deletion at the DNA level joins the proto-oncogene to a portion of the provirus, and the joint viral-oncogene transcript is incorporated into heterozygous virions along with a wild-type genome (from another provirus in the same cell).
3. An illegitimate recombination event during reverse transcription restores the 3′ end of the viral genome.

While this scheme seemed reasonable and still has a few supporters, studies with other models make it very clear that the postulated deletion at the DNA level is unnecessary: to be incorporated into proviruses, cellular sequences need only be packaged into virions; in at least some systems, coexpression of virus and the selectable gene (such as a drug resistance marker) is sufficient (287,706). Inclusion of a packaging signal increases the frequency, but detectable levels of recombinants are found even in its absence. Another model might more closely mimic the natural case which is likely to start with a tumor induced by promoter insertion (713). In this case (Fig. 25), frequent capture of the marker gene into the provirus was seen in cells infected with the progeny virus, and the frequency of capture was directly related to the frequency of readthrough transcripts (see above). No events attributable to DNA-level deletion could be seen. Furthermore, in all these systems, many of the proviruses lacked essential regions such

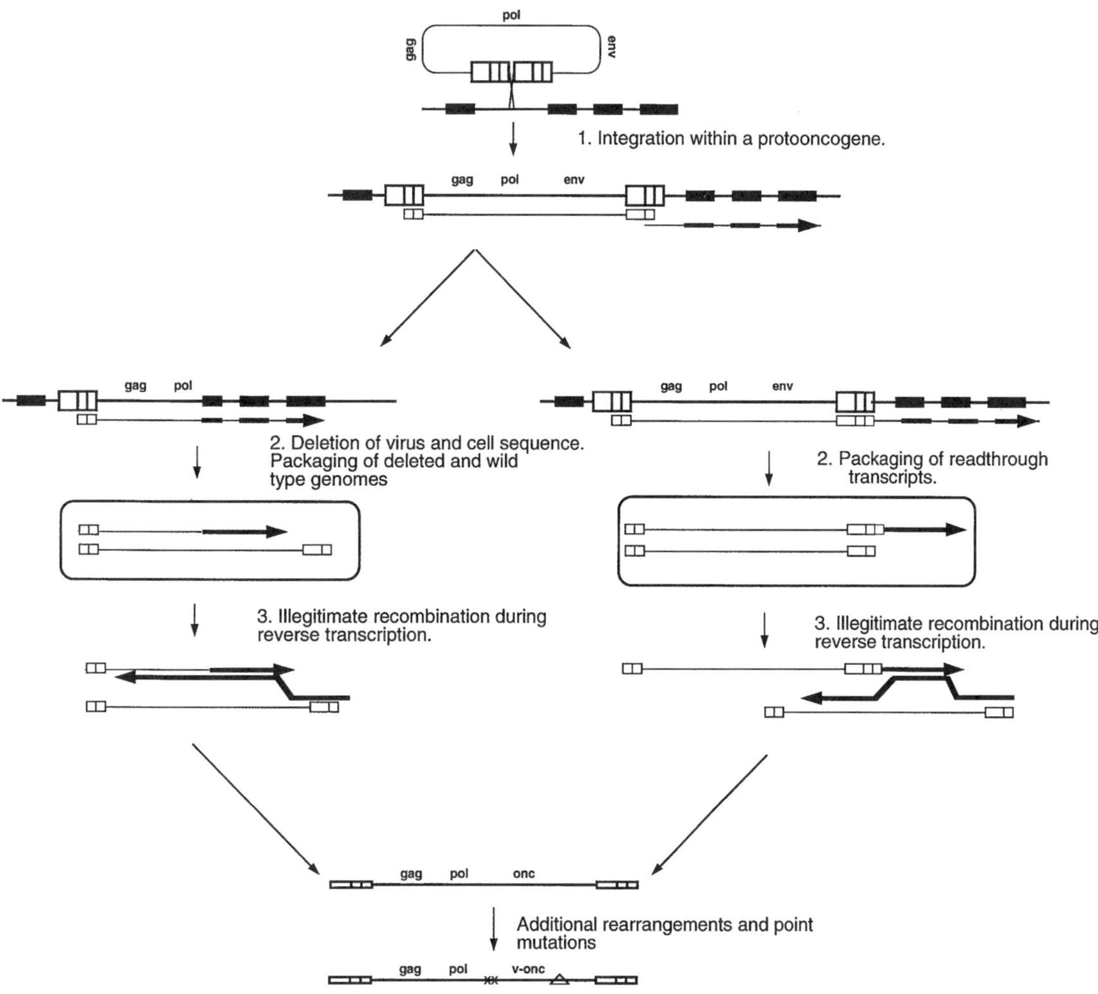

FIG. 25. Possible mechanisms for oncogene capture by retroviruses. One scheme (*left*) (715) differs from the other (304) by the requirement for a specific deletion in DNA of the infected cell which fuses viral and cellular sequences. As shown (*right*), this fusion can also be accomplished by reverse transcriptase acting on readthrough transcripts. In either case, important additional mutations and rearrangements probably occur during subsequent virus replication.

as packaging signals, implying that the rearrangements leading to their loss took place in the newly infected cell.

Thus, the current picture of oncogene capture is that it is entirely a consequence of nonhomologous recombination during reverse transcription, and the limiting step may be the frequency of copackaging of viral and cellular RNA. Note that the final structure of the oncogene containing virus genome may also be determined by additional rearrangements, as well as point mutations within the oncogene itself, so that the structure observed in viruses need not exactly reflect the initial events.

Endogenous Viruses

A particularly noteworthy and unique feature of retroviruses is their presence as inherited elements in the germline of many (perhaps all) vertebrates, where they behave as reasonably stable Mendelian genes (135,551,698, 799). It has been estimated that as much as 5% to 10% of the mammalian genome may consist of elements introduced by mechanisms involving reverse transcription (733). Perhaps 10% of these are identifiable on structural grounds as proviruslike; i.e., they contain LTRs flanked by short direct repeats and primer binding sites which flank internal coding regions with detectable relationship to *gag* and *pol.* The remainder comprise a variety of "retroelements," including movable elements such as the long interspersed repeat sequences (or LINES) whose motion is directed by reverse transcription machinery encoded within the element (55,241,333), as well as families of sequences such as processed pseudogenes and the abundant alu elements in human DNA which have been moved by processes that include reverse transcription but which do not themselves encode such activity (333). For the most part, this discussion will be confined to the proviruslike elements, in particular, the endogenous proviruses; i.e., those related to known viruses. The remainder of proviruslike retrotransposable elements such as copia and similar elements in flies (518), Ty 1 and other yeast transposons (56), and some plant elements (270), seem to use mechanisms quite similar to those of retroviruses for their motion and genetic association with the host. Indeed, recent evidence indicates that at least one of these—the gypsy element of Drosophila—is a true endogenous retrovirus (388). Elements defined as retrotransposons usually differ in the apparent lack of an extracellular or virion phase and the absence of an *env* gene. Neither of these characteristics is definitive in determining that an identified proviruslike element is not a true virus. An extracellular phase may be very transient and difficult to detect. Indeed, if many of the endogenous proviruses now known were first identified solely by their appearance in the genome without reference to known viruses, they, too, would be classed as retrotransposons. Also, as discussed in a previous section, *env* gene products can be readily supplied by other proviruses (even unrelated ones) in the same cell.

Distribution

It is convenient to divide endogenous proviruses into two types: ancient and modern. Ancient proviruses have been found in all vertebrate species examined, and the human genome contains a large number (66,799). In general, these proviruses were inserted into the germline of an ancestor of the modern species and are therefore present in the same location in the genome of all individuals, often of related species as well. For example, their species distribution has allowed the insertion of some specific human proviruses to be dated to more than 60 million years ago (661). By and large, these elements can be considered to be molecular fossils, of little present functional significance, but very informative regarding the evolutionary history of the *Retroviridae*. Indeed, retroviruses are the only known viruses to have left a fossil record, and we all carry a part of it in our DNA. Like true fossils, ancient endogenous proviruses have suffered abuse of the hands of time: none are known to yield infectious virions. A few [most notably one-celled HERV-K which is related to the B- and D-type virus groups (59, 523)] encode sufficient virion proteins to yield visible particles. Sequence analysis of ancient proviruses reveals the accumulation of numerous mutations since their insertion. These include frameshifts and premature stop codons in coding regions. Most notably, the two LTRs, which must have been identical at the time of integration, frequently demonstrate considerable divergence in sequence (611,661,739).

Modern endogenous proviruses (those closely related to and potentially encoding infectious viruses) are sporadically distributed among species. When present, they are unevenly distributed among individuals within a species. They have clearly become associated with the germline subsequent to speciation and are continuing to show reinsertion into it in modern times. Chickens, for example, contain from one to about four or so proviruses closely related to ALSV and residing at a variety of sites in the genome, whereas related species such as turkeys or quail or even other members of the same genus (*Gallus*) do not contain any related proiruses. In humans, no active proviruses have been detected, despite large numbers of ancient proviruses. Infectious endogenous viruses, including the baboon endogenous virus, are present in other primates.

In mice, four classes of elements totaling 500 to 1,000 members have been identified as endogenous proviruses, although two of them are not known to encode virus genomes. The VL30 (for "viruslike 30S") proviruses encode RNAs which apparently are not themselves competent but are efficiently encapsidated in MLV virions (337,384) and contaminate MLV based vectors at high frequency. They may be an ancient element devolved to the status of a passenger virus, only capable of replicating by using endogenous or exogenous MLV as its helper. The intracisternal A particles (IAPs) are present in more than 500 copies in the mouse genome (413). They are often expressed as large numbers of intracellular particles with active reverse transcriptase, but have never been seen in virions and mostly lack *env* genes, leading to the suggestion that they are retrotransposons limited to intracellular reverse transcription and integration. In support of this, intracellular "transposition" of an IAP vector construct has been detected with a sensitive assay (299). Thus, intracellular replication *can* occur with these elements, although this does not mean that it *does* occur. The same sort of assay also detects intracellular replication of an *env*-defective MLV, but at a very low level ($<10^{-6}$) relative to intercellular transmission of nondefective virus (727). The issue is further complicated by the presence of a subset of IAP sequences that contain *env* genes (614,615).

The two endogenous proviruses known to encode infectious virus are the B-type, closely related to exogenous MMTV, and the C-type, related to exogenous MLV. The former is present in 0 to 4 copies per mouse (68,405,424); the latter about 50 to 60 (231,700). The C-type proviruses can be further subdivided into groups according to properties of the envelope gene (and other linked sequence characteristics) into four classes: ecotropic, xenotropic, polytropic, and modified polytropic (699), with the latter three about tenfold more abundant (and somewhat more ancient) than the ecotropic class (309,407). Similar but less well studied groups of endogenous proviruses are also present in cats (291).

Detailed analysis of the distribution of the C-type proviruses in mice, including assignment of over 120 different proviruses in different inbred strains to specific chromosomal locations (231), has revealed the following general principles.

1. Proviruses are stably present at many locations on all chromosomes.
2. No two inbred strains are alike in their composition of proviruses and only a few are found in all strains.
3. Different proviruses within a group are very closely related to one another, differing principally by their chromosomal location, and seem to be separated by only a few viral replication cycles.
4. In most cases, acquisition of new germline proviruses is quite rare. In the course of inbreeding, a rate of about one new insertion in 3,500 generations was estimated (231). The frequency of appearance of new proviruses seems to correlate with the extent of virus replication in the animal, implying that any apparent "transposition" is a result of virus replication. In particular, when there is a high level of virus replication in newborn female mice in the absence of maternal antibody, rates of provirus insertion into the progeny as high as one provirus in ten animals can be seen (457,686).
5. Similarly, the rate of proviral loss is quite low. In several different types of experiments, values of about 4×10^{-6} have been estimated (231,657). The most common mechanism of loss seems to be recombination between the two LTRs, leaving a solo LTR behind. No evidence whatever has ever been presented for specific excision or intracellular transposition or any other mechanism of movement of proviruses except through the normal replication pathway.

Properties of Endogenous Viruses

Endogenous proviruses are not simply germline copies of exogenous infectious agents. Rather, they have a number of special features, some of which seem to be specific adaptations to the endogenous lifestyle, others of which are apparently enforced by residence in the host germline.

First, endogenous proviruses are usually transcriptionally silent. This effect is an epigenetic one. If expression of a nondefective endogenous provirus is induced and it infects new cells, it is often capable of directing at least moderate levels of transcription. As with silencing of other genes (50,442), this effect involves suppression of expression by extensive methylation of C residues in CpG dinucleotides (142) and possibly other mechanisms (such as heritable modifications in chromatin structure) as well. Agents that reverse DNA methylation, such as 5-azacytidine, can efficiently induce expression, at least transiently, of endogenous proviruses. The silencing of these proviruses is not a feature unique to them: proviruses of exogenous viruses, like Moloney MLV, are also affected the same way after introduction into the germline of mice (343,344). Presumably, this reflects a host mechanism for dealing with foreign DNA. Often, the transcriptional repression is very strong but not absolute. Under conditions where the endogenous provirus is capable of replicating in the host animal, even a very low level of expression can provide sufficient virus to cause a generalized infection. Some animals (such as AKR mice) usually become viremic at an early age. Not all endogenous proviruses are strongly transcriptionally suppressed. In chickens, for example, some defective proviruses are expressed at levels 100-fold higher than others. This phenomenon may be due to position effects of adjoining chromatin structure on the provirus (143). In some mice, coordinate expression of a number of C-type proviruses at different locations is controlled by a single genetic locus (designated *Gv-1*) (437).

Second, endogenous proviruses are often defective, typically differing from the canonical wild-type virus by deletions or point mutations that render them incapable of yielding infectious virus. One endogenous MLV provirus, for example, contains a mutation near the beginning of *gag* that prevents myristylation (145), and most endogenous ALV proviruses do not yield replicating virus.

Third, many endogenous viruses, although replication competent, are often unable to replicate in the animals in which they are found due to the mismatch in receptor ("xenotropism"), discussed above.

Fourth, with one exception, viruses induced from endogenous proviruses are nonpathogenic and often less efficient in replication than their exogenous relatives. This difference presumably represents an adaptation to the endogenous lifestyle; viruses of significant pathogenicity would reduce the reproductive potential of their host and be counterselected. The difference in replicative rate is largely attributable to sequence differences in the LTR, as shown by genetic exchange studies with both MLV and ALV (311,745). In particular, endogenous provirus LTRs have significantly reduced enhancer activity as compared to their pathogenic cousins (156,283).

Effects on the Host

Beneficial Effects

When endogenous retroviruses were first described, it became popular to propose ways in which they or related elements might provide some essential service to the host organism. The variability in these elements and the breeding of individuals lacking certain types of proviruses altogether laid these ideas to rest, at least in their simplest form. It became more popular to consider endogenous proviruses as a type of selfish element, derived from exogenous infectious viruses and specifically adapted to residence in the germline. It is now clear that the presence of some endogenous proviruses confers a beneficial effect relevant to pathogenesis by exogenous viruses. Expression of *env* proteins can induce interference to infection by viruses that use the same receptor. Several genetic loci, such as *Fv4* (247,447,710) and *Rmcf* (86) in mice and similar loci in chickens. were originally identified as conferring resistance to infection with MLV or ALV and later shown to be endogenous proviruses. Another possible beneficial effect is the induction of immunological tolerance to virus antigens, thus minimizing pathogenic effects due to immunological killing of infected cells (152). Consistent with selection for endogenous *env* gene expression is the observation that many of the endogenous ALVs are defective in *gag* and *pol* expression, yet express reasonably high levels of *env* gene product (129).

A related sort of effect is seen with the expression of the superantigen encoded by the *sag* gene of endogenous MMTV (1,134,557). As with expression of any antigen in developing animals, expression of an MMTV *sag* protein causes deletion of the subset of T helper cells that react with it. The effect of endogenous *sag* is much more dramatic, however. All T cells whose TCR gene is rearranged to express a specific subset of V_β chains are deleted by or soon after birth. Furthermore, B cells expressing *sag* genes of endogenous proviruses give a very strong V_β-specific stimulation of T cells in a mixed lymphocyte reaction. Since different MMTV proviruses in different inbred strains at different chromosomal locations react with different subsets of V_β chains (68,134), the genetics of this system can appear to be quite complex. Indeed, endogenous superantigen genes were initially recognized by their immunological effects and termed *mls* (for "minor lymphocyte stimulating") loci, and presented a significant puzzle to immunologists until their association with endogenous MMTV proviruses was recognized (118,230,329,557).

The deletion of reactive T-cell subsets by expression of endogenous MMTV, while it might at first seem disadvantageous, probably confers an advantage on the animal that carries it, since such an animal will not be able to respond to the *sag* protein of the homologous infecting MMTV and will be infected only inefficiently, if at all. There is good experimental evidence for this protection, both with natural endogenous proviruses (301,639) and exogenous MMTV *sag* genes introduced as transgenes (257). In both cases, the presence of an expressed *sag* gene from early in life prevents infection by MMTV of the same *sag* specificity. A consequence of this resistance is to put a strong selective pressure for variation in *sag* specificity on the exogenous MMTV, leading to the large variety of V_β specificities observed today. An analogous pressure to vary in specificity may well account for the polymorphism in receptor usage seen in ALSV, MLV, and FeLV.

Pathogenic Effects

In general, endogenous viruses, when expressed, do not exert pathogenic effects on their host. Chickens viremic with the endogenous virus, known as RAV-0, are indistiguishable from normal birds. The only endogenous virus known to directly cause disease is MMTV, in which inheritance of an endogenous provirus gives rise to a significant incidence of mammary carcinoma even in mice foster-nursed on virus-free strains to eliminate maternal transmission.

A more complex example was created when some strains of mice were inbred to select for high incidence of spontaneous leukemia. Mice of strains such as AKR and C58 typically die at about 1 year of age with a T-cell lymphoma induced by a retrovirus derived from proviruses endogenous to the particular strains. While the ultimate molecular event in oncogenesis involves insertional activation of one or more oncogenes by new proviruses (see above), a complex set of rearrangements is involved in the generation of the oncogenic viruses themselves (702) (Fig. 26). First, an ecotropic provirus is expressed and replicates widely in the animal, starting around the time of birth. Second, recombinants of this virus with a xenotropic provirus appear which have acquired the LTR of the xenotropic parent. Third, a specific rearrangement occurs in the LTR to duplicate (sometimes triplicate) the U_3 enhancer region. Fourth, recombination with a polytropic provirus alters the envelope SU (gp70) coding region and gives the polytropic host range to the virus. Finally, the new virus (an MCF virus) infects target T cells, and a set of events including integration of proviral DNA adjacent to a proto-oncogene and stimulation of cell growth by interaction of the *env* gene product with a growth factor receptor leads to transformation into a tumor cell.

These specific events occur synchronously (as can be detected using specific DNA probes) in virtually every mouse of affected strains. There is no imaginable basis by which the specific recombination events and rearrangements can be directed to occur; rather, the products seen most are those selected out of a very large pool of randomly occurring events.

The selective forces involved can be inferred. The LTR alterations presumably act to enhance the ability of the virus to replicate in the target tissue. The *env* alteration provides the recombinant virus with the ability to use a different receptor, but the more likely selected function is its ability to directly stimulate growth of infected cells. In all of this genetic contortion it is striking that the only genomic variation seen is that selected for, although initial recombination events must have generated a very large number of different combinations. Comparison of nucleotide sequence of recombinant and parental viruses reveals very little random sequence variation. This example points up the critical role of selective forces in shaping retroviral genomes. Such forces play a major role in the remarkable genetic variation observed in the course of HIV infection (Chapter 61 in *Fields Virology*, 3rd ed.).

Mutagenic Effects

The insertion of an endogenous provirus into a new location on the genome is a potentially mutagenic event.

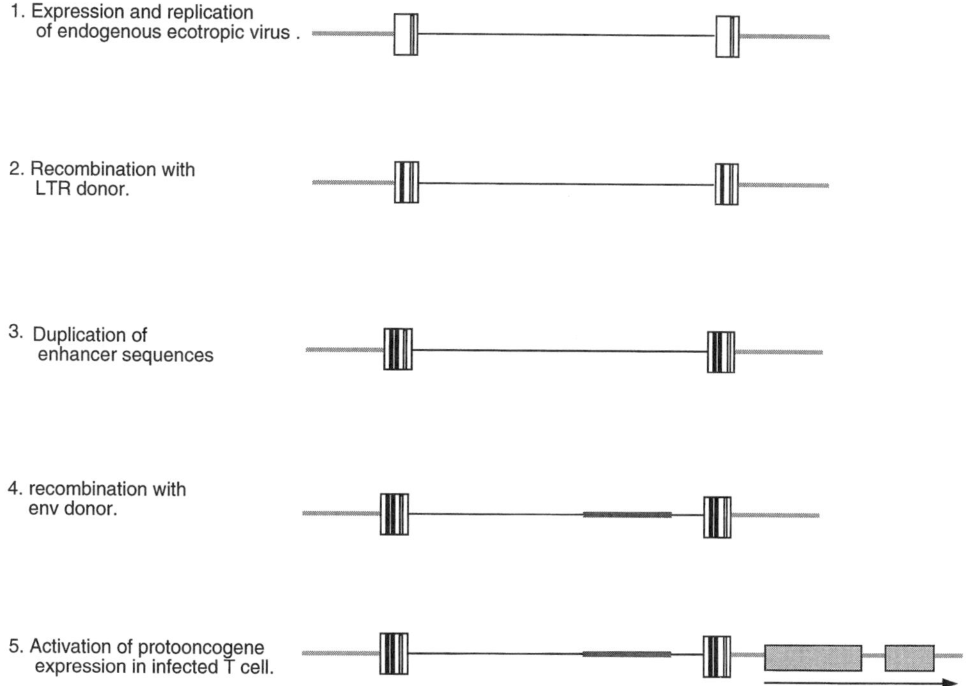

FIG. 26. Virological events during lymphomagenesis in AKR mice. It is important to keep in mind that each of the events itself is relatively rare, but the specific products are selected by their increased replication ability or effects on the target cell.

Considering the number and variety of endogenous proviruses in some species such as mice, it would not be surprising to find that some are associated with known mutations, and genetic analyses have supplied several examples. Two well-known mutations of mice, *d* (dilute), and *hr* (hairless) have each been found to be inseparable from specific endogenous proviruses in standard genetic experiments. In both cases, reversion of the mutation was accompanied by loss of the provirus, leaving behind a solo LTR (144,701) and providing proof of the causal relationship. In both cases, the provirus has been used as a tag from which flanking DNA corresponding to part of the gene could be cloned and analyzed (96,491). In a related example, introduction of exogenous proviruses into the mouse germline (by infection of early embryos) has also been found to be mutagenic. For example, one such provirus (known as *Mov-13*) was found to cause death of the homozygous embryo in a midembryonic stage due to its insertion into a gene whose product becomes essential to the survival of the embryo at that point—the gene encoding α-1 collagen (31,294,410).

Most of the mutations caused by insertion of endogenous proviruses are recessive, due to loss of expression of the affected gene. In principle, dominant effects, analogous to oncogene activation, might also be expected but are rarely observed, probably reflecting the lack of transcriptional activity of proviruses passed through the germline. Although no active endogenous viruses are known to have activated expression of a host gene, one interesting example of such activation by a more ancient provirus has been reported. In this example, expression of a duplicated copy of the mouse C4 complement gene [*Slp*, whose product is known as sex-limited protein (757)] is activated and made responsive to stimulation by androgen (460) due to the presence upstream of an ancient, deleted C-type provirus whose LTR provides the necessary hormone-responsive enhancer (687).

Closer to home is the interesting case of human salivary ß-amylase. In most species the ß-amylase gene is expressed only in the pancreas. In humans, however, it is expressed also in salivary glands, and its presence in saliva lends a sweet taste to starchy foods. When the genes for this enzyme were examined in detail, it was found that the human gene had suffered both a reduplication and the insertion, ahead of one of the copies, of an endogenous provirus (644) and that the LTR of this provirus is responsible for the salivary gland-specific expression (740). Thus, endogenous proviruses can exert evolutionary effects through the expression of neighboring genes, although only a very few examples have been observed to date.

Other Effects

Even aside from their direct role as mutagens, the number, variety, and antiquity of endogenous proviruses imply that they may have played important roles in shaping the vertebrate genome throughout evolution (76,135,173). It is possible that they contribute directly not only to their own expression and reinsertion, but also to that of unrelated sequence as well. Some defective endogenous retrovirus-related elements such as VL30 RNA are efficiently packaged into infectious virions by proteins of the unrelated MLV, and at least one small RNA related to repeated elements in the genome is copackaged into ASLV virions and reverse transcribed (116,286). Model systems for retroviral-assisted pseudogene formation, in which copackaging, reverse transcription, and integration of a selectable marker gene can be detected and studied (175,449), reveal complex rearrangements in the progeny not resembling the "natural" pseudogene structure (176,435). It is now considered more likely that the reverse transcription apparatus encoded by LINE elements is responsible for the vast bulk of non-LTR containing reverse transcripts, such as SINES (alu sequences and the like) and pseudogenes (333,350).

REFERENCES

1. Acha-Orbea H, Palmer E. Mls—a retrovirus exploits the immune system. *Immunol Today* 1991;12:356–361.
2. Adam MA, Miller AD. Identification of a signal in a murine retrovirus that is sufficient for packaging of nonretroviral RNA into virions. *J Virol* 1988;62:3802–3806.
3. Adam MA, Ramesh N, Miller AD, Osborne WRA. Internal initiation of translation in retroviral vectors carrying picornavirus 5′ nontranslated regions. *J Virol* 1991;65:4985–4990.
4. Ahmed YF, Gilmartin GM, Hanly SM, Nevins JR, Greene WC. The HTLV-1 rex response element mediates a novel form of mRNA polyadenylation. *Cell* 1991;64:727–738.
5. Aiyar A, Cobrinik D, Ge Z, Kung H-J, Leis J. Interaction between retroviral U5 RNA and the TΨC loop of the tRNATrp primer is required for efficient initiation of reverse transcription. *J Virol* 1992; 66:24604–2472.
6. Aiyar A, Ge Z, Leis J. A specific orientation of RNA secondary structures is required for initiation of reverse transcription. *J Virol* 1994; 68:611–618.
7. Akgun E, Ziegler M, Grez M. Determinants of retrovirus gene expression in embryonal carcinoma cells. *J Virol* 1991;65:382–388.
8. Akroyd J, Fincham VJ, Green AR, Levantis P, Searle S, Wyke JA. Transcription of Rous sarcoma proviruses in rat cells is determined by chromosomal position effects that fluctuate and can operate over long distances. *Oncogene* 1987;1:347–355.
9. Albritton LM, Kim JW, Tseng L, Cunningham JM. Envelope-binding domain in the cationic amino acid transporter determines the host range of ecotropic murine retroviruses. *J Virol* 1993;67:2091–2096.
10. Albritton LM, Tseng L, Scadden D, Cunningham JM. A putative murine ecotropic retrovirus receptor gene encodes a multiple membrane-spanning protein and confers susceptibility to virus infection. *Cell* 1989;57:659–666.
11. Alizon M, Dragic T. CD26 antigen and HIV fusion? *Science* 1994; 264:1161–1162.
12. Allain B, Lapadattapolsky M, Berlioz C, Darlix JL. Transactivation of the minus-strand DNA transfer by nucleocapsid protein during reverse transcription of the retroviral genome. *EMBO J* 1994;13:973–981.
13. Anderson DJ, Lee P, Levine KL, Sang J, Shah SA, Yang OO, Shank PR, Linial ML. Molecular cloning and characterization of the RNA packaging-defective retrovirus SE21Q1b. *J Virol* 1992;66:204–216.
14. Archer TK, Lefebvre P, Wolford RG, Hager GL. Transcription factor loading on the MMTV promoter: a bimodal mechanism for promoter activation. *Science* 1992;255:1573–1576.
15. Arima N, Molitor JA, Smith MR, Kim JH, Daitoku Y, Greene WC. Human T-cell leukemia virus type I tax induces expression of the rel-related family of κB enhancer-binding proteins: evidence for a pretranslational component of regulation. *J Virol* 1991;65:6892–6899.
16. Armstrong AP, Franklin AA, Uittenbogaard MN, Giebler HA, Nyborg JK. Pleiotropic effect of the human T-cell leukemia virus Tax protein on the DNA binding activity of eukaryotic transcription factors. *Proc Natl Acad Sci USA* 1993;90:7303–7307.
17. Aronoff R, Hajjar AM, Linial ML. Avian retroviral RNA encapsidation: reexamination of functional 5′ RNA sequences and the role of nucleocapsid cys-his motifs. *J Virol* 1993;67:178–188.
18. Aronoff R, Linial M. Specificity of retroviral RNA packaging. *J Virol* 1991;65:71–80.
19. Arrigo S, Yun M, Beemon K. Cis-acting regulatory elements within gag genes of avian retroviruses. *Mol Cell Biol* 1987;7:388–392.
20. Arthur LO, Bess JWJ, Sowder RCI, Benveniste RE. Mann, DL. Chermann, J-C, Henderson LE. Cellular proteins bound to immunodeficiency viruses: implications for pathogenesis for vaccines. *Science* 1992;258:1935–1938.
21. Aurigemma RE, Comstock RD, Smith RE. Persistent viral DNA synthesis associated with an avian osteopetrosis-inducing virus. *Virology* 1989;171:626–629.
22. Awang G, Sen D. Mode of dimerization of HIV-1 genomic RNA. *Biochemistry* 1993;32:11453–11457.
23. Aziz DC, Hanna Z, Jolicoeur P. Severe immunodeficiency disease induced by a defective murine leukaemia virus. *Nature* 1989;338:505–508.
24. Ball JK, Diggelmann H, Dekaban GA, Grossi GF, Semmler R, Waight PA, Fletcher RF. Alterations in the U3 region of the long terminal repeat of an infectious thymotropic type B retrovirus. *J Virol* 1988;62:2985–2993.19;
25. Ballard DW, Bohnlein E, Lowenthal JW, Wano Y, Franza BR, Greene WC. HTLV-I tax induces cellular proteins that activate κB element in the IL-2 receptor a gene. *Science* 1988;241:1652–1655.
26. Baltimore D. RNA-dependent DNA polymerase in virions of RNA tumour viruses. *Nature* 1970;226:1209–1211.
27. Ban J, Portetelle D, Altaner C, Horion B, Milan D, Krchnak V, Burny A, Kettmann R. Isolation and characterization of a 2.3-kilobase-pair cDNA fragment encoding the binding domain of the bovine leukemia virus cell receptor. *J Virol* 1993;67:1050–1057.
28. Bar-Shira A, Panet A, Honigman A. An RNA secondary structure juxtaposes two remote genetic signals for human T-cell leukemia virus type I RNA 3′-end processing. *J Virol* 1991;65:5165–5173.
29. Barillari G, Gendelman R, Gallo RC, Ensoli B. The tat protein of human immunodeficiency virus type 1, a growth factor for AIDS Kaposi sarcoma and cytokine-activated vascular cells, induces adehsion of the same cell types by using integrin receptors recognizing the RGD amino acid sequence. *Proc Natl Acad Sci USA* 1993;90:7941–7945.
30. Barker CS, Bear SE, Keler T, Copeland NG, Gilbert DJ, Jenkins NA, Yeung RS, Tsichlis PN. Activation of the prolactin receptor gene by promoter insertion in a Moloney murine leukemia virus-induced rat thymoma. *J Virol* 1992;66:6763–6768.
31. Barker DD, Wu H, Hartung S, Breindl M, Jaenisch R. Retrovirus-induced insertional mutagenesis: mechanism of collagen mutation in Mov13 mice. *Mol Cell Biol* 1991;11:5154–5163.
32. Barklis E, Mulligan RC, Jaenisch. R. Chromosomal position or virus mutation permits retrovirus expression in embryonal carcinoma cells. *Cell* 1986;47:391–399.
33. Bates P, Young JAT, Varmus HE. A receptor for subgroup A Rous sarcoma virus is related to the low densith lipoprotein receptor. *Cell* 1993;74:1043–1051.
34. Batschelet E, Domingo E, Weissmann C. The proportion of revertant and mutant phage in a growing population, as a function of mutation and growth rate. *Gene* 1976;1:27–32.
35. Bebenek K, Kunkel TA. THe fidelity of retroviral reverse transcriptases. In: Skalka AM, Goff SP, eds. *Reverse transcriptase.* Cold Spring Harbor, NY: Cold Spring Harbor Laboratory Press, 1993;85–102.
36. Bedinger P, Moriarty A, von Borstell II RC, Donovan NJ, Steimer KS, Littman DR. Internalizaton of the human immunodeficiency virus does not require the cytoplasmic domain of CD4. *Nature* 1988;334:162–165.
37. Belcourt MF, Farabaugh PJ. Ribosomal frameshifting in the yeast retrotransposon Ty: tRNAs induce slippage on a 7 nucleotide minimal site. *Cell* 1990;62:339–352.
38. Ben-David Y, Giddens EB, Bernstein A. Identification and mapping of a common proviral integration site *Fli-1* in erythroleukemia cells induced by Friend murine leukemia virus. *Proc Natl Acad Sci USA* 1990;87:1332–1336.
39. Ben-David Y, Giddens EB, Letwin K, Bernstein A. Erythroleukemia induction by Friend murine leukemia virus—insertional activation of

a new member of the ets gene family, Fli-1, closely linked to c-ets-1. *Genes Dev* 1991;5:908–918.

40. Ben-David Y, Lavigeur A, Cheong GY, Bernstein A. Insertional inactivation of the p53 gene during Friend leukemia: a new strategy for identifying tumor suppressor genes. *New Biologist* 1990;2:1015–1023.
41. Bender MA, Palmer TD, Gelinas RE, Miller AD. Evidence that the packaging signal of Moloney murine leukemia virus extends into the gag region. *J Virol* 1987;61:1639–1646.
42. Benson RE, Sanfridson A, Ottinger JS, Doyle C, Cullen BR. Down-regulation of cell-surface CD4 expression by simian imunodeficiency virus nef prevents viral superinfection. *J Exp Med* 1993;177:1561–1566.
43. Berberich SL, Stoltzfus CM. Mutations in the regions of the Rous sarcoma virus 3′ splice sites: implications for regulation of alternative splicing. *J Virol* 1991;65:2640–2646.
44. Bernstein HB, Tucker SP, Hunter E, Schutzbach JS, Compans RW. Human immunodeficiency virus type 1 envelope glycoprotein is modified by O-linked oligosaccharides. *J Virol* 1994;68:463–468.
45. Berwin B, Barklis E. Retrovirus-mediated insertion of expressed and non-expressed genes at identical chromosomal locations. *Nucl Acids Res* 1993;21:2399–2407.
46. Bess JWJ, Powell PJ, Issaq HJ, Schumack LJ, Grimes MK, Henderson LE, Arthur LO. Tightly bound zinc in human immunodeficiency virus type 1, human T-cell leukemia virus type I, and other retroviruses. *J Virol* 1992;66:840–847.
47. Bestwick RK, Hankins WD, Kabat D. Roles of helper and defective retroviral genomes in murine erythroleukemia: studies of spleen focus-forming virus in the absence of helper. *J Virol* 1985;56:660–664.
48. Beutner U, Kraus E, Kitamura D, Rajewsky K, Huber BT. B cells are essential for murine mammary tumor virus transmission, but not for presentation of endogenous superantigens. *J Exp Med* 179:1457–1466.
49. Beutner U, Rudy C, Huber BT. Molecular characterization of Mls-1. *Intern Rev Immunol* 1992;8:279–288.
50. Bird A. The essentials of DNA methylation. *Cell* 70:5–8.
51. Biswas DK, Ahlers CM, Dezube BJ, Pardee AB. Cooperative inhibition of NF-κB and tat-induced superactivation of human immunodeficiency virus type 1 long terminal repeat.*Proc Natl Acad Sci USA* 1993;90:11044–11048.
52. Bizub D, Weber IT, Cameron CE, Leis JP, Skalka AM. A range of catalytic efficiencies with avian retroviral protease subunits genetically linked to form single polypeptide chains. *J Biol Chem* 1991;266:4951–4958.
53. Blum HE, Harris JD, Ventura P, Staskus K, Retzel E, Haase AT. Synthesis in cell culture of the gapped linear duplex DNA of th slow virus visna. *Virology* 1985;142:270–277.
54. Bodor J, Svoboda J. The LTR, v-src, LTR provirus generated in the mammalian genome by src mRNA reverse transcription and integration. *J Virol* 1989;63:1015–1018.
55. Boeke JD. Retrotransposons. In: Domingo E, Holland JJ, Ahlquist P, eds. *RNA Genetics*, volume II. Retroviruses, viroids, and RNA recombination. Boca Raton, FL: CRC Press, 1988;59–103.
56. Boeke JD, Garfinkel DJ, Styles CA, Fink GR. Ty elements transpose through an RNA intermediate. *Cell* 1985;40:491–500.
57. Boerkoel CF, Kung H-J. Transcriptional interaction between retroviral long terminal repeats (LTRs): mechanism of 5′ LTR suppression and 3′ LTR promoter activation of c-*myc* in avian B-cell lymphomas.*J Virol* 1992;66:4814–4823.
58. Bohnlein E, Siekevitz M, Ballard DW, Lowenthal JW, Rimsky L, Bogerd H, Hoffman J, Wano Y, Franza BR, Greene WC. Stimulation of the human immunodeficiency virus type 1 enhancer by the human T-cell leukemia virus type I tax gene product involves the action of inducible cellular proteins. *J Virol* 1989;63:1578–1586.
59. Boller K, Konig H, Sauter M, Mueller-Lantzsch N, Lower R, Lower J, Kurth R. Evidence that HERV-K is the endogenous retrovirus sequence that codes for the human teratocarcinoma-derived retrovirus HTDV. *Virology* 1993;196:349–353.
60. Bolognesi DP, Montelaro RC, Frank H, Schafer W. Assembly of type C oncornaviruses: a model. *Science* 1978;199:183–186.
61. Bonnerot C, Legouy E, Choulika A, Nicolas J-F. Capture of a cellular transcriptional unit by a retrovirus: mode of provirus activation in embryonal carcinoma cells. *J Virol* 1992;66:4982–4991.
62. Bosselut R, Lim F, Romond P-C, Frampton J, Brady J, Ghysdael J. Myb protein binds to multiple sites in the human T cell lymphotropic virus type 1 long terminal repeat and transactivates LTR-mediated expression. *Virology* 1992;186:764–769.
63. Boucheix C, Benoit P, Frachet P, Billard M, Worthington RE, Gagnon J, Uzan G. Molecular cloning of the CD9 antigen. A new family of cell surface proteins. *J Biol Chem* 1991;266:117–22.
64. Bova CA, Olsen JC, Swanstrom R. The avian retrovirus env gene family: molecular analysis of host range and antigenic variants. *J Virol* 1988;62:75–83.
65. Bowerman B, Brown PO, Bishop JM, Varmus HE. A nucleoprotein complex mediates the integration of retroviral DNA. *Genes Dev* 1989;3:469–478.
66. Boyce-Jacino MT, O'Donoghue K, Faras AJ. Multiple complex families of endogenous retroviruses are highly conserved in the genus *Gallus*. *J Virol* 1992;66:4919–4929.
67. Boyer JC, Bebenek K, Kunkel TA. Unequal human immunodeficiency virus type 1 reverse transcriptase error rates with RNA and DNA templates. *Proc Natl Acad Sci USA* 1992;89:6919–6923.
68. Brandt-Carlson C, Butel JS, Wheeler D. Phylogenetic and structural analyses of MMTV LTR ORF sequences of exogenous and endogenous origins. *Virology* 1993;193:171–185.
69. Bray M, Prasad S, Dubay JW, Hunter E, Jeang K-T, Rekosh D, Hammarskljold M-L. A small element from the Mason-Pfizer monkey virus genome makes human immunodeficiency virus type 1 expression and replication Rev-independent. *Proc Natl Acad Sci USA* 1994; 91:1256–1260.
70. Brierley I, Boursnell MEG, Binns MM, Bilimoria B, Blok VC, Brown TDK, Inglis SC. An efficient ribosomal frameshifting signal in the polymerase-encoding region of the coronavirus IBV. *EMBO J* 1987;6:3779–3787.
71. Brightman BK, Rein A, Trepp DJ, Fan H. An enhancer variant of Moloney murine leukemia virus defective in leukemogenesis does not generate detectable mink cell focus-inducing virus in vitro. *Proc Natl Acad Sci USA* 1991;88:2264–2268.
72. Broder CC, Berger EA. CD4 molecules with a diversity of mutations encompassing the CDR3 region efficiently support human immunodeficiency virus type 1 envelope glycoprotein-mediated cell fusion. *J Virol* 1993;67:913–926.
73. Broder CC, Nussbaum O, Gutheil WG, Bachovchin WW, Berger EA. CD26 antigen and HIV fusion? *Science* 1994;264:1156–1159.
74. Brody BA, Rhee SS, Hunter E. Postassembly cleavage of a retroviral glycoprotein cytoplasmic domain removes a necessary incoporation signal and activates fusion activity. *J Virol* 1994;68:4620–4627.
75. Brody BA, Rhee SS, Sommerfelt MA, Hunter E. A viral protease-mediated cleavage of the transmembrane glycoprotein of Mason-Pfizer monkey virus can be suppressed by mutations within the matrix protein. *Proc Natl Acad Sci USA* 1992;89:3443–3447.
76. Brosius J. Retroposons—seeds of evolution. *Science* 1991;251:753.
77. Brown DW, Robinson HL. Role of RAV-0 genes in the permissive replication of subgroup E avian leukosis viruses on line 15B ev1 CEF. *Virology* 1988;162:239–242.
78. Murphy FA, Fauquet CM, Bishop DHL, Ghabrial SA, Jarvis AW, Martelli GP, Mayo MA, Summers MD. Virus taxonomy: Sixth report of the International Committee on the Taxonomy of Viruses. New York: Springer-Verlag, 1995.
79. Brown PH, Tiley LS, Cullen BR. Efficient polyadenylation with the human immunodeficiency virus type 1 long terminal repeat requires flanking U3-specific sequences. *J Virol* 1991;65:3340–3343.
80. Brown PO. Integration of retroviral DNA. In: Swanstrom R, Vogt PK, eds. *Retroviruses: strategies of replication*. New York: Springer-Verlag, 1990;19–48.
81. Brown PO, Bowerman B, Varmus HE, Bishop JM. Correct integration of retroviral DNA in vitro. *Cell* 1987;49:347–356.
82. Brown PO, Bowerman B, Varmus HE, Bishop JM. Retroviral integration: structure of the initial covalent product and its precursor, and a role for the IN protein. *Proc Natl Acad Sci USA* 1989;86:2525–2529.
83. Bukrinsky MI, Haggerty S, Dempsey MP, Sharova N, Adzhubei A, Spitz L, Lewis P, Goldfarb D, Emerman M, Stevenson M. A nuclear localization signal within HIV-1 matrix protein that governs infection of non-dividing cells. *Nature* 1993;365:666–669.
84. Bukrinsky MI, Sharova N, Dempsey MP, Stanwick TL, Bukrinskaya AG, Haggerty S, Stevenson M. Active nuclear import of human immunodeficiency virus type 1 preintegration complexes. *Proc Natl Acad Sci USA* 1992;89:6580–6584.
85. Bukrinsky MI, Sharova N, McDonald TL, Pushkarskaya T, Tarpley WG, Stevenson M. Association of integrase, matrix and reverse transcriptase antigens of human immunodeficiency virus type 1 with viral

nucleic acids following acute infection. *Proc Natl Acad Sci USA* 1993;90:6125–6129.
86. Buller RS, Ahmed A, Portis JL. Identification of two forms of an endogenous murine retroviral env gene linked to the Rmcf locus. *J Virol* 1987;61:29–34.
87. Buonaguro L, Barillari G, Chang HK, Bohan CA, Kao V, Morgan R, Gallo RC, Ensoli B. Effects of the human immunodeficiency virus type 1 tat protein on the expression of inflammatory cytokines. *J Virol* 1992;66:7159–7167.
88. Buonaguro L, Buonaguro FM, Giraldo G, Ensoli B. The human immunodeficiency virus type 1 tat protein transactives tumor necrosis factor beta gene expression through a TAR-like structure. *J Virol* 1994;68:2677–2682.
89. Burns DPW, Temin HM. High rates of frameshift mutations within homo-oligomeric runs during a single cycle of retroviiral replication. *J Virol* 1994;68:4196–4203.
90. Burns JC, Friedmann T, Driever W, Burrascano M, Yee J-K. Vesicular stomatitis virus G glycoprotein pseudotyped retroviral vectors: concentration to very high titer and efficient gene transfer into mammalian and nonmammalian cells. *Proc Natl Acad Sci USA* 1993;90:8033–8037.
91. Bushman FD, Craigie R. Activities of human immunodeficiency virus (HIV) integration protein in vitro: specific cleavage and integration of HIV DNA. *Proc Natl Acad Sci USA* 1991;88:1339–1343.
92. Bushman FD, Craigie R. Integration of human immunodeficiency virus DNA: adduct interference analysis of required DNA sites. *Proc Natl Acad Sci USA* 1992;89:3458–3462.
93. Bushman FD, Engelman A, Palmer I, Wingfield P, Craigie R. Domains of the integrase protein of human immunodeficiency virus type 1 responsible for polynucleotidyl transfer and zinc binding. *Proc Natl Acad Sci USA* 1993;90:3428–3432.
94. Bushman FD, Fujiwara T, Craigie R. Retroviral DNA integration directed by HIV integration protein in vitro. *Science* 1990;249:1555–1558.
95. Bushman FD, Wang B. Rous sarcoma virus integrase protein: mapping functions for catalysis and substrate binding. *J Virol* 1994;68:2215–2223.
96. Cachon-Gonzales MB, Fenner S, Coffin JM, Moran C, Best S, Stoye JP. Structure and expression of the hairless gene of mice. *Proc Natl Acad Sci USA* 1994;91:7717–7721.
97. Callabaut C, Jacotot E, Krust B, Hovanessian AG. Response: CD26 antigen and HIV fusion? *Science* 1994;264:1162–1164.
98. Callebaut C, Krust B, Jacotot E, Hovanessian AG. T cell activation antigen, CD26, as a cofactor for entry of HIV in CD4+ acells. *Science* 1993;262:2045–2050.
99. Camerini D, Planelles V, Chen ISY. CD26 antigen and HIV fusion? *Science* 1994;264:1160–1161.
100. Cameron CE, Grinde B, Jacques P, Jentoft J, Leis J, Wlodawer A, Weber IT. Comparison of the substrate-binding pockets of the Rous sarcoma virus and human immunodeficiency virus type 1 proteases. *J Biol Chem* 1993;268:11711–11720.
101. Cameron CE, Grinde B, Jentoft J, Leis J, Weber IT, Copeland TD, Wlodawer A. Mechanism of inhibition of the retroviral protease by a Rous sarcoma virus peptide substrate representing the cleavage site between the gag p2 and p10 proteins. *J Biol Chem* 1992;267:23735–23741.
102. Cameron CE, Ridky TW, Shulenin S, Leis J, Weber IT, Copeland T, Wlodawer A, Burstein H, Bizub-Bender D, Skalka AM. Mutational analysis of the substrate binding pockets of the Rous sarcoma virus and human immunodeficiency virus-1 proteases. *J Biol Chem* 1994;269:11170–11177.
103. Capon DJ, Ward RHR. THe CD4-gp120 interaction and AIDS pathogenesis. *Ann Rev Immunol* 1991;9:649–678.
104. Carlberg K, Beemon K. Proposed gag-encoded transcriptional activator is not necessary for Rous sarcoma virus replication or transformation. *J Virol* 1988;62:4003–4008.
105. Carlberg K, Ryden TA, Beemon K. Localization and footprinting of an enhancer within the avian sarcoma virus gag gene. *J Virol* 1988;62:1617–1624.
106. Caron C, Rousset R, Beraud C, Monocllin V, Egly JM, Jalinot P. Functional and biochemical interaction of the HTLV-I tax1 transactivator with TBP. *EMBO J* 1993;12:4269–4278.
107. Carr CM, Kim PS. A spring-loaded mechanism for the conformational change of influenza hemagglutinin. *Cell* 1993;73:823–832.
108. Chakrabarti S, Mizukami T, Franchini G, Moss B. Synthesis, oligomerization, and biological activity of the human immunodeficiency virus type 2 envelope glycoprotein expressed by a recombinant vaccinia virus. *Virology* 1990;178:134–142.
109. Challita P-M, Kohn DB. Lack of expression from a retroviral vector after transduction of murine hematopoietic stem cells is associated with methylation in vivo. *Proc Natl Acad Sci USA* 1994;91:2567–2571.
110. Champoux JJ. Roles of RNase H in reverse transcription. In: Skalka AM, Goff SP, eds. *Reverse transcriptase*. Cold Spring Harbor, NY: Cold Spring Harbor Laboratory Press, 1993;103–118.
111. Chang W, Hubbard SC, Friedel C, Ruley HE. Enrichment of insertional mutants following retrovirus gene trap selection. *Virology* 1993; 193:737–747.
112. Charneau P, Alizon M, Clavel F. A second origin of DNA plus-strand synthesis is required for optimal human immunodeficiency virus replication. *J Virol* 1992;66:2814–2820.
113. Charneau P, Clavel F. A single-stranded gap in human immunodeficiency virus unintegrated linear DNA defined by a central copy of the polypurine tract. *J Virol* 1991;65:2415–2521.
114. Chattopadhyay SK, Sengupta DN, Fredrickson TN, Morse HCI, Hartley JW. Characteristics and contributions of defective, ecotropic, and mink cell focus-inducing viruses involved in a retrovirus-induced immunodeficiency syndrome of mice. *J Virol* 1991;65:4232–4241.
115. Chen M-Y, Maldarelli F, Karczewski MK, Willey RL, Strebel K. Human immunodeficiency virus type 1 vpu protein induces degradation of CD4 in vitro: the cytoplasmic domain of CD4 contributes to Vpu sensitivity. *J Virol* 1993;67:3877–3884.
116. Chen P-J, Cywinski A, Taylor JM. Reverse transcription of 7S L RNA by an avian retrovirus. *J Virol* 1985;54:278–284.
117. Cheng-Mayer C, Shioda T, Levy JA. Host range, replicative, and cytopathic properties of human immunodeficiency virus type 1 are determined by very few amino acid changes in *tat* and gp120. *J Virol* 1991;65:6931–6941.
118. Choi Y, Kappler JW, Marrack P. A superantigen encoded in the open reading frame of the 3′ long terminal repeat of mouse mammary tumour virus. *Nature* 1991;350:203–207.
119. Choi Y, Marrack P, Kappler J. Structural analysis of a mouse mammary tumor virus superantigen. *J Exp Med* 1992;175:847–852.
120. Chow SA, Vincent KA, Ellison V, Brown PO. Reversal of integration and DNA splicing mediated by integrase of human immunodeficiency virus. *Science* 1992;255:723–726.
121. Chowdhury M, Taylor JP, Chang C-F, Rappaport J, Khalili K. Evidence that a sequence similar to TAR is important for induction of the JC virus late promoter by human immunodeficiency virus type 1 tat. *J Virol* 1992;66:7355–7361.
122. Chowdhury M, Taylor JP, Tada H, Rappaport J, Wong-Staal F, Amini S, Khalili K. Regulation of the human neurotropic virus promoter by JCV-T antigen and HIV-1 *tat* protein. *Oncogene* 1990;5:1737–1742.
123. Chung S-W, Wolff L, Ruscetti SK. Transmembrane domain of the envelope gene of a polycythemia-inducing retrovirus determines erythropoietin-independent growth. *Proc Natl Acad Sci USA* 1989;86:7957–7960.
124. Ciminale V, Pavlakis GN, Derse D, Cunningham CP, Felber BK. Complex splicing in the human T-cell leukemia virus (HTLV) family of retroviruses: novel mRNAs and proteins produced by HTLV type I. *J Virol* 1992;66:1737–1745.
125. Clayton LK, Hussey RE, Steinbrich R, Ramachandran H, Husain Y, Reinherz EL. Substitution of murine for human CD4 residues identifies amino acids critical for HIV-gp120 binding. *Nature* 1988;335:363–366.
126. Cobrinik D, Aiyar A, Ge Z, Katzman M, Huang H, Leis J. Overlapping retrovirus U5 sequence elements are required for efficient integration and initiation of reverse transcription. *J Virol* 1991;65:3864–3872.
127. Cobrinik D, Soskey L, Leis J. A retroviral RNA secondary structure required for efficient initiation of reverse transcription. *J Virol* 1988; 62:3622–3630.
128. Coffin JM. Structure, replication, and recombination of retrovirus genomes: some unifying hypotheses. *J Gen Virol* 1979;42:1–26.
129. Coffin JM. Endogenous viruses. In: Weiss R, Teich N, Varmus H, Coffin J, eds. *RNA tumor viruses*. Cold Spring Harbor, NY: Cold Spring Harbor Laboratory, 1982;1109–1204.
130. Coffin JM. Genetic variation in AIDS viruses. *Cell* 1986;46:1–4.
131. Coffin JM. Genetic variation in retroviruses. In: Kurstak E, Marusyk RG, Murphy FA, Regenmortel MHVV, eds. *Applied virology research*, volume 2. Virus variability, epidemiology, and control. New York: Plenum Press, 1990;11–33.
132. Coffin JM. Genetic diversity and evolution of retroviruses. *Curr Topics Microbiol Immunol* 1992;176:143–164.
133. Coffin JM. Structure and Classification of Retroviruses. In: Levy JA, ed. *The Retroviridae*. New York: Plenum Press, 1992;19–50.

134. Coffin JM. Superantigens and endogenous retroviruses: a confluence of puzzles. *Science* 1992;255:411–413.
135. Coffin JM. Reverse transcription and evolution. In: Goff S, Skalka AM, eds. *Reverse transcriptase*. Cold Spring Harbor, NY: Cold Spring Harbor Laboratory Press, 1993;445–479.
136. Coffin JM, Billeter MA. A physical map of the Rous sarcoma virus genome. *J Mol Biol* 1976;100:293–318.
137. Coffin JM, Haseltine WA. Terminal redundancy and the origin of replication of Rous sarcoma virus RNA. *Proc Natl Acad Sci USA* 1977;74:1908–1912.
138. Coffin JM, Moore C. Determination of 3′ end processing in retroelements. *Trends Genet Sci* 1990;6:276–277.
139. Coffin JM, Tsichlis PN, Barker CS, Voynow S. Variation in avian retrovirus genomes. *Ann NY Acad Sci* 1980;354:410–425.
140. Colicelli J, Goff SP. Isolation of a recombinant murine leukemia virus utilizing a new primer tRNA. *J Virol* 1986;57:37–45.
141. Colicelli J, Goff SP. Sequence and spacing requirements of a retrovirus integration site. *J Mol Biol* 1988;199:47–59.
142. Conklin KF, Coffin JM, Robinson HL, Groudine M, Eisenman R. Role of methylation in the induced and spontaneous expression of the avian endogenous virus ev-1: DNA structure and gene products. *Mol Cell Biol* 1982;2:638–652.
143. Conklin KF, Groudine M. Varied interactions between proviruses and adjacent host chromatin. *Mol Cell Biol* 1986;6:3999–4007.
144. Copeland NG, Hutchinson KW, Jenkins NA Excision of the DBA ecotropic provirus in dilute coat-color revertants of mice occurs by homologous recombination involving the viral LTRs. *Cell* 1983;33:379–387.
145. Copeland NG, Jenkins NA, Nexo B, Schultz AM, Rein A, Middelsen T, Jorgensen P. Poorly expressed endogenous ecotropic provirus of DBA/2 mice encodes a mutant Pr65gag protein that is not myristylated. *J Virol* 1988;62:479–487.
146. Corbin A, Prats AC, Darlix J-L, Sitbon M. A nonstructural gag-encoded glycoprotein precursor is necessary for efficient spreading and pathogenesis of murine leukemia viruses. *J Virol* 1994;68:3857–3867.
147. Craigie R. Hotspots and warm spots: integration specificity of retroelements. *Trends Genet* 1992;8:187–190.
148. Craigie R, Fujiwara T, Bushman F. The IN protein of Moloney murine leukemia virus processes the viral DNA ends and accomplishes their integration in vitro. *Cell* 1990;62:829–837.
149. Craigie R, Mizuuchi K, Bushman FD, Engelman A. A rapid in vitro assay for HIV DNA integration. *Nucl Acids Res* 1991;19:2729–2738.
150. Craven RC, Bennett RP, Wills JW. Role of the avian retroviral protease in the activation of reverse transcriptase during virion assembly. *J Virol* 1991;65:6205–6217.
151. Craven RC, Leure-duPree AE, Erdie CR, Wilson CB, Wills JW. Necessity of the spacer peptide between CA and NC in the Rous sarcoma virus gag protein. *J Virol* 1993;67:6246–6252.
152. Crittenden LB, McMahon S, Halpern MS, Fadly AM. Embryonic infection with the endogenous avian leukosis virus Rous-associated virus-0 alters responses to exogenous avian leukosis virus infection. *J Virol* 1987;61:722–725.
153. Cullen BR. Human immunodeficiency virus as a prototypic complex retrovirus. *J Virol* 1991;65:1053–1056.
154. Cullen BR. Regulation of HIV-1 gene expression. *Faseb J* 1991; 5:2361–2368.
155. Cullen BR. Does HIV-1 tat induce a change in viral initiation rates? *Cell* 1993;73:417–420.
156. Cullen BR, Skalka AM, Ju G. Endogenous avian retroviruses contain deficient promoter and leader sequences. *Proc Natl Acad Sci* 1983; 80:2946–2950.
157. Dalgleish AG, Beverly PCL, Clapham PR, Crawford DH, Greaves MF, Weiss RA. The CD4 (T4) antigen is an essential component of the receptor for the AIDS retrovirus. *Nature* 1984;312:763–767.
158. Darlix J-L, Gabus C, Allain B. Analytical study of avian reticuloendotheliosis virus dimeric RNA generated in vivo and in vitro. *J Virol* 1992;66:7245–7261.
159. David YB, Prideaux VR, Chow V, Benchimol S, Bernstein A. Inactivation of the p53 oncogene by internal deletion or retroviral integration in erythroleukemic cell lines induced by Friend leukemia virus. *Oncogene* 1988;3:179–186.
160. Davis NL, Rueckert RR. Properties of a ribonucleoprotein particle isolated from Nonidet P-40-treated Rous sarcoma virus.*J Virol* 1972;10:1010–1020.
161. de Mareuil J, Brichacek B, Salaun D, Chermann J-C, Hirsch I. The human immunodeficiency virus (HIV) *gag* gene product p18 is responsible for enhanced fusogenicity and host range tropism of the highly cytopathic HIV-1-NDK strain. *J Virol* 1992;66:6797–6801.
162. Delling U, Reid, LS, Barnett RW, Ma MYX, Climie S, Sumner-Smith M, Sonenberg N. Conserved nucleotides in the TAR RNA stem of human immunodeficiency virus type 1 are critical for tat binding and *trans* activation: model for TAR RNA tertiary structure. *J Virol* 1992; 66:3018–3025.
163. DeStefano JJ, Mallaber LM, Rodriguez-Rodriguez L, Fay PJ, Bambara RA. Requirements for strand transfer between internal regions of heteropolymer templates by human immunodeficiency virus reverse transcriptase. *J Virol* 1992;66:6370–6378.
164. DeZazzo JD, Kilpatrick JE, Imperiale MJ. Involvement of long terminal repeat U3 sequences overlapping the transcription control region in human immunodeficiency virus type 1 mRNA 3′ end formation. *Mol Cell Biol* 1991;11:1624–1630.
165. DeZazzo JD, Scott JM, Imperiale MJ. Relative roles of signals upstream of AAUAAA and promoter proximity in regulation of human immunodeficiency type 1 mRNA 3′ end formation. *Mol Cell Biol* 1992;12:5555–5562.
166. Dickson C, Eisenmann R, Fan H, Hunter E, Teich N. Protein biosynthesis and assembly. In: Weiss RA, Teich N, Varmus HE, Coffin JM, eds. *RNA tumor viruses*. Cold Spring Harbor, NY: Cold Spring Harbor Laboratory Press, 1982;513–648.
167. Dingwall C, Ernberg I, Gait MJ, Green SM, Heaphy S, Karn J, Lowe AD, Singh M, Skinner MA. HIV-1 *tat* protein stimulates transcription by binding to a U-rich bulge in the stem of the TAR RNA structure. *EMBO J* 1990;9:4145–4154.
168. Dingwall C, Ernberg I, Gait MJ, Green SM, Heaphy S, Karn J, Lowe AD, Singh M, Skinner MA, Valerio R. Human immunodeficiency virus 1 tat protein binds trans-activation-responsive region (TAR) RNA in vitro. *Proc Natl Acad Sci USA* 1989;86:6925–6929.
169. Donahue PR, Quackenbush SL, Gallo MV, DeNoronha CMC, Overbaugh J, Hoover EA, Mullins JI. Viral genetic determinants of T-cell killing and immunodeficiency disease induction by the feline leukemia virus FeLV-FAIDS. *J Virol* 1991;65:4461–4469.
170. Donehower LA. Analysis of mutant Moloney murine leukemia viruses containing linker insertion mutations in the 3′ region of pol. *J Virol* 1988;62:3958–3964.
171. Dong J, Dubay JW, Perez LG, Hunter E. Mutations within the proteolytic cleavage site of the Rous sarcoma virus glycoprotein define a requirement for dibasic residues for intracellular cleavage. *J Virol* 1992;66:865–874.
172. Dong J, Roth MG, Hunter E. A chimeric avian retrovirus containing the influeza virus hemagglutinin gene has an expanded host range. *J Virol* 1992;66:7374–7382.
173. Doolittle RF, Feng DF, McClure MA, Johnson MS. Retrovirus Phylogeny and evolution. In: Swanstrom R, Vogt PK, eds. *Retroviruses: strategies of replication*. New York: Springer-Verlag, 1990;1–18.
174. Dorfman T, Mammano F, Haseltine WA, Gottlinger HG. Role of the matrix protein in the virion association of the human immunodeficiency virus type 1 envelope glycoprotein. *J Virol* 1994;68:1689–1696.
175. Dornburg R, Temin HM. Retroviral vector system for the study of cDNA gene formation. *Mol Cell Biol* 1988;8:2328–2334.
176. Dornburg R, Temin HM. cDNA genes formed after infection with retroviral vector particles lack the hallmarks of natural processed pseudogenes. *Mol Cell Biol* 1990;10:68–74.
177. Dornburg R, Temin HM. Presence of a retroviral encapsidation sequence in nonretroviral RNA increases the efficiency of formation of cDNA genes. *J Virol* 1990;64:886–889.
178. Dorner AJ, Coffin JM. Determinants for receptor interaction and cell killing on the avian retrovirus glycoprotein gp85. *Cell* 1986;45:365–374.
179. Dorner AJ, Stoye JP, Coffin JM. Molecular basis of host range variation in avian retroviruses. *J Virol* 1985;53:32–39.
180. Dougherty JP, Temin HM. A promoterless retroviral vector indicates that there are sequences in U3 required for 3′ processing. *Proc Natl Acad Sci USA* 1987;84:1197–1201.
181. Dougherty JP, Temin. HM. High mutation rate of a spleen necrosis virus-based retrovirus vector. *Mol Cell Biol* 1986;6:4387–4395.
182. Drelich M, Wilhelm R, Mous J. Identification of amino acid residues critical for endonuclease and integration activities of HIV-1 IN protein in vitro. *Virology* 1992;188:459–468.
183. Dudley JP. Mouse mammary tumor proviruses from a T-cell lymphoma are associated with the retroposon L1Md. *J Virol* 1988;62:472–478.
184. Dunn MM, Olsen JC, Swanstrom R. Characterization of unintegrated retroviral DNA with long terminal repeat-associated cell-derived in-

serts. *J Virol* 1992;66:5735–5743.
185. Dupraz P, Oertle S, Meric C, Damay P, Spahr P-F. Point mutations in the proximal cys-his box of Rous sarcoma virus nucleocapsid protein. *J Virol* 1990;64:4978–4987.
186. Dupraz P, Spahr P-F. Specificity of Rous sarcoma virus nucleocapsid protein in genomic RNA packaging. *J Virol* 1992;66:4662–4670.
187. Dupraz P, Spahr P-F. Analysis of deletions and thermosensitive mutations in Rous sarcoma virus gag protein p10. *J Virol* 1993; 67:3826–3834.
188. Earl PL, Doms RW, Moss B. Oligomeric structure of the human immunodeficiency virus type 1 envelope glycoprotein. *Proc Natl Acad Sci USA* 1990;87:648–652.
189. Eiden MV, Farrell K, Wilson CA. Glycosylation-dependent invactivation of the ecotropic murine leukemia virus receptor. *J Virol* 1994;68:626–631.
190. Einfeld D, Hunter E. Oligomeric structure of a prototype retrovirus glycoprotein. *Proc Natl Acad Sci USA* 1988;85:8688–8692.
191. Einfeld DA, Hunter E. Expression of the TM protein of Rous sarecoma virus in the absence of SU shows that this domain is capable of oligomerization and intracellular transport. *J Virol* 1994;68:2513–2520.
192. Elder JH, Lerner DL, Hasselkus-Light CS, Fontenot DJ, Hunter E, Luciw PA, Montelaro RC, Phillips TR. Distinct subsets of retroviruses encode dUTPase. *J Virol* 1992;66:1791–1794.
193. Ellison V, Brown PO. A stable complex between integrase and viral DNA ends mediates human immunodeficiency virus integration in vitro. *Proc Natl Acad Sci USA* 1994;91:7316–7320.
194. Emi N, Friedmann T, Yee J-K. Pseudotype formation of murine leukemia virus with the G protein of vesicular stomatitis virus. *J Virol* 1991;65:1202–1207.
195. Engelman A, Craigie R. Identification of conserved amino acid residues critical for human immunodeficiency virus type 1 integrase function in vitro. *J Virol* 1992;66:6361–6369.
196. Engelman A, Mizuuchi K, Craigie R. HIV-1 DNA integration: mechanism of viral DNA cleavage and DNA strand transfer. *Cell* 1991;67:1211–1222.
197. Ensoli B, Barillari G, Salahuddin SZ, Gallo RC, Wong-Staal F. Tat protein of HIV-1 stimulates growth of cells derived from Kaposi's sarcoma lesions of AIDS patients. *Nature* 1990;345:84–86.
198. Ensoli B, Buonaguro L, Barillari G, Fiorelli V, Gendelman R, Morgan RA, Wingfield P, Gallo RC. Release, uptake and effects of extracellular human immunodeficiency virus type 1 tat protein on cell growth and viral transactivation. *J Virol* 1993;67:277–287.
199. Ensoli B, Lusso P, Schachter F, Josephs SF, Rappaport J, Negro F, Gallo RC, Wong-Staal F. Human herpes virus-6 increases HIV-1 expression in co-infected T cells via nuclear factors binding to the HIV-1 enhancer. *EMBO J* 1989;8:3019–3028.
200. Erdie CR, Wills JW. Myristylation of Rous sarcoma virus gag protein does not prevent replication in avian cells. *J Virol* 1990;64:5204–5208.
201. Evans LA, Levy JA. The heterogeneity and pathogenicity of HIV. In: Morrow WNW, Haigwood NL, eds. *HIV molecular organization, pathogenecity and treatment*. Amsterdam: Elsevier Science Publishers B.V, 1993;29–73.
202. Facke M, Janetzko A, Shoeman RL, Krausslich H-G. A large deletion in the matrix domain of the human immunodeficiency virus *gag* gene redirects virus particle assembly from the plasma membrane to the endoplasmic reticulum. *J Virol* 1993;67:4972–4980.
203. Falk H, Mador N, Udi P, A Honigman A. Two cis-acting signals control ribosomal framesh between human T-cell leukemia virus type II gag and pro genes. *J Virol* 1993;67:6273–6277.
204. Fan H. Retroviruses and their role in cancer. In: Levy JA, ed. *The Retroviridae*. New York: Plenum Press, 1994;313–349.
205. Fan H, Mittal S, Chute H, Chao E, Pattengale PK. Rearrangements and insertions in the Moloney murine leukemia virus long terminal repeat alter biological properties in vivo and in vitro. *J Virol* 1986;60:204–214.
206. Farnet CM, Haseltine WA. Integration of human immunodeficiency virus type 1 DNA in vitro. *Proc Natl Acad Sci USA* 1990;87:4164–4168.
207. Farnet CM, Haseltine WA. Circularization of human immunodeficiency virus type 1 DNA in vitro. *J Virol* 1991;65:6942–6952.
208. Farnet CM, Haseltine WA. Determination of viral proteins present in the human immunodeficiency virus type 1 preintegration complex. *J Virol* 1991;65:1910–1915.
209. Fauci AS. The human immunodeficiency virus: infectivity and mechanisms of pathogenesis. *Science* 1988;239:617–622.
210. Fauci AS. Immunopathogenesis of HIV infection. *J AIDS* 1993;6:655–662.
211. Fauci AS. Multifactorial nature of human immunodeficiency virus disease: implications for therapy. *Science* 1993;262:1011–1018.
212. Feinberg MB, Jarrett RF, Aldovini A, Gallo RC, Wong-Staal F. HTLV-III expression and production involve complex regulation at the levels of splicing and translation of viral RNA. *Cell* 1986;46:807–817.
213. Feinstein SC, Ross SR, Yamamoto KR. Chromosomal position effects determine transcriptional potential of integrated mammary tumor virus DNA. *J Mol Biol* 1982;156:549–566.
214. Felsenstein KM, Goff SP. Expression of the gag-pol fusion protein of Moloney murine leukemia virus without gag protein does not induce virion formation or proteolytic processing. *J Virol* 1988;62:2179–2182.
215. Felsenstein KM, Goff SP. Mutational analysis of the *gag-pol* junction of Moloney murine leukemia virus: requirements for expression of the *gag-pol* fusion protein. *J Virol* 1992;66:6601–6608.
216. Feng S, Holland EC. HIV-1 tat trans-activation requires the loop sequence within tar. *Nature* 1988;334:165–167.
217. Feng Y-X, Levin JG, Hatfield DL, Schaefer TS, Gorelick RJ, Rein A. Suppression of UAA and UGA termination codons in mutant murine leukemia viruses. *J Virol* 1989;63:2870–2873.
218. Feng Y-X, Yan H, Rein A, Levin JG. Bipartite signal for read-through suppression in murine leukemia virus mRNA: an eight-nucleotide purine-rich sequence immediately fownstream of the *gag* termination codon followed by an RNA pseudoknot. *J Virol* 1992;66:5127–5132.
219. Feuchter AE, Mager DL. SV40 large T antigen trans-activates the long terminal repeats of a large family of human endogenous retrovirus-like sequences. *Virology* 1992;187:242–250.
220. Feuer G, Taketo M, Hanecak RC, Fan H. Two blocks in Moloney murine leukemia virus expression in undifferentiated F9 embryonal carcinoma cells as determined by transient expression assays. *J Virol* 1989;63:2317–2324.
221. Fincham VJ, Wyke JA. Differences between cellular integration sites of transcribed and nontranscribed Rous sarcoma proviruses. *J Virol* 1991;65:461–463.
222. Finston WI, Champoux JJ. RNA-primed initiation of Moloney murine leukemia virus plus strands by reverse transcriptase in vitro. *J Virol* 1984;51:26–33.
223. Fitzgerald ML, Grandgenett DP. Retroviral integration: in vitro host site selection by avian integrase. *J Virol* 1994;68:4314–4321.
224. Fitzgerald ML, Vora AC, Zeh WG, Grandgenett DP. Concerted integration of viral DNA termini by purified avian myeloblastosis virus integrase. *J Virol* 1992;66:6257–6263.
225. Flubacher M, Bear SJ, Tsichlis PN. Replacement of IL-2 generated mitogenic signals by an MCF or xenotropic virus-induced IL-9 dependent autocrine loop. Implications for MCF virus-induced leukemogenesis. *J Virol* In Press.
226. Flugel RM. Spumaviruses - A group of complex retroviruses - Review. *J AIDS* 1991;4:739–750.
227. Flugel RM, Rethwilm A, Maurer B, Darai G. Nucleotide sequence of the env gene and its flanking regions of the human spumaretrovirus reveals two novel genes. *EMBO J* 1987;6:2077–2084.
228. Frankel AD, Pabo CO. Cellular uptake of the tat protein from human immunodeficiency virus. *Cell* 1988;55:1189–1193.
229. Frankel W, Potter TA, Rosenberg N, Lenz J, Rajan TV. Retroviral insertional mutagenesis of a target allele in a heterozygous murine ceml line. *Proc Natl Acad Sci USA* 1985;82:6600–6604.
230. Frankel WN, Rudy C, Coffin JM, Huber BT. Linkage of Mls genes to endogenous mammary tumour viruses of inbred mice. *Nature* 1991;349:526–528.
231. Frankel WN, Stoye JP, Taylor BA, Coffin JM. A linkage map of endogenous murine leukemia proviruses. *Genetics* 1990;124:221–236.
232. Freed EO, Myers DJ, Risser R. Characterization of the fusion domain of the human immunodeficiency virus type1 envelope glycoprotein gp41. *Proc Natl Acad Sci USA* 1990;87:4650–4654.
233. Frost SDW, McLean AR. Germinal centre destruction as a major pathway of HIV pathogenesis. *J AIDS* 1994;7:236–244.
234. Fu S, Phillips N, Jentoft J, Tuazon PT, Traugh JA, Leis J. Site-specific phosphorylation of avian retrovirus nucleocapsid protein pp12 regulates binding to RNA. *J Biol Chem* 1985;260:9941–9947.
235. Fu T-B, Taylor J. When retroviral reverse trasnscriptases reach the end of their RNA templates. *J Virol* 1992;66:4271–4278.
236. Fu W, Rein A. Maturation of dimeric viral RNA of Moloney murine leukemia virus. *J Virol* 1993;67:5443–5449.
237. Fujii M, Niki T, Mori T, Matsuda T, Matsui M, Nomura N, Seiki M. HTLV-1 Tax induces expression of various immediate early serum responsive genes. *Oncogene* 1991;6:887–894.

238. Fujii M, Tsuchiya H, Chuhjo T, Akizawa T, Seiki M. Interaction of HTLV-1 Tax1 with p67SRF causes the aberrant induction of cellular immediate early genes through CArG boxes. *Genes Dev* 1992;6:2066–2076.
239. Fujisawa J-i, Toita M, Yoshimura T, Yoshida M. The indirect association of human T-cell leukemia virus *tax* protein with DNA results in transcriptional activation. *J Virol* 1991;65:4525–4528.
240. Fujiwara T, Mizuuchi K. Retroviral DNA integration: structure of an integration intermediate. *Cell* 1988;54:497–504.
241. Gabriel A, Boeke JD. Retrotransposon reverse transcription. In: Skalks AM, Goff SP, eds. *Reverse transcriptase.* Cold Spring Harbor, NY: Cold Spring Harbor Laboratory Press, 1993;275–329.
242. Gallaher WR. Detection of a fusion peptide sequence in the transmembrane protein of human immunodeficiency virus. *Cell* 1987;50:327–328.
243. Gallis B, Linial M, Eisenmann R. An avian oncovirus mutant deficient in genomic RNA: characterization of the packaged RNA as cellular messenger RNA. *Virology* 1979;94:146–161.
244. Gao F, Yue L, White AT, Pappas PG, Barchue J, Hanson AP, Greene BM, Sharp PM, Shaw GM, Hahn BH. Human infection by genetically diverse SIV_{SM}-related HIV-2 in West Africa. *Nature* 1992;358:495–499.
245. Garcia JV, Miller AD. Serine phosphorylation-independent downregulation of cell-surface CD4 by nef. *Nature* 1991;350:508–511.
246. Garcia JV, Miller AD. Serine-phosphorylation-independent downregulation of cell-surface CD4 by *nef. Nature* 1992;350:508–511.
247. Gardner MB, Kozak CA, O'Brien SJ. The Lake Casitas wild mouse: evolving genetic resistance to retroviral disease. *Trends Genet* 1991; 7:22–27.
248. Gautsch J, Wilson MC. Delayed de novo methylation in teratocarcinoma suggests additional tissue-specific mechanisms for controlling gene expression. *Nature* 1993;301:32–37.
249. Gelinas C, Temin HM. Nondefective spleen necrosis virus-derived vectors define the upper size limit for packaging reticuloendotheliosis viruses. *Proc Natl Acad Sci USA* 1986;83:9211–9215.
250. Ghatta IR, Sanes JR, Majors JE. The encephalomyocarditis virus internal ribosome entry site allows efficient coexpression of two genes from a recombinant provirus in cultured cells and in embryos. *Mol Cell Biol* 1991;11:5848–5859.
251. Gilboa E, Mitra SW, Goff S, Baltimore D. A detailed model of reverse transcription and tests of crucial aspects. *Cell* 1979;18:93–100.
252. Goff SJ, Skalka AM, eds. *Reverse transcriptase.* Cold Spring Harbor, NY: Cold Spring Harbor Laboratory Press, 1993.
253. Gojobori T, Moriyama EN, Kimura M. Molecular clock of viral evolution, and the neutral theory. *Proc Natl Acad Sci USA* 1990;87:10015–10018.
254. Goldfarb MP, Weinberg RA. Generation of novel, biologically active Harvey sarcoma viruses via apparent illegitimate recombination. *J Virol* 1981;38:136–150.
255. Golding H, Blumenthal R, Manischewitz J, Littman DR, Dimitrov DS. Cell fusion mediated by interaction of a hybrid CD4-CD8 molecule with the human immunodeficiency virus type 1 envelope glycoprotein does occur after a long lag time. *J Virol* 1993;67:6469–6475.
256. Golemis EA, Speck NA, Hopkins N. Alignment of U3 region sequences of mammalian type C viruses: identification of highly conserved motifs and implications for enhancer design. *J Virol* 1990;64:534–542.
257. Golovkina TV, Chervonsky A, Dudley JP, Ross SR. Transgenic mouse mammary tumor virus superantigen expression prevents viral infection. *Cell* 1992;69:637–645.
258. Gontarek RR, McNally MT, Beemon K. Mutation of an RSV intronic element abolishes both U11/U12 snRNP binding and negative regulation of splicing. *Genes Dev* 1993;7:1926–1936.
259. Goodenow MM, Hayward WS. 5′ long terminal repeats of myc-associated proviruses appear structurally intact but are functionally impaired in tumors induced by avian leukosis viruses. *J Virol* 1987; 61:2489–2498.
260. Goodrich DW, Duesberg PH. Evidence that retroviral transduction is mediated by DNA, not by RNA. *Proc Natl Acad Sci USA* 1990; 87:3604–3608.
261. Goodwin RG, Rottman FM, Callaghan T, Kung H-J, Maroney PA, Nilsen TW. c-erbB activation in avian leukosis virus-induced erythroblastosis: multiple epidermal growth factor receptor mRNAs are generated by alternative RNA processing. *Mol Cell Biol* 1986;6:3128–3133.
262. Gopalakrishnan V, Peliska JA, Benkovic SJ. Human immunodeficiency virus type 1 reverse transcriptase: spatial and temporal relationshp between the polymerase and RNase H activities. *Proc Natl Acad Sci USA* 1992;89:10763–10767.
263. Gopinathan KP, Weymouth LA, Kunkel TA, Loeb LA. Mutagenesis in vitro by DNA polymerase from an RNA tumor virus. *Nature* 1979;278:857–859.
264. Gorelick RJ, Henderson LE, Hanser JP, Rein A. Point mutants of Moloney murine leukemia virus that fail to package viral RNA: evidence for specific RNA recognition by a "zinc finger-like" protein sequence. *Proc Natl Acad Sci USA* 1988;85:8420–8424.
265. Gorelick RJ, Nigida SMJ, Arthur LO, Henderson LE, Rein A. Roles of nucleocapsid cysteine arrays in retroviral assembly and replication: possible mechanisms in RNA encapsidation. In: *Advances in molecular biology and targeted treatment for AIDS.* New York: Plenum Press, 1990.
266. Gorman CM, Rigby PWJ, Lane DP. Negative regulation of viral enhancers in undifferentiated embryonic stem cells. *Cell* 1985;42:519–526.
267. Gottlinger HG, Dorfman T, Sodroski JG, Haseltine WA. Effect of mutations affecting the p6 *gag* protein on human immunodeficiency virus particle release. *Proc Natl Acad Sci USA* 1991;88:3195–3199.
268. Goudsmit J, Back NK, Nara PL. Genomic diversity and antigenic variation of HIV-1: links between pathogenesis, epidemiology and vaccine development. *Faseb J* 1991;5:2427–36.
269. Graeble MA, Churcher MJ, Lowe AD, Gait MJ, Karn J. Human immunodeficiency virus type 1 transactivator protein, tat, stimulates transcriptional read-through of distal terminator sequences in vitro. *Proc Natl Acad Sci USA* 1993;90:6184–6188.
270. Grandbastien M-A. Retroelements in higher plants. *Trends Genet* 1992;8:103–108.
271. Grandgenett DP, Mumm SR. Unraveling retrovirus integration. *Cell* 1990;60:3–4.
272. Graves BJ, Johnson PF, McKnight SL. Homologous recognition of a promoter domain common to the MSV LTR and the HSV tk gene. *Cell* 1986;44:565–576.
273. Gray KD, Roth MJ. Mutational analysis of the envelope gene of Moloney murine leukemia virus. *J Virol* 1993;67:3489–3496.
274. Green JE, Begley CG, Wagner DK, Waldmann TA, Jay G. transactivation of granulocyte-macrophage colony-stimulating factor and the interleukin-2 receptor in transgenic mice carrying the human T-lymphotropic virus type 1 tax gene. *Mol Cell Biol* 1989;9:4731–4737.
275. Green PL, Chen ISY. Molecular features of the human T-cell leukemia virus: mechanisms of transformation and malignancy. In: Levy JA, ed. *The Retroviridae.* New York: Plenum Press, 1994;277–312.
276. Greene WC, Bohulein E, Ballard DW. HIV-1, HTLV-I and T-cell growth: transcriptional strategies and suprises. *Immunol Today* 1989;10:272–278.
277. Greene WC, Leonard WJ, Wano Y, Svetlik PB, Peffer NJ, Sodroski JG, Rosen CA, Goh WC, Haseltine WA. Trans-activator gene of HTLV-II induces IL-2 receptor and IL-2 cellular gene expression. *Science* 1986;232:877–881.
278. Grez M, Zornig M, Nowock J, Ziegler M. A single point mutation activates the Moloney murine leukemia virus long terminal repeat in embryonal stem cells. *J Virol* 1991;65:4691–4698.
279. Grinde B, Cameron CE, Leis J, Weber IT, Wlodawer A, Burstein H, Skalka AM. Analysis of substrate interactions of the Rous sarcoma virus wild type and mutant proteases and human immunodeficiency virus-1 protease using a set of systematically altered peptide substrates. *J Biol Chem* 1992;267:9491–8.
280. Guntaka RV. Transcription termination and polyadenylation in retroviruses. *Microbiol Rev* 1993;57:511–521.
281. Günzberg WH, Heinemann F, Wintersperger S, Mlethke T, Wagner H, Salmons B. Endogenous superantigen expression controlled by a novel promoter in the MMTV lon terminal repeat. *Nature* 1993;364:154–158.
282. Günzberg WH, Salmons B. Factors controlling the expression of mouse mammary tumour virus. *Biochem J* 1992;283:625–632.
283. Habel DE, Dohrer KL, Conklin KF. Functional and defective components of avian endogenous virus long terminal repeat enhancer sequences. *J Virol* 1993;67:1545–1554.
284. Hackett PB, Petersen RB, Hensel CH, Albericio F, Gunderson SI, Palmenberg AC, Barany G. Synthesis in vitro of a seven amino acid peptide encoded in the leader RNA of Rous sarcoma virus. *J Mol Biol* 1986;190:45–58.
285. Hajihosseini M, Iavachev L, Price J. Evidence that retroviruses integrate into post-replication host DNA. *EMBO J* 1993;12:4969–4974.
286. Hajjar AM, Linial ML. Characterization of a unique retroviral recombinant containing 7S L sequences. *J Virol* 1993;67:7677–7679.

287. Hajjar AM, Linial ML. A model system for nonhomologous recombination between retroviral and cellular RNA. *J Virol* 1993;67:3845–3853.
288. Hallberg B, Schmidt J, Luz A, Pedersen FS, Grundstrom T. SL3-3 enhancer factor 1 transcriptional activators are required for tumor formation by SL3-3 murine leukemia virus. *J Virol* 1991;65:4177–4181.
289. Hanecak R, Pattengale PK, Fan H. Deletion of a GC-rich region flanking the enhancer element within the long terminal repeat sequences alters the disease specificity of Moloney murine leukemia virus. *J Virol* 1991;65:5357–5363.
290. Hansen M, Jelinek L, Whiting S, Barklis E. Transport and assembly of gag proteins into Moloney murine leukemia virus. *J Virol* 1990;64:5306–5316.
291. Hardy WD. Feline oncoretroviruses. In: Levy JA, ed. *The Retroviridae*. New York: Plenum Press, 1993;109–168.
292. Harris JD, Blum H, Scott J, Traynor B, Ventura P, Haase A. Slow virus visna: reproduction in vitro of virus from extrachromosomal DNA. *Proc Natl Acad Sci USA* 1984;81:7212–7215.
293. Hartley JW, Frederickson TN, Yetter RA, Makino M, Morse HCI. Retrovirus-induced murine acquired immunodeficiency syndrome: natural history of infection and differing susceptibility of inbred mouse strains. *J Virol* 1989;63:1223–1231.
294. Hartung S, Jaenisch R, Breindl M. Retrovirus insertion inactivates mouse a1(I) collagen gene by blocking initiation of transcription. *Nature* 1986;320:365–367.
295. Haseltine W. Replication and pathogenesis of the AIDS virus. *J AIDS* 1988;1:217–240.
296. Hatfield DL, Levin JG, Rein A, Oroszlan S. Translational suppression in retroviral gene expression. *Adv Virus Res* 1992;41:193–239.
297. Hayashi T, Shioda T, Iwakura Y, Shibuta H. RNA packaging signal of human immunodeficiency virus type 1. *Virology* 1992;188:590–599.
298. Hayward WS, Neel BG, Astrin SM. Activation of a cellular onc gene by promoter insertion in ALV-induced lymphomas. *Nature* 1981;290:475–480.
299. Heidmann O, Heidmann T. Retrotranposition of a mouse IAP sequence tagged with an indicator gene. *Cell* 1991;64:159–170.
300. Heinzinger NK, Bukrinsky MI, Haggerty SA, Ragland AM, Kewalramani V, Lee M-A, Gendelman HE, Ratner L, Stevenson M, Emerman M. The Vpr protein of human immunodeficiency virus type 1 influences nuclear localization of viral nucleic acids in nondividing host cells. *Proc Natl Acad Sci USA* 1994;91:7311–7315.
301. Held W, Shakhov AN, Waanders G, Scarpellino L, Luethy R, Kraehenbuhl J-P, MacDonald HR, Acha-Orbea H. An exogenous mouse mammary tumor virus with properties of Mls-1a (Mtv-7). *J Exp Med* 1992;175:1623–1634.
302. Held W, Waanders GA, Shakhov AN, Scarpellino L, Acha-Orbea H, MacDonald HR. Superantigen-induced immune stimulation amplifies mouse mammary tumor virus infection and allows virus transmission. *Cell* 1993;74:529–540.
303. Herman SA, Coffin JM. Differential transcription from the long terminal repeats of integrated avian leukosis virus DNA. *J Virol* 1986;60:497–505.
304. Herman SA, Coffin JM. Efficient packaging of readthrough RNA in ALV: implications for oncogene transduction. *Science* 1987; 236:845–848.
305. Hirai H, Suzuki T, Fujisawa J-i, Inoue J-i, Yoshida M. Tax protein of human T-cell leukemia virus type 1 binds to the ankyrin motifs of inhibitory factor κB and induces nuclear translocation of transcription factor NF-κB proteins for transcriptional activation. *Proc Natl Acad Sci USA* 1994;91:3584–3588.
306. Hizi A, Henderson LE, Copeland TD, Sowder RC, Hixson CV, Oroszlan S. Characterization of mouse mammary tumor virus gag-pro gene products and ribosomal frameshift site by protein sequencing. *Proc Natl Acad Sci USA* 1987;84:7041–7045.
307. Ho DD, Pomerantz RJ, Kaplan JC. Pathogenesis of infection with human immunodeficiency virus. *N Engl J Med* 1987;317:278–286.
308. Hoeben RC, Migchielsen AAJ, van der Jagt RCM, van Ormondt H, van der Eb AJ. Inactivation of the Moloney murine leukemia virus long terminal repeat in murine fibroblast cell lines is associated with methylation and dependent on its chromosomal position. *J Virol* 1991; 65:904–912.
309. Hoggan MD, O'Neill RR, Kozak CA. Nonecotropic murine leukemia viruses in BALB/c and NFS/N mice: characterization of the BALB/c Bxv-1 provirus and the single NFS endogenous xenotrope. *J Virol* 1986;60:980–986.
310. Hoggan MD, Willey RL, Strebel K, Martin MA, Repaske R. Genetic recombination of human immunodeficiency virus. *J Virol* 1989; 63:1455–1459.
311. Holland CA, Thomas CY, Chattopadhyay SK, Koehne C, O'Donnell PV. Influence of enhancer sequences on thymotropism and leukemogenicity of mink cell focus-forming viruses. *J Virol* 1989;63: 1284–1292.
312. Holland JJ, ed. Genetic Diversity of RNA Viruses. *Curr Top Microbiol Immunol* 1992;176 Springer-Verlag, Berlin
313. Honigman A, Wolf D, Yaish S, Falk H, Panet A. *cis* acting RNA sequences control the gag-pol translation readthreough in murine leukemia virus. *Virology* 1991;183:313–319.
314. Horvat RT, Wood C, Josephs SF, Balachandran N. Transactivation of the human immunodeficiency virus promoter by human herpesvirus 6 (HHV-6) strains GS and Z-29 in primary human T lymphocytes and identification of transactivating HHV-6(GS) gene fragments. *J Virol* 1991;65:2895–2902.
315. Hosie MJ, Willett BJ, Dunsford TH, Jarrett O, Neil JC. A monoclonal antibody which blocks infection with feline immunodeficiency virus identifies a possible non-CD4 receptor. *J Virol* 1993;67:1667–1671.
316. Hostomsky Z, Hudson GO, Rahmati S, Hostomska Z. RNase-D, a reported new activity associated with HIV-1 reverse transcriptase, displays the same cleavage specificity as Escherichi coli RNase III. *Nucl Acids Res* 1992;20:5819–5824.
317. Hostomsky Z, Hughes SH, Goff SP, LeGrice SFJ. Redesignation of the RNase D activity associated with retroviral reverse transcriptase as RNase H*. *J Virol* 1994;68:1970–1971.
318. Howcroft TK, Strebel K, Martin MA, Singer DS. Repression of MHC Class I gene promoter activity by two-exon tat of HIV. *Science* 1993;260:1320–1322.
319. Hu W-S, Pathek VK, Temin HM. Role of reverse transcriptase in retroviral recombination. In: Skalka AM, Goff SP, eds. *Reverse transcriptase*. Cold Spring Harbor, NY: Cold Spring Harbor Laboratory Press, 1993;251–274.
320. Hu W-S, Temin HM. Retroviral recombination and reverse transcription. *Science* 1990;250:1227–1233.
321. Hu W-S, Temin HM. Effect of gamma radiation on retroviral recombination. *J Virol* 1992;66:4457–4463.
322. Hu W. S, Temin HM. Genetic consequences of packaging two RNA genomes in one retroviral particle: pseudodiploidy and high rate of genetic recombination. *Proc Natl Acad Sci USA* 1990;87:1556–1560.
323. Huang M, Jolicoeur P. Characterization of the *gag*/fusion protein encoded by the defective duplan retrovirus inducing murine acquired immunodeficiency syndrome. *J Virol* 1990;64:5764–5772.
324. Huang M, Simard C, Jolicoeur P. Immunodeficiency and clonal growth of target cells induced by helper-free defective retrovirus. *Science* 1989;246:1614–1617.
325. Huang M, Simard C, Jolicoeur P. Susceptibility of inbred strains of mice to murine AIDS (MAIDS) correlates with target cell expansion and high expression of defective MAIDS virus. *J Virol* 1992;66:2398–2406.
326. Huber HE, McCoy JM, Seehra JS, Richardson CC. Human immunodeficiency virus 1 reverse transcriptase. Template binding, processivity, strand displacement synthesis, and template switching. *J Biol Chem* 1989;264:4669–4678.
327. Huber HE, Richardson CC. Processing of the primer for plus strand DNA synthesis by human immunodeficiency virus 1 reverse transcriptase. *J Biol Chem* 1990;265:10565–10573.
328. Hubner A, Kruhoffer Grosse, F Krauss, G. Fidelity of human immunodeficiency virus type 1 reverse transcriptase in copying natural RNA. *J Mol Biol* 1992;223:595–600.
329. Hugin AW, Vacchio MS, Morse HCI. A virus-encoded "superantigen" in a retrovirus-induced immunodeficiency syndrome of mice. *Science* 1991;252:424–427.
330. Hunter E, Hill E, Hardwick M, Bhown A, Schwartz DE, Tizard R. Complete sequence of the Rous sarcoma virus env gene: identification of structural and functional regions of its products. *J Virol* 1983;46:920–936.
331. Hunter K, Housman D, Hopkins N. Isolation and characterization of irradiation fusion hybrids from mouse chromosome 1 for mapping Rmc-1, a gene encoding a cellular receptor for MCF class murine retroviruses. *Somat Cell Mol Genet* 1991;17:169–183.
332. Hunter R, Swanstrom R. Retrovirus envelope glycoproteins. In: Swanstrom R, Vogt PK, eds. *Retroviruses: strategies of replication*. New York: Springer-Verlag, 1990;187–253.
333. Hutchison III CA, Hardies SC, Loeb DD, Shehee WR, Edgell MH. LINES and related retroposons: long interspersed repeated sequences

in the eucaryotic genome. In: Howe M, Berg D, eds. *Mobile DNA*. Washington: ASM, 1989;593–617.
334. Hwang JV, Gilboa E. Expression of genes introduced into cells by retroviral infection is more efficient than that of genes introduced into cells by DNA transfection. *J Virol* 1984;50:417–424.
335. Ikeda H, Sugimura H. Fv–4 resistance gene: a truncated endogenous murine leukemia virus with ecotropic interference properties. *J Virol* 1989;63:5405–5412.
336. Innes CL, Smith PB, Langenbach R, Tindall KR, Boone LR. Cationic liposomes (lipofectin) mediate retroviral infection in the absence of specific receptors. *J Virol* 1990;64:957–961.
337. Itin A, Keshet. E. A novel retroviruslike family in mouse DNA. *J Virol* 1986;59:301–307.
338. Iwasaki K, Temin HM. The efficiency of RNA 3′ end formation is determined by the distance between the cap site and the poly(A) site in spleen necrosis virus. *Genes Dev* 1990;4:2299–2307.
339. Iwasaki K, Temin HM. The U3 region is not necessary for 3′ end formation of spleen necrosis virus RNA. *J Virol* 1990;64:6329–6334.
340. Jacks T. Translational suppression in gene expression in retroviruses and retrotransposons. In: Swanstrom R, Vogt PK, eds. *Retroviruses. Strategies of Replication*. New York: Springer-Verlag, 1990;93–124.
341. Jacks T, Townsley K, Varmus HE, Majors J. Two efficient ribosomal frameshifing events are required for synthesis of mouse mammary tumor virus gag-related polyproteins. *Proc Natl Acad Sci USA* 1987;84:4298–4302.
342. Jacobo-Molina-A, Ding J, Nanni RG, Clark ADJ, Lu X, Tantillo C, Williams RL, Kamer G, Ferris AL, Clark P, Hizi A, Hughes SH, Arnold E. Crystal structure of human immunodeficiency virus type 1 reverse transcriptase complexed with double-stranded DNA at 3.0Å resolution shows bent DNA. *Proc Natl Acad Sci USA* 1993;90:6320–6324.
343. Jahner D, Jaenisch R. Chromosomal position and specific demethylation in enhancer sequences of germ line-transmitted retroviral genomes during mouse development. *Mol Cell Biol* 1985;5:2212–2220.
344. Jahner D, Jaenisch R. Retrovirus-induced de novo methylation of flanking host sequences corrrelates with gene inactivity. *Nature* 1985; 315:594–597.
345. Janeway C. MIs: makes a little sense. *Nature* 1991;349:459–461.
346. Jasin M, Page KA, Littman DR. Glycosylphosphatidylinositol-anchored CD4/Thy-1 chimeric molecules serve as human immunodeficiency virus receptor in human, but not mouse, cells and are modulated by gangliosides. *J Virol* 1991;65:440–444.
347. Jaskolski M, Miller M, Rao JK, Leis J, Wlodawer A. Structure of the aspartic protease from Rous sarcoma retrovirus refined at 2-A resolution. *Biochemistry* 1990;29:5889–5898.
348. Jeang K-T, Chun R, Lin NH, Gatignol A, Glabe CG, Fan H. In vitro and in vivo binding of human immunodeficiency virus type 1 tat protein and Sp1 transcription factor. *J Virol* 1993;67:6224–6233.
349. Jenkins NA, Copeland NG, Taylor BA, Lee BK. Dilute (*d*) coat colour mutation of DBA/2J mice is associated with the site of integration of an ecotropic MuLV genome. *Nature* 1981;293:370–374.
350. Jensen S, Heidmann T. An indicator gene for detection of germline retrotransposition in transgenic drosophila demosntrates RNA-mediated transposition of the LINE-1 element. *EMBO J* 1991;10:1927–1937.
351. Ji JP, Hoffmann JS, Loeb L. Mutagenicity and pausing of HIV reverse transcriptase during HIV plus-strand DNA synthesis. *Nucl Acids Res* 1994;22:47–52.
352. Johann SV, Gibbons JJ, O'Hara B. GLVR1, a receptor for gibbon ape leukemia virus, is homologous to a phosphate permease of *Neurospora crassa* and is expressed at high levels in the brain and thymus. *J Virol* 1992;66:1635–1640.
353. Johann SV, van Zeijl M, Cekleniak J, O'Hara B. Definition of a domain of GLVR1 which is necessary for infection by gibbon ape leukemia virus and which is highly polymorphic between species. *J Virol* 1993;67:6733–6736.
354. Jolicoeur P, Rassart E. Effect of *Fv-1* gene product on synthesis of linear and supercoiled viral DNA in cells infected with murine leukemia virus. *J Virol* 1980;33:183–195.
355. Joliot V, Boroughs K, Lasserre F, Crochet J, Dambrine G, Smith RE, Perbal B. Pathogenic potential of myeloblastosis-associated virus: implication of ENV proteins for osteopetrosis induction. *Virology* 1993;195:812–819.
356. Jones JS, Allan RW, Temin HM. Alteration of location of dimer linkage sequence in retroviral RNA: little effect on replication or homologous recombination. *J Virol* 1993;67:3151–3158.
357. Jones JS, Allan RW, Temin HM. One retroviral RNA is sufficient for synthesis of viral DNA. *J Virol* 1994;68:207–216.
358. Jonsson CB, Roth MJ. Role of the his-cys finger of Moloney murine leukemia virus integrase protein in integration and disintegration. *J Virol* 1993;67:5562–5571.
359. Jorgensen ECB, Pedersen FS, Jorgensen P. Matrix protein of Akv murine leukemia virus: genetic mapping of regions essential for particle formation. *J Virol* 1992;66:4479–4487.
360. Junghans RP, Boone LR, Skalka AM. Retroviral DNA H structures: displacement-assimilation model of recombination. *Cell* 1982;30:53–62.
361. Kamine J, Subramanian T, Chinnadurai G. Activation of a heterologous promoter by human immunodeficiency virus type 1 tat requires Sp1 and is distinct from the mode of activation by acidic transcriptional activators. *J Virol* 1993;67:6828–6834.
362. Kamps CA, Lin Y-C, Wong PKY. Oligomerization and transport of the envelope protein of Moloney murine leukemia virus-TB and of *ts*1, a neurovirulent temperature-sensitive mutant of MoMuLV-TB. *Virology* 1991;184:687–694.
363. Kanagawa O, Nussrallah BA, Wiebenga ME, Murphy KM, Morse HI, Carbone FR. Murine AIDS superantigen reactivity of the T cells bearing Vb5 T cell antigen receptor. *J Immunol* 1992;149:9–16.
364. Kane SE, Beemon K. Precise localization of m6A in Rous sarcoma virus RNA reveals clustering of methylation sites: implications for RNA processing. *Mol Cell Biol* 1985;5:2298–2306.
365. Karacostas V, Wolffe EJ, Nagashima K, Gonda MA, Moss B. Overexpression of the HIV-1 gag-pol polyprotein results in intracellular activation of HIV-1 protease and inhibition of assembly and budding of virus-like particles. *Virology* 1993;193:661–671.
366. Kashanchi F, Piras G, Radonovich MF, Duvall JF, Fattaey A, Chiang C-M, Roeder RG, Brady JN Direct interaction of human TFIUID with the HIV-1 transactivator tat. *Nature* 1994;367:295–299.
367. Kati WM, Johnson KA, Jerva LF, Anderson KS. Mechanism and fidelity of HIV reverse transcriptase. *J Biol Chem* 1992;267:25988–25997.
368. Katoh I, Yasunaga T, Yoshinaka Y. Bovine leukemia virus RNA sequences involved in dimerization and specific *gag* proteinb inding: close relation to the packaging sites of avian, murine and human retroviruses. *J Virol* 1993;67:1830–1839.
369. Katoh I, Yoshinaka Y, Rein A, Shibuya M, Odaka T, Oroszlan S. Murine leukemia virus maturation: protease region required for conversion from "immature" to "mature" core form and for virus infectivity.*Virology* 1985;145:280–292.
370. Katz RA, Cullen BR, Malavarca R, Skalka AM. Role of the avian retrovirus mRNA leader in expression: evidence for novel translational control. *Mol Cell Biol* 1986;6:372–379.
371. Katz RA, Kotler M, Skalka AM. cis-Acting intron mutations that affect the efficiency of avian retroviral RNA splicing: implications for mechanisms of control. *J Virol* 1988;62:2686–2695.
372. Katz RA, Mack JPG, Merkel G, Kulkosky J, Ge Z, Leis J, Skalka AM. Requirement for a conserved serine in both processing and joining activities of retroviral integrase. *Proc Natl Acad Sci USA* 1992;89: 6741–6745.
373. Katz RA, Merkel G, Kulkosky J, Leis J, Skalka AM. The avian retroviral IN protein is both necessary and sufficient for integrative recombination in vitro. *Cell* 1990;63:87–95.
374. Katz RA, Skalka AM. A C-terminal domain in the avian sarcoma-leukosis virus pol gene product is not essential for viral replication. *J Virol* 1988;62:528–533.
375. Katz RA, Skalka AM. Control of retroviral RNA splicing through maintenance of suboptimal processing signals. *Mol Cell Biol* 1990;10:696–704.
376. Katz RA, Skalka AM. The retroviral enzymes. *Ann Rev Biochem* 1994;63:133–173.
377. Katz RA, Terry RW, Skalka AM. A conserved cis-acting sequence in the 5′ leader of avian sarcoma virus RNA is required for packaging. *J Virol* 1986;59:163–167.
378. Katzman M, Katz RA, Skalka AM, Leis J. The avian retroviral integration protein cleaves the terminal sequences of linear viral DNA at the in vivo sites of integration. *J Virol* 1989;63:5319–5327.
379. Kavanaugh MP, Miller DG, Zhang W, Law W, Kozak SL, Kabat D, Miller AD. Cell-surface receptors for gibbon ape leukemia virus and amphotropic murine retrovirus are inducible sodium-dependent phosphate symporters. *Proc Natl Acad Sci USA* 1994;91:7071–7075.
380. Kawaguchi Y, Miyazawa T, Horimoto T, Itagaki S, Fukasawa M, Takahashi E, Mikami T. Activation of feline immunodeficiency virus long terminal repeat by feline herpesvirus type 1. *Virology* 1991;184:449–454.
381. Keller A, Garrett ED, Cullen BR. The Bel-1 protein of human foamy virus activates human immunodeficiency virus type 1 gene expression

via a novel DNA target site. *J Virol* 1992;66:3946–3949.
382. Keller A, Partin KM, Löchelt M, Bannert H, Flugel RM, Cullen BR. Characterization of the transcriptional trans activator of human foamy retrovirus. *J Virol* 1991;65:2589–2594.
383. Kempler GFB, Berwin B, Nanassy O, Barklis E. Characterization of the Moloney murine leukemia virus stem cell-specific repressor binding sites. *Virology* 1993;193:690–699.
384. Keshet E, Schiff R, Itin A. Mouse retrotransposons: a cellular reservoir of long terminal repeat (LTR) elements with diverse transcriptional specificities. *Adv Cancer Res* 1991;56:215–251.
385. Deleted in proofs.
386. Keshet E, Temin HM. Cell killing by spleen necrosis virus is correlated with a transient accumulation of spleen necrosis virus DNA. *J Virol* 1979;31:376–388.
387. Kestler HWI, Ringler DJ, Mori K, Panicali DL, Sehgal PK, Daniel MD, Desrosiers RC. Importance of the nef gene for maintenance of high virus loads and for development of AIDS. *Cell* 1991;65:651–662.
388. Kim A, Terzian C, Santamaria P, Pelisson A, Prud'homme N, Bucheton A. Retroviruses in inveterebrates: the gypsy retrotransposon is apparently an infectious retrovirus of *Drosophila melanogaster*. *Proc Natl Acad Sci USA* 1994;91:1285–1289.
389. Kim JW, Closs EI, Albritton LM, Cunningham JM. Transport of cationic amino acids by the mouse ecotropic retrovirus receptor. *Nature* 1991;352:725–728.
390. Kimpton J, Emerman M. Detection of replication-competent and pseudotyped human immunodeficiency virus with a sensitive cell line on the basis of activation of an integrated b-galactosidase gene. *J Virol* 1992;66:2232–2239.
391. King W, Patel MD, Lobel LI, Goff SP, Nguyen-Huu MC. Insertion mutagenesis of embryonal carcinoma cells by retroviruses. *Science* 1985;228:554–558.
392. Kitado H, Chen I. S-Y, Shah NP, Cann AJ, Shimotohno K, Fan. H. U3 sequences from HTLV-I and -II LTRs confer px protein response to a murine leukemia virus LTR. *Science* 1987;235:901–904.
393. Kitado H, Fan H. Chromatin structure of recombinant Moloney murine leukemia virus proviral DNAs that contain *tax*-responsive sequences from human T-cell lymphotropic virus type II in the presence and absence of *tax*. *J Virol* 1989;63:3072–3079.
394. Kitajima I, Shnohara T, Bilakovics J, Brown DA, Xu X, Nerenberg M. Ablation of transplanted HTLV-1 tax-transformed tumors in mice by antisense inhibition of NF-κB. *Science* 1992;258:1792–1795.
395. Kitamura Y, Lee YMH, Coffin JM. Nonrandom integration of retroviral DNA *in vitro*: effect of CpG methylation. *Proc Natl Acad Sci USA* 1992;89:5532–5536.
396. Klatzman D, Champagne E, Chamaret S, Gruest J, Guetard D, Hercend T, Gluckman J-C, Montagnier L. T-lymphocyte T4 molecule behaves as the receptor for human retrovirus LAV. *Nature* 1984;312:767–768.
397. Klotman ME, Kim S, Buchbinder A, DeRossi A, Baltimore D, Wong-Staal F. Kinetics of expression of multiply spliced RNA in human immunodeficiency virus type 1 infection of lymphocytes and monocytes. *Proc Natl Acad Sci USA* 1991;88:5011–5015.
398. Knight AM, Harrison GB, Pease RJ, Robinson PJ, Dyson PJ. Biochemical analysis of the mouse mammary tumor virus long terminal repeat product. Evidence for the molecular structure of an endogenous superantigen. *Eur J Immunol* 1992;22:879–882.
399. Kohl NE, Emini EA, Schleif WA, Davis LJ, Heimbach JC, Dixon RAF, Scolnick EM, Sigal IS. Active human immunodeficiency virus protease is required for viral infectivity. *Proc Natl Acad Sci USA* 1988;85:4686–4690.
400. Kohlstaedt LA, Wang J, Friedman JM, Rice PA, Steitz TA. Crystal structure at 3.5 Å resolution of HIV-1 reverse transcriptase complexed with an inhibitor. *Science* 1992;256:1783–1790.
401. Kohlstaedt LA, Wang J, Rice PA, Friedman JM, Steitz TA. The structure of HIV-I reverse transcriptase. In: Skalka AM, Goff SP, eds. *Reverse transcriptase*. Cold Spring Harbor, NY: Cold Spring Harbor Laboratory Press, 1993;223–250.
402. Koppe B, Menendez-Arias L, Oroszlan S. Expression and purification of the mouse mammary tumor virus *gag-pro* transframe protein p30 and characterization of its dUTPase activity. *J Virol* 1994;68:2313–2319.
403. Kotler M, Katz RA, Dahno W, Leis J, Skalka AM. Synthetic peptides as substrates and inhibitors of a retroviral protease. *Proc Natl Acad Sci USA* 1988;85:4185–4189.
404. Kotler M, Katz RA, Skalka AM. Activity of avian retroviral protease expressed in Escherichia coli. *J Virol* 1988;62:2696–2700.
405. Kozak C, Peters G, Pauley R, Morris V, Michalides R, Dudley J, Green M, Davisson M, Prakash O, Vaidya A, Hilgers J, Verstraeten A, Hynes N, Diggelmann H, Peterson D, Cohen JC, Dickson C, Sarkar N, Nusse R, Varmus H, Callahan R. A standardized nomenclature for endogenous mouse mammary tumor viruses. *J Virol* 1987;61:1651–1654.
406. Kozak CA. Susceptibility of wild mouse cells to exogenous infection with xenotropic leukemia viruses: control by a single dominant locus on chromosome 1. *J Virol* 1985;55:690–695.
407. Kozak CA, O'Neill RR. Diverse wild mouse origins of xenotropic, mink-cell focus-forming, and two types of ecotropic proviral genes. *J Virol* 1987;61:3082–3088.
408. Kozak CA, Ruscetti S. Retroviruses in rodents. In: Levy JA, ed. *The Retroviridae*. New York: Plenum Press, 1992;1:405–481.
409. Kramer RA, Schaber MD, Skalka AM, Ganguly K, Wong-Staal F, Reddy EP. HTLV-III gag protein is processed in yeast cells by the virus pol-protease. *Science* 1986;231:1580–1583.
410. Kratochwil K, von der Mark K, Kollar EJ, Jaenisch R, Mooslehner K, Schwarz M, Haase K, Gmachl I, Harbers K. Retrovirus-induced insertional mutation in Mov13 mice affects collagen I expression in a tissue-specific manner. *Cell* 1989;57:807–816.
411. Krausslich H-G. Specific inhibitor of human immunodeficiency virus proteinase prevents cytotoxic effects of a single-chain proteinase dimer and restores particle formation. *J Virol* 1992;66:567–572.
412. Kristal BS, Reinhart TA, Hoover EA, Mullins JI. Interference with superinfection with cell killing and determination of host range and growth kinetics mediated by feline leukemia virus surface glycoproteins. *J Virol* 1993;67:4142–4153.
413. Kuff EL, Leuders KK. The intracisternal A-particle gene family: structure and functional aspects. *Adv Cancer Res* 1988;51:183–276.
414. Kulkosky J, Jones KS, Katz RA, Mack JPG, Skalka AM. Residues critical for retroviral integrative recombination in a region that is highly conserved among retroviral/retrotransposon integrases and bacterial insertion sequences transposases. *Mol Cell Biol* 1992;12:2331–2338.
415. Kung H-J, Boerkoel C, Carter TH. Retroviral mutagenesis of cellular oncogenes: a review with insights into the mechanisms of insertional activation. *Curr Top Micro Immun* 1991,171:1–25.
416. LaFemina RL, Schneider CL, Robbins HL, Callahan PL, LeGrow K, Roth E, Schleif WA, Emini EA. Requirement of active human immunodeficiency virus type 1 integrase enzyme for productive infection of human T-lymphoid cells. *J Virol* 1992;66:7414–7419.
417. Lafon M, Lafage M, Martinez-Arends A, Ramirez R, Vuillier A, Charron D, Lotteau V, Scott-Algara D. Evidence for a viral superantigen in humans. *Nature* 1992;358:507–510.
418. Landau NR, Littman DR. Packaging system for rapid production of murine leukemia virus vectors with variable tropism. *J Virol* 1992;66:5110–5113.
419. Landau NR, Page KA, Littman DR. Pseudotyping with human T-cell leukemia virus type I broadens the human immunodeficiency virus host range. *J Virol* 1991;65:162–169.
420. Lanza F, Wolf D, Fox CF, Kieffer N, Seyer JM, Fried VA, Coughlin SR, Phillips DR, Jennings LK. cDNA cloning and expression of platelet p24/CD9. Evidence for a new family of multiple membrane-spanning proteins. *J Biol Chem* 1991;266:10638–10645.
421. Laurence J, Hodtsev AS, Posnett DN. Superantigen implicated in dependence of HIV-1 replication in T cells on TCR Vb expression. *Nature* 1992;358:255–255.
422. Lazo PA, Klein-Szanto AJP, Tsichlis P. T-cell lymphoma lines derived from rat thymomas induced by Moloney murine leukemia virus: phenotypic diversity and its implications. *J Virol* 1990;64:3948–3959.
423. Lee AH, Lee KJ, Kim S, Sung YC. Transactivation of human immunodeficiency virus type 1 long terminal repeat-direct3ed gene expression by the human foamy virus *bel1* protein requires a specific DNA sequence. *J Virol* 1992;66:3236–3240.
424. Lee BK, Eicher EM. Segregation patterns of endogenous mouse mammary tumor viruses in five recombinant inbred strain sets. *J Virol* 1990;64:4568–4572.
425. Lee KJ, Lee AH, Sung YC. Multiple positive and negative *cis*-acting elements that mediate transactivation by bel1 in the long terminal repeat of human foamy virus. *J Virol* 1993;67:2317–2326.
426. Lee YMH, Coffin JM. Efficient autointegration of avian retrovirus DNA in vitro. *J Virol* 1990;64:5958–5965.
427. Lee YMH, Coffin JM. Relationship of avian retrovirus DNA synthesis to integration in vitro. *Mol Cell Biol* 1991;11:1419–1430.
428. Lefebvre P, Berard DS, Cordingley MG, Hager GL. Two regions of the mouse mammary tumor virus long terminal repeat regulate the activity of its promoter in mammary cell lines. *Mol Cell Biol*

1991;11:2529–2537.
429. Leis J, Aiyar A, Cobrinik D. Regulation of initiation of reverse transcription of retroviruses. In: Skalka AM, Goff SP, eds. *Reverse transcriptase*. Cold Spring Harbor, NY: Cold Spring Harbor Laboratory Press, 1993;33–48.
430. Leis J, Baltimore D, Bishop JM, Coffin J, Fleissner E, Goff SP, Oroszlan S, Robinson H, Skalka AM, Temin HM, Vogt V. Standardized and simplified nomenclature for proteins common to all retroviruses. *J Virol* 1988;62:1808–1809.
431. Leis J, Weber I, Wlodawer A, Skalka AM. Structure-function analysis of the Rous sarcoma virus-specific proteinase. *ASM News* 1990;56:77–81.
432. Lenburg ME, Landau NR. Vpu-induced degradation of CD4: requirement for specific amino acid residues in the cytoplasmic domain of CD4. *J Virol* 1993;67:7238–7245.
433. Levantis P, Gillespie DAF, Hart K, Bissell MJ, Wyke JA. Control of expression of an integrated Rous sarcoma provirus in rat cells: role of 5′ genomic duplications reveals unexpected patterns of gene transcription and its regulation. *J Virol* 1986;57:907–916.
434. Levin JG, Hatfield DL, Oroszlan S, Rein A. Mechanisms of translational suppression used in the biosynthesis of reverse transcriptase. In: Skalka AM, Goff SP, eds. *Reverse transcriptase*. Cold Spring Harbor, NY: Cold Spring Harbor Laboratory Press, 1993;5–32.
435. Levine KL, Steiner B, Johnson K, Aronoff R, Quinton TJ, Linial ML. Unusual features of integrated cDNAs generated by infection with genome-free retroviruses. *Mol Cell Biol* 1990;10:1891–1900.
436. Levrero M, Balsano C, Natoli G, Avantaggiati ML, Elfassi E. Hepatitis B virus X protein transactivates the long terminal repeats of human immunodeficiency virus types 1 and 2. *J Virol* 1990;64:3082–3086.
437. Levy DE, Lerner RA, Wilson MC. The Gv-1 locus coordinately regulates the expression of multiple endogenous murine retroviruses. *Cell* 1985;41:289–299.
438. Levy DN, Fernandes LS, Williams WV, Weiner DB. Induction of cell differentiation by human immunodeficiency virus 1 *vpr*. *Cell* 1993;72:541–550.
439. Levy JA. The Retroviridae. Plenum Press, New York
440. Lewis P, Hensel M, Emerman M. Human immunodeficiency virus infection of cells arrested in the cell cycle. *EMBO J* 1992;11:3053–3058.
441. Lewis PF, Emerman M. Passage through mitosis is required for oncorentroviruses but not for the human immunodeficiency virus. *J Virol* 1994;68:510–516.
442. Li E, Beard C, Jaenisch R. Role for DNA methylation in genomic imprinting. *Nature* 1993;366:362–365.
443. Li J-P, Baltimore D. Mechanism of leukemogenesis induced by mink cell focus-forming murine leukemia viruses. *J Virol* 1991;65:2408–2414.
444. Li JP, D'Andrea AD, Lodish HF, Baltimore D. Activation of cell growth by binding of Friend spleen focus-forming virus gp55 glycoprotein to the erythropoietin receptor. *Nature* 1990;343:762–764.
445. Li Y, Golemis E, Hartley JW, Hopkins N. Disease specificity of nondefective Friend and Moloney murine leukemia viruses is controlled by a small number of nucleotides. *J Virol* 1987;61:693–700.
446. Lifson JD, Reyes GR, McGrath MS, Stein BS, Engelman EG. AIDS retrovirus induced cytopathology: giant cell formation and involvement of CD4 antigen. *Science* 1986;232:1123–1127.
447. Limhoco TI, Dickie P, Ikeda H, Silver J. Transgenic *Fv*-4 mice resistant to Friend virus. *J Virol* 1993;67:4163–4168.
448. Lindholm PF, Reid RL, Brady JN. Extracellular Tax_1 protein simulates tumor necrosis factor-b and immunoglobulin kappa light chain expression in lymphoid cells. *J Virol* 1992;66:1294–1302.
449. Linial M. Creation of a processed pseudogene by retroviral infection. *Cell* 1987;49:93–102.
450. Linial ML, Miller AD. Retroviral RNA packaging: sequence requirements and implications. In: Swanstrom R, Vogt PK, eds.*Retroviruses: strategies of replication*. New York: Springer-Verlag, 1990;125–152.
451. Linney E, Neill SD, Prestridge DS. Retroviral vector gene expression in F9 embryonal carcinoma cells. *J Virol* 1987;61:3248–3253.
452. Lobel LI, Goff SP. Reverse transcription of retroviral genomes: mutations in the terminal repeat sequences. *J Virol* 1985;53:447–455.
453. Lobel LI, Patel M, King W, Nguyen-Huu MC, Goff SP. Construction and recovery of viable retroviral genomes carrying a bacterial suppressor transfer RNA gene. *Science* 1985;228:329–332.
454. Löchelt M, Aboud M, Flugel RM. Increase in the basal transcriptional activity of the human foamy virus internal promoter by the homologous long terminal repeat promoter in cis. *Nucl Acids Res* 1993; 21:4226–4230.
455. Löchelt M, Flugel RM, Aboud M. The human foamy virus internal promoter directs the expression of the functional Bel 1 transactivator and Bet protein early after infection. *J Virol* 1994;68:638–645.
456. Löchelt M, Muranyi W, Flugel RM. Human foamy virus genome possesses an internal, Bel-1-dependent and functional promoter. *Proc Natl Acad Sci USA* 1993;90:7317–7321.
457. Lock LF, Keshet E, Gilbert DJ, Jenkins NA, Copeland NG. Studies of the mechanism of spontaneous germline ecotropic provirus acquisition in mice. *EMBO J* 1988;7:4169–4177.
458. Loh PC. Spumaviruses. In: Levy JA, ed. *The Retroviridae*. New York: Plenum Press, 1993;361–398.
459. Loh TP, Sievert LL, Scott RW. Negative regulation of retrovirus expression in embryonal carcinoma cells mediated by an intragenic domain. *J Virol* 1988;62:4086–4095.
460. Loreni F, Stavenhagen J, Kalff M, Robins DM. A complex androgen-responsive enhancer resides 2 kilobases upstream of the mouse Slp gene. *Mol Cell Biol* 1988;8:2350–2360.
461. Lori F, di Marzo Veronese F, DeVico AL, Lusso P, Reitz JSJ, Gallo RC. Viral DNA carried by human immunodeficiency virus type 1 virions. *J Virol* 1992;66:5067–5074.
462. LoSardo JE, Boral AL, Lenz J. Relative importance of elements within the SL3-3 virus enhancer for T-cell specificity. *J Virol* 1990;64:1756–1763.
463. LoSardo JE, Cupelli LA, Short MK, Berman JW, Lenz J. Differences in activities of murine retroviral long terminal repeats in cytotoxic T lymphocytes and T-lymphoma cells. *J Virol* 1989;63:1087–1094.
464. Low KG, Chu H-M, Tan Y, Schwartz PM, Daniels GM, Melner MH, Comb MJ. Novel interactions between human T-cell leukemia virus type I tax and activating transcription factor 3 at a cyclic AMP-responsive element. *Mol Cell Biol* 1994;14:4958–4974.
465. Lu X, Welsh TM, Peterline BM. The human immunodeficiency virus type 1 long terminal repeat specifies two different transcription complexes, only one of which is regulated by tat. *J Virol* 1993;67:1752–1760.
466. Luciw PA, Leung NJ. Mechanisms of retrovirus replication. In: Levy JA, ed. *The Retroviridae*. New York: Plenum Press, 1992.
467. Lund AH, Duch M, Lovmand J, Jorgensen P, Pedersen FS. Mutated primer binding sites interacting with different tRNAs allow efficient murine leu kemia virus replication. *J Virol* 1993;67:7125–7130.
468. Luo G, Sharmeen L, Taylor J. Specificities involved in the initiation of retrovirus plus-strand DNA. *J Virol* 1990;64:592–597.
469. Mador N, Panet A, Honigman A. Translation of gag, pro, and pol gene products of human T-cell leukemia virus type 2. *J Virol* 1989;63:2400–2404.
470. Majors J. The structure and function of retroviral long terminal repeats. In: Swanstrom R, Vogt PK, eds. *Retroviruses. Strategies of Replication*. New York: Springer-Verlag, 1990;49–92.
471. Manley NR, O'Connell M, Sun W, Speck NA, Hopkins N. Two factors that bind to highly conserved sequences in mammalian type C retroviral enhancers. *J Virol* 1993;67:1967–1975.
472. Mann RS, Mulligan RC, Baltimore D. Construction of a retrovirus packaging mutant and its use to produce helper-free defective retrovirus. *Cell* 1983;32:871–879.
473. Mansky LM, Temin HM. Lower mutation rate of bovine leukemia virus relative to that of spleen necrosis virus. *J Virol* 1994;68:494–499.
474. Marcuzzi A, Weinberger J, Weinberger OK. Transcellular activation of the human immunodeficiency virus type 1 long terminal repeat in cocultured lymphocytes. *J Virol* 1992;66:4228–4232.
475. Markovitz DM, Kenney S, Kamine J, Smith MS, Davis M, Huang E-S, Rosen C, Pagano JS. Disparate effects of two herpesviruses immediate-early gene trans-activators on the HIV-1 LTR. *Virology* 1989;173:750–754.
476. Marquet R, Baudin F, Gabus C, Darlix JL, Mougul M, Ehresman C, Ehresman B. Dimerization by human immunodeficiency virus (type1) RNA: stimulation by cations and possible mechanism. *Nucl Acids Res* 1991;19:2349–2357.
477. Marriott SJ, Boros I, Duvall JF, Brady JN. Indirect binding of human T-cell leukemia virus type I tax, to a responsive element in the viral long terminal repeat. *Mol Cell Biol* 1989;9:4152–4160.
478. Marriott SJ, Lindholm PF, Brown KM, Gitlin SD, Duvall JF, Radonovich MF, Brady JN. A 36-kilodalton cellular transcription factor mediates an indirect interaction of human T-cell leukemia/lymphoma virus type I TAX_1 with a responsive element in the viral long terminal repeat. *Mol Cell Biol* 1990;10:4192–4201.
479. Maruyama M, Shibuya H, Harada H, Hatakeyama M, Seiki M, Fujita

T, Inoue J, Yoshida M, Taniguchi. T. Evidence for aberrant activation of the interleukin-2 autocrine loop by HTLV-1-encoded p40x and T3/Ti complex triggering. *Cell* 1987;48:343–350.
480. Mateu MG, Martinez MA, Rocha E, Andreu D, Parejo J, Giralt E, Sobrino F, Domingo E. Implications of a quasispecies genome structure: effect of frequent, naturally occurring amino acid substitutions on the antigenicity of foot-and-mouth disease virus. *Proc Natl Acad Sci USA* 1989;86:5883–5887.
481. Maurer B, Bannert H, Darai G, Flugel RM. Analysis of the primary structure of the long terminal repeat and the gag and pol genes of the human spumaretrovirus. *J Virol* 1988;62:1590–1597.
482. McClure MA. Evolution of retroposons by acquisition or deletion of retrovirus-like genes. *Mol Biol Evol* 1992;8:835–856.
483. McClure MA. Evolutionary history of reverse transcriptase. In: Skalka AM, Goff SP, eds. *Reverse Transcriptase*. Cold Spring Harbor, NY: Cold Spring Harbor Laboratory Press, 1993;425–444.
484. McClure MA, Johnson MS, Doolittle RF. Relocation of a protease-like gene segment between two retroviruses. *Proc Natl Acad Sci USA* 1987;84:2693–2697.
485. McClure MO, Sommerfelt MA, Marsh M, Weiss RA. The pH independence of mammalian retrovirus infection. *J Gen Virol* 1990;71:767–.
486. McCune JM, Rabin LB, Feinberg MB, Lieberman M, Kosek JC, Reyes GR, Weissman IL. Endoproteolytic cleavage of gp160 is required for activation of human immunodeficiency virus. *Cell* 1988;53:55–67.
487. Mcgeoch DJ. Protein sequence comparisons show that the pseudoproteases encoded by poxviruses and certain retroviruses belong to the deoxyuridine triphosphatase family. *Nucl Acids Res* 1990;18:4105–4110.
488. McGuire KL, Curtiss VE, Larson EL, Haseltine WA. Influence of human T-cell leukemia virus type I *tax* and *rex* on interleukin-2 gene expression. *J Virol* 1993;67:1590–1599.
489. McNally MT, Beemon K. Intronic sequences and 3′ splice sites control Rous sarcoma virus RNA splicing. *J Virol* 1992;66:6–11.
490. McNally MT, Gontarek RR, Beemon K. Characterization of Rous sarcoma virus intronic sequences that negatively regulate splicing. *Virology* 1991;185:99–108.
491. Mercer JA, Seperack PK, Strobel MC, Copeland NG, Jenkins NA. Novel myosin heavy chain encoded by murine dilute coat colour locus. *Nature* 1991;349:709–713.
492. Mergia A, Luciw PA. Replication and regulation of foamy viruses. *Virology* 1991;184:475–482.
493. Mergia A, Pratt-Lowe E, Shaw KES, Renshaw-Gegg LW, Luciw PA. cis-acting regulatory regions in the long terminal repeat of simian foamy virus type 1. *J Virol* 1992;66:251–257.
494. Mergia A, Shaw KES, Pratt-Lowe E, Barry PA, Luciw PA. Simian foamy virus type 1 is a retrovirus which encodes a transcriptional transactivator. *J Virol* 1990;64:3598–3604.
495. Mergia A, Shaw KES, Pratt-Lowe E, Barry PA, Luciw PA. Identification of the simian foamy virus transcriptional transactivator gene (taf). *J Virol* 1991;65:2903–2909.
496. Meric C, Darlix JL, Spahr P-F. It is Rous sarcoma virus p12 and not p19 that binds tightly to Rous sarcoma virus RNA. *J Mol Biol* 1986;173:531–538.
497. Meric C, Goff SP. Characterization of Moloney murine leukemia virus mutants with single-amino-acid substitutions in the cys-his box of the nucleocapsid protein. *J Virol* 1989;63:1558–1568.
498. Meric C, Gouilloud E, Spahr. P-F. Mutations in Rous sarcoma virus nucleocapsid protein p12 (NC): deletions of Cys-His boxes. *J Virol* 1988;62:3228–3333.
499. Meric C, Spahr P-F. Rous sarcoma virus nucleic acid binding protein p12 is necessary for viral 70S RNA dimer formation and packaging. *J Virol* 1986;60:450–459.
500. Miles BD, Robinson HL. High frequency transduction of c-erbB in avian leukosis virus-induced erythroblastosis. *J Virol* 1985;54:295–303.
501. Miller CL, Garner R, Paetkau V. An activation-dependent, T-lymphocyte-specific transcriptional activator in the mouse mammary tumor virus *env* gene.*Mol Cell Biol* 1992;12:3262–3272.
502. Miller DG, Edwards RH, Miller AD. Cloning of the cellular receptor for amphotropic murine retroviruses reveals homology to that for gibbon ape leukemia virus. *Proc Natl Acad Sci USA* 1994;91:78–82.
503. Miller DG, Miller AD. Inhibitors of retrovirus infection are secreted by several hamster cell lines and are also present in hamster sera. *J Virol* 1993;67:5346–5352.
504. Miller JT, Stoltzfus CM. Two distant upstream regions containing *cis*-acting signals regulating splicing facilitate 3′-end processing of avian sarcoma virus RNA. *J Virol* 1992;66:4242–4251.
505. Miller M, Jaskolski M, Mohana Rao JK, Leis J, Wlodawer A. Crystal structure of a retroviral protease proves relationship to aspartic protease family. *Nature* 1989;337:576–579.
506. Mizutani S, Temin HM. Incorporation of noncomplementary deoxyribonucleotides at high frequencies by ribodeoxyvirus DNA polymerases and Escherichia coli DNA polymerase I. *Biochemistry* 1976;15:1510–1516.
507. Mizuuchi K, Adzuma K. Inversion of the phosphate chirality at the target site of Mu DNA strand transfer: evidence for a one-step transesterification mechanism. *Cell* 1991;66:129–140.
508. Mok E, Golovkina TV, Ross SR. A mouse mammary tumor virus mammary gland enhancer confers tissue-specific but not lactation-dependent expression in transgenic mice. *J Virol* 1992;66:7529–7532.
509. Monk RJ, Malik FG, Stokesberry D, Evans LH. Direct determination of the point mutation rate of a murine retrovirus.*J Virol* 1992; 66:3683–3689.
510. Moore JP, Jamieson BA, Weiss RA, Sattentau QJ. The HIV-cell fusion reaction. In: Bentz J, ed. *Viral fusion mechanisms*. Boca Raton, FL: CRC Press, 1992;233.
511. Moore JP, Sattentau QJ, Wyatt R, Sodroski J. Probing the structure of the human immunodeficiency virus surface glycoprotein gp120 with a panel of monoclonal antibodies. *J Virol* 1994;68:469–484.
512. Moore JP, Sweet RW. THe HIV gp120-CD4 association: a target for pharmocological or immunological intervention? *Persp Drug Disc Design* 1993;1:235–250.
513. Moore R, Dixon M, Smith R, Peters G, Dickson C. Complete nucleotide sequence of a milk-transmitted mouse mammary tumor virus: two frameshift suppression events are required for translation of gag and pol. *J Virol* 1987;61:480–490.
514. Morgan RA, Nussbaum O, Muenchau DD, Shu L, Couture L, Anderson WF. Analysis of the functional and host range-determining regions of the murine ecotropic and amphotropic retrovirus envelope proteins. *J Virol* 1993;67:4712–4721.
515. Mori K, Ringler DJ, Kodama T, Desrosiers RC. Complex determinants of macrophage tropism in *env* of simian immunodeficiency virus. *J Virol* 1992;66:2067–2075.
516. Morrison HL, Dai HY, Pedersen FS, Lenz J. Analysis of the significance of two single-base-pair differences in the SL3-3 and Akv virus long terminal repeats. *J Virol* 1991;65:1019–1022.
517. Mougel M, Tounekti N, Darlix JL, Paoletti J, Ehresmann B, Ehresmann C. Conformational analysis of the 5′ leader and the GAG initiation site of Mo-MuLV RNA and allosteric transitions induced by dimerization. *Nucl Acids Res* 1993;21:4677–4684.
518. Mount SM, Rubin GM. Complete nucleotide sequence of the Drosophila transposable element copia: homology between copia and retroviral proteins. *Mol Cell Biol* 1985;5:1630–1638.
519. Mous J, Heimer EP, LeGrice SFJ. Processing protease and reverse transcriptase from human immunodeficiency virus type I polyprotein in Escherichia coli. *J Virol* 1988;62:1433–1436.
520. Moustakas A, Sonstegard TS, Hackett PB. Alterations of the three short open reading frames in the Rous sarcoma virus leader RNA modulate viral replication and gene expression. *J Virol* 1993;67:4337–4349.
521. Moustakas A, Sonstegard TS, Hackett PB. Effects of the open reading frames in the Rous sarcoma virus leader RNA on translation. *J Virol* 1993;67:4350–4357.
522. Mowat M, Cheng A, Kimura N, Bernstein A, Benchimol S. Rearrangement of the cellular p53 gene in erythroleukaemic cells transformed by Friend virus. *Nature* 1985;314:633–636.
523. Mueller-Lantzsch N, Sauter M, Weiskircher A, Kramer K, Beat B, Buck M, Grässer F. The human endogenous retroviral K10 (HER-K10) encodes for a full length gag homologous 73 KD protein and a functional protease. *AIDS Res Human Retroviruses* 1993;9:343–350.
524. Muller B, Jones KS, Merkel GW, Skalka AM. Rapid solution assays for retroviral integration reactions and their use in kinetic analyses of wild-type and mutant Rous sarcoma virus integrases. *Proc Natl Acad Sci USA* 1993;90:11633–11637.
525. Mulligan MJ, Yamshchikov GV, Ritter GDJ, Gao F, Jin MJ, Nail CD, Spies CP, Hahn BH, Compans RW. Cytoplasmic domain truncation enhances fusion activity by the exterior glycoprotein complex of human immunodeficiency virus type 2 in selected cell types. *J Virol* 1992; 66:3971–3975.
526. Mullins JI, Chen CS, Hoover EA. Disease-specific and tissue- specific production of unintegrated feline leukaemia virus variant DNA in feline AIDS. *Nature* 1986;319:333–336.
527. Murphy JE, De Los Santos T, Goff SP. Mutation analysis of the se-

quences at the termini of the Moloney murine leukemia virus DNA required for integration. *Virology* 1993;195:432–440.
528. Murphy JE, Goff SP. Construction and analysis of deletion mutations in the U5 region of Moloney murine leukemia virus: effects on RNA packaging and reverse transcription. *J Virol* 1988;63:319–327.
529. Murphy JE, Goff SP. A mutation at one end of Moloney murine leukemia virus DNA blocks cleavage of both ends by the viral integrase in vivo. *J Virol* 1992;66:5092–5095.
530. Myers G, Korber B, Wain-Hobson S, Smith RF, Pavlakis GN. Human retroviruses and AIDS 1993. Los Alamos National Laboratory, Los Alamos, New Mexico
531. Myers G, Pavlakis GN. Evolutionary potential of complex retroviruses. In: Levy J, ed. *The Retroviridae*. New York: Plenum Press, 1992;51–106.
532. Nagy K, Young M, Baboonian C, Merson J, Whittle P, Oroszlan S. Antiviral activity of human immunodeficiency virus type 1 protease inhibitors in a single cycle of infection: evidence for a role of protease in the early phase. *J Virol* 1994;68:757–765.
533. Nam SH, Copeland TD, Hatanaka M, Oroszlan S. Characterization of ribosomal frameshifting for expression of *pol* gene products of human T-cell leukemia virus type 1. *J Virol* 1993;67:196–203.
534. Nerenberg M, Hinrichs SH, Reynolds RK, Khoury G, Jay G. The tat gene of human T-lymphotropic virus type 1 induces mesenchymal tumors in transgenic mice. *Science* 1987;237:1324–1329.
535. Nermut MV, Grief C, Hashmi S, Hockley DJ. Further evidence of icosahedral symmetry in human and simian immunodeficiency viruses. *AIDS Res Hum Retroviruses* 1993;9:929–938.
536. Nermut MV, Hockley DJ, Jowett JBM, Jones IM, Garreau M, Thomas D. Fullerene-like organization of HIV gag-protein shell in virus-like particles produced by recombinant baculovirus. *Virology* 1994; 198:288–296.
537. Netzer K-O, Rethwilm A, Maurer B, ter Meulen V. Identification of the major immunogenic structural proteins of human foamy virus. *J Gen Virol* 1990;71:1237–1241.
538. Netzer K-O, Schliephake A, Maurer B, Watanabe R, Aguzzi A, Rethwilm A. Identification of pol-related gene products of human foamy virus. *Virology* 1993;192:336–338.
539. Niederman TMJ, Hastings WR, Luria S, Bandres JC, Ratner L. HIV-1 nef protein inhibits the recruitment of AP-1 DNA-binding activity in human T-cells. *Virology* 1993;194:338–344.
540. Niki M, Ohtani K, Nakamura M, Sugamura K. Multistep regulation of enhancer activity of the 21-base-pair element of human T-cell leukemia virus type 1. *J Virol* 1992;66:4348–4357.
541. Nimer SD, Gasson JC, Hu K, Smalberg I, Williams JL, Chen ISY, Rosenblatt JD. Activation of the GM-CSF promoter by HTLV-I and -II tax proteins. *Oncogene* 1989;4:671–676.
542. Niwa O. Suppression of the hypomethylated Moloney leukemia virus genome in undifferentiated teratocarcinoma cells and inefficiency of transformation by a bacterial gene under control of the long terminal repeat. *Mol Cell Biol* 1985;5:2325–2331.
543. Norton PA, Coffin JM. Bacterial b-galactosidase as a marker of Rous Sarcoma Virus gene expression and replication. *Mol Cell Biol* 1985;5:281–290.
544. Norton PA, Coffin JM. Characterization of Rous sarcoma virus sequences essential for viral gene expression. *J Virol* 1987;61:1171–1179.
545. Nowak MA, Anderson RM, McLean AR, Wolfs TFW, Goudsmit J, May RM. Antigenic diversity thresholds and the development of AIDS. *Science* 1991;254:963–969.
546. O'Hara B, Johann SV, Klinger HP, Blair DG, Robinson H, Dunn KJ, Sass P, Vitek SM, Robbins T. Characterization of a human gene conferring sensitivity to infection by gibbon ape leukemia virus. *Cell Growth Differ* 1990;1:119–127.
547. Olsen JC, Bova-Hill C, Grandgenett DP, Quinn TP, Manfredi JP, Swanstrom R. Rearrangements in unintegrated retroviral DNA are complex and are the result of multiple genetic determinants. *J Virol* 1990;64:5475–5484.
548. Olshevsky U, Helseth E, Furman C, Li J, Haseltine W, Sodroski J. Identification of individual human immunodeficiency virus type 1 gp120 amino acids important for CD4 receptor binding. *J Virol* 1990; 64:5701–5707.
549. Omer CA. Mechanism of release of the avian retrovirus RNATrp primer molecule from viral DNA by ribonuclease H during reverse transcription. *Cell* 1982;30:797–805.
550. Omer CA, Pogue-Geile K, Guntaka R, Staskis KA, Faras AJ. Involvement of directly repeated sequences in the generation of deletions of of the avian sarcoma virus src gene. *J Virol* 1983;47:380–382.
551. Ono M. Molecular biology of type A endogenous retrovirus. *Kitasato Arch Exp Med* 1990;63:77–90.
552. Oroszlan S, Luftig RB. Retroviral proteinases. In: Swanstrom R, Vogt PK, eds. *Retroviruses: strategies of replication*. New York: Springer-Verlag, 1990;153–186.
553. Oswald M, von-der-Helm K. Fibronectin is a non-viral substrate for the HIV proteinase. *Febs Lett* 1991;292:298–300.
554. Ott D, Rein A. Basis for receptor specificity of nonecotropic murine leukemia virus surface glycoprotein $gp70^{SU}$. *J Virol* 1992;66:4632–4638.
555. Ou C-Y, Boone LR, Koh CK, Tennant RW, Yang WK. Nucleotide sequences of gag-pol regions that determine the *Fv-1* host range property of BALB/c N-tropic and B-tropic murine leukemia viruses. *J Virol* 1983;48:779–784.
556. Overbaugh J, Donahue PR, Quackenbush SL, Hoover EA, Mullins JI. Molecular cloning of a feline leukemia virus that induces fatal immunodeficiency disease in cats. *Science* 1988;239:906–910.
557. Palmer E. Infectious origin of superantigens. *Curr Biol* 1991;1:74–76.
558. Panganiban AT, Fiore D. Ordered interstrand and intrastrand DNA transfer during reverse transcription. *Science* 1988;241:1064–1069.
559. Panganiban AT, Temin HM. The terminal nucleotides of retrovirus DNA are required for integration but not virus production. *Nature* 1983;306:155–160.
560. Panganiban AT, Temin HM. Circles with two tandem LTR's are precursors to integrated retrovirus DNA. *Cell* 1984;36:673–679.
561. Panganiban AT, Temin HM. The retrovirus *pol* gene encodes a product required for DNA integration: identification of a retrovirus *int* locus. *Proc Natl Acad Sci USA* 1984;81:7885–7889.
562. Park J, Morrow CD. Mutations in the protease gene of human immunodeficiency virus type 1 affect release and stability of virus particle. *Virology* 1993;194:843–850.
563. Parkin NT, Chamorro M, Varmus HE. Human immunodeficiency virus type 1 *gag-pol* frameshifting is dependent on downstream mRNA secondary structure: demonstration by expression in vivo. *J Virol* 1992;66:5147–5151.
564. Parvin JD, Moscona A, Pan WT, Leider JM, Palese P. Measurement of the mutation rates of animal viruses: influenza A and poliovirus type 1. *J Virol* 1986;59:377–383.
565. Patel PH, Preston BD. Marked infidelity of human immunodeficiency virus type 1 reverse transcriptase at RNA and DNA template ends.*Proc Natl Acad Sci USA* 1994;91:549–553.
566. Pathak VK, Temin HM. Broad spectrum of in vivo forward mutations, hypermutations and mutational hotspots in a retroviral shuttle vector after a single replication cycle: deletions and deletions with insertions. *Proc Natl Acad Sci USA* 1990;87:6024–6028.
567. Pathak VK, Temin HM. Broad spectrum of in vivo forward mutations, hypermutations, and mutational hotspots in a retroviral shuttle vector after a single replication cycle: substitutions, frameshifts and hypermutations. *Proc Natl Acad Sci USA* 1990;87:6019–6023.
568. Pathak VK, Temin HM. 5-azacytidine and RNA secondary structure increase the retrovirus mutation rate. *J Virol* 1992;66:3093–3100.
569. Patientce C, McKnight A, Clapham PR, Boyde MT, Weiss RA, Schulz TF. CD26 antigen and HIV fusion? *Science* 1994;264:1159–1160.
570. Payne LN. Biology of avian retroviruses. In: Levy JA, ed. *The Retroviridae*. New York: Plenum Press, 1992;299–376.
571. Pearl LH, Taylor WR. Sequence specificity of retroviral proteases. *Nature* 1987;328:482–483.
572. Pedersen FS, Paludan K, Dai HY, Duch M, Jorgensen P, Kjeldgaard NO, Hallberg B, Grundstrom T, Schmidt J, Luz A. The murine leukemia virus LTR in oncogenesis: effect of point mutations and chromosomal integration sites. *Radiat Environ Biophys* 1991;30:195–197.
573. Pedersen K, Lovmand S, Cecilie E, Jorgensen B, Pedersen FS, Jorgensen P. Efficient replication and expression of murine leukemia virus with major deletions in the enhancer region of U3. *Virology* 1992; 187:821–824.
574. Peliska JA, Benkovic SJ. Mechanism of DNA strand transfer reactions catalyzed by HIV-1 reverse transcriptase. *Science* 1992;258:1112–1118.
575. Pelletier H, Sawaya MR, Kumar A, Wilson SH, Kraut J. Structures of ternary complexes of rat DNA polymerase b, a DNA template primer, and ddCTP. *Science* 1994;264:1891–1903.
576. Pepinsky RB, Mattaliano RJ, Vogt VM. Structure and processing of the p2 region of avian sarcoma and leukemia virus gag precursor polyproteins. *J Virol* 1986;58:50–58.
577. Pepinsky RB, Vogt VM. Fine-structure analyses of lipid-protein and

protein-protein interactions of gag protein p19 of the avian sarcoma and leukemia viruses by cyanogen bromide mapping. *J Virol* 1984;52:145–153.

578. Perez LG, Davis GL, Hunter E. Mutants of the Rous sarcoma virus envelope glycoprotein that lack the transmembrane anchor and cytoplasmic domains: analysis of intracellular transport and assembly into virions. *J Virol* 1987;61:2981–2988.
579. Perez LG, Hunter E. Mutations within proteolytic cleavage site of the Rous sarcoma virus glycoprotein that block processing to gp85 and gp37. *J Virol* 1987;61:1609–1614.
580. Peters GG, Hu J. Reverse transcriptase as the major determinant for selective packaging of tRNA's into avian sarcoma virus particles. *J Virol* 1980;36:692–700.
581. Petersen RB, Moustakas A, Hackett PB. A mutation in the short 5′-proximal open reading frame on Rous sarcoma virus RNA alters virus production. *J Virol* 1989;63:4787–4796.
582. Pierce J, Fee BE, Toohey MG, Peterson DO. A mouse mammary tumor virus promoter element near the transcription initiation site. *J Virol* 1993;67:415–424.
583. Pillemer EA, Kooistra DA, Witte ON, Weissman IL. Monoclonal antibody to the amino-terminal L sequence of murine leukemia virus glycosylated gag poly proteins demonstrates their unusual orientation in the cell membrane. *J Virol* 1986;57:413–421.
584. Pina B, Bruggemeier U, Beato M. Nucleosome positioning modulates accessibility of regulatory proteins to the mouse mammary tumor virus promoter. *Cell* 1990;60:719–731.
585. Pincus T, Hartley JW, Rowe WP. A major genetic locus affecting resistance to infection with murine leukemia viruses. IV. Dose-response relationships in *Fv-1* sensitive and resistant cell cultures. *Virology* 1975;65:333–342.
586. Pinter A, Honnen WJ. O-linked glycosylation of retroviral envelope gene products. *J Virol* 1988;62:1016–1021.
587. Portis JL, Atee FJ, Evans LH. Infectious entry of murine retroviruses into mouse cells: evidence of a postadsorption step inhibited by acidic pH. *J Virol* 1985;55:806–812.
588. Poulet FM, Bowser PR, Casey JW. Retroviruses of fish, reptiles, and molluscs. In: Levy JA, ed. *The Retroviridae*. New York: Plenum Press, 1994;1–38.
589. Poulin L, Evans LA, Tang S, Barboza A, Legg H, Littman DR, Levy JA. Several CD4 domains can play a role in human immunodeficiency virus infection of cells. *J Virol* 1991;65:4893–4901.
590. Power MD, Marx PA, Bryant ML, Gardner MB, Barr PJ, Luciw PA. Nucleotide sequence of SRV-1, a type D simian acquired immune deficiency syndrome retrovirus. *Science* 1986;231:1567–1572.
591. Prasad VR. Genetic analysis of reverse transcriptase structure and function. In: Skalka AM, Goff SP, eds. *Reverse transcriptase*. Cold Spring Harbor, NY: Cold Spring Harbor Laboratory Press, 1993;135–162.
592. Prats A-C, Roy C, Wang P, Erard M, Housset V, Gabus C, Paoletti C, Darlix J-L. *cis* elements and *trans*-acting factors involved in dimer formation of murine leukemia virus RNA. *J Virol* 1990;64:774–783.
593. Preston BD, Poiesz BJ, Loeb LA. Fidelity of HIV-1 reverse transcriptase. *Science* 1988;242:1168–1171.
594. Prince VE, Rigby PWJ. Derivatives of Moloney murine sarcoma virus capable of being transcribed in embryonal carcinoma stem cells have gained a kfunctional Sp1 binding site. *J Virol* 1991;65:1803–1811.
595. Proudfoot NJ. Poly(A) signals. *Cell* 1991;64:671–674.
596. Pruss D, Bushman FD, Wolffe AP. Human immunodeficiency virus integrase directs integration to sites of severe DNA distortion within the nucleosome core. *Proc Natl Acad Sci USA* 1994;91:5913–5917.
597. Pryciak PM, Sil A, Varmus HE. Retroviral integration into minichromosomes in vitro. *EMBO J* 1992;11:291–303.
598. Pryciak PM, Varmus HE. Fv-1 restriction and its effects on murine leukemia virus integration in vivo and in vitro. *J Virol* 1992;66:5959–5966.
599. Pryciak PM, Varmus HE. Nucleosomes, DNA-binding proteins, and DNA sequence modulate retroviral integration target site selection. *Cell* 1992;69:769–780.
600. Pulsinelli G, Temin H. Characterization of large deletions occuring during a single round of retrovirus vector replication: novel Deletion mechanism involving errors in strand transfer. *J Virol* 1991;65:4786–4797.
601. Deleted in proofs.
602. Pyra H, Boni J, Schupbach J. Ultrasensitive retrovirus detection by a reverse transcriptase assay based on product enhancement. *Proc Natl Acad Sci USA* 1994;91:1544–1548.
603. Raines MA, Maihle NJ, Moscovici C, Crittenden L, Kung H-J. Mechanism of *c-erbB* transduction: newly released transducing viruses retain poly(A) tracts of *erbB* transcripts and encode C-terminally intact *erbB* proteins. *J Virol* 1988;62:2437–2443.
604. Ramsey CA, Panganiban AT. Replication of the retroviral terminal repeat sequence during in vivo reverse transcription. *J Virol* 1993;67:4114–4121.
605. Randolph CA, Champoux JJ. The majority of simian immunodeficiency virus/mne circle junctions result from ligation of unintegrated viral DNA ends that are aberrant for integration. *Virology* 1993;194:851–854.
606. Rattray AJ, Champoux JJ. The role of Moloney murine leukemia virus Rnase H. activity in the formation of plus-strand primers.*J. Virol* 1987;61:2843–2851.
607. Reddy S, DeGregori JV, Melchner HV, Ruley HE. Retrovirus promoter-trap vector to induce *lacZ* gene fusions in mammalian cells. *J Virol* 1991;65:1507–1515.
608. Rein A. Interference grouping of murine leukemia viruses: a distinct receptor for MCF-recombinant viruses in mouse cells. *Virology* 1982;120:251–257.
609. Rein A, McClure MR, Rice NR, Luftig RB, Schultz AM. Myristylation site in Pr65gag is essential for virus particle formation by Moloney murine leukemia virus. *Proc Natl Acad Sci USA* 1986;83:7246–7250.
610. Reinhard TA, Ghosh AK, Hoover EA, Mullins JI. Distinct superinfection interference properties yet similar receptor utilization by cytopathic and noncytopathic feline leukemia viruses. *J Virol* 1993;67:5153–5162.
611. Repaske R, Steele PE, O'Neill RR, Rabson AB, Martin MA. Nucleotide sequence of a full-length human endogenous retroviral segment. *J Virol* 1985;54:764–772.
612. Rethwilm A, Erlwein O, Baunach G, Maurer B, ter Meulen V. The transcriptional transactivator of human foamy virus maps to the bel 1 genomic region. *Proc Natl Acad Sci USA* 1991;88:941–945.
613. Rethwilm A, Mori K, Maurer B, ter Meulen V. Transacting transcriptional activation of human spumaretrovirus LTR infected cells.*Virology* 1990;175:568–571.
614. Reuss FU. Expression of intracisternal A-particle-related retroviral element-encoded envelope proteins detected in cell lines. *J Virol* 1992;66:1915–1923.
615. Reuss FU, Schaller HC. cDNA sequence and genomic characterization of intracisternal A-particle-related retroviral elements containing an envelope gene. *J Virol* 1991;65:5702–5709.
616. Rhee SS, Huner E. Myristylation is required for intracellular transport but not for assembly of D-type retrovirus capsids. *J Virol* 1987;61:1045–1053.
617. Rhee SS, Hunter E. A single amino acid substitution within the matrix protein of a type D retrovirus converts its morphogenesis to that of a type C retrovirus. *Cell* 1990;63:77–86.
618. Rhee SS, Hunter E. Structural role of the matrix protein of type D retroviruses in gag polyprotein stability and capsid assembly. *J Virol* 1990;64:4383–4389.
619. Ricchetti M, Buc H. Reverse transcriptases and genomic variability: the accuracy of DNA replication is enzyme specific and sequence dependent. *EMBO J* 1990;9:1583–1594.
620. Richardson NE, Brown NR, Hussey RE, Vaid A, Matthews TJ, Bolonesi DP, Reinherz EL. Binding site for human immunodeficiency virus coat protein gp120 is located in the NH2-terminal region of T4 (CD4) and requires the intact variable- region-like domain. *Proc Natl Acad Sci USA* 1988;85:6102–6106.
621. Rimsky L, Hauber J, Dukovich M, Malim MH, Langlois A, Cullen BR, Greene WC. Functional replacement of the HIV-1 rev protein by the HTLV-1 rex protein. *Nature* 1988;335:738–740.
622. Ritter GDJ, Mulligan MJ, Lydy SL, Compans RW. Cell fusion activity of the simian immunodeficiency virus envelope protein is modulated by the intracytoplasmid domain. *Virology* 1993;197:255–264.
623. Roberts JD, Bebenek K, Kunkel TA. The accuracy of reverse transcriptase from HIV-1. *Science* 1988;242:1171–1173.
624. Roberts JD, Preston BD, Johnston LA, Soni A, Loeb LA, Kunkel T. Fidelity of two retroviral reverse transcriptases during DNA-dependent DNA synthesis in vitro. *Mol Cell Biol* 1989;9:469–476.
625. Roberts MM, Oroszlan S. The action of retroviral protease in various phases of virus replication. In: Pearl L, ed. *Retroviral proteases, control of maturation and morphogenesis*. London: Macmillan Press, 1991;131–139.

626. Robinson HL, Blais BM, Tsichlis PN, Coffin JM. At least two regions of the viral genome determine the oncogenic potential of avian leukosis viruses. *Proc Natl Acad Sci USA* 1982;79:1225–1229.
627. Robinson HL, Foster RG, Blais BP, Reinsch SS, Newstein M, Shank PR. 5′ avian leukosis virus sequences and osteopetrotic potential. *Virology* 1992;190:866–871.
628. Robinson HL, Gagnon GC. Patterns of proviral insertion in avian leukosis virus-induced lymphomas. *J Virol* 1986;57:28–36.
629. Robinson HL, Reinsch SS, Shank PR. Sequences near the 5′ long terminal repeat of avian leukosis viruses determine the ability to induce osteopetrosis. *J Virol* 1986;59:45–49.
630. Roe T-Y, Reynolds TC, Yu G, Brown PO. Integration of murine leukemia virus DNA depends on mitosis. *EMBO J* 1993;12:2099–2108.
631. Rohdewohld H, Weiher H, Reik W, Jaenisch R, Breindl M. Retrovirus integration and chromatin structure: Moloney murine leukemia proviral integration sites map near DNase I-hypersensitive sites. *J Virol* 1987;61:336–343.
632. Rommelaere J, Donis-Keller H, Hopkins N. RNA sequencing provides evidence for allelism of determinants of the N-, B-, or NB-tropism of murine leukemia viruses. *Cell* 1979;16:43–50.
633. Rosen CA. Regulation of HIV gene expression by RNA-protein interactions. *Trends Genet* 1991;7:9–15.
634. Rosenberg ZF, Fauci AS. Immunology of AIDS: approaches to understanding the immunopathogenesis of HIV infection. *Ric Clin Lab* 1989;19:189–209.
635. Ross SR, Hsu C-LL, Choi Y, Mok E, Dudley JP. Negative regulation in correct tissue-specific expression of mouse mammary tumor virus in transgenic mice. *Mol Cell Biol* 1990;10:5822–5829.
636. Roy C, Tounekti N, Mougel M, Darlix JL, Paoletti C, Ehresman C, Ehresman B, Paoletti J. An analytical study of the dimerization of in vitro generated RNA of Moloney murine leukemia virus MoMuLV. *Nucl Acids Res* 1990;18:7287–7292.
637. Ruben S, Poteat H, Tan T-H, Kawakami K, Roeder R, Haseltine W, Rosen CA. Cellular transcription factors and regulation of IL-2 receptor gene expression by HTLV-I tax gene product. *Science* 1988; 241:89–92.
638. Rubinstein E, Benoit P, Billard M, Plaisance S, Prenant M, Uzan G, Boucheix C. Organization of the human CD9 gene. *Genomics* 1993;16:132–138.
639. Rudy CK, Kraus E, Palmer E, Huber BT. Mls-1-like superantigen in the MA/MyJ mouse is encoded by a new mammary tumor provirus that is distinct from Mtv-7. *J Exp Med* 1992;175:1613–1622.
640. Ryden TA, Beemon K. Avian retroviral long terminal repeats bind CCAAT/enhancer-binding protein. *Mol Cell Biol* 1989;9:1155–1164.
641. Ryden TA, de Mars M, Beemon K. Mutation of the C/EBP bindings sites in the Rous sarcoma virus long terminal repeat and gag enhancers. *J Virol* 1993;67:2862–2870.
642. Sakai H, Kawamura M, Sakuragi J-i, Sakuragi S, Shibata R, Ishimoto A, Ono N, Ueda S, Adachi A. Integration is essential for efficient gene expression of human immunodeficiency virus type 1. *J Virol* 1993; 67:1169–1174.
643. Sakai K, Ma X, Gordienko I, Volsky DJ. Recombinational analysis of a natural noncytopathic human immunodeficiency virus type 1 (HIV-2) isolate: role of the *vif* gene in HIV-1 infection kinetics and cytopathicity. *J Virol* 1991;65:5765–5773.
644. Samuelson LC, Wiebauer K, Snow CM, Meisler MH. Retroviral and pseudogene insertion sites reveal the lineage of human salivary and pancreatic amylase genes from a single gene during primate evolution. *Mol Cell Biol* 1990;10:2513–2520.
645. Sarma PS, Gruber J. Human T-cell lymphotropic viruses in human diseases. *J Natl Cancer Inst* 1990;82:1100–1106.
646. Scadden DT, Fuller B, Cunningham JM. Human cells infected with retrovirus vectors acquire an endogenous murine provirus. *J Virol* 1990;64:424–427.
647. Schackleford GM, Varmus HE. construction of a clonable, infectious, and tumorigenic mouse mammary tumor virus provirus and a derivative genetic vector. *Proc Natl Acad Sci USA* 1988;85:9655–9659.
648. Scherdin U, Rhodes K, Breindl M. Transcriptionally active genome regions are preferred targets for retrovirus integration. *J Virol* 1990; 64:907–912.
649. Schiff LA, Nibert ML, Fields BN. Characterization of a zinc blotting technique: evidence that a retroviral gag protein binds zinc. *Proc Natl Acad Sci USA* 1988;85:4195–4199.
650. Schinazi R, Larder B, Mellors J. Mutations in HIV-I reverse transcriptase and protease associated with drug resistance. *Intl Antiviral News* 1994; 2:72–75.
651. Schubert U, Strebel K. Differential activities of the human immunodeficiency virus type 1-encoded vpu protein are regulated by phosphorylation and occur in different cellular compartments. *J Virol* 1994; 68:2260–2271.
652. Schultz AM, Rein A. Unmyristylated Moloney murine leukemia virus Pr65gag is excluded from virus assembly and maturation events. *J Virol* 1989;63:2370–2372.
653. Schwaller M, Smith GE, Skehel JJ, Wiley DC. Studies with crosslinking reagents on the oligomeric structure of the *env* glycoprotein of HIV. *Virology* 1989;172:367–369.
654. Sears RC, Sealy L. Characterization of nuclear proteins that bind the EFII enhancer sequence in the Rous sarcoma virus long terminal repeat. *J Virol* 1992;66:6338–6352.
655. Sears RC, Sealy L. Multiple forms of C/EBPb bind the EfII enhancer sequence in the Rous sarcoma virus long terminal repeat. *Mol Cell Biol* 1994;14:4855–4871.
656. Seeler J-S, Muchardt C, Podar M, Gaynor RB. Regulatory elements involved in tax-mediated transactivation of the HTLV-I LTR. *Virology* 1993;196:442–450.
657. Seperack PK, Strobel MC, Corrow DJ, Jenkins NA, Copeland NG. Somatic and germ-line reverse mutation rates of the retrovirus-induced dilute coat-color mutation of DBA mice. *Proc Natl Acad Sci USA* 1988;85:189–192.
658. Sharp PA, Marciniak RA. HIV TAR: an RNA enhancer? *Cell* 1989;59:229–230.
659. Sheldon M, Ratnasabapathy R, Hernandez N. Characterization of the inducer of short transcripts, a human immunodeficiency virus type 1 transcriptional element that activates the synthesis of short RNAs. *Mol Cell Biol* 1993;13:1251–1263.
660. Sherman PA, Dickson ML, Fyfe JA. Human immunodeficiency virus type 1 integration protein: DNA sequence reuqirements for cleaving and joining reactions. *J Virol* 1992;66:3593–3601.
661. Shih A, Coutavas EE, Rush MG. Evolutionary implications of primate endogenous retroviruses. *Virology* 1991;182:495–502.
662. Shih C-C, Stoye JP, Coffin JM. Highly preferred targets for retrovirus integration. *Cell* 1988;53:531–537.
663. Shimotohno K, Takano M, Teruuchi T, Miwa. M. Requirement of multiple copies of a 21-nucleotide sequence in the U3 regions of human T-cell leukemia virus type I and type II long terminal repeats for trans-acting activation of transcription. *Proc Natl Acad Sci USA* 1986;83:8112–8116.
664. Shimotohno K, Temin HM. Spontaneous variation and synthesis in the U3 region of the long terminal repeat of an avian retrovirus. *J Virol* 1982;41:163–171.
665. Shin C-G, Taddeo B, Haseltine WA, Farnet CM. Genetic analysis of the human immunodeficiency virus type 1 integrase protein. *J Virol* 1994;68:1633–1642.
666. Shoemaker CS, Goff S, Gilboa E, Paskind M, Mitra SW, Baltimore D. Structure of a cloned circular Moloney murine leukemia virus molecule containing an inverted segment: implications for retrovirus integration. *Proc Natl Acad Sci USA* 1980;77:3932–3936.
667. Shoeman RL, Honer B, Stoller TJ, Kesselmeier C, Miedel MC, Traub P, Graves MC. Human immunodeficiency virus type 1 protease cleaves the intermediate filamentproteins vimentin, desmin, and glial fibrillary acidic protein. *Proc Natl Acad Sci USA* 1990;87:6336–6340.
668. Short MK, Okenquist SA, Lenz. J. Correlation of leukemogenic potential of murine retroviruses with transcriptional tissue preference of the viral long terminal repeats. *J Virol* 1987;61:1067–1072.
669. Showers MO, DeMartino JC, Saito Y, D'Andrea AD. Fusion of erythropoietin receptor and the Friend spleen focus-forming virus gp55 glycoprotein transforms a factor-dependent hematopoietic cell line. *Mol Cell Biol* 1993;13:739–748.
670. Siddiqui A, Gaynor R, Srinivasan A, Mapoles J, Farr RW. trans-activation of viral enhancers including long terminal repeat of the human immunodeficiency virus by the hepatitis B virus X protein. *Virology* 1989;169:479–484.
671. Simard C, Huang M, Jolicoeur P. Murine AIDS is initiated in the lymph nodes draining the site of inoculation, and the infected B cells influence T cells located at distance, in noninfected organs.*J Virol* 1994;68:1903–1912.
672. Skalka AM, Leis J. Retroviral DNA integration. *Bioessays* 1984;1:206–210.

673. Skalka AM. Integrative recombination in retroviruses. In: Kutcherlapati R, Smith GR, eds. *Genetic Recombination*. Washington, DC: ASM, 1988;701–724.
674. Skalka AM. Retroviral proteases: first glimpses at the anatomy of a processing machine. *Cell* 1989;56:911–913.
675. Smith TF, Srinivasan A, Schochetman G, Marcus M, Myers G. The phylogenetic history of immunodeficiency viruses. *Nature* 1988; 333:573–575.
676. Sodroski J, Goh W. C, Rosen C, Campbell K, Haseltine. WA. Role of the HTLV-III/LAV envelope in syncytium formation and cytopathicity. *Nature* 1986;322:470–474.
677. Somasundaran M, Robinson HL. Unexpectedly high levels of HIV-1 RNA and protein synthesis in a cytocidal infection. *Science* 1988; 242:1554–1557.
678. Sonigo P, Alizon M, Staskus K, Klatzmann D, Cole S, Danos O, Retzel E, Tiollais P, Haase A, Wain-Hobson S. Nucleotide sequence of the visna lentivirus: relationship to the AIDS virus. *Cell* 1985;42: 369–382.
679. Sonigo P, Barker C, Hunter E, Wain-Hobson S. Nucleotide sequence of Mason-Pfizer monkey virus: an immunosuppressive D-type retrovirus. *Cell* 1986;45:375–385.
680. Sorge J, Ricci W, Hughes SH. cis-Acting packaging locus in the 115-nucleotide direct repeat of Rous sarcoma virus. *J Virol* 1983;48:667–675.
681. Sousa R, Chung YJ, Rose JP, Wang B-C. Crystal structure of bacteriophage T7 RNA polymerase at 3.3 Å resolution. *Nature* 1993; 364:593–599.
682. South TL, Blake PR, Sowder RCI, Arthur LO, Henderson LE, Summers MF. The nucleocapsid protein isolated from HIV-1 particles binds zinc and forms retroviral-type zinc fingers. *Biochemistry* 1990;29: 7786–7789.
683. Sova P, Volsky DJ. Efficiency ov viral DNA synthesis during infection of permissive and nonpermissive cells with vif-negative human immunodeficiency virus type 1. *J Virol* 1993;67:6322–6326.
684. Speck NA, Baltimore D. Six distinct nuclear factors interact with the 75-base-pair repeat of the Moloney murine leukemia virus enhancer. *Mol Cell Biol* 1987;7:1101–1110.
685. Speck NA, Renjifo B, Hopkins N. Point mutations in the Moloney murine leukemia virus enhancer identify a lymphoid-specific viral core motif and 1,3-phorbol myristate acetate-inducible element. *J Virol* 1990;64:543–550.
686. Spence SE, Gilbert DJ, Swing DA, Copeland NG, Jenkins NA. Spontaneous germ line virus infection and retroviral insertional mutagenesis in eighteen transgenic Srev lines of mice. *Mol Cell Biol* 1989;9:177–184.
687. Stavenhagen JB, Robins DM. An ancient provirus has imposed androgen regulation on the adjacent mouse sex-limited protein gene. *Cell* 1988;55:247–254.
688. Steeg CM, Vogt VM. RNA-binding properties of the matrix protein (p19gag) of avian sarcoma and leukemia viruses. *J Virol* 1990;64:847–855.
689. Steele PE, Martin MA, Rabson AB, Bryan T, O'Brien SJ. Amplification and chromosomal dispersion of human endogenous retroviral sequences. *J Virol* 1986;59:545–550.
690. Stein BS, Gowda SD, Lifson JD, Pennhallow RC, Bensch KG, Engelman EG. pH-independent HIV entry into CD4-positive T cells via virus envelope fusion to the plasma membrane. *Cell* 1987;49:659–669.
691. Steinhauer DA, de la Torre, JC. Meier, E Holland, JJ. Extreme heterogeneity in populations of vesicular stomatitis virus. *J Virol* 1989;63:2072–2080.
692. Stewart L, Schatz G, Vogt VM. Properties of avian retrovirus particles defective in viral protease. *J Virol* 1990;64:5076–5092.
693. Stewart L, Vogt VM. *trans*-acting viral protease is necessary and sufficient for activation of avian leukosis virus reverse transcriptase. *J Virol* 1991;65:6218–6231.
694. Stewart L, Vogt VM. Reverse transcriptase and protease activities of avian leukosis virus gag-pol fusion proteins expressed in insect cells. *J Virol* 1993;67:7582–7596.
695. Stewart MA, Forrest D, McFarlane R, Onions D, Wilkie N, Neil JC. Conservation of the c-myc coding sequence in transduced feline v-myc genes. *Virology* 1986;154:121–134.
696. Stocking CR, Kollek R, Bergholz U and Ostertag W. Point mutations in the U3 regon of the long terminal repeat of Moloney murine leukemia virus determine disease specificity of the myeloproliferative sarcoma virus. *Virology* 1986;153:145–149.
697. Stoltzfus CM, Dane RW. Accumulation of spliced avian retroviral RNA is inhibited in S-adenosyl methionine depleted chicken embryo fibroblasts. *J Virol* 1982;42:918–931.
698. Stoye JP, Coffin JM. Endogenous viruses. In: Weiss R, Teich N, Varmus H, Coffin J, eds. *RNA tumor viruses*. Cold Spring Harbor, NY: Cold Spring Harbor Laboratory Press, 1985;357–404.
699. Stoye JP, Coffin JM. The four classes of endogenous murine leukemia virus: structural relationships and potential for recombination. *J Virol* 1987;61:2659–2669.
700. Stoye JP, Coffin JM. Polymorphism of murine endogenous proviruses revealed by using virus class-specific oligonucleotide probes. *J Virol* 1988;62:168–175.
701. Stoye JP, Fenner S, Greenoak GE, Moran C, Coffin JM. Role of endogenous retroviruses as mutagens: the hairless mutation of mice. *Cell* 1988;54:383–391.
702. Stoye JP, Moroni C, Coffin J. Virological events leading to spontaneous AKR thymomas. *J Virol* 1991;65:1273–1285.
703. Strambio-de-Castilla C, Hunter E. Mutational analysis of the major homology region of Mason-Pfizer monkey virus by use of saturation mutagenesis.*J Virol* 1992;66:7021–7032.
704. Strauss EG, Rice CM, Strauss JH. Sequence coding for the alphavirus nonstructural proteins is interrupted by an opal codon. *Proc Natl Acad Sci USA* 1983;80:5271–5275.
705. Stuhlmann H, Berg P. Homologous recombination of copackaged retrovirus RNAs during reverse transcription. *J Virol* 1992;66:2378–2388.
706. Stuhlmann H, Dieckmann M, Berg P. Transduction of cellular neo mRNA by retrovirus-mediated recombination. *J Virol* 1990; 64:5783–5796.
707. Sugamura K, Hinuma Y. Human retroviruses HTLV-I and HTLV-II. In: Levy JA, ed. *The Retroviridae*. New York: Plenum Press, 1992;399–436.
708. Sun W, O'Connell M, Speck NA. Characterization of a protein that binds multiple sequences in mammalian type C retrovirus enhancers. *J Virol* 1993;67:1976–1986.
709. Sundquist WI, Heaphy S. Evidence for interstrand quadraplex formation in the dimerization of human immunodeficiency virus 1 genomic RNA. *1993* 1993;90:3393–3397.
710. Suzuki S. *Fv-4,* a new gene affecting the splenomegaly induction by Friend leukemia virus. *J Exp Med* 1975;45:473–478.
711. Suzuki T, Fujisawa J-i, Toita M, Yoshida M. The trans-activator tax of human T-cell leukemia virus type 1 (HTLV-1) interacts with cAMP-responsive element (CRE) binding and CRE modulator proteins that bind to the 21-base-pair enhancer of HTLV-1. *Proc Natl Acad Sci USA* 1993;90:610–614.
712. Swain A, Coffin JM. Polyadenylation at correct sites in genome RNA is not required for retrovirus replication or genome encapsidation. *J Virol* 1989;63:3301–3306.
713. Swain A, Coffin JM. Mechanism of transduction by retroviruses. *Science* 1992;255:841–845.
714. Swain A, Coffin JM. Influence of sequences in the long terminal repeat and flanking cell DNA on polyadenylation of retroviral transcripts. *J Virol* 1993;67:6265–6269.
715. Swanstrom R, Parker RC, Varmus HE, Bishop JM. Transduction of a cellular oncogene: the genesis of Rous sarcoma virus. *Proc Natl Acad Sci* 1983;80:2519–2523.
716. Swanstrom R, Vogt PK. *Retroviruses: strategies of replication*. Berlin: Springer-Verlag, 1990.
717. Szurek PF, Floyd E, Yuen PH, Wong PKY. Site-directed mutagenesis of the codon for Ile-25 in gPr80env alters the neurovirulence of ts1, a mutant of Moloney murine leukemia virus TB. *J Virol* 1990;64: 5241–5249.
718. Tada H, Rappaport J, Lashgari M, Amini S, Wong-Staal F, Khalili K. Trans-activation of the JC virus late promoter by the tat protein of type 1 human immunodeficiency virus in glial cells. *Proc Natl Acad Sci USA* 1990;87:3479–3483.
719. Tailor CS, Takeuchi Y, O'Hara B, Johann SV, Weiss RA, Collins MKL. Mutation of amino acids within the gibbon ape leukemia virus (GALV) receptor differentially affects feline leukemia virus subgruop B, simian sarcoma-associated virus and GALV infections. *J Virol* 1993; 67:6737–6741.
720. Taketo M, Tanaka M. A cellular enhancer of retrovirus gene expression in embryonal carcinoma cells. *Proc Natl Acad Sci USA* 1987;84:3748–3752.
721. Takeuchi Y, Simpson G, Vile RG, Weiss RA, Collins MKL. Retrovi-

ral pseudotypes produced by rescue of a Moloney murine leukemia virus vector by C-type, but not D-type, retroviruses. *Virology* 1992;186:792–794.
722. Takeuchi Y, Vile RG, Simpson G, O'Hara B, Collins MKL, Weiss RA. Feline leukemia virus supgroup B uses the same cell surface receptor as gibbon ape leukemia virus. *J Virol* 1992;66:1219–1222.
723. Tanaka A, Takahashi C, Yamaoka S, Nosaka T, Maki M, Hatanaka M. Oncogenic transformation by the tax gene of human T-cell leukemia virus type I in vitro. *Proc Natl Acad Sci USA* 1990;87:1071–1075.
724. Tanese N, Goff SP. Domain structure of the Moloney murine leukemia virus reverse transcriptase: mutational analysis and separate expression of the DNA polymerase and RNase H activities. *Proc Natl Acad Sci USA* 1988;85:1777–1781.
725. Tanese N, Telesnitsky A, Goff SP. Abortive reverse transcription by mutants of Moloney murine leukemia virus deficient in the reverse transcriptase-associated RNase H function. *J Virol* 1991;65:4387–4397.
726. Tao J, Frankel AD. Specific binding of arginine to TAR RNA. *Proc Natl Acad Sci USA* 1992;89:2723–2726.
727. Tchenio T, Heidmann T. High-frequency intracellular transposition of a defective mammalian provirus detected by an in situ colorimetric assay. *J Virol* 1992;66:1571–1578.
728. Teich N. Taxonomy of retroviruses. In: Weiss R, Teich N, Varmus H, Coffin J, eds. *RNA tumor viruses*. Cold Spring Harbor, NY: Cold Spring Harbor Laboratory, 1985;1–16.
729. Teich N, Weiss RA, Martin GR, Lowy DR. Virus infection of murine teratocarcinoma stem cell lines. *Cell* 1977;12:973–982.
730. Telenitsky A, Blain SW, Goff SP. Defects in Moloney murine leukemia virus replication caused by a reverse transcriptase mutation modeled on the structure of *Escherichia coli* RNase H. *J Virol* 1992;66:615–622.
731. Telesnitsky A, Goff SP. Two defective forms of reverse transcriptase can complement to restore retroviral infectivity. *EMBO J* 1993;12:4433–4438.
732. Temin HM. Nature of the provirus of Rous sarcoma. *Natl Cancer Inst Monogr* 1964;17:557–570.
733. Temin HM. Reverse transcription in the eukaryotic genome: retroviruses, pararetroviruses, retro transposons, and retrotranscripts. *Mol Biol Evol* 1985;6:455–468.
734. Temin HM. Sex and recombination in retroviruses. *Trends Genet* 1991;7:71–74.
735. Temin HM. Retrovirus variation and reverse transcription: abnormal strand transfers result in retrovirus genetic variation. *Proc Natl Acad Sci USA* 1993;90:6900–6903.
736. Temin HM, Mizutani S. RNA-dependent DNA polymerase in virions of Rous sarcoma virus. *Nature* 1970;226:1211–1213.
737. Tevethia MJ, Spector DJ. Heterologous transactivation among viruses. *Prog Med Virol* 1989;36:120–190.
738. Thomas E, Overbaugh J. Delayed cytopathicity of a feline leukemia virus variant is due to four mutations in the transmembrane protein gene. *J Virol* 1993;67:5724–5732.
739. Tikhonenko AT, Poloskaya AV, Vassetzky NSI, Golovkina TV, Gudkov AV. Long terminal repeats of dwarf hamster endogenous retrovirus are highly diverged and do not maintain efficient transcription. *Virology* 1991;181:367–370.
740. Ting C-N, Rosenberg MP, Snow CM, Samuelson LC, Meisler MH. Endogenous retroviral sequences are required for tissue-specific expression of a human salivary amylase gene. *Genes Dev* 1992; 6:1457–1465.
741. Toyoshima H, Itoh M, Inoue J-i, Seiki M, Takaku F, Yoshida M. Secondary structure of the human T-cell leukemia virus type 1 rex-responsive element is essential for rex regulation of RNA processing and transport of unspliced RNAs. *J Virol* 1990;64:2825–2832.
742. Trono D. Partial reverse transcripts in virions from human immunodeficiency and murine leukemia viruses. *J Virol* 1992;66:4893–4900.
743. Tsichlis PN, Bear SE. Infection by mink cell focus-forming viruses confers interleukin 2 (IL-2) independence to an IL-2-dependent rat T-cell lymphoma line. *Proc Natl Acad Sci USA* 1991;88:4611–4615.
744. Tsichlis PN, Coffin JM. Recombinants between endogenous and exogenous avian tumor viruses: role of the c region and other portions of the genome in the control of replication and transformation. *J Virol* 1980;33:238–249.
745. Tsichlis PN, Coffin JM. Recombinants between endogenous and exogenous avian tumor viruses: role of the C region and other portions of the genome in the control of replication and transformation. *J Virol* 1980;33:238–249.
746. Tsichlis PN, Lazo PA. Virus-host interactions and the pathogenesis of murine and human oncogenic retroviruses. *Curr Top Microbiol Immunol* 1991;171:95–179.
747. Tsichlis PN, Strauss PG, Lohse MA. Concerted DNA rearrangements in Moloney murine leukemia virus-induced thymomas: a potential synergistic relationship in oncogenesis. *J Virol* 1985;56:258–267.
748. Tsuchiya H, Fujii M, Niki T, Tokuhara M, Matsui M, Seiki M. Human T-cell leukemia virus type 1 tax activates transcription of the human fra-I gene through mutliple cis elements responsive to transmembrane signals. *J Virol* 1993;67:7001–7007.
749. Tsukiyama T, Niwa O, Yokoro K. Mechanism of suppression of the long terminal repeat of Moloney leukemia virus in mouse embryonal carcinoma cells. *Mol Cell Biol* 1989;9:4670–4676.
750. Tsukiyama T, Niwa O, Yokoro K. Characterization of the negative regulatory element of the 5′ noncoding region of Moloney murine leukemia virus in mouse embryonal carcinoma cells. *Virology* 1990;177:772–776.
751. Tsukiyami T, Niwa O, Yokoro K. Analysis of the binding proteins and activity of the long terminal repeat of Moloney murine leukemia virus during differentiation of mouse embryonal carcinoma cells. *J Virol* 1991;65:2979–2986.
752. Tucker SP, Srinivas RV, Compans RW. Molecular domains involved in oligomerization of the Friend murine leukemia virus envelope glycoprotein. *Virology* 1991;185:710–720.
753. Twu J-S, Chu K, Robinson WS. Hepatitis B virus X gene activated κB-like enhancer sequences in the long terminal repeat of human immunodeficiency virus 1. *Proc Natl Acad Sci USA* 1989;86:5168–5172.
754. Twu J-S, Robinson WS. Hepatitis B virus X gene can transactivate heterologous viral sequences. *Proc Natl Acad Sci USA* 1989;86:2046–2050.
755. Unger RE, Stout MW, Luciw PA. Simian immunodeficiency virus (SIVmac) exhibits complex splicing for *tat*, *rev* and *env* mRNA. *Virology* 1991;182:177–185.
756. Van Beveren C, Coffin JM, Hughes S. Appendixes. In: Weiss R, Teich N, Varmus H, Coffin J, eds. *RNA tumor viruses*. Cold Spring Harbor, NY: Cold Spring Harbor Laboratory, 1985;559–1222.
757. van den Berg CW, Demant P, Aerts PC, Van Dijk H. Slp is an essential component of an EDTA-resistant activation pathway of mouse complement. *Proc Natl Acad Sci USA* 1992;89:10711–10715.
758. van Gent DC, Oude Groeneger AAM, Plasterk RHA. Mutational analysis of the integrase protein of human immunodeficiency virus type 2. *Proc Natl Acad Sci USA* 1992;89:9598–9602.
759. van Gent DC, Oude Groeneger AAM, Plasterk RHA. Identification of amino acids in HIV-2 integrase involved in site-specific hydrolysis and alcoholysis of viral DNA termini. *Nucl Acids Res* 1993;21:3373–3377.
760. van Gent DC, Vink C, Oude Groeneger AAM, Plasterk RHA. Complementation between HIV integrase proteins mutated in different domains. *EMBO J* 1993;12:3261–3267.
761. Van Lint C, Burny A, Verdin E. The intragenic enhancer of human immunodeficiency virus type 1 contains functional AP-1 binding sites. *J Virol* 1991;65:7066–7072.
762. Van Lint C, Ghysdael J, Paras PJ, Burny A, Verdin E. A transcriptional regulatory element is associated with a nuclease-hypersensitive site in the *pol* gene of human immunodeficiency virus type 1. *J Virol* 1994;68:2632–2648.
763. van Lohuizen M, Berns A. Tumorigenesis by slow-transforming retroviruses-an update. *Biochim Biophys Acta* 1990;1032:213–215.
764. van Lohuizen M, Verbeek S, Krimpenfort P, Domen J, Saris C, Radaszkiewicz T, Berns A. Predisposition to lymphomagenesis in pim-1 transgenic mice: cooperation with c-myc and N-myc in murine leukemia virus-induced tumors. *Cell* 1989;56:673–682.
765. van Zeijl M, Johann SV, Closs E, Cunningham J, Eddy R, Shows TB, O'Hara B. A human amphotropic retrovirus receptor is a second member of the gibbon ape leukemia virus receptor family. *Proc Natl Acad Sci USA* 1994;91:1168–1172.
766. Varela-Echavarria A, Garvey N, Preston BD, Dougherty JP. Comparison of Moloney murine leukemia virus mutation rate with the fidelity of its reverse transcriptase in vitro. *J Biol Chem* 1992;267: 24681–24688.
767. Varmus H, Brown P. Retroviruses. In: Howe M, Berg D, eds. *Mobile DNA*. Washington: ASM, 1989;53–108.
768. Varmus HE, Quintrell NE, Ortiz S. Retroviruses as mutagens: insertion and excision of a non-transforming provirus alters expression of a resident transforming provirus. *Cell* 1981;25:23–26.
769. Vartanian J-P, Meyerhans A, Asjo B, Wain-Hobson S. Selection, recombination, and G—A hypermutation of human immunodeficiency

virus type 1 genomes. *J Virol* 1991;65:1779–1788.
770. Verdin E, Becker N, Bex F, Droogmans L, Burny A. Identification and characterization of an enhancer in the coding region of the genome of human immunodeficiency virus type 1. *Proc Natl Acad Sci USA* 1990;87:4874–4878.
771. Verdin E, Paras P, Vanlint C. Chromatin disruption in the promoter of human immunodeficiency virus type 1 during transcriptional activation. *EMBO J* 1993;12:4900.
772. Vijaya S, Steffen DL, Robinson HL. Acceptor sites for retroviral integrations map near DNase I-hypersensitive sites in chromatin. *J Virol* 1986;60:683–692.
773. Vile RG, Ali M, Hunter E, McClure MO. Identification of a generalized packaging sequence for D-type retroviruses and generation of a D-type retroviral vector. *Virology* 1992;189:786–791.
774. Vink C, van Gent DC, Elgersma Y, Plasterk RHA. Human immunodeficiency virus integrase protein requires a subterminal position of its viral DNA recognition sequence for efficient cleavage. *J Virol* 1991;65:4636–4644.
775. Vink C, Yeheskiely E, van der Marel GA, van Boom JH, Plasterk RA. Site-specific hydrolysis and alcoholysis of human immunodeficiency virus DNA termini mediated by the viral integrase protein. *Nucl Acids Res* 1991;19:6691–6698.
776. Vogel Hinrichs SH, Reynolds RK, Luciw PA, Jay G. The HIV tat gene induces dermal lesions resembling Kaposi's sarcoma in transgenic mice. *Nature* 1988;335:606–611.
777. von Melchner H, DeGregori JV, Rayburn H, Reddy S, Friedel C, Ruley HE. Selective disruption of genes expressed in totipotent embryonal stem cells. *Genes Dev* 1992;6:929–927.
778. von Schwedler U, Kornbluth RS, Trono D. The nuclear localization signal of the matrix protein of human immunodeficiency virus type 1 allows the establishment of infection in macrophages and quiescent T lyphocytes. *Proc Natl Acad Sci USA* 1994;91:6992–6996.
779. von Schwedler U, Song J, Aiken C, Trono D. *vif* is crucial for human immunodeficiency virus type 1 proviral DNA synthesis in infected cells. *J Virol* 1993;67:4945–4955.
780. Voynow SL, Coffin JM. Evolutionary variants of Rous sarcoma virus: large deletion mutants do not result from homologous recombination. *J Virol* 1985;55:67–78.
781. Wagaman PC, Hasselkus-Light CS, Henson M, Lerner DL, Phillips TR, Elder JH. Molecular cloning and charaterization of deoxyuridine triphophatase from feline immunodeficiency virus (FIV). *Virology* 1993;196:451–457.
782. Wang H, Kavanaugh MP, North RA, Kabat D. Cell-surface receptor for ecotropic murine retroviruses is a basic amino-acid transporter. *Nature* 1991;352:729–731.
783. Watanabe S, Temin HM. Encapsidation sequences for spleen necrosis virus, an avian retrovirus, are between the 5′ long terminal repeat and the start of the gag gene. *Proc Natl Acad Sci USA* 1982;79:5986–5990.
784. Weber IT, Miller M, Jaskolski M, Leis J, Skalka AM, Wlodawer A. Molecular modeling of the HIV-1 protease and its substrate binding site. *Science* 1989;243:928–931.
785. Weichs an der Glon C, Ashe M, Eggermont J, Proudfoot MJ. Tat-dependent occlusion of the HIV poly(A) site. *EMBO J* 1993;12:2119–2128.
786. Weichs an der Glon C, Monks J, Proudfoot NJ. Occlusion of the HIV pol(A) site. *Genes Dev* 1991;5:244–253.
787. Weiher H, Barklis E, Ostertag W, Jaenisch R. Two distinct sequence elements mediate retroviral gene expression in embryonal carcinoma cells. *J Virol* 1987;61:218–225.
788. Weiss CD, Levy JA, White JM. Oligomeric organization of gp120 on infectious human immunodeficiency virus type 1 particles. *J Virol* 1990;64:5674–5677.
789. Weiss R, Teich N, Varmus H, Coffin J. *RNA Tumor viruses*. Cold Spring Harbor Laboratory, Cold Spring Harbor, N.Y.
790. Weiss R, Teich N, Varmus H, Coffin J. *RNA Tumor viruses*. Cold Spring Harbor Laboratory, Cold Spring Harbor, N.Y.
791. Weiss RA. Experimental biology and assay of retroviruses. In: Weiss R, Teich N, Varmus H, Coffin J, eds. *RNA tumor viruses*. Cold Spring Harbor, NY: Cold Spring Harbor Laboratory, 1982;209–260.
792. Weiss RA. Experimental biology and assay of RNA tumor viruses. In: Weiss R, Teich N, Varmus H, Coffin J, eds. *RNA tumor viruses*. Cold Spring Harbor, NY: Cold Spring Harbor Laboratory, 1982;209–260.
793. Weiss RA. Human T-cell retroviruses. In: Weiss R, Teich N, Varmus H, Coffin J, eds. *RNA tumor viruses*. Cold Spring Harbor, NY: Cold Spring Harbor Laboratory, 1982;405–486.
794. Weiss RA. Cellular receptors and viral glycoproteins involved in retrovirus entry. In: Levy JA, ed. *The Retroviridae*. New York: Plenum Press, 1993;1–72.
795. Weldon RAJ, Wills JW. Characterization of a small (25-kilodalton) derivative of the Rous sarcoma virus gag protein competent for particle release. *J Virol* 1993;67:5550–5561.
796. Weller SK, Temin HM. Cell killing by avian leukosis viruses. *J Virol* 1981;39:713–721.
797. Westendorp MO, Li-Weber M, Frank RW, Krammer PH. Human immunodeficiency virus type 1 tat upregulates interleukin-2 secretion in activated T cells. *J Virol* 1994;68:4177–4185.
798. White JM, Littman DR. Viral receptors of the immunoglobulin superfamily. *Cell* 1989;56:725–728.
799. Wilkenson DA, Mager D, Leong J-A. Endogenous human retroviruses. In: Levy JA, ed. *The Retroviridae*. New York: Plenum Press, 1994;465–536.
800. Willett BJ, Hosie MJ, Jarrett O, Neil JC. Identification of a putative cellular receptor for feline immunodeficiency virus as the feline homologue of CD9. *Immunology* 1994;81:228–233.
801. Williams JL, Cann AJ, Leff T, P. S-C, Chen ISY. Studies of heterologous promoter trans-activation by the HTLV-II *tax* protein.*Nucl Acids Res* 1989;17:5737–5750.
802. Williams KJ, Loeb LA. Retroviral reverse transcriptases: error frequencies and mutagenesis. *Curr Topics Microbiol Immunol* 1992;176:165–180.
803. Wills JW, Craven RC. Form, function and use of retroviral gag proteins. *AIDS* 1991;5:639–654.
804. Wills NM, Gesteland RF, Atkins JF. Evidence that a downstream pseudoknot is required for translational read-through of the Moloney murine leukemia virus *gag* stop codon. *Proc Natl Acad Sci USA* 1991;88:6991–6995.
805. Wilson CA, Eiden MV. Viral and cellular factors governing hamster cell infection by murine and gibbon ape leukemia viruses. *J Virol* 1991;66:5975–5982.
806. Winslow GM, Scherer MT, Kappler JW, Marrack P. Detection and biochemical characterization of the mouse mammary tumor virus 7 superantigen (Mls-1[a]). *Cell* 1992;71:719–730.
807. Withers-Ward ES, Kitamura Y, Barnes JP, Coffin JM. Widespread distribution of targets for retrovirus DNA integration in vivo. *Genes Dev* 1994;8:1473–1487.
808. Wlodawer A, Miller M, Jaskolski M, Sathyanarayana BK, Baldwin E, Weber IT, Selk LM, Clawson L, Schneider J, Kent SBH. Conserved folding in retroviral proteases: crystal structure of a synthetic HIV-1 protease. *Science* 1989;245:616–621.
809. Wolff L, Koller R. Regions of the Moloney murine leukemia virus genome specifically related to induction of promonocytic tumors. *J Virol* 1990;64:155–160.
810. Wolff L, Ruscetti S. The spleen focus-forming virus (SFFV) envelope gene, when introduced into mice in the absence of other SFFV genes, induces acute erythroleukemia. *J Virol* 1988;62:2158–2163.
811. Wyatt R, Thali M, Tilley S, Pinter A, Posner M, Ho D, Robinson J, Sodroski J. Relationship of the human immunodeficiency virus type 1 gp120 third variable loop to a component of the CD4 binding site in the fourth conserved region. *J Virol* 1992;66:6997–7004.
812. Xiong Y, Eickbush TH. Origin and evolution of retroelements based on their reverse transcriptase sequences. *EMBO J* 1990;9:3353–3362.
813. Yadav PN, Yadav JS, Arnold E, Modak MJ. A computer-assisted analysis of conserved residues in the three-dimensional structures of the polymerase domains of Escherichia coli DNA polymerase I and HIV-1 reverse transcriptase. *J Biol Chem* 1994;269:716–720.
814. Yanagawa S-i, Kakimi K, Tanaka H, Murakami A, Nakagawa Y, Kubo Y, Yamada Y, Hiai H, Kuribayashi K, Masuda T, Ishimoto A. Mouse mammary tumor virus with rearranged long terminal repeats causes murine lymphomas. *J Virol* 1993;67:112–118.
815. Yang Y, Tojo A, Watanabe N, Amanuma H. Oligomerization of Friend spleen focus-forming virus (SFFV) env glycoproteins. *Virology* 1990;177:312–316.
816. Yazdanbakhsh K, Park CG, Winslow GM, Choi Y. Direct evidence for the role of COOH terminus of mouse mammary tumor virus superantigen in determining T cell receptor Vb specificity. *J Exp Med* 1993;178:737–741.
817. York DF, Vigne R, Verwoerd DW, Querat G. Nucleotide sequence of the Jaagsiekte retrovirus, an exogenous and endogenous type D and B

retrovirus of sheep and goats. *J Virol* 1992;66:4930–4939.
818. Yoshinaka Y, Katoh I, Copeland TD, Oroszlan SJ. Murine leukemia virus protease is encoded by the gag-pol gene and is synthesized through suppression of an amber termination codon. *Proc Natl Acad Sci USA* 1985;82:1618–1622.
819. Young JAT, Bates P, Varmus HE. Isolation of a chicken gene that confers suscpetibility to infection by subgroup A avian leukosis and sarcoma viruses. *J Virol* 1993;67:1811–1816.
820. Young JAT, Bates P, Willert K, Varmus HE. Efficient incorporation of human CD4 protein into avian leukosis virus particles. *Science* 1990;250:1421–1423.
821. Yu X, Yuan X, Matsuda Z, Lee T-H, Essex M. The matrix protein of human immunodeficiency virus type 1 is required for incorporation of viral envelope protein into mature virions. *J Virol* 1992;66:4966–4971.
822. Yu X, Yuan X, McLane MF, Lee T-H, Essex M. Mutations in the cytoplasmic domain of human immunodeficiency virus type 1 transmembrane protein impair the incorporation of env proteins into mature virions. *J Virol* 1993;67:213–221.
823. Zachow KR, Conklin KF. CArG, CCAAT, and CCAAT-like protein binding sites in avian retrovirus long terminal repeat enhancers. *J Virol* 1992;66:1959–1970.
824. Zhang J, Temin HM. Rate and mechanism of nonhomologous recombination during a single cycle of retroviral replication. *Science* 1993;259:234–238.
825. Zhang J, Temin HM. Retrovirus recombination depends on the length of sequence identity and is not error prone. *J Virol* 1994;68: 2409–2414.
826. Zhao L-J, Giam C-Z. Interaction of the human T-cell lymphotropic virus type I (HTLV-I) transcriptional activator tax with cellular factors that bind specifically to the 21-base pair repeats in the HTLV-I enhancer. *Proc Natl Acad Sci USA* 1991;88:11445–11449.
827. Zhao L-J, Giam C-Z. Human T-cell lymphotropic virus type I (HTLV-1) transcriptional activator, Tax, enhances CREB binding to HTLV-I 21-base-pair repeats by protein-protein interaction. *Proc Natl Acad Sci USA* 1992;89:7070–7074.
828. Zhou W, Parent LJ, Wills JW, Resh MD. Identification of a membrane-binding domain within the amino-terminal region of human immunodeficiency virus type 1 gag protein which interacts with acidic phospholipids. *J Virol* 1994;68:2556–2569.
829. Zhu J, Cunningham JM. Minus-strand DNA is present within murine type C ecotropic retroviruses prior to infection. *J Virol* 1993;67:2385–2388.
830. Zimmermann K, Dobrovnik M, Ballaun C, Bevec D, Hauber J, Bohnlein E. *trans*-activation of the HIV-1 LTR by the HIV-1 tat and HTLV-1 tax proteins is mediated by different *cis*-acting sequences. *Virology* 1991;182:874–878.
831. Zingler K, Littmann DR. Truncation of the cytoplasmic domain of the simian immunodeficiency virus envelope glycoprotein increases env incorporation into particles and fusogenicity and infectivity. *J Virol* 1993;67:2824–2831.
832. Zon LI, Moreau J-F, Koo J-W, Mathey-Prevot B, D'Andrea AD. The erythropoietin receptor transmembrane region is necessary for activation by the Friend spleen focus-forming virus gp55 glycoprotein. *Mol Cell Biol* 1992;12:2949–2957.

Fundamental Virology, Third Edition
edited by B.N. Fields, D.M. Knipe, P.M. Howley, et al.
Lippincott - Raven Publishers, Philadelphia © 1996

CHAPTER 27

Human Immunodeficiency Viruses and Their Replication

Paul A. Luciw

DISCOVERY OF PRIMATE LENTIVIRUSES

Knowledge gained over the last century has established that retroviruses cause a wide variety of diseases in many avian and mammalian species, including nonhuman primates (228); however, it was not until 1980 that a retrovirus, human T-cell leukemia virus type I (HTLV-I), was discovered and subsequently linked to both neoplasia and neuropathology in humans (see Chapter 59 in *Fields Virology*, 3rd ed.) (580,816). The epidemic of acquired immunodeficiency syndrome (AIDS), first recognized as a clinical entity in 1981, set the stage for the discovery of yet another human retrovirus (90,248). Hallmarks of this fatal disease are immunologic abnormalities, which are often accompanied by opportunistic infections, neurologic disorders, and unusual forms of cancer (see Chapter 61 in *Fields Virology*, 3rd ed.) (70,414). Epidemiologic studies implicated an infectious agent that was transmitted during sexual intercourse, through intravenous drug abuse, by therapies utilizing blood and blood products, and vertically from mother to child (339).

In 1983, Barre-Sinoussi, Chermann, and Montagnier (at the Pasteur Institute in Paris) isolated a retrovirus from lymph node cells of a patient with lymphadenopathy; accordingly, this virus was designated lymphadenopathy-associated virus (LAV) (40a,772). Lymphadenopathy-associated virus was shown to replicate and cause cytopathology in cultures of human peripheral blood lymphocytes. A year after the report on LAV, Gallo and co-workers (at the National Institutes of Health, Bethesda, MD) described the isolation of a cytopathic T-lymphotropic retrovirus, designated human T-lymphotropic virus type III (HTLV-III), from peripheral blood lymphocytes of patients with acquired immunodeficiency syndrome (AIDS) (221). Although this HTLV-III isolate was later shown to be a con-

P. A. Luciw: Department of Medical Pathology, School of Medicine, University of California, Davis, California 95616.

taminant of virus from the Pasteur Institute (72,97,772), subsequent isolates of HTLV-III were unique (262). Also, in 1984, Levy and co-workers (at the University of California, San Francisco) cultured an AIDS-associated retrovirus (ARV) from peripheral blood mononuclear cells of an AIDS patient (416). Molecular cloning and sequence analysis revealed that the genomes of LAV (771) and HTLV-III (511,599) were nearly identical, exhibiting 1% to 2% divergence. In contrast, the genome of the ARV-2 isolate displayed up to 15% differences in nucleotide sequences in pairwise alignments with the genomes of either LAV or HTLV-III (595,634). Electron microscopy demonstrated that these viruses from AIDS patients were morphologically similar to members of the lentivirus genus of the family *Retroviridae* (Fig. 1). (246). Furthermore, comparison of genome sequences also supported the notion that the human retroviruses associated with AIDS were related, albeit distantly, to lentiviruses of other animals (Fig. 2). (113,694)

The discovery of additional distinct lentiviruses in nonhuman primates as well as in humans has provided important insight into the biologic significance and evolutionary relationships of these viruses (229,292). In 1985, at the New England Regional Primate Research Center (Southboro, MA), a lentivirus was isolated from captive Asian macaques with an AIDS-like disease (141). Because of morphologic similarity (Fig. 1) and serologic cross-reactivity with HTLV-III, this macaque virus was designated simian T-lymphotropic virus type III (STLV-III). The finding that STLV-III caused fatal immunodeficiency in experimentally infected macaques strongly supported the idea that human lentiviruses are causative agents of AIDS (409). Additional lentiviruses have since been isolated from several monkey species in the wild throughout several regions in Africa (292); in the indigenous or natural host, these viruses do not produce disease (229). Importantly, a unique lentivirus discovered by Clavel, Montagnier, and their co-workers in 1986 has been linked to AIDS in West Africa (110,259). The designations human immunodeficiency virus (HIV) and simian immunodeficiency virus (SIV) have been adopted for the lentiviruses of humans and nonhuman primates, respectively (119). Human immunodeficiency virus type 1 refers to the genetically related viruses found in several regions of Africa, Asia, Europe, and both North and South America (345,570); HIV-2 is the distinct virus prevalent in certain West African countries (594). Although both of these viruses cause AIDS, individuals infected with HIV-2 exhibit a longer period of clinical latency and lower morbidity (187). The importance of the nonhuman primate lentiviruses is underscored by the fact that several HIV-2 isolates of West African origin are nearly indistinguishable at the nucleotide sequence level from certain strains of SIV (225).

SCOPE OF THIS CHAPTER

This chapter reviews the molecular biology of primate lentiviruses by focusing on the genetic organization of their genomes, the structure and function of their gene products, and the mechanisms regulating viral replication in cell culture. An objective of this chapter is to provide a basis for investigations aimed at elucidating mechanisms of HIV-1 and SIV pathogenesis on a molecular level. In addition, an

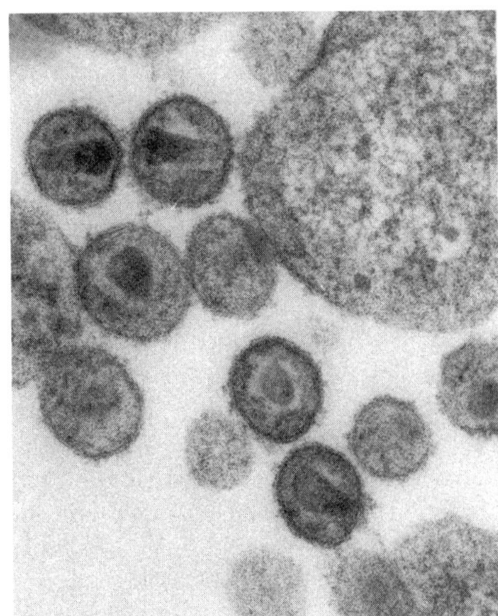
A

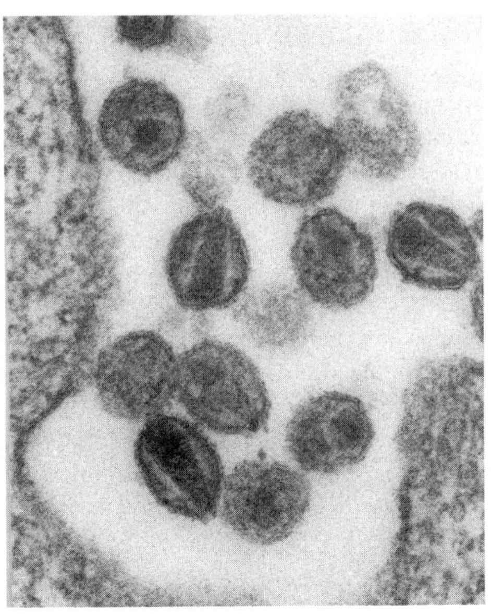
B

FIG. 1. Ultrastructure of primate lentiviruses. Electron microscopy of extracellular particles of HIV-1 (**A**) and SIV_{MAC} (**B**) reveals virions, about 110 nm in diameter, with a cone-shaped nucleoid surrounded by a lipid bilayer membrane, which contains envelope glycoprotein spikes (×100,000).

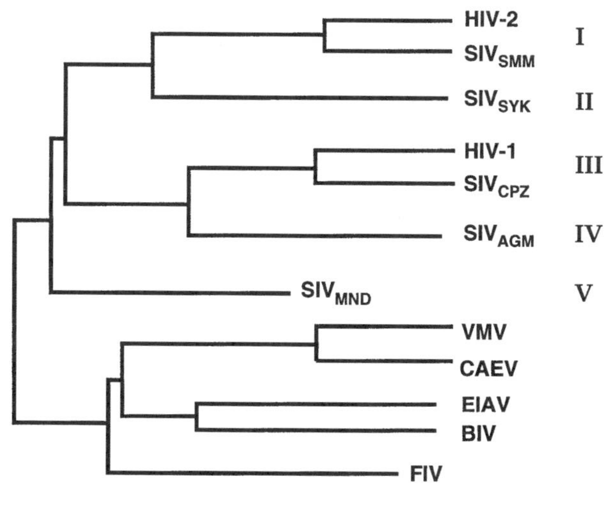

FIG. 2. Phylogenetic relationships of lentiviruses.Representative lentiviruses are compared using *pol* gene nucleotide sequences for establishing phylogenetic relationships. Five groups of primate lentiviruses are shown: HIV-1, HIV-2, SIV from sooty mangabey monkey (SIV_{SMM}), Sykes monkey (SIV_{SYK}), chimpanzee (SIV_{CPZ}), African green monkey (SIV_{AGM}), and mandrill (SIV_{MND}). Nonprimate lentiviruses are VMV, CAEV, EIAV, BIV, and FIV. The scale indicates percentage difference innucleotide sequences in the *pol* gene. The branching order of the primate lentiviruses is controversial. Adapted from Myers and Pavlakis (517).

understanding of viral replication mechanisms, coupled with information on the structure of virions and virion subunits, is essential for developing antiviral drugs and vaccines that block infection and/or prevent or delay manifestation of disease in infected individuals. Because an animal model to analyze both HIV-1 infection and disease is not yet available, infections of nonhuman primates with selected strains of SIV and HIV-2 are being exploited for identifying viral determinants of pathogenesis and for evaluating antiviral chemotherapies and immunization protocols (229). Accordingly, this review also encompasses many aspects of the molecular biology of SIV and HIV-2.

Several thousand papers have been published on the replication of primate lentiviruses; therefore, citations are generally limited to original papers, representative papers published in the last 3 to 4 years, and review articles on specific topics. Mechanisms of replication of the diverse members of the *Retroviridae* are described by Coffin in Chapter 26, and clinical features of HIV infection are presented by Hirsch and Curran in Chapter 61 in *Fields Virology*, 3rd ed.; overlap with these latter two chapters has been minimized. This chapter integrates knowledge of HIV and SIV replication with information on other retroviruses and discusses the implications of the molecular biology of these viruses for elucidating the virus-host relationship. Additional material on the molecular biology of both HIV and SIV can be found in the volumes edited by Cullen (136) and Morrow and Haigwood (508). The series entitled *Retroviridae* also reviews lentiviruses in other animal species including feline immunodeficiency virus (FIV) in cats, visna-maedi virus (VMV) in sheep, caprine arthritis-encephalitis virus (CAEV) in goats, bovine immunodeficiency-like virus (BIV) in cattle, and equine infectious anemia virus (EIAV) in horses (413–415). Although basic research has provided a wealth of information on the replication of retroviruses, this chapter will point out many gaps in knowledge that serve as opportunities for future discoveries on HIV and SIV.

CLASSIFICATION OF PRIMATE LENTIVIRUSES

Both HIV and SIV are genetically related members of the lentivirus genus of the *Retroviridae* family. Lentivirus isolates from humans are grouped into one of two types, designated HIV-1 and HIV-2, on the basis of serologic properties and sequence analysis of molecularly cloned viral genomes (517). Independent isolates of each type display the greatest sequence variation in the *env* gene, which encodes the glycoprotein in the virion membrane (519). A classification scheme, based on *env* gene sequences, recognizes nine subtypes (clades) of HIV-1 (A through I) (518, 792); HIV-2 isolates have been classified into five subtypes (A through E) (225). Accordingly, viral diversification, or speciation, is a feature of HIV-1 and HIV-2 phylogeny. Within each subtype, there is a high degree of variability (i.e., intrasubtype diversification). Viral speciation is a major concern because of implications for antiviral vaccine development and the possibility for increased virulence. Although mutation appears to be the major factor responsible for viral variation, recombination has been postulated to occur in individuals infected with viruses from different clades of HIV-1 (430) and HIV-2 (225). In addition, viral trafficking, i.e., the "transfer of viruses to new host populations" (509), is an ecologic driving force of the AIDS pandemic. Some areas of the world harbor predominantly a single subtype, whereas two or more subtypes may be prevalent in certain other populations (430). Molecular epidemiology surveys indicate that the pattern of global variation and distribution is due to viral migration rather than to viral mutation (518).

Genetically distinct lentiviruses have been obtained from several nonhuman primate species including African green monkeys (SIV_{AGM}), sooty mangabeys (SIV_{SMM}), mandrills (SIV_{MND}), and sykes (SIV_{SYK}) (300). In addition, a lentivirus designated SIV_{CPZ} has been found in chimpanzees (569). These unique strains of SIV are endemic to the respective African monkey species and appear not to produce disease in the natural host (229). Sequence analysis reveals that the genomes of SIV_{SMM} and HIV-2 exhibit a high degree of homology (290), and SIV_{CPZ} is most closely related to HIV-1 (317). Phylogenetic relationships based on sequences in the conserved *pol* gene of these primate lentiviruses are illustrated in Fig. 2 (292). Simian immunodeficiency virus has also been isolated from several Asian macaque species held in captivity in primate research facilities in the United States; these viruses include SIV from rhesus macaques (SIV_{MAC}), nemestrina macaques (SIV_{MNE}), and stump-tailed macaques (SIV_{STM}) (229). Each of these viruses causes a fatal AIDS-like disease in several macaque species (390). Because the various macaque viruses are all closely related to SIV_{SMM}, it appears that captive macaques were inadvertently infected either by cohousing Asian macaques and African monkeys (i.e., sooty mangabeys) and/or through experimental inoculation of macaques with fluids and other materials from African monkeys. Despite extensive sequence diversity, a unifying feature of human and nonhuman primate lentiviruses is that the cell receptor is the CD4 antigen, a differentiation marker on the surface of T-helper lymphocytes (784). A compendium of HIV and SIV sequence information, based on extensive analysis of numerous viral clones, is found in the database prepared by Myers et al. (519). At present, the origins of HIV-1 and HIV-2 infections remain an enigma, although it is possible that both of these viruses arose from zoonotic transmissions between nonhuman primates and humans (188).

VIRAL GENOME STRUCTURE AND ORGANIZATION

Infectious virions of HIV and SIV contain two identical copies of single-stranded RNA, about 9.2 kb long, that have positive polarity with respect to translation. In the early stages of infection, the virion RNA genome is converted into double-stranded linear DNA by the process of reverse transcription [via viral-encoded reverse transcriptase (RT)], which involves two strand-transfer steps to synthesize linear viral DNA with long terminal repeats (LTRs) flanking viral genes (Fig. 3). This linear viral DNA is integrated into the host cell genome to produce the provirus. Accordingly, HIV and SIV, like other retroviruses, have two genomic forms: single-stranded RNA in the extracellular phase of the viral life cycle (i.e., virions) and double-stranded DNA (i.e., provirus) within the cell. Genomic viral RNA is synthesized by cellular RNA polymerase II from proviral DNA and thus contains a cap structure at the 5′ end and a poly-A tail at the 3′ end. Generic features of both the RNA and DNA forms of retroviral genomes are described in detail in Chapter 26 in this volume.

Both HIV and SIV encode precursor polypeptides for virion proteins as well as several additional open reading frames (Fig. 2 ; and Table 1). The *gag* gene encodes the precursor for virion capsid proteins, the *pol* gene encodes the precursor for several virion enzymes [protease (PR), RT, RNase H, and integrase (IN)], and the *env* gene encodes the precursor for envelope glycoprotein (Env gp). The transcriptional transactivator (*tat*) and regulator of viral expression (*rev*) genes are each encoded by two overlapping exons and produce small nonvirion proteins which are essential for viral replication. Both HIV and SIV also encode several genes which are nonessential (i.e., dispensable) for viral replication in tissue culture cells. Nonessential genes, also designated "accessory" or "auxiliary" genes, encoded by HIV-1 are *vif, vpr, vpu,* and *nef*; HIV-2 and SIV encode *vif, vpx* and/or *vpr,* and *nef.* Pro-

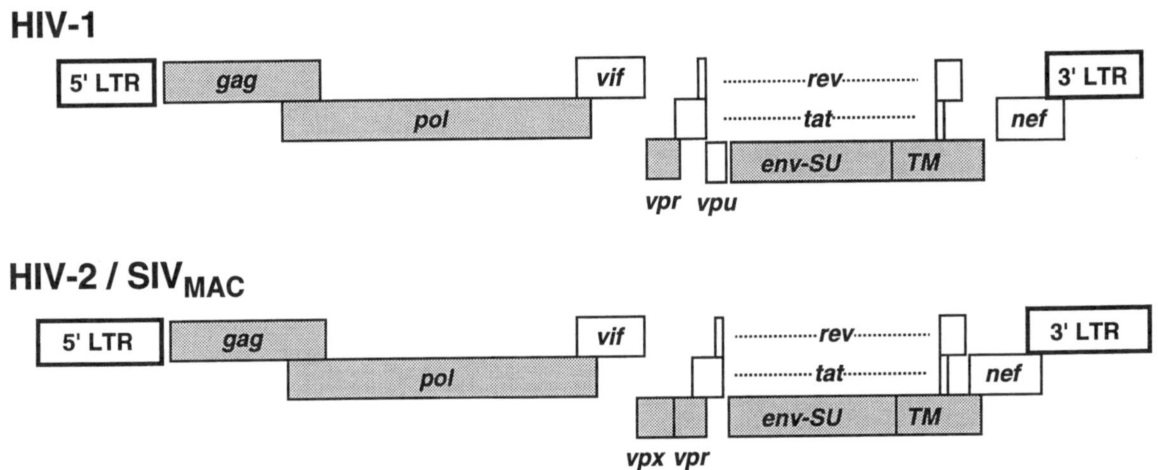

FIG. 3. Genome organization of primate lentiviruses. The linear double-stranded proviral DNA forms of HIV-1 and HIV-2/SIV_{MAC} show similar patterns of genomic organization. Structural genes (*gag, pol,* and *env*) are heavily shaded. Accessory genes, including essential regulatory genes (*tat* and *rev*), and nonessential genes (*nef, vif, vpu, vpx,* and *vpr*), are lightly shaded. (For viral gene nomenclature, see Table 1.) The 5′ and 3′ LTRs flanking viral genes are shown as open boxes. *Vpu* is found exclusively in HIV-1, whereas *vpx* is found only in HIV-2 and certain strains of SIV. *Vpr* is encoded by both HIV-1 and HIV-2/SIV_{MAC} (see text).

TABLE 1. *Genes and proteins of primate lentiviruses*

Gene[a]	Dispensable for Replication	Protein	Function	Localization
gag	No	$Pr55^{gag}$	Polyprotein precursor for virion core proteins MA (p17), CA (p24), NC (p9), p7	Virion nucleocapsid
pol	No	$Pr160^{gag\text{-}pol}$	Polyprotein precursor for virion enzymes PR: p10, RT and RNAse-H: p51/66, IN: p32	Virion (nucleocapsid?)
vif	Yes[b]	p23	Viral infectivity factor, function unresolved	Cell cytoplasm
vpx[c]	Yes	p16	Virion protein, function unresolved	Virion
vpr	Yes	p15	Virion protein, function unresolved	Virion
tat	No	p14	Transcriptional transactivator, binds TAR and cell factor(s) (initiation and elongation of viral transcripts)	Primarily in cell nucleus
rev	No	p19	Posttranscriptional transactivator, binds RRE and cell factor(s) (splicing and/or transport and translation of viral mRNA)	Primarily in cell nucleus
vpu[d]	Yes	p16	Influences virus release, augments turnover of CD4 antigen	Integral cell membrane protein
env	No	gp160	Precursor for envelope glycoprotein: SU (gp120): CD4 receptor binding, TM (gp41): membrane fusion	Virion envelope, plasma membrane
nef	Yes	p27	Negative effector?, downregulates CD4 receptor, influences T-cell activation, enhances virion infectivity	Cell cytoplasm, plasma membrane

[a] Gene nomenclature is based on that in Gallo et al. (222).
[b] Dependent on cell type.
[c] Encoded only by HIV-2 and several SIV strains (see text).
[d] Encoded only by HIV-1 and SIV_{CPZ}.
MA, matrix protein; CA, capsid protein; NC, nucleocapsid protein; PR, protease; RT, reverse transcriptase; IN, integrase; SU, surface glycoprotein; TM, transmembrane protein; TAR, tat-response element; RRE, rev-response element; HIV, human immunodeficiency virus; SIV, simian immunodeficiency virus.

teins produced from the *vpr* and *vpx* genes are assembled into virions, whereas the *vif, vpu,* and *nef* gene products appear not to be packaged into virions.

As a group, the proviral genomes of primate lentiviruses display a high content of adenosine deoxyribonucleotide (A) residues (38% to 39%) and a low frequency of cytidine deoxyribonucleotide (C) residues (16% to 19%). The preference for A residues is a feature of all lentiviruses (517). The generation of a A-rich viral genome may be due to an enzymatic property of RT (see section on reverse transcriptase) and/or selective pressure during evolution of the virus group. For example, the strong bias in the *env* gene for the triplet AAT and related codons favors serine, threonine, and asparagine; this bias appears to lead to the creation of new N-linked glycosylation sites, which could enable the virus to escape from the host immune response (62). A consequence of the high frequency of A-rich triplets is that HIV and SIV codon usage differs dramatically from that of cellular genes (52,388).

VIRION STRUCTURE

Models for the structure of HIV and SIV virions are based on a combination of high-resolution electron microscopy of viral particles and both biochemical and immunochemical analyses of virion components (233,514). By these approaches, extracellular particles produced by cells infected with HIV-1, HIV-2, and the various SIV strains are very similar in morphology and composition. Virions have a spherical shape, are about 110 nm in diameter, and consist of a lipid bilayer membrane or envelope that surrounds the cone-shaped nucleocapsid (Figs. 1 and 4). Although the overall shape of virions is spherical, computer simulation of shadowed replicas and photomicrographs produced by scanning electron microscopy suggest that virions are icosadeltahedrons (465,527). Each nucleocapsid is about 100 nm in the long dimension and spans the entire diameter of the virion. Analysis of electron micrographs by computer-imaging techniques reveals a nucleocapsid with a wide free end, 40 to 60 nm across, and a narrow end, about 20 nm in width. This narrow end of the nucleocapsid appears to be connected to the lipid bilayer through a proteinaceous structure designated the core (capsid)-envelope link (CEL) (298). The region between the viral envelope and the electron-dense nucleocapsid has been termed the paranucleoid region, core shell, or lateral body (823). The composition of both the CEL and the paranucleoid region remains to be determined. Although spherical virions with one nucleocapsid are the major forms released from infected cells, a small percentage of extracellular particles contains two or more spherical capsids, and tubular capsids, 40 to 200 nm long, have also been ob-

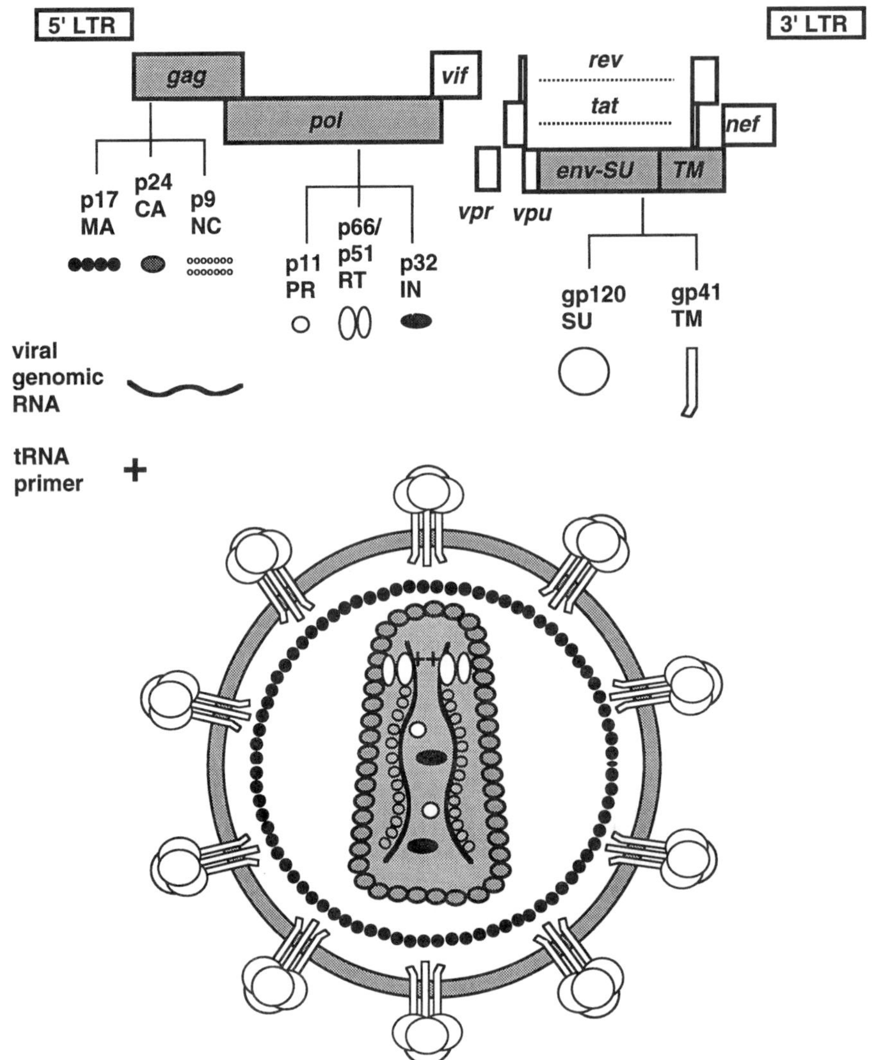

FIG. 4. Virion structure. Precursor polyproteins Gag-Pol (Pr160), Gag (Pr55), and Env (gp160) are enzymatically processed to yield mature virion proteins. Gag-Pol and Gag undergo several cleavage steps mediated by the viral aspartic PR to produce eight smaller proteins (see text). Env is cleaved once by a cellular PR, producing the SU gp120 and TM gp41. Typical lentivirus particles are spherical, about 110 nm in diameter, and consist of a lipid bilayer membrane surrounding a conical nucleocapsid (see text). The top portion of the figure shows the HIV-1 genome, with virion structural genes heavily shaded and accessory genes lightly shaded. Symbols representing the various virion proteins are indicated. Env gp are represented as trimers (see text). Exact positions of the proteins PR, RT, and IN in the viral core have not been elucidated. Of the accessory proteins encoded by HIV and SIV, only Vpr and Vpx are packaged into virions; the precise locations of these two proteins in virus particles have not been determined. Neither the other accessory proteins (Vif, Vpu, and Nef) nor the regulatory proteins (Tat, Rev) have been detected in virion particles.

served (233). The significance of these apparently aberrant forms is not known.

Biochemical and immunochemical analysis has demonstrated that the nucleocapsid within each mature virion is composed of two molecules of the viral single-stranded RNA genome encapsulated by proteins proteolytically processed from the Gag precursor polypeptide (Fig. 4) (see Chapter 26). These *gag* gene products are the matrix protein (MA), which is presumably located between the nucleocapsid and the virion envelope (membrane); the major capsid protein (CA), which forms the capsid shell (5 nm thick); and the nucleocapsid protein (NC), which binds tightly to the viral RNA genome. The 5′ ends of the two molecules of genomic single-stranded viral RNA (~9.2 kb) in the virion appear to interact with each other either through hydrogen bonds between molecules aligned in the same polarity or through short antiparallel arrangements (462); NC protein may also link the two viral genomes

(627). A transfer RNA^{lys} molecule is positioned near the 5′ end of each genomic RNA strand and serves as the primer for initiation of negative strand viral DNA synthesis by RT. A very small proportion of extracellular virus particles also contains viral DNA produced by incomplete reverse transcription which takes place after release of virions from cells (746). Viral enzymes derived from the *pol* gene precursor polypeptide are also packaged into virions; these enzymes are PR, RT, and IN. Reverse transcriptase is a heterodimer composed of p51 and p66 subunits (see section on reverse transcriptase). An additional enzymatic activity, ribonuclease specific to RNA in RNA-DNA hybrids (RNase H), is an independent domain in the RT heterodimer. Reverse transcriptase (and RNase H) and IN are presumably contained within the nucleocapsid in mature virions. In addition, Vpr (and Vpx for SIV and HIV-2) are small proteins associated with nucleocapsids; the precise relationship of these two accessory proteins with other capsid proteins is not well defined (see section on viral accessory proteins).

The membrane (or envelope) of extracellular HIV and SIV particles contains approximately 72 knobs (or spikes) that show triangular symmetry (233,552) (Figs. 1 and 4). Each knob is 9 to 10 nm long and has an ovoid distal end, 14 to 15 nm in diameter, linked to the lipid membrane by a 7- to 8-nm stalk. The knob is thought to contain four heterodimers of the Env gp, and each heterodimer is composed of a surface (SU) subunit that interacts with the transmembrane (TM) subunit through noncovalent bonds (see section on envelope glycoprotein). Human immunodeficiency virus type 1 SU and TM subunits are designated gp120 and gp41, respectively, on the basis of electrophoretic mobility under denaturing conditions in polyacrylamide gels. The counterpart subunits for HIV-2 and SIV Env gp are gp130 and gp41. The SU of the primate lentiviruses is extensively glycosylated ~50 kd of protein backbone and ~70 kd of carbohydrate side groups) and contains a domain that recognizes and binds the cell receptor; this receptor for HIV and SIV is the CD4 antigen, a cell surface protein on T-helper lymphocytes (see section on virion attachment and entry).

Analysis of the HIV-1 membrane reveals a phospholipid profile and fluidity significantly different from those of the plasma membrane of the cell in which virus is grown (8,9). Molar ratios of cholesterol to phospholipid in the viral envelope are about 2.5 times that of the host cell surface membrane. Spin resonance spectroscopy reveals low fluidity and a high degree of order in the viral membrane. Sequestration of specific lipids may occur during the budding process in which viral proteins select specific regions in the cell plasma membrane for release of nucleocapsids. An understanding of the role(s) of virion membrane lipids for both particle stability and entry into cells will be essential for developing lipophilic agents (e.g., the spermicide nonoxynol-9) which disrupt viral membranes and thereby block infectivity (483).

Biochemical and immunochemical analyses have shown that membranes of HIV and SIV particles also contain several cellular proteins acquired during the budding stage of virion assembly as nascent viral particles egress through the cell membrane (see section on virion assembly). These cellular proteins include β_2-microglobulin, the α and β chains of human lymphocyte DR antigen [major histocompatibility complex (MHC) class II], cyclophilins, as well as smaller amounts of other cell proteins (16,210,286, 463,737) (see section on virion assembly). Whether cellular proteins incorporated into virions play roles in viral replication and/or pathogenesis has not been resolved. Additional studies are also essential to establish the precise location of these cell membrane proteins in virions, including their physical relationship to other viral proteins.

VIRION PROTEINS

Capsid (Core) Proteins

The capsid functions to package genomic viral RNA into virions and participates in uncoating and possibly other early steps in the viral replication cycle (see Chapter 26) (318). Cleavage of the *gag* gene polyprotein $Pr55^{gag}$ by the viral coded PR produces the mature capsid proteins MA (p17), CA (p24), p2, NC (p7), p1, and P6 (Fig. 4) (288, 634); this proteolytic processing in infected cells is linked to virion morphogenesis (see section on virion assembly). Accordingly, a major aim of investigations on capsid proteins is to identify structural features (or domains) governing the function of each protein in assembly of intracellular particles and maturation of virions. This section integrates studies on the biochemical, immunochemical, and physical properties of mature capsid proteins with the results of genetic analyses. Expression vectors in genetically engineered microorganisms (e.g., bacterial and yeast cells as well as mammalian and insect cells) have been used to produce each capsid protein as well as the Gag polyprotein for structural studies. *In vitro* (cell-free) systems with defined components have also recently been employed to examine both binding of capsid proteins to viral RNA and assembly of the nucleocapsid. The two-hybrid system developed for analyzing intermolecular protein interactions in yeast cells has been utilized to examine multimerization of individual capsid proteins and interactions with other virion proteins (202). Moreover, cellular proteins that interact with viral proteins (including regulatory and accessory proteins) at all stages of the viral life cycle can be identified in the two-hybrid system. Although studies with expression vectors, virion components, and the yeast two-hybrid system have provided valuable insight into structural features and functions of viral capsid proteins, not all of the results obtained from these approaches have been validated through analysis of replication-competent and mutant viruses in tissue culture cells.

MA

Matrix protein is encoded by sequences at the 5′ end of the *gag* gene and is cleaved from the N-terminus of $Pr55^{gag}$ by viral PR (Fig. 4). Mature MA contains about 130 amino acids, displays a molecular weight of 17 to 18 k, and is located in the matrix between the virion capsid and envelope (Fig. 4) (233). The methionine at the N-terminus of the Gag polyprotein is removed shortly after translation, and myristic acid is covalently attached to the penultimate glycine residue by the host cell enzyme N-myristyl transferase (654). This posttranslational modification is important for targeting the Gag polyprotein to the cell plasma membrane. Mutations altering the penultimate glycine in Pr55gag block myristylation and abolish infectivity (73,249); however, these mutations do not prevent formation of capsids (i.e., viruslike particles, VLPs), within the cell (236) (see section on virion assembly). When fused to heterologous proteins, the first 31 residues of HIV-1 MA target the fusion protein to membranes (831); this region of MA has a high proportion of basic amino acids which are thought to be part of a topogenic signal(s) for membrane association as well as nuclear localization (see below). Matrix protein is also posttranslationally modified by phosphorylation of serine and tyrosine residues (220,288); roles for these modifications of MA in viral replication have not been established.

Mutagenesis studies have implicated additional functions for MA in the viral life cycle. Deletions in the first two-thirds of HIV-1 MA do not affect either processing of the gag polyprotein or virion assembly and release; however, incorporation of Env gp into virions is impaired, and the mutant virus particles are not infectious (168,820). Several of these MA mutants have a dominant negative effect on production of infectious virions in cells, harboring both wild-type and mutant viral genomes (820). Mutations at the carboxyl terminus of MA have no significant effect on the levels of Env gp recruited into particles or on virus replication, and HIV-1 Env gp assembles with hybrid (chimeric) particles that contain HIV-1 MA and the CA and NC proteins of VMV (168). The location of MA in virions between the envelope and the capsid, taken together with the analysis of viral mutants and hybrid Gag precursor polyprotein, implicates an interaction between MA and the TM subunit (gp41) of the Env gp; this interaction may occur during virion morphogenesis (426) (see section on virion assembly). Specific structural features in both the MA protein and gp41 mediating the (potential) interaction of these viral proteins remain to be identified.

Results of genetic studies suggest that MA also functions during early stages of viral infection. Cells transfected with HIV-1 genomes with small deletion mutations at the C-terminus of MA produce noninfectious virions containing processed Gag proteins and Env gp (820); additional investigations will be required to determine the replication step at which these mutant viruses are blocked. Inspection of amino acids in MA of both HIV and SIV reveals a stretch of basic amino acids between Lys18 and Lys32; this sequence resembles the nuclear localization signals (NLSs) of many nucleophilic proteins of viral and cellular origin (206). The corresponding region of MA of SIV and HIV-2 is also enriched in basic amino acids. Human immunodeficiency virus type 1 mutants in MA, produced by substituting Lys26 and Lys27 with threonine residues, infect proliferating cells but not cells arrested in the G2 phase of the cell cycle, whereas wild-type virus replicates efficiently in both growing and resting cells (77,284). These experiments support a model in which MA is part of the viral nucleoprotein complex (i.e., preintegration complex) produced after reverse transcription and mediates nuclear import of this complex; presumably, the basic region between Lys18 and Lys32 is a domain in MA that interacts with a cellular NLS receptor (711).

CA

Capsid protein is released from the central portion of the gag polyprotein by two cleavages mediated by viral PR (Fig. 4). The mature form of CA contains about 240 amino acids, exhibits a molecular weight of 24 to 27 k, has a high degree of hydrophobicity, and is the major subunit of the capsid shell. Although CA of HIV-1 is phosphorylated on a serine residue(s), the role of this posttranslational modification in viral replication has not been established (288). Physicochemical and immunochemical analyses on the structure of CA have been performed to obtain insight into the packing arrangement of protein subunits in the capsid (587); however, high-resolution electron density maps of CA are not yet available. An intriguing observation is that CA, as well as the polyprotein precursor $Pr55^{gag}$, binds the cellular proteins cyclophilins A and B, which are ubiquitous prolyl isomerases and function to mediate the correct assembly of other proteins (436); this interaction appears to be important for particle morphogenesis (see section on virion assembly).

A cell-free system to analyze capsid assembly is based on CA purified from genetically engineered *Escherichia coli* (180). In this system with defined components, CA self-associates to produce dimers and oligomeric complexes as large as dodecamers; alteration of ionic strength and pH conditions reveals that CA also has the capacity to assemble into long rodlike structures which can be disassembled into small irregular spheres. The relevance of these structures to the normal pathway of virion morphogenesis and/or to the uncoating process is not known. Specific interactions of CA with the viral RNA genome, other Gag proteins, as well as posttranslational modifications are presumed to play important roles in CA function during capsid assembly.

Genetic studies with mutations in the HIV-1 gag polyprotein have identified several functional domains in the CA protein. Small deletions in the C-terminal half of CA reduced viral particle formation (169,769); this includes deletions within the major homology region, a stretch of about 29 amino acids that is conserved in the CA proteins of diverse retroviruses (453,801). These mutations appear to affect precursor nucleocapsid assembly rather than particle release because cell-associated virus structures are not detected; in addition, proteolytic processing of the Gag precursor is impaired (169). Although deletions in the N-terminal domain of CA inhibited virus replication, these mutants assembled and released viral particles with the same efficiency as wild-type virus. With the use of the yeast two-hybrid system for detecting protein-protein interactions, sequences in the C-terminal portion of CA were implicated in homomeric multimerization of the Gag precursor (209). Taken together, these findings indicate that a domain governing nucleocapsid assembly is located in the C-terminal portion of CA (see section on virion assembly).

Several *gag* mutants with nonoverlapping mutations in the CA domain complement each other for production of infectious virions, and selected CA mutants blocked in Gag assembly interfere with assembly of wild-type Gag polyproteins (769). Analysis of complementation groups of different *gag* mutations is a means for mapping surfaces of CA involved in capsid formation. Characterization of *trans*-dominant mutants with a negative phenotype may identify CA peptides that inhibit virus assembly (534). Additional genetic studies, together with elucidation of the structure of mature CA protein and the further development of *in vitro* systems for analyzing Gag polyprotein processing and assembly (88), will yield significant insights into the role of CA as well as other domains of Gag in virion morphogenesis. The potential for CA to function in an early step in viral replication, such as uncoating of virions immediately after entry into the cell cytoplasm, requires further exploration.

NC

Protease-mediated cleavage from sequences in the C-terminal portion of the Gag polyprotein produces NC which contains about 70 amino acids and shows a molecular weight of 7 to 9 k (Fig. 4). This basic, hydrophilic protein binds genomic viral RNA in the nucleocapsid; each molecule of NC is estimated to cover 4 to 6 nucleotides (356). Accordingly, NC as part of the Gag polyprotein may function to condense the viral RNA genome for packaging into capsids during virion morphogenesis (see section on virion assembly). Genetic analysis has demonstrated that cells transfected with cloned viral genomes, containing deletion and point mutations within NC, release particles with capsids lacking viral RNA (i.e., VLPs); thus, NC is not reequired for production of VLPs (5,247).

Nucleocapsid protein of HIV and SIV has two copies of a cysteine-histidine motif (each motif represented as Cys-X_2-Cys-X_4-His-Cys) that is similar to the metal-binding finger domains of several proteins that interact with nucleic acids (80). However, this motif may be important for mediating intermolecular interactions between NC molecules during nucleocapsid assembly. Nuclear magnetic resonance spectroscopy reveals that NC has a globular central domain consisting of the two cysteine-histidine motifs and the flanking basic residues; this central domain is surrounded by flexible amino- and carboxy-terminal sequences (504). Zinc is coordinated to the metal finger motifs of NC in mature viral particles (695), and circular dichroism (CD) studies on synthetic peptides corresponding to the cysteine-histidine motifs show that binding of zinc ions induces conformational changes in this protein (719,721). Genetic studies with site-specific and deletion mutations in the viral genome have demonstrated that the first cysteine-histidine motif is required for packaging, whereas the second motif is dispensable (53). In *in vitro* binding studies with NC mutant proteins, one cysteine-histidine motif was found to be essential for efficient and specific interaction with viral RNA. Furthermore, the NC domain of HIV-1 Gag proteins purified from genetically engineered *E. coli* recognizes the major packaging site (*psi*) which is located 5′ from the *gag* gene (53,54,627) (see section on virion assembly). Chemically synthesized NC also binds RNA containing the psi element (142). A predicted stem-loop structure within psi is recognized by NC (627). The interaction of NC with psi on two genomic viral RNA molecules may influence the formation and/or stability of the dimeric form of viral RNA in virions (53, 143,462). Taken together, these studies show that the cysteine-histidine structural feature is important for the following NC functions: (a) interaction with psi on genomic viral RNA, (b) formation and/or stabilization of dimers of the viral RNA genome, and (c) multimerization of NC during nucleocapsid assembly. In addition, the cysteine-histidine motif appears to influence the annealing of the $tRNA^{lys}$ primer to the viral RNA genome (143,782,783). Because NC is an intrinsic component of the nucleocapsid, this protein is thought to play a role in virion uncoating during entry and may influence the reverse transcription process that takes place in a nucleoprotein complex containing viral RNA as well as capsid proteins.

The Viral Enzymes

The viral enzymes PR, RT (and RNase H), and IN are produced by cleavage of the Gag-Pol polyprotein $Pr160^{gag\text{-}pol}$ during virion morphogenesis (see section on virion assembly and Chapter 26 in this volume). This posttranslational processing is mediated by the PR domain in the Gag-Pol polyprotein which therefore represents a zymogen. Genetic and biochemical analyses of these viral enzymes have been performed in mammalian cells transfected with site-specific viral mutants as well as in mammalian, insect and microbial cells harboring vectors for expression of wild type, and mutant alleles of these enzymes. Large quantities of enzymatically active PR, RT (and RNAse H), and IN have been produced in genetically engineered microbial cells to examine enzymologic properties as well as for analysis of protein structure. In addition, HIV-1 PR has been chemically synthesized and shown to be enzymatically active (487,804). High-resolution electron density maps of PR, RT, and RNase H crystals have been deduced to provide detailed structures. Accordingly, knowledge of catalytic mechanisms of these viral enzymes, including active site interactions with substrates, can be integrated into the structure of each protein.

PR

The mature form of PR is 99 amino acids in length and displays a molecular weight of 10 k (359). In a proposed model for processing, the Gag-Pol polyprotein (i.e., zymogen) dimerizes in the infected cell, and the mature PR dimer is released by an autocatalytic cleavage that occurs either in *cis* (intramolecular) or in *trans* (intermolecular) (151,385,689). For steric reasons, PR is believed to cleave itself out of the precursor by a *trans* mechanism (801). After release, the fully active PR dimer targets other sites in the viral polyproteins; altogether, four and seven cleavage sites are recognized in the Gag and Gag-Pol polyproteins, respectively. Retroviral proteases are related to cellular aspartyl proteases because the sequence Asp-Thr/Ser-Gly is conserved in the active site of both the viral and cellular enzymes (428). This resemblance suggested that the active form of the retroviral PR may be a dimer, and detailed structural and biochemical studies have supported this notion (see below). Site-specific mutagenesis demonstrated that PR is required for replication; noninfectious particles, containing uncleaved Gag and Gag-Pol polyproteins, are produced if this enzyme is inactivated by mutation (377,427,454).

A detailed understanding of structure-function relationships of retroviral proteases has emerged from a combination of x-ray crystallography, analysis of site-specific mutants, and tests of substrate specificity (803). Proteases of HIV-1, HIV-2, SIV, and the avian sarcoma-leukemia virus have been crystallized, and high-resolution electron density maps reveal that these proteins possess similar structural features (478,481,524,738,803,830). The structure of the retroviral PR monomer is similar to a single domain of bilobal cellular aspartyl proteases and has several β-strands, a long α-helix, and a partial α-helix (see Fig. 9 in Chapter 26). In the dimer, N- and C-termini of both PR monomers are intertwined to produce a four-stranded antiparallel β-sheet. The conserved active site in each monomer is positioned in a loop which forms part of the catalytic site. A binding cleft for substrate is positioned above the active site; this cleft can accommodate a substrate seven amino acids in length. The entrance to the binding cleft is covered by two large flaps in the dimer, and these flaps move to permit substrate (or inhibitor) to enter.

The specificity of HIV-1 PR has been explored by comparing sequences of cleavage sites in the Gag and Gag-Pol polyproteins, testing synthetic peptide substrates *in vitro*, and analyzing site-specific mutations in viral polyprotein substrates as well as the PR catalytic site (152,573,803). Results of these studies demonstrate that PR has specificity for more than one cleavage site; accordingly, both primary sequence and structure of the substrate influence enzyme activity. Cleavage specificity is governed by four amino acids upstream and three amino acids downstream of the scissile bond in the target substrate. The amino acid upstream of the cleavage site is always hydrophobic and unbranched at the β-carbon. Three cleavage sites in the HIV-1 Gag-Pol polyprotein fit the consensus sequence Ser-Thr-Xa-Ya-Phe/Tyr-Pro-Z (cleavage between the Phe/Tyr-Pro linkage), and the other cleavages occur at Leu-Ala, Met-Met, Phe-Leu, and Leu-Phe linkages. Analysis of oligopeptide substrates for HIV-1 PR has provided information for designing inhibitors of this critical viral enzyme (361,751). Moreover, knowledge of the tertiary structure of PR has been used to identify compounds that inhibit enzymatic activity; these include symmetric inhibitors, inhibitors of enzyme dimerization, and compounds with steric complementarity to the active site (386,392,803) (see section on molecular therapeutics).

RT and RNase H

Reverse transcriptase is an RNA-dependent DNA polymerase which synthesizes DNA from RNA as well as DNA templates; in addition, this enzyme, first discovered in retroviruses, requires oligonucleotide primers bound to template (27,490,682,728). RNAse H functions in reverse transcription by degrading the RNA moiety of RNA/DNA hybrids and thereby uncovering the template for viral DNA synthesis. Additionally, RNase H appears to generate oligoribonucleotide primers for the reverse transcription process. Genetic analysis of HIV-1 clones with site-specific mutations has demonstrated that RT and RNase H are indispensable for viral replication (577,585a). Both enzymes are processed in two steps from the Gag-Pol polyprotein ($Pr160^{gag\text{-}pol}$) by PR during virion assembly (Fig. 4) (400). First, p66 is cleaved from this polyprotein and forms a homodimer. Subsequently, one subunit of p66 in this homodimer is cleaved by PR near the C-terminus to yield a heterodimer composed of p51 and p66 (Fig. 5); accordingly, the N-termini of p51 and p66 are identical. The heterodimer and the p66 homodimer both display RT and RNase H activities whereas p51 alone is not active although this subunit contains the RT sequence (305,399).

Methods of x-ray crystallography have been used to deduce electron density maps of HIV-1 RT purified from ge-

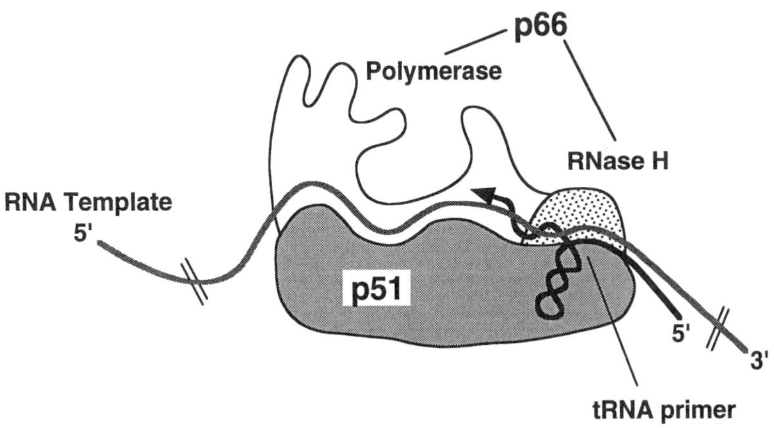

Polymerase domain

palm

RNase H domain

1 asp110 asp185,186 440 560

N- -C

fingers thumb connection

p51

p66

FIG. 5. RT domains. RT is processed by viral PR from the Gag-Pol precursor to yield a homodimer of p66; a portion of the C-terminus is subsequently cleaved from one p66 subunit to produce a heterodimer of p66/p51. Because the three-dimensional structure of p66 has been compared to that of a clenched right hand, specific subdomains are designated as palm, thumb, and fingers (379). A short connection subdomain joins the RT domain to the RNase H domain. Residues in RT that function in catalysis are Asp^{110} and $Asp^{185,186}$. The viral RNA template and the tRNA primer are positioned within the palm.

netically engineered *E. coli*, and the deduced structure provides a basis for elucidating the enzymatic mechanism of reverse transcription. A map at 3.5 Å has been derived for crystals of the RT heterodimer complexed with the nonnucleotide inhibitor nevirapine (378), and a map at 3.0-Å resolution has been produced for a ternary complex of the RT heterodimer, a monoclonal Fab fragment, and a DNA template-primer (326). The p51 and p66 subunits interact in a head-to-tail configuration to produce the heterodimer; such an asymmetric structure is unusual with very few precedents. Four subdomains in the p66 subunit of the heterodimer are arranged side-by-side to yield an elongated RT domain that has dimensions of 110 by 30 by 45 Å, and a connection subdomain in p66 joins the RT and RNAse H domains. In this model, the structure of the RT domain in the heterodimer resembles a right hand; accordingly, the subdomains are identified as fingers, palm, and thumb (Fig. 5). The catalytic site for RT lies in a cleft in the palm of the p66 subunit and contains the sequence Tyr183-Met184-Asp185-Asp186 which is highly conserved in retroviral RTs as well as in other DNA polymerases. Asp^{110} is also required for catalytic activity and is brought into the active site in the folded protein. Two α-helices in the thumb of the p66 subunit together with the palm subdomain act as a clamp to position the template-primer relative to the polymerase active site. The 3′ hydroxyl of the primer terminus is held close to the catalytically essential Asp110-Asp185-Asp186 residues; thus, this terminus is in a position for nucleophilic attack on the α-phosphate of an incoming nucleoside triphosphate. In the RT heterodimer, the p66 subunit has a subdomain which binds the template strand; the connection subdomain in p66 both interacts with the template-primer and links the RT and RNAse H domains. The 3′ end of the template strand contacts the active site of the RNAse H domain; backbone phosphates of the template interact with the two divalent metal ions bound to RNAse H (see below). About 20 base pairs of A-form DNA-RNA hybrid are accommodated between the primer terminus and the RNAse H active site.

Reverse transcription follows an ordered catalytic mechanism, as deduced from *in vitro* studies with various template/primers (13,225). DNA synthesis on the viral RNA template is accompanied by cleavage of this template by RNAse H at a position 16 to 18 nucleotides 3′ from the site of synthesis (Fig. 5). In the RT heterodimer, the conformation of p51 is very different from that of p66, although the smaller subunit is derived entirely from the larger subunit (378). No cleft is evident in the p51 subunit, and amino acid residues in p51 corresponding to the catalytic site of p66 are buried and thus not available for catalysis. In the proposed model for the RT heterodimer, a subdomain in p51 contacts the duplex formed between the $tRNA^{lys}$ primer and viral RNA template; in addition, p51 may interact with other sequences in the $tRNA^{lys}$ primer (33,34).

Reverse transcriptase plays a major role in the generation of diversity of retroviruses (118). Several error mechanisms have been ascribed to polymerases in general (45). First, direct misincorporation of a noncomplementary nucleotide produces a single-base substitution error. Second, slippage of the two DNA strands that may occur at repetitive sequences to generate either a deletion (unpaired nucleotide(s) in the template strand) or addition (unpaired nucleotide(s) in the primer strand); each of these events may involve one or more nucleotides. Third, frameshifts are caused by misincorporation followed by misalignment of the template-primer. Fourth, base substitutions can also result from dislocation mutagenesis which follows the following pathway: slippage, correct incorporation, and realignment. Each of these mechanisms can yield single-nucleotide mutations and may also operate over large distances to produce changes involving many nucleotides. The fidelity of RTs for a variety of retroviruses has been determined *in vitro* by measuring misincorporation (i.e., error) in reactions with defined RNA or DNA templates (800). These templates include synthetic polynucleotides as well as single-stranded circular phage DNA (e.g., bacteriophage M13). Mistakes occurring at the initial insertion step or the extension step are distinguished by steady-state kinetic analysis. The misincorporation rate for HIV-1 RT ranges from 1 per 1,700 to 4,000 nucleotides (44,586,610,611), whereas this rate for other retroviruses is lower (e.g., error rates are 1 per 9,000 to 17,000 nucleotides for avian myeloblastosis virus and 1 per 30,000 nucleotides for murine leukemia virus) (611). Because retroviral polymerases do not have 3′-5′ exonuclease activity for proofreading, these enzymes as a group have a much higher misincorporation rate than cellular DNA polymerases. For example, *E. coli* DNA polymerase, an enzyme with proofreading capacity, exhibits a misincorporation rate of 1 per 100,000 nucleotides (42). Parameters thought to contribute to polymerization error are base hydrogen bonding, base stacking, geometry of the base pair, and interactions of specific amino acids in the enzyme with template and/or primer. Accordingly, information of the structure of HIV-1 RT will provide insight on molecular basis for errors in reverse transcription. For the 9.5-kb HIV-1 genome, the *in vivo* error rate is estimated to be about one to three misincorporations per replication cycle (45,800).

Through these various error mechanisms as well as recombination events, RT is responsible for the production of viral sequence diversity in infected individuals. Antiviral immune responses and other factors (e.g., cell tropism and cytopathicity) provide selective pressure for accumulation of viral variants, designated quasispecies, in the host (773). In addition, reverse transcription errors are the source of variants resistant to antiviral drugs targeted to RT as well as other genes (606). The extensive G to A hypermutation, primarily at GpA sites, observed for HIV-1 (756) and SIV (333) appears to be the consequence of misincorporation of deoxythymidine monophosphate (dTmP) opposite template polyriboguanosine (poly-rG) and/or dislocation mutagenesis. It appears that this pattern can be influenced by an imbalance in the dCTP precursor pool during reverse transcription (757). Mutational and selective constraints have yet to be defined to account for this pattern of mutation; moreover, the basis for the limit to the drive toward an extreme base composition is not known.

The HIV-1 RNAse H domain has been expressed in genetically engineered *E. coli*, purified, and crystallized, and an electron density map at 2.4-Å resolution has been produced (144,306). Viral RNAse H is folded into a five-stranded mixed β-sheet flanked by an asymmetric distribution of four α-helices. This structure is very similar to that of *E. coli* RNAse H, although the viral and bacterial enzymes exhibit low sequence homology (144). Four acidic amino acid residues in HIV-1 RNase H, shown to be required for catalysis by site-directed mutagenesis, bind divalent metal ions. These four amino acids are highly conserved in the RNAse H of retroviruses and bacteria. The isolated RNAse H domain of HIV-1 RT is not catalytically active; however, addition of the p51 subunit of RT restores activity to this the RNAse H domain *in vitro* (305). The role of RNAse H in retroviral reverse transcription is described in the section on the viral DNA synthesis (also see ref. 95).

IN

Integrase is a viral enzyme possessing both DNA cleavage and joining (or strand-transfer) activities (see Chapter 26) (243). Accordingly, IN mediates covalent linkage of linear double-stranded viral DNA (i.e., substrate) into the host cell genome (i.e., target). The C-terminus of the HIV and SIV Gag-Pol precursor polypeptide is proteolytically processed by PR to produce the 32-kd IN (Fig. 4). Genetic studies of viral mutants have shown that this viral enzyme is required for replication in T cell lines as well as cultures of primary lymphocytes and macrophages (391, 709,802). A domain near the N-terminus of HIV and SIV IN exhibits high sequence conservation with IN of other retroviruses and retrotransposons. This domain contains pairs of histidine and cysteine residues (for HIV-1: His12, His16, Cys40, Cys43) that adopt a structure similar to the metal-finger motif in several DNA-binding proteins (83). A large central domain of IN has acidic amino acid residues (Asp64, Asp116, and Glu152) that are highly conserved in the INs of retroviruses, retrotransposons, and transposases of some bacterial transposable elements. The three-dimensional structure of the catalytic portion of HIV-1 IN (residues 50 to 212 out of 288) has been determined by x-ray crystallography (174). These studies reveal a core domain consisting of a central five-stranded β-sheet and six helices; the overall topology is similar to that of RNAse H and the core domain of bacteriophage Mu transposase. The catalytic site resembles that of polynucleotide transferases, which transfer chains of nucleotides in several biologic systems. Residues Asp64, Asp116, and Glu152 in the active site are believed to bind positively charged metal ions such as magnesium. Two monomers of IN interact through two contacts to produce a dimer. Interestingly, the subunits in the dimer do not recognize each other through a potential leucine zipper coiled coil involving residues 151 to 172 (342). Dimerization of IN may be essential not only for integration but also for assembly of this viral enzyme into virus particles. Each molecule of IN contains a single catalytic site but is thought to have separate binding sites for the ends of linear viral DNA (i.e., substrate or donor DNA) and double-stranded host cell DNA (i.e., target or acceptor DNA) (764).

Cell-free systems with purified IN and defined DNA substrates have been used to analyze the details of the enzymatic steps in retroviral integration (184,671). Substrates are short double-stranded oligonucleotides corresponding to termini (i.e., attachment, or *att* sites) of linear viral DNA. Integrase has the following activities in these *in vitro* systems: (a) cleavage of a deoxythymidylate-deoxythymidylate (TT) dinucleotide from the 3′ ends of double-stranded *att* site substrates (12 to 15 base oligonucleotides), (b) cleavage of double-stranded target DNA to produce a staggered 5-b overhang, and (c) strand-transfer activity in which the recessed 3′ end in the substrate DNA is joined to the 5′ phosphoryl end in the target DNA (see Fig. 17 in Chapter 26). The cleavage step is separable from the joining step, and thus joining to target DNA does not utilize the energy in the pre-existing phosphodiester bond of viral DNA substrate. In a one-step transesterification mechanism, the recessed 3′ ends of viral DNA are coupled to phosphodiester bonds in both strands of target DNA by a direct nucleophilic attack of water (184,765). Presumably, the conserved acidic amino acids in the catalytic site (Asp64, Asp116, and Glu152) bind two divalent metal ions, one of which activates a water molecule for nucleophilic attack on a phosphorus atom in the DNA backbone, while the second divalent metal ion stabilizes the transition state (174). Retroviruses and retrotransposons contain a highly conserved CpA dinucleotide adjacent to the 3′ processing site in the linear DNA substrate; in addition, IN recognizes an *att* site composed of six to nine nucleotides at each terminus of linear viral DNA (299,672,766). HIV-1 IN also mediates the reverse reaction (i.e., disintegration) *in vitro* with a synthetic DNA substrate designed to mimic the product of the joining reaction before repair (107). The observation that the integration reaction is reversible *in vitro* is consistent with ideas that IN is an enzyme and that the mechanism of integration involves a concerted mechanism and not a protein-DNA intermediate. Studies *in vitro* with deletion and site-specific point mutations in IN show that Asp64, Asp116, and Asp152 are essential for all three reactions (754). This finding, taken together with the three-dimensional model for the catalytic domain of IN (174), indicates that this viral enzyme carries out all three reactions with a single active site.

The Envelope Glycoprotein

The Env gp molecules on the surface of HIV and SIV particles bind to CD4 receptors, located on the plasma membrane of $CD4^+$ T-lymphocytes, monocytes, macrophages, and dendritic cells, and thereby attach virions to cells (784). After this attachment step, Env gp mediates uptake of virions into cells by fusion of viral and cellular membranes (see section on virion attachment and entry) (421). In addition, the viral glycoprotein is responsible for induction of syncytia in tissue culture cells (i.e., multinucleated cells) and is a major target for antiviral immune responses in the infected host.

For HIV-1, a spliced viral transcript encodes both Vpu and the precursor polyprotein for the Env gp (see Fig. 14). This bicistronic viral messenger RNA (mRNA) is synthesized in the late stage of viral replication and its expression is dependent on the posttranscriptional function of the viral *rev* gene. Both HIV-2 and SIV lack the *vpu* gene and thus produce monocistronic spliced transcripts for the Env gp.

The *env* genes of HIV and SIV are predicted to encode proteins of 850 to 880 amino acids; extensive glycosylation of the Env precursor polyprotein during synthesis produces gp160 which is the major form of the *env* gene product detected in infected cells (7,613). Intracellular cleavage of gp160 yields the N-terminal subunit (gp120), which is about 550 amino acids long, and the C-terminal subunit (gp41), which is about 350 amino acids long (763) (Fig. 6). Gp120, also designated the SU subunit, is a highly glycosylated, hydrophilic protein positioned on the external surface of virion membranes as well as plasma membranes of infected cells. This subunit has a binding domain for the

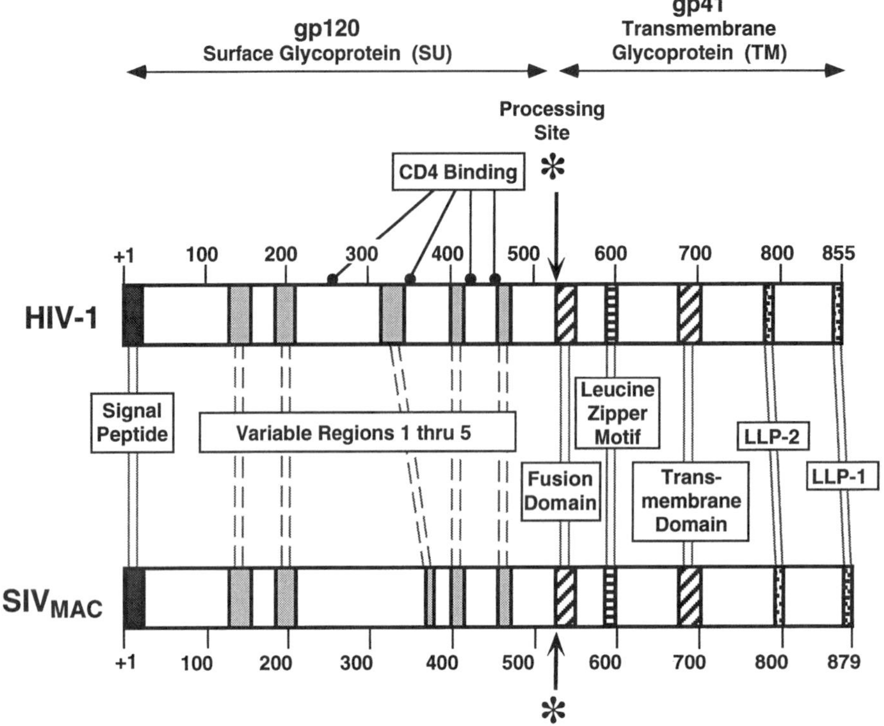

FIG. 6. Functional domains in the Env gp of HIV-1 and SIV_{MAC}. A signal peptide at the N-terminus of the Env precursor gp160 directs ribosomes translating the nascent protein to the endoplasmic reticulum; an intracellular PR removes this signal peptide during Env gp biogenesis. In addition, gp160 is cleaved at a processing site (*arrow* and *asterisk*) by a cellular protease to produce gp120 (SU subunit) and gp41 (TM subunit). The gp120 subunit contains five hypervariable regions (lightly shaded) and six conserved regions. Several sites in HIV-1 gp120, shown as *filled circles,* have been implicated in binding to the CD4 receptor (Thr^{257}, Trp^{427}, Asp^{368}/Glu^{370} and Asp^{457}). The third variable region, designated V3, is an immunodominant epitope; heterogeneity in the V3 sequence as well as other regions of gp120 influences differential cell tropism of HIV-1 isolates. The gp41 subunit contains a hydrophobic N-terminal fusion peptide (required for early steps in viral entry) (designated F), a sequence predicted to adopt a helical structure similar to the leucine zipper motif (designated Z), a hydrophobic transmembrane domain (anchor for the glycoprotein in membranes) (designated TM), and two amphipathic α-helices (designated LLP-1 and LLP-2). The leucine zipper region plays a role in oligomerization of Env gp and has been proposed to function in fusion of viral and cell membranes during virion entry (see Fig. 11).

CD4 receptor (382,395). Gp41, also designated the TM subunit, is relatively hydrophobic and traverses once the lipid bilayer membranes of both virions and cells; thus, gp41 is classified as a type 1 integral membrane protein. In the mature form, the Env gp is a heterodimer consisting of the gp120 and gp41 subunits held together by noncovalent bonds involving several points of contact between both subunits (176,655). Mutagenesis studies of HIV-1 have demonstrated that amino acid residues in both the N-terminus and C-terminus of gp120 are critical for maintaining the association of this subunit with gp41 (285,382,502, 797). The precise structural feature(s) in the gp41 subunit contacting the gp120 subunit has not been defined, although it has been proposed that a disulfide loop in gp41 may form a knob that fits into a pocket in the folded gp120 subunit (84,655). Through noncovalent interactions, Env gp heterodimers are organized into an oligomeric complex. Biochemical studies indicate that the Env gp oligomer may be a dimer, trimer, or tetramer (175,176,602,782); high-resolution electron microscopy of virions reveals trimeric symmetry of the Env gp spikes (233). The oligomeric arrangement of the Env gp may be important for multivalent interaction with the CD4 receptor during virion attachment to the cell surface (499,501); however, additional studies are required to fully define the role of Env gp oligomerization in viral replication.

SU Domain (gp120)

Several features in gp120 are evident from inspection of the predicted amino acid sequences (519). Human immunodeficiency virus type 1 gp120 contains 24 potential sites for N-linked glycosylation (Asn-X-Ser/Thr); about 13 of the 24 glycosylation motifs are conserved in different viral isolates. Analysis of HIV-1 Env gp produced in genetically engineered rodent cells has demonstrated that 17 of 24 potential glycosylation sites are modified with carbohydrate side chains (407,489). Because of extensive glycosylation of Env gp, models based on computer algorithms suggest that very few regions of the peptide backbone of gp120 protrude from the carbohydrate mass (523). These extensive and heterogeneous patterns of glycosylation have precluded crystallization of gp120 for structure analysis by x-ray diffraction. Results of genetic analyses aimed at determining the function of glycosylation sites are discussed below in the section on synthesis and processing of Env gp. The predicted sequence of HIV-1 gp120 shows 18 cysteine residues. Because these cysteine residues are highly conserved in the glycoproteins of diverse HIV-1 and HIV-2/SIV strains, disulfide bonds are presumed to play critical roles in the structure and function of this viral protein. A model for the gp120 subunit, based in part on biochemical analysis, shows nine intrachain disulfide bonds (Fig. 7) (310,407). This disulfide bonding pattern delineates gp120 into several functional regions which include a conformation-dependent domain for recognition of the CD4 receptor. A characteristic of *env* genes of independent HIV-1, HIV-2, and SIV isolates is extensive sequence heterogeneity (519); the nature of this sequence variation and a model for gp120, demonstrating five variable domains interspersed with conserved regions, are discussed below (Fig. 6). Regions of HIV-1 gp120 interacting with the CD4 receptor have been deduced from a combination of genetic, biochemical, and immunochemical approaches (501). Analysis of site-specific mutations in the env gene demonstrates that a limited number of conserved amino acids in different regions of gp120 are required for efficient binding to CD4 (382,395,545,722) (Fig. 6). In addition, revertants of site-specific mutations cause distal changes in the gp120 sequence (502). This finding highlights the importance of the interaction of noncontiguous regions in producing the functional conformation of gp120. Antibodies to gp120 that block CD4 binding either recognize a linear epitope around residue 430 (395,477) or are directed to a discontinuous epitope(s) (130,294,296,733). These diverse approaches support a model in which both primary sequence (i.e., conserved amino acids in noncontiguous regions) and conformational features (i.e., discontinuous epitopes) of the Env gp produce a configuration that recognizes the CD4 receptor in a selective fashion and with high affinity. The probable contact amino acids on gp120 are in two hydrophobic regions centered around Thr257 and Trp427 and in two hydrophilic regions around Asp368/Glu370 and Asp457 (545) (Fig. 6). Trp427 is presumed to lie at the base of the binding cleft. X-ray crystallographic analysis of a complex of CD4/gp120, to define the precise location and features of the high affinity binding site on both proteins, are confounded by the high degree of glycosylation of this subunit. Accordingly, a continuation of biochemical, genetic, and immunochemical analyses will be necessary to obtain more information on the precise structural features that are essential for the interaction of CD4 and gp120 and to deduce a three-dimensional model for gp120 (66,502,503).

Gp120 from laboratory strains of HIV-1 and soluble CD4, both produced in genetically engineered mammalian cells, interact with a high affinity constant ($K_d = 4 \times$

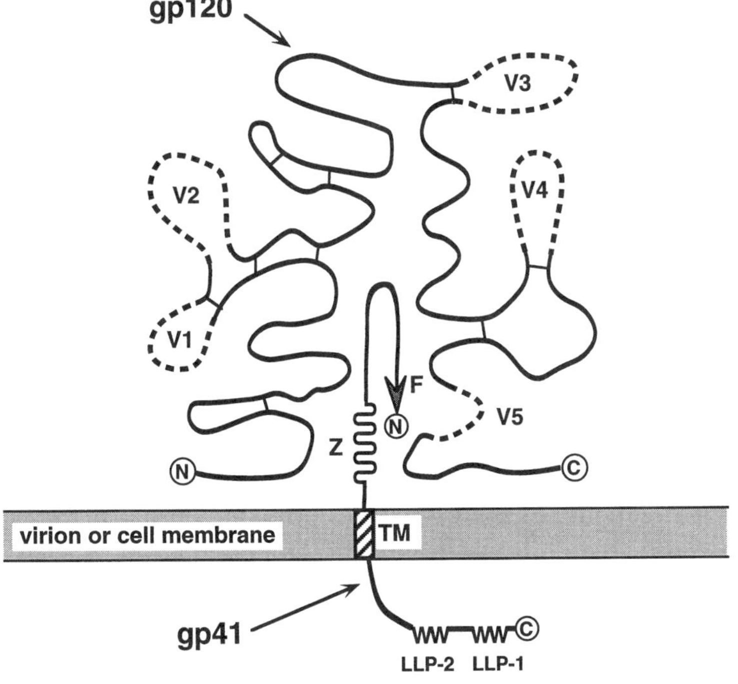

FIG. 7. Predicted folding pattern of HIV-1 Env gp. The predicted folding pattern of the gp120 subunit, as well as the positions of disulfide bonds (connecting lines in the loops), have been established by enzymatic digestion techniques combined with peptide mapping and sequencing (274, 447). Nine disulfide bonds have been identified; these linkages are indicated by short lines in the figure; disulfide-bonded domains are numbered with Roman numerals (*I* thru *V*). Hypervariable domains are designated *V1* thru *V5* and are drawn as broken lines (497). The predicted folding pattern and membrane association of the gp41 subunit are also shown (219). Regions labeled in the gp41 subunit are: F, fusion peptide (*large arrow* at N-terminus); Z, leucine zipperlike region (shown as a *helix*); TM, transmembrane domain (*cross-hatched box*); LLP-1 and LLP-2, lentivirus lytic peptides 1 and 2 (shown as *WW*). Amino and carboxyl termini are labeled *N* and *C*, respectively, for both subunits. The structures of the gp120 and gp41 subunits presented in this figure do not account for potential interactions between both subunits.

10^{-9}) (395,493). However, binding of CD4 to gp120 is influenced by oligomerization in a fashion that depends on the viral isolate. The gp120 molecules from primary isolates, solubilized by detergent treatment of virions, exhibit ten- to 30-fold reduced binding to CD4 compared to laboratory strains of virus (19,68). This difference in binding appears to be due to differences in the tertiary or quarternary structure of gp120 on virions (177,749). A consequence is that much higher concentrations of soluble CD4 are required to neutralize primary HIV-1 isolates than tissue culture adapted viral strains (139). The SU subunit of HIV-2 and SIV binds ten- to 300-fold less well than gp120 from laboratory strains of HIV-1 (322,495); the significance of these differences in affinities for CD4 between the human and non-human lentiviruses is not known.

TM Domain (gp41)

The gp41 sequence contains four potential glycosylation sites and three cysteine residues. About 20 amino acids at the N-terminus of gp41 are hydrophobic and define the fusion peptide (amino acids 512–527) which is required for fusion of the virion membrane with the cell plasma membrane during the entry step in viral replication (212, 303,382). A second hydrophobic domain (residues 684–705) spans virion and cell membranes and thereby enables gp41 to serve as an anchor for the Env gp heterodimer (Figs. 6 and 7) (219). The region between these two hydrophobic domains is external to the membrane (i.e., an ectodomain) and contains a highly conserved sequence (Leu553 to Leu590) predicted to be similar to the leucine zipper motif implicated in protein/protein interactions of a variety of viral and cellular proteins (155,794). Mutations in this leucine motif of gp41 block viral infectivity and cell fusion; however, synthesis, oligomer formation, transport, and proteolytic processing of the Env gp are not affected (84,101,172, also see 575). Mutation of either Cys598 or Cys604 abolishes viral infectivity (722). During viral entry, a portion of the gp41 ectodomain (including the leucine zipper) has been proposed to adopt a coiled-coil conformation which facilitates insertion of the gp41 fusion peptide into the target cell membrane (see section on viral entry) (794).

Analysis of gp41 by computer algorithms predicts two amphipathic α-helical regions in the cytoplasmic domain, one between Tyr768 to Arg788 and another at the C-terminus between Arg826 to Leu854 (Figs. 6 and 7) (482, 759). Synthetic peptides representing these two regions induce the formation of pores in plasma membranes of tissue culture cells and cause cell lysis (102,484,699); accordingly, these regions have been designated the lentivirus lytic peptides (LLP-1 for Arg826 to Leu854 and LLP-2 for Tyr768 to Arg788) (Fig. 7). Amphipathic α-helices are also a feature of the cytoplasmic domains of the TM subunits of other primate and animal lentiviruses (482). Interestingly, both LLP-1 and LLP-2 of HIV-1 bind to purified calmodulin, which is an integral component of many signal transduction pathways (485,699). The significance of these *in vitro* studies, showing cell lytic activity and calmodulin binding capacity of synthetic peptides representing env-TM sequences, to viral replication and cytopathicity is not known.

Simian immunodeficiency virus and HIV-2 encode a TM subunit, gp41, that possesses sequence homology to the HIV-1 counterpart; thus, the TM glycoproteins of these viruses exhibit similar structural features. Passage of SIV in cultures of human T-lymphoid tumor cell lines (e.g., HUT-78) selects for variant viruses which contain a truncated TM subunit (gp28) (93,94,291). The truncation is due to an in-frame TAG nonsense codon (stop codon) in the (SIV_{MAC}) *env* reading frame at position 736 or 737; this position corresponds to the splice acceptor site for the second coding exons of *tat* and *rev*, which are encoded in the remaining two translation frames (Fig. 2). Clones of SIV and HIV-2 with this premature stop codon are fully infectious in cultures of primary lymphocytes and T-cell lines (555). Moreover, clones of SIV with a truncated TM subunit infect rhesus monkeys, and virus recovered from these animals contains a full-length TM subunit, gp41 (376,440). This restoration of a coding sequence for the premature stop codon in the SIV TM subunit is correlated with viral persistence and fatal disease (440). Additional studies are required to determine the precise role(s) of the cytoplasmic domain of gp41 in viral load and pathogenesis.

Synthesis and Processing of Env gp

Synthesis and processing of the Env gp occur in the secretory pathway which is also utilized for producing cellular secreted and membrane proteins as well as glycoproteins of other enveloped viruses (166,600). The hydrophobic signal peptide, composed of 28 to 30 amino acids, at the N-terminus of the Env gp precursor attaches the nascent polyprotein to the endoplasmic reticulum (ER) and is cleaved by a host cell endoprotease after translocation into the lumen of the ER. During cotranslational transfer of the nascent precursor into the ER, cellular enzymes add oligosaccharide moieties, predominantly mannose residues esterified into long chains, to yield the Env gp precursor that has a molecular mass of 160 kd. Because a stop-transfer sequence is located near the C-terminus of gp160 (at about amino acid residues 687–699 in the region encoding the gp41 subunit), this precursor protein remains anchored in lipid membranes of the cell (218). Formation of intramolecular disulfide bonds in the ER lumen produces a folded monomer of gp160 (200), and these monomers associate into oligomeric complexes that have been reported by various investigators to consist of two, three, or four molecules of gp160 (176). At this intracellular stage of synthesis, Env gp is competent to bind the CD4 receptor. Oligomers are transported from the ER to the Golgi apparatus where mannose residues are trimmed from the carbohydrate side chains and other residues (i.e., N-acetylglucosamine, galactose, and fucose) are added to produce viral glycoproteins containing both complex and hybrid carbohydrate side chains (600). TM and the cytoplasmic tail of gp41 (Fig. 7) contain determinants of the amount of Env gp expressed on infected cell membranes (632,785). These experiments involved replacing the HIV-1 TM domain and cytoplasmic tail of gp41 with a glycophospholipid that served as an anchor for membrane proteins; such chimeric Env gp molecules exhibited a higher level of cell surface expression than native gp160. Studies on variants of SIV_{MAC} demonstrate that a small number of amino acid residues in both the gp120 and gp41 domains influence the amount of Env gp inserted into virion and infected cell membranes (389).

Glycosylation of the Env gp precursor appears to affect intracellular processing and influences both functional and immunologic properties of the mature glycoprotein (166). This posttranslational modification is presumed to govern transport of the glycoprotein precursor through various intracellular compartments. Consistent with this notion is the observation that treatment of infected cells with glycosidase inhibitors blocks Env gp maturation (and inhibits viral replication) (600). Certain carbohydrate modifications may be needed for proper folding of the Env precursor and may be removed at later steps in processing. Deglycosylated forms produced by endoglycosidase treatment of Env gp retain some capacity to bind CD4 (201,420,507). Studies on site-specific mutations in *env* have revealed that some but not all asparagine-linked carbohydrate modifications are essential for viral replication. Viruses with individual mutation of asparagine residues at five of the 24 potential glycosylation sites do not replicate, whereas mutation in the remaining 19 potential glycosylation sites does not affect viral infectivity (403). Interestingly, these critical sites (Asn88, Asn141, Asn197, Asn262, and Asn276) are located in the N-terminal half of gp120. Analysis of a mutation at Asn262 showed that the mutant Env gp assembled into noninfectious virions; because the mutant virus retained the abilty to bind CD4 and induce syncytia, it was proposed that this region of the Env gp was critical for a replication step after virion attachment to the cell receptor. In another study, mutation of two highly conserved glycosylation sites in gp41 (Asn611 and Asn616) reduced both infectivity and syncytium formation (84). Thus, glycosylation of certain sites on both subunits is important for different steps in viral replication. Carbohydrates can alter the recognition of Env gp by the immune system, in some cases, by occluding peptide epitopes, and thus provide a mechanism for virus to escape antiviral immune responses (62). Taken together, these findings demonstrate that several properties of the *env* gene product are governed by posttranslational glycosylation. Additional studies are required to determine whether carbohydrate modifications alter independent functional domains and/or produce changes in conformation of the viral glycoprotein. In addition to N-linked glycosylation, Env gp has been reported to contain O-linked carbohydrates which modify serine and threonine residues (268). Unresolved issues are the precise extent of O-linked glycosylation and and importance of this type of modification for viral replication. SIV infection of macaques offers opportunities to explore the potential role of both N- and O-linked glycosylation of Env gp in pathogenesis (92,550)

Cleavage of the gp160 precursor into the gp120 and gp41 subunits takes place in the Golgi apparatus and appears to be mediated by the host cell endoprotease furin, a mammalian subtilisin in the serine protease family (40,263,706). This cellular endoprotease acts on each chain in the Env gp oligomer immediately after a stretch of basic amino acids which define the C-terminus of gp120. The cleavage process presumably induces a conformational change that buries the hydrophobic fusion peptide of gp41 within the oligomer (303,639). After proteolytic cleavage of the gp160 precursor, the processed Env gp oligomers are directed out of the Golgi compartment and into the plasma membrane presumably by a cellular vesicle transport system (338). Mutational alteration of the basic residues at this cleavage site produce an uncleaved gp160, which folds and oligomerizes, and these oligomers insert into the cell plasma membrane and assemble into virions; however, virions with uncleaved gp160 are noninfectious (470). Presumably, the fusion peptide in the uncleaved gp160 is constrained from functioning in the entry step of viral replication. In some cell types infected with wild-type viruses, up to 90% of the mature Env gp oligomers inserted into the cell plasma membrane are targeted to lysosomes and degraded (176,795). Because the association of the gp120 and gp41 subunits is weak in the mature viral glycoprotein, the gp120 subunit is released into the extracellular medium (285,344). The significance of both the rapid turnover of the Env gp and release of the gp120 subunit into the medium is not known.

Sequence Variation in the env *Gene*

Comparison of the sequences of *env* genes from numerous HIV-1 isolates reveals a pattern of five variable regions interspersed with conserved regions for the gp120 subunit; in contrast, the gp41 subunit shows less heterogeneity and is thus more highly conserved (Fig. 6) (519, 702). This sequence variation consists of nucleotide changes, which produce amino acid substitutions, as well as small deletions and insertions (404). Up to 25% of the amino acids encoded by *env* may vary in HIV-1 strains from geographically separated locations (interpatient variation), although other regions of the genomes of diverse viral isolates are relatively conserved. Currently, molecular epidemiology surveys, based on *env* sequences of numerous

HIV-1 isolates, reveal a minimum of nine distinct viral subtypes (or clades) in the AIDS pandemic (519). Because of rapid spread of HIV-1 infections, two or more subtypes are prevalent in some areas of the world (518). Sequence analysis of HIV-2 and SIV isolates also supports a model with five variable regions interspersed with conserved sequences in the gp120 subunit of these two viruses (Fig. 6) (81). Sequencing of independent HIV-2 isolates reveals the existence of at least five distinct and roughly equidistant subtypes (225). Selection of *env* variants occurs in HIV-infected individuals and SIV-infected monkeys to produce populations of closely related but distinct viral genomes designated quasispecies (i.e., intrapatient variants) (81,773). Sequence variation in *env* of HIV and SIV has significant implications not only for antiviral immune responses but also for additional functions mediated by the Env gp such as binding to CD4, cell tropism, and cytopathicity.

Immunochemical analysis of the HIV and SIV Env gp has demonstrated that both linear and conformational determinants influence the antigenic structure of the glycoprotein (81,523). Essentially all neutralizing antibody activity in HIV-1 infected hosts is directed against the Env gp. Although the majority of this activity is specific for the V3 loop in gp120, the V2 and C4 domains as well as epitopes in gp41 also contain targets for neutralizing antibodies (98,126). In addition, immunogenicity of Env gp is influenced by conformational epitopes, including the quarternary structure of the oligomers (71,178,294,296,332, 500,704,733). Neutralizing antibodies may either block binding of gp120 to the CD4 receptor or may inhibit virion entry at a step after attachment. Investigations on the mechanism of virus neutralization, taken together with the analyses of *env* gene mutants and chimeric viruses (see below), support the notion that sequence heterogeneity may be under at least two selective pressures: (a) escape from antibodies which neutralize the virus and (b) adaptation to infection of different cell types. It is also tenable that other *env* gene phenotypes, such as cytopathicity and fusogenicity, also confer a selective advantage to the virus in the infected host.

The V3 Domain of Env gp

The V3 sequence, a variable domain in the gp120 subunit of HIV-1, contains 35 amino acids arranged in a disulfide loop involving Cys301 and Cys336 (Fig. 8) (480,496). This domain plays an important role in governing several biologic properties of the virus (i.e., cell tropism, cytopathicity, and fusogenicity), and deletions in the V3 loop abrogate viral infectivity (see below). In infected individuals, B-cell as well as T-cell responses are directed to the V3 loop (554,609,724,832); this region has been designated the principal neutralizing determinant because virus neutralizing antibodies are elicited by peptides encompassing the V3 sequence (87a,183,188,331). Because synthetic peptides representing V3 sequences block HIV-1 infection of primary lymphocytes and macrophages, this region of the Env gp is presumed to play an important role in the early stages of infection, perhaps by interfering with the interaction of gp120 and CD4 and/or a distinct coreceptor (812). Alternatively, synthetic peptides representing V3 may inhibit infection by interfering with a postentry step in viral infection (683).

Comparison of numerous HIV-1 isolates reveals relatively conserved subdomains as well as variable subdomains in the V3 loop (519).The Gly317-Pro318-Gly319-Arg320-Ala321-Phe322 motif, located at the top (or crown) of the loop, is highly conserved among HIV-1 isolates, and sequences near the Cys300 and Cys336 at the bottom of the loop also show relatively little variability (Fig. 8). Sec-

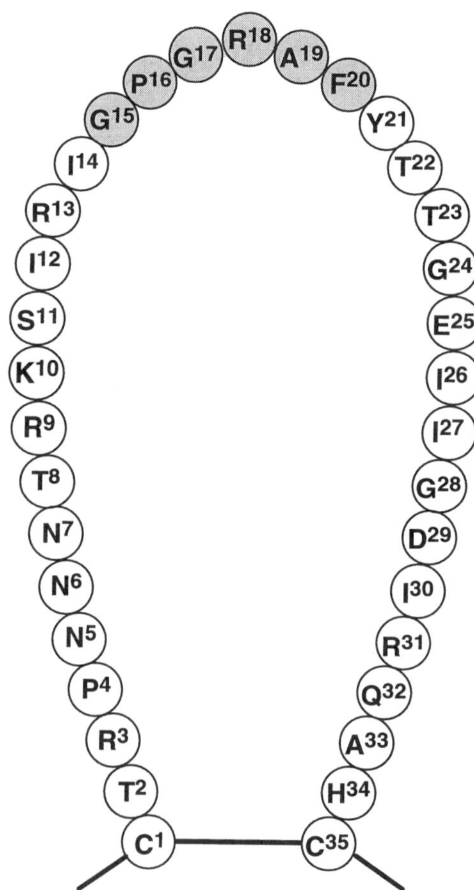

FIG. 8. V3 loop of HIV-1 Env. The one-letter amino acid code is used to identify residues in the consensus sequence for the V3 loop of HIV-1 (524). $G^{15}P^{16}G^{17}R^{18}A^{19}F^{20}$ at the crown of the loop (*shaded*) is a stretch of relatively conserved amino acids; sequence variation is observed in amino acids on either side of the crown. The sequence $N^6N^7T^8$, a (potential) signal for glycosylation, is found in the V3 sequence of many HIV-1 strains. T-cell line-tropic, SI viruses generally have a basic amino acid at one or more of the following positions: 11, 24, 25, 32. Macrophage-tropic, NSI viruses have either an acidic amino acid or alanine at position 25 (480).

ondary structure predictions and nuclear magnetic resonance analysis suggest that the HIV-1 V3 loop folds into a β-hairpin structure (β-strand type II, β-turn, β-strand, β-helix) (394), and x-ray diffraction studies on crystals of V3 peptides complexed with Fab fragments of virus neutralizing antibodies reveal that the Gly317-Pro318-Gly319-Arg320-Ala321-Phe322 motif adopts an S-shaped double turn (237,607). Many strains of HIV-1 and HIV-2/SIV have a potential glycosylation site in the V3 loop (Asn305-Asn306-Thr307 for HIV-1) (Fig. 8). Analysis of several HIV-1 isolates in clade B shows that the majority of sequence changes are in the ten to 12 amino acids on either side of the conserved crown of the loop; these changes are not random because substitutions are conservative and usually involve amino acids with similar chemical properties (480). These comparisons have also demonstrated that the identities of V3 amino acids are coordinately influenced at certain positions exhibiting variation in this domain. Studies on additional HIV-1 isolates in clade B as well as members of other clades are necessary to develop a full understanding of the extent of sequence variability in the V3 loop. Immunochemical analysis on the Env gp of SIV_{MAC} show that peptides representing the V3 loop of this virus do not elicit neutralizing antibodies (332); thus, it appears that the V3 loops of HIV-1 and SIV may not have equivalent immunochemical properties.

Viral Phenotypes Influenced by the V3 Domain

Analysis of viruses with site-specific mutations in *env* has implicated functions of conserved amino acids in the V3 loop of HIV-1. Proviruses with mutations altering single residues in Gly317-Pro318-Gly319 are not infectious (Fig. 8); however, gp160 of each mutant provirus exhibits normal patterns of synthesis and processing and retains the capacity to bind the CD4 receptor (254,321,553). Although deletion of this tripeptide also abolishes infectivity, this mutation has little effect on processing of gp160, binding to CD4, or release of virus particles (254). Some but not all mutants in the Gly317-Pro318-Gly319 motif do not induce syncytia (148,213). Substitution of the wild-type tripeptide with Gly-His-Gly or Gly-Pro-Gly does not abrogate virus infectivity; however, these mutant viruses are not efficiently neutralized by polyclonal and monoclonal antibodies which neutralize wild-type virus (254). Mutation of Pro313 also influences replication capacity in different T-cell lines (321,717). Mutation of the conserved Cys336 at the base of the V3 loop precludes cleavage of gp160 and binding to the CD4 receptor. Presumably, this latter mutation alters the overall tertiary structure of the Env gp so that the conformation of the binding site for CD4 is also altered. In addition, some mutations in the V3 sequence abrogate syncytium formation and also destabilize the association between gp120 and gp41 (213,285,798). Alteration of the conserved Arg320 at the top of the V3 loop causes a change in a potential cleavage site in the Env gp for a cellular protease which is hypothesized to play a role in viral entry into cells (see section on virion attachment and entry) (213). In summary, these genetic studies demonstrate that the conserved amino acids in the V3 loop influence several properties of HIV-1 that are controlled by the *env* gene.

Mutations (i.e., substitutions and small deletions) in V3 do not affect the ability of gp120 to interact with the CD4 receptor (see above); however, several investigations have shown that V3 sequences play an important role in syncytium formation and cell tropism (104,480,501,798). Analysis of natural variants of HIV-1, coupled with studies on point mutations introduced into V3 in molecular clones of virus, indicates that basic amino acids in one or more of positions 306, 321, 322 and 328 (positions 11, 24, 25, and 32 in Fig. 8) confer a syncytium-inducing (SI) phenotype, whereas hydrophobic amino acids in these positions correlate with a nonsyncytium-inducing (NSI) phenotype (105,148,207,480). V3 loops of SI and NSI variants generally have an overall charge of +5 and +3, respectively. Accordingly, both the position of charged amino acid residues and the overall charge influence the phenotype. Analysis of chimeric viruses constructed from two viruses with distinct replication phenotypes reveals that changes in the V3 loop converts an NSI, slowly replicating virus into an SI, rapidly replicating virus (667,676,787). Determinants of cell tropism have also been elucidated by examining natural viral variants and viral clones with site-specific mutations. Inspection of sequences of viral variants isolated from infected individuals indicates that T-cell line-tropic viruses generally have a nonacidic amino acid or alanine at position 328 (position 32 in Fig. 8), whereas macrophage-tropic viruses have either an acidic amino acid or alanine at this position (149,207,480). Site-specific mutations in V3 of a T-cell line-tropic virus confer macrophage tropism and simultaneously abolish ability to infect some T-cell lines; however, mutations in V3 of a macrophage-tropic virus reduce replication capacity in macrophages but do not confer the ability to propagate in T-cell lines (676). It has been proposed that the V3 domain from different HIV-1 isolates can assume two distinct conformations that determine preferential tropism for lymphocytes or macrophages (179).

Sequences in Env gp outside of the V3 loop can also influence phenotypes assigned to this domain. Analysis of chimeric viruses constructed from (certain) HIV-1 clones with distinct phenotypes indicates that a functional interaction between V3 and the second conserved region is important for infectivity as well as syncytium formation and cell tropism (380,712,798; also see 12,215,718). Immunochemical studies with monoclonal antibodies to epitopes in wild-type and mutant forms of Env gp implicate an intramolecular interaction between the V3 loop and amino acids in the fourth conserved region (of gp120) which constitutes part of the CD4 binding site (395,797,

810). These results support a model in which an interplay of the V3 loop with other regions in gp120 governs the conformation of this subunit and thereby controls biologic properties of the virus. With respect to the role of the V3 loop in the viral replication cycle, it is speculated that the interaction of Env gp with target cell moieties exposes the V3 loop (112,497,576). Subsequently, this proposed conformational change alters the relationship of both env subunits and thereby uncovers the fusion peptide at the N-terminus of the gp41 subunit to initiate fusion of virion and cell membranes (see section on virion attachment and entry; also Fig. 11) (501,790).

The mechanism(s) by which the V3 domain and other regions of the Env gp controls cell tropism has not been established. This domain may affect the interaction of Env gp with cell surface molecules other than the CD4 receptor (784). Thus, HIV and SIV may require a second receptor (or accessory molecule) to mediate virion entry into cells after the initial binding event (see section on virion attachment and entry). The V3 domain may play a role in viral entry by serving as a target for a cell surface protease; Arg320 is the site of cleavage by a trypsinlike enzyme(s) that may be located in the plasma membrane or in endosomes (112,278,366,515). Accordingly, the presence of an appropriate proteinase is presumed to be a determinant for viral infectivity in certain cell types. Additional studies are required to evaluate the significance of the proteolytic susceptibility of the V3 loop for viral entry and replication fully (656).

In vitro phenotypes controlled by Env gp may have significance for understanding viral transmission and pathogenesis. The population of viruses in individuals early after infection displays relatively little sequence heterogeneity and consists chiefly of macrophage-tropic, NSI viruses (615,831). In contrast, T-cell line-tropic, SI viruses are recovered from individuals who have progressed into AIDS (127,479,653,729). Thus, it is hypothesized that macrophage-tropic, NSI quasispecies are either more readily transmitted between individuals or that there is a strong selection for a limited virus population with these properties early after infection. T-cell line-tropic, SI viral variants are presumed to arise from the early quasispecies and are associated with fatal immunodeficiency in the host. Studies with molecular clones of HIV and SIV in nonhuman primate models will be necessary to test the biologic significance of these and other viral phenotypes controlled by the *env* gene (440) (see section below on molecular pathogenesis).

Variable Regions in HIV-2/SIV Env gp

Models for HIV-2 and SIV Env gp, based on comparisons of sequences of numerous clones of each virus, also show five variable regions (Fig. 6) (81,519). Both viruses have a V3 domain containing 34 amino acids between two cysteine residues; very little sequence homology is observed between the V3 loops of these viruses and HIV-1. Neutralizing antibodies are directed against the V3 loop of HIV-2 in infected humans (57); however, epitopes other than this domain are major immunological determinants for SIV neutralization in infected macaques (331,332). Synthetic peptides to the SIV V4 region elicit neutralizing antibodies (741), and this domain is also part of a conformational epitope recognized by monoclonal antibodies that neutralize virus (362). Analysis of SIV with site-specific mutations in *env* demonstrates that V3 sequences as well as several regions outside of this domain govern macrophage tropism *in vitro* (371,376,505). Depending on the SIV *env* allele, the restriction to viral replication in macrophages may be either at entry, at reverse transcription, or after viral DNA synthesis (30,506). Because HIV-2 and selected strains of SIV cause an AIDS-like disease in nonhuman primates, the significance of changes in *env* gene sequences (and *in vitro* properties governed by *env*) can be evaluated with respect to cell tropism and pathogenic potential in nonhuman primate models (333,464,550) (see section below on molecular pathogenesis).

VIRAL ACCESSORY GENES

The accessory genes of HIV (*vif, vpr, vpu,* and *nef*) and HIV-2/SIV (*vif, vpx* and/or *vpr,* and *nef*) are exclusive of the *gag, pol,* and *env* genes and the viral transactivators *tat* and *rev* (Fig. 3 and Table 1) (238). Vpr and Vpx are assembled into virions. Although several investigators have not detected Vif, Vpu, and Nef in virions, it is possible that small amounts of these accessory proteins may be incorporated into virus particles; this important issue may not be readily resolved because of the inherent difficulties in purifying retroviral particles from cell components. Roles for HIV and SIV accessory genes have been investigated by analyzing cells infected with cloned viral mutants, transfected with cloned proviral mutant DNA, and/or transfected with plasmid vectors which express viral accessory genes. Although numerous reports have been published on the functions of accessory genes through analysis of viral variants and viruses with site-specific mutations, not all of these genetic studies are in agreement. Discrepancies in descriptions of the phenotypes for these genes may be attributed to differences in virus strains, multiplicity of infection, cell types selected for analysis, and methods for measuring viral replication; accordingly, this section focuses largely on consistent observations for each viral accessory gene. The dispensable nature of these genes (based on studies of viral replication in tissue culture systems) suggests that tissue culture systems may not accurately reflect the cell types and physiologic conditions which are important for regulating viral replication *in vivo* (i.e., in the infected host); however, both antibodies and cell-mediated immune responses to the accessory proteins have been detected in hosts infected with HIV and SIV. Accordingly, accessory genes are expressed *in vivo* and are presumed to play important roles in the virus-host relationship (see section on molecular pathogenesis).

vif

All lentiviruses, with the exception of EIAV, encode the *vif gene*, which is located immediately downstream of the *pol* gene (Fig. 3) (536a). In the primate lentiviruses, a 5-kb singly spliced Vif transcript, which is dependent on *rev* function, is produced late in the viral replication cycle (Fig. 14) (230,658). Human immunodeficiency virus type 1 Vif is synthesized from an open reading frame encoding 193 amino acids, has a molecular weight of 23 to 27 k (346, 402,693), and accumulates in the cytosol and cytoplasmic membrane fractions of infected cells (245). Although Vif has not been detected in virions (204,713), additional studies with sensitive methods (e.g., immunoprecipitation with high-titer antibodies to Vif) are required to determine whether virions may contain small amounts of this viral protein. Potential sites for N-linked glycosylation are absent; the Vif proteins of HIV-1 isolates have two highly conserved cysteine residues and SIV/HIV-2 isolates have four to five conserved cysteine residues (519). The predicted Vif proteins contain three short sequences (Ser23 to Tyr40, 103–115, and Ser144 to Leu150) which are conserved in the primate lentiviruses; the latter two sequences are required for complementation of *vif* function in *trans* (245).

To explore the role of *vif* in viral replication, deletion mutations in this gene have been constructed by genetic engineering methods and analyzed in several tissue culture systems. Early studies on the requirement of the HIV-1 and HIV-2 *vif* gene for viral replication revealed that infectivity of cell-free mutant virus is reduced up to 1,000-fold relative to wild-type virus in certain $CD4^+$ T-cell lines (H9), although cell-to-cell transmission of mutant virus appears to be similar to wild-type virus (204,713). In recent studies, the HIV-1 *vif* gene was shown to be required for replication in certain T-cell lines (i.e., CEM and H9) and peripheral blood lymphocytes but not in other T-cell lines (i.e., SupT1, C8166, and Jurkat), and additional analysis demonstrated that the cell type used to produce *vif*-defective virus governs infectivity (190,218). For example, SupT1 cells yield *vif*-defective virus particles which are capable of initiating infection in many cell lines (e.g., CEM and H9); however, *vif*-defective virus particles produced in CEM cells do not intitiate infection in several cell lines permissive for wild-type virus. Taken together, these genetic studies on Vif function indicate that cells nonrestrictive to *vif*-defective virus contain a host factor or factors that complement the function of this viral gene and thereby supports viral replication.

Investigations aimed at defining the step(s) in the viral life cycle controlled by *vif* have focused on early stages of infection in T-cell lines restricting replication of mutant virus. Virions produced by *vif*-defective genomes enter restrictive cells and initiate reverse transcription; however, full-length linear molecules of double-stranded viral DNA are not produced (697,770). Mutagenesis of *vif* did not affect the amount of genomic viral RNA packaged into mutant virus particles. Accordingly, Vif may govern uncoating and/or internalization of virus in restrictive cells and thereby enable the reverse transcription process to produce complete linear viral DNA molecules. One report demonstrated that *vif*-defective virions were deficient in Env gp (258,628). Another report, utilizing electron microscopy for ultrastructure analysis, demonstrated that *vif* mutant virions produced in CEM cells (nonpermissive for vif-defective virus) contained a nucleocapsid core with aberrant morphology (298). These latter findings, suggesting a role for Vif in virion maturation, are consistent with a phenotype that is manifested at an early step of viral replication. Additional studies on *vif* gene function are required to define a functional domain(s) in Vif, characterize the precise early step of viral replication controlled by Vif, determine whether Vif modifies virion proteins and/or virion structure, and identify the putative cellular factor which appears to complement *vif*-defective virus.

vpr

Primate lentiviruses encode either one or two small open reading frames, each about 100 codons in length, between the end of *vif* and the beginning of *tat* (Fig. 3). Human immunodeficiency virus type 1, SIV_{CPZ}, SIV_{SYK}, SIV_{AGM}, and SIV_{MND} each contain only *vpr,* whereas HIV-2 and several other SIV strains (e.g., SIV_{MAC} and SIV_{SMM}) contain both *vpr* and *vpx* (519). Comparisons of *vpr* and vpx amino acid sequences reveal that these genes are homologous; thus, *vpx* may be a duplication of *vpr.* Accordingly, the *vpx* gene of SIV_{AGM} has been reclassified as *vpr* (742). Based on analysis of evolutionary distances of these viruses, the proposed duplication to produce *vpr* and *vpx* may have occurred at the time that the HIV-2 group and the other primate lentiviruses diverged from a common ancestral virus (743). Speculation is that Vpx plays a role in adaptation of the virus to the primate host species. Nonprimate lentiviruses (i.e., FIV, VMV, CAEV, BIV, and EIAV) appear not to encode genes related in sequence to *vpr* and *vpx*; however, these viruses contain small open reading frames which may be functional counterparts of *vpr* and/or *vpx* (113).

Vpr is a 15-kd virion-associated protein (122,287,466, 806) translated from a singly spliced mRNA, which is dependent on *rev* function and thus accumulates at late times in infection (Fig. 14) (230,658). Genetic studies on *vpr*-deficient mutants have been performed with several primate lentiviruses in various cell types with the goal of defining a phenotype for the *vpr* gene. Human immunodeficiency virus type 1, HIV-2, and SIV_{MAC} with mutations in *vpr* replicate with kinetics similar to wild-type viruses in CD4$^+$ T-cell lines and cultured primary lymphocytes (153,239,240,537,673). However, at low multiplicity of infection, HIV-1 *vpr* mutants replicate poorly in T-cell lines when compared to wild-type viruses. Vpr mutations in macrophage-tropic strains of HIV-1 and HIV-2 restrict viral replication in primary macrophage cultures (239,279,284), although macrophage-tropic strains of SIV_{MAC} with lesions in *vpr* replicate efficiently in macaque macrophages (29, 240). The differences in these assessments on the role of *vpr* in macrophage infection *in vitro* are difficult to interpret because the phenotype of cell tropism is influenced by more than one viral gene; furthermore, the molecular mechanism(s) mediating macrophage tropism is not well defined. Additional genetic studies have implicated a role for Vpr in nuclear localization of the preintegration complex (284). A macrophage-tropic clone of HIV-1 with mutations in the NLS of both the MA gag protein and Vpr is attenuated for nuclear localization of viral nucleic acids early after infection in nondividing cells. In addition, virus replication in macrophages is reduced in cells infected with this double mutant compared to wild-type virus. Thus, the primate lentiviruses appear to have redundant nucleophilic determinants, in MA and Vpr, that mediate efficient nuclear import of the preintegration complex in nondividing cells.

In mature virions, Vpr is associated with the nucleocapsid and is as abundant as Gag-related proteins (287,818). In infected cells, Vpr has been localized to cytoplasmic membranes and the nucleus (637). Studies on Vpr in COS cells transfected with a plasmid expression vector showed that this viral protein is largely a monomer but a small proportion was dimeric (567). Whether Vpr forms multimeric complexes in cells infected with virus remains to be established. Expression of $Pr55^{gag}$ and Vpr in transfected mammalian cells, in the absence of all other viral proteins, is sufficient for incorporation of Vpr into viral capsids and export from the cell into the culture medium (433). Although capsids are assembled from proteins encoded by gag genes with deletions in the p9 (NC) and p6 proline-rich proteins at the C-terminus of the Gag precursor, Vpr is unable to associate with these mutant capsids. Thus, Vpr, either as a monomer or as an oligomer, presumably interacts with the C-terminal domain of the Gag polyprotein during virion assembly. The precise location of Vpr in the mature nucleocapsid is not known. Because Vpr is assembled into virions, this protein may play a role in an early step in the viral life cycle such as nuclear localization of the viral preintegration complex (see section on integration of viral DNA).

Comparisons of predicted amino acid sequences of *vpr* genes reveals conservation of a single cysteine residue (Cys74 in HIV-1 and Cys76 in HIV-2/SIV), a potential amphipathic α-helical loop near the N-terminus of the protein, and an arginine-rich C-terminal region. Mutation of Cys74 of HIV-1 Vpr produces an unstable protein that is degraded in cells (567). Deletion mutations in the C-terminal domain preclude both localization of Vpr to the nucleus and incorporation into virions; however, point mutations that alter several arginine residues in Vpr do not prevent assembly of mutant proteins into virions (433). Whether these arginine residues are part of a nuclear localization sequence is unknown.

Vpr has been implicated in regulation of viral and cellular gene expression. In transient expression assays in cell lines, a plasmid expressing the HIV-1 *vpr* gene was shown to activate the homologous LTR as well as promoters of several heterologous viruses (122). The magnitude of this activation response is relatively small (i.e., three- to tenfold); accordingly, additional studies are necessary to confirm these findings and to map the *cis*-acting target element(s) for Vpr in these disparate promoters. Another investigation suggesting a nuclear role for Vpr is based on analysis of stable transfectants of a human muscle tumor cell line harboring the HIV-1 *vpr* gene (416). In these transfected cells, Vpr appears to block cell proliferation and induces expression (i.e., differentiation) of muscle cell markers. A speculation based on this finding in the muscle tumor cell line is that Vpr may influence viral gene expression by altering a host cell regulatory mechanism which thereby enhances viral replication.

vpx

Vpx is a 14- to 16-kd protein (351) presumed to be translated from a singly spliced, *rev*-dependent transcript produced late in infection (Fig. 14). In mature virions, the number of molecules of Vpx is approximately equal to the number of molecules of the major Gag protein CA (287, 351,817). Genetic analysis of viruses with mutations in *vpx* has not demonstrated a clear phenotype for this gene. Several studies on molecular clones of HIV-2 and SIV harboring mutations in *vpx* have reported that the mutant viruses replicate like wild-type viruses in T-cell lines and in cultures of primary T-lymphocytes and macrophages (314,461, 673). However, other studies with *vpx* mutations revealed that this gene is required for efficient viral replication in primary T-lymphocytes (260,352) and macrophages (240,819). These apparent discrepancies may be the consequence of differences in experimental conditions (e.g., viral strain, multiplicity of infection, cell culture conditions, and method for viral assay) utilized by different researchers.

The mechanism by which Vpx is assembled into virions has been investigated. Immunofluorescence studies with monospecific antibodies in infected cells demonstrate that HIV-2 Vpx is located at the inner surface of the cell plasma membrane (818). This pattern of subcellular compartmentation is believed to require another viral protein(s) because immunofluorescence analysis of COS cells transfected with a plasmid expressing only Vpx reveals that, under these conditions, Vpx is dispersed throughout the cytoplasm. In addition, packaging of HIV-2 Vpx into virions requires another protein(s) from the homologous virus because cells cotransfected with an HIV-1 proviral clone and an HIV-2 *vpx* expression plasmid assemble and release HIV-1 particles lacking Vpx (353). To determine directly the relationship of Vpx to other virion proteins, purified *vpx* mutant SIV_{MAC} particles were disrupted in nonionic detergents, and nucleocapsids were banded by density gradient centrifugation (822). Immunoblot analysis demonstrated that these nucleocapsids contained CA, NC, RT, and IN as well as Vpx; the majority of MA and Env gp (both SU and TM subunits) were absent. To investigate the requirements for incorporation of Vpx into virions, cells transfected with plasmids expressing the HIV-1 Gag precursor ($Pr55^{gag}$) and Vpx were examined by coimmunoprecipitation for formation of complexes between these two viral proteins (304,822). In these experiments, Vpx was observed to associate with the Gag precursor in the cell cytoplasm, and this complex was subsequently exported through the cell plasma membrane into the culture medium (304,822). In addition, Vpx coimmunoprecipitated with the processed CA protein. Thus, these studies in transfected cells show that the *pol* and *env* gene products are not required for assembly of Vpx into virions. The precise location in the cell for the binding of Vpx to $Pr55^{gag}$ and the potential involvement of cellular structures and cellular factors in this interaction are issues which remain to be explored. In addition, analysis of site-specific mutants of both Vpx and Gag may reveal a specific domain on each protein essential for the interaction between these two viral proteins. A strong amphipathic helix near the N-terminus of Vpx has been predicted by computerized secondary structure analysis (353); whether this potential structural feature plays a role in interactions between Vpx and Gag in assembly of viral particles is not known. As proposed for Vpr, Vpx may function at an early step(s) in viral infection because both of these accessory proteins are assembled into virions.

vpu

Human immunodeficiency virus type 1 and the related SIV_{CPZ} contain *vpu*, which is a small gene located 3′ from the first coding exons of *tat* and *rev* (Fig. 3); HIV-2 and other lentiviruses lack *vpu* (519). The reading frame for this gene encodes a protein 81 amino acids in length (120). About 50 to 80 bp of the 3′ end of *vpu* overlap the 5′ end of *env* in a different open reading frame; the extent of this overlap varies with different HIV-1 strains. A singly spliced viral mRNA, dependent on rev function and produced at late times in infection, utilizes a splice acceptor immediately upstream from the initiation codon for *vpu* (Fig. 13) (656a). This transcript is bicistronic because it is translated into both Vpu (16-kd protein) (714) and the precursor for the Env gp (i.e., gp160). Inspection of the predicted Vpu sequence reveals that 27 amino acids at the N-terminus are hydrophobic and that the remainder of the protein is hydrophilic. Biochemical studies show Vpu to be an amphipathic integral membrane protein (type I) which self-associates to form oligomeric complexes; the hydrophobic N-terminus is presumed to function as a membrane anchor domain (715). Immunofluorescence analysis of infected cells demonstrates that Vpu is localized in the perinuclear region, and thus this viral protein is associated with the ER and/or Golgi system. Analysis of Vpu by nuclear magnetic resonance and CD spectroscopy supports a model in which the hydrophilic portion of this protein forms two amphipathic α helices of opposite net charge (652). In infected cells, Vpu is phosphorylated at Ser52 and Ser56 by the cellular casein kinase 2; oligomerization of Vpu is not dependent on phosphorylation (651).

Genetic analysis of *vpu* function has been performed on HIV-1 isolates, with premature stop codons, obtained from infected individuals, and on strains of HIV-1 containing *vpu* lesions produced by site-directed mutagenesis. Although *vpu*-deficient viruses replicate in $CD4^+$ T-cell lines and in both primary T-lymphocytes and macrophages, these mutants are severalfold less efficient at releasing virions into the extracellular medium than viruses with an intact *vpu* gene (374,730). Electron microscopy of cells infected with *vpu* mutants revealed large accumulations of virions in intracytoplasmic vesicles relative to cells infected with virus encoding wild-type vpu. In addition, *vpu*-deficient viruses are more cytopathic in $CD4^+$ T-cells compared to wild-type virus.

Biochemical analysis of infected cells and cells transfected with plasmids expressing Vpu and CD4 (i.e., the cell surface receptor for virion attachment) demonstrated that Vpu causes rapid and selective degradation of CD4 in the ER; the intracellular half-life of CD4 is reduced from 6 hours to 12 minutes (796). In infected cells, CD4 is also trapped by the Env gp in the ER and produces gp160-CD4 complexes (132,796). Consequently, surface expression of CD4 is downregulated, and cleavage of gp160 to gp120 and gp41 is reduced. By decreasing the stability of intracellular CD4, Vpu reduces formation of gp160-CD4 complexes and thereby increases the rate of gp160 processing. The significance of this indirect effect of Vpu on gp160 processing, with respect to viral replication and/or cytopathicity, is not clear.

The relationship of the two phenotypes attributed to *vpu* (i.e., enhancement of virion release and destabilization of CD4) has been investigated (651). HeLa cells, which lack CD4, were cotransfected with a plasmid containing a *vpu*-deficient HIV-1 provirus and a plasmid expressing either wild-type or mutant Vpu. In these experiments, *vpu* augmented export of virions in cotransfected cells; thus, enhancement of virion release by Vpu does not require CD4 (814). To determine the role of phosphorylation in the two *vpu* phenotypes, mutant proviruses were constructed with mutations in Ser52 and Ser56 to preclude phosphorylation. These mutant proviruses did not induce decay of CD4; however, enhancement of virus release was found to be partially dependent on phosphorylation. To examine the role of intracellular compartmentalization on *Vpu* functions, protein traffic from the ER to the Golgi system in infected cells was inhibited with the fungal metabolite brefeldin. This inhibitor blocked the augmenting effect of Vpu on virion release but did not affect degradation of CD4. Thus, the two phenotypes governed by *vpu,* release of virions and decay of CD4, are independent.

A sequence(s) in CD4 that targets this T-cell surface protein for degradation by Vpu was identified by analyzing two types of chimeric proteins in transient expression assays in transfected cells: (a) chimeric proteins produced by exchanging the cytoplasmic tails of CD4 and CD8 (which is not sensitive to degradation by Vpu) (764) and (b) chimeric proteins constructed by fusing the gp120 subunit of HIV-1 *env* with the transmembrane domain and portions of the cytoplasmic domain of CD4 (799). Both of these studies demonstrated that the cytoplasmic tail of CD4 is both required and sufficient for degradation induced by by Vpu. Analysis of deletion mutations in the HIV-1 Env/CD4 chimeric proteins reveals that the sequence Leu414-Ser415-Lys416-Lys417-Thr418 in the tail of CD4 is required for this degradation; this sequence overlaps the target for CD4 interaction with $p56^{lck}$ kinase (764). However, another study suggests that the determinant in CD4 for Vpu-induced degradation is located between Glu418 and Pro425; this study also indicates that Vpu does not require either the phosphorylation sites (i.e., Ser410 and Ser417) or the target site for $p56^{lck}$ (405). In a proposed model, sequences in the hydrophilic portion of Vpu are presumed to interact with the cytoplasmic tail of CD4 and, subsequently, a cellular protease(s) recognizes and degrades this complex. Issues remaining to be investigated are the precise amino acid sequences and potential structural features in the CD4 tail that are required for Vpu degradation, whether phosphorylation of the two serine residues in the cytoplasmic tail of CD4 plays a role in this process, and identification of the cellular protease(s) that mediates Vpu-induced decay of CD4. In addition, it is not known if Vpu interacts with and degrades cellular proteins sharing sequences and/or structural features with the CD4 cytoplasmic tail.

Vpu displays structural similarities with the influenza virus M2 protein, an ion channel protein that modulates the pH of trans-Golgi compartments and thereby protects the viral hemagglutinin from prematurely changing conformation in the secretory pathway (see Chapter 21) (577). A provocative observation is that HIV-1 Vpu enhances the release of capsids produced by the *gag* gene of divergent retroviruses, including HIV-2, VMV, and Moloney murine leukemia virus (250). Whether Vpu has functional similarity to M2 protein in maturation of retroviral proteins and particle assembly remains to be resolved.

nef

All primate lentiviruses encode *nef*; however, a related counterpart is absent from the genomes of the other animal lentiviruses. The *nef* gene of HIV-1 and SIV_{CPZ} extends from the 3′ end of *env* into the U3 domain of the 3′ LTR (Fig. 3). In HIV-2 and several other SIV strains (e.g., SIV_{MAC}, SIV_{SMM}, and SIV_{AGM}), the 5′ end of *nef* overlaps the 3′ end of *env* in a different translation frame (Fig. 3). Human immunodeficiency virus type 1 and HIV-2/SIV *nef* genes encode about 210 and 260 amino acids, respectively (519). Nef is translated from two multiply spliced early transcripts that are independent of the posttranscriptional function of Rev; one transcript is monocistronic, and the other is bicistronic and encodes both Rev and Nef (Fig. 14) (124,608,656a). The 5′ ends of most HIV-1 *nef* genes have two initiation codons corresponding to Met1 and Met20; both initiation codons are utilized to produce two forms of Nef in infected cells (345). Variation in Nef sequence (both intra- and interpatient heterogeneity), together with differences in posttranslational modifications (see below), account for the observation that Nef proteins in cells infected with diverse virus isolates exhibit size heterogeneity in the 27 to 34 k range as measured by electrophoresis in polyacrylamide gels under denaturing conditions (7,750).

Structural Features of nef

Comparison of the predicted amino acid sequences of HIV and SIV Nef reveals both conserved and variable regions (519,678). The 5′ end of all HIV and SIV *nef* open reading frames encode Met1-Gly2. In infected cells, Met1 is removed, and myristic acid is covalently linked to Gly2 at the new N-terminus (256) (Fig. 9). Immunochemical and biochemical analysis of infected cells and cells transfected with expression plasmids show that Nef is predominantly localized to the cytoplasm and the inner surface of the cell plasma membrane (533,821). Mutation of Gly2 in Nef prevents myristylation and precludes attachment of the mutant protein to the cell membrane (533,821); this mutant Nef protein does not downregulate cell surface CD4 (4) (see below). In infected cells, Thr15 of Nef is a target for phosphorylation by protein kinase C (256,257); however, Thr15 is not highly conserved. Other threonine as well as serine residues in Nef may also be phosphorylated by this kinase (e.g., Thr80) (31). Early biochemical studies attributed guanosine triphosphate (GTP)-binding and GTPase activities to Nef, similar to the Ras protein (256, 525); however, these findings as well as Nef autophosphorylation activity have not been confirmed (271,345, 805). Co-immunoprecipitation studies with anti-Nef antibodies shows that Nef of HIV-1 and SIV_{MAC} associates with a cellular serine kinase in infected T-cells (640,641); this finding may provide insight into a potential role for Nef in cell activation (see below). Sequence comparisons of in-

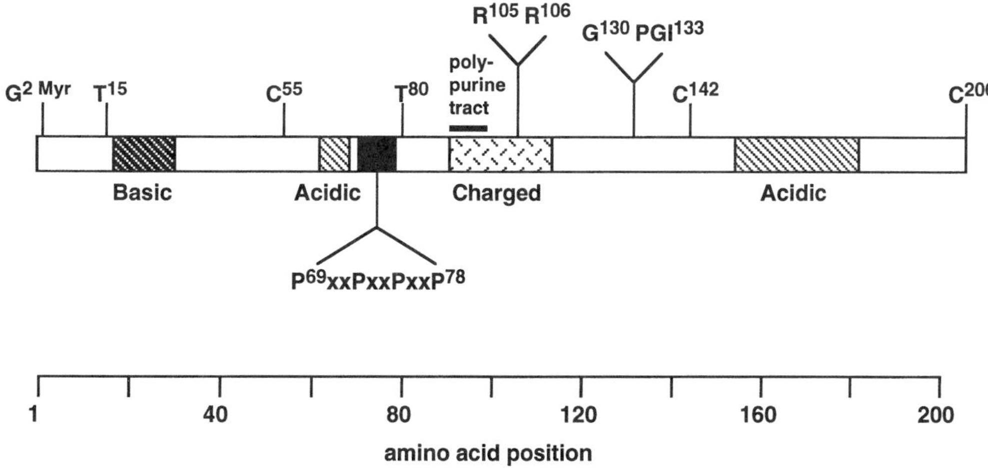

FIG. 9. Features of HIV-1 Nef. The consensus sequence for HIV-1 Nef contains 206 amino acids. The one-letter code is used to designate amino acid residues, and superscript numerals represent position in the sequence. This figure displays features highly conserved in the Nef sequences of HIV-1, HIV-2, and SIV. Nef is initiated with a methionine at position 1. Glycine at position 2 is myristylated ($G^{2,Myr}$) in the mature protein. Threonines at positions 15 and 80 may be phosphorylated by protein kinase C; however, serines at other positions may also be phosphorylated. A region near the amino terminus of Nef is enriched in basic amino acids (shown as a *heavily cross-hatched box*). Cysteine residues at positions 55, 142, and 206 are highly conserved and form intramolecular disulfide bonds. Two regions with a high abundance of acidic residues (i.e., aspartic and glutamic acid residues) are shown as *lightly cross-hatched boxes.* The *dark box* contains the proline-rich motif which resembles regions of proteins interacting with SH-3 domains on cellular signal transduction proteins involved. The polypurine tract also encodes a conserved stretch of six amino acids. A region of about 30 amino acids in the central portion of Nef (*stippled box*) contains a high proportion of charged residues and is therefore predicted to lie on the surface of the folded protein. Two conserved arginine residues (R^{105}, R^{106}) within this central charged region have been shown to be important for association of Nef with a cellular serine kinase, CD4 downregulation, and high virus load in SIV-infected rhesus monkeys (see text). Four amino acids that form a predicted surface turn ($G^{130}PGI^{133}$) are highly conserved in Nef of primate lentiviruses.

dependent HIV-1 isolates reveal a two- to 14-amino acid insertion in Nef at residue 24. These insertions are generally characterized by charged amino acids and include one or more proline residues. Nef of HIV-2 and several SIV strains contain about 12 glycine residues within the first 45 residues; thus, the N-terminus of the Nef proteins of these viruses has the potential to adopt many conformations because glycine does not have a side group to restrict rotation around the peptide bond. A highly conserved acidic region is located at position 62 to 65 of HIV-1 Nef and is found in the corresponding region of HIV-2 and SIV Nef; this region is predicted to be on the surface of the folded protein because of its hydrophilicity (Fig. 9). The region between positions 69 and 80 of HIV-1 Nef contains repeats of the sequence Pro-Xa-Xa-Pro (Fig. 9); only one copy of the motif is found in the corresponding location for HIV-2 and SIV Nef. This proline motif is a feature of proteins that interact with the *src*-homology region-3 (SH3) domain of several cellular proteins mediating signal transduction (629). The central portion of Nef is conserved in the primate lentiviruses (Fig. 9); this region consists of a stretch of about 30 amino acids, the majority of which have charged side groups and are predicted to lie on the surface of the folded protein. A sequence for six codons, located in the middle of Nef within this conserved region, overlaps the polypurine tract (i.e., a 16-nucleotide stretch of A and G residues) (Fig. 9). This part of Nef is highly conserved because the polypurine tract functions as a primer in reverse transcription (see section on viral DNA synthesis). The sequence Gly130-Pro131-Gly132-Ile133/Val133 is present in Nef of many HIV-1 isolates and is predicted to form a β-turn (Fig. 9). Nef proteins of HIV and SIV have a leucine repeat in the middle and an acidic domain near the C-terminus (Fig. 9) (633); these features are reminiscent of leucine zippers and activation domains of the acidic class of transcription factors which bind DNA. However, a role for Nef in transcriptional regulation has not been conclusively demonstrated (see below).

Structural features of Nef may be influenced by formation of intramolecular disulfide bonds. Most strains of HIV-1 contain three conserved cysteine residues (Cys55, Cys142, and Cys205) (Fig. 9) (519) which are reported to form intrachain disulfide bonds (826). Human immunodeficiency virus type 2 and the related SIV_{MAC} and SIV_{SMM} strains all contain Cys55 as well as two to three additional cysteine residues at nonconserved locations; HIV-1 Nef expressed in bacterial and insect cell vectors forms dimers and higher oligomers that are covalently linked by interchain disulfide bonds (367). The extent of Nef oligomerization in infected cells and its significance for function are not known. Whether formation of (potential) intrachain (and/or potential interchain) disulfide bonds influences Nef localization within the cell and/or function is not known.

Amino acid sequence analogies have been noted between Nef and many diverse proteins including *Bacillus* subtilis tetracycline-resistant determinant, yeast DNA binding protein RAP-1, scorpion neurotoxin, class II human lymphocyte antigen (HLA) β chain, mammalian G proteins (see below), and adenosine triphosphate (ATP)-specific protein kinases (678). The significance of any of these analogies with respect to Nef function *in vitro* (i.e., in infected tissue culture cells) and/or *in vivo* is not known. A speculation is that autoimmune responses observed in AIDS patients may be due to molecular mimicry between Nef and the HLA class II β chain.

Function(s) of Nef

Numerous studies have addressed the function(s) of HIV and SIV Nef; however, many of these reports are either disputed or unconfirmed (137). Differences in viral strains (i.e., Nef sequence heterogeneity) and experimental methodologies (e.g., cell type, cell activation state, and multiplicity of infection) for analyzing Nef properties probably account for many of these discrepancies. It is possible that Nef is multifunctional and exerts pleiotropic effects (e.g., enhancement of virion infectivity, activation of T cells, CD4 downregulation, and association with a cellular serine kinase). Nonetheless, the importance of Nef in pathogenesis has been clearly demonstrated in the SIV_{MAC} model because deletion of this gene from a pathogenic molecular clone of SIV_{MAC} produces an attenuated virus which is nonpathogenic in juvenile and adult macaques (365).

Genetic analysis of Nef has been based on examining replication properties of viral mutants in tissue culture systems (137). In initial reports, *nef*-deficient mutants of HIV-1 (i.e., deletion mutations and premature stop codons in *nef*) replicated to higher levels in T-cell cultures than viruses with an intact *nef* open reading frame; accordingly, *nef* was inferred to be a negative regulator that maintains viral latency in the infected individual. However, several studies on *nef*-deficient viruses in other HIV-1 strains as well as in SIV_{MAC} demonstrated that this gene either had no effect or had a positive effect on viral replication. Recent reports have examined the function of HIV-1 Nef by testing the effects of cell culture conditions (i.e., cell activation state) on viral replication (108,150,486,698). Viruses with intact *nef* genes do not require activation of peripheral blood lymphocytes with mitogens to replicate efficiently, whereas *nef*-deficient viruses exhibit reduced replicative capacity in resting lymphocytes. Moreover, the *nef* gene augments virus production in cultures of primary macrophages (486). These positive effects of Nef are dependent on the initial multiplicity of infection (MOI) and the timing of T-cell activation; Nef displays the greatest effect on viral replication (13- to 15-fold) at the lowest MOI. A mechanism by which Nef, a (presumably) nonvirion protein, modifies viral infectivity remains to be determined. These observations on Nef augmentation of viral replication in unstimulated cells, taken together with the ability of Nef to enhance virion infectivity, may have significance in an infected individual because the majority of T-lymphocytes *in vivo* are quiescent (76,825).

Expression of HIV and SIV *nef* genes in $CD4^+$ T cells is associated with reduction of the level of CD4 receptor on the cell surface (Fig. 10). This downregulation was observed in $CD4^+$ T cells and nonlymphoid cell lines (genetically engineered to express the CD4 receptor) after transfection with Nef expression plasmids or infection with a heterologous virus vector (i.e., vaccinia virus or murine leukemia virus) expressing Nef (227,256,603,635). Mutants in Gly2 of Nef are not myristylated and do not downregulate CD4 (4); thus, membrane targeting of Nef is a requisite for this function. CD4 downregulation mediated by Nef was shown to be posttranscriptional because levels of CD4 mRNA were not affected (227). In a transgenic mouse system, Nef expression was also linked to a reduction of CD4 on the surface of thymocytes (684). The target sequence for Nef-mediated downregulation in the cytoplasmic domain of the CD4 receptor was identified through studies with chimeric proteins and site-specific mutations. The region of CD4 between Cys397 to Glu416, proximal to the membrane spanning domain, was sufficient for Nef-mediated downregulation (4). Furthermore, the dileucine motif (Leu413-Leu414) within this region was critical; this motif is also found in a number of cell surface molecules that are endocytosed (4). Three observations indicate that this function of Nef does not involve the (intracellular) phosphotyrosine kinase $p56^{lck}$: (a) $p56^{lck}$ requires Cys420 and Cys422 in CD4 for downregulation (243); (b) nonlymphoid cells lack $p56^{lck}$, yet Nef downregulates CD4 in these cells (4); and (c) CD8 on the cell surface is not downregulated by Nef yet this cell protein binds $p56^{lck}$ (4). Assays aimed at analyzing internalization of cell surface proteins revealed that the rate of CD4 endocytosis is greatly increased in the presence of Nef (4). Whether Nef interacts directly or indirectly (through a host cell factor) with the cytoplasmic domain of CD4 to trigger endocytosis has not been resolved (47). CD4 removal from the cell surface may render a cell less permissive for reinfection by depleting the viral receptor. It is proposed that early expression of Nef downregulates CD4 and thereby prevents multiple rounds of superinfection and cytopathology (4,47a). In this fashion, a quiescent infected cell will maintain a latent provirus until an activation signal (e.g., cytokines, antigen triggering the T-cell receptor) stimulates transcriptional and posttranscriptional events that lead to synthesis of high

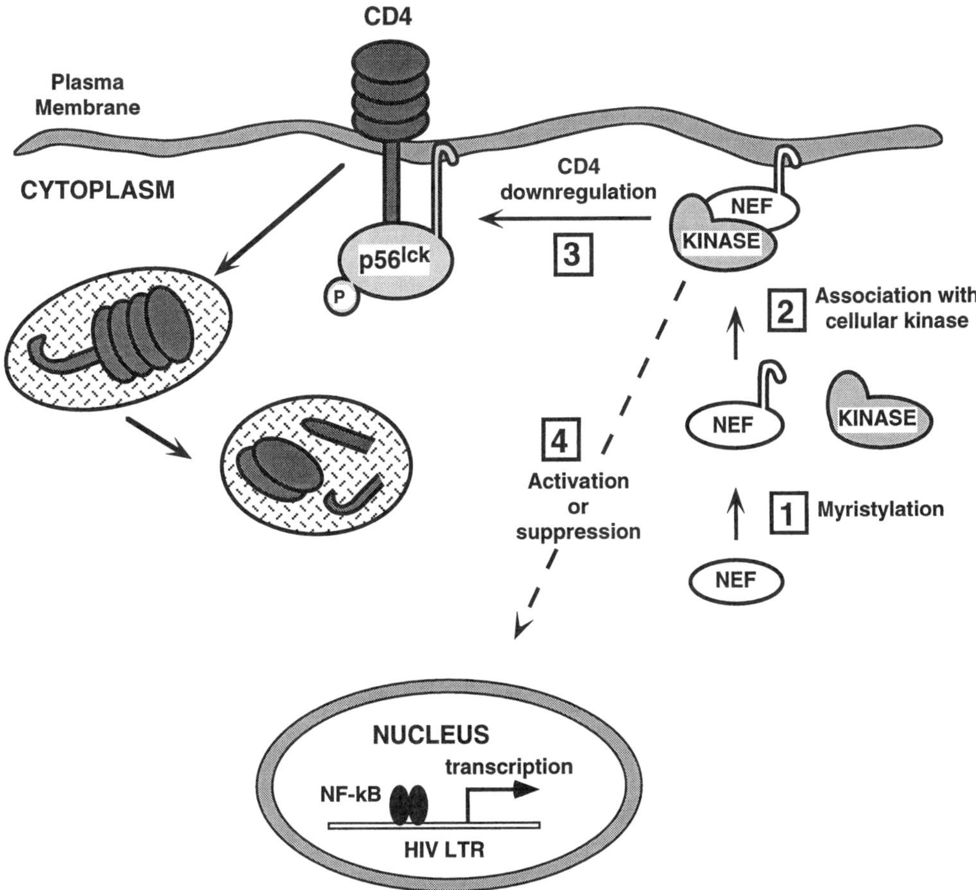

FIG. 10. Intracellular functions attributed to Nef. *(1)* Nef is posttranslationally myristylated; the hook represents myristic acid covalently linked to the glycine residue at position 2 in Nef. *(2)* Myristylation enables Nef to attach to cell membranes and interact with a cellular serine kinase (labeled *K*) and perhaps other cellular proteins. *(3)* The cytoplasmic form of Nef downregulates cell surface expression of CD4 by enhancing endocytosis (i.e., engulfment into endosomes through a process not involving the cellular tyrosine kinase $p56^{lck}$, which is a myristylated phosphoprotein (labeled *p*) also associated with the cell membrane. Endosomes are shown as stippled ovals. The cytoplasmic domain of CD4 contains the target sequences for Nef-mediated downregulation. *(4)* By an unknown mechanism, Nef increases the activity of the cellular enhancer factor, NF-κB, and perhaps other cellular transcription factors, thereby augmenting expression directed by the viral LTR. Nef has also been reported to suppress viral gene expression (see text).

levels of viral RNA and protein and subsequent virion assembly. Interestingly, Env gp and Vpu also downregulate CD4 in T cells in tissue culture; thus, it appears that the primate lentiviruses have redundant functions with respect to reducing the levels of the CD4 receptor on the cell surface (708,796). In addition to preventing superinfection of a cell, downregulation of CD4 may lead to T-cell dysfunction and subsequent immunodeficiency in the infected host.

Several studies have implicated Nef in controlling viral and cellular transcription (137). Transient expression assays in lymphoid and monocytic tumor cell lines and in transformed fibroblasts (COS monkey cell line) revealed that Nef suppressed LTR-directed gene expression (3,531, 748). The significance of this finding has not been determined through studies on permissive cells infected with replication-competent virus. Nef of some HIV-1 strains was shown to inhibit mitogen induction of the cellular transcription factors NF-κB and AP-1, whereas levels of other transcription factors that regulate the LTR were not affected (32) (Fig. 10). Accordingly, Nef suppressed induction of HIV-1 as well as the cellular interleukin-2 gene, both of which are transcriptionally controlled by NF-κB (443,532). The mechanism(s) by which Nef mediates these transcriptional effects has not been defined, and the role of these (potential) transcriptional effects of Nef on viral infection and disease progression has not been established.

A potential role for Nef in regulating the cell activation state has been investigated in tissue culture systems (i.e., *in vitro*) and extended to evaluations in a nonhuman primate model (i.e., *in vivo*). In one report, cell activation was associated with localization of Nef to cell membranes, whereas suppression of cell signaling events was observed when the interaction of Nef with cell membranes was precluded (43,146). The observation that HIV-1 and SIV Nef associate with a cellular serine kinase in T cells supports the notion that this viral protein may influence a signal transduction cascade (640) (Fig. 10). Mutagenesis studies on HIV-1 and SIV have demonstrated that the conserved central region of Nef is important both for interaction with the cellular serine kinase and for CD4 downregulation (641). With respect to significance in an infected host, Nef-mediated alterations in T-cell activation pathways may cause dysfunction of T-lymphocytes and immunosuppression; accordingly, levels of virus will increase in the infected individual. A molecular clone of SIV (SIVmac239nef$^{RR/LL}$), with point mutations in Nef (Arg137-Arg138 mutated to Leu137-Leu138) that inhibit the ability of this viral protein to associate with the cellular serine kinase and downregulate CD4, has been evaluated in rhesus macaques (642) (Fig. 10). Animals inoculated with this mutant clone exhibit low virus load at the early stage of infection and then show high virus load. This increase in levels of virus in the animal is associated with reversion of the point mutations and restoration of the Nef phenotypes (i.e., kinase association and CD4 downregulation). Additional studies in this nonhuman primate model are required to determine whether the same or different domains of Nef control the various *in vitro* functions of Nef (i.e., kinase association, CD4 downregulation, cell activation state, and enhancement of virion activity) and which of these functions are important *in vivo* for maintaining high virus load and for triggering immunodeficiency.

STEPS IN VIRAL REPLICATION

The replication cycle of primate lentiviruses is divided into an early and a late phase, each phase consists of sequential steps, and several of these steps involve specific interactions of viral proteins and nucleic acids with host cell factors (see Chapter 26). The early phase begins with attachment of a virion to a cell surface receptor and continues to formation of a provirus integrated into the host cell genome. After binding of the Env gp in the HIV (and SIV) particle to the CD4 receptor on the cell surface, viral and cellular membranes fuse, and the nucleocapsid is released into the cytoplasm. The viral enzymes RT (and RNase H) and IN remain associated with this uncoated nucleocapsid, and the process of reverse transcription produces linear double-stranded viral DNA which is maintained in a nucleoprotein complex that is subsequently transported into the nucleus. In a reaction mediated by IN, the ends of the linear double-stranded viral DNA are covalently joined to host cell DNA to produce the integrated provirus.

The late phase of replication begins with transcription and processing of viral RNA from the integrated proviral template and ends with release of progeny virions from the cell. In the provirus, the 5′ LTR encodes *cis*-acting sequences for cellular factors that control initiation of viral transcription by RNA polymerase II, and the 3′ LTR contains signals for processing the 3′ ends of all viral transcripts (by attachment of poly-A tails). The viral transcriptional transactivator Tat acts through a signal at the 5′ end of newly initiated viral transcripts to augment initiation and/or elongation by the host cell RNA polymerase II complex. Unspliced, singly spliced, and multiply spliced viral transripts are transported to the cytoplasm and translated into various viral proteins. The viral transactivator Rev recognizes a *cis*-acting element in the full-length viral transcript and controls the ratio of spliced and unspliced RNA; in addition, Rev may control both transport of viral RNA from the nucleus and translation on cytoplasmic polysomes. A portion of full-length viral transcripts interacts with virion precursor polypeptides (i.e., Gag and Gag-Pol) to produce immature nucleocapsids at the cell plasma membrane. These nucleocapsids acquire an envelope by extrusion (i.e., budding) through areas of the plasma membrane containing the Env gp. In newly released extracellular particles, the Gag and Gag-Pol polyproteins are proteolytically processed by the viral PR to yield fully infectious virions.

Virion Attachment and Entry

Interaction with the CD4 Receptor

Attachment of HIV (and SIV) to cells is mediated by an interaction between the extracellular domain (gp120) of the Env gp and the CD4 antigen which is a cellular glycoprotein (55 kd) located on the plasma membranes of T-helper lymphocytes and macrophages (784). This cell surface glycoprotein not only functions to bind virus but also triggers a conformational change in Env gp and thereby facilitates fusion of viral and cellular membranes during the entry process. CD4 was first implicated to be the viral receptor through experiments demonstrating that monoclonal antibodies to this cell surface protein inhibited viral infection in T-cell cultures (140,373). Analysis of extracts of T-cell cultures exposed to HIV-1 revealed that antibodies to gp120 coimmunoprecipitated CD4/gp120 complexes (308,474). Transfection of CD4-negative human cells, such as HeLa, with a complementary DNA (cDNA) clone expressing CD4, renders them permissive for HIV-1 binding, syncytium formation, and infection (444). Moreover, pseudotype and syncytial assays show receptor interference for the primate lentiviruses HIV-1, HIV-2, and SIV (276,309). Studies with antibodies to CD4 that block virion binding demonstrate also show that these viruses utilize the CD4 cell-surface protein for attachment (638).

CD4 is a member of the immunoglobulin superfamily and functions as a coreceptor in the antigen- and class II

MHC-dependent interactions that initiate T-cell activation (86). The *src*-related tyrosine kinase $p56^{lck}$ is noncovalently associated with the cytoplasmic tail of CD4 in T cells and mediates signal transduction. A detailed structure of CD4 has been obtained by integrating electron density maps, produced by x-ray crystallography (626,775), with the results of investigations on site-specific mutants and characterization of epitopes identified with monoclonal antibodies (274,501). The N-terminal extracellular portion of CD4 consists of four immunoglobulinlike domains (D1 to D4) and appears to be a rodlike molecule, about 125 Å in length. This extracellular portion is anchored to the cell surface by a transmembrane region followed by a short cytoplasmic tail at the C-terminus.

The region of the receptor that binds gp120 has been determined through biochemical and genetic analyses coupled with inhibition studies with monoclonal antibodies recognizing specific epitopes on CD4 (274,501). Many of these studies have been performed with soluble monomers of CD4 (i.e., extracellular domain lacking the transmembrane hydrophobic region and the cytoplasmic tail) and gp120 purified in a relatively native configuration from genetically engineered mammalian cells. A limitation to the use of these soluble proteins is that the role of multimers of CD4 and/or gp120 can not be evaluated. Nonetheless, a high-affinity binding site for gp120 has been localized to the first immunoglobulin-like domain (D1) of CD4 (66,69,680); this domain is analogous to the complementarity-determining region 2 of an antibody light-chain variable domain and consists primarily of a stretch of about 20 amino acids (residues 40–60) arranged into three β-strands to produce a ridge that is presumed to bind a cleft in gp120 (see below). Mutagenesis studies (15), as well as investigations on synthetic peptides mimicking portions of CD4 (344,422) and antibodies specific for epitopes outside the D1 region (747), indicate that regions outside this ridge also influence interactions with the viral glycoprotein. The soluble form of CD4 blocks replication of virus in tissue culture cells (687); accordingly, knowledge of the structural features of CD4 involved in binding to gp120 may form the basis for novel antiviral inhibitors which compete for attachment of virions to the CD4 receptor on the surface of permissive cells (86) or cause premature dissociation of the gp120/gp41 complex (734).

Entry

The mechanism(s) of virion entry (i.e., internalization) and the role of CD4 in this process has been studied in tissue culture systems employing T-cell lines (784). An initial investigation demonstrated that HIV-1 entry was pH-dependent because treatment of permissive T cells with ammonium chloride blocked infection this finding implicated receptor-mediated endocytosis as an entry pathway (444). However, other studies report that exposure of T-cells to lysosomotropic agents does not inhibit virus infection (469). Electron microscopy has revealed fusion of the viral and cell plasma membranes after exposure of tissue culture cells to virus (705). The consensus is that the major pathway for HIV internalization in most cells is direct fusion of viral and cell plasma membranes by a process independent of pH. Uptake of virus by endocytosis may be more important in cells in the macrophage lineage where CD4 endocytosis is more efficient than in T cells. To focus on the role of CD4 in entry, human (i.e., HeLa) and murine (i.e., NIH3T3) monolayer cell lines have been transfected with the cloned human CD4 gene (444). $CD4^+$ HeLa cells are infectable and form multinucleate syncytia, whereas $CD4^+$ NIH3T3 cells do not support viral replication; the latter cells bind but do not internalize virus. Accordingly, it is hypothesized that a second receptor (also termed the accessory receptor) mediates virus entry after attachment and that this receptor is species specific (i.e., found on human cells but not on murine cells) (784). Additional studies are required to identify the putative postattachment receptor and to investigate its interaction with gp120 as well as gp41.

To elucidate features of CD4 that play a role in viral internalization, HeLa cells were genetically engineered to express deletion mutants of CD4, chimeric CD4/CD8 molecules (46), chimeric CD4/thy-1 molecules (330), and a glycolipid-anchored form of CD4 (161). Substitutions as well as deletions in the cytoplasmic tail of CD4 do not prevent infection or inhibit the capacity to form syncytia; thus, the interaction of the cytoplasmic tail of CD4 with the $p56^{lck}$ kinase is not required for fusion and entry. However, hybrid proteins with the external domain of CD4 and the cytoplasmic tail of CD8 were less efficient than native CD4 at mediating viral infection. Taken together, these findings demonstrate that fusion of viral and cell plasma membranes does not require CD4-mediated signal transduction, the extracellular domain of CD4 is sufficient for infectivity, and the C-terminal domain of CD4 facilitates the entry process.

In the model for virion entry into lymphoid cells, it is thought that conformational changes are induced in both the external (i.e., gp120) and transmembrane (i.e., gp41) subunits of the Env gp oligomer after virions attach to cell surface CD4 (Fig. 11); in addition, the conformation of CD4 may also be altered (501). Structural changes in gp120 have been explored by measuring binding of monoclonal antibodies to specific epitopes and evaluating sensitivity to cleavage by exogenous protease (639,735). These experimental approaches indicate that exposure of the V3 loop is enhanced after gp120 interacts with CD4. The interaction of CD4 with the Env gp oligomer promotes dissociation of the gp120 subunit from the gp41 subunit in virion membranes (494). Additional studies with monoclonal antibodies to gp41 also suggest that previously occult epitopes on gp41 are also uncovered after CD4 binds the oligomer; however, this display of gp41 epitopes may

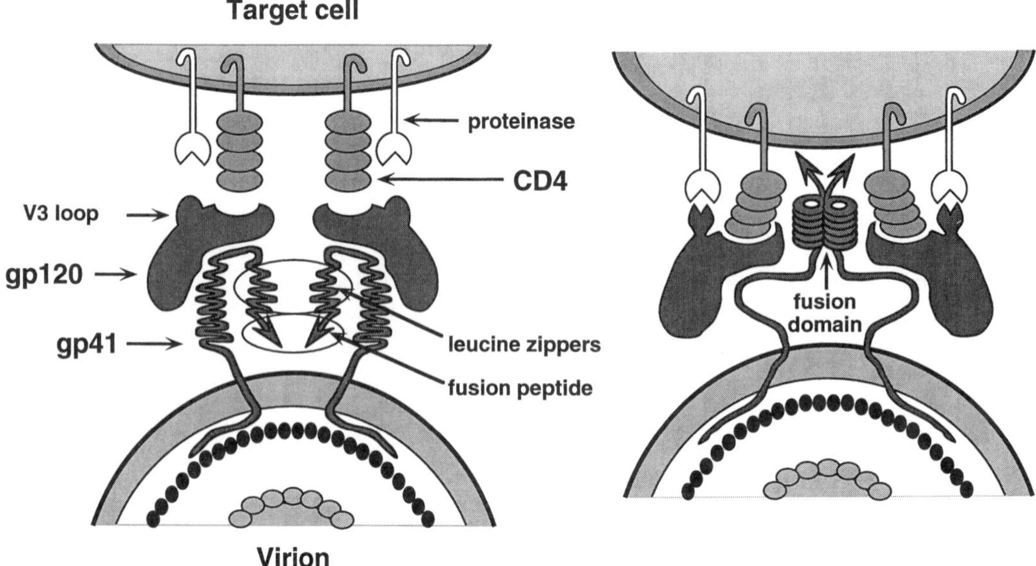

FIG. 11. Model for virion attachment and entry. This speculative model considers a role for both structural transitions in the Env gp and a cell surface protease during virion attachment and entry. **A:** Native state. The Env gp in the virion membrane is assumed to be dimeric, although trimeric and/or tetrameric forms have been reported (see text). Noncovalent protein-protein interactions maintain the gp120/gp41 heterodimer and are responsible for the association of this heterodimer into an oligomeric complex. Leucine zipper sequences in the gp41 subunit in each chain of the oligomeric complex are held apart in the native state. **B:** Fusogenic state. Each gp120 subunit of the Env gp binds to the CD4 receptor on the cell surface; the role of multimerization of CD4 in binding and subsequent entry steps remains to be defined. Presumably, a cellular protease in the plasma membrane recognizes and cleaves sequences in the V3 loop after attachment. Either the binding to CD4 or proteolytic cleavage of gp120 induces a conformational change in the Env gp, thereby exposing the hydrophobic fusion peptide at the N-terminus of the gp41 subunit. After Env gp binds to CD4, it is thought that the leucine zippers in the gp41 subunits interact to form stable coiled coils; this conformational change in the gp41 subunits is presumed to produce a fusion domain and thereby to facilitate insertion of the fusion peptide into the cell membrane. Insertion of the fusion peptide into the cell plasma membrane initiates fusion of viral and cell membranes.

simply reflect gp120 shedding (49,498,734). Syncytium-inducing strains of HIV-1 more readily release gp120 after exposure to soluble CD4 compared to NSI strains; thus, NSI variants, representing the majority of primary viral isolates, are more resistant to neutralization by soluble forms of CD4 than SI variants (277,499).

In a proposed model, structural changes in the Env gp oligomer on virions result in exposure of the hydrophobic fusion peptide at the N-terminus of gp41 (between amino acid residues 517 and 527) (501). The region between Leu553 and Leu590 is similar to a leucine zipper and has been proposed to adopt a coiled-coil conformation which plays a role in fusion (794). After exposure by dissociation of gp120, the fusion peptide inserts into the lipid bilayer of the cell plasma membrane and thereby initiates fusion of the viral and cell membranes (Fig. 11) (67). In the influenza virus system, a pH-dependent conformational change includes formation of a coiled-coil within a hydrophobic domain of the transmembrane subunit of the hemagglutinin protein in the viral envelope (87). A speculation is that attachment of Env gp to the CD4 receptor on the cell surface causes the receptor to condense, and thus the virion and cell membranes are brought into close proximity to allow for fusion (790). It is not known if dimerization and/or clustering of CD4 molecules on the cell plasma membrane play roles in viral attachment and entry.

Cell surface proteins other than CD4 may also induce conformational changes in Env gp during attachment and entry and thereby influence these steps in the early stage of viral replication. First, a cell membrane protease in the trypsin family cleaves the V3 loop in gp120 (112,366,515) (see section on the envelope glycoprotein; Fig. 8). It is proposed that this cleavage event produces a structural change in the Env gp oligomers in the virion surface. Subsequently, the hydrophobic N-terminus of gp41 is uncovered to mediate fusion of viral and cell membranes. Although the V3 loops of HIV-2, SIV, and some HIV-1 isolates lack the Gly-Pro-Gly-Arg-Ala-Phe cleavage site (Fig. 8), all have a potential cleavage site for a chymotrypsinlike enzyme immediately C-terminal. Second, inhibition of a cellular protein disulfide isomerase, localized in plasma membranes, blocks HIV infection in tissue culture cells (625). Presumably, this enzyme reduces critical disulfide bonds in the Env gp after virions attach to the CD4 receptor. Reductive cleavage of disulfide bonds is expected to alter the conformation of Env gp and thus expose the fusion peptide at the N-terminus of gp41. Both of these intriguing possibilities merit further investigation.

Alternative Receptors

Several alternative cell surface receptors have been implicated in virion attachment and entry into certain neuronal and colorectal tumor cell lines as well as several other cell types, all lacking detectable CD4 (784). Human immunodeficiency virus type 1 gp120 binds to galactosyl ceramide which is found on glial cell lines as well as colon cell lines (55,269,429), and monoclonal antibodies specific for this glycolipid block viral infection in these cells (193,811). Because virus is confined to macrophage-related microglial cells (and perhaps endothelial cells) in the brain of infected individuals, the significance of galactosyl ceramide in mediating interactions of virus with neuronal and colorectal cells is uncertain. Fc and complement receptors offer another potential mechanism for viral attachment and internalization. Virions coated with antibody molecules can bind to Fc receptors which are found in a variety of cell types (475,725). Although complement receptors have been implicated in HIV-1 infection independent of either antibody or CD4 (64), further studies are required to confirm this potential entry mechanism. It is not known whether this (potential) opsonization in infected individuals is significant with respect to permitting viral entry into cells lacking CD4 and/or enhancing infectivity in $CD4^+$ cells. Leukocyte function-associated antigen 1 is a cell adhesion molecule which may also play a role in viral entry because antibodies to this cell surface protein inhibit syncytium formation (289). Additional studies in tissue culture systems with antibodies to a member of the integrin family of transmembrane glycoproteins also suggest that this class of cell surface molecules, which involved in cell-matrix and cell-cell adhesion, may contribute to induction of syncytia by virus (539). Because syncytium induction is a complex process which is not completely understood, the significance of experiments implicating various cell surface molecules for HIV and SIV infection remains to be determined. It is most likely that these adhesion molecules facilitate cell-cell contact during infection rather than functioning as true receptors which, by definition, must interact with a virion surface component.

Viral DNA Synthesis by Reverse Transcription

The model for lentivirus reverse transcription is similar to that proposed for other retroviruses and has been elucidated by analysis in cell-free systems (i.e., in reactions with defined components) and by identification of viral DNA synthesis in the early stage of viral infection in tissue culture cells (see Chapter 26) (570,788). Reverse transcription requires two strand-transfers (or jumps) to convert the dimeric single-stranded viral RNA into a molecule of double-stranded linear DNA. The minimal requirements for this step in retroviral replication are (a) two copies of the viral single-stranded RNA genome (i.e., template), (b) the host $tRNA^{lys}$ primer that is bound near the 5′ end of the viral RNA template (i.e., primer binding site, PBS), and (c) the heterodimeric protein (i.e., p51 and p66) that has both RT and RNAse H activity. Nineteen nucleotides at the 3′CCA end of the $tRNA^{lys}$ primer are hydrogen bonded to the PBS, and the interaction of this primer with the PBS is presumed to be influenced by secondary structure in the leader region of viral RNA (49b).

Viral particles are partially uncoated immediately after entry into cells to produce a large nucleoprotein complex which is presumed to resemble the virion capsid (63a,194). Accordingly, capsid proteins and the architecture of this complex may influence the efficiency of reverse transcription. Reverse transcriptase is activated by a signal, as yet unidentified, in the cytoplasm to initiate DNA synthesis from the $tRNA^{lys}$ primer, which is bound to the genomic viral RNA template. The first DNA strand (i.e., minus strand) is extended by RT to the 5′ end of the viral RNA genome; this product is designated minus strand strong-stop DNA (Fig. 12). Subsequently, RNAse H activity in the p51/p66 heterodimer (Fig. 5) degrades RNA in the DNA/RNA hybrid and thereby exposes newly synthesized strong-stop DNA sequences which are complementary to the short repeat (R) at the 5′ end of the viral RNA genome (158,644). In the first strand-transfer reaction, the exposed strong-stop DNA hybridizes to complementary sequences in R at the 3′ end of the viral RNA genome (Fig. 12). This first strand transfer event appears to involve both viral RNA molecules in the nucleoprotein complex (i.e., intermolecular translocation), although there is support for an intramolecular transfer (158,313,441). After the strand transfer, minus strand DNA synthesis continues on the viral RNA template through U3 and into the viral genome. For synthesis of plus strand DNA, lentiviruses utilize two primers: a polypurine track (PPT) which borders the U3 domain in the 3′ LTR and a central PPT (cPPT) located at the end of the *pol* gene sequence (99,592), whereas most other retroviruses initiate plus strand DNA synthesis only at a PPT near the 3′ LTR (Fig. 12). Plus strand DNA is elongated to the end of the minus strand DNA template to produce plus strand strong-stop DNA. A second strand transfer is required to continue plus strand DNA synthesis; this strand transfer is

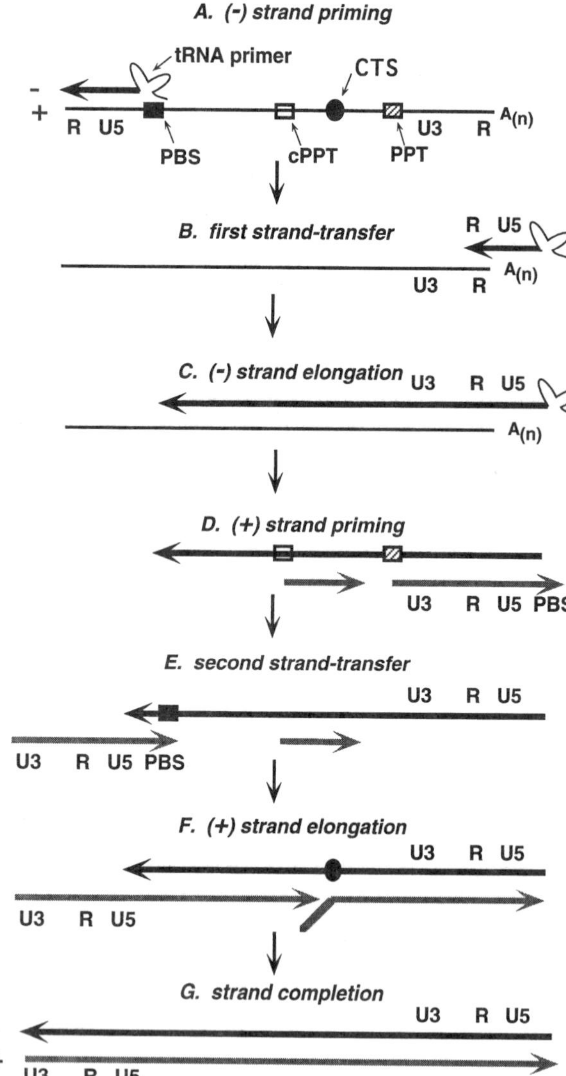

FIG. 12. Mechanism of reverse transcription. The viral RNA genome has plus strand polarity and is shown as a *thin solid line,* designated *(+).* The poly-A tail on viral RNA is labeled $A_{(n)}$. Synthesis of the first strand of DNA [negative polarity, represented as a *heavy solid line* and designated *(-)*] is primed by a tRNA^{lys} molecule hydrogen bonded to the PBS of the template RNA. Synthesis of the second DNA strand is primed by a short viral RNA fragment generated by RNase H activity at the PPT that borders U3 and a central PPT (cPPT) located in viral sequences near the 3′ end of the *pol* gene. The second DNA strand has plus strand polarity and is represented as a *heavy gray line.* A CTS, which functions to terminate plus strand viral DNA in step F, is located near the 3′ end of the *pol* gene and is shown as a *solid circle;* about 100 nucleotides of plus strand DNA is displaced to allow for termination at the CTS (100). To simplify the pictorial representation of lentivirus reverse transcription, important *cis*-acting signals (PBS, CTS, cPPT, and PPT) are shown only in key steps. Details of the reverse transcription process are presented in the text of this chapter and in Chapter 26.

an intramolecular event in which the tRNA^{lys} PBS in plus strand viral DNA is base paired with the 3′ end of the minus strand strong stop DNA (556,570). After this second strand transfer, RT continues to elongate plus strand DNA to the central termination signal (CTS) located immediately 3′ to the cPPT (100); in addition, plus strand DNA initiated at the cPPT is elongated to the end of the minus strand DNA template (Fig. 12). About 100 nucleotides of plus strand DNA, between the cPPT and CTS, is displaced. Cellular enzymes are presumed to remove the displaced sequences and seal the plus strand to yield double-stranded linear DNA which has an LTR at each end; the ends of the duplex form of viral DNA are thought to be blunt. Although many features of reverse transcription are shared between lentiviruses and other retroviruses, it appears that internal initiation of plus strand DNA (at the cPPT) is unique to lentiviruses. Cell cycle and/or cell activation events may influence the efficiency of the reverse transcription step in viral replication by regulating the amount of precursor nucleotides within the cell (224).

Integration of Viral DNA

The steps in retroviral integration have been established by characterizing integrative intermediates of viral DNA and analyzing viral-host cell DNA junctions in infected cells (see Chapter 26) (243,789). Integrase mutants of HIV-1 do not integrate and no infectious virus is produced (291). Importantly, cell-free integration systems with defined components have been used to produce a detailed model for this critical step in viral replication (see section on IN). The starting point for the model for viral integration is the viral nucleoprotein complex in the cell cytoplasm. Although the structure as well as the precise composition of this preintegration complex is not known, linear double-stranded viral DNA, produced by reverse transcription in the cytoplasm, and IN are minimal requirements. Biochemical analysis suggests that MA protein is also part of this complex (78) (see section on MA). Translocation of the preintegration complex into the nucleus, thought to be mediated by nuclear localization signals in MA and/or Vpr, is an active transport process of the host cell that requires ATP but is independent of cell division (220,284,418). Accordingly, it has been proposed that primate lentiviruses replicate in nonproliferating, terminally differentiated cells in contrast to other retroviruses (e.g., murine leukemia virus, avian sarcoma-leukemia virus) which require passage of cells through mitosis for efficient virus production (417,614). This view has been challenged by investigations of HIV-1 infection in primary macrophages where it appears that the cell activation state coinciding with the G1/S phase of the cell cycle, and not cell DNA synthesis or mitosis itself, governs viral replication (418,653).

Retroviruses show very little target sequence specificity for integration; however, it appears that integration is not random (755,675). Inspection of cell sequences flanking proviruses reveals that HIV-1 preferentially inserts into or near two classes of repeated DNA elements in the human genome, the L1 and *Alu* elements (707). These elements are termed retroposons and share properties with retroviruses. Thus, preferential integration of HIV-1 into L1 and *Alu* elements may reflect local chromatin structure which is more susceptible to acquisition of a retroposon. Analysis of retroviral integration *in vitro* with target DNA in the form of minichromosomes and nucleosome cores has shown that structural features of chromatin appear to influence target site selection by the integration complex; IN either prefers to interact with DNA segments having a wide major groove or utilizes pre-existing deformations of DNA within the nucleosome core (588–590). The role of cellular proteins in HIV integration has been investigated in the yeast two-hybrid system; using this approach, IN interactor 1 (Ini1) has been cloned from a cDNA library from a human macrophage cell line (343). Integrase interactor 1 not only binds to HIV IN but also stimulates its DNA joining activity *in vitro*. The sequence of Ini1 is similar to that of the yeast transcription factor SNF5, which is part of a large complex thought to recognize chromatin structure and thereby govern expression of several genes. Accordingly, in a proposed model, Ini1 may make certain stretches of DNA more accessible by loosening or opening up the tightly bound chromatin structure. Whether other cellular proteins participate in integration is not known, and the mechanism for attachment of the preintegration complex to host cell DNA in nuclear chromatin remains to be defined.

Steps for the cleavage-joining reactions in the integration process are described in detail above in the section on IN (83,174). Because the IN reaction leaves single-stranded gaps and two mismatched nucleotides at each 5′ terminus, host cell enzymes are presumed to remove the mismatched nucleotides, fill in the gaps, and nick seal (i.e., ligate) the remaining ends.

Supercoiled forms of HIV-1 DNA have been detected in infected cells (438,688,788); however, the significance of these supercoils for viral replication is unclear. Nucleoprotein complexes containing circular forms of HIV-1 DNA lack integrase and are, therefore, incapable of integrating (79). Moreover, biochemical analysis of the mechanism of retroviral integration shows that linear viral DNA, and not circular viral DNA, is the intermediate for the covalently linked provirus in the cell genome (see Chapter 26).

Regulation of Viral Gene Expression

Control of HIV (and SIV) RNA synthesis is complex and involves the interplay of *cis*-acting viral elements, viral transactivators, and several cellular proteins. These interactions regulate basal levels and induce high levels of viral gene expression (12a,336). The site of viral RNA synthesis is the cell nucleus, and the template is proviral DNA integrated into the host cell genome. Retroviral LTRs are divided into domains (U3, R, and U5) which have distinct functions in transcription (Fig. 12) (see Chapter 26). The U3 domain of HIV and SIV contains basal promoter elements, including a TATAA box for initiation by host cell RNA polymerase II and sites for binding the cellular transcription factor SP1. Additional *cis*-acting sequences in this domain are recognized by several cellular factors which modulate transcription by influencing the rate of initiation. Initiation of viral RNA takes place at the U3/R border of the 5′ LTR, and the 5′ ends of these viral transcripts (designated +1) are posttranscriptionally capped with 7-methylguanosine by cellular enzymes. The viral transactivator Tat functions through a *cis*-acting sequence, designated the tat-response element (TAR), which is located in R from +19 to +43 (see below). R-U5 (+1 to +185) is in the leader sequence of the full-length viral transcript as well as all spliced transcripts. The 3′ ends of viral transcripts are defined by the R/U5 border in the 3′ LTR (Fig. 14); signals in U3 and R (AAUAAA and a downstream GU-rich element) are recognized by cellular functions which add poly-A tails to the 3′ ends of viral transcripts. Although both LTRs are identical in sequence, retroviruses have mechanisms by which the 5′ LTR is used for initiation and the 3′ LTR signals addition of poly-A tails (103,160).

Full-length HIV and SIV transcripts serve three roles: (a) genomic RNA in progeny virions assembled at the plasma membrane, (b) mRNA for translation of Gag and Gag-Pol polyproteins in the cytoplasm, and (c) precursors for over 30 alternatively spliced mRNAs that are also translated in the cytoplasm to produce Env gp as well as accessory proteins (Fig. 14) (128,529,566). The ratio of spliced to unspliced viral mRNA is controlled by the regulator of viral gene expression, Rev, which is a small viral-encoded transactivator protein functioning through a Rev-response element (RRE) in the full-length transcript (see below). In addition, the *rev* gene product appears to facilitate transport of viral RNA from the nucleus to the cytoplasm and may influence translation of viral mRNA on cytoplasmic polysomes. In the model for regulation of viral gene expression, the early phase is characterized by synthesis of multiply spliced viral mRNAs that encode Tat, Rev, and Nef (i.e., nonvirion proteins); the late phase features preferential accumulation of unspliced and singly spliced viral mRNAs encoding proteins incorporated into virions (134,583). Importantly, cell activation signals regulate HIV and SIV transcription through *cis*-acting elements in the LTR and are presumed to be important in controlling viral replication in the infected host (522). Thus, knowledge of the molecular mechanisms governing viral gene expression will provide insight into clinical latency and pathogenesis and may point to specific targets for antiviral therapy.

Transcriptional Promoter Elements in the LTR

Cis-acting elements regulating basal and inducible transcription in the HIV and SIV LTR have been inferred by sequence analysis and directly identified by several methods. Plasmids containing the LTR linked to reporter genes, such as bacterial chloramphenicol acetyl transferase or firefly luciferase, have been tested in transient expression assays in a variety of cell types; this approach permits analysis of site-specific mutations in the HIV and SIV LTR to assess the role of specific sequences in viral RNA synthesis. Cellular factors interacting with transcriptional control elements in the LTR have been identified through electrophoretic mobility shift assays, DNAse footprinting, and in *in vitro* transcription systems reconstituted with LTR templates and defined factors. Importantly, several studies have evaluated the roles of *cis*-acting transcription elements on viral replication in tissue culture systems by constructing and testing mutant proviruses. The biologic significance of the results of detailed studies on molecular mechanisms regulating HIV and SIV transcription is based on observations in the murine and avian leukosis viruses; in these latter retroviruses, *cis*-acting regulatory elements in the U3 region of the LTR govern the pathogenic potential of the virus by modulating viral gene expression in a cell-type and tissue-specific manner (189).

Core Promoter

The U3 region of the HIV and SIV LTR encodes a TATAA box (–28 to –24) for binding a complex of cellular factors designated TFIID (Fig. 13); this complex is composed of the 38-kd TATAA-binding protein (TBP), additional cellular factors termed TBP-associated factors or TAFs, and the multisubunit RNA polymerase II (336). Viral transcripts are initiated 22 bp downstream from the TATAA box (599,634,771). Mutation of this promoter element in the HIV LTR greatly reduces basal promoter activity (49b, 521) and severely impairs viral replication in tissue culture cells (431). Sites for binding the cellular transcription factor SP1 are positioned immediately 5′ to the TATAA box

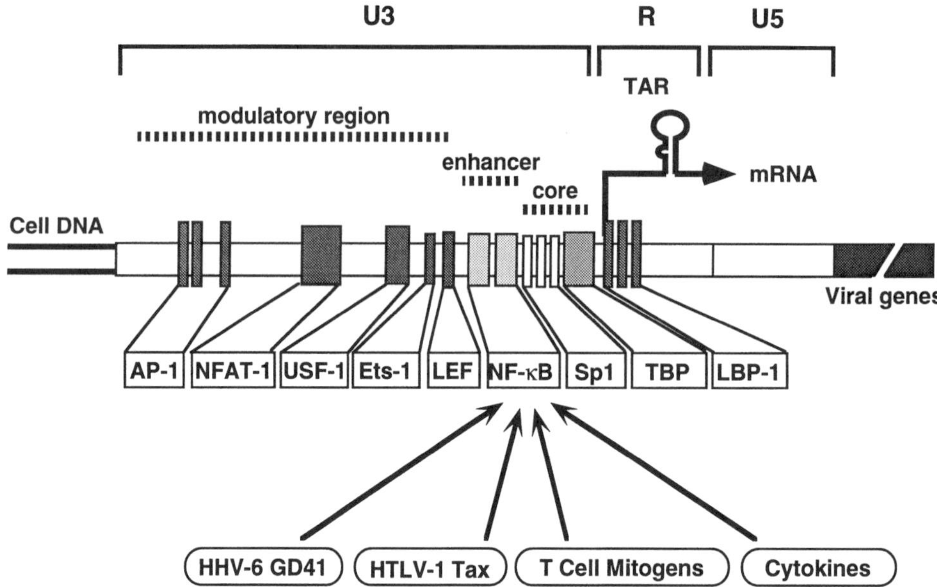

FIG. 13. Transcription elements in the HIV-1 LTR. The viral LTR is divided into three functionally distinct regions designated U3 (–453 to +1), R (+1 to +98), and U5 (+99 to +185). Three transcriptional domains constitute the viral promoter in the U3 region: the core or basal domain, the enhancer, and the modulatory domain. Transcription initiates at the U3/R border in the 5′ LTR; the core promoter contains binding sites for cellular transcription factors including TATAA-binding protein (TBP) (–28 to –24) and SP-1 (three sites between –78 to –45). The enhancer of HIV-1 consists of a duplication of the 10 bp NF-κB site (–104 to –81). Regulatory proteins of heterologous viruses (e.g., Tax of HTLV-I and GD41 of HHV-6), T-cell mitogens, and cytokines enhance HIV transcription and replication through activation/induction of NF-κB as well as other cellular transcription factors. The modulatory domain in the U3 portion of the LTR contains binding sites for several cellular factors, including AP-1 (–350 to –293), NFAT-1 (–256 to –218), USF-1 (–166 to –161), Ets-1 (–149 to –141), and LEF (–136 to –125) (571). Sequences located 3′ of the TATA box in the LTR are recognized by LBP-1 (–17 to +21). The R region also encodes the RNA sequence which forms the *trans*-activation response element, TAR (+19 to +43) (see Fig. 16). The R/U5 boundary at +98 in the 3′ LTR defines the 3′ end of all viral transcripts; this boundary is determined by the polyadenylation signal AAUAAA in R (+73 to +78).

(334). Human immunodeficiency virus type 1 has three consensus SP1 binding sites between –78 and –45 (Fig. 13); HIV-2 has four sites, and strains of SIV have from two to four sites (292). Taken together, sequences encompassing the TATAA box and the SP1 sites constitute the core promoter. Mutation of SP1 sites in the HIV-1 LTR diminishes both basal promoter activity (270,334) and viral replication in some but not all cell types (563,619). Additional studies have demonstrated that SP1 sites, as well as the spacing between the SP1 sites and the TATAA box, are important for efficient Tat-mediated transactivation of gene expression directed by the viral LTR (49b,315). The immediate vicinity of the TATAA box contains binding sites for the leader binding protein-1 (LBP-1), also designated upstream binding protein-1. This cellular factor is a 63 to 68-kd protein and has a high-affinity site located in sequences surrounding the site of transcription initiation (–16 to +27) and a low-affinity site encompassing the TATAA box (–38 to –16) (335,808). Analysis of viral RNA synthesis in cell-free systems that measure initiation from HIV-1 LTR template DNA has demonstrated that LBP-1 binding to the strong site activates transcription, whereas binding to the weak site inhibits transcription (335,357,815). Activation of cells with phorbol myristic acid (PMA) produces a change in the local chromatin structure near the LBP-1 binding site (762); accordingly, additional studies with chromatin-assembled DNA templates will be important for elucidating the function of LBP-1 in viral transcription.

Sequences near the HIV-1 RNA initiation site also contain at least three different regulatory elements: (a) inducer of short transcripts (IST), (b) HIV-1 initiator or SSR, and (c) TAR. The IST encompasses nucleotides –1 to +80 and mediates the synthesis of short transcripts, ranging from 55 to 60 nucleotides, in the absence of the viral transactivator Tat (598,668). The HIV-1 initiator or SSR element is not well characterized but appears to play a role in basal transcription (829). Additional studies are necessary to identify cellular factors interacting with IST and SSR and to determine the role of these elements in viral replication. The TAR is a stem-loop structure between nucleotides +19 and +43, and functions in Tat-mediated transactivation (see below).

Enhancer Elements

Immediately upstream of the core promoter, HIV and SIV contain one or more copies of the 10-bp sequence (GGGACTTTCC) recognized by the enhancer factor NF-κB (Fig. 13). This cellular transcription factor is a heterodimer comprised of a 50-kd DNA binding subunit and a 65-kd activation subunit (522). Human immunodeficiency virus type 1 has a duplication of the 10-bp NF-κB binding site between –104 and –81, and mutagenesis studies demonstrate that both sites are functional. Human immunodeficiency virus type 2 has one conserved NF-κB site and a second site which differs from the consensus; the latter appears to be nonfunctional. Various strains of SIV contain either one or two conserved NF-κB sites (292). In uninfected cells, NF-κB controls transcription of several cellular genes such as immunoglobulin κ light chain, class I and class II MHC, IL-2, IL-2 receptor α, β-interferon, and tumor necrosis factor-α (TNF-α) and also regulates transcription directed by the promoters of several viruses (e.g., SV 40 viral and HTLV) (25). Mutation analysis in HIV-1 reveals that the NF-κB sites are not required for efficient viral replication in cultures of activated primary T-lymphocytes (406). In SIV, duplication of the NF-κB site is associated with enhanced replicative capacity of virus in T-cell lines (164).

Regulation of HIV and SIV transcription by NF-κB may be significant in the infected host because T-lymphocytes and macrophages are generally quiescent (825). In unstimulated cells, NF-κB is complexed in the cytoplasm with an inhibitor (IkB, 105 kd) which blocks DNA-binding activity. Cell activation signals lead to dissociation of this complex, thereby permitting NF-κB to translocate into the nucleus (522). Tissue culture cells treated with PMA and mitogens (e.g., the plant lectin phytohemagglutinin) show increased levels of NF-κB and support relatively high levels of transcription directed by the viral LTR (56,360, 520,601,679,739). Cytokines such as TNF-α and IL-1 also augment NF-κB activity in cultured T cells and macrophages and thereby enhance both LTR-directed transcription and viral replication (114,205,546,581). The redox potential of a cell also influences NF-κB activity; antioxidants (e.g., N-acetyl cysteine) inhibit NF-κB activation and suppress transcription directed by the HIV LTR as well as viral replication in tissue culture cells (341,700). Conditions eliciting intracellular stress and DNA damage also affect the function of NF-κB; heat shock activates HIV proviruses in latently infected cell lines (701), and ultraviolet light treatment of stably transfected cells enhances LTR-directed transcription (753). Experiments in tissue culture systems have shown that regulatory gene products of a variety of heterologous viruses [e.g., Epstein-Barr virus (EBV), cytomegalovirus (CMV), human herpesvirus type 6 (HHV-6), adenovirus, HTLV-I, human hepadnavirus (HBV), and spumaviruses) augment levels of cellular transcription factors which activate the HIV LTR (12a,41,234,596). For example, in transient expression assays, ORF-A of the GD41 gene of HHV-6, a T-lymphotropic herpesvirus, activates HIV-1 transcription through the NF-κB sites in the LTR (831). Taken together, these findings indicate that several diverse agents (or cofactors) have the potential to govern the activity of NF-κB, and perhaps other cellular regulatory proteins, and thereby enhance HIV transcription and replication in cells harboring latent or quiescent HIV proviruses. Because many of these studies have only measured LTR-directed transcription in established cell lines, additional investigations are needed to explore the significance of various stimuli (e.g., cytokines, heat shock, ultraviolet irradiation, and regulatory proteins of heterologous viruses) for augmenting viral replication in primary T-lymphocytes and macrophages.

Modulatory Region

Cis-acting transcriptional regulatory elements, located in the U3 region upstream from the NF-κB sites, also influence viral gene expression directed by the viral LTR (12a,522). This portion of the LTR is designated the modulatory region (Fig. 13). Although HIV-1 LTR sequences from nucleotides –201 to –130 do not appreciably contribute to promoter activity *in vitro* or in transfected cells, analysis of site-specific mutations demonstrates that this region is important for viral replication in primary lymphocytes and some T-cell lines (369). Cellular factors with the capacity to bind sequences in this region, which is relatively well conserved among HIV-1 isolates, are LEF, Ets-1, and USF (Fig. 13). LEF is a lymphocyte-specific high mobility group protein, found in T cells, that induces a bend in DNA and thus may govern the local structure of the promoter (241,777). Ets-1 is also a T-cell-specific factor (669). The region between nucleotides –159 and –173 contains a consensus binding site for members of the helix-loop-helix family of transcriptional factors, including USF-1 (516). USF has been implicated in positive and negative roles with respect to HIV-1 gene expression (369,432). Because the effects of LEF, Ets-1, and USF sites are manifest only in replication assays that reflect transcriptional activity of the integrated provirus, it is presumed that these cellular factors produce a complex that maintains the promoter in an open configuration or counters (potential) repressive effects of cellular sequences flanking the proviral integration site in the host genome (336).

Sequences in U3 from the border of the LTR at nucleotide –453 to nucleotide–201 encode the 3′ portion of the *nef* gene and contain potential recognition sites for several cellular transcription factors, including AP-1 and NFAT-1 (Fig. 13). However, the sites for these factors are not highly conserved among viral isolates. Direct binding studies and analysis of LTR mutants in transient expression assays have not clearly demonstrated roles for these cellular factors in viral transcription (12a). Experiments in transient expression assays have implicated a negative regulatory element (NRE) in HIV-1 U3 between –420 and –167 (431, 616). However, the suppressive effect of this NRE is small, and a factor(s) which mediates this effect has not been identified. Genetic experiments with a pathogenic molecular clone of SIV (SIVmac239) have also investigated the role of the 5′ portion of U3 sequences (320,370). Extensive mutation analysis of the SIV LTR demonstrates that the region of U3 between –494 and –230 does not contain important *cis*-acting elements which mediate either positive or negative regulation of LTR-directed transcription in the context of replicating virus. Because the 3′ half of the SIV *nef* gene overlaps U3, proviruses were constructed with mutations in the third base of every codon in this portion of *nef* so as not to change the translation frame for this viral protein. One provirus had 99 mutations and another had 101 mutations in the portion of U3 (–494 to –230) overlapping the *nef* gene. Both mutant viruses express *nef*, replicate efficiently in tissue culture cells, and produce high virus load and immunosuppressive disease in experimentally infected rhesus macaques. Thus, SIV sequences in U3 located 5′ from –230 (i.e., encompassing the 3′ portion of the *nef* gene) are dispensable for both viral replication and pathogenesis. The picture that emerges is that proviral forms of HIV and SIV display low transcriptional capacity due to the low basal activity of the LTR and that NF-κB plays a key role in the inducible activation of viral transcription in both T-lymphocytes and macrophages. Although the LTR appears not to influence cell tropism *in vitro* in tissue culture cells (582), it is possible that regulation of viral gene expression in a cell type-specific fashion *in vivo* may be mediated at the transcriptional level (828). The significance of various regulatory sequences in the LTR, with respect to virus burden and pathogenesis, remains to be established by evaluating SIV LTR mutants and recombinants between SIV and HIV-1 (SHIV) in nonhuman primates (see section on molecular pathogenesis).

***Regulation by* tat *and* rev**

Human immunodeficiency virus and SIV encode regulatory genes, designated *tat* (17,691) and *rev* (195,692), which control viral gene expression at the transcriptional and posttranscriptional levels, respectively (134). Tat augments levels of viral RNA by increasing transcriptional initiation and/or elongation (572), and Rev regulates splicing and transport of viral RNA from the nucleus to the cytoplasm (564). Tat and Rev are synthesized at both early and late stages of the viral replication cycle and are not packaged into virions. Mutagenesis studies have revealed that each of these viral regulatory proteins is required for viral replication. For function, Tat requires a *cis*-acting RNA element, designated the TAR, located at the 5′ end of viral transcripts; this element forms a stem-loop structure (616, 663) (see below). The RRE is a complex stem-loop structure positioned within RNA sequences encoding the *env* gene (398,618). Both Tat and Rev have large effects on viral gene expression; accordingly, an understanding of the molecular mechanisms by which these transactivators function may provide insight into viral activation in the infected host and will serve as a basis for developing novel antiviral drugs.

tat

The first coding exon of *tat* contains 72 codons and is located in the central region of the viral genome between *vpr* and *env;* the second coding exon overlaps the translation frame for both *rev* and the TM subunit in *env* (Fig. 3) (17,691). Differences in position of translational stop codons in the second coding exon account for the range of sizes of Tat proteins (86 to 130 amino acids) found in different HIV and SIV isolates (519). Tat, a 14- to 15-kd protein, is translated from a group of multiply spliced monocistronic transcripts (Fig. 14); these transcripts are synthesized early in infection and are not dependent on *rev* function. The role of this transactivator is to increase steady-state levels of viral transcripts containing the TAR stem-loop RNA structure at the 5′ end of all viral transcripts greatly (see below).

A proposed model, encompassing the first 72 amino acids of Tat, shows five domains (Fig. 15) (571). First, the N-terminal domain (Met1 to Ala21) contains several acidic amino acids and is predicted to form an amphipathic α-helix. Proline residues in this region are presumed to introduce turns in this structural element. The N-terminal domain, together with the following two domains, constitutes the Tat activation domain (282,387,597). Second, the second domain (Cys22 to Cys37) consists of a cluster of seven cysteine residues. In a defined *in vitro* system, purified Tat binds divalent metal ions and forms homopolymeric complexes (211); however, the significance of these findings for Tat function *in vivo* is not known. Third, a core domain (Phe38 to Tyr47) is highly conserved in the Tat proteins of HIV and SIV (as well as EIAV). This domain influences binding of Tat to TAR RNA (see below). Fourth, a basic domain (Gly48 to Arg57) contains the nuclear targeting signal Gly48-Arg49-Lys50-Lys51-Arg52, which functions as an NLS when placed on a heterologous protein (282; also see 716). Immunocytochemical analysis shows that Tat accumulates in the nucleolus as well as in the nucleus (281,681); however, the significance of the nucleolus in Tat transactivation has not been determined and is likely to be an artifact of high levels of expression in COS cells. Synthetic peptides within the basic domain (Gly48 to Arg57) bind TAR oligonucleotides, and a peptide encompassing the core domain as well as the basic domain (Phe37 to Arg57) demonstrates high selectivity for wild-type TAR and discriminates mutant forms of TAR (778). Results of transient expression assays in mammalian cells also suggested that the basic domain was involved in

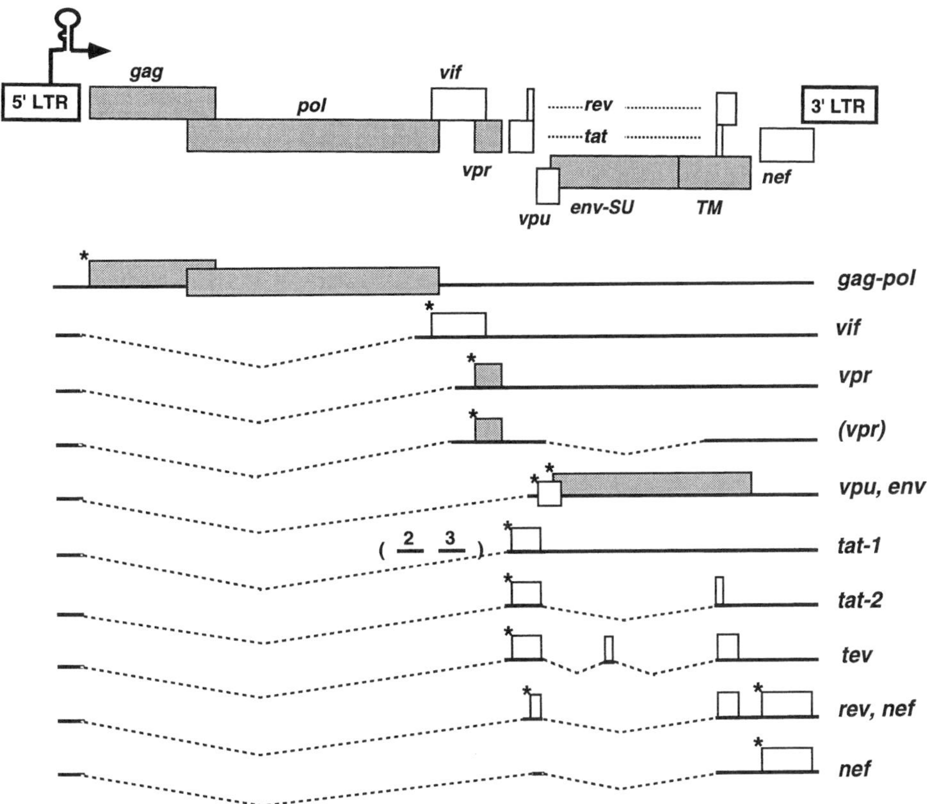

FIG. 14. Splicing pattern of HIV-1 transcripts. The *gag* and *gag-pol* mRNAs are unspliced; mRNAs for the early genes *tat, rev,* and *nef* are doubly spliced; mRNAs for the late genes *vpu, env, vif,* and *vpr* are singly spliced. Also shown are two minor mRNA species which encode novel regulatory genes: *tev/tnv* and the first exon of *tat.* Shown in parentheses are additional transcripts generated by alternative splicing; these include doublyspliced *vif* and *vpr* messages and messages which include miniexons 2 and 3. Redrawn from Pavlakis et al. (566).

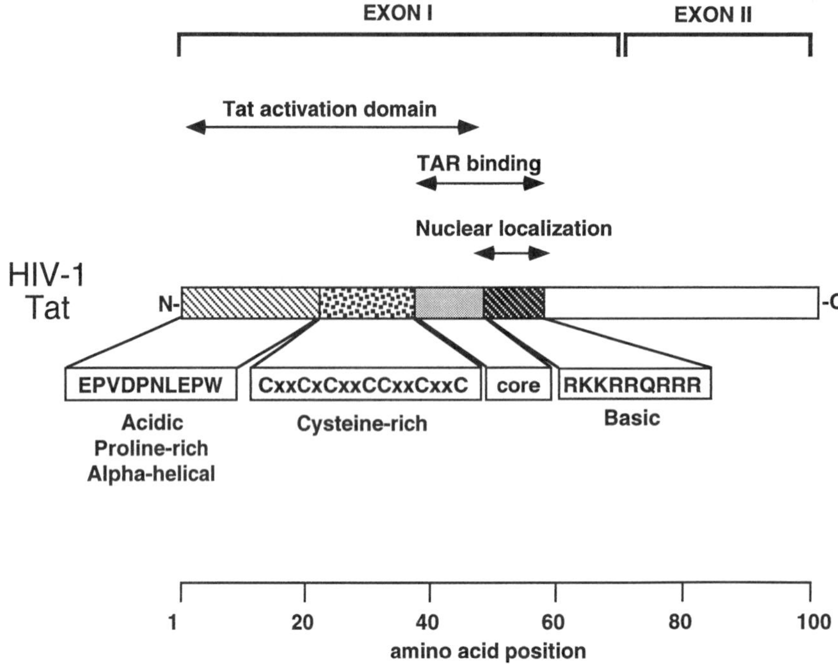

FIG. 15. Functional domains of Tat. Tat proteins of primate lentiviruses are encoded on doubly spliced mRNAs; the first coding exons are critical, and the second exons are dispensable for transactivation in transient expression assays. Four potentially important structural domains have been identified: (i) the N-terminal acidic domain, also rich in proline, resembles the activation domains of some transcription factors; (ii) the central cysteine-rich domain binds metal ions and has been implicated in *in vitro* dimerization of Tat; (iii) a core domain that influences binding of Tat to TAR RNA; and (iv) a basic domain that is essential for nuclear localization, TAR binding, and transactivation.

RNA binding (664). The core domain either also binds directly to TAR or influences the structure of the basic region of Tat and thereby affects TAR binding. Mutations that truncate or substitute neutral amino acids for lysine and arginine residues in the basic domain have a dominant negative phenotype (491). It is hypothesized that Tat dominant mutants may sequester cellular factors required for transactivation. Fifth, a C-terminal domain of Tat (Ala58 to Gly72) appears to contribute to both nuclear localization and RNA binding activity of this viral transactivator (572). A role for sequences in the second coding exon of *tat* has not been defined; Tat proteins missing this region are fully functional in various transactivation and binding assays (572). Tat is not phosphorylated in infected cells; however, the activity of this viral regulatory protein is influenced by levels of cellular protein kinase C (328). Presumably, this kinase regulates a cellular factor which participates in transactivation mediated by Tat.

TAR

The TAR for HIV-1 has been mapped to a stem-loop (or hairpin) structure at the 5′ end of viral transcripts (+19 to +43) (Fig. 16), whereas TAR for HIV-2 and SIV is encompassed within the first 130 nucleotides of viral transcripts and appears to fold into a structure with two stem-loops (49b,604). Because of its location in the R domain of the LTR, TAR is present in both viral RNA and DNA. The stem in HIV-1 TAR consists of 24 nucleotide pairs with a six-nucleotide loop (+30 to +35) and a small pyrimidine-rich bulge (+23 to +25) (199,512,623). The calculated free energy for TAR is –27 kcal, and there is evidence that this structure is stable within cells (333). In transient expression assays, optimal transactivation by TAR is obtained if this element is positioned immediately 3′ to the promoter; in addition, TAR functions only in the sense orientation (571). Transactivation requires the TATAA box and appears to involve additional sequences in the transcriptional promoter in the LTR including SP1 sequences; however, the enhancer and other upstream regulatory elements are dispensable (49b,544). In addition, TAR responds to Tat when positioned downstream from some but not all heterologous promoters (433).

Interactions of Tat with TAR RNA have been examined in both transient expression assays, which measure transactivation activity, and in binding systems, which utilize synthetic oligonucleotides and either Tat protein produced in genetically engineered bacteria or synthetic peptides (572). Transactivation requires the stem structure in TAR;

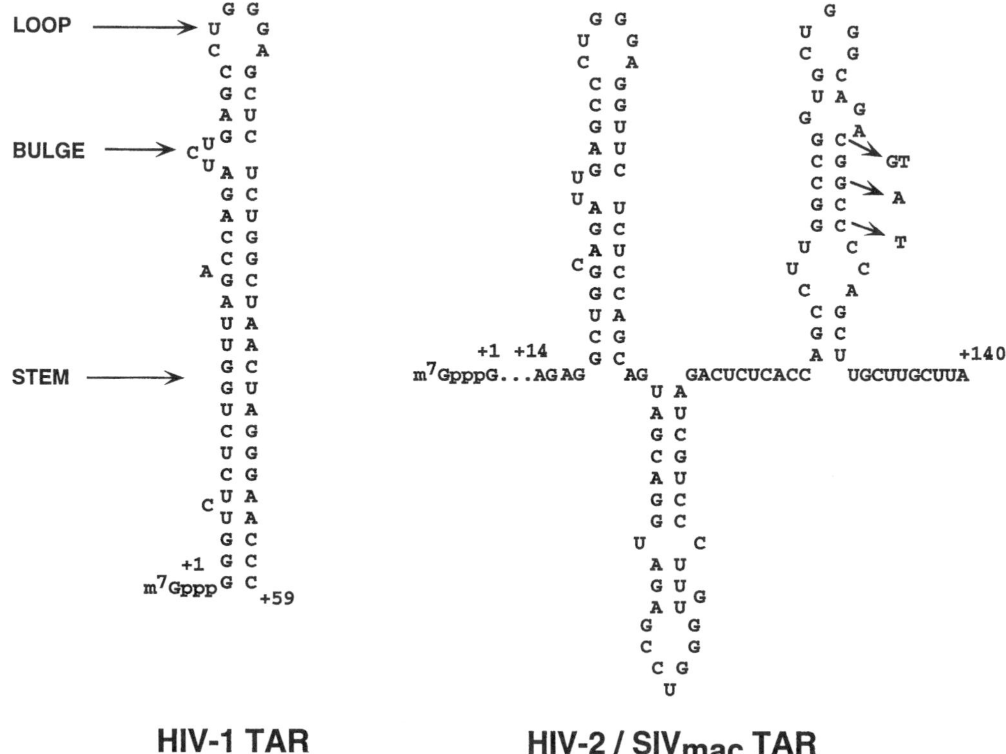

FIG. 16. Structure of the TAR. The transactivation-response elements (TAR) are conserved RNA secondary structures found in the R-U5 portion of the LTRs of HIV-1 and HIV-2/SIV. HIV-1 TAR is a single stem-loop of 59 nucleotides; the HIV-2/SIV TAR is more complex, consisting of as many as three stem-loops. Important structural features in the HIV-1 TAR are (i) the stem; (ii) the pyrimidine-rich bulge, which is essential for TAT binding; and (iii) the loop, which may bind cellular factors (see text). *Arrows* indicate base changes from HIV-2 (ROD) which occur in SIV_{MAC} (clone 239). Adapted from Feng and Holland (199).

analysis of compensatory mutations in the stem reveals that the primary nucleotide sequence can be altered without loss of activity as long as base-pairing is maintained (663). Mutations in the loop at the top of the stem or in the bulge inhibit the response to Tat in transfected cells (623), and the spacing between the bulge and loop is important (49c,199,226,663). The minimal TAR for transactivation is defined by nucleotides +19 to +43 (663). Binding of Tat to TAR RNA has been measured in both gel mobility-shift assays and filter binding assays (163). Tat forms a one-to-one complex with TAR and recognizes the pyrimidine-rich bulge (+23 to +25) below the apex of the stem-loop (778). In a proposed model, a bend in TAR at the bulged residues causes a local distortion widening the major groove in the double-stranded RNA stem (38). Tat interacts with the first nucleotide (U_{23}) in the bulge and with two nucleotide pairs on either side of the bulge (G_{26} to C_{39} and A_{27} to U_{38} above the bulge, and A_{22} to U_{40} and G_{21} to C_{41} below the bulge) (109). Binding studies utilizing chemically modified RNA substrates support a model in which Tat forms serveral hydrogen bonds to dispersed sites in the nucleotides located within the major groove of TAR RNA (265).

Cellular Factors Interacting with Tat and TAR

Cellular proteins which interact with Tat and/or TAR RNA were implicated from studies on the species specificity of Tat-mediated transactivation. In transient expression assays, Tat was found not to function efficiently in Chinese hamster ovary (CHO) cells (275). Analysis of hybrid CHO cell lines containing various human chromosomes revealed that human chromosome 12 was required to restore Tat-mediated transactivation (530). Several approaches have been utilized to search for specific cellular proteins that interact directly with Tat and TAR (572). A cDNA expression library from human cells was screened with biotinylated HIV-1 Tat, and a clone encoding a 50-kd protein, designated TBP-1, was isolated (526). Because TBP-1 inhibits Tat-mediated transactivation in transient expression assays, the relevance of this cellular protein remains unclear. Other members of the TBP-1 family of proteins also interact with Tat (538). These proteins are related to the 26S protease from human erythrocytes (173); the significance of this relationship is not known. In another approach, affinity chromatography using immobilized

HIV-1 Tat yielded a cellular 36-kd protein which displays transactivation function when coinjected into rat cells with a plasmid containing the LTR (156). Interactions of Tat with a cellular protein(s) has also been implicated in a cell-free transcription system; the antitermination activity ATN/TFIIS acts synergistically with HIV-1 Tat to stimulate elongation of transcripts initiated on the LTR (358). In this *in vitro* system, elongation factor TFIIF substitutes for Tat and stimulates transcription; accordingly, it has been proposed that Tat mimics the activity of this cellular factor (see below). Several cellular proteins that recognize TAR have been identified. TAR RNA binding protein-1 (TRP-1), also designated TRP-185, is a heterodimer of two proteins, 185 and 90 kd; the former subunit interacts with the loop (669,809). Sequences in the TAR loop are also recognized by human lupus antigen Ku (340) and a 68 kd protein in HeLa cells (459). The TAR bulge binds TRP-2, which is composed of at least four subunits of 70 to 100 kd (669,809). Proteins interacting with the double-stranded stem in TAR are PKR (469,624), SBP (621), and TRBP (232). Although the aforementioned results implicate several cellular factors that interact with Tat and TAR, the significance for transactivation remains to be established by a combination of approaches which (a) demonstrate that these proteins bind directly with Tat and/or TAR, (b) are active in reconstituted cell-free transcription systems, and (c) mediate transactivation in transient expression assays in intact cells (10).

Mechanism of tat Transactivation

Models for Tat-mediated transactivation are based on results obtained from transient expression assays and binding studies (see above) as well as on experiments measuring the effects of Tat in cell-free transcription systems utilizing the HIV-1 LTR as template (572,670). Although many studies have shown that Tat increases the steady-state levels of viral transcripts initiated in the LTR, a consensus for a single mechanism for transactivation has not emerged. This lack of agreement may be due to experimental differences (e.g., cell type, reporter gene, basal promoter level, transfection conditions, and level of Tat) or may mean that Tat functions to regulate viral gene expression through more than one mechanism. In one model, Tat is proposed to function primarily at the level of transcription initiation by increasing the number of stable RNA polymerase II initiation complexes (Fig. 17); this model assumes that formation of competent initiation complexes on the LTR is relatively inefficient in the absence of Tat (280,396,459). Accordingly, TAR may function as an RNA enhancer element (666). In another model, Tat is proposed to bind nascent TAR RNA and promote elongation of newly initiated viral transcripts, either by preventing premature termination (348) or by augmenting the processivity of transcription complexes (50,196,396,460) (Fig. 17). A role for Tat in elongation is supported by the results of tethering experiments in which chimeric proteins produced by fusing Tat to other nucleic acid binding proteins are capable of transactivation through the cognate target element for the nucleic acid binding protein (664,696). Thus, the function of TAR is to position Tat in the proximity of the transcription initiation complex. A variation of this model is that the low level of LTR-directed basal RNA synthesis is due to a repression mechanism which causes pausing and/or premature termination of transcription by the RNA polymerase II complex (51). Consequently, Tat may function as an attenuator to obviate this block to elongation of transcription. These mechanisms are not mutually exclusive; the relative contribution of each mechanism to HIV gene expression may be related to the level of participating cellular factors in different cell types. Accordingly, when Tat is absent, initiation directed by the LTR is low and transcriptional elongation is inefficient. Initiation complexes lose stability with increased distance from the promoter (polarity of transactivation), and short transcripts from +1 to about +50 accumulate in the cell. In the presence of Tat, both initiation and elongation are increased. This latter model highlights both potential functions of Tat, i.e., binding of Tat to TAR facilitates formation of an initiation complex at the promoter and stabilization of the elongation complex. The level of Tat may determine the mechanism of action if the initiation and elongation steps in RNA pol II transcription have different threshold sensitivities for Tat. Studies to establish the precise mechanism(s) of transcriptional transactivation by Tat will require identification of cellular factors which interact with Tat and TAR and mediate transcriptional transactivation.

Interestingly, support for a role for Tat in elongation of viral transcripts has also been obtained through analysis of viral RNA in lymphocytes from HIV-infected individuals (1). Short viral transcripts +1 to about +50 nucleotides are relatively abundant in asymptomatic individuals, whereas long viral transcripts predominate in individuals with AIDS. This observation suggests that Tat transactivation plays a critical role in regulating virus load in AIDS patients and that antiviral therapies directed against Tat (and TAR) may inhibit disease progression (312,717).

Effects of Tat on Expression of Heterologous Viral and Cellular Genes

Several investigators have shown that HIV-1 Tat activates promoters of certain heterologous viruses, influences expression of selected cellular genes, and stimulates cell growth. The promoters of the human polyomavirus and the human papilloma virus (HPV) are activated by Tat (723, 740). Expression of several cytokines, including transforming growth factor-β, TNF, IL-2, and IL-6 is activated by Tat through unique targets in the promoters for these genes (63,636,643,786). In contrast, the promoter of the

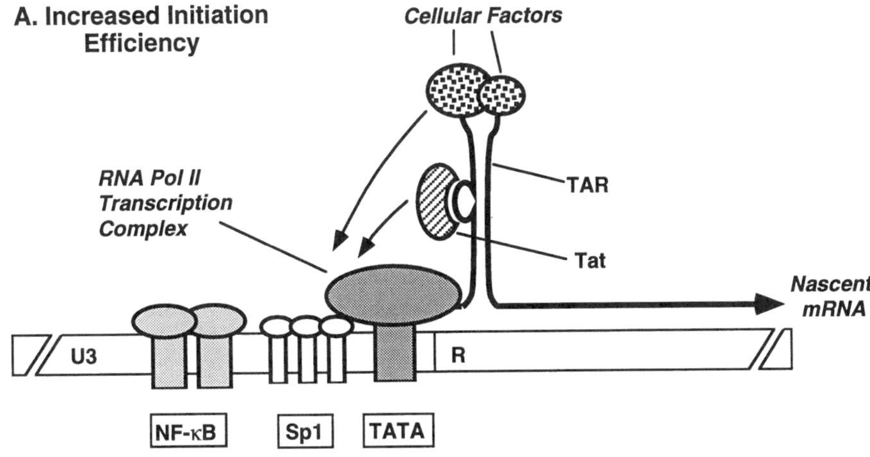

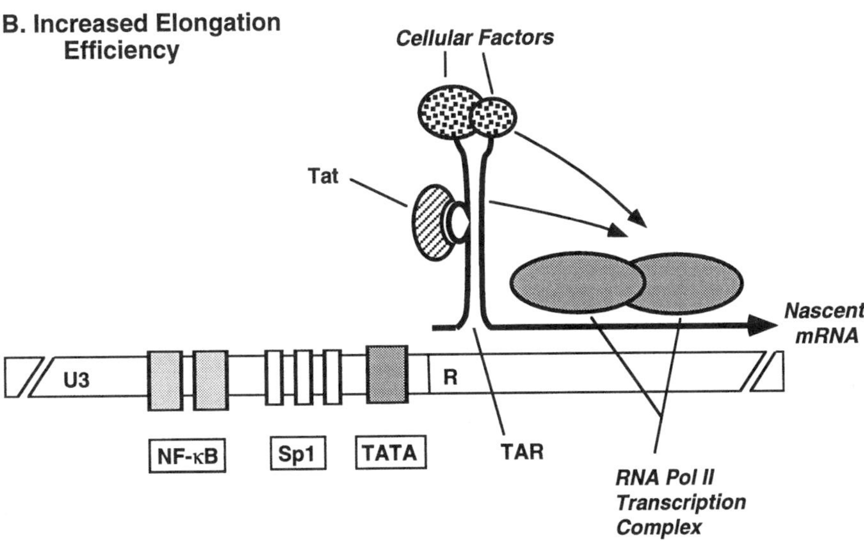

FIG. 17. Model for Tat transactivation. Tat may mediate transcriptional transactivation by a dual mechanism. In (**A**), Tat and TAR function at the level of transcription initiation to increase the number of stable RNA polymerase II initiation complexes formed at the transcription start site. In this model, TAR is proposed to be an RNA enhancer, recruiting cellular factors that promote the formation of stable initiation complexes. In (**B**), Tat functions by promoting elongation, either by preventing premature termination or by increasing the processivity of RNA polymerase II transcription complexes.

MHC class I gene is repressed by Tat (307). Additional observations in complex biologic systems also show that this viral transactivator influences cell proliferation and differentiation. Human immunodeficiency virus type 1-infected cells and cells transfected with *tat* expression plasmids secrete this viral protein into the culture medium (211). Exogenous Tat, produced in mammalian cells or in genetically engineered *E. coli*, is endocytosed by intact culture cells and transported to the nucleus where expression directed by the LTR is activated (186,455). In T-lymphocyte cultures, exogenous Tat suppresses antigen-induced but not lectin-induced cell proliferation (767). This viral protein stimulates growth of Kaposi's sarcoma (KS) cells and promotes adhesion of both KS cells and normal vascular cells (37). Moreover, KS-like dermal lesions have been observed in some *tat*-transgenic mice (185,768). The sequence Arg82-Gly83-Asp84 in the second coding exon of Tat is a recognition site for binding to cellular integrins; accordingly, this sequence may enable the viral transactivator to mediate effects on growth and differentiation of cells. (Neither HIV-2 nor SIV Tat proteins contain the Arg-Gly-Asp motif; interestingly, KS lesions have not been associated with these viruses.) Taken together, these findings support the notion that Tat activates and/or suppresses expression of cellular genes in infected and uninfected cells and thus contributes to cytokine dysregulation in individuals infected with HIV. In addition, an autocrine pathway induced by Tat may interrupt HIV latency, augment viral gene expression and replication, increase viral load, and thereby initiate and/or escalate disease in infected individuals. The mechanism(s) underlying these pleiotypic effects and its relationship to the mechanism of viral transactivation through TAR remains to be determined.

rev

The two coding exons for the HIV/SIV *rev* gene are joined by splicing to produce a monocistronic transcript, early in the viral replication cycle, for a protein about 110 amino acids in length (Fig. 18) (195,617). Each known lentivirus encodes a small gene which exhibits sequence relatedness to HIV/SIV *rev.* In the infected cell, Rev binds to the RRE in viral transcripts and shifts the balance from multiply spliced transcripts (encoding Tat, Rev, and Nef in the early stage of replication) to both singly spliced and unspliced transcripts (encoding viral structural proteins in the late stage of replication) (135). Accordingly, this viral transactivator mediates temporal regulation of viral gene expression (368,375,584). Because mammalian cell splicing mechanisms are coupled to transport of mRNA from the site of synthesis in the nucleus to the cytoplasm, Rev also appears to influence transport of viral transcripts containing RRE. In addition, Rev has been implicated in recruiting viral transcripts into polysomes and may thereby influence translation. Cellular factors which interact with Rev and RRE have been identified; thus, studies on posttranscriptional transactivation are adding to an understanding of basic mechanisms which regulate processing and nucleocytoplasmic transport of mRNA in eucaryotic cells.

The Rev of HIV-1 is 116 amino acids in length and is a 19-kd phosphoprotein. Immunocytochemical analysis reveals that this regulatory protein is localized in both the nucleus and nucleolus (120,138). Serine residues proximal to the C-terminus of Rev are phosphorylated by a nuclear protein kinase (448); however, the role of this modification is not clear because mutations in these serine residues do not affect the known Rev functions. Perhaps phosphorylation may be essential for a novel, as yet unidentified, function of Rev.

Results of numerous studies investigating Rev structure and function have been integrated into a model which shows several domains and takes into account the four essental properties of Rev: nuclear localization, RNA binding, protein oligomerization, and posttranscriptional transactivation (Fig. 18) (564). The region between Arg35 and Arg51 contains a high proportion of arginine residues and thus defines a basic domain. Situated within this region is a nuclear localization signal (Asn40-Arg41-Arg42-Arg43-Arg44-Trp45); mutagenesis analysis reveals that this region is essential for Rev activity in cells (301,761; also see

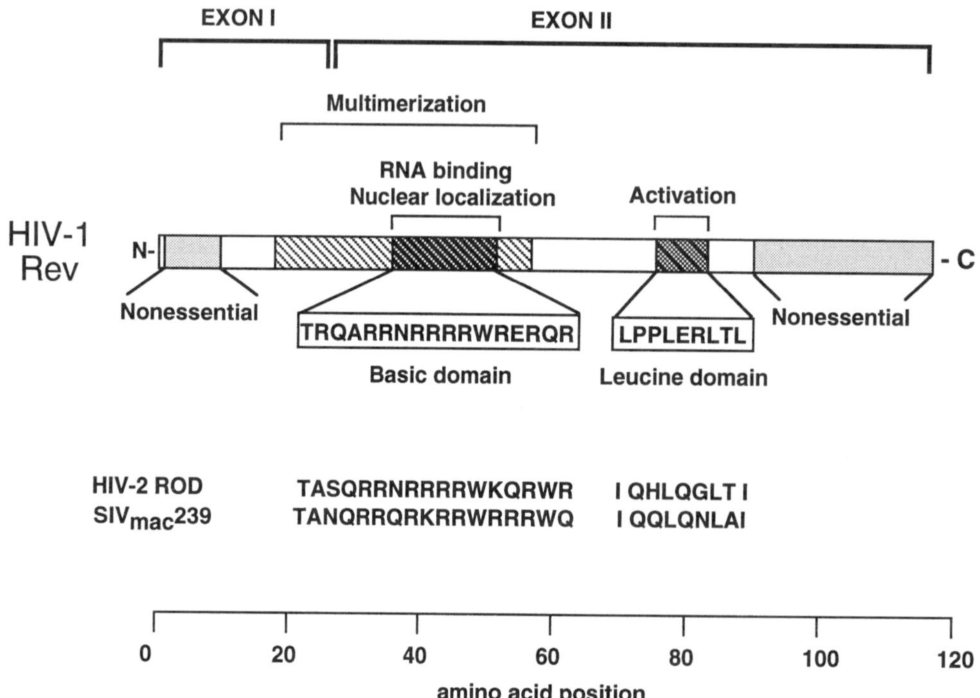

FIG. 18. Functional domains of Rev. Three important functional domains have been identified in HIV-1 Rev: (i) a basic domain responsible for nuclear/nucleolar localization and RRE binding; (ii) a multimerization domain, overlapping the nuclear localization domain; and (iii) a leucine-rich activation domain. Some mutations in the conserved leucine residues produce *trans*-dominant inhibitors of Rev. Similar functional domains have been identified in HIV-2 and also appear in SIV_{MAC}. Rev proteins of primate lentiviruses are encoded on doubly spliced mRNAs. Redrawn from Luciw and Shacklett (439).

300). *In vitro* binding studies have demonstrated that Rev interacts directly with the RRE RNA and that this interaction requires amino acids in the basic domain (825a); a peptide encompassing this basic domain binds the RRE with specificity similar to that of full-length Rev (372). Amino acids near the basic domain but distinct from amino acids of the RRE binding site mediate oligomerization of Rev which forms a tetramer in solution (533,825a). Oligomerization does not require RRE RNA substrate but depends on protein-protein interactions between Rev monomers. Genetic studies have shown that Rev mutants which bind RRE but fail to oligomerize are not functional for transactivation (825a). Taken together, these findings support a model in which multiple copies of Rev must be bound to the RRE for the transactivation response (see below). A leucine-rich region between Leu75 and Leu84 constitutes the activation domain (302). The tetrapeptide motif in this domain is conserved in the *rev* genes of several other lentiviruses as well as in the *rex* gene which is the posttranscriptional transactivator of HTLV (58,302). To investigate the domain structure of Rev, chimeric proteins have been produced by exchanging the activation domains of the Rev proteins of distant lentiviruses such as HIV-1 and VMV (737). Because these chimeric Rev proteins are functional, it appears that the posttranscriptional transactivators of distant lentiviruses function through the same pathway and that the activation domains interact with a common cellular factor. Mutations in the activation domain do not affect nuclear localization, RNA binding, or oligomerization of Rev; however, these mutants have an inhibitory effect on wild-type Rev in cotransfected cells (447,452,564). This transdominant negative phenotype is presumed to be due to the formation of nonfunctional oligomeric complexes of Rev.

RRE

The RRE of HIV-1 is an RNA sequence (234 nucleotides) located within the *env* gene immediately 3′ of sequences encoding the junction of the gp120 and gp41 subunits (261,398,446,618). Secondary structure predictions based on computer algorithms, coupled with analysis of site specific mutants, indicate that this region can be

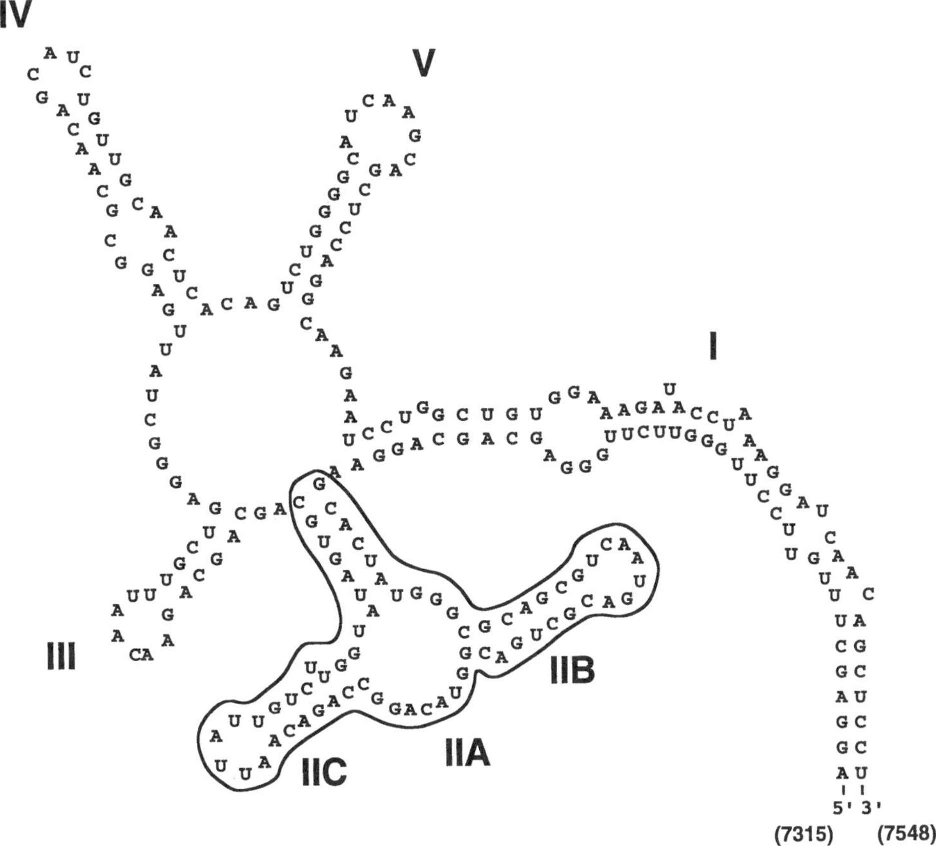

FIG. 19. Structure of the RRE. The RRE of HIV-1 is a complex mRNA secondary structure located within the *env* gene immediately 3′ of the SU/TM junction. The structure consists of five stem-loops. A portion of stem-loop II (*lightly shaded*) contains sequences required for Rev binding and transactivation. In addition, the long stem (I) is important for Rev function. Stem-loops III, IV, V, and a portion of stem-loop I are important for interaction with HTLV-I Rex. Secondary structure is shown as predicted by Malim et al. (449) for HIV-1$_{HXB-2}$; sequences are numbered according to Myers et al. (519).

folded into four stem-loops and a long stem with a total free energy of –110 kcal (Fig. 19). Analysis of RRE mutants in transient expression assays in tissue culture cells reveals that the long stem as well as stem-loop II compose the minimal target region for transactivation by Rev (145, 316,541). Transient expression assays also demonstrate that the RRE functions in a position-independent fashion; in its native state in the viral genome, RRE is within a potential intron but retains function if inserted into an upstream or downstream exon in plasmid expression vectors (446,448). However, the function of RRE is orientation dependent because this element must be encoded in the sense strand of mRNA. In HIV-2 and SIV, the RRE is also situated in *env* sequences 3′ of the junction of the gp120 and gp41 subunits (231). Because of extensive primary sequence differences, the predicted structure of HIV-2/SIV RRE (219 nucleotides) exhibits a different arrangement of stem-loops than the HIV-1 element (49d). These sequence and structure differences have implications with respect to function because HIV-1 Rev is active on RRE of HIV-2 whereas Rev of HIV-2 is not active on HIV-1 RRE.

Human immunodeficiency virus type 1 Rev has been purified from genetically engineered bacteria and shown to bind with high affinity to a 13-nucleotide sequence in stem-loop II RNA in the RRE (737). This high-affinity site is predicted to form an asymmetric bulged duplex (or "bubble") structure (Fig. 19) (283), and mutagenesis studies reveal that the Rev-RRE interaction is influenced by both the primary sequence and secondary structure in the target (372). At low Rev concentrations, a single protein-RRE complex, involving the high-affinity binding site, is produced (129,323,451). After this site is occupied, additional Rev molecules bind to low-affinity sites in the remainder of the RRE. Protein-protein interactions between Rev monomers are essential for binding to the low-affinity sites. The importance of the single binding site in stem-loop II is highlighted by the fact that mutations in this part of the RRE preclude both Rev binding *in vitro* and transactivation in transfected cells (737). Experiments with chimeric proteins produced by fusing Rev with the R17 bacteriophage RNA binding protein revealed that the RRE was dispensable for cytoplasmic expression of unspliced transcripts if these transcripts contained the target for the phage RNA binding protein (472). Also, transactivation in this system requires that multiple copies of the chimeric protein bind the target. Taken together, these observations demonstrate that multimerization of Rev is essential for the function of this viral transactivator.

Cellular Factors Interacting with Rev and RRE

The observation that Rev does not function in murine cells indicates that human cells express a factor essential for posttranscriptional transactivation (745). To determine a role(s) for cellular factors in posttranscriptional transactivation, cellular proteins interacting with Rev and RRE RNA have been identified and characterized. Rev attached to an affinity column binds a 38-kd nuclear protein from HeLa cells (191). This protein was shown to be a previously known protein, designated B23, which functions to translocate ribosomal components across the nuclear membrane by shuttling between the nucleus and cytoplasm in both directions (61). Localization of Rev to the nucleolus might be explained by the fact that B23 is found largely in this intranuclear compartment. Analysis in an *in vitro* system reveals that Rev dissociates from B23 after binding to RRE. Thus, in a proposed model, Rev binds RRE-containing transcripts in the nucleus and mediates transport to the cytoplasm where B23 attaches to and dissociates Rev from the transcript. Subsequently, the Rev-B23 complex is shuttled back into the nucleus where the cycle is completed when Rev binds RRE-containing transcripts and releases B23. Other binding studies have shown that Rev associates with the nuclear scaffold in stably transfected cells and interacts with a 110-kd protein which appears to function in mRNA transport. The yeast two-hybrid system has been used to isolate a mouse cell cDNA encoding a protein, designated YL-2, that interacts with the basic domain of Rev and potentiates activity of Rev in cotransfected cells (442). Because YL-2 is related to a cellular splicing factor, these findings support a role for Rev in splicing viral transcripts. A cellular factor recognizing the HIV-1 RRE has been identified in nuclear extracts of HeLa cells by gel mobility shift assay. This factor is a 56-kd protein, designated NF_{RRE}, which binds stem-loop II *in vitro* (752). A ternary complex is produced in a mixture of RRE, Rev, and NF_{RRE}, although this cellular protein can bind RRE RNA independently. In a proposed model, NF_{RRE} may act together with Rev to inhibit splicing and/or promote nucleocytoplasmic transport of RRE-containing transcripts. Interestingly, TRBP has also been shown to recognize not only TAR but also the RRE (232,561). Additional investigations are required to determine which of these cellular proteins are significant for viral transactivation and to establish their precise function(s) in the pathway of viral as well as cellular mRNA synthesis.

Mechanism of Rev Transactivation

The mechanism of Rev transactivation is not fully resolved because this viral regulatory protein functions through several cellular posttranscriptional mechanisms (i.e., mRNA splicing, stability, and nucleocytoplasmic transport) which are complex and not well understood. Nonetheless, numerous studies have provided insights on several posttranscriptional steps in viral gene expression regulated by Rev (Fig. 20) (564,566). First, Rev may influence the efficiency of splicing of transcripts containing the RRE (182,198,264,446). In this model, Rev binds to viral transcripts in the nucleus and increases their half-life. Thus, in

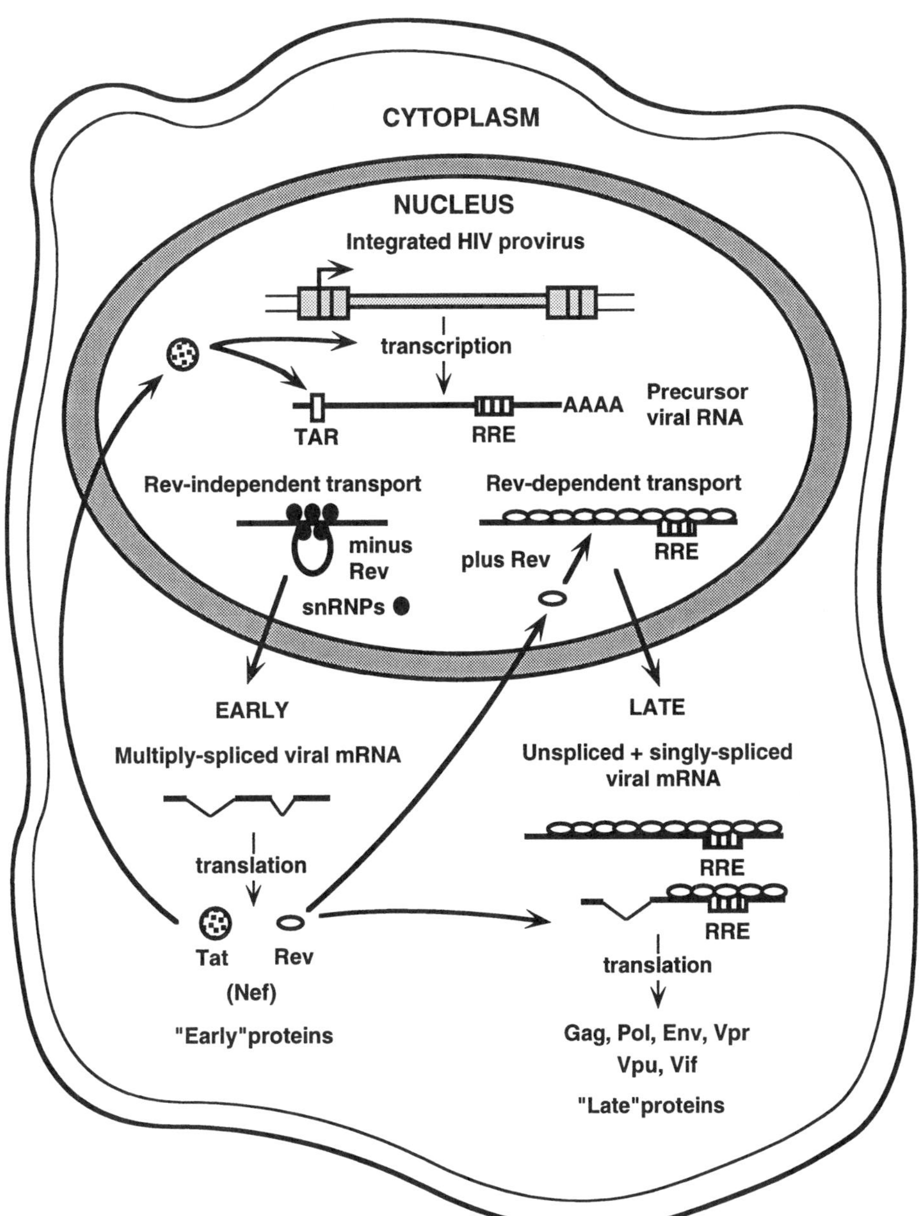

FIG. 20. Model for Rev transactivation. Rev is proposed to act as a key regulator and "molecular chaperone" for viral mRNAs. In the nucleus, integrated proviral DNA is transcribed by cellular RNA polymerase II; initially the level of viral-specific transcription is low in the absence of the transcriptional transactivator Tat. Viral transcripts in the nucleus may follow one of two pathways: A Rev-independent transport pathway, in which association with spliceosome components leads to removal of introns prior to export; or a Rev-dependent transport pathway, which leads to direct export of unspliced mRNA to the cytoplasm. The latter pathway requires an RRE. Early in infection, before a significant amount of Rev protein has accumulated, the splicing pathway predominates. Multiply spliced messages are translated in the cytoplasm, and virion regulatory proteins Tat and Rev shuttle back to the nucleus as directed by basic nuclear localization domains in each protein. Later in the infectious cycle, increased levels of Tat in the nucleus lead to increased transcription of viral mRNAs. As Rev accumulates in the nucleus, more RRE-containing viral transcripts associate with Rev and are shuttled via the Rev-dependent pathway out of the nucleus. In the cytoplasm, these unspliced and singly spliced messages are translated to produce virion structural polyproteins (Gag, Gag-Pol, and Env) and accessory proteins (Vpu, Vif, Vpr, and Vpx). The mechanistic details of these two alternative pathways with regard to spliceosome association, the role of nuclear matrix proteins, and the precise way in which Rev functions to escort unspliced mRNA out of the nucleus, are not yet understood. The stoichiometry of the Rev-RRE association remains unclear, although Rev is believed to multimerize on the RRE. Redrawn from Luciw and Shacklett (439).

the absence of Rev, unspliced and singly spliced viral transcripts are either degraded or spliced further to produce multiply spliced transcripts which are translated in the cytoplasm into Tat, Rev, and Nef. Accordingly, these regulatory proteins predominate in the early stage of viral replication. It is possible that Rev influences splicing by dissociating spliceosome components from precursor viral transcripts (96,432). Second, Rev may regulate stability of viral transcripts containing *cis*-acting repressive sequences (CRS) which are also designated instability elements (INS) (198,618). The CRS/INS are found in transcripts for *gag, pol,* and *env* and function by promoting degradation of these transcripts (117,660–662). The interaction of Rev with RRE in a transcript apparently overcomes the inhibitory effects of these elements. Third, Rev may function to facilitate transport of RRE-containing transcripts from the nucleus to the cytoplasm (182,446). Because relatively little is known about transport in mammalian cells, experimental approaches for defining the potential role of Rev in this process are very limited. Fourth, in the cytoplasm, Rev facilitates the association of RRE-containing RNA with polysomes and may also affect other steps in translation (14). As a "chaperone" for RRE-containing RNA, Rev escorts these transcripts from the nucleus to the cytoplasm and into polysomes. The weight of evidence suggests that this viral transactivator is a multifunctional protein that influences major steps in the pathway for synthesis and utilization of viral transcripts (139,397).

Although the mechanism(s) of Rev function is not well understood, this posttranscriptional transactivator is hypothesized to play a critical role in viral latency and activation (584). In cells harboring latent HIV-1 genomes, multiply spliced viral transcripts encoding early gene products are produced (i.e., Tat, Rev, and Nef). After induction, Rev regulates a shift to expression of unspliced and singly spliced viral transcripts encoding precursor polypeptides for virion proteins (i.e., Gag, Pol, and Env). In this model, differential viral gene expression is controlled by intracellular levels of Rev (197a).

Virion Assembly

Overview of Assembly

Although many aspects of virion formation have been explored in the HIV system, the basic model for primate lentivirus assembly draws primarily on investigations of other retroviruses (see Chapter 26). Approaches to elucidate retroviral morphogenesis have depended on electron microscopy of infected cells and virions as well as biochemical analysis of viral macromolecules in cells infected with wild-type and mutant viruses. In addition, many features of virion formation, including proteolytic processing, have been examined in bacteria, yeast, insect, and mammalian cells containing expression vectors designed to produce various viral proteins (and precursor proteins). Importantly, information on the molecular mechanisms governing virion assembly can be used as a basis for developing novel antiviral drugs.

The first event in virion assembly is presumed to be an interaction between Gag polyproteins ($Pr55^{gag}$) (233,801). This interaction may initiate in the cell cytosol to produce an assembly intermediate which is not visible by electron microscopy or at the plasma membrane where electron-dense crescents are observed. These electron-dense crescents are nucleoprotein complexes composed of $Pr55^{gag}$, as well as the Gag-Pol polyprotein ($Pr160^{gag\text{-}pol}$), and genomic viral RNA. A *cis*-acting packaging element (*psi*), also designated the encapsidation (E) element, is located near the 5′ end of the full-length viral transcript and is presumed to be recognized by the NC domain in the Gag polyprotein (see below). Oligomers of Env gp are inserted into the cell plasma membrane, and the MA domain of $Pr55^{gag}$ is presumed to interact with the cytoplasmic tail of the TM subunit in the Env oligomer during virion assembly. The viral nucleoprotein complex extrudes, or buds, through the plasma membrane to produce a virion with a nucleocapsid surrounded by a lipid bilayer membrane which contains oligomers of Env gp . Selected host cell proteins are also incorporated into mataure virus particles.

A major feature of retroviral morphogenesis is proteolytic processing of Gag, Gag-Pol, and Env-gp precursor proteins. Cleavage of the HIV and SIV Gag-containing polyproteins during assembly is mediated by the PR domain in $Pr160^{gag\text{-}pol}$ to produce a mature nucleocapsid composed of fully processed *gag* (MA, CA, NC, p6, p1, and p2) and *pol* gene products (PR, RT, RNase H, and IN) as well as two molecules of the viral single-stranded RNA genome. Mutations inactivating the viral protease, contained within $Pr160^{gag\text{-}pol}$, abolish infectivity (377,454). Although extracellular virions with uncleaved polyprotein precursors have been detected, it appears that proteolytic processing is initiated in an intracellular compartment before virion release (349,350). Thus, in the final stages for production of infectious virus particles, the following events occur in a coordinated fashion: (a) association of the Gag and Gag/Pol precursors, both with the plasma membrane of the infected cell, each other, and the viral RNA genome; (b) activation of the viral protease and processing of the polyprotein precursors; and (c) final assembly and budding.

Role of $Pr55^{gag}$ *in Assembly*

Assembly of viral particles is largely orchestrated by the Gag precursor $Pr55^{gag}$ (233,801). $Pr55^{gag}$ synthesized in both prokaryotic and eukaryotic expression systems assembles into VLPs which are similar in size (80- to 90-nm diameter) and morphology to the immature nucleocapsids in newly released virions (154,180,236); accordingly, the Gag polyprotein does not require viral RNA or any other viral gene product for assembly. High-resolution electron microscopy of mature virions reveals that $Pr55^{gag}$ is arranged in the form of a near-spherical "fullerene-like" icosahedral shell (527,528). In a proposed model, the icosahedral shell consists of closely packed rings of rodlike subunits of $Pr55^{gag}$; the rods are presumed to be tapered at the proximal end to allow for close packing (527,528).

Several functional domains in the Gag precursor play critical roles in the assembly of the nucleocapsid. Mutagenesis studies have shown that both myristylation of Gly2 and a stretch of basic amino acids (Arg15 to Lys30) at the N-terminus of MA are essential for targeting $Pr55^{gag}$ to the plasma membrane (73,249,831). The first 30 amino acids at the N-terminus of MA contains a high proportion of basic residues; thus, this region is thought to bind to acidic phospholipids on the cytoplasmic leaflet of the plasma membrane. In an *in vitro* approach to investigate viral nucleocapsid assembly, $Pr55^{gag}$ produced in a transcription/translation system was demonstrated to associate with cellular membranes isolated from a T-cell line; this cell-free system was used to show that deletion mutations in the NC domain of the gag precursor were also important for membrane association (579). Extensive mutagenesis of the MA domain in the Gag precursor has revealed that the association of the nucleocapsid with the Env gp (inserted into the cell plasma membrane) is influenced by many sequences throughout the first 100 amino acids of MA (168,169,820). About 20 amino acids at the carboxyl terminus of the MA domain are dispensable for virion assembly and infectivity. Accordingly, it appears that a three-dimensional structure of the first 100 amino acids of MA, rather than sequence specificity, governs the interaction of Env gp with MA in the Gag precursor. Mutagenesis studies on the CA domain in the Gag precursor suggest that CA contains two functional domains; mutations in the carboxy-terminal portion of CA block viral particle assembly, whereas mutations in the amino-terminal portion do not prevent assembly, but these particles are noninfectious (168,169). A genetic assay for protein multimerization in yeast also indicates that sequences in the carboxy-terminal portion of CA are required for homomeric interactions between Gag polyprotein molecules (209,210,435). Vpr as well as Vpx (of HIV-2/SIV) bind to capsid particles containing $Pr55^{gag}$ in mammalian cells transfected with plasmids expressing these viral proteins (353,433,822); accordingly, the Gag polyprotein is presumed to mediate incorporation of these viral accessory protein(s) into virions. Structural features of the Gag precursor mediating interactions with Vpr and Vpx have not been identified. Information on the structure and function of targeting signals and assembly domains in the Gag precursor, as well as the Gag-Pol precursor, is important for discovering and testing novel antiviral inhibitors. For example, viral replication is blocked in tissue culture cells treated with analogs of myristic acid that are incorporated into MA (74). Presumably, these analogs prevent targeting of the Gag precursor to the plasma membrane.

Incorporation of $Pr160^{gag\text{-}pol}$ into Virions

The mechanism for incorporation of $Pr160^{gag\text{-}pol}$ into virions has been investigated in infected cells as well as in cells transfected with various expression vectors (801). Although $Pr160^{gag\text{-}pol}$ contains Gag domains involved in particle assembly as well as PR, the Gag-Pol polyprotein is not sufficient for production of capsids. The Gag-Pol polyprotein exhibits autoprocessing into several gag and pol proteins in a full-defined *in vitro* system (88,384). In a vector system expressing large amounts of $Pr160^{gag\text{-}pol}$ in mammalin cells, the precursor was efficiently cleaved but particles were not assembled (354,559). However, when both Gag-containing polyproteins were expressed from separate vectors in the same cell, $Pr55^{gag}$ was cleaved in *trans* by the protease embedded in $Pr160^{gag\text{-}pol}$ (689). This observation suggests that $Pr160^{gag\text{-}pol}$ forms a complex with $Pr55^{gag}$ in the cytoplasm, perhaps immediately after translation; subsequently, the complex is transported to the site of capsid assembly and budding. Alternatively, because $Pr160^{gag\text{-}pol}$ has the acceptor site (i.e., Gly2) for myristylation, this viral polyprotein is presumed to be myristylated and to associate with $Pr55^{gag}$ at sites of budding in the plasma membrane. The ratio of $Pr55^{gag}$ to $Pr160^{gag\text{-}pol}$ in infected cells is about 9:1; this is a reflection of the efficiency of ribosomal frameshifting from the *gag* to the *pol* translation frame (324). The importance of the Gag polyprotein in capsid assembly is highlighted by the finding that an unmyristylated mutant form of $Pr160^{gag\text{-}pol}$ was incorporated into viral particles when coexpressed with a myristylated form of $Pr55^{gag}$ (560). A domain(s) in $Pr55^{gag}$ and $Pr160^{gag\text{-}pol}$ mediating the interaction of these polyproteins remains to be identified, and the site in the cell at which both polyproteins associate has not been defined.

Encapsidation of Genomic Viral RNA and tRNA Primer

Incorporation of genomic viral RNA into virions requires the *cis*-acting psi packaging element which encompasses about 120 nucleotides located 3′ of the 5′ LTR (5,111,424,437,627). Secondary structure in this region of HIV-1 has been analyzed by a combination of approaches including biochemical and enzymatic probes, algorithms for free energy minimization of predicted secondary structure, and phylogenetic comparisons of viral variants (272, 273). Accordingly, the predicted structure for psi of HIV-1 depends on base-pairing between sequences 5′ of the major splice donor (i.e., sequences common to both genomic and subgenomic viral transcripts) and sequences 3′ of the splice donor (i.e., sequences unique to unspliced viral RNA). A proportion of full-length viral transcripts enters the virion assembly pathway, whereas spliced transcripts lacking psi remain in the cytoplasm. A combination of genetic analysis of viral mutants and biochemical analysis (by methods that measure binding of Pr55^{gag} and mature Gag proteins to viral RNA) reveals that other sequences in the viral genome are also important for efficient packaging; this includes sequences extending into the 5′ end of the *gag* gene (434,437) and sequences within the *env* gene encompassing RRE (562,605). Each virion contains two single-stranded viral RNA genomes which can be isolated as a stable dimer from disrupted viral particles. The NC domain of the Gag polyprotein may facilitate formation of this dimeric RNA complex by interacting with psi and perhaps other sequences in the full-length viral transcript in the early steps of nucleocapsid assembly (54,143,462). *In vitro* studies with HIV-1 RNA fragments have demonstrated that dimerization involves sequences both 5′ and 3′ to the splice donor (685). Additional studies are needed to determine both the precise sequence requirements and the relationship of dimerization of the viral RNA genome to packaging. Viruses with mutations in NC do not package viral RNA (5). Human immunodeficiency virus type 1 with a defective protease has a dimer structure that differs from that found in wild-type virus; this finding suggests that dimer formation is probably mediated by mature NC and not by the Gag precursor (216). Incorporation of the tRNA^{lys} primer (i.e., minus strand primer) into virions appears to involve the RT domain in Pr160$^{gag-pol}$, although the NC protein may also play a role in encapsidating this primer (annealing of the tRNA^{lys} primer to the viral RNA genome (35,143,782,783).

As yet, cell-free *in vitro* systems recapitulating retroviral morphogenesis have not been established (88). Such systems will be essential for identifying intermediates, analyzing the order of assembly of viral components, and determining whether cellular factors play a role(s) in virion formation. Moreover, information on the mechanism of packaging will be essential for the development of novel antiviral therapies (see section on molecular therapeutics) and for the design of gene transfer vectors based on HIV and SIV (585).

Virus Budding and Incorporation of Cell Proteins into Virions

General features of retroviral extrusion through the cell plasma membrane are similar to those described for other enveloped viruses. The nature of the site for HIV and SIV release on the inner surface of the cell plasma membrane has not been characterized. Either nucleocapsids containing Pr55^{gag,} Pr160$^{gag-pol}$, and viral RNA are formed in the cytoplasm and then migrate to a budding site in the plasma membrane or the interactions of these virion components is initiated at the site of budding on the plasma membrane to produce the nucleocapsid. In either case, a critical role is mediated by the Env gp oligomer; the cytoplasmic tail of the gp41 subunit in this complex is available for interaction with the newly formed viral nucleocapsid in the cytoplasm. Analysis of HIV-1 mutants in T cell lines suggests that efficient release of virions requires an association between the cytoplasmic domain of the gp41 subunit and the MA domain in the Gag precursor (168,169,426, 820). Direct binding of the MA domain and gp41 has not been demonstrated. Because many gaps remain with respect to understanding the molecular interactions of Pr55^{gag}, Pr160gag-pol, and viral RNA with each other during assembly, the precise mechanism of budding for HIV and SIV is not known

In polarized epithelial cells, HIV-1 is released on the basolateral rather than the apical side (551). Particles lacking viral Env gp are extruded in a nonpolarized fashion; thus, *gag* gene products do not contain topogenic signals for targeting specifically to the basolateral side. However, expression of Env gp in *trans* polarizes budding of viral particles lacking the viral glycoprotein (426). Analysis of *env* gene mutants revealed that the gp41 subunit, and not the gp120 subunit, contained the targeting signal governing this pattern of release. In addition, nonpolarized release in epithelial cells is exhibited by mutants in the MA domain of the Gag precursor; these mutants contain lesions in the amino terminus of MA and are defective for incorporation of Env gp (168,169). Thus, an interaction between gp41 in the Env gp oligomer and the MA domain of the Gag precursor is essential for basolateral release. Human immunodeficiency virus type 1 has been reported to infect colonic and cervical epithelial cell lines productively (192, 727); accordingly, polarized virion budding is presumed to influence cell-to-cell spread of virus in the infected host.

Biochemical studies of purified HIV-1 particles show that a select set of host cell proteins assemble into mature viruses. These include the membrane proteins β-2 microglobulin, HLA-α and HLA-β chains DR (16,286), decay-accelerating factor (CD55) (463), as well as the cytoplasmic protein cyclophilin-A (210,736). The mechanism(s) for selecting and incorporating cell membrane proteins into the virion envelope has not been not defined. Cyclophilin-A is presumed to be assembled into virions through an interaction with Gag (436). Functions for cellular proteins in virus particles have been proposed. β-2 microglobulin and HLA may influence virion attachment and entry into T-lymphocytes, macrophages, and other cell types (16). Viral particles containing CD55, an inhibitor of activation of the complement system on the surface of mammalian cells, are resistant to complement (463); accordingly, CD55 may protect the virus from complement lysis in the infected host. The cyclophilins are molecular chaperones (i.e., peptidyl-prolyl *cis*-trans isomerases) that govern the folding of other proteins; thus, cyclophilin-A may influence the structural maturation of virion component(s) during assembly (436). It is also possible that virion-associated cyclophilin-A plays a role in an early step of viral replication, such as uncoating. Interestingly, the immunosuppressive drug cyclosporin-A has an antiviral effect (355, 774) thought to be due to interference with the interaction of Gag and cyclophilins (436). Additional studies are required to determine the role of cell type on the pattern of cell membrane incorporation into virions and to localize cell proteins relative to viral proteins in mature virions.

MOLECULAR THERAPEUTICS

Knowledge of the molecular mechanisms in each step of the HIV/SIV life cycle provides a compelling and essential basis for discovering and developing antiviral drugs and therapies aimed at blocking viral replication and preventing or delaying disease. Importantly, a firm understanding of the viral pathogenic process in infected individuals is also required for optimal development of treatment modalities (see Chapter 61 in *Fields Virology*, 3rd ed.). The need for antiviral agents is highlighted by the lack of an effective anti-HIV vaccine in the foreseeable future (60) and the limited possibility of altering human behavior to control the HIV pandemic (456,547). The current awareness of lentiviral replication mechanisms also indicates that HIV poses several difficult challenges for the development of effective antiviral therapies. First, integration of the provirus into the cell genome enables retroviruses to establish an intimate relationship with the cell and thereby behave like a host genetic element (see section on viral integration). Accordingly, integrated proviruses are passed on to progeny cells during mitosis and can remain latent (or can be expressed at low levels). Although anti-HIV drugs inhibit replicating virus and reduce virus load in infected individuals, proviruses within quiescent cells will not be affected. Second, HIV is disseminated in T-lymphocytes and monocyte/macrophages throughout the body, including the central nervous system (see Chapter 61 in *Fields Virology*, 3rd ed.). Thus, delivery of an antiviral agent to all target cells harboring virus is a formidable obstacle. Third, reverse transcription is a replication process that does not include proofreading to eliminate misincorporated nucleotides; consequently, variant viral genomes are readily generated during viral replication (see section on reverse transcriptase). Moreover, the reverse transcription process, coupled with the diploid nature of the viral RNA genome in each virion, produces a high rate of recombinant viruses. Thus, sequence variation due to RT errors and recombination renders lentiviruses elusive targets for chemotherapy because both of these mechanisms generate drug-resistant viral variants that undergo selection in the treated host. Taken together, these difficulties in the development of effective therapeutic agents against HIV continue to provide strong impetus for the search for novel approaches to the treatment of HIV infection and AIDS.

Agents directed at inhibiting HIV are categorized as conventional antivirals, such as chemical inhibitors of the viral enzyme RT (e.g., nucleoside analogs) (488), or genetic antivirals, which are administered as DNA or RNA into cells and block an intracellular step in viral replication either directly or by expression of RNA or proteins (171) (Table 2). Many conventional antivirals have been identified by screening methods which often depend on serendipity (Table 2); however, information on the three-dimensional structure of viral proteins provides a basis for molecular modeling for rational drug design in the development of novel chemical inhibitors. For this approach, high-resolution electron density maps are available for PR, RT, RNAse H, and the catalytic domain of IN (see section above on viral enzymes). As an example, the principles of rational drug design were used to identify a transition-state (peptidomimetic) analog of a characteristic HIV PR cleavage site (386,751). Three of eight cleavage sites in the Gag and Gag-Pol precursors are at Phe-Pro or Tyr-Pro (573). Because endopeptidase cleavage at an amino-terminal proline is very rare in mammalian cells, interference at this site is likely to be highly selective for the viral enzyme rather than cellular aspartic proteases. Saquinavir (Ro 31-8959) mimics the Phe-Pro and Tyr-Pro site and stabilizes viral PR in an inactive configuration, resembling the transition state; this compound inhibits both HIV-1 and HIV-2 *in vitro* at concentrations several orders of magnitude less than those required to cause cytotoxicity (131,612). Accordingly, clinical trials of saquinavir have been initiated in HIV-1-infected individuals (758).

Genetic antivirals encompass a variety of agents and approaches and include applications of the concept of intracellular immunization (28,171,197). First, antisense sequences are (synthetic) oligonucleotides that hybridize to target RNA and inhibit translation and/or induce degradation of hybridized RNAs (791). Second, ribozymes are catalytic antisense molecules that bind and cleave target RNA (620,791). Third, RNA decoys mimic and compete with viral RNAs that have critical functions in the viral replication cycle (717). Fourth, transdominant proteins are mutant forms of viral proteins that compete with the native protein molecules within infected cells (26,450,451,568, 744). Fifth, single-chain antibodies, directed to an antiviral protein, are transfected into cells via a gene-transfer vector and bind to and inhibit the function of the targeted viral protein (457). Although the majority of both conventional and genetic antiviral strategies block one or more steps in viral replication, agents that destroy infected cells within an individual also provide a means to reduce viral burden. For example, immunotoxins such as conjugates between CD4 and bacterial toxins (20,48) or between anti-Env antibody and the plant lectin ricin (574) have been shown to bind and destroy cultured lymphoid cells expressing the HIV-1 Env gp. Table 2 lists steps in viral replication targeted for antiviral therapies and gives selected examples of both conventional and genetic antivirals; this list demonstrates the diversity of approaches to a large number of viral targets.

Several of the genetic antivirals described above also serve as the basis for intracellular immunization methods

TABLE 2. *Molecular targets for antiviral chemotherapies*

Step in viral replication	Target for inhibitor	Examples (citation)
1. Mature virions	Viral membrane	Nonoxynol-9 (483)
2. Virion binding binding	Sites on Env gp involved in CD4	Soluble CD4 (86,687)
	CD4, gp120, and/or other surface molecules	Polyanions such as dextran sulfate and aurintricarboxilic acid (147), synthetic Env peptides (812)
3. Entry of virions into cells	Fusion function of Env-TM subunit (gp41)	Selected plant lectins (147,793)
	Cell surface protease that cleaves V3 loop of gp120	Trypsin inhibitors (366)
	Disulfide-isomerase in cell plasma membrane	Sulfhydryl inhibitors (625)
4. Uncoating	?	Bicyclams (147)
5. Viral DNA synthesis in the cytoplasm	RT binding site for dNTP precursors	Nucleoside analogs (AZT, DDI, DDC) (147,488)
	RT sites other than the substrate site	Nonnucleosides (benzodiazepines, TIBO; dipyridodiazepinones, nevirapine) (147,488)
	Cellular pool of dNTP substrates	Hydroxyurea (224)
	Viral RNA genome (i.e., template)	Ribozymes and antisense oligonucleotides (171,620,791)
	RNase H	?
6. Entry of preintegration complex into the nucleus	Nuclear localization site in MA domain of gag	Competitor peptides representing nuclear localization signal (255)
7. Integration	IN	Transdominant IN mutants (342), nucleoside analogs (AZT) (467), topoisomerase inhibitors (201a)
8. Viral RNA synthesis	TAR	TAR decoys (425,717), ribozymes, and antisense oligonucleotides (171,620,791)
	tat	Ro 5-3335 (311,312), transdominant *tat* mutants (26,491,568)
	NF-kB	Pentoxifylline (56)
9. Splicing and transport of viral RNA from nucleus to cytoplasm	RRE and/or rev	RRE decoys, ribozymes, and antisense oligonucleotides (171,620,791)
	rev	Transdominant *rev* mutants (452)
	Viral RNA	Ribozymes and antisense oligonucleotides (171,620,791)
10. Translation of viral mRNA	Frameshift site in *gag-pol* transcript	Ribozymes and antisense oligonucleotides (171,620,791)
11. Posttranslation modification of viral proteins	Glycosylation of Env gp by cellular glycosidases	Castanospermine and deoxynojirimycin (600)
	Myristylation of Gag and/or Nef	Myristic acid analogs (74)
12. Proteolytic cleavage of viral	PR	Transition state mimics: saquinavir (711,758), C2-symmetric (361) and dimerization (23) inhibitors (also see 152), transdominant PR (22)
	Cleavage of gp160	Brefeldin (alters compartmentation of gp160) (600), single-chain antibody to gp160 (457)
13. Virion assembly	Nucleocapsid assembly	Transdominant *gag* (744) and *env* (214) mutants, competitor Gag peptides (534), cyclosporin A (436,774)
	Viral RNA genome	Ribozymes and antisense oligonucleotides (171,620,791,813)

TM, transmembrane protein; RT, reverse transcriptase; dNTP, deoxynucleoside triphosphate; IN, integrase; AZT, zidovudine; DDI, didanosine; DDC, zalcitabine; TIBO; MA, matrix protein; TAR, tat-response element; RRE, rev-response element; PR, protease.

aimed at protecting cells in a population from HIV infection and cytopathology. An understanding of the replication mechanisms of HIV as well as other viruses has been essential for the development of these novel treatment modalities. For example, $CD4^+$ T cells have been genetically engineered to express the herpes simplex virus thymidine kinase (HSV TK) gene under the control of the HIV-1 LTR (89). Human immunodeficiency virus type 1, when infecting the cell, produces Tat which acts through the TAR to augment LTR-directed expression of HSV TK. Acyclovir is a nucleoside analog efficiently phosphorylated by HSV TK. If a culture of cells engineered to express HSV TK under the control of the LTR is treated with acyclovir, then this analog is converted into a nucleoside triphosphate and incorporated into elongating DNA by cellular DNA polymerases; DNA synthesis stops because of chain termination and cells eventually die. Thus, infected cells within a population are removed. For an additional example, LTR-regulated expression of inhibitors of translation in mammalian cells, such as poliovirus protein 2A (720) or diphtheria-A toxin (273), provides a means to eliminate cells infected with HIV (720). As a population, cells harboring a HIV-inducible suicide gene are protected from infection and cytopathology. Accordingly, these strategies provide the conceptual basis for therapeutic intervention in which the immune system of HIV-infected patients is reconstituted with a population of engineered bone marrow precursor cells (i.e., sources of progenitors for CD4$^+$ lymphocytes and monocytes) resistant to viral infection and spread in the recipient. A major challenge to the clinical application of suicide strategies as well as genetic antivirals (e.g., antisense oligonucleotides, ribozymes, and trans-dominant viral mutants) is that efficient delivery systems for gene therapy must be developed for *in vivo* use.

The interferons are a diverse group of cytokines that have antiviral properties and constitute one of the host defense responses to viral infection (see Chapter 13 in *Fields Virology,* 3rd ed.). Human immunodeficiency virus replication in cultures of primary lymphocytes and macrophages and in monocyte and T-cell lines has been shown to be inhibited by exogenous interferons (24,115,677). The step(s) in the viral life cycle affected by various interferons appears to be cell-type dependent; viral replication may be inhibited at an early step (e.g., proviral formation in primary macrophages and the CEMX174 hybrid T-cell/B-cell line) or a late step (e.g., assembly and release of infectious virions into the medium in primary lymphocytes and the majority of T-cell lines). The time of addition of interferon to a cell culture, whether before or after establishment of infection, also influences the step at which viral replication is affected (235,381). Although the interferon-induced cellular enzymes 2′,5′:-oligoadenylate synthetase (650) and double-stranded RNA-dependent protein kinase (665) have been implicated as mediators of anti-HIV activity, additional studies are required to elucidate the precise molecular mechanism(s) of inhibition. In AIDS patients, interferons (e.g., α-interferon) (170) as well as other cytokines (e.g., IL-2) (648) may find therapeutic utility in restoring or augmenting CD4$^+$ T-cell functions and thereby halting or delaying disease progression. Because certain cytokines have been shown to activate HIV transcription and replication in tissue culture systems (see section on viral RNA synthesis), there is a need to obtain more information on the role of cellular factors in viral gene regulation before clinical evaluation of cytokine therapies.

The potential for preventing disease on a long-term basis by treating HIV-infected individuals with antiviral drugs remains to be determined. However, studies with the RT inhibitor azidothymidine (AZT, or zidovudine) clearly demonstrate that this drug significantly reduces viral load and confers clinical benefits at least on a short-term basis in patients in both early and advanced stages of infection (125,488). Moreover, AZT reduces transmission of virus from an infected mother to the newborn infant (65). Further consideration needs to be given to treatment modalities based on alternating or combining inhibitors affecting different steps in viral replication (488) as well as to protocols involving both antiviral agents and reconstitution of general immune responsiveness of the host (648). Although major obstacles remain (i.e., proviral integration, widespread dissemination of virus in T-lymphocytes and monocytes/macrophages, and high viral mutation rate), progress in antiviral therapies will be enhanced by further studies aimed at elucidating both HIV replication mechanisms and the structure of viral proteins. Moreover, novel approaches offered by genetic antivirals (see above) will depend on key developments in other fields such as the technology of gene transfer vectors (513) and the biology of hematopoietic stem cells (732).

MOLECULAR BIOLOGY OF VIRAL PATHOGENESIS

The expanding knowledge of HIV and SIV replication mechanisms has been coupled with information from *in vivo* observations to obtain an understanding of the virus-host relationship, including the role of these viruses in T-lymphocyte and macrophage dysfunction, disease progression, and fatal immunodeficiency. Proof that HIV causes AIDS, defined by criteria established by the Centers for Disease Control (Atlanta, GA), rests on numerous observations (75,90,548). A wealth of epidemiologic evidence strongly implicates this lentivirus as the etiologic agent (339,578); the most cogent demonstration is that hemophiliacs receiving blood factors contaminated with HIV-1 develop fatal AIDS (106). Importantly, molecular epidemiology shows the evolution of viral variants in various geographical regions of the AIDS pandemic (225,430,519). In tissue culture systems, HIV and SIV are cytopathic for CD4$^+$ T-lymphocytes and cause dysfunction of macrophages (414). Although the mechanism(s) of cytopathology is not well defined, these *in vitro* findings recapitulate events *in vivo* in the respective susceptible host (i.e., depletion of CD4$^+$ T-lymphocytes and macrophage dysfunction) (see Chapter 61 in *Fields Virology*, 3rd ed.) (390). It is possible that apoptosis (i.e., programmed cell death) triggered by these viruses may account for loss of T cells (11). Replicating virus in infected individuals is readily detected in peripheral blood (414,647) and lymph nodes (181,557) throughout the course of infection, and increased viral load (or burden) is correlated with disease progression (127,630). Fatal disease occurs because the virus outpaces the immune system and causes progressive depletion of CD4$^+$ cells (297,779). Although the majority of HIV-1-infected individuals succumb to AIDS within 5 to 10 years after infection (549), a small proportion of individuals harboring virus have remained healthy for more than a decade (401,423,649). Some of these long-term nonprogressors, also designated long-term survivors, exhibit low virus load (85,558). In one instance, a long-term nonprogressor was shown to contain virus with a deletion in the *nef* gene (342). Interestingly, deletion of the nef gene from a pathogenic molecular clone of SIV produces a virus with attenuated virulence properties in adult rhesus macaques (21,365).

A relationship between viral phenotype and pathogenesis has been demonstrated. Nonsyncytium-inducing variants of HIV-1 that replicate slowly are the predominant viruses found early after infection and in long-term nonprogressors (615), whereas SI variants that replicate rapidly are generally isolated from individuals with disease (127, 479,653,729). Syncytium-induction is a property controlled largely by the *env* gene whereas replication rate may be influenced by *cis*-acting regulatory elements (e.g., transcriptional promoter elements) as well as various viral genes.

Treatment of infected individuals with anti-viral drugs, such as the RT inhibitor AZT, reduces virus load and confers short-term clinical benefit (125,488). In addition, AZT decreases the rate of HIV-1 transmission from mother to infant (65). Viral resistance to RT inhibitors in treated patients is correlated with an increased virus load and disease progression (606); this finding is genetic proof that HIV-1 is linked to (the phenotype) AIDS.

Studies on molecular clones of SIV in macaques (293, 364,365), infection of non-human primates with HIV-2 (39), and HIV-1 infection in SCID-hu mice (6,329,347) also provide persuasive evidence for a causative role for primate lentiviruses in AIDS. These animal models are highly manipulatable *in vivo* systems for investigating the functions of viral genes in all stages of infection and pathogenesis. Although cofactors such as infectious opportunistic agents and/or recreational drugs may influence disease progression (549), HIV is necessary and appears to be sufficient for AIDS (123).

An animal model to explore directly both HIV-1 infection and pathogenesis is not yet available; however, because the simian and human lentiviruses are genetically related, infection of nonhuman primates with selected strains and molecular clones of SIV has been utilized to examine the roles of viral genes in the acute and chronic stages of infection and in pathogenesis (229). Such studies are based on the analysis of molecularly cloned viruses with site-specific mutations and recombinant viruses constructed from molecularly cloned viruses that differ in phenotype. Elegant *in vivo* experiments on site-specific mutations in the *nef* gene of the pathogenic clone SIVmac239 reveal that this gene influences virus load and contributes to fatal immunodeficiency in rhesus macaques (365,641). Investigations in macaques inoculated with recombinant viruses constructed from pathogenic and nonpathogenic clones of SIV_{MAC} (SIVmac239 and SIVmac1A11, respectively) demonstrate that two or more genes control virus load and disease development (465,535). In addition, recombinant viruses, designated SHIV, have been constructed by substituting genes of SIVmac239 with counterpart genes of HIV-1 (674). Although HIV-1 produces only a transient infection in macaques (2), SHIV containing the HIV-1 *env* gene on the background of the molecular clone SIVmac239 establishes a persistent infection in these primates (419, 440,631). Thus, the SIVmac239 molecular clone serves as a vector to explore roles of HIV-1 genes *in vivo*. Serial passage of HIV-2 in macaques and baboons has produced viral variants that cause depletion of $CD4^+$ cells and an AIDS-like disease in these species (39); however, a molecular clone of HIV-1 that causes a fatal AIDS-like disease in non-human primates has not yet been isolated. Thus, the SIV_{MAC} model remains the most useful system for exploring viral determinants of acute infection, persistence, immune system dysfunction, and pathogenesis.

Mice are resistant to infection with the primate lentiviruses; however, two mouse model systems have been utilized to examine various aspects of viral infection and pathogenesis: transgenic mice produced by introducing one or more viral genes into the mouse germ line (266), and mice with severe combined immunodeficiency syndrome (SCID) containing xenografts of human tissues or organs (SCID-hu mice) (347). The transgenic mouse system is a means to study individual viral genes or combinations of viral genes without the complications of replicating virus. Examples of transgenic mouse lines constructed for investigating the roles of HIV-1 genes in pathogenesis are a line containing a transgene in which the viral LTRs flank the *tat* gene and a line with a *nef* transgene under the control of a T-cell promoter. The former develop a skin lesion similar to Kaposi's sarcoma late in life (768), and the latter show a reduction in thymocytes and downregulation of CD4 antigen on the surfaces of T cells early in life (684). SCID-hu mice engrafted with either human fetal hematolymphoid organs or peripheral blood lymphocytes from adults support HIV-1 replication (471,510); in addition, depletion of human $CD4^+$ T cells is observed in infected mice (6). Accordingly, this animal model has been used to examine HIV-1 phenotypes and gene functions. For example, SI variants replicate to high levels and cause depletion of human thymocytes in SCID-hu mice, whereas NSI viruses replicate to low levels and do not deplete human thymocytes in these animals (347). In another example, an HIV-1 clone with an intact *nef* gene replicates and causes depletion of human T cells in the implant, whereas a derivative of this clone with a deletion in the *nef* gene exhibits low virus load and does not deplete human T cells in the implant (329). This latter experiment complements the finding that the molecular clone SIV-mac239/nef-deletion persists at a low level without causing disease in juvenile and adult macaques (365). Thus, both the SCID-hu mouse model and SIV infection of nonhuman primates will be important for analyzing viral determinants of pathogenesis as well as for evaluating antiviral therapies and vaccines.

REPRISE AND PROSPECTS

Elucidation of molecular mechanisms of HIV and SIV replication has proceeded rapidly for the last decade. This enormous activity is due to the seriousness of the AIDS pandemic and to the power of molecular biologic tools and expertise that provide the means to make fundamental and exciting discoveries on many aspects of these viruses *in vitro* (i.e., in cell-free and tissue culture systems) and *in vivo* (i.e., in the susceptible host). Examples of relevant technological developments are new applications of polymerase chain reaction (PCR) amplification methods for detecting and cloning viral sequences (703), the yeast two-hybrid system for examining protein-protein interactions (202,343,436), and recombinatorial phage-display cloning methods for analyzing protein structure and developing antiviral inhibitors (36,159,408). Awareness that SIV is widespread in many nonhuman primates in the wild adds a sense of fascination; the diversity of these viruses with respect to genetic and biologic properties has captured the curiosity of virologists and those interested in evolutionary biology. Although studies on the molecular mechanisms of other retroviruses have guided many investigations on HIV and SIV, novel findings on the primate lentiviruses now outpace other retroviruses. For example, high-resolution three-dimensional structures have been derived for the HIV-1 enzymes PR, RT, and RNase H and the catalytic domain of IN. The cell receptor, CD4 antigen, was the first retroviral receptor to be identified and characterized in detail with respect to its interaction with the viral Env gp. Additionally, investigations on the viral transactivators Tat and Rev are providing new insights into regulation of gene expression in eukaryotic cells. Although much has been learned about the molecular biology of HIV and SIV, important aspects of the cell biology (e.g., subcellular compartmentation and trafficking of viral components) have begun to make significant impact on our knowledge of these viruses. The discipline of structural biology is expected to identify and characterize functional domains of viral proteins as well as cellular proteins that interact with them further. A combination of genetic and biochemical approaches will continue to generate new information on HIV and SIV replication mechanisms; an important goal is to establish cell-free systems, with defined viral (and cellular) components, that recapitulate each step in viral replication. Such *in vitro* systems are a foundation for identifying novel antiviral strategies as well as improving the efficacy of existing antiviral drugs. Animal models will be essential for determining the roles of viral genes in pathogenesis as well as evaluating both conventional and new approaches to antiviral vaccines and treatment modalities.

Expanded explorations on the molecular, cellular, and structural biology of HIV and SIV will be key to understanding the mechanisms by which these viruses initiate infection, persist, and cause fatal AIDS in the susceptible host. Because of the complexity of the immunodeficiency disease process, an understanding of the structures and functions of viral proteins and *cis*-acting elements must be integrated with information on the distribution, regulation, and immunology of the virus in the infected host (see Chapter 61 in *Fields Virology*, 3rd ed.). "We learn more and more about viruses, but not much about how they cause disease," said Harold S. Ginsberg in 1992. Accordingly, the challenge for the future is to utilize the increasing knowledge of molecular mechanisms of HIV and SIV replication to elucidate all aspects of the virus-host relationship fully and thereby to understand mechanisms of pathogenesis in AIDS further. "Science baits laws with stars to catch telescopes," wrote e. e. cummings in *New Poems 16.*

REFERENCES

1. Adams M, Sharmeen L, Kimpton J, Romeo JM, Garcia JV, Peterlin BM, Groudine M, and Emerman M. Cellular latency in human immunodeficiency virus infected individuals with high CD4 levels can be detected by the presence of promoter-proximal transcripts. *Proc Natl Acad Sci* 1994;91:3862–3866.
2. Agy MB, Frumkin L, Corey L, Coombs RW, Wolinski SM, Kochler J, Morton WR, Katze MG. Infection of Macaca nemestrina by human immunodeficiency virus type 1. *Science* 1992; 257:103–106.
3. Ahmad N, Venkatesan S. Nef protein of HIV-1 is a transcriptional repressor of HIV-1 LTR. *Science* 1988;241:1481–1485.
4. Aiken C, Konner J, Landau NR, Lenburg ME, Trono D. Nef induces CD4 endocytosis: requirement for a critical dileucine motif in the membrane-proximal CD4 cytoplasmic domain. *Cell* 1994;76:853–864.
5. Aldovini A, Young RA. Mutations of RNA and protein sequences involved in HIV-1 packaging in production of noninfectious virus. *J Virol* 1990;64:1920–1926.
6. Aldrovandi GM, Feuer G, Gao L, Jamieson B, Kristeve M, Chen IS, Zack JA. The SCID-hu mouse as a model for HIV-1 infection. *Nature* 1993;363:732–736.
7. Allan JS, Coligan JE, Barin F, McLane MF, Sodroski JG, Rosen CA, Haseltine WA, Lee TH, Essex M. Major glycoprotein antigens that induce antibodies in AIDS patients are encoded by HTLV-III. *Science* 1985;228:1091–1094.
8. Aloia RC, Jensen FC, Curtain CC, Mobley PW, Gordon LM. Lipid composition and fluidity of the human immunodeficiency virus. *Proc Natl Acad Sci USA* 1988;85:900–904.
9. Aloia RC, Tian H, Jensen FC. Lipid composition and fluidity of the human immunodeficiency virus envelope and host cell plasma membranes. *Proc Natl Acad Sci USA* 1993;90:5181–5185.
10. Alonso A, Cujec TP, Peterlin BM. Effects of human chromosome 12 on interactions between Tat and TAR of human immunodeficiency virus type 1. *J Virol* 1994;68:6505–6513.
11. Ameisen JC. Programmed cell death (apoptosis) and AIDS. In: Montagnier L, Gougeon M-L, eds. *New concepts in AIDS pathogenesis.* New York: Marcel Dekker; 1993:153–168.
12. Andeweg AC, Leeflang P, Osterhaus ADME, Bosch ML. Both the V2 and V3 regions of the human immunodeficiency virus type 1 surface glycoprotein functionally interact with other envelope regions in syncytium formation. *J Virol* 1993;67:3232–3239.

12a. Antoni BA, Stein SB, Rabson AB. Regulation of human immunodeficiency virus infection: implication for pathogenesis. *Adv Virus Res* 1994;43:53–145.

13. Arnold E, Jacobo-Molina A, Nanni RG, Williams RL, Lu X, Ding J, Clark AD, Zhang A, Ferris AL, Clark P, Hizi A, Hughes SH. Structure of HIV-1 reverse transcriptase/DNA complex at 7 Å resolution showing active site locations. *Nature* 1992;357:85–89.

14. Arrigo S, Chen ISY. Rev is necessary for translation but not cytoplasmic accumulation of HIV-1 vif, vpr, and env/vpu mRNAs. *Genes Dev* 1991;5:808–819.
15. Arthos J, Deen K, Chaikin M, Fornwald J, Sathe G, Sattentau Q, Clapham P, Weiss R, McDougal J, Pietropaolo C, Axel R, Truaneh A, Maddon P, Sweet R. Identification of the residues in human CD4 critical for the binding of HIV. *Cell* 1989;57:469–481.
16. Arthur LO, Bess JW, Sowder RC, Benveniste RE, Mann DL, Chermann J, Henderson LE. Cellular proteins bound to immunodeficiency viruses: implications for pathogenesis and vaccines. *Science* 1992;258:1935–1938.
17. Arya SK, Guo C, Josephs SF, Wong-Staal F. Trans-activator gene of human T-lymphotropic virus type III (HTLV-III). *Science* 1985;229:69–73.
18. Ashkenazi A, Presta LG, Marsters SA, Camerato TR, Rosenthal KA, Fendly BM, Capon DJ. Mapping the CD4 binding site for human immunodeficiency virus by alanine-scanning mutagenesis. *Proc Natl Acad Sci USA* 1990;87:1990–1994.
19. Ashkenazi A, Smith DH, Marsters SA, Riddle L, Gregory TJ, Ho DD, Capon DJ. Resistance of primary isolates of human immunodeficiency virus type 1 to soluble CD4 is independent of CD4-gp120 binding affinity. *Proc Natl Acad Sci USA* 1991;88:7056–7060.
20. Aullo P, Alcami J, Popoff MR, Klatzmann DR, Murphy JR, Boquet P. A recombinant diptheria toxin related human CD4 fusion protein specifically kills HIV infected cells which express gp120 but selects fusion toxin resistant cells which carry HIV. *EMBO J* 1992;11:575–578.
21. Baba TW, Jeong YS, Pennick D, Bronson R, Greene MF, Ruprecht RM. Pathogenicity of live attenuated SIV after mucosal infection of neonatal macaques. *Science* 1995;267:1820–1825.
22. Babe LM, Pichuantes S, Craik CS. Inhibition of HIV protease activity by heterodimer formation. *Biochemistry* 1991;30:106–111.
23. Babe LM, Rose J, Craik CS. Synthetic "interface" peptides alter dimeric assembly of the HIV 1 and 2 proteases. *Protein Sci* 1992;1:1244–1253.
24. Baca-Regen L, Heinzinger N, Stevenson M, Gendelman HA. Alpha interferon-induced antiretroviral activities: restriction of viral nucleic acid synthesis and progeny virion production in human immunodeficiency virus type 1-infected monocytes. *J Virol* 1995;68:7559–7565.
25. Baeurele PA, Henkel T. Function and activation of NF-κB in the immune system. *Annu Rev Immunol* 1994;12:141–179.
26. Bahner I, Zhou C, Yu XJ, Hao QL, Guatelli JC, Kohn DB. Comparison of transdominant inhibitory mutant human immunodeficiency virus type 1 genes expressed by retroviral vectors in human T lymphocytes. *J Virol* 1993;67:3199–3207.
27. Baltimore D. RNA-dependent DNA polymerase in virions of RNA tumor viruses. *Nature* 1970;226:1209–1211.
28. Baltimore D. Gene therapy: intracellular immunization. *Nature* 1988;335:395–396.
29. Banapour B, Marthas ML, Munn RJ, Luciw PA. In vitro macrophage tropism of pathogenic and nonpathogenic molecular clones of simian immunodeficiency virus (SIVmac). *Virology* 1991;183:12–19.
30. Banapour B, Marthas ML, Ramos RA, Lohman BL, Unger RE, Gardner MB, Pedersen NC, Luciw PA. Identification of viral determinants of macrophage tropism for simian immunodeficiency virus SIVmac. *J Virol* 1991;65:5798–5805.
31. Bandres JC, Luria S, Ratner L. Regulation of human immunodeficiency virus nef protein by phosphorylation. *Virology* 1994;201:157–161.
32. Bandres JC, Ratner L. Human immunodeficiency virus type 1 nef protein down-regulates transcription factors NF-kappa B and AP-1 in human T cells in vitro after T-cell receptor stimulation. *J Virol* 1994;68:3243–3249.
33. Barat C, Lullien V, Schatz O, Keith G, Nugeyre MT, Gruninger-Leitch F, Barre-Sinnoussi F, LeGrice SFJ, Darlix JL. HIV-1 reverse transcriptase specifically interacts with the anticodon domain of its cognate primer tRNA. *EMBO J* 1989;8:3279–3285.
34. Barat C, LeGrice SFJ, Darlix JL. Interaction of HIV-1 reverse transcriptase with a synthetic form of its replication primer, tRNAlys3. *Nucleic Acids Res* 1991;19:751–757.
35. Barat C, Schatz O, Le Grice S, Darliz J-L. Analysis of the interactions of HIV-1 replication primer tRNAlys,3 with nucleocapsid protein and reverse transcriptase. *J Mol Biol* 1993;231:185–190.
36. Barbas CF, Dunlop HDN, Sawyer L, Cababa D, Hendry RM, Nara PL, Burton DR. In vitro evolution of a neutralizing human antibody to human immunodeficiency virus type 1 to enhance affinity and broaden cross-reactivity. *Proc Natl Acad Sci USA* 1994;91:3809–3813.
37. Barillari G, Gendelman R, Gallo RC, Ensoli B. The tat protein of human immunodeficiency virus type 1, a growth factor for AIDS Kaposi sarcoma and cytokine-activated vascular cells, induces adhesion of the same cell types by using integrin receptors recognizing the RGD amino acid sequence. *Proc Natl Acad Sci USA* 1993;90:7941–7945.
38. Barnett RW, Delling U, Kuperman R, Sonenberg N, Sumner-Smith M. Rotational symmetry in ribonucleotide strand requirements for binding of HIV-1 tat protein to TAR RNA. *Nucleic Acids Res* 1993;21:151–154.
39. Barnett SW, Murthy KK, Herndier BG, Levy JA. An AIDS-like condition induced in baboons by HIV-2. *Science* 1994;266:642–646.
40. Barr PJ, Mason OB, Landsberg KE, Wong PA, Kiefer MC, Brake AJ. cDNA and gene structure for a human subtilisin-like protease with cleavage specificity for paired basic amino acid residues. *DNA Cell Biol* 1991;10:319–328.
40a. Barre-Sinousi F, Chermann JC, Rey F, et al. Isolation of a T-lymphotropic retrovirus from a patient at risk for acquired immunodeficiency syndrome (AIDS). *Science* 1983;220:868–871.
41. Barry PA, Pratt-Lowe E, Unger RE, Luciw PA. Cellular factors regulate transactivation of the human immunodeficiency virus type 1. *J Virol* 1991;64:1392–1399.
42. Battula N, Loeb LA. The infidelity of avian myeloblastosis virus deoxyribonucleic acid polymerase in polynucleotide replication. *J Biol Chem* 1974;249:4086.
43. Baur AS, Sawai ET, Dazin P, Fantl WJ, Cheng-Mayer C, Peterlin BM. HIV-1 nef leads to inhibition or activation of T cells depending on its intracellular localization. *Immunity* 1994;1:373–384.
44. Bebenek K, Abbotts J, Roberts JD, Wilson SH, Kunkel TA. Specificity and mechanism of error-prone replication by human immunodeficiency virus-1 reverse transcriptase. *J Biol Chem* 1989;264:16948–16956.
45. Bebenek K, Kunkel TA. The fidelity of retroviral reverse transcriptases. In: Skalka AM, Goff SP, eds. *Reverse transcriptase.* Cold Spring Harbor, NY: Cold Spring Harbor Laboratory Press; 1993:85–102.
46. Bedinger P, Moriarty A, von Borstel RC, Donovan NJ, Steimer KS, Littman DR. Internalization of the human immunodeficiency virus does not require the cytoplasmic domain of CD4. *Nature* 1988;334:162–165.
47. Benichou S, Bomsel M, Bodeus M, Durand H, Doute M, Letourneur F, Camonis J, Benarous R. Physical interaction of the HIV-1 nef protein with B-COP, a component of non-clathrin-coated vesicles essential for membrane traffic. *J Biol Chem* 1994;269:30073–30076.
47a. Benson RE, Sanfridson A, Ottinger JS, Doyle C, Cullen B. Downregulation of cell-surface CD4 expression by simian immunodeficiency virus Nef prevents viral superinfection. *J Exp Med* 1993;177:1561–1566.
48. Berger EA, Clouse KA, Chaudhary VK, Chakrabarti S, FitzGerald DJ, Pastan I, Moss B. CD4-Pseudomonas exotoxin hybrid protein blocks the spread of human immunoderficiency virus infection in vitro and is active against cells expressing the envelope glycoproteins from diverse primate immunodeficiency retroviruses. *Proc Natl Acad Sci USA* 1989;86:9539–9543.
49. Berger EA, Lifson JD, Eiden LE. Stimulation of glycoprotein gp120 dissociation from the envelope glycoprotein complex of human immunodeficiency virus type 1 by soluble CD4 and CD4 peptide derivatives: implications for the role of the complementarity-determining region 3-like region in membrane fusion. *Proc Natl Acad Sci USA* 1991;88:8082–8086.
49a. Berkhout B, Gatignol A, Silver J, Jeang KT. Efficient trans-activation by the HIV-2 Tat protein requires a duplicated TAR RNA structure. *Nucleic Acids Res* 1990;18:1839–1846.
49b. Berkhout B, Jeang KT. Functional roles for the TATA promoter and enhancers in basal and Tat-induced expression of the human immunodeficiency virus type 1 long terminal repeat. *J Virol* 1992;66:139–149.
49c. Berkhout B, Jeang KT. Detailed mutational analysis of TAR RNA: critical spacing between the bulge and loop recognition domains. *Nucleic Acids Res* 1991;19:6169–6176.
49d. Berkhout B, Schoneveld I. Secondary structure of the HIV-2 leader RNA comprising the tRNA-primer binding site. *Nucleic Acids Res* 1993;21:1171–1178.
50. Berkhout B, Silverman RH, Jeang K-T. Tat trans-activates the human immunodeficiency virus through a nascent RNA target. *Cell* 1989;59:273–282.
51. Berkhout B, Gatignol A, Rabson A, Jeang K-T. TAR-independent activation of the HIV-1 LTR: evidence that tat requires specific regions of the promoter. *Cell* 1990;62:757–767.

52. Berkhout B, van Hemert FJ. The unusual nucleotide content of the HIV RNA genome results in a biased amino acid composition of HIV proteins. *Nucleic Acids Res* 1994;9:1705–1711.
53. Berkowitz RD, Luban J, Goff SP. Specific binding of human immunodeficiency virus type 1 gag polyprotein and nucleocapsid protein to viral RNAs detected by RNA mobility shift assays. *J Virol* 1993;67:7190–7200.
54. Berkowitz RD, Goff SP. Analysis of binding elements in the human immunodeficiency virus type 1 genomic RNA and nucleocapsid protein. *Virology* 1994;202:233–246.
55. Bhat S, Spitalnik SL, Gonzalez-Scarano F, Silberberg DH. Galactosyl ceramide or a derivative is an essential component of the neural receptor for human immunodeficiency virus type 1 envelope glycoprotein gp120. *Proc Natl Acad Sci USA* 1991;88:7131–7134.
56. Biswas DK, Ahlers CM, Dezube BJ, Pardee AB. Pentoxifylline and other protein kinase C inhibitors down-regulate HIV-1 LTR NF-κB induced gene expression. *Mol Med* 1994;1:31–43.
57. Bjorling B, Broliden K, Bernardi D, Utter G, Thorstensson R, Chiodi F, Norrby E. Hyperimmune antisera against synthetic peptides representing the glycoprotein of human immunodeficiency virus type 2 can mediate neutralization and antibody-dependent cytotoxic activity. *Proc Natl Acad Sci USA* 1991;88:6082–6086.
58. Bogerd HP, Huckaby GL, Ahmed YF, Hanly SM, Greene WC. The type 1 human T-cell leukemia virus (HTLV-I) rex trans-activator binds directly to the HTLV-I rex and the type 1 human immunodeficiency virus rev RNA response elements. *Proc Natl Acad Sci USA* 1991;88:5704–5708.
59. Bohnlein E, Siekevitz M, Ballard DW, Lowenthal JW, Rimsky L, Bogerd H, Hoffman J, Wano Y, Franza BR, Greene WC. Stimulation of the human immunodeficiency virus type 1 enhancer by the human T-cell leukemia virus type 1 tax gene product involves the action of inducible cell proteins. *J Virol* 1989;63:1578–1586.
60. Bolognesi DP. Human immunodeficiency virus vaccines. *Adv Virus Res* 1993;42:103–148.
61. Borer RA, Lehner CF, Eppenberger HM, Nigg EA. Major nucleolar proteins shuttle between nucleus and cytoplasm. *Cell* 1989;56:379–390.
62. Bosch ML, Andeweg AC, Schipper R, Kenter M. Insertion of N-linked glycosylation sites in the variable regions of the human immunodeficiency virus type 1 surface glycoprotein through AAT triplet reiteration. *J Virol* 1994;68:7566–7569.
63. Boungorno L, Barillari G, Chang HK, Bohan CA, Kao V, Morgan R, Gallo RC, Ensoli B. Effects of the human immunodeficiency virus type 1 tat protein on the expression of inflammatory cytokines. *J Virol* 1992;66:7159–7167.
63a. Bowerman B, Brown PO, Bishop JM, Varmus HE. A nucleoprotein complex mediates the integration of retroviral DNA. *Genes Dev* 1989;3:469–478.
64. Boyer V, Desgranges C, Trabaud MA, Fischer E, Kazatchkine MD. Complement mediates human immunodeficiency virus type 1 infection of a human T cell line in a CD4- and antibody-independent fashion. *J Exp Med* 1991;173:1151–1158.
65. Boyer PJ, Dillon M, Navaie M, Deveikis A, Keller M, O'Rourke S, Bryson YJ. Factors predictive of maternal-fetal transmission of HIV-1—preliminery analysis of zidovudine given during pregnancy and/or delivery. *JAMA* 1994;271:1925–1930.
66. Brand D, Srinivasan K, Sodroski J. Determinants of human immunodeficiency virus type 1 entry in the CDR2 loop of the CD4 glycoprotein. *J Virol* 1995;69:166–171.
67. Brasseur R, Cornet B, Burny A, Vandenbranden M, Ruysschaert JM. Mode of insertion into a lipid membrane of the N-terminal HIV gp41 peptide segment. *AIDS Res Hum Retroviruses* 1988;4:83–90.
68. Brightly DW, Rosenberg M, Chen ISY, Ivey-Hoyle M. Envelope proteins from clinical isolates of human immunodeficiency virus type 1 that are refractory to neutralization by soluble CD4 possess high affinity for the CD4 receptor. *Proc Natl Acad Sci USA* 1991;88:7802–7806.
69. Broder CC, Berger EA. CD4 molecules with a diversity of mutations encompassing the CDR3 region efficiently support HIV-1 envelope glycoprotein-mediated cell fusion. J Virol 1993;67:913–926.
70. Broder S, Merigan TC, Bolognesi D. *Textbook of AIDS medicine.* Baltimore: Williams and Wilkins; 1994a.
71. Broder CC, Earl PL, Long D, Abedon ST, Moss B, Doms RW. Antigenic implication of human immunodeficiency virus type 1 envelope quarternary structure: oligomer-specific and -sensitive monoclonal antibodies. *Proc Natl Acad Sci USA* 1994;91:11699–11703.
72. Bryant ML, Yamamoti J, Luciw P, Munn R, Marx P, Higgins J, Pedersen N, Levine A, Gardner MB. Molecular comparison of retroviruses associated with human and simian AIDS. *Hematol Oncol* 1985;3:187–197.
73. Bryant M, Ratner L. Myristoylation-dependent replication and assembly of human immunodeficiency virus I. *Proc Natl Acad Sci USA* 1990;87:523–527.
74. Bryant ML, Ratner L, Duronio RJ, Kishore NS, Devadas B, Adams SP, Gordon JI. Incorporation of 12-methoxydodecanoate into the human immunodeficiency 1 gag polyprotein precursor inhibits its proteolytic processing and virus production in a chronically infected human lymphoid cell line. *Proc Natl Acad Sci USA* 1991;88:2055–2059.
75. Buehler JW. The surveillance definition for AIDS. *Am J Public Health* 1992;82:1462–1464.
76. Bukrinsky MI, Stanwick TL, Dempsey MP, Stevenson M. Quiescent T lymphocytes as an inducible virus reservoir in HIV-1 infection. *Science* 1991;254:423–427.
77. Bukrinsky MI, Haggerty S, Dempsey MP, Sharova N, Adzhubei A, Spitz L, Lewis P, Goldfarb D, Emerman M, Stevenson M. A nuclear localization signal within HIV-1 matrix protein that governs infection of non-dividing cells. *Nature* 1993;365:666–669.
78. Bukrinsky MI, Sharova N, McDonald TL, Pushkarskaya T, Tarpley WG, Stevenson M. Association of integrase, matrix, and reverse transcriptase antigens of human immunodeficiency virus type 1 with viral nucleic acids following acute infection. *Proc Natl Acad Sci USA* 1993;90:6125–6129.
79. Bukrinsky MI, Sharova N, Stevenson M. Human immunodeficiency virus type 1 2-LTR circles reside in a nucleoprotein complex which is different from the preintegration complex. *J Virol* 1993;67:6863–6865.
80. Burke CJ, Sanyal G, Bruner MW, Ryan JA, LaFemina RL, Robbins HL, Zeft AS, Middaugh CR, Cordingley MG. Structural implications of spectroscopic characterization of a putative zinc-finger peptide from HIV-1 integrase. *J Biol Chem* 1992;267:9639–9644.
81. Burns DPW, Desrosiers RC. Envelope sequence variation, neutralizing antibodies, and primate lentivirus persistence. *Curr Top Microbiol Immunol* 1994;188:185–219.
82. Bushman FD, Craigie R. Activities of human immunodeficiency virus (HIV) integration protein in vitro: specific cleavage and integration of HIV DNA. *Proc Natl Acad Sci USA* 1990;88:1339–1343.
83. Bushman FD, Engelman A, Palmer I, Wingfield P, Craigie R. Domains of the integrase protein of human immunodeficiency virus type 1 responsible for polynucleotidyl transfer and zinc binding. *Proc Natl Acad Sci USA* 1993;90:3428–3432.
84. Cao J, Bergeron L, Helseth ETM, Repke H, Sodroski J. Effects of amino acid changes in the extracellular domain of the human immunodeficiency virus type 1 gp41 envelope glycoprotein. *J Virol* 1993;67:2747–2755.
85. Cao YL, Qin L, Zhang L, Safrit J, Ho DD. Virologic and immunolgic characterization of long-term survivors of human immunodeficiency virus type 1 infection. *New Engl J Med* 1995;332:201–208.
86. Capon DJ, Ward RHR. The CD4-gp120 interaction and AIDS pathogenesis. *Annu Rev Immunol* 1991;9:649–678.
87. Carr CM, Kim PS. A spring-loaded mechanism for the conformational change of influenza hemagglutinin. *Cell* 1993;73:823–832.
87a. Carrow EW, Vujcic LK, Glass WL, et al. High prevalence of antibodies to the gp120 V3 region principal neutralizing determinant of HIV-1MN in sera from Africa and the Americas. *AIDS Res Hum Retroviruses* 1991;7:831–838.
88. Carter C, Zybarth G. Processing of retroviral gag polyproteins: an in vitro approach. In: Kuo LC, Shafer JA, eds.*Retroviral proteases.* San Diego: Academic Press; 1994:227–253.
89. Caruso M, Klatazmann D. Selective killing of CD4+ cells harboring a human immunodeficiency virus-inducible suicide gene prevents viral spread in an infected cell population. *Proc Natl Acad Sci USA* 1992;89:182–186.
90. Centers for Disease Control. Centers for Disease Control Task Force on Kaposi's sarcoma and opportunistic infections. *N Engl J Med* 1982;306:248–252.
91. Centers for Disease Control. 1993 revised classification system for HIV infection and expanded surveillance case definition for acquired immunodeficiency syndrome (AIDS) among adolescents and adults. *MMWR Morb Mortal Wkly Rep* 1992;32:1–19.
92. Chackerian B, Morton WR, Overbaugh J. Persistence of simian immunodeficiency virus Mne variants upon transmission. *J Virol* 1994;68:4080–4085.

93. Chakrabarti L, Guyader M, Alizon M, Daniel MD, Desrosiers RC, Tiollais P, Sonigo P. Molecular cloning and nucleotide sequence of simian immunodeficiency virus from macaques. *Nature* 1987;328:543–547.
94. Chakrabarti L, Emerman M, Tiollais P, Sonigo P. The cytoplasmic domain of simian immunodeficiency virus transmembrane protein modulates infectivity. *J Virol* 1989;63:4395–4403.
95. Champoux JJ. Roles of ribonuclease H in reverse transcription. In: Skalka AM, Goff SP, eds. *Reverse transcriptase.* Cold Spring Harbor, NY: Cold Spring Harbor Laboratory Press; 1993:103–117.
96. Chang DD, Sharp PA. Regulation by HIV rev depends upon recognition of splice sites. *Cell* 1989;59:789–795.
97. Chang SYP, Bowman BH, Weiss JB, Garcia RE, White TJ. The origin of HIV-1 isolate HTLV-IIIB. *Nature* 1993;363:466–469.
98. Chanh T, Dreesman GR, Kanda P, Linette GP, Sparrow JT, Ho DD, Kennedy RC. *EMBO J* 1986;5:3065–3071.
99. Charneau P, Alizon M, Clavel F. A second origin of DNA plus-strand synthesis is required for optimal human immunodeficiency virus replication. *J Virol* 1992;66:2814–2820.
100. Charneau P, Mirambeau G, Roux P, Paulos S, Buc H, Clavel F. HIV-1 reverse transcription: a termination step at the center of the genome. *J Mol Biol* 1995;241:651–662.
101. Chen S, Lee S-L, Lee C-N, McIntosh WR, Lee TH. Mutational analysis of the leucine zipper-like motif of the human immunodeficiency virus type 1 envelope transmembrane glycoprotein. *J Virol* 1993;67: 3615–3619.
102. Chernomordik L, Chanturiya AN, Suss-Toby E, Nora E, Zimmerberg J. An amphipathic peptide from the C-terminal region of the human immunodeficiency virus envelope glycoprotein causes pore formation in membranes. *J Virol* 1994;68:7115–7123.
103. Cherrington J, Ganem D. Regulation of polyadenylation in human immunodeficiency virus (HIV): contributions of promoter proximity and upstream sequences. *EMBO J* 1992;11:1513–1524.
104. Chesebro B, Nishio J, Perryman S, Cann A, O'Brien W, Chen ISY, Wehrly K. Identification of human immunodeficiency virus envelope gene sequences influencing viral entry into CD4-positive HeLa cells, T-leukemia cells, and macrophages. *J Virol* 1991;65:5782–5789.
105. Chesebro B, Wehrly K, Nishio J, Perryman S. Macrophage-tropic human immunodeficiency virus isolates from different patients exhibit unusual V3 envelope sequence homogeneity: definition of critical amino acids involved in cell tropism. *J Virol* 1992;66:6547–6554.
106. Chorba TL, Evatt BL. Transfusion-associated HIV-1 infection. In: Madhok R, Forbes CD, Evatt BL, eds. *Blood, blood products, and HIV.* London: Chapman and Hall Medical; 1994:39–60.
107. Chow SA, Vincent KA, Ellison V, Brown PO. Reversal of integration and DNA splicing mediated by integrase of human immunodeficiency virus. *Science* 1992;255:723–726.
108. Chowers MY, Spina CA, Kwoh TJ, Fitch NJ, Richman DD, Guatelli JC. Optimal infectivity in vitro of human immunodeficiency virus type 1 requires an intact nef gene. *J Virol* 1994;68:2906–2914.
109. Churcher MJ, Lamont C, Hamy F, et al. High affinity binding of TAR RNA by the human immunodeficiency type 1 tat protein requires base-pairs in the RNA stem and amino acid residues flanking the basic region. *J Mol Biol* 1993;230:90–110.
110. Clavel F, Guétard D, Brun-Vézinet F, Chamaret S, Rey MA, Santos-Ferreira MO, Laurent AG, Dauguet C, Katlama C, Rouzioux C. Isolation of a new human retrovirus from West African patients with AIDS. *Science* 1986;233:343–346.
111. Clavel F, Orenstein JM. A mutant of human immunodeficiency virus with reduced RNA packaging and abnormal particle morphology. *J Virol* 1990;64:5230–5234.
112. Clements GJ, Price-Jones MJ, Stephens PE, Sutton C, Schulz TF, Clapham PR, McKeating JA, McClure MO, Thomason S, Marsh M, Kay J, Weiss RA, Moore JP. The V3 loops of the HIV-1 and HIV-2 surface glycoproteins contain proteolytic cleavage sites: a possible function in viral fusion? *AIDS Res Hum Retroviruses* 1991;7:3–16.
113. Clements JE, Wong-Staal F. Molecular biology of lentiviruses. *Semin Virol* 1992;3:137–146.
114. Clouse K, Powell D, Washington I, Poli G, Strebel K, Farrar W, Barstad P, Kovacs J, Fauci AS, Folks TM. Monokine regulation in human immunodeficiency virus–1 expression in a chronically infected human T cell clone. *J Immunol* 1989;142:431–438.
115. Coccia EM, Krust B, Hovanessian AG. Specific inhibition of viral protein synthesis in HIV-infected cells in response to interferon treatment. *J Biol Chem* 1994;269:23087–23094.
116. Cochrane AW, Perkins A, Rosen CA. Identification of sequences important in the nucleolar localization of human immunodeficiency virus rev: relevance of nucleolar localization to function. *J Virol* 1990;64: 881–885.
117. Cochrane AW, Jones KS, Beidas S, Dillon PJ, Skalka AM, Rosen CA. Identification and characterization of intragenic sequences which repress human immunodeficiency virus structural gene expression. *J Virol* 1991;65:5305–5313.
118. Coffin JM. Genetic diversity and evolution of retroviruses. *Curr Top Microbiol Immunol* 1992;176:143–164.
119. Coffin JM, Haase A, Levy JA, et al. Human immunodeficiency viruses [letter]. *Science* 1986;232:697.
120. Cohen EA, Terwilliger EF, Sodroski JG, Haseltine WA. Identification of a protein encoded by the vpu gene of HIV-1. *Nature* 1988;334: 532–534.
121. Cohen EA, Dehni G, Sodroski JG, Haseltine WA. Human immunodeficiency virus vpr product is a virion-associated regulatory protein. *J Virol* 1990;64:3097–3099.
122. Cohen EA, Terwilliger EF, Jalinoos Y, Provix J, Sodroski JG, Haseltine WA. Identification of HIV-1 vpr product and function. *J Acquir Immune Defic Syndr* 1990;3:11–18.
123. Cohen J. The Duesberg phenomenon. *Science* 1995;266:1642–1649.
124. Colombini S, Arya S, Reitz MS, Jagodzinski L, Beaver B, Wong-Staal F. Structure of simian immunodeficiency virus regulatory genes. *Proc Natl Acad Sci USA* 1989;86:4813–4817.
125. Concorde Coordinating Committee. Concorde: MRC/ANRS randomized double-blind controlled trial of immediate and deferred zidovudine in symptom-free HIV infection. *Lancet* 1993;343:871–881.
126. Connick E, Schooley RT. HIV-1 specific immune responses. In: Morrow WJW, Haigwood N, eds.*HIV–molecular organization, pathogenicity and treatment.* Amsterdam: Elsevier; 1993:75–98.
127. Connor RI, Mohri H, Cao Y, Ho DD. Increased viral burden and cytopathicity correlate temporally with CD4+ T-lymphocyte decline and clinical progression in human immunodeficiency virus type 1-infected individuals. *J Virol* 1993;67:1772–1777.
128. Contag CH, Dewhurst S, Viglianti GA, Mullins JI. Simian immunodeficiency virus (SIV) from Old World monkeys. In: Gallo RC, Jay G, eds.*The human retroviruses.* San Diego: Academic Press; 1991:245–276.
129. Cook KS, Fisk GJ, Hauber J, Usman N, Daly TJ, Rusche JR. Characterization of HIV-1 REV protein: binding stoichiometry and minimal RNA substrate. *Nucleic Acids Res* 1991;19:1577–1583.
130. Cordell J, Moore JP, Dean CJ, Klasse PJ, Weiss RA, McKeating JA. Rat monoclonal antibodies to nonoverlapping epitopes of human immunodeficiency virus type 1 gp120 block CD4 binding. *Virology* 1991; 185:72–79.
131. Craig JC, Duncan IB, Hockley D, Greif C, Roberts NA, Mills JS. Antiviral properties of Ro 31-8959, an inhibitor of human immunodeficiency virus (HIV) proteinase. *Antiviral Res* 1991;16:295–305.
132. Crise B, Rose JK. Human immunodeficiency virus type 1 glycoprotein precursor retains a CD4-p56lck complex in the endoplasmic reticulum. *J Virol* 1992;66:2296–2301.
133. Cullen BR, Hauber J, Campbell K, Sodroski JG, Haseltine WA, Rosen CA. Subcellular localization of the human immunodeficiency virus trans-acting art gene product. *J Virol* 1988;62:2498–2501.
134. Cullen BR, Greene WC. Regulatory pathways governing HIV-1 replication. *Cell* 1989;58:423–426.
135. Cullen BR. Regulation of HIV-1 gene expression. *FASEB J* 1991;5: 2361–2368.
136. Cullen BR. *Human retroviruses.* Oxford: IRL Press; 1993:220.
137. Cullen BR. The role of nef in the replication cycle of the human and simian immunodeficiency viruses. *Virology* 1994;205:1–6.
138. Daar ES, Ho DD. Variable HIV sensitivity to neutralization by recombinant soluble CD4. In: Koff WC, Wong-Staal F, Kennedy RC, eds. *AIDS research reviews.* New York: Marcel Dekker; 1992: 249–258.
139. D'Agostino D, Felber BK, Harrison JE, Pavlakis GN. The rev protein of human immunodeficiency virus type 1 promotes polysomal association and translation of gag/pol and vpu/env mRNAs. *Mol Cell Biol* 1992;12:1375–1386.
140. Dalgliesh AG, Beverley PCL, Clapham PR, Crawford DH, Greaves MF, Weiss RA. The CD4 (T4) antigen is an essential component of the receptor for the AIDS retrovirus. *Nature* 1984;312:763–766.
141. Daniel MD, Letvin NL, King NW, et al. Isolation of T-cell tropic HTLV-III-like retrovirus from macaques. *Science* 1985;228:1201–1204.

142. Dannull J, Surovoy A, Jung G, Moelling K. Specific binding of HIV-1 nucleocapsid protein to PSI RNA in vitro requires N-terminal zinc finger and flanking basic amino acid residues. *EMBO J* 1994;13:1525–1533.
143. Darlix JL, Gabus C, Nugeyre MT, Clavel F, Barré-Sinoussi F. Cis elements and trans factors involved in the RNA dimerization of the human immunodeficiency virus HIV-1. *J Mol Biol* 1990;216:689–699.
144. Davies JF, Hostomska Z, Hostomsky Z, Jordan SR, Matthews DA. Crystal structure of the ribonuclease H domain of HIV-1 reverse transcriptase. *Science* 1991;252:88–95.
145. Dayton AI, Sodroski JG, Rosen CA, Goh WC, Haseltine WA. The trans-activator gene of the human T cell lymphotropic virus type III is required for replication. *Cell* 1986;44:941–947.
146. De SK, Marsh JW. HIV-1 nef inhibits a common activation pathway in NIH-3T3 cells. J Biol Chem 1994;269:6656–6660.
147. De Clercq E. Anti-HIV agents interfering with the initial stages of the HIV replication cycle. In: Morrow WJW, Haigwood NL, eds. *HIV–molecular organization, pathogenicity and treatment.* Amsterdam: Elsevier; 1993:267–292.
148. de Jong JJ, de Ronde A, Keulen W, Tersmette M, Goudsmit J. Minimal requirements for the human immunodeficiency virus type 1 domain to support the syncytium-inducing phenotype. Analysis by single amino acid substitution. *J Virol* 1992;66:6777–6780.
149. de Jong JJ, Goudsmit J, Keulen W, Klaver B, Krone W, Tersmette M, DeRonde A. Human immunodeficiency virus type 1 clones chimeric for the envelope V3 domain differ in syncytium formation and replication capacity. *J Virol* 1992;63:273–280.
150. De Ronde A, Klaver B, Keulen W, Smit L, Goudsmit J. Natural HIV-1 nef accelerates virus replication in primary human lymphocytes. *Virology* 1993;188:391–395.
151. Debouck C, Gorniak JG, Strickler JE, Meek TD, Metcalf BW, Rosenberg M. Human immunodeficiency virus protease expressed in *Escherichia coli* exhibits autoprocessing and specific maturation of the gag precursor. *Proc Natl Acad Sci USA* 1987;84:8903–8906.
152. Debouck C. The HIV-1 protease as a therapeutic target for AIDS. *AIDS Res Hum Retroviruses* 1992;8:153–164.
153. Dedera D, Hu W, Vander Heyden N, Ratner L. Viral protein R of human immunodeficiency virus types 1 and 2 is dispensable for replication and cytopathogenicity in lymphoid cells. *J Virol* 1989;63:3205–3208.
154. Delchambre M, Gheysen D, Thines D, Thiriart C, Jacobs E, Verdin E, Horth M, Burny A, Bex F. The gag precursor of simian immunodeficiency virus assembles into virus-like particles. *EMBO J* 1989;8:2653–2660.
155. Delwart EL, Mosalios G. Retroviral envelope glycoproteins contain a "leucine zipper"-like repeat. *AIDS Res Hum Retroviruses* 1990;6:703–706.
156. Desai K, Loewenstein PM, Green M. Isolation of a cellular protein that binds to the human immunodeficiency virus Tat protein and can potentiate transactivation of the viral promoter [published erratum appears in *Proc Natl Acad Sci USA* 1991;88:11589]. *Proc Natl Acad Sci USA* 1991;88:8875–8879.
157. Desrosiers RC. HIV with multiple gene deletions as a live attenuated vaccine for AIDS. *AIDS Res Hum Retroviruses* 1992;8:411–421.
158. DeStefano JJ, Buiser RG, Mallaber LM, Myers TW, Bambara RA, Fay PJ. Polymerization and RNase H activities of the reverse transcriptases from avian myeloblastosis, human immunodeficiency, and Moloney murine leukemia viruses are functionally uncoupled. *J Biol Chem* 1991;266:7423–7431.
159. Devlin JJ, Panganiban LC, Devlin PE. Random peptide libraries: a source of specific protein binding molecules. *Science* 1990;249:404–406.
160. DeZazzo JD, Scott JM, Imperiale MJ. Relative roles of signals upstream of AAUAAA and promoter proximity in regulation of human immunodeficiency virus type 1 mRNA 3′ end formation.*Mol Cell Biol* 1992;12:5555–5562.
161. Diamond DC, Fineberg R, Chaudhuri S, Sleckman BP, Burakoff SJ. Human immunodeficiency virus infection is efficiently mediated by a glycolipid-anchored form of CD4. *Proc Natl Acad Sci USA* 1990;87:5001–5005.
162. Dillon PJ, Nelbock P, Perkins A, Rosen CA. Function of human immunodeficiency virus types 1 and 2 rev proteins is dependent on their ability to interact with a structured region present in env gene mRNA. *J Virol* 1990;64:4428–4437.
163. Dingwall C, Ernberg I, Gait MJ, Green SM, Heaphy S, Karn J, Lowe AD, Singh M, Skinner MA. HIV-1 tat protein stimulates transcription by binding to a U-rich bulge in the stem of the TAR RNA structure. *EMBO J* 1990;9:4145–4153.
164. Dollard SC, Gummuluru S, Tsang S, Fultz PN, Dewhurst S. Enhanced responsiveness to nuclear factor kappa B contributes to the unique phenotype of simian immunodeficiency virus variant SIV_{smm}PBj14. *J Virol* 1994;68:7800–7809.
165. Doms RW, Earl PL, Chakrabarti S, Moss B. Human immunodeficiency virus types 1 and 2 and simian immunodeficiency virus env proteins possess a functionally conserved assembly domain. *J Virol* 1990;64:3537–3540.
166. Doms RW, Lamb RA, Rose JK, Helenius A. Folding and assembly of viral membrane proteins. *Virology* 1993;193:545–562.
167. Dorfman T, Luban J, Goff SP, Haseltine WA, Gottlinger HG. Mapping of functionally important residues of a cysteine-histidine box in the human immunodeficiency virus type 1 nucleocapsid protein. *J Virol* 1993;67:6159–6169.
168. Dorfman T, Mammano F, Haseltine WA, Gottlinger HG. Role of the matrix protein in the virion association of the human immunodeficiency virus type 1 envelope glycoprotein. *J Virol* 1994;68:1689–1696.
169. Dorfman T, Bukovsky A, Ohagaen A, Hoglund S, Gottlinger HG. Functional domains of the capsid protein of human immunodeficiency virus type 1. *J Virol* 1994;68:8180–8187.
170. Dorr RT. Interferon-alpha in malignant and viral diseases. *Drugs* 1993;45:177–211.
171. Dropulic B, Jeang K-T. Gene therapy for human immunodeficiency virus infection: genetic antiviral strategies and targets for intervention. *Hum Gene Ther* 1994;5:927–939.
172. Dubay JW, Roberts SJ, Hahn BH, Hunter E. Truncation of the human immunodeficiency viryus type 1 transmembrane glycoprotein cytoplasmic domain blocks virus infectivity. *J Virol* 1992;66:6616–6625.
173. Dubiel W, Ferrell K, Rechsteiner M. Tat-binding protein 7 is a subunit of the 26S protease. *Biol Chem Hoppe Seyler* 1994;375:237–240.
174. Dyda F, Hickman AD, Jenkins TM, Engelman A, Craigie R, Davies DR. Crystal structure of the catalytic domain of HIV-1 integrase: similarity to other polynucleotide transferases. *Science* 1994;266:1981–1986.
175. Earl PL, Doms RW, Moss B. Oligomeric structure of the human immunodeficiency virus type 1 envelope glycoprotein. *Proc Natl Acad Sci USA* 1990;87:648–652.
176. Earl PL, Moss B, Doms RW. Folding, interaction with GRP78-BiP, assembly, and transport of the human immunodeficiency virus type 1 envelope protein. *J Virol* 1991;65:2047–2055.
177. Earl PL, Doms RW, Moss B. Multimeric CD4 binding exhibited by human and simian immunodeficiency virus envelope protein dimers. *J Virol* 1992;66:5610–5614.
178. Earl PL, Broder CA, Long D, Lee SA, Peterson J, Chakrabarti S, Doms RW, Moss B. Native oligomeric human immunodeficiency virus type 1 envelope glycoprotein elicits diverse monoclonal antibody reactivities. *J Virol* 1994;68:3015–3026.
179. Ebenbichler C, Westervelt P, Carillo A, Henkel T, Johnson D, Ratner L. Structure-function relationships of the HIV-1 envelope V3 loop tropism determinant: evidence for two distinct conformations. AIDS 1993;7:639–646.
180. Ehrlich LS, Agresta BE, Carter CA. Assembly of recombinant human immunodeficiency virus type 1 capsid in vitro. *J Virol* 1992;66:4874–4883.
181. Embretson J, Zupancic M, Ribas JL, Burke A, Racz P, Tenner-Racz K, Haase AT. Massive cohort infection of helper T lymphocytes and macrophages by HIV during the incubation period of AIDS. *Nature* 1993;362:359–362.
182. Emerman M, Vazeux R, Peden K. The rev gene product of the human immunodeficiency virus affects envelope-specific RNA localization. *Cell* 1989;57:1155–1165.
183. Emini EA, Schleif WA, Nunberg JH, Conley AJ, Eda Y, Tokiyoshi S, Putney SD. Prevention of HIV-1 infection in chimpanzees by gp120 V3 domain-specific monoclonal antibody. *Nature* 1992;355:728–730.
184. Engelman A, Mizuuchi K, Craigie R. HIV-1 DNA integration: mechanism of viral DNA cleavage and DNA strand transfer. *Cell* 1991;67:1211–1221.
185. Ensoli B, Garillari G, Salahuddin SZ, Gallo RC, Wong-Staal F. Tat protein of HIV-1 stimulates growth of cells derived from Kaposi's sarcoma lesions of AIDS patients. *Nature* 1990;3:84–86.
186. Ensoli B, Buonaguro L, Barillari G, Fiorelli V, Gendelman R, Morgan

RA, Wingfield P, Gallo RC. Release, uptake, and effects of extracellular human immunodeficiency virus type 1 tat protein on cell growth and viral transactivation. *J Virol* 1993;67:277–287.

187. Essex M, Kanki PJ. Human immunodeficiency virus type 2 (HIV-2). In: Broder S, Merigan TC, Bolognesi D, eds. *Textbook of AIDS medicine.* Baltimore: Williams and Wilkins; 1994:873–886.

188. Ewald PW. *Evolution of infectious disease.* Oxford: Oxford University Press; 1994:119–157.

189. Fan H. Influences of the long terminal repeats on retrovirus pathogenicity. *Semin Virol* 1990;1:165–174.

190. Fan L, Peden K. Cell-free transmission of vif mutants of HIV-1. *Virology* 1992;190:19–29.

191. Fankhauser C, Izaurralde E, Adachi Y, Wingfield P, Laemmli U. Specific complex of human immunodeficiency virus type 1 rev and nucleolar B23 proteins: dissociation by the rev response element. *Mol Cell Biol* 1991;11:2567–2575.

192. Fantini J, Yahi N, Baghdiguian S, Chermann J-C. Human colonic epithelial cells productively infected with human immunodeficiency virus show impaired differentiation and altered secretion. *J Virol* 1992;66: 580–585.

193. Fantini J, Cook DG, Nathanson N, Spitalnik SL, Gonzalez-Scarano F. Infection of colonic epithelial cell lines by type 1 human immunodeficiency virus is associated with cell surface expression of galactosylceramide, a potential alternative gp120 receptor. *Proc Natl Acad Sci USA* 1993;90:2700–2704.

194. Farnet CM, Haseltine WA. Determination of viral proteins present in the human immunodeficiency virus type-1 preintegration complex. J Virol 1991;65:1910–1915.

195. Feinberg MB, Jarrett RF, Aldovini A, Gallo RC, Wong-Staal F. HTLV-III expression and production involve complex regulation at the levels of splicing and translation of viral RNA. *Cell* 1986;46:807–817.

196. Feinberg MB, Baltimore D, Frankel AD. The role of tat in the human immunodeficiency virus life cycle indicates a primary effect on transcriptional elongation. *Proc Natl Acad Sci USA* 1991;88:4045–4049.

197. Feinberg MB, Trono D. Intracellular immunization: trans-dominant mutants of HIV gene products as tools for the study and interruption of viral replication. *AIDS Res Hum Retroviruses* 1992;8:1013–1022.

197a. Felber BK, Drysdale CM, Pavlakis GN. Feedback regulation of human immunodeficiency virus type 1 expression by the rev protein. *J Virol* 1990;64:3734–3741.

198. Felber BK, Hadzopoulou-Cladaras M, Cladaras C, Copeland T, Pavlakis GN. Rev protein of HIV-1 affects the stability and transport of viral mRNA. *Proc Natl Acad Sci USA* 1989;86:1495–1499.

199. Feng S, Holland EC. HIV-1 tat trans-activation requires the loop sequence within TAR. *Nature* 1988;334:165–167.

200. Fennie C, Lasky LA. Model for intracellular folding of the human immunodeficiency virus type 1 gp120. *J Virol* 1989;63:639–646.

201. Fenouillet E, Gluckman JC, Bahraoui E. Role of N-linked glycans of envelope glycoproteins in infectivity of human immunodeficiency virus type 1. *J Virol* 1990;64:2841–2848.

201a. Fesen MR, Kohn KW, Leteurtre F, Pommier Y. Inhibitors of human immunodeficiency virus integrase. *Proc Natl Acad Sci USA* 1993;90: 2399–2403.

202. Fields S, Song O. A novel genetic system to detect protein-protein interactions. *Nature* 1989;340:245–246.

203. Fisher AG, Feinberg MB, Josephs SF, Harper ME, Marselle LM, Reyes G, Gonda MA, Aldovini A, Debouck C, Gallo RC, Wong-Staal F. The trans-activator gene of HTLV-III is essential for virus replication.*Nature* 1986;320:367–371.

204. Fisher AG, Ensoli B, Ivanoff I, Chamberlain M, Petteway S, Ratner L, Gallo RC, Wong-Staal F. The sor gene of HIV-1 is required for efficient virus transmission in vitro. *Science* 1987;237:888–893.

205. Folks TM, Justement J, Kinter A, Dinarello CA, Fauci AS. Cytokine-induced expression of HIV-1 in a chronically infected promonocyte cell line. *Science* 1987;238:800–802.

206. Forbes D. Structure and function of the nuclear pore complex. *Annu Rev Cell Biol* 1992;8:495–527.

207. Fouchier RAM, Groenink M, Koostra NA, Tersmette M, Huisman HG, Miedema F, Schuitemaker H. Phenotype-associated sequence variation in the third variable domain of the human immunodeficiency virus type 1 gp120 molecule. *J Virol* 1992;63:3810–3821.

208. Franchini G, Rusche JR, O'Keefe TJ, Wong-Staal F. The human immunodeficiency virus type 2 (HIV-2) contains a novel gene encoding a 16-kDa protein associated with mature virions. *AIDS Res Hum Retroviruses* 1988;4:243–250.

209. Franke EK, Yuan HEH, Luban J. Specific incorporation of cyclophilin A into HIV-1 virions. *Nature* 1994;372:359–362.

210. Franke EK, Yuan HEH, Bossolt KL, Goff SP, Luban J. Specificity and sequence requirements for interactions between various retroviral gag proteins. *J Virol* 1994;68:5300–5305.

211. Frankel AD, Chen L, Cotter RJ, Pabo CO. Dimerization of the tat protein from human immunodeficiency virus: a cysteine-rich peptide mimics the normal metal-linked dimer interface. *Proc Natl Acad Sci USA* 1988;85:6297–6300.

212. Freed EO, Myers DJ, Risser R. Characterization of the fusion domain of the human immunodeficiency virus type 1 envelope glycoprotein gp41. *Proc Natl Acad Sci USA* 1990;87:4650–4654.

213. Freed EO, Myers DJ, Risser R. Identification of the principal neutralizing determinant of human immunodeficiency virus type 1 as a fusion domain. *J Virol* 1991;65:190–194.

214. Freed EO, Delwart EL, Bichschacher GJ, Panganiban AT. A mutation in the human immunodeficiency virus type 1 transmembrane glycoprotein gp41 dominantly interferes with fusion and infectivity. *Proc Natl Acad Sci USA* 1992;89:70–74.

215. Freed EO, Martin MA. Evidence for a functional interaction between the V1/V2 and C4 domains of human immunodeficiency virus type 1 envelope glycoprotein. *J Virol* 1994;68:2503–2512.

216. Fu W, Gorelick RJ, Rein A. Characterization of human immunodeficiency virus type 1 dimeric RNA from wild-type and protease-defective virions. *J Virol* 1994;68:5013–5018.

217. Gabuzda D, Olshevsky U, Bertani P, Haseltine W, Sodroski J. Identification of membrane anchorage domains of the HIV-1 gp160 envelope glycoprotein precursor. *J Acquir Immune Defic Syndr* 1991;4:34–40.

218. Gabuzda DH, Lawrence K, Langhoff E, Terwilliger E, Dorfman T, Haseltine WA, Sodroski J. Role of vif in replication of human immunodeficiency virus type 1 in CD4+ lymphocytes. *J Virol* 1992;66: 6489–6495.

219. Gallaher WR, Ball JR, Garry RF, Griffin MC, Montelaro RC. A general model for the transmembrane proteins of HIV and other retroviruses. *AIDS Res Hum Retroviruses* 1989;5:431–440.

220. Gallay P, Swingler S, Aiken C, Trono D. HIV-1 infection of nondividing cells: C-terminal tyrosine phosphorylation of the viral matrix protein is a key regulator. *Cell* 1995;80:379–388.

221. Gallo RC, Salahuddin SZ, Popovic M, et al. Frequent detection and isolation of cytopathic retroviruses (HTLV-III) from patients with AIDS and at risk for AIDS. *Science* 1984;224:500–503.

222. Gallo RC, Wong-Staal F, Montagnier L, Haseltine WA, Yoshida M. HIV-HTLV gene nomenclature. *Nature* 1987;333:504.

223. Gao F, Yue L, White AT, Pappas PG, Barchue J, Hanson AP, Greene BM, Sharp PM, Shaw GM, Hahn BH. Human infection by genetically diverse SIVsm-related HIV-2 in West Africa. *Nature* 1992;358: 495–499.

224. Gao W-Y, Cara A, Gallo RC, Lori F. Low levels of deoxynucleotides in peripheral blood lymphocytes: a strategy to inhibit human immunodeficiency virus type 1 replication. *Proc Natl Acad Sci USA* 1993;90: 8925–8928.

225. Gao F, Yue L, Robertson DL, Hill SC, Hui H, Biggar RJ, Neequaye AE, Whelan TM, Ho DD, Shaw GM, Sharp PM, Hahn BH. Genetic diversity of human immunodeficiency virus type 2: evidence for distinct sequence subtypes with differences in virus biology. *J Virol* 1994; 68:7433–7447.

226. Garcia JA, Harrich D, Pearson L, Mitsuyasu R, Gaynor RB. Functional domains required for tat-induced transcriptional activation of the HIV-1 long terminal repeat. *EMBO J* 1988;7:3143–3147.

227. Garcia JV, Miller AD. Serine phosphorylation independent downregulation of cell-surface CD4 by nef. *Nature* 1991;350:508–511.

228. Gardner MB. Historical background. In: Stephenson JR, ed. *Molecular biology of RNA tumor viruses.* New York: Academic Press; 1980: 1–46.

229. Gardner MB, Endres M, Barry PA. The simian retroviruses: SIV and SRV. In: Levy JA, ed.*The retroviridae.* New York: Plenum Press; 1994: 133–276.

230. Garrett ED, Tiley LS, Cullen BR. Rev activates expression of the human immunodeficiency virus type 1 vif and vpr gene products. *J Virol* 1991; 65:1653–1657.

231. Garrett ED, Cullen BR. Comparative analysis of rev function in human immunodeficiency virus types 1 and 2. *J Virol* 1992;66:4288–4294.

232. Gatignol A, Kumar A, Rabson A, Jeang K-T. Identification of cellular proteins that bind to the HIV-1 TAR RNA. *Proc Natl Acad Sci USA* 1989;83:9759–9763.

233. Gelderblom HR. Assembly and morphology of HIV: potential effect of structure on viral function. *AIDS* 1991;5:617–638.
234. Gendelman HE, Phelps W, Feigenbaum L, Ostrove JM, Adachi A, Howley PM, Khoury G, Ginsberg HS, Martin MA. Transactivation of the human immunodeficiency virus long terminal repeat by DNA viruses. *Proc Natl Acad Sci USA* 1986;83:9759–9763.
235. Gendelman HE, Baca LM, Turpin J, Kalter DC, Hansen A, Orenstein JM, Dieffenbach CW, Friedman RM, Meltzer MS. Regulation of HIV replication in infected monocytes by IFN-alpha. Mechanisms for viral restriction. *J Immunol* 1990;145:2669–2676.
236. Gheysen D, Jacobs E, De Foresta F, Thiriart C, Francotte M, Thines D, De Wilde M. Assembly and release of HIV-1 precursor Pr55 gag virus-like particles from recombinant baculovirus-infected insect cells. *Cell* 1989;59:103–112.
237. Ghiara JB, Stura EA, Stanfield RL, Profy AT, Wilson IA. Crystal structure of the principal neutralization site of HIV-1. *Science* 1994;264:82–85.
238. Gibbs JS, Desrosiers RC. Auxiliary proteins of the primate immunodeficiency viruses. In: Cullen BR, ed. *Human retroviruses.* Oxford: IRL Press; 1993:137–158.
239. Gibbs JS, Regier DA, Desrosiers RC, Construction and in vitro properties of HIV-1 mutants with deletions in "nonessential" genes. *AIDS Res Hum Retroviruses* 1994;10:343–350.
240. Gibbs JS, Regier DA, Desrosiers RC. Construction and in vitro properties of HIV-1 mutants with deletions in "nonessential" genes. *AIDS Res Hum Retroviruses* 1994;10:607–616.
241. Giese K, Cox J, Groschedl R. The HMG domain of lymphoid enhancer factor 1 bends DNA and facilitates assembly of functional nucleoprotein structures. *Cell* 1992;69:185–196.
242. Glaichenhaus N, Shastri N, Littman DR, Turner JM. Requirement for association of p56lck with CD4 in antigen-specific signal transduction in T cells. *Cell* 1991;64:511–520.
243. Goff SP. Genetics of retroviral integration. *Annu Rev Genet* 1992;26:527–544.
244. Golub EI, Li G, Volsky DJ. Differences in the basal activity of the long terminal repeat determine different replicative capacities of two closely related human immunodeficiency virus type 1 isolates. *J Virol* 1990;64:3654–3660.
245. Goncalves J, Jallepalli P, Gabuzda DH. Subcellular localization of the vif protein of human immunondeficiency virus type 1. *J Virol* 1994;68:704–712.
246. Gonda MA, Wong-Staal F, Gallo RC, Clements JE, Narayan O, Gilden RV. Sequence homolgy and morphologic similarity of HTLV-III and visna virus, a pathogenic lentivirus. *Science* 1985;227:173–177.
247. Gorelick RJ, Nigida SM, Bess JW, Arthur LO, Henderson LE, Rein A. Noninfectious human immunodeficiency virus type 1 mutants deficient in genomic RNA. *J Virol* 1990;64:3207–3211.
248. Gottlieb MS, Schroff R, Schanker H, Weisman JD, Fan PT, Wolf RA, Saxon A. *Pneumocystis carinii* pneumonia and mucosal candidiasis in previously healthy homosexual men. *N Engl J Med* 1981;305:1425–1430.
249. Gottlinger HG, Sodroski JG, Haseltine WA. Role of capsid precursor processing and myristoylation in morphogenesis and infectivity of human immunodeficiency virus type 1. *Proc Natl Acad Sci USA* 1989;86:5781–5785.
250. Gottlinger HG, Dorfman T, Cohen EA, Haseltine WA. Vpu protein of human immunodeficiency virus type 1 enhances the release of capsids produced by gag gene constructs of widely divergent retroviruses. *Proc Natl Acad Sci USA* 1993;90:7381–7385.
251. Goudsmit J, Debouck C, Meloen RH, Smit L, Bakker M, Asher DM, Wolff AV, Gibbs CJ, Gajdusek DC. Human immunodeficiency virus type 1 neutralizing epitope with conserved architecture elicits early type-specific antibodies in experimentally infected chimpanzees. *Proc Natl Acad Sci USA* 1988;85:4478–4482.
252. Graves MC, Lim JJ, Heimer EP, Kramer RA. An 11-kDa form of human immunodeficiency virus protease expressed in *Escherichia coli* is sufficient for enzymatic activity. *Proc Natl Acad Sci USA* 1988;85:2449–2453.
253. Gregory T, Hoxie J, Watanabe C, Spellman M. Structure and function in recombinant HIV-1 gp120 and speculation about the disulfide bonding in the gp120 homologs of HIV-2 and SIV. *Adv Exp Med Biol* 1991;303:1–4.
254. Grimalia RJ, Fuller BA, Rennert PD, Nelson MB, Hammarskjold ML, Potts B, Murray M, Putney SD, Gray G. Mutations in the principal neutralization domain of the human immunodeficiency virus type 1 affect syncytium formation, virus infectivity, growth kinetics, and neutralization. *J Virol* 1992;66:1875–1883.
255. Gulizia J, Dempsey MP, Sharova N, Bukrinsky MI, Spitz L, Goldfarb D, Stevenson M. Reduced nuclear import of human immunodeficiency virus type 1 preintegration complexes in the presence of a prototypic nuclear targeting signal. *J Virol* 1994;68:2021–2025.
256. Guy B, Keiny M-P, Rivière Y. HIV F/3′-orf encodes a phosphorylated GTP-binding protein resembling an oncogene product. *Nature* 1987;330:266–269.
257. Guy B, Rivière Y, Dott K, Regnault A, Kieny MP. Mutational analysis of the HIV nef protein. *Virology* 1990;176:413.
258. Guy B, Geist M, Dott K, Spehner D, Kieny M-P, Lecocq J-P. A specific inhibitor of cysteine proteases impairs a vif-dependent modification of human immunodeficiency virus type 1 env protein. *J Virol* 1991;65:1325–1331.
259. Guyader M, Emerman M, Sonigo P, Clavel F, Montagnier L, Alizon M. Genome organization and transactivation of HIV-2. *Nature* 1987;326:662–669.
260. Guyader M, Emerman M, Montagnier L, Peden K. Vpx mutants of HIV-2 are infectious in established cell lines but display a severe defect in peripheral blood lymphocytes. *EMBO J* 1989;8:1169–1175.
261. Hadzopoulou-Cladaras M, Felber BK, Cladaras C, Athanassopoulos A, Tse A, Pavlakis GN. The rev protein of HIV-1 affects viral mRNA and protein expression via a cis-acting sequence in the env region. *J Virol* 1989;63:1265–1274.
262. Hahn BH, Gonda MA, Shaw GM, Popovic M, Hoxie JA, Gallo RC, Wong-Staal F. Genomic diversity of the acquired immune deficiency syndrome virus HTLV-III: different viruses exhibit greatest divergence in their envelope genes. *Proc Natl Acad Sci USA* 1985;82:4813–4817.
263. Hallenberger S, Bosch V, Angliker H, Shaw E, Klenk H-D, Garten W. Inhibition of furin-mediated cleavage activation of HIV-1 glycoprotein gp160. *Nature* 1992;360:358–361.
264. Hammarskjold M-L, Heimer J, Hammarskjold B, Sanwan I, Albert L, Rekosh D. Regulation of human immunodeficiency virus env expression by the rev gene product. *J Virol* 1989;63:1959–1966.
265. Hamy F, Asseline U, Grasby J, et al. Hydrogen-bonding contacts in teh major groove are required for human immunodeficiency virus type 1 tat protein recognition of TAR RNA. *J Mol Biol* 1993;230:111–123.
266. Hanahan D. Transgenic mice as probes into complex systems. *Science* 1989;246:1265–1275.
267. Hanly SM, Rimsky LT, Malim MH, Kim JH, Hauber J, Duc Dodon M, Le S-Y, Maizel JV, Cullen BR, Greene WC. Comparative analysis of the HTLV-1 rex and HIV-1 rev trans-regulatory proteins and their RNA response elements. *Genes Dev* 1989;3:1534–1544.
268. Hansen JE, Clausen H, Hu SL, Nielsen JO, Olofsson S. An O-linked carbohydrate neutralization epitope of HIV-1 gp120 is expressed by HIV-1 env gene recombinant vaccinia virus. *Arch Virol* 1992;126:11–20.
269. Harouse JM, Bhat S, Spitalnik SL, Laughlin M, Stefano K, Silberberg DH, Gonzalez-Scarano F. Inhibition of entry of HIV-1 in neural cell lines by antibodies against galactosyl ceramide. *Science* 1991;253:320–323.
270. Harrich D, Garcia J, Wu F, Mitsuyasu R, Gonzalez J, Gaynor R. Role of Sp1-binding domains in in vivo transcriptional regulation of the human immunodeficiency virus type 1 long terminal repeat. *J Virol* 1989;63:2585–2591.
271. Harris M, Hislop S, Patsilinacos P, Neil JC. In vivo derived HIV-1 nef gene products are heterogeneous and lack detectable nucleotide binding activity. *AIDS Res Hum Retroviruses* 1992;8:537–543.
272. Harrison GP, Lever AML. The human immunodeficiency virus type 1 packaging and major splice donor region have a conserved stable secondary structure. *J Virol* 1992;66:4144–4153.
273. Harrison GS, Long CJ, Curiel TJ, Maxwell F, Maxwell IH. Inhibition of human immunodeficiency virus-1 production resulting from transduction with a retrovirus containing a HIV regulated diphtheria toxin a chain gene. *Hum Gene Ther* 1992;3:461–469.
274. Harrison SC. CD4: the receptor for HIV. In: Wimmer E, ed. *Cellular receptors for animal viruses.* Cold Spring Harbor, NY: Cold Spring Harbor Press; 1994:33–48.
275. Hart C, Ou C-Y, Galphin J, Moore J, Bacheler LT, Wasmuth JJ, Petteway SR, Schochetman G. Human chromosome 12 is required for elevated HIV-1 expression in human-hamster hybrid cells. *Science* 1989;246:488–491.
276. Hart AR, Cloyd MW. Interference patterns of human immunodeficiency viruses 1 and 2. *Virology* 1990;177:1–10.
277. Hart TK, Kirsh R, Ellens H, Sweet RW, Lambert DM, Petteway SE,

Leary J, Bugelski PJ. Binding of soluble CD4 proteins to human immunodeficiency virus type 1 and infected cells induces release of envelope glycoprotein gp120. *Proc Natl Acad Sci USA* 1991;88:2189–2193.
278. Hattori T, Koito A, Takasuki K, Kido H, Katunuma N. Involvement of a tryptase-related cellular protease(s) in human immunodeficiency virus type 1 infection. *FEBS Lett* 1989;248:48–52.
279. Hattori N, Michaels F, Fargnoli K, Marcon L, Gallo RC, Franchini G. The human immunodeficiency virus type 2 vpr gene is essential for productive infection of human macrophages. *Proc Natl Acad Sci USA* 1990;87:8080–8084.
280. Hauber J, Perkins A, Heimer EP, Cullen BR. Trans-activation of human immunodeficiency virus gene expression is mediated by nuclear events. *Proc Natl Acad Sci USA* 1987;84:6364–6368.
281. Hauber J, Cullen BR. Mutational analysis of the trans-activation-responsive region of the human immunodeficiency virus type 1 long terminal repeat. *J Virol* 1988;62:673–679.
282. Hauber J, Malim MH, Cullen BR. Mutational analysis of the conserved basic domain of human immunodeficiency virus tat protein. *J Virol* 1989;63:1181–1187.
283. Heaphy S, Finch JT, Gait MJ, Karn J, Singh M. Human immunodeficiency virus type 1 regulator of virion expression, rev, forms nucleoprotein filaments after binding to a purine-rich "bubble" located within the rev-responsive region of viral mRNAs. *Proc Natl Acad Sci USA* 1991;88:7366–7370.
284. Heinzinger NK, Bukrinsky MI, Haggerty SA, Ragland AM, Kewalramani V, Lee M-A, Gendelman HE, Ratner L, Stevenson M, Emerman M. The vpr protein of human immunodeficiency virus type 1 influences nuclear localization of viral nucleic acids in nondividing host cells. *Proc Natl Acad Sci USA* 1994;91:7311–7315.
285. Helseth E, Olshevsky U, Furman U, Sodroski J. Human immunodeficiency virus type 1 glycoprotein regions important for association with the gp41 transmembrane glycoprotein. *J Virol* 1991;65:2119–2123.
286. Henderson LE, Sowder R, Copeland TD, Oroszlan S, Arthur LO, Robey WG, Fischinger PJ. Direct identification of class II histocompatibility DR proteins in preparations of human cell lymphotropic virus type III. *J Virol* 1987;61:629–632.
287. Henderson LE, Sowder RC, Copeland TD, Benveniste RE, Oroszlan S. Isolation and characterization of a novel protein (X-ORF product) from SIV and HIV-2. *Science* 1988;241:199–201.
288. Henderson LE, Bowers MA, Sowder RC, Serabyn SA, Johnson DG, Bess JW, Arthur LO, Bryant DK, Fenselau C. Gag proteins of the highly replicative MN strain of human immunodeficiency virus type 1: posttranslational modifications, proteolytic processing, and complete amino acid sequences. *J Virol* 1992;66:1856–1865.
289. Hildreth JEK, Orentas RJ. Involvement of a leukocyte adhesion receptor (LFA-1) in HIV-induced syncytium formation. *Science* 1989; 244:1075–1078.
290. Hirsch VM, Olmsted RA, Murphy-Corb M, Purcell RH, Johnson PR. An African primate lentivirus (SIVsm) closely related to HIV-2. *Nature* 1989;339:389–392.
291. Hirsch VM, Edmonson P, Murphey-Corb M, Arbeille B, Johnson PR, Mullins JI. SIV adaptation to human cells. *Nature* 1989;341:573–574.
292. Hirsch VM, Myers G, Johnson PR. Genetic diversity and phylogeny of primate lentiviruses. In: Morrow WJW, Haigwood NL, eds. *HIV–molecular organization, pathogenicity and treatment.* Amsterdam: Elsevier; 1993:221–240.
293. Hirsch VM, Dapolito G, Johnson PR, Elkine WR, London WT, Montali RJ, Goldstein S, Brown C. Induction of AIDS by simian immunodeficiency virus from an African green monkey: species-specific variation in pathogenicity correlates with the extent of in vivo replication. *J Virol* 1995;69:955–967.
294. Ho DD, Fung MSC, Cao X, Li L, Sun C, Chang TW, Sun N-C. Another discontinuous epitope on glycoprotein gp 120 that is important in human immunodeficiency virus type 1 neutralization is identified by a monoclonal antibody. *Proc Natl Acad Sci USA* 1991;88:8949–8953.
295. Ho DD, Kaplan JC, Rackauskas IE, Gurney ME. Second conserved domain of gp120 is important for HIV infectivity and antibody neutralization. *Science* 1988;239:1021–1023.
296. Ho DD, McKeating JA, Li XL, Moudgil T, Daar ES, Sun NC, Robinson JE. Conformational epitope on gp120 important in CD4 binding and human immunodeficiency virus type 1 neutralization identified by a human monoclonal antibody. *J Virol* 1991;65:489–493.
297. Ho DD, Neuman AU, Perelson AS, Whcn W, Leonard JM, Markowitz M. Rapid turnover of plasma virions and CD4 lymphocytes in HIV-1 infection. *Nature* 1995;373:123–126.
298. Hoglund S, Ofverstedt L-G, Nilsson A, Lundquist P, Gelderblom H, Ozel M, Skoglund U. Spatial visualization of the maturing HIV-1 core and its linkage to the envelope. *AIDS Res Hum Retroviruses* 1992;8:1–7.
299. Hong T, Murphy E, Groarke J, Drlica K. Human immunodeficiency virus type 1 DNA integration: fine structure target analysis using synthetic oligonucleotides. *J Virol* 1993;67:1127–1131.
300. Hope TJ, Huang XJ, McDonald D, Parslow TG. Steroid-receptor fusion of the human immunodeficiency virus type 1 rev transactivator: mapping cryptic functions of the arginine-rich motif. *Proc Natl Acad Sci USA* 1990;87:7787–7791.
301. Hope TJ, McDonald D, Huang XJ, Low J, Parslow TG. Mutational analysis of the human immunodeficiency virus type 1 rev transactivator: essential residues near the amino terminus. *J Virol* 1990;64:5360–5366.
302. Hope TJ, Bond BL, McDonald D, Klein NP, Parslow TG. Effector domains of human immunodeficiency virus type 1 rev and human T-cell leukemia virus type 1 rex are functionally interchangeable and share an essential peptide motif. *J Virol* 1991;65:6001–6007.
303. Horth M, Lambrecht B, Khim MC, Bex F, Thiriart C, Ruysschaert JM, Burny A, Brasseur R. Theoretical and functional analysis of the SIV fusion peptide. *EMBO J* 1991;10:2747–2755.
304. Horton R, Spearman P, Ratner L. HIV-1 viral protein X association with the gag p27 capsid protein. Virology 1994;199:453–457.
305. Hostomsky Z, Hostomska Z, Hudson GO, Woomaw EW, Nodes BR. Reconstitution in vitro of RNase H activity by using purified N-terminal and C-terminal domains of human immunodeficiency virus type 1 reverse transcriptase. *Proc Natl Acad Sci USA* 1991;88:1148–1152.
306. Hostomsky Z, Hostomska Z, Matthews DA. Ribonucleases H. In: Roberts RJ, Lloyd RS, Linn SN, eds. *Nucleases II.* Cold Spring Harbor, NY: Cold Spring Harbor Laboratory Press, 1993.
307. Howcroft TK, Strebel K, Martin MA, Singer DS. Repression of MHC class I gene promoter activity by two-exon tat. *Science* 1993;260:1319–1322.
308. Hoxie JA, Flaherty LE, Haggarty BS, Rackowski JL. Infection of T4 lymphocytes by HTLV-III does not require expression of the OKT4 epitope. *J Immunol* 1986;136:361–363.
309. Hoxie J, Haggarty B, Bonser S, Packowski J, Shan H, Kanki P. Biological characterization of a simian immunodeficiency virus-like retrovirus (HTLV-IV): evidence for CD4-associated molecules required for infection. *J Virol* 1988;62:2557–2568.
310. Hoxie JA. Hypothetical assignment of intrachain disulfide bonds for HIV-2 and SIV envelope glycoproteins. *AIDS Res Hum Retroviruses* 1991;6:495–499.
311. *Hsu MC, Schutt AD, Holly M, Slice L, Sherman MI, Richman DD, Potash MJ, Volsky DJ. Inhibition of HIV replication in acute and chronic infections in vitro by a tat antagonist. *Science* 1991;254:1799–1802.
312. Hsu M-C, Dhingra U, Earley JV, Holly M, Keith D, Nalin CM, Richou AR, Schutt AD, Tam SY, Potash MJ, Volsky DJ, Richman DD. Inhibition of type 1 human immunodeficiency virus replication by a tat antagonist to which the virus remains sensitive after prolonged exposure in vitro. *Proc Natl Acad Sci USA* 1993;90:6395–6399.
313. Hu WS, Temin HM. Retroviral recombination and reverse transcription. *Science* 1990;250:1227–1233.
314. Hu W, Vander Heyden N, Ratner L. Analysis of the function of viral protein X (VPX) of HIV-2. *Virology* 1989;173:624–630.
315. Huang LM, Jeang KT. Increased spacing between Sp1 and TATAA renders human immunodeficiency virus type 1 replication defective: implication for Tat function. *J Virol* 1993;67:6937–6944.
316. Huang XJ, Hope TJ, Bond BL, McDonald D, Grahl K, Parslow TG. Minimal rev-response element for type 1 human immunodeficiency virus. *J Virol* 1991;65:2131–2134.
317. Huet T, Cheynier R, Meyerhans A, Roelants G, Wain-Hobson S. Genetic organization of a chimpanzee lentivirus related to HIV-1. *Nature* 1990;345:356–359.
318. Hunter E. Macromolecular interactions in the assembly of HIV and other retroviruses. *Sem Virol* 1994;5:71–83.
319. Hwang SS, Boyle TJ, Lyerly HK, Cullen BR. Identification of the envelope V3 loop as the primary determinant of cell tropism in HIV-1. *Science* 1991;253:71–74.
320. Ilyinskii PO, Daniel MD, Simon MA, Lackner AA, Desrosiers RC. The role of upstream U3 sequences in the pathogenesis of simian immunodeficiency virus-induced AIDS in rhesus monkeys. *J Virol* 1994; 68:5933–5944.

321. Ivanoff LA, Dubay JW, Morris JF, et al. V3 loop region of the HIV-1 gp120 envelope protein is essential for virus infectivity. *Virology* 1992; 187:423–432.
322. Ivey-Hoyle M, Culp JS, Chaikin MA, Hellmig BD, Matthews TJ, Sweet RW, Rosenberg M. Envelope glycoproteins from biologically diverse isolates of immunodeficiency viruses have widely different affinities for CD4. *Proc Natl Acad Sci USA* 1991;88:512–516.
323. Iwai S, Pritchard C, Mann DA, Karn J, Gait MJ. Recognition of the high affinity binding site in rev-response element RNA by the human immunodeficiency virus type-1 rev protein. *Nucleic Acids Res* 1992; 20:6465–6472.
324. Jacks T, Power MD, Masiarz FR, Luciw PA, Barr PJ, Varmus HE. Characterization of ribosomal frameshifting in HIV-1 gag-pol expression. *Nature* 1988;331:280–283.
325. Jacobo-Molina A, Arnold E. HIV reverse transcriptase structure-function relationships. *Biochemistry* 1991;30:6351–6356.
326. Jacobo-Molina A, Ding J, Nanni RG, Clark AD, Lu X, Tantillo C, Williams RL, Kamer G, Ferris AL, Clark P, Hizi A, Hughes SH, Arnold E. Crystal structure of human immunodeficiency virus type 1 reverse transcriptase complexed with double-stranded DNA at 3.0A resolution shows bent DNA. *Proc Natl Acad Sci USA* 1993;90:6320–6324.
327. Jakobovits A, Smith DH, Jakobovits EB, Capon DJ. A discrete element 3′ of human immunodeficiency virus HIV-1 and HIV-2 mRNA initiation sites mediates transcriptional activation by an HIV trans-activator. *Mol Cell Biol* 1988;8:2555–2561.
328. Jakobovits A, Rosenthal A, Capon DJ. Trans-activation of HIV-1 LTR-directed gene expression by tat requires protein kinase C. *EMBO J* 1990;9:1165–1170.
329. Jamieson BD, Aldrovandi GM, Planelles V, Jowett JB, Gao L, Bloch LM, Chen IS, Zack JA. Requirement of human immunodeficiency virus type 1 nef for in vivo replication and pathogenicity. *J Virol* 1994;68: 3478–3485.
330. Jasin M, Page KA, Littman DR. Glycosylphosphatidylinositol-anchored CD4/Thy-1 chimeric molecules serve as human immunodeficiency virus receptors in human but not mouse cells and are modulated by gangliosides. *J Virol* 1991;65:440–444.
331. Javaherian K, Langlois AJ, McDonal C, Ross KL, Eckler LI, Jellis CL, Profy AT, Rusche JR, Bolognesi DP, Putney SD. Principal neutralizing determinant of the human immunodeficiency virus type 1 envelope protein. *Proc Natl Acad Sci USA* 1989;86:6768–6772.
332. Javaherian K, Langlois AJ, Montefiori DC, Kent KA, Ryan KA, Wyman PD, Stott J, Bolognesi DP, Murphey-Corb M, LaRosa GJ. Studies of the conformation-dependent neutralizing epitopes of simian immunodeficiency virus envelope protein. *J Virol* 1994;68:2624–2631.
333. Johnson PR, Hamm TE, Goldstein S, Kitov S, Hirsch VM. The genetic fate of molecularly cloned SIV in experimentally infected macaques. *Virology* 1991;185:217.
334. Jones KA, Kadonaga JT, Luciw PA, Tijan R. Activation of the AIDS retrovirus promoter by the cellular transcription factor, Sp1. *Science* 1986;232:755–759.
335. Jones KA, Luciw PA, Duchange N. Structural arrangements of transcription control elements within the 5′-untranslated leader regions of the HIV-1 and HIV-2 promoters. *Genes Dev* 1988;2:1101–1114.
336. Jones KA, Peterlin BM. Control of RNA initiation and elongation at the HIV-1 promoter. *Annu Rev Biochem* 1994;63:717–743.
337. Jones KS, Coleman J, Merkel GW, Laue TM, Skalka AM. Retroviral integrase functions as a multimer and can turn over catalytically. J Biol Chem 1992;267:16037–16040.
338. Jones TA, Blaug G, Hansen M, Barklis E. Assembly of gag-B-galactosidase proteins into retrovirus particles. *J Virol* 1990;64:2265–2279.
339. Jones WK, Curran JW. Epidemiology of AIDS and HIV infection in industrialized countries. In: Broder S, Merigan TC, Bolognesi D, eds. *Textbook of AIDS medicine.* Baltimore: Williams and Wilkins; 1994: 91–108.
340. Kaczmarski W, Khan SA. Lupus autoantigen Ku protein binds HIV-1 TAR RNA in vitro. *Biochem Biophys Res Communications* 1993;196: 935–942.
341. Kalebic T, Kinter A, Poli G, Anderson ME, Meister A, Fauci AS. Suppression of human immunodeficiency virus expression in chronically infected monocytic cells by glutathione, glutathione ester, and N-acetylcysteine. *Proc Natl Acad Sci USA* 1991;88:986–990.
342. Kalpana GV, Goff SP. Genetic analysis of homomeric interactions of human immunodeficiency virus type 1 integrase using the yeast two-hybrid system. *Proc Natl Acad Sci USA* 1993;90:10593–10597.
343. Kalpana GV, Marmon S, Wang W, Crabtree GR, Goff SP. Binding and stimulation of HIV-1 integrase by a human homolog of yeast transcription factor SNF5. *Science* 1994;266:2002–2006.
344. Kalyanaraman VS, Rodriguez V, Veronese F, Rahman R, Lusso P, DeVico AL, Copeland T, Oroszlan S, Gallo RC, Sarngadharan MG. Characterization of the secreted, native gp120 and gp160 of the human immunodeficiency virus type 1. *AIDS Res Hum Retroviruses* 1990;6: 371–380.
345. Kaminchik J, Margalit R, Yaish S, Drummer H, Amit B, Sarver N, Gorecki M, Panet A. Cellular distribution of HIV type 1 nef protein: identification of domains in nef required for association with membrane and detergent-insoluble cellular matrix. *AIDS Res Hum Retroviruses* 1994;10:1003–1010.
346. Kan NC, Franchini G, Wong-Staal F, DuBois GC, Robey WG, Lautenberger JA, Papas TS. Identification of HTLV-III/LAV sor gene product and detection of antibodies in human sera. *Science* 1986;231:1553–1555.
347. Kaneshima H, Su L, Bonyhadi ML, Connor RI, Ho DD, McCune JM. Rapid-high syncytium-inducing isolates of human immunodeficiency virus type 1 induce cytopathicity in the human thymus of the SCID-hu mouse. *J Virol* 1994;68:8188–8192.
348. Kao S-Y, Calman AF, Luciw PA, Peterlin BM. Anti-termination of transcription within the long terminal repeat of HIV-1 by tat gene product. *Nature* 1987;330:489–493.
349. Kaplan AH, Swanstrom R. Human immunodeficiency virus type 1 gag proteins are processed in two cellular compartments. *Proc Natl Acad Sci USA* 1991;88:4528–4532.
350. Kaplan AH, Manchester M, Swanstrom R. The activity of the protease of human immunodeficiency virus type 1 is initiated at the membrane of infected cells before the release of viral proteins and is required for release to occur with maximun efficiency. *J Virol* 1994;68:6782–6786.
351. Kappes JC, Morrow CD, Lee S-W, Jameson BA, Kent SB, Hood LE, Shaw GM, Hahn BH. Identification of a novel retroviral gene unique to human immunodeficiency virus type 2 and simian immunodeficiency virus SIVmac. *J Virol* 1988;62:3501–3505.
352. Kappes JC, Conway JA, Lee SW, Shaw GM, Hahn BH. Human immunodeficiency virus type 2 vpx protein augments viral infectivity. *Virology* 1991;184:197–209.
353. Kappes JC, Parkin JS, Conway JA, Kim J, Brouillette CG, Shaw GM, Hahn BH. Intracellular transport and virion incorporation of vpx requires interaction with other virus type-specific components. *Virology* 1993;193:222–233.
354. Karacostas V, Nagashima K, Gonda M, Moss B. Human immunodeficiency virus-like particles produced by a vaccinia virus expression vector. *Proc Natl Acad Sci USA* 1989;86:8964–8967.
355. Karpas A, Lowdell M, Jacobson SK, Hill F. Inhibition of human immunodeficiency virus and growth of infected T cells by the immunosuppressive drugs cyclosporin A and FK506. *Proc Natl Acad Sci USA* 1992;89:8351–8355.
356. Karpel RL, Henderson LE, Oroszlan S. Interactions of retroviral structural proteins with single-stranded nucleic acids. *J Biol Chem* 1987; 262:4961–4967.
357. Kato H, Horikoshi M, Roeder RG. Repression of HIV-1 transcription by a cellular protein. *Science* 1991;251:1476–1479.
358. Kato H, Sumimoto M, Pognonec P, Chen CH, Rosen CA, Roeder RG. HIV-1 tat acts as a processivity factor in vitro in conjunction with cellular elongation factors. *Genes Dev* 1992;6:655–666.
359. Katz RA, Skalka AM. The retroviral enzymes. *Annu Rev Biochem* 1994; 63:133–173.
360. Kaufman JD, Valandra G, Rodriquez G, Bushar G, Giri C, Norcross MA. Phorbol ester enhances human immunodeficiency virus-promoted gene expression and acts on a repeated 10-base pair functional enhancer element. *Mol Cell Biol* 1987;7:3759–3766.
361. Kempf DJ. Design of symmetry-based, peptidomimetic inhibitors of human immunodeficiency virus protease. In: Kuo LC, Shafer JA, eds.*Retroviral proteases.* San Diego: Academic Press; 1994:334–354.
362. Kent RA, Rud E, Corcoran T, Powell C, Thiriart C, Collignon C, Stott EJ. Identification of two neutralizing and 8 non-neutralizing epitopes on simian immunodeficiency virus envelope using monoclonal antibodies. *AIDS Res Hum Retroviruses* 1992;8:1147–1151.
363. Kestler HW, Li Y, Naidu YM, Butler CV, Ochs MF, Jaenel G, King NW, Daniel MD, Desrosiers RC. Comparison of simian immunodeficiency virus isolates. *Nature* 1988;331:619–622.
364. Kestler HW, Kodama T, Ringler DJ, Marthas M, Pedersen NC, Lackner A, Regier D, Sehgal P, Daniel M, King N, Desrosiers RC. Induc-

tion of AIDS in rhesus monkeys by molecularly cloned simian immunodeficiency virus. *Science* 1990;248:1109–1112.

365. Kestler HW, Ringler DJ, Mori K, Panicall D, Sehgal P, Daniel M, Desrosiers RC. Importance of the nef gene for maintenance of high virus loads and for development of AIDS. *Cell* 1991;65:651–662.
366. Kido H, Fukutomi A, Katunuma N. A novel membrane-bound serine esterase in human T4+ lymphocytes immunologically reactive with antibody inhibiting syncytia induce by HIV-1: purification and characterization. *J Biol Chem* 1990;265:21,979.
367. Kienzle N, Freund J, Kalbitzer HR, Mueller-Lantzsch N. Oligermerization of the Nef protein from human immunodeficiency virus (HIV) type 1. *Eur J Biochem* 1993;214:451–457.
368. Kim S, Byrn R, Groopman J, Baltimore D. Temporal aspects of DNA and RNA synthesis during human immunodeficiency virus infection: evidence for differential gene expression. *J Virol* 1989;63:3708–3713.
369. Kim JY, Gonzalez-Scarano F, Zeichner SL, Alwine JC. Replication of type 1 human immunodeficiency viruses containing linker scanning mutations in the –201 to –130 region of the long terminal repeat. *J Virol* 1993;67:1658–1662.
370. Kirchoff F, Mori K, Desrosiers RC. The "V3" domain is a determinant of simian immunodeficiency virus cell tropism. *J Virol* 1994;68: 3682–3692.
371. Kirchoff F, Greenough RC, Brettler DB, Sullivan JL, Desrosiers RC. Absence of intact nef sequences in a long-term survivor with nonprogressive HIV-1 infection. *New Engl J Med* 1995;332:228–232.
372. Kjems J, Brown M, Chang DD, Sharp PA. Structural analysis of the interaction between the human immunodeficiency virus rev protein and the rev response element. *Proc Natl Acad Sci USA* 1991;88:683–687.
373. Klatzmann D, Champagne E, Chamaret S, Gruest J, Guétard D, Hercend T, Gluckman JC, Montagnier L. T-lymphocyte T4 molecule behaves as the receptor for human retrovirus LAV. *Nature* 1984;312:767–768.
374. Klimkait T, Strebel K, Hoggan MD, Martin MA, Orenstein JM. The human immunodeficiency virus type 1-specific protein vpu is required for efficient virus maturation and release. *J Virol* 1990;64:621–629.
375. Klotman ME, Kim S, Buchbinder A, DeRossi A, Baltimore D, Wong-Staal F. Kinetics of expression of multiply spliced RNA in early human immunodeficiency virus type 1 infection of lymphocytes and monocytes. *Proc Natl Acad Sci USA* 1991;88:5011–5015.
376. Kodama T, Wooley DP, Naidu YM, Kestler HW 3, Daniel MD, Li Y, Desrosiers RC. Significance of premature stop codons in env of simian immunodeficiency virus. *J Virol* 1989;63:4709–4714.
377. Kohl NE, Emini EA, Schleif WA, Davis LJ, Heimbach JC, Dixon RA, Scolnick EM, Sigal IS. Active human immunodeficiency virus protease is required for viral infectivity. *Proc Natl Acad Sci USA* 1988; 85:4686–4690.
378. Kohlstaedt LA, Wang JM, Friedman JM, Rice PA, Steitz TA. Crystal structure at 3.5 Å resolution of HIV-1 reverse transcriptase complexed with an inhibitor. *Science* 1992;256:1783–1790.
379. Kohlstaedt LA, Wang JM, Friedman JM, Rice PA, Steitz TA. The structure of HIV-1 reverse transcriptase. In: Skalka AM, Goff SP, eds.*Reverse Transcriptase.* Cold Spring Harbor: Cold Spring Harbor Laboratory Press; 1993:223–249.
380. Koito A, Harrowe G, Levy JA, Cheng-Mayer C. Functional role of the V1/V2 region of human immunodeficiency virus type 1 envelope glycoprotein gp120 in infection of primary macrophages and soluble CD4 neutralization. *J Virol* 1994;68:2253–2259.
381. Kornbluth RS, Oh PS, Munis JR, Cleveland PH, Richman DD. Interferons and bacterial lipopolysaccharide protect macrophages from productive infection: evidence for differential gene expression. *Clin Immunol Immunopathol* 1990;54:200–219.
382. Kowalski M, Potz J, Basiripour L, Dorfman T, Goh WC, Terwilliger E, Dayton A, Rosen C, Haseltine W, Sodroski J. Functional regions of the envelope protein of human immunodeficiency virus type 1. *Science* 1987;237:1351–1355.
383. Koyanagi Y, O'Brien WA, Zhao JQ, Golde DW, Gasson JC, Chen ISY. Cytokines alter production of HIV-1 from primary mononuclear phagocytes. *Science* 1988;241:1673–1675.
384. Kramer RA, Schaber MD, Skalka AM, Ganguly K, Wong-Staal F, Reddy EP. HTLV-III in gag protein is processed in yeast cells by the virus pol-protease. *Science* 1986;231:1580–1584.
385. Krausslich H-G, Schneider H, Zybarth G, Carter CA, Wimmer E. Processing of in vitro-synthesized gag precursor proteins of human immunodeficiency virus (HIV) type 1 by HIV proteinase generated in *Escherichia coli. J Virol* 1988;62:4393–4397.
386. Kuo LC, Shafer JA. *Retroviral proteases.* San Diego: Academic Press; 1994:432.
387. Kuppuswamy M, Subramanian T, Srinivasan A, Chinnadurai G. Multiple functional domains of tat, the trans-activator of HIV-1, defined by mutational analysis. *Nucleic Acids Res* 1989;17:3551–3561.
388. Kypr J, Mrazek J, Reich J. Nucleotide composition bias and CpG dinucleotide content in the genomes of HIV and HTLV 1/2. *Biochim Biophys Acta* 1989;1009:280.
389. LaBranche CC, Sauter MM, Haggarty BS, et al. Biological, molecular, and structural analysis of a cytopathic variant from a molecularly cloned simian immunodeficiency virus. *J Virol* 1994;68:5509–5522.
390. Lackner A. Pathology of simian immunodeficiency virus. *Curr Top Microbiol Immunol* 1994;188:35–64.
391. LaFemina RL, Callahan PL, Cordingley MG. Substrate specificity of recombinant human immunodeficiency virus integrase protein. *J Virol* 1991;65:5624–5630.
392. Lam PYS, Jadhav PK, Eyermann CJ, et al. Rationale design of potent, bioavailable, nonapeptide cyclic ureas as HIV protease inhibitors. *Science* 1994;263:380–384.
393. Larder B. Inhibitors of HIV reverse transcriptase as antiviral agents and drug resistance. In: Skalka AM, Goff SP, eds. *Reverse transcriptase.* Cold Spring Harbor, NY: Cold Spring Harbor Laboratory Press; 1993:205–222.
394. LaRosa GJ, Davide JP, Weinhold K, et al. Conserved sequence and structural elements in the HIV-1 principal neutralizing determinant [published erratum appears in *Science* 1991;251:811]. *Science* 1990; 249:932–935.
395. Lasky LA, Nakamura G, Smith DH, Fennie C, Shimasaki C, Patzer E, Berman P, Gregory T, Capon DJ. Delineation of a region of the human immunodeficiency virus type 1 gp120 glycoprotein critical for interaction with the CD4 receptor. *Cell* 1987;50:975–985.
396. Laspia MF, Rice AP, Mathews MB. HIV-1 tat protein increases transcriptional initiation and stabilizes elongation. *Cell* 1989;59:283–292.
397. Lawrence JB, Cochrane AW, Johnson CV, Perkins A, Rosen CA. The HIV-1 rev protein: a model system for coupled RNA transport and translation. *New Biol* 1991;3:1220–1232.
398. Le S-Y, Malim MH, Cullen BR, Maizel JV. A highly conserved RNA folding region coincident with the rev response element of primate immunodeficiency viruses. *Nucleic Acids Res* 1990;18:1613–1623.
399. Le Grice SFJ, Naas T, Wohlgensingerr B, Schatz O. Subunit-selective mutagenesis indicates minimal polymerase activity in heterodimer-associated p51 HIV-1 reverse transcriptase. *EMBO J* 1991;10:3905–3911.
400. Le Grice SFJ. Human immunodeficiency virus reverse transcriptase. In: Skalka AM, Goff SP, eds. *Reverse transcriptase.* Cold Spring Harbor, NY: Cold Spring Harbor Laboratory Press; 1993:163–191.
401. Learmont J, Tindall B, Evans L, Cunningham A, Cunningham P, Wells J, Penny R, Kaldor J, Cooper DA. Long-term symptom-less HIV-1 infection in recipients of blood products from a single donor. *Lancet* 1992;340:863–867.
402. Lee TH, Coligan JE, Allan JS, McLane MF, Groopman JE, Essex M. A new HTLV-III/Lav protein encoded by a gene found in cytopathic retroviruses. *Science* 1986;231:1546–1549.
403. Lee W-R, Syu W-J, Du B, Maatsuda M, Tan S, Wolf A, Essex M, Lee T-H. Nonrandom distribution of gp120 N-linked glycosylation sites important for infectivity of human immunodeficiency virus type 1. *Proc Natl Acad Sci USA* 1993;89:2213–2217.
404. Leigh-Brown AJ. Sequence variability of human immunodeficiency viruses: pattern and process in viral evolution. *AIDS* 1991;5S:35–42.
405. Lenburg ME, Landau NR. Vpu-induced degradation of CD4: requirement for specific amino acid residues in the cytoplasmic domain of CD4. *J Virol* 1993;67:7238–7245.
406. Leonard J, Parrott C, Buckler-White AJ, Turner W, Ross EK, Martin MA, Rabson AB. The NF-κB binding sites in the human immunodeficiency virus type 1 long terminal repeat are not required for virus infectivity. *J Virol* 1989;63:4919–4924.
407. Leonard CK, Spellman MW, Riddle L, Harris RJ, Thomas JN, Gregory TJ. Assignment of intrachain disulfide bonds and characterization of potential glycosylation sites of the type 1 recombinant human immunodeficiency virus envelope glycoprotein (gp120) expressed in Chinese hamster ovary cells. *J Biol Chem* 1990;265:10373–10382.
408. Lerner RA, Kang AS, Bain JD, Burton DR, Barbas CF. Antibodies without immunization. *Science* 1992;258:1313–1314.
409. Letvin NL, Daniel MD, Sehgal PK, Desrosiers RC, Hunt RD, Waldron LM, MacKey JJ, Schmidt DK, Chalifoux LV, King NW. Induction of

AIDS-like disease in macaque monkeys with T-cell tropic retrovirus STLV-III. *Science* 1985;230:71–73.

410. Levy DN, Fernandes LS, Williams WV, Weiner DB. Induction of cell differentiation by human immunodeficiency virus 1 vpr. *Cell* 1993;72: 541–550.
411. Levy DN, Refaeli Y, MacGregor RR, Weiner DB. Serum Vpr regulates productive infection and latency of human immunodeficiency virus type 1. *Proc Natl Acad Sci USA* 1994;91:10873–10877.
412. Levy DN, Refaeli Y, Weiner DB. Extracellular Vpr protein increases cellular permissiveness to human immunodeficiency virus replication and reactivates virus from latency. *J Virol* 1995;69:1243–1252.
413. Levy JA. The retroviridae. In: Fraenkel-Conrat H, Wagner RR, eds. *The viruses.* Vol. 2. New York: Plenum; 1993:489.
414. Levy JA. *HIV and AIDS pathogenesis.* Washington, DC: ASM Press; 1994:359.
415. See ref. 413.
416. Levy JA, Hoffman AD, Kramer SM, Landis JA, Shimabukuro JM, Oshiro LS. Isolation of lymphocytopathic retroviruses from San Francisco patients with AIDS. *Science* 1984;225:840–842.
417. Lewis PF, Emerman M. Passage through mitosis is required for oncoretroviruses by not for the human immunodeficiency virus. *J Virol* 1994;68:510–516.
418. Lewis P, Hensel M, Emerman M. Human immunodeficiency virus infection of cells arrested in the cell cycle. *EMBO J* 1994;11: 3053–3058.
419. Li J, Lord CI, Haseltine WA, Letvien NL, Sodroski J. Infection of cynomologous monkeys with a chimeric HIV-1/SIVmac virus that expresses the HIV-1 envelope glycoproteins. *J Acquir Immune Defic Syndr* 1992;5:639–646.
420. Li Y, Luo L, Rasool N, Kang CY. Glycosylation is necessary for the correct folding of human immunodeficiency virus gp120 in CD4 binding. *J Virol* 1993;67:584–588.
421. Lifson JD, Feinberg MB, Wong-Staal F, Reyes GR, Rabin L, Banapour B, Chakrabarti S, Moss B, Steimer KS, Engelman EG. Induction of CD4-dependent cell fusion by the HTLV-III/LAV envelope glycoprotein. *Nature* 1986;323:725–728.
422. Lifson JD, Hwang KM, Nara PL, Fraser B, Padgett M, Dunlop MN, Eiden L. Synthetic CD4 peptide derivatives that inhibit HIV infection and cytopathicity. *Science* 1988;241:712–716.
423. Lifson AR, Buchbinder S, Sheppard HS. Long-term human immunodeficiency virus infection in asymptomatic homosexual and bisexual men with normal CD4+ lymphocyte counts: immunologic and virologic characteristics. *J Infect Dis* 1991;163:959–965.
424. Linial ML, Miller AD. Retroviral RNA packaging: sequence requirements and implications. *Curr Top Microbiol Immunol* 1990;157: 125–152.
425. Lisziewicz J, Sun D, Smythe J, Lusso P, Lori F, Louie A, Markham P, Rossi J, Reitz M, Gallo RC. Inhibition of human immunodeficiency virus type 1 replication by regulated expression of a polymeric tat activation response RNA decoy as a strategy for gene therapy in AIDS. *Proc Natl Acad Sci USA* 1993;90:8000–8004.
426. Lodge R, Gottlinger H, Gabuzda D, Cohen EA, Lemay G. The intracytoplasmic domain of gp41 mediates polarized budding of human immunodeficiency virus type 1 in MDCK cells. *J Virol* 1994;68:4857–4861.
427. Loeb DD, Swanstrom R, Everitt L, Manchester M, Stamper SE, Hutchison CA. Complete mutagenesis of the HIV-1 protease. *Nature* 1989; 340:397–400.
428. Loeb DD, Hutchison CA, Edgell MH, Farmerie WG, Swanstrom R. Mutational analysis of human immunodeficiency virus type 1 protease suggests functional homology with aspartic proteases. *J Virol* 1989; 63:111–121.
429. Long D, Berson JF, Cook DG, Doms RW. Characterization of human immunodeficiency virus type 1 gp120 binding to liposomes containing galactosylceramide. *J Virol* 1994;68:5890–5898.
430. Louwagie J, Janssens W, Mascola J, Heyndrickx L, Hegerich P, Groen G, MiCutchan FE, Burke DS. Genetic diversity of the envelope glycoprotein from human immunodeficiency virus type 1 isolates of African origin. *J Virol* 1995;69:263–271.
431. Lu Y, Stenzel M, Sodroski JG, Haseltine WA. Effects of long terminal repeat mutations on human immunodeficiency virus type 1 replication. *J Virol* 1989;63:4115–4119.
432. Lu X, Heimer J, Rekosh D, Hammarskjold M-L. U1 small nuclear RNA plays a direct role in the formation of a rev-regulated human immunodeficiency virus env mRNA that remains unspliced. *Proc Natl Acad Sci USA* 1990;87:7598–7602.
433. Lu X, Welsh TM, Peterlin BM. The human immunodeficiency virus type 1 long terminal repeat specifies two different transcription complexes, only one of which is regulated by tat. *J Virol* 1993;67:1752–1760.
434. Luban J, Goff SP. Binding of human immunodeficiency virus type 1 (HIV-1) RNA to recombinant HIV-1 gag polyprotein. *J Virol* 1991;65: 3203–3212.
435. Luban J, Alin KB, Bossolt KL, Humaran T, Goff SP. Genetic assay for multimerization of retroviral gag polyproteins. *J Virol* 1992;66: 5157–5160.
436. Luban J, Bossolt KL, Franke EK, Kalpana GV, Goff SP. Human immunodeficiency virus type 1 gag protein binds to the cyclophilins A and B. *Cell* 1993;73:1067–1078.
437. Luban J, Goff SP. Mutational analysis of cis-acting packaging signals in human immunodeficiency virus type 1 RNA. *J Virol* 1994;68:3784–3793.
438. Luciw PA, Oppermann H, Bishop JM, Varmus HE. Integration and expression of several molecular forms of Rous sarcoma virus. *Mol Cell Biol* 1983;4:1260–1268.
439. Luciw PA, Shacklett BL. Molecular biology of the human and simian immunodeficiency viruses. In: Morrow WJW, Haigwood NL, eds. *HIV—molecular organization, pathogenicity and treatment.* Amsterdam: Elsevier; 1993:123–219.
440. Luciw PA, Pratt-Lowe E, Shaw KES, Levy JA, Cheng-Mayer C. Infection of rhesus macaques with T-cell line-tropic and macrophage-tropic clones of simian/human immunodeficiency viruses (SHIV). *Proc Natl Acad Sci USA* 1995;92:[in press].
441. Luo GX, Taylor J. Templae switching by reverse transcriptase during DNA synthesis. *J Virol* 1990;64:4321–4328.
442. Luo Y, Yu H, Peterlin BM. Cellular protein modulates effects of human immunodeficiency virus type 1. *Rev J Virol* 1994;68:3850–3856.
443. Luria S, Chambers I, Berg P. Expression of the type 1 human immunodeficiency virus nef protein in T cells prevents antigen receptor-mediated induction of interleukin-2 mRNA. *Proc Natl Acad Sci USA* 1991; 88:5326–5330.
444. Maddon PJ, McDougal JS, Clapham PR, Dalgliesh AG, Jamal S, Weiss RA, Axel R. HIV infection does not require endocytosis of its receptor, CD4. *Cell* 1988;54:865–874.
445. Malim MH, Hauber J, Fenrick R, Cullen BR. Immunodeficiency virus rev trans-activator modulates the expression of the viral regulatory genes. *Nature* 1988;335:181–183.
446. Malim MH, Hauber J, Le SY, Maizel JV, Cullen BR. The HIV-1 rev trans-activator acts through a structured target sequence to activate nuclear export of unspliced viral mRNA. *Nature* 1989;338:254–257.
447. Malim MH, Bohnlein S, Hauber J, Cullen BR. Functional dissection of the HIV-1 rev trans-activator-derivation of a trans-dominant repressor of rev function. *Cell* 1989;58:205–214.
448. Malim MH, Bohnlein S, Fenrick R, Le S-Y, Maizel JV, Cullen BR. Functional comparison of the rev trans-activators encoded by different primate immunodeficiency virus species. *Proc Natl Acad Sci USA* 1989;86:8222–8226.
449. Malim MH, Tiley LS, McCarn DF, Rusche JR, Hauber J, Cullen BR. HIV-1 structural gene expression requires binding of the rev trans-activator to its RNA target sequence. *Cell* 1990;60:675–683.
450. Malim MH, McCarn DF, Tiley LS, Cullen BR. Mutational definition of the human immunodeficiency virus type 1 rev activation domain. *J Virol* 1991;65:4248–4254.
451. Malim MH, Cullen BR. HIV-1 structural gene expression requires the binding of multiple rev monomers to the viral RRE: implications for HIV-1 latency. *Cell* 1991;65:241–248.
452. Malim MH, Freimuth J, Liu J, Boyle TJ, Lyerly HK, Cullen BR. Stable expression of transdominant rev protein in human T cells inhibits human immunodeficiency virus replication. *J Exp Med* 1992;176: 1197–1201.
453. Mammano F, Ohagen A, Hoglund S, Gottlinger HG. Role of the major homology region of human immunodeficiency virus type 1 in virion morphogenesis. *J Virol* 1994;68:4927–4936.
454. Manchester M, Everitt L, Loeb DD, Hutchison CA, Swanstrom R. Identification of temperature-sensitive mutants of human immunodeficiency virus type 1 protease through saturation mutagenesis. Amino acid requirements for temperature sensisitivty. *J Biol Chem* 1994;269: 7689–7695.

455. Mann DA, Frankel AD. Endocytosis and targeting of exogenous HIV-1 Tat protein. *EMBO J* 1991;10:1733–1739.
456. Mann JM, Tarantola DJM, Netter TW. *AIDS in the world.* Cambridge, MA: Harvard University Press; 1992.
457. Marasco WA, Haseltine WA, Chen S-Y. Design, intracellular expression, and activity of a human anti-human immunodeficiency virus type 1 gp120 single-chain antibody. *Proc Natl Acad Sci USA* 1993;90:7889–7893.
458. Marciniak RA, Garcia-Blanco MA, Sharp PA. Identification and characterization of a HeLa nuclear protein that specifically binds to the trans-activation-response (TAR) element of human immunodeficiency virus. *Proc Natl Acad Sci USA* 1990;87:3624–3628.
459. Marciniak RA, Calnan BJ, Frankel AD, Sharp PA. HIV-1 tat protein trans-activates transcription in vitro. *Cell* 1990;63:791–802.
460. Marciniak RA, Sharp PA. HIV-1 tat protein promotes formation of more-processive elongation complexes. *EMBO J* 1991;10:4189–4196.
461. Marcon L, Michaels F, Hattori N, Fargnoli K, Gallo RC, Franchini G. Dispensable role of the human immunodeficiency virus type 2 vpx protein in viral replication. *J Virol* 1991;65:3938–3942.
462. Marquet R, Baudin F, Gabus C, Darlix JL, Mougel M, Ehresman C, Ehresman B. Dimerization of human immunodeficiency virus (type 1) RNA: stimulation by cations and possible mechanism. *Nucleic Acids Res* 1991;19:2349–2357.
463. Marschang P, Sodroski J, Wurzner R, Dierich MP. Decay-accelerating factor (CD55) protects human immunodeficiency virus type 1 from inactivation by human complement. *Eur J Immunol* 1995;25:285–290.
464. Marthas ML, Ramos RA, Lohman BA, Van Rompay K, Unger RE, Miller CJ, Banapour B, Pedersen NC, Luciw PA. Viral determinants of simian immunodeficiency virus (SIV) virulence in rhesus macaques assessed by using attenuated and pathogenic molecular clones of SIVmac. *J Virol* 1993;67:6047–6055.
465. Marx PA, Munn RJ, Joy KI. Computer emulation of thin section electron microscopy predicts an envelope-associated icosadeltahedral capsid for human immunodeficiency virus. *Lab Invest* 1988;58:112–118.
466. Matsuda Z, Chou MJ, Matsuda M, Huang JH, Chen YM, Redfield R, Mayer K, Essex M, Lee TH. Human immunodeficiency virus type 1 has an additional coding sequence in the central region of the genome. *Proc Natl Acad Sci USA* 1988;85:6968–6972.
467. Mazumder A, Cooney D, Agbaria R, Gupta M, Pommier Y. Inhibition of human immunodeficiency virus type 1 integrase by 3′-azido-3′-deoxythymidylate. *Proc Natl Acad Sci USA* 1994;91:5771–5775.
468. McClure MO, Marsh M, Weiss RA. HIV infection of CD4 bearing cells occurs by a pH-independent mechanism. *EMBO J* 1988;7:513–518.
469. McCormack S, Thomis D, Samuel C. Mechanism of ionterferon action: indentification of an RNA-binding domain within the N-terminal region of the human RNA dependent P1/eIF-2 alpha protein kinase. *Virology* 1992;188:47–56.
470. McCune JM, Rabin LB, Feinberg MB, Lieberman M, Kosek JC, Reyes GR, Weissman IL. Endoproteolytic cleavage of gp160 is required for activation of human immunodeficiency virus. *Cell* 1988;53:55–67.
471. McCune JM, Namikawa R, Kaneshima H, Schultz LD, Lieberman M, Weismann IL. The SCID-hu mouse model for the analysis of human hematolymphoid differentiation and function. *Science* 1988;241:1632–1639.
472. McDonald D, Hope TJ, Parslow TG. Posttranscriptional regulation by the human immunodeficiency virus type 1 Rev and human T-cell leukemia virus type I Rex proteins through a heterologous RNA binding site. *J Virol* 1992;66:7232–7238.
473. McDougal JS, Kennedy MS, Sligh JM, Cort SP, Mawle AC, Nicholson JKA. Binding of HTLV-III/LAV to T4+ T cells by a complex of the 110K viral protein and the T4 molecule. *Science* 1986;231:382–385.
474. McDougal JS, Nicholson JKA, Cross GD, Cort SP, Kennedy MS, Mawle AC. Binding of the human retrovirus HTLV-III/ARV/HIV to the CD4 (T4) molecule: conformation dependence, epitope mapping, antibody inhibition, and potential for idiotypic mimicry. *J Immunol* 1986;137:2937–2944.
475. McKeating JA, Griffiths PD, Weiss RA. HIV susceptibility conferred to human fibroblasts by cytomegalovirus-induced Fc receptor. *Nature* 1990;343:659–661.
476. McKeating JA, McKnight A, McMoore JP. Differential loss of envelope glycoprotein from virions of human immunodeficiency virus type 1 isolates: effect on infectivity and neutralisation. *J Virol* 1991;65: 852–860.
477. McKeating JA, Moore JP, Ferguson M, Graham S, Almond JW, Evans DJ, Weiss RA. Monoclonal antibodies to the C4 region of human immunodeficiency virus type 1 gp120: use in topological analysis of a CD4 biding site. *AIDS Res Hum Retroviruses* 1992;8:451–459.
478. McKeever BM, Navia MA, Fitzgerald PMD, Springer JP, Leu C-T, Heimbach JC, Herber WK, Sigal IS, Darke PL. Crystallization of the aspartyl protease from the human immunodeficiency virus, HIV-1. *J Biol Chem* 1989;264:1919–1921.
479. McNearney T, Hornickova Z, Markham R, Birdwell A, Arens M, Saah A, Ratner L. Relationship of human immunodeficiency virus type 1 sequence heterogeneity to stage of disease. *Proc Natl Acad Sci USA* 1992;89:10247–10251.
480. Milich L, Margolin B, Swanstrom R. V3 loop of the human immunodeficiency virus type 1 env protein: interpreting sequence variability. *J Virol* 1993;67:1919–1921.
481. Miller M, Schneider J, Sathyanrayana BK, Toth MY, Marshall GR, Clawson L, Selk L, Kent SBH, Wlodawer A. Structure of complex of synthetic HIV-1 protease with a substrate-based inhibitor at 2.3 Å resolution. *Science* 1989;246:1149–1152.
482. Miller MA, Garry RF, Jaynes JM, Montelaro RC. A structural correlation between lentivirus transmembrane proteins and natural cytolytic peptides. *AIDS Res Hum Retroviruses* 1991;7:511–519.
483. Miller CJ, Alexander NJ, Gettie A, Hendrickx AG, Marx PA. The effect of contraceptives containing nonxynol-9 on the genital transmission of simian immunodeficiency virus in rhesus macaques. *Fertil Steril* 1992;57:1126–1128.
484. Miller MA, Cloyd MW, Liebmann J, Rinaldo CR, Islam KR, Wang SZS, Mietzner TA, Montelaro RC. Alterations in cell membrane permeability by the lentivirus lytic peptide (LLP-1) of HIV-1 transmembrane protein. *Virology* 1993;196:89–100.
485. Miller MA, Mietzner TA, Cloyd MW, Robey WG, Montelaro RC. Identification of calmodulin-binding and inhibitory peptide domain in the HIV-1 transmembrane glycoprotein. *AIDS Res Hum Retroviruses* 1993; 9:1057–1065.
486. Miller MD, Warmerdam MT, Gaston I, Greene WC, Feinberg MB. The human immunodeficiency virus-1 nef gene product: a positive factor for viral infection and replication in primary lymphocytes and macrophages. *J Exp Med* 1994;179:101–113.
487. Milton RCL, Milton SCF, Kent SBH. Total chemical synthesis of a D-enzyme: the enantiomers of HIV-1 protease show demonstration of reciprocal chiral substrate specificity. *Science* 1992;256:1445–1448.
488. Mitsuyasu H, Yarochan R. Development of antiretroviral therapy for AIDS and related disorders. In: Broder S, Merigan TC, Bolognesi D, eds. *Textbook of AIDS medicine.* Baltimore: Williams and Wilkins; 1994:721–742.
489. Mizuochi T, Matthews TJ, Kato M, Hamako J, Titani K, Solomon J, Feizi T. Diversity of oligosaccharide structure on the envelope glycoprotein gp120 of human immunodeficiency virus 1 from the lymphoblastoid cell line H9. Presence of complex-type oligosaccharides with bisecting N-acetylglucosamine residues. *J Biol Chem* 1990;265: 8519–8524.
490. Mizutani S, Temin HM. RNA-dependent DNA polymerase in virions of Rous sarcoma virus. *Nature* 1970;226:1211–1213.
491. Modesti N, Garcia J, DeBouck C, Peterlin BM, Gaynor R. Transdominant tat mutants with alterations in the basic domain inhibit HIV-1 gene expression. *New Biol* 1991;3:759–768.
492. Modrow S, Hahn BH, Shaw GM, Gallo RC, Wong-Staal F Wolf H. Computer-assisted analysis of envelope protein sequences of seven human immunodeficiency virus isolates: prediction of antigenic epitopes in conserved and variable sequences. *J Virol* 1987;61:570–578.
493. Moebius U, Clayton LK, Abraham S, Harrison SC, Reinherz EL. The HIV gp120 bonding site on CD4: delineation by quantitative equilibrium and kinetic binding studies of mutants in conjunction with a high resolution CD4 atomic structure. *J Exp Med* 1992;176:507–517.
494. Moore JP, McKeating JA, Weiss RA, Sattentau Q. Dissociation of gp120 from HIV-1 virions induced by soluble CD4. *Science* 1990;250: 1139–1142.
495. Moore JP. Simple methods for monitoring HIV-1 and HIV-2 gp120 binding to sCD4 by ELISA: HIV-2 has a 25-fold lower affinity than HIV-1 for sCD4. *AIDS* 1990;3:297–305.
496. Moore JP, Nara PL. The role of the V3 loop of gp120 in HIV infection. *AIDS* 1991;5(suppl. 2):S21–S33.
497. Moore JP, McKeating JM, Norton WA, Sattentau QJ. Direct mea-

surement of soluble CD4 binding to human immunodeficiency virus type 1: gp120 dissociation and its implications for virus-cell binding and fusion reactions and their neutralization by soluble CD4. J Virol 1991;65:1133–1140.
498. Moore JP, Klasse PJ. Thermodynamic and kinetic analysis of sCD4-binding to HIV-1 virions and of gp120 dissociation. *AIDS Res Hum Retroviruses* 1992;8:443–450.
499. Moore JP, McKeating JA, Huang Y, Ashkenazi A, Ho DD. Virions of primary human immunodeficiency virus type 1 isolates resistant to soluble CD4 (sCD4) neutralization differ in sCD4 binding and glycoprotein gp120 retention from sCD4-sensitive isolates. *J Virol* 1992;66: 235–243.
500. Moore JP, Ho DD. Antibodies to discontinuous or conformationally sensitive epitopes on the gp120 glycoprotein of human immunodeficiency virus type 1 are highly prevalent in sera of infected individuals. *J Virol* 1992;67:863–875.
501. Moore JP, Jameson BA, Weiss RA, Sattentau QJ. The HIV-cell fusion reaction. In: Bentz J, ed. *Viral fusion mechanisms.* Boca Raton, FL: CRC Press; 1993.
502. Moore JP, Willey RL, Lewis GK, Robinson J, Sodroski J. Immunological evidence for interactions between the first, second, and fifth conserved domains of the gp120 surface glycoprotein of human immunodeficiency virus type 1. *J Virol* 1994;68:6836–6847.
503. Moore JP, Sattentau QJ, Wyatt R, Sodroski J. Probing the structure of the human immunodeficiency virus surface glycoprotein gp120 with a panel of monoclonal antibodies. *J Virol* 1994;68:469–484.
504. Morellet N, de Rocquigny H, Mely Y, et al. Conformational behaviour of the active and inactive forms of the nucleocapsid NCp7 of HIV-1 studied by 1H NMR. *J Mol Biol* 1994;235:287–301.
505. Mori K, Ringler DJ, Kodama T, Desrosiers RC. Complex determinants of macrophage tropism in env of simian immunodeficiency virus. *J Virol* 1992;66:2067–2075.
506. Mori K, Ringler DJ, Desrosiers RC. Restricted replication of simian immunodeficiency virus strain 239 in macrophages is determined by env but is not due to restricted entry. *J Virol* 1993;67:2807–2814.
507. Morikawa Y, Moore JP, Wilkinson AJ, Jones IM. Reduction in CD4 binding affinity associated with removal of a single glycosylation site in the external glycoprotein of HIV-1. *Virology* 1991;180:853–856.
508. Morrow WJW, Haigwood NL. *HIV molecular organization, pathogenicity and treatment.* Amsterdam: Elsevier; 1993:386.
509. Morse SS. The viruses of the future? Emerging viruses and evolution. In: Morse SS, ed. *Evolutionary biology of viruses.* New York: Raven Press; 1994:325–335.
510. Mosier DE, Gulizia RJ, Baird SM, Wilson DB. Transfer of a functional human immune system to mice with severe combined immunodeficiency. *Nature* 1988;335:256–259.
511. Muesing MA, Smith DH, Cabradilla CD, Benton CV, Lasky LA, Capon DJ. Nucleic acid structure and expression of the human AIDS/lymphadenopathy retrovirus. *Nature* 1985;313:450–458.
512. Muesing MA, Smith DH, Capon DJ. Regulation of mRNA accumulation by a human immunodefiency virus trans-activator protein. *Cell* 1987;48:691–701.
513. Mulligan RC. The basic science of gene therapy. Science 1993;260: 926–932.
514. Munn R, Marx PA, Yamamoto JK, Gardner MB. Ultrastructural comparison of the retroviruses associated with human and simian acquired immunodeficiency syndromes. *Lab Invest* 1985;53:194–199.
515. Murakami T, Hattori T, Takatsuki K. A principal nuetralizing domain of human immunodeficiency virus type 1 interacts with proteinase-like molecule(s) at the surface of Molt-4 clone 8 cells. *Biochim Biophys Acta* 1991;1079:279.
516. Murre C, McGaw SP, Baltimore D. A new DNA binding and dimerization motif in immunoglobulin enhancer binding, daughterless, MyoD, and myc proteins. *Cell* 1990;56:777–783.
517. Myers G, Pavlakis GN. Evolutionary potential of complex retroviruses. In: Levy JA, ed. *The retroviridae.* New York: Plenum Press; 1992: 51–105.
518. Myers G. HIV: between past and future. *AIDS Res Hum Retroviruses* 1994;10:1317–1324.
519. Myers G, Korber B, Wain-Hobson S, Jeang KT, Henderson LE, Pavlakis GN. *Human retroviruses and AIDS. A compilation and analysis of nucleic acid and amino acid sequences.* Los Alamos, NM: Los Alamos National Laboratory; 1994.
520. Nabel GJ, Baltimore D. An inducible transcription factor activates expression of human immunodeficiency virus in T cells. *Nature* 1987; 325:711–713.
521. Nabel GJ, Rice SA, Knipe DM, Baltimore D. Alternative mechanisms for activation of human immunodeficiency virus enhancer in T cells. *Science* 1988;239:1299–1302.
522. Nabel GJ. The role of cellular transcription factors in the regulation of human immunodeficiency virus gene expression. In: Cullen BR, ed. *Human retroviruses.* Oxford: IRL Press; 1993:49–73.
523. Nara PL, Garrity RR, Goudsmit J. Neutralization of HIV-1: a paradox of humoral proportions. *FASEB J* 1991;5:2437–2455.
524. Navia MA, Fitzgerald PMD, McKeever BM. Three dimensional structure of aspartyl protease from human immunodeficiency virus HIV-1. *Nature* 1989;337:615–620.
525. Nebreda AR, Bryan T, Segade F, Wingfield P, Venkatesan S, Santos E. Biochemical and biological comparison of HIV-1 nef and ras gene products. *Virology* 1991;181:151–159.
526. Nelbock P, Dillon R, Perkins A, Rosen CA. A cDNA for a protein that interacts with the human immunodeficiency virus tat transactivator. *Science* 1990;248:1650–1653.
527. Nermut MV, Grief C, Hashmi S, Hockley DJ. Further evidence of icosahedral symmetry in human and simian immunodeficiency virus. *AIDS Res Hum Retroviruses* 1993;9:929–938.
528. Nermut MV, Hockley DJ, Jowett JB, Jones IM, Garreau M, Thomas D. Fullerene-like organization of HIV gag-protein shell in a virus-like paerticles produced by recombinant baculovirus. *Virology* 1994;198: 288–296.
529. Neumann M, Harrison JSM, Hadziyannis E, Erfle V, Felber BK, Pavlakis GN. Splicing variability in HIV type 1 revealed by quantitative RNA polymerase chain reaction. *AIDS Res Hum Retroviruses* 1994; 10:1531–1542.
530. Newstein M, Stanbridge EJ, Casey G, Shank PR. Human chromosome 12 encodes a species-specific factor which increases human immunodeficiency virus type 1 tat-mediated trans-activation in rodent cells. *J Virol* 1990;64:4565–4567.
531. Niederman TMJ, Thielan GJ, Ratner L. Human immunodeficiency virus type 1 negative factor is a transcriptional silencer. *Proc Natl Acad Sci USA* 1989;86:1128–1132.
532. Niederman TMJ, Garcia JV, Hastings WR, Luria S, Ratner L. Human immunodeficiency virus type 1 nef protein inhibits NF-κB induction in human T cells. *J Virol* 1992;66:6213–6219.
533. Niederman TMJ, Randall-Hastings W, Ratner L. Myristoylation-enhanced binding of the HIV-1 nef protein to T cell skeletal matrix. *Virology* 1993;197:420–425.
534. Niedrig M, Gelderblom HR, Pauli G, Marx J, Bickhard H, Wolf H, Modrow S. Inhibition of infectious human immunodeficiency virus type 1 by gag protein-derived peptides. J Gen Virol 1994;75:1469–1474.
535. Novembre F, Johnson P, Lewis M, Anderson D, Klumpp S, McClure H, Hirsch V. Multiple determinants contribute to pathogenicity of the acutely lethal simian immunodeficiency virus SIVsmmPBj variant. *J Virol* 1993;67:2466–2474.
536. Oberste MS, Gonda MA. Conservation of amino-acid sequence motifs in lentivirus vif proteins. *Virus Genes* 1992;6:95–102.
536a. O'Brien WA, Koyangi Y, Namazie A, et al. HIV-1 tropism for mononuclearphagocytes can be determined by regions outside the CD4-binding domain. *Nature* 1990;348:69–73.
537. Ogawa K, Shibata R, Kiyomasu T, Higuchi I, Kishida Y, Ishimoto A, Adachi A. Mutational analysis of the human immunodeficiency virus vpr open reading frame. *J Virol* 1989;63:4110–4114.
538. Ohana B, Moore PA, Ruben SM, Southgate CD, Green MR, Rosen CA. The type 1 human immunodeficiency virus tat binding protein is a transcriptional activator belonging to an additional family of evolutionarily conserved genes. *Proc Natl Acad Sci USA* 1993;90:138–142.
539. Ohta H, Tsurudome M, Masumura H, et al. Molecular and biological characterization of fusion regulatory proteins (FRPs): anti-FRP mAbs induced HIV-mediated cell fusion via an integrin system. *EMBO J* 1994;13:2044–2055.
540. Olsen HS, Beidas S, Dillon P, Rosen CA, Cochrane AW. Mutational analysis of the HIV-1 rev protein and its target sequence, the rev response element. *J Acquir Immune Defic Syndr* 1991;4:558–567.
541. Olsen HS, Cochrane AW, Dillon PJ, Nalin CM, Rosen CA. Interaction of the human immunodeficiency virus type 1 rev protein with a structured region in env mRNA is dependent on multimer formation

mediated through a basic stretch of amino acids. *Genes Dev* 1990;4: 1357–1364.
542. Olsen HS, Cochrane AW, Rosen C. Interaction of cellular factors with intragenic *cis*-acting repressive sequences within the HIV genome. *Virology* 1992;191:709–715.
543. Olsen HS, Nelbock P, Cochrane AW, Rosen CA. Secondary structure is the major determinant for interaction of HIV rev protein with RNA. *Science* 1990;247:845–848.
544. Olsen HS, Rosen CA. Contribution of the TATA motif to Tat-mediated transcriptional activation of human immunodeficiency virus gene expression. *J Virol* 1992;66:5594–5597.
545. Olshevsky U, Helseth E, Furman C, Li J, Haseltine W, Sodroski J. Identification of individual human immunodeficiency virus type 1 gp120 amino acids important for CD4 receptor binding. *J Virol* 1990;64:5701–5707.
546. Osborn L, Kunkel S, Nabel GJ. Tumor necrosis factor a and interleukin-1 stimulate the human immunodeficiency virus enhancer by activation of the nuclear factor kB. *Proc Natl Acad Sci USA* 1989;86:2336–2340.
547. Osborn JE. Public health, HIV, and AIDS. In: Broder S, Merigan TC, Bolognesi D, eds. *Textbook of AIDS medicine.* Baltimore: Williams and Wilkins; 1994:133–146.
548. Osmond DH. Classification and staging of HIV disease. In: Cohen PT, Sande MA, Volberding PA, eds. *The AIDS knowledge base.* Boston: Little, Brown and Company; 1994a:1.1–1.15.
549. Osmond DH. HIV disease progression from infection to CDC-defined AIDS: incubation period, cofactors, and laboratory markers. In: Cohen PT, Sande MA, Volberding PA, eds. *The AIDS knowledge base.* Boston: Little, Brown and Company; 1994b:1.7-1–1.7-19.
550. Overbaugh J, Rudensky LM. Alterations in potential sites for glycosylation predominate during evolution of the simian immunodeficiency virus envelope gene in macaques. *J Virol* 1992;66:5937–5948.
551. Owens R, Dubay JW, Hunter E, Compans RW. Human immunodeficiency virus envelope protein determines the site of virus release in polarized epithelial cells. *Proc Natl Acad Sci USA* 1991;88:3987–3991.
552. Ozel M, Pauli G, Gelderblom HR. The organization of the envelope projections on the surface of HIV. *Arch Virol* 1988;100:255–266.
553. Page KA, Stearns SM, Littman DR. Analysis of mutations in the V3 domain of gp160 that affect fusion and infectivity. *J Virol* 1992;66: 524–533.
554. Palker TJ, Clark ME, Langlois AJ, Matthews TJ, Weinhold KJ, Randall RR, Bolognesi DP, Haynes BF. Type-specific neutralization of the human immunodeficiency virus with antibodies to env-encoded peptides. *Proc Natl Acad Sci USA* 1988;85:1932–1936.
555. Pancino G, Ellerbrok H, Sitbon M, Sonigo P. Conserved framework of envelope glycoproteins among lentiviruses. *Curr Top Microbiol Immunol* 1994;188:77–105.
556. Panganiban AT, Fiore D. Ordered interstrand and intrastrand DNA transfer during reverse transcription. *Science* 1988;241:1064–1069.
557. Pantaleo G, Grazioni C, Demarest JF, Butini L, Montroni M, Fox CH, Orenstein JM, Kotler DP, Fauci AS. HIV infection is active and progressive in lymphoid tissue during clinically latent stage of disease. *Nature* 1993;362:355–358.
558. Pantaleo G, Menzo S, Vaccarezza M, Grazioni C, Cohen OJ, Demarest JF, et al. Studies in subjects with long-term nonprogressive human immunodeficiency virus infection. *N Engl J Med* 1995;332:209–216.
559. Park J, Morrow CD. Overexpression of the gag-pol precursor from human immunodeficiency virus type 1 proviral genomes results in efficient proteolytic processing in the absence of virion production. *J Virol* 1991;65:5111–5117.
560. Park J, Morrow CD. The non-myristylated Pr160gag-pol polyprotein of human immunodeficiency virus type 1 interacts with Pr55gag and is incorporated into virus-like particles. *J Virol* 1992;66:6304–6313.
561. Park H, Davis MV, Langland JO, Chang H, Nam Y, Tartaglia J, et al. TAR RNA-binding protein is an inhibitor of interferon-induced protein kinase PKR. *Proc Natl Acad Sci USA* 1994;91:4713–4717.
562. Parolin C, Dorfman T, Palu G, Gottlinger H, Sodroski J. Analysis in human immunodeficiency virus type 1 vectors of cis-acting sequences that affect gene transfer into human lymphocytes. *J Virol* 1994;68: 3888–3895.
563. Parrott C, Seidner T, Duh E, Leonard J, Theodore TS, Buckler-White A, Martin MA, Rabson AB. Variable role of the long terminal repeat Sp-1 sites in human immunodeficiency virus replication in T lymphocytes. *J Virol* 1991;65:1414–1419.
564. Parslow TG. Post-transcriptional regulation of human retroviral gene expression. In: Cullen BR, ed. *Human retroviruses.* Oxford: IRL Press; 1993:101–136.
565. Pauza CD, Price TM. HIV infection of T cells and monocytes proceeds via receptor-mediated endocytosis. *J Cell Biol* 1988;107:959–968.
566. Pavlakis GN, Schwartz S, D'Agostino D, Felber B. Structure, splicing, and regulation of expression of HIV-1: a model for the general organization of lentiviruses and other complex retroviruses. In: Koff WC, Kennedy RC, Wong-Staal F, eds. *AIDS research reviews.* New York: Marcel Dekker; 1992:41–63.
567. Paxton W, Connor RI, LAndau NR. Incorporation of Vpr into human immunodeficiency virus type 1 virions:requirement for the p6 region of gag and mutational analysis. *J Virol* 1993;67:7229–7237.
568. Pearson L, Garcia J, Wu F, Modesti N, Nelson J, Gaynor RA. Transdominant tat mutant that inhibits tat-induced gene expression from the human immunodeficiency virus long terminal repeat. *Proc Natl Acad Sci USA* 1990;87:5079–5083.
569. Peeters M, Honore C, Huet T, Bedjaba L, Ossari S, Bussi P, Cooper RW, Delaporte E. Isolation and partial characterization of an HIV-related virus occurring naturally in chimpanzees in Gabon. *AIDS* 1989; 3:625–630.
570. Peliska JA, Benkovic SJ. Mechanism of DNA strand transfer reactions catalyzed by HIV-1 reverse transcriptase. *Science* 1992;258:1112–1118.
571. Peterlin BM, Luciw PA, Barr PJ, Walker MD. Elevated levels of mRNA can account for the trans-activation of human immunodeficiency virus. *Proc Natl Acad Sci USA* 1986;83:9734–9738.
572. Peterlin BM, Adams M, Alonso A, Baur A, Ghosh S, Lu X, Luo L. Tat trans-activator. In: Cullen BR, ed. *Human retroviruses.* Oxford: IRL Press; 1993:75–100.
573. Pettit SC, Simsic J, Loeb DD, Everitt L, Hutchison CA, Swanstrom R. Analysis of retroviral protease cleavage sites reveals two types of cleavage sites and the structural requirements of the P1 amino acid. *J Biol Chem* 1991;266:14539–14547.
574. Pincus SH, Cole RL, Hersh E, Lake D, Masuho Y, Durda PJ, McClure J. In vitro efficacy of anti-HIV immunotoxins targeted by various antibodies to the envelope protein. *J Immunol* 1991;146:4315–4324.
575. Pinter A, Honnen WJ, Tilley SA, Bona C, Zaghouani H, Gorny MK, Zolla-Pazner S. Oligomeric structure of gp41, the transmembrane protein of human immunodeficiency virus type 1. *J Virol* 1989;63:2674–2679.
576. Pinter A, Honnen WJ, Tilley SA. Conformational changes affecting the V3 and CD4-binding domains of human immunodeficiency virus type 1 gp120 associated with env processing and with binding of ligands to these sites. *J Virol* 1993;67:5692–5697.
577. Pinto LH, Holsinger LJ, Lamb RA. Influenza virus M2 protein has ion channel activity. *Cell* 1992;69:517–528.
578. Piot P, Laga M. Epidemiology of AIDS in the developing world. In: Broder S, Merigan TC, Bolognesi D, eds. *Textbook of AIDS medicine.* Baltimore: Williams and Wilkins; 1994:109–132.
579. Platt EJ, Haffar OK. Characterization of human immunodeficiency virus type 1 Pr55gag membrane association in a cell-free system: requirement for a C-terminal domain. *Proc Natl Acad Sci USA* 1994;91: 4594–4598.
580. Poiesz BJ, Ruscetti RW, Gazdar AF, Bunn PA, Minna JD, Gallo RC. Detection and isolation of type c retrovirus particles from fresh and cultured lymphocytes of a patient with cutaneous T cell lymphoma. *Proc Natl Acad Sci USA* 1980;77:7415–7418.
581. Poli G, Fauci AS. The effect of cytokines and pharmacologic agents on chronic HIV infection. *AIDS Res Hum Retroviruses* 1992;8: 191–197.
582. Pomerantz RJ, Feinberg MB, Andino R, Baltimore D. The long terminal repeat is not a major determinant of the cellular tropism of human immunodeficiency virus type 1. *J Virol* 1991;65:1414–1419.
583. Pomerantz RJ, Bagasra O, Baltimore D. Cellular latency of human immunodeficiency virus type 1. *Curr Opin Immunol* 1992;4:475–480.
584. Pomerantz RJ, Seshamma T, Trono D. Efficient replication of human immunodeficiency virus type 1 requires a threshold level of rev: potential implications for latency. *J Virol* 1992;66:1809–1813.
585. Poznansky M, Lever A, Bergeron L, Haseltine W, Sodroski J. Gene transfer into human lymphocytes by a defective human immunodeficiency virus type 1 vector. *J Virol* 1991;65:532–536.
585a. Prasad VR, Goff SP. Linker insertion mutagenesis of human immunodeficiency virus reverse transcriptase expressed in bacteria: definition of the minimal polymerase domain. *Proc Natl Acad Sci USA* 1989; 86:3104–3108.

586. Preston BD, Poiez BJ, Loeb L. Fidelity of HIV-1 reverse transcriptase. *Science* 1988;242:1168–1171.
587. Prongay AJ, Smith TJ, Rossman MG, Ehrlich LS, Carter CA, McClure J. Preparation and crystallization of a human immunodeficiency virus p24-Fab complex. *Proc Natl Acad Sci USA* 1990;87:9980–9984.
588. Pruss, D, Bushman FD, Wolffe AP. Human immunodeficiency integrase directs integration to sites of severe DNA distortion within the nucleosome core. *Proc Natl Acad Sci USA* 1994;91:5913–5917.
589. Pruss D, Reeves R, Bushman FD, Wolffe AP. The influence of DNA and nucleosome structure on integration events directed by HIV integrase. *J Biol Chem* 1994;269:25031–25041.
590. Pryciak PA, Varmus HE. Nucleosomes, DNA-binding proteins, and DNA sequence modulate retroviral integration target site selection. *Cell* 1992;69:769–780.
591. Pryciak PM, Sil A, Varmus HE. Retroviral integration into minichromosomes in vitro. *EMBO J* 1992;11:291–303.
592. Pullen KA, Champoux JJ. Plus-strand origin for human immunodeficiency virus type 1: implications for integration. *J Virol* 1990;64:6274–6277.
593. Pullen KA, Ishimoto LK, Champoux JJ. Incomplete removal of the RNA primer for minus-strand DNA synthesis by human immunodeficiency virus type 1 reverse transcriptase. *J Virol* 1992;66:367–373.
594. Quinn TC. Population migration and the spread of types 1 and 2 human immunodeficiency viruses. *Proc Natl Acad Sci USA* 1994;91:2407–2414.
595. Rabson AB, Martin MA. Molecular organization of the AIDS retrovirus. *Cell* 1985;40:477–480.
596. Rando RF, Pellett PE, Luciw PA, Bohan CA, Srinivasan A. Transactivation of human immunodeficiency virus by herpesviruses. *Oncogene* 1987;1:13–18.
597. Rappaport J, Lee S-J, Khalili K, Wong-Staal F. The acidic amino-terminal region of the HIV-1 tat protein constitutes an essential activating domain. *New Biol* 1989;1:101–110.
598. Ratnasabapathy R, Sheldon M, Hohal L, Hernandez N. The HIV-1 long terminal repeat contains an unusual element that induces the synthesis of short RNAs from various mRNA and snRNA promoters. *Genes Dev* 1990;4:2061–2074.
599. Ratner L, Haseltine WA, Patarca R, Livak KJ, Starcich B, Josephs SF, Doran ER, Rafalski JA, Whitehorn EA, Baumeister K. Complete nucleotide sequence of the AIDS virus, HTLV-III. *Nature* 1985;313:277–284.
600. Ratner L. Glucosidase inhibitors for treatment of HIV-1 infection. *AIDS Res Hum Retroviruses* 1992;8:165–173.
601. Renjifo B, Speck NA, Winandy S, Hopkins N, Li Y. Cis-acting elements in the U3 region of a simian immunodeficiency virus. *J Virol* 1990;64:3130–3134.
602. Rey M, Krust B, Laurent AG, Montagnier L, Hovanessian AG. Characterization of human immunodeficiency virus type 2 envelope glycoproteins: dimerization of the glycoprotein precursor during processing. *J Virol* 1989;63:647–658.
603. Rhee SS, Marsh JW. Human immunodeficiency virus type 1 nef-induced down-modulation of CD4 is due to rapid internalization and degradation of surface CD4. *J Virol* 1994;68:5156–5163.
604. Rhim H, Rice AP. TAR RNA binding properties and relative transactivation activities of human immunodeficiency virus type 1 and 2 tat proteins. *J Virol* 1993;67:1110–1121.
605. Richardson JH, Child LA, Lever AML. Packaging of human immunodeficiency virus type 1 RNA requires cis-acting sequences outside the 5′ leader region. *J Virol* 1993;67:3997–4005.
606. Richman DD. Viral resistance to antiretroviral therapy. In: Broder S, Merigan TC, Bolognesi D, eds. *Textbook of AIDS medicine.* Baltimore: Williams and Wilkins; 1994:795–805.
607. Rini JM, Stanfield RL, Stura EA, Salinas PA, Profy AT, Wilson IA. Crystal structure of a human immunodeficiency virus type 1 neutralizing antibody, 50.1, in complex with its V3 loop peptide antigen. *Proc Natl Acad Sci USA* 1993;90:6325–6329.
608. Robert-Guroff M, Popovic M, Gartner S, Markham P, Gallo RC, Reitz MS. Structure and expression of tat-, rev-, and nef-specific transcripts of human immunodeficiency virus type 1 in infected lymphocytes and macrophages. *J Virol* 1990;64:3391–3398.
609. Robert-Guroff M, Louie A, Myagkikh M, Michaels F, Kieny MP, White-Scharf ME, Potts B, Grogg D, Reitz MS. Alteration of V3 loop context within the envelope of human immunodeficiency virus type 1 enhances neutralization. *J Virol* 1994;68:3459–3466.
610. Roberts JD, Bebenek K, Kunkel TA. The accuracy of reverse transcriptase from HIV-1. *Science* 1988;242:1171.
611. Roberts JD, Preston BD, Johnston LA, Soni A, Loeb LA, Kunkel TA. Fidelity of two retroviral reverse transcriptases during DNA-dependent DNA synthesis in vitro. *Mol Cell Biol* 1989;9:469–476.
612. Roberts NA, Martin JA, Kinchington D, et al. Rational design of peptide-based HIV proteinase inhibitors. *Science* 1990;248:358–361.
613. Robey WG, Safai B, Oroszlan S, Arthur LO, Gonda MA, Gallo RC, Fischinger PJ. Characterization of envelope and core structural gene products of HTVL-III with sera from AIDS patients. *Science* 1985;228:593–595.
614. Roe T, Reynolds TC, Yu G, Brown PO. Integration of murine leukemia virus DNA depends on mitosis. *EMBO J* 1993;40:477–480.
615. Roos MT, Lange JM, De Goede RE, Coutinho RA, Schellekens PT, Miedema F, Tersmette M. Viral phenotype and immune response in primary human immunodeficiency virus type 1 infection. *J Infect Dis* 1992;165:427–432.
616. Rosen CA, Sodroski JG, Haseltine WA. The location of cis-acting regulatory sequences in the human T cell lymphotropic virus type III (HTLV-III/LAV) long terminal repeat. *Cell* 1985;41:813–823.
617. Rosen CA, Sodroski JG, Goh WC, Dayton AI, Lippke J, Haseltine WA. Post-transcriptional regulation accounts for the trans-activation of the human T-lymphotropic virus type III. *Nature* 1986;319:555–559.
618. Rosen CA, Terwilliger E, Dayton A, Sodroski JG, Haseltine WA. Intragenic cis-acting art gene-responsive sequences of the human immunodeficiency virus. *Proc Natl Acad Sci USA* 1988;85:2071–2075.
619. Ross EK, Buckler-White AJ, Rabson AB, Englund G, Martin MA. Contribution of NF-κB and Sp1 binding motifs to the replicative capacity of human immunodeficiency virus type 1: distinct patterns of viral growth are determined by T-cell types. *J Virol* 1991;65:4350–4358.
620. Rossi JJ, Sarver N. Catalytic antisense RNA (ribozymes): their potential and use as anti-HIV-1 therapeutic agents. *Adv Exp Med Biol* 1992;312:95–109.
621. Rounseville MP, Kumar A. Binding of a host cell nuclear protein to the stem region of human immunodeficiency virus type 1 *trans*-activation-responsive RNA. *J Virol* 1992;66:1688–1694.
622. Roy S, Parkin NT, Rosen C, Itovitch J, Sonenberg N. Structural requirements for trans activation of human immunodeficiency virus type 1 long terminal repeat-directed gene expression by tat: importance of base pairing, loop sequence, and bulges in the tat-responsive sequence. *J Virol* 1990;64:1402–1406.
623. Roy S, Delling U, Chen C-H, Rosen CA, Sonenberg N. A bulge structure in HIV-1 TAR RNA is required for tat binding and tat-mediated trans-activation. *Genes Dev* 1990;4:1365–1373.
624. Roy S, Agy M, Horvenessian A, Sonenberg N, Katze M. The integrity of the stem structure of human immunodeficiency virus type 1 Tat-responsive sequence RNA is required for interaction with the interferon-induced 68,000Mr protein kinase. *J Virol* 1991;65:632–640.
625. Ryser H, Levy EM, Mandel R, DiSciullo GJ. Inhibition of human immunodeficiency virus infection by agents that interfere with thiol-disulfide interchange upon virus-receptor interaction. *Proc Natl Acad Sci USA* 1994;91:4559–4563.
626. Ryu SE, Kwong PD, Truneh A, Porter TG, Arthos J, Rosenberg M, Dai XP, Xuong NH, Axel R, Sweet RW. Crystal structure of an HIV-binding recombinant fragment of human CD4. *Nature* 1990;348:419–426.
627. Sakaguchi K, Zambrano N, Baldwin ET, Shapiro BA, Erickson JW, Omichinski JG, Clore GM, Gronenborn AM, Appella E. Identification of a binding site for the human immunodeficiency virus type 1 nucleocapsid protein. *Proc Natl Acad Sci USA* 1993;90:5219–5223.
628. Sakai H, Shibata R, Sakuragi J-I, Sakuragi S, Kawamura M, Adachi A. Cell-dependent requirement of human immunodeficiency virus type 1 vif protein for maturation of virus particles. *J Virol* 1993;67:1663–1666.
629. Saksela K, Cheng G, Baltimore D. Proline-rich (PxxP) motifs in HIV-1 Nef bind to SH3 domains of a subset of Src kinases and are required for the enhanced growth of Nef+ viruses but not for down-regulation of CD4. *EMBO J* 1995;14:484–491.
630. Saksela K, Stevens C, Rubenstein P, Baltimore D. Human immunodeficiency virus type 1 mRNA expression in peripheral blood cells predicts disease progression independently of the numbers of CD4+ lymphocytes. *Proc Natl Acad Sci USA* 1994;91:1104–1108.
631. Sakuragi S, Shibata R, Mukai R, Komatsu T, Fukasawa M, Sakai H, Sakuragi J, Kawamura M, Ibuki K, Hayami M. Infection of macaque

monkeys with a chimeric human and simian immunodeficiency virus. *J Gen Virol* 1992;73:2983–2987.
632. Salzwedel K, Johnston PB, Roberts SJ, Dubay JW, Hunter E. Expression and characterization of glycophospholipid-anchored human immunodeficiency virus type 1 envelope glycoproteins. *J Virol* 1993;67: 5279–5288.
633. Samuel KP, Hodge DR, Chen Y-M, Papas T. Nef proteins of the human immunodeficiency viruses (HIV-1 and HIV-2) and simian immunodeficiency virus (SIV) are structurally similar to leucine zipper transcriptional activation factors. *AIDS Res Hum Retroviruses* 1991;7: 697–706.
634. Sanchez-Pescador R, Power MD, Barr PJ, Steimer KS, Stempien MM, Brown-Shimer SL, Gee WW, Renard A, Randolph A, Levy JA, Luciw PA. Nucleotide sequence and expression of an AIDS-associated retrovirus (ARV-2). *Science* 1985;227:484–492.
635. Sanfridson A, Cullen BR, Doyle C. The simian immunodeficiency virus nef protein promotes degradation of CD4 in human T cells. *J Biol Chem* 1994;269:3917–3920.
636. Sastry KJ, Reddy HR, Pandita R, Tatpal K, Aggarawal BB. HIV-1 tat gene induces tumor necrosis factor-beta (lymphotoxin) in a human B-lymphoblastoid cell line. *J Biol Chem* 1990;265:20091–20093.
637. Sato A, Igarashi H, Adachi A, Hayami M. Identification and localization of vpr gene product of human immunodeficiency virus type 1. *Virus Genes* 1990;4:303–312.
638. Sattentau QJ, Clapham PR, Weiss RA, Beverley PCL, Montagnier L, Alhalabi MF, Gluckman JC, Klatzman D. The human and simian immunodeficiency viruses HIV-1, HIV-2, and SIV interact with similar epitopes on their cellular receptor. *AIDS* 1988;2:101–105.
639. Sattentau QJ, Moore JP. Conformational changes in the human immunodeficiency virus envelope glycoproteins by soluble CD4 binding. *J Exp Med* 1991;174:407–415.
640. Sawai ET, Baur A, Struble H, Peterlin BM, Levy JA, Cheng-Mayer C. Human immunodeficiency virus type 1 nef associates with a cellular serine kinase in T lymphocytes. *Proc Natl Acad Sci USA* 1994;91:1539–1543.
641. Sawai ET, Khan I, Cheng-Mayer C, Luciw PA. Simian immunodeficiency virus (SIV) nef interaction with a cellular serine kinase is important for high virus load in infected primates. *[in preparation]* 1995a
642. Sawai ET, Baur AS, Peterlin BM, Levy JA, Cheng-Mayer C. A conserved domain and membrane targeting of nef from HIV and SIV are required for association with a cellular serine kinase. *J Biol Chem [in press]* 1995;270:15307–15314.
643. Scala F, Ruocco MR, Ambrosion C, Mallardo M, Giordano V, Baldassarre E, Dragonetti E, Quinto I, Venuta S. The expression of the interleukin 6 gene is induced by the human immunodeficiency virus TAT protein. *J Exp Med* 1994;179:961–971.
644. Schatz O, Mous J, LeGrice SF. HIV-1 RT-associated ribonuclease H displays both endonuclease and 3′-5′ exonuclease activity. *EMBO J* 1990;9:1171–1176.
645. Schawaller M, Smith GE, Skehel JJ, Wiley DC. Studies with crosslinking reagents on the oligomeric structure of the env glycoprotein of HIV. *Virology* 1989;172:367–369.
646. Schneider J, Kaaden O, Copeland TD, Oroszlan S, Hunsmann G. Shedding and interspecies type seroreactivity of the envelope glycopolypeptide gp120 of the human immunodeficiency virus. *J Gen Virol* 1986;67:2533–2538.
647. Schnittman SM, Psallidopoulos MC, Lane HC, Thompson L, Baseler M, Massari F, Fox CH, Salzman NP, Fauci AS. The reservoir for HIV-1 in human peripheral blood is a T cell that maintains expression of CD4. *Science* 1989;245:305–308.
648. Schooley RT. Immune-based therapies for HIV-1 infection. In: Broder S, Merigan TC, Bolognesi D, eds. *Textbook of AIDS medicine.* Baltimore: Williams and Wilkins; 1994:713–720.
649. Schrager LK, Young JM, Fowler MG, Mathieson BJ, Vermund SH. Long-term survivors of HIV-1 infection: definition and research challenges. *AIDS* 1994;8(suppl. 1):S95–S108.
650. Schroeder HC, Wenger R, Kuchino Y, Muller WEG. Modulation of nuclear matrix-associated 2′,5′-oligoadenylate metabolism and ribonuclease L activity in H9 cells by human immunodeficiency virus. *J Biol Chem* 1989;264:5669–5673.
651. Schubert U, Strebel K. Differential activities of the human immunodeficiency virus type 1-encoded vpu protein are regulated by phosphorylation and occur in different cellular compartments. *J Virol* 1994;68: 2260–2271.
652. Schubert U, Henklein P, Boldyreff B, Wingender E, Strebel K, Portsmann T. The human immunodeficiency virus type 1 encoded vpu protein is phosphorylated by casein kinase-2 (CK-2) at positions Ser52 and Ser56 within a predicted a-helix-turn-a-helix motif. *J Mol Biol* 1994;236:16–25.
653. Schuitemaker H, Koostra NA, Fouchier RAM, Hooibrink B, Miedema F. Productive HIV-1 infection of macrophages restricted to the cell fraction with proliferative capacity. *EMBO J* 1994;13:5929–5936.
654. Schultz AM, Henderson LE, Oroszlan S. Fatty acylation of proteins. *Annu Rev Cell Biol* 1988;4:611–647.
655. Schulz TF, Jameson BA, Lopalco L, Siccardi AG, Weiss RA, Moore JP. Conserved structural features in the interaction between retroviral surface and transmembrane glycoproteins. *AIDS Res Hum Retroviruses* 1992;9:1571–1580.
656. Schulz TF, Reeves JD, Hoad JG, Tailor C, Stephens P, Clements G, Ortlepp S, Page KA, Moore JP, Weiss RA. Effect of mutations in the V3 loop on HIV-1 gp120 on infectivity and susceptibility to proteolytic cleavage. *AIDS Res Hum Retroviruses* 1993;9:159–166.
656a. Schwartz S, Felber BK, Benko DM, Fenyo EM, Pavlakis GM. Cloning and functional analysis of multiply spliced mRNA species of human immunodeficiency virus type 1. *J Virol* 1990;64:2519–2529.
657. Schwartz S, Felber BK, Fenyo EM, Pavlakis GN. Env and vpu proteins of human immunodeficiency virus type 1 are produced from multiple bicistronic mRNAs. *J Virol* 1990;64:5448–5464.
658. Schwartz S, Felber BK, Pavlakis GN. Expression of human immunodeficiency virus type 1 vif and vpr mRNAs is rev-dependent and regulated by splicing. *Virology* 1991;183:677–686.
659. Schwartz O, Arenzana-Seisdedos F, Heard J-M, Danos O. Activation pathways and human immunodeficiency virus type 1 replication are not altered in CD4+ T cells expressing the nef protein. *AIDS Res Hum Retroviruses* 1992;8:545–551.
660. Schwartz S, Felber BK, Pavlakis GN. Distinct RNA sequences in the gag region of human immunodeficiency virus type 1 decrease RNA stability and inhibit expression in the absence of rev protein. *J Virol* 1992;66:150–159.
661. Schwartz S, Campbell M, Nasioulas G, Harrison J, Felber BK, Pavlakis GN. Mutational inactivation of an inhibitory sequence in human immunodeficiency virus type 1 results in rev-independent gag expression. *J Virol* 1992;66:7176–7182.
662. Schwartz S, Felber BK, Pavlakis GN. Mechanism of translation of monocistronic and multicistronic human immunodeficiency virus type 1 mRNAs. *Mol Cell Biol* 1992;12:207–219.
663. Selby MJ, Bain ES, Luciw PA, Peterlin BM. Structure, sequence, and position of the stem-loop in tar determine transcriptional elongation by tat through the HIV-1 long terminal repeat. *Genes Dev* 1989;3:547–558.
664. Selby MJ, Peterlin BM. Trans-activation by HIV-1 tat via a heterologous RNA binding protein. *Cell* 1990;62:769–776.
665. SenGupta DN, Silverman RH. Activation of interferon-regulated, dsRNA-dependent enzymes by HIV-1 leader RNA. *Nucleic Acids Res* 1989;17:969–978.
666. Sharp PA, Marciniak RA. HIV TAR: an RNA enhancer? *Cell* 1989; 59:229–230.
667. Sharpless NE, O'Brien WA, Verdin E, Kufta CV, Chen ISY, Dubois-Dalcq M. Human immunodeficiency virus type 1 tropism for brain microglial cells is determined by a region of the env glycoprotein that also controls macrophage tropism. *J Virol* 1992;66:2588–2593.
668. Sheldon M, Ratnasbapathy R, Hernandez N. Characterization of the inducer of short transcripts, a human immunodeficiency virus type 1 transcriptional element that activates the synthesis of short RNAs. *Mol Cell Biol* 1993;13:1251–1263.
669. Sheline CT, Milocco LH, Jones KA. Two distinct transcription factors recognize loop and bulge residues of the HIV-1 TAR RNA hairpin. *Genes Dev* 1991;5:2508–2520.
670. Sheridan PL, Sheline CT, Miloco SH, Jones KA. Tat and the HIV-1 promoter: a model for RNA-mediated regulation of transcription. *Semin Virol* 1993;4:69–80.
671. Sherman PA, Fyfe JA. Human immunodeficiency virus integration protein expressed in *Escherichia coli* possesses selective DNA cleaving ability. *Proc Natl Acad Sci USA* 1990;87:5119–5123.
672. Sherman P, Dickson ML, Fyfe JA. Human immunodeficiency virus type 1 integration protein: DNA sequence requirements for cleaving and joining reactions. *J Virol* 1992;66:3693–3602.
673. Shibata R, Miura T, Hayami M, Ogawa K, Sakai H, Kiyomasu T, Ishimoto A, Adachi A. Mutational analysis of the human immunodeficiency virus type 2 (HIV-2) genome in relation to HIV-1 and simian immunodeficiency virus SIVagm. *J Virol* 1990;64:742–747.

674. Shibata R, Kawamura M, Sakai H, Hayami M, Ishimoto A, Adachi A. Generation of a chimeric human and simian immunodeficiency virus infectious to monkey peripheral blood mononuclear cells. *J Virol* 1991; 65:3514–3520.
675. Shih CC, Stoye JP, Coffin JM. Highly preferred targets for retrovirus integration. *Cell* 1988;53:531–537.
676. Shioda T, Levy JA, Cheng-Mayer C. Small amino acid changes in the V3 hypervariable region of gp120 can affect the T-cell-line and macrophage tropism of human immunodeficiency virus type 1. *Proc Natl Acad Sci USA* 1992;89:9434–9438.
677. Shirazi Y, Pitha PM. Alpha interferon inhibits early stages of the human immunodeficiency virus type 1 replication cycle. *J Virol* 1992;66:1321–1328.
678. Shugars DC, Smith MS, Glueck DG, Nantermet PV, Seillier-Moiseiwitsch F, Swanstrom R. Analysis of human immunodeficiency virus type 1 nef gene sequences present in vivo. J Virol 1993;67:4639–4650.
679. Siekevitz M, Josephs SF, Dukovich M, Peffer N, Wong-Staal F, Greene WC. Activation of the HIV-1 LTR by T cell mitogens and the trans-activator protein of HTLV-I. *Science* 1987;238:1575–1578.
680. Simon JHM, Somoza C, Schockmel GA, Collin M, Davis SJ, Williams AF, James W. A rat CD4 mutant containing the gp120-binding site mediates human immunodeficiency virus type 1 infection. *J Exp Med* 1993;177:949–954.
681. Siomi H, Shida H, Maki M, Hatanaka M. Effects of a highly basic region of human immunodeficiency virus Tat protein on nucleolar localization. *J Virol* 1990;64:1803–1807.
682. Skalka AM, Goff SP. *Reverse transcriptase.* Cold Spring Harbor, NY: Cold Spring Harbor Laboratory Press; 1993:492.
683. Skinner MA, Langlois AJ, McDanal CB, McDougal JS, Bolognesi DP, Matthews TJ. Neutralizing antibodies to an immunodominant envelope sequence do not prevent gp120 binding to CD4. *J Virol* 1988;62: 4195–4200.
684. Skowronski J, Parks D, Mariani R. Altered T cell activation and development in transgenic mice expressing the HIV-1 nef gene. *EMBO J* 1993;12:703–713.
685. Skripkin E, Paillart J-C, Marquet R, Ehresmann B, Ehresmann C. Identification of the primary site of the human immunodeficiency virus type 1 RNA dimerization in vitro. *Proc Natl Acad Sci USA* 1994;91:4945–4949.
686. Smerdon SJ, Jager J, Wang J, Kohlstaedt LA, Chirino AJ, Friedman JM, Rice PA, Steitz TA. Structure of the binding site for nonnucleoside inhibitors of the reverse transcriptase of human immunodeficiency virus type 1. *Proc Natl Acad Sci USA* 1994;91:3911–3915.
687. Smith AJ, Srinivasakumar N, Hammarskjold ML, Rekosh D. Requirements for incorporation of Pr160gag-pol from human immunodeficiency virus type 1 into virus-like particles. *J Virol* 1993;67:2266–2275.
688. Smith DH, Byrn RA, Marsters MA, Gregory T, Groopman JA, Capon DJ. Blocking of HIV-1 infectivity by a soluble, secreted form of the CD4 antigen. *Science* 1987;238:1704–1707.
689. Smith JS, Kim S, Roth MJ. Analysis of long terminal repeat circle junctions of human immunodeficiency virus type 1. *J Virol* 1990;64:6268–6290.
690. Sodroski JG, Rosen CA, Haseltine WA. Trans-cting transcriptional activation of the long terminal repeat of human T-lymphotropic virus in infected cells. *Science* 1984;225:381–385.
691. Sodroski JG, Patarca R, Rosen CA, Haseltine WA. Location of the trans-cting region on the genome of human T-cell lymphotropic virus type III. *Science* 1985;229:74–77.
692. Sodroski JG, Goh WC, Rosen CA, Dayton A, Terwilliger E, Haseltine WA. A second post-transcriptional trans-activator gene required for HTLV-III replication. *Nature* 1986;321:412–417.
693. Sodroski JG, Goh WC, Rosen CA, Tartar A, Portetelle D, Burny A, Haseltine WA. Replication and cytopathic potential of HTLV-III/LAV with sor gene deletions. *Science* 1986;231:1549–1553.
694. Sonigo P, Alizon M, Staskus K, et al. Nucleotide sequence of the visna lentivirus: relationship to the AIDS virus. *Cell* 1985;42:369–382.
695. South TL, Blake PR, Sowdwe RCI, Arthur LO, Henderson LE, Summers MF. The nucleocapsid protein isolated from HIV-1 particles binds zinc and forms retroviral-type zinc fingers. *Biochemistry* 1990;29:7786–7789.
696. Southgate C, Zapp M, Green M. Activation of transcription by HIV-1 tat protein tethered to nascent RNA through another protein. *Nature* 1990;345:640–642.
697. Sova p, Volsky DJ. Efficiency of viral DNA synthesis during infection of permissive and nonpermissive cells with vif-negative human immunodefiency virus type 1. *J Virol* 1993;67:6322–6326.
698. Spina CA, Kwoh TJ, Chowers MY, Guatelli JC, Richman DD. The importance of nef in the induction of human immunodeficiency virus type 1 replication from primary quiescent CD4 lymphocytes. *J Exp Med* 1994;179:115–123.
699. Srinivas SK, Srinivas RV, Anantharamaiah GM, Segrest JP. Membrane interactions of synthetic peptides corresponding to amphipathic helical segments of the human immunodeficiency virus type-1 envelope glycoprotein. *J Biol Chem* 1992;267:7121–7127.
700. Staal FJ, Roederer M, Herzenberg LA, Herzenberg LA. Intracellular thiols regulate activation of nuclear factor kappa B and transcription of human immunodeficiency virus. *Proc Natl Acad Sci USA* 1990;87: 9943–9947.
701. Stanley SK, Bressler PB, Poli G, Fauci AS. Heat shock induction of HIV production from chronically infected promonocytic and T cell lines. *J Immunol* 1990;145:1120–1126.
702. Starcich BR, Hahn BH, Shaw GM, McNeely PD, Modrow S, Wolf H, Parks ES, Parks WP, Josephs SF, Gallo RC, Wong-Staal F. Identification and characterization of conserved and variable regions in the envelope gene of HTLV-III/LAV, the retrovirus of AIDS. *Cell* 1986; 45:637–648.
703. Staskus KA, Embretson JE, Retzel EF, Beneke J, Haase AT. PCR in situ: new technologies with single-cell resolution for the detection and investigation of viral latency and persistence. In: Clewley JP, ed. *The polymerase chain reaction (PCR) for human viral diagnosis.* Boca Raton: CRC Press; 1995:23–39.
704. Steimer KS, Klasse PJ, McKeating JA. HIV-1 neutralization directed to epitopes other than linear V3 determinants. *AIDS* 1991;5S:135–143.
705. Stein BS, Gowda SD, Lifson JD, Penhallow RC, Bensch KG, Engleman EG. pH-independent HIV entry into CD4-positive T cells via virus envelope fusion to the plasma membrane. *Cell* 1987;49:659–668.
706. Stein BS, Engelman EG. Intracellular processing of the gp160 HIV-1 envelope precursor. Endoproteolytic cleavage occurs in a cis or medial component of the Golgi complex. *J Biol Chem* 1990;265:2640–2649.
707. Stevens SW, Griffith JD. Human immunodeficiency virus type 1 may preferentially integrate into chromatin occupied by L1Hs repetitive elements. *Proc Natl Acad Sci USA* 1994;91:5557–5561.
708. Stevenson M, Meier C, Mann A, Chapman N, Wasiak A. Envelope glycoprotein of HIV induces interference and cytolysis resistance in CD4+ cells: mechanism for persistence in AIDS. *Cell* 1988;53: 483–496.
709. Stevenson M, Haggerty S, Lamonica CA, Meier CM, Welch S-K, Wasiak AJ. Integration is not necessary for expression of human immunodeficiency virus type 1 protein products. *J Virol* 1990;64:2421–2425.
710. Stevenson M, Stanwick TL, Dempsey MP, Lamonica CA. HIV-1 replication is controlled at the level of T-cell activation and proviral integration. *EMBO J* 1990;9:1551–1560.
711. Stevenson M, Gendelman HE. Cellular and viral determinants that regulate HIV-1 infection in macrophages. *J Leuk Biol* 1994;56:278–288.
712. Stomatatos L, Cheng-Mayer C. Evidence that the structural conformation of envelope gp120 affects human immunodeficiency virus type 1 infectivity, host range, and syncytium-forming ability. *J Virol* 1993; 67:5635–5639.
713. Strebel K, Daugherty D, Cohen D, Folks T, Martin M. The HIV "A" (sor) gene product is essential for virus infectivity. *Nature* 1987;328: 728–730.
714. Strebel K, Klimkait T, Martin MA. A novel gene of HIV-1, vpu, and its 16 kilodalton product. *Science* 1988;241:1221–1223.
715. Strebel K, Klimkait T, Maldarelli F, Martin MA. Molecular and biochemical analyses of human immunodeficiency virus type 1 vpu protein. *J Virol* 1989;63:3784–3791.
716. Subramanian T, Kuppuswamy M, Venkatesh L, Srinivasan A, Chinnadurai G. Functional substitution of the basic domain of the HIV-1 trans-activator, tat, with the basic domain of the functionally heterologous rev. *Virology* 1990;176:178–183.
717. Sullenberger BA, Gallardo HF, Ungers GE, Gilboa E. Overexpression of TAR sequences renders cells resistant to HIV replication. *Cell* 1990; 63:601–608.
718. Sullivan N, Thali M, Furman C, Ho DD, Sodroski J. Effect of amino acid changes in the V1/V2 region of the human immunodeficiency virus type 1 gp120 glycoprotein on subunit association, syncytium formation, and recognition by a neutralizing antibody. *J Virol* 1993;67: 3674–3679.
719. Summers MF, Henderson LE, Chance MR, Bess JW, South TL, Blake

PR, Sagi I, Perez-Alvarado G, Sowder RC, Hare DR, Arthur LO. Nucleocapsid zinc fingers detected in retroviruses: EXAFS studies of intact viruses and the solution-state structure of the nucleocapsid protein from HIV-1. *Protein Sci* 1992;1:563–574.

720. Sun X-H, Baltimore D. Human immunodeficiency virus tat-activated expression of poliovirus protein 2A inhibits mRNA translation. *Proc Natl Acad Sci USA* 1989;86:2143–2146.
721. Surovoy A, Dannuli J, Moelling K, Jung G. Conformational and nucleic acid binding studies on the synthetic nucleocapsid protein of HIV-1. *J Mol Biol* 1993;229:94–104.
722. Syu W-J, Lee W-R, Du B, Yu Q-C, Essex M, Lee T-H. Role of the conserved gp41 cysteine residues in the processing of human immunodeficiency virus envelope precursor and viral infectivity. *J Virol* 1991; 65:6349–6352.
723. Tada HJ, Rappaport J, Lashgary M, Amini S, Wong-Staal F, Khalili K. Trans-activation of the JC virus late promoter by the tat protein of type 1 human immunodeficiency virus in glial cells. *Proc Natl Acad Sci USA* 1990;87:3479.
724. Takahashi H, Cohen J, Hosmalin A, Cease KB, Houghten R, Cornette JC, DeLisi C, Moss B, Germain RN, Berzofsky JA. An immunodominant epitope of the human immunodeficiency virus envelope glycoprotein gp160 recognized by class I major histocompatibility complex molecule-restricted murine cytotoxic T lymphocytes. *Proc Natl Acad Sci USA* 1988;85:3105–3109.
725. Takeda A, Tuazon CU, Ennis FA. Two receptors are required for antibody-dependent enhancement of human immunodeficiency virus the 1 infection. *J Virol* 1990;64:5605.
726. Takeuchi Y, Akutsu M, Murayama K, Shimizu N, Hoshino H. Host-range mutant of human immunodeficiency virus type 1: modification of cell tropism by a single point mutation at the neutralization epitope in the env gene. *J Virol* 1991;66:757–765.
727. Tan ST, Pearce-Pratt R, Phillips DM. Productive infection of a cervical epithelial cell line with human immunodeficiency virus: implications for sexual transmission. *J Virol* 1993;67:6447–6452.
728. Temin HM. The DNA provirus hypothesis. *Science* 1976;192:1075–1080.
729. Tersmette M, Gruters RA, DeWolf F, DeGoede REY, Lange JMA, Schellekens PTA, Goudsmit J, Huisman HG, Miedema F. Evidence for a role of virulent human immunodeficiency virus (HIV) variants in the pathogenesis of acquired immunodeficiency syndrome: studies on sequential HIV isolates. *J Virol* 1989;63:2118–2125.
730. Terwilliger E, Cohen EA, Lu YC, Sodroski JG, Haseltine WA. Functional role of human immunodeficiency virus type 1 vpu. *Proc Natl Acad Sci USA* 1989;86:5163–5167.
731. Terwilliger E, Langhoff E, Gabuzda D, Haseltine WA. Allelic variation in the effects of the nef gene on replication of HIV-1. *Proc Natl Acad Sci USA* 1991;88:10971–10975.
732. Testa NG, Molineux G. *Haemopoiesis—a practical approach.* Oxford: IRL Press; 1993:293.
733. Thali M, Furman C, Ho DD, Robinson J, Tilley S, Pinter A, Sodroski J. Discontinuous, conserved neutralization epitopes overlapping the CD4-binding region of human immunodeficiency virus type 1 gp120 envelope glycoprotein. *J Virol* 1992;66:5635–5641.
734. Thali M, Fur,man C, Helseth E, Repke H, Sodroski J. Lack of correlation between soluble CD$-induced shedding of the human immunodeficiency virus type 1 exterior envelope glycoprotein and subsequent membrane fusion events. *J Virol* 1992;66:5516–5524.
735. Thali M, Moore JP, Furman C, Charles M, Ho DD, Robinson J, Sodroski J. Characterization of conserved human immunodeficiency virus type 1 gp120 neutralization epitopes exposed upon gp120-CD4 binding. *J Virol* 1993;67:3978–3988.
736. Thali M, Bukovsky A, Kondo E, Rosenwirth B, Walsh CT, Sodroski J, Gottlinger HG. Functional association of cyclophilin A with HIV-1 virions. *Nature* 1994;372:363–365.
737. Tiley LS, Brown PH, Cullen BR. Does the human immunodeficiency virus tat trans-activator contain a discrete activation domain? *Virology* 1990;178:560–567.
738. Tong L, Pav S, Pargellis C, Do F, Lamarre D, Anderson PC. Crystal structure of human immunodeficiency virus type 2 protease in complex with a reduced amide inhibitor and comparison with HIV-1 protease structures. *Proc Natl Acad Sci USA* 1993;90:8387–8391.
739. Tong-Starksen S, Luciw PA, Peterlin BM. Human immunodeficiency virus long terminal repeat responds to T-cell activation signals. *Proc Natl Acad Sci USA* 1987;84:6845–6849.
740. Tornesello MC, Buonaguro FM, Beth-Giraldo E, Giraldo G. Human immunodeficiency virus type 1 tat gene enhances human papillomavirus early gene expression. *Intervirology* 1993;36:57–64.
741. Torres JV, Malley A, Banapour B, Anderson DE, Axthelm MK, Gardner MB, Benjamini E. An epitope on the surface envelope glycoprotein (gp130) of simian immunodeficiency virus (SIV_{mac}) involved in viral neutralization and T cell activation. *AIDS Res Hum Retroviruses* 1993;9:423–430.
742. Tristem M, Marshall C, Karpas A, Petrik J, Hill F. Origin of vpx in lentiviruses. *Nature* 1990;347:341–342.
743. *Tristem M, Marshall C, Karpas A, Hill F. Evolution of the primate lentiviruses: evidence from vpx and vpr. *EMBO J* 1992;11:3405–3412.
744. Trono D, Feinberg MB, Baltimore D. Gag mutants can dominantly interfere with the replication of the wild-type virus. *Cell* 1989;59:113–120.
745. Trono D, Baltimore D. A human cell factor is essential for HIV-1 rev action. *EMBO J* 1990;9:4155–4160.
746. Trono D. Partial reverse transcripts in virions from human immunodeficiency and murine leukemia viruses. *J Virol* 1992;66:4893–4900.
747. Truneh A, Buck D, Cassatt R, Juszczak R, Kassis S, Ryu S-E, et al. A region in domain 1 of CD4 distinct from the primary gp120 binding site is involved in HIV infection and virus mediated fusion. *J Biol Chem* 1991;266:5942–5948.
748. Tsunetsugu-Yokota Y, Matsuda S, Maekawa M, Saito T, Takemori T, Takebe Y. Constitutive expression of the nef gene suppresses human immunodeficiency virus type 1 (HIV-1) replication in monocytic cell lines. *Virology* 1992;191:960–963.
749. Turner S, Tizard R, DeMarinis J, Pepinsky RB, Zullo J, Schooley R, Fisher R. Resistance of primary isolates of human immunodeficiency virus type 1 to neutralization by soluble CD4 is not due to lower affinity with the viral envelope glycoprotein gp120. *Proc Natl Acad Sci USA* 1992;89:1335–1339.
750. Unger RE, Marthas ML, Pratt-Lowe E, Padrid PA, Luciw PA. The nef gene of simian immunodeficiency virus SIVmac1A11. *J Virol* 1992; 66:5432–5442.
751. Vaca JP. Design of tight-binding human immunodeficiency virus type 1 protease inhibitors. In: Kuo LC, Shafer JA, eds. *Retroviral proteases.* San Diego: Academic Press; 1994:311–334.
752. Vaishnav YN, Vaishnav M, Wong-Staal F. Identification and characterization of a nuclear factor that specifically binds to the rev-response element (RRE) of human immunodeficiency virus type 1 (HIV-1). *New Biol* 1991;3:142–150.
753. Valerie K, Rosenberg M. Chromatin structure implicated in activation of HIV-1 gene expression by ultraviolet light. *New Biol* 1990;2:712–718.
754. van Gent DC, Elgersma Y, Bolk MWJ, Vink C, Plasterk RHA. DNA binding properties of the integrase proteins of human immunodeficiency virus types 1 and 2. *Nucleic Acids Res* 1991;19:3821–3827.
755. Varmus H, Brown P. Retroviruses. In: Berg DE, Howe MM, eds. *Mobile DNA.* Washington, DC: American Society for Microbiology; 1989: 53–108.
756. Vartanian J-P, Meyerhans A, Asjo B, Wain-Hobson S. Selection, recombination, and G to A hypermutation of human immunodeficiency virus type 1 genomes. *J Virol* 1991;65:1779–1788.
757. Vartanian J-P, Meyerhans A, Sala M, Wain-Hobson S. G to A hypermutation of the human immunodeficiency virus type 1 genome: evidence for dCTP pool imbalance during reverse transcription. *Proc Natl Acad Sci USA* 1994;91:3092–3096.
758. Vella S. Update on a proteinase inhibitor. *AIDS* 1994;8(suppl.3): S25–S29.
759. Venable RM, Pastor RW, Brooks BR, Carson FW. Theoretically determined three-dimensional structures for amphipathic segments of HIV-1 gp41 envelope protein. *AIDS Res Hum Retroviruses* 1989;5:7–22.
760. Venkatesh LK, Chinnadurai G. Mutants in a conserved region near the carboxy-terminus of HIV-1 rev identify functionally important residues and exhibit a dominant negative phenotype. *Virology* 1990;178:327–330.
761. Venkatesh LK, Mohammed S, Chinnadurai G. Functional domains of the HIV-1 rev gene required for trans-regulation and subcellular localization. *Virology* 1990;176:39–47.
762. Verdin E, Paras JP, Van Lint C. Chromatin disruption in the promoter of human immunodeficiency virus type 1 during transcriptional activation. *EMBO J* 1993;12:3249–3259.
763. Veronese FD, DeVico AL, Copeland TD, Oroszlan S, Gallo RC, Sarngadharan MG. Characterization of gp41 as the transmembrane protein coded by the HTLV-III/LAV envelope gene. *Science* 1985;229: 1402–1405.

764. Vincent KA, Ellison V, Chow SA, Brown PO. Characterization of human immunodeficiency virus type 1 integrase expressed in *Escherichia coli* and analysis of variants with amino-terminal mutations. *J Virol* 1993;67:425–437.
765. Vink C, Yeheskiely E, van der Marel GA, van Boom JH, Plasterk RHA. Site-specific hydrolysis and alcoholysis of human immunodeficiency virus DNA termini mediated by the viral integrase protein. *Nucleic Acids Res* 1991;19:6691–6698.
766. Vink C, van Gent DC, Elgersma Y, Plasterk RHA. Human immunodeficiency virus integrase protein requires a subterminal position of its viral DNA recognition sequence for efficient cleavage. *J Virol* 1991; 65:4636–4644.
767. Viscidi RP, Mayur K, Lederman HM, Frankel AD. Inhibition of antigen-induced lymphocyte proliferation by tat protein from HIV-1. *Science* 1989;246:1606–1608.
768. Vogel J, Hinrichs SH, Reynolds RK, Luciw PA, Jay G. The HIV tat gene induces dermal lesions resembling Kaposi's sarcoma in transgenic mice. *Nature* 1988;335:606–611.
769. von Poblotzki A, Wagner R, Niedrig M, Wanner G, Wolf H, Modrow S. Identification of a region in the Pr55gag-polyprotein essential for HIV-1 particle formation. *Virology* 1993;193:981–985.
770. von Schwedler U, Song J, Aiken C, Trono D. Vif is crucial for human immunodeficiency virus type 1 proviral DNA synthesis in infected cells. *J Virol* 1993;67:4945–4955.
771. Wain-Hobson S, Sonigo P, Danos O, Cole S, Alizon M. Nucleotide sequence of the AIDS virus, LAV. *Cell* 1985;40:9–17.
772. Wain-Hobson S, Vartanian JP, Henry M, Chenciner N, Cheynier R, Delassus S, Martins LP, Sala M, Nugeyre MT, Guetard D. LAV revisited: origins of the early HIV-1 isolates from Institut Pasteur. *Science* 1991;252:961–965.
773. Wain-Hobson S. Human immunodeficiency virus type 1 quasispecies in vivo and ex vivo. *Curr Top Microbiol Immunol* 1992;176:181–193.
774. Wainberg MA, Dascal A, Blain N, Fitz-Gibbon L, Boulerice F, Numazaki K, Tremblay M. The effect of cyclosporine A on infection of susceptible cells by human immunodeficiency virus type 1. *Blood* 1988; 72:1904–1910.
775. Wang JH, Yan YW, Garrett TP, Liu JH, Rodgers DW, Garlick RL, Tarr GE, Husain Y, Reinherz EL, Harrison SC. Atomic structure of a fragment of human CD4 containing two immunoglobulin-like domains. *Nature* 1990;348:411–418.
776. Waterman ML, Jones KA. Purification of TCF-1 alpha, a T-cell-specific transcription factor that activates the T-cell receptor C alpha gene enhancer in a context-dependent manner. *New Biol* 1990;2:621–636.
777. Waterman ML, Fischer WH, Jones KA. A thymus-specific member of the HMG protein family regulates the human T cell receptor C alpha enhancer. *Genes Dev* 1991;5:656–659.
778. Weeks KM, Ampe C, Schultz SC, Steitz TA, Crothers DM. Fragments of the HIV-1 tat protein specifically bind TAR RNA. *Science* 1990; 249:1281–1285.
779. Wei A, Ghosh SK, Taylor MS, Johnson VA, Emini EA, Deutsch P, Lifson JD, Bonhoeffer S, Nowak MA, Hahn BA, Saag MS, Shaw GM. Viral dynamics in human immunodeficiency virus type 1 infection. *Nature* 1995;373:117–122.
780. Weichselbraun I, Farrington K, Rusche JR, Bohnlein E, Hauber J. Definition of the human immunodeficiency virus type 1 rev and human T-cell leukemia virus type 1 rex protein activation domain by functional exchange. *J Virol* 1992;66:2583–2587.
781. Weiss CD, White JM. Characterization of stable Chinese hamster ovary cells expressing wild-type, secreted, and glycosylphosphatidylinositol-anchored human immunodeficiency virus type 1 envelope glycoprotein. *J Virol* 1993;67:7060–7066.
782. Weiss CD, Levy JA, White JM. Oligomeric structure of gp120 on infectious human immunodeficiency virus type 1 particles. *J Virol* 1990; 64:5674–5677.
783. Weiss S, König B, Morikawa Y, Jones I. Recombinant HIV-1 nucleocapsid protein p15 produced as a fusion protein with glutathione S-transferase in *Escherichia coli* mediates dimerization and enhances reverse transcription of retroviral RNA. *Gene* 1992;121:203–212.
784. Weiss S, König B, Müller H-J, Seidel H, Goody RS. Synthetic human tRNAlys3 and natural bovine tRNAlys3 interact with HIV-1 reverse transcriptase and serve as specific primers for retroviral cDNA synthesis. *Gene* 1992;111:183–197.
785. Weiss RA. Cellular receptors and viral glycoproteins involved in retrovirus entry. In: Levy JA, ed. *The retroviridae.* New York: Plenum Press; 1993:1–108.
786. Westendorp MO, Li-Weber M, Frank RW, Krammer PH. Human immunodeficiency virus type 1 tat upregulates interleukin-2 secretion in activated T cells. *J Virol* 1994;68:4177–4185.
787. Westervelt P, Trowbridge DB, Epstein LG, Li Y, Hahn BH, Shaw GM, Price RW, Ratner L. Macrophage tropism determinants of human immunodeficiency virus type 1 in vivo. *J Virol* 1992;66:2577–2582.
788. Whitcomb JM, Kumar R, Hughes SH. Sequence of the circle junction of human immunodeficiency virus type 1: implications for reverse transcription and integration. *J Virol* 1990;64:4903–4906.
789. Whitcomb JM, Hughes SM. Retroviral reverse transcription and integration: progress and problems. *Annu Rev Cell Biol* 1992;8:275–306.
790. White JM. Membrane fusion. *Science* 1992;258:917–924.
791. Whitton JL. Antisense treatment of viral infection. *Adv Virus Res* 1994; 44:267–303.
792. WHO Network for HIV Isolation and Characterization. HIV type 1 variation in World Health Organization-sponsored vaccine evaluation sites: genetic screening, sequence analysis, and preliminary biological characterization of selected viral strains. *AIDS Res Hum Retroviruses* 1994;10:1327–1343.
793. Wild C, Oas T, McDanal C, Bolognesi D, Matthews T. A synthetic peptide inhibitor of human immunodeficiency virus replication: correlation between solution structure and viral inhibition. *Proc Natl Acad Sci USA* 1992;89:10537–10541.
794. Wild C, Dubay JW, Greenwall T, Baird T, Oas TG, McDanal C, Hunter E, Matthews T. Propensity for a leucine zipper-like domain of human immunodeficiency virus type 1 gp41 to form oligomers correlates with a role in virus-induced fusion rather than assembly of the glycoprotein complex. *Proc Natl Acad Sci USA* 1994;91:12676–12680.
795. Willey RL, Bonifacino JS, Potts BJ, Martin MA, Klausner RD. Biosynthesis, cleavage, and degradation of the human immunodeficiency virus 1 envelope glycoprotein gp160. *Proc Natl Acad Sci USA* 1988; 85:9580–9584.
796. Willey RL, Maldarellli F, Martin MA, Strebel K. Human immunodeficiency virus type 1 vpu protein induces rapid degradation of CD4. *J Virol* 1992;66:7193–7200.
797. Willey RL, Martin MA. Association of human immunodeficiency virus type 1 envelope glycoprotein with particles depends on interactions between the third variable and conserved regions of gp120. *J Virol* 1993; 67:3639–3643.
798. Willey RL, Theodore TS, Martin MA. Amino acid substitutions in the human immunodeficiency virus type 1 gp120 V3 loop that change viral tropism also alter physical and functional properties of the virion envelope. *J Virol* 1994;68:4409–4419.
799. Willey RL, Buckler-White A, Strebel K. Sequences present in the cytoplasmic domain of CD4 are necessary and sufficient to confer sensitivity to the human immunodeficiency virus type 1 vpu protein. *J Virol* 1994;68:1207–1212.
800. Williams KJ, Loeb LA. Retroviral reverse transcriptases: error frequencies and mutagenesis. *Curr Top Microbiol Immunol* 1992;176:165–180.
801. Wills NM, Craven RC. Form, function, and use of retroviral gag proteins. *AIDS* 1991;5:639–654.
802. Wiskerchen M, Muesing MA. Human immunodeficiency virus type 1 integrase: effects of mutations on viral ability to integrate, direct viral gene expression from unintegrataed viral DNA templates, and sustain viral propagation in primary cells. *J Virol* 1995;69:376–386.
803. Wlodawer A, Erickson JW. Structure-based inhibitors of HIV-1 protease. *Annu Rev Biochem* 1993;62:543–585.
804. Wlodawer A, Miller M, Jaskolski M, Sathyanarayana BK, Baldwin E, Weber IT, Selk LM, Clawson L, Schneider J, Kent SBH. Conserved folding in retroviral proteases: crystal structure of a synthetic HIV-1 protease. *Science* 1989;245:616–621.
805. Wolber V, Rensland H, Brandmeier B, Sagemann M, Hoffmann R, Kalbitzer HR, Wittingshofer A. Expression, purification and biochemical characterisation of the human immunodeficiency virus 1 nef gene product. *Eur J Biochem* 1992;205:1115–1121.
806. Wong-Staal F, Chanda PK, Ghrayeb J. Human immunodeficiency virus: the eighth gene. *AIDS Res Hum Retroviruses* 1987;3:33–39.
807. Wong-Staal F. Cellular factors in HIV transactivation. In: Koff WC, Kennedy RC, Wong-Staal F, eds.*AIDS research reviews.* New York: Marcel Dekker; 1992:29–40.
808. Wu FK, Garcia JA, Harrich D, Gaynor RB. Purification of the human immunodeficiency virus type 1 enhancer and TAR binding proteins, EBP-1 and UBP-1. *EMBO J* 1988;7:2117–2129.

809. Wu FK, Garcia JA, Sigman D, Gaynor RB. Tat regulates binding of the human immunodeficiency virus trans-activating region RNA loop-binding protein. *Genes Dev* 1991;5:2128–2140.
810. Wyatt R, Thali M, Tilley S, Pinter A, Posner M, Ho D, Robinson J, Sodroski J. Relationship of the human immunodeficiency virus type 1 gp120 third variable loop to a component of the CD4 binding site in the fourth conserved region. *J Virol* 1992;66:6997–7004.
811. Yahi N, Baghdiguian S, Moreau H, Fantini J. Galactosyl ceramide (or a closely related molecule) is the recptor human immunodeficiency virus type 1 on human colon epithelial HT29 cells. *J Virol* 1992;66: 4848–4854.
812. Yahi N, Fantini J, Mabrouk K, Tamalet C, De Micco P, Van Rietschoten J, Rochat H, Sabatier JM. Multibranched V3 peptides inhibit human immunodeficiency virus type 1 infection in human lymphocytes and macrophages. *J Virol* 1994;68:5714–5720.
813. Yamada O, Kraus G, Leavitt MC, Yu N, Wong-Staal F. Activity and cleavage site specificity of an anti-HIV-1 hairpin ribozyme in human T cells. *Virology* 1994;205:121–126.
814. Yao XJ, Gottlinger H, Haseltine WA, Cohen EA. Envelope glycoprotein and CD4 independence of vpu-facilitated human immunodeficiency virus type 1 capsid export. *J Virol* 1992;66:5119–5126.
815. Yoon J-B, Li G, Roeder RG. Characterization of a family of related cellular transcription factors which can modulate human immunodeficiency virus type 1 transcription in vitro. *Mol Cell Biol* 1994;14: 1776–1785.
816. Yoshida M, Miyoshi I, Hinuma Y. Isolation and characterization of retrovirus from cell lines of human adult T-cell leukemia and its implication in the disease. *Proc Natl Acad Sci USA* 1982;79:2031–2035.
817. Yu XF, Ito S, Essex M, Lee TH. A naturally immunogenic virion-associated protein specific for HIV-2 and SIV. *Nature* 1988;335:262–265.
818. Yu XF, Matsuda M, Essex M, Lee TH. Open reading frame vpr of simian immunodeficiency virus encodes a virion-associated protein. *J Virol* 1990;64:5688–5693.
819. Yu XF, Yu QC, Essex M, Lee TH. The vpx gene of simian immunodeficiency virus facilitates efficiency viral replication in fresh lymphocytes and macrophages. *J Virol* 1991;65:5088–5091.
820. Yu XF, Yuan X, Matsuda Z, Lee TH, Essex M. The matrix protein of human immunodeficiency virus type 1 is required for incorporation of viral envelope protein into mature virions. *J Virol* 1992;66:4966–4971.
821. Yu G, Felsted RL. Effect of myristoylation on p27nef subcellular distribution and suppression of HIV-LTR transcription. *Virology* 1992; 187:46–55.
822. Yu XF, Yuan X, McLane MF, Lee TH, Essex M. Mutations in the cytoplasmic domain of human immunodeficiency virus type 1 transmembrane protein impair the incorporation of env proteins into mature virions. J Virol 1993;67:213–221.
823. Yu OC, Matsuda Z, Yu X, Ito S, Essex M, Lee T-H. An electron-lucent region within the virion distinguishes HIV-1 from HIV-2 and simian immunodeficiency viruses. *AIDS Res Hum Retroviruses* 1994;10: 757–761.
824. Yuan X, Matsuda M, Essex M, Lee TH. Human immunodeficiency virus vpr gene encodes a virion-associated protein. *AIDS Res Hum Retroviruses* 1990;6:1265–1271.
825. Zack JA, Arrigo SJ. HIV-1 entry into primary lymphocytes: molecular analysis reveals a labile, latent viral structure. *Cell* 1990;61:213–222.
825a. Zapp ML, Green MR. Sequence-specific RNA binding by the HIV-1 rev protein. *Nature* 1989;342:714–716.
826. Zazopoulos E, Haseltine WA. Disulfide bond formation in the human immunodeficiency virus type 1 nef protein. *J Virol* 1993;67:1676–1680.
827. Zazopoulos E, Haseltine WA. Effect of nef alleles on replication of human immunodeficiency virus type 1. *Virology* 1993;194:20–27.
828. Zeichner SL, Hirka G, Andrews PW, Alwine JC. Differentiation-dependent human immunodeficiency virus long terminal repeat regulatory elements active in human teratocarcinoma cells. *J Virol* 1992;66: 2268–2273.
829. Zenzie-Gregory B, Sheridan P, Jones KA, Smale ST. HIV-1 core promoter lacks a simple initiator element but contains a bipartite activator at the transcription start site. *J Biol Chem* 1993;268:15823–15832.
830. Zhao B, Winborne E, Minnich MD, Culp JS, Debouck C, Abdel-Meguid SA. Three-dimensional structure of a simian immunodeficiency virus protease/inhibitor complex. Implication for the design of human immunodeficiency virus type 1 and 2 protease inhibitors. *Biochemistry* 1993;32:13054–13060.
831. Zhou W, Parent LJ, Wills JW, Resh MD. Identification of a membrane-binding domain within the amino-terminal region of human immunodeficiency virus type 1 gag protein which interacts with acidic phospholipids. *J Virol* 1994;68:2556–2569.
832. Zolla-Pazner S, Goudsmit J, Nara P. Characterization of human neutralizing antibodies derived from HIV-1 infected individuals. *Semin Virol* 1992;3:203–211.
833. Zweig M, Samuel KP, Showalter SD, Bladen SV, DuBois GC, Lautenberger JA, Hodge DR, Papas TS. Heterogeneity of nef proteins in cells infected with human immunodeficiency virus type 1. *Virology* 1990;179:504–507.

Fundamental Virology, Third Edition
edited by B.N. Fields, D.M. Knipe, P.M. Howley, et al.
Lippincott - Raven Publishers, Philadelphia © 1996

CHAPTER 28

Polyomavirinae: The Viruses and Their Replication

Charles N. Cole

DISCOVERY AND CLASSIFICATION

The *Polyomavirinae* are a subfamily of the *Papovaviridae,* a family of small, nonenveloped viruses with icosahedral capsids. The family name is derived from the names of three prototypical members: rabbit papilloma virus (pa), mouse polyoma virus (po), and simian virus 40 (SV40), originally called vacuolating virus (va). Their genomes are single molecules of covalently closed, superhelical, double-stranded DNA that are replicated in the nucleus. Thorough investigations over the past 20 years indicate that SV40 and mouse polyoma virus differ from the papillomaviruses in having smaller capsids (diameters of 45 nm versus 55 nm), smaller genomes (approximately 5,000 bp versus approximately 8,000 bp), and a different genomic organization. On this basis the *Polyomavirinae* are considered a distinct subfamily of the *Papovaviridae* rather than a distinct genus. In this chapter polyoma is used to refer to mouse polyoma virus; and the group of viruses is referred to as polyomaviruses.

Twelve members of the *polyomavirinae* have now been identified. All have capsids that are the same size and are constructed from three viral capsid proteins. All have genomes of approximately 5,000 bp and display a similar genomic organization. Many regions of their genomes are highly conserved, demonstrating that the *polyomavirinae* are descended from a common ancestor. Different family members infect several species of mammals—including humans, other primates, rodents, and rabbits—as well as birds. Table 1 lists the members of the *polyomavirinae,* gives their natural hosts, and provides some details about their genomes and gene products. Most of these viruses display a narrow host range and do not productively infect other species. However, infection of cells in which these viruses do not grow productively often leads to the malignant transformation of infected cells. Some of these viruses also induce tumor formation in newborn hamsters.

These small DNA tumor viruses, particularly SV40 and polyoma, have been subjects of intensive studies by virologists and molecular biologists since their discovery more

C. N. Cole: Department of Biochemistry, Dartmouth Medical School, 7200 Vail Building, Hanover, NH 03755

TABLE 1. *Viruses of the polyomavirinae family*

Virus	Host	Genome Size (bp)	Virus-Encoded Proteins (Amino Acids) Large T Ag	Middle T Ag	Small T Ag	VP1	VP2	VP3	Agno-protein
Simian virus 40 (SV 40)	Rhesus monkey	5243	708		174	364	352	234	62
JC virus (JCV)	Human	5130	688		172	354	344	225	71
BK virus (BKV)	Human	5133	695		172	362	351	232	66
Lymphotropic Papovirus (LPV)	African green monkey	5270	679		189	368	356	237	
Bovine polyomavirus (BPyV)	Cattle	4967	586		124	365	353	232	
Hamster polyomavirus (HaPV)	Hamster	5366	751	401	194	373	346	221	
Polyoma virus (PyV)	Mouse	5392	785	432	195	385	319	204	
Kirsten virus (KV)	Mouse	4754	646		158	373	321	222	
Rabbit polyomavirus (RKV)	Rabbit	NR	NR	NR	NR	NR	NR	NR	
Rat polyomavirus (RPV)	Rat	NR	NR	NR	NR	NR	NR	NR	
Simian Agent 12 (SA12)	Baboon	NR	699		172	NR	NR	NR	
Budgerigar fledgling disease virus	Parakeet	4980	554		145	343	341	235	

NR, Not reported.

than 30 years ago. Polyoma was the first family member to be discovered. In his studies on the transmission of murine leukemia Ludwig Gross noted that extracts from infected animals could be used to transmit leukemia (123). However, some inoculated animals developed salivary gland tumors rather than leukemia. His further studies showed that the two agents could be separated on the basis of differences in sedimentation or filtration [the murine leukemia virus (MLV) was larger], as well as heat inactivation. (The agent that induced salivary gland tumors was insensitive to treatment at 65°C while MLV was completely inactivated.) The virus that caused the salivary gland tumors was subsequently named mouse polyoma virus because of its ability to cause a variety of different types of tumors in newborn mice (280). Infection of adult mice with polyoma virus does not usually result in tumorigenesis.

SV40 is one of several viruses identified by screening for viruses in the secondary rhesus monkey kidney cell cultures used for production of poliovirus vaccines. Although SV40 does not induce a visible cytopathic effect in rhesus monkey kidney cells, Sweet and Hilleman (280a) noted a pronounced cytopathic effect when African green monkey kidney cells were infected with extracts from the rhesus kidney cell cultures. Soon afterward, it was discovered that tumors were induced by injection of SV40 into newborn hamsters (85,113). Many lots of poliovirus vaccine were contaminated with live SV40 virus, raising the concern that this virus, which is oncogenic for newborn hamsters, might also be oncogenic for humans. Fortunately, studies to follow the incidence of cancer in those who were inadvertently inoculated with SV40 during poliovirus vaccination indicate clearly that SV40 does not cause tumors in humans (214).

Two polyomaviruses of humans have been described. JC virus (JCV) was isolated in 1971 (223) by inoculating human fetal brain cells with extracts of diseased brain tissue from patients with progressive multifocal leukoencephalopathy (PML). BK virus (BKV) was isolated in the same year from the urine of an immunosuppressed renal transplant recipient (104). The genomes of both of these human viruses show closest homology to SV40. A majority of people worldwide become infected with and acquire antibodies to these viruses during childhood with no apparent disease manifestations. These viruses are then thought to lie dormant in cells of a subset of infected persons. Once extremely rare, the incidence of PML has increased dramatically in recent years as a consequence of the immunosuppression associated with infection by HIV.

Although most polyomaviruses replicate primarily in epithelial or fibroblastic cells, zur Hausen and Gissmann (332) discovered a B-lymphotropic virus antigenically related to SV40 in a B-lymphoblastoid cell line derived from an African green monkey. Since this virus grows only in B-lymphoblastoid cells, it has been named lymphotropic papovavirus (LPV). Serologic studies suggest that humans and most other primates harbor viruses antigenically related to LPV. This virus does not induce tumors in newborn hamsters but will transform cultured hamster and mouse cells.

Hamster polyomavirus (HaPV) was isolated originally from a spontaneously occurring hair follicle epithelioma of a Syrian hamster (118). These tumors resemble papillomavirus-induced tumors in being highly keratinized and having virus particles exclusively in the differentiated cell layer. When injected into newborn hamsters, this virus causes leukemias and lymphomas, a tumor spectrum quite distinct in the polyomavirus subfamily. On the basis of nucleic acid sequence homology and genome organization (66,117), this virus is more closely related to polyoma than to SV40 or the human polyomaviruses. Other mammalian polyomaviruses include bovine polyomavirus (ByPV) (256), which is sometimes found in fetal calf serum, a rabbit polyomavirus (RKV) (130), a rat polyomavirus (RPV) (313), and Kilham virus, an additional polyomavirus of mice (202).

The only known avian polyomavirus was identified in 1986 (182). Budgerigar fledgling disease virus (BFDV) is classified as a polyomavirus on the basis of nucleic acid homology to other polyomaviruses as well as on morphologic and serologic criteria. It is also more closely related to polyoma than to SV40 (134).

POLYOMAVIRUSES AS PARADIGMS

The initial impetus for studies of these viruses was their oncogenic potential. However, these viruses are easy to cultivate in tissue culture and to purify, making them suitable model systems for diverse studies in molecular biology. Because of the small size of their genomes, physical maps of their genomes were generated as soon as restriction endonucleases became available in the 1970s, and they were among the first DNA genomes to be completely sequenced (92,237,268,269). Over the past 30 years these viruses have been used to examine many fundamental questions in eukaryotic molecular biology.

Among the major discoveries resulting from work on these viruses are the structure of supercoiled DNA, the identification of eukaryotic origins of DNA replication, the discovery of enhancers and elucidation of the organization of promoters involved in transcriptional regulation, the discovery of alternative splicing, and detailed understanding of the mechanisms of negative and positive regulation of gene expression. Because viral DNA is organized into chromatin and because replication of viral DNA involves only a single viral gene product and uses almost all the proteins used for cellular DNA replication, replication of SV40 origin-containing DNA *in vitro* has served as a model for understanding eukaryotic chromosomal DNA replication. Studies to understand the oncogenic potential of these viruses have provided fundamental insights into cell cycle regulation, oncogenes, and tumor suppressor genes. Much of our understanding of the polyomaviruses derives from studies on SV40 and polyoma, and this chapter will focus primarily on these two viruses.

VIRION STRUCTURE

The capsids of the polyomaviruses contain three virus-encoded proteins, VP1, VP2, and VP3 surrounding a single molecule of viral DNA (Fig. 1A) complexed with cellular histones H2A, H2B, H3, and H4 in the form of chromatin (Fig. 1B). The sizes of the capsid proteins of SV40 and polyoma are shown in Table 1. These proteins are arranged to form a T = 7 icosahedral capsid (Fig. 1C) containing 360 molecules of the major capsid protein, VP1, and approximately 30–60 molecules of each of the minor capsid proteins VP2 and VP3. The virus particle is 88% protein and 12% DNA and has a sedimentation coefficient of 240S in sucrose density gradients. Because they lack envelopes, the virus particles are resistant to lipid solvents. They are also relatively resistant to heat inactivation. Virions have a density of 1.34 g/ml in CsCl equilibrium density gradients, whereas empty capsids have a density of 1.29 g/ml.

Although originally thought to contain 60 hexameric (hexons) and 12 pentameric (pentons) capsomeres, it is now clear that these viruses contain 72 pentameric capsomers that each contain five molecules of VP1 (184,235). In fact, polyoma VP1 produced in *E. coli* will self-assemble into pentameric capsomeres, and under appropriate conditions, these molecules will further assemble into virus-like empty capsids and other structures (246). Since icosahedral viruses have both sixfold and fivefold axes of symmetry, the structure of the polyoma capsids presents a puzzle, since these capsids contain capsomeres that have only fivefold axes of symmetry. The high-resolution structure of the SV40 capsid (184) indicates that the C-terminal arm of VP1 does not form part of the capsomere structure itself but instead makes contacts with neighboring pentamers (Fig. 1D). It is the flexible geometry of these contacts that appears to allow the pentamers to fit together in such a way that an icosahedral capsid is formed. Calcium ions are required for virion stability (28,42) and are thought to stabilize pentamer/pentamer interactions (184). There are no disulfide bonds within the SV40 pentamer subunits, but disulfide bonds may exist between adjacent pentamers, since reducing agents are required to disassemble virus particles.

The sequences of the minor capsid proteins, VP2 and VP3, are overlapping; VP2 contains the entire VP3 sequence at its C-terminus and an additional sequence of approximately 115 amino acids at its N-terminus. The N-terminus of SV40 and polyoma VP2 is myristylated (282). The location of these proteins within the capsid cannot be discerned from available crystallographic data. Each of the 72 capsomeres of the virus particle contains five molecules of VP1 and displays additional electron density within the central axis of each pentamer. This density is probably VP2 or VP3. There is now considerable evidence that VP1 interacts with and forms complexes with VP2 and VP3, with sequences near the carboxy-terminus of VP2 and VP3 playing an important role in mediating this interaction (11). The most likely structure is one with pentamers containing a molecule of either VP2 or VP3 oriented with their amino termini extending outward from the virion.

Preparations of polyomaviruses contain three kinds of virus particles. Infectious virus particles contain a single molecule of viral DNA in association with four cellular histones. The DNA and histones are arranged as chromatin, and the histone/viral DNA complex is often referred to as the viral mini-chromosome. This mini-chromosome has

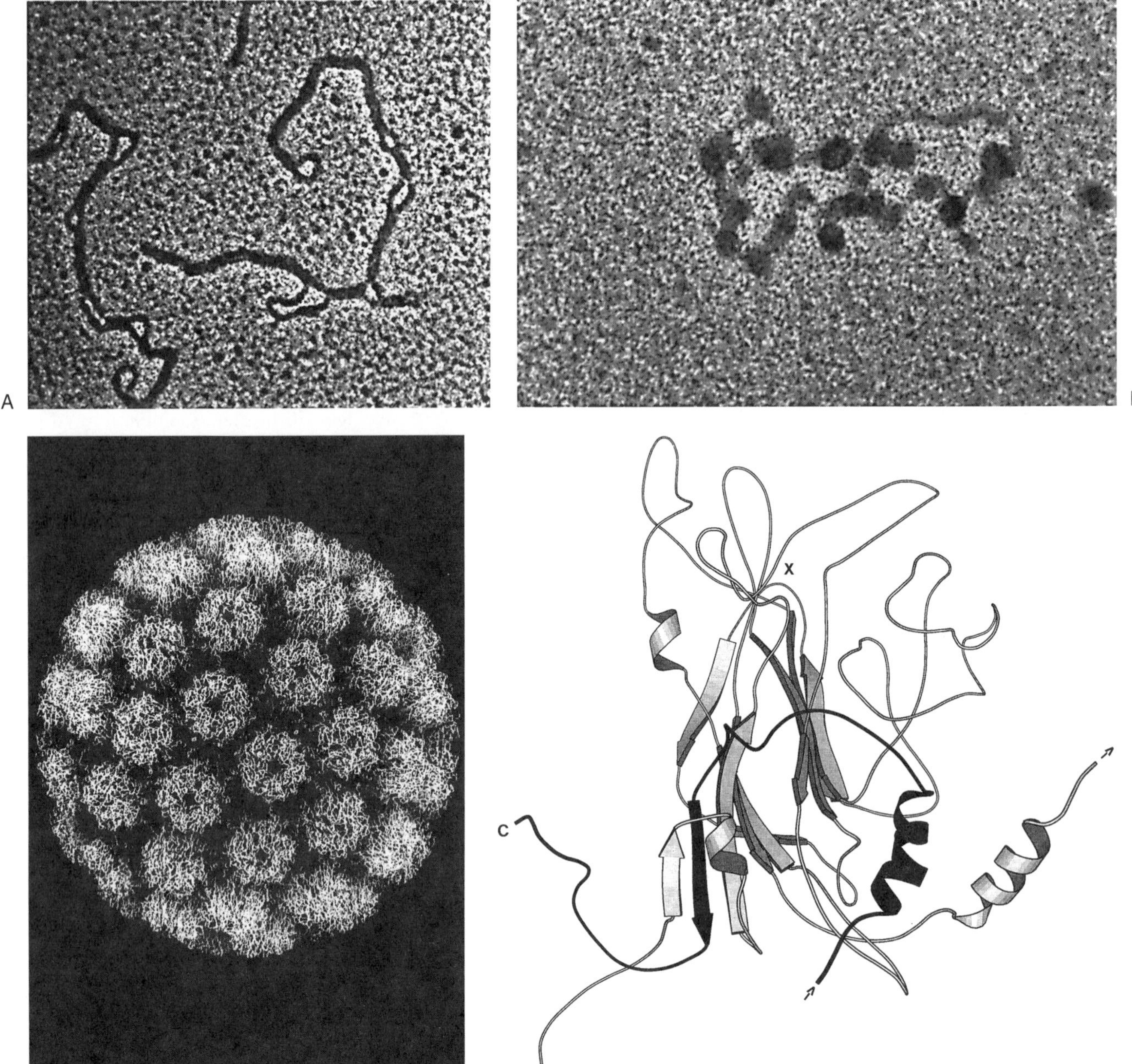

FIG 1. Viral DNA, minichromosomes, and virions. **(A)** Electron micrograph of supercoiled SV40 DNA molecules. **(B)** Electron micrograph of an SV40 minichromosome. Note that this particular minichromosome displays a nucleosome-free region. About 20% of minichromosomes have a nucleosome-free region surrounding the regulatory region at the origin of DNA replication. **(C)** Computer graphic representation of the structure of the SV40 virion. Note that the shell is made up of 72 pentamers of VP1. Twelve of these lie on icosahedral fivefold axes and are surrounded by five other pentamers. The remaining 60 pentamers, such as the ones near the center of this diagram, are surrounded by six other pentamers. The pentamers are linked by extended C-terminal arms of VP1 molecules of tile Subunits. **(D)** A ribbon diagram of a VP1 subunit of the mouse polyomavirus virion. N refers to the location of residue 15 of the VP1 polypeptide; the first 14 residues are thought to be disordered. C is the C-terminus of an invading arm (darkly shaded) of another VP1 molecule. At the right of the figure is an arrow showing that this VP1 molecule goes off to the right, where it interacts with another VP1 molecule. The interactions of the C-terminal arms of VP1 molecules can be pairwise with each of two VP1 molecules interacting with the C-terminal arm of the other (as in this figure), or the incoming C- terminal arm of VP1 can come from a third VP1 subunit (see Chapter 3 for more details on protein-protein interactions in the capsids of the polyomaviruses). X is the site of the mouse polyomavirus receptor-binding pocket. (Panels A and B courtesy of Dr. Jack Griffith, University of North Carolina; Panels C and D courtesy of Dr. Steven Harrison, Harvard University.)

been used as a model system to study the effects of chromatin structure on DNA replication and transcription. Most viral DNA molecules contain a full complement of nucleosomes, but some lack nucleosomes over viral regulatory sequences. Virion mini-chromosomes lack histone H1, whereas the form of the mini-chromosome found within infected cells contains it (306). Virion preparations also contain empty capsids as well as pseudovirions, capsids containing cellular rather than viral DNA. Passage of polyomaviruses at high multiplicities of infection leads to the generation of defective viral genomes containing deletions, duplications, and rearrangements of viral genetic information, often with duplications of the viral origin of DNA replication. If they are within appropriate size limits, these defective viral genomes can be encapsidated.

GENOME ORGANIZATION

The genomes of all polyomaviruses are divided into early and late regions (Fig. 2). The early region is, by definition, that portion of the genome transcribed and expressed early after the virus enters the cells, and it continues to be expressed at late times after infection, after the onset of viral DNA replication. The late region of the genome is expressed efficiently only after viral DNA replication begins, though low levels of transcription of the late region occur early after infection as well. The first maps of the polyoma and SV40 genomes were divided into 100 map units with the unique Eco RI site in both the SV40 and polyoma genomes defined as 0/100 map units. Since the complete nucleotide sequences of these genomes have been determined, use of nucleotide positions rather than map units permits more precise description of specific sites within the genome. Nucleotide position 1/5243 for SV40 is the center of a 27-bp palindrome located at the origin of viral DNA replication (ORE in Fig. 3). The numbering system used for SV40 is that of Buchman and colleagues (31). The numbering system used for polyoma virus is that of Griffin and colleagues (121). For polyoma, nucleotide position 1/5295 corresponds to the center of the Hpa II cleavage site at the junction of the fifth and third largest of eight fragments of polyoma DNA produced by cleavage with Hpa II. This site is also close to the polyoma origin of DNA replication.

Two lines of evidence indicate that polyomavirus genomes contain a single unique origin of DNA replication. In one of the earliest uses of restriction endonucleases as tools to address important problems in molecular biology, Danna and Nathans (61) exposed SV40-infected BSC- 1 monkey cells to pulses of 3H-thymidine and, at various subsequent times, isolated viral DNA molecules that had completed replication. They then hybridized this labeled DNA to fragments of SV40 DNA generated with restriction endonucleases Hind II + III. In mature viral DNA molecules, label should first appear at sites near the terminus for DNA replication, whereas the origin region should become labeled later, since molecules just initiating DNA replication at the time of addition of label require a longer time to be completed. The region of the genome around map position 17 (nt 2622) became labeled first, whereas that around position 67 (nt 0/5243) was labeled last and designated as the origin region.

This location for the origin was also revealed by electron microscopy. Replicating SV40 DNA molecules were cleaved with Eco RI, which cuts SV40 DNA once (nt 1782), and examined by electron microscopy. The unique cleavage site allows mapping of the sites at which replication bubbles are initiated; these studies indicated that replication began one-third of the way around the genome from the Eco RI site. By performing similar analyses with viral DNA molecules lacking portions of the early or late region, it was shown that there is no unique site for the termination of DNA replication. Rather, bidirectional replication proceeds away from the origin with the replication forks meeting at a site approximately 180° from the initiation site. Detailed mutational analyses to define the origin more precisely will be discussed later.

The promoters and enhancers for transcription are located close to the origin of replication (Figs. 2 and 3). Together the promoters, enhancers, and origin are referred to as the *viral regulatory region*. Transcription extends bidirectionally from initiation sites near the origin, with early and late mRNAs being transcribed from opposite strands of the viral genome. Thus, the early region extends from the origin to a site approximately halfway around the genome. The early region encodes the viral regulatory proteins, the tumor or T antigens, so called because they can be detected with antisera derived from animals bearing tumors induced by these viruses or by cells transformed by these viruses. SV40 and the other primate polyomaviruses encode two T antigens—designated large T and small t antigens on the basis of size. Polyoma and closely related viruses encode three T antigens—large T, middle (or medium) T, and small t. All the T antigens of each virus share N-terminal sequences and contain different C-terminal regions. The mRNAs encoding them are produced by alternative splicing from a common pre-mRNA. Fig. 2 shows the common and unique regions of SV40 and polyoma T antigens.

The late regions of these viral genomes encode the three capsid proteins VP1, VP2, and VP3. As with the early mRNAs, late mRNAs are generated from a common pre-mRNA by alternative splicing. The coding regions of VP2 and VP3 overlap, with VP3 sequences being a subset of VP2 sequences. The coding region for the N-terminus of VP1 overlaps that for the C-termini of VP2 and VP3, with translation of VP1 being from a different reading frame than those of VP2 and VP3. SV40 and the human polyomaviruses encode a fourth late protein, called the agnoprotein (Table 1). This 62–71 amino acid protein is encoded by the leader region of some species of late viral mRNA and accumulates in the perinuclear region during the late

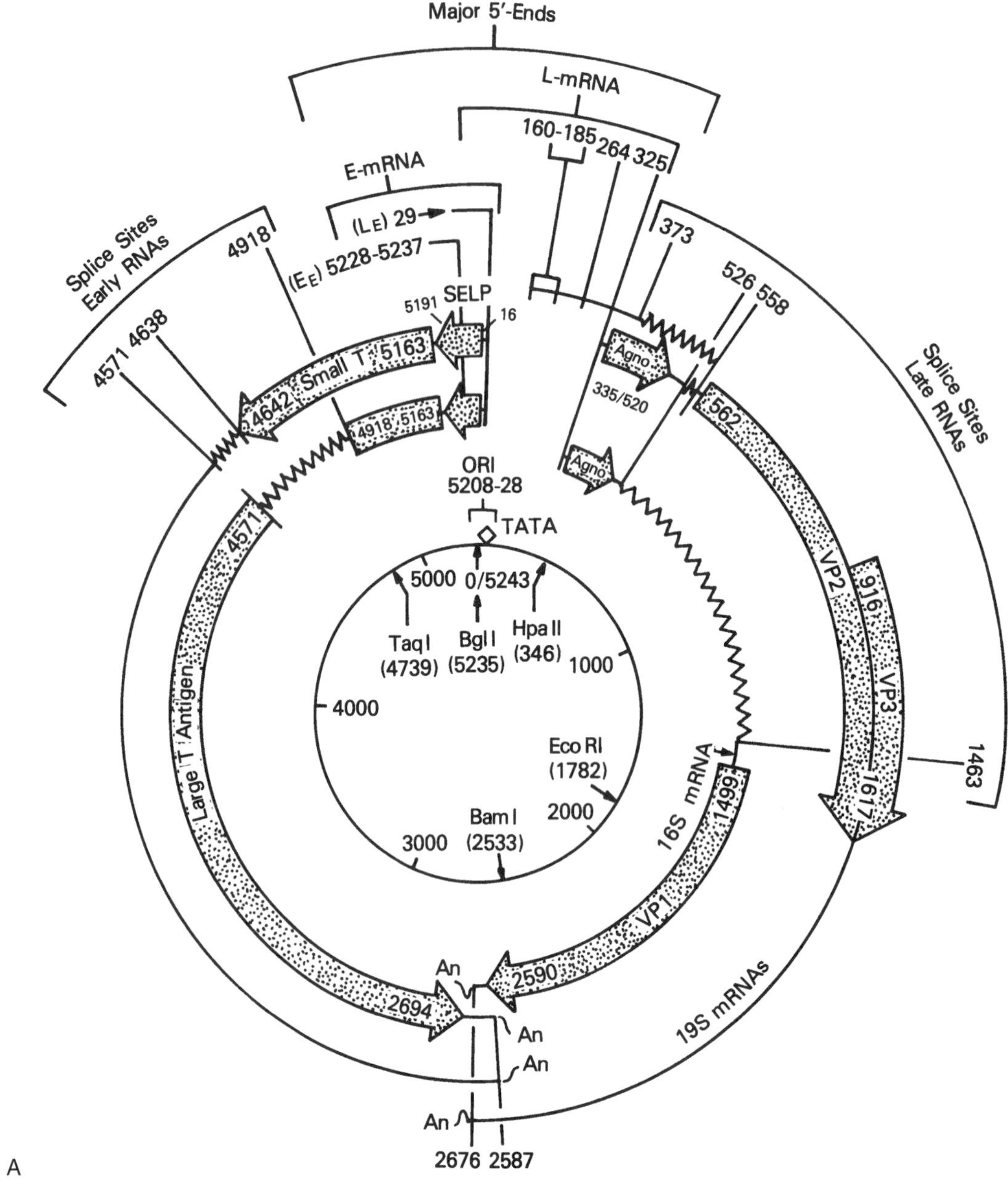

FIG. 2. Genomic organization of SV40 and mouse polyomavirus. **(A)** SV40. The origin of replication and transcriptional regulatory region is at the top. The early region extends counterclockwise and the late region extends clockwise from the top. The regions encoding viral proteins are shaded. Also shown are the nucleotide positions for the 5′> and 3′ ends of viral mRNAs and the positions of introns. (From ref. 26, with permission.) **(B)** Mouse polyomavirus. The 0/100 map unit position is at the top and corresponds to the unique Eco RI site. The origin is located counterclockwise approximately one-third of the way around the circle. The early region is located in the top half of the figure, and the late region in the bottom half. Also shown are the positions where each of the viral gene products are encoded, the nucleotide numbers of the 5′> and 3′ ends of the viral mRNAs, and the nucleotide positions of the introns. (From ref. 120, with permission.)

phase of the infection cycle (143,222). It was originally named "agnoprotein" because its function was unknown. It appears to facilitate the localization of the major capsid protein to the nucleus and may enhance the efficiency with which the virus spreads from cell to cell (36,239). Mutants that do not produce the agnoprotein yield fewer progeny virions and produce small plaques (9,208). Pseudorevertants of these mutants map to the major capsid protein, VP1, suggesting a possible interaction between the agnoprotein and VP1. Additional small, open reading frames that could encode other proteins are found in other species of late mRNA as well as in some species of early-strand

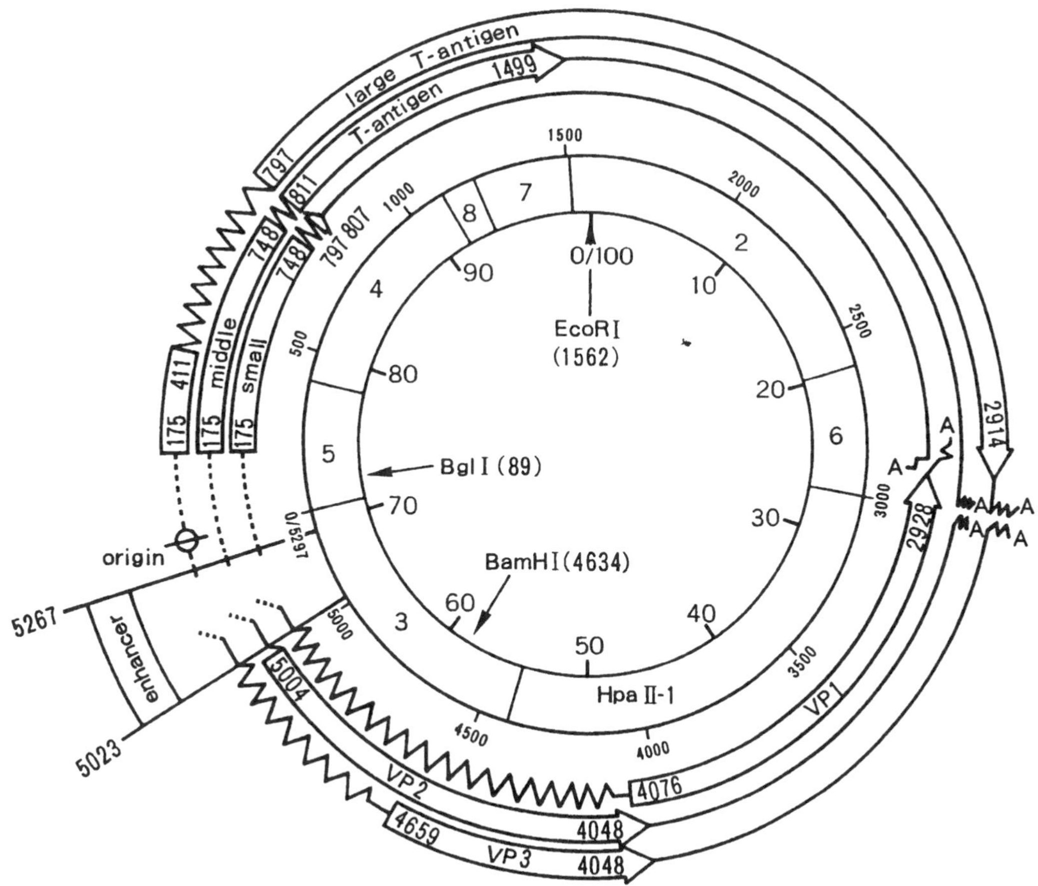

FIG. 2. *Continued.*

mRNA present late after infection (161). Thus, many SV40 late mRNA species are bi-cistronic.

VIRAL GENETICS

Genetic analyses have played a central role in our understanding of the biology and replication cycle of the polyomaviruses (for a review, see ref. 295). The first mutants to be isolated were host range mutants of polyoma whose growth was restricted to certain mouse cell lines and temperature-sensitive (ts) mutants of SV40 and polyoma. The availability of restriction endonucleases led to the construction of a wide variety of deletion and substitution mutants during the 1970s and 1980s. Site-directed mutagenesis and PCR-based approaches have also been used to produce mutants of the polyomaviruses. Today it is relatively easy to construct any desired point, deletion, or substitution mutant of these viruses.

For SV40, temperature-sensitive mutants were divided into five complementation groups—A, B, C, BC, and D—based on their ability to complement one another to produce plaques at the nonpermissive temperature. We now know that the product of the A gene is large T antigen. VP2 and VP3 are both altered by mutations in gene D, while mutants in groups B, BC, and C produce altered VP1. Intracistronic complementation is often seen in genes that encode proteins that assemble into multimers or in multifunctional proteins where separate domains perform distinct roles. Intracistronic complementation is observed between VP1 mutants in groups B and C but not between those in the BC group and either group B or group C. It has also been observed between mutants of large T antigen that affect the host range of the virus and mutants that affect replicative activities of the protein (296). Temperature-sensitive mutants of polyomavirus have been sorted into similar complementation groups.

By infecting cells at the permissive temperature and shifting them to the nonpermissive temperature at various times after infection, it was shown that the products of genes A and D were required early after infection, whereas the products of genes B and C were required only at late stages of infection. It is not surprising that large T antigen is required early after infection, since it is required for viral DNA replication. Studies that suggest a role for VP2 and VP3 in adsorption or uncoating of the virus explain the finding that the D gene product is also required for early events of the infection cycle.

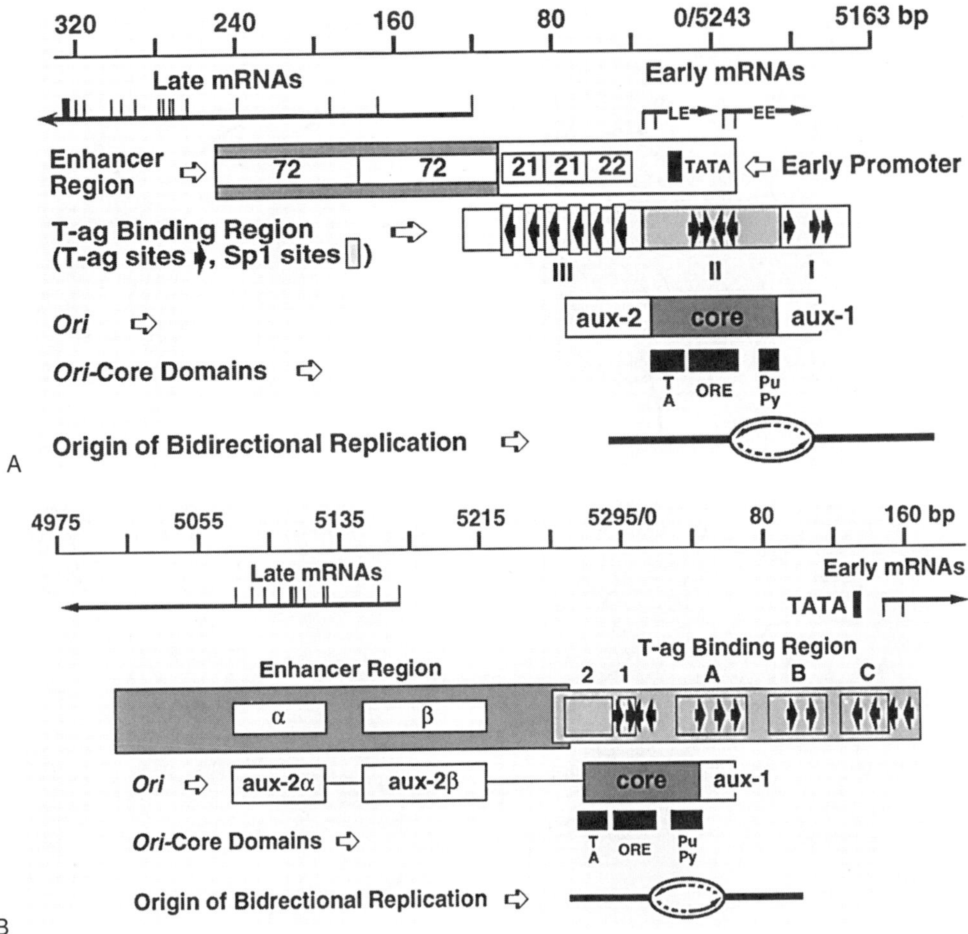

FIG. 3. The regulatory regions of SV40 and polyoma. Shown are the location of the core origin of viral DNA replication and the location of auxiliary sequences (aux) that enhance viral DNA replication. Replication of polyoma requires either aux-2α or aux-2β. Within each core origin region are located a TA tract, with a strand bias for Ts and As, the palindromic region that serves as an origin recognition element (ORE), and a region with a purine (PU)/pyrimidine (PY) strand bias, within which bidirectional DNA replication is initiated. Also shown are sites in which large T antigen binds to viral DNA; pentanucleotides involved directly in binding are indicated by arrowhead s Po that point the direction of the pentanucleotide, 5′GAGGC3′. Sites within this region at which transcription is initiated for production of viral early and late mRNAs are also indicated. The sites marked EE for SV40 are those used early after infection; those marked LE are used to produce early mRNAs following the onset of viral DNA replication. Additional sites for production of both SV40 and polyoma late mRNAs are located further downstream for both SV40 and Py. For SV40 the locations that serve as enhancers and of the three nearly perfect 21-bp GC-rich repeats of the two 72-bp repeats are also shown. Within these repeats are six sites to which transcription factor Sp I can bind. (Figure courtesy of Dr. Mel DePamphilis, Roche Institute of Molecular Biology, Nutley, NJ.)

The sites of mutation of representative members of each complementation group were mapped by marker rescue using restriction endonuclease fragments of viral DNA (171). This involved producing partial heteroduplexes containing a circular strand of a particular ts mutant and a linear restriction endonuclease fragment. Successful marker rescue depends on the wild-type restriction endonuclease fragment containing the information that is mutated in the mutant genome. In this way the SV40 A gene was mapped to the early region and the SV40 B, C, BC, and D groups were mapped to the late region of the genome. Similar mutants of polyoma were mapped to analogous regions of the polyoma genome.

Although these studies were able to identify viral genes and their positions on the physical map of the viral genome, they did not indicate which of the viral gene products were encoded by each gene. Deletion mutants were constructed using restriction endonucleases, and this permitted assignment of some viral proteins to viral genes, since it was easy to determine which viral gene product had been shortened by the deletion mutation. Many of these mutants could be propagated only in the presence of wild-type helper virus. Mutants of SV40 with deletions at the unique Eco RI or BamH I sites produce internally deleted or truncated VP1 polypeptides, indicating that the gene for VP1 spans the Eco RI and Bam HI sites (49). This is the region of the genome where marker rescue studies placed mutations in the B, BC, and C complementation groups.

Mutants with deletions at the unique Hae II site affected only VP2 (49). These mutants were able to form very small plaques and hence could be propagated without helper virus. Because these very-slow-growing mutants could complement tsD mutants at the nonpermissive temperature, they were assigned to a new complementation group E, with VP3 assigned to complementation group D. SV40 mutants with deletions in the early region were also constructed and found to produce large T antigens of reduced size, thus proving that the large T antigen was the product of the A gene, rather than a cellular protein induced by SV40 infection (172).

A set of mutants with deletions located randomly around the SV40 genome was produced by treating SV40 DNA with DNAse in the presence of Mn2+. This treatment causes double-strand breaks in the circular DNA. By treating the resulting linear molecules with nuclease S1, small amounts of DNA were removed from each end of the linear molecule. After transfection into mammalian cells, these linear molecules become recircularized and carry deletions at the site of linearization. Those molecules with deletions of nonessential information retain the ability to induce plaque formation. In this way, three regions of the viral genome were identified that could accommodate deletions without impairing the plaque-forming ability of the virus (264). These sites are located in the agnogene region, in the region encoding the 3′ untranslated regions of the early and late mRNAs, and in the early region between map positions 54 and 59 (nts 4,900–4,600, approximately).

Normal-sized plaques were produced by the 54/59 region deletion mutants, but these viruses were unable to transform nonpermissive cells as efficiently as wild-type SV40, suggesting that a gene product needed for transformation was encoded in this region. No tsA mutants (affecting large T antigen) had been mapped to this region. Analysis of T antigens encoded by mutants with deletions in the 54/59 region showed that the deletions did not affect large T antigen, although size measurements of large T suggested that most or all of the early region was required to encode it. Studies by a number of investigators subsequently showed that the SV40 early region encodes both small and large T antigens (17,59,225), and that small t antigen was reduced in size by mutations in the 54/59 region. Alternative splicing allows the production of mRNAs that encode these two early gene products; there is not sufficient genetic information in the SV40 early region to encode distinct polypeptides of 84 kD and 17kD.

Viable deletion mutants of polyoma virus have also been isolated and characterized. Host range mutants of polyoma were selected that grew poorly or not at all in mouse 3T3 cells but replicated efficiently in primary cells or in cells transformed by retroviruses or by polyoma itself (14). Because these mutants display this host range phenotype and are also defective for transformation, they are called "hrt" mutants. Marker rescue experiments indicated that the "hrt" mutants map to the analogous region of the polyoma genome as the SV40 54/59 deletion mutants. Complementation is observed between polyoma tsA and "hrt" mutants to permit cell transformation at a temperature restrictive for tsA mutants. The "hrt" mutants are known to encode truncated forms of both small and middle T antigens. The role of these proteins in transformation will be discussed later.

BIOLOGICAL PROPERTIES OF THE POLYOMAVIRUSES

Although SV40 and polyoma are able to infect a wide range of mammalian cell lines and cultures, the response to infection can be productive or nonproductive. In some types of cells, viral DNA replication occurs, followed by the assembly of progeny virions and, ultimately, the death of the cell. This is called a productive infection, and cells in which infection is productive are said to be permissive for growth of the virus. Monkey cells are permissive for growth of SV40, and mouse cells are permissive for growth of polyoma. Some polyomaviruses show an extremely narrow host range for productive infection. For example, JCV replicates only in human fetal glial cells.

Nonproductive infections result when viral DNA replication cannot take place in the infected cell. For viral DNA replication to occur, SV40 or polyoma must produce adequate levels of large T antigen, the only viral protein needed for replication, and this protein must interact with cellular replication factors and the viral origin of DNA replication. An inability of SV40 large T antigen to interact productively with host replication factors in rodent cells is the reason that SV40 infection of rodent cells is unproductive. A similar inability of the polyoma large T antigen to interact productively with simian or human host cell factors is responsible for the inability of polyoma to replicate in primate cells.

In nonproductively infected cells, the infection cycle begins normally. The viral early mRNAs are produced and the viral T antigens can be detected. The early proteins exert a variety of effects on the host cell. Since the polyomaviruses require the host cell DNA replication machinery, they replicate only during S phase, and T antigen stimulates cells to move through the cell cycle into S phase. When expressed in cells that are nonpermissive for viral replication, the viral T antigens also cause cells to acquire the properties of transformed cells. Such cells will proliferate in semisolid media or in the absence of a high concentration of fetal calf serum. Usually, these transformed properties are manifested for only a few days. Frequently, the viral genome, which does not replicate, is lost from the cells, and they return to normal growth. Such cells are said to have been "abortively" transformed. At a low frequency, the viral DNA becomes integrated into the host cell genome and is subsequently inherited as if it were a cellular gene. Integration appears to occur randomly with respect to sites in both the viral and host cell genomes. If the arrangement of viral DNA sequences following integration permits continued expression of SV40 large T antigen or polyoma middle T antigen, then the transformed phenotype will be expressed permanently, and the cell is said to have been transformed by the virus.

Most studies of transformation by SV40 have been conducted in mouse, rat, and hamster cells, but primate cells can be transformed by SV40 if replication is prevented by mutation of the viral origin of replication or the viral T-antigen gene. The Cos-1 and Cos-7 monkey kidney cell lines express wild-type SV40 large T antigen and contain an integrated copy of SV40 DNA carrying a deletion of sequences within the origin of DNA replication that are essential for DNA replication (114). For polyoma, rat cells are normally used to study viral transformation. Although mouse cells are permissive for replication of polyoma virus, they can also be transformed by polyoma if the virus carries a mutation affecting large T antigen or its origin of replication or if viral DNA becomes integrated into the mouse cell genome in a way that interrupts the early coding region and prevents continued synthesis of full-length large T antigen. Although multiple regions spanning most of the SV40 large T antigen are required for transformation, the critical protein for transformation by polyoma is the middle T antigen. Although these viral gene products must be present continuously to maintain the transformed phenotype, mutation of cellular genes can eliminate the need for maintenance of the viral transforming region (232,233).

In addition to being able to cause malignant transformation of cultured cells, these viruses can immortalize primary cells that would normally undergo a limited number of cell doublings. Immortalization is a function of the large T antigens of SV40 and polyoma. Studies of immortalization of primary mouse embryo fibroblasts by mutants of SV40 T antigen suggest that some mutants that can neither immortalize primary cells nor transform established cell lines have the ability to extend the normal lifespan of the primary mouse cells (S Conzen and CN Cole, unpublished). Although the lifespan of human fibroblasts and epithelial cells can be extended by wild-type SV40 T antigen, their immortalization is very rare and likely requires mutation of cellular genes (263,305).

THE REPLICATION CYCLES OF SV40 AND POLYOMA

Productive infection of cells by polyomaviruses can be divided into early and late stages. The early stage begins with attachment of virus to cells (Fig. 4) and continues until the beginning of viral DNA replication. Thus, the early stage is marked by adsorption and penetration of the viri-

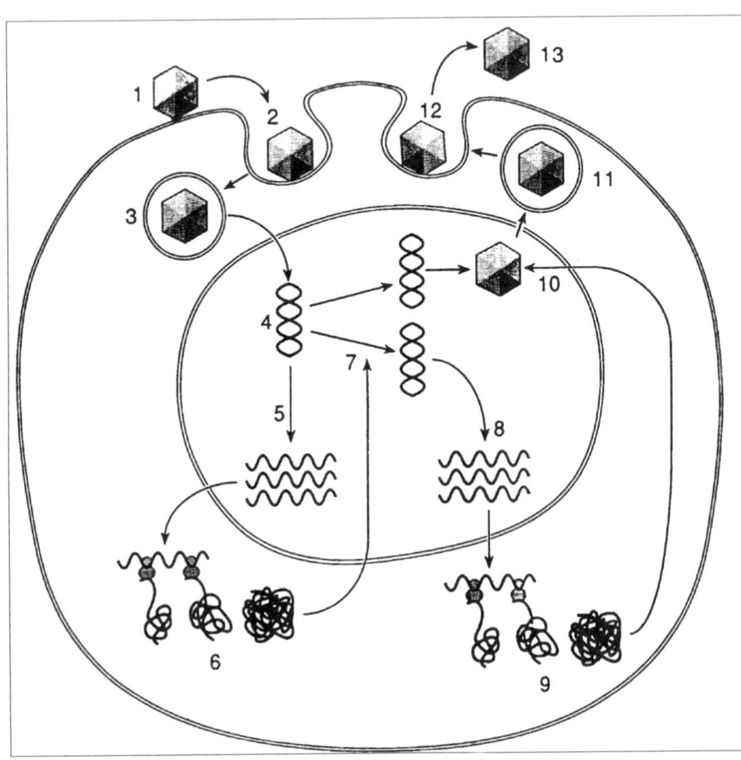

FIG. 4. Replication cycle of polyomaviruses. Steps in the replication cycle are indicated by numbers as follows: 1, adsorption of virions to the cell Surface, 2, entry by endocytosis; 3, transport to the cell nucleus (route and mechanism not yet known); 4, uncoating; 5, transcription to produce early= region mRNAs; 6, translation to produce early proteins (T antigens); 7, viral DNA replication; 8, transcription to produce late region mRNAs, 9, translation to produce late proteins (capsid proteins); 10, assembly of progeny virions in the nucleus; 11, entry of virions into cytoplasmic vesicles (mechanism unknown); 12, release of virions from the cell by fusion of membrane vesicles with the plasma membrane; 13, released virion. Virions are most likely also released from cells at cell death, when virions have an opportunity to leak out of the nucleus. In nonpermissive cells the first six steps occur normally, but viral DNA replication cannot occur and subsequent events do not take place.

on and its migration to the nucleus where the viral genome is uncoated and made available for transcription. During the early phase of infection, the viral early proteins, the T antigens, are produced, and they affect the host cell by stimulating the production of enzymes required for cellular DNA replication, thereby preparing the cell for replication of viral DNA. These viral early proteins also stimulate resting cells to reenter the cell cycle. The late stage of infection extends from the onset of viral DNA replication to the end of the infection cycle and involves the replication of viral DNA, expression of the viral late genes encoding the capsid proteins, the assembly of progeny virus particles in the nucleus of the infected cells, and the release of virus, possibly at the time of the death of the cell.

The time course of infection by SV40 and polyoma depends primarily on two parameters. Infection proceeds more rapidly at higher multiplicities of infection than at lower multiplicities. This most likely reflects the fact that critical levels of the viral early proteins are attained earlier when more viral genomes are available for transcription to produce the viral early mRNAs. The other critical parameter is the growth state of the host cell. At the same multiplicity of infection, viral DNA replication begins sooner and the production of progeny virions is completed more rapidly when cells are growing exponentially than when cells are confluent and in a G0 state at the time of infection. Figure 5 illustrates the time course for the expression of polyoma T antigen, induction of cellular enzymes, replication of viral DNA, and production of progeny virions.

Adsorption and Receptors

Early studies of the initial events of polyomavirus infection were complicated by the fact that virus preparations used contained not only infectious virus particles, but empty capsids and pseudovirions as well. Less ambiguous results have been obtained by using purified virion preparations. Norkin and colleagues found that binding of SV40 virions to LLC-MK2 rhesus monkey kidney cells was inhibited by antibodies to MHC class I antigens but not by antibodies to other cell surface proteins (29). Little binding was observed to two human lymphoblastoid cell lines that do not express MHC class I molecules because of failure to express either β_2- microglobulin or HLA class I molecules. Binding did occur when class I expression was restored by transfection of cells with plasmids encoding β_2-microglobulin or HLA class I molecules. These results suggest a role for class I MHC antigens in SV40 adsorption.

Different results were obtained by Compans and colleagues, who studied the binding of SV40 virions to polarized monkey kidney cells (Vero C1008) (12). Binding was restricted to the apical surface of these polarized cells, while HLA expression was detected at both the apical and basolateral surface. Perhaps MHC class I molecules are required for adsorption of SV40 virions but are not sufficient, with other cell surface molecules also being part of the receptor. In nonpolarized cells, binding sites for SV40 were distributed uniformly over the cell surface. Different subclones of C1008 Vero cells bound different numbers of virions, and the abundance of binding sites was cell cycle regulated. Those cells that bound the greatest number of virions contained approximately 10^5 receptors per cell.

Competition experiments indicate that SV40 and polyoma receptors are distinct (12). This was also inferred from earlier studies that showed that binding of polyoma was sensitive to treatment of the cells with sialidase, whereas binding of SV40 was not. Little is known about the identity of the polyomavirus receptor. In contrast, more is known

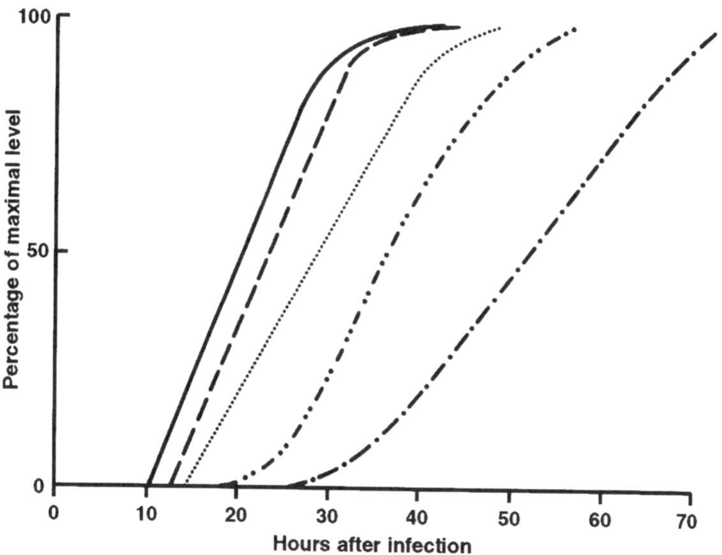

FIG. 5. Idealized time course of central events during productive infection of mouse cells by mouse polyoma virus. Events following infection of monkey cells by simian virus 40 follow a similar course. Parameters shown are expressed as a percentage of maximal level reached. T antigen (solid line) was measured by percentage of cells showing positive immunofluorescence; the actual rate of accumulation of T antigen may be slower, since T-antigen positive cells continued to synthesize T antigen over a long period. Cellular enzyme production (dashed line) is expressed as stimulation of enzyme activities compared with activity levels present in uninfected cells. Cellular DNA synthesis (dotted line), viral DNA and RNA (double dot- and dashed line), and infectious virus (single dot- and dashed line) are expressed as the percentage of the total final yields of these macromolecules attained at particular times, not as the rate of production at any particular time. (From ref. 295, with permission.)

about the interaction of polyoma virions with cells than is known about that of SV40 with cells. Antibodies to VP1 block adsorption of polyoma virions (20). Small and large plaque strains of polyoma produce different VP1s, suggesting that plaque morphology may reflect different efficiencies of interaction of the virus particles with their receptors. When examined by isoelectric focusing (20), polyoma virions are seen to contain six distinct isoforms of VP1 (A–F) that differ in posttranslational modifications. Empty capsids lack one of these species (E) and do not compete with virions for binding to mouse kidney cells, suggesting that species E plays a central role in specific virus adsorption (21). Empty capsids can still bind to guinea pig erythrocytes and are internalized and degraded in lysosomes, rather than being transported to the nucleus for productive infection. Additional evidence for a role for VP1 in virus binding comes from binding studies with capsid-like structures that self-assemble from polyoma VP1 expressed in and purified from bacteria. The fact that the binding of these "capsids" to quiescent cells stimulates modest induction of c-*fos* and c-*myc* (331) suggests that these "capsids" may be interacting with a growth factor receptor or related protein.

The VP2 polypeptide of polyoma and SV40 is myristylated at its N-terminus (282), and this modification plays a role in the early events of infection. Deletion mutants of SV40 that produce no VP2 are weakly viable. Plaques formed by these mutants are extremely slow to appear and enlarge slowly, suggesting that these mutants produce a dramatically reduced yield of infectious progeny (49). However, the yield of physical virus particles is nearly normal, suggesting that these progeny are defective in some early event of the infection cycle, perhaps adsorption. Mutation of polyoma VP2 so that it can no longer be myristylated still permits its incorporation into virions, but infectivity is reduced 15–20 times and the infection cycle is prolonged by several hours (245). These results suggest that the myristic acid moiety of VP2 may interact with cellular membranes, whereas VP1 probably interacts with specific receptor polypeptides.

Virion Entry and Uncoating

How polyomaviruses gain entry to the nucleus is uncertain. The vast majority of particles in a virus preparation do not initiate infection successfully, so there is the problem that the particles observed during the early stages of infection are likely to be primarily those that do not initiate an infectious cycle. Soon after infection, virus particles can be detected by electron microscopy in pinocytic vesicles within the cytoplasm (139,200). These vesicles may ferry virus particles to the nucleus, and the fusion of these vesicles with the nuclear envelope has been observed by electron microscopy (122,219). Virus particles have also been observed within the endoplasmic reticulum (152), so the ER may also provide a pathway to the nuclear envelope.

The nuclear pore complex can serve as an entry portal for virus particles, since SV40 virions injected into the cytoplasm enter the nucleus and begin production of SV40 early mRNA (45). This process is blocked by wheat germ agglutinin or by monoclonal antibodies to nuclear pore complex proteins (321), treatments known to block the movement of karyophilic proteins through the nuclear pore complex. Large T antigen is normally not detected until 10–12 hours after infection but can be detected within 3–4 hours after microinjection of virus particles into the cytoplasm. This time differential may reflect slow steps in penetration and the escape of virions from membrane vesicles into the cytoplasm prior to entry through the nuclear pore complex. Alternatively, it may mean that the nuclear pore complex is not generally a route of entry into the nucleus for SV40 or polyoma virions, with the pathway to the nucleus involving movement of virus particles within the endoplasmic reticulum or other membrane-bound structures. Since virus capsid proteins contain nuclear localization signals used to direct the proteins to the nucleus prior to virion assembly, these same nuclear localization signals may be able to mediate the entry into the nucleus of microinjected virions, even if this route is not normally used.

Most studies indicate that virions are uncoated within the nucleus, since intact virions can be detected within the nucleus prior to expression of the viral T antigen (6,139, 191). However, in some studies, partially disassembled virus structures were seen in the cytoplasm and were presumed to be uncoating intermediates (101). Whatever the route of entry, the viral minichromosome is transcribed, replicated, and subsequently encapsidated within the nucleus.

Transcription and Processing of Early Viral mRNAs

Once virions have been uncoated, the viral minichromosome can be used as a template for transcription by RNA polymerase II to produce early viral mRNAs, derived from the early region of the viral genome. Production of mRNAs derived from the late strand of viral DNA is very inefficient early after infection. Maximal production of the viral late mRNAs requires both an activity of large T antigen and the onset of viral DNA replication. Viral late mRNAs will be discussed later.

For many years studies on patterns of transcription of polyomavirus genomes were hampered by the very low levels of the viral early mRNAs, which comprise only 0.01–0.02% of the total cellular RNA synthesized prior to viral DNA replication. During the initial 10–15 hours of an SV40 or polyoma infection, the total production of early mRNA likely represents not more than a few hundred molecules, in comparison to the 300,000 molecules of total mRNA found in a eukaryotic cell (for a review, see ref.

295). From the application of more sophisticated and sensitive mapping techniques, it is now known that transcription initiates within the viral regulatory region and extends at least 180° around the viral genome (163), and it may extend further. Functional viral early mRNAs are produced by alternative splicing of this pre-mRNA (17). For SV40, alternative splicing generates two species of early mRNA. mRNAs encoding large T antigen result from excision of an intron that extends from nucleotide 4918 to 4571, whereas excision of an intron extending from nucleotide 4638 to the same splice acceptor at 4571 yields mRNAs encoding small t antigen. Thus, the SV40 early pre-mRNA contains two splice donor sites and a common splice acceptor. All these mRNAs are polyadenylated at the same site at nucleotide 2694. Recent studies suggest that infected cells also contain a third alternatively spliced early SV40 mRNA (326).

For polyoma, a similar early pre-mRNA is spliced to generate three polyoma early mRNAs (149,150). These mRNAs share 5′ ends located in the viral origin region and 3′ ends located at nucleotide 2930; they differ with respect to sequences excised by splicing. The introns for the mRNAs for the polyoma large, middle, and small T antigens extend from nucleotides 409 to 794, 746 to 808, and 746 to 794, respectively. Thus, there are two splice donors and two splice acceptors. Low levels of polyoma early mRNAs that use a weak polyadenylation site at nucleotide 1525 have also been detected (135,298).

Other polyomaviruses show similar patterns of early transcription. For those that produce two T antigens, including JCV and BKV, two early mRNAs are generated by alternative splicing; for those that produce large, middle, and small T antigens, the three mRNAs are produced by alternative splicing from a common precursor, analogous to the situation with polyoma.

Early Promoters and Enhancers

The sites at which SV40 early mRNAs are initiated have been mapped in three ways—by nuclease S1 protection using end-labeled DNA probes, by primer extension of short labeled oligonucleotides, and by analysis of the nucleotide sequences attached to 5′-capped oligonucleotides produced following enzymatic digestion of purified labeled viral early mRNAs. All three approaches can produce artifactual results, particularly if the RNAs to be analyzed are not abundant (as viral early mRNAs are) or if they have significant secondary structure. Taken together, these experiments suggest that there are multiple clustered start sites for the SV40 early mRNAs, positioned around nucleotides 5237 and 5231 (Fig. 3) (107,126,292).

Additional species of SV40 early mRNA are also detected that contain 5′ ends at nucleotides 34 through 28 (107). Molecules using these upstream starts can be detected relatively early after infection, but these sites become the primary initiation sites after viral DNA replication begins (100). This shift from early early to late early start sites requires both T antigen and viral DNA replication. When all T-antigen binding sites in the origin region are mutated, and a second origin is introduced at another site, the shift from early early to late early sites still occurs (32). These 5′ extended early mRNAs have the potential to encode an additional viral polypeptide of 23 amino acids (Fig. 2) (161), though deletion mutants unable to produce this protein behave like wild type. Translation from these longer early mRNAs, where the initiation codon for the T antigens is not the first one, is inefficient (32).

The promoters that govern the production of the SV40 early mRNAs were among the first to be studied in detail and dissected by exhaustive mutagenesis. The critical features of the SV40 regulatory region are shown in Fig. 3. Three sites (I, II, and III) for binding of SV40 large T antigen are located in this region. Upstream from the start sites for transcription is located an AT- rich region that contains a TATA-box-like element, TATTTAT. Analysis of deletion and point mutants affecting the SV40 early TATA box and sequences between it and the normal transcriptional start sites demonstrated for the first time that TATA boxes function to direct transcription initiation to a site approximately 30 nucleotides downstream (53). Detailed point mutagenesis of this region indicates that it actually contains two TATA boxes that specify initiation at the two clusters, around nucleotides 5237 and 5231 (226).

Upstream from the AT-rich region is a GC-rich cluster that contains two repeats of a 21-bp sequence and a homologous 22-bp sequence (often referred to as the three 21-bp repeats). Altogether, this region contains six copies of a GC-rich sequence (CCGCCC) present 40–103 nucleotides upstream of the RNA initiation sites. Lying further upstream are two copies of a 72- bp repeat. A variety of *in vivo* and *in vitro* studies have been performed in many laboratories to determine the importance for viral early transcription of these various regulatory region elements. Together they indicate that the 21-bp elements are important promoter elements but are not absolutely essential for early transcription (16,88,100,131). Studies of the SV40 promoter indicate that transcription factor Sp1 binds to these sites *in vitro* and is absolutely required for transcription *in vitro* from the SV40 early promoter (78,79).

The 72-bp repeat elements function as enhancers for the SV40 early promoter. In fact, it was the study of these repeats that led to the discovery of eukaryotic transcriptional enhancers. These elements act to increase transcription initiation in an orientation-independent mechanism when located either upstream or downstream from transcriptional start sites, and with little dependence on distance for their enhancing effects. Many different transcription factor binding sites are located within the SV40 and polyoma enhancers, but it has not yet been possible to identify which factors actually interact with them *in vivo* (for review, see ref. 144). The SV40 enhancer is quite strong and functions well in cells of a variety of species and types; it has been widely used as an enhancer to drive expression of heterologous genes. The BKV enhancer consists of three repeats of a 68-bp sequence, with the middle copy containing an internal 18-bp deletion, and also includes another sequence element located adjacent to the repeat element closest to the BKV late genes (71). JCV contains two copies of a 98-bp sequence that functions as an enhancer and that, like SV40 and BKV enhancers, is located on the late side of the origin or replication (158).

The organization of the regulatory region of polyoma virus is similar to that of SV40 (Fig. 3) but has some important differences. Similar approaches were used to map the major sites of transcription initiation to sequences between nucleotides 147 and 158 (55,148). As in the case of SV40, 5′ sites mapping further upstream were detected during the late phase of polyoma infection (55,91,148). In contrast to their production by transcription initiation during SV40 infection, these 5′-extended polyoma virus transcripts appear to arise by processing of long viral pre-mRNAs that themselves arise by transcription that continues more than once around the complete polyoma virus genome (1). Also in contrast to the late early SV40 transcripts, these polyoma transcripts are uncapped and primarily nuclear.

The polyoma virus regulatory region contains an enhancer element that has a more complex organization than does the SV40 enhancer. The polyoma enhancer contains a single copy of two different adjacent enhancers, called A and B or a and b (138), which can function independently and which confer different cell type specificities. Extensive studies have been conducted of polyoma virus variants selected for efficient growth on either differentiated or undifferentiated embryonal carcinoma (EC) cells; these viral variants contain point mutations within the enhancers (60,205,206). Since enhancers contain multiple sites to which transcription factors can bind productively (144), these genetic studies suggest that there has been a selection for mutants that acquire binding sites for transcription factors present in EC cells.

Regulation of Early Transcription

The early promoters of all polyomaviruses are autoregulated by their large T antigens. This was shown first for SV40. In cells infected with tsA mutants affecting the large T antigen, early mRNAs are overproduced and the rate of synthesis of large T antigen is elevated in parallel (164, 238,289). A similar result has been obtained for polyoma (47). Autoregulation of viral early transcription has been reproduced *in vitro* and shown to be dependent on T-antigen binding sites (128,242,243), indicating that large T antigen autoregulates synthesis of its own mRNA. The binding of large T antigen to viral DNA blocks the assembly of a functional transcription complex, thereby repressing early transcription (217). The location of the binding sites that are the target for this repression is different for SV40 and polyoma. The SV40 binding sites I and II (Fig. 3A) are located downstream of, or overlap, the early mRNA start sites used early after infection, respectively (243), while the binding sites critical for autoregulation by polyoma T antigen (Fig. 3B) are located upstream of the early transcription initiation sites (56,90).

Most studies on autoregulation by large T antigen were performed under conditions where T antigen levels were relatively high. Early after infection, when T antigen levels are much lower, T-antigen binding sites are unoccupied. Under these conditions, large T antigen stimulates transcription from its own promoter by a mechanism that does not require direct binding of T antigen to viral DNA (315).

Processing of Viral Early pre-mRNAs

Alternative splicing is used to generate multiple species of early mRNA from a single class of mRNA precursor (Fig. 2). Because of the advantages of working with viral systems, alternative splicing of SV40 pre-mRNA has been extensively analyzed, both *in vivo* and *in vitro*. as a model system for alternative splicing. The very small size of the intron excised to produce the SV40 small t mRNA (66 nucleotides) helped to define the relationship between minimum intron size and efficiency of excision (102). Splicing to produce large T mRNA uses multiple alternative branch-point sites, whereas splicing to produce small t mRNA uses only the most distal of these sites (220). The use of alternative lariat branch sites is probably central to alternative splicing in other systems. The ratio of small t/large T mRNA produced varies from cell line to cell line, with the ratio being 10–20 times higher in human 293 cells than in many other mammalian cell lines (102). Since the ratio of large T to small t splicing *in vitro* in HeLa cell extracts is 100:1 (221), extracts supplemented with fractions from a human 293 cell extract were used to identify and purify a factor from 293 cells that promoted use of the small t intron. This resulted in the discovery of ASF, a protein factor that controls alternative splicing in SV40 (105) and that appears to play an important role in alternative splicing in many cellular gene systems.

For polyoma, the 48-nucleotide intron excised to produce small t mRNA is the smallest known mammalian intron and is below the minimum size thought to be required for excision. Excision of this intron involves two branch point sites, one only four nucleotides upstream of the small t 3′ splice acceptor site, and requires an intact 3′ splice site for middle T antigen, located 14 nucleotides downstream of the small t 3′ splice acceptor site (106). This indicates that the two 3′ splice sites somehow cooperate to permit excision of the tiny small t intron. Such splice site cooperation could be important for alternative splicing of cellular pre-mRNAs as well.

Synthesis and Functions of the Viral T Antigens

Large T Antigen

The large T antigens of the polyomaviruses are complex multifunctional proteins that have multiple enzymatic activities, interact with several cellular proteins, and perform several different roles during infection. The structures and activities of different large T antigens have been compared and reviewed recently (228). Considerably more is known about SV40 large T antigen than about the other large T antigens. Although this portion of this chapter focuses on SV40 large T antigen, key differences between SV40 and polyoma will be highlighted.

The sizes of the T antigens of several polyomavirinae are listed in Table 1. The functional organization of SV40 and polyoma large T antigens is shown in Fig. 6. All the large T proteins are closely related and contain identical or similar sequences over most of their lengths. Polyoma large T contains a region of 154 amino acids following amino acid 82 with no homology to SV40 large T sequences; the viral DNA encoding this portion of large T is the portion of the polyoma genome that also encodes, in a different reading frame, the unique portion of middle T antigen. Conversely, SV40, JCV, and BKV large T antigen contain approximately 70 amino acids at their carboxy-termini that have no homology with any sequences in polyoma large T. This carboxy-terminal domain of large T is involved in host range and adenovirus helper function (see later).

The SV40 large T protein is modified posttranslationally in several ways. It contains two clusters of phosphorylation sites (249,288) (Fig. 6) and is modified by O-glycosylation (141,253), acylation (167), adenylation (25),

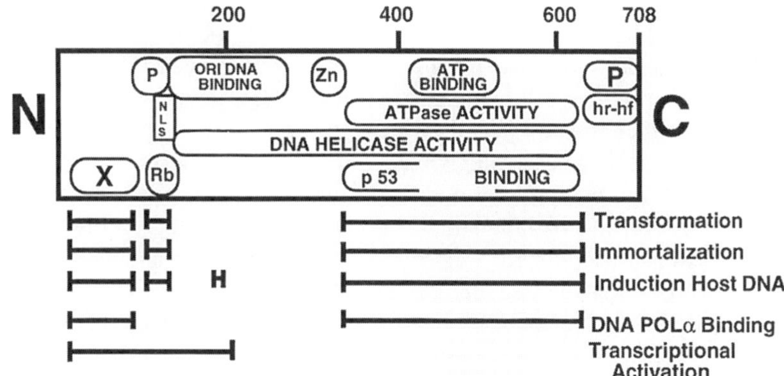

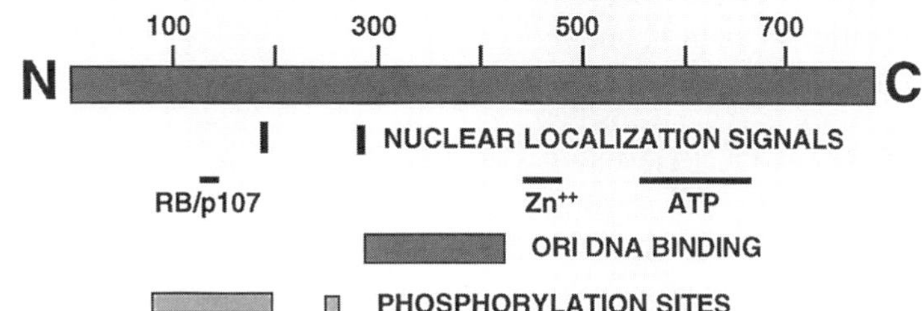

FIG. 6. The domain structures of SV40 and mouse polyomavirus large T antigens. **(A)** SV40. Shown are the domains of large T antigen required for nuclear localization (NLS) and for binding ATP, Zn^{2+}, and for binding to the viral origin of DNA replication. The regions of T antigen required for ATPase activity, helicase activity, host-range helper function activity (hr-hf), as well as for binding to plO5Rb/p107/p130 (indicated by Rb) and to p53 are shown. The region denoted X is required for transformation and immortalization, but the precise role of this region is not known. Also shown are the regions required for transformation of most rodent cell lines, for immortalization of primary mouse fibroblasts, for binding to DNA polymerase, for induction of host DNA synthesis, and for transcriptional activation of the SV40 late promoter and many simple modular promoters by SV40 serve as nuclear for binding to the large T antigen. **(B)** Mouse polyoma virus. Shown are the sequences that serve as nuclear localization signals (NLS) and sequences required for binding ATP, Zn^{2+}, for binding to the polyoma virus organ of DNA replication and to p105Rb/p107/p130. Sites of phosphorylation are also shown. (Panel B courtesy of Dr. Brian Schaffhausen, Tufts University Medical School, Boston, MA.)

poly(ADP)-ribosylation (116), and amino-terminal acetylation (207). Little is known about the functions of any of these modifications except for phosphorylation, which is known to play a major role in controlling the activities and functions of the protein. More than 95% of SV40 large T antigen is located in the nucleus, although a small percentage is found at the plasma membrane. Nuclear T antigen exists free in the nucleoplasm and is also associated with chromatin and nuclear matrix (278). It is estimated that lytically infected cells contain between 5×10^5 and 10^6 molecules of large T antigen.

Detailed studies of the sequences of SV40 large T antigen required for nuclear localization led to the definition of amino acids 127–133 (KKKRKVD) as the first identified nuclear localization signal (145,174). Polyoma large T antigen contains two nuclear localization signals, one at a place analogous to the site of the SV40 nuclear localization signal, the other in a region without homology to SV40 T antigen (241).

SV40 large T antigen is a DNA binding protein that interacts specifically with several pentameric sequences (GAGGC) at the SV40 origin of DNA replication. It also binds, but with considerably lower affinity, to both single- and double-stranded DNA in a sequence-independent manner. The organization of these T-antigen binding sites is shown in Fig. 3. Binding of T antigen to site I permits T antigen to autoregulate production of SV40 early mRNAs (74,128,242). Binding at site II plays a central role in initiation of viral DNA replication. A role for binding site III has not been determined, but the affinity for site III is much lower than that for sites I and II. Binding site II is the preferred site for binding of SV40 large T antigen in the presence of physiologic levels of ATP (22,64). In the presence of ATP, binding of polyoma T antigen to the polyoma regulatory region is also stimulated, and binding occurs over an extended portion of the origin (190).

A second activity of large T antigen is DNA helicase activity (62,276). SV40 large T antigen can bind and hydrolyze ATP, and this activity is stimulated by single-stranded DNA (110,294). Helicase activity is crucial for viral DNA replication and requires functional ATPase activity and the ability to bind to SV40 DNA. *In vitro*, T antigen displays helicase activity with substrates lacking specific T antigen binding sites. T antigen also possesses RNA helicase activity (248), but the importance of this activity is unknown. There have also been reports that T antigen has (or is tightly associated with) topoisomerase I activity and that T antigen is able to promote strand reannealing and DNA looping (193,196,251). The domains of T antigen required for some of these enzymatic activities are shown in Fig. 6. Based on its primary sequence and structural modeling, the DNA binding domain of large T antigen does not resemble any known DNA binding domain, and hence there is no information available about its structure. The ATP-binding/ATPase domain of large T shows substantial similarity to many known ATP binding and hydrolyzing proteins. Located between the DNA binding domain and the ATP binding/ATPase domain is a zinc finger domain important for the overall structure and function of large T antigen (188). Attempts to crystallize T antigen have so far been unsuccessful, perhaps because of heterogeneity resulting from nonuniform posttranslational modification.

T antigen forms complexes with several cellular proteins important for DNA replication, including the largest subunit of DNA polymerase α and the single-stranded DNA binding protein, RPA (76,77,204). Complexes are also formed between T antigen and several proteins important for regulation of cell growth. Through sequences located between amino acids 102 and 115, T antigen interacts with the tumor suppressor protein, p105^{Rb}, as well as with related cellular proteins, p107 and p130 (65,80,89,127). A sequence very similar to T antigen amino acids 102–115 is found in the adenovirus E1A and the human papillomavirus E7 proteins, which also interact with p105^{Rb} and related proteins. In fact, the adenovirus E1A gene remains functional when its p105^{Rb} binding domain is replaced with the p105^{Rb} binding region of SV40 large T antigen (212). SV40 large T antigen has also been reported to interact with an unidentified cellular protein of 185 kD (168), but the significance of this interaction and the identity of the 185-kD protein are unknown. The tumor suppressor protein, p53, forms complexes with SV40 large T antigen (173, 186) through portions of a large, conformationally sensitive domain located in the carboxy-terminal half of the protein (165,252,327). Although p53 also interacts with the JCV and BKV T antigens, it does not form complexes with the polyoma large T antigen, and this is one of the most important differences between SV40 and polyoma large T antigens. Recent evidence indicates that T antigen also forms complexes with the p300 protein (Eckner, Morgan, and Livingston, submitted), which also binds to the adenovirus E1A protein (84,314,320). SV40 T antigen also interacts with cellular transcription factors, including the TATA box binding protein, TBP (125), AP2, and TEF-1.

Small t Antigen

The small t antigens of most polyomaviruses are cysteine-rich proteins of 172–195 amino acids (Table 1) whose amino-terminal 82 amino acids are shared with large T antigen. For those polyomaviruses that encode middle T antigens, almost all of the sequence of small t antigen is contained within the middle T antigen of the same virus. Small t antigen is located in both the nucleus and the cytoplasm (87,330). Studies with mutants of SV40 and polyoma that contain deletions within the region of small t not shared with large T antigen indicate that small t antigen is dispensable for the lytic cycle of these viruses in cultured cells (264,277). The small t antigens associate with two cellular proteins of 36 kd and 63 kd, the regulatory and catalytic subunits of protein phosphatase 2A (224). This interaction is thought to inactivate the mitogen-activated protein kinase ERK1 and the mitogen-activated protein kinase MEK1, resulting in growth stimulation (272).

In some studies, SV40 small t antigen expressed from plasmids has been shown to be capable of activating transcription from the SV40 late and other viral promoters (18, 189). However, when small t was expressed during a viral infection, transcriptional activation was not observed (230), making it uncertain whether small t antigens play a role in regulation of transcription. SV40 small t antigen has been reported to bind zinc ions through its cysteine clusters (304), but the function and importance of this interaction are uncertain, since the small t antigens of the bovine and parakeet viruses contain only a single cysteine and would therefore be unable to bind zinc ions.

Middle T Antigen

T antigens of medium size (middle T) are encoded by polyoma and closely related viruses, but not by SV40 or the human polyomaviruses. The middle T protein is located primarily at the plasma membrane (140,260), where it associates with several cellular proteins. These include c-*src*, the cellular homologue of the transforming protein v-*src* of Rous sarcoma virus (54), and related *src*-family kinase members. This association leads to the activation of the tyrosine kinase activity of c-*src*. Also complexed with middle T antigen and c-*src* is the phosphoinositide-3 kinase (3), *Shc*, a two-protein complex that lies upstream of ras in the signal transduction pathway (34,73), and the same two subunits of protein phosphatase 2A that also associate with the small t antigens of the polyomaviruses (224, 311). Mutants of polyoma virus unable to produce middle T antigen are defective for replication, persistence, transformation, and tumor induction in mice (97).

While formation of complexes with *src*-family kinases is important for cellular transformation, a role for middle T antigen during lytic infection is not known. Since polyoma large T antigen does not interact with p53, and the interaction of SV40 large T with p53 is thought to be central to its ability to move cells into S phase, thereby creating an optimal environment for viral DNA replication, an interesting possibility is that the polyoma middle T/cellular protein interactions perform a related function for polyoma infection. In this view, middle T antigen plays the role of an activated growth factor receptor, and its interactions with *src*-family members, PI-3 kinase, and other proteins of the signal transduction pathway would stimulate signal transduction, ultimately impacting on cell cycle progression and cell growth.

Another t Antigen?

Recent studies suggest that there is yet another t antigen encoded in the SV40 genome, expressed from a third species of SV40 early mRNA that is spliced between nucleotides 4425 and 3679 (326). This alternatively spliced mRNA encodes a 17-kD t antigen containing the first 131 amino acids of large T antigen and 4 amino acids encoded downstream of the splice junction in an alternate reading frame from that used to produce large T antigen. Only limited studies have been done on this protein, but it and its mRNA were detected both in lytically infected monkey cells and in SV40-transformed rat cell lines. In addition, rat F111 cells expressing this protein displayed some properties of transformed cells, similar to what is seen when minimal levels of large T are expressed in nonpermissive cells. Further studies will be necessary to understand fully the properties and importance of this viral gene product.

Preparation for Viral DNA Replication

Soon after T antigen is detected within cells, the levels of many cellular enzymes increase (Fig. 5), and this requires functional T antigen (reviewed in ref. 295). This probably reflects the ability of T antigen to stimulate expression from a variety of viral and cellular promoters (111, 125,240). Diverse mechanisms underlie T antigen's ability to alter host cell gene expression. For example, when T antigen forms complexes with p105Rb and related proteins, other host cell proteins, including transcription factors of the E2F family, are released and can reach important promoter targets (4,5,40). Many genes involved in growth regulation and cell cycle progression, including *c-fos* and *c-myc*, contain E2F binding sites.

The regulation of other cellular genes is probably affected by binding of p53 to T antigen, since p53 is a DNA binding protein (7,159) with both transcriptional activating (154,325) and repressing activities (112). p53 binding to DNA is inhibited by T antigen (8), abrogating p53's ability to stimulate gene expression. Among the targets of p53 transcriptional activation is *WAF1/CIP1*, a potent inhibitor of the cell cycle, because of its ability to block the kinase activity of some cdk/cyclin complexes. It is reasonable to expect that *WAF1/CIP1* expression would be reduced in the presence of T antigen, resulting in cell cycle progression because of the action of cdk/cyclin complexes (86, 129,319).

T antigen stimulates resting cells to enter the cell cycle and replicate their DNA, thereby creating an environment optimal for viral DNA replication. Multiple distinct subregions of large T are able independently to stimulate cellular DNA replication and probably cooperate to provide maximal stimulation (75,273,293). T antigen also stimulates the expression of ribosomal RNA synthesis (273).

Viral DNA Replication

Replication of viral DNA within infected cells requires a functional origin of DNA replication, large T antigen with its DNA binding and helicase activities intact, and the set of cellular proteins involved in replicative DNA synthesis. Viral DNA is in the form of a minichromosome, and its replication occurs in the nucleus, like cellular DNA replication. By other parameters, including the enzymes used and the geometry of the replication fork, replication of SV40 and polyoma DNA are sufficiently similar to cellular DNA replication that the viral systems have been exploited as powerful tools to understand cellular DNA replication. Excellent *in vitro* systems have been developed (183,281,316) that permit accurate and efficient replication of viral DNA using purified cellular proteins, viral DNA containing a functional origin, and purified large T antigen. Viral DNA replication differs from cellular DNA replication in that the SV40 DNA origin can fire multiple times during S phase, whereas cellular DNA replication is tightly controlled to prevent any region of the genome from being replicated more than once in a single cell cycle.

The Viral Origin of DNA Replication

The region of DNA required for initiation of viral DNA replication (*ORI*) has been defined genetically for each of the polyomaviruses by the construction and analysis of detailed sets of deletion and point substitution mutations and by the analysis of evolutionary variants of these viruses that carry deletions of sequences not required for replication (for reviews, see refs. 68 and 69). These studies indicate that both SV40 and polyoma contain a core of sequences absolutely required for viral DNA replication, and auxiliary sequences on either side of the core which act to enhance initiation of DNA replication within the viral core (Fig. 3). For SV40, the core origin of 64 bp extends from nucleotides 5209 to 29, and the auxiliary sequences extend from nucleotides 5164 to 5208 and 30 to 72.

The major difference between SV40 and polyoma in the organization of their *ORI* regions is that polyoma replication requires promoter/enhancer sequences on the late side of the *ORI* core, whereas SV40 replication does not. Critical sequences, shown in Fig. 3, include the core, auxiliary sequences on the early side of *ORI* (aux-1), and enhancer sequences. Replication of DNA-containing polyoma sequences requires core and either aux-2a or aux-2b. The polyoma core covers 66 bp. Although quite different in sequence, the auxiliary sequences of polyoma will substitute for those of SV40, and the SV40 72-bp enhancer can be used in place of the polyoma a and b enhancers. The SV40 21-bp repeats, which play a role as auxiliary origin elements, cannot substitute for the polyoma enhancer elements. Replication of SV40 in monkey cells and polyoma virus in mouse cells absolutely requires the presence of the appropriate core *ORI* and large T antigen (15). Interestingly, the SV40 core *ORI* contains start sites for early mRNA, whereas polyoma early mRNAs are initiated at sites that lie outside of the polyoma core *ORI*.

The features shared by all polyomavirinae *ORI* regions are an inverted repeat of 14 bp on the early gene side of *ORI* (PU/PY in Fig. 3), a GC-rich palindrome of 23–34 bp in the center (ORE in Fig. 3), and an AT sequence of 15–20 bp on the late side (T/A in Fig. 3). That all polyomavirinae core origins contain these elements suggests that common structures play a role in binding proteins and in initiation of viral DNA replication. SV40, BKV, and JCV contain a second nearby palindrome of differing size, but it is not required for viral DNA replication and is absent from the polyoma genome.

Initiation of Viral DNA Replication

In the absence of ATP, T antigen binds as a tetramer to binding site II in the origin region and protects approximately 35 bp of DNA from DNAse (244). In the presence of ATP, T antigen undergoes a conformational shift, permitting the assembly of a bilobed double hexamer of T antigen at the origin of DNA replication (22,63,64,197). Although a single monomer of T antigen can bind to a GAGGC pentanucleotide (dark arrowheads in Fig. 3), clearly not all 12 molecules of T antigen will be bound to a pentanucleotide in the assembled dodecamer/DNA structure. Some of the T antigen subunits can be thought of as playing an allosteric role, modifying the conformation of other subunits to permit proper binding and assembly of the double hexamer (24). ATP hydrolysis is not required for the assembly of this structure on the DNA.

A variety of physical and chemical analyses indicate that the bound T antigen double hexamer catalyzes the local unwinding of part of the early palindrome (PU/PY in Fig. 3) and the distortion or untwisting of the TA-element (22, 262). Once this initial unwinding has occurred, T antigen no longer binds specifically to viral DNA, since the *ORI* region is no longer double-stranded (2). T antigen then associates with replication protein A (RPA), a three-subunit, single-stranded DNA binding protein that keeps unwound regions single stranded, and this permits more extensive unwinding of the DNA through T antigen's helicase activity (62,317). In its ability to unwind double-stranded DNA from an internal site, T antigen is unusual among DNA helicases (198,276).

The association of the viral *ORI*, the T antigen double-hexamer, and RPA is referred to as the preinitiation complex. Recruitment of DNA polymerase α-primase into this complex converts it to an initiation complex. It is this step in viral DNA replication that defines the very limited host range for members of the *polyomavirinae.* SV40 DNA can be replicated in extracts prepared from murine cells if these are supplemented with DNA polymerase α-primase from monkey cells (216) Specific protein contacts form between T antigen, RPA, and DNA polymerase α-primase (52,76, 77,103,267), and the nature of these interactions at the viral origin of replication limits productive complex formation with SV40 and polyoma virus T antigens to primate and murine DNA polymerase α-primases, respectively (254). Polymerase α-primase synthesizes short primers, which are then extended into short DNA fragments (33,133,199).

DNA Synthesis at the Replication Fork

Following initiation, DNA synthesis moves bidirectionally away from the *ORI*. Replication fork movement is facilitated by the helicase activity of T antigen, which translocates along the DNA. The cellular replication factor C (RFC) binds the 3′ end of nascent DNA strands and facilitates the binding of proliferating cell nuclear antigen (PCNA) and DNA polymerase δ (177,303). This complex extends both the continuous leading strand (301,302) and the Okazaki fragments on the lagging strand (310). The actions of RNase H, a 5′-3′-exonuclease (MFI) and DNA ligase I are required for removal of primers and ligation of Okazaki fragments, yielding covalently closed circular form I DNA (309).

Termination of Replication and Separation of Daughter Molecules

Replicative intermediates have been isolated by sucrose gradient centrifugation and analyzed biochemically, electrophoretically, and by electron microscopy. Preparations of replicative intermediates are enriched in molecules that are almost completely replicated. Replication moves bidirectionally around the circular chromosome, with both replication forks advancing at approximately the same rate. These replication intermediates contain three DNA loops—two daughter loops and the unreplicated parental DNA. Molecules that are 85–95% complete are enriched in replicative intermediate pools, suggesting that separation of daughter molecules is a slow step in DNA replication (283,284,286). Topoisomerase II is required for this separation (322). There is no specific termination site for viral DNA replication. Deletion analysis indicates that termination can occur within any sequence located approximately 180° from the origin.

Regulation of Viral DNA Replication by Phosphorylation

The activity of T antigen in viral DNA replication is regulated by phosphorylation of the protein. Both SV40 and polyoma large T antigens carry phosphorylation sites near their DNA binding domains in the amino-terminal half of each protein. An underphosphorylated population of T antigen isolated from infected cells, or T antigen that is dephosphorylated enzymatically, is more active in DNA binding and viral DNA replication *in vitro* (210). While phosphorylation of threonine-124 of SV40 T antigen is essential for unwinding of viral origin DNA and replication (209), dephosphorylation (210,307,308) or substitution (255) of adjacent serine residues enhances origin unwinding and replication. Interestingly, the phosphorylation site at thr-124 is a substrate for the cyclin-dependent kinases (203), whereas protein phosphatase 2A is active in dephosphorylation of the nearby serines whose phosphorylation inhibits viral DNA replication *in vitro* (250). This suggests that the replication activity of T antigen may be regulated by the cell cycle so that T is maximally active for replication during S phase and inactive for replication at earlier stages of the viral infection.

Viral Late Gene Expression

The onset of viral DNA replication brings about a shift in the pattern of viral transcription. The start sites for production of the viral early mRNAs shift to upstream positions (LE in Fig. 3A), and there is a dramatic increase in transcription to produce the viral late mRNAs. Infected cells contain up to 200,000 molecules of viral DNA, and more than half of this may be encapsidated into progeny virions. With 360 molecules of VP1 per virus particle, a minimum of 3.6×10^8 molecules of VP1 must be produced. Assuming a normal loading of ribosomes on viral late mRNA and a normal rate of translation to produce VP1, each molecule of mRNA encoding VP1 could generate 5,000 to 10,000 molecules of VP1 in an infected cell. Thus, more than 30,000 molecules of message encoding VP1 are required. This is several hundred times greater than the abundance of viral early mRNA, making studies of the synthesis and structure of viral late mRNAs relatively easy.

Late transcription of SV40 and polyoma has been studied in the greatest detail. In both systems, two sets of late mRNAs are produced, originally referred to as 16S and 19S late RNAs based on velocity sedimentation in sucrose gradients. The 16S late mRNAs are considerably more abundant than the 19S late mRNAs. These late mRNAs are all derived by transcription that begins near the origin and extends around the opposite strand from that transcribed to produce the early mRNAs (Fig. 2). The start sites for SV40 late mRNAs are heterogeneous and map to many positions (Fig. 3A) between nucleotides 120 and 482, though the start at nucleotide 325 is the most abundant (108,109, 236). All viral late 16S mRNAs contain a leader sequence spliced to a second exon that extends from nucleotide 1464 to a polyadenylation site at nucleotide 2674 (Fig. 2A). All lack sequences between 527 and 1464. Thus, the 16S late mRNAs differ in the structure of their leader regions, with some containing an additional splice and a rare group containing a duplication of sequences in the leader region. The 19S class of late mRNAs are also heterogeneous both at their 5′ ends and in the structure of their leader regions. A minor class is unspliced. All the others contain a leader spliced to sequences extending from nucleotide 558 to the same polyadenylation site used for 16S late mRNAs at nucleotide 2674. All SV40 late mRNAs are derived by processing of precursors which begin at heterogeneous initiation sites and extend most of the way around the viral genome.

Polyoma virus late transcription differs from SV40 in that most viral late pre-mRNAs are extremely large, resulting from transcription completely around the viral genome multiple times (1,181). The mRNAs produced from these giant transcripts contain a single copy of the coding region for late proteins spliced to multiple copies of the leader region by leader-to-leader splicing (147). As with SV40, the start sites for late transcription are heterogeneous (Fig. 3B); they extend from nucleotide 5075 to 5170, with about 90% of them lying in the region from nucleotide 5077 to 5101 (55,58,93,299). The late promoters of SV40 and polyoma lack TATA boxes, which likely explains the heterogeneity of initiation sites for viral late transcription. Maximal rates of SV40 late transcription depend on sequences within the 21-bp GC repeats as well as within the 72-bp enhancers (156,201).

The two primary mechanisms involved in the production of high levels of the viral late mRNAs are amplification of templates for late transcription by viral DNA replication and activation of transcription from the viral late promoter by large T antigen. Studies with origin-defective viral DNA molecules indicate that genome amplification is not required for activation of the late promoter by large T antigen (27,155). The region of large T antigen involved in activation of the late promoter maps to the amino-terminal half of the protein (328). Transcriptional activation by T antigen requires sequences within the 72-bp enhancer region, including a unique sequence element located at the junction of the two 72-bp repeats (124). One factor able to bind to these sequences is TEF-1, which is able to form a complex with large T antigen (125,157). The structure of replicated minichromosomes likely also plays a key role in activation of late transcription (285).

Synthesis of Late Proteins, Assembly and Release of Progeny Virions

Within the cytoplasm of infected cells, viral 16S late mRNAs are translated to produce VP1, and viral 19S late mRNAs are translated to produce VP2 and VP3. Because the sequence of VP3 is contained entirely within that of VP2, VP3 could be derived from VP2 by proteolytic processing. However, deletion mutants lacking the initiation codon for VP2 still produce VP3, indicating that VP3 can be synthesized independently of synthesis of VP2 (49). All the SV40 late mRNAs are polycistronic and contain open reading frames upstream from the capsid protein genes. For some late mRNAs, the upstream ORF encodes the agnoprotein. Studies of translation of SV40 late mRNAs (257–259) have dramatically increased our understanding of how leaky scanning permits internal AUGs to function efficiently as initiation codons (169,170).

Virion proteins VP1, VP2, and VP3 are synthesized in the cytoplasm of infected cells and transported to the nucleus for assembly into virions. When all three capsid proteins are produced, all three enter the nucleus efficiently. Detailed studies of these proteins indicate that sequences near the amino-terminus of VP1 and near the carboxy-termini of VP2 and VP3 are essential for this nuclear accumulation. Within the N-terminus of both SV40 and polyoma VP1 are nuclear localization signals (38,213,318). While sequences required for localization of VP2 and VP3 to the nucleus have also been identified, it is likely that these sequences permit the formation of complexes with VP1, and it is these complexes that are transported to the nucleus by virtue of the VP1 nuclear localization signal (11,67,95,153). Within the nucleus these proteins are found in the same ratio as within the viral capsid (185).

Viral chromatin is not required for the assembly of capsids or capsidlike structures. Purified polyoma VP1 will self-assemble into pentamers and capsidlike structures (246), whereas coexpression of polyoma VP1, VP2, and VP3 in insect cells from baculovirus vectors results in the assembly of capsidlike structures containing all three capsid proteins in approximately the same relative ratio as is found in mature virions (95). Analysis of the process of encapsidation in polyoma-infected cells suggests either that empty capsids are precursors to virions or that capsomers condense on viral chromatin to assemble a virus capsid (324). 75–90S chromatin structures containing viral DNA are detected as well as 200S previrions and 240S virions. Although intracellular chromatin contains histone H1, virion chromatin does not, but the precise time or mechanism of removal of histone H1 is not known. Analysis of immunoprecipitates of insect cells expressing one or more polyoma virus capsid proteins indicates that cellular histones associate with capsid proteins and that VP2 or VP3 may be the direct mediator of capsid protein/histone interactions (95). Polyoma VP1 and SV40 VP2 and VP3 have been shown to bind to DNA in a sequence-independent manner (37,44). The importance of these interactions for virion assembly or stability is uncertain, since capsid protein/histone interactions could serve the same roles in assembly or stability as would capsid protein/DNA interactions.

SV40 and other polyomaviruses do not cause cell lysis. It was previously thought that exit of SV40 from infected cells occurred when dying cells disintegrated or ruptured. This may in fact be an important mechanism for release of virions from infected cells. However, studies of virus- infected epithelial cells lines indicate that SV40 is released from the cell surface, primarily from the apical surface of polarized epithelial cells and uniformly from the surface of nonpolarized epithelial cells (43). By electron microscopy, SV40 was seen in cytoplasmic smooth membrane vesicles and release of SV40 was inhibited by monensin, an inhibitor of intracellular vesicular transport. How and where these vesicles form and the fraction of total infectious progeny released in this manner are not known.

The Host Range/Helper Function of SV40 Large T Antigen

Most monkey cell lines infected by human adenoviruses produce low yields of viral progeny, with the block to productive infection occurring very late in the adenovirus infection cycle (for reviews, see refs. 166 and 295). This block can be overcome by the extreme carboxy-terminus of SV40 large T antigen (48,229), which is unrelated to any portions of polyoma large T antigen. Mutants of SV40 whose T antigens lack a normal carboxy terminus are defective in this helper function and show a restricted growth range among different African green monkey kidney cell lines (50,51,194,227). Human adenoviruses and host range mutants of SV40 both grow well in some monkey kidney cells lines (e.g., Vero) and grow poorly in others (e.g., CV-1). This activity of large T antigen is referred to as the host range/helper function (hr/hf). It is a distinct and separable activity of large T, since a VP1 fusion protein containing the normal carboxy-terminus of large T can provide hr/hf activity (297).

Mutants lacking the hr/hf domain of large T show defects that affect late gene expression. The levels of viral late mRNA and capsid protein are reduced to about one-third of their normal levels, but virion production is reduced much more dramatically (275). Agnoprotein is not produced (162,275), since the 5′ ends of late viral mRNAs in mutant-infected cells map to normally used minor sites downstream of the agnoprotein initiation codon (275). Provision of agnoprotein in trans permits hr/hf mutants to form plaques but does not increase late mRNA or VP1 production. This suggests that agnoprotein permits more efficient use of the capsid proteins available in mutant-infected cells. This is consistent with studies that suggested a role for the agnoprotein either in transport of VP1 to the nucleus or in assembly of progeny virions (10,36,239).

Evolutionary Variants and Defective Viral Genomes

As with most animal viruses, passage of polyomavirinae at high multiplicities of infection leads to the generation of defective viral genomes and the accumulation of those that have selective growth advantages. These evolutionary variants contain deletions, inversions, and rearrangements, and often contain multiple origins of DNA replication (119,178). The only sequences retained by all defective viruses is the viral origin region, indicating that it is the only cis-acting sequence required for propagation of the origin-containing DNA. This has led to the development of many eukaryotic expression vectors containing an SV40 origin of replication. These molecules will replicate in monkey kidney cells, which provide the SV40 T antigen constitutively (e.g., Cos cells) because of the integration into the cellular genome of an origin- defective SV40 genome encoding wild-type T antigen.

Those variants with more than one origin replicate more efficiently than wild-type virus and become enriched. Defective viruses require the presence of wild-type virus for their propagation, since they often contain no intact viral genes, but nondefective variants have been constructed that contain two functional origins (192). Virus stocks maintained by high multiplicity passage can become contaminated with these evolutionary variants, but this can be prevented by frequent reisolation of wild-type virus through plaque purification.

TRANSFORMATION AND IMMORTALIZATION BY SV40

The polyomaviruses are also known as small DNA tumor viruses. The ability of these viruses to transform established cell lines, to immortalize primary cell cultures, and to induce tumors in animals resulted in the widespread study of these viruses as a window into understanding the molecular basis of oncogenesis and growth regulation. Transformed cells grow readily into tumors when injected into syngeneic or immunocompromised hosts.

Early studies indicated that genetic information in the viral early region was necessary and sufficient for transformation (for a review, see refs. 99 and 295). One of the first experiments performed after the development of Southern hybridization was an analysis of the patterns of integration of viral DNA in SV40-transformed cells. Whereas these studies demonstrated that there were no unique or preferred sites of integration in either the viral or host genomes, a functional viral early region was present in all transformed lines analyzed (23,160,247). For SV40, studies using tsA mutants readily demonstrated that T antigen was required for both the initiation and maintenance of the transformed state (30,195,287).

For polyoma, large T antigen is required for transformation (72,82,98), but subsequent studies indicated that T-antigen expression is needed only for initiation but not for the maintenance of the transformed state. The viral DNA in transformed cells is usually integrated so as to interrupt the early region, blocking continued expression of large T antigen but permitting continued synthesis of both small and middle T antigens (13,175). Studies of polyoma hr-t mutants (host range and transformation) (14), which encode a wild-type large T antigen, demonstrated that other early region gene products besides large T antigen were required for transformation by polyoma virus (83,94). It is now known that a cDNA encoding just middle T antigen is sufficient to transform established cell lines (231,300). The requirement for polyoma large T antigen to initiate transformation following viral infection may reflect modest amplification of the genome by replication. Subsequent integration of the viral DNA often interrupts the large T-antigen coding region, thereby eliminating the replicative activities of polyoma large T antigen while leaving the middle T-antigen coding region intact. Levels of middle T antigen sufficient for transformation may be achieved by transfection, thus eliminating the need for large T antigen. In transformation, the polyoma virus middle T antigen resembles an activated growth factor receptor. Its activity requires association with the plasma membrane, where it interacts with many of the same cellular signal transduction proteins as do growth factor receptors (for a review, see ref. 151), presumably leading to constitutive activation of the cell cycle. Since middle T antigen can transform cells that contain a disruption of the *c-src* gene, the interaction between middle T and pp60c-src cannot be essential, but middle T continues to interact with other *src* family kinases and PI-3 kinase in cells lacking *c-src* (290).

Mutational analyses of transformation by SV40 large T antigen demonstrate that only the amino-terminal portion of large T (amino acids 1-121) is required to transform mouse C3H 10T1/2 cells, while sequences extending over much of large T antigen are required to transform REF-52 rat fibroblasts and many other established rodent cell lines (274). Transformation by T antigen can be separated from its role in viral DNA replication (115,146) and does not require any of the activities involved in viral DNA replication (e.g., DNA binding, ATP binding, ATPase activity, helicase activity) or host range/helper function. SV40 mutants defective in transformation of REF-52 cells map to three regions of large T antigen (329)—the amino terminus (amino acids 1–82) and the domains that bind p105^{Rb} (41,146) and p53. SV40 small t antigen is clearly required for transformation under many conditions (264,266), but sufficiently high levels of expression of large T (19), the growth conditions of the infected cells (261), or contributions to transformation from the cellular genetic background probably account for the lack of a uniform requirement for small t antigen for SV40 transformation. Under circumstances where SV40 small t antigen is required for transformation, mutants of small t defective for interaction with and inhibition of protein phosphatase 2A are defective for transformation, suggesting that the interaction with PP2A is central to small t's effect on cellular growth properties (215).

The role of the SV40 large T amino terminus in cellular transformation is not known. Mutants that express only amino acids 83–708 are defective for transformation, as are several deletion, insertion, and point mutations within the first 82 amino acids. Mutants of SV40 that express a T antigen lacking residues 1–82 can be complemented for transformation by SV40 small t antigen (211). The amino terminal region of large T shows some similarity to one of the regions through which the adenovirus E1A protein binds the cellular p300 protein (312,314). Furthermore, mutants of E1A unable to bind p105^{Rb} can be complemented for transformation by mutants of SV40 large T antigen with mutations within the amino-terminal domain, whereas mutants of E1A unable to bind p300 can be complemented by mutants of T antigen unable to bind p105^{Rb} (320). This suggests that the amino terminus of T antigen might bind p300 or provide an activity that controls p300 function within cells. However, more detailed complementation studies show that the ability of large T antigen to complement mutants of E1A defective for p300 binding was substantially reduced by deletions within either the amino-terminal domain or the p105^{Rb} binding site of T antigen. Similarly, the ability of large T antigen to complement mutants of E1A defective for p105^{Rb} binding was also substantially reduced by mutations within the same two transformation-related domains (Zhu J, Cole CN, manuscript in preparation). This suggests that the interactions between cellular proteins and the first 120 amino acids of SV40 large T antigen is complex. Recently, it was reported that p300 forms complexes with SV40 large T antigen but that the sequences within T antigen required for interaction with p300 have not yet been fully defined (Eckner, Morgan, and Livingston, submitted).

Binding of T antigen to p105^{Rb} and related cellular proteins ("pocket proteins") is thought to disrupt their functions. A clue to the importance of p105^{Rb} in cellular growth control is its interaction with the E2F family of transcription factors (218). This group of proteins appears to bind as heterodimers to sequence-specific sites present in the promoter–enhancer regions of genes known to have important roles in cellular growth control, including c-*myc* and c-*fos* (137). Different E2F family members are thought to associate with different p105^{Rb}-related proteins at unique points in the cell cycle (81,180). For example, in G_0, hypophosphorylated p105^{Rb} interacts with E2F, preventing E2F from interacting with target genes (39,136). By binding to the hypophosphorylated form of p105^{Rb}, T antigen causes the release of E2F, and this activates E2F- regulated genes, leading to cell cycle progression (40). p107 complexes with different members of the E2F family, and these complexes are most abundant at the G_1/S boundary (81). T antigen's binding to p107 releases E2F proteins, and the genes activated by these E2F family members may encode proteins that trigger initiation of DNA synthesis. p130-E2F complexes exhibit a third cell cycle pattern of expression and, in lymphocytes and mouse fibroblasts, are found predominantly at the G_0/G_1 border (46).

Yet another p105^{Rb}/p107/p130 mechanism for cell cycle control was identified by the discovery that these proteins, in association with E2Fs, can bind to the cyclins and cyclin-dependent (cdk) kinases. The association of "pocket protein"/E2F complexes with cyclin/cdk kinases appears to be specific for particular cyclin/cdk complexes and to be cell cycle regulated, although the functional consequences of such associations have yet to be elucidated (35, 46,70,179,265)

The ability to immortalize primary rodent cells is another property of the large T antigens of polyoma and SV40 that has also been studied in detail. The domains of polyoma and SV40 large T antigens required for immortalization have also been analyzed but with conflicting results. Some studies demonstrate that the ability of SV40 large T antigen to bind p53 is required for immortalization of primary mouse cells (291,327); others indicate that the amino-terminal 147 amino acids of large T can immortalize primary mouse or rat cells (270,271). Studies with a temperature-sensitive mutant of SV40 T antigen indicate, however, that T antigen is required continuously to maintain immortalization, since cultures shifted to the nonpermissive temperature ceased proliferation (142).

Critical to evaluation of the immortalization potential of these viruses and their individual T antigens are the assays used to measure immortalization. If immortalization of primary cultures requires multiple genetic or epigenetic alterations (e.g., the formation of a complex between SV40 T antigen and p53), mutant viral T antigens may provide some or most of the required functions, and a single additional cellular mutation could result in the full spectrum of genetic changes leading to immortalization. In some assays no attempts were made to distinguish between extension of normal lifespan, with eventual senescence, and true immortalization. In other assays populations of cells still proliferating at a time when control cells had senesced were serially subcultured and considered to be immortal if the culture survived for several months. However, the probability is high that cellular mutations will occur that contribute to immortalization of cells expressing mutant T antigens, particularly if mutation of one cellular gene in one cell in the population is sufficient to cause its immortalization and eventual predominance in the culture.

There is solid evidence that mutation of p53 plays a central role in immortalization. Mutation of the p53 gene was seen in all spontaneously immortalized mouse embryo fibroblast cell lines examined (132). It is easy to see that mutation of p53 or its functional inactivation by complex formation with SV40 large T antigen could increase the frequency of cellular mutations, since p53 loss of function is known to be associated with gene amplification and genomic instability (187,323). In fact, inactivation of p53 by complex formation with SV40 T antigen is likely to be at least partially responsible for the known genomic instability associated with expression of SV40 T antigen (234,279).

Careful analysis of immortalization by mutant SV40 large T antigens suggests that all three domains required for transformation are required for immortalization of primary mouse embryo fibroblasts (Conzen S, Cole CN, manuscript in preparation). Substantial lifespan extension was seen in cultures expressing mutant T antigens defective for the amino-terminal transformation function or unable to bind $p105^{Rb}$, and a small minority of colonies expressing these mutant T antigens became immortalized on continued subculturing. In contrast, neither lifespan extension nor immortalization was achieved with mutants of T antigen unable to bind p53.

Fewer studies have been performed to understand immortalization by polyoma large T antigen, which does not bind p53. An essential role in immortalization for the $p105^{Rb}$-binding activity of polyoma large T antigen has been demonstrated (96,176), and it is likely that additional activities are also required.

REFERENCES

1. Acheson NH. Polyoma giant RNAs contain tandem repeats of the nucleotide sequence of the entire viral genome. *Proc Natl Acad Sci USA* 1978;75:4754–4758.
2. Auborn KJ, Markowitz RB, Wang E, Yu YT, Prives C. Simian virus 40 (SV40) T antigen binds specifically to double-stranded DNA but not to single-stranded DNA or DNA/RNA hybrids containing the SV40 regulatory sequences. *J Virol* 1988;62:2204–2208.
3. Auger KR, Carpenter CL, Shoelson SE, Piwnica-Worms H, Cantley LC. Polyoma virus middle T antigen–pp60c–src complex associates with purified phosphatidylinositol 3-kinase in vitro. *J Biol Chem* 1992; 267:5408–5415.
4. Bagchi S, Weinmann R, Raychaudhuri P. The retinoblastoma protein copurifies with E2F-1, an E1A-regulated inhibitor of the transcription factor E2F. *Cell* 1991;65:1063–1072.
5. Bandara LR, Adamczewski JP, Hunt T, La Thangue NB. Cyclin A and the retinoblastoma gene product complex with a common transcription factor. *Nature* 1991;352:249–251.
6. Barbanti-Brodano G, Swetly P, Koprowski H. Early events in the infection of permissive cells with simian virus 40: adsorption, penetration, and uncoating. *J Virol* 1970;6:78–86.
7. Bargonetti J, Friedman PN, Kern SE, Vogelstein B, Prives C. Wild-type but not mutant p53 immunopurified proteins bind to sequences adjacent to the SV40 origin of replication. *Cell* 1991;65:1083–1091.
8. Bargonetti J, Reynisdóttir I, Friedman PN, Prives C. Site-specific binding of wild-type p53 to cellular DNA is inhibited by SV40 T antigen and mutant p53. *Genes Dev* 1992;6:1886–1898.
9. Barkan A, Mertz JE. DNA sequence analysis of simian virus 40 mutants with deletions mapping in the leader region of the late mRNAs: mutants with deletions similar in size and position exhibit varied phenotypes. *J Virol* 1981;37:730–737.
10. Barkan A, Welch RC, Mertz JE. Missense mutations in the VP1 gene of simian virus 40 that compensate for defects caused by deletions in the viral genome. *J Virol* 1987;61:3190–3198.
11. Barouch DH, Harrison SC. The interaction between the major and minor coat proteins of polyomavirus. *J Virol* 1994;68:3982–3989.
12. Basak S, Turner H, Compans RW. Expression of SV40 receptors on apical surfaces of polarized epithelial cells. *Virology* 1992;190:393–402.
13. Basilico C, Zouzias D, Della-Valle G, et al. Integration and excision of polyoma virus genomes. *Cold Spring Harbor Symp Quant Biol* 1979; 44:611–620.
14. Benjamin TL. Host-range mutants of polyoma virus. *Proc Natl Acad Sci USA* 1970;67:394–401.
15. Bennett ER, Naujokas M, Hassell JA. Requirements for species-specific papovavirus DNA replication. *J Virol* 1989;63:5371–5385.
16. Benoist C, Chambon P. In vivo sequence requirements of the SV40 early promoter region. *Nature* 1981;290:304–310.
17. Berk AJ, Sharp PA. Spliced early mRNAs of simian virus 40. *Proc Natl Acad Sci USA* 1978;75:1274–1278.
18. Bikel I, Loeken MR. Involvement of simian virus 40 (SV40) small t antigen in transactivation of SV40 early and late promoters. *J Virol* 1992;66:1489–1494.
19. Bikel I, Montano X, Agha M, Brown M, McCormack M, Boltax J, Livingston D. SV40 small t antigen enhances the transformation activity of limiting concentrations of SV40 large T antigen. *Cell* 1987;48: 321–330.
20. Bolen JB, Anders DG, Trempy J, Consigli RA. Difference in the subpopulations of the structural proteins of polyoma virions and capsids: biological functions of the VP1 species. *J Virol* 1981;37:80–91.
21. Bolen JB, Consigli RA. Differential adsorption of polyoma virions and

capsids to mouse kidney cells and guinea pig erythrocytes. *J Virol* 1979;32:679–683.

22. Borowiec JA, Hurwitz J. ATP stimulates the binding of simian virus 40 (SV40) large tumor antigen to the SV40 origin of replication. *Proc Natl Acad Sci USA* 1988;85:64–68.
23. Botchan M, Topp WC, Sambrook J. The arrangement of simian virus 40 sequences in the DNA of transformed cells. *Cell* 1976;9:269–287.
24. Bradley MK. Activation of ATPase activity of simian virus 40 large T antigen by the covalent affinity analog of ATP, fluorosulfonylbenzoyl 5′-adenosine. *J Virol* 1990;64:4939–4947.
25. Bradley MK, Hudson J, Villanueva MS, Livingston DM. Specific in vitro adenylation of simian virus 40 large tumor antigen. *Proc Natl Acad Sci USA* 1984;81:6574–6578.
26. Brady JN, Pipas J. DNA viruses: SV40. In: O'Brien SJ, ed. *Genetic maps: locus maps for complex genomes.* Cold Spring Harbor, NY: Cold Spring Harbor Laboratory; 1993:1.88–1.104.
27. Brady JN, Bolen JB, Radonovich M, Salzman N, Khoury G. Stimulation of simian virus 40 late gene expression by simian virus 40 tumor antigen. *Proc Natl Acad Sci USA* 1984;81:2040–2044.
28. Brady JN, Wihnston VD, Consigli RA. Dissociation of polyoma virus by chelation of calcium ions found associated with purified virions. *J Virol* 1977;23:717–724.
29. Breau WC, Atwood WJ, Norkin LC. Class I major histocompatibility proteins are an essential component of the simian virus 40 receptor. *J Virol* 1992;66:2037–2045.
30. Brugge JS, Butel JS. Involvement of the simian virus 40 gene A function in maintenance of transformation. *J Virol* 1975;15:619–635.
31. Buchman AR, Burnett L, Berg P. 1981. The SV40 nucleotide sequence. In: Tooze, J., ed. *DNA Tumor Viruses* Cold Spring Harbor, NY: Cold Spring Harbor Laboratory; 1981:799–841.
32. Buchman AR, Fromm M, Berg P. Complex regulation of SV40 early-region transcription from different overlapping promoters. *Mol Cell Biol* 1984;4:1900–1914.
33. Bullock PA, Seo YS, Hurwitz J. Initiation of simian virus 40 DNA synthesis in vitro. *Mol Cell Biol* 1991;11:2350–2361.
34. Campbell KS, Ogris E, Burke B, et al. Polyoma middle T antigen interacts with SHC via the NPTY motif in middle T. *Proc Natl Acad Sci USA* 1994;91:6344–6348.
35. Cao L, Faha B, Dembski M, Tsai L-H, Harlow E, Dyson N. Independent binding of the retinoblastoma protein and p107 to the transcription factor E2F. *Nature* 1992;355:176–179.
36. Carswell S, Alwine JC. Simian virus 40 agnoprotein facilitates perinuclear–nuclear localization of VP1, the major capsid protein. *J Virol* 1986;60:1055–1061.
37. Chang D, Cai X, Consigli RA. Characterization of the DNA binding properties of polyomavirus capsid protein. *J Virol* 1993;67:6327–6331.
38. Chang D, Hayes DI, Brady JN, Consigli RA. The use of additive and subtractive approaches to examine the nuclear localization sequence of the polyomavirus major capsid protein VP1. *Virology* 1992;189:821–827.
39. Chellappan S, Hiebert S, Mudryj M, Horowitz JM, Nevins JR. The E2F cellular transcription factor is a target for the RB protein. *Cell* 1991;89:4549–4553.
40. Chellappan S, Kraus VB, Kroger B, et al. Adenovirus E1A, simian virus 40 tumor antigen, and human papillomavirus E7 protein share the capacity to disrupt the interaction between transcription factor E2F and the retinoblastoma gene product. *Proc Natl Acad Sci USA* 1992;89:4549–4553.
41. Chen S, Paucha E. Identification of a region of simian virus 40 large T antigen required for cell transformation. *J Virol* 1990;64:3350–3357.
42. Christiansen G, Landers T, Griffith J, Berg P. Characterization of components released by alkali disruption of simian virus 40. *J Virol* 1977;21:1079–1084.
43. Clayson ET, Brando LV, Compans RW. Release of simian virus 40 virions from epithelial cells is polarized and occurs without cell lysis. *J Virol* 1989;63:2278–2288.
44. Clever J, Dean DA, Kasamatsu H. Identification of a DNA binding domain in simian virus 40 capsid proteins VP2 and VP3. *J Biol Chem* 1993;268:20877–20883.
45. Clever J, Yamada M, Kasamatsu H. Import of simian virus 40 virions through nuclear pore complexes. *Proc Natl Acad Sci USA* 1991;88:7333–7337.
46. Cobrinik D, Whyte P, Peeper D, Jacks T, Weinberg R. Cell cycle-specific association of E2F with the p130 E1A-binding protein. *Genes Dev* 1993;7:2392–2404.
47. Cogen B. Virus-specific early RNA in 3T6 cells infected by a tsA mutant of polyoma virus. *Virology* 1978;85:222–230.
48. Cole CN, Crawford LV, Berg P. Simian virus 40 mutants with deletions at the 3′ end of the early region are defective in adenovirus helper function. *J Virol* 1979;30:683–691.
49. Cole CN, Landers T, Goff SP, Manteuil-Brutlag S, Berg P. Physical and genetic characterization of deletion mutants of simian virus 40 constructed in vitro. *J Virol* 1977;24:277–294.
50. Cole CN, Stacy TP. Biological properties of simian virus 40 host range mutants lacking the COOH-terminus of large T antigen. *Virology* 1987;161:170–180.
51. Cole CN, Tornow J, Clark R, Tjian R. Properties of the simian virus 40 (SV40) large T antigens encoded by SV40 mutants with deletions in gene A. *J Virol* 1986;57:539–546.
52. Collins KL, Kelly TJ. Effects of T antigen and replication protein A on the initiation of DNA synthesis by DNA polymerase α-primase. *Mol Cell Biol* 1991;11:2108–2115.
53. Corden J, Wasylyk B, Buchwalder A, Sassone-Corsi P, Kedinger C, Chambon P. Promoter sequences of eukaryotic promoters. *Science* 1980;209:1406–1414.
54. Courtneidge SA, Smith AE. Polyoma virus transforming protein associates with the product of the c-src cellular gene. *Nature* 1983;303:435–439.
55. Cowie A, Jat P, Kamen R. Determination of sequences at the capped 5′-ends of polyomavirus early region transcripts synthesized in vivo and in vitro demonstrates an unusual microheterogeneity. *J Mol Biol* 1982;159:225–255.
56. Cowie A, Kamen R. Multiple binding sites for polymavirus large T antigen within regulatory sequences of polyomavirus DNA. *J Virol* 1984;52:750–760.
57. Cowie A, Kamen R. Guanine nucleotide contacts within viral DNA sequences bound by polyomavirus large T antigen. *J Virol* 1986;57:505–514.
58. Cowie A, Tyndall C, Kamen R. Sequences at the capped 5′-ends of polyomavirus late region mRNAs: an example of extreme heterogeneity. *Nucl Acids Res* 1981;9:6305–6322.
59. Crawford L, Cole CN, Smith AE, et al. The organization and expression of simian virus 40's early genes. *Proc Natl Acad Sci USA* 1978;75:117–122.
60. Dandolo L, Blangy D, Kamen R. Regulation of polyoma virus transcription in murine embryonal carcinoma cells. *J Virol* 1983;47:55–64.
61. Danna K, Nathans D. Bidirectional replication of simian virus 40 DNA. *Proc Nat Acad Sci USA* 1972;69:2391–2395.
62. Dean FB, Bullock P, Murakami Y, Wobbe CR, Weissbach L, Hurwitz J. Simian virus 40 (SV40) DNA replication: SV40 large T antigen unwinds DNA containing an SV40 origin of DNA replication. *Proc Natl Acad Sci USA* 1987;84:16–20.
63. Dean FB, Dodson M, Echols H, Hurwitz J. ATP-dependent formation of a specialized nucleoprotein structure by simian virus 40 (SV40) large tumor antigen at the SV40 replication origin. *Proc Natl Acad Sci USA* 1987;84:8981–8985.
64. Deb SP, Tegtmeyer P. ATP enhances the binding of simian virus 40 large T antigen to the origin of replication. *J Virol* 1987;61:3649–3654.
65. DeCaprio JA, Ludlow JW, Figge J, et al. SV40 large tumor antigen forms a specific complex with the product of the retinoblastoma susceptibility gene. *Cell* 1988;54:275–283.
66. Delmas V, Bastien C, Scherneck S, Feunteun J. A new member of the polyomavirus family: the hamster papovavirus. Complete nucleotide sequence and transformation properties. *EMBO J* 1985;4:1279–1286.
67. Delos SE, Montross L, Moreland RB, Garcea RL. Expression of the polyomavirus VP2 and VP3 proteins in insect cells: coexpression with the major capsid protein VP1 alters VP2/3 localization. *Virology* 1993;194:393–398.
68. DePamphilis M. Eukaryotic DNA replication: anatomy of an origin. *Annu Rev Biochem* 1993;62:29–63.
69. DePamphilis ML, Bradley MK. 1986. Replication of SV40 and polyoma virus chromosomes. In: Salzman N, ed. *The papovaviridae,* vol.1. New York: Plenum; 1986:99–246.
70. Devoto S, Mudryj M, Pines J, Hunter T, Nevins J. A cyclin A–protein kinase complex possesses sequence-specific DNA binding-activity: p33 cdk2 is a component of the E2F–cyclin A complex. *Cell* 1992;68:167–176.
71. Deyerle KD, Cassill JA, Subramani S. Analysis of the early regulatory region of the human papovavirus BK. *Virology* 1987;158:181–193.

72. Di Mayorca G, Callender J, Marin G, Giordano R. Temperature-sensitive mutants of polyoma virus. *Virology* 1969;38:126–133.
73. Dilworth SM, Brewster CEP, Jones MD, Lanfrancone L, Pelicci G, Pelicci PG. Transformation by polyoma virus middle T-antigen involves the binding and tyrosine phosphorylation of Shc. *Nature* 1994; 367:87–90.
74. DiMaio D, Nathans D. Regulatory mutants of simian virus 40: effect of mutations at a T antigen binding site on DNA replication and expression of viral genes. *J Mol Biol* 1982;156:531–548.
75. Dobbelsein M, Arthur AK, Dehde S, van Zee K, Dickmanns A, Fanning E. Intracistronic complementation reveals a new function of SV40 T antigen that co-operates with Rb and p53 binding to stimulate DNA synthesis in quiescent cells. *Oncogene* 1992;7:837–847.
76. Dornreiter I, Erdile LF, Gilbert IU, von Winkler D, Kelly TJ, Fanning E. Interaction of DNA polymerase alpha-primase with cellular replication protein A and SV40 T antigen. *EMBO J* 1992;11:769–776.
77. Dornreiter I, Höss A, Arthur AK, Fanning E. SV40 T antigen binds directly to the large subunit of purified DNA polymerase alpha. *EMBO J* 1990;9:3329–3336.
78. Dynan WS, Tjian R. Isolation of transcription factors that discriminate between different promoters recognized by RNA polymerase II. *Cell* 1983;32:669–680.
79. Dynan WS, Tjian R. The promoter-specific transcription factor Sp1 binds to upstream sequence in the SV40 early promoter. *Cell* 1983;35: 79–87.
80. Dyson N, Buchkovich K, Whyte P, Harlow E. The cellular 107K protein that binds to adenovirus E1A also associates with the large T antigens of SV40 and JC virus. *Cell* 1989;58:249–255.
81. Dyson N, Dembski A, Fattaey A, Ngwu C, Ewen M, Helin K. Analysis of the p107-associated proteins: p107 associates with a form of E2F that differs from pRb-associated E2F-1. *J Virol* 1993;67:7641–7647.
82. Eckhart W. Complementation and transformation by temperature-sensitive mutants of polyoma virus. *Virology* 1969;38:120–125.
83. Eckhart W. Complementation between temperature-sensitive and host range nontransforming mutants of polyoma virus. *Virology* 1977;77: 589–597.
84. Eckner R, Ewen ME, Newsome D, et al. Molecular cloning and functional analysis of the adenovirus E1A-associated 300-kD protein (p300) reveal a protein with properties of a transcriptional adaptor. *Genes Dev* 1994;8:869–884.
85. Eddy BE, Borman GS, Grubbs GE, Young RD. Identification of the oncogenic substance in rhesus monkey kidney cell cultures as simian virus 40. *Virology* 1962;17:65–75.
86. El-Deiry WS, Tokino T, Velculescu VE, et al. WAF1, a potential mediator of p53 tumor suppression. *Cell* 1993;75:817–825.
87. Ellman M, Bikel I, Figge J, Roberts T, Schlossman R, Livingston DM. Localization of the simian virus 40 small t antigen in the nucleus and cytoplasm of monkey and mouse cells. *J Virol* 1984;50:623–628.
88. Everett RD, Baty D, Chambon P. The repeated GC-rich motifs upstream from the TATA box are important elements of the SV40 early promoter. *Nucl Acids Res* 1983;11:2447–2464.
89. Ewen ME, Ludlow JW, Marsilio E, et al. An N-terminal transformation-governing sequence of SV40 large T antigen contributes to the binding of both p110Rb and a second cellular protein, p120. *Cell* 1989; 58:257–267.
90. Farmerie WG, Folk WR. Regulation of polyoma virus transcription by large T antigen. *Proc Natl Acad Sci USA* 1984;81:6919–6923.
91. Fenton RG, Basilico C. Changes in the topography of early region transcription during polyoma virus lytic infection. *Proc Natl Acad Sci USA* 1982;79:7142–7146.
92. Fiers W, Contreras R, Haegeman G, et al. The complete nucleotide sequence of SV40 DNA. *Nature* 1978;273:113–120.
93. Flavell AJ, Cowie A, Arrand JR, Kamen R. Localization of three major capped 5′ ends of polyoma late mRNAs within a single tetranucleotide sequence in the viral genome. *J Virol* 1980;33:902–908.
94. Fluck MM, Staneloni RJ, Benjamin TL. Hr-t and ts-a: two early gene functions in polyoma virus DNA. *Virology* 1977;77:610–624.
95. Forstova J, Krauzewica N, Wallace S, et al. Cooperation of structural proteins during late events in the life cycle of polyomavirus. *J Virol* 1993;67:1405–1413.
96. Freund R, Bronson RT, Benjamin TL. Separation of immortalization from tumor induction with polyoma large T mutants that fail to bind the retinoblastoma gene product. *Oncogene* 1992;7:1979–1987.
97. Freund R, Sotnikov A, Bronson RT, Benjamin TL. Polyoma virus middle T is essential for virus replication and persistence as well as for tumor induction in mice. *Virology* 1992;191:716–723.
98. Fried M. Cell-transforming ability of a temperature-sensitive mutant of polyoma virus. *Proc Natl Acad Sci USA* 1965;53:486–491.
99. Fried M, Prives C. The biology of simian virus 40 and polyomavirus. In: Botchan M, Grodzicker T, Sharp P, eds. *Cancer cells 4: DNA tumor viruses: control of gene expression and replication.* Cold Spring Harbor, NY: Cold Spring Harbor Laboratory; 1986:1–16.
100. Fromm M, Berg P. Deletion mapping of DNA regions required for SV40 early region promoter function in vivo. *J Mol Appl Genet* 1982; 1:457–481.
101. Frost E, Borgaux P. Decapsidation of polyoma virus: identification of subviral species. *Virology* 1975;68:245–255.
102. Fu X-Y, Manley JL. Factors influencing alternative splice site utilization in vivo. *Mol Cell Biol* 1987;7:738–748.
103. Gannon JV, Lane DP. Interactions between SV40 T antigen and DNA polymerase α. *New Biol* 1990;2:84–92.
104. Gardner SD, Field AM, Coleman DV, Hulme B. New human papovavirus (B.K.) isolated from urine after renal transplantation. *Lancet* 1971;i:1253–1257.
105. Ge H, Manley JL. A protein factor, ASF, controls cell-specific alternative splicing of SV40 early pre-mRNA in vitro. *Cell* 1990;62:25–34.
106. Ge H, Noble J, Colgan J, Manley JL. Polyoma virus small tumor antigen pre-mRNA splicing requires cooperation between 3′ splice sites. *Proc Natl Acad Sci USA* 1990;87:3338–3342.
107. Ghosh PK, Lebowitz P. Simian virus 40 early mRNAs contain multiple 5′-termini upstream and downstream from a Hogness-Goldberg sequence: a shift in 5′-termini during the lytic cycle is mediated by large T antigen. *J Virol* 1981;40:224–240.
108. Ghosh PK, Reddy VB, Swinscoe J, Choudary PV, Lebowitz P, Weissman SM. The 5′ terminal leader sequence of late 16S mRNA from cells infected with siman virus 40. *J Biol Chem* 1978;253:3643–3647.
109. Ghosh PK, Reddy VB, Swinscoe J, Lebowitz P, Weissman SM. The heterogeneity and 5′ terminal structures of the late RNAs of simian virus 40. *J Mol Biol* 1978;126:813–846.
110. Giacherio D, Hager LP. A poly(dT) stimulated ATPase activity associated with simian virus 40 large T antigen. *J Biol Chem* 1979;254: 8113–8120.
111. Gilinger G, Alwine JC. Transcriptional activation by simian virus 40 large T antigen: requirements for simple promoter structures containing either TATA or initiator elements with variable upstream factor binding sites. *J Virol* 1993;11:6682–6688.
112. Ginsberg D, Mechta F, Yaniv M, Oren M. Wild-type p53 can downmodulate the activity of various promoters. *Proc Natl Acad Sci USA* 1991;88:9979–9983.
113. Girardi AJ, Sweet BH, Slotnick VB, Hilleman MR. Development of tumors in hamsters inoculated in the neo-natal period with vacuolating virus, SV40. *Proc Soc Exp Biol Med* 1962;109:649–660.
114. Gluzman Y. SV40 transformed cells support the replication of early SV40 mutants. *Cell* 1981;23:175–182.
115. Gluzman Y, Ahrens B. SV40 early mutants that are defective for viral synthesis but competent for transformation of cultured rat and simian cells. *Virology* 1982;123:78–92.
116. Goldman ND, Brown M, Khoury G. Modification of SV40 T-antigen by poly ADP-ribosylation. *Cell* 1981;24:567–572.
117. Goutebroze L, Feunteun J. Transformation by hamster polyomavirus: identification and functional analysis of the early genes. *J Virol* 1992; 66:2495–2504.
118. Graffi A, Schramm T, Graffi I, Bierwolf D, Bender E. Virus-associated skin tumors of the Syrian hamster: preliminary note. *J Natl Cancer Inst* 1968;40:867–873.
119. Griffin B, Fried M. Amplification of a specific region of the polyoma virus genome. *Nature* 1975;256:175–179.
120. Griffin BE, Ito Y. DNA viruses: polyomavirus. In: O'Brien SJ, ed. *Genetic maps: locus maps of complex genomes.* Cold Spring Harbor, NY: Cold Spring Harbor Laboratory; 1993:1.105–1.113.
121. Griffin BE, Soeda E, Barrell BG, Staden R. 1981. Sequence and analysis of polyoma virus DNA. In: Tooze J, ed. *DNA tumor viruses.* Cold Spring Harbor, NY: Cold Spring Harbor Laboratory; 1981:843–913.
122. Griffith GR, Marriott SJ, Rintoul DA, Consigli RA. Early events in polyomavirus infection: fusion of mono-pinocytotic vesicles containing virions with mouse kidney cell nuclei. *Virus Res* 1988;10:41–52.
123. Gross L. A filterable agent, recovered from Ak leukemic extracts, caus-

ing salivary gland carcinomas in C3H mice. *Proc Soc Exp Biol Med* 1953;83:414–421.
124. Gruda MC, Alwine JC. Simian virus 40 (SV40) T antigen transcriptional activation mediated through the Oct/SPH region of the SV40 late promoter. *J Virol* 1991;65:3553–3558.
125. Gruda MC, Zabolotny JM, Xiao JH, Davidson I, Alwine JC. Transcriptional activation by simian virus 40 large T antigen: interactions with multiple components of the transcription complex. *Mol Cell Biol* 1993;13:961–969.
126. Haegeman G, Fiers W. Characterization of the 5′-terminal cap structures of early simian virus-40 RNA. *J Virol* 1980;35:955–961.
127. Hannon GJ, Demetrick D, Beach D. Isolation of the RB-related p130 through its interaction with CDK2 and cyclins. *Genes Dev* 1993;7: 2378–2391.
128. Hansen U, Tenen DG, Livingston DM, Sharp PA. T antigen repression of SV40 early transcription from two promoters. *Cell* 1981;27:603–612.
129. Harper JW, Adami GR, Wei N, Keyomarsi K, Elledge SJ. The p21 Cdk-interacting protein Cip1 is a potent inhibitor of G1 cyclin-dependent kinases. *Cell* 1993;75:805–816.
130. Hartley JW, Rowe WP. New papovavirus contaminating Shope papillomata. *Science* 1964;143:258–260.
131. Hartzell SW, Yamaguchi J, Subramanian KN. SV40 deletion mutants lacking the 21-bp repeated sequences are viable, but have non-complementable deficiencies. *Nucl Acids Res* 1983;11:1601–1616.
132. Harvey DM, Levine AJ. p53 alteration is a common event in the spontaneous immortalization of primary Balb/c murine embryo fibroblasts. *Genes Dev* 1991;5:2375–2385.
133. Hay RT, DePamphilis ML. Initiation of simian virus 40 DNA replication in vivo: location and structure of 5′-ends of DNA synthesized in the ori region. *Cell* 1982;28:767–779.
134. Hayes JL, Consigli RA. Phosphorylation of the budgerigar fledgling disease virus major capsid protein VP1. *J Virol* 1992;66:4551–4555.
135. Heiser WC, Eckhart W. Polyoma virus early and late mRNAs in productively infected mouse 3T6 cells. *J Virol* 1982;44:175–188.
136. Helin K, Lees J, Vidal M, Dyson N, Harlow E, Fattaey A. cDNA encoding a pRb-binding protein with properties of the transcription factor E2F. *Cell* 1992;70:337–350.
137. Helin K, Wu C, Fattaey A, Lees J, Dynlacht B, Ngwu C, Harlow E. Heterodimerization of the transcription factors E2F and DP-1 leads to cooperative transactivation. *Genes Dev* 1993;1850–1861.
138. Herbomel P, Bourachot B, Yaniv M. Two distinct enhancers with different cell specificities coexist in the regulatory region of polyoma. *Cell* 1984;39:653–662.
139. Hummeler K, Tomassini N, Sokol F. Morphological aspects of the uptake of simian virus 40 by permissive cells. *J Virol* 1970;6:87–93.
140. Ito Y, Brocklehurst JR, Dulbecco R. Virus-specific proteins in the plasma membrane of cells lytically infected or transformed by polyoma virus. *Proc Natl Acad Sci USA* 1977;74:4666–4670.
141. Jarvis DL, Butel JS. Modification of simian virus 40 large tumor antigen by glycosylation. *Virology* 1985;141:173–189.
142. Jat PS, Sharp PA. Cell lines established by a temperature-sensitive simian virus 40 large-T-antigen gene are growth restricted at the nonpermissive temperature. *Mol Cell Biol* 1989;9:1672–1681.
143. Jay G, Nomura S, Anderson CW, Khoury G. Identification of the SV40 agnogene product: a DNA binding protein. *Nature* 1981;291:346–349.
144. Jones NC, Rigby PWJ, Ziff EB. Trans-acting protein factors and the regulation of eukaryotic transcription: lessons from studies on DNA tumor viruses. *Genes Dev* 1988;2:267–281.
145. Kalderon D, Roberts B, Richardson WD, Smith AE. A short amino acid sequence able to specify nuclear localization. *Cell* 1984;39: 499–509.
146. Kalderon D, Smith AE. In vitro mutagenesis of a putative DNA binding domain of SV40 large-T antigen. *Virology* 1984;139:109–137.
147. Kamen R, Favaloro J, Parker J. Topography of the three late mRNAs of polyomavirus which encode the virion proteins. *J Virol* 1980;33: 637–651.
148. Kamen R, Jat P, Treisman R, Favaloro J. Termini of polyoma virus early region transcripts synthesized in vivo by wild-type virus and viable deletion mutants. *J Mol Biol* 1982;159:189–224.
149. Kamen R, Lindstrom DM, Shure H, Old RW. Virus-specific RNA in cells productively infected or transformed by polyoma virus. *Cold Spring Harbor Symp Quant Biol* 1975;39:187–198.
150. Kamen R, Shure H. Topography of polyoma virus messenger RNA molecules. *Cell* 1976;7:361–371.
151. Kaplan DR, Pallas DC, Morgan W, Schaffhausen B, Roberts TM. Mechanisms of transformation by polyoma virus middle T antigen. *Biochim Biophys Acta* 1989;948:345–364.
152. Kartenbeck J, Stukenbrok H, Helenius A. Endocytosis of simian virus 40 into the endoplasmic reticulum. *J Cell Biol* 1989;109:2721–2729.
153. Kasamatsu H, Nehorayan A. VP1 affects the intracellular localization of VP3 polypeptide during simian virus 40 infection. *Proc Natl Acad Sci USA* 1979;76:2808–2812.
154. Kastan MB, Onyekwere O, Sidransky D, Vogelstein B, Craig RW. Participation of p53 protein in the cellular response to DNA damage. *Cancer Res* 1991;51:6304–6311.
155. Keller JM, Alwine JC. Activation of the SV40 late promoter: direct effects of T antigen in the absence of viral DNA replication. *Cell* 1984; 36:381–389.
156. Keller JM, Alwine JC. Analysis of an activatable promoter: sequences in the simian virus 40 late promoter required for T-antigen-mediated trans activation. *Mol Cell Biol* 1985;5:1859–1869.
157. Kelly JJ, Wildeman AG. Role of the SV40 enhancer in the early to late shift in viral transcription. *Nucl Acids Res* 1991;19:6799–6804.
158. Kenney S, Natarajan V, Strike D, Khoury G, Salzman NP. JC virus enhancer-promoter active in human brain cells. *Science* 1984;226:1337–1339.
159. Kern SE, Kinzler KW, Bruskin A, et al. Identification of p53 as a sequence-specific DNA- binding protein. *Science* 1991;252:1708–1711.
160. Ketner G, Kelly TJ. Integrated simian virus 40 sequences in transformed cell DNA: analysis using restriction endonucleases. *Proc Natl Acad Sci USA* 1976;73:1102–1106.
161. Khalili K, Brady J, Khoury G. Translational regulation of SV40 early mRNA defines a new viral protein. *Cell* 1987;48:639–645.
162. Khalili K, Brady J, Pipas JM, Spence SL, Sadofsky M, Khoury G. Carboxyl-terminal mutants of the large tumor antigen of simian virus 40: a role for the early protein late in the lytic cycle. *Proc Natl Acad Sci USA* 1988;85:354–358.
163. Khoury G, Howley P, Nathans D, Martin M. Posttranscriptional selection of simian virus 40-specific RNA. *J Virol* 1975;15:433–437.
164. Khoury G, May E. Regulation of early and late simian virus 40 transcription: overproduction of early viral RNA in the absence of a functional T antigen. *J Virol* 1977;23:167–176.
165. Kierstead TD, Tevethia MJ. Association of p53 binding and immortalization of primary C57BL/6 mouse embryo fibroblasts by using simian virus 40 T-antigen mutants bearing internal overlapping deletion mutations. *J Virol* 1993;67:1817–1829.
166. Klessig DF. Adenovirus-simian virus 40 interactions. In: Ginsberg H, ed. *The adenoviruses.* New York: Plenum; 1984:399–449.
167. Klockmann U, Deppert W. Acylated simian virus 40 large T-antigen: a new subclass associated with a detergent resistant lamina of the plasma membrane. *EMBO J* 1983;2:1151–1157.
168. Kohrman DC, Imperiale MJ. Simian virus 40 large T antigen stably complexes with a 185-kilodalton host protein. *J Virol* 1992;66:1752–1760.
169. Kozak M. Mechanism of mRNA recognition by eukaryotic ribosomes during initiation of protein synthesis. *Curr Top Microbiol Immunol* 1981;93:81–123.
170. Kozak M. Point mutations define a sequence flanking the AUG initiator codon that modulates translation by eukaryotic ribosomes. *Cell* 1986;44:283–292.
171. Lai C, Nathans D. A map of temperature-sensitive mutants of simian virus 40. *Virology* 1975;66:70–81.
172. Lai C-J, Nathans D. Deletion mutants of simian virus 40 generated by enzymatic excision of DNA segments from the viral genome. *J Mol Biol* 1974;89:179–193.
173. Lane DP, Crawford LV. T antigen is bound to a host protein in SV40-transformed cells. *Nature* 1979;278:261–263.
174. Lanford RE, Butel JS. Construction and characterization of an SV40 mutant defective in nuclear transport of T antigen. *Cell* 1984;37:801–813.
175. Lania L, Hayday A, Bjursell G, Gandini-Attardi D, Fried M. Organization and expression of integrated polyoma virus DNA in transformed rodent cells. *Cold Spring Harbor Symp Quant Biol* 1979;44:597–603.
176. Larose A, Dyson N, Sullivan M, Harlow E, Bastin M. Polyomavirus large T mutants affected in retinoblastoma protein binding are defective in immortalization. *J Virol* 1991;65:2308–2313.
177. Lee SH, Kwong AD, Pan ZQ, Hurwitz J. Studies on the activator 1 protein complex, an accessory factor for proliferating cell nuclear antigen-dependent DNA polymerase delta. *J Biol Chem* 1991;266:594–602.

178. Lee TNH, Brockman WW, Nathans D. Evolutionary variants of SV40: cloned substituted variants containing multiple initiation sites for DNA replication. *Virology* 1975;66:53–69.
179. Lees E, Faha B, Dulic V, Reed S, Harlow E. Cyclin E/cdk 2 and cyclin A/cdk2 kinases associate with p107 and E2F in a temporally distinct manner. *Genes Dev* 1992;6:1874–1885.
180. Lees JA, Saito M, Vidal M, et al. The retinoblastoma protein binds to a family of E2F transcription factors. *Mol Cell Biol* 1993;13:7813–7825.
181. Legon S, Flavell A, Cowie A, Kamen R. Amplification of the leader sequences of "late" polyomavirus mRNAs. *Cell* 1979;16:373–388.
182. Lehn H, Müller H. Cloning and characterization of budgerigar fledgling disease virus, an avian polyomavirus. *Virology* 1986;151:362–370.
183. Li JJ, Kelly TJ. Simian virus 40 DNA replication in vitro. *Proc Natl Acad Sci USA* 1984;81:6973–6977.
184. Liddington RC, Yan Y, Moulai J, Sahli R, Benjamin TL, Harrison SC. Structure of simian virus 40 at 3.8-Å resolution. *Nature* 1991;354: 278–294.
185. Lin W, Hata T, Kasamatsu H. Subcellular distribution of viral structural proteins during simian virus 40 infection. *J Virol* 1984;50:363–371.
186. Linzer DIH, Levine AJ. Characterization of a 54 Kdalton cellular SV40 tumor antigen present in SV40 transformed cells and uninfected embryonal carcinoma cells. *Cell* 1979;17:43–52.
187. Livingstone LR, White A, Sprouse J, Livanos E, Jacks T, Tlsty TD. Altered cell cycle arrest and gene amplification potential accompany loss of wild-type p53. *Cell* 1992;70:923–935.
188. Loeber G, Parsons R, Tegtmeyer P. A genetic analysis of the zinc finger of SV40 large T antigen. *Curr Top Microbiol Immunol* 1989;144: 21–29.
189. Loeken MR, Bikel I, Livingston DM, Brady J. Trans-activation of RNA polymerase II and III promoters by SV40 small t antigen. *Cell* 1988; 55:1171–1177.
190. Lorimer HE, Wang EH, Prives C. The DNA-binding properties of polyomavirus large T antigen are altered by ATP and other nucleotides. *J Virol* 1991;65:687–699.
191. Mackay RL, Consigli RA. Early events in polyoma virus infection: attachment, penetration, and nuclear entry. *J Virol* 1976;19:620–636.
192. Magnusson TG, Nilsson M-G. Viable polyoma virus variant with two origins of DNA replication. *Virology* 1982;119:12–21.
193. Mann K. Topoisomerase activity is associated with purified SV40 T antigen. *Nucl Acids Res* 1993;21:1697–1704.
194. Manos MM, Gluzman Y. Genetic and biochemical analysis of transformation-competent, replication-defective simian virus 40 large T antigen mutants. *J Virol* 1985;53:120–127.
195. Martin RG, Chou JY. Simian virus 40 functions required for the establishment and maintenance of malignant transformation. *J Virol* 1975; 15:599–612.
196. Marton A, Jean D, Delbecchi L, Simmons DT, Bourgaux P. Topoisomerase activity associated with SV40 large tumor antigen. *Nucl Acids Res* 1990;21:1689–1695.
197. Mastrangelo IA, Hough PVC, Wall JS, Dodson M, Dean FB, Hurwitz J. ATP-dependent assembly of double hexamers of SV40 T antigen at the viral origin of DNA replication. *Nature* 1989;338:658–662.
198. Matson SW, Kaiser-Rogers KA. DNA helicases. *Annu Rev Biochem* 1990;59:289–329.
199. Matsumoto T, Eki T, Hurwitz J. Studies on the initiation and elongation reactions in the simian virus 40 DNA replication system. *Proc Natl Acad Sci USA* 1990;87:9712–9716.
200. Mattern CF, Takemoto KK, Daniel WA. Replication of polyoma virus in mouse embryo cells: electron microscopic observations. *Virology* 1966;30:242–256.
201. May E, Omilli F, Ernoult-Lange E, Zenke M, Chambon P. The sequence motifs that are involved in SV40 enhancer function also control SV40 late promoter activity. *Nucl Acids Res* 1987;15:2445–2461.
202. Mayer M, Dorries K. Nucleotide sequence and genome organization of the murine polyomavirus, Kilham strain. *Virology* 1991;181:469–480.
203. McVey D, Brizuela L, Mohr I, Marshak D, Gluzman Y, Beach D. Phosphorylation of large tumour antigen by cdc2 stimulates SV40 DNA replication. *Nature* 1989;341:503–507.
204. Melendy T, Stillman B. An interaction between replication protein A and SV40 T antigen appears essential for primasome assembly during SV40 DNA replication. *J Biol Chem* 1993;268:3389–3395.
205. Melin F, Pinon H, Kress C, Blangy D. Isolation of polyomavirus mutants multiadapted to murine embryonal carcinoma cells. *J Virol* 1985; 28:992–996.
206. Melin F, Pinon H, Reiss C, Kress C, Montreau N, Blangy D. Common features of polyomavirus mutants selected on PCC4 embryonal carcinoma cells. *EMBO J* 1985;4:1799–1803.
207. Mellor A, Smith AE. Characterization of the amino terminal tryptic peptide of simian virus 40 small-t and large-T antigens. *J Virol* 1978; 28:992–996.
208. Mertz JE, Berg P. Viable deletion mutants of simian virus 40: selective isolation by means of a restriction endonuclease from Haemophilus parainfluenzae. *Proc Nat Acad Sci USA* 1974;71:4879–4883.
209. Moarefi IF, Small D, Gilbert I, et al. Mutation of the cyclin-dependent kinase phosphorylation site in simian virus 40 (SV40) large T antigen specifically blocks SV40 origin DNA unwinding. *J Virol* 1993;67: 4992–5002.
210. Mohr IJ, Stillman B, Gluzman Y. Regulation of SV40 DNA replication by phosphorylation of T antigen. *EMBO J* 1987;6:153–160.
211. Montano X, Millikan R, Milhaven JM, et al. Simian virus 40 small tumor antigen and an amino-terminal domain of large tumor antigen share a common transforming function. *Proc Nat Acad Sci USA* 1990; 97:7448–7452.
212. Moran E. A region of SV40 large T antigen can substitute for a transforming domain of the adenovirus E1A products. *Nature* 1988;334: 168–170.
213. Moreland RB, Garcea RL. Characterization of a nuclear localization sequence in the polyomavirus capsid protein VP1. *Virology* 1991;185: 513–518.
214. Mortimer EA, Lepow ML, Gold E, Robbins FC, Burton GJ, Fraumeni JF. Long-term follow-up of persons inadvertently inoculated with SV40 as neonates. *N Engl J Med* 1981;305:1517–1518.
215. Mungre S, Enderle K, Turk B, Porras A, Wu Y-Q, Mumby MC, Rundell K. Mutations which affect the inhibition of protein phosphatase 2A by simian virus 40 small-t antigen in vitro decrease viral transformation. *J Virol* 1994;68:1675–1681.
216. Murakami Y, Wobbe CR, Weissbach L, Dean FB, Hurwitz J. Role of DNA polymerase α and DNA primase in simian virus 40 DNA replication in vitro. *Proc Natl Acad Sci USA* 1986;83:2869–2873.
217. Myers RM, Rio DC, Robbins AK, Tjian R. SV40 gene expression is modulated by the cooperative binding of T antigen to DNA. *Cell* 1981; 25:373–384.
218. Nevins JR. E2F: A link between the Rb tumor suppressor protein and viral oncoproteins. *Science* 1992;258:424–429.
219. Nishimura T, Kawai N, Kawai M, Notake K, Ichihara I. Fusion of SV40-induced endocytic vacuoles with the nuclear membrane. *Cell Struct Funct* 1986;11:135–141.
220. Noble JCS, Pan Z-Q, Prives C, Manley JL. Splicing of SV40 early pre-mRNA to large T and small t mRNAs utilizes different patterns of lariat branch sites. *Cell* 1987;50:227–236.
221. Noble JCS, Prives C, Manley JL. In vitro splicing of simian virus 40 early pre-mRNA. *Nucl Acids Res* 1986;14:1219–1235.
222. Nomura S, Khoury G, Jay G. Subcellular localization of the simina virus 40 agnoprotein. *J Virol* 1983;45:428–433.
223. Padgett BL, Walker DL, ZuRhein GM, Eckroade RJ, Dessel BH. Cultivation of a papova- like virus from human brain with progressive multifocal leukoencephalopathy. *Lancet* 1971;i:1257–1260.
224. Pallas DC, Shahrink LK, Martin BL, et al. Polyoma small and middle T antigens and SV40 small t antigen form stable complexes with protein phosphatase 2A. *Cell* 1990;60:167–176.
225. Paucha E, Mellor A, Harvey R, Smith AE, Hewick RM, Waterfield MD. Large and small tumor antigens from SV40 have identical amino termini mapping at 0.65 map units. *Proc Natl Acad Sci USA* 1978;75: 2165–2169.
226. Pauly M, Treger M, Westhof E, Chambon P. The initiation accuracy of SV40 early transcription is determined by the functional domains of two TATA elements. *Nucl Acids Res* 1992;20:975–982.
227. Pipas JM. Mutations near the carboxyl terminus of the simian virus 40 large T antigen alter viral host range. *J Virol* 1985;54:569–575.
228. Pipas JM. Common and unique features of the T antigens encoded by the polyomavirus group. *J Virol* 1992;66:3979–3985.
229. Polvino-Bodnar M, Cole CN. Construction and characterization of viable deletion mutants of simian virus 40 lacking sequences near the 3′ end of the early region. *J Virol* 1982;43:489–502.
230. Rajan P, Dhamankar V, Rundell K, Thimmappaya B. Simian virus 40 small-t does not transactivate polymerase II promoters in virus infections. *J Virol* 1991;65:6553–6561.
231. Rassoulzadegan M, Cowie A, Carr A, Glaichenhaus N, Kamen R, Cuzin

F. The role of individual polyoma virus early proteins in oncogenic transformation. *Nature* 1982;300:713–718.
232. Rassoulzadegan M, Perbal B, Cuzin F. Growth control in simian virus 40-transformed rat cells-temperature independent expression of transformed phenotype in tsA transformants derived by agar selection. *J Virol* 1978;28:1–5.
233. Rassoulzadegan M, Seif R, Cuzin F. Conditions leading to establishment of N (A gene dependent) and A (A gene independent) transformed states after polyoma virus infection of rat fibroblasts. *J Virol* 1978;28: 421–426.
234. Ray FA, Peabody DS, Cooper JL, Cram LS, Kraemer PM. SV40 T antigen alone drives karyotype instability that precedes neoplastic transformation of human diploid fibroblasts. *J Cell Biochem* 1990;42:13–31.
235. Rayment I, Baker TS, Caspar DL, Murakami WT. Polyoma virus capsid structure at 22.5 Å resolution. *Nature* 1982;295:110–115.
236. Reddy VB, Ghosh PK, Lebowitz P, Weissman SM. Gaps and duplicated sequence in the leaders of SV40 16S RNA. *Nucl Acids Res* 1978; 5:4195–4213.
237. Reddy VB, Thimmappaya B, Dhar R, et al. The genome of simian virus 40. *Science* 1978;200:494–502.
238. Reed SI, Stark GR, Alwine JC. Autoregulation of SV40 gene A by T antigen. *Proc Natl Acad Sci USA* 1976;73:3083–3088.
239. Resnick J, Shenk T. Simian virus 40 agnoprotein facilitates normal nuclear location of the major capsid polypeptide and cell-to-cell spread of virus. *J Virol* 1986;60:1098–1106.
240. Rice PW, Cole CN. Efficient transcriptional activation of many simple modular promoters by simian virus 40 large T antigen. *J Virol* 1993; 67:6689–6697.
241. Richardson WD, Roberts BL, Smith AE. Nuclear location signals in polyoma virus large-T. *Cell* 1986;44:77–85.
242. Rio DC, Robbins A, Myers R, Tjian R. Regulation of SV40 early transcription in vitro by a purified tumor antigen. *Proc Natl Acad Sci USA* 1980;77:5706–5710.
243. Rio DC, Tjian R. SV40 T antigen binding site mutations that affect autoregulation. *Cell* 1983;32:1227–1240.
244. Ryder K, Vakalopoulou E, Mertz R, et al. Seventeen base pairs of region I encode a novel tripartite binding signal for SV40 T antigen. *Cell* 1985;42:539–548.
245. Sahli R, Freund R, Dubensky T, Garcea R, Bronson R, Benjamin T. Defect in entry and altered pathogenicity of a polyoma virus mutant blocked in VP2 myristylation. *Virology* 1993;192:142–153.
246. Salunke DM, Caspar DLD, Garcea RL. Self-assembly of purified polyomavirus capsid protein VP1. *Cell* 1986;46:895–904.
247. Sambrook J, Westphal H, Srinivasan PR, Dulbecco R. The integrated state of viral DNA in SV40-transformed cells. *Proc Natl Acad Sci USA* 1968;60:1288–1295.
248. Scheffner M, Knippers R, Stahl H. RNA unwinding activity of SV40 large T antigen. *Cell* 1989;57:955–963.
249. Scheidtmann K-H, Echle B, Walter G. Simian virus 40 large T antigen is phosphorylated at multiple sites clustered in two separate regions. *J Virol* 1982;44:116–133.
250. Scheidtmann KH, Virshup DM, Kelly TJ. Protein phosphatase 2A dephosphorylates simian virus 40 large T antigen specifically at residues involved in regulation of DNA-binding activity. *J Virol* 1991;65:2098–2101.
251. Schiedner G, Wessel R, Scheffner M, Stahl H. Renaturation and DNA looping promoted by SV40 large tumour antigen. *EMBO J* 1990;9: 2937–2943.
252. Schmieg FI, Simmons DT. Characterization of the in vitro interaction between SV40 T antigen and p53: mapping the p53 binding site. *Virology* 1988;164:132–140.
253. Schmitt MK, Mann K. Glycosylation of simian virus 40 T antigen and localization of glycosylated T antigen in the nuclear matrix. *Virology* 1987;156:268–281.
254. Schneider C, Weisshart K, Guarino LA, Dornreiter I, Fanning E. Species-specific functional interactions of DNA polymerase α-primase with simian virus 40 (SV40) T antigen require SV40 origin DNA. *Mol Cell Biol* 1994;14:3176–3185.
255. Schneider J, Fanning E. Mutations in the phosphorylation sites of simian virus 40 (SV40) T antigen alter its origin DNA-binding specificity for sites I or II and affect SV40 DNA replication activity. *J Virol* 1988; 62:1598–1605.
256. Schuurman R, Sol C, van der Noordaa J. The complete nucleotide sequence of bovine polyomavirus. *J Gen Virol* 1990;71:1723–1735.
257. Sedman S, Gelembiuk GW, Mertz JE. Translation initiation at a downstream AGU occurs with increased efficiency when the upstream AUG is located very close to the 5′ cap. *J Virol* 1990;64:453–457.
258. Sedman S, Good PJ, Mertz JE. Leader-encoded open reading frames modulate both the absolute and relative rates of synthesis of the virion proteins of simian virus 40. *J Virol* 1989;63:3884–3993.
259. Sedman SA, Mertz JE. Mechanisms of synthesis of virion proteins from the functionally bigenic late mRNAs of simian virus 40. *J Virol* 1988; 62:954–961.
260. Segawa K, Ito Y. Differential subcellular localization of in vivo-phosphorylated and nonphosphorylated middle-sized tumor antigen of polyma virus and its relationship to middle- sized tumor antigen phosphorylating activity in vitro. *Proc Natl Acad Sci USA* 1982;79:6812–6816.
261. Seif R, Martin RG. Simian virus 40 small t antigen is not required for the maintenance of transformation but may act as a promoter (cocarcinogen) during establishment of transformation in resting rat cells. *J Virol* 1979;32:979–988.
262. SenGupta DJ, Borowiec JA. Strand and face: the topography of interactions between the SV40 origin of replication and T-antigen during the initiation of replication. *EMBO J* 1994;12:982–992.
263. Shay JW, Wright WE. Quantitation of the frequency of immortalization of normal human diploid fibroblasts by SV40 large T-antigen. *Exp Cell Res* 1989;184:109–118.
264. Shenk TE, Carbon J, Berg P. Construction and analysis of viable deletion mutants of simian virus 40. *J Virol* 1976;18:664–672.
265. Shirodkar S, Ewen M, DeCaprio J, Morgan J, Morgan D, Chittenden T. The transcription factor E2F interacts with retinoblastoma product and a p107–cyclin A complex in a cell cycle- regulated manner. *Cell* 1992;68:157–166.
266. Sleigh MJ, Topp WC, Hanich R, Sambrook JF. Mutants of SV40 with an altered small t protein are reduced in their ability to transform cells. *Cell* 1978;14:79–88.
267. Smale ST, Tjian R. T-antigen-DNA polymerase α complex implicated in simian virus 40 DNA replication. *Mol Cell Biol* 1986;6:4077–4087.
268. Soeda E, Arrand JR, Smolar N, Girffin BE. Sequences from the early region of polyoma virus DNA containing the viral replication origin and encoding small, middle and (part of) large T-antigens. *Cell* 1979; 17:357–370.
269. Soeda E, Arrand JR, Smolar N, Walsh J, Griffin BE. Coding potential and regulatory signals of the polyoma virus genome. *Nature* 1980;283: 445–453.
270. Sompayrac L, Danna KJ. The SV40 sequences between 0.169 and 0.423 map units are not essential to immortalize early passage rat embryo cells. *Mol Cell Biol* 1985;5:1191–1194.
271. Sompayrac L, Danna KJ. The amino-terminal 147 amino acids of SV40 large T antigen transform secondary rat embryo fibroblasts. *Virology* 1991;181:412–415.
272. Sontag E, Federov S, Kamibayashi C, Robbins D, Cobb M, Mumby M. The interaction of SV40 small tumor antigen with protein phosphatase 2A stimulates the map kinase pathway and induces cell proliferation. *Cell* 1993;75:887–897.
273. Soprano KJ, Galanti N, Jonak GK, et al. Mutational analysis of simian virus 40 T antigen: stimulation of cellular DNA synthesis and activation of rRNA genes by mutants with deletions in the T-antigen gene. *Mol Cell Biol* 1983;3:214–219.
274. Srinivasan A, Peden KW, Pipas JM. The large tumor antigen of simian virus 40 encodes at least two distinct transforming functions. *J Virol* 1989;63:5459–5463.
275. Stacy T, Chamberlain M, Cole CN. Simian virus 40 host range/helper function mutations cause multiple defects in viral late gene expression. *J Virol* 1989;63:5280–5215.
276. Stahl H, Dröge P, Knippers R. DNA helicase activity of SV40 large tumor antigen. *EMBO J* 1986;5:1939–1944.
277. Staneloni RJ, M FM, Benjamin TL. Host range selection of transformation-defective hr-t mutants of polyoma virus. *Virology* 1977;77: 598–609.
278. Staufenbiel M, Deppert W. Different structural systems of the nucleus are targets for SV40 large T antigen. *Cell* 1983;33:173–181.
279. Stewart N, Bacchetti S. Expression of SV40 large T antigen, but not small t antigen, is required for the induction of chromosomal alterations in transformed human cells. *Virology* 1991;180:49–57.
280. Stewart SE, Eddy BE, Borgese NG. Neoplasms in mice inoculated with a tumor agent carried in tissue culture. *J Natl Cancer Inst* 1958;20: 1223–1243.

280a. Sweet BH, Hilleman MR. The vacuolating virus, SV40. *Proc Soc Exp Biol Med* 1960;105:420–427.

281. Stillman BW, Gluzman Y. Replication and supercoiling of simian virus 40 DNA in cell extracts from human cells. *Mol Cell Biol* 1985;5:2051–2060.

282. Streuli CH, Griffin BE. Myristic acid is coupled to a structural protein of polyoma virus and SV40. *Nature* 1987;326:619–622.

283. Sundin O, Varshavsky A. Terminal stages of SV40 DNA replication proceed via multiple intertwined catenated dimers. *Cell* 1980;21:103–114.

284. Sundin O, Varshavsky A. Arrest of segregation leads to accumulation of highly intertwined catenated dimers: dissection of the final stages of SV40 DNA replication. *Cell* 1981;25:659–669.

285. Tack LC, Beard P. Both trans-acting factors and chromatin structure are involved in regulation of transcription from the early and late promoters in simian virus 40 chromosomes. *J Virol* 1985;54:207–218.

286. Tapper DP, DePamphilis ML. Discontinuous DNA replication: accumulation of simian virus 40 DNA at specific stages in its replication. *J Mol Biol* 1978;120:401–422.

287. Tegtmeyer P. Function of simian virus 40 gene A in transforming infection. *J Virol* 1975;15:613–618.

288. Tegtmeyer P, Rundell K, Collins JK. Modification of simian virus 40 protein A. *J Virol* 1977;21:647–657.

289. Tegtmeyer P, Schwartz M, Collins JK, Rundell K. Regulation of tumor antigen synthesis by simian virus 40 gene A. *J Virol* 1975;16:168–178.

290. Thomas JE, Aguzzi A, Soriano P, Wagner EF, Brugge JS. Induction of tumor formation and cell transformation by polyoma middle T antigen in the absence of src. *Oncogene* 1993;8:2521–2529.

291. Thompson DL, Kalderon D, Smith A, Tevethia M. Dissociation of Rb-binding and anchorage-independent growth from immortalization and tumorigenicity using SV40 mutants producing N-terminally truncated large T antigens. *Virology* 1990;178:15–34.

292. Thompson JA, Radonovich MF, Salzman NP. Characterization of the 5′ terminal structure of SV40 early mRNAs. *J Virol* 1979;31:437–446.

293. Tjian R, Fey G, Graessmann A. Biological activity of purified SV40 T-antigen proteins. *Proc Natl Acad Sci USA* 1978;75:1279–1283.

294. Tjian R, Robbins A. Enzymatic activities associated with a purified SV40 T-antigen related protein. *Proc Natl Acad Sci USA* 1979;76:610–615.

295. Tooze J, ed. *DNA tumor viruses*, 2nd ed. Cold Spring Harbor, NY: Cold Spring Harbor Laboratory; 1981.

296. Tornow J, Cole CN. Intracistronic complementation in the simian virus 40 A gene. *Proc Natl Acad Sci USA* 1983;80:6312–6316.

297. Tornow J, Polvino-Bodnar M, Santangelo G, Cole CN. Two separable functional domains of simian virus 40 large T antigen: carboxyl-terminal region of simian virus 40 large T antigen is required for efficient capsid protein synthesis. *J Virol* 1985;53:415–424.

298. Treisman R, Cowie A, Favaloro J, Jat P, Kamen R. The structure of the spliced mRNAs encoding polyoma virus early region proteins. *J Mol Appl Genet* 1981;1:83–92.

299. Treisman R, Kamen R. Structure of polyoma virus late nuclear RNA. *J Mol Biol* 1981;148:273–301.

300. Treisman RH, Novack V, Favaloro J, Kamen R. Transformation of rat cells by an altered polyoma virus genome expressing only the middle T protein. *Nature* 1981;292:595–600.

301. Tsurimoto T, Melendy T, Stillman B. Sequential initiation of lagging and leading strand synthesis by two different polymerase complexes at the SV40 DNA replication origin. *Nature* 1990;346:534–539.

302. Tsurimoto T, Stillman B. Replication factors required for SV40 DNA replication in vitro. I. DNA structure-specific recognition of a primer–template junction by eukaryotic DNA polymerases and their accessory proteins. *J Biol Chem* 1991;266:1950–1960.

303. Tsurimoto T, Stillman B. Replication factors required for SV40 DNA replication in vitro. II. Switching of DNA polymerase alpha and delta during initiation of leading and lagging strand synthesis. *J Biol Chem* 1991;266:1961–1968.

304. Turk B, Porras A, Mumby MC, Rundell K. Simian virus 40 small-t antigen binds two zinc ions. *J Virol* 1993;67:3671–3673.

305. van der Haegen BA, Shay JW. Immortalization of human mammary epithelial cells by SV40 large T-antigen involves a two-step mechanism. *In Vitro Cell Devel Biol* 1993;29:180–182.

306. Varshavsky AJ, Bakayev VV, Chumackov PM, Georgiev GP. Minichromosome of simian virus 40: Presence of H1. *Nucl Acids Res* 1976;3:2101–2113.

307. Virshup DM, Kauffman MG, Kelly TJ. Activation of SV40 DNA replication in vitro by cellular protein phosphatase 2A. *EMBO J* 1989;8:3891–3898.

308. Virshup DM, Russo A, Kelly TJ. Mechanism of activation of simian virus 40 DNA replication by protein phosphatase 2A. *Mol Cell Biol* 1992;12:4883–4895.

309. Waga S, Bauer G, Stillman B. Reconstitution of complete SV40 DNA replication with purified replication factors. *J Biol Chem* 1994;269:10923–10934.

310. Waga S, Stillman B. Anatomy of a DNA replication fork revealed by reconstitution of SV40 DNA replication in vitro. *Nature* 1994;369:207–212.

311. Walter G, Ruediger R, Slaughter C, Mumby M. Association of protein phosphatase 2A with polyoma virus medium tumor antigen. *Proc Natl Acad Sci USA* 1990;87:2521–2525.

312. Wang HG, Rikitake Y, Carter MC, et al. Identification of specific adenovirus E1A N- terminal residues critical to the binding of cellular proteins and to the control of cell growth. *J Virol* 1993;67:476–488.

313. Ward JM, Lock A, Collins JMG, Gonda MA, Reynolds CW. Papovaviral sialoadenitis in athymic nude rats. *Lab Anim* 1984;18:84–89.

314. Whyte P, Williamson NM, Harlow E. Cellular targets for transformation by the adenovirus E1A proteins. *Cell* 1989;56:67–75.

315. Wildeman AG. Transactivation of both early and late simian virus 40 promoters by large tumor antigen does not require nuclear localization of the protein. *Proc Natl Acad Sci USA* 1989;86:2123–2127.

316. Wobbe CR, Dean F, Weissbach L, Hurwitz J. In vitro replication of duplex circular DNA containing the simian virus 40 DNA origin site. *Proc Nat Acad Sci USA* 1985;82:5710–5714.

317. Wold M, Li J, Kelly T. Initiation of simian virus 40 DNA replication in vitro: large-tumor-antigen- and origin-dependent unwinding of the template. *Proc Natl Acad Sci USA* 1987;84:3643–3647.

318. Wychowski C, Benichou D, Girard M. A domain of SV40 capsid polypeptide VP1 that specifies migration into the cell nucleus. *EMBO J* 1986;5:2569–2576.

319. Xiong Y, Hannon GJ, Zhang H, Casso D, Kobayashi R, Beach D. p21 is a universal inhibitor of cyclin kinases. *Nature* 1993;366:701–704.

320. Yaciuk P, Carter MC, Pipas JM, Moran E. Simian virus 40 large T antigen expresses a biological activity complementary to the p300-associated transforming function of the adenovirus E1A gene products. *Mol Cell Biol* 1991;11:2116–2124.

321. Yamada M, Kasamatsu H. Role of nuclear pore complex in simian virus 40 nuclear targeting. *J Virol* 1993;67:119–130.

322. Yang L, Wold MS, Li JJ, Kelly TJ, Liu LF. Roles of DNA topoisomerases in simian virus 40 DNA replication in vitro. *Proc Natl Acad Sci USA* 1987;84:950–954.

323. Yin Y, Tainsky MA, Bischoff FZ, Strong LC, Wahl GM. Wild-type p53 restores cell cycle control and inhibits gene amplification in cells with mutant p53 alleles. *Cell* 1992;70:937–948.

324. Yuen LKC, Consigli RA. Identification and protein analysis of polyomavirus assembly intermediates from infected primary mouse embryo cells. *Virology* 1985;144:127–138.

325. Zambetti GP, Bargonetti J, Walker K, Prives C, Levine AJ. Wild-type p53 mediates positive regulation of gene expression through a specific DNA sequence element. *Genes Dev* 1992;6:1143–1152.

326. Zerrahn J, Knippschild U, Winkler T, Deppert W. Independent expression of the transforming amino-terminal domain of SV40 large T an alternatively spliced third SV40 early mRNA. *EMBO J* 1993;12:4739–4746.

327. Zhu J, Rice PW, Abate M, Cole CN. The ability of SV40 large T antigen to immortalize primary mouse embryo fibroblasts co-segregates with its ability to bind p53. *J Virol* 1991;65:6872–6880.

328. Zhu J, Rice PW, Chamberlain M, Cole CN. Mapping the transcriptional transactivation function of SV40 large T antigen. *J Virol* 1991;65:2778–2790.

329. Zhu J, Rice PW, Gorsch L, Abate M, Cole CN. Transformation of a continuous rat embryo fibroblast cell line requires three separate domains of simian virus 40 large T antigen. *J Virol* 1992;66:2780–2791.

330. Zhu Z, Veldman GM, Cowie A, Carr A, Schaffhausen B, Kamen R. Construction and functional characterization of polyomavirus genomes that separately encode the three early proteins. *J Virol* 1984;51:170–180.

331. Zullo J, Stiles CD, Garcea RL. Regulation of c-myc and c-fos mRNA levels by polyomavirus: distinct roles for the capsid protein VP1 and the viral early proteins. *Proc Natl Acad Sci USA* 1987;84:1210–1214.

332. zur Hausen H, Gissmann L. Lymphotropic papovaviruses isolated from African green monkey and human cells. *Med Microbiol Immunol* 1979;167:137–153.

Fundamental Virology, Third Edition
edited by B.N. Fields, D.M. Knipe, P.M. Howley, et al.
Lippincott - Raven Publishers, Philadelphia © 1996

CHAPTER 29

Papillomavirinae: The Viruses and Their Replication

Peter M. Howley

GENERAL DEFINITION AND PROPERTIES

The papillomaviruses are a group of small DNA viruses which induce warts (or papillomas) in a variety of higher vertebrates, including man. The viral nature of human warts was first indicated over 80 years ago by Ciuffo (41), who demonstrated transmission of common warts using cell-free filtrates. The first papillomavirus was described in 1933 when Richard Shope (254) recognized the cottontail rabbit papillomavirus (CRPV) as the etiologic agent responsible for cutaneous papillomatosis in the cottontail rabbit. This group of viruses has remained refractory to standard virologic study because, until recently, a tissue culture system for the propagation of any of the papillomaviruses in the laboratory did not exist. It was not until the late 1970s, when the first papillomavirus genome was successfully cloned in bacteria, that investigators had reagents that were sufficiently standardized to begin a detailed analysis of the molecular biology of this group of viruses. In addition to standardizing reagents, the molecular cloning of the papillomavirus genomes provided adequate viral genetic material to permit the sequencing of the genomes of a number of papillomaviruses and to initiate a systematic mutational analysis and definition of the genes encoded by this group of viruses. The study of the papillomaviruses was spurred during the 1980s by the development of *in vitro* transformation assays which permitted the analysis of the viral functions involved in the induction of cellular proliferation. The bovine papillomavirus type 1 (BPV-1) initially served as the prototype for studies on various aspects of the molecular biology of the papillomaviruses and, therefore, a major focus of this chapter will be BPV-1. During the past decade, the recognition that specific human papillomaviruses (HPVs) are closely linked with certain human cancers, most notably human cervical carcinoma, has focused interest on the specific subgroup of HPVs which are associated with genital lesions. Consequently, information has recently been generated on the molecular biology of certain of the genital-

P. M. Howley: Department of Pathology, Harvard Medical School, Boston, Massachusetts 02115.

associated HPVs, and a discussion of the molecular biology of these viruses will be included in this chapter.

CLASSIFICATION

Historically, the papillomaviruses were grouped together with the polyomaviruses to form the papovavirus family. The term *papovavirus* is derived from the first two letters of the virus first grouped together to form this family of viruses: rabbit *pa*pillomavirus, mouse *po*lyomavirus, and simian *va*cuolating virus (SV40). The properties shared by these viruses include small size, a nonenveloped virion, an icosahedral capsid, a double-stranded circular DNA genome, and the nucleus as a site of multiplication. The papillomavirus particles are slightly larger than those of the polyomaviruses, measuring 55 nm in diameter instead of 45 nm, and on this basis the family was divided into two genera. More recent studies on the biology and genomic organization of these two groups of viruses has indicated that these should not be grouped together as different genera of the same family. Indeed, based on the fundamental differences in the genomic organization and the biology of the papillomas and the polyomaviruses, these two groups of viruses are now considered as individual subfamilies of the papovaviruses.

Members of the papillomavirus and polyomavirus subfamilies can readily be distinguished by differences in the size in the virions (55 nm versus 40 nm), and by the size of the viral genomes (8,000 bp versus 5,000 bp). Also, there are subfamily-specific antigens characteristic of the papillomaviruses or the polyomaviruses which permit distinction of these viruses (137). Furthermore, there is sufficient nucleic acid conservation among the individual members of each of the subfamilies such that they can individually cross-hybridize with other members of the subfamilies under hybridization conditions of reduced stringency (158). Under similar nonstringent hybridization conditions, the genomes of the polyomaviruses will not cross-hybridize with the genomes of the papillomaviruses.

The papillomaviruses are widespread in nature and have been recognized primarily in higher vertebrates. Viruses have been characterized from humans, cattle, rabbits, horses, dogs, sheep, elk, deer, nonhuman primates, the harvest mouse, and the multimammate mouse (*Mastomys natalensis*). Papillomavirus antigens have been identified in a variety of other mammals including other nonhuman primates as recently reviewed (268). Papillomaviruses have also been described in some avian species, namely, the parrot and the chafinch. In general, the papillomaviruses are highly species-specific, and there are no examples of a papillomavirus from one species causing a productive infection in a second species. Most animal papillomaviruses are associated with purely squamous epithelial proliferative lesions (warts) which can be cutaneous or can involve the mucosal squamous epithelium from the oral pharynx, the esophagus, or the genital tract. Most of the papillomaviruses have a specific cellular tropism for squamous epithelial cells. Expression of the productive functions necessary for virion replication appear to be limited to terminally differentiated squamous epithelial cells.

There is a subgroup of papillomaviruses which induce benign fibropapillomas in which there is a proliferative dermal fibroblastic component as well as the proliferative squamous epithelial component. The best studied of this group of papillomaviruses is BPV-1. Other papillomaviruses, which also induce fibropapillomas, include BPV-2, the European elk papillomavirus (EEPV), the deer papillomavirus (DPV), and the reindeer papillomavirus. Interest in this subgroup of papillomaviruses stems from the fact that, in addition to their ability to induce fibropapillomas in their natural host, they also have the ability to induce fibroblastic tumors in other species such as hamsters (85). Also, members of this subgroup of papillomaviruses can readily transform a variety of rodent cells in culture. The transformation capacity of BPV in tissue culture was first demonstrated in the early 1960s (21,23,278). Quantitative transformation focus assays were developed for BPV-1 in the late 1970s, which provided a biological assay to define and study the viral functions involved in the induction of cellular proliferation (69). While most studies have been carried out with BPV-1 transformation of these rodent cells, it is notable that the EEPV and the DPV have also been demonstrated to have these transformation properties.

To date, over 70 different human papillomaviruses have been described. Since serologic reagents are not generally available to distinguish each of these types, they are not referred to as serotypes. The classification of viral types is based on the species of origin and the extent and degree of relatedness of the viral genomes. For most species, only a single papillomavirus type has been described, but this likely is due to the fact that extensive comparative studies have not yet been carried out for most animal species at a molecular level. However, six distinct bovine papillomaviruses have now been described (196).

The papillomavirus DNA isolates from one species are classified according to their sequence homology. As noted above, there are over 70 HPV types that are now recognized. The initial classification of a specific type was based on the extent of homology of the DNA genomes utilizing liquid hybridization techniques under stringent reassociation conditions (44). Classification of types is now carried out by comparison of the nucleotide sequence of specific regions of the genome. Based on criteria adopted by the Papillomavirus Nomenclature Committee, the nucleotide sequences of the E6, E7, and L1 open reading frames (ORFs) of a new type should not exceed 90% of the corresponding sequences of the genomes of known HPV types (57). A closer relationship is considered a subtype of the type with the highest degree of homology. The 70 human papillomavirus types that have been characterized are listed in Table 1, along with the clinical entities with which some of them have been associated. In some cases, extensive analyses have not yet been carried out which establish with any assurance the specific clinical associations.

VIRION STRUCTURE

Papillomaviruses are small, nonenveloped, icosahedral DNA viruses that replicate in the nucleus of squamous epithelial cells. The papillomavirus particles have a sedimentation coefficient (S_{20},W) of 300. The papillomavirus particles are 52 to 55 nm in diameter (Fig. 1). The virion particles consist of a single molecule of double-stranded circular DNA approximately 8,000 bp in size, contained within a spherical protein coat, or capsid, composed of 72 capsomeres. The virus particles have a density in cesium chloride of 1.34 g/ml (48). Fine structural analysis by cryoelectron microscopy on three-dimensional image reconstruction techniques have revealed that the viruses consist of 72 pentameric capsomeres arranged on a T=7 surface lattice (10). Like the polyomavirus capsids, the capsomeres exist in two states, one capable of making contact with six neighbors, as observed in the 60 hexavalent capsomeres, and the other with 5 neighbors in the 12 pentavalent capsomeres (Fig. 2).

The capsid consists of two structural proteins. The major capsid protein (L1) has a molecular weight of approximately 55 kd (77,97,213) and represents approximately 80% of the total viral protein. A minor protein (L2) has a molecular size of approximately 70 kd. Both of these proteins are virally encoded. In addition, analysis of proteins in the virus particle reveals that the viral DNA is associated with cellular histones to form a chromatin-like complex (77,213). Several groups have recently produced viruslike particles (VLPs) from different papillomaviruses by expressing L1 alone or the combination of L1 and L2 using either vaccinia virus or baculovirus expression systems (112,148,228,312). Although not required for assembly, L2 is incorporated into VLPs when coexpressed in mammalian or insect cells with L1. The morphology of VLPs containing only L1 appears identical to intact virus particles by cryoelectron microscopy (111). It is, therefore, presently unclear what functional role L2 has in the capsid.

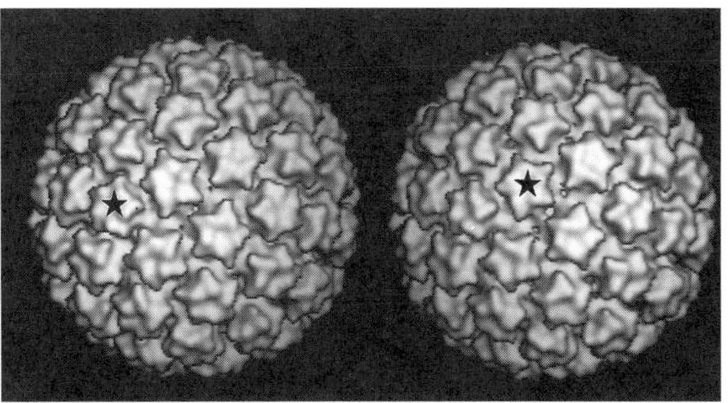

FIG. 2. Surface-shaded displays of papillomavirus three-dimensional image reconstructions of a cryoelectron microscopic analysis of virion particle structures. The individual capsomers can exist in either a pentavalent (*starred, left*) or hexavalent (*starred, right* state. From Baker et al. (10), with permission.

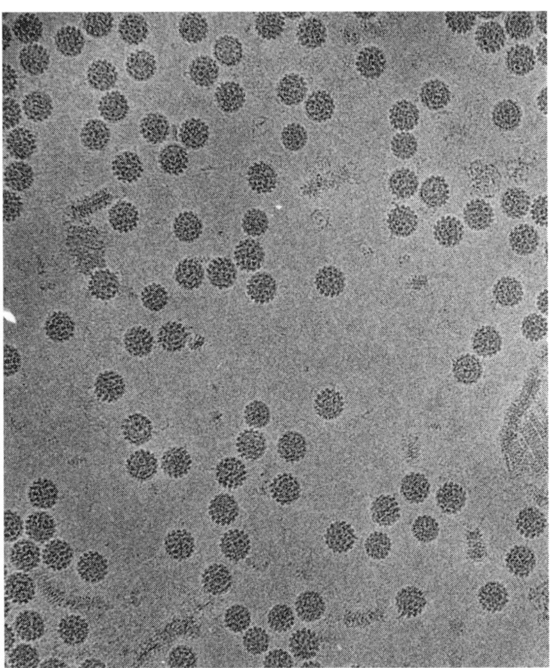

FIG. 1. BPV-1 virion particles (55 nm in diameter). From Baker et al. (10), with permission.

Full papillomavirus particles contain a double-stranded circular DNA viral genome of approximately 8,000 bp. The guanosine cytosine content of most papillomavirus genomes is approximately 42%. The DNA constitutes approximately 12% of the virion by weight, accounting for the density in cesium chloride of approximately 1.34 g/ml (48).

THE GENOME STRUCTURE AND ORGANIZATION

To date, at least 7 animal papillomaviruses and over 22 human papillomavirus genomes have been sequenced in their entirety (Table 2). In addition, a variety of other human and animal papillomavirus genomes have been partially sequenced, and these sequences are available through GenBank. The genomic organization of each of the papillomaviruses is remarkably similar. The genomic map of BPV-1 DNA is shown in Fig. 3. One characteristic of the genomic organization of BPV-1 which is shared with all of the other papillomaviruses is that all of the ORFs are located on one strand of the viral DNA, indicating that all of the viral genes are located on one strand (35). Analysis of the RNAs encoded by the papillomaviruses have indicated that, indeed, only one strand serves as a template for transcription.

TABLE 1. *The human papillomaviruses*

HPV	Location	Isolated from	Associated with	References
1	Cutaneous	Verruca plantaris	Verruca plantaris	(83)
2	Cutaneous Verruca plantaris	Verruca vulgaris	Verruca vulgaris	(200)
3	Cutaneous	Verruca plana	Verruca plana	(152,201)
4	Cutaneous Verruca plantaris	Verruca vulgaris Verruca plantaris	Verruca vulgaris	(97,121)
5	Cutaneous[a]	Pityriasis versicolor-like macular lesions of epidermodysplasia verruciformis (EV)	EV (benign) EV (squamous cell carcinoma)	(151,201)
6	Genital mucosa[b]	Condyloma acuminatum	Condyloma acuminatum Laryngeal papilloma Buschke-Lowenstein tumors	(98,247)
7	Cutaneous	Butcher's wart	Butcher's wart	(197,202)
8	Cutaneous[a]	Macular lesions (EV)	EV (benign) EV (squamous cell carcinomas)	(86,199)
9	Cutaneous	EV lesions	EV (benign)	(199)
10	Cutaneous[b]	Verruca plana	Verruca plana	(152,201)
11	Genital mucosa[b]	Laryngeal papilloma	Condyloma acuminatum Laryngeal papilloma	(52,96)
12	Cutaneous	Macular lesions (EV)	EV (benign)	(152,201)
13	Oral mucosa	Focal epithelial hyperplasia (FEH) (morbus heck)	FEH	(214)
14	Cutaneous[a]	Flat warts (EV) EV (squamous cell carcinoma	EV (benign)	(150)
15	Cutaneous	Flat warts (EV)	EV (benign)	(150)
16	Genital mucosa[a]	Cervical carcinoma	CIN cervical carcinoma	(68,249)
17	Cutaneous[a]	Macular lesions (EV) EV (squamous cell carcinoma)	EV (benign)	(150)
18	Genital mucosa[a]	Cervical carcinoma	CIN Cervical carcinoma	(24,46)
19	Cutaneous	Macular lesions (EV)	EV (benign)	(150)
20	Cutaneous[a]	Flat warts (EV)	EV (benign) EV (squamous cell carcinoma)	(150)
21	Cutaneous	Flat warts (EV)	EV (benign)	(150)
22	Cutaneous	Macular lesions (EV)	EV (benign)	(150)
23	Cutaneous	Macular lesions (EV)	EV (benign)	(150)
24	Cutaneous	Macular lesions (EV)	EV (benign)	(150)
25	Cutaneous	Macular lesions (EV)	EV (benign)	(92)
26	Cutaneous	Verruca vulgaris (immunosuppressed patient)		(204)
27	Cutaneous	Verruca (immunosuppressed patient)		(203)
28	Cutaneous	Verruca plana		(81)
29	Cutaneous	Verruca vulgaris		(78)
30	Genital mucosa[a]	Laryngeal carcinoma	CIN	(139)
31	Genital mucosa[a]	CIN 1	CIN Cervical carcinoma	(167)
32	Oral mucosa	FEH	FEH Oral papilloma	(17)
33	Genital mucosa[a]	Cervical carcinoma	CIN Cervical carcinoma	(15,47)
34	Genital mucosa; cutaneous	Bowen's disease	CIN (genital)	(144)

(continued)

TABLE 1. *Continued*

HPV	Location	Isolated from	Associated with	References
35	Genital mucosa[a]	Cervical adenocarcinoma	CIN Cervical carcinoma	(170)
36	Cutaneous	Actinic keratosis	EV (benign)	(144)
37	Cutaneous	Keratoacanthoma		(240)
38	Cutaneous[a]	Malignant melanoma		(240)
39	Genital mucosa[a]	Penile bowenoid papulosis (PIN)	CIN Cervical carcinoma	(16)
40	Genital mucosa	PIN	CIN	(54)
41	Cutaneous[a]	Disseminated warts	Cutaneous squamous cell carcinoma	(105)
42	Genital mucosa	Vulvar papilloma	CIN	(16)
43	Genital mucosa	Vulvar hyperplasia	CIN	(169)
44	Genital mucosa	Vulvar condyloma	CIN	(169)
45	Genital mucosa[a]	CIN	CIN Cervical carcinoma	(188)
46	Cutaneous	Macular lesions (Hodgkin's disease patient)	EV (benign)	(108)
47	Cutaneous	Macular and verrucae lesions (EV)	EV (benign)	(1)
48	Cutaneous[a]	Cutaneous squamous cell carcinoma (transplant patient)		(184)
49	Cutaneous	Verruca plana (immunosuppressed patient)		(82)
50	Cutaneous	EV (benign)		(80)
51	Genital mucosa[a]	CIN 1	CIN Cervical carcinoma	(195)
52	Genital mucosa[a]	CIN	CIN Cervical carcinoma	(253)
53	Genital mucosa	Normal cervical mucosa		(91)
54	Genital mucosa	Condyloma acuminatum		(79)
55	Genital mucosa	Bowenoid papulosis		(79)
56	Genital mucosa[a]	CIN 1	CIN	(168)
57	Oral and genital mucosa; cutaneous	Inverted papilloma of the maxillary sinus	CIN Verruca vulgaris	(54)
58	Genital mucosa[a]	Cervical carcinoma	CIN	(178)
59	Genital mucosa	Vulvar IN		T. Matsukura, unpublished.
60	Cutaneous	Epidermoid cyst	Plantar warts	(178)
61	Genital mucosa	Vulvar IN + CIN	CIN	S. Beaudenon, and T. Matsukura, unpublished.
62	Genital mucosa	Vulvar IN	CIN	T. Matsukura, unpublished
63	Cutaneous	Plantar Wart		(73)
64	Genital mucosa	Vulvar IN		T. Matsukura, unpublished.
65	Cutaneous mucosa	Pigmented wart		(73)
66	Genital mucosa	Cervical carcinoma		(274)
67	Genital mucosa	Vulvar IN		T. Matsukura, unpublished.
68	Genital mucosa	Genital lesion		M. Longue, unpublished.
69	Genital mucosa	CIN		T. Matsukura, unpublished.
70	Genital mucosa	Vulvar papilloma		(16)

[a]These HPV types have been found to be associated with clinical lesions that may have a risk for malignant progression.
[b]On rare occasions these HPV types have also been found to be associated with carcinomas.
CIN, cervical intraepithelial neoplasia; EV, epidermodysplasia verruciformis; FEH, focal epithelial hyperplasia; PIN, penile bowenoid papulosis or penile intraepithelial neoplasia.

TABLE 2. *Sequenced papillomavirus genomes*[a]

Virus	Host	Types of lesions	Reference
Animal papillomaviruses:			
BPV-1	Cattle	Cutaneous fibropapillomas	(35)
BPV-2	Cattle	Cutaneous fibropapillomas	[b]
BPV-4	Cattle	Alimentary tract papillomas	(209)
CRPV	Rabbit	Cutaneous papillomas	(94)
DPV	Deer	Fibropapillomas	(107)
MnPV	Mastomys natalensis	Cutaneous epithelioma	(271)
EEPV	Elk	Fibropapillomas	(76)
Human papillomaviruses:			
HPV-1	Human	Plantar warts	(51)
HPV-2	Human	Verruca vulgaris	(125)
HPV-4	Human	Verruca vulgaris, Verruca plantaris	(73)
HPV-5	Human	Macular lesions in EV	(311)
HPV-6	Human	Condyloma acuminata and laryngeal papillomas	(247)
HPV-8	Human	Macular lesions in EV	(86)
HPV-11	Human	Condyloma acuminata and laryngeal papillomas	(52)
HPV-13	Human	Focal epithelial hyperplasia	(287)
HPV-16	Human	Genital intraepithelial neoplasia	(249)
HPV-18	Human	Genital intraepithelial neoplasia	(46)
HPV-31	Human	Genital intraepithelial neoplasia	(99)
HPV-33	Human	Genital intraepithelial neoplasia	(47)
HPV-35	Human	CIN, cervical carcinoma	(175)
HPV-39	Human	CIN, cervical carcinoma	(291)
HPV-41	Human	Cutaneous squamous cell carcinoma	(126)
HPV-42	Human	CIN	(217)
HPV-47	Human	EV (benign)	(149)
HPV-51	Human	CIN	(172)
HPV-57	Human	CIN	(125)
HPV-58	Human	CIN	(147)
HPV-63	Human		(73)
HPV-65	Human		(73)

[a]Additional sequences of partial and complete papillomavirus genomes can be found in GenBank.
[b]Unpublished, but sequence is available through GenBank by D. E. Groff, R. Mitra, and W. D. Lancaster (Accession #M20219, M19551).
EV, epidermodysplasia verruciformis.

The coding strand for each of the papillomaviruses contains approximately ten designated translational ORFs that have been classified as either the *early* (E) or *late* (L) ORFs, based on their location in the genome. The early ORFs of the BPV-1 genome are located within the fragment of the BPV-1 genome which is sufficient for inducing cellular transformation, i.e., the 69% subgenomic transforming fragment (171). It is this segment of the viral genome which is expressed in nonproductively infected cells and in transformed cells (120). The analogous region of each of the other papillomavirus genomes is also referred to as the *early region*. The L ORFs are expressed only in productively infected cells (3,8). The position, size, and function of many of the ORFs are well conserved among the various papillomaviruses that have been sequenced and studied in detail thus far (7). The function of the individuals ORFs whose functions have been well characterized are described in more detail in the appropriate sections of this chapter.

There is a region in each of the papillomavirus genomes in which there are no ORFs. The region varies slightly in size among the different papillomavirus genomes; in the BPV-1 genome, it is approximately one kilobase in size. This region has been referred to by several terms, including the *long control region* (LCR), the *upstream regulatory region* (the URR), and the *noncoding region*. The term *LCR* will be used in this chapter because of the many *cis*-acting genetic elements that map to this region.

The organization of all of the papillomavirus genomes is quite similar. Figure 4 presents a map of the genomes of BPV-1, CRPV, and some of the human papillomaviruses that have been sequenced. As noted above, all of the ORFs

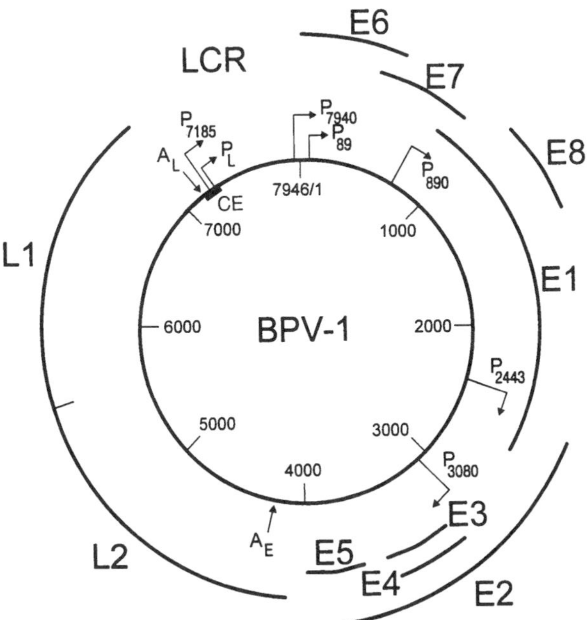

FIG. 3. BPV-1 genomic map. The numbers (*circled*) indicate the nucleotide positions. The individual ORFs of the early (E) and late (L) regions are depicted as areas outside the double-stranded circular genome. Only one strand is transcribed, and transcription occurs in the clockwise direction. Early promoters are indicated (*arrow labeled* P_n), where *n* is the approximate nucleotide position of the RNA initiation site. P_L is the late promoter whose initiation sites map between nucleotides 7214 and 7256. *LCR* designates the long control region which contains the origin of DNA replication (nt 7911 to 22) and the constitutive (CE) transcriptional enhancer (nt 7162 to 7275). The early (A_E) and late (A_L) poly(A) sites are located at nt 4203 and 7175, respectively. Adapted from Ustav et al. (285), Vande Pol and Howley (288), and Yang et al. (304).

are located on one strand. In addition, the size and location of many of the ORFs are well conserved among the papillomaviruses.

VIRUS REPLICATION

The papillomaviruses are highly species-specific and induce squamous epithelial tumors and fibroepithelial tumors in their natural hosts. The viruses also have a specific tropism for squamous epithelial cells. The productive infection of cells by the papillomaviruses can be divided into early and late stages. These stages are linked to the differentiation state of the epithelial cell. Histologically, the lesions induced by papillomaviruses share a number of features. In general, there is thickening of the epidermis (acanthosis) and hyperkeratosis. Usually, there is some degree of papillomatosis. Keratohyalin granules are often prominent in the granular layer of a keratinized epithelium, and occasionally, basophilic intranuclear inclusions can be detected in the cells of the upper layer of the epidermis. These histologic features reflect the biological properties of the papillomaviruses; most likely these morphological changes are induced by specific viral gene products. The specific tropism of the papillomaviruses for squamous epithelial cells is evidenced by the restriction of the viral replication functions such as vegetative viral DNA synthesis, the production of viral capsid proteins and the assembly of virions, to the most terminally differentiated keratinocytes. The replication cycle for the papillomavirus is depicted in Fig. 5 and is discussed in more detail below.

Since the basal cell is the only cell in the squamous epithelium capable of dividing, the virus must infect the basal cell in order to induce a lesion which can persist. By *in situ* hybridization it has been demonstrated that the viral DNA is indeed present within the basal cells and the parabasal cells of a papilloma (246). Furthermore, using probes to the early gene regions of the papillomaviruses, viral transcripts have been detected in the basal cells of the epidermis (264). Late gene expression, synthesis of capsid proteins, vegetative viral DNA synthesis, and assembly of virions, however, occur only in terminally differentiated squamous epithelial cells. The link of viral DNA replication to the state of differentiation of the squamous epithelial cell is illustrated in Fig. 6B which shows an *in situ* hybridization of a BPV-1–induced fibropapilloma. Using a BPV-1 DNA probe which can hybridize to denatured viral DNA, positive hybridization can readily be detected in the nuclei of cells in the middle and upper layers of the squamous epithelium but not within the cells of the lower portion of the epithelium nor within the fibroblasts in the dermis. This represents a relatively short exposure and is only sensitive enough to detect viral DNA in those cells in which there has been significant amplification of the viral DNA (i.e., vegetative viral DNA replication). Thus, vegetative viral DNA synthesis occurs only in differentiating keratinocytes and is therefore linked to the differentiation program of the squamous epithelial cell.

The late viral genes which encode the viral capsid proteins are expressed only in the terminally differentiated epithelial cells of the wart. Using antisera to the major capsid protein, antigen can be detected only in the most superficial cells of the wart (137). The regulation of expression of the genes encoding the capsid protein is at the level of transcription, as demonstrated in the *in situ* hybridization analysis depicted in Fig. 6C, in which a probe complementary to the late viral mRNAs localizes the transcripts to the most terminally differentiated keratinocytes within the wart.

Attachment, Entry, and Uncoating

Little is known concerning virus attachment, receptors, virion entry, or uncoating. Difficulties in propagating the viruses in the laboratory and generating infectious virus *in vitro* have hampered studies of these early events. Binding studies with radiolabeled virion particles have revealed that

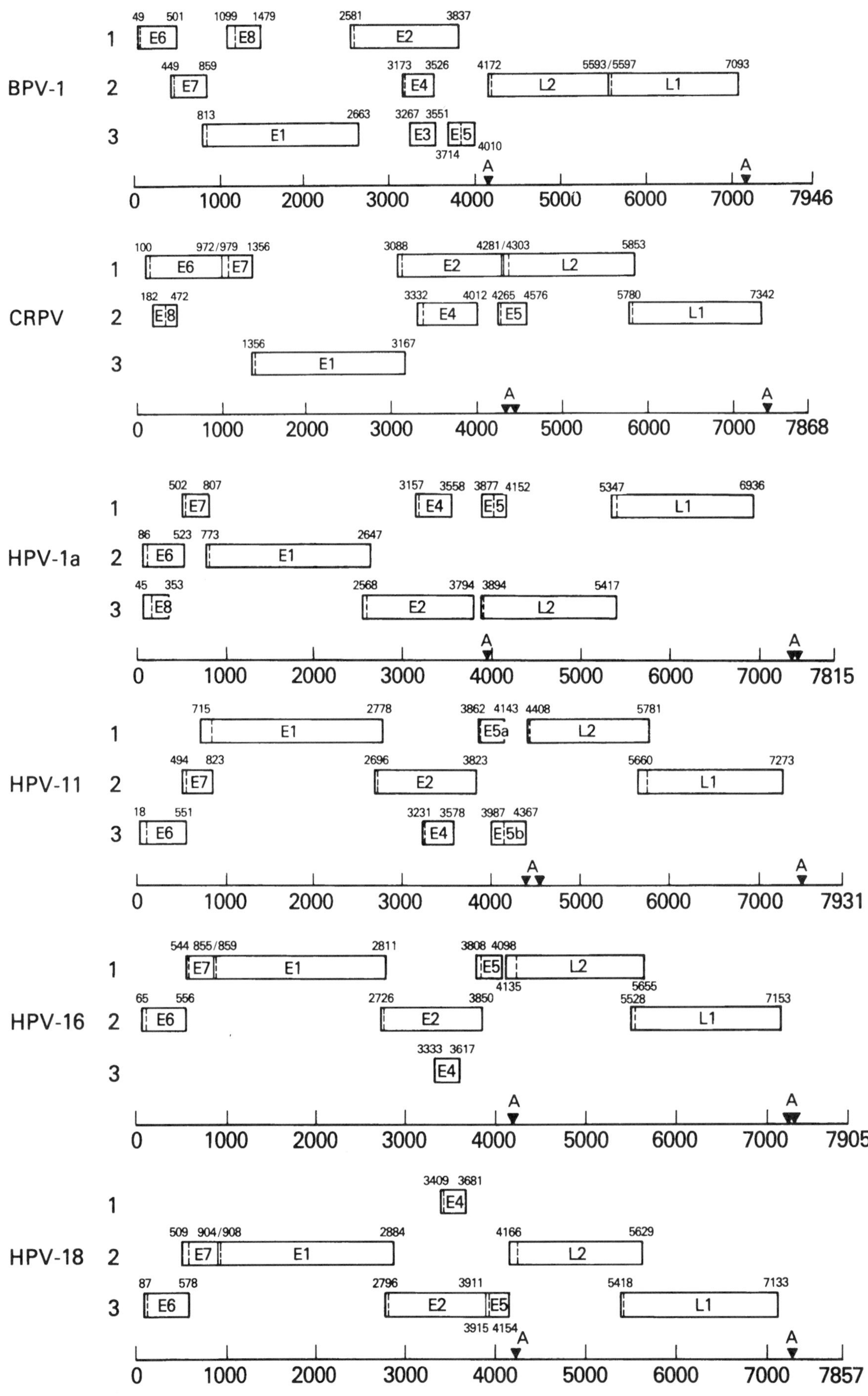
BPV-1
1
2
3
49 501 E6
1099 1479 E8
2581 3837 E2
449 859 E7
3173 3526 E4
4172 5593/5597 7093 L2 L1
813 2663 E1
3267 3551 E3 E5
3714 4010
A A
0 1000 2000 3000 4000 5000 6000 7000 7946
CRPV
1
2
3
100 972/979 1356 E6 E7
3088 4281/4303 5853 E2 L2
182 472 E8
3332 4012 4265 4576 E4 E5
5780 7342 L1
1356 3167 E1
A A
0 1000 2000 3000 4000 5000 6000 7000 7868
HPV-1a
1
2
3
502 807 E7
3157 3558 3877 4152 E4 E5
5347 6936 L1
86 523 773 2647 E6 E1
45 353 E8
2568 3794 3894 5417 E2 L2
A A
0 1000 2000 3000 4000 5000 6000 7000 7815
HPV-11
1
2
3
715 2778 E1
3862 4143 4408 5781 E5a L2
494 823 E7
2696 3823 E2
5660 7273 L1
18 551 E6
3231 3578 E4
3987 4367 E5b
A A
0 1000 2000 3000 4000 5000 6000 7000 7931
HPV-16
1
2
3
544 855/859 2811 E7 E1
3808 4098 E5 L2
4135 5655
65 556 E6
2726 3850 E2
5528 7153 L1
3333 3617 E4
A A
0 1000 2000 3000 4000 5000 6000 7000 7905
HPV-18
1
2
3
3409 3681 E4
509 904/908 2884 E7 E1
4166 5629 L2
87 578 E6
2796 3911 E2 E5
5418 7133 L1
3915 4154
A A
0 1000 2000 3000 4000 5000 6000 7000 7857

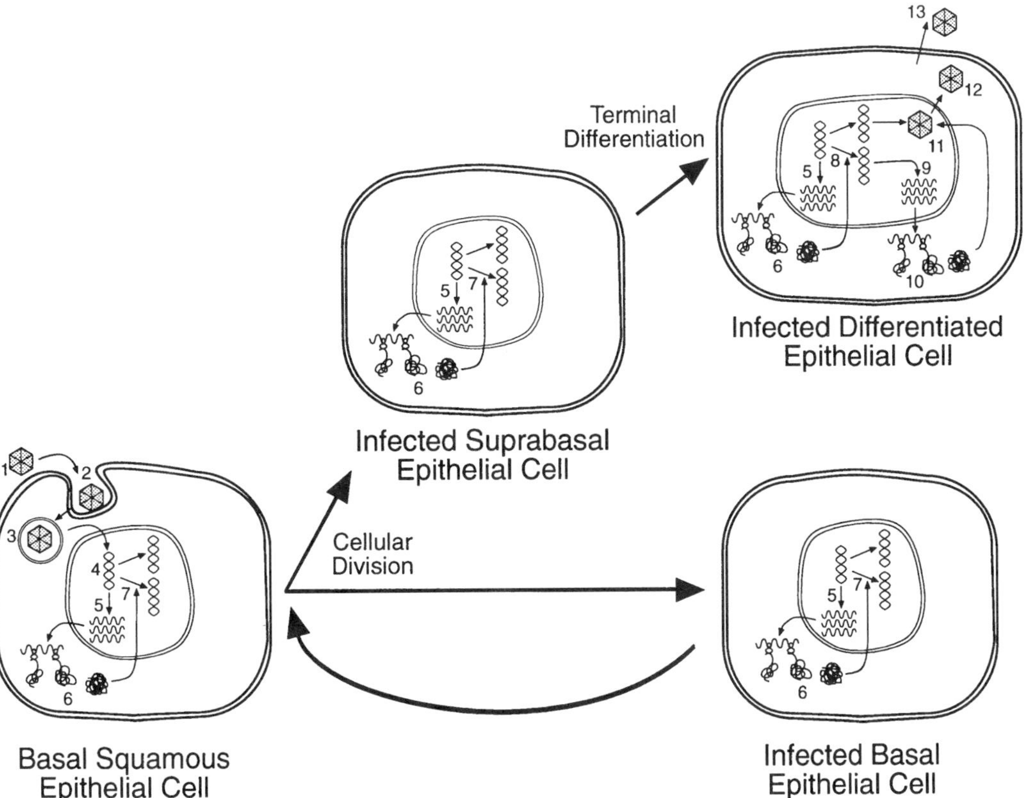

FIG. 5. Replication cycle of a papillomavirus. In order to establish a wart or papilloma, the virus must infect a basal epithelial cell. Our knowledge is quite limited about the initial steps in the replication cycle such as attachment (*1*), uptake (*2*), endocytosis (*3*), and transport to the nucleus and uncoating of the viral DNA (*4*). Early region transcription (*5*), translation of the early proteins (*6*), and steady state viral DNA replication (*7*) all occur in the basal cell and in the infected suprabasal epithelial cell. Events in the viral life cycle leading to the production of virion particles occur in the differentiated keratinocyte: vegetative viral DNA replication (*8*), transcription of the late region (*9*), production of the capsid proteins L1 and L2 (*10*), assembly of the virion particles (*11*), nuclear breakdown (*12*) and release of virus (*13*).

the papillomaviruses can bind a wide variety of cell types in addition to the normal host cell, the squamous epithelial cell (223). Therefore the specific tropism of these viruses for keratinocytes does not appear to be due to a cell type-specific receptor. This observation is consistent with studies with the BPV-1 which have shown that virus can infect and transform a wide variety of cells including rodent fibroblasts. Binding to the cell surface appears to depend upon the major capsid protein (L1) since VLPs consisting only of the L1 protein are capable of inhibiting virus attachment (223). The cell surface receptor for the papillomaviruses has not yet been identified. The same cell surface molecule(s) do, however, react with both BPV-1 and HPV-16 VLPs, suggesting that all of the papillomaviruses likely bind a widely expressed and evolutionarily conserved cell surface receptor (223). There have been no published studies on the mechanisms by which the papillomavirus enters the cell, gains entry to the nucleus, or uncoats its DNA. It is presumed that mechanisms are similar to those used by the polyomaviruses.

Transcription of the Early Region

The replicative phase of the papillomavirus life cycle has been difficult to study because of its link to the terminal differentiation program of the epithelium. BPV-1 has been the best studied of the papillomaviruses with regard

FIG. 4. The genomic organization of a series of animal and human papillomaviruses deduced from the primary DNA sequences. Each of the genomes has been linearized for ease of presentation at a site just upstream of the E6 ORF corresponding to the Hpa I site of BPV-1. As with BPV-1, all of the ORFs of the papillomavirus genomes are located on one strand, and only that strand is transcribed. (The references for the primary sequence citations are listed in Table 2.) The map for HPV-16 includes the correction for the frame shift in the E1 ORF and for the E5 ORF. From Baker (6), with permission.

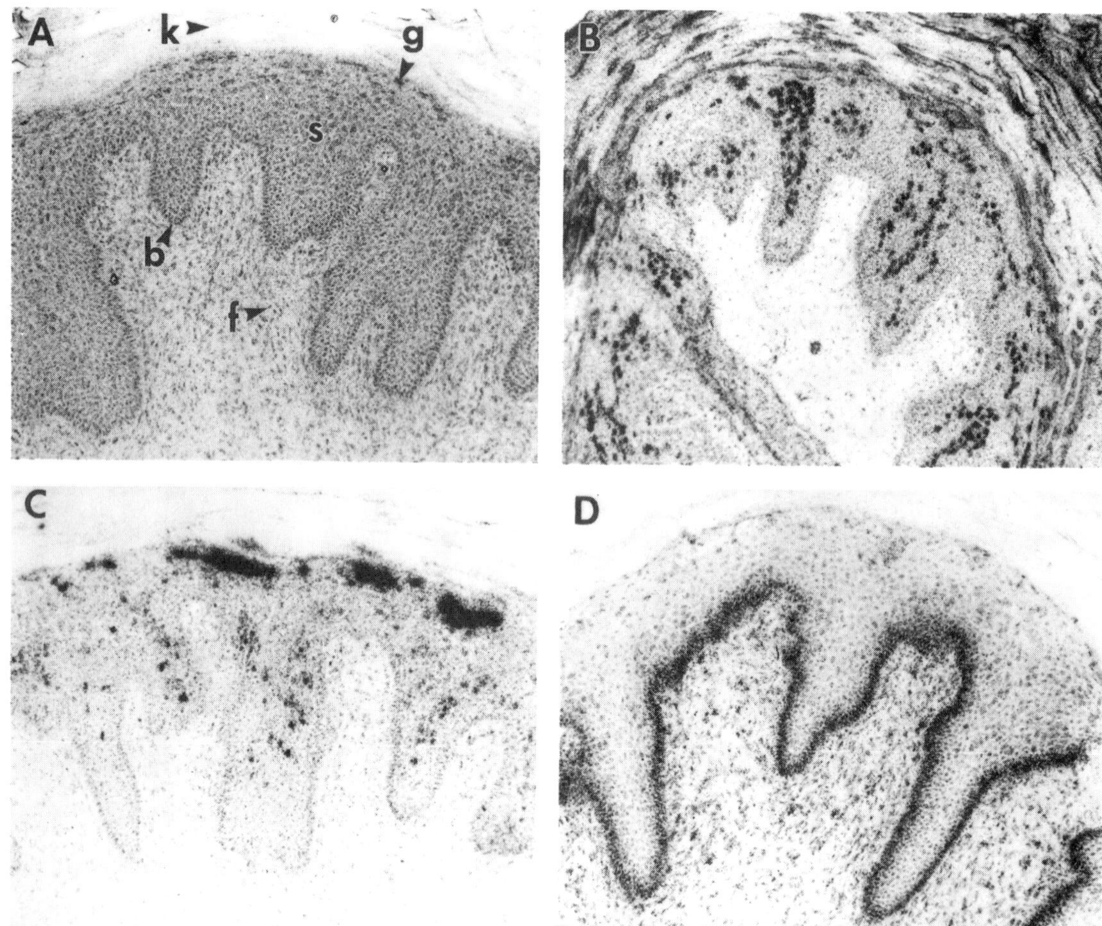

FIG. 6. *In situ* hybridization of a bovine fibropapilloma induced by BPV-1, revealing the linkage of the papillomavirus replication cycle to the squamous epithelial cell differentiation program. **A:** Control section with a sense probe and no denaturation: keratinized horn (*k*), granular layer (*g*), basal layer (*b*), and dermal fibroma (*f*). **B:** Hybridization of a sense probe to a denatured section revealing cells with a high copy number of viral DNA molecules due to vegetative viral DNA replication. **C:** Hybridization with a probe complementary to the L1 ORF indicating late gene expression in the more differentiated epithelial cells of the epithelium. **D:** Hybridization with a probe complementary to a spliced message specific for the E5 transcript revealing maximal expression in the basal cells and some expression in the dermal fibroblasts. From Barksdale and Baker (14), with permission.

to its transcriptional program. These studies have been carried out using rodent cells transformed by BPV-1 as a source of transcripts in nonproductively infected cells and using viral RNAs extracted from infected wart tissues. In recent years, studies have been extended to some of the HPVs associated with genital tract lesions such as HPV-11, HPV-16, and HPV-18 by using HPV-positive clinical lesions, xenograft tissue in nude mice, and cervical carcinoma cell lines. This section will focus primarily on BPV-1 but will include relevant information on the HPVs.

Early Promoters and Enhancers (cis Elements)

Papillomavirus transcription is complex due to the presence of multiple promoters, alternate and multiple splice patterns, and the differential production of mRNA species in different cells. Transcription has been best analyzed for BPV-1. In BPV-1, seven different transcriptional promoters have been identified (2,8,39,261). These are indicated on the circular genomic map of BPV-1 (Fig. 3). Over 20 different mRNA species have been identified in BPV-1 transformed cells and in the productively infected cells of a fibropapilloma (6). The various RNA species and the genes they might possibly encode are depicted in Fig. 7. The viral RNA species are often at a very low abundance, making it likely that additional RNA species exist. The majority of the RNA species utilize the polyadenylation site (A_e) at nucleotide 4180, downstream of the early genes. These RNA species principally encode viral factors involved in viral plasmid replication, regulation of viral transcription, and cellular transformation (120). This polyadenylation site is the only one utilized in transformed rodent cells. A second set of RNAs, which are expressed in productively infected cells, utilize the polyadenylation signal

(A_l) at nucleotide 7156 which is positioned downstream of the L1 and L2 ORFs. RNAs utilizing this late polyadenylation site are found neither in transformed cells nor in the nonproductively infected cells of a BPV-1 fibropapilloma; they are found in only the terminally differentiated keratinocytes of such a lesion. The control of late transcription is complex and involves at least three regulatory mechanisms, as discussed below.

Six different transcriptional promoters for BPV-1 are active in transformed cells. These promoters are also active in productively infected fibropapillomas (8). The nomenclature which has been generally adopted for use in the papillomavirus field designates the promoters (P) with a subscript indicating the nucleotide position of the 5' end of the most abundant RNA species expressed from the designated promoter. Thus, for BPV-1, P_{89} is the promoter that has its major transcription start at nucleotide 89. Another promoter, P_{2443}, gives rise to RNA species with 5' prime ends heterogeneously mapping in the vicinity of nucleotide 2443. The six promoters active in transformed cells which have been mapped to date are P_{89}, P_{890}, P_{2443}, P_{3080}, P_{7185}, and P_{7940} (2,8,39,261,262). One additional promoter, P_{7250}, is referred to as the major late promoter, (P_L), and is principally active in productively infected keratinocytes, giving rise to the late mRNA species containing the E4 and L1 ORFs (8).

A variety of mRNA and promoter mapping studies have been carried out on the genital-tract-associated papillomaviruses including HPV-11, HPV-16, and HPV-18. A transcription map of HPV-16 is shown in Fig. 8. Only one promoter (P_{97}) has been well studied. An important difference in the structures of the E6 and E7 mRNAs and in the manner by which they are expressed distinguishes the "high-risk" and "low-risk" HPVs. As discussed in detail in Chapter 66 in *Fields Virology*, 3rd ed., the HPVs associated with the genital tract lesions can be divided into "high-risk" (HPV types 16 and 18) and "low-risk" (HPV types 6 and 11) categories based on the risk of malignant progression of the lesions which they cause. For the "high-risk" HPVs such as HPV-16 and HPV-18, a single promoter (P_{97} for HPV-16 or P_{105} for HPV-18) directs the synthesis of mRNAs with E6 and E7 intact (Fig. 8, Species A) or with splices in the E6 gene (Species B and C). The A species could be translated into E6 but not E7, since there is insufficient spacing for translation reinitiation. The B and C species of mRNAs splice the 5' end of the E6 ORF into a reading frame with stop codons providing sufficient spacing for translation reinitiation for the E7 ORF and are, therefore, likely to represent the E7 mRNAs. In contrast, the E6 and E7 genes of the "low-risk" HPVs such as HPV-6 and HPV-11 are expressed from two independent promoters.

Regulation of Transcription (cis elements)

Papillomavirus transcription is tightly regulated in infected cells. This is evident in part from the differential expression of different viral RNAs in the cells of the wart at the different levels in the epithelium (14,40). The bovine papillomavirus transformed cells have provided a useful model for examining transcriptional regulation in the nonproductively infected cell, in which transcription is also tightly regulated. The genomes of the papillomaviruses contain multiple *cis* regulatory elements and encode several transcriptional factors that modulate viral gene expression.

The LCR of the papillomaviruses contains enhancer elements which are responsive to cellular factors as well as to viral encoded transcriptional regulatory factors. Each of the viral LCRs that have been studied in detail have been shown to contain constitutive enhancer elements which have some tissue or cell type specificity. This was first established for HPV-16 and HPV-18 (49,275) and has also been shown for HPV-11 (38). In the BPV-1 genome, a constitutive enhancer has been mapped between nucleotide 7162 and 7275, which is active in a variety of cell types in the absence of any viral gene products (288,289). It is thought that these constitutive enhancer elements are essential for the initial expression of the viral genes after virus infection and that they may be important in the maintenance of viral latency.

E2 Regulatory Proteins

The papillomavirus E2 gene products are important regulators of viral transcription and replication. The E2 gene product of BPV-1 was first described as a transcriptional activator (259) capable of activating viral transcription through E2-responsive elements located within the viral genome (257). The E2 proteins are relatively well conserved among the papillomaviruses in two domains: a sequence-specific DNA binding and dimerization domain located in the carboxy terminal region of the protein and a transactivating domain that is located within the amino terminal half of the protein (95,179). These two domains are separated by an internal hinge region which is not well conserved in size or in amino acid composition among different papillomaviruses. The E2 proteins bind the consensus sequence $ACCN_6GGT$ (4,165) and can regulate transcription from promoters containing E2 binding sites (115,124,256). E2 binds $ACCN_6GGT$ motifs as a dimer; the dimerization domain has been localized to the carboxy terminus of the E2 protein (180).

The E2 proteins have been best studied in the BPV system where three species have been identified (Fig. 9). The full-length protein (E2TA) can function as a transactivator or a repressor depending on the position of the E2 binding sites within the enhancer/promoter region (Fig. 10). The two shorter forms of E2 called *E2TR* and *E8/E2* have been described as repressors because they can inhibit the transactivation function of the full-length E2TA (156,157). The shorter E2 proteins contain the DNA binding and dimerization domains of the C terminus but lack the transactivation domain. E2TR and E8/E2 can inhibit the transcriptional transactivating function of the full-length polypeptide

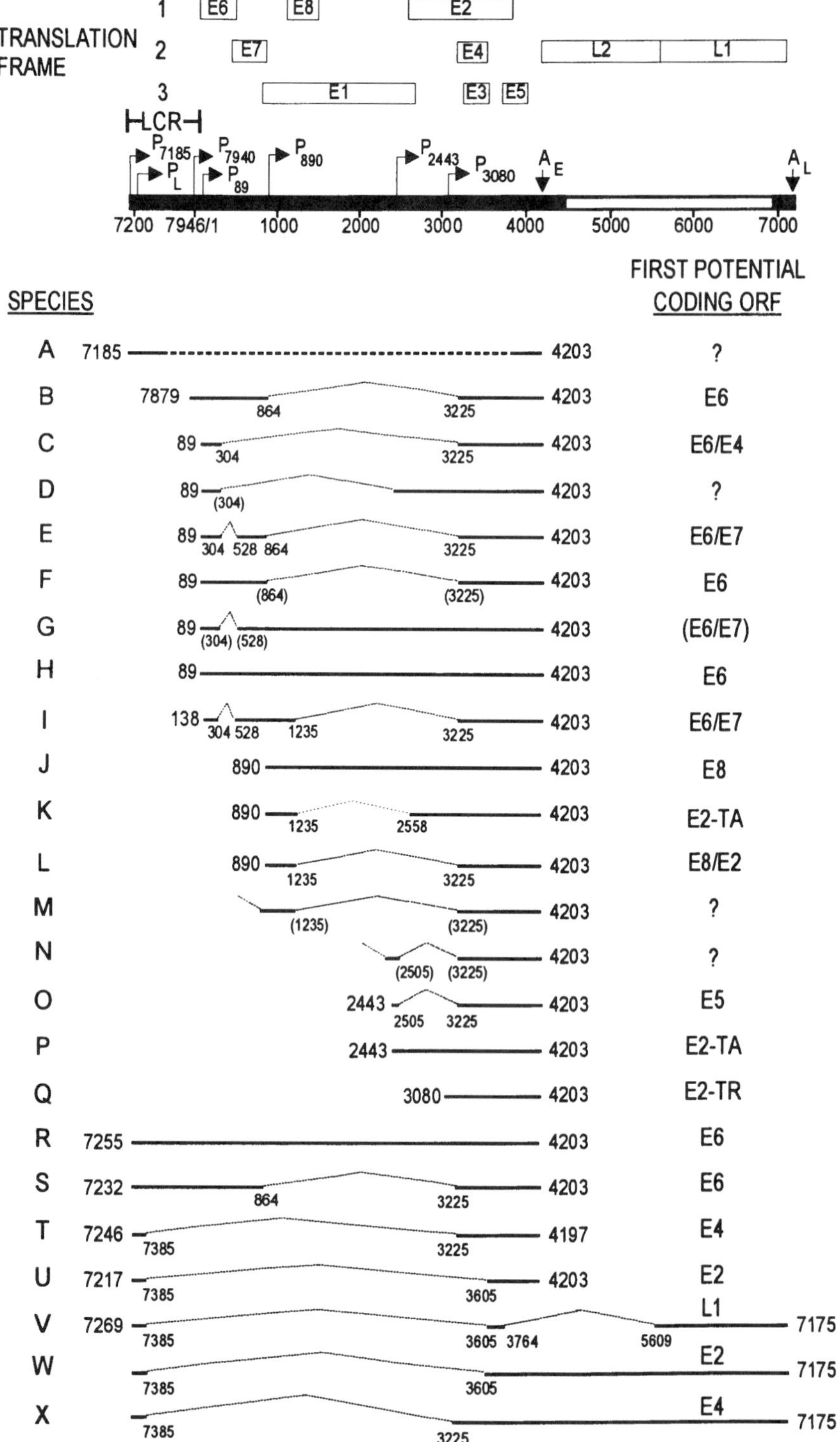

FIG. 7. Transcription map of BPV-1. The genomic map (*top*) indicates the 69% transforming region containing the LCR and the early region ORFs (*black line*). The known promoters and poly(A) sites are indicated. The structures of the species A to Q were determined from RNA species in BPV-1–transformed mouse C127 cells by cDNA cloning, electron microscopy, nuclease protection, PCR, and primer extension (2,8,39,261,262,279,286,306). The 5' most ORF which is likely to be encoded by the mRNA species is indicated (*right*). Structures of additional mRNAs that appear to be unique for productively infected BPV-1 fibropapillomas (species R-X) were determined by RT-PCR and cDNA cloning (6,8). Although E2 and E4 are the first significant ORFs for the W and X species, these mRNAs could also encode the L2 capsid protein. From Baker (6), with permission.

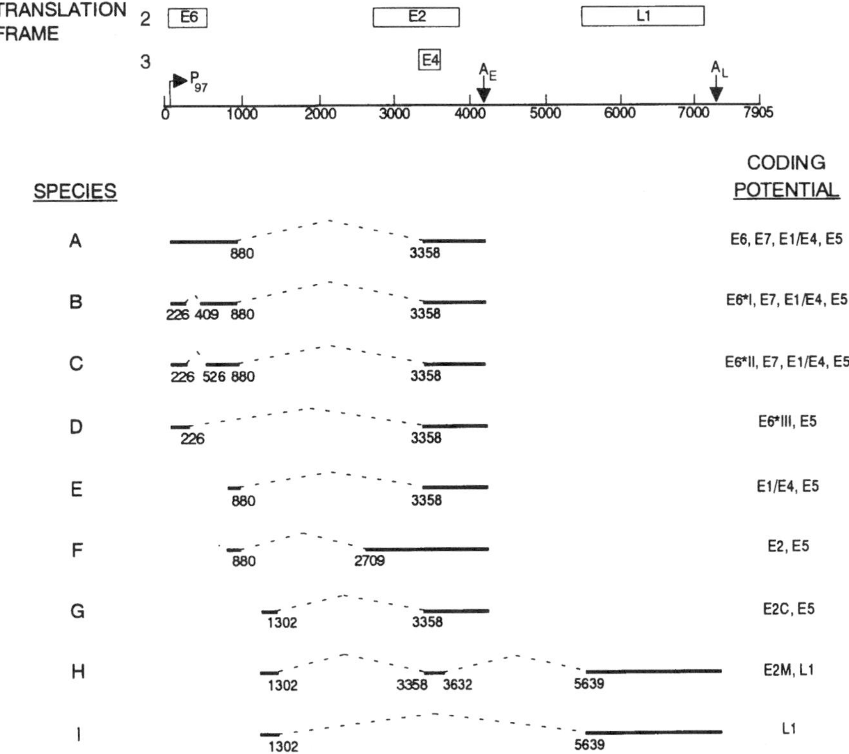

FIG. 8. Transcription map of HPV-16. The genomic map of HPV-16 is shown (*top*). The structures of viral mRNAs transcribed from full genomes are shown. Viral/cellular fusion transcripts have been detected in HPV-16–positive cancers and cancer cell lines; however, these are not included in this summary. Species A to D are transcribed from the only documented HPV-16 promoter, P_{97} (66,225,255). The 5' ends of the species E to I have not yet been mapped (66,252). The numbers below each mRNA species indicate the splice junctions. The coding potential for each mRNA is indicated (*right*). From Baker (6), with permission.

by competing for its cognate DNA binding sites and by forming inactive heterodimers with the full-length transactivator protein (180). The crystal structure of the dimeric DNA binding domain of BPV E2 complexed to DNA has been determined and reveals a previously unobserved structure for a DNA binding protein which is a dimeric antiparallel fl barrel (119).

Mutations in the BPV-1 ORF are pleiotropic, disrupting transformation, replication, and transcriptional regulation functions (59,61,106,122,220). Studies have shown that all of the early region viral gene expression is under the control of the viral E2 gene product through E2-responsive elements located within the viral LCR (4,114,123,166,219, 257,259,270). The LCR contains 12 of the 17 E2 binding sites (ATTN$_6$GGT) in the BPV1 genome (165). Pairs of the E2 binding sites are sufficient to function as an E2-dependent enhancer (4,116,256). The major E2-dependent enhancer in the LCR, E2RE1, contains four high-affinity E2 binding sites (257) and is located upstream of the P_{89} promoter complex. In addition to the P_{89} promoter, E2 can activate all of the early region promoters (P_{890}, P_{2443}, and P_{3080}) through the E2-dependent enhancer elements in the LCR (4,114,123, 166,219,257,259,270). Despite this dependence on a common E2-responsive element in the LCR, the various promoters have different sensitivities to the E2 transactivator (270). These differences are most likely due to specific promoter elements in the vicinities of the individual promoters. For instance, there is an SP1 site upstream of the P_{2443} promoter that is necessary for both its basal and E2 transactivated expression (258).

E2 transcriptional regulation has also been well studied for the genital-tract-associated HPVs. The binding of E2 to its cognate sites within the LCR of the HPV genomes results in the modulation of viral promoter activity. The E6 and E7 transforming genes of HPV-16 and HPV-18 are transcribed from a major promoter (P_{97} and P_{105}, respectively) contained within the LCR of their respective genomes (255,277). Analyses of promoter activity in human epithelial cells have shown that the P_{97} promoter of HPV-16 and the P_{105} promoter of HPV-18 possess a basal activity which can be repressed by full-length E2 gene products (20,227,276,277). There are four E2-binding sites within the LCRs of the HPV-16 and HPV-18 genomes. Two E2-binding sites are located immediately adjacent to the TATA box of the P_{97} and P_{105} promoters of HPV-16 and HPV-18, respectively. The basal activity is dependent on the ker-

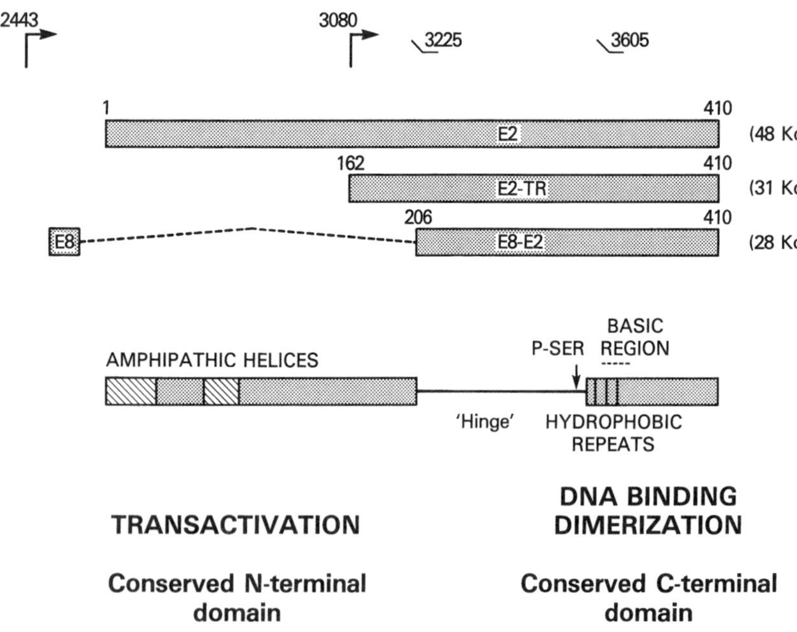

FIG. 9. Structure of the BPV-1 E2 gene products. The structures of the three known proteins encoded by the BPV-1 E2 ORF are indicated. The 48-kd full-length E2 transactivator can be expressed from an unspliced message from P_{2443} or from a spliced mRNA from upstream promoters by utilizing a splice acceptor at nT 2558. The 31-kd and 28-kd forms of the repressor are expressed from P_{3080} and as an E8/E2 fusion by a spliced mRNA as shown, respectively. The transactivation domain consists of a region of approximately 200 amino acids at the N-terminal region of the full-length E2 protein which is relatively well conserved among papillomaviruses. This region, which contains predicted amphipathic helices and is acidic, is only present in the full length form of E2. The 85 C-terminal amino acids are also conserved and comprise the DNA-binding and dimerization domain. The basic region and the hydrophobic repeats are indicated. From McBride et al. (180), with permission.

atinocyte-dependent enhancer contained within the LCR, and E2 repression occurs through binding to its cognate $ACCN_6GGT$ sites in very close proximity to the TATA boxes of each promoter, most probably interfering with the assembly of the preinitiation transcription complex (20,49,227,269,277). In the context of the full genome, the modulation of HPV gene expression by E2 is even more pronounced. In human keratinocyte immortalization assays dependent upon the expression of the E6 and E7 genes, E2 can decrease the efficiency of HPV-16 immortalization. E2 can also suppress the growth of HPV-positive cervical cancer cell lines through mechanisms that in part involve the transcriptional repression of the viral E6 and E7 genes (135,277). Wild-type HPV-16 DNA immortalizes prima-

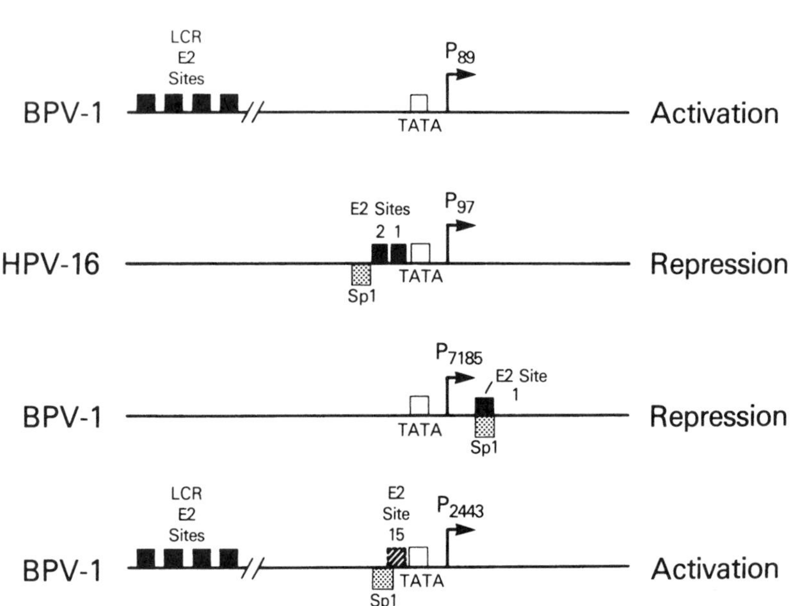

FIG. 10. Papillomavirus E2 transcriptional regulation. Transcriptional activation or repression by E2 depends upon the position and proximity of the E2 binding sites in the promoter with respect to the promoter. The E2 binding sites in the LCR are positioned several hundred bp upstream of P_{89}, which is activated by E2 (114,257). The HPV-16 P_{97} promoter (and the analogous P_{105} promoter in HPV-18), which directs the synthesis of the E6 and E7 viral oncoproteins, is repressed by E2 through occupancy of E2 binding sites which overlap an SP1 site and are in close proximity to the TATA box (20,227,272,276). It is thought that through displacement of SP1 from the promoter and interference with the formation of the transcription preinitiation complex, E2 may repress transcription. In a similar manner, E2 may repress the BPV-1 P_{7185} promoter by interfering with SP1 binding to a site downstream of the promoter (260,288). The P_{2443} promoter has a low affinity E2 binding site overlapping an SP1 site but does not appear to play an important role in its regulation. The promoter is activated by E2 binding sites located 2.5 kilobases upstream in the LCR (123,258). From McBride et al. (180), with permission.

ry human keratinocytes with a low efficiency, whereas genomes with mutations of the E2 gene are approximately tenfold more efficient (226). These results are consistent with the model that at certain concentrations E2 can repress the viral P_{97} promoter through binding to the E2 DNA binding sites in the proximity of the promoter. This model, however, cannot fully account for the inhibition of cellular immortalization by E2. Mutation of the E2-binding sites adjacent to the P_{97} promoter only partially alleviate E2-mediated repression, suggesting that additional mechanisms of E2-mediated repression may also be involved. Such a mechanism may also involve the E1 protein, since mutations in the E1 gene also result in an increase in the HPV-16 immortalization efficiency (226). In contrast to the repression mediated by E2 on the expression of the high-risk HPVs, in BPV-1, the full-length E2TA stimulates the expression of a number of genes including E2 itself and E5 (123,219).

In addition to its role in transcriptional regulation, E2 is also necessary for efficient viral DNA replication together with the viral E1 protein (37,56,284). The role of E2 in DNA replication may be as an auxiliary protein. This has been best studied in the BPV system where E2 has been shown to complex with E1 and strengthen the affinity of E1 for binding to the origin of DNA replication (182,284).

Late Gene Expression

The viral late functions, such as vegetative viral DNA synthesis, capsid protein synthesis, and virion assembly, occur exclusively in differentiated keratinocytes. Transcription of the late genes in BPV-1 was initially studied by analyzing the late ORF-specific RNAs by filter hybridization analysis (3,75,120) and by direct cDNA cloning of the viral mRNAs (8). The L1- and L2-specific mRNAs are polyadenylated at nucleotide 7175, a site not utilized in transformed cells and therefore considered the late poly(A) site (A_L) (8,75). In addition, the late RNAs are transcribed from a wart-specific or late promoter (P_L) at nt7250, a promoter that is apparently only active in terminally differentiated keratinocytes of a wart. Both the L1 and L2 genes are expressed from messenger RNAs transcribed from P_L and polyadenylated at A_L (14). The P_L promoter is regulated, at least in part, by cellular factors. In addition to the L1 and L2 mRNA species, this promoter also gives rise to other viral mRNAs which are specific for terminally differentiated cells. For instance, one abundant mRNA from P_L gives rise to the message that appears to encode the E4 protein (8). E4 proteins are abundant cytoplasmic proteins found predominantly in the differentiated keratinocytes of the wart (63). Thus, although E4 is located in the early region of the viral genomes (Figs. 3,4), it is a late protein by virtue of the promoter from which it is expressed.

BPV-1 late gene expression is also regulated from elements located within the late region. One system that has been useful for the study of papillomavirus late gene regulation has been the analysis of BPV-1 transformed mouse fibroblasts. Although no cytoplasmic mRNAs are expressed from the late region in these cells, nuclear run-off analysis has indicated that there is significant transcription of the late region (9). Thus, in fibroblasts and in nondifferentiated keratinocytes, there are mechanisms which either prevent pre-mRNAs transcribed from the early promoters from being polyadenylated at the late poly(A) site or destabilize any late mRNAs that are made. Transcriptional termination (or pausing) may significantly influence poly(A) site choice in BPV-1–infected cells (9,88). There is a negative element in the late 3' UTR (untranslated region) that acts at a posttranscriptional level to decrease the steady-state levels of late mRNAs in infected cells. This element does not function by destabilization of cytoplasmic mRNA but rather appears to inhibit the efficiency of utilization of the late poly(A) site (89). The element is conserved among the different papillomaviruses, including HPV-16, and has homology to 5' splice sites. Since there is no 3' splice site between this inhibitory element and the late poly(A) site, it is thought to interfere with 3' terminal exon definition (89). By inhibiting the use of the late poly(A) site, the use of the upstream early poly(A) site is preferred, resulting in mRNAs that encode only the early proteins.

Several models have been proposed for how transcriptional termination might be regulated during keratinocyte differentiation in a wart (9). One possibility is that cellular factors required for transcriptional pausing are titrated out during the process of vegetative DNA amplification, thereby permitting more efficient transcription of the late region. This is consistent with the observation that late region mRNAs become abundant only after the onset of vegetative DNA replication. An alternative model is that termination, or antitermination, factors might be part of the initial transcription complex that assembles at various promoters. If an antitermination factor were bound to the transcription complex only at the late promoter and not at the early promoter, then expression of the L1 and L2 ORFs would be linked to use of the late wart-specific promoter. Further experiments will be necessary to identify the cellular factors involved in papillomavirus late gene regulation.

The E4 Proteins

The E4 ORF of the papillomaviruses is located in the early region (see Figs. 3,4), yet it appears to be expressed as a late gene with a role in productive infection. It entirely overlaps the E2 ORF in a different reading frame and, therefore, encodes a protein with an entirely different amino acid sequence. In general, the E4 gene is not highly conserved among the papillomaviruses. Transcripts that could encode E4 proteins have been described both for BPV-1 and the HPVs (see Figs. 8,9). A viral transcript formed by splicing a few codons from the beginning of E1 to E4 appears to be the major RNA in HPV-induced lesions (190). In papillomas the E4 proteins are expressed primarily in the differentiating layers of the epithelium and the E1^E4 protein has been co-localized with L1 (26,28,50,206). Studies with the HPVs have revealed that there are multiple species of E4 proteins in infected cells. Although the E4 proteins are expressed at high levels in infected tissues, their precise role in the viral life cycle is unclear. E4 proteins are not found in the virion particles. Mutational analysis of the E4 gene in BPV-1 showed that E4 was not essential for viral transformation or viral DNA replication (122,192).

Recent reports have shown that the E4 proteins are associated with the keratin cytoskeleton of cultured epithelial cells (64,222,224). The HPV-16 E4 protein has been shown to induce the collapse of the cytokeratin network (64,222), suggesting that it may function to aid the virus in its egress from the cell. A dramatic collapse of the keratin intermediate-filaments has not however, been observed with cells expressing HPV-1 E4, raising some question as to the generality of this effect (222,224).

Studies with HPV-1 E4 have revealed multiple protein doublets of 10/11 kd, 16/17 kd, 21/23 kd, and 32/34 kd in size (26, 63). The 17-kd polypeptide is thought to represent the E1^E4 product, and although the cleavage sites are not yet known, the smaller species are thought to arise from N-terminal proteolytic processing. It is thought that the larger species represent multimers of the E4 proteins (63,65). The different E4 species appear to be differentially expressed in warts. The 17-kd protein is first detected in the suprabasal layer, and the other species appear progressively and concomitantly with epithelial cell differentiation (26). The 10/11-kd species are most abundant in the most differentiated superficial levels. It is thought that the different forms of the E4 proteins may have different functions.

Virus Assembly and Release

There is little known about the papillomavirus assembly or the release. Virus particles are observed in the granular layer of the epithelium and not at lower levels. The virus is not believed to be cytolytic, and the release of the virion particles occurs into the cornified layers of a keratinized epithelium.

VIRAL DNA REPLICATION

For the papillomaviruses, there are two modes of viral DNA replication. The first likely occurs in the cells of the lower portion of the epidermis, including the basal cells, as well as in the dermal fibroblasts in fibropapillomas, such as those induced by BPV-1. In these cells, the viral DNA is apparently maintained as a stable multicopy plasmid. The viral genomes replicate an average of once per cell cycle during S-phase in synchrony with the host cell chromosome (93) and may be faithfully partitioned to the daughter cells. This type of DNA replication ensures a persistent and latent infection in the stem cells of the epidermis. The second type of DNA replication is vegetative DNA replication, which occurs in the more differentiated epithelial cells of the papilloma. In the differentiated squamous epithelial cells, in which there is no cellular DNA synthesis, one observes a burst of viral DNA synthesis, generating the genomes that are packaged into progeny virions.

Plasmid Replication

In BPV-1 transformed cells, the viral genome can exist as a stable multicopy plasmid (159). This system has provided a model for studying nonvegetative viral DNA replication in mammalian cells, and it has been assumed that DNA replication in these rodent cells may be analogous to the stable plasmid replication seen in the dermal fibroblast and in the nondifferentiated keratinocytes of a fibropapilloma. Plasmid DNA replication in virally transformed rodent cells can be divided into two phases: establishment and maintenance. There is an initial establishment period in which the viral genome is amplified from a low copy number to a moderate copy number of approximately 50 to 400 copies per cell. During the subsequent maintenance phase, this copy number of the viral plasmids remains relatively constant for many cell generations. BPV-1 transformed rodent cells have been a useful model system to analyze these two phases of DNA replication.

Origin of DNA Replication

The origin of BPV-1 DNA replication was initially localized to the LCR by electronmicroscopic analysis of replicative intermediates isolated from BPV-1 transformed rodent cells harboring extrachromosomal viral DNA (293). BPV DNA replication requires the origin of DNA replication in cis and the viral E1 and E2 proteins in trans. *In vivo*, the minimal origin of replication (nT 7911-27) contains an A+T-rich region, the E1 binding site that includes a region of dyad symmetry (DSR), and an E2 binding site (Fig. 11) (285). The origins of DNA replication have been mapped for a number of the human papillomaviruses and contain similar ATR and DSR domains adjacent to E2 binding sites (37,56). Furthermore, in transient replication assays the viral E1 and E2 proteins can function in trans on replication origins from heterologous papillomavirus genomes (37). BPV-1 origin-dependent DNA replication can be achieved *in vitro* with cell extracts with high levels of E1 alone and no E2 (251,304).

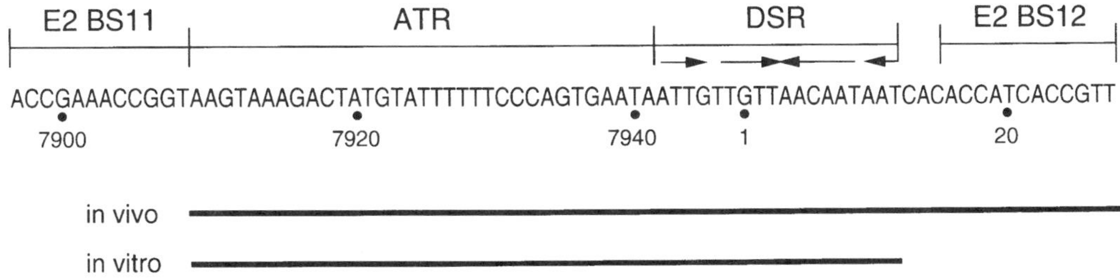

FIG. 11. The BPV-1 origin of DNA replication. The minimal origin for *in vivo* DNA replication (nt 7911-27) requires the A/T-rich region (*ATR*), the dyad symmetry repeats (*DSR*), and a binding site for the E2 protein. For *in vitro* DNA replication there is a requirement only for the ATR and DSR.

The E1 Protein

Genetic studies have revealed that stable BPV-1 plasmid replication is dependent upon the expression of the viral E1 and E2 genes. Initial studies of mutated BPV-1 genomes implicated both E1 and E2 as essential for stable viral DNA replication (59,61,106,220,233) and for transient viral DNA replication (284). However, E1 appears to be the only factor that is directly involved in plasmid replication; E2 may have an auxiliary role in DNA replication, as discussed below. The E1 ORF is the largest ORF in the papillomavirus genome and is relatively well conserved among all of the papillomaviruses. The E1 proteins share structural similarities with the large T antigen of simian virus (SV40), an essential replication protein of that virus. The similarity includes regions of tumor antigen (TAg) with ATPase, helicase, and nucleotide-binding activities (42,251). These similarities suggest a common function for these proteins in the initiation of viral DNA replication. The BPV-1 E1 protein is a 68-kd nuclear phosphoprotein that binds specifically to the origin of replication (280,285,300). By itself, E1 binds the origin with weak affinity; the additional binding of E2 to adjacent sites increases this affinity (182,285).

The papillomavirus E1 proteins can bind and hydrolyze ATP (25,251,305), and BPV-1 has been shown to have ATP-dependent helicase activity (251,305). E1 also interacts with the p180 subunit of the cellular polymerase α-primase and presumably thereby recruits the cellular DNA replication initiation machinery to the viral replication origin (208). In addition to the 68-kd protein encoded by the full gene, the 5' end of E1 encodes a protein with an apparent molecular weight of 23,000 (279). Although the protein has been detected by BPV-1 transformed cells, no function has yet been ascribed to this N-terminal E1 protein (131).

The Role of the E2 Proteins

As noted above, papillomavirus DNA replication also requires the viral E2 protein (284,304). A requirement for E2 in origin-dependent DNA replication has been shown for the HPVs as well as BPV-1 in transient assays. Although not essential for origin-dependent DNA replication *in vitro*, E2 greatly stimulates the ability of E1 to initiate DNA replication (304). The actual role of E2 in stimulating E1-dependent DNA replication is not yet known. E2 interacts with E1 (22,173,182) and greatly enhances the ability of E1 to bind the replication origin (182,250). E2 can relieve nucleosome-mediated repression of papillomavirus DNA replication *in vitro* (163). Also, E2 may stimulate DNA replication by recruiting host replication factors to the origin, such as the host cell single-stranded DNA binding protein RPA (replication protein A), which can bind to acid transactivation domains similar to that found in E2 (164). A recent study has shown that the role of E2 may be in the assembly of the preinitiation complex at the origin, but that E2 plays no role in the replication process (174).

Model for Plasmid Replication

In the initial stages of a papillomavirus infection in the cell, the viral genome may undergo amplification for a short period of time. There is then a transition to a maintenance stage in which the plasmids replicate on an average of once per cell cycle in a manner analogous to chromosomal DNA replication. It was initially thought that stable plasmid DNA replication might involve the "marking" of the daughter viral genomes so that they did not return to the replicating pool until the next cell cycle and that each viral plasmid molecule replicated only once per cell cycle. An alternative model in which BPV-1 plasmids replicate on average once per cell generation does not rely upon plasmid "marking" (93). At this point the evidence strongly supports the second model in which the plasmid replication is not restricted to once per cell cycle but occurs an average of once per cell cycle (189). However, precise mechanisms involved in regulating plasmid maintenance, as well as vegetative DNA replication, have not yet been elucidated.

Vegetative DNA Replication

As noted above, papillomavirus DNA must also undergo vegetative replication to generate the genomes to be packaged in virions, a process which normally occurs only in the more terminally differentiated epithelial cells of a papilloma. The mechanisms regulating the switch from plasmid maintenance to vegetative viral DNA replication are not known. The switch may involve the presence or absence of controlling cellular factors in differentiating keratinocytes. In addition or alternatively, the relative levels of viral factors such as E1 or E2 (or their modification) may change in terminally differentiating keratinocytes. A high level of BPV-1 DNA amplification has recently been demonstrated in growth-arrested transformed cells (32). In this study, it was shown that stable plasmid replication occurs in normal proliferating cells; however, viral DNA amplified to a high level only in cells which were growth-arrested. Both E1 and E2 have been implicated as having a role in this DNA amplification (32,33).

VIRAL TRANSFORMATION

The BPV-1 Transformation

Despite the fact that there is no efficient *in vitro* tissue culture system for the propagation of the papillomaviruses, cellular transformation by certain of these viruses has permitted researchers to study the viral functions involved in the induction of cellular proliferation by these viruses. The best studied of the transforming papillomaviruses has been BPV-1. Morphologic transformation in tissue culture was first described for BPV in the early 1960s (21,23,278). In the late 1970s, a focus assay was developed using established cell lines to study BPV-1 transformation (69). In general, investigators have relied upon mouse C127 cells and NIH 3T3 cells for these transformation studies, although a variety of other rodent cells including hamster and rat cells are susceptible to BPV-1–mediated transformation. Transformation of mouse C127 cells by BPV-1 causes alterations in morphology, loss of contact inhibition, anchorage independence, and tumorigenicity in nude mice (69).

One interesting characteristic of BPV-1 transformed rodent cells is that the viral DNA is often maintained as a multicopy plasmid (159), and that integration of the viral genome is not required for either the initiation or maintenance of the transformed state. However, transformation is dependent upon the continued expression of viral DNA, as evidenced by the loss of the transformed phenotype in mouse cells which have been "cured" of the viral DNA by treatment with interferon (283). The extrachromosomal state of the viral genome, however, is not a prerequisite for this transformation, since replication-defective viruses are capable of transforming cells (233). In cells in which the extrachromosomal DNA is maintained, it is likely that a high gene dosage resulting from the viral plasmid replication is necessary for efficient transformation by the wild-type genome. Cells harboring BPV-1 mutants which maintain a low copy number of extrachromosomal DNA have been reported to have a nontransformed phenotype (19).

The initial studies with cloned BPV-1 DNA indicated that the entire viral genome was not required for transformation of rodent cells (171). A specific 69% subgenomic fragment was sufficient for inducing cellular transformation; hybrid plasmids containing this 69% fragment could be maintained as stable plasmids within transformed cells. Genetic studies have mapped the BPV-1 transforming genes to the E5, E6, and E7 ORFs as discussed below. These early studies establishing that the BPV-1 genome could be maintained as a stable plasmid in rodent cells led to the development of BPV-1–based plasmid vectors for mammalian cells (62,231,232).

BPV-1 encodes three independent transforming proteins, encoded by the E6 (242,306), E7, and E5 ORFs (106,243, 307). Each of these genes is contained within the 69% transforming fragment and each is expressed in transformed cells. The E5 gene is the major transforming gene of BPV-1 expressed in transformed cells (59,106,220,242). This gene is highly conserved among those papillomaviruses that induce fibropapillomas in their natural host and are also able to induce fibroblastic tumors in hamsters. The E5 gene is believed to be responsible for the proliferation of dermal fibroblasts in fibropapillomas.

The BPV-1 E5 Oncoprotein

The BPV-1 E5 ORF encodes a 44 amino acid protein (245) and is sufficient for the transformation of certain established rodent cells in culture (58,106,220,243) The amino acid sequence of E5 and its predicted structure is shown in Fig. 12. Structurally it is composed of two protein domains; a very hydrophobic segment at its amino terminus which is responsible for localization of E5 in membrane fractions of BPV-1 transformed cells (245), and a carboxy-terminal hydrophilic domain. The E5 protein is associated with intracellular membranes and is thought to be an integral membrane protein (29). In transformed cells the E5 protein exists largely as a homodimer and is localized primarily in the Golgi apparatus and in the endoplasmic reticulum (30,31). E5 does not appear to be secreted and the hydrophilic C-terminus appears to extend into the lumen of the Golgi (30). The E5 protein contains two cysteine residues that are required for transformation; these residues appear to be involved in dimer formation (29,127).

The E5 protein does not possess intrinsic enzymatic activity, and it is likely that it functions by altering the activity of cellular membrane proteins involved in proliferation. The first direct evidence that BPV-1 E5 can affect the activity and metabolism of growth factor receptors came from experiments that showed that E5 can cooperate with exogenously introduced EGF receptor or CSF1 receptor in the transformation of NIH3T3 cells (176). Subsequently, it has been established that the primary endogenous target

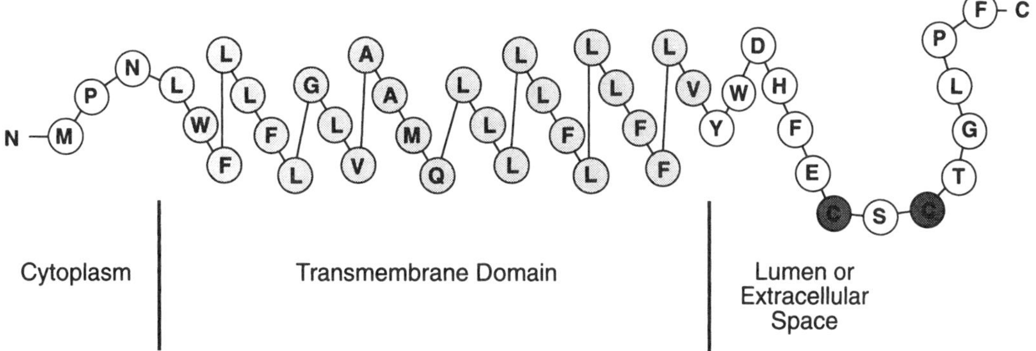

FIG. 12. Amino acid sequence and predicted secondary structure of the BPV-1 E5 oncoprotein. The 27 hydrophobic amino acids comprising the first two-thirds of the 44 amino acid protein (*shaded lightly)* are predicted to be in alpha-helical conformation. The 14 carboxy-terminal residues are predicted to be in a nonhelical configuration. The cysteine residues, which are highly conserved among the E5 proteins of the transforming ungulate papillomaviruses are indicated *(shaded darkly)*. These cysteines mediate dimer formation (29,127).

for the BPV-1 E5 protein in fibroblasts is the β receptor for the platelet-derived growth factor (PDGF) (212). There is an increase in the level of tyrosine phosphorylation of the endogenous PDGF receptor in the E5 transformed rodent cells and bovine fibroblasts (210, 212), and a prominent form of the tyrosine phosphorylated PDGF receptor in the transformed cells is an intracellular, membrane-associated, premature form of the receptor. Similarly, gene transfer experiments in heterologous cell types demonstrate that BPV-1 E5 transformation can be mediated by the PDGF β receptor but not by a variety of other growth factor receptors (103,194). The mechanism of E5 activation of the PDGF β receptor appears to involve complex formation, resulting in receptor activation in a ligand-independent manner (210,211). This is depicted in Fig. 13 which shows a model suggesting that the individual subunits of the E5 dimer may each bind a molecule of the PDGF β receptor, resulting in receptor dimerization and activation (60). The activated receptor would then initiate a cascade of intracellular events, resulting in mitogenic stimulation.

BPV-1 E5 has also been shown to complex with several other cellular proteins. The first cellular protein shown to bind the BPV-1 E5 protein was the 16-kd transmembrane channel-forming subunit of the vacuolar H^+-ATPase (102,104), an abundant cellular protein located in the membranes of intracytoplasmic membranes and plasma membranes. The possible functional significance of the binding of E5 to this 16-kd protein is unclear. It has been proposed that E5 binding to the 16-kd protein might affect its ATPase activity or affect gap-junctional communication, since the 16-kd protein may also be a major component of gap junctions (84,100). However, there is not yet any biochemical evidence that E5 effects the ATPase or gap junction activity of the 16-kd protein. Alternatively, there is also evidence that E5, the PDGF receptor, and the 16-kd ATPase subunit can form a ternary complex, making it possible that the role of 16-kd protein may be to mediate or stabilize the binding of E5 to the PDGF fl receptor (101).

The carboxyl-terminal hydrophilic domain of BPV-1 E5 has been shown to interact with a 125-kd cellular protein related to α-adaptin (45). Although the importance of this interaction has not yet been established, since α-adaptin may be involved in receptor-mediated endocytosis, some of the effects of the E5 protein on growth factor receptor metabolism could be mediated through its interaction with α-adaptin.

The E6 and E7 Genes of BPV-1

The BPV-1 E6 and E7 genes also encode proteins with transforming activities. In mouse cells, the full transformed phenotype requires the expression of the E6 and E7 genes as well as E5 (191,233). Furthermore, the BPV-1 E6 ORF when expressed from a strong heterologous promoter is sufficient for transformation of C127 cells (242). The E6 and E7 proteins are themselves structurally related and are conserved, at least in part, among all of the papillomaviruses. In BPV-1, cDNAs have been isolated and characterized which could encode the full-length E6 protein, the full-length E7 protein, as well as E6 fusion protein in which the amino terminal half of E6 could be fused to portions of downstream genes (306) (see Fig. 7). The E6 and E7 genes of all the papillomaviruses encode proteins with conserved structural motifs. They contain domains of almost identically spaced CYS-X-X-CYS motifs (four in E6 and two in the carboxy-terminal portion of E7). It has been postulated that the E6 and E7 genes may have arisen from

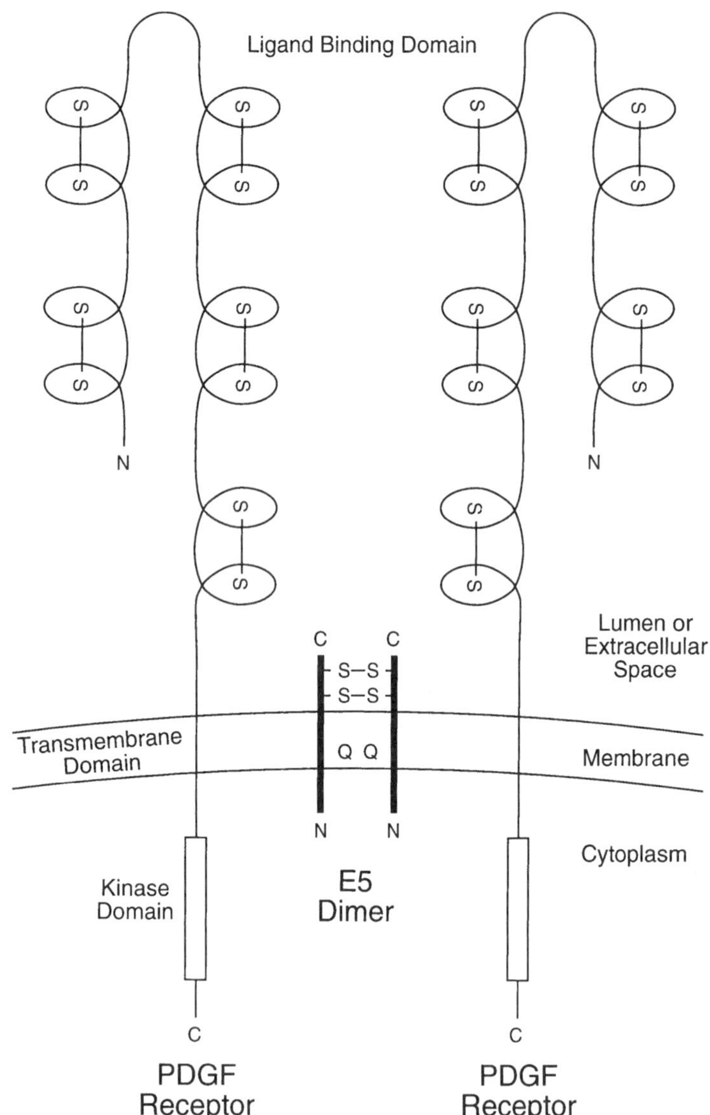

FIG. 13. A model of the E5 oncoprotein/PDGF receptor complex in transformed cells. The disulfide-linked E5 dimer is shown complexed to two PDGF receptor monomers, resulting in receptor dimerization, activation, and tyrosine phosphorylation. *Q* in the E5 protein represents the glutamine residue in transmembrane domain which is essential for transformation. Not depicted is the 16-kd transmembrane channel-forming subunit of the vacuolar H^+-ATPase which was the first cellular protein shown to bind E5 (102). Adapted from DiMaio et al. (60).

duplication events involving a 39 codon core sequence containing one of these motifs (46). The CYS-X-X-CYS motifs found in a number of nucleic acid binding proteins are characteristic of zinc binding proteins. The E6 and E7 proteins of several papillomaviruses have now been shown to bind zinc (13,109,110). The E6 protein of BPV-1 is present at low levels in stably transformed cells, and cell fractionation experiments have revealed that the BPV-1 E6 protein is located in the nucleus and in nonnuclear membranes of transformed cells (5).

The product of the E7 ORF has been detected in BPV-1 transformed cells (136). Although the E7 gene by itself is not able to induce foci in transformation assays, its integrity is necessary for the fully transformed phenotype as assayed by anchorage independence and by tumorigenicity (191). A recent study has shown that BPV-1 E6 and E7 are under the negative regulatory control of the viral E1 and E2 genes, and in the absence of these gene products, E6 and E7 together are potent oncogenes that are sufficient for the full transformation of C127 cells (290).

In addition to genes which are directly involved in BPV-1 transformation (E5, E6, and E7), the early region also encodes genes which are indirectly involved in transformation. This includes genes encoded by the E1 and E2 ORF which are involved in the transcriptional regulation and replication of the virus. Initial studies with BPV-1 E2 mutants suggested they were defective for transformation (59,106,220). This defect is not because of the loss of a direct transforming activity of E2, but rather because E2 is necessary for sufficient expression of the viral oncoproteins to manifest a transformed phenotype. Mutations in the E1 ORF may result in a higher transforming activity due to the increase in the transcriptional activity of the mutant viral genome (155,241). BPV-1 E1 has been shown to repress the E2 transactivated expression of the virus (229).

HPV Transformation and Immortalization

Transformation studies with the papillomaviruses have not been limited to BPV-1. Assays have also been developed for the transforming functions encoded by those HPVs associated with anogenital cancers in humans (18,140,177, 216,218,282,308). The HPV-16 and HPV-18 genomes are not as efficient as BPV-1 at inducing transformation of established rodent cells; however, alternative assays have been developed. DNA cotransfection with a second dominant selective marker such as the neomycin resistance gene and biochemical selection for the transfected cells permits HPV-16 and HPV-18 to transform established rodent cells (308). More informative have been assays employing primary rodent cells and primary human fibroblast and keratinocyte cultures (67,141,177,216,218,244,294). In these assays, those high-risk HPVs associated with clinical lesions that are at an increased risk for malignant progression such as HPV-16 and HPV-18 are transformation-positive, whereas the lower-risk viruses such as HPV-6 and HPV-11 are not (244,266). These assays have permitted the mapping of the viral genes directly involved in cellular transformation to the E6 and E7 ORFs of HPV-16 and HPV-18, which are discussed in detail below.

In established rodent cells, such as the NIH3T3 cells, the E7 ORF scores as the major transforming gene for HPV-16 and HPV-18 (216,273,292,295,309). HPV-16 and HPV-18 by themselves are not able to transform primary rat fibroblasts or baby rat kidney cells (154,216). However, the E7 gene can cooperate with an activated *ras* oncogene to fully transform primary rat cells (16,165,195,216,266). It is of note that the DNAs of the HPVs associated with lesions at a risk for malignant progression (i.e., HPV-16, -18, -31, and -33) display transforming properties in rodent cells, whereas DNAs from the genital HPVs not associated with this risk for malignant progression (i.e., HPV-6 and HPV-11) do not (266).

The DNAs of the high-risk HPVs can also be distinguished from the DNAs of the low-risk HPVs by their ability to immortalize primary cultures of human fibroblasts, human foreskin keratinocytes, and human cervical epithelial cells (67,218,244,294,301). The resulting cell lines are neither anchorage-dependent nor tumorigenic in nude mice, but they do display altered growth properties and are resistant in the response to signals for terminal differentiation (67,143,218,244). All of the genital tract HPV types, (i.e., HPV-6, -11, -16, and -18) are capable of transiently inducing cellular proliferation. Only the cancer-associated HPVs, however, are able to extend the life span and give rise to immortalized cell lines which are refractory to differentiation signals (244).

The HPV E5 Oncoprotein

The E5 ORF is highly conserved among those papillomaviruses that induce fibropapillomas and which readily transform rodent fibroblasts in tissue culture, including BPV-1, DPV, and EEPV (153), suggesting that E5 may play an important role in the tropism for fibroblasts observed *in vivo* for these viruses. Many of the other papillomaviruses that induce purely epithelial papillomas (such as CRPV and the HPVs) also contain E5 genes with the potential to encode short hydrophobic peptides. The structural similarity of these potential peptides to the BPV-1 E5 protein has prompted studies of the potential transforming activities of these putative genes. Studies have shown the HPV-6 E5 gene can induce some transformation alterations in NIH3T3 cells (36); however, the role HPV-6 E5 is unclear since the HPV-6 DNA is often not retained in transformed mouse cells (183). Transforming activity has also been demonstrated for the HPV-16 E5 gene. There have been several studies which have shown that HPV-16 E5 can induce some transformed alterations in mouse fibroblast lines or mouse epidermal keratinocytes (161,162,267), increase the proliferative capacity of human keratinocytes (265), and stimulate cellular DNA synthesis in human keratinocytes in a manner that is potentiated by EGF (267). The biochemical mechanisms by which the E5 genes of the epitheliotropic papillomaviruses exert their growth stimulatory effects have not yet been elaborated. It should be noted, however, that in most HPV-positive cancers, the E5 gene is not expressed, suggesting that if the E5 gene does stimulate cell proliferation *in vivo*, it presumably functions in benign papillomas and not in the cancers. It might participate, however, in the initiation of the carcinogenic process.

The HPV E7 Oncoprotein

The E7 protein encoded by the high-risk HPVs is a small nuclear protein of about 100 amino acids, has been shown to bind zinc, and is phosphorylated by casein kinase II (CK II) (186). Insight into its mechanism of action came initially from the recognition that E7 has functional similarities with the adenovirus (Ad) 12S E1A product (216). Like Ad E1A, E7 can transactivate the Ad E2 promoter (216), induce DNA synthesis in quiescent cells (234), and cooperate with an activated *ras* oncogene to transform primary rodent cells (177, 216).

In addition to these functional similarities, the HPV-16 E7 shares amino acid sequence similarity with portions of the AdE1A proteins and the SV40 large TAg (Fig. 14). The conserved regions in all of these oncoproteins bind cellular proteins, one of which is the product of the retinoblastoma tumor suppressor gene pRB (55,71,298). Complex formation with pRB involves conserved region 2 of the Ad E1A protein and the corresponding region in the E7 protein and in SV40 large TAg (55,186,299).

The retinoblastoma protein is a member of a family of cellular proteins which also include p107 and p130 which are homologous in their binding "pockets" for E7, Ad E1A, and SV40 TAg. The retinoblastoma protein is the most ex-

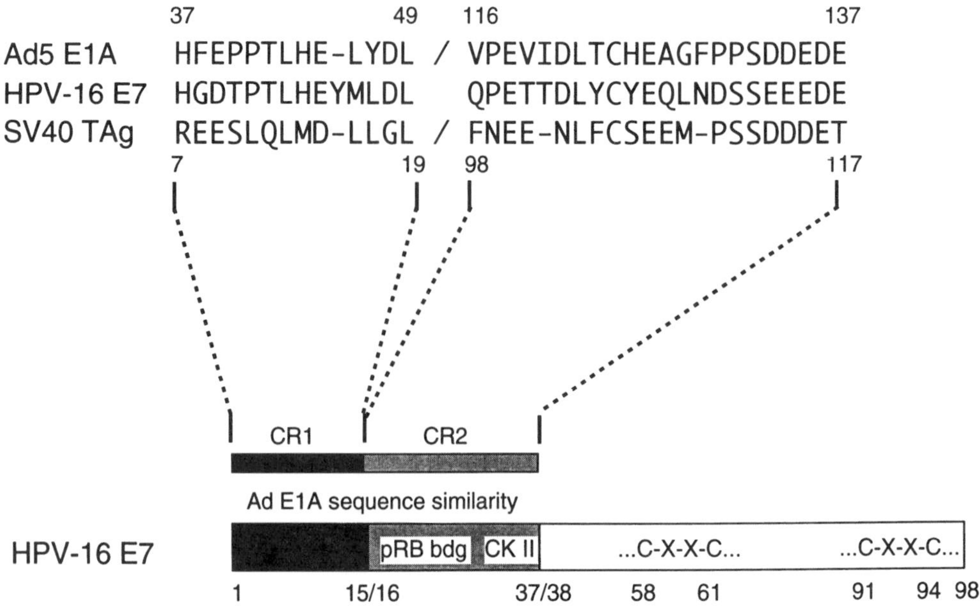

FIG. 14. Amino acid sequence similarity between portions of conserved regions 1 and 2 (*CR1, CR2*) of the Ad5 E1A proteins and the regions of SV40 large T antigen with the amino terminal 38 amino acids of HPV-16 E7. CR2 contains the pRB binding site and the casein kinase II (*CKII*) phosphorylation site of HPV-16 E7. From Scheffner et al. (238), with permission.

tensively studied member of this family of proteins. Its phosphorylation state is regulated through the cell cycle, being hypophosphorylated in G0 and G1 and phosphorylated during S, E2, and M. pRB becomes phosphorylated at multiple serine residues by one or more cyclin-dependent kinases (cdk) at the G1/S boundary and remains phosphorylated until late M when it becomes hypophosphorylated again through the action of a specific phosphatase (Fig. 15). Since pRB acts as a negative regulator of cell growth at the G1/S border, it follows that the hypophosphorylated form represents the active form with respect to its ability to inhibit cell cycle progression. HPV-16 E7, like SV40 TAg, binds preferentially to the hypophosphorylated form of pRB, consistent with the model that this interaction results in the functional inactivation of pRB and permits progression of the cell into S phase of the cell cycle (43). This property of the viral oncoproteins to complex pRB would appear to account, at least in part, for their ability to induce DNA synthesis.

Like Ad E1A and SV40 TAg, the HPV E7 proteins can also bind the cellular proteins p107 and p130 (70) through a region conserved among this group of proteins referred to as the *pocket*, which is also involved in binding specific cellular proteins. The consequence of the binding of the viral oncoproteins to pRB, p107, and p130 may be the displacement of these cellular proteins. This has been demonstrated in that binding of the E2F transcription factors, which normally interact with the pocket proteins, is significantly reduced in the presence of the viral oncoproteins (193). However, the model is not as simple as it may seem. Whereas Ad E1A disrupts the pRB/E2F complex and the p107/E2F complex with similar efficiency, the HPV E7 disrupts the pRB/E2F complex more efficiently than the p107/E2F complex (34,205). Like Ad E1A, HPV-16 E7 is also able to form a complex with cyclin A. This interaction may not be direct, however, and may be mediated by p107 (70,281).

Genetic studies of HPV-16 E7 have revealed that an intact and high-affinity pRB binding site is necessary for transformation of rodent cells. The amino terminal sequences of E7, which are similar to CR1 and Ad E1A, may also affect cellular transformation independent of the pRB binding. In the N-terminus Ad E1A and SV40 TAg there are amino sequences involved in complexing the cellular protein, p300. Similar sequences are not found in E7, and binding of E7 to p300 has not been detected, although mutational studies have demonstrated the importance of the amino terminus of E7 in its transformation functions (27,53,72,215).

The E7 proteins from the high-risk and low-risk HPV types differ in a number of biochemical and biological properties. The E7 proteins from the low-risk HPV types 6 and 11 bind pRB with about a tenfold lower efficiency than the E7 proteins of HPV types 16 and 18 (90,187). Furthermore, the E7 proteins of the low-risk HPVs function very inefficiently in cellular transformation assays with an activated ras oncogene and are phosphorylated by CK II at a lower rate (12). Studies with chimeric E7 proteins containing domains of high-risk and low-risk HPV E7 proteins showed that the difference in transformation efficiency in rodent cells was due to the pRB-binding site (118). Sequence comparison of the pRB binding sites re-

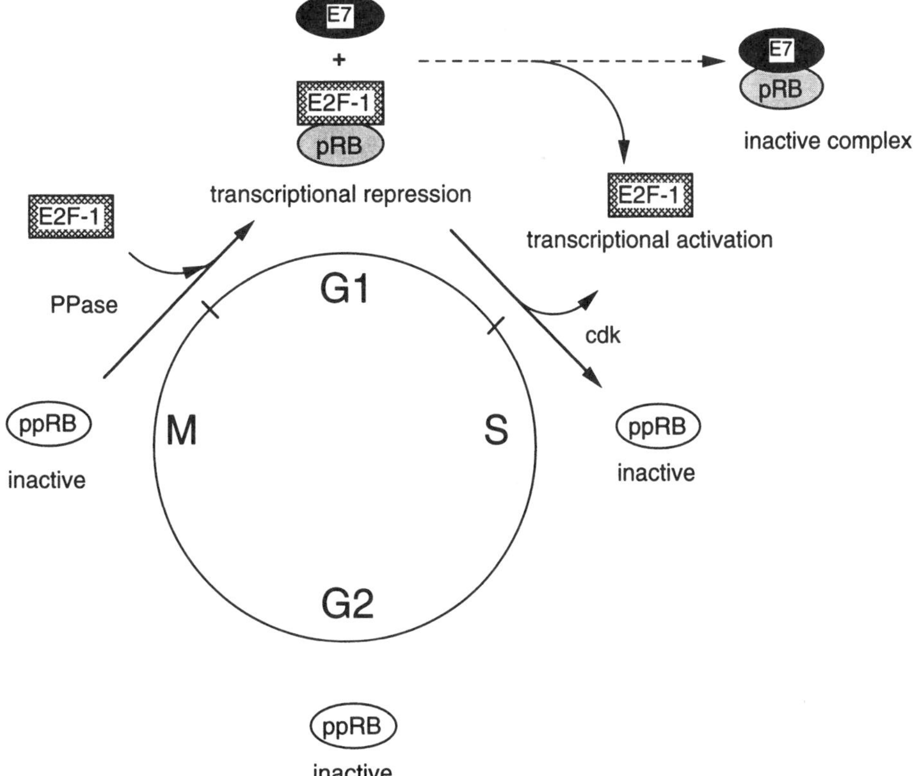

FIG. 15. E7 abrogates the cell cycle regulation mediated by pRB (as well as the related proteins p107 and p130) by complex formation. During the cell cycle, pRB is differentially phosphorylated, and the underphosphorylated form is detected only in the G0/G1 phase. This underphosphorylated form is the active form of pRB, acting as a negative regulator of the cell cycle. During the transition to the S-phase, pRB is phosphorylated by cyclin-dependent kinases (cdk), resulting in the inactivation of its cell cycle regulatory functions. The cellular transcription factor E2F-1 is preferentially bound to the under-phosphorylated form of pRB, and in complex with pRB cannot activate transcription. Phosphorylation of pRB or complex formation with E7 results in the release of E2F-1, allowing it to function as a transcriptional activator of cellular genes involved in cellular DNA synthesis and progression into the S phase of the cell cycle. From Scheffner et al. (238), with permission.

vealed a single consistent amino acid sequence difference between the high-risk and the low-risk E7 proteins: an aspartic acid residue (Asp 21 in HPV-16 E7) corresponding to a glycine residue in the low-risk E7 sequence (Gly 22 in HPV-6 E7) (Fig. 16). Substitution of this residue in the respective E7 genes revealed that this single amino acid residue was the principle determinant responsible for the difference in pRB binding affinity and in the transforming capacity of the low-risk and the high-risk E7 proteins (118,230).

Mutations in the carboxy terminal half of E7 that interfere with E7 function have also generally affected the intracellular stability of the protein (72,215). Nevertheless, several studies have suggested some specific contributions of the carboxy terminus to E7 functions. E7 can exist as a dimer, and the C-terminal half is important in mediating dimerization (154). Also, it has been shown that binding of E7 to pRB abrogates the nonspecific DNA-binding properties of pRB and that carboxy terminal E7 sequences, in addition to the pRB-binding site, may be necessary for this activity (263). The ability of E7 to disrupt the complex between the cellular transcription factor E2F and pRB also involves sequences in the carboxy terminus of E7 in addition to the pRB-binding site (129,302). Although the ability of E7 to transform rodent cells requires an intact pRB binding domain, a mutant of HPV-16 E7 that is defective for pRB binding is competent for the immortalization of primary human genital keratinocytes, indicating that other properties of E7 are involved in this function. One mutation that rendered E7 incompetent for immortalization of human keratinocyte was located in the carboxy terminus (138).

The HPV E6 Oncoprotein

Although HPV-16 E7 is sufficient for the transformation of established rodent cells such as NIH3T3 cells and for cooperation with *ras* in the transformation of baby rat

HPV- 6 E7	... P V G L H C Y E Q L N D ...
HPV-11 E7	... P V G L H C Y E Q L E D ...
HPV-16 E7	... T T D L Y C Y E Q L N D ...
HPV-18 E7	... P V D L L C H E Q L S D ...
HPV-31 E7	... A T D L H C Y E Q L P S ...
HPV-33 E7	... P T D L Y C Y E Q L S D ...
HPV-39 E7	... P V D L V C H E Q L G E ...
HPV-51 E7	... E I D L Q C Y E Q F D S ...
HPV-58 E7	... P T D L F C Y E Q L C D ...
ME180 HPV E7	... P V D L V C H E Q L G D ...

FIG. 16. Amino acid sequence comparison of the pRB binding sites of the E7 proteins of the low-risk (HPV-6 and -11) and the high-risk (HPV-16, -18, -31, -33, -39, -51, -58, and ME180) genital-tract-associated HPVs. The core of the binding site is the LXCXE sequence. A single amino acid difference (*arrowhead*), an aspartic acid in E7 proteins of the high-risk viruses, and a glycine in the low-risk viruses, is responsible for a difference in binding affinity for pRB and transforming activity in rodent cells (118).

kidney cells, E6 together with E7 is required for the efficient immortalization of primary human fibroblasts or keratinocytes (117,185,294). E6 together with E7 can extend the life span of human keratinocytes and lead to the outgrowth of clones with an immortalized phenotype which is resistant to challenges to terminal differentiation. This property is dependent upon the full-length E6 gene; mutational analysis has shown that the HPV-16 E6 gene, encoding a truncated form of E6, cannot provide this function (185).

The HPV E6 proteins are approximately 150 amino acids in size and contain four Cys-X-X-Cys motifs which presumably are involved in binding zinc (13,109,110). The E6 proteins from the low-risk and high-risk HPVs appear to have similar transcriptional activation properties, as has been shown using "minimal" promoters containing a TATA box element only (248). Since the E6 proteins of the low-risk HPVs have little or no transformation activity in any of the assays tested, this transactivation property of the E6 proteins is probably not linked mechanistically to the transforming functions of the E6 proteins of the high-risk HPVs.

The small DNA tumor viruses have also evolved mechanisms to complex and functionally inactivate the tumor suppressor gene product p53. Like the SV40 TAg and the Ad5 E1B 55K protein, the E6 proteins encoded by the high-risk HPVs can complex with p53; however, an interaction between the low-risk HPV E6 proteins and p53 has not been detected (296). p53 functions as a sequence-specific transcriptional transactivator (87,145), and this function is necessary for its activity in regulating cell growth and in tumor growth suppression. One downstream target of p53 has been identified, called p21 (also WAF1 or CIP1), which functions as an inhibitor of cdks (74,113,303). p53 can transcriptionally activate the p21 gene, leading to a cell cycle arrest in many cells due to the inhibition of the cdks. The viral oncoproteins SV40 TAg, AdE1B 55K, and the high-risk E6 proteins can efficiently abrogate the transcriptional transactivation activity of p53 (181,310). Thus, like SV40 TAg and Ad E1B, HPV E6 can interfere with the negative cell cycle regulatory function of p53.

Although SV40 TAg, the Ad5 E1B 55-kd protein, and the high-risk HPV E6 proteins can all complex p53, the consequence of these interactions are different with respect to the stability of the p53 protein. In SV40 and adenovirus-transformed cells, levels of p53 are usually quite high, and the half-life of p53 is increased (198,221). In contrast, the levels of p53 in HPV-infected cells are low compared to uninfected primary host cells (130,237). Unlike TAg and the E1B 55-kd protein, which may inactivate p53 at least in part by sequestering it into stable complexes, the E6 proteins of the high-risk HPVs inactivate p53 by inducing its degradation. This was first demonstrated by *in vitro* studies which showed that the high-risk HPV E6 proteins can facilitate the rapid degradation of p53 through the ubiquitin-dependent proteolytic system (239). The low-risk HPV E6 proteins, which do not bind detectable amounts of p53, have no effect on p53 stability *in vitro* (239). The binding of E6 to p53 is mediated by a 100-kd cellular protein designated E6-associated protein (E6AP) (132). E6AP binds stably to the high-risk HPV E6 proteins in the absence of p53 but not to p53 in the absence of E6. Since both E6 and E6-AP are necessary to detect complex formation with p53, it is not yet possible to determine whether E6, E6-AP, or both actually contact p53. The cDNA encoding E6-AP has been cloned (133), and database searches showed that it is a member of a family of proteins of generally unknown function but which share significant homology with the carboxy-terminal region of E6AP (134). Biochemical studies have demonstrated that E6AP has a direct role in the E6-dependent ubiquitination of p53 and that it functions as a ubiquitin-protein ligase (Fig. 17) (236). The ubiquitin proteolysis system is a major pathway for the intracellular degradation of proteins. Ubiquitin is a 76 amino acid protein which is highly conserved among all eukaryotic organisms. The hallmark of the pathway is the covalent linkage of the carboxy terminus of ubiquitin to the epsilon-amino group of lysine side chains, forming a stable isopeptide bond. Additional ubiquitin moieties can be linked sequentially to each other via a lysine residue at amino acid 48 of ubiquitin. Multiubiquitinated proteins can then be specifically recognized and degraded by the 26S proteasome, a large multisubunit complex of proteins. Three classes of proteins are involved in targeting a substrate protein for ubiquitination: the E1 ubiquitin-activating enzyme, the E2 ubiquitin-conjugating enzymes, and the E3 ubiquitin protein ligases (reviewed in 123a) (see Fig. 17). The E1 ubiquitin-activating enzyme catalyzes the formation of a high-energy thioester between the carboxy terminus of ubiquitin and the thiol group of a cysteine side chain on the E1 enzyme itself in a reac-

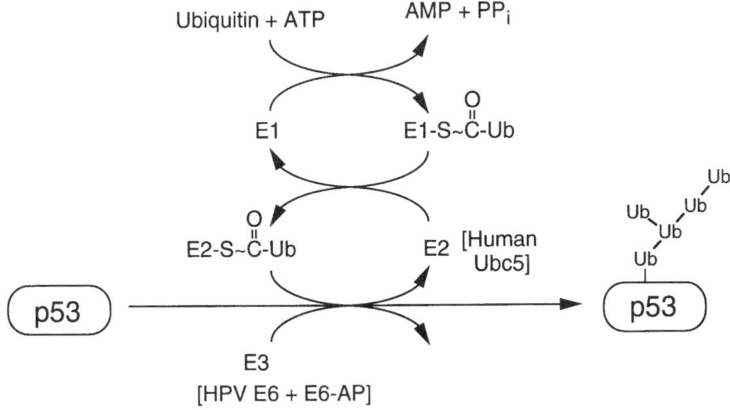

FIG. 17. A. HPV E6-mediated ubiquitination of p53. The E6 protein binds to the cellular protein E6-AP and the complex together functions as an E3 (ubiquitin protein ligase) in facilitating the ubiquitination of p53 (236). The ubiquitination of a protein involves three cellular activities: E1 (ubiquitin activating enzyme), E2 (ubiquitin conjugating enzyme), and E3. In the case of the E6/E6AP mediated ubiquitination of p53, a specific E2 called *human Ubc5* is involved in the process (235).

tion that requires ATP. The activated ubiquitin is then transferred to a thiol group of the active site cysteine of an E2 ubiquitin-conjugating enzyme (ubc). There are several distinct E2 enzymes (at least 10 in *S. cerevisiae*), all of which appear to be recognized by a single type of E1 enzyme. In the current model the E2 enzymes then directly catalyze the formation of a stable isopeptide bond between ubiquitin and a lysine on the substrate protein. The substrate specificity in this model is provided by the E2 itself which in some cases requires the activity of an E3 ubiquitin protein ligase. The E3 proteins until now have been poorly characterized and have been simply defined as activities that are required in addition to an E2 enzyme to ubiquitinate a specific substrate.

The E6/E6-AP-dependent ubiquitination of p53 has been reconstituted *in vitro* with purified and/or recombinant proteins (236). In addition to ATP, ubiquitin, and the cellular E1 enzyme, there is a requirement for a specific cellular E2, the activity of which is provided by human UBC5 (235). The E3 activity is provided by the complex of E6 and E6AP. Furthermore, recent studies have shown that E6AP can form a thioester with ubiquitin (134), suggesting that E6AP may be involved directly in the ubiquitination of p53 and that the order of transfer of ubiquitin to p53 is from E1 to E2 to E6AP in the form of a high-energy thioester and, finally, to p53 through a stable isopeptide linkage.

In addition to the *in vitro* evidence that the high-risk HPV E6 proteins accelerate the degradation of p53, E6 also appears to affect the stability of intracellular p53 *in vivo* (11,130,160,237). Levels of p53 in E6-immortalized cells or in HPV-positive cervical carcinoma cells are, on average, two- to threefold decreased compared to primary cells (237). The half-life of p53 is reduced from 3 hr to 20 min in human keratinocytes expressing E6 (130). In uninfected cells, intracellular p53 levels increase significantly in response to DNA damage induced by γ irradiation or other agents (142). The higher levels of p53 are thought to result in growth arrest or apoptosis of the treated cells, which may be a cell defense mechanism that would allow for the DNA damage to be repaired prior to the initiation of a new round of DNA replication. E6-expressing cells, however, do not manifest a p53-mediated cellular response to DNA damage (146), indicating the ability of E6 to promote the degradation of p53 and prevent the steady level of p53 to rise above a certain threshold level (Fig. 18). Under DNA-damaging conditions, the E6-stimulated degradation of p53 would presumably abrogate the negative growth regulatory effect of p53 and, as such, is thought to contribute to genomic instability (297).

What are the roles of the E6 and E7 oncoproteins in the normal life cycle of an HPV infection? It is likely that they play critical roles in providing a cellular environment for the replication of the viral DNA. The viral E1 and E2 proteins are necessary for the initiation of viral DNA replication, but the virus is otherwise totally dependent on host cell factors including DNA polymerase α, thymidine kinase, PCNA, etc. for the replication of its DNA. These are proteins that are normally expressed only in S phase during cellular DNA replication in cycling cells. Vegetative DNA replication for the papillomaviruses, however, occurs only in the more differentiated cells of the epithelium that are no longer cycling (see Figs. 5,6). Thus, the papillomaviruses have evolved a mechanism similar to that of the polyomaviruses and the adenoviruses, to activate the genes necessary for the replication of their own DNA in otherwise quiescent cells. These viruses may do so through the E7 proteins and their ability to release the E2F transcription factors by binding the pocket proteins, including pRB (see Fig. 15). However, the high-risk HPV E7 proteins, when expressed in the absence of E6, apparently induce increased levels of p53, resulting in either a G1-mediated cell cycle arrest or apoptosis, depending upon the cell type (128,207). E6, by promoting the degradation of p53, resulting in a lower steady-state level in the cell, counters this activity of E7 and permits progression into the S phase of the cell cycle and the E7-dependent activation of the cellular DNA replication genes required for viral DNA replication.

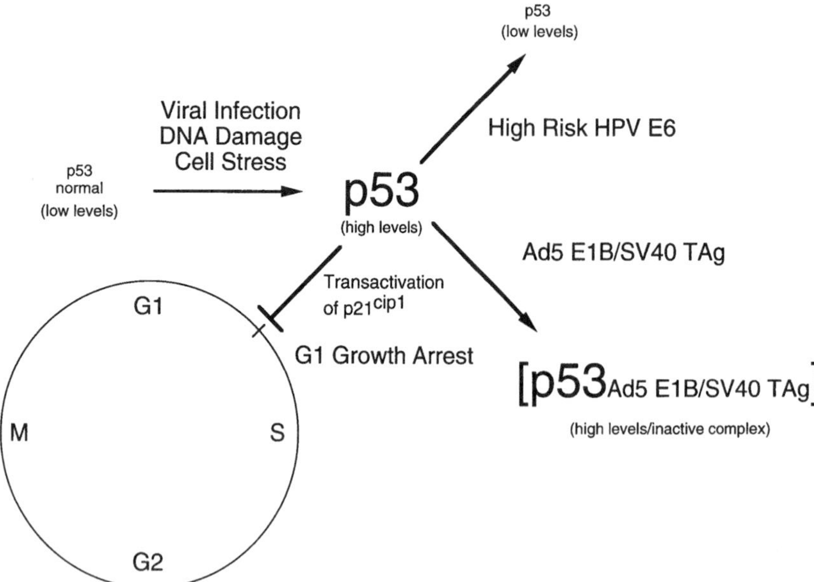

FIG. 18. The level of p53 in primary cells is generally low. DNA-damaging agents, viral infection, and expression of E7 increase the level of p53. Elevated levels of p53 can lead to either apoptosis or a cell cycle checkpoint arrest in G1. Viral oncoproteins may interfere with this negative growth regulatory function of p53, either by sequestering p53 into a stable but nonfunctional complex (such as with SV40 TAg or the Ad5 55-kd E1B protein) or by enhanced degradation as observed with the high risk HPV E6 proteins.

REFERENCES

1. Adachi A, Yasue H, Ohashi M, et al. A novel type of human papillomavirus DNA from the lesion of epidermodysplasia verruciformis. *Japan J Cancer Res* 1986;77:978–984.
2. Ahola H, Stenlund A, Moreno-Lopez J, et al. Promoters and processing sites within the transforming region of bovine papillomavirus type 1. *J Virol* 1987;61:2240–2244.
3. Amtmann E, Sauer G. Bovine papilloma virus transcription: Polyadenylated RNA species and assessment of the direction of transcription. *J Virol* 1982;43:59–66.
4. Androphy EJ, Lowy DR, Schiller JT. Bovine papillomavirus E2 trans-acting gene product binds to specific sites in papillomavirus DNA. *Nature* 1987;325:70–73.
5. Androphy EJ, Schiller JT, Lowy DR. Identification of the protein encoded by the E6 transforming gene of bovine papillomavirus. *Science* 1985;230:442–445.
6. Baker CC. The genomes of the papillomaviruses. In: SJ O'Brien, ed. *Genetic maps: locus maps of complex genomes.* Cold Spring Harbor, NY: Cold Spring Harbor Laboratory Press, 1993;1.134–1.146.
7. Baker CC. Sequence analysis of papillomavirus genomes. In: NP Salzman, PM Howley, eds. *The Papillomaviruses, vol. 2.* New York: Plenum Press, 1987;321–385.
8. Baker CC, Howley PM. Differential promoter utilization by the papillomavirus in transformed cells and productively infected wart tissues. *EMBO* 1987;6:1027–1035.
9. Baker CC, Noe JS. Transcriptional termination between bovine papillomavirus type 1 (BPV-1) early and late polyadenylation sites blocks late transcription in BPV-1 transformed cells. *J Virol* 1989;63:3529–3534.
10. Baker TS, Newcomb WW, Olson NH, et al. Structures of bovine and human papillomaviruses—analysis by cryoelectron microscopy and three-dimensional image reconstruction. *Biophys J* 1991;60:1445–1456.
11. Band V, DeCaprio JA, Delmolino L, et al. Loss of p53 protein in human papillomavirus type 16 E6-immortalized human mammary epithelial cells. *J Virol* 1991;65:6671–6676.
12. Barbosa MS, Edmonds C, Fisher C, et al. The region of the HPV E7 oncoprotein homologous to adenovirus E1a and SV40 large T antigen contains separate domains for RB binding and casein kinase II. *EMBO J* 1990;9:153–160.
13. Barbosa MS, Lowy DR, Schiller JT. Papillomavirus polypeptides E6 and E7 are zinc binding proteins. *J Virol* 1989;63:1404–1407.
14. Barksdale SK, Baker CC. Differentiation-specific expression from the bovine papillomavirus type 1 P_{2443} and late promoters. *J Virol* 1993;67:5605–5616.
15. Beaudenon S, Kremsdorf D, Croissant O, et al. A novel type of human papillomavirus associated with genital neoplasias. *Nature* 1986;321:246–249.
16. Beaudenon S, Kremsdorf D, Obalek S, et al. Plurality of genital human papillomaviruses: Characterization of two new types with distinct biological properties. *Virology* 1987;161:374–384.
17. Beaudenon S, Praetorius F, Kremsdorf D, et al. A new type of human papillomavirus associated with oral focal epithelial hyperplasia. *J Invest Derm* 1987;88:130–135.
18. Bedell MA, Jones KH, Laimins LA. The E6-E7 region of human papillomavirus 18 is sufficient for transformation of NIH 3T3 and Rat-1 cells. *J Virol* 1987;61:3635–3640.
19. Berg LJ, Singh K, Botchan M. Complementation of a bovine papilloma virus low-copy-number mutant: Evidence for a temporal requirement of the complementing gene. *Mol Cell Biol* 1986;6:859–869.
20. Bernard BA, Bailly C, Lenoir MC, et al. The human papillomavirus type 18 (HPV18) E2 gene product is a repressor of the HPV18 regulatory region in human keratinocytes. *J Virol* 1989;63:4317–4324.
21. Black PH, Hartley JW, W.P. R, et al. Transformation of bovine tissue culture cells by bovine papilloma virus. *Nature* 1963;199:1016–1018.
22. Blitz IL, Laimins LA. The 68-kilodalton E1 protein of bovine papillomavirus is a DNA-binding phosphoprotein which associates with the E2 transcriptional activator in vitro. *J Virol* 1991;65:649–656.
23. Boiron M, Levy JP, Thomas M, et al. Some properties of bovine papilloma virus. *Nature* 1964;201:423–424.
24. Boshart M, Gissman L, Ikenberg H, et al. A new type of papillomavirus DNA, its presence in genital cancer biopsies and in cell lines derived from cervical cancer. *EMBO J* 1984;3:1151–1157.
25. Bream GL, Ohmstede C-A, Phelps WC. Characterization of human papillomavirus type 11 E1 and E2 proteins expressed in insect cells. *J Virol* 1993;67:2655–2663.
26. Breitburd F, Croissant O, Orth G. Expression of human papillomavirus type 1 E4 gene products in warts. In: BM Steinberg, JL Brandsma, LB Taichman, eds. *Papillomaviruses: cancer cells.* New York: Cold Spring Harbor Laboratory Press, 1987.
27. Brokaw JL, Yee CL, Munger K. A mutational analysis of the amino terminal domain of the human papillomavirus type 16 E7 oncoprotein. *Virology* 1994; 205:603–607.
28. Brown DR, Fan L, Jones J, et al. Colocalization of human papillomavirus type 11 E1/E4 and L1 proteins in human foreskin implants grown in athymic mice. *Virology* 1994;201:46–54.
29. Burkhardt A, DiMaio D, Schlegel R. Genetic and biochemical definition of the bovine papillomavirus E5 transforming protein. *EMBO J* 1987;6:2381–2385.
30. Burkhardt A, Willingham M, Gay C, et al. The E5 oncoprotein of bovine

papillomavirus is oriented asymmetrically in Golgi and plasma membranes. *Virology* 1989;170:334–339.

31. Burnett S, Jareborg N, DiMaio D. Localization of bovine papillomavirus type 1 E5 protein to transformed keratinocytes and permissive differentiated cells in fibropapilloma tissue. *Proc Natl Acad Sci USA* 1992;89:5665–5669.
32. Burnett S, Kiessling U, Pettersson U. Loss of bovine papillomavirus DNA replication control in growth-arrested transformed cells. *J Virol* 1989;63:2215–2225.
33. Burnett S, Strom AC, Jareborg N, et al. Induction of bovine papillomavirus E2 gene expression and early region transcription by cell growth arrest: Correlation with viral DNA amplification and evidence for differential promoter induction. *J Virol* 1990;64:5529–5541.
34. Chellappan S, Kraus VB, Kroger B, et al. Adenovirus E1A, simian virus 40 tumor antigen and human papillomavirus E7 protein share the capacity to disrupt the interaction between transcription factor E2F and the retinoblastoma gene product. *Proc Natl Acad Sci USA* 1992;89: 4549–4553.
35. Chen EY, Howley PM, Levinson AD, et al. The primary structure and genetic organization of the bovine papillomavirus type 1 genome. *Nature* 1982;299:529–534.
36. Chen S-L, Mounts P. Transforming activity of E5a protein of human papillomavirus type 6 in NIH3T3 and C127 cells. *J Virol* 1990;64:3226–3233.
37. Chiang C-M, Ustav M, Stenlund A, et al. Viral E1 and E2 proteins support replication of homologous and heterologous papillomavirus origins. *Proc Natl Acad Sci USA* 1992;89:5799–5803.
38. Chin MT, Broker TR, Chow LT. Identification of a novel constitutive enhancer element and an associated binding protein: Implications for human papillomavirus type 11 enhancer regulation. *J Virol* 1989;63: 2967–2976.
39. Choe J, Vaillancourt P, Stenlund A, et al. Bovine papillomavirus type 1 encodes two forms of a transcriptional repressor: Structural and functional analysis of new viral cDNAs. *J Virol* 1989;63:1743–1755.
40. Chow LT, Hirochika H, Nasseri M, et al. Human papilloma virus gene expression. In: BM Steinberg, JL Brandsma, LB Taichman, eds. *Papillomaviruses, vol. 5*. New York: Cold Spring Harbor Press, 1987; 55–72.
41. Ciuffo G. Imnfesto positivo con filtrato di verruca volgare. *Giorn Ital Mal Venereol* 1907;48:12–17.
42. Clertant P, Seif I. A common function for polyoma virus large-T and papillomavirus E1 proteins. *Nature* 1984;311:276–279.
43. Cobrinik D, Dowdy SF, Hinds PW, et al. The retinoblastoma protein and the regulation of cell cycling. *TIBS* 1992;17:312–315.
44. Coggin JR, zur Hausen H. Workshop on papillomaviruses and cancer. *Cancer Res* 1979;39:545–546.
45. Cohen BD, Goldstein DJ, Rutledge L, et al. Transformation-specific interaction of the bovine papillomavirus E5 oncoprotein with the platelet-derived growth factor receptor transmembrane domain and the epidermal growth factor receptor cytoplasmic domain. *J Virol* 1993;67: 5303–5311.
46. Cole ST, Danos O. Nucleotide sequence and comparative analysis of the human papillomavirus type 18 genome. *J Mol Biol* 1987;159: 599–608.
47. Cole ST, Streeck RE. Genome organization and nucleotide sequence of human papillomavirus type 33 which is associated with cervical cancer. *J Virol* 1986;58:991–995.
48. Crawford LV, Crawford EM. A comparative study of polyoma and papilloma viruses. *Virology* 1963;21:258–263.
49. Cripe TP, Haugen TH, Turk JP, et al. Transcriptional regulation of the human papillomavirus-16 E6-E7 promoter by a keratinocyte-dependent enhancer, and by viral E2 trans-activator and repressor gene products: Implications for cervical carcinogenesis. *EMBO J* 1987;6: 3745–3753.
50. Crum CP, Barber S, Symbula M, et al. Coexpression of the human papillomavirus type 16 E4 and L1 open reading frames in early cervical neoplasia. *Virology* 1990;178:238–246.
51. Danos O, Katinka M, Yaniv M. Human papillomavirus la complete DNA sequence: A novel type of genome organization among papovaviridae. *EMBO J* 1982;1:231–236.
52. Dartmann K, Schwarz E, Gissmann L, et al. The nucleotide sequence and genome organization of human papillomavirus type 11. *Virology* 1986;151:124–130.
53. Davies RC, Vousden KH. Functional analysis of human papillomavirus type 16 E7 by complemention with adenovirus E1A mutants. *J Gen Virol* 1992;73:2135–2139.
54. de Villiers E-M, Hirsch-Benham A, von Knebel Doeberitz C, et al. Two newly identified human papillomavirus types (HPV 40 and HPV 57) isolated from mucosal lesions. *J Virol* 1989;171:248–252.
55. DeCaprio JA, Ludlow JW, Figge J, et al. SV40 large tumor antigen forms a specific complex with the product of the retinoblastoma suscriptibility gene. *Cell* 1988;54:275–283.
56. Del Vecchio AM, Romanczuk H, Howley PM, et al. Transient replication of human papillomavirus DNAs. *J Virol* 1992;66:5949–5958.
57. Delius H, Hofmann B. Primer-directed sequencing of human papillomavirus types. *Curr Top Microbiol Immunol* 1994;186:13–32.
58. DiMaio D, Guralski D, Schiller JT. Translation of open reading frame E5 of bovine papillomavirus is required for its transforming activity. *Proc Natl Acad Sci USA* 1986;83:1797–1801.
59. DiMaio D, Metherall J, Neary K, et al. Nonsense mutation in open reading frame E2 of bovine papillomavirus DNA. *J Virol* 1986;57:475–480.
60. DiMaio D, Petti L, Hwang E-S. The E5 transforming proteins of the papillomaviruses. *Seminars in Virology* (in press) 1994.
61. DiMaio D, Settleman J. Bovine papillomavirus mutant temperature defective for transformation, replication and transactivation. *EMBO J* 1988;7:1197–1204.
62. DiMaio D, Treisman RH, Maniatis T. Bovine papillomavirus vector that propagates as a plasmid in both mouse and bacterial cells. *Proc Natl Acad Sci USA* 1982;77:4030–4034.
63. Doorbar J, Campbell D, Grand RJA, et al. Identification of the human papillomavirus-1a E4 gene products. *EMBO J* 1986;5:355–362.
64. Doorbar J, Ely S, Sterling J, et al. Specific interaction between HPV-16 E1-E4 and cytokeratins results in collapse of the epithelial cell intermediate filament network. *Nature* 1991;352:824–827.
65. Doorbar J, Evans HS, Coneron I, et al. Analysis of HPV-1 E4 gene expression using epitope-defined antibodies. *EMBO J* 1988;7:825–833.
66. Doorbar J, Parton A, Hartley K, et al. Detection of novel splicing patterns in a HPV16-containing keratinocyte cell line. *Virology* 1990;178:254–262.
67. Durst M, Dzarlieva-Petruskova RT, Boukamp P, et al. Molecular and cytogenetic analysis of immortalized human primary keratinocytes obtained after transfection with human papillomavirus type 16 DNA. *Oncogene* 1987a;1:251–256.
68. Durst M, Gissmann L, Ikenberg H, et al. A papillomavirus DNA from a cervical carcinoma and its prevalence in cancer biopsy samples from different geographic regions. *Proc Natl Acad Sci USA* 1983;80: 3812–3815.
69. Dvoretzky I, Shober R, Chattopadhyay SK, et al. A quantitative in vitro focus assay for bovine papilloma virus. *Virology* 1980;103:369–375.
70. Dyson N, Guida P, Munger K, et al. Homologous sequences in adenovirus E1A and human papillomavirus E7 proteins mediate interaction with the same set of cellular proteins. *J Virol* 1992;66:6893–6902.
71. Dyson N, Howley PM, Munger K, et al. The human papillomavirus-16 E7 oncoprotein is able to bind the retinoblastoma gene product. *Science* 1989;243:934–937.
72. Edmonds C, Vousden KH. A point mutational analysis of human papillomavirus type 16 E7 protein. *J Virol* 1989;63:2650–2656.
73. Egawa K, Delius H, Matsukura T, et al. Two novel types of human papillomavirus, HPV 63 and HPV 65: Comparisons of their clinical and histological features and DNA sequences to other HPV types. *Virology* 1993;194:789–799.
74. El-Diery WS, Tokino T, Velculescu VE, et al. WAF1, a potential mediator of p53 tumor suppression. *Cell* 1993;75:817–825.
75. Engel LW, Heilman CA, Howley PM. Transcriptional organization of the bovine papillomavirus type 1. *J Virol* 1983;47:516–528.
76. Eriksson A, Ahola H, Pettersson U, et al. Genome of the European elk papillomavirus (EEPV). *Virus Genes* 1988;1:123–133.
77. Favre M, Breitburd F, Croissant O, et al. Structural polypeptides of rabbit, bovine, and human papilloma viruses. *J Virol* 1975;15: 1239–1247.
78. Favre M, Croissant O, Orth G. Human papillomavirus type 29 (HPV-29), an HPV type cross-hybridizing with HPV-2 and with HPV-3-related types. *J Virol* 1989;63:4906.
79. Favre M, Kremsdorf D, Jablonska S, et al. Two new human papillomavirus types (HPV54 and 55) characterized from genital tumors illustrate the plurality of genital HPVs. *Int J Cancer* 1990;45:40–46.
80. Favre M, Obalek S, Jablonska S, et al. Human papillomavirus (HPV) type 50, a type associated with epidermodysplasia verruciformis (EV) and only weakly related to other EV-specific HPVs. *J Virol* 1989;63:4910.
81. Favre M, Obalek S, Jablonska S, et al. Human papillomavirus type 28

(HPV-28), an HPV-3-related type associated with skin warts. *J Virol* 1989;63:4905.
82. Favre M, Obalek S, Jablonska S, et al. Human papillomavirus type 49, a type isolated from flat warts of renal transplant patients. *J Virol* 1989;63:4909.
83. Favre M, Orth G, Croissant O, et al. Human papillomavirus DNA: Physical map. *Proc Natl Acad Sci USA* 1975;72:4810–4814.
84. Finbow ME, Pitts JD, Goldstein DJ, et al. The E5 oncoprotein target: A 16-kDa channel-forming protein with diverse functions. *Molec Carcinogenesis* 1991;4:441–444.
85. Friedman JC, Levy JP, Lasneret J, et al. Induction de fibromes souscutanes chez le hamster dore par inoculation d'extraits acellulaires de papillomes bovins. *C R Acad Sci* 1963;257:2328–2331.
86. Fuchs PG, Iftner T, Weninger J, et al. Epidermodysplasia verruci formis-associated human papillomavirus 8: Genomic sequence and comparative analysis. *J Virol* 1986;58:626–634.
87. Funk WD, Pak DT, Karas RH, et al. A transcriptionally active DNA-binding site for human p53 protein complexes. *Mol Cell Biol* 1992; 12:2866–2871.
88. Furth PA, Baker CC. An element in the bovine papillomavirus late 3' untranslated region reduces polyadenylated cytoplasmic RNA levels. *J Virol* 1991;65:5806–5812.
89. Furth PA, Choe W-T, Rex JH, et al. Sequences homologous to 5' splice sites are required for the inhibitory activity of papillomavirus late 3' untranslated regions. *Mol and Cell Biol* 1994;14:5278–5289.
90. Gage JR, Meyers C, Wettstein FO. The E7 proteins of the nononcogenic human papillomavirus type 6b (HPV-6b) and of the oncogenic HPV-16 differ in retinoblastoma protein binding and other properties. *J Virol* 1990;64:723–730.
91. Gallahan D, Muller M, Schneider A, et al. Human papillomavirus type 53. *J Virol* 1989;63:4911–4912.
92. Gassenmaier A, Lammel M, Pfister H. Molecular cloning and characterization of the DNAs of human papillomaviruses 19, 20 and 25 from a patient with epidermodysplasia verruciformis. *J Virol* 1984;52: 1019–1023.
93. Gilbert DM, Cohen SN. Bovine papilloma virus plasmids replicate randomly in mouse fibroblasts throughout S phase of the cell cycle. *Cell* 1987;50:59–68.
94. Giri I, Danos O, Yaniv M. Genomic structure of the cottontail rabbit (Shope) papillomavirus. *Proc Natl Acad Sci USA* 1985;82:1580–1584.
95. Giri I, Yaniv M. Study of the E2 gene product of the cottontail rabbit papillomavirus reveals a common mechanism of trans-activation among the papillomaviruses. *J Virol* 1988;62:1573–1581.
96. Gissmann L, Diehl V, Schulz-Coulon H-J, et al. Molecular cloning and characterization of human papilloma virus DNA derived from a laryngeal papilloma. *J Virol* 1982;44:393–400.
97. Gissmann L, Pfister H, zur Hausen H. Human papilloma virus (HPV): Characterization of 4 different isolates. *Virology* 1977;76:569–580.
98. Gissmann L, zur Hausen H. Partial characterization of viral DNA from human genital warts (condylomata acuminata). *Int J Cancer* 1980; 25:605–609.
99. Goldsborough MD, DiSilvestra D, Temple GF, et al. Nucleotide sequence of human papillomavirus type 31: A cervical neoplasia associated virus. *Virology* 1989;171:306–311.
100. Goldstein D, Kulke R, DiMaio D, et al. A glutamine residue in the membrane-associating domain of the BPV-1 E5 oncoprotein mediates its binding to a transmembrane component of the vacuolar H^+-ATPase. *J Virol* 1992;66:405–413.
101. Goldstein DJ, Andersson T, Sparkowski JJ, et al. The BPV-1 E5 protein, the 16 kDa membrane pore-forming protein and the PDGF receptor exist in a complex that is dependent on hydrophobic transmembrane interactions. *EMBO J* 1992;11:4851–4859.
102. Goldstein DJ, Finbow ME, Andersson T, et al. Bovine papillomavirus E5 oncoprotein binds to the 16K component of the vacuolar H^+-ATPase. *Nature* 1991;352:347–349.
103. Goldstein DJ, Li W, Wang L-M, et al. The bovine papillomavirus type 1 E5 transforming protein specifically binds and activates the b-type receptor for the platelet-derived growth factor but not other related tyrosine kinase-containing receptors to induce cellular transformation. *J Virol* 1994;68:4432–4441.
104. Goldstein DJ, Schlegel R. The E5 oncoprotein of bovine papillomavirus binds to a 16 kd cellular protein. *EMBO J* 1990;9:137–146.
105. Grimmel M, de Villiers EM, Neumann C, et al. Characterization of a new human papillomavirus (HPV 41) from disseminated warts and detection of its DNA in some skin carcinomas. *Int J Cancer* 1988;41:5–9.
106. Groff DE, Lancaster WD. Genetic analysis of the 3' early region transformation and replication functions of bovine papillomavirus type 1. *Virology* 1986;150:221–230.
107. Groff DE, Lancaster WD. Molecular cloning and nucleotide sequence of deer papillomavirus. *J Virol* 1985;56:85–91.
108. Gross G, Ellinger K, Roussaki A, et al. Epidermodysplasia verruciformis in a patient with Hodgkin's disease: Characterization of a new papillomavirus type and interferon treatment. *J Invest Derm* 1988; 91:43–48.
109. Grossman SR, Laimins LA. E6 protein of human papillomavirus type 18 binds zinc. *Oncogene* 1989;4:1089–1093.
110. Grossman SR, Mora R, Laimins LA. Intracellular localization and DNA-binding properties of human papillomavirus type 18 E6 protein expressed with a baculovirus vector. *J Virol* 1989;63:366–374.
111. Hagensee ME, Olson NH, Baker TS, et al. Three-dimensional structure of vaccinia virus-produced human papillomavirus type 1 capsids. *J Virol* 1994;68:4503–4505.
112. Hagensee ME, Yaegashi N, Galloway DA. Self-assembly of human papillomavirus type 1 capsids by expression of the L1 protein alone or by coexpression of the L1 and L2 capsid proteins. *J Virol* 1993;67: 315–322.
113. Harper JW, Adami GR, Wei N, et al. The p21 cdk-interacting protein Cip1 is a potent inhibitor of G1 cyclin dependent kinases. *Cell* 1993;75:805–816.
114. Haugen TH, Cripe TP, Ginder GD, et al. Trans-activation of an upstream early gene promoter of bovine papilloma virus-1 by a product of the viral E2 gene. *EMBO J* 1987;6:145–152.
115. Haugen TH, Turek LP, Mercurio FM, et al. Sequence specific and general transactivation by the BPV-1 E2 transactivator require an N-terminal amphipathic helix-containing E2 domain. *EMBO J* 1988;7: 4245–4253.
116. Hawley-Nelson P, Androphy EJ, Lowy DR, et al. The specific DNA recognition sequences of the bovine papillomavirus E2 protein is an E2-dependent enhancer. *EMBO J* 1988;7:525–531.
117. Hawley-Nelson P, Vousden KH, Hubbert NL, et al. HPV16 E6 and E7 proteins cooperate to immortalize human foreskin keratinocytes. *EMBO J* 1989;8:3905–3910.
118. Heck DV, Yee CL, Howley PM, et al. Efficiency of binding the retinoblastoma protein correlates with the transforming capacity of the E7 oncoproteins of the human papillomaviruses. *Proc Natl Acad Sci USA* 1992;89:4442–4446.
119. Hedge RS, Rossman SR, Laimins LA, et al. Crystal structure at 1.7A of the bovine papillomavirus-1 E2 DNA-binding domain bound to its DNA target. *Nature* 1992;359:505–512.
120. Heilman CA, Engel L, Lowy DR, et al. Virus-specific transcription in bovine papillomavirus-transformed mouse cells. *Virology* 1982;119: 22–34.
121. Heilman CA, Law MF, Israel MA, et al. Cloning of human papillomavirus genomic DNAs and analysis of homologous polynucleotide sequences. *J Virol* 1980;36:395–407.
122. Hermonat PL, Howley PM. Mutational analysis of the 3' open reading frames and the splice junction at nucleotide 3225 of bovine papillomavirus type 1. *J Virol* 1987;61:3889–3895.
123. Hermonat PL, Spalholz BA, Howley PM. The bovine papillomavirus P_{2443} promoter is E2 *trans*-responsive: Evidence for E2 autoregulation. *EMBO J* 1988;7:2815–2822.
123a. Hershko A, Ciechanover A. The ubiquitin system for protein degradation. *Annu Rev Biochem* 1992;61:761–807.
124. Hirochika H, Hirochika R, Broker TR, et al. Functional mapping of the human papillomavirus type 11 transcriptional enhancer and its interaction with the trans-acting E2 protein. *Genes Dev* 1988;2:54–67.
125. Hirsch-Benham A, Delius H, de Villiers E-M. A comparative sequence analysis of two human papillomavirus (HPV) types 2a and 57. *Virus Res* 1990;18:81–98.
126. Hirt L, Hirsch-Benham A, de Villiers E-M. Nucleotide sequence of human papillomavirus (HPV) type 41: An unusual HPV type without a typical E2 binding site consensus sequence. *Virus Res* 1991;18: 179–190.
127. Horwitz BH, Burkardt AL, Schlegel R, et al. The 44 amino acid E5 transforming protein of bovine papillomavirus requires a hydrophobic core and specific carboxyl-terminal amino acids. *Mol Cell Biol* 1988; 8:4071–4078.
128. Howes KA, Ransom N, Papermaster DF, et al. Apoptosis or retinoblastoma: Alternative fates of photoreceptors expressing the HPV-16 E7 gene in the presence or absence of p53. *Genes Dev* 1994;8:1300–1310.
129. Huang PS, Patrick DR, Edwards G, et al. Protein domains governing

interactions between E2F, the retinoblastoma gene product, and human papillomavirus type 16 E7 protein. *Mol Cell Biol* 1993;13:953–960.
130. Hubbert NL, Sedman SA, Schiller JT. Human papillomavirus type 16 E6 increases the degradation rate of p53 in human keratinocytes. *J Virol* 1992;66:6237–6241.
131. Hubert WG, Lambert PF. The 23-kilodalton phosphoprotein of bovine papillomavirus type 1 is non-essential for stable plasmid replication in murine C127 cells. *J Virol* 1993;67:2932–2937.
132. Huibregtse JM, Scheffner M, Howley PM. A cellular protein mediates association of p53 with the E6 oncoprotein of human papillomavirus types 16 or 18. *EMBO J* 1991;10:4125–4135.
133. Huibregtse JM, Scheffner M, Howley PM. Cloning and expression of the cDNA for E6-AP: A protein that mediates the interaction of the human papillomavirus E6 oncoprotein with p53. *Mol Cell Biol* 1993;13:775–784.
134. Huibregtse JM, Scheffner M, Howley PM. E6-AP directs the HPV E6-dependent inactivation of p53 and represents a family of structurally and functionally related proteins. 1994 Cold Spring Harbor Symposia on Quantitative Biology 1994;59:237–245.
135. Hwang ES, Riese DJ, Settleman J, et al. Inhibition of cervical carcinoma cell line proliferation by the introduction of a bovine papillomavirus regulatory gene. *J Virol* 1993;67:3720–3729.
136. Jareborg N, Alderborn A, Burnett S. Identification and Genetic Definition of a bovine papillomavirus type 1 E7 protein and absence of a low-copy-number phenotype exhibited by E5, E6, or E7 viral mutants. *J Virol* 1992;66:4957–4965.
137. Jenson AB, Rosenthal JD, Olson C, et al. Immunologic relatedness of papillomaviruses from different species. *J Natl Cancer Inst* 1980;64:495–500.
138. Jewers RJ, Hildebrandt P, Ludlow JW, et al. Regions of human papillomavirus type 16 E7 oncoprotein required for immortalization of human keratinocytes. *J Virol* 1992;66:1329–1335.
139. Kahn T, Schwarz E, zur Hausen H. Molecular cloning and characterization of the DNA of a new human papillomavirus (HPV30) from a laryngeal carcinoma. *Int J Cancer* 1986;37:61–65.
140. Kanda T, Watanabe S, Yoshiike K. Human papillomavirus type 16 transformation of rate 3Y1 cells. *Jpn J Cancer Res* 1987;78:103–108.
141. Kanda T, Watanabe S, Yoshiiko K. Immortalization of primary rat cells by human papillomavirus type 16 subgenomic fragments controlled by the SV40 promoter. *Virology* 1988;165:321–325.
142. Kastan MB, Zhan Q, El Deiry W-S, et al. A mammalian cell cycle checkpoint pathway utilizing p53 and GADD 45 is defective in ataxia-telangiectasia. *Cell* 1992;71:587–597.
143. Kaur P, McDougall JK. Characterization of primary human keratinocytes transformed by human papillomavirus type 18. *J Virol* 1988;62:1917–1924.
144. Kawashima M, Jablonska S, Favre M, et al. Characterization of a new type of human papillomavirus found in a lesion of Bowen's disease of the skin. *J Virol* 1986;57:688–692.
145. Kern SE, Pietenpol JA, Thiagalingam S, et al. Oncogenic forms of p53 inhibit p53-regulated gene expression. *Science* 1992;256:827–830.
146. Kessis TD, Slebos RJ, Nelson WG, et al. Human papillomavirus 16 E6 expression disrupts the p53-mediated cellular response to DNA damage. *Proc Natl Acad Sci USA* 1993;90:3988–3992.
147. Kirii Y, Iwamoto S, Matsukura T. Human papillomavirus type 58 DNA sequence. *Virology* 1991;185:424–427.
148. Kirnbauer R, Booy F, Cheng N, et al. Papillomavirus L1 major capsid protein self-assembles into virus-like particles that are highly immunogenic. *Proc Natl Acad Sci USA* 1992;89:12180–12184.
149. Kiyono T, Adachi A, Ishibashi M. Genome organization and taxonomic position of human papillomavirus type 47 inferred from its DNA sequence. *Virology* 1990;177:401–405.
150. Kremsdorf D, Favre M, Jablonska S, et al. Molecular cloning and characterization of the genomes of nine newly recognized human papillomavirus types associated with epidermodysplasia verruciformis. *J Virol* 1984;52:1013–1018.
151. Kremsdorf D, Jablonska S, Favre M, et al. Biochemical characterization of two types of human papillomaviruses associated with epidermodysplasia verruciformis. *J Virol* 1982;43:436–447.
152. Kremsdorf D, Jablonska S, Favre M, et al. Human papillomaviruses associated with epidermodysplasia verruciformis. II. Molecular cloning and biochemical characterization of human papillomavirus 2a, 8, 10 and 12 genomes. *J Virol* 1983;48:340–351.
153. Kulke R, DiMaio D. Biological activities of the E5 protein of the deer papillomavirus in mouse C127 cells: morphologic transformation, induction of cellular DNA synthesis and activation of the PDGF receptor. *J Virol* 1991;65:4943–4949.
154. Laimins LA, Bedell MA, Jones KH, et al. Transformation of NIH-3T3 and primary rat embryo fibroblasts by human papillomavirus type 16 and type 18. In: BM Steinberg, JL Brandsma, LB Taichman, eds. *Papillomaviruses, vol. 5.* New York: Cold Spring Harbor Press, 1987; 201–207.
155. Lambert PF, Howley PM. Bovine papillomavirus type 1 E1 replication-defective mutants are altered in their transcriptional regulation. *J Virol* 1988;62:4009–4015.
156. Lambert PF, Hubbert NL, Howley PM, et al. Genetic assignment of the multiple E2 gene products in bovine papillomavirus transformed cells. *J Virol* 1989;63:3151–3154.
157. Lambert PF, Spalholz BA, Howley PM. A transcriptional repressor encoded by BPV-1 shares a common carboxy terminal domain with the E2 transactivator. *Cell* 1987;50:68–78.
158. Law M-F, Lancaster WD, Howley PM. Conserved sequences among the genomes of papillomaviruses. *J Virol* 1979;32:199–207.
159. Law M-F, Lowy DR, Dvoretzky I, et al. Mouse cells transformed by bovine papillomavirus contain only extrachromosomal viral DNA sequences. *Proc Natl Acad Sci USA* 1981;78:2727–2731.
160. Lechner MS, Mack DH, Finicle AB, et al. Human papillomavirus E6 proteins bind p53 *in vivo* and abrogate p53-mediated repression of transcription. *EMBO J* 1992;11:3045–3052.
161. Leechanachai P, Banks L, Moreau F, et al. The E5 gene from human papillomavirus type 16 is an oncogene which enhances growth factor-mediated signal transduction to the nucleus. *Oncogene* 1992;7:17–25.
162. Leptak C, Ramon y Cajal S, Kulke R, et al. Tumorigenic transformation of mouse keratinocytes by the E5 genes of human papillomavirus type 16 and bovine papillomavirus type 1. *J Virol* 1991;65:7078–7083.
163. Li R, Botchan MR. Acidic transcription factors alleviate nucleosome-mediated repression of BPV-1 DNA replication. *Proc Natl Acad Sci USA* 1994;91:7051–7055.
164. Li R, Botchan MR. The acidic transcriptional activation domains of VP16 and p53 bind the cellular replication protein A and stimulate in vitro BPV-1 DNA replication. *Cell* 1993;73:1207–1221.
165. Li R, Knight J, Bream G, et al. Specific recognition nucleotides and their context determine the affinity of E2 protein for 17 binding sites in the BPV-1 genome. *Genes & Develop.* 1989;3:510–526.
166. Li R, Knight JD, Jackson SP, et al. Direct interaction between Sp1 and the BPV enhancer E2 protein mediates synergistic activation of transcription. *Cell* 1991;65:493–505.
167. Lorincz AT, Lancaster WD, Temple GF. Cloning and characterization of the DNA of a new human papillomavirus from a woman with dysplasia of the uterine cervix. *J Virol* 1986;58:225–229.
168. Lorincz AT, Quinn AP, Goldsborough MD, et al. Human papillomavirus type 56: A new virus detected in cervical cancers. *J Gen Virol* 1989;70:3099–3104.
169. Lorincz AT, Quinn AP, Goldsborough MD, et al. Cloning and partial sequencing of two new human papillomavirus types associated with condylomas and low grade cervical neoplasia. *J Virol* 1989;63: 2829–2834.
170. Lorincz AT, Quinn AP, Lancaster WD, et al. A new type of papillomavirus associated with cancer of the uterine cervix. *Virology* 1987; 159:187–190.
171. Lowy DR, Dvoretzky I, Shober R, et al. In vitro tumorigenic transformation by a defined sub-genomic fragment of bovine papilloma virus DNA. *Nature* 1980;287:72–74.
172. Lungu O, Crum CP, Silverstein S. Biologic properties and nucleotide sequence analysis of human papillomavirus type 51. *J Virol* 1991; 65:4216–4225.
173. Lusky M, Fontane E. Formation of the complex of bovine papillomavirus E1 and E2 proteins is modulated by E2 phosphorylation and depends upon sequences within the carboxyl terminus of E1. *Proc Natl Acad Sci USA* 1991;88:6363–6367.
174. Lusky M, Hurwitz J, Seo YS. The bovine papillomavirus E2 protein modulates the assembly of but is not stably maintained in a replication-competent multimeric E1-replication origin complex. *Proc Natl Acad Sci USA* 1994;91:8895–8899.
175. Marich JE, Pontsler AV, Rice SM, et al. The phylogenetic relationship and complete nucleotide sequence of human papillomavirus type 35. *Virology* 1992;186:770–776.
176. Martin P, Vass W, Schiller JT, et al. The bovine papillomavirus E5 transforming protein can stimulate the transforming activity of EGF and CSF-1 receptors. *Cell* 1989;59:21–23.

177. Matlashewski G, Schneider J, Banks L, et al. Human papillomavirus type 16 DNA cooperates with activated ras in transforming primary cells. *EMBO J* 1987;**6**:1741–1746.
178. Matsukura T, Sugase M. Molecular cloning of a novel papillomavirus (type 58) from an invasive cervical carcinoma. *Virology* 1990;**177**: 833–836.
179. McBride AA, Byrne JC, Howley PM. E2 polypeptides encoded by bovine papillomavirus I form dimers through the carboxyl-terminal DNA binding domain: transactivation is mediated through the conserved amino-terminal domain. *Proc Natl Acad Sci USA* 1989;**86**: 510–514.
180. McBride AA, Romanczuk H, Howley PM. The papillomavirus E2 regulatory proteins. *J Biol Chem* 1991;**266**:18411–18414.
181. Mietz JA, Unger T, Huibregtse JM, et al. The transcriptional transactivation function of wild-type p53 is inhibited by SV40 large T-antigen and by HPV-16 oncoprotein. *EMBO J* 1992;**11**:5013–5020.
182. Mohr IJ, Clark R, Sun S, et al. Targeting the E1 replication protein to the papillomavirus origin of replication by complex formation with the E2 transactivator. *Science* 1990;**250**:1694–1699.
183. Morgan D, Pecoraro G, Rosenberg I, et al. Human papillomavirus type 6b DNA required for initiation but not maintenance of transformation of C127 mouse cells. *J Virol* 1990;**64**:969–976.
184. Muller M, Kelly G, Fiedler M, et al. Human papillomavirus type 48. *J Virol* 1989;**63**:4907–4908.
185. Munger K, Phelps WC, Bubb V, et al. The E6 and E7 genes of the human papillomavirus together are necessary and sufficient for transformation of primary human keratinocytes. *J Virol* 1989;**63**:4417–4421.
186. Munger K, Scheffner M, Huibregtse JM, et al. Interactions of HPV E6 and E7 with tumor suppressor gene products. *Cancer Surv* 1992; **12**:197–217.
187. Munger K, Werness BA, Dyson N, et al. Complex formation of human papillomavirus E7 proteins with the retinoblastoma tumor suppressor gene product. *EMBO J* 1989;**8**:4099–4105.
188. Naghashfar ZS, Lorincz AT, Buscema J, et al. Characterization of human papillomavirus type 45, a new type 18-related virus of the genital tract. *J Gen Virol* 1987;**68**:3073–3079.
189. Nallaseth FS, DePamphilis ML. Papillomavirus contains cis-acting sequences that can suppress but not replicate origins of DNA replication. *J Virol* 1994;**68**:3051–3064.
190. Nasseri M, Hirochika R, Broker TR, et al. A human papillomavirus type 11 transcript encoding an E1^E4 protein. *Virology* 1987;**159**:433–439.
191. Neary K, DiMaio D. Open reading frames E6 and E7 of bovine papillomavirus type 1 are both required for full transformation of mouse C127 cells. *J Virol* 1989;**63**:259–266.
192. Neary K, Horwitz BH, DiMaio D. Mutational analysis of open reading frame E4 of bovine papillomavirus type 1. *J Virol* 1987;**61**: 1248–1252.
193. Nevins JR. A link between the Rb tumor suppressor protein and viral oncoproteins. *Science* 1992;**258**:424–429.
194. Nilson LA, DiMaio D. Platelet-derived growth factor receptor can mediate tumorigenic transformation by the bovine papillomavirus E5 protein. *Mol Cell Biol* 1993;**13**:4137–4145.
195. Nuovo GJ, Crum CP, de Villiers E-M, et al. Isolation of a novel human papillomavirus (Type 15) from a cervical condyloma. *J Virol* 1988;**62**:1452–1455.
196. Olson C. Animal papillomas: Historical perspectives. In: NP Salzman, PM Howley, eds.*The Papillomaviruses, vol. 2.* New York: Plenum Press, 1987;39–66.
197. Oltersdorf T, Campo MS, Favre M, et al. Molecular cloning and characterization of human papillomavirus type 7 DNA. *Virology* 1986; **149**:247–250.
198. Oren M, Maltzman W, Levine AJ. Post-translational regulation of the 54K cellular tumor antigen in normal and transformed cells. *Mol Cell Biol* 1981;**1**:101–110.
199. Orth G, Favre M, Breitburd F, et al. Epidermodysplasia verruciformis: A model for the role of papilloma viruses in human cancer. *Cold Spring Harbor Conf Cell Prolif* 1980;**7**:259–282.
200. Orth G, Favre M, Croissant O. Characterization of a new type of human papillomavirus that causes skin warts. *J Virol* 1977;**24**:108–120.
201. Orth G, Jablonska S, Favre M, et al. Characterization of two types of human papillomaviruses in lesion of epidermodysplasia verruciformis. *Proc Natl Acad Sci USA* 1978;**75**:1537–1541.
202. Orth G, Jablonska S, Favre M, et al. Identification of papillomaviruses in butcher's warts. *J Invest Dermatol* 1981;**76**:97–102.
203. Ostrow RS, Zachow K, Weber D, et al. Presence and possible involvement of HPV DNA in premalignant and malignant tumors. In: PM Howley, TR Broker, eds. *Papillomaviruses: molecular and clinical aspects.* 1985;101–122.
204. Ostrow RS, Zachow KR, Thompson O, et al. Molecular cloning and characterization of a unique type of human papillomavirus from an immune deficient patient. *J Invest Dermatol* 1984;**82**:362–366.
205. Pagano M, Durst M, Joswig S, et al. Binding of the human E2F transcription factor to the retinoblastoma protein but not to cyclin A is abolished in HPV-16-immortalized cells. *Oncogene* 1992;**7**:1681–1686.
206. Palefsky JM, Winkler B, Rabanus JP, et al. Characterization of *in vivo* expression of the human papillomavirus type 16 E4 protein in cervical biopsy tissues. *J Clin Invest* 1991;**87**:2132–2141.
207. Pan H, Griep AE. Altered cell cycle regulation in the lens of HPV-16 E6 and E7 transgenic mice: Implications for tumor suppressor gene function in development. *Genes Dev* 1994;**8**:1285–1299.
208. Park P, Copeland W, Yang L, et al. The cellular DNA polymerase α-primase is required for papillomavirus DNA replication and associates with the viral E1 helicase. *Proc Natl Acad Sci USA* 1994;**91**:8700–8704.
209. Patel KR, Smith KT, Campo MS. Bovine papillomavirus type 4 (BPV-4) genome sequence. *J Gen Virol* 1987;**68**:2117–2128.
210. Petti L, DiMaio D. Specific interaction between the bovine papillomavirus E5 transforming protein and the b receptor for platelet-derived growth factor in stably transformed and acutely transfected cells. *J Virol* 1994;**68**:3582–3592.
211. Petti L, DiMaio D. Stable association between the bovine papillomavirus E5 transforming protein and activated platelet-derived growth factor receptor in transformed mouse cells. *Proc Natl Acad Sci USA* 1992; **89**:6736–6740.
212. Petti L, Nilson L, DiMaio D. Activation of the platelet-derived growth factor receptor by the bovine papillomavirus E5 protein. *EMBO J* 1991;**10**:845–855.
213. Pfister H, Gissman L, zur Hausen H. Partial characterization of proteins of human papilloma viruses (HPV) 1–3. *Virology* 1977; **83**:131–137.
214. Pfister H, Hettich I, Runne U, et al. Characterization of human papillomavirus type 13 from lesions of focal epithelial hyperplasia Heck. *J Virol* 1983;**47**:363–366.
215. Phelps WC, Munger K, Yee CL, et al. Structure-function analysis of the human papillomavirus type 16 E7 oncoprotein. *J Virol* 1992;**66**: 2418–2427.
216. Phelps WC, Yee CL, Munger K, et al. The human papillomavirus type 16 E7 gene encodes transactivation and transformation functions similar to adenovirus E1a. *Cell* 1988;**53**:339–347.
217. Philip W, Honore N, Sapp M, et al. Human papillomavirus type 42: New sequences, conserved genome organization. *Virology* 1992; **186**:331–334.
218. Pirisi L, Yasumoto S, Feller M, et al. Transformation of human fibroblasts and keratinocytes with human papillomavirus type 16 DNA. *J Virol* 1987;**61**:1061–1066.
219. Prakash SS, Horwitz BH, Zibello T, et al. Bovine papillomavirus E2 gene regulates expression of the viral E5 transforming gene. *J Virol* 1988;**62**:3608–3613.
220. Rabson MS, Yee C, Yang Y-C, et al. Bovine papillomavirus type 1 3' early region transformation and plasmid maintenance functions. *J Virol* 1986;**60**:626–634.
221. Reich NC, Oren M, Levine AJ. Two distinct mechanisms regulate the levels of a cellular tumor antigen. *Mol Cell Biol* 1983;**3**:2134–2150.
222. Roberts S, Ashmole I, Johnson GD, et al. Cutaneous and mucosal human papillomavirus E4 proteins form intermediate filament-like structures in epithelial cells. *Virology* 1993;**197**:176–187.
223. Roden RBS, Kirnbauer R, Jenson AB, et al. Interaction of papillomaviruses with the cell surface. *J Virol* 1994;**68**:7260–7266.
224. Rogel-Gaillard C, Pehau-Arnaudet G, Breitburd F, et al. Cytopathic effect in human papillomavirus type 1-induced inclusion warts: in vitro analysis of the contribution of two forms of the viral E4 protein. *J Invest Dermatol* 1993;**101**:843–851.
225. Rohlfs M, Winkenbach S, Meyer S, et al. Viral transcription in human keratinocyte cell lines immortalized by human papillomavirus type-16. *Virology* 1991;**183**:331–342.
226. Romanczuk H, Howley PM. Disruption of either the E1 or E2 regulatory gene of human papillomavirus type 16 increases viral immortalization capacity. *Proc Natl Acad Sci USA* 1992;**89**:3153–3163.
227. Romanczuk H, Thierry F, Howley PM. Mutational analysis of *cis*-el-

ements involved in E2 modulation of human papillomavirus type 16 P_{97} and Type 18 P_{105} promoters. *J Virol* 1990;64:2849–2859.
228. Rose RC, Bonnez W, Reichman RC, et al. Expression of human papillomavirus type 11 L1 protein in insect cells: in vivo and in vitro assembly of virus like particles. *J Virol* 1993;67:1936–1944.
229. Sandler AB, Vande Pol SB, Spalholz BA. Repression of the bovine papillomavirus type 1 transcription by the E1 replication protein. *J Virol* 1993;67:5079–5087.
230. Sang B-C, Barbosa MS. Single amino acid substitutions in "low risk" human papillomavirus (HPV) type 6 E7 protein enhance features characteristic of the "high risk" HPV E7 oncoproteins. *Proc Natl Acad Sci USA* 1992;89:8063–8067.
231. Sarver N, Byrne JC, Howley PM. Transformation and replication in mouse cells of a bovine papillomavirus-pML2 plasmid vector that can be rescued in bacteria. *Proc Natl Acad Sci USA* 1982;79:7147–7151.
232. Sarver N, Gruss P, Law M-F, et al. Bovine papilloma virus deoxyribonucleic acid: A novel eucaryotic cloning vector. *Mol Cell Biol* 1981;1:486–496.
233. Sarver N, Rabson MS, Yang YC, et al. Localization and analysis of bovine papillomavirus type 1 transforming functions. *J Virol* 1984;52:377–388.
234. Sato H, Furuno A, Yoshiike K. Expression of human papillomavirus type 16 E7 gene induces DNA synthesis of rat 3Y1 cells. *Virology* 1989;168:195–199.
235. Scheffner M, Huibregtse JM, Howley PM. Identification of a human ubiquitin-conjugating enzyme that mediates the E6-AP-dependent ubiquitination of p53. *Proc Natl Acad Sci USA* 1994;91:8797–8801.
236. Scheffner M, Huibregtse JM, Vierstra RD, et al. The HPV-16 E6 and E6-AP complex functions as a ubiquitin-protein ligase in the ubiquination of p53. *Cell* 1993;75:495–505.
237. Scheffner M, Munger K, Byrne JC, et al. The state of the p53 and retinoblastoma genes in human cervical carcinoma cell lines. *Proc Natl Acad Sci USA* 1991;88:5523–5527.
238. Scheffner M, Romanczuk H, Munger K, et al. Functions of human papillomavirus proteins. *Curr Top Microbiol Immunol* 1994;186:83–99.
239. Scheffner M, Werness BA, Huibregtse JM, et al. The E6 oncoprotein encoded by human papillomavirus 16 and 18 promotes the degradation of p53. *Cell* 1990;63:1129–1136.
240. Scheurlen W, Gissmann L, Gross G, et al. Molecular cloning of two new HPV types (HPV-37 and HPV-38) from a keratoacanthoma and a malignant melanoma. *Int J Cancer* 1986;37:505–510.
241. Schiller JT, Kleiner E, Androphy EJ, et al. Identification of bovine papillomavirus E1 mutants with increased transforming and transcriptional activity. *J Virol* 1989;63:1775–1782.
242. Schiller JT, Vass WC, Lowy DR. Identification of a second transforming region in bovine papillomavirus DNA. *Proc Natl Acad Sci USA* 1984;81:7880–7884.
243. Schiller JT, Vass WC, Vousdan KH, et al. The E5 open reading frame of bovine papillomavirus type 1 encodes a transforming gene. *J Virol* 1986;57:1–6.
244. Schlegel R, Phelps WC, Zhang Y-L, et al. Quantitative keratinocyte assay detects two biological activities of human papillomavirus DNA and identifies viral types associated with cervical carcinoma. *EMBO J* 1988;7:3181–3187.
245. Schlegel R, Wade-Glass M, Rabson M, et al. The E5 transforming gene of bovine papillomavirus encodes a small, hydrophobic polypeptide. *Science* 1986;233:464–467.
246. Schneider A, Oltersdorf T, Schneider V, et al. Distribution of human papillomavirus 16 genome in cervical neoplasia by molecular in situ hybridization of tissue sections. *Int J Cancer* 1987;39:717–721.
247. Schwarz E, Durst M, Demankowski C, et al. DNA sequence and genome organization of genital human papillomavirus type 6b. *EMBO J* 1983;2:2341–2348.
248. Sedman SA, Barbosa MS, Vass WC, et al. The full-length E6 protein of human papillomavirus type 16 has transforming and trans-activating activities and cooperates with E7 to immortalize keratinocytes in culture. *J Virol* 1991;65:4860–4866.
249. Seedorf K, Krammer G, Durst M, et al. Human papillomavirus type 16 DNA sequence. *Virology* 1985;145:181.
250. Seo Y-S, Muller F, Lusky M, et al. Bovine papillomavirus-encoded E2 protein enhances binding of E1 protein to the BPV replication origin. *Proc Natl Acad Sci USA* 1993;90:2865–2869.
251. Seo Y-S, Muller F, Lusky M, et al. Bovine papilloma virus-encoded E1 protein contains multiple activities required for BPV DNA replication. *Proc Natl Acad Sci USA* 1993;90:702–706.
252. Sherman L, Alloul N, Golan I, et al. Expression and splicing patterns of human papillomavirus type-16 mRNAs in pre-cancerous lesions and carcinomas of the cervix, in human keratinocytes immortalized by HPV 16, and in cell lines established from cervical cancers. *Int J Cancer* 1992;50:356–364.
253. Shimoda K, Lorincz AT, Temple GF, et al. Human papillomavirus type 52: A new virus associated with cervical neoplasia. *J Gen Virol* 1988;69:2925–2928.
254. Shope RE, Hurst EW. Infectious papillomatosis of rabbits; with a note on the histopathology. *J Exp Med* 1933;58:607–624.
255. Smotkin D, Wettstein FO. Transcription of human papillomavirus type 16 early genes in cervical cancer and a cervical cancer derived cell line and identification of the E7 protein. *Proc Natl Acad Sci USA* 1986;83:4680–4684.
256. Spalholz BA, Byrne JC, Howley PM. Evidence for cooperativity between E2 binding sites in E2 *trans*-regulation of bovine papillomavirus type 1. *J Virol* 1988;62:3143–3150.
257. Spalholz BA, Lambert PF, Yee CL, et al. Bovine papillomavirus transcriptional regulation: Localization of the E2-responsive elements of the long control region. *J Virol* 1987;61:2128–2137.
258. Spalholz BA, Vande Pol SB, Howley PM. Characterization of the cis elements involved in the basal and E2 transactivated expression of the bovine papillomavirus P_{2443} promoter. *J Virol* 1991;65:743–753.
259. Spalholz BA, Yang Y-C, Howley PM. Transactivation of a bovine papilloma virus transcriptional regulatory element by the E2 gene product. *Cell* 1985;42:183–191.
260. Stenlund A, Botchan MR. The E2 trans-activator can act as a repressor by interfering with a cellular transcription factor. *Genes Dev* 1990;4:123–136.
261. Stenlund A, Bream GL, Botchan MR. A promoter with an internal regulatory domain is part of the origin of replication in BPV-1. *Science* 1987;236:1666–1671.
262. Stenlund A, Zabielski J, Ahola H, et al. Messenger RNAs from the transforming region of bovine papilloma virus type 1. *J Mol Biol* 1985;182:541–554.
263. Stirdivant SM, Huber HE, Patrick DR, et al. Human papillomavirus type 16 E7 protein inhibits DNA binding by the retinoblastoma gene product. *Mol Cell Biol* 1992;12:1905–1914.
264. Stoler MH, Broker TR. In situ hybridization detection of human papilloma virus DNA and messenger RNA in genital condylomas and a cervical carcinoma. *Human Pathol* 1986;17:1250–1258.
265. Storey A, Greenfield I, Banks L, et al. Lack of immortalizing activity of a human papillomavirus type 16 variant DNA with a mutation in the E2 gene isolated from normal human cervical keratinocytes. *Oncogene* 1992;7:459–465.
266. Storey A, Pim D, Murray A, et al. Comparison of the in vitro transforming activities of human papillomavirus types. *EMBO J* 1988;6:1815–1820.
267. Straight SW, Hinkle PM, Jewers RJ, et al. The E5 oncoprotein of human papillomavirus type 16 transforms fibroblasts and effects downregulation of the epidermal growth factor receptor in keratinocytes. *J Virol* 1993;67:4521–4532.
268. Sundberg JP. Papillomavirus infections in animals. In: K Syrjanen, LL Gissmann, LG Koss, eds.*Papillomaviruses and human disease.* Berlin: Springer-Verlag, 1987;40–103.
269. Swift FV, Bhat K, Younghusband HB, et al. Characterization of a cell type-specific enhancer found in the human papilloma virus type 18 genome. *EMBO J* 1987;6:1339–1344.
270. Szymanski P, Stenlund A. Regulation of early gene expression from the bovine papillomavirus genome in transiently transfected C127 cells. *J Virol* 1991;65:5710–5720.
271. Tan CH, Tachezy R, van Ranst M, et al. The Mastomys natalensis papillomavirus: Nucleotide sequence, genome organization, and phylogenetic relationship of a rodent papillomavirus involved in tumorigenesis of cutaneous epithelia. *Virology* 1994;198:534–541.
272. Tan S-H, Gloss B, Bernard H-U. During negative regulation of the human papillomavirus-16 E6 promoter, the viral E2 protein can displace Sp1 from a proximal promoter element. *Nucl Acids Res* 1992;20:251–256.
273. Tanaka A, Noda T, Yajima H, et al. Identification of a transforming gene of human papillomavirus type 16. *J Virol* 1989;63:1465–1469.
274. Tawheed AR, Banderson S, Favre M, et al. Characterization of human

papillomavirus type 66 from an invasive carcinoma of the uterine cervix. *J Clin Microbiol* 1991;29:2656–2660.
275. Thierry F, Garcia-Carranca A, Yaniv M. Elements that control the transcription of genital papillomavirus type 18. *Cancer Cells* 1987;5:23–32.
276. Thierry F, Howley PM. Functional analysis of E2 mediated repression of the HPV-18 P_{105} promoter. *The New Biologist* 1991;3:90–100.
277. Thierry F, Yaniv M. The BPV1-E2 trans-acting protein can be either an activator or a repressor of the HPV 18 regulatory region. *EMBO J* 1987;6:3391–3397.
278. Thomas M, Boiron M, Tanzer J, et al. In vitro transformation of mice cells by bovine papilloma virus. *Nature* 1964;202:709–710.
279. Thorner L, Bucay N, Choe J, et al. The product of the bovine papillomavirus type 1 modulator gene (M) is a phosphoprotein. *J Virol* 1988;62:2474–2482.
280. Thorner LK, Lim DA, Botchan MR. DNA-binding domain of bovine papillomavirus type 1 E1 helicase: Structural and functional aspects. *J Virol* 1993;67:6000–6014.
281. Tommasino M, Adamczewski JP, Carlotti F, et al. HPV16 E7 protein associates with the protein kinase p33CDK2 and cyclin A. *Oncogene* 1993;8:195–202.
282. Tsunokawa Y, Takebe N, Kasamatsu T, et al. Transforming activity of human papillomavirus type 16 DNA sequences in a cervical cancer. *Proc Natl Acad Sci USA* 1986;83:2200–2203.
283. Turek LP, Byrne JC, Lowy DR, et al. Interferon induces morphologic reversion with elimination of extrachromosomal viral genomes in bovine papillomavirus-transformed mouse cells. *Proc Natl Acad Sci USA* 1982;79:7914–7918.
284. Ustav M, Stenlund A. Transient replication of BPV-1 requires two viral polypepetides encoded by the E1 and E2 open reading frames. *EMBO J* 1991;10:449–457.
285. Ustav M, Ustav E, Szymanski P, et al. Identification of the origin of replication of bovine papillomavirus and characterization of the viral origin recognition factor E1. *EMBO J* 1991;10:4321–4329.
286. Vaillancourt P, Nottoli T, Choe J, et al. The E2 transactivator of bovine papillomavirus type 1 is expressed from multiple promoters. *J Virol* 1990;64:3927–3937.
287. van Ranst M, Fuse A, Fiten P, et al. Human papillomavirus type 13 and pygmy chimpanzee papillomavirus type 1: Comparison of the genome organizations. *Virology* 1992;190:587–596.
288. Vande Pol SB, Howley PM. A bovine papilloma virus constitutive enhancer is negatively regulated by the E2 repressor through competitive binding. *J Virol* 1990;64:5420–5429.
289. Vande Pol SB, Howley PM. The bovine papillomavirus constitutive enhancer is essential for viral transformation, DNA replication, and the maintenace of latency. *J Virol* 1992;66:2346–2358.
290. Vande Pol SB, Howley PM. Negative regulation of the bovine papillomavirus E5, E6, and E7 oncogenes by the viral E1 and E2 genes. *J Virol,* 1995;69:395–402.
291. Volpers C, Streeck RE. Genome organization and nucleotide sequence of human papillomavirus type 39. *Virology* 1991;181:419–423.
292. Vousden KH, Doninger J, DiPaolo JA, et al. The E7 open reading frame of human papillomavirus type 16 encodes a transforming gene. *Oncogene Res* 1988;3:167–175.
293. Waldeck S, Rosl F, Zentgraf H. Origin of replication in episomal bovine papilloma virus type 1 DNA isolated from transformed cells. *EMBO J* 1984;3:2173–2178.
294. Watanabe S, Kanda T, Yoshiike K. Human papillomavirus type 16 transformation of primary human embryonic fibroblasts requires expression of open reading frames E6 and E7. *J Virol* 1989;63:965–969.
295. Watanabe S, Yoshiike K. Transformation of rat 3Y1 cells by human papillomavirus type 18 DNA. *Int J Cancer* 1988;41:896–900.
296. Werness BA, Levine AJ, Howley PM. Association of human papillomavirus types 16 and 18 E6 proteins with p53. *Science* 1990;248:76–79.
297. White A, Livanos EM, Tlsty TD. Differential disruption of genomic integrity and cell cycle regulation in normal human fibroblasts by the HPV oncoproteins. *Genes Dev* 1994;8:666–677.
298. Whyte P, Buchovich KJ, Horowitz JM, et al. Association between an oncogene and an anti-oncogene: The adenovirus E1a proteins bind to the retinoblastoma gene product. *Nature* 1988;334:124–129.
299. Whyte P, Williamson NM, Harlow E. Cellular targets for transformation by the adenovirus E1A proteins. *Cell* 1989;56:67–75.
300. Wilson VG, Ludes-Meyers J. A bovine papillomavirus E1-related protein binds specifically to bovine papillomavirus DNA. *J Virol* 1991;65:5314–5322.
301. Woodworth CD, Bowden PE, Doninger J, et al. Characterization of normal human exocervical epithelial cells immortalized in vitro by papillomavirus types 16 and 18 DNA. *Cancer Res* 1988;48:4620–4628.
302. Wu EW, Clemens KE, Heck DV, et al. The human papillomavirus E7 oncoprotein and the cellular transcription factor E2F bind to separate sites on the retinoblastoma tumor suppressor protein. *J Virol* 1993;67:2402–2407.
303. Xiong Y, Hannon GJ, Zhang H, et al. p21 is a universal inhibitor of cyclin kinases. *Nature* 1993;366:701–704.
304. Yang L, Li R, Mohr IJ, et al. Activation of BPV-1 replication in vitro by the transcription factor E2. *Science* 1991;353:628–632.
305. Yang L, Mohr I, Fouts E, et al. The E1 protein of bovine papilloma virus 1 is an ATP-dependent DNA helicase. *Proc Natl Acad Sci USA* 1993;90:5086–5090.
306. Yang Y-C, Okayama H, Howley PM. Bovine papillomavirus contains multiple transforming genes. *Proc Natl Acad Sci USA* 1985;82:1030–1034.
307. Yang Y-C, Spalholz BA, Rabson MS, et al. Dissociation of transforming and transactivating functions for bovine papillomavirus type 1. *Nature* 1985;318:575–577.
308. Yasumoto S, Burkhardt AL, Doninger J, et al. Human papillomavirus type 16-induced malignant transformation of NIH 3T3 cells. *J Virol* 1986;57:572–577.
309. Yatsudo M, Okamoto Y, Hakura A. Functional dissociation of transforming genes of human papillomavirus type 16. *Virology* 1988;166:594–597.
310. Yew PR, Berk A. Inhibition of p53 transactivation required for transformation by adenovirus early 1B protein. *Nature* 1992;357:82–85.
311. Zachow KR, Ostrow RS,yFaras AJ. Nucleotide sequence and genome organization of human papillomavirus type 5. *Virology* 1987;158:251–254.
312. Zhou J, Sun XY, Stenzel DJ, et al. Expression of vaccinia recombinant HPV 16 L1 and L2 ORF proteins in epithelial cells is sufficient for assembly of HPV virion-like particles. *Virology* 1991;185:251–257.

Fundamental Virology, Third Edition
edited by B.N. Fields, D.M. Knipe, P.M. Howley, et al.
Lippincott - Raven Publishers, Philadelphia © 1996

CHAPTER 30

Adenoviridae: The Viruses and Their Replication

Thomas Shenk

Adenoviruses were first isolated and characterized as distinct viral agents by two groups of investigators who were searching for the etiologic agents of acute respiratory infections. In 1953, Rowe and colleagues (389) observed the spontaneous degeneration of primary cell cultures derived from human adenoids. The pathogenic changes proved to result from the replication of previously unidentified viruses present in the adenoid tissues. In 1954, Hilleman and Werner (212) were studying an epidemic of respiratory disease in army recruits, and they isolated agents from respiratory secretions that induced cytopathic changes in cultures of human cells. The viruses discovered by the two groups were soon shown to be related (233), and they were initially called *adenoid degeneration* (AD), *respiratory illness* (RI), *adenoidal-pharyngeal-conjunctival* (APC) or *acute respiratory disease* (ARD) agents. In 1956, the agents were named *adenoviruses*, after the original source of tissue (adenoid) in which the prototype viral strain was discovered (116). Epidemiologic studies confirmed that adenoviruses were the cause of a large number of acute febrile respiratory syndromes among military recruits (1,105,151). It soon became clear, however, that adenoviruses were not the etiologic agents of the common cold; they are responsible for only a small portion of acute respiratory morbidity in the general population and about 5% to 10% of respiratory illness in children. Besides respiratory disease, adenoviruses cause epidemic conjunctivitis (242), and they have been associated with a variety of additional clinical syndromes—perhaps most notably, infantile gastroenteritis (136,510).

During the first years after their discovery, many viruses belonging to the same general group were isolated from humans, and agents with similar properties were obtained from a variety of animal species, including monkeys (234) and mice (189). Today well over 100 members of the adenovirus group have been identified which infect a wide range of mammalian and avian hosts. All of these viruses contain a linear, double-stranded DNA genome encapsidated in an icosahedral protein shell measuring 70 to 100 nm in diameter.

In 1962, Trentin and his colleagues (456) made a seminal discovery, showing that human adenovirus type 12 in-

T. Shenk: Howard Hughes Medical Institute, Department of Molecular Biology, Lewis Thomas Laboratory, Princeton University, Princeton, New Jersey 08544-1014.

duced malignant tumors following inoculation into newborn hamsters. This was the first time that a human virus was shown to sponsor oncogenesis. So far, no epidemiological evidence has been reported linking adenoviruses with malignant disease in the human; although there is one report of adenovirus-related RNA in neurogenic tumors (235), extensive searches have generally failed to find adenovirus nucleic acids in human tumors (170,295). Nevertheless, the ability to induce tumors in animals and to transform cultured cells has established adenovirus as an important model system for probing the mysteries of oncogenesis.

As the interest in adenoviruses as tumor viruses intensified, their virtues as an experimental system became evident. The prototype human adenoviruses are easily propagated to produce high titer stocks, and they initiate highly synchronous infections of established cell lines. Further, the viral genome is readily manipulated, facilitating the study of adenovirus gene functions by mutational analysis. Studies of adenovirus-infected cells have made numerous contributions to our understanding of viral and cellular gene expression and regulation, DNA replication, cell cycle control, and cellular growth regulation.

Perhaps the signal contribution to modern biology of the adenovirus system has been to host the discovery of mRNA splicing. Studies on the biogenesis of viral mRNA first demonstrated that many mRNAs were produced from a large nuclear transcript (17,480), and subsequent analysis of the structure of adenovirus mRNAs revealed the existence of introns (31,77). Today, the utility of adenovirus as a vector for gene therapy is the subject of intense exploration. This chapter will overview the structure of the adenovirus particle, the adenovirus replication cycle in human cells, its ability to oncogenically transform cells, and its interactions with host cells and host organisms. Details of adenovirus transformation are also considered in Chapter 9, and a discussion of the pathogenesis, clinical syndromes, epidemiology, techniques for diagnosis, modes of treatment, and the utility of adenoviruses as vectors will follow in Chapter 68 in *Fields Virology*, 3rd ed.

CLASSIFICATION

The adenoviruses constitute the *Adenoviridae* family of viruses, which is divided into two genera, *Mastadenovirus* and *Aviadenovirus* (341). Whereas the *Aviadenovirus* genus is limited to viruses of birds, the *Mastadenovirus* genus includes human, simian, bovine, equine, porcine, ovine, canine, and opossum viruses. Although there is antigenic cross-reactivity among members within each genera due to conserved epitopes located on the hexon protein of the virion (340), there is no known antigen common to all adenoviruses.

So far, 47 human adenovirus serotypes (Table 1) have been distinguished on the basis of their resistance to neutralization by antisera to other known adenovirus serotypes. Type-specific neutralization results predominantly from antibody binding to epitopes on the virion hexon protein and the terminal knob portion of the fiber protein (340,453). The various serotypes are classified into six subgroups (Table 1) based on their ability to agglutinate red blood cells (209,211,384). The central shaft of the viral fiber protein is responsible for binding to erythrocytes, and the hemagglutination reaction of an adenovirus is inhibited by antisera specific for viruses of the same type but not by antisera to viruses of different types. A variety of additional classification schemes have been explored, including subgroupings based on oncogenicity in rodents (167), relatedness of tumor antigens (305), electrophoretic mobility of virion proteins (474), and genome homologies revealed by base composition (361), cross-hybridization (145), or digestion with restriction endonucleases (475). The various schemes produce reasonably concordant groupings (Table 1), suggesting that the widely utilized classification based on hemagglutination is a reasonable standard.

TABLE 1. *Classification schemes for human adenoviruses (mastadenovirus H)*

Subgroup	Hemagglutination groups		Serotypes	Oncogenic potential		Percentage of G–C in DNA
				Tumors in animals	Transformation in tissue culture	
A	IV	(little or no agglutination)	12,18,31	High	+	48–49
B	I	(complete agglutination of monkey erythrocytes)	3,7,11,14,16, 21,34,35	Moderate	+	50–52
C	III	(partial agglutination of rat erythrocytes)	1,2,5,6	Low or none	+	57–59
D	II	(complete agglutination of rat erythrocytes)	8,9,10,13, 15,17,19,20,22–30, 32,33,36–39,42–47	Low or none (mammary tumors)	+	57–61
E	III		4	Low or none	+	57–59
F	III		40,41	Unknown		

Modified from Baum (26).

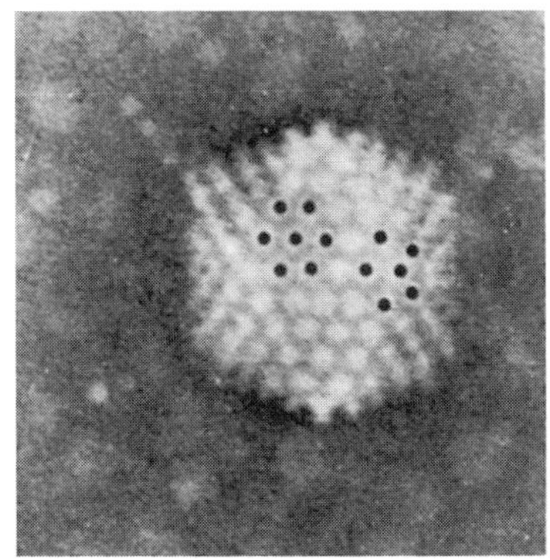
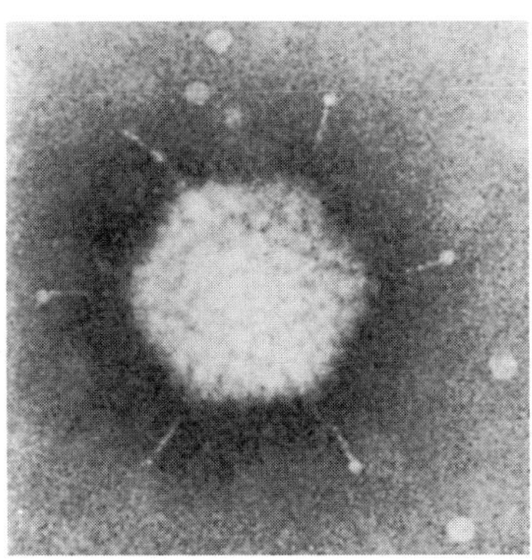
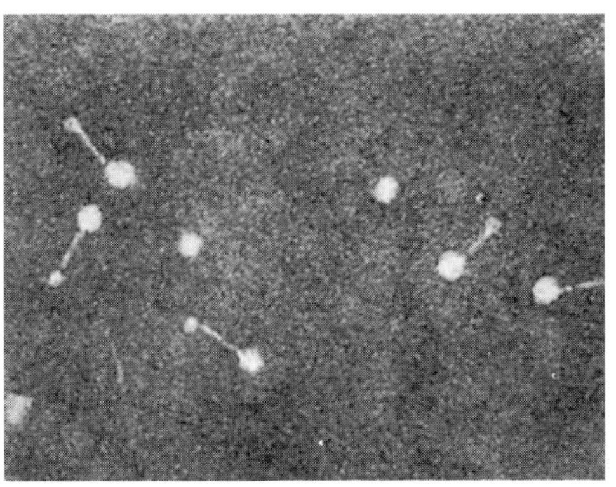

A, B C

FIG. 1. Adenovirus type 5. **A:** The virion is an icosahedron. One of the 240 hexon capsomeres surrounded by 6 hexons and one of the 12 penton capsomeres surrounded by 5 hexons are marked (*dots*). **B:** Six of the twelve fibers that are present on each virus particle are shown projecting from penton capsomeres located at the verticies of the icosahedral capsid. **C:** Free penton capsomeres containing penton base and fiber. (×285,000.) From Valentine and Pereira (459), with permission.

VIRION STRUCTURE

Adenoviruses are icosahedral particles (20 triangular surfaces and 12 vertices) that are 70 to 100 nm in diameter (219) (Fig. 1). The particles (virions) contain DNA (13% of mass), protein (87% of mass), no membrane or lipid, and trace amounts of carbohydrate since the virion fiber protein is modified by addition of glucosamine (169,237, 238). Virions consist of a protein shell surrounding a DNA-containing core. The protein shell (capsid) is composed of 252 subunits (capsomeres), of which 240 are hexons and 12 are pentons (154). As suggested by their names, penton and hexon subunits are surrounded by five and six neighbors, respectively. Each penton contains a base, which forms part of the surface of the capsid, and a projecting fiber whose length varies among different serotypes (338,339).

Virion Polypeptides and DNA

Most of the structural studies of adenoviruses have focused on the closely related adenoviruses type 2 and 5 (Ad2 and Ad5). Electrophoretic analyses of purified virions disrupted with sodium dodecylsulfate was employed initially to identify structural polypeptides (238,296,465). Comparison of electrophoretic results with genomic open reading frames (ORFs) suggests there are probably 11 virion proteins. These proteins are numbered by convention (297), with no polypeptide I since the moiety originally designated I proved to be a mixture of aggregated smaller molecules.

The outer shell of the virion or capsid is comprised of seven known polypeptides. Polypeptide II (967 amino acids) is the most abundant virion constituent. The hexon protein is comprised of three tightly associated molecules of polypeptide II (221), and this trimeric protein is often referred to as the *hexon capsomere*. Polypeptides VI (217 amino acids), VIII (134 amino acids), and IX (139 amino acids) are associated with the hexon protein after various isolation procedures (125). All three polypeptides likely stabilize the hexon capsomere lattice, and polypeptides VI and VIII probably bridge between the capsid and core components of the virion. Five copies of polypeptide III (571 amino acids) associate to form the penton base protein (465), which is found at each vertex of the icosahedral particle. Polypeptide IIIa (566 amino acids) is associated with hexon units that surround the penton after pyridine dissociation of virions, and it probably links adjacent facets of the capsid and serves a bridging function between hexons and polypeptide VII of the core (124,125). Polypeptide IV (582 amino acids) forms the trimeric fiber protein (465) which projects from the penton base at each vertex of the icosahedron. The combination of penton base plus fiber is called the *penton capsomere.* All human adenoviruses examined to date encode a single fiber protein with the exception of Ad40 and Ad41 which encode two fiber proteins and incorporate both polypeptides into their virions (254). Since the fiber interacts with a cellular receptor protein, these viruses might recognize two independent receptors. Hexon and penton capsomeres are the major components on the surface of the virion, and their constituents, polypeptides II, III, and IV, contain tyrosine residues that are exposed on the virion surface and can be labeled by iodination of intact particles (124).

The core of the virion contains four known proteins and the viral genome. Polypeptides V (368 amino acids), VII (174 amino acids), and mu (19 amino acids) are basic, arginine-rich constituents of cores prepared by disruption of virions (223,391), and all three proteins contact the viral DNA (5,66). The function of the mu protein is unknown (223). Polypeptide VII is the major core protein, and it probably serves as a histonelike center around which the viral DNA is wrapped (66,312). Polypeptide V can bind to a penton base (124) and it might bridge between the core and capsid, positioning one relative to the other. The fourth protein present in the core is the so-called terminal protein (671 amino acids), which is covalently attached to the 5' ends of the viral DNA. It was first identified indirectly by its ability to mediate circularization of the viral DNA through a protease-sensitive, noncovalent interaction (382). The circularizing agent was shown to be a 55-kd protein covalently attached to each 5' end of the viral DNA (378), and the protein was subsequently visualized on viral DNA by avidin-biotin labeling (45). The linkage between DNA and protein is a phosphodiester bond formed between the β-hydroxyl group of a serine residue residue 562 of terminal protein) and the 5' hydroxyl of the terminal deoxycytosine residue (100,420). The protein is not evident in electrophoretic analyses of virion proteins since it is present in only two copies per virion. The terminal protein serves as a primer for DNA replication (54,58,120,280,441), and it mediates attachment of the viral genome to the nuclear matrix (38,141,398).

Ad2 DNA was the first adenovirus genome to be completely sequenced (381), and its sequence includes a total of 35,937 bp. The sequence of Ad5 DNA was completed more recently (80), and portions of many other adenovirus genomes also have been sequenced. The Ad5 genome is about 95% identical to the Ad2 sequence, except in the fiber coding region, where a substantial portion of the differences underlying type specificity reside. Adenovirus DNA has inverted terminal repeat sequences ranging in size from about 100 to 140 bp, depending on the serotype (11,146,425,454,502). The inverted repeats enable single strands of viral DNA to circularize by base-pairing of their terminal sequences, and the resulting base-paired panhandles are thought to be important for replication of the viral DNA.

Capsid and Core Three-Dimensional Structure

X-ray crystallography, electron microscopy, and combinations of the two methods have been used to generate a fairly refined picture of the adenovirus capsid [reviewed in (430)]. The x-ray structure of the major capsid protein, the hexon, has been determined to a resolution of 2.9 Å (12,380). The hexon protein is a trimer comprised of three hexon polypeptides which are extensively interwoven. As is the case for polioviruses (216) and rhinoviruses (386), a β-structure stabilizes the association of the three subunits into a larger structure consisting of two domains: a triangular top facing the outside of the capsid and a hexagonal base with a central cavity. Group-of-nine hexons ("ninemers"), which can be isolated from each triangular face of the virion by treatment with 10% pyridine (362), provided insight to how interactions between hexons are stabilized. The structure of these ninemers was examined by subtracting an array of nine projected x-ray–determined hexons from scanning transmission electron microscopic images (142). The resulting two-dimensional difference image revealed density from a minor protein component extending along the hexon-hexon interfaces. Biochemical analysis had previously identified this component as polypeptide IX (465), a minor capsid constituent involved in stabilization of the capsid (85).

The three-dimensional structure of the complete adenovirus particle was determined to 35-Å resolution by image reconstruction from cryoelectron micrographs (429). A density map of the virion was generated from multiple images of the particle in different orientations. The reconstruction process relied on the known icosahedral symmetry of the virion to align the individual images. This work provided the first detailed visualization of the vertex proteins, including the penton base and its associated protruding fiber; it confirmed the earlier placement of protein IX; and it located minor capsid polypeptides at the edges of triangular facets, bridging hexons in adjacent facets. The three-dimensional structure of the virion has been refined by subtracting 240 copies of the crystallographic hexon from the cryoelectron microscopic image reconstruction (430). The difference map revealed more precisely the positions of several capsid proteins (Fig. 2). A less abundant stabilizing protein is located at each specialized position in the hexon assemblage. The penton complex fills the large gaps at the vertices, polypeptide IX stabilizes hexon-hexon contacts within a facet, polypeptide IIIa joins hexons of adjacent facets, and polypeptide VI anchors the ring of peripentonal hexons on the inside surface of the capsid and connects the capsid to the core.

The fiber protein which projects from each vertex of the particle is comprised of three copies of polypeptide IV (390,465). The amino-terminal 40 residues of each subunit are embedded in the penton base (101,482). A central extended shaft is comprised of repeating motifs about 15 amino acids in length, and the number of repeat units differs among adenovirus serotypes (172). The carboxy-terminal 180-residue segment of each subunit comprising the fiber protein contributes to formation of a terminal bulb (101). A three-dimensional model of the fiber shaft has been proposed (434), predicting that its constituent polypeptide chains form a left-handed, triple helical structure comprised of short β-strands interspersed with extended loops that follow the overall helical path. This model fits reasonably well with the cryoelectron microscopic results (430).

The adenovirus core is composed of the linear, double-stranded DNA and four virus-coded proteins. Unlike cellular chromatin which yields discrete mononucleosomal DNA fragments of about 146 bp upon digestion with micrococcal nuclease, the adenovirus core yields a heterogeneous population of DNA fragments ranging from about 50 to 300 bp in length (47,90,312,444,468,469), and these fragments are associated with polypeptide VII (312,468). Although the reason for the variability in DNA length is not understood, it nevertheless seems clear that adenovirus

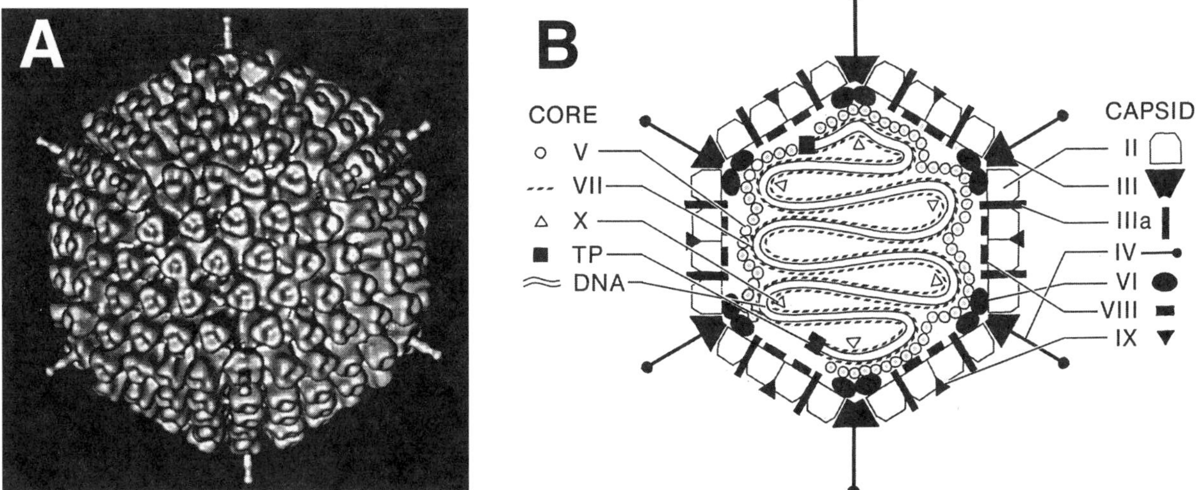

FIG. 2. Models of the adenovirus virion. **A:** A three-dimensional image reconstruction of the intact adenovirus particle viewed along an icosahedral three-fold axis. **B:** A stylized section of the adenovirus particle based on current understanding of its polypeptide components and DNA. No real section of the icosahedral virion would contain all components. Virion constituents are designated by their polypeptide numbers with the exception of the terminal protein (*TP*). From Stewart et al. (429), with permission.

chromatin is organized into unit particles comprised of DNA and polypeptide VII (312,468). The notion that adenovirus chromatin is arranged in a repeating unit is consistent with electron microscopic visualization of particles within the adenovirus core (46,329,469) and with x-ray scattering patterns obtained with cores (102). Digestion of cores with limiting amounts of nuclease fails to produce a ladderlike series of DNA fragments upon electrophoresis, as would be predicted for an oligomeric structure, and this might be due to the fragile and unstable nature of the core after release from virions (330,469).

The higher level organization of the core remains obscure. Psoralen crosslinking experiments of viral DNA within partially disassembled virions indicates that the linear molecule is organized into eight supercoiled domains (503), in agreement with earlier electron microscopic studies which suggested that the core is organized into 8 to 12 domains (46,336). It is intriguing to note that the eight-domain organization places most of the viral transcription units into independent supercoiled domains (503).

GENOME ORGANIZATION

All adenovirus genomes that have been examined to date have the same general organization, i.e., the genes encoding specific functions are located at the same position on the viral chromosome, (with the exception of the VA RNA gene which is positioned differently on the avian CELO virus genome than on the chromosomes of mammalian adenoviruses (270). As noted above, the genome consists of a single linear, double-stranded DNA molecule with relatively short terminal repeats whose inverted structure plays a role in replication of the DNA. The genome contains two origins for DNA replication. The two origins are identical, and one is present in each terminal repeat. The genome also includes a *cis*-acting packaging sequence (163,180,198) which must be located within several hundred base pairs of an end of the chromosome to direct the interaction of the viral DNA with its encapsidating proteins (197).

The viral chromosome carries five early transcription units (E1A, E1B, E2, E3, and E4), two delayed early units (IX and IVa2), and one late unit (major late) which is processed to generate five families of late mRNAs (L1 to L5), all of which are transcribed by RNA polymerase II [reviewed in (356)] (Fig. 3). Families of mRNAs were initially localized on the adenovirus map by hybridization to restriction fragments of the viral DNA (e.g., 408), and transcription units were defined by ultraviolet (UV) mapping experiments (33,499). The chromosome also carries one or two (depending on the serotype) VA genes transcribed by RNA polymerase III. By convention, the map is drawn with the E1A gene at the left end (Fig. 3). Both strands of the viral DNA are transcribed with the so-called rightward reading strand on the conventional map coding for the E1A, E1B, IX, major late, VA RNA, and E3 units and the leftward reading strand coding the E4, E2, and IVa2 units.

Little is known about the functional and evolutionary considerations that have led to the current organization of the transcription units on the viral chromosome. Their arrangement might serve a timing function. Viral cores have an organized structure (46,336,503), and they are converted from a more compact to a more open structure as the early phase of infection progresses (503). Perhaps RNA

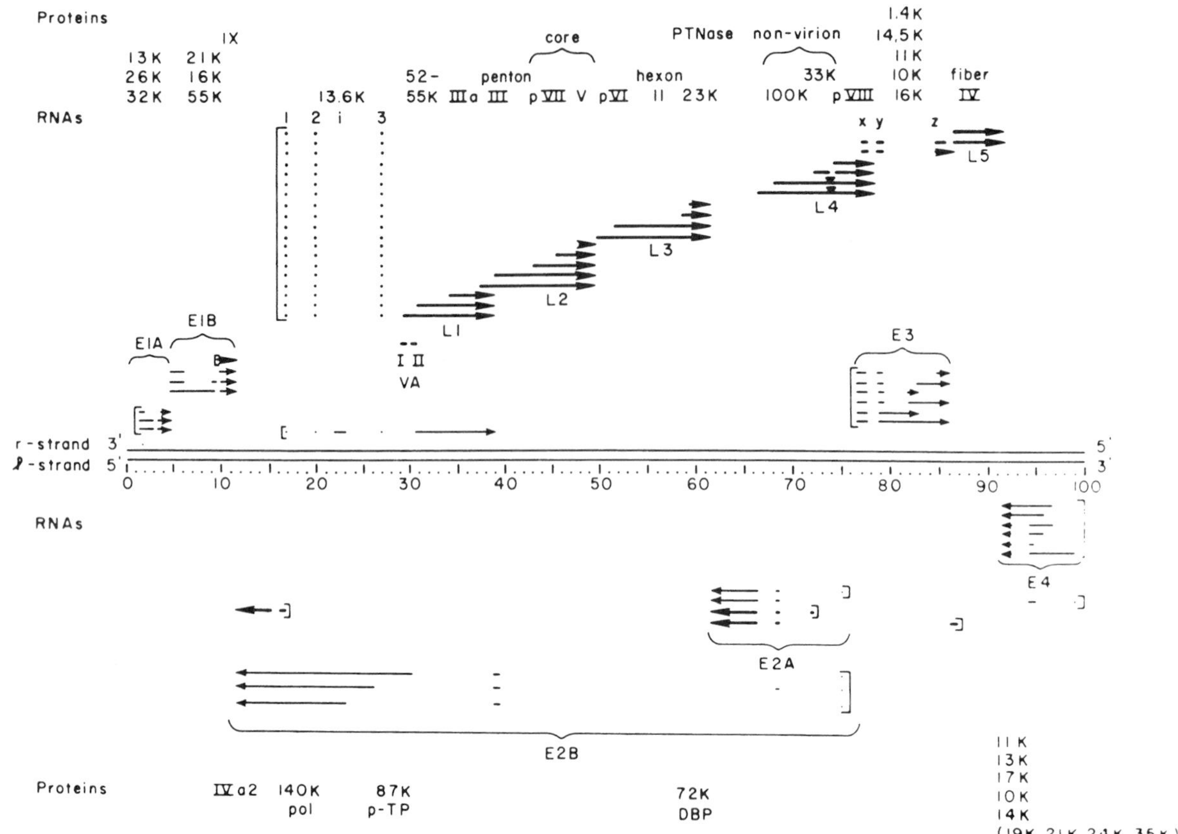

FIG. 3. Transcription and translation map of adenovirus type 2. The early mRNAs are designated E (*thin lines*). Late mRNAs are designated L (*heavy lines*). Most late mRNAs originate at 16.3 map units and contain the tripartite leader whose components are labeled *1, 2,* and *3.* Some of the late mRNAs also contain a fourth leader segment *i*, which codes for a 14-kd polypeptide. *Heavy lines* also designate mRNA species derived from early transcription units that are synthesized at elevated levels late after infection. Polypeptides are identified by the conventional numbering system (*Roman numerals*) for the virion structural components and by size (*kd*) for nonstructural polypeptides. From Broker, ref. 44, with permission.

polymerase initially interacts with promoters exposed at the ends of the chromosome, and transcription of the terminal units drives further opening of the core structure. This proposal predicts that the terminal E1A and E4 transcription units would be the first to be expressed, and this is the case. Thus, the location of a transcription unit on the chromosome could influence the order in which it is expressed relative to other units during the early phase of the infectious cycle.

The activation of transcription units can also be influenced by their location relative to other units on the chromosome. Insertion of a strong transcriptional termination sequence between the E1A and E1B units inhibits activation of the E1B unit, suggesting that transcriptional readthrough from E1A to E1B contributes to the activation of E1B (130). One might also imagine that transcription units embedded within the major late unit would be influenced in *cis* by its very active expression late after infection. However, different viral molecules could serve as templates for early and late transcription units, avoiding the complications that might arise from overlapping transcription units. Further, initiation is rate-limiting, so even if opposing and overlapping units were transcribed from the same template, transcription complexes would seldom meet on the chromosome.

Each of the adenovirus genes transcribed by RNA polymerase II gives rise to multiple mRNAs that are differentiated by alternative splicing and in some cases by the use of different poly(A) sites. The structures of mRNAs were defined by a variety of procedures, including electron microscopic heteroduplex analysis (76,78) and *S1* nuclease mapping (33), and 5' ends were precisely localized by primer extension analysis (527). As mentioned earlier, the detailed analysis of adenovirus mRNA structure led to the discovery of splicing (31,77) (Fig. 4).

Some of the protein products generated from the same transcription unit are partially related in their sequence, e.g., the two major polypeptides encoded by the E1A unit; others have no sequence in common, e.g., the two major E1B-coded polypeptides. Unfortunately, no consistent ter-

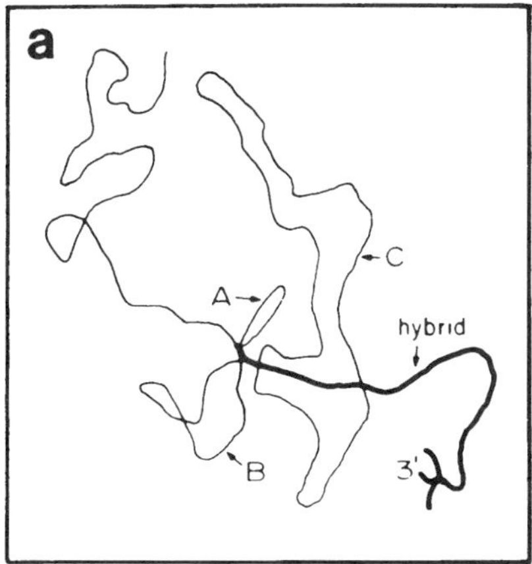

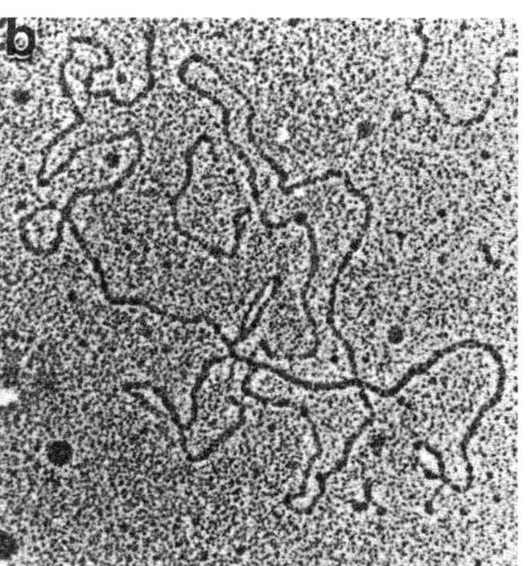

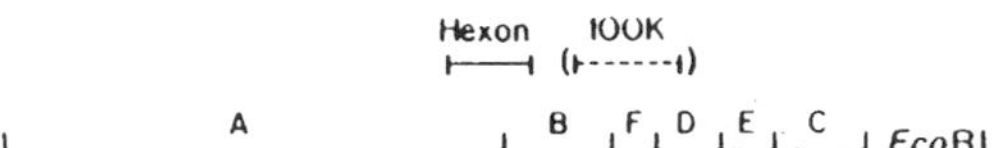

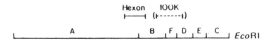

FIG. 4. Detection of adenovirus hexon transcripts processed by splicing near their 5' ends. When hexon mRNA was hybridized to viral DNA, it formed a duplex structure with three short DNA loops (labeled *A, B,* and *C*) which correspond to introns removed from the mRNA by splicing. **A:** diagram of duplex molecules; **B:** electron micrograph of the RNA:DNA hybrid. The position of the hexon and 100-kd protein coding regions are displayed on the EcoR1 cleavage map of adenovirus type 2 (*bottom of figure*). From Berget et al. (31), with permission.

minology has been adopted for naming viral proteins: the E1A proteins are named for the sedimentation coefficient of the mRNAs which encode them; E1B, E3, and E4 proteins are designated by their molecular mass; E2 proteins are named for their functions; and the structural proteins encoded by the late transcription unit are termed II-IX. The various historical names of viral polypeptides are used in this chapter, generally preceded by the name of the transcription unit or family of late mRNAs which encodes them, e.g., E1A 13S, E4-34kd, L5-IV.

Many of the individual adenovirus transcription units encode a series of polypeptides with related functions. As will be discussed in detail below, the E1A unit encodes two proteins that activate transcription and induce the host cell to enter the S phase of the cell cycle; E1B encodes two proteins that cooperate with E1A products to induce cell growth; E2 encodes three different proteins, all of which function directly in DNA replication; E3 encodes products that modulate the response of the host to the adenovirus infection; and the late family of mRNAs are concerned with the production and assembly of capsid components. Only the E4 unit encodes an apparently disparate set of functions, mediating transcriptional regulation, mRNA transport, and DNA replication. Thus, evolution has selected for a virus with many of its related functions assembled into groups. One might speculate that an ancestral adenovirus encoding fewer gene products has evolved to generate a modern virus in which many of the ancestral genes have given rise to groups of more specialized products that remain functionally related. The grouping might also be driven in part by the advantage of using a single transcriptional control element to regulate the expression of multiple polypeptides which are needed simultaneously to execute a function such as DNA replication. Further, it might be useful to closely group the coding regions for products that interact physically and/or functionally to reduce the frequency with which they can be separated by recombination. Otherwise, recombinant variants might be generated at high frequency which could be defective because they encode products unable to function well together.

GENETIC SYSTEM AND RECOMBINATION

Adenovirus mutants have proven invaluable as reagents for the study of viral physiological processes [reviewed in (414,515)]. Most genetic studies have utilized Ad2 or Ad5. Initially, physical and chemical mutagens were employed to generate variants that could be recognized on the basis of their growth phenotype. Subsequently, mutants have been produced by the targeted manipulation of viral DNA. The ability to propagate mutant viruses is currently the only limitation to their production. Many variants have proven to be viable, often growing more poorly than their wild-type parent, but nevertheless growing sufficiently well for the production of useful virus stocks. Other mutants have been selected for conditional growth. These include temperature-sensitive as well as host range variants. Host range mutants have played an especially important role in adenovirus genetics. E1A and E1B mutants can be propagated in 293 cells, a human embryonic kidney cell line that contains and expresses the Ad5 E1A and E1B genes (4,164); then the physiological consequences of the mutations can be analyzed in standard laboratory host cells such as HeLa cells. Similarly, complementing cell lines have been developed for propagation of viruses with mutations in their E2 and E4 genes (261,485).

Recombination between adenoviruses contributes to the evolution and diversity of viral serotypes. It is well established that recombination occurs among adenoviruses of the same subgroup. Homologous recombination occurs with high efficiency during growth in coinfected cultured cells (156,438,496), and comparison of field strains with disparate neutralization (hexon) and hemagglutination (fiber) serology (210,211) indicates that such exchanges also occur in nature. Further, recombination between viruses of different subgroups appears to have given rise to Ad4, the sole member of subgroup E adenoviruses. Ad4 exhibits sequence similarity in its E1A (0 to 5 map units) and E2 regions (62 to 66 map units) to group B adenoviruses (258,451), and its hexon gene (51 to 61 map units) is immunologically related to group B viruses (342). However, the Ad4 fiber gene sequence (173) and immunological characteristics (340) are most similar to group C viruses. Thus, the simplest model for the origination of Ad4 posits a single recombinatorial crossover event somewhere between the E2 coding region of a group B virus and the fiber coding region of a group C virus (173).

Efficient adenovirus recombination requires that viral DNA replication occur (514,516), but the virus does not appear to encode any gene products that function specifically to facilitate recombination (121). Rather, it appears that single strands of DNA produced during the viral replication process, which can pair with each other or invade duplex DNAs, are the driving force behind recombination (137).

Although adenoviruses have never unambiguously been shown to integrate into the chromosomes of permissive host cells, integration does occur during the process of transformation. No specific motifs have been identified in the cellular DNA sequences at adenovirus integration sites, although patchy homologies between viral sequences and cellular target sites suggest that partial homologies probably influence the integration sites. Cellular proteins that mediate such recombination events have been partially purified (445).

REPLICATIVE CYCLE

Studies of the adenovirus replicative cycle have focused primarily on the closely related Ad2 and Ad5 viruses. These serotypes have been favored because they are easily grown in the laboratory, and an extensive collection of Ad2 and Ad5 mutant viruses have been developed. When other serotypes have been studied, their growth strategies have proven similar to the paradigm established for the prototypes. Most studies of adenovirus growth have been performed by infection of HeLa or KB cells at fairly high multiplicities of infection (>10 plaque-forming units per cell). High multiplicities of infection have been used so that all cells in the culture are synchronously infected, allowing the ordered series of biochemical events during the infectious cycle to be observed in a timewise fashion. HeLa and KB cells have been favored as hosts since they are easily propagated in large quantities, and the viruses grow in them rapidly and to high yield. These tumor cells support more rapid viral growth than "normal" human diploid fibroblasts where the replication cycle is substantially prolonged.

The replication cycle is divided by convention into two phases which are separated by the onset of viral DNA replication (Fig. 5). Early events are those which commence as soon as the infecting virus interacts with the host cell. These include adsorption, penetration, transcription, and translation of an early set of genes. Early viral gene products mediate viral gene expression and DNA replication, induce cell cycle progression, block apoptosis, and antagonize a variety of host antiviral measures. In HeLa cells infected at a multiplicity of 10 plaque-forming units per cell, the early phase lasts for about 5 to 6 hours, after which viral DNA replication is first detected. Concomitant with the onset of viral DNA replication, the late phase of the cycle begins with expression of a new set of "late" viral genes and assembly of progeny virions. The infectious cycle is completed after 20 to 24 hours in HeLa cells. At the end of the cycle, approximately 10^4 progeny virus particles per cell have been produced, along with the synthesis of a substantial excess of virion proteins and DNA that are not assembled into virions (168). Cells infected at high multiplicity seldom divide (220), so at the completion of the replication cycle, the DNA and protein content of the infected cell has increased about twofold.

While *early* and *late* are convenient terms for description of events that occur during the replication cycle, the functional distinction between early and late events is often blurred. For example, early genes continue to be expressed at late times after infection, and the promoter controlling expression of the major late transcription unit directs a low level of transcription early after infection. The viral genes encoding proteins IVa2 and IX begin to be expressed at an intermediate time (356) and thus form a "delayed early" category.

Adsorption and Entry

Attachment of Ad2 to cells is mediated by the fiber protein (290). The distal, carboxy-terminal domain of the fiber protein terminates in a knob that is presumed to bind to the cellular receptor (101). Soluble fiber protein and some fiber-specific antibodies can block virus attachment and infection (98,355,358,500). The identity of the cellular receptor remains a mystery, although three cellular membrane polypeptides have been captured on a penton-fiber affinity matrix and shown to inhibit Ad2 attachment to cells (206). Since the amino acid sequence of the knob region varies considerably among serotypes, it is possible that different adenovirus serotypes bind to different cellular receptor proteins. Attachment occurs efficiently at 0°C, but subsequent steps in the infectious process require energy and are inhibited at low temperatures (63).

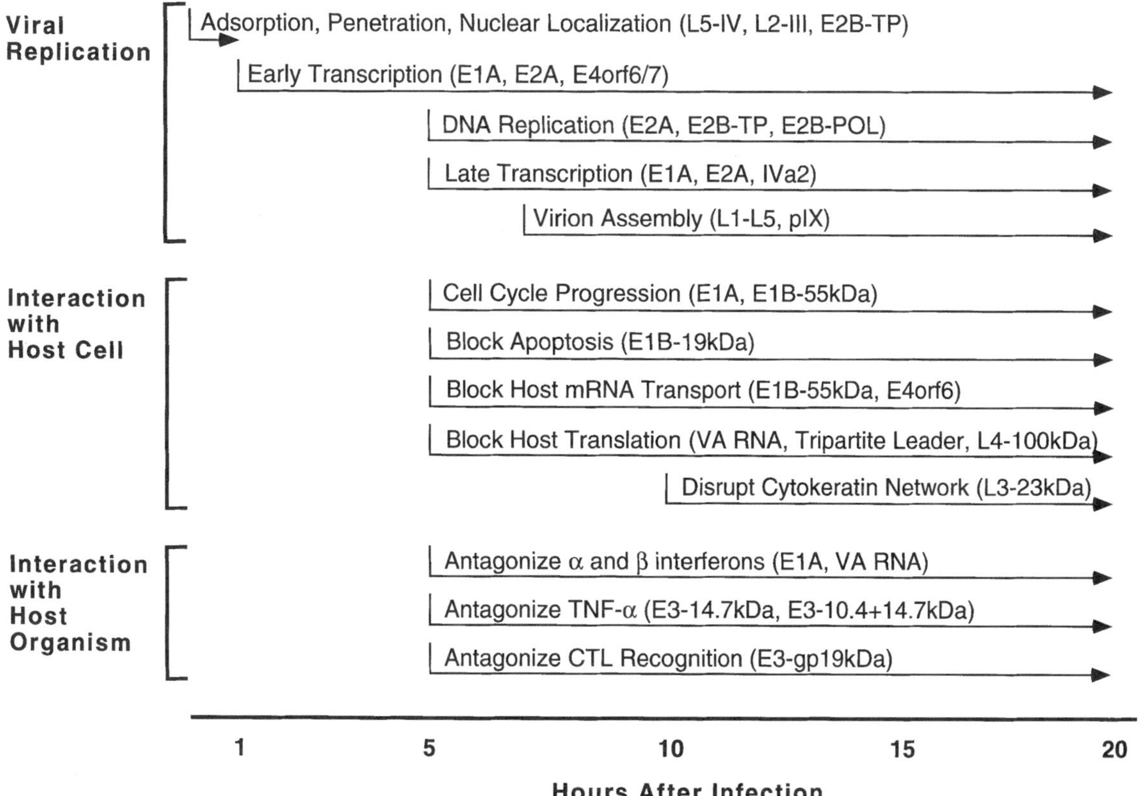

FIG. 5. Diagram displaying the relative timing of the main events that occur during the adenovirus growth cycle. Events are separated into three groups: viral replication, host cell interactions, and host organism interactions. The viral gene products that are known to mediate each event are listed (*in parentheses*). Times for the stages of viral replication and for interactions with the host cell are as occur during growth in HeLa cells which have been infected at a multiplicity of 10 pfu/cell. HeLa cells die at 20 to 24 hours after infection. Times for interactions with the host organism are artificial and meant to place the interactive events in a time frame relative to events that occur during the viral replication cycle. Viral replication within an infected host animal proceeds more slowly than in HeLa cells.

The observation that Ad2 binds to some cells but does not efficiently enter them (417) raised the possibility that a second protein-protein recognition event might be needed for internalization, and a second interaction has been identified. The penton base protein binds to specific members of a family of heterodimeric cell surface receptors, termed *integrins* (494). The $\alpha_v\beta_3$ and $\alpha_v\beta_5$ vitronectin-binding integrins and perhaps other integrin family members interact with the viral protein. The interaction occurs through an arg-gly-asp (RGD) sequence present in each of the five polypeptide III molecules that comprise the penton base, the same sequence motif present in a number of extracellular adhesion molecules that bind to integrins. RGD-containing peptides can block internalization of Ad2 (494), and Ad5 mutants with mutations that alter the RGD sequence in polypeptide III exhibit a delay in the onset of viral DNA synthesis under certain conditions, consistent with a reduced efficiency of internalization (20).

Interestingly, the growth of cells on matrices that bind to α_v integrins significantly reduced their susceptibility to Ad2 infection (494), suggesting that cells with similar fiber and penton receptor repertoires located in different tissue environments might exhibit marked differences in susceptibility to adenovirus infection. The penton-integrin interaction is not sufficient for viral binding to cells since soluble fiber or fiber antibodies can completely block adsorption. Thus, adsorption and internalization is a two-component process requiring the interaction of both fiber and penton proteins with their cognate cellular targets.

After adsorption, Ad2-receptor complexes diffuse into coated pits and they are internalized by receptor-mediated endocytosis (61,62,134,467). The process is triggered by the penton-integrin interaction, since purified penton but not fiber protein is rapidly internalized by cultured cells (494). The mechanism by which the interaction induces endocytosis is not clear, but the process must be induced locally because the adsorption of RGD-deficient virus cannot be complemented by wild-type virus (20).

The internalization process of adsorbed virus is remarkably efficient (166). Eighty to eighty-five percent of virus that binds to the surface of a susceptible cell is internalized, and penetration occurs rapidly with half of the adsorbed virus moving to endosomes within 10 minutes. About 90% of the virus within endosomes successfully moves to the cytosol with a half time of about 5 minutes. The movement to the cytosol is somehow triggered by the acidic pH of the endosome (351,405,437), and the penton base is believed to play an essential role in the process. The rapidity of the movement to the cytosol suggests that the virus escapes from the endosomal unit known as the *early endosome* prior to formation of a lysosome (310). The endosome appears to be substantially disrupted since DNA-protein complexes that enter the endosome together with virus particles—but not physically attached to virions—also escape to the cytosol with high efficiency (512). Virus particles are transported across the cytoplasm to the nucleus by a process that probably involves microtubules (94,293). *In vitro* studies have shown that virions can attach to microtubules through the hexon, and some drugs that interact with microtubules can inhibit adenovirus infection (94). About 40 minutes after penetration, virus particles can be seen at nuclear pore complexes by electron microscopy, suggesting that DNA release occurs at the nuclear membrane. After 120 minutes, about 40% of internalized particles have released their DNA free from hexon proteins, although it is not known what portion of this released DNA is localized to the nucleus.

During the internalization process, there is a sequential disassembly of the virion (166). The disassembly occurs by selective dissociation and proteolytic degradation of virion constituents. First, the proteins at the vertices of the particle are lost. Polypeptide IV, which trimerizes to form the fiber; polypeptide IIIa, located near the peripentonal hexons linking adjacent facets of the capsid; and polypeptide III, which forms the pentameric penton base, are substantially lost by 15 minutes after penetration of the cell. Polypeptides IV and IIIa dissociate somewhat more rapidly than III, and this could reflect a need to free the penton base for a role in escape from the endosome. Polypeptide VIII dissociates from the particle as the penton capsomeres are lost, and, shortly after entry of the particle to the cytosol, polypeptide VI is degraded. The internal location of polypeptide VI suggests that it might be degraded by the virus-coded, DNA-dependent protease (299,450,483). Both polypeptides VI and VIII bridge from the DNA core to the capsid, and their loss should prepare the particle for release of its DNA. Somewhat later polypeptide IX, which stabilizes hexon facets, exits from the infecting particle; finally, the DNA-containing core is freed from hexons. Thus, the virion, which is very stable outside the cell, is dismantled by an ordered elimination of structural proteins upon entry to the cell so that it can deliver its DNA to the nucleus.

The nature of the events that signal disassembly are mostly unknown. Acidification induces exit of the infecting particle from the endosome to the cytosol, but disassembly of the vertex constituents can occur in the absence of acidification (166). It seems likely that the signal for the initial dissociation of the fiber from the penton is a determining event that initiates a cascade in which each sequential disassembly event is driven by changes in the structure of the virion that result from previous events.

The components and structure of the DNA-protein complex that reach the nucleus have not been described, although it is known that protein V is removed from the core before it enters the nucleus (166). The viral DNA has been reported to convert into a structure that can be digested with a low concentration of DNase I to generate DNA fragments that form a ladder pattern similar to that of cellular chromatin upon electrophoresis (444), suggesting that cellular histone proteins might replace polypeptide VII, the major core constituent, before the infecting DNA is transcribed. However, using sensitive assays, others have been unable to detect nucleosomal particles comprised of viral DNA and cellular histones (468). It is possible that the viral genome is expressed as a chromosomal structure containing viral basic proteins rather than cellular histones.

When the viral DNA reaches the nucleus, it associates with the nuclear matrix through its terminal protein (38,141,398). DNAs from mutant Ad5 variants with lesions in their terminal protein genes fail to become tightly associated with the nuclear matrix, and they also fail to be efficiently transcribed (398). This correlation between nuclear matrix association and activation of transcription suggests the two events are functionally interrelated. Apparently, the terminal protein, which arrives with the infecting genome, is the first viral gene product that functions within the nucleus to initiate the program of viral gene expression.

Activation of Early Viral Genes

There are three main goals of early adenovirus gene expression. The first is to induce the host cell to enter the S phase of the cell cycle, providing an optimal environment for viral replication. As will be discussed later (see "Interactions with the Host" section), both E1A and E1B gene products play roles in this process. The second is to set up viral systems that protect the infected cell from various antiviral defenses of the host organism. The E3 and VA RNA genes contribute to these defenses, and these will also be discussed later (see "Virus-Host Interactions" section). The third is to synthesize viral gene products needed for viral DNA replication. All three of these goals depend on the transcriptional activation of the viral genome, and the principal activating proteins of adenovirus are encoded by the E1A gene [reviewed in (332,412)].

E1A is the first viral transcription unit to be expressed after the viral chromosome reaches the nucleus. Transcription of the E1A unit is controlled by a constitutively active promoter that includes a duplicated enhancer element (198,200). The E1A unit encodes two mRNAs during the early phase of infection (Fig. 6). Three additional E1A mRNA species accumulate later in the infectious cycle (428,458), but no definitive function has been described

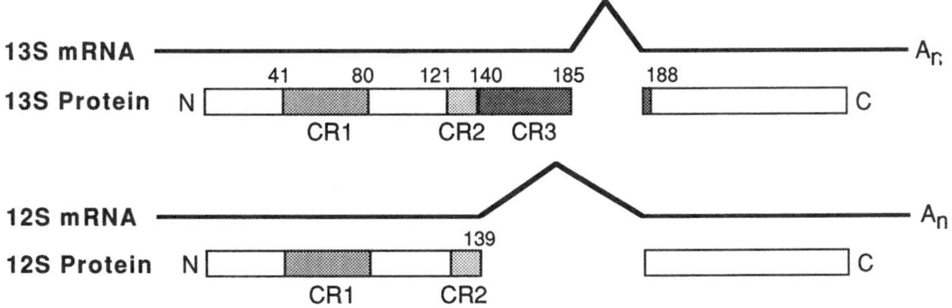

CR1
41-PTLHELYDLDVTAPEDPNEEAVSQIFPDSVMLAVQEGIDL-80

CR2
121-DLTCHEAGFPPSDDEDEEG-139

CR3
140-EEFVLDYVEHPGHG**C**RS**C**HYHRRNTGDPDIM**C**SL**C**YMRTCGMFVYSPVS-188

FIG. 6. Diagram of E1A mRNAs, the polypeptides which they encode, and the amino acid sequence of conserved domains. 13S and 12S mRNA exons are represented (*lines*), as are introns (*caret symbols*) and poly(A) sequences (A_n). The polypeptides encoded by the 13S and 12S mRNAs are designated (*rectangles*). Domains in the proteins that are conserved among adenovirus serotypes are identified as conserved regions 1 to 3 (*CR1, CR2, CR3*), and the amino acids at the boundaries of conserved regions are indicated (*above rectangles*). The amino acid sequences of the conserved domains are displayed at the bottom of the figure, and the cysteines comprising the zinc finger motif in CR3 are marked (*enlarged bold letters*).

for their products. The two early mRNAs contain identical 5' and 3' ends, differ internally due to differential splicing, and encode proteins that are identical except for an additional 46 amino acid segment that is present in the larger polypeptide. The two polypeptides are commonly referred to as the 12S and 13S E1A proteins, names derived from the sedimentation coefficients of their corresponding mRNAs. The primary E1A translation products undergo extensive phosphorylation (185) but, so far, the functional consequences of the modifications are not clear.

When the amino acid sequences of the E1A proteins encoded by a variety of human serotypes are compared, the E1A products prove to be constructed of three conserved regions (CR1, CR2, and CR3) (Fig. 6) separated by less highly conserved domains (110,317,332,412,466). The E1A proteins do not exhibit sequence-specific DNA binding (64,132); rather, they bind to cellular proteins and modulate their function. The three conserved regions in the E1A proteins mark domains that play major roles in protein-protein interactions.

The ability of E1A proteins to activate expression of the other adenovirus transcription units was discovered when viral mutants were examined that carried either a single base pair substitution (32) or a deletion (245,413) within the E1A coding region. These mutant viruses, which could be grown in transformed human cells expressing the E1A and E1B genes (293 cells) (4,164), failed to accumulate early viral mRNAs with normal kinetics when grown in HeLa cells that were unable to complement the defect. Compared to wild-type virus, they accumulated early mRNAs after a long delay and then only very slowly. Subsequent analysis demonstrated that E1A proteins acted to dramatically increase the rate of transcription (331). Since the E1A proteins can activate other viral genes in *trans*, they are often referred to as *trans*-activators.

Initially, attempts to identify a promoter element that was responsive to the E1A proteins proved frustrating. In a number of studies, it was possible to mutate any specific DNA element without blocking the ability of E1A to activate test promoters, provided that basal transcription was not eliminated (199,255,325). The reason for the promiscuous activation is now clear. The E1A proteins can activate transcription by binding to a variety of different cellular transcription factors and regulatory proteins (Table 2). The E1A proteins can interact directly with auxiliary factors that mediate basal transcription, with activating proteins that bind to upstream promoter and enhancer elements, and with regulatory subunits that influence the activity of DNA-binding factors. All three conserved domains of E1A mediate protein-protein interactions that activate transcription.

E1A proteins can activate transcription through the TATA motif (171,418,504), an element found 25 to 30 bp upstream of transcriptional initiation sites in many viral and cellular genes. The activation is believed to result in part from the ability of the 13S E1A protein to bind directly to the TATA-binding protein (TBP) (217,272), which is the DNA-binding subunit of the auxiliary transcription factor

TABLE 2. *Cellular E1A binding partners*[a]

E1A-Binding protein	Binds to		Other proteins in complex
	12S E1A	13S E1A	
pRB	+	+	None known
p107	+	+	cyclin A, cdk2/cyclin E, cdk2
p130	+	+	cyclin A, cdk2/cyclin E, cdk2
p300	+	+	None known
TBP	−	+	TAFs
ATF-2	−	+	None known
YY1	+	+	None known

[a] This listing contains only those cellular proteins that have been shown to bind directly to E1A proteins, and it is limited to binding partners for which there is evidence that the interaction is biologically relevant.

pRB, retinoblastoma susceptibility protein; cdk2, cyclin-dependent kinase 2; TBP, TATA-binding proteins; TAFs, TBP-associated factors.

IID (TFIID). TFIID is the first auxiliary transcription factor to interact with TATA-containing promoters, and it plays a central role in the assembly of a preinitiation complex at the appropriate position on a template DNA (367). The interaction between TBP and E1A is mediated by the 13S-specific CR3 domain of the E1A protein (272), and this helps to explain why the 13S E1A protein activates transcription through TATA motifs more efficiently than the 12S protein. Part of the mechanism by which the 13S E1A protein activates transcription as a result of binding to TBP appears to involve the cellular tumor suppressor protein, p53, which can bind to TFIID (288,406) and repress transcription (406). The two proteins bind to overlapping domains on TBP, and the E1A protein can displace p53 from TBP and relieve p53-mediated repression (218). The 12S E1A protein can also activate through a TATA motif (267), but more weakly than the 13S protein. This activation event involves Dr1, a factor that can bind to TBP and inhibit transcription (236). The 12S protein can bind directly to Dr1, preventing it from associating with TBP (266). Thus, both 12S and 13S proteins can activate transcription through TATA motifs by binding to cellular factors and relieving transcriptional repression. It is likely that additional activating events occur as a result of E1A interactions with cellular factors at the TATA motif. For example, the 13S E1A protein might serve as an adaptor, bridging between TBP and other transcriptional activating proteins. One curious aspect of activation through TATA motifs is that the E1A proteins appear to activate through some, but not all, TATA sequences (353,418). Perhaps different classes of TFIID complexes assemble at variants of the TATA element and some respond to the presence of E1A while others do not.

In addition to interactions with the basal transcriptional machinery, E1A proteins can activate transcription through the binding sites for factors that bind well upstream of the basal promoter. The best studied activation event occurs through E2F binding sites. This transcription factor is named for the first promoter where it was found to bind, the adenovirus E2 promoter (263); it also binds within the enhancer domain of the E1A gene. E2F forms a complex with the cellular retinoblastoma tumor suppressor protein, pRB (19,21,69,74); and pRB inhibits transcriptional activation by E2F (68,95,179,208). The 12S and 13S E1A proteins activate transcription through E2F binding sites (264,505,517), and both E1A proteins bind to pRB (492) through CR1 and CR2 (109,114), dissociating it from E2F (18,19,21,23,69). The strong correlation between the dissociation of the E2F-pRB complex and transcriptional activation by E1A proteins argues that the two events are causally related and leads to the proposal that E1A activates transcription by dissociating pRB from E2F. Indeed, the E2F1 cDNA has been cloned (202,247,407), and pRB has been shown to directly inhibit its function (135,201). The activation of E2F by E1A not only impacts on the expression of viral genes, but it also influences the expression of cellular genes, dramatically effecting cell cycle progression of the infected host cell. This will be discussed in detail below (see "Activation of the Host Cell" section).

E1A can activate transcription through its ability to bind directly to several additional cellular factors. All of the early adenovirus promoters, with the exception of the E1B promoter, contain binding sites for the ATF (also termed CREB) family of transcription factors. E1A activates transcription through ATF binding sites (286), presumably in part because the 13S E1A protein can bind to ATF-2 (2,283,286,287). The interaction occurs between the DNA binding domain of ATF-2 and CR3 of E1A. Different subdomains of CR3 mediate E1A binding to ATF-2 and TBP, raising the possibility that E1A functions as an adaptor, bridging between ATF-2 bound at an upstream site and TBP (272). Such a bridging function might stabilize the initiation complex at the promoter.

E1A proteins can also activate through YY1 recognition sites (415). In the absence of E1A protein, YY1 often represses the activity of promoters that contain its binding site. The E1A proteins bind to YY1 through an amino terminal sequence of E1A, and, in the case of the 13S E1A protein, the interaction is stabilized by CR3 (277). E1A binding correlates with the relief of repression mediated by YY1. As yet, the mechanism by which E1A binding alters the function of YY1 is unknown. The YY1-E1A interaction was initially identified through study of the adeno-associated virus P5 promoter which is strongly activated by E1A through a YY1 recognition site (415). The E1A-responsive enhancer domain of the E1A promoter also contains a YY1 binding site, and YY1 has been implicated in the activation of the Ad12 major late promoter (529).

E1A proteins have been reported to both activate and repress through the transcription factor AP1. These activities involve both an amino terminal domain and the CR1 domain of E1A (118,148,346). E1A can repress the action of AP1 through its consensus binding sites in the collagenase and stromelysin genes (346), and enhance the ability of AP1 to activate through the ATF sites (to which AP1 can bind) of adenovirus early genes (117,184). The AP1 transcription factor is a dimeric molecule which is generally comprised of fos and jun family members, although jun family members can also heterodimerize with members of the ATF family (30). E1A repression appears to result from a block to DNA binding by c-fos/c-jun heterodimers (177), and activation by E1A through AP1 occurs, at least in part, by an indirect mechanism. E1A proteins cooperate with cyclic AMP to induce the level of AP1, which activates transcription of viral genes (323). The E1A-mediated induction of AP1 activity results from transcriptional activation of the c-fos and junB genes through their TATA motifs (260), and the c-jun gene has been shown to be induced through an ATF-like element in its promoter, which probably responds to a c-jun/ATF-2 heterodimer (461). So, E1A can activate transcription indirectly through AP1 binding sites by inducing AP1 activity, and E1A can activate through ATF binding sites either directly by binding to ATF-2 or indirectly by inducing AP1 activity.

Expression of the adenovirus VA RNA genes is also induced by the E1A proteins (214,215,513). These genes are transcribed by RNA polymerase III, and it appears that the activity of transcription factor IIIC is altered by the E1A proteins.

Besides the E1A proteins, two additional early gene products have been shown to activate adenovirus promoters. The E4-17 kd polypeptide binds to E2F, with two molecules of the E4 protein apparently bridging between a pair of E2F proteins and causing them to bind cooperatively to the pair of E2F binding sites within the E2 promoter (183,231,328,372). The E2 DNA-binding protein can activate a variety of early adenovirus promoters as well as the major late promoter (60,318), but as yet the mechanism by which it stimulates transcription is unclear.

Individual adenovirus early promoters are activated by multiple mechanisms. For example, the E2 promoter can be activated by E1A through its TATA motif, ATF site, or E2F sites. In addition, the E4-17 kd protein contributes to activation at the E2F sites, and the E2 promoter can also be induced by the E2 DNA-binding protein. Perhaps the activation pathways function with different efficiencies within various cell types, and these apparently redundant activation mechanisms have evolved to insure that early viral promoters will be efficiently activated in a variety of different cell types as the virus spreads within its infected host.

Generally, adenovirus early genes remain active throughout the viral replication cycle, although the rate at which they are transcribed slowly declines. In part, the decline results from cell death. However, there are three known down-regulatory events, and in each case it appears that a viral protein, which accumulates in response to an activation event, subsequently acts to inhibit continued transcriptional stimulation mediated by one or more early promoter elements. First, the E1A proteins can repress the activity of a variety of known enhancers, including the enhancer residing upstream of the E1A gene itself (40,470). This down-regulatory function correlates with the ability of the E1A protein to bind to a cellular protein, p300 (383,426,478); the consequences of this protein-protein interaction will be considered further below (see "Activation of the Host Cell" section). Second, whereas the E2 DNA-binding protein activates some promoters, it appears to inhibit transcription from the E4 promoter (181,335) by an unknown mechanism. Third, the induction of AP1 activity by E1A and cyclic AMP is transient (323). It is antagonized by the E4-14 kd polypeptide which accumulates in response to the E1A-mediated activation of the E4 gene (322). Mutant viruses unable to produce the E4-14 kd protein are viable, but more cytopathic than the wild-type virus, suggesting that it is important to counterbalance the E1A-mediated induction of AP1 to preserve integrity of the host cell for the replicating virus. The E4-14 kd protein binds to protein phosphatase 2A, and phosphatase activity is essential for the down-regulatory event (260).

Once early mRNAs have been synthesized, they are translated on polysomes together with cellular mRNAs. Initially, they do not appear to enjoy any competitive advantage, but as the infection enters the late phase, cellular mRNAs are excluded from polysomes (see "Activation of Late Gene Expression" and "Host Cell Shutoff" sections).

Activation of the Host Cell

Adenovirus infection has long been known to induce quiescent cells to enter the S phase of the cell cycle, creating an environment optimally conducive to viral replication [reviewed in (454)]. Modulation of the cell cycle is primarily a function of the E1A proteins, and the modulation is mediated by their CR1, CR2, and nonconserved amino-terminal domains. The key to understanding how E1A proteins manipulate cell cycle regulation came from the observation that a set of cellular proteins can be coimmunoprecipitated with the E1A proteins (157,186,224,506). The main coprecipitating cellular polypeptides have molecular weights of about 33, 60, 80, 90, 105, 107, 130, 300, and 400 kd (Fig. 7); and several of these polypeptides have been shown to interact directly with E1A proteins (Table 2). The 105-kd moiety was the first to be identified (492). It is the retinoblastoma tumor suppressor protein pRB which, as discussed above, regulates the ability of the E2F cellular transcription factor to activate transcription. The human papillomavirus E7 protein and the large T antigens of papovaviruses also bind to pRB (97a, 111), underscoring its importance as a target of oncoproteins encoded by DNA tumor viruses. Although E2F sites are present in the adenovirus E1A and E2 promoters and contribute to their activation, neither the papovaviruses nor papillomavirus-

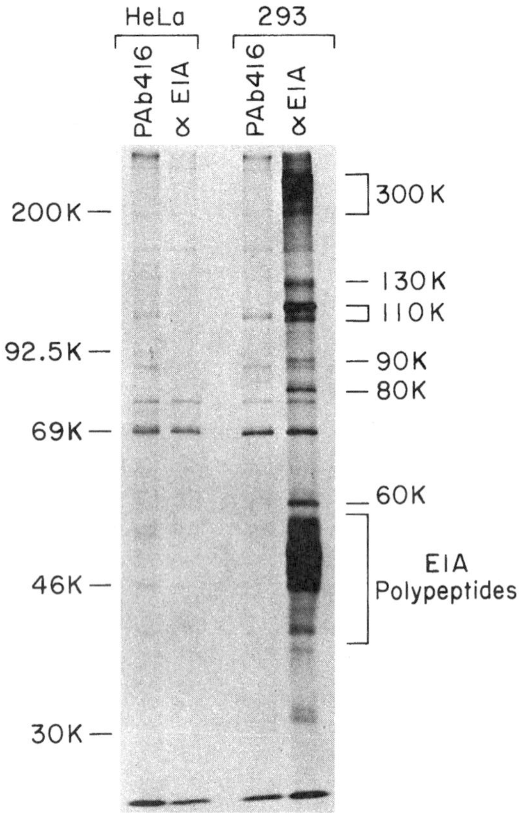

FIG. 7. Coimmunoprecipitation of cellular polypeptides with E1A proteins. Radioactively labeled polypeptides were immunoprecipitated from lysates of HeLa or 293 cells with a monoclonal antibody specific for SV40 T antigen (PAb416) or a mixture of antibodies specific for E1A proteins (αE1A), and immune complexes were analyzed by electrophoresis. The size (kd) of marker proteins is indicated (*left*) and the size of polypeptides coimmunoprecipitated with E1A proteins as well as the location of E1A polypeptides is indicated (*right of autoradiogram*).

es have E2F binding sites on their genome. Therefore, tumor virus oncoproteins must target pRB for reasons other than to influence E2F function at their own promoters.

pRB has been shown in a number of assays to inhibit cell cycle progression, causing cells to arrest in mid-to-late G_1 (162,213), and growth arrest correlates with the ability of pRB to bind to E2F and block E2F-dependent transcriptional activation (208,368,369). pRB activity is regulated by phosphorylation. pRB can be isolated in association with cyclin and cyclin-dependent kinase (cdk), and this complex might, in part, represent an enzyme-substrate intermediate in which the cdk phosphorylates serine and threonine residues on pRB (228,256,498). Hyperphosphorylation of pRB inhibits the formation of the pRB-E2F complex (69,202,403), and it reverses the ability of pRB to arrest cell growth (213). These observations are consistent with the fact that hyperphosphorylated pRB and free E2F accumulate as growing cells reach late G_1 and enter S. The strong correlation between the ability of pRB to bind E2F and to block progression from G_1 to S has led to the belief that pRB regulates cell cycle progression through its interactions with E2F. However, it is now clear that pRB is one of several proteins that regulate E2F activity, and pRB-E2F complexes persist at the G_1/S boundary of the cell cycle, even though free E2F accumulates at this time (403).

Free E2F can activate transcription, and ectopic expression of E2F induces quiescent cells to enter the S phase (243). E2F likely activates a series of genes important for S phase and cell growth that contain E2F binding sites in their promoters, including dihydrofolate reductase (DHFR), thymidine kinase, thymidylate synthetase, ribonucleotide reductase, DNA polymerase α, c-myb, cdc2, c-myc, and N-myc [reviewed in (332)]. Indeed, E2F binding sites have been shown to be essential for the activation of the DHFR gene; a single E2F binding element is sufficient for activation of transcription at the G_1/S boundary (419).

The CR2 domain is primarily responsible for the interaction of E1A proteins with pRB, and the N-terminal portion of the CR1 domain plays an auxiliary role, stabilizing the interaction (207,224,477,493). CR2 contains the sequence Leu-X-Cys-X-Glu (where *X* is any amino acid), which is also present in other viral and cellular proteins that bind to pRB. The E1A binding domain on pRB is often referred to as the *pocket*, and it is composed of two essential sequences, termed *A* and *B*, separated by a spacer region (227,246). E2F also binds to the pocket domain, so E1A binding competes with E2F for access to the pRB pocket. As a result, E2F can be liberated from pRB by E1A, and the free E2F would be expected to activate cellular genes that facilitate cell cycle progression. This accomplishes a common goal of DNA tumor viruses: to move the infected cell into the S phase, creating an environment favorable for efficient viral DNA replication. Mutant E1A proteins that fail to bind pRB are unable to activate transcription through E2F or modulate cellular growth.

Two additional cellular pRB family members have been identified that interact with E2F: p107 (127) and p130 (182,278). p107 and p130 are more closely related in sequence to each other than to pRB, but as pRB family members, they contain "pocket" domains to which E2F can bind. Complexes containing E2F and p107 are found in the G_1 and S phase, and complexes with p130 and E2F are evident during G_0 and early G_1. Like pRB, p107 has been shown to inhibit E2F-dependent gene expression (403,518,525) and inhibit cellular proliferation, blocking cell cycle progression in late G_1 (525). Presumably, p130 will prove to do the same.

The spacer region separating the A and B elements of the pocket is larger in p107 and p130 than in pRB, and the enlarged spacer serves as a site for direct binding of cyclins (126,128). p107 forms two complexes that contain E2F, cyclin, and a cdk: an E2F-p107-cyclinA-cdk2 complex that is found predominantly in S phase (103,416), and an E2F-p107-cyclinE-cdk2 complex found in the G_1 phase

(273). p130 also is present in complexes containing E2F and cyclinA or cyclinE *in vivo* (84,129,278). The p107 and p130 cyclin-cdk complexes might represent enzyme-substrate intermediates. However, the complexes appear fairly stable, and they include a substantial portion of the nuclear pool of p107 and p130 molecules. An enzyme would be expected to interact only briefly and sequentially with many molecules of its substrate. Thus, the complex might be more than an intermediate in the reaction mediating phosphorylation of pRB family members. Perhaps pRB family members direct the activity of the kinase and its regulatory subunit to specific target substrates.

As is the case for pRB, E1A can bind to p107 and p130 [these proteins can be coimmunoprecipitated with E1A (Fig. 7)], dissociating E2F. As for pRB, the interaction is mediated by the E1A CR2 domain, and E1A-p107 binding is stabilized by interactions within the CR1 domain (478). E1A does not disrupt the interactions of p107 or p130 with cyclins and cdks. Thus, E1A can be found in complexes that include kinase activities (157,207,259) and, further, the phosphorylation state of p107 and p130 in these complexes is substantially elevated (24). Whether these complexes represent nonfunctional byproducts of the release of E2F or whether the complexes that include E1A, p107 or p130, cyclin and cdk (p60 and p33) (Fig. 7) perform an active, E1A-modulated function is unknown.

Not only are there multiple pRB family members present in E2F complexes, but there are also multiple E2Fs. E2F DNA-binding activity is composed of a group of polypeptides including E2F1, E2F2, and E2F3 (202,274) and DP-1 (158); and at least a portion of the E2F activity results from heterodimer formation between various E2F family members (22,203,232). Further, it appears that different E2F family members are likely to associate with specific pRB family members; for example, E2F1 does not interact with p107.

In sum, the E1A proteins disrupt a series of complexes that contain different pRB family members, multiple E2F subunits, and cyclins with associated kinases. These complexes normally regulate cell cycle progression, and their disruption by E1A deregulates normal cell cycle control, allowing quiescent cells to begin DNA synthesis subsequent to infection.

The E1A proteins also bind to a 300-kd cellular protein, p300 (Fig. 7). Its binding site on E1A includes the poorly conserved N-terminus and the C-terminal half of CR1 (478). E1A mutants lacking CR2 and unable to bind pRB family members can nevertheless stimulate cellular DNA synthesis, provided the p300 binding site remains intact. Thus, the E1A proteins contain two independent domains, either of which can stimulate cells to progress from G_1 to S (225,282,522). Besides stimulating entry into S phase, the p300 binding domain is responsible for the ability of E1A proteins to repress enhancer function (383,426,478). E1A can repress the activity of a variety of enhancer elements, including elements controlling genes associated with the differentiation state of cells (204,427,484). p300 has been reported to bind directly to DNA, recognizing NFκB/H2TF1-like sites (379). A p300 cDNA has been cloned (112), and it encodes a 2,414-residue polypeptide that contains domains reminiscent of transcriptional coactivators. The E1A-associated p300 is closely related in its amino acid sequence to the CRE-binding protein (CBP) (79). These two proteins appear to be members of a family of transcriptional regulatory proteins (10), and it is possible that E1A proteins bind to multiple family members. The p300 whose cDNA has been cloned contributes to transcriptional activation, probably by bridging from the enhancer domain to the initiation complex assembled at the promoter (112). E1A binding blocks the ability of p300 to activate transcription, and overexpression of p300 can relieve E1A-mediated repression. Some of the genes that are regulated by p300 might encode products required for cell cycle regulation. This would explain how p300 binding by E1A could deregulate the cell cycle and induce quiescent cells to enter the S phase.

p400 is the largest protein that is known to be complexed with E1A. It binds to the same domains on E1A as does p300, and comparison of peptide maps indicates that the two proteins are related in their amino acid sequences (24); so p300 and p400 will probably prove to be functionally related.

Two separate regions of the E1A proteins can induce cells to move from G_1 to S through two independent mechanisms: binding to pRB family members and binding to p300. Presumably, E1A proteins simultaneously employ two mechanisms because together they function more effectively than either alone or because there may be cells in which one pathway works more efficiently than the other. Further, although the E1A domains mediating binding to either pRB family members or p300 can induce DNA synthesis in quiescent cells, both regions are required to pass the G_2/M checkpoint and progress to mitosis (224,225, 477,522). This could be an important function in animal hosts where infections are likely to proceed much more slowly than in cultured cells. Without the ability to divide, cells that pass completely through S would be stuck in G_2; if the cells can divide, they can continue cycling and return to the S phase where viral replication is most efficient. Possibly, the simultaneous inhibition of both the pRB family and p300 by E1A binding somehow enables the infected cell to pass through the G_2/M checkpoint. Alternatively, E1A proteins might alter the function of these regulatory proteins rather than simply inhibit them. E1A has been reported to bind simultaneously to a pRB family member and p300 (24), and this new complex might influence passage through mitosis. It is also possible that E1A proteins bind to additional, as yet unidentified, proteins to antagonize this checkpoint.

Like the E1A proteins, the E1B-55 kd protein also modulates cell cycle progression. The E1B protein targets the cellular p53 tumor suppressor protein which, like pRB family members, regulates progression from G_1 to S. p53 is the most commonly mutated gene in human tumors, and the loss of p53 function or its alteration by mutation can contribute to tumor progression [reviewed in (276,473,519)]. Consistent with its role in tumorigenesis, high-level expression of p53 blocks cell cycle progression at the G_1/S boundary [reviewed in (354)]. p53 is a sequence-specific DNA-binding protein that can activate transcription when it binds to p53 response elements, and it can repress a variety of genes that lack a binding site for p53 (reviewed in 519).

It appears that p53 normally functions as a component of a G_1 checkpoint that is induced by DNA damage (104,190,284). Recent work indicates that DNA damage can transiently induce p53 levels; p53, in turn, can block cell cycle progression and contribute to the activation of genes known to be induced by DNA damage. Specifically, p53 can induce expression of the WAF1 (also termed *p21*) gene (115) whose product is able to arrest cell cycle progression (187), and it can induce the GADD 45 gene which is activated by DNA damage (251). High levels of p53 can also result in apoptosis (81,292,410,511), and this will be considered below (see "Inhibition of Apoptosis" section). Thus, the current view of p53 is that it transcriptionally activates genes that prevent entry into S phase. It is not yet clear whether transcriptional repression by p53 also contributes to this process.

The Ad5 E1B-55 kd protein binds to p53 within infected cells (394), and it can block transcriptional activation by p53. Analysis of mutant E1B-55 kd proteins has revealed a strict correlation between the ability of E1B to block p53-mediated transcriptional activation and its ability to cooperate with E1A in the oncogenic transformation of cells (507). As will be discussed below (see "Oncogenesis" section), the E1A and E1B proteins of adenovirus are oncoproteins, and they cooperate to transform cultured cells. Oncogenic transformation is a manifestation of the ability of these proteins to interfere with the normal function of tumor suppressor proteins. Thus, the correlation between a block to transcriptional activation and transforming activity of the viral protein suggests that the E1B-55 kd protein antagonizes the ability of p53 to influence cell cycle progression by blocking its activation function. The E1B-55 kd protein binds to the N-terminal, acidic transcriptional activation domain of p53 (248), and this suggests that the viral protein might simply mask the activation domain. Steric hindrance might contribute to the ability of E1B-55 kd to block activation by p53, but it is not the entire mechanism. The E1B-55 kd protein can inhibit transcription if it is artificially anchored to a promoter, indicating that it can actively repress transcription (508). Apparently, the interaction of E1B-55 kd with p53 serves to bring the viral protein to the promoter, where it not only blocks the activation function of p53 but it also actively represses transcription. So, like the E1A proteins, the E1B-55 kd protein binds to a tumor suppressor protein, antagonizes its normal activity, and helps to deregulate cell cycle progression.

In contrast to the E1A proteins, expression of the E1B-55 kd protein alone is not sufficient to stimulate quiescent cells to enter the S phase of the cell cycle. This is consistent with the observation that deletion of both p53 alleles does not directly lead to the loss of regulated cell division (107). Nevertheless, p53 is targeted by a variety of tumor virus oncoproteins: SV40 large T antigen binds to p53, blocks its DNA-binding activity, and interferes with its ability to activate transcripton (25,131); and human papilloma virus type 16 E6 protein complexes with p53 and cooperates with a cellular protein to promote its degradation (399). Presumably, then, the E1B-55 kd protein collaborates with the E1A proteins to more effectively activate quiescent cells. Further, since the E1A proteins somehow stabilize p53, increasing its steady state level (291), the E1B proteins must contribute to the activation of quiescent cells by preventing the implementation of a cell cycle block by elevated levels of p53. In a similar vein, as will be discussed below (see "Inhibition of Apoptosis" section), the E1B-55 kd protein helps to prevent cells from undergoing apoptosis in response to E1A activities.

It is noteworthy that the Ad12 homologue of the Ad5 E1B-55 kd protein, the Ad12 E1B-54 kd protein, can inhibit p53-mediated transcriptional activation and cooperate with E1A to transform cells (507), even though it shows no indication of p53 binding (165,521). Also, the Ad5 E1B protein can sequester a substantial portion of the cell's p53 in a discrete cytoplasmic structure outside of the nucleus (520), while the Ad12 protein cannot (521); this has been taken as further evidence that the Ad12 protein does not bind to p53. The Ad12 E1B protein might bind weakly to p53, and the resulting complex might not be sufficiently stable to be detected by the assays that have been employed. p53 is stabilized in Ad12 transformed cells, and the stabilization is dependent on expression of the E1B-54 kd protein (462). Stabilization is often observed for mutant p53, but the p53 present in the Ad12 transformed cells is wild type. Further, while the Ad5 E1B protein inhibited transformation by *myc* plus *ras*, the Ad12 E1B protein, like mutant p53, enhanced the number of foci produced by *myc* plus *ras*. If, indeed, there is no direct interaction between the Ad12 E1B protein and p53, these observations suggest that the Ad12 protein somehow causes wild-type p53 to be modified, perhaps by association with another cellular protein, so that it behaves in some respects like mutant p53.

Inhibition of Apoptosis

As discussed above, expression of the E1A proteins in quiescent cells can efficiently induce cellular DNA synthesis and transient cell proliferation. However, E1A expression is not sufficient to induce long-term growth of primary cells. This is because in addition to inducing proliferation, E1A proteins trigger apoptosis. Apoptosis, which is also termed *programmed cell death*, is associated with well-defined morphological changes of the nucleus and DNA fragmentation; these characteristics distinguish it from necrosis, which is cell death characterized by extensive cytoplasmic destruction that is induced by an inhospitable environment or injury [reviewed in (488,495)].

E1A-induced apoptosis involves the induction of p53 (97,291). High-level expression of p53 can block cell cycle progression [reviewed in (354)] or it can induce apoptosis (81,292,410,511). E1A proteins stabilize p53, causing it to accumulate within the nucleus (291), and both Ad5 E1B proteins can block apoptosis induced by the E1A proteins (370). The Ad5 E1B-55 kd protein probably blocks apoptosis as a result of its ability to bind to p53 and alter its function. Either the E1B-19 kd protein or the cellular Bcl-2 proto-oncoprotein can block E1A-induced apoptosis more efficiently than the larger E1B protein (370). So far, the mechanism by which p53 induces apoptosis, as well as the mechanism by which the smaller E1B protein blocks apoptosis, remain a mystery. The E1B-19 kd and Bcl-2 proteins each can block p53-mediated transcriptional repression (411). So the correlation between the ability of these proteins to block apoptosis and to alleviate p53-mediated repression raises the possibility that p53 might mediate apoptosis, at least in part, by repressing transcription.

Adenovirus mutants that fail to produce a functional E1B-19 kd protein induce extensive degradation of host cell and viral DNA, enhanced cytopathic effect, and reduced viral yield when grown in cultured cells (359,439,491). Presumably, the inhibitory effects of premature cell death on viral propagation would be even more severe in natural infections where the viral growth cycle is slower than in cultured HeLa cells. Thus, apoptosis is an example of a cellular response to infection that has the potential to inhibit viral growth and block its spread within the infected organism. However, adenoviruses have evolved to encode a gene product that effectively blocks the cellular defense.

Adenovirus is not the only virus to carry a gene that can block the onset of apoptosis. The Epstein-Barr virus LMP-1 protein induces expression of the cellular Bcl-2 protein which blocks apoptosis (205). African swine fever virus (327), herpes simplex virus (75), and baculovirus (83) have also been reported to encode proteins likely to inhibit apoptosis.

Viral DNA Replication

As E2 gene products accumulate and the infected cell enters the S phase of the cell cycle, the stage is set for viral DNA replication [reviewed in (55,195,431)]. Ad2 or Ad5 DNA replication begins about 5 hr after infection of HeLa cells at a multiplicity of 10 plaque-forming units per cell, and it continues until the host cell dies (Fig. 5).

The inverted terminal repeats of the viral chromosome serve as replication origins. *In vivo* studies support a model in which adenovirus DNA replication takes place in two stages (271) (Fig. 8). First, synthesis is initiated at either terminus of the linear DNA and proceeds in a continuous fashion to the other end of the genome. Only one of the two DNA strands serves as a template for the synthesis, so the products of the replication are a duplex consisting of a daughter and parental strand plus a displaced single strand of DNA. In the second stage of the replication process, a complement to the displaced single strand is synthesized. The single-stranded template circularizes through annealing of its self-complementary termini, and the resulting duplex "panhandle" has the same structure as the termini of the duplex viral genome. This structure allows it to be recognized by the same initiation machinery that operates in the first stage of replication, and complementary strand synthesis generates a second completed duplex consisting of one parental and one daughter strand. Adenovirus DNA was the first eukaryotic template to be replicated *in vitro* (56), and increasingly defined cell-free systems have allowed the analysis of replication in considerable detail.

Cis-acting sequences comprising the replication origins are located within the inverted terminal repeats of the viral chromosome. Three functional domains have been defined within the terminal 51bp of the repeats. Domain A consists of the first 18 bp of the viral DNA, and it comprises a minimal origin of replication. This domain is required for replication but it supports only limited replication on its own (59,269,442,460). A cellular protein, termed *ORP-A*, binds to the first 12 bp of the genome within the minimal origin (385), but this protein does not appear essential for replication. The sequence between base pairs 9 to 18 (5'-ATAATATACC-3') is conserved among different human adenovirus serotypes, and a complex of two viral protein binds here (72,320,446): the preterminal protein and the DNA polymerase. The E2-coded terminal protein is synthesized as an 80-kd polypeptide [preterminal protein (pTP)] that is active in initiation of DNA replication (54,432), and, as discussed above, is found covalently attached to the 5' ends of the viral chromosome. It is subsequently processed by proteolysis during assembly of virions to generate a 55-kd fragment [terminal protein (TP)] that is covalently attached to the genome (57), but it appears that the entire protein with its single cleaved peptide bond remains associated with the genomic termini (398). The E2-coded polymerase is a 140-kd protein with biochemical properties distinct from other known DNA polymerases (120,133). It contains both 5' to 3' polymerase activity and 3' to 5' exonuclease activity that probably serves a proofreading function during polymerization (133). Preterminal protein and the polymerase form a heterodimeric complex in solution (120,279,433,446) so they would be expected to bind to the origin as a unit.

Domain B consists of base pairs 19 to 39, and domain C includes base pairs 40 to 51. These two elements are not absolutely required for adenovirus DNA replication, but they substantially enhance the efficiency of the initiation

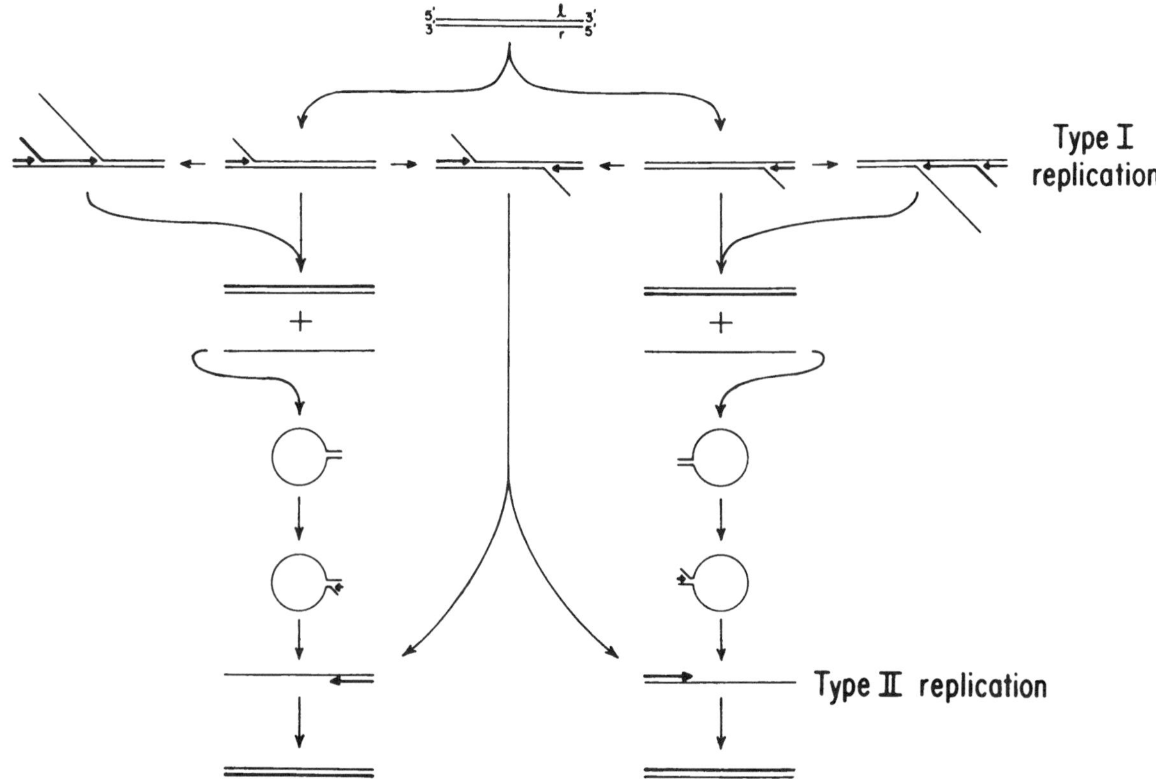

FIG. 8 A model for adenovirus DNA replication. Replication begins at either end of the duplex adenovirus DNA molecule, displacing one of the parental strands and subsequently replicating the displaced strand. Type I replication refers to replication on duplex templates, and type II replication refers to replication initiated on a single-stranded molecule. From Lechner and Kelly, ref. 271, with permission.

reaction. Cellular factors bind to these elements: nuclear factor I (NFI) binds to domain B and nuclear factor III (NFIII) binds to domain C. NFI interacts with the preterminal protein-polymerase complex, stabilizing it at the origin (41,72,320,321). The binding of NFI at domain B appears to be stimulated by the third viral protein that functions in DNA replication, the E2-coded single-stranded DNA-bending protein (82,435). The mechanism by which NFIII stimulates initiation remains uncertain; NFIII binding induces DNA bending, and this might be involved in its function (472). It is interesting to note that NFI and NFIII are also transcription factors. NFI is a member of the CTF family of transcriptional activators (244), and NFIII is also known as Oct 1, a member of the family of proteins that contain POU DNA-binding domains (345,366).

There is flexibility in the requirement for the B and C domains in the replication origin. A mutant Ad5 virus lacking domain C grows well, with sequences adjacent to the inverted terminal repeat compensating for the lack of an NFIII binding site (193). Ad4 lacks domain B, the binding site for NFI, and its replication appears to require only domain A (188,194). Possibly, the Ad4 terminal protein-polymerase complex has an intrinsically higher affinity for its binding site in the minimal origin than is the case for the complex encoded by other adenoviruses, and this could bypass the need for the stabilizing function of NFI.

A cloned adenovirus terminal sequence will support initiation of replication if it is cleaved to produce an end near the normal terminus (441). However, a natural template with a molecule of terminal protein attached at its 5' end is replicated at much higher efficiency. The activity of the cloned template can be substantially increased if it is treated with a 5' to 3' exonuclease. Thus, the production of a single-stranded 3' terminus can compensate for the absence of an attached molecule at the 5' terminus, and this suggests that the terminal protein bound to the template DNA normally plays some role in opening the terminal duplex. The preterminal protein-polymerase complex can bind in a sequence-specific fashion to the double-stranded form of the A domain of the replication origin (446), and it also can bind to single-stranded DNA (72,253). A sequential interaction of the terminal protein-polymerase complex, first with the duplex terminus and subsequently with a single strand, also could help to unwind the terminal duplex DNA.

The preterminal protein serves as a primer for DNA replication (378), preserving the integrity of the viral chromosome's terminal sequence during multiple rounds of DNA replication. The priming reaction begins with the formation of an ester bond between the β-OH of a serine residue in the preterminal protein and the α-phosphoryl group of dCMP, the first residue at the 5' end of the DNA chain (54,280). The preterminal protein-dCMP interaction requires the presence of polymerase and it is template-dependent (58,280,441), suggesting that it occurs after the preterminal protein-polymerase complex is properly positioned on the template. The 3'-OH group of the preterminal protein-dCMP complex then serves to prime synthesis of the nascent strand by the DNA polymerase.

Chain elongation requires two virus E2-coded proteins, the polymerase and single-stranded DNA-binding protein, and a cellular protein, nuclear factor II (NFII). The virus-coded DNA-binding protein is a 59-kd phosphoprotein that migrates in SDS-polyacrylamide gels with an apparent molecular weight of 72 kd (it is often referred to as the 72-kd protein). It binds tightly and cooperatively in a sequence-independent fashion to single-stranded DNA (463). The role of the DNA-binding protein in DNA replication was first revealed by analysis of an adenovirus temperature-sensitive variant carrying a mutation within the E2 gene; its ability to replicate DNA is exquisitely temperature-dependent (464). In the presence of the DNA-binding protein, which coats single-stranded replication intermediates, the polymerase is highly processive (133,285), and this is probably the basis for its requirement in chain elongation. The highly processive nature of the polymerase likely enables it to travel the entire length of the chromosome after an initiation event at the terminus. NFII copurifies with a cellular DNA topoisomerase activity (326), and mammalian topoisomerase I will substitute for it in an *in vitro* replication reaction. NFII does not significantly enhance the synthesis of nascent chains up to 9,000 nucleotides in length, so it must be needed to overcome a DNA structural problem that arises only after extensive replication.

In sum, a set of polypeptides have been identified that mediate the initiation of adenovirus DNA replication (preterminal protein, polymerase, NFI, and NFIII) and chain elongation (polymerase, DNA-binding protein, and NFII). These polypeptides, together with a template containing an adenovirus replication origin, are sufficient to reconstitute the complete viral DNA replication reaction *in vitro*. The adenovirus E4 gene also encodes one or more products required for efficient DNA replication (43,178,486), but their role in the process is probably indirect and remains obscure.

Activation of Late Gene Expression and Host Cell Shutoff

As for most DNA viruses, adenovirus late genes begin to be expressed efficiently at the onset of viral DNA replication (Fig. 5). As described earlier, the adenovirus late coding regions are organized into a single large transcription unit whose primary transcript is about 29,000 nucleotides in length (123,333). This transcript is processed by differential poly(A) site utilization and splicing to generate at least 18 distinct mRNAs (Fig. 3). These mRNAs have been grouped into five families, termed L1 to L5, based on the utilization of common poly(A) addition sites (78,333,526). Expression of this large family of late mRNAs is controlled by the major late promoter. This promoter exhibits a low level of activity early after infection, and it becomes several hundredfold more active on a per DNA molecule basis at late times (409). There appear to be at least two distinct components that contribute to activation of the major late promoter: a *cis*-acting change in the viral chromosome and induction of at least one new virus-coded *trans*-acting factor.

The time-dependent *cis*-acting modification of the adenovirus chromosome was revealed by the sequential infection of cells with two closely related adenovirus strains whose late products could be distinguished (448). Expression from the second virus to infect was initially restricted to early polypeptides, even though the first virus to reach the nucleus was actively expressing its late products. The second viral chromosome did not express late gene products until it had completed the early phase and had begun DNA replication. Thus, gene products from the first virus did not act in *trans* to initiate late expression from the second virus.

There are several possible explanations for this observation. It might be necessary to alter the constituents of the viral chromatin during the process of DNA replication to activate the major late promoter. Unreplicated viral chromatin appears to be associated with a different set of proteins than replicated chromatin (66,99). Viral DNA might remain associated with virus-coded core proteins until cellular histones bind during replication. Such an exchange could be required to activate the major late promoter. However, this proposal seems somewhat unlikely. Adenovirus DNA does not appear to be associated with histones (468), and papovaviruses exhibit a DNA replication-coupled early-to-late switch even though they package their DNA in cellular histones. Perhaps the replication process allows transcription factors to gain access to the promoter. Histones or histonelike proteins are displaced and then reassemble on DNA during replication, and this could provide an opportunity for transcription factors to compete for binding to the DNA. Alternatively, a newly arriving viral core might simply require a period of time to decondense and become available to the transcription machinery. Chromosomal domains could become accessible to the transcriptional machinery in a defined order, and the promoter responsible for most late gene expression might become accessible only after a delay. The major late promoter is active at a low level early after infection, but this might reflect early activation by a relatively minor subset of the total infecting population of viral chromosomes. It is also possible that the viral chromosome might establish a compartmentalized environment in the nucleus, and it might take a period of time to recruit cellular factors to its defined environment that are needed for late gene expression. Discrete viral centers at which replication and transcription occur have been observed in the infected nucleus by electron microscopy (302,319).

Whatever the underlying mechanism, *in vivo* footprint analysis has revealed that a transcription factor, termed USF or MLTF (52,313,397), is present, bound to its upstream site in the major late promoter after the onset of DNA replication but not before (455). This factor activates transcription when it is bound to its recognition sequence in the major late promoter. However, a point mutation which blocks binding of USF/MLTF does not prevent the normal activation of the major late promoter (373,374). So if it normally plays an important role in the late activation process, there must be redundant activators that can work in its absence or USF/MLTF might be brought to the promoter through a protein-protein interaction in the absence of its binding site. Nevertheless, the demonstration that a factor gains access to the major late promoter only after the onset of DNA replication suggests that the time-dependent *cis*-acting modification to viral chromatin likely involves changes that permit access of transcription factors to the template.

The second component that contributes to activation of the major late promoter is a virus-coded transcription factor. In addition to the upstream region with USF/MLTF, CAAT, and TATA binding motifs; the region between +85 and +120, downstream of the initiation site, contributes to activation of the promoter (274,300). This downstream domain contains binding sites for at least two factors which can cooperate with the upstream USF/MLTF element to activate transcription (314,315). One of these factors is coded by the adenovirus IVa2 gene (457). The IVa2 gene has been classified as a delayed early gene since it is activated somewhat later than the standard early genes. Its promoter also contains important downstream elements (50,70,249), and these elements include the binding site for a factor that represses IVa2 transcription (71). The mechanism by which repression is relieved is not yet known, but one would anticipate that an early gene product plays a role. One can build a model in which a cascade of events leads to activation of the major late promoter: early transcription is activated; an early gene product relieves repression of the IVa2 promoter; and the IVa2 protein binds downstream of the major late start site and contributes to its induction. This model is attractive because the sequential nature of these events can serve as a timer, and this clock, together with the requirement for DNA replication, delays the synthesis of the late mRNAs until their products are needed for virus assembly.

Accumulation of late mRNA not only requires activation of the major late promoter, but it also depends on the alleviation of premature termination by the polymerase as it traverses the 29,000-bp late unit. Premature termination occurs at two positions within the unit. At late times, 80% of the RNA chains initiated at the major late promoter terminate before elongation has proceeded 500 nucleotides (122,140,404). This termination primarily mitigates against late mRNA accumulation, since prematurely terminated transcripts with 3' ends in this region are not detected early after infection (140). The second termination event occurs early after infection when molecules of polymerase do not transcribe beyond the end of the L2 coding region (334); later the polymerase transcribes through all of the late coding regions to the end of the genome (139). The molecular basis for the termination at early but not late times remains a mystery. Among other possibilities, it could result from *cis*-acting changes in the structure of adenovirus chromatin.

When DNA replication begins and the full spectrum of late mRNAs are synthesized, the cytoplasmic accumulation of cellular mRNAs is blocked (29). Synthesis and processing of cellular transcripts continue, but the mRNAs fail to accumulate in the cytoplasm, suggesting that their transport is blocked. The block to cellular mRNA accumulation is mediated by the E1B-55 kd polypeptide (15,360) and the E4-34 kd polypeptide (178,486), which exist in a complex (393). The same E1B and E4 proteins are required for efficient cytoplasmic accumulation of viral mRNAs late after infection (16,42,178,360,486). Thus, a complex that includes two viral proteins both inhibits accumulation of cellular mRNAs and facilitates cytoplasmic accumulation of viral mRNAs. Analysis of late viral mRNA metabolism in cells infected with a mutant virus unable to synthesize the E1B-55 kd protein indicated that a block occurred immediately prior to movement of the mRNA from the nuclear compartment to the cytoplasm (275), consistent with the nuclear location of the E1B and E4 proteins (91,395). As discussed above, the E1B-55 kd protein also binds to p53, but there is no evidence that p53 plays a role in the mRNA transport function mediated by the complex formed between the E1B and E4 proteins.

The discrimination between host and viral mRNAs is not based on the identity of individual mRNAs. Cellular genes expressed from recombinant viral chromosomes accumulate in the cytoplasm late after infection as if they were viral mRNAs (147,199). Immunoelectron microscopy revealed that the E1B-E4 protein complex is localized within and surrounding intranuclear inclusions believed to represent viral transcription and replication centers (348). Together, these observations suggest a model in which the E1B-E4 complex relocalizes a cellular factor required for transport of mRNAs from their site of synthesis to the nuclear pore (348). This proposal can explain the simultaneous inhibition of host and activation of viral mRNA transport, since a transport factor would be moved from the many sites of host transcription and processing to the viral centers. So far, this model has not been validated by identification of the putative transport factor.

In addition to their facilitated transport from the nucleus, viral mRNAs are preferentially translated when they reach the cytoplasm late after infection [reviewed in (524)].

At late times, when viral mRNAs constitute about 20% of the total cytoplasmic pool (440), they are translated to the exclusion of host mRNAs (14,509). Host mRNAs are not degraded; if they are extracted from the infected cell, they can be translated with normal efficiency in a cell-free extract (447). The translational block is not dependent on the inhibition of host cell mRNA accumulation in the cytoplasm. In contrast to most host mRNAs, β-tubulin mRNA continues to accumulate in the cytoplasm, but it is not translated (316). Also, 2-aminopurine can prevent the inhibition of translation without relieving the block to accumulation of host mRNAs (229). There are several regulatory components that cooperate to facilitate selective translation of viral mRNAs late after infection.

The first regulatory component involves the cellular protein kinase R (PKR), which is activated by double-stranded RNA that accumulates within adenovirus-infected cells (301,344). After activation, it can phosphorylate eIF-2α, inactivating the initiation factor and blocking translation. Translation of host cell mRNAs is not inhibited after infection of cells deficient in PKR activity (229,343), suggesting that the kinase, and presumably inactivation of eIF-2α, are key components of the block to host cell translation. This proposal requires that viral and host translation reside in different functional compartments so that neither are inhibited by activated kinase. The adenovirus-coded VA RNAs, which will be discussed in detail below ("Viral Antagonists of α and β Interferon" section), inhibit activation of the cellular PKR kinase. These small RNAs have been shown to copurify with viral mRNAs (303) and, as a result, they might protect viral but not cellular protein synthesis, providing a functional compartmentalization.

Inactivation of eIF-4F also contributes to selective translation in adenovirus-infected cells (230). This initiation factor binds to the cap of mRNAs and facilitates scanning of the 40S ribosome from the cap to the AUG through its intrinsic helicase activity. eIF-4F is normally activated by phosphorylation, and it becomes substantially inactivated by dephosphorylation late after adenovirus infection when cellular mRNAs are not translated. The five families of mRNAs encoded by the adenovirus major late transcription unit all contain the same 200-nucleotide-long 5' noncoding region, which has been termed the *tripartite leader sequence* (31). This 5' noncoding region is important for translation of mRNAs late but not early after infection (34,289), and it substantially lacks secondary structure (106). Thus, it has been postulated (230) that tripartite leader-containing adenovirus mRNAs can continue to be translated late after infection because the 40S ribosome can scan from cap to AUG without the need for a helicase as eIF-4F activity becomes limiting. In contrast, most cellular mRNAs are no longer translated in the absence of eIF-4F because they require the helicase to permit scanning through the more extensive secondary structure at their 5' ends.

Finally, there is a selective activation of late viral protein synthesis by the 100-kd protein encoded by the L4 family of late mRNAs (196). A mutant virus expressing defective L4-100 kd protein fails to efficiently translate its late mRNAs, but it is nevertheless able to block host cell translation. The L4-100 kd protein can bind to mRNA (3), suggesting it may function at the polysome to facilitate viral translation.

Virus Assembly and Release from the Cell

The replication of viral DNA coupled with the production of large quantities of the adenovirus structural polypeptides sets the stage for virus assembly. Trimeric hexon capsomeres are rapidly assembled from monomers after their synthesis in the cytoplasm (222). Assembly of the hexon requires the participation of a second late viral protein (the L4-100 kd protein), the same protein that stimulates late viral translation. Biochemical experiments indicate that a multimeric complex of the L4-100 kd protein binds to hexon monomers (53), and genetic analyses have demonstrated that mutations in the L4-100 kd protein can block assembly of the hexon capsomere (281,347). Apparently the L4-coded protein acts as a scaffold to facilitate assembly of trimers, but the mechanism underlying the process is unknown. Penton capsomeres consisting of a pentameric penton base and trimeric fiber assemble somewhat more slowly in the cytoplasm (222). Pulse-chase experiments indicate that the penton base and fiber assemble independently, and they join to form a complete penton capsomere (222,471). After their production, hexon and penton capsomeres accumulate in the nucleus where assembly of the virion occurs.

Mutations within a variety of viral genes can interfere with the assembly process. As would be expected, alterations in structural polypeptides that comprise the hexon or penton base can prevent accumulation of mature virions as well as subassemblies such as empty capsids (92,113,392). Alterations in the polypeptide forming the fiber, polypeptide IIIa, the E2-coded DNA polymerase, or the L1-coded 52-/55-kd protein can prevent assembly of virions and lead to the accumulation of incomplete capsid-like particles (67,92,93,113,192,392). Finally, a mutant virus with a defective L3-coded protease accumulates noninfectious virionlike particles with a series of unprocessed polypeptides (479).

Studies of mutant viruses combined with analysis of the kinetics with which polypeptides are incorporated into capsids and mature virions have provided a rough outline of the adenovirus assembly process [reviewed in (357)]. Assembly appears to begin with the formation of an empty capsid (357,436) and, subsequently, a viral DNA molecule enters the capsid. The DNA-capsid recognition event is mediated by the packaging sequence, a *cis*-acting DNA element that is centered about 260 bp from the left end of the viral chromosome (163,180,197,449). Presumably one or more proteins bind at the packaging sequence and mediate the interaction between DNA and capsid, but they remain unidentified. The packaging element can function only when it is positioned near an end of the viral chromosome, raising the possibility that proteins interacting at the terminal replication origin might contribute to its function. This could explain the ability of mutations in the virus-coded DNA polymerase to interfere with assembly (92). No empty capsids accumulate in cells infected with a mutant adenovirus containing a partially defective packaging sequence, even though all of the constituents of the capsid accumulate (191,198). This suggests that capsid assembly does not occur spontaneously in the absence of an interacting DNA; rather, it probably is initiated in association with a viral DNA, an association mediated by the viral packaging sequence. Encapsidation of the chromosome is polar, beginning with the left end of the viral DNA (449), as one might anticipate given the location of the packaging sequence. The mechanism by which the viral DNA enters the capsid is controversial (92,191,481). Some experiments favor the possibility that viral DNA replication and encapsidation are coupled, i.e., that the viral DNA enters the capsid as it is replicated; other work suggests that ongoing viral DNA replication is not required and that replication and encapsidation occur in separate nuclear compartments. The L1-coded 52-/55-kd protein, which is comprised of two differentially phosphorylated forms of a 48-kd precursor polypeptide, facilitates the encapsidation process (191,192). It is thought to act as a scaffold protein since it is present in all known assembly intermediates but not in mature virions. The L3-coded proteinase, a cysteine proteinase that requires DNA and a fragment of another viral polypeptide as cofactors (299,450,483), functions late in the assembly process. It cleaves at least four virion constituents, generating the mature VI, VII, and VIII and terminal protein moieties. These cleavages stabilize the structure of the particle and render it infectious.

The assembly process has long been known to be substantially inhibited when the amount of arginine becomes limiting, even though all of the major structural proteins are made (39,363,387). Several of the capsid polypeptides are substantially under-acetylated when arginine is limiting (13), and the polypeptides apparently fail to assemble in the absence of the modification.

There appear to be at least two active viral systems which facilitate the escape and spread of progeny virus, and both involve disruption of intermediate filaments, which are components of the cytoskeleton. Vimentin is cleaved very rapidly after infection, reportedly both by a reaction initiated by the adsorption process that does not require viral gene expression (27) and in response to the E1B-19 kd polypeptide (489,490). As a result, the extended vimentin system collapses into the perinuclear region (98,523). Late in the infectious cycle, the viral proteinase cleaves the cellular cytokeratin K18 (73). The cleavage event occurs at amino acid 74 of the cytokeratin, creating a "headless" protein that is not able to polymerize and form filaments; rather, it accumulates in cytoplasmic clumps. A normal intermediate filament system helps to maintain the mechanical integrity of cells, and perturbations to the network would be expected to make the infected cell more susceptible to lysis and release of progeny virus.

INTERACTIONS WITH THE HOST

Adenovirus can maintain a long-term association with its human host, persisting for years after an initial infection. Adenovirus DNA has been identified by *in situ* hybridization in 1 in 10^4 to 10^6 peripheral lymphocytes of apparently normal individuals (152), and lymphocytes could certainly be the source of virus in the original isolations of adenoviruses from adenoid tissue. Thus, adenovirus probably persists in lymphocytes in the human. The virus encodes three gene products that should facilitate persistence by antagonizing antiviral responses of the host: E1A proteins and VA RNAs which inhibit the cellular response to α and β interferons; and E3-coded products that protect infected cells from lysis by cytotoxic T lymphocytes and tumor necrosis factor. It is intriguing to note that these protective viral products are expressed from the three adenovirus genes most likely to be active in lymphocytes. The E1A promoter and the polymerase III-transcribed VA RNAs are constitutively expressed in all cells that have been tested, and the E3 promoter is unique among adenovirus control elements in that it contains several binding sites for the lymphoid-specific NFκB transcription factor, facilitating constitutive expression in lymphoid cells (497). Additionally, it appears that these protective viral genes might be further induced in response to the production of interleukin 6, a cytokine that is induced at the site of adenovirus infection in experimental animals (153), as a host protective measure. Transcription of the E2 gene is strongly activated by interleukin 6 through the cellular transcription factor NF-IL6 (423,424); the resulting E2-coded DNA binding protein, in turn, can activate the E1A and E3 promoters (60). Thus, the production of a cytokine by the host as a protective response likely induces a viral countermeasure.

Viral Antagonists of α and β Interferons

The adenovirus E1A protein and virus-associated RNAs both afford protection from α and β interferons. E1A proteins inhibit the antiviral affect of interferon (6) by blocking the activation of interferon response genes (176,375). This appears to result from inhibition of the DNA-binding activity of ISGF3, a cellular transcription factor that is activated when α or β interferon binds to its receptor. This E1A function maps to its CR1 domain. Although the CR1 domain is involved with binding of pRB family members, p300 and YY1, the mapping does not precisely correlate with the binding requirements for any known E1A target. So the immediate cellular target of E1A that mediates the effect on ISGF3 binding has not been identified.

The small, abundant VA RNAs [reviewed in (304)] were named when their viral origin was still uncertain (376). Different adenovirus serotypes have been reported to encode one or two VA RNAs; Ad2 and Ad5 encode two species, termed VA RNA_I and VA RNA_{II}. The RNAs are each about 160 nucleotides in length, they are GC-rich, they adopt stable secondary structures that are important for their function, and they are transcribed by RNA polymerase III (138,174,364,487). VA RNA synthesis begins during the early phase of the infectious cycle and dramatically accelerates during the late phase. VA RNA_I accumulates to ~10^8 molecules per infected HeLa cell, roughly the abundance of ribosomes, and VA RNA_{II} reaches ~10^7 molecules per cell.

The first hint to the function of VA RNAs came from analysis of a mutant Ad5 virus in which the VA RNAI gene was deleted (447). The virus grew poorly, and its defect was traced to inefficient protein synthesis during the late phase of infection. Additional work identified a defect in polypeptide chain initiation (377,401) resulting from phosphorylation of the eukaryotic initiation factor 2 (eIF-2) (377,400).

eIF-2, which includes three subunits (α, β, and τ), is central to one of the best-studied pathways of translational control. Early in the initiation process, eIF-2 binds to GTP and the initiator tRNA to form a ternary complex, which then interacts with the 40S ribosomal subunit. Subsequent steps in the initiation process involving additional factors result in the binding of mRNA and the 60S ribosomal subunit. As the round of initiation is completed, eIF-2 leaves the ribosome, lacking the initiator tRNA and complexed with GDP instead of GTP. For eIF-2•GDP to participate in another round of initiation, its GDP must be replaced with GTP in a reaction catalyzed by eIF-2B (also termed guanosine nucleotide exchange factor). The exchange reaction is inhibited by phosphorylation of the α-subunit of eIF-2. eIF-2α(P) forms a tight complex with eIF-2B, preventing it from cycling and catalyzing the exchange reaction. As a result, initiation is brought to a halt when about one-third to one-half of eIF-2 is phosphorylated, trapping all available eIF-2B. The degree of eIF-2α phosphorylation is greatly increased during the late phase of infection with the VA RNA_I-minus mutant (344), and protein synthesis in extracts of these infected cells can be restored to normal levels by addition of eIF-2 or eIF-2B (377,400). Further, the mutant phenotype is abrogated in cells that contain mutant forms of the eIF-2α subunit (96) or lack the eIF-2α kinase (344,400). Thus, the reduction in late translation results from the phosphorylation and inactivation of eIF-2.

Phosphorylation of eIF-2 is mediated by a kinase known as protein kinase R [(PKR) (also known as the dsRNA-activated inhibitor kinase, the P1 kinase, p68 kinase, P1/eIF-2α kinase, or interferon-inducible kinase)]. Synthesis of a latent or inactive form of PKR is induced by interferon, and the latent enzyme is then activated by double-stranded RNA. Activation of PKR is accomplished by autophosphorylation, which appears to require that two molecules of the enzyme interact with a molecule of double-stranded RNA (262) that is >30 bp in length (298) Short, double-stranded RNA molecules (<30 bp) can block activation, presumably because only one molecule of kinase can bind, and the autophosphorylation step cannot proceed. Activated PKR can phosphorylate the eIF-2α subunit; this leads to sequestration of eIF-2B and the cessation of translation. VA RNA_I can bind to PKR (252,308) and block its activation (257,343). The current structural model for VA RNA_I (Fig. 9A) suggests that it consists of two extended duplex domains (apical and terminal) with a central domain that is comprised of several stem-loop structures and includes two complementary tetranucleotides that are conserved in VA RNAs encoded by a variety of adenovirus serotypes (294). Mutational analysis indicates that the central domain is critical for VA RNA_I function (143,150,307,308,352) and the apical stem is also involved in the interaction of VA RNA_I with PKR (149,150, 307,309).

In an adenovirus-infected cell, PKR is probably activated by double-stranded RNA produced as a result of the symmetrical transcription of the viral chromosome (301,344). Activation of endogenous latent kinase is presumably responsible for the inhibition of translation observed in cells infected with VA RNA_I-minus mutants (304). The level of latent PKR is induced by a factor of 5 to 10 in response to interferon. Thus, one might anticipate that VA RNA_I could antagonize the antiviral effect of interferon by blocking activation of PKR, and this has proven to be the case (257). Whereas α-interferon inhibits growth of VA RNA_I-minus virus, it has no effect on wild-type Ad5 (Fig. 9B). In sum, VA RNA_I, and presumably VA RNA_{II} as well, blocks activation of latent PKR, and this function would seem critical for optimal viral growth in an infected host capable of responding to viral infection by the production of interferon.

Two additional viruses encode RNAs with the potential to control PKR: Epstein-Barr virus encodes EBER RNAs, which can partially substitute for VA RNAs in adenovirus-infected cells (37); and human immunodeficiency virus encodes RNAs carrying TAR elements, which can block activation of DAI *in vitro* (175).

Viral Antagonists of CTLs and TNF-α

Upon intranasal inoculation of cotton rats (*Sigmodon hispidus*) or mice with Ad5, viral gene expression is detected in the epithelial cells of the bronchi and bronchioles of the lung and nasal mucosa. The pulmonary histopathology is very similar to that observed in the human [reviewed in (155)], and disease proceeds in two stages (365). The

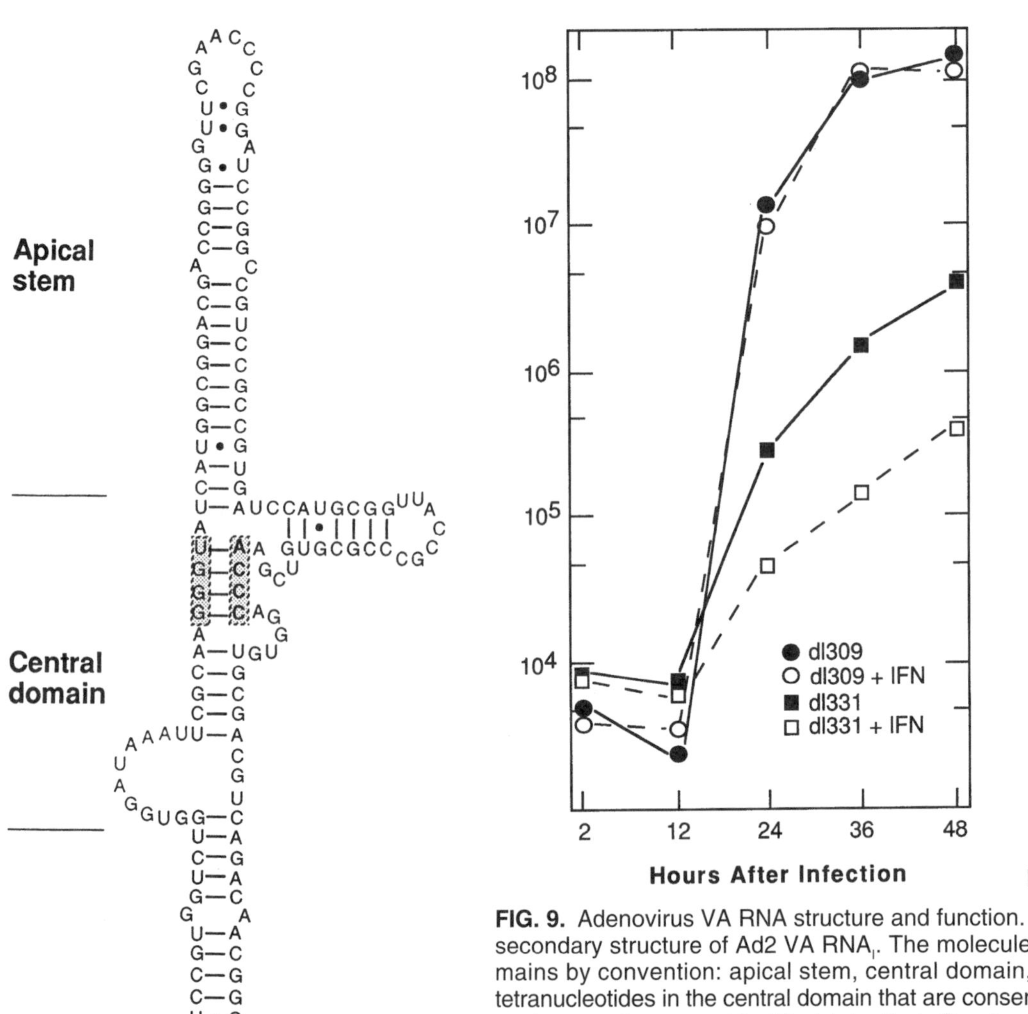

FIG. 9. Adenovirus VA RNA structure and function. **A:** Current model for the secondary structure of Ad2 VA RNA_I. The molecule is divided into three domains by convention: apical stem, central domain, terminal stem. The two tetranucleotides in the central domain that are conserved among different adenovirus serotypes are identified (*shading*). Courtesy of M. B. Mathews, Cold Spring Harbor Laboratory. **B:** VA RNA_I antagonizes the antiviral effects of α-interferon. Cells were infected with a virus that contained a wild-type VA RNA_I gene (*dl309, circles*) or a mutant with a deletion blocking expression of VA RNA_I (*dl331, squares*), and either treated (*open symbols*) or not treated (*solid symbols*) with α-interferon. The adenovirus with the wild-type VA RNA_I gene was resistant to α-interferon while the mutant virus was inhibited. From Kitajewski et al. (257), with permission.

earlier stage consists of a mild-to-moderate damage to bronchiolar epithelial cells and a diffuse cellular infiltration of peribronchiolar and alveolar regions. At this time, tumor necrosis factor (TNF-α), interleukin 1, and interleukin 6 appear in infected lung tissue (153). The later stage of the disease consists almost entirely of an infiltration of lymphocytes. A normal early phase ensues but little late response is observed when athymic nude mice are infected (153), consistent with the interpretation that the late phase results from infiltration of virus-specific cytotoxic T lymphocytes (CTLs). Thus, the two stages of pathogenesis appear to result from an initial nonspecific host response to viral infection that includes synthesis of cytokines followed by an immunologically specific CTL response. Adenovirus E3-coded proteins, which are not needed for efficient viral growth in cultured cells, combat at least a portion of these host antiviral measures [reviewed in (159,501)], inhibiting cytolysis of infected cells by CTLs or TNF-α.

Recognition and lysis of a virus-infected cell by CTLs requires that viral peptide antigens be displayed on the infected cell's surface in a complex with a major histocompatibility complex (MHC) class I antigen. Ad2 or Ad5-infected cells expressing the E3-coded 19-kd glycoprotein (E3-gp19 kd) are considerably less sensitive to CTL-mediated lysis than cells infected with mutant viruses unable to express the E3 protein (7,49). The E3-gp19 kd protein is a transmembrane protein of 142 amino acids, after cleavage and removal of its N-terminal signal sequence. It resides in the membrane of the endoplasmic reticulum where it is held through the action of a retention signal that has been mapped to its C-terminal domain (349). The lumenal domain of the E3-gp19 kd protein binds directly to the peptide-binding domain of MHC class I antigens (49,268), and this interaction is believed to cause retention of class I antigen in the endoplasmic reticulum. A reduced level of class I antigen on the cell surface should protect against premature lysis of the infected cell by CTLs. Such protection can be inferred from experiments with cotton rats. Pulmonary infection with a mutant virus, unable to express the E3-gp19 kd protein, induced a markedly increased late phase inflammatory response (155). Thus, the E3 protein must normally ameliorate the late phase response by protecting against cytolysis by CTLs.

TNF-α is a cytokine that is secreted by activated macrophages and lymphocytes. It exerts cytotoxic or cytostatic effects on many tumor cells, and it can induce the lysis of cells infected with some viruses. Whereas uninfected cells or cells infected with wild-type adenovirus are not affected, cells infected with mutant adenoviruses lacking the E3 coding region are lysed by TNF-α (160). The CR1 domain of the E1A proteins is responsible for inducing susceptibility to TNF-α (108). This is the same E1A domain that has been reported to antagonize the activation of interferon-responsive genes, and the mapping does not precisely correlate with the binding requirements for any known E1A target. Either the E3-14.7 kd protein (160) or the complex formed between the E3-14.5 kd and E3-10.4 kd proteins can prevent cytolysis by TNF-α (161). It appears that the E3-14.7 kd protein inhibits the release of arachidonic acid following the stimulation of phospholipase A2 by TNF-α (528).

TNF-α can induce apoptosis, and cytolysis is the consequence of cell death. The E1B-19 kd protein, as discussed above, can block the onset of apoptosis, and like the E3-coded proteins, it can protect against apoptosis and the induction of cytolysis by TNF-α. So adenovirus encodes three proteins that can protect against the antiviral effects of TNF-α in human cells: E1B-19 kd protein, the E3-14.7 kd protein, and the E3-14.5 kd/E3-10.4 kd protein complex. Why the virus carries three apparently redundant systems is a matter for speculation. Perhaps the different viral products function with different degrees of efficiency in the variety of cell types where the virus replicates in an infected host. Alternatively, the three products might each address antiviral pathways at different levels, and the antagonism of TNF-α could represent a point of convergence. If this is true, each of the proteins might encode additional, unrelated functions, besides their TNF-α related activity.

This appears to be the case for the E3-14.5 kd/E3-10.4 kd complex which, in addition to its effect on TNF-α sensitivity, reduces the level of epidermal growth factor receptor on the surface of infected cells (51,452) The E3-10.4 kd protein contains a small domain with sequence similarity to epidermal growth factor, so it is possible that the reduction might reflect activation of the receptor with subsequent internalization. If this proves to be the case, the E3 protein might help to activate quiescent cells through the epidermal growth factor response pathway. However, it remains possible that the E3 products simply reduce the level of the receptor and block the ability of the infected cell to respond to the factor. The viral strategy behind such an activity is not yet apparent.

The adenoviruses are not alone in their ability to counteract CTLs and TNF-α. Shope fibroma virus encodes a soluble TNF-binding protein that competes and blocks binding of the cytokine to its normal receptor (421), Epstein-Barr virus produces a homologue of interleukin 10 that blocks the synthesis of cytokines such as TNF-α (226), and human cytomegalovirus blocks cell surface accumulation of MHC class I antigen (48).

ONCOGENESIS

All human adenoviruses that have been tested are able to oncogenically transform cultured rodent cells. The transformants undergo morphological changes, and they tend to grow as dense, multilayered colonies. They can display various phenotypic hallmarks of oncogenic transformation, including growth in reduced serum, anchorage-independent growth, and growth to higher densities than normal. Only a subset of adenovirus serotypes can directly induce the formation of tumors within rats and hamsters (Table 1). Group A adenoviruses (e.g., Ad12, Ad18) are highly oncogenic, producing tumors in most animals within 4 months; group B viruses (e.g., Ad3, Ad7) are weakly oncogenic, inducing tumors within 4 to 18 months; at least one group D virus (Ad9) is oncogenic, efficiently inducing mammary tumors within 3 to 5 months; and group C and E viruses are not known to be tumorigenic.

Transformation

Three main lines of evidence initially demonstrated that the E1A and E1B genes mediate transformation [reviewed in (454)]. First, virus-transformed cells always contain and express these genes; second, transfection of cultured cells with cloned E1A and E1B genes leads to transformation; and, third, mutant viruses with alterations to these genes can be defective for transformation. As discussed above, the E1A and E1B proteins can manipulate cellular growth regulation. In a lytically infected human cell, E1A proteins induce quiescent cells to enter the S phase of the cell cycle and E1B proteins block the onset of apoptosis, creating an environment optimally conducive to viral replication. In the context of a rodent cell, these same events contribute to oncogenic transformation.

One can construct a detailed model for how the E1A and E1B-55 kd oncoproteins deregulate the $G_0/G_1/S$ cell cycle checkpoint through their interactions with pRB family members, p300 and p53. However, although E1A proteins have been shown to drive cells past the G_2/M checkpoint into mitosis, the mechanism by which this checkpoint is antagonized remains a mystery. Mutations in the E1A proteins that interfere with the binding of p300 or pRB family members can block the ability of E1A proteins to induce mitosis (224). Perhaps these cellular proteins, either in their normal complexes or as a result of their interaction with E1A proteins, influence both the $G_0/G_1/S$ and G_2/M cell cycle checkpoints. Alternatively, the E1A proteins interact with additional, as yet unidentified proteins that control the G_2/M checkpoint.

Tumorigenesis

Rat cells transformed by the highly oncogenic Ad12 generally produce tumors in newborn syngeneic rats, while most cells transformed by the nononcogenic adenoviruses, such as Ad2 or Ad5, are not tumorigenic. However, some Ad2-transformed rat cell lines can induce tumors in immunosuppressed rats (144), suggesting that cells transformed by nononcogenic adenoviruses are rejected by the host's immune system. Newborn hamsters, which have not yet developed a thymus-dependent immune response, are susceptible to tumor production by Ad2-transformed cells. While thymectomized hamsters remain susceptible, normal animals become resistant to tumor challenge at the age of about 3 weeks when their thymic response has developed (87), suggesting that the thymus-dependent cellular immune system plays a key role in the rejection.

The thymus-dependent CTL response leads to lysis of transformed cells. As mentioned earlier, CTLs target tumor cells by recognizing foreign antigens displayed on their cell surface in the context of major histocompatibility complex (MHC) class I antigens. In the case of Ad5 transformed cells, the E1A protein itself serves as a cell surface target (28), and several specific E1A-encoded epitopes have been identified that serve as targets of class I-restricted CTLs (250,388).

The Ad12 E1A proteins inhibit the CTL response that blocks tumor formation by cells transformed with nononcogenic adenoviruses (36). Ad12-transformed cells contain reduced quantities of MHC class I antigen on their surface (402), and, after introduction of an MHC class I gene that cannot be down-regulated by the Ad12 E1A proteins, Ad12 transformants are unable to induce tumors (443). Thus, the Ad12 E1A proteins repress expression of the class I antigen, and this prevents recognition by CTLs. The repression occurs at the level of transcription, and it appears that the activity of at least two factors which interact with the class I gene control region are altered: an activator (NFκB-like) is decreased (306,337) and a repressor (retinoic acid receptor-like) is increased (265) in cells transformed with Ad12 as compared to Ad5 .

Not all Ad12 transformants have proven resistant to CTL lysis (311), and some Ad12 transformed lines have become more tumorigenic (i.e., fewer cells can induce a tumor) when they were modified to express higher levels of class I antigen (422). Thus, although we can conclude that Ad12 E1A proteins reduce MHC class I levels in some transformed cells, facilitating escape from class I-restricted CTL-mediated destruction, the correlation is not reliable. This indicates that there are additional immune mechanisms that govern adenovirus tumor formation.

Many Ad2 transformed cells are markedly less efficient than Ad12 transformants in the formation of tumors in athymic nude mice which lack class I-restricted CTLs. Nude mice contain natural killer (NK) cells, a class of immunologically nonspecific cytotoxic lymphocytes. Ad2 transformants are more susceptible to NK cell-mediated lysis than their Ad12 counterparts (86,371). Analysis of cells transformed by mutant viruses, recombinant Ad5/Ad12 viruses, or transfected genes demonstrated that susceptibility or resistance to NK killing is governed by the E1A gene (88,89,396), and addition of the Ad5 E1A gene to a highly tumorigenic sarcoma cell line induced susceptibility to lysis by NK cells, blocking its ability to form tumors (476).

Non-E1A-mediated mechanisms can also influence tumorigenesis. As discussed above, the E3-19 kd glycoprotein can influence class I MHC levels. So, the E3 gene might also contribute to the differential CTL sensitivity of Ad12 versus Ad2 or Ad5 transformants. While the E3 gene is often present in Ad12 transformants, it is generally not present in Ad2 or Ad5 transformed cells. Although there are exceptions, the majority of rodent cells transformed with Ad2 or Ad5 contain only portions of the viral genome; only the E1A and E1B genes are consistently present. In a similar vein, additional E3-coded proteins, if present, might afford protection from TNF-α. The notion that additional genes play auxiliary roles in tumorigenesis is consistent with the observation that substitution of the Ad12 E1A and/or E1B genes into the corresponding Ad5 domain did not equip the resulting hybrid viruses to induce tumors in newborn rats or hamsters (35,396).

In contrast with group A adenoviruses such as Ad12, which induces tumors at the site of injection, Ad9, a group D virus, can induce mammary tumors. Subcutaneous injection of newborn Wistar-Furth rats produces mammary tumors in females but not males (9). The tumors are predominantly benign fibroadenomas, but phyllodeslike tumors and malignant sarcomas are also produced; all of the tumors are estrogen-dependent (8,240). When hybrid viruses were constructed between Ad9 and a closely related group D virus that fails to induce mammary tumors, the Ad9 E4 gene was found to be required for mammary tumorigenesis. Presumably, the E4 gene product acts in concert with the Ad9 E1A and E1B oncoproteins to induce tumor formation. The Ad9 E4 gene can oncogenically transform rat cells in the absence of the E1A and E1B genes, in contrast to the Ad5 E4 gene which does not function as an independent oncogene (241).

The 12.5-kd oncoprotein is encoded by ORF 1 of the Ad9 E4 gene (239). The mechanisms by which Ad9 targets mammary fibroblasts and by which its E4 protein subverts cellular growth control are not yet known.

Oncogenesis in Rodents But Not Humans

Adenoviruses transform rodent cells much more efficiently than human cells, and the oncogenic adenoviruses induce tumors in rodents while extensive screens have failed to correlate the presence of adenoviruses with oncogenesis in humans. Initially it seemed that the more efficient oncogenic transformation of rodent cells might result from the fact that they are generally less permissive for growth of human adenoviruses, and, as a result, they could more readily survive an infection and yield transformants. However, cloned E1A and E1B genes exhibit the same host cell preferences for transformation as does whole virus, arguing that differences in permissivity to viral growth are not a primary determining factor in transformation efficiency.

Perhaps rodent and human cells both contain regulatory proteins that actively mitigate against transformation or tumorigenesis but are not targeted by the adenovirus oncoproteins. E1A plus E1B proteins might allow limited cell cycle progression, and changes in the expression of additional genes might then permit long-term, robust growth of the transformed cell and endow it with tumorigenic properties. Conceivably, an epigenetic event such as methylation might activate or repress expression of these putative regulatory genes more rapidly in rodent than human cells; this could explain the difference in the efficiency with which the two cell types are transformed and become tumorigenic. If such alterations are necessary during the transformation process, they probably occur rapidly and with high efficiency in rodent cells; abortive transformants which might arise in the absence of the putative secondary events are rare.

It is also possible that the viral oncoproteins interact differently with their targets in rodent as compared to human cells. Differences in the affinity of protein-protein interactions could influence how completely the viral oncoproteins subvert normal function of their targets. Alternatively, to achieve transformation or a tumorigenic phenotype, it might be necessary to alter the function of additional regulatory proteins in human cells as compared to rodent cells. Although the rodent would be expected to contain genes homologous to each of the human genes that regulate cell growth, it is possible that some respond somewhat differently to upstream signaling proteins or differ in details of their interactions with downstream targets. Subtle differences could influence the ability of auxiliary regulatory proteins to control cellular growth in the presence of adenovirus oncoproteins that have blocked the function of a subset of regulators.

PERSPECTIVES

Much is known about the adenovirus growth cycle; yet, much remains a mystery, including the identity of the viral receptor, the structure of the viral core both within the virion and as viral DNA is transcribed, the nature of the recognition signal that mediates packaging of the viral DNA into a capsid, and the basis for host restrictions and tissue tropisms in infected organisms. The tools are all in place, and there is no doubt that these and many other questions will be resolved in the near future.

Adenoviruses have brought much more to the table than a system of viral replication—they have also contributed to a molecular understanding of many fundamental cellular processes. For example, studies of adenoviruses have yielded important insights to the complex mechanisms by which cells control their growth. Early adenovirus gene products have helped to identify many important cell cycle regulatory proteins, and they continue to serve as probes with which to manipulate these cellular factors and reveal their normal functions.

Adenoviruses no longer serve simply as systems for study. Rather, they are now poised to serve as vectors for gene therapy [reviewed in (324)]. One can readily substitute a therapeutic gene for the E1A and E1B genes and grow the resulting recombinant adenovirus to high titer in 293 cells which complement the E1A/E1B defect. Animal studies have demonstrated that these recombinants can effectively and efficiently deliver genes to target organs *in vivo*. Cystic fibrosis is an especially attractive target disease for genetic intervention with recombinant adenoviruses since the lung epithelial cells, where a functional cystic fibrosis transmembrane conductance regulator protein is needed, are a natural target for adenovirus infection. If the virus can be tamed so that it delivers and expresses therapeutic genes without attendant pathology [reviewed in (119)], we will undoubtedly find ourselves "sleeping with the enemy" in the near future.

REFERENCES

1. Commission on Acute Respiratory Disease. Experimental transmission of minor respiratory illness to human volunteers by filter-passing agents. Demonstration of two types of illness characterized by long and short incubation periods and different clinical features. *J Clin Invest* 1947;26:957–973.
2. Abdel-Hafiz HA, Chen CY, Marcell T, Kroll DJ, Hoeffler JP. Structural determinants outside of the leucine zipper influence the interactions of CREB and ATF-2: interaction of CREB with ATF-2 blocks E1a-ATF-2 complex formation. *Oncogene* 1993;8:1161–1174.
3. Adam SA, Dreyfuss G. Adenovirus proteins associated with mRNA and hnRNA in infected Hela cells. *J Virol* 1987;61:3276–3283.
4. Aiello L, Guilfoyle R, Huebner K, Weinmann R. Adenovirus 5 DNA sequences present and RNA sequences transcribed in transformed human embryo kidney cells. *Virology* 1979;94:460–469.
5. Anderson CW, Young ME, Flint SJ. Characterization of the adenovirus 2 virion protein, mu. *Virology* 1989;172:506–512.
6. Anderson K, Fennie E. Adenovirus early region 1A modulation of interferon antiviral activity. *J Virol* 1987;61:787–795.

7. Andersson M, McMichael A, Peterson PA. Reduced allorecognition of adenovirus 2 infected cells. *J Immunol* 1987;138:3960–3966.
8. Ankerst J, Jonsson N. Adenovirus type 9-induced tumorigenesis in the rat mammary gland related to sex hormonal state. *J Natl Cancer Inst* 1989;81:294–298.
9. Ankerst J, Jonsson N, Kjellen L, Norrby E, Sjogren HO. Induction of mammary fibroadenomas in rats by adenovirus type 9. *Int J Cancer* 1974;13:286–290.
10. Arany Z, Sellers WR, Livingston DM, Eckner R. E1A-associated p300 and CREB-associated CBP belong to a conserved family of coactivators. *Cell* 1994;77:799–800.
11. Arrand JR, Roberts RJ. The nucleotide sequences at the termini of adenovirus 2 DNA. *J Mol Biol* 1973;128:577–594.
12. Athappilly FK, Murali R, Rux JJ, Cai Z, Burnett RM. The refined crystal structure of hexon, the major coat protein of adenovirus type 2, at 2.9Å resolution. *J Mol Biol* 1994;242:430–455.
13. Auborn KJ, Rouse H. The mechanism of the arginine dependent function in adenovirus type-2 infected cells: late viral polypeptides are under-acetylated. *Virus Res* 1984;1:615–624.
14. Babich A, Feldman CT, Nevins JR, Darnell JE, Weinberger C. Effect of adenovirus on metabolism of specific host mRNAs: transport control and specific translational discrimination. *Mol Cell Biol* 1983;3:1212–1221.
15. Babiss LE, Ginsberg HS. Adenovirus type 5 early region 1b gene product is required for efficient shutoff of host protein synthesis. *J Virol* 1984;50:202–212.
16. Babiss LE, Ginsberg HS, Darnell JE Jr. Adenovirus E1B proteins are required for accumulation of late viral mRNA and for effects on cellular mRNA translation and transport. *Mol Cell Biol* 1985;5:2552–2558.
17. Bachenheimer S, Darnell JE. Adenovirus 2 mRNA is transcribed as part of a high molecular weight precursor RNA. *Proc Natl Acad Sci USA* 1975;72:4445–4449.
18. Bagchi S, Raychaudhuri P, Nevins JR. Adenovirus E1A proteins can dissociate heteromeric complexes involving the E2F transcription factor: a novel mechanism for E1A trans-activation. *Cell* 1990;62:659–669.
19. Bagchi S, Weinmann R, Raychaudhuri P. The retinoblastoma protein copurifies with E2F-I, an E1A-regulated inhibitor of the transcription factor E2F. *Cell* 1991;65:1063–1072.
20. Bai M, Harfe B, Freimuth P. Mutations that alter an Arg-Gly-Asp (RGD) sequence in the adenovirus type 2 penton base protein abolish its cell-rounding activity and delay virus reproduction in flat cells. *J Virol* 1993;67:5198–5205.
21. Bandara LR, Adamczewski JP, Hunt T, La Thangue NB. Cyclin A and the retinoblastoma gene product complex with a common transcription factor. *Nature* 1991;352:249–251.
22. Bandara LR, Buck VM, Zamanian M, Johnston LH, La Thangue NB. Functional interaction between DP-1 and E2F-1 in the cell cycle-regulating transcription factor DRTF1/E2F. *EMBO J* 1993;12:4317–4324.
23. Bandara LR, LaThangue NB. Adenovirus E1A prevents the retinoblastoma gene product from complexing with a cellular transcription factor. *Nature* 1991;351:494–497.
24. Barbeau D, Charbonneau R, Whalen SG, Bayley ST, Branton PE. Functional interactions within adenovirus E1A protein complexes. *Oncogene* 1994;9:359–373.
25. Bargonetti J, Reynisdottir I, Friedman PN, Prives C. Site-specific binding of wild-type p53 to cellular DNA is inhibited by SV40 T antigen and mutant p53. *Genes Dev* 1992;6:1886–1898.
26. Baum SG. Adenoviridae. In: Mandel GL, Douglas RG, Bennet JE, eds. *Principles and practice of infectious diseases*, 2nd ed. New York: John Wiley and Sons, 1984:1353–1361.
27. Belin M, Boulanger P. Processing of vimentin occurs during the early stages of adenovirus infection. *J Virol* 1987;61:2559–2566.
28. Bellgrau D, Walker TA, Cook JL. Recognition of adenovirus E1A gene products on immortalized cell surfaces by cytotoxic T lymphocytes. *J Virol* 1988;62:1513–1519.
29. Beltz GA, Flint SJ. Inhibition of Hela cell protein synthesis during adenovirus infection. *J Mol Biol* 1979;131:353–373.
30. Benbrook DM, Jones NC. Heterodimer formation between CREB and Jun proteins. *Oncogene* 1990;5:295–302.
31. Berget SM, Moore C, Sharp PA. Spliced segments at the 5' terminus of adenovirus 2 late mRNA. *Proc Natl Acad Sci USA* 1977;74:3171–3175.
32. Berk AJ, Lee F, Harrison T, Williams J, Sharp PA. Pre-early adenovirus 5 gene product regulates synthesis of early viral messenger RNAs. *Cell* 1979;17:935–944.
33. Berk AJ, Sharp PA. Sizing and mapping of early adenovirus mRNAs by gel electrophoresis of S1 endonuclease digested hybrids. *Cell* 1977;12:721–732.
34. Berkner KE, Sharp PA. Effect of tripartite leader on synthesis of a non-viral protein in a adenovirus 5' recombinant. *Nucl Acids Res* 1985;13:841–857.
35. Bernards R, deLeeuw MGW, Vaessen M, Houweling A, van der Eb AJ. Oncogenicity by adenovirus is not determined by the transforming region only. *J Virol* 1984;50:847–853.
36. Bernards R, Schrier PI, Houweling A, Bos JV, van der Eb AJ, Zijlstra M, Melief CJM. Tumorigenicity of cells transformed by adenovirus type 12 by evasion of T-cell immunity. *Nature* 1983;305:776–779.
37. Bhat RA, Thimmappaya B. Construction and analysis of additional adenovirus substitution mutants confirm the complementation of VAI RNA function by two small RNAs encoded by Epstein-Barr virus. *J Virol* 1985;56:750–756.
38. Bodnar JW, Hanson PI, Polvino-Bodnar M, Zempsky W, Ward DC. The terminal regions of adenovirus and minute virus of mice DNAs are preferentially associated with the nuclear matrix in infected cells. *J Virol* 1989;63:4344–4353.
39. Bonifas V, Schlesinger RW. Nutritional requirement for plaque production by adenovirus. *Fed Proc Fed Am Soc Exp Biol* 1959;18:500.
40. Borelli E, Hen R, Chambon P. Adenovirus-2 E1A products repress enhancer-induced stimulation of transcription. *Nature* 1984;312:608–612.
41. Bosher J, Robinson EC, Hay RT. Interactions between the adenovirus type 2 DNA polymerase and the DNA binding domain of nuclear factor I. *New Biol* 1990;2:1083–1090.
42. Bridge E, Ketner G. Interaction of adenoviral E4 and E1B products in late gene expression. *Virology* 1990;174:345–353.
43. Bridge E, Medghalchi S, Ubol S, Leesong M, Ketner G. Adenovirus early region 4 and viral DNA synthesis. *Virology* 1993;193:794–801.
44. Broker TR. Animal virus RNA processing. In: Apirion D, ed. *Processing of RNA*. Boca Raton, Florida: CRC press, 1984:181–212.
45. Broker TR, Chow L. In: Tooze J, ed. *DNA tumor viruses: molecular biology of tumor viruses*, 2nd ed., revised ed. Cold Spring Harbor, New York: Cold Spring Harbor Laboratory, 1981:408–409.
46. Brown DT, Westphal M, Burlingham BT, Winterhof U, Doerfler W. Structure and composition of adenovirus type 2 core. *J Virol* 1975;16:366–387.
47. Brown M, Weber J. Virion core-like organization of intranuclear adenovirus chromatin late in infection. *J Virol* 1980;107:306–310.
48. Browne H, Smith G, Beck S, Minson T. A complex between the MHC class I homologue encoded by human cytomegalovirus and beta 2 microglobulin. *Nature* 1990;347:770–772.
49. Burget H-G, Kvist S. The E3/19K protein of adenovirus type 2 binds to the domains of histocompatibility antigens required for CTL recognition. *EMBO J* 1987;6:2019–2026.
50. Carcamo J, Maldonado E, Cortes P, Ahn M-H, Ilho-Ha, Kasai Y, Flint SJ, Reinberg D. A TATA-like sequence located downstream of the transcription site is required for expression of an RNA polymerase II transcribed gene. *Genes Dev* 1990;4:1611–1622.
51. Carlin CR, Tollefson AE, Brady HA, Hoffman BL, Wold WSM. Epidermal growth factor receptor is down-regulated by a 10,400 MW protein encoded by the E3 region of adenovirus. *Cell* 1989;57:135–144.
52. Carthew RW, Chodosh LA, Sharp PA. An RNA polymerase II transcription factor binds to an upstream element in the adenovirus major late promoter. *Cell* 1985;43:439–448.
53. Cepko CL, Sharp PA. Assembly of adenovirus major capsid protein is mediated by a non-virion proten. *Cell* 1982;3:407–414.
54. Challberg MD, Desiderio SV, Kelly TJ Jr. Adenovirus DNA replication *in vitro*: characterization of a protein covalently linked to nascent DNA strands. *Proc Natl Acad Sci USA* 1980;77:5105–5109.
55. Challberg MD, Kelly JK. Animal virus DNA replication. *Annu Rev Biochem* 1989;58:671–717.
56. Challberg MD, Kelly TJ Jr. Adenovirus DNA replication *in vitro*. *Proc Natl Acad Sci USA* 1979;76:272–277.
57. Challberg MD, Kelly TJ Jr. Processing of the adenovirus terminal protein. *J Virol* 1981;38:272–277.
58. Challberg MD, Ostrove JM, Kelly TJ Jr. Initiation of adenovirus DNA replication: detection of covalent complexes between nucleotide and the 80-kilodalton terminal protein. *J Virol* 1982;41:265–270.
59. Challberg MD, Rawlins DR. Template requirements for the initiation of adenovirus DNA replication. *Proc Natl Acad Sci USA* 1984;81:100–104.
60. Chang L-S, Shenk T. The adenovirus DNA binding protein can stim-

ulate the rate of transcription directed by adenovirus and adeno-associated virus promoters. *J Virol* 1990;64:2103–2109.

61. Chardonnet Y, Dales S. Early events in the interaction of adenoviruses with HeLa cells. I Penetration of type 5 and intracellular release of the DNA genome. *Virology* 1970;40:462–477.
62. Chardonnet Y, Dales S. Early events in the interaction of adenoviruses with HeLa cells. II. Comparative observations on the penetration of types 1, 5, 7 and 12. *Virology* 1970;40:478–485.
63. Chardonnet Y, Dales S. Early events in the interaction of adenoviruses with HeLa cells. III. Relationship between an ATP-ase activity in nuclear envelopes and transfer of core material. A hypothesis. *Virology* 1972;48:342–359.
64. Chatterjee PK, Bruner M, Flint SJ, Harter ML. DNA-binding properties of an adenovirus 289R E1A protein. *EMBO J* 1988;7:835–841.
65. Chatterjee PK, Vayda ME, Flint SJ. Adenoviral protein VII packages intracellular viral DNA throughout the early phase of infection. *EMBO J* 1986;5:1633–1644.
66. Chatterjee PK, Vayda ME, Flint SJ. Identification of proteins and protein domains that contact DNA within adenovirus nucleoprotein cores by ultraviolet light crosslinking of oligonucleotides 32p-labelled *in vivo*. *J Mol Biol* 1986;188:23–37.
67. Chee-Sheung CC, Ginsberg HS. Characterization of a temperature-sensitive fiber mutant of type 5 adenovirus and effect of the mutation on virion assembly. *J Virol* 1982;42:932–950.
68. Chellappan S, Kraus VB, Kroger B, Munger K, Howley PM, Phelps WC, Nevins JR. Adenovirus E1A, simian virus 40 tumor antigen, and human papillomavirus E7 protein share the capacity to disrupt the interaction between transcription factor E2F and the retinoblastoma gene product. *Proc Natl Acad Sci USA* 1992;89:4549–4553.
69. Chellappan SP, Hiebert S, Mudryj M, Horowitz JM, Nevins JR. The E2F transcription factor is a cellular target for the RB protein. *Cell* 1991;65:1053–1061.
70. Chen H, Flint SJ. Mutational analysis of the adenovirus 2 IV_{a2}. *J Biol Chem* 1992;267:25457–25465.
71. Chen H, Vinnakota R, Flint SJ. Intragenic activating and repressing elements control transcription from the adenovirus IV_{a2} initiator. *Mol Cell Biol* 1994;14:676–685.
72. Chen M, Mermod N, Horwitz MS. Protein-protein interactions between adenovirus DNA polymerase and nuclear factor-I mediate formation of the DNA replication preinitiation complex. *J Biol Chem* 1990;265: 18634–18642.
73. Chen PH, Ornelles DA, Shenk T. The adenovirus L3 23-kilodalton proteinase cleaves the amino-terminal head domain from cytokeratin 18 and disrupts the cytokeratin network of HeLa cells. *J Virol* 1993;67: 3507–3514.
74. Chittenden T, Livingston DM, Kaelin WG Jr. The T/E1A-binding domain of the retinoblastoma product can interact selectively with a sequence-specific DNA-binding protein. *Cell* 1991;65:1073–1082.
75. Chou J, Roizman B. The $_{A1}$34.5 gene of herpes simplex virus 1 precludes neuroblastoma cells from triggering total shutoff of protein synthesis characteristic of programed cell death in neuronal cells. *Proc Natl Acad Sci USA* 1992;89:3266–3270.
76. Chow LT, Broker TR, Lewis JB. Complex splicing patterns of RNAs from the early regions of adenovirus 2. *J Mol Biol* 1979;134:265–303.
77. Chow LT, Gelinas RE, Broker TR, Roberts RJ. An amazing sequence arrangement at the 5' ends of adenovirus 2 messenger RNA. *Cell* 1977;12:1–8.
78. Chow LT, Roberts JM, Lewis JB, Broker TR. A map of cytoplasmic RNA transcripts from lytic adenovirus type 2, determined by electron microscopy of RNA:DNA hybrids. *Cell* 1977;11:819–836.
79. Chrivia JC, Kwok RPS, Lamb N, Hagiwara M, Montminy MR, Goodman RH. Phosphorylated CREB binds specifically to the nuclear protein CBP. *Nature* 1993;365:855–859.
80. Chroboczek J, Bieber F, Jacrot B. The sequence of the genome of adenovirus type 5 and its comparison with the genome of adenovirus type 2. *Virology* 1992;186:280–285.
81. Clarke AR, Purdie CA, Harrison DJ, Morris RG, Bird CC, Hooper ML, Wyllie AH. Thymocyte apoptosis induced by p53-dependent and -independent pathways. *Nature* 1993;362:849–852.
82. Cleat PH, Hay RT. Co-operative interactions between NFI and the adenovirus DNA binding protein at the adenovirus origin of replication. *EMBO J* 1989;8:1841–1848.
83. Clem RJ, Fechheimer M, Miller LK. Prevention of apoptosis by a baculovirus gene during infection of insect cells. *Science* 1991;254: 1388–1390.
84. Cobrinik D, Whyte P, Peeper DS, Jacks T, Weinberg R. Cell cycle-specific association of E2F with the p130 E1A-binding protein. *Genes Dev* 1993;7:2392–2404.
85. Colby W, Shenk T. Adenovirus type 5 virions can be assembled *in vivo* in the absence of polypeptide IX. *J Virol* 1981;39:977–980.
86. Cook JL, Hibbs JB, Lewis AM Jr. DNA virus transformed hamster cell-host effector cell interactions: level of resistance to cytolysis is correlated with tumorigenicity. *Int J Cancer* 1982;30:795–803.
87. Cook JL, Lewis AM. Host response to adenovirus 2-transformed hamster embryo cells. *Cancer Res* 1979;39:1455–1461.
88. Cook JL, May DL, Lewis AM Jr., Walker TA. Adenovirus E1A gene induction of susceptibility to lysis by natural killer cells and activated macrophages in infected rodent cells. *J Virol* 1987;61:3510–3520.
89. Cook JL, Walker TA, Lewis AM, Ruley HE, Graham FL, Pilder SH. Expression of the adenovirus E1A oncogene during cell transformation is sufficient to induce susceptibility to lysis by host inflammatory cells. *Proc Natl Acad Sci USA* 1986;83:6965–6969.
90. Corden J, Engelking HM, Pearson GD. Chromatin-like organization of the adenovirus chromosome. *Proc Natl Acad Sci USA* 1976;73: 401–404.
91. Cutt JR, Shenk T, Hearing P. Analysis of adenovirus early region 4-encoded polypeptides synthesized in productively infected cells. *J Virol* 1987;61:543–552.
92. D'Halluin J-C, Milleville M, Martin GR, Boulanger P. Morphogenesis of human adenovirus type 2 studied with fiber and penton base-defective temperature-sensitive mutants. *J Virol* 1980;33:88–99.
93. D'Halluin JC, Milleville M, Boulanger PA, Martin GR. Temperature-sensitive mutant of adenovirus type 2 blocked in virion assembly: accumulation of light intermediate particles. *J Virol* 1978;26:344–356.
94. Dales S, Chardonnet Y. Early events in the interaction of adenovirus with HeLa cells. IV. Association with microtubules and the nuclear pore complex during vectorial movement in the inoculum. *Virology* 1973;56:465–483.
95. Dalton S. Cell cycle regulation of the human cdc2 gene. *EMBO J* 1992;11:1797–1804.
96. Davies MV, Furtado M, Hershey JWB, Thimmappaya B, Kaufman RWJ. Complementation of adenovirus virus-associated RNA I gene deletion by expression of a mutant eukaryotic translation initiation factor. *Proc Natl Acad Sci USA* 1989;86:9163–9167.
97. Debbas M, White E. Wild-type p53 mediates apoptosis by E1A, which is inhibited by E1B. *Genes Dev* 1993;7:546–554.

97a. DeCaprio JA, Ludlow JW, Figge J, et al. SV40 large tumor antigen forms a complex with the product of retinoblastoma gene. *Cell* 1988;4: 275–283.

98. Defer C, Belin MT, Caillet-Boudin ML, Boulanger P. Human adenovirus-host cell interactions: comparative study with members of subgroups B and C. *J Virol* 1990;64:3661–3673.
99. Dery CV, Toth M, Brown M, Horvath J, Allaire S, Weber JM. The structure of adenovirus chromatin in infected cells. *J Gen Virol* 1985;66:2671–2684.
100. Desiderio SV, Kelly TJ Jr. Structure of the linkage between adenovirus DNA and the 55,000 molecular weight terminal protein. *J Mol Biol* 1981;145:319–337.
101. Devaux C, Caillet-Boudin ML, Jacrot B, Boulanger P. Crystallization, enzymatic cleavage and the polarity of the adenovirus type 2 fiber. *Virology* 1987;161:121–128.
102. Devaux C, Timmins PA, Berthet-Colominas C. Structural studies of adenovirus type 2 by neutron and X-ray scattering. *J Mol Biol* 1983;167:119–128.
103. Devoto SH, Mudryj M, Pines J, Hunter T, Nevins JR. A cyclin A-specific protein kinase complex possesses sequence-specific DNA binding activity: p33cdk2 is a component of the E2F-cyclin A complex. *Cell* 1992;68:167–176.
104. Diller L, Kassel J, Nelson CE, Gryka MA, Litwak G, Gebhardt M, Bressac B, Ozturk M, Baker SJ, Vogelstein B, Friend SH. p53 functions as a cell cycle control protein in osteosarcomas. *Mol Cell Biol* 1990;10:5772–5781.
105. Dingle J, Langmuir AD. Epidemiology of acute respiratory disease in military recruits. *Am Rev Respir Dis* 1968;97:1–65.
106. Dolph PJ, Huang J, Schneider RJ. Translation by the adenovirus tripartite leader: elements which determine independence from cap-binding protein complex. *J Virol* 1990;64:2669–2677.
107. Donehower LA, Harvey M, Slagle BL, McArthur MJ, Montgomery CA, Butel JS, Bradley A. Mice deficient for p53 are developmentally normal but susceptible to spontaneous tumors. *Nature* 1992;356: 215–221.

108. Duerksen-Hughes PJ, Hermiston TW, Wold WSM, Gooding LR. The amino-terminal portion of CD1 of the adenovirus E1 proteins is required to induce susceptibility to tumor necrosis factor cytolysis in adenovirus-infected mouse cells. *J Virol* 1991;65:1236–1244.
109. Dyson N, Guida P, McCall C, Harlow E. Adenovirus E1A makes two distinct contacts with the retinoblastoma protein. *J Virol* 1992;66: 4606–4611.
110. Dyson N, Guida P, Munger K, Harlow E. Homologous sequences in adenovirus E1A and human papillomavirus E7 proteins mediate interaction with the same set of cellular proteins. *J Virol* 1992;66:6893–6902.
111. Dyson N, Howley PM, Munger K, Harlow E. The human papilloma virus-16 E7 oncoprotein is able to bind to the retinoblastoma gene product. *Science* 1989;243:934–937.
112. Eckner R, Ewen ME, Newsome D, Gerdes M, DeCaprio JA, Bentley-Lawrence JB, Livingston D. Molecular cloning and functional analysis of the adenovirus E1A associated 300kD protein (p300) reveals a protein with properties of a transcriptional adaptor. *Genes Dev* 1994;8:869–884.
113. Edvardsson B, Ustacelebi S, Williams J, Philipson L. Assembly intermediates among adenovirus type 5 temperature-sensitive mutants. *J Virol* 1978;25:641–651.
114. Egan C, Jelsma TN, Howe JA, Bayley ST, Furguson B, Branton PE. Mapping of cellular protein binding sites on the products of early region 1A of human adenovirus type 5. *Mol Cell Biol* 1988;8:3955–3959.
115. El-Deiry WS, Tokino T, Velculescu VE, Levy DB, Parsons R, Trent JM, Lin D, Mercer WE, Kinzler KW, Vogelstein B. WAF1, a potential mediator of p53 tumor suppression. *Cell* 1993;75:817–825.
116. Enders JF, Bell JA, Dingle JH et al. "Adenoviruses": group name proposed for new respiratory-tract viruses. *Science* 1956;124:119–120.
117. Engel DA, Hardy S, Shenk T. Cyclic AMP acts in synergy with E1A protein to activate transcription of the adenovirus early genes E4 and E1A. *Genes Dev* 1988;2:1517–1528.
118. Engel DA, Muller U, Gedrich RW, Eubanks JS, Shenk T. Induction of c-fos mRNA and AP-1 DNA-binding activity by cAMP in cooperation with either the adenovirus 243- or the adenovirus 289-amino acid E1A protein. *Proc Natl Acad Sci USA* 1991;88:3957–3961.
119. Engelhardt JF, Ye X, Doranx B, Wilson JM. Ablation of E2A in recombinant adenoviruses improves transgene persistence and decreases inflammatory response in mouse liver. *Proc Natl Acad Sci USA* 1994;91:6196–6200.
120. Enomoto T, Lichy JH, Ikeda J, Hurwitz J. Adenovirus DNA replication *in vitro*: purification of the terminal protein in a functional form. *Proc Natl Acad Sci USA* 1981;78:6779–6783.
121. Epstein LH, Young CS. Adenovirus homologous recombination does not require expression of the immediate-early E1A gene. *J Virol* 1991;65:4475–4479.
122. Evans R, Weber J, Ziff E, Darnell JE. Premature termination during adenovirus transcription. *Nature* 1979;278:367.
123. Evans RM, Fraser N, Ziff E, Weber J, Wilson M, Darnell JE. The initiation sites for RNA transcription in Ad2 DNA. *Cell* 1977;12:733–739.
124. Everitt E, Lutter L, Philipson L. Structural proteins of adenoviruses. XII. Location and neighbor relationship among proteins of adenovirus type 2 as revealed by enzymatic iodination, immunoprecipitation and chemical cross-linking. *Virology* 1975;67:197–208.
125. Everitt E, Sundquist B, Pettersson U, Philipson L. Structural proteins of adenoviruses. X. Isolation and topography of low molecular weight antigens from the virion of adenovirus type 2. *Virology* 1973;62: 130–147.
126. Ewen M, Faha B, Harlow E, Livingston D. Interaction of p107 with cyclin A independent of complex formation with viral oncoproteins. *Science* 1992;255:85–87.
127. Ewen M, Xing Y, Lawrence JB, Livingston DM. Molecular cloning, chromosmal mapping, and expression of the cDNA for p107, a retinoblastoma gene product-related protein. *Cell* 1991;66:1155–1164.
128. Faha B, Ewen M, Tsai L-H, Livingston D, Harlow E. Interaction between human cyclin A and adenovirus E1A-associated p107 protein. *Science* 1992;255:87–90.
129. Faha B, Harlow E, Lees E. The adenovirus E1A-associated kinase consists of cyclin E-p33^{cdk2} and cyclin A-p33^{cdk2}. *J Virol* 1993;67: 2456–2465.
130. Falck-Pedersen E, Logan J, Shenk T, Darnell JE Jr. Transcription termination within the E1A gene of adenovirus induced by insertion of the mouse beta-major globin terminator element. *Cell* 1985;40:897–905.
131. Farmer G, Bargonetti J, Zhu H, Friedman P, Prywes R, Prives C. Wild-type p53 activates transcription *in vitro*. *Nature* 1992;358:83–86.
132. Ferguson B, Krippl B, Andrisani O, Jones N, Westphal H. E1A 13S and 12S mRNA products made in *Escherichia coli* both function as nucleus-localised transcription activators but do not directly bind DNA. *Mol Cell Biol* 1985;5:2653–2661.
133. Field J, Gronostajski RM, Hurwitz J. Properties of the adenovirus DNA polymerase. *J Biol Chem* 1984;259:9487–9495.
134. FitzGerald DJP, Padmanabhan R, Pastan I, Willingham MC. Adenovirus-induced release of epidermal growth factor and pseudomonas toxin into the cytosol of KB cells during receptor-mediated endocytosis. *Cell* 1983;32:607–617.
135. Flemington EK, Speck SH, Kaelin WG. E2F-1 mediated transactivation is inhibited by complex formation with the retinoblastoma susceptibility gene product. *Proc Natl Acad Sci USA* 1993;90:6914–6918.
136. Flewett TH, Bryden AS, Davies H, Morris CA. Virus particles in gastroenteritis? Epidemic viral enteritis in a long-stay children's ward. *Lancet* 1975;1:4–5.
137. Flint SJ, Berget SM, Sharp PA. Characterization of single-stranded viral DNA sequences present during replication of adenovirus types 2 and 5. *Cell* 1976;9:559–571.
138. Fowlkes DM, Shenk T. Transcriptional control regions of the adenovirus VAI RNA gene. *Cell* 1980;22:405–413.
139. Fraser NW, Nevins JR, Ziff E, Darnell JE. The major late adenovirus type-2 transcription unit: termination is downstream from the last poly(A) site. *J Mol Biol* 1979;129:643–646.
140. Fraser NW, Sehgal P, Darnell JE. Multiple discrete sites for premature RNA chain termination late in adenovirus-2 infection: enhancement by 5,6-dichloro-1-β-D-ribofuranosylbenzimidazole. *Proc Natl Acad Sci USA* 1979;76:2571–2575.
141. Fredman JN, Engler JA. Adenovirus precursor to terminal protein interacts with the nuclear matrix *in vivo* and *in vitro*. *J Virol* 1993;67:3384–3395.
142. Furcinitti PS, van Oostrum J, Burnett RM. Adenovirus polypeptide IX revealed as capsid cement by difference images from electron microscopy and crystallography. *EMBO J* 1989;8:3563–3570.
143. Furtado MR, Subramanian S, Bhat RA, Fowlkes DM, Safer B, Thimmappaya B. Functional dissection of adenovirus VAI RNA. *J Virol* 1989;63:3423–3434.
144. Gallimore PH. Tumor production in immunosuppressed rats with cells transformed *in vitro* by Ad2. *J Gen Virol* 1972;62:99–102.
145. Garon CF, Berry KW, Hierholzer JC, Rose JA. Mapping of base sequence heterologies between genomes from different adenovirus serotypes. *Virology* 1973;54:414–426.
146. Garon CF, Berry KW, Rose JA. A unique form of terminal redundancy in adenovirus DNA molecules. *Proc Natl Acad Sci USA* 1972;69:2391–2395.
147. Gaynor RB, Hillman D, Berk AJ. Adenovirus early region 1A protein activates transcription of a nonviral gene introduced into mammalian cells by infection or transfection. *Proc Natl Acad Sci USA* 1984;81:1193–1197.
148. Gedrich RW, Bayley ST, Engel DA. Induction of AP-1 DNA-binding activity and c-fos mRNA by the adenovirus 243R E1A protein and cyclic AMP requires domains necessary for transformation. *J Virol* 1992;66:5849–5859.
149. Ghadge GD, Malhotra P, Furtado MR, Dhar R, Thimmappaya B. *In vitro* analysis of virus-associated RNA I (VAI RNA): inhibition of the double-stranded RNA activated protein kinase PKR by VAI RNA mutants correlates with the *in vivo* phenotype and the structural integrity of the central domain. *J Virol* 1994;68:4137–4151.
150. Ghadge GD, Swaminathan S, Katze MG, Thimmappaya B. Binding of adenovirus VAI RNA to the interferon-induced 68-kDa protein kinase correlates with function. *Proc Natl Acad Sci USA* 1991;88:7140–7144.
151. Ginsberg HS, Gold E, Jordan WS Jr., Katz S, Badger GF, Dingle JH. Relation of the new respiratory agents to acute respiratory diseases. *Am J Public Health* 1955;45:915–922.
152. Ginsberg HS, Lundholm-Beauchamp U, Prince G. Adenovirus as a model of disease. In: Russell W, Almond J, ed. *Molecular basis of virus disease*. New York: Cambridge University Press, 1987:245–258.
153. Ginsberg HS, Moldawer LL, Sehgal PB, Redington M, Kilian PL, Chanock RM, Prince GA. A mouse model for investigating the molecular pathogenesis of adenovirus pneumonia. *Proc Natl Acad Sci USA* 1991;88:1651–1655.
154. Ginsberg HS, Pereira HG, Valentine RC, Wilcox WC. A proposed terminology for the adenovirus antigens and virion morphological subunits. *Virology* 1966;28:782–783.

155. Ginsberg HS, Prince G. Gene functions directing adenovirus pathogenesis. In: Notkins A, Oldstone M, eds. *Concepts in viral pathogenesis III*. New York: Springer-Verlag, 1989:275–281.
156. Ginsberg HS, Young CSH. The genetics of adenoviruses. In: Fraenkel-Conrat H, Wagner RR, eds. *Comprehensive virology*, vol. 9. New York: Plenum Press, 1977.
157. Giordano A, Lee JH, Scheppler JA, Herrmann C, Harlow E, Deuschle U, Beach D, Franza BR Jr. Cell cycle regulation of histone H1 kinase activity associated with the adenoviral protein E1A. *Science* 1991;253: 1271–1275.
158. Girling R, Partridge JF, Bandara LR, Burden N, Totty NF, Hsuan JJ, LaThangue NB. A new component of the transcription factor DRTF1/E2F. *Nature* 1993;362:83–87.
159. Gooding LR. Virus proteins that counteract host immune defenses. *Cell* 1992;71:5–7.
160. Gooding LR, Elmore LW, Tollefson AE, Brady HA, Wold WSM. A 14,700 MW protein from the E3 region of adenovirus inhibits cytolysis by tumor necrosis factor. *Cell* 1988;53:341–346.
161. Gooding LR, Ranheim TS, Tollefson AE, Aquino L, Duerksen-Hughes P, Horton TM, Wold WSM. The 10,400- and 14,500-dalton proteins encoded by region E3 of adenovirus function together to protect many but not all mouse cell lines against lysis by tumor necrosis factor. *J Virol* 1991;65:4114–4123.
162. Goodrich DW, Wang NP, Qian Y-W, Lee EY-HP, Lee WH. The retinoblastoma gene product regulates progression through the G1 phase of the cell-cycle. *Cell* 1991;67:293–302.
163. Grable M, Hearing P. cis and trans requirements for the selective packaging of adenovirus type 5 DNA. *J Virol* 1992;66:723–731.
164. Graham FL, Smiley J, Russell WC, Nairu R. Characteristics of a human cell line transformed by DNA from human adenovirus type 5. *J Gen Virol* 1977;36:59–72.
165. Grand RJ, Lecane PS, Roberts S, Grant ML, Lane DP, Young LS, Dawson CW, Gallimore PH. Overexpression of p53 and c-myc in human fetal cells transformed with adenovirus early region 1. *Virology* 1993;193:579–591.
166. Greber UF, Willetts M, Webster P, Helenius A. Stepwise dismantling of adenovirus 2 during entry into cells. *Cell* 1993;75:477–486.
167. Green M. Oncogenic viruses. *Ann Rev Biochem* 1970;39:701–756.
168. Green M, Daesch GE. Biochemical studies on adenovirus multiplication. I. Kinetics of nucleic acid and protein synthesis in suspension cultures. *Virology* 1961;13:169–176.
169. Green M, Piña M. Biochemical studies on adenovirus multiplication. IV. Isolation, purification, and chemical analysis of adenovirus. *Virology* 1963;20:199–207.
170. Green M, Wold WSM, Brackmann KH et al. Human adenovirus transforming genes: group relationships, integration, expression in transformed cells and analysis of human cancers and tonsils. In: Essex M, Todaro G, zurHausen H, eds. *Seventh Cold Spring Harbor Conference on Cell Proliferation Viruses in Naturally Occurring Tumors*. Cold Spring Harbor, New York: Cold Spring Harbor Laboratory, 1980:373–397.
171. Green MR, Treismann R, Maniatis T. Transcriptional activation of cloned human β-globin genes by viral immediate-early gene products. *Cell* 1983;35:137–148.
172. Green NM, Wrigley NG, Russel WC, Martin SR, McLachlan AD. Evidence for a repeating β-sheet structure in the adenovirus fibre. *EMBO J* 1983;8:1357–1365.
173. Gruber WC, Russell DJ, Tibbetts C. Fiber gene and genomic origin of human adenovirus type 4. *Virology* 1993;196:603–611.
174. Guilfoyle R, Weinmann R. The control region for adenovirus VA RNA transcription. *Proc Natl Acad Sci USA* 1981;78:3378–3382.
175. Gunnery S, Rice AP, Robertson HD, Mathews MB. Tat-responsive region RNA of human immunodeficiency virus 1 can prevent activation of the double-stranded RNA-activated protein kinase. *Proc Natl Acad Sci USA* 1990;87:8687–8691.
176. Gutch MJ, Reich NC. Repression of the interferon signal transduction pathway by the adenovirus E1A oncogene. *Proc Natl Acad Sci USA* 1991;88:7913–7917.
177. Hagmeyer BM, Konig H, Herr I, Offringa R, Zantema A, van der Eb A, Herrlich P, Angel P. Adenovirus E1A negatively and positively modulates transcription of AP-1 dependent genes by dimer-specific regulation of the DNA binding and transactivation activities of Jun. *EMBO J* 1993;12:3559–3572.
178. Halbert DN, Cutt JR, Shenk T. Adenovirus early region 4 encodes functions required for efficient DNA replication, late gene expression, and host cell shutoff. *J Virol* 1985;56:250–257.
179. Hamel PA, Gill RM, Phillips RA, Gallie BL. Regions controlling hyperphosphorylation and conformation of the retinoblastoma gene product are independent of domains required for transcriptional repression. *Oncogene* 1992;7:693–701.
180. Hammarskjold ML, Winberg G. Encapsidation of adenovirus 16 DNA is directed by a small DNA sequence at the left end of the genome. *Cell* 1980;20:787–795.
181. Handa H, Kingston RE, Sharp PA. Inhibition of adenovirus early region IV transcription *in vitro* by a purified DNA binding protein. *Nature* 1983;302:545–547.
182. Hannon GJ, Demetrick D, Beach D. Isolation of the Rb-related p130 through its interaction with CDK2 and cyclins. *Genes Dev* 1993;7: 2378–2391.
183. Hardy S, Engel D, Shenk T. An adenovirus early region 4 gene product is required for induction of the infection-specific form of cellular E2F activity. *Genes Dev* 1989;3:1062–1074.
184. Hardy S, Shenk T. Adenosine 3',5'-cyclic monophosphate response element and adenoviral control regions activated by E1A bind the same factor. *Proc Natl Acad Sci USA* 1988;85:4171–4175.
185. Harlow E, Franza BR, Schley C. Monoclonal antibodies specific for adenovirus early region 1A proteins: extensive heterogeneity in early region 1A products. *J Virol* 1985;55:533–546.
186. Harlow E, Whyte P, Franza BR, Schley C. Association of adenovirus early region 1A proteins with cellular polypeptides. *Mol Cell Biol* 1986;6:1579–1589.
187. Harper JW, Adami GR, Wei N, Keyomarsi K, Elledge SJ. The p21 Cdk-interacting protein Cip1 is a potent inhibitor of G1 cyclin-dependent kinases. *Cell* 1993;75:805–816.
188. Harris MP, Hay RT. DNA sequences required for the initiation of adenovirus type 4 DNA replication *in vitro*. *J Mol Biol* 1988;201:57–67.
189. Hartley JW, Rowe WP. A new mouse virus apparently related to the adenovirus group. *Virology* 1960;11:645–652.
190. Hartwell L. Defects in a cell cycle checkpoint may be responsible for the genomic instability of cancer cells. *Cell* 1992;71:543–546.
191. Hasson TB, Ornelles DA, Shenk T. Adenovirus L1 52- and 55-kilodalton proteins are present within assembling virions and colocalize with nuclear structures distinct from replication centers. *J Virol* 1992;66:6133–6142.
192. Hasson TB, Soloway PD, Ornelles DA, Doerfler W, Shenk T. Adenovirus L1 52- and 55-kilodalton proteins are required for assembly of virions. *J Virol* 1989;63:3612–3621.
193. Hatfield L, Hearing P. The NFIII/OCT-1 binding site stimulates adenovirus DNA replication *in vivo* and is functionally redundant with adjacent sequences. *J Virol* 1993;67:3931–3939.
194. Hay RT. The origin of adenovirus DNA replication: minimal DNA sequence requirement *in vivo*. *EMBO J* 1985;4:421–426.
195. Hay RT, Russell WC. Recognition mechanisms in the synthesis of animal virus DNA. *Biochem* J 1989;258:3–16.
196. Hayes BW, Telling GC, Myat MM, Williams JF, Flint SJ. The adenovirus L4 100 kilodalton protein is necessary for efficient translation of viral late mRNA species. *J Virol* 1990;64:2732–2742.
197. Hearing P, Samulski R, Wishart W, Shenk T. Identification of a repeated sequence element required for efficient encapsidation of the adenovirus type 5 chromosome. *J Virol* 1987;61:2555–2558.
198. Hearing P, Shenk T. The adenovirus type 5 E1A transcriptional control region contains a duplicated enhancer element. *Cell* 1983;33: 695–703.
199. Hearing P, Shenk T. Sequence-independent autoregulation of the adenovirus type 5 E1A transcription unit. *Mol Cell Biol* 1985;5:413–421.
200. Hearing P, Shenk T. Adenovirus 5 E1A enhancer contains two distinct domains: one is specific for E1A and the other modulates expression of all early units in *cis*. *Cell* 1986;45:229–236.
201. Helin K, Harlow E, Fattaey A. Inhibition of E2F-1 transactivation by direct binding of the retinoblastoma protein. *Mol Cell Biol* 1993;13: 6501–6508.
202. Helin K, Lees JA, Vidal M, Dyson N, Harlow E, Fattaey A. A cDNA encoding a pRB-binding protein with properties of the transcription factor E2F. *Cell* 1992;70:337–350.
203. Helin K, Wu C-L, Fattaey AR, Lees JA, Dynlacht BD, Ngwu C, Harlow E. Heterodimerization of the transcription factors E2F-1 and DP-1 leads to cooperative trans-activation. *Genes Dev* 1993;7:1850–1861.
204. Hen R, Borelli E, Chambon P. Repression of the immunoglobulin heavy

chain enhancer by the adenovirus-2 E1A products. *Science* 1985;230:1391–1394.
205. Henderson S, Rowe M, Gregory C, Croom-Carter D, Wang F, Longnecker R, Kieff E, Rickinson A. Induction of *bcl*-2 expression by Epstein-Barr virus latent membrane protein 1 protects infected B cells from programmed cell death. *Cell* 1991;65:1107–1115.
206. Hennache B, Boulanger P. Biochemical study of KB cell receptor for adenovirus. *Biochem J* 1977;166:237–247.
207. Hermann CH, Su L-K, Harlow E. Adenovirus E1A is associated with a serine/threonine protein kinase. *J Virol* 1991;65:5848–5859.
208. Hiebert SW, Chellappan SP, Horowitz JM, Nevins JR. The interaction of RB with E2F coincides with an inhibition of the transcriptional activity of E2F. *Genes Dev* 1992;6:177–185.
209. Hierholzer JC. Further subgrouping of the human adenoviruses by differential hemagglutination. *J Infect Dis* 1973;128:541–550.
210. Hierholzer JC, Torrence AE, Wright P. Generalized viral illness caused by an intermediate strain of adenovirus (21/H21+35). *J Infect Dis* 1980;141:281–288.
211. Hierholzer JC, Wigand R, Anderson LJ, Adrian T, Gold JWM. Adenoviruses from patients with AIDS: a plethora of serotypes and a description of five new serotypes of subgenus D (Types 43-47). *J Infect Dis* 1988;158:804–813.
212. Hilleman MR, Werner JH. Recovery of new agents from patients with acute respiratory illness. *Proc Soc Exp Biol Med* 1954;85:183–188.
213. Hinds PW, Mittnacht S, Dulic V, Arnold A, Reed SI, Weinberg RA. Regulation of retinoblastoma protein functions by ectopic expression of human cyclins. *Cell* 1992;70:993–1006.
214. Hoeffler WK, Kovelman R, Roeder RG. Activation of transcription factor IIIC by the adenovirus E1A protein. *Cell* 1988:907–920.
215. Hoeffler WK, Roeder RG. Enhancement of RNA polymerase III transcription by the E1A gene product of adenovirus. *Cell* 1985;41:955–963.
216. Hogle JM, Chow M, Filman DJ. Three-dimensional structure of poliovirus at 2.9Å resolution. *Science* 1985;229:1358–1365.
217. Horikoshi N, Maguire K, Kralli A, Maldonado E, Reinberg D, Weinmann R. Direct interaction between adenovirus E1A protein and the TATA box binding transcription factor IID. *Proc Natl Acad Sci USA* 1991;88:5124–5128.
218. Horikoshi N, Usheva A, Chen J, Levine A.J. L, Weinmann R, Shenk T. Two domains on p53 interact with the TATA-binding protein and the adenovirus 13S E1A protein disrupts the association, relieving p53-mediated transcriptional repression. *Mol Cell Biol* 1995;15:227–234.
219. Horne RW, Bonner S, Waterson AP, Wildy P. The icosahedral form of an adenovirus. *J Mol Biol* 1959;1:84–86.
220. Horwitz MS. Intermediates in the replication of type 2 adenovirus DNA. *Virology* 1971;8:675–683.
221. Horwitz MS, Maizel JV Jr., Scharff MD. Molecular weight of adenovirus type 2 hexon polypeptide. *J Virol* 1970;6:569–571.
222. Horwitz MS, Scharff MD, Maizel JV. Synthesis and assembly of adenovirus 2. I. Polypeptide synthesis, assembly of capsomers and morphogenesis of the virion. *Virology* 1969;39:682–694.
223. Hosakawa K, Sung MT. Isolation and characterization of an extremely basic protein from adenovirus type 5. *J Virol* 1976;17:924–934.
224. Howe JA, Bayley ST. Effects of Ad5 E1A mutant viruses on the cell cycle in relation to the binding of cellular proteins including the retinoblastoma protein and cyclin A. *Virology* 1992;186:15–24.
225. Howe JA, Mymryk JS, Egan C, Branton PE, Bayley ST. Retinoblastoma growth suppressor and a 300kDa protein appear to regulate cellular DNA synthesis. *Proc Natl Acad Sci USA* 1990;87:5883–5887.
226. Hsu DH, de Waal-Malefyt R, Fiorentino DF, Dang MN, Vieira P, de Vries J, Spits H, Mosmann TR, Moore KW. Expression of interleukin-10 activity by Epstein-Barr virus protein BCRF1. *Science* 1990;250: 830–832.
227. Hu Q, Dyson N, Harlow E. The regions of the retinoblastoma protein needed for binding to adenovirus E1A or SV40 large T antigen are common sites for mutations. *EMBO J* 1990;9:1147–1155.
228. Hu QJ, Lees JA, Buchkovich KJ, Harlow E. The retinoblastoma protein physically associates with the human cdc2 kinase. *Mol Cell Biol* 1992;12:971–980.
229. Huang J, Schneider RJ. Adenovirus inhibition of cellular protein synthesis is prevented by the drug 2-aminopurine. *Proc Natl Acad Sci USA* 1990;87:7115–7119.
230. Huang J, Schneider RJ. Adenovirus inhibition of cellular protein synthesis involves inactivation of cap binding protein. *Cell* 1991;65: 271–280.
231. Huang M-M, Hearing P. The adenovirus early region 4 open reading frame 6/7 protein regulates the DNA binding activity of the cellular transcription factor, E2F, through a direct complex. *Genes Dev* 1989;3:1699–1710.
232. Huber HE, Edwards G, Goodhart PJ, Patrick DR, Huang PS, Ivey-Hoyle M, Barnett SF, Oliff A, Heimbrook DC. Transcription factor E2F binds DNA as a heterodimer. *Proc Natl Acad Sci USA* 1993; 90:3525–3529.
233. Huebner RJ, Rowe WP, Ward TG, Parrott RH, Bell JA. Adenoidal-pharyngeal conjunctival agents. *N Engl J Med* 1954;251:1077–1086.
234. Hull RN, Minner JR, Mascoli CC. New viral agents recovered from tissue cultures of monkey cells. III. Additional agents both from cultures of monkey tissues and directly from tissues and excreta. *Am J Hyg* 1958;68:31.
235. Ibelgaufts H, Jones KW, Maitland N, Shaw NF. Adenovirus-related RNA sequences in human neurogenic tumors. *Acta Neurpathol* 1982;56:113–117.
236. Inostroza JS, Mermelstein FH, Ha I, Lane WS, Reinberg D. Dr1, a TATA-binding protein-associated phosphoprotein and inhibitor of Class II gene transcription. *Cell* 1992;70:477–489.
237. Ishibashi M, Maizel JV Jr. The polypeptides of adenovirus. V. Young virion, structural intermediates between top components and aged virions. *Virology* 1974;57:409–424.
238. Ishibashi M, Maizel JV Jr. The polypeptides of adenovirus. VI. Early and late glycoproteins. *Virology* 1974;58:345–361.
239. Javier R. Adenovirus type 9 E4 open reading frame 1 encodes a transforming protein required for the production of mammary tumors in rats. *J Virol* 1994;68:3917–3924.
240. Javier R, Raska K, MacDonald C, Shenk T. Human adenovirus type 9-induced rat mammary tumors. *J Virol* 1991;65:3192–3202.
241. Javier R, Raska K, Shenk T. The adenovirus type 9 E4 region is required for production of mammary tumors. *Science* 1992;257: 1267–1271.
242. Jawetz E. The story of shipyard eye. *Br Med J* 1959;1:873–878.
243. Johnson DG, Schwarz JK, Cress WD, Nevins JR. Expression of transcription factor E2F1 induces quiescent cells to enter S phase. *Nature* 1993;365:349–352.
244. Jones KA, Kadonaga JT, Rosenfeld PJ, Kelly TJ, Tjian R. A cellular DNA-binding protein that activates eukaryotic transcription and DNA replication. *Cell* 1987;48:79–89.
245. Jones N, Shenk T. An adenovirus type 5 early gene function regulates expression of other early viral genes. *Proc Natl Acad Sci USA* 1979;76:3665–3669.
246. Kaelin WG, Ewen ME, Livingston DM. Definition of the minimal simian virus 40 large T antigen- and adenovirus E1A-binding domain in the retinoblastoma gene product. *Mol Cell Biol* 1990;10:3761–3769.
247. Kaelin WG Jr., Krek W, Sellers WR, DeCaprio JA, Ajchanbaum F, Fuchs CS, Chittenden T, Li Y, Farnham PJ, Blanar MA, et al. Expression cloning of a cDNA encoding a retinoblastoma-binding protein with E2F-like properties. *Cell* 1992;70:351–364.
248. Kao CC, Yew PR, Berk AJ. Domains required for *in vitro* association between the cellular p53 and the adenovirus 2 E1B 55K proteins. *Virology* 1990;179:806–814.
249. Kasai Y, Chen H, Flint SJ. Anatomy of an unusual RNA polymerase II promoter containing a downstream TATA element. *Mol Cell Biol* 1992;12:2884–2897.
250. Kast WM, Offringa R, Peters PJ, Voordouw AC, Meloen RH, van der Eb AJ, Melief CJM. Eradication of adenovirus E1-induced tumors by E1A-specific cytotoxic T lymphocytes. *Cell* 1989;62:1513–1519.
251. Kastan MB, Zhan Q, El-Deiry WS, Carrier F, Jacks T, Walsh WV, Plunkett BS, Vogelstein B, Fornace AJ Jr. A mammalian cell cycle checkpoint pathway utilizing p53 and GADD45 is defective in ataxia-telangiectasia. *Cell* 1992;71:587–597.
252. Katze MG, DeCorato D, Safer B, Galabru J, Hovanessian AG. Adenovirus VAI RNA complexes with the 68,000 M_r protein kinase to regulate its autophosphorylation and activity. *EMBO J* 1987;6:689–697.
253. Kenny MK, Hurwitz J. Initiation of adenovirus DNA replication. II. Structural requirement using synthetic oligonucleotide adenovirus templates. *J Biol Chem* 1988;263:9809–9817.
254. Kidd AH, Chroboczek J, Cusack S, Ruigrok RW. Adenovirus type 40 virions contain two distinct fibers. *Virology* 1993;192:73–84.
255. Kingston RE, Kaufman RJ, Sharp PA. Regulation of transcription of the adenovirus EII promoter be E1A gene products: absence of sequence specificity. *Mol Cell Biol* 1984;4:1970–1977.
256. Kitagawa M, Saitoh S, Ogino H, Okabe T, Matsumoto H, Okuyama A, Tamai K, Ohba Y, Yasuda H, Nishimura S, Taya Y. cdc2-like ki-

nase is associated with the retinoblastoma protein. *Oncogene* 1992;7:1067–1074.

257. Kitajewski J, Schneider RJ, Safer B, Munemitsu SM, Samuel CE, Thimmappaya B, Shenk T. Adenovirus VAI RNA antagonizes the antiviral action of the interferon-induced eIF-2a kinase. *Cell* 1986;45:195–200.
258. Kitchingman GR. Sequence of the DNA-binding protein of a human subgroup E adenovirus (type 4): Comparisons with subgroup A (type 12), subgroup B (type 7), and subgroup C (type 5). *Virology* 1985;146: 90–101.
259. Kleinberger T, Shenk T. A protein kinase is present in a complex with adenovirus E1A proteins. *Proc Natl Acad Sci USA* 1991;88: 11143–11147.
260. Kleinberger T, Shenk T. Adenovirus E4 or f4 protein binds to protein phosphatase 2A, and the complex down regulates E1A-enhanced junB transcription. *J Virol* 1993;67:7556–7560.
261. Klessig DF, Brough DE, Cleghorn VG. Introducing stable integration, and controlled expression of a chimeric adenovirus gene whose product is toxic to the recipient human cell. *Mol Cell Biol* 1984;4:1354–1362.
262. Kostura M, Mathews MB. Purification and activation of the double-stranded RNA-dependent eIF-2 kinase DAI. *Mol Cell Biol* 1989;9: 1576–1586.
263. Kovesdi I, Reichel R, Nevins JR. Identification of a cellular factor involved in E1A transactivation. *Cell* 1986;45:219–228.
264. Kovesdi I, Reichel R, Nevins JR. Role of an adenovirus E2 promoter-binding factor in E1A-mediated coordinate gene control. *Proc Natl Acad Sci USA* 1987;84:2180–2184.
265. Kralli A, Ge R, Graeven U, Ricciardi RP, Weinmann R. Negative regulation of the major histocompatibility complex class I enhancer in adenovirus type 12-transformed cells via a retinoic acid response element. *J Virol* 1992;66:6979–6988.
266. Kraus VB, Inostroza JA, Yeung K, Reinberg D, Nevins J. Interaction of the DR1 inhibitory factor with the TATA binding protein is disrupted by adenovirus E1A. *Proc Natl Acad Sci USA* 1994;91: 6279–6282.
267. Kraus VB, Moran E, Nevins JR. Promoter-specific trans-activation by the adenovirus E1A12S product involves separate E1A domains. *Mol Cell Biol* 1992;12:4391–4399.
268. Kvist S, Ostberg L, Persson H, Philipson L, Peterson PA. Molecular association between transplantation antigens and cell surface antigen in adenovirus-transformed cell line. *Proc Natl Acad Sci USA* 1978;75:5674–5678.
269. Lally C, Dörper T, Gröger W, Antoine G, Winnacker E-L. A size analysis of the adenovirus replicon. *EMBO J* 1984;3:333-341.
270. Larsson S, Bellett A, Akusjarvi G. VA RNAs from avian and human adenoviruses: dramatic differences in length, sequence and gene location. *J Virol* 1986;58:600–609.
271. Lechner RL, Kelly TJ Jr. The structure of replicating adenovirus 2 DNA molecules. *Cell* 1977;12:1007–1020.
272. Lee WS, Kao CC, Bryant GO, Liu X, Berk AJ. Adenovirus E1A activation domain binds the basic repeat in the TATA box transcription factor. *Cell* 1991;67:365–376.
273. Lees E, Faha BF, Dulic V, Reed SI, Harlow E. Cyclin E/cdk2 and cyclin A/cdk2 kinases associate with p107 and E2F in a temporally distinct manner. *Genes Dev* 1992;6:1874–1885.
274. Leong K, Lee W, Berk AJ. High-level transcription from the adenovirus major late promoter requires downstream binding sites for late-phase-specific factors. *J Virol* 1990;64:51–60.
275. Leppard KN, Shenk T. The adenovirus E1B 55kd protein influences mRNA transport via an intranuclear effect on RNA metabolism. *EMBO J* 1989;8:2329–2336.
276. Levine AJ, Momand J, Finlay CA. The p53 tumor suppressor gene. *Nature* 1991;351:453–456.
277. Lewis BA, Tullis G, Seto E, Horikoshi N, Weinmann R, Shenk T. Adenovirus E1A proteins interact with the cellular YY1 transcription factor *J Virol* 1995;69:1628–1636.
278. Li Y, Graham C, Lacy S, Duncan AM, Whyte P. The adenovirus E1A-associated 130-kD protein is encoded by a member of the retinoblastoma gene family and physically interacts with cyclins A and E. *Genes Dev* 1993;7:2366–2377.
279. Lichy JH, Field J, Horwitz MS, Hurwitz J. Separation of the adenovirus terminal protein precursor from its associated DNA polymerase: role of both proteins in the initiation of adenovirus DNA replication. *Proc Natl Acad Sci USA* 1982;79:5225–5229.
280. Lichy JH, Horwitz MS, Hurwitz J. Formation of a covalent complex between the 80,000 dalton adenovirus terminal protein and 5' dCMP *in vitro*. *Proc Natl Acad Sci USA* 1981;79:2678–2682.
281. Liebowitz J, Horwitz MS. Synthesis and assembly of adenovirus polypeptides. III. Reversible inhibition of hexon assembly in adenovirus type 5 temperature-sensitive mutants. *Virology* 1975;66:10–24.
282. Lillie J, Loewenstein P, Green MR, Green M. Functional domains of adenovirus type 5 E1a proteins. *Cell* 1987;50:1091–1100.
283. Lillie JW, Green MR. Transcription activation by the adenovirus E1A protein. *Nature* 1989;338:39–44.
284. Lin D, Shields MT, Ullrich SJ, Appella E, Mercer WE. Growth arrest induced by wild-type p53 protein blocks cells prior to or near the restriction point in late G1 phase. *Proc Natl Acad Sci USA* 1992;89: 9210–9214.
285. Lindenbaum JO, Field J, Hurwitz J. The adenovirus DNA binding protein and adenovirus DNA polymerase interact to catalyze elongation of primed DNA templates. *J Biol Chem* 1986;261:10218–10227.
286. Liu F, Green MR. A specific member of the ATF transcription factor family can mediate transcription activation by the adenovirus E1A protein. *Cell* 1990;61:1217–1224.
287. Liu F, Green MR. Promoter targeting by adenovirus E1A through interactions with different cellular DNA-binding domains. *Nature* 1994;368:520–525.
288. Liu X, Miller CW, Koeffler PH, Berk AJ. The p53 activation domain binds the TATA box-binding polypeptide in Holo-TFIID, and a neighboring p53 domain inhibits transcription. *Mol Cell Biol* 1993;13: 3291–3300.
289. Logan J, Shenk T. Adenovirus tripartite leader sequence enhances translation of mRNAs late after infection. *Proc Natl Acad Sci USA* 1984;81:3655–3659.
290. Londberg-Holm K, Philipson L. Early events of virus-cell interactions in an adenovirus system. *J Virol* 1969;4:323-338.
291. Lowe SW, Ruley HE. Stabilization of the p53 tumor suppressor is induced by adenovirus 5 E1A and accompanies apoptosis. *Genes Dev* 1993;7:535–545.
292. Lowe SW, Schmitt E, Smith S, Osborne B, Jacks T. p53 is required for radiation-induced apoptosis in mouse thymocytes. *Nature* 1993;362:847–849.
293. Luftig RB, Weihing RR. Adenovirus binds to rat brain microtubules *in vitro*. *J Virol* 1975;16:696–706.
294. Ma Y, Mathews MB. Comparative analysis of the structure and function of adenovirus virus-associated RNAs. *Mol Cell Biol* 1993;67: 6005–6017.
295. Mackey JK, Rigden PM, Green M. Do highly oncogenic group A human adenoviruses cause human cancer? Analysis of human tumors for adenovirus 12 transforming DNA sequences. *Proc Natl Acad Sci USA* 1976;73:4675–4661.
296. Maizel JV Jr., White DO, Scharff MD. The polypeptides of adenovirus. I. Evidence for multiple protein components in the virion and a comparison of types 2, 7 and 12. *Virology* 1968;36:115–125.
297. Maizel JV Jr., White DO, Scharff MD. The polypeptides of adenovirus. II. Soluble proteins, cores, top components and structure of the virion. *Virology* 1968;36:126–136.
298. Manche L, Green SR, Schmedt C, Mathews MB. Interactions between double stranded RNA regulators and the protein kinase DAI. *Mol Cell Biol* 1992;12:5238–5248.
299. Mangel WF, McGrath WJ, Toledo DL, Anderson CW. Viral DNA and a viral peptide can act as cofactors of adenovirus virion proteinase activity. *Nature* 1993;361:274–275.
300. Mansour SL, Grodzicker T, Tjian R. Downstream sequences affect transcription initiation from the adenovirus major late promoter. *Mol Cell Biol* 1986;6:2684–2694.
301. Maran A, Mathews MB. Characterization of the double-stranded RNA implicated in the inhibition of protein synthesis in cells infected with a mutant adenovirus defective for VA RNA_I. *Virology* 1988;164: 106–113.
302. Martinez-Palomo A, Granboulan N. Electron microscopy of adenovirus 12 replication. II. High-resolution autoradiography of infected KB cells labeled with tritiated thymidine. *J Virol* 1967;1:1010–1018.
303. Mathews MB. Binding of adenovirus VA RNA to mRNA: a possible role in splicing. *Nature* 1980;285:575–577.
304. Mathews MB, Shenk T. Adenovirus virus-associated RNA and translation control. *J Virol* 1991;65:5657–5662.
305. McAllister RM, Nicholson MO, Reed G, Kern J, Gilden RV, Huebner RJ. Transformation of rodent cells by adenovirus 19 and other group D adenoviruses. *J Natl Cancer Inst* 1969;43:917–922.

306. Meijer I, Boot AJ, Mahabir G, Zantema A, van der Eb AJ. Reduced binding activity of transcription factor NF-kappa B accounts for MHC class I repression in adenovirus type 12 E 1-transformed cells. *Cell Immunol* 1992;145:56–65.
307. Mellits KH, Kostura M, Mathews MB. Interaction of adenovirus VA RNA_I, with the protein kinase DAI: nonequivalence of binding and function. *Cell* 1990;1961:843–852.
308. Mellits KH, Mathews MB. Effects of mutations in stem and loop regions on the structure and function of adenovirus VA RNA_I. *EMBO J* 1988;7:2849–2859.
309. Mellits KH, Pe'ery T, Mathews MB. Structure and function of adenovirus VA RNA: role of the apical stem. *J Virol* 1992;66:2369–2377.
310. Mellman I. The importance of being acidic: the role of acidification in intracellular membrane traffic. *J Exp Biol* 1992;39–45.
311. Mellow GH, Fohring B, Dougherty J, Gallimore PH, Raska K Jr. Tumorigenicity of adenovirus transformed rat cells and expression of class I major histocompatibility antigen. *Virology* 1984;134:460–465.
312. Mirza MA, Weber J. Structure of adenovirus chromatin. *Biochem Biophys Acta* 1982;696:76–86.
313. Miyamoto NG, Moncolin V, Egly JM, Cambon P. Specific interaction between a transcription factor and the upstream element of the adenovirus-2 major late promoter. *EMBO J* 1985;4:3563–3570.
314. Mondesert G, Kedinger C. Cooperation between upstream and downstream elements of the adenovirus major late promoter for maximal late phase-specific transcription. *Nucleic Acids Res* 1991;19:3221–3228.
315. Mondesert G, Tribouley C, Kedinger C. Identification of a novel downstream binding protein implicated in late-phase-specific activation of the adenovirus major late promotor. *Nucleic Acids Res* 1992;20: 3881–3889.
316. Moore M, Schaack J, Baim SB, Morimoto RI, Shenk T. Induced heat shock mRNAs escape the nucleocytoplasmic transport block in adenovirus infected Hela cells. *Mol Cell Biol* 1987;7:4505–4512.
317. Moran E, Mathews MB. Multiple functional domains in the adenovirus E1A gene. *Cell* 1987;48:177–178.
318. Morin N, Delsert C, Klessig DF. Mutations that affect phosphorylation of the adenovirus DNA-binding protein alter its ability to enhance its own synthesis. *J Virol* 1989;63:5228–5237.
319. Moyne G, Pichard E, Bernhard W. Localization of simian adenovirus 7 (SA 7) transcription and replication in lytic infection. An ultracytochemical and autoradiographical study. *J Gen Virol* 1978;40:77–92.
320. Mul YM, van der Vliet PC. Nuclear factor I enhances adenovirus DNA replication by increasing the stability of a preinitiation complex. *EMBO J* 1992;11:751–760.
321. Mul YM, Verrijzer CP, van der Vliet PC. Transcription factors NFI and NFIII/Oct-1 function independently, employing different mechanisms to enhance adenovirus DNA replication. *J Virol* 1990;64: 5510–5518.
322. Muller U, Kleinberger T, Shenk T. Adenovirus E4 or f4 protein reduces phosphorylation of c-Fos and E1A proteins while simultaneously reducing the level of AP-1. *J Virol* 1992;66:5867–5878.
323. Muller U, Roberts MP, Engel DA, Doerfler W, Shenk T. Induction of transcription factor AP-1 by adenovirus E1A protein and cAMP. *Genes Dev* 1989;3:1991–2002.
324. Mulligan RC. The basic science of gene therapy. *Science* 1993;260: 926–932.
325. Murthy SCS, Bhat GP, Thimmappaya B. Adenovirus EIIA early promoter; transcription control elements and induction by the viral preearly EIA gene, which appears to be sequence independent. *Proc Natl Acad Sci USA* 1985;82:2230–2234.
326. Nagata K, Guggenheimer RA, Hurwitz J. Adenovirus DNA replication *in vitro*: synthesis of full-length DNA with purified proteins. *Proc Natl Acad Sci USA* 1983;80:4266–4270.
327. Neilan JG, Lu Z, Afonso CL, Kutish GF, Sussman MD, Rock DL. An African swine fever virus gene with similarity to the proto-oncogene bcl-2 and the Epstein-Barr virus gene BHRF1. *J Virol* 1993;67: 4391–4394.
328. Neill SD, Hemstrom C, Virtanen A. An adenovirus E4 gene product trans-activates E2 transcription and stimulates stable E2F binding through a direct association with E2F. *Proc Natl Acad Sci USA* 1990;87:2008–2012.
329. Nermut MV. Structural elements in adenovirus cores: evidence for a "core shell" and linear structures in "relaxed" cores. *Arch Virol* 1979;62:101–113.
330. Nermut V. *The architecture of adenoviruses*. Ginsberg HS, ed. NY: Plenum Press, 1984:6–34.
331. Nevins JR. Mechanism of activation of early viral transcription by the adenovirus E1A gene product. *Cell* 1981;26:213–220.
332. Nevins JR. E2F: a link between the Rb tumor suppressor protein and viral oncoproteins. *Science* 1992;258:424–429.
333. Nevins JR, Darnell JE. Groups of adenovirus type 2 mRNAs derived from a large primary transcript: probable nuclear origin and possible common 3' ends. *J Virol* 1978;25:811–823.
334. Nevins JR, Wilson MC. Regulation of adenovirus-2 gene expression at the level of transcriptional termination and RNA processing. *Nature* 1981;290:113–118.
335. Nevins JR, Winkler JJ. Regulation of early adenovirus transcription: a protein product of early region 2 specifically represses region 4 transcription. *Proc Natl Acad Sci USA* 1980;77:1893–1897.
336. Newcomb WF, Boring JW, Brown JC. Ion etching of human adenovirus 2: structure of the core. *J Virol* 1984;51:52–56.
337. Nielsch U, Zimmer SG, Babiss LE. Changes in NF-kappa B and ISGF3 DNA binding activities are responsible for differences in MHC and beta-IFN gene expression in Ad5- versus Ad12-transformed cells. *EMBO J* 1991;10:4169–4175.
338. Norrby E. The relationship between soluble antigens and the virion of adenovirus type 3. I. Morphological characteristics. *Virology* 1966;28:236–248.
339. Norrby E. The relationship between soluble antigens and the virion of adenovirus type 3. III Immunologic characteristics. *Virology* 1969;37: 565–576.
340. Norrby E. The structural and functional diversity of adenovirus capsid components. *J Gen Virol* 1969;5:221–236.
341. Norrby E, Bartha A, Boulanger P. Adenoviridae. *Intervirology* 1976;7:117–125.
342. Norrby E, Wadell G. Immunological relationships between hexons of certain adenoviruses. *J Virol* 1969;4:663–670.
343. O'Malley RP, Duncan RF, Hershey JWB, Mathews MB. Modification of protein synthesis initiation factors and the shut-off of host protein synthesis in adenovirus infected cells. *Virology* 1989;168:112–118.
344. O'Malley RP, Mariano TM, Siekierka J, Merrick WC, Reichel PA, Mathews MB. The control of protein synthesis by adenovirus VA RNA. *Cancer Cells* 1986;4:291–301.
345. O'Neill EA, Fletcher C, Burrow CR, Heintz N, Roeder RG, Kelly TJ. Transcription factor OTF-1 is functionally identical to the DNA replication factor NF-III. *Science* 1988;241:1210–1213.
346. Offringa R, Gebel S, Van Dam H, Timmers M, Smits A, Zwart R, Stein JL, Bos A, van der Eb A, Herrlich P. A novel function of the transforming region of E1A: repression of AP1 activity. *Cell* 1990;62: 527–538.
347. Oosterom-Dragon EA, Ginsberg HS. Characterization of two temperature-sensitive mutants of type 5 adenovirus with mutations in the 100,000-dalton protein gene. *J Virol* 1981;40:491–500.
348. Ornelles D, Shenk T. Localization of the adenovirus early region 1B 55-kildalton protein during lytic infection: association with nuclear viral inclusions requires the early region 4 34-kilodalton protein. *J Virol* 1991;65:424–439.
349. Pääbo S, Bhat BM, Wold WSM, Peterson PA. A short sequence in the COOH-terminus makes an adenovirus membrane glycoprotein a resident of the endoplasmic reticulum. *Cell* 1987;50:311–317.
350. Pääbo S, Weber F, Kampe O, Schaffner W, Peterson PA. Association between transplantation antigens and a viral membrane protein synthesized from a mammalian expression vector. *Cell* 1983; 35:445–453.
351. Pasten I, Seth P, FitzGerald D, Willingham M. Adenovirus entry into cells; some new observations on an old problem. In: Notkins A, Oldstone M, ed. *Concepts in viral pathogenesis II*. New York: Springer-Verlag, 1986:141–146.
352. Pe'ery T, Mellits K, Mathews MB. Mutational analysis of the central domain of adenovirus virus-associated RNA mandates a revision of the proposed secondary structure. *J Virol* 1993;67:3534–3543.
353. Pei R, Berk AJ. Multiple transcription factor binding sites mediate adenovirus E1A transactivation. *J Virol* 1989;63:3499-3506.
354. Perry ME, Levine AJ. Tumor-suppressor p53 and the cell cycle. *Curr Opin Gen Dev* 1993;3:50–54.
355. Pettersson U, Philipson L, Hoglund SV. Structural proteins of adenovirus II. Purification and characterization of the adenovirus type 2 fiber antigen. *Virology* 1968;35:204–215.
356. Pettersson U, Roberts RJ. Adenovirus gene expression and replication: a historical review. *Cancer Cells* 1986;4:37–57.

357. Philipson L. Adenovirus assembly. In: Ginsberg H, ed.*The adenoviruses*. New York: Plenum Press, 1984.
358. Philipson L, Lonberg-Holm K, Petterson U. Virus-receptor interaction in an adenovirus system. *J Virol* 1968;2:1064–1075.
359. Pilder S, Logan J, Shenk T. Deletion of the gene encoding the adenovirus type 5 E1B-21K polypeptide leads to degradation of viral and host cell DNA. *J Virol* 1984;52:664–671.
360. Pilder S, Moore M, Logan J, Shenk T. The adenovirus E1B-55K transforming polypeptide modulates transport or cytoplasmic stabilization of viral and host cell mRNAs. *Mol Cell Biol* 1986;6:470–476.
361. Piña M, Green M. Biochemical studies on adenovirus multiplication. IX. Chemical and base composition analysis of 28 human adenoviruses. *Proc Natl Acad Sci USA* 1965;54:547–551.
362. Prage L, Pettersson U, Hoglund S, Londberg-Holm K, Philipson L. Structural proteins of adenoviruses. IV. Sequential degradation of the adenovirus type 2 virion. *Virology* 1970;42:341–358.
363. Prage L, Rouse HC. Effect of arginine starvation on macromolecular synthesis in infection with type 2 adenovirus. III. Immunofluorescence studies of the synthesis of the hexon and the major core antigen (AAP). *Virology* 1976;69:352–359.
364. Price R, Penman S. Transcription of the adenovirus genome by an α-amanitin-sensitive RNA polymerase in HeLa cells. *J Virol* 1972;9:621–626.
365. Prince GA, Porter DD, Jenson AB, Horswood RL, Chanock RM, Ginsberg HS. Pathogenesis of adenovirus type 5 pneumonia in cotton rats (Sigmodon hispidus). *J Virol* 1993;67:101–111.
366. Pruijn GJM, van der Vleit P, Dathan NA, Mattaj IW. Anti-OTF-1 antibodies inhibit NFIII stimulation of *in vitro* adenovirus DNA replication. *Nucleic Acids Res* 1989;17:1845–1863.
367. Pugh BF, Tjian R. Diverse transcriptional functions of the multisubunit eukaryotic TFIID complex. *J Biol Chem* 1992;267:3310–3321.
368. Qian Y, Luckey C, Horton L, Esser M, Templeton DJ. Biological function of the retinoblastoma protein requires distinct domains for hyperphosphorylation and transcription factor binding. *Mol Cell Biol* 1992;12:5363–5372.
369. Qin XQ, Chittenden T, Livingston DM, Kaelin WG. Identification of a growth suppression domain within the retinoblastoma gene product. *Genes Dev* 1992;6:953–964.
370. Rao L, Debbas M, Sabbatini P, Hockenbery D, Korsmeyer S, White E. The adenovirus E1A proteins induce apoptosis, which is inhibited by the E1B 19-kDa and Bcl-2 proteins. *Proc Natl Acad Sci USA* 1992;89:7742–7746.
371. Raska K Jr., Gallimore PH. An inverse relation of the oncogenic potential of adenovirus-transformed cells and their sensitivity to killing by syngeneic natural killer cells. *Virology* 1982;123:8–18.
372. Raychaudhuri P, Bagchi S, Neill SD, Nevins JR. Activation of the E2F transcription factor in adenovirus-infected cells involves E1A-dependent stimulation of DNA-binding activity and induction of cooperative binding mediated by an E4 gene product. *J Virol* 1990;64:2702–2710.
373. Reach M, Babiss L-E, Young C-S. The upstream factor-binding site is not essential for activation of transcription from the adenovirus major late promoter. *J Virol* 1990;64:5851–5860.
374. Reach M, Xu L-X, Young C-S. Transcription from the adenovirus major late promoter uses redundant activation elements. *EMBO J* 1991;10:3439–3446.
375. Reich N, Pine R, Levy D, Darnell JE. Transcription of interferon-stimulated genes is induced by adenovirus particles but is suppressed by E1A gene products. *J Virol* 1988;62:114–119.
376. Reich PR, Rose J, Forget B, Weissman SM. RNA of low molecular weight in KB cells infected with Ad2. *J Mol Biol* 1966;17:428–439.
377. Reichel PA, Merrick WC, Siekierka J, Mathews MB. Adenovirus VA RNA_I regulates the activity of a protein synthesis initiator factor. *Nature* 1985;313:196–200.
378. Rekosh DMK, Russell WC, Bellet AJD, Robinson AJ. Identification of a protein linked to the ends of adenovirus DNA. *Cell* 1977;11:283–295.
379. Rikitake Y, Moran E. DNA-binding properties of the E1A-associated 300-kilodalton protein. *Mol Cell Biol* 1992;12:2826–2836.
380. Roberts MM, White JL, Grutter MG, Burnett RM. Three-dimensional structure of the adenovirus major coat protein hexon. *Science* 1986;232:1148–1151.
381. Roberts RJ, O'Neill KE, Yen CT. DNA sequences from the adenovirus 2 genome. *J Biol Chem* 1984;259:13968–13975.
382. Robinson AJ, Younghusband HB, Bellett AJD. A circular DNA-protein complex from adenoviruses. *Virology* 1973;56:54–69.
383. Rochette-Egly C, Fromental C, Chambon P. General repression of enhanson activity by the adenovirus-2 E1A proteins. *Genes Dev* 1990;4:137–150.
384. Rosen I. A hemagglutination-inhibition technique for typing adenoviruses. *Am J Hyg* 1960;71:120–128.
385. Rosenfeld PJ, O'Neill E, Wides RJ, Kelly TJ. Sequence-specific interactions between cellular DNA-binding proteins and the adenovirus origin of DNA replication. *Mol Cell Biol* 1987;7:875–886.
386. Rossmann MG, Arnold E, Erickson JW, Frankenberger EA, Griffith JP, Hecht HJ, Johnson JE, Kamer G, Luo M, Mosser AG et al. Structure of a human common cold virus and functional relationship to other picornaviruses. *Nature* 1985;317:145–153.
387. Rouse HC, Schlesinger RW. An arginine-dependent step in the maturation of type 1 adenovirus. *Virology* 1967;33:513–522.
388. Routes JM, Bellgrau D, McGrory WJ, Bautista DS, Graham FL, Cook JL. Anti-adenovirus type 5 cytotoxic T lymphocytes: immunodominant epitopes are encoded by the E1A gene. *J Virol* 1991;65:1450–1457.
389. Rowe WP, Huebner RJ, Gilmore LK, Parrott RH, Ward TG. Isolation of a cytopathogenic agent from human adenoids undergoing spontaneous degeneration in tissue culture. *Proc Soc Exp Biol Med* 1953;84:570–573.
390. Ruigrok RWH, Barge A, Albiges-Rizo C, Dayan S. Structure of adenovirus fibre. II. Morphology of single fibres. *J Mol Biol* 1990;589–596.
391. Russell WC, Laver WG, Sanderson PJ. Internal components of adenovirus. *Nature* 1968;219:1127–1130.
392. Russell WC, Newman C, Williams JF. Characterization of temperature-sensitive mutants of adenovirus type 5-serology. *J Gen Virol* 1972;17:265–279.
393. Sarnow P, Hearing P, Anderson CW, Halbert DN, Shenk T, Levine AJ. Adenovirus early region 1B 58,000-dalton tumor antigen is physically associated with an early region 4 25,000-dalton protein in productively infected cells. *J Virol* 1984;49:692–700.
394. Sarnow P, Ho Y-S, Williams J, Levine AJ. Adenovirus E1b-58kd tumor antigen and SV40 large tumor antigen are physically associated with the same 54 kd cellular protein in transformed cells. *Cell* 1982a;28:387–394.
395. Sarnow P, Sullivan CA, Levine AJ. A monoclonal antibody detecting the adenovirus type 5 E1B-58kD tumor antigen: characterization of the E1b-58kD tumor antigen in adenovirus-infected and -transformed cells. *Virology* 1982;120:510–517.
396. Sawada Y, Fohring B, Shenk T, Raska K. Tumorigenicity of adenovirus-transformed cells: region E1A of adenovirus 12 confers resistance to natural killer cells. *Virology* 1985;147:413–421.
397. Sawadago M, Roeder RG. Interaction of a gene-specific transcription factor with the adenovirus major late promoter upstream of the TATA box region. *Cell* 1985;43:165–175.
398. Schaack J, Ho WY-W, Freimuth P, Shenk T. Adenovirus terminal protein mediates both nuclear matrix association and efficient transcription of adenovirus DNA. *Genes Dev* 1990;4:1197–1208.
399. Scheffner M, Werness BA, Huibregtse JM, Levine AJ, Howley PM. The E6 oncoprotein encoded by human papillomavirus types 16 and 18 promotes the degradation of p53. *Cell* 1990;63:1129–1136.
400. Schneider RJ, Safer B, Munemitsu SM, Samuel CE, Shenk T. Adenovirus VAI RNA prevents phosphorylation of the eukaryotic initiation factor 2 alpha subunit subsequent to infection. *Proc Natl Acad Sci USA* 1985;82:4321–4324.
401. Schneider RJ, Weinberger C, Shenk T. The adenovirus VAI RNA functions during initiation of translation in virus-infected cells. *Cell* 1984;37:291–298.
402. Schrier PI, Bernards R, Vaessen TJ, Houweling A, van der Eb AJ. Expression of class I major histocompatibility antigens switched off by highly oncogenic adenovirus 12 in transformed rat cells. *Nature* 1983;305:771–775.
403. Schwarz JK, Devoto SH, Smith EJ, Chellappan S, Jakoi L, Nevins JR. Interactions of the p107 and Rb proteins with E2F during the cell proliferatve response. *EMBO J* 1993;122:1013–1020.
404. Seiberg M, Kessler M, Levine AJ, Aloni Y. Human RNA polymerase II can prematurely terminate transcription of the adenovirus type 2 late transcription unit at a precise site that resembles a prokaryotic termination signal. *Virus-Genes* 1987;1:97–116.
405. Seth P, Fitzgerald DJ, Willingham MC, Pastan I. Role of a low-pH en-

vironment in adenovirus enhancement of the toxicity of a Pseudomonas exotoxin-epidermal growth factor conjugate. *J Virol* 1984;51:650–655.
406. Seto E, Usheva A, Zambetti GP, Momand J, Horikoshi N, Weinmann R, Levine AJ, Shenk T. Wild-type p53 binds to the TATA binding protein and interferes with transcription. *Proc Natl Acad Sci USA* 1992;89:12028–12032.
407. Shan B, Zhu X, Chen P-L, Durfee T, Yang Y, Sharp D, Lee W-H. Molecular cloning of cellular genes encoding retinoblastoma-associated proteins: identification of a gene with properties of the transcription factor E2F. *Mol Cell Biol* 1992;12:5620–5631.
408. Sharp PA, Gallimore PH, Flint SJ. Mapping of adenovirus 2 RNA sequences in lytically infected cells and transformed cell lines. *Cold Spring Harbor Symp Quant Biol* 1975;39:457–474.
409. Shaw AR, Ziff EB. Transcripts from the adenovirus-2 major late promoter yield a single early family of 3' coterminal mRNAs and five late families. *Cell* 1980;22:905–916.
410. Shaw P, Bovey R, Tardy S, Sahli R, Sordat B, Costa J. Induction of apoptosis by wild-type p53 in a human colon tumor-derived cell line. *Proc Natl Acad Sci USA* 1992;89:4495–4499.
411. Shen Y, Shenk T. Relief of p53-mediated transcriptonal repression by the adenovirus E1B-19kDa protein or the cellular Bcl-2 protein. *Proc Natl Acad Sci USA* 1994;91:8940–8944.
412. Shenk T, Flint SJ. Transcriptional and transforming activities of the adenovirus E1A proteins. *Adv Cancer Res* 1991;57:47–85.
413. Shenk T, Jones N, Colby W, Fowlkes D. Functional analysis of adenovirus type 5 host-range deletion mutants defective for transformation of rat embryo cells. *Cold Spring Harbor Symp Quant Biol* 1979;44:367–375.
414. Shenk T, Williams J. Genetic analysis of adenoviruses. *Curr Top Micro Immunol* 1984;111:1–39.
415. Shi Y, Seto E, Chang LS, Shenk T. Transcriptional repression by YY1, a human GLI-Kruppel-related protein, and relief of repression by adenovirus E1A protein. *Cell* 1991;67:377–388.
416. Shirodkar S, Ewen M, DeCaprio JA, Morgan D, Livingston D, Chittenden T. The transcription factor E2F interacts with the retinoblastoma product and a p107-cyclin A complex in a cell cycle-regulated manner. *Cell* 1992;68:157–166.
417. Silver L, Anderson CW. Interaction of human adenovirus serotype 2 with human lymphoid cells. *Virology* 1988;165:377–387.
418. Simon MC, Fisch TM, Benecke BJ, Nevins JR, Heintz N. Identification of multiple, functionally distinct TATA elements, one of which is the target in the hsp70 promoter for E1A regulation. *Cell* 1988;52:723–729.
419. Slansky JE, Li Y, Kaelin WG, Farnham PJ. A protein synthesis-dependent increase in E2F1 mRNA correlates with growth regulation of the dihydrofolate reductase promoter. *Mol Cell Biol* 1993;13:1610–1618.
420. Smart J, Stillman BW. Adenovirus terminal protein precursor. Partial amino acid sequence and the site of covalent linkage to virus DNA. *J Biol Chem* 1982;257:13499–13506.
421. Smith CA, Davis T, Wignall JM, Din WS, Farrah T, Upton C, McFadden G, Goodwin RG. T2 open reading frame from the Shope fibroma virus encodes a soluble form of the TNF receptor. *Biochem Biophys Res Commun* 1991;176:335–342.
422. Soddu S, Lewis AM Jr. Driving adenovirus type 12-transformed BALB/c mouse cells to express high levels of class I major histocompatibility complex proteins enhances, rather than abrogates, their tumorigenicity. *J Virol* 1992;66:2875–2884.
423. Spergel JM, Chen-Kiang S. Interleukin 6 enhances a cellular activity that functionally substitutes for E1A protein in transactivation. *Proc Natl Acad Sci USA* 1991;88:6472–6476.
424. Spergel JM, Hsu W, Akira S, Thimmappaya B, Kishimoto T, Chen-Kiang S. NF-IL6, a member of the C/EBP family, regulates E1A-responsive promoters in the absence of E1A. *J Virol* 1992;66:1021–1030.
425. Steenberg PH, Maat J, van Ormondt H, Sussanbach JS. The sequence at the termini of adenovirus type 5 DNA. *Nucleic Acids Res* 1977;4:4371–4389.
426. Stein RW, Corrigan M, Yaciuk P, Whelan J, Moran E. Analysis of E1A-mediated growth regulation functions: binding of the 300-kilodalton cellular product correlates with E1A enhancer repression function and DNA synthesis-inducing activity. *J Virol* 1990;64:4421–4427.
427. Stein RW, Ziff EB. Repression of insulin gene expression by adenovirus type 5 E1a proteins. *Mol Cell Biol* 1987;7:1164–1170.
428. Stephens C, Harlow E. Differential splicing yields novel adenovirus 5 E1A mRNAs that encode 30kd and 35kd proteins. *EMBO J* 1987;6:2027–2035.
429. Stewart PL, Burnett RM, Cyrklaff M, Fuller SD. Image reconstruction reveals the complex molecular organization of adenovirus. *Cell* 1991;67:145–154.
430. Stewart PL, Fuller SD, Burnett RM. Difference imaging of adenovirus: bridging the resolution gap between X-ray crystallography and electron microscopy. *EMBO J* 1993;12:2589–2599.
431. Stillman B. Initiation of eukaryotic DNA replication *in vitro*. *Ann Rev Cell Biol* 1989;5:197–245.
432. Stillman BW, Lewis JB, Chow LT, Mathews MB, Smart E. Identification of the gene and mRNA for the adenovirus terminal protein precursor. *Cell* 1981;23:497–508.
433. Stillman BW, Tamanoi F, Mathews MB. Purification of an adenovirus-coded DNA polymerase that is required for initiation of DNA replication. *Cell* 1982;31:613–623.
434. Stouten PF, Sander C, Ruigrok RW, Cusack S. New triple-helical model for the shaft of the adenovirus fibre. *J Mol Biol* 1992;226:1073–1084.
435. Stuiver MH, van der Vliet PC. The adenovirus DNA binding protein forms a multimeric protein complex with double-stranded DNA and enhances binding of nuclear factor I. *J Virol* 1990;64:379–386.
436. Sundquist B, Everitt E, Philipson L, Höglund S. Assembly of adenoviruses. *J Virol* 1973;11:449–459.
437. Svensson U. Role of vesicles during adenovirus 2 internalization into HeLa cells. *J Virol* 1985;55:442–449.
438. Takemori N. Genetic studies with tumorigenic adenoviruses. III. Recombination in adenovirus type 12. *Virology* 1972;47:157–167.
439. Takemori N, Cladaras C, Bhat B, Conley AJ, Wold WSM. cyt gene of adenovirus 2 and 5 is an oncogene for transforming function in early region E1B and encodes the E1B 19,000-molecular-weight polypeptide. *J Virol* 1984;52:793–805.
440. Tal JT, Craig EA, Raskas HJ. Sequence relationships between adenovirus 2 early RNA and viral RNA size classes synthesized at 18 hours after infection. *J Virol* 1975;15:137–144.
441. Tamanoi F, Stillman BW. Function of adenovirus terminal protein in the initiation of DNA replication. *Proc Natl Acad Sci USA* 1982;79:2221–2225.
442. Tamanoi F, Stillman BW. Initiation of adenovirus DNA replication *in vitro* requires a specific DNA sequence. *Proc Natl Acad Sci USA* 1983;80:6446–6450.
443. Tanaka K, Isselbacher KJ, Khoury G, Jay G. Reversal of tumorigenicity by expression of a major histocompatibility complex class I gene. *Science* 1985;228:226–230.
444. Tate V, Philipson J. Parental adenovirus DNA accumulates in nucleosome-like structures in infected cells. *Nucleic Acids Res* 1979;6:2769–2785.
445. Tatzelt J, Fechteler K, Langenbach P, Doerfler W. Fractionated nuclear extracts from hamster cells catalyze cell-free recombination at selective sequences between adenovirus DNA and a hamster preinsertion site. *Proc Natl Acad Sci USA* 1993;90:7356–7360.
446. Temperley SM, Hay RT. Recognition of the adenovirus type 2 origin of DNA replication by the virally encoded DNA polymerase and preterminal proteins. *EMBO J* 1992;11:761–768.
447. Thimmappaya B, Weinberger C, Schneider RJ, Shenk T. Adenovirus VA1 RNA is required for efficient translation of viral mRNA at late times after infection. *Cell* 1982;31:543–551.
448. Thomas GP, Mathews MB. DNA replication and the early to late transition in adenovirus infection. *Cell* 1980;22:523–533.
449. Tibbetts C. Viral DNA sequences from incomplete particles of human adenovirus type 7. *Cell* 1977;12:243–249.
450. Tihanyi K, Bourbonniere M, Houde A, Rancourt C, Weber JM. Isolation and properties of adenovirus type 2 proteinase. *J Biol Chem* 1993;268:1780–1785.
451. Tokunaga O, Yaegashi R, Lowe J, Dobbs L, Padmanabhan R. Sequence analysis in the E1 region of adenovirus type 4 DNA. *Virology* 1986;155:418–433.
452. Tollefson AE, Stewart AR, Yei SP, Saha SK, Wold WS. The 10,400- and 14,500-dalton proteins encoded by region E3 of adenovirus form a complex and function together to down-regulate the epidermal growth factor receptor. *J Virol* 1991;65:3095–3105.
453. Toogood CI, Crompton J, Hay RT. Antipeptide antisera define neutralizing epitopes on the adenovirus hexon. *J Gen Virol* 1992;73:1429–1435.
454. Tooze J. *DNA tumor viruses*, 2nd ed. Cold Spring Harbor, New York: Cold Spring Harbor Laboratory, 1981:943–1054.
455. Toth M, Doerfler W, Shenk T. Adenovirus DNA replication facilitates binding of the MLTF/USF transcription factor to the viral major

late promoter within infected cells. *Nucleic Acids Res* 1992; 20:5143–5148.
456. Trentin JJ, Yabe Y, Taylor G. The quest for human cancer viruses. *Science* 1962;137:835–849.
457. Tribouley C, Lutz P, Staub A, Kedinger C. The product of the adenovirus intermediate gene IVa2 is a transcriptional activator of the major late promoter. *J Virol* 1994;68:4450–4457.
458. Ulfendahl PJ, Linder S, Kreivi J-P, Nordquist K, Sevensson C, Hultberg H, Akusjarvi G. A novel adenovirus-2 E1A mRNA encoding a protein with transcription activation properties. *EMBO J* 1987;6: 2037–2044.
459. Valentine RG, Pereira HG. Antigens and structure of the adenovirus. *J Mol Biol* 1965;13:13–20.
460. van Bergen BGM, van der Ley PA, van Driel W, van Mansfield ADM, van der Vliet PC. Replication of origin containing adenovirus DNA fragments that do not carry the terminal protein. *Nucleic Acids Res* 1983;11:1975–1989.
461. van Dam H, Duyndam M, Rottier R, Bosch A, de Vries-Smits L, Herrlich P, Zantema A, Angel P, van der Eb AJ. Heterodimer formation of cJun and ATF-2 is responsible for induction of c-jun by the 243 amino acid adenovirus E1A protein. *EMBO J* 1993;12:479–487.
462. van den Heuvel SJ, van Laar T, The I, van der Eb AJ. Large E1B proteins of adenovirus types 5 and 12 have different effects on p53 and distinct roles in cell transformation. *J Virol* 1993;67:5226–5234.
463. van der Vliet PC, Levine AJ. DNA-binding proteins specific for cells infected by adenovirus. *Nature* 1973;246:170–174.
464. van der Vliet PC, Levine AJ, Ensinger MJ, Ginsberg HS. Thermolabile DNA-binding protein from cells infected with a temperature-sensitive mutant of adenovirus defective in viral DNA synthesis. *J Virol* 1975;15:348–354.
465. van Oostrum J, Burnett RM. The molecular composition of the adenovirus type 2 virion. *J Virol* 1985;56:439–448.
466. van Ormondt H, Maat J, Dijkema R. Comparison of nucleotide sequences of the early E1a regions for subgroups A, B and C of human adenoviruses. Gene 1980;12:63–76.
467. Varga MJ, Weibull C, Everitt E. Infectious entry pathway of adenovirus type 2. *J Virol* 1991;65:6061-6070.
468. Vayda ME, Flint SJ. Isolation and characterization of adenovirus core nucleoprotein subunits. *J Virol* 1987;16:3335–3339.
469. Vayda ME, Rogers AE, Flint SJ. The structure of nucleoprotein cores released from adenoviruses. *Nucleic Acids Res* 1983;11:441–460.
470. Velcich A, Ziff E. Adenovirus E1A proteins repress transcription from the SV40 early promoter. *Cell* 1985;40:705–716.
471. Velicer LF, Ginsberg HS. Synthesis, transport and morphogenesis of type 5 adenovirus capsid proteins. *J Virol* 1970;5:338–347.
472. Verrijzer CP, van Oosterhout AWM, van Weperen WW, van der Vliet PC. POU proteins bend DNA via the POU-specific domain. *EMBO J* 1991;10:3007–3014.
473. Vogelstein B, Kinzler KW. p53 function and dysfunction. *Cell* 1992;70:523–526.
474. Wadell G. Classification of human adenoviruses by SDS-polyacrylamide gel electrophoresis of structural proteins. *Intervirology* 1979;11:47–59.
475. Wadell G, Hammarskjold ML, Winberg G, Varsanyi TM, Sundell G. Genetic variability of adenoviruses. *Ann NY Acad Sci* 1980; 354:16–42.
476. Walker TA, Wilson BA, Lewis AM Jr., Cook JL. E1A oncogene induction of cytolytic susceptibility eliminates sarcoma cell tumorigenicity. *Proc Natl Acad Sci USA* 1991;88:6491–6495.
477. Wang H-G, Draetta G, Moran E. E1A induces phosphorylation of the retinoblastoma protein independently of direct physical association between the E1A and retinoblastoma products. *Mol Cell Biol* 1991; 11:4253–4265.
478. Wang H-G, Rikitake Y, Carter MC, Yaciuk P, Abraham SE, Zerler B, Moran E. Identification of specific adenovirus E1A N-terminal residues critical to the binding of cellular proteins and to the control of cell growth. *J Virol* 1993;67:476–488.
479. Weber J. Genetic analysis of adenovirus type 2. III. Temperature sensitivity of processing of viral proteins. *J Virol* 1976;17:462–471.
480. Weber J, Jelinek W, Darnell JE Jr. The definition of a large viral transcription unit late in Ad2 infection of HeLa cells: mapping of nascent RNA molecules labeled in isolated nuclei. *Cell* 1977;10:611–616.
481. Weber JM, Dery CV, Mirza MA, Horvath J. Adenovirus DNA synthesis is coupled to virus assembly. *Virology* 1985;140:351–359.
482. Weber JM, Talbot BG, Delorme L. The orientation of the adenovirus fiber and its anchor domain identified through molecular mimicry. *Virology* 1989;168:180–182.
483. Webster A, Hay RT, Kemp G. The adenovirus protease is activated by a virus-coded disulphide-linked peptide. *Cell* 1993;72:97–104.
484. Webster KA, Muscat GEO, Kedes L. Adenovirus E1A products suppress myogenic differentiation and inhibit transcription from muscle-specific promoters. *Nature* 1988;332:553–557.
485. Weinberg DH, Ketner G. A cell line that supports the growth of a defective early region 4 deletion mutant of human adenovirus type 2. *Proc Natl Acad Sci USA* 1983;80:5383–5386.
486. Weinberg DH, Ketner G. Adenoviral early region 4 is required for efficient viral DNA replication and for late gene expression. *J Virol* 1986;57:833–838.
487. Weinmann R, Raskas HJ, Roeder RG. Role of DNA-dependent RNA polymerases II and III in transcription of the adenovirus genome late in productive infection. *Proc Natl Acad Sci USA* 1974;71:3426–3430.
488. White E. Death-defying acts: a meeting review on apoptosis. *Genes Dev* 1993;7:2277–2284.
489. White E, Cipriani R. Specific disruption of intermediate filaments and the nuclear lamina by the 19-kD product of the adenovirus E1B oncogene. *Proc Natl Acad Sci USA* 1989;86:9886–9890.
490. White E, Cipriani R. Role of adenovirus E1B proteins in transformation: altered organization of intermediate filaments in transformed cells that express the 19-kildalton protein. *Mol Cell Biol* 1990;10:120–130.
491. White E, Grodzicker T, Stillman BW. Mutations in the gene encoding the adenovirus early region 1B 19,000 molecular weight tumor antigen cause the degradation of chromosomal DNA. *J Virol* 1984;42: 410–419.
492. Whyte P, Buchkovich KJ, Horowitz JM, Friend SH, Raybuck M, Weinberg RA, Harlow E. Association between an oncogene and an anti-oncogene: the adenovirus E1A proteins bind to the retinoblastoma gene product. *Nature* 1988;334:124–129.
493. Whyte P, Williamson NM, Harlow E. Cellular targets for transformation by the adenovirus E1A proteins. *Cell* 1989;56:67–75.
494. Wickham TJ, Mathias P, Cheresh DA, Nemerow GR. Integrins alpha v beta 3 and alpha v beta 5 promote adenovirus internalization but not virus attachment. *Cell* 1993;73:309–319.
495. Williams GT, Smith CA. Molecular regulation of apoptosis: genetic controls on cell death. *Cell* 1993;74:777–779.
496. Williams J, Grodzicker T, Sharp P, Sambrook S. Adenovirus recombination: physical mapping of crossover events. *Cell* 1975;4:113–119.
497. Williams JL, Garcia J, Harrich D, Pearson L, Wu F, Gaynor R. Lymphoid specific gene expression of the adenovirus early region 3 promoter is mediated by NF-κB binding motifs. *EMBO J* 1990; 9:4435–4442.
498. Williams RT, Carbonaro-Hall DA, Hall FL. Co-purification of p34cdc2/p58 cyclin A proline-directed protein kinase and the retinoblastoma tumor susceptibility gene product: interaction of an oncogenic: serine/threonine protein kinase with a tumor-suppressor protein. *Oncogene* 1992;7:423–432.
499. Wilson MC, Fraser NW, Darnell JE Jr. Mapping of RNA initiation sites by high doses of UV-irradiation: evidence for three independent promoters within the left-hand 11% of the Ad2 genome. *Virology* 1979;94:175–184.
500. Wohlfart CEG, Svensson UK, Everitt E. Interaction between HeLa cells and adenovirus type 2 virions neutralized by different sera. *J Virol* 1985;56:896–903.
501. Wold WS, Gooding LR. Region E3 of adenovirus: a cassette of genes involved in host immunosurveillance and virus-cell interactions. *Virology* 1991;184:1–8.
502. Wolfson J, Dressler D. Adenovirus DNA contains an inverted terminal repetition. *Proc Natl Acad Sci USA* 1972;69:3054–3057.
503. Wong ML, Hsu MT. Linear adenovirus DNA is organized into supercoiled domains in virus particles. *Nucleic Acids Res* 1989;17:3535–3550.
504. Wu L, Rosser DSE, Schmidt MG, Berk AJ. A TATA box implicated in E1a transcriptional activation of a simple adenovirus 2 promoter. *Nature* 1987;326:512–515.
505. Yee AS, Raychauduri P, Jakoi L, Nevins JR. The adenovirus-inducible factor E2F stimulates transcription after specific DNA binding. *Mol Cell Biol* 1989;9:578–585.
506. Yee S, Branton P. Detection of cellular proteins associated with human adenovirus type 5 early region 1A polypeptides. *Virology* 1985;147: 142–153.
507. Yew PR, Berk AJ. Inhibition of p53 transactivation required for transformation by adenovirus early 1B protein. *Nature* 1992;357:82–85.

508. Yew PR, Liu X, Berk AJ. Adenovirus E1B oncoprotein tethers a transcriptional repression domain to p53. *Genes Dev* 1994;8:190–202.
509. Yoder SS, Robberson BL, Leys EJ, Hook AG, Al-Ubaidi M, Yeung CY, Kellems RE, Berget SM. Control of cellular gene expression during adenovirus infection: induction and shut off of dihydrofolate reductase gene expression by adenovirus type 2. *Mol Cell Biol* 1983;3: 819–828.
510. Yolken RH. Gastroenteritis associated with enteric type adenovirus in hospitalized infants. *J Pediatr* 1982;101:21–26.
511. Yonish-Rouach E, Resnitzky D, Lotem J, Sachs L, Kimchi A, Oren M. Wild-type p53 induces apoptosis of myeloid leukaemic cells that is inhibited by interleukin-6. *Nature* 1991;352:345–347.
512. Yoshimura K, Rosenfeld MA, Seth P, Crystal RG. Adenovirus-mediated augmentation of cell transfection with unmodified plasmid vectors. *J Biol Chem* 1993;268:2300–2303.
513. Yoshinaga S, Dean N, Han M, Berk AJ. Adenovirus stimulation of transcription by RNA polymerase III: evidence for an E1A-dependent increase in transcription factor IIIC concentration. *EMBO J* 1986;5:343–354.
514. Young CSH, Cachianes G, Munz P, Silverstein S. Replication and recombination in adenovirus-infected cells are temporally and functionally related. *J Virol* 1984;51:571–577.
515. Young CSH, Shenk T, Ginsberg HS. The genetic system. In: *The viruses: adenovirus*. New York: Plenum, 1984:125–172.
516. Young CSH, Silverstein SJ. The kinetics of adenovirus recombination in homotypic and heterotypic genetic crosses. *Virology* 1980;101: 503–515.
517. Zamanian M, La Thangue NB. Adenovirus E1a prevents the retinoblastoma gene product from repressing the activity of a cellular transcription factor. *EMBO J* 1992;11:2603–2610.
518. Zamanian M, LaThangue NB. Transcriptional repression by the Rb-related protein p107. *Mol Biol Cell* 1993;4:389–396.
519. Zambetti GP, Levine AJ. A comparison of the biological activities of wild-type and mutant p53. *FASEB J* 1993;7:855–865.
520. Zantema A, Fransen JAM, Davis-Olivier A, Ramaekers FCS, Voojis GP, DeLeys B, van der Eb AJ. Localization of the E1B proteins of adenovirus 5 in transformed cells as revealed by interaction with monoclonal antibodies. *Virology* 1985;142:44–58.
521. Zantema A, Schrier PO, Davis-Olivier A, van Laar T, Vaessen RTMJ, van der Eb AJ. Adenovirus serotype determines association and localization of the large E1B tumor antigen with cellular tumor antigen p53 in transformed cells. *Mol Cell Biol* 1985;5:3084–3091.
522. Zerler B, Roberts RJ, Mathews MB, Moran E. Different functional domains of the adenovirus E1A gene are involved in regulation of host cell cycle products. *Mol Cell Biol* 1987;7:821–829.
523. Zhai Z, Wang X, Qian X. Nuclear matrix-intermediate filament system and its alteration in adenovirus infected HeLa cell. *Cell Biol Int Rep* 1988;12:99–108.
524. Zhang Y, Schneider R. Adenovirus inhibition of cellular protein synthesis and the specific translation of late viral mRNAs. *Seminars in Virology* 1993;4:229–236.
525. Zhu L, van de Heuvel S, Helin K, Fattaey A, Ewen M, Livingston D, Dyson N, Harlow E. Inhibition of cell proliferation by p107, a relative of the retinoblastoma protein. *Genes Dev* 1993;7:1111–1125.
526. Ziff E, Fraser NW. Adenovirus type 2 late mRNA: structural evidence for 3' coterminal species. *J Virol* 1978;25:897–906.
527. Ziff EB, Evans RM. Coincidence of the promoter and capped 5' terminus of RNA from the adenovirus 2 major late transcription unit. *Cell* 1978;15:1463–1476.
528. Zilli D, Voelkel-Johnson C, Skinner T, Laster SM. The adenovirus E3 region 14.7 kDa protein, heat and sodium arsinate inhibit the TNF-induced release of arachidonic acid. *Biochem Biophys Res Comm* 1992;188:177–183.
529. Zock C, Iselt A, Doerfler W. A unique mitigator sequence determines the species specificity of the major late promoter in adenovirus type 12 DNA. *J Virol* 1993;67:682–693.

Fundamental Virology, Third Edition
edited by B.N. Fields, D.M. Knipe, P.M. Howley, et al.
Lippincott - Raven Publishers, Philadelphia © 1996

CHAPTER 31

Parvoviridae: The Viruses and Their Replication

Kenneth I. Berns

The parvoviruses are among the smallest of the DNA animal viruses. The virion has a diameter of 18 to 26 nm and is composed entirely of protein and DNA. The family *Parvoviridae* (27) contains two subfamilies: the *Parvovirinae*, which infect vertebrates, and the *Densovirinae*, which infect insects. Each of the subfamilies contains three genera (Table 1). Because the host range of the *Densovirinae* lies outside the scope of this volume, the replication of these viruses is not reviewed. The interested reader is referred to several recent articles (12,13,81,94). The *Parvovirinae* have a wide distribution among warm-blooded animals, ranging from domestic fowl to humans (Table 1). *Dependoviruses,* also known as adeno-associated viruses (AAV), are unique among animal viruses in that, except under special conditions, they require coinfection with an unrelated helper virus, either adenovirus or a herpesvirus, for productive infection in cell culture (10,42,109,195). In spite of this distinction, there are many similarities between AAV and the members of the *Parvovirus* and *Erythrovirus* genera in genome structure, organization, and expression. It is the intent of this chapter to compare and contrast the replication of these three genera at the molecular and biological levels.

K. I. Berns: Department of Microbiology, Hearst Microbiology Research Center, Cornell University Medical College, New York, New York 10021.

THE VIRION

The parvovirus virion has a relatively simple structure composed of only three proteins and a linear, single-strand DNA molecule. The particle has icosahedral symmetry and a diameter of 18 to 26 nm (1,10,52,109,261,277). The crystal structure of canine parvovirus has been determined. There are 60 protein subunits, consisting primarily of VP2 (see below). The main structural motif is an 8-stranded ß barrel composed of VP2 (the major structural protein, composing 90% of the capsid), but much of the protein is present as large loops connecting the strands in the ß-barrel structure. The arrangement of the protein subunits is unusual; the capsid protein occupies a kite-shaped asymmetric unit around the fivefold axis of symmetry. Each kite-shaped wedge is bounded by a fivefold, a threefold, and two twofold axes of symmetry. In the intact virion there is a "tube" at each fivefold axis of symmetry formed by the association of five copies of two of the strands at the narrow end of the ß barrel (85,261).

The particle has a molecular weight (MW) of 5.5 to 6.2 $\times 10^6$. Approximately 50% of the mass is protein, and the

TABLE 1. *Parvoviridae*

Subfamily *Parvovirinae*
- Genus *Parvovirus*
  - Members
    - mice minute virus [J02275]
    - Aleutian mink disease [M20036]
    - bovine parvovirus [M14363]
    - canine parvovirus [M19296]
    - chicken parvovirus
    - feline panleukopenia [M75728]
    - feline parvovirus
    - goose parvovirus
    - HB parvovirus
    - H-1 parvovirus [X01457]
    - Kilham rat
    - lapine parvovirus
    - LuIII [M81888]
    - mink enteritis
    - canine minute virus
    - porcine parvovirus [D00623]
    - RA-1
    - raccoon parvovirus [M24005]
    - RT parvovirus
    - tumor virus X
- Genus *Erythrovirus*
  - Members
    - B19 virus [M13178,M24682]
- Genus *Dependovirus*
  - Members
    - adeno-associated type 1
    - adeno-associated type 2 [J01901]
    - adeno-associated type 3
    - adeno-associated type 4
    - adeno-associated type 5
    - avian adeno-associated
    - bovine adeno-associated
    - canine adeno-associated
    - equine adeno-associated
    - ovine adeno-associated

Subfamily *Densovirinae*
- Genus *Densovirus*
  - Members
    - *Junonia coenia* densovirus
    - *Galleria mellonella* densovirus
- Genus *Iteravirus*
  - Members
    - *Bombyx mori* densovirus [M15123, M60583, M60584]
- Genus *Contravirus*
  - Members
    - *Aedea aegypti* densovirus
    - *Aedes albopictus* densovirus
  - Tentative species in the genus
    - *Acheta domestica* densovirus
    - *Aedes pseudoscutellaris* densovirus
    - *Agraulis vanillae* densovirus
    - *Casphalia extranea* densovirus
    - *Diatraea saccharalis* densovirus
    - *Euxoa auxiliaris* densovirus
    - *Leuco rrhinia dubia* densovirus
    - *Lymantria dubia* densovirus
    - *Periplanata fuliginosa* densovirus
    - *Pieris rapae* densovirus
    - *Pseudaletia includens* densovirus
    - *Sibine fusca* densovirus
    - *Simulium vittatum* densovirus
  - Tentative species in the subfamily
    - parvo-like virus of crabs
    - hepatopancreatic parvo-like virus of shrimps

DNA sequence accession numbers are shown in brackets.

The table is derived from a report to the ICTV in 1994 by K. I. Berns, M. Bergoin, M. Bloom, M. Lederman, N. Muzyczka, G. Siegl, J. Tal, P. Tattersall (27).

remainder is DNA. Because of the relatively high DNA-to-protein ratio, the buoyant density of the intact virion in CsC1 is 1.39 to 1.42 g/cm^3 (27). The heavy buoyant density in CsC1 permits the ready separation of AAV from helper adenovirus in coinfections. Both heavier and lighter forms of the virion occur. The latter are particles containing DNA molecules with significant deletions, and these particles can function as defective interfering (DI) particles (61,74,75,83,179,204,253). The exact role of the heavier-than-normal particles (153) in infections is unknown. The encapsidated DNA molecules are indistinguishable from those of normal density particles and so, presumably, some of the coat protein molecules are missing. Whether the missing proteins constitute a specific set is unknown. Finally, the sedimentation coefficient of the virion in neutral sucrose gradients is 110 to 122 (27,69).

Possibly as a consequence of its structural simplicity, the virion is extremely resistant to inactivation. It is stable between pH 3 and 9 and at 56°C for 60 min. The virus can be inactivated by formalin, ß-propriolactone, hydroxylamine, and oxidizing agents (27).

The Genome

The genome is a linear, single-strand polydeoxynucleotide chain (69). Several parvovirus DNAs have been completely sequenced (8,9,34,55,78,81,201,208,231,242, 262). H-1 and MMV DNAs (autonomous parvoviruses) contain 5,176 and 5,084 bases, respectively; AAV2 DNA contains 4,680 bases. In general, autonomous parvoviruses encapsidate primarily strands of one polarity, that which is complementary to mRNA, while AAV encapsidates strands of both polarities with equal frequency (26,32,163, 214). However, the distinction is far from absolute. In the case of bovine parvovirus, approximately 20% to 30% of the encapsidated strands are of the same polarity as mRNA (218); in some hosts, LuIII encapsidates equal numbers of strands of both polarities (16,160).

All parvovirus genomes have palindromic sequences at both the 5′ and 3′ termini of the virion (antimessenger) strand. The palindromic sequence at the orientation of the 3′ end of the virion strand of most murine autonomous parvovirus DNAs is approximately 115 bases long (7,8,220). In the human parvovirus B19, these sequences appear to be more than 300 nucleotides in length (77,231). This segment of the DNA can fold back on itself to form a hairpin structure stabilized by hydrogen bonding between the self-complementary sequences (Fig. 1). In contrast to the sequence at the 5′ end of the virion strand, which can exist in either of two orientations, the 3′-terminal sequence is unique. (The relevance of this organization to the process of DNA replication is discussed later.) The hairpin structure that the 3′-terminal sequence can form (Fig. 1) is shaped like either a *Y* or a *T*. Of special interest has been the fact that almost every base pair formed in the short arms of the *Y* or *T* structure is a GC base pair.

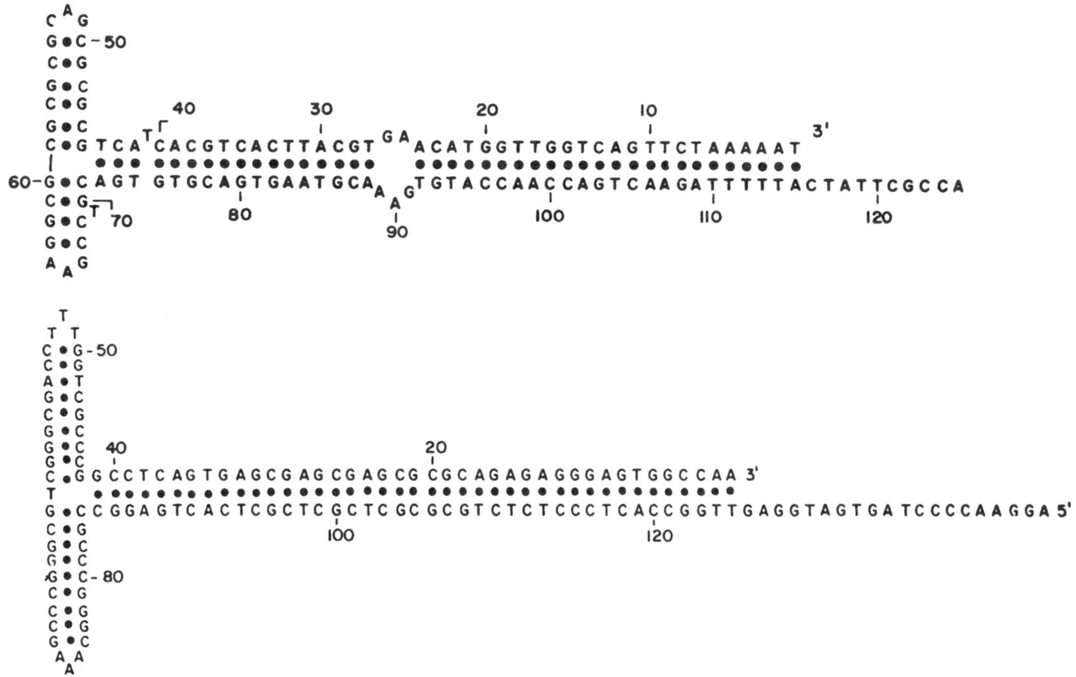

FIG. 1. Comparison of the 3′-terminal nucleotide sequences and hairpins of MMV DNA (*top*) and AAV-2 DNA (*bottom*). The sequences are shown in their most stable secondary structure.

The 5′-terminal sequence is also palindromic and can form a hairpin, but the sequence is completely unrelated to that at the 3′ end of the virion strand. [An apparent exception is the 5′ end of the human B19 genome, which appears to present a terminal repeat of the sequence at the 3′ end (231).] Both the MMV and the H1 DNA 5′ palindromes have been sequenced, and they are 207 and 245 bases long, respectively (9,208). Like the 3′-terminal sequence, the 5′-terminal sequence is not a perfect palindrome; again there are short, self-contained internal palindromic sequences. Unlike the 3′-terminal sequence, the 5′-terminal sequence is not unique. Two sequences that represent an inversion of the terminal segment of the genome are found with equal frequency. The two sequences are a consequence of the fact that the 5′-terminal hairpin is not perfectly symmetric. [A recent report has demonstrated that about 10% of the 3′-terminal sequences of the bovine parvovirus (BPV) genome also represent an inversion (233).]

In contrast to autonomous parvovirus DNA, the AAV genome has an inverted terminal repetition (29,93,129) of 145 nucleotides (158). The first 125 bases form a palindromic sequence that, when folded over to maximize self-base pairing, forms a *Y*- or *T*-shaped structure almost identical to that formed by the 3′-terminal sequence of the virion strand of autonomous parvovirus DNAs. As with the sequence in autonomous parvovirus DNA, the palindromic terminal sequence in AAV DNA may be characterized more accurately as two internal palindromes (nucleotides 42 to 84) flanked by a more extensive palindrome (nucleotides 1 to 41 and 85 to 125). In the folded configuration, only seven bases remain unpaired. Six are required to allow the internal palindromes to fold over, and the seventh separates the two internal palindromes. Unlike the autonomous 3′-terminal sequences, the palindromic part of the AAV terminal repetition is heterogeneous. There are two sequences that reflect an inversion of the terminal 125 bases. In this respect, the AAV terminal repetition is more similar to the autonomous parvovirus 5′-terminal palindromic sequence. Again, the inversion is manifested in two terminal sequences because the internal palindromes are not identical and result in the absence of perfect symmetry in the terminal repeat. Furthermore, there is an additional heterogeneity in the extreme terminal bases. The 5′-terminal sequence in AAV2 DNA is 5′ TTG. However, 50% of the molecules isolated from virions lack the first *T*, and a further 15% lack both *T*s (84).

Analysis by restriction enzyme mapping and DNA sequencing of the terminal sequences in the DNAs of viruses of both genera have played an important role in the current understanding of the mechanism of viral DNA replication. Because mature virion DNA is the end product of DNA replication, any model proposed must be able to account for the nucleotide sequence arrangements found in virion DNA. In return, the terminal sequence organization has offered valuable clues as to the correct mechanism. Thus, the similarities and differences in the terminal sequence arrangements in autonomous parvovirus DNA as compared to AAV DNA have led to a model of DNA replication similar to that for AAV DNA but with special constraints.

There are additional similarities in the internal organization of the genomes of both types of viruses (Fig. 2). All have two large open reading frames (ORFs) that do not overlap (9,55,67,208,231,242). In all, on the physical map, one extends approximately from map position (mp) 5 to 40, while the second extends approximately from mp 50 to 90. As is discussed later, the ORF in the right half of the genome appears to code for all the coat proteins, while that in the left half codes for nonstructural proteins.

Coat Proteins

Both autonomous parvovirus and AAV virions contain three coat proteins [with the exception of Aleutian disease virus (ADV), which contains only two coat proteins] (27). For different species of autonomous parvoviruses, the proteins have approximate MWs of 80,000 to 86,000 (VP-1), 64,000 to 75,000 (VP-2), and 60,000 to 62,000 (VP-3). Those of lapine parvovirus are significantly larger: 96,000 (VP-1), 85,000 (VP-2), and 75,000 (VP-3). Except for ADV, VP-3 is the major coat protein, representing about 80% of the total mass. Some virus preparations are lacking in VP-3, and its abundance appears to depend on the time during infection when the virions are isolated. Majaniemi et al. (160) have reported that, under some conditions, LuIII preparations may contain only VP-3 in detectable quantities. All AAV preparations characterized to date have three coat proteins: VP-1, 87,000; VP-2, 73,000; and VP-3, 62,000 (121–123,215). In neither genus is there any evidence for glycosylation of any of the coat proteins (122). Compositional analyses have been carried out on the coat proteins, and they appear to be relatively rich in acidic amino acids (123,215). AAV-2 coat proteins contain 10.6% glutamine and glutamic acid and 14% asparagine and aspartic acid. To date, however, the coat proteins have proved refractory to sequence analysis. They are very poorly soluble in aqueous solution, and the N terminus is blocked in the case of the AAV coat proteins (121). A theoretical VP-3 amino acid sequence has been derived from the AAV-2 DNA sequence, and the composition is in good agreement with that determined experimentally (242).

Possibly the most interesting property of the coat proteins has been revealed by comparative tryptic digestion. For both autonomous parvoviruses and AAV, all three coat proteins of a given virus yielded very similar tryptic peptides with only a few differences (121,249). These data suggested that almost all the amino acid sequences of all three coat proteins are determined by a common DNA sequence. *In vitro* translation experiments have supported this notion (see below).

In addition to the fact that all three parvovirus coat proteins appear to be coded for by the overlapping in-frame DNA sequences, a further complexity has been reported. In the cases of both AAV-1 and AAV-2, the coat proteins VP-1 and VP-3 can be further subdivided by polyacrylamide gel electrophoresis into several subspecies (171). The molecular basis for the difference in mobilities is unknown.

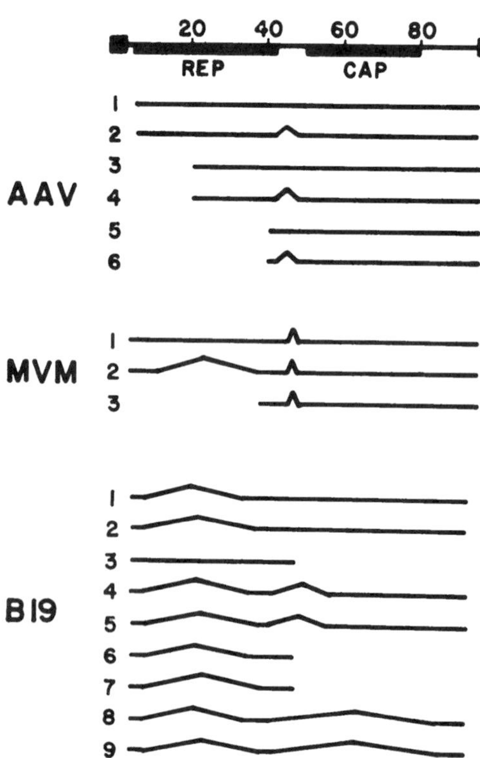

FIG. 2. Transcriptional maps of three parvoviruses. The generalized genetic map is shown (*top*), with terminal palindromes represented (*filled boxes at the ends*). In the cases of AAV and B19, these are terminal repeats. *Internal boxes* represent generalized ORFs. Rep encodes functions required for transcription and DNA replication. CAP encodes the coat proteins. AAV has three promoters, at mp 5, 19, and 40; MMV has two promoters, at mp 4 and 38; and B19 apparently has only one promoter, at mp 6. All AAV and MMV transcripts are coterminous at a polyadenylation signal near the right end of the genomes, whereas some B19 transcripts apparently are polyadenylated at a sequence near the middle of the genome.

PARVOVIRUS INFECTION

Parvoviruses replicate in the nuclei of infected cells. Of all the DNA viruses, they seem to be among the most dependent on cellular function. The autonomous parvoviruses require the cell to go through S phase in order to replicate (30,203,247,265,268). Unlike the polyomaviruses, the parvoviruses do not have the ability to stimulate or turn on host DNA synthesis in resting cells. AAV is even more dependent on the intracellular milieu for a productive infection to occur. In most cases, the cell must be infected by a helper adeno- or herpesvirus to be permissive for AAV replication (10,17,18,30,42,172,195) (Fig. 3). The nature

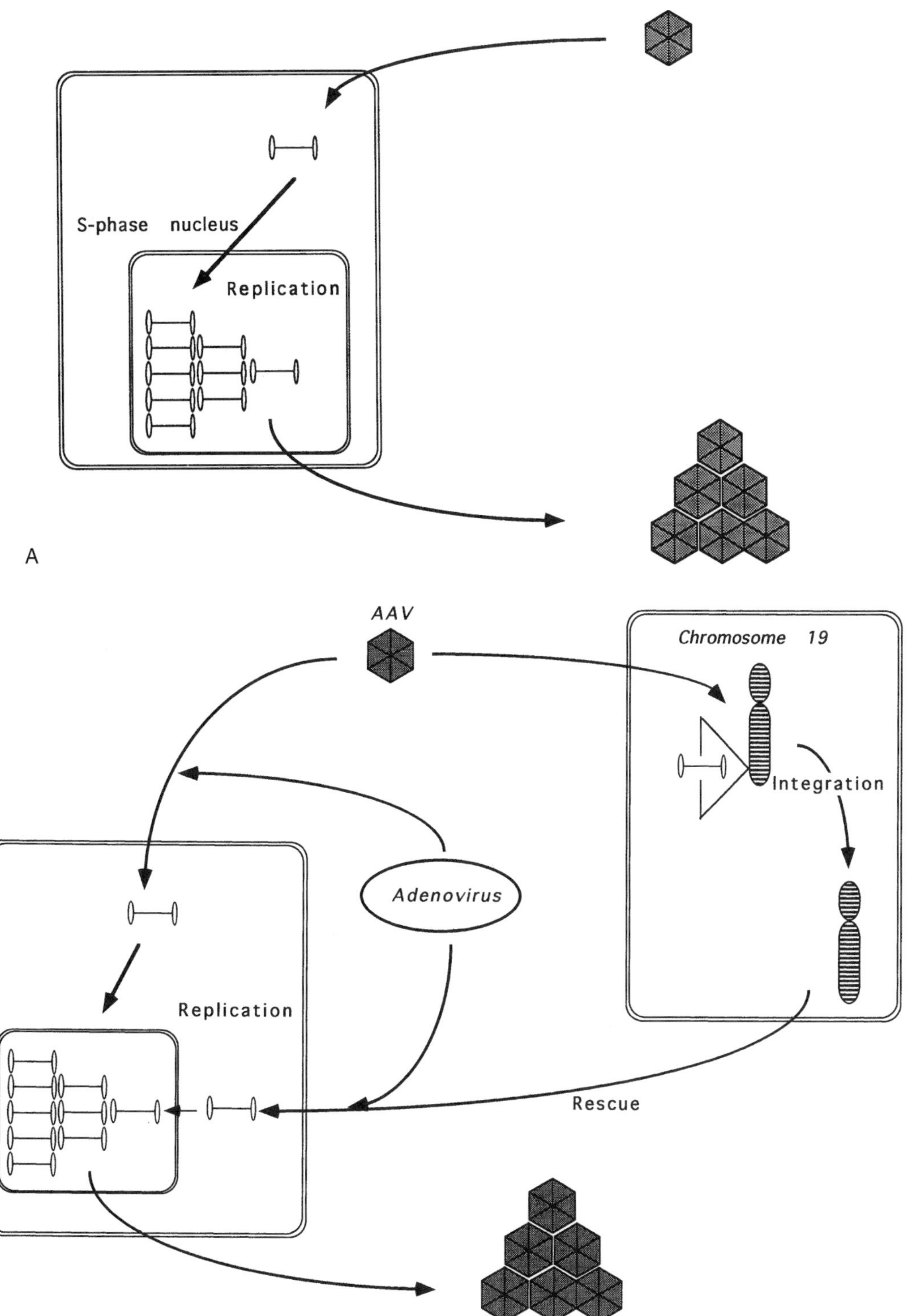

FIG. 3. A: Replication of the autonomous parvoviridae. After adsorption of the virus at the cell membrane of a variety of host cells, the virus enters and is transported to the nucleus. During S-phase the cellular replication machinery is recruited for viral replication leading to virus production and cell lysis. **B:** Replication of adeno-associated virus (AAV). Under nonpermissive conditions, AAV integrates into the q arm of human chromosome 19 where it remains silent until challenged by a helper virus, e.g., adenovirus. This leads to rescue of the integrated virus from the chromosome and induction of the lytic cycle. Under permissive conditions (i.e., in the presence of a helper virus, such as adenovirus), AAV replicates, resulting in host cell lysis.

of the helper factors is described below. There are cell lines that can be made permissive for AAV infection without helper coinfection by various chemical or physical treatments described in the section on helper functions. The autonomous parvoviruses are relatively host-specific, although some hamster viruses were first isolated as apparent contaminants from human tumor cells grown in the laboratory (101), and MMV will grow in both mouse and hamster transformed cell lines (67,248). In the case of MMV, tissue specificity has been observed for differentiated cells (38,169,240). The prototype strain MMV(p) replicates in murine fibroblasts but not in T lymphocytes. A second strain has been discovered, MMV(i), which displays the opposite phenotype.

With their great dependence on cellular functions and on helper virus coinfection in the case of AAV, there was a significant question for a long time as to the extent of parvovirus autoregulation of replication. With development of methods for detailed genetic analysis, it has been demonstrated that the parvoviruses tightly regulate their own replication in the cases of both the autonomous and dependoviruses. Genetic analysis has been greatly facilitated by determination of the entire genomic sequences for several of the parvoviruses and by the infectivity of the cloned duplex forms of the viral DNAs. Thus, it is possible to make and propagate any mutant in a bacterial plasmid vector and to then transfect the intact plasmid clone into cell culture. If the cells are permissive or infected by helper virus (in the case of AAV), the parvoviral genes are expressed, the inserted genome is rescued from the plasmid and replicated, and infectious virions are produced. In this way it has been possible to correlate specific phenotypes with specific mutations at fixed points along the genome and, thus, has been possible to develop meaningful genetic maps (107, 145,174,221,229,230,233).

AUTONOMOUS PARVOVIRUSES

Genetic Map

The autonomous parvovirus genome contains two large ORFs, both on the virion strand (27). The first covers much of the left half of the genome (approximately mp 6 to 42) and encodes two nonstructural (NS) proteins, NS-1 and NS-2. Any mutation within this ORF blocks viral replication and gene expression (206,208,233). The second large ORF occupies much of the right half of the genome (mp 45 to 90) and encodes the coat proteins (208,233). Up to three coat proteins have been detected in the virion. In the case of MMV, it appears that the smallest, VP-3, is generated in the intact capsid by proteolytic cleavage of VP-2 (249). The amino acid sequences of VP-1 and VP-2 are identical except for additional amino acids at the NH_2 terminus of VP-1. Synthesis of VP-1 thus appears to involve splicing to add a second (smaller) ORF to the 5′ end of the major ORF on the right side of the genome (138,177). Mutants altered in either the rep or coat protein ORFs can be complemented in *trans*. However, the palindromic sequences at both termini are required in *cis* for replication to occur.

Transcription

For most autonomous parvoviruses, three polyadenylated transcripts have been identified (27). Two of these are initiated downstream of a promoter (TATA box) at mp 4 and extend to a point near the right end of the genome near mp 95 to 96, where a polyadenylation signal (AATAAA) is located. Both of these transcripts have an intron removed between mp 46 and 48, and one has a large intron removed between mp 8 and 38 (177,196). Those two transcripts are translated to yield nonstructural proteins NS-1 and NS-2, respectively (44,66,148,208). A third polyadenylated transcript is initiated downstream from a promoter at mp 38, and the majority of this species is spliced between mp 46 and 48. The coat proteins VP-1 and VP-2 are translated from the spliced species. Significant variations of this general scheme occur. All transcripts of the autonomous human parvovirus B19 have been reported to initiate from the promoter at mp 6 (35,79,194). Positive regulation of p6 by upstream sequences has been noted (154). Potential internal polyadenylation signals (AATAAA) exist near the middle of several parvovirus genomes, and evidence has recently been obtained that this sequence is used for polyadenylation in the case of ADV and B19 virus (3,194). By the same token, instances of alternative splicing have been demonstrated (177). In some cases at least, this is considered to be significant, especially for the synthesis of the two coat proteins.

Another form of alternative splicing involves the p4 transcript. The large intron donor site regulates the relative amounts of splicing of the larger p4 mRNA (NS1) to form the smaller p4 mRNA (by removal of the large intron, NS2). However, sequences within the small intron (map position 44 to 46) play a primary role in the excision of the large intron. The role of the small intron in this process has been hypothesized to involve initial entry of the spliceasome (281). As an added factor, any frame shift mutations in either exon will block splicing (186).

There is evidence that the large nonstructural protein (NS1) transactivates both regulatory and structural gene expression and is, in turn, regulated by sequences upstream in the terminal hairpin (154). B19 NS1 transactivates the P6 promoter which controls transcription of all B19 genes (80). MMV NS1 has been demonstrated to transactivate the MMV p4 promoter (80) and also the p39 promoter (80). H1 NS1 transactivates the p38 promoter, and the target sequences in the p38 promoter have been identified (209). For full activity the following *cis*-active signals must be intact: (a) the tar sequence (transactivation region sensitive to NS1); (b) the SP-1 binding site (197); and (c) the

TATA box (209), although only the GC box and TATA are essential (2). The transactivation domain in NS1 has been mapped to the C-terminal 129 amino acid residues (103). Negative regulation of p39 has also been observed (135,91), which can be caused by a cellular repressor. Evidence for regulation by RNA processing and protein stability has also been found (228). Another form of regulation has been suggested to involve attenuation of transcription (22,136,137). Temporal regulation of gene expression has been reported for MMV (60), which is in accord with *trans*-activation of the p38 transcript by one or both of the NS proteins.

Regulation of gene expression is also a major factor in tissue specificity. The parvovirus H1 NS2 gene regulates gene expression in a manner dependent on cell type; it does so by virtue of sequences in the 3′ untranslated region (152). A different type of host range mutant is observed for MMV. Wild-type MMV productively infects murine fibroblasts but not T lymphocytes. A mutant MMVi has the reverse phenotype. Both strains are replicated for one round after transfection. The genetic difference is in the middle of the capsid gene and represents several missense mutations, which by extrapolation from the CPV crystal structure affect amino acids on the surface of the virion. However, simple binding to the cells is not apparently affected. Thus, the physiologic lesion is somewhere between entry and uncoating (11).

DNA Replication

Parvovirus DNA replication takes place in the nucleus. The location of replication, combined with the fact that the cell is required to go through S phase in order for replication to occur (62,203,247,265,268), suggests a very close relationship between viral and cellular replication. This notion is further buttressed by the involvement of cellular DNA polymerase(s) in the process. Inhibitor studies have clearly demonstrated that either cellular DNA polymerase α or δ is used (130,199). Sequence analysis of the termini of virion and replicative forms of minute virus of mice DNA suggests a modified rolling hairpin model for autonomous parvovirus DNA replication (6,250). Because all known DNA polymerases have a requirement for a primer with an available 3′ OH, in addition to a template, linear DNA genomes have had to evolve specialized terminal sequences to allow them to maintain the 5′-terminal sequences intact when the primer structure is resolved during replication. In the case of the autonomous parvovirus genome, the terminal sequences are palindromic. It is assumed that the 3′ terminus of the virion strand folds back on itself to serve as the primer to initiate DNA synthesis. Subsequently, the complement of the 5′ terminus (i.e., the 3′ terminus of the complementary strand) is assumed to hairpin to serve as a primer. In some cases, the covalent linkage of the hairpinned sequences to both parental and progeny strands must be resolved, and the palindromic sequence itself must serve as a template to be copied. This synthetic event may require a second DNA polymerase different from that used initially to copy the virion strand.

The model for autonomous parvovirus DNA replication is shown in Fig. 4. Any model of DNA replication must be able to account for the particular features found in mature virion DNA, as well as for intracellular structures (67,151, 205,234,268). In particular, the model in Fig. 4 is somewhat more complex than that illustrated below for AAV because the rodent parvoviruses, which have been studied in most detail, have terminal sequence arrangements, which indicate that the 5′-terminal sequence of the virion is inverted during replication but that the 3′-terminal sequence is not (7,8,67). There are some exceptions to this feature of the mature chromosome. Bovine parvovirus DNA shows evidence that both terminal sequences are inverted during replication, although the majority of the virion strands have the 3′-terminal sequence in the same orientation (233). The model for autonomous parvovirus DNA replication is made more complex by the observation that, in effect, with H1 virus there are two stages of replication (207). The first stage is the conversion of the virion single strand to a duplex replicative form (RF). This phase obviously initiates at the 3′ hairpin end of the parental virion strand. The second stage is the replication of progeny strands from the RF. In this case, initiation predominantly (if not exclusively) takes place at the right end of the RF or at the 5′ end of the parental strand. Two *cis*-acting sequences between nts 4,489 and 4,695 have been identified which are essential for DNA replication (219). Several cellular proteins which bind specifically to this region of the genome have been detected (219,246).

The model for replication has, therefore, the following features:

1. It is a single-strand displacement model of the type also currently favored for adenovirus DNA replication (147). Therefore, there is no discontinuous, lagging strand synthesis.
2. In both stages of DNA synthesis, terminal DNA sequences are used as primers. Therefore, there is no need for RNA or protein primers or a primase in the conventional sense. As described above, one or more cellular DNA polymerases have been implicated in the process, but no requirements for other cellular proteins (such as topoisomerase) have been demonstrated, although certain features of the model would involve a possible topoisomerase-like activity.
3. The model does predict site-specific cleavage of replicative intermediates. This is accomplished by NS1 in *in vitro* assays.

Replicative DNA has protein covalently linked to the 5′ termini, and even the 5′ terminus in virion DNA is blocked to a large degree (58,68,202). The terminal protein corresponds to NS-1 in the case of MMV (68). Possibly conflicting data have been published that material which cross-reacts with the terminal protein is present in uninfected cells

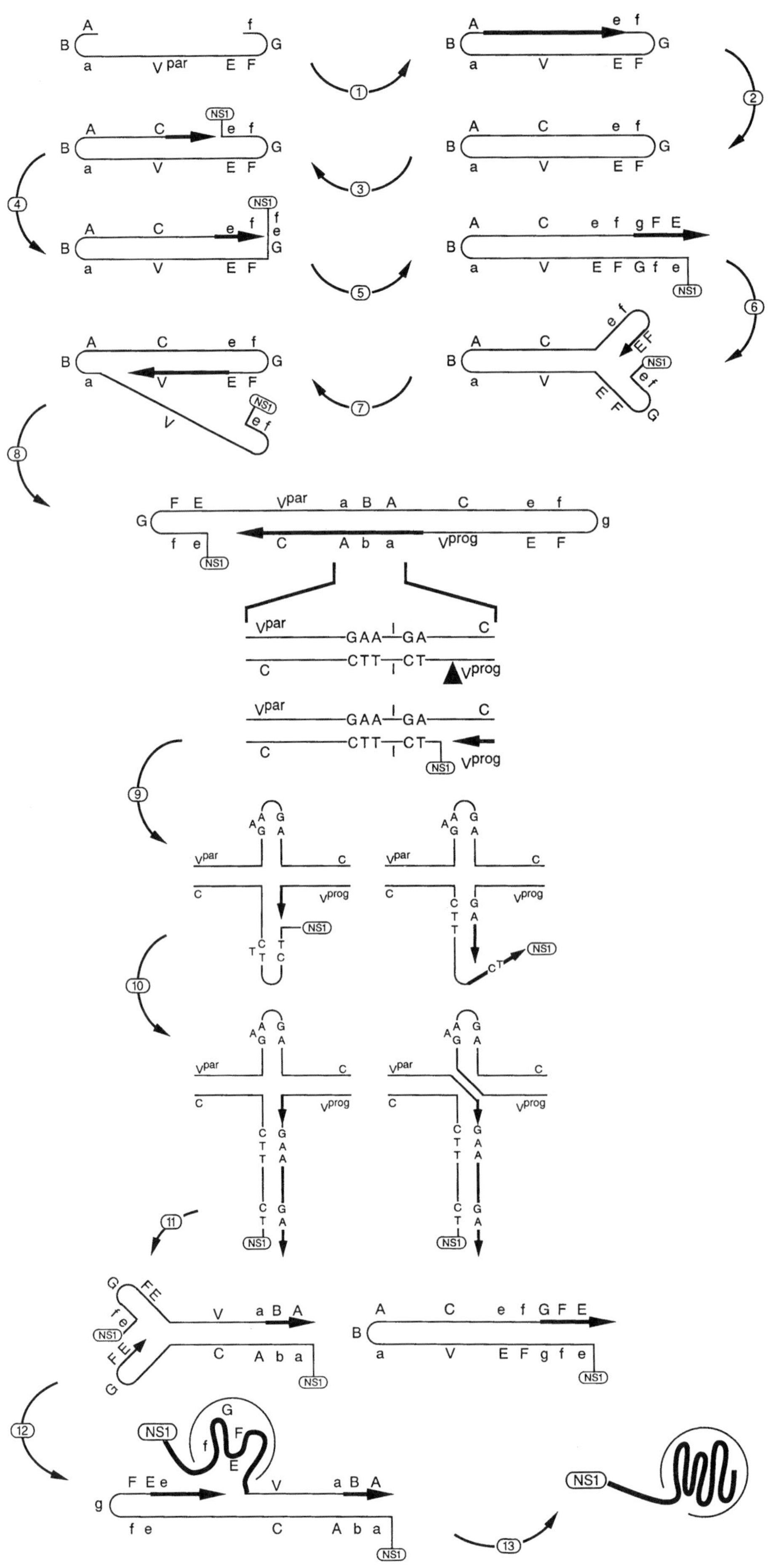

A f
B G
a V par E F
1
A e f
B G
a V E F
2
A C NS1 e f
B G
a V E F
A C e f
B G
a V E F
3
4
NS1
A C e f f e
B G
a V E F
5
A C e f g F E
B
a V E F G f e
NS1
6
A C e f
B G
a V E F
NS1
e f
V
7
A C e f
B
a V
E F
NS1
e f
E F
G
8
F E Vpar a B A C e f
G g
f e C A b a Vprog E F
NS1
Vpar C
GAA GA
CTT CT
C Vprog
Vpar C
GAA GA
CTT CT
C NS1 Vprog
9
10
11
12
13
Vpar
Vprog
C
NS1
A A G G A
C T T C T
C T T
G A A
G F E
f e
NS1
F E
G
V a B A
C A b a
NS1
A C e f G F E
B
a V E F g f e
NS1
NS1
G
f F
E
F E e V a B A
g
f e C A b a
NS1
NS1

(58). Alternatively, these two sets of data may indicate a complex between NS-1 and cellular protein; nucleolin was identified in one case (15). The role of the terminal protein(s) as either a site-specific DNase or a topoisomerase—or as a possible form of primer for DNA synthesis—remains to be elucidated (235,236). NS1 has ATPase and helicase activities which are dissociated by a mutation in the ATP binding domain (120,189).

Special features of the model for DNA replication must be invoked to account for the lack of inversion of the 3′-terminal palindrome of the virion strand. The first is formation of an obligatory dimer duplex replicative intermediate (step 8, Fig. 4). The second involves small, unpaired regions within the terminal palindromes (Fig. 1) which are considered to represent potential signal sequences for specific nicking of intermediates, including that of the dimer-length duplex replicative intermediate, so that the original 3′-terminal sequence of the virion strand can be recreated (step 10, Fig. 4). This step has now been achieved in *in vitro* assays (156). MMV NS1 is the nuclease and becomes covalently attached to the 5′ side of the nick. The dimer bridge is the stretched-out 3′ terminal hairpin in duplex form. The actual nick occurs on the strand complementary to the virion strand to the 5′ side of the sequence TC. In order to have only one orientation of the left end palindrome, there must then be synthesis from the 3′OH produced by the nick, and such synthesis is thought to lead to cruciform formation. Finally, cleavage of the cruciform in a specific orientation is invoked. Such cleavage is suggested to be by a cellular recombinase (S. Cotmore and P. Tattersall, personal communication; C. Astell, personal communication) which may correspond to a cellular protein found associated with the left end of BPV DNA (175). Capsid protein can recognize the 3′ hairpin and can package single strands (273,274). An unusual property of duplex replicative intermediates is that the 5′-terminal sequences of the viral strand are found to be 18 to 26 nucleotides longer than that found in virion DNA (68). It appears that mature virions contain DNA covalently linked at its 5′ end to NS-1, which is located on the outside of the capsid structure (steps 11 to 13, Fig. 4). The 18 to 26 nucleotides, as well as NS-1, potentially can be cleaved off this structure by nucleolytic activity, either in the extracellular milieu or upon entry into a new host cell.

An additional complexity of replication is the generation of variant genomes of two types. The first represents internal deletions of varying extents, and the second is a duplex hairpin molecule representing segments from either the left or right ends of the molecule (83,204). Virions containing the incomplete genomes can function as DI particles (204).

The determination of the details of replication awaits the development of an *in vitro* system for DNA replication (130,199). An assay for initiation of replication has already been reported and was the source of data concerning dimer resolution (65,156). The assay requires a cellular extract and NS1 produced by an expression vector. In some instances synthesis begins to approach full-length DNA. Detailed study of this system should yield significant information concerning the biochemistry of autonomous parvovirus replication.

Protein Synthesis

Little is known concerning the regulation of the synthesis of parvovirus proteins. Synthesis of the nonstructural protein appears to occur first, because the nonstructural protein transcripts appear earlier in the course of infection than the transcripts for the structural proteins and because one or both of the nonstructural proteins regulate gene expression (80,152,209). Both NS-1 and NS-2 are phosphorylated subsequent to translation (67): NS-1 self-associates prior to nuclear localization (190). The coat proteins are thought to be acetylated at the NH_2 termini (121). VP2 has also been reported to be phosphorylated (225). Synthesis of coat proteins is marked by two unusual phenomena. The first is that the major coat protein is produced by proteolytic cleavage of VP-2 (249). The ratio of VP-2 to VP-3 is dependent upon the time course of the infection. The second unusual phenomenon is that the initiator AUG (9) for VP-2 is several hundred bases downstream from the cap site. A possible advantage of this distance between cap site and initiation will be suggested below when AAV protein synthesis is considered. Finally, the available data suggest that alternative splice donor and/or acceptor sites are used to remove the intron between mp 46 and 48 to create the transcript from which VP-1 is translated (177). When expressed in insect cells, the coat proteins can self-assemble to form particles (39,59,125). In the case of B19, the unique region of VP-1 (the NH_2 terminus) appears to project from the surface of the particle (216). For efficient translation of MMV mRNA, NS2 is required (184,185).

Oncolysis

Several original parvovirus isolates were made from tumor cell lines (101). Hence, the question of the ability

FIG. 4. Model for MMV DNA replication. See text for details. ABa, 3′-terminal palindrome of virion strand; FGf, 5′-terminal palindrome of virion strand; e, 18- to 26-nucleotide sequence present in replicative intermediates but not in DNA of nuclease-treated virions; T, site-specific nucleases; V, virion strand; V^{par}, parental virion strand; V^{prog}, progeny virion strand; C, complementary strand; X, site of possible topoisomerase action. Modified from Cotmore et al., ref. 65, with permission.

of the autonomous parvoviruses to transform cells was considered, but no evidence for this has been found. Rather, the isolation of viruses from such cells would seem to reflect the requirement for cell division in order for viral replication to occur. Thus, transformation of nonpermissive murine cells by SV40 renders them permissive for productive infection by MMV (62). This observation may help to explain several instances wherein animals with autonomous parvovirus infection at birth had only 20% of the incidence of tumors found in uninfected siblings. Those animals that had severe manifestations of H1 infection had an even lower incidence of tumors (251). In other studies, Toolan's group also demonstrated that H1 infection inhibited tumor formation in hamsters caused either by treatment with dimethyl benzanthracene or by adenovirus infection (252). Similarly, Kilham rat virus infection inhibited the ability of Moloney leukemia virus to induce leukemia in rats (23). Thus, parvovirus infection can be seen to inhibit tumor formation, but whether this represents a direct inhibition of cellular transformation or simply reflects a lytic effect of parvovirus replication in permissive cells remains to be elucidated.

Susceptibility of transformed cells to cytotoxic replication of parvoviruses may represent cellular dedifferentiation. The model suggested is that the transformed state allows viral expression which, in turn, is cytotoxic. However, the mechanisms involved are more complex, as illustrated by the following experiment. Human cells transformed by human papillomavirus, SV40, hepatitis B virus, Epstein-Barr virus, and human T-cell lymphotrophic virus and containing the viral DNAs were susceptible to H-1 virus infection. However, transformed cells which had lost the DNA of the transforming virus were no longer as susceptible. Of six such lines, two were susceptible, two partially susceptible, and the other two were resistant (82). The levels of NS1 synthesis and parvovirus DNA replication are increased in transformed cells (63,64). Two rat cell lines FR3T3 and NRK differ in permissivity for MMV DNA replication; viral DNA replication is greater in FR3T3 cells. Transformation by the EJ Ha-ras-1 oncogene renders the FR3T3 cells, but not the NRK cells, susceptible to killing by MMV infection. The level of viral DNA replication in the transformed FR3T3 cells was unchanged, but early gene expression was enhanced, suggesting that both early gene expression and DNA replication are involved in the cytolytic effect (64).

Tissue Specificity and Cryptic Infection

Although dividing cells are required for the replication of autonomous parvoviruses, not all dividing cells are susceptible to, or permissive for, viral infection (178). It would appear that with cellular differentiation during the course of development, specific cellular functions that are required for a productive infection may be lost (248). In the case of MMV, susceptible cells have large numbers (10^5 or greater) of surface receptors. The nature of these receptors has not been defined; however, the receptors appear to contain sialic acid, since pretreatment of cells with neuraminidase prevents viral attachment (67). The surface receptors are missing from B lymphocytes. The prototype strain of MMV is restricted in T lymphocytes at the level of gene expression—although the virus enters the cell, as does MMV(i), for which the T cells are permissive (240). However, host range is extended by mutations in the capsid (4,11,89,90). In other cells, transcription can occur, but DNA replication is inhibited. The loss of permissiveness seen at the level of cell culture is reflected in the general resistance of older animals to infection in spite of the presence of rapidly dividing tissues.

The human B19 virus receptor has been identified as globoside (the blood group P antigen). The receptor is found on only a few cell types, and this helps to explain the narrow host range (primarily bone marrow erythrogenic precursors) of B19 virus (40,41). Some host range specificity has also been attributed to tissue-specific early gene expression (155). Both NS and capsid genes have been implicated in porcine parvovirus host range (263). In the case of MMV, NS2 has been specifically implicated in tissue specificity (184).

Parvoviruses frequently cause cryptic infections, which can imply a low level of viral multiplication in the intact host (67). At the level of cell culture, this seems best illustrated by the establishment of carrier cultures in which viral multiplication also occurs at a low level. The situation is due to a low fraction of cells that are permissive and is plastic in that viral variants frequently arise which can infect the majority cells in the culture (213). The reciprocal situation has also been observed in which resistant cells arise from clones of susceptible cells. The frequency of cryptic infection raises the question of whether the autonomous parvoviruses can cause latent infections by integration of the viral genome into cellular DNA as a provirus, as is seen with AAV. Evidence for persistence of B19 viral DNA *in vivo* has been reported (88). This point is undecided at present. No evidence for integration of viral DNA during viral replication can be detected, nor has integration been observed during abortive infections. However, MMV DNA cloned into bacterial plasmid vectors is infectious, and the genome can be rescued (albeit at a relatively low frequency) upon transfection into permissive cells (174).

DEPENDOVIRUSES

Replication

Until very recently, the outstanding feature of AAV replication in cell culture has been the requirement for coinfection of the cell by an unrelated helper virus, either an adenovirus or a herpesvirus (10,42,109,172,195). Two groups have reported that pretreatment of several cell lines with a variety of toxic agents has rendered the cells permissive for AAV replication in the absence of helper virus coinfection (226,278,279). Because of these results, the concept of AAV defectiveness has had to be reconsidered. How do the helper functions supplied by adenovirus or herpesvirus help AAV replication? Why is AAV successful as a virus in spite of its dependence on the helper viruses? The aim of this section is to understand the nature of the interaction between AAV and its helper viruses and also between AAV and the cell. The latter consideration is particularly emphasized by the frequency with which AAV is able to establish latent infections, both in the intact host and in cell culture. AAV has not been associated with disease in any of its wide range of hosts and, in particular, has never been demonstrated to function as a tumor virus, in spite of its propensity to integrate into the cellular genome as a rescuable provirus. The question of the consequences to the host of AAV latent infection remains unanswered, but the intriguing possibility that AAV infection may actually be beneficial to the host will be considered below. Finally, the closeness of the relationship between the dependoviruses and the autonomous parvoviruses will be considered in light of new concepts developed concerning AAV replication.

Helper Functions

Adenoviruses (Ad), herpes simplex virus (HSV) types I and II, cytomegalovirus, and pseudorabies virus all serve as complete helpers for AAV replication (10,17,42,109,172, 195). The helper host range is identical to the normal host range for the helper virus. Genetic analysis of helper functions has been most extensive for Ad, which has the best characterized genetic map (46,48,117–119,144,170,183, 200,210–212,222). Many of the identified Ad early functions serve as helper functions for AAV replication, but no Ad late helper functions have been found. Ad early region 1A (E1A) function is required for the other Ad early regions to be transcribed (25,124). Similarly, an E1A function is required for AAV transcripts to be detected by Northern blotting (118,212). Two E1A proteins with overlapping sequences of 289 and 243 amino acids, respectively, have been identified (24). The former can both activate and inhibit gene expression in *trans*, whereas the latter E1A product primarily inhibits gene expression in *trans*. Thus, although no definitive data have been published, it is assumed that the 289-amino-acid E1A protein is responsible for *trans*-activation of AAV gene expression.

E4 was originally identified as encoding a helper function specifically required for AAV DNA replication (210). The E4 product involved is a 35-kd protein that normally forms a complex with the 55-kd E1B protein during a productive Ad infection. In contrast to the original identification of E4 function as directed specifically toward AAV DNA replication, it now appears more likely that the E4 35-kd protein is involved, along with the E1B 55-kd protein, in regulation of gene expression, possibly at the level of transcript transport (222). In a general sense, E1B has also been identified as being required for AAV DNA replication to occur but does not seem to have a consistent effect on AAV transcript accumulation (118,144,211).

E2A encodes a 72-kd single-strand DNA-binding protein (DBP) which is required for Ad DNA synthesis but which is not required for AAV DNA replication (183), although particle formation is greatly inhibited by certain *E2A* gene mutations (119). The DBP stimulates transcription from AAV promoters (45,51). Another mechanism of action of the E2A protein from the point of view of AAV replication has been suggested to be at the level of transcript transport from nucleus to cytoplasm. Significantly, E2B, which produces the terminal protein, (243) and the DNA polymerase (244), both of which are directly involved in the process of Ad DNA replication, is not required for AAV replication. Two salient points are apparent from the available analyses of the Ad helper functions. The first is that all of the Ad helper functions affect gene expression in a general sense within the infected cell. The second is that none involve products that appear to interact directly with the AAV genome. Thus, AAV sequence specificity does not appear to be directly involved with the helper functions. Rather, the helper functions all appear aimed in a general sense at enhancing expression of certain classes of genes within the cell, regardless of whether or not the genes are of cellular or viral origin. This type of rather nonspecific effect makes it easier to understand why a variety of unrelated herpesviruses can also serve as complete helpers. Finally, Ad VAI RNA has been reported to facilitate the initiation of AAV protein synthesis (117,227).

Two detailed studies on the HSV-1 genes required for productive AAV infection have been reported. Unfortunately, the conclusions were somewhat different. The first study (176) identified the following HSV genes as providing helper functions for AAV: ICP4 transactivator, DNA polymerase, ICP8 single-stranded DNA binding protein, the origin binding protein, and two of the three units of the helicase-primase complex (UL5, UL8). The second study (271) identified ICP8 and all three components of the helicase-primase complex as helper functions. It should be noted that the latter study used a heterologous promoter; normally, ICP4 function is required for expression of HSV replication genes. The former study found that maximal AAV replication required all the genes identified but that only ICP4 was essential for AAV replication per se. Similarly, in the latter experiments, the third component of the helicase-primase complex (UL52) was not essential. A significant question is: Why, with Ad as a helper, are none of the Ad replication genes required (except E2A) while HSV replication genes are required? The difference may be a reflection that AAV DNA replication primarily uses cellular factors in an Ad coinfection, but apparently uses viral equivalents with HSV as a helper.

The notion of AAV replication reflecting the general classes of gene expression permitted by the intracellular milieu suggests that it ought to be possible to find cells in culture with a permissive intracellular milieu in the absence of a helper virus infection. Indeed, this has now been accomplished by two laboratories. Exposure of several cell lines to a variety of genotoxic agents renders the cells permissive for AAV productive infection in the absence of helper virus coinfection. Agents that have been successfully used include ultraviolet irradiation, cycloheximide, hydroxyurea, and several chemical carcinogens (226,278, 279). Thus, the rather unusual behavior of AAV in cell culture does not appear to reflect viral defectiveness but, instead, seems to indicate a rather selective program of replication. It does seem likely, though, that AAV active replication is tied closely to that of Ad *in vivo*.

Genetic Map

The AAV genetic map is quite similar to that of the autonomous parvoviruses and has been derived almost exclusively from studies of AAV2 (107,229,230,254). There is a large ORF (cap) on the right side of the genome (mp 50 to 90) which encodes the coat proteins of the virus (Fig. 2). Mutations within this region do not block DNA replication, but the accumulation of progeny single strands is inhibited, presumably because this requires encapsidation. There is also a large ORF in the left half of the genome (mp 5 to 40) which has been dubbed the *rep region* because any frame-shift mutation or significant deletion within the region blocks DNA replication. At least four proteins have been detected which correspond to four mRNAs that have been mapped to this region (173). Because these four proteins have the majority of their amino acid sequences in common, it has been difficult to assign specific functions to each. It is possible to selectively eliminate two of them (from the p5 transcripts), and this type of mutation completely blocks all AAV-directed transcript accumulation that can be detected by Northern blot analysis. In one instance, it has been possible to detect an inhibitory effect on heterologous promoters by the remaining two products, but this effect was dependent on the type of cells used for transfection (140). In another study the initiator AUG for the two smaller Rep proteins was altered to GGG (54). The resulting mutant was able to replicate DNA, but no mature single-stranded DNA was encapsidated. The phenotype was similar to that of a coat protein mutant. The significance is unclear; it was hypothesized that the mutation affected a specific step involved in separation of single strands from duplex replicative intermediates.

The larger Rep proteins play a critical regulatory role in every phase of the AAV life cycle. Under nonpermissive conditions (no helper) Rep 68/78 negatively regulates AAV gene expression and DNA replication and appears to be important for site-specific integration in the host cell genome to establish a latent infection (282). In the presence of helper, Rep 68/78 is a transactivator of AAV gene expression and essential for DNA replication and rescue of the viral genome from the integrated state. Other (smaller) ORFs exist near the middle of the genome, but the production of proteins corresponding to these regions is less clear. There has been a report of the synthesis of smaller polypeptides in *in vitro* translation systems programmed by RNAs from infected cells which may correspond to these ORFs (20,43). One ORF near the middle of the genome (mp 42 to 48) has been dubbed the *Lip region*, because a mutation in this region gives a low yield of infectious particles even though the level of DNA and structural protein synthesis is nearly at normal levels. It now appears that this sequence encodes the amino terminus of the largest coat protein, VP-1, which is a minor component of the capsid (20,259).

Mutations in either the rep or cap ORFs can be complemented, so it was originally assumed that they contained no critical *cis*-acting regulatory sequences. However, as described below, it now appears that the rep ORF contains one or two sequences that act as *cis*-active negative regulators of transcript accumulation for those RNAs that cover this region (141,258). The inverted terminal repeat (tr) of 145 bases is required in *cis* for both DNA replication and transcription. In addition to these functions, in normal productive infection the tr is required for (a) encapsidation, (b) integration of the genome during the establishment of a latent infection, and (c) rescue of the genome from the integrated state. All of these functions are detailed below.

It would appear that almost all viral genomes must encode a minimal number of functions in order to assure biological continuity. In the case of the smaller genomes, in particular, this requirement means that many of the genomic sequences must serve more than one function. Thus, in the case of AAV the left ORF encodes at least four proteins with overlapping sequences, and the right ORF encodes at least three coat proteins with overlapping sequences. The manner in which AAV achieves this is described below.

Transcription

The regulation of AAV transcription is especially complex because of the requirement for helper virus coinfection or for alteration of the internal cellular milieu by treatment with toxic agents (see above). In the case of Ad coinfection, at least four Ad early regions (E1A, E1B, E2A, E4) contribute to the synthesis, transport, and translation of AAV mRNAs (46,48,117–119,144,183,210–212,222). Thus, the initial question was the extent, if any, to which AAV regulated its own expression. It has now become clear that AAV highly regulates its expression and that of the helper virus. In a productive infection, six polyadenylated, capped RNAs are detectable by Northern blotting (96,98,146). In decreasing size, the mRNAs have apparent lengths of 4.2 kb, 3.9 kb, 3.6 kb, 3.3 kb, 2.6 kb, and

2.3 kb. The smallest RNA is the major species and is translated into all three-coat proteins. Three promoters at mp 5, 19, and 40 have been identified (97,159). All of the RNAs are 3′-coterminal and are polyadenylated downstream from an AATAAA signal at mp 96 (242). There is another AATAAA at mp 42, but transcripts corresponding to polyadenylation at this signal are not found reproducibly. The two largest transcripts (4.2 and 3.9 kb) are synthesized from the promoter at mp 5 (p5 promoter and transcripts), the next two (3.6 and 3.3 kb) are from the promoter at mp 19 (p19 promoter and transcripts), and the two smallest RNAs are from the promoter at mp40 (p40 promoter and transcripts). There is an intron at mp 42 to 46, and there are both spliced and unspliced mRNAs produced from all three promoters—hence the three pairs. Recent evidence detailed below suggests that an alternative splice acceptor may be used for at least one of the 2.3-kb RNAs (p40) (20). In the case of the p5 and p19 RNAs, the unspliced species is the major one, but the reverse is true of the p40 RNAs.

One or more of the *rep* gene products *trans*-activates transcription from all three AAV promoters. Even when the AAV genome is delivered to the cell in the double-strand form as a plasmid clone, an intact *rep* gene is required for the synthesis of large amounts of AAV RNA in the presence of an Ad infection (141,256). Nuclear run-on experiments have demonstrated that the *trans*-activation occurs at the level of initiation of transcription (141). In the absence of permissive intracellular conditions (e.g., no helper virus coinfection), it has not been possible to detect AAV transcripts by Northern blot analysis; however, *rep* gene products can be detected immunologically (173), and it is possible to detect *rep* gene function biologically (140,257). Under these conditions, AAV inhibits expression from a number of heterologous promoters (see below) and inhibits at least one, if not all, of its own promoters. The clearest evidence for this relates to the p40 promoter, which functions constitutively in many cells when removed from the context of the AAV genome. Data regarding effects on the p5 and p19 promoters are more preliminary, but p5 cat and p19 cat constructs function constitutively when transfected into human cells. This expression is inhibited by transfection with an intact AAV clone; in turn, the inhibition (mapped to the *rep* gene) can be reversed by Ad infection (19). When a heterologous constitutive promoter is substituted for the p5 promoter within the context of the AAV genome, it is no longer constitutive, although it still functions in the presence of an Ad infection (139). Thus, it seems likely that in the absence of a permissive intracellular milieu there is a low level of rep expression that represses normal levels of AAV gene expression and that the repression can be overcome specifically for AAV gene expression by coinfection with helper virus.

Chang and Shenk (51) mapped two *cis*-active sequences in the p5 promoter. The first corresponded to a sequence in the adenovirus major late promoter activated by E1A. The second was a tandemly repeated decamer (TTTTGCGACA) which binds the cellular YY1 factor. Initially, the latter was thought to explain the lack of permissivity in the absence of helper virus, because in the presence of adenovirus, constructs containing the p5 promoter were activated. However, Beaton et al. (19) demonstrated that if all the sequences upstream of the p5 promoter up to the boundary of the tr were present in the construct, the p5 promoter was active without an adenovirus helper. Addition of a *rep* gene construct did inhibit the latter construct, and the inhibition was reversed by adenovirus infection. Thus, the exact role of YY1 in AAV infection is uncertain. YY1 has been found to act as both an activator and an inhibitor (232). McCarty et al. (165) have demonstrated that *cis*-active sequences in p19 and p40 work coordinately to regulate transcription. Presumably, this reflects the ability of Rep 68 to bind to these regions (167).

The p5 and p19 transcripts usually represent only 10% to 20% of the AAV RNA detected by Northern blot analysis. However, in nuclear run-on experiments, the levels of initiation of transcription appear to be the same for all three AAV promoters (141). The discordance between transcript initiation and RNA accumulation appears to be due to *cis*-active negative regulation by one or more sequences contained within the structural region of the *rep* gene. When a mutant genome deleted between mp 10 and 37 is complemented by another with an intact *rep* gene, both p5 and p40 transcripts are made from the mp 10 to 37 deletion mutant. (No p19 transcript is made, because the p19 promoter is gone.) The p40 transcripts accumulate at a level equal to or greater than that of the complementer, but now the amount of p5 RNA from the mutant accumulates in an amount equal to that of p40 RNA, even though the complementer's p5 RNA is at the normal low level. Thus, there is a *cis*-active negative effect. Smaller deletion mutants have been used to map one such *cis*-active negative regulatory sequence to a region between mp 28 and 37 (141). The sequence effect is not specific for the p5 and/or p19 promoters. Placement of the sequence downstream from the p40 promoter has the same effect on the accumulation of transcripts from this promoter (M. A. Labow, unpublished data). Whether the *cis*-active negative regulatory sequence inhibits transcript elongation (e.g., attenuation) or adversely affects transcript stability is not known. Teleologically, this type of regulation makes some sense. Presumably, greater amounts of structural protein mRNA than of regulatory protein mRNA are required.

In summary, AAV transcription and gene expression are tightly controlled and autoregulated in both positive and negative fashions. The autoregulation has both *trans*- and *cis*-active components.

DNA Replication

AAV DNA replication occurs via a single-strand displacement mechanism (Fig. 5) that is quite similar to that

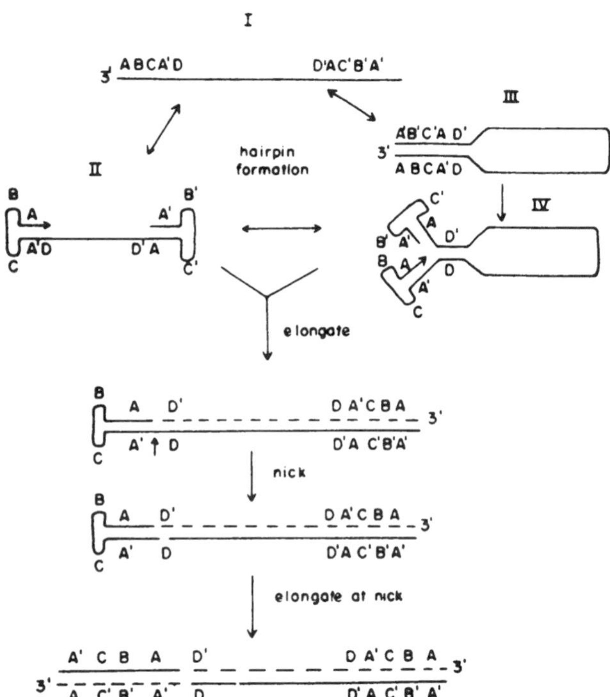

FIG. 5. Model for AAV DNA replication. See text for details. ABCA′D and D′AC′B′A′, inverted palindromic terminal repeats.

of the autonomous parvoviruses (104,105,245). The palindromic inverted terminal repeats (trs) at the 3′ ends of both strands can form a hairpin to serve as primers to initiate synthesis. No evidence has been found for RNA primers or for the equivalent of Okazaki fragments (the presence of which would be indicative of lagging strand synthesis). There are two major differences in comparison to autonomous parvovirus DNA replication. The first is the existence of the terminal repetition and the evidence for inversion of the terminal sequences at both the 3′ and 5′ termini of both complementary strands. These differences introduce a note of symmetry into the process and make a much simpler model possible. However, the possibility of base pairing between the complementary ends is introduced. The possible consequences of this are discussed below. The second major difference is that no evidence has been found for the covalent attachment of a protein to the 5′ termini of replicative intermediates. The model assumes that a protein does interact near the termini to resolve the covalent terminal hairpin structure of a replicative intermediate by making a nick in the parental strand at or near base 125, but there is no evidence *in vivo* for formation of a stable protein-DNA linkage resulting from this process, although such a linkage is seen *in vitro* (see below). The nicking reaction does not appear to be absolutely precise. The 5′-terminal sequence is TTG; however, 50% of virion strands are missing the terminal T, and another 15% are missing both Ts (84).

In the steps leading to initiation of the first round of DNA synthesis, two pathways are thought to be possible. One alternative would be for the 3′ terminus to simply hairpin on itself to directly form a primer. The other possibility is for the trs to base-pair to form an H-bond-stabilized single-strand circle. Two potential advantages would occur from this. The first would be that the substrate for the initiation of the first round of synthesis would be a double helix and would thus be conformationally equivalent to the duplex substrates used to initiate subsequent rounds of replication in a manner first suggested by Lechner and Kelly for Ad DNA synthesis (147). The second advantage is that formation of such a structure would help to explain the mechanism by which AAV DNA can repair significant deletions and mutations in one tr if the tr at the other end of the genome is intact (223,230,254). As described above, the cloned duplex form of the genome is infectious when transfected into Ad-infected cells. Clones with deletions of up to 113 bases at one end are still infectious, and the resulting progeny contain genomes with wild-type termini at both ends (223). The repair process is thought to involve (a) base pairing between the remaining complementary bases (32 base pairs in this case) and (b) use of the shorter 3′-terminal strand as a primer for repair synthesis, using the intact 5′-tr strand as the template. Mutants with deletions that extend beyond the bounds of the tr cannot be rescued. Thus, the tr not only helps to resolve the problem inherent to linear genomes of maintaining the integrity of the 5′-terminal sequences during the process of DNA replication, but it also confers the ability to repair terminal deletions of the type described above. There also exists the ability to correct mutant sequences in the tr (described below).

The outstanding feature of the tr is the *T*-shaped structure that is formed when the terminal 125 bases are folded on themselves to optimize base pairing. Deletion of an 11-base symmetrical sequence (at nucleotides 50 to 60) from both trs, thus effectively removing one of the cross-arms of the *T*, abolishes the ability (>99%) of a cloned genome to be rescued and replicated (223). Substitution of an alternative 8- or 12-base symmetrical sequence restores viability to the cloned genome (149). Therefore, at least in the region of the cross-arms, the conformation of the tr, when folded on itself, appears to take precedence over the actual sequence in terms of rescue and replication of the genome from the cloned state in a plasmid. Virions containing genomes with mutant trs of this type appear to be nearly as infectious as wild-type virus. No differences can be detected in either the rate or extent of DNA replication that occurs after viral infection under permissive conditions. Substitution of an asymmetric 12-base sequence does not significantly restore viability (37).

The viability of the symmetrical substitutions offers the possibility of testing, whether or not the ability of the trs to perfectly base-pair is a requirement for normal AAV DNA replication. The AAV tr in the virion genome can exist in either the flip or the flop orientation as a consequence of the inversion of the imperfectly symmetrical

125-base terminal sequence during replication. Thus, in a molecule with a tr in the flip orientation at one end and with a tr in the flop orientation at the other end, the trs are not perfectly complementary. Packaged genomes are equally divided between those whose trs at both ends are in the same orientation and those whose trs are of opposite orientation (157). This would be expected from the model as a consequence of the last round of replication prior to packaging. The real question is whether or not there must be gene conversion to produce identical trs before a subsequent round of replication can occur.

To test this, chimeric molecules with one wild-type tr and one mutant tr were tested. Under some conditions, the chimeras were able to undergo many rounds of replication and encapsidation without the occurrence of gene conversion. Thus, although the termini of AAV DNA definitely can interact during the process of replication, it does not appear to be a requirement that they be able to perfectly base-pair or even that they must interact in order for replication to occur (36).

As is the case with the autonomous parvoviruses, AAV DNA replication generates defective genomes. Again, structures representing internal deletions and hairpinned sequences from either end of the genome have been reported (74,75,105).

Although Ad functions are required normally for AAV DNA replication to occur, there is no evidence that any Ad gene functions interact directly in the actual process of AAV DNA synthesis. The three Ad gene products directly required for Ad DNA synthesis—the terminal protein, the Ad DNA polymerase, and the Ad single-strand DNA-binding protein—are not required for AAV DNA synthesis. On the other hand, AAV gene functions are required directly for AAV DNA synthesis to occur. The *rep* gene was so named because one or more of the products it produces is required in *trans* for AAV DNA replication to occur (107). The initial question was whether the requirement was indirect (i.e., for transcription) or direct. Three lines of evidence suggest that at least some of the interaction is direct. The first is a mutant that is the consequence of an 8-base frame-shift insertion at mp 42. This mutant accumulates about 20% to 40% of the normal level of AAV RNA under permissive conditions, but DNA synthesis is decreased 100-fold (107). The second piece of data suggesting a direct role for an AAV gene product comes from studies on AAV/SV40 hybrid genomes (139). The SV40 regulatory region (nucleotides 5,170 to 270), which contains all the known SV40 promoters and enhancers as well as the SV40 origin of DNA replication (ori), was substituted for the AAV p5 promoter and approximately 100 upstream bases (to the boundary of the tr). The cap site of the p5 transcript was left intact, and the SV40 sequence was oriented so that the early promoter could function. In HeLa cells, this construct was phenotypically indistinguishable from wild-type AAV; no transcripts were made in the absence of Ad infection, but in the presence of Ad the hybrid construct was infectious and produced virus. The critical property, however, was observed in COS7 cells (SV40-transformed monkey cells), which synthesize the SV40 T antigen constitutively. Normally, when a plasmid with an SV40 *ori* is transfected into COS7 cells, it replicates. However, the AAV/SV40 hybrid described above was greatly inhibited in its replication. The inhibition was relieved by mutation within the *rep* gene, but replication of the *rep* mutant could be inhibited in *trans* by the original hybrid construct. Interestingly, the hybrid did not inhibit replication of pSV2 *cat*, which contained an identical set of SV40 sequences; thus, the existence of a *cis*-active AAV target sequence was implied. The target sequence has been mapped to the AAV tr between nucleotides 55 and 125. Removal of the trs from the original hybrid genome allows it to replicate as well as pSV2 *cat*. These data clearly show that one or more rep proteins must be produced and be functional without helper virus.

Although these data were obtained with an AAV/SV40 hybrid, they suggest the possibility that AAV has the ability to actively inhibit its own DNA replication under non-permissive conditions, regardless of whether the replication is via an AAV mechanism or a heterologous (SV40) mechanism. Possible implications of this are discussed below in the section on latent infections.

A third line of evidence for a direct role for Rep 68/78 in DNA replication has been demonstrated in two assays of *in vitro* AAV DNA replication. The first assay utilizes a linear AAV duplex molecule which is covalently crosslinked at both ends by the hairpinned form of the tr. In this assay Rep 68/78 protein (produced in a baculovirus expression system) is required, as is an extract from cells infected by Ad (113,188). The major product is linear duplex AAV DNA containing one newly synthesized strand and extended ends (i.e., each strand contains a complete, linear itr). In this assay, extracts of uninfected HeLa cells do not support replication, even if supplemented with Rep 68/78. A second assay has been described (57,111,112,150, 270) in which the substrate is either linear duplex AAV DNA or the same DNA contained in a pBR322 vector. Again Rep 68/78 is required (produced in either a vaccinia/T7 expression system or as a maltose binding protein fusion in an *E. coli* expression system). However, in this assay an extract from uninfected HeLa cells does support replication. When the template is the plasmid construct, replication leads to rescue of the AAV insert from pBR322 and, in some instances, pBR322 gets replicated. The AAV product is like that described for the previous assay. If the substrate is the "no-end" DNA used in the first assay described, an Ad-infected HeLa cell extract is again required to support replication. Thus, with one substrate-template, only Rep 68/78 is required for one round of replication, but a change in the structure of the substrate-template requires an additional factor(s). Whether the additional factor(s) is an Ad gene product or a cellular product induced by Ad remains to be determined.

An important observation is that rescue of the AAV insert from the integrated state can be separated from the initiation of DNA replication in terms of template sequence requirement. Deletion of the terminal 55 bases from both trs in the plasmid construct prevents rescue of the AAV insert, but allows the entire construct to be replicated. The implications of this are that ori function does not require the entire itr and that a hairpin per se is not required for DNA replication (111). Requirements are Rep binding and the so-called terminal resolution site nicked by Rep (see Fig. 5).

The availability of Rep 68/78 from a variety of expression vectors has permitted more detailed genetic and biochemical analyses of the properties of the protein. The protein has been found to bind to the tr (5,114,116,164,192, 193,237,239) and is dependent on a dodecamer sequence $(GCTC)_3$ in the stem of the tr (115). Rep 68/78 is an ATPase, a helicase, and a site-specific DNase (115) which nicks after nt124 in the tr *in vitro*. The nicking reaction results in covalent linkage of Rep 68/78 to the 5′ PO_4 at the nick site (114–116,238). Rep 68/78 can also resolve covalently joined AAV ends *in vitro* (238,270).

Protein Synthesis

Four nonstructural proteins have been identified immunologically (173,260) and expressed *in vitro* (57,113, 150). Approximate MWs estimated from electrophoretic mobilities on SDS polyacrylamide gels are 78 kd, 68 kd, 52 kd, and 40 kd, respectively. The sizes would correspond well to the predicted translational products of the spliced and unspliced mRNAs from the p5 and p19 promoters. The amounts of the individual proteins correspond reasonably well to the relative accumulations of the individual transcripts, and mutations that delete the p5 promoter inhibit synthesis of the two larger nonstructural proteins but not that of the smaller ones. Although the data are not entirely consistent, the two larger species appear to be nuclear proteins, while the smaller ones from the p19 transcripts are also found in the cytoplasm (173). Although it is known that in the case of a helper virus coinfection the Ad-virus-associated RNAs are required for synthesis of the coat proteins, it is not known whether the virus-associated RNAs are also required for the levels of Rep protein synthesis necessary for productive infection. At present, there is no evidence that Rep protein synthesis precedes that of the structural proteins—although it would seem likely, since one or more of the *rep* gene functions is to *trans*-activate the p40 promoter. The existence of smaller Rep proteins has been suggested from *in vitro* translation of RNAs from infected cells (43). It remains to be determined from which RNAs they are translated and what, if any, role they play in replication.

The various functions ascribed to Rep 68/78 have been mapped along the primary structure of the proteins (166, 193,280). Amino acids 56 to 490 were important for DNA replication and for transactivation. Nicking at the terminal resolution site (nts 124 to 125) also required amino acids 134 to 490. Binding to the tr required amino acids 134 to 142 and 415 to 490 (166,193,280). Rep 68/78 has been implicated in the inhibition by AAV of oncogenesis; mutants involving amino acids 25 to 56, 73 to 74, 164 to 165, 257 to 346, had a significantly reduced capacity to inhibit E1A-ras transformation. A nuclear homing signal has been mapped to amino acids 483 to 519 (280).

All of the coat proteins, VP1-VP3, have been reported to be translated from the spliced 2.3-kb p40 mRNA species (119). This presents a potential problem because if the originally reported intron (nucleotides 1,907 to 2,208) is removed, the first AUG is nearly 700 bases downstream from the cap, and the ORF is only large enough to encode the smallest capsid protein VP-3, which is most abundant in the virion (27). It has now been reported that AAV has evolved two alternative strategies to overcome this conundrum. VP-2 translation apparently initiates at an ACG 195 bases upstream from the first AUG (21), while an alternative splice acceptor site at nucleotide 2,201 is used in place of the reported acceptor site at nucleotide 2,102 to generate a second 2.3-kb RNA with an AUG-initiated ORF large enough to encode VP-1 (20,259). Synthetic DNA constructs with this sequence have been transcribed to produce mRNAs which, when translated *in vivo* or *in vitro*, generate a product the size of VP-1. The assignments of initiator codons for all three coat proteins has been confirmed by site-directed mutagenesis (180).

Why the first AUG in the predominant mRNA should be so far downstream is uncertain. However, evidence has been presented that rep activity can inhibit gene expression from the p40 promoter at the level of protein translation when a vector with a reporter gene under control of the p40 promoter is cotransfected with a plasmid containing rep into Ad-infected cells (256). Thus, it is possible that the long leader sequence in the major p40 spliced RNA from which VP-2 and VP-3 are translated serves to protect against this type of inhibition.

Expression of the structural proteins in insect cells leads to spontaneous particle self-assembly. VP2 was specifically required for self-assembly, but either VP1 or VP3 could be omitted (217). However, all three were required for infectivity.

Latent Infection

In the absence of a helper virus coinfection, the AAV virion can penetrate the nucleus where the DNA is uncoated; but there is insufficient gene expression for transcripts to be readily detected by Northern blots. Under these conditions, the AAV genome is integrated into the cellular DNA to establish a latent infection that can be activated by subsequent helper virus infection (28). AAV latent infection was discovered by Hoggan et al. (110) in the course of a federal project to screen primary cell lots intended for vaccine production. Although there was no immunological evidence for AAV, upon challenge by Ad infection, 20% of the lots of African green monkey kidney cells and 1% to 2% of the lots of human embryonic kidney cells tested produced AAV. Thus, AAV latent infection *in vivo* appeared to be common. Although most human transmission of AAV appears to be horizontal, evidence for vertical germline transmission of avian AAV has been reported in chickens (73). Human cells in culture can be latently infected simply by infecting with a high multiplicity of AAV (250 tissue culture infectious doses per cell) in the absence of helper virus (31). The original mass culture remained positive for induction of AAV by Ad challenge for more than 50 passages. When the culture was cloned after the 39th passage, 30% of the clones were AAV-positive after challenge. At least 30% of the cells in positive clones produced AAV, a rate equivalent to that seen with an exogenous coinfection with AAV and Ad. These figures far exceed those seen with other DNA animal viruses and approach the efficiency of lambda bacteriophage lysogeny and rescue. One of the original clones was subcloned and followed for many passages. It remained positive for AAV for over 150 passages.

Initial characterization of the AAV DNA in latently infected clones showed that it was integrated into cellular DNA (56,102) and that the viral termini were present at the junctions with cell DNA (56). The viral DNA in most latently infected cell lines was integrated as a tandem array of several genome equivalents (31,56,143,168). The tandem array was characterized in several instances by restriction endonuclease digestion as head-to-tail. In a more recent study in which the junction between two integrated genomes was sequenced, the array was seen to be tail-to-tail (131), as might be expected if the tandem array were the consequence of replication during or before the process of integration. It is more difficult to account for the presence of a tandem array of head-to-tail orientation, except by a sequential process of recombination during or after the integration process or by a hitherto unseen rolling circle mode of replication. In the latter study (131) which was the first instance in which integrated sequences were cloned and sequenced, there was evidence for significant rearrangements of viral sequences near the junction with cellular sequence and deletion of some of the tr sequence. Since the DNA was cloned from cells which contained an AAV provirus that could be rescued by superinfection with Ad, the assumption is that among the tandem of genome equivalents there is one that is intact and thus able to produce infectious virus.

It seems likely that the rearrangements noted above occurred mainly during integration. However, a distinct difference has been noted between the stability of integrated sequences representing the unique sequence portion of the AAV genome in contrast to tr sequences. In one clone of latently infected cells which was followed for over 100 passages, the restriction patterns obtained, using enzymes which cut in the unique sequences, were unchanged, but significant changes were noted using Sma1, which cuts exclusively in the tr (48). Thus, it seems that the tr is a particularly unstable sequence in terms of being subject to rearrangement and recombination. This notion was supported by the observation that after 100 passages, free copies of the AAV genome were present, even though all of the detectable sequences had been integrated into high MW DNA in earlier passages (56).

Initially, it was concluded that integration was either random or occurred at a large number of possible sites, because restriction analysis found that junction fragments of viral and cellular DNA differed in size in every independently derived clone of latently infected cells. In a naive sense, if integration had been site-specific, the junction fragments would have been expected to have been uniform. With the availability of isolated junctions, it has now been demonstrated in several continuous lines of human cells that integration of AAV does occur at a specific site on chromosome 19q13.3-qter (131,133,134,224). Integration was by nonhomologous recombination, although there are 4 to 5 base homologies at the site of recombination and was associated with rearrangements and inversions of both viral DNA and cellular sequences. Although integration is site-specific, it is not specific at the individual nucleotide level, but occurs within a region of several hundred nucleotides (133,134,224).

The preintegration site has been isolated as an 8-kb fragment of which the first 4 kb have been sequenced (132). The sequenced region contains all of the known sites of recombination. A number of features of the sequence may be pertinent to the specificity of integration. The average base composition is 65% GC and is 82% GC in the first 900 bases, which are near the recombination sites. Toward the 3′ end of the determined sequence there is a tandem repeat of 10 copies of a 35mer minisatellite which occurs at 60 places in the human genome, all of which are on 19q (72). There are also a large number of direct repeats in which there is an 11/12 base match, but these are distributed unevenly. In the first 900 bases there is one such repeat for every 45 bases, and in the last 1,600 bases there is a repeat every 70 bases. In between, there is only one repeat every 220 bases, the frequency of which would be expected on a random basis. The first 900 bases fulfill the criteria for a CpG island, which could serve as a promoter for expression of the ORF found in the middle of the sequenced region. Evidence for transcription of this region has come from isolation of a cDNA containing part of the sequence and positive reverse transcription PCR assays for several cell lines and tissues.

The most interesting feature of the preintegration site may well be that it contains the AAV Rep sequence binding site (GCTC)3. Weitzman et al. (272) have demonstrated that Rep 68 can bind to a plasmid containing a fragment of the preintegration site in which the binding sequence is present. Rep can also serve as a bridge between the AAV tr and the preintegration site Rep binding site. Thus, the ability of Rep to link the tr to the preintegration site and the known ability of Rep to function as a nuclease and helicase suggest a mechanism for the site specificity of AAV integration. Support for Rep recognition of this site has been provided by the results of Kotin et al. (personal communication), who have found that Rep can utilize the binding site as an ori for DNA replication in the *in vitro* assay described above.

The critical question of whether the sequence of the preintegration site is the determining factor in the site specificity of AAV integration or whether higher order chromatin structure present in chromosome 19q is an important factor has been addressed by putting the preintegration sequence in an extrachromosomal Epstein-Barr virus-based shuttle vector. When human cells which contained 50 to 100 copies of the shuttle vector replicating in synchrony with the cellular genome were infected with wild-type AAV virions, the viral genome preferentially integrated into those shuttle vectors containing specific regions of the preintegration site. The critical region was a 500-base fragment which contained the Rep binding site. Interestingly, it was possible to dissociate the Rep binding site from the sequences at which recombination had occurred in the latently infected cell lines which had been characterized. Thus, it appears as though the critical factor is a recognition site, and the actual recombination takes place several hundred bases distal to the recognition site. Any higher order chromatin structure specific to chromosome 19q does not appear to be essential for site-specific integration (95).

Several questions arise with regard to the mechanisms involved in integration and rescue of the AAV genome. First, with respect to integration, the genome is converted from an initial unit-length, single-strand conformation in the virion to a duplex tandem state within the cellular DNA. Therefore, although no AAV replication can be readily identified, at some level it must occur. Second, the model for AAV DNA replication in Fig. 5 predicts formation of tail-to-tail tandems but does not predict formation of head-to-tail tandems. To produce these, some form of rolling circle mechanism would need to be hypothesized. Heterologous recombination between AAV and SV40 was studied as a possible model of the integration process (99,100). The salient result was that recombinants produced by coinfection of monkey cells with SV40 viruses and AAV virions had structures which resembled that of the integrated AAV DNA in latently infected cells: AAV terminal sequences were at one of the junctions with SV40 DNA, and the SV40 and AAV sequences were present as a head-to-tail tandem repeat. A simple deletion-substitution pattern with no preference for the AAV termini at junctions was seen in recombinants produced by cotransfection of monkey cells by duplex SV40 and AAV DNAs. These results would suggest that the ability of the trs to either hairpin or anneal with one another when the genome is single-stranded may play a significant role in the establishment of the latent state. Interestingly, some of the recombinants formed in the EBV shuttle vector model system also contain head-to-tail tandem junctions (C. Giraud and K.I. Berns, unpublished data).

The rescue of the AAV genome from plasmids has been viewed as a model for the rescue and replication process from the latent state. These studies have indicated that an intact *rep* gene and intact or reparable trs are required for rescue and replication (111,223,269). Because part of the tr (approximately 45 to 75 bases) had been deleted, both in the one case where the AAV termini at the junctions with cell DNA have been sequenced and in the AAV/SV40 recombinants that were characterized at this level, the tandem structure may be important to preserve an intact copy of the tr for rescue to occur.

Consequences of latent infection for the host are uncertain. Clearly, the process serves to maintain the AAV genome in a proviral state until it is activated and, thus, helps to assure the biological continuity of the virus.

AAV as a Vector

AAV has received a significant amount of interest as a potential vector for gene therapy (53,86,87,108,142,168, 181,182,187,198,255,257,264,282). There are a number of characteristics which contribute to the potential utility of the virus for this purpose. Among these are the ability to integrate to establish a stable latent state with high frequency (in 19% to 70% of exposed cells), the lack of association with any known disease, and the fact that vectors can be constructed with few, if any, viral genes that will express protein products on the surface of transformed cells. The ability to latently infect and transform nondividing cells has also been reported (276). Site specificity of integration is a potential advantage, although most latently infected cell lines have only one copy of the normal sequence disrupted by integration; the consequences of interrupting both copies would be deleterious. Also unknown is whether all transferred genes would be equally expressed from the specific AAV location on chromosome 19q.

The size limit for foreign DNA which can be packaged in an AAV virion is about 4 to 4.5 kb. Two general types of vectors have been made. The first deletes the capsid gene and substitutes the foreign DNA (about 2 kb). This type of vector retains the *rep* gene and presumably can integrate at a specific site. The second type of vector simply puts an AAV tr (plus a few more bases) on both ends of the foreign DNA. In some cases such a vector can integrate and transform with >50% frequency, but the incidence of site specificity is significantly reduced (283). Both with cells and in one animal system, AAV vectors have given promising results in terms of stable expression of the transformed gene (86,87). AAV as a vector is an area of intense interest.

Inhibition of Helper Virus Replication and Heterologous Gene Expression

In addition to AAV gene expression and replication being affected by Ad coinfection, the course of Ad infection is also significantly affected by AAV coinfection (47,49,50). The net effect is enhanced AAV gene expression and replication with greatly reduced Ad gene expression and replication. These effects are dependent on the ratio of the multiplicities of infection of the two viruses and the temporal order of addition. AAV is effectively autoinhibitory because, if the ratio of AAV to Ad becomes too great, the inhibition of Ad gene expression is sufficient that it no longer provides the helper functions needed for AAV replication. By the same token, if AAV infection is delayed until Ad DNA synthesis has begun, AAV replication occurs but that of Ad is not inhibited. The inhibition appears to require *rep* gene expression because DI particles of AAV, which have very large internal deletions but intact trs, cannot inhibit Ad replication (75). Nuclear run-on experiments have demonstrated inhibition by AAV coinfection of both E1 and E4 transcription (M. A. Labow, unpublished data). Whether the inhibition of Ad includes direct effects on DNA replication or simply a lack of early regulatory proteins is not known. Avian AAV coinfection of chicken cells has also been reported to inhibit replication of a helper herpesvirus, Marek's disease virus (17). This type of inhibition is not usually seen with other herpesvirus coinfections (14), probably because the herpesvirus infection is so rapid.

AAV has also been reported to inhibit permanent cellular transformation by selectable markers under the control of the SV40 early promoter, the HSV type II thymidine kinase promoter, and the inducible murine metallothionein promoter. The level of inhibition was greater than 95% and depended on an intact *rep* gene. Under the conditions employed, the cells were not killed by *rep* gene expression, nor was the ability of plasmid DNA to integrate into cellular DNA decreased. Thus, it appeared as though the inhibition was caused by repression of gene expression. This notion was supported by the observation that in cotransfections the AAV *rep* gene was able to inhibit CAT expression under control of the SV40 early promoter. These types of inhibitory effects have been observed in both murine and human cells (140,258), but the extent of inhibition was very dependent on the specific cells used.

Initially, infection by AAV in the absence of helper virus was not noted to affect the cell phenotype. More detailed studies of latently infected cells have indicated several phenotypic changes. These include decreased plating efficiency, evidence of cell cycle arrest, and increased sensitivity to a variety of genotoxic agents [e.g., UV, chemical carcinogens (266,267)]. Latent infection can also inhibit DNA amplification after exposure to genotoxic agents (275). AAV infection of the human leukemic cell line HL60 and immortalized keratinocyte lines led to reduced cell growth and reduced levels of detectable differentiation associated antigens, plus altered expression of c-myc, c-myb, and c-fos (128). Whether these consequences are mediated by exposure to the intact viral particle or are the results of Rep expression is not yet known.

Inhibition of Adenovirus and Herpesvirus Oncogenicity

Oncogenic serotypes of Ad induce tumors in newborn Syrian hamsters. Coinfection of AAV with the oncogenic Ad reduces the frequency of tumors and lengthens the induction time of those that do occur (127,162). AAV coinfection also inhibits the ability of Ad to transform cells. Ad-transformed hamster cells are also oncogenic in the newborn, and this oncogenicity was reduced by infection of the cells with AAV, although the growth rate in culture was unaltered (191). Therefore, it was possible to study at the molecular level the effects of AAV on the system. In the transformed cells, AAV infection did not affect the integrated Ad DNA sequences, nor did it have any apparent effect on the Ad E1A and E1B transcripts that accumulated. However, the amount of the major tumor antigen (E1B 55-kd protein) in the transformed cells was reduced by more than 80%. Thus, AAV infection of the transformed cells inhibited E1B expression. Because the accumulation of the E1B major transcript for the 55-kd protein was unaffected, it appeared that the inhibition was at the level of translation. De la Maza and Carter (76) reported that DI particles of AAV and DI genomes could inhibit Ad induction of tumors in hamsters. Because some of the preparations of defective particles contained only the termini of the AAV genome, it was suggested that the termini may be critical in the inhibition of Ad oncogenicity. On the other hand, the inhibition of E1B expression observed in AAV-infected transformed cells correlated well with the observed decrease in tumor formation, and the former would seem likely to be a *rep* gene function. Indeed, Carter's group has reported that *rep* can inhibit gene expression at the level of translation (258). This apparent divergence in mechanism awaits further experiments for clarification. AAV has also been reported to inhibit the oncogenicity of HSV-II-transformed cells after infection (33,70,71). Not only was the incidence of tumor induction decreased after cells were injected into hamsters, but the induction period was lengthened. Possibly the most interesting result was that even in those animals that did develop tumors there was no evidence of metastasis. AAV has also been shown to inhibit transformation by bovine papilloma virus, and this inhibition is attributable to Rep (106). Similarly Rep inhibits cell transformation by EJ Ha Ras (126).

The reports of AAV inhibition of oncogenicity have raised the question of whether AAV might inhibit oncogenicity in humans. More than 90% of adults are seropositive for AAV, suggesting that many carry AAV as a latent infection. According to the model hypothesized, latent virus might be activated by oncogenic virus infection and, once activated, inhibit oncogenesis. Several retrospective epidemiological studies have been done, studying patients suffering from cervical carcinoma. Interestingly, in each study the patients were markedly deficient in antibodies to AAV compared to a group of matched controls (92,161,241). These results raise the unusual possibility that a viral infection might, for once, be beneficial to the host.

SUMMARY AND CONCLUSION

In terms of physical structure, genetic map, autoregulation, and close dependence on the state of the intracellular milieu, the autonomous parvoviruses are indeed closely related to the dependoviruses. There are even significant stretches of homology at the amino acid sequence level in both structural and nonstructural proteins of viruses of both genera. In some ways there appears to be a continuous spectrum of properties among the viruses of both genera—for example, LuIII and bovine parvovirus encapsidate both strands, the 3′-terminal palindrome of the bovine parvovirus shows evidence of inversion during replication, and B19 contains terminal repeats. However, there remains a major difference at the biological level. It seems certain that latent infection plays a major role in the replication cycle of AAV, whereas it is unclear whether the autonomous parvoviruses can establish latency. They do frequently cause cryptic infections, and perhaps these are the biological equivalent of the AAV latent infection. The other major difference, of course, is that all of the autonomous viruses are serious pathogens in their normal hosts, while AAV has yet to be associated with disease. Since the second edition of this book, our knowledge of the mechanisms regulating parvovirus replication has greatly increased, but fundamental questions at the biological level remain to be answered.

REFERENCES

1. Agbandje M, McKenna R, Rossmann MG, Strassheim ML, Parrish CR. Structure determination of feline panleukopenia virus empty particles. *Proteins* 1993; 16:155.
2. Ahn JK, Pitluk ZW, Ward DC. The GC box and TATA transcription control elements in the P38 promoter of the minute virus of mice are necessary and sufficient for transactivation by the nonstructural protein NS1. *J Virol* 1992; 66:3776.
3. Alexandersen R, Bloom M, Perryman S. A detailed transcription map of Aleutian mink disease virus. *J Virol* 1988; 62:3684.
4. Antonietti J-P, Sali R, Beard P, Hirt B. Characterization of the cell type-specific determinant in the genome of minute virus of mice. *J Virol* 1988; 62:552.
5. Ashktorab H, Srivastava A. Identification of nuclear proteins that specifically interact with adeno-associated virus type 2 inverted terminal repeat hairpin DNA. *J Virol* 1989; 63:3034.
6. Astell CR, Chow MB, Ward DC. Sequence analysis of the termini of virion and replicative forms of minute virus of mice DNA suggests a modified rolling hairpin model for autonomous parvovirus DNA replication. *J Virol* 1985; 54:171.
7. Astell CR, Smith M, Chow MB, Ward DC. Sequence of the 3′-terminus of the genome from Kilham rat virus, a nondefective parvovirus. *Virology* 1979; 96:669.
8. Astell CR, Thomas H, Chow MB, Ward DC. Structure and replication of minute virus of mouse DNA. *Cold Spring Harbor Symp Quant Biol* 1982; 47:751.
9. Astell CR, Thomson M, Merchlinsky MJ, Ward DC. The complete DNA sequence of minute virus of mice, an autonomous parvovirus. *Nucleic Acids Res* 1984; 11:999.
10. Atchison RW, Casto BC, Hammon W. Adenovirus-associated defective virus particles. *Science* 1965; 149:754.
11. Ball-Goodrich LJ, Moir RD, Tattersall P. Parvoviral target cell specificity: acquisition of fibrotropism by a mutant of the lymphotropic strain of minute virus of mice involves multiple amino acid substitutions within the capsid. *Virology* 1991; 184:175.
12. Bando H, Choi H, Ito Y, Kawase S. Terminal structure of a Densovirus implies a hairpin transfer replication which is similar to the model for AAV. *Virology* 1990; 179:57.
13. Bando H, Kusuda J, Gojobori T, Maruyama T, Kawase S. Organization and nucleotide sequence of a densovirus genome imply a host-dependent evolution of the parvoviruses. *J Virol* 1987; 61:553.
14. Bantel Schaal U, zur Hausen H. Adeno-associated viruses inhibit SV40 DNA amplification and replication of herpes simplex virus in SV40-transformed hamster cells. *Virology* 1988; 164:64.
15. Barrijal S, Perros M, Gu Z, Avalosse BL, Belenguer P, Amalric F, Rommelaere J. Nucleolin forms of a specific complex with a fragment of the viral (minus) strand of minute virus of mice DNA. *Nucleic Acids Res* 1992; 20:5053.
16. Bates RC, Snyder CE, Banerjee PT, Mitra S. Autonomous parvovirus LuIII encapsidates equal amounts of plus and minus DNA strands. *J Virol* 1984; 49:319.
17. Bauer HJ, Monreal G. Herpesviruses provide helper functions for avian adeno-associated parvovirus. *J Gen Virol* 1986; 67:181.
18. Bauer HJ, Monreal G. Avian adeno-associated parvovirus and Marek's disease virus: Studies of viral interactions in chicken embryo fibroblasts. *Arch Virol* 1988; 98:271.
19. Beaton A, Palumbo P, Berns KI. Expression from the adeno-associated virus p5 and p19 promoters is negatively regulated in trans by the rep protein. *J Virol* 1989; 63:4450.
20. Becerra SP, Koczot F, Fabisch P, Rose JA. Synthesis of adeno-associated virus proteins requires both alternative mRNA splicing and alternative initiations from a single transcript. *J Virol* 1988; 62:2745.
21. Becerra SP, Rose JA, Hardy M, Baroudy BM, Anderson CW. Direct mapping of adeno-associated virus capsid proteins B and C: A possible ACG initiation codon. *Proc Natl Acad Sci USA* 1985; 82:7919.
22. Ben-Asher E, Aloni Y. Transcription of minute virus of mice, an autonomous parvovirus, may be regulated by attenuation. *J Virol* 1984; 52:266.
23. Bergs VV. Rat virus-mediated suppression of leukemia induction by Moloney virus in rats. *Cancer Res* 1969; 17:935.
24. Berk AJ. Adenovirus promoters and E1a transactivation. *Ann Rev Genet* 1986; 20:45.
25. Berk AJ, Lee F, Harrison T, Williams J, Sharp PA. Pre-early Ad5 gene product regulates synthesis of early viral mRNA's. *Cell* 1979; 17:935.
26. Berns KI, Adler S. Separation of two types of adeno-associated virus particles containing complementary polynucleotide chains. *J Virol* 1972; 9:394.
27. Berns KI, Bergoin M, Bloom M, Lederman M, Muzyczka N, Siegl G, Tal J, Tattersall P. *Parvoviridae*. VIth Report of International Committee on Taxonomy of Viruses. In: Murphy FA, Fauquet CM, Bishop DHL, et al., eds. *Virus Taxonomy,* Archives of Virology, suppl 10. 1994;166–178.
28. Berns KI, Cheung AKM, Ostrove JM, Lewis M. Adeno-associated virus latent infection. In: Mahy BWJ, Minson AC, Darby GK, eds. *Virus Persistence.* Cambridge: Cambridge University Press, 1982;249.
29. Berns KI, Kelly TJJ. Visualization of the inverted terminal repetition in adeno-associated virus DNA. *J Mol Biol* 1974; 82:267.
30. Berns KI, Labow MA. Parvovirus gene regulation. *J Gen Virol* 1987; 68:601.
31. Berns KI, Pinkerton TC, Thomas GF, Hoggan MD. Detection of adeno-associated virus (AAV) specific nucleotide sequences in DNA isolated from latently infected Detroit 6 cells. *Virology* 1975; 68:556.
32. Berns KI, Rose JA. Evidence for a single-stranded adenovirus-associated virus genome: isolation and separation of complementary single strands. *J Virol* 1979; 5:693.
33. Blacklow NR, Cukor G, Kibrick S, Quinman G. Interactions of adeno-

associated viruses with cells transformed by herpes simplex virus. In: Ward DC, Tattersall P, eds. *Replication of mammalian parvoviruses.* Cold Spring Harbor, NY: Cold Spring Harbor Laboratory Press, 1978;87.

34. Bloom ME, Kaaden OR, Huggans E, Cohn A, Wolfingbarger JB. Molecular comparisons of *in vivo*- and *in vitro*-derived strains of Aleutian disease of mink parvovirus. *J Virol* 1988; 62:132.
35. Blundell MC, Beard C, Astell CR. *In vitro* identification of a B19 parvovirus promoter. *Virology* 1987; 157:534.
36. Bohenzky A, Berns KI. Interactions between the termini of adeno-associated virus DNA. *J Mol Biol* 1989; 206:91.
37. Bohenzky RA, Lefebvre RB, Berns KI. Sequence and symmetry requirements within the internal palindromic sequences of the adeno-associated virus terminal repeat. *Virology* 1988; 166:316.
38. Bonnard GD, Manders EK, Campbell DAJ, Herberman RB, Collins MJ. Immunosuppressive activity of a subline of the mouse EL-4 lymphoma. Evidence for minute virus of mice causing the inhibition. *J Exp Med* 1976; 143:187.
39. Brown CS, Van Lent JW, Vlak JM, Spaan WJ. Assembly of empty capsids by using baculovirus recombinants expressing human parvovirus B19 structural proteins. *J Virol* 1991; 65:2702.
40. Brown KE, Gallinella G, Hibbs JR, Anderson SM, Young NS. Blood group P antigen (globoside) is the cellular receptor for B19 parvovirus. *Fifth Parvovirus Workshop.*Crystal River, FL: 1993.
41. Brown KE, Hibbs JR, Gallinella G, Anderson SM, Lehman ED, McCarthy P, Young NS. Resistance to parvovirus B19 infection due to lack of virus receptor (erythrocyte P antigen). *N Engl J Med* 1994; 330:1192.
42. Buller RML, Janik JE, Sebring ED, Rose JA. Herpes simplex virus types 1 and 2 help adeno-associated virus replication. *J Virol* 1981; 40:241.
43. Buller RML, Rose JA. Characterization of adenovirus-associated virus-induced polypeptides in KB cells. *J Virol* 1978; 25:331.
44. Carlson JO, Lynde-Mass MK, Shen ZD. A nonstructural protein of feline panleukopenia virus: expression in *Escherichia coli* and detection of multiple forms in infected cells. *J Virol* 1987; 61:621.
45. Carter BJ, Antoni BA, Klessig DF. Adenovirus containing a deletion of the early region 2A gene allows growth of adeno-associated virus with decreased efficiency. *Virology* 1992; 191:473.
46. Carter BJ, Laughlin CA. Adeno-associated virus defectiveness and the nature of the helper function. In: Berns KI, ed. *The Parvoviruses.* New York: Plenum Press, 1983;67.
47. Carter BJ, Laughlin CA, de la Maza LM, Myers M. Adeno-associated virus autointerference. *Virology* 1979; 92:449.
48. Carter BJ, Marcus-Sekura CJ, Laughlin CA, Ketner G. Properties of an adenovirus type 2 mutant, *Ad2dl807*, having a deletion near the right-hand genome terminus: failure to help AAV replication. *Virology* 1983; 126:505.
49. Casto BC, Atchison RW, Hammond W. Studies on the relationship between adeno-associated virus type 1 (AAV-1) and adenoviruses. I. Replication of AAV in certain cell cultures and its effects on helper adenovirus. *Virology* 1967; 32:52.
50. Casto BC, Goodheart CR. Inhibition of adenovirus transformation *in vitro* by AAV-1. *Proc Soc Exp Biol Med* 1972; 140:72.
51. Chang LS, Shenk T. The adenovirus DNA-binding protein stimulates the rate of transcription directed by adenovirus and adeno-associated virus promoters. *J Virol* 1990; 64:2103.
52. Chapman MS, Rossmann MG. Structure, sequence, and function correlations among parvoviruses. *Virology* 1993; 194:491.
53. Chatterjee S, Johnson PR, Wong KKJ. Dual-target inhibition of HIV-1 *in vitro* by means of an adeno-associated virus antisense vector. *Science* 1992; 258:1485.
54. Chejanovsky N, Carter BJ. Mutagenesis of an AUG codon in the adeno-associated virus rep gene: effects on viral DNA replication. *Virology* 1989; 173:120.
55. Chen KC, Shull BC, Moses EA, Lederman M, Stout ER, Bates RC. Complete nucleotide sequence and genome organization of bovine parvovirus. *J Virol* 1986; 60:1085.
56. Cheung A-M, Hoggan MD, Hauswirth WW, Berns KI. Integration of the adeno-associated virus genome into cellular DNA in latently infected human Detroit 6 cells. *J Virol* 1980; 33:739.
57. Chiorini JA, Weitzman MD, Owens RA, Urcelay E, Safer B, Kotin RM. Biologically active Rep proteins of adeno-associated virus type 2 produced as fusion proteins in *Escherichia coli. J Virol* 1994; 68:797.
58. Chow M, Bodnar JW, Polvino-Bodnar M, Ward D. Identification and characterization of a protein covalently bound to DNA of minute virus of mice. *J Virol* 1986; 57:1094.
59. Clemens DL, Wolfinbarger JB, Mori S, Berry BD, Hayes SF, Bloom ME. Expression of Aleutian mink disease parvovirus capsid proteins by a recombinant vaccinia virus: Self-assembly of capsid proteins into particles. *J Virol* 1992; 66:3077.
60. Clemens KE, Pintel DJ. The two transcription units of the autonomous parvovirus minute virus of mice are transcribed in a temporal order. *J Virol* 1988; 62:1448.
61. Clinton GM, Hayashi J. The parvovirus MVM: a comparison of heavy and light particle infectivity and their conversion *in vitro. Virology* 1976; 74:57.
62. Cornelis JJ, Becquart P, Duponchel N, et al. Transformation of human fibroblasts by ionizing radiation, a chemical carcinogen, or simian virus 40 correlates with an increase in susceptibility to the autonomous parvoviruses H-1 virus and minute virus of mice. *J Virol* 1988; 62:1679.
63. Cornelis JJ, Chen YQ, Spruyt N, Duponchel N, Cotmore SF, Tattersall P, Rommelaere J. Susceptibility of human cells to killing by the parvoviruses H-1 and minute virus of mice correlates with viral transcription. *J Virol* 1990; 64:2537.
64. Cornelis JJ, Spryt N, Spegelaere P, Guetta E, Darawshi T, Cotmore SF, Tal J, Rommelaere J. Sensitization of transformed rat fibroblasts to killing by parvovirus minute virus of mice correlates with an increase in viral gene expression. *J Virol* 1988; 62:3438.
65. Cotmore SF, Nuesch JP, Tattersall P. Asymmetric resolution of a parvovirus palindrome *in vitro. J Virol* 1993; 67:1579.
66. Cotmore SF, Sturzenbecker LJ, Tattersall P. The autonomous parvovirus MVM encodes two nonstructural proteins in addition to its capsid polypeptides. *Virology* 1983; 129:333.
67. Cotmore SF, Tattersall P. The autonomously replicating parvoviruses of vertebrates. *Adv Virus Res* 1987; 33:91.
68. Cotmore SF, Tattersall P. The NS-1 polypeptide of minute virus of mice is covalently attached to the 5′ termini of duplex replicative-form DNA and progeny single strands. *J Virol* 1988; 62:851.
69. Crawford LV, Follet EAC, Burdon MG, McGeoch DJ. The DNA of a minute virus of mice. *J Gen Virol* 1969; 4:37.
70. Cukor G, Blacklow NR, Hoggan MD, Berns KI. Biology of adeno-associated virus. In: Berns KI, ed. *The parvoviruses.* New York: Plenum Press, 1983;33.
71. Cukor G, Blacklow NR, Kibrick S, Swan IC. Effects of adeno-associated virus on cancer expression by herpesvirus-transformed hamster cells. *J Natl Cancer Inst* 1975; 55:957.
72. Das HK, Jackson CL, Miller DA, Leff T, Breslow JL. The human apolipoprotein C-11 gene sequence contains a novel chromosome 19-specific minisatellite in its third intron. *J Biol Chem* 1987; 262:4787.
73. Dawson, GJ Yates, VJ Chang, PW Wattanavijarin, W. Egg transmission of avian adeno-associated virus and CELO virus during experimental infections. *Am J Vet Res* 1981; 42:1833.
74. de la Maza LM, Carter BJ. DNA structure of incomplete adeno-associated virus particles. In: Ward D, Tattersall P, ed. *Replication of mammalian parvoviruses.* Cold Spring Harbor, NY: Cold Spring Harbor Laboratory Press, 1978;193.
75. de la Maza LM, Carter BJ. Molecular structure of adeno-associated virus variant DNA. *J Biol Chem* 1980; 255:3194.
76. de la Maza LM, Carter BJ. Inhibition of adenovirus oncogenicity by adeno-associated virus DNA. *J Natl Cancer Inst* 1981; 67:1323.
77. Deiss V, Tratschin JD, Weitz M, Siegl G. Cloning of the human parvovirus B19 genome and structural analysis of its palindromic termini. *Virology* 1990; 175:247.
78. Diffoot N, Chen KC, Bates RC, Lederman M. The complete nucleotide sequence of parvovirus LuIII and localization of a unique sequence possibly responsible for its encapsidation pattern. *Virology* 1993; 192:339.
79. Doerig C, Beard P, Hirt B. A transcriptional promoter of the human parvovirus B19 active *in vitro* and *in vivo. Virology* 1987; 157:539.
80. Doerig C, Hirt B, Antonietti JP, Beard P. Nonstructural protein of parvoviruses B19 and minute virus of mice controls transcription. *J Virol* 1990; 64:387.
81. Dumas B, Jourdan M, Pascaud AM, Bergoin M. Complete nucleotide sequence of the cloned infectious genome of Junonia coenia densovirus reveals an organization unique among parvoviruses. *Virology* 1992; 191:202.
82. Faisst S, Schlehofer JR, zur Hausen H. Transformation of human cells

by oncogenic virus supports permissiveness for H1 parvovirus propagation. *J Virol* 1989; 63:2152.
83. Faust EA, Ward DC. Incomplete genomes of the parvovirus minute virus of mice: selective conservation of genome termini, including the origin for DNA replication. *J Virol* 1979; 32:276.
84. Fife KH, Berns KI, Murray K. Structure and nucleotide sequence of the terminal regions of adeno-associated virus DNA. *Virology* 1977; 78:475.
85. Filman DJ, Hogle JM. Virology. Architecture with a difference. *Nature (UK)* 1991; 351:100.
86. Flotte TR, Afione SA, Solow R, Drumm ML, Markakis D, Guggino WB, Zeitlin PL, Carter BJ. Expression of the cystic fibrosis transmembrane conductance regulator from a novel adeno-associated virus promoter. *J Biol Chem* 1993; 268:3781.
87. Flotte TR, Solow R, Owens RA, Afione S, Zeitlin PL, Carter BJ. Gene expression from adeno-associated virus vectors in airway epithelial cells. *Am J Respir Cell Mol Biol* 1992; 7:349.
88. Foto F, Saag KG, Scharosch LL, Howard EJ, Naides SJ. Parvovirus B19-specific DNA in bone marrow from B19 arthropathy patients: evidence for B19 virus persistence. *J Infect Dis* 1993; 167:744.
89. Gardiner EM, Tattersall P. Mapping of the fibrotropic and lymphotropic host range determinants of the parvovirus minute virus of mice. *J Virol* 1988; 62:2605.
90. Gardiner EM, Tattersall P. Evidence that developmentally regulated control gene expression by a parvoviral allotropic determinant is particle mediated. *J Virol* 1988; 62:1713.
91. Gavin BJ, Ward DC. Positive and negative regulation of the minute virus of mice P38 promoter. *J Virol* 1990; 64:2057.
92. Georg-Fries B, Biederlack S, Wolf J, zur Hausen H. Analysis of proteins, helper dependence, and seroepidemiology of a new human parvovirus. *Virology* 1984; 134:64.
93. Gerry HW, Kelly TJJ, Berns KI. Arrangement of nucleotide sequences in adeno-associated virus DNA. *J Mol Biol* 1973; 79:201.
94. Giraud C, Devauchelle G, Bergoin M. The densovirus of Junonia coenia (JcDNV) as an insect cell expression vector. *Virology* 1992; 186:207.
95. Giraud C, Winocour E, Berns KI. Site-specific integration by adeno-associated virus is directed by a cellular DNA sequence. *Proc Natl Acad Sci USA* 1994; 91:in press.
96. Green MR, Roeder RG. Transcripts of the adenovirus-associated virus genome: mapping of the major RNA's. *J Virol* 1980; 36:79.
97. Green MR, Roeder RG. Definition of a novel promoter for the major adeno-associated virus mRNA. *Cell* 1980; 22:231.
98. Green MR, Straus SE, Roeder RG. Transcripts of the adenovirus-associated genome: multiple polyadenylated RNA's including a potential primary transcript. *J Virol* 1980; 35:560.
99. Grossman Z, Berns KI, Winocour E. Structure of simian virus 40—adeno-associated virus recombinant genomes. *J Virol* 1985; 56:457.
100. Grossman Z, Winocour E, Berns KI. Recombination between simian virus 40 and adeno-associated virus: virion coinfection compared to DNA cotransfection. *Virology* 1984; 134:125.
101. Hallauer C, Kronauer G, G S. Parvoviruses as contaminants of permanent human cell lines. I. Virus isolations from 1960–1970. *Arch Ges Virusforsch* 1971; 35:80.
102. Handa H, Shiroki K, Shimojo H. Establishment and characterization of KB cell lines latently infected with adeno-associated virus type 1. *Virology* 1977; 82:84.
103. Harris C, Astell CR. Transcriptional activation and SV40 T antigen responsive domains are present in NS1, the major nonstructural protein of the minute virus of mice. *Fifth Parvovirus Workshop*. Crystal River, FL: 1993
104. Hauswirth WW, Berns KI. Origin and termination of adeno-associated virus DNA replication. *Virology* 1977; 79:488.
105. Hauswirth WW, Berns KI. Adeno-associated virus DNA replication: non unit-length molecules. *Virology* 1979; 93:57.
106. Hermonat PL. Inhibition of bovine papillomavirus plasmid DNA replication by adeno-associated virus. *Virology* 1992; 189:329.
107. Hermonat PL, Labow MA, Wright R, Berns KI, Muzyczka N. Genetics of adeno-associated virus: isolation and preliminary characterization of adeno-associated virus type 2 mutants. *J Virol* 1984; 51:329.
108. Hermonat PL, Muzyczka N. Use of adeno-associated virus as a mammalian DNA cloning vector: transduction of neomycin resistance into mammalian tissue culture cells. *Proc Natl Acad Sci USA* 1984; 81:6466.
109. Hoggan MD, Blacklow NR, Rowe WP. Studies of small DNA viruses found in various adenovirus preparations: physical, biological and immunological characteristics. *Proc Natl Acad Sci USA* 1966; 55:1467.
110. Hoggan MD, Thomas GF, Thomas FB, Johnson FB. Continuous "carriage" of adenovirus associated virus genome in cell cultures in the absence of helper adenoviruses. *Proceedings of the Fourth Lepetit Colloquium*. Cocoyac, Mexico: Amsterdam: North-Holland, 1972.
111. Hong G, Ward P, Berns KI. *In vitro* replication of adeno-associated virus DNA. *Proc Natl Acad Sci USA* 1992; 89:4673.
112. Hong G, Ward P, Berns KI. Replicative intermediates in adeno-associated virus DNA replication *in vitro*. *J Virol* 1994; 68:2011.
113. Huang TH, Zhou X, McCarty D, Zolotukhin I, Muzyczka N. *In vitro* replication of adeno-associated virus DNA. *J Virol* 1994; 68:1128.
114. Im DS, Muzyczka N. Factors that bind to adeno-associated virus terminal repeats. *J Virol* 1989; 63:3095.
115. Im DS, Muzyczka N. The AAV origin binding protein Rep68 is an ATP-dependent site-specific endonuclease with DNA helicase activity. *Cell* 1990; 61:447.
116. Im DS, Muzyczka N. Partial purification of adeno-associated virus Rep78, Rep52, and Rep40 and their biochemical characterization. *J Virol* 1992; 66:1119.
117. Janik JE, Huston MM, Cho K, Rose JA. Requirement of adenovirus DNA-binding protein and VA-1 RNA for production of adeno-associated virus polypeptides. *J Cell Biochem (Suppl)* 1982; 6:209.
118. Janik JE, Huston MM, Rose JA. Locations of adenovirus genes required for the replication of adenovirus-associated virus. *Proc Natl Acad Sci USA* 1981; 78:1925.
119. Jay FT, Laughlin CA, Carter BJ. Eukaryotic translational control: adeno-associated virus protein synthesis is affected by a mutation in the adenovirus DNA binding protein. *Proc Natl Acad Sci USA* 1981; 78:2927.
120. Jindal HK, Yong CB, Wilson GM, Tam P, Astell CR. Mutations in the NTP-binding motif of minute virus of mice (MVM) NS-1 protein uncouple ATPase and DNA helicase functions. *J Biol Chem* 1994; 269:3283.
121. Johnson FB. Parvovirus proteins. In: Berns KI, ed. *The Parvoviruses*. New York: Plenum Press, 1983;239.
122. Johnson FB, Blacklow NR, Hoggan MD. Immunological reactivity of anti-sera prepared against the sodium dodecyl sulfate-treated structural polypeptides of adenovirus-associated virus. *J Virol* 1972; 9:1017.
123. Johnson FB, Ozer HL, Hoggan MD. Structural proteins of adenovirus-associated virus type 3. *J Virol* 1971; 9:860.
124. Jones N, Shenk T. An adenovirus type 5 early gene function regulates expression of other early viral genes. *Proc Natl Acad Sci USA* 1979; 76:3665.
125. Kajigaya S, Fujii H, Field A, Anderson S, Rosenfeld S, Anderson LJ, Shimada T, Young NS. Self-assembled B19 parvovirus capsids, produced in a baculovirus system, are antigenically and immunogenically similar to native virions. *Proc Natl Acad Sci USA* 1991; 88:4646.
126. Khleif SN, Myers T, Carter BJ, Trempe JP. Inhibition of cellular transformation by the adeno-associated virus rep gene. *Virology* 1991; 181:738.
127. Kirschstein RL, Smith KO, Peters EA. Inhibition of adenovirus 12 oncogenicity by adeno-associated virus. *Proc Soc Exp Biol Med* 1968; 128:670.
128. Klein-Bauernschmitt P, zur Hausen H, Schlehofer JR. Induction of differentiation-associated changes in established human cells by infection with adeno-associated virus type 2. *J Virol* 1992; 66:4191.
129. Koczot FJ, Carter BJ, Garon CF, Rose JA. Self-complementarity of terminal sequences within plus or minus strands of adenovirus-associated virus DNA. *Proc Natl Acad Sci USA* 1973; 70:215.
130. Kolleck R, Tseng BY, Goulian M. DNA polymerase requirements for parvovirus H1 DNA replication *in vitro*. *J Virol* 1982; 41:982.
131. Kotin RM, Berns KI. Organization of adeno-associated virus DNA in latently infected Detroit 6 cells. *Virology* 1989; 170:460.
132. Kotin RM, Linden RM, Berns KI. Characterization of a preferred site on human chromosome 19q for integration of adeno-associated virus DNA by non-homologous recombination. *EMBO J* 1992; 11:5071.
133. Kotin RM, Menninger JC, Ward DC, Berns KI. Mapping and direct visualization of a region-specific viral DNA integration site on chromosome 19q 13-qter. *Genomics* 1991; 10:831.
134. Kotin RM, Siniscalco M, Samulski RJ, Zhu XD, Hunter L, Laughlin CA, McLaughlin S, Muzyczka N, Rocchi M, Berns KI. Site-specific integration by adeno-associated virus. *Proc Natl Acad Sci USA* 1990; 87:2211.
135. Krauskopf A, Aloni Y. A cellular repressor regulates transcription ini-

tiation from the minute virus of mice P38 promoter. *Nucleic Acids Res* 1994; 22:828.

136. Krauskopf A, Bengal E, Aloni Y. The block to transcription elongation at the minute virus of mice attenuator is regulated by cellular elongation factors. *Mol Cell Biol* 1991; 11:3515.
137. Krauskopf A, Resnekov O, Aloni Y. A cis downstream element participates in regulation of *in vitro* transcription initiation from the P38 promoter of minute virus of mice. *J Virol* 1990; 64:354.
138. Labieniec-Pintel L, Pintel D. The minute virus of mice P_{39} transcription unit can encode both capsid proteins. *J Virol* 1986; 57:1163.
139. Labow MA, Berns KI. The adeno-associated virus rep gene inhibits replication of an adeno-associated virus/simian virus 40 hybrid gene genome in cos-7 cells. *J Virol* 1988; 62:1705.
140. Labow MA, Graf LHJ, Berns KI. Adeno-associated virus gene expression inhibits cellular transformation by heterologous genes. *Mol Cell Biol* 1987; 7:1320.
141. Labow MA, Hermonat PL, Berns KI. Positive and negative autoregulation of the adeno-associated virus type 2 genome. *J Virol* 1986; 60:251.
142. LaFace D, Hermonat P, Wakeland E, Peck A. Gene transfer into hematopoietic progenitor cells mediated by an adeno-associated virus vector. *Virology* 1988; 162:483.
143. Laughlin CA, Cardellichio CB, Coon HC. Latent infection of KB cells with adeno-associated virus type 2. *J Virol* 1986; 60:515.
144. Laughlin CA, Jones N, Carter BJ. Effect of deletions in adenovirus early region I genes upon replication of adeno-associated virus. *J Virol* 1982; 41:868.
145. Laughlin CA, Tratschin JD, Coon H, Carter BJ. Cloning of infectious adeno-associated virus genomes in bacterial plasmids. *Gene* 1983; 23:65.
146. Laughlin CA, Westphal H, Carter BJ. Spliced adenovirus-associated virus RNA. *Proc Natl Acad Sci USA* 1979; 76:5567.
147. Lechner RL, Kelly TJJ. The structure of replicating adenovirus 2 DNA molecules. *Cell* 1977; 12:1007.
148. Lederman M, Cotmore SF, Stout ER, Bates RC. Detection of bovine parvovirus proteins homologous to the nonstructural NS-1 proteins of other autonomous parvoviruses. *J Virol* 1987; 61:3612.
149. Lefebvre RB, Riva S, Berns KI. Conformation takes precedence over sequence in adeno-associated virus DNA replication. *Mol Cell Biol* 1984; 4:1416.
150. Leonard CJ, Berns KI. Cloning, expression, and partial purification of Rep78: An adeno-associated virus replication protein. *Virology* 1994; 200:566.
151. Li AT, Lavelle GC, Tennant RW. DNA replication of Kilham rat virus: characterization of intracellular forms of viral DNA extracted by guanidine hydrochloride. In: Ward DC, Tattersall P, eds. *Replication of mammalian parvoviruses.* Cold Spring Harbor, NY: Cold Spring Harbor Laboratory Press, 1978.
152. Li X, Rhode SL. The parvovirus H1 NS2 protein affects viral gene expression through sequences in the 3′ untranslated region. *Virology* 1993; 194:10.
153. Lipps BV, Mayor HD. Characterization of heavy particles of adeno-associated virus type 1. *J Gen Virol* 1982; 58:63.
154. Liu JM, Green SW, Hao YS, McDonagh KT, Young NS, Shimada T. Upstream sequences within the terminal hairpin positively regulate the P6 promoter of B19 parvovirus. *Virology* 1991; 185:39.
155. Liu JM, Green SW, Shimada T, Young NS. A block in full-length transcript maturation in cells nonpermissive for B19 parvovirus. *J Virol* 1992; 66:6989.
156. Liu Q, Yong CB, Astell CR. *In vitro* resolution of the dimer bridge of the minute virus of mice (MVM0 genome supports the modified rolling hairpin model for MVM replication). *Virology* 1994; 201:251.
157. Lusby E, Bohenzky R, Berns KI. The inverted terminal repetition in adeno-associated virus DNA: independence of orientation at either end of the genome. *J Virol* 1981; 37:1083.
158. Lusby E, Fife KH, Berns KI. Nucleotide sequence of the inverted terminal repetition in adeno-associated virus DNA. *J Virol* 1980; 34:402.
159. Lusby EW, Berns KI. Mapping of the 5′ termini of two adeno-associated virus 2 RNA's in the left half of the genome. *J Virol* 1982; 41:518.
160. Majaniemi I, Tratschin JD, Seigl G. A reassessment of the nucleic acid and protein components of parvovirus LuIII. In: *Abstracts, Vth International Congress of Virology*.1981.
161. Mayor HD, Drake S, Stahmann J, Mumford DM. Antibodies to adeno-associated satellite virus and herpes simplex in sera from cancer patients and normal adults. *Am J Obstet Gynecol* 1976; 126:100.
162. Mayor HD, Houlditch GS, Mumford DM. Influence of adeno-associated satellite virus on adenovirus-induced tumors in hamsters. *Nature (New Biol)* 1973; 241:44.
163. Mayor HD, Torikai K, Melnick J, Mandel M. Plus and minus single-stranded DNA separately encapsidated in adeno-associated satellite virions. *Science* 1969; 166:1280.
164. McCarty DH, Ryan JH, Zolotukhin I, Zhou X, Muzyczka N. Interaction of the adeno-associated virus Rep protein with a sequence within the A palindrome of the viral terminal repeat. *J Virol* 1994; 68:4998.
165. McCarty DM, Christensen M, Muzyczka N. Sequences required for coordinate induction of adeno-associated virus p19 and p40 promoters by Rep protein. *J Virol* 1991; 65:2936.
166. McCarty DM, Ni TH, Muzyczka N. Analysis of mutations in adeno-associated virus Rep protein *in vivo* and *in vitro*. *J Virol* 1992; 66:4050.
167. McCarty DM, Pereira DJ, Zolotukhin E, Zhou X, Ryan JH, Muzyczka N. Identification of linear DNA sequences that specifically bind the adeno-associated virus Rep protein. *J Virol* 1994; 68:4988.
168. McLaughlin SK, Collis P, Hermonat PL, Muzyczka N. Adeno-associated virus general transduction vectors: Analysis of proviral structures. *J Virol* 1988; 62:1963.
169. McMaster GK, Beard P, Engers HK, Hirt BJ. Characterization of an immunosuppressive parvovirus related to minute virus of mice. *J Virol* 1981; 38:317.
170. McPherson RA, Ginsberg HS, Rose JA. Adeno-associated virus helper activity of adenovirus DNA binding protein. *J Virol* 1983; 44:666.
171. McPherson RA, Rose JA. Structural proteins of adenovirus associated virus: Subspecies and their relatedness. *J Virol* 1983; 46:523.
172. McPherson RA, Rosenthal LJ. Human cytomegalovirus completely helps adeno-associated virus replication. *Virology* 1985; 147:217.
173. Mendelson E, Trempe JP, Carter BJ. Identification of the trans-acting rep proteins of adeno-associated virus by antibodies to a synthetic oligopeptide. *J Virol* 1986; 60:823.
174. Merchlinsky MJ, Tattersall P, Leary JJ, Cotmore SF, Gardiner EM, Ward DC. Construction of an infectious molecular clone of the autonomous parvovirus minute virus of mice. *J Virol* 1983; 47:227.
175. Metcalf JB, Bates RC, Lederman M. Interaction of virally coded protein and a cell cycle-regulated cellular protein with the bovine parvovirus left terminus ori. *J Virol* 1990; 64:5485.
176. Mishra L, Rose JA. Adeno-associated virus DNA synthesis. *Virology* 1990; 179:632.
177. Morgan WR, Ward DC. Three splicing patterns are used to excise the small intron common to all minute virus of mice RNAs. *J Virol* 1986; 60:1170.
178. Mousset S, Rommelaere J. Susceptibility to parvovirus minute virus of mice as a function of the degree of host cell termination: Little effect of simian virus 40 infection and phorbol ester treatment. *Virus Res* 1988; 9:107.
179. Muller HP, Gautschi M, Seigl G. Defective particles of parvovirus LuIII. In: Ward DC, Tattersall P, eds. *Replication of mammalian parvoviruses.* Cold Spring Harbor, NY: Cold Spring Harbor Laboratory Press, 1978;231.
180. Muralidhar S, Becerra SP, Rose JA. Site-directed mutagenesis of adeno-associated virus type 2 structural protein initiation codons: effects on regulation of synthesis and biological activity. *J Virol* 1994; 68:170.
181. Muro-Cacho CA, Samulski RJ, Kaplan D. Gene transfer in human lymphocytes using a vector based on adeno-associated virus. *J Immunother* 1992; 11:231.
182. Muzyczka N. Use of adeno-associated virus as a general transduction vector for mammalian cells. *Curr Top Microbiol Immunol* 1992; Heidelberg.
183. Myers MW, Laughlin CA, Jay FT, Carter BJ. Adenovirus helper function for growth of adeno-associated virus: Effect of temperature sensitive mutations in adenovirus early gene region. *J Virol* 1980; 35:65.
184. Naeger LK, Cater J, Pintel DJ. The small nonstructural protein (NS2) of the parvovirus minute virus of mice is required for efficient DNA replication and infectious virus production in a cell-type-specific manner. *J Virol* 1990; 64:6166.
185. Naeger LK, Salome N, Pintel DJ. NS2 is required for efficient translation of viral mRNA in minute virus of mice-infected murine cells. *J Virol* 1993; 67:1034.
186. Naeger LK, Schoborg RV, Zhao Q, Tullis GE, Pintel DJ. Nonsense mutations inhibit splicing of MVM RNA in *cis* when they interrupt the reading frame of either exon of the final spliced product. *Genes Dev* 1992; 6:1107.

187. Nahreini P, Woody MJ, Zhou SZ, Srivastava A. Versatile adeno-associated virus 2-based vectors for constructing recombinant virions. *Gene (Netherlands)* 1993; 124:257.
188. Ni TH, Zhou X, McCarty DM, Zolotukhin E, Muzyczka N. *In vitro* replication of adeno-associated virus DNA. *J Virol* 1994; 68:1128.
189. Nuesch JP, Cotmore SF, Tattersall P. Expression of functional parvoviral NS1 from recombinant vaccinia virus: Effects of mutations in the nucleotide-binding motif. *Virology* 1992; 191:406.
190. Nuesch JP, Tattersall P. Nuclear targeting of the parvoviral replicator molecule NS1: evidence for self-association prior to nuclear transport. *Virology* 1993; 196:637.
191. Ostrove JM, Duckworth DH, Berns KI. Inhibition of adenovirus-transformed cell oncogenicity by adeno-associated virus. *Virology* 1981; 113:521.
192. Owens RA, Trempe JP, Chejanovsky N, Carter BJ. Adeno-associated virus rep proteins produced in insect and mammalian expression systems: wild-type and dominant-negative mutant proteins bind to the viral replication origin. *Virology* 1991; 184:14.
193. Owens RA, Weitzman MD, Kyostio SR, Carter BJ. Identification of a DNA-binding domain in the amino terminus of adeno-associated virus Rep proteins. *J Virol* 1993; 67:997.
194. Ozawa K, Ayub J, Hao YS, Kurtzman G, Shimada T, Young N. Novel transcription map for the B19 (human) pathogenic parvovirus. *J Virol* 1987; 61:2395.
195. Parks WP, Melnick JL, Rongey R, Mayor HD. Physical assay and growth cycle studies of a defective adeno-satellite virus. *J Virol* 1967; 1:171.
196. Pintel D, Dadachani D, Astell CR, Ward DC. The genome of minute virus of mice, an autonomous parvovirus, encodes two overlapping transcription units. *Nucleic Acids Res* 1983; 11:1019.
197. Pitluk ZW, Ward DC. Unusual Sp1-GC box interaction in a parvovirus promoter. *J Virol* 1991; 65:22.
198. Ponnazhagan S, Nallari ML, Srivastava A. Suppression of human alpha-globin gene expression mediated by the recombinant adeno-associated virus 2-based antisense vectors. *J Exp Med* 1994; 179:733.
199. Pritchard C, Stout ER, Bates RC. Replication of parvoviral DNA. I. Characterization of a nuclear lysate system. *J Virol* 1981; 37:352.
200. Quinn CO, Kitchingman GR. Functional analysis of the adenovirus type 5 DNA-binding protein: Site-directed mutants which are defective for adeno-associated virus helper activity. *J Virol* 1986; 60:653.
201. Reed AP, Jones EV, Miller TJ. Nucleotide sequence and genome organization of canine parvovirus. *J Virol* 1988; 62:266.
202. Revie D, Tseng BY, Grafstrom RH, Goulian M. Covalent association of protein with replicative form DNA of parvovirus H-1. *Proc Natl Acad Sci USA* 1979; 76:5539.
203. Rhode SL. Replication of the parvovirus H-1. Kinetics in a parasynchronous cell system. *J Virol* 1973; 11:856.
204. Rhode SL. Defective interfering particles of parvovirus H-1. *J Virol* 1978; 27:347.
205. Rhode SL. Replication process of the parvovirus H-1. X Isolation of a mutant defective in replicative form DNA replication. *J Virol* 1978; 25:215.
206. Rhode SL. Trans-activation parvovirus p38 promoter by the 76K noncapsid protein. *J Virol* 1985; 55:886.
207. Rhode SL, Klaassen B. DNA sequence of the 5′-terminus containing the replication origin of parvovirus replicative form DNA. *J Virol* 1982; 41:990.
208. Rhode SL, Paradiso PR. Parvovirus genome: Nucleotide sequence of H1 and mapping of its genes by hybrid arrest translations. *J Virol* 1983; 45:173.
209. Rhode SL, Richard SM. Characterization of the *trans*-activation-responsive element of the parvovirus H-1 p38 promoter. *J Virol* 1987; 61:2807.
210. Richardson WD, Carter BJ, Westphal H. Vero cells injected with adenovirus type 2 mRNA produce authentic viral polypeptide patterns: early mRNA promotes growth of adenovirus-associated virus. *Proc Natl Acad Sci USA* 1980; 77:931.
211. Richardson WD, Westphal H. A cascade of adenovirus early function is required for expression of adeno-associated virus. *Cell* 1981; 27:133.
212. Richardson WD, Westphal H. Requirement for either early region 1a or early region 1b adenovirus gene products in the helper effect for adeno-associated virus. *J Virol* 1984; 51:404.
213. Ron D, Tal J. Coevolution of cells and virus as a mechanism for the persistence of lymphotropic minute virus of mice in L cells. *J Virol* 1985; 55:424.
214. Rose JA, Berns KI, Hoggan MD, Koczot FJ. Evidence for a single-stranded adenovirus-associated virus genome: Formation of a DNA density hybrid on release of viral DNA. *Proc Natl Acad Sci USA* 1969; 64:863.
215. Rose JA, Maizel JVJ, Inman JK, Shatkin AJ. Structural proteins of adenovirus-associated viruses. *J Virol* 1971; 8:766.
216. Rosenfeld SJ, Yoshimoto K, Kajigaya S, Anderson S, Young NS, Field A, Warrener P, Bansal G, Collett MS. Unique region of the minor capsid protein of human parvovirus B19 is exposed on the virion surface. *J Clin Invest* 1992; 90:2609.
217. Ruffing M, Zentgraf H, Kleinschmidt JA. Assembly of viruslike particles by recombinant structural proteins of adeno-associated virus type 2 in insect cells. *J Virol* 1992; 66:6922.
218. Saemundsen A. *Structure of the BPV genome*. Virginia Polytechnic University: MS Thesis, 1978.
219. Salvino R, Skiadopoulos M, Faust EA, Tam P, Shade RO, Astell CR. Two spatially distinct genetic elements constitute a bipartite DNA replication origin in the minute virus of mice genome. *J Virol* 1991; 65:1352.
220. Salzman LA, Fabisch P. Nucleotide sequence of the self-priming 3′ terminus of the single-stranded DNA extracted from the parvovirus KRV. *J Virol* 1970; 30:946.
221. Samulski RJ, Berns KI, Tan M, Muzyczka N. Cloning of adeno-associated virus into pBR322: Rescue of intact from the recombinant plasmid in human cells. *Proc Natl Acad Sci USA* 1982; 79:2007.
222. Samulski RJ, Shenk T. Adenovirus E1B 55-Mr polypeptide facilitates timely cytoplasmic accumulation of adeno-associated virus mRNAs. *J Virol* 1988; 62:206.
223. Samulski RJ, Srivastava A, Berns KI, Muzyczka N. Rescue of adeno-associated virus from recombinant plasmids: Gene correction within the terminal repeats of AAV. *Cell* 1983; 33:135.
224. Samulski RJ, Zhu X, Xiao X, Brook JD, Houseman DE, Epstein N, Hunter LA. Targeted integration of adeno-associated virus (AAV) into human chromosome 19. *EMBO J* 1991; 10:3941.
225. Santaren JF, Ramirez JC, Almendral JM. Protein species of the parvovirus minute virus of mice strain MVMp: involvement of phosphorylated VP-2 subtypes in viral morphogenesis. *J Virol* 1993; 67:5126.
226. Schlehofer JR, Ehrbar M, zur Hausen H. Vaccinia virus, herpes simplex virus, and carcinogens induce DNA amplification in a human cell line and support replication of a helpervirus dependent parvovirus. *Virology* 1986; 152:110.
227. Schneider RJ, Weinberger C, Shenk T. Adenovirus VAI RNA facilitates the initiation of translation in virus-infected cells. *Cell* 1984; 37:291.
228. Schoborg RV, Pintel DJ. Accumulation of MVM gene products is differentially regulated by transcription initiation, RNA processing and protein stability. *Virology* 1991; 181:22.
229. Senapathy P, Carter BJ. Molecular cloning of adeno-associated virus variant genomes and generation of infectious virus by recombination in mammalian cells. *J Biol Chem* 1984; 259:4661.
230. Senapathy P, Tratschin JD, Carter BJ. Replication of adeno-associated virus DNA. Complementation of naturally occurring *rep* mutants by a wild-type genome or an *ori* mutant and correction of terminal palindrome deletions. *J Mol Biol* 1984; 179:1.
231. Shade RO, Blundell MC, Cotmore SR, Tattersall P, Astell CR. Nucleotide sequence and genome organization of human parvovirus B19 isolated from the serum of a child during aplastic crisis. *J Virol* 1986; 58:921.
232. Shi Y, Seto E, Chang LS, Shenk T. Transcriptional repression by YY1, a human GLI-Kruppel-related protein, and relief of repression by adenovirus E1A protein. *Cell* 1991; 67:377.
233. Shull BC, Chen KC, Lederman M, Stout ER, Bates RC. Genomic clones of bovine parvovirus: Construction and effects of deletions and terminal sequence inversions on infectivity. *J Virol* 1988; 62:417.
234. Siegl G, Gautschi M. Purification and properties of replicative-form and replicative-intermediate DNA molecules of parvovirus LuIII. In: Ward DC, Tattersall P, eds. *Replication of mammalian parvoviruses*. Cold Spring Harbor, NY: Cold Spring Harbor Laboratory Press, 1978;315.
235. Skiadopoulos MH, Faust EA. Mutational analysis of conserved tyrosines in the NSl-1 protein of the parvovirus minute virus of mice. *Virology* 1993; 194:509.
236. Skiadopoulos MH, Salvino R, Leong WL, Faust EA. Characterization of linker insertion and point mutations in the NS-1 gene of minute virus of mice: Effects on DNA replication and transcriptional activation functions of NS-1. *Virology* 1992; 188:122.

237. Snyder RO, Im DS, Muzyczka N. Evidence for covalent attachment of the adeno-associated virus (AAV) rep protein to the ends of the AAV genome. *J Virol* 1990a; 64:6204.
238. Snyder RO, Im DS, Muzyczka N. *In vitro* resolution of covalently joined AAV chromosome ends. *Cell,* 1990b; 60:105.
239. Snyder RO, Im DS, Ni T, Xiao X, Samulski RJ, Muzyczka N. Features of the adeno-associated virus origin involved in substrate recognition by the viral Rep protein. *J Virol* 1993; 67:6096.
240. Spalholz BA, Tattersall P. Interaction of minute virus of mice with differentiated cells: Strain dependent target cell specificity is mediated by intracellular factors. *J Virol* 1983; 46:937.
241. Sprecher-Goldberger S, Thiry L, Lefebvre N, Dekegel D, DeHalleux F. Complement-fixation antibodies to adenovirus-associated viruses, adenoviruses, cytomegaloviruses and herpes simplex viruses in patients with tumors and in control individuals. *Am J Epidemiol* 1971; 94:351.
242. Srivastava A, Lusby EW, Berns KI. Nucleotide sequence and organization of the adeno-associated virus 2 genome. *J Virol* 1983; 45:555.
243. Stillman BW, Lewis JB, Chow LT, Mathews MB, Smart JE. Identification of the gene for the adenovirus terminal protein precursor. *Cell* 1981; 23:497.
244. Stillman BW, Tamonoi F, Matthews MB. Purification of an adenovirus-coded DNA polymerase that is required for initiation of DNA replication. *Cell* 1982; 31:613.
245. Straus SE, Sebring E, Rose JA. Concatamers of alternating plus and minus strands are intermediates in adenovirus-associated virus DNA synthesis. *Proc Natl Acad Sci USA* 1976; 73:742.
246. Tam P, Astell CR. Multiple cellular factors bind to cis-regulatory elements found inboard of the 5′ palindrome of minute virus of mice. *J Virol* 1994; 68:2840.
247. Tattersall P. Replication of the parvovirus MVM. I. Dependence of virus multiplication and plaque formation on cell growth. *J Virol* 1972; 10:586.
248. Tattersall P. Susceptibility to minute virus of mice as a function of host-cell differentiation. In: Ward DC, Tattersall P, eds. *Replication of mammalian parvoviruses.* Cold Spring Harbor, NY: Cold Spring Harbor Laboratory Press, 1978;131.
249. Tattersall P, Shatkin A, Ward D. Sequence overlap between the structural polypeptides of parvovirus MVM. *J Mol Biol* 1977; 11:375.
250. Tattersall P, Ward DC. Rolling hairpin model for replication of parvovirus and linear chromosomal DNA. *Nature* 1976; 263:106.
251. Toolan H. Maternal role in susceptibility of embryonic and newborn hamsters to H-1 parvovirus. In: Ward DC, Tattersall P, eds. *Replication of mammalian parvoviruses.* Cold Spring Harbor, NY: Cold Spring Harbor Laboratory Press, 1978;161.
252. Toolan HW, Rhode SL, Gierthy JF. Inhibition of 7,12-dimethylbenz (**a**) anthracene-induced tumors in Syrian hamsters by prior infection with H-1 parvovirus. *Cancer Res* 1982; 42:2552.
253. Torikai K, Ito M, Jordan LE, Mayor HD. Properties of light particles produced during growth of type 4 adeno-associated satellite virus. *J Virol* 1970; 6:363.
254. Tratschin JD, Miller IL, Carter BJ. Genetic analysis of adeno-associated virus: Properties of deletion mutants constructed *in vitro* and evidence for an adeno-associated virus replication function. *J Virol* 1984; 51:611.
255. Tratschin JD, Miller IL, Smith MG, Carter BJ. Adeno-associated virus vector for high-frequency integration, expression, and rescue of genes in mammalian cells. *Mol Cell Biol* 1985; 5:3251.
256. Tratschin JD, Tal J, Carter BJ. Negative and positive regulation in *trans* of gene expression from adeno-associated virus vectors in mammalian cells by a viral *rep* gene product. *Mol Cell Biol* 1986; 6:2884.
257. Tratschin JD, West MH, Sandbank T, Carter BJ. A human parvovirus, adeno-associated virus, as a eukaryotic vector: transient expression and encapsidation of the prokaryotic gene for chloramphenicol acetyltransferase. *Mol Cell Biol* 1984; 4:2072.
258. Trempe JP, Carter BJ. Regulation of adeno-associated virus gene expression in 293 cells: Control of mRNA abundance and translation. *J Virol* 1988; 62:68.
259. Trempe JP, Carter BJ. Alternative mRNA splicing is required for synthesis of adeno-associated virus VP-1 capsid protein. *J Virol* 1988; 62:3356.
260. Trempe JP, Mendelson E, Carter BJ. Characterization of adeno-associated virus rep proteins in human cells by antibodies raised against rep expressed in *Escherichia coli*. *Virology* 1987; 161:18.
261. Tsao J, Chapman MS, Agbandje M, Keller W, Smith K, Wu H, Luo M, Smith TJ, Rossmann MG, Compans RW, et al. The three-dimensional structure of canine parvovirus and its functional implications. *Science* 1991; 251:1456.
262. Vasudevacharya J, Basak S, Srinivas RV, Compans RW. The complete nucleotide sequence of an infectious clone of porcine parvovirus, strain NADL-2. *Virology* 1990; 178:611.
263. Vasudevacharya J, Compans RW. The NS and capsid genes determine the host range of porcine parvovirus. *Virology* 1992; 187:515.
264. Walsh CE, Liu JM, Xiao X, Young NS, Nienhuis AW, Samulski RJ. Regulated high level expression of a human τ-globin gene introduced into erythroid cells by an adeno-associated virus vector. *Proc Natl Acad Sci USA* 1992; 89:7257.
265. Walter S, Richards R, Armentrout RW. Cell cycle-dependent replication of the DNA of minute virus of mice, a parvovirus. *Biochim Biophys Acta* 1980; 607:420.
266. Walz C, Schlehofer JR. Modification of some biological properties of HeLa cells containing adeno-associated virus DNA integrated into chromosome 17. *J Virol* 1992; 66:2990.
267. Walz C, Schlehofer JR, Flentje M, Rudat V, zur Hausen H. Adeno-associated virus sensitizes HeLa cell tumors to gamma rays. *J Virol* 1992; 66:5651.
268. Ward DC, Dadchanji D. Replication of minute-virus-of-mice DNA. In: Ward DC, Tattersall P, eds. *Replication of mammalian parvoviruses.* Cold Spring Harbor, NY: Cold Spring Harbor Laboratory Press, 1978;197.
269. Ward P, Berns KI. *In vitro* rescue of an integrated hybrid adeno-associated virus/Simian virus 40 genome. *J Mol Biol* 1991; 218:791.
270. Ward P, Urcelay E, Kotin R, Safer B, Berns KI. Adeno-associated virus DNA replication *in vitro*: Activation by a maltose binding protein/Rep 68 fusion protein. *J Virol* 1994; 68:6029.
271. Weindler FW, Heilbronn R. A subset of herpes simplex virus replication genes provides helper functions for productive adeno-associated virus replication. *J Virol* 1991; 65:2476.
272. Weitzman MD, Kyostio SRM, Kotin RM, Owens RA. Adeno-associated virus (AAV) rep proteins mediate complete formation between AAV DNA and the human integration site. *Proc Natl Acad Sci USA* 1994; 91:5808.
273. Willwand K, Hirt B. The minute virus of mice capsid specifically recognizes the 3′ hairpin structure of the viral replicative-form DNA: Mapping of the binding site by hydroxyl radical footprinting. *J Virol* 1991; 65:4629.
274. Willwand K, Hirt B. The major capsid protein VP2 of minute virus of mice (MVM) can form particles which bind to the 3′-terminal hairpin of MVM replicative-form DNA and package single-stranded viral progeny DNA. *J Virol* 1993; 67:5660.
275. Winocour E, Puzis L, Etkins S, Koch T, Danovitch B, Mendelson E, Shaulian E, Karby S, Lavi S. Modulation of the cellular phenotype by integrated adeno-associated virus. *Virology* 1992; 190:316.
276. Wong KK, Podsakoff G, Lu D, Chatterjee S. High efficiency gene transfer into growth arrested cells utilizing an adeno-associated virus (AAV)-based vector. *Blood* 1993;82:302a.
277. Wu H, Rossmann MG. The canine parvovirus empty capsid structure. *J Mol Biol* 1993; 233:231.
278. Yakinoglu AO, Heilbronn R, Burkle A, Schlehofer JR, zur Hausen H. DNA amplification of adeno-associated virus as a response to cellular genotoxic stress. *Cancer Res* 1988; 48:3123.
279. Yakobson B, Koch T, Winocour E. Replication of adeno-associated virus in synchronized cells without the addition of a helper virus. *J Virol* 1987; 61:972.
280. Yang Q, Trempe JP. Analysis of the terminal repeat binding abilities of mutant adeno-associated virus replication proteins. *J Virol* 1993; 67:4442.
281. Zhao O, Schoborg RV, Pintel D. Efficient excision of the large intron from RNA encoding the nonstructural proteins of the autonomous parvovirus MVM requires the downstream small intron and is suppressed by the large intron nonconsensus donor. *Fifth Parvovirus Workshop.* Crystal River, FL: 1993.
282. Zhou SZ, Broxmeyer HE, Cooper S, Harrington MA, Srivastava A. Adeno-associated virus 2-mediated gene transfer in murine hematopoietic progenitor cells. *Exp Hematol* 1993; 21:928.
283. Zhu X. *Characterization of adeno-associated virus proviral structure in latently infected human cells.* University of Pittsburgh: PhD Thesis, 1993.

Fundamental Virology, Third Edition
edited by B.N. Fields, D.M. Knipe, P.M. Howley, et al.
Lippincott - Raven Publishers, Philadelphia © 1996

CHAPTER 32

Herpes Simplex Viruses and Their Replication

Bernard Roizman and Amy E. Sears

> . . . paradoxically it is much easier for people to adapt the observed facts dialectically to the ruling paradigm than to renounce the ruling paradigm in response to possible new interpretations of the facts.
>
> Carlo M. Cipolla in *Miasmas and Disease. Public Health and the Environment in the Pre-Industrial Age*. p. 6. Yale University Press, 1992.

B. Roizman: The Marjorie B. Kovler Viral Oncology Laboratories, The University of Chicago, Chicago, Illinois 60637.

A. E. Sears, Department of Microbiology and Immunology, Emory University, School of Medicine, Atlanta, Georgia 30322.

Herpes simplex viruses (HSV) were the first of the human herpesviruses to be discovered and are among the most intensively investigated of all viruses. Their attractions are their biologic properties, particularly their abilities to cause a variety of infections, to remain latent in their host for life, and to be reactivated to cause lesions at or near the site of initial infection. They serve as models and tools for the study of translocation of proteins, synaptic connections in the nervous system, membrane structure, gene regulation, and a myriad of other biologic problems, both general to viruses and specific to HSV.

For years, their size and complexity served as formidable obstacles to intensive research. More than 40 years passed from the time of their isolation until Schneweiss (597) demonstrated that there were in fact two serotypes, HSV-1 and HSV-2, whose formal designations under International Conference for Taxonomy of Viruses (ICTV): rules are now human herpesviruses 1 and 2 (556). Not until 1961 were practical plaque assays published (564), and only much later were the genome sizes and the extent of homology between these two viruses reported. This chapter recounts well established facts, but its main emphasis is on burning issues, the problems whose time has come.

Virology conserves three myths. The first is that research on a virus reaches its peak when the number of investigators approaches the number of nucleotides in its genome. This formula calls for 152,000+ investigators, one for each base pair (309,402). In orders of magnitude, we are close but not yet there. There are times when we think almost that many bodies will be needed to unravel all the mysteries of these viruses. The second myth is that virologists repeat the same experiment over and over again. As with all myths, there may be a grain of truth here. In wading through the mass of papers published in the past decade, it was instructive to see how many times the same phenomenon was published or rediscovered time and time again under the same or different name.

Lastly, the oft-made statement that science simplifies knowledge is patently a myth. The falsehood of this premise is attested by this chapter. What should have been a simple update of the chapter published in the second edition of *Field's Virology* unraveled complexities far beyond our expectations.

VIRION STRUCTURE

The HSV virion consists of four elements: (i) an electron-opaque core, (ii) an icosadeltahedral capsid surrounding the core, (iii) an amorphous tegument surrounding the capsid, and (iv) an outer envelope exhibiting spikes on its surface (557).

The dry masses of herpes simplex virions, full nucleocapsids, empty nucleocapsids, and cores were calculated from permeability of virions to electrons to be $13.33 \pm 2.56 \times 10^{-16}$ g, $7.55 \pm 1.11 \times 10^{-16}$ g, $5.22 \pm 1.10 \times 10^{-16}$ g, and $2.07 \pm 0.95 \times 10^{-16}$ g, respectively (350). The average mass ratios of the virion, full capsid, and core to DNA are 8.1, 4.6, and 1.25, respectively. The experimentally derived ratio of virus mass to DNA is 10.73 ± 0.96, from which it has been calculated that the virion contains 19.4×10^{-16} g of protein (242). A similar value was derived from counts of virions in purified virus preparations (R. W. Honess and B. Roizman, unpublished data), although the error in the these determinations was higher. This ratio was used in the calculation of the polypeptide content of HSV virions by Heine et al. (242).

Virion Polypeptides

Early studies on purified HSV-1 virions suggested that they contain 30+ proteins designated as virion polypeptides (VP) and given serial numbers (242,643). All of the virion proteins were made after infection, and no host proteins could be detected in purified virion preparations. Of the approximately 30 known and another ten suspected virion proteins (Table 1), at least 9 are on the surface of the virion (accessible to antibody) and at least ten are glycosylated. The glycoproteins are gB (VP7 and VP8.5), gC (VP8), gD (VP17 and VP18) and gE (VP12.3 and VP12.6), gG, gH, gI, gK, gL, and gM. Another small glycoprotein, given the designation gJ, was predicted by DNA sequence analyses. Virion envelopes also contain at least two (U_L20 and U_L34) and possibly more (U_L24 and U_L43??) nonglycosylated intrinsic membrane proteins. Stannard et al. (653) reported that spikes projecting from envelopes are, as was expected, the viral glycoproteins, and that the latter were nonrandomly distributed.

Gibson and Roizman (213,215) described three kinds of capsids, i.e., those that lack DNA and were never enveloped (type A), those that contain DNA and were never enveloped (type B), and those that contain DNA and were obtained by deenveloping intact virions (type C). In the current nomenclature, the term A capsid refers to capsids without an internal toroidal structure which is thought to act as scaffolding for packaging of DNA, those with scaffolding but without DNA are designated as B capsids, and those with DNA have been designated as the C capsids (Fig. 1).

The empty (A) capsids consist of four proteins, i.e., VP5 (U_L19), VP19C (U_L38), VP23 (U_L38), and a smaller M_r

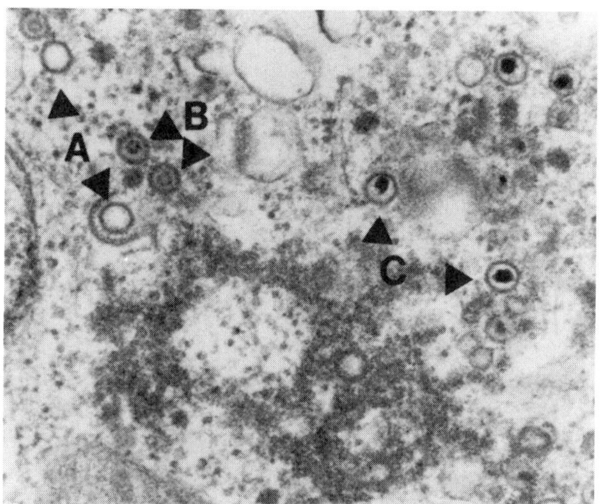

FIG. 1. This section of a nucleus of a Vero cell culture harvested 18 hr postinfection with HSV-1(F). **A:** Empty capsids devoid of scaffolding protein. **B:** Capsids containing internal proteins arranged as a ring and presumed to be the scaffolding proteins. **C:** Capsids containing DNA.

TABLE 1. *Herpes simplex virus genes, their products, and their functions*

Gene or Transcriptional Unit	Designation of Protein	Dispensable for Replication in Cell Culture	Regulation: Kinetic Group	Function of Gene Product
$\gamma_1 34.5$	ICP34.5	Y	γ_1	Deletion mutants fail to replicate in central nervous system of mice. In human cells in culture, the $\gamma_1 34.5^-$ viruses fail to preclude programmed cell death initiated by complete cessation of viral protein synthesis. Carboxyl terminus homologous to the corresponding domain of the GADD34 and MyD116 proteins.
ORF P	ORF P	Y	??	Open reading frame is antisense to $\gamma_1 34.5$. The gene is expressed under conditions in which ICP4 is nonfunctional.
ORF O	ORF O	Y	?	Open reading frame partially antisense to the $\gamma_1 34.5$ gene. The protein is expressed under conditions in which ICP4 is not functional.
$\alpha 0$	ICP0	Y	α	Nucleotidylylated, phosphoprotein, promiscuous *trans*-activator of genes introduced by transfection or infection; optimal activity requires presence of ICP4. Deletion mutants debilitated with respect to replication at low multiplicities of infection.
$U_L 1$	gL	N	γ	Forms complex with gH. Complex required for transport of both proteins to plasma membrane and for viral [illegible], mediated by gH. Contains *syn* locus.
$U_L 2$		Y	β	Uracil DNA glycosylase.
$U_L 3$		Y	Unknown	Unknown function. Identified in HSV-2 as a nuclear phosphoprotein. Protein has nuclear localization signal and is unglycosylated. Reported to localize to perinuclear region early and to the nucleus late in infection.
$U_L 4$		Y	γ_2	Identified three protein species: the M_r 60k species is found in virions and light particles (19).
$U_L 5$		N	β	Forms complex with $U_L 8$ and $U_L 52$ proteins.
$U_L 6$		N	Unknown	Virion protein; required for DNA cleavage-packaging.
$U_L 7$		N	Unknown	Unknown.
$U_L 8$		N	β	Forms complex with $U_L 5$ and $U_L 52$, acts as a primase and expresses helicase activity in the presence of $U_L 9$ protein. Stabilizes interaction between primers and DNA template.
$U_L 9$		N	γ(?)	Binds to origins of DNA synthesis in sequence-specific (origin) fashion; carries out helicase and ATPase activities.
$U_L 10$	gM	Y	γ	Glycoprotein present in virions and plasma membranes.
$U_L 11$		Y	γ(?)	Myristylated protein; necessary for efficient capsid envelopment and exocytosis.
$U_L 12$		Y	β	Alkaline exonuclease (DNAse)—involved in viral nucleic acid metabolism; reported to localize in nucleoli and in virally induced nuclear dense bodies and to bind to a sequence along with other unidentified proteins. Complex may be involved in cleavage-packaging of viral DNA.
$U_L 13$		Y	γ	Nuclear protein suspected of being a protein kinase; required for phosphorylation of ICP22, VHS.
$U_L 14$		N	Unknown	Unknown.
$U_L 15$		N	γ	ts mutant DNA+. Two exons; protein required for packaging of DNA.
$U_L 16$		Y	Unknown	Located within intron of $U_L 15$.
$U_L 17$		N	γ	Located within intron of $U_L 15$.
$U_L 18$	VP23	N	γ	Capsid protein required for capsid formation and cleavage-packaging of replicated viral DNA.
$U_L 19$	VP5, ICP5	N	γ_1	Major capsid protein.
$U_L 20$		Y	γ	Intrinsic membrane protein necessary for viral exocytosis, particularly in cells in which the Golgi apparatus is fragmented and dispersed.
$U_L 20.5$	$U_L 20.5$	Y	γ_2	Unknown.
$U_L 21$		Y	Unknown	Nucleotidylylated phosphoprotein; unknown function.
$U_L 22$	gH	N	γ_2	Forms complex with gL (see above). Appears to play a role in entry, egress, and cell-cell spread.

TABLE 1. *Continued.*

Gene or Transcriptional Unit	Designation of Protein	Dispensable for Replication in Cell Culture	Regulation: Kinetic Group	Function of Gene Product
U_L23	ICP36	Y	β	Thymidine kinase, more properly a nucleoside kinase.
U_L24		Y	γ	*syn⁻* locus; membrane-associated protein?
U_L25		N	γ	Virion protein reported to be required for cleavage-packaging of viral DNA.
U_L26		N	γ	Serine protease; substrates are U_L26 protein and $U_L26.5$ (ICP35 capsid proteins)—VP21, VP22a, and VP24 capsid proteins are all products of the self-cleavage of U_L26; VP21 suggested to be C-terminal portion of U_L26 after cleavage release of VP24. VP24 contains protease domain from N-terminal portion of U_L26.
$U_L26.5$	ICP35	N	γ	Substrate of U_L26 protease; the precursor, ICP35b,c is cleaved to ICP35e,f. The protein is unique to B capsids and forms its inner core or scaffolding. On packaging of DNA, VP22A is removed from capsid shell.
U_L27	gB, VP7	N	γ_1	Glycoprotein forms a dimer and induces neutralizing antibody. Required for viral entry. A syn locus maps to the carboxyl terminus.
U_L28	ICP18.5	N	γ	M_r 87- to 95-k protein required for DNA cleavage-packaging.
U_L29	ICP8	N	β	Binds singled-stranded DNA cooperatively, required for viral DNA replication. Mutants are DNA^-. Expression of early and late genes may be affected positively or negatively by the function of ICP8. ICP8 binds to single-stranded DNA and facilitates renaturation of complementary strands of DNA, homologous pairing, and strand transfer.
U_L30		N	β	DNA polymerase; forms complex with C-terminal 247 amino acids of U_L42 protein.
U_L31		N	γ_2	Nucleotidylylated phosphoprotein, cofractionates with nuclear matrix.
U_L32		N	γ_2	ts mutant is deficient in DNA packaging.
U_L33		N	Unknown	DNA packaging; necessary for assembly of capsids containing DNA.
U_L34		N	Unknown	Abundant nonglycosylated, membrane-associatad, virion protein phosphorylated by protein kinase U_S3.
U_L35	VP26	N	γ_2	Basic phosphorylated capsid protein.
U_L36	ICP1-2	N	γ_2	Virion tegument phosphoprotein. In cells infected with ts mutant at nonpermissive temperatures, DNA is not released from capsids at nuclear pores. Reported to form complex with M_r 140k protein that binds a sequence DNA. May be involved in cleavage and/or packaging of newly synthesized viral DNA.
U_L37	ICP32	N	γ	Cytoplasmic phosphoprotein; in presence of ICP8, it is transported to nucleus and associates with DNA, but phosphorylation is not dependent on ICP8.
U_L38	VP19C	N	γ_2	Capsid assembly protein, binds DNA, and may be involved in anchoring DNA in the capsid.
U_L39	ICP6	Y	β	Large subunit of ribonucleotide reductase. Autophosphorylates via unique N-terminus but does not transphosphorylate. HSV-2 homolog can be transphosphorylated.
U_L40		Y	β	Small subunit of ribonucleotide reductase.
U_L41	VHS	Y	γ	Causes nonspecific degradation of mRNA and shut off of macromolecular synthesis after infection. Exists as two phosphorylated species with M_r of 58 and 59.5 k.
U_L42		N	β	Double-stranded DNA-binding protein, binds to and increases processivity of DNA polymerase.
U_L43		Y	Unknown	Amino acid sequence predicts membrane-associated protein.
U_L44	gC, VP8	Y	γ_2	Glycoprotein involved in cell attachment; required for attachment to the apical surface of polarized MDCK cells.
U_L45		Y	γ_2	Encodes a M_r 18k protein of unknown function.
U_L46	VP11/12	Y	γ	Tegument phosphoprotein reported to modulate the activity of U_L48 (αTIF) protein.

TABLE 1. *Continued.*

Gene or Transcriptional Unit	Designation of Protein	Dispensable for Replication in Cell Culture	Regulation: Kinetic Group	Function of Gene Product
U_L47	VP13/14	Y	γ_2	Nucleotidylylated tegument phosphoprotein reported to modulate the activity of U_L48 (αTIF) protein.
U_L48	VP16, ICP25, αTIF	N	γ	Tegument protein, induces α genes by interacting with host proteins, including Oct-1. The complex binds to specific sequences with the consensus GyATGnTAATGArATTCyTTGnGGG-NC.
U_L49		N	γ	Nucleotidylylated mono(ADP-ribosyl)ated tegument phosphoprotein.
$U_L49.5$		N	γ_2	Sequence predicts a M_r 12k membrane-associated protein.
U_L50		Y	β	dUTPase.
U_L51		Y	γ	Unknown.
U_L52		N	β	Component of the helicase/primase complex.
U_L53	gK	N	γ	Glycoprotein required for efficient viral exocytosis; contains syn locus.
U_L54	α27, ICP27	N	α	Nucleotidylylated, multifunctional regulatory protein required for late gene expression. The protein negatively regulates early genes. It was reported to cause the redistribution of snRNPs and to inhibit RNA splicing.
U_L55		Y	Unknown	Unknown.
U_L56		Y	Unknown	Nuclear, virion-associated protein of unknown function.
α4	ICP4	N	α	Nucleotidylylated, poly(ADP-ribosyl)ated, phosphoprotein regulates positively most ß and γ genes and negatively itself, OFRF-P and the α0 genes. It binds to DNA in sequence-specific fashion.
U_S1	α22, ICP22	Y	α	Nucleotidylylated regulatory protein, phosphorylated by U_L13 protein kinase, required for optimal expression of ICP0 and of a subset of γ proteins.
$U_S1.5$	$U_S1.5$	Y	α	Gene 31 coterminal with U_S1. Not required for optimal expression of α0 or late (γ) genes.
U_S2		Y	Unknown	Unknown.
U_S3		Y	β	Protein kinase; major substrate is U_L34 protein.
U_S4	gG	Y	γ	Glycoprotein involved in entry, egress, and spread from cell to cell.
U_S5	gJ(?)	Y	Unknown	Sequence predicts glycoprotein.
U_S6	gD	N	γ_1	Glycoprotein required for postattachment entry of virus into cells.
U_S7	VP17/18, gI	Y	γ	gI and gE glycoproteins form complex for transport to plasma membrane and also to constitute a high-affinity Fc receptor. gI is required for basolateral spread of virus in polarized cells.
U_S8	gE	Y	γ_2	Fc receptor; involved in basolateral spread of virus in polarized cells.
$U_S8.5$		Y	ß or γ_1	Unknown.
U_S9		Y	Unknown	Tegument protein.
U_S10		Y	Unknown	Tegument protein.
U_S11		Y	γ_2	Abundant virion tegument protein binds to U_L34 mRNA in sequence and conformation specific fashion and acts as an antiattenuation factor; binds to the 60S ribosomal subunit and localizes in the nucleolus.
U_S12	α47, ICP47	Y	α	Blocks presentation of viral peptides to MHCI restricted cells.
Ori_STU	Ori_SRNA	Y	γ_2	RNA transcribed across S component origins of DNA synthesis. Most probably not translated and function is not known.
LATU	LATs	Y	?	Transcripts, some spliced, from the inverted repeat sequences flanking U_L sequences. The function of these transcripts is not known.

syn locus, mutation causes infected cells to fuse; tegument, a component of the virion located between the envelope and the capsid; Ori_STU, transcriptional unit comprising the domain of the S component origins of DNA synthesis; LATU, latency-associated transcriptional units; αTIF, α *trans*-inducing factor; HSV, herpes simplex virus; dUTP, deoxyuridine triphosphate; mRNA, messenger RNA.

12,000 protein described subsequently (105) and often referred to as VP26 (U_L35) (130,486). VP5 was estimated to be present in ratios of 850 to 1,000 per virion, i.e., approximately six per hexameric capsomere (242,557,703), but Schrag et al. (598) suggested that VP5 is a component of both pentameric and hexameric capsomeres. VP19C and VP5 appear to be linked by disulfide bond (758) and are present in approximately similar ratios per virion (242). Braun et al. (58) showed that VP19C, identified as the infected cell protein (ICP) number 32, bound to DNA and may be involved in anchoring the viral DNA in the capsid. The HSV-2 counterpart has also been mapped (752). The A-type capsids may be a decay product not in the pathway of virion maturation (618). B-type capsids differ from the A type in that they contain three additional proteins, i.e., VP21, VP22a, and VP24. VP22a corresponds to ICP35e-f, the product of the open reading frame $U_L26.5$ cleaved at the carboxyl terminus by the protease (370,371). VP21 corresponds to ICP35b (amino acids 248–610) of the protease Pra, the nascent U_L26 gene product. VP24 corresponds to the N-terminal domain (codons 1 through 247) of Pra designated as Prn, the smallest form of the protease encoded by the open reading frame U_L26. Pra is cleaved by the autologous protease between Ala^{247} and Ser^{248} and between Ala^{610} and Ser^{611} (144,368). The ICP35 family of proteins described by Braun et al. (60) plays a vital role in capsid assembly and encapsidation of viral DNA (205,507,538, 618). Gibson and Roizman (215) suggested that VP21 is an internal capsid protein. Newcomb and Brown (451,452) demonstrated that VP19 and VP23 are on the surface of the capsid and could form a network of fibers located between capsomeres (intercapsomeric fibers) and that VP22a is in the interior of the capsid and forms a ringlike structure which is quantitatively removed by 2.0 M guanidine hydrochloride.

Type C capsids were reported to contain a smaller protein—VP22—but not VP22a, and it has been suggested that the proteins are related (213,215). Sherman and Bachenheimer (618) and Rixon et al. (538) suggested that the VP22 found in the C-type capsids may not be related to VP22a. Depending on the procedure used for stripping the envelope, the C-type capsids may contain variable amounts of tegument proteins. Schrag et al. (598) have reported an elegant model of the HSV-1 capsid.

It should be noted that, in the interval between 1965 and 1974, a large number of articles dealt with the structure and morphogenesis of herpesvirus capsids; a list of citations and review of that literature was published by Roizman and Furlong (557). On the basis of electron microscopic appearance of the capsid and the core, capsids form eight groups (557). Although the various forms probably reflect different stages of capsid assembly, the reagents necessary to relate the various forms to specific proteins are only now becoming available.

The space between the undersurface of the envelope and the surface of the capsid was designated as the tegument (557); it contains the rest of the virion proteins. The most notable of the proteins associated with the space between the underside of the envelope and the capsid are the α-*trans*-inducing factor (αTIF; ICP25; VP16), VP11-12 (U_L46), VP13-14, (U_L47), the virion host shut off (VHS) protein (U_L41), the product of the U_S11 gene, and a very large protein (VP1-2) associated with a complex which binds to the terminal *a* sequence of the viral genome (33,96, 530,574,573,755). Extensive discussion of the various types of capsids and virions can be found elsewhere (557).

Recently, several publications have reported on the production of "light particles" devoid of DNA. These particles consist of enveloped tegumentlike structures (417,536, 678). They appear to contain, in addition, other nonstructural proteins previously associated with virions (e.g., ICP4) (750). Little is known of their synthesis beyond the facts that they do not appear to be uniform in size and that the capsid is not an essential trigger for envelopment or egress.

Viral DNA

Like other herpesvirus DNAs, the bulk of packaged HSV DNA is linear and double stranded (36,309,488). In the virion, HSV DNA is packaged in the form of a toroid (202). The ends of the genome are probably held together or are in close proximity inasmuch as a small fraction of the packaged DNA appears to be circular and a large fraction of the linear DNA circularizes rapidly in the absence of protein synthesis after it enters the nuclei of infected cells (489). DNA extracted from virions contains ribonucleotides, nicks, and gaps (42,197,736).

The HSV genome is approximately 150 kbp, with a G+C content of 68% for HSV-1 and 69% for HSV-2 (36,309,402). It consists of two covalently linked components, designated as L (long) and S (short) (Fig. 2). Each component consists of unique sequences bracketed by inverted repeats (604; 695). The repeats of the L component are designated *ab* and *b′a′*; those of the S component are *a′c′* and *ca* (Fig. 2) (706). The number of *a* sequence repeats at the L-S junction and at the L terminus is variable; the HSV genome can then be represented as

$$a_L a_n b\text{-}U_L\text{-}b'a'_m c'\text{-}U_S\text{-}ca_S$$

where a_L and a_S are terminal sequences with unique properties described below, and a_n and a_m are terminal *a* sequences directly repeated zero or more times (n) or present in one to many copies (m) (372,549,550,706,710). The structure of the *a* sequence is highly conserved but consists of a variable number of repeat elements. In the HSV-1(F) strain, the *a* sequence consists of a 20-bp direct repeat (DR1), a 65-bp unique sequence (U_b), a 12-bp direct repeat (DR2) present in 19 to 23 copies per *a* sequence, a 37-bp direct repeat (DR4) present in two to three copies, a 58-bp unique sequence (U_c), and a final copy of DR1 (436,438). The size of the *a* sequence varies from strain to

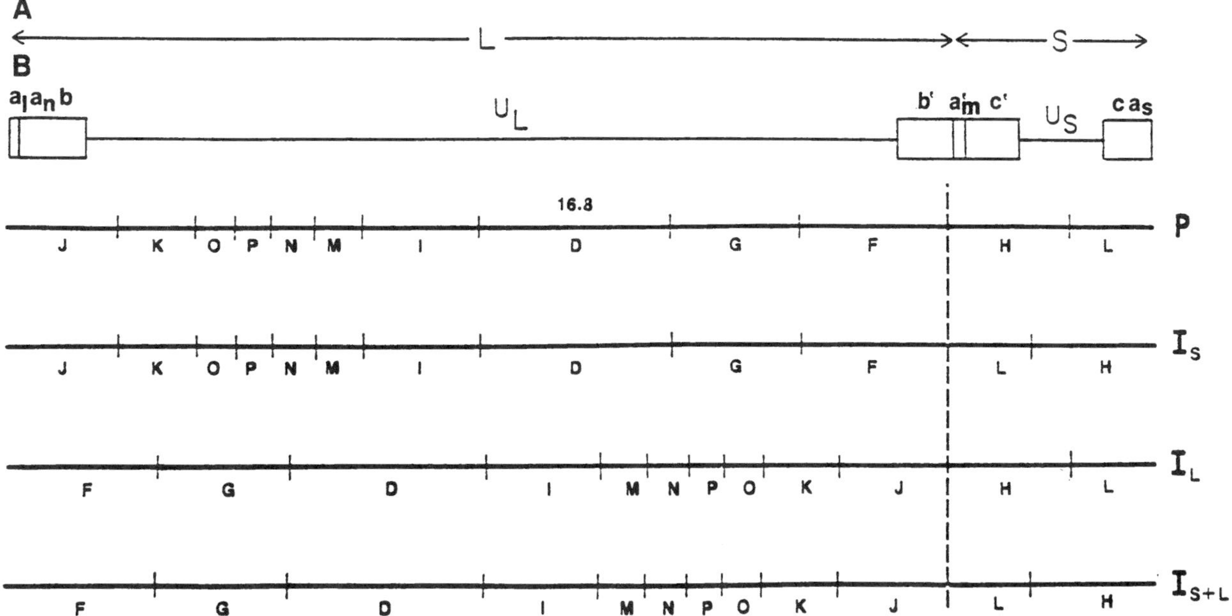

FIG. 2. Schematic representation of the arrangement of DNA sequences in the HSV genome. **A:** The domains of the L and S components are denoted by the *arrows*. The second line shows the unique sequences (*thin lines*) flanked by the inverted repeats (*boxes*). The letters above the second line designate the terminal *a* sequence of the L component (a_L), a variable (n) number of additional *a* sequences, the *b* sequence, the unique sequence of the L component (U_L), the repetitions of the *b* sequence and of a variable (m) number of *a* sequences (a_m), the inverted *c* sequence, the unique sequence of the S component (U_S), and finally the terminal a sequence (a_S) of the S component. **B:** The HindIII restriction endonuclease map of HSV-1(F) strain for the P, I_S, I_L, and I_{SL} isomers of the DNA. Note that, because HindIII does not cleave within the inverted repeat sequences, there are four terminal fragments and four fragments spanning the internal inverted repeats in concentrations of 0.5 and 0.25 M, respectively, relative to the concentration of the viral DNA.

strain, reflecting in part the number of copies of DR2 and DR4. The structure of the HSV-1(F) *a* sequence can be represented as

$$DR1\text{-}U_b\text{-}DR2_n\text{-}DR4_m\text{-}U_c\text{-}DR1$$

with adjacent *a* sequences sharing the intervening DR1. Linear virion DNA contains asymmetric terminal *a* sequence ends. The terminal *a* sequence of the L component (a_L) contains a truncated DR1 with 18 bp and one 3′ nucleotide extension, whereas the terminal *a* sequence of the S component (a_S) ends with a DR1 containing only 1 bp and one 3′ overhanging nucleotide (436). The two truncated DR1 sequences form one complete DR1 upon circularization.

The L and S components of HSV can invert relative to one another, yielding four linear isomers (Fig. 2) (135, 239). The isomers have been designated as P (prototype), I_L (inversion of the L component), I_S (inversion of the S component), and I_{SL} (inversion of both S and L components) (239,441,442).

The first evidence for the repetition of terminal sequences in inverted orientation was based on electron microscopic studies of denatured HSV-1 DNA allowed to self-anneal (615). Electron microscopic analyses of denatured molecules allowed to self-anneal and of partial denaturation profiles of HSV DNA revealed that the terminal repeats are also repeated internally and that the repeats flanking the L component differ from those of the S component in size and sequence arrangement (706). The demonstration that restriction endonucleases which cleave outside the inverted repeats yield four terminal 0.5 M fragments and four L-S component junction fragments that are 0.25 M (Fig. 2) (239), as well as analyses of the partial denaturation profiles of Wadsworth et al. (706), supported the conclusion that the L and S components can invert relative to each other.

The internal inverted repeat sequences are not essential for growth of the virus in cell culture; mutants from which portions of unique sequences and most of the internal inverted repeats have been deleted have been obtained in all four arrangements of HSV DNA (286,490). The genomes of these mutants do not invert; each is frozen in one arrangement of the L and S components, but all retain their viability in cell culture.

Other Constituents

Polyamines

The search for polyamines in the virion evolved from the observations that HSV capsid assembly requires the addition of arginine to the medium (388,683) and that the capsid does not contain highly basic proteins that would neutralize viral DNA for proper folding inside the capsid. Highly purified virions contain the polyamines spermidine and spermine in a nearly constant ratio of 1.6 ± 0.2:1 or approximately 70,000 molecules of spermidine and 40,000 molecules of spermine per virion (214,216). The polyamines appear to be tightly bound and cannot be exchanged with exogenously added labeled polyamines. Disruption of the envelope with nonionic detergents and urea removed the spermidine but not the spermine. The spermine contained in the virion is sufficient to neutralize approximately 40% of the DNA phosphate (214). Parenthetically, proteins have been noted in association with the toroidal structure (202) in the capsid. and a capsid protein has been reported to bind DNA (58).

The compartmentalization of spermine and spermidine may reflect the distribution of polyamines in the infected cell. It is of interest to note that, after infection, the conversion of ornithine to putrescine appears to be blocked, but the synthesis of spermine and spermidine does not appear to be affected (214).

Lipids

It has been assumed that HSV acquires the envelope lipids from its host. Little is known of the composition of the lipid in the envelopes. The hypothesis that it is determined by its host was supported by the observation that the buoyant density of the virus was host cell-dependent on serial passage of HSV-1 alternately in HEp-2 and chick embryo cells (644). Since the envelope is derived from cellular membranes, it has been assumed that the viral envelope and cellular membranes contain similar or identical lipids. Recent studies suggest that the virion lipids are similar to those of cytoplasmic membranes and different from those of nuclear membranes (700).

HSV Polymorphism

Intertypic Variation

Although the genetic maps of HSV-1 and HSV-2 are largely colinear, they differ in restriction endonuclease cleavage sites and in the apparent sizes of viral proteins. Thus, the initial locations of viral genes on the linear map of HSV genomes were based on analyses of HSV-1 × HSV-2 recombinants and took advantage of the intertypic differences in the sizes of the proteins and the locations of restriction endonuclease cleavage sites (390,441,442,508).

Intratypic Variation

The first evidence of intratypic polymorphism emerged from studies of virion structural proteins and indicated that nonglycosylated proteins vary sufficiently in electrophoretic mobility to be used as strain markers (482). Although specimens from epidemiologically related individuals appeared to yield similar electrophoretic profiles, the usefulness of virion proteins as markers for molecular epidemiologic studies was limited by the effort required to purify virions for such analyses.

At the DNA level, differences between HSV-1 strains appear to result from base substitutions, which may add or eliminate a restriction endonuclease cleavage site and, on occasion, change an amino acid, or variability in the number of repeated sequences present in a number of regions of the genome, e.g., $\gamma_1$34.5, U_S11, and so forth (95,540). The restriction endonuclease cleavage patterns of a given strain are relatively stable, while the number of repeats are not (64,238,555,692). Thus, no changes in restriction endonuclease patterns were noted in isolates from the same individual over an interval of 13 years or in genomes of an HSV-1 strain passaged serially numerous times in cell culture. However, restriction endonuclease site polymorphism was readily noted in isolates from epidemiologically unrelated individuals (232,570). On the basis of these properties, restriction endonuclease site polymorphism was used in several epidemiologic studies of HSV transmission in the human population (64,555,570), and blind restriction endonuclease analysis of virus isolates has been used to trace the spread of infection from patients to hospital personnel (62), from patient to patient (366), and from hospital personnel to patient (7,63). Recently, Sakaoka et al. (587) reported on clustering of divergent sites along geographically and racially distinct areas. The conclusion that "the evolution of HSV-1 may be host dependent" is an aggressive interpretation of the data; a more conservative interpretation is that random mutations were conserved and dispersed in different populations.

VIRAL REPLICATION

General Pattern of Replication

It is convenient to begin this section on viral replication with a bird's-eye view of the major events (Fig. 3).

To initiate infection, the virus must attach to cell surface receptors. Fusion of the envelope with the plasma membrane rapidly follows the initial attachment. The deenveloped capsid is then transported to the nuclear pores where DNA is released into the nucleus.

Transcription, replication of viral DNA, and assembly of new capsids take place in the nucleus (Figs. 3 and 4).

Viral DNA is transcribed throughout the reproductive cycle by host RNA polymerase II, but with the participation of viral factors at all stages of infection. The syn-

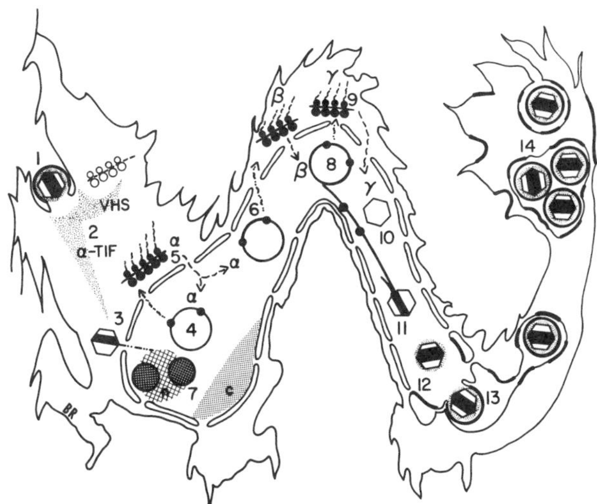

FIG. 3. Schematic representation of the replication of HSV in susceptible cells. **1:** The virus initiates infection by the fusion of the viral envelope with the plasma membrane following attachment to the cell surface. **2:** Fusion of the membranes releases two proteins from the virion. VHS shuts off protein synthesis (broken RNA in open polyribosomes). α-TIF is transported to the nucleus. **3:** The capsid is transported to the nuclear pore where viral DNA is released into the nucleus and immediately circularizes. **4:** The transcription of α genes by cellular enzymes is induced by α-TIF. **5:** The 5 αmRNAs are transported into the cytoplasm and translated (filled polyribosome); the proteins are transported into the nucleus. **6:** A new round of transcription results in the synthesis of β proteins. **7:** At this stage in the infection, the chromatin (*c*) is degraded and displaced toward the nuclear membrane, whereas the nucleoli (*round hatched structures*) become disaggregated. **8:** Viral DNA is replicated by a rolling circle mechanism, which yields head-to-tail concatemers of unit-length viral DNA. **9:** A new round of transcription/translation yields the γ proteins, consisting primarily of structural proteins of the virus. **10:** The capsid proteins form empty capsids. **11:** Unit-length viral DNA is cleaved from concatemers and packaged into the preformed capsids. **12:** Capsids containing viral DNA acquire a new protein. **13:** Viral glycoproteins and tegument proteins accumulate and form patches in cellular membranes. The capsids containing DNA and the additional protein attach to the underside of the membrane patches containing viral proteins and are enveloped. **14:** The enveloped proteins accumulate in the endoplasmic reticulum and are transported into the extracellular space.

thesis of viral gene products is tightly regulated (Fig. 5). Viral gene expression is coordinately regulated and sequentially ordered in a cascade fashion. The gene products studied to date form at least five groups on the basis of both transcriptional and posttranscriptional regulation (Fig. 5).

Several of the gene products are enzymes and DNA-binding proteins involved in viral DNA replication. The bulk of viral DNA is synthesized by a rolling circle mechanism, yielding concatemers which are cleaved into monomers and packaged into capsids.

Assembly occurs in stages; after packaging of DNA into preassembled capsids, the virus matures and acquires infectivity by budding through the inner lamella of the nuclear membrane (Figs. 3 and 6). In fully permissive tissue culture cells, the entire process takes approximately 18 to 20 hr.

Initial Stages of Infection

The available information on events preceding the transcription of viral genes is still fragmentary. The central issue is that two of the initial events, attachment to the cell surface and fusion of the viral envelope with the plasma membrane, must necessarily involve viral surface proteins. Of the ten known HSV membrane glycoproteins (gB, gC, gD, gE, gG, gH, gI, gK, gL, and gM) (19,192, 271,529,640), five (gC, gE, gG, gI, and gM) are dispensable in most cells in culture, both for entry into and egress from cells (22,242,375,376,377). The predicted product of the U_S5 gene, gJ, has not been shown to be present in virions, and its posttranslational modification or function has not been published.

Attachment

Both HSV-1 and HSV-2 are readily detected on the surface of cells, particularly juxtaposed to coated pits of cells exposed for a brief interval to infectious virus (Fig. 7). Attempts to find cultured mammalian cells lacking receptors have not been successful, leaving the species specificity of natural infection by this virus a mystery: only chimpanzees, other than humans, are "naturally" infected with this virus (399).

Numerous studies have been unable to assign the responsibility for attachment to the cell surface to a sole viral glycoprotein. The genes encoding each of the HSV-1 glycoproteins except gK have been deleted individually from the viral genome, and each of those viruses is able to attach to and infect nonpolarized epithelial cells (22,66,83, 190,289,290,375,376,575). The reason for this became apparent as data indicated that HSV can utilize more than one attachment pathway.

The apparent lack of a single viral protein-cell surface receptor interaction for HSV is in part due to the use for most of the attachment studies of nonpolarized, continuous cell lines. During the course of its normal life cycle, HSV must infect and replicate in two different cell types: epithelial cells and neurons. These cells are quite different from one another and also, *in vivo*, highly polarized; they sort membrane and secreted proteins to one surface or another. In the case of epithelial cells, membrane proteins are sorted to the apical or basal surfaces (77,547,548); in neurons, they are sorted to axons or dendrites (148,149). The virus therefore must be able to attach to and infect at least three very different types of membranes. For example, in nonpolarized cells, gC is dispensable for viral attachment and replication (242). The first evidence that gC was in-

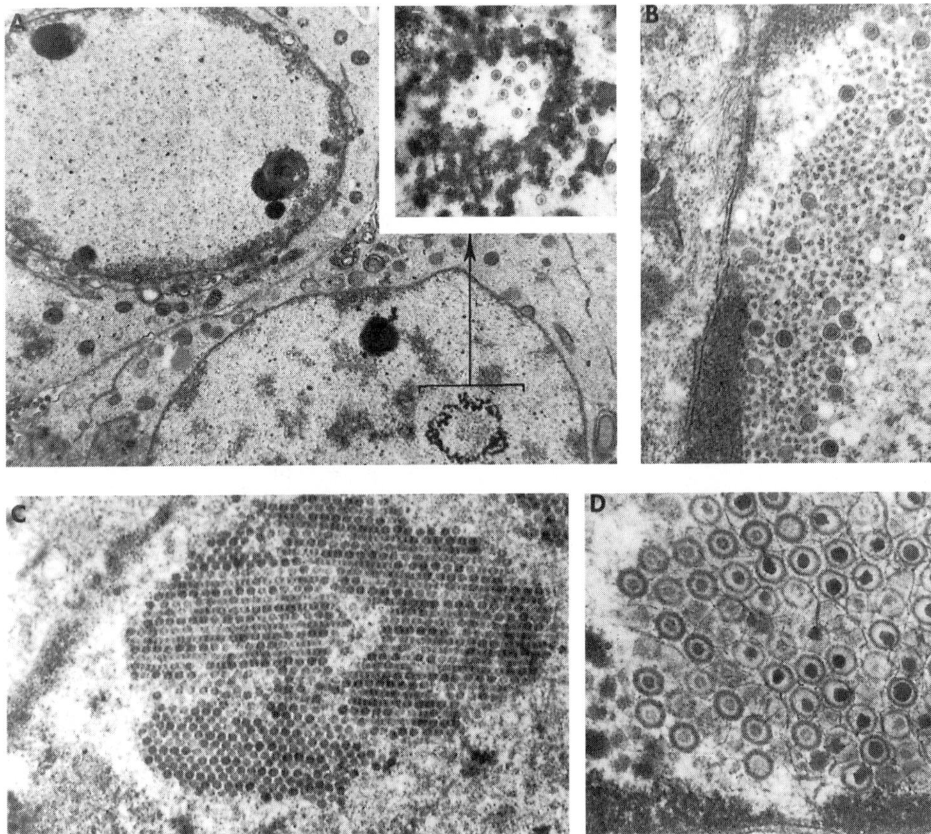

FIG. 4. Electron micrographs of the intracellular events in HSV-1 replication. **A:** Electron-opaque bodies (magnified in insert) showing sites of assembly of capsids. **B:** A region near the end of the nucleus showing accumulation of chromatin, small particles that appear to be capsid precursors and capsids. **C:** A paracrystalline array of capsids, both empty and containing DNA, frequently found in nuclei of infected cells. **D:** Capsids in nuclei of infected cells in various stages of packaging of viral DNA. Electron micrographs assembled from Roizman and Furlong (557) and Schwartz and Roizman (600,601), with permission.

volved in attachment was obtained in studies on the inhibition of attachment of HSV to BHK cells by the polycation neomycin (75). In this study, it was shown that the resistance of HSV-2 to inhibition of attachment by neomycin mapped to the gC-2 gene. It was later reported that deletion of the membrane anchor domain of gC decreased, but did not eliminate, the ability of purified virions to attach to nonpolarized cells in culture (245). When polarized epithelial cells were used in attachment assays, it was shown that gC is required for attachment of HSV virions to a receptor found on the apical surface of polarized cells, but not for attachment of the virus to the basal surfaces of the same cells (605). Polarized epithelial cells in culture therefore express more than one receptor for HSV, and those receptors are recognized by different viral proteins.

This attachment to different cell surface receptors mediated by different viral glycoproteins raises the possibility that nonpolarized cells may express more than one receptor on the same surface, and at least some of the myriad of apparently nonessential glycoproteins encoded by this virus may then be functionally redundant in infection of nonpolarized cells. In that case, removal of one glycoprotein at a time cannot completely eliminate attachment, since the viral protein required for attachment to another receptor would still be present. Viruses lacking gC, gG, and gE, or gE and gI, are able to attach to and infect nonpolarized cells (A. Sears and B. McGwire, unpublished data). Because gB is also a strong binder of heparan sulfate, it has been postulated (245) that gB is responsible for gC-independent attachment. However, this hypothesis depends upon heparan sulfate being the sole viral receptor, which no longer appears likely.

Spear et al. (206,622,747) have identified cell surface heparan sulfate as a major factor in binding of HSV to the cell surface; heparin is a potent inhibitor of HSV attachment, and removal of heparan sulfate from cells either enzymatically or by the use of mutant cell lines deficient in heparan sulfate synthesis reduces the levels of virus attachment to and infection of those cells by approximately 85%. However, in no case has removal of heparan sulfate or competition by heparin resulted in a complete loss of

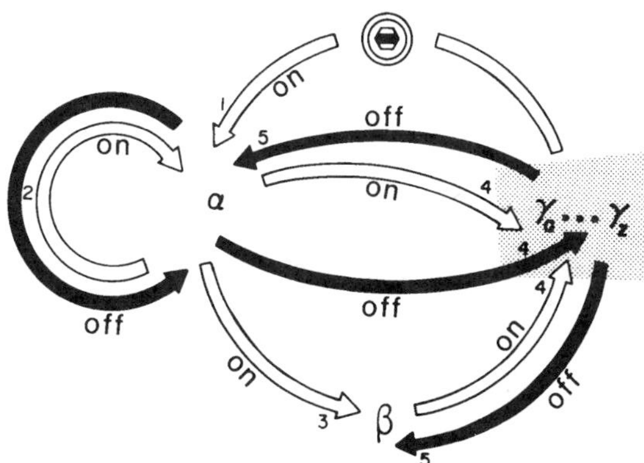

FIG. 5. Schematic representation of the regulation of HSV-1 gene expression. *Open* and *filled arrows* represent events in the reproductive cycle which turn gene expression "on" and "off," respectively. **1:** Turning on of α gene transcription by α-TIF, a γ protein packaged in the virion. **2:** Autoregulation of α gene expression. **3:** Turning on of β gene transcription. **4:** Turning on of γ gene transcription by α and β gene products through transactivation of γ genes, release of γ genes from repression, and replication of viral DNA. Note that γ genes differ with respect to the stringency of the requirement for DNA synthesis. The heterogeneity is shown as a continuum in which inhibitors of viral DNA synthesis are shown to have minimal effect on γ_a gene expression but totally preclude the expression of γ_z genes. **5:** Turn off of α and β gene expression by the products of γ genes late in infection.

attachment or infectivity. Two possible explanations for this are that heparan sulfate is one of multiple receptors, or that it acts as a cofactor, enhancing binding of gC or other glycoproteins to cell surface proteins.

Another factor that has been implicated in attachment of HSV is the basic fibroblast growth factor receptor (FGFR). Reports that the presence of this protein on the cell surface was required for viral attachment (301) could not be substantiated by other laboratories (431,445,621). However, inasmuch as there is more than one attachment pathway available to HSV in infection of nonpolarized cells, the possibility that FGFR is one of multiple receptors was not conclusively ruled out.

Recent experiments have identified the receptor for gC-dependent attachment of HSV-1 to the apical surface of polarized epithelial cells as the complement receptor CR1 (CD35; A. Sears, in preparation). This protein, closely related to the Epstein-Barr virus receptor CR2, has a wide tissue distribution *in vivo* (479). The role of heparan sulfate in the gC-CR1 interaction has yet to be determined.

Penetration into the Infected Cell

Attachment to the cell surface activates a process mediated by viral surface proteins which causes the fusion of the viral envelope and the cell plasma membrane. There is overwhelming acceptance of the hypothesis that productive infection results from the entry of virus mediated by fusion of the envelope and plasma membranes rather than that mediated by phagocytosis (440). There is also evidence that fusion of the envelope with membranes lining endocytic vesicles results in a nonproductive infection (70). The demonstration that virion envelope Fc receptors (i.e., glycoproteins gE and gI) (291) could be detected on cell surfaces following penetration in the absence of viral gene expression is consistent with this hypothesis (474).

Penetration may be a multistep event and involves more than one viral glycoprotein. In nonpolarized cells, the cumulative evidence indicates that (a) a ts mutant in HSV-1 gB attaches to but does not penetrate into cells (386), but infection does ensue, and progeny virus is made after chemically induced fusion of the envelope of the adsorbed virus to the plasma membrane (590,591); (b) HSV-1 gB^-, gD^- and gH^- recombinant viruses also attach to cell surfaces but do not penetrate (66,190,289); (c) cells expressing HSV-1 gD allow attachment and endocytosis of both HSV-1 and HSV-2, but fusion of the membranes and penetration do not ensue (70); and (d) viruses deleted in gL attach to but do not penetrate into cells; however, the gL^- virions also lack gH (575), so that the role of gL in penetration remains to be seen. In the case of polarized epithelial cells, gG, gE, and gI are required for a postattachment stage of entry, the nature of the step(s) requiring gG, gE, and gI remains to be determined (A. Sears, B. S. McGwire, and L. Tran, in preparation).

The role of gD in viral penetration deserves a further note. Clonal lines originally derived from BHK cells have been shown to vary with respect to susceptibility to infection by wild-type HSV-1. Mutant viruses capable of infecting the clonal lines have been isolated. A mutation in gD capable of conferring on recipient virus some but not all of the properties of the mutant virus mapped in the gD gene (572). In addition, cell lines expressing HSV-1 gD allow attachment and endocytosis of both HSV-1 and HSV-2, but fusion of the membranes and penetration do not ensue (70). It appears that the gD expressed in these cells is responsible at least in part for the observed resistance to infection, that viral mutants which overcome the resistance contain mutations in gD, and that transfer of the mutations in gD to a recipient virus overcomes in part the resistance to infection. The evidence favors additional mutations in the viral genome necessary to overcome resistance to infection completely and, by extension, argues that entry of virus into cells requires the participation of several gene products (56).

The transition from attached to penetrated virus (as measured by the loss of susceptibility to neutralization characteristic of virus still attached to the cell surface) is very rapid (265).

Release of Viral DNA

Upon entry into the cell, the capsids are transported to the nuclear pores (Figs. 3 and 7) (32,693). Release of viral

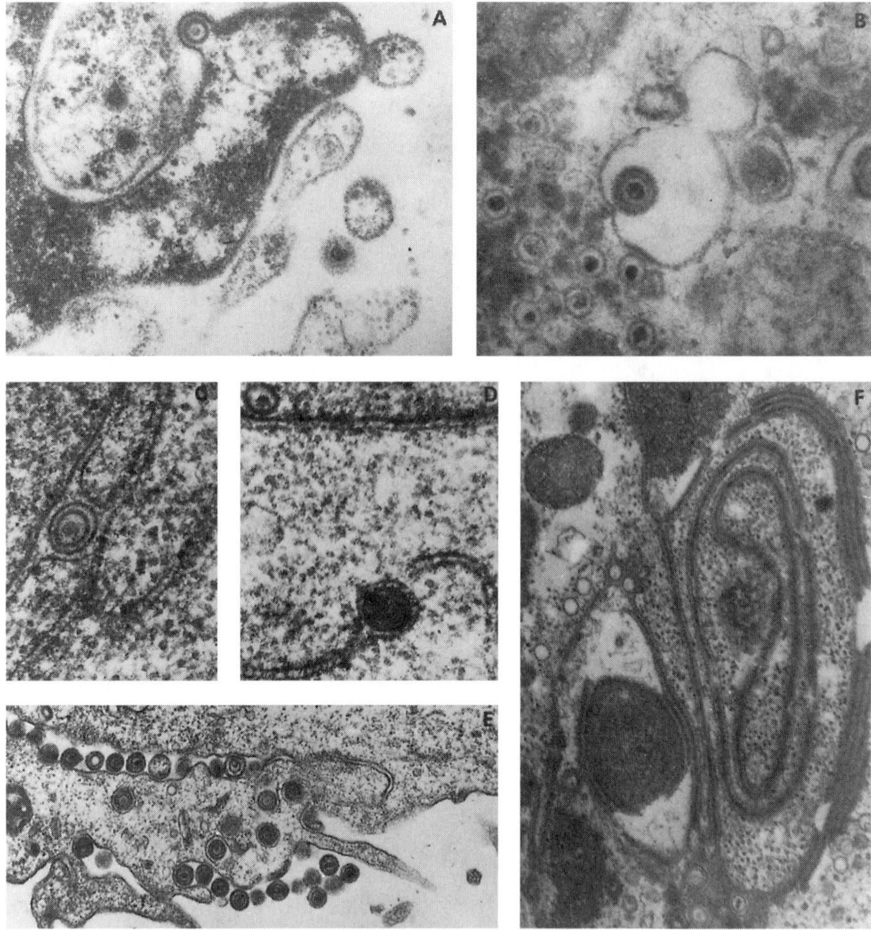

FIG. 6. Electron micrographs of the envelopment and egress of virus from infected cells. **A:** Envelopment of virus from a protrusion of the nucleus. Note that the nucleus contains marginated chromatin. The inner lamellae of the nuclear membrane contain electron-dense, slightly curved patches representing regions of the membrane at which envelopment takes place. Note the spikes projecting from the surface of the membrane of the capsid being enveloped. **B:** An enveloped capsid and numerous unenveloped capsids found late in infection in the cytoplasm of infected cells. Some of the capsids appear to be in the process of being either enveloped or deenveloped. **C:** Micrograph showing an enveloped capsid in the space between the inner and outer lamellae of the nuclear membrane connecting with the cisternae of the endoplasmic reticulum. **D:** An unenveloped capsid in the nucleus and an enveloped particle bulging in the cisternae of the endoplasmic reticulum. **E:** Cytoplasmic enveloped particles enclosed in vesicles or cisternae of the endoplasmic reticulum. **F:** Modified nuclear membranes folded upon themselves, frequently seen in cells late in infection. The structures formed by such membranes have been designated as "reduplicated membranes." Electron micrographs assembled from Roizman and Furlong (557) and J. Schwartz and B. Roizman (unpublished micrographs).

DNA into the nucleoplasm requires a viral function; thus, capsids of the ts mutant HSV-1(HFEM)*ts*B7 accumulate at nuclear pores and release viral DNA only after a shift down from nonpermissive to permissive temperature (32). Empty capsids are readily found at nuclear pores early in infection with wild-type viruses. The cellular cytoskeleton probably mediates the transport of herpesvirus capsids to the nuclear pores (124,331). Parental viral DNA accumulates in the nucleus.

Virion Components Required for Replication in Permissive Cells

Transfection of "permissive" cells with intact, deproteinated viral DNA yields infectious viral progeny (225,351,616). However, the specific activity of viral DNA is many orders of magnitude lower than that of virions, and the duration of the reproductive cycle is longer. Moreover, there is no certainty that the sequence of events in trans-

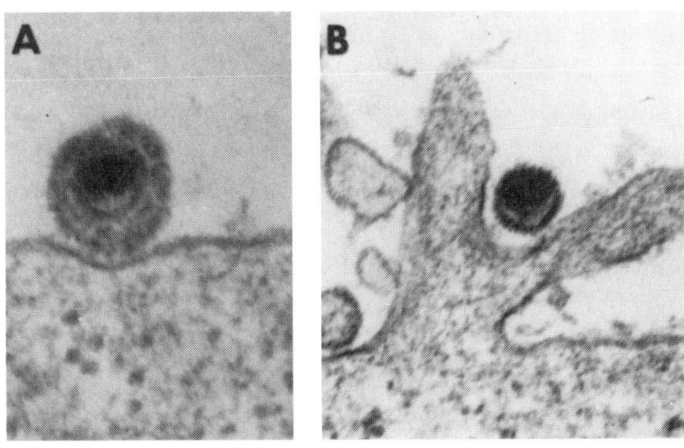

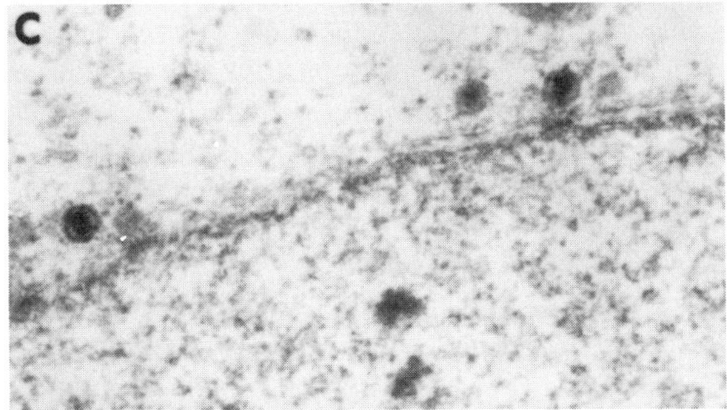

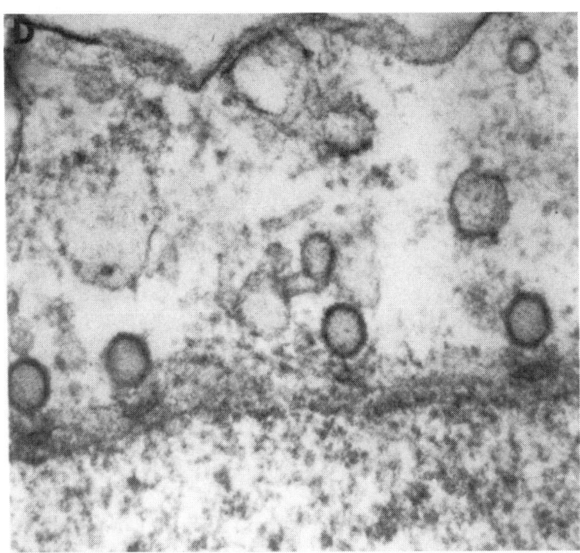

FIG. 7. Attachment and penetration of HSV-1 to cells in culture. **A,B:** Virions attached to plasma membrane. **C:** Capsids with DNA at nuclear pores in cells infected with HSV-1(HFEM)*ts*B7 maintained at the nonpermissive temperature (32). **D:** Empty capsids accumulating in cells late in infection with mutant HSV1(50B) late in infection (693). In cells infected with this mutant, virtually every pore contains a juxtaposed empty capsid.

fected cells resembles the viral reproductive cycle occurring in cells infected with competent virions.

The components of the virion other than its DNA appear to have several functions. The envelope obviously enables entry of the virus into cells, and the capsid acts a vector for the viral DNA. Several tegument proteins have been shown to have important functions for the initiation of viral replication, and surprises are yet to come. Among these are (i) VHS (U_L41), which is involved in the early shut off of host macromolecular synthesis (341,342,458,457,530, 554,595, 675–677); (ii) a protein designated by the Spear and Roizman (643) nomenclature as VP16 acts in *trans* to induce α genes, the first set of genes to be expressed (33,76, 481,498) (since the induction of α genes is a nuclear event, it is evident that at least some tegument components make their way into the nucleus); (iii) two proteins (U_L46, U_L47), which appear to modulate the function of VP16 (400a); (iv) a protein encoded by U_S11, which binds the 60S ribosomal subunit and whose function is not known (573); (v) a protein kinase (U_L13) whose function in newly infected cells is not known (520,521); and (vi) proteins which may facilitate attachment to the nuclear pore and the release of viral DNA (32).

Preston and Notarianni (506) reported adenosine diphosphate (ADP) ribosylation of VP23, a capsid protein (213,643). It has previously been reported that phosphate cycles on and off VP23, suggesting that a kinase associated with virions (although not necessarily viral in origin) phosphorylates and dephosphorylates VP23 and substrate proteins (360).

It should be noted that 38 of the 75 open reading frames encoding unique proteins are dispensable at least in some cell lines and that a large proportion of the dispensable open reading frames are structural proteins, particularly of the tegument and envelope. The functions of the dispensable proteins are particularly important inasmuch as they may eventually identify the cellular functions required for viral replication in specific tissues or cell types.

Viral Genes: Pattern of Expression and Characterization of Their Products

Timing and Requirements for Gene Product Synthesis

The transcription of viral DNA takes place in the nucleus. As would be expected, all viral proteins are synthesized in the cytoplasm. The number of abundant, i.e., readily detectable, polypeptides specified by HSV does not exceed 50 (116,262,442). McGeoch et al. (402,405) predicted 56 open reading frames in U_L, 12 in U_S, and one each in the repeats flanking the L and S components for a total of 72. However, as discussed later in this text, the definition of open reading frames in the viral genome was somewhat conservative, and the actual number is higher. Since the HSV-1 sequence was published, ten additional transcribed open reading frames have been reported. These are $\gamma_1$34.5, ORF, and ORF P, which map in the repeats flanking the L component and therefore present in two copies, U_L26.5, which encodes a capsid scaffolding protein, and the substrate of the protease, U_L49.5, U_L20.5, U_S1.5, and U_S8.5 (5,28,97,211,346, 370,371, P. L. Ward, K. Carter, and B. Roizman, in preparation.). The functions of ORF P, U_L49.5, and U_S8.5 are not known.

In cells productively infected with HSV, the regulation of viral gene expression schematically represented in Fig. 5 has three features: (i) HSV proteins form several groups whose synthesis is coordinately regulated in that they have similar requirements for and kinetics of synthesis, (ii) the absolute rate of synthesis and ultimate abundance of each protein may vary, and (iii) the protein groups are sequentially ordered in a cascade fashion (183,260–262,329,483).

The α genes are the first to be expressed. There are five α proteins, i.e., ICP0, ICP4, ICP22, ICP27, and ICP47 (Table 1). The α genes were initially defined as those that are expressed in the absence of viral protein synthesis. The α genes may be defined more precisely by the presence of the sequence 5′ NC GyATGnTAATGArATTCyTTGnGGG 3′ in one to several copies within several hundred base pairs upstream of the cap site (382).

The synthesis of α polypeptides reaches peak rates at approximately 2 to 4 hr postinfection, but α proteins continue to accumulate until late in infection at nonuniform rates (4,261). As discussed below, all α proteins, with the exception of α47, have been shown to have regulatory functions. and functional α proteins are required for the synthesis of subsequent polypeptide groups.

The ß genes are expressed at very low levels in the absence of competent α proteins, and their expression is enhanced rather than reduced in the presence of inhibitory concentrations of drugs that block viral DNA synthesis or in cells infected with tight DNA-ts mutants in ß genes. The $ß_1$ and $ß_2$ groups of polypeptides reach peak rates of synthesis at about 5 to 7 hr postinfection (260,261). The $ß_1$ proteins, exemplified by polypeptides ICP6 (the large component of the viral ribonucleotide reductase) (270) and ICP8 (the major DNA binding protein) (108) appear very early after infection and in the past have been mistaken for α proteins (102). They are differentiated from the latter by their requirement for functional α4 protein for their synthesis (260,261). The $ß_2$ polypeptides include the viral thymidine kinase (TK) and DNA polymerase. The appearance of ß proteins signals the onset of viral DNA synthesis, and most viral proteins involved in viral nucleic acid metabolism appear to be in the ß group.

The γ genes have been lumped for convenience into two groups, γ_1 and γ_2, although in reality they form a continuum differing in their timing and dependence on viral DNA synthesis for expression (108,113,300,625,707). The prototype γ_1 gene (e.g., the genes specifying glycoproteins B and D and γ34.5) is expressed relatively early in infection and is only minimally affected by inhibitors of DNA synthesis. The relatively abundant major capsid protein ICP5 (γ_1) is made both early and late in infection. In contrast, prototypic γ_2 proteins [e.g., gC (U_L44) and U_S11] are expressed late in infection and are not expressed in the presence of effective concentrations of inhibitors of viral DNA synthesis.

The γ_1 genes have also been designated as ßγ or leaky γ genes (114,253). The differentiation of ß genes into $ß_1$ and $ß_2$ and the variability in the requirements for the expression of genes are the major reasons for the designation of HSV genes as α, ß, and γ rather than immediate-early, early, and late (261).

Functional Organization of HSV Genomes

The sources of the data for the functional organization of the HSV-1 genome shown in Fig. 8 are useful to present for both historical and heuristic reasons. Globally, the key sources were the transcriptional maps painstakingly collected and defined by E. K. Wagner et al. (11,12,113,114, 151,200,230,253–255,707,709). These maps served as the basis for the interpretation of the nucleotide sequence data generated by McGeoch et al. (402,405), although in some instances transcriptional analyses and even translational analyses were ignored in favor of nucleotide sequences denoting putative transcriptional initiation sites or terminations. Identification of the proteins specified by the individual open reading frames is based on several sources. The framework and much of the initial mapping of the HSV genome is based on analyses of proteins and DNA sequence arrangements of HSV-1 × HSV-2 recombinants (390,441, 442,582) supplemented by (i) rescue of mutants by transfection of cells with intact mutant viral DNA and DNA fragments generated by restriction endonuclease digestion of wild-type genomes (319,320,444,476), (ii) transfer of a dominant or assayable marker from one genome to another with restriction endonuclease fragments (320,327,498, 499,582), (iii) expression of the gene product from purified messenger RNA (mRNA) or from a DNA fragment in a suitable system (108,247,355,383,498), and (iv) insertion in frame with the putative open reading frame of a "tag" consisting of a nucleotide sequence encoding a non-HSV epitope for a known monoclonal antibody (346,370, 371). The products of a large number of putative open reading frames have not been identified. The sequence-dependent, in contrast to transcription- or function-dependent, identification of open reading frames is conservative and does not take into account proven exceptions (e.g., the arbitrary rules would have excluded α0 as an open reading frame had its product not been known). Nevertheless, the overall organization of the genome is becoming apparent and can be summarized as follows:

1. The α genes map near the termini of the L and S components (11,299,300,383,441,448,508,714,715). The α0 and α4 genes map within the inverted repeats of the L and S components, respectively,and are therefore each present in two copies per wild-type genome. However, a single copy of each is sufficient inasmuch as I358, a HSV-1 mutant lacking most of the internal inverted repeat sequences, is viable (490). In the circular arrangement of viral DNA, the α genes form two clusters. The first consists of α genes 0, 4, and 22, whereas the second consists of α genes 47, 4, and 0. A key feature of these two clusters is that each con-

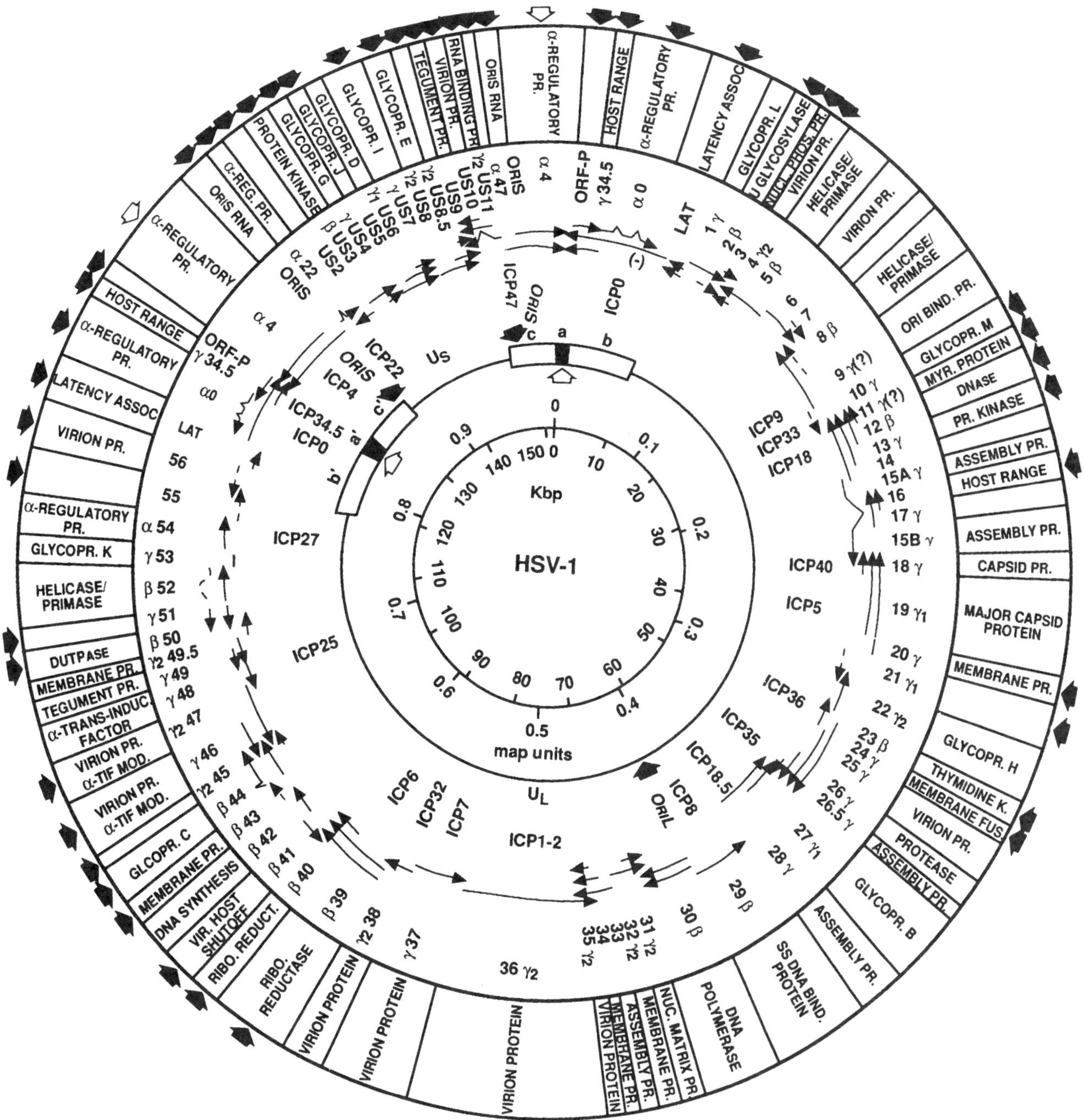

FIG. 8. Functional organization of the HSV-1 genome. The circles are described from inside out. **Circle 1:** Map units and kilobase pairs. **Circle 2:** Sequence arrangement of HSV genome. The letters *ab, U_L*, and *b′a′* identify the long (L) component consisting of the unique sequences U_L flanked by the inverted repeats. The letters *a′c′, U_S*, and *ca* identify the corresponding sequences of the short (S) component. The *open arrow* shows the sites of cleavage of concatemeric or circular DNA to yield linear DNA. Since the L and S components can invert relative to each other, the arrangement shown is that which would yield the "prototype arrangement" if linearization were to occur by cleavage of the DNA between map units 0 and 100. The *filled arrows* represent the three origins of viral DNA synthesis, one in the middle of U_L (OriL) and two (both designated OriS) within the inverted repeats flanking the S component. **Circle 3:** The transcriptional map of the HSV-1 genome. The map serves the purpose of identifying the direction of transcription, the approximate initiation and termination sites, and the families of 3′ coterminal transcripts. *Dashed lines* identify transcripts mapped imprecisely. The designation between the second and third ring identify proteins encoded by the transcripts according to their ICP number. The designations outside this ring identify the open reading frame (those mapping in U_S have the prefix US) number and the kinetic class (α, β, or γ) to which they belong. **Circle 4:** The known functions of the proteins specified by the open reading frames. The *filled arrows* identify open reading frames which can be deleted without affecting the ability of the virus to multiply in cells in culture. The *open arrows* identify the two copies of the α4 gene; only one copy of this gene can be deleted without affecting the capacity of the virus to multiply. The data for the circles 3 and 4 are derived from refs. 11, 28, 53, 81, 97, 113, 125, 150, 168, 200, 201, 230, 255, 280, 346, 371, 402, 402, 411, 416, 448, 481, 484, 523, 537, 540, 614, 707, and 716.

tains an origin of DNA synthesis (Ori_S) sandwiched between α4 and α22 or between α4 and α47. Notwithstanding the clustering, each α gene has its own promoter-regulatory region and transcription initiation and termination sites (381–383).

2. With few exceptions, ß and γ genes are scattered in the unique sequences of both the L and S components. The exceptions are the $\gamma_1$34.5 and ORF P genes located in the reiterated sequences flanking the L component between the terminal *a* sequence and the α0 gene (6,97,346). At present only two functional gene clusters are strikingly apparent, but their significance is uncertain. The ß genes specifying the DNA polymerase and the single-stranded DNA-binding protein flank the L component origin of DNA synthesis (Ori_L), and the γ genes specifying membrane glycoproteins D, E, G, and I and the presumed glycoprotein J map next to each other within the unique sequences of the S component (6,191,354, 535,561,582,640,716). Although there are several instances of apparent sharing of 5′ or 3′ gene domains (151,537,714), there is altogether little gene overlap (ORF P and γ_1 34.5) and only a few known instances of RNA splicing (e.g., α22, α47, α0, and U_L15) (707) relative to the frequency with which these events have been observed to occur in adenovirus and papovavirus genomes. In several instances, replacement of the genomic DNA sequences of spliced genes with the corresponding complementary DNA (cDNA) had no effect on viral replication in cells in culture (21).

A more detailed listing of the gene products is presented in Table 1. Each viral protein is designated by one of three criteria, i.e., by function, if it is precisely defined (e.g., tk, DNA polymerase, and so forth), by its first published designation, or by its open reading frame.

Synthesis and Processing of Viral Proteins

Viral proteins appear to be made on both free and bound polyribosomes. Most of the proteins examined to date appear to be processed extensively after synthesis (4,43,44,59, 74,167,213,263,384,483,506,521–523,639–641). Processing includes cleavage, phosphorylation, sulfation, glycosylation, myristylation, ADP-ribosylation, and nucleotidylylation. In some instances, the modifications in protein structure accompany the translocation of proteins across membranes (393). Current information concerning processing of proteins and the relationship of processing to function is detailed in the section on general properties and functions of viral proteins and in the section on viral glycoproteins.

With the exception of some glycoproteins, the extent to which processing is a requirement of virus growth rather than the consequence of an encounter between cellular or viral enzymes and molecules resembling natural substrates remains uncertain. N-linked glycosylation of gD is not necessary for its function (637). However, at least some glycosylation within the infected cells is necessary to produce infectious progeny (74,639).

HSV Protease

Gibson and Roizman (213,215) noted that protein VP22a present in empty B capsids was replaced in DNA-containing C capsids by a faster migrating protein with similar characteristics, designated as VP22. VP22a proteins were identified as members of a family of proteins designated as ICP35a-f (59,60). VP22a, corresponding to ICP35e,f are derived from ICP35c,d and form the toroidal structure which functions as a scaffolding for DNA packaging into capsids (451). As predicted from early studies (213), it is absent from DNA-containing capsids. McGeoch et al. (403) assigned ICP35 to the open reading frame U_L26 on the basis of the observation that a temperature-sensitive mutation affecting the processing of ICP35 and accumulation of mature capsids mapped in the U_L26 open reading frame (507). Numerous studies (144,368–371) indicate the following:

1. The domain of U_L26 contains two independently transcribed open reading frames, U_L26 and U_L26.5. U_L26 encodes a protein of 635 amino acids. The U_L26.5 promoter maps within the coding domain of U_L26. U_L26.5 encodes 329 amino acids; the methionine initiator codon of U_L26.5 corresponds to the methionine 306 of U_L26. Although they are transcribed separately and although the abundance of the U_L26.5 product is higher than that of U_L26, the product of U_L26.5 has the same amino acid sequence as the carboxy-terminal amino acids of U_L26.
2. U_L26 is a serine protease (368,369,371). U_L26 cleaves itself between Ala^{610} and Ser^{611} and between Ala^{247} and Ser^{248} (144,368,371). The amino-terminal 247-amino acid polypeptide of U_L26 retains its catalytic activity (368). The carboxy-terminal product of the cleavage of the product of the U_L26 gene between Ala^{247} and Ser^{248} exists in two states, i.e., with and without the 25 carboxy-terminal amino acids; these proteins correspond to the ICP35a,b of Braun et al. (58). The protease also cleaves the terminal 25 amino acids at the corresponding Ala-Ser site off the U_L26.5 product ICP35c,d to form ICP35e,f. Mutagenesis studies have identified two histidines and two glutamic acid residues essential for proteolytic activity, but a conserved cysteine was dispensable in the 247-amino acid polypeptide (368,369).
3. The unprocessed product of the U_L26.5 gene and ICP35e,f form double bands. The modification of the protein which causes double band formation maps in the domain of the U_L26.5 gene between codons 307 and 417 (368).
4. At this time, no other substrates have been identified for the U_L26 protease.

Protein Kinases

Herpes simplex virus appears to specify at least three protein kinases. One protein kinase activity has been associated with the large subunit of the ribonucleotide reductase (ICP6) of HSV-2 but not of HSV-1 (100). In contrast to the carboxy-terminal domain, the amino-terminal 300+-amino acid stretch of the HSV-2 polypeptide shares little homology with the corresponding domain of the HSV-1 homolog (455). Whereas the ribonucleotide reductase activity is associated with the conserved carboxy-terminal domain, the protein kinase activity maps near the amino terminus (9,109). The large subunit of ribonucleotide reductase appears to be a multifunctional protein. The observation that HSV-2 but not HSV-1 evolved this activity of the ribonucleotide reductase is surprising. The substrate specificity of this protein kinase activity is not known.

McGeoch and Davison (404) predicted that U_S3 encodes a protein kinase on the basis of its sequence, a conclusion verified by studies with antibody to the protein (193) and with deletion mutants in the gene (519). The major substrate of this enzyme is an intrinsic membrane protein exposed on the surface of infected cells and encoded by U_L34. The U_L34 protein contains the amino acid motif recognized by the U_S3 protein kinase; substitution of the serine or threonine residues within this motif with alanine resulted in a loss of phosphorylation, but the virus grew poorly, and revertants were readily detected. It is noteworthy that four prominent phosphoproteins unrelated to the U_L34 protein appear in lysates of cells infected with the U_S3^- mutant or with mutants in the amino acid motif recognized in U_L34 by the U_S3 kinase. The anti-U_L34 serum coprecipitates the four phosphoproteins along with the U_L34 protein. It would appear that the four phosphoproteins compensate in some fashion for the absence of phosphorylation by U_S3 kinase. Undoubtedly, the U_S3 kinase phosphorylates other as yet unidentified proteins (522,523).

Smith and Smith (636) and Chee et al. (88) reported that the sequence of the U_L13 open reading frame contains the signature common to other protein kinases and is shared among α, ß, and γ herpesviruses. Cunningham et al. (121) reported on the properties of a new kinase very similar to the protein kinase activity demonstrated in tegument-capsid structures described by LeMaster and Roizman (360) and ascribed the new kinase to the product of the U_L13 gene. Studies of U_L13^- and $U_L13^-U_S3^-$ mutants led to the conclusion that the U_L13 kinase affects the phosphorylation and processing of the gene products of α22 and U_L47 and the accumulation of the α0, U_L26 and $U_L26.5$ (protease and its substrate), and U_S11 gene products (520,521).

Although the U_L13 kinase is associated with structural proteins, the enzyme brought into the infected cell by the virion does not phosphorylate the newly synthesized ICP22 (520,521). The phenotype of the U_L13^- virus is similar to that described for the α22⁻ mutant by Sears et al. (603).

A large number of other proteins (e.g., ICP4 and gE; see Table 1 for a more detailed list) are phosphorylated in the course of the reproductive cycle. The kinases responsible for the phosphorylation and the role of phosphorylation in the functions of these gene products have not been elucidated.

ADP-Ribosylation

Preston and Notarianni (506) reported that ICP4 and VP23 are poly(ADP-ribosyl)ated in isolated nuclei, a significant finding that nevertheless left unanswered the question whether this reaction actually takes place in the infected cell. Blaho et al. (43) reported that antibody specific for poly(ADP-ribose) reacts with ICP4 extracted from cells late in infection, in effect answering the question in the affirmative. However, the poly(ADP-ribose) added to ICP4 was digested by poly(ADP-ribose) glycohydrolase, but only after denaturation of the protein. In contrast, poly(ADP-ribose) added to ICP4 in isolated nuclei was readily removed from the native protein by the glycohydrolase. The results indicate that ICP4 is poly(ADP-ribosyl)ated and suggest that, in the isolated nuclei, the poly(ADP-ribose) is added either by elongation of existing chains or to novel sites.

Nucleotidylylation of Viral Proteins

An initial report by Blaho and Roizman (46) showed that ICP4 is both guanylylated and adenylylated. The label transferred by α^{32}P-adenosine triphosphate (ATP) or α^{32}P-guanosine triphosphate is associated with the slowly migrating forms of the protein. Conclusive evidence for the nucleotidylylation emerged from transfer of a ^{3}H-labeled purine ring from ATP to ICP4. More recent studies revealed that an as yet unidentified late viral gene product was involved in the reaction and that the number of nucleotidylylated proteins is higher (43,44). The first to be identified in addition to ICP4 were ICP0, ICP22, and ICP27. ICP22 and ICP27 share the amino acid sequence Arg/ProArgAlaPro/SerArg which is also found in ICP4, ICP0, and the products of the HSV-1 genes U_L21, U_L31, U_L47, and U_L49. ICP0, ICP4, ICP22, and ICP27 are regulatory proteins; U_L21 is dispensable for growth in cultured cells (24); U_L31 cofractionates with the nuclear matrix (85); and U_L47 may interact stoichiometrically with αTIF (or VP16, the product of the U_L48 gene) (425). U_L49 is a virion protein which is labeled in cells with [^{32}P]orthophosphate and also is (ADP-ribosyl)ated. The genes encoding the additional nucleotidylylated proteins were identified initially by analyses of intertypic recombinant viruses and subsequently by analyses of the products encoded by the mapped genome domains and were shown to correspond to U_L21, U_L31, U_L47, and U_L49 proteins (44). Preliminary evidence indicates that casein kinase II nucleotidylylates at least ICP22 and that an additional late viral protein is required for nucleotidylylation of most of the remaining seven proteins (45,432). The significance of the nucleotidylylation is not known. In fact, there is no evidence that the identified "consensus" sequence is nucleotidylylated.

Modification of Membrane-Associated Proteins

Of the 11 predicted glycoproteins, at least ten have been studied in sufficient detail to demonstrate the presence of oligosaccharide chains. The 11th, gJ, remains elusive. In addition, it has been reported recently that the products of the U_L47 gene, VP13-14, contain O-linked polysaccharide chains and are phosphorylated (425).

At least one protein, the product of the U_L11 gene, has been shown to be myristylated (384). Deletion mutants in the U_L11 gene show an impairment in egress from infected cells (20).

Application of Genetic Techniques to the Identification of Gene Product Function: Genes Essential and Dispensable for Growth in Cell Cultures

Key to the identification of viral functions and mapping of viral genes encoding these functions are temperature sensitive (ts) and null mutants. Earlier studies identified approximately 30 complementation groups (723)—an extraordinary accomplishment in itself given the difficulties inherent in the selection and testing of the numerous mutants produced by many laboratories. The ts mutants have been enormously helpful in mapping genes. Nevertheless, this approach to identification and mapping of viral functions suffers from several problems, i.e., (i) the phenotypes of viruses containing extensive mutations in some nonessential genes cannot be readily differentiated from that of the wild-type parent, (ii) conditional lethal (e.g., ts) mutants produced by general mutagenesis of the viral genome may contain a large number of silent nonlethal mutations in both essential and nonessential genes, (iii) the phenotypes of mutations introduced into domains shared by more than one gene cannot be readily attributed to the malfunction of a specific gene product, and (iv) while the usefulness of ts mutants is in part dependent on their efficiency of plating at permissive and nonpermissive temperatures, tight mutants with high permissive/nonpermissive ratios may well contain more than one point mutation. Although the presence of multiple mutations in a single gene should not affect the mapping or identification of the gene function, it does present a problem in mapping the functional domains of the gene.

An alternative to the random or fragment-specific substitution of bases in DNA is site-specific deletion of the viral genome. A protocol for site-specific insertion/deletion of viral genes was first reported by Post and Roizman (499). It was based on selection of recombinants generated by double recombination through homologous flanking sequences between an intact viral DNA molecule and a DNA fragment containing an insertion or deletion and a selectable marker. The selectable marker used in these studies was the viral TK gene (*tk*) because (i) it can be deleted from the HSV genome without affecting the growth of virus in cell culture, (ii) a plasmid-borne *tk* gene can be altered so that it cannot recombine by double crossover to repair the deletion in the genomic *tk* gene, (iii) viruses carrying a functional *tk* gene can be selected against by plating viral progeny in the presence of nucleoside analogs phosphorylated by the viral TK (e.g., AraT), and (iv) viruses expressing the *tk* gene can be selected for by plating the virus in TK^- cells in medium containing methotrexate or aminopterin, which block the conversion of thymidine monophosphate (TMP) from deoxyuridine monophosphate (dUMP) by thymidylate synthetase and precludes the *de novo* pathway of TMP synthesis. This procedure permits the selection of viable mutants with deletions or insertions in genes which appear to be nonessential for growth in cells in culture. Other investigators adapted the double crossover protocol for selection of mutants with deletions in essential genes (66,137,363). In this protocol, the gene to be deleted was transfected into and expressed in cells in culture, and the host cell line, i.e., the cells expressing the gene, was then transfected with intact viral DNA and the mutated DNA fragment. The progeny of transfection were screened for deletion mutants that multiplied only in the vector cell line.

A still different protocol for insertional mutagenesis is based on the use of transposons (e.g., miniMu phage, Tn5) (285,558,718). Its principles were described first by Jenkins et al. (285), taking advantage of the random insertion of the DNA of phage miniMu into target plasmid DNAs. A miniMu phage was constructed containing a modified HSV-1 *tk* gene. Transposition of this miniMu into an HSV fragment is random and limited to one insert per plasmid copy. Transfection of intact tk^- viral DNA with an HSV DNA fragment containing random insertions of the modified miniMu would result in recombinants in which the miniMu, which is randomly inserted into the viral DNA fragment, would become recombined at the identical position in the viral genome. However, only the HSV genomes containing the miniMu sequences at nonessential sites multiplied in cells in culture.

Although the transposons have a technical advantage of reduced labor to produce the insertion mutants, they suffer from the fact that the target gene is not actually deleted and the site of insertion must be ascertained precisely since truncated genes may still yield a functional product.

The genes known to be dispensable for growth in cells in culture are listed in Table 1. It should be stressed that most of the dispensable genes are required for replication in experimental animal systems and that in no instance has a virus lacking a dispensable gene been isolated from human lesions (although viruses that fail to react with a specific monoclonal antibody are readily isolated).

Genes dispensable for viral replication in cells in culture fall into several groups whose products are involved in entry of HSV into cells, regulation of gene expression, posttranslational modification of proteins, exocytosis, inhibition of host response to infection, and spread of virus from cell to cell.

In a special category are deletion mutants whose ability to multiply is cell species-dependent. One example of such mutants is the $\alpha 22^-$ virus, which grows well in Vero and HEp-2 cell lines but not in human fibroblast strains or in rodent cell lines (603). In the nonpermissive cells, the virus fails to express γ_2 genes efficiently. Another example of a cell-specific gene is $\gamma_1 34.5$, which enables HSV to multiply in human cells but is dispensable in Vero, baby hamster kidney cells, and so forth. In the absence of the gene, there is total shut off of protein synthesis before significant amounts of virus are synthesized (91,93,94).

It could be predicted that viral genes which specify products whose functions are identical and interchangeable with those of cellular genes would be dispensable, at least in cells which express these functions. In this category are the *tk* gene, the genes specifying ribonucleotide reductase, and so forth. The observation that some virion proteins are dispensable for infection and replication of virus at least in cell culture was very puzzling for many years.

Although we cannot exclude the possibility that cells express proteins with similar functions which complement the deletion mutants, a more likely scenario is that cells in culture express many more genes than cells *in situ* in animal organs. The exceptions to date are the polarized epithelial cells and the neuroblastoma cells in which the gene products are sorted differently or which differ from other cells in culture with respect to the nature of the genes that are expressed. Herpes simplex virus may carry a set of genes which enables the virus to enter (e.g., the dispensable glycoproteins), multiply (e.g., $\gamma_1 34.5$), or egress (e.g., $U_L 20$) from a wide variety of human cells. Since these genes are not required for replication in all cells, there exists the formal possibility that functional analogs of the viral gene are encoded and may be expressed by cells. The obvious examples are the *tk*, ribonucleotide reductase, and so forth. The less obvious homologs are the cellular protein kinases which substitute for some of the viral enzymes, the homolog of $U_L 20$ which enables virus to egress from a variety of cell lines other than Vero cells, and so forth. The complexity involved in defining the function of 38 genes whose products are dispensable for viral replication in at least some cells in culture is significantly offset by the fact that these genes are excellent probes for analyses of cellular functions which at least in some instances complement the missing viral function.

Synthesis of Viral DNA

Temporal Pattern of Synthesis

A characteristic of herpesviruses not shared by other animal nuclear DNA viruses is that they specify a large number of enzymes involved in DNA synthesis. Although the sequence of events in viral DNA replication is roughly known, many details are still lacking. In HSV-infected cells, viral DNA synthesis is detected at about 3 hr postinfection and continues for at least another 12 hr (272,553,554,562). The DNA is made in the nucleus (Fig. 9). Earlier studies relied on incorporation of labeled thymidine into viral DNA—a procedure that yielded biased results inasmuch as the deoxynucleotide triphosphate pool increases and becomes saturated early in infection. Hence, the rate of viral DNA synthesis, as determined by the use of labeled deoxynucleosides, appears to be highest relatively early in infection. Analyses of viral DNA synthesis by hybridization with specific probes suggest that the bulk of viral DNA is made relatively late in infection (272).

Structure of Replicating DNA

At least in HSV-1-infected cells, only a small portion of total input (parental) viral DNA is replicated (278). The DNA labeled during a pulse lacks free ends, i.e., it consists of circles or head-to-tail concatemers (277,278). Labeled precursors become incorporated into molecules banding at a higher density which sediment at a faster rate than intact double-stranded DNA. In alkaline sucrose density gradients, the bulk of the labeled DNA bands at a position expected for small single-stranded fragments. Early after the onset of viral DNA synthesis, parental DNA, circles, and linear branched forms can be found in the DNA banding at the density of viral DNA. These are replaced late in the reproductive cycle by large, rapidly sedimenting bodies of tangled DNA. Available evidence suggests that, at least late in infection, herpesvirus DNAs replicate by a rolling circle mechanism (37,277). Attempts to find "theta" forms of replicating DNA early in infection have not been successful.

Origins of DNA Replication

The origins of DNA replication in the HSV genome were initially deduced from the structures of defective genomes (196,599) and have more recently been operationally defined as those sequences which must be present in a fragment of HSV DNA in order for it to be amplified in permissive cells transfected with the fragment and either transfected or infected with helper virus (437,704). By this definition, HSV-1 and presumably HSV-2 each contain three origins of DNA replication. Two of the origins map in the *c* reiterated sequence of the S component between

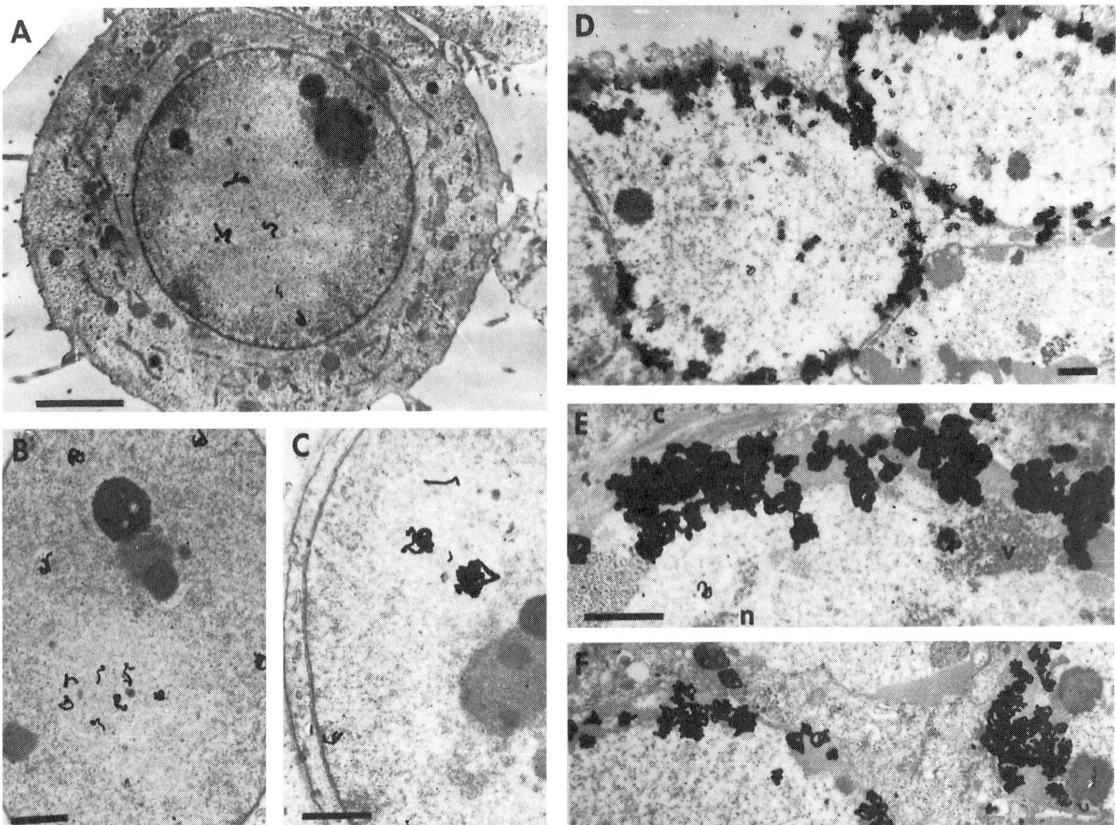

FIG. 9. Electron photomicrographs of thin-section autoradiography of HEp-2 cells infected with HSV. **A:** A 4-hr-infected cell pulse labeled for 15 minutes with ^{3}H-methyl thymidine prior to fixation. **B,C:** Enlargements of nuclei prepared as in part A. **D:** Portions of three nuclei of 18-hr-infected cells labeled with ^{3}H-methyl thymidine prior to infection. Unlabeled thymidine was present in the medium during and after infection. **E,F:** Electron micrographs of nuclei taken at higher magnifications. Note the aggregation of chromatin at the nuclear membrane. One of the cells in parts D and F did not synthesize DNA during the short labeling pulse. *n*, nucleus; *c*, cytoplasm; *v*, aggregation of virus-specific, electron-opaque material. From Roizman and Furlong (557) and J. Schwartz and B. Roizman (unpublished micrographs).

the promoters of α4 and α22 (ori_S1) or α4 and α47 (ori_S2) (29,131,437,665,666), whereas a third (ori_L) maps in the middle of the L component sandwiched between the promoters of the ß genes specifying the major DNA-binding protein (ICP8) and the DNA polymerase (373,725).

The L component origin consists of an A+T-rich sequence of 144 bp, forming a perfect palindrome (321,374, 527,725). Because of its extensive dyad symmetry, it tends to be unstable in DNA fragments cloned in *Escherichia coli.* The S component origin is smaller and contains a much shorter A+T-rich palindrome which is related to but lacks the complete dyad symmetry of ori_L. It has been suggested that the structure of ori_L enables bidirectional synthesis, whereas DNA synthesis initiated in ori_S would be unidirectional (725). The existence or necessity for bidirectional synthesis of DNA remains to be established. Earlier studies have shown that ori_L (494), and at least one ori_S (377) but not both (633) are dispensable. More recently, Igarashi et al. (272) deleted both ori_S sequences with little effect on viral yields or viral DNA accumulation in infected cells. The results indicate that none of the three origins is uniquely required for viral replication. Rather, if an origin is required for either initiation or maintenance of DNA synthesis, any one of the three origins will most likely suffice.

All three origins, ori_L and the two ori_S, are situated between transcription initiation sites. The locations of the origins suggest that initiation of DNA synthesis might be *trans*-activated or at least enhanced by the changes in the local environment of the DNA due to transcription initiation events. In addition, it has been reported that a transcript designated as Ori_SRNA originates in the domain of the α22 and α47 genes and runs across the origins coterminally with α4 mRNA (268). The Ori_SRNA is detected at 9 hr and later postinfection, i.e., coincident with peak rates and termination of synthesis of viral DNA. The initiation of Ori_SRNA transcription is imprecise. A sequence [N(GTGGGTGGG)$_2$(N $\leq$ 10)] overlapping with the site of initiation of synthesis of the bulk of Ori_SRNA binds a cellular protein with an unusual property; the protein binds in a sequence-specific fashion the above sequence in the context of single- or double-stranded DNA or RNA. The cognate sequence is not present elsewhere in the viral genome, and the DNA binding activity is present in a variety of tissues (400,571). The significance and function of the Ori_SRNA or of the cellular activity described above are unknown.

A central question that concerns the origins is why there are three of them. Also, it seems appropriate to wonder why the origins are distributed in both components and quasi-symmetrically, i.e., in the middle of the L component and in the sequences flanking the S component, when in fact one origin appears to suffice. The number and distribution of origins are particularly puzzling because the bulk of the DNA is replicated as a rolling circle, obviating the need for *de novo* initiation of viral DNA synthesis at an origin each time a copy of viral DNA is made.

Viral Proteins Involved in DNA Metabolism

Herpes simplex virus specifies a large array of proteins involved in nucleic acid metabolism and DNA synthesis. These proteins fall into two categories, i.e., proteins that are essential for viral origin-dependent amplification of DNA, and enzymes involved in nucleic acid metabolism [e.g., TK, ribonucleotide reductase, deoxyuridine triphosphatase (dUTPase), uracil glycosylase, and alkaline exonuclease] which for the most part appear not to be essential for viral growth in cells in culture. Contrary to early reports, topoisomerases do not appear to be encoded by HSV-1; the virus utilizes the p170 form of the host topoisomerase II (156).

Viral Proteins Essential for Origin-Dependent DNA Synthesis

Much of the initial evidence for viral proteins essential for DNA synthesis emerged from studies of the defects in DNA^- ts mutants. More recently, the genes whose products are essential for DNA synthesis were identified by transfecting cells with a plasmid containing an origin of DNA synthesis and various fragments of the HSV genome. These studies identified seven genes mapping in the L component (open reading frames U_L5, 8, 9, 29, 30, 42, and 52) required for viral origin-dependent DNA synthesis. The seven genes specify a DNA polymerase (U_L30) with an apparent molecular weight of 140,000 (86,103,237,259,305, 306,500), a single-strand specific DNA-binding protein designated as ICP8 (U_L29) with an apparent molecular weight of 124,000 (84,99,109,287,387,394,500,525,580, 745), a protein binding to as many as three sites at or near the origin of viral DNA synthesis (U_L9) (160,161,323,324, 465) with a translated molecular weight of 94,000 (161), a protein which binds to double-stranded DNA and confers processivity on the DNA polymerase (U_L42) with a molecular weight of 62,000 (389,402,475,745), and three additional proteins (U_L5, predicted molecular weight of 99,000; U_L8, predicted molecular weight of 80,000; and U_L52, predicted molecular weight of 114,000). These three proteins form a complex in which each protein is present in equimolar ratios and which functions as a primase and, in the presence of ICP8, also as a helicase (119).

The DNA polymerase, the product of U_L30 ($ß_2$) in particular has been the object of numerous studies because of its unusual sensitivity to a variety of compounds [e.g., phosphonoacetate (PAA) and phosphonoformate]. In HSV DNA polymerase, *ts* mutants have been described (87,103,516, 517), and some have been found to be resistant to a variety of drugs inhibitory to wild-type viruses, including PAA (87,287,517) and nucleoside analogs (e.g., acycloguanosine) (118). The DNA polymerase forms a complex (1:1) with the product of the U_L42 gene (244).

ICP8 ($ß_1$) has also been extensively investigated, particularly by Knipe et al. (204,217–219,352,353,525,526) and by Ruyechan (579,581,583). The protein has an apparent molecular weight of approximately 120,000. It has an affinity for and binds cooperatively to single-stranded DNA (518,580). The *ts* mutants in this gene fail to synthesize viral DNA at the nonpermissive temperature (108,219), as do deletion mutants in cells which do not provide ICP8 in *trans* (468). ICP8 appears to be essential in anchoring the polymerase to the replication complexes and appears to interact specifically with U_L9 origin-binding protein (52,204). The protein promotes renaturation of cDNA strands and strand transfer—a key function for a high level of recombination in infected cells (55,154).

Dimers of the origin-binding protein (U_L9) bind to the sequence CGTTCGCACTT or its derivatives at three unequal sites with decreasing affinity (159,324). The sites flank the AT-rich sequences in the origins. The binding is cooperative, and the order of binding to these sites (designated as I, II, and III) reflects the affinity of the protein for each site. The bound U_L9 protein binding to two sites loops and distorts the intervening AT-rich sequences (323). The looping is independent of the phasing of the binding sites. In contrast, the required distortion is dependent both on the position of the binding sites and the free energy of supercoiled DNA. The origin-binding protein has been shown to be an ATP- or dATP-dependent helicase, unwinding DNA in the 3′ to 5′ direction. The helicase activity is increased in both rate and extent by ICP8 (52).

Relevant to the role of U_L9 and the origins of HSV DNA replication is the report that transfection of cells with plasmids encoding six (U_L5, U_L8, U_L29, U_L30, U_L42, and U_L52) of the seven proteins essential for viral replication is sufficient to induce the amplification of SV40 DNA integrated into cellular chromosomes but not HSV DNA synthesis (241). The U_L9 gene was dispensable. While the results are convincing, the interpretation of these results is far from clear. The key issue is whether cellular proteins (or T-antigen) direct the viral gene products to specific sites analogous to the function attributed to U_L9 or whether the viral proteins produced changes in cellular chromosomes that resulted in cellular DNA synthesis. While the Heilbronn et al. (241) report could be construed as supporting the hypothesis that the U_L9 protein modifies the AT-rich sequence at the origin and enables the assembly of the proteins required for initiation of viral DNA synthesis, it casts a pall on the notion that the seven viral proteins listed above are all that is required for viral DNA synthesis. If, in fact, the six proteins encoded by U_L5, U_L8, U_L29, U_L30, U_L42, and U_L52 interact with cellular proteins to amplify the cellular DNA, the failure to replicate viral DNA *in vitro* by the seven proteins cannot be attributed solely to failure to find the right buffer composition for the reaction mixture.

Proteins Involved in Nucleic Acid Metabolism and Not Essential for Viral DNA Synthesis in Cells in Culture

Other proteins undoubtedly play a role in processing, cleavage, and packaging of the genomic viral DNA, as well as in the production of precursors of DNA synthesis; for example, as described below, the alkaline DNAse is not among the seven essential genes and appears to play no role in DNA synthesis (724), notwithstanding reports to the contrary (194,443).

Alkaline DNAse activity in infected cells was first reported in 1963 (306). The gene has been mapped by transient expression in oocytes (503) and by the use of HSV-1 × HSV-2 recombinants (26) to a site corresponding to U_L12 (402). The protein is encoded by a 2.3-kb mRNA (150) and has a predicted translated molecular weight of 67,503 (402) and an apparent molecular weight of 80,000 to 85,000 (27). It has been reported that U_L12 is required for egress of capsids from the nucleus (612). While it can be rationalized that the function is required to enable capsids to make their way through marginated chromosomes at the inner nuclear membrane, it would be useful to know whether the mutation on which this conclusion is based does not affect the expression of a U_L11 since deletion mutant in U_L11 had a similar phenotype (20). The alkaline DNAse accumulates in large amounts in nuclear dense bodies shown to contain proteins derived from the nucleolus (524).

Thymidine kinase is one of the most studied viral proteins. A unique characteristic of TK is that its substrate range is far greater than that of its host counterpart. Although it has been designated as a deoxypyrimidine kinase, it in fact phosphorylates purine pentosides and a wide diversity of nucleoside analogs that are not phosphorylated efficiently by cellular kinases (281,310,318). This characteristic of TK is the basis for the effectiveness of various nucleoside analogs in the treatment of experimental and natural herpesvirus infections. The observation that TK is essential for normal virus multiplication in experimental infections (186,687) but not in cell culture (311) is the basis of much of the probing of the HSV genome structure done in recent years (490,498,499). Mutants in the *tk* gene fall into several groups. Some fail to produce functional TK altogether, whereas others either make reduced amounts of enzyme or an enzyme with an altered substrate specificity which is resistant to the analog used in the selection process (126,186,502,674).

The ribonucleotide reductase encoded by HSV-1 consists of two proteins. The large subunit, ICP6 (260,261,270, 510) has an apparent molecular weight of 140,000 and a predicted translated molecular weight of 124,043 (402). The small subunit has an apparent molecular weight of 38,000 (16,510) and a predicted translated molecular weight of 38,017 (402). The two proteins are encoded by 3′ coterminal mRNAs of 5.0 kb for the large subunit and 1.2 kb for the small subunit (12). The two proteins are tightly associated in a $\alpha2\beta2$ complex (16,17,275), and both subunits are required for activity (16,191,264). Ribonucleotide reductase functions to reduce ribonucleotides to deoxyribonucleotides, creating a pool of substrates for DNA synthesis. The viral enzyme is not essential for growth in actively dividing cells maintained at 37°C (123,220). However, it is required for efficient viral growth and DNA replication in nondividing cells or in cells maintained at 39.5°C (221,510), indicating that, at 37°C, actively dividing cells can complement the viral function.

The uracil DNA glycosylase encoded by HSV presumably functions in DNA repair and proofreading. Uracil DNA glycosylase acts to correct insertion of dUTP and deamination of cytosine residues in DNA; the extremely high G+C content of HSV DNA makes this an important element of error correction in HSV DNA replication. The HSV-induced uracil DNA glycosylase has been identified by Caradonna and Cheng (78), and its coding domain was initially mapped to between 0.065 and 0.08 map units (79), corresponding to the U_L2 open reading frame (402). Subsequent *in vitro* translation experiments definitively identified U_L2 as the uracil DNA glycosylase gene (744). The protein has an apparent molecular weight of 39,000 (79) and a predicted translated molecular weight of 36,326 (402). The gene has been deleted and is nonessential for growth of the virus in culture (446).

Deoxyuridine triphosphate nucleotidohydrolase (dUTPase) acts to hydrolyze dUTP to dUMP, providing both a mechanism to prevent incorporation of dUTP into DNA and a pool of dUMP for conversion to dTMP by thymidylate synthetase. An HSV-encoded dUTPase has been identified (78,740); contrary to early reports (740), the purified enzyme is specific for the hydrolysis of dUTP (737). The viral gene has been mapped to 0.69 to 0.70 map units by transient expression (509), corresponding to the U_L50 open reading frame (402). In HSV-1(17)*ts*K, a mutant in ICP4 used for analyses of gene regulation, dUTPase activity appears to be lacking (127). The dUTPase gene has subsequently been shown to be nonessential for growth of the virus in tissue culture (187).

Assembly of Capsids

Capsids are assembled in the nucleus (Figs. 1 and 4). The steps in the assembly have not been defined, although capsids have been assembled *in vitro* from capsid proteins made in insect cells by baculoviruses carrying the genes specifying capsid proteins (691). Viral DNA is packaged into preformed capsids containing the protease specified by U_L26 and the scaffolding protein (the products of self-digestion of the U_L26 protein and of the substrate specified by $U_L26.5$ (144,368–371).

The assembly of full capsids containing DNA requires the participation of numerous noncapsid proteins as well. These include at the very least the products of U_L6, U_L15, U_L25, U_L28, U_L32, U_L33, U_L36, and U_L37 genes (8,10,96, 189,495,539,618,619,686).

Cleavage and Packaging of HSV DNA

Newly synthesized viral DNA is "processed" and packaged into preformed capsids. "Processing" involves amplification of *a* sequences and cleavage of viral DNA lacking free ends, i.e., in circular or head-to-tail concatemeric form. Associated with the processes of DNA replication, cleavage, and packaging is the isomerization of the DNA. There is considerable genetic evidence that cleavage and packaging of DNA are linked processes (344,345). The data on the isomerization of the DNA has come from three sources: (a) analyses of the termini of standard viral genomes (129,436,438), (b) analyses of termini of viral genomes containing insertions of additional *a* sequences (98,435,701), and (c) studies on amplicons (plasmids containing an origin of viral DNA synthesis and one or more *a* sequences which are amplified and packaged with the aid of a helper virus) (133,134,638,704,705).

The net result of the process of cleavage of standard genomes from concatemers is the generation of a free S component terminus consisting of one *a* sequence with a terminal DR1 sequence, containing only a single base pair and one 3′ nucleotide extension (436), and a free L component terminus, consisting of one to several directly repeated *a* sequences and ending in a DR1 containing 18 bp and one 3′ nucleotide extension. Upon circularization of the DNA following entry into cells, the two partial DR1 sequences together would form one complete DR1 shared by two *a* sequences. In the reverse process of linearization of viral DNA for packaging, cleavage of endless (circular or concatemeric) DNA occurs asymmetrically within a second DR1 distal from the *c* sequence and, in an ideal case, shared by two *a* sequences. Junctions containing a single *a* sequence are cleaved (134). The results of such studies have been interpreted to indicate either that the sequence *xay* is cleaved to yield *xa* and *y* and the *y* product is processively degraded along the DNA to the next *a* sequence or that the cleavage simultaneously yields both *xa* and *ay* by amplification of the *a* sequence during the cleavage process (133,134,701). Parenthetically, there is little doubt that DNA lacking a terminal *a* sequence could be degraded, but inasmuch as nearly 50% of the L-S component junctions are of the *bac* type, i.e., have a single *a* sequence, a hypothesis whose logical extension is that 50% of newly synthesized DNA is degraded during packaging, does not make biologic sense.

Deiss et al. (133) analyzed the cleavage and packaging of a series of amplicons. Those lacking the U_b sequence were amplified and packaged, but they acquired an intact *a* sequence from the helper virus. Those lacking the U_c sequence were not subject to cleavage-packaging. Furthermore, U_b and U_c contain domains conserved in several herpesviruses and which were designated *pac1* and *pac2*, respectively. The model (133) that best fits the data, presented here in a slightly modified form (Fig. 10), consists of several steps: (i) a cleavage-packaging protein attaches to the U_c sequence; (ii) a putative structure on the surface of the capsid complexes with a U_c bound protein sequence, loops the viral DNA, and scans from the bound *a* sequence (a_1) across the L component toward the end of the S component until it detects the first U_c domain of an *a* sequence in an identical orientation; (iii) in the juxtaposed *a* sequences, the DR1 sequence of one *a* is cleaved and the gap is repaired, resulting in the generation of an *a* sequence by the mechanism proposed by Szostak et al. (679) to explain recombinational events resulting in gene conversion; and (iv) cleavage then occurs in the DR1 shared by the two *a* sequences. In this model, the *a* sequences in the internal inverted repeats play no role in the packaging of the unit-length molecule, consistent with the observation that HSV-1 DNAs from which the internal inverted repeats are deleted do package effectively.

The packaging component of the model of Deiss et al. (133) predicted that the length of the packaged DNA be defined by the distance between two directly repeated *a* sequences. However, defective genomes consisting of 17+ direct reiterations of a unit consisting of an Ori_S and an *a* sequence are readily detected in virions of HSV stocks derived by serial passages at relatively high multiplicities (196). These observations are consistent with the hypothesis that, besides the scanning mode, there is a "head-full" recognition element which selects the juxtaposed *a* sequence once a threshold amount of DNA has been packaged. Shorter fragments of HSV DNA are packaged into capsids, but these capsids do not become enveloped (705). A hypothesis that may explain the apparent contradiction is that packaging aborts when the DNA reeled into the capsid is smaller than full length, but that the capsid does not disgorge the packaged fragment.

The viral proteins responsible for the cleavage-packaging event have not been identified. However, (a) Chou and

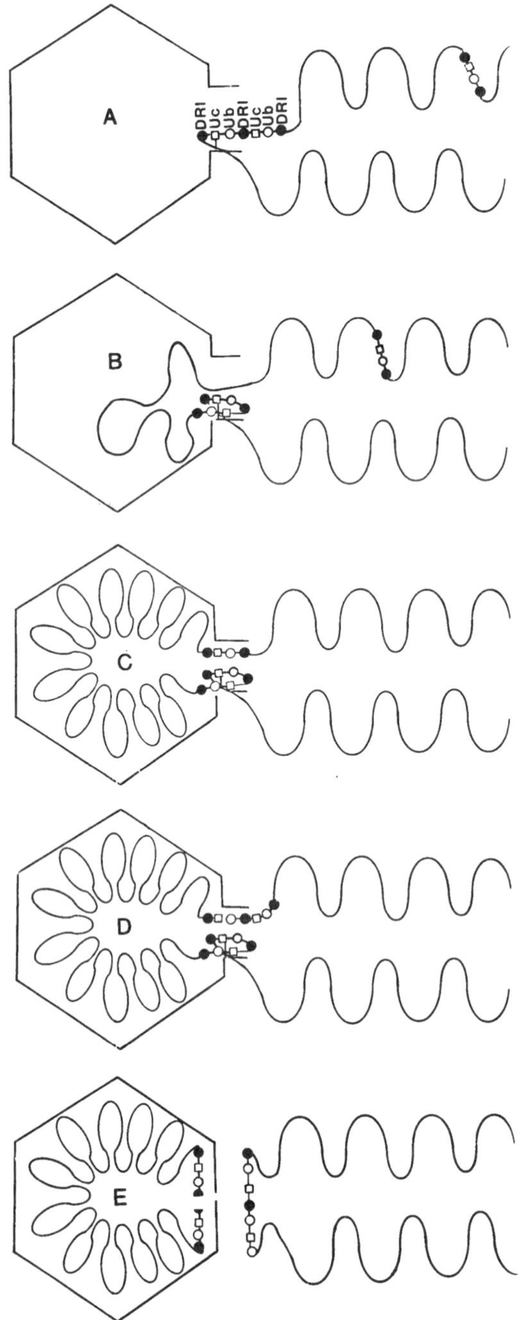

FIG. 10. Packaging of HSV-1 DNA. The current model developed by N. Frenkel and associates is described in Deiss et al. (133) and in the text. The model requires that (a) proteins attach to components of the *a* sequence, probably U_c; (b) empty capsids scan concatemeric DNA until contact is made in a specific orientation with the first protein-U_c sequence (capsid A); the DNA is than taken into the capsid B until a "head full" or contact is made with an *a* sequence whose nucleotide arrangement is in the same orientation (i.e., one genome equivalent in length away) is encountered (capsid C); the packaging signal requires nicking of both strands from signals on opposite sites of a DR1 sequence. In the absence of two adjacent *a* sequences (capsid D), the juxtaposition of the *a* sequences would results in duplication of the *a* sequence (capsid E), as described by Deiss et al. (133).

Roizman (96) have identified two viral proteins that form a sequence-specific complex with the portions of the U_c sequence containing *pac2*, (b) capsids contain a protein which binds viral DNA (VP19C or ICP32) (60), (c) Wolhtrab et al. (741) reported that the *a* sequence is specifically cleaved by virus-induced endonuclease, and (d) Table 1 lists numerous genes whose products appear to play a role in the cleavage-packaging of viral DNA.

Inversions of the L and S Components

The isomerization of HSV DNA resulting from the inversion of the L and S components relative to each other is an intriguing, tantalizing feature of the HSV genomes shared with only a few other herpesviruses.

In its circular form, the HSV genome forms two isomers, each containing two L-S component junctions. Cleavage of one circular isomeric form at the two junctions would yield the P and I_{SL} arrangements, whereas the corresponding cleavages of the other circular isomer would yield the I_S and I_L isomers. Generation of the I_S and I_L arrangements from the first circular isomeric form would require inversion of either the S or the L component through the inverted repeat sequences.

Fundamentally, there are several issues. First, inversion of covalently linked components is not a property of all herpesvirus genomes. Second, the physiologic function of the inversions is not clear inasmuch as genomes frozen in one orientation as a consequence of deletion of internal inverted repeats are viable (286,490). However, all wild-type isolates examined to date do contain the inverted repeat sequences, and viruses lacking internal inverted repeats have a reduced capacity for growth in animal tissues. Third, insertion of the junction between the L and S components, and especially of the 500-bp *a* sequence, results in additional inversions of DNA segments contained between inverted repeats of *a* sequences (98,435,436,438). Deletion analyses have shown that inversions are associated with the sequences DR2 and DR4; deletion of these sequences results in a gross reduction in the inversion frequency (98). Lastly, inversion of viral DNA segments flanked by other domains of the genome or inversion between repeated foreign DNA sequences was observed in some instances but not in others. In the case of fragments duplicated in different components of the HSV DNA, the segment of the genome flanked by the inverted repeats does not invert (435,498). DNA fragments flanked by inverted repeats contained in the same component do invert. In some instances, the inversions were accompanied by a high-frequency gene conversion (492). Weber et al. (717) reported that inversions of DNA segments flanked by inverted Tn5 transposon elements resulted from recombination events through homologous sequences and was not the consequence of a recombinational event mediated at a specific *cis*-acting site by *trans*-acting viral proteins. Thus, inversions of amplified DNA sequences flanked by inverted Tn5 sequences at least 600 bp or longer were noted in cells transfected with the genes specifying the seven proteins required for viral DNA synthesis. As in the case of amplicons containing two *a* sequences in an inverted orientation (436), inversions were not observed in the absence of DNA synthesis.

The central issue is not that DNA sequences flanked by inverted repeats tend to invert as a consequence of homologous recombination, but rather the frequency of such inversions and the specificity of the process in infected cells. The DNA extracted from a plaque generated by a single virus particle, presumed to be in one arrangement of DNA, contains all four isomers of HSV DNA in equimolar concentrations. In the case of DNA segments flanked by inverted repeats of nonjunction fragments, the fraction of the genomes showing inversions even after many serial passages is seldom more than a small fraction of the total. A more careful analysis by Dutch et al. (153,155) showed that DNA flanked by inverted *a* sequences had a higher rate of inversion than DNA flanked by non-*a* sequences of equal length. These studies also demonstrated that the sequence required for this process was a 95-bp sequence containing DR1 and U_c. Bruckner et al. (61) reported the partial purification of an activity mediating *in vitro* recombination between repeated copies of the *a* sequence and concluded that the recombination proceeds by a site-specific mechanism.

Viral Membrane Proteins, Virion Envelopment, and Egress

Appearance of Modified Nuclear Membranes in Infected Cells

The hallmark of infected cells late in infection is the appearance of reduplicated membranes and thick, concave or convex patches, particularly in nuclear membranes (Fig. 6) (111,166,356,401,439,454,600,601,624,743). Nuclear envelopment takes place at these patches. Because the enveloped virions do not contain detectable amounts of host membrane proteins, it is likely that the patches represent aggregations of viral membrane proteins, presumably including the viral glycoproteins on the outside surface and anchorage and tegument proteins on the inside surface.

Processing of HSV Glycoproteins

The general pattern of the biosynthesis of herpesvirus glycoproteins appears to follow that of eukaryotic cell glycosylated proteins (74,640). Specifically, nonglycosylated precursors of herpesvirus membrane proteins are synthesized on polyribosomes bound to the rough endoplasmic reticulum. Glycosylation includes translational and posttranslational events. Thus N-linked glycosylation is initiated by transfer of preformed glycans [(glucose)$_3$-mannose-(N-acetylglucosamine)$_2$] from a dolichol phosphate lipid carrier to asparagine residues in the sequence Asn-X-Thr/Ser (X can be almost any amino acid) of a nascent polypeptide (362,391,680). During transit through the Golgi apparatus, the oligosaccharide chains are trimmed by glucosidases, mannosidases, and so forth to yield a polymannosyl chain frequently referred to as the high-mannose glycans (267,649). The high-mannose glycans are frequently converted by glycosyl transferases to complex glycans which consist of a pentasaccharide core [(mannose)$_3$-(N-acetylglucosamine)$_2$] and a number of side chains (antennae) with the composition of sialic acid-galactose-glucosamine. Fucose, when present, is usually added to the completed side chains (267). O-linked glycosylation occurs less frequently than N-linked glycosylation (74,292,453,466,620,640); it is initiated by the transfer of N-acetylgalactosamine to the hydroxyl group of threonine or serine and is followed by the addition of galactosamine, N-acetylglucosamine, fucose, and sialic acid in the Golgi apparatus (38). The extent of glycosylation and the structure of complete glycans is affected by the conformation of protein around the glycosylation site, inasmuch as much as the structure affects access by enzymes involved in processing. Conformation of the protein may also explain the heterogeneity of glycans attached to a given protein.

Current information on the structure of the HSV glycans has been summarized in detail elsewhere (74,640). Thus N-linked high-mannose, O-linked, and complex heterogeneous glycans have all been reported to exist in glycoproteins specified by HSV-1. There is no evidence to date that processing of nonglycosylated precursors to the fully glycosylated stage requires the synthesis of virus-specified enzymes, although the data do not specifically exclude the possibility that the virus encodes at least some enzymes whose functions are similar to those of the host.

Nothing is known regarding the function and requirements for O-linked glycosylation of herpesvirus glycoproteins. N-linked glycosylation is required for infectivity, inasmuch as blocking of N-linked glycosylation by tunicamycin blocks the accumulation of glycosylated proteins and of enveloped virus (480,485,487). Conversion of high-mannose glycans into complex-type glycans appears to be required for the egress of the virus from the infected cell but not for infectivity (72,291,327,611).

There is considerable evidence that, after synthesis, the viral glycoproteins are transported to the plasma membrane and can be found in all cytoplasmic membranes of the cell. The viral glycoproteins in the cellular membranes are major targets of the immune response to the virus. Several authors reported that, when viral glycoproteins are specified by genes resident in the environment of the cells, maturation and transport occur faster than when the proteins are specified by genes resident in the viral genome and are expressed during infection (14). The difference may simply reflect timing and intracellular traffic congestion. In cells expressing both a glycoprotein gene resident in the cellular genome and the corresponding gene resident in the viral genome, the former gene is expressed earlier; the glycosylation and transport of this protein does not compete with that of abundant viral glycoproteins made later from transcripts of genes resident in the viral genome.

Properties of Minor Glycosylated and Nonglycosylated Membrane Proteins

McGeoch et al. (402) predicted that the products of U_S9, U_L10, U_L20, U_L43, and U_L53 are membrane proteins. U_L53 has been shown to encode glycoprotein K (271,529). U_L10 encodes an abundant dispensable glycoprotein (gM) associated with both cellular membranes and virion envelopes (19,22). Current studies (J. Baines and B. Roizman, in preparation) suggest that the protein contains several transmembrane domains and that at least two short loops accessible to proteases project from the surface of the cell. One of the loops contains a single glycosylation site which is apparently used. Deletion mutants are impaired in growth properties.

U_L20 is dispensable. In cells infected with wild-type virus, the product of the gene is made in relatively small amounts; it is highly hydrophobic, appears to be associated with membranes, and does not appear to be glycosylated (23,712). Preliminary observations suggest that it localizes in nuclear membranes, the intermediate compartment, and the Golgi but not in the plasma membrane. The properties of the deletion mutant are described below.

U_L43 presents a special problem in that the domain of the gene is poorly defined. The transcription initiation site is unknown, and the TATAA box is only a few nucleotides away from the putative translational initiation site. MacLean et al. (385) reported that the open reading frame as defined by McGeoch et al. (402) is dispensable without effect on viral replication. L. Kaplan, D. Barker, and B. Roizman (unpublished results) also deleted the sequences assigned to this open reading frame. Although a gene product could not be demonstrated by immunoblotting of lysates of cells infected with a mutant containing an in-frame insertion of an epitope, as has been done for other open reading frames (95,371), immunofluorescence studies demonstrated the presence of an antigen which appeared to be associated with Golgi (P. L. Ward and B. Roizman, in preparation).

In addition to these proteins, two others have recently been shown to associate with membranes. The product of the membrane protein encoded by U_L34 has been described above in connection with its phosphorylation by the U_S3 protein kinase. The association of the U_L24 gene product with membranes is based on the observation that deletions within the domain of the gene cause infected cells to fuse (280). Several lines of evidence suggest that HSV membrane proteins form specific complexes. The existence of one complex, the Fc receptor formed by glycoproteins E and I, can be deduced from the observation that monoclonal antibody to either precipitates both glycoproteins (288). Similarly, it has been reported that gH associates with gL (U_L1) (271).

Nothing is known of the mechanism by which viral glycoproteins enter nuclear membranes. One facile (and possibly incorrect) explanation is that viral membrane proteins are made on rough endoplasmic reticulum and are translocated laterally to the contiguous nuclear membrane.

Envelopment

Nuclear DNA-containing capsids attach to patches of modified inner lamella of the nuclear membrane and become enveloped in the process. The emphasis on "DNA-containing" stems from electron microscopic observations which show that envelopment of empty capsids occurs rarely, although there is no evidence that "full" capsids always contain a full-length molecule of HSV DNA (557). As discussed above, Vlazny et al. (705) demonstrated that capsids containing fragments of HSV DNA less than standard genome length are retained in the nucleus. A plausible explanation for this phenomenon is that capsids containing DNA become modified and acquire affinity for membrane tegument structures in nuclei. Conceivably, capsids totally lacking this putative modification are unable to bind to the underside of the thickened patches containing viral proteins in the nuclear membranes.

There is general agreement that the inner lamella is the site of initial envelopment (Fig. 6). However, even cursory examinations of thin sections of infected cells elicits the rediscovery that envelopment occurs in the cytoplasm since the cytoplasm abounds in capsids juxtaposed to patches of modified cytoplasmic membranes in the process of envelopment or deenvelopment. Stackpole (650) is the originator of the idea that the capsids become enveloped at the inner lamella, deenveloped at the outer lamella, reenveloped by the endoplasmic reticulum, and released in the extracellular environment either by envelopment at the plasma membrane or by fusion of vesicles carrying enveloped virus at the plasma membrane. The large number of particles seemingly undergoing cytoplasmic envelopment makes a strong case for this model. However, thin sections showing capsids being enveloped at the nuclear membranes are extremely rare, suggesting that the process of envelopment is very rapid. Since every capsid undergoing putative envelopment in the cytoplasm must have been enveloped and deenveloped in transit through the nuclear membranes, the disparity in the numbers of capsids being enveloped at nuclear and cytoplasmic membranes suggest that either (a) the rate of envelopment at the nuclear membranes is significantly faster than that at the cytoplasmic membranes or (b) that the capsids in juxtaposition to cytoplasmic membranes are artifacts and represent capsids that are deenveloped or arrested in their movement through membranes. The key question of whether the cytoplasmic, semienveloped capsids are in fact in the process of being enveloped rather than arrested, i.e., transient structures resulting from deenvelopment, cannot be answered by electron microscopic snapshots anymore than the plot of a motion picture can be deduced by viewing the prints of a few random frames. It is of interest to note that electron micrographs published in a recent study which strongly defended the serial envelopment hypothesis showed that some of the capsids undergoing cytoplasmic envelopment were in fact partially degraded, as evident from the protruding DNA (726).

One hypothesis for the abundance of partially enveloped structures at cytoplasmic membranes is that virions contained in the endoplasmic reticulum attach to receptors and reinfect the cell from within, but unlike capsids which become deenveloped at the plasma membrane during entry, the capsids entering the cytoplasm by fusion of envelopes with membranes of transport vesicles may not transported to the nuclear pore. Thus the capsids deenveloped at the plasma membrane are most likely transported to the nuclear pores by specific elements of the cytoskeleton. These connections may not be available for capsids entering cytoplasm through transport vesicles. It is of interest that cytoplasmic semienveloped capsids are prevalent in continuous cell lines but less frequent in infected primary human diploid cells (B. Roizman, unpublished observations).

Translocation of Virions Across the Cytoplasm to Extracellular Space

Following envelopment, virions accumulate in the space between the inner and outer lamellae of the nuclear membrane. In the cytoplasm, intact enveloped particles are usually seen inside structures bounded by membranes (128,601). This observation is not surprising, inasmuch as structures bounded by membranes with surface glycoproteins are not likely to fare well unprotected in the cytoplasm. In two-dimensional sections, these structures appear to be vesicles; in a few electron micrographs, tubular structures have been seen, and the probability exists that, in some cells, the cisternae of the endoplasmic reticulum extend to the plasma membrane (601).

There is a general expectation that (a) virions must pass through Golgi as they traverse the cytoplasm and (b) the sorting of virions is mediated by cellular machinery. Indeed, Johnson and Spear (291) proposed, in part on the basis of studies done with monensin, that virions are secreted via the Golgi apparatus following a pathway similar to that taken by secreted soluble proteins. Several observations suggest that the situation may be more complex and that HSV controls and at the very least directs traffic through the infected cell.

1. In cells defective in processing of high-mannose to mature glycoproteins, infectious virions accumulate but are not transported into the extracellular space (25,72).
2. Virions made in cells infected with mutants in gH and carrying the mutated glycoprotein are not translocated into the extracellular space. Conversely, the translocated virions do not contain gH (142). It is tempting to deduce from this study that the mutated gH exhibits a retention sequence which precludes the translocation of the virion.
3. Deletion mutants in U_L20 replicate and form plaques in human 143 cells but not in Vero or HEp-2 cells. In Vero cells, infectious virus accumulates in the space between the inner and outer lamellae, and transport vesicles normally associated with virion traffic across the cytoplasm are lacking. Moreover, U_L20^- virus does not form plaques, which is concordant with the absence of virions in the extracellular space (23). Additional studies showed that (a) purified virions were significantly enriched with high-mannose glycoproteins; (b) the total cell glycoproteins exhibited a ratio of fully processed to high-mannose glycoproteins similar to those of wild-type infected cells, and the processed glycoproteins contained terminal sialic acid, suggesting that the viral glycoproteins not associated with virions are sorted at least through the *trans*-Golgi; and (c) infected cell plasma membranes contained significantly less glycoprotein than cells infected with wild-type virus (15).
4. Virions in or associated with Golgi have been seen but rarely. In light of the large number of virions produced in the infected cells, either the processing through Golgi is extremely rapid, or Golgi enzymes are translocated from the Golgi to the transport vesicles. Part of the solution of this riddle may stem from the observation that the Golgi is fragmented and dispersed throughout the cytoplasm in infected Vero or HEp-2 cells (69). The evidence suggests that U_L20 is essential primarily in cells in which the Golgi is fragmented and that, in these cells, it enables the traffic from the outer nuclear membrane to the Golgi and from the *trans*-Golgi to the plasma membrane.
5. Both gE and gI, besides being required for entry through the membranes of polarized epithelial cells, also play an essential role in the basolateral spread of virus in those cells (A. Bloom, K. Rott, A. Sears, and B. S. McGwire, in preparation) and enhance cell-to-cell spread in nonpolarized cells (145). It is conceivable that the virus employs several different gene products to ensure full processing of glycoproteins and safe transport across the cytoplasm to the appropriate cell surface.

Superinfection of Infected Cells

In cells infected with a mutant mapping in the S component, but not in cells infected with wild-type virus, empty capsids accumulated in large numbers at the outer surface of the nuclear membrane, suggesting the possibility that HSV encodes a function to prevent reinfection of cells, particularly with virus which had been released from those cells (693). This function has been attributed to gD (70,71, 73). Indeed, cells expressing gD allow attachment, but the virus is endocytosed and destroyed. Treatment of the cells with antibody to gD renders the cells susceptible to infection. Mutants capable of infecting these cells were readily isolated and were found to map foremost in gD, but also in other, as yet unidentified viral genes (56,70,73).

A most interesting, and potentially significant, finding was that clones of baby hamster kidney cells were found to vary with respect to their ability to express α genes, and indications are that they vary with respect to their susceptibility to infection. Mutations which confer the capacity to infect these cells, albeit at variable efficiencies, also map at least in part in gD (572).

The model that best fits the available data is that cells may express at least two surface receptors for entry of HSV into cells. One receptor has an affinity for gD and is sequestered by wild-type gD. In cells expressing gD, this receptor becomes available to incoming wild-type virus by pretreatment of the cells with antibody to gD. The other, secondary receptor does not interact with wild-type gD but has a low affinity for a specific set of mutated gD molecules. While the secondary receptor can be sequestered by mutated gD, the affinity of the secondary receptor for the mutant gD molecules may be lower than that of wild-type gD for the primary receptor. This may explain why cell lines expressing mutant gD are infectable by both wild-type virus (primary receptor is present) and viruses carrying mutated gD (low affinity favors reequilibration of the mutant gD in the presence of viral particles attached to the surface of the cells). In the highly restrictive clonal cell lines which do not express gD, the primary receptor may be absent or mutated. A component of this model is that other viral proteins (e.g., gB, gH, and so forth) affect the interaction of gD with cellular proteins and hence are subject to selection of mutations which increase the efficiency of viral entry into cells (56).

REGULATION OF VIRAL GENE EXPRESSION

A key feature of productive HSV infection is that the genes comprising the five groups, α, $ß_1$, $ß_2$, γ_1, and γ_2 are tightly regulated with respect to both abundance and timing of their expression. Herpes simplex virus genes function within the environment of eukaryotic nuclei, and by necessity they contain signals that enable their transcription by cellular factors. In addition, viral genes contain response elements *trans*-activated by viral gene products. The focus of this section is on the components of virus-specific regulation of gene expression.

Structure of HSV mRNAs

Herpes simplex virus DNA is transcribed by RNA polymerase II (115). Viral mRNAs are capped, methylated, and polyadenylated, although nonpolyadenylated RNAs of the same sequences can be isolated (18,30,31,626,628,668). Internal methylation is readily apparent in RNA made early but not late in infection (30). Notwithstanding the efficient expression of HSV genes in the environment of higher eukaryotic cells, only four open reading frames, α0, α22, α47, and U_L15, have been shown to yield spliced mRNAs. Minor subsets of the transcripts of mapping to the gC (U_L44) and polymerase (U_L30) genes are also spliced (49,200). Transcripts sharing 5′ and particularly 3′ termini have been described (707). Attention has also been drawn to multiple initiation sites for the transcription of selected HSV genes (200,448,614,715,756), to the common observation that RNAs may extend beyond the usual polyadenylation site (12,97,255,528), and to selective and possibly regulated use of alternative polyadenylation signals (418). In contrast with the orderly transcription of intact cells are the random initiations experienced by more than one laboratory in nuclear run-off transcription assays late in infection (217,721).

The abundance and stability of the various HSV-1 mRNAs vary (198,199,626,628). In general, mRNAs of α and ß genes appear more stable than those of γ genes (742). Viral mRNAs may persist in the cell after their translation ceases (293,328,329).

Environment of the Viral Genes

The open reading frames identified to date are embedded for the most part in domains exhibiting both virus/host-common (e.g., TATAA and CAAT, SP1 response elements, and so forth) and virus-specific *cis*-acting sites. The absence of TATAA boxes from some transcriptional units has been noted (e.g., $\gamma_1 34.5$) (97).

Studies on the structure of HSV genes have focused on two specific objectives. The first concerned the minimal promoter domain and the *cis*-acting sites required for gene expression. The second objective was to identify the *cis*-acting sites which confer upon the reporter gene the capacity to be regulated as an authentic viral gene. Only a few HSV genes have been analyzed in sufficient detail to reveal and dissect the *cis*-acting signals embedded in them. The most thoroughly studied, and the one which has generated the most perplexing results, is the *tk* gene. The α4 gene (Fig. 11) and, to a much lesser extent, two representative γ_2 genes are also worthy of discussion, although in these instances the results are more novel than perplexing.

α4 Gene

In the α4 gene, the 5′ nontranscribed domain extending upstream from the cap site to nucleotide −110 is capable of imparting to a reporter gene the capacity to be transcribed efficiently in the absence of viral *trans*-activating factors (380,381,498). Other than the transcription

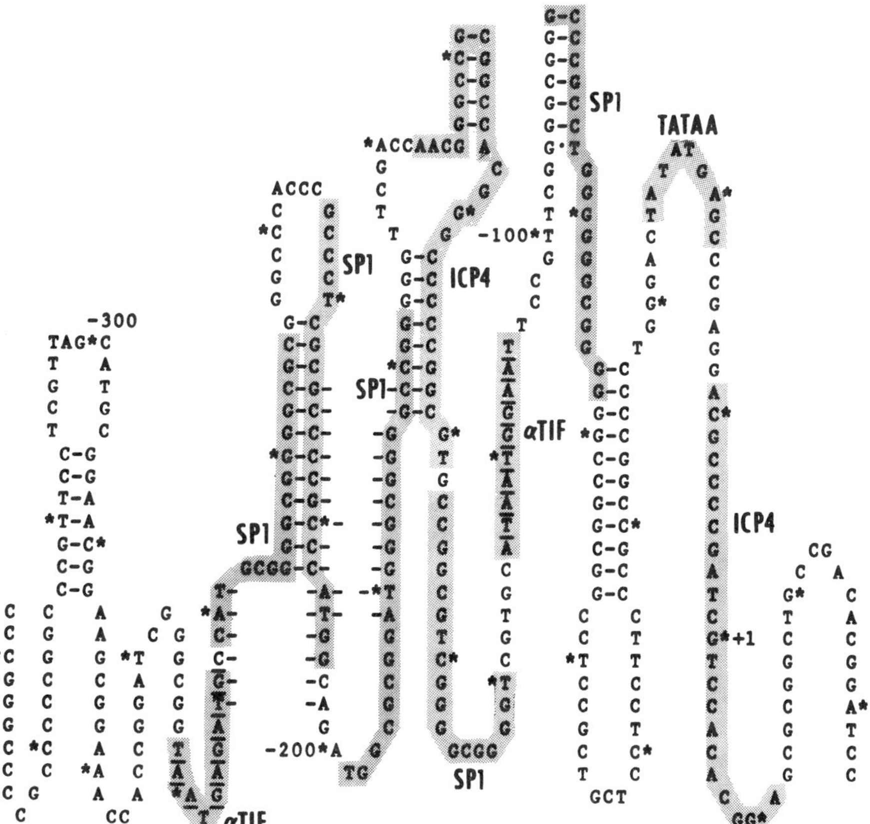

FIG. 11. Schematic representation of the structure of the 5′ nontranscribed domain of the α4 gene. The symbol * marks every tenth nucleotide. The transcription initiates at nucleotide −1. The shaded areas represent the binding sites of TATAA protein, SP1 and ICP4, and the host protein-αTIF complex. The sequence is shown in a form to emphasize the presence of the perfect G+C-rich inverted repeats that abound in this domain of the gene. In some instances, the G+C-rich regions can form alternate stem structures; these are identified by the dashed lines. Data from Mackem and Roizman (382).

initiation site, the *cis*-acting sites which affect expression have not been investigated in detail. The sequences upstream from nucleotide −110 confer on α and ß promoters a higher basal level of expression as well as the capacity to be induced as α genes by viral *trans*-activating factors (337). The higher level of expression conferred by the sequences upstream from nucleotide −110 is very likely due to the SP1 binding sites embedded in G+C-rich inverted repeats which abound in that region (296,297,337,380–382). At least one sequence which confers inducibility as an α gene is the *cis*-acting site for αTIF, i.e., 5′ NCGyATGnTAATGArATTCyTTGnGGG 3′ (337,380). Separation of minimal promoter and regulatory domains has been noted in other α genes (380–383). Although G+C-rich stretches are frequently found in α genes, SP1 binding sites have not been reported in other α genes (Fig. 11).

ICP4 binding sites have been reported in the promoter-regulatory domain, and across the transcription initiation site of the α4 gene, and in the promoter domain of the α0 gene, but not in the other α genes (176,335,336,447,559), and their significance is detailed below in the context of the function of ICP4.

In addition to the SP1 and αTIF-responsive *cis*-acting site, α gene promoters contain numerous other *cis*-acting sites for CAAT-box binding proteins, shown to affect gene expression markedly (648). The element GCGGAA, often present a variable number of times upstream of the transcribed domains of α genes, has been reported to bind proteins by footprint analysis (695). The role of these repeats and their environment may be deduced from the observation that the proteins purified by affinity to this sequence do not bind to a single isolated hexanucleotide (348). Complementary DNA clones that encode two distinct subunits of this transcription factor have been isolated; the amino acid sequence of one protein exhibits similarity to the Ets DNA binding domain (349). The function of the purine-rich GA element in the context of α gene promoters remains a mystery, however, inasmuch as mutations within this region of the α27 promoter in the context of the viral genome reduced the accumulation of α27-*tk* mRNA only slightly.

tk *Gene*

The initial analyses of the *tk* gene rested on the notion that it exemplified the structure of a typical eukaryotic gene. Only much later has there been an effort to understand its structure as a viral ß gene. The *tk* gene appears to have two transcription initiation sites, and mRNAs derived from both sites have been reported (141,505,756). The role of these sites in the expression of the gene in lytic infection is not clear. The 5′ nontranscribed region has been very thoroughly investigated by McKnight et al. (157, 410–414) and by Silverstein et al. (162,163,210). Initial studies identified a 110-nucleotide region upstream from the site of initiation of transcription that was minimally required for efficient expression in the absence of viral *trans*-activating factors. This promoter contains a "proximal" and two "distal" *cis*-acting sites, all of which are important for constitutive expression of the *tk* gene (756). In subsequent studies, a CCAAT box, two SP1 binding sites and an octamer motif (ATTGCAT) upstream of −116 were also identified (478). Mutations in all of these sites affect expression of the uninduced gene in rodent cells. However, sequence-specific mutagenesis of the octamer motif and of the SP1 binding sites indicated that they play a minimal role in the *trans*-activation of the *tk* gene by HSV infection, and attempts to define a *cis*-acting site that is virus-specific have not been successful (470,471). The protocol employed involved studies on linker scanning mutations. Inasmuch as the authors failed to identify mutations which affected *trans*-activation but not expression in the absence of viral *trans*-activating factors, they concluded that the viral factors which *trans*-activate viral ß gene expression act on a host factor and not directly on DNA.

More recent studies have focused on the interaction of ICP4 with domains of the *tk* gene and on mutations downstream from the *tk* cap site. Notwithstanding earlier reports that ICP4 does not bind to the *tk* gene directly (138,542), it is evident that the ICP4 does indeed bind to several domains of the gene, both upstream and downstream of the cap site, and that the binding sites include both sequences similar to the consensus binding site reported by Faber and Wilcox (177) and highly degenerate sites with little resemblance to the consensus (274,430). Of particular interest are the ICP4-binding sites in the transcribed noncoding domains and the role of the transcribed noncoding sequences on the regulation of *tk* as a ß gene, as discussed below.

γ *Genes*

The structure of the γ genes, and particularly of the γ_2 genes, is the least well understood and is likely to vary from gene to gene. Analyses of several genes, i.e., U_L44 (gC), U_L38, U_L49.5, and U_S11, suggest that the sequences required for efficient expression include the TATAA box and extend into the 5′ transcribed noncoding domain (114,141, 172,229,257,258,266,294,295,396,397,613,625). The sequences required for the γ_2 regulation of these genes also appear to include sequences downstream from the TATAA box. Embedded in transcribed domains as well as the 5′ nontranscribed sequences are ICP4 binding sites. Specific sequences which are required for efficient transcription have been reported within 5′ transcribed, noncoding domain of some but not all γ genes (229).

Trans-Activation of α Genes by αTIF

Background

The *trans*-activation of α genes by a putative structural component of the virus was first reported by Post et al. (498) in experiments designed to map the domains of the α gene promoters. In these experiments, transcription of a chimeric gene residing in L cells and consisting of the α4 gene transcribed noncoding and 5′ untranscribed domains fused to the transcribed noncoding and coding domains of the *tk* gene was induced by an infection with a tk^- virus in the presence of cycloheximide. Similar results obtained with other α gene promoters suggested the possibility that the inducing factor was brought into the infected cells with the virion (380–382). The compelling evidence that the *trans*-activator is a component of the virion which acts on a specific response element came from three sources. First, Mackem and Roizman (382) reported that all DNA fragments derived from α promoters that were capable of being induced contained a consensus sequence 5′ GyATGnTAATGArATTCyTTGnGGG (noncoding) common to the regulatory domains of all α genes. They also reported that some promoters contained G+C-rich, perfect inverted repeats; some of these stretches were later shown to contain SP1 binding sites. Kristie and Roizman (337) showed that homologs of the consensus sequence imparted on chimeric constructs the capacity to be induced by the virion component, whereas the G+C-rich domains affected the basal level of expression of chimeric genes. Second, Batterson and Roizman (33) reported that ultraviolet light-inactivated virus was capable of inducing α-*tk* chimeric genes and that the *trans*-activator was a component of the virion tegument inasmuch as HSV-1(HFEM)*ts*B7, a ts mutant, at the nonpermissive temperature induced the α-*tk* chimeric genes even though the capsids accumulated at the nuclear pore and failed to release viral DNA into the nucleus. Lastly, Campbell et al. (76) mapped the gene whose product induced the α chimeric gene to U_L48. The nucleotide sequences of the HSV-1(F) and HSV(17)syn$^+$ U_L48 genes were reported (125,481). The product of this gene had been previously recognized as VP16 present in 500 to 1,000 copies/virion (242,643) and also as ICP25 (261,262). The name by which this protein is known is subject to considerable confusion largely because Dalrymple et al. (125) gave it still another designation based presumably on its apparent molecular weight deduced from its migration in polyacrylamide gels. The designation given to it by Pellett et al. (481), αTIF, reflects its function as the *trans*-activator of α genes.

Properties of the αTIF Protein

A report from McKnight et al. (409), followed quickly by many others, established that αTIF forms complexes with DNA, but only in the presence of a cellular protein, Oct-1, or under conditions in which Oct-1 can participate in the reaction (212,460,461,504,648,695). Specifically, (i) *in vitro* synthesized, labeled VP16 was demonstrated to form a complex with unlabeled DNA in the presence of crude uninfected cell extracts or partially purified Oct-1 (212,409) and (ii) the complex was retarded by antibody to αTIF, but not by preimmune serum (212,409,504). The αTIF gene falls into the γ_1 kinetic class and is predicted to encode a maximum of 490 amino acids with a translated molecular weight of 54,000 and a minimum of 479 amino acids if the first in-frame methionine codon is not used or is used intermittently (125,481). The functional domains of αTIF have been mapped by several laboratories (2,3,226, 227,660,694) with the following results.

(i) Kristie and Sharp (338) demonstrated that, at very high concentrations, αTIF could form complexes with DNA in the absence of additional host factors.

(ii) The key domains of αTIF responsible for the formation of multiprotein complexes have been mapped to amino acid stretches. Region I consists of the most basic amino acid stretch located between residues 173 and 241, whereas region II lies between residues 378 and 389. Mutations in region I diminished the ability of the protein to interact with DNA in the absence or presence of the Oct-1 POU domain, suggesting that this region is required for and may be involved in binding DNA (660).

(iii) The αTIF interacts specifically with the Oct-1 POU domain, but it is incapable of interacting with a POU domain in which helix 2 of the Oct-2 protein has been substituted for the corresponding helix of Oct-1 (661). Point mutations within region II also completely destroyed αTIF-induced complex assembly (226). Region II appears to be the domain of αTIF responsible for specific interaction with the Oct-1 homeo domain (660). Mutational analysis also suggested that region II is primarily responsible for αTIF interaction with an additional host factor, although mutations in region I can also affect this interaction (660). Since αTIF-lacking region II has a reduced capacity to interact with DNA, it is conceivable that deletions in region II alter its conformation. Parenthetically, the entire carboxy-terminal acidic activating region of the protein is absent from the varicella-zoster virus homolog of the protein encoded by the ORF10 gene. The ORF10 protein has been reported to be incapable of forming multiprotein complexes with Oct-1 (406).

(iv) The domain responsible for the *trans*-activation of α genes is contained within the highly acidic carboxy-terminal 80 amino acids. Proteins which lack the *trans*-activation domain may retain their capacity to form complexes with Oct-1 and DNA. However, deletion of the domain responsible for complex formation obliterates the capacity of αTIF to *trans*-activate. The key evidence concerning the role of this region emerged from observations that a chimeric protein consisting of the DNA binding domain of the yeast transcriptional activator GAL4 fused to the C-terminal domain of the αTIF gene induced a reporter gene to a high level (586). Analysis of this region has revealed that a net negative charge contributed to, but was not sufficient for, transcriptional activation. The observations that transcriptional activators of several classes contained hydrophobic amino acids in patterns resembling that of αTIF, and that a phenylalanine residue at position 442 was markedly sensitive to mutation, suggested that the mechanism of transcriptional activation by αTIF could involve both ionic and specific hydrophobic interactions with target molecules (117).

(v) The identity of the interactive cellular protein has been the subject of numerous studies. Stringer et al. (669) reported that the αTIF activation domain bound strongly and selectively to the human and yeast TATA-box binding factor TFIID. Lin and Green (365) presented evidence that αTIF stimulates transcription in the absence of ATP by increasing the number of functional preinitiation complexes through recruitment of the general transcription factor TFIIB to the promoter. In both of the above studies, mutations in the αTIF activation domain which affected *trans*-activation also decreased the affinity of the protein for transcription factors (276,365,669). Liljelund et al. (364) also reported that αTIF alters the binding of the TATA box factor to its response element. Kelleher et al. (307) suggested that the direct target of αTIF is not a general transcription factor but rather another factor referred to as an "adapter" or a "coactivator," which forms a bridge to one of the general transcription factors. Recently, the gene for such a putative adapter molecule, ADA2, has been isolated from yeast on the basis of its interaction with the acidic activation domain of αTIF and was shown to respond to some but not all acidic activator domains (39).

(vi) Located in the tegument, αTIF is a structural protein. Weinheimer et al. (720) deleted the U_L48 gene encoding the αTIF using a cell line which produced the protein to propagate the mutant virus. In infected cells, the deletion mutant induced nearly normal levels of viral DNA synthesis and capsid production, but the amounts of encapsidated DNA were lower, and the electron microscopic studies suggested that it was defective in other steps of viral maturation.

(vii) Attempts to dissociate the function of αTIF as a *trans*-activator of α genes from its as yet undefined function as a structural protein have run aground. A mutant with an insertion into the U_L48 gene was shown to abolish the ability of the protein to induce recorder genes but did not render the virus noninfectious, although its ability to multiply and spread at low multiplicity of infection was impaired (3). The virus stocks were characterized by high virion/plaque-forming units (pfu) ratios. In infected cells, the expression of α0 and α27 were reduced four- to fivefold, whereas that of α4 was not. Interpretation of these results presents three problems. Foremost, the high virion/pfu ratios may also be interpreted to indicate that the structural function of αTIF was compromised, and therefore the two functions of αTIF have not been resolved. Second, the promoter domains of α genes contain numerous *cis*-acting sites for a variety of transcriptional factors, and each gene is readily expressed in cultured cells, albeit at a reduced basal level, in the absence of αTIF. High multiplicity of infection could compensate for weak promoter activity. Lastly, a perhaps more worrisome problem is that a very much reduced level of *trans*-activation, which might not be detectable in transfected cells, would be sufficient to enable the virus to multiply. In light of these issues, the report that the virus appeared to be avirulent in mice, was able to establish latency in the trigeminal ganglion, and reactivated upon explantation is subject to conflicting interpretations (657).

(viii) Lastly, observation that VHS binds to αTIF and precludes its interaction with its response element has been suggested to indicate that αTIF modulates the activity of VHS during infection (630). Since αTIF and VHS enter the cell with the infecting virus at the same time, but perform their functions in different compartments of the cell, such regulatory functions are probably of no great significance in the initial stages of infection. The association of the two protein does solve an important riddle, i.e., why the massive quantities of αTIF made late in infection fail to induce α gene expression at that time.

Oct-1 Protein

The Oct-1 protein was independently discovered in several laboratories studying a variety of biologic systems and has been designated as NF-A1, OBP100, OTF-1, NFIII, origin recognition protein C, αH1, and Oct-1 (189,334,467, 655,671,672). Oct-1 interacts with the octamer motif of the *cis*-regulatory element in the promoters of ubiquitously expressed sequences such as those encoding the small nuclear RNAs and the histone H2b genes (343,392). The identical octamer motifs present in cell-specific immunoglobulin promoters are thought to interact with the transcriptional factor Oct-2 (594). Oct-1 has been demonstrated as a stimulatory replication factor in the adenovirus system (467,514).

A nearly complete cDNA cloned by Sturm et al. (672) led to the recognition that Oct-1 is a homeobox protein which, along with two other mammalian transcription factors, Pit-1 and Oct-2, and the product of the nematode *Caenorhabditis elegans* Unc-86 gene, helped define a new class of homeobox proteins called POU proteins (for Pit, Oct, Unc) (246). These four proteins share a highly related 150- to 160-amino acid-long region of sequence designated as the POU domain. Polymerase chain reaction (PCR) amplification of human and rat brain cDNA and rat testes cDNA with degenerate primers to the POU domain led to the discovery of the Brn-1, Brn-2, Brn-3, and Tst-1 POU proteins (240). All of these known mammalian POU proteins have been implicated by *in situ* hybridization as playing a role in the development of the nervous system. A number of other proteins have recently been discovered with homology to the known POU proteins and many more are certain to follow.

The POU domain can be further subdivided into the homeobox-related POU-homeo and the unique POU-specific domains. The POU homeo domain is more highly conserved among the POU proteins than among other homeobox proteins. The homology between Oct-1 and Oct-2 in this region approaches 88%. The POU homeo domain is involved in both DNA binding and in directing the formation of the multiprotein-DNA complex with αTIF. Helix 2 of Oct-1 makes direct contact with αTIF. Thus, substitution of helix 2 of Oct-2 with that of Oct-1 imparts upon Oct-2 the ability to complex with αTIF (661). Conversely, substitution of helix 2 of Oct-1 with that of Oct-2 destroys the ability of Oct-1 to bind αTIF. The POU-specific region is 75 to 83 amino acids long and located in the amino-terminal half of the overall POU domain. The POU-specific domain is 99% conserved between Oct-1 and Oct-2. If the POU-specific domain does play a role in complex formation, it must be at amino acids conserved between Pit-1 and Oct-1 because replacing the POU-specific domain of Oct-1 with that of Pit-1 reduced did not eliminate αTIF complex formation (661). Its function is largely unknown, although in Oct-1 it appears to contribute to DNA binding (673).

The POU-homeo and POU-specific domains are separated by a short nonconserved 15- to 20-amino acid linker. The insertion of an additional six alanines into the linker had no effect on DNA binding, suggesting that the POU domain is a bipartite DNA-binding structure. Such flexibility in the DNA-binding domain might explain the many apparently dissimilar DNA sequences to which Oct-1 appears to bind. Sequences which are identical in only four of 14 residues have been shown to bind, yet simple point mutations within some binding sites can abolish binding (35).

α-TIF Response Element

As noted earlier in the text, several lines of evidence indicated that the consensus sequence identified by Mackem and Roizman (380,381) [5′ GyATGnTAATGArATT 3′] is the *cis*-site for induction of α genes by αTIF (337). Gaffney et al. (203) designated the *cis*-acting site as the sequence TAATGARAT. For reasons discussed below, the *cis*-acting site is longer and extends upstream.

Gel retardation studies using a 29-bp fragment from the α0 gene (29Rα0) and a 48-bp fragment from the α27 gene (48Rα27) revealed that the *cis*-acting site binds host proteins (334). Both DNase protection and methylation interference studies indicated that those purines whose methylation interfered with the binding of the largest of these host proteins, αH1, was part of the sequence ATGCTAAT, which resembled the consensus binding site ATGCAAAT of several proteins subsequently designated as Oct-1 (333). Extensive mutational analysis of the αTIF consensus site in the context of the viral genome (647,648) led to three conclusions. First, mutagenesis of the ATGCTAAT domain led to a loss of Oct-1 binding and reduced the level of expression of the reporter gene to 6% that of wild-type virus. Second, transversion of the sequence GATATG, which was designed to destroy the consensus octamer binding site but preserve the TAATGARAT sequence also abolished high-affinity binding of Oct-1 and αTIF, but, at vastly increased amounts of probe, trace amounts of bound Oct-1 could be detected, presumably reflecting direct binding of the protein to the TAATGARAT sequence. Consequently, in cells infected with this virus, the reporter gene expression was reduced to 20% to 30% of the level observed in cells infected with the wild-type parent. Further studies showed that expression of the reporter gene correlated with the respective Oct-1 binding levels. Third, mutations in the GARAT domain abolished the formation of αTIF/Oct-1/DNA complexes and reduced the expression of the reporter gene to 20% to 30% of the wild-type level but had no effect on the binding of Oct-1 to the DNA. The stringency of the requirement for a GARAT sequence varied, depending on the cell in which the virus carrying the mutation was tested. In some cell lines, the mutant promoter expressed the reporter gene at a level 80% that of the wild-type. On the basis of these findings, it was postulated that an unknown cellular factor is capable of interacting with Oct-1 to induce gene expression, albeit to a lesser degree than when Oct-1 is complexed with αTIF, that this factor does not require the GARAT sequence, and that certain cell lines possess more of this factor than others. Indeed, such cellular factors have recently been discovered (449).

Stable formation of a complex between purified Oct-1 and αTIF requires an additional factor designated as C1 (212,332,338). Highly purified C1 was reported to consist of a family of polypeptides derived from a common precursor by proteolytic cleavage (303, 332a, 738). C1 binds to αTIF rather than to the Oct-1/DNA complex. An additional factor designated as C2 and forming a complex designated by the same name has also been reported (332,338).

In summary, the multiprotein complex responsible for induction of the α genes is initiated by the Oct-1 binding to the octamer element within the αTIF *cis*-acting site. The αTIF/C1 factor complex binds to the Oct-1/DNA complex and is stabilized by specific interactions between the αTIF and the Oct-1 POU-homeo domain and between αTIF and the GARAT sequences of the *cis*-acting site. Although αTIF interacts directly with the POU-homeo domain, no interaction is detected in the absence of DNA, suggesting that this interaction is of low affinity. Likewise, the DNA-binding affinity of αTIF is extremely low and may require an interaction with Oct-1 *in vivo.* The nature of the interactions of the C2 factor with the complex involve most likely protein-protein interactions as the binding of this component does not affect the DNA footprint of the complex.

Regulatory Functions of α Gene Products

ICP0

The product of the α0 gene is predicted to be 80,000 in molecular weight, but in denaturing polyacrylamide gels it migrates with an apparent molecular weight of 110,000 to 124,000, depending on the type of cross-linking agent used and on the acrylamide concentration. ICP0 oligomerizes in solution (89,174). It has been reported that ICP0 is a virion protein (750), but definitive studies involving separation of "heavy" or complete particles from light particles which may package nonvirion proteins have not been done. The α0⁻ mutants are viable in cell culture, and ts mutants have not been reported (585,667). In transient expression systems, ICP0 has been reported to *trans*-activate transfected genes promiscuously by itself or in combination with ICP4 (169–171,208–210,462–464,484,526,589). Of the various experimental designs, the most convincing are those in which the *trans*-activation of target genes was done in conjunction with ICP4.

The transient expression studies suggest ICP0 enhances the function of ICP4, possibly by interacting with it, but this may be only one of its functions. ICP0 response elements have not been identified, although it has been reported that the protein interacts with a cellular protein (426). Deletion mutants in the α0 gene grow in cell culture, albeit more sluggishly than the wild-type virus, particularly at low multiplicities of infection (585,667).

Mutants in ICP0 have been studied in some detail by both Cai and Schaffer (67) and Chen and Silverstein (90).

Overall, it is difficult to argue with the basic conclusion by Chen and Silverstein that, while it is not essential for viral replication in some cells in culture, defects in this gene delay the expression of ß and γ genes and impair viral replication.

ICP4

ICP4 is a phosphoprotein essential for expression of viral genes and for viral replication. Long a subject of study, it has generated a rich, passionate, and at times confusing literature.

Properties of the Protein

This protein is predicted to be 132,835 in translated molecular weight (402). Most limited-passage HSV-1 isolates carry a ts lesion mapping in the α4 gene and 37°C/39°C ratios of plating efficiencies as high as 10^6 have been recorded [e.g., HSV-1(F)]. It is perhaps then not surprising that ts mutants in the α4 gene have been readily isolated by a number of laboratories (361). ICP4 forms three bands designated as 4a, 4b, and 4c in denaturing polyacrylamide gels (483). In cells infected with wild-type virus, the fastest migrating band (4a) has an apparent molecular weight of 160,000 and is readily detected in the cytoplasm after pulse labeling with radioactive precursors (261,262,483,735). It is also the only form accumulating in cells infected with certain α4 ts mutants and incubated at nonpermissive temperatures (320). The other bands (4b and 4c) have apparent molecular weights of 163,000 and 170,000, respectively, and accumulate in the nucleus (337,483). The accumulation of the slower migrating bands coincides with the translocation of the protein into the nucleus and labeling with inorganic ^{32}P added to the medium (183,483). ICP4a and ICP4c can be pulse labeled with ^{32}P during the reproductive cycle long after the synthesis of this protein ceases, suggesting that phosphate cycles during infection (735). As noted above, ICP4 is (ADP-ribosyl)ated (43,506). Adenosyl ribosylation could account for only a fraction of the labeling of ICP4 by inorganic phosphate from the medium (483), inasmuch as phosphate also cycles on and off ICP4, ICP22, and ICP27 (735), but ICP4 appears to be the only α protein to be poly(ADP) ribosylated. Recently, Blaho and Roizman (46) reported that ICP4 is also guanylylated and adenylylated.

ICP4 has been shown to bind to DNA directly in several types of assays (177,335,336,430). Initial studies identified a strong binding site with a consensus sequence of ATCGTCnnnnCnGnn (176). Subsequent studies have identified numerous binding sites which do not correspond to this consensus sequence (335,336,429,430). Studies by Michael and Roizman (428,429) have shown that two molecules of ICP4 bind to both consensus and nonconsensus sites.

The various forms of ICP4 differ with respect to their affinities for the binding sites (430). Papavassiliou et al. (472) reported that, whereas dephosphorylated forms of ICP4 could bind to α promoters, only phosphorylated forms bound to ß and γ gene promoters. Anatomic dissection of the α4 gene in the context of the viral genome has outlined several domains which play a key role in the function of this protein. Most of the key functions of the protein appear to be associated with the carboxyl half of the protein. Mutations in this region affect autoregulation, *trans*-activation of viral genes, intranuclear localization, interaction with DNA, and so forth (137–139,617,634). Wu and Wilcox (746) mapped the ICP4 sequence-specific DNA-binding site to residues 262–490.

To *trans*-activate transcription of HSV genes, ICP4 must interact with a component of the PolII transciptional complex. Indeed, interaction of ICP4 with TATA binding protein and TFIIB has been reported (632).

ICP4 has been reported to be packaged in virions (749). The uncertainty surrounding this datum rests on the concern that ICP4 detected in these assays may be packaged nonspecifically into "light" particles and is not a bona fide component of the virion. If both ICP4 and ICP0 were to be packaged in virions, we would expect to see transcription of genes in cells infected and maintained in the presence of cycloheximide. This is not the case.

The $\alpha4^-$ mutants express only α genes, but with time both ß and γ genes are also expressed. To multiply, $\alpha4^-$ mutants must be grown in cells capable of expressing ICP4 proteins from a copy of the α4 gene embedded in the cellular genome. The observation that ICP4 is required for the expression of genes expressed later in infection has been taken as an indication that ICP4 regulates genes positively. Evidence has also amassed indicating that ICP4 regulates specific α genes negatively. It is convenient to consider these manifestations of ICP4 function separately.

Role of ICP4 in the Expression of α Genes

As noted above, ts mutants in the α4 gene abound. In these viruses, both copies of the gene are mutated, as would be expected for the expression of the ts phenotype (320). The phenotypes of these mutants vary. At the nonpermissive temperature, some mutants express both α proteins and selected sets of proteins normally made later in infection (136,140). A most interesting set of ts mutants in the α4 gene overproduce α proteins at the nonpermissive temperature (136,501,713). There is convincing evidence that ICP4 turns off its own synthesis and that this autoregulation correlates with the binding of the protein to a *cis*-acting site across the transcription initiation site of the gene (138,427,437,542,713). Measurements of α RNAs accumulating in cells infected and maintained at permissive and nonpermissive temperatures indicate that the α genes subjected to repression are primarily α4 and α0 (J. Hubenthal-Voss and B. Roizman, unpublished observations).

In the case of the α0 gene, the binding site ATCGT-CactgCcGcc is at position −64 to −49 (335). The α4 gene 5′ untranscribed and transcribed, noncoding domain contains three ICP4 binding sites. Two nonconsensus binding sites map at positions −194 to −171 (site α4-1 distal) and −162 to −145 (α4-1 proximal), whereas one consensus site (α4-2) maps across the cap site (176,335,336,430,447). The function of these binding sites has been the subject of numerous and somewhat contradictory reports.

Roberts at al. (542) reported transient expression assays of the function of the ICP4 binding site at the transcription initiation site of ICP4 (α4-2 site). Deletion of the nucleotides ¯8 to +30 abolished both binding of DNA and "negative autoregulation." Detailed studies by Resnick et al. (532) showed that ICP4 *trans*-activated a mutated promoter lacking the binding site much more efficiently than the wild-type promoter containing an intact DNA binding site. In contrast, Everett and Orr (173) substituted a mutated promoter lacking the ICP4 binding site for the wild-type promoter in both copies of the viral genome. They reported that the wild-type and mutant viruses did not differ with respect to the accumulation of ICP0 throughout infection. Although the results would support our view that transient expression systems are useful to determine the requirements for expression but not for the regulation of viral genes, in this instance, analyses of the accumulation of mRNA rather than protein would have dealt more directly with the role of the ICP4 binding site on the expression of ICP0. A remarkable feature of HSV gene expression is that the accumulation of many viral proteins is fine tuned at several levels. In the case of ICP0, recent studies indicate that its accumulation is regulated by the products of both U_L13 and α22 (520).

As noted above, ICP4 binds to three sites within the 5′ untranscribed and transcribed, noncoding domains of the α4 gene. The purines whose methylation interferes with the binding of ICP4 to these sites have been mapped (429, 559,560). Mutagenesis of these three sites in a construct in which the 5′ untranscribed and the entire transcribed noncoding domain of the α4 gene were fused to the coding sequence of a reporter gene and inserted into the viral genome revealed the following. First, ICP4 did not bind to the mutated binding sites. Second, identical amounts of RNA of the chimeric gene were recovered from cells infected with viruses carrying wild-type and mutated α4-1 and α4-2 sites linked to the reporter gene and maintained under cycloheximide, indicating that the mutations did not affect either the *trans*-activation of the α4 promoters linked to the reporter gene or the stability of the RNAs. Third, mutations of the α4-2 sites resulted in a seven- to 18-fold increase in the amounts of reporter gene RNA recovered at 4 and 8 hr postinfection. Fourth, mutations of either the distal or proximal α4-1 binding sites alone or in combination did not affect the recovery of reporter gene RNA at these times postinfection, whereas the mutagenesis of all (α4-1 and α4-2) binding sites increased at least threefold the amount of RNA recovered compared to mutagenesis of the α4-2 site alone (427). These results strongly support the hypothesis that (i) ICP4 acts by binding to both consensus and nonconsensus sites on DNA, (ii) autoregulation of α4 gene expression is the consequence of binding of ICP4 to DNA, and (iii) the ICP4 binding sites upstream from the cap site can contribute to the repression of gene expression.

Koop et al. (325) inserted wild-type and mutated ICP4 binding sites into a variety of promoter domains. They noted that ICP4 binding sites upstream or downstream of TATA boxes effectively decreased gene expression from these promoters. An interesting finding was that DNA synthesis alleviated the apparent repression caused by the insertion of the wild-type ICP4 binding sites.

Role of ICP4 in the Expression of ß and γ Genes

ICP4-dependent activation of transcription of a ß gene embedded in the viral genome occurs from a much lower level of basal expression than that seen from an isolated gene introduced in cells and selected for *tk* expression. After *trans*-activation with ICP4, the level of *tk* gene expression is higher than that attained in cells transfected with the isolated *tk* gene but not as high as that seen in lytically infected cells. ICP4 DNA binding sites in the domain of the *tk* gene both upstream of the cap site and downstream from nucleotide +50 have been demonstrated by several groups (336,470–472). Studies by Halpern and Smiley (231) and by Mavromara-Nazos and Roizman (396) have failed to demonstrate a significant role of the binding to sites downstream of nucleotide +51. The report by Papavassiliou et al. (472) that the binding of ICP4 to ß and γ genes requires infected cell factors and is determined by both concentration and phosphorylation of ICP4 is consistent with and supports the conclusion that ICP4 interacts with cellular factors and both enhances and depends on them for the stability of its own binding to DNA.

In infected cells, ICP4 is required but is not sufficient for efficient and timely expression of γ genes, particularly $γ_2$ genes. DNA binding sites have been observed in both 5′ untranscribed and transcribed, noncoding domains of genes (428,430,684). The role of these sites is unclear.

Does ICP4 block ß and γ gene transcription, *trans*-activate it, or both? The evidence that, in cell-free systems, ICP4 increased the transcription of gD, a $γ_1$ gene (685), is not in itself impressive since (a) the ICP4 binding site tested was upstream from the reported minimal sequence required by gD to be regulated as a $γ_1$ gene (nucleotide −55) (148,150); (b) in transient expression systems, for what it is worth, late $γ_2$ gene expression was activated at low concentrations of the ICP4 gene but not by high concentrations of the gene; and (c) the amount of ICP4 used in that study was arbitrary, without relevance to either known positive or negative *trans*-activation. Subsequently, Tedder et al. (684) reported that ICP4 binding sites enhance the transcription of the gD gene *in vitro*. However, studies involving mutagenesis of the ICP4 binding sites within the domains of the gD gene failed to affect the transcription of that gene (631).

In other studies, some of the characteristics of γ_2 gene regulation, i.e., expression of the gene late in infection and sensitivity to inhibitory effects of PAA, were transferred to the *tk* gene by substitution of the *tk* cap site and a portion of the 5′ transcribed noncoding domain with the corresponding sequences of the U_L49.5 gene. The substituted γ_2 sequences contained two ICP4 binding sites; mutagenesis of one of these sites led to a loss in the capacity of the chimeric gene to be expressed late in infection (325). While the hypothesis that ICP4 does not act directly on the DNA cannot be dismissed out of hand, it seems likely that ICP4 acts in at least two ways, by stabilizing the assembly of TATA box-dependent transcriptional factors and by a local effect at the site of binding. Studies showing that ICP4 bends DNA have been reported (175). The jury evaluating the role of ICP4 binding sites in ß and γ genes is still out. While the data clearly indicate that functional ICP4 is required for the expression of these genes, the precise mechanism by which ICP4 acts remains elusive.

Much has been made of the observation that α4 and the equivalent gene product of pseudorabies virus induce not only herpesvirus genes but also adenovirus and cellular genes (e.g., ß globin), which are introduced into cells by transfection (41,180,228). More recent studies have shown that the *trans*-activation of adenovirus late gene expression by the ICP4 equivalent of pseudorabies virus is through enhanced formation of transcription initiation complexes (1). Although these studies are consistent with the hypothesis that ICP4 may stabilize specific sets of transcriptional factors on the DNA, they do not exclude the possibility that weak interactions with DNA may be necessary for *trans*-activation of gene expression.

ICP27

The realization that this nuclear phosphoprotein performs a plethora of diverse regulatory functions throughout infection emerged only in recent years (588). The key functions and properties of ICP27 may be summarized as follows.

1. ICP27 is an essential gene. Null mutants do not replicate; they exhibit a gross decrease in viral DNA synthesis and late gene expression. Some ts and nonsense mutants do not block viral DNA synthesis but do decrease the expression of late genes (419,584).
2. The key phenotype of ts mutants studied in some detail indicate that ICP27 exhibits both activator and repressor functions. Thus, in the absence of functional ICP27, α genes are overexpressed, whereas late genes are poorly expressed (398,533,584,635). As articulated by McCarthy et al. (398), the general conclusion is that ICP27 acts at a transcriptional level.

   More recent studies indicate that ICP27 does have a role in transcriptional regulation at two levels. First, ICP27 determines which polyadenylation signal is used where more than one signal is present (418). Second, it has been noted that the amount of spliced RNA is decreased in the presence of functional ICP27. Splicing studies support the hypothesis that ICP27 directly inhibits splicing by sequestering snRNPs (236).
3. Studies on ICP27 mutants have identified both repressor and activator domains which map in the carboxy-terminal half of the protein (235,419,433,534). Amino-terminal domains have been shown to be essential for optimal levels of DNA synthesis and for localization of ICP27 in the nucleus. One domain localized roughly between amino acids 138 and 152 bears a striking similarity to several cellular proteins which have been implicated in nuclear RNA processing and appears to be required for nucleolar localization (347).
4. ICP27 binds to single-stranded DNA-agarose columns and is eluted by high salt. In these studies, ICP27 copurified with an unidentified protein 110,000 in molecular weight. In addition, the predicted amino acid sequence has a potential metal binding domain, and *in vitro* the protein binds zinc. This domain is located at the carboxyl-terminal 105 amino acids. Conservation of this domain is essential for at least some of its regulatory effects (702).

In summary, ICP27 has been reported to perform an extraordinary range of regulatory feats which include selection of transcriptional termination sites, inhibition of RNA splicing, stimulation of DNA synthesis, and posttranscriptional destabilization of α mRNA (588). This is a remarkable range. In time, it may decrease as some of the manifestations now thought to be due to direct action of the protein are recognized to be the consequences of secondary events in a regulatory cascade.

ICP22

Inasmuch as this gene is dispensable (499), it has not been studied intensively. The studies done to date (603) indicated the following. The α22⁻ virus multiplied in Vero and HEp-2 cells as efficiently as the parent virus. In BHK and RAT-1 cell lines and in human embryonic lung cells, the plating efficiency of α22⁻ virus was reduced, and the yield was multiplicity dependent. Moreover, in these cells, the shut off of synthesis of ß proteins was delayed, and the expression of γ proteins and the number of capsids detected in infected cells was reduced. Deletion of α22 had no effect on viral DNA synthesis. The α22 protein is phosphorylated by several kinases, including U_L13 and U_S3 (520,521) and is adenylylated by protein kinase II *in vitro* (432). The phenotype of α22⁻ cannot be differentiated from that of U_L13^- in cells in culture (520).

ICP47

This gene is dispensable (377,395) and discussed in more detail in the section on viral modulation of host response to infection.

Other Regulatory Proteins

Recently, it was noted that U_S11, an abundant γ_2 protein, has two functions of potential importance in regulating gene expression (573,574). First, U_S11 protein was found to accumulate in nucleoli, and subsequent studies have shown that it binds to the 60S ribosomal component late in infection, in newly infected cells, and in cells constitutively expressing the protein (573). Second, U_S11 protein binds specifically in a conformation-dependent manner to the mRNA specified by the U_L34 open reading frame (574). Cells infected with a mutant deleted in the U_S11 gene accumulated a truncated form of the U_L34 mRNA. The truncated mRNA is not polyadenylated, but it is transported into the cytoplasm. The U_S11 protein binds to the truncated U_L34 mRNA at or near its terminus. The nucleotide sequence of the 3′ terminus of the truncated mRNA suggests that the truncation takes place immediately before a stem-loop structure. U_S11 protein may act as an antiterminator of transcription of the gene. Inasmuch as no cell line has so far exhibited a phenotype associated with the truncation, the function of the U_S11 protein in regulating U_L34 mRNA accumulation is not known.

Regulation of α Gene Expression

The induction of α genes by αTIF and the autorepression of the α4 gene by its product have been discussed in detail earlier in the text. Although the focus of much of the current studies on α gene regulation centers on these two phenomena, it seems important to stress several observations.

First, from the initial studies of α gene-chimeras (498), it has been observed that α genes are efficiently expressed in transfected cells or in cell lines carrying genes under α promoters. These studies indicate that α gene promoters contain *cis*-acting sites for other transcriptional factors in addition to the αTIF-specific response element. In fact, the studies by Spector et al. (647,648) indicate that cells vary in their capacity to express α genes in the absence of functional αTIF *cis*-acting elements.

The second issue concerns the turn off of synthesis of α proteins. Although the studies by Ackermann et al. (4) led to the conclusion that at least some of the α proteins continue to accumulate throughout infection, there is a significant decrease in α protein synthesis with the onset of synthesis of ß and γ proteins. However, late in infection, Oct-1 is still present, and there is significant accumulation of αTIF—a condition which should result in stimulation rather than shut off of α protein synthesis. Attempts to demonstrate alteration in Oct-1 (e.g., phosphorylation) late in infection have not been productive (F. C. Purves, D. Spector, and B. Roizman, unpublished data). One possible explanation is that the association of VHS with αTIF cited earlier in the text precludes the function of αTIF as a *trans*-activator.

The last point that should be made concerning α gene regulation is that the accumulation of some α proteins may be fine tuned at a posttranscriptional level. The observation that ICP0 accumulation in in HEp-2 and especially in BHK cells was affected by the U_L13 protein kinase raises the possibility that the actual amount of α proteins present at any time may be regulated at many different levels by numerous cellular and viral factors.

Regulation of ß Gene Expression

Exemplified by the *tk* gene, ß genes appear to have the capacity to be expressed in the context of the cellular genome in the absence of other viral gene products. There is a general agreement that the response elements required for the expression of ß genes consist of binding sites for cellular transcriptional factors, a TATA box, and a cap site (413,415), although removal of all known response elements does not appear to ablate entirely ß gene regulation (273,274). In the context of the viral genome introduced by infection into permissive cells, these genes require the expression of α genes and especially of functional ICP4. The level of expression of the *tk* gene in the presence of functional α proteins is higher than that seen in cells stably expressing the *tk* gene in the absence of other viral genes (337).

It would seem that *trans*-activation of ß genes in the context of the viral genome involves two functions, i.e., (i) release from a repressive state and (ii) *trans*-activation. Since neither occurs at the nonpermissive temperature in cells infected with ts mutants in the α4 gene, then at least one, the initial event, depends on ICP4. To aficionados of transient expression, it is worthwhile to point out that, in cells transfected and selected for *tk* activity, the ratio of induced to basal enzyme activity after *trans*-activation with virus is considerably lower than that obtained in cells transfected with a plasmid containing the *tk* gene and another marker and selected for the other, covalently linked marker (337). We interpret this to indicate that, for a constant ratio of *tk* genes per cell, the fraction of derepressed *tk* genes is higher in the cells selected for the TK enzyme activity but that α4 derepresses *tk* genes in both systems to about the same level.

There is little doubt that *trans*-activation of ß genes is enabled by ICP4. The wording of this statement itself betrays how little we know of the precise mechanism by which this takes place.

Regulation of γ Gene Expression

The key questions regarding γ gene regulation are why are the genes not expressed early in infection, what is the nature of *trans*-activators, and what is the identity of the *cis*-acting sites? The mechanism by which HSV regulates its late gene expression remains a formidable challenge and, as such, deserves special consideration. While the exact mechanism for γ gene regulation has yet to be elucidated, there is evidence for (a) involvement of α proteins, (b) inhibition of expression early in infection, (c) response elements in both 5′ transcribed noncoding and coding domains of γ genes, and (d) activation of gene expression after the onset of viral DNA synthesis.

Environmental Constraints on γ Genes for Proper Expression

Initial studies on HSV-1 γ genes were done in transient expression systems transfected with chimeric genes consisting of a reporter gene under the control of a γ gene promoter. In these studies [e.g., those of Dennis and Smiley (141) on γ_1VP5 or those of Silver and Roizman (625) on γ_2UL49.5 promoters], the expression of the γ-*tk* gene transfected into TK-negative cells was low compared to that of the *tk* under its natural promoter. The expression was increased by infection of the cells with a tk^- mutant. Silver and Roizman (625) found that, when cells were treated with the inhibitor of viral DNA synthesis, PAA, and infected with a tk^- mutant, the cells still produced high levels of TK activity. Thus the *tk* gene was regulated as a ß gene rather than as a γ gene. Similar results were obtained with cell lines that contain a stably integrated gC, a γ_2 gene (14). In the environment of the cellular genome, the gene was not expressed unless the cells were infected with HSV. In addition, gC was produced in cells infected with a ts DNA^- virus at the nonpermissive temperature or infected with wild-type virus and treated with PAA, indicating that it was regulated as a ß gene. These observations led to the conclusion that γ genes must be studied within the context of a viral genome in productive infection.

Dependence of γ Gene Expression on Viral DNA Synthesis

Two key hypotheses could explain the dependence of γ gene expression on DNA synthesis. The first is that γ genes are negatively regulated during the early phase of viral infection either (i) by the binding of *trans*-acting negative factors or (ii) by constraints placed on the late genes by a particular DNA secondary structure formed in the vicinity of the γ genes. Viral DNA replication could, in either case, relieve the block and allow full expression of γ genes. The second hypothesis predicts the production or modification of a *trans*-acting factor during viral DNA replication that allows the activation of γ gene expression.

To differentiate between these two hypotheses, cells infected for 6 hr with a tk^- mutant were placed in a medium containing PAA and superinfected with a virus containing its *tk* gene under the control of the promoter of the $\gamma_2$49.5 gene. Assays done on the doubly infected cells failed to detect the expression of the *tk* gene of the superinfecting virus and indicated that the expression of γ genes is tightly linked to viral DNA synthesis and the effect of DNA synthesis is mediated by a *cis*-acting function (397).

It has been reported that ICP8, the HSV-1 single-stranded DNA binding protein essential for viral DNA synthesis, may be a negative regulator of γ gene expression. In addition, ICP8 is an essential protein for viral DNA synthesis inasmuch as the mutant HSV-1 KOS1.1 *ts*18, at the nonpermissive temperature, blocked viral DNA synthesis but allowed low level expression of the γ_2 gC gene even in the presence of inhibitors of DNA synthesis (218). Conversely, certain mutants in the α27 gene downregulate γ gene expression in the face of DNA synthesis, as does an $\alpha22^-$ mutant in restrictive cells. Viral DNA synthesis is necessary but not sufficient for the proper expression of γ genes.

Role of α and ß Gene Products in γ Gene Regulation

The roles of α and ß genes in the expression of γ and, particularly, γ_2 genes may be summarized as follows. First, functional ICP4 is required but is not sufficient for the expression of γ genes. Second, in most instances tested, ICP0 enhances the capacity of ICP4 to *trans*-activate γ genes. The hypothesis that α0 regulates γ gene expression by regulating α27 which in turn affects the switch from ß to γ gene expression is interesting, but much more data will be needed to support it (90). Third, deletion of the ICP27 gene resulted in the downregulation of the expression of γ genes. In the case of specific mutants in this gene, γ gene expression was reduced, notwithstanding viral DNA synthesis (235,398,419,533,534,635). Fourth, the requirement for ICP22 for optimal gene expression is cell-type dependent (603). As noted above, ICP22 is posttranslationally modified by phosphorylation by U_L13 (521). In the absence of U_L13, γ_2 gene expression is phenotypically similar to that seen in cells infected with U_L13^- virus (520). Fifth, other than the ß genes required for viral DNA synthesis and the protein kinase described above, several others exert a profound effect on the expression of γ genes.

Cis-Acting Sites Involved in γ Gene Regulation

There is an incipient convergence of views on the structure of γ gene promoters and, particularly, of γ_2 promoters. One series of studies suggested that they consist simply of a specific TATA box (188,257,294). Indeed, Johnson and Everett (294) reported that the cap site and TATA box of the gene encoding U_S11 is all that is required for "fully efficient regulated activity." At the other extreme, Mavromara-Nazos and Roizman (396) demonstrated that a reporter gene driven by the 5′ nontranscribed sequences of the *tk* (ß) gene fused to the 5′ transcribed noncoding domain of the U_L49.5, a γ_2 gene, was regulated as a γ_1 gene, leading to the hypothesis that response elements of γ_2 genes are located in the 5′ transcribed noncoding domains and that fusion of the ß 5′ untranscribed domains to the γ_2 response elements yields promoters with characteristics of γ_1 genes. Similar results were obtained with a chimeric promoter consisting of the *tk* 5′ untranscribed region fused to the gC (a γ_2 gene) 5′ transcribed noncoding domain (258, 396,722). Homa et al. (258) concluded that the γ_2 response elements resided in the 5′ transcribed noncoding domains. The situation may be far more complex. Thus the U_L24, a γ gene, was expressed when its TATA box was replaced by that of the gC (γ_2) gene but not when it was replaced by the TATA box of the ß-*tk* gene or that of the $\gamma_2 U_S$11 gene (308). Linker-scanning mutations in the gC and gH promoters led to the conclusion that the genomic sites important for the expression of γ_2 genes are the TATA box, the cap site, and nucleotides +30 to +40 relative to the cap site (656). The mutations in the TATA box and cap site affected transcription, whereas the mutation in the 5′ transcribed noncoding domain had little effect on transcription but had a dramatic effect on the amount of reporter protein produced. Although all three mutations reduced the expression of the reporter gene, none altered its temporal regulation.

Clues to the possible function of the γ_2 5′ transcribed noncoding domains emerged from studies of a series of chimeric genes consisting of the gC (γ_2) 5′ untranscribed and transcribed noncoding domains fused to the chick ovalbumin coding sequences and inserted into the viral genome. The key observation is that some genes with deletions within the 5′ transcribed noncoding domain were expressed in the presence of cycloheximide, although the indicator gene was expressed to a higher level late in infection (R. King, M. Arsenakis, A. Poon, and B. Roizman, unpublished data).

At least two hypotheses could explain the observation that deletion mutants in the 5′ transcribed noncoding sequences of a γ_2 gene could be expressed as an α gene. First, the 5′ noncoding domain may contain a site that allows the binding of specific *trans*-acting proteins necessary for the regulation of γ_2 gene expression. A survey of several γ_2 genes of the published HSV-1 DNA sequence (402) failed to reveal a potential response element conserved in all or most γ_2 genes, although the data do not exclude the possibility that the shared response element is highly degenerate. Second, the 5′ transcribed noncoding domain may form a secondary structure which affects expression of the γ_2 genes. Specifically, analyses of the nucleotide sequences with the RNAFOLD program (Genetics Computer Group, Madison, WI) led to the conclusion that the γ_2 5′ transcribed noncoding domains could form long and apparently stable stem-loop structures, although evidence that they actually exist is lacking. Whereas the wild-type sequence is potentially capable of forming a stem 25 bp long, the mutation which confers upon the chimeric gene the capacity to be expressed as an α gene lost the ability to form a stable stem-loop structure (R. King, M. Arsenakis, A. Poon, and B. Roizman, unpublished data).

Irrespective of whether the secondary structure theory is supported by future studies, the available data suggest that the 5′ transcribed noncoding domains contain response elements for a repressor which precludes expression of γ_2 genes under conditions which enable the expression of α and ß genes and an activator which enables their expression in the absence of the repressor (R. King, M. Arsenakis, A. Poon and B. Roizman, unpublished data) (397). It is tempting to speculate that the function of DNA synthesis and of ICP8 is to disable the repressor and that the function of ICP27 relates to the function of the *trans*-activator. The available data also suggest that the repressor is absent or inactive in γ genes embedded in the context of nonviral episomes or cellular chromosomes.

Posttranscriptional Regulation

The evidence for posttranscriptional controls is based on reports that translocation of viral transcripts into the cytoplasm appears to be regulated (298,300,329). Specifically, the genetic complexity of the RNA accumulating in the nuclei of cells infected with HSV in the presence of cycloheximide and maintained in medium containing the drug was greater than that observed in the cytoplasm. In retrospect, the interpretation of the data is not clear. The failure to demonstrate RNA complementary to ß genes (e.g., to the *tk* gene) in nuclei of infected cells treated with cycloheximide (361) suggests that the transcripts accumulating in the nuclei might be random transcripts of the DNA rather than transcripts of specific genes belonging to the ß and γ groups.

The evidence for translational regulation is based on several observations. Specifically, the inhibition of host protein synthesis by structural components of the virion soon after infection (184,458,530) and the inhibition of α gene product synthesis by subsequent gene expression (183,262) are translational events, inasmuch as they occur in physically and chemically enucleated cells. A significant finding to emerge from the studies by Read and Frenkel (530) is that virion structural components exert an inhibitory effect on both host and α protein synthesis, inasmuch as mutants defective in the virion host shut-off function produce more α gene products than their wild-type parents.

Lastly, several studies reported here have ascribed posttranscriptional regulatory functions to ICP27 and to the two protein kinases encoded by U_L13 and U_S3, respectively.

HSV Gene Regulation: The Problem in Experimental Designs

The "gold standard" for the studies of viral gene regulation is the pattern of expression in productive infection of natural or reporter genes contained in the viral genome. Tests of modified *trans*- or *cis*-acting domains of individual genes are easier to perform and may, in some cases, be more meaningful if they can be done in the environment of the cell and in the presence of only a minimal amount of viral genetic information. However, the validity of such tests hinges on the extent to which they reproduce the regulation of the gene embedded in the viral genome and expressed in the course of viral infection. The expression of isolated α genes, in biochemically transformed cells or in transient expression systems, appears to mimic to some extent the regulation of the corresponding genes contained in viral genomes during productive infections (498). Notwithstanding the massive number of transfections which argue that ICP0 is a promiscuous *trans*-activator, supporting evidence from studies on deletion mutants in the α0 gene is not readily available. The transfection system apparently fails if more than two components of the regulatory pathway are introduced into the cell simultaneously, for example, the cotransfection of αTIF, α4, and the intended target gene of α4 (542). In the case of γ_2 genes, the transient expression system yields totally false results; viral genes permanently integrated in cellular genomes or transiently expressed after transfection are regulated as ß genes (14,47,625). The transfection system has given rise to a veritable cottage industry, but the data generated by it are not totally reliable. What is the evidence that viral genes other than those carrying the α *cis*-acting sites can be regulated in that system in a mode which resembles viral gene regulation? If γ_2 genes are regulated as ß genes, what is then the evidence that ß genes in transfected cells are regulated as bona fide ß genes?

FATE OF THE INFECTED CELL

Cells productively infected with herpesviruses do not survive. Almost from the beginning of the reproductive cycle, the infected cells undergo major structural and biochemical alterations that ultimately result in their destruction.

Structural Alterations

Changes in Host Chromatin

As described in detail elsewhere (557) and shown in Fig. 4, one of the earliest manifestations of productive infection is in the nucleolus; it becomes enlarged, displaced toward the nuclear membrane, and ultimately disaggregates or fragments. Concurrently, host chromosomes become marginated, and later in infection the nucleus becomes distorted and multilobed. The numerous protrusions and distortions have in the past been mistaken for amitotic division (302,602). Margination of the chromosomes may or may not be linked with the chromosome breakage reported by numerous investigators (557).

Virus-Induced Alteration of Cellular Membranes

Duplication and Folding of Intracellular Membranes

Changes in the appearance of cellular membranes and, in particular, of nuclear membranes is characteristic of cells late in infection. Deposition of material (tegument proteins?) on the inner surface facing the nucleoplasm or cytoplasm, but not in the space between inner and outer lamella or cisternae of the endoplasmic reticulum, results in the formation of thickened patches along the membranes. Ultimately, the patches in the nuclear membrane coalesce and fold upon themselves to give the impression of reduplicated membranes (Fig. 6) (111,166,356,401,439,454,500, 601,624,743).

Insertion of Viral Proteins into Cellular Membranes

The first inkling that herpesviruses modify cellular membranes was based on the observations that mutants differ from wild-type strains with respect to their effects on cells; while wild-type viruses usually cause cells to round up and clump together, some mutants cause cells to fuse into polykaryocytes (158,552). These observations led to the prediction that herpesviruses alter the structure and antigenicity of cellular membranes—a prediction fulfilled by the demonstration of altered structure and antigenic specificity (541,564,563,569) and the presence of viral glycoproteins in the cytoplasmic and plasma membranes of infected cells (243,567,568,642).

Polykaryocytosis

Both HSV-1 and HSV-2 cause infected cells to round up and cling to each other. Some viral mutants cause cells to fuse into polykaryocytes; this fusion may be cell type specific or cell type independent (158,252,582). Polykaryocytosis has been studied for several reasons: as a probe in the structure and function of cellular membranes, reflected in the "social behavior of cells;" as a tool for analyses of the functions of viral membrane proteins; and as a model of the initial interaction between HSV and susceptible cells that results in the fusion of the viral envelope and the plasma membrane (74,552,639,640). Cell fusion induced by HSV also requires conditions which favor processing of high-mannose glycans to complex glycans, but in this instance it is not clear whether complex glycans must be present only in viral glycoproteins present on the surface of the infected recruiter cells or on both the recruiter cells and on the uninfected cells to be recruited into polykaryocytes (74,639,640).

Polykaryocytosis can be viewed as an aberrant manifestation of the interaction of altered membrane domains of infected cells and unaltered membranes of juxtaposed cells (552). Genetic analyses have shown that mutations (*syn*) which confer the capacity to fuse cells map in at least five and possibly more loci within the viral genome (22,54, 132,367,386,491,493,576,582,748). These loci map within the domains of gB, gK, gL, U_L24, and U_L20. One interpretation of this observation is that the membrane proteins form complexes whose structure and conformation become altered by mutations in any of the component polypeptides and that the changes in the conformation are similar to those which occur in the envelopes of virions interacting with the plasma membrane (582).

Host Macromolecular Metabolism

Background

A characteristic of herpesvirus-infected cells is the rapid shut off of host macromolecular metabolism early in infection. Thus, host DNA synthesis is shut off (562), host protein synthesis declines very rapidly (554,674,676), and glycosylation of host proteins ceases (642).

Herpes simplex virus-induced host shut off occurs in two stages. The first stage, documented initially by Fenwick and Walker (184) and by Nishioka and Silverstein (456–458) involves structural proteins of the virus and does not require *de novo* protein synthesis. Thus, HSV shuts off host protein synthesis in physically or in chemically enucleated cells (183); the shut off was effected by density gradient-purified virus but not by purified virus inactivated by heating or neutralization with antibody. The shut off is faster and more effective in HSV-2- than in HSV-1-infected cells; this observation permitted the initial mapping of the genetic locus that confers upon HSV-1 × HSV-2 recombinants the accelerated shut off characteristic of HSV-2 (181). Isolation of *vhs* mutants which fail to shut off host polypeptide synthesis in HSV-infected cells (530) has demonstrated more conclusively that this function is due to a virally encoded protein.

The second stage requires *de novo* synthesis of proteins after infection (183,261,262,456,457,627). The shut off coincides with the onset of synthesis of ß proteins, but the experimental results do not exclude the possibility that the shut off is caused by γ rather than ß gene products.

Expression of the vhs *Gene and the Function of Its Product*

The *vhs* function was initially mapped to 0.52 to 0.59 by (181). Isolation of a mutant defective in this function (530) allowed further mapping of the gene responsible. Mapping studies by Oroskar and Read (469) and Kwong et al. (342) have identified sequences mapping from 0.604 to 0.606 on the viral genome (U_L41 open reading frame) as being responsible for the *vhs*⁻ phenotype of the mutants. U_L41 RNA is expressed as a γ_1 gene (200). The products of U_L41 are an abundant phosphoprotein of M_r 58,000 and a less abundant, more extensively phosphorylated protein of apparent M_r of 59,500. Only the faster migrating protein is found in virions (531). It has been reported that the U_L41 protein forms a complex with VP16 (630).

Early studies showed that virion components were responsible for destabilization and degradation of host mRNA (184). Further studies have led to the conclusion that the virion component responsible for both mRNA destabilization and degradation, *vhs*, is also responsible for a nondiscriminatory destabilization and degradation of viral α, ß, and γ mRNAs (182,342,469,595,670).

In cells infected with the *vhs*⁻ mutant, host protein synthesis is not shut off at least early in infection; α and ß protein synthesis are somewhat prolonged compared to wild type. Both of these effects have been shown to be due to a stabilization of host and viral mRNAs; in cells infected by *vhs*⁻ mutants, mRNAs are not degraded as rapidly as in cells infected by wild-type virus.

This function confers at least two advantages on the virus. First, it removes preexisting host mRNA from the pool of translatable messages, allowing the viral mRNAs to take over the pool rapidly. Second, destabilization of viral mRNAs allows a rapid transition from one regulatory class to the next. In the absence of the *vhs* function, α and ß proteins are produced beyond the time spans normally seen; the positive transcriptional controls discussed in the rest of this paper are not enough to ensure efficient α to ß and ß to γ transitions. Although the *vhs*⁻ mutation is not lethal, wild-type virus does have a growth advantage in tissue culture, indicating that efficient separation of the regulatory classes is helpful to the virus.

VIRULENCE

Virulence, if defined as the ability of the virus to cause disease, is composed of several parts. During infection of humans, viral disease includes primary and recurrent epithelial lesions as well as disseminated disease and encephalitis. In studies on the molecular basis of disease induced by HSV, the end point of the research objective—the disease—is often taken to be synonymous with the destruction of central nervous system (CNS) tissue. In healthy nonimmunocompromised humans, encephalitis occurs rarely but with catastrophic results (727). In experimental animals, it is frequently a major component of the disease. Neurogrowth, as measured solely by intracerebral inoculation of virus, is the most commonly measured aspect of virulence. Direct injection of virus into the CNS measures the capacity of the virus to destroy an amount of CNS tissue that will result in death before the immune system blocks further virus spread. Because in most instances destruction of the CNS and death are related to virus multiplication, in quantitative terms, the growth of the virus in the CNS is measured in terms of the amount of virus required to reach a specific level of mortality (50% of inoculated animals).

A second attribute of virulence is invasiveness—the capacity to reach a target organ from the portal of entry. To disseminate to the target organ, it may be necessary for the virus to multiply at peripheral sites. In experimental systems, virulence is composed of (a) peripheral multiplication, (b) invasion of the CNS, and (c) growth in the CNS. Peripheral growth and invasiveness into the CNS can be quantified by measuring the amount of virus recovered at the peripheral site and in the CNS as a function of the quantity of virus inoculated at a peripheral site, i.e. footpad, eye, ear, and so forth.

Several types of studies have been done to define genes required for neurovirulence or neuroinvasiveness. In the first, the doses lethal to 50% of inoculated animals (LD_{50} values) of wild-type isolates taken either from CNS tissues of encephalitic patients or from facial or genital isolates of the same patients or from normal individuals have been compared in mice inoculated by intracerebral or peripheral routes. Although the encephalitis isolates appear to have a slightly lower average LD_{50} as compared to nonencephalitis strains, no correlation was apparent between encephalitis in humans and neurogrowth or neuroinvasiveness in experimental animals (R. J. Whitley, personal communication). However, as a variant of this approach, infection of neurons in culture has indicated that virus isolates from encephalitis cases are better able to infect and be transported to the cell body than isolates from peripheral lesions (40).

Numerous studies have been undertaken to assess the relative abilities of various viral mutants to grow in the CNS after direct intracerebral inoculation. What has emerged from the accumulated data is that so-called neurovirulence factors are not the exception, but the rule. In almost all virus mutants tested to date, including a number of HSV-1 × HSV-2 recombinants, many ts viruses, a variety of spontaneous mutants, and a myriad of recombinant viruses with deletions in one or more genes, LD_{50} values following intracerebral inoculation ranged from 100- to 100,000-fold higher than those of the parental strains. Experiments utilizing nongenetically engineered viruses mapped loci implicated in "neurovirulence" to a number of regions, including the *tk* gene and sequences around the right terminus of the L component (82,185,282–284,322, 511,577,652,689,690). Recombinant viruses with elevated intracerebral LD_{50} values include those with deletions in the genes encoding ICP22, U_S2, U_S3 protein kinase, U_L13, U_L16, U_L24, gG, gJ, gE, gI, ribonucleotide reductase, TK, $\gamma_1 34.5$, U_L55 and U_L56 (E. Kern, B. Meignier, J. Baines, R. J. Whitley, A. Sears, and B. Roizman, unpublished results) (68,421,422,603,718,728). Most of the attenuated deletion mutants listed above also exhibit impaired growth in peripheral tissues (e.g., cornea), as assayed either by the occurrence of epithelial lesions or by quantitation of infectious virus in the tissues, indicating that the genes deleted are required generally for growth in differentiated and/or polarized cells, not specifically for growth in neurons or in CNS tissues (A. Sears, B. Meignier, and B. Roizman, unpublished results) (57,279,280). In fact, the only genes that have, to date, been shown not to be required for neurovirulence in the murine model are gC, U_S9, U_S10, U_S11, and $\alpha 47$ (R. J. Roller and B. Roizman unpublished data) (459). Since the function of $\alpha 47$ appears to be human cell specific (754), it is conceivable that, if the U_S9-11 gene cluster shares in this property, their function will be inapparent in the mouse.

The animal models have thus so far failed to identify any viral function uniquely required for neurogrowth.

LATENCY

The ability of HSV to remain latent in the human host for its lifetime is the unique and intellectually most challenging aspect of its biology. The virus enters sensory nerves innervating the cells of mucosal membranes (Fig. 12). In latently infected neurons, the viral genome acquires the characteristics of endless or circular DNA (195,424,543, 544). To our knowledge, the virus expresses no functions which are required for the establishment or maintenance of the latent state. In a fraction of neurons harboring latent HSV, the virus is periodically reactivated; infectious virus is carried back to peripheral tissues by axonal transport (110), usually to cells at or near the site of initial infection (81,122,222,551). Depending on the host immune response, the resulting lesion may vary considerably in severity, from barely visible vesicles to rather severe, debilitating lesions in immunosuppressed individuals. The clinical aspects of latent infection and reactivation are discussed in Chapter 73 in *Fields Virology*, 3rd ed. This section concerns the molecular biology of latency.

HSV Latency in Experimental Systems

Experimental Models

The most useful model systems are mice, guinea pigs, and rabbits. In the mouse, latent infection is readily established after eye, footpad, or ear inoculation, but the rate of spontaneous reactivation is extremely low (50,51,248,249, 662). Latent virus in the rabbit does reactivate spontaneously (450). The guinea pig shows recurrent lesions after vaginal infection with high doses of HSV-2 (651).

At the other extreme are latency models in cells cultured *in vitro*. Neurons are nonpermissive at the time they harbor the virus in the latent state, and in the ganglia their permissivity is transient. When placed in culture, neurons become permissive. Those that contain latent virus activate its multiplication. It has been reported that neurons retain virus in a latent state in the presence of nerve growth factor and, conversely, that the virus is activated when the growth factor is withdrawn (734). In this system, only latency-associated transcripts (LAT) are detectable (146). A protein encoded by an open reading frame within the stable LAT intron has been reported (147), but the detected

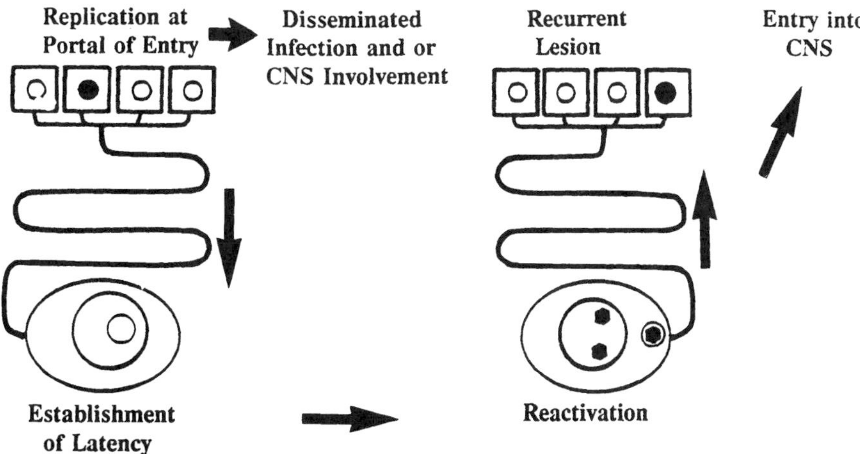

FIG. 12. Natural course of HSV infection *in vivo*. Virus first replicates in epithelial cells (*squares*) at the portal of entry and then moves through neurites (*curved lines*) to establish latent infections in neurons (*ovals*).

protein is several times the predicted size, and the report has never been confirmed. Attempts to detect proteins encoded by several of the major open reading frames 3′ of the $\gamma_1$34.5 gene but within the domain of the 8.3-kb LAT have been unsuccessful (346).

A number of laboratories have reported the maintenance of the viral genome by rendering cells nonpermissive by a variety of methods (106,107,578,729–733). To our jaundiced eye, the state of nonpermissivity induced in cells in culture by elevated temperatures, interferon, or antiviral drugs is not equivalent to the nonpermissive state of the neuronal cells *in vitro*.

The events transpiring in animal models can be divided into several stages. After infection with wild-type viruses, virus replication ensues in the tissues at or near the site of inoculation. Under normal conditions, this initial multiplication probably ensures contact with and entry into the sensory nerve endings. In animal models, perhaps because of the huge inoculating doses normally used, this step is not required; even mutants totally deficient in replication capability are able to establish latent infections following corneal inoculation (306). The capsid is transported to the neuronal nucleus (331,378). Studies on neurons cultured *in vitro* indicate that the viral capsids are transported by retrograde axonal transport involving microtubules; drugs that disrupt neuronal microtubule structures, or that are known to inhibit retrograde transport of certain compounds, also inhibit the ability of the virus to move from the peripheral endings of neurons to the nuclei (331). Electron microscopic studies indicate that, in neurons infected in cell culture, the viral particle being transported is the unenveloped capsid (378).

In animal models, there is a short period of viral replication in the ganglia at this stage (312,314–317,407,408, 512,513,698,697,711,739). It is not known whether this also occurs in humans or, as seems more likely, is an artifact due to the animal system, route of inoculation, induction of host *trans*-activating factors, or the large amount of virus used in the inoculum to attain a high percentage of latently infected ganglia. For example, corneal scarification, discussed below, has been reported to induce a number of host *trans*-activating factors and viral genes (ICP4, VP5) in neurons harboring latent virus (699).

In the second stage, at most 2 to 4 weeks after inoculation, no replicating virus can be detected in the sensory ganglia innervating the site of inoculation.

In the last stage, certain stimuli (e.g., physical or emotional stress, peripheral tissue damage or intake of certain hormones in humans, and both peripheral tissue damage and administration of drugs that stimulate prostaglandin synthesis in experimental animals), may result in activation of virus multiplication concurrent with axonal transport of the virus progeny, usually to a site at or near the portal of entry. Although the issue is still being debated, the sum of all of the data currently available strongly suggest that virus multiplication results in destruction of the neuronal cell.

Role of Viral Multiplication at the Portal of Entry in the Establishment of the Latent State

There are several important facets of latent infections which relate to the role of virus multiplication, both at the portal of entry and in the neurons harboring the virus.

(i) As noted above, HSV must have access to the nerve endings in order to establish latency, and therefore it could be expected that the greater the number of peripheral cells that become infected and support virus multiplication is, the larger the number of neurons which will harbor latent virus. The relevant phenomenon in humans is that the frequency of reactivations resulting in recrudescences of lesions is related to the severity of lesions caused by the first infection. In the model we have proposed below, the frequency of recurrences would be determined in part by the number of neurons harboring virus. However, in animal models, there is no absolute requirement for peripheral replication. Viruses lacking the essential α4 and α27 genes, totally unable to replicate in any cell *in vivo,* were shown to be able to establish and maintain latent infections (304,609).

(ii) Many years ago, it was proposed that the latent virus makes a "round trip," i.e., the reactivated virus reestablished the latent state by infecting the nerve endings of hitherto uninfected neurons (313). Currently, available evidence does not support this hypothesis. First, it is very difficult to superinfect in experimental systems ganglia harboring latent virus with a second, marked virus (423). Perhaps even more significant, the "round trip" does not appear to take place in humans even under conditions which would favor such a phenomenon. Thus, in a small number of individuals, mutations which were virulent and acyclovir resistant have arisen (164,477). Recurrent lesions which emerged after the mutant was eliminated with the aid of other drugs did not contain the acyclovir-resistant virus. While this phenomenon has been attributed to rapid elimination of the peripheral infected cells by the immune system (82), the observations may have more profound implications, inasmuch as induction of latent virus should eliminate it from the ganglion.

Viral Gene Expression in Latently Infected Neurons

Extensive studies on ganglia harboring latent HSV have been rewarded by an extreme paucity of evidence for viral gene expression. The only transcript detected to date is one designated optimistically as the LAT (663). This transcript is abundant and accumulates in the nuclei of neurons of latently infected animals and humans (330,545,659,663, 664). The transcript, originally identified as an abundant, nuclear RNA (664), is now believed to be a stable intron (178), approximately 2 kb in length, derived from a primary transcript of approximately 8.3 kb (434,757). The intron itself can be spliced (708,719). The LAT RNAs are conserved in a very similar form in HSV-2 (433). It is difficult to believe that these viruses have conserved throughout evolution the expression of a transcript with no function. However, the function of LAT during latency remains elusive. Confounding the issue are the accumulated masses of data indicating that LAT is not absolutely required for the establishment or maintenance of latency (251,608, 658) or for reactivation of latent virus (48,250,357,696). Experiments must, therefore, be designed to measure small differences in reactivation rates or frequencies, which can be difficult to reproduce. There have been conflicting reports on whether expression of LAT during latency affects reactivation (40,250,357,696); however, in the absence of data on the number of neurons actually harboring latent virus, it is difficult to determine whether this decrease is due to a true defect in reactivation or to an earlier defect in peripheral replication or establishment of latency.

Because it runs in part antisense to the 3′ terminus of the α0 gene, it has been postulated that the function of LAT is to preclude the expression of α0 (664). In this case, however, it would seem that the absence of LAT would allow expression of α0 mRNA, and this does not appear to be so. It has also been postulated that either the primary, polyadenylated transcript or the stable intron may encode a protein and that the function of LAT, whatever that may be, is carried out by this protein. To date, numerous attempts by several laboratories to identify a protein encoded by LAT and expressed during latency *in vivo* have failed. As noted above, antibodies to a peptide homologous to one open reading frame within the nonpolyadenylated intron sequence have been shown to react with a protein expressed during nonproductive infection of sensory neurons in culture; in this case, however, the reactivity of the antiserum with the sensory neurons *in situ* has not been reported (147).

Confounding a total rejection of viral functions expressed for the establishment or maintenance of the latent state is the recent discovery of an open reading frame antisense to $\gamma_1$34.5 and which expresses a protein (346). The transcript was detected late in infection and in nuclear run-off experiments, but not in cells treated with cycloheximide from the time after infection (53,751). However, tagged ORF P protein was detected in cells infected and maintained at 37°C in the present of cycloheximide and then allowed to make proteins in the presence of actinomycin D (346a). The transcript or the protein it encodes has been detected early in infection in cells infected with α4$^-$ virus (751) or at nonpermissive temperature with mutants in the α4 gene (346). Although attempts to detected the transcript in latently infected neurons have been unsuccessful (751), the results remain inconclusive since both the transcript and the protein are made in very small amounts, and Yeh and Schaffer (751) failed to detect the 8.3-kbp LAT which contains the sequences contained in the ORF P transcript. An added complication is that ORF P and the $\gamma_1$34.5 open reading frame are virtually superimposable, albeit antisense to each other. At least with respect to HSV-1, mutations in the $\gamma_1$34.5 gene result in a decreased incidence of latency and reactivation in the murine model (728).

One of the central issues of HSV latency is why there is no transcription in sensory neurons of promoters expressed in a wide variety of other cell types, particularly the α genes. As discussed in the section on regulation, transcription of α genes is induced by a combination of host and viral factors. It is conceivable then that the lack of α gene expression in sensory ganglia could be due to (a) a lack of the viral α *trans*-inducing factor, (b) a lack of host factor, (e.g., Oct-1) required for αTIF function, or (c) a direct repression of the α gene promoters, as opposed to a lack of induction. The first possibility was tested using a recombinant virus that contained an insertion of a second copy of the αTIF gene under the control of the mouse metallothionein promoter (604). Despite expression of the chimeric αTIF gene in latently infected sensory neurons, latent infection continued. Furthermore, expression of the same chimeric gene in transgenic mice also failed to prevent establishment or maintenance of latency. *In situ* hybridization with probes specific for Oct-1 demonstrated detectable levels of Oct-1 mRNA in the sensory neurons of the MT-αTIF transgenics (V. Hukkannen and B. Roizman, unpublished data), indicating that the presence of both αTIF and Oct-1 was insufficient to prevent latent infection. Similar studies on the α0 and α4 genes have also failed to block the establishment of latency (R. Fawl, V. Hukkannen, and B. Roizman, in preparation).

Another series of experiments has shown that corneal scarification, commonly used in inoculation, causes induction of transcription of several viral genes and host *trans*-activating factors (Oct-1, c-*jun,* and c-*fos*) (699). At this point, a cause-and-effect relationship between the upregulation of one of these factors and the reactivation of latent virus has not been defined, but is an intriguing possibility.

Copy Number of Viral DNA in Latently Infected Neurons

In a different category from LAT is the observation that, in trigeminal ganglia harboring latent virus, there are between 0.1 and 1 viral genome equivalents per cell genome (65,515,543,544). This datum poses an intriguing question. Heretofore, the number of neurons harboring virus was thought to be between 0.1% and 3% of the total neurons. A recent report by Rodahl and Stevens (546) indicates that, under some circumstances, the number can be higher; however, neurons account for only about 10% of the total cells in a sensory ganglion. Unless viral genomes are contained in every single cell in a ganglion, including glial cells, it is obvious that, as we first calculated a number of years ago (566), each latently infected neuron must contain more than one viral genome.

Two series of experiments indicate that, as we have previously predicted (566), replication of viral DNA by viral enzymes is not necessary for the establishment of latent infections and is not responsible for the attainment of the high copy number. In these experiments (629,645,646), mice were infected by the footpad route, and assays of LAT, lytic antigens, and DNA copy number were done. Although there was a correlation between levels of lytic replication and the level of viral DNA per LAT+ neuron, even those ganglia which showed little or no signs of viral antigen expression still had DNA copy numbers of greater than 20 viral genomes per LAT+ neuron. To account for the high number of viral genomes per cell harboring latent virus, it is then necessary to postulate that (i) more than one viral genome can enter a single neuron during the establishment of latency or (ii) viral genomes are amplified by the cellular machinery during latency (566). In support of the second of these hypotheses, a host-dependent origin of DNA replication embedded within the viral genome has been identified (607).

Viral Gene Expression Required for the Maintenance of the Latent State

To date, no gene or sequence tested, including several essential for lytic viral replication, has been identified as being essential for maintenance of latency. These include all of the genes discussed above under virulence, as well as α0, α4, α27 (358), α47, U_S9-11 (421,459), and U_L24 (606).

Reactivation of Virus from Latent State

The original operational definition of latent virus was dependent on reactivation, i.e., that virus which could be detected after incubation of intact ganglionic tissue with suitable susceptible cells, but not by inoculation of the susceptible cells with macerated ganglia (662). This definition obviously also included the ability to reactivate and replicate to detectable levels in the ganglia. More recently, the definition of latency has come to be extended to include viruses that can be detected in sensory ganglia several weeks after infection by *in situ* hybridization with probes for LAT or by assays of viral DNA in the ganglia (104,304,358,359,379,688). This has allowed assessment of the roles of specific genes in the establishment and maintenance of latency, separate from their roles in reactivation. However, the evidence that reactivatible latent virus can be found under some circumstances in the absence of detectable LAT (546) indicates that these assays must be interpreted cautiously. At the present time, the only technique for proving that a virus is incapable of establishing latent infections is the inability to detect viral DNA by PCR assays of DNA extracted from whole ganglia. It seems likely that a new standard will be the identification of neurons containing viral DNA by *in situ* PCR (165,654).

In humans, latent virus is reactivated after local stimuli such as injury to tissues innervated by neurons harboring latent virus or after systemic stimuli such as physical or emotional stress, menstruation, hormonal imbalance, and so forth which may reactivate virus simultaneously in neurons of diverse ganglia (e.g., trigeminal and sacral). In experimental systems, multiplication of latent virus has been induced by physical trauma to tissues innervated by the neurons harboring virus (13,233), by iontophoresis of epinephrine (339,340) or other drugs (223,234,623), by transient hyperthermia (592,593), and by corneal scarification (699). Recent studies showed that injection of cadmium resulted in reactivation of a HSV-1 in a fraction of neurons harboring latent virus and that this reactivation was not associated with induction of the metallothionein genes, inasmuch as other injections of other heavy metal solutions at much higher amounts failed to induce viral reactivation (179). The hypothesis advanced to explain these results is that cadmium specifically inactivates a repressor.

The molecular basis of reactivation and the order in which viral genes are induced is not known. Attempts to identify genes specifically required for reactivation have implicated two genes specifically in reactivation events, i.e., ICP0 and LAT, although in both cases conflicting data have been published (48,101,224,248,357,358,696). However, as outlined above in the case of LAT, the absence of data on the number of latently infected neurons from which the virus is being reactivated makes these data difficult to assess. Particularly in the case of ICP0 deletion mutants, the attenuation of virus growth in all tissues makes it very hard to say that the gene plays a reactivation-specific role. Virtually all the recombinant viruses described in the Virulence section of this chapter that are attenuated for growth in peripheral and CNS tissues also exhibit reduced reactivation rates by explant cultivation of latently infected sensory ganglia; this makes it difficult to attribute any reduced reactivation frequency to a reactivation-specific role.

Modeling the Latent State

The molecular basis of latency rests on answers to several key questions. First, since HSV readily multiplies in a variety of cells derived from human or animal tissues, why does lytic infection not ensue in neurons harboring latent virus? Second, at what stage in the reproductive cycle is viral multiplication arrested? Third, what is the origin of the high copy number of the viral genome in latently infected neurons? Fourth, why does the virus not reactivate from all neurons at the same time? Fifth, at what stage in the cascade of viral gene expression does replication of reactivating virus begin? Sixth, why is HSV-2 more readily reactivated from sacral ganglia, whereas HSV-1 is more readily activated from trigeminal or cervical ganglia?

Our model, presented largely for its heuristic value, is based primarily on the hypothesis that latency is required for the perpetuation of the virus in its natural host population and that the virus has evolved elaborate mechanisms to control the latent state. It is a further modification of a model that we have proposed in the past (58,565) and consists of several components as follows.

1. Inasmuch as expression of αTIF plus Oct-1 in latently infected neurons was not able to induce expression of lytic replication, it appears that the lack of expression of α genes must be due to either repression, a lack of other host proteins involved in α gene transcription, or both. Repression could be caused by a viral or host-encoded protein binding to some sequence in the vicinity of α gene promoters, or it could be due to a DNA conformation effect (the two are not mutually exclusive). Either of these modes of repression could potentially be relieved by replication of the viral genome. Obviously, a requirement for DNA replication for expression of HSV genes is not unique to this situation but is directly analogous to the regulation of γ genes.
2. The model also proposes that activation of virus multiplication is the consequence of the cumulative effect of stimuli to which each cell harboring latent virus responds independently. Specifically, the hypothesis envisions that both local and systemic stimuli cause the latent viral genome to be replicated by host enzymes, utilizing a previously mapped host origin of DNA replication present in the viral DNA (607), thereby causing the copy number to be increased. DNA replication could overcome repression by either diluting a repressor (a copy number effect) or changing the conformation of replicating DNA. As the stimuli encountered by each neuron would be different, the copy number of viral DNA in each latently infected cell would increase independently.
3. The increase in the DNA copy number may not be sufficient to ablate the block in virus multiplication. A second function, e.g., induction of a host transcription factor, could also be necessary. The combination of requirements for both release of repression (by DNA amplification) and addition of a positive factor could explain why latently infected neurons reactivate individually, rather than as a group.
4. Additional requirements for viral multiplication during reactivation fall under the heading of capacity for gene expression and are poorly defined. A common feature of this property is seen in the case of deletion mutants (e.g., α22⁻) (603) infecting nonpermissive or restrictive cells. Such mutants often exhibit a multiplicity dependence reflected in the failure of virus multiplication at low, but not at high, multiplicities of infection. The additional functions that may be required to achieve clinically detectable reactivation (or detectable amounts of infectious virus in experimental systems) may be those of a set of viral genes. The model proposes that (a) this set would include genes that are required as well as those that are dispensable for multiplication in cells in culture and (b) the expression rate and product abundance of these genes would determine whether infection is productive or abortive in the particular cell in which the virus is latent.

A similar model of HSV latency and reactivation has recently been proposed by Kosz-Vnenchak et al. (326). These authors have also proposed that DNA replication may be a critical stimulus during reactivation of latent virus. However, they propose that the replication is taking place via the normal viral enzymes, expressed as β genes. In their experiments, mice were infected with wild-type virus or a *tk*$^-$ viral mutant, and after the establishment of latency ganglia were explanted and incubated for 48 hr. *In situ* hybridization indicated that expression of viral α and ß genes was much lower in ganglia from mice infected with the *tk*$^-$ virus than in those containing wild-type virus. The authors have proposed that the ß genes involved in DNA replication, including *tk*$^+$, are required for the initial stages of reactivation by increasing the expression of α genes to a level sufficient to stimulate lytic infection. In this case, the high DNA copy number during latency is accounted for by sporadic low-level expression of the viral replication enzymes. The significance of this model stems from the authors' conclusions that DNA replication somehow stimulates α gene expression and further that the mechanisms of gene expression defined in cells in culture do not hold true for sensory neurons. In these experiments, very few sections were examined for each ganglion, and the number of LAT$^+$ neurons per ganglia were not assayed prior to explant, leaving open the possibility that the wild-type virus had spread within the ganglia and that the positive signals were coming not only from cells that had initially contained latent virus. The γ gene products in wild-type infected ganglia were not detectable under these conditions until 48 hr postexplant; there may have been a sensitivity problem in these studies in that experiments by a different group of investigators have shown that, following explant of wild-type latently infected ganglia, transcripts of all classes of genes (α, ß, and γ) can be detected by PCR analyses as early as 4 hr postexplant (143). This very quick induction of viral gene expression following explant indicates that, by 48 hr, considerable virus spread within the ganglion may have occurred.

SIGNIFICANCE OF THE NUCLEOTIDE DIVERGENCE OF HSV-1 AND HSV-2

It can be argued that the differences in the nucleotide sequences of HSV-1 and HSV-2 predicts appropriate differences in the cells lining the oral and genital mucous membranes. The data described above for gC-1 and gC-2 support this hypothesis. In addition, the differences in rates of reactivation of HSV-1 and HSV-2 from trigeminal and sacral dorsal root ganglia may also reflect differences in viral functions required for lytic growth in each cell type.

VIRAL MODULATION OF HOST DEFENSE MECHANISMS

Like other large DNA viruses, HSV appears to have evolved a variety of mechanisms of evasion of the host defense machinery. They include genes that prevent apoptosis in differentiated cells, that interact with antibody and complement, and that prevent induction of CD8$^+$ cytotoxic T cells.

The association of the $\gamma_1$34.5 gene with neurogrowth is an excellent example. As noted above, several laboratories have associated loss of the capacity to cause death by intracerebral inoculation of mice with mutations in sequences in the right end of the L component (282,283,577,681,682, 689,690). Independently of these studies, a gene mapping in the repeats of the L component and designated as $\gamma_1$34.5 (5,93,95,97) was found to play a major role in the ability of the virus to replicate in the CNS, inasmuch as mutants lacking a functional $\gamma_1$34.5 gene exhibited $>10^7$ pfu/LD_{50} by intracerebral inoculation even though the virus is capable of limited replication on intravaginal inoculation of mice (728). Because of these results, further studies were carried out in a variety of cell lines with the following results.

1. The $\gamma_1$34.5 gene is dispensable in some cell lines (e.g., Vero) and the ability of the mutant virus to replicate cannot be differentiated from that of the wild-type virus (5,91,97). In the human neuroblastoma cell line SK-N-SH and in other cell lines of neuronal origin, deletion mutants and stop codon mutants in the $\gamma_1$34.5 gene trigger a premature total shut off of all protein synthesis, thereby rendering the cell nonviable and drastically reducing viral yields (94). The stress response which leads to the termination of synthesis is triggered by an event associated with viral DNA synthesis, inasmuch as exposure of cells to PAA, an inhibitor of viral DNA synthesis, precludes the premature termination of translation (94). It is noteworthy that, although protein synthesis is shut off, viral DNA accumulates to near-normal levels, and the presence of γ_2 mRNA late in infection argues against a role of the viral *vhs* gene in this process.
2. The HSV-1(F) 263-amino acid $\gamma_1$34.5 protein consists of a 159-amino terminal domain, ten repeats of the amino acids AlaThrPro, and a 74-amino acid carboxy-terminal domain (95). The repeat varies from strain to strain (five to ten repeats) and may be considered a linker sequence (95). Cellular homologs identified to date include MyD116, a 657-amino acid protein expressed in myeloid leukemia cells induced to differentiate by interleukin-6, and GADD34, a protein induced by growth arrest and DNA damage. These proteins also contain a long amino-terminal domain separated from a shorter carboxy-terminal domain by a repeat—in these instances, a 38-amino acid sequence repeated 4.5 times. The carboxyl terminus of $\gamma_1$34.5 is homologous to those of Myd116 and GADD34. Analyses of viral mutants from which various domains of the $\gamma_1$34.5 gene had been deleted or rendered mute by the insertion of a stop codon have shown that the carboxyl terminus of $\gamma_1$34.5 protein, i.e., the sequence shared with Myd116 and GADD34, is required in order to preclude the stress response which leads to total shut off of protein synthesis (93).
3. A surprising finding was the discovery that in human foreskin fibroblasts, the shut off of protein synthesis was triggered by a deletion mutant lacking 1 kbp of

coding sequence but not by the stop codon mutants in human foreskin fibroblasts. This observation dispelled the notion that the $\gamma_1 34.5$ was required exclusively in neurons but raises the issue as to why stop codon mutants behave like null mutants in human neuroblastoma cells but not in human foreskin fibroblasts. One clue is based on the observation that very small amounts of $\gamma_1 34.5$ are required to block the stress response which leads to shut off of protein synthesis. It is conceivable that there is a low-level suppression of the stop codon in human fibroblasts or that the amounts required to block the stress response are lower in fibroblasts than in the neuroblastoma cells (92).

The studies on the $\gamma_1 34.5$ gene mutants indicate that at least some host cells respond to the stress of HSV infection triggered by the onset of viral DNA synthesis. This response results in complete shut off of protein synthesis and, invariably, from the point of view of viral infection, a premature death of the infected cell. It is an important response, inasmuch as it curtails the ability of the cell to support viral replication and spread throughout the body. The inability to prevent the cellular stress response which causes the infected cell to die may be responsible for the inability of the deletion mutant to multiply and cause pathology in the CNS of mice. The virus has evolved a gene, possibly borrowed in part from eukaryotic cells, to preclude this response in order to replicate in a variety of infected cells.

Herpes simplex virus type 1 expresses two types of IgG Fc receptors on the cell surface (34,152,473,474). One, a complex of gE and gI (288), binds monomeric IgG, while the other, consisting of gE alone, binds polymeric IgG (152). The Fc receptors appear to play a role in protecting infected cells or virions from the host antibody response. Cells in culture infected with wild-type virus are relatively resistant to antibody-dependent cellular cytotoxicity as compared to cells infected with a gE deletion mutant (152). Because of the multifunctional nature of gE and gI, studies to determine the role of this protection in pathogenesis will have to await specific mutants that eliminate the Fc binding functions without altering their functions in the viral entry into and egress from infected cells.

The second component of the immune system known to interact with HSV proteins is complement. The complement component C3b has been shown to bind to both gC1 and gC2 (420,610). Mapping of the domains of the viral glycoproteins involved in binding C3b has indicated that several domains are necessary, including one that has some homology to the cellular C3b receptor, CR1 (269). Here again, because of the role of gC in attachment to some cell surfaces, specific mutations that do not affect attachment to polarized cells must be made to determine whether gC fulfills its predicted role in protection of the virus from the host complement system *in vivo.*

Herpes simplex virus also appears to protect itself from a strong host $CD8^+$ cytotoxic T-cell response. It has been known for a number of years that, in humans, epithelial lesions caused by HSV contain a disproportionate number of $CD4^+$ and very few $CD8^+$ T cells (120,596). Experiments in which HSV-infected human fibroblasts were exposed to human anti-HSV cytotoxic T lymphocytes (CTL) indicated that infection actually caused the fibroblasts to become resistant to lysis (497). This resistance phenotype was mapped to the right end of the S component (80). Later studies have shown that expression of sequences containing the α47 gene causes a resistance to lysis by $CD8^+$ CTLs, due to retention of the MHC class I molecules in the cytoplasm and a lack of peptide presentation on the cell surface (754). ICP 47 has recently been shown to interact directly with the transporter associated with antigen processing (TAP), inhibiting peptide translocation into the ER in human but not murine cells (Zola, 247a).

Herpes simplex virus also appears to have a second function capable of inactivating $CD8^+$ T cells that come into contact with an HSV-infected fibroblast; in this case, the viral gene products responsible for this activity have not been identified (496,753).

CONCLUSIONS

We permit ourselves a personal comment. The field has grown enormously since the first attempt to understand the biology of the virus was undertaken by one of us more than a third of a century ago. The studies on HSV are at last entering a most exciting period largely because the words "structure and function" are beginning to have an operational meaning. As a field of endeavor, we are beginning to characterize the interaction of proteins among themselves and with viral nucleic acids. In addition, the host factors crucial to virus multiplication and, potentially, to latency are being sought out. The armamentarium for a major assault on the mysteries underlying the biology of these viruses is in place and reflects the contributions of many laboratories over many, many years.

At the same time, the field has grown too much. The task before us was far greater for not having followed Ludwig Wittgenstein's dictum that "Whereof one cannot speak, thereof one must be silent," but chapters of citations bereft of commentary are dull. We fear that the commentary of the future will be significantly restrained for lack of both space and capability to deal with the enormous windfall of information that only a genome of 150,000 bp can produce.

REFERENCES

1. Abmayr SM, Workman JL, Roeder RG. The pseudorabies immediate early protein stimulates in vitro transcription by facilitating TFIID: promoter interactions. *Genes Dev* 1988;2:542–553.
2. Ace CI, Dalrymple MA, Ramsay FH, Preston VG, Preston CM. Mutational analysis of the herpes simplex virus type 1 trans-inducing factor Vmw65. *J Gen Virol* 1988;69:2595–2605.
3. Ace CI, McKee TA, Ryan M, Cameron JM, Preston CM. Construction and characterization of a herpes simplex virus type 1 mutant unable to

transinduce immediate-early gene expression. *J Virol* 1989; 63:2260–2269.
4. Ackermann M, Braun DK, Pereira L, Roizman B. Characterization of HSV-1 α proteins 0, 4, and 27 with monoclonal antibodies. *J Virol* 1984;52:108–118.
5. Ackermann M, Chou J, Sarmiento M, Lerner RA, Roizman B. Identification by antibody to a synthetic peptide of a protein specified by a diploid gene located in the terminal repeats of the L component of herpes simplex virus genome. *J Virol* 1986;58:843–850.
6. Ackermann M, Longnecker R, Roizman B, Pereira L. Identification, properties, and gene location of a novel glycoprotein specified by herpes simplex virus 1. *Virology* 1986;150:207–220.
7. Adams G, Stover BH, Keenlyside RA. Nosocomial herpetic infection in a pediatric intensive care unit. *Am J Epidemiol* 1981;113:126–132.
8. Addison C, Rixon FJ, Preston VG. Herpes simplex virus type 1 UL28 gene product is important for the formation of mature capsids. *J Gen Virol* 1990;71:2377–2384.
9. Ali MA, Prakash SS, Jariwalla RJ. Localization of antigenic sites and intrinsic protein kinase domain within a 300 amino acid segment of the ribonucleotide reductase large subunit from herpes simplex virus type 2. *Virology* 1992;187:360–367.
10. Al-Kobaisi MF, Rixon FJ, McDougall I, Preston VG. The herpes simplex UL33 gene product is required for the assembly of full capsids. *Virology* 1991;180:380–388.
11. Anderson KP, Costa R, Holland L, Wagner E. Characterization of HSV-1 RNA present in the absence of de novo protein synthesis. *J Virol* 1980;34:9–27.
12. Anderson K, Frink R, Devi G, Gaylord B, Costa R, Wagner E. Detailed characterization of the mRNA mapping in the Hind III fragment K region of the HSV-1 genome. *J Virol* 1981;37:1011–1027.
13. Anderson WA, Margruder B, Kilbourne ED. Induced reactivation of herpes simplex virus in healed rabbit corneal lesions. *Proc Soc Exp Biol Med* 1961;107:628–632.
14. Arsenakis M, Foa-Tomasi L, Speziali V, Roizman B, Campadelli-Fiume G. Expression and regulation of glycoprotein C gene of herpes simplex virus 1 resident in a clonal L cell line. *J Virol* 1986;58: 367–376.
15. Avitabile E, Ward PL, Di Lazzaro C, Torrisi MR, Roizman B, Campadelli-Fiume G. The herpes simplex virus U_L20 protein compensates for the differential disruption of exocytosis of virions and membrane glycoproteins associated with the fragmentation of the Golgi apparatus. *J Virol* 1994;68:7397–7405.
16. Bacchetti S, Evelegh MJ, Muirhead B. Identification and separation of the two subunits of the herpes simplex virus ribonucleotide reductase. *J Virol* 1986;57:1177–1181.
17. Bacchetti S, Evelegh MJ, Muirhead B, Sartori CS, Huszar D. Immunological characterization of herpes simplex virus type 1 and 2 polypeptides involved in viral ribonucleotide reductase activity. *J Virol* 1984;49:591–593.
18. Bachenheimer SL, Roizman B. Ribonucleic acid synthesis in cells infected with herpes simplex virus. VI. Polyadenylic acid sequences in viral messenger ribonucleic acid. *J Virol* 1972;10:875–879.
19. Baines JD, Roizman B. The U_L10 gene of herpes simplex virus 1 encodes a novel viral glycoprotein, gM, which is present in the virion and plasma membrane of infected cells. *J Virol* 1993;67:1441–1452.
20. Baines JD, Roizman B. The U_L11 gene of herpes simplex virus 1 encodes a function that facilitates nucleocapsid envelopment and egress from cells. *J Virol* 1992;66:5168–5174.
21. Baines JD, Roizman B. The cDNA of U_L15, a highly conserved herpes simplex virus 1 gene effectively replaces the two exons of the wild type virus. *J Virol* 1992;66:5621–5626.
22. Baines JD, Roizman B. The open reading frames U_L3, U_L4, U_L10, and U_L16 are dispensable for the replication of herpes simplex virus 1 in cell culture. *J Virol* 1991;65:938–944.
23. Baines JD, Ward PL, Campadelli-Fiume G, Roizman B. The U_L20 gene of herpes simplex virus 1 encodes a function necessary for viral egress. *J Virol* 1991;65:6414–6424.
24. Baines JD, Koyama AH, Huang T, Roizman B. The U_L21 gene products of herpes simplex virus 1 are dispensable for growth in cultured cells. *J Virol* 1994;68:2929–2936.
25. Banfield BW, Tufaro F. Herpes simplex virus particles are unable to traverse the secretory pathway in the mouse L-cell mutant gro29. *J Virol* 1990;64:5716–5729.
26. Banks LM, Halliburton IW, Purifoy DJM, Killington RA, Powell KL. Studies on the herpes simplex virus alkaline nuclease: Detection of type-common and type-specific epitopes on the enzyme. *J Gen Virol* 1985;66:1–14.
27. Banks L, Purifoy DJM, Hurst PF, Killington RA, Powell KL. Herpes simplex virus nonstructural proteins. IV. Purification of the virus-induced deoxyribonuclease and characterization of the enzyme with monoclonal antibodies. *J Gen Virol* 1983;64:2249–2260.
28. Barker DE, Roizman B. The unique sequence of the herpes simplex virus 1 L component contains an additional translated open reading frame designated $U_L49.5$. *J Virol* 1992;66:562–566.
29. Barnett JW, Eppstein DA, Chan HW. Class I defective herpes simplex virus DNA as a molecular cloning vehicle in eucaryotic cells. *J Virol* 1983;48:384–395.
30. Bartkoski MJ Jr, Roizman B. Regulation of herpesvirus macromolecular synthesis. VII. Inhibition of internal methylation of mRNA late in infection. *Virology* 1978;85:146–156.
31. Bartkoski MJ Jr, Roizman B. RNA synthesis in cells infected with herpes simplex virus. XIII. Differences in the methylation patterns of viral RNA during the reproductive cycle. *J Virol* 1976;20:583–588.
32. Batterson W, Furlong D, Roizman B. Molecular genetics of herpes simplex virus. VII. Further characterization of a ts mutant defective in release of viral DNA and in other stages of viral reproductive cycle. *J Virol* 1983;45:397–407.
33. Batterson W, Roizman, B. Characterization of the herpes simplex virion-associated factor responsible for the induction of α genes. *J Virol* 1983;46:371–377.
34. Baucke RB, Spear PG. Membrane proteins specified by herpes simplex viruses. V. Identification of an Fc binding glycoprotein. *J Virol* 1979;32:779–789.
35. Baumruker T, Sturm R, Herr W. OBP100 binds remarkably degenerate octamer motifs through specific interactions with flanking sequences. *Genes Dev* 1988;2:1400–1413.
36. Becker Y, Dym H, Sarov I. Herpes simplex virus DNA. *Virology* 1968;36:184–192.
37. Ben-Porat T, Tokazewski S. Replication of herpesvirus DNA. II. Sedimentation characteristics of newly synthesized DNA. *Virology* 1977;79:292–301.
38. Berger EC, Buddecke E, Kamerling JP, Kobata A, Paulson JC, Vliegenthart JFC. Structure, biosynthesis and functions of glycoprotein glycans. *Experientia* 1982;38:1129–1162.
39. Berger SL, Pina B, Silverman N, Marcus GA, Agapite J, Regier JL, Triezenberg SJ, Guarente L. Genetic isolation of ADA2: a potential transcriptional adapter required for the function of certain acidic activation domains. *Cell* 1992;70:251–265.
40. Bergstrom T, Lycke EJ. Neuroinvasion by herpes simplex virus. An in vitro model for characterization of neurovirulent strains. *Gen Virol* 1990;71:405–410.
41. Berk AJ, Lee F, Harrison T, Williams J, Sharp PA. Pre-early adenovirus 5 gene product regulates synthesis of early viral messenger RNAs. *Cell* 1979;17:935–944.
42. Biswal N, Murray BK, Benyesh-Melnick M. Ribonucleotides in newly synthesized DNA of herpes simplex virus. Virology 1974;61:87–99.
43. Blaho JA, Michael N, Kang V, Aboul-Ela N, Smulson ME, Jacobson MK, Roizman B. Differences in the poly(ADP-ribosyl)ation patterns of ICP4, the herpes simplex virus major regulatory protein, in infected cells and in isolated nuclei. *J Virol* 1992;66:6398–6407.
44. Blaho, JA, Mitchell C, Roizman B. An amino acid sequence shared by the herpes simplex virus 1 αproteins 0, 4, 22, and 27 predicted the nucleotidylylation of proteins encoded by the U_L21, U_L31, U_L47, and U_L49 genes. *J Biol Chem* 1994;269:17401–17410.
45. Blaho JA, Mitchell C, Roizman B. Guanylylation and adenylylation of the α regulatory proteins of herpes simplex virus requires a viral or function. *J Virol* 1993:67:3891–3900.
46. Blaho JA, Roizman B. ICP4, the major regulatory protein of herpes simplex virus shares features comon to GTP binding proteins and is adenylated and guanylated. *J Virol* 1991;65:3759–3769.
47. Blair ED, Wagner EK. A single regulatory region modulates both cis activation and trans activation of the herpes simplex VP5 promoter in transient expression assays in vivo. *J Virol* 1986;60:460–469.
48. Block TM, Spivack JG, Steiner I, Deshmane S, McIntosh MT, Lirette RP, Fraser NW. A herpes simplex virus type 1 latency associated transcript mutant reactivates with normal kinetics from latent infection. *J Virol* 1990;64:3417–3426.
49. Bludau H, Freese UK. Analysis of the HSV-1 strain 17 DNA poly-

merase gene reveals the expression of four different classes of pol transcripts. *Virology* 1991;183:505–518.
50. Blue WT, Winland RD, Stobbs DG, Kirksey DF, Savage RE. Effects of adenosine monophosphate on the reactivation of latent herpes simplex virus type 1 infections of mice. *Antimicrob Agents Chemother* 1981;20:547–548.
51. Blyth WA, Hill TJ, Field HJ, Harbour DA. Reactivation of herpes simplex virus infection by ultraviolet light and possible involvement of prostaglandin. *J Gen Virol* 1976;33:547–550.
52. Boehmer PE, Dodson MS, Lehman IR. The herpes simplex virus type 1 origin binding protein. DNA helicase activity. *J Biol Chem* 1993; 268:1220–1225.
53. Bohenzky RA, Papavassiliou AG, Gelman IH, Silverstein S. Identification of a promoter mapping within the reiterated sequences that flank the herpes simplex virus type 1 U_L region. *J Virol* 1993;67:632–642.
54. Bond VC, Person S. Fine structure physical map locations of alterations that affect cell fusion in herpes simplex virus type 1. *Virology* 1984; 132:368–376.
55. Bortner C, Hernandez TR, Lehman IR. Griffith J. Herpes simplex virus 1 single strand DNA-binding protein (ICP8) will promote homologous pairing and strand transfer. *J Mol Biol* 1993;231:241–250.
56. Brandimarti R, Huang T, Roizman B, Campadelli-Fiume G. Mapping of herpes simplex virus 1 genes with mutations which overcome host restrictions to infection. *Proc Natl Acad Sci USA* 1994:91:5406–5410.
57. Brandt CR, Kintner RL, Pumfrey AM, Visalli RJ, Grau DR. The herpes simplex virus ribonucleotide reductase is required for ocular virulence. *J Gen Virol* 1991;72:2043–2049.
58. Braun DK, Batterson W, Roizman B. Identification and genetic mapping of a herpes simplex virus capsid protein which binds DNA. *J Virol* 1984:50:645–648.
59. Braun DK, Pereira L, Norrild B, Roizman B. Application of denatured, electro-phoretically separated, and immobilized lysates of herpes simplex virus-infected cells for detection of monoclonal antibodies and for studies of the properties of viral proteins. *J Virol* 1983:46:103–112.
60. Braun DK, Roizman B, Pereira L. Characterization of post-translational products of herpes simplex virus gene 35 proteins binding to the surfaces of full capsids but not empty capsids. *J Virol* 1984;49:142–153.
61. Bruckner RC, Durch RE, Zemelman BV, Mocarski ES, Lehman IR. Recombination *in vitro* between herpes simplex virus type 1 *a* sequences. *Proc Natl Acad Sci USA* 1992;89:10950–10954.
62. Buchman TG, Roizman B, Adams G, Stover H. Restriction endonuclease fingerprinting of herpes simplex virus DNA: a novel epidemiological tool applied to a nosocomial outbreak. *J Infect Dis* 1978; 138:488–498.
63. Buchman TG, Roizman B, Nahmias AJ. Demonstration of exogenous genital reinfection with herpes simplex virus type 2 by restriction endonuclease fingerprinting of viral DNA. *J Infect Dis* 1979;140:295–304.
64. Buchman TG, Simpson T, Nosal C, Roizman B, Nahmias AJ. The structure of herpes simplex virus DNA and its application to molecular epidemiology. *Ann NY Acad Sci* 1980;354:279–290.
65. Cabrera CV, Wohlenberg C, Openshaw H, Rey-Mendez M, Puga A, Notkins AL. Herpes simplex virus DNA sequences in the CNS of latently infected mice. *Nature* 1980;288:288–290.
66. Cai W, Gu B, Person S. Role of glycoprotein B of herpes simplex virus type 1 in viral entry and cell fusion. *J Virol* 1988;62:2596–2604.
67. Cai W, Schaffer PA. Herpes simplex virus type 1 ICP0 regulates expression of immediate-early, early, and late genes in productively infected cells. *J Virol* 1992;66:2904–2915.
68. Cameron JM, McDougall I, Marsden HS, Preston VG, Ryan DM, Subak-Sharpe JH. Ribonucleotide reductase encoded by herpes simplex virus is a determinant of the pathogenicity of the virus in mice and a valid antiviral target. *J Gen Virol* 1988;69:2607–2612.
69. Campadelli G, Brandimarti R, Di Lazzaro C, Ward PL, Roizman B, Torrisi MR. Fragmentation and dispersal of Golgi proteins and redistribution of glycoproteins and glycolipids processed through Golgi following infection with herpes simplex virus 1. *Proc Natl Acad Sci USA* 1993;90:2798–2802.
70. Campadelli-Fiume G, Arsenakis M, Farabegoli F, Roizman B. Entry of herpes simplex virus 1 in BJ cells that constitutively express viral glycoprotein D is by endocytosis and results in the degradation of the virus. *J Virol* 1988;62:159–167.
71. Campadelli-Fiume G, Farabegoli F, Di Gaeta S, Roizman B. Origin of unenveloped capsids in the cytoplasm of cells infected with herpes simplex virus 1. *J Virol* 1991;65:1589–1595.
72. Campadelli-Fiume G, Poletti L, Dall'Olio F, Serafini-Cessi F. Infectivity and glycoprotein processing of herpes simplex virus type 1 grown in a ricine-resistant cell line deficient in N-acetylglucosaminyl transferase 1. *J Virol* 1982;43:1061–1071.
73. Campadelli-Fiume G, Qi S, Avitabile E, Foa-Tomasi L, Brandimarti R, Roizman B. Glycoprotein D of herpes simplex virus encodes a domain which precludes penetration of cells expressing the glycoprotein by superinfecting herpes simplex virus. *J Virol* 1990;64:6070–6079.
74. Campadelli-Fiume G, Serafini-Cessi F. Processing of the oligosaccharide chains of herpes simplex virus type 1 glycoproteins. In: Roizman B, ed.*The herpesviruses*. vol. 3. New York: Plenum Press; 1985:357–382.
75. Campadelli-Fiume G, Stirpe D, Boscaro A, Avitabile E, Foa-Tomasi L, Barker D, Roizman B. Glycoprotein C-dependent attachment of herpes simplex virus to susceptible cells leading to productive infection. *Virology* 1990;178:213–222.
76. Campbell MEM, Palfreyman JW, Preston CM. Identification of the herpes simplex virus DNA sequences which encodes a *trans*-acting polypeptide responsible for the stimulation of the immediate early transcription. *J Mol Biol* 1984;180:1–19.
77. Caplan M, Matlin KS. Sorting of membrane and secretory proteins in polarized epithelial cells. In: Matlin KS, Valentich JD, eds. *Functional epithelial cells in culture*. New York: Alan R. Liss; 1989:71–127.
78. Caradonna SJ, Cheng YC. Induction of uracil-DNA-glycosylase and dUTP hydrolase in herpes simplex virus infected human cells. *J Biol Chem* 1981;256:9834–9837.
79. Caradonna S, Worrad D, Lirette R. Isolation of a herpes simplex virus cDNA encoding the DNA repair enzyme uracil-DNA glycosylase. *J Virol* 1987;61:3040–3047.
80. Carter VC, Jennings SR, Rice PL, Tevethia SS. Mapping of a herpes simplex virus type 2-encoded function that affects the susceptibility of herpes simplex virus-infected cells to lysis by herpes simplex virus-specific cytotoxic T lymphocytes. *J Virol* 1984;49:766–771.
81. Carton CA, Kilbourne ED. Activation of latent herpes simplex by trigeminal sensory-root section. *N Engl J Med* 1952;246:172–176.
82. Centifanto-Fitzgerald YM, Vanell ED, Kaufman HE. Initial herpes simplex virus type 1 infection prevents ganglionic superinfection by other strains. *Infect Immun* 1982;35:1125–1132.
83. Centifanto-Fitzgerald YM, Yamaguchi T, Kaufman HE, Tognon M, Roizman B. Ocular disease pattern induced by herpes simplex virus is genetically determined by a specific region of viral DNA. *J Exp Med* 1982;155:475–489.
84. Challberg MD. A method for identifying the viral genes required for herpesvirus DNA replication. *Proc Natl Acad Sci USA* 1986;83:9094–9098.
85. Chang YE, Roizman B. The product of U_L31 gene of herpes simplex virus 1 is a nuclear phosphoprotein which partitions with the nuclear matrix. *J Virol* 1993;67:6348–6356.
86. Chartrand P, Crumpacker CS, Schaffer PA, Wilkie NM. Physical and genetic analysis of the herpes simplex virus DNA polymerase locus. *Virology* 1980;103:311–325.
87. Chartrand P, Stow ND, Timbury MC, Wilkie NM. Physical mapping of Paar mutations of herpes simplex virus type 1 and type 2 by intertypic marker rescue. *J Virol* 1979;31:265–276.
88. Chee MS, Lawrence GL, Barrell BG. Alpha-, beta- and gamma herpesviruses encode a putative phosphotransferase. *J Gen Virol* 1989; 70:1151–1160.
89. Chen J, Panagiotidis C, Silverstein S. Multimerization of ICP0, a herpes simplex virus immediate early protein. *J Virol* 1991;66:5598–5602.
90. Chen J, Silverstein S. Herpes simplex viruses with mutations in the gene encoding ICP0 are defective in gene expression.*J Virol* 1992;66:2916–2927.
91. Chou J, Kern ER, Whitley RJ, Roizman B. Mapping of herpes simplex virus 1 neurovirulence to $\gamma_1$34.5, a gene nonessential for growth in cell culture. *Science* 1990;252:1262–1266.
92. Chou J, Poon APW, Johnson J, Roizman B. Differential response of human cells to deletions and stop codons in the $\gamma_1$34.5 gene of herpes simplex virus. *J Virol* 1994:68:8304–8311.
93. Chou J, Roizman B. Herpes simplex virus 1 $\gamma_1$34.5 gene function which blocks the host response to infection maps in the homologous domain of the genes expressed during growth arrest and DNA damage. *Proc Natl Acad Sci USA* 1994:91:5247–5251.
94. Chou J, Roizman B. The $\gamma_1$34.5 gene of herpes simplex virus 1 precludes neuroblastoma cells from triggering total shutoff of protein synthesis characteristic of programmed cell death in neuronal cells. *Proc Natl Acad Sci USA* 1992;89:3266–3270.

95. Chou J, Roizman B. The herpes simplex virus 1 gene for ICP34.5, which maps in inverted repeats, is conserved in several limited passage isolates but not in strain 17synÿ2D. *J Virol* 1990;64:1014–1020.
96. Chou J, Roizman B. Characterization of DNA sequence common and DNA sequence specific proteins binding to the cis-acting sites for the cleavage of the terminal a sequence of herpes simplex virus 1 genome. *J Virol* 1989;63:1059–1068.
97. Chou J, Roizman B. The terminal *a* sequence of the herpes simplex virus genome contains the promoter of a gene located in the repeat sequences of the L component.*J Virol* 1986:57:629–637.
98. Chou J, Roizman B. The isomerization of the herpes simplex virus 1 genome: identification of the cis-acting and recombination sites within the domain of the *a* sequence. *Cell* 1985;41:803–811.
99. Chu CT, Parris DS, Dixon RAF, Farber FE, Schaffer PA. Hydroxylamine mutagenesis of HSV DNA and DNA fragments: introduction of mutations into selected regions of the viral genome. *Virology* 1979;98:168–181.
100. Chung TD, Wymer JP, Smith CC, Kulka M, Aurelian L. Protein kinase activity associated with the large subunit of herpes simplex virus type 2 ribonucleotide reductase (ICP10) *J Virol* 1989;63:3389–3398.
101. Clements GB, Stow ND. A herpes simplex virus type 1 mutant containing a deletion within immediate early gene 1 is latency competent in mice. *J Gen Virol* 1989;70:2501–2506.
102. Clements JB, Watson RJ, Wilkie NM. Temporal regulation of herpes simplex virus type 1 transcription: location of transcripts in the viral genome. *Cell* 1977;12:275–285.
103. Coen DM, Aschman DP, Gelep PT, Retondo MJ, Weller SK, Schaffer PA. Fine mapping and molecular cloning of mutations in the herpes simplex virus DNA polymerase locus. *J Virol* 1984;49:236–247.
104. Coen DM, Kosz-Vnenchak M, Jacobson JG, Leib DA, Bogard CL, Schaffer PA, Tyler KL, Knipe DM. Thymidine kinase-negative herpes simplex virus mutants establish latency in mouse trigeminal ganglia but do not reactivate. *Proc Natl Acad Sci USA* 1989;86:4736–4740.
105. Cohen GH, Ponce de Leon M, Deggelman H, Lawrence WC, Vernon SK, Eisenberg RJ. Structural analysis of the capsid polypeptides of herpes simplex virus types 1 and 2. *J Virol* 1980;34:421–531.
106. Colberg-Poley AM, Isom HC, Rapp F. Involvement of an early cytomegalovirus function in reactivation of quiescent herpes simplex virus type 2. *J Virol* 1981;37:1051–1059.
107. Colberg-Poley AM, Isom HC, Rapp F. Reactivation of herpes simplex virus type 2 from a quiescent state by human cytomegalovirus. *Proc Natl Acad Sci USA* 1979;76:5948–5951.
108. Conley AF, Knipe DM, Jones PC, Roizman B. Molecular genetics of herpes simplex virus. VII. Characterization of a temperature-sensitive mutant produced by in vitro mutagenesis and defective in DNA synthesis and accumulation of polypeptides. *J Virol* 1981;37:191–206.
109. Conner J, Macfarlane J, Lankinen H, Marsden H. The unique N terminus of the herpes simplex virus type 1 large subunit is not required for ribonucleotide reductase activity. *J Gen Virol* 1992;73:103–112.
110. Cook ML, Stevens JG. Pathogenesis of herpetic neuritis and ganglionitis in mice: evidence of intra-axonal transport of infection. *Infect Immun* 1973;7:272–288.
111. Cook ML, Stevens JG. Replication of varicella-zoster virus in cell cultures. An ultrastructural study. J Ultrastruct Res 1970;32:334–350.
112. Costa RH, Cohen G, Eisenberg R, Long D, Wagner EK. Direct demonstration that the abundant 6-kilobase herpes simplex virus type 1 mRNA mapping between 0.23 and 0.27 map units encodes the major capsid protein VP5. *J Virol* 1984;49:287–292.
113. Costa RH, Devi BG, Anderson KP, Gaylord BH, Wagner EK. Characterization of a major late herpes simplex virus type 1 mRNA. *J Virol* 1981;38:483–496.
114. Costa RH, Draper KG, Devi-Rao G, Thompson RL, Wagner EK. Virus-induced modification of the host cell is required for expression of the bacterial chloramphenicol acetyltransferase gene controlled by a late herpes simplex virus promoter (VP5). *J Virol* 1985;56:19–30.
115. Costanzo F, Campadelli-Fiume G, Foa-Tomas L, Cassai E. Evidence that herpes simplex virus DNA is transcribed by cellular RNA polymerase II. *J Virol* l977;21:996–1001.
116. Courtney RJ, Powell KL. Immunological and biochemical characterization of polypeptides induced by herpes simplex virus types 1 and 2. In: de The G, et al, eds. *Oncogenesis and herpesviruses II*. Lyon: International Agency for Research on Cancer; 1975:63.
117. Cress WD, Triezenberg SJ. Critical structural elements of the VP16 transcriptional activation domain. *Science* 1991;51:87–90.
118. Crumpacker CS, Chartrand P, Subak-Sharpe JH, Wilkie NM. Resistance of herpes simplex virus to acycloguanosine—genetic and physical analysis. *Virology* 1980;05:171–184.
119. Crute JW, Tsurumi T, Zhu L, Weller SK, Olivo PD, Challberg MD, Mocarski ES, Lehman IR. Herpes simplex virus 1 helicase-primase: a complex of three herpes encoded gene products. *Proc Natl Acad Sci USA* 1989;86:2186–2189.
120. Cunningham AL, Turner RR, Miller C, Para MF, Merigan TC. Evolution of recurrent herpes simplex virus lesions and immunohistologic study. *J Clin Invest* 1985;75:226–233.
121. Cunningham C, Davison AJ, Dolan A, Frame MC, McGeoch DJ, Meredith DM, Moss HWM, Orr AC. The UL13 virion protein of herpes simplex virus 1 is phosphorylated by a novel virus-induced protein kinase. *J Gen Virol* 1992;73:303–311.
122. Cushing H. Surgical aspects of major neuralgia of trigeminal nerve: report of 20 cases of operation upon the gasserian ganglion with anatomic and physiologic notes on the consequence of its removal. *JAMA* 1905;4:1002–1008.
123. Daikoku T, Yamamoto N, Maeno K, Nishiyama Y. Role of viral ribonucleotide reductase in the increase of dTTP pool size in herpes simplex virus-infected Vero cells. *J Gen Virol* 1991;72:1441–1444.
124. Dales S, Chardonnet Y. Early events in the interaction of adenoviruses with HeLa cells. IV. Association with microtubles and the nuclear pore complex during vectorial movement of the inoculum. *Virology* 1973;56:465–483.
125. Dalrymple MA, McGeoch DJ, Davison AJ, Preston CM. DNA sequence of the herpes simplex virus type 1 gene whose product is responsible for transcriptional activation of immediate early promoters. *Nucleic Acids Res* 1985;13:7865–7879.
126. Darby G, Field HJ, Salisbury SA. Altered substrate specificity of herpes simplex virus thymidine kinase confers acyclovir resistance. *Nature* 1981;289:81–83.
127. Dargan DJ, Subak-Sharpe JH. Isolation and characterization of revertants from fourteen herpes simplex virus type 1 (strain 17) temperature sensitive mutants. *J Gen Virol* 1984;65:477–491.
128. Darlington RW, Moss LH III. The envelope of herpesvirus. *Prog Med Virol* 1969;11:16–45.
129. Davison AJ, Wilkie NM. Nucleotide sequences of the joint between the L and S segments of herpes simplex virus types 1 and 2. *J Gen Virol* 1981;55:315–331.
130. Davison MD, Rixon FJ, Davison, AJ. Identification of genes encoding two capsid proteins (VP24 and VP26) of herpes simplex virus type 1. *J Gen Virol* 1992;73:2709–2713.
131. Deb S, Doelberg M. A 67-base-pair segment from the ori-S region of herpes simplex virus type 1 encodes origin function. *J Virol* 1988;62:2516–2519.
132. Debroy C, Pederson N, Person S. Nucleotide sequence of a herpes simplex virus type 1 gene that causes cell fusion. *Virology* 1985;145:36–48.
133. Deiss LP, Chou J, Frenkel N. Functional domains within the a sequence involved in the cleavage-packaging of herpes simplex virus DNA. *J Virol* 1986;59:605–618.
134. Deiss LP, Frenkel N. Herpes simplex virus amplicon: cleavage of concatemeric DNA is linked to packaging and involves amplification of the terminally reiterated a sequence. *J Virol* 1986;57:933–941.
135. Delius H, Clements JB. A partial denaturation map of herpes simplex virus type 1 DNA: evidence for inversions of the unique DNA regions. *J Gen Virol* 1976;33:125–133.
136. DeLuca NA, Courtney MA, Schaffer PA. Temperature-sensitive mutants in herpes simplex virus type 1 ICP4 permissive for early gene expression. *J Virol* 1984;52:767–776.
137. DeLuca NA, McCarthy AM, Schaffer PA. Isolation and characterization of deletion mutants of herpes simplex virus type 1 in the gene encoding the immediate-early regulatory protein ICP4. *J Virol* 1985;56:558–570.
138. DeLuca NA, Schaffer PA. Physical and functional domains of the herpes simplex virus transcriptional regulatory protein ICP4. *J Virol* 1988;62:732–743.
139. DeLuca NA, Schaffer PA. Activities of herpes simplex virus type 1 (HSV-1) ICP4 genes specifying nonsense peptides. *Nucleic Acids Res* 1987;15:4491–4511.
140. DeLuca NA, Schaffer PA. Activation of immediate-early, early, and late promoters by temperature-sensitive and wild-type forms of herpes simplex virus type 1 protein ICP4. *Mol Cell Biol* 1985;5:1997–2008.
141. Dennis D, Smiley JR. Transactivation of a late herpes simplex virus promoter. *Mol Cell Biol* 1984;4:544–551.

142. Desai PJ, Schaffer PA, Minson AC. Excretion of non-infectious virus particles lacking glycoprotein H by a temperature-sensitive mutant of herpes simplex virus type 1: evidence that gH is essential for virion infectivity. *J Gen Virol* 1988;69:1147–1156.
143. Devi-Rao GB, Bloom DC, Stevens JG, Wagner EK. Herpes simplex virus type 1 DNA replication and gene expression during explant-induced reactivation of latently infected murine sensory ganglia. *J Virol* 1994;68:1271–1282.
144. DiIanni CL, Drier DA, Deckman IC, McCann PJ III, Liu F, Roizman B, Colonno RJ, Cordingley MG. Identification of the herpes simplex virus 1 protease cleavage sites. *J Biol Chem* 1993;368:2048–2051.
145. Dingwell KS, Brunetti CR, Hendricks RL, Tang Q, Tang M, Rainbow AJ, Johnson DC. Herpes simplex virus glycoprotein E and I facilitate cell to cell spread in vivo and across junctions of cultured cells. *J Virol* 1994;68:834–845.
146. Doerig, C, Pizer LI, Wilcox CL. Detection of the latency associated transcript in neuronal cultures during the latent infection with herpes simplex virus type 1. *Virology* 1991;183:423–426.
147. Doerig C, Pizer LI, Wilcox CL. An antigen encoded by the latency associated transcript in neuronal cell cultures latently infected with herpes simplex virus type 1. *J Virol* 1991;65:2724–2727.
148. Dotti CG, Banker GA, Binder LI. The expression and distribution of the microtubule-associated proteins tau and microtubule associated protein 2 in hippocampal neurons in the rat in situ and in cell culture. *Neuroscience* 1987;23:121–130.
149. Dotti CG, Simons K. Polarized sorting of viral glycoproteins to the axons and dendrites of hippocampal neurons in culture. *Cell* 1990;62:63–72.
150. Draper KG, Devi-Rao G, Costa RH, Blair ED, Thompson RL, Wagner EK. Characterization of the genes encoding herpes simplex virus type 1 and type 2 alkaline exonucleases and overlapping proteins. *J Virol* 1986;57:1023–1036.
151. Draper KG, Frink RJ, Wagner EK. Detailed characterization of an apparently unspliced ß herpes simplex virus type 1 gene mapping in the interior of another. *J Virol* 1982;44:1123–1128.
152. Dubin G, Socolof E, Frank I, Friedman HM. Herpes simplex virus type 1 Fc receptor protects infected cells from antibody-dependent cellular cytotoxicity.*J Virol* 1991;65:7046–7050.
153. Dutch RE, Bruckner RC, Mocarski ES, Lehman IR. Herpes simplex virus type 1 recombination: role of DNA replication and viral *a* sequences. *J Virol* 1992;66:277–285.
154. Dutch RE, Lehman IR. Renaturation of complementary DNA strands by herpes simplex virus type 1 ICP8. *J Virol* 1993;67:6945–6949.
155. Dutch RE, Zemelman BV, Lehman IR. Herpes simplex virus type 1 recombination: the U_c-DR1 region is required for high-level *a*-sequence mediated recombination. *J Virol* 1944;68:3733–3741.
156. Ebert SN, Subramanian D, Shtrom SS, Chung IK, Parris D, and Muller MT. Association between the p170 form of human topoisomerase II and progeny viral DNA in cells infected with herpes simplex virus type 1. *J Virol* 1994;68:1010–1020.
157. Eisenberg SP, Coen DM, McKnight SL. Promoter domains required for expression of plasmid-borne copies of the herpes simplex virus thymidine kinase gene in virus-infected mouse fibroblasts and microinjected frog oocytes. *Mol Cell Biol* 1985;5:1940–1947.
158. Ejercito PM, Kieff ED, Roizman B. Characterization of herpes simplex virus strains differing in their effect on social behavior of infected cells. *J Gen Virol* 1968;2:357–364.
159. Elias P, Gustafsson CM, Hammarsten O. The origin binding protein of herpes simplex virus 1 binds cooperatively to the viral origin of replication Ori_S. *J Biol Chem* 1990;265:17167–17173.
160. Elias P, Lehman IR. Interaction of origin binding protein with an origin of replication of herpes simplex virus 1. *Proc Natl Acad Sci USA* 1988;85:2959–2963.
161. Elias P, O'Donnell ME, Mocarski ED, Lehman IR. A DNA binding protein specific for an origin of replication of herpes simplex virus type 1. *Proc Natl Acad Sci USA* 1986;83:6322–6326.
162. Elkareh AA, Murphy AJM, Fichter T, Efstradiatis A, Silverstein S. "Transactivation" control signals in the promoter of the herpesvirus thymidine kinase gene. *Proc Natl Acad Sci USA* 1985;82:1002–1006.
163. Elkareh A, Silverstein S, Smiley J. Control of expression of the herpes simplex virus thymidine kinase gene in biochemically transformed cells. *J Gen Virol* 1984;65:19–36.
164. Ellis MN, Keller PM, Fyfe JA, Martin JL, Rooney JF, Straus SE, Lehrman SN, Barry DW. Clinical isolate of herpes simplex virus type 2 that induces a thymidine kinase with an altered substrate specificity. *Antimicrob Agents Chemother* 1987;31:1117–1125.
165. Embretson J, Staskus K, Retzel EF, Haase AT, Bitterman P. PCR amplification of viral DNA and viral host cell mRNAs in situ. In: *The polymerase chain reaction [in press]*.
166. Epstein MA. Observations on the mode of release of herpes virus from infected HeLa cells. *J Cell Biol* 1962;12:589–597.
167. Erickson JS, Kaplan AS. Synthesis of proteins in cells infected with herpesvirus. IX. Sulfated proteins. *Virology* 1973;55:94–102.
168. Everett RD. DNA sequence elements required for regulated expression of the HSV-1 glycoprotein D gene lie within 83 bp of the RNA capsites. *Nucleic Acids Res* 1983;11:6647–6666.
169. Everett RD. A detailed mutational analysis of Vmw110, a trans-acting transcriptional activator encoded by herpes simplex virus type 1. *EMBO J* 1987;6:2069–2076.
170. Everett RD. A detailed analysis of an HSV-1 early promoter: sequences involved in trans-activation by immediate-early gene products are not early-gene specific. *Nucleic Acids Res* 1984;12:3037–3055.
171. Everett RD. Transactivation of transcription by herpes virus products: requirement for two HSV-1 immediate-early polypeptides for maximum activity. *EMBO J* 1984;3:3135–3141.
172. Everett RD, Dunlop M. Trans-activation of plasmid-borne promoters by adenovirus and several herpes group viruses. *Nucleic Acids Res* 1984;12:5969.
173. Everett RD, Orr A. The Vmw175 binding site in the IE-1 promoter has no apparent role in the expression of Vmw110 during herpes simplex virus type 1 infection. *Virology* 1991;180:509–517.
174. Everett RD, Orr A, Elliott M. High level expression and purification of herpes simplex virus type 1 immediate early polypeptide VmW110. *Nucleic Acid Res* 1991;19:6155–6161.
175. Everett RJ, DiDonato J, Elliott M, Muller M. Herpes simplex virus type 1 polypeptide ICP4 binds DNA. *Nucleic Acids Res* 1992;20:1229–1233.
176. Faber SW, Wilcox KW. Association of herpes simplex virus regulatory protein ICP4 with sequences spanning the ICP4 gene transcription initiation site. Nucleic Acids Res 1988;16:555–570.
177. Faber SW, Wilcox KW. Association of the herpes simplex virus regulatory protein ICP4 with specific nucleotide sequences in DNA. *Nucleic Acids Res* 1986;14:6067–6083.
178. Farrell MJ, Dobson AT, Feldman LT. Herpes simplex virus latency associated transcript is a stable intron. *Proc Natl Acad Sci USA* 1991;88:790–794.
179. Fawl RL, Roizman, B. Induction of reactivation of herpes simplex virus in murine sensory ganglia in vivo by cadmium. *J Virol* 1993;67:7025–7031.
180. Feldman LT, Imperiale MJ, Nevins JR. Activation of early adenovirus transcription by the herpesvirus immediate early gene: evidence for a common cellular control factor. *Proc Natl Acad Sci USA* 1982; 79:4952–4956.
181. Fenwick M, Morse LS, Roizman B. Anatomy of herpes simplex virus DNA. XI. Apparent clustering of functions effecting rapid inhibition of host DNA and protein synthesis. *J Virol* 1979;29:825–827.
182. Fenwick ML, Owen SA. On the control of immediate early (α) mRNA survival in cells infected with herpes simplex virus. *J Gen Virol* 1988;69:2869–2877.
183. Fenwick ML, Roizman B. Regulation of herpesvirus macromolecular synthesis. VI. Synthesis and modification of viral polypeptides in enucleated cells. *J Virol* 1977;22:720–725.
184. Fenwick ML, Walker MJ. Suppression of the synthesis of cellular macromolecules by herpes simplex virus. *J Gen Virol* 1978;41:37–5l.
185. Field HJ, Darby G. Pathogenicity in mice of herpes simplex viruses which are resistant to acyclovir in vitro and in vivo. *Antimicrob Agents Chemother* 1980;17:209–216.
186. Field HJ, Wildy P. The pathogenicity of thymidine kinase-deficient mutants of herpes simplex virus in mice. *J Hyg* 1978;81:267–277.
187. Fisher FB, Preston VG. Isolation and characterisation of herpes simplex virus type 1 mutants which fail to induce dUTPase activity. *Virology* 1986;148:190–197.
188. Flanagan WM, Papavassiliou AG, Rice M, Hecht LB,Silverstein S, Wagner EK. Analysis of the herpes simplex virus type 1 promoter controlling the expression of U_L38, a true late gene involved in capsid assembly. *J Virol* 1991:65:769–786.
189. Fletcher C, Heintz N, Roeder RG. Purification and characterization of OTF-1, a transcription factor regulating cell cycle expression of a human histone H2b gene. *Cell* 1987;51:773–781.

190. Forrester A, Farrell H, Wilkinson G, Kaye J, Davis-Poynter N, Minson T. Construction and properties of a mutant of herpes simplex virus type 1 with glycoprotein H coding sequences deleted. *J Virol* 1992; 66:341–348.
191. Frame MC, Marsden HS, Dutia BM. The ribonucleotide reductase induced by herpes simplex virus type 1 involves minimally a complex of two polypeptides (136K and 38K). *J Gen Virol* 1985;66:1581–1587.
192. Frame MC, Marsden HS, McGeoch DJ. Novel herpes simplex virus type 1 glycoproteins identified by antiserum against a synthetic oligopeptide from the predicted product of gene US4. *J Gen Virol* 1986;67:745–751.
193. Frame MC, Purves FC, McGeoch DJ, Marsden HS, Leader DP. Identification of the herpes simplex virus protein kinase as the product of viral gene US3. *J Gen Virol* 1987;68:2699–2704.
194. Francke B, Moss H, Timbury MC, Hay J. Alkaline DNAse activity in cells infected with a temperature-sensitive mutant of herpes simplex virus type 2. *J Virol* 1978;26:209–213.
195. Fraser JW, Deatly AM, Mellerick MI, Muggeridge JI, Spivack JG. Molecular biology of latent HSV-1. In: Lopez C, Roizman B, eds. *Human herpesvirus infections: pathogenesis, diagnosis, and treatment.* New York: Raven Press; 1986:39–54.
196. Frenkel N, Locker H, Batterson W, Hayward G, Roizman B. Anatomy of herpes simplex DNA. VI. Defective DNA originates from the S component. *J Virol* 1976;20:527–531.
197. Frenkel N, Roizman B. Separation of the herpesvirus deoxyribonucleic acid on sedimentation in alkaline gradients. *J Virol* 1972;10:565–572.
198. Frenkel N, Roizman B. Ribonucleic acid synthesis in cells infected with herpes simplex virus: control of transcription and of RNA abundance. *Proc Natl Acad Sci USA* 1972;69:2654–2659.
199. Frenkel N, Silverstein NS, Cassai E, Roizman B. RNA synthesis in cells infected with herpes simplex virus. VII. Control of transcription and of transcript abundancies of unique and common sequences of herpes simplex 1 and 2. *J Virol* 1973;11:886–892.
200. Frink RJ, Anderson KP, Wagner EK. Herpes simplex virus type 1 Hind III fragment L encodes spliced and complementary mRNA species. *J Virol* 1981;39:559–572.
201. Frink RJ, Eisenberg R, Cohen G, Wagner EK. Detailed analysis of the portion of the herpes simplex virus type 1 genome encoding glycoprotein C. *J Virol* 1983;45:634–647.
201a. Fruth K, Kwangseog A, Djaballah H, Sempe P, van Edert PM, Tampe P, Peterson PA, Yang Y. A viral inhibitor of peptide transporters for antigen presentation. *Nature* 1995;375:415–418.
202. Furlong D, Swift H, Roizman B. Arrangement of herpes-virus deoxyribonucleic acid in the core. *J Virol* 1972;10:1071–1074.
203. Gaffney DF, McLauchlan J, Whitton JL, Clements JB. A modular system for the assay of transcription regulatory signals: the sequence TAATGARAT is required for herpes simplex virus immediate early gene activation. *Nucleic Acids Res* 1985;13:7847–7863.
204. Gao M, Knipe DM. Potential role for herpes simplex virus ICP8 DNA replication protein in stimulation of late gene expression. *J Virol* 1991;65:2666–2675.
205. Gao M, Matisick-Kumar L, Hurlburt W, DiTusa SF, Newcomb WW, Brown JC, McCann PJ III, Deckman I, Colonno RJ. The protease of herpes simplex virus is essential for functional capsid formation and viral growth. *J Virol* 1994;68:3702–3712.
206. Gatzke L, Meadows H, Gruenheid S, Tufaro F. Isolation of mutant mouse L cell lines defective in the binding of herpes simplex virus to the cell surface. In: *Abstracts of the 16th International Herpesvirus Workshop.* 1991:56.
207. Gelman IH, Silverstein S. Dissection of immediate-early gene promoters from herpes simplex virus: sequences that respond to the virus transcriptional activators. *J Virol* 1987;61:3167–3172.
208. Gelman IH, Silverstein S. Herpes simplex virus immediate-early promoters are responsive to virus and cell *trans*-acting factors. *J Virol* 1987;61:2286–2296.
209. Gelman IH, Silverstein S. Co-ordinate regulation of herpes simplex virus gene expression is mediated by the functional interaction of two immediate early gene products. *J Mol Biol* 1986;191:395–409.
210. Gelman IH, Silverstein S. Identification of immediate early genes from herpes simplex virus that transactivate the virus thymidine kinase gene. *Proc Natl Acad Sci USA* 1985; 82:5265–5269.
211. Georgopoulou U, Michaelidou A, Roizman B, Mavromara-Nazos P. Identification of a new transcriptional unit that yields a gene product within the unique sequences of the short component of the herpes simplex virus 1 genome. *J Virol* 1993;67:3961–3968.
212. Gerster T, Roeder RG. A herpesvirus trans-activating protein interacts with transcription factor OTF-1 and other cellular proteins. *Proc Natl Acad Sci USA* 1988;85:6347–6351.
213. Gibson W, Roizman B. Protein specified by herpes simplex virus. X. Staining and radiolabeling properties of B capsids and virion proteins in polyacrylamide gels. *J Virol* 1974;113:155–165.
214. Gibson W, Roizman B. The structural and metabolic involvement of polyamines with herpes simplex virus. In: Russell DH, ed. *Polyamines in normal and neoplastic growth.* New York: Raven Press; 1973:123–135.
215. Gibson W, Roizman B. Proteins specified by herpes simplex virus. VIII. Characterization and composition of multiple capsid forms of subtypes 1 and 2. *J Virol* 1972;10:1044–1052.
216. Gibson W, Roizman B. Compartmentalization of spermine and spermidine in the herpes simplex virion. *Proc Natl Acad Sci USA* 1971; 68:2818–2821.
217. Godowski PJ, Knipe DM. Transcriptional control of herpesvirus gene expression: gene functions required for positive and negative regulation. *Proc Natl Acad Sci USA* 1986;83:256–260.
218. Godowski PJ, Knipe DM. Identification of a herpes simplex virus that represses late gene expression from parental viral genomes. *J Virol* 1985;55:357–365.
219. Godowski PJ, Knipe DM. Mutations in the major DNA-binding protein gene of herpes simplex virus type 1 result in increased levels of viral gene expression. *J Virol* 1983;47:478–486.
220. Goldstein DJ, Weller SK. Herpes simplex virus type 1-induced ribonucleotide reductase activity is dispensable for virus growth and DNA synthesis: isolation and characterization of an ICP6 lacZ insertion mutant. *J Virol* 1988;62:196–205.
221. Goldstein DJ, Weller SK. Factor(s) present in herpes simplex virus type 1-infected cells can compensate for the loss of the large subunit of the viral ribonucleotide reductase: characterization of and ICP6 deletion mutant. *Virology* 1988;166:41–51.
222. Goodpasture EW. Herpetic infections with special reference to involvement of the nervous system. *Medicine (Baltimore)* 1929; 8:223–243.
223. Gordon YJ, Arullo-Cruz TP, Romanowski E, Ruziczka L, Balouris C, Oren J, Cheng K, Kim S. The development of an improved murine iontophoresis reactivation model for the study of HSV-1 latency. *Invest Ophthalmol Vis Sci* 1986;27:1230–1234.
224. Gordon YJ, McKnight JLC, Ostrove JM, Romanowski E, Araullo-Cruz T. Host species and strain differences affect the ability of an HSV-1 ICP0 deletion mutant to establish latency and spontaneously reactivate in vivo. *Virology* 1990;178:469–477.
225. Graham FL, Velihaisen G, Wilkie NM. Infectious herpes virus DNA. *Nature* 1973;245:265–266.
226. Greaves R, O'Hare P. Separation of requirements for protein-DNA complex assembly from those for functional activity in the herpes simplex virus regulatory protein Vmw65. *J Virol* 1989;63:1641–1650.
227. Greaves RF, O'Hare P. Structural requirements in the herpes simplex virus type 1 transactivator Vmw65 for interaction with the cellular octamer-binding protein and target TAATGARAT sequences. *J Virol* 1990;64:2716–2724.
228. Green MR, Maniatis T, Melton DA. Human ß-globin pre-mRNA synthesized in vitro is accurately spliced in Xenopus oocyte nuclei. *Cell* 1983;32:681–694.
229. Guzowski J, Wagner EK. Mutational analysis of the herpes simplex virus type 1 strict late U_L38 promoter/leader reveals two regions critical in transcriptional regulation. *J Virol* 1993;67:5098–5108.
230. Hall LM, Draper KG, Fluck RJ, Carter RH, Wagner EK. Herpes simplex virus mRNA species mapping in EcoRI fragment I. *J Virol* 1982;43:594–607.
231. Halpern ME, Smiley JR. Effects of deletions on expression of the herpes simplex virus thymidine kinase gene from the intact viral genome: the amino terminus of the enzyme is dispensable for catalytic activity. *J Virol* 1984;50:733–738.
232. Hammer SM, Buchman TG, D'Angelo LJ, Karchmer AW, Roizman B, Hirsch MS. Temporal cluster of herpes simplex encephalitis: investigation by restriction endonuclease cleavage of viral DNA. *J Infect Dis* 1980;141:436–440.
233. Harbour DA, Hill TJ, Blyth WA. Recurrent herpes simplex in the mouse: inflammation in the skin and activation of virus in the ganglia following peripheral stimuli. *J Gen Virol* 1983;64:1491–1498.
234. Hardwick J, Romanowski E, Arullo-Cruz T, Gordon YJ. Timolol pro-

motes reactivation of latent HSV-1 in the mouse iontophoresis model. *Invest Ophthalmol Vis Sci* 1987;28:580–584.

235. Hardwicke MA, Vaughan PJ, Sekulovich RE, O'Conner R, Sandri-Goldin RM. The regions important for the activator and repressor functions of herpes simplex virus type 1 α protein ICP27 map to the C-terminal half of the molecule. *J Virol* 1989;63:4590–4602.
236. Hardy WR, Hardwicke MA, Sandri-Goldin RM. The HSV-1 regulatory protein ICP27 appears to impair host cell splicing. In: *Abstracts of the 17th International Herpesvirus Workshop, Edinburgh.* 1992:105.
237. Hay J, Subak-Sharpe JH. Mutants of herpes simplex virus types 1 and 2 that are resistant to phosphonoacetic acid induce altered DNA polymerase activities in infected cells. *J Gen Virol* 1976;31:145–148.
238. Hayward GS, Frenkel N, Roizman B. The anatomy of herpes simplex virus DNA: strain differences and heterogeneity in the locations of restriction endonuclease cleavage sites. *Proc Natl Acad Sci USA* 1975;72:1768–1772.
239. Hayward GS, Jacob RJ, Wadsworth SC, Roizman B. Anatomy of herpes simplex virus DNA: evidence for four populations of molecules that differ in the relative orientations of their long and short segments. *Proc Natl Acad Sci USA* 1975;72:4243–4247.
240. He X, Treacy MN, Simmons DM, Ingraham HA, Swanson LW, Rosenfeld MG. Expression of a large family of POU-domain regulatory genes in mammalian brain development. *Nature* 1989;340:35–42.
241. Heilbronn R, Weller SK, zur Hausen H. Herpes simplex virus type 1 mutants for the origin-binding protein induce DNA amplification in the absence of viral replication. *Virology* 1990;179:478–481.
242. Heine JW, Honess RW, Cassai E, Roizman B. Proteins specified by herpes simplex virus. XII. The virion polypeptides of type 1 strains.*J Virol* 1974;14:640–651.
243. Heine UI. Intranuclear viruses. In: Busch H, ed. *The cell nucleus.* New York: Academic Press; 1974:489.
244. Hernandez TR, Lehman IR. Functional interaction between the herpes simplex 1 DNA polymerase and UL42 protein. *J Biol Chem* 1990;265:11227–11232.
245. Herold BC, WuDunn D, Soltys N, Spear PG. Glycoprotein C of herpes simplex virus type 1 plays a principal role in the adsorption of virus to cells and in infectivity. *J Virol* 1991;65:1090–1098.
246. Herr W, Sturm RA, Clerc RG, Corcoran LM, Baltimore D, Sharp PA, Ingraham HA, Rosenfeld MG, Finney M, Ruvkun G, Horvitz HR. The POU domain: a large conserved region in the mammaian *pit*-1, *oct*-1, *oct*-2, and *Caenorhabditis elegans unc*-86 gene products. *Genes Dev* 1988;2:1513–1516.
247. Herz C, Roizman B. The α promoter regulator-ovalbumin chimeric gene resident in human cells is regulated like the authentic α4 gene after infection with herpes simplex virus 1 mutants in α4 gene. *Cell* 1983;33:145–151.
247a. Hill A, Jugovic P, York I, Russ G, Bennink J, Yewdell J, Ploegh H, Johnson D. Herpes simplex virus turns off the TAP to evade host immunity. *Nature* 1995;375:411–415.
248. Hill TJ, Blyth WA, Harbour DA. Recurrent herpes simplex in mice: topical treatment with acyclovir cream. *Antiviral Res* 1982;2:135–146.
249. Hill TJ, Blyth WA, Harbour DA. Trauma to the skin causes recurrence of herpes simplex in the mouse. *J Gen Virol* 1978;39:21–28.
250. Hill JM, Sederati F, Javier RT, Wagner EK, Stevens JG. Herpes simplex virus latent phase transcription facilitates in vivo reactivation. *Virology* 1990;174:117–125.
251. Ho DY, Mocarski ES. Herpes simplex virus latent RNA (LAT) is not required for latent infection in the mouse. *Proc Natl Acad Sci USA* 1989;86:7596–7600.
252. Hoggan MD, Roizman B. The isolation and properties of a variant of herpes simplex producing multinucleated giant cells in monolayer cultures in the presence of antibody. *Am J Hyg* 1959;70:208–219.
253. Holland LE, Anderson KP, Shipman C, Wagner EK. Viral DNA synthesis is required for the efficient expression of specific herpes virus type 1 mRNA species. *Virology* 1980;101:50–53.
254. Holland LE, Anderson KP, Stringer JR, Wagner EK. Isolation and localization of herpes simplex virus type 1 mRNA abundant before viral DNA synthesis. *J Virol* 1979;31:447–462.
255. Holland LE, Sandri-Goldin RM, Goldin AL, Glorioso JC, Levine M. Transcriptional and genetic analyses of the herpes simplex virus type 1 genome: coordinates 0.29 to 0.45. *J Virol* 1984;49:947–959.
256. Holmes AM, Wietstock SM, Ruyechan WT. Identification and characterization of a DNA primase activity present in herpes simplex virus type 1-infected HeLa cells. *J Virol* 1988;62:1038–1045.
257. Homa FL, Glorioso JC, Levine M. A specific 15-bp TATA box promoter element is required for expression of a herpes simplex virus type 1 late gene. *Genes Dev* 1988;2:40–53.
258. Homa FL, Otal TM, Glorioso JC, Levine M. Transcriptional control signals of a herpes simplex virus type 1 late (γ_2) gene lie within bases −34 to −124 relative to the 5′ terminus of the mRNA. *Mol Cell Biol* 1986;6:3652–3666.
259. Honess RW, Purifoy DJM, Young D, Gopal R, Cammack N, O'Hare P. Single mutations at many sites within the DNA polymerase locus of herpes simplex viruses can confer hypersensitivity to aphidocolin and resistance to phosphonoacetic acid. *J Gen Virol* 1984:65:1–17.
260. Honess RW, Roizman B. Regulation of herpesvirus macro-molecular synthesis: sequential transition of polypeptide synthesis requires functional viral polypeptides. *Proc Natl Acad Sci USA* 1975;72:1276–1280.
261. Honess RW, Roizman B. Regulation of herpesvirus macro-molecular synthesis. I. Cascade regulation of the synthesis of three groups of viral proteins. *J Virol* 1974;14:8–19.
262. Honess RW, Roizman B. Proteins specified by herpes simplex virus. XI. Identification and relative molar rates of synthesis of structural and non-structural herpesvirus polypeptides in infected cells. *J Virol* 1973;12:1346–1365.
263. Hope RG, Marsden HS. Processing of glycoproteins induced by herpes simplex virus type 1: sulphation and nature of the oligosaccharide linkages. *J Gen Virol* 1983;64:1943–1953.
264. Huang A, Jacobi G, Haj-Ahmad Y, Bacchetti S. Expression of the HSV-2 ribonucleotide reductase subunits in adenovirus vectors or stably transformed cells: restoration of enzymatic activity by reassociation of enzyme subunits in the absence of other HSV proteins. *Virology* 1988;163:462–470.
265. Huang AS, Wagner RR. Penetration of herpes simplex virus into human epidermoid cells. *Proc Soc Exp Biol Med* 1966;116:863–869.
266. Huang C-J, Goodart SA, Rice MK, Guzowski JF, Wagner EK. Mutational analysis of sequence downstream of the TATA box of the herpes simplex virus type 1 major promoter capsid protein (VP5/U_L19) promoter. *J Virol* 1993;67:5109–5116.
267. Hubbard SC, Ivatt RJ. Synthesis and processing of asparagine-linked oligosaccharides. *Annu Rev Biochem* 1981;50:555–583.
268. Hubenthal-Voss J, Starr L, Roizman B. The herpes simplex virus origins of DNA synthesis in the S component are each contained in a transcribed open reading frame. *J Virol* 1987;61:3349–3355.
269. Hung S-L, Srinivasan S, Friedman HM, Eisenberg RJ, Cohen GH. Structural basis of C3b binding by glycoprotein C of herpes simplex virus. *J Virol* 1992;66:4013–4027.
270. Huszar D, Bacchetti S. Partial purification and characterization of the ribonucleotide reductase induced by herpes simplex virus infection of mammalian cells. *J Virol* 1981;37:580–588.
271. Hutchinson L, Browne H, Wargent V, Davis-Poynter N, Primorac S, Goldsmith K, Minson A, Johnson DC. A novel herpes simplex virus glycoprotein, gL, forms a complex with glycoprotein H (gH) and affects normal folding and surface expression of gH. *J Virol* 1992;6:2240–2250.
272. Igarashi K, Fawl R, Roller R, Roizman B. Construction and properties of a recombinant herpes simplex virus 1 lacking both S component origins of DNA synthesis. *J Virol* 1993;67:2123–2132.
273. Imbalzano AN, Coen DM, DeLuca NA. Herpes simplex virus transactivator ICP4 operationally substitutes for the cellular transcription factor Sp1 for efficient expression of the viral thymidine kinase gene. *J Virol* 1991;65:565–574.
274. Imbalzano AN, Shapard AA, DeLuca NA. Functional relevance of specific interactions between herpes simplex virus type 1 ICP4 and sequences from the promoter-regulatory domain of the viral thymidine kinase gene. *J Virol* 1990;64:2620–2631.
275. Ingemarson R, Lankinen H. The herpes simplex virus type 1 ribonucleotide reductase is a tight complex of the type α2ß2 composed of 40K and 140K proteins, of which the latter shows multiple forms due to proteolysis. *Virology* 1987;56:417–422.
276. Ingles CJ, Shales M, Cress WD, Triezenberg SJ, Greenblatt J. Reduced binding of TFIID to transcriptionally compromised mutants of VP16. *Nature* 1991;51:588–590.
277. Jacob RJ, Morse LS, Roizman B. Anatomy of herpes simplex virus DNA. XIII. Accumulation of head to tail concatemers in nuclei of infected cells and their role in the generation of the four isomeric arrangements of viral DNA. *J Virol* 1979;9:448–457.
278. Jacob RJ, Roizman B. Anatomy of herpes simplex virus DNA. VIII. Properties of the replicating DNA. *J Virol* 1977;23:394–411.

279. Jacobson JG, Leib DA, Goldstein DJ, Bogard CC, Schaffer PA, Weller SK, Coen DM. A herpes simplex virus ribonucleotide reductase deletion mutant is defective for productive, acute, and reactivatible infections of mice and for replication in mouse cells. *Virology* 1989;3:276–283.
280. Jacobson JG, Martin SL, Coen DM. A conserved open reading frame that overlaps the herpes simplex virus thymidine kinase gene is important for viral growth in cell culture. *J Virol* 1989;63:1839–1843.
281. Jamieson AT, Subak-Sharpe JH. Biochemical studies on the herpes simplex virus-specified deoxypyrimidine kinase activity. *J Gen Virol* 1974;4:481–492.
282. Javier RT, Izumi KM, Stevens JG. Localization of a herpes simplex virus neurovirulence gene dissociated from high-titer virus replication in the brain. *J Virol* 1988;2:1381–1387.
283. Javier RT, Stevens JG, Dissette VB, Wagner EK. A herpes simplex virus transcript abundant in latently infected neurons is dispensable for establishment of the latent state. *Virology* 1988;66:254–257.
284. Javier RT, Thompson RL, Stevens JG. Genetic and biological analyses of a herpes simplex virus intertypic recombinant reduced specifically for neurovirulence. *J Virol* 1987;65:1978–1984.
285. Jenkins FJ, Casadaban M, Roizman B. Application of the mini Mu phage for target sequence specific insertional mutagenesis of the herpes simplex virus genome. *Proc Natl Acad Sci USA* 1985;82:4773–4777.
286. Jenkins FJ, Roizman B. Herpes simplex virus recombinants with noninverting genomes frozen in different isomeric arrangements are capable of independent replication. *J Virol* 1986;59:494–499.
287. Jofrie JT, Schaffer PA, Parris DS. Genetics of resistance to phosphonoacetic acid in strain KOS of herpes simplex type 1. *J Virol* 1977;23:833–836.
288. Johnson DC, Frame MC, Ligas MW, Cross AM, Stow ND. Herpes simplex virus immunoglobulin G Fc receptor activity depends on a complex of two viral glycoproteins, gE and gI. *J Virol* 1988;62: 1347–1354.
289. Johnson DC, Ligas MW. Herpes simplex viruses lacking glycoprotein D are unable to inhibit virus penetration: quantitative evidence for virus-specific cell surface receptors. *J Virol* 1988;62:4605–4612.
290. Johnson DC, McDermott MR, Chrisp C, Glorioso JC. Pathogenicity in mice of herpes simplex virus type 2 mutants unable to express glycoprotein C. *J Virol* 1986;58:36–42.
291. Johnson DC, Spear PG. Monensin inhibits the processing of herpes simplex virus glycoproteins, their transport to the cell surface, and the egress of virions from infected cells. *J Virol* 1982;43:1102–1112.
292. Johnson DC, Spear PG. O-linked oligosaccharides are acquired by herpes simplex virus glycoproteins in the Golgi apparatus. *Cell* 1983;32:987–997.
293. Johnson DC, Spear PG. Evidence for translational regulation of herpes simplex virus type 1 gD expression. *J Virol* 1984;51:389–394.
294. Johnson PA, Everett RD. The control of herpes simplex virus type-1 late gene trancription: a `TATA-box'/cap-site region is sufficient for fully efficient regulated activity. *Nucleic Acids Res* 1986;14:8247–8264.
295. Johnson PA, MacLean C, Marsden HS, Dalziel RG, Everett RD. The product of gene US11 of herpes simplex virus type 1 is expressed as a true late gene. *J Gen Virol* 1986;67:871–883.
296. Jones KA, Tijan R. Sp1 binds to promoter sequences and activates herpes simplex virus `immediate-early' gene transcription in vitro. *Nature* 1985;317:179–185.
297. Jones KA, Yamamoto KR, Tjian R. Two distinct transcription factors bind to the HSV thymidine kinase promoter in vitro. *Cell* 1985; 42:559–572.
298. Jones N, Shenk T. An adenovirus 5 early gene product function regulates expression of other early viral genes. *Proc Natl Acad Sci USA* 1979;76:3665–3669.
299. Jones PC, Hayward GS, Roizman B. Anatomy of herpes simplex virus DNA. VII. αRNA is homologous to noncontiguous sites in the L and S components of viral DNA. *J Virol* 1977;21:268–278.
300. Jones PC, Roizman B. Regulation of herpesvirus macromolecular synthesis: VIII. The transcription program consists of three phases during which both extent of transcription and accumulation of RNA in the cytoplasm are regulated. *J Virol* 1979;31:299–314.
301. Kaner RA, Baird A, Mansukhani A, Basilico C, Summers B, Florkiewicz R, Hajjar D. Fibroblast growth factor receptor is a portal of cellular entry for herpes simplex virus type 1. *Science* 1990; 248:1410–1413.
302. Kaplan AS, Ben-Porat T. The effect of pseudorabies virus on the nucleic acid metabolism and on the nuclei of rabbit kidney cells. *Virology* 1959;8:352–366.
303. Katan M, Haigh A, Verrijzer CP, van der Vliet PC, O'Hare P. Characterization of a cellular factor which interacts functionally with Oct-1 in the assembly of a multicomponent transcription complex. *Nucleic Acids Res* 1990;18:6871–6880.
304. Katz JP, Bodin ET, Coen DM. Quantitative polymerase chain reaction analysis of herpes simplex virus DNA in ganglia of mice infected with replication incompetent mutants. *J Virol* 1990;64:4288–4295.
305. Keir HM. Virus-induced enzymes in mammalian cells infected with DNA-viruses. In: Crawford LV, Stoker MGP, eds. *The molecular biology of viruses.* Cambridge: Cambridge University Press; 1968:67–99.
306. Keir HM, Gold E. Deoxyribonucleic acid nucleotidyltransferase and deoxyribonuclease from cultured cells infected with herpes simplex virus. *Biochim Biophys Acta* 1963;72:263–276.
307. Kelleher RJ III, Flanagan PM, Kornberg RD. A novel mediator between activator proteins and the RNA polymerase II transcription apparatus. *Cell* 1990;61:1209–1215.
308. Kibler PK, Duncan J, Keith BD, Hupel T, Smiley JR. Regulation of herpes simplex virus true late gene expression: sequences downstream from the U_S11 TATA box inhibit expression from an unreplicated template. *J Virol* 1991;65:6749–6760.
309. Kieff ED, Bachenheimer SL, Roizman B. Size, composition and structure of the DNA of subtypes 1 and 2 herpes simplex virus. *J Virol* 1971;8:125–129.
310. Kit S, Dubbs DR. Properties of deoxythymidine kinase partially purified from noninfected and virus-infected mouse fibroblast cells. *Virology* 1965;26:16–27.
311. Kit S, Dubbs DR. Acquisition of thymidine kinase activity by herpes simplex infected mouse fibroblast cells. *Biochem Biophys Res Commun* 1963;11:55–59.
312. Klein RJ. Effect of immune serum on the establishment of herpes simplex virus infection in trigeminal ganglia of hairless mice. *J Gen Virol* 1980;49:401–405.
313. Klein RJ. Pathogenetic mechanisms of recurrent herpes simplex viral infections. *Arch Virol* 1976;51:1–13.
314. Klein RJ, Friedman AE, Yellin PB. Orofacial herpes simplex virus infection in hairless mice: latent virus in trigeminal ganglia after topical antiviral treatment. *Infect Immun* 1978;20:130–135.
315. Klein RJ, Friedman-Kien AE, Brady E. Latent herpes simplex virus infection in ganglia of mice after primary infection and re-inoculation at a distant site. *Arch Virol* 1978;57:161–166.
316. Klein RJ, Friedman-Kien AE, DeStefano E. Latent herpes simplex virus infections in sensory ganglia of hairless mice prevented by acycloguanosine. *Antimicrob Agents Chemother* 1979;15:723–729.
317. Klein RJ, Friedman-Kien AE, Fondak AA, Buimovici-Klein E. Immune response and latent infection after topical treatment of herpes simplex virus infection in hairless mice. *Infect Immun* 1977;16:842–848.
318. Klemperer HG, Haynes GR, Sheddon WIH, Watson DH. A virus-specific thymidine kinase in BHK 2l cells infected with herpes simplex virus. *Virology* 1967;31:120–128.
319. Knipe DM, Ruyechan WT, Roizman B. Molecular genetics of herpes simplex virus. III. Fine mapping of a genetic locus determining resistance to phosphonoacetate by two methods of marker transfer. *J Virol* 1979;29:698–704.
320. Knipe DM, Ruyechan WT, Roizman B, Halliburton IW. Molecular genetics of herpes simplex virus. Demonstration of regions of obligatory and non-obligatory identity in diploid regions of the genome by sequence replacement and insertion. *Proc Natl Acad Sci USA* 1978; 75:3896–3900.
321. Knopf CW, Spies B, Kaerner HC. The DNA replication origins of herpes simplex virus type 1 strain Angelotti. *Nucleic Acid Res* 1986;14:8655–8667.
322. Koch H-G, Rosen A, Ernst F, Becker Y, Darai G. Determination of the nucleotide sequence flanking the deletion (0.762 to 0.789 map units) in the genome of and intraperitoneally avirulent HSV-1 strain HFEM. *Virus Res* 1987;7:105–115.
323. Koff A, Schwedes JF, Tegtmeyer P. Herpes simplex virus origin binding protein (U_L9) loops and distorts the viral rteplication origin. *J Virol* 1991;65:3284–3292.
324. Koff A, Tegtmeyer P. Characterization of major recognition sequences for a herpes simplex virus type 1 origin-binding protein. *J Virol* 1988;2:4096–4103.
325. Koop KE, Duncan J, Smiley JR. Binding sites for the herpes simplex

virus immediate early protein ICP4 impose an increased dependence on viral DNA replication on simple model promoters located in the viral genome. *J Virol* 1993;67:7254–7263.

326. Kosz-Vnenchak M, Jacobson J, Coen DM, Knipe DM. Evidence for a novel regulatory pathway for herpes simplex virus gene expression in trigeminal ganglion neurons. *J Virol* 1993;67:5383–5393.
327. Kousoulas KG, Pellett PE, Pereira L, Roizman B. Mutations affecting conformation or sequence of neutralizing epitopes identified by reactivity of viable plaques segregate from syn and ts domains of HSV-1(F) gB gene. *Virology* 1984;35:379–395.
328. Kozak M, Roizman B. RNA synthesis in cells infected with herpes simplex virus. IX. Evidence for accumulation of abundant symmetric transcripts in nuclei. *J Virol* 1975;15:36–40.
329. Kozak M, Roizman B. Regulation of herpesvirus macromolecular synthesis: nuclear retention of non-translated viral RNA sequences. *Proc Natl Acad Sci USA* 1974;71;4322–4326.
330. Krause PR, Croen KD, Straus SE, Ostrove JM. Detection and preliminary characterization of herpes simplex virus type 1 transcripts in latently infected human trigeminal ganglia. *J Virol* 1988;62:4819–4823.
331. Kristensson K, Lycke E, Roytta M, Svennerholm B, Vahlne A. Neuritic transport of herpes simplex virus in rat sensory neurons in vitro. Effects of substances interacting with microtubular function and axonal flow [Nocodazone, Taxol and Erythro-9-3-(2-hydroxynonyl)adenine]. *J Gen Virol* 1986;67:2023–2028.
332. Kristie TM, LeBowitz JH, Sharp PA. The octamer-binding proteins form multi-protein-DNA complexes with the HSV αTIF regulatory protein. *EMBO J* 1989;8:4229–4238.
332a. Kristie TM, Pomerantz JL, Twomey TC, Parent SA, Sharp PA. The cellular C1 factor of herpes simplex virus enhancer complex is a family of polypeptides. *J Biol Chem* 1995;270:4387–4394.
333. Kristie TM, Roizman B. Differentiation and DNA contact points of host proteins binding at the *cis* site for virion-mediated induction of α genes of herpes simplex virus 1. *J Virol* 1988;62:1145–1157.
334. Kristie TM, Roizman B. Host cell proteins bind to the cis-acting site required for virion-mediated induction of herpes simplex virus 1 α genes. *Proc Natl Acad Sci USA* 1987;84:71–75.
335. Kristie TM, Roizman B. DNA-binding site of major regulatory protein α4 specifically associated with the promoter-regulatory domains of α genes of herpes simplex virus type 1. *Proc Natl Acad Sci USA* 1986; 83:4700–4704.
336. Kristie TM, Roizman B. α4, the major regulatory protein of herpes simplex virus type 1, is stably and specifically associated with promoter-regulatory domains of α genes and of selected other viral genes. *Proc Natl Acad Sci USA* 1986;83:3218–3222.
337. Kristie TM, Roizman B. Separation of sequences defining basal expression from those conferring α gene recognition within the regulatory domains of herpes simplex virus 1 α genes. *Proc Natl Acad Sci USA* 1984;81:4065–4069.
338. Kristie TM, Sharp PA. Interactions of the Oct-1 POU subdomains with specific DNA sequences and with the HSV α-trans-activator protein. *Genes Dev* 1990;4:2383–2396.
339. Kwon BS, Gangarosa LP, Burch KD, deBack J, Hill JM. Induction of ocular herpes simplex virus shedding by iontophoresis of epinephrine into rabbit cornea. *Invest Ophthamol Vis Sci* 1981;21:442–449.
340. Kwon BS, Gangarosa LP, Green K, Hill JM. Kinetics of ocular herpes simplex virus shedding induced by epinephrine iontophoresis. *Invest Ophthalmol Vis Sci* 1982;22:818–821.
341. Kwong AD, Frenkel N. Herpes simplex virus-infected cells contain a function(s) that destabilizes both host and viral mRNAs. *Proc Natl Acad Sci USA* 1987;84:1926–1930.
342. Kwong AD, Kruper JA, Frenkel N. Herpes simplex virus virion host shutoff function. *J Virol* 1988;62:912–921.
343. LaBella F, Sive HL, Roeder RG, Heintz N. Cell-cycle regulation of a human histone H2b gene is mediated by the H2b subtype-specific consensus element. *Genes Dev* 1988;2:32–39.
344. Ladin BF, Blankenship ML, Ben-Porat T. Replication of herpesvirus DNA. V. The maturation of concatemeric DNA of pseudorabies virus to genome length is related to capsid formation. *J Virol* 1980;33:1151–1164.
345. Ladin BF, Ihara S, Hampl H, Ben-Porat T. Pathway of assembly of herpesvirus capsids: an analysis using DNA temperature sensitive mutants of pseudorabies virus. *Virology* 1982;116:544–561.
346. Lagunoff M, Roizman B. Expression of a herpes simplex virus 1 open reading frame antisense to the $\gamma_1 34.5$ gene and transcribed by an RNA 3′ coterminal with the unspliced latency associated transcript. *J Virol* 1994;68:6021–6028.
346a. Lagunoff M, Roizman B. The regulation of synthesis and properties of the protein product of the open reading frame P of the herpes simplex virus 1 genome. *J Virol* 1995;69:3615–3623.
347. Lam V, Eamon W, Rice S. Identification of functional regions in the amino-terminal half of the HSV-1 regulatory protein ICP27. In: *Abstracts of the 17th International Herpesvirus Workshop.* 1992:85.
348. LaMarco KL, McKnight SL. Purification of a set of cellular polypeptides that bind to the purine-rich cis-regulatory element of herpes simplex virus immediate early genes. *Genes Dev* 1989;3:1372–1383.
349. LaMarco KL, Thompson CC, Byers BP, Walton EM, McKnight SL. Identification of Ets- and notch-related subunits in GA binding protein. *Science* 1991;253:789–791.
350. Lampert F, Bahr GF, Rabson AS. Herpes simplex virus: dry mass. *Science* 1969;166:1163–1165.
351. Lando D, Ryhiner ML. Pouvoir infectieux du DNA d'herpes virus hominis en culture cellulaire. *C R Acad Sci* 1969;269:527.
352. Lee CK, Knipe DM. An immunoassay for the study of DNA-binding activities of herpes simplex virus protein ICP8. *J Virol* 1985;54:731–738.
353. Lee CK, Knipe DM. Thermolabile in vivo DNA-binding activity associated with a protein encoded by mutants of herpes simplex virus type 1. *J Virol* 1983;46:909–919.
354. Lee GTY, Para MF, Spear PG. Location of the structural genes for glycoproteins gD and gE and for other polypeptides in the S component of herpes simplex virus type 1 DNA. *J Virol* 1982;43:41–49.
355. Lee GTY, Pogue-Geile KL, Pereira L, Spear PG. Expression of herpes simplex virus glycoprotein C from a DNA fragment inserted into the thymidine gene of this virus. *Proc Natl Acad Sci USA* 1982; 79:6612–6616.
356. Leetsma JE, Bornstein MB, Sheppard RD, Feldman LA. Ultrastructural aspects of herpes simplex virus infection in organized cultures of mammalian nervous tissue. *Lab Invest* 1969;20:70–78.
357. Leib DA, Bogard CL, Kosz-Vnenchak M, Hicks KA, Coen DM, Knipe DM, Schaffer PA. A deletion mutant of the latency associated transcript of herpes simplex virus type 1 reactivates from the latent state with reduced frequency. *J Virol* 1989;63:2893–2900.
358. Leib DA, Coen DM, Bogard CL, Hicks KA, Yager DR, Knipe DM, Schaffer PA. Immediate-early gene mutants define different stages in the establishment and reactivation of herpes simplex virus latency.*J Virol* 1989;63:759–768.
359. Leist TP, Sandri-Goldin RM, Stevens JG. Latent infections in spinal ganglia with thymidine kinase deficient herpes simplex virus. *J Virol* 1989;63:4976–4978.
360. LeMaster S, Roizman B. Herpes simplex virus phosphoproteins: II. Characterization of the virion protein kinase and of the polypeptides phosphorylated in the virion. *J Virol* 1980;35:798–811.
361. Leung W-C, Dimock K, Smiley J, Bacchetti S. HSV thymidine kinase transcripts are absent from both nucleus and cytoplasm during infection in the presence of cycloheximide. *J Virol* 1980;36:361–365.
362. Li E, Tabas I, Kornfeld S. The synthesis of complex type oligosaccharides. I. Structure of the lipid-linked oligosaccharide precursor of the complex type oligosaccharides of the vesicular stomatitis virus G protein. *J Biol Chem* 1978;253:7762–7770.
363. Ligas MW, Johnson DC. A herpes simplex virus mutant in which glycoprotein D sequences are replaced by ß-galactosidase sequences binds to but is unable to penetrate into cells. *J Virol* 1988;62:1486–1494.
364. Liljelund P, Ingles CJ, Greenblatt J. Altered promoter binding of the TATA box-binding factor induced by the transcriptional activation domain of VP16 and suppressed by TFIIA. *Mol Gen Genet* 1993;241:694–699.
365. Lin Y-S, Green MR. Mechanism of action of an acidic transcriptional activator in vitro. *Cell* 1991;64:971–981.
366. Linneman CC, Buchman TG, Light IJ, Ballard JL, Roizman B. Transmission of herpes simplex virus type 1 in a nursery for the newborn: identification of viral isolates by DNA "fingerprinting." *Lancet* 1978;1:964–966.
367. Little SP, Schaffer PA. Expression of the syncytial (syn) phenotype in HSV-12, strain KOS: genetic and phenotypic studies of mutants in two syn loci. *Virology* 1981;112:686–702.
368. Liu F, Roizman B. Characterization of the protease and of other products of the amino terminus proximal cleavage of the herpes simplex virus 1 $U_L 26$ protein. *J Virol* 1993;67:1441–1452.
369. Liu F, Roizman, B. Differentiation of multiple domains in the herpes

simplex virus 1 protease encoded by the U_L26 gene. *Proc Natl Acad Sci USA* 1992;89:2076–2080.
370. Liu F, Roizman B. The herpes simplex virus 1 gene encoding a protease also contains within its coding domain the gene encoding the more abundant substrate. *J Virol* 1991;65:5149–5156.
371. Liu F, Roizman B. The promoter, transcriptional unit and coding sequence of herpes simplex virus 1 family 35 proteins are contained within and in frame with the U_L26 open reading frame. *J Virol* 1991;65:206–212.
372. Locker H, Frenkel N. Bam HI, Kpn I and Sal I restriction enzyme maps of the DNAs of herpes simplex virus strains Justin and F: occurrence of heterogeneities in defined regions of the viral DNA. *J Virol* 1979; 32:424–441.
373. Locker H, Frenkel H, Halliburton I. Structure and expression of class II defective herpes simplex virus genomes encoding infected cell polypeptide number 8. *J Virol* 1982;43:574–593.
374. Lockshon D, Galloway DA. Cloning and characterization of oriL2, a large palindromic DNA replication origin of herpes simplex virus type 2. *J Virol* 1986;62:513–521.
375. Longnecker R, Chatterjee S, Whitley RJ, Roizman B. Identification of a novel herpes simplex virus 1 glycoprotein gene within a gene cluster dispensable for growth in cell culture. *Proc Natl Acad Sci USA* 1987;84:4303–4307.
376. Longnecker R, Roizman B. Clustering of genes dispensable for growth in cell culture in the small component of the herpes simplex virus 1 genome. *Science* 1987;236:573–576.
377. Longnecker R, Roizman B. Generation of an inverting herpes simplex virus 1 mutant lacking the L-S junction a sequences, an origin of DNA synthesis, and several genes including those specifying glycoprotein E and the α47 gene. *J Virol* 1986;58:583–591.
378. Lycke E, Kristensson K, Svennerholm B, Vahlne A, Ziegler R. Uptake and transport of herpes simplex virus in neurites of rat dorsal root ganglia cells in culture. *J Gen Virol* 1984;65:55–64.
379. Lynas C, Laycock KA, Cook SD, Hill TJ, Blyth WA, Maitland NJ. Detection of herpes simplex virus type 1 gene expression in latently and productively infected mouse ganglia using the polymerase chain reaction. *J Gen Virol* 1989;70:2345–2355.
380. Mackem S, Roizman B. Regulation of α genes of herpes simplex virus: the α27 gene promoter-thymidine kinase chimera is positively regulated in converted L cells. *J Virol* 1982;43:1015–1023.
381. Mackem S, Roizman B. Differentiation between α promoter and regulator regions of herpes simplex virus1: the functional domains and sequence of a movable α regulator. *Proc Natl Acad Sci USA* 1982; 79:4917–4921.
382. Mackem S, Roizman B. Structural features of the herpes simplex virus α gene 4, 0, and 27 promoter-regulatory sequences which confer α regulation on chimeric thymidine kinase genes. *J Virol* 1982;44:939–949.
383. Mackem S, Roizman B. Regulation of herpesvirus macromolecular synthesis: transcription-initiation sites and domains of α genes. *Proc Natl Acad Sci USA* 1980;77:7122–7126.
384. MacLean CA, Dolan A, Jamieson FE, McGeoch DJ. The myristylated virion proteins of herpes simplex virus type 1: investigation of their role in the virus life cycle. *J Gen Virol* 1992;73:539–547.
385. MacLean CA, Efstathiou S, Elliott ML, Jamieson FE, McGeoch DJ. Investigation of herpes simplex virus type 1 genes encoding multiply inserted membrane proteins. *J Gen Virol* 1991;72:897–906.
386. Manservigi R, Spear PG, Buchan A. Cell fusion induced by herpes simplex virus is promoted and suppressed by different viral glycoproteins. *Proc Natl Acad Sci USA* 1977;74:3913–3917.
387. Marchetti ME, Smith CA, Schaffer PA. A temperature-sensitive mutation in a herpes simplex virus type 1 gene required for viral DNA synthesis maps to coordinates 0.609 through 0.614 in UL. *J Virol* 1988;62:715–721.
388. Mark GE, Kaplan AS. Synthesis of protein in cells infected with herpesvirus. VII. Lack of migration of structural viral proteins to the nucleus of arginine-deprived cells. *Virology* 1971;45:53–60.
389. Marsden HS, Campbell MEM, Haarr L, Frame MC, Parris DS, Murphy M, Hope RG, Muller MT, Preston CM. The 65,000-Mr DNA-binding and virion trans-inducing proteins of herpes simplex virus type 1. *J Virol* 1987;61:2428–2437.
390. Marsden HS, Stow ND, Preston VG, Timbury MC, Wilkie NM. Physical mapping of herpes simplex virus induced polypeptides. *J Virol* 1978;28:624–642.
391. Marshall R. Glycoproteins. *Annu Rev Biochem* 1978;41:673–702.
392. Mattaj IW, Lienhard S, Jiricny J, DeRobertis EM. An enhancer-like sequence within the Xenopus U2 gene promoter facilitates the formation of stable transcription complexes. *Nature* 1985;316:163–167.
393. Matthews TJ, Cohen GH, Eisenberg RJ. Synthesis and processing of glycoprotein D of herpes simplex virus types 1 and 2 in an *in vitro* system. *J Virol* 1983;48:521–533.
394. Matz B, Subak-Sharpe JH, Preston VG. Physical mapping of temperature-sensitive mutations of herpes simplex virus type 1 using cloned restriction endonuclease fragments. *J Gen Virol* 1983;64:2261–2270.
395. Mavromara-Nazos P, Ackermann M, Roizman B. Construction and properties of a viable herpes simplex virus 1 recombinant lacking the coding sequences of the α47 gene. *J Virol* 1986;60:807–812.
396. Mavromara-Nazos P, Roizman B. Delineation of regulatory domains of early (ß) and late (γ_2) genes by construction of chimeric genes expressed in herpes simplex virus 1 genomes. *Proc Natl Acad Sci USA* 1989;86:4071–4079.
397. Mavromara-Nazos P, Roizman B. Activation of herpes simplex virus 1 γ_2 genes by viral DNA replication. *Virology* 1987;161:593–598.
398. McCarthy AM, McMahan L, Schaffer PA. Herpes simplex virus type 1 ICP27 deletion mutants exhibit altered patterns of transcription and are DNA deficient. *J Virol* 1989;63:18–27.
399. McClure HM, Swanson RB, Kalter SS, Lester TL. Natural genital herpesvirus hominis infections in chimpanzees (*Pan trogloditis* and *Pan iseus*). *Lab Animal Sci* 1980;30:895–901.
400. McCormick L, Roller RJ, Roizman B. Characterization of a herpes simplex virus sequence which binds a cellular protein as either a single stranded or double stranded DNA or RNA. *J Virol* 1992;66:3435–3447.
401. McCracken RM, Clarke JK. A thin section study of the morphogenesis of Aujeszky's disease virus in synchronously infected cell cultures. *Arch Ges Virusforsch* 1971;34:189–201.
402. McGeoch DJ, Dalrymple MA, Davison AJ, Dolan A, Frame MC, McNab D, Perry LJ, Scott JE, Taylor P. The complete DNA sequence of the long unique region in the genome of herpes simplex virus type 1. *J Gen Virol* 1988;69:1531–1574.
403. McGeoch DJ, Dalrymple MA, Dolan A. Structures of herpes simplex virus type 1 genes required for replication of virus DNA. *J Virol* 1988;62:444–453.
404. McGeoch DJ, Davison AJ. Alpha herpesviruses possess a gene homologous to the protein kinase family of eukaryotes and retroviruses. *Nucleic Acids Res* 1986;14:1765–1777.
405. McGeoch DJ, Dolan A, Donald S, Rixon FJ. Sequence determination and genetic content of the short unique region in the genome of herpes simplex virus type 1. *J Mol Biol* 1985;181:1–13.
406. McKee TA, Disney GH, Everett RD, Preston CM. Control of expression of the varicella-zoster virus major immediate early gene. *J Gen Virol* 1990;71:897–906.
407. McKendall RR. Efficacy of herpes simplex virus type 1 immunisation in protecting against acute and latent infections by herpes simplex virus type 2 in mice. *Infect Immun* 1977;16:717–719.
408. McKendall RR, Klassen T, Baringer JR. Host defenses in herpes simplex infections of the nervous system: effect of antibody on disease and viral spread. *Infect Immun* 1979;23:305–311.
409. McKnight JLC, Kristie TM, Roizman B. Binding of the virion protein mediating α gene induction in herpes simplex virus 1-infected cells to its cis site requires cellular proteins. *Proc Natl Acad Sci USA* 1987;84:7061–7065.
410. McKnight SL. Functional relationships between transcriptional control signals of the thymidine kinase gene of herpes simplex virus. *Cell* 1982;31:355–365.
411. McKnight SL. The nucleotide sequence and transcript map of the herpes simplex virus thymidine kinase gene. *Nucleic Acids Res* 1980; 8:5949–5964.
412. McKnight SL, Gavis ER, Kingsbury R. Analysis of transcriptional regulatory signals of the HSV thymidine kinase gene: identification of an upstream control region. *Cell* 1981;25:385–398.
413. McKnight S, Kingsbury RC. Transcriptional control signals of a eukaryotic protein-coding gene. *Science* 1982;217:316–324.
414. McKnight SL, Kingsbury RC, Spence A, Smith M. The distal transcription signals of the herpesvirus tk gene share a common hexanucleotide control sequence. *Cell* 1984;37:253–262.
415. McKnight SL, Tijan R. Trancriptional selectivity of viral genes in mammalian cells. *Cell* 1986;46:795–805.
416. McLauchlan J, Clements JB. A 3′ co-terminus of two early herpes simplex virus type 1 mRNAs. *Nucleic Acids Res* 1982;10:501–512.

417. McLauchlan J, Rixon FJ. Characterization of enveloped tegument structures (L particles) produced by alpha herpesviruses: integrity of the tegument does not depend on the presence of capsid or envelope. *J Gen Virol* 1992;73:269–276.
418. McLauchlan J, Simpson S, Clements JB. Herpes simplex virus induces a processing factor that stimulates poly(A) site usage. *Cell* 1989; 59:1093–1105.
419. McMahan L, Schaffer PA. The repressing and enhancing functions of the herpes simplex virus regulatory protein ICP27 map to the C-terminal regions and are required to modulate viral gene expression very early in infection. *J Virol* 1990;64:3471–3485.
420. McNearny TA, Odell C, Holers VM, Spear PG, Atkinson JP. Herpes simplex virus glycoproteins gC-1 and gC-2 bind to the third component of complement and provide protection against complement-mediated neutralization of viral infectivity. *J Exp Med* 1987; 166:1525–1535.
421. Meignier B, Longnecker R, Mavromara-Nazos P, Sears A, Roizman B. Virulence of and establishment of latency by genetically engineered mutants of herpes simplex virus 1. *Virology* 1987;162:251–254.
422. Meignier B, Longnecker R, Roizman B. *In vivo* behavior of genetically engineered herpes simplex viruses R7017 and 7020: construction and evaluation in rodents. *J Infect Dis* 1988;158:602–614.
423. Meignier B, Norrild B, Roizman B. Colonization of murine ganglia by a superinfecting strain of herpes simplex virus. *Infect Immun* 1983; 41:702–708.
424. Mellerick DM, Fraser NW. Physical state of the latent herpes simplex virus genome in a mouse model system: evidence suggesting an episomal state. *Virology* 1987;158:265–275.
425. Meredith DM, Lindsay JA, Halliburton IW, Whittaker GR. Post-translational modification of the tegument proteins (VP13 and VP14) of herpes simplex virus type 1 by glycosylation and phosphorylation. *J Gen Virol* 1991;72:2771–2775.
426. Meredith M, Orr A, Everett R. Herpes simplex virus type 1 immediate early protein vmw110 binds strongly and specifically to a 135-kDa cellular protein. *Virology* 1994;200:457–469.
427. Michael N, Roizman B. Repression of the herpes simplex virus 1 α4 gene by its gene product takes place within the context of the viral genome and is associated with all three identified cognate sites. *Proc Natl Acad Sci USA* 1993;90:2286–2290.
428. Michael N, Roizman B. Determination of the number of protein monomers binding to DNA with Fab fragments of monoclonal antibodies to the protein. *Methods Mol Cell Biol* 1990;1:203–211.
429. Michael N, Roizman B. The binding of herpes simplex virus major regulatory protein to viral DNA. *Proc Natl Acad Sci USA* 1989; 86:9808–9817.
430. Michael N, Spector D, Mavromara-Nazos P, Kristie TM, Roizman B. The DNA-binding properties of the major regulatory protein α4 of herpes simplex viruses. *Science* 1988;239:1531–1534.
431. Mirda DP, Navarro D, Paz P, Lee PL, Pereira L, Williams LT. The fibroblast growth factor receptor is not required for herpes simplex virus type 1 infection. *J Virol* 1992;66:448–457.
432. Mitchell C, Blaho JD, Roizman B. Casein kinase II specifically nucleotidylylates *in vitro* the amino acid sequence of the protein encoded by the α22 gene of the herpes simplex virus 1. *Proc Natl Acad Sci USA* 1994;91:11864–11868.
433. Mitchell WJ, Deshmane SL, Dolan A, McGeoch DJ, Fraser NW. Characterization of herpes simplex virus type 2 transcription during latent infection of mouse trigeminal ganglia. *J Virol* 1990;64:5342–5348.
434. Mitchell WJ, Lirette RP, Fraser NW. Mapping of low abundance latency associated RNA in the trigeminal ganglia of mice latently infected with herpes simplex virus type 1. *J Gen Virol* 1990;71:125–132.
435. Mocarski ES, Post LE, Roizman B. Molecular engineering of the herpes simplex virus genome: insertion of a second L-S junction into the genome causes additional genome inversions. *Cell* 1980;22:243–255.
436. Mocarski ES, Roizman B. The structure and role of the herpes simplex virus DNA termini in inversion, circularization and generation of virion DNA. *Cell* 1982;31:89–97.
437. Mocarski ES, Roizman B. Herpesvirus-dependent amplification and inversion of a cell-associated viral thymidine kinase gene flanked by viral *a* sequences and linked to an origin of viral DNA replication. *Proc Natl Acad Sci USA* 1982;79:5626–5630.
438. Mocarski ES, Roizman B. Site specific inversion sequence of herpes simplex virus genome: domain and structural features. *Proc Natl Acad Sci USA* 1981;78:7047–7051.
439. Morgan C, Rose HM, Holden M, Jones EP. Electron microscopic observations on the development of herpes simplex virus. *J Exp Med* 1959;110:643–656.
440. Morgan C, Rose HM, Mednis B. Electron microscopy of herpes simplex virus. I. Entry. *J Virol* 1968;2:507–516.
441. Morse LS, Buchman TG, Roizman B, Schaffer PA. Anatomy of herpes simplex virus DNA. IX. Apparent exclusion of some parental DNA arrangements in the generation of intertypic (HSV-1 × HSV-2) recombinants. *J Virol* 1977;24:231–248.
442. Morse LS, Pereira L, Roizman B, Schaffer PA. Anatomy of HSV DNA. XI. Mapping of viral genes by analysis of polypeptides and functions specified by HSV-1 × HSV-2 recombinants. *J Virol* 1978;26:389–410.
443. Moss H. The herpes simplex virus type 2 alkaline DNase activity is essential for replication and growth. *J Gen Virol* 1986;67:1173–1178.
444. Moss H, Chartrand P, Timbury MC, Hay J. Mutant of herpes simplex virus type 2 with temperature sensitive lesions affecting virion thermostability and DNase activity: identification of the lethal mutation and physical mapping of the nuc⁻ lesion. *J Virol* 1979;32:140–146.
445. Muggeridge M, Cohen GH, Eisenberg RJ. Herpes simplex virus infection can occur without involvement of the fibroblast growth factor receptor. *J Virol* 1992;66:824–830.
446. Mullaney J, Moss HW, McGeoch DJ. Gene UL2 of herpes simplex virus type 1 encodes a uracil-DNA glycosylase. *J Gen Virol* 1989; 70:449–454.
447. Muller MT. Binding of the herpes simplex virus immediate-early gene product ICP4 to its own transcription start site. *J Virol* 1987;61:858–865.
448. Murchie MJ, McGeoch DJ. DNA sequence analysis of an immediate-early gene region of the herpes simplex virus type 1 genome (map coordinates 0.950–0.978). *J Gen Virol* 1982;62:1–15.
449. Murphy S, Yoon J-B, Gerster T, Roeder RG. Oct-1 and Oct-2 potentiate functional interactions of a transcription factor with the proximal sequence element of small nuclear RNA genes. *Mol Cell Biol* 1992;12:3247–3261.
450. Nesburn AB, Elliot JM, Leibowitz HM. Spontaneous reactivation of experimental herpes simplex keratitis in rabbits. Arch Ophthalmol 1967;78:523–529.
451. Newcomb WW, Brown JC. Structure of the herpes simplex virus capsid: effects of extraction with guanidine hydrochloride and partial reconstitution of extracted capsids. *J Virol* 1991;65:613–620.
452. Newcomb WW, Brown JC. Use of Ar⁻ plasma to localize structural proteins in the capsid of herpes simplex virus type 1. *J Virol* 1989;63:4697–4702.
453. Nieman H, Klenk HD. Coronavirus glycoprotein E1, a new type of viral glycoprotein. *J Mol Biol* 1981;153:993–1010.
454. Nii S, Morgan C, Rose HM, Hsu KC. Electron microscopy of herpes simplex virus. IV. Studies with ferritin conjugated antibodies. *J Virol* 1968;2:1172–1184.
455. Nikas I, McLauchlan J, Davison AJ, Taylor WR, Clements JB. Structural features of ribonucelotide reductase. *Protein Struct Funct Genet* 1986;1:376–384.
456. Nishioka Y, Silverstein S. Requirement of protein synthesis for the degradation of host mRNA in Friend erythroleukemia cells infected with herpes simplex virus type 1. *J Virol* 1978;27:619–627.
457. Nishioka Y, Silverstein S. Alterations in the protein synthetic apparatus of Friend erythroleukemia cells infected with vesicular stomatitis virus or herpes simplex virus. *J Virol* 1978;25:422–426.
458. Nishioka Y, Silverstein S. Degradation of cellular mRNA during infection by herpes simplex virus. *Proc Natl Acad Sci USA* 1977; 74:2370–2374.
459. Nishiyama Y, Kurachi R, Daikoku T, Umene K. The US 9, 10, 11, and 12 genes of herpes simplex virus type 1 are of no importance for its neurovirulence and latency in mice. *Virology* 1993;194:419–423.
460. O'Hare P, Goding CR. Herpes simplex virus regulatory elements and the immunoglobulin octamer domain bind a common factor and are both targets for virion transactivation. *Cell* 1988;52:435–445.
461. O'Hare P, Goding CR, Haigh A. Direct combinatorial interaction between a herpes simplex virus regulatory protein and a cellular octamer-binding factor mediates specific induction of virus immediate-early gene expression. *EMBO J* 1988;7:4231–4238.
462. O'Hare P, Hayward GS. Evidence for a direct role for both the 175,000 and 110,000-molecular weight immediate-early proteins of herpes simplex virus in the transactivation of delayed-early promoters. *J Virol* 1985;53:751–760.
463. O'Hare P, Hayward GS. Three trans-acting regulatory proteins of her-

pes simplex virus modulate immediate-early gene expression in a pathway involving positive and negative feedback regulation. *J Virol* 1985;56:723–733.
464. O'Hare P, Mosca JD, Hayward GS. Multiple trans-acting proteins of herpes simplex virus that have different target promoter specificities and exhibit both positive and negative regulatory functions. *Cancer Cells* 1986;4:175–188.
465. Olivo PD, Nelson NJ, Challberg MD. Herpes simplex type 1 gene products required for DNA replication: identification and overexpression.*J Virol* 1989;63:196–204.
466. Olofsson S, Blomberg J, Lycke E. O-linked glycosidic carbohydrate-peptide linkages of herpes simplex virus glycoproteins. *Arch Virol* 1981;70:321–329.
467. O'Neill EA, Kelly TJ. Purification and characterization of nuclear factor III (origin recognition protein C), a sequence-specific DNA binding protein required for efficient initiation of adenovirus DNA replication. *J Biol Chem* 1988;263:931–937.
468. Orberg P, Schaffer PA. Expression of herpes simplex virus type 1 major DNA-binding protein, ICP8, in transformed cell lines: complementation of deletion mutants and inhibition of wild-type virus. *J Virol* 1987;61:1136–1146.
469. Oroskar AA, Read GS. A mutant of herpes simplex virus type 1 exhibits increased stability of immediate-early (α) mRNAs. *J Virol* 1987;61:604–606.
470. Papavassiliou AG, Silverstein SJ. Characterization of DNA-protein complex formation in nuclear extracts with a sequence from the herpes simplex virus thymidine kinase gene. *J Biol Chem* 1990;265:1648–1657.
471. Papavassiliou AG, Silverstein SJ. Interaction of cell and virus proteins with DNA sequences encompassing the promoter/regulatory and leader regions of the herpes simplex virus thymidine kinase gene. *J Biol Chem* 1990;265:9402–9412.
472. Papavassiliou AG, Wilcox KW, Silverstein SJ. The interaction of ICP4 with cell/infected-cell factors and its state of phosphorylation modulate differential recognition of leader sequences in herpes simplex virus DNA. *EMBO J* 1991;10:397–406.
473. Para MF, Baucke RB, Spear PG. Glycoprotein gE of herpes simplex virus type 1: effects of anti-gE on virion infectivity and on virus-induced Fc-binding receptors. *J Virol* 1982;41:129–136.
474. Para M, Baucke R, Spear PG. IG-G (FC)-binding receptors on virions of HSV-1 and transfer of these receptors to the cell surface. *J Virol* 1980;34:512–520.
475. Parris DS, Cross A, Haarr L, Orr A, Frame MC, Murphy M, McGeoch DJ, Marsden HS. Identification of the gene encoding the 65-kilodalton DNA-binding protein of herpes simplex virus type 1. *J Virol* 1988;62:818–825.
476. Parris DS, Dixon RAF, Schaffer PA. Physical mapping of herpes simplex virus type 1 ts mutants by marker rescue: correlation of the physical and genetic maps. *Virology* 1980;100:275–287.
477. Parris D, Harrington JE. Herpes simplex virus variants resistant to high concentrations of acyclovir exist in clinical isolates. *Antimicrob Agents Chemother* 1982;22:71–77.
478. Parslow TG, Jones SD, Bond B, Yamamoto K. The immunoglobulin octanucleotide: independent activity and selective interaction with enhancers. *Science* 1987;235:1498–1501.
479. Paul MS, Aegerter M, O'Brien SE, Kurtz CB, Wess JH. The murine complement receptor family. Analysis of mCRY gene products and their homology to human CR1. *J Immunol* 1989;142:582–589.
480. Peake ML, Nystrom P, Pizer LI. Herpesvirus glycoprotein synthesis and insertion into plasma membranes. *J Virol* 1982;42:678– 690.
481. Pellett PE, McKnight JLC, Jenkins FJ, Roizman B. Nucleotide sequence and predicted amino acid sequence of a protein encoded in a small herpes simplex virus DNA fragment capable of trans-inducing α genes. *Proc Natl Acad Sci USA* 1985;82:5870–5874.
482. Pereira L, Cassai E, Honess RW, Roizman B, Terni M, Nahmias A. Variability in the structural polypeptides of herpes simplex virus 1 strains: potential applications in molecular epidemiology. *Infect Immun* 1976;13:211–220.
483. Pereira L, Wolff M, Fenwick M, Roizman B. Regulation of herpesvirus synthesis. V. Properties of a polypeptides specified by HSV-1 and HSV-2. *Virology* 1977;77:733–749.
484. Perry LJ, Rixon FJ, Everett RD, Frame MC, McGeoch DJ. Characterization of the IE110 gene of herpes simplex virus type 1. *J Gen Virol* 1986;67:2365–2380.
485. Person S, Kousoulas KC, Knowles RW, Read GS, Holland TC, Keller PM, Warner SC. Glycoprotein processing in mutants of HSV-1 that induce cell fusion. *Virology* 1982;117:293–306.
486. Person S, Laquerre S, Desai P, Hempel J. Herpes simplex virus type 1 capsid protein, VP21, originates within the U_L26 open reading frame. *J Gen Virol* 1993;74:2269–2273.
487. Pizer LI, Cohen GH, Eisenberg RJ. Effect of tunicamycin on herpes simplex virus glycoproteins and infectious virus production. *J Virol* 1980;34:142–153.
488. Plummer G, Goodheart CR, Henson D, Bowling CP. A comparative study of the DNA density and behavior in tissue cultures of fourteen different herpesviruses. *Virology* 1969;39:134–137.
489. Poffenberger KL, Roizman B. Studies on non-inverting genome of a viable herpes simplex virus 1. Presence of head-to-tail linkages in packaged genomes and requirements for circularization after infection. *J Virol* 1985;53:589–595.
490. Poffenberger KL, Tabares E, Roizman B. Characterization of a viable, non-inverting herpes simplex virus 1 genome derived by insertion of sequences at the L-S component junction. *Proc Natl Acad Sci USA* 1983;80:2690–2694.
491. Pogue-Geile KL, Lee GT-Y, Shapira SK, Spear PG. Fine mapping of mutations in the fusion-inducing MP strain of herpes simplex virus type 1. *Virology* 1984;136:100–109.
492. Pogue-Geile KL, Lee GT-Y, Spear PG. Novel rearrangements of herpes simplex virus DNA sequences resulting from duplication of a sequence within the unique region of the L component. *J Virol* 1985;53:456–461.
493. Pogue-Geile KL, Spear PG. The single base pair substitution responsible for the syn phenotype of herpes simplex virus type 1, strain MP. *Virology* 1987;157:67–74.
494. Polvino-Bodnar M, Orberg PK, Schaffer PA. Herpes simplex virus type 1 oriL is not required for virus replication or for the establishment and reactivation of latent infection in mice. *J Virol* 1987;61:3528–3535.
495. Poon APW, Roizman B. Characterization of a *ts* mutant of the U_L15 open reading frame of HSV-1. *J Virol* 1993;67:4497–4503.
496. Posavad CM, Newton JJ, Rosenthal KL. Inhibition of human CTL-mediated lysis by fibroblasts infected with herpes simplex virus. *J Immunol* 1993;151:4865–4873.
497. Posavad CM, Rosenthal KL. Herpes simplex virus-infected human fibroblasts are resistant to and inhibit cytotoxic T lymphocyte activity. *J Virol* 1992;66:6264–6272.
498. Post LE, Mackem S, Roizman B. Regulation of α genes of herpes simplex virus: expression of chimeric genes produced by fusion of thymidine kinase with α gene promoters. *Cell* 1981;24:555–565.
499. Post LE, Roizman B. A generalized technique for deletion of specific genes in large genomes: α gene 22 of herpes simplex virus 1 is not essential for growth. *Cell* 1981;25:227–232.
500. Powell K, Purifoy D. Nonstructural proteins of herpes simplex virus. I. Purification of the induced DNA polymerase.*J Virol* 1977;24:618–626.
501. Preston CM. Control of herpes simplex virus type 1 mRNA synthesis in cells infected with wild type virus or the temperature sensitive mutant tsK. *J Virol* 1979;29:275–284.
502. Preston CM. Cell-free synthesis of herpes-simplex virus coded pyrimidine deoxyribonucleotide kinase enzyme. *J Virol* 1977;23:455–460.
503. Preston CM, Cordingley MG. mRNA- and DNA-directed synthesis of herpes simplex virus-coded exonuclease in *Xenopus laevis* oocytes. *J Virol* 1982;43:386–394.
504. Preston CM, Frame MC, Campbell MEM. A complex formed between cell components and an HSV structural polypeptide binds to a viral immediate early gene regulatory DNA sequence. *Cell* 1988;52:425–434.
505. Preston CM, McGeoch DJ. Identification and mapping of two polypeptides encoded within the herpes simplex virus type 1 thymidine kinase gene. *J Virol* 1981;38:593–605.
506. Preston CM, Notarianni EL. Poly (ADP-ribosyl)ation of a herpes simplex virus immediate early polypeptide. *Virology* 1983;131:492–501.
507. Preston VG, Coates JAV, Rixon FJ. Identification and characterization of a herpes simplex virus gene product required for encapsidation of virus DNA. *J Virol* 1983;45:1056–1064.
508. Preston VG, Davison AJ, Marsden HS, Timbury MC, Subak-Sharpe JH, Wilkie NM. Recombinants between herpes simplex virus types 1 and 2: analyses of genome structures and expression of immediate-early polypeptides. *J Virol* 1978;28:499–517.
509. Preston VG, Fisher FB. Identification of the herpes simplex virus type 1 gene encoding the dUTPase. *Virology* 1984;138:58–68.
510. Preston VG, Palfreyman JW, Dutia BM. Identification of a herpes sim-

plex virus type 1 polypeptide which is a component of the virus-induced ribonucleotide reductase. *J Gen Virol* 1984;65:1457–1466.
511. Price RW, Kahn A. Resistance of peripheral autonomic neurons to in vivo productive infection by herpes simplex virus mutants deficient in thymidine kinase activity. *Infect Immun* 1981;43:571–580.
512. Price RW, Schmitz J. Route of infection, systemic host resistance, and integrity of ganglionic axons influence acute and latent herpes simplex virus infection of the superior cervical ganglion. *Infect Immun* 1979; 23:373–383.
513. Price RW, Walz MA, Wohlenberg C, Notkins AL. Latent infection of sensory ganglia with herpes simplex virus: efficacy of immunization. *Science* 1975;188:938–940.
514. Pruijn GJ, van Driel W, van der Vliet PC. Nuclear factor III, a novel sequence-specific DNA-binding protein from HeLa cells stimulating adenovirus DNA replication. *Nature* 1986;322:656–659.
515. Puga A, Rosenthal JD, Openshaw H, Notkins AL. Herpes simplex virus DNA and mRNA sequences in acutely and chronically infected trigeminal ganglia of infected mice. *Virology* 1978;89:102–111.
516. Purifoy DJM, Powell KL. Temperature-sensitive mutants in two distinct complementation groups of herpes simplex virus type 1 specify thermolabile DNA polymerase. *J Gen Virol* 1981;54:219–222.
517. Purifoy DJM, Powell KL. DNA-binding proteins induced by herpes simplex virus type 2 in HEp-2 cells. *J Virol* 1976;19:717–731.
518. Purifoy DJM, Powell KL. Herpes simplex virus DNA polymerase as the site of phosphonoacetate sensitivity: temperature sensitive mutants. *J Virol* 1977;24:470–477.
519. Purves FC, Longnecker RM, Leader DP, Roizman B. The herpes simplex virus 1 protein kinase is encoded by open reading frame US3 which is not essential for virus growth in cell culture. *J Virol* 1987;61:2896–2901.
520. Purves FC, Ogle WO, Roizman B. The processing of the herpes simplex virus regulatory protein α22 mediated by the U_L13 protein kinase determines the accumulation of a subset of α and mRNAs and proteins in infected cells. *Proc Natl Acad Sci USA* 1993;90:6701–6705.
521. Purves FC, Roizman B. The U_L13 gene of herpes simplex virus 1 encodes the functions for posttranslational processing associated with phosphorylation of the regulatory protein α22. *Proc Natl Acad Sci USA* 1992;89:7310–7314.
522. Purves FC, Spector D, Roizman B. U_L34, the target of the herpes simplex virus U_S3 protein kinase, is a membrane protein which in its unphosphorylated state associates with novel phosphoproteins. *J Virol* 1992;66:4295–4303.
523. Purves FC, Spector D, Roizman B. The herpes simplex virus 1 protein kinase encoded by U_S3 gene mediates posttranslational modification of the phosphoprotein encoded by the U_L34 gene. *J Virol* 1991;65:5757–5764.
524. Puvion-Dutilleul F, Pichard E. Viral alkaline nuclease in intranuclear dense bodies induced by herpes simplex infection. *Biol Cell* 1986;58:15–22.
525. Quinlan MP, Chen LB, Knipe DM. The intranuclear location of a herpes simplex virus DNA-binding protein is determined by the status of viral DNA replication. *Cell* 1984;36:857–868.
526. Quinlan MP, Knipe DM. Stimulation of expression of a herpes simplex virus DNA-binding protein by two viral functions. *Mol Cell Biol* 1985;5:957–963.
527. Quinn JP, McGeoch DJ. DNA sequence of the region in the genome of herpes simplex virus type 1 containing the gene for DNA polymerase and the major DNA binding protein. *Nucleic Acids Res* 1985; 13:8143–8163.
528. Rafield LF, Knipe DM. Characterization of the major RNAs transcribed from the genes for glycoprotein B and DNA-binding protein ICP8 of herpes simplex virus type 1. *J Virol* 1985;49:960–969.
529. Ramaswamy R, Holland TC. In vitro characterization of the HSV-1 UL53 gene product. *Virology* 1992;186:579–587.
530. Read GS, Frenkel N. Herpes simplex virus mutants defective in the virion associated shut-off of host polypeptide synthesis and exhibiting abnormal synthesis of α (immediate early) viral polypeptides. *J Virol* 1983;46:498–512.
531. Read GS, Karr BM, Knight K. Isolation of a herpes simplex virus type 1 mutant with a deletion in the virion host shutoff gene and identification of multiple forms of the *vhs* (U_L41) polypetide 1993; 67:7149–7160.
532. Resnick J, Boyd BA, Haffey ML. DNA binding by the herpes simplex virus type 1 ICP4 protein is necessary for efficient down regulation of the ICP0 promoter. *J Virol* 1989;63:2497–2503.
533. Rice SA, Knipe DM. Genetic evidence for two distinct transactivation functions of the herpes simplex virus α protein ICP27. *J Virol* 1990;64:1704–1715.
534. Rice SA, Su L, Knipe DM. Herpes simplex α ICP27 protein possesses separable positive and negative regulatory activities. *J Virol* 1989;63:3399–3407.
535. Richman DD, Buckmaster A, Bell S, Hodgman C, Minson AC. Identification of a new glycoprotein of herpes simplex virus type 1 and genetic mapping of the gene that codes for it. *J Virol* 1986;57:647–655.
536. Rixon FJ, Addison C, McLauchlan J. Assembly of enveloped tegument structures (L particles) can occur independently of virion maturation in herpes simplex virus type 1-infected cells. *J Gen Virol* 1992;73:277–284.
537. Rixon FJ, Clements JB. Detailed structural analysis of two spliced HSV-1 immediate-early mRNAs. *Nucleic Acids Res* 1982;10: 2244–2256.
538. Rixon FJ, Cross AM, Addison C, Preston VG. The products of herpes simplex virus type 1 gene UL26 which are involved in DNA packaging are strongly associated with empty but not with full capsids. *J Gen Virol* 1988;69:2879–2891.
539. Rixon FJ, Davison MD, Davison AJ. Identification of the genes encoding two capsid proteins of herpes simplex virus type 1 by direct amino acid sequencing. *J Gen Virol* 1990;71:1211–1214.
540. Rixon FJ, McGeoch DJ. A 3′ co-terminal family of mRNAs from the herpes simplex virus type 1 short region: two overlapping reading frames encode unrelated polypeptides one of which has a highly reiterated amino acid sequence. *Nucleic Acids Res* 1984;12:2473–2487.
541. Roane PR Jr, Roizman B. Studies of the determinant antigens of viable cells. II. Demonstration of altered antigenic reactivity of HEp-2 cells infected with herpes simplex virus. *Virology* 1964;22:1–8.
542. Roberts MS, Boundy A, O'Hare P, Pizzorno MC, Ciufo DM, Hayward GD. Direct correlation between a negative autoregulatory response element at the cap site of the herpes simplex virus type 1 IE175 (α4) promoter and a specific binding site for the IE175 (ICP4) protein. *J Virol* 1988;62:4307–4320.
543. Rock DL, Fraser NW. Latent herpes simplex virus type 1 DNA contains two copies of the virion DNA joint region. *J Virol* 1985; 55:849–852.
544. Rock DL, Fraser NW. Detection of HSV-1 genome in central nervous system of latently infected mice. *Nature* 1983;302:523–525.
545. Rock DL, Nesburn AB, Ghiasi H, Ong J, Lewis TL, Lokensgard JR, Wechsler SL. Detection of latency-related viral RNAs in trigeminal ganglia of rabbits latently infected with herpes simplex virus type 1. *J Virol* 1987;62:3820–3826.
546. Rodahl E, Stevens JG. Differential accumulation of herpes simplex virus type 1 latency associated transcripts in sensory and autonomic ganglia. *Virology* 1992;189:385–388.
547. Rodriguez-Boulan EJ, Pendergast M. Polarized distribution of viral envelope glyocproteins in the plasma membrane of infected epithelial cells. *Cell* 1980;20:45–54.
548. Rodriguez-Boulan E, Sabatini DD. Asymmetric budding of viruses in epithelia monolayers: a model system for study of epithelial polarity. *Proc Natl Acad Sci USA* 1978;75:5071–5075.
549. Roizman B. The structure and isomerization of herpes simplex virus genomes. *Cell* 1979;16:481–494.
550. Roizman B. The organization of the herpes simplex virus genomes. *Annu Rev Genet* 1979;13:25–57.
551. Roizman B. An inquiry into the mechanisms of recurrent herpes infection of man. In: Pollard M, ed. *Perspectives in virology IV.*New York: Harper-Row; 1966:283–304.
552. Roizman B. Polykaryocytosis. *Cold Spring Harb Symp Quant Biol* 1962;27:327–340.
553. Roizman B, Aurelian L, Roane PR Jr. The multiplication of herpes simplex virus. I. The programming of viral DNA duplication in HEp-2 cells. *Virology* 1963;21:482–498.
554. Roizman B, Borman GS, Kamali-Rousta M. Macromolecular synthesis in cells infected with herpes simplex virus. *Nature* 1965;206:1374–1375.
555. Roizman B, Buchman TG. The molecular epidemiology of herpes simplex viruses. *Hosp Pract* 1979;14:95–104.
556. Roizman B, Carmichael LE, Deinhardt F, de The G, Nahmias AJ, Plowright W, Rapp F, Sheldrick P, Takahashi M, Wolf K. Herpesviridae: definition, provisional nomenclature and taxonomy. *Intervirology* 1981;16:201–217.
557. Roizman B, Furlong D. The replication of herpesviruses. In: Fraenkel-

Conrat H, Wagner RR, eds. *Comprehensive virology*. vol. 3. New York: Plenum Press; 1974:229–403.
558. Roizman B, Jenkins FJ. Genetic engineering of novel genomes of large DNA viruses. *Science* 1985;129:1208–1218.
559. Roizman B, Kristie T, McKnight JLC, Michael N, Mavromara-Nazos P, Spector D. The trans-activation of herpes simplex virus gene expression: comparison of two factors and their cis sites. *Biochimie* 1988;70:1031–1043.
560. Roizman B, Kristie T, Michael N, McKnight JLC, Mavromara-Nazos P, Spector D. The *trans*-activation of viral gene expression in herpes simplex virus infected cells. In: De Palo G, Rilke F, zur Hausen H, eds. *Herpes and papilloma viruses, their role in the carcinogenesis of the lower genital tract II*. vol. 46. New York: Raven Press; 1988: 21–40.
561. Roizman B, Norrild B, Chan C, Pereira L. Identification of a herpes simplex virus 2 glycoprotein lacking a known type 1 counterpart. *Virology* 1984;133:242–247.
562. Roizman B, Roane PR Jr. Multiplication of herpes simplex virus. II. The relation between protein synthesis and the duplication of viral DNA in infected HEp-2 cells. *Virology* 1964;22:262–269.
563. Roizman B, Roane PR Jr. Studies on the determinant antigens of viable cells. I. A method, and its application in tissue culture studies, for enumeration of killed cells, based on the failure of virus multiplication following injury by cytotoxic antibody and complement.*J Immunol* 1961;87:714–727.
564. Roizman B, Roane PR Jr. A physical difference between two strains of herpes simplex virus apparent on sedimentation in cesium chloride. *Virology* 1961;15:75–79.
565. Roizman B, Sears AE. Herpes simplex viruses and their replication. In: Fields BN, Knipe DM, Chanock RM, Hirsch MS, Melnick JL, Monath TP, Roizman B, eds. *Virology*. 2nd ed. New York: Raven Press; 1990:1795–1841.
566. Roizman B, Sears AE. An inquiry into the mechanism of herpes simplex virus latency. *Annu Rev Microbiol* 1987;41:543–571.
567. Roizman B, Spear PG. Herpesvirus antigens on cell membranes detected by centrifugation of membrane-antibody complexes. *Science* 1971;171:298–300.
568. Roizman B, Spear PG. The role of herpesvirus glycoproteins in the modification of membranes of infected cells. Proceedings of the Miami Winter Symposia, January 18–22, 1971. In: Ribbons DW, Woessner JF, Schultz J, eds. *Nucleic acid-protein interactions and nucleic acid synthesis in viral infection*. vol. 2. Amsterdam: North Holland Publishing; 1971:435–455.
569. Roizman B, Spring SB. Alteration in immunologic specificity of cells infected with cytolytic viruses. In: Trentin JJ, ed. *Proceedings of the Conference on Cross Reacting Antigens*. Baltimore: Williams and Wilkins; 1967:85–96.
570. Roizman B, Tognon M. Restriction endonuclease patterns of herpes simplex virus DNA: application to diagnosis and molecular epidemiology. *Curr Top Microbiol Immunol* 1983;104:275–286.
571. Roller RJ, McCormick AL, Roizman B. Cellular proteins specifically bind single- and double-stranded DNA and RNA from the initiation site of a transcript which crosses the origin of DNA replication of herpes simplex virus 1. *Proc Natl Acad Sci USA* 1989;86:6518–6522.
572. Roller RJ, Roizman B. A herpes simplex virus-1 U_S11-expressing cell line is resistant to herpes simplex virus infection at a step in viral entry mediated by glycoprotein D. *J Virol* 1994;68:2830–2839.
573. Roller RJ, Roizman B. The herpes simplex virus 1 RNA binding protein U_S11 is a virion component and associates with ribosomal 60S subunits. *J Virol* 1992;66:3624–3632.
574. Roller RJ, Roizman B. The herpes simplex virus 1 RNA binding protein U_S11 negatively regulates the accumulation of a truncated viral mRNA. *J Virol* 1991;65:5873–5879.
575. Roop C, Hutchinson L, Johnson DC. A mutant herpes simplex virus type 1 unable to express glycoprotein 1 cannot enter cells, and its particles lack glycoprotein H. *J Virol* 1993;67:2285–2297.
576. Romanelli MG, Cattozzo EM, Faggioli L, Tognon M. Fine mapping and characterization of the syn 6 locus in the herpes simplex virus type 1 genome. *J Gen Virol* 1991;72:1991–1995.
577. Rosen A, Ernst F, Koch H-G, Gederblom H, Darai G, Hadar J, Tabor E, Ben-Hur T, Becker Y. Replacement of the deletion in the genome (0.762–0.789) of avirulent HSV-1 HFEM using cloned MluI DNA fragment (0.7615–0.796) of virulent HSV-1 F leads to generation of virulent intratypic recombinant. *Virus Res* 1986;5:157–175.
578. Russell J, Preston CM. An in vitro latency system for herpes simplex virus type 2. *J Gen Virol* 1986;67:397–403.
579. Ruyechan WT. N-ethylmaleimide inhibition of the DNA-binding activity of the herpes simplex virus type 1 major DNA-binding protein. *J Virol* 1988;62:810–817.
580. Ruyechan WT. The major herpes simplex virus DNA-binding protein holds single-stranded DNA in an extended conformation. *J Virol* 1983;46:661–666.
581. Ruyechan WT, Chytil A, Fisher CM. In vitro characterization of a thermolabile herpes simplex virus DNA-binding protein. *J Virol* 1986;59:31–36.
582. Ruyechan WT, Morse LS, Knipe DM, Roizman B. Molecular genetics of herpes simplex virus. II. Mapping of the major viral glycoproteins and of the genetic loci specifying the social behavior of infected cells. *J Virol* 1979;29:677–697.
583. Ruyechan WT, Weir AC. Interaction with nucleic acids and stimulation of the viral DNA polymerase by the herpes simplex virus type 1 major DNA-binding protein. *J Virol* 1984;52:727–733.
584. Sacks WR, Greene CC, Ashman DP, Schaffer PA. Herpes simplex virus type 1 ICP27 is an essential regulatory protein. *J Virol* 1985;55:796–805.
585. Sacks WR, Schaffer PA. Deletion mutants in the gene encoding the herpes simplex virus type 1 immediate-early protein ICP0 exhibit impaired growth in cell culture. *J Virol* 1987;61:829–839.
586. Sadowski I, Ma J, Triezenberg S, Ptashne M. GAL4-VP16 is an unusually potent transcriptional activator. *Nature* 1988;335:563–564.
587. Sakaoka H, Kurita K, Iida Y, Takada S, Umene K, Kim YT, Ren C-S, Nahmias AJ. Quantitative analsysis of genomic polymorphism of herpes simplex virus type 1 from six countries: studies of molecular evolution and molecular epidemiology of the virus. *J Gen Virol* 1994; 74:513–527.
588. Sandri-Goldin. Properties of an HSV-1 regulatory protein that appears to impair host cell splicing. *Infect Agents Dis* 1994;3;59–67.
589. Sandri-Goldin RM, Sekulovich RE, Leary K. The alpha protein ICP0 does not appear to play a major role in the regulation of herpes simplex virus gene expression during infection in tissue culture. *Nucleic Acids Res* 1987;15:905–919.
590. Sarmiento M, Haffey M, Spear PG. Membrane proteins specified by herpes simplex viruses. III. Role of glycoprotein VP7(B2) in virion infectivity. *J Virol* 1979;29:1149–1158.
591. Sarmiento M, Spear PG. Membrane proteins specified by herpes simplex virus. IV. Conformation of the virion glycoprotein designated VP7 (B2). *J Virol* 1979;29:1159–1167.
592. Sawtell NM, Thompson RL. Rapid in vivo reactivation of herpes simplex virus in latently infected murine ganglionic neurons after transient hypothermia. *J Virol* 1992;66:2150–2156.
593. Sawtell NM, Thompson RL. Herpes simplex virus type 1 latency associated transcription unit promotes anatomical site-dependent establishment and reactivation from latency. *J Virol* 1992;66:2157–2169.
594. Scheidereit C, Heguy A, Roeder RG. Identification and purification of a human lymphoid-specific octamer-binding protein (OTF-2) that activates transcription of an immunoglobulin promoter in vivo. *Cell* 1987;51:783–793.
595. Schek N, Bachenheimer SL. Degradation of cellular mRNAs induced by a virion-associated factor during herpes simplex virus infection of Vero cells. *J Virol* 1985;55:601–610.
596. Schmid DS, Rouse BT. The role of T cell immunity in control of herpes simplex virus. In: Rouse BT, ed. *Herpes simplex virus: pathogesis, immunobiology, and control*. Berlin: Springer-Verlag; 1992:57–74.
597. Schneweiss KE. Serologische Untersuchungen zur Typendifferenzierung des Herpesvirus hominis. *Z Immuno-Forsch* 1962;124:24–28.
598. Schrag JD, Prasad BVV, Rixon RJ, Chiu W. Three dimensional structure of the HSV-1 nucleocapsid. *Cell* 1989;56:651–660.
599. Schroder CH, Stegmann B, Lauppe HF, Kaerner HC. An universal defective genotype derived from herpes simplex virus strain ANG. *Intervirology* 1975/76;6:270–284.
600. Schwartz J, Roizman B. Similarities and differences in the development of laboratory strains of herpes simplex virus in HEp-2 cells: electron microscopy. *J Virol* 1969;4:879–889.
601. Schwartz J, Roizman B. Concerning the egress of herpes simplex virus from infected cells: electron and light microscope observations. *Virology* 1969;38:42–49.
602. Scott TF, Burgoon CF, Coriell LL, Blank M. The growth curve of the

virus of herpes simplex in rabbit corneal cells grown in tissue culture with parallel observations on the development of the intranuclear inclusion body. *J Immunol* 1953;71:385–396.
603. Sears AE, Halliburton IW, Meignier B, Silver S, Roizman B. Herpes simplex virus 1 mutant deleted in the α22 gene: growth and gene expression in permissive and restrictive cells and establishment of latency in mice. *J Virol* 1985;55:338–346.
604. Sears AE, Hukkanen V, Labow MA, Levine AJ, Roizman B. Expression of the herpes simplex virus 1 αtransinducing factor (VP16) does not induce reactivation of latent virus or prevent the establishment of latency in mice. *J Virol* 1991;65:2929–2935.
605. Sears AE, McGwire BS, Roizman B. Infection of polarized MDCK cells with herpes simplex virus 1: two asymmetrically distributed cell receptors interact with different viral proteins. *Proc Natl Acad Sci USA* 1991;88:5087–5091.
606. Sears AE, Meignier B, Roizman B. Establishment of latency in mice by herpes simplex virus 1 recombinants carrying insertions affecting the regulation of the thymidine kinase gene. *J Virol* 1985;55:410–416.
607. Sears AE, Roizman B. Amplification by host factors of a sequence contained within the herpes simplex virus 1 genome. *Proc Natl Acad Sci USA* 1990;87:9441–9445.
608. Sederati F, Izumi KM, Wagner EK, Stevens JG. Herpes simplex virus type 1 latency associated transcription plays no role in establishment or maintenance of a latent infection in murine sensory neurons. *J Virol* 1989;63:4455–4458.
609. Sedarati F, Margolis TP, and Stevens JG. Latent infection can be established with drastically restricted transcription and replication of the HSV-1 genome. *Virology* 1993;192:687–691.
610. Seidel-Dugan C, Ponce de Leon M, Friedman HM, Fries LF, Frank MM, Cohen GH, Eisenberg RJ. C3b receptor activity on transfected cells expressing glycoprotein C of herpes simplex types 1 and 2. *J Virol* 1988;62:4027–4036.
611. Serafini-Cessi F, Dall'Olio F, Scannavini M, Campadelli-Fiume G. Processing of herpes simplex virus 1 glycans in cells defective in glycosyl transferase of the Golgi system: relationship to cell fusion and virion egress. *Virology* 1983;131:59–70.
612. Shao L, Rapp LM, Weller SK. Herpes simplex virus 1 alkaline nuclease is required for efficient egress of capsids from the nucleus. *Virology* 1993;196:146–162.
613. Shapira M, Homa FL, GLorioso JC, Levine M. Regulaton of the herpes simplex virus type 1 late (gamma 2) glycoprotein C gene: sequences between base pairs ⁻34 to ⁻29 control transient expression and responsiveness to transactivation by the products of the immediate early (alpha) 4 and 0 genes. *Nucleic Acids Res* 1987;5:3097–3111.
614. Sharp JA, Wagner MJ, Summers WC. Transcription of herpes simplex virus gene in vivo: overlap of a late promoter with the 3′ end of the early thymidine kinase gene. *J Virol* 1983;45:10–17.
615. Sheldrick P, Berthelot N. Inverted repetitions in the chromosome of herpes simplex virus. *Cold Spring Harb Symp Quant Biol* 1975; 39:667–678.
616. Sheldrick P, Laithier M, Larria D, Ryhinder ML. Infectious DNA from herpes simplex virus: infectivity of double and single-stranded molecules. *Proc Natl Acad Sci USA* 1973;70:3621–3625.
617. Shepard AA, Imbalzano AN, DeLuca NA. Separation of primary structural components conferring autoregulation, transactivation, and DNA-binding properties to the herpes simplex virus transcriptional regulatory protein ICP4. *J Virol* 1989;63:3714–3728.
618. Sherman G, Bachenheimer SL. Characterization of intranuclear capsids made by *ts* morphogenic mutants of HSV-1. *Virology* 1988; 163:471–480.
619. Sherman G, Bachenheimer SL. DNA processing in temperature sensitive morphogenic mutants of HSV-1. *Virology* 1987;158:427–430.
620. Shida H, Dales S. Biogenesis of vaccinia: carbohydrate of the hemagglutinin molecule. *Virology* 1981;111:56–76.
621. Shieh MT, Spear PG. Fibroblast growth factor receptor: does it have a role in the binding of herpes simplex virus? *Science* 1991;253: 208–210.
622. Shieh MT, WuDunn D, Montgomery RI, Esko JD, Spear PG. Cell surface receptors for herpes simplex virus are heparan sulfate proteoglycans. *J Cell Biol* 1992;116:1273–1281.
623. Shimomura Y, Gangarosa LP, Kataoka M, Hill JM. HSV-1 shedding by iontophoresis of 6-hydroxydopamine followed by topical epinephrine. *Invest Ophthalmol Vis Sci* 1983;24:1588–1594.
624. Shipkey FH, Erlandson RA, Bailey RB, Babcock VI, Southam CM. Virus biographies. II. Growth of herpes simplex virus in tissue culture. *Exp Mol Pathol* 1967:6:39–67.
625. Silver S, Roizman B. γ_2-thymidine kinase chimeras are identically transcribed but regulated as γ_2 genes in herpes simplex virus genomes and as b genes in cell genomes. *Mol Cell Biol* 1985;5:518–528.
626. Silverstein S, Bachenheimer SL, Frenkel N, Roizman B. Relationship between post-transcriptional adenylation of herpes virus RNA and messenger RNA abundance. *Proc Natl Acad Sci USA* 1973;70:2101–2105.
627. Silverstein S, Engelhardt EL. Alterations in the protein synthetic apparatus of cells infected with herpes simplex virus. *Virology* 1979; 95:324–342.
628. Silverstein S, Millette R, Jones P, Roizman B. RNA synthesis in cells infected with herpes simplex virus. XII. Sequence complexity and properties of RNA differing in extent of adenylation. *J Virol* 1976; 18:977–991.
629. Simmons A, Slobedman B, Arthur J, Efstathiou S. Two patterns of persistence of herpes simplex virus DNA sequences in the nervous system of latently infected mice. *J Gen Virol* 1992;73:1287–1291.
630. Smibert CA, Popova B, Xiao P, Capone JP, Smiley JR. Herpes simplex virus VP16 forms a complex with the virion host shutoff protein vhs. *J Virol* 1994;68:2339–2346.
631. Smiley JR, Johnson DC, Pizer LI, Everett RD. The ICP4 binding sites in the herpes simplex virus type 1 glycoprotein D (gD) promoter are not essential for efficient gD transcription during virus infection. *J Virol* 1992;66:623–631.
632. Smith CA, Bates P, Rivera-Gonzales R, Gu B, DeLuca NA. ICP4, the major transcriptional regulatory protein of herpes simplex virus type 1 forms a tripartite complex with TATA-binding protein and TFIIB. *J Virol* 1993;67:4676–4687.
633. Smith CA, Marchetti ME, Edmonsin P, Schaffer PA. Herpes simplex virus type 2 mutants with deletions in the intergenic region between ICP4 and ICP22/47: identification of nonessential *cis*-acting elements in the context of the viral genome. *J Virol* 1989;63:2036–2047.
634. Smith CA, Schaffer PA. Mutants defective in herpes simplex virus type 2 ICP4: isolation and preliminary characterization. *J Virol* 1987; 61:1092–1097.
635. Smith IL, Hardwicke MA, Sandri-Goldin RM. Evidence that the herpes simplex virus immediate early protein ICP27 acts post-transcriptionally during infection to regulate gene expression. *Virology* 1992; 186:74–85.
636. Smith RF, Smith TF. Identification of new protein kinase-related genes in three herpesviruses, herpes simplex virus, varicella-zoster virus, and Epstein-Barr virus. *J Virol* 1989;63:450–455.
637. Sodora DL, Cohen GH, Muggeridge MI, Eisenberg RJ. Absence of asparagine-linked oligosaccharides from glycoprotein D of herpes simplex virus type 1 results in structurally altered but biologically active protein. *J Virol* 1991;65:4424–4431.
638. Spaete RR, Frenkel N. The herpes simplex virus amplicon: analysis of cis-acting replicaiton functions. *Proc Natl Acad Sci USA* 1985; 82:694–698.
639. Spear PG. Antigenic structure of herpes simplex viruses. In: van Regenmortel MHV, Neurath AR, eds. *Immunochemistry of viruses. The basis for serodiagnosis and vaccines*. Amsterdam: Elsevier Science Publishers; 1985:425–446.
640. Spear PG. Glycoproteins specified by herpes simplex viruses. In: Roizman B, ed. *The herpesviruses*. vol. 3. New York: Plenum Press; 1985:315–356.
641. Spear PG. Membrane proteins specified by herpes simplex virus. I. Identification of four glycoprotein precursors and their products in type 1-infected cells. *J Virol* 1976;17:991–1008.
642. Spear PG, Keller JM, Roizman B. The proteins specified by herpes simplex virus. II. Viral glycoproteins associated with cellular membranes. *J Virol* 1970;5:123–131.
643. Spear PG, Roizman B. Proteins specified by herpes simplex virus. V. Purification and structural proteins of the herpes virion. *J Virol* 1972;9:143–159.
644. Spear PG, Roizman B. Buoyant density of herpes simplex virus in solutions of cesium chloride. *Nature* 1967;214:713–714.
645. Speck PG, Simmons A. Synchronous appearance of antigen-positive and latently infected neurons in spinal ganglia of mice infected with a virulent strain of herpes simplex virus. *J Gen Virol* 1992;73:1281–1285.
646. Speck PG, Simmons A. Divergent molecular pathways of productive and latent infection with a virulent strain of herpes simplex virus type 1. *J Virol* 1991;65:4001–4005.

647. Spector D, Purves F, Roizman B. Role of α-transinducing factor (VP16) in the induction of α genes within the context of viral genomes. *J Virol* 1991;65:3504–3513.
648. Spector D, Purves F, Roizman B. Mutational analysis of the promoter region of the α27 gene of herpes simplex virus 1 within the context of the viral genome. *Proc Natl Acad Sci USA* 1990;87:5268–5272.
649. Spiro RG, Spiro MJ. Studies on the synthesis and processing of the asparagine linked carbohydrate units of glycoproteins. *Philos Trans R Soc Lond B Biol Sci* 1982;300:117–127.
650. Stackpole CW. Herpes-type virus of the frog renal adenocarcinoma. I. Virus development in tumor transplants maintained at low temperature. *J Virol* 1969;4:75–93.
651. Stanberry LR, Kern ER, Richards JT, Abott TH, Overall JC. Genital herpes in guinea pigs: pathogenesis of the primary infection and description of recurrent disease. *J Infect Dis* 1982;146:397–404.
652. Stanberry LR, Kit S, Myers MG. Thymidine kinase-deficient herpes simplex virus type 2 genital infection in guinea pigs. *J Virol* 1985;55:322–328.
653. Stannard LM, Fuller AO, Spear PG. Herpes simplex virus glycoproteins associated with different morphological entities projecting from the virion envelope. *J Gen Virol* 1987;68:715–725.
654. Staskus KA, Couch L, Bitterman P, Retzel EF, Zupancic M, List J, Haase AT. *In situ* amplification of visna virus DNA in tissue sections reveals a reservoir of latently infected cells. *Microb Pathog* 1991; 11:67–76.
655. Staudt LM, Singh H, Sen R, Wirth T, Sharp PA, Baltimore D. A lymphoid-specific protein binding to the octamer motif of immunoglobulin genes. *Nature* 1986;323:640–643.
656. Steffy KR, Weir JP. Mutational analysis of two herpes simplex virus type 1 late promoters. *J Virol* 1991;65:6454–6460.
657. Steiner I, Spivack JG, Deshmane SL, Ace CI, Preston CM, Fraser NW. A herpes simplex virus type 1 mutant containing a nontransinducing Vmw65 protein establishes latent infection in vivo in the absence of viral replication and reactivates efficiently from explanted trigeminal ganglia. *J Virol* 1990;64:1630–1638.
658. Steiner I, Spivack JG, Lirette RP, Brown SM, MacLean AR, Subak-Sharpe JH, Fraser NW. Herpes simplex virus type 1 latency associated transcripts are evidently not essential for latent infection. *EMBO J* 1989;8:505–511.
659. Steiner I, Spivack JG, O'Boyle DR, Lavi E, Fraser NW. Latent herpes simplex virus type 1 transcription in human trigeminal ganglia. *J Virol* 1988;62:3493–3496.
660. Stern S, Herr W. The herpes simplex virus trans-activator VP16 recognizes the Oct-1 homeo domain: evidence for a homeo domain recognition subdomain. *Genes Dev* 1991;5:2555–2566.
661. Stern S, Tanaka M, Herr W. The Oct-1 homeodomain directs formation of a multiprotein-DNA complex with the HSV transactivator VP16. *Nature* 1989;341:624–630.
662. Stevens JG, Cook ML. Latent herpes simplex virus in spinal ganglia of mice. *Science* 1971;173:843–845.
663. Stevens JG, Haarr L, Porter DD, Cook ML, Wagner EK. Prominence of the herpes simplex virus latency-associated transcript in trigeminal ganglia from seropositive humans. *J Infect Dis* 1988;158:117–123.
664. Stevens JG, Wagner EK, Devi-Rao GB, Cook ML, Feldman LT. RNA complementary to a herpesvirus α gene mRNA is prominent in latently infected neurons. *Science* 1987;235:1056–1059.
665. Stow ND. Localization of an origin of DNA replication within the TRs/IRs repeated region of the herpes simplex virus type 1 genome. *EMBO J* 1982;1:863–867.
666. Stow ND, McMonagle EC. Characterization of the TRs/IRs origin of DNA replication of herpes simplex virus type 1. *Virology* 1983;130:427–438.
667. Stow ND, Stow EC. Isolation and characterization of a herpes simplex virus type 1 mutant containing a deletion within the gene encoding the immediate early polypeptide Vmw110. *J Gen Virol* 1986;67:2571–2585.
668. Stringer J, Holland JL, Swanstrom R, Pivo K, Wagner E. Quantitation of herpes simplex virus type 1 RNA in infected HeLa cells. *J Virol* 1977;21:889–901.
669. Stringer KF, Ingles CJ, Greenblatt J. Direct and selective binding of an acidic transcriptional activation domain to the TATA-box factor TFIID. *Nature* 1990;345:783–786.
670. Strom T, Frenkel N. Effects of herpes simplex virus on mRNA stability. *J Virol* 1987;61:2198–2207.
671. Sturm R, Baumruker T, Franza BR Jr, Herr W. A 100-kD HeLa cell octamer binding protein (OBP100) interacts differently with two separate octamer-related sequences within the SV40 enhancer. *Genes Dev* 1987;1:1147–1160.
672. Sturm R, Das G, Herr W. The ubiquitous octamer binding protein Oct-1 contains a POU domain with a homeobox subdomain. *Genes Dev* 1988;2:1582–1599.
673. Sturm RA, Herr W. The POU domain is a bipartite DNA-binding structure. *Nature* 1988;336:601–604.
674. Summers WP, Wagner M, Summers WC. Possible peptide chain termination mutants in thymidine kinase gene of a mammalian virus, herpes simplex virus. *Proc Natl Acad Sci USA* 1975;72:4081–4084.
675. Sydiskis RJ, Roizman B. The sedimentation profiles of cytoplasmic polyribosomes in mammalian cells productively and abortively infected with herpes simplex virus. *Virology* 1968;34:562–565.
676. Sydiskis RJ, Roizman B. The disaggregation of host polyribosomes in productive and abortive infection with herpes simplex virus. *Virology* 1966;32:678–686.
677. Sydiskis RJ, Roizman B. Polysomes and protein synthesis in cells infected with a DNA virus. *Science* 1966;153:76–78.
678. Szilagyi JF, Cunningham C. Identification and characterization of a novel non-infectious herpes simplex virus-related particle. *J Gen Virol* 1991;72:661–668.
679. Szostak JK, Orr-Weavwe TL, Rothstein RJ, Stahl FW. The double-strand-break repair model for recombination. *Cell* 1983;33:25–35.
680. Tabas I, Schlesinger S, Kornfeld S. Processing of high mannose oligosaccharides to form complex type oligosaccharides on the newly synthesized polypeptides of the vesicular stomatitis virus G protein and IgG heavy chain. *J Biol Chem* 1978;253:716–722.
681. Taha MY, Clements GB, Brown SM. A variant of herpes simplex virus type 2 strain HG52 with a 1.5 Kb deletion in R_L between 0 to 0.02 and 0.81 to 0.83 map units is non-neurovirulent in mice. *J Gen Virol* 1989;70:705–716.
682. Taha MY, Clements GB, Brown SM. The herpes simplex virus type 2 (HG52) variant JH2604 has a 1488 bp deletion which eliminates neurovirulence in mice. *J Gen Virol* 1989;70:3073–3078.
683. Tankersley RW. Amino acid requirements of herpes simplex virus in human cells. *J Bacteriol* 1964;87:609–613.
684. Tedder DG, Everett RD, Wilcox KW, Beard P, Pizer LI. ICP4 binding sites in the promoter and coding regions of the herpes simplex virus gD gene contribute to activation of in vitro transcription by ICP4. *J Virol* 1989;63:2510–2520.
685. Tedder DG, Pizer LI. Role for DNA-protein interaction in activation of the herpes simplex virus glycoprotein D gene. *J Virol* 1988;62:4661–4672.
686. Tengelsen LA, Pederson NE, Shaver PR, Wathen MW, Homa FL. Herpes simplex virus type 1 cleavage and encapsidation require the product of the U_L28 gene; isolation and characterization of two U_L28 deletion mutants. *J Virol* 1993;67:3470–3480.
687. Tenser RB, Dunston ME. Herpes simplex virus thymidine kinase expression in infection of the trigeminal ganglion. *Virology* 1979;99:417–422.
688. Tenser RB, Hay KA, Edris WA. Latency associated transcript but not reactivatible virus is present in sensory ganglion neurons after inoculation of thymidine kinase negative mutants of herpes simplex virus type 1. *J Virol* 1989;63:2861–2865.
689. Thompson RL, Cook ML, Devi-Rao GB, Wagner EK, Stevens JG. Functional and molecular analyses of the avirulent wild-type herpes simplex virus type 1 strain KOS. *J Virol* 1986;51:203–211.
690. Thompson RL, Devi-Rao GV, Stevens JG, Wagner EK. Rescue of a herpes simplex virus type 1 neurovirulence function with a cloned DNA fragment. *J Virol* 1985;55:504–508.
691. Thomsen DR, Roof LI, Homa FL. Assembly of herpes simplex virus (HSV) intermediate capsids in insect cells infected with recombinant baculoviruses expressing capsid proteins. *J Virol* 1994;68:2442–2457.
692. Tognon M, Cassai E, Rotola A, Roizman B. The heterogeneous regions in herpes simplex virus 1 DNA. *Microbiologica* 1983;6:191–198.
693. Tognon M, Furlong D, Conley AJ, Roizman B. Molecular genetics of herpes simplex virus. V. Characterization of a mutant defective in ability to form plaques at low temperatures and in a viral function which prevents accumulation of coreless capsids at nuclear pores late in infection. *J Virol* 1981;40:870–880.
694. Triezenberg SJ, Kingsbury RC, McKnight SL. Functional dissection of VP16, the trans-activator of herpes simplex virus immediate early gene expression. *Genes Dev* 1988;2:718–729.

695. Triezenberg SJ, LaMarco KL, McKnight SL. Evidence of DNA:protein interactions that mediate HSV-1 immediate early gene activation by VP16. *Genes* 1988;2:730–742.
696. Trousdale MD, Steiner I, Spivack JG, Deshmane SL, Brown SM, MacLean AR, Subak-Sharpe, JH, Fraser NW. *In vivo* and *in vitro* reactivation impairment of a herpes simplex virus type 1 latency associated transcript variant in a rabbit eye model. *J Virol* 1991; 65:6989–6993.
697. Tullo AB, Shimeld C, Blyth WA, Hill TJ, Easty DL. Ocular infection with HSV in non-immune and immune mice. *Arch Ophthalmol* 1983;101:961–964.
698. Tullo AB, Shimeld C, Blyth WA, Hill TJ, Easty DL. Spread of virus and distribution of latent infection following ocular herpes simplex in the non-immune and immune mouse. *J Gen Virol* 1982;63:95–101.
699. Valyi-Nagy T, Deshmane S, Dillner A, Fraser NW. Induction of cellular transcription factors in trigeminal ganglia of mice by corneal scarification, herpes simplex virus type 1 infection, and explantation of trigeminal ganglia. *J Virol* 1991;65:4142–4152.
700. van Genderen IL, Bradimarti R, Torrisi MR, Campadelli G, van Meer G. The phospholipid composition of extracellular herpes simplex virions differs from that of host cell nuclei. *Virology* 1994;200:831–832.
701. Varmuza SL, Smiley JR. Signals for site-specific cleavage of herpes simplex virus DNA: maturation involves two separate cleavage events at sites distal to the recognition site. *Cell* 1985;41:792–802.
702. Vaughan PJ, Thibault KJ, Hardwicke MA, Sandri-Goldin RM. The herpes simplex virus immediate early protein ICP27 encodes a potential metal binding domain and binds zinc in vitro. *Virology* 1992;189:377–384.
703. Vernon SK, Ponce de Leon M, Cohen GH, Eisenberg RJ, Rubins BA. Morphological components of herpesvirus. III. Localization of herpes simplex virus type 1 nucleocapsid polypeptides by immune electron microscopy. *J Gen Virol* 1981;54:39–46.
704. Vlazny DA, Frenkel N. Replication of herpes simplex virus DNA: location of replication recognition signals within defective virus genomes. *Proc Natl Acad Sci* USA 1981;78:742–746.
705. Vlazny DA, Kwong A, Frenkel N. Site specific cleavage packaging of herpes simplex virus DNA and the selective maturation of nucleocapsids containing full length viral DNA. *Proc Natl Acad Sci USA* 1982; 79:1423–1427.
706. Wadsworth S, Jacob RJ, Roizman B. Anatomy of herpes simplex virus DNA. II. Size, composition, and arrangement of inverted terminal repetitions. *J Virol* 1975;15:1487–1497.
707. Wagner EK. Individual HSV transcripts: characterization of specific genes. In: Roizman B, ed. *The herpesviruses*. vol. 3. New York: Plenum Press; 1985:45–104.
708. Wagner EK, Flanagan M, Devi-Rao G, Zhang Y-F, Hill JM, Anderson KP, Stevens JG. The herpes simplex virus latency-associated transcript is spliced during the latent phase of infection. *J Virol* 1988; 62:4577–4585.
709. Wagner EK, Roizman B. RNA synthesis in cells infected with herpes simplex virus. I. The patterns of RNA synthesis in productively infected cells. *J Virol* 1969;4:36–46.
710. Wagner HM, Summers WC. Structure of the joint region and the termini of the DNA of herpes simplex virus type 1. *J Virol* 1978; 27:374–387.
711. Walz MA, Yamamoto H, Notkins AL. Immunologic response restricts the number of cells in sensory ganglia infected with herpes simplex. *Nature* 1976;264:554–556.
712. Ward PL, Campadelli-Fiume G, Avitabile E, Roizman B. The localization and putative function of the U_L20 membrane protein in cells infected with herpes simplex virus 1. *J Virol* 1994;68:7406–7011.
713. Watson RJ, Clements JB. A herpes simplex virus type 1 function continuously required for early and late virus RNA synthesis. *Nature* 1980;285:329–330.
714. Watson RJ, Sullivan M, Vande Woude GF. Structures of two spliced herpes simplex virus type 1 immediate-early mRNA's which map at the junctions of the unique and reiterated regions of the virus DNA S component. *J Virol* 1981;37:431–444.
715. Watson RJ, Vande Woude GF. DNA sequence of an immediate-early gene (IE mRNA-5) of herpes simplex virus 1. *Nucleic Acids Res* 1982;10:979–991.
716. Watson RJ, Weis JH, Salstrom JS, Enquist LW. Herpes simplex virus type 1 glycoprotein D gene: nucleotide sequence and expression in *Escherichia coli*. *Science* 1982;218:381–383.
717. Weber PC, Challberg MD, Nelson NJ, Levine J, Glorioso JC. Inversion events in the HSV-1 genome are directly mediated by the viral DNA replication machinery and lack sequence specificity. *Cell* 1988;54:369–381.
718. Weber PC, Levine M, Glorioso JC. Rapid identification of nonessential genes of herpes simplex virus type 1 by Tn5 mutagenesis. *Science* 1987;236:576–579.
719. Wechsler SL, Nesburn A, Watson R, Slanina SM, Ghiasi H. Fine mapping of the latency-related gene of herpes simplex virus type 1: alternative splicing produces distinct latency-related RNAs containing open reading frames. *J Virol* 1988;62:4051–4058.
720. Weinheimer SP, Boyd BA, Durham SK, Resnick JL, O'Boyle II DR. Deletion of the VP16 open reading frame of herpes simplex virus type 1. *J Virol* 1992;66:258–269.
721. Weinheimer SP, McKnight SL. Transcriptional and post-transcriptional controls establish the cascade of herpes simplex virus protein synthesis. *J Mol Biol* 1987;195:819–833.
722. Weir JP, Narayanan PR. The use of B-galactosidase as a marker gene to define the regulatory sequences of the herpes simplex virus type 1 glycoprotein C in recombinant herpesvirus. *Nucleic Acids Res* 1988; 16:10267–10282.
723. Weller SK, Aschman DP, Sacks WR, Coen DM, Schaffer PA. Genetic analysis of temperature sensitive mutants of HSV-1: the combined use of complementation and physical mapping for cistron assignment. *Virology* 1983;130:290–305.
724. Weller SK, Seghatoleslami RM, Shao L, Rowse D, Charmichael EP. The herpes simplex virus type 1 alkaline nuclease is not essential for viral DNA synthesis: isolation and characterization of a lacZ insertion mutant. *J Gen Virol* 1990;71:2941–2952.
725. Weller SK, Spadoro A, Schaffer JE, Murray AW, Maxam AM, Schaffer PA. Cloning, sequencing, and functional analysis of oriL, a herpes simplex virus type 1 origin of DNA synthesis. *Mol Cell Biol* 1985; 5:930–942.
726. Whealy ME, Card JP, Meade RP, Robbins AK, Enquist LW. Effect of brefeldin A on alpha herpesvirus membrane protein glycosylation and virus egress. *J Virol* 1991;65:1066–1081.
727. Whitley RJ. Epidemiology of herpes simplex viruses. In: Roizman B, ed. *The herpesviruses*. vol. 3. New York: Plenum Press; 1985:1–44.
728. Whitley RJ, Kern E, Chattopadhay S, Chou J, Roizman B. Replication, establishment of latency, and induced reactivation of herpes simplex virus $\gamma_1 34.5$ deletion mutants in rodent models. *J Clin Invest* 1993; 91:2837–2843.
729. Wigdahl BL, Isom HC, deClerq E, Rapp F. Activation of herpes simplex virus (HSV) type 1 genome by temperature sensitive mutants of HSV type 2. *Virology* 1982;116:468–479.
730. Wigdahl BL, Isom HC, Rapp F. Repression and activation of the genome of herpes simplex viruses in human cells. *Proc Natl Acad Sci USA* 1981;78:6522–6526.
731. Wigdahl BL, Scheck AC, De Clerq E, Rapp F. High efficiency latency and reactivation of herpes simplex virus in human cells. *Science* 1982;217:1145–1146.
732. Wigdahl BL, Scheck AC, Ziegler RJ, De Clercq E, Rapp F. Analysis of the herpes simplex virus genome during in vitro latency in human diploid fibroblasts and rat sensory neurons. *J Virol* 1984;49:205–213.
733. Wigdahl BL, Ziegler RJ, Sneve M, Rapp F. Herpes simplex virus latency and reactivation in isolated rat sensory neurons. *Virology* 1983;127:159–167.
734. Wilcox CL, Johnson EM Jr. Nerve growth factor deprivation results in the reactivation of latent herpes simplex virus in vitro. *J Virol* 1987;61:2311–2315.
735. Wilcox KW, Kohn A, Sklyanskaya E, Roizman B. Herpes simplex virus phosphoproteins. I. Phosphate cycles on and off some viral polypeptides and can alter their affinity for DNA. *J Virol* 1980;33:167–182.
736. Wilkie NM. The synthesis and substructure of herpesvirus DNA: the distribution of alkali labile single strand interruptions in HSV-1 DNA. *J Gen Virol* 1973;21:453–467.
737. Williams MV, Parris DS. Characterization of a herpes simplex virus type 2 deoxyuridine triphosphate nucleotidohydrolase and mapping of a gene conferring type specificity for the enzyme. Virology 1987; 156:282–292.
738. Wilson AC, LaMarco K, Peterson MG, Herr W. The VP16 accessory protein HCF is a family of polypeptides processed from a large precursor protein. *Cell* 1993;74:115–125.
739. Wohlenberg CR, Walz MA, Notkins AL. Efficacy of phosphonoacetic acid on herpes simplex virus infection of sensory ganglia. *Infect Immun* 1976;13:1519–1521.

740. Wohlrab F, Francke B. Deoxypyrimidine triphosphatase activity specific for cells infected with herpes simplex virus type 1. *Proc Natl Acad Sci USA* 1980;77:1872–1876.
741. Wohltrab F, Chatterjee S, Wells RD. The herpes simplex virus 1 segment inversion site is specifically cleaved by a virus-induced nuclear endonuclease. *Proc Natl Acad Sci USA* 1991;88:6432–6436.
742. Wolf H, Roizman B. The regulation of (structural) polypeptide synthesis in herpes simplex virus types 1 and 2 infected cells. In: de-The G, et al., eds. *Oncogenesis and herpesviruses III*. Lyon: International Agency for Research on Cancer; 1978:327.
743. Wolf K, Darlington RW. Channel catfish virus: a new herpesvirus of ictalurid fish. *J Virol* 1971;8:525–533.
744. Worrad DM, Caradonna S. Identification of the coding sequence for herpes simplex virus uracil-DNA glycosylase. *J Virol* 1988;62:4774–4777.
745. Wu CA, Nelson NJ, McGeoch DJ, Challberg MD. Identification of herpes simplex virus type 1 genes required for origin-dependent DNA synthesis. *J Virol* 1988;62:435–443.
746. Wu CL, Wilcox KW. Codons 262 to 490 from the herpes simplex virus ICP4 gene are sufficient to encode a sequence-specific DNA binding protein. *Nucleic Acids Res* 1990;18:531–538.
747. Wudunn D, Spear PG. Initial interaction of herpes simplex virus with cells is binding to heparan sulfate. *J Virol* 1989;63:52–58.
748. Yamamoto S, Kabuta H. Genetic analysis of polykaryocytosis by herpes simplex virus. III. Complementation and recombination between non-fusing mutants and construction of a linkage map with regard to the fusion function. *Kurume Med J* 1977;24:163.
749. Yao F, Courtney R. A major transcriptional regulatory protein (ICP4) of herpes simplex virus type 1 is associated with purified virions. *J Virol* 1989;63:3338–3344.
750. Yao F, Courtney RJ. Association of ICP0 but not of ICP27 with purified virions of herpes simplex virus type 1. *J Virol* 1992;66:2709–2716.
751. Yeh L, Schaffer PA. A novel class of transcripts expressed with late kinetics in the absence of ICP4 spans the junction between the long and short segments of the herpes simplex virus type 1 genome. *J Virol* 1993;67:7373–7382.
752. Yei S, Chowdhury SI, Bhat BM, Conley AJ, Wold WSM, Batterson W. Identification and characterization of the herpes simplex virus type 2 gene encoding the essential capsid protein ICP32/VP19c. *J Virol* 1990;64:1124–1134.
753. York I, Johnson DC. Direct contact with herpes simplex virus-infected cells results in inhibition of lymphokine-activated killer cells because of cell-to-cell spread of virus. *J Infect Dis* 1993;168:1127–1132.
754. York IA, Roop C, Andrews DW, Riddell SR, Graham FL, Johnson DC. A cytosolic herpes simplex virus protein inhibits antigen presentation to $CD8^-$ T lymphocytes. *Cell* 1994;77:525–535.
755. Zhang Y, McKnight JLC. Herpes simpex virus type 1 U_L46 and U_L47 deletion mutants lack VP11 and VP12 or VP13 and VP14, respectively, and exhibit altered viral thymidine kinase expression. *J Virol* 1993;67:1482–1492.
756. Zipser D, Lipsich L, Kwoh J. Mapping functional domains in the promoter region of the herpes thymidine kinase gene. *Proc Natl Acad Sci USA* 1981;78:6276–6280.
757. Zwaagstra JC, Ghiasi H, Slanina SM, Nesburn AB, Wheatly SC, Lillycrop K, Wood J, Latchman DS, Patel L, Wechsler SL. Activity of herpes simplex virus type 1 latency associated trancript in neuron derived cells: evidence for neuron specificity and for a large LAT transcript. *J Virol* 1990;64:5019–5028.
758. Zweig M, Heilman CJ, Hampar B. Identification of disulfide-linked protein complexes in the nucleocapsid of herpes simplex virus type 2. *Virology* 1979;94:442–450.

Fundamental Virology, Third Edition
edited by B.N. Fields, D.M. Knipe, P.M. Howley, et al.
Lippincott - Raven Publishers, Philadelphia © 1996

CHAPTER 33

Epstein-Barr Virus and Its Replication

Elliott Kieff

CLASSIFICATION

Epstein-Barr virus (EBV) is the most extensively studied gamma herpes virus and the only human virus in the gamma herpes virus subfamily. Two EBV types circulate in most human populations. The two types differ in only a few genes, but there are significant and consistent genetic differences between the two alleles characteristic of the few genes that are type specific. Some of the differences are important in biologic activity. The two types were originally referred to as types A and B; however, a change to types 1 and 2 is suggested to make the nomenclature similar to that for herpes simplex virus types. Throughout this review the term isolate will refer to a virus or virus-infected cell that has had only a few passages in culture; strain will refer to highly passaged virus or virus-infected cells. Strains frequently differ from new isolates because of the deletion of one or more genes.

The gamma herpes virus subfamily includes both the *lymphocryptovirus* and *rhadinovirus* genera. EBV is the prototype *lymphocryptovirus*. Herpes virus saimiri (HVS) and Marek's disease virus (MDV) are prototypes for the *rhadinovirus* genus.

Lymphocryptoviruses have only been found in Old World primates. Humans are the exclusive natural host for EBV; each of the other Old World primate species is infected with a closely related lymphocryptovirus (45,110,155,189, 190,192,249,250,253,352,364,465,481,480,522,523). In the past, lymphocryptoviruses of nonhuman old world primates have simply been named by the species in which they are naturally endemic, e.g. Herpes virus pan is the chimpanzee lymphocryptovirus.

The lymphocryptoviruses are similar to each other in genome structure and gene organization (Fig. 1). Their genomes are colinearly homologous with the EBV genome (109,110,192,249,250,253,364,518,555). The structural and nonstructural proteins of one primate lymphocryptovirus are frequently antigenically related to the proteins of other lymphocryptoviruses (189,190,192,366,465,480, 481,522,523).

Consistent with the taxonomic relationship between the lymphocryptoviruses and the rhadinoviruses, the lymphocryptovirus genomes are more closely related to the genome of HVS than to the genomes of the alpha or beta human herpes viruses (24,120,121,467). The similarity includes the herpes virus early and late lytic infection genes and the location of highly specific genes characteristic of

E. Kieff: Departments of Medicine and Microbiology, and Molecular Genetics, Harvard Medical School, Boston, Massachusetts 02115.

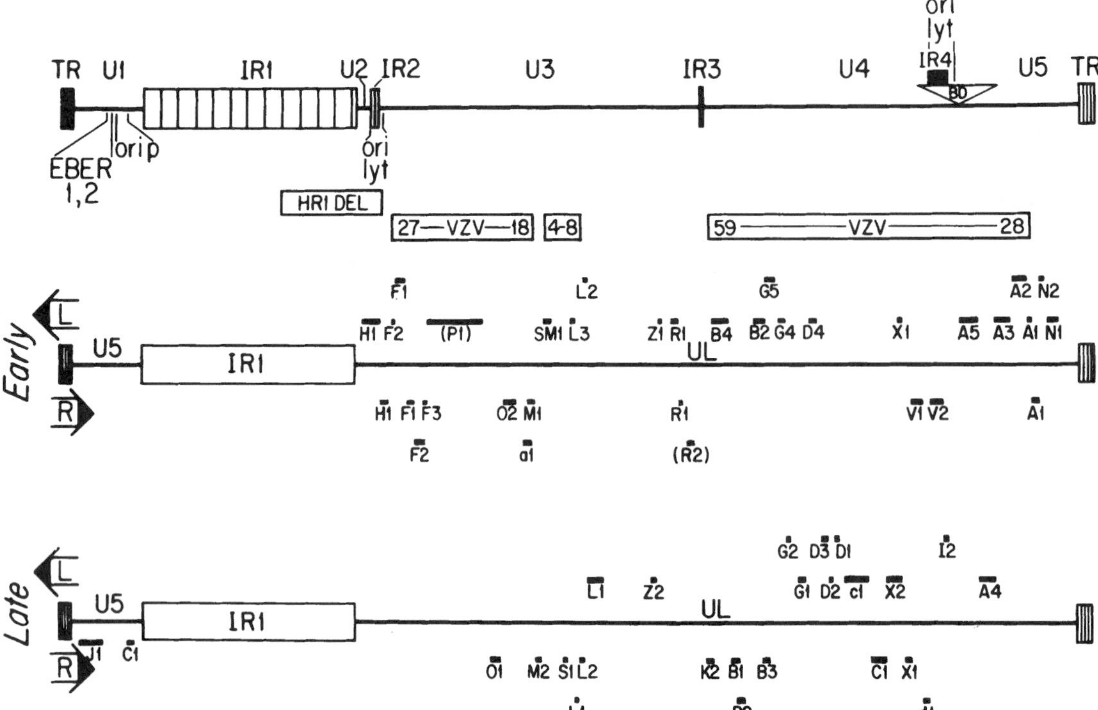

FIG. 1. Schematic depiction of the linear EBV genome. Terminal repeat (TR), internal repeat (IR1-4), and largely unique sequence domains (U1-U5) as well as other features are shown in proportion to their overall size. EBER-1 and -2 indicate the location of the DNA encoding the EBV small, nonpolyadenylated RNAs. Ori lyt and ori P indicate *cis*-acting DNA domains that permit EBV DNA to replicate in lytically or latently infected cells, respectively. HR1 DEL and BD indicate DNA segments deleted from the EBV DNA in the P3HR-1 or B95-8 strains, respectively. The line above BD is the segment of EBV DNA which is deleted at the point indicated by the apex of the BD triangle. The rightward ori lyt is part of the DNA for which the B95-8 strain is deleted. VZV 18 to 27, 4 to 8, and 28 to 59 are large segments of VZV ORFs that are homologous to the indicated regions of EBV DNA. The middle and lower part of the diagram depict known early (middle) and late (lower) mRNAs named according to their order going rightward or leftward in the *Bam*HI fragment, which includes the start of the ORF. Leftward RNAs are shown above the genome and rightward RNAs below. For example, O1 below the bottom line indicates a late mRNA transcribed rightward through the *Bam*HI O fragment first rightward reading frame.

the gamma herpes virus subfamily or the *lymphocryptovirus* or *rhadinovirus* genera. Large blocks of early and late lytic virus genes serve analogous roles in EBV and HVS. There is little nucleotide sequence homology or antigenic cross-reactivity between the lymphocryptoviruses and the rhadinoviruses. The genes that are not involved in lytic infection do not at this point seem similar, although they may have similar functions.

The gamma herpes virus classification was initially established not on the basis of similarity in genome organization but on the basis of similarity in biologic properties. The gamma herpes viruses establish latent infection in lymphocytes and are associated with cell proliferation. Although some members of the beta herpes virus group, such as HHV-6 and HHV-7, also may establish latent infection in lymphocytes, they do not cause the proliferation of latently infected cells. EBV, HVS, and MDV are associated with malignancies in their naturally endemic species or in related species. Indeed, much of the interest in the gamma herpes viruses is because of their association with cell proliferation and cancer. EBV can directly cause a polyclonal B lymphoproliferative disease in immune deficient humans and is an etiologic agent in monoclonal B lymphoproliferative disease, Burkitt's lymphoma (BL), Hodgkin's disease (HD), unusual T-cell lymphomas, and nasopharyngeal carcinoma (NPC) (Fig. 2; for reviews see Chapter 75 in *Fields Virology,* 3rd ed. and refs. 143–145,329,339,729).

Although historically important in their segregation as a distinctive subfamily, the biologic distinctions between beta and gamma herpes viruses may blur with more detailed studies of larger numbers of lymphotropic beta and gamma herpesviruses. If HHV-6 and -7 can persist in lymphocytes without expanding the pool of latently infected cells, then some EBV-, HVS-, or MDV-related agents may have acquired similar capabilities, lost or modified those genes that cause cell proliferation, and retained their close relationship to other gamma herpes viruses because of their similarity among the larger number of virus genes needed

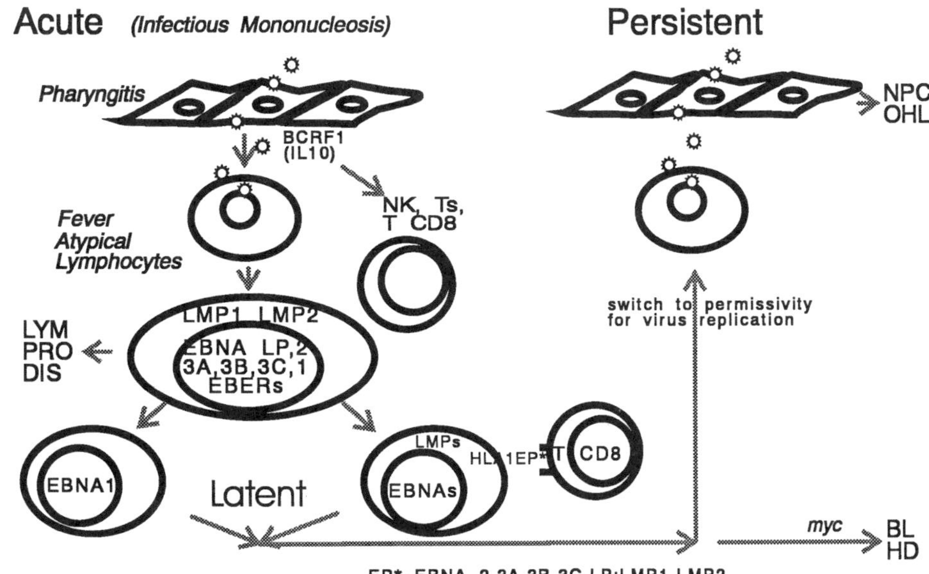

FIG. 2. Schematic depiction of events in EBV infection *in vivo* (see next chapter). EBV initiates infection on the oropharyngeal epithelium where it can produce symptomatic pharyngitis. B-lymphocyes are then infected as they traffic in close proximity to or through the oropharyngeal epithelium. EBV BCRF1 is a close homologue of human IL-10 and probably blunts the NK and CD8+ T-lymphocyte response, enabling EBV to better infect B-lymphocytes. In acutely infected B-lymphocytes EBV expresses all of the EBNAs, LMPs, and EBERS. In immune-suppressed patients, an acute lymphoproliferative disease may emerge. In acute lymphoproliferative disease, the EBNAs, LMPs, and EBERS are expressed. An EBV-specific CD8+ T-lymphocyte response is demonstrable with convalescence from primary infection and is presumed to account for the decrease in EBV-infected B-lymphocytes from 1 in 10 to 1 in 10^5 to 10^6 after acute EBV infection. CD8+ T-lymphocytes recognize determinants from EBNA 2, 3A, 3B, 3C, or LP or from LMP1 or LMP2 in the context of specific class 1 MHC molecules. B-lymphocytes expressing these proteins must continue to be present because a high level of EBV immune CD8+ T-lymphocytes persists for life. EBV also persists in some lymphocytes expressing only EBNA 1 and EBERs or EBNA 1 and LMP2 and EBERs. Because CD8+ responses to EBNA 1 are rare, EBNA 1-expressing lymphocytes would maintain EBV episome and escape immune destruction. Lymphocytes probably carry virus to other organs and to epithelial surfaces, including the oropharynx. Persistent replication in the oropharynx is dependent on activation of lytic infection in B-lymphocytes. Years after primary EBV infection, EBV is closely associated with emergence of BL, HD, and NPC. These tumors initiate from a clone of EBV-infected cells.

for lytic infection. Alternatively, a lymphotropic virus may be discovered that has greater similarity to beta herpes viruses in nucleotide sequence and genome organization but has evolved to use cell proliferation as part of its strategy for latent persistence. This putative agent would likely be placed into the beta herpes virus sub-family.

Taxonomists have renamed EBV human herpes virus 4. Although most authors still use the name EBV, the National Library of Medicine and some other indexing services have adopted the human herpes virus 4 nomenclature, and those doing bibliographical searches should be prepared to use both entries.

VIRUS STRUCTURE

Like other herpes viruses, EBV has a toroid-shaped protein core that is wrapped with DNA, a nucleocapsid with 162 capsomeres, a protein tegument between the nucleocapsid and the envelope, and an outer envelope with external glycoprotein spikes (138,139,143–147,507). The major EBV capsid proteins are 160, 47, and 28 kd, similar in size to the major herpes simplex virus (HSV)-1 capsid proteins (138,139). Like other herpes viruses, EBV has a number of minor virion proteins. The most abundant EBV envelope and tegument proteins are 350/220 and 152 kd, respectively, different in size from the major envelope and tegument proteins of HSV-1 (138,139,455,513,515, 650–653). EBV also differs from most other herpes viruses in the predominance of a single glycoprotein in the outer envelope.

GENOME STRUCTURE

The EBV genome is a linear, double-stranded, 172-kbp DNA composed of 60 mol% guanine or cytosine (24,109, 201–203,301,318,510,511,680) (Fig. 1). The characteristic features of EBV and other lymphocryptovirus genomes include (a) a single overall format and gene arrangement (109,115,201–203,518); (b) reiterated 0.5-kbp terminal direct repeats (TR) (203,238,336); and (3) reiterated 3-kbp internal direct repeats (IR1) (74,75,201,202,241) that divide the genome into short and long largely unique sequence domains (US and UL). The largely unique sequence domains are replete with perfect or imperfect tandem repeat elements, many of which encode for repeat domains in proteins. One region (DL) near the left end of the UL domain consists of multiple, highly conserved, G-C rich, 125-bp repeats and 2 kbp of adjacent unique DNA, all of which have extensive homology to another region (DR) near the right end of the EBV DNA UL domain. DR consists of multiple highly conserved, G-C rich, 102-bp tandem repeats and 1 kbp of nearby unique DNA (113,239, 247,249,250,253,281,300,354).

The surprising exception to the rule that lymphocryptovirus genomes are colinearly homologous is the nonhomology among their TR sequences. Each genome has an analogous TR of similar length. The EBV TR does not hybridize to the HV papio or HV pan TR even under reduced stringency (249,250,253).

The major DNA repeat elements are important landmarks on the EBV genome maps (Fig. 1). Although each EBV strain tends to maintain a characteristic modal number of repeats through sequential passage (57,115,247,519), isolates frequently differ from each other in their tandem repeat reiteration frequency. Thus, repeat reiteration frequency measured by the DNA length between unique restriction endonuclease sites that bracket a repeat element is a useful indicator of differences among isolates or strains and is not an invariant characteristic of EBV DNA. Some of the repeats encode protein, and differences in protein size on immunoblot are also useful isolate- or strain-specific markers. Furthermore, each time EBV establishes latent infection in a proliferating cell, the genome persists as an episome that has a number of repeats of TR characteristic of that infected cell. Each progeny cell in that infected cell lineage will tend to have the same number of TRs (63,610). The number of TRs in latently infected cells are therefore useful in determining whether latently infected cells arose from a common progenitor. BL or NPC tumors can thereby be shown to usually arise from a single EBV-infected cell (57,519).

EBV was the first herpes virus whose genome was completely cloned into *Escherichia coli* (109). The DNA of this strain is an unusual deletion derivative (518). A number of additional clone libraries are now available (22,58,162,504, 518,604).

EBV was also the first herpes virus to be completely sequenced (22,235,492). Subsequently, the alpha herpes virus genomes, varicella zoster virus (VZV) and HSV, the beta herpes virus genomes cytomegalovirus (CMV) and HHV6, and the rhadinovirus genome HVS were cloned and completely or partially sequenced (82,120–122,343,418,466, 467). Despite enormous differences in base composition, ranging from 43% guanine plus cytosine for VZV to 71% guanine plus cytosine for HSV, comparison of these herpes virus DNA sequences shows large regions with colinear, albeit distant, homology at the predicted protein level (Fig. 1, Table 1) (24,66,98,120–122,343,496). For example, three large EBV DNA segments are distantly, but colinearly, homologous to VZV DNA segments encoding open reading frames (ORFs) 4 to 8, 18 to 27, and 28 to 59 (120,121) (Fig. 1). The relatively conserved domains encode for EBV genes that function in lytic infection. One large conserved domain, homologous to VZV ORFs 28 to 59 (Fig. 1), encodes the major EBV DNA-binding protein, the DNA polymerase, the glycoprotein (gp)110, the TK, and the gp85 genes. A second short conserved domain, ho-

TABLE 1. *Epstein-Barr virus open reading frames expressed in lytic injection.*

Reading frame	Proposed function
Early	
Leftward	
SM1	44-kd nuclear protein; promiscuous transactivator
L3	32-kd dUTPase by homology to HSV
Z1	38-kd immediate nuclear protein; transactivator of lytic replication
R1	immediate transactivator
G5	alkaline exonuclease by homology to HSV
X1	thymidine kinase; homologous to HSV
A5	DNA polymerase; homologous to HSV
A2	p135; major protein; homologous to HSV major DNA binding protein
N1	D1LMP
Rightward	
H1	17-kd nuclear and cytoplasmic membrane; bcl-2 homology
O1	140-kd ribonucleotide reductase subunit by homology to HSV
a1	38-kd ribonucleotide reductase subunit by homology to HSV
M1	47-kd nuclear protein; transcription or DNA replication
Late	
Leftward	
L1	gp350/220; ligand for CD21; virion and plasma membrane protein
D3	gp35; virion and plasma membrane protein
c1	p150; major capsid protein by homology to HSV
X2	gp85; membrane fusion; virion and plasma membrane protein homologous to HSV gH
I2	gp, p28
A4	gp110; major nonvirion endoplasmic reticulum glycoprotein; HSV gB homolog
Rightward	
J1	140-kd protein (tegument or membrane, 5NXTS(?))
X2	Basic core protein by homology to VZV

mologous to VZV ORFs 4 to 8 (Fig. 1) encodes for a major EBV early *trans*-activator, BSMLF1. A third domain, homologous to VZV ORFs 18 to 27 (Fig. 1), encodes for the EBV ribonucleotide reductase genes. Although the VZV ORFs are numbered sequentially in the VZV genome, the EBV DNA segments corresponding to these VZV DNA segments are arranged in the EBV genome so that the segment corresponding to VZV ORFs 28 to 59 precedes and is separated from the segment corresponding to VZV ORFs 18 to 27. The DNA segment corresponding to VZV ORFs 4 to 8 is inverted relative to the other segments and placed between them. HVS has the same inversion and interposition relative to VZV and is colinearly homologous with EBV throughout these regions, indicative of a closer evolutionary relatedness. The distant colinear homology to HVS is interrupted by EBV DNA segments, which function in latent B lyphocyte infection (466–467). Similarities notwithstanding, base pair conservation even in the homologous domains is inadequate to permit EBV DNA to hybridize to HSV, VZV, CMV, or even HVS DNAs. Furthermore, antigenic cross-reactivity between EBV and other human herpes viruses is rare, even among proteins encoded by the most conserved genes.

Because the EBV genome was sequenced from an EBV DNA *Bam*HI fragment cloned library, ORFs, genes, or sites for transcription or RNA processing are frequently referenced to specific *Bam*HI fragments (24,156). Thus, the EBV DNA polymerase gene is frequently referred to as *BALF3* for *Bam*HI, A fragment, third leftward ORF. Exons of spliced messenger RNAs (mRNAs) are also frequently designated by the *Bam*HI DNA fragment that encodes the exon.

In contrast to the similarity among herpes viruses in most early and late lytic infection genes, the EBV genes expressed in latent B lymphocyte infection and a few EBV genes expressed in lytic infection have no detectable homology to other herpes virus genes and may have arisen in part from cell DNA. Consistent with this hypothesis is the finding that irregular GGGGCAGGA repeat motifs, which are part of the latent infection cycle nuclear protein 1 gene [Epstein-Barr nuclear antigen (*EBNA*) 1] (254,262,264) are also interspersed in cell DNA (248,251,252). Furthermore, monoclonal antibodies to the irregular gly and ala repeating peptide motifs of EBNA1 cross-react with cell protein(s) (395). Moreover, two cell proteins can specifically bind to the EBNA1 cognate DNA sequence (698). Thus, at least parts of EBNA1 may have arisen from cell DNA.

Further evidence for the premise that some EBV DNA sequences have been recently (on an evolutionary time scale of 10^7 years) appropriated from the primate genome is the finding that a few other EBV, HVS, or CMV genes expressed in lytic infection have interesting homology to cell genes and are not homologous to genes of other herpes viruses. The EBV immediate early gene BZLF1 is closely related to the *jun*/*fos* family of transcriptional activators (490). The EBV early gene BHRF1 has distant but colinear and functional homology to the *Bcl-2* gene, which is at the chromosomal break point of giant follicular lymphomas (90,494). The EBV late gene *BCRF1* is the most striking example of a recent acquisition from the cell gene pool. *BCRF1* is nearly identical to human IL10 in primary amino acid sequence (441,442). Both CMV and HVS have genes that are homologous to cellular G protein coupled receptors (466). EBV does not have such a gene but induces expression of two cellular G protein–coupled peptide receptors in latently infected cells, thereby accomplishing the same result using a different strategy (42). These receptors may be important in the growth or movement of virus-infected cells.

EBV types 1 and 2 are found in the oropharynx of humans in most populations (4,20,111,115,191,214,247,333, 550,565,579,581,600,681,718,725). The type 1 and 2 EBV genomes are nearly identical (318,366,473,510) except for the genes that encode for nuclear proteins (EBNA) LP, 2, 3A, 3B, and 3C in latently infected cells. The type 1 and 2 alleles for EBNA 2, 3A, 3B, and 3C differ in predicted primary amino acid sequence by 47%, 16%, 20%, and 28%, respectively (4,111,565). EBNA LP differs less overall (111,302). The differences among the EBER genes of EBV isolates segregate less clearly into type as opposed to strain-specific differences (52). Aside from these genes, the genomes appear to have few differences beyond those which characterize individual EBV strains. The genes encoding (latent) integral membrane proteins (LMP1 or 2) that are expressed in latent infection vary somewhat among isolates from different geographic locale, but the differences between types 1 and 2 are minor and not of biologic significance (63,279,562). The restriction endonuclease sites in most of the genomes of type 1 and type 2 EBV strains are remarkably similar (109,112,115,162,247,510,511,518, 553,554,725). Thus, EBV 1 and 2 are much more closely related to each other overall than are HSV 1 and 2.

The differences between EBV 1 and 2 in the EBNA 2, 3A, 3B, and 3C genes are reflected in type-specific and type-common EBNA epitopes for antibody (109,263,550, 579,581) and cytotoxic T-lymphocyte responses (see Chapter 75). Because EBV 1 is more common in developed societies, most EBV immune human sera from developed societies react preferentially or exclusively with the EBV 1 EBNA 2, 3A, 3B, and 3C; African sera are more evenly split in their serologic reactivity (550,600,718,725).

Although almost half of the African BLs are infected with EBV 2, recovery of EBV 2 from peripheral blood lymphocytes in developed countries is unusual (54,579,600, 718,725). However, EBV 2 DNA can frequently be detected in oropharyngeal secretions from people living in developed countries (600,718). The failure to recover EBV 2 from the peripheral blood of normal people in developed countries may be due to the less aggressive growth of EBV 2–infected lymphocytes *in vitro* (532) and the overall inefficiency in recovering EBV-infected lymphocytes from peripheral blood.

STAGES OF INFECTION

In vitro, EBV infection is largely restricted to primate B-lymphocytes. Most human peripheral blood B-lymphocytes are susceptible to EBV infection. Tonsillar lymphocytes and fetal chord blood lymphocytes also have been used, chord blood being the most efficient cell for virus isolation. New virus isolates can be obtained by infecting human B-lymphocytes with virus from throat washings. Most adults have been infected by EBV and intermittently shed virus in saliva. Previously infected people also have latently infected B-lymphocytes in their peripheral blood and lymphatic organs. Approximately 1 in 10^5 to 10^6 of their peripheral blood B-lymphocytes are already EBV infected. An advantage of chord blood lymphocytes is that transplacental transmission of EBV infection has not been described so that an EBV recovered in *in vitro*–infected chord blood B-lymphocytes is almost certainly from the innoculum and not from the cell preparation. The virus usually does not replicate in recently infected B-lymphocytes. Instead, infected B-lymphocytes become stably latently infected. In latently infected B-lymphocytes, EBV expresses six different nuclear proteins or EBNAs, two different integral membrane proteins or LMPs, and two small nonpolyadenylated RNAs or EBERs (Figs. 2 and 3). These viral gene products maintain the latent infection and cause the previously resting B-lymphocyte to continuously proliferate (409). Thus, latent means nonlytic infection but is otherwise a misnomer in that the cell is patently affected by EBV infection. The effect on cell growth is immediate and efficient (256,622), with most cells entering DNA synthesis 48 to 72 hr after EBV infection (7,453,454). Aside from the presence of the EBV genome and its gene products, the EBV-infected proliferating B-lymphocytes are similar to lymphocytes proliferating in response to antigen, mitogen, or interleukin (IL)-4 and anti-CD40 in their expression of a similar repertoire of activation-associated proteins, their secretion of immunoglobulin, and their adherence to each other (28,339,719) (see Chapter 75 in *Fields Virology,* 3rd ed.). At least 10% of the infected B-lymphocytes are capable of long-term growth *in vitro* as lymphoblastoid cell lines (LCLs) (256,622,665). The target cell amenable to *in vitro* conversion to an LCL is usually a resting B-lymphocyte before infection (339). Early progenitor B-lymphocytes, pre B-lymphocytes, and cells with fully rearranged and mature immunoglobulin genes can be transformed by EBV *in vitro* (148). EBV-infected B-lymphocytes continue to differentiate and undergo class switching in culture when exposed to interleukin (IL)-4 (188,296).

Approximately 1 in 10^5 to 10^6 of the B-lymphocytes purified from the peripheral blood of previously infected people are latently infected with EBV. These latently infected B-lymphocytes will either proliferate into long-term lymphoblastoid cell lines *in vitro* or will undergo an initial activation of lytic infection and the resultant virus will transform other primary B-lymphocytes in the initial culture (531,598,665). All cell lines that grow out *in vitro* from the peripheral blood of normal humans are EBV-infected B-lymphocytes, and using this peripheral blood is a convenient way to isolate an individual's endogenous virus. LCL outgrowth is also the simplest means for establishing immortal cell lines from individual humans for chemical, biological, or genetic analyses.

The standard conditions for growing infected primary B-lymphocytes are to plate 10^5 T cell–depleted lymphocytes per 0.1 ml of RPMI 1640 medium supplemented with 10% fetal calf serum in each well of a 96-well microtiter tray. The putatively infected cells are fed every 2 weeks, and the assay is scored after 6 to 8 weeks to avoid potential artifacts of transiently stimulated cells. By 6 to 8 weeks a single infectious EBV should have initiated at least one focus of actively proliferating cell clump(s). The isolation yield can be improved by culturing the infected cells on microwells of human diploid fibroblasts, which act as a feeder layer. EBV infection is usually confirmed by polymerase chain reaction (PCR) of putatively infected cell DNA using primers specific for EBV DNA.

Most non–EBV-infected continuous B-lymphocyte cell lines are derived from BLs, and many can be infected with EBV *in vitro* (64,65). The growth of BL cells *in vitro* is attributed to constitutive *c-myc* expression (108,399,408) and to less well characterized changes on chromosome 1 (36). The BL cell is similar to a germinal center B-lymphocyte even though it proliferates rapidly in culture. BL cells grow as single cells and do not express activation markers or adhesion proteins characteristic of antigen- or mitogen-activated normal B-lymphocytes or EBV-infected LCLs. The efficiency of infection of BL cells *in vitro* is not as high as with primary human B-lymphocytes. Most frequently, only a single EBV protein, EBNA 1, is expressed in these cells (404–406,689). This form of latent infection is similar to that in many EBV-infected BL cells *in vivo* (14,216,549,558) (see Chapter 75 in *Fields Virology,* 3rd ed.). The viral genome remains as an episome for many cell divisions. In some cells the viral genome eventually integrates into cell DNA and episomes completely disappear (255,291,358,416). Other cells maintain both integrated and episomal EBV DNA. In some BL cell lines the same EBV genes are expressed as in latently infected primary B-lymphocytes (42,64,65). These infected cells then express the same repertoire of B-lymphocyte activation markers and adhesion proteins as EBV-infected primary B-lymphocytes, and the cells grow in clumps and reach a higher cell density (64,65,613,614). Cells of some other B- or even T-lymphocyte malignancies can be grown as continuous cells lines; some express low levels of EBV receptors. When infected *in vitro,* these infected cells usually express only EBNA 1 (668).

With a frequency characteristic of each EBV-infected B-lymphocyte line, some progeny, latently infected proliferating cells growing *in vitro*, spontaneously become permissive for virus replication. Primary human B-lymphocyte derived cell lines (LCLs) vary from zero to a few percent of cells permissive for EBV replication. *In vitro*–infected BL cell lines tend to be even less permissive for virus replication. Permissively infected cells exhibit the characteristic cytopathic changes associated with herpes virus replication, including formation of an intranuclear inclusion, margination of nuclear chromatin, assembly of capsids within the nucleus near the nuclear membrane, budding of virus through the nuclear mem-

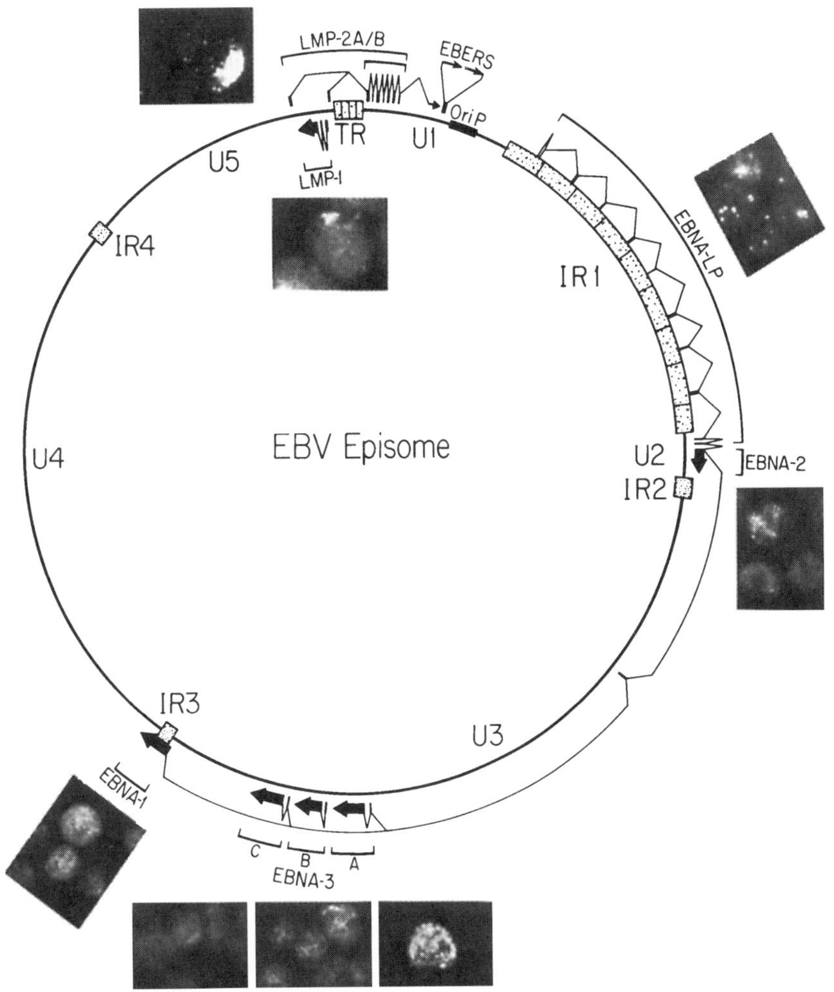

FIG. 3. EBV episome, transcripts, and mRNAs in latent B-lymphocyte infection. Largely unique (U1-U5) and highly repetitive internal (IR1-4) or terminal repeat (TR) DNA domains, the episomal replicon (ori P), and the location of exons encoding EBV nuclear proteins (EBNA 1, 2, 3A, 3B, 3C, and LP) or membrane proteins (LMP 1, 2A, or 2B) genes are indicated by dark boxes. The EBNA mRNAs are spliced as indicated from a single transcript. Transcription is shown as it initiates from the IR1 or *Bam* W promoter. In many infected lymphocytes, after expression of EBNA LP and EBNA 2, the *Bam* C promoter in U1 dominates EBNA transcription. Alternative splicing then leads to expression of all of the EBNAs. EBNA LP is encoded by repeating exons (W1 or W2) from IR1 and two short unique exons from U2. Translation of EBNA LP is dependent on a splice between the exons downstream of the W or C promoters and the first W1 exon. RNAs with a 5-base shorter first W1 exon have an ATG at the beginning of the EBNA LP ORF and express EBNA LP. RNAs with a longer W1 exon are not translated to EBNA LP. These RNAs have a long untranslated leader and EBNA 2, 3A, 3B, 3C, or EBNA 1 from alternative ORFs at the end of their respective mRNAs. The EBERs are two small, nonpolyadenylated, palindromic RNA transcribed in latent or productive infection. LMP 1 and LMP 2B are transcribed in opposite directions under control of promoters separated by only 200 bp. The promoters share the same EBNA 2 response element and are both upregulated by EBNA 2 expression. Another clockwise promoter downstream of the LMP 1 polyadenylation site transcribes LMP 2A. LMP2A also has an upstream EBNA 2 response element. Shown with inserts is the characteristic localization of each of the EBNAs or LMP in a typical latently infected cell line by immune microscopy using monospecific antibody. Each of the EBNA LP, 1, 2, 3A, 3B, and 3C, and LMP 1 and LMP2A have been shown by immune microscopy with specific antisera to be expressed in each latently infected and growth-transformed cell.

brane, and formation of cytoplasmic vesicles (143–147, 507). Cell macromolecular synthesis is inhibited early in EBV replication (207).

The frequency with which latently infected lymphocytes become permissive for virus replication can sometimes be influenced by specific culture conditions. Transient exposure to drugs that inhibit host macromolecular synthesis, cell starvation, arginine-free media, inhibition of DNA methylation with azacytidine, exposure to butyric acid that can also inhibit DNA methylation, treatment with phorbol ester to stimulate protein kinase C, exposure to ionophores that increase intracellular free calcium, or cross-linking of surface immunoglobulin (sIg) increases the permissivity of some EBV-infected cells for lytic infection (30,119,280, 355,462,556,636–639,708,709,727,728). Except for one cell line that reproducibly responds very well to sIg cross-linking, azacytidine or butyrate and phorbal ester have the most reproducible effects on increasing permissivity.

The identification of the EBV-encoded BZLF1 protein as a key immediate early transactivator of EBV lytic infection has led to the use of Z expression from heterologous vectors in transient transfection as a relatively reproducible inducer of lytic replication (100,101,211). Even with this strategy, not all cells in which Z is expressed go on to lytic infection (101).

Induction of virus replication in latently infected B-lymphocytes remains the only way to study EBV replication. Fortunately, latently infected cell lines have been developed in which a substantial fraction of the cells can be rendered permissive for lytic virus infection (105,106,267, 432,433,636–638). These cell lines are essential for studying lytic infection *in vitro* and are the best source of EBV for studying virus structure or cell infection. Lytic EBV infection in B-lymphocytes in culture requires 2 to 3 days for maximum late gene expression. Maximum titers of extracellular virus are reached at 4 to 5 days.

The most reliable method for achieving passage of virus is to cultivate 5×10^4 induced and lethally irradiated infected cells with 10^5 primary B-lymphocytes in a microwell (432,433). Under such conditions, the irradiated cells provide feeder effects for the newly infected primary B-lymphocytes.

Because epithelial cells are fully permissive for lytic EBV infection *in vivo* (468), considerable effort had been made to adopt EBV to growth in epithelial organ cultures or transformed epithelial cell lines. These cultures can be infected, but the infection is inefficient and little virus results (596,602). The efficiency of infection has been slightly improved by expression of the B-lymphocyte EBV receptor in an epithelial cell line (368). More cells are infected, but the infection is largely abortive. Epithelial cell cultures have not been useful for isolation or propagation of virus or for detailed biochemical studies of EBV infection. Therefore, most of our knowledge of latent or lytic EBV infection *in vitro* is based on infection of B-lymphocytes.

Adsorption

The *in vitro* host range restriction of lymphocryptoviruses to preplasmacytoid B-lymphocytes is partly due to a restriction in high-level expression of a cell surface protein, CD21, which is the EBV receptor and the receptor for the C3d component of complement. The fact that B-lymphocyte infectability correlates with the stages of B-lymphocyte development in which type 2 complement receptors (CR2, now renamed CD21) are expressed initially led to the expectation that CD21 was the EBV receptor (293,309). Blockade of virus infection by monoclonal antibodies to the 140-kd CD21 glycoprotein provided a direct link between receptor activity and CD21 (160,172,461,649,694). Purified CD21 was then shown to bind to EBV (464,591), and CD21 expressed on heterologous cells conferred the ability to adsorb EBV (5,368). CD21 is a member of the immunoglobulin superfamily and consists of 15 or 16 imperfect repeats of a 60-amino acid domain, a transmembrane domain, and a short carboxy-terminal cytoplasmic domain (444,695). Each extracellular repeat domain has two cysteines at either end that, if disulfide bonded, would result in the rest of the domain forming an intrachain loop. The mRNA is multiply spliced, and the mRNA from some cells is deleted for an exon encoding for one of the extracellular repeat domains (44,444,695). The EBV binding activity maps to one or both of the two amino-terminal repeat domains (413,439,443,463).

The major EBV outer envelope glycoprotein, gp350/220, is the CD21 ligand (459,460,643,697). gp350 and gp220 are translated from abundant late replication cycle EBV mRNAs, which are transcribed from the same gene (34, 289). The mRNA for gp350 is not spliced, whereas the mRNA for gp220 is spliced in frame (34,40,699). CD21 is the only B-lymphocyte surface protein that binds to gp350/220 (643). The gp350/220–CD21 binding affinity is 1.2×10^{-8} M (644). Soluble gp350/220 can saturate B-lymphocyte receptors and block virus infection, indicating an essential role for the gp350/220–CD21 interaction in virus adsorption (459,644). A gp350/220 peptide sequence EDPGFFNVEI is similar to the peptide sequence EDPGKQLYNVEA, through which the C3d component of complement binds to CD21 (351,460,643). Synthetic peptides containing the LYNVEA C3d sequence block C3d binding to CD21 (351), whereas peptides containing the EDPGFFNVEA sequence block EBV infection (460,463). In addition, mutant gp350/220 molecules, which delete the two amino acids VE from the EDPGFFNVE peptide, no longer have ligand activity for CD21 (644).

The identification of the EBV receptor on epithelial cells or cell lines has been more difficult. Virus adsorbs to epithelial cells or to transformed epithelial cell lines much less efficiently than to B-lymphocytes. One of the most efficient anti-CD21 monoclonal antibodies reacts with a 200-kd glycoprotein on normal or transformed epithelial cells, consistent with the possibility that this glycoprotein (which is antigenically similar to CD21) is the epithelial EBV receptor (599,716). However, hybridization of CD21 complementary DNA (cDNA) to northern blots of RNA from two transformed epithelial cell lines that express the 200-kd protein and adsorb small amounts of EBV shows a small amount of CD21 homologous RNA indistinguishable in size from B-lymphocyte CD21 mRNA (42,44). Furthermore, the 200-kd glycoprotein did not interact with gp350. In fact, gp350 bound to an inert matrix could retrieve a molecule from the surface of epithelial cells that is identical in size to CD21 (44). Cloning and sequencing of the CD21 homologous RNA from an epithelial cell line showed it to be identical to B-lymphocyte CD21 (44). The RNA begins at least 50 bp upstream of the previously identified CD21 start site, but a similar start site is used in some lymphocyte cell lines. CD21 RNA also has been detected in NPC cells, adding additional support to the likelihood that CD21 is the epithelial cell EBV receptor (411). However, CD21 has not as yet been shown to be the EBV receptor on normal epithelial cells. Proof of low-level CD21 expression in normal epithelial tissues is complicated by the presence of small numbers of B-lymphocytes in epithelium as evidenced by Ig mRNA.

Efficient penetration into cells may require CD21 not only for adsorption but CD21 accessory molecules and other cellular proteins to facilitate uptake, fusion, or entry. There is little direct evidence on these points and some evidence against a role for CD21 beyond adsorption (412). Receptor transfers by membrane fusion initially indicated that the host range might be extended (678). However, high-level expression of CD21 on mouse L cells, human T or erythroleukemia cells, or transformed human epithelial cells results in inefficient EBV infection as measured by latent or lytic cycle EBV gene expression (5, 341, 368; Wang and Kieff, unpublished observations). Whether infection aborts at the steps of penetration, uncoating, nuclear translocation, or virus gene expression is uncertain.

Penetration and Uncoating

EBV infection of primary B-lymphocytes *in vitro* involves binding of CD21 to the B-lymphocyte plasma membrane, aggregation of CD21 in the plasma membrane and coaggregation of sIg, and internalization of EBV into cytoplasmic vesicles (68,462,643). The virus envelope then fuses with the vesicle membrane, releasing the nucleocapsid and tegument into the cytoplasm. Infection of BL cells is somewhat different in that EBV binding to the plasma membrane does not result in as significant a patching of CD21 and sIg and the envelope fuses with the plasma membrane, releasing the nucleocapsid and tegument into the plasma membrane. The difference between the primary B-lymphocytes and the BL cells is likely to be due to the cytoskeletal abnormalities of the tumor cells. Studies of gp350/220-coated microbeads and primary B-lymphocytes or BL cells indicate multiple similarities in the processing of the gp350/220-coated beads and virus, including the release of beads free into the cytoplasm. The simplest, although not necessarily correct, interpretation of the release of gp350/220-coated beads into the cytoplasm is that gp350/220 not only mediates EBV absorption but also has a role in penetration (643). gp350/220 has a potentially amphipathic domain that could participate in the fusion of the EBV envelope with vesicle membranes (34,643). However, the natural process of penetration is likely to involve other glycoproteins as well.

A second EBV glycoprotein, gp85, has been more directly implicated in fusion of the EBV envelope with the vesicle membrane. Genetic and biochemical evidence indicates that HSV-1 gH is important in HSV-1 envelope and cell membrane fusion and in nucleocapsid exocytosis into the cytoplasm (207). The EBV gp85 primary amino acid sequence is similar to that of HSV-1 gH (246,477). Furthermore, monoclonal antibodies to gp85 do not affect virus adsorption but do inhibit the fusion of the EBV envelope and the cell membranes (434).

Early Intracellular Events in Infection

Little is known about EBV capsid dissolution, genome transport to the cell nucleus, or DNA circularization. By analogy with other DNA viruses that replicate in the nucleus, the cytoskeleton is likely to mediate EBV capsid transport to the nucleus or to nuclear pores (107). A protein similar in size to the HSV virion tegumentary protein, Vmw65 or VP 16, has been detected in purified EBV preparations and remains associated with cells after infection (7). HSV VP16 is a key transinducer of immediate early lytic virus gene expression (see Chapter 72). The EBV BPLF1 ORF is colinearly 30% homologous to the HSV VP16 ORF. EBV BPLF1 could therefore be a transactivator of lytic EBV replication in epithelial cells. BPLF1 may be unnecessary for activation of latent gene expression in B-lymphocytes because the first latent infection promoter, Wp, has high constitutive activity in B-lymphocytes (674).

The first evidence that a herpes virus genome circularizes in infected cells arose from the observation that almost all EBV-infected cells contain multiple copies of covalently closed circular EBV episomes (3,381). In primary B-lymphocytes, EBV genome circularization precedes or coincides with earliest virus gene expression (7,290). By 16 hr postinfection, each infected cell contains one EBV episome. Whether the putative protein that recognizes EBV TR for cleavage and packaging (see section on Late Gene Expression, etc.) facilitates the joining of the linear virion EBV TRs to produce genome circularization is not known. Studies with inhibitors of cell DNA, RNA, and protein synthesis suggest that active cell macromolecular synthesis is required for a step before or including genome circularization.

Cell transcription factors probably determine if latent or lytic infection ensues after the genome enters the nucleus and circularizes. As indicated above, EBV's narrow host range is only partly determined by the presence of CD21 receptors. Although, transfection of HSV, VZV, or CMV DNA into cells permissive for virus replication results in lytic infection and release of infectious virus, attempts to similarly extend the EBV host range by DNA transfection have so far been unrewarding, consistent with the notion that EBV gene expression is tightly regulated by cell transcription factors. In one series of experiments with placental fibroblasts, transfection of EBV DNA resulted in release of virus as detected by primary B-lymphocyte growth transformation (428). This experiment suggests that diploid fibroblasts might be permissive for lytic cycle EBV transcription if the EBV genome could be efficiently placed in these cells. Transfection of cells that permit transfection-mediated lytic infection with other human herpes virus DNAs, microinjection of EBV DNA, or receptor transplantation followed by infection have not so far resulted in latent or lytic EBV infection, even though some EBV gene expression has been observed (212,218,677,678). B-lymphocyte– or epithelial cell–specific transcriptional factors are probably necessary for efficient latent or lytic cycle EBV promoter activity.

Latent Infection in B-lymphocytes and Growth Transformation

As described briefly above, the usual outcome of B-lymphocyte infection with EBV is persistent latent infection; expression of EBNAs, LMPs, and EBERs; immediate and efficient entry of the infected cell into the cell cycle; and continuous cell proliferation. An early event in EBV infection may activate the B-lymphocyte before viral DNA transcription. Cross-linking of CD21 on primary B-lymphocytes has costimulatory effects on antigen-activated cells and may be part or all of this early B-lymphocyte activating effect (412). CD21 is part of a complex that includes CD19 and associated molecules. Some of the associated molecules have intracellular tyrosine kinase signaling domains similar to the sIg-associated signal-transducing cell surface proteins. Probably as a consequence of multivalent CD21 engagement, infected cells begin to synthesize RNA, enlarge, express activation and adhesion molecules, clump, and secrete Ig soon after exposure to virus (7,18,290,643,665). How much of this is due to multivalent CD21 ligation or other surface phenomena has not been sorted out. CD21 ligation appears to have a nonspecific effect on increasing expression from transfected DNA, and this may be related to these activating effects (497,548). CD21 also has been implicated as an alpha interferon receptor, raising the possibility that alpha interferon might block EBV adsorption (126).

Initial data suggest that as the EBV-infected cells begin to proliferate, cell growth is critically dependent on high serum concentration, feeder layers, or cell density. In all of these characteristics, EBV-infected cells parallel B-lymphocytes stimulated to proliferate by antigen and cytokines or by IgM cross-linking and IL-2 and IL-4, or by CD40 ligation and IL-4 (18,28,290,339). The initial dependence of EBV-infected cells on cell concentration, high serum concentration, or feeder layers is evidence indicating that EBV-infected B-lymphocytes produce and are dependent on autostimulatory substances that promote cell growth *in vitro* (210). Autocrine-dependent growth was first described for cells derived from certain sarcomas and may be a route to cell immortality and malignancy (125). Autocrine growth stimulation may contribute to the proliferation of B-lymphocytes latently infected with EBV *in vitro* as well as to EBV transformation of B-lymphocytes *in vivo*.

In contrast to B-lymphocytes stimulated to proliferate for a short time *in vitro* by polyclonal B-cell activators or by surface IgM cross-linking, EBV-transformed B-lymphocytes proliferate indefinitely in cell culture (507,508, 622). The similarity between EBV and antigen-driven proliferating B cells and the distinct difference between LCLs transformed by EBV- and non–EBV-infected BL cells in the repertoire of cell genes expressed is compatible with the hypothesis that EBV transforms primary B-lymphocytes, at least in part, through autocrine growth mechanisms that maintain differentiated, antigen-driven, B-lymphocyte characteristics as opposed to the less differentiated characteristics of a *c-myc* translocated BL cell. Non–EBV-infected BL cells also express CD10, a marker of acute lymphocytic leukemia or normal germinal center B-lymphocytes, and do not express CD23, CD48 (163), or other activation markers, or high levels of adhesion molecules as is characteristic of EBV-induced primary B-lymphocyte proliferation.

With continued cell proliferation, EBV-infected LCLs do not undergo a crisis but gradually evolve into more rapidly growing cells, which are somewhat less dependent on high cell concentration, high serum concentration, or other growth conditions. This change in growth is associated with stabilization of chromosome telomeres and constitutive expression of telomerase (99). The established LCLs tend to secrete less immunoglobulin (452).

Viral Gene Expression In Latent Infection

Because EBV-infected lymphocytes are growth transformed by the virus, they can be grown indefinitely in culture and are amenable to detailed biochemical analyses, including investigation of the mechanism of latent genome persistence and of cell growth transformation. In defining the genes expressed in latent infection, it is important to recognize that most latently infected cell lines contain a small number of progeny cells that are spontaneously permissive for virus replication. Because virus gene expression in lytic infection is at a much higher level than in latent infection, lytic cycle RNAs or proteins are frequently detected among RNA or proteins from predominantly latently infected B cell lines. Proof that an EBV gene is expressed in latent infection requires the demonstration that the specific virus gene product is in the vast majority of latently infected cells in a culture. For the most part, this has been accomplished by defining an mRNA sequence whose abundance is not affected by the number of permissively infected cells (159,254,334,335,509,551,674,675), deriving monospecific antibody from the putative protein product using heterologous antigen expression systems (260–265,389,499–501,690) or peptide synthesis (135–137), and using the antibody to identify the protein in latently infected cells by immune microscopy (260,261,263, 265,389,499,501,690).

As indicated in Fig. 2, at least 11 EBV genes are expressed in latent infection. Two of these encode small, nonpolyadenylated RNAs (EBER 1 and EBER 2), six encode nuclear proteins (EBNA 1, 2, 3A, 3B, 3C, and LP), and three encode integral membrane proteins (LMP1, 2A, and 2B). The EBV-encoded nuclear proteins in latent infection are all called EBNAs because the presence of viral proteins was first detected by immunofluorescence microscopy using serum from normal seropositive adults. LMP is simply an acronym for latent infection membrane protein. Some authors have used different nomenclatures for EBNA 3A, 3B, 3C and LMP2A and 2B; naming EBNA 3B and 3C as EBNA4, 5, or 6, depending on the author, and LMP2A and 2B as TP1 and TP2, because the mRNAs cross the ends of the linear EBV genome (339,427).

The time course of events in *in vitro* primary B-lymphocyte infection vary somewhat depending on the virus innoculum and the physiologic state of the lymphocytes. The EBV genome circularizes in the infected cell nucleus within 12 to 16 h of infection (Figs. 3 and 4). At about the same time, the Wp promoter initiates rightward transcription (7,11,290,453,454,539). As is true of the other herpes viruses, EBV does not encode an RNA polymerase and uses host cell RNA polymerase II for transcription of viral mRNAs. By nuclear run-on or transient transfection analysis, the initial EBNA promoter, Wp, or the second EBNA promoter, Cp, are the strongest EBV pol II promoters in latent EBV infection. The Wp promoter is in the 3-kbp, *Bam* W or IR1, long internal direct repeat element (560, 561,674). The promoter in the left-most repeat appears to dominate because most of the EBNA cDNAs have multiple exons from downstream copies of the W repeat (46–50,560,611). The simplest model that would provide a basis for the domination of a leftward or the leftmost W repeat element is to ascribe the dominance to the effect of a positive transcriptional element in the upstream U1 region (Fig. 2).

The first mRNAs have a less than 40-base first exon, W0, which ends with an AT. This is spliced to alternating 66-base W1 and 132-base W2 exons derived from successive reiterations of the long internal repeats and then to two short and one long exon (7,560,690). The 5′ exons of half of the RNAs are spliced from the 40-base W0 exon to an alternate acceptor site 5 base into the first 66-base W1 exon. This alternate splice provides a G, which completes an ATG initiation codon for translation of an ORF encoded by the repeating 66-base and 132-base exons and by two ensuing short exons (Y1, Y2) from the unique DNA (*Bam* Y) downstream of the long internal repeat (Fig. 3). Translation of the alternately spliced RNA initiates at the W0W1 ATG and stops near the end of the Y2 exon. The resulting protein is a nuclear protein. Because this EBNA is encoded in the leader of the EBNA mRNAs, it is usually called EBNA leader protein or EBNA LP.

The Y2 exon is spliced to a long exon, YH, from the right end of *Bam* Y and the left end of *Bam* H (Fig. 3), which has a translational initiation codon followed by a long ORF. This exon encodes for a nuclear protein, EBNA 2. Although those mRNAs with an ATG at the beginning of the EBNA LP ORF produce EBNA LP as their major translation product, those without an ATG at the beginning of the EBNA LP ORF do not translate EBNA LP and efficiently translate EBNA 2 (7,536, 537,690). Efficient translation of the EBNA 2 ORF, despite the long untranslated leader, is surprising because 40S ribosomal subunits associate with mRNA at the 5′ cap and movement along the RNA is associated with some instability (345). Addition of the 60S subunit just before translational initiation constitutes the complete ribosome, which can move more stably along the RNA. Thus, translation of the 5′ EBNA LP ORF would be expected to yield more efficient translation of the 3′ ORF than would occur if a 40S subunit had to transverse the entire leader domain. Either the 40S subunit is more stable than the complete ribosome in this system or EBNA LP translation destabilizes the ribosome mRNA complex, or the EBNA 2 mRNA, with the long untranslated leader, can internally initiate translation at the beginning of the EBNA 2 ORF.

EBNA LP and EBNA 2 are the first viral proteins expressed in B-lymphocyte infection. They reach the level that is maintained in transformed B-lymphocytes (LCLs) by 24 to 32 hr postinfection (7). EBNA 2, in turn, transactivates cell genes (including CD23, CD21, and *c-fgr*) and viral genes (including LMP1 and LMP2) (Fig. 4) (1,7,97, 150,151,340,667,686–688,690,691,724,726). Upregulation of these genes is readily observed after transient or stable expression of EBNA 2 in non–EBV-infected BL cells but is less efficient in cells other than human B-lymphocytes. In fact, *c-fgr* upregulation is more impressive in BL cells than in primary B-lymphocyte infection. *c-myc* expression is also up regulated in primary B-lymphocyte infection at the time that EBNA LP and EBNA 2 are expressed (7).

Probably as a consequence of EBNA 2 effects on EBNA 2 response elements in the unique DNA in the EBV *Bam* C fragment (which is upstream of the long internal repeat), EBNA promoter usage in many or most cells moves upstream from Wp to the nearby Cp latency promoter (7,629, 702,703). One EBNA 2 response element is immediately

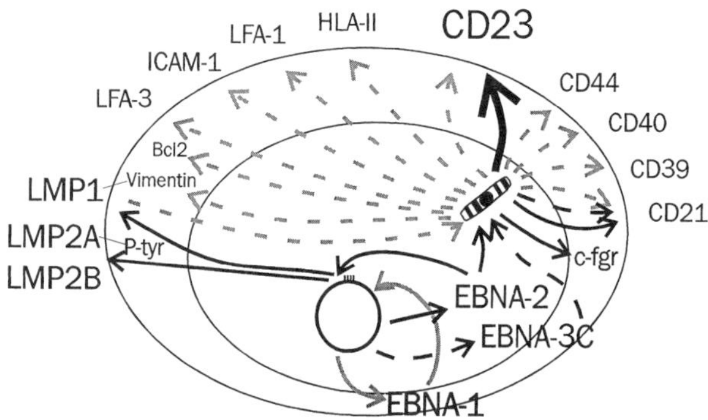

FIG. 4. Expression of EBNAs or LMPs in non–EBV-infected BL lymphoblasts provides indications of the biochemical mechanisms by which these proteins effect changes in primary B-lymphocyte growth. EBNA 1 binds to ori P and enables the EBV episome to be replicated during S phase. EBNA 2 specifically *trans*-activates CD23, CD21, and cfgr as well as the LMP promoters. EBNA 2 response elements also have been described upstream of the first EBNA promoter and may be involved in promoter switching. EBNA 3C upregulates CD21 mRNA and protein but has no effect on CD23. LMP 1 patches in the cell plasma membrane where it is associated with (and induces higher level expression of) vimentin. LMP1 has dramatic effects on BL cell growth. LMP1 functionally activates homotypic adhesion, upregulates ICAM 1, LFA 1, and LFA 3, upregulates bcl2, raises the lymphocyte intracellular free calcium, and induces activation markers, including transferrin receptor, HLA-II, CD21, CD23, CD39, CD40, and CD44. LMP also downregulates CD10 and alters lymphoblast TGF beta responsiveness. Each of these effects is cell line-dependent, and most effects are not evident in all BL cell lines. For example, BL cell lines vary in their level of adhesion molecule expression, and LMP1 effects on adhesion molecule expression are most evident in cell lines in which adhesion molecule expression is low before LMP1 expression. LMP2A associates with LMP1 and with fyn and lyn. LMP2A is serine/threonine and tyrosine phosphorylated. LMP2A blocks calcium mobilization in response to cross-linking of B-lymphocyte surface molecules, including Ig.

upstream of the Cp promoter (628,702,703), whereas two other elements are downstream of the promoter (306,682). Two short exons downstream of the Cp promoter, C1 and C2, replace the short W0 exon in the EBNA mRNAs in those cells in which the switch has occurred. As with the W0 exon, the C2 exon ends in an AT, which is spliced to either the 66 or 61 b internal repeat derived exon, W1 or WO1, resulting in an mRNA leader incapable or capable, respectively, of translating EBNA LP. All resulting C- or W-initiated mRNAs have the same repeating 66- and 132-base exons in their leader (46–50,307,537,560,611). A potential splice donor near the beginning of the EBNA 2–encoding YH exon but preceding the EBNA 2 initiation codon is now activated in some transcripts, and some RNA molecules are spliced to a far downstream acceptor in *Bam* U. The U exon is then spliced to any of four alternate acceptor sites of exons that begin the ORFs that encode for EBNA 3A, EBNA 3B, EBNA 3C, or EBNA 1 (Fig. 3). Now, as a consequence of alternate 5′ splicing of Cp- or Wp-initiated RNAs, some mRNAs will still encode EBNA LP. Furthermore, as a consequence of partial use of the alternate donor site between the beginning of the YH exon and the EBNA 2 initiation codon, some RNAs that do not encode for LP or EBNA 2 will be spliced to a U exon and now encode EBNA 3A, 3B, 3C, or 1 instead of EBNA 2. Alternate acceptor site usage after the U donor site will now determine which RNAs will encode EBNA 3A, 3B, 3C, or 1 (Fig. 3).

The simplest model to explain the turn-on of EBNA 3A, 3B, 3C, and 1 expression is that activation of the Cp promoter, or enhancement of Wp activity in those cells that do not switch, leads to higher level transcription of the EBNA mRNAs. Many transcripts now pass the polyadenylation site downstream of EBNA 2 and extend to EBNA 3A, 3B, 3C, or 1. The processing of the transcripts is then determined by their polyadenylation site. Because both EBNA 2 and EBNA 1 activate EBNA transcription, the question arises as to what limits the Cp or activated Wp promoters that they do not simply continue to increase in response to higher EBNA 2 and 1 levels. The likely candidates are the EBNA 3s, although there is as yet no direct experimental evidence on this point.

By 32 hr all EBNAs and LMP RNAs are expressed (7). The LMP1 mRNA is the most abundant EBV mRNA in latent infection, more abundant than the EBNA or LMP2 mRNAs. LMP2A and 2B are probably regulated in a manner similar to that in which LMP1 is regulated because their promoters are also EBNA 2 responsive. In fact, the LMP2B promoter shares regulatory elements with the LMP1 promoter. The LMP2B promoter regulates rightward transcription, whereas the LMP1 promoter is a leftward promoter (Fig. 3). The transcriptional start sites are only 266 bp apart. The LMP2A promoter is 3 kbp downstream of the LMP1 promoter. LMP2A regulation occurs largely through a separate EBNA 2 response element upstream of its promoter (724,726). The temporal onset of LMP2 A and B expression in acute infection has not been documented due to the poor quality of specific antisera. Contemporaneous with the expression of the other EBNAs and of LMP1, CD21 and CD23 are further upregulated. Cell DNA synthesis follows the onset of LMP1 expression. EBNA 1 acts at the EBV episome origin, presumably to enable initiation of EBV episome replication and amplification (see below). By 48 hr of primary human B-lymphocyte infection *in vitro*, all of the EBNA proteins, LMP1, and probably LMP2 A and B are near the level that is maintained consistently through latent infection (7,11,290,453,454,538,539).

EBER expression lags by approximately 24 hr and does not reach substantial levels until 70 hr postinfection (7). The EBER 1 and EBER 2 RNAs are mostly transcribed by cell RNA polymerase III, although polymerase II also may be involved (21,273,274,299,365,541). The EBERs not only contain the typical intragenic control regions common to small RNAs but possess upstream Sp1, ATF, and

TATA box elements characteristic of pol II transcriptional sites (273). These sequences appear to be required for optimal EBER transcription. The EBERs are the most heavily transcribed and most abundant EBV RNAs in latent infection (21,274,334,335,509,561,664).

The delineation of EBNA transcripts through Northern blot, S1 nuclease mapping, cDNA cloning analysis, and recent nuclear run-on studies now explains the initial observations that the EBV genome is extensively transcribed in latent infection and that only a highly restricted complexity of RNA is processed into cytoplasmic polyadenylated polyribosomal mRNA (114,237,239,334,335,487). In particular, nuclear run-on studies indicate that almost the entire strand of the genome 3′ to the Wp promoter is transcribed (Fig. 3) (560). Although the rate of transcription drops off somewhat downstream of EBNA 2 and again downstream of EBNA 1, transcription, per se, does not regulate EBV DNA R strand gene expression.

As indicated above, polyadenylation may determine splicing and thereby regulate expression of the six EBNA mRNAs from the same promoter. However, EBV *cis* or *trans* factors could also play a role in splice choice. None have been defined. The possible role for the EBERs or EBNAs (particularly EBNA LP) as *trans*-acting factors is considered below. One potential *cis*-acting factor in the EBNA RNAs that could affect RNA transcription or processing is the tandem copies of IR1 or *Bam* W. Each IR1 or *Bam* W copy contains a 500-bp nearly perfect palindrome that is transcribed into a repetitive part of all EBNA transcripts (81). Analysis of a defective EBV DNA sequence has shown that this palindrome can serve as a site for DNA inversion within the EBV genome (302). Thus, the IR1 palindrome is formed within DNA under physiologic conditions and could well impart considerable secondary structure to the EBNA primary transcripts.

Latent Infection Proteins

EBNA LP

EBNA LP is encoded by the leader of each of the EBNA mRNAs and is translated from those mRNAs when the first and second exons are spliced so as to create the EBNA LP initiation codon. As expected from the observation that the IR1 repeat number varies among EBV isolates, and IR1 contributes 66 and 132 bp alternating W1 and W2 EBNA LP encoding exons, the EBNA LP size also varies among EBV isolates (135,161,690). The distinctive features of the EBNA LP primary amino acid sequence are alternating 22– or 44–amino acid repeat domains and a 45–amino acid long unique carboxy-terminal domain. The 44–amino acid repeat domain has two runs of basic amino acids, arg-arg-his-arg or arg-arg-val-arg-arg-arg, which could be nuclear translocation signals. EBNA LP ends with seven acidic residues. EBNA LP is phosphorylated at multiple sites, probably at least once within each repeating exon because the number of isoelectric forms varies with the exon repeat number (500). EBNA LP is presumably phosphorylated on serine residues because each 44–amino acid repeat domain has a near consensus casein kinase II type serine phosphorylation site. EBNA LP is more associated with the nuclear matrix fraction than are the other EBNAs (500).

EBNA LP has an unusual and intriguing nuclear localization as shown by immunofluorescence microscopy (500, 690). Some EBNA LP is diffusely spread through the nucleus, but the rest is concentrated in a few small nuclear granules. These granules are frequently in a curved linear array, suggesting that they could be associated with a continuous linear structure. *In situ* RNA cytohybridizations with EBV IR1 DNA reveal similar linear intranuclear structures that could be EBV transcripts (357). EBNA LP could associate with EBV nuclear RNA and play a role in EBV RNA transcription or processing.

Molecular genetic analyses of EBNA LP have focused on the last two exons which are within the DNA fragment deleted from the nontransforming P3HR-1 EBV strain. Deletion of these two exons from EBV recombinants and plating of infected primary B-lymphocytes in soft agarose over fibroblast feeder layers resulted in a modest reduction in transformed cell outgrowth (228). However, only one colony was grown out as an LCL in culture. When this LCL was transfected with a full-length EBNA LP expression vector with a positive selection marker, LCLs expressing wild-type EBNA LP grew out more rapidly under selective conditions than did LCLs that did not express EBNA LP (8). Attempts to derive similar deletion mutants using an LCL outgrowth assay without fibroblast feeder layers were repeatedly unsuccessful (403). One additional deletion mutant was recovered by culture of newly infected cells on fibroblast feeder layers. Because the deletion mutation could have a *cis* effect on EBNA RNA processing, a nonsense mutation at the beginning of the Y1 exon was also evaluated (403). EBNA LP nonsense mutant recombinants were also deficient in cell growth transformation. Despite growth of the infected primary B-lymphocytes on fibroblast feeder layers, the frequency of nonsense recombinant transformed LCLs was 5% that of wild-type. Of the four that were obtained, only one grew out within the normal outgrowth period, and that LCL proved to be infected with a revertant (403). Induction of lytic virus infection in the few LCLs that were infected with deletion or nonsense EBNA LP recombinants and infection of primary B-lymphocytes with the virus from these cells confirmed an unusual dependence on fibroblast feeder layers. The initiation of multiple simultaneous transforming events could also enable the outgrowth of LCLs infected with the mutant EBNA LP EBV recombinants, presumably by cross-feeding among the multiple mutant EBV-infected cells (403). The mutant virus-infected LCLs also tended to be more highly differentiated toward immunoglobulin secretion. There was no difference in latent EBV gene expression other than the absence of EBNA LP. The activity of fibroblast feeder layers in partially compensating for the EBNA LP deletion or nonsense mutation and the enhanced differentiation toward immunoglobulin secretion is most compatible with the hypothesis that wild-type EBNA LP directly or indirectly upregulates expression of an autocrine factor(s) critical for LCL outgrowth. Recent results of transient transfection of EBNA LP and EBNA 2 into primary B-lymphocytes costimulated with gp 350 indicate that EBNA LP and 2 together can induce G0 to G1 transition as marked by cyclin D2 induction (596). Where EBNA LP enters into a pathway leading to cyclin D2 induction is uncertain. EBNA LP may complement EBNA 2 in inducing *c-myc* or another early response gene, which may regulate a growth factor or may itself be sufficient for G1 entry. Association of EBNA LP with Rb and p53 has been suggested based on *in vitro* biochemical interactions and colocalization of EBNA LP with Rb as detected with one monoclonal antibody but not with others (305,635). Somewhat inconsistent with such effects, expression of EBNA LP in transgeneic mice had no discernible effect on development or tumor incidence. Animals died of heart failure, which is likely to be a toxic effect of EBNA LP expression or of the insertion site (285).

EBNA 2

EBNA 2 was long suspected to be important in the ability of EBV to growth transform B-lymphocytes because of the isolation 25 years ago of the laboratory mutant P3HR-1 (267). This mutant EBV is transformation incompetent (419,432). A small deletion was identified in the P3HR-1 EBV genome by subtractive hybridization and Southern blot analysis and was shown to encode EBNA 2 and the last two exons of EBNA LP (111,135–137,263, 690). The few transformation-competent recombinants that arose after infection of latently infected cells with the P3HR-1 EBV had part of the P3HR-1 genome but were wild-type at the deletion site (603).

More recently, the deletion in the P3HR-1 EBV genome has been useful for assaying the ability of specifically mutated EBV DNA fragments to rescue transforming virus from P3HR-1 cells. Transfection of P3HR-1 cells with *Escherichia coli*–cloned wild type EBV DNA fragments that span the P3HR-1 deletion and induction of lytic replication result in homologous recombination and restoration of the deleted DNA segment in one in 10^5 progeny virus. Only virus with the restored wild-type EBV DNA segment can transform primary B-lymphocytes into LCLs (95,228). Transfection with cloned EBV DNA fragments deleted for part of the EBNA 2 ORF or containing a stop codon one third of the way into the ORF repeatedly failed to restore transforming ability, whereas wild-type DNA restored transforming ability. Recombinant EBVs with a nontransforming EBNA 2 mutation could be recovered by infection of EBV-negative BL cells (404). Thus, genetic analyses proved that EBNA 2 is essential for primary B-lymphocyte growth transformation.

EBNA 2 differs more extensively between EBV types 1 and 2 than any other protein (6,111,565), and genetic analyses indPC IM rE Ir EBNA 2 is the principal determinant of the biological difference that enables type 1 EBV strains to transform primary B-lymphocytes with greater efficiency than can type 2 EBV strains (95). Unlike infection of primary B-lymphocytes with type 1 EBV strains, which results in efficient and rapid outgrowth of LCLs, infection with type 2 strains results in inefficient and slow outgrowth of LCLs (532). The P3HR-1 EBV is a type 2 strain, but when recombined with the type 1 EBNA 2, the recombinants transform cells with high efficiency (95,228). The infected primary B-lymphocytes proliferate as rapidly as do lymphocytes transformed by type 1 strains. Recombination of cloned type 2 EBNA 2 DNA into the site of the P3HR-1 deletion restores the normal poorly transforming phenotype of type 2 EBV strains (95).

The type 1 EBNA 2 is predicted from DNA sequence to consist of 484 amino acids, whereas the type 2 EBNA 2 is 443 amino acids, primarily because of a shorter repeat domain (111,263,457,552). EBNA 2 is acidic overall and consists of several obvious domains (Fig. 5): (a) an amino-terminal domain of 57 amino acids that are negatively charged and highly conserved between EBV-1 and EBV-2; (b) a polyproline domain of between 10 and 40 proline residues; (c) 46 amino acids, among which nine of ten charged amino acids and 30 of 46 overall are identical between EBV types 1 and 2; (d) 149 amino acids, of which few of the 16 charged amino acids and only 52 of 149 amino acids overall are identical between EBV-1 and -2; (e) 46 amino acids, of which only three are charged and 26 of 46 overall are identical; (f) a 26–amino acid imperfect arg-gly repeat, of which 13 amino acids are arg or lys; and (g) 121 carboxy-terminal amino acids, of which 19 of 24 charged amino acids and 84 of 121 amino acids overall are identical between EBV-1 and -2. Of the carboxy-terminal 62 amino acids, 16 are acidic. The arg-gly repeat region or a lys-arg-pro-arg sequence near the carboxy-terminus are each adequate for nuclear localization. Deletion of either sequence is compatible with continued nuclear translocation (94). Deletion of both results in cytoplasmic localization (94).

EBNA 2 localizes to large nuclear granules (263,500). In nuclear fractionation, EBNA 2 is associated with nucleoplasmic, chromatin and nuclear matrix fractions (500). Like EBNA LP and EBNA 1, EBNA 2 is phosphorylated on serine and threonine residues. EBNA 2 undergoes significant posttranslational modification other than phosphorylation because the size of the nascent protein is 10 kd smaller than that of stable intranuclear EBNA 2 (690). The basis for this modification is not known. Fractionation of EBNA 2 extracted under physiologic salt conditions on sucrose gradients yields complexes the size of tetramers and a much larger complex (213,575).

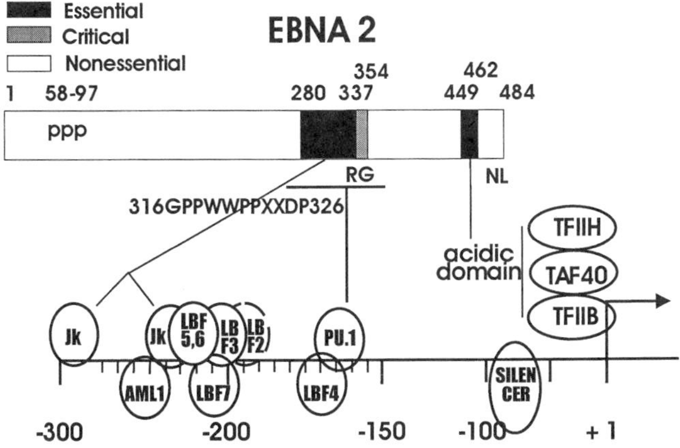

FIG. 5. Model of EBNA 2 interactions with the LMP1 response element binding factors and basal transcription factors in B-lymphocytes. EBNA 2 is an acidic transcriptional transactivator that requires sequence-specific DNA binding proteins to bring it to response elements. The acidic domain can interact with TFIIB, TAF40, and TFIIH in stimulating transcription. Whether it does so sequentially or as a complex of more than one EBNA 2 molecule interacting simultaneously with more than one factor remains to be determined. Jk is a sequence-specific DNA binding protein that recognizes a GTGGGAA sequence present in all EBNA 2 response elements and interacts with a 27–amino acid domain of EBNA 2 that includes a key GPPWWPP sequence. The LMP1 promoter has Jk binding sites at –298/–290 and –223/–213. Truncation of the –298/–290 Jk site by the –234/–40 promoter construct has surprisingly little effect on EBNA 2 responsiveness in these and previous experiments. Further truncation to –214 removes the more proximal Jk site (–223/–213) as well as the AML1 site (–230/–224) and encroaches on the LBF5 and six sites (–216/–209). The effect on EBNA 2 responsiveness is similar to that of a specific mutation in the Jk binding site, i.e., a 50% reduction in EBNA-2 responsiveness. Initial studies indicate that EBNA 2 also can interact with PU.1, and this interaction is likely to account for the B-lymphocyte specificity of EBNA 2 responsiveness of the LMP1 promoter. Mutation of the PU.1 binding site (–169/–158) has a profound effect on EBNA 2 responsiveness, whereas mutation of the overlapping LBF4 site (–175/–158) has no affect. However, the PU.1 site is only active in the context of upstream sequence, including the LBF3, 5, 6, and 7 (–211/–203) binding sites. Which of these latter factors acts in conjunction with PU.1 to convey EBNA 2 responsivness is not known. An epithelial cell–specific factor LBF2, which recognizes a –215/–192 site is also identified in this study. LBF2 could account for the importance of the –214/–144 LMP-1 promoter sequence in the promoter's high basal activity in epithelial cells.

EBNA 2 is a specific transactivator of cell and viral gene expression. This property was first described in studies delineating the mechanism for upregulated expression of CD23 (687). CD23 is a B-lymphocyte–specific surface protein expressed at high abundance on EBV-transformed or antigen-activated primary B-lymphocytes (128,290,338). EBNA 2 transactivation of CD23 occurs in part through an EBNA 2–responsive element upstream of the CD23 promoter (686,688). Subsequently, EBNA 2 also has been shown to upregulate CD21 (686) and *c-fgr* (340) mRNA levels, to upregulate LMP1 (1,150,151,194,667,691) and LMP2 (724,726) mRNA levels and to upregulate a *cis*-acting elements upstream and downstream of the *Bam* C promoter (306,629,682). Curiously, EBNA 2 also transactivates the human immunodeficiency virus (HIV) LTR, and the transactivation is dependent on the NFkB sites in the LTR, despite the absence of NFkB sites in other EBNA-responsive *cis*-acting elements (570).

The EBNA 2 response elements upstream of the CD23, LMP1, LMP2, and *Bam* C promoters also have been partially defined by deletional analysis of EBV promoter constructs or of the response element positioned near a heterologous promoter (306,667,688,724,726). For the most part, EBNA 2 responsiveness is independent of the position and orientation of the response element, relative to the promoter. Some EBNA 2 response elements, such as the LMP1 promoter EBNA 2 response element, are only responsive to EBNA 2 in the context of human B-lymphocytes (151,691). Others, such as the *Bam* C EBNA 2 response element, are responsive in Jurkat, a human T-cell line.

Molecular genetic analyses of EBNA 2 using as the phenotypic markers either rescue of transforming virus from P3HR-1–infected cells or transactivation of the LMP1 promoter in transient transfection of BL cells have in general established a correlation between the sequences in EBNA 2 that are important for transformation and those important for transactivation of the LMP1 promoter (94) (Fig. 5). Surprisingly, two substantial deletions are compatible with full *trans*-forming and *trans*-activating activity. One is the deletion of amino acids 195 to 230, a relatively proline-rich region not well conserved between EBV types 1 and 2. Extension of this deletion to include amino acids 143 to 230 reduced transformation and transactivating activity by approximately 50%. Deletion of amino acids 112 to 141, which are also well conserved between EBV 1 and 2, had a more pronounced effect in decreasing transforming and transactivating activity but still resulted in virus capable of transforming cells with low efficiency. Thus, amino acids 112 to 230 are unlikely to mediate a key EBNA 2 interaction, and the effects of this mutation are more likely to be due to effects on protein folding. The other innocuous deletion was of amino acids 463 to 483, which are well conserved between EBV 1 and 2. This latter deletion enhances transforming and transactivating activity. Other domains also could be deleted without completely blocking transformation, and in most instances the transformed cell grew nearly as well as those transformed by wild-type virus. This includes amino acids 2 to 95 (Yalamanchili, Harada, and Kieff, unpublished data) and amino acids 357 to 420 (662). The 337 to 357 RG oligomer domain can be deleted with a 90% reduction in transforming efficiency, but the cells transformed by the mutant recombinant virus grow almost as well as those transformed by wild-type virus (662). Deletion of the RG domain increases transactivation of the LMP1 promoter tenfold in transient transfection assays. Only three regions appear to be stringently required for transformation, and at least two (and probably all three) are essential for transactivation. The essential regions are amino acids 95 to 110, 280 to 337, and 425 to 462. As described below, region 425 to 462 is essential because of its acidic transactivating characteristics, whereas region 280 to 337 is essential in its interaction with DNA sequence-specific binding proteins that bring EBNA 2 to its response elements. The role of region 95 to 110 is uncertain. Similar genetic analyses of adenovirus E1A or Simian virus 40 T-antigen readily yielded mutants that were deficient in transformation but retained wild-type transactivating efficiency. These transformation-negative mutations in E1A or T defined regions of E1A or T that interact with Rb or p53 or other cell proteins likely to be involved in cell growth transformation as opposed to specific gene transactivation.

Detailed analysis of the acidic EBNA 2 domain from 425 to 462, which is essential for transformation and for LMP1 transactivation, indicates that this domain is similar in many respects to the prototype VP16 acidic domain (92,93,94). Although there is little colinear homology, both domains are rich in glutamic and aspartic acid. Fusion of the EBNA 2 acidic domain to the DNA binding domain of gal 4 and transfection into BL cells along with CAT reporter constructs having multiple GAL 4 binding sites upstream of a basal promoter results in strong activation of the reporter constructs (93,383). The GAL 4–EBNA 2 fusion was more active in BL cells than in CHO cells, whereas gal 4–VP16 fusions were more active in CHO cells than in BL cells, providing another indication of EBNA 2 selectivity for transactivation in B-lymphocytes (93). In GAL 4 fusions, the 14 EBNA 2 amino acids from 449 to 462 had about 25% of the activity of the entire VP16 transactivating domain, and the terminal 105 EBNA 2 amino acids from 379 to 483 had greater activity than did VP16 (93). Part of the VP16 acidic domain can substitute for part of the EBNA 2 acidic domain in the context of a reconstituted EBV recombinant (92). The EBNA 2 acidic domain shares with VP16 an affinity for TFIIB, TAF40, TFIIH, and RPA70, but has much less affinity for TBP (663). Unlike VP16, which appears to have two separable domains, all of the interactions of the EBNA 2 acidic domain are inhibited by mutation of a single W to S or T (92,663). Thus, the critical role of the 425 to 462 domain is as an acidic domain that recruits TFIIB, TAF40, and TFIIH to EBNA 2–responsive promoters.

Despite numerous attempts to demonstrate specific interaction with response elements, EBNA 2 does not interact directly with any of the known response elements. Each response element has at least two components required for EBNA 2 effect. The smallest response element is 50 bp of the *Bam* C promoter between –331 and –380 relative to the transcriptional start site (306). The minimal LMP1 response element is 87 bp between –147 and –234 relative to the LMP1 promoter transcriptional start site (150,151,308,667), whereas the minimal LMP2A response element is 81 bp between –177 and –258 relative to the LMP2A transcriptional start site (382,724). Further deletions from each of the response elements significantly impairs EBNA 2 responsiveness.

All three response elements have been intensively investigated by electrophoretic mobility shift assays using nuclear extracts from EBV-infected lymphocytes or BL cells. Each has at least two major DNA binding activities. The Cp EBNA 2 response element appears to be the simplest, with only two gel shift activities. One gel shift traces to an oligonucleotide that includes a GTGGGAA component that is present twice in the minimal EBNA 2–responsive LMP2A promoter and once in the minimal EBNA 2–responsive LMP1 promoter (306,382,724). A faint super shift of an LMP2A element that included the oligonucleotide was detected when EBNA 2 was added to nuclear extracts (724). The super shift was further shifted by EBNA 2–specific antibody. Recombinant proteins composed of glutathione transferase fused to EBNA 2 amino acids 242 to 425 also can supershift the Cp element (382). Even as small a part of EBNA 2 as amino acids 310 to 337 could affinity purify a 63-kd nuclear protein that reproduced the gel shift activity (707). The EBNA 2 28–amino acid core p63 binding domain is a region of EBNA 2 that differs between EBV types 1 and 2. In fact the only common sequence between the two types is GPPWWPPXXDPI/A (707). Mutation of the WW to FF in the context of small fusion proteins destroys their ability to interact with the p63 (707). The sequence of p63 identified it to be the previously cloned J kappa recombination signal sequence binding protein (15,221,259). The original identification, cloning, and naming of the J kappa protein is in retrospect an artifact of the recombinant heptamere sequence used for screening of proteins that recognize the J kappa heptamere (415). The heptamere GTG was placed next to the GGATCC of a *Bam*HI site, resulting in GTGGGAT, close to the consensus GTGGGAA Jk recognition site that is part of all EBNA 2 response elements. The Jk protein is expressed in all cells and is conserved in evolution. A homozygous Jk null mutation is lethal to drosophila embryonic development, and another drosophila mutation, hairless, is a neural developmental defect (184,578). Thus, Jk is probably a key adapter protein that brings transcriptional regulatory proteins to specific cell genes. Search of the nucleotide sequence database shows matches of GTGGGAA to many cell genes, including CD23, the prototype EBNA 2–responsive cell gene. *c-fgr* has a consensus Jk site in an intron before the exon that encodes *c-fgr*. A promoter in this exon just downstream of the Jk site regulates *c-fgr* transcription in LCLs (225). Other cell genes with Jk sites are likely to be activated by EBNA 2 in transformed cells and have been previously undetected.

The –234/–147 LMP1 EBNA 2 response element includes at least seven interactive sites, only one of which is the JK site (221). In fact, Jk interaction is only part of

EBNA 2 responsiveness. Mutation of the Jk site or a mutation in the critical EBNA 2 $W_{319}W_{320}$ residues that interact with J kappa reduces EBNA 2 responsiveness by only 50%. In contrast, WW to SS mutated EBNA 2 is completely inactive in transactivating the simpler Cp promoter (662). PU.1 is the second DNA binding protein important in EBNA 2 responsiveness of the LMP1 promoter. Mutation of the PU.1 binding site in the promoter is in fact more destructive of EBNA 2 responsiveness than mutation of the Jk binding site. The EBNA 2 domain that interacts with PU.1 appears to include the site that interacts with Jk. Deletional analysis of the LMP1 EBNA 2 responsive element shows a third site near –214 that also contributes to EBNA 2 responsiveness. The protein that interacts with this site has not been identified.

The current model (Fig. 5) for EBNA 2 role in growth transformation is that EBNA 2 is brought to promoters by an interaction with the GTGGGAA binding protein Jk and by interaction with other DNA binding proteins such as PU.1 in the case of the LMP1 promoter. PU.1 is likely to be an important factor not only for the LMP1 promoter but for many of the B-lymphocyte–specific genes that are activated by EBNA 2 because PU.1 is generally important in B-lymphocyte–specific gene transcription. Even the EBV Cp promoter—the upstream element of which is EBNA 2 responsive—has two candidate PU.1 sites downstream of the promoter (and upstream of the Wp promoter). These two sites have a much greater effect on the EBNA 2 responsiveness of the Cp promoter than does its upstream element (646). The RG domain of EBNA 2 can interact with histones, potentially facilitating the interaction of EBNA 2 with DNA. Probably several EBNA 2 molecules associated with a single response element coordinately recruit the basal transcription factors TFIIB, TAF40, and TFIIH, resulting in transcriptional activation. The RG and 462 to 482 terminal basic domains probably modulate the activity of the acidic transactivating domain because deletion of either domain results in increased transactivation. Clearly, the readily dispensable region between the RG domain and the acidic domain is a necessary spacer separating the critical acidic domain from the domain that interacts with sequence-specific DNA binding proteins. The interactions of the 280 to 337 region with DNA binding proteins may extend beyond Jk and PU.1. Whether the seemingly essential 97 to 110 domain is important for transactivation or for some other transformation-related function remains to be established.

EBNA 3A, 3B, and 3C

EBNA 3A, 3B, and 3C are encoded by three genes that are likely to have had a common origin (261,265,307,312, 327,417,499,501,533,565,590), although they are sufficiently different that a statistical analysis judges them to be borderline unrelated (315). The three genes are tandemly placed in the EBV genome. Each protein is encoded by a short exon and a long exon, which are at the 3′ end of their respective mRNAs (46–50,261,265,307,312,499, 501,533,580,601). The intron separating the two exons is short. The RNAs all initiate at the Cp or Wp EBNA promoters and are spliced through the W1W2Y1Y2Y3 and U1 exons to an acceptor site on one of the three short exons. The EBNA3 mRNAs are the least abundant EBNA mRNAs, with only a few molecules in each latently infected cell. The predicted primary amino acid sequences of EBV 1 EBNA 3A, 3B, and 3C are 944, 938, and 992 amino acids, respectively, whereas those of EBV 2 EBNA 3A, 3B, and 3C are 925, 946, and 1,069 amino acids, respectively. The EBV 1 and 2 EBNA 3A, 3B, or 3C alleles are 84%, 80%, or 72% identical in predicted primary amino acid sequence (565). EBNA 3A, 3B, and 3C each have different repeating polypeptide domains near their carboxy-termini, representing unusual examples of protein domain amplification among genes that are themselves likely to have amplified from a common progenitor and then diverged.

The EBNA 3 proteins are remarkably hydrophilic, with up to 20% charged amino acids. Each has several localized concentrations of arginine or lysine that could be responsible for nuclear localization. Each also has heptad repeats of leucine, isoleucine, or valine that could facilitate hydrophobic homo- or heterodimerization. Despite the low abundance of the three mRNAs in latently infected cells, EBNA 3A, 3B, and 3C accumulate in nuclei and localize to large intranuclear clumps that fill the nucleus, sparing only the nucleolus (261,499–501). Their distribution resembles that of EBNA 2 but is independent of EBNA 2 expression. In biochemical fractionation, EBNA 3A, 3B, and 3C are found in the nuclear matrix, chromatin, and nucleoplasmic fractions (500).

Because of the similarities between EBNA 3A, 3B, and 3C, these proteins are likely to have similar roles in EBV latent infection or cell growth transformation. The finding that EBNA 3C can upregulate CD21 mRNA and protein in non–EBV-infected BL cells *in vitro* provides the first indication of a relevant biochemical activity for these proteins (686). Subsequently, 3C expression in Raji cells has been shown to upregulate LMP1 expression (12,13). An unusual property of some clones of the EBV-infected BL cell line, Raji, is that LMP1 expression is lower than in most other EBNA 2–expressing cell lines. Expression of 3C returns expression to the same level as LCLs and maintains expression in G1-arrested cells (120). Effects on cell gene expression have not as yet been documented for EBNA 3A or 3B, nor have the mechanisms of upregulation of CD21 mRNA by EBNA 3C been delineated.

Molecular genetic analyses of EBNA 3A, 3B, and 3C began with an evaluation of the role of type-specific differences in B-lymphocyte growth transformation. EBV recombinants were made in a type 2 EBV background using the P3HR-1 strain. The P3HR-1–deleted DNA segment was replaced with type 1 EBNA LP– and EBNA 2–encoding DNA, and the P3HR-1 EBV type 2 EBNA 3 genes were either left intact or replaced with type 1 EBNA 3 genes. The EBNA 3 gene type did not affect the ability of the recombinants to initiate growth transformation, to be maintained as episomes, or to replicate in response to induction of permissive infection (658). Cells transformed by EBV recombinants that were type 1 or 2 at the EBNA 3 loci grew indistinguishably. Thus, the type specificity of EBNA 3 genes does not appear to affect virus infection of B-lymphocytes *in vitro*.

Initial experiments with type-specific changes in EBNA 3 also served as prototype controls for deriving EBV recombinants with specific mutations in each of the EBNA 3 genes. Insertion of an amber nonsense codon after codon 109 of the 938 EBNA 3B codons resulted in an EBV recombinant that was fully competent for transformation of primary B-lymphocytes *in vitro* or for lytic infection (659). In contrast, insertion of an amber nonsense codon after codon 302 of EBNA 3A or after codon 365 of EBNA3C resulted in EBV recombinants that were unable to transform primary B-lymphocytes (660). The mutant EBNA 3C recombinants could transform primary B-lymphocytes when the lymphocytes were also infected with the P3HR-1 EBV, which would provide wild-type EBNA 3A, 3B, and 3C in *trans*. Surprisingly, the mutant EBNA 3A recombinants were unable to stably transform primary B-lymphocytes even when B-lymphocytes were initially coinfected with P3HR-1 EBV. The simplest interpretation of these data is that the 302–amino acid EBNA 3A mutant polypeptide has a dominant negative effect on B-lymphocyte growth transformation as a consequence of an essential interaction with a cell protein that is present in limited abundance. The finding that EBNA 3B is nonessential for any aspect of lymphocyte infection *in vitro* but that EBNA 3A and 3C are essential for virus-mediated lymphocyte growth transformation is surprising because this is a family of large and distantly related proteins, each of which has epitopes that are frequently recognized by human immune cytotoxic T-lymphocytes (see Chapter 75) so that their immune recognition should provide strong selection against expression unless that expression is essential for virus infection *in vivo*.

EBNA 1

EBNA 1 was initially described as an EBV nuclear neoantigen that associated with chromosomes during mitosis (396,482,528). EBNA 1 is the only EBNA that associates with chromosomes during mitosis (219,482,500,528). EBNA 1 is entirely encoded by the 2-kb terminal exon of the EBNA transcript (244,254,262,628) (Fig. 2). The full EBNA 1 mRNA is 3.5 kb. The cloning of EBNA 1 mRNA cDNAs established the linkage between the EBNA 1-encoding exon and the U, YH, Y2, Y1, W2, W1, and W0 exons transcribed from the Wp or Cp promoter (Fig. 3) (46–50,560,611,620).

Based on predicted primary amino acid sequence, EBNA 1 from a typical EBV strain consists of 641 amino acids. The protein has a high proline content, is charged, and migrates on denaturing polyacrylamide gels with an apparent size of 76 kd. EBNA 1 has four obvious domains: (a) an amino-terminus of 89 amino acids that is rich in basic amino acids; (b) the next 239 amino acids, which are a copolymer of glycine and alanine that could form beta sheets and might participate in intramolecular interactions; (c) a short domain rich in basic amino acids; and (d) a long hydrophilic domain from 459 to 607 that has sequence-specific DNA binding and dimerization activity (16,585). Mutational analysis of the 459 to 607 DNA binding and dimerization domain has not clearly separated these activities (72). The failure to separate subdomains within the 459 to 607 domain is consistent with its relative protease resistance. Both properties are frequently due to highly folded and interactive tertiary structures. Mutations in the region from 459 to around 500 affect nonspecific and sequence-specific DNA binding, whereas mutations between 501 and 532 or between 554 and 598 affect dimerization with less affect on DNA binding (72). Although primary sequence algorithms suggest that the EBNA 1 ori P binding domain may contain a basic helix-loop-helix motif similar to that of several other sequence-specific DNA binding proteins (294), the lack of effect of proline substitutions at sites that should then disrupt DNA binding and dimerization is somewhat inconsistent with the model (72). These mutational and biochemical analyses of EBNA 1 indicate that there is considerable interaction among regions within the 459 to 607 domain and probably interactions of this domain with other parts of EBNA 1.

EBNA 1 is phosphorylated on serine residues in at least two separable domains in the carboxy half of the molecule (243,500,505,612). The role of these phosphorylations in EBNA 1 function has not as yet been established through genetic analyses.

Although most of the EBNAs bind to DNA cellulose, only EBNA 1 has sequence-specific DNA binding properties. The specific EBNA 1 cognate sequence is a partial palindrome: TAGGATAGCATATGCTACCCAGATCCAG (17,310,332,527). EBNA 1 has a high affinity for its cognate sequence, and interaction with the cognate sequence can be demonstrated even after protein denaturation and renaturation on cellulose nitrate, providing some evidence that even monomers of EBNA 1 may have some specific affinity for the cognate sequence. EBNA 1 cognate sequences are located at three sites in the EBV genome (310, 527,621). The highest affinity site consists of 20 tandem direct 30-bp repeats. This site is about 7 kbp from the left end of the genome. The second highest affinity site is 1 kbp to the right of the first and consists of two cognate sequences in dyad symmetry and two in tandem. A third site is in the *Bam*HI Q fragment, about 10 kb downstream of the EBNA 2–encoding exon (527). This site is composed of two divergent tandem repeats of the 30-bp recognition element and is important in the negative regulation of an alternative promoter for EBNA 1 transcription, the FQ promoter (559).

EBNA 1 enables covalently closed circular DNA molecules carrying the 20 tandem direct 30-bp repeats and the dyad symmetry sites to initially replicate and persist as an episome in primate cells. The two sites comprise an EBNA 1-dependent, *cis*-acting factor for EBV episome maintenance and replication in latently infected cells (81,421,530, 623,711–714). The EBV DNA segment containing both sites is therefore designated ori P, for plasmid DNA replication origin. The family of repeats and dyad symmetry components of ori P, as with EBNA 1, are necessary for efficient function in episome persistence or episome replication in latent infection. The intervening sequence between the family of repeats and dyad symmetry sites is not critical (397,530,542,706).

The 30-bp repeat component has an EBNA 1–dependent enhancer effect on heterologous promoters or on neighboring EBV promoters in transient transfection assays (529, 626). The transcriptional activating activity is also important in enhancing episome replication (130,506,712). The family of repeats arrests and functionally terminates the replication of episomes containing ori P in cells expressing EBNA 1 (132,185,502). Surprisingly, only seven to eight copies of the 30-bp repeats are adequate for full activity of the family of repeats in transcription or DNA replication. The role of the family of repeats in arresting DNA replication function is probably less important for replication because tandem duplication of the dyad symmetry element can partially relieve the requirement for the family of repeats in episome maintenance, replication, or enhancer activity.

The dyad repeat is stringently required for episome replication. The dyad can partially denature, forming bubble and cruciform structures as measured by nuclease sensitivity (488,700). Such structures are characteristic of other efficient origins. Denaturation of the dyad is effected by EBNA 1 binding (242). Gel analyses confirm that the dyad is the origin of bidirectional episomal plasmid replication in EBNA 1–expressing primate cells (185). Without the family of repeats, the dyad is inefficient in enhancing transcription or in enhancing plasmid replication or persistence in the presence of EBNA 1. At least one cellular cofactor that is required for the dyad component of ori P to be maintained as a plasmid is not required for enhancer activity because EBNA 1 binding to the family of repeats enhances transcription in mouse or human cells, whereas EBNA 1 promotes episome replication only in primate cells. The barrier to replication in rodent cells can be bypassed by providing an origin sequence active in rodent cells; such vectors still are dependent on the enhancer function of EBNA 1 and the family of repeats (346). Two human cell proteins that interact specifically with the EBV ori P element have been identified and their cDNAs sequenced (723).

The interaction of EBNA 1 with the repeats and the dyad sites is cooperative (435,527), indicating facilitative alteration of the template or an intrinsic stabilizing interaction between EBNA 1 molecules. Electron microscopic and biochemical studies indicate that EBNA 1 binding to the family of repeats and the dyad symmetry results in high-order structures that lead to bending of the DNA, distortion of the duplex, and looping out of the sequence between the family of repeats and the dyad symmetry element (173–176,422,488,621). The formation of this macromolecular complex is almost certainly due to the abilities of EBNA 1 to recognize both sites and to form homopolymers after associating with the templates (173–176,206,234). Initial deletional analyses indicate that the key EBNA 1 amino acids for bridging the family of repeats and the dyad are amino acids 336 to 450, that is, on the amino terminal side of the dimerization domain (234).

The carboxy-terminus of EBNA 1 appears to determine its subnuclear distribution, its ability to persist in cells, and its ability to associate with chromosomes during mitosis (712). The carboxy-terminal EBNA 1 domain may interact with a specific chromosomal protein. EBNA 1 does not interact with a specific chromosome site but is homogeneously distributed on chromosomes (219,233,482,500, 528). This property is likely to be important in the segregation of episomes into progeny nuclei and could be important in interphase as well. Part of the EBNA 1 in cells is associated with the nuclear matrix fraction. Consistent with a physiologic role for this EBNA 1 is the observation that degradation of matrix associated-DNA leaves the EBNA 1 cognate sites in ori P still associated with the nuclear matrix (297).

EBNA 1 could play an additional role in latent infection if EBNA 1 response elements exist in the cell genome (479,723). The two cell proteins that bind to ori P with a high affinity (723) could be important in DNA replication or transcription. If EBNA 1 and ori P are related to a family of nuclear protein transactivators and their cognate sequence elements, EBNA 1 also may affect cell gene transcription. In single gene transfer experiments, however, EBNA 1 has not induced changes in cell growth or in expression of cellular genes known to be affected by EBV infection.

EBNA 1 is expressed in the absence of the other EBNAs in some BL cells, in Hodgkin's cells, in nasopharyngeal carcinoma, and in some infected B-lymphocytes in normal human peripheral blood (see section on other types of latent infection). These EBNA mRNAs initiate at a promoter near 62.3 kb in the EBV genome map and are spliced to the same U and K exons as the LCL type EBNA 1 mRNA (558).

EBNA 1 is also the only EBNA that continues to be made as cells are activated to lytic infection (254,547, 692,693). Cp/Wp activity usually ceases in lytic infection and the downstream Fp promoter is activated (547,558,571). The lytic mRNA initiates near 62.2 kb and otherwise has the same sequence as the RNA that initiates at 62.3 (558) in latent infection in some lymphocytes *in vivo* in BL lines, in HD, and in NPC. Some of the regulatory elements of the promoters for these RNAs have been delineated (559). The two EBNA 1 binding sites in the EBV genome that are not part of the ori P element are between these two transcriptional start sites and may have a role in negative regulation of these promoters (559).

LMP1

Despite a weak promoter for the LMP1 gene relative to the EBNA 2 promoter as shown by nuclear run-on assays, latently infected cells have almost 10 times as much LMP1 mRNA as EBNA mRNA, indicating that LMP1 mRNA is more stable (159,561). The reverse is probably true of the relative stability of the LMP1 and EBNA proteins. The LMP1 transcript has two short introns (29,159). From inspection of the predicted primary amino acid sequence, LMP1 was almost certain to be an integral membrane protein with at least three domains: (a) a 20–amino acid hydrophilic amino-terminus lacking the characteristics of a signal peptide; (b) six markedly hydrophobic 20 amino acid, alpha helical, transmembrane segments, separated by five reverse turns, each eight to ten amino acids in length; and (c) a 200–amino acid carboxy-terminus, rich in acidic residues. LMP1, translated *in vitro*, posttranslationally inserts into canine pancreatic microsomal membranes, confirming the activity of the three putative membrane insertion sequences, each consisting of two highly hydrophobic transmembrane segments joined by a short reverse turn (374,376). Although LMP1 has little primary sequence homology to other proteins, aspects of its three-domain organization are similar to those of other integral membrane proteins, such as erythrocyte membrane band three or the *mas* oncogene.

Immunofluorescence microscopy (260,376) and membrane fractionation (401) indicate that about half of the LMP1 molecules are in the plasma membrane. LMP1 is less abundant in other cytoplasmic membranes. Live cell protease cleavage studies are consistent with a proposed model for LMP1 in the plasma membrane that positions both the N- and C-termini on the cytoplasmic side of the plasma membrane and only three short reverse turn domains on the extracellular side of the plasma membrane (376).

LMP1 is phosphorylated on serine and threonine residues at a ratio of 6:1 and is not phosphorylated on tyrosines (25, 402,445,447). Half or more of LMP1 is associated with the cytoplasmic cytoskeleton as defined by resistance to extraction with buffers supplemented with nonionic detergents (376,402,445,447). Because the LMP1 patches in the plasma membrane colocalize with a patch of the intermediate filament protein, vimentin, and is not extractable with nonionic detergent it is likely that a significant fraction of the LMP1 in the plasma membrane of LCLs is associated with the cytoskeleton (374). The half-life of LMP1 has been measured in different ways, using time to reach stable levels of ^{35}S methionine label accumulation in LMP1 in various cell fractions or by determining the rate of decay of LMP1 in the presence of protein synthesis inhibitors. The data do not entirely agree, possibly due to effects of the protein synthesis inhibitors in increasing total LMP1 turnover in those studies using inhibitors or to differences in the cell lines studied (25,374,402,445,447). Nascent, nonionic, detergent-soluble LMP1 has a relatively short half-life and is converted to a tightly cytoskeletally associated form (374,402,445,447). The cytoskeletal form is phosphorylated, whereas the soluble LMP1 is not phosphorylated (374,402,445,447). The half-life of the soluble form is less than 2 hr, whereas the half-life of the cytoskeletal form of LMP1 is on the order of 3 to 15 hr. LMP1 is cleaved near the beginning of the carboxy-terminal cytoplasmic domain, resulting in a soluble C-terminal domain of about 25 kd (445,447). The principal serine and threonine phosphorylation sites are near each other in the 25-kd cleavage product (447).

LMP1 forms discrete patches in the plasma membrane that are often further organized into a single caplike structure (260,376). Unlike growth factor receptors, which form patches and caps in response to ligand binding, LMP1 constitutively forms patches in LCL plasma membranes in the absence of exogenous growth factors (374,376). A vimentin cap colocalizes with the LMP1 cap in the latently infected B-lymphocyte plasma membrane and is a key cytoskeletal element that directly or indirectly links LMP1 to the cytoskeleton (374). Vimentin is not ordinarily found in a plasma membrane patch, and is drawn into this patch by LMP1. Vimentin relocalizes into perinuclear rings and coils when cells are treated with colcemid, and LMP1 then relocalizes to the vimentin rings (374). Thus, once LMP1 is associated with vimentin, vimentin can further direct LMP1's localization. By expressing LMP1 in a human B-lymphoblast cell that does not express vimentin, LMP1 could be shown to form patches and caps in the plasma membrane without any associated vimentin (373). Patch formation also does not require another EBV protein because LMP1 forms patches in human B-lymphocytes in the absence of any other EBV proteins (375,685) (Fig. 6). Thus, LMP1 appears to interact with itself or with other cell membrane proteins to form large, noncovalently linked membrane complexes. No clear evidence for LMP1 association with itself or with cell proteins other than vimentin has emerged until recently, despite intensive biochemical investigation.

In single gene transfer experiments under heterologous promoters, LMP1 has transforming effects in continuous rodent fibroblast cell lines (26,446,6834,684). In Rat-1 or NIH 3T3 cells, LMP1 alters cell morphology and enables cells to grow in medium supplemented with low serum (684). In Rat-1 cells, LMP1 also causes loss of contact inhibition so that cells heap up in monolayer culture (684). LMP1 also caused Rat-1 or Balb/c 3T3 cells to lose their

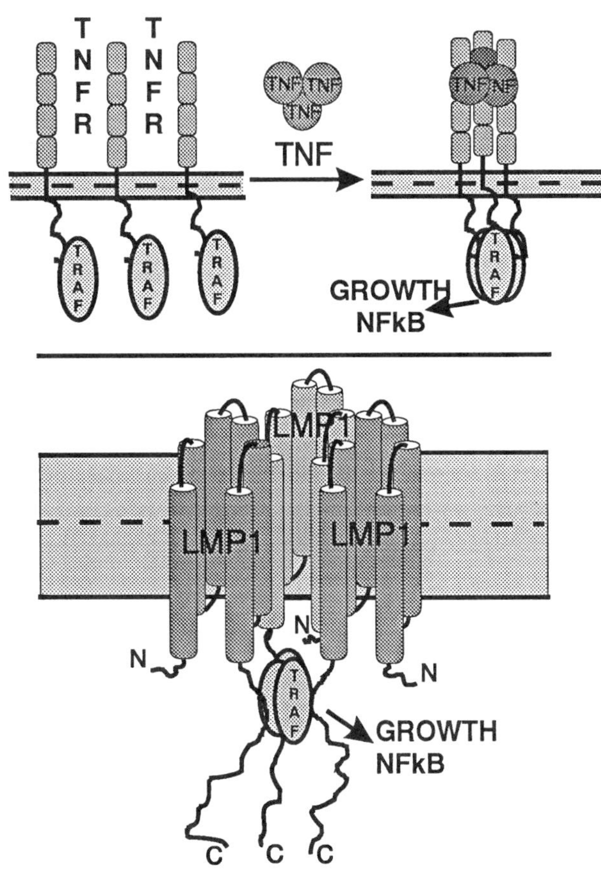

FIG. 6. Schematic diagram of the mechanisms by which LMP1 affects cell growth and death pathways in EBV latent and lytic infection. The six hydrophobic transmembrane domains of LMP1 enable it to aggregate in the plasma membrane (as a consequence of posttranslational membrane insertion into ER membranes and membrane bulk flow into other cytoplamsic membrane) (374,375, 376). The amino-terminal cytoplasmic domain is fully dispensible for primary B-lymphocyte growth transformation, and its primary function is in cytoplasmic tethering of the first transmembrane domain (295). The transmembrane domains and the carboxy-terminus are essential for primary B-lymphocyte growth transformation (321,322). The first 44 amino acids of the transmembrane domain interact with a human ring finger and extended coiled-coil domain protein (LMP-associated protein 1 or LAP1) (449), which is homologous to the recently discovered murine tumor necrosis factor receptor (TNFR)-associated factors (TRAFs) (543). LMP1 associates in B-lymphoblasts with LAP1 and with an EBV-induced cell protein EBI6 (449). EBI6 is the human homologue of the murine TRAF1 (449). The murine TRAFs associate with a region of the p80 TNFR implicated in cell growth and NFkB activation (542). LAP1 binds directly not only to the LMP1 carboxy-terminal cytoplasmic domain but also to the p80 TNFR, CD40, and lymphotoxin beta receptor cytoplasmic domains. LAP1 binds less well to the FAS and p60 TNFR cytoplasmic domains. EBI6 also binds to the p80 and p60 TNFR cytoplasmic domains. Thus, the LMP1 cytoplasmic domain interacts with and associates with cell proteins that are mediators of cytoplasmic signaling from the family of TNFRs. LMP1 aggregated in the plasma membrane with its aggregated cytoplasmic domans hanging into the cytoplasm and engaging LAP1 and EBI6 constitutively activates the TNFR growth, death, and NFkB signaling pathways, thereby causing constitutive cell growth, inhibiting apoptosis, and activating NFkB (449). In doing so, the LMP1 aggregates may interact with TNFR family-TRAF aggregates to form large complexes. Constitutive activation of this pathway by LMP1 is likely to be fundamentally important in EBV-associated lymphoproliferative diseases, HD, and NPC, situations in which LMP1 is expressed in the EBV-infected cells and in which cell growth is dependent on TNFR activation.

anchorage dependence so that they clone with a high efficiency in soft agar (26,683,684). Rat-1 cells expressing LMP1 are tumorigenic in nude mice, whereas control Rat-1 cells are nontumorigenic (684). The growth of Balb/c 3T3 cells in soft agar correlates quantitatively with the extent of LMP1 expression up to the levels ordinarily expressed in LCLs (683). Expression at higher levels results in toxicity. The hypothesis has been put forward that the toxic effects of LMP1 expressed at high levels may be a useful surrogate marker for LMP1 biological activity in rodent fibroblasts, and experiments have shown a correlation among LMP1 mutants between transforming activity when expressed at levels similar to those expressed in LCLs and toxicity when expressed at higher levels (413). Transforming activity at moderate levels and toxicity at high levels may be partly related phenomena, although many proteins are toxic to cells when expressed at supraphysiologic levels.

LMP1 also dramatically alters the growth of EBV-negative BL lymphoblasts when expressed stably or transiently at the appropriate level in such cells after gene transfer (43, 373,375,414,445,497,548,685,686,721). In fact, LMP1 induces many of the changes usually associated with EBV infection of primary B-lymphocytes or with antigen activation of primary B-lymphocytes, including cell clumping; increased villous projections; increased vimentin expression; increased cell surface expression of CD23, CD39, CD40, CD44, and class II major histocompatibility complex (MHC); increased IL-10 expression; decreased expression of CD10; and increased expression of the cell adhesion molecules LFA-1, ICAM-1, and LFA-3 (Fig. 4). LMP1 also has been shown to protect B-lymphocytes from apoptosis (215,258, 548). The effects mediated in part through the induction of *bcl-2* by LMP1 (258,414,548) and probably also through induction of A20 (258). LMP1 not only increases plasma membrane expression of adhesion molecules, but also functionally activates adhesion and induces higher levels of LFA-1 mRNA (685,686). Each of these effects in BL cells is dependent on the particular cell background. Some BL cell types already express high levels of adhesion molecules. In such cells LMP1 has no effect on adhesion molecule expression.

LMP1 induction of B-lymphocyte adhesion molecules may be important because of at least two potentially countervailing effects on latently infected cell growth. First, homotypic B-lymphocyte adhesion facilitates cell growth because proliferating B-lymphocytes secrete autocrine growth factors. When growing at low cell density, EBV-infected cells adhere to fibroblast feeders and similar heterotypic interaction may occur *in vivo*. Proliferation may depend on such interactions. Second, *in vivo* the elimination of EBV-transformed B-lymphocytes is important to host survival and to the perpetuation of latent and lytic virus infection. Increased conjugate formation with T cells (686) as a consequence of EBV LMP1 induction of LFA-3 (and to a lesser extent LFA-1) results in increased efficiency of immune T-cell surveillance and increased elimination of latently infected growth-transformed B-lymphocytes.

LMP1 expression also alters the growth of multipotent hematopoietic stem cells and epithelial cells (123,152,153, 278,701). The effects of LMP1 on epithelial cell growth were first demonstrated by expressing LMP1 in the skin of transgenic mice, in which LMP1 induces epidermal hyperplasia and alters keratin gene expression (701). Similar effects were observed when LMP1 was expressed in immortalized human keratinocytes grown in monolayer or stratified air–liquid interface raft cultures (123,152). In monolayer cultures, LMP1 alters keratinocyte morphology and cytokeratin expression (152). In stratified air–liquid interface raft cultures of immortalized epithelial cell lines, LMP1 inhibits cell differentiation (123).

Studies with LMP1 deletion mutants in rodent fibroblasts or in BL cells have yielded some consistent results that highlight the importance of parts of the proteins, whereas other results have not been consistent in various assay systems. The importance of the amino-terminus and first four transmembrane domains as a single entity is evident from the complete inability of the D1LMP1 to alter the growth of Rat-1 or Balb/c 3T3 cells (27,683,685) or BL lymphoblasts. Addition of the 25–amino acid amino terminal domain to the last two transmembrane and cytoplasmic domains does not restore the ability of this truncated LMP1 to cause anchorage-independent growth of Balb/c 3T3 cells (27,413). Complete deletion of just the amino-terminal cytoplasmic and first transmembrane domains markedly reduces or eliminates anchorage-independent growth in Balb/c 3T3 cells and eliminates most of LMP1 BL cell–activating activity (27,375,413). Deletion of the free cytoplasmic amino-terminus up to the arginine, which should serve to retain the amino-terminus of the first transmembrane domain on the cytoplamic side of the membrane, resulted in only a slight change in the tightness of LMP1 plasma membrane patching, slightly smaller cell clumps, downregulated expression of CD10, and upregulation of most activation and adhesion markers except for CD40 and CD23 (430). Deletion of the amino-terminus and first transmembrane domain or of the middle two transmembrane domains had more pronounced effects on LMP1 localization and on gene expression in BL cells (375). Deletions of the carboxy-terminus have not yielded consistent results and have been reported to have no effect on anchorage-independent growth in soft agar but important effects on contact inhibition (446). These data suggest that the cytoplasmic amino-terminus is not responsible for LMP1-activating effects in BL lymphoblasts, that the transmembrane domains are critical to LMP1 effects probably because of their importance in LMP1 aggregation in the plasma membrane, and that the carboxy-terminal cytoplasmic domain may be important.

The isolation of EBV recombinants specifically mutated in LMP1 has enabled a biologically more relevant analysis of the role of LMP1 and of domains of LMP1 in primary B-lymphocyte growth transformation (295,316,321). As expected from the profound effects of LMP1 expression in rodent fibroblasts, LMP1 is essential for primary B-lymphocyte growth transformation (316). EBV recombinants with mutations in LMP1 that result in expression of proteins that are deleted for the amino-terminal cytoplasmic domain and the first transmembrane domain or larger deletions of the amino-terminus result in a non-transforming phenotype (316). Surprisingly, this effect is not due to specific interactions mediated by the amino-terminal cytoplasmic domain (295). Deletions of any part of the amino-terminal cytoplasmic domain result in no more than a 90% effect on transforming efficiency and the transformed cells grow well (397). In contrast, deletion of all of the carboxy-terminal cytoplasmic domain results in a complete loss of the ability to transform primary B-lymphocytes, and deletion of all but the first 44 amino acids of the carboxy-terminus results in the inability to transform primary B-lymphocytes without fibroblast feeder layers (321). Primary B-lymphocytes infected with EBV recombinants that express the LMP1 amino-terminal cytoplasmic domain, the transmembrane domains, and the first 44 amino acids grow well on human diploid fibroblast feeder layers. Thus, there appears to be two components to the carboxy-terminus in that the first 44 amino acids are stringently required for growth on or off fibroblast feeders, whereas deletion of the rest of the carboxy-terminus can be transcomplemented with fibroblasts. Taken together, these results are most consistent with a model in which LMP1 patching in the plasma membrane mediated by the LMP1 transmembrane domains enables the cytoplasmic domain to interact with growth factor receptor pathway(s) so as to mimic constitutive activation of that pathway. The growth factor pathway through which LMP1 acts was identified when a yeast two hybrid screen for proteins that interact with the LMP1 C-terminal cytoplasmic domain revealed a novel cytoplasmic protein related to putative tumor necrosis factor growth signaling proteins (Fig. 2, ref 449). The LMP1 interactive protein also interacts with the CD40 cytoplasmic domain and with the lymphotoxin beta cytoplasmic domain, probably explaining the similarity between LMP1 and CD40 cross-linking effects on B-lymphocyte growth. LMP1 also induces higher level expression and associates with TRAF1, another tumor necrosis factor receptor associated proteins.

Downstream of LMP1 and its immediate cytoplasmic interactions with signaling pathways, LMP1 induces transcription of a number of cell genes. The best studied of these are CD23 and A20. In inducing CD23 transcription, LMP1 turns on an internal promoter in an intron of the CD23 gene

(687). The mRNA and even the protein are slightly different from the mRNA and protein expressed in the absence of LMP1. EBNA 2 synergizes with LMP1 in the dramatic upregulation of the internal promoter, although EBNA 2 alone cannot turn on this promoter and only increases the activity of the normal 5 promoter (349,687,691). At least part of LMP1's transcriptional activating effects are mediated by NF-κB activation (230,350,548). A20 is a putative antiapoptosis gene the expression of which is upregulated by LMP1 (350). Upregulation of the A20 promoter by LMP1 is dependent on the NfkB element upstream of the A20 promoter (350). An effect of LMP1 expression on the function of the HIV promoter is also dependent on the NFkB site in that promoter (230). Expression of LMP1 in primary cells alters NFkB activity before inducing bcl-2 expression, as is most compatible with an indirect effect on bcl-2 expression (548). Thus, downstream of the plasma membrane, one important mechanism of LMP1 effects on cell gene expression is through NFkB activation.

LMP1 and EBNA 1 are the only latent infection-associated genes that are also transcribed in lytic EBV infection. In some latently infected BL cells such as Raji, LMP1 expression can be increased by treatment with activators of protein kinase C, which increase early lytic cycle promoter activity (51,355,546). The promoter for full-length LMP1 expression in lytic infection has not been characterized. In late lytic infection, a promoter in the third LMP1 exon transcribes part of the LMP1 ORF, which encodes for the last two transmembrane domains and the entire LMP1 cytoplasmic domain (283). This smaller integral membrane protein, referred to as D1LMP1, localizes to cytoplasmic membranes and does not have the transforming or cell-activating properties of LMP1 and does not associate with vimentin or other cytoskeletal elements (374,375,401,683–685). The detection of LMP1 but not D1LMP1 in preparations of purified virus suggests that LMP1 may be incorporated into virions (401). This raises the possibility that virion-associated LMP1 could have an effect on the growth of newly infected cells.

LMP2A and 2B

The cloning and sequencing of cDNAs (356,564) to 2.3- and 2.0-kb cytoplasmic polyadenylated RNAs (334,335,509) from latently infected cells provided the first evidence for two integral membrane proteins, which were designated LMP2A and 2B or TP1 and 2 (177,389). The LMP2A and B RNAs are transcribed from the same strand as the EBNA RNAs and opposite to the strand from which LMP1 is transcribed. The LMP2A transcriptional start site is 3 kbp downstream of the LMP1 polyadenylation site, whereas the LMP2B start site is 0.2 kbp upstream of the LMP1 transcriptional start site (356,564). The LMP2A transcript is antisense to the LMP1 transcript and vice versa. The first LMP2A exon maps downstream of the LMP1 mRNA, and the second exon maps upstream of the LMP1 initiation site so that the LMP1 and LMP2A mRNAs are not antisense to each other. The LMP2B promoter and the LMP1 promoter are a bidirectional promoter unit separated by 200 bp. The EBNA 2 response element, which maps from –147 to –234 relative to the LMP1 transcriptional initiation site, extends into the LMP2B promoter. LMP2B transcription is therefore regulated by the LMP1 EBNA-2 response element. A separate EBNA 2 response element upstream of the LMP2A promoter regulates LMP2A transcription (724). The LMP2 transcripts cross the TR into U1, giving rise to the designation of LMP2A and B as terminal proteins, or TP1 and 2. By nuclear run-on, the level of transcription is low (561).

The first exons of LMP2A and 2B are the only unique exons; all other exons are shared by LMP2A and 2B. The first LMP2A exon is predicted to encode for a 119–amino acid hydrophilic amino-terminal cytoplasmic domain. The first LMP2B exon is short and lacks a methionine codon. Translation of LMP2B initiates at a methionine codon at the beginning of the common second exon, at the start of the first transmembrane sequence (356,564). The remaining LMP2A and 2B exons are predicted to encode 12 hydrophobic integral membrane sequences separated by short reverse turns and a 27–amino acid hydrophylic carboxy-terminal domain. Monospecific antisera generated against LMP2 fusion proteins have been used to demonstrate the presence of LMP2 in the plasma membrane of latently infected B-lymphocytes where it colocalizes with LMP1 (388,389).

LMP2A is a substrate for B-lymphocyte src family tyrosine kinases, and antiphosphotyrosine antibodies colocalize with LMP2 antibodies to sites in the plasma membrane, indicating that LMP2A is one of the few localized stably tyrosine phosphorylated proteins in BL cells or LCLs (388). LMP2A also associates with a 70-kd tyrosine phosphorylated cellular protein (60,388). Deletion analysis indicates that the first 167 amino acids of LMP2A, the amino-terminal cytoplasmic domain, and the first two transmembrane domains retain the ability to associate with and act as a substrate for a tyrosine kinase (60,388). Among src family tyrosine kinases, LMP2A exhibits specificity for fyn and lyn (60).

LMP2 expression in BL lymphoblasts blocks the normal calcium mobilization that should follow the crosslinking of sIgM, CD19, or class II MHC, indicating that LMP2 can modulate transmembrane signal transduction (426). The simplest model that would account for this activity is that LMP2 interaction with fyn and lyn in the lymphocyte plasma membrane results in an abrogation of the effects of signal transduction on plasma membrane tyrosine kinases and an inhibition of activation of phospholipase C. Membrane-associated tyrosine kinases are known to phosphorylate phospholipase C and to mediate *trans* membrane effects on calcium mobilization.

Despite the interaction of LMP2A with B-lymphocyte tyrosine kinases, LMP2A and 2B do not positively or negatively affect the processes by which EBV transforms primary B-lymphocytes. Each part of LMP2A and 2B has been specifically mutated in an EBV recombinant without an effect of the mutation on the efficiency of primary B-lymphocyte growth transformation, on the growth of the transformed B-lymphocytes under standard culture conditions, in reduced serum, or in soft agarose, on the growth of the transformed lymphocytes in SCID mice, or on the lytic infection that ensues when latently infected lymphocytes are treated with chemical inducers of lytic infection (331,390–392). The specific mutations include insertion of an amber nonsense codon after codon 19 of the LMP2A ORF, which results in ablation of expression of LMP2A and continued expression of LMP2B (390); insertion of an amber nonsense codon after DNA encoding the first five transmembrane domains, which results in ablation of expression in the last seven transmembrane domains and the carboxy-terminus of LMP2A and 2B (391); and deletion of DNA encoding the first five transmembrane domains of LMP2A and 2B, which results in truncation of LMP2A after the amino-terminal cytoplasmic domain and the addition of missense codons (392).

The effects of LMP2A on src family kinases and on *trans* membrane signaling–mediated calcium mobilization enable LMP2A to block the switch from latent infection to lytic infection in B-lymphocytes (425). Cross-linking of sIg on primary B-lymphocytes transformed by LMP2 null mutant recombinants results in activation of lyn, syk, and phospholipase C gamma 2 and transient tyrosine phosphorylation of these effectors (424). In lymphocytes transformed by wild-type virus, LMP2A is constitutively phosphorylated, lyn activity is reduced by 50%, and syk and phospholipase C gamma 2 are constitutively phosphorylated. A significant fraction of the lyn and syk in these cells is associated with LMP2A. Syk is the 70-kd phosphoprotein originally identified in LMP2A immunoprecipitates. The phosphorylation of these molecules does not change with sIg cross-linking. The constitutive phophorylation of these effectors and their failure to respond to sIg crosslinking is most compatible with their being phosphorylated on a negative regulatory site. Negative phosphorylation of src family tyrosine kinases is known to be important in their return to basal activity after stimulation. The constitutive phosphorylation of syk and phospholipase-C PLC gamma 2 and their inactivity is probably the result of constitutive stimulation by LMP2A and feedback-negative regulation of these effectors as well as the src family kinase. The amino-terminus of LMP2A has a YXXL(N)$_7$YXXL sequence, which in the B- or T-cell receptor–associated molecules binds syk or zap70, and a YEEI sequence, which in the receptor-associated molecules binds lyn or other src family tyrosine kinases (607,696). When these LMP2A domains are fused to the CD8 surface and transmembrane domains and CD8 is cross-linked on the surface transfected cells, the LMP2A domains can transmit an increase in intracellular free calcium and are then refractory to further stimulation, as is characteristic of sIg cross-linking (33). Cross-linking of sIg on null mutant recombinant infected LCLs results in normal signal transduction and in activation of lytic EBV infection, whereas cross-linking of sIg on wild-type recombinant infected cells results in little or no change in lytic infection (424–426). The LMP2A block can be by passed by pharmacologically activating protein kinase C and raising intracellular free calcium. Treatment of wild-type or mutant transformed cells with TPA and calcium ionophore results in similar inductions of lytic infection (425).

LMP2A appears to be expressed in most latently infected cells *in vivo* and *in vitro* (389,512,545). The simplest model for its role in latent infection is that the multiple transmembrane domains result in constitutive aggregation of LMP2A and approximation of the tyrosine kinase interactive motifs in the N-terminal cytoplasmic domain, mimicking a constitutively cross-linked, constitutively activated, and constitutively desensitized receptor. PI3 kinase is constitutively activated in LCLs, providing some correlative evidence for constitutive activity of LMP2 in these cells (630). Because LMP2B lacks the amino-terminal receptor homology domain, LMP2B may act to simply modulate the aggregating effects of LMP2A. By depleting lyn activity and by desensitizing the cell, LMP2A prevents many of the normal activating signals that are generated at the B-lymphocyte plasma membrane by sIg, CD19, or class II ligation from triggering B-cell activation and concomitant activation of the EBV lytic cycle. This prevents reactivation of lytic infection as latently infected cells traffic through the peripheral blood, lymphatic tissue, spleen, or bone marrow, where B-cell receptor activation would activate a response element upstream of the Z immediate early transactivator (598) turning on lytic infection. In the peripheral blood or lymphatic tissues, reactivation would result in neutralization of released virus by antibody and cytotoxic T-cell attack on infected cells (Fig. 2; see also Chapter 75 in *Fields Virology,* 3rd ed.). A corollary to this model is that epithelial cells must secrete some cytokine or engage B-lymphocytes so as to transmit a transmembrane signal through JAK kinase or G protein signalling receptors that function independently of the src family cascade and are thereby able to induce lytic reactivation in the appropriate setting.

EBERs

The EBV-encoded, small nonpolyadenylated RNAs are by far the most abundant EBV RNAs in latently infected cells. Estimates of abundance place the EBERs at 10^7 copies per cell (21,274,275,672). Most of the EBERs localize to the cell nucleus, where they are complexed with cellular La protein (275,365). La protein complexes are recognized by specific antisera from patients with systemic lupus erythematosus. The EBER RNAs have stable secondary structures so that in purified RNA or RNA-La protein complexes the RNAs are extensively intramolecularly base paired (204). La protein is associated with the 3′ terminus of the EBER RNAs (204). EBER 1 and 2 have extensive primary sequence similarity to adenovirus VA1 and VA2 and cell U6 small RNAs, both of which also form similar secondary structures and complex with La protein (204,541). EBERs also associate with the L22 ribosomal protein and the novel nuclear protein EAP and localize with these proteins to the nucleus (475,655,656). EAP binds to an EBER stem loop structure (657).

EBER 1 and 2, adenovirus VA, and U6 cell RNA, may have similar primary sequences, secondary structures, and association with La protein in order to accomplish similar functions. Based on the known functions of VA and U6 RNAs, two alternative roles have been proposed for the EBERs. In adenovirus infection, VA1 RNA acts in the cytoplasm to directly inhibit activation of an interferon-induced protein kinase, which would phosphorylate protein synthesis initiation factor eIF-2 alpha and block translation (574). In fact, EBER 1 and 2, provided by transfection into cells or substituted into the adenovirus genome, could partially complement the replication of adenovirus with null mutations in VA1 and 2 (37,38). However, purified EBERs had less effect on eIF-2 alpha kinase activity in *in vitro* assays at levels at which VA1 RNA was inhibitory, and most of the EBERs are not in the cytoplasm to interact with the kinase (88,89,275,586). On the other hand, small nuclear RNAs have long been suspected to be involved in RNA splicing, partly because of base complementarity to splice sites. U6 RNA has been shown to be base paired to U4 RNA in particles required for *in vitro* RNA splicing. Six of seven identical nucleotides to those through which U6 binds to U4 could be provided by a single strand loop in EBER 2 RNA (204).

The recent observation that EBER expression in initial infection and growth transformation of primary B-lymphocytes is delayed until after EBNA and LMP gene expression and the onset of cell DNA synthesis (7) is somewhat incompatible with the proposals that the EBERs function to block interferon inhibition or to enable the splicing of the multiple spliced EBNA and LMP mRNAs. In either scenario, the EBERS could be expected to be expressed as the earliest or one of the earliest EBV RNAs. Nevertheless, the earliest events in primary B-lymphocyte infection are interferon-sensitive, and a role for the EBERs in blocking eIF-2 kinase remains a possibility (187,204).

EBV recombinants specifically deleted for the EBER genes have been compared with wild-type recombinants in primary B-lymphocyte infection (633). EBER-deleted virus was as able to initiate primary B-lymphocyte infection and growth transformation as wild-type virus. No difference could be found in the growth of LCLs infected with EBER-deleted or control virus or in the permissivity of these cells for lytic virus infection after induction.

To further evaluate the hypothesis that the EBERs could have an antiinterferon effect, the effect of the EBERs on the replication of an interferon-sensitive virus in EBV-transformed human B-lymphocytes was determined (632). VSV replication in human lymphocytes is sensitive to interferon. However, interferon had a similar inhibitory effect on VSV replication in LCLs infected with EBER-deleted or wild-type virus (632). Furthermore, lytic wild-type EBV replication could not be inhibited by high levels of interferon, and replication in cells infected with EBER-deleted virus was not more sensitive to interferon. Furthermore, even low doses of interferon are inhibitory of initial B-lymphocyte growth transformation by EBV. EBER-deleted virus was not more sensitive to the antitransforming effects of interferon (632). Neither EBER-deleted nor wild-type EBV-infected LCLs were sensitive to potential antiproliferative effects of interferon. Thus, the role of the EBERs in EBV infection is not obvious from studies in B-lymphocytes *in vitro* and may emerge from testing of the mutants in other cell types or in an *in vivo* model.

Other Viral Gene Products in Latent Infection: BHRF1 and BARF0 RNAs

BHRF1 (23,494) and BARF0 (56,199,268,606) RNAs have been identified in what are otherwise strictly latently infected cells. They have not been shown to be associated with polyribosomes, and proteins have not been detected in typical latently infected cells. BHRF1 RNA and protein are abundantly expressed early in lytic infection, when BHRF1 appears to have an antiapoptotic effect (23, 257,494). Because BHRF1 is the next rightward ORF after EBNA 2 and part of many EBNA transcripts, RNAs with intact BHRF1 ORFs are not necessarily mRNAs that will be translated into protein in latently infected cells. In fact, monoclonal antibody to BHRF1 has been used to attempt to detect the BHRF1-encoded protein in latently infected cells and in newly infected primary B-lymphocytes (7). The only positive outcome of such investigations has been the finding that Raji cells transiently express BHRF1 protein when refed with fresh medium after growth arrest in spent medium (G. Pearson, personal communication). Although this is an important example of anomalous expression of an early lytic gene in latent infection without description of latency, LCLs do not express BHRF1 protein after similar manipulations. Furthermore, nonsense mutation or deletion of the BHRF1 ORF does not affect the ability of the resultant recombinants to establish latent infection in primary B-lymphocytes *in vitro* or to growth transform the cells into actively growing LCLs (407).

BARF0 is a highly spliced RNA that was originally identified from cDNA cloning of nasopharyngeal carcinoma RNA (268,605). The RNA is in all latently infected cells that have been studied so far, including LCLs (56). Humans have antibody to a polypeptide that can be translated *in vitro* from the cDNA (199). However, no protein has been detected in any latently infected cell. Furthermore, deletion of the entire BARF0 RNA-encoding region of the EBV genome has no discernible effect on the ability of the resultant recombinants to establish latent infection in primary B-lymphocytes or to growth transform the cells into LCLs (534).

Role of Latent Infection Proteins in Cell Growth Transformation

The end result of the mechanisms by which EBV transforms primary B-lymphocytes into LCLs is a cell that is nearly indistinguishable from B-lymphocytes proliferating in response to antigen, mitogens, or cytokines in its growth and appearance and in display of molecular markers of adhesion, activation, and differentiation (18,163,290,654). The major characteristics that distinguish LCLs from activated normal B-lymphocytes are latent viral infection, immortality, and autonomy. Two general models need to be considered for the mechanisms that could result in the striking similarity between LCLs and B-lymphocytes proliferating in response to physiologic stimuli. One is that virus gene expression activates cytokine–cytokine receptor pathways that are similar to the normal physiologic pathways and the cell is responding in a similar way to similar effectors. The other is that the virus throws a very small number of key switches (perhaps c-myc, Rb, and p53) in the resting B-lymphocyte, and the panoply of effects seen reflects activation effects preprogrammed in resting B-lymphocytes.

At this moment in EBV research, most evidence favors the first model over the second. However, the two models are not mutually exclusive, and even under the first model the virus must aid and abet normal processes for passing through check points in cell proliferation.

The following observations support the first model: (a) Molecular genetic EBV-recombinant analyses described in detail elsewhere in this chapter indicate that although most of the EBV genome is not important for B-lymphocyte growth transformation, a complex repertoire of EBV genes—EBNA LP, 2, 3A, and 3C, as well as LMP1 and possibly also EBNA 1—are critical or essential for growth transformation. (b) EBNA 2, EBNA 3C, and LMP1 directly affect expression of cell genes involved in normal B-lymphocyte activation pathways. (c) Non–EBV-infected BL cells and LCLs are both immortalized B-lymphocytes at a similar stage of differentiation, but LCLs resemble normal proliferating B-lymphocytes, whereas BL cells resemble germinal center B-lymphocytes (see Chapter 75 in *Fields Virology,* 3rd ed.). Infection of BL cells with EBV or the key transformation-associated EBV gene products in BL cells induces an LCL-like growth phenotype in the BL cells. (d) LCLs at early passage differ from BL cells in being dependent on feeder layers or cell density for continued proliferation. IL-5, IL-6, IL-10, and tumor necrosis factor (TNF) beta are important components of this autocrine growth factor dependence (31,59,149,277, 348,441,458,569,616,645,666).

Although much needs to be done to clarify these mechanisms and to demonstrate that they are sufficient to cause LCL outgrowth, there is little evidence to suggest that the EBNAs or LMP1 are functionally equivalent to SV40 T, adeno E1A, HPV E6 and E7, or Ras in directly abrogating cell cycle check point controls. For example, the EBNAs have so far failed to complement Ras in primary fibroblast growth transformation assays, and although LMP1 has Ras-like effects on immortalized fibroblasts, LMP1 cannot complement E1A or T in primary fibroblast transformations (F. Wang and E. Kieff, unpublished observations). The limited interactions that have been detected between EBNA LP and Rb or p53 (635) are not likely to have the dominant effects on these proteins that characterize SV40 T. Furthermore, although LCLs can support the growth of E1A- or E2-deficient adenovirus mutants, this may be nonspecifically due to the active growth of these cells rather than a direct complementation of an the E1A deficiency with an EBNA that has precisely the same effects as E1A (272). A heuristic argument can be made that unlike the transforming genes of smaller tumor viruses, which are only expressed in natural infection as part of lytic virus infection, the EBV-transforming genes have a much more sophisticated role in causing controlled expansion of a reservoir of latently infected cells in natural infection of normal humans. Abrogation of a major cell check point would result in shortened survival of the host and thereby limit virus persistence in almost all individuals.

Because in non–EBV-infected BL cells lines EBV induces changes in cell growth and gene expression similar to those induced in primary B-lymphocytes (see section on Stages of Infection and Chapter 75 in *Fields Virology,* 3rd ed.), single gene transfer into BL cells has been used to identify the role of specific EBNA or LMP genes in these effects. The identification of a specific gene effect in BL cells opens the way for intensive biochemical analysis of the mechanism of the effect because these cells can be grown in large numbers. However, the utility of BL cells for investigating EBV effects has limitations, the most obvious of which is that BL cells are immortalized and may constitutively express cell genes involves in immortalization that would otherwise be induced by EBV infection in primary B-lymphocytes. Also, EBV may alter cell growth by effecting changes in protein function without affecting level of expression. From a more practical experimental perspective, EBNA 3A and EBNA LP have no obvious effect on BL cell growth or gene expression despite their critical roles in primary B-lymphocyte growth transformation. The failure to detect an effect of these or other EBV genes could be a consequence of the limited repertoire of antibodies to B-lymphocyte proteins, and an effect on gene expression in BL cells, once detected by differential display or similar technology, can be readily investigated.

cDNA clones of cell mRNAs expressed at higher abundance in EBV-infected BL cells than in noninfected cells have been identified (42,43,163,450,452,715). Several are known or novel genes whose expression is probably affected as a consequence of EBV's activating and differentiating effects and are not likely to be pleotropic effectors of changes in cell growth. These include vimentin, MARCKs, serglycin, CD48, and a novel evolutionarily conserved actin-bundling protein (42,43,163,451,715). Others are novel genes that are likely to be components of the effect of EBV on cell growth, differentiation, or motility. These include a calcium calmodulin-dependent protein kinase and two G protein–coupled peptide receptors similar to IL-8 or thrombin receptors (43,450). The calcium calmodulin-dependent kinase could be related to a previously identified effect of calcium levels on the initiation of B-lymphocyte growth transformation (452). Cytokine and cytokine receptors are also likely to be identified in such screens because expression of cytokines or their receptor genes is frequently regulated. CD23 is markedly upregulated with latent EBV infection (41,624), and soluble CD23 has been implicated as a B-cell growth factor (634), although other experiments using recombinant soluble CD23 have not shown growth-stimulating effects (673).

Persistence of EBV DNA in Latently Infected Cells

Circular episomal EBV DNA is detectable in the nuclei of acutely *in vitro*–infected peripheral blood lymphocytes within 24 hr (7,290). Initial circularization appears to require ongoing infected cell DNA synthesis, transcription, and protein synthesis. Probably only DNA repair synthesis is required for circularization because significant nucleotide incorporation does not occur until after 48 hr postinfection (598), and S phase synthesis is not required for initial circularization of other herpes virus DNAs in acute infection. Stable latently infected proliferating lymphocytes characteristically contain multiple copies of EBV episomes (3,316,381,472,625). Most LCLs have 10 or fewer episomes per cell. The range of episome copy number in BL cells is greater. Raji cells have 50 episomes per cell and the number has been stable over 30 years in culture (2, 17). EBV episomes are replicated early in S phase by cell DNA polymerase (2,224,231). The initial amplification of circular EBV DNA is at the start of latent infection and requires early S phase DNA synthesis because before S phase each infected cell has only one episome (7,625). Plasmids containing ori P increase over several days in cells in which EBNA 1 is expressed (625). This appears to be due to EBNA 1 effects on S phase ori P dependent DNA replication. Although EBNA 1 abundance does not appear to correlate with episome copy number (615), other EBV or cell proteins have not been identified that affect episome copy number (420).

The EBV genome usually persists in cells as an episome, but also can persist by integrating into chromosomal DNA or as both integrated and episomal DNA (127,255,291, 292,358,416,642). Only a few such cell lines have been studied in detail. In one, an EBV-infected human Burkitt tumor cell line, EBV DNA has persisted by integrating on chromosome one at IP35 (255,358). In this instance, the entire EBV genome is integrated into A-T–rich cell DNA through the EBV TR (416). More than 15 kbps of cell DNA were deleted by the integration of the 170-kb complete EBV genome. The entire block of cell DNA with the integrated EBV genome appears to have been duplicated as an inverted DNA domain (358). The second example is a latently infected fetal human lymphoblastoid cell line that contains an EBV genome integrated at chromosome 4q25 (255,416). In this instance, integration appears to occur through a site within U1 (252,255), and there are several EBV genomes integrated in tandem (292). In the first instance, integration through TR favors the hypothesis that integration occurred before circularization, whereas in the second instance integration through U1 and the finding of tandem repeats of the EBV genome are more consistent with the hypothesis that integration occurred after initial circularization. Integration occurs frequently when BL cells are infected with EBV *in vitro* and passaged for years (291). In contrast, infection of BL cells with EBV and selection for cells infected with virus almost always results in cell lines containing episomal EBV DNA (404–406, 689). The genome persists as an episome in such cell lines for at least 1 year. Clearly, integration is not chromosome site-specific or a regular feature of EBV infection or EBV-induced cell growth transformation. However, because LMP 2 is not required for cell transformation, linear EBV DNA may be able to transform cells and persist solely as integrated DNA.

Episomal DNA is likely to be necessary for lytic cycle EBV DNA replication because lytic herpes virus DNA replication characteristically proceeds from a circular DNA intermediate (see Chapter 75 in *Fields Virology,* 3rd ed.), and circular EBV DNA copy number increases in lytic infection (587). Lytic EBV replication has not been observed in cells that contain only integrated EBV DNA, and virus has not been recovered from such cells in response to known inducers of virus replication.

In the establishment of latent infection, the EBV genome undergoes progressive methylation (337,353,498,557) and associates with chromosomal proteins so as to give a nucleosomal pattern after micrococcal nuclease digestion (140,588). Regulatory domains involved in maintaining latent infection such as ori P tend to be undermethylated (154). Extensive methylation of parts of the genome not expressed in latent infection probably helps to maintain latency by inhibiting lytic gene expression (298,337,436,471, 557,636). Treatment of latently infected cells with drugs that reduce DNA methylation increases the frequency of spontaneous activation to lytic infection (35,556).

Other Types of Latent Infection

Infection of non–EBV-infected BL cells *in vitro* only infrequently results in the expression of the same set of EBV genes expressed in LCLs. Typically, only EBNA 1 and the EBERs are expressed in BL cells (399,404–406,539). As described in the ensuing chapter, BL cells infected with EBV *in vivo* also frequently express only EBNA 1 and the EBERs, whereas HD cells and NPC cells characteristically express EBNA 1, EBERs, LMP1, and LMP2 (55,56,62, 114,124,236,491,520,535,672). Primary B-lymphocytes *in vivo* may exhibit all three types of latent infection: EBNA 1 and EBERs; EBNA1, EBERs, and LMP1 or LMP2; or all EBNAs, LMPs, and EBERs (51,535). Moreover, a few LCLs growing *in vitro* express lower amounts of the other EBNAs and the LMPs. When explanted to culture, some BL cells that express EBNA 1 and EBERs switch to the full repertoire of LCL-associated EBV gene expression *in vitro* (14,215,216,470,549). In cells expressing EBNA 1 without other EBNAs, the Cp and Wp promoters are not active, and transcription of EBNA 1 is under control of a far downstream promoter at approximately 62.3 in the EBV genome (558,571,606). A glucocorticoid response element upstream of the Cp promoter may be a component of the differential activity of the Cp promoter in B-lymphocytes under different growth conditions (595). In the absence of EBNA 2 expression, LMP1 also can be expressed from a second promoter, in this case upstream from the LCL LMP1 promoter (300). The various forms of latency also can be seen in hybrids between LCLs and somatic cells (194). Clearly, host cell factors determine the pattern of EBV gene expression in latency (328).

Induction of Lytic Infection

Lytic EBV infection is usually studied using chemical inducers of cell permissivity for lytic virus gene expression (35,280,394,526,556,727,728). The various inducers of lytic virus infection give characteristic, although not precisely reproducible, levels of permissivity for each latently infected B cell line. Autocrine TGF beta may participate in lytic reactivation because antibody to TGF beta reduces the responsiveness of cells to lytic induction (133). Phorbol esters are among the most reproducible and most broadly applicable inducers. The effect is probably mediated by protein kinase C activation of jun-fos interaction with AP-1 sites upstream of the immediate early virus genes (19, 156–158,168,355,363). The rapidity of activation of lytic infection by phorbol ester is most consistent with the induction of an activator or inactivation of repressor of lytic infection. Among latently infected lymphoblast cell lines, marmoset lymphoblasts tend to be more inducible than adult human lymphoblasts, whereas neonatal human lymphoblasts are least inducible for virus replication (191,427, 432,433). A few cell lines can be consistently induced to permit lytic virus replication in 10% of the cells (267,433, 637,638). One cell line, Akata, can be induced to 20% to 50% lytic infection by cross-linking sIg and is currently the best source for large quantities of EBV (637,638). sIg cross-linking induces activation of phospholipase C. Phospholipase C releases IP3 and diacylglycerol, which in turn mobilize free calcium and activate protein kinase C. Both second messengers are components of the activation pathway, as evidenced by the induction of lytic infection in response to phorbol ester and calcium ionophore, which directly activate protein kinase C and increase intracellular free calcium, by-passing a requirement for sIg cross-linking (104,105,416). As described above, TPA response elements upstream of the BZLF1 and BRLF1 promoters are probably mediators of the protein kinase C response. Curiously, an sIg cross-linking response element maps just upstream of a TPA element (589) and is a candidate effector site for the calcium-dependent pathway. The calcium effect could be mediated by a calcium calmodulin-dependent protein kinase similar to that induced by LMP1 in latent infection (448).

The realization that the unique response of Akata cells is due in large measure to the lack of expression of LMP2A in this cell line and the otherwise dominant effect of LMP2A in blocking sIg cross-linking induced second messenger release has led to the use of calcium ionophore and TPA together to induce lytic infection in typical latently infected cells (425). An alternative strategy is to use EBV recombinants with specific null mutations in the LMP2A gene (425). Such mutations in LMP2A result in virus recombinants that can be reactivated to lytic infection by sIg cross-linking.

A second approach to studying virus replication is to induce lytic replication by superinfection of Raji cells with a defective EBV from the P3HR-1 cell line (39,455,456, 513,515,517). Raji is an EBV-infected, Burkitt tumor–derived cell line that has an unusually high EBV episome copy number and is unusually responsive to induction. The endogenous Raji genome has at least two deletions and is defective for EBV DNA replication and late gene expression (235,503). Thus, EBV late genes are only expressed in Raji cells after superinfection. The P3HR-1 virus is probably uniquely efficient in inducing lytic infection in Raji because of the presence of defective virus in P3HR-1 preparations (429,431,540). An abundant class of defective virions from P3HR-1 cells contains rearranged DNA molecules in which one or more key immediate early *trans*-activator(s) are downstream of the latent infection cycle IR1 EBNA gene promoter and are therefore expressed in latently infected cell lines after superinfection (82,247,303, 431,525,540). The induction of productive infection after high multiplicity superinfection tends to be more rapid and synchronous than after chemical induction. In either case, maximal late virus protein expression and enveloped virus release requires 48 to 72 hr.

After induction, cells that have become permissive for virus replication undergo cytopathic changes characteristic of herpes viruses, including margination of nuclear chromatin (319), synthesis of viral DNA at the center of the nucleus, assembly of nucleocapsids at the nuclear periphery, nucleation of nucleocapsids, envelopment of virus by budding through the inner nuclear membrane, and inhibition of host macromolecular synthesis (193). Virus gene expression follows a temporal and sequential order (156, 638). Some virus genes are expressed independently of new protein synthesis, early after induction, and are classified as immediate early. Early lytic virus genes are expressed slightly later, and their expression is not affected by inhibition of viral DNA synthesis. Late proteins are expressed late but are formally categorized as late based on a marked reduction in expression after inhibition of viral DNA synthesis.

Immediate Early Genes and Their Targets

In herpes simplex infection, immediate early genes are defined by their transcription after infection in the presence of complete inhibition of protein synthesis (see Chapter 75 in *Fields Virology,* 3rd ed.). A virion tegument protein contributes to the activation of HSV immediate early gene expression. The definition needs to be operationally modified for EBV because the host range of virus infection *in vitro* does not include epithelial cells, which are the permissive cell for lytic infection *in vivo,* and B-lymphocytes are initially nonpermissive. Primary B-lymphocytes are also killed by exposure to inhibitors of protein synthesis for time intervals adequate to observe initial latent gene expression. Although it cannot be rigorously determined whether EBV gene expression in acute latent lymphocyte infection is independent of new virus protein synthesis, the constitutive activity of the Wp EBNA promoter and the very early expression of EBNA LP and EBNA 2 are compatible with the notion that EBNA LP and EBNA 2 are equivalent to immediate early virus gene products in latent infection of primary B-lymphocytes. Similarly, the constitutive activity of the *Bam* FQ EBNA 1 promoter in BL cells (559) and the sole expression of EBNA 1 in infected BL cells are consistent with EBNA 1 being an immediate early gene in that type of latent infection.

Because lytic EBV infection *in vitro* can only be studied by induction of permissivity in cells that are already latently infected, the definition of immediate early genes again must be modified to acknowledge the potential direct or indirect effects of EBV latent infection genes on transcription from the activated EBV genome or the potential effect of the induction on the transcription of specific EBV genes. Moreover, chemical induction may fail to activate an immediate early gene whose transcription is dependent on a virion *trans*-acting factor.

P3HR-1 superinfection of Raji cells and inhibition of protein synthesis with anisomycin results in accumulation of 1 kb BZLF1 and 0.8 kb BRLF1 mRNAs (39). A 4.1-kb RNA from the right end of the genome also was detected. Induction of lytic infection in Akata cells by sIg cross-linking and incubation of the cells in the presence of moderately high concentrations of cycloheximide (50 μg/ml) results in expression of the 1-kb BZLF1 and 2.8-kb BRLF1 mRNAs, confirming that these are immediate early gene products (638). Cycloheximide-resistant 1.6- and 4.1-kb mRNAs from *Bam* M and A, respectively, were also detected in induced Akata cells but were much less abundant in cycloheximide-treated cells than in nontreated cells and in the absence of cycloheximide were most abundant 4 to 12 hr postinduction, whereas the BZLF1 and BRLF1 RNAs are expressed earlier. Although they are expressed before most early genes and are less dependent on viral protein synthesis for their expression, the M and A mRNAs also clearly differ from the Z and R RNAs in being inhibited by cycloheximide treatment and are likely to be very early but not immediate early genes. The *Bam* A RNA is encoded by BALF2, which is the EBV homologue to HSV-1 ICP8. ICP8 is also expressed early in HSV infection and is less sensitive to viral protein synthesis inhibition than are other early virus genes. Another immediate early gene is encoded from the EBV BI'LF4 ORF (410).

The BZLF1, BRLF1, and BI'LF4-encoded proteins are transactivators of early EBV lytic gene expression (Fig. 7). The size of Z varies slightly amoung EBV strains, but the sequence differences have not been described (489). The first indication of transactivation came from a cotransfection experiment in which a DNA fragment encoding (in retrospect) BRLF1 was transfected into cells alone with a DNA fragment containing a responsive early EBV gene, resulting in expression of the early gene (641). Transfection of a DNA fragment from highly defective P3HR-1 DNA that expresses BZLF1 under control of the Wp promoter or BZLF1 expression from heterologous vectors was then found to broadly activate lytic infection (217,302–304,431). BZLF1 and BRLF1 were found to coordinately

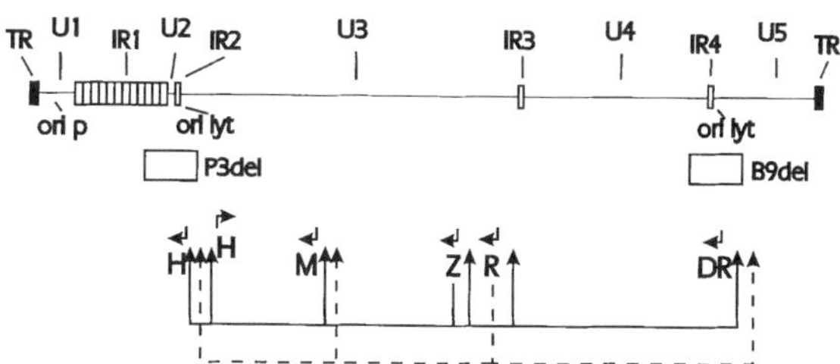

FIG. 7. Immediate early lytic EBV infection. The promoter for BZLF1 is presumably activated by a physiologic change in the cell, such as TPA-initiated, increased protein kinase C–mediated activation of upstream AP-1 sites. BZLF1 is transcribed (1) and translated. Z protein targets to the nucleus where it binds directly to AP-1–like sites upstream of BZLF1 and BRLF1, increasing BZLF1 transcription and turning on BRLF1 (2). Z and R proteins act coordinately to turn on the first early genes BHLF1, BHRF1, DR, BSMLF1, and BALF2 (not shown). R *trans*-activates an enhancer element usually upstream of the Z recognition sequences.

upregulate expression from an early EBV promoter (71,76), and BI′LF4 was discovered to be host cell restricted in transactivating effects (410), completing an initial view of immediate early regulation of early transcription in B-lymphocytes *in vitro*. Consistent with these observations is the more recent finding that BZLF1 expression is evident in advancing margins of lesions of EBV lytic infection in epithelium (717).

Two key early promoter regulatory elements within DL and DR (duplication left and right) are coordinately regulated by BZLF1 and BRLF1. DL and DR encode abundant early RNAs (113,179,281,288,476) and include the origins for lytic viral DNA replication within their promoter upstream regulatory domains (229). BZLF1 and BRLF1 synergistically upregulate the bidirectional BHLF1 and BHRF1 promoters of DL in transient transfection assays (71,76, 77,102,200). The *cis*-acting domains of DR and DL have been partially dissected (71,76,77,232). The promoter proximal domain (–79 to –221 bp relative to BHLF1) confers strong promoter activity in HeLa or Vero cells only when BZLF1 is provided in *trans*. A DNA segment 639 bp upstream of the BHLF1 cap site is a position- and orientation-independent enhancer composed of one component with constitutive activity in fibroblasts or epithelial lines but not in B-lymphocytes and a second component, TTGTCCCGTGGACAATGTCC, which is responsive to BRLF1.

The BSMLF1 and BMRF1 promoters are also coordinately regulated by BZLF1 and BRLF1 (271,516). BZLF probably interacts with a TPA response element upstream of the BSMLF1 promoter (387). Direct BRLF1 binding to the BMRF1 promoter upstream DNA is also essential for BSMRF1 coordinate upregulation by the BZLF1 and BRLF1 *trans*-activators (516). BRLF1 *trans*-activation of the BSMLF1 early promoter is probably mediated by activation of an upstream sequence CCGTGGAGAATGTC similar to the activation sequence in HLF1 (323). *In vitro*–translated R binds directly to this R response element by interacting with two sequences, catGTCCCtctatcatGGCGCagac, within the element (222,223,226,400).

BZLF1 also functionally and physically interacts with NF-κB (227). NF-κB is an important mediator of LMP1 effects in latent infection. BZLF1 also can downregulate the EBNA Cp promoter, perhaps facilitating the transition from latent to lytic infection (324,593). This would be biologically important if a latent gene product is an antagonist to lytic infection but could be coincidental. The action of BZLF1 on Cp appears to be indirect involving the upregulation of fos by BZLF1 (166,593) and downmodulation of the positive effect of a glucocorticoid response element, which maps about 1 kbp upstream of Cp (347,593,595). Glucocorticoid withdrawal from culture media increases permissivity for replication, providing support for a mutual antagonism between Cp function and lytic reactivation (593). The putative antagonism of lytic reactivation by a product of the Cp promoter is unlikely to be mediated by EBNA 1 because EBNA 1 continues to be expressed under control of a downstream promoter after the switch to lytic infection. The activation of the downstream EBNA 1 promoter in response to Z expression places lytic EBNA 1 expression in the early lytic infection gene class along with LMP1 (360,402,547).

The initial activation of Z may be partially dependent on two 7-bp dyad symmetry elements separated by 27 bp. These sequences are similar to a reversible silencing element in adeno-associated virus and may contribute to the lack of Z expression during latency (440). Other negative regulatory elements are more upstream from the transcriptional initiation site (577).

BZLF1 probably autoregulates its own promoter and the BRLF1 promoter. The Z promoter has several AP1-like upstream elements that appear to mediate activation in the presence of low levels of Z and repression in the presence of high levels of Z (185,425,648). Differences in Z binding to the various upstream sites could be mediated by changes in Z abundance during lytic infection.

BZLF1 mRNA is spliced and consists of three exons (161,370). The first exon encodes amino acids 1 to 167, which includes the transactivating domain (78,169,369). This domain activates GAL 4 response elements when fused to a GAL 4 DNA binding domain (169,430) and stabilizes TFIID association with TATA elements through protein–protein interactions (370). Replacement of the activating domain with the VPI6 acidic domain results in enhanced transactivation of early promoters (32). The second exon encodes amino acids 168 to 202, which includes a strongly basic domain that has homology to a conserved region of the c-fos/c-jun family of transcriptional modulators (158). This domain confers the ability to interact with AP1-related sites in DNA (70,171,370) and also targets Z to the nucleus (423). A BZLF1 intron also has homology to the 3′ untranslated c-fos sequences, suggesting that BZLF1 relatively recently evolved from c-fos (158). The third exon encodes for amino acids 203 to 245, which includes a near perfect leucine or isoleucine heptad repeat capable of coiled coil dimer formation (70,158,168,344, 370). Like c-fos, BZLF1 is a dimerizing DNA binding protein that binds directly to AP1-related sequences (158,165–167,171,369–371). Dimerization may facilitate high-affinity interaction with templates having two potential Z response elements (69). The similarity between TPA response elements, AP1 sites, and Z recognition sites may partially explain the activity of TPA in activating lytic EBV infection (371,567).

BZLF1 is partially phosphorylated on serine 336, and the phosphorylation regulates DNA binding activity (104, 342). BZLF1 appears to associate with p53 and may thereby inhibit an effect of p53 on inducing apoptosis in response to lytic EBV DNA replication (722). Later in lytic infection Z becomes associated with virion structural proteins, although Z has not been detected in virus (317). The interaction of structural proteins with Z could serve to downmodulate Z activity late in infection so as to minimize the toxic effects of high-level Z expression.

The R transactivator is a DNA sequence-specific acidic transactivator that has distant homology to c-myb (226,227, 400). c-myb can interact synergistically with Z in transactivating the BMRF1 promoter in lymphocytes (326). Two domains of R when fused to the GAL 4 DNA binding domain have transactivating activity for GAL 4 response elements upstream of basal promoters (232). Amino acids 416 to 519 are weakly activating and are required only for activation in B-lymphocytes, whereas the carboxy-terminal amino acids, 520 to 605, are a potent acidic transactivating domain similar to VP 16.

A minor immediate early RNA is an unusual spliced product of the BRLF1 and BZLF1 bicistronic transcript. This RNA has the beginning of the BRLF1 ORF spliced in frame to the DNA binding and dimerization domains of BZLF1. Because the BRLF1 activation domain is at the C-terminus and the BZLF1 activation domain is at the N-terminus, the spliced product lacks an activation domain and when dimerized with Z could be a repressor of Z activity (183).

Early Lytic Infection

EBV early replicative cycle genes are operationally differentiated from late genes by their persistent synthesis in the presence of inhibitors of viral DNA synthesis. By this criteria, at least 30 EBV mRNAs are early mRNAs, and almost 30EBV mRNAs are late mRNAs (Fig. 1, Table 1) (24,29,39,40,156,157,196,197,235,282–284,288,289,563, 583). Early and late mRNAs are intermingled through most of the EBV genome. Frequently, different promoters initiate nested transcripts that begin with different ORFs and terminate at the same polyadenylation site so that the longer mRNAs include all of the shorter mRNAs (563). Some early and late genes are spliced, whereas others are not. In most instances in which genes have been studied in some detail, minor spliced mRNA species have been discerned.

Because of the difficulty in doing genetic studies of EBV replicative functions, proteins encoded by EBV early or late genes have been identified or assigned functions from analysis of the protein predicted from mRNA or DNA ORF sequences (24,29,34,98,103,196,197,235,398,583), from the temporal class and abundance of the RNA or the size of its *in vitro* translation products (96,141,287–289,581, 582), or from functional assays of *in vitro*–translated or *in vivo*–expressed proteins (34,178,249,286,289,313,314, 380,384,386,485,544,618). The function of many EBV genes has been first suggested from comparison of predicted primary amino acid sequences of mRNA or DNA ORFs with the primary amino acid sequences of other herpes virus proteins of known function. Subsequent studies have confirmed these tentative functional assignments and frequently have shown unique as well as common attributes of the EBV proteins.

Aside from the two immediate early proteins BZLF1 and BRLF1, two EBV early proteins, BSMLF1 (possibly a delayed immediate early) and BMRF1, also may *trans*-activate expression of other early EBV genes. Both are moderately abundant early nuclear proteins (85,86,705,709). BMRF1 activity has only been demonstrated by cotransfection with SV40 promoter-driven chloramphenicol acetyl transferase reporter gene DNA into BHK cells (478). BSMLF1 is a promiscuous *trans*-activator of gene expression that acts synergistically with BZLF1 or BRLF1 in inducing higher level expression in transient expression assays (325,372,478,704). Because the level of CAT reporter gene mRNA in such assays frequently does not increase in cells cotransfected with BSMLF1-expressing plasmids, BSMLF1 may be a *trans*-activator of translation (325). However, a direct effect on translation has not been demonstrated.

Two other very abundant EBV early proteins have been mapped to specific EBV DNA sequences. One is the 135-kd BALF2 protein (287). This protein has primary amino acid sequence homology to the major HSV DNA binding protein ICP8 and is important in DNA replication (164). BHRF1 encodes another abundant EBV early protein (494). Surprisingly, the mRNA for HRF1 is only moderately abundant (23,494). HRF1 encodes an 18-kd protein that consists of a predicted hydrophobic signal peptide, a 135–amino acid hydrophilic domain that contains two potential *N*-linked glycosylation sites, a hydrophobic potential transmembrane domain, and a short carboxy-terminal cytoplasmic domain (494). Immunofluorescence microscopy with specific sera shows nuclear membrane and cytoplasmic localization compatible with endoplasmic reticulum residence (494). HRF1 has extensive, colinear homology with bcl-2, a cell gene activated by the chromosome 14/18 translocation in human follicular B cell lymphomas (90). bcl-2 is important in lymphocyte survival because of its role in preventing apoptotic death. BHRF1 is therefore likely to prevent apoptotic cell death in lytic EBV infection. Because lytic infection in LCLs commences in a latently infected cell in which bcl-2 and A20 (another antiapoptosis protein) are expressed at high levels, LCLs, or their *in vivo* counterpart, may not be the cells in which BHRF1 is critical to lytic infection. BHRF1 is likely to be more critical to lytic infection in B-lymphocytes, which previously expressed only EBNA1 and EBERs and lack Bcl2, or to lytic infection in epithelial cells.

Because of the relationship to bcl-2, its abundant expression in lytic infection, and the possibility of low-level expression in latent infection, mutations have been made in BHRF1 and mutant EBV recombinants constructed (362, 407). EBV recombinants with a stop codon early in BHRF1 or with deletions of the BHRF1 ORF were fully able to initiate or maintain cell growth transformation, and the mutant recombinant transformed cells grew, as did wild-type recombinant transformed cells (472), even under adverse conditions of low serum or low cell concentration, which favor apoptasis. They also were able to enter lytic cycle

EBV infection and produce virus (362,407), even in low serum or at low cell concentration (407). As described in the section on other types of latent infection and in Chapter 75, some EBV-infected lymphocytes *in vivo* express only EBNA 1 and the EBERs. When such infected cells proceed to lytic infection, they are more likely to be subject to apoptotic cell death as a consequence of p53 expression in response to viral DNA replication in lytic infection. BHRF1 can protect non–EBV-infected BL cells from apoptosis-inducing stresses, such as infection with an E1b-deficient adenovirus mutant (257,646).

Among the EBV early genes identified by their homology to early genes of other herpes viruses are several that are linked to DNA replication. These include DNA polymerase (BALF5), major DNA binding protein (BALF2), ribonucleotide reductase (BORF2 and BaRF1), thymidine kinase (BXLF1) and alkaline exonuclease (BGLF5). These genes are distributed through the long EBV US domain (Fig. 1).

The EBV DNA polymerase has been extensively purified and is 117 kd (314). Partially purified DNA polymerase is associated with several other EBV nuclear proteins, including the 50-kd protein encoded by BMRF1 (80,330,368). The catalytic subunit expressed in insect cells from a baculovirus vector has 3′ to 5′ proofreading exonuclease activity in addition to DNA polymerase activity (671). Reconstitution of baculovirus-expressed polymerase with a 2:1 molar ratio of baculovirus-expressed BMRF1 enhances the double-strand 3′ to 5′ exonuclease activity fourfold (670) and the DNA polymerase activity tenfold (669). BMRF1 also markedly enhances polymerase processivity and results in full-length products (669). Unlike cellular polymerases, the EBV polymerase is active in 100 mM ammonium sulfate or in KCl and is sensitive to inhibition by ethyl maleimide, phosphonoacetic acid, phosphonoformic acid, arabinofuranosylthymine, and acycloguanosine triphosphate (9,73,79,91,110–119,220,276–279,483,484,486,627). The EBV DNA polymerase also differs from HSV DNA polymerase in its salt sensitivity and its relative resistance to amphidicolin. The EBV DNA polymerase only weakly distinguishes between 2-amino-purine nucleoside triphosphate and dATP, so that 2-amino-purine is preferentially incorporated into viral as opposed to cell DNA (218).

The EBV, VZV, and HSV ribonucleotide reductases are extensively colinearly homologous (196). The large HSV subunit is 140 kd, whereas the EBV subunit is 85 kd, and VZV is predicted to be 87-kd (196). The EBV, VZV, and HSV large and small subunits have domains that are common to ribonucleotide reductases from a wide variety of prokaryotic and eukaryotic species (67,469). The HSV-1 large subunit has an amino-terminal domain that is absent from other ribonucleotide reductases. Surprisingly, the EBV ribonucleotide reductase large subunit is confined to multiple, discrete regions in the cytoplasm of productively infected cells (205). The large subunit is a delayed early protein that accumulates in cells approximately 4 hr after the MRF1 nuclear early protein (205,495). Acetone fixation of cells destroys the immunologic reactivity of the large subunit, suggesting that it may be a major component of the restricted early antigen complex originally identified by EBV immune human sera from African patients with BL (see Chapter 75).

The existence of an EBV-specific deoxy pyrimidine kinase was inferred from the coinduction of EBV lytic replication and a new kinase activity in P3HR-1 cells (131,617). An EBV-induced dTK and dCK activity elutes as a single peak on DEAE-cellulose, away from the cellular dTK and dCK activities. Gene transfer of EBV DNA or of the EBV DNA fragment with homology to HSV TK into TK-negative mammalian cells or into TK-negative *E. coli* restores TK activity, mapping the kinase activity to this DNA fragment (384–386). The EBV TK is similar to HSV TK in accepting a broad range of nucleoside or nucleoside analogues as substrates and has some antigenic cross-reactivity with the HSV TK (385).

Despite initial evidence to the contrary, EBV DNA polymerase is not required for viral DNA replication associated with episome establishment (531,598). Episome establishment and B-lymphocyte transformation are not inhibited by infection in the presence of acycloguanosine. With induction of lytic EBV replication, episome copy number increases, and labeled thymidine, initially incorporated into circular EBV DNA, chases into very large, presumably linear concatemeric EBV DNA (87,597,603).

The DL and DR EBV DNA segments (Fig. 1), which have strong early promoter activity also function as origins of lytic infection viral DNA replication (229,572,573). Using transient transfection to activate lytic EBV replication cotransfection of either of these EBV DNA segments on plasmids results in replication of the plasmid DNAs in the lytically infected cells (229). An essential 1.4-kb core element replicates with considerably reduced efficiency (229). The core 1.4-kb origin has numerous inverted repeat elements that could serve as the definitive origin. Even within the 1.4-kb segment, at least three domains contribute to the replicative activity. One domain can be functionally replaced in part by the CMV immediate early promoter and enhancer, suggesting that this domain is important for lytic origin function because of transcriptional enhancement (130). Interestingly, when cells containing lytic origin plasmids are induced to lytic EBV infection, plasmid copy number increases slightly and linear concatemers are produced (568,587,592). Thus, the EBV lytic origin region, although as yet not defined as a sequence, is considerably larger than the HSV origin but appears to function similarly as an amplicon. As expected, inhibition of EBV DNA polymerase inhibits EBV lytic origin activity in lytically infected cells. Other lymphocryptoviruses have DL and DR elements (249,250,253,364). The HV Papio element has transcriptional activation and DNA replication sites that are similar to those in EBV (555).

Considerable progress has been made in constituting an *in vivo* minimal replication system similar to the Challberg

HSV system (164). The components include BALF5, which encodes the core DNA polymerase; BALF2, the single-strand DNA binding protein; BMRF1, the processivity factor; BSLF1 and BBLF4, the primase and helicase complex; BBLF2/3, a spliced primase-helicase complex component; and BKRF3, the uracil DNA glycosylase. A surprising aspect of the EBV DNA replication genes is the extent to which they are dependent on complementing expression of the BZLF1, BRLF1, and BSMLF1 transactivators. EBV DNA replication also requires topoisomerase 1 and 2 activity (320). The redistribution of BZLF1 and BMRF1 gene products to the same nuclear site may identify the site at which viral DNA is replicated (640).

Late Gene Expression, Cleavage, Packaging, Envelopment, and Egress

With the initial sequencing of TR from integrated EBV DNA (416), similarities between TR and the HSV *a* sequence were apparent (438,608–610). Both EBV TR and the HSV *a* sequence are directly repeated at both ends of the DNA. Both sequences are high in G-C composition, with runs of three to five Cs. In both HSV and EBV, there is a short extra oligonucleotide segment (DR1) in which the repeat joins the unique DNA. The EBV DR1 oligonucleotide GCATGGGGGG brackets the multiple TR copies (416,438). In HSV-1, DR1 is the site of cleavage of head full-length DNA for packaging into virions (608–610). There seems to be little sequence specificity for DR1 because the nonhomologous oligonucleotides of HSV-1, HSV-2, and CMV are interchangeable (610). Instead, HSV or CMV cleavage and packaging appear to occur 33 b 3′ to a conserved sequence of $C(G)_{6\text{-}8}$ TGT G/T $(T)_3$ NG C/G $(G)_6$ G/C T/C. A similar sequence C $(G)_5$ $TGT(T)_2CCT$-$(G)_5CC$ occurs 25b before the EBV DR1 and may be the key cleavage packaging recognition sequence. Plasmids containing the EBV latent infection episome origin, the EBV lytic origin, and the EBV terminal repeat sequence are stably maintained as an episome in latently infected cells, are replicated into linear concatemers by EBV DNA polymerase when lytic infection is induced, and are cleaved and packaged into infectious virions (228,229,689).

Those EBV late genes whose functions are partially known or can be predicted from homology to other known herpes virus genes are mostly structural viral proteins or proteins that modify the infected cells so as to permit virus envelopment or egress. Late genes with enzymatic or regulatory functions have not as yet been defined. The viral glycoprotein genes that have been identified so far are all late genes. These have been more intensively studied than other late genes because of their potential importance in antibody-mediated immunity to virus infection. Current knowledge of nonglycoprotein late genes is limited. Among the nonglycoprotein late genes, the major nucleocapsid protein is almost certainly encoded by cLF1. cLF1 is homologous to the HSV major capsid protein gene (418). Furthermore, cLF1 encodes a 4.5-kb late mRNA that, translated *in vitro*, yields a 150-kd protein identical in size to the major EBV nucleocapsid protein (287). NRF1 probably encodes the major virion external nonglycoprotein. The principal nonglycosylated protein extractable from virus with nonionic detergents is 140 kd. NRF1 encodes a 4.1-kb late mRNA that translates a 140-kd protein and is therefore presumed to encode the external virus component (287). Because the BNRF1 predicted sequence has no major hydrophobic domains, it is more likely to encode a tegument protein than to encode for an integral membrane protein. Another late RNA includes an ORF BXRF1 homologous to the VZV basic virion core protein gene (120,179). BXRF1 is therefore tentatively designated as an EBV basic core protein gene. BXRF1 encodes a 2.4-kb late mRNA, but its product has not been identified (121).

The known EBV glycoprotein genes are BLLF1 (gp350/220), BALF4 (gp110), BXLF2 (gp85), ILF2 (gp55/80), and BDLF3 (gp42) (34,209,218,245,246,289, 398,477). A sixth glycoprotein has been detected in cells using a monoclonal antibody but has not been mapped (566). BILF expressed from a recombinant vaccinia virus is 55 to 80 kd; a similar size glycoprotein has been identified in EBV lytically infected cells and in virus using antiserum from recombinant vaccinia virus–infected rabbits or EBV-immunized mice (398). The BILF2 glycoprotein corresponds to an abundant glycoprotein in purified virus preparations. However, antibody to BILF2 fails to neutralize virus infectivity (398). BDLF3 cDNA cloning and sequencing confirm that BDLF3 mRNA is not spliced (245). The predicted amino acid sequence includes a hydrophobic amino terminus, consistent with a signal peptide, a hydrophilic external domain with nine potential *N*-linked glycosylation sites, and a single hydrophobic transmembrane domain. Translation of the 0.9-kb mRNA *in vitro* yields a 32-kd protein, which is processed by dog pancreas microsomes to 60 kd (245). Surprisingly, antisera to bacterially expressed DLF3 identify only a 35- to 36-kd protein in lymphocytes late in EBV replication (245). The antisera appear to react with the plasma membrane of cells replicating virus and with virus (245).

Because HSV-1 gB is a major virion and infected cell surface glycoprotein, the homologous EBV BALF4 ORF is of interest (142,209,496). Both BALF4 and HSV-1 gB are predicted from their primary amino acid sequence to have amino-terminal cleavable signal peptides, external domains of 650 amino acids with four potential *N*-linked glycosylation sites, three potential membrane-spanning domains, and a 150–amino acid cytoplasmic domain (496). BALF4 encodes two late mRNAs: an abundant 3-kb RNA that is not spliced and a scarce 1.8-kb RNA (209). The BALF4-encoded 3-kb mRNA translates a 93-kd protein that is glycosylated to 110 kd (208,209). gp110 is one of the most abundant late EBV proteins (142,208,209). The glycosylation is probably both *N*- and *O*-linked because *N*-glycanase, which cleaves *N*-linked oligosaccharides, only reduces gp110 to 105 kd (208,209). Endoglycosidase H has a similar effect on almost all of the gp110. Because endoglycosidase H only cleaves *N*-linked oligosaccharides that have not been modified in the Golgi, this suggests that most of the gp110 is not Golgi processed. A small fraction of gp110 is processed to a form compatible with a size of 120 kd. This 120-kd glycoprotein is not endo H sensitive and is partially incorporated into virus (Alfieri and Kieff, unpublished data). The intracellular distribution of gp110 closely parallels that of an endoplasmic reticulum resident protein (208,209). Immune light or electron microscopy localize gp110 to the inner and outer nuclear membrane and to cytoplasmic membranes frequently surrounding enveloped virus, but not to the Golgi or plasma membrane. Furthermore, despite gp110's high abundance in the inner nuclear membrane through which nucleocapsids bud to acquire their initial envelope, gp110 cannot be demonstrated in enveloped intracellular or extracellular virus by immune electron microscopy (208,209). Protein or sugar stains of polyacrylamide gel–separated proteins from purified virus indicates that the Golgi-processed gp110 is not a major virion glycoprotein. Thus, EBV gp110 is an abundant viral glycoprotein present in the cell nuclear and endoplasmic reticulum, and a small amount of gp110 is a minor structural protein of the virion envelope. The small amount of gp110 in virus is likely to be acquired as the virus buds through the nuclear membrane and to be processed during virus residence in endoplasmic reticular vesicles that interact with the Golgi.

In contrast to gp110, gp85 and gp350/220 are processed much more efficiently through the Golgi and are found on the virus and in the plasma membrane of lytically infected cells (34,138,139,208,209,245,246,270,398,455,456, 474–477,513–515,619,651–653).

Gp85 is encoded by XLF2, which has significant colinear homology to HSV-1 gH (246,477), whereas gp350/220 is encoded by LLF, which has only a small region of distant homology to HSV-1 gC (34). Gp85 is a relatively minor virus component that appears to be important in fusion between virus and cell membranes (434), whereas gp350/220 is the dominant external virus glycoprotein that mediates virus binding to the B-lymphocyte receptor CD21. Interestingly, gp85 expressed in NIH 3T3 cells localizes to the internal cytoplasmic and nuclear membranes, whereas gp350/220 localizes to the Golgi and plasma membrane (209,699). This suggests that gp85 may require another virus protein to bring it to plasma membrane of lytically infected lymphocytes, where it is characteristically found, whereas gp350/220 possesses the necessary signals for efficient Golgi transport, processing, and plasma membrane insertion. In fact, gp85 does complex with the BKRF2-encoded protein, which seems to enable its transport to the plasma membrane (710).

Although gp350/220 is the most abundant viral protein in the lytically infected cell plasma membrane and in the virus envelope, only small amounts of gp350/220 accumulate in the lytically infected cell nuclear membrane (208,664). EBV gp350/220 is extensively *N*- and *O*-glycosylated (208,584,651–653,699). gp350 nascent protein has an apparent size of 135 kd and 37 potential *N*-linked glycosylation sites, whereas mature glycoprotein has an apparent size of 350 kd. Approximately half of the glycosylation is due to *N*-linkage as opposed to *O*-linkage. All of the *N*-linked oligosaccharides are complexed with tri- and tetra-antenary chains, whereas most *O*-linked chains are probably di- or tri-*N*-acetyl lactosamine (584). This high degree of *N*- and *O*-linked complex oligosaccharides is further evidence of extensive Golgi processing.

The question arises as to why gp350/220 is expressed in large amounts in the infected cell plasma membrane, where it renders the infected cell susceptible to antibody- and complement-mediated cytolysis. Early electron microscopic observations of lymphocytes lytically infected with EBV describe the presence of nucleocapsids in the cytoplasm. Although this could occur as a consequence of nuclear disruption late in infection, an alternative hypothesis is that virus acquires an initial envelope as it buds through the nuclear membrane, that deenvelopment occurs within cytoplasm vesicles resulting in release of nucleocapsids into the cytoplasm, and that reenvelopment occurs at the plasma membrane, where the virus acquires a definitive envelope rich in gp350/220 and gp85. Consistent with this hypothesis, the plasma membrane and virus are both rich in gp350/220, whereas the nuclear membrane is relatively deficient in gp350/220 (208). Even if this hypothesis is correct, late in infection, as the cell deteriorates, partially enveloped or initially enveloped virus may be released in substantial quantity. The relative importance of these various pathways in natural infection is difficult to evaluate. A second possible role for high-level plasma membrane gp350/220 expression is in saturation of lymphocyte CD21 so that virus can be completely released from and not reabsorbed to the lytically infected cells.

gp350/220 is not only the most abundant viral protein in the lytically infected cell plasma membrane, but it is also one of the most abundant late viral proteins and the most abundant protein on the outer surface of the virus. Most of the human EBV neutralizing antibody response is directed to gp350/220 (653). gp350/220 is therefore an essential component of any prospective EBV vaccine that would aim to engender neutralizing antibody (448,474). Significant parts or all of the protein have been expressed in *E. coli, Sacchromyces cerevesiae,* and in mammalian cells (34,463,576,644,699,720). A recombinant VZV, which expresses gp350/220 under control of the gpI promoter has been constructed (393). gp350 is expressed in large amounts in recombinant VZV-infected cells and is incorporated into the infected cell plasma membrane and into the VZV outer envelope (393). gp350/220 also has been inserted into vaccinia under control of the p8.5 promoter (448). Injection with purified EBV gp350/220 or infection with vaccinia expressing gp350/220 protects cottontop tamarins against a lethal, lymphomagenic EBV challenge (448).

BCRF1 is a close colinear homologue to the human IL-10 gene, with nearly 90% colinear identity in amino acid sequence (277,676). BCRF1 is expressed only late in EBV replication despite its EBV genomic map location in the middle of the EBNA regulatory domain between ori P and the Cp promoter. IL-10 has B-cell growth factor activity (442), and some data have suggested that BCRF1 could also be important in latent EBV infection (437). However, nonsense or deletion mutations involving BCRF1 have no effect on the ability of the recombinants to initiate growth transformation, to maintain wild-type latent infection and transformed cell growth, or to lytically infect B-lymphocytes *in vitro* (631). Thus, the role of BCRF1 must be in lytic infection *in vivo.* The principal role of IL-10 in mice and humans appears to be as a negative regulator of macrophage and NK cell functions that otherwise positively regulate TH1 cytotoxic T-lymphocytes (442,679). BCRF1 has most of the activities of human IL-10 *in vitro.* Thus, BCRF1 is likely to act to blunt the initial NK and T-cell cytotoxic response to EBV infection in epithelial cells and B-lymphocytes. One important aspect of that response is the release of gamma interferon. The initiation of cell growth transformation is quite sensitive to interferon (187,631). Mixture of lytically induced EBV recombinant transformed LCLs with normal peripheral blood mononuclear cells results in low interferon release and efficient transformation of primary B-lymphocytes, whereas mixture of lytically induced BCRF1 null mutant recombinant infected LCLs with peripheral blood mononuclear cells results in high-level gamma interferon release and inefficient transformation of the primary B-lymphocytes (631). A second aspect of the NK and CD8 cytotoxic responses is the direct cytotoxic component. Blunting of that component is likely to be important in both primary EBV infection and in reactivation of lytic infection in B-lymphocytes. To create an analogous murine model for this possible effect, infection of C57BL/6 mice with vaccinia virus expressing beta galactosidase was compared with infection of C57BL/6 mice with vaccinia virus expressing murine IL-10. The NK- and vaccinia-specific CD8 cytotoxic responses were lower in the mice infected with vaccinia virus expressing IL-10, although there was no difference in virulence or in antibody response (341). Thus, virally expressed IL-10 appears to have an effect on the initial interferon and NK and CD8 cytotoxic responses and may have a local effect on these responses to reactivated infection.

Defective Virus

The complete EBV genome is usually maintained as an episome in latently infected lymphocytes. However, in several instances incomplete genomes have been detected in latently infected BL cells growing in culture or in subclones of such cells. The three best studied examples are the EBV genomes in the Raji, P3HR-1, and Daudi cell lines (53,82–84,129,235,247,300,302–304,311,333,521). The Raji cell has 50 EBV episomes, each of which is deleted for EBNA-3C and for the ALF2, 135-kd major DNA binding protein. Presumably as a consequence of the deletion of the ALF2 gene, the EBV genome in Raji cells is not replicated by the Raji DNA polymerase when lytic infection is induced. Thus, no late genes are expressed, and the infection is abortive.

P3HR-1 and Daudi have fewer EBV episomes and are deleted for a single EBV DNA fragment, which includes some of the IR1 copies, all of U2 and IR2, and several hundred base pairs of U3. The P3HR-1 EBV deletion appears to have arisen on *in vitro* passage because the parent cell lines from which P3HR-1 was cloned are not deleted (4, 333,524). The EBV genome in P3HR-1 can be induced to full lytic infection, and virus is produced. Thus, EBV genes deleted from P3HR-1, including EBNA-LP, EBNA-2, and HLF1, are not necessary for EBV replication in B-lymphoma cells *in vitro.* The resulting P3HR-1 or Daudi virus cannot transform normal B-lymphocytes. These cell lines carrying minimally defective EBV genomes are useful for EBV recombinant genetic studies. Interestingly, similar defectives arise during EBV infection in epithelial cells *in vivo* (493,601)

Aside from the minimally defective EBV DNA, some of the EBV DNA in P3HR-1 cells is more highly defective. There are several formats for the more highly defective DNAs, and several biologic properties have been delineated (82,84,247,302–304,647). Analysis of P3HR-1 subclones shows that most cells contain only the minimally defective EBV P3HR-1 genome (266). A rare cell contains both the minimally and more defective P3HR-1 EBV genomes. The more defective genome is spread by cell-to-cell transmission after spontaneous reactivation of lytic EBV replication (429). Reactivation occurs in cells containing the defective genomes because the defective genomes contain the ZLF1 immediate early *trans*-activator under control of the IR1 latent infection cycle promoter (217,302–304). One population of defective genomes appears to be centered about the IR1 palindrome (Fig. 8) (302–304). In the highly defective DNA, the left arm of the palindrome is a complete inverted copy of the right arm. IR1 sequence to the right of the palindrome now also extends from the palindrome in the opposite direction into a new linkage with the right end of *Bam* Z (Fig. 8). A second new linkage joins the left of *Bam* Z to *Bam* S and *Bam* M. A third new linkage joins *Bam* M to *Bam* B1, W1I1, and *Bam*HI A. *Bam*HI A is presumably linked to TR. The lytic ori in *Bam*HI B1 and the TR presumably provide the necessary *cis*-acting lytic EBV DNA replication origin and packaging signals, respectively, to assure the perpetuation of the defective virus in the P3HR-1 cell cultures. The specific and selective inclusion of BZLF1, BSMLF1, and BI′LF1 probably arises because of the adjuvant effect of these transactivators on lytic EBV replication.

GENETICS

Attempts at constructing EBV recombinants have been hindered by the limitation in the EBV host range to human B-lymphocytes and the nonpermissivity of these cells for EBV replication. The development of strategies for switching latently infected cells into lytic infection partially bypassed this major experimental barrier. The first attempt at developing an EBV-recombinant genetics system was to infect replication-defective Raji cells with P3HR-1 EBV stocks and to plate the resultant virus onto primary B-lymphocytes (83,180). Raji EBV cannot replicate its own genome in lytic infection because of deletion of the major single-strand DNA binding protein BALF2 but has EBNA LP and EBNA 2. Raji is also deleted for EBNA 3C, which in later experiments turned out to be a critical gene for B-lymphocyte growth transformation. The P3HR-1 EBV genome is lytic replication competent but transformation incompetent because of the deletion of a DNA segment that encodes EBNA LP and EBNA 2. Infection of Raji cells with P3HR-1 EBV induced partial permissivity for lytic infection, and replication of both genomes ensued. Because Raji cells contain 50 episome copies of EBV DNA, most of the transforming virus that came out of the infection should have been simply Raji EBV genomes. However, the few transforming viruses that were obtained in transformed primary B-lymphocyte–derived cell lines were in fact recombinants between Raji and P3HR-1 that had restored to P3HR-1 the deleted EBNA LP and EBNA 2 encoding fragment, providing initial genetic evidence that EBNA 2 or EBNA LP is essential for transformation (603). In retrospect, these results also provided an indication that the Raji EBV genome was itself transformation incompetent because of the EBNA 3C deletion.

More recently, the entire EBV genome has become amenable to molecular genetic manipulation. In these experiments specifically mutated recombinant EBV genomes have been obtained by transfecting latently infected lymphocytes with specifically mutated recombinant EBV DNA fragments that had been cloned and amplified in *E. coli.* When virus replication is then induced, the mutant EBV DNA fragment undergoes homologous recombination with replicating viral DNA (94,95,228,295,321,322,361,362, 390–392,403–407,534,631,633,658–661,689). Parental and recombinant virus can be used to infect primary B-lymphocytes or non–EBV-infected BL cells. Because primary B-lymphocytes are dependent on EBV infection for their ability to grow *in vitro*, a mutation in an essential transforming gene will therefore only be recoverable in primary B-lymphocytes if the cells are coinfected with a defective EBV that is wild-type for the mutated gene. The usual source for defective EBV is P3HR-1 cells because 10^7 to 10^8 P3HR-1 EBV can be obtained from 10^7 P3HR-1 cells, and the virus is wild-type except for the EBNA LP and EBNA 2 deletion. BL cells are not dependent on EBV for their growth in culture and therefore are potentially useful for recovery of recombinants with mutations in essential transforming genes (404–407,689). BL cells or primary B-lymphocytes can be made dependent on latent recombinant virus infection for their growth by including a positive selection marker such as SV 40 promoter driven hygromycin phosphotransferase in the transfected recombinant EBV DNA. Only cells infected with a recombinant virus will then grow out under selective conditions (689). Alternatively, cells infected with the replication-competent but transformation-defective P3HR-1 EBV can be used as the source of a parent virus strain. When these cells are transfected with a wild-type EBV DNA fragment spanning the EBNA LP and EBNA 2 deletion, a small number of transformation-competent wild-type EBNA LP and EBNA 2 P3HR-1 EBV recombinants result, and these can be specifically isolated in primary B-lymphocyte–derived cell lines

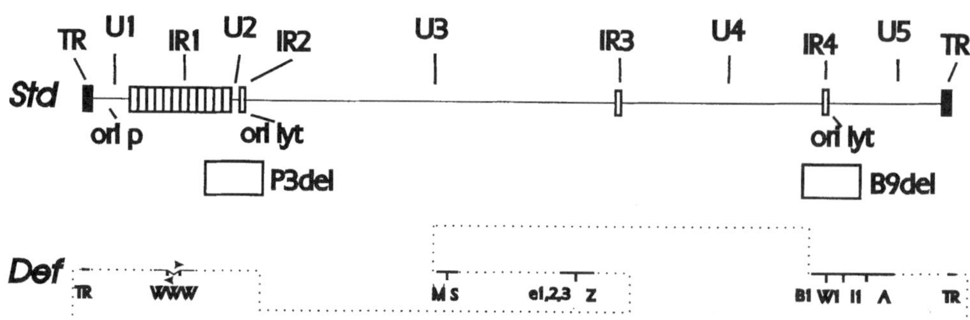

FIG. 8. Schematic depiction of the standard EBV genome and of a major P3HR-1 maximally defective EBV genome. The central feature of this defective genome is an extension of the IR1 palindrome so that the sequence extending leftward from the center of the palindrome is identical to the IR1 sequence that extends rightward from the palindrome in the standard EBV genome. IR1 sequences downstream of the IR1 promoter are recombined with the *Bam* Z sequence so that BZLF1 is now downstream of the IR1 promoter. Much of the DNA between BZLF1 and BSM1 has been deleted. Another recombinational event has linked *Bam* M to a large DNA segment from the right end of the genome, which includes a lytic replication origin. The linkage between that segment and the TR sequences necessary for cleavage and packaging of EBV DNA is uncertain.

because only the recombinants will be able to cause LCL outgrowth (94,95,228).

Because transfection with a wild-type EBV cosmid DNA fragment that includes EBNA LP and EBNA 2 into P3HR-1 EBV-infected cells, and induction of lytic infection results in homologous recombination between the transfected DNA and the replicating P3HR-1 EBV with restoration of transforming ability to recombinant P3HR-1 genomes, the effect on transformation of specific EBNA LP or EBNA 2 mutations or of EBNA LP or EBNA 2 type specific differences could be determined (94,95,229,403). Recombination with a wild-type EBV cosmid DNA fragment that spans the deletion uniformly resulted in transforming recombinants. The number of transforming recombinants was sufficiently consistent in a clonal transformation assay that differences in transformation efficiency could be detected. Some deletion mutations within the EBNA 2 ORF or a stop codon mutation resulted in no transforming recombinants, demonstrating that EBNA 2 is essential for lymphocyte growth transformation. EBNA 2 was also shown to account for the differences between high transforming type 1 EBV strains and low transforming type 2 EBV strains (95). Analysis of the phenotype of 11 linker insertions and 15 deletion mutations within the EBNA 2 ORF showed four separable domains that are essential for transformation of primary B-lymphocytes (94). Subsequent experiments with EBV recombinants containing EBNA LP deletion or stop codon insertion mutants showed a markedly reduced B-lymphocyte transformation efficiency.

The use of large cosmid DNA fragments for EBNA 2/EBNA LP marker rescue of transformation in P3HR-1 cells enables specific mutations to be made in genes mapping near EBNA LP and EBNA 2, and within the flanking EBV DNA that was common to both the P3HR-1 and transfected EBV DNA fragment. Deletion of the EBER-encoding DNA or mutation of the BHLF1 or BHRF1 ORFs in the cosmid fragment resulted in many of the transforming recombinants also being mutant in the adjacent gene (407,633). The resultant mutant EBV recombinants could transform cells with wild-type efficiency and replicate after induction.

Because the EBERs map about 30 kbp away from EBNA 2, the high frequency with which the EBER deletion is incorporated into virus recombinants is therefore surprising. This led to the observation that almost as high a frequency of incorporation into recombinant EBV genomes occurs with a second, non-linked, cosmid EBV DNA fragment transfected along with the EBNA LP and EBNA 2 cosmid DNA into P3HR-1 cells (658–660). Using this strategy of frequent second-site homologous recombination, mutations could be made on any EBV DNA fragment, and after transfection into P3HR-1 cells, harvest of the resultant virus, and clonal infection of primary B-lymphocytes, a substantial fraction of the resultant LCLs were infected with mutant recombinants (Fig. 8). The frequency of incorporation of nonlinked second EBV DNA fragments is directly related to transfected fragment size. Fragments of 4 to 5 kbp are incorporated with about 20% to 30% of the efficiency of 40-kbp fragments, the latter being incorporated into about 10% of the EBV DNAs that recombine with the EBNA LP/2 positive selection marker.

An alternative strategy for constructing EBV recombinants that is particularly useful for making deletions in the EBV genome is to use the P3HR-1 cells as a lytic infection transcomplementing and packaging cell line for cosmid cloned EBV DNA fragments (534,660). Approximately 50% of all transforming recombinants from cells transfected with five overlapping EBV DNA fragments representative of the entire genome consist of the transfected cosmid DNAs without any evidence for recombination with the P3HR-1 EBV genome. For studying the transforming EBV genes, only two and a half cosmid-cloned EBV DNA fragments are necessary because all of the genes between EBNA 1 and LMP1 are unnecessary, and most of the noncoding exons and introns of the EBNA mRNAs can be deleted.

Although positive selection for transformation from the transformation-negative P3HR-1 parent strain is useful for making specifically mutated recombinants in most genes, there are limitations. First, non-transforming mutations in EBNA LP or EBNA 2 cannot be recovered, and the effect of the mutation can only be ascertained by the failure to generate a transforming recombinant from P3HR-1. Second, nontransforming mutations in other essential or critical transforming genes, such as EBNA 3A or 3C or LMP1, can only be recovered when the wild-type EBNA 3A or 3C or LMP1 is provided to the infected primary B-lymphocyte. In many instances, this can be conveniently provided by nonrecombinant P3HR-1 virus because 10^7 to 10^8 nonrecombinant P3HR-1 EBV are in each preparation along with 50 to 200 recombinant EBVs. Because the virus preparation is used to infect 10^7 primary B-lymphocytes, 50% of the primary B-lymphocytes are initially infected with nonrecombinant P3HR-1. Any cell infected with a recombinant therefore has a 50% chance of being initially coinfected with a nonrecombinant. Both genomes can stably persist in cells and function, although the nonrecombinant P3HR-1 genome is usually lost over a few months in continuous culture of the resultant LCL if it is not required to maintain cell growth transformation. The characterization of the mutant recombinant genome in coinfected cells is more complicated than in cells infected with a single EBV genome. Furthermore, in some instances the mutant recombinant can be lost due to the mutant recombinant with its mutated EBNA 3A, EBNA 3C, or LMP1 gene and its wild-type EBNA 2 and EBNA LP genes undergoing secondary recombination with the coinfecting P3HR-1 EBV, which is wild-type for EBNA 3A, EBNA 3C, or LMP1 but deleted for EBNA 2 and EBNA LP. The secondary recombinant genome could be fully wild-type. Third, the efficiency of second site recombination is about 10%. Many recombinants are obtained that are not mutant at the second site, and these must be identified.

To circumvent some of these problems, non–EBV-infected BL cells or primary B-lymphocytes can be used for the isolation and recovery of mutant EBV recombinants carrying a selectable marker (361,362,404–406,689). BL cells can be infected with EBV *in vitro*. The efficiency of infection varies among BL cell lines and is frequently 10% of the efficiency of primary B-lymphocyte infection. The EBV genome is almost always maintained in such cells as a multicopy episome. Frequently, only EBNA 1 is expressed in the infected BL cells, although the full panoply of EBNAs and LMPs can be expressed in some clones of the same cell line. EBV recombinants carrying positive selection cassettes as a consequence of recombination between transfected cloned EBV DNA with the positive selection cassette and replicating EBV genomes can be specifically identified and recovered by positive selection of infected BL cells. In some BL cell backgrounds, cells latently infected with mutant recombinants can be induced to lytic EBV replication by transfection with an expression cassette for the Z immediate ealy EBV transactivator and treatment with TPA. Mutant recombinant virus can be recovered from such cells. Because the BL cells are not dependent on EBV for their ability to grow *in vitro,* mutant recombinants deleted for an essential transforming gene can be transferred to non–EBV-infected BL cells and marker rescue experiments similar to those done in P3HR-1 cells can be conducted. The major limitation of this strategy has been the difficulty in recovering substantial quantities of virus from most infected BL cells even of the more permissive cell backgrounds.

INHIBITORS OF VIRUS REPLICATION

Lytic EBV infection *in vitro* or *in vivo* (see next chapter) are sensitive to acycloguanosine, which is a substrate for the EBV deoxynucleoside kinase (9,10,378,384) and an inhibitor of EBV DNA polymerase. Other inhibitors of herpes virus DNA polymerases, such as phosphonoacetic acid and phosphonoformic acid, are also useful for *in vitro* inhibition of EBV DNA polymerase in B-lymphocytes so as to distinguish early from late gene expression (119,377,379,531). Because of the limited clinical benefit in inhibiting EBV replication with acycloguanosine in patients with acute primary EBV infection (see Chapter 75 in *Fields Virology,* 3rd ed.), there has been little effort to develop other specific inhibitors of lytic EBV infection.

BRIEF SUMMARY

In the first decade of EBV research, a great deal was learned about the biology of EBV infection *in vitro* and *in vivo.* The virus was found to cause infectious mononucleosis and to be latent in most adult humans. The tight linkage to BL and NPC, the rare occurrence of lymphoproliferative disease in young children and renal transplant recipients, and the ability of the virus to efficiently immortalize lymphocytes *in vitro* and to induce lymphomas in marmosets provided a biologic framework indicating that EBV can be oncogenic under unusual circumstances. The discovery of cell lines that could be induced to replicate EBV led in the second decade to increasingly sophisticated biochemical analyses. The structure of the genome was determined, the genome cloned, the transcriptional program in latent and lytic infection worked out, and the entire genome sequenced. A nuclear tumor antigen became a protein on a polyacrylamide gel and then more than one protein. Studies of transcripts in "latently" infected and growth-transformed B-lymphocytes indicated a complicated repertoire. By the beginning of the third decade, the genes encoding the EBV latent infection nuclear and membrane proteins and the proteins themselves were being identified in rapid succession. Six EBNAs, two LMPs, and two EBERs were involved in latent infection or cell growth transformation. Studies of the biologic activity of these genes after single gene transfer showed EBNA 1 and ori P to be sufficient for plasmid episome maintenance and only one LMP to have significant transforming activity in rodent fibroblasts. Gene transfer into BL cells and recombinant EBV molecular genetics emerged as important experimental paradigms with the demonstration that activities of the EBNAs and LMPs were more evident in human B-lymphocytes, a natural target of EBV infection.

In the past 5 years knowledge of the roles of specific EBNAs and LMPs in maintaining latent infection and growth transformation and of the various types of latent infection has advanced rapidly. The interaction of EBNAs and LMPs with B-lymphocyte proteins is offering increasing understanding of the mechanisms of EBV-induced and normal B-lymphocyte activation and proliferation. The past decade also has seen an explosion in knowledge of immediate early and early regulators of lytic EBV infection. New strategies for manipulating the transition to lytic infection have facilitated EBV molecular genetic and biochemical research. These discoveries and those described in the next chapter on the molecular pathogenesis of EBV infection *in vivo* open new possibilities for contemporary approaches at understanding, preventing, diagnosing, and treating EBV infection and its associated malignancies.

REFERENCES

1. Abbot SD, Rowe M, Cadwallader K, et al. Epstein-Barr virus nuclear antigen 2 induces expression of the virus-encoded latent membrane protein. *J Virol* 1990;64:2126–2134.
2. Adams A. Replication of latent Epstein-Barr virus genomes in Raji cells. *J Virol* 1987;61:1743–1746.
3. Adams A, Lindahl T. Epstein-Barr virus genomes with properties of circular DNA molecules in carrier cells. *Proc Natl Acad Sci USA* 1975; 72:1477–1481.
4. Adldinger HK, Delius H, Freese UK, Clarke J, Bornkamm GW. A putative transforming gene of the Jijoye virus differs from that of Epstein-Barr virus prototypes. *Virology* 1985;141:221–234.

5. Ahearn JM, Hayward SD, Hickey JC, Fearon DT. Epstein-Barr virus (Epstein-Barr virus) infection of murine L cells expressing recombinant human Epstein-Barr virus/C3d receptor. *Proc Natl Acad Sci USA* 1988;85:9307–9311.
6. Aitken C, Sengupta SK, Aedes C, Moss DJ, Sculley TB. Heterogeneity within the Epstein-Barr virus nuclear antigen 2 gene in different strains of Epstein-Barr virus. *J Gen Virol* 1994;75:95–100.
7. Alfieri C, Birkenbach M, Kieff E. Early events in Epstein-Barr virus infection of human B-lymphocytes. *Virology* 1991;181:595–608.
8. Allan GJ, Inman GJ, Parker BD, Rowe DT, Farrell PJ. Cell growth effects of Epstein-Barr virus leader protein. *J Gen Virol* 1992;73:1547–1551.
9. Allaudeen HS. Distinctive properties of DNA polymerase induced by herpes simplex virus type-1 and Epstein-Barr virus. *Antiviral Res* 1985; 5:1–12.
10. Allaudeen HS, Descamps J, Sehgal RK. Mode of action of acyclovir triphosphate on herpes viral and cellular DNA polymerase. *Antiviral Res* 1982;2:123–133.
11. Allday MJ, Crawford DH, Griffin BE. Epstein-Barr virus latent gene expression during the initiation of B cell immortalization. *J Gen Virol* 1989;70:1755–1764.
12. Allday MJ, Crawford DH, Thomas JA. Epstein-Barr virus (Epstein-Barr virus) nuclear antigen 6 induces expression of the Epstein-Barr virus latent membrane protein and an activated phenotype in Raji cells. *J Gen Virol* 1993:74:661–669.
13. Allday MJ, Farrell PJ. Epstein-Barr virus nuclear antigen EBNA3C/6 expression maintains the level of latent membrane protein 1 in G1-arrested cells. *J Virol* 1994;68:3491–3498.
14. Altiok E, Minarovits J, Hu LF, Contreras-Brodin B, Klein G, Ernberg I. Host-cell-phenotype-dependent control of the BCR2/BWR1 promoter complex regulates the expression of Epstein-Barr virus nuclear antigens 2-6. *Proc Natl Acad Sci USA* 1992;89:905–909.
15. Amakawa R, Jing W, Ozawa K, et al. Human Jk recombination signal binding protein gene (IGKJRB): comparison with its mouse homologue. *Genomics* 1993;17:306–315.
16. Ambinder RF, Mullen MA, Chang YN, Hayward GS, Hayward SD. Functional domains of Epstein-Barr virus nuclear antigen EBNA-1. *J Virol* 1991;65:1466–1478.
17. Ambinder RF, Shah WA, Rawlins DR, Hayward GS, Hayward SD. Definition of the sequence requirements for binding of the EBNA-1 protein to its palindromic target sites in Epstein-Barr virus DNA. *J Virol* 1990;64:2369–2379.
18. Amen P, Lewin N, Nordstrum M, Klein G. Epstein-Barr virus-activation of human B lymphocytes. *Curr Top Microbiol Immunol* 1986;132: 266–271.
19. Angel P, Imagawa M, Chiu R, et al. Phobol ester-inducible genes contain a common cis element recognized by a TPA-modulated trans acting factor. *Cell* 1987;49:729–739.
20. Apolloni A, Sculley TB. Detection of A-type and B-type Epstein-Barr virus in throat washings and lymphocytes. *Virology* 1994;202:978–981.
21. Arrand JR, Rymo L. Characterization of the major Epstein-Barr virus-specific RNA in Burkitt lymphoma-derived cells. *J Virol* 1982;41: 376–389.
22. Arrand JR, Rymo L, Walsh JE, Bjorck E, Lindahl T, Griffin BE. Molecular cloning of the complete Epstein-Barr virus genome as a set of overlapping restriction endonuclease fragments. *Nucleic Acids Res* 1981;9:2999–3014.
23. Austin PJ, Flemington E, Yandava CN, Strominger JL, Speck SH. Complex transcription of the Epstein-Barr virus BamHI fragment H rightward open reading frame 1 (BHRF1) in latently and lytically infected B lymphocytes. *Proc Natl Acad Sci USA* 1988;85:3678–3682.
24. Baer R, Bankier AT, Biggin MD, et al. DNA sequence and expression of the B95-8 Epstein-Barr virus genome. *Nature* 1984;310:207–211.
25. Baichwal VR, Sugden B. Post-translational processing of an Epstein-Barr virus-encoded membrane protein expressed in cells transformed by Epstein-Barr virus. *J Virol* 1987;61:866–875.
26. Baichwal VR, Sugden B. Transformation of Balb/3T3 cells by the BNLF-1 gene of Epstein-Barr virus. *Oncogene* 1988;2:461–467.
27. Baichwal VR, Sugden B. The multiple membrane-spanning segments of the BNLF-1 oncogene from Epstein-Barr virus are required for transformation. *Oncogene* 1989;4:67–74.
28. Banchereau J, de Paoli P, Valle A, Garcia E, Rousset F. Long-term human B cell lines dependent on interleukin-4 and antibody to CD40. *Science* 1991;251:70–72.
29. Bankier AT, Deininger PL, Satchwell SC, Baer R, Griffin BE. DNA sequence analysis of the EcoRI Dhet fragment of B95-8 Epstein-Barr virus containing the terminal repeat sequences. *Mol Biol Med* 1983;1: 425–446.
30. Bauer G, Hofler P, zur Hausen H. Epstein-Barr virus induction by a serum factor. I. Induction and cooperation with additonal inducers. *Virology* 1982;121:184–194.
31. Baumann MA, Paul CC. Interleukin-5 is an autocrine growth factor for Epstein-Barr virus-transformed B lymphocytes. *Blood* 1992;79:1763–1767.
32. Baumann R, Grogan E, Ptashne M, Miller G. Changing Epstein-Barr viral ZEBRA protein into a more powerful activator enhances its capacity to disrupt latency. *Proc Natl Acad Sci USA* 1993;90:4436–4440.
33. Beaufils P, Choquet D, Mamoun RZ, Malissen B. The (YXXL/I)2 signalling motif found in the cytoplasmic segments of the bovine leukaemia virus envelope protein and Epstein-Barr virus latent membrane protein 2A can elicit early and late lymphocyte activation events. *EMBO J* 1993;12:5105–1512.
34. Beisel C, Tanner J, Matsuo T, Thorley-Lawson D, Kezdy F, Kieff E. Two major outer envelope glycoproteins of Epstein-Barr virus are encoded by the same gene. *J Virol* 1985;54:665–674.
35. Ben-Sasson SA, Klein G. Activation of the Epstein-Barr virus genome by 5-aza-cytidine in latently infected human lymphoid lines. *Int J Cancer* 1981;28:131–135.
36. Berger R, Bernheim A. Cytogenetics of Burkitt's lymphoma-leukemia: a review. In: *Burkitt's lymphoma, a human cancer model.* Lyon, France: IARC Scientific Publication #60; 1985:65–80.
37. Bhat R, Thimmappaya B. Two small RNAs encoded by Epstein-Barr virus can be functionally substituted for the virus-associated RNAs in the lytic growth of adenovirus 5. *Proc Natl Acad Sci USA* 1983;80: 4789–4793.
38. Bhat R, Thimmappaya B. Construction and analysis of additional adenovirus substitution mutants confirm the complementation of VA1 RNA function by two small RNAs encoded by Epstein-Barr virus. *J Virol* 1985;56:750–756.
39. Biggin M, Bodescot M, Perricaudet M, Farrell P. Epstein-Barr virus gene expression in P3HR1-superinfected Raji cells. *J Virol* 1987;61: 3120–3132.
40. Biggin M, Farrell PJ, Barrell BG. Transcription and DNA sequence of the Bam HI L fragment of B95-8 Epstein-Barr virus. *EMBO J* 1984;3: 1083–1090.
41. Billaud M, Busson P, Huang D, et al. Epstein-Barr virus (Epstein-Barr virus)-containing nasopharyngeal carcinoma cells express the B-cell activation antigen Blast2/CD23 and low levels of the Epstein-Barr virus receptor CR2. *J Virol* 1989;63:4121–4128.
42. Birkenbach M, Josefsen K, Yalamanchili R, Lenoir G, Kieff E. Epstein-Barr virus-induced genes: first lymphocyte-specific G protein-coupled peptide receptors. *J Virol* 1993;67:2209–2220.
43. Birkenbach M, Liebowitz D, Wang F, Sample J, Kieff E. Epstein-Barr virus infection or the latent infection membrane protein induces vimentin expression. *J Virol* 1989;63:4079–4084.
44. Birkenbach M, Tong X, Bradbury L, Tedder T, Kieff E. Characterization of an Epstein-Barr virus receptor on human epithelial cells. *J Exp Med* 1991;176:1405–1414.
45. Bocker JF, Tiedmann KH, Bornkamm GW, Bornkamm GW, zur Hausen H. Characterization of an Epstein-Barr virus-like virus from African green monkey lymphoblasts. *Virology* 1980;101:291–295.
46. Bodescot M, Brison O, Perricaudet M. An Epstein-Barr virus transcription unit is at least 84 kilobases long. *Nucleic Acids Res* 1986;14: 2611–2620.
47. Bodescot M, Chambraud B, Farrell P, Perricaudet M. Spliced RNA from the IRI-U2 region of Epstein-Barr virus: presence of an opening reading frame for a repetitive polypeptide. *EMBO J* 1984;3:1913–1917.
48. Bodescot M, Perricaudet M. Clustered alternative splicing sites in Epstein-Barr virus RNAs. *Nucleic Acids Res* 1987;15:5887.
49. Bodescot M, Perricaudet M. Epstein-Barr virus mRNAs produced by alternative splicing. *Nucleic Acids Res* 1986;17:7130–7134.
50. Bodescot M, Perricaudet M, Farrell PJ. A promoter for the highly spliced EBNA family of RNAs of Epstein-Barr virus. *J Virol* 1987;61:3424–3430.
51. Boos H, Berger R, Kuklik-Roos C, Iftner T, Mueller-Lantzsch N. Enhancement of Epstein-Barr virus membrane protein (LMP) expression by serum, TPA or n-butyrate in latently infected Raji cells. *Virology* 1987;159:20–30.

52. Bornkamm G, Delius H, Zimber U, Hudewentz J, Epstein MA. Comparison of Epstein-Barr virus strains of different origin by analysis of the viral DNAs. *J Virol* 1980;35:603–618.
53. Bornkamm GW, Hudewentz J, Freese UK, Zimber U. Deletion of the nontransforming Epstein-Barr virus strain P3HR-1 causes fusion of the large internal repeat to the subregion. *J Virol* 1982;43:952–968.
54. Bornkamm GW, Von Knebel-Doebertiz M, Lenoir GM. No evidence for differences in the Epstein-Barr virus genome carried in Burkitt lymphoma cells and nonmalignant lymphoblastoid cells from the same patients. *Proc Natl Acad Sci USA* 1984;81:4930–4940.
55. Brooks L, Yao QY, Rickinson AB, Young LS. Epstein-Barr virus latent gene transcription in nasopharyngeal carcinoma cells: coexpression of EBNA1, LMP1, and LMP2 transcripts. *J Virol* 1992;66:2689–2697.
56. Brooks LA, Lear AL, Young LS, Rickinson AB. Transcripts from the Epstein-Barr virus BamHI A fragment are detectable in all three forms of virus latency. *J Virol* 1993;67:3182–3190.
57. Brown N, Liu C, Garcia CR, Wang YF, Griffith A, Sparkes RS, Calame KL. Clonal origins of lymphoproliferative disease induced by Epstein-Barr virus. *J Virol* 1986;58:975–978.
58. Buell G, Reisman D, Kintner C, Crouse G, Sugden B. Cloning overlapping DNA fragments from the B95-8 strain of Epstein-Barr virus reveals a site of homology to the internal repetition. *J Virol* 1981;40:977–982.
59. Burdin N, Peronne C, Banchereau J, Rousset F. Epstein-Barr virus transformation induces B lymphocytes to produce human interleukin 10. *J Exp Med* 1993;177:295–304.
60. Burkhardt A, Bolen J, Kieff E, Longnecker R. An Epstein-Barr virus transformation associated membrane protein interacts with src family tyrosine kinases. *J Virol* 1992;66:5161–5167.
61. Busson P, Edwards RI, Tursz T, Raab-Traub N. Sequence polymorphism in the Epstein-Barr virus latent membrane protein 2 (LMP2) gene. *J Gen Virol* 1995;76:139–145.
62. Busson P, McCoy R, Sadler R, Gilligan K, Tursz T, Raab-Traub N. Consistent transcription of the Epstein-Barr virus LMP2 gene in nasopharyngeal carcinoma. *J Virol* 1992;66:3257–3262.
63. Busson P, Zhang Q, Guillon JM, et al. Elevated expression of ICAM1 (CD54) and minimal expression of LFA3 (CD58) in Epstein-Barr-virus-positive nasopharyngeal carcinoma cells. *Int J Cancer* 1992;50:863–867.
64. Calendar A, Billaud M, Aubry J, Banchereau J, Vuillaume M, Lenoir G. Epstein-Barr virus induces expression of B cell activation markers on in vitro infection of Epstein-Barr virus negative B lymphoma cells. *Proc Natl Acad Sci USA* 1987;84:8060–8064.
65. Calender A, Cordier M, Billaud M, Lenoir GM. Modulation of cellular gene expression in B lymphoma cells following in vitro infection by Epstein-Barr virus (Epstein-Barr virus). *Int J Cancer* 1990;46: 658–663.
66. Cameron KR, Staminger T, Craxton M, Bodmer W, Honess RW, Fleckstein B. The 160,000-Mr virion protein encoded at the right end of the herpesvirus Saimiri genome is homologous to the 140,000-Mr membrane antigen encoded at the left end of the Epstein-Barr genome. *J Virol* 1987;61:2063–2070.
67. Caras IW, Levinson BB, Fabry M, Williams SR, Martin DW Jr. Cloned mouse ribonucleotide reductase subunit M1 cDNA reveals amino acid sequence homology with Escherichia coli and herpes virus ribonucleotide reductases. *J Biol Chem* 1985;260:7015–7022.
68. Carel JC, Myones BL, Frazier B, Holers VM. Structural requirements for C3d,g/Epstein-Barr virus receptor (CR2/CD21) ligand binding, internalization, and viral infection. *J Biol Chem* 1990;265:12293–12299.
69. Carey M, Kolman J, Katz DA, Gradoville L, Barberis L, Miller G. Transcriptional synergy by the Epstein-Barr virus transactivator ZEBRA. *J Virol* 1992;66:4803–4813.
70. Chang YN, Dong DL, Hayward GS, Hayward SD. The Epstein-Barr virus Zta transactivator: a member of the bZIP family with unique DNA-binding specificity and a dimerization domain that lacks the characteristic heptad leucine zipper motif. *J Virol* 1990;64:3358–3369.
71. Chavier P, Gruffat H, Chevallier-Greco A, Buisson M, Sergeant A. The Epstein-Barr virus (Epstein-Barr virus) early promoter DR contains a cis-acting element responsive to the Epstein-Barr virus transactivator EB1 and an enhancer with constitutive and inducible activities. *J Virol* 1989;63:607–614.
72. Chen MR, Middeldorp JM, Hayward SD. Separation of the complex DNA binding domain of EBNA-1 into DNA recognition and dimerization subdomains of novel structure. *J Virol* 1993;67:4875–4885.
73. Cheng YC, Huang ES, Lin JC. Unique spectrum of activity of 9-[(1,3-dihydroxy-2-propoxy)methyl]-guanine against herpesviruses in vitro and its mode of action against herpes simplex virus type 1. *Proc Natl Acad Sci USA* 1983;80:2767–2770.
74. Cheung A, Kieff E. Long internal direct repeat in Epstein-Barr virus DNAs. *J Virol* 1982;44:286–294.
75. Cheung A, Kieff E. Epstein-Barr virus DNA. X. A direct repeat within the internal direct repeat of Epstein-Barr virus DNA. *J Virol* 1981; 40:501–507.
76. Chevallier-Greco A, Gruffat H, Manet E, Calender A, Sergeant A. The Epstein-Barr virus (Epstein-Barr virus) DR enhancer contains two functionally different domains: domain A is constitutive and cell specific, domain B transactivated by the Epstein-Barr virus early protein R. *J Virol* 1989;63:615–623.
77. Chevallier-Greco A, Manet E, Chavrier P, Mosnier C, Daillie J, Sergeant A. Both Epstein-Barr virus (Epstein-Barr virus)-encoded transacting factors, EB1 and EB2, are required to activate transcription from an Epstein-Barr virus early promoter. *EMBO J* 1986;5:3243–3250.
78. Chi T, Carey M. The ZEBRA activation domain: modular organization and mechanism of action. *Mol Cell Biol* 1993;13:7045–7055.
79. Chiou JF, Cheng YC. Interaction of Epstein-Barr virus DNA polymerase and 5′-triphosphates of several antiviral nucleoside analogs. *Antimicrob Agents Chemother* 1985;27:416–418.
80. Chiou JF, Li JK, Cheng YC. Demonstration of a stimulatory protein for virus-specified DNA polymerase in phorbol ester-treated Epstein-Barr virus-carrying cells. *Proc Natl Acad Sci USA* 1985;82:5728–5731.
81. Chittenden T, Lupton S, Levine AJ. Functional limits of oriP, the Epstein-Barr plasmid origin of replication. *J Virol* 1989;63:3016–3025.
82. Cho MS, Bornkamm GW, zur Hausen H. Structure of defective DNA molecules in Epstein-Barr virus preparations from P3HR-1 cells. *J Virol* 1984;51:199–207.
83. Cho MS, Fresen KO, zur Hausen H. Multiplicity dependent biological and biochemical properties of Epstein-Barr virus (Epstein-Barr virus) rescued from non-producer lines after superinfection with P3HR1 Epstein-Barr virus. *Int J Cancer* 1980;26:357–363.
84. Cho MS, Gissman L, Hayward SD. Epstein-Barr virus (P3HR-1) defective DNA codes for components of both the early antigen and viral capsid antigen complexes. *Virology* 1984;137:9–19.
85. Cho MS, Jeang KT, Hayward SD. Localization of the coding region for an Epstein-Barr virus EA and inducible expression of this 60 kd nuclear protein in transfected fibroblast cell lines. *J Virol* 1985;56: 852–859.
86. Cho MS, Milman G, Hayward SD. A second Epstein-Barr virus EA gene in BamHI fragment M encoded a 48- to 50-kilodalton nuclear protein. *J Virol* 1985;56:860–866.
87. Cho MS, Tran VM. A concatenated form of Epstein-Barr viral DNA in lymphoblastoid cell lines induced by transfection with BZLF1. *Virology* 1993;194:838–842.
88. Clarke PA, Schwemmle M, Schickinger J, Hilse K, Clemens MJ. Binding of Epstein-Barr virus small RNA EBER-1 to the double-stranded RNA-activated protein kinase DAI. *Nucleic Acids Res* 1991;19:243–248.
89. Clarke PA, Sharp NA, Clemens MJ. Translational control by the Epstein-Barr virus small RNA EBER-1. Reversal of the double-stranded RNA-induced inhibition of protein synthesis in reticulocyte lysates. *Eur J Biochem* 1990;193:635–641.
90. Cleary ML, Smith SD, Sklar J. Cloning and structural analysis of cDNAs for bcl-2 and hybrid bcl-2/immunoglobulin transcript resulting from the t(14-18) translocation. *Cell* 1986;47:19–28.
91. Clough W, McMahon J. Characterization of the Epstein-Barr virion-associated DNA polymerase as isolated from superinfected and drug-stimulated cells. *Biochim Biophys Acta* 1981;656:76–85.
92. Cohen J. A region of herpes simplex virus VP16 can substitute for a transforming domain of Epstein-Barr virus nuclear protein 2. *Proc Natl Acad Sci USA* 1992;89:8030–8034
93. Cohen J, Kieff E. An Epstein-Barr virus nuclear protein 2 domain essential for transformation is a direct transcriptional activator. *J Virol* 1991;65:5880–5885.
94. Cohen J, Wang, F, Kieff E. Epstein-Barr virus nuclear protein-2 mutations define essential domains for transformation and transactivation. *J Virol* 1991;65:2545–2554.
95. Cohen J, Wang F, Mannick J, Kieff E. Epstein-Barr virus nuclear protein 2 is a key determinant of lymphocyte transformation. *Proc Natl Acad Sci USA* 1989;86:9558–9562.
96. Cohen LK, Speck SH, Roberts BE, Strominger JL. Identification and mapping of polypeptides encoded by the P3HR-1 strain of Epstein-Barr virus. *Proc Natl Acad Sci USA* 1984;81:4183–4187.

97. Cordier M, Calender A, Billaud M, et al. Stable transfection of Epstein-Barr virus (Epstein-Barr virus) nuclear antigen 2 in lymphoma cells containing the Epstein-Barr virus P3HR1 genome induces expression of B-cell activation molecules CD21 and CD23. *J Virol* 1990;64:1002–1013.
98. Costa RH, Draper KG, Kelly TJ, Wagner EK. An unusual HSV-1 transcript with sequence homology to Epstein-Barr virus DNA. *J Virol* 1985;54:317–328.
99. Counter CM, Botelho FM, Wang P, Harley CB, Bacchetti S. Stabilization of short telomeres and telomerase activity accompany immortalization of Epstein-Barr virus-transformed human B lymphocytes. *J Virol* 1994;68:3410–3414.
100. Countryman J, Jenson H, Seibl R, Wolf H, Miller G. Polymorphic proteins encoded within BZF1 of defective and standard Epstein-Barr viruses disrupt latency. *J. Virol* 1987;12:3672–3679.
101. Countryman JU, Miller G. Activation of expression of latent Epstein-Barr herpes virus after gene transfer with a small clones subfragment of heterogeneous viral DNA. *Proc Natl Acad Sci* 1985;82:4085–4089.
102. Cox MA, Leahy J, Hardwick JM. An enhancer within the divergent promoter of Epstein-Barr virus responds synergistically to the R and Z transactivators. *J Virol* 1990;64:313–321.
103. Cranage MP, Smith GL, Bell SE, et al. Identification and expression of a human cytomegalovirus glycoprotein with homology to the Epstein-Barr virus BXLF2 product, varicella-zoster virus gpIII, and herpes simplex virus type 1 glycoprotein H. *J Virol* 1988;62:1416–1422.
104. Daibata M, Humphreys RE, Sairenji T. Phosphorylation of the Epstein-Barr virus BZLF1 immediate-early gene product ZEBRA. *Virology* 1992;188:916–920.
105. Daibata M, Humphreys RE, Takada K, Sairenji T. Activation of latent Epstein-Barr virus via anti-IgG-triggered, second messenger pathways in the Burkitt's lymphoma cell line Akata. *J Immunol* 1990;144:4788–4793.
106. Daibata M, Sairenji T. Epstein-Barr virus (Epstein-Barr virus) replication and expressions of EA-D (BMRF1 gene product), virus-specific deoxyribonuclease, and DNA polymerase in Epstein-Barr virus-activated Akata cells. *Virology* 1993;196:900–904.
107. Dales S, Chardonet Y. Early events in the interaction of adenoviruses with Hela cells. IV. Association with microtubules and the nuclear pore complex during vectorial movement of the inoculum.*Virology* 1973; 56:465–483.
108. Dalla-Favara R, Martinetti S, Gallo R, Erikson J, Croce C. Translocation and rearrangements of the c-myc oncogene locus in human undifferentiated B Cell lymphomas. *Science* 1983;219:963–967.
109. Dambaugh T, Beisel C, Hummel M, et al. Epstein-Barr virus DNA. VII. Molecular cloning and detailed mapping of Epstein-Barr virus (B95-8) DNA. *Proc Natl Acad Sci USA* 1980;77:2999–3003.
110. Dambaugh T, Heller M, Raab-Traub N, et al. DNAs of Epstein-Barr virus and herpes virus Papio. In: Nahamias A, ed. *The Human Herpes Viruses*. New York: Elsevier; 1980:85–90.
111. Dambaugh T, Hennessy K, Chamnaukit L, Kieff E. U2 region of Epstein-Barr virus DNA may encode Epstein-Barr virus nuclear antigen 2. *Proc Natl Acad Sci USA* 1984:81:7632–7636.
112. Dambaugh T, Hennessy K, Fennewald S, Kieff E. The Epstein-Barr virus genome and its expression in latent infection. In: Epstein MA, Achong BG, eds. *The Epstein-Barr virus: recent advances*. London: William Heinemann; 1986:13–45.
113. Dambaugh T, Kieff E. Two related tandem direct repeat sequences in Epstein-Barr virus DNA. *J Virol* 1982;44:823–833.
114. Dambaugh T, Kieff E, Nkrumah FK, Biggar RJ. Epstein-Barr virus RNA. IV. Viral RNA in Burkitt tumor tissue. *Cell* 1979;16:313–322.
115. Dambaugh T, Raab-Traub N, Heller M, et al. Variations among isolates of Epstein-Barr virus. *Ann N Y Acad Sci* 1980;354:309–325.
116. Dambaugh T, Wang F, Hennessy K, Rickinson A, Kieff E. Expression of the Epstein-Barr virus nuclear protein 2 in rodent cells. *J Virol* 1986; 59:453–462.
117. Datta AK, Colby BM, Shaw JE, Pagano JS. Acyclovir inhibition of Epstein-Barr virus replication. *Proc Natl Acad Sci USA* 1980;77:5163–5166.
118. Datta AK, Feighny RJ, Pagano JS. Induction of Epstein-Barr virus-associated DNA polymerase by 12-0-tetradecanoylphorbol-13-acetate. Purification and characterization. *J Biol Chem* 1980;255:5120–5125.
119. Datta AK, Hood RE. Mechanism of inhibition of Epstein-Barr virus replication by phosphonoformic acid. *Virology* 1981;114:52–59.
120. Davison AJ, Scott JE. The complete DNA sequence of varicella-zoster virus. *J Gen Virol* 1986;67:1759–1816.
121. Davison AJ, Taylor P. Genetic relations between varicella-zoster virus and Epstein-Barr virus. *J Gen Virol* 1987;68:1067–1079.
122. Davison AJ, Wilkie NM. Nucleotide sequences of the joint between L and S segments of herpes simplex virus type 1 and 2. *J Gen Virol* 1981; 55:315–331.
123. Dawson CW, Rickinson AB, Young LS. Epstein-Barr virus latent membrane protein inhibits human epithelial cell differentiation. *Nature* 1990; 344:777–780.
124. Deacon EM, Pallesen G, Niedobitek G, et al. Epstein-Barr virus and Hodgkin's disease: transcriptional analysis of virus latency in the malignant cells. *J Exp Med* 1993;177:339–349.
125. Delarco J, Todaro G. Growth factor from murine sarcoma virus-transformed cells. *Proc Natl Acad Sci USA* 1987;75:4001–4005.
126. Delcayre AX, Lotz M, Lernhardt W. Inhibition of Epstein-Barr virus-mediated capping of CD21/CR2 by alpha interferon (IFN-alpha): immediate antiviral activity of IFN-alpha during the early phase of infection. *J Virol* 1993;67:2918–2921.
127. Delecluse HJ, Bartnizke S, Hammerschmidt W, Bullerdiek J, Bornkamm GW. Episomal and integrated copies of Epstein-Barr virus coexist in Burkitt lymphoma cell lines. *J Virol* 1993;67:1292–1299.
128. Delespesse C, Sarfati M, Peteman R. Influence of recombinant IL-4, IFN alpha, and IFN gamma on the production of human IgE-binding factor (soluble CD23). *J Immunol* 1989;142:134–138.
129. Delius H, Bornkamm GW. Heterogeneity of Epstein-Barr virus. III. Comparison of a transforming and a nontransforming virus by partial denaturation mapping of their DNAs. *J Virol* 1978;27:81–89.
130. DePamphilis ML. Transcriptional elements as components of eucaryotic origin of DNA replication. *Cell* 1988;52:635–638.
131. de Turenne-Tessier M, Ooka T, De The G, Daillie J. Characterization of an Epstein-Barr virus-induced thymidine kinase. *J Gen Virol* 1986; 3:1105–1112.
132. Dhar V, Schildkraut CL. Role of EBNA-1 in arresting replication forks at the Epstein-Barr virus oriP family of tandem repeats. *Mol Cell Biol* 1991;11:6268–6278.
133. di Renzo L, Altiok A, Klein G, Klein E. Endogenous TGF-beta contributes to the induction of the Epstein-Barr virus lytic cycle in two Burkitt lymphoma cell lines. *Int J Cancer* 1994;57:914–919.
134. Diala ES, Koffman RM. Epstein-Barr HR-1 virion DNA is very highly methylated. *J Virol* 1983;45:482–483.
135. Dillner J, Kallin B, Alexander H, et al. An Epstein-Barr virus (Epstein-Barr virus)-determined nuclear antigen (EBNA5) partly encoded by the transformation-associated Bam WYH region of Epstein-Barr virus DNA: preferential expression in lymphoblastoid cell lines. *Proc Natl Acad Sci USA* 1986;83:6641–6645.
136. Dillner J, Kallin B, Klein G, Jornvall H, Alexander H, Lerner R. Antibodies against synthetic peptides react with the second Epstein-Barr virus nuclear antigen. *EMBO J* 1985;4:1813–1818.
137. Dillner J, Sternas L, Kallin B, et al. Antibodies against a synthetic peptide identify the Epstein-Barr virus–determined nuclear antigen. *Proc Natl Acad Sci USA* 1984;81:4652–4656.
138. Dolyniuk M, Pritchett R, Kieff ED. Proteins of Epstein-Barr virus. I. Analysis of the polypeptides of purified enveloped Epstein-Barr virus. *J Virol* 1976;17:935–949.
139. Dolyniuk M, Wolff E, Kieff ED. Proteins of Epstein-Barr virus. II. Electrophoretic analysis of the polypeptides of the nucleocapsid and the glucosamine- and polysaccharide-containing components of enveloped virus. *J Virol* 1976;18:289–297.
140. Dyson PJ, Farrell PJ. Chromatin structure of Epstein-Barr virus. *J Gen Virol* 1985;66:1931–1940.
141. Edson CM, Cohen LK, Henle W, Strominger JL. An unusually high-titer human anti-Epstein Barr virus (Epstein-Barr virus) serum and its use in the study of Epstein-Barr virus-specific proteins synthesized in vitro and in vivo. *J Immunol* 1983;130:919–924.
142. Emini EA, Luka J, Armstrong ME, Keller PM, Ellis RW, Pearson GR. Identification of an Epstein-Barr virus glycoprotein which is antigenically homologous to the varicella-zoster virus glycoprotein II and the herpes simplex virus glycoprotein B. *Virology* 1987;157:552–555.
143. Epstein M, Achong B. *The Epstein-Barr virus. Recent advances*. London: Heinemann; 1986.
144. Epstein M, Achong B. *The Epstein-Barr virus*. London: Springer-Verlag; 1979.
145. Epstein M, Achong B. The Epstein Barr virus. *Ann Rev Microbiol* 1973; 27:413–436.
146. Epstein M, Achong B, Barr Y. Morphological and biological studies

on a virus in cultured lymphoblasts from Burkitt's lymphoma. *J Exp Med* 1965;121:761–770.
147. Epstein M, Achong B, Barr Y. Virus particles in cultured lymphoblasts from Burkitt's lymphoma. *Lancet* 1964;1:702–703.
148. Ernberg I, Falk K, Hansson M. Progenitor and pre-B lymphocytes transformed by Epstein-Barr virus. *Int J Cancer* 1987;39:190–197.
149. Estrov Z, Kurzrock R, Pocsik E, et al. Lymphotoxin is an autocrine growth factor for Epstein-Barr virus–infected B cell lines. *J Exp Med* 1993;177:763–74.
150. Fahraeus R, Jansson A, Ricksten A, Sjoblom A, Rymo L. Epstein-Barr virus-encoded nuclear antigen 2 activates the viral latent membrane protein promoter by modulating the activity of a negative regulatory element. *Proc Natl Acad Sci USA* 1990;87:7390–7394.
151. Fahraeus R, Jansson A, Sjoblom A, Nilsson T, Klein G, Rymo L. Cell phenotype-dependent control of Epstein-Barr virus latent membrane protein 1 gene regulatory sequences. *Virology* 1993;195:71–80.
152. Fahraeus R, Rymo L, Rhim JS, Klein G. Morphological transformation of human keratinocytes expressing the LMP gene of Epstein-Barr virus. *Nature* 1990;345:447–449.
153. Fairbairn LJ, Stewart JP, Hampson IN, Arrand JR, Dexter TM. Expression of Epstein-Barr virus latent membrane protein influences self-renewal and differentiation in a multipotential murine haemopoietic 'stem cell' line. *J Gen Virol* 1993:247–54.
154. Falk K, Ernberg I. An origin of DNA replication (oriP) in highly methylated episomal Epstein-Barr virus DNA localizes to a 4.5-kb unmethylated region. *Virology* 1993;195:608–615.
155. Falk L, Deinhardt F, Nonoyama M, Wolfe C, Bergholz C. Properties of a baboon lymphotropic herpesvirus related to related to Epstein-Barr virus. *Int J Cancer* 1976;18:798–807.
156. Farrell PJ. Epstein-Barr virus. In: O'Brien SJ, ed. *Genetic maps*. New York: Cold Spring Harbor Press; 1992:120–133.
157. Farrell PJ, Bankier A, Seguin C, Deininger P, Barrell BG. Latent and lytic cycle promoters of the Epstein-Barr virus. *EMBO J* 1983;2:1331–1338.
158. Farrell PJ, Rowe DT, Rooney CM, Kouzarides T. Epstein-Barr virus BZLF1 transactivator specfically binds to a consensus AP-1 site and is related to c-fos. *EMBO J* 1989;8:127–133.
159. Fennewald S, van Santen V, Kieff E. The nucleotide sequence of a messenger RNA transcribed in latent growth transforming virus infection indicates that it may encode a membrane protein. *J Virol* 1984; 51:411–419.
160. Fingeroth JD, Weis JJ, Tedder TF, Strominger JL, Biro PA, Fearson DT. Epstein-Barr virus receptor of human B lymphocytes is the C3d receptor CR2. *Proc Natl Acad Sci USA* 1984;81:4510–4516.
161. Finke J, Rowe M, Kallin B, et al. Monoclonal and polyclonal antibodies against Epstein-Barr virus nuclear antigen 5 (EBNA-5) detect multiple protein species in Burkitt's lymphoma and lymphoblastoid cell lines. *J Virol* 1987;61:3870–3878.
162. Fischer DK, Miller G, Gradoville L, et al. Genome of a mononucleosis Epstein-Barr virus contains DNA fragments previously regarded to be unique to Burkitt's lymphoma isolates. *Cell* 1981;24:543–553.
163. Fisher RC, Thorley-Lawson DA. Characterization of the Epstein-Barr virus-inducible gene encoding the human leukocyte adhesion and activation antigen BLAST-1 (CD48). *Mol Cell Biol* 1991;11:1614–1623.
164. Fixman ED Hayward GS Hayward SD. Trans-acting requirements for replication of Epstein-Barr virus ori- Lyt. *J Virol* 1992;66:5030–5039.
165. Flemington E, Speck SH. Autoregulation of Epstein-Barr virus putative lytic switch gene BZLF1. *J Virol* 1990;64:1227–1232.
166. Flemington E, Speck SH. Epstein-Barr virus BZLF1 trans activator induces the promoter of a cellular cognate gene, c-fos. *J Virol* 1990;64: 4549–4552.
167. Flemington E, Speck SH. Evidence for coiled-coil dimer formation by an Epstein-Barr virus transactivator that lacks a heptad repeat of leucine residues. *Proc Natl Acad Sci USA* 1990;87:9459–9463.
168. Flemington E, Speck SH. Identification of phorbol ester response elements in the promoter of Epstein-Barr virus putative lytic switch gene BZLF1. *J Virol* 1990;64:1217–1226.
169. Flemington EK, Borras AM, Lytle JP, Speck SH. Characterization of the Epstein-Barr virus BZLF1 protein transactivation domain. *J Virol* 1992;66:922–929.
170. Flemington EK, Goldfeld A, Speck S. Efficient transcription of the Epstein-Barr virus-immediate early Z and R genes requires protein synthesis. *J Virol* 1991;65:7073–7077.
171. Flemington EK, Lytle JP, Cayrol C, Borras AM, Speck SH. DNA-binding-defective mutants of the Epstein-Barr virus lytic switch activator Zta transactivate with altered specificities. *Mol Cell Biol* 1994;14: 3041–52.
172. Frade R, Barel M, Ehlin-Henriksson B, Klein G. gp140, the C3d receptor of human B lymphocytes, is also the Epstein-Barr virus receptor. *Proc Natl Acad Sci USA* 1985;82:1490–1493.
173. Frappier L, Goldsmith K, Bendell L. Stabilization of the EBNA1 protein on the Epstein-Barr virus latent origin of DNA replication by a DNA looping mechanism. *J Biol Chem* 1994;269:1057–62.
174. Frappier L, O'Donnell M. EBNA1 distorts oriP, the Epstein-Barr virus latent replication origin. *J Virol* 1992;66:1786–1790.
175. Frappier L, O'Donnell M. Epstein-Barr nuclear antigen 1 mediates a DNA loop within the latent replication origin of Epstein-Barr virus. *Proc Natl Acad Sci USA* 1991;88:10875–10879.
176. Frappier L, O'Donnell M. Overproduction, purification, and characterization of EBNA1, the origin binding protein of Epstein-Barr virus. *J Biol Chem* 1991;266:7819–7826.
177. Frech B, Zimber-Strobl U, Suentzenich KO, et al. Identification of Epstein-Barr virus terminal protein 1 (TP1) in extracts of four lymphoid cell lines, expression in insect cells, and detection of antibodies in human sera. *J Virol* 1990;64:2759–2767.
178. Freemer CS, Bertoni G, Takagi S, Sairenji T. A novel early antigen associated with Epstein-Barr virus productive cycle. *Virology* 1993;194: 387–92.
179. Freese UK, Laux G, Hudenwentz J, Schwarz E, Bornkamm G. Two distant clusters of partially homologous small repeats of Epstein-Barr virus are transcribed upon induction of an abortive or lytic cycle the virus. *J Virol* 1983;48:731–743.
180. Fresen KO, Cho MS, Gissman L, zur Hausen H. NC 37-R1 EB-virus: a possible recombinant between intracellular NC 37 viral DNA and superinfecting P3HR-1 Epstein-Barr virus. *Intervirology* 1979;12:303–310.
181. Furnari FB, Adams MD, Pagano JS. Unconventional processing of the 3′ termini of the Epstein-Barr virus DNA polymerase mRNA. *Proc Natl Acad Sci USA* 1993;90:378–382.
182. Furnari FB, Adams MD, Pagano JS. Regulation of the Epstein-Barr virus DNA polymerase gene. *J Virol* 1992;66:2837–2845.
183. Furnari FB, Zacny V, Quinlivan EB, Kenney S, Pagano JS. RAZ, an Epstein-Barr virus transdominant repressor that modulates the viral reactivation mechanism. *J Virol* 1994;68:1827–1836.
184. Furukawa T, Maruyama N, Kawaichi M, Honjo T. The drosophila homologue of the recombination signal-binding protein regulates peripheral nervous system development. *Cell* 1992;69:1191–1197.
185. Gahn TA, Schildkraut CL. The Epstein-Barr virus origin of plasmid replication, oriP, contains both the initiation and termination sites of DNA replication. *Cell* 1989;58:527–535.
186. Gahn TA, Sugden B. Marked, transient inhibition of expression of the Epstein-Barr virus latent membrane protein gene in Burkitt's lymphoma cell lines by electroporation. *J Virol* 1993;67:6379–86.
187. Garner JG, Hirsch MS, Schooley RT. Prevention of Epstein-Barr virus-induced B-cell outgrowth by interferon alpha. *Infect Immun* 1984;43: 920–924.
188. Gauchat JF, Gascan H, de Waal Malefyt R, de Vries JE. Regulation of germ-line epsilon transcription and induction of epsilon switching in cloned Epstein-Barr virus-transformed and malignant human B cell lines by cytokines and CD4+ T cells. *J Immunol* 1992;148:2291–2299.
189. Gerber P, Birch SM. Complement-fixing antibodies in sera of human and non-human primates to viral antigens derived from Burkitt's lymphoma cells. *Prot Natl Acad Sci USA* 1967;58:478–484.
190. Gerber P, Kalter SS, Schidlovsky G, Peterson WD Jr, Daniel MD. Biologic and antigenic characteristics of Epstein-Barr virus–related herpes-viruses of chimpanzees and baboons. *Int J Cancer* 1977;20:448–459.
191. Gerber P, Nkrumah F, Pritchett R, Kieff ED. Comparative studies of Epstein-Barr virus strains from Ghana and the United States. *Int J Cancer* 1976;17:71–81.
192. Gerber P, Pritchett R, Kieff ED. Antigens and DNA of a chimpanzee agent related to Epstein-Barr virus. *J Virol* 1976;19:1090–1100.
193. Gergely L, Klein G, Ernberg I. Host cell macromolecular synthesis in cells containing Epstein-Barr virus induced early antigens, studied by combined immunofluorescence and radioautography. *Virology* 1971; 45:22–29.
194. Ghosh D, Kieff E. cis-Acting regulatory elements near the Epstein-Barr virus latent infection membrane protein transcriptional start site. *J Virol* 1990;64:1855–1858.
195. Gibbons DL, Rowe M, Cope AP, Feldmann M, Brennan FM. Lym-

photoxin acts as an autocrine growth factor for Epstein-Barr virus-transformed B cells and differentiated Burkitt lymphoma cell lines. *J Immunol* 1994;24:1879–1885.
196. Gibson T, Stockwell P, Ginsburg M, Barrell B. Homology between two Epstein-Barr virus early genes and HSV ribonucleotide reductase and 38k genes. *Nucleic Acids Res* 1984;12:5087–5099.
197. Gibson TJ, Barrell BG, Farrell PJ. Coding content and expression of the Epstein-Barr virus B95-8 genome in the region from base 62,248 to base 82,920. *Virology* 1986;152:136–148.
198. Gilligan K, Sato H, Rajadurai P, et al. Novel transcription from the Epstein-Barr virus terminal EcoRI fragment, DIJhet, in a nasopharyngeal carcinoma. *J Virol* 1990;64:4948–4956.
199. Gilligan KJ, Rajadurai P, Lin JC, et al. Expression of the Epstein-Barr virus BamHI A fragment in nasopharyngeal carcinoma: evidence for a viral protein expressed in vivo. *J Virol* 1991;65:6252–6259.
200. Giot JF, Mikaelian I, Buisson M, Manet E, Joab I, Nicolas JC, Sergeant A. Transcriptional interference between the Epstein-Barr virus transcription factors EB1 and R: both DNA-binding and activation domains of EB1 are required. *Nucleic Acids Res* 1991;19:1251–1258.
201. Given D, Kieff E. DNA of Epstein-Barr virus. VI. Mapping of the internal tandem reiteration. *J Virol* 1979;31:315–324.
202. Given D, Kieff E. DNA of Epstein-Barr virus. IV. Linkage map for restriction enzyme fragments of the B95-8 and W91 strains of Epstein-Barr virus. *J Virol* 1978;21:524–542.
203. Given D, Yee D, Griem K, Kieff E. DNA of Epstein-Barr virus: V. Direct repeats of the ends of Epstein-Barr virus DNA. *J Virol* 1979;30: 852–862.
204. Glickman J, Howe G, Steitz J. Structural analyses of EBER 1 and EBER 2 ribonucleoprotein particles present in Epstein-Barr virus infected cells. *J Virol* 1988;62:902–911.
205. Goldschmidts W, Luka J, Pearson GR. A restricted component of the Epstein-Barr virus early antigen complex is structurally related to ribonucleotide reductase. *Virology* 1987;157:220–226.
206. Goldsmith K, Bendell L, Frappier L. Identification of EBNA1 amino acid sequences required for the interaction of the functional elements of the Epstein-Barr virus latent origin of DNA replication. *J Virol* 1993; 67:3418–3426.
207. Gompels U, Minson A. The properties and sequence of glycoprotein H of herpes simplex virus type 1. *Virology* 1986;153:230–247.
208. Gong M, Kieff E. Intracellular trafficking of two major Epstein-Barr virus glycoproteins, gp350/220 and gp110. *J Virol* 1990;64:1507–1516.
209. Gong M, Ooka T, Matsuo T, Kieff E. The Epstein-Barr virus glycoprotein gene homologous to HSVgB. *J Virol* 1987;61:499–508.
210. Gordon J, Ley SC, Melamed MD, Aman P, Hughes-Jones NC. Soluble factor requirements for the autostimulatory growth of B lymphoblasts immortalized by Epstein-Barr virus. *J Exp Med* 1984;159:1554–1559.
211. Gradoville L, Grogan E, Taylor N, Miller G. Differences in the extent of activation of Epstein-Barr virus replicative gene expression among four nonproducer cell lines stably transformed by oriP/BZLF1 plasmids. *Virology* 1990;178:345–354.
212. Graessmann A, Wolf H, Bornkamm GW. Expression of Epstein-Barr virus genes in different cell types after microinjection of viral DNA. *Proc Natl Acad Sci USA* 1980;77:433–436.
213. Grasser FA, Haiss P, Gottel S, Mueller-Lantzsch N. Biochemical characterization of Epstein-Barr virus nuclear antigen 2A. *J Virol* 1991;65: 3779–3788.
214. Gratama JW, Lennette ET, Lonnqvist B, et al. Detection of multiple Eptein-Barr viral strains in allogeneic bone marrow transplant recipients. *J Med Virol* 1992;37:39–47.
215. Gregory CD, Dive C, Henderson S, et al. Activation of Epstein-Barr virus latent genes protects human B cells from death by apoptosis. *Nature* 1991;349:612–614.
216. Gregory CD, Rowe DM, Rickinson AB. Different Epstein-Barr virus (Epstein-Barr virus) B cell interactions in phenotypically distinct clones of a Burkitt lymphoma cell line. *J Gen Virol* 1990;71:1481–1495.
217. Grogan E, Jenson H, Countryman J, Heston L, Gradoville L, Miller G. Transfection of a rearranged viral DNA fragment, WZhet, stably converts latent Epstein-Barr viral infection to productive infection in lymphoid cells. *Proc Natl Acad Sci USA* 1987;84:1332–1336.
218. Grogan EA, Miller G, Henle W, Rabson M, Shedd D, Niederman JC. Expression of Epstein-Barr viral early antigen in monolayer tissue cultures after transfection with viral DNA and DNA fragments. *J Virol* 1981;40:861–869.
219. Grogan EA, Summers WP, Dowling S, Shedd D, Gradoville L, Miller G. Two Epstein-Barr viral nuclear neoantigens distinguished by gene transfer, serology, and chromosome binding. *Proc Natl Acad Sci USA* 1983;80:7650–7653.
220. Grossberger D, Clough W. Characterization of purified Epstein-Barr virus induced deoxyribonucleic acid polymerase: nucleotide turnover, processiveness, and phosphonoacetic acid sensitivity. *Biochemistry* 1981;20:4049–4055.
221. Grossman S, Johannsen E, Tong X, Yalamanchili R, Kieff E. The Epstein-Barr virus nuclear protein 2 transactivator is directed to response elements by the Jk recombination signal binding protein. *PNAS* 1994; 91:7568–7572.
222. Gruffat H, Duran N, Buisson M, Wild F, Buckland R, Sergeant A. Characterization of an R-binding site mediating the R-induced activation of the Epstein-Barr virus BMLF1 promoter. *J Virol* 1992;66:46–52.
223. Gruffat H, Manet E, Rigolet A, Sergeant A. The enhancer factor R of Epstein-Barr virus (Epstein-Barr virus) is a sequence-specific DNA binding protein. *Nucleic Acids Res* 1990;18:6835–6843.
224. Gussander E, Adams A. Electron microscopic evidence for replication of circular Epstein-Barr virus genomes in latently infected Raji cells. *J Virol* 1984;52:549–556.
225. Gutkind JS, Link DC, Katamine S, et al. A novel c-fgr exon utilized in Epstein-Barr virus-infected B lymphocytes but not in normal monocytes. *Mol Cell Biol* 1991;11:1500–1507.
226. Gruffat H, Sergeant A. Characterization of the DNA-binding site repertoire for the Epstein-Barr virus transcription factor R. *Nucleic Acids Res* 1994;22:1172–1178.
227. Gutsch DE, Holley GE, Zhang Q, et al. The bZIP transactivator of Epstein-Barr virus, BZLF1, functionally and physically interacts with the p65 subunit of NF-kappa B. *Mol Cell Biol* 1994;14:1939–1948.
228. Hammerschmidt W, Sugden B. Genetic analysis of immortalizing functions of Epstein-Barr virus in human B lymphocytes. *Nature* 1989;340: 393–397.
229. Hammerschmidt W, Sugden B. Identification and characterization of orilyt, a lytic origin of DNA replication of Epstein-Barr virus. *Cell* 1988;55:427–433.
230. Hammerskjold M, Simurda M. Epstein-Barr virus latent membrane protein transactivates the human imminodeficiency virus type 1 long terminal repeat through induction of NF-kB activity. *J Virol* 1992; 66:6496–6501
231. Hampar B, Tanaka A, Nonoyama M, Derge JG. Replication of the resident repressed Epstein-Barr virus genome during the early S phase (S-1 period) of nonproducer Raji cells. *Proc Natl Acad Sci USA* 1974;71: 631–635.
232. Hardwick JM, Lieberman PM, Hayward SD. A new Epstein-Barr virus transactivator, R, induces expression of a cytoplasmic early antigen. *J Virol* 1988;62:2274–2284.
233. Harris A, Young BD, Griffin BE. Random association of Epstein-Barr virus genomes with host cell metaphase chromosomes in Burkitt's lymphoma-derived cell lines. *J Virol* 1985;56:328–332.
234. Harrison S, Fisenne K, Hearing J. Sequence requirements of the Epstein-Barr virus latent origin of DNA replication. *J Virol* 1994;68:1913–25.
235. Hatfull G, Bankier AT, Barrell BG, Farrell PJ. Sequence analysis of Raji Epstein-Barr virus DNA. *Virology* 1988;164:334–340.
236. Hatzubai A, Anafi M, Masucci MG, et al. Down-regulation of the Epstein-Barr virus-encoded membrane protein (LMP) in Burkitt lymphomas. *Int J Cancer* 1987;40:358–364.
237. Hayward D, Pritchett R, Orellana T, King W, Kieff E. The DNA of Epstein-Barr virus fragments produced by restriction enzymes: homologous DNA and RNA in lymphoblastoid cells. ICN-UCLA Symposia on Molecular and Cellular Biology, In: Baltimore D, Huang A, Fox CF, eds.*Animal viruses*. Vol. 4. New York: Academic; 1976:619–640.
238. Hayward SD, Kieff E. DNA of Epstein-Barr virus. II. Comparison of the molecular weights of restriction endonuclease fragments of the DNA of Epstein-Barr virus strains and identification of end fragments of the B95-8 strain. *J Virol* 1977;23:421–429.
239. Hayward SD, Kieff ED. Epstein-Barr virus–specific RNA. I. Analysis of viral RNA in cellular extracts and in the polyribosomal fraction of permissive and nonpermissive lymphoblastoid cell lines. *J Virol* 1976; 18:518–525.
240. Hayward SD, Lazarowitz S, Hayward GS. Organization of the Epstein-Barr virus DNA molecule. II. Fine mapping of the boundaries of the internal repeat cluster of B95-8 and identification of additional small tandem repeats adjacent to the HR-1 deletion. *J Virol* 1982;43:201–212.
241. Hayward SD, Nogee DL, Hayward GS. Organization of repeated re-

gions within the Epstein-Barr virus DNA molecule. *J Virol* 1980;33: 507–521.
242. Hearing J, Mulhaupt Y, Harper S. Interaction of Epstein-Barr virus nuclear antigen 1 with the viral latent origin of replication. *J Virol* 1992;66:694–705.
243. Hearing JC, Levine AJ. The Epstein-Barr virus nuclear antigen (BamHI K antigen) is a single-stranded DNA binding phosphoprotein. *Virology* 1985;145:105–116.
244. Hearing JC, Nicolas JC, Levine AJ. Identification of Epstein-Barr virus sequences that encode a nuclear antigen expressed in latently infected lymphocytes. *Proc Natl Acad Sci USA* 1984;81:4183–4187.
245. Heineman T. Glycoproteins of Epstein-Barr virus [Ph.D. Dissertation]. Chicago: The University of Chicago, 1988.
246. Heineman T, Gong M, Sample J, Kieff E. Identification of the Epstein-Barr virus gp85 gene. *J Virol* 1988;62:1101–1107.
247. Heller M, Dambaugh T, Kieff E. Epstein-Barr virus DNA: IX. Variation among viral DNAs from producer and nonproducer infected cells. *J Virol* 1981;38:632–648.
248. Heller M, Flemington E, Kieff E, Deininger P. Repeat arrays in cell DNA related to Epstein-Barr virus IR3 repeat. *Mol Cell Biol* 1985;5: 457–465.
249. Heller M, Gerber P, Kieff E. The DNA of herpes virus pan, a third member of the Epstein-Barr virus-herpes virus Papio group. *J Virol* 1982;41:931–939.
250. Heller M, Gerber P, Kieff E. Herpes virus Papio DNA is similar in organization to Epstein-Barr virus DNA. *J Virol* 1981;37:698–709.
251. Heller M, Henderson A, Kieff E. The IR3 repeat sequence in Epstein-Barr virus DNA has homology to a repeat sequence in human and mouse cell DNA: sites of homology on human chromosomes. *Proc Natl Acad Sci USA* 1982;79:5916–5920.
252. Heller M, Henderson A, Ripley S, van Santen V, Kieff E. The IR3 repeat in Epstein-Barr virus DNA has homology to cell DNA, encodes part of a messenger RNA in Epstein-Barr virus transformed cells but does not mediate integration of Epstein-Barr virus DNA. In: Prasad U et al., eds. *Nasopharyngeal carcinoma: current concepts*, Vol. 6. Kuala Lumpur: University of Malaya; 1983:177–202.
253. Heller M, Kieff E. Colinearity between the DNAs of Epstein-Barr virus and herpes virus Papio. *J Virol* 1981;37:821–826.
254. Heller M, van Santen V, Kieff E. A simple repeat sequence in Epstein-Barr virus DNA is transcribed in latent and productive infection. *J Virol* 1982;44:311–320.
255. Henderson A, Ripley S, Heller M, Kieff E. Human chromosome association of Epstein-Barr virus DNA in a Burkitt tumor cell line and in lymphocytes growth transformed in vitro. *Proc Natl Acad Sci USA* 1983;80:1987–1991.
256. Henderson E, Miller G, Robinson J, Heston L. Efficiency of transformation of lymphocytes by Epstein-Barr virus. *Virology* 1977;76:152–163.
257. Henderson S, Huen D, Rowe M, Dawson C, Johnson G, Rickinson A. Epstein-Barr virus-coded BHRF1 protein, a viral homologue of Bcl-2, protects human B cells from programmed cell death. *Proc Natl Acad Sci USA* 1993;90:8479–8483.
258. Henderson S, Rowe M, Gregory C, et al. Induction of *bcl-2* expression by Epstein-Barr virus latent membrane protein-1 protects infected B cells from programmed cell death. *Cell* 1991;65:1071–1115.
259. Henkel T, Ling PD, Hayward SD, Peterson MG. Mediation of Epstein-Barr virus EBNA2 transactivation by recombination signal-binding protein J kappa. *Science* 1994;265:92–95.
260. Hennessy K, Fennewald S, Hummel M, Cole T, Kieff E. A membrane protein encoded by Epstein-Barr virus in latent growth-transforming infection. *Proc Natl Acad Sci USA* 1984;81:7201–7211.
261. Hennessy K, Fennewald S, Kieff E. A third virus nuclear protein in lymphoblasts immortalized by Epstein-Barr virus. *Proc Natl Acad Sci USA* 1985;81:5944–5948.
262. Hennessy K, Heller M, van Santen V, Kieff E. Simple repeat array in Epstein-Barr virus DNA encodes part of the Epstein-Barr nuclear antigen. *Science* 1983;220:1396–1398.
263. Hennessy K, Kieff E. A second nuclear protein is encoded by Epstein-Barr virus in latent infection. *Science* 1985;227:1238–1240.
264. Hennessy K, Kieff E. One of two Epstein-Barr virus nuclear antigens contains a glycine-alanine copolymer domain. *Proc Natl Acad Sci USA* 1983;80:5665–5669.
265. Hennessy K, Wang F, Woodland Bushman E, Kieff E. Definitive identification of a member of the Epstein-Barr virus nuclear protein 3 family. *Proc Natl Acad Sci USA* 1986;83:5693–5697.
266. Heston L, Rabson M, Brown N, Miller G. New Epstein-Barr virus variants from cellular subclones of P3J-HR-1 Burkitt lymphoma. *Nature* 1982;295:160–163.
267. Hinuma Y, Konn M, Yamaguchi J, Wudarski D, Blakeslee J, Grace J. Immunofluorescence and herpes type virus particles in the P3HR-1 Burkitt lymphoma gene. *J Virol* 1967;1:1045–1051.
268. Hitt MM, Allday M, Hara T, et al. Epstein-Barr virus gene expression in an NPC-related tumour. *EMBO J* 1989;8:2639–2651.
269. Hockenbery D, Nunex G, Milliman C, Schreiber RD, Korsmeyer SJ. Bcl2 is an inner mitochondrial membrane protein that blocks programmed cell death. *Nature (London)* 1990;348:334–336.
270. Hoffman GJ, Lazarowitz SG, Hayward SD. Monoclonal antibody against a 250,000 dalton glycoprotein of Epstein-Barr virus identifies a membrane antigen and a neutralizing antigen. *Proc Natl Acad Sci USA* 1980;77:2979–2983.
271. Holley-Guthrie EA, Quinlivan EB, Mar EC, Kenney S. The Epstein-Barr virus (Epstein-Barr virus) BMRF1 promoter for early antigen (EA-D) is regulated by the Epstein-Barr virus transactivators, BRLF1 and BZLF1, in a cell-specific manner. *J Virol* 1990;64:3753–3759.
272. Horvath J, Faxing C, Weber JM. Complementation of adenovirus early region 1a and 2a mutants by Epstein-Barr virus immortalized lymphoblastoid cell lines. *Virology* 1991;184:141–148.
273. Howe JG, Shu MD. Upstream basal promoter element important for exclusive RNA polymerase III transcription of the EBER 2 gene. *Mol Cell Biol* 1993;13:2655–2665.
274. Howe JG, Shu MD. Epstein-Barr virus small RNA (EBER) genes: unique transcription units that combine RNA polymerase II and III promoter elements. *Cell* 1989;57:825–834.
275. Howe JG, Steitz JA. Localization of Epstein-Barr virus encoded small RNAs by in situ hybridization. *Proc Natl Acad Sci USA* 1986;83: 9006–9010.
276. Hsieh DJ, Camiolo SM, Yates JL. Constitutive binding of EBNA1 protein to the Epstein-Barr virus replication origin, oriP, with distortion of DNA structure during latent infection. *EMBO J* 1993;12:4933–4944.
277. Hsu DH, de Waal-Malefyt R, Fiorentino DF, et al. Expression of interleukin-10 activity by Epstein-Barr virus protein BCRF1. *Science* 1990;250:830–832.
278. Hu LF, Chen F, Zheng X, et al. Clonability and tumorigenicity of human epithelial cells expressing the Epstein-Barr virus encoded membrane protein LMP1. *Oncogene* 1993;8:1575–1583.
279. Hu LF, Zabarovsky ER, Chen F, Cao SL, Ernberg I, Klein G, Winberg G. Isolation and sequencing of the Epstein-Barr virus BNLF-1 gene (LMP1) from a Chinese nasopharyngeal carcinoma. *J Gen Virol* 1991; 72:2399–2409.
280. Hudewentz J, Bornkamm GW, zur Hausen H. Effect of the diterpene ester TPA on Epstein-Barr virus antigen and DNA synthesis in producer and nonproducer cell lines. *Virology* 1980;100:175–178.
281. Hudewentz J, Delius H, Freese U, Zimbar U, Bornkamm G. Two distant regions of the Epstein-Barr virus genome with sequence homologies have the same orientation and involve small tandem repeats. *EMBO J* 1982;1:21–26.
282. Hudson GS, Bankier AT, Satchwell SC, Barrell BG. The short unique region of the B95-8 Epstein-Barr virus genome. *Virology* 1985;147: 81–98.
283. Hudson GS, Farrell PJ, Barrell BG. Two related but differentially expressed potential membrane proteins encoded by the EcoRI Dhet region of Epstein-Barr virus B95-8. *J Virol* 1985;53:528–535.
284. Hudson GS, Gibson TJ, Barrell BG. The BamHI F region of the B95-8 Epstein-Barr virus genome. *Virology* 1985;147:99–109.
285. Huen DS, Fox A, Kumar P, Searle PF. Dilated heart failure in transgenic mice expressing the Epstein-Barr virus nuclear antigen-leader protein. *J Gen Virol* 1993;74:1381–1391.
286. Hummel M, Arsenakis M, Marchini A, Lee L, Roizman B, Kieff E. Herpes simplex virus expressing Epstein-Barr virus nuclear antigen. *Virology* 1986;148:337–348.
287. Hummel M, Kieff E. Mapping of polypeptides encoded by the Epstein-Barr virus genome in productive infection. *Proc Natl Acad Sci USA* 1982;79:5698–5702.
288. Hummel M, Kieff E. Epstein-Barr virus RNA. VIII. Viral RNA in permissively infected B95-8 cells. *J Virol* 1982;43:262–272.
289. Hummel M, Thorley-Lawson D, Kieff E. An Epstein-Barr virus DNA fragment encodes messages for the major envelope glycoproteins (gp350/300 and gp220/200). *J Virol* 1984;49:413–417.

290. Hurley E, Thorley-Lawson D. B cell activation and the establishment of Epstein-Barr virus latency. *J Exp Med* 1989;168:2059–2075.
291. Hurley EA, Agger S, McNeil JA, et al. When Epstein-Barr virus persistently infects B-cell lines, it frequently integrates. *J Virol* 1991;65: 1245–1254.
292. Hurley EA, Klaman LD, Agger S, Lawrence JB, Thorley-Lawson DA. The prototypical Epstein-Barr virus-transformed lymphoblastoid cell line IB4 is an unusual variant containing integrated but no episomal viral DNA. *J Virol* 1991;65:3958–3963.
293. Hutt-Fletcher LM, Fowler E, Lambris JD, Feingey RJ, Simmons JG, Ross GD. Studies of the Epstein-Barr virus receptor found on Raji cells. II. A comparison of lymphocyte binding sites for Epstein-Barr and C3d. *J Immunol* 1983;130:1309–1312.
294. Inoue N, Harada S, Honma T, Kitamura T, Yanagi K. The domain of Epstein-Barr virus nuclear antigen 1 essential for binding to oriP region has a sequence fitted for the hypothetical basic-helix-loop-helix structure. *Virology* 1991;182:84–93.
295. Izumi KM, Kaye KM, Kieff ED. Epstein-Barr virus recombinant molecular genetic analysis of the LMP1 amino-terminal cytoplasmic domain reveals a probable structural role, with no component essential for primary B-lymphocyte growth transformation. *J Virol* 1994;68: 4369–4376.
296. Jabara H, Schneider L, Shapira S, et al. Induction of germ-line and mature C_3 transcripts in human B cell stimulated with rlL-4 and Epstein-Barr virus. *J Immunol* 1990;145:3468–3473.
297. Jankelevich S, Kolman JL, Bodnar JW, Miller G. A nuclear matrix attachment region organizes the Epstein-Barr viral plasmid in Raji cells into a single DNA domain. *EMBO J* 1992;11:1165–1176.
298. Jansson A, Masucci M, Rymo L. Methylation of discrete sites within the enhancer region regulates the activity of the Epstein-Barr virus BamHI W promoter in Burkitt lymphoma lines. *J Virol* 1992;66:62–69.
299. Jat P, Arrand JR. In vitro transcription of two Epstein-Barr virus specified small RNA molecules. *Nucleic Acids Res* 1982;10:3407–3425.
300. Jeang KT, Hayward SD. Organization of the Epstein-Barr virus DNA molecule: III. Location of the P3HR-I deletion junction and characterization of the Not I repeat units that form part of the template for an abundant 12-0-tetradecanoylphorbol-13-acetate-induced m RNA transcript. *J Virol* 1983;48:135–148.
301. Jehn U, Lindahl T, Klein G. Fate of virus DNA in the abortive infection of human lymphoid cell lines by Epstein-Barr virus. *J Gen Virol* 1972;16:409–412.
302. Jenson HB, Farrell PJ, Miller G. Sequences of the Epstein-Barr virus (Epstein-Barr virus) large internal repeat form the center of a 16-kilobase-pair palindrome of Epstein-Barr virus (P3HR-1) heterogeneous DNA. *J Virol* 1987;61:1495–1506.
303. Jenson HB, Miller G. Polymorphisms of the region of the Epstein-Barr virus genome which disrupts latency. *Virology* 1988;185:549–564.
304. Jenson HB, Rabson MS, Miller G. Palindromic structure and polypeptide expression of 36 kilobase pairs of heterogeneous Epstein-Barr virus (P3HR-1) DNA. *J Virol* 1986;58:475–486.
305. Jiang WQ, Szekely L, Wendel-Hansen V, Ringertz N, Klein G, Rosen A. Co-localization of the retinoblastoma protein and the Epstein-Barr virus-encoded nuclear antigen EBNA-5. *Exp Cell Res* 1991;197: 314–318.
306. Jin XW, Speck SH. Identification of critical cis elements involved in mediating Epstein-Barr virus nuclear antigen 2-dependent activity of an enhancer located upstream of the viral BamHI C promoter. *J Virol* 1992;66:2846–2852.
307. Joab I, Rowe DT, Bodescot M, Nicolas JC, Farrell PJ, Perricaudet M. Mapping of the gene coding for Epstein-Barr virus–determined nuclear antigen EBNA 3 and its transient overexpression in a human cell line by using an adenovirus expression vector. *J Virol* 1987;61:3340–3344.
308. Johannsen E, Koh E, Mosialos G, Tong X, Kieff E, Grossman E. Epstein-Barr virus nuclear protein 2 transactivation of the latent membrane protein 1 promoter is mediated by Jk and PU.1. *J Virol* 1994;69: 253–262.
309. Jondal M, Klein G, Oldstone M, Bokish V, Yefenof E. Surface markers on human B and T lymphocytes VIII: association between complement and Epstein-Barr virus receptors on human lymphoid cells. *Scand J Immunol* 1976;5:401–410.
310. Jones CH, Hayward SD, Rawlins DR. Interaction of the lymphocyte-derived Epstein-Barr virus nuclear antigen EBNA-1 with its DNA-binding sites. *J Virol* 1989;63:101–110.
311. Jones M, Foster L, Sheedy T, Griffin BE. The EB virus genome in Daudi Burkitts lymphoma cells has a deletion similar to that observed in a non-transforming strain (P3HR1) of the virus. *EMBO J* 1984;3: 813–821.
312. Kallin B, Dillner J, Ernberg I, et al. Four virally determined nuclear antigens are expressed in Epstein-Barr virus–transformed cells. *Proc Natl Acad Sci USA* 1986;83:1499–1503.
313. Kallin B, Luka J, Klein G. Immunochemical characterization of Epstein-Barr virus–associated early and late antigens in n-butyrate-treated P3HR-1 cells. *J Virol* 1979;32:710–716.
314. Kallin B, Sternas L, Saemundsen AK, et al. Purification of Epstein-Barr virus DNA polymerase from P3HR-1 cells. *J Virol* 1985;54: 561–568.
315. Karlin S, Blaisdell BE, Schachtel GA. Contrasts in codon usage of latent versus productive genes of Epstein-Barr virus: data and hypotheses. *J Virol* 1990;64:4264–4273.
316. Kaschka-Dierich C, Adams A, Lindahl T, et al. Intracellular forms of Epstein-Barr virus DNA in human tumor cells in vivo. *Nature* 1976; 260:302–306.
317. Katz DA, Baumann RP, Sun R, Kolman JL, Taylor N, Miller G. Viral proteins associated with the Epstein-Barr virus transactivator, ZEBRA. *Proc Natl Acad Sci USA* 1992;89:378–382.
318. Kawai Y, Nonoyama M, Pagano J. Reassociation kinetics for Epstein-Barr virus DNA. Non-homology to mammalian DNA and homology to viral DNA in various diseases. *J Virol* 1973;12:1006–1012.
319. Kawanishi M. Epstein-Barr virus induces fragmentation of chromosomal DNA during lytic infection. *J Virol* 1993;67:7654–7658.
320. Kawanishi M. Topoisomerase I and II activities are required for Epstein-Barr virus replication. *J Gen Virol* 1993:2263–2268.
321. Kaye K, Izumi K, Mosialos G, Kieff E. The Epstein-Barr virus LMP1 cytoplasmic carboxy terminus is essential for B lymphocyte transformation; fibroblast co-cultivation complements a critical function within the terminal 155 residues. *J Virol* 1995;69:675–683.
322. Kaye KM, Izumi KM, Kieff E. Epstein-Barr virus latent membrane protein 1 is essential for B-lymphocyte growth transformation. *Proc Natl Acad Sci USA* 1993;90:9150–9154.
323. Kenney S, Holley-Guthrie E, Mar E, Smith M. The Epstein-Barr virus BMLF1 promoter contains an enhancer element responsive to the BZLF1 and BRLF1 trans activator. *J Virol* 1989;63:3878–3883.
324. Kenney S, Kamine J, Holley-Guthrie E, Lin J, Mar E, Pagano J. The Epstein-Barr virus (Epstein-Barr virus) BZLF-1 immediate-early gene product differentially affects latent versus productive Epstein-Barr virus promoters. *J Virol* 1989;63:1729–1736.
325. Kenney S, Kamine J, Holley-Guthrie E, et al. The Epstein-Barr virus immediate-early gene product, BMLF1 acts in trans by a posttranscriptional mechanism which is reporter gene dependent. *J Virol* 1989; 63:3870–3877.
326. Kenney SC, Holley-Guthrie E, Quinlivan EB, et al. The cellular oncogene c-myb can interact synergistically with the Epstein-Barr virus BZLF1 transactivator in lymphoid cells. *Mol Cell Biol* 1992;12:136–146.
327. Kerdiles B, Walls D, Triki H, Perricaudet M, Joab I. cDNA cloning and transient expression of the Epstein-Barr virus-determined nuclear antigen EBNA3B in human cells and identification of novel transcripts from its coding region. *J Virol* 1990;64:1812–1816.
328. Kerr BM, Lear AL, Rowe M, et al. Three transcriptionally distinct forms of Epstein-Barr virus latency in somatic cell hybrids: cell phenotype dependence of virus promoter usage. *Virology* 1992;187:189–201.
329. Kieff E, Wang F, Birkenbach M, et al. Molecular biology of lymphocyte transformation by Epstein-Barr virus. In: Brugge J, Curran T, Harlow E, McCormack F, eds. *Origins of human cancer: a comprehensive review.* Cold Spring Harbor, NY:Cold Spring Harbor Press; 1991: 563–576.
330. Kiehl A, Dorsky DI. Cooperation of Epstein-Barr virus DNA polymerase and EA-D(BMRF1) in vitro and colocalization in nuclei of infected cells. *Virology* 1991;184:330–340.
331. Kim OJ, Yates JL. Mutants of Epstein-Barr virus with a selective marker disrupting the TP gene transform B cells and replicate normally in culture. *J Virol* 1993;67:7634–7640.
332. Kimball AS, Milman G, Tuulius TD. High-resolution footprints of the DNA-binding domain of Epstein-Barr virus nuclear antigen 1. *Mol Cell Biol* 1989;9:2738–2742.
333. King W, Dambaugh T, Heller M, Dowling J, Kieff E. Epstein-Barr virus DNA. XII. A variable region of the Epstein-Barr virus genome is included in the P3HR-1 deletion. *J Virol* 1982;43:979–986.
334. King W, Thomas-Powell AL, Raab-Traub N, Hawke M, Kieff E. Ep-

stein-Barr virus RNA. V. Viral RNA in a restringently infected, growth-transformed cell line. *J Virol* 1980;36:506–518.

335. King W, van Santen V, Kieff E. Epstein-Barr virus RNA. VI. Viral RNA in restringently and abortively infected Raji cells. *J Virol* 1981; 38:649–660.
336. Kintner C, Sugden B. The structure of the termini of the DNA of Epstein-Barr virus. *Cell* 1979;17:661–671.
337. Kintner C, Sugden B. Conservation and progressive methylation of Epstein-Barr virus DNA sequences in transformed cells. *J Virol* 1981;38: 305–316.
338. Kintner C, Sugden B. Identification of antigenic determinants unique to the surfaces of cells transformed by Epstein-Barr virus. *Nature (London)* 1981;294:458–460.
339. Klein G. *Advances in viral onocolgy*. New York: Raven; 1987.
340. Knutson JC. The level of c-fgr RNA is increased by EBNA-2, an Epstein-Barr virus gene required for B-cell immortalization. *J Virol* 1990; 64:2530–2536.
341. Koizumi S, Zhang XK, Imai S, Sugiura M, Usui N, Osato T. Infection of the HTLV-I-harbouring T-lymphoblastoid line MT-2 by Epstein-Barr virus. *Virology* 1992;188:859–863.
342. Kolman JL, Taylor N, Marshak DR, Miller G. Serine-173 of the Epstein-Barr virus ZEBRA protein is required for DNA binding and is a target for casein kinase II phosphorylation. *Proc Natl Acad Sci USA* 1993;90:10115–9.
343. Kouzarides T, Bankier AT, Satchwell SC, Weston K, Tomlinson P, Barrell BG. Large-scale rearrangement of homologous regions in the genomes of HCMV and Epstein-Barr virus. *Virology* 1987;157:397–413.
344. Kouzarides T, Packham G, Cook A, Farrell PJ. The BZLF1 protein of Epstein-Barr virus has a coiled coil dimerisation domain without a heptad leucine repeat but with homology to the C/EBP leucine zipper. *Oncogene* 1991;6:195–204.
345. Kozak M. The scanning model for translation: an update. *J Cell Biol* 1989;108:229–241.
346. Krysan PJ, Calos MP. Epstein-Barr virus-based vectors that replicate in rodent cells. *Gene* 1993;136:137–143.
347. Kupfer SR, Summers WC. Identification of a glucocorticoid-responsive element in Epstein-Barr virus. *J Virol* 1990;64:1984–1990.
348. Kurilla M, Swaminathan S, Welsh R, Kieff E, Brutkiewicz E. Effects of virally expressed interleukin-10 on vaccinia virus infection in mice. *J Virol* 1993;67:7623–7628.
349. Lacy J, Rudnick H. Transcriptional regulation of the human IgE receptor (Fc epsilon RII/CD23) by Epstein-Barr virus. Identification of Epstein-Barr virus-responsive regulatory elements in intron 1. *J Immunol* 1992;148:1554–1560.
350. Laherty C, Hu H, Opipari A, Wang F, Dixit V. The Epstein-Barr virus LMP1 gene product induces A20 zinc finger protein expression by activating nuclear factor, kB. *J Biol Chem* 1992;267:24157–24160.
351. Lambris JD, Ganu S, Hirani S, Muller-Eberhard HJ. Mapping of the C3d receptor (CR2) binding site and a neoantigen site in the C3d domain of the third component of complement. *Proc Natl Acad Sci USA* 1985;82:4236–4239.
352. Landon JC, Ellis LB, Zene VH, Frabrizio DPA. Herpes-type virus in cultured leukocytes from chipanzees. *J Natl Cancer Inst* 1968;40: 181–192.
353. Larocca D, Clough W. Hypomethylation of Epstein-Barr virus DNA in the nonproducer B-cell line Epstein-Barr virus. *J Virol* 1982;43: 1129–1131.
354. Laux G, Freese UK, Bornkamm GW. Structure and evolution of two related transcription units of Epstein-Barr virus carrying small tandem repeats. *J Virol* 1985;56:987–995.
355. Laux G, Freese UK, Fischer R, Polack A, Kofler E, Bornkamm GW. TPA-inducible Epstein-Barr virus genes in Raji cells and their regulation. *Virology* 1988;162:503–507.
356. Laux G, Perricaudet M, Farrell PJ. A spliced Epstein-Barr virus gene expressed in latently transformed lymphocytes is created by circularisation of the linear viral genome. *EMBO J* 1988;7:769–774.
357. Lawrence JB, Singer RH, Marselle LM. Highly localized tracks of specific transcripts within interphase nuclei visualized by in situ hybridization. *Cell* 1989;57:493–502.
358. Lawrence JB, Villnave CA, Singer RH. Sensitive, high-resolution chromatin and chromosome mapping in situ: presence and orientation of two closely intergrated copies of Epstein-Barr virus in a lymphoma line. *Cell* 1988;52:51–61.
359. Le RA, Berebbi M, Moukaddem M, Perricaudet M, Joab I. Identification of a short amino acid sequence essential for efficient nuclear targeting of the Epstein-Barr virus nuclear antigen 3A. *J Virol* 1993;67: 1716–1720.
360. Lear AL, Rowe M, Kurilla MG, Lee S, Henderson S, Kieff E, Rickinson AB. The Epstein-Barr virus (Epstein-Barr virus) nuclear antigen 1 *Bam*HI F promoter is activated on entry of Epstein-Barr virus-transformed B cells into the lytic cycle. *J Virol* 1991;66:7461–7468.
361. Lee MA, Kim OJ, Yates JL. Targeted gene disruption in Epstein-Barr virus. *Virology* 1992;189:253–265.
362. Lee MA, Yates JL. BHRF1 of Epstein-Barr virus, which is homologous to human proto-oncogene bcl2, is not essential for transformation of B cells or for virus replication in vitro. *J Virol* 1992;66:1899–1906.
363. Lee W, Mitchell P, Tjian R. Purified transcription factor AP-1 interacts with TPA-inducible enhancer elements. *Cell* 1987;49:741–752.
364. Lee Y, Tanaka A, Lau RY, Nonoyama M, Rabin H. Linkage map of the fragments of herpesvirus Papio DNA. *J Virol* 1981;37:710–720.
365. Lerner MR, Andrews NC, Miller G, Steitz JA. Two small RNAs encoded by Epstein-Barr virus and complexed with protein are precipitated by antibodies from patients with systemic lupus erythematosus. *Proc Natl Acad Sci USA* 1981;78:805–809.
366. Levy JA, Levy SB, Hirshaut Y, Kafuko G, Prince A. Presence of Epstein-Barr virus antibodies in sera from wild chimpanzees. *Nature* 1971; 233:559–560.
367. Li JS, Zhou BS, Dutschman GE, Grill SP, Tan RS, Cheng YC. Association of Epstein-Barr virus early antigen diffuse component and virus-specified DNA polymerase activity. *J Virol* 1987;61:2947–2949.
368. Li QX, Young LS, Niedobitek G, et al. Epstein-Barr virus infection and replication in a human epithelial cell system. *Nature* 1992;356: 347–350.
369. Lieberman PM, Berk AJ. The Zta trans-activator protein stabilizes TFIID association with promoter DNA by direct protein-protein interaction. *Genes Dev* 1991;5:2441–2454.
370. Lieberman PM, Berk AJ. In vitro transcriptional activation, dimerization, and DNA-binding specificity of the Epstein-Barr virus Zta protein. *J Virol* 1990;64:2560–2568.
371. Lieberman PM, Hardwick JM, Sample J, Hayward GS, Hayward SD. The zta transactivator involved in induction of lytic cycle gene expression in Epstein-Barr virus-infected lymphocytes binds to both AP-1 and ZRE sites in target promoter and enhancer regions. *J Virol* 1990; 64:1143–1155.
372. Lieberman PM, O'Hare P, Hayward GS, Hayward SD. Promiscuous trans activation of gene expression by an Epstein-Barr virus-encoded early protein. *J Virol* 1986;60:140–148.
373. Liebowitz D, Kieff E. The Epstein-Barr virus latent membrane protein (LMP) induction of B cell activation antigens and membrane patch formation does not require vimentin. *J Virol* 1989;63:4051–4054.
374. Liebowitz D, Kopan R, Fuchs E, Sample J, Kieff E. An Epstein-Barr virus transforming protein associates with vimentin in lymphocytes. *Mol Cell Biol* 1987;7:2299–2308.
375. Liebowitz D, Mannick J, Takada K, Kieff E. Phenotypes of Epstein-Barr virus LMP1 deletion mutants indicate transmembrane and amino terminal cytoplasmic domains necessary for effect in B-lymphoma cells. *J Virol* 1992;66:4612–4616.
376. Liebowitz D, Wang D, Kieff E. Orientation and patching of the latent infection membrane protein encoded by Epstein-Barr virus. *J Virol* 1986;58:233–237.
377. Lin JC, DeClercq E, Pagano JS. Novel acyclic adenosine analogs inhibits Epstein-Barr virus replication. *Antimicrob Agents Chemother* 1987;31:1431–1433.
378. Lin JC, Machida H. Comparision of two bromovinyl nucleoside analogs, 1-beta-D-arabinofuranosyl-E-5-(2-bromovinyl) uracil and E-5-(2-bromovinyl)uracil and E-5-(2-bromovinyl)-2′-deoxyuridine, with acyclovir in inhibition of Epstein-Barr virus replication. *Antimicrob Agents Chemother* 1988;32:1068–1072.
379. Lin JC, Zhang ZX, Smith MC, Biron K, Pagano JS. Anti-human immunodeficiency virus agent 3′-azido-3′-deoxythymidine inhibits replication of Epstein-Barr virus. *Antimicrob Agents Chemother* 1988;32: 265–267.
380. Lin SF, Lin SW, Hsu TY, Liu MY, Chen JY, Yang CS. Functional analysis of the amino terminus of Epstein-Barr virus deoxyribonuclease. *Virology* 1994;199:223–227.
381. Lindahl T, Adams A, Bjursell G, Bornkamm GW, Kaschka-Dierich C, Jehn U. Covalently closed circular duplex DNA of Epstein-Barr virus in a human lymphoid cell line. *J Mol Biol* 1976;102:511–530.

382. Ling PD, Rawlins DR, Hayward SD. The Epstein-Barr virus immortalizing protein EBNA-2 is targeted to DNA by a cellular enhancer-binding protein. *Proc Natl Acad Sci USA* 1993;90:9237–9241.
383. Ling PD, Ryon JJ, Hayward SD. EBNA-2 of herpesvirus Papio diverges significantly from the type A and type B EBNA-2 proteins of Epstein-Barr virus but retains an efficient transactivation domain with a conserved hydrophobic motif. *J Virol* 1993;67:2990–3003.
384. Littler E, Arrand JR. Characterization of the Epstein-Barr virus encoded thymidine kinase expressed in heterologous eucaryotic and procaryotic systems. *J Virol* 1988;62:3892–3895.
385. Littler E, Halliburton IW, Powell KL, Snowden BW, Arrand JR. Immunological conservation between Epstein-Barr virus and herpes simplex virus. *J Gen Virol* 1988;69:2021–2031.
386. Littler E, Zeuthen J, McBride AA, Sorensen E, Powell KL, Walsh-Arrand JR. Identification of an Epstein-Barr virus–coded thymidine kinase. *EMBO J* 1986;5:1959–1966.
387. Liu Q, Summers WC. Identification of the 12-O-tetradecanoylphorbol-13-acetate-responsive enhancer of the MS gene of the Epstein-Barr virus. *J Biol Chem* 1992;267:12049–12054.
388. Longnecker R, Druker B, Roberts T, Kieff E. An Epstein-Barr virus protein associated with cell growth transformation interacts with a tyrosine kinase. *J Virol* 1991;65:3681–3692.
389. Longnecker R, Kieff E. A second Epstein-Barr virus membrane protein (LMP2) is expressed in latent infection and colocalizes with LMP1. *J Virol* 1990;64:2319–2326.
390. Longnecker R, Miller C, Maio X-Q, Marchini A, Kieff E. The only domain which distinguishes Epstein-Barr virus latent membrane protein 2A (LMP2A) from LMP2B is dispensable for lymphocyte infection and growth transformation *in vitro*. *J Virol* 1992;66:6461–6469.
391. Longnecker R, Miller CL, Miao XQ, Tomkinson B, Kieff E. The last seven transmembrane and carboxy-terminal cytoplasmic domains of Epstein-Barr virus latent membrane protein 2 (LMP2) are dispensable for lymphocyte infection and growth transformation in vitro. *J Virol* 1993;67:2006–2013.
392. Longnecker R, Miller CL, Tomkinson B, Miao XQ, Kieff E. Deletion of DNA encoding the first five transmembrane domains of Epstein-Barr virus latent membrane proteins 2A and 2B. *J Virol* 1993;67: 5068–74.
393. Lowe RS, Keller PM, Keech BJ, et al. Varicella-Zoster virus: a recombinant vaccine vector. *Proc Natl Acad Sci USA* 1987;84:3896–3900.
394. Luka J, Kallin B, Klein G. Induction of the Epstein-Barr virus (Epstein-Barr virus) cycle in latently infected cells by n-butyrate. *Virology* 1979; 94:228–231.
395. Luka J, Kreofsky T, Pearson GR, Hennessy K, Kieff E. (1984): Partial purification and characterization of a cellular protein crossreacting with the 72K EBNA. *J Virol* 52:833–838.
396. Luka J, Siegert W, Klein G. Solubilization of the Epstein-Barr virus–determined nuclear antigen and its characterization as a DNA-binding protein. *J Virol* 1977;22:1–8.
397. Lupton S, Levine AJ. Mapping genetic elements of Epstein-Barr virus that facilitate extrachromosomal persistence of Epstein-Barr virus-derived plasmids in human cells. *Mol Cell Biol* 1985;5:2533–2542.
398. Mackett M, Conway MJ, Arrand JR, Haddad RS, Hutt-Fletcher LM. Characterization and expression of a glycoprotein encoded by the Epstein-Barr virus BamHI I fragment. *J Virol* 1990;64:2545–2552.
399. Magrath I. The pathogenesis of Burkitt's lymphoma. *Adv Cancer Res* 1990;55:133–270.
400. Manet E, Rigolet A, Gruffat H, Giot JF, Sergeant A. Domains of the Epstein-Barr virus (Epstein-Barr virus) transcription factor R required for dimerization, DNA binding and activation. *Nucleic Acids Res* 1991; 19:2661–2667.
401. Mann KP, Staunton D, Thorley-Lawson D. Epstein-Barr virus–encoded protein found in plasma membranes of transformed cells. *J Virol* 1985; 55:710–720.
402. Mann KP, Thorley-Lawson D. Posttranslational processing of the Epstein-Barr virus encoded p63/LMP protein. *J Virol* 1987;61:2100–2108.
403. Mannick JB, Cohen JI, Birkenbach M, Marchini A, Kieff E. The Epstein-Barr virus nuclear protein encoded by the leader of the EBNA RNAs (EBNA-LP) is important in B-lymphocyte transformation. *J Virol* 1991;65:6826–6837.
404. Marchini A, Cohen J, Wang F, Kieff E. A selectable marker allows investigation of a non-transforming Epstein-Barr virus mutant. *J Virol* 1992;66:3214–3219.
405. Marchini A, Kieff E, Longnecker R. Marker rescue of a transformation-negative Epstein-Barr virus recombinant from an infected Burkitt lymphoma cell line: a method useful for analysis of genes essential for transformation. *J Virol* 1993;67:606–609.
406. Marchini A, Longnecker R, Kieff E. Epstein-Barr virus (Epstein-Barr virus) negative B-lymphoma cell lines for clonal isolation and replication of Epstein-Barr virus recombinants. *J Virol* 1992;66:4972–4981.
407. Marchini A, Tomkinson B, Cohen J, Kieff, E. BHRF1, the Epstein-Barr virus gene with homology to Bc12, is dispensable for B-lymphocyte transformation and virus replication. *J Virol* 1991;65:5991–6000.
408. Marcu K, Bossone S, Patel A. Myc function and regulation. *Ann Rev Biochem* 1992;61:809–860.
409. Mark W, Sugden B. Transformation of lymphocytes by Epstein-Barr virus requires only one-fourth of the viral genome. *Virology* 1982;122: 431–443.
410. Marschall M, Schwarzmann F, Leser U, Oker B, Alliger P, Mairhofer H, Wolf H. The BI′LF4 *trans*-activator of Epstein-Barr virus is modulated by type and differentiation of the host cell. *Virology* 1991;181: 172–179.
411. Martin D, Yuryev A, Kalli KR, Fearon DT, Ahearn JM. Determination of the structural basis for selective binding of Epstein-Barr virus to human complement receptor type 2. *J Exp Med* 1991;174:1299–1311.
412. Martin DR, Marlowe RL, Ahearn JM. Determination of the role for CD21 during Epstein-Barr virus infection of B-lymphoblastoid cells. *J Virol* 1994;68:4716–4726.
413. Martin J, Sugden B. Transformation by the oncogenic latent membrane protein correlates with its rapid turnover, membrane localization, and cytoskeletal association. *J Virol* 1991;65:3246–3258.
414. Martin JM, Veis D, Korsmeyer SJ, Sugden B. Latent membrane protein of Epstein-Barr virus induces cellular phenotypes independently of expression of Bcl-2. *J Virol* 1993;67:5269–5278.
415. Matsunami N, Hamaguchi Y, Yamamoto Y, et al. A protein binding to the Jk recombination sequence of immunoglobulin genes contains a sequence related to the integrase motif. *Nature* 1989;342:934–937.
416. Matsuo T, Heller M, Petti L, Oshiro E, Kieff E. Persistence of the entire Epstein-Barr virus genome integrated into human lymphocyte DNA. *Science* 1984;226:1322–1325.
417. Maunders MJ, Petti L, Rowe M. Precipitation of the Epstein-Barr virus protein EBNA 2 by an EBNA 3c-specific monoclonal antibody. *J Gen Virol* 1994;:769–778.
418. Mcgeoch DJ, Dalrymple MA, Davison AJ, et al. The complete DNA sequence of the long unique region in the genome of herpes simplex virus type 1. *J Gen Virol* 1988;69:1531–1574.
419. Menezes J, Leibold W, Klein G. Biological differences between Epstein-Barr virus (Epstein-Barr virus) strains with regard to lymphocyte transforming ability, superinfection and antigen induction. *Exp Cell Res* 1975;92:478–484.
420. Metzenberg S. Levels of Epstein-Barr virus DNA in lymphoblastoid cell lines are correlated with frequencies of spontaneous lytic growth but not with levels of expression of EBNA-1, EBNA-2, or latent membrane protein. *J Virol* 1990;64:437–444.
421. Middleton T, Sugden B. Retention of plasmid DNA in mammalian cells is enhanced by binding of the Epstein-Barr virus replication protein EBNA1. *J Virol* 1994;68:4067–4071.
422. Middleton T, Sugden B. EBNA1 can link the enhancer element to the initiator element of the Epstein-Barr virus plasmid origin of DNA replication. *J Virol* 1992;66:489–495.
423. Mikaelian I, Drouet E, Marechal V, Denoyel G, Nicolas JC, Sergeant A. The DNA-binding domain of two bZIP transcription factors, the Epstein-Barr virus switch gene product EB1 and Jun, is a bipartite nuclear targeting sequence. *J Virol* 1993;67:734–742.
424. Miller C, Burkhardt A, Lee J, et al. Integral membrane protein 2 (LMP2) of Epstein-Barr virus regulates reactivation from latency through dominant negative effects on protein tyrosine kinases. *Immunity* 1995;2: (in press).
425. Miller CL, Lee JH, Kieff E, Longnecker R. An integral membrane protein (LMP2) blocks reactivation of Epstein-Barr virus from latency following surface immunoglobulin crosslinking. *Proc Natl Acad Sci USA* 1994;91:772–776.
426. Miller CL, Longnecker R, Kieff E. Epstein-Barr virus latent membrane protein 2A blocks calcium mobilization in B lymphocytes. *J Virol* 1993; 67:3087–3094.
427. Miller G. The Epstein-Barr virus. In: Fields B, Knipe D, eds.*Fields virology*. New York: Raven; 1989.
428. Miller G, Grogan E, Heston L, Robinson J, Smith D. Epstein-Barr viral

DNA: infectivity for human placental cells. *Science* 1981;212:452–455.

429. Miller G, Heston L, Countryman J. P3HR-1 Epstein-Barr virus with heterogeneous DNA is an independent replicon maintained by cell-to-cell spread. *J Virol* 1985;54:45–52.
430. Miller G, Himmelfarb H, Heston L, et al. Comparing regions of the Epstein-Barr virus ZEBRA protein which function as transcriptional activating sequences in Saccharomyces cerevisiae and in B cells. *J Virol* 1993;67:7472–7481.
431. Miller G, Rabson M, Heston L. Epstein-Barr virus with heterogenous DNA disrupts latency. *J Virol* 1984;50:174–182.
432. Miller G, Robinson J, Heston L, Lipman M. Differences between laboratory strains of Epstein-Barr virus based on immortalization, abortive infection, and interference. *Proc Natl Acad Sci USA* 1974;71:4006–4010.
433. Miller G, Shope T, Lisco H, Stitt D, Lipman M. Epstein-Barr virus: transformation, cytopathic changes, and viral antigens in squirrel monkey and marmoset leukocytes. *Proc Natl Acad Sci USA* 1972;69:383–387.
434. Miller N, Hutt-Fletcher L. A monoclonal antibody to glycoprotein gp85 inhibits fusion but not attachment of Epstein-Barr virus. *J Virol* 1988; 62:2366–2372.
435. Milman G, Hwang ES. Epstein-Barr virus nuclear antigen forms a complex that binds with high concentration dependence to a single DNA-binding site. *J Virol* 1987;61:465–471.
436. Minarovits J, Hu LF, Minarovits KS, Klein G, Ernberg I. Sequence-specific methylation inhibits the activity of the Epstein- Barr virus LMP 1 and BCR2 enhancer-promoter regions. *Virology* 1994;200:661–667.
437. Miyazaki I, Cheung RK, Dosch HM. Viral interleukin 10 is critical for the induction of B cell growth transformation by Epstein-Barr virus. *J Exp Med* 1993;178:439–447.
438. Mocarski ES, Roizman B. Structure and role of the herpes simplex virus DNA termini in inversion, circularization and generation of viron DNA. *Cell* 1982;31:89–97.
439. Molina H, Brenner C, Jacobi S, Gorka J, Carel JC, Kinoshita T, Holers VM. Analysis of Epstein-Barr virus-binding sites on complement receptor 2 (CR2/CD21) using human-mouse chimeras and peptides. At least two distinct sites are necessary for ligand-receptor interaction.*J Biol Chem* 1991;266:12173–12179.
440. Montalvo EA, Shi Y, Shenk TE, Levine AJ. Negative regulation of the BZLF1 promoter of Epstein-Barr virus. *J Virol* 1991;65:3647–3655.
441. Moore KW, O'Garra A, de Waal M, Vieira P, Mosmann TR. Interleukin-10. *Annu Rev Immunol* 1993;11:165–190.
442. Moore KW, Vieira P, Fiorentino DF, Trounstine ML, Khan TA, Mosmann TR. Homology of cytokine synthesis inhibitory factor (IL-10) to the Epstein-Barr virus gene BCRFI [published erratum appears in Science 1990;250:494]. *Science* 1990;248:1230–1234.
443. Moore MD, Cannon MJ, Sewall A, Finlayson M, Okimoto M, Nemerow GR. Inhibition of Epstein-Barr virus infection in vitro and in vivo by soluble CR2 (CD21) containing two short consensus repeats. *J Virol* 1991;65:3559–3565.
444. Moore MD, Cooper NR, Tack BF, Nemerow G. Molecular cloning of the cDNA encoding the Epstein-Barr virus/C3d receptor (complement receptor type 2) of human B lymphocytes. *Proc Natl Acad Sci USA* 1987;84:9194–9198.
445. Moorthy R, Thorley-Lawson DA. Processing of the Epstein-Barr virus–encoded latent membrane protein p63/LMP. *J Virol* 1990;64:829–837.
446. Moorthy RK, Thorley-Lawson D. All three domains of the Epstein-Barr virus-encoded latent membrane protein LMP-1 are required for transformation of rat-1 fibroblasts. *J Virol* 1993;67:1638–1646.
447. Moorthy RK, Thorley-Lawson D. Biochemical, genetic, and functional analyses of the phosphorylation sites on the Epstein-Barr virus-encoded oncogenic latent membrane protein LMP-1. *J Virol* 1993;67: 2637–45.
448. Morgan AJ, Mackett M, Finerty S, Arrand JR, Scullion FT, Epstein MA. Recombinant vaccinia virus expressing Epstein-Barr virus glycoprotein gp340 protects cottontop tamarins against EB virus-induced malignant lymphomas. *J Med Virol* 1988;25:189–195.
449. Mosialos G, Birkenbach M, VanArsdale T, Ware C, Yalamanchili R, Kieff E. The Epstein-Barr virus transforming protein LMP1 engages signaling proteins for the tumor necrosis factor receptor family. *Cell* 1995;80:389–399.
450. Mosialos G, Hanissian SH, Jawahar S, Vara L, Kieff E, Chatila TA. A Ca2-/calmodulin-dependent protein kinase, CaM kinase-Gr, expressed after transformation of primary human B lymphocytes by Epstein-Barr virus (Epstein-Barr virus) is induced by the Epstein-Barr virus oncogene LMP1. *J Virol* 1994;68:1697–705.
451. Mosialos G, Yamashiro S, Baughman RW, et al. Epstein-Barr virus infection induces expression in B lymphocytes of a novel gene encoding an evolutionarily conserved 55-kD actin-bundling protein. *J Virol* 1994; 68:7320–7328.
452. Moss DJ, Burrows SR, Parsons PG. Calcium concentration defines two stages in transformation of lymphocytes by Epstein-Barr virus. *Int J Cancer* 1984;33:89–97.
453. Moss DJ, Rickinson AB, Wallace LE, Epstein MA. Sequential appearance of Epstein-Barr virus nuclear and lymphocyte-detected membrane antigens in B cell transformation. *Nature* 1981;291:664–666.
454. Moss DJ, Sculley T, Pope J. Induction of Epstein-Barr virus nuclear antigens. *J Virol* 1986;58:988–990.
455. Mueller-Lantzsch N, Georg B, Yamamoto N, zur Hausen H. Epstein-Barr virus-induced proteins. II. Analysis of surface polypeptides from Epstein-Barr virus-producing and superinfected cells by immunoprecipitation. *Virology* 1980;102:401–411.
456. Mueller-Lantzsch N, Georg B, Yamamoto N, zur Hausen H. Epstein-Barr virus-induced proteins. III. Analysis of polypeptides from P3HR-1-Epstein-Barr virus-superinfected NC37 cells by immunoprecipitation. *Virology* 1980;102:231–233.
457. Mueller-Lantzsch N, Lenoir GM, Sauter M, et al. Identification of the coding region for a second Epstein-Barr virus nuclear antigen (EBNA 2) by transfection of cloned DNA fragments. *EMBO J* 1985;4:1805–1811.
458. Nakagomi H, Dolcetti R, Bejarano MT, Pisa P, Kiessling R, Masucci MG. The Epstein-Barr virus latent membrane protein-1 (LMP1) induces interleukin-10 production in Burkitt lymphoma lines. *Int J Cancer* 1994;57:240–244.
459. Nemerow G, Houghton R, Moore M, Cooper N. Identification of an epitope in the major envelope protein of Epstein-Barr virus that mediates viral binding to the B lymphocyte Epstein-Barr virus receptor (CR2). *Cell* 1989;56:369–377.
460. Nemerow G, Mold C, Keivens-Schwend V, Tollefson V, Cooper NR. Identification of gp350 as the viral glycoprotein mediating attachment of Epstein-Barr virus (Epstein-Barr virus) to the Epstein-Barr virus/C3d receptor of B cells: sequence homology of gp350 and C3 complement fragment C3d. *J Virol* 1987;61:1416–1420.
461. Nemerow G, Wolfert R, McNaughton M, Cooper N. Identification and Characterization of the Epstein-Barr virus receptor on human B lymphocytes and its relationship to the C3d complement receptor CR2. *J Virol* 1985;55:347–351.
462. Nemerow GR, Cooper NR. Early events in infection of human B lymphocytes by Epstein-Barr virus: internalization process. *Virology* 1984; 132:186–196.
463. Nemerow GR, Mullen JJ 3d, Dickson PW, Cooper NR. Soluble recombinant CR2 (CD21) inhibits Epstein-Barr virus infection. *J Virol* 1990; 64:1348–1352.
464. Nemerow GR, Siaw MFE, Cooper NR. Purification of the Epstein-Barr virus/C3d complement receptor of human B lymphocytes: antigenic and functional properties of the purified protein. *J Virol* 1986;58:709–712.
465. Neubauer RH, Rabin H, Strnad BC, Nonoyama M, Nelson-Rees WA. Establishment of a lymphoblastoid cell line and isolation of an Epstein-Barr-related virus of gorilla origin. *J Virol* 1979;31:845–853.
466. Nicholas J, Cameron K, Honess R. Herpes virus Saimiri encodes homologues of G protein coupled receptors and cyclins. *Nature* 1992;355: 362–365.
467. Nicholas J, Cameron KR, Coleman H, Newman C, Honess RW. Analysis of nucleotide sequence of the rightmost 43 kbp of herpesvirus Saimiri (HVS) L-DNA: general conservation of genetic organization between HVS and Epstein-Barr virus. *Virology* 1992;188:296–310.
468. Niedobitek G, Young LS, Lau R, et al. Epstein-Barr virus infection in oral hairy leukoplakia: virus replication in the absence of a detectable latent phase. *J Gen Virol* 1991;72:3035–3046.
469. Nikas I, McLauchlan J, Davison AJ, Taylor WR, Clements JB. Structural features of ribonucleotide reductase. *Proteins* 1986;1:376–384.
470. Nilsson T, Sjoblom A, Masucci MG, Rymo L. Viral and cellular factors influence the activity of the Epstein-Barr virus BCR2 and BWR1 promoters in cells of different phenotype. *Virology* 1993;193:774–785.
471. Nonkwelo CB, Long WK. Regulation of Epstein-Barr virus BamHI-H divergent promoter by DNA methylation. *Virology* 1993;197:205–15.
472. Nonoyama M, Pagano JS. Separation of Epstein-Barr virus DNA from large chromosomal DNA in non-virus-producing cells. *Nature* 1972;333: 41–45.
473. Nonoyama M, Pagano JS. Homology between Epstein-Barr viruses

DNA and viral-DNA from Burkitt's lymphoma and nasopharyngeal carcinoma determined by DNA-DNA reassociation kinetics. *Nature* 1973;242:44–47.

474. North JR, Morgan AJ, Epstein MA. Observations on the EB virus envelope and the virus-determined membrane antigen (MA) polypeptides. *Int J Cancer* 1980;26:231–240.
475. Nucifora G, Begy CR, Erickson P, Drabkin HA, Rowley JD. The 3;21 translocation in myelodysplasia results in a fusion transcript between the AML1 gene and the gene for EAP, a highly conserved protein associated with the Epstein-Barr virus small RNA EBER 1. *Proc Natl Acad Sci USA* 1993;90:7784–7788.
476. Nuebling CM, Mueller-Lantzsch N. Identification of the gene product encoded by the PstI repeats (IR4) of the Epstein-Barr virus genome. *Virology* 1991;185:519–523.
477. Oba DE, Hutt-Fletcher LM. Induction of antibodies to the Epstein-Barr virus glycoprotein gp85 with a synthetic peptide corresponding to a sequence in the BXLF2 open reading frame. *J Virol* 1988;62:1108–1114.
478. Oguro MO, Shimizu N, Ono Y, Takada K. Both the rightward and leftward open reading frames within the BamHI M DNA fragment of Epstein-Barr virus act as trans-activators of gene expression. *J Virol* 1987; 61:3310–3313.
479. Oh SJ, Chittenden T, Levine AJ. Identification of cellular factors that bind specifically to the Epstein-Barr virus origin of DNA replication. *J Virol* 1991;65:514–519.
480. Ohno S, Luka J, Falk L, Klein G. Detection of a nuclear, EBNA-type antigen in apparently EBNA negative herpesvirus Papio (HVP)-transformed lymphoid lines by the acid fixed nuclear binding technique. *Int J Cancer* 1979;20:941–946.
481. Ohno S, Luka J, Falk LA, Klein G. Serological reactivities of human and baboon sera against EBNA and herpesvirus Papio-determined nuclear antigen. *Eur J Cancer* 1978;14:955–960.
482. Ohno S, Luka J, Lindahl T, Klein G. Identification of a purified complement-fixing antigen as the Epstein-Barr virus determined nuclear antigen (EBNA) by its binding to metaphase chromosomes. *Proc Natl Acad Sci USA* 1977;74:1605–1609.
483. Ooka T, Calender A. Effects of arabinofuranosylthymine on Epstein-Barr virus replication. *Virology* 1980;104:219–223.
484. Ooka T, Calender A, de Turenne M, Daillie J. Effect of arabinofuranosylthymine on the replication of Epstein-Barr virus and relationship with a new induced thymidine kinase activity. *J Virol* 1983;46:187–195.
485. Ooka T, de Turenne M, de The G, Daillie J. Epstein-Barr virus-specific DNase activity in nonproducer Raji cells after treatment with 12-o-tetradecanoylphorbol-13-acetate and sodium butyrate. *J Virol* 1984; 49:626–628.
486. Ooka T, Lenoir GM, Decaussin G, Bornkamm GW, Daillie J. Epstein-Barr virus-specific DNA polymerase in virus-nonproducer Raji cells. *J Virol* 1986;58:671–675.
487. Orellana T, Kieff E. Epstein-Barr virus specific RNA. II. Analysis of polyadenylated viral RNA in restringent, abortive and productive infection. *J Virol* 1977;22:321–330.
488. Orlowski R, Miller G. Single-stranded structures are present within plasmids containing the Epstein-Barr virus latent origin of replication. *J Virol* 1991;65:677–686.
489. Packham G, Brimmell M, Cook D, Sinclair AJ, Farrell PJ. Strain variation in Epstein-Barr virus immediate early genes. *Virology* 1993;192: 541–50.
490. Packham G, Economou A, Rooney CM, Rowe DT, Farrell PJ. Structure and function of the Epstein-Barr virus BZLF1 protein. *J Virol* 1990; 64:2110–2116.
491. Pallesen G, Hamilton-Dutoit SJ, Rowe M, Young LS. Expression of Epstein-Barr virus latent gene products in tumour cells of Hodgkin's disease. *Lancet* 1991;337:320–322.
492. Parker BD, Bankier A, Satchwell S, Barrell B, Farrell PJ. Sequence and transcription of Raji Epstein-Barr virus DNA spanning the B95-8 deletion region. *Virology* 1990;179:339–346.
493. Patton DF, Shirley P, Raab-Traub N, Resnick L, Sixbey JW. Defective viral DNA in Epstein-Barr virus–associated oral hairy leukoplakia. *J Virol* 1990;64:397–400.
494. Pearson G, Luka J, Petti L, et al. Identification of an Epstein-Barr virus early gene encoding for a second component of the restricted early antigen complex. *Virology* 1987;160:151–161.
495. Pearson GR, Vroman B, Chase B, Sculley T, Hummel M, Kieff E. Identification of polypeptide components of the Epstein-Barr virus early antigen complex with monoclonal antibodies. *J Virol* 1983;47:193–201.
496. Pellett PE, Biggin MD, Barrell B, Roizman B. Epstein-Barr virus genome may encode a protein showing significant amino acid and predicted secondary structure homology with glycoprotein B of herpes simplex virus 1. *J Virol* 1985;56:807–813.
497. Peng M, Lundgren E. Transient expression of the Epstein-Barr virus LMP1 gene in human primary B cells induces cellular activation and DNA synthesis. *Oncogene* 1992;7:1775–1782
498. Perlmann C, Saemundsen AK, Klein G. A fraction of Epstein-Barr virus virion DNA is methylated in and around the EcoRI-J fragment. *Virology* 1982;123:217–221.
499. Petti L, Kieff E. A sixth Epstein-Barr virus nuclear protein (EBNA3B) is expressed in latently infected growth transformed lymphocytes. *J Virol* 1988;62:2173–2178.
500. Petti L, Sample C, Kieff E. Subnuclear localization and phosphorylation of Epstein-Barr virus latent infection nuclear proteins. *Virology* 1990;176:563–574.
501. Petti L, Sample J, Wang F, Kieff E. A fifth Epstein-Barr virus nuclear protein (EBNA3C) is expressed in latently infected growth transformed lymphocytes. *J Virol* 1988;62:1330–1338.
502. Platt TH, Tcherepanova IY, Schildkraut CL. Effect of number and position of EBNA-1 binding sites in Epstein-Barr virus oriP on the sites of initiation, barrier formation, and termination of replication. *J Virol* 1993; 67:1739–1745.
503. Polack A, Delius H, Zimber U, Bornkamm GW. Two deletions in the Epstein-Barr virus genome of the Burkitt lymphoma nonproducer line, Raji. *Virology* 1984;133:146–157.
504. Polack A, Hartl G, Zimber U, et al. A complete set of overlapping cosmid clones of M-ABA virus derived from asopharyngeal carcimona and its similarity to other Epstein-Barr virus isolates. *Gene* 1984;27:279–288.
505. Polvino-Bodnar M, Kiso J, Schaffer PA. Mutational analysis of Epstein-Barr virus nuclear antigen 1 (EBNA 1). *Nucleic Acids Res* 1988; 16:3415–3434.
506. Polvino-Bodnar M, Schaffer PA. DNA binding activity is required for EBNA-1-dependent transcriptional activation and DNA replication. *Virology* 1992;187:591–603.
507. Pope JH, Achong B, Epstein M. Cultivation and pure structure of virus bearing lymphoblasts from 2nd N.G. Burkitt lymphoma. *Int J Cancer* 1968;3:171–182.
508. Pope JH, Horne MK, Scott W. Transformation of fetal human leucocytes in vitro by filtrates of a human leukemic cell line containing herpes-like virus. *Int J Cancer* 1968;3:857–866.
509. Powell ALT, King W, Kieff E. Epstein-Barr virus specific RNA. III. Mapping of the DNA encoding viral specific RNA in restringently infected cells. *J Virol* 1979;29:261–274.
510. Pritchett R, Pedersen M, Kieff E. Complexity of Epstein-Barr virus homologous DNA in continuous lymphoblastoid cell lines. *Virology* 1976; 74:227–231.
511. Pritchett RF, Hayward SD, Kieff ED. DNA of Epstein-Barr virus. I. Comparative studies of the DNA of Epstein-Barr virus from HR-1 and B95-8 cells: size, structure and relatedness. *J Virol* 1975;15:556–569.
512. Qu L, Rowe DT. Epstein-Barr virus latent gene expression in uncultured peripheral blood lymphocytes. *J Virol* 1992;66:3715–3724.
513. Qualtiere L, Pearson G. Epstein-Barr virus induced membrane antigens: immunochemical characterization of Triton X-100 solubilized viral membrane antigens from Epstein-Barr virus-superinfected Raji cells. *Int J Cancer* 1979;23:808–817.
514. Qualtiere LF, Decoteau JF, Nasr-El-Din E. Epitope mapping of the major Epstein-Barr virus outer envelope glycoprotein gp350/220. *J Gen Virol* 1987;68:535–543.
515. Qualtiere LF, Pearson GR. Radioimmune precipitation study comparing the Epstein-Barr virus–superinfected Raji cells to those expressed on cells in a B-95 virus-transformed producer culture activated with tumor-promoting agent (TPA). *Virology* 1980;102:360–369.
516. Quinlivan EB, Holley GE, Norris M, Gutsch D, Bachenheimer SL, Kenney SC. Direct BRLF1 binding is required for cooperative BZLF1/BRLF1 activation of the Epstein-Barr virus early promoter, BMRF1 [corrected and republished with original paging, article originally printed in *Nucleic Acids Res* 1993:21:1999–2007]. *Nucleic Acids Res* 1993;21:1999–2007.
517. Quinlivan EB, Holley-Guthrie E, Mar EC, Smith MS, Kenney S. The Epstein-Barr virus BRLF1 immediate-early gene product transactivates the human immunodeficiency virus type 1 long terminal repeat by a mechanism which is enhancer independent. *J Virol* 1990;64:1817–1820.
518. Raab-Traub N, Dambaugh T, Kieff E. DNA of Epstein-Barr virus VIII.

B95-8, the previous prototype is an unusual deletion derivative. *Cell* 1980;22:257–267.
519. Raab-Traub N, Flynn K. The structure of the termini of the Epstein-Barr virus as a marker of clonal cellular proliferation. *Cell* 1986;47:883–889.
520. Raab-Traub N, Hood R, Yang CS, Henry B, Pagano JS. Epstein-Barr virus transcription in nasopharyngeal carcinoma. *J Virol* 1983;48: 580–590.
521. Raab-Traub N, Pritchett R, Kieff E. DNA of Epstein-Barr virus. III. Identification of restriction enzyme fragments which contain DNA sequences which differ among strains of Epstein-Barr virus. *J Virol* 1978; 27:388–398.
522. Rabin H, Neubauer R, Hopkins F, Nonoyama M. Further characterization of a herpes-virus-positive orangutan cell line and comparative aspects of in vitro transformation with lymphotropic Old World primate herpesvirus. *Int J Cancer* 1978;21:762–767.
523. Rabin H, Strnad BC, Neubauer RH, Brown AM, Hopkins RF, Nelson-Rees R. Comparisons of nuclear antigens of Epstein-Barr virus (Epstein-Barr virus) and Epstein-Barr virus-like simian viruses. *J Gen Virol* 1980;48:265–272.
524. Rabson M, Gradoville L, Heston L, Miller G. Non-immortalizing P3J-HR-1 Epstein-Barr virus: a deletion mutant of its transforming parent, Jijoye. *J Virol* 1982;44:834–844.
525. Rabson M, Heston L, Miller G. Identification of rare Epstein-Barr virus varient that enhances antigen expression in Raji cells. *Proc Natl Acad Sci USA* 1983;80:2762–2766.
526. Ragona G, Ernberg I, Klein G. Induction and biological characterization of the Epstein-Barr virus. *Virology* 1980;101:553–557.
527. Rawlins DR, Milman G, Hayward SD, Hayward GS. Sequence specific DNA binding of the Epstein-Barr virus nuclear antigen (EBNA-1) to clustered sites in the plasmid maintainance region. *Cell* 1985;42:859–868.
528. Reedman B, Klein G. Cellular localization of an Epstein-Barr virus–associated complement-fixing antigen in producer and nonproducer lymphoblastoid cell lines. *Int J Cancer* 1973;11:499–520.
529. Reisman D, Sugden B. Trans activation of an Epstein-Barr viral transcriptional enhancer by the Epstein-Barr viral nuclear antigen 1. *Mol Cell Biol* 1986;6:3838–3846.
530. Reisman D, Yates J, Sugden B. A putative origin of replication of plasmids derived from Epstein-Barr virus is composed of two cis-acting components. *Mol Cell Biol* 1985;5:1822–1832.
531. Rickinson AB, Epstein MA. Sensitivity of the transforming and replicative functions of Epstein-Barr virus to inhibition by phosphoacetate. *J Gen Virol* 1978;40:409–420.
532. Rickinson AB, Young LS, Rowe M. Influence of the Epstein-Barr virus nuclear antigen EBNA 2 on the growth phenotype of virus-transformed B cells. *J Virol* 1987;61:1310–1317.
533. Ricksten A, Kallin B, Alexander H, et al. BamHI E region of the Epstein-Barr virus genome encodes three transformation-associated nuclear proteins. *Proc Natl Acad Sci USA* 1988;85:995–999.
534. Robertson ES, Tomkinson B, Kieff E. An Epstein-Barr virus with a 58-kilobase-pair deletion that includes BARF0 transforms B lymphocytes in vitro. *J Virol* 1994;68:1449–1458.
535. Rochford R, Hobbs MV, Garnier JL, Cooper NR, Cannon MJ. Plasmacytoid differentiation of Epstein-Barr virus-transformed B cells in vivo is associated with reduced expression of viral latent genes. *Proc Natl Acad Sci USA* 1993;90:352–356.
536. Rogers RP, Strominger JL, Speck SH. Epstein-Barr virus in B lymphocytes: viral gene expression and function in latency. *Adv Cancer Res* 1992;58:1–26.
537. Rogers RP, Woisetschlaeger M, Speck SH. Alternative splicing dictates translational start in Epstein-Barr virus transcripts. *EMBO J* 1990; 9:2273–2277.
538. Rooney C, Brimmell M, Buschle M, Allan G, Farrell PJ, Kolman JL. Host cell and EBNA-2 regulation of Epstein-Barr virus latent-cycle promoter activity in B lymphocytes. *J Virol* 1992;66:496–504.
539. Rooney C, Howe JG, Speck SH, Miller G. Influences of Burkitt's lymphoma and primary B cells on latent gene expression by the nonimmortalizing P3J-HR-1 strain of Epstein-Barr virus. *J Virol* 1989;63: 1531–1539.
540. Rooney C, Taylor N, Countryman J, Jenson H, Kolman J, Miller G. Genome rearrangements activate the Epstein-Barr virus gene whose product disrupts latency. *Proc Natl Acad Sci USA* 1988;85:9801–9805.
541. Rosa MD, Gottlieb M, Lerner MR, Steitz JA. Striking similarities are exhibited by two small Epstein-Barr virus encoded ribonucleic acids and the adenovirus associates ribonucleic acids VA_1 and VA_2. *Mol Cell Biol* 1981;1:785–796.
542. Roth G, Curiel T, Lacy J. Epstein-Barr viral nuclear antigen 1 antisense oligodeoxynucleotide inhibits proliferation of Epstein-Barr virus-immortalized B cells. *Blood* 1994;84:582–587.
543. Rothe M, Wong S, Henzel W, Goedel D. A novel family of putative signal transducers associate with the cytoplasmic domain of the 75 kDa tumor necrosis factor receptor. *Cell* 1994;78:681–692
544. Roubal J, Kallin B, Luka J, Klein G. Early DNA-binding polypeptides of Epstein-Barr virus. *Virology* 1981;113:285–292.
545. Rowe DT, Hall L, Joab I, Laux G. Identification of the Epstein-Barr virus terminal protein gene products in latently infected lymphocytes. *J Virol* 1990;64:2866–2875.
546. Rowe M, Evans H, Young L, Hennessy K, Kieff E, Rickinson A. Monoclonal antibodies to the latent membrane protein of Epstein-Barr virus reveal heterogeneity of the protein and inducible expression in virus-transformed cells. *J Gen Virol* 1987;68:1575–1586.
547. Rowe M, Lear AL, Croom-Carter D, Davies AH, Rickinson AB. Three pathways of Epstein-Barr virus gene activation from EBNA1-positive latency in B lymphocytes. *J Virol* 1992;66:122–131.
548. Rowe M, Peng-Pilon M, Huen D, et al. Upregulation of bcl-2 by the Epstein-Barr virus latent membrane protein LMP1: a B-cell specific response that is delayed relative to NF-kB activation and to induction of cell surface markers. *J Virol* 1994;68:5602–5612.
549. Rowe M, Rowe DT, Gregory CD, et al. Differences in B cell growth phenotype reflect novel patterns of Epstein-Barr virus latent gene expression in Burkitt's lyphoma cells. *EMBO J* 1987;6:2743–2751.
550. Rowe M, Young LS, Cadwallader K, Petti L, Kieff E, Rickinson AB. Distinction between Epstein-Barr virus type-A (EBNA-2A) and type-B (EBNA-2B) isolates extends to the EBNA-3 family of nuclear proteins. *J Virol* 1989;63:1031–1039.
551. Rymo L. Identification of transcribed regions of Epstein-Barr virus DNA in Burkitt lymphoma-derived cells. *J Virol* 1979;32:8–18.
552. Rymo L, Klein G, Ricksten A. Expression of a second Epstein-Barr virus-determined nuclear antigen in mouse cells after gene transfer with a cloned fragment of the viral genome. *Proc Natl Acad Sci USA* 1985; 82:3435–3459.
553. Rymo L, Lindahl T, Adams A. Sites of sequence variability in Epstein-Barr virus DNA from different sources. *Proc Natl Acad Sci USA* 1979; 76:2794–2798.
554. Rymo L, Lindahl T, Povey S, Klein G. Analysis of restriction endonuclease fragments of intracellular Epstein-Barr virus DNA and isoenzymes indicate a common origin of the Raji. *Virology* 1981;115:115–124.
555. Ryon JJ, Fixman ED, Houchens C, et al. The lytic origin of herpesvirus Papio is highly homologous to Epstein- Barr virus ori-Lyt: evolutionary conservation of transcriptional activation and replication signals. *J Virol* 1993;67:4006–4016.
556. Saemundsen AK, Kallin B, Klein G. Effect of n-butyrate on cellular and viral DNA synthesis in cells latently infected with Epstein-Barr virus. *Virology* 1980;107:557–561.
557. Saemundsen AK, Perlman G, Klein G. Intracellular Epstein-Barr virus DNA is methylated in and around the EcoRI-J fragment in both producer and nonproducer cell lines. *Virology* 1983;126:701–706.
558. Sample J, Brooks L, Sample C, Young L, Rowe M, Rickinson A, Kieff E. Restricted Epstein-Barr virus protein expression in Burkitt lymphoma is due to a different Epstein-Barr nuclear antigen 1 transcriptional initiation site. *Proc Natl Acad Sci USA* 1991;88:6343–6347.
559. Sample J, Henson EB, Sample C. The Epstein-Barr virus nuclear protein 1 promoter active in type I latency is autoregulated. *J Virol* 1992; 66:4654–4661.
560. Sample J, Hummel M, Braun D, Birkenbach M, Kieff E. Nucleotide sequences of mRNAs encoding Epstein-Barr virus nuclear proteins: a probable transcriptional intiation site. *Proc Natl Acad Sci USA* 1986; 83:5096–5100.
561. Sample J, Kieff E. Transcription of the Epstein-Barr virus genome during latentcy in growth-transformed lymphocytes. *J Virol* 1990;64: 1667–1674.
562. Sample J, Kieff EF, Kieff ED. Epstein-Barr virus types 1 and 2 have nearly identical LMP-1 transforming genes. *J Gen Virol* 1994;75:2741–2746.
563. Sample J, Lancz G, Nonoyama M. Mapping of genes in BamH1 fragment M of Epstein-Barr virus DNA that may determine the fate of viral infection. *J Virol* 1986;57:145–154.
564. Sample J, Liebowitz D, Kieff E. Two related Epstein-Barr virus membrane proteins are encoded by separate genes. *J Virol* 1989;63:933–937.
565. Sample J, Young L, Martin B, et al. Epstein-Barr virus type 1 and 2 differ in their EBNA-3A, EBNA-3B and EBNA-3C genes. *J Virol* 1990;

64:4084–4092.
566. Sanchez-Pinel A, Bernad J, Rives H, Lapchine L, Icart J, Didier J. Identification of a novel Epstein-Barr virus-induced membrane glycoprotein of 43 kDa with H667 MAb. *Virology* 1991;180:31–40.
567. Sato H, Takeshita H, Furukawa M, Seiki M. Epstein-Barr virus BZLF1 transactivator is a negative regulator of Jun. *J Virol* 1992;66:4732–4736.
568. Sato H, Takimoto T, Tanaka S, Tanaka J, Raab-Traub N. Concatameric replication of Epstein-Barr virus: structure of the termini in virus-producer and newly transformed cell lines. *J Virol* 1990;64:5295–5300.
569. Scala G, Quinto I, Ruocco MR, et al. Expression of an exogenous interleukin 6 gene in human Epstein Barr virus B cells confers growth advantage and in vivo tumorigenicity. *J Exp Med* 1990;172:61–68.
570. Scala G, Quinto I, Ruocco MR, et al. Epstein-Barr virus nuclear antigen 2 transactivates the long terminal repeat of human immunodeficiency virus type 1. *J Virol* 1993;67:2853–2861.
571. Schaefer BC, Woisetschlaeger M, Strominger JL, Speck SH. Exclusive expression of Epstein-Barr virus nuclear antigen 1 in Burkitt lymphoma arises from a third promoter, distinct from the promoters used in latently infected lymphocytes. *Proc Natl Acad Sci USA* 1991;88: 6550–6554.
572. Schepers A, Pich D, Hammerschmidt W. A transcription factor with homology to the AP-1 family links RNA transcription and DNA replication in the lytic cycle of Epstein-Barr virus. *EMBO J* 1993;12:3921–3929.
573. Schepers A, Pich D, Mankertz J, Hammerschmidt W. cis-acting elements in the lytic origin of DNA replication of Epstein-Barr virus. *J Virol* 1993;67:4237–4245.
574. Schneider R, Safer B, Munemitsu S, Samuel C, Shenk T. Adenovirus VA1 RNA prevents phosphorylation of the eukaryotic initiation factor 2 alpha subunit subsequent to infection. *Proc Natl Acad Sci USA* 1985;82:4321–4325.
575. Schubach WH, Horvath G, Spoth B, Hearing JC. Expression of Epstein-Barr virus nuclear antigen 2 in insect cells from a baculovirus vector. *Virology* 1991;185:428–431.
576. Schultz LD, Tanner J, Hofmann K, et al. Expression in yeast of the gene encoding the major envelope glycoproteins (gp350/gp220) of Epstein-Barr virus. *Gene* 1987;54:113–123.
577. Schwarzmann F, Prang N, Reichelt B, et al. Negatively cis-acting elements in the distal part of the promoter of Epstein-Barr virus transactivator gene BZLF1. *J Gen Virol* 1994;75:1999–2006.
578. Schweisguth F, Psokany J. Suppressor of hairless the Drosophila homolog of the mouse recombination signal binding protein gene controls sensory organ fates. *Cell* 1992;69:1199–1212
579. Sculley TB, Sculley DG, Pope JH, Bornkamm GW, Lenoir GM, Rickinson AB.Epstein-Barr virus nuclear antigens 1 and 2 in Burkitt lymphoma cell lines containing either 'A'- or 'B'-type virus. *Intervirology* 1988;29:77–85.
580. Sculley TB, Walker PJ, Moss DJ, Pope JH. Identification of multiple Epstein-Barr virus-induced nuclear antigens with sera from patients with rheumatoid arthritis. *J Virol* 1984;52:88–93.
581. Seibl R, Motz M, Wolf H. Strain-specific transcription and translation of the BamHI Z area of Epstein-Barr virus. *J Virol* 1986;60:902–909.
582. Seibl R, Wolf H. Mapping of Epstein-Barr virus proteins on the genome by translation of hybrid-selected RNA from induced Raji cells. *Virology* 1985;141:1–13.
583. Sequin C, Farrell PJ, Barrell BG. DNA sequence and transcription of the BamHI fragment B region of B95-8 Epstein-Barr virus. *Mol Biol Med* 1983;1:425–445.
584. Serafini-Cessi F, Malagolini N, Nanni M, et al. Characterization of N- and O-linked oligasaccharides of glycoprotein 350 from Epstein-Barr virus. *Virology* 1989;170:1–10.
585. Shah WA, Ambinder RF, Hayward GS, Hayward SD. Binding of EBNA-1 to DNA creates a protease-resistant domain that encompasses the DNA recognition and dimerization functions. *J Virol* 1992;66: 3355–3362.
586. Sharp TV, Schwemmle M, Jeffrey I, et al. Comparative analysis of the regulation of the interferon-inducible protein kinase PKR by Epstein-Barr virus RNAs EBER-1 and EBER-2 and adenovirus VAI RNA. *Nucleic Acids Res* 1993;21:4483–4490.
587. Shaw JE. The circular intracellular form of Epstein-Barr virus DNA is amplified by the virus associated DNA polymerase. *J Virol* 1985;53: 1012–1015.
588. Shaw JE, Levinger LF, Carter CW. Nucleosomal structure of Epstein-Barr virus DNA in transformed cell lines. *J Virol* 1979;29:657–665.
589. Shimizu N, Takada K. Analysis of the BZLF1 promoter of Epstein-Barr virus: identification of an anti-immunoglobulin response sequence. *J Virol* 1993;67:3240–3245.
590. Shimizu N, Yamaki M, Sakuma S, Ono Y, Takada K. Three Epstein-Barr virus (Epstein-Barr virus)-determined nuclear antigens induced by the BamHI E region of Epstein-Barr virus DNA. *Int J Cancer* 1988; 41:744–751.
591. Siaw MFW, Nemerow GR, Cooper NR. Biochemical and analysis of the Epstein-Barr virus/C3d receptor (CR2). *J Immunol* 1986;136:4146–4151.
592. Siegel PJ, Clough W, Strominger JL. Sedimentation characterization of newly synthesized Epstein-Barr viral DNA in superinfected cells. *J Virol* 1981;38:880–885.
593. Sinclair AJ, Brimmell M, Farrell PJ. Reciprocal antagonism of steroid hormones and BZLF1 in switch between Epstein-Barr virus latent and productive cycle gene expression. *J Virol* 1992;66:70–77.
594. Sinclair AJ, Brimmell M, Shanahan F, Farrell PJ. Pathways of activation of the Epstein-Barr virus productive cycle. *Virology* 1991;65:2237–2244.
595. Sinclair AJ, Jacquemin MG, Brooks L, et al. Reduced signal transduction through glucocorticoid receptor in Burkitt's lymphoma cell lines. *Virology* 1994;199:339–353.
596. Sinclair AJ, Palmero I, Peters G, Farrell PJ. EBNA-2 and EBNA-LP cooperate to cause G0 to G1 transition during immortalization of resting human B lymphocytes by Epstein-Barr virus. *EMBO J* 1994;13: 3321–8.
597. Sinha SK, Todd SC, Hedrick JA, Speiser CL, Lambris JD, Tsoukas CD. Characterization of the Epstein-Barr virus/C3d receptor on the human Jurkat T cell line: evidence for a novel transcript. *J Immunol* 1993;150:5311–20.
598. Sixbey J, Pagano JS. Epstein-Barr virus transformation of human B lymphocytes despite inhibition of viral polymerase. *J Virol* 1985;53: 299–301.
599. Sixbey JW, Davis DS, Young LS, Hutt-Fletcher L, Tedder TF, Rickinson AB. Human epithelial cell expression of an Epstein-Barr virus receptor. *J Gen Virol* 1987;68:805–811.
600. Sixbey JW, Shirley P, Chesney PJ, Buntin DM, Resnick L. Detection of a second widespread strain of Epstein-Barr virus. *Lancet* 1989;2: 761–765.
601. Sixbey JW, Shirley P, Sloas M, Raab-Traub N, Israele V. A transformation-incompetent, nuclear antigen 2-deleted Epstein-Barr virus associated with replicative infection. *J Infect Dis* 1991;163:1008–1015.
602. Sixbey JW, Vesterinen EH, Nedrud JG, Raab-Traub N, Walton LA, Pagano JS. Replication of Epstein-Barr virus in human epithelial cells infected in vitro. *Nature* 1983;306:480–483.
603. Skare J, Farley J, Strominger JL, Fresen O, Cho MS, zur Hausen H. Transformation by Epstein-Barr virus requires DNA sequences in the region of BamHI fragments Y and H. *J Virol* 1985;55:286–297.
604. Skare J, Strominger J. Cloning and mapping of BamHI endonuclease fragments from the transforming B95-8 strain of Epstein-Barr virus. *Proc Natl Acad Sci USA* 1980;77:3860–3864.
605. Smith PR, Gao Y, Karran L, Jones MD, Snudden D, Griffin BE. Complex nature of the major viral polyadenylated transcripts in Epstein-Barr virus-associated tumors. *J Virol* 1993;67:3217–3225.
606. Smith PR, Griffin BE. Transcription of the Epstein-Barr virus gene EBNA-1 from different promoters in nasopharyngeal carcinoma and B-lymphoblastoid cells. *J Virol* 1992;66:706–714.
607. Songyang Z, Shoelson S, Chadhuri M, et al. SH2 domains recognize specific phosphopeptide sequences. *Cell* 1993;71:767–778.
608. Spaete RR, Frenkel N. The herpes simplex virus amplicon: analyses of cis-acting repilcation functions. *Proc Natl Acad Sci USA* 1985;82: 694–698.
609. Spaete RR, Frenkel N. The herpes simplex virus amplicon: a new eukaryotic defective-virus cloning-amplifying vector. *Cell* 1982;30: 295–304.
610. Spaete RR, Mocarski ES. The a sequence of the cytomegalovirus genome functions as a cleavage/packing signal for herpes simplex virus defective genomes. *J Virol* 1985;54:817–824.
611. Speck S, Strominger J. Analysis of the transcript encoding the latent Epstein-Barr virus nuclear antigen 1: a potentially polycistronic message generated by long range splicing of several exons. *Proc Natl Acad Sci USA* 1985;82:8305–8309.
612. Spelsberg TC, Sculley TB, Pikler GM, Gilbert JA, Pearson GR. Evidence for two classes of chromatin-associated Epstein-Barr virus–determined nuclear antigen. *J Virol* 1982;43:555–565.
613. Steinitz M, Klein G. Epstein-Barr virus (Epstein-Barr virus)-induced

change in the saturation sensitivity and serum dependence of established, Epstein-Barr virus-negative lymphoma lines in vitro. *Virology* 1976;70:570–573.
614. Steinitz M, Klein G. Comparison between growth characteristics of an Epstein-Barr virus (Epstein-Barr virus)-genome-negative lymphoma line and its Epstein-Barr virus-converted subline in vitro. *Proc Natl Acad Sci USA* 1975;72:3518–3520.
615. Sternas L, Middleton T, Sugden B. The average number of molecules of Epstein-Barr nuclear antigen 1 per cell does not correlate with the average number of Epstein-Barr virus (Epstein-Barr virus) DNA molecules per cell among different clones of Epstein-Barr virus-immortalized cells. *J Virol* 1990;64:2407–2410.
616. Stewart JP, Behm FG, Arrand JR, Rooney CM. Differential expression of viral and human interleukin-10 (IL-10) by primary B cell tumors and B cell lines. *Virology* 1994;200:724–732.
617. Stinchcombe T, Clough W. Epstein-Barr virus induces a unique pyrimidine deoxynucleoside kinase activity in superinfected and virus-producer B cell lines. *Biochemistry* 1985;24:2027–2033.
618. Stolzenberg MC, Ooka T. Purification and properties of Epstein-Barr virus DNase expressed in Escherichia coli. *J Virol* 1990;64:96–104.
619. Strnad BC, Neubauer RH, Rabin H, Mazur RA. Correlation between Epstein-Barr virus membrane antigen and three large cell surface glycoproteins. *J Virol* 1979;32:885–894.
620. Strnad BC, Schuster TC, Hopkins RF 3d, Neubauer RH, Rabin H. Identification of an Epstein-Barr virus nuclear antigen by fluoroimmunoelectrophoresis and radioimmunoelectrophoresis. *J Virol* 1981;38: 996–1004.
621. Su W, Middleton T, Sugden B, Echols H. DNA looping between the origin of replication of Epstein-Barr virus and its enhancer site: stabilization of an origin complex with Epstein-Barr nuclear antigen 1. *Proc Natl Acad Sci USA* 1991;88:10870–10874.
622. Sudgen B, Mark W. Clonal transformation of adult human leukocytes by Epstein-Barr virus. *J Virol* 1977;23:503–508.
623. Sugden B, Marsh K, Yates J. A vector that replicates as a plasmid and can be efficiently selected in B-lymphoblasts transformed by Epstein-Barr virus. *Mol Cell Biol* 1985;5:410–413.
624. Sugden B, Metzenberg S. Characterisation of an antigen whose cell surface expression is induced by infection with Epstein-Barr virus. *J Virol* 1982;46:800–807.
625. Sugden B, Phelps M, Domoradzki J. Epstein-Barr virus DNA is amplified in transformed lymphocytes. *J Virol* 1979;31:590–595.
626. Sudgen B, Warren N. A promoter of Epstein-Barr virus that can function during latent infection can be transactivated by EBNA-1, a viral protein required for viral DNA replication during latent infection. *J Virol* 1989;63:2644–2649.
627. Summers WC, Klein G. Inhibition of Epstein-Barr virus DNA synthesis and late gene expression by phosphonoacetic acid. *J Virol* 1976;18: 151–155.
628. Summers WP, Grogan EA, Sheed D, Robert M, Liu CR, Miller G. Stable expression in mouse cells of nuclear neoantigen after transfer of a 3.4 megadalton cloned fragment of Epstein-Barr virus DNA. *Proc Natl Acad Sci USA* 1982;79:5688–5692.
629. Sung NS, Kenney S, Gutsch D, Pagano JS. EBNA-2 transactivates a lymphoid-specific enhancer in the BamHI C promoter of Epstein-Barr virus. *J Virol* 1991;65:2164–2169.
630. Suzuki Y, Ohsugi K, Ono Y. Epstein-Barr virus increases phosphoinositide kinase activities in human B cells. *J Immunol* 1992;149: 207–213.
631. Swaminathan S, Hesselton R, Sullivan J, Kieff E. Epstein-Barr virus recombinants with specifically mutated BCRF1 genes. *J Virol* 1993; 67:7406–7413.
632. Swaminathan S, Huneycutt B, Reiss, C, Kieff E. Epstein-Barr virus encoded small RNAs (EBERs) do not modulate interferon effects in Epstein-Barr virus infected lymphocytes. *J Virol* 1992;66:5133–5136.
633. Swaminathan S, Tomkinson B, Kieff E. Recombinant Epstein-Barr virus deleted for small RNA (EBER) genes transforms lymphocytes and replicates in vitro. *Proc Natl Acad Sci USA* 1991;88:1546–1550.
634. Swendeman S, Thorley-Lawson DA. The activation antigen BLAST-2, when shed, is an autocrine BCGF for normal and transformed B cells. *EMBO J* 1987;6:1637–1642.
635. Szekely L, Selivanova G, Magnusson KP, Klein G, Wiman KG. EBNA-5, an Epstein-Barr virus–encoded nuclear antigen, binds to the retinoblastoma and p53 proteins. *Proc Natl Acad Sci USA* 1993;90:5455–9.
636. Szyf M, Eliasson L, Mann V, Klein G, Razin A. Cellular and viral DNA hypomethylation associated with induction of Epstein-Barr virus lytic cycle. *Proc Natl Acad Sci USA* 1985;82:8090–8094.
637. Takada K. Cross-linking of cell surface immunoglobulin induces Epstein-Barr virus in Burkitt lymphoma lines. *Int J Cancer* 1984;33:27–32.
638. Takada K, Ono Y. Synchronous and sequential activation of latently infected Epstein-Barr virus genomes. *J Virol* 1989;63:445–449.
639. Takada K, Shimizu N, Sakuma S, Ono Y. Trans activation of the latent Epstein-Barr virus (Epstein-Barr virus) genome after transfection of the Epstein-Barr virus DNA fragment. *J Virol* 1986;57:1016–1022.
640. Takagi S, Takada K, Sairenji T. Formation of intranuclear replication compartments of Epstein-Barr virus with redistribution of BZLF1 and BMRF1 gene products. *Virology* 1991;185:309–315.
641. Takaki K, Polack A, Bornkamm GW. Expression of a nuclear and a cytoplasmic Epstein-Barr virus early antigen after DNA transfer: cooperation of two distant parts of the genome for expression of the cytoplasmic antigen. *Proc Natl Acad Sci USA* 1984;81:4178–4182.
642. Tanaka A, Nonoyama M. Latent DNA of Epstein-Barr virus. Separation from high molecular weight cell DNA in a neutral glycerol gradient. *Proc Natl Acad Sci USA* 1974;11:4658–4661.
643. Tanner J, Weis J, Fearon D, Whang Y, Kieff E. Epstein-Barr virus gp350/220 binding to the B lymphocyte C3d receptor mediates adsorption, capping, and endocytosis. *Cell* 1987;50:203–213.
644. Tanner J, Whang Y, Sears A, Kieff E. Soluble gp350/220 and deletion mutant glycoproteins block Epstein-Barr virus adsorption to lymphocytes. *J Virol* 1988;62:4452–4464.
645. Tanner JE, Tosato G. Regulation of B-cell growth and immunoglobulin gene transcription by interleukin-6. *Blood* 1992;79:452–459.
646. Tarodi B, Subramanian T, Chinnadurai G. Epstein-Barr virus BHRF1 protein protects against cell death induced by DNA-damaging agents and heterologous viral infection. *Virology* 1994;201:404–407.
647. Taylor N, Countryman J, Rooney C, Katz D, Miller G. Expression of the BZLF1 latency-disrupting gene differs in standard and defective Epstein-Barr viruses. *J Virol* 1989;63:1721–1728.
648. Taylor N, Flemington E, Kolman JL, Baumann RP, Speck SH, Miller G. ZEBRA and a Fos-GCN4 chimeric protein differ in their DNA-binding specificities for sites in the Epstein-Barr virus BZLF1 promoter. *J Virol* 1991;65:4033–4041.
649. Tedder T, Weis J, Clement L, Fearon D, Cooper M. Role of receptors for complement in the induction of polyclonal B cell proliferation and differentiation. *J Clin Immunol* 1986;6:65–73.
650. Thorley-Lawson, D. Characterization of cross reacting antigens on the Epstein-Barr virus envelope and plasma membrane of producer cells. *Cell* 1979;16:33–42.
651. Thorley-Lawson DA, Edson CM. Polypeptides of the Epstein-Barr virus membrane antigen complex. *J Virol* 1979;32:458–467.
652. Thorley-Lawson DA, Geilinger K. Monoclonal antibodies against the major glycoprotein (gp 350/220) of Epstein-Barr virus neutralize infectivity. *Proc Natl Acad Sci USA* 1980;77:5307–5311.
653. Thorley-Lawson DA, Poodry CA. Identification and isolation of the main component (gp350-gp220) of Epstein-Barr virus responsible for generating neutralizing antibodies in vivo. *J Virol* 1982;43:730–736.
654. Thorley-Lawson DA, Schooley RT, Bhan AK, Nadler LM. Epstein-Barr virus superinduces a new human B cell differentiation antigen (B-last 1) expressed on transformed lymphocytes. *Cell* 1982;30:415–425.
655. Toczyski DP, Matera AG, Ward DC, Steitz JA. The Epstein-Barr virus (Epstein-Barr virus) small RNA EBER1 binds and relocalizes ribosomal protein L22 in Epstein-Barr virus-infected human B lymphocytes. *Proc Natl Acad Sci USA* 1994;91:3463–3467.
656. Toczyski DP, Steitz JA. EAP, a highly conserved cellular protein associated with Epstein-Barr virus small RNAs (EBERs). *EMBO J* 1991; 10:459–466.
657. Toczyski DP, Steitz JA. The cellular RNA-binding protein EAP recognizes a conserved stem-loop in the Epstein-Barr virus small RNA EBER 1. *Mol Cell Biol* 1993;13:703–710.
658. Tomkinson B, Kieff E. Second site homologous recombination in Epstein-Barr virus: insertion of type 1 EBNA-3 genes in place of type 2 has no effect on *in vitro* infection. *J Virol* 1992;66:780–789.
659. Tomkinson B, Kieff E. Use of second site homologous recombination to demonstrate that Epstein-Barr virus nuclear protein EBNA-3B is not important for B-lymphocyte infection or growth transformation *in vitro*. *J Virol* 1992;66:2893–2903.
660. Tomkinson B, Robertson E, Kieff E. Epstein Barr virus nuclear proteins (EBNA) 3A and 3C are essential for B lymphocyte growth transformation. *J Virol* (in press).
661. Tomkinson B, Robertson E, Yalamanchili R, Longnecker R, Kieff E. Epstein-Barr virus recombinants from overlapping cosmid fragments.

J Virol 1993;67:7298–7306.

662. Tong X, Yalamanchili R, Harada S, Kieff E. The EBNA-2 arginine-glycine domain is critical but not essential for B-lymphocyte growth transformation; the rest of region 3 lacks essential interactive domains. *J Virol* 1994;68:6188–6197.
663. Tong X, Wang F, Thut C, Kieff E. The Epstein-Barr virus nuclear protein 2 acidic domain can interact with TFIIB, TAF40, and RPA70, but not with TBP. *J Virol* 1995;69:585–588.
664. Torrisi MR, Cirone M, Pavan A, et al. Localization of Epstein-Barr virus envelope glycoproteins on the inner nuclear membrane of virus-producing cells. *J Virol* 1989;63:828–832.
665. Tosato G, Blaese RM. Epstein-Barr virus infection and immunoregulation in man. *Adv Immunol* 1985;37:99–149.
666. Tosato G, Tanner J, Jones KD, Revel M, Pike SE. Identification of interleukin-6 as an autocrine growth factor for Epstein-Barr virus-immortalized B cells. *J Virol* 1990;64:3033–3041.
667. Tsang S, Wang F, Izumi KM, Kieff E. Delineation of the cis acting element mediating EBNA2 transactivation of latent infection membrane protein expression. *J Virol* 1991;65:6765–6771.
668. Tsoukas CD, Lambris JD. Expression of Epstein-Barr virus/C3d receptors on T cells: biological significance. *Immunol Today* 1993;14: 56–59.
669. Tsurumi T, Daikoku T, Kurachi R, Nishiyama Y. Functional interaction between Epstein-Barr virus DNA polymerase catalytic subunit and its accessory subunit in vitro. *J Virol* 1993;67:7648–7653.
670. Tsurumi T, Daikoku T, Nishiyama Y. Further characterization of the interaction between the Epstein-Barr virus DNA polymerase catalytic subunit and its accessory subunit with regard to the 3′-to-5′ exonucleolytic activity and stability of initiation complex at primer terminus. *J Virol* 1994;68:3354–3363.
671. Tsurumi T, Kobayashi A, Tamai K, Daikoku T, Kurachi R, Nishiyama Y. Functional expression and characterization of the Epstein-Barr virus DNA polymerase catalytic subunit. *J Virol* 1993;67:4651–4658.
672. Tugwood JD, Lau WH, O SK, et al. Epstein-Barr virus–specific transcription in normal and malignant nasopharyngeal biopsies and in lymphocytes from healthy donors and infectious mononucleosis patients. *J Gen Virol* 1987;68:1081–1091.
673. Uchibayashi N, Kikutani H, Barsumian E, et al. Recombinant soluble Fc-epsilon receptor II (Fc_ERII/CD23) has IgE binding activity but no B cell growth promoting activity. *J Immunol* 1989;142:3901–3908.
674. van Santen V, Cheung A, Hummel M, Kieff E. RNA encoded by the IR1-U2 region of Epstein-Barr virus DNA in latently infected growth transformed cells. *J Virol* 1983;46:424–433.
675. van Santen V, Cheung A, Kieff E. Epstein-Barr virus (Epstein-Barr virus) RNA. VII. Viral cytoplasmic RNA in a restringently infected cell line transformed in vitro by Epstein-Barr virus. *Proc Natl Acad Sci USA* 1981;78:1930–1934.
676. Vieira P, de Waal-Malefyt R, Dang MN, et al. Isolation and expression of human cytokine synthesis inhibitory factor cDNA clones: homology to Epstein-Barr virus open reading frame BCRFI. *Proc Natl Acad Sci USA* 1991;88:1172–1176.
677. Volsky D, Gross T, Sinangil F, et al. Expression of Epstein-Barr virus (Epstein-Barr virus) DNA and cloned DNA fragments in human lymphocytes following Sendai virus envelope-mediated gene transfer. *Proc Natl Acad Sci USA* 1984;81:5926–5930.
678. Volsky DJ, Shapiro IM, Klein G. Transfer of Epstein-Barr virus receptor-negative cells permits virus penetration and antigen expression. *Proc Natl Acad Sci USA* 1980;77:5453–5457.
679. de Waal-Malefyt R, Haanen J, Spits H, et al. Interleukin 10 (IL-10) and viral IL-10 strongly reduce antigen-specific human T cell proliferation by diminishing the antigen-presenting capacity of monocytes via downregulation of class II major histocompatibility complex expression. *J Exp Med* 1991;174:915–924.
680. Wagner EK, Roizman B, Savage T, et al. Characterization of the DNA of herpesviruses associated with Lucke's adenocarcinoma of the frog and Burkitt's Lymphoma of man. *Virology* 1970;42:257–261.
681. Walling DM, Edmiston SN, Sixbey JW, Abdel-Hamid M, Resnick L, Raab-Traub N. Coinfection with multiple strains of the Epstein-Barr virus in human immunodeficiency virus-associated hairy leukoplakia. *Proc Natl Acad Sci USA* 1992;89:6560–6564.
682. Walls D, Perricaudet M. Novel downstream elements upregulate transcription initiated from an Epstein-Barr virus latent promoter. *EMBO J* 1991;10:143–151.
683. Wang D, Liebowitz D, Kieff E. The truncated form of the Epstein-Barr virus latent-infection membrane protein expressed in virus replication does not transform rodent fibroblasts. *J Virol* 1988;62:2337–2346.
684. Wang D, Liebowitz D, Kieff E. An Epstein-Barr virus membrane protein expressed in immortalized lymphocytes transforms established rodent cells. *Cell* 1985;43:831–840.
685. Wang D, Liebowitz D, Wang F, et al. Epstein-Barr virus latent infection membrane protein alters the human B-lymphocyte phenotype: deletion of the amino terminus abolishes activity. *J Virol* 1988;62:4173–4184.
686. Wang F, Gregory C, Sample C, et al. Epstein-Barr virus latent infection membrane and nuclear proteins 2 and 3C are effectors of phenotypic changes in B lymphocytes: EBNA 2 and LMP cooperatively induce CD23. *J Virol* 1990;64:2309–2318.
687. Wang F, Gregory CD, Rowe M, et al. Epstein-Barr virus nuclear antigen 2 specifically induces expression of the B-cell activation antigen CD23. *Proc Natl Acad Sci USA* 1987;83:3452–3457.
688. Wang F, Kikutani H, Tsang S, Kishimoto T, Kieff E. Epstein-Barr virus nuclear protein 2 transactivates a cis-acting CD23 DNA element. *J Virol* 1991;65:4101–4106.
689. Wang F, Marchini A, Kieff E. Epstein-Barr virus recombinants: use of positive selection markers to rescue mutants in Epstein-Barr virus negative B lymphoma cells. *J Virol* 1991;65:1701–1709.
690. Wang F, Petti L, Braun D, Seung S, Kieff E. A bicistronic Epstein-Barr virus mRNA encodes two nuclear proteins in latently infected, growth transformed lymphocytes. *J Virol* 1987;61:945–954.
691. Wang F, Tsang S, Kurilla M, Cohen J, Kieff E. Epstein-Barr virus nuclear antigen 2 transactivates the latent membrane protein (LMP1). *J Virol* 1990;64:3407–3416.
692. Weigel R, Fischer DK, Heston L, Miller G. Constitutive expression of Epstein-Barr virus-encoded RNAs and nuclear antigen during latency and after induction of Epstein-Barr virus replication. *J Virol* 1985;53: 254–259.
693. Weigel R, Miller G. Major EB virus specific cytoplasmic transcripts in a cellular clone of the HR-1 Burkitt lymphoma line during latency and after induction of viral replicative cycle by phorbol esters. *Virology* 1983;125:287–298.
694. Weis JJ, Tedder TF, Fearon DT. Identification of a 145,000 M_r membrane protein as the C3d receptor (CR2) of human B lymphocytes. *Proc Natl Acad Sci USA* 1984;81:881–885.
695. Weis JJ, Toothaker LE, Smith JA, Weis JH, Fearon DT. Structure of the human B lymphocyte receptor for C3d and the Epstein-Barr virus and relatedness to other members of the family of C3/C4 binding proteins. *J Exp Med* 1988;167:1047–1066.
696. Weiss A, Littman D. Signal transduction by lymphocyte antigen receptors. *Cell* 1994;76:263–274.
697. Wells A, Koide N, Klein G. Two large virion envelope glycoproteins mediate Epstein-Barr virus binding to receptor-positive cells. *J Virol* 1982;41:286–297.
698. Wen LT, Lai PK, Bradley G, Tanaka A, Nonoyama M. Interaction of Epstein-Barr viral (Epstein-Barr virus) origin of replication (oriP) with EBNA-1 and cellular anti-EBNA-1 proteins. *Virology* 1990;178: 293–296.
699. Whang Y, Silberklang M, Morgan A, et al. Expression of the Epstein-Barr virus gp350/220 gene in rodent and primate cells. *J Virol* 1987; 61:1796–1807.
700. Williams DL, Kowalski D. Easily unwound DNA sequences and hairpin structures in the Epstein-Barr virus origin of plasmid replication. *J Virol* 1993;67:2707–2715.
701. Wilson JB, Weinberg W, Johnson R, Yuspa S, Levine AJ. Expression of the BNLF-1 oncogene of Epstein-Barr virus in the skin of transgenic mice induces hyperplasia and aberrant expression of keratin 6. *Cell* 1990;61:1315–1327.
702. Woisetschlaeger M, Jin XW, Yandava CN, Furmanski LA, Strominger JL, Speck SH. Role for the Epstein-Barr virus nuclear antigen 2 in viral promoter switching during initial stages of infection. *Proc Natl Acad Sci USA* 1991;88:3942–3946.
703. Woisetschlaeger M, Yandava CN, Furmanski LA, Strominger JL, Speck SH. Promoter switching in Epstein-Barr virus during the initial stages of infection of B lymphocytes. *Proc Natl Acad Sci USA* 1990;87:1725–1729.
704. Wong K, Levine A. Identification and mapping of Epstein-Barr virus early antigens and demonstration of a viral gene activator that functions in trans. *J Virol* 1986;60:149–156.
705. Wong KM, Levine AJ. Characterization of proteins encoded by the Epstein-Barr virus transactivator gene BMLF1. *Virology* 1989;168: 387–393.

706. Wysokenski D, Yates J. Multiple EBNA-1-binding sites are required to form an EBNA-1-dependent enhancer and to activate a minimal replicative origin within ori P of Epstein-Barr virus. *J Virol* 1989;63: 2657–2666.
707. Yalamanchili R, Tong X, Grossman S, Johansen E, Mosialos G, Kieff E. Genetic and biochemical evidence that EBNA 2 interaction with a 63 kd cellular GTG-binding protein is essential for B lymphocyte growth transformation by Epstein-Barr virus. *Virology* 1994;204: 634–641
708. Yamamoto N, Mueller-Lantzch N, zur Hausen H. Effect of actinomycin D and cycloheximide on Epstein-Barr virus early antigen induction in lymphoblastoid cells. *J Gen Virol* 1980;51:255–261.
709. Yamamoto N, zur Hausen H. Inducing Epstein-Barr virus early antigens by intercalating chemicals in B95-8 cells. *Virology* 1981;115:390–394.
710. Yaswen LR, Stephens EB, Davenport LC, Hutt FL. Epstein-Barr virus glycoprotein gp85 associates with the BKRF2 gene product and is incompletely processed as a recombinant protein. *Virology* 1993;195: 387–96.
711. Yates J, Warren N, Reisman D, Sugden B. A cis-acting element from the Epstein-Barr viral genome that permits stable replication of recombinant plasmids in latently infected cells. *Proc Natl Acad Sci USA* 1984; 81:3806–3810.
712. Yates JL, Camiolo SM. Dissection of DNA replication and enhancer activation function of Epstein-Barr virus nuclear antigen 1. *Cancer Cells* 1988;6:197–205.
713. Yates JL, Guan N. Epstein-Barr virus-derived plasmids replicate only once per cell cycle and are not amplified after entry into cells. *J Virol* 1991;65:483–488.
714. Yates JL, Warren N, Sugden B. Stable replication of plasmids derived from Epstein-Barr virus in various mammalian cells. *Nature* 1985;313: 812–815.
715. Yokoyama S, Staunton D, Fisher R, Amiot M, Fortin JJ, Thorley-Lawson DA. Expression of the Blast-1 activation/adhesion molecule and its identification as CD48. *J Immunol* 1991;146:2192–2200.
716. Young LS, Clark D, Sixbey JW, Rickinson AB. Epstein-Barr virus receptors on human pharyngeal epithelium. *Lancet* 1986;8475:240–242.
717. Young LS, Lau R, Rowe M, et al. Differentiation-associated expression of the Epstein-Barr virus BZLF1 transactivator protein in oral hairy leukoplakia. *J Virol* 1991;65:2868–2874.
718. Young LS, Yao QY, Rooney CM, et al. New type B isolates of Epstein-Barr virus from Burkitt's lymphoma and from normal individuals in endemic areas. *J Gen Virol* 1987;68:2853–2862.
719. Zhang K, Clark EA, Saxon A. CD40 stimulation provides an IFN-gamma-independent and IL-4-dependent differentiation signal directly to human B cells for IgE production. *J Immunol* 1991;146:1836–1842.
720. Zhang PF, Marcus SC. Conformation-dependent recognition of baculovirus-expressed Epstein-Barr virus gp350 by a panel of monoclonal antibodies. *J Gen Virol* 1993;74:2171–2179.
721. Zhang Q, Brooks L, Busson P, et al. Epstein-Barr virus (Epstein-Barr virus) latent membrane protein 1 increases HLA class II expression in an Epstein-Barr virus-negative B cell line. *Eur J Immunol* 1994;24: 1467–1470.
722. Zhang Q, Gutsch D, Kenney S. Functional and physical interaction between p53 and BZLF1: implications for Epstein-Barr virus latency. *Mol Cell Biol* 1994;14:1929–1938.
723. Zhang S, Nonoyama M. The cellular proteins that bind specifically to the Epstein-Barr virus origin of plasmid DNA replication belong to a gene family. *Proc Natl Acad Sci USA* 1994;91:2843–2847.
724. Zimber SU, Kremmer E, Grasser F, Marschall G, Laux G, Bornkamm GW. The Epstein-Barr virus nuclear antigen 2 interacts with an EBNA2 responsive cis-element of the terminal protein 1 gene promoter. *EMBO J* 1993;12:167–175.
725. Zimber U, Aldinger HK, Lenoir GM, et al. Geographic prevalence of two Epstein-Barr virus types. *Virology* 1986;154:56–66.
726. Zimber-Strobl U, Suentzenich KO, Laux G, et al. Epstein-Barr virus nuclear antigen 2 activates transcription of the terminal protein gene. *J Virol* 1991;65:415–423.
727. zur Hausen H, Bornkamm GW, Schmidt R, Hecker E. Tumor initators and promoters in the induction of Epstein-Barr virus. *Proc Natl Acad Sci USA* 1979;76:782–785.
728. zur Hausen H, O'Neill FJ, Freese UK, Hecher E. (1978): Persisting oncogenic herpesvirus induced by tumor promoter TPA. *Nature*, 1978; 272:373–375.
729. zur Hausen H, Schulte-Holthausen H, Klein G, et al. EB-virus DNA in biopsies of Burkitt tumors and anaplastic carcinomas of the nasopharynx. *Nature* 1970;228:1056–1057.

Fundamental Virology, Third Edition
edited by B.N. Fields, D.M. Knipe, P.M. Howley, et al.
Lippincott - Raven Publishers, Philadelphia © 1996

CHAPTER 34

Poxviridae: The Viruses and Their Replication

Bernard Moss

The *Poxviridae* comprise a large family of complex DNA viruses that replicate in the cytoplasm of vertebrate and invertebrate cells. The most notorious member, variola virus, caused smallpox and consequently had a profound impact on human history. Smallpox was finally eradicated in 1977, nearly two centuries after the introduction of prophylactic inoculations with cowpox and vaccinia virus. Vaccination contributed to present concepts of infectious disease and immunity. Moreover, vaccinia virus was the first animal virus seen microscopically, grown in tissue culture, accurately titered, physically purified, and chemically analyzed. A simplified view of virus particles, as packets of nucleic acid, was revised following the discovery of RNA synthetic activity in purified vaccinia viri-

B. Moss: Laboratory of Viral Diseases, National Institute of Allergy and Infectious Diseases, National Institutes of Health, Bethesda, Maryland 20892.

ons. This finding stimulated investigations that led to the discovery of transcriptase and reverse transcriptase activities in RNA viruses and to the elucidation of structural features of viral and eukaryotic mRNA, including the 5′ cap and 3′ poly(A) tail.

Research on poxviruses did not languish following the eradication of smallpox. Recombinant DNA technology eliminated an obstacle to working with these large viruses, and remarkable progress has been made in our understanding of virus replication (Fig. 1). Furthermore, the development of vaccinia virus as a live recombinant expression vector provided a new tool for immunologists and biochemists as well as an alternative approach to the development of vaccines against a variety of infectious agents. While past studies with poxviruses contributed to an understanding of viral pathogenesis, discoveries of virus-encoded proteins that affect cell growth and modulate immune defense mechanisms provide new insights into virus/host relationships.

CLASSIFICATION

The general properties of *Poxviridae* include (a) a large complex virion containing enzymes that synthesize mRNA, (b) a genome composed of a single linear double-stranded DNA molecule of 130–300 kilobase pairs (kbp) with a hairpin loop at each end, and (c) a cytoplasmic site of replication. The *Poxviridae* are divided into two subfamilies, *Chordopoxvirinae* and *Entomopoxvirinae,* based on vertebrate and insect host range (Table 1). African swine fever virus shares some properties with

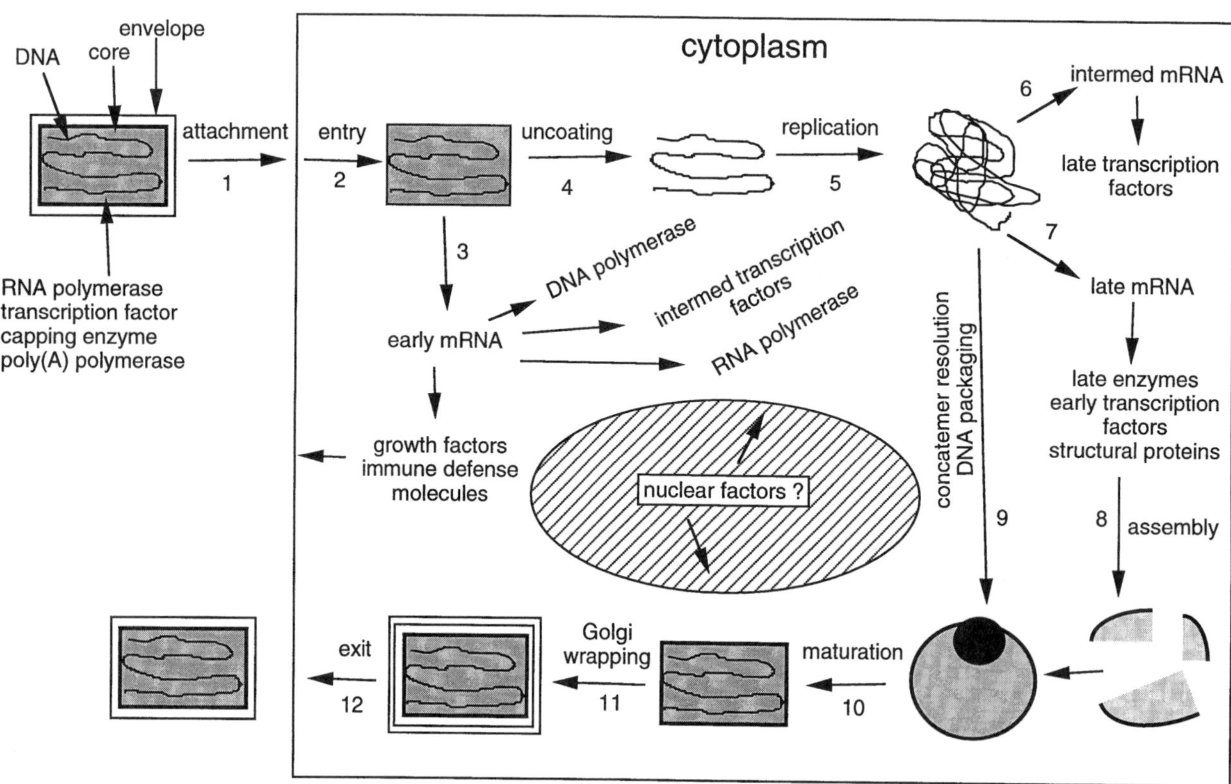

FIG. 1. The replication cycle of vaccinia virus. Virions, containing a double-stranded DNA genome, enzymes and transcription factors, attach to cells *[1]* and fuse with the cell membrane, releasing cores into the cytoplasm *[2]*. The cores synthesize early mRNAs that are translated into a variety of proteins, including growth factors, immune defense molecules, enzymes, and factors for DNA replication and intermediate transcription *[3]*. Uncoating occurs *[4]* and the DNA is replicated to form concatemeric molecules *[5]*. Intermediate genes in the progeny DNA are transcribed and the mRNAs are translated to form late transcription factors *[6]*. The late genes are transcribed and the mRNAs are translated to form virion structural proteins, enzymes, and early transcription factors *[7]*. Assembly begins with the formation of discrete membrane structures *[8]*. The concatemeric DNA intermediates are resolved into unit genomes and packaged in immature virions *[9]*. Maturation proceeds to the formation of infectious intracellular mature virions *[10]*. The virions are wrapped by modified Golgi membranes and transported to the periphery of the cell *[11]*. Fusion of the wrapped virions with the plasma membrane results in release of extracellular enveloped virus *[12]*. Although replication occurs entirely in the cytoplasm, nuclear factors may be involved in transcription and assembly.

TABLE 1. *Family* Poxviridae

Subfamilies	Genera	Member viruses	Features
Chordopoxvirinae (vertebrate poxviruses)	*Orthopoxvirus*	Camelpox, cowpox, ectromelia, monkeypox racoonpox, skunkpox, taterapox, Uasin Gishu[a], vaccinia[b], variola, volepox	Brick-shaped virion, DNA ~200 kbp, G+C ~36%, wide to narrow host range, variola (smallpox), vaccinia (smallpox vaccine)
	Parapoxvirus	Auzduk disease[a], chamois contagious ecthyma[a], orf[b], pseudocowpox, parapox of deer, sealpox[a]	Ovoid virion, DNA~140 kbp, G+C ~64%, mainly ungulates
	Avipoxvirus	Canarypox, fowlpox[b], juncopox, mynahpox, pigeonpox, psittacinepox, quailpox, peacockpox[a], penguinpox[a], sparrowpox, starlingpox, turkeypox	Brick-shaped virion, DNA~260 kbp, G+C ~35%, birds, arthropod tranmission
	Capripoxvirus	goatpox, lumpy skin disease, sheeppox[b]	Brick-shaped virion, DNA ~150 kbp, ungulates, arthropod transmission
	Leporipoxvirus	Hare fibroma, myxoma[b], rabbit fibroma, squirrel fibroma	Brick-shaped virion, DNA ~160 kbp, G+C ~40%, leporids and squirrels, localized tumors, arthropod tranmission
	Suipoxvirus	Swinepox	Brick-shaped virion, DNA ~170 kbp, narrow host range
	Molluscipoxvirus	Molluscum contagiosum	Brick-shaped virion, DNA ~180 kbp, G+C ~60%, human host, localized tumors, spread by contact
	Yatapoxvirus	Tanapox, Yaba monkey tumor[b]	Brick-shaped virion, DNA ~145 kbp, G+C ~33%, primates and rodents (?)
Entomopoxvirinae (insect poxviruses)	*Entomopoxvirus* A	*Melolontha melolontha*[b]	Ovoid virions, DNA ~260–370 kbp, Coleoptera
	Entomopoxvirus B	*Amsacta moorei*[b]	Ovoid virions, DNA~225 kbp, G+C ~18.5%, Lepidoptera and Orthoptera
	Entomopoxvirus C	Chironimus luridus[b]	Brick-shaped virions, DNA ~250–380 kbp, Diptera

[a] Probable member of genus.
[b] Prototypal member.

poxviruses, including the ones listed earlier, but is morphologically distinct and separately classified.

The *Chordopoxvirinae* consist of eight genera: *Orthopoxvirus*, *Parapoxvirus*, *Avipoxvirus*, *Capripoxvirus*, *Leporipoxvirus*, *Suipoxvirus*, *Molluscipoxvirus*, and *Yatapoxvirus*. Members of a genus are genetically and antigenically related and have a similar morphology and host range. There is, however, intergenus serologic cross-reactivity for at least the group-specific NP antigen (522,576) and generally members of one genus can reactivate heat-treated poxviruses of another (176,217). While no intergenus cross hybridizations of DNA were detected under stringent conditions (>75% identity) (160), some underlying gene order and sequence similarities with orthopoxviruses have been shown for capripoxviruses (189), avipoxviruses (347,525), leporipoxviruses (544), and parapoxviruses (178).

The similarity in restriction endonuclease maps of several *Orthopoxvirus* genomes (160,314) has been substantiated by the more than 90% sequence identity of genes of vaccinia and variola viruses (201,323,473). Vaccinia virus has no known host and the question of its origin, from an independent species or hybrid strain, has not been resolved (38). Orthopoxviruses indigenous to the Americas (e.g., raccoon poxvirus and volepox virus) are genetically divergent from geographically separated members of the genus (284). Biological descriptions of individual orthopoxviruses and references to that literature are available (175), but there is less comparative information for members of the other poxvirus genera. The restriction endonuclease patterns of several avipoxviruses are similar (466), although that of quailpox appears divergent (199). Members of the *Capripoxvirus* genus are closely related as judged by immunologic analysis (280) and genome structure (190), whereas parapoxviruses appear to be genetically diverse (187,434). Restriction endonuclease analysis revealed two principal variants of molluscum contagiosum virus (123,420).

The *Entomopoxvirinae* have been divided into three genera based on the insect host of isolation (18). Only the prototypal members are listed in Table 1. Genetic information regarding these viruses remains scant. A distant relationship between insect and vertebrate poxviruses is supported by comparisons of the DNA sequences of genes encoding the thymidine kinase (210,311b), DNA polymerase (368), and nucleoside triphosphate phosphohydrolase I (NPH I) (214,588).

VIRION STRUCTURE

Morphology

The virions of poxviruses are larger than those of other animal viruses and are just discernible by light microscopy (81,396). The ultrastructural appearance of the particles varies according to the preparation methods. Vaccinia virions appear as smooth, rounded rectangles of about 350 by 270 nm by cryoelectron microscopy of unfixed and unstained vitrified samples (Fig. 2) (145). A 30-nm-membrane-delimited surface layer surrounds a homogenous core. Using the same technique, cores produced by treating virions with a detergent, reducing agent, and deoxyribonuclease appear to be studded with 20-nm spikes. Two types of particles, C and M, are seen by negative staining of whole virions (371,565). The C form has a smooth exterior similar to the particles viewed by cryoelectron microscopy, whereas the M form has a beaded appearance also noted by freeze-etching (332). Incubation of virions with a nonionic detergent converts the M forms to C forms, but otherwise the morphology appears unchanged (152). Further treatment with a reducing agent leads to removal of the outer coat, suggesting that disulfide-bonded proteins hold the latter together. Negatively stained images of cores,

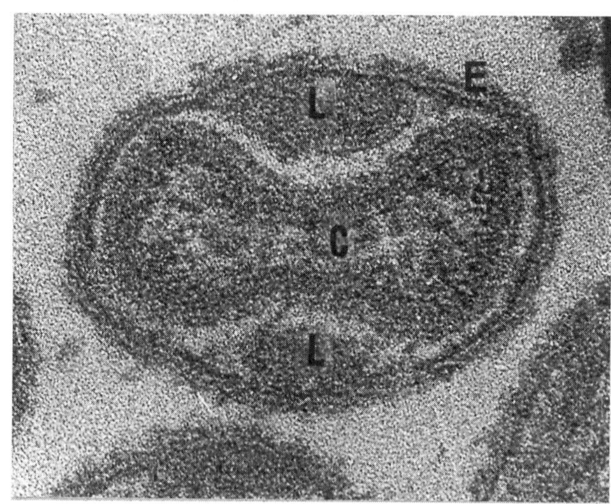

FIG. 3. Electron-microscopic image of a thin-sectioned intracellular mature vaccinia virus particle showing core *(C)*, lateral bodies *(L)*, and external membrane *(E)*. From Pogo and Dales (416), with permission.

isolated by treatment with a detergent and reducing agent, have a rectangular shape; the wall of the core appears to be composed of an outer layer of cylindrical subunits 10 nm in length and 5 nm in diameter and an inner 5-nm-thick smooth layer (152). Trypsin-sensitive structures called lateral bodies remain associated with cores prepared in this manner. In fixed and stained thin sections of virions, the core frequently appears dumbbell-shaped with the lateral bodies in the concavities (Fig. 3) (122a). Recently, it was suggested that the surface tubules and the dumbbell-shaped cores result from nonisotropic drying, as they are most clearly seen in dehydrated samples (145). Even so, an underlying stucture must contribute to the acquisition of this highly characteristic appearance. Cylindrical elements that may take an S-shape or more complex flowerlike structures, presumably representing nucleoprotein, have been visualized within poxvirus cores (240b,411b).

The preceding descriptions apply to infectious particles isolated from cells and originally called *intracellular naked virions* but recently termed *intracellular mature virions* (IMV). Extracellular enveloped virions (EEV), isolated from the tissue culture medium, contain an additional lipoprotein envelope and have a lower buoyant density than IMV (400).

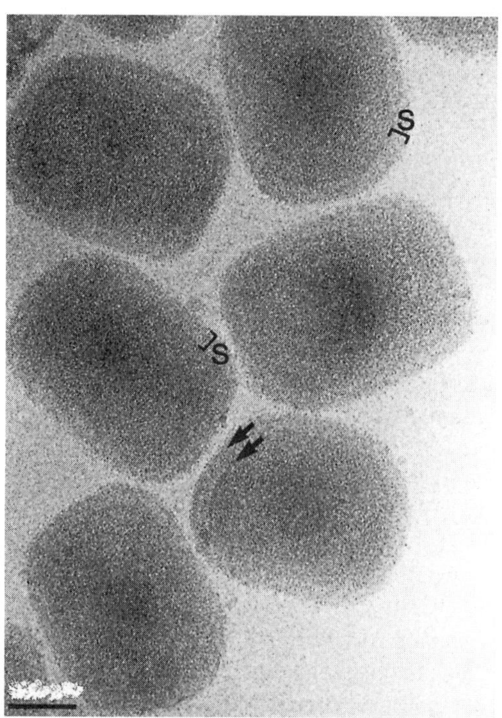

FIG. 2. Vitrified suspension of intracellular mature vaccinia virus particles observed by cryoelectron microscopy. The 30-nm-thick surface domain *(S)* is delimited by arrows. From Dubochet et al. (145), with permission.

Chemical Composition

The large size of poxvirus virions facilitated their isolation by low-speed centrifugation and permitted accurate chemical determinations to be made at a relatively early date (498). The principal components of vaccinia virions are protein, lipid, and DNA. These account for 90%, 5%, and 3.2%, respectively, of the dry weight (590), which has been estimated to be about 5×10^{-15} g (497). In contrast, about one-third of fowlpox virions is lipid (310). The lipid components of vaccinia virions are predominantly cholesterol and phospholipids (512), whereas fowlpox virions also contain squalene and cholesterol esters (310). Carbohydrate is present in the EEV as a constituent of glycoproteins. Spermine and spermidine (295) and trace amounts of RNA (443) also have been found in vaccinia virions, but their significance has not been established.

Genome

Poxviruses have linear double-stranded DNA genomes that vary from about 130 kbp in parapoxviruses (334,434) to about 300 kbp in avipoxviruses (240a). Inverted terminal repetitions (ITRs), which are identical but oppositely oriented sequences at the two ends of the genome (Fig. 4), are present in all poxviruses examined: orthopoxviruses (160,185,284,323,573), parapoxviruses (182,334), leporipoxviruses (132), and capripoxviruses (190). The lengths of the ITRs are variable even within a genus. For example, the ITRs of vaccinia virus Western Reserve (WR) and Copenhagen are 10.5–12 kbp, respectively, whereas that of variola virus is only 725 bp (323).

The two strands of vaccinia virus DNA are connected by hairpin loops to form a covalently continuous polynucleotide chain (35,198), as depicted in Fig. 4. The loops are A+T-rich, cannot form a completely base-paired structure, and exist in two forms that are inverted and complementary in sequence (35). Similar telomeric structures have been found in variola (323), cowpox (412) and Shope fibroma viruses (131) and are undoubtedly characteristic of the entire family. A similar structure is also present at the ends of the genome of African swine fever virus (205).

A region of less than 100 bp, adjacent to the hairpin loops, is highly conserved and contains sequences required for the resolution of replicating concatemeric forms of DNA (135,339). In orthopoxviruses this region is followed by sets of short, tandemly repeated sequences separated by a highly homologous 250–350-bp segment (35,323,394, 413,574). Sequence similarities suggest that various-length repeats evolved by unequal crossing-over events (34). The precise number of repeats varies in different species and even in individual isolates from the same passaged strain (360). Parapoxviruses also contain sets of short tandem repeats (182), whereas in leporipoxviruses there are related, tandemly repeated, imperfect palindromes (537).

Since the ITRs may include coding regions, some genes are present at both ends of the genome. In the orthopoxviruses, the terminal regions are hypervariable and contain transpositions and deletions (17,158,365,415). In Shope fibroma virus, closely related genes that probably arose by duplication are present within the ITRs (542). The variability of the end regions of poxvirus genomes is consistent with the presence there of large blocks of genes that are nonessential for replication in tissue culture (288,410).

The complete genomic sequence of the Copenhagen strain of vaccinia virus (201) and most of the sequence of the laboratory prototype WR strain (151) are available. The

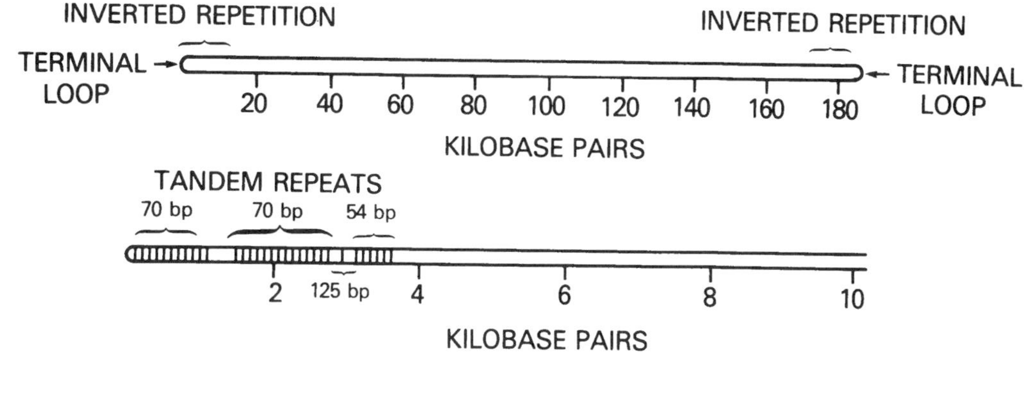

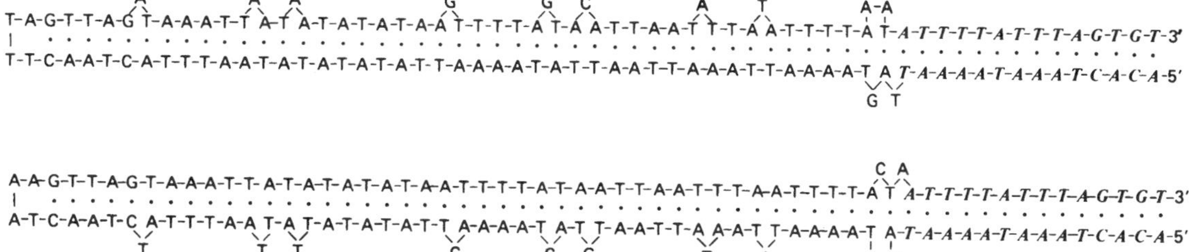

FIG. 4. Structural features of vaccinia viral DNA. A representation of the entire linear double-stranded DNA genome and an expansion of the 10,000-bp inverted terminal repetition are shown in the upper part. The nucleotide sequences of the inverted and complementary forms of the terminal loops are below. From Moss et al. (361) and Baroudy (35), with permission.

sequences are similar except for some small transpositions and deletions, primarily near the ends of the genome. The Copenhagen sequence is 191,636 bp and contains about 185 unique, nonoverlapping, open reading frames (ORFs) of more than 65 amino acids. The ORFs within the left 70,000 bp nearly all point leftward and the remainder, with the exception of a large block between 108,000 and 130,000 bp and a few small blocks, mostly point rightward.

The convention adopted for naming orthopoxvirus ORFs consists of using the *Hin*dIII restriction endonuclease DNA fragment letter, followed by the ORF number (from left to right) within the fragment, and L or R, depending on the direction of the ORF (448). For example, the B1R ORF is the leftmost one that starts in the *Hin*dIII B fragment and is read toward the right end of the genome. An exception to this rule was made for the *Hin*dIII C fragment; the ORFs were numbered from right to left to avoid starting at the end of the genome, which is highly variable (201). In papers published prior to this convention, the ORFs in the *Hin*dIII C fragment may be numbered from left to right. There are also strain differences in the precise number of ORFs caused by small nucleotide changes or terminal transpositions. For consistency in this chapter the Copenhagen nomenclature is used even though most references are to the WR strain. The order of the 16 *Hin*dIII fragments of vaccinia virus and the ORFs with known roles or significant homologies are shown in Fig. 5.

The complete and nearly identical DNA sequences of the India (473) and Bangladesh (323) strains of variola virus are also available. The 186,102-bp genome contains a short ITR of 725 bp and 187 major ORFs of which 150 have more than 90% sequence identity to those of vaccinia

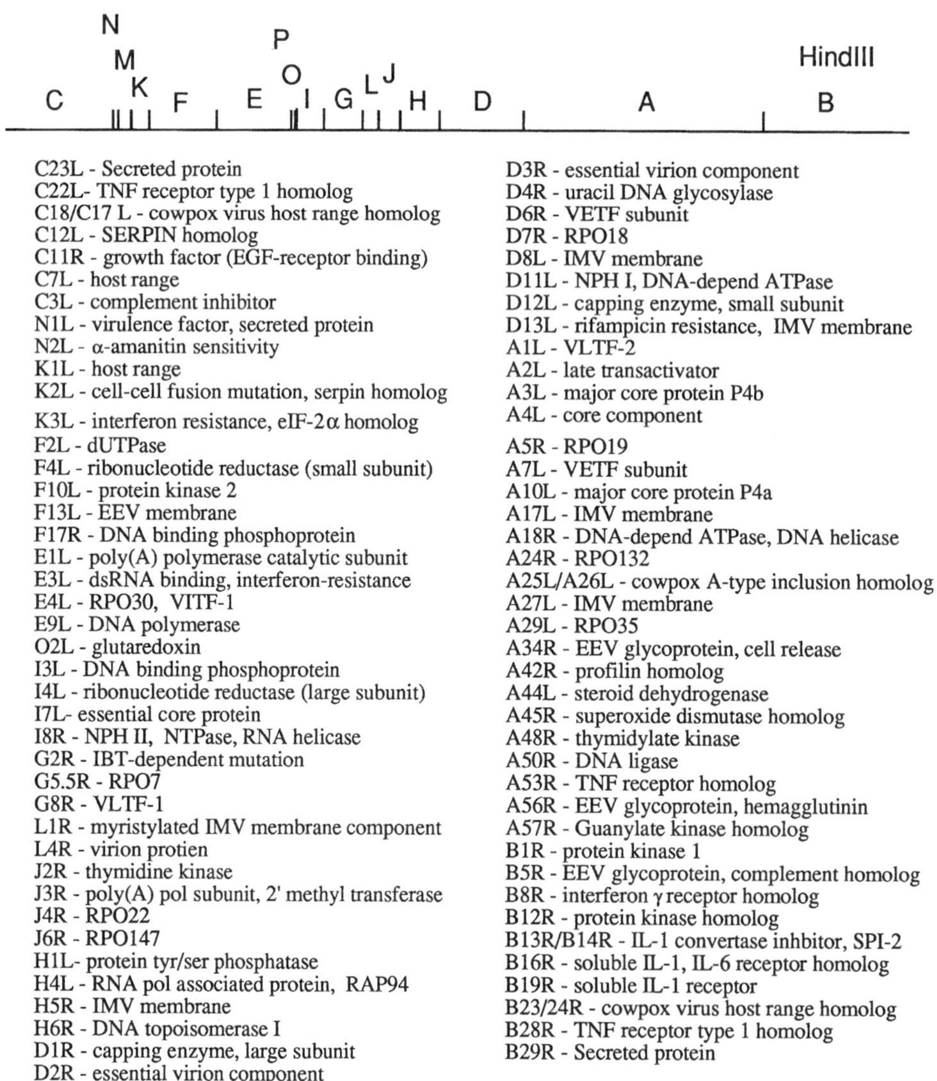

FIG. 5. Arrangement of vaccinia virus genes. A *Hin*dIII restriction endonuclease map of the vaccinia virus genome is shown. Below are listed ORFs encoding proteins for which there is functional information or a significant homology. The first and last three ORFs in the list are present in the inverted terminal repetition.

virus. Variola virus has some ORFs that are truncated and others that are elongated, relative to those of vaccinia virus. The terminal region encodes several potentially novel proteins, including one that predicts a 214-kd polypeptide with transmembrane domains.

Sequence information for other poxviruses is still fragmentary, although more than 65 kbp have been reported for fowlpox virus (347,525).

Polypeptides

Consistent with their large and complex structures, poxvirus virions are composed of many polypeptides. About 30 bands are readily resolved by polyacrylamide gel electrophoresis of vaccinia virus virions that have been purified from infected cells and disrupted with sodium dodecyl sulfate and a reducing agent (458). Considerably more polypeptides can be detected by two-dimensional analysis (161,379). Several polypeptides have been localized near the exterior of IMV by one or more of the following procedures: surface-specific labeling, sensitivity to proteases, extraction with nonionic detergents, and reactivity with virus neutralizing antibodies (208,380,426,437,458,513, 523). Proteins that meet the criteria of IMV membrane-association to varying degrees are listed in Table 2. Some of these proteins may form a disulfide-bonded matrix, as both a reducing agent and a nonionic detergent are required for their release (244). There is evidence for physical association of the A17L polypeptide with the trimeric A27L protein. The L1R polypeptide is myristylated, presumably for its association with the IMV membrane. One of the membrane-associated polypeptides, not yet genetically identified, contains N-acetylglucosamine but apparently lacks other types of sugar residues (186).

After IMV are treated with a nonionic detergent and reducing agent, the core can be recovered by low-speed centrifugation. Disruption of the core with deoxycholate releases a soluble enzyme fraction, described in the next section, from insoluble structural proteins (390). Four of the latter proteins, encoded by the F17R, L4R, A3L, and A10L ORFs, account for about 70% of the viral core by weight (Table 2). The F17R 11-kd phosphoprotein and the L4R 25-kd polypeptide bind DNA and may correspond to nucleoprotein constituents (244). The mature G7L, L4R, A3L, A10L, and A12L proteins are smaller than predicted

TABLE 2. *Nonenzymatic virion components*

	ORF	kd	Properties	References
Intracellular mature virus membrane-associated	I5L	8.7	Hydrophobic, basic	523
	L1R	27.3	Myristylated, hydrophobic	179,181,426
	H3L	37.5	Hydrophobic	102,523
	H5R	22.3	Surface location, hydrophobic C-terminal, NA	207,208
	D8L	35.3	Surface location, hydrophobic C-terminal, cell-membrane binding, NE, plaque size/virulence	101,292,312,376
	D13L	61.9	Rifampicin resistance	26,344,505,524,549,596
	A13L	7.7	Oligomeric	523
	A14L	10.0	Oligomeric, hydrophobic	523
	A17L	23.0	Putative signal peptide, dimer, assoc. with A27 product, NA	245,435
	A27L	12.6	Surface location, fusion protein, NE, id, required for EEV, NA	203,293,437,442
Intracellular mature virus internal	F17R	11.3	Phosphoprotein, DNA binding, id	262,572,595
	I7L	49.0	Homology to topoisomerase II, ts	260
	G7L	41.9	Proteolytic process, ts	523
	L4R	28.5	Proteolytic process, DNA binding	548,562,583,584
	D2L	16.9	ts	147
	D3R	28.0	ts	147
	A3L	72.6	Proteolytic process from P4b	448,548
	A4L	30.8		136
	A10L	102.3	Proteolytic process from P4a	523,546,551
	A12L	20.5	Proteolytic process	523
Extracellular enveloped virus-specific	F13L	41.8	Nonglycosylated, required for EEV, id, IMCBH sensitivity	54,227,462
	A34R	19.5	Hydrophobic N-terminus, N-glycosylated, EEV release, id, lectin homology	56,146
	A36R	25.1	Hydrophobic N-terminus, deletion reduced EEV	392
	A56R	34.8	N- and O-glycosylated, hemagglutinin, NE, mutation causes cell fusion	401,470,474,475
	B5R	35.1	Glycosylated, required for EEV, SCR motif	154,155,250a,321,575

ts, ts mutant; id, inducer-dependent mutant; NA, neutralizing antibody; NE, nonessential; SCR, short consensus repeat.

from the ORFs (Table 2) and have undergone cleavage (356) at AlaGly↓Ala/Ser motifs (551).

EEV have a lower buoyant density than IMV because of the additional lipid membrane and contain several unique glycosylated and nonglycosylated proteins (400,403). The major nonglycosylated, acylated protein maps to ORF F13L and glycoproteins are encoded by the A56R, B5R, and A34R ORFs (Table 2). The hemagglutinin (A56R) is both N- and O-glycosylated. The B5R product, like the HA, is a type 1 integral membrane glycoprotein, whereas the glycoprotein encoded by the A34R ORF is predicted to be type 2. The product of the A36R ORF is also associated with the EEV membrane.

Enzymes

Infectious poxvirus particles contain a transcription system (266,367) that can carry out *in vitro* synthesis of mRNAs that are polyadenylylated (264), capped, and methylated (559). A large number of virus-encoded enzymes and factors are packaged in the virus particle (Table 3). The RNA polymerase, RNA polymerase-associated protein of 94-kd (RAP94), vaccinia early transcription factor (VETF), capping and methylating enzymes, and poly(A) polymerase are clearly involved in the synthesis and modification of mRNA. Other enzymes, such as NPH I and II, DNA topoisomerase, protein kinases, nicking-joining enzyme, and glutaredoxin, may also be involved in transcription, although roles in virus assembly and DNA packaging or release are also possible.

VIRUS ENTRY

Studies of poxvirus entry into cells is complicated by the existence of IMV and EEV forms, both of which are infectious. The existence of a unique viral attachment protein or cell surface receptor has not been established. Electron micrographs demonstrate vaccinia IMV particles fused to membranes within vacuoles formed by surface invagination (119) and to the plasma membrane (95). With polarized cells there may be a preference for entry through basolateral surfaces (436). Vaccinia virus plaque formation is

TABLE 3. *Virion enzymes and factors*

Enzymes/factors	ORF	kd	Properties	References
RNA polymerase			Multisubunit enzyme	33,507
RPO147	J6R	147	Yeast RPB1 homolog	75
RPO132	A24R	132	Yeast RPB2 homolog	11,229
RPO35	A29L	35		13
RPO30	E4L	30	Eukaryotic TFIIS homolog	1,79
RPO22	J4R	22		75
RPO19	A5R	19		6
RPO18	D7R	18		3,425
RPO7	G5.5R	7	Yeast RPB10 homolog	12
RNA pol assoc. protein	H4L	94	RAP94, early promoter-specificity factor	2,5,259
Early transcription factor			VETF, early promoter-binding, DNA-dependent ATPase	74,78,80
	A7L	82		193
	D6R	74		72,73,193,299
Poly(A) polymerase			Primer-dependent	357
	E1L	55	Catalytic subunit	188,194,196
	J3R	39	Stimulatory subunit	195
Capping enzyme			RNA triphosphatase, guanyltransferase, guanine-7-methyltransferase, termination factor	319,485,554
	D1R	97	Covalent GMP, catalytic activities	211,350,477
	D12L	33	Stimulates guanine-7-methyltransferase	112,222,375
RNA (nucleoside-2′) methyltransferase	J3R	39	Also poly(A) polymerase subunit	31,463
DNA-dependent ATPase	D11L	72	NPH I	76,389,390,439
RNA/DNA-dependent NTPase	I8R	77	NPH II, RNA helicase	285,389,390,480, 481
DNA-dependent ATPase	A18R		DNA helicase	37b,491
DNA topoisomerase I	H6R	37	Sequence-specific nicking	37a,283,478,483, 486,487,544
Nicking-joining enzyme	?	~50	Cleaves supercoiled DNA and joins ends, single-strand DNase	294,338,428,452
Protein kinase 1	B1R	34	Ser/Thr	29,30a,303,430
Protein kinase 2	F10L	52	Ser/Thr	281,282,302
Glutaredoxin	O2L	12	Thioltransferase, dehydroascorbate reductase	7

not inhibited by lysomotropic agents (251), and some virion proteins remain associated with fractions containing the plasma membrane (251,317), implicating the latter as a site of fusion. Support for a neutral pH, temperature-dependent process was obtained by directly monitoring the cell fusion of vaccinia IMV using a fluorescence assay (140).

Roles for several vaccinia IMV membrane proteins in cell surface attachment and/or penetration have been suggested on the basis of antibody neutralization (380,438, 513). Treatment of vaccinia virus with proteases increases infectivity, apparently by enhancing cell penetration (243). An IMV envelope protein encoded by the D8L ORF (Table 2) binds with high affinity and in a saturatable manner to the surfaces of a variety of cells (292,312). However, intact D8L protein is not required for virus infectivity, nor can saturating amounts of the exogenous protein efficiently prevent virus plaque formation. Three proteins—a 54-kd tubule component (513), the product of the A27L ORF (437,440), and the product of the A17L ORF (245, 435)—have been implicated in vaccinia virus/cell fusion and penetration.

The wide host range of vaccinia virus suggests that the putative cellular receptor(s) must be highly conserved. A vaccinia virus–encoded protein, vaccinia growth factor (VGF), is capable of binding to the cellular EGF receptor (515,535). It was proposed—based on infectivity blocking studies with epidermal growth factor (EGF), synthetic EGF-derived peptides, and antibody to EGF receptor—that the latter serves as a portal of entry for vaccinia virus (157,318). The general importance of the EGF receptor for vaccinia virus entry is doubtful, since infectivity was incompletely blocked by EGF, wild-type virus can infect cells lacking the EGF receptor, and vaccinia virus mutants that do not make VGF can readily infect cells (82). As will be discussed, secreted VGF enhances virus replication by stimulating cell proliferation.

EEV are important for spread of infection in animals and cultured cells (64,402). The entry of EEV presents an interesting topological problem, since fusion with the cell membrane should release cytoplasmic IMV that are still membrane enclosed. Fluorescent assays indicated that EEV fusion occurs at neutral pH and is more rapid than that of IMV (140). EEV infectivity is not neutralized by antibody to IMV, suggesting that different proteins are involved (404). A single amino acid change in a lectin homology motif in the A34R ORF influences the release of progeny EEV from the external surface of infected cells (56); whether the protein is involved in entry as well as dissemination must await further studies.

Cells infected with viruses that have mutations in the hemagglutinin (241) or in the SERPIN homolog encoded by the K2L ORF (534) form syncytia, although the mechanisms are not understood. Oddly, in view of the evidence for fusion of IMV and EEV with the cell membrane at neutral pH, briefly lowering the pH of the medium induces vaccinia virus–infected cells to fuse (140,204).

UNCOATING

Following entry into the cytoplasm, virus cores synthesize mRNA and then undergo a second uncoating step. Electron microscopic images suggest that the nucleoprotein complex passes out through breaches in the core wall (117). Susceptibility of the virion DNA to treatment with deoxyribonuclease has been used as biochemical evidence of uncoating (253). Prevention of the uncoating process by inhibitors of transcription or translation indicates a requirement for either a viral-induced or -encoded protein. A putative 23-kd uncoating protein with trypsinlike activity has been partially purified from infected cell extracts (405), but genetic studies would be needed to establish its role. Viral particles in intermediate stages of disassembly have been isolated and the polypeptide compositions have been analyzed (228,457,591).

GENE EXPRESSION

Regulation of Early-Stage Transcription

The early transcription system is packaged within the core of the infectious poxvirus particle (266,367), providing a mechanism for the synthesis of viral mRNA in the cytoplasm (Fig. 1). Vaccinia virus early mRNA is detected within 20 minutes and accumulates to maximal levels within 1–2 hours under conditions of synchronous infec-

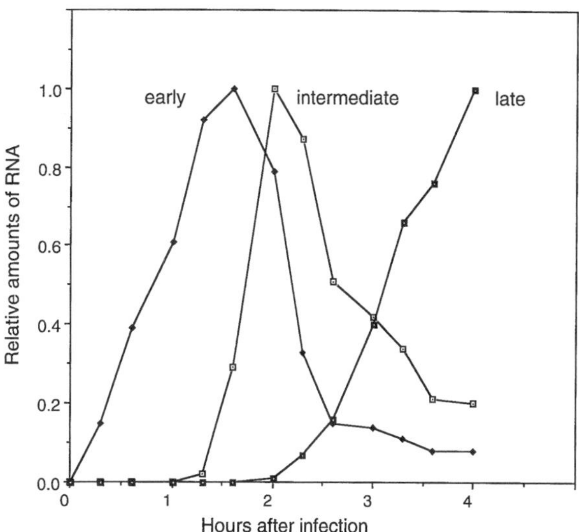

FIG. 6. Steady-state levels of representative early, intermediate, and late-stage mRNAs in vaccinia virus–infected cells. Total RNA was isolated from infected HeLa cells at various times after infection and hybridized to antisense RNA probes specific for the 5′-ends of mRNAs encoded by the C11R (early), G8R (intermediate) or F17R (late) ORFs (27). After ribonuclease digestion, the protected probe fragments were analyzed by polyacrylamide gel electrophoresis and the radioactivity quantitated. The numbers were normalized to the peak value in each case.

tion (Fig. 6) (27). The rapid decline of steady-state early mRNA levels is consistent with an enhanced rate of mRNA degradation after virus infection (378,469). In the presence of protein synthesis inhibitors, uncoating of the core is prevented (254), and greatly increased amounts of RNA are made (27,577), suggesting that core disassembly may negatively regulate early transcription. By comparison, DNA replication inhibitors do not prevent uncoating and, at most, prolong early transcription to a limited extent. RNA/DNA hybridization studies revealed that about half of the vaccinia virus genome is transcribed prior to DNA replication (61,271,378,388). The early genes encode proteins involved in viral DNA replication, intermediate gene expression, and host interactions.

Early-Stage Promoters and Termination Signal

The transcription of early genes is regulated by A/T-rich sequences located immediately upstream of the RNA start sites (104,246,563). Saturation mutagenesis of a vaccinia virus early promoter defined a critical core region, from –13 to –27, in which many single-base substitutions have a drastic effect on expression (Fig. 7) (127a). Adenylate residues were needed at positions –13, –18 to –20, and –25 to –27 and were optimal at certain other positions; a G-residue was needed at –21; and a T either at –22 or at –23. A consensus core sequence 5′-AAAAAATGAAAAAA/TA-3′ is close to the optimal one defined by mutagenesis. Transcription initiation occurs with a purine, predominantly 12–17 nucleotides downstream of the core region (127a). The intervening DNA, between the core and RNA start site, appears to have a spacer role. There are no evident sequence requirements upstream of the core or downstream of the initiation site.

Structural and functional studies suggest that promoter sequences are conserved between poxvirus genera (66). Such conservation is consistent with "nongenetic reactivation" of a heat-inactivated poxvirus by coinfection with a second poxvirus from another genus (176,217). Evidently, the heat-killed poxvirus provides the template and the second poxvirus provides the enzymes for transcription.

The 3′-ends of vaccinia virus early mRNAs occur 20–50 bp downstream of the sequence TTTTTNT (586,587). Such sequences are present near the ends of most viral early genes but only rarely in their middle. *In vivo* studies suggested that the efficiency of termination is about 80% (149), although in some cases it is much less (298), probably because of RNA secondary structure (308). T-rich sequences have been noted near the ends of genes in *Leporipoxvirus* (537), *Avipoxvirus* (530), *Capripoxvirus* (191), and *Molluscipoxvirus* (53), suggesting a similar role in termination.

VACCINIA VIRUS PROMOTER SEQUENCES

	CORE	INITIATOR
EARLY	AAAA T GAAA TA	A/G
INTER	T TT AAA AA	TAAA
LATE	A/T-rich	TAAATG/A
	-30 -20 -10	+1

FIG. 7. Early, intermediate, and late-stage promoter sequences. Nucleotides that have a strong positive effect on transcription are shown. Positions of nucleotides in the nontemplate strand are relative to the RNA start site (+1). From Moss (354), with permission.

Enzymes and Factors for Early-Stage Transcription

Soluble extracts of vaccinia virus virions are competent to transcribe an early promoter template *in vitro* (202,444, 445) and provide a source of materials for characterization of the relevant enzymes and factors. The virion RNA polymerase is eukaryotic-like with regard to its size and subunit complexity (33). The subunits, ranging from 7- to 147-kd, are encoded by at least eight viral genes (Table 3). Although there is a single copy of each large subunit, the stoichiometry of the small subunits has not been accurately determined, partly because of polypeptide heterogeneity caused by alternative transcriptional and translational start sites. RPO147 and RPO132 are homologous to the corresponding subunits of cellular RNA polymerases and more closely resemble those of eukaryotes and archaebacteria (20% to 30% amino acid identity over the entire proteins) than those of eubacteria (11,75,398). The vaccinia virus RPO30 subunit is approximately 23% identical in sequence to eukaryotic transcription elongation factor SII (TFIIS) over a 180-amino-acid region (1); RPO7, the smallest vaccinia virus RNA polymerase subunit, is about 23% identical in amino acid sequence with the smallest subunit of yeast RNA polymerase II (12). Synthesis of the RNA polymerase subunits begins during the early phase of infection and continues throughout the cycle. The functions of the individual subunits have not yet been determined. Temperature-sensitive mutants, with defects mapped to RNA polymerase subunits, have been isolated (156,229,231,528). Such mutants, propagated at the permissive temperature, generally contain active RNA polymerase so that early gene expression occurs even when cells are infected at the nonpermissive temperature. The phenotype of these mutants is a block in late gene expression, probably caused by aberrant assembly of the multisubunit enzyme.

An additional polypeptide is stably associated with approximately 40% of RNA polymerase molecules in vaccinia virions (2,5,137,259). The RNA polymerase-associated polypeptide of 94-kd (RAP94) is specifically required for transcription of early promoter templates in conjunction with the vaccinia virus early transcription factor (VETF). RNA polymerase lacking RAP94 can transcribe single-stranded DNA nonspecifically or double-stranded, intermediate and late promoter templates with the corresponding transcription factors. It seems likely that RAP94 interacts with VETF, although this has not been demonstrated. Synthesis of RAP94 occurs late, at the time of virion assembly, consistent with its exclusive role as a virion-associated early transcription factor. In this regard, RAP94 differs from the RNA polymerase subunits, which must be synthesized early in the infectious cycle for intermediate and late transcription.

Transcriptional activity can be reconstituted *in vitro* with RNA polymerase and VETF, a heterodimer of 82- and 70-kd subunits (72,80,193). VETF, like RAP94, is synthesized only at late times after infection. The protein binds to the core region of early promoters and to DNA downstream of the RNA start site, thereby altering the conformation of the DNA (73,74). Single nucleotide substitutions, in the core sequence of the promoter, that decrease transcription also abrogate specific DNA binding (585a). Complexes of VETF and RNA polymerase have been detected, suggesting that VETF may recruit RNA polymerase to the promoter (24,77,300). A DNA-dependent ATPase activity, associated with the small subunit of VETF, is not required for promoter binding but is essential for transcription, possibly via a promoter clearance mechanism (70,78,299). The elongation complex has a 3′ RNase activity that permits resumption of transcription by stalled polymerase (213b). A role for RPO30 has been suggested, since this RNase activity is similar to one exhibited by eukaryotic RNA polymerase II in the presence of TFIIS, an RPO30 homolog.

RNA polymerase and VETF can reconstitute the transcription initiation and elongation activities of soluble virion extracts but not the transcription termination activity. Complementation studies revealed that termination was restored by addition of a protein identified as the viral enzyme that caps the 5′-ends of mRNA (307,482). Although the termination signal was first noted by its DNA sequence TTTTTNT (587), subsequent *in vitro* studies indicated that it is recognized in RNA as a UUUUUNU (488). Capping enzyme is able to bind to RNA polymerase as well as to RNA, and its role in termination is independent of its ability to modify the 5′-end of the message (77,212,213a,309). A requirement for hydrolyzable ATP in termination or RNA release has been demonstrated (212). As indicated later, the UUUUUNU termination signal also could provide the uridylate residues needed for interaction with the catalytic subunit of the poly(A) polymerase (196), although the latter enzyme is not needed for termination.

Early viral transcripts made *in vivo* or *in vitro* are capped (559) and polyadenylylated (264), as are typical eukaryotic mRNAs. RNA synthesized by virus cores contains a cap I structure that consists of a terminal 7-methylguanosine connected via a triphosphate bridge to a 2′-O-methylribonucleoside. The following are the first three steps in cap formation: (a) removal of the terminal phosphate of the triphosphate end of the nascent RNA to form a pp(5′)N-terminus, (b) transfer of a GMP residue from GTP to form G(5′)ppp(5′)N-, and (c) methylation of the N^7 to produce m^7G(5′)ppp(5′)N-. They are catalyzed by a virus-encoded 127-kd heterodimeric capping enzyme (319,320,350,375, 554). The fourth step in cap formation, ribose methylation of the penultimate nucleoside to form m^7G(5′)ppp(5′)Nm, is carried out by a separate viral enzyme (31,463). Capping occurs during transcription when the nascent RNA chains are approximately 30 nucleotides long (213a). The capping enzyme large subunit forms a covalent lysyl-GMP intermediate (111,374,453,485). The RNA triphosphatase and guanylyltransferase activities reside in a 55- to 59-kd N-terminal fragment of the large subunit (211,477,486), whereas a complex of the C-terminal part of the large subunit and the small subunit contains the N^7-methyltransferase activity (112,222). The nucleoside 2′-methyltransferase exists both as a 39-kd protein and as a subunit of the poly(A) polymerase (463).

The N^7-methylguanosine component of the cap is required for mRNA stability and for binding of vaccinia virus mRNA to ribosomes, whereas the role of ribose methylation has not been determined (369). Viral mRNAs synthesized *in vivo* have additional methylations that are catalyzed by cellular enzymes (60).

The enzyme that catalyzes poly(A) tail formation is a heterodimer of virus-encoded 55- and 39-kd subunits called VP55 and VP39, respectively (188,357). The two subunits have been separated and have been expressed independently by recombinant methods. VP55 binds to uridylate-rich sequences near the end of the RNA chain and catalyzes the processive addition of 30–35 adenylate residues before changing to a slow and nonprocessive mechanism (194, 196). VP39 binds to poly(A) and stimulates VP55 to semiprocessively add additional adenylate residues (194). VP39 is present in a fivefold molar excess over VP55 and therefore exists in both heterodimeric and monomeric forms, both of which have methyltransferase activity (463). Mutated forms of VP39 that lack methyltransferase but retain adenylyltransferase stimulatory activity *in vitro* have been studied (464).

The finding that the capping enzyme is involved in termination and that one of the cap methyltransferases is a subunit of the poly(A) polymerase is intriguing. Whether the association of such disparate functions in the same enzymes provides a specific advantage or merely represents an economical use of proteins is unknown.

The minimal components for synthesis of correctly initiated, terminated, capped, and polyadenylylated mRNA have been defined by *in vitro* reconstitution assays. However, additional enzymes may be needed within the virus core. For example, the ATP requirement for RNA extrusion (265,555) suggests a role for NPH II, which has RNA-dependent RNA nucleoside triphosphatase (389,390) and RNA helicase (480,481) activities. Transcription studies with the noninfectious particles formed by vaccinia virus conditional lethal mutants with substitutions in the NPH II gene (171,172a) may help to determine the role of this enzyme.

The products of ORF D11L (NPH 1) (76,389,390,439) and A18R (37b) exhibit DNA-dependent ATPase activities. A prediction that the latter is a DNA helicase (285) has been substantiated (37b). Studies with ts mutants suggested that both enzymes have roles in early and late transcription (139, 258,491). Eukaryotic DNA type 1 topoisomerase has been shown to regulate RNA polymerase II transcription *in vitro*. Vaccinia virus particles contain an essential DNA type 1 topoisomerase that structurally and functionally resembles its eukaryotic counterpart, except for the small size of the viral enzyme and its resistance to camptothecin and specificity for binding and cleaving duplex DNA at 5′(C/T) CCTT sequences (37a,283,471,479,483,484,486,487). The enzyme catalyzes a reversible, site-specific strand cleavage and resealing reaction with duplex DNA involving a transient 3′-phosphotyrosyl linkage between the DNA and Tyr-274 of the enzyme. Shope fibroma virus encodes a protein of the same length as the vaccinia virus topoisomerase and with 61% sequence identity (544).

In view of the regulatory role of protein kinases in eukaryotic gene expression, a similar role may be invoked for the virion-associated serine/threonine protein kinases (281,282). Vaccinia virus genes encoding two separate protein kinases have been identified (29,302,303,430). As will be discussed later, conditional lethal mutants of the ORF B1R–encoded protein kinase are defective in DNA replication.

Regulation of Intermediate-Stage Transcription

DNA replication precedes a profound shift in viral gene expression. Amino acid pulse-labeling (359,407) and DNA transfection and transcription (558) studies indicated the existence of an intermediate class of genes that are expressed after DNA replication but before expression of the more abundant late genes (Fig. 1). Experiments with metabolic inhibitors further suggested that synthesis of certain viral early proteins is required for intermediate gene transcription. After synchronous infections, intermediate mRNAs are detected at 100 min, the time of peak early mRNA accumulation (Fig. 6). Intermediate mRNAs reach their peak values soon after and then decline in quantity. Only five vaccinia virus genes belonging to the intermediate class have been identified thus far (272,558). Three of the genes (A1L, A2L, and G8R) encode transactivators of late gene expression (272), one (I8R) is NPH II, and one (I3L) that has both early and intermediate promoters is a DNA binding protein that interacts with ribonucleotide reductase (126). Evidence for additional intermediate genes has been obtained (597).

The requirement for DNA replication in intermediate gene expression may be explained if the genome within the infecting particle is inaccessible to the newly synthesized RNA polymerase and transcription factors. This hypothesis is consistent with transfection experiments showing that the deproteinized DNA of purified virus particles can serve as a template for intermediate and late transcription in the absence of DNA replication (272). The proposed inaccessibility of the parental DNA could be due to remaining virion proteins or to putative repressor proteins. There are, of course, numerous other possibilities.

Intermediate-Stage Promoters

Mutagenesis of intermediate promoters indicated two important regions: a 14-bp core element separated by 10 or 11 bp from a 4-bp initiator element (25,226). The intermediate core element, which resembles that of early promoters in A/T richness but not in sequence, cannot tolerate substitutions at positions –13 to –15; –17 to –19; –21, –22, and –26 (Fig. 7). The A-residues at –18 and –19 and the T-residue at –21 are particularly important. The tetranucleotide TAAA serves as an initiator element of intermediate promoters but is not required as such for early promoters (25). Intermediate-stage RNAs are initiated within the A triplet, but as discussed later, they contain additional A residues incorporated by a polymerase slippage mechanism. With a slight adjustment, early and intermediate promoter motifs can be accommodated within a single, dual-function, synthetic early/intermediate promoter (27). It seems likely that such dual promoters also occur naturally.

Analysis of some intermediate mRNAs by agarose gel electrophoresis indicated diffuse bands, equal to and longer than the coding regions, suggesting preferred sites of 3′-end formation that did not correlate with early gene transcriptional termination signals (27). No further information is available regarding intermediate transcription termination or processing.

Enzymes and Factors for Intermediate-Stage Transcription

Intermediate promoter templates can be transcribed by extracts prepared from cells infected with vaccinia virus in the presence of an inhibitor of DNA replication (558). Initial experiments indicated a requirement for two factors in addition to the viral RNA polymerase. One factor was identified as the virus-encoded capping enzyme (557). Subsequent studies showed that the transcriptional role of capping enzyme occurs at, or immediately after, initiation by a mechanism that does not require RNA guanylylation (220). The RNA polymerase and capping enzyme from virions could substitute for the corresponding enzymes from infected cell extracts, indicating that no modification is required. The second partially purified intermediate transcription factor, which was not present in virion extracts, formed a template-committed complex and conferred

$KMnO_4$ sensitivity to T-residues at the site of transcription initiation, suggesting sequence-specific binding and strand separation (556). Further studies revealed two vaccinia virus intermediate transcription factors, VITF-1 and VITF-2 (446). VITF-1 is encoded by the gene for RPO30, a viral RNA polymerase subunit with homology to eukaryotic transcription elongation factor TFIIS. Thus, free VITF-1 is required even though the polypeptide is present in a complexed form as an RNA polymerase subunit. The finding that VITF-2 is a cellular protein (447) was unexpected in view of the viral origin of other transcription components. VITF-2 activity can be extracted from the nucleus of uninfected HeLa cells but is distributed within the cytoplasmic and nuclear fractions of infected cells. Whether vaccinia virus infection induces transit of VITF-2 out of the nucleus, activates a cryptic cytoplasmic factor, or depends on newly synthesized VITF-2 has not yet been determined. The latter two possibilities would be consistent with previous enucleation experiments (235). A relationship between VITF-2 activity and certain examples of host range restriction has been suggested (447).

Regulation of Late-Stage Transcription

The transcription of late genes follows that of intermediate genes (Fig. 1). In HeLa cells, late-stage RNA is detected at about 140 minutes after synchronous infection with vaccinia virus and continues for about 48 hours (Fig. 6). The persistent synthesis of late proteins reflects continued transcription, since the half-life of late mRNAs has been estimated to be 30 minutes or less (378,469). Many late proteins, including the major virion components, accumulate in large amounts during this long period. Other late proteins include the factors that are specifically required for transcription of early genes, such as VETF and RAP94, as well as certain other virion enzymes (Table 3). Although distributed throughout the genome, the late-stage genes cluster in the central region (45).

Late-Stage Promoters and RNA Processing Signal

Late-stage promoters also may be considered in terms of three regions: a core sequence of about 20 bp with some consecutive T or A residues, separated by a region of about 6 bp from a highly conserved TAAAT element within which transcription inititiates (Fig. 7) (47,127b,449,564). A very strong late promoter was made by employing exclusively T-residues for the core sequence. Any mutations within TAAAT severely decreased transcription. The intermediate (TAAA) and late (TAAAT) initiator elements are obviously similar to each other. The late promoter TAAAT sequence is usually followed by G or A. In the former case, the TAAATG transcription initiation squence and the ATG translation initiation codon overlap. The seeming absence of an untranslated RNA leader in this situation was puzzling, until it was found that late mRNAs have a 5′ capped, heterogeneous-length, poly(A) sequence that is evidently synthesized by RNA polymerase slippage (4,48,397,468, 518). Poly(A) leaders are also present on mRNAs of certain early genes that have a TAAAT initiation site (3,247) as well as on intermediate mRNAs (27), suggesting that slippage on an AAA sequence is an intrinsic property of the viral RNA polymerase.

Most late transcripts are long and heterogeneous, lacking defined 3′-ends (115,316). The early termination signal is not recognized by the late transcription system; consequently, TTTTTNT is frequently present within the coding region of late genes. Terminal heterogeneity, combined with transcription from both DNA strands, explains the ability of late transcripts to self-anneal or anneal with early transcripts to form ribonuclease-resistant hybrids *in vitro* (63,105,552). Double-stranded RNAs could be deleterious, and it is reasonable to consider that poxviruses may have special mechanisms to minimize their accumulation or effect. Whether the vaccinia virus–encoded double-stranded RNA binding protein (96) or RNA helicase (480) has such a role is unknown. The 5′ poly(A) leader could compensate for the complementary RNA by providing a single-stranded binding site for initiation factors and the 40s ribosomal subunit, which would then move unimpeded by antisense RNA to the first AUG where ribosome assembly and translation occur.

An exception to the general 3′ heterogeneity of late mRNAs has been noted (15). The cowpox virus late mRNA encoding the A-type inclusion protein has a 3′-end corresponding to a precise site in the DNA template. The DNA sequence at this position encodes an RNA *cis*-acting signal for RNA 3′-end formation, which can function independently of either the nature of the promoter or the RNA polymerase responsible for generating the primary RNA. There is evidence for induction or activation, late in infection, of a specific endoribonuclease that cleaves this RNA, which is then polyadenylylated. The number of late mRNAs that are processed in this manner remains to be determined. Therefore, poxviruses employ at least two mechanisms of RNA 3′-end formation. The first, operative at early times in viral replication, terminates transcription downstream of an RNA signal, whereas the second, operative at late times, involves RNA site–specific cleavage.

Enzymes and Factors for Late-Stage Transcription

Templates containing late promoters can be transcribed by extracts prepared from cells at the late stage of vaccinia virus infection (467,581). Activity was reconstituted with three fractions, one of which contained RNA polymerase (582). The identification of factors was facilitated by a reverse genetic approach using transfected DNA (272). By systematic screening of cloned DNA fragments, ORFs A1L,

A2L, and G8R encoding proteins of 16.9, 26.3, and 20.9 kd were found to be necessary and sufficient for transactivation of a transfected late promoter reporter gene in vaccinia virus–infected cells that were blocked in DNA replication. Each of these transactivator genes is regulated by an intermediate promoter, consistent with a cascade model of transcriptional regulation. Biochemical studies confirmed that the products of the G8R and A1L ORFs are vaccinia virus late transcription factors (274,579,580); the names VLTF-1 and VLTF-2 have been used. In addition, temperature-sensitive and repressible mutations of A1L and G8R, respectively, block late gene expression under nonpermissive conditions (90,597). Evidence that the A2L product functions directly as a transcription factor has not yet been obtained. However, one or two additional late transcription factors, which appear to be early gene products, have been partially purified (291,579). *In vitro* studies indicate that ORFs A1L and A2L encode zinc binding proteins (273).

Additional proteins appear to be involved in the transcription of late genes *in vivo*, although their roles have not been well defined. Temperature-sensitive mutants with defects in the A18R gene exhibit unregulated transcription from regions of the genome that are normally quiescent at late times (39). The resulting phenotype is related to increased double-stranded RNA activation of the endogenous cellular 2-5A endoribonucleolytic pathway. A similar effect is caused by the drug isatin-fl-thiosemicarbazone (IBT) (114,578). The IBT-resistant mutations map to RNA polymerase subunit RPO130 (107), whereas the IBT-dependent mutants map to ORF G2R, which encodes a protein of unknown function (333).

Several observations point to a role of RNA polymerase II during vaccinia virus infection (118). These include inhibitory effects on virus replication of (a) α-amanitin (235, 490), a drug that specifically inhibits cellular RNA polymerase II at low concentrations but does not affect the vaccinia viral polymerase (33,116,373,507); (b) enucleation (408,422); and (c) UV treatment of cells prior to infection (235). Additional studies demonstrated α-amanitin-sensitive transcription of poxvirus DNA by extracts of uninfected HeLa cells (16) and showed that cellular RNA polymerase II (or at least the large subunit) exits the nucleus during vaccinia virus infection, associates with virosomal regions of the cytoplasm, and may even be packaged in virus particles (351,570,571). A higher yield of vaccinia virus from HeLa and L cells, compared to myoblasts, was attributed to the relative RNA polymerase II contents of these cells (571). Nevertheless, interpretation of these data is difficult, since the major reported effects of α-amanitin, enucleation, and UV treatment are on poxvirus assembly rather than on gene expression; the transcripts made *in vitro* by HeLa cell extracts were not well characterized or shown to be expressed *in vivo*; only nuclear matrix-bound proteins were used as comparisons to suggest that RNA polymerase II was specifically transported out of the nucleus; and other explanations for virus yield differences in the three cell types are possible. At this time, there is no compelling evidence that RNA polymerase II has a direct role in the transcription of poxviral genes.

Posttranscriptional Regulation of Gene Expression

Transcriptional initiation is an important mechanism for the temporal regulation of poxvirus gene expression. As discussed earlier, accelerated mRNA degradation also facilitates rapid changes in viral RNA populations (Fig. 6). The basis for mRNA instability, however, is not understood.

No evidence of RNA splicing has been found, despite the analysis of many vaccinia virus transcripts. Apparently, the cytoplasmic location of viral transcription precludes use of the cellular, splicing apparatus in the nucleus. Infrequent instances of RNA self-splicing, as occurs with some bacteriophage, has not been excluded.

Translational regulation has been suggested (233,325, 465,553), but molecular mechanisms have not been determined. A translational role for the 5′ poly(A) leader remains an intriguing possibility.

DNA REPLICATION

General Features

An exclusively cytoplasmic location for DNA replication has been described only for poxviruses and African swine fever virus. Discrete cytoplasmic foci of replication, termed *factory areas,* have been discerned by autoradiographic and microscopic procedures (87,122a,219,268). Reports of poxviral DNA synthesis in the nuclei of infected cells have been discounted (345) and vaccinia virus DNA replication occurs in enucleated cells (408,422). The cytoplasmic location of DNA synthesis suggests that poxviruses use viral counterparts of cellular proteins for replication. A recent review of poxvirus DNA replication is available (532).

Enzymes Involved in DNA Precursor Metabolism

Poxviruses encode several enzymes involved in the synthesis of deoxyribonucleotides, evidently to enhance DNA replication in cells with suboptimal precursor pools. These enzymes include a thymidine kinase (144,234,560), thymidylate kinase (502), ribonucleotide reductase (494,527), and dUTPase (71). ORFs present in the genomes of *Orthopoxvirus* (159,237,561), *Avipoxvirus* (69), *Suipoxvirus* (172b), *Capripoxvirus* (191), *Leporipoxvirus* (250b), and *Entomopoxvirinae* (210,311b) predict homologous thymidine kinases with molecular weights on the order of 20–25 kd. Avipoxvirus and entomopoxvirus thymidine kinase genes differ most from those of orthopoxviruses. The poxvirus thymidine kinases are related in sequence to corresponding eukaryotic enzymes (35% to 70% amino acid identity) but not to the pyrimidine kinase of herpesviruses. The thymidine kinase gene is regulated by an early promoter, as befits its role in increasing precursors for DNA replication (563). The vaccinia virus enzyme exists as a tetramer (51,279) and is susceptible to feedback inhibition by dTDP or dTTP (52). The putative ATP and Mg^{2+} binding domains have been identified by site-directed mutagenesis (50,52). Although the thymidine kinase gene is not required for vaccinia virus growth in tissue culture cells, deletion mutants are significantly attenuated *in vivo* (85). Fowlpox virus also contains an ORF that has homology to human deoxcytidine kinase (286).

Thymidylate kinase catalyzes the next step in TMP metabolism. The vaccinia virus gene was identified by the 42% similarity of its predicted 23.2-kd polypeptide with the *Saccharomyces cerevisiae* thymidylate kinase (502). Moreover, the viral gene was capable of complementing a yeast thymidylate kinase–deficient mutant (239). The protein is expressed early in infection but is not required for replication in tissue culture.

The synthesis of ribonucleotide reductase, an enzyme that converts ribonucleoside diphosphates to deoxyribonucleoside diphosphates, is induced soon after vaccinia virus infection (493). Both the small catalytic subunit and the large regulatory subunit are virus-encoded and closely resemble their eukaryotic counterparts both structurally (70% to 80% identity) and functionally (232,492,494,527). Catalytic activity is inhibited by hydroxyurea, and drug-resistant mutants have direct tandem repeats of the gene encoding the catalytic subunit (496). Inactivation of the gene encoding the large subunit abolished induced ribonucleotide reductase activity in tissue culture cells without affecting replication of the mutant virus (103). However, the mutant virus was mildly attenuated in a mouse model.

Vaccinia virus also encodes a functional dUTPase, which could produce dUMP, an intermediate in the biosynthesis of TTP, as well as minimize dUTP incorporation into DNA (71). The protein is synthesized early in infection (495) and is nonessential for virus replication (410).

DNA Synthesis

The timing of DNA synthesis varies with different members of the poxvirus family and to some extent with the multiplicity of infection and cell type. In cells synchronously infected with vaccinia virus, DNA replication begins 1–2 hours after infection and results in the generation of about 10,000 genome copies per cell of which half are ultimately packaged into virions (255,454). Initiation of DNA replication is detected within 4–6 hours with orf virus (23) and within 12–16 hours with fowlpox virus (423). Studies depending on thymidine incorporation, instead of hybridization for DNA quantitation, generally underestimate the length of the replication period, probably because of a decline in the activity of the viral thymidine kinase used for incorporation of the radioactive precursor via the salvage pathway.

Some studies indicate that replication begins at each end of the genome following the introduction of terminal nicks (417,419) and suggest a strand displacement mechanism as well as the existence of small DNA fragments covalently linked to RNA primers (162–164,418). However, these preliminary observations have not been followed up using modern technology.

Efforts to locate a specific poxvirus origin sequence, using a plasmid replication assay, have been unsuccessful. Surprisingly, any circular plasmid can replicate in cells infected with Shope fibroma (133) and vaccinia virus (340). This result has led to speculation that poxviruses, unlike nuclear DNA viruses, do not require specific origin sequences because of the absence of competing cellular DNA in the cytoplasmic factory areas as well as the use of novel replication mechanisms.

A large number of virus-encoded enzymes essential for DNA replication might be anticipated. Thus far, four complementation groups of temperature-sensitive (*ts*) mutants, which express vaccinia virus early proteins but are impaired in DNA synthesis, have been found (108,109,329b). Not surprisingly, one of the groups contains mutations in the DNA polymerase (109,511,533). The vaccinia virus DNA polymerase has a molecular weight of about 110 kd, has an associated 3′ exonuclease activity (94), and shares sequence similarities with other eukaryotic and viral DNA polymerases (150,256). Certain codon substitutions confer resistance to inhibitors of DNA synthesis, providing information regarding the active site of the polymerase (129, 150,520,521,533). Transcription of the vaccinia virus DNA polymerase gene is stringently regulated (327). The predicted amino acid sequence of the fowlpox virus DNA polymerase gene is 32% identical to that of vaccinia virus.

The second DNA^- complementation group maps to the D5R ORF, which encodes a 90-kd protein (167,450). A detailed characterization of one of those mutants indicated that viral DNA synthesis was virtually undetectable during nonpermissive infections and incorporation of thymidine ceased rapidly when cultures were shifted to the nonpermissive temperature in the midst of replication (168). Some mutants were impaired in marker rescue at the nonpermissive temperature. Examination of the DNA sequence of the D5R ORF revealed an ATP/GTP binding motif (206), but no function for this protein has been determined.

The third DNA^- complementation group was mapped to the B1R ORF (429), which encodes a serine/threonine protein kinase that is expressed early in infection, and is packaged in virions (29,30a,303,430). The connection between protein kinase activity and DNA replication remains unknown.

A fourth *ts* mutant impaired in DNA replication at the nonpermissive temperature (329b) was mapped to ORF

D4R (343b), which is transcribed early in infection (298). Curiously, the D4R transcript co-terminates with the downstream D5R transcript. Both the D4R ORF and its Shope fibroma virus homolog encode functional uracil DNA glycosylases (517,545). Since these enzymes remove uracil residues that have been introduced into DNA, either through misincorporation of dUTP or through the deamination of cytosine, a DNA$^-$ phenotype is surprising. The possibility that the uracil DNA glycosylase forms part of a multienzyme repair or replication complex has been suggested (343b).

The vaccinia virus–encoded ATP-dependent DNA ligase is not essential for viral replication in tissue culture, although it affects virulence as well as sensitivity to DNA damaging agents (106,275,276). Whether DNA ligation is unnecessary for replication or this requirement is fulfilled by an enzyme of cellular origin is unknown. The viral gene can functionally substitute for the *S. cerevisiae* enzyme (275). Sequence comparisons with other ligases indicate a conserved region around the active site lysine.

Concatemer Resolution

The replication of the poxvirus genome involves the formation of concatemeric intermediates and their resolution into unit-length molecules (36,361,364). The concatemer junction consists of a precise duplex copy of the hairpin loop (Fig. 4) present at the ends of mature DNA genomes (338). Studies with temperature-sensitive mutants of vaccinia virus, as well as specific inhibitors, indicate that concatemeric forms of DNA accumulate when late gene expression is prevented (130,342). Resolution occurs upon allowing late proteins to be made, providing evidence that the concatemers are replicative intermediates.

Circular plasmids, containing vaccinia virus (339) or Shope fibroma virus (135) concatemer junctions, are converted into linear molecules with hairpin termini when transfected into poxvirus-infected cells. Using this assay, the structural and sequence requirements for resolution of concatemer junctions were determined by site-directed mutagenesis (134,330,336,341). The minimal requirement for resolution is two copies of the sequence T_6-N_{7-9}-T/C-AAA-T/A present in an inverted repeat orientation on either side of an extended double-stranded copy of the hairpin loop. Structurally and functionally similar resolution sequences are present in at least four poxvirus genera: *Orthopoxvirus*, *Leporipoxvirus*, *Capripoxvirus*, and *Avipoxvirus* (Fig. 8). The sequence of the intervening region, destined to form the hairpin loop, is not highly conserved but must be palindromic and less than 200 bp long. Further studies suggested that resolution is accomplished either by conservative site-specific recombination and oriented branch migration (337) or by nicking and sealing of an extruded cruciform structure (330).

Data showing that all conditional lethal mutants of vaccinia virus blocked in late gene expression are also defective in concatemer resolution at the nonpermissive temperature were interpreted as indicating a requirement for a specific viral late protein (130,342). However, the resolution sequence contains a consensus late promoter (compare Figs. 7 and 8) and functions as such (395,516). Therefore, it is unclear whether transcription of the resolution sequence, the translation product of a late transcript, or both are required. There are at least two viral late proteins that might be components of a resolvase. A 50-kd homodimeric DNase (452) with nicking/joining activity (294, 338,428) is present in vaccinia virus cores. The nicking/joining reaction requires no energy cofactor and generates 3′ P and 5′ OH termini. When the purified enzyme was incubated with concatemeric junction fragments, however, cleavage occurred at the apex of the cruciform instead of at the base, which would be necessary for telomere formation (338). The DNA type I topoisomerase, which has nicking/sealing activity, is also a candidate resolvase. The phenotype of a conditional lethal mutant with a lesion in the small subunit of the capping enzyme has led to a suggestion that this protein might also be involved in resolution (89).

Homologous Recombination

High rates of recombination occur within poxvirus-infected cells (174). Natural recombination has apparently occurred between two leporipoxviruses—Shope fibroma virus, which produces benign fibromas in rabbits and myxoma virus, the agent of myxomatosis—to form malignant rabbit fibroma virus (57), and it has occurred between individual capripoxviruses (192). Recombination between the terminal sequences of poxvirus DNA may explain variations in the number of tandem repeats (360), translocations of rabbitpox (365), cowpox (17,415), monkeypox (158) and vaccinia (288) virus DNA, and mirror image deletions (329a).

Recombination can also occur between virus-derived genomic DNA and transfected subgenomic DNA fragments (372,455a) or recombinant plasmids (560). Recombina-

```
VA  TTTTTTTCTAG  ACAC   TAAAT  --hairpin loop--  ATTTA   GTGT  CTAGAAAAAAA
CP  TTTTTTTCTAG  ACAC   TAAAT  --hairpin loop--  ATTTA   GTGT  CTAGAAAAAAA
RP  TTTTTTTCTAG  ACAT   TAAAT  --hairpin loop--  ATTTA   ATGT  CTAGAAAAAAA
SF  TTTTTTTCTAG  GGTTA  TAAAT  --hairpin loop--  ATTTA  TAACC  CTAGAAAAAAA
```

FIG. 8. Concatemer resolution sequence. *VA*, vaccinia; *CP*, cowpox; *RP*, rabbitpox; *SF*, Shope fibroma. The boxed regions represent conserved sequences necessary for concatemer resolution. Extended palindromic hairpin loop sequences are indicated.

tion has been exploited to map and construct mutations and to insert genes for expression (353).

Viral genomes rapidly eliminate direct repeats with the formation of intra- and intermolecular recombination products (28). Single- and double-crossover products, resulting from recombination between transfected plasmids and viral genomes (510), and inter- and intramolecular plasmid or bacteriophage DNA recombinants (166,393) have been detected in poxvirus-infected cells. Recombination does not require late gene products, and there appears to be a strong connection between recombination and replication (335). Using lambda phage DNA transfected into Shope fibroma virus–infected cells, heteroduplex formation was shown to coincide with the onset of both replication and recombination, suggesting that poxviruses make no clear biochemical distinction between these processes (177). Evidence for DNA strand exchange catalyzed by proteins from vaccinia virus–infected cells has been reported, and a derivative of the T4-replication-primed recombination model has been suggested (592).

DNA Replication Model

Although large information gaps exist in our understanding of poxvirus DNA replication, the unique terminal structure of the poxvirus genome, the presence of concatemer junctions in replicating DNA, and the absence of a defined replication origin suggest a self-priming replication model (35,364) similar to that proposed for replication of single-stranded parvovirus DNA (526). As depicted in Fig. 9, a hypothetical nick occurring at one or both ends of the genome provides a free 3′-end for priming. The replicated DNA strand then folds back on itself and copies the remainder of the genome. Concatemer junctions form by replication through the hairpin; very large branched concatemers can arise by initiating new rounds of replication before resolution occurs. Recombinational strand invasion may further contribute to the formation of complex multibranched molecules that cannot penetrate agarose gels during pulse-field electrophoresis (130,342). After the onset of late-stage transcription, unit-length

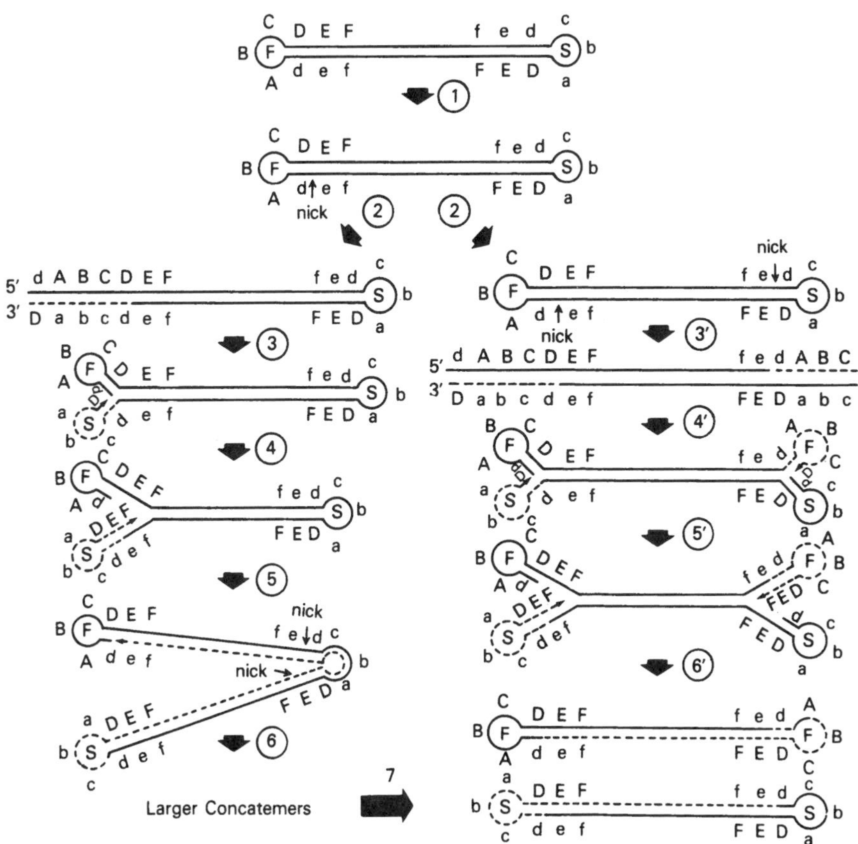

FIG. 9. Self-priming model for vaccinia virus DNA replication. *F* and *S* refer to the difference in electrophoretic mobilities of the two alternative inverted and complementary hairpin sequences present at the ends of the vaccinia virus genome (35). The scheme on the left assumes that the replication of a single DNA molecule is inititiated at one end and continues to the other end to form a concatemer, whereas on the right initiation occurs at both ends without concatemer formation. From Moss et al. (361), with permission.

genomes are resolved and the incompletely base-paired terminal loops, with inverted and complementary sequences, are regenerated.

VIRION ASSEMBLY, MATURATION, AND RELEASE

Intracellular Mature Virions

The initial stages of assembly occur in circumscribed, granular, electron-dense areas of the cytoplasm. The first morphologically distinct structure is a crescent (or cupulae in three dimensions) consisting of a membrane with a brushlike border of spicules on the convex surface and granular material adjacent to the concave side (Fig. 10, panel 1). Both biochemical analyses and the inability to detect continuity with cellular organelles contributed to the idea that the viral membranes are formed *de novo* (121), a proposal that is now challenged. Though no cellular protein markers are detectable on the viral membranes themselves, recent electron-microscopic images suggest that the latter are derived from cisternae of the intermediate compartment between the endoplasmic reticulum and the Golgi stacks (504). The new observations further suggest that a tightly opposed double membrane is formed by a wrapping mechanism.

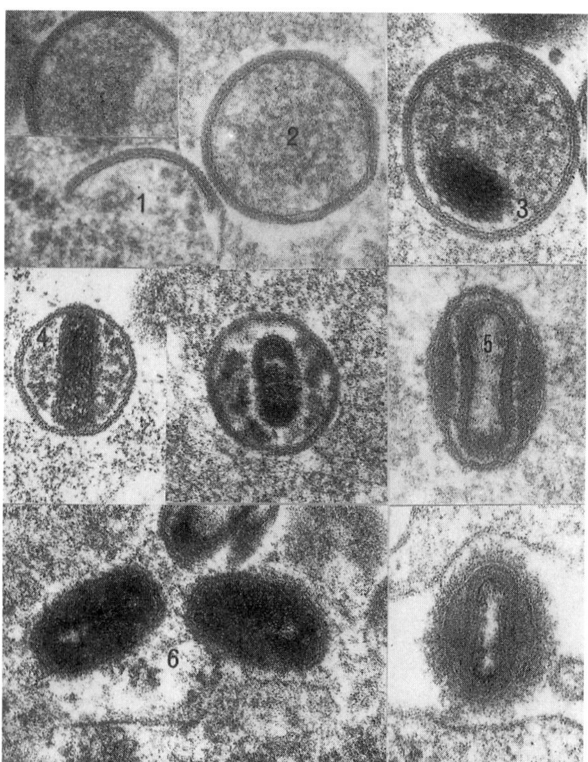

FIG. 10. Electron micrographs of developing and mature virus particles in thin section. Courtesy of P. Grimley.

Morphogenesis studies are facilitated by use of the macrolide antibiotic rifampicin. In the presence of this drug the viral membranes that accumulate are irregularly shaped, lack spicules, and fail to progress in maturation (Fig. 11) (358,370). These irregular membranes and the material they enclose have been called membrane-limited domains or rifampicin bodies. Within minutes after removal of the drug, the membranes become spicule coated and assume a regular convex shape (Fig. 11) even in the presence of inhibitors of RNA or protein synthesis, suggesting that rifampicin directly interferes with assembly (209). The gene responsible for resistance of mutant viruses to rifampicin has been mapped to the D13L ORF, which encodes a 65-kd protein (26,524) that rapidly associates with the rifampicin body membranes after drug removal (344,505, 549). Moreover, when expression of the D13L ORF is repressed, morphogenesis of the viral envelope is blocked at the same stage as occurs with rifampicin (596). The 65-kd protein appears to concentrate on the inner side of the membrane, perhaps serving as a scaffold.

In subsequent stages of development, the immature virions appear circular (or spherical in three dimensions) with

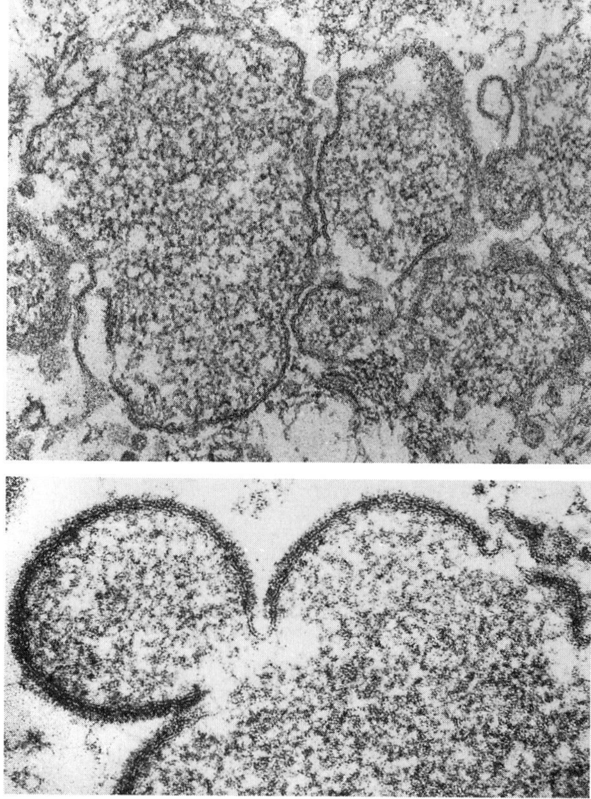

FIG. 11. Electron micrographs of thin sections of cell infected in the presence of rifampicin. Upper panel, HeLa cells were fixed and sectioned at 8 hours after infection in the presence of rifampicin. Lower panel, cells were fixed and sectioned at 10 minutes after removal of rifampicin. From Grimley (209), with permission.

a dense nucleoprotein mass embedded in a granular matrix (Fig. 10, panels 2 and 3). Occasional electron-microscopic pictures suggest that the nucleoprotein enters the immature envelopes just before they are completely sealed (349). Further steps in maturation are inferred from selected images (Fig. 10, panels 4–6). Several vaccinia virus major core proteins have been detected within the immature particles visualized by immunoelectron microscopy (549) or isolated by sedimentation (456,550).

Pulse-chase experiments indicated that at least several late polypeptides, including major structural proteins of the core, undergo proteolytic processing (356) at Ala-Gly↓Ala/Ser motifs (297). Processing of the major structural proteins is coupled to virus assembly, since their cleavage is prevented by rifampicin (270). Conditional lethal mutants can provide tools to study the roles of individual gene products in viral maturation (120,489). Specific repression of the D13L ORF mimics the effects of rifampicin and prevents both viral assembly and protein cleavages (596). Repression of synthesis of the F17R core phosphoprotein halts assembly at an immature particle stage and prevents proteolytic processing (595). Under nonpermissive conditions, *ts* mutants with substitutions in the I7L ORF appear to be blocked at a slightly earlier stage, but protein processing was not examined (260).

The question of how the transcription apparatus becomes enclosed within the core of the assembling virus particle is intriguing. It seems unlikely that each enzyme has its own targeting signal. A clue to possible mechanisms was obtained by examining the noninfectious virus particles that formed when synthesis of RAP94, the RNA polymerase–associated protein that confers specificity for early promoters, was repressed. Such particles contained an apparently complete set of structural proteins as well as the early transcription factor VETF but lacked the viral RNA polymerase, poly(A) polymerase, capping enzyme, topoisomerase, NPH I, and NPH II (593). Such a specific effect could result if the mechanism of entry involved the formation of a protein complex that associates with promoter-bound VETF through RAP94. Reduced amounts of RNA polymerase in virions formed with a ts mutation in VETF (301) is consistent with the latter model.

Occluded Virions

The IMV of some *Chordopoxvirinae* (e.g., cowpox, ectromelia, and fowlpox) become occluded in a dense protein matrix within the cytoplasm. These have been referred to as A-type inclusions to differentiate them from the sites of virus replication and assembly, which are sometimes called B-type inclusions (267). Presumably, the A-type inclusions are released into the environment following degeneration of infected cells. The major A-type inclusion protein of cowpox virus is a 160-kd species that may represent up to 4% of the total cell protein at late times after infection (184,399). Although vaccinia virus does not form A-type inclusions, a smaller homologous protein is made (398) because of a frame-shift mutation (14,128).

The virions of *Entomopoxvirinae* are also occluded to protect them from the environment (46). Following ingestion by a new larval host, infectious particles may be released in the alkaline pH of the gut. The sequences of homologous occlusion proteins, called spheroidin or spherulin, have been deduced from the ORFs of several entomopoxviruses (30,214,216,455b). These proteins are cysteine-rich, approximately 100 kd, and lack homology to fusolin, the abundant 50-kd spindle-body protein of entomopoxviruses (122b,585b), the A-type inclusion protein of chordopoxviruses, or the polyhedrin protein of baculoviruses.

Extracellular Enveloped Virions

Mature vaccinia virus particles move out of the assembly areas to the cell periphery, where they may become wrapped by additional membranes, derived from the trans-Golgi or early endosomal network, that contain viral proteins destined to be in the EEV (224,242,349,461,531). Studies with mutant vaccinia viruses indicate that wrapping requires expression of at least one IMV membrane protein, the A27L 14-kd protein, and two EEV membrane proteins, the F13L 37-kd protein and the B5R 42-kd glycoprotein (54,155,442,462,575). These wrapped particles, containing two inner membranes apparently derived from the intermediate compartment and two outer membranes from the Golgi, are transported along actin-containing microfilaments to the plasma membrane (223,225,514). Fusion of the wrapped particles with the plasma membrane results in the externalization of the virus, with loss of the outermost Golgi-derived membrane. Only a portion of the externalized virus is found in the medium as EEV, whereas the rest adheres to the cell surface and is called cell-associated enveloped virions (CEV). The ratio of adherent to released virions varies in different vaccinia virus strains and cell hosts (401), largely because of a single amino acid difference in the putative lectin-binding domain of the membrane protein encoded by the A34R ORF (56). The extracellular virus is important for virus dissemination (64,402). On cell monolayers the adherent virus can mediate efficient cell-to-cell spread and plaque formation, whereas the EEV provide long-range dissemination (55).

An avipoxvirus homolog of the 37-kd EEV protein determines fowlpox virus plaque size and EEV formation (88).

VIRUS/HOST INTERACTIONS

Inhibitory Effects on Host Macromolecular Synthesis

Infection of tissue culture cells with vaccinia virus or other orthopoxviruses results in profound cytopathic effects (19), changes in membrane permeability (91), and inhibition of DNA, RNA, and protein synthesis. Although these phenomena have been studied intensively, the absence of combined genetic and biochemical studies has made interpretations difficult. The effects on protein synthesis are dramatic (165,352). It seems likely that several factors may lead to the switch from host to viral protein synthesis and that the relative contribution of each factor may depend on the virus multiplicity, cell type, time of analysis, and use of metabolic inhibitors. Some experiments suggested that inhibition of host protein synthesis can occur in the absence of viral gene expression, thereby implicating a protein in the vaccinia virus particle. Candidate inhibitors are the surface tubules (324), the F17R phosphoprotein (411a), and the B1R protein kinase (43). There is indirect evidence that some viral early proteins may allow translation of viral mRNA to continue under conditions that inhibit host protein synthesis (42,355). Other studies have suggested that viral transcription is necessary for inhibition of host protein synthesis. For example, high concentration of actinomycin D and cordycepin prevented shutoff (20). Also in some cell lines, inhibition of host protein synthesis remains incomplete until after DNA replication. An inhibitory role for small poly(A)-containing RNA molecules was originally proposed to explain the effects on host protein synthesis that occur in the presence of actinomycin D or by ultraviolet-irradiated virus (197,451). Subsequent reports, however, suggested that small poly(A)-containing RNAs may have an inhibitory role during normal virus infection (21,22,86).

After several hours, most of the mRNA present in the cytoplasm of infected cells is viral, accounting for the predominance of viral protein synthesis (61,113). The shift in mRNA species may result from the short half-life of all mRNAs following vaccinia virus infection (44,469), coupled with very active virus transcription and inhibition of host RNA synthesis and transport. Rapid degradation of actin and tubulin mRNAs has been demonstrated (432).

Viral inhibitory effects on the synthesis or processing of cellular RNAs have been reported (44,252,469). This inhibition involves all classes of RNA and requires vaccinia virus protein synthesis; heated or ultraviolet irradiated virus was ineffective (406). This inhibition may be explained in part by the transit of RNA polymerase II activity from the nucleus to the cytoplasm (351,571). However, whole cell extracts of vaccinia virus–infected cells were no longer able to transcribe genes regulated by eukaryotic RNA polymerase II promoters, although they were still able to transcribe a polymerase III promoter (424). Nuclear DNA replication also is inhibited (218,257,278), possibly by vaccinia virus–encoded deoxyribonucleases (138).

Under certain nonpermissive conditions of vaccinia virus infection, the inhibition of host macromolecular synthesis may be accelerated. These situations include the restricted growth of vaccinia virus in Chinese hamster ovary cells, which can be overcome by the cowpox virus gene encoding a 77-kd protein (143,236,506); host range mutants of rabbitpox virus (363); and vaccinia virus K1L deletion mutants (142,200,409,519).

Stimulatory Effects on Cell Growth

Many poxviruses induce hyperplastic responses and even tumors in the skin of infected animals. These effects are pronounced with fowlpox virus (100), Shope fibroma virus (476), Yaba virus (40,377), and molluscum contagiosum virus (421). The hyperplasia can be explained by the secretion of poxvirus-encoded growth factors (269). The vaccinia virus growth factor (VGF) is a homolog of epidermal growth factor (EGF) and of transforming growth factor α (58). The processed, glycosylated, and secreted VGF is capable of binding to the EGF receptor, stimulating its autophosphorylation and inducing anchorage-independent cell growth (97,277,515,535). In tissue culture cells there was little or no difference in infectivity or yield between a mutant virus with deletions in both copies of the VGF gene and the parental virus (82). *In vivo* experiments revealed, however, that infection with wild-type virus resulted in a rapid proliferation of ectodermal and entodermal cells of the chicken embryo chorioallantoic membrane, whereas this did not occur with the mutant (83). Moreover, the proliferating cells had no viral antigens, confirming that this was a response to the secreted VGF and preceded virus infection. The absence of VGF resulted in diminished mutant virus replication *in vivo* and attenuated pathogenicity. A gene with 89% amino acid identity to VGF is present in variola virus (322). Leporipoxviruses (Shope fibroma virus, myxoma virus, and malignant rabbit fibroma viruses) also have stuctural and functional homologs of EGF that are nonessential for replication in tissue culture but contribute to virulence (98,99,304,305,381, 382,538).

A gene encoding a polypeptide with homology (22% to 27% amino acid identity) to mammalian vascular endothelial growth factor is present in the orf virus genome and is transcribed early in infection (311a). Although the activity of the viral protein has not yet been reported, it could account for the extensive capillary proliferation seen in histologic sections of infected skin.

Viral Defense Molecules

Host immune responses can halt the spread of viruses and eliminate them. Immediately after virus invasion, nonspecifc mechanisms involving interferons, complement, and natural killer cells predominate; subsequently, cytotoxic T cells and antibodies become important (84). In response to these selective pressures many viruses have developed ways of evading or impeding the immune response (328,500). None of the poxviruses undergo a latent phase.

TABLE 4. *Poxvirus immune defense molecules*

System	Target	Virus	Gene	Homolog	Properties	References
Complement	C4B and C3B	Vaccinia Variola	C3L (VCP), D15L	C4B binding protein	4 SCRs, secreted, binds and inhibits C4B and C3B, virulence factor	249,287,331
	?	Vaccinia Variola	B5R B6R	complement control proteins	4 SCRs, EEV class I membrane glycoprotein, for virus egress	154,155, 250a,575
Interferon	PKR	Vaccinia Variola	K3L C3L	eIF-2α	Binds PKR, inhibits phosphorylation of eIF-2α, INF resistance	41,92,125
	dsRNA	Vaccinia Variola	E3L E3L	PKR	Binds ds RNA, nuclear local, inhibits activation of PKR, INF resistance	96,124,589
	IFN-γ	Myx, SFV Vaccinia Variola	T7 B8R B8R	IFN-γ receptor	Secreted, binds and inhibits IFN-γ	543
IL-1	ICE	Coxpox Vaccinia Variola	crmA B14R (SPI-2) B12R	SERPIN	Prevents proteolytic activation of IL-1β, inhibits inflammatory response, inhibits apoptosis	346,384,385, 427
	IL-1β	Vaccinia Variola Cowpox	B16R, disrupted	IL-1 receptor	Secreted glycoprotein, binds and inhibits IL-1β, variable virulence	9,509
	?	Vaccinia Variola	B19R B17R	IL-1 and IL-6 receptor	Cell surface, no binding IL-1 or IL-6	501,536
TNF	TNF-α, TNF-β	Myx, SFV Vaccinia Variola	T2 G2R Truncated	TNF receptor	Secreted, binds and inhibits TNF-α, TNF-β, virulence factor	499,540
?	Serine proteases	Myx	Serp 1	SERPIN	Inhibits variety of serine proteases, anti-inflammatory, virulence factor	313,541
?	?	Vaccinia Variola	A44L A47L	Steroid dehydrog	Pregnenolone to progesterone, virulence factor	348

Myx, myxoma; SFV, Shope fibroma; IFN, interferon.

Instead poxviruses encode multiple proteins that interfere with the induction or activity of complement and the principal cytokines (Table 4). Some of these inhibitory proteins, like the growth factors described in a previous section, are secreted from infected cells. The general names *virokine* (289) and *viroceptor* (540) were suggested for such viral proteins and the strategy of engaging host molecules has been likened to the "star wars" defense policy (32).

Complement Regulatory Protein

One of the major secreted proteins of vaccinia virus consists largely of four, tandem, inexact copies of a 60-amino-acid sequence known as the short consensus repeat (SCR) (289). SCRs are present in a number of proteins that regulate complement activation, and those of vaccinia virus are most closely related to the SCRs in the human C4B binding protein. This vaccinia viral protein, called VCP, inhibits the classical and alternative pathways of complement activation through its ability to bind and inactivate both C4B and C3B (287,331). The concentration of VCP in the medium of vaccinia virus–infected tissue culture cells is sufficient to prevent complement-enhanced neutralization of vaccinia virus by antibody (249). A deletion mutant of vaccinia virus, specifically lacking the VCP gene, replicates well in tissue culture yet is severely attenuated when inoculated into animals. The skin lesions caused by the mutant virus are smaller and heal more rapidly than those produced by the parental virus (249). The variola genome encodes a closely related protein (322) that presumably had a role in the clinical manifestations of smallpox.

Vaccinia virus encodes another protein with four SCRs, the product of the B5R ORF, that has a 29% amino acid identity with VCP (154,250a). Unlike VCP, the B5R protein has a C-terminal hydrophobic domain that anchors it in the surface membrane of the infected cell and of the extracellular virus. As discussed in a preceding section, mutants with deletions of the B5R ORF produce low amounts of EEV and small plaques. Anticomplement activity has not been demonstrated for this protein or for several other orthopoxvirus proteins with low homology to complement regulators (59).

Inhibitors of Interferon and the Interferon Transduction Pathway

Types I and II interferons contribute to host defenses against poxviruses (183,238,263,460). Nevertheless, vaccinia virus is relatively resistant to interferon in a number of different cell lines and can even rescue interferon-sensitive RNA viruses, such as vesicular stomatitis virus and encephalomyocarditis virus (566,568). Two modes of interferon regulation have been well characterized: The double-stranded RNA-dependent protein kinase (PKR) pathway leads to the phosphorylation of the eukaryotic translation initiation factor eIF-2α and inhibition of protein synthesis; the 2-5A/RNase L pathway leads to degradation of mRNA and rRNA. Evidence for vaccinia viral inhibition of both pathways has been obtained (8,383,431, 433,567).

Two vaccinia viral genes encode inhibitors of the PKR pathway. Sequence analysis (201) indicated that the 88-amino-acid product of the vaccinia virus K3L gene is a homolog of the N-terminus of eIF-2α, the target of PKR. Transient expression of K3L reduced PKR activity and eIF-2α phosphorylation (125). *In vitro* studies further indicate that the viral gene product directly prevents phosphorylation of eIF-2α, as well as autophosphorylation of the kinase, by acting as a tightly binding pseudosubstrate of the kinase (92). Deletion of the K3L gene can increase the interferon sensitivity of vaccinia virus (41). The 25-kd product of the vaccinia virus E3L gene is a double-stranded RNA-binding protein that inhibits activation of PKR (96,124). The C-terminus of the viral protein is homologous to the double-stranded RNA-binding domain within the N-terminus of PKR (326). The E3L gene product was independently identified as a unique vaccinia virus protein that localized within the nucleus, in addition to the cytoplasm (589). Whether this localization signifies an additional role of the E3L gene product is unknown.

Enhanced degradation of mRNA and rRNA, apparently mediated through the 2–5A/RNase L pathway as a result of aberrant transcription, is caused by mutation of the A18R gene (39). As already discussed, the A18R protein is a DNA-dependent ATPase as well as a helicase (37b,285).

Poxviruses also employ secretory proteins for protection against interferons. Virus-encoded 35-kd proteins, homologous to the ligand-binding domain of the cellular receptor for interferon γ, are secreted from cells infected with myxoma and Shope fibroma virus (543). These leporipoxvirus proteins bind and inhibit rabbit interferon γ. The orthopoxviruses, vaccinia and variola, also encode interferon γ receptor homologs (322,472), though their biological activities have not yet been established.

IL-1 and TNF Receptor Homologs

The cytokines IL-1α and IL-1β participate in the early response to viral infections and regulate the inflammatory process. Vaccinia and cowpox viruses encode an IL-1 receptor homolog that is secreted from infected cells and binds and inhibits IL-1β (9,501,509). Vaccinia virus mutants with deletions in the corresponding B16R ORF replicate normally in tissue culture cells but have altered pathogenicity. Curiously, the presence of several frame-shift mutations in the variola homolog would preclude synthesis of an active protein, so that it was unnecessary for smallpox (323). Certain defense mechanisms may be less important for poxviruses, such as variola virus, that produce a rapid and fulminant disease than for poxviruses that have a milder and more prolonged interaction with the host.

The protein encoded by the B19R ORF of vaccinia virus, which is present in both cell surface and secreted forms, also has similarity to the IL-1 and IL-6 receptors (501,536). Since no interaction with IL-1 or IL-6 could be detected, the B19R product might bind to another, still untested, homologous ligand.

Members of the SERPIN Superfamily

Poxviruses have developed a second strategy to prevent IL-1 action. Cowpox virus produces pocks on the chicken chorioallantoic membrane that commonly appear red; the occurrence of rare white pocks was recognized as a viral mutational event that allowed an influx of inflammatory cells (141,173,547). The culmination of investigations of mutant cowpox virus genomes led to the finding that the gene primarily responsible for inhibiting the inflammatory response in this system has sequence motifs common to the serine protease inhibitor (SERPIN) superfamily (384, 414). *In vitro* studies demonstrated that this protein, called SPI-2 or crmA, inhibits the the IL-1β converting enzyme, which is an atypical cysteine proteinase (427). By preventing the intracellular conversion of the inactive precursor of IL-1β to the active secreted form, crmA could function as a viral anti-inflammatory molecule. crmA also has been shown to prevent IL-1β-converting enzyme-induced apoptosis of fibroblasts, suggesting a second role for this viral protein (346). There is also evidence for a third role, regulation of arachidonic acid metabolism (385). Deletion of the SPI-2 gene attenuates both cowpox and rabbitpox viruses (529)

Orthopoxviruses encode two additional members of the SERPIN superfamily, known as SPI-1 and SPI-3 (65,290, 503). The targets of these two proteins are unknown; however, deletion of the SPI-3 gene induces fusion of infected cells (296,534,598). Deletion of the SPI-1 gene causes host range effects and a white pock phenotype for rabbitpox virus but not cowpox virus (10).

A related but distinct SERPIN homolog (SERP 1), present in the genome of myxoma virus and malignant rabbit fibroma virus, is secreted from infected cells, interferes with inflammation, and is required for virus-induced disease in rabbits (313,541). SERP 1 inhibits a number of serine proteases (plasmin, urokinase, tissue plasminogen activator, and a member of the complement cascade) *in vitro*, but whether any or all of these are significant biological targets *in vivo* is unknown (306).

MUTANTS

Viral mutants have contributed substantially to the studies of poxvirus replication described in the previous sections. The purpose here is to indicate approaches to the generation of poxvirus mutants that may have further applicability. Poxviruses contain many genes, particularly in the near terminal regions of the genome, that are not required for replication *in vitro*; consequently, deletion mutants have arisen spontaneously during tissue culture passage (366). Recombinant DNA technology, however, has allowed investigators to insertionally inactivate or "knock out" specific genes using color screening (93,386) or antibiotic selection (68,169,180) to isolate the desired mutants. Both spontaneous and targeted deletion mutants have been effectively used to study viral gene function, as described in previous sections. Nevertheless, the procedure is limited to genes that are not essential for replication. Recently, a cell line that was stably transfected with a vaccinia virus gene was shown to complement a vaccinia virus mutant (519). Extensions of this approach could greatly increase the number of genes that can be studied by deletion mutagenesis.

Transient dominant selection, a versatile procedure employing single-site recombination by circular plasmids, permits the isolation of replication-competent vaccinia virus mutants without regard to the number of bases added, deleted, or changed or the resulting phenotype (170). This procedure can be coupled with reverse guanine phosphoribosyltransferase selection (248).

Conditional lethal *ts* mutants of vaccinia virus have been isolated after random mutagenesis and both classical genetic and marker rescue techniques used to determine their map locations (110). However, only about 40 complementation groups were obtained by such procedures. This number should soon be increased using the recently demonstrated ability to produce temperature-sensitive mutations in specific genes (221).

A novel inducer-dependent class of conditional mutants of vaccinia virus was developed using the repressor and operator elements of the *E. coli lac* operon (441,594). This procedure has been applied to several intermediate and late-stage genes encoding structural proteins and transcription factors (146,442,569,593,595-597).

EXPRESSION VECTORS

Several attributes of poxviruses have led to their extensive use as expression vectors (353). These include relative ease of formation and isolation of recombinant viruses, capacity for large amounts of DNA, relatively high expression, and wide host range. Foreign DNA has been inserted into orthopoxvirus or avipoxvirus genomes by homologous recombination (67,315,387) or *in vitro* ligation (343a,459). Color screening, antibiotic selection, plaque size, DNA hybridization, and antibody binding all can be used for recombinant virus isolation (148). Expression has been achieved either by using poxvirus promoters or by employing bacteriophage RNA polymerases and cognate promoters (153). Recombinant viruses have been used for the synthesis of proteins *in vivo* or *in vitro* and as vaccine candidates.

CONCLUSIONS

The poxviruses are among the largest and most genetically complex of all animal viruses. Unlike most other DNA viruses, they replicate in the cytoplasm and encode many proteins that permit transcription and replication to occur outside of the nucleus. Gene expression is stringently regulated by a cascade mechanism, and virion assembly is a complex process that involves the formation of multiple membranes. Viral proteins are used for evasion of the host immune defense system.

REFERENCES

1. Ahn B-Y, Gershon PD, Jones EV, Moss B. Identification of *rpo30*, a vaccinia virus RNA polymerase gene with structural similarity to a eukaryotic transcription factor. *Mol Cell Biol* 1990;10:5433–5441.
2. Ahn B-Y, Gershon PD, Moss B. RNA-polymerase associated protein RAP94 confers promoter specificity for initiating transcription of vaccinia virus early stage genes. *J Biol Chem* 1994;269:7552–7557.
3. Ahn B-Y, Jones EV, Moss B. Identification of the vaccinia virus gene encoding an 18-kilodalton subunit of RNA polymerase and demonstration of a 5′ poly(A) leader on its early transcript. *J Virol* 1990;64:3019–3024.
4. Ahn B-Y, Moss B. Capped poly(A) leader of variable lengths at the 5′ ends of vaccinia virus late mRNAs. *J Virol* 1989;63:226–232.
5. Ahn B-Y, Moss B. RNA polymerase-associated transcription specificity factor encoded by vaccinia virus. *Proc Natl Acad Sci USA* 1992;89:3536–3540.
6. Ahn B-Y, Rosel J, Cole NB, Moss B. Identification and expression of *rpo19*, a vaccinia virus gene encoding a 19-kilodalton DNA-dependent RNA polymerase subunit. *J Virol* 1992;66:971–982.
7. Ahn BY, Moss B. Glutaredoxin homolog encoded by vaccinia virus is a virion-associated enzyme with thioltransferase and dehydroascorbate reductase activities. *Proc Natl Acad Sci USA* 1992;89:7060–7064.
8. Akkaraju GR, Whitaker-Dowling P, Youngner JS, Jagus R. Vaccinia specific kinase inhibitory factor prevents translational inhibition by double-stranded RNA in rabbit reticulocyte lysate. *J Biol Chem* 1989;264:10321–10325.
9. Alcami A, Smith GL. A soluble receptor for interleukin-1b encoded by vaccinia virus—a novel mechanism of virus modulation of the host response to infection. *Cell* 1992;71:153–167.
10. Ali AN, Turner PC, Brooks MA, Moyer RW. The SPI-1 gene of rabbitpox virus determines host range and is required for hemorrhagic pock formation. *Virology* 1994;202:305–314.
11. Amegadzie B, Holmes M, Cole NB, Jones EV, Earl PL, Moss B. Identification, sequence, and expression of the gene encoding the second-largest subunit of the vaccinia virus RNA polymerase. *Virology* 1991;180:88–98.
12. Amegadzie BY, Ahn BY, Moss B. Characterization of a 7-kilodalton subunit of vaccinia virus DNA-dependent RNA polymerase with structural similarities to the smallest subunit of eukaryotic RNA polymerase-II. *J Virol* 1992;66:3003–3010.
13. Amegadzie BY, Cole N, Ahn BY, Moss B. Identification, sequence and expression of the gene encoding a Mr 35,000 subunit of the vac-

cinia virus DNA-dependent RNA polymerase. *J Biol Chem* 1991;266: 13712–13718.
14. Amegadzie BY, Sisler JR, Moss B. Frame-shift mutations within the vaccinia virus A-type inclusion protein gene. *Virology* 1992;186: 777–782.
15. Antczak JB, Patel DD, Ray CA, Ink BS, Pickup DJ. Site-specific RNA cleavage generates the 3′ end of a poxvirus late mRNA. *Proc Natl Acad Sci USA* 1992;89:12033–12037.
16. Archard LC, Johnson K, Malcolm AD. Specific transcription of orthopox virus DNA by HeLa cell RNA polymerase II. *Febs Lett* 1985; 192:53–56.
17. Archard LC, Mackett M, Barnes DE, Dumbell KR. The genome structure of cowpox virus white pock variants. *J Gen Virol* 1984;65:875–876.
18. Arif BM. The entomopoxviruses. In: Kurstak E, ed. *Viruses of invertebrates*. New York: Dekker; 1991:179–195.
19. Bablanian R, Baxt B, Sonnabend JA, Esteban M. Studies on the mechanisms of vaccinia virus cytopathic effects. II. Early cell rounding is associated with virus polypeptide synthesis. *J Gen Virol* 1978;39: 403–413.
20. Bablanian R, Coppola G, Scribani S, Esteban M. Inhibition of protein synthesis by vaccinia virus. IV. The role of low-molecular-weight viral RNA in the inhibition of protein synthesis. *Virology* 1981;112:13–24.
21. Bablanian R, Goswami SK, Esteban M, Banerjee AK, Merrick WC. Mechanism of selective translation of vaccinia virus mRNAs: differential role of poly(A) and initiation factors in the translation of viral and cellular mRNAs [published erratum appears in *J Virol* 1991;65: 5655]. *J Virol* 1991;65:4449–4460.
22. Bablanian R, Scribani S, Esteban M. Amplification of polyadenylated nontranslated small RNA sequences (POLADS) during superinfection correlates with the inhibition of viral and cellular protein synthesis. *Cell Mol Biol Res* 1993;39:243–255.
23. Balassu TC, Robinson AJ. Orf virus replication in bovine testis cells: kinetics of viral DNA, polypeptides, and infectious virus production and analysis of virion polypeptides. *Arch Virol* 1987;97:267–278.
24. Baldick CJ Jr, Cassetti MC, Harris N, Moss B. Ordered assembly of a functional pre-initiation transcription complex, containing vaccinia virus early transcription factor and RNA polymerase, on an immobilized template. *J Virol* 1994;68:6052–6056.
25. Baldick CJ Jr, Keck JG, Moss B. Mutational analysis of the core, spacer and initiator regions of vaccinia virus intermediate class promoters. *J Virol* 1992;66:4710–4719.
26. Baldick CJ Jr, Moss B. Resistance of vaccinia virus to rifampicin conferred by a single nucleotide substitution near the predicted NH_2 terminus of a gene encoding an M_r 62,000 polypeptide. *Virology* 1987; 156:138–145.
27. Baldick CJ Jr, Moss B. Characterization and temporal regulation of mRNAs encoded by vaccinia virus intermediate stage genes. *J Virol* 1993;67:3515–3527.
28. Ball LA. High frequency recombination in vaccinia virus DNA. *J Virol* 1987;61:1788–1795.
29. Banham A, Smith GL. Vaccinia virus gene B1R encodes a 34-kDa serine/threonine protein kinase that localizes in cytoplasmic factories and is packaged into virions. *Virology* 1992;191:803–812.
30a. Banham AH, Leader DP, Smith GL. Phosphorylation of ribosomal proteins by the vaccinia virus B1R protein kinase. *FEBS* 1993;321:27–31.
30b. Banville M, Duma SF, Trifiro S, Arif B, Richarson C. The predicted amino acid sequence of the spheroidin protein of Amsacta moorei entomopoxvirus. Lack of homology between major occlusion body proteins of different poxviruses. *J Gen Virol* 1992;73:559–566.
31. Barbosa E, Moss B. mRNA(nucleoside-2′-)-methyltransferase from vaccinia virus. Characteristics and substrate specificity. *J Biol Chem* 1978;253:7698–7702.
32. Barinaga M. Viruses launch their own "star wars." *Nature* 1992;258: 1730–1731.
33. Baroudy BM, Moss B. Purification and characterization of a DNA-dependent RNA polymerase from vaccinia virions. *J Biol Chem* 1980; 255:4372–4380.
34. Baroudy BM, Moss B. Sequence homologies of diverse length tandem repetitions near ends of vaccinia virus genome suggest unequal crossing over. *Nucleic Acids Res* 1982;10:5673–5679.
35. Baroudy BM, Venkatesan S, Moss B. Incompletely base-paired flip-flop terminal loops link the two DNA strands of the vaccinia virus genome into one uninterrupted polynucleotide chain. *Cell* 1982;28: 315–324.
36. Baroudy BM, Venkatesan S, Moss B. Structure and replication of vaccinia virus telomeres. *Cold Spring Harbor Symp Quant Biol* 1982;47: 723–729.
37a. Bauer WR, Ressner EC, Kates J, Patzke J. A DNA nicking-closing enzyme encapsidated in vaccinia virus: partial purification and properties. *Proc Natl Acad Sci USA* 1977;74:1841–1845.
37b. Bayliss CD, Simpson DA, Condit, RC. Personal communication.
38. Baxby D. The origins of vaccinia virus. *J Infect Dis* 1977;136:453–455.
39. Bayliss CD, Condit RC. Temperature-sensitive mutants in the vaccinia virus A18R gene increases double-stranded RNA synthesis as a result of aberrant viral transcription. *Virology* 1993;194:254–262.
40. Bearcroft WCC, Jamieson MF. An outbreak of subcutaneous tumors in Rhesus monkeys. *Nature* 1958;182:195–196.
41. Beattie E, Tartaglia J, Paoletti E. Vaccinia virus encoded eIF-2α homolog abrogates the antiviral effect of interferon. *Virology* 1991;183: 419–422.
42. Beaud G, Dru A. Protein synthesis in vaccinia virus-infected cells in the presence of amino acid analogs: a translational control mechanism. *Virology* 1980;100:10–21.
43. Beaud G, Sharif A, Topamass A, Leader DP. Ribosomal-protein S2/SA kinase purified from HeLa cells infected with vaccinia virus corresponds to the B1R protein-kinase and phosphorylates *in vitro* the viral ssDNA-binding protein. *J Gen Virol* 1994;75:283–293.
44. Becker Y, Joklik WK. Messenger RNA in cells infected with vaccinia virus. *Proc Natl Acad Sci USA* 1964;51:577–585.
45. Belle Isle H, Venkatesan S, Moss B. Cell-free translation of early and late mRNAs selected by hybridization to cloned DNA fragments derived from the left 14 million to 72 million daltons of the vaccinia virus genome. *Virology* 1981;112:306–317.
46. Bergoin M, Devauchelle G, Vago C. Electron microscopy study of Melolontha poxvirus: The fine structure of occluded virions. *Virology* 1971;43:453–467.
47. Bertholet C, Stocco P, Van Meir E, Wittek R. Functional analysis of the 5′ flanking sequence of a vaccinia virus late gene. *EMBO J* 1986; 5:1951–1957.
48. Bertholet C, Van Meir E, ten Heggeler-Bordier B, Wittek R. Vaccinia virus produces late mRNAs by discontinuous synthesis. *Cell* 1987;50: 153–162.
49. Binns MM, Boursnell ME, Skinner MA. Gene translocations in poxviruses: the fowlpox virus thymidine kinase gene is flanked by 15 bp direct repeats and occupies the locus which in vaccinia virus is occupied by the ribonucleotide reductase large subunit gene. *Virus Res* 1992;24:161–72.
50. Black ME, Hruby DE. Identification of the ATP-binding domain of vaccinia virus thymidine kinase. *J Biol Chem* 1990;265:17584–17592.
51. Black ME, Hruby DE. Quaternary structure of vaccinia virus thymidine kinase. *Biochem Biophys Res Commun* 1990;169:1080–1086.
52. Black ME, Hruby DE. Site-directed mutagenesis of a conserved domain in vaccinia virus thymidine kinase—evidence for a potential role in magnesium binding. *J Biol Chem* 1992;267:6801–6806.
53. Blake NW, Porter CD, Archard LC. Characterization of a molluscum contagiosum virus homolog of the vaccinia virus p37K major envelope antigen. *J Virol* 1991;65:3583–3589.
54. Blasco R, Moss B. Extracellular vaccinia virus formation and cell-to-cell virus transmission are prevented by deletion of the gene encoding the 37,000 Dalton outer envelope protein. *J Virol* 1991;65:5910–5920.
55. Blasco R, Moss B. Role of cell-associated enveloped vaccinia virus in cell-to-cell spread. *J Virol* 1992;66:4170–4179.
56. Blasco R, Sisler JR, Moss B. Dissociation of progeny vaccinia virus from the cell membrane is regulated by a viral envelope glycoprotein: effect of a pont mutation in the lectin homology domain of the A34R gene. *J Virol* 1993;3319–3325.
57. Block W, Upton C, McFadden G. Tumorigenic poxviruses: genomic organization of malignant rabbit virus, a recombinant between Shope fibroma virus and myxoma virus. *Virology* 1985;140:113–124.
58. Blomquist MCL, Hunt LT, Barker WC. Vaccinia virus 19-kilodalton protein: relationship to several mammalian proteins, including two growth factors. *Proc Natl Acad Sci USA* 1984;81:7363–7367.
59. Bloom DC, Edwards KM, Hager C, Moyer RW. Identification and characterization of two nonessential regions of the rabbitpox virus genome involved in virulence. *J Virol* 1991;65:1530–1542.
60. Boone RF, Moss B. Methylated 5′ terminal sequences of vaccinia virus mRNA species made *in vivo* at early and late times after infection. *Virology* 1977;79:67–80.

61. Boone RF, Moss B. Sequence complexity and relative abundance of vaccinia virus mRNA's synthesized *in vivo* and *in vitro*. *J Virol* 1978; 26:554–569.
63. Boone RF, Parr RP, Moss B. Intermolecular duplexes formed from polyadenylated vaccinia virus RNA. *J Virol* 1979;30:365–374.
64. Boulter EA, Appleyard G. Differences between extracellular and intracellular forms of poxvirus and their implications. *Prog Med Virol* 1973;16:86–108.
65. Boursnell MEG, Foulds IJ, Campbell JI, Binns MM. Non-essential genes in the vaccinia virus HindIII K fragment: a gene related to serine protease inhibitors and a gene related to the 37K vaccinia virus major envelope antigen. *J Gen Virol* 1988;69:2995–3003.
66. Boyle DB. Quantitative assessment of poxvirus promoters in fowlpox and vaccinia virus recombinants. *Virus Genes* 1992;6:281–290.
67. Boyle DB, Coupar BEH. Construction of recombinant fowlpox viruses as vectors for poultry vaccines. *Virus Res* 1988;10:343–356.
68. Boyle DB, Coupar BEH. A dominant selectable marker for the construction of recombinant poxviruses. *Gene* 1988;65:123–128.
69. Boyle DB, Coupar BH. Identification and cloning of the fowlpox virus thymidine kinase gene using vaccinia virus. *J Gen Virol* 1986;67:1591–1600.
70. Broyles SS. A role for ATP hydrolysis in vaccinia virus early gene transcription. *J Biol Chem* 1991;266:15545–15548.
71. Broyles SS. Vaccinia virus encodes a functional dUTPase. *Virology* 1993;195:863–865.
72. Broyles SS, Fesler BS. Vaccinia virus gene encoding a component of the viral early transcription factor. *J Virol* 1990;64:1523–1529.
73. Broyles SS, Li J. The small subunit of the vaccinia virus early transcription factor contacts the transcription promoter DNA. *J Virol* 1993; 67:5677–5680.
74. Broyles SS, Li J, Moss B. Promoter DNA contacts made by the vaccinia virus early transcription factor. *J Biol Chem* 1991;266:15539–15544.
75. Broyles SS, Moss B. Homology between RNA polymerases of poxviruses, prokaryotes, and eukaryotes: nucleotide sequence and transcriptional analysis of vaccinia virus genes encoding 147-kDa and 22-kDa subunits. *Proc Natl Acad Sci USA* 1986;83:3141–3145.
76. Broyles SS, Moss B. Identification of the vaccinia virus gene encoding nucleoside triphosphate phosphohydrolase I, a DNA-dependent ATPase. *J Virol* 1987;61:1738–1742.
77. Broyles SS, Moss B. Sedimentation of an RNA polymerase complex from vaccinia virus that specifically initiates and terminates transcription. *Mol Cell Biol* 1987;7:7–14.
78. Broyles SS, Moss B. DNA-dependent ATPase activity associated with vaccinia virus early transcription factor. *J Biol Chem* 1988;263:10761–10765.
79. Broyles SS, Pennington MJ. Vaccinia virus gene encoding a 30-kilodalton subunit of the viral DNA-dependent RNA polymerase. *J Virol* 1990;64:5376–5382.
80. Broyles SS, Yuen L, Shuman S, Moss B. Purification of a factor required for transcription of vaccinia virus early genes. *J Biol Chem* 1988; 263:10754–10760.
81. Buist JB. The life-history of the micro-organisms associated with variola and vaccinia. *Proc R Soc Edinburgh* 1886;13:603–615.
82. Buller RM, Chakrabarti S, Cooper JA, Twardzik DR, Moss B. Deletion of the vaccinia virus growth factor gene reduces virus virulence. *J Virol* 1988;62:866–877.
83. Buller RML, Chakrabarti S, Moss B, Frederickson T. Cell proliferative response to vaccinia virus is mediated by VGF. *Virology* 1988; 164:182–192.
84. Buller RML, Palumbo GJ. Poxvirus pathogenesis. *Microbiol Rev* 1991; 55:80–122.
85. Buller RML, Smith GL, Cremer K, Notkins AL, Moss B. Decreased virulence of recombinant vaccinia virus expression vectors is associated with a thymidine kinase-negative phenotype. *Nature* 1985;317: 813–815.
86. Cacoullos N, Bablanian R. Polyadenylated RNA sequences produced in vaccinia virus-infected cells under aberrant conditions inhibit protein synthesis *in vitro*. *Virology* 1991;184:747–751.
87. Cairns J. The initiation of vaccinia infection. *Virology* 1960;11:603–623.
88. Calvert JG, Ogawa R, Yanagida N, Nazerian K. Identification and functional analysis of the fowlpox virus homolog of the vaccinia virus p37K major envelope antigen gene. *Virology* 1992;191:783–792.
89. Carpenter MS, DeLange AM. A temperature-sensitive lesion in the small subunit of the vaccinia virus-encoded mRNA capping enzyme causes a defect in viral telomere resolution. *J Virol* 1991;65:4042–4050.
90. Carpenter MS, DeLange AM. Identification of a temperature-sensitive mutant of vaccinia virus defective in late but not intermediate gene expression. *Virology* 1992;188:233–244.
91. Carrasco L, Esteban M. Modification of membrane permeability in vaccinia virus infected cells. *Virology* 1982;117:62–69.
92. Carroll K, Elroy-Stein O, Moss B, Jagus R. Recombinant vaccinia virus K3L gene product prevents activation of double-stranded RNA-dependent, initiation factor 2 alpha-specific protein kinase. *J Biol Chem* 1993;268:12837–12842.
93. Chakrabarti S, Brechling K, Moss B. Vaccinia virus expression vector: coexpression of β-galactosidase provides visual screening of recombinant virus plaques. *Mol Cell Biol* 1985;5:3403–3409.
94. Challberg MD, Englund PT. Purification and properties of the deoxyribonucleic acid polymerase induced by vaccinia virus. *J Biol Chem* 1979;254:7812–7819.
95. Chang A, Metz DH. Further investigations on the mode of entry of vaccinia virus into cells. *J Gen Virol* 1976;32:275–282.
96. Chang HW, Watson JC, Jacobs BL. The E3L gene of vaccinia virus encodes an inhibitor of the interferon-induced, double-stranded RNA-dependent protein kinase. *Proc Natl Acad Sci USA* 1992;89:4825–4829.
97. Chang W, Lim JG, Hellstrom I, Gentry LE. Characterization of vaccinia virus growth factor biosynthetic pathway with an antipeptide antiserum. *J Virol* 1988;62:1080–1083.
98. Chang W, Macaulay C, Hu S-L, Tam JP, McFadden G. Tumorigenic poxviruses: characterization of the expression of an epidermal growth factor related gene in Shope fibroma virus. *Virology* 1990;179:926–930.
99. Chang W, Upton C, Hu SL, Purchio AF, McFadden G. The genome of Shope fibroma virus, a tumorigenic poxvirus, contains a growth factor gene with sequence similarity to those encoding epidermal growth factor and transforming growth factor alpha. *Mol Cell Biol* 1987;7:535–540.
100. Cheevers WP, O'Callaghan DJ, Randall CC. Biosynthesis of host and viral deoxyribonucleic acid during hyperplastic fowlpox-infection *in vivo*. *J Virol* 1968;2:421–429.
101. Chernos VI, Vovk TS, Ivanova ON, Antonova TP, Loparev VN. [Insertion mutants of the vaccinia virus. The effect of inactivating E7R and D8L genes on the biological properties of the virus].*Mol Gen Mikrobiol Virusol* 1993;2:30–34.
102. Chertov OY, Telezhinskaya IN, Zaitseva, EV, Golubeva TB, Zinov'ev VV, Ovechkina LG, et al. Amino acid sequence determination of vaccinia virus immunodominant protein p35 and identification of the gene. *Biomed Sci* 1991;2:151–154.
103. Child SJ, Palumbo G, Buller RM, Hruby D. Insertional inactivation of the large subunit of ribonucleotide reductase encoded by vaccinia virus is associated with reduced virulence *in vivo*. *Virology* 1990;174: 625–629.
104. Cochran MA, Puckett C, Moss B. In vitro mutagenesis of the promoter region for a vaccinia virus gene: evidence for tandem early and late regulatory signals. *J Virol* 1985;54:30–37.
105. Colby C, Jurale C, Kates JR. Mechanism of synthesis of vaccinia virus double-stranded ribonucleic acid *in vivo* and *in vitro*. *J Virol* 1971;7: 71–76.
106. Colinas RJ, Goebel SJ, Davis SW, Johnson GP, Norton EK, Paoletti E. A DNA ligase gene in the Copenhagen strain of vaccinia virus is nonessential for viral replication and recombination. *Virology* 1990; 179:267–275.
107. Condit RC, Easterly R, Pacha RF, Fathi Z, Meis RJ. A vaccinia virus isatin-fl-thiosemicarbazone resistance mutation maps in the viral gene encoding the 132-kDa subunit of RNA polymerase. *Virology* 1991; 185:857–861.
108. Condit RC, Motyczka A. Isolation and preliminary characterization of temperature-sensitive mutants of vaccinia virus. *Virology* 1981;113: 224–241.
109. Condit RC, Motyczka A, Spizz G. Isolation, characterization, and physical mapping of temperature-sensitive mutants of vaccinia virus. *Virology* 1983;128:429–443.
110. Condit RC, Niles EG. Orthopoxvirus genetics. *Curr Topics Microbiol Immunol* 1990;169:1–40.
111. Cong P, Shuman S. Covalent catalysis in nucleotidyl transfer. A KTDG motif essential for enzyme-GMP complex formation by mRNA capping enzyme is conserved at the active sites of RNA and DNA ligases. *J Biol Chem* 1993;268:7256–7260.
112. Cong PJ, Shuman S. Methyltransferase and subunit association do-

mains of vaccinia virus messenger RNA capping enzyme. *J Biol Chem* 1992;267:16424–16429.

113. Cooper JA, Moss B. In vitro translation of immediate early, early, and late classes of RNA from vaccinia virus infected cells. *Virology* 1979; 96:368–380.

114. Cooper JA, Moss B, Katz E. Inhibition of vaccinia virus late protein synthesis by isatin-β-thiosemicarbazone. Characterization and *in vitro* translation of viral mRNA. *Virology* 1970;96:381–392.

115. Cooper JA, Wittek R, Moss B. Extension of the transcriptional and translational map of the left end of the vaccinia virus genome to 21 kilobase pairs. *J Virol* 1981;39:733–745.

116. Costanzo F, Fiume L, La Placa M, Mannini-Palenzona A, Novello F, Stirpe F. Ribonucleic acid polymerase induced by vaccinia virus: lack of inhibition by rifampicin and alpha-amanitin. *J Virol* 1970;5:266–269.

117. Dales S. Relation between penetration of vaccinia, release of viral DNA, and initiation of genetic functions. *Perspect Virol* 1965;4:47–71.

118. Dales S. Reciprocity in the interactions between the poxviruses and their host cells. *Ann Rev Microbiol* 1990;44:173–192.

119. Dales S, Kajioka R. The cycle of multiplication of vaccinia virus in Earle's strain L cells. I. Uptake and penetration. *Virology* 1964;24: 278–294.

120. Dales S, Milovanovitch V, Pogo BGT, Weintraub SB, Huima T, Wilton S, McFadden G. Biogenesis of vaccinia: isolation of conditional lethal mutants and electron microscopic characterization of their phenotypically expressed defects. *Virology* 1978;84:403–428.

121. Dales S, Mosbach EH. Vaccinia as a model for membrane biogenesis. *Virology* 1968;35:564–583.

122a. Dales S, Siminovitch L. The development of vaccinia virus in Earles L strain cells as examined by electron microscopy. *J Biophys Biochem Cytol* 1961;10:475–503.

122b. Dall D, Sriskantha A, Vera A, Lai-Fook J, Symonds T. A gene encoding a highly expressed spindle body protein of Heliothis armigera entomopoxvirus. *J Gen Virol* 1993;74:1811–1818.

123. Darai G, Reisner H, Scholz J, Schnitzler P, Lorbacher de Ruiz H. Analysis of the genome of molluscum contagiosum virus by restriction endonuclease analysis and molecular cloning. *J Med Virol* 1986; 18:29–39.

124. Davies MV, Chang H-W, Jacobs BL, Kaufman RJ. The E3L and K3L vaccinia virus gene products stimulate translation through inhibition of the double-stranded RNA-dependent protein kinase by different mechanisms. *J Virol* 1993;67:1688–1692.

125. Davies MV, Elroy-Stein O, Jagus R, Moss B, Kaufman RJ. The vaccinia virus K3L gene product potentiates translation by inhibiting double-stranded-RNA-activated protein kinase and phosphorylation of the alpha subunit of eukaryotic initiation factor 2. *J Virol* 1992;66:1943–1950.

126. Davis RE, Mathews CK. Acidic-C terminus of vaccinia virus DNA-Binding protein interacts with ribonucleotide reductase. *Proc Natl Acad Sci USA* 1993;90:745–749.

127a. Davison AJ, Moss B. The structure of vaccinia virus early promoters. *J Mol Biol* 1989;210:749–769.

127b. Davison AJ, Moss B. The structure of vaccinia virus late promoters. *J Mol Biol* 1989;210:771–784.

128. De Carlos A, Paez E. Isolation and characterization of mutants of vaccinia virus with a modified 94-kDa inclusion protein. *Virology* 1991; 185:768–778.

129. DeFilippes FM. Site of the base change in the vaccinia virus DNA polymerase gene which confers aphidicolin resistance. *J Virol* 1989;63: 4060–4063.

130. DeLange AM. Identification of temperature-sensitive mutants of vaccinia virus that are defective in conversion of concatemeric replicative intermediates to the mature linear DNA genome. *J Virol* 1989;63:2437–2444.

131. DeLange AM, Futcher B, Morgan R, McFadden G. Cloning of the vaccinia virus telomere in a yeast plasmid vector. *Gene* 1984;27:13–21.

132. Delange AM, Macaulay C, Block W, Mueller T, McFadden G. Tumorigenic poxviruses: construction of the composite physical map of the Shope fibroma virus genome. *J Virol* 1984;50:408–416.

133. DeLange AM, McFadden G. Sequence-nonspecific replication of transfected plasmid DNA in poxvirus-infected cells. *Proc Natl Acad Sci USA* 1986;83:614–618.

134. DeLange AM, McFadden G. Efficient resolution of replicated poxvirus telomeres to native hairpin structures requires two inverted symmetrical copies of a core target DNA sequence. *J Virol* 1987;61:1957–1963.

135. Delange AM, Reddy M, Scraba D, Upton C, McFadden G. Replication and resolution of cloned poxvirus telomeres *in vivo* generates linear minichromosomes with intact viral hairpin termini. *J Virol* 1986; 59:249–259.

136. Demkowicz WE, Maa JS, Esteban M. Identification and characterization of vaccinia virus genes encoding proteins that are highly antigenic in animals and are immunodominant in vaccinated humans. *J Virol* 1992;66:386–398.

137. Deng S, Shuman S. A role for the H4 subunit of vaccinia RNA polymerase in transcription initiation at a viral early promoter. *J Biol Chem* 1994;269:14323–14329.

138. Des Gouttes Olgiati D, Pogo BG, Dales S. Biogenesis of vaccinia: specific inhibition of rapidly labeled host DNA in vaccinia inoculated cells. *Virology* 1976;71:325–35.

139. Diaz-Guerra M, Esteban M. Vaccinia virus nucleoside triphosphate phosphohydrolase I controls early and late gene expression by regulating the rate of transcription. *J Virol* 1993;67:7561–7572.

140. Doms RW, Blumenthal R, Moss B. Fusion of intra- and extracellular forms of vaccinia virus with the cell membrane. *J Virol* 1990;64:4884–4892.

141. Downie AW, Haddock DW. A variant of cowpox virus. *Lancet* 1952; i:1049–1050.

142. Drillien R, Koehren F, Kirn A. Host range deletion mutant of vaccinia virus defective in human cells. *Virology* 1981;111:488–499.

143. Drillien RD, Spehner D, Kirn A. Host range restriction of vaccinia virus in Chinese hamster ovary cells: relationship to shutoff of protein synthesis. *J Virol* 1978;28:843–850.

144. Dubbs DR, Kit S. Isolation and properties of vaccinia mutants deficient in thymidine kinase-inducing activity. *Virology* 1964;22:214–225.

145. Dubochet J, Adrian M, Richter K, Garces J, Wittek R. Structure of intracellular mature vaccinia virus observed by cryoelectron microscopy. *J Virol* 1994;68:1935–1941.

146. Duncan SA, Smith GL. Identification and characterization of an extracellular envelope glycoprotein affecting vaccinia virus egress. *J Virol* 1992;66:1610–1621.

147. Dyster LM, Niles EG. Genetic and biochemical characterization of vaccinia virus genes D2L and D3R which encode virion structural proteins. *Virology* 1991;182:455–467.

148. Earl PL, Moss B. Generation of recombinant vaccinia viruses. In: Ausubel FM, Brent R, Kingston RE, Moore DD, Seidman JG, Smith JA, Struhl K, ed. *Currrent protocols in molecular biology*. New York: Greene Publishing Associates and Wiley Interscience; 1991:16.17.1–16.17.16.

149. Earl PL, Hügen AW, Moss B. Removal of cryptic poxvirus transcription termination signals from the human immunodeficiency virus type 1 envelope gene enhances expression and immunogenicity of a recombinant vaccinia virus. *J Virol* 1990;64:2448–2451.

150. Earl PL, Jones EV, Moss B. Homology between DNA polymerases of poxviruses, herpesviruses, and adenoviruses: nucleotide sequence of the vaccinia virus DNA polymerase gene. *Proc Natl Acad Sci USA* 1986;83:3659–3663.

151. Earl PL, Moss B. Vaccinia virus. In: O'Brien SJ, ed. *Genetic maps: locus maps of complex genomes*, Cold Spring Harbor, NY: Cold Spring Harbor Laboratory Press; 1993:1.157–1.165.

152. Easterbrook KB. Controlled degradation of vaccinia virions *in vitro*: an electron microscopic study. *J Ultrastruct Res* 1966;14:484–496.

153. Elroy-Stein O, Moss B. Gene expression using the vaccinia virus/T7 RNA polymerase hybrid system. In: Ausubel FM, et al., eds. *Current protocols in molecular biology*. New York: Greene Publishing Associates and Wiley Interscience; 1992:16.19.1–16.19.12.

154. Engelstad M, Howard ST, Smith GL. A constitutively expressed vaccinia gene encodes a 42-kDa glycoprotein related to complement control factors that forms part of the extracellular virus envelope. *Virology* 1992;188:801–810.

155. Engelstad M, Smith GL. The vaccinia virus 42-kDa envelope protein is required for the envelopment and egress of extracellular virus and for virus virulence. *Virology* 1993;194:627–637.

156. Ensinger MJ. Phenotypic characterization of temperature-sensitive mutants of vaccinia virus with mutations in a 125,000-Mr subunit of the virion-associated DNA-dependent RNA polymerase. *J Virol* 1987;61: 1842–1850.

157. Eppstein DA, Marsh YV, Schreiber AB, Newman SR, Todaro GJ, Nestor JJ. Epidermal growth factor receptor occupancy inhibits vaccinia virus infection. *Nature* 1985;266:550–552.

158. Esposito JJ, Cabradilla CD, Nakano JH, Obijeski JF. Intragenomic sequence transposition in monkeypox virus. *Virology* 1981;109:231–240.
159. Esposito JJ, Knight JC. Nucleotide sequence of the thymidine kinase gene region of monkeypox and variola viruses. *Virology* 1984;135: 561–567.
160. Esposito JJ, Knight JC. Orthopoxvirus DNA: a comparison of restriction profiles and maps. *Virology* 1985;143:230–251.
161. Essani K, Dales S. Biogenesis of vaccinia: evidence for more than 100 polypeptides in the virion. *Virology* 1979;95:385–394.
162. Esteban M, Flores L, Holowczak JA. Model for vaccinia virus DNA replication. *Virology* 1977;83:467–473.
163. Esteban M, Flores L, Holowczak JA. Topography of vaccinia virus DNA. *Virology* 1977;82:163–181.
164. Esteban M, Holowczak JA. Replication of vaccinia DNA in mouse L cells. I. *In vivo* DNA synthesis. *Virology* 1977;78:57–75.
165. Esteban M, Metz DH. Early virus protein synthesis in vaccinia virus-infected cells. *J Gen Virol* 1973;19:201–216.
166. Evans DH, Stuart D, McFadden G. High levels of genetic recombination among cotransfected plasmid DNAs in poxvirus-infected mammalian cells. *J Virol* 1988;62:367–375.
167. Evans E, Traktman P. Molecular genetic analysis of a vaccinia virus gene with an essential role in DNA replication. *J Virol* 1987;61:3152–3162.
168. Evans E, Traktman P. Characterization of vaccinia virus DNA replication mutants with lesions in the D5 gene. *Chromosoma* 1992;102: S72–S82.
169. Falkner FG, Moss B. *Escherichia coli gpt* gene provides dominant selection for vaccinia virus open reading frame expression vectors. *J Virol* 1988;62:1849–1854.
170. Falkner FG, Moss B. Transient dominant selection of recombinant vaccinia viruses. *J Virol* 1990;64:3108–3111.
171. Fathi Z, Condit RC. Genetic and molecular biological characterization of a vaccinia virus temperature sensitive complementation group affecting a virion component. *Virology* 1991;181:258–272.
172a. Fathi Z, Condit RC. Phenotypic characterization of a vaccinia virus temperature-sensitive complementation group affecting a virion component. *Virology* 1991;181:272–276.
172b. Feller JA, Massung RF, Turner PC, Gibbs EPJ, Bockamp EO, Beloso A, et al. Isolation and molecular characterization of the swinepox virus thymidine kinase gene. *Virology* 1991;183:578–585
173. Fenner F. The biological characters of several strains of vaccinia, cowpox and rabbitpox viruses. *Virology* 1958;5:502–529.
174. Fenner F, Comben BM. Genetic studies with mammalian poxviruses. I. Demonstration of recombination between two strains of poxviruses. *Virology* 1958;5:530–548.
175. Fenner F, Wittek R, Dumbell KR. *The orthopoxviruses*. San Diego: Academic Press; 1988:1–432.
176. Fenner F, Woodroofe GM. The reactivation of poxviruses. II. The range of reactivating viruses. *Virology* 1960;11:185–201.
177. Fisher C, Parks RJ, Lauzon ML, Evans DH. Heteroduplex DNA formation is associated with replication and recombination in poxvirus-infected cells. *Genetics* 1991;129:7–18.
178. Fleming SB, Blok J, Fraser KM, Mercer AA, Robinson AJ. Conservation of gene structure and arrangement between vaccinia virus and orf virus. *Virology* 1993;195:175–184.
179. Franke CA, Reynolds PL, Hruby DE. Fatty acid acylation of vaccinia virus proteins. *J Virol* 1989;63:4285–4291.
180. Franke CA, Rice CM, Strauss JH, Hruby DE. Neomycin resistance as a dominant selectable marker for selection and isolation of vaccinia virus recombinants. *Mol Cell Biol* 1985;5:1918–1924.
181. Franke CA, Wilson EM, Hruby MD. Use of a cell-free system to identify the vaccinia virus L1R gene product as the major late myristylated virion protein M25. *J Virol* 1990;64:5988–5996.
182. Fraser KM, Hill DF, Mercer AA, Robinson AJ. Sequence analysis of the inverted terminal repetition in the genome of parapoxvirus, Orf virus. *Virology* 1990;176:379–389.
183. Friedman RM, Baron S, Buckler CE, Steinmuller RI. The role of antibody, delayed hypersensitivity, and interferon production in recovery of guinea pigs from primary infection with vaccinia virus. *J Exp Med* 1962;116:347–356.
184. Funahashi S, Sato T, Shida H. Cloning and characterization of the gene encoding the major protein of the A-type inclusion body of cowpox virus. *J Gen Virol* 1988;69:35–47.
185. Garon CF, Barbosa E, Moss B. Visualization of an inverted terminal repetition in vaccinia virus DNA. *Proc Natl Acad Sci USA* 1978;75: 4863–4867.
186. Garon CF, Moss B. Glycoprotein synthesis in cells infected with vaccinia virus: II. A glycoprotein component of the virion. *Virology* 1971; 46:223–246.
187. Gassmann U, Wyler R, Wittek R. Analysis of parapoxvirus genomes. *Arch Virol* 1985;83:17–31.
188. Gershon PD, Ahn BY, Garfield M, Moss B. Poly(A) polymerase and a dissociable polyadenylation stimulatory factor encoded by vaccinia virus. *Cell* 1991;66:1269–1278.
189. Gershon PD, Ansell DM, Black DN. A comparison of the genome organization of capripoxvirus with that of the orthopoxviruses. *J Virol* 1989;63:4703–4708.
190. Gershon PD, Black DN. A comparison of the genomes of capripoxvirus isolates of sheep, goats, and cattle. *Virology* 1988;164:341–349.
191. Gershon PD, Black DN. The nucleotide sequence around the capripoxvirus thymidine kinase gene reveals a gene shared specifically with leporipoxviruses. *J Gen Virol* 1989;70:525–533.
192. Gershon PD, Kitching RP, Hammond JM, Black DN. Poxvirus genetic recombination during natural virus transmission. *J Gen Virol* 1989; 70:485–489.
193. Gershon PD, Moss B. Early transcription factor subunits are encoded by vaccinia virus late genes. *Proc Natl Acad Sci USA* 1990;87:4401–4405.
194. Gershon PD, Moss B. Transition from rapid processive to slow nonprocessive polyadenylation by vaccinia virus poly(A) polymerase catalytic subunit is regulated by the net length of the poly(A) tail. *Genes Develop* 1992;6:1575–1586.
195. Gershon PD, Moss B. Stimulation of poly(A) tail elongation by the VP39 subunit of the vaccinia virus-encoded poly(A) polymerase. *J Biol Chem* 1993;268:2203–2210.
196. Gershon PD, Moss B. Uridylate-containing RNA sequences determine specificity for binding and polyadenylation by the catalytic subunit of vaccinia virus poly(A) polymerase. *EMBO J* 1994;12:4705–4714.
197. Gershowitz A, Moss B. Abortive transcription products of vaccinia virus are guanylylated, methylated, and polyadenylylated. *J Virol* 1979; 31:849–853.
198. Geshelin P, Berns KI. Characterization and localization of the naturally occurring cross-links in vaccinia virus DNA. *J Mol Biol* 1974;88: 785–796.
199. Ghildyal N, Schnitzlein WM, Tripathy DN. Genetic and antigenic differences between fowlpox and quailpox viruses. *Arch Virol* 1989;106: 85–92.
200. Gillard S, Spehner D, Drillien R, Kirn A. Localization and sequence of a vaccinia virus gene required for multiplication in human cells. *Proc Natl Acad Sci USA* 1986;83:5573–5577.
201. Goebel SJ, Johnson GP, Perkus ME, Davis SW, Winslow JP, Paoletti E. The complete DNA sequence of vaccinia virus. *Virology* 1990;179: 247–266.
202. Golini F, Kates JR. A soluble transcription system derived from purified vaccinia virions. *J Virol* 1985;53:205–213.
203. Gong S, Lai C, Dallo S, Esteban M. A single point mutation of Ala-25 to Asp in the 14,000-Mr envelope protein of vaccinia virus induces a size change that leads to the small plaque size phenotype of the virus. *J Virol* 1989;63:4507–4514.
204. Gong SC, Lai CF, Esteban M. Vaccinia virus induces cell fusion at acid pH and this activity is mediated by the N-terminus of the 14-kDa virus envelope protein. *Virology* 1990;178:81–91.
205. Gonzalez A, Talavera A, Almendral JM, Vinuela E. Hairpin loop structure of African swine fever virus DNA. *Nucleic Acids Res* 1986;14: 6835–6844.
206. Gorbalenya AE, Koonin EV. Viral proteins containing the purine NTP-binding sequence pattern. *Nucleic Acids Res* 1989;17:8413–8440.
207. Gordon J, Kovala T, Dales S. Molecular characterization of a prominent antigen of the vaccinia virus envelope. *Virology* 1988;167: 361–369.
208. Gordon J, Mohandas A, Wilton S, Dales S. A prominent antigenic surface polypeptide involved in the biogenesis and function of the vaccinia virus envelope. *Virology* 1991;181:671–686.
209. Grimley PM, Rosenblum EN, Mims SJ, Moss B. Interruption by rifampin of an early stage in vaccinia virus morphogenesis: accumulation of membranes which are precursors of virus envelopes. *J Virol* 1970;6:519–533.
210. Gruidl ME, Hall RL, Moyer RW. Mapping and molecular characteri-

zation of a functional thymidine kinase from Amsacta moorei entomopoxvirus. *Virology* 1992;186:507–516.
211. Guo P, Moss B. Interaction and mutual stabilization of the two subunits of vaccinia virus mRNA capping enzyme coexpressed in Escherichia coli. *Proc Natl Acad Sci USA* 1990;87:4023–4027.
212. Hagler J, Luo Y, Shuman S. Factor-dependent transcription termination by vaccinia RNA polymerase–kinetic coupling and requirement for ATP hydrolysis. *J Biol Chem* 1994;269:50–60.
213a. Hagler J, Shuman S. A freeze-frame view of eukaryotic transcription during elongation and capping of nascent mRNA. *Science* 1992;255: 983–986.
213b. Hagler J, Shuman S. Nascent RNA cleavage by purified ternary complexes of vaccinia RNA polymerase. *J Biol Chem* 1994;268:2166–2173
214. Hall RL, Moyer RW. Identification, cloning, and sequencing of a fragment of Amsacta moorei entomopoxvirus DNA containing the spheroidin gene and three vaccinia virus-related open reading frames. *J Virol* 1991;65:6516–6527.
216. Hall RL, Moyer RW. Identification of an amsacta spheroidin-like protein within the occlusion bodies of Choristoneura entomopox viruses. *Virology* 1993;192:179–187.
217. Hanafusa H, Hanafusa T, Kamahora J. Transformation phenomena in the pox group viruses. II. Transformation between several members of the pox group. *Biken J* 1959;2:85–91.
218. Hanafusa T. Alteration of nucleic acid metabolism by active and inactive forms of vaccinia virus. *Biken J* 1960;3:313–327.
219. Harford C, Hamlin A, Riders E. Electron microscopic autoradiography of DNA synthesis in cell infected with vaccinia virus. *Exp Cell Res* 1966;42:50–57.
220. Harris N, Rosales R, Moss B. Transcription initiation factor activity of vaccinia virus capping enzyme is independent of mRNA guanylylation. *Proc Natl Acad Sci USA* 1993;90:2860–2864.
221. Hassett DE, Condit DE. Targeted construction of temperature-sensitive mutations in vaccinia virus by replacing clustered charged residues with alanine. *Proc Natl Acad Sci USA* 1994;91:4554–4558.
222. Higman MA, Bourgeois N, Niles EG. The vaccinia virus messenger RNA (guanine-N^7-)-methyltransferase requires both subunits of the messenger RNA capping enzyme for activity. *J Biol Chem* 1992;267: 16430–16437.
223. Hiller G, Jungwirth C, Weber K. Fluorescence microscopical analysis of the life cycle of vaccinia virus in chick embryo fibroblasts. Virus-cytoskeleton interactions. *Exp Cell Res* 1981;132:81–87.
224. Hiller G, Weber K. Golgi-derived membranes that contain an acylated viral polypeptide are used for vaccinia virus envelopment. *J Virol* 1985;55:651–659.
225. Hiller G, Weber K, Schneider L, Parajsz C, Jungwirth C. Interaction of assembled progeny pox viruses with the cellular cytoskeleton. *Virology* 1979;98:142–153.
226. Hirschmann P, Vos JC, Stunnenberg HG. Mutational analysis of a vaccinia virus intermediate promoter *in vivo* and *in vitro*. *J Virol* 1990;64: 6063–6069.
227. Hirt P, Hiller G, Wittek R. Localization and fine structure of a vaccinia virus gene encoding an envelope antigen. *J Virol* 1986;58:757–764.
228. Holowczak JA. Uncoating of poxviruses. 1. Detection and characterization of subviral particles in the uncoating process. *Virology* 1972;50: 216–232.
229. Hooda-Dhingra U, Patel DD, Pickup DJ, Condit RC. Fine structure mapping and phenotypic analysis of five temperature sensitive mutations in the second largest subunit of vaccinia virus DNA-dependent RNA polymerase. *Virology* 1990;174:60–69.
231. Hooda-Dhingra U, Thompson CL, Condit RC. Detailed phenotypic characterization of five temperature-sensitive mutants in the 22- and 147-kilodalton subunits of vaccinia virus DNA-dependent RNA polymerase. *J Virol* 1989;63:714–729.
232. Howell ML, Sanders-Loehr J, Loehr TM, Roseman NA, Mathews CK, Slabaugh MB. Cloning of the vaccinia virus ribonucleotide reductase small subunit gene. Characterization of the gene product expressed in *Escherichia coli*. *J Biol Chem* 1992;267:1705–1711.
233. Hruby DE, Ball LA. Control of expression of the vaccinia virus thymidine kinase gene. *J Virol* 1981;40:456–464.
234. Hruby DE, Ball LA. Mapping and identification of the vaccinia virus thymidine kinase gene. *J Virol* 1982;43:403–409.
235. Hruby DE, Guarino LA, Kates JR. Vaccinia virus replication. I. Requirement for the host cell nucleus. *J Virol* 1979;29:705–715.
236. Hruby DE, Lynn DL, Condit RC, Kates JR. Cellular differences in the molecular mechanisms of vaccinia virus host range restriction. *J Gen Virol* 1980;47:485–488.
237. Hruby DE, Maki RA, Miller DB, Ball LA. Fine structure analysis and nucleotide sequence of the vaccinia virus thymidine kinase gene. *Proc Natl Acad Sci USA* 1983;80:3411–3415.
238. Huang S, Hendriks W, Althage A, Hemmi S, Bluethmann H, Kamijo, et al. Immune response in mice that lack the interferon-g receptor. *Science* 1993;259:1742–1745.
239. Hughes SJ, Johnston LH, Decarlos A, Smith GL. Vaccinia virus encodes an active thymidylate kinase that complements a cdc8 mutant of *Saccharomyces cerevisiae*. *J Biol Chem* 1991;266:20103–20109.
240a. Hyde JM, Gafford LG, Randall CC. Molecular weight determination of fowlpox virus DNA by electron microscopy. *Virology* 1967;33: 112–120.
240b. Hyde JM, Peters D. The organization of nucleoprotein within fowlpox virus. *J. Ultrastruct Res* 1971;35:626–641
241. Ichihashi Y, Dales S. Biogenesis of poxviruses: interrelationship between hemagglutinin production and polykaryocytosis. *Virology* 1971; 46:533–543.
242. Ichihashi Y, Matsumoto S, Dales S. Biogenesis of poxviruses: role of A-type inclusions and host cell membranes in virus dissemination. *Virology* 1971;46:507–532.
243. Ichihashi Y, Oie M. Proteolytic activation of vaccinia virus for penetration phase of infection. *Virology* 1982;116:297–305.
244. Ichihashi Y, Oie M, Tsuruhara T. Location of DNA-binding proteins and disulfide-linked proteins in vaccinia virus structural elements. *J Virol* 1984;50:929–938.
245. Ichihashi Y, Takahashi T, Oie M. Identification of a vaccinia virus penetration protein. *Virology* 1994;202:834–843.
246. Ink BS, Pickup DJ. Transcription of poxvirus early gene is regulated both by a short promoter element and by a transcriptional termination signal controlling transcriptional interference. *J Virol* 1989;63:4632–4644.
247. Ink BS, Pickup DJ. Vaccinia virus directs the synthesis of early mRNAs containing 5′ poly(A) sequences. *Proc Natl Acad Sci USA* 1990;87: 1536–1540.
248. Isaacs SN, Kotwal GJ, Moss B. Reverse guanine phosphoribosyltransferase selection of recombinant vaccinia viruses. *Virology* 1990; 178:626–630.
249. Isaacs SN, Kotwal GJ, Moss B. Vaccinia virus complement-control protein prevents antibody-dependent complement-enhanced neutralization of infectivity and contributes to virulence. *Proc Natl Acad Sci USA* 1992;89:628–632.
250a. Isaacs SN, Wolffe EJ, Payne LG, Moss B. Characterization of a vaccinia virus-encoded 42-kilodalton class I membrane glycoprotein component of the extracellular virus envelope. *J Virol* 1992;66:7217–7224.
250b. Jackson RJ, Bults H G. The myxoma virus thymidine kinase gene: sequence and transcriptional mapping. *J Gen Virol* 1992;73:323–328.
251. Janeczko RA, Rodriguez JF, Esteban M. Studies on the mechanism of entry of vaccinia virus into animal cells. *Arch Virol* 1987;92:135–150.
252. Jefferts ER, Holowczak. RNA synthesis in vaccinia-infected L cells: inhibition of ribosome formation and maturation. *Virology* 1971;46: 730–744.
253. Joklik WK. The intracellular uncoating of poxvirus DNA. I. The fate of radioactively labeled rabbitpox virus. *J Mol Biol* 1964;8:263–276.
254. Joklik WK. The intracellular uncoating of poxvirus DNA. II. The molecular basis of the uncoating process. *J Mol Biol* 1964;8:277–288.
255. Joklik WK, Becker Y. The replication and coating of vaccinia DNA. *J Mol Biol* 1964;10:452–474.
256. Jones EV, Moss B. Mapping of the vaccinia virus DNA polymerase gene by marker rescue and cell-free translation of selected mRNA. *J Virol* 1984;49:72–77.
257. Jungwirth C, Launer J. Effect of poxvirus infection on host cell DNA synthesis. *J Virol* 1968;2:401–408.
258. Kahn JS, Esteban JS. Identification of the point mutations in two vaccinia virus nucleoside triphosphate phosphohydrolase I temperature sensitive mutants and role of DNA-dependent ATPase enzyme in virus gene expression. *Virology* 1990;174:459–471.
259. Kane EM, Shuman S. Temperature-sensitive mutations in the vaccinia Virus-H4 gene encoding a component of the virion RNA polymerase. *J Virol* 1992;66:5752–5762.
260. Kane EM, Shuman S. Vaccinia virus morphogenesis is blocked by a temperature sensitive mutation in the I7 gene that encodes a virion component. *J Virol* 1993;67:2689–2698.

262. Kao S-Y, Ressner E, Kates J, Bauer WR. Purification and characterization of a superhelix binding protein from vaccinia virus. *Virology* 1981;111:500–508.
263. Karupiah G, Frederickson TN, Holmes KL, Khairallah LH, Buller RML. Importance of interferons in recovery from mousepox. *J Virol* 1993;67:4214–4226.
264. Kates J, Beeson J. Ribonucleic acid synthesis in vaccinia virus. II. Synthesis of polyriboadenylic acid. *J Mol Biol* 1970;50:19–23.
265. Kates J, Beeson J. Ribonucleic acid synthesis in vaccinia virus. I. The mechanism of synthesis and release of RNA in vaccinia cores. *J Mol Biol* 1970;50:1–18.
266. Kates JR, McAuslan BR. Poxvirus DNA-dependent RNA polymerase. *Proc Natl Acad Sci USA* 1967;58:134–141.
267. Kato S, Kamahora J. The significance of the inclusion formation of poxvirus group and herpes simplex virus. *Symp Cell Chem* 1962;12: 47–90.
268. Kato S, Kameyama S, Kamahora J. Autoradiography with tritium-labeled thymidine of pox virus and human amnion cell system in tissue culture. *Biken J* 1960;3:135–138.
269. Kato S, Ono K, Miyamoto H, Mantani M. Virus-host cell interactions in rabbit fibrosarcoma produced by Shope fibroma virus. *Biken J* 1966; 9:51–61.
270. Katz E, Moss B. Formation of a vaccinia virus structural polypeptide from a higher molecular weight precursor: inhibition by rifampicin. *Proc Natl Acad Sci USA* 1970;6:677–684.
271. Kaverin NV, Varich NL, Surgay VV, Chernos VI. A quantitative estimation of poxvirus genome fraction transcribed as early and late mRNA. *Virology* 1975;65:112–119.
272. Keck JG, Baldick CJ Jr, Moss B. Role of DNA replication in vaccinia virus gene expression: a naked template is required for transcription of three late transactivator genes. *Cell* 1990;61:801–809.
273. Keck JG, Feigenbaum F, Moss B. Mutational analysis of a predicted zinc binding motif in the 26-kDa protein encoded by the vaccinia virus A2L gene: correlation of zinc binding with late transcriptional activity. *J Virol* 1993;67:5740–5748.
274. Keck JG, Kovacs GR, Moss B. Overexpression, purification and late transcription factor activity of the 17-kDa protein encoded by the vaccinia virus A1L gene. *J Virol* 1993;67:5740–5748.
275. Kerr SM, Johnston LH, Odell M, Duncan SA, Law KM, Smith GL. Vaccinia DNA ligase complements *Saccharomyces cerevisiae* Cdc9, localizes in cytoplasmic factories and affects virulence and virus sensitivity to DNA damaging agents. *EMBO J* 1991;10:4343–4350.
276. Kerr SM, Smith GL. Vaccinia virus encodes a polypeptide with DNA ligase activity. *Nucleic Acids Res* 1989;17:9039–9050.
277. King CS, Cooper JA, Moss B, Twardzik DR. Vaccinia virus growth factor stimulates tyrosine protein kinase activity of A431 cell epidermal growth factor receptors. *Mol Cell Biol* 1986;6:332–336.
278. Kit S, Dubbs DR. Biochemistry of vaccinia-infected mouse fibroblasts (strain L-M). Effects on nucleic acid and protein synthesis. *Virology* 1962;18:274–285.
279. Kit S, Jorgensen GN, Liau A, Zaslavsky V. Purification of vaccinia virus induced thymidine kinase activity from (^{35}S)methionine-labeled cells. *Virology* 1977;77:661–676.
280. Kitching RP, Hammond JM, Black DN. Studies on the major common precipitating antigen of capripoxvirus. *J Gen Virol* 1986;67:139–148.
281. Kleiman JH, Moss B. Characterization of a protein kinase and two phosphate acceptor proteins from vaccinia virions. *J Biol Chem* 1975; 250:2430–2437.
282. Kleiman JH, Moss B. Purification of a protein kinase and two phosphate acceptor proteins from vaccinia virions. *J Biol Chem* 1975;250: 2420–2429.
283. Klemperer N, Traktman P. Biochemical analysis of mutant alleles of the vaccinia virus topoisomerase I carrying targeted substitutions in a highly conserved domain. *J Biol Chem* 1993;268:15887–15899.
284. Knight JC, Goldsmith CS, Tamin A, Regnery RL, Regnery DC, Esposito JJ. Further analyses of the orthopoxviruses volepox virus and raccoon poxvirus. *Virology* 1992;190:423–433.
285. Koonin EV, Senkevich TG. Vaccinia virus encodes four putative DNA and/or RNA helicases distantly related to each other. *J Gen Virol* 1992; 73:989–993.
286. Koonin EV, Senkevich TG. Fowlpox virus encodes a protein related to human deoxycytidine kinase: further evidence for independent acquisition of genes for enzymes of nucleotide metabolism by different viruses. *Virus Genes* 1993;7:289–295.
287. Kotwal GJ, Isaacs SN, McKenzie R, Frank MM, Moss B. Inhibition of the complement cascade by the major secretory protein of vaccinia virus. *Science* 1990;250:827–830.
288. Kotwal GJ, Moss B. Analysis of a large cluster of nonessential genes deleted from a vaccinia virus terminal transposition mutant. *Virology* 1988;167:524–537.
289. Kotwal GJ, Moss B. Vaccinia virus encodes a secretory polypeptide structurally related to complement control proteins. *Nature* 1988;335: 176–178.
290. Kotwal GJ, Moss B. Vaccinia virus encodes two proteins that are structurally related to members of the plasma serine protease inhibitor superfamily. *J Virol* 1989;63:600–606.
291. Kovacs GR, Rosales R, Keck JG, Moss B. Modulation of the cascade model for regulation of vaccinia virus gene expression: purification of a prereplicative, late-stage-specific transcription factor. *J Virol* 1994; 68:3443–3447.
292. Lai C, Gong S, Esteban M. The 32-kilodalton envelope protein of vaccinia virus synthesized in *Escherichia coli* binds with specificity to cell surfaces. *J Virol* 1991;65:499–504.
293. Lai CF, Gong SC, Esteban M. Structural and functional properties of the 14-kDa envelope protein of vaccinia virus synthesized in *Escherichia coli*. *J Biol Chem* 1990;265:22174–22180.
294. Lakritz N, Fogelsong PD, Reddy M, Baum S, Hurwitz J, Bauer WR. A vaccinia virus DNase preparation which cross-links superhelical DNA. *J Virol* 1985;53:935–943.
295. Lanzer W, Holowczak JA. Polyamines in vaccinia virions and polypeptides released from viral cores by acid extraction. *J Virol* 1975;16:1254–1264.
296. Law KM, Smith GL. A vaccinia serine protease inhibitor which prevents virus-induced cell fusion. *J Gen Virol* 1992;73:549–557.
297. Lee P, Hruby DE. Proteolytic cleavage of vaccinia virus virion proteins. Mutational analysis of the specificity determinants. *J Biol Chem* 1994;269:8616–8622.
298. Lee-Chen GJ, Bourgeois N, Davidson K, Condit RC, Niles EG. Structure of the transcription initiation and termination sequences of seven early genes in the vaccinia virus *Hin*dIII D fragment. *Virology* 1988; 163:64–79.
299. Li J, Broyles SS. The DNA-dependent ATPase activity of vaccinia virus early gene transcription factor is essential for its transcription activation function. *J Biol Chem* 1993;268:20016–20021.
300. Li J, Broyles SS. Recruitment of vaccinia virus RNA polymerase to an early gene promoter by the viral early transcription factor. *J Biol Chem* 1993;268:2773–2780.
301. Li J, Pennington MJ, Broyles SS. Temperature-sensitive mutations in the gene encoding the small subunit of the vaccinia virus early transcription factor impair promoter binding, transcription activation, and packaging of multiple virion components. *J Virol* 1994;68:2605–2614.
302. Lin S, Broyles SS. VPK2: a second essential serine/threonine protein kinase encoded by vaccinia virus. *Proc Natl Acad Sci USA* 1994;in press.
303. Lin S, Chen W, Broyles SS. The vaccinia virus B1R gene product is a serine/threonine protein kinase. *J Virol* 1992;66:2717–2723.
304. Lin Y-Z, Caporaso G, Chang P-Y, Ke X-H, Tam JP. Synthesis of a biologically active tumor growth factor from the predicted sequence of Shope fibroma virus. *Biochemistry* 1988;27:5640–5645.
305. Lin YZ, Ke XH, Tam JP. Synthesis and structure-activity study of myxoma virus growth factor. *Biochemistry* 1991;30:3310–3314.
306. Lomas DA, Evans DL, Upton C, Mcfadden G, Carrell RW. Inhibition of plasmin, urokinase, tissue plasminogen activator, and C1S by a myxoma virus serine proteinase inhibitor. *J Biol Chem* 1993;268:516–521.
307. Luo Y, Hagler J, Shuman S. Discrete functional stages of vaccinia virus early transcription during a single round of RNA synthesis *in vitro*. *J Biol Chem* 1991;266:13303–13310.
308. Luo Y, Shuman S. Antitermination of vaccinia virus early transcription: possible role of RNA secondary structure. *Virology* 1991;185: 432–436.
309. Luo Y, Shuman S. RNA binding properties of vaccinia virus capping enzyme. *J Biol Chem* 1993;268:21253–21262.
310. Lyles DS, Randall CC, Gafford LG, White HB Jr. Cellular fatty acids during fowlpox virus infection of three different host systems. *Virology* 1976;70:227–229.
311a. Lyttle DJ, K.M.F, Fleming SB, Mercer AA, Robinson AJ. Homologs of vascular endothelial growth factor are encoded by the poxvirus orf virus. *J Virol* 1994;68:84–92.

311b. Lytvyn V, Fortin Y, Banville M, Arif B, Richardson C. Comparison of the thymidine kinase genes from 3 entomopoxviruses. J Gen Virol 1992;73:3235–3240.
312. Maa J-S, Rodriguez JF, Esteban M. Structural and functional characterization of a cell surface binding protein of vaccinia virus. *J Biol Chem* 1990;265:1569–1577.
313. Macen JL, Upton C, Nation N, McFadden G. SERP1, a serine protease inhibitor encoded by myxoma virus, is a secreted glycoprotein that interferes with inflammation. *Virology* 1993;195:348–363.
314. Mackett M, Archard LC. Conservation and variation in orthopoxvirus genome structure. *J Gen Virol* 1979;45:683–701.
315. Mackett M, Smith GL, Moss B. Vaccinia virus: a selectable eukaryotic cloning and expression vector. *Proc Natl Acad Sci USA* 1982;79: 7415–7419.
316. Mahr A, Roberts BE. Arrangement of late RNAs transcribed from a 7.1 kilobase *Eco* R1 vaccinia virus DNA fragment. *J Virol* 1984;49: 510–520.
317. Mallon V, Holowczak JA. Vaccinia virus antigens on the plasma membrane of infected cells. 1. Viral antigens transferred from infecting virus particles and synthesized after infection. *Virology* 1985;141:201–220.
318. Marsh YV, Eppstein DA. Vaccinia virus and the EGF receptor: a portal of entry for infectivity? *J Cell Biochem* 1987;34:239–245.
319. Martin SA, Moss B. Modification of RNA by mRNA guanylyltransferase and mRNA (guanine-7-)methyl-transferase from vaccinia virions. *J Biol Chem* 1975;250:9330–9335.
320. Martin SA, Moss B. mRNA guanylyltransferase and mRNA (guanine-7)methyltransferase from vaccinia virions. Donor and acceptor substrate activities. *J Biol Chem* 1976;251:7313–7321.
321. Martinez-Pomares L, Stern RJ, Moyer RW. The ps/hr gene (B5R open reading frame homolog) of rabbitpox virus controls pock color, is a component of the extracellular enveloped virus, and is secreted into the medium. *J Virol* 1993;67:5450–5462.
322. Massung RF, Esposito JJ, Liu L-L, Qi J, Utterback TR, Knight JC, et al. Potential virulence determinants in terminal regions of variola smallpox virus genome. *Nature* 1993;366:748–751.
323. Massung RF, Liu L-I, Qi J, Knight JC, Yuran TE, Kerlavage AR, et al. Analysis of the complete genome of smallpox variola major virus strain Bangladesh-1975. *Virology* 1994;215–240.
324. Mbuy GN, Morris RE, Bubel HC. Inhibition of cellular protein synthesis by vaccinia virus surface tubules. *Virology* 1982;116:137–147.
325. McAuslan BR. Control of induced thymidine kinase activity in the poxvirus infected cell. *Virology* 1963;20:162–168.
326. McCormack SJ, Thomas DC, Samuel CE. Mechanism of interferon action: identification of a RNA binding domain within the N-terminal region of human RNA-dependent P1/eIF-2α protein kinase. *Virology* 1992;188:47–56.
327. McDonald WF, Crozel-Goudot V, Traktman P. Transient expression of the vaccinia virus DNA polymerase is an intrinsic feature of the early phase of infection and is unlinked to DNA replication and late gene expression. *J Virol* 1992;66:534–547.
328. McFadden G. DNA viruses that affect cytokine networks. In: Agarwal BB, Puri RK, ed. *Human cytokines: their role in health and disease.* Blackwell Press; 1994;in press.
329a. McFadden G, Dales S. Biogenesis of poxviruses: mirror image deletions in vaccinia virus DNA. *Cell* 1979;18:101–108.
329b. McFadden G, Dales S. Biogenesis of poxviruses: preliminary characterization of conditional lethal mutants of vaccinia virus defective in DNA synthesis. *Virology* 1980;103:68–79.
330. McFadden G, Stuart D, Upton C, Dickie P, Morgan AR. Replication and resolution of poxvirus telomeres. *Cancer Cells* 1988;6:77–85.
331. McKenzie R, Kotwal GJ, Moss B, Hammer CH, Frank MM. Regulation of complement activity by vaccinia virus complement-control protein. *J Infect Dis* 1992;166:1245–1250.
332. Medzon EL, Bauer H. Structural features of vaccinia virus revealed by negative staining. *Virology* 1970;40:860–867.
333. Meis RJ, Condit RC. Genetic and molecular biological characterization of a vaccinia virus gene which renders the virus dependent on isatin-fl-thiosemicarbazone (IBT). *Virology* 1991;182:442–454.
334. Menna A, Wittek R, Bachmann PA, Mayr A, Wyler R. Physical characterization of a stomatitis papulosa virus genome: a cleavage map for the restriction endonucleases HindIII and EcoRI. *Arch Virol* 1979;59: 145–156.
335. Merchlinsky M. Intramolecular homologous recombination in cells infected with temperature-sensitive mutants of vaccinia virus. *J Virol* 1989;63:2030–2035.
336. Merchlinsky M. Mutational analysis of the resolution sequence of vaccinia virus DNA—essential sequence consists of 2 separate AT-rich regions highly conserved among poxviruses. *J Virol* 1990;64:5029–5035.
337. Merchlinsky M. Resolution of poxvirus telomeres: processing of vaccinia virus concatemer junctions by conservative strand exchange. *J Virol* 1990;64:3437–3446.
338. Merchlinsky M, Garon C, Moss B. Molecular cloning and sequence of the concatemer junction from vaccinia virus replicative DNA: viral nuclease cleavage sites in cruciform structures. *J Mol Biol* 1988;199: 399–413.
339. Merchlinsky M, Moss B. Resolution of linear minichromosomes with hairpin ends from circular plasmids containing vaccinia virus concatemer junctions. *Cell* 1986;45:879–884.
340. Merchlinsky M, Moss B. Sequence-independent replication and sequence-specific resolution of plasmids containing the vaccinia virus concatemer junction: requirements for early and late trans-acting factors. In: Kelly T, Stillman B, eds. *Cancer cells 6/eukaryotic DNA replication*. Cold Spring Harbor, NY: Cold Spring Harbor Laboratory; 1988: 87–93.
341. Merchlinsky M, Moss B. Nucleotide sequence required for resolution of the concatemer junction of vaccinia virus DNA. *J Virol* 1989;63: 1595–1603
342. Merchlinsky M, Moss B. Resolution of vaccinia virus DNA concatemer junctions requires late gene expression. *J Virol* 1989;63:1595–1603.
343a. Merchlinsky M, Moss B. Introduction of foreign DNA into the vaccinia virus genome by *in vitro* ligation—recombination-independent selectable cloning vectors. *Virology* 1992;190:522–526.
343b. Millns AK, Carpenter MS, DeLange, AM. The vaccinia virus-encoded uracil DNA glycosylase has an essential role in viral DNA replication. *Virology* 1994;198:504–513.
344. Miner JN, Hruby DE. Rifampicin prevents virosome localization of L65, an essential vaccinia virus polypeptide. *Virology* 1989;170: 227–237.
345. Minnegan H, Moyer RW. Intracellular localization of rabbit poxvirus nucleic acid within infected cells as determined by *in situ* hybridization. *J Virol* 1985;55:634–643.
346. Miura M, Zhu H, Rotello R, Hartweig EA, Yuan J. Induction of apoptosis in fibroblasts by IL-1fl-converting enzyme, a mammalian homolog of the *C. elegans* cell death gene ced-3. *Cell* 1993;75:653–660.
347. Mockett B, Binns MM, Boursnell MEG, Skinner MA. Comparison of the locations of homologous fowlpox and vaccinia virus genes reveals major genome reorganization. *J Gen Virol* 1992;73:2661–2668.
348. Moore JB, Smith GL. Steroid hormone synthesis by a vaccinia enzyme—a new type of virus virulence factor. *EMBO J* 1992;11:1973–1980.
349. Morgan C. Vaccinia virus reexamined: development and release. *Virology* 1976;73:43–58.
350. Morgan JR, Cohen LK, Roberts BE. Identification of the DNA sequences encoding the large subunit of the mRNA capping enzyme of vaccinia virus. *J Virol* 1984;52:206–214.
351. Morrison DK, Moyer RW. Detection of a subunit of cellular Pol II within highly purified preparations of RNA polymerase isolated from poxvirus virions. *Cell* 1986;44:587–596.
352. Moss B. Inhibition of HeLa cell protein synthesis by the vaccinia virion. *J Virol* 1968;2:1028–1037.
353. Moss B. Vaccinia virus: a tool for research and vaccine development. *Science* 1991;252:1662–1667.
354. Moss B. Vaccinia virus transcription. In: Conaway R, Conaway J, eds. *Transcription*. New York: Raven Press; 1994:185–206.
355. Moss B, Filler R. Irreversible effects of cycloheximide during the early period of vaccinia virus replication. *J Virol* 1970;5:99–108.
356. Moss B, Rosenblum EN. Protein cleavage and poxvirus morphogenesis: tryptic peptide analysis of core precursors accumulated by blocking assembly with rifampicin. *J Mol Biol* 1973;81:267–269.
357. Moss B, Rosenblum EN, Gershowitz A. Characterization of a polyriboadenylate polymerase from vaccinia virions. *J Biol Chem* 1975;250: 4722–4729.
358. Moss B, Rosenblum EN, Katz E, Grimley PM. Rifampicin: a specific inhibitor of vaccinia virus assembly. *Nature* 1969;224:1280–1284.
359. Moss B, Salzman NP. Sequential protein synthesis following vaccinia virus infection. *J Virol* 1968;2:1016–1027.
360. Moss B, Winters E, Cooper N. Instability and reiteration of DNA sequences within the vaccinia virus genome. *Proc Natl Acad Sci USA* 1981;78:1614–1618.

361. Moss B, Winters E, Jones EV. Replication of vaccinia virus. In: Cozzarelli N, ed. *Mechanics of DNA replication and recombination.* New York: A. Liss; 1983:449–461.
363. Moyer RW, Brown GD, Graves RL. The white pock mutants of rabbit poxvirus. II. The early white pock (m) host range (hr) mutants of rabbit poxvirus uncouple transcription and translation in non-permissive cells. *Virology* 1980;106:234–249.
364. Moyer RW, Graves RL. The mechanism of cytoplasmic orthopoxvirus DNA replication. *Cell* 1981;27:391–401.
365. Moyer RW, Graves RL, Rothe CT. The white pock (m) mutants of rabbit poxvirus. III. The terminal DNA sequence duplication and transposition in rabbit poxvirus. *Cell* 1980;22:545–553.
366. Moyer RW, Rothe CT. The white pock mutants of rabbit poxvirus I. Spontaneous host range mutants contain deletions. *Virology* 1980;102: 119–132.
367. Munyon WE, Paoletti E, Grace JT Jr. RNA polymerase activity in purified infectious vaccinia virus. *Proc Natl Acad Sci USA* 1967;58:2280–2288.
368. Mustafa A, Yuen L. Identification and sequencing of the *Choristoneura biennis* entomopoxvirus DNA polymerase gene. *DNA Seq* 1991;2: 39–45.
369. Muthukrishnan S, Moss B, Cooper JA, Maxwell ES. Influence of 5′ terminal cap structure on the initiation of translation of vaccinia virus mRNA. *J Biol Chem* 1978;253:1710–1715.
370. Nagayama A, Pogo BGT, Dales S. Biogenesis of vaccinia: separation of early stages from maturation by means of rifampicin. *Virology* 1970; 40:1039–1051.
371. Nagington J, Horne RW. Morphological studies of orf and vaccinia viruses. *Virology* 1962;16:248–260.
372. Nakano E, Panicali D, Paoletti E. Molecular genetics of vaccinia virus: demonstration of marker rescue. *Proc Natl Acad Sci USA* 1982;79: 1593–1596.
373. Nevins JR, Joklik WK. Isolation and properties of the vaccinia virus DNA-dependent RNA polymerase. *J Biol Chem* 1977;252:6930–6938.
374. Niles EG, Christen L. Identification of the vaccinia virus mRNA guanyltransferase active site lysine. *J Biol Chem* 1993;268:24986–24989.
375. Niles EG, Lee-Chen G-J, Shuman S, Moss B, Broyles SS. Vaccinia virus gene D12L encodes the small subunit of the viral mRNA capping enzyme. *Virology* 1989;172:513–522.
376. Niles EG, Seto J. Vaccinia virus gene D8 encodes a virion transmembrane protein. *J Virol* 1988;62:3772–3778.
377. Niven JSF, Armstrong JA, Andrewes CH, Pereira HG, Valentine RC. Subcutaneous "growths" in monkeys produced by a poxvirus. *J Pathol Bacteriol* 1961;81:1–10.
378. Oda K, Joklik WK. Hybridization and sedimentation studies on "early" and "late" vaccinia messenger RNA. *J Mol Biol* 1967;27:395–419.
379. Oie M, Ichihashi Y. Characterization of vaccinia polypeptides. *Virology* 1981;113:263–276.
380. Oie M, Ichihashi Y. Modification of vaccinia virus penetration proteins analyzed by monoclonal antibodies. *Virology* 1987;157:449–459.
381. Opgenorth A, Nation N, Graham K, McFadden G. Transforming growth factor a, Shope fibroma growth factor, and vaccinia growth factor can replace myxoma growth factor in the induction of myxomatosis in rabbits. *Virology* 1993;192:701–709.
382. Opgenorth A, Strayer D, Upton C, McFadden G. Deletion of the growth factor gene related to EGF and TGFa reduces virulence of malignant rabbit fibroma virus. *Virology* 1992;186:175–191.
383. Paez E, Esteban M. Nature and mode of action of vaccinia virus products that block activation of the interferon-mediated ppp(A2′p)$_n$A-synthetase. *Virology* 1984;134:29–39.
384. Palumbo G, Pickup DJ, Frederickson TN, McIntyre LJ, Buller RML. Inhibition of an inflammatory response is mediated by 38-kDa protein of cowpox virus. *Virology* 1989;172:262–273.
385. Palumbo GJ, Glasgow WC, Buller RML. Poxvirus-induced alteration of arachidonate metabolism. *Proc Natl Acad Sci USA* 1993;90:2020–2024.
386. Panicali D, Grzelecki A, Huang C. Vaccinia virus vectors utilizing the β-galactosidase assay for rapid selection of recombinant viruses and measurement of gene expression. *Gene* 1986;47:193–199.
387. Panicali D, Paoletti E. Construction of poxviruses as cloning vectors: insertion of the thymidine kinase gene from herpes simplex virus into the DNA of infectious vaccinia virus. *Proc Natl Acad Sci USA* 1982; 79:4927–4931.
388. Paoletti E, Grady LJ. Transcriptional complexity of vaccinia virus *in vivo* and *in vitro*. *J Virol* 1977;23:608–615.
389. Paoletti E, Moss B. Two nucleic acid-dependent nucleoside triphosphate phosphohydrolases from vaccinia virus. Nucleotide substrate and polynucleotide cofactor specificities. *J Biol Chem* 1974;249:3281–3286.
390. Paoletti E, Rosemond-Hornbeak H, Moss B. Two nucleic acid-dependent nucleoside triphosphate phosphohydrolases from vaccinia virus: Purification and characterization. *J Biol Chem* 1974;249:3273–3280.
392. Parkinson JE, Smith GL. Vaccinia virus gene A36R encodes a Mr 43-50 K protein on the surface of extracellular enveloped virus. *Virology* 1994;in press:
393. Parks RJ, Evans DH. Effect of marker distance and orientation on recombinant formation in poxvirus-infected cells. *J Virol* 1991;65:1263–1272.
394. Parsons BL, Pickup DJ. Tandemly repeated sequences are present at the ends of the DNA of raccoonpox virus. *Virology* 1987;161:45–53.
395. Parsons BL, Pickup DJ. Transcription of orthopoxvirus telomeres at late times during infection. *Virology* 1990;175:69–80.
396. Paschen E. Was wissen wir uber den vakzineerreger? *Muncher medizinische wochenschrift* 1906;53:2391–2392.
397. Patel DD, Pickup DJ. Messenger RNAs of a strongly expressed late gene of cowpox virus contains a 5′-terminal poly(A) leader. *EMBO J* 1987;6:3787–3794.
398. Patel DD, Pickup DJ. The second-largest subunit of the poxvirus RNA polymerase is similar to the corresponding subunits of procaryotic and eucaryotic RNA polymerases. *J Virol* 1989;63:1076–1086.
399. Patel DD, Pickup DJ, Joklik WK. Isolation of cowpox virus A-type inclusions and characterization of their major protein component. *Virology* 1986;149:174–189.
400. Payne L. Polypeptide composition of extracellular enveloped vaccinia virus. *J Virol* 1978;27:28–37.
401. Payne LG. Identification of the vaccinia hemagglutinin polypeptide from a cell system yielding large amounts of extracellular enveloped virus.*J Virol* 1979;31:147–155.
402. Payne LG. Significance of extracellular virus in the *in vitro* and *in vivo* dissemination of vaccinia virus. *J Gen Virol* 1980;50:89–100.
403. Payne LG. Characterization of vaccinia virus glycoproteins by monoclonal antibody preparations. *Virology* 1992;187:251–260.
404. Payne LG, Norrby E. Adsorption and penetration of enveloped and naked vaccinia virus particles. *J Virol* 1978;27:19–27.
405. Pedley CB, Cooper RJ. The assay, purification and properties of vaccinia virus-induced uncoating protein. *J Gen Virol* 1987;68:1021–1028.
406. Pedley S, Cooper RJ. The inhibition of HeLa cell RNA synthesis following infection with vaccinia virus. *J Gen Virol* 1984;65:1687–1697.
407. Pennington TH. Vaccinia virus polypeptide synthesis: sequential appearance and stability of pre- and post-replicative polypeptides. *J Gen Virol* 1974;25:433–444.
408. Pennington TH, Follett EA. Vaccinia virus replication in enucleated BSC-1 cells: particle production and synthesis of viral DNA and proteins. *J Virol* 1974;13:488–493.
409. Perkus ME, Goebel SJ, Davis SW, Johnson GP, Limbach K, Norton EK, Paoletti E. Vaccinia virus host range genes. *Virology* 1990;179: 276–286.
410. Perkus ME, Goebel SJ, Davis SW, Johnson GP, Norton EK, Paoletti E. Deletion of 55 open reading frames from the termini of vaccinia virus. *Virology* 1991;180:406–410.
411a. Person-Fernandez A, Beaud G. Purification and characterization of a protein synthesis inhibitor associated with vaccinia virus. *J Biol Chem* 1986;261:8283–8289.
411b. Peters D., Müller G. The fine structure of the DNA containing core of vaccinia virus. *Virology* 1963;21:266–269.
412. Pickup DJ, Bastia D, Joklik WK. Cloning of the terminal loop of vaccinia virus DNA. *Virology* 1983;124:215–217.
413. Pickup DJ, Bastia D, Stone HO, Joklik WK. Sequence of terminal regions of cowpox virus DNA: arrangement of repeated and unique sequence elements. *Proc Natl Acad Sci USA* 1982;79:7112–7116.
414. Pickup DJ, Ink B, Hu W, Ray CA, Joklik WK. Hemorrhage in lesions caused by cowpox virus is induced by a viral protein that is related to plasma protein inhibitors of serine proteases. *Proc Natl Acad Sci USA* 1986;83:7698–7702.
415. Pickup DJ, Ink BS, Parsons BL, Hu W, Joklik WK. Spontaneous deletions and duplications of sequences in the genome of cowpox virus. *Proc Natl Acad Sci USA* 1984;81:6817–6821.
416. Pogo BGT, Dales S. Two deoxyribonuclease activities within purified vaccinia virus. *Proc Natl Acad Sci USA* 1969;63:820–827.
417. Pogo BGT, O'Shea M, Freimuth P. Initiation and termination of vaccinia virus DNA replication. *Virology* 1981;108:241–248.

418. Pogo BGT, O'Shea MT. The mode of replication of vaccinia virus DNA. *Virology* 1978;86:1–8.
419. Pogo BT. Changes in parental vaccinia virus DNA after viral penetration into cells. *Virology* 1980;101:520–524.
420. Porter CD, Archard LC. Characterisation by restriction mapping of three subtypes of molluscum contagiosum virus. *J Med Virol* 1992; 38:1–6.
421. Postlethwaite R. Molluscum contagiosum. A review. *Arch Environ Health* 1970;21:432–452.
422. Prescott DM, Kates J, Kirkpatrick JB. Replication of vaccinia virus DNA in enucleated L-cells. *J Mol Biol* 1971;59:505–508.
423. Prideaux CT, Boyle DB. Fowlpox virus polypeptides: sequential appearance and virion associated polypeptides. *Arch Virol* 1987;96: 185–199.
424. Puckett C, Moss B. Selective transcription of vaccinia virus genes in template dependent soluble extracts of infected cells. *Cell* 1983;35: 441–448.
425. Quick SD, Broyles SS. Vaccinia virus gene D7R encodes a 20,000-Dalton subunit of the viral DNA-dependent RNA polymerase. *Virology* 1990;178:603–605.
426. Ravanello MP, Hruby DE. Characterization of the vaccinia virus L1R myristylprotein as a component of the intracellular virion envelope. *J Gen Virol* 1994;75:1479–1483.
427. Ray CA, Black RA, Kronheim SR, Greenstreet TA, Sleath PR, Salvesen GS, Pickup DJ. Viral inhibition of inflammation—cowpox virus encodes an inhibitor of the interleukin-1fl converting enzyme. *Cell* 1992; 69:597–604.
428. Reddy MK, Bauer WR. Activation of the vaccinia virus nicking-joining enzyme by trypsinization. *J Biol Chem* 1989;264:443–449.
429. Rempel RE, Anderson MK, Evans E, Traktman P. Temperature-sensitive vaccinia virus mutants identify a gene with an essential role in viral replication. *J Virol* 1990;64:574–583.
430. Rempel RE, Traktman P. Vaccinia virus-B1 kinase—phenotypic analysis of temperature-sensitive mutants and enzymatic characterization of recombinant proteins. *J Virol* 1992;66:4413–4426.
431. Rice AP, Kerr IM. Interferon-mediated, double-stranded RNA-dependent protein kinase is inhibited in extracts from vaccinia virus-infected cells. *J Virol* 1984;50:229–236.
432. Rice AP, Roberts BE. Vaccinia virus induces cellular mRNA degradation. *J Virol* 1983;47:529–539.
433. Rice AP, Roberts WK, Kerr IM. 2-5A accumulates to high levels in interferon-treated, vaccinia virus-infected cells in the absence of any inhibition of virus replication. *J Virol* 1984;50:220–228.
434. Robinson AJ, Barns G, Fraser K, Carpenter E, Mercer AA. Conservation and variation in orf virus genomes. *Virology* 1987;157:13–23.
435. Rodriguez D, Rodriguez J-R, Esteban M. The vaccinia virus 14-kilodalton fusion protein forms a stable complex with the processed protein encoded by the vaccinia virus A17L gene. *J Virol* 1993;67:3435–3440.
436. Rodriguez D, Rodriguez J-R, Ojakian GK, Esteban M. Vaccinia virus preferentially enters polarized epithelial cell through the basolateral surface. *J Virol* 1991;65:494–498.
437. Rodriguez JF, Esteban M. Mapping and nucleotide sequence of the vaccinia virus gene that encodes a 14-kilodalton fusion protein. *J Virol* 1987;61:3550–3554.
438. Rodriguez JF, Janeczko R, Esteban M. Isolation and characterization of neutralizing monoclonal antibodies to vaccinia virus. *J Virol* 1985; 56:482–488.
439. Rodriguez JF, Kahn JS, Esteban M. Molecular cloning, encoding sequence, and expression of vaccinia virus nucleic acid-dependent nucleoside triphosphatase gene. *Proc Natl Acad Sci USA* 1986;83:9566–9570.
440. Rodriguez JF, Paez E, Esteban M. A 14,000-Mr envelope protein of vaccinia virus is involved in cell fusion and forms covalently linked trimers. *J Virol* 1987;61:395–404.
441. Rodriguez JF, Smith GL. Inducible gene expression from vaccinia virus. *Virology* 1990;177:239–250.
442. Rodriguez JF, Smith GL. IPTG-dependent vaccinia virus: identification of a virus protein enabling virion envelopment by Golgi membrane and egress. *Nucleic Acids Res* 1990;18:5347–5351.
443. Roening G, Holowczak JA. Evidence for the presence of RNA in the purified virions of vaccinia virus. *J Virol* 1974;14:704–708.
444. Rohrmann G, Moss B. Transcription of vaccinia virus early genes by a template-dependent soluble extract of purified virions. *J Virol* 1985; 56:349–355.
445. Rohrmann G, Yuen L, Moss B. Transcription of vaccinia virus early genes by enzymes isolated from vaccinia virions terminates downstream of a regulatory sequence. *Cell* 1986;46:1029–1035.
446. Rosales R, Harris N, Ahn B-Y, Moss B. Purification and identification of a vaccinia virus-encoded intermediate stage promoter-specific transcription factor that has homology to eukaryotic transcription factor SII (TFIIS) and an additional role as a viral RNA polymerase subunit. *J Biol Chem* 1994;269;14260–14267.
447. Rosales R, Sutter G, Moss B. A cellular factor is required for transcription of vaccinia viral intermediate stage genes. *Proc Natl Acad Sci USA* 1994;91:3794–3798.
448. Rosel J, Moss B. Transcriptional and translational mapping and nucleotide sequence analysis of a vaccinia virus gene encoding the precursor of the major core polypeptide 4b. *J Virol* 1985;56:830–838.
449. Rosel JL, Earl PL, Weir JP, Moss B. Conserved TAAATG sequence at the transcriptional and translational initiation sites of vaccinia virus late genes deduced by structural and functional analysis of the HindIII H genome fragment. *J Virol* 1986;60:436–439.
450. Roseman NA, Hruby DE. Nucleotide sequence and transcript organization of a region of the vaccinia virus genome which encodes a constitutively expressed gene required for DNA replication. *J Virol* 1987; 61:1398–1406.
451. Rosemond-Hornbeak H, Moss B. Inhibition of host protein synthesis by vaccinia virus: fate of cell mRNA and synthesis of small poly(A)-rich polyribonucleotides in the presence of actinomycin D. *J Virol* 1975; 16:34–42.
452. Rosemond-Hornbeak H, Paoletti E, Moss B. Single-stranded deoxyribonucleic acid-specific nuclease from vaccinia virus. Purification and characterization.*J Biol Chem* 1974;249:3287–3291.
453. Roth MJ, Hurwitz J. RNA capping by the vaccinia virus guanylyltransferase. Structure of enzyme-guanylate intermediate. *J Biol Chem* 1984;259:13488–13494.
454. Salzman NP. The rate of formation of vaccinia deoxyribonucleic acid and vaccinia virus. *Virology* 1960;10:150–152.
455a. Sam CK, Dumbell KR. Expression of poxvirus DNA in coinfected cells and marker rescue of thermosensitive mutants by subgenomic fragments of DNA. *Ann Virol* 1981;132E:135–150.
455b. Sanz P, Veyrunes J-C, Cousserans F, Bergoin M. Cloning and sequencing of the spherulin gene, the occlusion body major polypeptide of the *Melolontha melolontha* entomopoxvirus (*Mm*EPV). *Virology* 1994;202:449–457.
456. Sarov I, Joklik W. Isolation and characterization of intermediates in vaccinia virus morphogenesis. *Virology* 1973;52:223–233.
457. Sarov I, Joklik WK. Characterization of intermediates in the uncoating of vaccinia virus DNA. *Virology* 1972;50:593–602.
458. Sarov I, Joklik WK. Studies on the nature and location of the capsid polypeptides of vaccinia virions. *Virology* 1972;50:579–592.
459. Scheiflinger F, Dorner F, Falkner FG. Construction of chimeric vaccinia viruses by molecular cloning and packaging. *Proc Natl Acad Sci USA* 1992;89:9977–9981.
460. Schellekens H, de Reus A, Bolhuis R, Fountoulakis M, Schein C, Ecsodi J, Nagata S, Weissmann C. Comparative antiviral efficiency of leukocyte and bacterially produced human α-interferon in rhesus monkeys. *Nature* 1981;292:775–776.
461. Schmelz M, Sodeik B, Ericsson M, Wolffe EJ, Shida H, Hiller G, Griffiths G. Assembly of vaccinia virus: the second wrapping cisterna is derived from the trans Golgi network. *J Virol* 1994;68:130–147.
462. Schmutz C, Payne LG, Gubser J, Wittek R. A mutation in the gene encoding the vaccinia virus 37,000-M(r) protein confers resistance to an inhibitor of virus envelopment and release. *J Virol* 1991;65:3435–3442.
463. Schnierle BS, Gershon PD, Moss B. Cap-specific mRNA (nucleoside-$O^{2'}$-)-methyltransferase and poly(A) polymerase stimulatory activities of vaccinia virus are mediated by a single protein. *Proc Natl Acad Sci USA* 1992;89:2897–2901.
464. Schnierle BS, Gershon PD, Moss B. Mutational analysis of a multifunctional protein, with mRNA 5′ cap-specific (nucleoside-2′-O-)-methyltransferase and 3′ adenylyltransferase stimulatory activities, encoded by vaccinia virus. *J Biol Chem* 1994;in press:
465. Schnierle BS, Moss B. Vaccinia virus-mediated inhibition of host protein synthesis involves neither degradation nor underphosphorylation of components of the cap-bindng eukaryotic translation initiation factor complex eIF-4F. *Virology* 1992;188:931–933.
466. Schnitzlein WM, Ghildyal N, Tripathy DN. Genomic and antigenic characterization of avipoxviruses. *Virus Res* 1988;10:65–75.
467. Schwer B, Stunnenberg HG. Vaccinia virus late transcripts generated *in vitro* have a poly(A) head. *EMBO J* 1988;7:1183–1190.

468. Schwer B, Visca P, Vos JC, Stunnenberg HG. Discontinuous transcription or RNA processing of vaccinia virus late messengers results in a 5′ poly(A) leader. *Cell* 1987;50:163–169.
469. Sebring ED, Salzman NP. Metabolic properties of early and late vaccinia messenger ribonucleic acid. *J Virol* 1967;1:550–575.
470. Seki M, Oie M, Ichihashi Y, Shida H. Hemadsorbtion and fusion inhibition activities of hemagglutinin analyzed by vaccinia virus mutants. *Virology* 1990;175:372–384.
471. Shaffer R, Traktman P. Vaccinia virus encapsidates a novel topoisomerase with the properties of a eucaryotic type I enzyme. *J Biol Chem* 1987;262:9309–9315.
472. Shchelkunov SN, Blinov VM, Sandakhchiev LS. Genes of variola and vaccinia viruses necessary to overcome the host protective mechanisms. *FEBS Lett* 1993;319:80–83.
473. Shchelkunov SN, Resenchuk SM, Totmenin AV, Blinov VM, Marennikova SS, Sandakhchiev LS. Comparison of the genetic maps of variola and vaccinia viruses. *FEBS Lett* 1993;327:321–324.
474. Shida H. Nucleotide sequence of the vaccinia virus hemagglutinin gene. *Virology* 1986;150:451–462.
475. Shida H, Dales S. Biogenesis of vaccinia: carbohydrate of the hemagglutinin molecule. *Virology* 1981;111:56–72.
476. Shope RE. A filtrable virus causing a tumor-like condition in rabbits and its relationship to virus myxomatosum. *J Exp Med* 1932;56:803–822.
477. Shuman S. Catalytic activity of vaccinia messenger RNA capping enzyme subunits coexpressed in *Escherichia coli*. *J Biol Chem* 1990;265: 11960–11966.
478. Shuman S. Site-specific DNA cleavage by vaccinia virus DNA topoisomerase-I—role of nucleotide sequence and DNA secondary structure. *J Biol Chem* 1991;266:1796–1803.
479. Shuman S. DNA strand transfer reactions catalyzed by vaccinia topoisomerase-I. *J Biol Chem* 1992;267:8620–8627.
480. Shuman S. Vaccinia virus RNA helicase—an essential enzyme related to the DE-H family of RNA-Dependent NTPases. *Proc Natl Acad Sci USA* 1992;89:10935–10939.
481. Shuman S. Vaccinia virus RNA helicase. Directionality and substrate specificity. *J Biol Chem* 1993;268:11798–802.
482. Shuman S, Broyles SS, Moss B. Purification and characterization of a transcription termination factor from vaccinia virions. *J Biol Chem* 1987;262:12372–12380.
483. Shuman S, Golder M, Moss B. Characterization of vaccinia virus DNA topoisomerase I expressed in *Escherichia coli*. *J Biol Chem* 1988;263: 16401–16407.
484. Shuman S, Golder M, Moss B. Insertional mutagenesis of the vaccinia virus gene encoding a type I DNA topoisomerase: evidence that the gene is essential for virus growth. *Virology* 1989;170:302–306.
485. Shuman S, Hurwitz J. Mechanism of mRNA capping by vaccinia virus guanylyltransferase: characterization of an enzyme-guanylate intermediate. *Proc Natl Acad Sci USA* 1981;78:187–191.
486. Shuman S, Kane EM, Morham SG. Mapping the active-site tyrosine of vaccinia virus DNA topoisomerase gene. *Proc Natl Acad Sci USA* 1989;86:9793–9797.
487. Shuman S, Moss B. Identification of a vaccinia virus gene encoding a type I DNA topoisomerase. *Proc Natl Acad Sci USA* 1987;84:7478–7482.
488. Shuman S, Moss B. Bromouridine triphosphate inhibits transcription termination and mRNA release by vaccinia virions. *J Biol Chem* 1989; 264:21356–21360.
489. Silver M, Dales S. Biogenesis of vaccinia: interrelationship between post-translational cleavage, virus assembly and maturation. *Virology* 1982;117:341–356.
490. Silver M, McFadden G, Wilton S, Dales S. Biogenesis of poxviruses: role for the DNA dependent RNA polymerase II of the host during expression of late function. *Proc Natl Acad Sci USA* 1979;76:4122–4125.
491. Simpson DA, Condit RC. The vaccinia virus A18R protein plays a role in viral transcription during both the early and late phase of infection. *J Virol* 1994;68:3642–3649.
492. Slabaugh MB, Davis RE, Roseman NA, Mathews CK. Vaccinia virus ribonucleotide reductase expression and isolation of the recombinant large subunit. *J Biol Chem* 1993;268:17803–17810.
493. Slabaugh MB, Johnson TL, Mathews CK. Vaccinia virus induces ribonucleotide reductase in primate cells. *J Virol* 1984;52:507–514.
494. Slabaugh MB, Roseman N, Davis R, Mathews C. Vaccinia virus-encoded ribonucleotide reductase: sequence conservation of the gene for the small subunit and its amplification in hydroxyurea-resistant mutants. *J Virol* 1988;62:519–527.
495. Slabaugh MB, Roseman NA. Retroviral protease-like gene in the vaccinia virus genome. *Proc Natl Acad Sci USA* 1989;86:4152–4155.
496. Slabaugh MB, Roseman NA, Mathews CK. Amplification of the ribonucleotide reductase small subunit gene: analysis of novel joints and the mechanism of gene duplication in vaccinia virus. *Nucleic Acids Res* 1989;17:7073–7088.
497. Smadel Je, Hoagland CL. Estimation of the purity of preparation of elementary bodies of vaccinia. *J Exp Med* 1939;70:379–385.
498. Smadel JE, Hoagland CL. Elementary bodies of vaccinia. *Bacteriol Rev* 1942;6:79–110.
499. Smith CA, Davis T, Wignall JM, Din WS, Farrah T, Upton C, Mcfadden G, Goodwin RG. T2 open reading frame from the Shope fibroma virus encodes a soluble form of the TNF receptor. *Biochem Biophys Res Commun* 1991;176:335–342.
500. Smith GL. Virus strategies for evasion of the host response to infection. *Trends Microbiol* 1994;2:81–88.
501. Smith GL, Chan YS. Two vaccinia virus proteins structurally related to the interleukin-1 receptor and the immunoglobulin superfamily. *J Gen Virol* 1991;72:511–518.
502. Smith GL, de Carlos A, Chan YS. Vaccinia virus encodes a thymidylate kinase gene: sequence and transcriptional mapping. *Nucleic Acids Res* 1989;17:7581–7590.
503. Smith GL, Howard ST, Chan TS. Vaccinia virus encodes a family of genes with homology to serine protease inhibitors. *J Gen Virol* 1989; 70:2333–2343.
504. Sodeik B, Doms RW, Ericsson M, Hiller G, Machamer CE, van't Hof W, van Meer G, Moss B, Griffiths G. Assembly of vaccinia virus: role of the intermediate compartment between the endoplasmic reticulum and the Golgi stacks. *J Cell Biol* 1993;121:521–541.
505. Sodeik B, Griffiths G, Ericsson M, Moss B, Doms RW. Assembly of vaccinia virus: effects of rifampin on the intracellular distribution of viral protein p65. *J Virol* 1994;68:1103–1114.
506. Spehner D, Gillard S, Drillien R, Kirn A. A cowpox virus gene required for multiplication in Chinese hamster ovary cells. *J Virol* 1988;62: 1297–1304.
507. Spencer E, Shuman S, Hurwitz J. Purification and properties of vaccinia virus DNA-dependent RNA polymerase. *J Biol Chem* 1980;255: 5388–5395.
509. Spriggs MK, Hruby DE, Maliszewski CR, Pickup DJ, Sims JE, Buller RML, VanSlyke J. Vaccinia and cowpox viruses encode a novel secreted interleukin-1-binding protein. *Cell* 1992;71:145–152.
510. Spyropoulos DD, Roberts BE, Panicali DL, Cohen LK. Delineation of the viral products of recombination in vaccinia virus-infected cells. *J Virol* 1988;62:1046–1054.
511. Sridhar P, Condit RC. Selection for temperature-sensitive mutations in specific vaccinia virus genes: isolation and characterization of a virus mutant which encodes a phosphonoacetic acid-resistant, temperature-sensitive DNA polymerase. *Virology* 1983;128:444–457.
512. Stern W, Dales S. Biogenesis of vaccinia: concerning the origin of the envelope phospholipids. *Virology* 1974;62:293–306.
513. Stern W, Dales S. Biogenesis of vaccinia. Isolation and characterization of a surface component that elicits antibody suppressing infectivity and cell-cell fusion. *Virology* 1976;75:232–241.
514. Stokes GV. High-voltage electron microscope study of the release of vaccinia virus from whole cells. *J Virol* 1976;18:636–643.
515. Stroobant P, Rice AP, Gullick WJ, Cheng DJ, Kerr IM, Waterfield MD. Purification and characterization of vaccinia virus growth factor. *Cell* 1985;42:383–393.
516. Stuart D, Graham K, Schreiber M, Macaulay C, McFadden G. The target DNA sequence for resolution of poxvirus replicative intermediates is an active late promoter. *J Virol* 1991;65:61–70.
517. Stuart DT, Upton C, Higman MA, Niles EG, McFadden G. A poxvirus-encoded uracil DNA glycosylase is essential for virus viability. *J Virol* 1993;67:2503–2512.
518. Stunnenberg HG, de Magistris L, Schwer B. The generation of poly(A) heads on vaccinia late mRNA: a proposal of a slippage mechanism. In: Cech TR, ed. *Molecular biology of RNA*. New York: Alan R. Liss; 1989:199–208.
519. Sutter G, Ramsey-Ewing A, Rosales R, Moss B. Stable expression of the vaccinia virus K1L gene in rabbit cells complements the host range defect of a vaccinia virus mutant. *J Virol* 1994;68:4109–4116.
520. Taddie JA, Traktman P. Genetic characterization of the vaccinia virus DNA polymerase: identification of point mutations conferring altered drug sensitivities and reduced fidelity. *J Virol* 1991;65:869–879.
521. Taddie JA, Traktman P. Genetic characterization of the vaccinia virus

DNA polymerase: cytosine arabinoside resistance requires a variable lesion conferring phosphonoacetate resistance in conjunction with an invariant mutation localized to the 3′–5′ exonuclease domain. *J Virol* 1993;67:4323–4336.

522. Takahashi M, Kameyama S, Kato S, Kahamora J. The immunological relationship of the poxvirus group. *Biken J* 1959;2:27–29.
523. Takahashi T, Oie M, Ichihashi Y. N-terminal amino acid sequences of vaccinia virus structural proteins. *Virology* 1994;202:844–852.
524. Tartaglia J, Piccini A, Paoletti E. Vaccinia virus rifampicin-resistance locus specifies a late 63,000 Da gene product. *Virology* 1986;150:45–54.
525. Tartaglia J, Winslow J, Soebel S, Johnson GP, Taylor J, Paoletti E. Nucleotide sequence analysis of a 10.5 kbp *Hind*III fragment of fowlpox virus: relatedness to the central portion of the vaccinia virus *Hind*III D region. *J Gen Virol* 1990;71:1517–1524.
526. Tattersall P, Ward DC. Rolling hairpin model for replication of parvovirus and linear chromosomal DNA. *Nature* 1976;263:106–109.
527. Tengelsen LA, Slabaugh MB, Bibler JK, Hruby DE. Nucleotide sequence and molecular genetic analysis of the large subunit of ribonucleotide reductase encoded by vaccinia virus. *Virology* 1988;164: 121–131.
528. Thompson CL, Hooda-Dhingra U, Condit RC. Fine structure mapping of five temperature-sensitive mutants in the 22- and 147-kilodalton subunits of vaccinia virus DNA-dependent RNA polymerase. *J Virol* 1991;63:705–713.
529. Thompson JP, Turner PC, Ali AN, Crenshaw BC, Moyer RW. The effects of serpin gene mutations on the distinctive pathobiology of cowpox and rabbitpox virus following intranasal inoculation of balb/c mice. *Virology* 1993;197:328–338.
530. Tomley F, Binns M, Campbell J, Boursnell M. Sequence analysis of an 11.2 kilobase, near-terminal, *Bam*HI fragment of fowlpox virus. *J Gen Virol* 1988;69:1025–1040.
531. Tooze J, Hollinshead M, Reis B, Radsak K, Kern H. Progeny vaccinia and human cytomegalovirus particles utilize early endosomal cisternae for their envelopes. *Eur J Cell Biol* 1993;60:163–178.
532. Traktman P. Molecular genetic and biochemical analysis of poxvirus DNA replication. *Semin Virol* 1991;2:291–304.
533. Traktman P, Kelvin M, Pacheco S. Molecular genetic analysis of vaccinia virus DNA polymerase mutants. *J Virol* 1989;63:841–846.
534. Turner PC, Moyer RW. An orthopoxvirus serpin-like gene controls the ability of infected cells to fuse. *J Virol* 1992;66:2076–2085.
535. Twardzik DR, Brown JP, Ranchalis JE, Todaro GJ, Moss B. Vaccinia virus-infected cells release a novel polypeptide functionally related to transforming and epidermal growth factors. *Proc Natl Acad Sci USA* 1985;82:5300–5304.
536. Ueda Y, Morikawa S, Matsuura Y. Identification and nucleotide sequence of the gene encoding a surface antigen induced by vaccinia virus. *Virology* 1990;177:588–594.
537. Upton C, DeLange AM, McFadden G. Tumorigenic poxviruses: genomic organization and DNA sequence of the telomeric region of the Shope fibroma virus genome. *Virology* 1987;160:20–30.
538. Upton C, Macen JL, McFadden G. Mapping and sequencing of a gene from myxoma virus that is related to those encoding epidermal growth factor and transforming growth factor α. *J Virol* 1987;61:1271–1275.
540. Upton C, Macen JL, Schreiber M, McFadden G. Myxoma virus expresses a secreted protein with homology to the tumor necrosis factor receptor gene family that contributes to viral virulence. *Virology* 1991; 184:370–382.
541. Upton C, Macen JL, Wishart DS, McFadden G. Myxoma virus and malignant rabbit fibroma virus encode a serpin-like protein important for virus virulence. *Virology* 1990;179:618–631.
542. Upton C, McFadden G. Tumorigenic poxviruses: analysis of viral DNA sequences implicated in the tumorigenicity of Shope fibroma virus and malignant rabbit virus. *Virology* 1986;152:308–321.
543. Upton C, Mossman K, McFadden G. Encoding of a homolog of the IFN-γ receptor by myxoma virus. *Science* 1992;258:1369–1373.
544. Upton C, Opgenorth A, Traktman P, McFadden G. Identification and DNA sequence of the Shope fibroma virus DNA topoisomerase gene. *Virology* 1990;176:439–447.
545. Upton C, Stuart DT, McFadden G. Identification of a poxvirus gene encoding a uracil DNA glycosylase. *Proc Natl Acad Sci USA* 1993;90: 4518–4522.
546. Van Meir E, Wittek R. Fine structure of the vaccinia virus gene encoding the precursor of the major core protein 4a. *Arch Virol* 1988;102: 19–27.
547. Van Tongeren HAE. Spontaneous mutation of cowpox-virus by means of egg passage. *Arch Gesamte Virusforsch* 1952;5:35–52.
548. VanSlyke JK, Franke CA, Hruby DE. Proteolytic maturation of vaccinia virus core proteins—identification of a conserved motif at the N termini of the 4b and 25K virion proteins. *J Gen Virol* 1991;72:411–416.
549. VanSlyke JK, Hruby DE. Immunolocalization of vaccinia virus structural proteins during virion formation. *Virology* 1994;198:624–635.
550. VanSlyke JK, Lee P, Wilson EM, Hruby DE. Isolation and analysis of vaccinia virus previrions. *Virus Genes* 1993;7:311–324.
551. VanSlyke JK, Whitehead SS, Wilson EM, Hruby DE. The multistep proteolytic maturation pathway utilized by vaccinia virus P4a protein: a degenerate conserved cleavage motif within core proteins. *Virology* 1991;183:467–478.
552. Varich NL, Sychova IV, Kaverin NV, Antonova TP, Chernos VI. Transcription of both DNA strands of vaccinia virus genome *in vivo*. *Virology* 1979;96:412–430.
553. Vassef A, Ben-Hamida F, Dru A, Beaud G. Translational control of early protein synthesis at the late stage of vaccinia virus infection. *Virology* 1982;118:45–53.
554. Venkatesan S, Gershowitz A, Moss B. Modification of the 5′-end of mRNA: Association of RNA triphosphatase with the RNA guanylyltransferase-RNA (guanine-7)methyltransferase complex from vaccinia virus. *J Biol Chem* 1980;255:903–908.
555. Veomett GE, Kates JR. ATP requirement for extrusion of RNA from vaccinia cores and release of RNA from nuclei *in vitro*. In: Fox CF, Robinson WS, eds. *ICN-UCLA symposium on molecular biology*. New York: Academic Press; 1973:127–142.
556. Vos JC, Saskar M, Stunnenberg HG. Promoter melting by a stage-specific vaccinia virus transcription factor is independent of the presence of RNA polymerase. *Cell* 1991;65:105–114.
557. Vos JC, Sasker M, Stunnenberg HG. Vaccinia virus capping enzyme is a transcription initiation factor. *EMBO J* 1991;10:2553–2558.
558. Vos JC, Stunnenberg HG. Derepression of a novel class of vaccinia virus genes upon DNA replication. *EMBO J* 1988;7:3487–3492.
559. Wei CM, Moss B. Methylated nucleotides block 5′-terminus of vaccinia virus mRNA. *Proc Natl Acad Sci USA* 1975;72:318–322.
560. Weir JP, Bajszar G, Moss B. Mapping of the vaccinia virus thymidine kinase gene by marker rescue and by cell-free translation of selected mRNA. *Proc Natl Acad Sci USA* 1982;79:1210–1214.
561. Weir JP, Moss B. Nucleotide sequence of the vaccinia virus thymidine kinase gene and the nature of spontaneous frameshift mutations. *J Virol* 1983;46:530–537.
562. Weir JP, Moss B. Use of a bacterial expression vector to identify the gene encoding a major core protein of vaccinia virus. *J Virol* 1985;56: 534–540.
563. Weir JP, Moss B. Determination of the promoter region of an early vaccinia virus gene encoding thymidine kinase. *Virology* 1987;158: 206–210.
564. Weir JP, Moss B. Determination of the transcriptional regulatory region of a vaccinia virus late gene. *J Virol* 1987;61:75–80.
565. Westwood JCN, Harris WJ, Zwartouw HT, Titmus DHJ, Appleyard G. Studies on the structure of vaccinia virus. *J Gen Microbiol* 1964; 34:67–78.
566. Whitaker-Dowling P, Youngner JS. Vaccinia rescue of VSV from interferon-induced resistance: reversal of translation block and inhibition of protein kinase activity. *Virology* 1983;131:128–136.
567. Whitaker-Dowling P, Youngner JS. Characterization of a specific kinase inhibitory factor produced by vaccinia virus which inhibits the interferon-induced protein kinase. *Virology* 1984;137:171–181.
568. Whitaker-Dowling P, Youngner JS. Vaccinia-mediated rescue of encephalomyocarditis virus from the inhibitory effects of interferon. *Virology* 1986;152:50–57.
569. Wilcock D, Smith GL. Vaccinia virus core protein VP8 is required for virus infectivity, but not for core protein processing or for INV and EEV formation. *Virology* 1994;202:294–304.
570. Wilton S, Dales S. Influence of RNA polymerase II upon vaccinia virus-related translation examined by means of α-amanitin. *Virus Res* 1986; 5:323–341.
571. Wilton S, Dales S. Relationship between RNA polymerase II and efficiency of vaccinia virus replication. *J Virol* 1989;63:1540–1548.
572. Wittek R, Hanggi M, Hiller G. Mapping of a gene coding for a major late structural polypeptide on the vaccinia virus genome. *J Virol* 1984; 49:371–378.
573. Wittek R, Menna A, Muller K, Schumperli D, Bosley PG, Wyler R.

Inverted terminal repeats in rabbits poxvirus and vaccinia virus DNA. *J Virol* 1978;28:171–181.
574. Wittek R, Moss B. Tandem repeats within the inverted terminal repetition of vaccinia virus DNA. *Cell* 1980;21:277–284.
575. Wolffe EJ, Isaacs SN, Moss B. Deletion of the vaccinia virus B5R gene encoding a 42-kilodalton membrane glycoprotein inhibits extracellular virus envelope formation and dissemination. *J Virol* 1993;67:4732–4741.
576. Woodroofe GM, Fenner G. Serological relationships within the poxvirus group: an antigen common to all members of the group. *Virology* 1962; 16:334–341.
577. Woodson B. Vaccinia mRNA synthesis under conditions which prevent uncoating. *Biochem Biophys Res Commun* 1967;27:169–175.
578. Woodson B, Joklik WK. The inhibition of vaccinia virus multiplication by isatin-fl-thiosemicarbazone. *Proc Natl Acad Sci USA* 1965;54: 946–953.
579. Wright CF, Coroneos AM. Purification of the late transcription system of vaccinia virus: identification of a novel transcription factor. *J Virol* 1993;67:7264–7270.
580. Wright CF, Keck JG, Moss B. A transcription factor for expression of vaccinia virus late genes is encoded by an intermediate gene. *J Virol* 1991;65:3715–3720.
581. Wright CF, Moss B. *In vitro* synthesis of vaccinia virus late mRNA containing a 5′ poly(A) leader sequence. *Proc Natl Acad Sci USA* 1987; 84:8883–8887.
582. Wright CF, Moss B. Identification of factors specific for transcription of the late class of vaccinia virus genes. *J Virol* 1989;63:4224–4233.
583. Yang WP, Bauer WR. Purification and characterization of vaccinia virus structural protein VP8. *Virology* 1988;167:578–584.
584. Yang WP, Kao SY, Bauer WR. Biosynthesis and post-translational cleavage of vaccinia virus structural protein VP8. *Virology* 1988;167: 585–590.
585a. Yuen L, Davison AJ, Moss B. Early promoter-binding factor from vaccinia virions. *Proc Natl Acad Sci USA* 1987;84:6069–6073.
585b. Yuen L, Donne J, Arif B, Richardson C. Identification and sequencing of the spheroidin gene of *Choristoneura biennis* entomopoxvirus. *Virology* 1990;175:427–433.
586. Yuen L, Moss B. Multiple 3′ ends of mRNA encoding vaccinia virus growth factor occur within a series of repeated sequences downstream of T clusters. *J Virol* 1986;60:320–323.
587. Yuen L, Moss B. Oligonucleotide sequence signaling transcriptional termination of vaccinia virus early genes. *Proc Natl Acad Sci USA* 1987;84:6417–6421.
588. Yuen L, Noiseux M, Gomes M. DNA sequence of the nucleoside triphosphate phosphohydrolase I (NPH I) of the *Choristoneura biennis* entomopoxvirus. *Virology* 1991;182:403–406.
589. Yuwen H, Cox JH, Yewdell JW, Bennink JR, Moss B. Nuclear localization of a double-stranded RNA-binding protein encoded by the vaccinia virus E3L gene. *Virology* 1993;195:732–744.
590. Zartouw HT. The chemical composition of vaccinia virus. *J Gen Microbiol* 1964;34:115–123.
591. Zaslavsky V. Uncoating of vaccinia virus. *J Virol* 1985;55:352–356.
592. Zhang WD, Evans DH. DNA strand exchange catalyzed by proteins from vaccinia virus-infected cells. *J Virol* 1993;67:204–212.
593. Zhang Y, Ahn B-Y, Moss B. Targeting of a multicomponent transcription apparatus into assembling vaccinia virus particles requires RAP94, an RNA polymerase-associated protein. *J Virol* 1994;68:1360–1370.
594. Zhang Y, Moss B. Inducer-dependent conditional-lethal mutant animal viruses. *Proc Natl Acad Sci USA* 1991;88:1511–1515.
595. Zhang Y, Moss B. Vaccinia virus morphogenesis is interrupted when expression of the gene encoding an 11-kDa phosphoprotein is prevented by the *Escherichia coli lac* repressor. *J Virol* 1991;65:6101–6110.
596. Zhang Y, Moss B. Immature viral envelope formation is interrupted at the same stage by *lac* operator-mediated repression of the vaccinia virus D13L gene and by the drug rifampicin. *Virology* 1992;187: 643–653.
597. Zhang YF, Keck JG, Moss B. Transcription of viral late genes is dependent on expression of the viral intermediate gene G8R in cells infected with an inducible conditional-lethal mutant vaccinia virus. *J Virol* 1992;66:6470–6479.
598. Zhou J, Sun XY, Fernando GJP, Frazer IH. The vaccinia virus K2L gene encodes a serine protease inhibitor which inhibits cell-cell fusion. *Virology* 1992;189:678–686.

Fundamental Virology, Third Edition
edited by B.N. Fields, D.M. Knipe, P.M. Howley, et al.
Lippincott - Raven Publishers, Philadelphia © 1996

CHAPTER 35

Hepadnaviridae and Their Replication

Don Ganem

Of the many viral causes of human hepatitis few are of greater global importance than hepatitis B virus (HBV). Over 250 million people worldwide are persistently infected with HBV, and of these a significant minority develop severe pathologic consequences, including chronic hepatic insufficiency, cirrhosis, and hepatocellular carcinoma (HCC) (20,21,113,115,270,295,327,358,408). For many years, insight into the fundamentals of virus replication was impeded by a number of experimental factors. Chief among these were the narrow host range of the virus, which precluded transmission to convenient animal hosts, and the apparent absence of cell lines that supported infection by virus particles. As a result, early investigations were limited to descriptive study of the virion and its associated antigens and to characterization of the serologic responses to viral infection (147). These studies nonetheless had enormous impact, profoundly advancing our understanding of the epidemiology, transmission, and natural history of HBV infection (see Chapter 86 in *Fields Virology*, 3rd ed.). However, progress in understanding the molecular basis of viral replication became possible only in the late 1970s, with the advent of techniques for the molecular cloning of the viral genome and the discovery of natural animal models of viral infection (106,108, 236,240,244,285,382,404,444,446). These advances have transformed our understanding of the life cycle of these remarkable viruses and, in the process, spawned a landmark event in the history of medicine: the development of the first successful recombinant vaccine for a human infectious disease (245).

D. Ganem: Howard Hughes Medical Institute, Departments of Microbiology and Immunology and Medicine, University of California Medical Center, San Francisco, California 94143-0502.

HBV BIOLOGY: AN OVERVIEW

A detailed account of the epidemiologic, clinical, and serologic features of HBV infection can be found in Chapter 86 in *Fields Virology*, 3rd ed. Here we highlight only those general features of HBV biology that are pertinent to an understanding of the viral replication cycle.

Primary HBV infection of susceptible adults results from sexual contact with an infected host or from parenteral exposure to virus-containing blood or blood products (113, 142,187). Primary infection may be asymptomatic or may result in varying degrees of acute liver injury (acute hepatitis). Although such hepatitis can be severe, the vast majority of adult individuals will resolve the primary infection (147,149,172,327). Host immune responses to viral antigens result in the clearance of infected cells from the liver and removal of virions from the bloodstream; lasting immunity to reinfection typically results (147). However, approximately 5% of infected adults will not resolve the primary infection; these individuals go on to a persistent infection characterized by active viral replication in hepatocytes and varying (but often substantial) degrees of viremia (147,149,187). As in primary infection, the clinical manifestations of the persistent infection vary greatly; many patients are relatively symptom-free ("asymptomatic carriers"), while others have varying grades of chronic liver injury and inflammation (chronic hepatitis B). Although it is the minority outcome of primary infection, persistence is important for several reasons. First, most of the morbidity and mortality of hepatitis B virus infection results from the persistent infection. Some subsets of symptomatic patients (e.g., those with severe chronic active hepatitis B) have 5-year survival rates of less than 50%, with most deaths resulting from liver failure and its complications (113,187). Chronic hepatitis B patients who survive 25 to 30 years of viral persistence also have a markedly increased risk of developing HCC (20,21), a dreaded malignancy which is usually fatal. Finally, the asymptomatic carriers are the major epidemiologic reservoir of infection; it is principally from them that spread of HBV to susceptible hosts occurs.

The reasons why some individuals resolve HBV infection while others do not remain poorly understood. Much correlative clinical evidence suggests that variations in host immune responses are a critical variable. For example, individuals with overt deficits in cell-mediated immunity (transplant recipients, patients with acquired immunodeficiency virus, and so forth) are more likely to become chronic carriers than are fully immunocompetent hosts. The most biologically important example of this phenomenon is seen in the vertical transmission of HBV from pregnant mother to newborn baby (22,207,277,395). Hepatitis B virus transmission to babies usually takes place at the time of delivery, when the newborn is exposed to large quantities of viremic maternal blood during passage through the birth canal. The cellular immune system of the neonate is known to be incompletely developed at birth; in this context, 80% to 90% of HBV exposures result in persistent infections (22,395). It is important to emphasize, however, that most *adults* who develop chronic infection are not grossly immunodeficient. Most were clinically well until their encounter with HBV, having successfully handled the usual viremic infections of childhood, received live viral vaccines without incident, and so forth. What if any subtler immunologic defects such carriers may harbor is unknown at present.

Although there is general agreement on the importance of host immune factors in controlling HBV infection, detailed understanding of which effector arms of the response are critical and which viral targets they recognize is lacking (67,89,172,255,290). Although preexposure prophylaxis with antiviral antibody can confer some protection against the development of infection, there is little evidence that humoral immunity plays a major role in the clearance of established infection. Clinical data (see above) and analogy with other chronic viral infections have led most workers to emphasize the importance of cell-mediated immune responses, particularly those involving cytotoxic T-lymphocytes (CTLs). This notion seemed to be supported by early studies (254,255), indicating the frequent presence in infected individuals of circulating lymphocytes cytotoxic for autologous infected hepatocytes. Since this toxicity appeared to be blocked by antibodies to the viral nucleocapsid (core), it was presumed to derive from CTLs directed against the viral core protein. In retrospect, this observation is puzzling, since we now know that CTLs recognize peptide fragments of antigen that are usually not cross-reactive with antibodies to the cognate protein. Be that as it may, more recent studies using modern immunologic methods have now unequivocally demonstrated the presence of CD8-positive, class I major histocompatibility complex (MHC)-restricted CTLs directed against HBV nucleocapsid protein in the peripheral blood of patients with acute, resolving hepatitis B (23,128). In line with contemporary views, the recognition target for these cells is a small peptide from the N-terminal region of the core protein. Interestingly, such cells are barely detectable in the blood of patients with chronic HBV infection, consistent with the notion that the inability to generate such cells may predispose to persistent infection (289). However, it cannot be excluded that their absence from the blood in chronic infection may simply be due to their sequestration elsewhere (e.g., within the liver itself). More recently, CTLs against envelope glycoprotein determinants have also been detected; surprisingly, these are often CD4-positive, class II MHC-restricted (290). The relative importance of the latter in the termination of infection is unclear. Definitive resolution of the immunologic basis of viral clearance will probably require the development of models of HBV infection in immunologically well-understood inbred animals (e.g., mice); this would permit adoptive transfer and T-cell subset manipulation studies that can delineate the contributions of these and other candidate effectors.

VIRION STRUCTURE

Particle Types

Hepatitis B virus is unusual among animal viruses in that infected cells produce multiple types of virus-related particles (19,79,122,123,182,333). Electron microscopy (EM) of partially purified preparations of HBV shows three types of particles: (a) 42 to 47-nm double-shelled particles

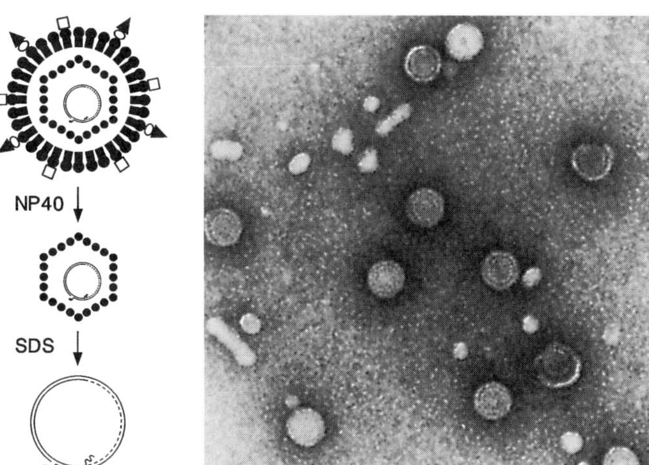

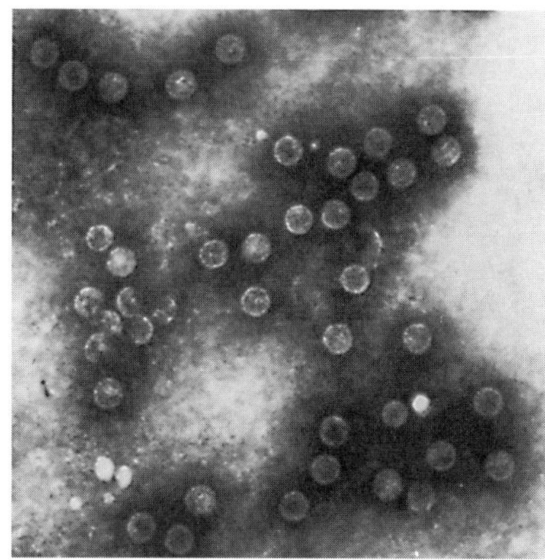

FIG. 1. The structure of hepadnaviral virions and subviral particles. ***Left:*** Schematic depiction of virion *(top)*, core particle *(middle)*, and virion DNA *(bottom)*. The outer envelope of the virion contains three related surface glycoproteins (*L, M,* and *S*); the inner nucleocapsid contains a single capsid protein *(C)*. The viral DNA contains a terminal protein *(oval)* attached to the (–) strand and a short RNA *(wavy line)* attached to the (+) strand. Dashes indicate single-stranded gap region on virion DNA. **Center:** Electron micrograph of HBV particles, including virions, 20-nm spheres, and filaments. **Right:** Electron micrograph of virion cores produced by detergent (NP40) treatment of virions. (Experiment by June Almeida.)

(known as Dane particles, after their discoverer), (b) 20-nm spheres, usually present in a 10,000- to 1,000,000-fold excess over Dane particles, and (c) smaller quantities of filaments of 20-nm diameter and variable length (Fig. 1). All three forms have a common antigen on their surface, termed hepatitis B surface antigen (HBsAg), which is present in enormous quantity in the serum of infected hosts, with concentrations ranging from 50 to 300 μg/mL. Most of the circulating pool of HBsAg is composed of 20-nm spheres, which can reach titers as high as 10^{13}/mL. This unprecedented amount of viral antigen in the circulation allows physicians to use direct viral antigen detection as a sensitive diagnostic test for HBV infection (147).

The Dane particle is the infectious virion of HBV; titers of Dane particles in the blood can range from less that 10^4/mL to greater than 10^9/mL. Its outer shell is a lipoprotein envelope containing the viral surface glycoproteins, originally detected serologically as HBsAg (27). These are the determinants against which neutralizing antibody (anti-HBs) is directed (147). The envelope can be removed by treatment with nonionic detergents, liberating an inner core particle or nucleocapsid of 25 to 27 nm. The major structural protein of the core is the C protein, a 21-kd basic phosphoprotein that was also originally detected serologically and is still frequently called hepatitis B core antigen (HBcAg) (7,36,148,152,333). Within the core is the viral DNA (331,397) and a polymerase (*P*) activity (176,321, 332) now known to be centrally involved in viral genomic replication. Purified cores also contain a protein kinase activity detected by its ability to phosphorylate C protein *in vitro* (3,99,124); since recombinant C and P proteins do not possess kinase activity, it is thought that this enzyme is of host origin (351). Whether it is truly an internal capsid component or merely tightly associated with cores from without is still uncertain, as is its biologic role (although there is no doubt that C protein is phosphorylated *in vivo* (229,314,337,338,478,482).

The 20-nm spheres and filaments are composed exclusively of HBsAg and host-derived lipid (approximately 30% by weight); their principal lipids include phospholipids, cholesterol, cholesterol esters, and triglycerides (2, 119,296,297). These particles lack nucleic acid altogether and hence are noninfectious. Nonetheless, in pure form these particles are highly immunogenic and efficiently induce a neutralizing anti-HBs antibody response. [Such 20-nm spheres, as purified from the serum of HBV carriers, in fact served as the initial form of HBV vaccine prior to the development of recombinant HBsAg preparations (187)]. Natural infection thus presents the seeming paradox of efficient progression despite the accompanying production of highly immunogenic particles that can elicit neutralizing host responses. How this comes about is not known; one school of thought is that the excess HBsAg particles function to adsorb neutralizing antibody and thus help shield virions from host defenses.

Viral Genome

Hepatitis B virus virion DNA is a relaxed circular, partially duplex species of 3.2 kb whose circularity is maintained by 5′ cohesive ends (Fig. 2) (331,345,401). This molecule has an unusual structure in that its two DNA strands are not perfectly symmetric. The viral minus strand

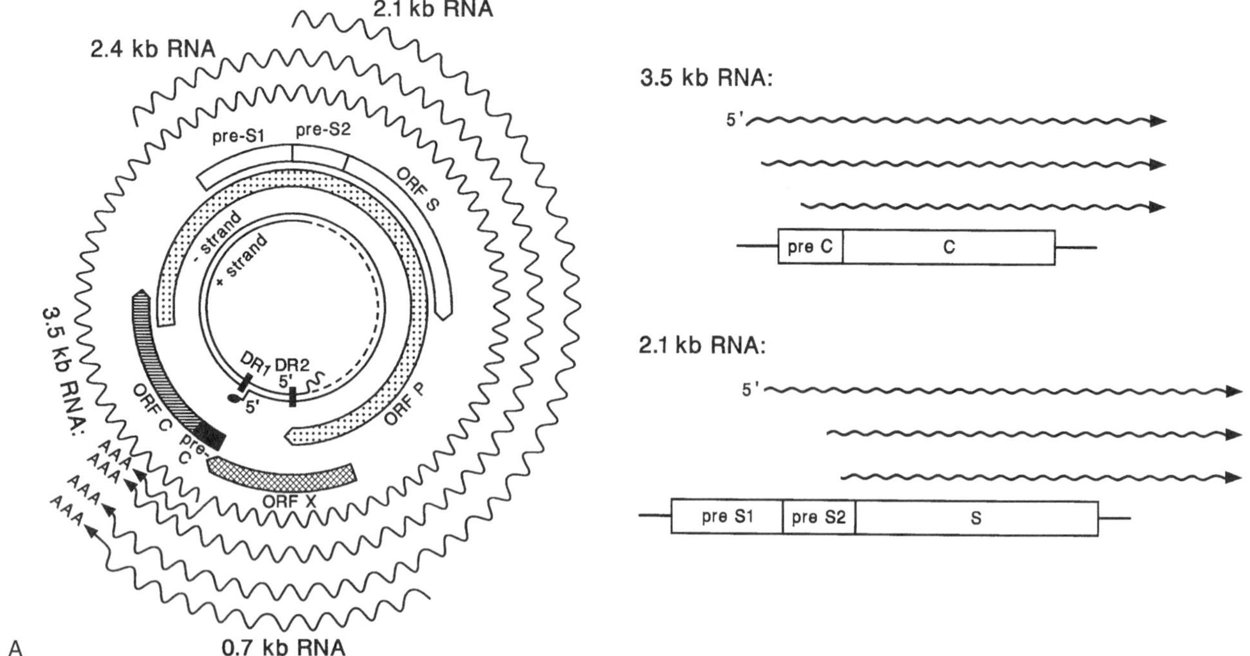

FIG. 2. A: HBV coding organization. *Inner circle* represents virion DNA, with dashes signifying the single-stranded genomic region; the locations of DR1 and DR2 sequence elements are as indicated. Boxes denote viral coding regions, with *arrows* indicating direction of translation. Outermost *wavy lines* depict the viral RNAs identified in infected cells, with *arrows* indicating direction of transcription. **B:** Fine structure of the 5′ ends of the pre-C/C transcripts *(top)* and pre-S2/S transcripts *(bottom)* relative to their respective ORFs.

is unit length and has protein covalently linked to its 5′ end (125). The existence of this terminal protein emerged from the observation that, unless the viral genome was first treated with proteases, the DNA was extracted from the aqueous phase by phenol. Restriction endonuclease digestion of viral DNA indicated that only terminal DNA fragments bearing the minus strand 5′ end displayed this behavior, indicating linkage of the protein to the 5′ end of minus strand DNA. By contrast, the plus strand is less than unit length and bears a capped oligoribonucleotide at its 5′ end (214,362,466). The presence of these terminal structures explained the long-standing observation that neither 5′ end of viral DNA could be phosphorylated by polynucleotide kinase. Importantly, the positions of the 5′ ends of both strands map to the regions of short (11 nucleotide) direct repeats (DRs) in viral DNA. The 5′ end of minus strand DNA maps within the repeat termed DR1, while plus strand DNA begins within DR2 (214,252,362,466) (Fig. 1). As will be detailed below, these repeats are importantly involved in priming the synthesis of their respective DNA strands.

Virion DNA thus contains a single-stranded region or gap of fixed polarity but variable length. Early studies showed that cores purified from extracellular virions contain a polymerase activity that could fill in this gap, generating a fully duplex genome (176,331,332,401). In this so-called endogenous polymerase reaction, the 3′ hydroxyl of plus strand DNA serves as the primer, with synthesis extending along minus strand DNA templates. The newly synthesized product is therefore entirely of plus strand polarity, an asymmetry that did not seem simply consistent with a semiconservative DNA replication scheme (see below).

Molecular cloning of HBV DNA extracted from Dane particles reveals a coding organization that is highly compact, i.e., every nucleotide in the genome is within a coding region, and over half of the sequence is translated in more than one frame (108,446). As shown in Fig. 2, four open reading frames (ORFs) are present in the DNA. The viral polymerase, the central enzymatic activity in genomic replication, is encoded by the *P* gene; ORF *P* also encodes the terminal protein found on minus strand DNA (see below) (28). The *C* (core) region encodes the structural protein of the nucleocapsid (285), while ORF *S*/pre-S encodes the viral surface glycoproteins (444).

The DNA sequence of HBV revealed a number of surprises not anticipated from earlier biologic studies. Chief among these was the existence of ORF *X*, a coding region whose product had not previously been detected. The product of ORF *X* is a poorly understood regulatory protein that enhances the expression of heterologous and homologous genes in *trans,* at least in transient assays performed in cultured cells (339). How it does so is a matter of great debate and will be considered in a later section.

Moreover, the coding organization of the genes for the known virion structural components (HBsAg and HBcAg) turned out to be more complex than expected. Early studies (122,297) had suggested that the major component of HBsAg was a protein of 24 kd (30% to 50% of which is glycosylated, appearing as a second species of 27 kd), but small quantities of larger polypeptides often copurified with this material (330). The coding region for HBsAg was identified by aligning the translated HBV DNA sequence with the known N-terminal amino acid sequence of the 24-kd chain (444). This immediately revealed that the reading frame for this antigen was open for some 400 nucleotides 5′ to its AUG and that this upstream (or pre-S) ORF could be divided into two subregions (pre-S1 and pre-S2) by the presence of two in-frame initiation codons (Fig. 2). This coding organization provided an explanation for most of the larger polypeptides earlier observed in HBsAg preparations. The largest of these so-called pre-S proteins is the 39-kd L protein, which is the product of initiation at the first AUG of the ORF (139,468). Initiation at the second AUG generates the 31-kd M protein (227,228,293,396). Both proteins share the common C-terminal S domain and differ principally by the length and structure of their N-terminal (pre-S) extensions. Classic HBsAg, which contains only the S domain, is now more commonly referred to as the S protein.

Both pre-S-encoded polypeptides are quantitatively minor components of the circulating pool of S-related antigens, with the M protein accounting for approximately 5% to 15% of the total and the L protein typically representing only 1% to 2% or less of the total. Each protein exists in two isomeric forms, differing only by the presence or absence of S-domain glycosylation; in addition, the M protein contains an additional N-linked oligosaccharide on its pre-S2-specific domain (139,396). The L protein contains a further posttranslational modification, a myristic acid group in amide linkage to its amino-terminal glycine residue (291). Despite their low abundance, L chains are now thought to play a key role in viral assembly and infectivity (see below).

Careful fractionation studies have established that the three envelope glycoproteins are not distributed uniformly among the various HBV particle types (139). Subviral 20-nm particles are composed predominantly of the S protein, with variable quantities of M polypeptides and few or no L chains. By contrast, Dane particles are substantially enriched for L chains. Since L chains are thought to carry the receptor recognition domain (183,267,305) (T. Ishikawa and D. Ganem, unpublished data), this enrichment may prevent the more numerous 20-nm particles from competing effectively with virions for cell surface receptors.

The coding organization of the core protein gene revealed a similar surprise; again, two in-frame AUGs were found in the core ORF, and the classic HBcAg was found to be the product of initiation from the more internal start codon. Initiation at the upstream AUG gives rise to a C-related protein that is not incorporated into virions but instead is independently secreted from cells, accumulating in serum as an immunologically distinct antigen known as HBeAg (38,118,246,264,274,282,337,338,390,465). The function of HBeAg remains enigmatic (but see below).

CLASSIFICATION

Hepadnavirus Family

These unusual structural features clearly set HBV apart from the other families of animal DNA viruses. However, it is now clear that HBV is but one member of a family of related viruses now known as hepadnaviruses (for *hepa*totropic *DNA* viruses). The first of the nonhuman hepadnaviruses to be discovered was the woodchuck hepatitis virus (WHV). In a colony of captive woodchucks (*Marmota monax*) veterinarians noted the frequent presence of chronic active hepatitis and hepatoma at necropsy (405); these features recalled the histopathology of advanced HBV infection. A search for HBV-like particles in serum from these animals ultimately culminated in the identification of WHV (404), a novel virus that is morphologically indistinguishable from HBV and whose genome shares approximately 60% nucleotide sequence identity with its human counterpart (107,186). A series of similar viruses have now been recovered from a variety of animal species, including the Beechey ground squirrel, *Spermophilis beecheyi* (ground squirrel hepatitis virus, GSHV) (114,116, 239,240,363), and the Pekin duck, *Anas domesticus* (duck HBV, DHBV) (237,238,244). Less well characterized viruses have been recovered from wild herons, domestic geese, marsupials, and other hosts (387,418). All of these viruses share the following common properties: (a) enveloped virions bearing 3- to 3.3-kb relaxed circular, partially duplex DNA; (b) virion-associated polymerases that can repair the gap in the virion DNA template; (c) the production of excess subviral lipoprotein particles composed of envelope proteins; (d) narrow host range, growing only in species close to the natural host (12,200,237,312,364,387, 430); and (e) production of persistent infections displaying pronounced (but not absolute) hepatotropism.

Table 1 summarizes the biologic properties of the major hepadnaviruses characterized thus far. Although the similarities between these viruses outnumber the differences, there are important distinctions to be made among the individual members of this virus family. In general, the avian viruses are the most divergent, i.e., their viral genomes are smaller than those of the mammalian viruses and share little primary nucleotide sequence homology with them (235). Most avian viruses encode only two envelope proteins (L and S) rather than three (101,315,354), and all lack the *X* coding region altogether. Although they too display a very narrow host range, their hepatotropism is much less marked; viral antigens and replicative intermediates can be readily

TABLE 1. *Hepadnavirus family*

	HBV	WHV	GSHV	DHBV
Genome	3.2 kb	3.3 kb	3.3 kb	3.0 kb
ORFs	*S,C,P,X*	*S,C,P,X*	*S,C,P,X*	*S,C,P*
Hosts	Humans Chimps	Wood-chucks	Ground squirrels Wood-chucks Chipmunks	Ducks Geese
Replication	Liver Kidney Pancreas WBC	Liver Kidney Pancreas WBC Other	Liver	Liver Kidney Pancreas Spleen Other?
Diseases	ACS Hepatitis Cirrhosis HCC	ACS Hepatitis HCC	ACS Hepatitis HCC	ACS Hepatitis

ACS, asymptomatic carrier state; HCC, hepatocellular carcinoma; ORF, open reading frame; WBC, white blood cells; HBV, hepatitis B virus; WHV, woodchuck hepatitis virus; GSHV, ground squirrel hepatitis virus; DHBV, duck hepatitis B virus.

detected in several extrahepatic sites including the pancreas, kidney, and spleen (132–135,151,171). In animals infected *in ovo,* the yolk sac is a major site of viral replication (409). [Extrahepatic infection is also well-documented for HBV (205,213,306,336,385,480) and WHV (191–193,275), but the titers of their extrahepatic viral DNAs are generally much lower than those observed for DHBV]. Nonetheless, even in the avian viruses the liver remains the predominant site of virus production. Most strikingly, DHBV infection, while it can be associated with mild grades of hepatitis, is not strongly linked to the development of HCC. While in some DHBV-infected flocks in China and Japan the occasional occurrence of hepatoma has been noted (161,279,481), infected birds in United States and European flocks (and laboratory-held animals) virtually never develop this lesion. It seems likely that the Asian flocks have been exposed to environmental cofactors that may accelerate HCC formation; dietary aflatoxin, a potent hepatic carcinogen, is often mentioned in this connection (78).

VIRAL REPLICATION

Overview of the Life Cycle

The peculiar asymmetries of virion DNA (and of the endogenous polymerase reaction) provided early clues that a mechanism other than semiconservative DNA synthesis is involved in hepadnaviral genomic replication. With the availability of hepadnavirus-infected animals, it became possible to examine intrahepatic replicative intermediates, and these were found to bear even more striking asymmetries. Infected liver cells were noted to harbor large quantities of minus strands of less than unit length, most of which were not associated with plus strands (243,460). In 1982, Summers and Mason (400) reported seminal experiments that culminated in the landmark discovery that viral DNA replication proceeds not by conventional semiconservative DNA synthesis but by reverse transcription of an RNA intermediate. To do this, they first prepared subviral particles from DHBV-infected liver. These particles, unlike mature virions, incorporate labeled deoxynucleotides (dNTPs) into *both* plus and minus strands of viral DNA, exactly as would be anticipated for authentic replication intermediates. Electron microscopy of the particles proved them to be immature (unenveloped) cores that were cytoplasmic in location; thus, most viral replication was occurring outside of the nucleus, the principal site of host DNA synthesis. Importantly, the synthesis of minus strand DNA was resistant to actinomycin D, implying its template was not DNA; plus strand synthesis was sensitive to this compound. In addition, a portion of newly made minus DNA was found in the form of RNA-DNA hybrids. These observations suggested that minus strand DNA was made from an RNA template, while the plus strand was copied from a DNA template, by inference the minus strand DNA whose RNA template had been removed.

These experiments predicted the existence of a full-length, unspliced RNA that would serve as the template for reverse transcription, and such an RNA was indeed identified soon thereafter (39,93,257). This RNA [sometimes called pregenomic RNA (or pgRNA) to denote its role in genomic replication] had all the hallmarks of a transcript produced by host RNA polymerase II. However if pol II, a nuclear enzyme, is the enzyme responsible for genomic RNA synthesis, then clearly a subpopulation of nuclear viral DNA molecules must exist to serve as its template. This function is provided by a nuclear pool of viral DNA consisting of ten to 20 molecules/cell of unit-length, covalently closed circular (ccc) DNA (243,460). As the presumed viral transcriptional template, these molecules thus play in the hepadnaviral life cycle the role analogous to that of integrated proviral DNA in retroviral replication; however, in hepadnaviral infection these molecules remain episomal. Presumably, incoming viral DNA is transported to the nucleus and converted to the cccDNA form. Consistent with this, cccDNA is the first novel virus-specific DNA species to appear following DHBV infection, preceding the accumulation of viral RNA (242).

These observations allow formulation of a model for the overall viral replicative cycle (Fig. 3). Following receptor binding, virions deliver their nucleocapsids to the cytoplasm. These then translocate to the nucleus, where their genomic DNA is matured to the cccDNA form. This DNA is then transcribed by host RNA polymerase II, and the resulting RNAs are translated to give rise to the *P, C,* pre-*S/S,* and (in mammalian viruses) *X* gene products. Viral

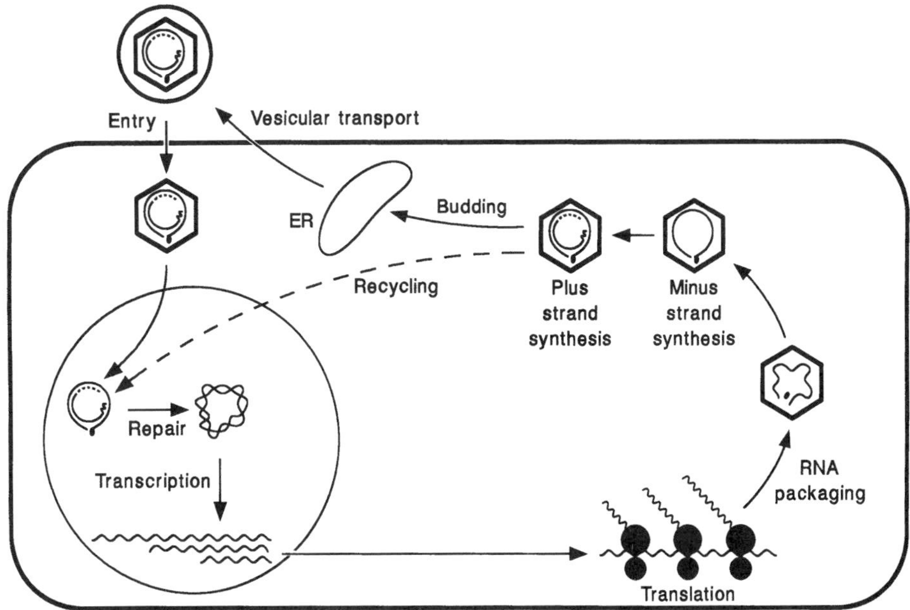

FIG. 3. Hepadnaviral life cycle. See text for details.

pgRNAs are selectively encapsidated within core particles in the cytoplasm (92,400), together with the *P* gene product. Within this structure, viral DNA synthesis is initiated; following minus strand synthesis (and concomitant degradation of the RNA template), plus strand DNA synthesis occurs. Upon completion of genomic DNA synthesis, progeny cores bud into intracellular membranes [generally the endoplasmic reticulum (ER) or proximal Golgi (175,335)] to acquire their glycoprotein envelope. Enveloped virions are then secreted via the constitutive pathway of vesicular transport (180).

Many of the events schematized in Fig. 3 can now be studied in cell culture. Although there are still no established cell lines that support infection with hepadnaviral virions (see below), primary duck hepatocytes freshly explanted directly from liver will support DHBV infection (109,316,436); these are terminally differentiated epithelial cell cultures that engage in little or no cell division and cannot be passaged *in vitro.* Similar preparations of mammalian hepatocytes appear to support cognate mammalian virus infections, but much less efficiently (127,272,329). The reason(s) for this inefficiency are unclear, but a recent report indicates that HBV infection of primary human hepatocytes can be significantly enhanced by the addition of low concentrations of polyethylene glycol to the medium (126), suggesting that inefficient membrane fusion may be the cause.

Although they will not support virion infection, several well-differentiated and immortalized hepatoma cell lines (e.g., human HepG2 or HuH7 and chick LMH cells) will support viral replication and release if cloned viral DNA is delivered by transfection (1,44,73,110,150,174,313,317, 369,406,432,474). Presumably, the incoming transfected DNA functionally substitutes for cccDNA upon arrival in the nucleus, allowing the generation of the correct panel of viral transcripts; these can then encode the viral gene products necessary to complete all subsequent stages of the life cycle. Transfection of hepatocytes can also occur *in vivo* following the direct inoculation of cloned hepadnaviral DNA into the liver of susceptible hosts (388,389, 464). In this case, progeny virions elaborated by the transfected cells establish a hepatic and systemic infection exactly as observed after virion inoculation.

The availability of these systems has revolutionized hepadnavirology, as it has made possible mutational analysis of virtually all steps in replication. These systems have also revealed an important additional feature of the replication cycle, i.e., its noncytocidal nature. Primary cells supporting productive infection generally show no cytopathic effects and are morphologically normal (126,127, 272,316,436); transfected hepatoma cells bearing viral replicative intermediates display growth rates identical to those of their untransfected parents. These facts agree well with clinical observations that many densely infected HBV carriers have minimal or no hepatocellular injury and support the contention that such injury *in vivo* is largely the result of host immune or inflammatory responses (67).

Attachment and Entry

Little is known about the earliest events in the viral life cycle. Enveloped virions must first bind to the cell surface, a reaction that has been best studied for DHBV. Binding occurs at 4°C, but at this temperature no productive infection takes place. The binding reaction appears to have two components: a low-affinity, nonsaturable one and a high-affinity, saturable component (183). On warming to 37°C, entry and infection take place. Following binding, a viral-host membrane fusion event must occur. For some viruses (e.g., influenza) this occurs following endocytosis and is triggered by a pH-dependent fusion reaction governed by an envelope glycoprotein; in other viruses [e.g., paramyxoviruses, herpes simplex virus (HSV) and human immunodeficiency virus (HIV)], this is mediated by a pH-independent fusion reaction occurring at the cell surface (232,233,461,462). Experiments using inhibitors of endosomal acidification to examine the pH-dependence of DHBV entry have come to divergent conclusions (273,328), but the weight of evidence favors a pH-independent mechanism (328).

The kinetics of DHBV uptake are odd, i.e., binding and/or entry appear to be slow, such that for maximal infection to occur the cells must remain exposed to the inoculum for up to 16 hours (316); the basis of this is not understood. As for several other viruses, entry is blocked by suramin (294) and by some but not all antienvelope monoclonal antibodies (mAbs) (60,61,199,201,484).

Much work has been done on the viral envelope components that may be involved in host cell interactions. For HBV and DHBV several lines of evidence indicate that pre-S proteins participate in cellular receptor binding. Binding of HBV virions to human liver plasma membranes fractions is blocked by mAbs to the HBV pre-S1 domain; antipeptide antibodies to pre-S1 block adherence of HBV particles to HepG2 cells (267,305). Unfortunately, neither experiment is able to relate binding to productive infection. However, using recombinant DHBV envelope proteins, it has been shown that S proteins do not block DHBV infection of primary duck hepatocytes, while recombinant preparations containing both pre-S and S proteins do (183).

Another strong inference that pre-S determinants play a key role in receptor interactions comes from genetic studies of viral host range determinants. Hepadnaviruses typically are restricted to species close to that of their natural host, and the block to efficient cross-species infection is at the level of virus entry (110,378). For example, the heron hepatitis B virus (HHBV) grows in herons but not in ducks or chicks, despite its substantial sequence homology with DHBV (387). If the normal entry pathway is bypassed by transfecting HHBV DNA into heterologous cells, replication proceeds normally (T. Ishikawa, D. Loeb, and D. Ganem, unpublished data). Consistent with this, pseudotyping experiments show that HHBV can be enveloped by DHBV envelope proteins and that these pseudotyped virions can efficiently infect duck hepatocytes. Examination of HHBV/DHBV chimeric envelope proteins in such pseudotyping assays indicates that pre-S determinants play the dominant role in host range determination (T. Ishikawa and D. Ganem, unpublished data).

Much less is known of the identities of the cellular receptors for hepadnaviruses. A large variety of proteins have been identified that bind to HBV envelope glycoproteins or to peptides derived from them. Some of these are serum derived, others are found in or on hepatocytes. A major problem has been that none of these molecules has been convincingly tied to infectivity. Mehdi et al. (247) have produced excellent evidence that HBV S determinants bind the serum protein apolipoprotein H, but this molecule is not an integral transmembrane protein of the hepatocyte and its role in infection is uncertain. Conceivably, it might play a role in delivery of virus from the periphery to the liver. Similarly, HBV S particles (HBsAg) have been shown to bind the phospholipid-binding protein endonexin 2 (140). The biologic significance of this remains unclear, as the observed interaction may simply reflect the known ability of endonexin 2 to bind phospholipids, which are abundant in HBsAg lipoprotein. Since this binding does not involve pre-S determinants, it is unlikely to be the sole important component of attachment; it might, however, play some role in a postbinding membrane fusion event. Other studies have suggested possible roles for yet other molecules in viral entry (35,105,189,190,268,269,298–300,307), but no evidence that these proteins play a role in permissive infection has yet been forthcoming. In the DHBV system, a 180-kd surface glycoprotein has been identified that binds specifically to the pre-S region of the viral L protein; this binding is blocked by neutralizing antiviral mAbs (195). The species distribution of this protein mirrors the known host range of DHBV, and all pre-S mutations that block this binding destroy DHBV infectivity without affecting viral assembly (162). These properties are consistent with a role for this protein in viral entry. However, the protein is expressed on a wide variety of cell types in the animal, including many (e.g., fibroblasts) that are not thought to be susceptible to infection. Thus, though this protein may well be involved the entry pathway, it is unlikely to be its sole component.

The remaining (postreceptor) steps in viral entry are even less well characterized (318,319). Cytoplasmic cores must be delivered to the nucleus, where their DNA is converted to the superhelical form. Nothing is known of this delivery process. Conceivably, microtubule-based motor systems (447) might play a role in their transport, but it is also possible that passive diffusion suffices to achieve this in vivo. Although microtubule assembly inhibitors profoundly depress viral replication (D. Kedes and D. Ganem, unpublished data), the exact mechanism of this inhibition remains uncertain. How and in what form the genome is deposited in the nucleus likewise has not been investigated. Some speculate that core particles themselves are transported to the nucleus, since C protein is known to harbor nuclear localization signals in its C-terminus (88,284,477) and the diameter of viral cores (approximately 25 to 27 nm) is just at the limit of the functional nuclear pore size, as determined by experimental measurements in *Xenopus* oocytes

(100). However, the possibility that cores disassemble in the cytosol (perhaps at the nuclear pore?) and deliver their DNA to the nucleus in nonparticulate form, as posited for adenovirus (301), has by no means been excluded.

Once in the nucleus, viral DNA is repaired to the cccDNA form. This requires repair of the single-stranded gap, removal of the 5′ terminal structures (RNA and P protein), and covalent ligation of the strands. Several lines of evidence suggest that this process can be carried out largely by host machinery. The cloning of Dane particle DNA in *Escherichia coli* results in the recovery of fully duplex, superhelical plasmid clones in which the terminal structures have been correctly processed: thus, there would seem to be no need for viral machinery to effect these steps. Consistent with this notion, DHBV cccDNA formation in primary hepatocytes is not inhibited by phosphonoformate, a known inhibitor of P protein polymerase activity (185), but the biochemical mechanisms by which the terminal structures are removed remain largely unknown.

Viral Transcription

Figure 2 shows a schematic representation of the major viral transcripts that have been identified in either virus-infected liver or transfected hepatoma cells in culture. (The figure depicts the transcripts of the mammalian hepadnaviruses only.) Formally, two major classes of transcripts exist: genomic and subgenomic. Each class contains multiple individual members, specialized for the translation of different viral gene products. All are capped, unspliced, and polyadenylated at a common position within the *C* gene. [Spliced derivatives of HBV genomic RNAs have been observed in some transfected cells (52,399,407,419, 469), but their role in the life cycle remains uncertain, since mutational inactivation of the splice sites does not impair HBV replication in these cells and the infectivity of genomic RNA (158) strongly implies that unspliced genomic RNA is competent to carry out all its essential coding and template functions.] The subgenomic RNAs function exclusively as the messenger RNAs (mRNAs) for translation of the envelope and X proteins. The genomic RNAs are bifunctional, serving as both the templates for viral DNA synthesis and as the messages for ORF pre-*C*, *C*, and *P* translation.

Cis-*Acting Elements*

The transcriptional map summarized in Fig. 2 implies that the different classes of viral transcripts must arise from the use of several promoters. At present, four separate viral promoters have been identified (5,346,373,479), driving expression of (a) genomic (pre-*C* and *C*) RNAs (146,177, 337,472); (b) L protein mRNA (46,47,322,324); (c) M and S protein mRNAs (37,46,81,94,323,325,326,381); and (d) X protein mRNA (380,429,473). These promoters are referred to as the genomic, pre-*S*1, *S*, and *X* promoters, respectively. The existence of each of these promoter elements was demonstrated by showing that 100 to 300-nucleotide sequence blocks excised from the regions 5′ to each of the viral transcripts can drive reporter gene expression. Deletion mutagenesis of each such promoter region has helped define the key elements required for efficient transcription; the protein factors that bind to these elements are now being enumerated (346,373,479).

The mammalian virus L and X mRNAs appear to be weakly expressed in liver tissue *in vivo,* and their corresponding RNAs were not detected until transfection systems became available. Why their transcripts are so much better expressed in transfected hepatoma lines is unclear. This disparity is greatest for the X mRNA, which is often quite abundant in transfected HepG2 cells but has yet to be detected in virus-infected liver. When the X promoter is used to drive reporter gene expression in transient assays, it behaves as a strong promoter (5). Thus, additional factors must operate *in vivo* to determine the steady-state level of X mRNA, but the nature of these controls is unknown. No such discordancy exists for the L mRNA; in transient assays, the isolated pre-*S*1 promoter is the weakest of the viral promoters (5,322). This accords well with the low levels of L protein produced in infection and suggests that control of L expression is determined largely at the level of transcriptional initiation.

The viral *S* promoter appears to be constitutively active in a wide variety of cell types, as judged by transfection assays in culture (84,392,394). However, it is still more active in differentiated hepatic cell lines (46,81,94), and studies of transgenic mice receiving HBV DNA suggest that *S* transcripts retain significant preferential expression in the liver *in vivo* (6,10,69,97,98). This may reflect the influence of a liver-specific enhancer element (enhancer II, see below) located elsewhere on the genome (479,492). The expression of the genomic and pre-*S*1 transcripts is even more strongly liver specific, i.e., these RNAs are expressed poorly or not at all in most nonhepatic cell lines (146,223,322, 398,485,486). For the pre-*S*1 promoter, this tissue specificity derives from its dependence upon hepatocyte nuclear factor 1 (HNF-1), a transcription factor found only in differentiated hepatocytes (75). Similarly, for the genomic promoter, liver-specificity results at least in part from the binding of liver-enriched transcription factors (e.g., HNF-4 and C/EBP) to an upstream regulatory region (also termed *EnII*; see below) (131,222,223). In addition, some data suggest that the genomic promoter may be under negative regulation in nonhepatic cells (131). Much evidence indicates that the liver specificity of genomic transcription is an important determinant of viral hepatotropism. For example, transfection of WHV DNA into mouse 3T3 cells (fibroblasts) results in no viral replication; however, if the genomic promoter/enhancer of WHV is replaced by the broadly active cytomegalovirus promoter, replication in 3T3 is efficiently observed (361).

Searches for enhancer elements on the genome have been conducted by looking for genomic regions that activate heterologous promoters driving reporter gene expression. To date, two genomic regions [designated enhancer I (*EnI*) and enhancer II (*EnII*)] have been identified in this fashion (53,54,374,420,455,476,487). Both can activate heterologous promoters in a position- and orientation-independent manner, and both display markedly increased activity in hepatocytic lines; in fact, *EnII* appears to be active only in hepatic cells (476,479,486), while En I retains some activity in nonhepatic cells (420,448). Each is located just 5′ to a known viral promoter; *EnI* is adjacent to the *X* promoter, and *EnII* adjoins the genomic promoter. Since the enhancers strongly activate these promoters, they can also be thought of as upstream activator elements of these promoters. This semantic distinction from their enhancer function serves no important purpose, however, since both activities reflect similar biochemical mechanisms, i.e., the binding of transcription factors that affect the activity of initiation complexes assembled near start sites. (Given the small size and circularity of HBV DNA and the fact that eukaryotic transcription factors can exert their effects over many kilobases, it is no surprise that viral promoters other than the ones in which these elements are embedded can be influenced by them.) The functional relevance of these enhancer elements *in vivo* is supported by the fact that the two DNase hypersensitive sites found in HBV-transgenic mice map to these two enhancers (91).

The best characterized of these, *EnI,* is an approximately 200-nucleotide element located within the body of the genomic and envelope transcripts, between the *S* and *X* coding regions. In transient transfection assays it appears that *EnI,* both in artificial reporter constructs and in its native context in HBV DNA, can upregulate all of the major viral promoters (5,154). Numerous factors binding to this region have been identified, some of which are found in many nonhepatic tissues (e.g., NF-1, AP-1, NFkB, and EF-C) and others of which appear to be relatively or absolutely liver specific (e.g., C/EBP, HNF-4, and HBLF) (117,131, 165,281,383,431). In most cases, this identification has been inferred from the nucleotide sequences protected in DNase footprinting experiments using *EnI* DNA probes in nuclear extracts of hepatoma cell lines, but for some factors this binding has been directly demonstrated with purified proteins.

EnII is also thought to play an important role in viral gene regulation *in vivo.* Apart from its key role as a liver-specific activator of the genomic promoter, *EnII* also upregulates the *S* promoter by about tenfold, both in artificial reporter constructs and in its native genomic context (222,487,492); interestingly, it appears to have little effect on the pre-*S*1 promoter (492). In addition, recent experiments suggest that, in WHV-induced hepatic tumors, in which integrated viral sequences transcriptionally activate nearby N-*myc* genes, it is *EnII* sequences that likely play the dominant role in upregulation (K. Ueda and D. Ganem, unpublished data). A full accounting of the factors binding to *EnII* has not yet been completed. The enhancer can be dissected into two subregions, each of which is necessary for full activity (487,492); at least two liver-enriched factors (C/EBP and HNF-4) are thought to operate through *EnII* (131,222,417).

Deletion of either *EnI* or *En* II in the context of the intact viral genome strongly diminishes viral transcription (222). Mutational assessment of the roles of the individual *En*-binding factors on viral replication *in vivo* has been hampered by the likely functional redundancy of the activating proteins and by the fact that the entire enhancer is overlapped by the *P* coding region, severely constraining mutagenesis. Serious attempts to determine the role of individual enhancer-binding factors in viral replication have been made only in DHBV, which appears to contain a single viral enhancer located upstream of the core promoter (76,218,356). DNase footprinting of this region reveals approximately seven binding sites for transcription factors; inspection of the binding site sequences suggests the involvement of C/EBP, HNF-1, HNF-3, and possibly GATA-binding protein family members (219). Point mutations engineered into the putative HNF-1 and HNF-3 sites (and designed not to alter the P protein sequence) produced only modest impairment of viral transcription in transiently transfected hepatoma cells (two- to fourfold), and replication of virus in primary hepatocytes was similarly affected. (The most drastic effect seen was a tenfold impairment of replication in double mutants affecting both of the two HNF-3 sites.) This suggests the evolution of significant functional redundancy in the system. Interestingly, one virus with a lesion in a single HNF-3 site proved incapable of pancreatic infection in the whole animal, although it produced an essentially normal hepatic infection (219). This interesting and unexpected result reaffirms that transcriptional factors can strongly influence tissue tropism (see above) and indicates that redundancy does not extend to every aspect of transcriptional control.

X Protein and Transcriptional Control

In transient expression assays, cotransfection of a reporter gene with an X protein expression vector reproducibly results in a three- to tenfold upregulation of reporter gene expression (72,155,166,210,371,379,386,412, 437–439,488,490). This holds true in a large number of cell lines (both of liver and nonliver origin) and for a huge array of target promoters, both viral and cellular. Viral *cis*-acting elements include the HSV *tk* promoter, the simian vacuolating virus 40 (SV40) early promoter, the HIV long terminal repeat (LTR), and the HBV *EnI* (224,371,386, 437–440,488). Conflicting reports exist about the impact of *X* mutations on hepadnaviral transcripts in the context of the whole viral genome; some report no effect (26), while others report three- to eightfold decreases in X-deficient mutants (72,260,494). The HBV pre-*S* and *S* promoters show little (260) or no (487,492) upregulation by X when examined in reporter constructs in isolation from other genomic regions of HBV, suggesting that any effects of X on subgenomic RNA levels may be due to enhancer upregulation. Among the cellular promoters that are activated are those for α-globin, MHC class I, c-*myc,* and others (11, 155,339,490). Even pol III promoters (e.g., VA RNA and transfer RNA) can be stimulated by X activity (8).

How the X protein achieves this broad-spectrum activation has proven to be one of the most difficult problems in hepadnavirus research. There are many technical reasons for this, i.e., the magnitude of the X effect (in cell culture, at least) is modest and variable; X protein levels in transfected cells are very low, in part due to a short half-life (376); and the protein is poorly immunogenic, resulting in a paucity of good antisera. In addition, it appears that several related proteins can be translated from ORF *X* by initiation at several in-frame AUG codons and that the functional properties of these various X-related proteins may differ (198,260). All of these problems have frustrated attempts to detect, purify, and quantify the X protein as well as to set up *in vitro* assays to determine its function. Nonetheless, there is general agreement on a few central facts as follows: (a) the protein is required for infection in the whole animal [woodchucks inoculated with WHV *X* (–) mutants do not develop WHV infection (51,494)]; (b) target gene regulation by X protein operates at the level of primary transcription (11,72,379); (c) the X protein does not bind to DNA (479), excluding a simple role as a direct, sequence-specific nuclear transcription factor; and (d) no single DNA sequence is common to all targets activated by X (339). These facts strongly suggest that X-induced transcriptional upregulation likely operates indirectly, by somehow altering the activity of numerous cellular transcription factors. The identities of some of these factors can be inferred from deletion analyses in which the *cis*-acting sequences that mediate X-induced activation are mapped. As expected, in different promoters different factors are implicated. Some of these include AP-1, AP-2, NFkB, EF-C, ATF-2, CREB, and TF IIIC (95,234,266,339,370,437).

Unfortunately, consensus about X biology ends here. There is even controversy about the subcellular localization of X protein. Most investigators find the protein largely restricted to the cytoplasm (178,208), while others find it in both cytoplasm and nucleus (155) or in a perinuclear location (64,210,380). Broadly speaking, thoughts about how X might activate transcription factors take several forms:

1. X expression might lead to a posttranslational modification of the factors (e.g., by phosphorylation), activating their DNA binding and/or transactivation functions.
2. X might engage in protein-protein interactions with components of the transcriptional apparatus to stimulate transcriptional activity.

Evidence exists to support both schools of thought, and given the broad spectrum of X action there is no reason why both might not be applicable. In favor of the idea that X can activate protein kinase cascades, it has been reported that X expression leads to an activation of protein kinase C (PKC) and that this activation is required for stimulation of AP-1-dependent gene expression (179,225), but others contend that X expression does not activate PKC and that in any case activation of NFkB can proceed in the absence of PKC activation (226). Still others suggest that activation of some transcription factors by X proceeds via upregulation of the MAP kinase pathway (77). X itself does not possess protein kinase activity.

Another posttranslational modification that could be controlled by X is proteolytic cleavage. Very limited sequence homology between X and serine protease inhibitors has been noted (413), and binding of X protein to tryptase TL2 has been documented *in vitro* (411). It has been suggested that X might inhibit this or a related protease *in vivo;* if so, such inhibition could be a locus for transcription factor regulation. However, the inhibitory activity of X for tryptase TL2 is extremely weak (411), and the biologic relevance of this activity remains unclear.

Regarding direct protein-protein interactions, it has been shown *in vitro* that, while neither ATF2 nor CREB binds to HBV *EnI* DNA in the absence of X, addition of a partially purified X protein preparation led to efficient DNA binding, as judged by electrophoretic mobility shift assays (234). These complexes could be supershifted by anti-X antibodies, implying that X protein was part of the complex. *In vivo* evidence that X determinants can potentially interact with transcription components has also been proffered. When the X coding region is fused to a heterologous DNA binding protein, it can strongly activate promoters bearing the cognate DNA recognition element (370,443). This implies that domains of X, once brought into proximity with assembling transcription complexes, can interact with them to upregulate them.

Other Regulatory Factors

Recently it has been shown that an additional product of viral gene expression can also affect host and viral gene expression: artificially truncated M proteins (203,204,265). Such C-terminally deleted chains are not produced during a typical productive infection but can be encoded by rare integrated, rearranged copies of viral DNA in which the host-viral DNA junction falls within the *S* gene; such structures are sometimes found in hepatomas. Cotransfection of such rearranged HBV genomes with reporter genes leads to upregulation of reporter gene expression. As for X protein, the activation is weak but promiscuous, affecting a wide array of target promoters. How the activation comes about is uncertain; since the truncated M protein is translocated into the ER lumen, it seems likely that signal transduction pathways are being activated inappropriately, much as would be the case in dysregulated growth factor production. Presumably, multiple transcription factors are upregulated as a result. What if anything this means for hepadnaviral infection *in vivo* is unknown; it is difficult to envision a role for such products in vegetative replication, but they might contribute to growth deregulation on the pathway to hepatoma formation in occasional individuals.

Hepatitis B virus transcription is also under hormonal control. A glucocorticoid response element has been identified within the *S* coding region (433,434), and viral transcription in HBV transgenic mice is enhanced by glucocorticoids (98). This accords well with the upregulation of viremia often observed in HBV carrier patients taking glucocorticoids (359), but the latter effect may also be influenced by the immunoregulatory properties of steroids. Androgens also upregulate HBV transcripts and HBsAg levels in transgenic mice (98), a fact that might contribute to the known male-predominance of HBV infection and disease (113). However, many other factors also influence the sex ratio of chronic hepatitis B (113).

Processing of Viral RNA

The coterminal nature of the viral transcripts (Fig. 2) poses an interesting problem in the control of RNA processing. The subgenomic mRNAs are efficiently polyadenylated at a signal within ORF *C*. The genomic RNAs, being terminally redundant, contain two copies of this signal, one at the 5′ end of the transcript and one at the 3′ end. Thus, the transcriptional apparatus, which efficiently polyadenylates at this signal in subgenomic RNAs, must bypass this signal on its first pass in genomic RNA synthesis but use it efficiently on its second pass (342). (This problem is formally analogous to that encountered in retroviral genomic RNA synthesis, in which polyadenylation signals embedded in the 5′ LTR are ignored, but identical signals in the 3′ LTR are utilized.).

Exactly how this problem is solved is still somewhat unclear, but the elements of the solution have at least been identified. Canonical eukaryotic poly(A) signals include the hexanucleotide AAUAAA, a cleavage site approximately 30 nucleotides downstream and a U- or GU-rich element 3′ to the cleavage site. Mammalian hepadnaviruses have polyadenylation signals that resemble these canonical signals, with one exception: the hexanucleotide is UAUAAA, which is known to function inefficiently (375). To bolster the efficiency of hepadnaviral RNA processing, the virus has evolved additional elements upstream of the hexanucleotide that enhance its utilization (342). Some of these elements are 5′ to the genomic RNA cap site, such that full processing efficiency is only possible at the 3′ end of the RNA. However, beyond this it appears that the proximity of the 5′ processing site to the 5′ end of the RNA somehow drastically curtails its use (59). The mechanism by which cap-site proximity exerts this effect is unknown, but similar effects have been observed in cellular and retroviral transcripts (58,163,459).

Several additional, less well-characterized *cis*-acting elements on the hepadnaviral genome affect other steps in viral RNA production. RNA sequences within the HBV *X* coding region appear to play some role in the cytoplasmic accumulation of stable viral transcripts, i.e., deletions in this region reduce the abundance of mature, stable viral mRNAs in the cytoplasm without influencing the rate of primary transcription or the stability of the final transcript. Since nuclear accumulation of the deleted viral RNAs is normal, it has been suggested that this element might govern the transport of viral transcripts from nucleus to cytoplasm (156,159). More recently, sequences at the 5′ end of DHBV pgRNA have also been shown to be necessary for the selective accumulation of this transcript; unlike the HBV sequences just mentioned, deletion of this element reduces pgRNA levels but has no impact on the accumulation of subgenomic RNA (Chang, Hirsch, and D. Ganem, unpublished data) (157). Interestingly, the phenotype of such deletions can be reverted by deletion of a second genomic region located downstream. It has been suggested that this element may operate by suppressing premature termination of pgRNA transcription (157).

Translation of Viral Gene Products

Hepadnaviruses have evolved multiple transcriptional and translational strategies to maximize the use of the limited coding capacity of their DNA. As previously noted, they make extensive use of overlapping reading frames (Fig. 2) and, within reading frames, often employ multiple initiation codons to generate structurally related but functionally distinct proteins. Because of their 5′ scanning mechanism, host ribosomes tend not to initiate internally with efficiency (194). Thus, in order to guarantee host ribosomes access to these internal AUG codons, the viral transcriptional program has deployed separate mRNA 5′ ends just 5′ to each AUG. In the case of the envelope proteins, it does this by (a) employing a dedicated promoter for the pre-*S1* mRNAs and (b) having the *S* promoter display microheterogeneity of its start sites, such that some begin 5′ to the pre-*S2* AUG and others just 3′ to it (41,392). In this way, both M and S proteins can be translated as the first ORF of a transcript (Fig. 2).

Heterogeneous start sites also characterize the genomic promoter, whose RNA 5′ ends bracket the pre-*C* AUG; those initiating 3′ to this AUG encode the C protein, the structural protein of the nucleocapsid (93,257,466). Those initiating 5′ to this AUG encode the pre-C protein, which enjoys an altogether different biosynthetic fate. The pre-*C* region encodes a signal sequence that targets this protein to the secretory pathway (118,282,352,390,456). Following cleavage of the signal in the ER lumen, the protein is transported through the vesicular transport system, undergoing cleavage of its basic C-terminal region in the process (168,169,353,390,454). Ultimately, it is secreted from the cell, accumulating in the extracellular medium (or serum) as the 17-kd protein known serologically as e antigen (HBeAg). [In a minority of chains, ER translocation is aborted following signal peptide cleavage and the resulting products returned to the cytosol (118) or even the nucleus (284). The significance of these unusual events is uncertain.] In addition, HBeAg immunoreactivity can be found on the plasma membrane (350,353), although the mechanism of this cell surface localization is not well defined.

The function of HBeAg is still unknown. Clearly it is dispensable for replication *in vitro,* since nonsense or frameshift mutations within pre-C grow normally in cultured cells or in experimental animals (43,55,349,355,425). However, clinical studies of human HBV carriers show that, during persistent infection, mutant viruses with pre-*C* lesions spontaneously and regularly arise, often coming to prevail over the wild-type virus after years of carriage (24, 30,31,129,278,343,427). [Transmission of these mutants confirms that they are infectious (137).] This implies that there is some selection against HBeAg expression, most likely due to host immunologic attack on HBeAg-bearing cells. However, this raises a deeper paradox, as yet unresolved, i.e., if there is a selection against pre-C products at the level of the individual host, how has the pre-*C* gene been maintained in the population? No entirely satisfactory answer to this question has yet been forthcoming. One attractive proposal is that HBeAg may serve as a neonatal tolerogen (248); if so, transmission of pre-*C*(+) viruses would be selected for, since the induction of neonatal immune tolerance to HBV would promote the development of chronic infection following vertical transmission.

The genomic RNA species that initiates downstream of the pre-*C* AUG is the mRNA not only for C protein but also for P protein. Evidence that this is so comes from several sources as follows: (a) it is the only identified viral transcript bearing the entire *P* coding region; (b) no viral transcripts have been identified with 5′ ends just upstream of the *P* AUG, despite vigorous and sensitive searches; and (c) the fact that DHBV pregenomic RNAs are infectious (158) implies that P protein must have been translated from the input RNA. Examination of the coding organization of pgRNA shows that ORF *P* overlaps the 3′ end of the upstream *C* gene; the *P* frame is +1 relative to the *C* frame. This organization recalls that of most retroviruses, whose upstream core gene (*gag*) overlaps the downstream polymerase gene (*pol*); there, however, the *pol* reading frame is –1 relative to that of *gag.* In these cases, the retroviral polymerase is expressed as a gag-pol polyprotein by ribosomal frameshifting in the overlap region (164). Early assumptions that this would prove true for hepadnaviruses seemed to be supported by the finding of proteins with both C and P immunoreactivity in HBV-infected liver samples (467). However, genetic studies with DHBV soon established that polymerase is made by *de novo* initiation at the *P* initiator rather than by *C-P* frameshifting. These studies showed that (a) mutation of the first *P* AUG to ACG inactivates DNA synthesis, (b) stop codons in ORF *P* located 3′ but not 5′ to this AUG inactivate polymerase activity, and (c) *C* gene frameshift mutations do not impair the production of P protein (48,50,348). Subsequent experiments in HBV confirmed these findings and indicated that missense mutations in the *P* AUG could be reverted by construction of a new, in-frame AUG either upstream or downstream of the original mutation (170,341).

The mechanism by which ribosomes gain access to this internal AUG remains incompletely understood. Attempts to identify internal ribosome entry sequences of the type that function in picornavirus translation (167,288) have failed, and *P* gene translation appears fully sensitive to poliovirus superinfection (48), suggesting that, like most translational events, it is cap dependent. Recent genetic studies have led to a model that invokes modified ribosomal scanning from the 5′ end of genomic RNA. In this model, ribosomes scan from the cap site until they reach a small ORF overlapping the *C* gene; translation of this ORF allows bypass of a strong out-of-frame AUG that would otherwise occlude scanning further downstream, thereby allowing ribosomes that originated at the 5′ end of the transcript access to the internal *P* AUG (102). *In vitro* translation studies of *P* expression are also consistent with ribosomal scanning (216).

Genomic Replication

Viral Polymerase

Early inferences that ORF *P* encoded the viral polymerase were strongly supported by the subsequent recognition of homology between this coding region and those of retroviral reverse transcriptases (RTs) (421). Consistent with this assignment, mutations in ORF *P* inactivate viral DNA synthesis *in vivo* (but see below) (17). Since then many other RTs have been identified from numerous sources, and all of them share sequence relatedness to the hepadnaviral enzyme. Sequence alignments between these coding regions reveal the existence of subregions with obvious homology to either other polymerases (including RNA-directed RNA and DNA polymerases) or to *E. coli* RNase H (49,320,357); mutational analysis of retroviral *pol* genes has confirmed that these regions indeed represent functional domains that control the polymerase and RNase H activities of the retroviral enzyme (414). In hepadnaviruses the polymerase homology domain is centrally located, while the RNase H homology region is located near the C-terminus of the chain (Fig. 4).

In retroviruses and many retrotransposons, polymerase genes also encode protease and endonuclease activities, and recognizable homologies also characterize each of these

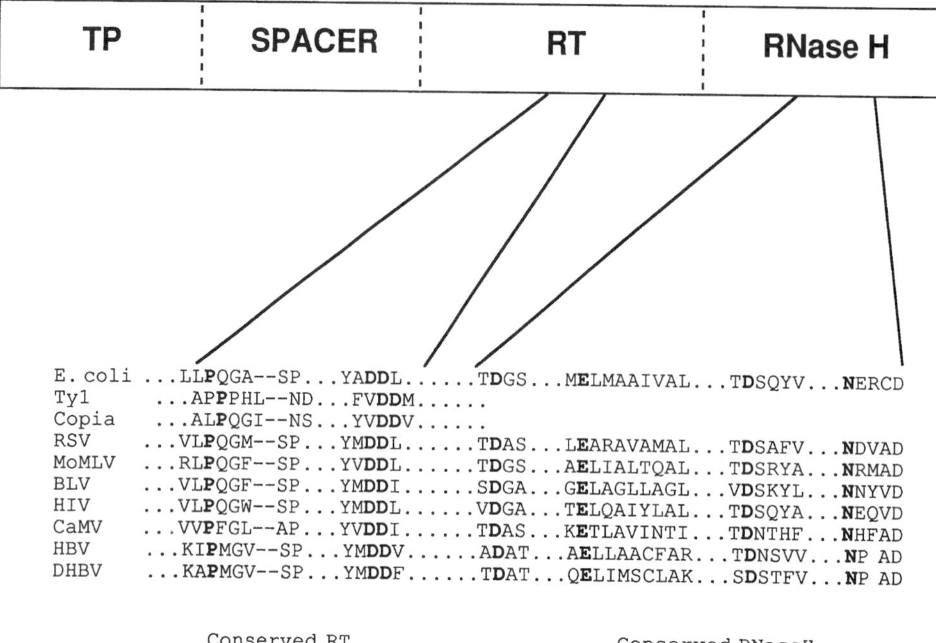

FIG. 4. Domain structure of the hepadnaviral P protein. ***Top:*** Functional domains of P protein. *TP,* terminal protein; *RT,* reverse transcriptase; *RNaseH,* ribonuclease H. ***Bottom:*** Amino acid sequence alignments with other RNA-dependent DNA polymerases; these homologies form the basis of the assignment of the RT and RNaseH domains.Reproduced from ref. 221a with permission.

activities. As might have been anticipated from the absence of a known role for integration in the growth of hepadnaviruses, no homology exists between ORF *P* and the integrases of other reverse-transcribing elements. Similarly, no functional protease homologies have been located within ORF *P* (or ORF *C*) (262), consistent with the fact that P protein is not processed from a C-P polyprotein.

The functional significance of these homology-based domain assignments has been validated *in vivo* through the study of mutant viral genomes bearing lesions in these regions. Hepatitis B virus and DHBV mutants bearing point mutations in highly conserved residues within the polymerase homology region are defective for viral DNA synthesis, while certain mutations within the RNaseH homology region allow minus strand but not plus strand DNA synthesis (49,96,320). Just upstream of the polymerase homology region, overlapping the pre-*S* coding domain, is a region of the *P* gene that is apparently unrelated to other polymerases and displays high sequence divergence between different hepadnaviral *P* genes. This region appears to define an inessential portion of the P protein, as it tolerates a wide variety of substitution, deletion, and insertion mutations (14,49,211,320); 5′ to it is the region implicated in the covalent linkage of the P protein to the viral DNA (see below). For this reason the inessential region is often referred to as a "spacer" or "tether" region, connecting the terminal protein domain to the polymerase and RNaseH domains (Fig. 4).

Owing to the low abundance of P protein *in vivo,* it has been difficult to visualize P chains in virions. Attempts to do so by immunoblotting (230) or activity gel analysis (18, 271) have often revealed species of less than full length, raising the question of posttranslational proteolytic processing. However, by engineering a protein kinase recognition site into the P chain, Bartenschlager et al. (16) were able to visualize encapsidated P chains directly following their phosphorylation *in vitro;* these studies strongly suggest that cores bear full-length P chains.

RNA Encapsidation

The classic experiments of Summers and Mason (400) indicated that virtually all reverse transcription takes place within subviral core particles, thereby implying that encapsidation of the genomic RNA template must represent the initial step in the genomic replication pathway. As noted previously, hepadnaviral genomic RNAs display microheterogeneities at their 5′ ends (93,257,474) (Fig. 1), with the longer ones encoding pre-C proteins and the smallest one encoding the C and P gene products. Importantly, only the smallest genomic mRNA is encapsidated into the core particle (92), indicating that only this transcript (also called pgRNA) can serve as a template for reverse transcription; host RNAs, subgenomic RNAs, and even the closely similar longer genomic transcripts are excluded. (The basis of this fine discrimination among such closely related RNAs will be discussed below.)

The viral proteins required for encapsidation have been identified by examining the impact of mutations in individual genes on pgRNA packaging. As expected, C protein is required to form the structure into which the RNA is packaged (143,206). Capsid formation is independent of pgRNA packaging, i.e., isolated *C* gene expression in heterologous hosts results in the accumulation of morphologically normal capsids devoid of pgRNA (25,71,251, 282,337,465,489,491). However, such recombinant capsids do contain host and viral RNAs, which are encapsidated by virtue of the sequence-nonspecific nucleic acid-binding properties of the arginine-rich C terminal domain of the molecule (25,111,138,261,347). Assembly of capsids proceeds via C protein dimer intermediates (491); these dimers form very rapidly, perhaps even while the chains are still nascent (45). Once the dimers have reached a threshold concentration (estimated at 1 μM), a sharply cooperative assembly reaction takes place to yield intact capsids (368). In general, no higher-order assembly intermediates can be detected *in vivo* between dimers and capsids. As a result, it is not yet known exactly where in this sequence the addition of RNA occurs.

In addition to the expected requirement for C protein, *P* gene products are also necessary for RNA packaging (15, 56,62,63,143). Null mutations in ORF *P* result in cores with reduced or undetectable levels of pgRNA. The P protein function required for packaging can be dissociated from those involved in DNA synthesis, i.e., many missense mutations that ablate polymerase or RNase H activity are still competent for RNA encapsidation (15,49,320,340). This requirement for polymerase gene products in packaging of viral RNA is strikingly different from the retroviral case, in which only gag proteins are required.

How does the encapsidation machinery select the proper RNA? First, a *cis*-acting packaging signal is present on the encapsidated message. The location of this packaging signal was determined by assaying for the ability of hepadnaviral sequences to confer encapsidation upon heterologous transcripts to which they were fused. In HBV, a small (approximately 100 nucleotides) region from the 5′ end of pgRNA (termed ϵ) suffices to allow packaging (63, 173,304). In DHBV, the corresponding ϵ region is necessary but not sufficient; a second, noncontiguous region downstream of ϵ is also required (40,144). Because of the terminal redundancy in pgRNA, the ϵ element is present at both ends of the RNA. Interestingly, however, only the 5′ copy is functional for RNA packaging: deletion of the 3′ element has no impact on viral replication (144). How this position dependence comes about is unknown, but it helps to explain the exclusion of subgenomic RNAs from encapsidation (92,144). Although all viral transcripts harbor ϵ sequences at their 3′ ends, only the genomic RNAs harbor them in the 5′ (functional) position (Fig. 2); thus, subgenomic RNAs cannot be encapsidated.

Examination of the nucleotide sequence within ϵ reveals a series of inverted repeats that can be predicted to fold into an RNA stem-loop structure (Fig. 5). This predicted stem-loop is phylogenetically conserved among all he-

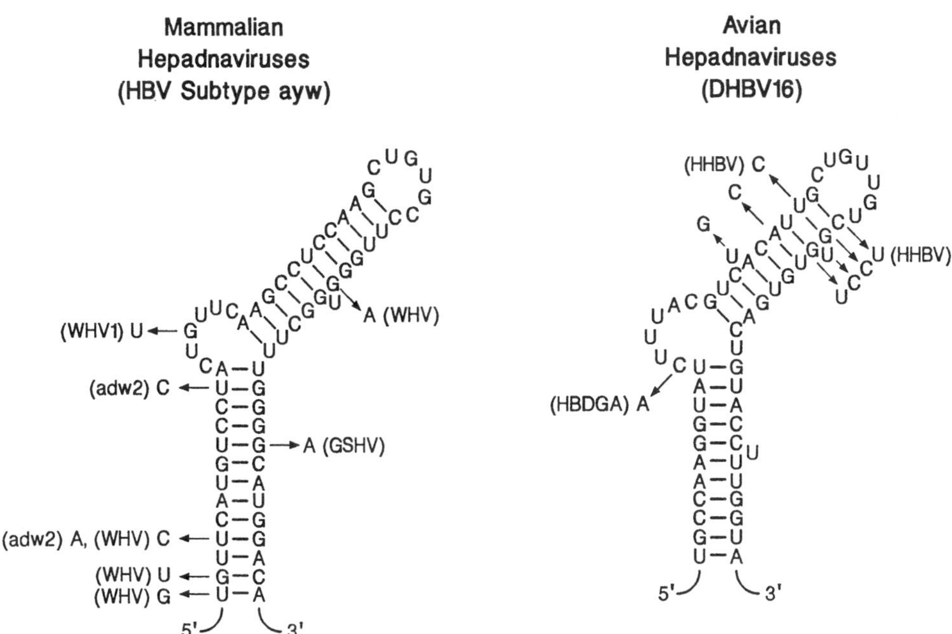

FIG. 5. Phylogenetic conservation of the ϵ stem-loop structure in mammalian *(left)* and avian *(right)* hepadnaviruses. The positions of base changes in naturally occurring isolates are indicated. Reproduced from ref. 304 with permission.

padnaviruses despite significant primary sequence variation in this region among the avian viruses (173). Consistent with this finding, direct biochemical analyses confirm that such a structure actually exists in HBV RNA (184,304). Mutational analyses have probed the regions of this structure that are functionally important for encapsidation (184, 304,426). The lower stem (and limited portions of the upper stem) must be base-paired, but primary sequence in this region is not critical; the six-nucleotide bulge must be present, but again primary sequence within the bulge can be radically altered without impairment of encapsidation. By contrast, the specific nucleotide sequence in the loop is critical, and base changes there are poorly tolerated by the packaging machinery.

What is the ϵ recognition apparatus? The simplest model compatible with the genetic analysis of encapsidation is that P protein directly recognizes this element. However, formally the genetic factors are compatible with recognition by a C-P complex or even by a C protein modified by its interaction with P. Very recently, direct *in vitro* evidence for P protein were found, i.e., RNA interactions have been obtained. Duck HBV P protein synthesized *in vitro* can bind in vitro ϵ RNA transcripts in a variety of assay formats (303). That this binding relates importantly to RNA encapsidation is supported by analysis of ϵ mutations, i.e, all mutations that block P-ϵ interaction *in vitro* block RNA encapsidation *in vivo.* However, there is one important discordancy: loop mutations, known to impair packaging strongly *in vivo,* do not abolish P-ϵ binding (303). This might mean that additional factors, possibly of host origin, may be binding to the loop region.

The next step in RNA encapsidation is presumably the association of the P-ϵ complex with assembling core subunits. Detailed understanding of these events is lacking; no direct evidence for noncovalent P-C interactions has yet been produced. However, some information has emerged from the study of structure-function relations within the DHBV P protein. By examining truncation mutants of DHBV P produced by *in vitro* translation, it has been shown that the C-terminus of P protein is dispensable for ϵ binding; however, this same region is important for RNA encapsidation *in vivo* (143,303). This is compatible with a model in which the N-terminal two thirds of P protein binds pgRNA, leaving the C-terminus free to interact (directly or indirectly) with assembling C polypeptides. A full accounting of the mechanism of the encapsidation reaction will likely require development of an *in vitro* system in which the entire process is faithfully reproduced. The first step in this direction has recently been achieved: the assembly of "empty" HBV cores from C proteins produced in cell-free translation systems (217).

Thus, in hepadnaviruses the encapsidation of polymerase occurs by a fundamentally different mechanism from that employed by retroviruses. The latter produce pol as a gag-pol polyprotein, which is assembled into capsids by interactions of its gag domain with assembling gag polypeptides. In hepadnaviruses, P protein is bound to the RNA that is destined for encapsidation, and noncovalent interactions, direct or indirect, between C and P must exist to package the P-RNA complex. Such interactions, however, must depend upon a unique property of the P-ϵ complex not shared by free P, since DHBV genomes bearing ϵ deletions produce cores that lack P protein as well as RNA (13).

An interesting feature of the encapsidation reaction is the marked preference of the P gene product for its own mRNA during the packaging process. This is most easily demonstrated by cotransfecting a wild-type and P-minus mutant genome (which is differentially marked with a new restriction enzyme site). Examination of either the encapsidated RNA or the subsequently reverse transcribed viral DNA reveals that the mRNA encoding the wild-type (functional) P gene product is preferentially encapsidated (90% to 95% of the encapsidated genomes are of wild-type origin) (143,206). The mechanistic basis of this *cis* preference is unclear. One possibility is that the nascent P polypeptide cotranslationally binds to its own mRNA. Alternatively, the levels of functional P gene product might be limiting, so that concentrations of P sufficient for encapsidation are found only in the vicinity of its own mRNA. In the case of HBV, the inefficiency in *trans* complementation can be at least partially overcome by deleting the encapsidation signal (15).

As noted earlier, the 5′ ends of the genomic RNAs are heterogeneous, but only the shortest of these RNAs is encapsidated. Since all these transcripts contain an intact 5′ copy of ϵ, how is packaging of all but the shortest pregenomic RNA suppressed? Examination of the 5′ termini of these RNAs indicates that all but the shortest transcript (i.e., pgRNA) include the upstream, in-frame pre-*C* AUG (Fig. 2). Mutations that inactivate this AUG allow these longer messages to be encapsidated (263). These and other data suggest that translation through this region suppresses recognition of ϵ. Since the shortest genomic RNA (pgRNA) lacks the pre-*C* AUG, such ribosome-mediated suppression cannot operate on this RNA.

Viral DNA Synthesis

The complex mechanism by which single-stranded pgRNA is converted to partially duplex virion DNA is now understood in considerable detail. The current view of this reaction is summarized schematically in Fig. 6. Pregenomic RNA is terminally redundant; its approximately 200-nucleotide redundancies (termed R) include the ϵ stem-loop and a short sequence of 11 to 12 nucleotides termed DR1. Another copy of this same sequence is present near the 3′ portion of the unique region of the RNA and is termed DR2. The potential importance of the DR sequences in viral DNA synthesis was recognized when the 5′ termini of the viral DNA strands were precisely mapped. This showed that the 5′ end of minus strand DNA mapped with-

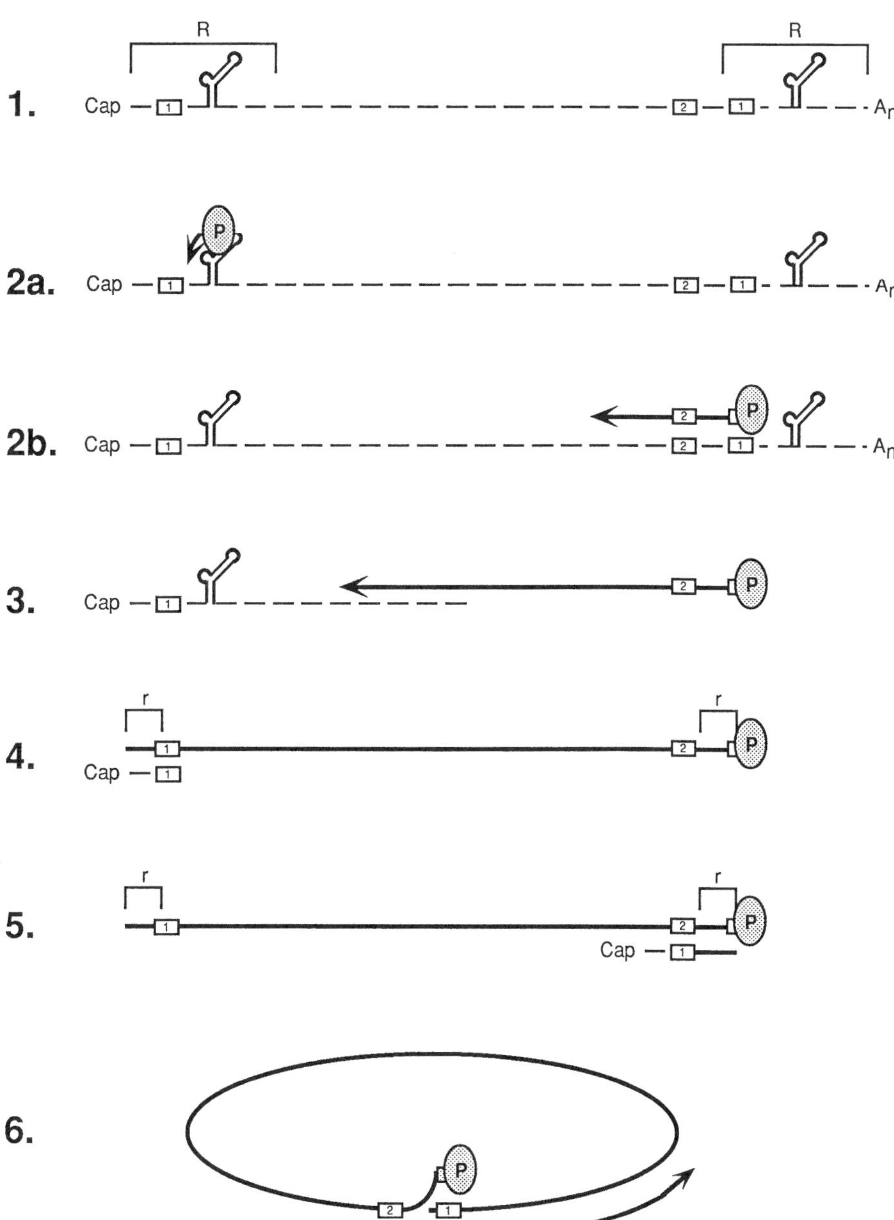

FIG. 6. The hepadnaviral reverse transcription pathway. Pregenomic RNA *(dashed line, step 1)* is capped, is polyadenylated, and has a large terminal redundancy *("R")*. The locations of direct repeats 1 and 2 (DR1 and DR2) are shown as correspondingly numbered boxes, and the ϵ stem-loops are indicated. Pregenomic RNA packaging into cores is initiated by the interaction of P protein with the 5′ copy of ϵ. P initiates reverse transcription at the 5′ stem-loop and extends minus strand DNA *(solid line)* for three to four nucleotides *(step 2a)*. Then P and the covalently attached nascent DNA are transferred to the 3′ copy of DR1 *(step 2b)*, and the DNA is extended. During minus strand elongation, pgRNA is degraded by the RNaseH activity of P *(step 3)*. When P reaches the 5′ end of the template, its RNaseH activity leaves an RNA oligomer consisting of r plus DR1 sequences *(step 4)*. This RNA oligomer is translocated and annealed to DR2, where it primes plus strand DNA synthesis *(lower solid line, step 5)*. Plus strand elongation proceeds to the 5′ end of the minus DNA template, including the sequences denoted as r. Since complementary r sequences are found at the 3′ end of minus DNA, a second homology-mediated template transfer can now circularize the genome. The plus strand is then extended for a variable length *(step 6)* to yield mature viral DNA. Reproduced from ref. 416 with permission.

in DR1, while the 5′ end of plus strand DNA mapped to DR2 (74,214,252,362,466). The exact roles of these sequences will be discussed below.

Minus Strand Synthesis

Minus strand DNA has protein covalently linked to its 5′ end (114,125,253). This so-called terminal protein was identified as the viral P protein by showing that viral DNA could be immunoprecipitated by anti-P protein antibodies (28). Limited proteolysis suggested that the N-terminal portion of the P polypeptide contained the site of covalent attachment to the DNA (14); this has now been directly confirmed for DHBV, in which the DNA-protein joint has been precisely mapped to tyrosine 96 of the P protein (493).

Because of the fact that all nascent chains of viral DNA are found linked to P protein (253), virologists have for years suspected that the protein serves as the primer for DNA synthesis, in a fashion formally analogous to that established for the terminal protein of adenovirus DNA (42). This inference has now been directly sustained by a seminal experiment of Wang and Seeger (451). When DHBV P protein was translated *in vitro* (from a mRNA that also contains DR1 and ϵ sequences) in the presence of ^{32}P-dNTPs, P protein chains become labeled by covalent attachment of nucleotides. The majority product consists of P chains bearing approximately four nucleotides; longer products (up to several hundred nucleotides) are present at a very low level. That this reaction reflects authentic viral DNA synthesis is indicated by the following: it is inhibited by mutations in the RT active site, and the sequence of bases at the DNA-protein joint corresponds exactly to the authentic 5′ end of viral DNA. In this reaction, the priming hydroxyl group is supplied not by a 3′OH of DNA or RNA but by the side chain of a tyrosine residue on P protein (457,493).

If P protein is the primer, what is the template for initiation? Because the 5′ ends of viral DNA map to DR1, it was long assumed that the priming of minus strand DNA was templated by DR1 sequences. However, analyses of the DHBV priming reaction both *in vitro* (450) and in recombinant yeast expressing functional P protein (415,416) indicate that this is not the case. Rather, minus strand initiation is templated by sequences within the bulge of the ϵ stem-loop. The involvement of ϵ sequences in the RT reaction had previously gone unrecognized because *in vivo* ϵ is required for the RNA packaging step that is a necessary precondition for DNA synthesis. Thus, defects in DNA synthesis due to ϵ mutations could be accounted for entirely by their RNA packaging defect. However, both of the recombinant P protein expression systems mentioned above were designed to be independent of DHBV packaging functions. Under these conditions, the involvement of ϵ in reverse transcription rapidly became apparent; deletions of ϵ (or even of the ϵ bulge alone) completely ablated all viral reverse transcription, a result that is not compatible with earlier models of initiation within DR1.

A full account of the experimental details that established the priming scheme shown in Fig. 6 is beyond the scope of the present discussion (416,450). To summarize the principal conclusions, initiation occurs within the ϵ bulge, generating an initial P-lined oligonucleotide of approximately four nucleotides, exactly as observed *in vitro* by Wang and Seeger (451). Because the sequences of the ϵ bulge and DR1 share four nucleotides of identity, this DNA product can be transferred and annealed to DR1 sequences in a homology-mediated reaction; from this position the DNA can be extended by traditional elongation mechanisms.

Because both DR1 and ϵ are represented twice in the pgRNA template, it was important to determine which copies of these sequences participate in the above reactions. The fact that the 3′ copy of ϵ can be deleted from pgRNA without blocking viral replication (365,366) suggests that it is the 5′ copy of ϵ that is functional in initiation. This inference has been directly validated by genetically marking the 5′ ϵ with a sequence polymorphism and showing that this sequence marker appears at the 5′ end of minus strand DNA after one round of reverse transcription (416,450). Analogous genetic tagging experiments indicate that it is the 3′ copy of DR1 to which the nascent P-linked primer is transferred (366). Once this transfer reaction is complete, elongation can continue to the end of the template, generating a minus DNA that is actually terminally redundant by approximately eight nucleotides (this shorter redundancy is known as r) (214,362,466).

Plus Strand Synthesis

As minus strand elongation proceeds, degradation of the newly copied pgRNA template occurs concomitantly (249, 400). As in other reverse transcription reactions, this function is attributed to the RNase H activity of the viral polymerase itself (320). Once minus strand DNA synthesis is complete, plus strand DNA synthesis can begin. Plus strands are initiated at DR2 (214,252,362,466) (Fig. 6), in a reaction primed by short (approximately 15 to 18 nucleotides), capped oligoribonucleotides derived from the 5′ of pgRNA; these include DR1 sequences and the six-nucleotide 5′ of DR1 (214). The viral RNaseH is also presumed to be responsible for the cleavages that generate these primers. How its cleavage sites are chosen has recently been clarified. The simplest model, that sequence-specific endonucleolytic cleavages are made at the border of DR1, is excluded by the finding that nucleotide changes around the cleavage sites do not perturb either the specificity or efficiency of cleavage (367,393). However, the specificity of cleavage can be altered by insertions or deletions between the 5′ end of pregenome and the 3′ end of DR1 (221). In these mutants the sites of cleavage remain a constant distance (15 to 18 nucleotides) from the 5′ end of the pregenome, indicating that the cleavages are somehow positioned by measurement from the end of the template and that the actual scission event is independent of the sequence at the cleavage site. A model for how this distance might be measured has been proposed (221).

The site of primer generation is many hundreds of bases away from the actual site of plus strand DNA synthesis at DR2. Thus, the newly cleaved RNA primer must be translocated to the 5′ end of minus DNA to base-pair with DR2 for plus strand DNA synthesis to begin (214,362,466). Because the primer is being moved from a region (including r and DR1) where it is base-paired for 18 nucleotides to a region (DR2) where it is base-paired for only 12 nucleotides (Fig. 6), this primer translocation is not thought to occur passively. Most likely it is actively facilitated by proteins, but how this is accomplished remains a matter for study. It is known that the primer translocation step is very sensitive to mutations not only within the primer but also to lesions just outside (3′) of it (393). This implies that the machinery involved in primer translocation interacts not only with the RNA primer but with a larger local region of the genome.

Occasionally, primers that are cleaved fail to be translocated. This occurs at a low rate with wild-type virus genomes (approximately 1% to 5% of primers) but can be greatly increased by *cis*-acting mutations in and around DR1. Typically, primers that are cleaved but not translocated are instead extended from their original position. The result is fully duplex linear viral DNA (393). Such molecules are found in all stocks of hepadnaviral particles. They are unlikely to be infectious, since in effect they represent complementary DNA copies of pgRNA and, as such, lack appropriate transcriptional signals. However, because they are linear, their termini may be recombinogenic *in vivo*; it is thus conceivable that they could participate in genomic integration into host DNA in infected cells (see below).

Once the RNA primer is translocated, synthesis of plus strand DNA can begin. Elongation proceeds to the 5′ end of minus DNA, at which point the template is exhausted and an intramolecular strand transfer is required to complete plus strand synthesis (Fig. 6). This transfer is facilitated by the short redundancy (r) in the minus DNA template: annealing of the 3′ end of plus strand DNA (via its r homology) to the 3′ end of the minus strand circularizes the genome and allows continuation of plus strand synthesis. Thus, the strand transfer allows plus strand synthesis access to the rest of the minus DNA template. However, in most cases synthesis does not proceed to completion [the exception is DHBV, where 80% of the plus strands complete elongation (215,252)]. The result is the characteristic single-stranded gap that is a hallmark of hepadnaviral DNA.

Why plus strand synthesis terminates prematurely is unknown but might be related to the process of viral budding. The DNA of cytoplasmic core particles is substantially more immature than that of extracellular virions, suggesting that maturation of viral DNA synthesis is somehow coupled with envelopment and secretion of the Dane particle. Presumably a structural change or posttranslational modification of the core particle occurs concomitant with reaching a certain stage in plus strand DNA synthesis (483). If so, such a change could be the signal for core envelopment. Once budded into the ER the core particle would no longer have access to cytoplasmic dNTP pools, and further DNA synthesis would be arrested. However, this model, while attractive, remains largely conjectural at present.

Viral Assembly and Release

20-nm Subviral Particle Assembly

As noted previously, 20-nm particles contain predominantly the S protein with variable amounts of M and only trace quantities of L subunits (139). With the advent of cloned HBV DNA, it became possible to express the surface protein coding regions in heterologous cells, leading to the discovery that cells expressing only the S polypeptide can assemble and secrete morphologically normal 20-nm particles. Thus, all of the viral information necessary for this assembly process resides in the S domain (202,220). Since the cellular machinery for particle assembly is present in virtually all vertebrate cells, expression of 20-nm particles is possible in a wide variety of expression systems (84,256,394,397). This has made possible the development of recombinant HBV vaccine, the first recombinant DNA-based vaccine licensed for any human infectious disease. Interestingly, yeast cells (the source of one widely used recombinant vaccine) do not secrete HBsAg at all; however, lipoprotein arrays of S subunits do form intracellularly, and such complexes can be readily extracted and purified to yield immunogenic particulate antigen (445,449).

How do 20-nm particles form? Early EM studies of infected liver revealed that particles accumulated within cisternae of the endoplasmic reticulum (120), an observation also confirmed by more recent EM studies of transfected, HBsAg-positive cells in culture (286). Consistent with this is the fact that intracellular particles contain asparagine-linked high-mannose oligosaccharide chains, a modification specific to the ER (287). Subsequent biochemical studies have more precisely localized the major site of particle assembly to the recently recognized "intermediate compartment" between the ER and Golgi (160). Assembled complexes are then transported through the constitutive secretory pathway (180), traversing the Golgi complex, where their carbohydrates are processed to the complex, endoglycosidase H-resistant form (287). Transport of the complexes through the Golgi appears to be the rate-limiting step in export, since all intracellular glycosylated S protein is found in the endoglycosidase H-sensitive form, whereas all extracellular HBsAg is endoglycosidase H resistant.

The S proteins are synthesized in the ER as integral membrane proteins. Their transmembrane topology has been determined in cell-free translation systems supplemented with ER-derived microsomal vesicles (86,87). In these experiments, nascent S protein chains were found to be oriented in the bilayer such that both N- and C-termini are in the vesicle lumen. To do so, the polypeptide must span the bilayer at least twice; many current models, based largely upon theoretical considerations, envision up to four transmembrane passages (Fig. 7). (The S coding region contains three hydrophobic domains, encompassing residues 4–28, 80–100, and 164–221. Hydrophobicity and other considerations lead most to assume that all of these regions are located within the bilayer, generating the model of Fig. 7. However, the disposition of the C-terminal hydrophobic domain has not been experimentally assessed.) This topology is achieved through the conjoint action of two topogenic signal sequences (87). Signal I corresponds to the first hydrophobic domain and initiates chain translocation across the bilayer. It resembles conventional N-terminal signal sequences except for the fact that it remains uncleaved during translocation. Signal II (corresponding to the second hydrophobic domain) is a complex topogenic element that serves both as a stop-transfer sequence (anchoring the protein in the bilayer with residue 80 facing the cytoplasm and residue 100 facing the ER lumen) and a signal sequence directing the translocation of the distal region of the molecule. Thus, the region between signals I and II is a cytoplasmic loop; downstream of signal II is the major glycosylation site of S as well as its immunodominant surface epitope.

This transmembrane form of S is only a transient intermediate in particle formation; *in vivo* this form is rapidly converted to a series of defined assembly intermediates (160,384), and *S* gene mutations affecting one or more of these steps have been examined (33,311). Following membrane insertion, the transmembrane monomers aggregate and in the process exclude host proteins from the complex. The aggregate then buds or exudes into the lumen. It is likely that this extrusion event differs from more conventional viral budding reactions in that the "budded" product lacks a typical unit membrane structure. In addition, its lipid content is much lower than that of most cellular membranes, suggesting that that the host lipids may be substantially reorganized in the process.

This pathway shares many features with morphogenetic events postulated for the envelope proteins of other animal viruses: transmembrane insertion, aggregation, and

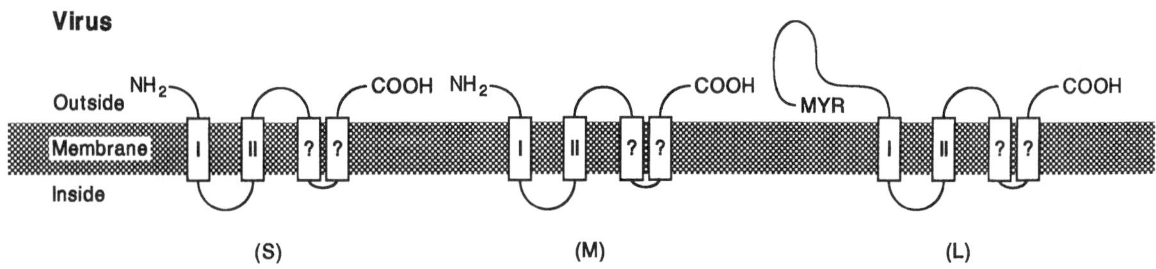

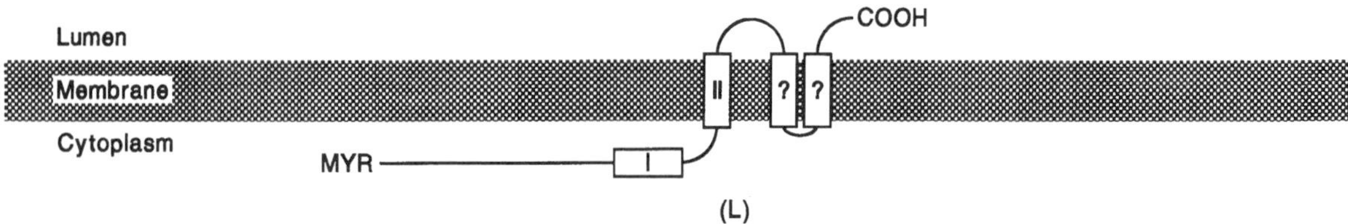

FIG. 7. ***Top:*** Proposed transmembrane topologies for the HBV L, M, and S glycoproteins in the virion envelope. Rectangles denote S protein hydrophobic domains. [The transmembrane orientations of domains I and II have been addressed experimentally (87), while the disposition of the remaining hydrophobic regions remains speculative.) The ψ symbol denotes N-linked carbohydrate. Myr, myristate group in amide linkage to glycine at the N-terminus of L protein (291). ***Bottom:*** transmembrane structure of L protein in endoplasmic reticulum membranes. Virtually all pre-S sequences reside on the cytoplasmic face of the membrane. (The disposition of S hydrophobic region I is here drawn as cytoplasmic, but this has not been directly validated.)

budding. However, in other viruses, envelope glycoproteins remain anchored in the bilayer until interactions with other viral components (e.g., nucleocapsid or matrix proteins) trigger the budding step, thus ensuring that envelope proteins can exit the cell only by enveloping the nucleocapsid. The distinctive feature of the S protein is that it can carry out the entire assembly sequence without the involvement of other viral proteins. The result is the release of the subviral particles containing only envelope proteins—the distinctive signature of hepadnaviral infection.

In authentic viral infection, 20-nm particles frequently contain M as well as S subunits (139,396). In such particles, the pre-S2-encoded domain of the M protein is exposed on the particle surface, as demonstrated by their reactivity with antisera specific for the pre-S2 domain of M and by their ability to bind polymers of human albumin, another activity [of uncertain function (302)] mediated by this domain (227,228,293). Given its surface location on the particle, in the transmembrane form of the M protein, this domain would be predicted to be translocated into the ER lumen (Fig. 7); this prediction has been experimentally confirmed (85). Since the pre-S2-encoded domain lacks a definable signal sequence, its transmembrane transport must be mediated by the downstream signals in the S domain.

As noted above, natural 20-nm particles often lack detectable L protein (139). The reason for the low abundance of L chains on 20-nm particles became clear in experiments in which L chains were deliberately overexpressed in cultured cells. In cells infected with vaccinia vectors that express only the L protein, the L chains themselves are not secreted, despite the presence of an intact S domain that contains all the information required for the export process (57). This suggests that the pre-S1-specific domain of L contains signals that promote intracellular retention of the chain and that can override the secretory elements in the S domain. When L chains are overexpressed in the presence of M and S chains, the ability of the M and S chains to be secreted is inhibited in a dose-dependent fashion (57,66,292,391); all three chains accumulate intracellularly in the endoplasmic reticulum (65). Presumably this reflects the formation of mixed arrays of L-M-S aggregates, with the inhibitory influence of the pre-S1 domain now being conferred upon the entire aggregate. The inhibitory effect of L overexpression on S particle secretion explains the underrepresentation of L chains on 20-nm spherical particles.

The region of the pre-S1 domain that specifies the intracellular retention of HBsAg has been mapped by deletion mutagenesis to the extreme N-terminus of the molecule (197). The observation that this region of the L protein is myristylated (291) raised the possibility that ER retention might be mediated by the affinity of this hydrophobic fatty acid for the ER membrane. However, point mutations that ablate myristylation have either no (197) or only incomplete (310) effects on secretion inhibition, indicating that primary sequences in this region of the chain are important for this activity.

Interestingly, in EM studies of the hepatocytes of transgenic mice overexpressing L chains (relative to S), 20-nm *filamentous* forms were frequently observed in the ER (65). This accords well with earlier observations that filaments are relatively enriched in L subunits (139) and suggests that filaments may arise in regions of the ER where local concentrations of L protein are high.

The overaccumulation of retained L-M-S aggregates can be injurious to the host cell. Hepatocytes of transgenic mice bearing such aggregates develop characteristic cytopathic effects (in the relative absence of inflammation) (65,68). Morphologically similar cells are sometimes seen in biopsy specimens from acute hepatitis B, indicating that deregulation of envelope protein expression may make some contribution to the process of liver cell injury. The magnitude of this contribution, however, is uncertain, and is likely to be modest.

Dane Particle Assembly

Conceptual views of Dane particle assembly have been heavily influenced by the emerging understanding of the morphogenesis of subviral particles. Since all three HBV envelope proteins are expressed primarily on intracellular membranes rather than on the cell surface (283), virion formation almost certainly proceeds on intracellular membranes; EM studies consistent with this have been reported (175,335). However, an important distinguishing feature of virions is that they contain relatively large quantities of L proteins; in the best preparations the ratio of L:M:S subunits was approximately 1:1:4 (139). Since these preparations are often contaminated to some extent with 20-nm particles, it is possible that the ratio of L:S chains in authentic virions is still higher. It is unlikely that subviral particles bearing this much L protein could be efficiently released; this suggests that distinctive mechanisms are involved in virion assembly and export.

To define the components required for Dane particle production, permissive HepG2 cells have been transfected with missense or amber mutants selectively affecting the HBV L, M, or S proteins and the ability of these mutants to support virion release assayed (32,442). Several conclusions have emerged from these studies as follows: (a) unlike retroviruses, whose nucleocapsids can still bud out of cells in the absence of envelope proteins, no HBV cores are exported if the synthesis of all envelope proteins is aborted by nonsense mutations in S and (b) both L and S proteins are required for virion formation and release. Although one report to the contrary has appeared (442), the weight of evidence is that M proteins are not required for assembly (32,101,344). Missense mutants affecting L myristylation also remain competent for virion assembly and secretion, though the unmyristylated virions are noninfectious (32,231,403). Myristylation may thus play its most important role in the adsorption, entry, or uncoating of virions on the next cycle of infection.

How the L protein contributes to viral assembly remains incompletely understood, but recent studies of its transmembrane topology have suggested new possibilities. Studies of HBV virion architecture reveal that the L protein-specific, pre-S1-encoded domain is largely or entirely exposed on the virion surface. For example, it is accessible to exogenous proteases as well as to pre-S1-specific mAbs (139,196,280). This would suggest that, like pre-S2, the pre-S1 domain in the transmembrane form of L should reside in the ER lumen. However, recent biochemical studies (280) (V. Bruss and W. Gerlich, unpublished data) indicate that this is not the case; rather, the majority of pre-S-encoded L protein sequences are on the *cytoplasmic* surface in the nascent L chain (Fig. 7). This helps to explain how L might make a distinctive contribution to virion morphogenesis, since it now can present unique determinants to nucleocapsids on the cytoplasmic face of the ER membrane. In one attractive model, capsid-L protein interactions trigger the budding event that incorporates the mature core into the viral envelope; this might explain the enrichment for L chains on virions. This notion is supported by mutational studies examining the impact of pre-*S* deletions on virion assembly: only those mutants whose pre-S sequences appeared to be cytoplasmically disposed were assembly competent (34).

However, we are left with an important remaining problem: how does the pre-S1 domain wind up on the surface of the budded virion? One possibility is that nascent L chains exist in two orientations in the ER: one with luminal pre-S sequences and another with cytoplasmic pre-S domains. However, virtually all detected nascent chains have only cytoplasmic pre-S determinants. A more likely scenario is that, during or following budding, the components of the envelope undergo a dramatic rearrangement, perhaps analogous to that envisioned during 20-nm subviral particle formation.

Nuclear Delivery of Progeny Core DNA: cccDNA Amplification

Consideration of the overall replication cycle schematized in Fig. 3 indicates that the central genomic species involved in viral persistence is nuclear cccDNA. Only cells that harbor this form of the genome can continue to produce virus. Since hepadnavirus-infected cells are not killed by viral replication, they may continue to grow and divide. (Admittedly, under normal circumstances hepatocytes divide infrequently, but proliferation is often greatly enhanced during chronic infection as a result of inflammation and regeneration.) Thus, a mechanism must exist to assure the inheritance of this molecule by both daughter cells following mitosis. In principle, this problem could be solved by equipping cccDNA with an origin for semiconservative DNA replication, since all the machinery for this process exists in the nucleus; this would be analogous to the solution adopted by Epstein-Barr virus or papillomaviruses, for example. However, density labeling studies (435) indicate that this is not the case, i.e., all nuclear cccDNA is derived instead from the reverse transcription pathway. Since reverse transcription is a cytoplasmic process, this implies that mechanisms exist to deliver cytoplasmic reverse transcription products into the nucleus. For DHBV, such mechanisms operate efficiently early in the establishment of infection in primary duck hepatocytes, leading to the rapid accumulation of ten to 20 or more copies of cccDNA per nucleus.

One possible way to account for the maintenance of a pool of cccDNA would be to imagine that it was replenished by reinfection of cells by extracellular virus. However, it was soon shown that cccDNA amplification could be observed in the presence of inhibitors of viral entry like suramin or neutralizing antibody; it can also occur following transfection of viral DNA into cells that lack the surface receptor, thereby eliminating the possibility of reinfection by released progeny virus (470). Thus, cytoplasmic cores bearing virion DNA can be shunted to the nucleus by an intracellular pathway. A key step forward was made with the recognition that DHBV mutants lacking the L envelope protein displayed greatly enhanced amplification, resulting in the nuclear accumulation of very high levels of cccDNA (402,403). This implies that the subcellular trafficking of cores is regulated; early in infection, when envelope proteins are at low levels, nuclear delivery is favored, allowing accumulation of cccDNA. This ensures that this infected cell will be stably colonized. Late in infection, with rising levels of envelope proteins, export of cores as virions is favored, allowing horizontal spread of infection to surrounding cells. Thus, in this view, nuclear transit is the default pathway; as envelope proteins accumulate, cores are shunted from this pathway to the export pathway. Detailed mutational studies of DHBV L protein are consistent with this view, i.e., virtually all mutations that block virion assembly display enhanced nuclear cccDNA accumulation (209).

HEPADNAVIRUSES AND HCC

Biology and Epidemiology

The first indication that chronic HBV infection might predispose to the development of HCC emerged from studies of the global epidemiology of HBV carriage (408). These studies revealed that HBV is not uniformly distributed among human populations. In Western Europe and North America, only about 0.1% to 0.5% of individuals are HBsAg carriers; however, in Southeast Asia and Sub-Saharan Africa, fully 5% to 15% of all human beings are chronically infected with HBV. (The remaining areas of the globe have an intermediate prevalence of HBV carriage.) This pattern of HBV infection strikingly reproduces the known global epidemiology of HCC, a rare tumor in the West but long known as a leading malignancy in Africa and Asia. This remarkable finding prompted deeper investigation of this linkage.

In HBV-endemic countries, the majority of HCCs arise in patients who are HBV carriers (113,408); even though carriers may represent 5% to 10% of the population, they represent 50% to 80% of HCC patients in those societies. However, the strongest evidence for an association comes from careful prospective studies done in Taiwan, in which large numbers of carriers and noncarriers were followed prospectively and deaths due to HCC recorded. These classic studies demonstrated that chronic HBV carriage is associated with a 100-fold increase in risk for HCC development relative to noncarriers (20,21). This massive increment in cancer risk makes HBV one of the most important environmental risk factors in human cancer epidemiology.

In Taiwan (as in all HBV-endemic areas) most infections are acquired in the first decade of life (113). Thus, in these studies most of the individuals being followed had been carrying HBV for many decades. From the age distribution of Taiwanese hepatoma cases, it can be inferred that most tumors arise after 30 or more years of persistent infection; relatively few cases occur in children and adolescents. Extrapolation of this estimate of relative cancer risk to Western societies should be done with caution. In the West, most infections are acquired in young adulthood, as a result of sexual transmission. Although few doubt that such infections present some increase in cancer risk, it is unclear if the magnitude of this risk is as large as in the HBV-endemic zones, if only because those infected as adults have fewer years at risk. Nonetheless, even in the West, the interval between acquisition of infection and HCC development remains at least 30 years (113). Further support for the oncogenic potential of hepadnavirus infection came with the discovery of the animal hepadnaviruses. Studies of WHV infection showed that this virus is even more potent than HBV as a hepatic carcinogen; nearly 100% of woodchucks infected from birth with WHV will develop HCC, starting at around 18 to 24 months of age (121,308,309,360). (Uninfected woodchucks housed in similar circumstances do not develop HCC at all.) The oncogenic drive accompanying this infection is so strong that many animals develop multiple independent hepatic tumors; even animals who experienced only a transient WHV infection (as judged by the presence of antisurface and anticore antibodies in the absence of surface antigenemia) display an increased prevalence of HCC (121). Ground squirrels infected with GSHV also display an increased prevalence of HCC, though the tumors are less frequent and arise much later in life (241,428). [Since WHV and GSHV can both be grown in woodchucks (4,364), it has been possible to compare the oncogenic potential of both infections in a common host. These studies (360) show that the biology just described is determined largely by the viral genome and not by the host, i.e., GSHV-infected woodchucks display HCC later in life and at a lower frequency than WHV-infected woodchucks.) As noted earlier, studies of the oncogenic potential of DHBV have not clearly established it as a tumor virus. Although some DHBV-infected flocks in Asia have shown HCC, most laboratory-raised animals or commercial flocks in America and Europe have not exhibited HCC (78), suggesting that environmental factors (or cofactors) may account for a large portion of the oncogenesis in Asia. There is no evidence that Asian DHBV isolates (441) differ from Western isolates in their oncogenicity.

Viral DNA and HCC

When the tumors themselves are examined, some can be shown to produce one or more viral antigens; most, however, have extinguished all HBV gene expression. Nonetheless, the majority (85% or more) of such tumors harbor integrated viral DNA, often multiple copies per cell (29,372). The HBV integrants are usually highly rearranged, with deletions, inversions, and sequence reiterations all commonly observed (90,259); as expected, most of these rearrangements ablate viral gene expression. [WHV insertions in woodchuck HCC in general display similar features (104,258,276).] In addition, alterations of host DNA often accompany these integrations. Described alterations include deletions or repeats at the host-viral junctions (334, 424,471), as well as (more rarely) large chromosomal translocations, some of which harbor HBV sequences at the interchromosomal junctions (141,422); the latter might have arisen by homologous recombination between viral integrants on two different chromosomes. Importantly, tumors are clonal with respect to these integrants, i.e., every cell in the tumor harbors an identical complement of HBV insertions. This implies that the integration event(s) accompanied or preceded the clonal expansion of the cells. Although much has been made of this fact, by itself, it puts HBV DNA at the "scene of the crime" but does not establish "criminality."

How HBV integration is achieved is poorly understood (377). As opposed to its role in retroviral replication, integration is not an obligatory part of the hepadnaviral replication cycle; it is therefore not surprising that hepadnaviruses have no virus-encoded integration machinery. Presumably, HBV DNA is assimilated into the nucleus by host mechanisms; it is reasonable to suppose that this event proceeds by pathways akin to those operating in the integration of exogenous DNA in transfected mammalian cells in culture. Such mechanisms likely include topoisomerase I (452) and other components. A major problem that has confounded study of the integration process has been that the integrants characterized in HCC specimens are many years removed from the primary insertion event. During this (approximately 30-year) interval, there has presumably been strong immunologic selection against viral antigen expression; this is the likely force that has selected the genomic rearrangements described above.

Models for HBV Oncogenesis

The molecular mechanisms by which hepadnaviruses predispose to malignancy remain much debated. Models for how they might act fall into two broad categories, direct and indirect, according to the roles they posit for the viral genome and its products (112).

Direct Models

These posit that viral DNA makes a direct genetic contribution to the lesion, either by providing *cis*-acting sequences that activate or suppress host growth-regulating genes or by elaborating *trans*-acting factors that alter the program of cellular growth control. The most straightforward version of such a model is that a viral protein might act as a dominant oncogene, much as *src* does in Rous sarcoma virus infection. However, hepadnaviruses carry no gene directly transduced from the host cell genome; in addition, hepadnaviral infection does not lead to growth transformation or immortalization of primary cells *in vitro* [though there is one report (145) of further growth deregulation induced by HBV X expression in a cell line already immortalized by SV40 T antigen]. These facts, together with the prolonged incubation period between infection and tumor development, make it extremely unlikely that the direct action of a single viral gene product can itself transform a hepatocyte. However, they by no means exclude more subtle roles for viral proteins that might help initiate the loss of growth control, even if they cannot by themselves complete this process.

In this connection, two viral proteins are frequently mentioned. Primary attention has centered on the X protein, for two reasons: (a) it is absent in the avian viruses, the only hepadnaviruses not reliably associated with HCC; and (b) as a known regulator of gene expression, it is easy to envision models for how such a molecule might affect cellular growth; indeed, many known oncogenes affect the two processes—transcription and signal transduction—postulated to be targeted by X. The debate about X protein's role in oncogenesis has been powerfully influenced by reports that transgenic mice bearing X sequences under the control of the *X* promoter and *EnI* develop HCC (181,188). However, this effect has not been observed by other groups using X vectors driven by heterologous promoters (208), nor has comparable oncogenesis been seen in transgenic mice bearing other subgenomic or genomic HBV DNAs containing the *X* region (10,69,97,98). The reasons for these divergent results are obscure; possibilities include differing levels of X expression (188) or the different genetic backgrounds of the mice used in different laboratories. One thing is certain: if X protein plays a role in oncogenesis, it cannot be necessary for the entire life of the transformed cell, since most advanced tumors have extinguished all HBV gene expression. Thus, if X is required early in oncogenesis to promote cell proliferation, this requirement must be obviated at later stages of transformation.

The second viral gene product for which a role in oncogenesis has been postulated is the truncated M protein generated by occasional viral integration events (203,204,265). The roles ascribed to such factors are formally identical to those outlined above for X protein, but experimental evidence supporting such roles is nil at present, i.e., many tumors cannot encode such M variants, expression of these proteins in cultured cells does not lead to loss of growth control, and no transgenic mice expressing them have been reported. Finally, rare viral integration events might be able to generate other novel proteins by fusion of viral coding sequences with adjacent cellular genes, analogous to the generation of novel oncogenic proteins by chromosomal translocations in other settings (70). One potential example of this surfaced in a human HCC in which integration fused the HBV pre-*S* region to a retinoic acid receptor locus (80,82). However, such events have been documented extremely rarely and are unlikely to make a major contribution to oncogenesis.

Most advocates of direct models favor the notion that viral sequences act primarily by contributing *cis*-acting sequences rather than coding regions (153). There is unequivocal evidence for this mode of oncogenesis in WHV infection. In this system, approximately 40% of all HCCs contain an integrated WHV genome within several kilobases of one of two N-*myc* loci; these N-*myc* rearrangements are readily detected by Southern blotting (104,136). N-*myc* genes are normally silent in adult liver but are transcriptionally activated by enhancer elements within WHV DNA (458); recent studies implicate *EnII* in this effect (K. Ueda and D. Ganem, unpublished data). Interestingly, N-*myc* transcription is elevated very early in the oncogenic process, even in premalignant nodules (475) and is observed in virtually 100% of cases. Very recently, Fourel al. (103) have noted that, among many tumors that lack N-*myc* rearrangements by conventional Southern blotting, integrated WHV DNA can be found in *cis* to N-*myc2* by pulse-field gel analysis, at distances up to 150 to 180 kb away from the activated locus. This raises the possibility that all WHV-associated HCC may be initiated by insertional activation of N-*myc,* with many cases resulting from previously unsuspected long-range activation events.

Efforts to make a similar case for insertional activation by HBV have been largely unsuccessful. Dozens of human hepatomas have been screened for common sites of HBV integration (259,423). This is done by cloning HBV integrants and using the flanking host DNA as a probe to screen

Southern blots of DNA from other HCCs for rearrangements of the corresponding host sequences. (This is the same strategy that was employed productively for WHV). To date, no rearrangements common to many human HCCs have been detected, although a few isolated examples of insertions near interesting loci have been reported (80,82, 452), these insertions appear to be unique to the tumors in which they were discovered. Similarly, deliberate screens of human HCCs for rearrangements of known host oncogenes have likewise been largely unrevealing. It should be emphasized, however, that all of these methods can only detect rearrangements within a few kilobases of the locus being probed; if long-range activation events of the type identified for WHV (103) operate in HBV infection, then insertional activation could still contribute importantly to the biology of HBV-induced HCC.

Indirect Models

The relative paucity of experimental support for direct models of HBV-induced oncogenesis has led to the growing popularity of indirect models. By "indirect" is meant that HBV genes and their products make no direct genetic contribution to the transforming event. Rather, HBV-induced liver injury (itself largely the result of host immune or inflammatory responses to infection) triggers a series of stereotypic host responses that lead to liver cell regeneration. Vertebrate hosts are known to respond to liver cell injury of many types with a brisk regenerative response (although the mechanism of this response is poorly understood). Under conditions of chronic regeneration, the increase in the number of cells undergoing DNA synthesis results in an increased probability of mutation; this in turn increases the probability that one or more of the resulting mutant cells will have a proliferative advantage. As this cell is preferentially expanded, the same logic predicts an increased opportunity for additional mutational lesions in its progeny (and so on, until full growth transformation is achieved). Thus, HBV's role in this is to act as an agent of liver injury; everything else is contributed by the host.

Supporters of indirect models point to several compelling facts, chief among which is that most conditions in human medicine that lead to chronic liver injury are associated to one degree or another with a risk for HCC development: alcoholic cirrhosis (212), α_1-antitrypsin deficiency, Wilson's disease, and so forth. Even hepatitis C virus, an RNA virus with no known DNA intermediate, can predispose to HCC during chronic infection (250). (However, none of these precedents can fully explain why the magnitude of the HCC risk appears to be so much larger in chronic HBV infection). Strong experimental support for this hypothesis is derived from elegant recent experiments in transgenic mice (68). These mice were engineered to overexpress the HBV L protein relative to the M and S proteins. As noted above, when L is overexpressed, it leads to retention of all the viral envelope glycoproteins in the ER (57,66,292,391); this retention is toxic to cells, although the mechanism of this toxicity is unknown. Initially the mice display only hepatocellular necrosis and accompanying regeneration. Over time, however, malignant HCC arises with great regularity.

Confusion has arisen over the proper interpretation of this experiment, in part because it uses HBV gene products to drive the liver injury. Strictly speaking, the experiment demonstrates that chronic liver injury promotes regeneration and HCC. While it raises the possibility that viral envelope glycoproteins might contribute to this injury, it does not establish that this form of injury is relevant to clinical hepatitis B. Indeed, it appears unlikely that this would be so, i.e., most evidence favors the notion that the bulk of the injury is produced by immune attack on infected hepatocytes (67), and the relative levels of L protein in natural infection are nowhere near those achieved in the HCC-prone transgenic mice. Although hepatocytes with the histologic appearance of cells undergoing damage from deregulated envelope expression can be seen in hepatitis B, in most cases they are not numerous enough to explain the regenerative drive that accompanies this infection.

The notion that hepatocellular injury is an important factor in the HCC risk of viral infection, if sustained, has enormous implications. It suggests that a better understanding of the immunologic mechanisms of liver cell injury (and of the inflammatory mediators that amplify it and trigger the regenerative response) could allow the development of therapeutics that would blunt these responses. Such agents would then be expected not only to ameliorate clinical hepatitis but also to reduce the attendant risk of HCC.

REFERENCES

1. Acs G, Sells MA, Purcell RH, Price P, Engle R, Shapiro M, Popper H. Hepatitis B virus produced by transfected Hep G2 cells causes hepatitis in chimpanzees. *Proc Natl Acad Sci USA* 1987;84:4641–4644.
2. Aggerback LP, Peterson DL. Electron microscopic and solution X-ray scattering observations on the structure of hepatitis B surface antigen. *Virology* 1985;141:155–161.
3. Albin C, Robinson WS. Protein kinase activity in hepatitis B virions. *J Virol* 1980;34:297–302.
4. Aldrich CE, Coates L, Wu T-T, Newbold J, Tennant BC, Summers J, Seeger C, Mason WS. *In vitro* infection of woodchuck hepatocytes with woodchuck hepatitis virus and with ground squirrel hepatitis virus. *Virology* 1989;172:247–252.
5. Antonucci T, Rutter W. Hepatitis B virus promoters are regulated by the HBV enhancer in a tissue-specific manner. *J Virol* 1989;63:579–585.
6. Araki K, Miyazaki J-I, Hino O, Tomita N, Chisaka O, Matsubara K, Yamamura K. Expression and replication of hepatitis B virus genome in transgenic mice. *Proc Natl Acad Sci USA* 1989;86:207–211.
7. Argos P, Fuller SD. A model for the hepatitis B virus core protein: prediction of antigenic sites and relationship to RNA virus capsid proteins. *EMBO J* 1988;7:819–824.
8. Aufiero B, Schneider R. The hepatitis B virus X gene product *trans*-activates both RNA polymerase II and III promoters. *EMBO J* 1990; 9:497–504.
9. Ayola B, Kanda P, Lanford RE. High level expression and phosphorylation of hepatitis B virus polymerase in insect cells with recombinant baculoviruses. *Virology* 1993;194:370–373.
10. Babinet C, Farza H, Morello D, Hadchouel M, Pourcel C. Specific ex-

pression of hepatitis B surface antigen in transgenic mice. *Science* 1985; 230:1160–1163.
11. Balsano C, Avantaggiati, Natoli G, DeMarzio E, Will H, Perricaudet M, Levrero M. Full-length and truncated versions of the HBV X protein (pX) transactivate the c-*myc* proto-oncogene at the transcriptional level. *Biochem Biophys Res Commun* 1991;176:985–992.
12. Barker LF, Chisari FV, McGrath PP, et al. Transmission of type B viral hepatitis to chimpanzees. *J Infect Dis* 1973;127:648–662.
13. Bartenschlager R, Schaller H. Hepadnaviral assembly is initiated by polymerase binding to the encapsidation signal in the viral RNA genome. *EMBO J* 1992;11:3413–3420.
14. Bartenschlager R, Schaller H. The amino terminal domain of the hepadnaviral P gene encodes the terminal protein (genome-linked protein) believed to prime reverse transcription. *EMBO J* 1988;7:4185–4192.
15. Bartenschlager R, Junker-Niepmann M, Schaller H. The P gene product of the hepatitis B virus is required as a structural component for genomic RNA encapsidation. *J Virol* 1990;64:5324–5332.
16. Bartenschlager R, Kuhn C, Schaller H. Expression of the P protein of the human hepatitis B virus in a vaccinia system and detection of the nucleocapsid-associated P gene product by radiolabeling at newly introduced phosphorylation sites. *Nucleic Acids Res* 1991;20:195–202.
17. Bavand M, Feitelson M, Laub O. The hepatitis B virus-associated reverse transcriptase is encoded by the viral *pol* gene. *J Virol* 1989;63: 1019–1021.
18. Bavand MR, Laub O. Two proteins with reverse transcriptase activities associated with hepatitis B virus-like particles. *J Virol* 1988;62: 626–628.
19. Bayer M, Blumberg B, Werner B. Particles associated with Australia antigen in the sera of patients with leukemia, Down's syndrome and hepatitis. *Nature* 1968;218:1057–1059.
20. Beasley RP. Hepatitis B virus—the major etiology of hepatocellular carcinoma. *Cancer* 1988;61:1942–1956.
21. Beasley RP, Huang L-Y, Lin C, Chien C. Hepatocellular carcinoma and HBV: a prospective study of 22,707 men in Taiwan. *Lancet* 1981; 2:1129–1133.
22. Beasley RP, Trepo C, Stevens CE, Szmuness W. The e antigen and vertical transmission of hepatitis B surface antigen. *Am J Epidemiol* 1977;105:94–98.
23. Bertoletti A, Ferrari C, Fiaccadori F, Penna FA, Margolskee R, Schlicht HJ, Fowler P, Guilhot S, Chisari FV. HLA class I-restricted human cytotoxic T cells recognize endogenously synthesized hepatitis B virus nucleocapsid antigen. *Proc Natl Acad Sci USA* 1991;88:10445–10449.
24. Bhat R, Ulrich P, Vyas GN. Molecular characterization of a new variant of hepatitis B virus in a persistently infected homosexual man. *Hepatology* 1990;11:271–276.
25. Birnbaum F, Nassal M. Hepatitis B virus nucleocapsid assembly: primary structure requirements in the core protein. *J Virol* 1990;64:3319–3330.
26. Blum HE, Zhang Z-S, Galun E, von Weizsacker F, Garner B, Liang TJ, Wands JR. Hepatitis B virus X protein is not central to the viral life cycle in vitro. *J Virol* 1992;66:1223–1227.
27. Blumberg BS, Alter HJ, Visnich S. A "new" antigen in leukemia sera. *JAMA* 1965;191:541–546.
28. Bosch V, Bartenschlager R, Radziwill G, Schaller H. The duck hepatitis B virus P-gene codes for protein strongly associated with the 5′-end of the viral DNA minus strand. *Virology* 1988;166:475–485.
29. Brechot C, Pourcel C, Louise A, Rain B, Tiollais P. Presence of integrated hepatitis B virus DNA sequences in cellular DNA in human hepatocellular carcinoma. *Nature* 1980;286:533–535.
30. Brunetto M, Stemmler M, Schodel F, Will H, Ottobrelli A, Rizetto M, Verme G, Bonino F. Identification of HBV variants which cannot produce precore-derived HBeAg and may be responsible for severe hepatitis. *Ital J Gastroenterol* 1989;21:151–154.
31. Brunetto MR, Giarin MM, Oliveri F, Chaberge E, Baldi M, Alfarano A, Serra A, Saracco G, Verme G, Will H, Bonino F. Wild-type and e antigen-minus hepatitis B viruses and course of chronic hepatitis. *Proc Natl Acad Sci USA* 1991;88:4186–4190.
32. Bruss V, Ganem D. The role of envelope proteins in hepatitis B virus assembly. *Proc Natl Acad Sci USA* 1991;88:1059–1063.
33. Bruss V, Ganem D. Mutational analysis of hepatitis B surface antigen particle assembly and secretion. *J Virol* 1991;65:3813–3820.
34. Bruss V, Thomssen R. Mapping a region of the large envelope protein required for hepatitis B virion maturation. *J Virol* 1994;68:1643–1650.
35. Budkowska A, Quan C, Groh F, Bedossa P, Dubreuil P, Bouvet JP, Pillot J. Hepatitis B virus (HBV) binding factor in human serum: candidate for a soluble form of hepatocyte HBV receptor. *J Virol* 1993; 67:4316–4322.
36. Budkowska A, Shih JWK, Gerin JL. Immunochemistry and polypeptide composition of hepatitis B core antigen (HBcAg). *J Immunol* 1977; 118:1300–1350.
37. Bulla G, Siddiqui A. The hepatitis B virus enhancer modulates transcription of the hepatitis B virus surface antigen gene from an internal location. *J Virol* 1988;62:1437–1441.
38. Burrell CJ, Mackay P, Greenway PJ, Hofschneider PH, Murray K. Expression in *Escherichia coli* of hepatitis B virus DNA sequences cloned in plasmid pBR322. *Nature* 1979;279:43–47.
39. Buscher M, Reiser W, Will H, Schaller H. Transcripts and the putative RNA pregenome of duck hepatitis B virus: implications for reverse transcription. *Cell* 1985;40:717–724.
40. Calvert J, Summers J. Two regions of an avian hepadnavirus RNA pregenome are required in cis for RNA encapsidation. *J Virol* 1994; 68:2084–2090.
41. Cattaneo R, Will H, Hernandez N, Schaller H. Signals regulating hepatitis B surface antigen transcription. *Nature* 1983;305:336–338.
42. Challberg M, Kelly T. Animal virus DNA replication. *Annu Rev Biochem* 1989;58:671–717.
43. Chang C, Enders G, Sprengel R, Peters N, Varmus HE, Ganem D. Expression of the precore region of an avian hepatitis B virus is not required for viral replication. *J Virol* 1987;61:3322–3325.
44. Chang C, Jeng K-S, Hu C-P, Lo SJ, Su T-S, Ting L-P, Chou C-K, Han S-H, Pfaff E, Salfeld J, Schaller H. Production of hepatitis B virus *in vitro* by transient expression of cloned HBV DNA in a hepatoma cell line. *EMBO J* 1987;6:675–680.
45. Chang C, Zhou S, Ganem D, Standring D. Phenotypic mixing between different hepadnaviral nucleocapsid subunits reveals that C protein dimerization is cis-preferential. *J Virol* [in press].
46. Chang H-K, Ting L-P. The surface gene promoter of the human hepatitis B virus displays a preference for differentiated hepatocytes. *Virology* 1989;170:176–183.
47. Chang H-K, Wang B-Y, Yuh C-H, Wsei C-L, Ting L-P. A liver-specific nuclear factor interacts with the promoter region of the large surface protein gene of human hepatitis B virus. *Mol Cell Biol* 1989;9: 5189–5197.
48. Chang L-J, Ganem D, Varmus HE. Mechanism of translation of the hepadnaviral polymerase (P) gene. *Proc Natl Acad Sci USA* 1990;87: 5158–5162.
49. Chang L-J, Hirsch R, Ganem D, Varmus H. Effects of insertional and point mutations on the functions of the duck hepatitis B virus polymerase. *J Virol* 1990;64:5553–5558.
50. Chang L-J, Pryciak P, Ganem D, Varmus HE. Biosynthesis of the reverse transcriptase of the hepatitis B viruses involves de novo translational initiation not ribosomal frameshifting. *Nature* 1989;337:364–368.
51. Chen H, Kaneko S, Girones R, Anderson R, Hornbuckle W, Tennant B, Cote P, Gerin J, Purcell R, Miller R. The woodchuck hepatitis X gene is important for the establishment of virus infection in woodchucks. *J Virol* 1993;67:1218–1226.
52. Chen P, Chen C, Sung J, Chen D. Identification of a doubly spliced viral transcript joining the separated domains for putative protease and reverse transcriptase of hepatitis B virus. *J Virol* 1989;63:4165–4171.
53. Chen S-T, La Porte P, Yee J-K. Mutational analysis of hepatitis B virus enhancer 2. *Virology* 1993;196:652–659.
54. Chen S-T, Su H, Yee J-K. Repression of liver-specific hepatitis B virus enhancer 2 activity by adenovirus E1A proteins. *J Virol* 1992;66:7452–7460.
55. Chen Y, Kew MC, Hornbuckle WE, Tennant BC, Cote PJ, Gerin JL, Purcell RH, Miller RH. The precore gene of the woodchuck hepatitis virus genome is not essential for viral replication in the natural host. *J Virol* 1992;66:5682–5684.
56. Chen Y, Robinson WS, Marion PL. Naturally occurring point mutation in the C terminus of the polymerase gene prevents duck hepatitis B virus RNA packaging. *J Virol* 1992;66:1282–1287.
57. Cheng C, Smith K, Moss B. Hepatitis B virus large surface protein is not secreted but is immunogenic when selectively expressed by recombinant vaccinia virus. *J Virol* 1986;60:337–344.
58. Cherrington J, Ganem D. Regulation of polyadenylation in human immunodeficiency virus (HIV): contributions of promoter proximity and upstream sequences. *EMBO J* 1992;11:1513–1524.

59. Cherrington J, Russnak R, Ganem D. Upstream sequences and cap proximity in the regulation of polyadenylation in ground squirrel hepatitis virus *J Virol* 1992;66:7589–7595.
60. Cheung RC, Robinson WS, Marion PL, Greenberg HB. Epitope mapping of neutralizing monoclonal antibodies against duck hepatitis B virus. *J Virol* 1989; 63:2445–2451.
61. Cheung RC, Trujillo DE, Robinson WS, Greenberg HB, Marion PL. Epitope-specific antibody response to the surface antigen of duck hepatitis B virus infected ducks. *Virology* 1990;176:546–552.
62. Chiang P-W, Hu C-P, Su T-S, Lo SJ, Chu M-HH, Schaller H, Chang C. Encapsidation of truncated human hepatitis B virus genomes through trans-complementation of the core protein and polymerase. *Virology* 1990;176:355–361.
63. Chiang P-W, Jeng K-S, Hu C-P, Chang C. Characterization of a *cis* element required for packaging and replication of the human hepatitis B virus. *Virology* 1992;186:701–711.
64. Chisaka O, Araki K, Ochiya T, Tsurimoto T, Hiranyawasitte W, Yanihara N, Matsubara K. Purification of hepatitis B virus X product synthesized in *E. coli* and its detection in a human hepatoblastoma cell line producing hepatitis B virus. *Gene* 1987;60:183–189.
65. Chisari F, Filippi P, Buras J, MacLachlan A, Popper H, Pinkert C, Palmiter R, Brinster R. Structural and pathological effects of synthesis of hepatitis B virus large envelope polypeptide in transgenic mice. *Proc Natl Acad Sci USA* 1987;84:6909–6913.
66. Chisari F, Fillipi P, MacLachlan A, Milich D, Riggs M, Lee S, Palmiter R, Pinkert C, Brinster R. Expression of hepatitis B large envelope polypeptide inhibits hepatitis B surface antigen secretion in transgenic mice. *J Virol* 1986;60:880–887.
67. Chisari FV, Ferrari C, Mondelli M. Hepatitis B virus structure and biology. *Microb Pathog* 1989;6:311–314.
68. Chisari FV, Klopchin K, Moriyama T, Pasquinelli C, Dunsford HA, Brinster RL, Palmiter RD. Molecular pathogenesis of hepatocellular carcinoma in hepatitis B virus transgenic mice. *Cell* 1989;59:1145–1156.
69. Choo K-B, Liew L-N, Chong KY, Lu R, Cheng WTK. Transgenome transcription and replication in the liver and extrahepatic tissues of a human hepatitis B virus transgenic mouse. *Virology* 1991;182:785–792.
70. Clark S, McLaughlin J, Crist W, et al. Unique forms of the abl tyrosine kinase distinguish Ph-positive CML from Ph-positive ALL. *Science* 1987;235:85–88.
71. Cohen B, Richmond J. Electron microscopy of hepatitis B core antigen synthesized in *E. coli*. *Nature* 1982;296:677–678.
72. Colgrove R, Simon G, Ganem D. Transcriptional activation of homologous and heterologous genes by the hepatitis B virus X gene product in cells permissive for viral replication. *J Virol* 1989;63:4019–4026.
73. Condreay LD, Aldrich CE, Coates L, Mason WS, Wu T-T. Efficient duck hepatitis B virus production by an avian liver tumor cell line. *J Virol* 1990;64:3249–3258.
74. Condreay LD, Wu T-T, Aldrich CE, Delaney MA, Summers J, Seeger C, Mason WS. Replication of DHBV genomes with mutations at the sites of initiation of minus- and plus-strand DNA synthesis. *Virology* 1992;88:208–216.
75. Courtois G, Baumheuter S, Crabtree G. Purified hepatocyte nuclear factor 1 interacts with a family of hepatocyte-specific promoters. *Proc Natl Acad Sci USA* 1988;85:7937–7941.
76. Crescenzo-Chaigne J, Pillot, Lilienbaum A, Leurero M, Elfassi E. Identification of a strong enhancer element upstream from the pregenomic RNA start site of the duck hepatitis B virus genome. *J Virol* 1991; 65:3882–3883.
77. Cross J, Wen P, Rutter WJ. Transactivation by hepatitis B virus X protein is promiscuous and dependent on mitogen-activated cellular serine/threonine kinases. *Proc Natl Acad Sci USA* 1993;90:8078–8082.
78. Cullen J, Marion P, Sherman G, Xin H, Newbold J. Hepatic neoplasms in aflatoxin B1-treated congenitally DHBV-infected and virus-free Pekin ducks. In: Hollinger FB, Lemon S, Margolis H, eds. *Viral hepatitis and liver disease*. 1991:601–604.
79. Dane DS, Cameron CH, Briggs M. Virus-like particles in serum of patients with Australia antigen-associated hepatitis. *Lancet* 1970;1: 695–698.
80. Dejean A, Bouguelleret L, Grzeschick K, Tiollais P. Hepatitis B virus DNA integration in a sequence homologous to v-erbA and steroid receptor genes in a hepatocellular carcinoma. *Nature* 1986;322:70–72.
81. De-Medina T, Faktor O, Shaul Y. The S promoter of hepatitis B virus is regulated by positive and negative elements. *Mol Cell Biol* 1988;8: 2449–2455.
82. deThe H, Marchio A, Tiollais P, Dejean A. A novel steroid/thyroid hormone receptor-related gene inappropriately expressed in a hepatocellular carcinoma. *Nature* 1987;330:667–670.
83. Dikstein R, Faktor O, Ben-Levy R, Shaul Y. Functional organization of the hepatitis B virus enhancer. *Mol Cell Biol* 1990;10:3683–3689.
84. Dubois MF, Pourcel C, Rousset S, Chany C, Tiollais P. Excretion of hepatitis B surface antigen particles from mouse cells transformed with cloned viral DNA. *Proc Natl Acad Sci USA* 1980;77:4549–4553.
85. Eble B, Lingappa V, Ganem D. The N-terminal (preS2) domain of a hepatitis B virus surface glycoprotein is translocated across membranes by downstream signal sequences. *J Virol* 1990;64:1414–1419.
86. Eble BE, Lingappa VR, Ganem D. Hepatitis B surface antigen: an unusual secreted protein initially synthesized as a transmembrane polypeptide. *Mol Cell Biol* 1986;6:1454–1463.
87. Eble BE, Macrae DR, Lingappa VR, Ganem D. Multiple topogenic sequences determine the transmembrane orientation hepatitis B surface antigen. *Mol Cell Biol* 1987;7:3591–3601.
88. Eckhardt SG, Milich DR, McLachlan A. Hepatitis B virus core antigen has two nuclear localization sequences in the arginine-rich carboxyl terminus. *J Virol* 1991;65:575–582.
89. Eddleston A. Overview of HBV pathogenesis. In: Hollinger F, Lemon S, Margolis H, eds. *Viral hepatitis and liver disease*. Baltimore: Williams and Wilkins; 1991:234–237.
90. Edman J, Gray P, Valenzuela P, Rall L, Rutter W. Integration of hepatitis B virus sequences and their expression in a human hepatoma cell. *Nature* 1980;286:535–538.
91. El-Ghor N, Burk R. DNase I hypersensitive site maps to the HBV enhancer. *Virology* 1989;172:478–488.
92. Enders GH, Ganem D, Varmus HE. 5′-Terminal sequences influence the segregation of ground squirrel hepatitis virus RNAs into polyribosomes and viral core particles. *J Virol* 1987;61:35–41.
93. Enders GH, Ganem D, Varmus HE. Mapping the major transcripts of ground squirrel hepatitis virus: the presumptive template for reverse transcriptase is terminally redundant. *Cell* 1985;42:297–308.
94. Faktor O, DeMedina T, Shaul Y. Regulation of hepatitis B virus S gene promoter in transfected cell lines. *Virology* 1988;162:362–368.
95. Faktor O, Shaul Y. The identification of hepatitis B virus X gene responsive elements reveals functional similarity of X and HTLV-1 tax. *Oncogene* 1990;5:867–872.
96. Faruqi AF, Roychoudhury S, Greenberg R, Israel J, Shih C. Replication-defective missense mutations within the terminal protein and spacer/intron regions of the polymerase gene of human hepatitis B virus. *Virology* 1991;183:764–768.
97. Farza H, Hadchouel M, Scotto J, Tiollais P, Babinet C, Pourcel C. Replication and gene expression of hepatitis B virus in a transgenic mouse that contains the complete viral genome. *J Virol* 1988;62:4144–4152.
98. Farza H, Salmon A, Hadchouel M, Tiollais P, Pourcel C. Hepatitis B surface antigen expression is regulated by sex steroids and glucocorticoids in transgenic mice. *Proc Natl Acad Sci USA* 1987;84:1187–1191.
99. Feitelson MA, Marion P, Robinson W. Core particles of hepatitis B virus and ground squirrel hepatitis virus II: characterization of the protein kinase reaction associated with ground squirrel hepatitis virus and hepatitis B virus. *J Virol* 1986;43:741–748.
100. Feldherr C, Kallenbach E, Schultz N. Movement of a karyophilic protein through the nuclear pores of oocytes. *J Cell Biol* 1984;99:2216–2222.
101. Fernholz D, Wildner G, Will H. Minor envelope proteins of duck hepatitis B virus are initiated at internal pre-S AUG condons but are not essential for infectivity. *Virology* 1993;197:64–73.
102. Fouillot N, Tlouzeau S, Rossignol J, Jean-Jean O. Translation of the hepatitis B virus P gene by ribosomal scanning as an alternative to internal initiation. *J Virol* 1993;67:4886–4895.
103. Fourel G, Couturier J, Wei Y, Apiou F, Tiollais P, Buendia MA. Evidence for long-range oncogene activation by hepadnavirus insertion. *EMBO J* [in press].
104. Fourel G, Trepo C, Bougueleret L, Henglein B, Ponzetto A, Tiollais P, Buendia M-A. Frequent activation of N-myc genes by hepadnavirus insertion in woodchuck liver tumours. *Nature* 1990;347:294–298.
105. Franco A, Paroli M, Tresta U, Benvenuto R, Peschle C, Balsano F, Barnaba V. Transferrin receptor mediates uptake and presentation of hepatitis B envelope antigen by T lymphocytes. *J Exp Med* 1992;175: 1095–1105.
106. Fujiyama A, Miyanohara A, Nozaki C, Yoneyama T, Ohtomo N, Mat-

subara K. Cloning and structural analyses of hepatitis B virus DNAs, subtype adr. *Nucleic Acids Res* 1983;11:4601–4610.
107. Galibert F, Chen T-N, Mandart E. Nucleotide sequence of a cloned woodchuck hepatitis virus genome: comparison with the hepatitis B virus sequence. *J Virol* 1982;41:51–65.
108. Galibert F, Mandart E, Fitoussi F, Tiollais P, Charnay P. Nucleotide sequence of hepatitis B virus genome (subtype ayw) cloned in *E. coli. Nature* 1979;281:646–650.
109. Galle PR, Schlicht H, Kuhn C, Schaller H. Replication of duck hepatitis B virus in primary duck hepatocytes and its dependence on the state of differentiation of the host cell. *Hepatology* 1989;10:459–465.
110. Galle PR, Schlicht HJ, Fisher M, Schaller H. Production of infectious duck hepatitis B virus in a human hepatoma cell line. *J Virol* 1988;62: 1736–1740.
111. Gallina A, Bonelli F, Zentilin L, Rindi G, Muttini M, Milanesi G. A recombinant hepatitis B core antigen polypeptide with the protamine-like domain deleted self-assembles into capsid particles but fails to bind nucleic acids. *J Virol* 1989;63:4645–4652.
112. Ganem D. Of marmots and men. *Nature* 1990;347:230–232.
113. Ganem D. Persistent infection of humans with hepatitis B virus: mechanisms and consequences. *Rev Infect Dis* 1982;4:1026–1047.
114. Ganem D, Greenbaum L, Varmus HE. Virion DNA of ground squirrel hepatitis virus: structural analysis and molecular cloning. *J Virol* 1982;44:374–383.
115. Ganem D, Varmus HE. The molecular biology of the hepatitis B viruses. *Annu Rev Biochem* 1987;56:651–694.
116. Ganem D, Weiser B, Barchuk A, Brown RJ, Varmus HE. Biological characterization of acute infection with ground squirrel hepatitis virus. *J Virol* 1982;44:366–373.
117. Garcia A, Ostapchuk P, Hearing P. Functional interaction of nuclear factors EF-C, HNF-4 and RXR-alpha with hepatitis B virus enhancer I. *J Virol* 1993;67:3940–3950.
118. Garcia P, Ou J, Rutter WJ, Walter P. Targeting of precore protein of hepatitis B virus to the endoplasmic reticulum membrane: after signal peptide cleavage translocation can be aborted and the product released into the cytoplasm. *J Cell Biol* 1988;106:1093–1104.
119. Gavilanes F, Gonzales-Ros A, Peterson D. Structure of hepatitis B surface antigen: characterization of the lipid components and their association with the viral proteins. *J Biol Chem* 1982;257:7770–7777.
120. Gerber M, Hadziyannis S, Vissoulis C, Schaffner F, Paronetto F, Popper H. Electron microscopy and immunoelectron microscopy of cytoplasmic hepatitis B antigen in hepatocytes. *Am J Pathol* 1974;75: 489–502.
121. Gerin J, Cote P, Korba B, Miller R, Purcell R, Tennant B. Hepatitis B virus and liver cancer: the woodchuck as an experimental model of hepadnavirus-induced liver cancer. In: Hollinger B, Lemon S, Margolis H, eds. *Viral hepatitis and liver disease*. 1991:556–559.
122. Gerin J, Purcell R, Hoggan M, Holland P, Chanock R. Biophysical properties of Australian antigen. *J Virol* 1980;36:787–795.
123. Gerin JL, Holland PW, Purcell RH. Australia antigen: large-scale purification from human serum and biochemical studies of its proteins. *J Virol* 1971;7:569–576.
124. Gerlich W, Goldmann U, Muller R, Stibbe W, Wolff W. Specificity and localization of the hepatitis B virus-associated protein kinase. *J Virol* 1982;42:761–766.
125. Gerlich W, Robinson WS. Hepatitis B virus contains protein attached to the 5′ end of its complete strand. *Cell* 1980;21:801–811.
126. Gripon P, Diot C, Guguen-Guillouzo C. Reproducible high-level infection of cultured adult hepatocytes by hepatitis B virus: effect of polyethylene glycol on adsorption and penetration. *Virology* 1993;192: 534–540.
127. Gripon P, Diot C, Theze N, Fourel I, Loreal O, Brechot C, Guguen-Guillouzo C. Hepatitis B virus infection of adult human hepatocytes cultured in the presence of dimethyl sulfoxide. *J Virol* 1988;62:4136–4143.
128. Guilhot S, Fowler P, Portillo G, Margolskee RF, Ferrari C, Bertoletti A, Chisari FV. Hepatitis B virus (HBV)-specific cytotoxic T-cell response in humans: production of target cells by stable expression of HBV-encoded proteins in immortalized human B-cell lines. *J Virol* 1992;66:2670–2678.
129. Gunther S, Meisel H, Reip A, Miska S, Kruger DH, Will H. Frequent and rapid emergence of mutated pre-C sequences in HBV from e-antigen positive carriers who seroconvert to anti-HBe during interferon treatment. *Virology* 1992;187:271–279.
130. Guo W, Bell KD, Ou J-H. Characterization of the hepatitis B virus EnhI enhancer and X promoter complex. *J Virol* 1991;65:6686–6692.
131. Guo W, Chen M, Yen TSB, Ou J-H. Hepatocyte-specific expression of the hepatitis B virus core promoter depends on both positive and negative regulation. *Mol Cell Biol* 1993;13:443–448.
132. Halpern M, Egan J, McMahon S, Ewert D. Duck hepatitis B virus is tropic for exocrine cells of the pancreas. *Virology* 1985;146:157–161.
133. Halpern M, England J, Deery D, Petcu D, Mason W, Molnar K. Viral nucleic acid synthesis and antigen accumulation in pancreas and kidney of Pekin ducks infected with duck hepatitis B virus. *Proc Natl Acad Sci USA* 1983;80:4865–4869.
134. Halpern M, McMahon S, Mason W, O'Connell A. Viral antigen expression in the pancreas of DHBV-infected embryos and young ducks. *Virology* 1986;150:276–282.
135. Halpern MS, England JM, Flores L, Egan J, Newbold J, Mason WS. Individual cells in tissues of DHBV-infected ducks express antigens crossreactive with those on virus surface antigen particles and immature viral cores. *Virology* 1984;37:408–413.
136. Hansen LJ, Tennant BC, Seeger C, Ganem D. Differential activation of *myc* gene family members in hepatic carcinogenesis by closely related hepatitis B virus. *Mol Cell Biol* 1993;13:659–667.
137. Hasegawa K, Huang J, Wands JR, Obata H, Liang TJ. Association of hepatitis B viral precore mutations with fulminant hepatitis B in Japan. *Virology* 1991;185:460–463.
138. Hatton T, Zhou S, Standring DN. RNA- and DNA-binding activities in hepatitis B virus capsid protein: a model for their roles in viral replication. *J Virol* 1992;66:5232–5241.
139. Heermann KH, Goldmann U, Schwartz W, Seyffarth T, Baumgarten H, Gerlich WH. Large surface proteins of hepatitis B virus containing the pre-S sequence. *J Virol* 1984;52:396–402.
140. Hertogs K, Leenders WPJ, Depla E, De Brun WCC, Meheus L, Raymackers J, Moshoge H, Yap SH. Endonexin II, present on human liver plasma membranes, is a specific binding protein of small hepatitis B virus (HBV) envelope protein. *Virology* 1993;197:549–557.
141. Hino O, Shows T, Rogler C. Hepatitis B virus integration site in hepatocellular carcinoma at chromosome 17:18 translocation. *Proc Natl Acad Sci USA* 1986;83:8338–8342.
142. Hirsch R, Colgrove R, Ganem D. Replication of duck hepatitis B virus in two differentiated human hepatoma cell lines after transfection with cloned viral DNA. *Virology* 1988;167:136–142.
143. Hirsch R, Lavine J, Chang L, Varmus H, Ganem D. Polymerase gene products of hepatitis B viruses are required for genomic RNA packaging as well as for reverse transcription. *Nature* 1990;344:552–555.
144. Hirsch R, Loeb DD, Pollack JR, Ganem D. *Cis*-acting sequences required for encapsidation of duck hepatitis B virus pregenomic RNA. *J Virol* 1991;65:3309–3316.
145. Hohne M, Schaefer S, Seifer M, Feitelson MA, Paul D, Gerlich WH. Malignant transformation of immortalized transgenic hepatocytes after transfection with hepatitis B virus DNA. *EMBO J* 1990;9:1137–1145.
146. Honigwachs J, Faktor O, Dikstein R, Shaul Y, Laub O. Liver-specific expression of hepatitis B virus is determined by the combined action of the core gene promoter and the enhancer. *J Virol* 1989;63: 919–927.
147. Hoofnagle JH. Serologic markers of hepatitis B virus infection. *Annu Rev Med* 1981;32:1–11.
148. Hoofnagle JH, Gerety RJ, Barker LF. Antibody to hepatitis B virus core in man. *Lancet* 1973;2:869.
149. Hoofnagle JH, Seeff LB, Bales ZB, Gerety RJ, Tabor E. Serologic responses in HB. In: Vyas GN, Cohen SN, Schmid R, eds. *Viral hepatitis*. Philadelphia: Franklin Institute Press; 1978:219–244.
150. Horwich AL, Furtak K, Pugh J, Summers J. Synthesis of hepadnavirus particles that contain replication-defective duck hepatitis B virus genomes in cultured HuH7 cells. *J Virol* 1990;64:642–650.
151. Hosoda K, Omata M, Uchiumi K, Imazeki F, Yokosuka O, Ito Y, Okuda K, Ohto M. Extrahepatic replication of duck hepatitis B virus: more than expected. *Hepatology* 1990;11:44–48.
152. Hruska JF, Robinson WS. The proteins of hepatitis B Dane particle cores. *J Med Virol* 1977;1:119–131.
153. Hsu T, Moroy T, Etiemble J, Louise A, Trepo C, Tiollais P, Buendia M-A. Activation of c-*myc* by woodchuck hepatitis virus insertion in hepatocellular carcinoma. *Cell* 1988;55:627–635.
154. Hu K, Siddiqui A. Regulation of the hepatitis B virus gene expression by the enhancer element I. *Virology* 1991;181:721–726.
155. Hu K, Vierling J, Siddiqui A. Trans-activation of HLA-DR gene by

hepatitis B virus X gene product. *Proc Natl Acad Sci USA* 1990;87: 7140–7144
156. Huang J, Liang J. A novel hepatitis B virus (HBV) genetic element with rev response element-like properties that is essential for expression of HBV gene products. *Mol Cell Biol* 1993;13:7476–7486.
157. Huang M, Summers J. Pet, a small sequence distal to the pregenome cap site, is required for the expression of the duck hepatitis B virus pregenome. *J Virol* 1994;68:1564–1572.
158. Huang M, Summers J. Infection initiated by the RNA pregenome of a DNA virus. *J Virol* 1991;65:5435–5439.
159. Huang Z-M, Yen TSB. Hepatitis B virus RNA element that facilitates accumulation of surface gene transcripts in the cytoplasm. *J Virol* 1994; 68:3193–3199.
160. Huovila A-PJ, Eder AM, Fuller SD. Hepatitis B surface antigen assembles in a post-ER, pre-Golgi compartment. *J Cell Biol* 1992;118: 1305–1320.
161. Imazeki F, Yaginuma K, Mmata M, Okuda K, Kobayashi M, Koike K. Integrated structures of duck hepatitis B virus DNA in hepatocellular carcinoma. *J Virol* 1988;62:861–865.
162. Ishikawa T, Kuroki K, Lenhoff R, Summers J, Ganem D. Analysis of the binding of a host cell glycoprotein to the preS protein of duck hepatitis B virus. *Virology* [in press].
163. Iwasaki K, Temin HM. The efficiency of RNA 3′-end formation is determined by the distance between the cap site and the poly(A) site in spleen necrosis virus. *Genes Dev* 1990;4:2299–2307.
164. Jacks T, Varmus HE. Expression of Rous sarcoma virus pol gene by ribosomal frameshifting. *Science* 1985;230:1237–1242.
165. Jameel S, Siddiqui A. The human hepatitis B virus enhancer requires *trans*-acting cellular factor(s) for activity. *Mol Cell Biol* 1986;6: 710–715.
166. Jameel S, Siddiqui A, Maguire HF, Rao KVS. Hepatitis B virus X protein produced in *E. coli* is biologically functional. *J Virol* 1990;64: 3963–3966.
167. Jang SK, Davies MV, Kaufman RJ, Wimmer E. Initiation of protein synthesis by internal entry of ribosomes into the 5′ nontranslated region of encephalomyocarditis virus RNA in vivo. *J Virol* 1989;63: 1651–1660.
168. Jean-Jean O, Levrero M, Will H, Perricaudet M, Rossignol J-M. Expression mechanism of the hepatitis B virus (HBV) C gene and biosynthesis of HBe antigen. *Virology* 1989;170:99–106.
169. Jean-Jean O, Salhi S, Carlier D, Elie C, DeRecondo A-M, Rossignol J-M. Biosynthesis of hepatitis B virus e antigen: directed mutagenesis of the putative aspartyl protease site. *J Virol* 1989;63:5497–5500.
170. Jean-Jean O, Weimer T, de Recondo AM, Will H, Rossignol JM. Internal entry of ribosomes and ribosomal scanning involved in hepatitis B virus P gene expression. *J Virol* 1989b;63:5451–5454.
171. Jilbert AR, Freiman JS, Gowans EJ, Holmes M, Cossart YE, Burrell CJ. Duck hepatitis B virus in liver, spleen, and pancreas: analysis by *in situ* and southern blot hybridization. *Virology* 1987;158:330–338.
172. Jilbert AR, Wu T-T, England JM, Hall PM, Carp NZ, O'Connell AP, Mason WS. Rapid resolution of duck hepatitis B virus infections occurs after massive hepatocellular involvement. *J Virol* 1992;66:1377–1388.
173. Junker-Niepmann M, Bartenschlager R, Schaller H. A short *cis*-acting sequence is required for hepatitis B virus pregenome encapsidation and sufficient for packaging of foreign RNA. *EMBO J* 1990;9: 3389–3396.
174. Junker-Niepmann M, Galle P, Schaller H. Expression and replication of the hepatitis B virus genome under foreign promoter control. *Nucleic Acids Res* 1987;15:10117–10132.
175. Kamimura T, Yoshikawa A, Ichida F, Sasaki H. Electron microscopic studies of Dane particles in hepatocytes with special references to intracellular development of Dane particles and their relation with the HBeAg in the serum. *Hepatology* 1981;1:392–397.
176. Kaplan PM, Greenman RL, Gerin JL, Purcell RH, Robinson WS. DNA polymerase associated with human hepatitis B antigen. *J Virol* 1973; 12:995–1005.
177. Karpen S, Banerjee R, Zelent A, Price P, Acs G. Identification of protein-binding sites in the hepatitis B virus enhancer and core promoter domains. *Mol Cell Biol* 1988;8:5159–5165.
178. Katayama K, Hayashi N, Sasaki Y, Kasahara A, Ueda K, Fusamoto H, Sato N, Chisaka O, Matsubara K, Takenobu K. Detection of hepatitis B virus X gene protein and antibody in type B chronic liver disease. *Gastroenterology* 1989;97:990–998.
179. Kekulé AS, Lauer U, Weiss L, Luber B, Hofschneider P. Hepatitis B virus transactivator HBx uses a tumor promoter signalling pathway. *Nature* 1993;361:742–745.
180. Kelly R. Pathways of protein secretion in eukaryotes. *Science* 1985; 230:25–31.
181. Kim C-H, Koike K, Saito I, Miyamura T, Jay G. HBx gene of hepatitis B virus induces liver cancer in transgenic mice. *Nature* 1991;353: 317–320.
182. Kim CY, Tilles JG. Purification and biophysical characterization of hepatitis B antigen. *J Clin Invest* 1970;52:1176.
183. Klingmüller U, Schaller H. Hepadnavirus infection requires interaction between the viral pre-S domain and a specific hepatocellular receptor. *J Virol* 1993;67:7414–7422.
184. Knaus T, Nassal M. The encapsidation signal on the hepatitis B virus RNA pregenome forms a stem-loop structure that is critical for its function. *Nucleic Acids Res* 1993;21:3967–3965.
185. Köck J, Schilicht H-J. Analysis of the earliest steps of hepadnavirus replication genome repair after infectious entry into hepatocytes does not depend on viral polymerase activity. *J Virol* 1993;67:4867–4874.
186. Kodama K, Ogasawara N, Yoshikawa H, Murakami S. Nucleotide sequence of a cloned woodchuck hepatitis virus genome: evolutional relationship between hepadnaviruses. *J Virol* 1985;56:978–986.
187. Koff RS, Galambos JT. Viral hepatitis. In: Schiff L, Schiff ER, eds. *Diseases of the liver*. Philadelphia: JB Lippincott; 1987:457–581.
188. Koike K, Moriya K, Iino S, Yotsuyanagi H, Endo Y, Miyamura T, Kurokawa K. High-level expression of hepatitis B virus HBx gene and hepatocarcinogenesis in transgenic mice. *Hepatology* 1994;19: 810–819.
189. Komai K, Kaplan M, Peeples ME. The Vero cell receptor for the hepatitis B virus small S protein is a sialoglycoprotein. *Virology* 1988;163: 629–634.
190. Komai K, Peeples ME. Physiology and function of the Vero cell receptor for the hepatitis B virus small S protein. *Virology* 1990;177: 332–338.
191. Korba BE, Gowans EJ, Wells FV, Tennant BC, Clarke R, Gerin JL. Systemic distribution of woodchuck hepatitis virus in the tissues of experimentally infected woodchucks. *Virology* 1988;165:172–181.
192. Korba BE, Wells F, Tennant B, Yoakum G, Purcell R, Gerin J. Hepadnavirus infection of peripheral blood lymphocytes in vivo: woodchuck and chimpanzee models of viral hepatitis. *J Virol* 1986;58:1–8.
193. Korba BE, Wells F, Tennant BC, Cote PJ, Gerin JL. Lymphoid cells in the spleens of woodchuck hepatitis virus-infected woodchucks are a site of active viral replication. *J Virol* 1987;61:1318–1324.
194. Kozak M. The scanning model for translation: an update. *J Cell Biol* 1989;108:229–241.
195. Kuroki K, Cheung R, Marion P, Ganem D. A cell surface protein that binds avian hepatitis B virus particles. *J Virol* 1994;68:2091–2096.
196. Kuroki K, Floreani M, Mimms L, Ganem D. Epitope mapping of the pre S1 domain of the hepatitis B virus large surface antigen. *Virology* 1990;176:620–624.
197. Kuroki K, Russnak R, Ganem D. A novel N-terminal amino acid sequence required for retention of a hepatitis B virus glycoprotein in the endoplasmic reticulum. *Mol Cell Biol* 1989;9:4459–4466.
198. Kwee L, Lucito R, Aufiero B, Schneider RJ. Alternate translation initiation on hepatitis B virus X mRNA produces multiple polypeptides that differentially transactivate class II and III promoters. *J Virol* 1992; 66:4382–4389.
199. Lambert V, Chassot S, Kay A, Trepo C, Cova L. *In vivo* neutralization of duck hepatitis B virus by antibodies specific to the N-terminal portion of pre-S protein. *Virology* 1991;185:446–450.
200. Lambert V, Cova L, Chevallier P, Mehrotra R, Trepo C. Natural and experimental infection of wild mallard ducks with duck hepatitis B virus. *J Gen Virol* 1991;72:417–420.
201. Lambert V, Fernholz D, Sprengel R, Fourel I, Deleange G, Wildner G, Peyret C, Trepo C, Cova L, Will H. Virus-neutralizing monoclonal antibody to a conserved epitope on the duck hepatitis B virus pre-s protein. *J Virol* 1990;64:1290–1297.
202. Laub O, Rall LB, Truett M, Shaul Y, Standring DN, Valenzuela P, Rutter WJ. Synthesis of hepatitis B surface antigen in mammalian cells: expression of the entire gene and the coding region. *J Virol* 1983;48: 271–280.
203. Lauer U, Weiss L, Hofschneider PH, Kekule AS. The hepatitis B virus pre-S/S t transactivator is generated by 3′ truncations within a defined region of the S gene. *J Virol* 1992;66:5284–5289.

204. Lauer U, Weiss L, Lipp M, Hofschneider P, Kekule A. The hepatitis B virus preS2/S-t transactivator utilizes AP-1 and other transcription factors for transactivation. *Hepatology* 1994;19:23–31.
205. Laure F, Zagury D, Saimot AG, Gallo RC, Hahn BH, Brechot C. Hepatitis B virus DNA sequences in lymphoid cells from patients with AIDS and AIDS-related complex. *Science* 1985;229:561–563.
206. Lavine J, Hirsch R, Ganem D. A system for studying the selective encapsidation of hepadnaviral RNA. *J Virol* 1989;63:4257–4263.
207. Lee AKY, Ip HMH, Wong VCW. Mechanisms of maternal-fetal transmission of hepatitis B virus. *J Infect Dis* 1978;138:688–671.
208. Lee T-H, Finegold MJ, Shen R-F, DeMayo JL, Woo SLC, Butel JS. Hepatitis B virus transactivator X protein is not tumorigenic in transgenic mice. *J Virol* 1990;64:5939–5947.
209. Lenhoff R, Summers J. Coordinate regulation of replication and virus assembly by the large envelope protein of an avian hepadnavirus. *J Virol* [in press].
210. Levrero M, Jean-Jean O, Balsano C, Will H, Perricaudet M. Hepatitis B virus X gene expression in human cells and anti-HBx antibody detection in chronic HBV infection. *Virology* 1990;174:299–304.
211. Li J-S, Cova L, Buckland R, Lambert V, Deleage G, Trepo C. Duck hepatitis B virus can tolerate insertion, deletion, and partial frameshift mutation in the distal pre-S region. *J Virol* 1989;63:4965–4968.
212. Lieber CS, Garro A, Leo MA, Mak KM, Worner T. Alcohol and cancer. *Hepatology* 1986;6:1005–1009.
213. Lie-Injo LE, Balasegaram M, Lopez CG, et al. Hepatitis B virus DNA in liver and white blood cells of patients with hepatoma. *DNA* 1983;2: 301.
214. Lien JM, Aldrich CE, Mason WS. Evidence that a capped oligoribonucleotide is the primer for duck hepatitis B virus plus-strand DNA synthesis. *J Virol* 1986;57:229–236.
215. Lien JM, Petcu D, Aldrich C, Mason W. Initiation and termination of duck hepatitis B virus DNA synthesis during virus maturation. *J Virol* 1987;61:3832–3840.
216. Lin C-G, Lo SJ. Evidence for involvement of a ribosomal leaky scanning mechanism in the translation of the hepatitis B virus *pol* gene from the viral pregenome RNA. *Virology* 1992;188:342–352.
217. Lingappa J, Martin R, Wong M-L, Ganem D, Welch W, Lingappa V. A eukaryotic cytosolic chaperonin is associated with a high molecular weight intermediate in the assembly of hepatitis B virus capsid, a multimeric particle. *J Cell Biol* 1994;125:99–111.
218. Liu C, Condreay L, Burch J, Mason W. Characterization of the core promoter and enhancer of duck hepatitis B virus. *Virology* 1991;184: 242–252.
219. Liu C, Mason W, Burch J. Identification of factor-binding sites in the duck hepatitis B virus enhancer and in vivo effects of enhancer mutations. *J Virol* 1994;68:2286–2296.
220. Liu CC, Yansura D, Levinson A. Direct expression of hepatitis B surface antigen in monkey cells from an SV40 vector. *DNA* 1982;1: 213–221.
221. Loeb DD, Hirsch RC, Ganem D. Sequence-independent RNA cleavages generate the primers for plus strand DNA synthesis in hepatitis B viruses: implications for other reverse transcribing elements. *EMBO J* 1991;10:3533–3540.
221a. Loeb D, Ganem D. Reverse transcription pathway of the hepatitis B virus. In: Skalpa A, Goff S, eds. *Reverse trascriptase.* Cold Spring Harbor, NY: Cold Spring Harbor Press; 1995:329–355.
222. Lopez-Cabrera M, Letovsky J, Hu K-Q, Siddiqui A. Transcriptional factor C/EBP binds to and transactivates the enhancer element II of the hepatitis B virus. *Virology* 1991;183:825–829.
223. Lopez-Cabrera M, Letovsky J, Hu K-Q, Siddiqui A. Multiple liver-specific factors bind to the hepatitis B virus core/pregenomic promoter: trans-activation and repression by CCAAT/enhancer binding protein. *Proc Natl Acad Sci USA* 1990;87:5069–5073.
224. Luber B, Burgelt E, Fromental C, Kanno M, Koch W. Multiple simian virus 40 enhancer elements mediate the *trans*-activating function of the X protein of hepatitis B virus. *Virology* 1991;184:808–813.
225. Luber B, Lauer U, Weiss L, Hohne M, Hofschneider P, Kekule A. The hepatitis B virus transactivator HBx causes elevation of diacylglycerol and activation of protein kinase C. *Res Virol* 1993;144:311–321.
226. Lucito R, Schneider RJ. Hepatitis B virus X protein activates transcription factor NF-KB without a requirement for protein kinase C. *J Virol* 1992;66:983–991.
227. Machida A, Kishimoto S, Ohnuma H, Baba K, Ito Y, Miyamoto H, Funatsu G, Oda K. A polypeptide containing 55 amino acid residues coded by the pre-S region of hepatitis B virus deoxyribonucleic bears the receptor for polymerized human as well as chimpanzee albumins. *Gastroenterology* 1984;86:910–918.
228. Machida, A., S. Kishimoto, H. Ohumura, H. Miyamoto, K. Baba, K. Oda, T. Nakamura, and Y. Miyakawa. 1983. A hepatitis B surface antigen poly–peptide (P31) with the receptor for polymerized human as well as chimpanzee albumins. *Gastroenterology* 85:268–274.
229. Machida A, Ohnuma H, Tsuda F, Yoshikawa A, Hoshi Y, Tanaka T, Kishimoto S, Akahane Y, Miyakawa Y, Mayumi M. Phosphorylation in the carboxyl-terminal domain of the capsid protein of hepatitis B virus: evaluation with a monoclonal antibody. *J Virol* 1991;65:6024–6030.
230. Mack DH, Bloch W, Nath N, Sninsky JJ. Hepatitis B virus particles contain a polypeptide encoded by the largest open reading frame: a putative reverse transcriptase. *J Virol* 1988;62:4786–4790.
231. Macrae D, Bruss V, Ganem D. Myristylation of duck hepatitis B virus envelope protein is essential for infectivity but not for virus assembly. *Virology* 1991;181:359–363.
232. Maddon PJ, Dalgleish J, McDougal AG, Clapham JS, Weiss PR, Axel R. The T4 gene encodes the AIDS virus receptor and is expressed in the immune system and the brain. *Cell* 1986;47:333–348.
233. Maddon PJ, McDougal JS, Clapham PR, Dalgleish AG, Jamal S, Weiss R, Axel R. HIV infection does not require endocytosis of its receptor, CD4. *Cell* 1989;54:865–874.
234. Maguire H, Hoeffler J, Siddiqui A. HBV X protein alters the DNA binding specificity of CREB and ATF-2 by protein-protein interactions. *Science* 1991;252:842–844.
235. Mandart E, Kay A, Galibert F. Nucleotide sequence of a cloned duck hepatitis B virus genome: comparison with woodchuck and human hepatitis B virus sequences. *J Virol* 1984;49:782–792.
236. Marion PL. Use of animal models to study hepatitis B viruses. *Prog Med Virol* 1988;35:43–75.
237. Marion PL, Cullen JM, Azcarraga RR, Van Davelar MJ,Robinson SW. Experimental transmission of duck hepatitis B virus to Pekin ducks and to domestic geese. *Hepatology* 1987;7:724–731.
238. Marion PL, Knight SS, Ho BK, Guo YY, Robinson WS, Popper H. Liver disease associated with duck hepatitis B virus infection of domestic ducks. *Proc Natl Acad Sci USA* 1984;81:898–902.
239. Marion PL, Knight SS, Salazar FH, Popper H, Robinson WS. Ground squirrel hepatitis virus infection. *Hepatology* 1983;3:519–527.
240. Marion PL, Oshiro L, Regnery DC, Scullard GH, Robinson WS. A virus in Beechey ground squirrels that is related to hepatitis B virus of man. *Proc Natl Acad Sci USA* 1980;77:2941–2945.
241. Marion PL, Van Davelaar MJ, Knight SS, Salazar FH, Garcia G, Popper H, Robinson WS. Hepatocellular carcinoma in ground squirrels persistently infected with ground squirrel hepatitis virus. *Proc Natl Acad Sci USA* 1986;83:4543–4546.
242. Mason W, Halpern M, England J, Seal G, Egan J, Coates L, Aldrich C, Summers J. Experimental transmission of duck hepatitis B virus. *Virology* 1983;131:375–384.
243. Mason WS, Aldrich C, Summers J, Taylor JM. Asymmetric replication of duck hepatitis B virus DNA in liver cells: free minus strand DNA. *Proc Natl Acad Sci USA* 1982;79:3007–4001.
244. Mason WS, Seal G, Summers J. Virus of Pekin ducks with structural and biological relatedness to human hepatitis B virus. *J Virol* 1980;36: 829–836.
245. McAller WJ, Buynak EB, Maigetter RZ, Wampler DE, Miller WJ, Hilleman MR. Human hepatitis B vaccine from recombinant genes. *Nature* 1984;307:178–180.
246. McLachlan A, Milich DR, Raney AK, Riggs MG, Hughes JL, Sorge J, Chisari FV. Expression of hepatitis B virus surface and core antigens: influences of pre-S and precore sequences. *J Virol* 1987;61: 683–692.
247. Mehdi H, Kaplan M, Anlar F, Yang X, Bayer R, Sutherland K, Peeples M. Hepatitis B virus surface antigen binds apolipoprotein H. *J Virol* 1994;68:2415–2424.
248. Milich DR, Jones JE, Hughes JL, Price J, Rnaey AK, McLachlan A. Is a function of the secreted hepatitis B e antigen to induce immunologic tolerance in utero? *Proc Natl Acad Sci USA* 1990;87:6599–6603.
249. Miller RH, Marion PL, Robinson WS. Hepatitis B virus DNA-RNA hybrid molecules in particles from infected liver are converted to viral DNA molecules during an endogenous DNA polymerase reaction. *Virology* 1984;139:64–72.
250. Miyamura T, Saito I, Yoneyama T, Takeuchi K, Ohbayashi A, Watanabe Y, Choo Q, Houghton M, Kuo G. Role of hepatitis C virus in hepatocellular carcinoma. In: Hollinger FB, Lemon S, Margolis H, eds.

Viral hepatitis and liver disease. 1991:559–562.

251. Miyanohara A, Imamura T, Araki M, Sugawara K, Ohtomo N, Matsubara K. Expression of hepatitis B virus core antigen gene in *Saccharomyces cerevisiae:* synthesis of two polypeptides translated from different initiation codons. *J Virol* 1986;59:176–180.
252. Molnar-Kimber KL, Summers J, Mason WS. Mapping of the cohesive overlap of duck hepatitis B virus DNA and of the site of initiation of reverse transcriptions. *J Virol* 1984;51:181–191.
253. Molnar-Kimber KL, Summers J, Taylor JM, Mason WS. Protein covalently bound to minus-strand DNA intermediates on duck hepatitis B virus. *J Virol* 1983;51:181–191.
254. Mondelli M, Bortolotti F, Pontisso P, Rondanelli E, Williams R, Realdi G, Alberti A, Eddleston A. Definition of hepatitis B virus-specific target antigens recognized by cytotoxic T cells in acute HBV infection. *Clin Exp Immunol* 1987;68:242–250.
255. Mondelli M, Vergani G, Alberti A, Vergani D, Portmann B, Eddleston A, Williams R. Specificity of T lymphocyte cytotoxicity to autologous hepatocytes in chronic hepatitis B virus infection: evidence that T cells are directed against HBV core antigen expressed on hepatocytes. *J Immunol* 1982;129:2773–2777.
256. Moriarty AM, Hoyer BH, Shih JWK, Gerin JL, Hamer DH. Expression of the hepatitis B virus surface antigen gene in cell culture by using a simian virus 40 vector. *Proc Natl Acad Sci USA* 1981;78:2606–2610.
257. Moroy T, Etiemble J, Trepo C, Tiollais P, Buendia M-A. Transcription of woodchuck hepatitis virus in the chronically infected liver. *EMBO J* 1985;4:1507–1514.
258. Moroy T, Marchio A, Etiemble J, Trepo C, Tiollais P, Buendia M-A. Rearrangement and enhanced expression of c-*myc* in hepatocellular carcinoma of hepatitis virus infected woodchucks. *Nature* 1986;324: 276–279.
259. Nagaya T, Nakamura T, Tokino T, Tsurimoto T, Imai M, Mayumi T, Kamino K, Yamamura K, Matsubara K. The mode of hepatitis B virus DNA integration in chromosomes of human hepatocellular carcinoma. *Genes Dev* 1987;1:773–782.
260. Nakatake H, Chisaka O, Yamamoto S, Matsubara K, Koshy R. Effect of X protein on transactivation of hepatitis B virus promoters and on viral replication. *Virology* 1993;195:305–314.
261. Nassal M. The arginine-rich domain of the hepatitis B virus core protein is required for pregenome encapsidation and productive viral positive-strand DNA synthesis but not for virus assembly. *J Virol* 1992; 66:4107–4116.
262. Nassal M, Galle PR, Schaller H. Proteaselike sequence in hepatitis B virus core antigen is not required for e antigen generation and may not be part of an aspartic acid-type protease. *J Virol* 1989;63:2598–2604.
263. Nassal M, Junker-Niepmann M, Schaller H. Translational inactivation of RNA function: discrimination against a subset of genomic transcripts during HBV nucleocapsid assembly. *Cell* 1990;63:1357–1363.
264. Nassal M, Rieger A. An intramolecular disulfide bridge between Cys-7 and Cys-61 determines the structure of the secretory core gene product (e antigen) of hepatitis B virus. *J Virol* 1993;67:4307–4315.
265. Natoli G, Avantaggiati ML, Balsano C, DeMarzio E, Collepardo D, Elfassi E, Levrero M. Characterization of the hepatitis B virus preS/S region encoded transcriptional transactivator. *Virology* 1992;187: 663–670.
266. Natoli G, Avantaggiati M, Chirillo P, Costanzo A, Artini M, Balsano C, Levrero M. Induction of the DNA binding activity of c-*jun*/c-*fos* heterodimers by the hepatitis B virus transactivator pX. *Mol Cell Biol* 1994;14:989–998.
267. Neurath AR, Kent SBH, Strick N, Parker K. Identification and chemical synthesis of a host cell receptor binding site on hepatitis B virus. *Cell* 1986;46:429–436.
268. Neurath AR, Strick N, Li Y. Cells transfected with human IL6 cDNA acquire binding sites for the hepatitis B virus envelope protein. *J Exp Med* 1992;176:1561–1569
269. Neurath AR, Strick N, Sproul P, Ralph HE, Valinsky J. Detection of receptors for hepatitis B virus on cells of extrahepatic origin. *Virology* 1990;176:448–457.
270. Obata H, Hayashi N, Motoike Y. A prospective study of development of hepatocellular carcinoma from liver cirrhosis with persistent hepatitis B virus infection. *Int J Cancer* 1980;25:741.
271. Oberhaus SM, Newbold JE. Detection of DNA polymerase activities associated with purified duck hepatitis B virus core particles by using an activity gel assay. *J Virol* 1993;67:6558–6566.
272. Ochiya T, Tsurimoto T, Ueda K, Okubo K, Shiozawa M, Matsubara K. An in vitro system for infection with hepatitis B virus that uses primary human fetal hepatocytes. *Proc Natl Acad Sci USA* 1989;86:1875–1879.
273. Offensperger W-B, Offensperger S, Walter E, Blum HE, Gerok W. Inhibition of duck hepatitis B virus infection by lysosomotropic agents. *Virology* 1991;183:415–418.
274. Ogata N, Miller RH, Ishak KG, Purcell RH. The complete nucleotide sequence of a pre-core mutant of hepatitis B virus implicated in fulminant hepatitis and its biological characterization in chimpanzees. *Virology* 1993;194:263–276.
275. Ogston CW, Schechter EM, Humes CA, Pranikoff MB. Extrahepatic replication of woodchuck hepatitis virus in chronic infection. *Virology* 1989;169:9–14.
276. Ogston W, Jonak G, Rogler C, Astrin S, Summers J. Cloning and structural analysis of integrated woodchuck hepatitis virus sequences from hepatocellular carcinomas of woodchucks. *Cell* 1982;29:385–394.
277. Okada K, Kamiyama I, Inomata M, Imai M, Miyakawa Y, Mayumi M. E antigen and anti-e in the serum of asymptomatic carrier mothers as indicators of positive and negative transmission of hepatitis B virus to their infants. *N Engl J Med* 1976;294:746–749.
278. Okamoto H, Yotsumoto S, Akahane Y, Yamanaka T, Miyazaki Y, Sugai Y, Tsuda F, Tanaka T, Miyakawa Y, Mayumi M. Hepatitis B viruses with precore region defects prevail in persistently infected hosts along with seroconversion to the antibody against e antigen. *J Virol* 1990;64:1298–1303.
279. Omata M, Uchiumi K, Ito Y, Yokosuka O, Mori J, Terao K, et al. Duck hepatitis B virus and liver diseases. *Gastroenterology* 1983;85:260–267.
280. Ostapchuck P, Hearing P, Ganem D. A dramatic shift in the transmembrane topology of a viral envelope glycoprotein accompanies hepatitis B viral morphogenesis. *EMBO J* 1994;13:1048–1057.
281. Ostapchuk P, Scheirle G, Hearing P. Binding of nuclear factor EF-C to a functional domain of the hepatitis B virus enhancer region. *Mol Cell Biol* 1989;9:2787–2797.
282. Ou J, Laub O, Rutter WJ. Hepatitis B gene function: the precore region targets core protein to the endoplasmic reticulum. *Proc Natl Acad Sci USA* 1986;83:1578–1582.
283. Ou J-H, Rutter WJ. Regulation of secretion of the hepatitis B virus major surface antigen by the preS-1 protein. *J Virol* 1987;61:782–786.
284. Ou J, Yeh C, Yen TSB. Transport of hepatitis B virus precore protein into the nucleus after cleavage of its signal peptide. *J Virol* 1989;63: 5238–5243.
285. Pasek M, Goto T, Gilbert W, Zink B, Schaller H, Mackay P, Ledbetter G, Murray K. Hepatitis B virus genes and their expression in *E. coli. Nature* 1978;282:575–579.
286. Patzer EJ, Nakamura GR, Simonsen GC, Levinson AD, Brands R. Intracellular assembly and packaging of hepatitis B surface antigen particles occur in the endoplasmic reticulum. *J Virol* 1986;58:884–892.
287. Patzer EJ, Nakamura GR, Yaffe A. Intracellular transport and secretion of hepatitis B surface antigen in mammalian cells. *J Virol* 1984; 51:346–353.
288. Pelletier J, Sonenberg N. Internal initiation of translation of eukaryotic mRNA directed by a sequence derived from poliovirus RNA. *Nature* 1988;334:320–325.
289. Penna A, Chisari F, Bertoletti A, Missale G, Fowler P, Giuberti T, Fiaccadori F, Ferrari C. Cytotoxic T lymphocytes recognize an HLA A2-restricted epitope within the hepatitis B virus nucleocapsid antigen. *J Exp Med* 1991;174:1565–1570.
290. Penna A, Fowler P, Bertoletti A, Guilhot S, Moss B, Margolskee RF, Cavalli A, Valli A, Fiaccadori F, Chisari FV, Ferrari C. Hepatitis B virus (HBV)-specific cytotoxic T-cell (CTL) response in humans: characterization of HLA class II-restricted CTLs that recognize endogenously synthesized HBV envelope antigens. *J Virol* 1992;66:1193–1198.
291. Persing DH, Varmus HE, Ganem D. The preS1 protein of hepatitis B virus is acylated at its amino terminus with myristic acid. *J Virol* 1987; 61:1672–1677.
292. Persing D, Varmus H, Ganem D. Inhibition of secretion of hepatitis B surface antigen by a related presurface polypeptide. *Science* 1986;234: 1388–1392.
293. Persing DH, Varmus HE, Ganem D. A frameshift mutation in the pre-S region of the human hepatitis B virus genome allows production of surface antigen particles but eliminates binding to polymerized albumin. *Proc Natl Acad Sci USA* 1985;82:3440–3444.
294. Petcu DJ, Aldrich CE, Coates L, Taylor JM, Mason WS. Suramin inhibits in vitro infection by duck hepatitis B virus, Rous sarcoma virus, and hepatitis delta virus. *Virology* 1988;167:385–392.
295. Peters RL. Viral hepatitis: a pathologic spectrum. *Am J Med Sci* 1975;

270:17.
296. Peterson D. The structure of hepatitis B surface antigen and its antigenic sites. *Bioessays* 1987;6:258–262.
297. Peterson DL. Isolation and characterization of the major protein and glycoprotein of hepatitis B surface antigen. *J Biol Chem* 1981;256: 6975–6983.
298. Petit M, Capel F, Dubanchet S, Mabit H. PreS1-specific binding proteins as potential receptors for hepatitis B virus in human hepatocytes. *Virology* 1992;187:211–222.
299. Petit M-A, Capel F, Dubanchet S, Mabit H. PreS1-specific binding proteins as potential receptors for hepatitis B virus in human hepatocytes. *Virology* 1992;187:211–222.
300. Petit MA, Dubanchet S, Capel F, Voet P, Dauguet C, Hauser P. HepG2 cell binding activities of different hepatitis B virus isolates: inhibitory effect of anti-HBs and anti-preS1 (21-47). *Virology* 1991;180:483–491.
301. Phillipson L, Lonborg-Holm K, Pettersson U. Virus-receptor interaction in an adenovirus system. *J Virol* 1968;2:1064–1075.
302. Pohl CJ, Cote PJ, Purcell RH, Gerin JL. Failure to detect polyalbumin-binding sites on the woodchuck hepatitis virus surface antigen: implications for the pathogenesis of hepatitis B virus in humans. *J Virol* 1986;60:943–949.
303. Pollack J, Ganem D. Site-specific RNA binding by a hepatitis B virus reverse transcriptase initiates two reactions: RNA packaging and DNA synthesis. *J Virol* [in press].
304. Pollack J, Ganem D. An RNA stem-loop structure directs hepatitis B virus genomic RNA encapsidation. *J Virol* 1993;67:3254–3263.
305. Pontisso P, Petit MA, Vankowski M, Peeples M. Human liver plasma membrane contains receptors for the hepatitis B virus pre S1 region and, via polymerized human serum albumin, for the pre S2 region. *J Virol* 1989;63:1981–1988.
306. Pontisso P, Poon MC, Tiollais P, Brechot C. Detection of hepatitis B virus DNA in mononuclear blood cells. *BMJ* 1984;288:1563–1566.
307. Pontisso P, Ruoletto M, Tiribelli C, Gerlich W, Ruol A, Alberti A. The preS1 domain of hepatitis B virus and IgA cross-react in their binding to the hepatocyte surface. *J Gen Virol* 1992;73:2041–2045.
308. Popper H, Roth L, Purcell RH, Tennant BC, Gerin JL. Hepatocarcinogenicity of the woodchuck hepatitis virus. *Proc Natl Acad Sci USA* 1987;84:866–870.
309. Popper H, Shih JW-K, Gerin JL, Wong DC, Hoyer BH, London WT, Sly DL, Purcell RH. Woodchuck hepatitis and hepatocellular carcinoma: correlation of histologic with virologic observations. *Hepatology* 1981;1:91–98.
310. Prange R, Clemen A, Streeck RF. Myristylation is involved in intracellular retention of hepatitis B virus envelope proteins. *J Virol* 1991; 65:3919–3923.
311. Prange R, Nagel N, Streeck RE. Deletions in the hepatitis B virus small envelope protein: effect on assembly and secretion of surface antigen particles. *J Virol* 1992;66:5832–5841.
312. Pugh J, Simmons H. Duck hepatitis B virus infection of Muscovy duck hepatocytes and nature of virus resistance in vivo. *J Virol* 1994;68: 2487–2494.
313. Pugh J, Yaginuma K, Koike K, Summers J. HBV particles produced by transient expression of DHBV DNA in a human hepatoma cell line are infections *in vitro*. *J Virol* 1988;62:3513–3516.
314. Pugh J, Zweidler A, Summers J. Characterization of the major duck hepatitis B virus core particle protein. *J Virol* 1989;63:1371–1376.
315. Pugh JC, Sninsky JJ, Summers JW, Schaeffer E. Characterization of a pre-S polypeptide on the surfaces of infectious avian hepadnavirus particles. *J Virol* 1987;61:1384–1390.
316. Pugh JC, Summers JW. Infection and uptake of duck hepatitis B virus by duck hepatocytes maintained in the presence of dimethyl sulfoxide. *Virology* 1989;172:564–572.
317. Pugh JC, Yaginuma K, Koike K, Summers J. Duck hepatitis B virus (DHBV) particles produced by transient expression of DHBV DNA in a human hepatoma cell line are infectious is vitro. *J Virol* 1988;62: 3513–3516.
318. Qiao M, Gowans EJ, Burrell CJ. Intracellular factors, but not virus receptor levels, influence the age-related outcome of DHBV infection of ducks. *Virology* 1992;186:517–523.
319. Qiao M, MacNaughton T, Gowans E. Adsorption and penetration of hepatitis B virus in a nonpermissive cell line. *Virology* 1994;201: 356–363.
320. Radziwill G, Tucker W, Schaller H. Mutational analysis of the hepatitis B virus P gene product: domain structure and Rnase H activity. *J Virol* 1990;64:613–620.
321. Radziwill G, Zentgraf H, Schaller H, Bosch V. The duck hepatitis B virus DNA polymerase is tightly associated with the viral core structure and unable to switch to an exogenous template. *Virology* 1988; 163:123–132.
322. Raney A, Milich D, Easton A, McLachlan A. Differentiation-specific transcriptional regulation of the hepatitis B virus large surface antigen gene in human hepatoma cell lines. *J Virol* 1990;64:2360–2368.
323. Raney A, Milich D, McLachlan A. Characterization of the major surface antigen gene transcriptional regulatory elements in differentiated hepatoma cell lines. *J Virol* 1989;63:3919–3925.
324. Raney AK, Easton AJ, Milich DR, McLachlan A. Promoter-specific transactivation of hepatitis B virus transcription by a glutamine-and proline-rich domain of hepatocyte nuclear factor 1. *J Virol* 1991;65: 5774–5781.
325. Raney AK, Le HB, McLachlan A. Regulation of transcription from the hepatitis B virus major surface antigen promoter by the Sp1 transcription factor. *J Virol* 1992;66:6912–6921.
326. Raney AK, Milich DR, McLachlan A. Complex regulation of transcription from the hepatitis B virus major surface antigen promoter in human hepatoma cell lines. *J Virol* 1991;65:4805–4811.
327. Redeker AG. Viral hepatitis: clinical aspects. *Am J Med Sci* 1975;270: 9–16.
328. Rigg RJ, Schaller H. Duck hepatitis B virus infection of hepatocytes is not dependent on low pH. *J Virol* 1992;66:2829–2836.
329. Rijntjes PJM, Moshage HJ, Yap SH. *In vitro* infection of primary cultures of cryopreserved adult human hepatocytes with hepatitis B virus.*Virus Res* 1988;10:95–110.
330. Robinson WS. The genome of hepatitis B virus. *Annu Rev Microbiol* 1977;31:357–377.
331. Robinson WS, Clayton DA, Greenman RL. DNA of a human hepatitis B virus candidate. *J Virol* 1974;14:384–391.
332. Robinson WS, Greenman R. DNA polymerase in the core of the human hepatitis B virus candidate. *J Virol* 1974;13:1231–1236.
333. Robinson WS, Lutwick LI. The virus of hepatitis type B. *N Engl J Med* 1976;295:1168–1175.
334. Rogler C, Sherman M, Yu CJ, Shafritz D. Deletion in chromosome 11p associated with a hepatitis B integration site in hepatocellular carcinoma. *Science* 1985;230:319–322.
335. Roingeard P, Lu S, Sureau C, Freschlin M, Arbeille B, Essex M, Romet-Lemonne J. Immunocytochemical and electron microscopic study of hepatitis B virus antigen and complete particle production in hepatitis B virus DNA transfected HepG2 cells. *Hepatology* 1990;11:277–285.
336. Romet-Lemonne JL, McLane MF, Elfassi E, et al. Hepatitis B virus infection in cultured human lymphoblastoid cells. *Science* 1983;21:667.
337. Roossinck MJ, Jameel S, Loukin SH, Siddiqui A. Expression of hepatitis B viral core region in mammalian cells. *Mol Cell Biol* 1986;6: 1393–1400.
338. Roossinck MJ, Siddiqui A. In vivo phosphorylation and protein analysis of hepatitis B virus core antigen. *J Virol* 1987;61:955–961.
339. Rossner M. Review: hepatitis B virus X gene product: a promiscuous transcriptional activator. *J Med Virol* 1992;36:101–117.
340. Roychoudhury A, Faruqui F, Shih C. Pregenomic RNA encapsidation analysis of eleven missense and nonsense polymerase mutants of human hepatitis B virus. *J Virol* 1991;65:3617–3624.
341. Roychoudhury S, Shih C. *Cis* rescue of a mutated reverse transcriptase gene of human hepatitis B virus by creation of an internal ATG. *J Virol* 1990;64:1063–1069.
342. Russnak R, Ganem D. Sequences 5′ to the poly A signal mediate differential poly A site use in hepatitis B viruses. *Genes Dev* 1990;4: 764–776.
343. Santantonio T, Jung M-C, Miska S, Pastore G, Pape GR, Will H. Prevalence and type of pre-C HBV mutants in anti-HBe positive carriers with chronic liver disease in a highly endemic area. *Virology* 1991;183: 840–844.
344. Santantonio T, Jung M-C, Schneider R, Fernholz D, Milella M, Monno L, Pastore G, Pape GR, Will H. Hepatitis B virus genomes that cannot synthesize pre-S2 proteins occur frequently and as dominant virus populations in chronic carriers in Italy. *Virology* 1992;188:948–952.
345. Sattler F, Robinson WS. Hepatitis B viral DNA molecules have cohesive ends. *J Virol* 1979;32:226–233.
346. Schaller H, Fischer M. Transcriptional control of hepadnavirus gene expression. *Curr Top Microbiol Immunol* 1991;168:21–39.
347. Schlicht H, Bartenschlager R, Schaller H. The duck hepatitis B virus

core protein contains a highly phosphorylated C-terminus that is essential for replication but not for RNA packaging. *J Virol* 1989;63: 2995–3000.
348. Schlicht H, Radziwill G, Schaller H. Synthesis and encapsidation of duck hepatitis B virus reverse transcriptase do not require formation of core-polymerase fusion proteins. *Cell* 1989;56:85–92.
349. Schlicht H, Salfeld J, Schaller H. The duck hepatitis B virus pre-C region encodes a signal sequence which is essential for synthesis and secretion of processed core proteins but not for virus formation. *J Virol* 1987;61:3701–3709.
350. Schlicht H, Schaller H. The secretory core protein of human hepatitis B virus is expressed on the cell surface. *J Virol* 1989;63:5399–5404.
351. Schlicht HJ, Schaller H. Analysis of hepatitis B virus gene functions in tissue culture and in vivo. *Curr Top Microbiol Immunol* 1989;144: 253–263.
352. Schlicht H-J, Wasenauer G. The quaternary structure, antigenicity, and aggregational behavior of the secretory core protein of human hepatitis B virus are determined by its signal sequence. *J Virol* 1991;65:6817–6825.
353. Schlicht HJ. Biosynthesis of the secretory core protein of duck hepatitis B virus: intracellular transport, proteolytic processing, and membrane expression of the precore protein. *J Virol* 1991;65:3489–3495.
354. Schlicht HJ, Kuhn C, Guhr B, Mattaliano RJ, Schaller H. Biochemical and immunological characterization of the duck hepatitis B virus envelope proteins. *J Virol* 1987;61:2280–2285.
355. Schneider, R., D. Fernholz, G. Wildner, and H. Will. 1991. Mechanism, kinetics, and role of duck hepatitis B virus e–antigen expression in vivo. *Virology* 182: 503–512.
356. Schneider R, Will H. Regulatory sequences of duck hepatitis B virus C gene transcription. *J Virol* 1991;65:5693–5701.
357. Schodel F, Weimer T, Will H, Sprengel R. Amino acid sequence similarity between retroviral and *E. coli* RNase H and hepadnaviral gene products. *AIDS Res Hum Retroviruses* 1988;4:9–11.
358. Schweitzer IL, Dunn AEF, Peters RL, Spears RL. Viral hepatitis in neonates and infants. *Am J Med* 1973;55:762.
359. Scullard G, Smith C, Merigan T, Robinson W, Gregory P. Effects of immunosuppressive therapy on viral markers in chronic active hepatitis B. *Gastroenterology* 1981;81:987–991.
360. Seeger C, Baldwin B, Hornbuckle WE, Yeager AE, Tennant BC, Cote P, Ferrell L, Ganem D, Varmus HE. Woodchuck hepatitis virus is a more efficient oncogenic agent than ground squirrel hepatitis virus in a common host. *J Virol* 1991;65:1673–1679.
361. Seeger C, Baldwin B, Tennant BC. Expression of infectious woodchuck hepatitis virus in murine and avian fibroblasts. *J Virol* 1989;63: 4665–4669.
362. Seeger C, Ganem D, Varmus HE. Genetic and biochemical evidence for the hepatitis B virus replication strategy. *Science* 1986;232:477–485.
363. Seeger C, Ganem D, Varmus HE. Nucleotide sequence of an infectious molecularly cloned genome of ground squirrel hepatitis virus. *J Virol* 1984;51:367–375.
364. Seeger C, Marion PL, Ganem D, Varmus HE. *In vitro* recombinants of ground squirrel and woodchuck hepatitis viral DNAs produce infectious virus in squirrels. *J Virol* 1987;61:3241–3247.
365. Seeger C, Maragos J. Identification of a signal necessary for initiation of reverse transcription of the hepadnavirus genome. *J Virol* 1991;65: 5190–5195.
366. Seeger C, Maragos J. Identification and characterization of the woodchuck hepatitis virus origin of DNA replication. *J Virol* 1990;64:16–23.
367. Seeger C, Maragos J. Molecular analysis of the function of direct repeats and a polypurine tract for plus-strand DNA priming in woodchuck hepatitis virus. *J Virol* 1989;63:1907–1915.
368. Seifer M, Zhou S, Standring DN. A micromolar pool of antigenically distinct precursors is required to initiate cooperative assembly of hepatitis B virus capsids in *Xenopus* oocytes. *J Virol* 1993;67:249–257.
369. Sells MA, Chen ML, Acs G. Production of hepatitis B virus particles in Hep G2 cells transfected with cloned hepatitis B virus DNA. *Proc Natl Acad Sci USA* 1987;84:1005–1009.
370. Seto E, Mitchell PJ, Yen TSB. Transactivation by the hepatitis B virus X protein depends on AP-2 and other transcription factors. *Nature* 1990; 344:72–74.
371. Seto E, Yen TSB, Peterlin BM, Ou J. Trans-activation of the human immunodeficiency virus long terminal repeat by the hepatitis B virus X protein. *Proc Natl Acad Sci USA* 1988;85:8286–8290.
372. Shafritz D, Shouval D, Sherman H, Hadziyannis S, Kew M. Integration of hepatitis B virus DNA into the genome of liver cell in chronic liver disease and hepatocellular carcinoma. *N Engl J Med* 1981;305: 1067–1073.
373. Shaul Y. Regulation of hepadnavirus transcription. In: McLachlan A, ed.*Molecular biology of hepatitis B viruses.* Boca Raton, FL: CRC Press; 1991:193–211.
374. Shaul Y, Rutter WJ, Laub O. A human hepatitis B viral enhancer element. *EMBO J* 1985;4:427–430.
375. Sheets M, Ogg S, Wickens MP. Point mutations in AAUAAA and the poly (A) addition site: effects on the accuracy and efficiency of cleavage and polyadenylation in vitro. *Nucleic Acids Res* 1990;18:5799–5805.
376. Shek N, Bartenschlager R, Kuhn C, Schaller H. Phosphorylation and rapid turnover of hepatitis B virus X protein expressed in HepG2 cells from a recombinant vaccinia virus. *Oncogene* 1991;6:1735–1744.
377. Shih C, Burke K, Chou MJ, Zeldis JB, Yang CS, Lee CS, Isselbacher KJ, Wands JR, Goodman HM. Tight clustering of human hepatitis B virus integration sites in hepatomas near a triple-stranded region. *J Virol* 1987;61:3491–3498.
378. Shih C, Li L-S, Roychoudhury S, Ho M-H. *In vitro* propagation of human hepatitis B virus in rat hepatoma cell line. *Proc Natl Acad Sci USA* 1989;86:6323–6327.
379. Siddiqui A, Gaynor R, Srinivasan A, Mapoles J, Farr R. Transactivation of viral enhancers including the long terminal repeat of HIV by the hepatitis B virus X protein. *Virology* 1989;169:479–484.
380. Siddiqui A, Jameel S, Mapoles J. Expression of the hepatitis B virus X gene in mammalian cells. *Proc Natl Acad Sci USA* 1987;84:2513–2517.
381. Siddiqui A, Jameel S, Mapoles J. Transcriptional control elements of hepatitis B surface antigen gene. *Proc Natl Acad Sci USA* 1986;83: 566–570.
382. Siddiqui A, Sattler FR, Robinson WS. Restriction endonuclease cleavage map and location of unique features of the DNA of hepatitis B virus subtype adw_2. *Proc Natl Acad Sci USA* 1979;76:4664–4668.
383. Siegrist CA, Durnad B, Emery P, David E, Hearing P, Mach B, Reith W. RFX1 is identical to enhancer factor C and functions as a transactivator of the hepatitis B virus enhancer. *Mol Cell Biol* 1993;13:6375–6384.
384. Simon K, Lingappa V, Ganem D. Secreted hepatitis B surface antigen polypeptides are derived from a transmembrane precursor. *J Cell Biol* 1988;107:2163–2168.
385. Sing GK, Prior S, Fernan A, Cooksley G. Hepatitis B virus differentially suppresses myelopoiesis and displays tropism for immature hematopoietic cells. *J Virol* 1993;67:3454–3460.
386. Spandau DF, Lee C. *Trans*-activation of viral enhancers by the hepatitis B virus X protein. *J Virol* 1988;62:427–434.
387. Sprengel R, Kaleta EF, Will H. Isolation and characterization of a hepatitis B virus endemic in herons. *J Virol* 1988;62:3832–3839.
388. Sprengel R, Kuhn C, Manso C, Will H. Cloned duck hepatitis B virus DNA is infectious in Pekin ducks. *J Virol* 1984;52:932–937.
389. Sprengel R, Varmus HE, Ganem D. Homologous recombination between hepadnaviral genomes following in vivo DNA transfection: implications for studies of viral infectivity. *Virology* 1987;159:454–456.
390. Standring D, Ou J, Masiarz F, Rutter WJ. A signal peptide encoded within the precore region of hepatitis B virus directs the secretion of a heterogeneous population of e antigens in *Xenopus* oocytes. *Proc Natl Acad Sci USA* 1988;85:8405–8409.
391. Standring DN, Ou J, Rutter WJ. Assembly of viral particles in Xenopus oocytes: pre-surface-antigens regulate secretion of the hepatitis B viral surface envelope particle. *Proc Natl Acad Sci USA* 1986;83:9338–9342.
392. Standring DN, Rutter WJ, Varmus HE, Ganem D. Transcription of the hepatitis B surface antigen gene in cultured murine cells initiates within the presurface region. *J Virol* 1984;50:563–571.
393. Staprans S, Loeb D, Ganem D. Mutations affecting hepadnavirus plus-strand DNA synthesis dissociate primer cleavage from translocation and reveal the origin of linear viral DNA. *J Virol* 1991;65:1255–1262.
394. Stenlund A, Lamy D, Moreno-Lopez J, Ahola H, Pettersson U, Tiollais P. Secretion of the hepatitis B virus surface antigen from mouse cells using an extra-chromosomal eukaryotic vector. *EMBO J* 1983;2: 669–673.
395. Stevens CE, Neurath RA, Beasley RP, Szmuness W. Vertical transmission of hepatitis B antigen in Taiwan. *N Engl J Med* 1975;292: 771–774.
396. Stibbe W, Gerlich W. Structural relationships between minor and major proteins of hepatitis B surface antigen. *J Virol* 1983;46:626–628.

397. Stratowa C, Doehmer J, Wang Y, Hofschneider PH. Recombinant retroviral DNA yielding high expression of hepatitis B surface antigen. *EMBO J* 1982;1:1573–1578.
398. Su H, Yee JK. Regulation of hepatitis B virus gene expression by its two enhancers. *Proc Natl Acad Sci USA* 1992;89:2708–2712.
399. Su T, Lai C, Huand J, Lin L, Yauk Y, Chang C, Lo S, Han S. Hepatitis B virus transcript produced by RNA splicing. *J Virol* 1989;63:4011–4018.
400. Summers J, Mason WS. Replication of the genome of a hepatitis B-like virus by reverse transcription of an RNA intermediate. *Cell* 1982; 29:403–415.
401. Summers J, O'Connell A, Millman I. Genome of hepatitis B virus: restriction enzyme cleavage and structure of DNA extracted from Dane particles. *Proc Natl Acad Sci USA* 1975;72:4597–4601.
402. Summers J, Smith P, Horwich A. Hepadnavirus envelope proteins regulate covalently closed circular DNA amplification. *J Virol* 1990;64: 2819–2824.
403. Summers J, Smith P, Huang M, Yu M. Morphogenetic and regulatory effects of mutations in the envelope proteins of an avian hepadnavirus. *J Virol* 1991;65:1310–1317.
404. Summers J, Smolec JM, Snyder R. A virus similar to human hepatitis B virus associated with hepatitis and hepatoma in woodchucks. *Proc Natl Acad Sci USA* 1978;75:4533–4537.
405. Summers J, Smolec JM, Werner BG, et al. Hepatitis B virus and woodchuck hepatitis virus are members of a novel class of DNA viruses. In: *Viruses in naturally occurring tumors. Cold Spring Harbor Conference on Cell Proliferation VII*. Cold Spring Harbor, NY: Cold Spring Harbor Press; 1980:459–470.
406. Sureau C, Romet-Lemonne J-L, Mullins JI, Essex M. Production of hepatitis B virus by a differentiated human hepatoma cell line after transfection with cloned circular HBV DNA. *Cell* 1986;47:37–47.
407. Suzuki T, Masui N, Kajino K, Saito I, Miyamura T. Detection and mapping of spliced RNA from a human hepatoma cell line transfected with the hepatitis B virus genome. *Proc Natl Acad Sci USA* 1989;86:8422–8426.
408. Szmuness W. Hepatocellular carcinoma and the hepatitis B virus: evidence for a causal association. *Prog Med Virol* 1978;24:40.
409. Tagawa M, Robinson WS, Marion RL. Duck hepatitis B virus replicates in the yolk sac of developing embryos. *J Virol* 1987;61:2273–2279.
411. Takada S, H. Kido, A. Fukutomi, T. Mori, and K. Koike 1994. Interaction of hepatitis B virus X protein with a serine protease, tryptase TL2 as an inhibitor. *Oncogene* 9:341–348.
412. Takada S, Koike K. Trans-activation function of a 3′ truncated X gene-cell fusion product from integrated hepatitis B virus DNA in chronic hepatitis tissues. *Proc Natl Acad Sci USA* 1990;87:5628–5632.
413. Takada S, Koike K. X protein of hepatitis B virus resembles a serine protease inhibitor. *Jpn J Cancer Res* 1990;81:1191–1194.
414. Tanese N, Goff SP. Domain structure of the Moloney murine leukemia virus reverse transcriptase: mutational analysis and separate expression of the DNA polymerase and RNase H activities. *Proc Natl Acad Sci USA* 1988;85:1777–1781.
415. Tavis JE, Ganem D. Expression of functional hepatitis B virus polymerase in yeast reveals it to be the sole viral protein required for correct initiation of reverse transcription. *Proc Natl Acad Sci USA* 1993; 90:4107–4111.
416. Tavis J, Perri S, Ganem D. Hepadnaviral reverse transcription initiates within the RNA stem-loop of the viral encapsidation signal and employs a novel strand transfer. *J Virol* [in press].
417. Tay N, Chan S-H, Ren E-C. Identification and cloning of a novel heterogeneous nuclear ribonucleoprotein C-like protein that functions as a transcriptional activator of the hepatitis B virus enhancer II. *J Virol* 1992;66:6841–6848.
418. Tennant BC, Mrosovsky N, McLean K, Cote PJ, Korba BE, Engle RE, Gerin JL, Wright J, Michener GR, Uhl E, King JM. hepatocellular carcinoma in Richardson's ground squirrels (*Spermophilus richardsonii*): evidence for association with hepatitis B-like virus infection. *Hepatology* 1992;13:1215–1221.
419. Terre S, Petit M-A, Brehcot C. Defective hepatitis B virus particles are generated by packaging and reverse transcription of spliced viral RNAs in vivo. *J Virol* 1991;65:5539–5543.
420. Tognoni A, Cattaneo R, Serfling E, Schaffner W. A novel expression selection approach allows precise mapping of the hepatitis B virus enhancer. *Nucleic Acids Res* 1985;13:7457–7471.
421. Toh H, Hayashida H, Miyata T. Sequence homology between retroviral reverse transcriptase and putative polymerase of hepatitis B virus and cauliflower mosaic virus. *Nature* 1983;305:827–829.
422. Tokino T, Fukushiga S, Nakamura T, Nagaya T, Murotsu T, Shiga K, Aoki N, Matsubara K. Chromosomal translocation and inverted duplication associated with integrated hepatitis B virus in hepatocellular carcinoma. *J Virol* 1987;61:3848–3854.
423. Tokino T, Matsubara K. Chromosomal sites for hepatitis B virus integration in human hepatocellular carcinoma. *J Virol* 1991;65:6761–6764.
424. Tokino T, Tamura H, Hori N, Matsubara K. Chromosome deletions associated with hepatitis B virus integration. *Virology* 1991;185: 879–882.
425. Tong S, Diot C, Gripon P, Li J, Vitvitski L, Trepo C, Guguen-Guillouzo C. In vitro replication competence of a cloned hepatitis B virus variant with a nonsense mutation in the distal pre-C region. *Virology* 1991;181:733–737.
426. Tong S-P, Li J-S, Vitvitski L, Kay A, Trépo C. Evidence for a base-paired region of hepatitis B virus progenome encapsidation signal which influences the patterns of precore mutations abolishing HBe protein expression. *J Virol* 1993;67:5651–5655.
427. Tran A, Kremsdorf D, Capel F, Housset C, Douguet C, Petit M-A, Brechot C. Emergence of and takeover by hepatitis B virus (HBV) with rearrangements in the pre-S/S and pre-C/C genes. *J Virol* 1991;65: 3566–3574.
428. Transy C, Fourel G, Robinson WS, Tiollais P, Marion PL, Buendia M-A. Frequent amplification of c-*myc* in ground squirrel liver tumors associated with past or ongoing infection with a hepadnavirus. *Proc Natl Acad Sci USA* 1992;89:3874–3878.
429. Treinin M, Laub O. Identification of a promoter element located upstream from the hepatitis B virus X gene. *Mol Cell Biol* 1987;7:545–550.
430. Trueba D, Phelan M, Nelson J, et al. Transmission of ground squirrel hepatitis virus to homologous and heterologous hosts. *Hepatology* 1985; 5:435–439.
431. Trujillo MA, Letovsky J, Maguire HF, Lopez-Cabrera M, Siddiqui A. Functional analysis of a liver-specific enhancer of the hepatitis B virus. *Proc Natl Acad Sci USA* 1991;88:3797–3801.
432. Tsurimoto T, Fujiyama A, Matsubara K. Stable expression and replication of hepatitis B virus genome in an integrated state in a human hepatoma cell line transfected with the cloned viral DNA. *Proc Natl Acad Sci USA* 1987;84:444–448.
433. Tur-Kaspa R, Burk R, Shaul Y, Shafritz D. Hepatitis B virus DNA contains a glucocorticoid response element. *Proc Natl Acad Sci USA* 1986; 83:1627–1631.
434. Tur-Kaspa R, Shaul Y, Moore D, Burk R, Okret S, Poellinger L, Shafritz D. The glucocorticoid receptor recognizes a specific nucleotide sequence in hepatitis B virus DNA causing increased activity of the HBV enhancer. *Virology* 1988;167:630–633.
435. Tuttleman J, Pourcel C, Summers J. Formation of the pool of covalently closed circular viral DNA in hepadnavirus-infected cells. *Cell* 1986;47:451–460.
436. Tuttleman J, Pugh JC, Summers JW. In vitro experimental infection of primary duck hepatocyte cultures with duck hepatitis B virus. *J Virol* 1986;58:17–25.
437. Twu J-S, Chu K, Robinson WS. Hepatitis B virus X gene activates kB-like enhancer sequences in the long terminal repeat of human immunodeficiency virus 1. *Proc Natl Acad Sci USA* 1989;86:5168–5172.
438. Twu J-S, Robinson WS. Hepatitis B virus X gene can transactivate heterologous viral sequences. *Proc Natl Acad Sci USA* 1989;86:2046–2050.
439. Twu J-S, Schloemer RH. Transcriptional *trans*-activating function of hepatitis B virus. *J Virol* 1987;61:3448–3453.
440. Twu JS, Rosen CA, Haseltine WA, Robinson WS. Identification of a region within the human immunodeficiency virus type 1 long terminal repeat that is essential for transactivation by the hepatitis B virus gene X. *J Virol* 1989;63:2857–2860.
441. Uchida M, Esumi M, Shikata T. Molecular cloning and sequence analysis of duck hepatitis B virus genomes of a new variant isolated from Shanghai ducks. *Virology* 1989;173:600–606.
442. Ueda K, Tsurimoto T, Matsubara K. Three envelope proteins of hepatitis B virus: large S, middle S, and major S proteins needed for the formation of Dane particles. *J Virol* 1991;65:3251–3259.
443. Unger T, Shaul Y. The X protein of the hepatitis B virus acts as a transcription factor when targeted to its responsive element. *EMBO J* 1990; 9:1889–1895.
444. Valenzuela P, Gray P, Quiroga M, Zaldivar J, Goodman HM, Rutter WJ. Nucleotide sequence of the gene coding for the major protein of

hepatitis B virus surface antigen. *Nature* 1979;280:815–819.
445. Valenzuela P, Medina A, Rutter WJ. Synthesis and assembly of hepatitis B virus surface antigen particles in yeast. *Nature* 1982;298:347–351.
446. Valenzuela P, Quiroga M, Zaldivar J, Gray P, Rutter WJ. The nucleotide sequence of the hepatitis B viral genome and the identification of the major viral genes. *Mol Cell Biol* 1980;:57–70.
447. Vallee R, Shpetner H. Motor proteins of cytoplasmic microtubules. *Annu Rev Biochem* 1990;59:909–932.
448. Vannice J, Levinson A. Properties of the human hepatitis B virus enhancer—position effects and cell-type nonspecificity. *J Virol* 1988;62: 1305–1313.
449. Wampler DE, Lehman ED, Boger J, McAleer WJ, Scolnick EM. Multiple chemical forms of hepatitis B surface antigen produced in yeast. *Proc Natl Acad Sci USA* 1985;82:6830–6834.
450. Wang G-H, Seeger C. Novel mechanism for reverse transcription in hepatitis B viruses. *J Virol* 1993;67:6507–6512.
451. Wang G-H, Seeger C. The reverse transcriptase of hepatitis B virus acts as a protein primer for viral DNA synthesis. *Cell* 1992;71:663–670.
452. Wang H-P, Rogler CE. Topoisomerase I-mediated integration of hepadnavirus DNA in vitro. *J Virol* 1991;65:2381–2392.
453. Wang J, Chenivesse X, Henglein B, Brechot C. Hepatitis B virus integration in a cyclin A gene in a hepatocellular carcinoma. *Nature* 1990; 343:555–557.
454. Wang J, Lee AS, Ou J-H. Proteolytic conversion of hepatitis B virus e antigen precursor to end product occurs in a postendoplasmic reticulum compartment. *J Virol* 1991;65:5080–5083.
455. Wang Y, Chen P, Wu X, Sun A-L, Wang H, Zhu Y-A, Li Z-P. A new enhancer element, ENII, identified in the X gene of hepatitis B virus. *J Virol* 1990;64:3977–3981.
456. Wasenauer G, Kock J, Schlicht H-J. A cysteine and a hydrophobic sequence in the noncleaved portion of the pre-C leader peptide determine the biophysical properties of the secretory core protein (HBe protein) of human hepatitis B virus. *J Virol* 1992;66:5338–5346.
457. Weber M, Bronsema V, Bartos H, Bosserhoff A, Bartenschlager R, Schaller H. Hepadnavirus P protein utilizes a tyrosine residue in the TP domain to prime reverse transcription. *J Virol* 1994;68:2994–2999.
458. Wei Y, Fourel G, Ponzetto A, Silvestro M, Tiollais P, Buendia M-A. Hepadnavirus integration: mechanisms of activation of the N-*myc* 2 retrotransposon in woodchuck liver tumors. *J Virol* 1992;66:5265–5276.
459. Weichs an der Glon C, Monks J, Proudfoot NJ. Occlusion of the HIV poly(A) site. *Genes Dev* 1991;5:244–253.
460. Weiser B, Ganem D, Seeger C, Varmus HE. Closed circular viral DNA and asymmetrical heterogeneous forms in livers from animals infected with ground squirrel hepatitis virus. *J Virol* 1983;48:1–9.
461. White J. Membrane fusion. *Science* 1992;258:917–924.
462. White J. Viral and cellular membrane fusion proteins. *Annu Rev Physiol* 1990;52:675–697.
463. Wildner G, Fernholz D, Sprengel R, Schneider R, Will H. Characterization of infectious and defective cloned avian hepadnavirus genomes. *Virology* 1991;185:345–353.
464. Will H, Cattaneo R, Koch HG, Darai G, Schaller H. Cloned HBV DNA causes hepatitis in chimpanzees. *Nature* 1982;299:740–742.
465. Will H, Cattaneo R, Pfaff E, Kuhn C, Roggendorf M, Schaller H. Expression of hepatitis B antigens with a simian virus 40 infection. *J Virol* 1984;50:335–342.
466. Will H, Reiser W, Weimer T, Pfaff E, Buscher M, Sprengel R, Cattaneo R, Schaller H. Replication strategy of human hepatitis B virus. *J Virol* 1987;61:904–911.
467. Will H, Salfeld J, Pfaff E, Manso C, Theimann L, Schaller H. Putative reverse transcriptase intermediates of human hepatitis B virus in primary liver carcinomas. *Science* 1986;231:594–596.
468. Wong DT, Nath N, Sninsky JJ. Identification of hepatitis B virus polypeptides encoded by the entire pre-S open reading frame. *J Virol* 1985; 55:223–231.
469. Wu H-L, Chen P-J, Tu S-J, Lin M-H, Lai M-Y, Chen D-S. Characterization and genetic analysis of alternatively spliced transcripts of hepatitis B virus in infected human liver tissues and transfected HepG2 cells. *J Virol* 1991;65:1680–1686.
470. Wu T-T, Coates L, Aldrich CE, Summers J, Mason WS. In hepatocytes infected with hepatitis B virus, the template for viral RNA synthesis is amplified by an intracellular pathway. *Virology* 1990;175:255–261.
471. Yaginuma K, Koboyashi M, Yoshida E, Koike K. Hepatitis B virus integration in hepatocellular carcinoma DNA: duplication of cellular flanking sequences at the integration site. *Proc Natl Acad Sci USA* 1985;82: 4458–4462.
472. Yaginuma K, Koike K. Identification of a promoter region for 3.6-kilobase mRNA of hepatitis B virus and specific cellular binding protein. *J Virol* 1989;63:2914–2920.
473. Yaginuma K, Nadamura I, Takada S, Koike K. A transcription initiation site for the hepatitis B virus X gene is directed by the promoter-binding protein. *J Virol* 1993;67:2559–2565.
474. Yaginuma K, Shirakata Y, Kobayashi M, Koike K. Hepatitis B virus (HBV) particles are produced in a cell culture system by transient expression of transfected HBV DNA. *Proc Natl Acad Sci USA* 1987;84: 2678–2682.
475. Yang D, Alt E, Rogler CE. Coordinate expression of N-*myc* 2 and insulin-like growth factor II in pre-cancerous altered hepatic foci in woodchuck hepatitis virus carriers[1]. *Cancer Res* 1993;53:2020–2027.
476. Yee J. A liver-specific enhancer in the core promoter region of human hepatitis B virus. *Science* 1989;246:658–670.
477. Yeh C-T, Liaw Y-F, Ou J-H. The arginine-rich domain of hepatitis B virus precore and core protein contains a signal for nuclear transport. *J Virol* 1990;64:6141–6147.
478. Yeh C-T, Ou J-H. Phosphorylation of hepatitis B virus precore and core proteins. *J Virol* 1991;65:2327–2331.
479. Yen TSB. Regulation of hepatitis B virus gene expression. *Semin Virol* 1993;4:33–42.
480. Yoffe B, Noonan C, Melnick J, Hollinger FB. Hepatitis B virus DNA in mononuclear cells and analysis of cell subsets for the presence of replicative intermediates of viral DNA. *J Infect Dis* 1986;153:471–477.
481. Yokosuka O, Omata M, Zhou Y-Z, Imazeki F, Okuda K. Duck hepatitis B virus DNA in liver and serum of Chinese ducks: integration of viral DNA in a hepatocellular carcinoma. *Proc Natl Acad Sci USA* 1985; 82:5180–5184.
482. Yu M, Summers J. Phosphorylation of the duck hepatitis B virus capsid protein associated with conformational changes in the C-terminus. *J Virol* 1994;68:2965–2969.
483. Yu M, Summers J. A domain of the hepadnavirus capsid protein is specifically required for DNA maturation and virus assembly. *J Virol* 1991; 65:2511–2517.
484. Yuasa S, Cheung RC, Pham Q, Robinson WS, Marion PL. Peptide mapping of neutralizing and nonneutralizing epitopes of duck hepatitis B virus pre-S polypeptide. *Virology* 1991;181:14–21.
485. Yuh C-H, Chang Y-L, Ting L-P. Transcriptional regulation of precore and pregenomic RNAs of hepatitis B virus. *J Virol* 1992;66:4073–4084.
486. Yuh C-H, Ting L-P. Differentiated liver cell specificity of the second enhancer of hepatitis B virus. *J Virol* 1993;67:142–149.
487. Yuh C-H, Ting L-P. The genome of hepatitis B virus contains a second enhancer: cooperation of two elements within this enhancer is required for its function. *J Virol* 1990;64:4281–4287.
488. Zahm P, Hofschneider PH, Koshy R. HBV X-ORF encodes a transactivator: a potential factor in viral hepatocarcinogenesis. *Oncogene* 1988; 3:169–177.
489. Zhou S, Standring DN. Cys residues of the hepatitis B virus capsid protein are not essential for the assembly of viral core particles but can influence their stability. *J Virol* 1992;66:5393–5398.
490. Zhou D, Taraboulos A, Ou J, Yen TSB. Activation of class I MHC gene expression by hepatitis B virus *J Virol* 1990;64:4025–4028.
491. Zhou S, Yang SQ, Standring DN. Characterization of hepatitis B virus capsid particle assembly in *Xenopus* oocytes. *J Virol* 1992;66:3086–3092.
492. Zhou D, Yen TSB. Differential regulation of the hepatitis B viral surface gene promoters by a second viral enhancer. *J Biol Chem* 1990;265: 20731–20734.
493. Zoulim F, Seeger C. Reverse transcription in hepatitis B viruses is primed by a tyrosine residue of the polymerase. *J Virol* 1994;68:6–13.
494. Zoulim F, Saputelli J, Seeger C. Woodchuck hepatitis virus X protein is required for viral infection in vivo. *J Virol* 1994;68:2026–2030.

Fundamental Virology, Third Edition
edited by B.N. Fields, D.M. Knipe, P.M. Howley, et al.
Lippincott - Raven Publishers, Philadelphia © 1996

CHAPTER 36

Hepatitis Delta Virus and Its Replication

John M. Taylor

CLASSIFICATION

Hepatitis delta virus (HDV) is a "subviral" agent (20). A complete cycle of replication of HDV depends upon the presence of an helper hepadnavirus to provide envelope proteins. In nature, HDV has been found only in humans, where its replication is supported by coinfection with one specific hepadnavirus, human hepatitis B virus (HBV). Experimentally, HDV can be transmitted to primates in the presence of HBV (65). It can also be transmitted to woodchucks; however, since woodchucks are resistant to HBV infection, HDV replication in this host requires coinfection with another hepadnavirus, woodchuck hepatitis virus (WHV) (63). The genome of HDV shows no apparent sequence relationships to those of the hepadnaviruses (80). Thus, HDV is not a defective-interfering species of HBV, and should be considered as a satellite virus (20). In summary, HDV can be described as a natural subviral satellite of HBV.

As described below, significant similarities exist between HDV and certain subviral agents of plants, especially the viroids, in terms of genome structure and the mechanism of genome replication (5,21,77). Nevertheless, the differences are sufficiently great that, as presented in Chapter 2, the International Committee on Taxonomy of Viruses has approved a floating genus for HDV: Deltavirus.

J.M. Taylor: Fox Chase Cancer Center, Philadelphia, Pennsylvania 19111-2497.

VIRION STRUCTURE

As represented in Fig. 1, HDV, when examined by electron microcopy after negative-staining, appears roughly spherical with an average diameter, variously estimated to be 36 to 43 nm (30,63,67). Thus, HDV is roughly similar in size to the helper virus, HBV (23). In one careful study of HDV, although the average diameter was 36 nm, the particles were heterogeneous in size, ranging largely from 28 to 39 nm (30). The significance of this size heterogeneity is not yet known. Negative staining has also revealed a nucleocapsid structure within undisrupted HDV (67). And, when HDV is disrupted by a mixture of nonionic detergent and dithiothreitol, this nucleocapsid is released. After negative staining, it is revealed as a roughly spherical particle of a diameter around 19 nm (68). By contrast, the isolated nucleocapsids of HBV have a diameter of about 27 nm and, as explained below, are of a different protein composition than HDV nucleocapsids (23).

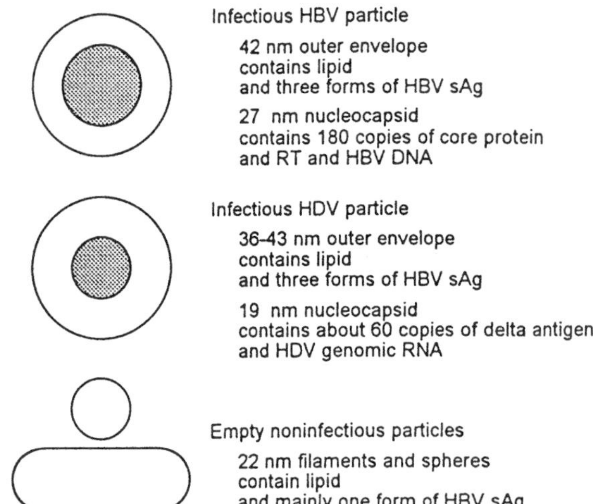

FIG. 1. Representation of the three types of particles found in the serum of an HDV infected animal. Sera will typically contain not only HDV but also relatively small amounts of the infectious helper virus particles, the Dane particles, and a 10- to 100-fold excess of the empty, noninfectious, surface antigen particles which are present both as spherical and filamentous forms. From Ganem (23), with permission.

The outer surface or envelope proteins of the HDV are provided entirely by the helper hepadnavirus (3). This envelope contains some lipids and all three envelope proteins of the hepadnavirus, referred to here as sAg. In the case of HBV, these sAg carry the well-known hepatitis B surface antigenic markers (HBsAgs) used in diagnostic virology. The sAg can make intermolecular disulfide linkages, which explains the need for a reducing agent such as dithiothreitol for disruption of the viral envelope (23,68). The number of molecules of each of the sAg per HDV virion is not known.

The hepadnavirus envelope proteins have been studied in great detail. These are a nested set of proteins made from a single ORF; have a common C-terminus; and are designated according to their size as large, middle, and small, or sAg-L, sAg-M, and sAg-S. These three proteins each exist in two major electrophoretic forms, with the differences being due to posttranslational modifications, mainly glycosylation (23). All three proteins, especially the smallest, appear to be produced in vast excess of the amounts needed for virion assembly, and the excess is secreted from cells as noninfectious, rodlike and spherical particles, 22 nm in diameter. These particles contain almost exclusively sAg-S. In HDV the ratio of sAg-L:sAg-M:sAg-S has been reported as 1:5:95 (3), but these proportions might have been distorted by the presence in the serum of relatively large amounts of the empty, noninfectious sAg-S particles (68).

The nucleocapsid structure of HDV is composed of the 1.7-kb single-stranded viral RNA genome and multiple copies of delta antigen, the only HDV-encoded protein. As discussed subsequently, there are two forms of the delta antigen which differ in that the larger form is 19-aa longer at the C-terminus than the small form. These are referred to as large and small delta antigen, or δAg-L and δAg-S, respectively. The total number of copies of these delta antigens appears to be around 60 per RNA genome, which is consistent with the possibility that the HDV nucleocapsid has icosahedral symmetry (68). It is believed, but not established, that there is only one copy of the RNA genome per virion.

GENOME STRUCTURE AND ORGANIZATION

Fig. 2 is a representation of the three RNAs involved in the replication of HDV. The first, indicated at the left side of the figure, is the single-stranded RNA genome. It is unique relative to those of other RNA viruses, for several reasons: (a) At 1.7 kb, it is the smallest; (b) the conformation is not linear but circular (80); (c) the RNA has the potential to fold on itself, with about 70% of all bases paired, to form an unbranched, rodlike structure (80); and (d) the genomic RNA, as well as its complement, the antigenome, can function as a ribozyme to carry out self-cleavage (71) and self-ligation reactions (70).

Inside the infected liver cell there are not only about 300,000 copies of this genomic RNA, but also numerous copies of two other RNAs (14). In particular (Fig. 2), there are about 50,000 copies the complement of genomic RNA, the antigenome (14), and there are around 600 copies of a third RNA, which is also complementary to the genome but is only 800 b long and is polyadenylated (32). This latter RNA is the mRNA for synthesis of the delta antigen (δAg). Thus, since the coding is on the strand complementary to the RNA genome, HDV is a negative-stranded agent.

The genomic and antigenomic RNAs of HDV have the ability to function as ribozymes. As indicated in Fig. 2, the RNAs each possess a unique site at which self-cleavage can occur. Such self-cleavage is readily demonstrated in the test tube. When certain HDV RNA sequences are heated in the presence of magnesium ions, cleavage is efficient and specific (39,71). The cleavage reaction in the test tube is a transesterification and produces a 5′OH and a 2′,3′-cyclic-monophosphate terminus (39). The RNA sequences at and around the cleavage sites on the genome and antigenome are very similar but not identical. They are related to each other because of the rodlike structure of the genome (39). Detailed biochemical studies have determined which sequences in the vicinity of the self-cleavage sites are essential for cleavage; 84 nucleotides 3′ of the cleavage site are needed but only 1 nucleotide is needed on the 5′ side (1,39,58,61,66,79,82). This minimum contiguous domain on the genomic RNA apparently folds into a four-stem structure that includes a pseudoknot (59,61). The overall sequence and structure of this ribozyme is quite different from other known ribozymes, especially the so-called hammerhead and hairpin ribozymes found on the RNAs of cer-

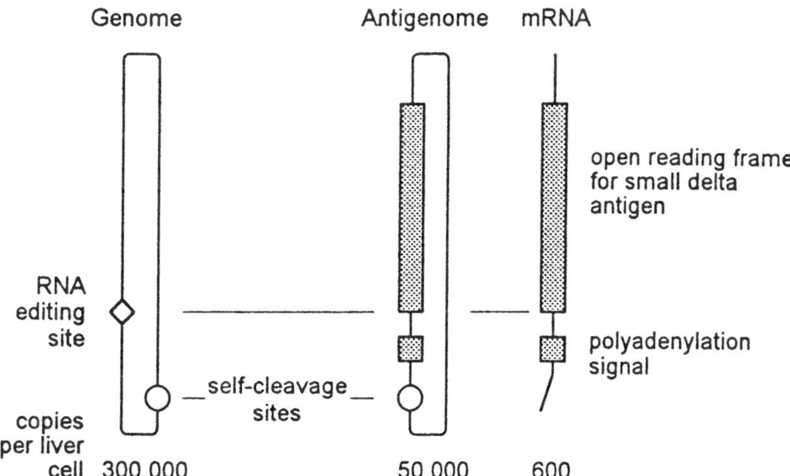

FIG. 2. The three RNAs of HDV. As indicated, these RNAs undergo RNA-editing, self-cleavage and self-ligation, and polyadenylation. The potential for genomic and antigenomic HDV RNAs to form an unbranched rodlike structure is indicated; folding of the plant viroid RNAs into an unbranched, rodlike structure is also possible (5). Adapted from Lazinski and Taylor (41).

tain subviral plant agents (5). The HDV genomic ribozyme has been divided into two domains: one that acts as an enzyme and the other that acts as the substrate (6,60,82).

In addition to carrying out the self-cleavage reactions, the genomic and antigenomic RNAs can also undergo self-ligation reactions at the same site. The self-ligation reaction was first studied using purified fragments of HDV RNA, and the data supported the hypothesis that in the test tube the side of the rodlike RNA opposite to the ligation site acts as a backbone or guide to bring the termini together (70). Subsequent studies of the ligation reaction within transfected cells have required a modification of the interpretation of the test tube experiments: the side of the rod opposite the ligation site is not required for self-ligation but stabilizes the ligated RNA against further self-cleavage (D. Lazinski, unpublished data).

As discussed below, HDV shows significant similarities to certain pathogenic RNAs of plants. These similarities are not only at the level of genome structure and organization but also at the level of genome replication.

STAGES OF THE VIRUS LIFE CYCLE

In the following sections, the life cycle of HDV may be broken down into seven separate stages.

Attachment, Entry, and Uncoating

The pre-S1 domain of sAg-L is essential for infectivity (73), presumably at the stage of attachment to a specific cell surface receptor. In this respect, HDV probably behaves just like its helper virus, HBV (56); however, as with HBV, the receptor has not yet been identified. For HBV, a possible candidate is the IL6 receptor (57).

Nothing is known about the subsequent two steps of entry and uncoating, either for HDV or for HBV.

Transcription

With most negative-stranded RNA viruses, it is possible to conceptually separate early RNA-directed RNA synthesis, which produces the positive-stranded mRNAs, from RNA-directed RNA synthesis to produce transcripts that are used as templates for genome replication. For HDV, the synthesis of the mRNA and the production of the template for replication of the genome are, in contrast, mechanistically linked together.

The HDV RNA genome is replicated by RNA-directed RNA synthesis in the nucleus of an infected cell (78). Indirect studies with both isolated nuclei and nuclear homogenates suggest that the host RNA polymerase II carries out HDV RNA synthesis and that the delta antigen is not required (22,48). The RNA products obtained with the nuclear homogenates are largely unit-length and circular (22), just as detected in a natural infection (14). *In vitro* transcription studies have also been carried out with purified RNA polymerase II and individual basal Pol II transcription factors, and provide strong evidence that the HDV RNA is synthesized by the host RNA polymerase II. HDV RNA will act as a template for Pol II only when the TATAA-binding protein, TBP, and the factor TFIIB are also present. These factors are necessary, but it is not yet clear whether they are sufficient. Addition of other basal transcription factors can augment the synthesis to the level achieved with a complete nuclear homogenate (T.-B. Fu, T. Shenk, A. Usheva, unpublished data).

To study HDV RNA-directed RNA synthesis or genome replication *in vivo*, cells can be transfected with cDNA (38, 75) or *in vitro* synthesized cRNA (24). Such transfection studies have established that the small form of the delta antigen, δAg-S, is essential for genome replication (38). The other form of the delta antigen, δAg-L, is nonfunctional in this respect. In fact, it acts as a dominant negative inhibitor of the replication supported by δAg-S (13,26). However, as discussed below, the δAg-L protein is essential for virus assembly (10,67). The actual role of δAg-S in RNA replication and the mechanism by which δAg-L inhibits replication are not known for certain. Some studies suggest that δAg-S is required to transport and/or retain HDV RNA in the nucleus, the site of genome replication (48). In addition, by binding in a ribonucleoprotein, the δAg appear to protect the HDV RNAs from degradation by the host cell (40).

In order to better understand the functions of δAg-S and δAg-L in HDV replication, transfected cell extracts have been examined for the presence of virus-specific RNPs containing these proteins. Inside the cell, δAg-S has been found to participate in the formation of RNPs, containing either genomic or antigenomic HDV RNAs (68). At present, the functional significance of these structures is unclear.

Fig. 3 shows a simplified model of HDV genome replication (33). It resembles the so-called double-rolling-circle models that have been proposed for some of the plant viroids (5) but is different in that it has to explain not only the synthesis of the complementary RNA, but also the synthesis of the 800-b mRNA for the delta antigen. With negative-stranded RNA viruses, RNA transcription is regulated; initially, mRNA(s) is transcribed from the incoming minus-strand genome and later, after the translation of the mRNA to make essential replication proteins, there is a switch in the mode of RNA-directed RNA synthesis to facilitate replication of the RNA genome (33). According to the model presented in Fig. 3, HDV synthesis, though clearly regulated, differs conceptually from the scheme employed by more well-studied negative-strand RNA viruses. In Step 1, transcription of genomic RNA into antigenomic RNA is proposed to initiate at a unique site, corresponding to the mapped 5′-end of the mRNA. In Step 2 this nascent RNA is processed to produce the polyadenylated mRNA for the delta antigen (32). Steps 1 and 2 are thus consistent with the mRNA that has been observed. As indicated in Step 3, the fate of the downstream fragment of such processing is proposed to differ from that of a typical mRNA precursor in that self-cleavage allows it to escape otherwise certain and prompt degradation. This proposal is supported by HDV cDNA transfections studies (33,34). The continuing transcript is proposed to proceed around the template for at least another 1.7 kb, so that at least two RNAs, one mRNA, and one antigenome can be made from a single initiation event. Two mechanisms are considered to ultimately suppress polyadenylation of the continuing RNA transcript. First, the structure of this RNA, unlike that of the 800-b mRNA, includes the polyadenylation signal in the rodlike structure. Second, as the amount of newly translated δAg increases it can, by its ability to bind rodlike HDV RNA, even further suppress polyadenylation. Evidence for these two mechanisms again comes from cDNA transfection studies (33,34). In these two ways the elongating antigenomic transcript can undergo a second self-cleavage, (as indicated in Step 4) to release a unit-length antigenomic RNA, which can then go on (in Step 5) to self-ligate to form a closed-circular RNA. The subsequent replication of such antigenomic circles into genomic circles is not shown in the Fig. 3, but is expected to be similar but simpler in that there is no longer a polyadenylation site to affect RNA elongation. No data exist for the site(s) at which the transcription of genomic RNA is initiated.

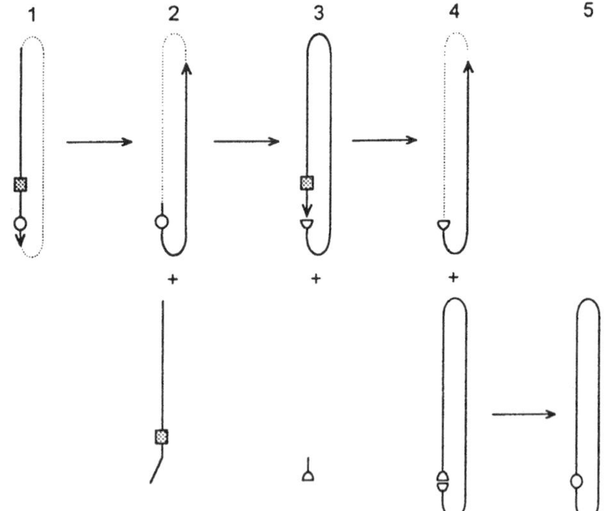

FIG. 3. A model for HDV RNA synthesis. The genomic RNA template (*dotted line*), acts as template for the initiation of RNA-directed RNA synthesis at a site near to the top of the rodlike structure. Transcription passes through the polyadenylation signal (*shaded box*) and the self-cleavage domain (*open circle*). This RNA is processed by polyadenylation to release the mRNA; the continuing RNA fragment undergoes self-cleavage. The temporal order of these two processing events is not known, but somehow the continuing RNA fragment avoids prompt degradation. Elongation then continues at least 1.7 kb further. Polyadenylation is suppressed by both the rodlike structure and the binding to this structure of the delta antigen, and the molecule undergoes a second self-cleavage. This releases a unit-length linear antigenomic RNA which proceeds to self-ligate. In subsequent steps not shown, the circular genome acts as template for the synthesis of new genomic RNA. Adapted from Hsieh et al. (33).

Translation

There are two electrophoretic forms of the delta antigen; they share a common N-terminus and differ at the C-terminus (81). HDV might appear to produce just one mRNA species, the 800-b polyadenlyated mRNA described above. However, as will be discussed in more detail below, there is a specific posttranscriptional RNA-editing event that allows HDV to make two mRNAs and two proteins instead of just one. This event indirectly changes the UAG amber termination codon of the 195-aa δAg-S to UGG, allowing the incorporation of tryptophan and a total C-terminal extension of 19 aa, to make the 214 aa δAg-L. As intimated earlier, these two proteins have essential roles in the HDV life cycle. Fig. 4 is a representation of the two proteins, so as to allow a discussion of their shared and unique features.

δAg-S and δAg-L can both be phosphorylated at one or more of nine potential serine residues (11), although there is some indication that the large form might be able to undergo, albeit in insect cells, a sixfold greater level of phosphorylation (35). The proteins are not glycosylated (11). As indicated in Fig. 5, a number of the properties of the delta antigens can be assigned to regions on the primary amino acid sequence. Between amino acids 13 and 48 is a domain that is predicted to fold into a coiled coil structure and which actually facilitates intermolecular binding between monomers of the delta antigen (42,83). Almost adjacent to this domain is a region between amino acids 69 and 88 that contains two subregions essential for the nuclear localization of the delta antigens (83). Beyond this, the region 97 to 143 is involved in the RNA binding. This RNA-binding region has been further subdivided; similar to certain other RNA-binding proteins, it contains two essential arginine-rich sequences separated by a nonessential intermediate region (45,46). The RNA-binding ability of the delta antigens, as studied using recombinant proteins, is specific for HDV RNAs (11,46) but also dependent upon the rodlike structure (12). Transfection studies support the interpretation that inside the cell, the delta antigens specifically recognize rodlike HDV RNA (40). Between amino acids 145 and 195 is a domain rich in proline and glycine. Although both forms of the delta antigen share this domain, mutation studies support the interpretation that when the extra 19 amino acids at the C-terminus of δAg-L are present, the adjacent proline-glycine rich region is altered in its structure in a way that contributes to the functionality of the protein in virus assembly (39). However, 33 amino acids of this sequence (163 to 195) can be deleted from δAg-L without interfering with the packaging ability (44), as discussed below.

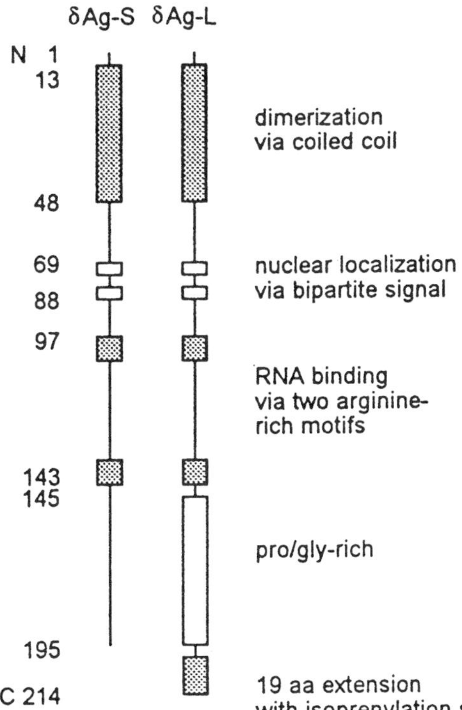

FIG. 4. Representation of the two forms of the delta antigen. Adapted from Lazinski and Taylor (41) and explained in the text.

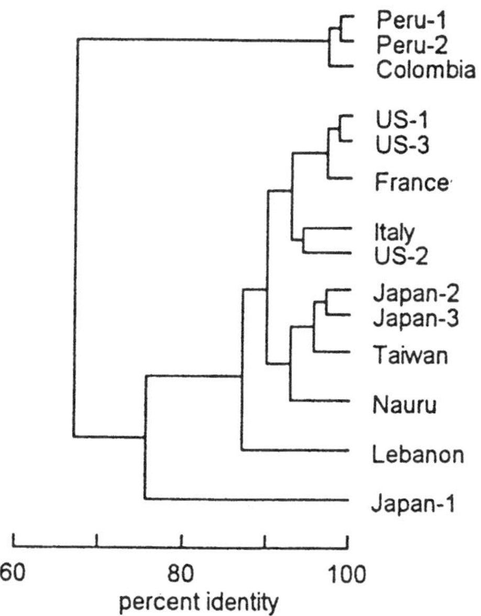

FIG. 5. Sequence relationship between 14 naturally occurring isolates of HDV. Adapted from Casey et al. (8).

Despite their many similarities, functional differences between δAg-S and δAg-L can be readily demonstrated inside cells. Only δAg-S is able to support genome replication (38). Only δAg-L is able to direct the assembly of virus particles (10,67). Only δAg-L undergoes posttranslational modification by isoprenylation; this occurs in the unique 19-aa extension at the cysteine 4 aa in from the C-terminus (25) and may actually be a geranyl geranylation (44). This modification is necessary but not sufficient for the ability of δAg-L to direct particle assembly (25,44).

RNA Editing

As discussed above, though the two forms of the delta antigen each have unique roles to play in the life cycle of

HDV, only δAg-S is encoded within the (infectious) viral genome. In transfected cells only δAg-S is made initially, with δAg-L appearing subsequently as a consequence of a specific posttranscriptional RNA editing event that changes the coding capacity of the viral genome.

Editing of the HDV genome at position 1012 was first discovered in transfected cells (13), and is now realized to occur in infected chimpanzees (74) and infected woodchucks (84). In Fig. 2, the RNA editing is indicated as occurring on the genomic RNA (7,47,85). More recent evidence, however, argues that the change is actually on the antigenomic RNA (8a). This means that an adenosine is changed to a nucleotide which behaves as a guanosine. For this adenosine to be edited it must be in the context of the rodlike structure. Certain bases and predicted base-pairs in the vicinity are essential (7). A phylogenetic study of the sequence of many different HDV isolates, discussed below, shows that in some cases the predicted pairing around the editing site is fundamentally different from the isolates that were first studied (8).

The activity responsible for HDV RNA editing seems to be double-stranded RNA-activated adenosine deaminase, DRADA (8a). This enzyme is present in the nucleus of animal cells and acts on RNAs with double-stranded structure by deaminating adenosine to inosine (9b). If the modified RNA substrate is subsequently copied by two rounds of RNA-directed RNA synthesis, the change is equivalent to replacing adenosine with guanosine. Such post-transcriptional RNA editing by DRADA also occurs for RNAs of other viruses and also for certain cellular RNAs (9,9a,9b).

While much evidence supports the interpretation that an RNA-editing event is essential to the life cycle of HDV, allowing the synthesis of two forms of the delta antigen, there is a contrary report. For one HDV isolate, associated with a particularly virulent infection, there seemed to be evidence of only one form of the delta antigen (76).

Assembly

HDV requires the envelope proteins of a helper hepadnavirus, HBV or WHV, so the assembly pathway significantly overlaps that of the helper virus. During an initial HDV infection of a human or woodchuck already chronically infected by HBV or WHV, respectively, when the amount of HDV released into the serum reaches a peak, there is a corresponding major drop-off in the production of the hepadnavirus. After a period of several days, this is followed by a major drop in HDV levels (63,64). The levels of hepadnavirus then increase but do not return to their original level.

While HDV does infect cultured hepatocytes (15,72,73, 78), there are as yet no data from assembly studies employing coinfections of hepatocyte cultures. As an alternative, assembly of HDV is typically studied in cells cotransfected with hepadnavirus DNA sequences (10,67,72, 73). Such studies must be interpreted with caution, however, because the proportions of the three sAg thus provided may not reflect those that would be achieved inside an infected hepatocyte. Nonetheless, some important information has been obtained about the role of the hepadnavirus in HDV replication. Transfection studies have shown that, in addition to the sAg of the hepadnavirus, δAg-L must also be present if delta antigen-containing particles are to be released (10). The δAg-S, if present, is "copackaged," but it does not have to be present for viruslike particles to be formed (10,67). Similarly, the HDV genomic RNA is packaged, if present. A range of nonreplicating deleted forms of HDV can also be assembled into viruslike particles (40). Reconstruction experiments have shown that the delta antigens specifically bind to HDV rodlike RNA (12). The same requirement apparently applies inside transfected cells, because for an HDV-related RNA to be assembled into viruslike particles it must be able to fold, at least to some extent, into the rodlike structure (40). No single origin-of-assembly sequence is revealed by such transfection studies. Also even nonreplicating antigenomic RNAs, as long as they contain the rodlike structure, can be assembled. However, for some reason, these findings made in the absence of genome replication do not fully reflect reality, because when genome replication takes place, the strand specificity of RNA packaging becomes very high; only about 1 part in 10,000 of the packaged RNA is antigenomic (H. J. Netter, unpublished data).

Using nuclei from cells undergoing HDV genome replication, it is possible to elute HDV-specific RNP structures made up of delta antigens and genomic RNA, and delta antigens complexed with antigenomic RNA (68). These RNP contain somewhat less δAg per RNA than RNP isolated from virions (68). It is not yet understood how the presence of the large form of the delta antigen, coupled with the presence of the hepadnavirus sAgs, can lead to the specific migration of genomic RNP to the cytoplasm and the assembly there, so as to ultimately make and release new virion particles. Presumably, the phenomenon is basically similar to the assembly of HBV particles. A speculation is that the genomic RNA, already assembled into an RNP involving both δAg-S and δAg-L, makes an interaction with sAg molecules already inserted into the membranes of the endoplasmic reticulum. In a series of as yet unclarified steps, this is followed by posttranslational modifications of the sAg, and the passage of assembled particles onto the Golgi apparatus, the trans-Golgi network, and ultimately the release of the particles from the cell, without any direct toxicity. However, such steps have not yet been delineated for HDV, and only to a limited extent for HBV (23).

HDV particles can be assembled in transfected cells using only the sAg-S (67,73), but sAg-L also has to be present for the particles to be infectious (72,73). This requirement for sAg-L is predicted by studies with HBV, where the N-terminal domain on sAg-L appears to contain the ligand for the binding of HBV to a cell-surface receptor (56).

Release

As discussed below, cultured cells can release as much as 2×10^7 HDV RNA-containing particles per mL (67), which can account for approximately 20% of total genomic RNA synthesis (40). This value is high, but not in comparison to the > 90% release from transfected cultured cells of sAg into empty, noninfectious particles (53,67). These values can be compared with the values for particles in the serum of an infected animal at the peak of an acute HDV infection; the HDV titer can be up to 10^{12} RNA-containing particles per mL (53,64). This apparent efficiency is less impressive when it is pointed out that the serum contains a further 10- to 100-fold excess of empty sAg-containing particles (68).

Production, Evolution, and Fate of Defective Viruses

The HDV genome undergoes small but measurable levels of mutation during passage through a host (19,43). As with the plant viroids, there are strong constraints for the maintenance of the viral sequence, especially to maintain the rodlike structure (5,37). The sequences of 14 independent isolates of HDV have been compared (8). As represented in the dendrogram of Fig. 5, there are variations of up to 40% between the nucleotide sequences. Within the HDV sequence, some regions are more conserved than others; these are largely in the self-cleavage domains and certain critical parts of the delta antigen ORF (8).

In addition to the natural variants of HDV, deliberately altered HDV genomes have been produced using transfection of modified cDNAs. These cDNAs can be delivered as multimers inserted into an expression vector (38) or even as circularized monomers (75) and lead to the production of variant HDVs. Chimeras made between HDV genomes of different sequence (and also as specific mutants with up to several nucleotides changed) are often able to replicate, provided that the rodlike structure is conserved (7). In contrast, transfection of cells with cDNA constructs with large deletions (>500 bp), even when the predicted rodlike structure is conserved, fail to produce replicating genomes, even when δAg-S is provided in trans (40). Nevertheless, the nascent DNA-directed RNA transcripts can often undergo self-cleavage, self-ligation, delta antigen binding, and, in the presence of sAg, assembly and release into viruslike particles (40). Genome insertions of as much has 50 nucleotides of foreign sequence at the bottom of the rodlike structure in some, but not all, cases produces replication-competent species. However, the level of replication corresponds to no more than 2% relative to that of an unmodified HDV RNA (55). Thus, structural features of the HDV genome other than rodlike folding and known ribozyme activities are essential for RNA replication. The nature of these features is not yet defined.

EFFECT ON HOST CELL

The pathogenic effects of HDV infections are considered by Dr. Purcell in the next chapter. Nevertheless, some comments here are appropriate. Careful studies have been made of sections from HDV-infected liver. Cytotoxic effects of hepatic infection upon hepatocytes have been described (27,51,52), but it is hard to discriminate in such studies between the roles of the HDV infection, the associated hepadnavirus infection, and the immune responses of the host to both infections upon the hepatic cells.

It has been reported that the expression in cultured cells from transfected cDNA of δAg-S, but not δAg-L, can have a cytotoxic effect (16,17). This was indicated by a reduction in the rate of growth of the cells, and in some cases by cell death, apparently by apoptosis. In contrast, a study of the expression of either δAg-S or δAg-L in the livers of transgenic mice failed to show any direct cytopathic effect (28). Also, when single cells transfected with cDNA or even the complete HDV genome were followed during several subsequent cell divisions, no specific apoptosis was detected. The cells apparently continued to express δAg and to divide. Apparently, HDV infections are not always cytopathic or even cytostatic (2). In studies of mice and woodchuck hepatocytes infected with HDV in the absence of helper hepadnavirus, as discussed below, infected cells disappeared at a rate of around 50% per 3 to 5 days (2,54). The basis for this disappearance is not yet known. However, in SCID mice lacking T and B cells and in neonatal mice the rate of disappearance was essentially the same (54).

FEATURES OF GROWTH

In nature, HDV is found to infect only humans. However, in the presence of its helper virus, HBV, HDV can be experimentally transmitted to chimpanzees (65). Alternatively, the helper virus can be substituted with WHV, allowing experimental transmission of HDV to the woodchuck (63). These different infections are typically classified according to the associated helper virus infection as either superinfections or coinfections.

If the host animal has a prior chronic infection with helper virus, the HDV infection is called a *superinfection*. Such superinfections are very efficient for HDV spread, and at the peak of an acute infection, the amount of HDV in the serum can exceed 10^{12} RNA-containing particles per mL (53,64). The titer of an HDV superinfection peaks at around 2 to 5 weeks postinoculation; this peak is quite transient, often spanning less than 1 to 2 weeks (63,64). Considering the high titers of HDV that can be found in the bloodstream, it is a puzzling observation that HDV infections are strictly limited to the liver (51), inasmuch as hepadnaviruses have a small but real capacity for infection of extrahepatic cells of the pancreas, kidney, bone marrow, and spleen (49,51).

In contrast to a superinfection, if the host has not been previously exposed to either virus and both viruses are administered at once, the infection is referred to as a *coinfection*. The probability of achieving a productive coinfection does depend upon both the absolute and relative amounts of the two inoculated viruses. If there is not enough helper virus to achieve a sufficient level of coinfection of hepatocytes, HDV genome replication will be initiated but progeny virions will not be released. Such helper-independent infections have been observed in recipient livers of transplant patients (18). In these patients, originally infected with HDV and HBV, HBV infections had spontaneously ceased, and continued HDV production by the original liver was supported by HBV envelope proteins, possibly expressed from rare integrated HBV sequences. Such an infection has also been called "latent" (27) or "autonomous." (18) A model for this phenomenon can be set up in the woodchuck; a helper-independent HDV infection can be rescued by the administration of helper virus 7, even 33 days after the original HDV infection (53). In this context, one has to bear in mind that even in typical superinfections and coinfections, there may be some hepatocytes that initially undergo helper-independent HDV infection, with probably a subsequent conversion to a productive infection as a result of infection by the helper hepadnavirus.

Experimentally, HDV can be used to infect a mouse; the virus genome replicates in hepatocytes in a helper-independent manner and virus particles are thus not released (54). HDV can also be used to infected monolayer cultures of primary human, chimpanzee, or woodchuck hepatocytes (15,73,78). HDV will not infect established cell lines, presumably because of the lack of appropriate receptor molecules. However, when administered in the presence of a cationic lipid formulation, virus can be essentially transfected in, and as many as 10% of such cells go on to demonstrate helper-independent genome replication. In fact, such replication can persist for as long as 6 weeks, even when the cells undergo numerous rounds of cell division (2).

As with other RNA viruses (3), it is possible to assemble a cDNA copy of HDV in a eukaryotic expression vector and to use this to transfect cultured cells. This leads to helper-independent HDV genome replication (38). Even a circularized cDNA monomer is sufficient for such transfection (73). Furthermore, in the presence of helper virus sequences, it is possible to achieve assembly of HDV RNA into viruslike particles (10,67) or even infectious particles (72,73). HDV cDNA has also been used to transfect chimpanzees (74) and woodchucks (80). Also, since such studies were carried out in the presence of helper virus, the transfections led to a full HDV viremia.

Attempts have been made to transfect cells with HDV cRNA. This has failed, presumably because the small delta antigen, which is essential for genome replcation, was not present (24). In contrast, if the recipient cell was one in which the small delta antigen was already expressed, then HDV genome replication was initiated (24). Alternatively, if the HDV RNA was already complexed with the delta antigens in an RNP structure, this RNP could be used to successfully transfect a cell that did not contain the delta antigen (2).

As discussed above, it is also possible to study aspects of HDV genome replication using HDV RNA added to nuclear homogenates or even to purified RNA polymerase II and certain specific basal transcription factors.

INHIBITORS OF REPLICATION

As discussed earlier, HDV genome replication by itself leads to the synthesis of an inhibitor of replication. That is, the editing of the HDV RNA leads to the synthesis of δAg-L. This species, as studied in reconstruction experiments in cotransfected cells, has a dominant negative effect on genome replication (13,26). Such inhibition requires interactions with δAg-S, using the common coiled coil region (41,42). Nevertheless, beyond this, it is not yet understood how δAg-L is able to so act.

Of more practical interest, certain drugs have been claimed to act on HDV infections. Suramin is able to block the infection of primary hepatocytes by HDV (62). This drug is also known to act on infections by hepadnaviruses and retroviruses (62). Apparently it acts at an early step, presumably by preventing virus penetration or uncoating. In addition, α-amanitin has been shown to block HDV replication (48), and this is consistent with the interpretation that genome replication requires the host RNA polymerase II (22,48). Using cultured woodchuck hepatocytes infected with HDV, certain other potential inhibitors of HDV replication have been tested. Ribovirin behaved as a reversible inhibitor (15). Interferons have been tested on cells transfected with HDV cDNA; apparently, they have no direct effect (36). In contrast, for infected patients, interferon therapy does have an effect, but it is considered to be most likely an indirect one, possibly via an effect on the helper hepadnavirus and/or on the immune reponse to the infections (31).

GENETICS

As discussed in an earlier section, there is considerable sequence heterogeneity among natural isolates of HDV. It is possible to use HDV cDNAs and make replication-competent chimeras between such variants (7). Moreover, it is clear that sequence heterogeneity occurs even in infected individuals (80). Also, a limited number of variations can be made on the genome of a given isolate without losing replication competence. However, at this time, there is no evidence for or against recombination events between HDV genomes.

RELATIONSHIP TO CERTAIN PATHOGENIC RNAs OF PLANTS

It has become progressively clearer that the structure and replication of the HDV genome show significant similarities to certain much-studied pathogenic RNAs of plants. One class of such agents is known as *viroids* (20). About 24 of these have been isolated. They each have single-stranded RNA genomes and the analogies to HDV include: (a) the RNA has a circular conformation; (b) the RNA will fold by base-pairing into an unbranched, rodlike structure; (c) one of these RNAs has been shown to possess a unique site at which self-cleavage occurs; (d) evidence has been obtained with plant nuclear extracts that at least one of these viroids is replicated by the host RNA polymerase II (69); and (e) also revealed is a patch of relationship to 7S L RNA, the 300-b RNA component of the signal recognition particle (29,51). There are, however, also real differences: (a) the viroids are much smaller, being 250 to 350 b in length; (b) the viroids have no known requirement for a helper virus; and (c) the viroids have no known protein coding ability. In addition to the viroids there are other classes of plant pathogens that show some similarities to HDV. These are certain of the plant "satellite RNAs" and a subclass of these, the "virusoids," which also have a circular RNA genome (20). These satellite RNAs differ from HDV in that the helper virus provides the polymerase needed for genome replication, but they are similar to HDV in that they depend upon a helper virus for particle assembly. Yet another class of plant agents, the "satellite viruses," includes members with similarities to HDV in that they encode a protein that facilitates particle assembly (20).

REFERENCES

1. Belinsky M, Dinter-Gottlieb G. Characterizing the self-cleavage of a 135 nucleotide ribozyme from genomic hepatitis delta virus. Gerin JL, Purcell RH, Rizzetto M., eds. *The hepatitis delta virus*.New York: Wiley-Liss, 1991;265–274.
2. Bichko V, Netter HJ, Taylor J. Introduction of hepatitis delta virus into animal cell lines via cationic liposomes. *J Virol* 1994; 68:5247–5252.
3. Bonino F, Heermann KF, Rizzetto M, et al. Hepatitis delta virus: protein composition of delta antigen and its hepatitis B-derived envelope. *J Virol* 1986;58:945–950.
4. Boyer J-C, Haenni A-L. Infectious transcripts and cDNA clones of RNA viruses. *Virology* 1994;198:415–426.
5. Branch AD, Levine BL, Robertson HD. The brotherhood of circular RNA pathogens: viroids, circular satellites, and the delta agent. *Sem Virol* 1990;1:143–152.
6. Branch AD, Robertson HD. Efficient trans cleavage and a common structural motif for ribozymes of the human hepatitis δ agent. *PNAS* 1991;88:10163–10167.
7. Casey JL, Bergmann KF, Brown TL, et al. Structural requirements for RNA editing in hepatitis delta virus: evidence for a uridine-to-cytidine editing mechanism. *PNAS* 1992;89:7149–7153.
8. Casey JL, Brown TL, Colan EJ, et al. A genotype of hepatitis D virus that occurs in northern South America. *PNAS* 1993;90:9016–9020.

8a. Casey JL, Gerin JL. Hepatitis data virus: RNA editing and genotype variations. In: Dinter-Gottlieb G, ed. *The unique hepatitis delta virus*. Heidelberg: Springer-Verlag, 1995;111–124.

9. Cattaneo R. Different types of messenger RNA editing. *Ann Rev Genet* 1991;25:71–88.

9a. Cattaneo R. Biased (adenosine to iosine) hypermutation of animal RNA virus genomes. *Current Biol* 1994;4:895–899.

9b. Cattaneo R. RNA duplexes guide base conversions. *Current Biol* 1994;4:134–136.

10. Chang FL, Chen PJ, Tu SJ, et al. The large form of hepatitis δ antigen is crucial for assembly of hepatitis δ virus. *PNAS* 1991;88:8490–8494.
11. Chang M-F, Baker SC, Soe LH, et al. Human hepatitis delta antigen is a nuclear phosphoprotein with RNA-binding activity. *J Virol* 1988;62:2403–2410.
12. Chao M, Hsieh S-Y, Taylor J. The antigen of hepatitis delta virus: examination of in vitro RNA-binding specificity. *J Virol* 1991;65:4057–4062.
13. Chao M, Hsieh S-Y, Taylor J. Role of two forms of the hepatitis delta virus antigen: evidence for a mechanism of self-limiting genome replication. *J Virol* 1990;64:5066–5069.
14. Chen P-J, Kalpana G, Goldberg J, et al. Structure and replication of the genome of hepatitis delta virus. *PNAS* 1986;83:8774–8778.
15. Choi S-S, Rasshofer R, Roggendorf M. Inhibition of hepatitis delta virus RNA replication in primary woodchuck hepatocytes. *Antiviral Res* 1989;12:213–222.
16. Cole SM, Gowans EJ, Macnaughton TB, et al. Direct evidence for cytotoxicity associated with expression of hepatitis delta virus antigen. *Hepatol* 1991;13:845–851.
17. Cole SM, Macnaughton TB, Gowans EJ. Differential roles for HDAg-p24 and -p27 in HDV pathogenesis. In: Hadziyannis SJ, Taylor JM, Bonino F, eds.*Hepatitis delta virus*: molecular biology, pathogenesis, and clinical aspects. New York: Wiley Liss, 1993;131–138.
18. David E, Pucci A, Rahier J, et al. Histopathology of recurrent delta hepatitis in liver transplant. In: Hadziyannis SJ, Taylor JM, Bonino F, eds. *Hepatitis delta virus*: molecular biology, pathogenesis and clinical aspects. New York: Wiley-Liss, 1993;419–424.
19. Dény P, Zignego AL, Rascalou N, et al. Nucleotide sequence analysis of three different hepatitis delta viruses isolated from a woodchuck and humans. *J Gen Virol* 1991;72:735–739.
20. Diener TO, Prusiner SB. The recognition of subviral pathogens. In: Maramorosch K, McKelvey, JJ Jr, eds., Subviral pathogens of plants and animals: viroids and prions. Orlando: Academic Press, 1985;3–18.
21. Elena SF, Dopazo J, Flores R, et al. Phylogeny of viroids, viroidlike satellite RNAs, and viroidlike domain of hepatitis δ virus RNA. *PNAS* 1991;88:5631–5634.
22. Fu T-B, Taylor J. The RNAs of hepatitis delta virus are copied by RNA polymerase II in nuclear homogenates. *J Virol* 1993;67:6965–6972.
23. Ganem D. Assembly of hepadnaviral virions and subviral particles. In: Mason WS, Seeger C, eds. Hepadnaviruses: molecular biology and pathogenesis. *Curr Topics Microbiol and Immunol* 1991;168:61–83.
24. Glenn JS, Taylor JM., White JM. In vitro-synthesized hepatitis delta virus RNA initiates genome replication in cultured cells. *J Virol* 1990;64:3104–3107.
25. Glenn JS, Watson JA, Havel CM, et al. Identification of a prenylation site in delta virus large antigen. *Science* 1992;256:1331–1333.
26. Glenn JS, White JM. Trans-dominant inhibition of human hepatitis delta virus genome replication. *J Virol* 1991;65:2357–2361.
27. Gowans EJ, Bonino F. *Hepatitis delta virus pathogenicity*. In: Hadziyannis SJ, Taylor JM, Bonino F, eds. Hepatitis delta virus: molecular biology, pathogenesis, and clinical aspects. New York: Wiley Liss, 1993;125–131.
28. Guilhot S, Huang S-N, Xia YP, et al. Expression of the hepatitis delta virus large and small antigens in transgenic mice. *J Virol* 1994;68:1052–1058.
29. Haas B, Klanner A, Ramm K, et al. The 7S RNA from tomato leaf tissue resembles a signal recognition particle RNA and exhibits a remarkable sequence complementarity to viroids. *EMBO J* 1988;7:4063–4074.
30. He L-F, Ford E, Purcell RH, et al. The size of the hepatitis delta agent. *J Med Virol* 1989;27:31–33.
31. Hoofnagle JH, Di Bisceglie AM. *Therapy of chronic delta hepatitis*: overview. In: Hadziyannis SJ, Taylor JM, Bonino F, eds. Hepatitis delta virus: molecular biology, pathogenesis and clinical aspects. New York: Wiley-Liss, 1993;337–343.
32. Hsieh S-Y, Chao M, Coates L, et al. Hepatitis delta virus genome replication: a polyadenylated mRNA for delta antigen. *J Virol* 1990;64:3192–3198.
33. Hsieh S-Y, JM Taylor JM. Regulation of polyadenylation of HDV antigenomic RNA. *J Virol* 1991;65:6438–6446.
34. Hsieh S-Y, Yang P-Y, Ou JT, et al. Polyadenylation of the mRNA of hepatitis delta virus is dependent upon the structure of the nascent RNA and regulated by the small or large delta antigen. *Nucl Acids Res* 1994;22:391–396.
35. Hwang SB, Lee S.-Z, Lai MMC. Hepatitis delta antigen expressed by recombinant baculoviruses: comparison of biochemical properties and post-translational modifications between large and small forms. *Virology* 1992;190:413–422.

36. Ilan Y., Klein A, Taylor J, et al. Resistance of hepatitis delta virus replication to interferon-α treatment in transfected human cells. *J Inf Dis* 1992;166:1164–1166.
37. Keese P, Visvader JE, Symons RH. *Sequence variability in plant viroid RNAs*. In Domingo E, Holland JJ, Ahlquist P, eds. RNA genetics, vol III, Variability of RNA genomes. Boca Raton: CRC Press, 1988;71–98
38. Kuo MY-P, Chao M, Taylor J. Initiation of replication of human hepatitis delta virus genome from cloned DNA: role of the delta antigen. *J Virol* 1989;63:1945–1950.
39. Kuo MY-P, Sharmeen L, Dinter-Gottlieb G, et al. Characterization of self-cleaving RNA sequences on the genome and antigenome of human hepatitis delta virus. *J Virol* 1988;62:1855–1861.
40. Lazinski DW, Taylor JM. Expression of hepatitis delta virus RNA deletions: cis and trans requirements for self-cleavage, ligation and RNA packaging. *J Virol* 1994;68:2879–2888.
41. Lazinski DW, Taylor JM. Recent developments in hepatitis delta virus research. *Adv Virus Res* 1994;43:187–231.
42. Lazinski DW, Taylor JM. Relating structure to function in the hepatitis delta virus antigen. *J Virol* 1993;67:2672–2680.
43. Lee C-M, Bih F-Y, Chao Y-C, et al. Evolution of hepatitis delta virus RNA during chronic infection. *Virology* 1992;188:265–273.
44. Lee C-Z, Chen P-J, Lai MMC, et al. Isoprenylation of large hepatitis delta antigen is necessary but not sufficient for hepatitis delta virus assembly. *Virology* 1994;199:169–175.
45. Lee C-Z, Lin J-H, McKnight K, et al. RNA binding activity of hepatitis delta antigen involves two arginine-rich motifs and is required for hepatitis delta virus RNA replication. *J Virol* 1993;67:2221–2227.
46. Lin J-H, Chang M-F, Baker SC, et al. Characterization of hepatitis delta antigen: specific binding to hepatitis delta virus RNA. *J Virol* 1990;64:4051–4058.
47. Luo G, Chao M, Hsieh S-Y, et al. A specific base transition occurs on replicating hepatitis delta virus RNA. *J Virol* 1990;64:1021–1027.
48. Macnaughton TB, Gowans EJ, McNamara SP, et al. Hepatitis δ antigen is necessary for access of hepatitis δ virus RNA to the cell transcriptional machinery but is not part of the transcriptional complex. *Virology* 1991;184:387–390.
49. Mason WS, Taylor JM, Seal G, et al. The duck hepatitis B virus, an HBV-like virus of domestic ducks. In: Alter H, Maynard J, Szmuness W, eds. *Viral Hepatitis*. 1989; 107–116.
50. Negro F, Gerin JL, Purcell RH, et al. Basis of hepatitis delta virus disease. *Nature* 1989;341:111.
51. Negro F, Korba BE, Forzani B, et al. Hepatitis delta virus (HDV) and woodchuck hepatitis virus (WHV) nucleic acids in tissues of HDV-infected chronic WHV carrier woodchucks. *J Virol* 1989;63:1612–1618.
52. Negro F, Pacchioni D, Bussolati G, et al. *Pathobiology of hepatitis delta virus infection at the cell level*. In: Hadziyannis SJ, Taylor JM, Bonino F, eds. Hepatitis delta virus: molecular biology, pathogenesis and clinical aspects. New York: Wiley-Liss, 1993;155–160.
53. Netter HJ, Gerin JL, Tennant BD, et al. Apparent helper-independent infection of woodchucks by hepatitis delta virus and subsequent rescue with woodchuck hepatitis virus. *J Virol* 1994; 68:5344–5350.
54. Netter HJ, Kajino K, Taylor JM. Experimental transmission of human hepatitis delta virus to the laboratory mouse. *J Virol* 1993;67:3357–3362.
55. Netter HJ, Lazinski DW, Taylor JM. Hepatitis delta virus as a vector for RNA sequences. In: Vos J-MH, ed. *Viruses in human gene therapy*. Durham: Carolina Academic Press, 1995;33–54.
56. Neurath AR, Kent SBH, Strick N, et al. Identification and chemical synthesis of a host cell receptor binding site on hepatitis B virus. *Cell* 1986;46:429–436.
57. Neurath AR, Strick N, Sproul P. Search for hepatitis B virus cell receptors reveals binding sites for interleukin 6 on the virus envelope protein. *J Exp Med* 1992;175:461–469.
58. Nishikawa S, Suh Y-A, Kumar PKR, et al. Identification of important bases for the self-cleavage activity at two single-stranded regions of genomic HDV ribozyme. *Nucl Acids Res* 1992;20:41–42.
59. Perrotta AT, Been MD. Assessment of disparate structural features in three models of the hepatitis delta virus ribozyme. *Nucl Acids Res* 1993;21:3959–3965.
60. Perotta AT, Been MD. Cleavage of oligoribonucleotides by a ribozyme derived from the hepatitis delta virus RNA sequence. *Biochemistry* 1992;31:16–21.
61. Perotta AT, Been MD. A pseudoknot stucture required for efficient self-cleavage of hepatitis delta virus RNA. *Nature* 1991;350:434–436.
62. Petcu DJ, Aldrich CE, Coates L, et al. Suramin effectively inhibits in vitro infection by duck hepatitis B virus, Rous sarcoma virus, and hepatitis delta virus. *Virology* 1988;167:385–392.
63. Ponzetto A, Cote PJ, Popper H, et al. Transmission of the hepatitis B virus-associated δ agent to the eastern woodchuck. *PNAS* 1984;81:2208–2212.
64. Ponzetto A, Negro F, Gerin JL, et al. Experimental hepatitis delta virus infection in the animal model. In: Gerin JL, Purcell, RH, Rizzetto M, eds. *The hepatitis delta virus*. New York: Wiley Liss, 1989;147–157.
65. Rizzetto M, Canese MG, London WT, et al. Transmission of hepatitis B virus-associated delta antigen to chimpanzees. *J Inf Dis* 1980; 141:590–602.
66. Rosenstein SP, Been MD. Evidence that genomic and antigenomic RNA self-cleaving elements from hepatitis delta virus have similar secondary structure. *Nucl Acids Res* 1991:19:5409–5416.
67. Ryu W-S, Bayer M, Taylor JM. Assembly of hepatitis delta virus particles. J Virol 1992;66:2310–2315.
68. Ryu W-S, Netter HJ, Bayer M, et al. The ribonucleoprotein complexes of hepatitis delta virus. *J Virol* 1993;67:3281–3287.
69. Schindler I-M, Mühlbach H-P. Involvement of nuclear DNA-dependent RNA polymerase in potato spindle tuber viroid replication: a reevaluation. *Plant Science* 1992;84:221–229.
70. Sharmeen L, Kuo MY, Taylor J. Self-ligating RNA sequences on the antigenome of human hepatitis delta virus. *J Virol* 1989;63:1428–1430.
71. Sharmeen L, Kuo MY-P, Dinter-Gottlieb G, et al. The antigenomic RNA of human hepatitis delta virus can undergo self-cleavage. *J Virol* 1988;62:2674–2679.
72. Sureau, C, Guerra B, Lanford RE. Role of the large hepatitis B envelope protein in infectivity of the hepatitis delta virion. *J Virol* 1993;67:366–372.
73. Sureau C, Moriarty AM, Thornton GB, et al. Production of infectious hepatitis delta virus in vitro and neutralization with antibodies directed against hepatitis B virus pre-S antigens. *J Virol* 1992; 66:1241–1245.
74. Sureau C, Taylor J, Chao M, et al. A cloned DNA copy of hepatitis delta virus is infectious in the chimpanzee. *J Virol* 1989;63:4292–4297.
75. Tai F-P, Chen P-J, Chen D-S. Hepatitis delta virus cDNA can be used in transfection experiments to initiate viral RNA replication. *Virology* 1993;197:137–142.
76. Tang JR, Hantz O, Vitvitski L, et al. Discovery of a novel point mutation changing the HDAg expression of a hepatitis delta virus isolate from Central African Republic. *J Gen Virol* 1993;74:1827–1835.
77. Taylor J, Chao M, Kuo M, et al. Human hepatitis delta: Unique or not unique. In: Brinton M, Heinz FX, eds. *New aspects of positive-strand RNA viruses*. Washington: ASM Publications, 1990;20–24.
78. Taylor J, Mason W, Summers J, et al. Replication of human hepatitis delta virus in primary cultures of woodchuck hepatocytes. *J Virol* 1987;61:2891–2895.
79. Thill G, Vasseur M, Tanner NK. Structural and sequence elements required for the self-cleaving activity of the hepatitis delta virus ribozyme. *Biochemistry* 1993;32:4245–4262.
80. Wang K-S, Choo Q-L, Weiner A, et al. Structure, sequence, and expression of the hepatitis delta (δ) viral genome. *Nature* 1986; 323:508–514.
81. Weiner AJ, Choo Q-L, Wang K-S, et al. A single antigenomic open reading frame of the hepatitis delta virus encodes the epitope(s) of both hepatitis delta antigen polypeptides $p24^{\delta}$ and $p27^{\delta}$. *J Virol* 1988;62:594–599.
82. Wu HN, Wang YJ, Hung CF, et al. Sequence and structure of the catalytic RNA of hepatitis delta virus genomic RNA. *J Mol Biol* 1992;223:233–245.
83. Xia Y-P, Yeh C-T, Ou J-H, et al. Characterization of nuclear targeting signal of hepatitis delta antigen: nuclear transport as a protein complex. *J Virol* 1992;66:914–921.
84. Yang A, Karayiannis P., Cheng D, et al. Emergence of the long form of hepatitis delta virus antigen in transfected cells after intrahepatic transfection and during natural infection. In: Hadziyannis SJ, Taylor JM, Bonino F, eds. *Hepatitis delta virus: molecular biology, pathogenesis and clinical aspects*. New York: Wiley-Liss, 1993;145–148.
85. Zheng H, Fu T-B, Lazinski D, et al. Post-transcriptional modification of genomic RNA of human hepatitis delta virus. *J Virol* 1992;66:4693–4697.

Fundamental Virology, Third Edition
edited by B.N. Fields, D.M. Knipe, P.M. Howley, et al.
Lippincott - Raven Publishers, Philadelphia © 1996

CHAPTER 37

Prions

Stanley B. Prusiner

Prions cause a group of human and animal neurodegenerative diseases which are now classified together because their etiology and pathogenesis involve modification of the prion protein (PrP) (452,447). Prion diseases are manifest as infectious, genetic, and sporadic disorders (Table 1). These diseases can be transmitted among mammals by the infectious particle designated *prion* (456). Despite intensive searches over the past three decades, no nucleic acid has been found within prions (9,11,270,482); yet, a modified isoform of the host-encoded PrP designated *PrP*^{Sc} is essential for infectivity (80,452,463,464,475). In fact, considerable experimental data argue that prions are composed exclusively of PrP^{Sc}. Earlier terms used to describe the prion diseases include *transmissible encephalopathies, spongiform encephalopathies* and *slow virus diseases* (196,197,515).

The quartet of human (Hu) prion diseases are frequently referred to as *kuru, Creutzfeldt-Jakob disease* (CJD), *Gerstmann-Sträussler-Scheinker* (GSS) *disease*, and *fatal familial insomnia* (FFI). Kuru was the first of the human prion diseases to be transmitted to experimental animals, and it has often been suggested that kuru spread among the Fore people of Papua New Guinea by ritualistic cannibalism (197,199). The experimental and presumed human-to-human transmission of kuru led to the belief that prion diseases are infectious disorders caused by unusual viruses similar to those causing scrapie in sheep and goats. Yet, a paradox was presented by the occurrence of CJD in families, first reported almost 70 years ago (305,374), which appeared to be a genetic disease. The significance of familial CJD remained unappreciated until mutations in the protein coding region of the PrP gene on the short arm of chromosome 20 were discovered (259,448,523). The earlier finding that brain extracts from patients who had died of familial prion diseases inoculated into experimental animals often transmit disease posed a conundrum that was resolved with the genetic linkage of these diseases to mutations of the PrP gene (358,453,539).

The most common form of prion disease in humans is sporadic CJD. Many attempts to show that the sporadic prion diseases are caused by infection have been unsuccessful (63,121,248,344). The discovery that inherited prion diseases are caused by germline mutation of the PrP gene raised the possibility that sporadic forms of these diseases might result from a somatic mutation (453). The discovery that PrP^{Sc} is formed from the cellular isoform of the prion protein, PrP^{C}, by a posttranslational process (53) and that overexpression of wild-type (wt) PrP transgenes produces spongiform degeneration and infectivity *de novo* (559) has raised the possibility that sporadic prion diseases result from the spontaneous conversion of PrP^{C} into PrP^{Sc}.

CJD has a world-wide incidence of ~1 case per 10^6 population annually (370). Less than 1% of CJD cases are infectious, and most of those appear to be iatrogenic. Between 10% and 15% of prion disease cases are inherited, while the remaining cases are sporadic. Kuru was once the most common cause of death among New Guinea women in the Fore region of the Highlands (199,201,202) but has virtually disappeared with the cessation of ritualistic cannibalism (13). Patients with CJD frequently present with dementia, but ~10% of patients exhibit cerebellar dysfunction initially. Patients with either kuru or GSS usually present with ataxia, while those with FFI manifest insomnia and autonomic dysfunction (59,262,373).

PrP^{CJD} has been found in the brains of most patients who died of prion disease. The term *PrP*^{CJD} is preferred by some investigators when referring to the abnormal isoform of HuPrP in human brain. Here, *PrP*^{Sc} is used interchangeably with *PrP*^{CJD}. PrP^{Sc} is always used after human CJD prions have been passaged into an experimental animal since the nascent PrP^{Sc} molecules are produced from host PrP^{C}, and the PrP^{CJD} in the inoculum serves only to initiate the

S. B. Prusiner: Departments of Neurology and Biochemistry and Biophysics, University of California, San Francisco, California 94143.

TABLE 1. *Human prion diseases*

Disease	Etiology
Kuru	Infection
Creutzfeldt-Jakob disease	
Iatrogenic	Infection
Sporadic	Unknown
Familial	PrP mutation
Gerstmann-Sträussler-Scheinker disease	PrP mutation
Fatal Familial Insomnia	PrP mutation

process. In the brains of some patients with inherited prion diseases as well as transgenic (Tg) mice expressing mouse (Mo) PrP with the human GSS point mutation (Pro→Leu), detection of PrP^{Sc} has been problematic despite clinical and neuropathologic hallmarks of neurodegeneration (265,267). Of note, horizontal transmission of neurodegeneration from the brains of patients with inherited prion diseases to inoculated rodents has been less frequent than with sporadic cases (539). Whether this distinction between transmissible and nontransmissible inherited prion diseases will persist is unclear. Tg mice expressing a chimeric Hu/Mo PrP gene have been found to be highly susceptible to Hu prions from sporadic and iatrogenic CJD cases (544). These Tg(MHu2M) mice should make the use of apes and monkeys for the study of human prion diseases unnecessary and allow for tailoring the PrP^{C} translated from the transgene to match the sequence of the PrP^{CJD} in the inoculum. The use of mice expressing MHu2MPrP transgenes with mutations may enhance the ability to transmit cases of the inherited human prion diseases.

Scrapie is the most common natural prion disease of animals. An investigation into the etiology of scrapie followed the vaccination of sheep for looping ill virus with formalin-treated extracts of ovine lymphoid tissue unknowingly contaminated with scrapie prions (241). Two years later, more than 1,500 sheep developed scrapie from this vaccine. While the transmissibility of experimental scrapie became well established, the spread of natural scrapie within and among flocks of sheep remained puzzling. Parry (426,428) argued that host genes were responsible for the development of scrapie in sheep. He was convinced that natural scrapie is a genetic disease which could be eradicated by proper breeding protocols. He considered its transmission by inoculation to be of importance primarily for laboratory studies and communicable infection to be of little consequence in nature. Other investigators viewed natural scrapie as an infectious disease and argued that host genetics only modulates susceptibility to an endemic infectious agent (154).

The offal of scrapied sheep in Great Britain is thought to be responsible for the current epidemic of bovine spongiform encephalopathy (BSE) or mad cow disease (568). Prions in the offal from scrapie-infected sheep appear to have survived the rendering process which produced meat and bone meal (MBM). The MBM was fed to cattle as a nutritional supplement. After BSE was recognized, MBM produced from domestic animal offal was banned from further use. Since 1986, when BSE was first recognized, >150,000 cattle have died of BSE. Whether humans will develop CJD after consuming beef from cattle with BSE prions is of considerable concern.

Prions differ from all other known infectious pathogens in several respects. First, prions do not contain a nucleic acid genome which codes for their progeny. Viruses, viroids, bacteria, fungi, and parasites all have nucleic acid genomes that code for their progeny. Second, the only known component of the prion is a modified protein that is encoded by a cellular gene. Third, the major, and possibly the only, component of the prion is PrP^{Sc} which is a pathogenic conformer of PrP^{C}.

The fundamental event in prion diseases seems to be a conformational change in PrP. The cellular isoform of the prion protein (PrP^{C}) has been identified in all mammals and birds examined to date; it is anchored to the external surface of cells by a glycolipid moiety, and its function is unknown (527). All attempts to identify a posttranslational chemical modification that distinguishes PrP^{Sc} from PrP^{C} have been unsuccessful to date (526). PrP^{C} contains ~45% α-helix and is virtually devoid of β-sheet (424). Conversion to PrP^{Sc} creates a protein which contains ~30% α-helix and 45% β-sheet. The mechanism by which PrP^{C} is converted into PrP^{Sc} remains unknown, but PrP^{C} appears to bind to PrP^{Sc} to form an intermediate complex during the formation of nascent PrP^{Sc}. Tg mouse studies have demonstrated that PrP^{Sc} in the inoculum interacts preferentially with homotypic PrP^{C} during the propagation of prions (477,507).

As our knowledge of the prion diseases increases and more is learned about the molecular and genetic characteristics of prion proteins, these disorders will undoubtedly undergo modification with respect to their classification. Indeed, the discovery of the PrP and the identification of pathogenic PrP gene mutations have already forced us to view these illnesses from perspectives not previously imagined.

DEVELOPMENT OF THE PRION CONCEPT

Hypotheses on the Nature of the Scrapie Agent

The published literature contains a fascinating record of the structural hypotheses for the scrapie agent proposed first to explain the unusual features of the disease and later those of the infectious agent (Table 2). Among the earliest hypotheses was the notion that scrapie was a disease of muscle caused by the parasite *Sarcosporidia* (341,342). With the successful transmission of scrapie to animals, the hypothesis that scrapie is caused by a "filterable" virus became popular (125,576). With the findings of Tikvah Alper and her colleagues (9,11) that scrapie infectivity resists inactivation by UV and ionizing radiation, a myriad of hypotheses on the chemical nature of the scrapie agent emerged. Among the

TABLE 2. *Hypothetical structures for the scrapie agent*

1. Sarcosporidia-like parasite
2. "Filterable" virus
3. Small DNA virus
4. Replicating protein
5. Replicating abnormal polysaccharide with membranes
6. DNA subvirus controlled by a transmissible linkage substance
7. Provirus consisting of recessive genes generating RNA particles
8. Naked nucleic acid similar to plant viroids
9. Unconventional virus
10. Aggregated conventional virus with unusual properties
11. Replicating polysaccharide
12. Nucleoprotein complex
13. Nucleic acid surrounded by a polysaccharide coat
14. Spiroplasma-like organism
15. Multicomponent system with one component quite small
16. Membrane-bound DNA
17. Virino (viroidlike DNA complexed with host proteins)
18. Filamentous animal virus (SAF)
19. Aluminum-silicate amyloid complex
20. Computer virus

hypothetical structures proposed were a small DNA virus (296), a replicating protein (242,331,332,434), replicating abnormal polysaccharide with membranes (209,273), a DNA subvirus controlled by a transmissible linkage substance (1,3), a provirus consisting of recessive genes generating RNA particles (427,428), and a naked nucleic acid similar to plant viroids (157). While subsequent investigations showed the viroid suggestion to be incorrect (158), other studies argue that the scrapie agent may be composed only of a protein that adopts an abnormal conformation (317,424, 526), as previously proposed (242).

The term *unconventional virus* was proposed, but no structural details were ever given with respect to how these unconventional virions differ from the conventional viral particles (197); some investigators have suggested that this term obscured the ignorance that continued to shroud the infectious scrapie agent (429). Other suggestions included: aggregated conventional virus with unusual properties (493), replicating polysaccharide (176), nucleoprotein complex (329), nucleic acid surrounded by a polysaccharide coat (2,397,514), spiroplasma-like organism (31,269), multicomponent system with one component quite small (272, 521), membrane-bound DNA (355), virino (viroidlike DNA complexed with host proteins), filamentous animal virus (SAF) (376), aluminum-silicate amyloid complex, and even a computer virus (195).

Despite so much evidence to the contrary, some investigators persist with the belief that scrapie is caused by a virus (8,104,293,419,489,578).

Bioassays of Prion Infectivity

The experimental transmission of scrapie from sheep (241) to mice (102) gave investigators a more convenient laboratory model, which yielded considerable information on the nature of the unusual infectious pathogen that causes scrapie (9–11,209,382,434). Yet progress was slow because quantitation of infectivity in a single sample required holding 60 mice for 1 year before accurate scoring could be accomplished (102).

Attempts to develop a more economical bioassay by relating the titer to incubation times in mice were unsuccessful (170,275); however, some investigators used incubation times to characterize different "strains" of scrapie agent, while others determined the kinetics of prion replication in rodents (147,148,298,299). Yet these investigators refrained from trying to establish quantitative bioassays for prions based on incubation times, despite the successful application of such an approach for the measurement of picorna and other viruses three decades earlier (203). After scrapie incubation times were reported to be ~50% shorter in Syrian hamsters compared to mice (354), studies were undertaken to determine if the incubation times in hamsters could be related to the titer of the inoculated sample. Once it was found that there was an excellent correlation between the length of the incubation time and the dose of inoculated prions, a more rapid and economical bioassay was developed (459,467). This improved bioassay for the scrapie agent in Syrian golden hamsters accelerated purification of the infectious particles by a factor of nearly 100.

Bioassays for human prions were initially performed in apes and monkeys (214,215). Over the last 30 years, 300 cases of CJD, kuru, and GSS have been transmitted to a variety of apes and monkeys (68). The scarcity of these primates, the expense of their long-term care, and the increasing ethical objections to such experiments have limited investigations. It has been stated that >90% of cases clinically and neuropathologically diagnosed as CJD transmit to nonhuman primates after prolonged incubation times (68).

Because of the "species barrier" (431), the initial passage of prions from humans to rodents requires prolonged incubation times with relatively few animals developing illness (215,311,348,392,455,539). Subsequent passage in the same species occurs with high frequency and shortened incubation times. The molecular basis for the species barrier between Syrian hamster (SHa) and mouse was found to reside in the sequence of the PrP gene using Tg mice (506); SHaPrP differs from MoPrP at 16 of 254 amino acid residues (29,337). Construction of chimeric SHa/Mo transgenes identified a region within PrP where the substitution of five amino acids found in SHaPrP lying between residues 94 and 188 rendered the chimeric PrP designated MH2M eligible for conversion to PrP^{Sc} by SHa prions (507). Substitution of only two SHaPrP residues in the chimeric PrP designated MHM2 did not permit conversion into PrP^{Sc} by SHa prions (508) (M. Scott and S. B. Prusiner, unpublished data).

Since mice expressing HuPrP transgenes have not given abbreviated incubation times similar to Tg(SHaPrP) mice when inoculated with homologous prions, a chimeric Hu/Mo transgene was constructed, designated MHu2M. Human PrP differs from mouse PrP at 28 of 254 positions (322), while chimeric MHu2MPrP differs from mouse PrP at nine residues. The MHu2M transgenes are susceptible to human prions and exhibit abbreviated incubation times (544).

Purification of Scrapie Infectivity

Many investigators attempted to purify the scrapie agent for several decades but with relatively little success. The slow, cumbersome, and tedious bioassays in sheep and later mice, greatly limited the number of samples that could be analyzed. Since the ease of purifying any biologically active macromolecule is directly related to the rapidity of the assay, it is not surprising that little progress was made with sheep and goats, where only very limited numbers of samples could be analyzed and incubation times exceeding 18 months were required (270,429).

Experimental transmission of scrapie to mice allowed many more samples to be analyzed, but a year was required to complete the measurement of scrapie infectivity by endpoint titration using 60 animals to evaluate one sample (102). Although some properties of the scrapie agent were determined using this rather cumbersome mouse bioassay, it was difficult to develop an effective purification protocol when the interval between execution of the experiment and the availability of the results was nearly a year (454, 474,513). The resistance of scrapie infectivity to nondenaturing detergents, nucleases, proteases, and glycosidases (271,274,275,383), as well as the sedimentation properties of the scrapie agent was determined using endpoint titrations in mice (469,470). Attempts to purify infectivity were complicated by the apparent size and charge heterogeneity of scrapie infectivity, which was interpreted to be a consequence of hydrophobic interactions (471).

The transmission of scrapie to Syrian hamsters by an inoculum previously passaged in rats produced disease in about 70 days (354). These shorter incubation times, coupled with the development of standard curves relating the length of the incubation time to the size of the inoculated dose, permitted much more rapid quantitation of specimens (459,467). This methodological advance made possible the development of protocols for the significant enrichment of scrapie infectivity, using a series of detergent extractions; limited digestions with proteases and nucleases; and differential centrifugation (467), followed first by agarose gel electrophoresis (465) and later by sucrose gradient centrifugation (457,475).

The Prion Concept

Once an effective protocol was developed for preparation of partially purified fractions of scrapie agent from hamster brain, it became possible to demonstrate that those procedures which modify or hydrolyze proteins produce a diminution in scrapie infectivity (456,476). At the same time, tests done in search of a scrapie-specific nucleic acid were unable to demonstrate any dependence of infectivity on a polynucleotide (456), in agreement with earlier studies reporting the extreme resistance of infectivity to UV irradiation at 254 nm (9).

Based on these findings, it seemed likely that the infectious pathogen capable of transmitting scrapie was neither a virus nor a viroid. For this reason the term *prion* was introduced to distinguish the **pro**teinaceous **in**fectious particles that cause scrapie, CJD, GSS, and kuru from both viroids and viruses (456). Hypotheses for the structure of the infectious prion particle included: (a) proteins surrounding a nucleic acid encoding them (a virus), (b) proteins associated with a small polynucleotide, and (c) proteins devoid of nucleic acid (456). Mechanisms postulated for the replication of infectious prion particles ranged from those used by viruses to the synthesis of polypeptides in the absence of nucleic acid template to posttranslational modifications of cellular proteins. Subsequent discoveries have narrowed hypotheses for both prion structure and the mechanism of replication.

Considerable evidence has accumulated over the past decade supporting the prion hypothesis (452). Furthermore, the replication of prions and their mode of pathogenesis also appear to be without precedent. After a decade of severe criticism and serious doubt, the prion concept is now enjoying considerable acceptance.

Search for a Scrapie-Specific Nucleic Acid

The search for a scrapie-specific nucleic acid has been intense, thorough, and comprehensive, yet it has been unrewarding. The challenge to find a scrapie-specific polynucleotide was initiated by investigators who found that scrapie agent infectivity is highly resistant to UV and ionizing radiation (9–11). Their results prompted speculation that the scrapie pathogen might be devoid of nucleic acid—a postulate initially dismissed by many scientists. Although some investigators have argued that the interpretation of these data was flawed (489–492), they and others have failed to demonstrate the putative scrapie nucleic acid.

Based on the resistance of the scrapie agent to both UV and ionizing radiation, the possibility was raised that the scrapie agent might contain a small polynucleotide similar in size and properties to viroids of plants (157). Subsequently, evidence for a putative DNA-like viroid was published (345,346,355), but the findings could not be confirmed (465) and the properties of the scrapie agent were found to be incompatible with those of viroids (158). Besides ultraviolet irradiation, reagents specifically modifying or damaging nucleic acids such as nucleases, psoralens, hydroxylamine and Zn^{2+} ions were found not to alter scrapie infectivity in homogenates (456), microsomal fractions (456), purified prion rod preparations, or detergent-lipid-protein complexes (33–35,187,367,401).

Attempts to find a scrapie-specific polynucleotide using physical techniques such as polyacrylamide gel electrophoresis were as unsuccessful as molecular cloning approaches. Subtractive hybridization studies identified several cellular genes, the expression of which is increased in scrapie, but no unique sequence could be identified (155, 168,556). Extensively purified fractions were analyzed for a scrapie-specific nucleic acid using a specially developed technique designated return refocusing gel electrophoresis, but none was found (380). These studies argue that if such a molecule exists, its size is 80 nucleotides or less; larger nucleic acids were excluded as components essential for infectivity (292,482). Data from UV inactivation studies of scrapie prions argue that if prions contain a scrapie-specific nucleic acid, then it cannot exceed 6 bases in length for a single-stranded molecule (33). If such a nucleic acid were double-stranded, then it might contain as many as 40 bp. This larger estimate for a double-stranded polynucleotide takes into account the possibility of repair to UV damaged molecules after their injection into rodents for bioassay. Attempts to use these highly enriched fractions to identify a scrapie-specific nucleic acid by molecular cloning were also unsuccessful (408).

In spite of these studies, some investigators continue to champion the idea that scrapie is caused by a "virus." (103, 293) A few argue that the scrapie virus is similar to a retrovirus (7,351,393,517–520), while others argue that the scrapie virus induces amyloid deposition in brain (57,159, 160) or that scrapie is caused by a larger pathogen, similar to spiroplasma bacterium (30,31). Still others contend that elongated protein polymers covered by DNA are the etiologic agents in scrapie (394,396,398–400). DNA molecules like the D-loop DNA of mitochondria have also been suggested as the cause of scrapie (5,6).

The search for a component within the prion particle other than PrP has focused largely on a nucleic acid because some properties of prions are similar to those of viruses, and a polynucleotide would most readily explain different isolates or "strains" of infectivity (75,143,151,300). Specific scrapie isolates characterized by distinct incubation times retain this property when repeatedly passaged in mice or hamsters (75,143,151,300). Although available data do not permit exclusion of a scrapie-specific polynucleotide, its existence seems unlikely. The possibility that prions might contain noncovalently bound cofactors such as peptides, oligosaccharides, fatty acids, sterols, or inorganic compounds deserves consideration.

DISCOVERY OF THE PRION PROTEIN

Copurification of Prion Infectivity and PrP^{Sc}

Data from several studies suggested that scrapie infectivity might depend upon protein (106,107,271,383), while others had demonstrated that infectivity was resistant to protease digestion (355). Only after an effective protocol was developed for enriching fractions ~100-fold for scrapie infectivity with respect to cellular protein (465,467) could the dependence of scrapie infectivity on protein be established (476). Studies with partially purified fractions prepared from SHa brain showed loss of infectivity as a function of the concentration of protease and the time of digestion; these results demonstrated that a polypeptide is required for propagation of the infectious scrapie pathogen (476).

Once the dependence of prion infectivity upon protein was clear, the search for a scrapie-specific protein intensified. While the insolubility of scrapie infectivity made purification problematic, we took advantage of this property along with its relative resistance to degradation by proteases to extend the degree of purification (467,476). In subcellular fractions from hamster brain enriched for scrapie infectivity, a protease-resistant polypepetide of 27 to 30 kD, later designated PrP 27-30, was identified; it was absent from controls (49,366,457). Radioiodination of partially purified fractions revealed a protein unique to preparations from scrapie-infected brains (49,457). The existence of this protein was rapidly confirmed (162).

Determination of the N-terminal Sequence of PrP 27-30

Purification of PrP 27-30 to homogeneity allowed determination of its NH_2-terminal amino acid sequence (466). These studies were particularly difficult because multiple signals were found in each cycle of the Edman degradation. Whether multiple proteins were present in these "purified fractions" or a single protein with a ragged NH_2-terminus was present was resolved only after data from five different preparations were compared. When the signals in each cycle were grouped according to their intensities of strong, intermediate, and weak, it became clear that a single protein with a ragged NH_2-terminus was being sequenced. Determination of a single, unique sequence for the NH_2-terminus of PrP 27-30 permitted the synthesis of an isocoding mixture of oligonucleotides that was subsequently used to identify incomplete PrP cDNA clones from hamster (410) and mouse (105). cDNA clones encoding the entire open reading frames (ORFs) of SHa and Mo PrP were subsequently recovered (29,337).

PrP is encoded by a chromosomal gene and not by a nucleic acid within the infectious scrapie prion particle (29, 410). Levels of PrP mRNA remain unchanged throughout the course of scrapie infection—an unpredicted observation which led to the identification of the normal PrP gene product, a protein of 33 to 35 kd, designated PrP^C (29,410). PrP^C is protease-sensitive and soluble in nondenaturing detergents while PrP 27-30 is the protease-resistant core of a 33- to 35-kD disease-specific protein, designated PrP^{Sc}, which is insoluble in detergents (381). Progress in the study of prions was greatly accelerated by the discovery of PrP and determination of its N-terminal sequence (49,457,466). Indeed, all of the elegant molecular genetic studies in humans and animals as well as many highly informative transgenetic investigations have their origin in the purification of PrP 27-30 (475) and the determination of its N-terminal sequence (466).

Prions Contain PrP^{Sc}

There is evidence to argue that PrP^{Sc} is an essential component of the infectious prion particle (Table 3). Attempts to find a second component of the prion particle have been unsuccessful to date; indeed, many lines of investigation have converged to contend that prions are composed largely, if not entirely, of PrP^{Sc} molecules. Although some investigators think that PrP^{Sc} is merely a pathologic product of scrapie infection and that PrP^{Sc} coincidentally purifies with the "scrapie virus" (5–7,57,351,393,518,520), such views are not supported by the data. No infective fractions containing <1 PrP^{Sc} molecule per ID_{50} unit have been found, arguing that PrP^{Sc} is required for infectivity. Some investigators report that PrP^{Sc} accumulation in hamsters occurs after the synthesis of many infective units (127,128), but these results have been refuted (283). In another study, the kinetics of PrP^{Sc} and infectivity production in mice inoculated with mouse passaged CJD prions were similar in brain but were thought to be different in salivary gland (502). The discrepancies between PrP^{Sc} and infectivity levels in the above studies appear to be due to comparisons of infectivity in crude homogenates with PrP^{Sc} concentrations in purified fractions. Other investigators claim to have dissociated scrapie infectivity from PrP 27-30 in brains of Syrian hamsters treated with amphotericin B and inoculated with the 263K isolate, but not if they were inoculated with the 139H isolate; also, no dissociation was seen with mice inoculated with 139a prions (578). To date, no confirmation of these studies with amphotericin has been published.

The covalent structure of PrP^{Sc} remains uncertain because purified fractions contain ~10^5 PrP 27-30 molecules per ID_{50} unit, the infectious dose at which 50% of the animals develop scrapie (49,366,457). If <1% of the PrP^{Sc} molecules contained an amino acid substitution or posttranslational modification that conferred scrapie infectivity, our methods would not detect such a change (526).

TABLE 3. *Evidence that PrP^{Sc} is a major and essential component of the infectious prion*

1. Copurification of PrP 27-30 and scrapie infectivity by biochemical methods. Concentration of PrP 27-30 is proportional to prion titer (49,255,283,366,457,501,549).
2. Kinetics of proteolytic digestion of PrP 27-30 and infectivity are similar (49,366,457).
3. Copurification of PrP^{Sc} and infectivity by immunoaffinity chromatography. α-PrP antisera neutralization of infectivity (188,191).
4. PrP^{Sc} detected only in clones of cultured cells producing infectivity (82,369,538).
5. PrP amyloid plaques are specific for prion diseases of animals and humans (36,136,312,484). Deposition of PrP amyloid is controlled, at least in part, by the PrP sequence (477).
6. PrP^{Sc} (or PrP^{CJD}) is specific for prion diseases of animals and humans (47,66,510). Deposition of PrP^{Sc} precedes spongiform degeneration and reactive gliosis (85,138, 251,534).
7. Genetic linkage between the PrP gene and scrapie incubation times in mice with short and long incubation times (86,87,278,479) encoding PrP molecules differing at residues 108 and 189 (560). The length of the incubation time is determined by the level of PrP expression and the PrP sequence (85).
8. Expression of SHaPrP in Tg(SHaPrP) mice renders them susceptible to SHa prions (506). The primary structure of PrP^{Sc} in the inoculum governs the neuropathology and prion synthesis (477). Expression of a chimeric PrP in Tg(MHu2M) mice renders them susceptible to Hu prions (544).
9. Genetic linkage between PrP gene point mutations at codons 102, 178, 198, or 200 and the development of inherited prion diseases in humans was demonstrated (163,192,259,439). Genetic linkage was also established between the mutation insert of six additional octarepeats and familial CJD (444).
10. Mice expressing MoPrP transgenes with the P102L point mutation of GSS spontaneously develop neurologic dysfunction, spongiform brain degeneration, and astrocytic gliosis (267). Serial transmission of neurodegeneration was initiated with brain extracts from these Tg mice.
11. Ablation of the PrP gene in mice prevents scrapie and propagation of prions after intracerebral inoculation of prions (80,463).
12. Mice expressing chimeric Mo/SHaPrP transgenes produce "artificial" prions with novel properties (507).
13. Overexpression of MoPrP-B and SHaPrP produces spongiform degeneration, myopathy and peripheral neuropathy in older transgenic mice; serial transmission of neurodegeneration was initiated with brain extracts (559).

PrP Gene Structure and Organization

The entire ORF of all known mammalian and avian PrP genes is contained within a single exon (Fig. 1) (29,193, 259,560). This feature of the PrP gene eliminates the possibility that PrP^{Sc} arises from alternative RNA splicing (29,560,561); however, mechanisms such as RNA editing or protein splicing remain a possibility (45,287). The two exons of the SHaPrP gene are separated by a 10-kbp intron: exon 1 encodes a portion of the 5′ untranslated leader sequence while exon 2 encodes the ORF and 3′ untranslated region (29). The Mo and sheep PrP gene is comprised of three exons with exon 3 analogous to exon 2 of the hamster (558,561). The promoters of both the SHa and MoPrP genes contain multiple copies of G-C-rich repeats and are

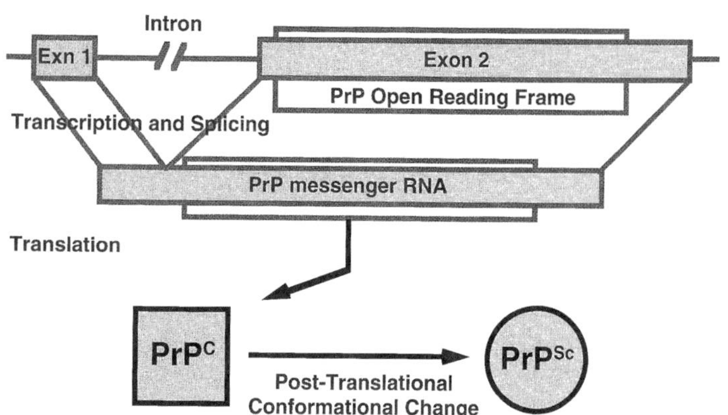

FIG. 1. Structure and organization of the chromosomal PrP gene. In all mammals examined the entire ORF is contained within a single exon. The 5′ untranslated region of the PrP mRNA is derived from either one or two additional exons (29,478,558,561). Only one PrP mRNA has been detected. PrP^{Sc} is thought to be derived from PrP^{C} by a posttranslational process (29,53,54,98,535). The amino acid sequence of PrP^{Sc} is identical to that predicted from the translated sequence of the DNA encoding the PrP gene (29,526), and no unique posttranslational chemical modifications have been identified that might distinguish PrP^{Sc} from PrP^{C}. Thus, it seems likely that PrP^{C} undergoes a conformational change as it is converted to PrP^{Sc}. From Prusiner (450), with permission.

devoid of TATA boxes. These G-C nonamers represent a motif which may function as a canonical binding site for the transcription factor Sp1 (371).

Mapping PrP genes to the short arm of Hu chromosome 20 and the homologous region of Mo chromosome 2 argues for the existence of PrP genes prior to the speciation of mammals (523). Hybridization studies demonstrated <0.002 PrP gene sequences per ID_{50} unit in purified prion fractions, indicating that a nucleic acid encoding PrP^{Sc} is not a component of the infectious prion particle (410). This is a major feature that distinguishes prions from viruses, including those retroviruses that carry cellular oncogenes and from satellite viruses that derive their coat proteins from other viruses previously infecting plant cells.

Expression of the PrP Gene

Although PrP mRNA is constitutively expressed in the brains of adult animals (105,410), it is highly regulated during development. In the septum, levels of PrP mRNA and choline acetyltransferase were found to increase in parallel during development (386). In other brain regions, PrP gene expression occurred at an earlier age. *In situ* hybridization studies show that the highest levels of PrP mRNA are found in neurons (321).

Since no antibodies are currently available that distinguish PrP^{C} from PrP^{Sc}, and vice versa, PrP^{C} is generally measured in tissues from uninfected control animals where no PrP^{Sc} is found. PrP^{Sc} must be measured in tissues of in-

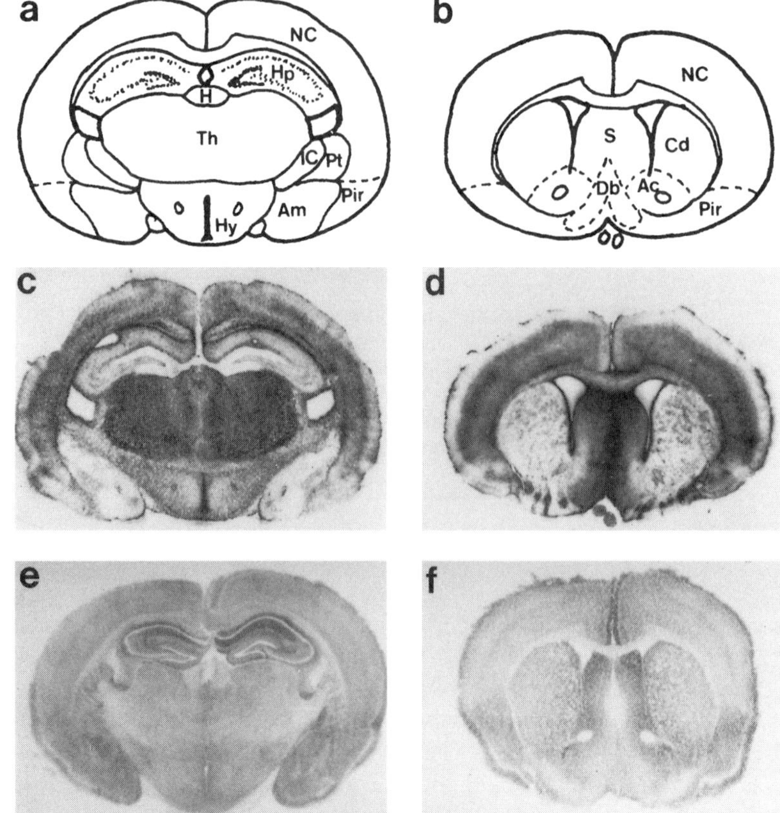

FIG. 2. Histoblots of Syrian hamster brain immunostained for PrP^{C} or PrP^{Sc}. Coronal sections through the hippocampus-thalamus (**a**,**c**,**e**) and the septum-caudate (**b**,**d**,**f**). Brain sections of a Syrian hamster clinically ill after inoculation with Sc237 prions (**c**,**d**) and an uninfected control animal (**e**,**f**). Immunostaining for PrP^{Sc} shown in **c**,**d**; for PrP^{C} in **e**,**f**. Ac, nucleus accumbens; Am, amygdala; Cd, caudate nucleus; Db, diagonal band of Broca; H, habenula; Hp, hippocampus; Hy, hypothalamus; IC, internal capsule; NC, neocortex; Th, thalamus. From Taraboulos et al. (534), with permission.

fected animals, but after PrP^{C} has been hydrolyzed by digestion with a proteolytic enzyme. PrP^{C} expression in brain was defined by standard immunohistochemistry (137) and by histoblotting in the brains of uninfected controls (Fig. 2) (534). Immunostaining of PrP^{C} in the SHa brain was most intense in the stratum radiatum and stratum oriens of the CA1 region of the hippocampus and was virtually absent from the granule cell layer of the dentate gyrus and the pyramidal cell layer throughout Ammon's horn. PrP^{Sc} staining was minimal in these regions, which were intensely stained for PrP^{C}. A similar relationship between PrP^{C} and PrP^{Sc} was found in the amygdala. In contrast, PrP^{Sc} accumulated in the medial habenular nucleus, the medial septal nuclei, and the diagonal band of Broca; these areas were virtually devoid of PrP^{C}. In the white matter, bundles of myelinated axons contained PrP^{Sc} but were devoid of PrP^{C}. These findings suggest that prions are transported along axons, in agreement with earlier findings in which scrapie infectivity was found to migrate in a pattern consistent with retrograde transport (182,283,295). While the rate of PrP^{Sc} synthesis appears to be a function of the level of PrP^{C} expression in Tg mice, the level to which PrP^{Sc} accumulates appears to be independent of PrP^{C} concentration (477).

PrP AMYLOID

The discovery of PrP 27-30 in fractions enriched for scrapie infectivity was accompanied by the identification of rod-shaped particles in the same fractions (457,475). Both by rotary shadowing and negative staining, the fine structure of these rod-shaped particles failed to reveal any regular substructure. Indeed, it was the irregular ultrastructure of the prion rods that differentiated them from viruses, which have regular, distinct structures (573), and made them indistinguishable ultrastructurally from many purified amyloids (111). This analogy was extended when the prion rods were found to display the tinctorial properties of amyloids (475). These findings were followed by the demonstration that amyloid plaques in the brains of humans and other animals with prion diseases contain PrP, as determined by immunoreactivity and amino acid sequencing (36,136,312,484,529).

The formation of prion rods requires limited proteolysis in the presence of detergent (368). Thus, the prion rods in fractions enriched for scrapie infectivity are largely, if not entirely, artifacts of the purification protocol. Solubilization of PrP 27-30 into liposomes with retention of infectivity (189) demonstrated that large PrP polymers are not required for infectivity and permitted the immunoaffinity copurification of PrP^{Sc} and infectivity (188, 191). In scrapie-infected mouse neuroblastoma cells, immunocytochemical studies demonstrated PrP^{Sc} confined largely to secondary lysosomes; there was no ultrastructural evidence for polymers of PrP (369). In Tg(SHaPrP) mice inoculated with SHa prions, numerous amyloid plaques were found but none were observed if these mice were inoculated with Mo prions, indicating that amyloid formation is not an obligatory feature of prion diseases (460,477). In accord with these Tg(SHaPrP) mouse studies, PrP plaques are consistently found in some inherited prion diseases (207) but absent in others (261).

Ultrastructural Studies in Search of a Virus

For many years, investigators searched for viruslike particles in brain sections from scrapie-infected sheep and rodents as well as humans who had died of CJD. In spite of the small target size of the infectious pathogen based on inactivation by ionizing radiation (11), there were many candidate structures reported (55,282,289,324,553). Among these were tubulovesicular structures within postsynaptic evaginations which seemed to be composed of arrays of spherical particles (26,27,130,333,334). These particles were relatively infrequent and could not be found in the brains of Syrian hamsters with clinical signs of scrapie (25). Other structures such as filamentous particles composed of protein internally and DNA on the exterior have also been reported (394–396,398,399), but these findings have not been confirmed.

Scrapie-Associated Fibrils

In crude extracts prepared from brain tissue of rodents with scrapie, as well as humans with CJD, fibrillar structures composed of two or four helically wound subfilaments were found (377,378). The crossing of these subfilaments occurred at specific intervals, and the distinctive ultrastructure of these fibers, designated *scrapie-associated fibrils* (SAF), permitted them to be distinguished from both intermediate filaments and amyloids (379). The regular substructure of SAF prompted some investigators to propose that these particles might be the first example of a filamentous animal virus and that this virus causes scrapie (376). Subsequently, some investigators argued that SAF are amyloid, despite earlier data to the contrary and the well-documented, irregular ultrastructure of purified amyloids (162,293,375,522). The term *SAF* has been inappropriately used as a synonym for *prion rods*, leading to the conclusion that SAF are composed of PrP (156).

SYNTHESIS, PROCESSING AND DEGRADATION OF PrP^{C}

Metabolic Labeling Studies

In cultured cells, PrP^{C} was found to be synthesized and degraded rapidly (96). In contrast, PrP^{Sc} is synthesized slowly by a posttranslational process (Fig. 3) (53,54,98). These observations are consistent with earlier findings, showing that PrP^{Sc} accumulates in the brains of scrapie-infected animals, while PrP mRNA levels remain unchanged (105,410).

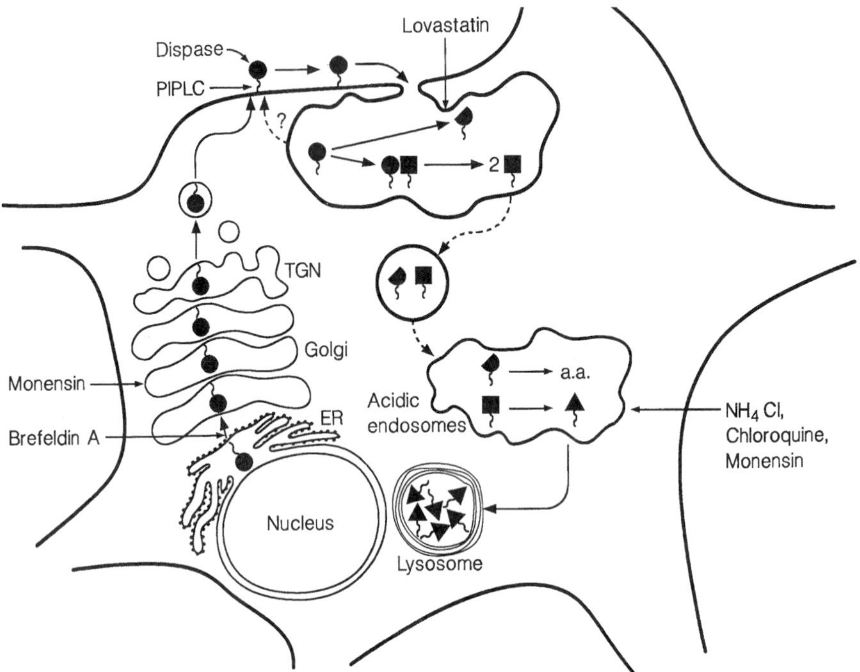

FIG. 3. Pathways of prion protein synthesis and degradation in cultured cells. PrP^{Sc} is indicated *(squares)*, as is PrP^{C} *(circles)*. Prior to becoming protease-resistant, the PrP^{Sc} precursor transits through the plasma membrane and is sensitive to dispase or PIPLC added to the medium. PrP^{Sc} formation probably occurs in a compartment accessible from the plasma membrane, such as caveolae or early endosomes, both of which are nonacidic compartments. The synthesis of nascent PrP^{Sc} seems to require the interaction of PrP^{C} with existing PrP^{Sc}. In cultured cells, but not brain, the N-terminus of PrP^{Sc} is trimmed to form PrP 27-30; PrP^{Sc} then accumulates primarily in secondary lysosomes. The inhibition of PrP^{Sc} synthesis by Brefeldin A demonstrates that the endoplasmic reticulum (ER)-Golgi is not competent for its synthesis and that transport of PrP down the secretory pathway is required for the formation of PrP^{Sc}.

PrP^{C} Transits the Golgi and Is Anchored to the Cell Surface

PrP^{C} transits through the Golgi apparatus where its Asn-linked oligosaccharides are modified and sialylated (Fig. 3) (50,173,246,352,487). Since about 30% of the glycosyl phosphatylinositol (GPI) anchors of PrP are also sialylated (525), we presume that sialylation of the anchors occurs in the Golgi. PrP^{C} is presumably transported within secretory vesicles to the external cell surface, where it is attached by a GPI anchor (Fig. 4) (497,525,527). Like other GPI anchored proteins, mouse PrP^{C} seems to return into the cell through caveolae (291,537,579). Some investigators argue that chicken PrP, which is ≈30% homologous with mammalian PrPs, is internalized into mouse neuroblastoma cells through clathrin-coated pits and that a small proportion of chicken PrP^{C} which is endocytosed is degraded, while the majority is recycled to the cell surface, along with other proteins and lipids of the endosome (511). In pulse-chase experiments, PrP^{C} in uninfected cells is rapidly labeled by a radioactive amino acid tracer and appears to be degraded in about 6 hours (53,96). Within a cholesterol-rich, nonacidic compartment, GPI-anchored PrP^{C} is partially degraded and is subsequently hydrolyzed in an acidic endosomal compartment.

Cell-Free Translation Studies

A transmembrane form of PrP—which spans the bilayer twice at the transmembrane (TM) and amphipathic helix domains—and a secretory form of PrP were identified by cell-free translation studies (338,580). The stop transfer effector (STE) domain controls the topogenesis of PrP (134). That PrP contains both a TM domain and a GPI anchor poses a topologic conundrum. It seems likely that membrane-dependent events feature in the synthesis of PrP^{Sc}, especially since brefeldin A, which selectively destroys the Golgi stacks (166), prevents PrP^{Sc} synthesis in scrapie-infected cultured cells (535). For many years, the association of scrapie infectivity with membrane fractions has been appreciated (209); indeed, hydrophobic interactions are thought to account for many of the physical properties displayed by infectious prion particles (191,467).

Function of PrP^{C} Is Unknown

The discovery that PrP^{C} is bound to the cell surface by a GPI anchor suggested that this protein might function as a receptor or adhesion molecule; alternatively, it might function upon release as a trophic factor (527). The report

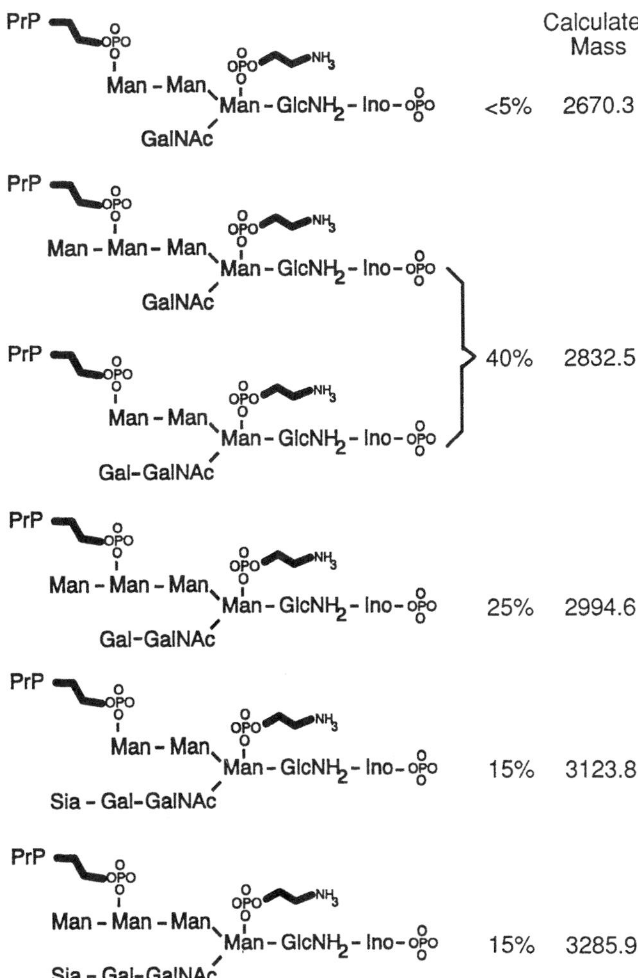

FIG. 4. Glycoinositol phospholipid anchors of the prion protein. The proposed GPI anchor structures were determined for SHaPrP^{Sc} by mass spectrometry (525). The percentages indicate an estimate of the approximate relative abundance of each glycoform. The calculated masses are based on the average molecular weight of each element in the GPI and include the mass of the C-terminal PrP (K12) peptide which accounts for 1312.5 mass units of the total. From Stahl et al. (525), with permission.

that a PrP-like molecule from chickens may have acetylcholine receptor-inducing activity (ARIA) (249) prompted considerable interest, but subsequent studies showed that ARIA is distinct from chicken PrP and that it is a member of the neu ligand family (174). Studies of Chk PrP expressed in mouse neuroblastoma cells have shown recycling of the protein between a presumed endosomal compartment and the cell surface, raising the possibility that PrP^{C} functions as a receptor for the internalization of a ligand which binds to PrP^{C} in the extracellular space (512). Furthermore, PrP^{C} does not seem to be essential, at least in young Prnp$^{0/0}$ mice, since disruption of the PrP gene has not caused any detectable abnormalities in the nervous, musculoskeletal, or lymphoreticular systems at 24 months of age (81). These results argue that scrapie and the other prion diseases do not result from an inhibition of PrP^{C} function due to PrP^{Sc}, but rather from the accumulation of PrP^{Sc} which interferes with some as-yet-undefined cellular process. It has been suggested that PrP^{C} functions in the mitogenesis of lymphocytes (93), but the lymphocytes of Prnp$^{0/0}$ mice were shown to exhibit normal responses to a number of mitogens (81). Electrophysiological studies of hippocampal slices from Prnp$^{0/0}$ mice suggest that the absence of PrP^{C} may alter synapse formation (118), while other studies have shown PrP^{C} is transported down the axons of both the central and peripheral nervous systems (51). By immunocytochemistry, PrP^{C} has been localized to the neuromuscular junction (21)

FORMATION OF PrP^{Sc}

Whether PrP^{C} is the substrate for PrP^{Sc} formation or a restricted subset of PrP molecules are precursors for PrP^{Sc} remains to be established. Several experimental results argue that PrP molecules destined to become PrP^{Sc} exit to the cell surface, as does PrP^{C} (527) prior to its conversion into PrP^{Sc} (54,98,535).

Like other GPI-anchored proteins, PrP^{C} appears to reenter the cell through a subcellular compartment bounded by cholesterol-rich, detergent-insoluble membranes (537) which might be caveolae or early endosomes (Fig. 3) (18, 291). Within this cholesterol-rich, nonacidic compartment, GPI-anchored PrP^{C} seems to be either converted into PrP^{Sc} or partially degraded (537). The partially degraded fragment of PrP^{C} appears to be the same as the protein previously designated PrP^{C}-II in partially purified fractions prepared from Syrian hamster brain (246,425). After denaturation PrP^{Sc}, like PrP^{C}, can be released from the cell membranes by digestion with phospatidylinositol-specific phospholipase C, suggesting that PrP^{Sc} is tethered only by the GPI anchor (52). In scrapie-infected cultured cells, PrP^{Sc} is trimmed at the N-terminus to form PrP 27-30 in an acidic compartment (99,535). Whether this acidic compartment is endosomal or lysosomal where PrP 27-30 accumulates (370) remains to be determined. In contrast to cultured cells, the N-terminal trimming of PrP^{Sc} is minimal in brain, where little PrP 27-30 is found (368). Deleting the GPI addition signal resulted in greatly diminished synthesis of PrP^{Sc} (488). In contrast to PrP^{C}, PrP^{Sc} accumulates primarily within cells, where it is deposited in cytoplasmic vesicles, many of which appear to be secondary lysosomes (54, 99,369,535,538).

Although most of the difference in the mass of PrP 27-30 predicted from the amino acid sequence and observed after posttranslational modification is due to complex-type oligosaccharides (246), these sugar chains are not required for PrP^{Sc} synthesis in scrapie-infected cultured cells, based on experiments with the Asn-linked glycosylation inhibitor tunicamycin and site-directed mutagenesis studies (536).

Search for a Chemical Modification

The discovery that the entire ORF of the PrP gene is contained within a single exon, first in Syrian hamsters and later in humans and other animals, argued that PrP^{Sc} is not generated by alternative splicing (29,238,239,259,558,562). This prompted us to search for a posttranslational chemical modification to explain the differences in the properties of these two PrP isoforms (526). PrP^{Sc} was analyzed by mass spectrometry and gas phase sequencing in order to identify any amino acid substitutions or posttranslational chemical modifications. The amino acid sequence was the same as that deduced from the translated ORF of the PrP gene, and no candidate posttranslational chemical modifications that might differentiate PrP^{C} from PrP^{Sc} were found (526). These findings forced consideration of the possibility that conformation distinguishes the two PrP isoforms.

Secondary Structure Prediction Studies

By comparing the amino acid sequences of 11 mammalian and 1 avian prion proteins, structural analyses predicted four α-helical regions (205,268). These prediction studies of $SHaPrP^{C}$ and $SHaPrP^{Sc}$ (residues 23 to 231) were performed using a neural network algorithm (315,446). Class-dependent (α/α, α/β, β/β) and naive predictions were performed. The α/α class contains proteins that are composed largely of α-helices. Similarly, β/β class contains proteins that are mostly β-sheets. Interestingly, the four putative α-helical domains of PrP (205) showed both strong helix preference in the α/α class prediction and strong β-sheet preference in the β/β class prediction. These results are consistent with the hypothesis that these domains undergo conformational changes from α-helices to β-sheets during the formation of PrP^{Sc}. Further support for this hypothesis comes from structural investigations of synthetic PrP peptides.

Structures of Purified PrP^{C} and PrP^{Sc}

To gather evidence for or against the hypothesis that a conformational change features in PrP^{Sc} synthesis, we purified both PrP^{C} and PrP^{Sc} using nondenaturing procedures and determined the secondary structure of each (424,549). Fourier transform infrared (FTIR) spectroscopy demonstrated that PrP^{C} has a high α-helix content (42%) and no β-sheet (3%), findings that were confirmed by circular dichroism measurements (Fig. 5) (424). In contrast, the β-sheet content of PrP^{Sc} was 43% and the α-helix 30% as measured by FTIR (424) and CD spectroscopy (499). As determined in earlier studies, N-terminally truncated PrP^{Sc} derived by limited proteolysis and designated PrP 27-30 has an even higher β-sheet content (54%) and a lower α-helix (21%) (100,204,500).

Neither purified PrP^{C} nor PrP^{Sc} formed aggregates detectable by electron microscopy, while PrP 27-30 polymerized into rod-shaped amyloids (Fig. 6). While these findings argue that the conversion of α-helices into β-sheets underlies the formation of PrP^{Sc}, we cannot eliminate the possibility that an undetected chemical modification of a small fraction of PrP^{Sc} initiates this process. Since PrP^{Sc} seems to be the only component of the "infectious" prion particle, it is likely that this conformational transition is a fundamental event in the propagation of prions. In support of the foregoing statement is the finding that denaturation of PrP 27-30 under conditions which reduced scrapie infectivity resulted in a concomitant diminution of β-sheet content (204,500).

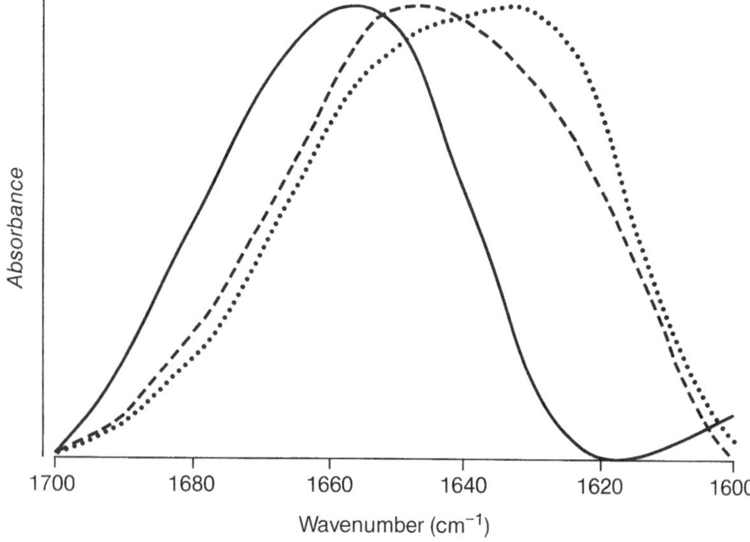

FIG. 5. Fourier transform infrared spectroscopy of prion proteins. The amide I′ band (1,700–1,600 cm^{-1}) of transmission FTIR spectra of PrP^{C} *(solid line)*, PrP^{Sc} *(dashed line)* and PrP 27-30 *(dotted line)*. These proteins were suspended in a buffer in D_2O containing 0.15 M NaCl/10 mM sodium phosphate, pD 7.5 (uncorrected)/0.12% ZW. The spectra are scaled independently to be full scale on the ordinate axis (absorbance). From Pan et al. (424), with permission.

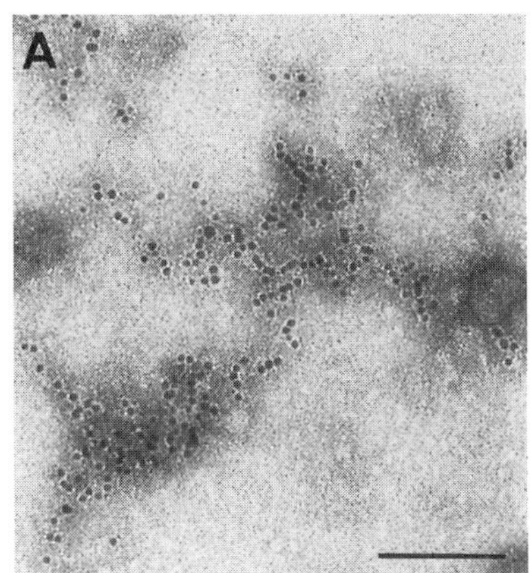

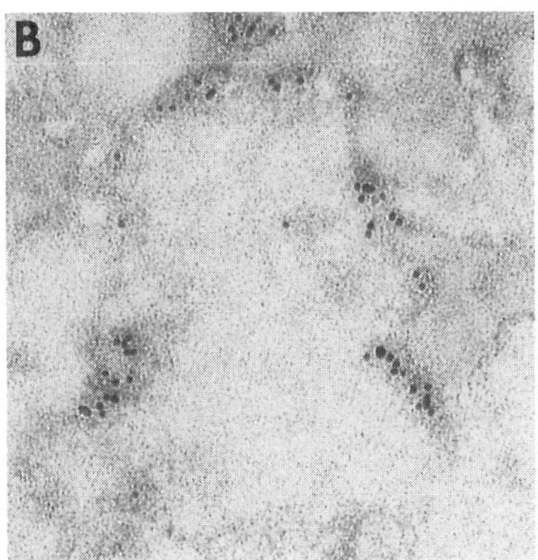

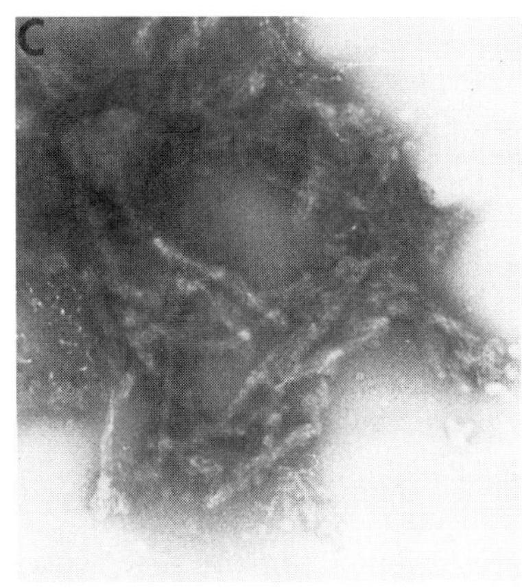

FIG. 6. Electron micrographs of negatively stained and immunogold-labeled prion proteins. **A**: PrP^C. **B**: PrP^{Sc}. **C**: Prion rods composed of PrP 27-30 were negatively stained with uranyl acetate. Bar = 100 nm. From Pan et al. (424), with permission.

Synthetic PrP Peptides

Peptides corresponding to the four putative α-helical domains of SHaPrP were synthesized and, contrary to predictions, three of the four spontaneously formed amyloids when dispersed into water, as shown by electron microscopy and Congo red staining (205). By infrared spectroscopy, these amyloid peptides were found to exhibit a secondary structure comprised largely of β-sheets, but formed α-helices when dissolved in hexafluoroisopropanol. The first of the predicted helices is the 14-residue peptide corresponding to residues 109 to 122; this peptide, designated H1, and the overlapping 15-residue sequence 113 to 127 both form amyloid. The most highly amyloidogenic peptide is the sequence AGAAAAGA corresponding to PrP residues 113 to 120. This peptide is in a region of PrP that is conserved across all known species. Two other predicted α-helices corresponding to residues 178 to 191 (H3) and 202 to 218 (H4) form amyloid and exhibit considerable β-sheet structure when synthesized as peptides.

Synthetic peptides corresponding to the region of PrP around H1 have been produced, and their propensity to polymerize studied (226,509,530). Peptides differing by only a single amino acid at position 129 exhibit slowed amyloidogenesis when mixed together (119). Polymers of one synthetic PrP peptide have been found to be neurotoxic for primary cultures of hippocampal neurons (179).

On the Mechanism of Conversion of PrP^C into PrP^{Sc}

We and others have suggested that the conversion of PrP^C into PrP^{Sc} may proceed through a metastable or partially unfolded intermediate designated PrP*, but no physical evidence for the existence of PrP* has been obtained to date (112,281). Intermediates in the refolding of PrP 27-30 from a denatured state have been identified by fluorescence and CD spectroscopy (498). Experiments with immunoprecipitated [^{35}S-] Met-labeled $SHaPrP^C$ and a >50-fold excess of purified SHaPrP 27-30 have been interpreted as showing that PrP^C becomes protease-resistant under these conditions (317). Results of experiments with radiolabeled $MoPrP^C$ as a negative control were subsequently reported. Whether the protease resistance is due to the conversion of PrP^C into PrP^{Sc} under these conditions or results from the binding of PrP^C to PrP 27-30, which is quite hydrophobic, remains to be established. When the ratios of PrP^C and PrP^{Sc} approximate those found *in vivo*, no evidence for the conversion of PrP^C into PrP^{Sc} could be found in cell-free systems (480).

In studies with synthetic peptides, the hydrophobicity of H1 that seems to be primarily responsible for its amyloid-forming properties could be offset by extending the N-terminus to include more hydrophilic residues. Longer peptides that contain Lys-104 and Lys-106 were found to be more stable as α-helices or coils than in the β-sheet conformation (404). Ambiguity between coils and α-helices is reminiscent of many nascent helices that have been characterized by 2D-NMR spectroscopy that form α-helices in appropriate solvents only if an intrinsic helical propensity exists in the sequence. Both the SHaPrP peptide designated 104H1, which extends from residue 104 to 122, and peptide H2, which extends from residue 129 to 141, provide substrates for monitoring conditions that might induce conformational changes that mirror those in PrP. H2 is the only one of the four predicted helices that does not adopt β-sheet structure when synthesized as a peptide (205).

Noncovalent interactions between peptides from different regions of a protein have been shown to enhance secondary structure when the same structures in isolated peptides have only marginal stability. We found that H2 and 104H1 both adopt β-sheet structure when mixed with H1 in solvents favoring β-sheets (404). Thus, interactions in solution are able to promote the conformational transitions. As little as 1% (wt/wt) H1 in the mixture was sufficient to promote the conformational change detected in H2 or 104H1. We also compared the efficiency of the interactions between SHa 104H1 and H1 peptides corresponding to SHa and MoPrP sequences. The SHa and Mo sequences differ by two amino acids: SHaPrP encodes Met at residues 109 and 112 while MoPrP encodes Leu and Val, respectively. In 30% acetonitrile, β-sheet is favored for both SHaH1 and MoH1, whereas SHa104H1 is largely coil. Over 24 h, homotypic interactions between the two SHa peptides resulted in the progressive conversion of coils into β-sheets, while heterotypic interactions in a mixture of the SHa and Mo PrP peptides produced substantially less β-sheet (404). These findings support the concept that the species barrier for prions is due, at least in part, to inefficient interactions between PrP molecules having slightly different amino acid sequences.

The foregoing experiments with synthetic PrP peptides appear to model some of the unique characteristics of the conformational transition underlying the conversion of PrP^{C} to PrP^{Sc}. Although it is a formal possibility that an as yet undetected covalent modification triggers a conformational transition in PrP^{C} *in vivo* to form PrP^{Sc}, these synthetic PrP peptides seem to undergo a similar transition *in vitro* without any apparent chemical change.

While the highly conserved H1 region of PrP is likely to be critical to the replication of prions, we have no information about its structure in PrP^{Sc}. Considerable data argue that PrP^{Sc} in inoculated prions either induces a conformational transition in PrP^{C} (template assistance) or interacts with low abundance intermediates (equilibrium assembly). Rarely does PrP^{C} seem to transform spontaneously into PrP^{Sc}, as is probably the case in sporadic CJD. While the energy barriers to these conformational isomerizations may be relatively modest in isolated peptides, the equivalent processes for intact proteins *in vivo* may require the intervention of molecular chaperones. The concentration dependence of aggregation may further modify the details of the conformational interconversion of PrP^{C} to PrP^{Sc}. The enthalpic advantage of aggregation may be outweighed by adverse entropic factors at low concentration.

No Evidence that Amyloid Formation Features in PrP^{Sc} Synthesis

Some investigators have suggested that scrapie agent multiplication proceeds through a crystallization process involving PrP amyloid formation (194,195,198,281). While the high β-sheet content of PrP^{Sc} is clearly a hallmark distinguishing it from PrP^{C} (424,499), there is no evidence that PrP^{Sc} forms paracrystalline arrays, as is the case for amyloids, which also have a high β-sheet content (218,219). Purified infectious prions isolated from scrapie-infected SHa brains with a cocktail of protease-inhibitors contain only PrP^{Sc} which exists as amorphous aggregates; only if PrP^{Sc} molecules are exposed to detergents and limited proteolysis do they then polymerize into prion rods exhibiting the ultrastructural and tinctorial features of amyloid (368,424) (see Fig. 6C). Furthermore, dispersion of prion rods into detergent-lipid-protein complexes results in a 10- to 100-fold increase in scrapie titer, and no rods could be identified in these fractions by electron microscopy (189). Consistent with these findings is the absence or rarity of amyloid plaques in many prion diseases, as well as the inability to identify any amyloid-like polymers in cultured cells chronically synthesizing prions (368,477).

While prion replication seems to be "seeded" by PrP^{Sc}, it can also be initiated by specific mutations, the specificity of which has been dramatically demonstrated recently when the change from a glutamate to a lysine at codon 200 produces PrP^{CJD} and CJD in humans (192), while the same glutamate to lysine mutation at codon 219 is a common polymorphism occurring in ~6% of the Japanese population (310).

The specificity of prion replication has also been demonstrated in studies of Tg mice in which the prion produced is homologous to that in the inoculum (477). While $SHaPrP^{C}$ or a chimeric $SHa/MoPrP^{C}$ can be converted into the corresponding PrP^{Sc} molecules in Tg mice inoculated with SHa prions (507), the situation with Hu prions is different (544). Tg(HuPrP) mice rarely produce $HuPrP^{Sc}$ after inoculation with Hu prions, while Tg(MHu2M) mice expressing a chimeric $Hu/MoPrP^{C}$ readily produce chimeric PrP^{Sc} after inoculation with Hu prions. These results argue that a species-specific factor(s), possibly a chaperone-like molecule, is involved in the conversion of PrP^{C} into PrP^{Sc} (544).

It is difficult to reconcile the foregoing data with the conjecture that crystallization is the mechanism by which prions replicate. A crystallization phenomenon producing multidimensional, macroscopic ordered arrays implies a high degree of at least two-dimensional—and more commonly, three-dimensional—organization in aggregates of PrP^{Sc}. It seems more likely that a complex process involving chaperones or similar molecules is responsible for transforming PrP^{C} into PrP^{Sc}.

PRION PROPAGATION

Mechanism of Prion Formation

Although the search for a scrapie-specific nucleic acid continues to be unrewarding, some investigators steadfastly cling to the notion that this putative polynucleotide drives prion replication. If prions are found to contain a scrapie-specific nucleic acid, then such a molecule would be expected to direct scrapie agent replication using a strategy similar to that employed by viruses. In the absence of any chemical or physical evidence for a scrapie-specific polynucleotide, it seems reasonable to consider some alternative

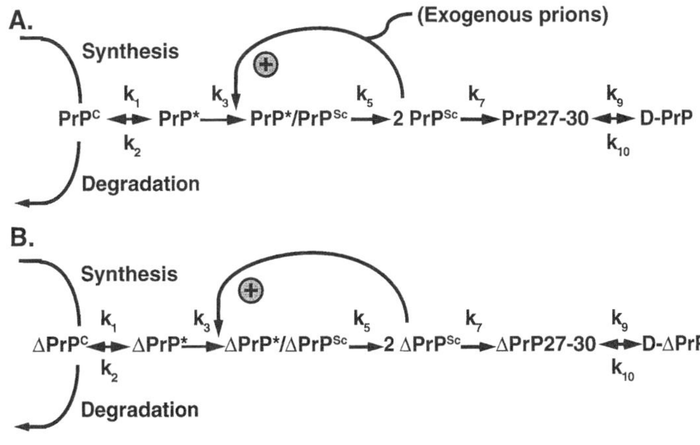

FIG. 7. Models for the replication of prions. **A**: Proposed scheme for the replication of prions in sporadic and infectious prion diseases. wtPrP^C is synthesized and degraded as part of the normal metabolism of many cells. Stochastic fluctuations in the structure of PrP^C can create (k_1) a rare, partially unfolded, monomeric structure, PrP*, that is an intermediate in the formation of PrP^{Sc} but can revert (k_2) to PrP^C or be degraded prior to its conversion (k_3) into PrP^{Sc}. Normally, the concentration of PrP* is small and PrP^{Sc} formation is insignificant. In infectious prion diseases, exogenous prions enter the cell and stimulate conversion of PrP* into PrP^{Sc}. In the absence of exogenous prions, the concentration of PrP^{Sc} may eventually reach a threshold level in sporadic prion diseases, after which a positive feedback loop would stimulate the formation of PrP^{Sc}. Limited proteolysis of the N-terminus of PrP^{Sc} produces (k_5) PrP 27-30, which can also be generated in scrapie-infected cells from a recombinant vector encoding PrP truncated at the N-terminus (488). Denaturation (k_7) of PrP^{Sc} or PrP 27-30 renders these molecules protease-sensitive and abolishes scrapie infectivity; attempts to renature (k_8) these PrP^{Sc} or PrP 27-30 have been unsuccessful to date (464,475). **B**: Scheme for the replication of prions in genetic prion diseases. Mutant (Δ) PrP^C is synthesized and degraded as part of the normal metabolism of many cells. Stochastic fluctuations in the structure of ΔPrP^C are increased compared to wtPrP^C, which creates (k_1) a partially unfolded, monomeric structure, ΔPrP*, that is an intermediate in the formation of ΔPrP^{Sc}, but can revert (k_2) to ΔPrP^C or be degraded prior to its conversion (k_3) into ΔPrP^{Sc}. Limited proteolysis of the N-terminus of ΔPrP^{Sc} produces (k_5) ΔPrP 27-30 which in some cases may be less protease resistant than wtPrP 27-30 (387,472). Adapted from Cohen et al. (112).

mechanisms that might feature in prion biosynthesis. The multiplication of prion infectivity is an exponential process in which the posttranslational conversion of PrP^C or a precursor to PrP^{Sc} appears to be obligatory (53).

Stochastic fluctuations in the structure of PrP^C can create a rare, partially unfolded, monomeric structure, PrP*, that is an intermediate in the formation of PrP^{Sc} but can revert to PrP^C or be degraded prior to its conversion into PrP^{Sc} (Fig. 7) (112). Normally, the concentration of PrP* is small and PrP^{Sc} formation is insignificant. Whether PrP^{Sc} formation involves oligomerization remains uncertain, since the insolubility of PrP^{Sc} has precluded analysis of its physical state (189).

Infection with exogenous prions containing PrP^{Sc} would act as a template to promote the conversion of PrP* into PrP^{Sc} (Fig. 7A). The insolubility of PrP^{Sc} would make this process irreversible and drive the formation of PrP* as well as PrP^{Sc} by mass action. Examples of prion infection include: kuru caused by ritualistic cannibalism among the Fore people of New Guinea, CJD in young adults treated with prion-contaminated growth hormone extracted from human pituitaries, and "mad cow" disease caused by prion-contaminated meat and bone meal prepared from the offal of scrapied sheep or "mad" cattle (72,197,566).

The sporadic prion diseases might result from the relatively infrequent accumulation of sufficient PrP* to produce PrP^{Sc} (Fig. 7A). In support of this proposal are Tg mice overexpressing wtPrP that develop spongiform degeneration and produce prions in their brains (559). Alternatively, somatic mutations might destabilize PrP^C, which would promote its conversion into PrP*. Of note is the common polymorphism encoding Met or Val at codon 129 in human PrP which lies at the N-terminal border of second putative α-helix in PrP^C (Fig. 7B) but may be at the end of a β-strand in PrP^{Sc} (268). That homozygosity at codon 129 appears to predispose to sporadic CJD (421) is intriguing, since this residue could lie at a multimeric interface joining two β-sheets. Studies of other proteins suggest that Met-Met and Val-Val are common neighbors in β-sheets and Met-Val are distinctly unusual (335). Patients with the D178N mutation present with insomnia if codon 129 encodes Met on the mutant allele or dementia if codon 129 encodes Val (232,387).

Homophilic Interactions

The results of transgenetic studies argue that PrP^{Sc} combines with PrP* or PrP^C to form a transient complex which is subsequently transformed into two molecules of PrP^{Sc} (Fig. 7A) (477). In the next cycle, two PrP^{Sc} molecules combine with two PrP* or PrP^C molecules, giving rise to two complexes that dissociate to combine with four PrP* or PrP^C molecules to create an exponential process. As noted above, PrP^{Sc} formation may involve oligomerization, but the insolubility of PrP^{Sc} has precluded determination of the stoichiometry (189). If prion biosynthesis simply involved amplification of posttranslationally modified PrP molecules, we might expect Tg(SHaPrP) mice to produce both SHa and Mo prions after inoculation with either prion, since these mice produce SHa and MoPrP^C. Yet Tg(SHaPrP) mice synthesize only those prions present in the inoculum (see below, Figs. 8E,8F). These results contend that the incoming prion and PrP^{Sc} interact with the homologous PrP* or PrP^C substrate to replicate more of the same prions (Fig. 7A).

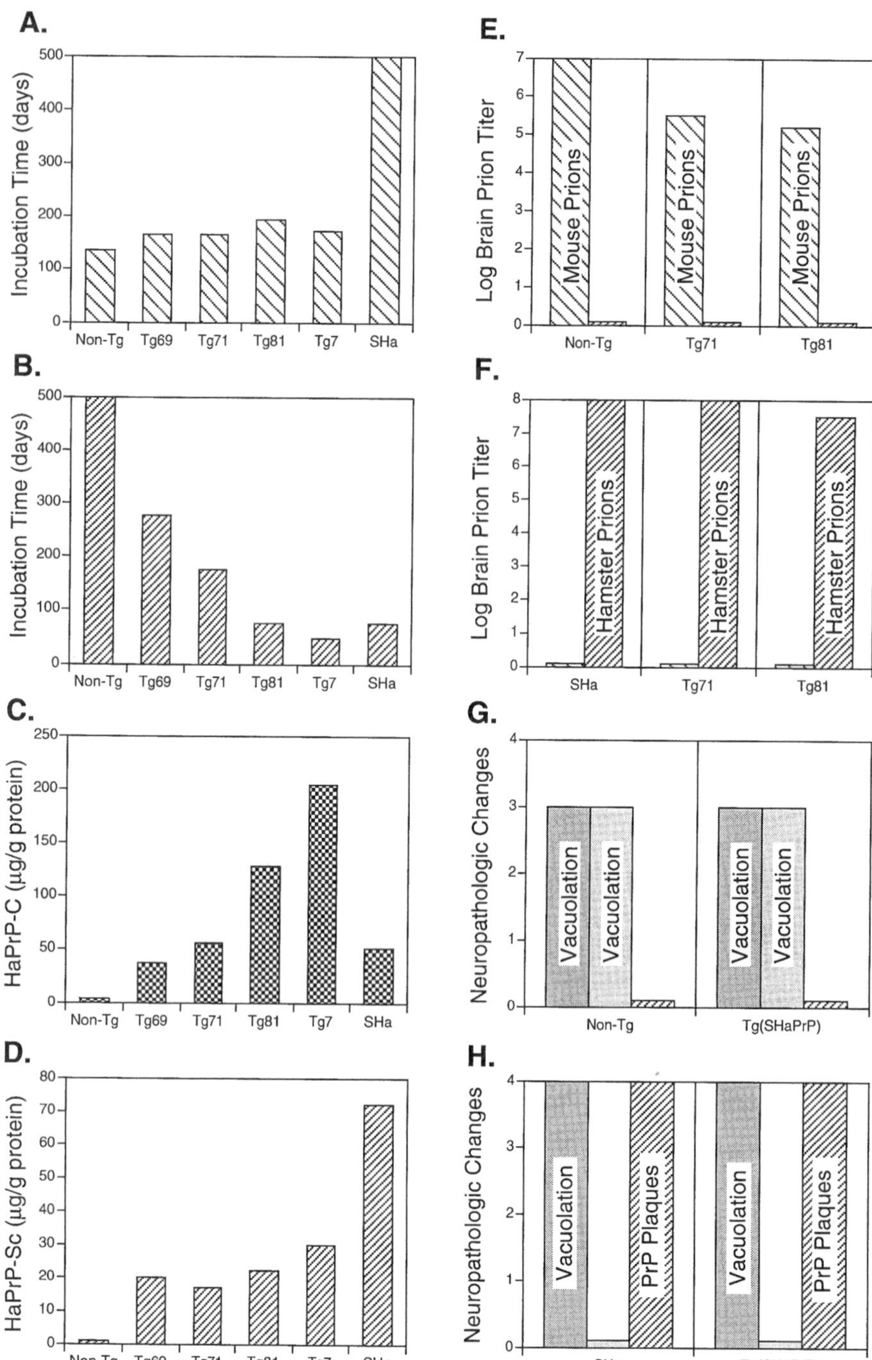

FIG. 8. Transgenic *(Tg)* mice expressing Syrian hamster *(SHa)* prion protein exhibit species-specific scrapie incubation times, infectious prion synthesis, and neuropathology (477). **A**: Scrapie incubation times in nontransgenic mice (Non-Tg) and four lines of Tg mice expressing SHaPrP and Syrian hamsters inoculated intracerebrally with ~10^6 ID_{50} units of Chandler Mo prions serially passaged in Swiss mice. The four lines of Tg mice have different numbers of transgene copies: Tg69 and 71 mice have two to four copies of the SHaPrP transgene, whereas Tg81 have 30 to 50, and Tg7 mice have >60. Incubation times are measured in days from inoculation to onset of neurologic dysfunction. **B**: Scrapie incubation times in mice and hamsters inoculated with ~10^7 ID_{50} units of Sc237 prions serially passaged in Syrian hamsters and as described in **A**. **C**: Brain $SHaPrP^C$ levels in Tg mice and hamsters. $SHaPrP^C$ levels were quantitated by an enzyme-linked immunoassay. **D**: Brain $SHaPrP^{Sc}$ in Tg mice and hamsters. Animals were killed after exhibiting clinical signs of scrapie. $SHaPrP^{Sc}$ levels were determined by immunoassay. **E**: Prion titers in brains of clinically ill animals after inoculation with Mo prions. Brain extracts from Non-Tg, Tg71, and Tg81 mice were bioassayed for prions in mice *(left)* and hamsters *(right)*. **F**: Prion titers in brains of clinically ill animals after inoculation with SHa prions. Brain extracts from Syrian hamsters as well as Tg71 and Tg81 mice were bioassayed for prions in mice *(left)* and hamsters *(right)*. **G**: Neuropathology in non-Tg mice and Tg(SHaPrP) mice with clinical signs of scrapie after inoculation with Mo prions. Vacuolation in grey *(left)* and white matter *(center)*; PrP amyloid plaques *(right)*. Vacuolation score: 0, none; 1, rare; 2, modest; 3, moderate; 4, intense. **H**: Neuropathology in Syrian hamsters and transgenic mice inoculated with SHa prions. Degree of vacuolation and frequency of PrP amyloid plaques as described in **G**. Adapted from Prusiner (452).

Additional evidence in support of the proposed model for prion replication comes from Tg(Mo/SHaPrP) mice expressing chimeric Mo/SHaPrP^{C} (507). The chimeric Mo/SHaPrP gene was constructed by substituting the SHaPrP sequence for MoPrP from codon 94 to 188; within this domain, there are five amino acid substitutions which distinguish Mo from SHaPrP. When inoculated with either Mo or SHa prions, these Tg(Mo/SHaPrP) mice develop scrapie after ~140 days. The chimeric Tg mice produce Mo/SHaPrP^{Sc} and Mo/SHa prions after inoculation with SHa prions and probably Mo prions as well. Evidence for chimeric Mo/SHa prions comes from the development of scrapie in Tg(Mo/SHaPrP) mice ~70 days after inoculation with brain extracts from Tg(Mo/SHaPrP) mice containing the chimeric prions.

Because mice expressing HuPrP transgenes have not had abbreviated incubation times similar to Tg(SHaPrP) mice when inoculated with homologous prions, we constructed a chimeric Hu/Mo transgene similar to the chimeric MH2M transgene. Hu PrP differs from mouse PrP at 28 of 254 positions (322) while chimeric Hu/Mo PrP, designated MHu2M differs at nine residues. Mice expressing the MHu2M transgene are susceptible to human prions and exhibit abbreviated incubation times (544). These findings argue that a mouse-specific factor that we have provisionally designated protein X forms a ternary complex with PrP^{C} and PrP^{Sc} during the formation of nascent PrP^{Sc}. Failure to produce nascent PrP^{Sc} when PrP^{Sc} is mixed with PrP^{C} (480) also argues for another factor such as protein X. Whether protein X is a chaperone that is involved in catalyzing the conformational changes that feature in the formation of PrP^{Sc} (424) remains to be established. One possible explanation for the difference in susceptibility of Tg(MHu2M) and Tg(HuPrP) mice to Hu prions is that mouse chaperones catalyzing the refolding of PrP^{C} into PrP^{Sc} can readily interact with the MHu2MPrP^{C}/HuPrP^{CJD} complex but not with HuPrP^{C}/HuPrP^{CJD}.

Conversion of PrP α-Helices into β-Sheets

Numerous investigations described above have been directed toward finding a scrapie-specific nucleic acid, without success. In contrast, there has been a compelling convergence of data implicating PrP^{Sc} in the transmission and pathogenesis of prion diseases (Table 3). Since studies of PrP^{Sc} failed to reveal a candidate posttranslational chemical modification that might distinguish it from PrP^{C} (526), the secondary structures of PrP^{C} and PrP^{Sc} were determined (424). As described above, FTIR spectroscopy demonstrated that PrP^{C} has a high α-helix and low β-sheet content (Fig. 5), findings that were confirmed by circular dichroism measurements (424). In contrast, the β-sheet content of PrP^{Sc} was more than 40% and the α-helix 30%, as measured by FTIR. While these findings argue that the conversion of α-helices into β-sheets underlies the formation of both PrP^{Sc} and infectious prions, we cannot eliminate the possibility that an undetected chemical modification of a small fraction of PrP^{Sc} initiates this process.

Mutant PrP Molecules

In humans carrying point mutations or inserts in their PrP genes (see below, Fig. 12), mutant PrP^{C} molecules might spontaneously convert into PrP^{Sc} (Fig. 7B). Eleven point mutations and seven inserts composed of different numbers of tandemly repeated octamers in PrP have been found to segregate with inherited prion diseases; five mutations have occurred in families sufficiently large to establish genetic linkage (163,192,259,439,444). Ten of eleven point mutations are located within or near the four α-helices, and thus cluster around a central core region that is essential for the structural stability of PrP^{C} (268). The tertiary structural model of PrP^{C} was generated by examining all plausible spatial arrangements for the α-helices in this putative four-helix bundle protein. Presumably, PrP mutations (Δ) destabilize ΔPrP^{C} and promote its conversion into ΔPrP* which results in the formation of ΔPrP^{Sc} (Fig. 7B) (112). While the initial stochastic event may be inefficient, once it happens the process becomes autocatalytic.

The proposed mechanism for prion replication in genetic prion diseases is consistent with individuals harboring germline mutations who do not develop CNS dysfunction for decades and with studies on Tg(MoPrP-P101L)H mice that spontaneously develop CNS degeneration (266,267). Whether all GSS and familial CJD cases contain infectious prions or some represent inborn errors of PrP metabolism in which neither PrP^{Sc} nor prion infectivity accumulates is unknown; however, transmission of inherited human prion diseases to rodents is less frequent than for sporadic CJD (539). In contrast, transmission of inherited prion diseases to apes and monkeys seems to be only slightly less efficient than from sporadic or iatrogenic cases (68). The use of Tg(MHu2M) mice will allow more detailed studies of the transmission of the inherited human prion diseases to experimental hosts and should clarify some of the unresolved issues surrounding mutant PrP^{C} molecules and their role in the pathogenesis of CNS degeneration (544).

TRANSGENETICS AND GENE TARGETING

While transgenetic studies have yielded a wealth of new knowledge about infectious, genetic and sporadic prion diseases, the laborious production of Tg mice limits the number of studies that can be performed. The relatively long gestation period of mice, coupled with the need to do microinjections of fertilized embryos, prevents the creation of the very large numbers of different Tg mice that would yield the greatest amount of new information. However, it is important to stress that transgenetic studies can readily yield an incomplete and sometimes erroneous interpretation of the data if the number of lines of mice examined expressing a particular construct is inadequate. Defining an adequate number of lines is difficult, but certainly, comparisons of lines expressing high and low levels of a given PrP transgene have proved to be quite helpful (265,477).

Species Barriers for Transmission of Prion Diseases

The passage of prions between species is a stochastic process characterized by prolonged incubation times (430, 431,434). Prions synthesized *de novo* reflect the sequence of the host PrP gene and not that of the PrP^{Sc} molecules in the inoculum (48). On subsequent passage in a homologous host, the incubation time shortens to that recorded for all subsequent passages, and it becomes a nonstochastic process. The species barrier concept is of practical importance in assessing the risk for humans of developing CJD after consumption of scrapie-infected lamb or BSE-infected beef (238,256,461,567,568).

To test the hypothesis that differences in PrP gene sequences might be responsible for the species barrier, Tg mice expressing SHaPrP were constructed (477,506). The PrP genes of Syrian hamsters and mice encode proteins differing at 16 positions. Incubation times in four lines of Tg(SHaPrP) mice inoculated with Mo prions were prolonged compared to those observed for non-Tg, control mice (Fig. 8A). Inoculation of Tg(SHaPrP) mice with SHa prions demonstrated abrogation of the species barrier, resulting in abbreviated incubation times due to a nonstochastic process (Fig. 8B) (477,506). The length of the incubation time after inoculation with SHa prions was inversely proportional to the level of $SHaPrP^{C}$ in the brains of Tg(SHaPrP) mice (Figs. 8B,8C) (477). $SHaPrP^{Sc}$ levels in the brains of clinically ill mice were similar in all four Tg(SHaPrP) lines inoculated with SHa prions (Fig. 8D). Bioassays of brain extracts from clinically ill Tg(SHaPrP) mice inoculated with Mo prions revealed that only Mo prions and no SHa prions were produced (Fig. 8E). Conversely, inoculation of Tg(SHaPrP) mice with SHa prions led to only the synthesis of SHa prions (Fig. 8F). Thus, the *de novo* synthesis of prions is species-specific and reflects the genetic origin of the inoculated prions. Similarly, the neuropathology of Tg(SHaPrP) mice is determined by the genetic origin of prion inoculum. Mo prions injected into Tg(SHaPrP) mice produced a neuropathology characteristic of mice with scrapie. A moderate degree of vacuolation in both the gray and white matter was found while amyloid plaques were rarely detected (Fig. 8G) (Table 4). Inoculation of Tg(SHaPrP) mice with SHa prions produced intense vacuolation of the gray matter, sparing of the white matter, and numerous SHaPrP amyloid plaques characteristic of Syrian hamsters with scrapie (Fig. 8H).

Overexpression of wt PrP Transgenes

During transgenetic studies, we discovered that uninoculated older mice harboring high copy numbers of wt PrP transgenes derived from Syrian hamsters, sheep, and PrP-B mice spontaneously developed truncal ataxia, hind-limb paralysis, and tremors (558). These Tg mice exhibited a profound necrotizing myopathy involving skeletal muscle, a demyelinating polyneuropathy, and focal vacuolation of the CNS. Development of disease was dependent on transgene dosage. For example, $Tg(SHaPrP^{+/+})7$ mice homozygous for the SHaPrP transgene array regularly developed disease between 400 and 600 days of age, while hemizy-

TABLE 4. *Species-specific prion inocula determine the distribution of spongiform change and deposition of PrP amyloid plaques in transgenic mice*

	SHa prions					Mo prions			
		Spongiform change[a]		PrP plaques[b]			Spongiform changes[a]		PrP plaques[b]
Animal	n	Grey	Wht	Frequency	Diameter[c]	n	Grey	Wht	Frequency
Non-Tg		N.D.		N.D.		10	+	+	–
Tg 69	6	+	–	Numerous	6.5±3.1 (389)	2	+	+	–
Tg 71	5	+	–	Numerous	8.1±3.6 (345)	2	+	+	–
Tg 81	7	+	–	Numerous	8.3±3.0 (439)	3	+	+	Few
Tg 7	3	+[d]	–	Numerous	14.0±8.3 (19)	4	+	+	–
SHa	3	+	–	Numerous	5.7±2.7 (247)		N.D.		N.D.

[a] Spongiform change evaluated in hippocampus, thalamus, cerebral cortex and brainstem for grey matter and the deep cerebellum for white matter.

[b] Plaques in the subcallosal region were stained with SHaPrP mAb 13A5, anti-PrP rabbit antisera R073 and trichrome stain.

[c] Mean diameter of PrP plaques given in microns±standard error with the number of observations in parentheses.

[d] Focal: confirmed to the dorsal nucleus of the raphe.

From Prusiner (451). n, number of brains examined; N.D., not determined; +, present; –, not found.

gous Tg(SHaPrP$^{+/0}$)7 mice also developed disease, but after >650 days.

Attempts to demonstrate PrP^{Sc} in either muscle or brain were unsuccessful, but transmission of disease with brain extracts from Tg(SHaPrP$^{+/+}$)7 mice inoculated into Syrian hamsters did occur. These Syrian hamsters had PrP^{Sc}, as detected by immunoblotting and spongiform degeneration (D. Groth and S. B. Prusiner, unpublished data). Serial passage with brain extracts from these animals to recipients was observed. *De novo* synthesis of prions in Tg(SHaPrP$^{+/+}$)7 mice overexpressing wtSHaPrP^{C} provides support for the hypothesis that sporadic CJD does not result from infection but rather is a consequence of the spontaneous, although rare, conversion of PrP^{C} into PrP^{Sc}. Alternatively, a somatic mutation in which mutant SHaPrP^{C} is spontaneously converted into PrP^{Sc} as in the inherited prion diseases could also explain sporadic CJD. These findings, as well as those described below for Tg(MoPrP-P101L) mice, argue that prions are devoid of foreign nucleic acid, in accord with many earlier studies that use other experimental approaches, as described above.

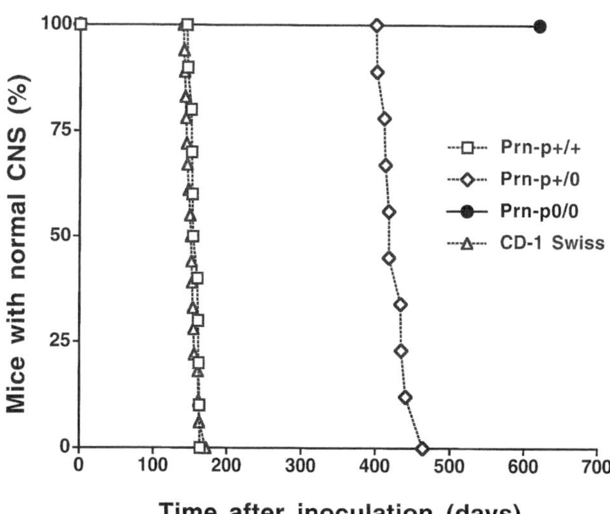

FIG. 9. Incubation times in PrP gene ablated Prnp$^{0/+}$ and Prnp$^{0/0}$ mice as well as wt Prnp$^{+/+}$ and CD-1 mice inoculated with RML mouse prions. The RML prions were heated and irradiated at 254 nm prior to intracerebral inoculation into CD-1 Swiss mice *(open triangles)*, Prn-p$^{+/+}$ mice *(open squares)*, Prn-p$^{0/+}$ mice *(open diamonds)* or Prn-p$^{0/0}$ mice *(filled circle)*.

Ablation of the PrP Gene

Unexpectedly, ablation of the PrP gene in Tg (Prnp$^{0/0}$) mice has not affected the development of these animals (81). In fact, they are healthy at almost 2 years of age. Prnp$^{0/0}$ mice are resistant to prions (Fig. 9) and do not propagate scrapie infectivity (80,463). Prnp$^{0/0}$ mice were sacrificed 5, 60, 120, and 315 days after inoculation with RML prions passaged in CD-1 Swiss mice. Except for residual infectivity from the inoculum detected at 5 days after inoculation, no infectivity was detected in the brains of Prnp$^{0/0}$ mice (Table 5).

Prnp$^{0/0}$ mice crossed with Tg(SHaPrP) mice were rendered susceptible to SHa prions but remained resistant to Mo prions (80,463). Since the absence of PrP^{C} expression does not provoke disease, it is likely that scrapie and other prion diseases are a consequence of PrP^{Sc} accumulation rather than an inhibition of PrP^{C} function (81).

Mice heterozygous (Prnp$^{0/+}$) for ablation of the PrP gene had prolonged incubation times when inoculated with Mo prions (Fig. 9) (463). The Prnp$^{0/+}$ mice developed signs of neurologic dysfunction at 400 to 460 days after inocula-

TABLE 5. *Prion titers in brains of Prn-p$^{0/0}$ and Prn-p$^{0/+}$ mice*

	Time of sacrifice after inoculation with RML scrapie prions				
	5 days	60 days	120 days	315 days	500 days
Mouse	Log scrapie prion titers (ID_{50} units/ml±SE)[a]				
Prn-p$^{+/+}$	<1	3.9±0.4	6.4±0.3		
	<1	4.8±0.3	7.1±0.1		
	<1	4.6±0.2	6.6±0.2		
Prn-p$^{+/0}$	<1	<1	5.1±0.2		
	0.6±0.7	<1	5.2±0.6		
	1.2±0.1[b]	3.4±0.2	2.8±0.1		
Prn-p$^{0/0}$	<1[c]	<1	<1	<1	<1
	<1[d]	<1	<1	<1	<1
		<1		<1	
				<1	

[a] Titers are for 10% (w/v) brain homogenates. Log titers of <1 reflect no signs of CNS dysfunction in CD-1 mice for >250 days after inoculation except as noted.
[b] 3/9 mice developed scrapie between 208 and 210 days after inoculation.
[c] 2/9 mice developed scrapie between 208 and 225 days after inoculation.
[d] 2/10 mice developed scrapie between 208 and 225 days after inoculation.

tion. These findings are in accord with studies on Tg(SHaPrP) mice in which increased SHaPrP expression was accompanied by diminished incubation times (Fig. 8B) (477).

Since $Prnp^{0/0}$ mice do not express PrP^C, we reasoned that they might more readily produce α-PrP antibodies. $Prnp^{0/0}$ mice immunized with Mo or SHa prion rods produced α-PrP antisera which bound Mo, SHa, and Hu PrP (463). These findings contrast with earlier studies in which α-MoPrP antibodies could not be produced in mice, presumably because the mice had been rendered tolerant by the presence of $MoPrP^C$ (28,288,486). That $Prnp^{0/0}$ mice readily produce α-PrP antibodies is consistent with the hypothesis that the lack of an immune response in prion diseases is due to the fact that PrP^C and PrP^{Sc} share many epitopes. Whether $Prnp^{0/0}$ mice produce α-PrP antibodies that specifically recognize conformational-dependent epitopes present on PrP^{Sc} but absent from PrP^C remains to be determined.

Modeling of GSS in Tg(MoPrP-P101L) Mice

The codon 102 point mutation found in GSS patients was introduced into the MoPrP gene, and Tg(MoPrP-P101L)H mice were created expressing high (H) levels of the mutant transgene product. The two lines of Tg(MoPrP-P101L)H mice designated 174 and 87 spontaneously developed CNS degeneration, characterized by clinical signs indistinguishable from experimental murine scrapie and neuropathology consisting of widespread spongiform morphology, astrocytic gliosis, (267) and PrP amyloid plaques (Fig. 10) (265). By inference, these results contend that PrP gene mutations cause GSS, familial CJD, and FFI.

Brain extracts prepared from spontaneously ill Tg (MoPrP-P101L)H mice transmitted CNS degeneration to Tg196 mice and some Syrian hamsters (265). The Tg196 mice express low levels of the mutant transgene product and do not develop spontaneous disease. Many Tg196 mice

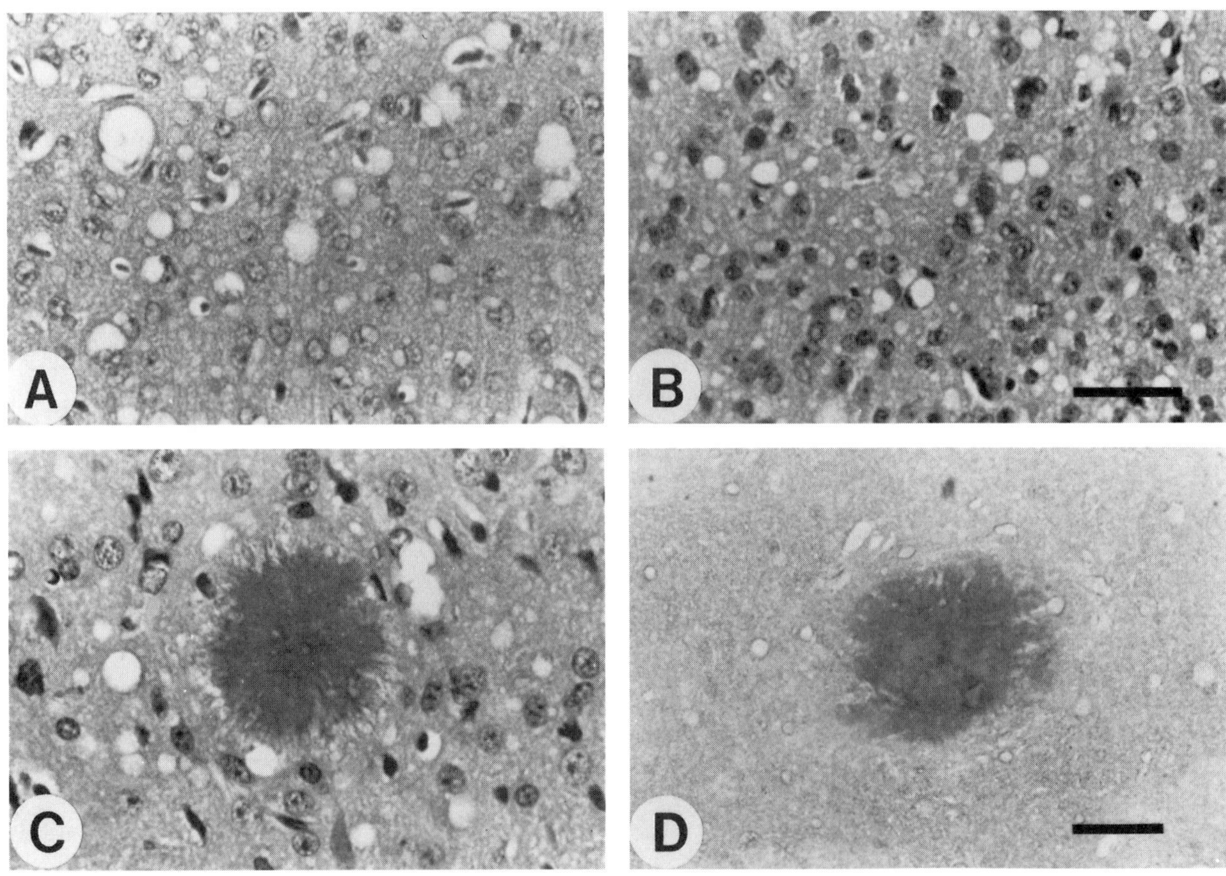

FIG. 10. Neuropathology of Tg(MoPrP-P101L) mice developing neurodegeneration spontaneously. The mice harbor transgenes carrying the PrP point mutation found in GSS(P102L) of humans. **A**: Vacuolation in cerebral cortex of a Swiss CD-1 mouse that exhibited signs of neurologic dysfunction at 138 days after intracerebral inoculation with ~10^6 ID_{50} units of RML scrapie prions. **B**: Vacuolation in cerebral cortex of a Tg(MoPrP-P101L) mouse that exhibited signs of neurologic dysfunction at 252 days of age. **C**: Kuru-type PrP amyloid plaque stained with periodic acid Schiff in the caudate nucleus of a Tg(MoPrP-P101L) mouse that exhibited signs of neurologic dysfunction. **D**: PrP amyloid plaques stained with α-PrP antiserum (RO73) in the caudate nucleus of a Tg(MoPrP-P101L) mouse that exhibited signs of neurologic dysfunction. Bar in **B** also applies to **A** (Bar=50 μm). Bar in **D** also applies to **C** (Bar = 25 μm. From Prusiner (449), with permission.

and some Syrian hamsters developed CNS degeneration between 200 and 700 days after inoculation, while inoculated CD-1 Swiss mice remained well. Serial transmission of CNS degeneration in Tg196 mice required about 1 year, while serial transmission in Syrian hamsters occurred after ~75 days (265). Although brain extracts prepared from Tg (MoPrP-P101L)H mice transmitted CNS degeneration to some inoculated recipients, little or no PrP^{Sc} was detected by immunoassays after limited proteolysis. Undetectable or low levels of PrP^{Sc} in the brains of these Tg(MoPrP-P101L)H mice are consistent with the results of these transmission experiments, which suggest low titers of infectious prions. Though no PrP^{Sc} was detected in the brains of inoculated Tg196 mice exhibiting neurologic dysfunction by immunoassays after limited proteolysis, PrP amyloid plaques as well as spongiform degeneration were frequently found. The neurodegeneration found in inoculated Tg196 mice seems likely to result from a modification of mutant PrP^{C} that is initiated by mutant PrP^{Sc} present in the brain extracts prepared from ill Tg(MoPrP-P101L)H mice. In support of this explanation are the findings in some of the inherited human prion diseases, as described above, where neither protease-resistant PrP (69,372) nor transmission to experimental rodents could be demonstrated (539). Furthermore, transmission of disease from Tg(MoPrP-P101L)H mice to Tg196 mice, but not to Swiss mice, is consistent with earlier findings which demonstrate that homotypic interactions between PrP^{C} and PrP^{Sc} markedly enhance the formation of PrP^{Sc}. Why Syrian hamsters are more permissive than Swiss mice for prion transmission from Tg (MoPrP-P101L)H mice is unknown. Presumbably, transmission to hamsters reflects the differences in tertiary structure between the two substrates $SHaPrP^{C}$ and $MoPrP^{C}$.

In other studies, modifying the expression of mutant and wtPrP genes in transgenic (Tg) mice permitted experimental manipulation of the pathogenesis of both inherited and infectious prion diseases. Although overexpression of the wtPrP-A transgene ~eightfold was not deleterious to the mice, it did shorten scrapie incubation times from ~145 days to ~45 days after inoculation with Mo scrapie prions (G. C. Telling, T. Haga, D. Foster, et al., unpublished data). In contrast, overexpression at the same level of a PrP-A transgene, mutated at codon 101, produced spontaneous, fatal neurodegeneration between 150 and 300 days of age in two new lines of Tg(MoPrP-P101L) mice designated 2866 and 2247. Genetic crosses of Tg(MoPrP-P101L)2866 mice with gene targeted mice lacking both PrP alleles ($Prnp^{0/0}$) produced animals with a highly synchronous onset of illness between 150 and 160 days of age. The Tg(MoPrP-P101L)2866/$Prnp^{0/0}$ mice had numerous PrP plaques and widespread spongiform degeneration in contrast to the Tg2866 and 2247 mice that exhibited spongiform degeneration, but only a few PrP amyloid plaques. Another line of mice designated Tg2862 overexpress the mutant transgene ~32-fold and develop fatal neurodegeneration between 200 and 400 days of age. Tg2862 mice exhibited the most severe spongiform degeneration and had numerous, large PrP amyloid plaques. While mutant PrP^{C}(P101L) clearly produces neurodegeneration, $wtPrP^{C}$ profoundly modifies both the age of onset of illness and the neuropathology for a given level of transgene expression. These findings and those from other studies (544) suggest that mutant and wtPrP interact, perhaps through a chaperone-like protein, to modify the pathogenesis of the dominantly inherited prion diseases.

NON-PrP PROTEINS MODIFY PRION DISEASES

The results of the foregoing investigations indicate that PrP transgenes modulate virtually all aspects of scrapie, including prion propagation, incubation time length, synthesis of PrP^{Sc}, the species barrier, and neuropathologic lesions. However, evidence for the role of proteins not encoded by PrP genes is now beginning to emerge. In particular, studies with Tg(MHu2M) and Tg(HuPrP) mice argue for the existence of a species-specific factor that has provisionally been designated protein X (544). Investigations of prion strains in congenic mice suggest that a gene linked to, but separate from, PrP profoundly modifies the neuropathology of scrapie (85). We have provisionally designated this gene product as protein Y.

Protein X and the Transmission of Prions

Attempts to abrogate the prion species barrier between humans and mice by using an approach similar to that described for the abrogation of the species barrier between Syrian hamsters and mice were unsuccessful. Mice expressing HuPrP transgenes did not develop signs of CNS dysfunction more rapidly or frequently than non-Tg controls (544).

The successful breaking of the species barrier between humans and mice has its origins in a set of studies with Tg mice expressing chimeric PrP genes derived from SHa and Mo PrP genes (508). One SHa/MoPrP gene, designated MH2M PrP, contains five amino acid substitutions encoded by SHaPrP, while another construct designated MHM2 PrP has two substitutions. Tg(MH2M PrP) mice were susceptible to both SHa or Mo prions, whereas three lines expressing MHM2 PrP were resistant to SHa prions (507). The brains of Tg(MH2M PrP) mice dying of scrapie contained chimeric PrP^{Sc} and prions with an artificial host range favoring propagation in mice that express the corresponding chimeric PrP and were also transmissible, at reduced efficiency, to non-Tg mice and hamsters. These findings provided additional genetic evidence for homophilic interactions between PrP^{Sc} in the inoculum and PrP^{C} synthesized by the host.

With the recognition that Tg(HuPrP) mice were not suitable recipients for the transmission of Hu prions, we constructed Tg(MHu2M) mice analogous to the Tg(MH2M) mice described above. HuPrP differs from MoPrP at 28 of 254 positions (322), while chimeric MHu2MPrP differs at nine residues. The mice expressing the MHu2M transgene are susceptible to human prions and exhibit abbreviated incubation times of ~200 days (544). In these initial studies, the chimeric MHu2M transgene encoded a Met at codon 129, and all three of the patients were homozygous for Met at this residue. Two of the cases were sporadic CJD, and the third was an iatrogenic case which occurred after treatment with pituitary derived human growth hormone (HGH). Whether it will be necessary to match the PrP genotype of the Tg(MHu2M) mouse with that of the CJD patient from whom the inoculum was derived or some variations in sequence can be tolerated remains to be established.

From Tg(SHaPrP) mouse studies, prion propagation is thought to involve the formation of a complex between PrP^{Sc} and the homotypic substrate PrP^{C} (477). Attempts to mix PrP^{Sc} with PrP^{C} have failed to produce nascent PrP^{Sc} (480), raising the possibility that proteins such as chaperones might be involved in catalyzing the conformational changes that feature in the formation of PrP^{Sc} (424). One explanation for the difference in susceptibility of Tg(MHu2M) and Tg (HuPrP) mice to Hu prions in mice may be that mouse chaperones catalyzing the refolding of PrP^{C} into PrP^{Sc} can readily interact with the $MHu2MPrP^{C}/HuPrP^{CJD}$ complex but not with $HuPrP^{C}/HuPrP^{CJD}$. The identification of protein X is an important avenue of research since isolation of this protein or complex of proteins would presumably facilitate studies of PrP^{Sc} formation. To date, attempts to isolate specific proteins that bind to PrP have been disappointing (409). Whether or not identification of protein X will require isolation of a ternary complex composed of PrP^{C}, PrP^{Sc}, and protein X remains to be determined.

The sensitivity of Tg(MHu2M) mice to Hu prions suggests that a similar approach to the construction of Tg mice susceptible to BSE and scrapie sheep prions may prove fruitful. The BSE epidemic has led to considerable concern about the safety for humans consuming beef and dairy products. Although epidemiologic studies over the past two decades argue that humans do not contract CJD from scrapie-infected sheep products (63,121,248), it is unknown whether any of the seven amino acid substitutions that distinguish bovine from sheep PrP render bovine prions permissive in humans (461). Whether Tg(MHu2M) mice are susceptible to bovine or sheep prions is unknown.

Protein Y and the Neuropathology of Prion Disease

Concurrent with many of the transgenic mouse studies already described, four lines of congenic mice were produced by crossing the PrP gene of the ILn/J mouse onto the C57BL background. The four lines of congenic mice were derived by backcrosses through 20 generations which are designated: B6.I-4 for B6.I-*B2m*^{a}, B6.I-1 for B6.I-*Prnp*^{b}, B6.I-2 for B6.I-*Il-1a*^{d} *Prnp*^{b}, B6.I-3 for B6.I-*B2m*^{a} *Prnp*^{b} (84). Neuropathologic examination of B6.I-1, B6.I-2, I/LnJ, and VM/Dk mice inoculated with 87V prions showed numerous PrP amyloid plaques, in accord with an earlier report on VM/Dk mice (77). In B6.I-1 mice, intense spongiform degeneration, gliosis, and PrP immunostaining were found in the ventral posterior lateral (VPL) nucleus of the thalamus, the habenula, and the raphe nuclei of the brainstem (85). These same regions showed intense immunoreactivity for PrP^{Sc} on histoblots. Unexpectedly, B6.I-2 and ILn/J mice exhibited only mild vacuolation of the thalamus and brainstem. These findings suggest that a locus near *Prnp* influences the deposition of PrP^{Sc} (and thus vacuolation) in the thalamus, the habenula, and raphe nuclei. We have provisionally designated the product of this gene protein Y.

Identification of the gene that encodes protein Y, distinct from but near *Prnp,* will be important. The gene *Y* product appears to control, at least in part, neuronal vacuolation and presumably PrP^{Sc} deposition in mice inoculated with scrapie prions. Isolation of protein Y should be helpful in dissecting the molecular events that feature in the pathogenesis of the prion diseases.

PRION DIVERSITY

The diversity of scrapie prions was first appreciated in goats inoculated with “hyper” and “drowsy” isolates (436). Scrapie isolates or “strains” from goats with a drowsy syndrome transmitted a similar syndrome to inoculated recipients, whereas those from goats with a “hyper” or ataxic syndrome transmitted an ataxic form of scrapie to recipient goats. Subsequently, studies in mice demonstrated the existence of many scrapie “strains,” where extracts prepared from the brains of mice inoculated with a particular preparation of prions produced a similar disease in inoculated recipients (75,143,151,294). While the clinical signs of scrapie for different prion isolates in mice tended to be similar, the isolates could be distinguished by the incubation times, the distribution of CNS vacuolation that they produced, and whether or not amyloid plaques formed.

That such isolates could be propagated through multiple passages in mice suggested that the scrapie pathogen has a nucleic acid genome which is copied and passed onto nascent prions (75,150). However, no evidence for scrapie-specific nucleic acid encoding information which specifies the incubation time and the distribution of neuropathological lesions has emerged from considerable efforts using a variety of experimental approaches, as described above. In striking contrast, mice expressing PrP transgenes have demonstrated that the level of PrP expression is inversely related to the incubation time (477). Furthermore, the distribution of CNS vacuolation and attendant gliosis are a consequence of pattern of PrP^{Sc} deposition which can be altered by both PrP genes and non-PrP genes (477).

These observations, taken together, begin to build an argument for PrP^{Sc} as the informational molecule in which prion "strain-specific" information is encrypted. Deciphering the mechanism by which PrP^{Sc} carries information for prion diversity and passes it on the nascent prions is a challenging goal. Whether PrP^{Sc} can adopt multiple conformations, each of which produces prions exhibiting distinct incubation times and patterns of PrP^{Sc} deposition, remains to be determined (112).

Prion Strains and Variations in Patterns of Disease

For many years, studies of experimental scrapie were performed exclusively with sheep and goats. The disease was first transmitted by intraocular inoculation (125) and later by intracerebral, oral, subcutaneous, intramuscular, and intravenous injections of brain extracts from sheep developing scrapie. Incubation periods of 1 to 3 years were common, and often many of the inoculated animals failed to develop disease (153,243,244). Different breeds of sheep exhibited markedly different susceptibilities to scrapie prions inoculated subcutaneously, suggesting that the genetic background might influence host permissiveness (240).

The lengths of the incubation times have been used to distinguish prion strains inoculated into sheep, goats, mice, and hamsters. Dickinson and his colleagues (142,149,150) developed a system for "strain typing," by which mice with genetically determined short and long incubation times were used in combination with the F1 cross. For example, C57BL mice exhibited short incubation times of ~150 days when inoculated with either the Me7 or Chandler isolates; VM mice inoculated with these same isolates had prolonged incubation times of ~300 days. The mouse gene controlling incubation times was labeled *Sinc*, and long incubation times were said to be a dominant trait because of prolonged incubation times in F1 mice. Prion strains were categorized into two groups, based upon their incubation times: (a) those causing disease more rapidly in "short" incubation time C57BL mice and (b) those causing disease more rapidly in "long" incubation time VM mice. Noteworthy are the 22a and 87V prion strains that can be passaged in VM mice while maintaining their distinct characteristics.

PrP Gene Dosage Controls the Length of the Scrapie Incubation Time

More than a decade of study was required to unravel the mechanism responsible for the "dominance" of long incubation times; not unexpectedly, long incubation times were found not to be dominant traits. Instead, the apparent dominance of long incubation times is due to a gene dosage effect (85).

Our own studies began with the identification of a widely available mouse strain with long incubation times. ILn/J mice inoculated with RML prions were found to have incubation times exceeding 200 days (302), a finding that was confirmed by others (90). Once molecular clones of the PrP gene were available, we asked whether or not the PrP genes of short and long mice segregate with incubation times. A restriction fragment-length polymorphism (RFLP) of the PrP gene was used to follow the segregation of MoPrP genes (*Prnp*) from short NZW or C57BL mice with long ILn/J mice in F1 and F2 crosses. This approach permitted the demonstration of genetic linkage between a *Prnp* and a gene modulating incubation times (*Prn-i*) (87). Other investigators have confirmed the genetic linkage, and one group has shown that the incubation-time gene *Sinc* is also linked to PrP (278,479). It now seems likely that the genes for PrP, *Prn-i*, and *Sinc* are all congruent; the term *Sinc* is no longer used (407). The PrP sequences of NZW with short and long scrapie incubation times, respectively, differ at codons 108 (L→F) and 189 (T→V) (560).

Although the amino acid substitutions in PrP that distinguish $Prnp^a$ from $Prnp^b$ mice argued for the congruency of *Prnp* and *Prn-i*, experiments with $Prnp^a$ mice expressing $Prnp^b$ transgenes demonstrated a "paradoxical" shortening of incubation times (561). We had predicted that these Tg mice would exhibit a prolongation of the incubation time after inoculation with RML prions based on ($Prnp^a \times Prnp^b$) F1 mice, which do exhibit long incubation times. We described those findings as "paradoxical shortening" because we and others had believed for many years that long incubation times are dominant traits (87,150). From studies of congenic and transgenic mice expressing different numbers of the *a* and *b* alleles of *Prnp* (Table 6), we now realize that these findings were not paradoxical; indeed, they result from increased PrP gene dosage (85). When the RML isolate was inoculated into congenic and transgenic mice, increasing the number of copies of the *a* allele was found to be the major determinant in reducing the incubation time; however, increasing the number of copies of the *b* allele also reduced the incubation time, but not to the same extent as that seen with the *a* allele (Table 6).

The discovery that incubation times are controlled by the relative dosage of $Prnp^a$ and $Prnp^b$ alleles was foreshadowed by studies of Tg(SHaPrP) mice, in which the length of the incubation time after inoculation with SHa prions was inversely proportional to the transgene product, $SHaPrP^C$ (477). Not only does the PrP gene dose determine the length of the incubation time, but also the passage history of the inoculum, particularly in $Prnp^b$ mice (Table 7). The PrP^{Sc} allotype in the inoculum produced the shortest incubation times when it was the same as that of PrP^C in the host (88). The term *allotype* is used to describe allelic variants of PrP. To address the issue of whether gene products other than PrP might be responsible for these findings, we inoculated B6 and B6.I-4 mice carrying $Prnp^{a/a}$ as well as I/Ln and B6.I-2 mice (84,85) with RML prions passaged in mice homozygous for either the *a* or *b* allele of *Prnp* (Table 7). CD-1 and NZW/LacJ mice produce prions con-

TABLE 6. *MoPrP-A expression is a major determinant of incubation times in mice inoculated with the RML scrapie prions. The RML prions were passaged in CD-1 (Prnp^a) mice prior to inoculation.*

Mice	Prnp Genotype (copies)	Prnp Transgenes (copies)	Alleles		Incubation Time[a] (days±SEM)	(n)
			a	*b*		
$Prnp^{0/0}$	0/0		0	0	>600	4
$Prnp^{+/0}$	a/0		1	0	426±18	9[a]
B6.I-1	b/b		0	2	360±16	7[b]
B6.I-2	b/b		0	2	379±8	10[b]
B6.I-3	b/b		0	2	404±10	20
(B6 × B6.I-1)F1	a/b		1	1	268±4	7
B6.I-1 × Tg(MoPrP-$B^{0/0}$)15	a/b		1	1	255±7	11[c]
B6.I-1 × Tg(MoPrP-$B^{0/0}$)15	a/b		1	1	274±3	9[d]
B6.I-1 × Tg(MoPrP-$B^{+/0}$)15	a/b	bbb/0	1	4	166±2	11[c]
B6.I-1 × Tg(MoPrP-$B^{+/0}$)15	a/b	bbb/0	1	4	162±3	8[d]
C57BL/6J (B6)	a/a		2	0	143±4	8
B6.I-4	a/a		2	0	144±5	8
non-Tg(MoPrP-$B^{0/0}$)15	a/a		2	0	130±3	10
Tg(MoPrP-$B^{+/0}$)15	a/a	bbb/0	2	3	115±2	18
Tg(MoPrP-$B^{+/+}$)15	a/a	bbb/bbb	2	6	111±5	5
Tg(MoPrP-$B^{+/0}$)94	a/a	>30b	2	>30	75±2	15[e]
Tg(MoPrP-$A^{+/0}$)B4053	a/a	>30a	>30	0	50±2	16

[a] Data from Prusiner et al. (463).
[b] Data from Carlson, et al. (84).
[c] The homozygous Tg(MoPrP-$B^{+/+}$)15 mice were maintained as a distinct subline selected for transgene homozygosity two generations removed from the (B6 x LT/Sv)F2 founder. Hemizygous Tg(MoPrP-$B^{+/0}$)15 mice were produced by crossing the Tg(MoPrP-$B^{+/+}$)15 line with B6 mice.
[d] Tg(MoPrP-$B^{+/0}$)15 mice were maintained by repeated back-crossing to B6 mice.
[e] Data from Westaway et al. (561).

taining PrP^{Sc}-A encoded by *$Prnp^a$*, while I/LnJ mice produce PrP^{Sc}-B prions. The incubation times in the congenic mice reflected the PrP allotype rather than other factors acquired during prion passage. The effect of the allotype barrier was small when measured in *$Prnp^{a/a}$* mice but was clearly demonstrable in *$Prnp^{b/b}$* mice. B6.I-2 congenic mice inoculated with prions from I/Ln mice had an incubation time of 237±8 days compared to times of 360±16 days and 404±4 days for mice inoculated with prions passaged in CD-1 and NZW mice, respectively. Thus, previous passage of prions in *$Prnp^b$* mice shortened the incubation time by ~40% when assayed in *$Prnp^b$* mice, compared to those inoculated with prions passaged in *$Prnp^a$* mice (88).

TABLE 7. *Mismatching of PrP allotypes between PrP^{Sc} in the inoculum and PrP^C in the inoculated host extends prion incubation times in congenic mice*

Mice	Inoculum Donor		Inoculum	Recipient Host	
	Genotype	Donor	Genotype	Incubation Time	(n)
C57BL/6J (B6)	a/a	CD-1	a/a	143±4	(8)
B6.I-4	a/a	NZW	a/a	144±5	(8)
B6.I-4	a/a	I/Ln	b/b	150±6	(6)
B6.I-2	b/b	CD-1	a/a	360±16	(8)
B6.I-2	b/b	NZW	a/a	404±4	(20)
B6.I-2	b/b	I/Ln	b/b	237±8	(17)
I/LnJ[a]	b/b	CD-1	a/a	314±13	(11)
I/LnJ	b/b	NZW	a/a	283±21	(8)
I/LnJ	b/b	I/Ln	b/b	193±6	(16)

[a] I/LnJ results previously reported [Carlson et al. (85)].

Overdominance

The phenomenon of *overdominance*, in which incubation times in F1 hybrids are longer than those of either parent (147), contributed to the confusion surrounding control of scrapie incubation times. When the 22A scrapie isolate was inoculated into B6, B6.I-1, and (B6×B6.I-1)F1, overdominance was observed: the scrapie incubation time in B6 mice was 405±2 days; in B6.I mice, 194±10 days; and in (B6×B6.I-1)F1 mice, 508±14 days (Table 8). Shorter incubation times were observed in Tg(MoPrP-B)15 mice, which were either homozygous or hemizygous for the *$Prnp^b$* transgene. Hemizygous Tg(MoPrP-$B^{+/0}$)15 mice exhibited a scrapie incubation time of 395±12 days, while the homozygous mice had an incubation time of 286±15 days.

As with the results with the RML isolate (Table 6), the findings with the 22A isolate can be explained on the basis of gene dosage; however, the relative effects of the *a* and *b* alleles differ in two respects. First, the *b* allele is the major determinant of the scrapie incubation time with the 22A isolate, not the *a* allele. Second, increasing the number of

TABLE 8. *MoPrP^{C}-A inhibits the synthesis of 22A scrapie prions. The 22A prions were passaged in B6.I1 mice prior to inoculation.*

Mice	Prnp Genotype (copies)	Prnp Transgenes (copies)	Alleles *a*	Alleles *b*	Incubation Time (days±SEM)	(n)
B6.I-1	b/b		0	2	194±10	7
(B6×B6.I-1)F1	a/b		1	1	508±14	7
C57BL/6J (B6)	a/a		2	0	405±2	8
non-Tg(MoPrP-B$^{0/0}$)15	a/a		2	0	378±8	3[a]
Tg(MoPrP-B$^{+/0}$)15	a/a	bbb/0	2	3	318±14	15[a]
Tg(MoPrP-B$^{+/0}$)15	a/a	bbb/0	2	3	395+12	6[b]
Tg(MoPrP-B$^{+/+}$)15	a/a	bbb/bbb	2	6	266±1	6[a]
Tg(MoPrP-B$^{+/+}$)15	a/a	bbb/bbb	2	6	286±15	5[b]

[a] The homozygous Tg(MoPrP-B$^{+/+}$)15 mice were maintained as a distinct subline selected for transgene homozygosity two generations removed from the (B6×LT/Sv)F2 founder. Hemizygous Tg(MoPrP-B$^{+/0}$)15 mice were produced by crossing the Tg(MoPrP-B$^{+/+}$)15 line with B6 mice.

[b] Tg(MoPrP-B$^{+/0}$)15 mice were maintained by repeated backcrossing to B6 mice.

copies of the *a* allele does not diminish the incubation but prolongs it: the *a* allele is inhibitory with the 22A isolate (Table 8). With the 87V prion isolate the inhibitory effect of the *Prnp^{a}* allele is even more pronounced, since only a few *Prnp^{a}* and (*Prnp^{a}*×*Prnp^{b}*)F1 mice develop scrapie after >600 days postinoculation (85).

The most interesting feature of the incubation time profile for 22A is the overdominance of the *a* allele of *Prnp* in prolonging incubation period. On the basis of overdominance, Dickinson and Outram put forth the replication site hypothesis, postulating that dimers of the *Sinc* gene product feature in the replication of the scrapie agent (152). The results in Table 8 are compatible with the interpretation that the target for PrP^{Sc} may be a PrP^{C} dimer or multimer. The assumptions under this model are that PrP^{C}-B dimers are more readily converted to PrP^{Sc} than are PrP^{C}-A dimers and that PrP^{C}-A:PrP^{C}-B heterodimers are even more resistant to conversion to PrP^{Sc} than PrP^{C}-A dimers. Increasing the ratio of PrP-B to PrP-A would lead to shorter incubation times by favoring the formation of PrP^{C}-B homodimers (Table 8). A similar mechanism may account for the relative paucity of individuals heterozygous for the Met/Val polymorphism at codon 129 of the human PrP gene in spontaneous and iatrogenic CJD (421). Alternatively, PrP^{C}-PrP^{Sc} interaction can be broken down to two distinct aspects: binding affinity and efficacy of conversion to PrP^{Sc}. If PrP-A has a higher affinity for 22A PrP^{Sc} than does PrP^{C}-B, but is inefficiently converted to PrP^{Sc}, the exceptionally long incubation time of *Prnp$^{a/b}$* heterozygotes might reflect reduction in the supply of 22A prions available for interaction with the PrP^{C}-B product of the single *Prnp^{b}* allele. Additionally, PrP^{C}-A may inhibit the interaction of 22A PrP^{Sc} with PrP^{C}-B, leading to prolongation of the incubation time. This interpretation is supported by prolonged incubation times in Tg(SHaPrP) mice inoculated with mouse prions in which SHaPrP^{C} is thought to inhibit the binding of MoPrP^{Sc} to the substrate MoPrP^{C} (477).

Patterns of PrP^{Sc} Deposition

Besides measurements of the length of the incubation time, profiles of spongiform degeneration have also been used to characterize different prion strains (181,183). With the development of a new procedure for *in situ* detection of PrP^{Sc}, designated *histoblotting* (534), it became possible to localize and quantify PrP^{Sc} as well as to determine whether or not "strains" produce different, reproducible patterns of PrP^{Sc} accumulation (Fig. 2) (138,251).

Histoblotting overcame two obstacles that plagued PrP^{Sc} detection in brain by standard immunohistochemical techniques: the presence of PrP^{C} and weak antigenicity of PrP^{Sc} (137). The histoblot is made by pressing 10-μm thick cryostat sections of fresh-frozen brain tissue to nitrocellulose paper. To localize protease-resistant PrP^{Sc} in brain, the histoblot is digested with proteinase K to eliminate PrP^{C}, followed by denaturation of the undigested PrP^{Sc} to enhance binding of PrP antibodies. Immunohistochemical staining yields a far more intense, specific, and reproducible PrP signal than can be achieved by immunohistochemistry on standard tissue sections. The intensity of immunostaining correlates well with neurochemical estimates of PrP^{Sc} concentration in homogenates of dissected brain regions. PrP^{C} can be localized in histoblots of uninfected, normal brains by eliminating the proteinase K digestion step.

Comparisons of PrP^{Sc} accumulation on histoblots with histologic sections showed that PrP^{Sc} deposition preceded vacuolation, and only those regions with PrP^{Sc} underwent degeneration. Microdissection of individual brain regions confirmed the conclusions of the histoblot studies: those regions with high levels of PrP 27-30 had intense vacuolation (92). Thus, we concluded that the deposition of PrP^{Sc} is responsible for the neuropathologic changes found in the prion diseases.

While studies with both mice and Syrian hamsters established that each isolate has a specific signature as defined by a specific pattern of PrP^{Sc} accumulation in the brain (85,138,251), comparisons must be done on an isogenic background (265,507). When a single strain is inoculated into mice expressing different PrP genes, variations in the patterns of PrP^{Sc} accumulation were found to be equally as great as those seen between two strains. Based upon the initial studies which were performed in animals of a single genotype, we suggested that PrP^{Sc} synthesis occurs in specific populations of cells for a given distinct prion isolate (251).

Prion Strain-Specific Patterns of PrP^{Sc} Accumulation

RML prions were inoculated into B6 or B6.I-1 congenic mice (84,85) and the patterns of PrP^{Sc} accumulation compared (Fig. 11A–F). First, the PrP^{Sc} signal was more widely and uniformly distributed in the neocortex and hippocampus in B6 mice (compare Figs. 11B,E). Second, the PrP^{Sc} signal was more intense in the granule cell layer of the cerebellum in B6.I-1 mice (compare Figs. 11F,C). Third, PrP^{Sc} accumulated preferentially in the cerebellar white matter in B6 mice but not in B6.I-1 mice (compare Figs. 11C,F). Other differences were also apparent, such as the weaker signal in the hypothalamus in B6.I-1 mice. These findings argue that the distribution of PrP^{Sc} is influenced by both the infecting prion and the amino acid sequence of PrP^{C}.

Although the distributions of PrP^{Sc} in the brains of B6.I-1 mice inoculated with RML or 22A were similar, clear differences were observed (Figs. 11D–I). For example, there was a more intense PrP^{Sc} signal in the molecular layer of the dentate gyrus with RML than with 22A, while the hypothalamus contained more PrP^{Sc} in the mice inoculated with 22A prions. The full thickness of the neocortex exhibited PrP^{Sc} immunostaining with 22A prions but only the inner layers stained with RML reminiscent of the differences between Sc237 and 139H scrapie in Syrian hamsters (251). The PrP^{Sc} signal was more intense in the corpus callosum and in the white matter tracts coursing through the caudate nucleus with RML than with 22A (compare Figs. 11D,G). With both RML and 22A prions, there was intense PrP^{Sc} staining in the granule cell layer of the cerebellum, as well as the notable absence of PrP amyloid plaques.

PrP^{Sc} deposition with 87V prions was markedly different from that with 22A and RML (Figs. 11D–L) and was most intense in the thalamus, particularly in the VPL nucleus (Fig. 11K), in the habenula (Fig. 11K), and in the locus coeruleus and raphe nuclei of the brainstem (Fig. 11L). Little or no PrP^{Sc} accumulated in the neocortex, the hippocampus, or the hypothalamus. Relatively weak PrP^{Sc} signals were distributed diffusely in the brainstem tegmentum, inferior colliculi, and amygdala. Multiple, intense, punctate signals were located beneath the corpus callosum overlying the hippocampus (Fig. 11K), in the caudate nucleus, and in the cerebellum. Histological sections stained with the periodic acid Schiff method or α-PrP antibodies indicated that the punctate PrP^{Sc} signals were kuru-type and primitive PrP amyloid plaques. Much of the intense PrP^{Sc} signal in the raphe nuclei and locus coeruleus of the brainstem (Fig. 11L) seems to be perivascular PrP plaques.

The development of Tg(MHu2M) mice that are highly susceptible to human prions offers a new approach to studying prion diversity. The patterns of PrP^{Sc} in Tg(MHu2M) mice were remarkably similar for three inocula from humans who had died of CJD (544). Two of the inocula were from patients who had died of sporadic CJD, while the third was from a patient previously treated with HGH who had died of iatrogenic CJD. Whether these three inocula all represent the same "strain" of prions or whether differences among them are obscured by passage in the Tg(MHu2M) mice remains to be established. Interestingly, PrP^{CJD} accumulation was much greater in the cerebral cortex than in the cerebellum with sporadic prion disease, while the distribution was reversed in iatrogenic CJD that followed HGH treatment (140). Studies in rodents have shown that prion strains produce different patterns of PrP^{Sc} accumulation (138,251), which can be dramatically changed by the expression of different PrP sequences (85,507). Transmission studies of Hu prions from patients dying of CJD or kuru to nonhuman primates suggested that each case might be due to a different "strain." (215)

Of note, mice inoculated intracerebrally or intraperitoneally with Mo prions accumulated PrP^{Sc} in follicular dendritic cells but did not when inoculated with human prions (307,365,390,391).

PrP Gene Specific Patterns of PrP^{Sc} Accumulation

Studies of a single isolate inoculated into mice expressing different PrP genes demonstrated that the PrP^{Sc} accumulation could be profoundly altered by factors other than strain of prions. For example, the RML isolate inoculated into CD-1 mice and Tg196 mice expressing the mutant MoPrP-P101L transgene produces very different patterns of PrP^{Sc} accumulation. The deposition of PrP^{Sc} in the brains of CD-1 mice is widespread throughout the neocortex, hippocampus, thalamus, and caudate, whereas it is restricted to the thalamus and hypothalamus of Tg196 mice. In both CD-1 and Tg196 mice, RML prions seem to be propagated. The CD-1 mice have incubation times of ~140 days, and extracts from their brains bioassayed in CD-1 mice have similar incubation times. The Tg196 mice have incubation times of ~245 days after inoculation with RML prions, while extracts from their brains bioassayed in CD-1 mice have incubation times of ~140 days (266). The prolonged incubation times in Tg196 mice indicate that the restricted pattern of PrP^{Sc} is not due to an abbreviated illness in which the spread of PrP^{Sc} to other regions did not have sufficient time to occur. The mechanism by which PrP^{Sc} accumulation remains restricted to a few specific

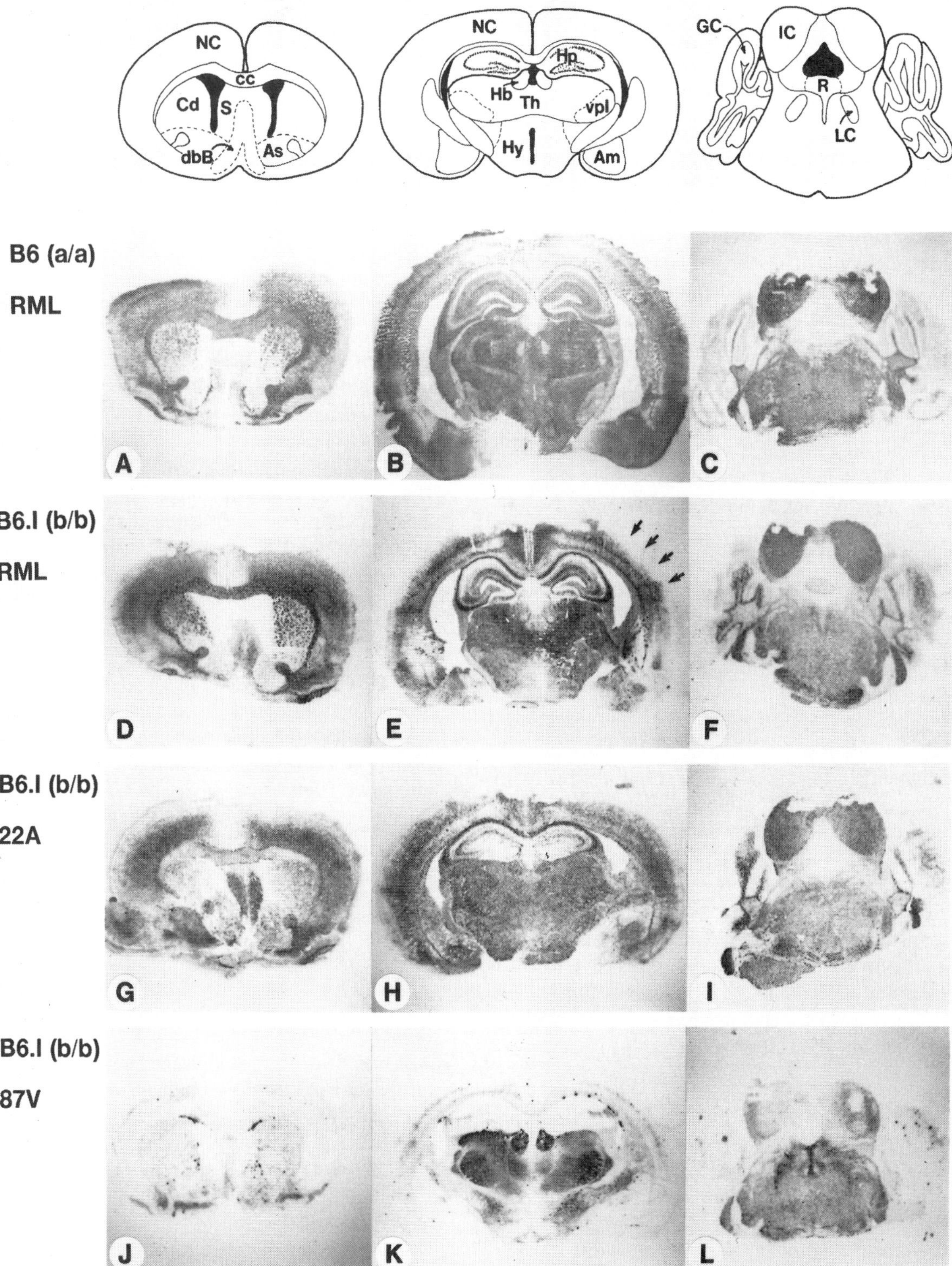

FIG. 11. Histoblots showing the PrP^{Sc} in coronal sections of mouse brain. Sections cut at three levels: caudate and septal nuclei *(left column)*; hippocampus and thalamus *(middle column)*; inferior colliculus *(right column)*. **A–C**: B6; and **D–F**: B6.I-1 mice were inoculated with RML prions and sacrificed after showing clinical signs of scrapie. **G–I**: 22A; and **J–L**: 87V prions were inoculated into B6.I-1 mice. The outer surface of the cerebral cortex is indicated (**E**, *arrows*). Am, amygdala; As, accumbens septi; cc, corpus callosum; Cd, caudate nucleus; dbB, diagonal band of Broca; GC, granule cell layer of the cerebellar cortex; Hb, habenula; Hp, hippocampus; Hy, hypothalamus; IC, inferior colliculus; LC, locus coeruleus; NC, neocortex; R, raphe nuclei of the brainstem; S, septal nuclei; Th, thalamus; vpl, ventral posterior lateral nucleus of the thalamus. From Carlson et al. (85), with permission.

areas of the brain, as in the case of RML in Tg196 mice and 87V in B6.I-1 mice, is unknown. How the protein Y modulates the accumulation of PrP^{Sc}, and thus neuronal vacuolation, remains to be established.

Mutations, Mixtures, and Cloning

The isolation of scrapie "strains" in mice has been performed generally with extracts prepared from the brains of scrapied sheep (89,150). Cloning of new "strains" has been performed by limiting dilution in mice. While many strains were isolated, most of the studies were performed with only a few strains, and passaging was limited. For example, Me7 prions were passaged in C57BL mice (145) which were later shown to have the *Prnp*[a] allele, while 22a and 87V were passaged in VM mice which have the *Prnp*[b] allele. Of course, the *a* and *b* alleles of *Prnp* encode MoPrP molecules that differ at codons 108 and 189 (560); propagation of the Me7 strain was much more rapid in the *Prnp*[a] mouse than the *Prnp*[b] mouse and vice versa for 22a and 87V in the *Prnp*[b] mouse (77,149). In other words, propagation of a particular strain was restricted by the PrP sequence in the host. It is noteworthy that a number of new "strains" have been isolated by passage of murine isolates into hamsters (294,301), in which the PrP genes differ at 16 positions (29,337).

Although "mutation" of scrapie isolates or "strains" was reported, virtually nothing is known about the molecules that participate in this process. Low dilution of 87A was reported to give rise to 7D prions, while passage at high dilution preserved the 87A properties (76). Dickinson thought that it was important to prepare inocula from the smallest regions of individual brains in order to minimize contamination with other strains or mutants (151).

The construction of Tg(MH2MPrP) mice that are susceptible to both Mo and hamster prions has provided a new tool for the study of strains. The Tg(MH2MPrP) mice produce artificial prions which infect Syrian hamsters as well as non-Tg and Tg(MH2MPrP) mice (507).

Molecular Basis of Prion Strains

The mechanism by which isolate-specific information is carried by prions remains enigmatic; indeed, explaining the molecular basis of prion diversity seems to be a formidable challenge. For many years some investigators argued that scrapie is caused by a viruslike particle which contains a scrapie-specific nucleic acid that encodes the information expressed by each isolate (75). To date, no such polynucleotide has been identified using a wide variety of techniques, including measurements of the nucleic acids in purified preparations. An alternative hypothesis has been suggested—that PrP^{Sc} alone is capable of transmitting disease but the characteristics of PrP^{Sc} might be modified by a cellular RNA (555). This accessory cellular RNA is postulated to induce its own synthesis upon transmission from one host to another, but there is no experimental evidence to support its existence. In fact, recent studies comparing the resistance of two prion strains show that each exhibits the same resistance to inactivation by irradiation at 254 nm (S. Prusiner, A. Serban, and J. Cleaver, unpublished data).

Two additional hypotheses not involving a nucleic acid have been offered to explain distinct prion isolates: a non-nucleic acid second component might create prion diversity or posttranslational modification of PrP^{Sc} might be responsible for the different properties of distinct prion isolates (452). Whether the PrP^{Sc} modification is chemical or only conformational remains to be established, but no candidate chemical modifications have been identified (526). Structural studies of GPI anchors of two SHa isolates have failed to reveal any differences; interestingly, about 40% of the anchor glycans have sialic acid residues (Fig. 4) (525). A portion of the PrP^{C} GPI anchors also have sialic acid residues; PrP is the first protein found to have sialic acid residues attached to GPI anchors.

The finding that the pattern of PrP^{Sc} accumulation in the CNS is characteristic for a particular strain offers a mechanism for the propagation of distinct prion isolates (138, 251). In this model, a different set of cells would propagate each isolate. Whether different Asn-linked CHOs function to target PrP^{Sc} of a distinct isolate to a particular set of cells in which the same Asn-linked CHOs will be coupled to PrP^{C} prior to its conversion to PrP^{Sc} remains to be established. The great diversity of Asn-linked CHOs makes them potential candidates for carrying isolate-specific information (453). Though this hypothesis is attractive, it must be noted that PrP^{Sc} synthesis in scrapie-infected cells occurs in the presence of tunicamycin which inhibits Asn-linked glycosylation and with PrP molecules mutated at the Asn-linked glycosylation consensus sites (538). Although the structures of Asn-linked CHOs have been analyzed for PrP^{Sc} of one isolate (173), no data are available for PrP^{Sc} of other isolates or for PrP^{C}. The large number of Asn-linked CHOs found attached to the PrP 27-30 of Sc237 prions purified from Syrian hamster would seem to argue that Asn-linked CHOs being responsible for strain variation is less likely, but experimental data addressing this point are still needed. Furthermore, as described above, a single prion strain can produce markedly different patterns of PrP^{Sc} in mice expressing different PrP allotypes (507).

Are Prion Strains Different PrP^{Sc} Conformers?

Multiple prion isolates might be explained by distinct PrP^{Sc} conformers that act as templates for the folding of *de novo* synthesized PrP^{Sc} molecules during prion "replication." (Fig. 7) Although it is clear that passage history can be responsible for the prolongation of incubation time when prions are passed between mice expressing different PrP allotypes (88) or between species (477), many scrapie strains show distinct incubation times in the same inbred host (78).

In recent studies we inoculated three strains of prions into congenic and Tg mice harboring various numbers of the *a* and *b* alleles of *Prnp* (85). The number of *Prnp*^{a} genes was the major determinant of incubation times in mice inoculated with the RML prion isolate and was inversely related to the length of the incubation time (Table 6) . In contrast, the *Prnp*^{a} allele prevented scrapie in mice inoculated with 87V prions. *Prnp*^{b} genes were permissive for 87V prions and shortened incubation times in most mice inoculated with 22A prions (Table 8). Experiments with the 87V isolate suggest that a genetic locus encoding protein Y, distinct from *Prnp*, controls the deposition of PrP^{Sc} and the attendant neuropathology. While each prion isolate produced distinguishable patterns of PrP^{Sc} accumulation in brain, a comparison showed that RML and 22A prions in congenic *Prnp*^{b} mice yielded similar patterns while RML prions produced very different patterns in *Prnp*^{a} and *Prnp*^{b} congenic mice. Thus, both the PrP genotype and prion isolate modify the distribution of PrP^{Sc} and the length of the incubation time. These findings suggest that prion strain-specified properties result from different affinities of PrP^{Sc} in the inocula for PrP^C-A and PrP^C-B allotypes encoded by the host.

Although the proposal for multiple PrP^{Sc} conformers is rather unorthodox, we already know that PrP can assume at least two profoundly different conformations: PrP^C and PrP^{Sc} (424). Of note, two different isolates from mink dying of transmissible mink encephalopathy exhibit different sensitivities of PrP^{Sc} to proteolytic digestion, supporting the suggestion that isolate-specific information might be carried by PrP^{Sc} (42,43,353). How many conformations PrP^{Sc} can assume is unknown. The molecular weight of a PrP^{Sc} homodimer is consistent with the ionizing radiation target size of 55,000±9,000 daltons, as determined for infectious prion particles independent of their polymeric form (35). If prions are oligomers of PrP^{Sc}, which seems likely, this offers another level of complexity which, in turn, generates additional diversity.

SCRAPIE

Experimental Scrapie

For many years, studies of experimental scrapie were performed exclusively with sheep and goats. The disease was first transmitted by intraocular inoculation (125) and later by intracerebral, oral, subcutaneous, intramuscular, and intravenous injections of brain extracts from sheep developing scrapie. Incubation periods of 1 to 3 years were common, and often many of the inoculated animals failed to develop disease (153,243,244). Different breeds of sheep exhibited markedly different susceptibilities to scrapie prions inoculated subcutaneously, suggesting that the genetic background might influence host permissiveness (240).

A crucial methodologic advance in experimental studies of scrapie was created by the demonstration that scrapie could be transmitted to mice (101,102), which could be used for endpoint titrations of particular samples. In addition, pathogenesis experiments directed at elucidating factors governing incubation times and neuropathological lesions were performed (150,171,182).

Natural Scrapie in Sheep and Goats

Even though scrapie was recognized as a distinct disorder of sheep with respect to its clinical manifestations as early as 1738, the disease remained enigmatic even with respect to its pathology for more than two centuries (426). Some veterinarians thought that scrapie was a disease of muscle caused by parasites, while others thought that it was a dystrophic process (342). An investigation into the etiology of scrapie followed the vaccination of sheep for looping ill virus with formalin-treated extracts of ovine lymphoid tissue unknowingly contaminated with scrapie prions (241). Two years later, more than 1,500 sheep developed scrapie from this vaccine.

Communicability

Scrapie of sheep and goats appears to be unique among the prion diseases in that it seems to be readily communicable within flocks. While the transmissibility of scrapie seems to be well established, the mechanism of the natural spread of scrapie among sheep is so puzzling that it bears close scrutiny. The placenta has been implicated as one source of prions accounting for the horizontal spread of scrapie within flocks (411,432,433,435). Whether or not this view is correct remains to be established. In Iceland, scrapied flocks of sheep were destroyed and the pastures left vacant for several years; however, reintroduction of sheep from flocks known to be free of scrapie for many years eventually resulted in scrapie (423). The source of the scrapie prions that attacked the sheep from flocks without a history of scrapie is unknown.

Genetics of Sheep

Parry (426,428) argued that host genes were responsible for the development of scrapie in sheep. He was convinced that natural scrapie is a genetic disease which could be eradicated by proper breeding protocols. He considered its transmission by inoculation to be of importance primarily for laboratory studies and communicable infection to be of little consequence in nature. Other investigators viewed natural scrapie as an infectious disease and argued that host genetics only modulates susceptibility to an endemic infectious agent (154). The incubation time gene for experimental scrapie in Cheviot sheep, called *Sip*, is said to be linked to a PrP gene RFLP (277); however, the null-hypothesis of nonlinkage has yet to be tested; this is important, especially in view of earlier studies which argue that susceptibility of sheep to scrapie is governed by a recessive gene (426,428).

Polymorphisms at codons 136 and 171 of the PrP gene in sheep that produce amino acid substitutions have been studied with respect to the occurrence of scrapie in sheep (108). In Romanov and Ile-de-France breeds of sheep, a polymorphism in the PrP ORF was found at codon 136 (A→V), which seems to correlate with scrapie (328). Sheep homozygous or heterozygous for Val at codon 136 were susceptible to scrapie while those that were homozygous for Ala were resistant. Unexpectedly, only one of 74 scrapied autochthonous sheep had a Val at codon 136; these sheep were from three breeds denoted *Lacaune, Manech* and *Presalpes* (326).

In Suffolk sheep, a polymorphism in the PrP ORF was found at codon 171 (Q→R) (236,237). Studies of natural scrapie in the USA have shown that ~85% of the afflicted sheep are of the Suffolk breed. Only those Suffolk sheep homozygous for Gln (Q) at codon 171 were found with scrapie, although healthy controls with QQ, QR, and RR genotypes were found (562). These results argue that susceptibility in Suffolk sheep is governed by the PrP codon 171 polymorphism. Whether the PrP codon 171 or 136 polymorphisms in Cheviot sheep have the same profound influence on susceptibility to scrapie as has been found for codon 171 in Suffolks is unknown (235,276).

BOVINE SPONGIFORM ENCEPHALOPATHY

Epidemic of Mad Cow Disease

Beginning in 1986, an epidemic of a previously unknown disease appeared in cattle in Great Britain (557). This disease was initially named *bovine spongiform encephalopathy* (BSE) but is frequently called *mad cow disease*. BSE was shown to be a prion disease by demonstration of protease-resistant PrP in brains of ill cattle (256,461). Based mainly on epidemiologic evidence, it has been proposed that BSE represents a massive, common-source epidemic which has caused more than 150,000 cases to date. In Britain, cattle, particularly dairy cows, were routinely fed MBM as a nutritional supplement (135,565,567–569). The MBM was prepared by rendering the offal of sheep and cattle using a process that involved steam treatment and hydrocardon solvent extraction. The extraction process produced protein- and fat-rich fractions; the protein or greaves fraction contained about 1% fat, from which the MBM was prepared. In the late 1970s, the price of tallow prepared from the fat fraction fell, making it no longer profitable to use hydrocarbons in the rendering process. The resulting MBM contained about 14% fat, and it is postulated that the high lipid content protected scrapie prions in the sheep offal from being completely inactivated by steam.

Since 1988, the practice of using dietary protein supplements derived from rendered sheep or cattle offal for domestic animals has been forbidden in the U.K. Curiously, almost half of the BSE cases have occurred in herds where only a single affected animal has been found; several cases of BSE in a single herd are infrequent (135,565, 569). Whether the distribution of BSE cases within herds will change as the epidemic progresses and BSE will disappear with the cessation of feeding rendered MBM are uncertain.

Crossing a Species Barrier

Assuming the above postulate is correct, only sheep prions were present initially in the contaminated MBM. Since the species barrier depends, at least in part, on the amino acid sequences of PrP in the donor host and recipient, the similarity between bovine and sheep PrP was probably an important factor in initiating the BSE epidemic. Bovine PrP differs from sheep PrP at seven or eight residues, depending on the breed of sheep (238,461). As the BSE epidemic expanded, infected bovine offal began to be rendered into MBM which contained bovine prions.

Transmission of BSE to Experimental Animals

Brain extracts from BSE cattle have transmitted disease to mice, cattle, sheep, and pigs after intracerebral inoculation (74,132,133,185). Transmissions to mice and sheep suggest that cattle preferentially propagate a single "strain" of prions. Seven BSE brains all produced similar incubation times as measured in each of three strains of inbred mice (74).

Of particular importance to the BSE epidemic is the recent transmission of BSE to the nonhuman primate marmoset after intracerebral inoculation followed by a prolonged incubation period (24). The potential parallels with kuru of humans, confined to the Fore region of New Guinea (197,199), are worthy of consideration. Once the most common cause of death among women and children, kuru has almost disappeared with the cessation of ritualistic cannibalism (12). While it seems likely that kuru was transmitted orally, as proposed for BSE among cattle, some investigators argue that routes other than oral routes were important since oral transmission of kuru prions to apes and monkeys has been difficult to demonstrate (197,210).

There is no example of zoonotic transmission of prions from animals to humans, based on many epidemiological studies which have attempted to implicate scrapie prions from sheep as a cause of CJD (121,248,344). Whether BSE poses any risk to humans is unknown, but two farmers with BSE-afflicted cattle have died of CJD during the past year (19,20,503).

Oral Transmission of BSE Prions

Besides BSE, four other animal diseases appear to have arisen from the oral consumption of prions. It has been suggested that an outbreak of transmissible mink encephalopathy in 1985 arose from the use for feed of a cow with a sporadic case of BSE (353). The source of prions in chronic wasting disease of captive mule deer and elk is unclear (571,572). The prion-contaminated MBM thought to be the cause of BSE is also hypothesized to be the cause of feline spongiform encephalopathy (FSE) and exotic ungulate encephalopathy. FSE has been found in almost 30 domestic cats in Great Britain as well as in a puma and a cheetah (575). Three cases of FSE in domestic cats have been transmitted to laboratory mice, and PrP^{Sc} has been identified in their brains by immunoblotting (438). Prion disease has been found in the brains of the nyala, greater kudu, eland, gembok, and Arabian oryx in British zoos; all of these animals are exotic ungulates. Five of eight greater kudu born into a herd maintained in a London zoo since 1987 have developed prion disease. Except for the first case, none of the other four kudu were exposed to feeds containing ruminant-derived MBM (303). Brain extracts prepared from a nyala and a greater kudu have been transmitted to mice (126,304). PrP of the greater kudu differs from the bovine protein at four residues; Arabian oryx PrP differs from the sheep PrP at only one residue (443).

HUMAN PRION DISEASES

Clinical Manifestations of Human Prion Diseases

The human prion diseases are manifest as infectious, inherited, and sporadic disorders and are often referred to as kuru, CJD, GSS and FFI, depending upon the clinical and neuropathological findings (Table 1).

Infectious forms of prion diseases result from the horizontal transmission of infectious prions, as occurs in iatrogenic CJD and kuru. Inherited forms, notably GSS, familial CJD, and FFI comprise 10% to 15% of all cases of prion disease. A mutation in the ORF or protein coding region of the PrP gene has been found in all reported kindreds with inherited human prion disease (41,163,165,192, 222,229,230,233,259,306,308,373,439,444). Sporadic forms of prion disease comprise most cases of CJD and possibly some cases of GSS (361). How prions arise in patients with sporadic forms is unknown; hypotheses include horizontal transmission from humans or animals (197), somatic mutation of the PrP gene ORF, and spontaneous conversion of PrP^{C} into PrP^{Sc} (261,453). Numerous attempts to establish an infectious link between sporadic CJD and a preexisting prion disease in animals or humans have been unrewarding (46,63,121,248,344).

Diagnosis of Human Prion Diseases

Human prion disease should be considered in any patient who develops a progressive subacute or chronic decline in cognitive or motor function. Typically adults between 40 and 70 years of age, patients often exhibit clinical features helpful in providing a premorbid diagnosis of prion disease, particularly sporadic CJD (62,494). There is as yet no specific diagnostic test for prion disease in the cerebrospinal fluid. A definitive diagnosis of human prion disease, which is invariably fatal, can often be made from the examination of brain tissue. Over the past 4 years, knowledge of the molecular genetics of prion diseases has made it possible to diagnose inherited prion disease in living patients using peripheral tissues.

A broad spectrum of neuropathological features in human prion diseases precludes a precise neuropathological definition. The classic neuropathological features of human prion disease include spongiform degeneration, gliosis, and neuronal loss in the absence of an inflammatory reaction. When present, amyloid plaques which stain with α-PrP antibodies are diagnostic.

The presence of protease-resistant PrP (PrP^{Sc} or PrP^{CJD}) in the infectious and sporadic forms and most of the inherited forms of these diseases implicates prions in their pathogenesis. The absence of PrP^{CJD} in a biopsy specimen may simply reflect regional variations in the concentration of the protein (510). In some patients with inherited prion disease, PrP^{Sc} is barely detectable or undetectable (70,336, 347,372); this situation seems to be mimicked in transgenic mice which express a mutant PrP gene and spontaneously develop neurologic illness indistinguishable from experimental murine scrapie (265,267).

In humans and Tg mice which have no detectable protease-resistant PrP but express mutant PrP, neurodegeneration may, at least in part, be caused by abnormal metabolism of mutant PrP. Because molecular genetic analyses of PrP genes in patients with unusual dementing illnesses are readily performed, the diagnosis of inherited prion disease can often be established where there was either little or no neuropathology (115), atypical neurodegenerative disease (372), or misdiagnosed neurodegenerative disease (22,254), including Alzheimer's disease.

Although horizontal transmission of neurodegeneration to experimental hosts was for a time the "gold standard" of prion disease, it can no longer be used as such. Some investigators have reported that transmission of the inherited prion diseases from humans to experimental animals is frequently negative when using rodents, despite the presence of a pathogenic mutation in the PrP gene (539), while others state that this is not the case with apes and monkeys as hosts (71). The discovery that Tg(MHu2M) mice are susceptible to Hu prions (544) promises to make feasible transmission studies which were not practical in apes and monkeys (68).

The hallmark common to all of the prion diseases whether sporadic, dominantly inherited, or acquired by infection is that they involve the aberrant metabolism of the prion protein (452). Making a definitive diagnosis of human prion disease can be rapidly accomplished if PrP^{Sc} can be detected immunologically. Frequently PrP^{Sc} can be detected by either dot blot method or Western immunoblot analysis of brain homogenates in which samples are subjected to limited proteolysis to remove PrP^{C} prior to immunostaining (47,48,66,510). The dot blot method exploits enhancement of PrP^{Sc} immunoreactivity following denaturation in the chaotropic salt, guanidinium chloride. Because of regional variations in PrP^{Sc} concentration, methods using homogenates prepared from small brain regions can give false-negative results. Alternatively, PrP^{Sc} may be detected *in situ* in cryostat sections bound to nitrocellulose membranes followed by limited proteolysis to remove PrP^{C} and guanidinium treatment to denature PrP^{Sc}, thus enhancing its avidity for α-PrP antibodies (Figs. 2,11) (534). Denaturation of PrP^{Sc} *in situ* prior to immunostaining has also been accomplished by autoclaving fixed tissue sections (309).

In the familial forms of the prion diseases, molecular genetic analyses of PrP can be diagnostic and can be performed on DNA extracted from blood leucocytes *ante mortem*. Unfortunately, such testing is of little value in the diagnosis of the sporadic or infectious forms of prion disease. Although the first missense PrP mutation was discovered when the two PrP alleles of a patient with GSS were cloned from a genomic library and sequenced (259), all subsequent novel missense and insertional mutations have been identified in PrP ORFs amplified by polymerase chain reaction (PCR) and sequenced. The 759 base pairs encoding the 253 amino acids of PrP reside in a single exon of the PrP gene, providing an ideal situation for the use of PCR. Amplified PrP ORFs can be screened for known mutations using one of several methods, the most reliable of which is allele-specific oligonucleotide hybridization. If known mutations are absent, novel mutations may be found when the PrP ORF is sequenced.

When PrP amyloid plaques in brain are present, they are diagnostic for prion disease, as noted above. Unfortunately, they are thought to be present in only ~10% of CJD cases, and by definition all cases of GSS. The amyloid plaques in CJD are compact (kuru plaques). Those in GSS are either multicentric (diffuse) or compact. The amyloid plaques in prion diseases contain PrP (312,484,485). The multicentric amyloid plaques which are pathognomonic for GSS may be difficult to distinguish from the neuritic plaques of Alzheimer's disease except by immunohistology (207,279,406). In the GSS kindreds the diagnosis of Alzheimer's disease was excluded because the amyloid plaques failed to stain with β-amyloid antiserum but stained with PrP antiserum. In subsequent studies, missense mutations were found in the PrP genes of these kindreds.

In summary, the diagnosis of prion or prion protein disease may be made in patients on the basis of: (a) the presence of PrP^{Sc}, (b) mutant PrP genotype, or (c) appropriate immunohistology and should not be excluded in patients with atypical neurodegenerative diseases until one or preferably two of these examinations have been performed (113, 115,325).

INFECTIOUS PRION DISEASES

Kuru

For many decades kuru devastated the lives of the Fore Highlanders of Papua New Guinea (197). The high incidence of the disease among women left a society of motherless children raised by their fathers (Table 9). It was unusual in the Fore region to see an older woman. With the cessation of traditional warfare, older men are now found. Many of these older men have had a succession of wives, each dying of kuru after leaving several children. Because contamination during ritualistic cannibalism appears to have been the mode of spread of kuru among the Fore people, and since cannibalism had ceased by 1960 in the Fore region, the patients now developing kuru presumably were exposed to the kuru agent more than three decades ago (14,197). In many cases, histories from patients and their families of the episode in which they cannibalized the remains of a near relative who had died of kuru have been obtained, presumably providing the source of infection. That the kuru prions could remain apparently quiescent in these patients for periods of two decades and then manifest in the form of a fatal neurological disease is support-

TABLE 9. *Infectious prion diseases of humans*[a]

Diseases	No. Cases
A. Kuru (1957–1982)	
1. Adult females	1739
2. Adult males	248
3. Children and adolescents	597
Total	2584
B. Iatrogenic Creutzfeldt-Jakob disease	
1. Depth electrodes	2
2. Corneal transplants	1
3. Human pituitary growth hormone	55
4. Human pituitary gonadotropin	5
5. Dura mater grafts	11
6. Neurosurgical procedures	4
Total	78

[a] References cited in text.

ed by incubation periods of over 7.5 years in some monkeys inoculated with kuru agent (210).

The uniform clinical presentation of kuru is remarkable (12,201,202,257,258,516,581). In one study, the prodromal symptoms and onset of the disease were similar in all of the patients investigated (462). Even the time interval between the prodromal symptoms of headache and joint pain and the onset of difficulty walking was always 6 to 12 weeks. In most cases the disease progressed to death within 12 months, and all patients were dead within 2 years of onset. The average duration of illness for the 15 patients was 16 months (462). Invariably, signs of cerebellar dysfunction dominated the clinical picture. All patients remained ambulatory with the aid of a stick for more than half of the clinical phase of their illness. These clinical characteristics were similar to those reported for adult patients at the peak of the kuru epidemic (13,201,202,257, 258,516).

Incubation Periods that Exceed Three Decades

No individual born in the South Fore after 1959 (when cannibalism ceased) has developed kuru (12,14). Kuru has progressively disappeared, first among children and thereafter among adolescents. The number of deaths in adult females has decreased steadily, and adult male deaths have remained almost invariant. Each year the youngest new patients are older than those of the previous year.

Of several hundred kuru orphans born since 1957 to mothers who later died of kuru, none has yet developed the disease. Thus, the many children with kuru seen in the 1950s were not infected prenatally, perinatally, or neonatally by their mothers in spite of evidence for prions in the placenta and colostrum of a pregnant woman who died of CJD (532). Attempts to demonstrate consistent transmission of prion disease from mother to offspring in experimental animals have been unsuccessful (16,350,388, 433,531).

While patients currently afflicted with kuru exhibit greatly prolonged incubation periods, children with kuru who were observed 30 years ago provide some information on the minimum incubation period. The youngest patient with kuru was 4 years old at the onset of the disease and died at age 5, but it is not known at what age young children were infected. CJD accidentally transmitted to humans has required only 18 months after intracerebral or intraoptic inoculation (39,167) to manifest. An incubation period of 18 months has also been found in chimpanzees inoculated intracerebrally with kuru prions.

The regular disappearance of kuru is inconsistent with the existence of any natural reservoirs for kuru besides humans. Indeed, there is no evidence for animal or insect reservoirs. Thus, patients dying of kuru over the past decade seem to have incubation periods exceeding two or even three decades (12,314,462).

Transmission by Cannibalism

Considerable evidence implicates ritualistic cannibalism as the mode of transmission for kuru among the Fore and neighboring tribes (197). Oral transmission of kuru to monkeys has been documented (210). Proposed transmission routes through laceration of the skin and rubbing of the eyes were suggested when early experiments on oral transmission to apes and monkeys failed (197), but documentation of non-oral transmission remains to be established. The experimental results from oral transmission of scrapie to hamsters suggest that insufficient doses of kuru prions were used in those protocols (458).

Origin of Kuru

It has been suggested that kuru began at the turn of the century as a spontaneous case of CJD that was propagated by ritualistic cannibalism (14,197). It is not known whether the Fore peoples and their immediate neighbors provide an especially permissive genetic background on which kuru prions multiply. Sequencing of the ORF of the PrP gene from three kuru patients failed to reveal any mutations (224). Noteworthy is a case of CJD outside the kuru region in Papua New Guinea.

Transmission to Animals

Kuru has been regularly transmitted after intracerebral inoculation to apes and monkeys (197,199,215). The prolonged incubation periods in experimental animals are sim-

TABLE 10. *Incubation periods and durations of illness in Creutzfeldt-Jakob disease and kuru*

Host	Incubation period (mos.)[a]		Duration of illness (mos.)	
	CJD	Kuru	CJD	Kuru
Natural				
Humans	18–360	60–360	1–55	3–12
Experimental				
Apes	11–71	10–82	1–6	1–15
Monkeys	4–73	8–92	1–27	1–23
Sheep				
Goats	36–48	39		2–6
Ferrets		18–71		
Mink		45		
Domestic cats	19–30	2–6		
Guinea pigs	7–16	1		
Hamsters	5–18			
Mice	3–20	1–2		

[a] Incubation period is the time in months from prion exposure to the onset of illness. Data compiled from (14,213,215, 244,348,349,360,542; W. J. Hadlow, unpublished data).

From Prusiner et al. (473), with permission.

ilar to those observed with CJD and GSS (Table 10). Oral transmission of kuru to apes and monkeys has been difficult (197), but recent studies have demonstrated transmission to monkeys (210). Presumably, the difficulties in transmitting human kuru prions orally to apes and monkeys are due to the inefficiency of this route (458) and possibly the crossing of a species barrier. While the number of differences in PrP amino acid sequences between humans and nonhuman primates are small, a few specific amino acid changes could have a profound effect on susceptibility (504).

Immunologic Studies

Using PrP antiserum, the kuru or amyloid plaques showed intense immunostaining (485). It has been estimated that ~70% of patients dying of kuru have PrP amyloid plaques (313). Protease-resistant immunoreactive proteins were reported in the brain extract of one of two patients (50%) who died of kuru (65). Presumably, these two patients are included in a larger series where three of four kuru patients (75%) were found to have the abnormal isoform of the prion protein (66).

Iatrogenic Creutzfeldt-Jakob Disease

Accidental transmission of CJD to humans appears to have occurred by corneal transplantation (167), contaminated EEG electrode implantation (39), and surgical operations using contaminated instruments or apparatus (Table 9) (129,318,362,570). A cornea unknowingly removed from a donor with CJD was transplanted to an apparently healthy recipient who developed CJD after a prolonged incubation period. Corneas of animals have significant levels of prions (83), making this scenario seem quite probable. The same improperly decontaminated EEG electrodes which caused CJD in two young patients with intractable epilepsy were found to cause CJD in a chimpanzee 18 months after their experimental implantation (40,212).

Surgical procedures may have resulted in accidental inoculation of patients with prions during their operations (72,197,570), presumably because some instrument or apparatus in the operating theater became contaminated when a CJD patient underwent surgery. Although the epidemiology of these studies is highly suggestive, no proof for such episodes exists.

Since 1988, eleven cases of CJD after implantation of dura mater grafts have been recorded (72,356,364,384,405, 412,546,574). All of the grafts were thought to have been acquired from a single manufacturer whose preparative procedures were inadequate to inactivate human prions (72). One case of CJD occurred after repair of an eardrum perforation with a pericardium graft (533).

Thirty cases of CJD in physicians and health care workers have been reported (38); however, no occupational link has been established (481). Whether any of these cases represent infectious prion diseases contracted during care of patients with CJD or processing of specimens from these patients remains uncertain.

Human Growth Hormone Therapy

The possibility of transmission of CJD from contaminated human grown hormone (HGH) preparations derived from human pituitaries has been raised by the occurrence of fatal cerebellar disorders with dementia in >55 patients ranging in age from 10 to 41 years (Table 9) (61,72,79,180). While one case of spontaneous CJD in a 20-year-old woman has been reported (61,217,420), CJD in patients under 40 years of age is very rare. These patients received injections of HGH every 2 to 4 days for 4 to 12 years (17, 44,124,172,217,316,343,357,403,445,547). Interestingly, most of the patients presented with cerebellar syndromes which progressed over periods varying from 6 to 18 months. Some patients became demented during the terminal phases of their illnesses. This clinical course resembles kuru more than ataxic CJD in some respects (462). Assuming these patients developed CJD from injections of prion-contaminated HGH preparations, the possible incubation periods range from 4 to 30 years (72). The longest incubation periods are similar to those (20 to 30 years) associated with recent cases of kuru (200,314,462). Many patients received several common lots of HGH at various times during their prolonged therapies, but no single lot was administered to all the American patients. An aliquot of one lot of HGH has been reported to transmit CNS disease to a squirrel monkey after a prolonged incubation period (211). How many lots of the HGH might have been contaminated with prions is unknown.

Although CJD is a rare disease with an annual incidence of approximately 1 per million population (362), it is reasonable to assume that CJD is present with a proportional frequency among people who have died. About 1% of the population dies each year, and most CJD patients die within 1 year of developing symptoms. Thus, we estimate that 1 per 10^4 people who have died had CJD. Since 10,000 human pituitaries were typically processed in a single HGH preparation, the possibility of hormone preparations contaminated with CJD prions is not remote (60,64,67).

The concentration of CJD prions within infected human pituitaries is unknown; it is interesting that widespread degenerative changes have been observed in both the hypothalamus and pituitary of sheep with scrapie (32). The forebrains from scrapie-infected mice have been added to human pituitary suspensions to determine whether prions and HGH copurify (340). Bioassays in mice suggest that prions and HGH do not copurify with currently used protocols (543). Although these results seem reassuring, especially for patients treated with HGH over much of the last decade, the relatively low titers of the murine scrapie prions used in these studies may not have provided an adequate test (61). The extremely small size and charge heterogeneity exhibited by scrapie (11,50,468,471,475) and presumably CJD prions (37,47) may complicate procedures designed to separate pituitary hormones from these slow infectious pathogens. Though additional investigations argue for the efficacy of inactivating prions in HGH fractions prepared from human pituitaries using 6 M urea (440), it seems doubtful that such protocols will be used for purifying HGH since recombinant HGH is available.

Molecular genetic studies have shown that most patients developing iatrogenic CJD after receiving pituitary-derived HGH are homozygous for either met or val at codon 129 of the PrP gene (64,117,141). Homozygosity at the codon 129 polymorphism has also been shown to predispose individuals to sporadic CJD (421). Interestingly, valine homozygosity seems to be overrepresented in these HGH cases, compared to the general population.

Five cases of CJD have occurred in women receiving human pituitary gonadotropin (109,111,250).

INHERITED PRION DISEASES

Familial Prion Disease

The recognition that ≈10% of CJD cases are familial (131,186,280,305,358–360,374,496,528) posed a perplexing problem once it was established that CJD is transmissible (214,216). Equally puzzling was the transmission of GSS to nonhuman primates (214,216,358) and mice (542), since most cases of GSS are familial (206). Like sheep scrapie, the relative contributions of genetic and infectious etiologies in the human prion diseases remained a conundrum until molecular clones of the PrP gene became available to probe the inherited aspects of these disorders.

PrP Mutations and Genetic Linkage

The discovery of the PrP gene and its linkage to scrapie incubation times in mice (87) raised the possibility that mutation might feature in the hereditary human prion diseases. A proline (P)→leucine (L) mutation at codon 102 was shown to be linked genetically to development of GSS with a logarithm of odds (LOD) score exceeding 3 (Fig. 12) (259). This mutation may be due to the deamination of a methylated CpG in a germline PrP gene resulting in the substitution of a thymine (T) for cytosine (C). The P102L mutation has been found in ten different families in nine different countries, including the original GSS family (165, 233,234,319,320). Patients with GSS who have a proline (P)→leucine (L) substitution at PrP codon 105 have also been reported (308).

Patients with a dementing or telencephalic form of GSS have a mutation at codon 117. These patients, as well as some in other families with PrP mutations at codons 198

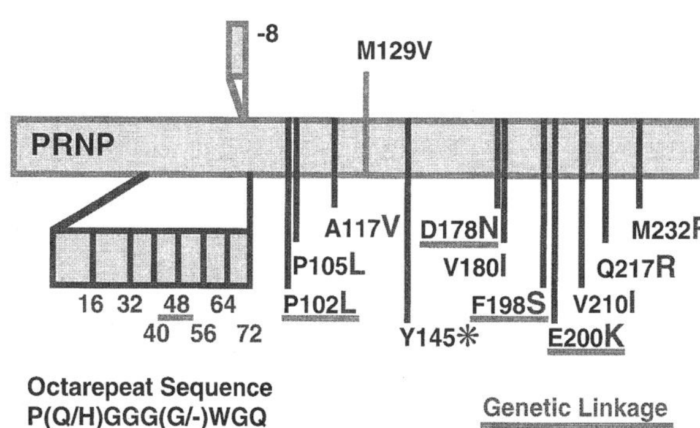

FIG. 12. Human prion protein gene (PRNP). The open reading frame (ORF) is denoted *(large gray rectangle)*. Human PRNP wild-type coding polymorphisms are shown *(above the rectangle)*; as are mutations that segregate with the inherited prion diseases *(below the rectangle)*. The wild-type human PrP gene contains five octarepeats [P(Q/H)GGG(G/-)WGQ] from codons 51 to 91 (322). Deletion of a single octarepeat at codon 81 or 82 is not associated with prion disease (327,478, 554); whether this deletion alters the phenotypic characteristics of a prion disease is unknown. There are common polymorphisms at codons 117 (Ala→Ala) and 129 (Met→Val); homozygosity for Met or Val at codon 129 appears to increase susceptibility to sporadic CJD (421). Octarepeat inserts of 16, 32, 40, 48, 56, 64, and 72 amino acids at codons 67, 75, or 83 are designated *(small rectangle below the ORF)*. These inserts segregate with familial CJD, and genetic linkage has been demonstrated where sufficient specimens from family members are available (114,115,123,224,227,417,418,422). Point mutations are expressed by the wild-type amino acid letter code and the codon number, followed by the mutant residue, e.g., P102L. These point mutations segregate with the inherited prion diseases, and significant genetic linkage *(underlined mutations)* has been demonstrated where sufficient specimens from family members are available. Mutations at codons 102 (Pro→Leu), 117 (Ala→Val), 198 (Phe→Ser), and 217 (Gln→Arg) are found in patients with GSS (165,221,224, 231,233,259,262–264,541). Point mutations at codons 178 (Asp→Asn), 200 (Glu→Lys) and 210 (Val→Ile) are found in patients with familial CJD (190,222,230,261,483). Point mutations at codons 198 (Phe→Ser) and 217 (Gln→Arg) are found in patients with GSS who have PrP amyloid plaques and neurofibrillary tangles (163,260). Additional point mutations at codons 145 (Tyr→Stop), 105 (Pro→Leu), 180 (Val→Ile) and 232 (Met→Arg) have been recently reported (306,308). Single letter code for amino acids is as follows: A, Ala; D, Asp; E, Glu; F, Phe; I, Ile; K, Lys; L, Leu; M, Met; N, Asn; P, Pro; Q, Gln; R, Arg; S, Ser; T, Thr; and V, Val; Y, Tyr.

and 217, were once thought to have familial Alzheimer's disease, but are now known to have prion diseases on the basis of PrP immunostaining of amyloid plaques and PrP gene mutations (175,207,208,406). Patients with the codon 198 mutation have numerous neurofibrillary tangles that stain with antibodies to tau (τ) protein and have amyloid plaques (175,207,208,406) that are composed largely of a PrP fragment extending from residues 58 to 150 (529). A genetic linkage study of this family produced a LOD score exceeding 6 (163). The neuropathology of two patients of Swedish ancestry with the codon 217 mutation (279) was similar to that of patients with the codon 198 mutation.

An insert of 144 bp at codon 53 containing six octarepeats has been described in patients with CJD from four families, all residing in southern England (Fig. 12) (113–115,123,416–418,444). This mutation must have arisen through a complex series of events, since the human PrP gene contains only five octarepeats, indicating that a single recombination event could not have created the insert. Genealogic investigations have shown that all four families are related, arguing for a single founder born more than two centuries ago (123). The LOD score for this extended pedigree exceeds 11. Studies from several laboratories have demonstrated that two, four, five, six, seven, eight, or nine octarepeats in addition to the normal five are found in individuals with inherited CJD (58,114,115,227,415,417, 418), whereas deletion of one octarepeat has been identified without the neurologic disease (327,422,554).

For many years the unusually high incidence of CJD among Israeli Jews of Libyan origin was thought to be due to the consumption of lightly cooked sheep brain or eyeballs (15,253,285,286,402,582). Recent studies have shown that some Libyan and Tunisian Jews in families with CJD have a PrP gene point mutation at codon 200, resulting in a glutamate (E)→lysine (K) substitution (190,231,261). One patient was homozygous for the E200K mutation, but her clinical presentation was similar to that of heterozygotes (261), arguing that familial prion diseases are true autosomal dominant disorders like Huntington's disease (563). The E200K mutation has also been found in Slovaks originating from Orava in North Central Czechoslovakia (231), in a cluster of familial cases in Chile (228), and in a large German family living in the U.S. (41) Some investigators have argued that the E200K mutation originated in a Sephardic Jew whose descendants migrated from Spain and Portugal at the time of the Inquisition (228). It is more likely that the E200K mutation has arisen independently multiple times by the deamidation of a methylated CpG (as described above), the codon 102 mutation (259,261). In support of this hypothesis are historical records of Libyan and Tunisian Jews, indicating that they are descended from Jews living on the island of Jerba off the southern coast of Tunisia, where Jews first settled around 500 B.C. and not from Sephardim (550).

Many families with CJD have been found to have a point mutation at codon 178, resulting in an aspartate (D)→asparagine (N) substitution (69,178,225,230,245). In these patients, as well as those with the E200K mutation, PrP amyloid plaques are rare; the neuropathologic changes generally consist of widespread spongiform degeneration. Recently a new prion disease called fatal familial insomnia (FFI) which presents with insomnia was described in three Italian families with the D178N mutation (339,372). The neuropathology in these patients with FFI is restricted to selected nuclei of the thalamus. It is unclear whether all patients with the D178N mutation or only a subset present with sleep disturbances. It has been proposed that the allele with the D178N mutation encodes an M at position 129 in FFI while a V is encoded at position 129 in familial CJD (232). The discovery that FFI is an inherited prion disease clearly widens the clinical spectrum of these disorders and raises the possibility that many other degenerative diseases of unknown etiology may be caused by prions (284,372). The D178N mutation has been linked to the development of prion disease with a LOD score exceeding 5 (439). Studies of PrP^{Sc} in FFI and familial CJD caused by the D178N mutation show that after limited proteolysis the M_r of the FFI PrP^{Sc} is ~1 kD smaller (387). Whether this difference in protease resistance reflects distinct conformations of PrP^{Sc}—which give rise to the different clinical and neuropathologic manifestations of these inherited prion diseases—remains to be established.

Like the E200K and D178N(V129) mutations, a valine (V)→isoleucine (I) mutation at PrP codon 210 produces CJD with classic symptoms and signs (441,483). It appears that this V210I mutation is also incompletely penetrant.

Other point mutations at codons 145 and possibly 232 also segregate with inherited prion diseases (58,165,260, 263,306,308,310,548). One patient with a prolonged neurologic illness spanning almost two decades who had PrP amyloid plaques was found to have an amber mutation of the PrP gene resulting in a stop codon at residue 145 (306, 310). Staining of the plaques with α-PrP peptide antisera suggested that they might be composed exclusively of the truncated PrP molecules. That a PrP peptide ending at residue 145 polymerizes in amyloid filaments is to be expected since an earlier study noted above showed that the major PrP peptide in plaques from patients with the F198S mutation was an 11-kD PrP peptide beginning at codon 58 and ending at ~150 (529). Furthermore, synthetic PrP peptides adjacent to and including residues 109 to 122 readily polymerize into rod-shaped structures with the tinctorial properties of amyloid (119,179,205,226).

One view of the PrP gene mutations has been that they render individuals susceptible to a common "virus" (4,103, 293). In this scenario, the putative scrapie virus is thought to persist within a world-wide reservoir of humans, animals, or insects without causing detectable illness. Yet $1/10^6$ individuals develop sporadic CJD and die from a lethal "infection" while ~100% of people with PrP point mutations or inserts appear to eventually develop neurologic dysfunction. That germline mutations found in the PrP genes of patients and at-risk individuals are the cause of familial prion diseases is supported by experiments with Tg(MoPrP-P101L) mice, described above (262,265,266). These and other Tg mouse studies also argue that sporadic CJD might arise from either a somatic mutation of the PrP gene or the spontaneous conversion of PrP^C to PrP^{CJD} (452,559).

PrP GENE POLYMORPHISMS

Polymorphisms at Codons 129 and 219

At PrP codon 129, an amino acid polymorphism for the M→V has been identified (Fig. 10) (414). This polymorphism appears able to influence prion disease expression, not only in inherited forms, but also in iatrogenic and sporadic forms of prion disease. A second polymorphism resulting in an amino acid substitution at codon 219 (E→K) has been reported in the Japanese population, in which the K allele occurs with a frequency of 6% (310).

Does Homozygosity at Codon 129 Predispose to CJD?

Studies of Caucasian patients with sporadic CJD have shown that most are homozygous for M or V at codon 129 (421). This contrasts with the general population, in which frequencies for the codon 129 polymorphism in Caucasians are 12% V/V, 37% M/M, and 51% M/V (117). In contrast, the frequency of the V allele in the Japanese population is much lower (164,385), and heterozygosity at codon 129 (M/V) is more frequent (18%) in CJD patients than in the general population, where the polymorphism frequencies are 0% V/V, 92% M/M, and 8% M/V (540).

While no specific mutations have been identified in the PrP gene of patients with sporadic CJD (224), homozygosity at codon 129 in sporadic CJD (247,421) is consistent with the results of Tg mouse studies. The finding that homozygosity at codon 129 predisposes to CJD supports a model of prion production which favors PrP interactions between homologous proteins, as appears to occur in Tg mice expressing SHaPrP inoculated with either hamster prions or mouse prions (452,477,506), as well as Tg mice expressing a chimeric mouse/hamster PrP transgene inoculated with "artificial" prions (507).

Codon 129 May Influence Iatrogenic CJD

Susceptibility to infection may be partially determined by the PrP codon 129 genotype (117), analogous in principle to the incubation-time alleles in mice (87,117). In sixteen patients (fifteen Caucasian, one Afro-American) from the U.K., U.S., and France with iatrogenic CJD from contaminated growth hormone extracts, eight (50%) were V/V, five (31%) were M/M, and three (19%) were M/V (64,72,79,117,141,180,223). Thus, a disproportionate number of patients with iatrogenic CJD were homozygous for valine at PrP codon 129, and heterozygosity at codon 129 may provide partial protection. Whether these associations are strongly significant awaits statistical analysis of larger samples. Thousands of children who received pituitary growth hormone extracts are still at risk for the development of CJD. Fortunately, the use of genetically engineered growth hormone will eliminate this form of iatrogenic CJD.

Approximately 15% of patients with sporadic CJD develop ataxia as an early sign, accompanied by dementia (73). Most, but not all, patients with ataxia have compact "kuru" plaques in the cerebellum (437). Patients with ataxia and compact plaques exhibit a protracted clinical course which may last up to 3 years. The molecular basis for the differences between CJD of shorter and longer duration have not yet been fully elucidated; however, some preliminary analyses have suggested that patients with protracted, atypical clinical courses are more likely to be heterozygous at codon 129 (116,164).

Codon 129 and Inherited Prion Diseases

Homozygosity at codon 129 has been reported to be associated with an earlier age of onset in the inherited prion disease caused by the 6 octarepeat insert, but not by the E200K mutation in Libyan Jews (23,192). As noted above, the FFI phenotype is found in patients with the D178N mutation who encode a met at codon 129 on the mutant allele while those with dementing illness (familial CJD) encode a V at 129 (232). Homozygosity for either M or V at codon 129 is thought to be associated with an earlier age of onset for the D178N mutation.

THERAPEUTIC APPROACHES TO PRION DISEASES

There is no known effective therapy for treating or preventing CJD. There are no well-documented cases of patients with CJD showing recovery, either spontaneously or after therapy—with one possible exception (349). Since people at risk for inherited prion diseases can now be identified decades before neurologic dysfunction is evident, the development of an effective therapy is imperative.

Prenatal Screening for PrP Mutations

The inherited prion diseases can be prevented by genetic counseling coupled with prenatal DNA screening, but such testing presents ethical problems. For example, during the child-bearing years, the parents are generally symptom-free and may not want to know their own genotype. The apparent incomplete penetrance of some of the inherited prion diseases makes predicting the future for an asymptomatic individual uncertain (192,261).

Gene Therapy and Anti-Sense Oligonucleotides

Unexpectedly, ablation of the PrP gene in Tg($Prnp^{0/0}$) mice has not affected the development of these animals, and they remain healthy at almost 2 years of age (81). Since $Prnp^{0/0}$ mice are resistant to prions (Fig. 9) and do not propagate scrapie infectivity (Table 5) (80,463), gene therapy or antisense oligonucleotides might ultimately provide an effective therapeutic approach. Mice heterozygous ($Prnp^{0/+}$) for ablation of the PrP gene had prolonged incubation times when inoculated with mouse prions (463). This finding is in accord with studies on Tg(SHaPrP) mice in which increased SHaPrP expression was accompanied by diminished incubation times (477).

Inhibitors of β-Sheet Formation

Because the absence of PrP^C expression does not provoke disease, it seems reasonable to conclude that scrapie and other prion diseases are a consequence of PrP^{Sc} accumulation rather than an inhibition of PrP^C function (81). The function of PrP^C remains unknown to date. These findings suggest that perhaps the most effective therapy may evolve from the development of drugs which block the conversion of PrP^C into PrP^{Sc}. Since the fundamental event in both the formation of PrP^{Sc} and the propagation of prions seems to be the unfolding of α-helices and their refolding into β-sheets (Fig. 5) (424), drugs targeting this structural transformation would seem likely to be efficacious (112,268).

Whether dyes like Congo red which bind to PrP amyloid will be effective *in vivo* in preventing the formation of PrP^{Sc} remains to be established. Congo red has been reported to inhibit the formation of PrP^{Sc} in cultured scrapie-infected mouse neuroblastoma cells (95), but the mechanism by which this occurs is unknown.

Dextran sulfates have been used to delay the onset of scrapie in laboratory animals, but these compounds are effective only when given before or concurrently with the prion inoculum (146,161,169). The inhibitory effect of dextrans on prion propagation has been most pronounced in animals inoculated intraperitoneally. The mechanism by which the sulphated sugar polymers inhibit PrP^{Sc} formation in cultured cells and animals is unknown (97).

Injection of corticosteroids into mice at the time of intraperitoneal inoculation of prions results in the prolongation of the incubation time (413). These results suggested that the immune system played a role in the initial phase of scrapie infection, as did studies with spleenless mice (144). Studies with nude mice and, more recently, SCID mice showed that incubation times were unaltered compared to controls after intracerebral inoculation; in contrast, SCID mice were not susceptible to prions inoculated intraperitoneally (307). Studies of lymphoid tissues after intraperitoneal inoculation of prions show that the follicular dendritic cells which are thought to function in the presentation of antigens selectively accumulate PrP^{Sc}. In SCID mice no focal accumulations of PrP^{Sc} in lymphoid tissues have been found (307).

Chronic treatment with the antifungal drug amphotericin started at the time of intracerebral inoculation prolonged incubation periods in mice and hamsters (91,442,578). Administration of amphotericin to humans with CJD had no therapeutic effect (363); likewise, the antiviral drug amantidine was also ineffective (56,252,545). HPA-23 is an inhibitor of viral glycoprotein synthesis. When given to scrapie-infected animals around the time of inoculation, but not later, it profoundly extends the length of the incubation period (297). The effects in human CJD are uncertain (94). Interferon has been used in experimental scrapie of rodents, but the incubation times were unaltered (177,290,577).

Though antibodies have been raised against the scrapie prion protein and these cross-react with prion proteins in CJD human brains (47), passive immunization or even vaccination would seem to be of little value. CJD and scrapie both progress in the absence of any immune response to the offending prions; however, neutralization of scrapie prion infectivity was accomplished when the infectious particles were dispersed into detergent-lipid-protein complexes (187).

Resistant Breeds of Sheep

With the discovery that QQ homozygosity at codon 171 appears to render Suffolk sheep susceptible to scrapie while QR and RR sheep are resistant (562), the notion of breeding scrapie-resistant sheep as proposed by Parry seems quite reasonable (428). The screening procedures for identifying QQ sheep are simple and could be applied in an agricultural setting with ease.

CONCLUSIONS

Prions Are Not Viruses

The study of prions has taken several unexpected directions over the past few years. The discovery that prion diseases in humans are uniquely both genetic and infectious has greatly strengthened and extended the prion concept. To date, 18 different mutations in the human PrP gene, all resulting in nonconservative substitutions, have been found to either be linked genetically to or segregate with the inherited prion diseases (Fig. 12). Yet, the transmissible prion particle is composed largely, if not entirely, of an abnormal isoform of the prion protein designated PrP^{Sc} (452). These findings argue that prion diseases should be considered pseudoinfections, since the particles transmitting disease appear to be devoid of a foreign nucleic acid and thus differ from all known microorganisms as well as viruses and viroids. Because much information, especially about scrapie of rodents, has been derived using experimental protocols adapted from virology, we continue to use terms such as *infection, incubation period, transmissibility*, and *endpoint titration* in studies of prion diseases.

Do Prions Exist in Lower Organisms?

In *S. cerevisiae*, ure2 and [URA3] mutants were described that can grow on ureidosuccinate under conditions of nitrogen repression such as glutamic acid and ammonia (323). Mutants of URE2 exhibit Mendelian inheritance, whereas [URE3] is cytoplasmically inherited (564). The [URE3] phenotype can be induced by UV irradiation and by overexpression of ure2p, the gene product of ure2; deletion of ure2 abolishes [URE3]. The function of ure2p is unknown, but it has substantial homology with glutathione-S-transferase; attempts to demonstrate this enzymic activity with purified ure2p have not been successful (120). Whether the [URE3] protein is a posttranslationally modified form of ure2p which acts upon unmodified ure2p to produce more of itself remains to be established.

Another possible yeast prion is the [PSI] phenotype (564). [PSI] is a non-Mendelian inherited trait that can be induced by expression of the PNM2 gene (122). Both [PSI] and [URE3] can be cured by exposure of the yeast to 3 mM GdnHCl. The mechanism responsible for abolishing [PSI] and [URE3] with a low concentration of GdnHCl is unknown. In the filamentous fungus *Podospora anserina*, the het-s locus controls the vegetative incompatibility; conversion from the S^s to the s state seems to be a posttranslational, autocatalytic process (139).

If any of the above cited examples can be shown to function in a manner similar to prions in animals, many new, more rapid and economical approaches to prion diseases should be forthcoming.

Common Neurodegenerative Diseases

The knowledge accrued from the study of prion diseases may provide an effective strategy for defining the etiologies and dissecting the molecular pathogenesis of the more common neurodegenerative disorders such as Alzheimer's disease, Parkinson's disease, and amyotrophic lateral sclerosis (ALS). Advances in the molecular genetics of Alzheimer's disease and ALS suggest that, like the prion diseases, an important subset is caused by mutations that result in nonconservative amino acid substitutions in proteins expressed in the CNS (220,330,389,495,505,524, 551,552).

Future Studies

Tg mice expressing foreign or mutant PrP genes now permit virtually all facets of prion diseases to be studied and have created a framework for future investigations. Furthermore, the structure and organization of the PrP gene suggested that PrP^{Sc} is derived from PrP^{C} or a precursor by a posttranslational process. Studies with scrapie-infected cultured cells have provided much evidence that the conversion of PrP^{C} to PrP^{Sc} occurs within a subcellular compartment bounded by cholesterol-rich membranes (Fig. 3). The molecular mechanism of PrP^{Sc} formation remains to be elucidated.

The study of prion biology and diseases has emerged as a new area of biomedical investigation. While prion biology has its roots in virology, neurology, and neuropathology; its relationships to the disciplines of molecular and cell biology as well as protein chemistry are evident. Learning how prions multiply and cause disease will open up new vistas in biochemistry and genetics.

REFERENCES

1. Adams DH. The nature of the scrapie agent: a review of recent progress. *Pathol Biol* 1970;18:559–577.
2. Adams DH, Caspary EA. Nature of the scrapie virus. *Br Med J* 1967; 3:173.
3. Adams DH, Field EJ. The infective process in scrapie. *Lancet* 1968;2: 714–716.
4. Aiken JM, Marsh RF. The search for scrapie agent nucleic acid. *Microbiol Rev* 1990;54:242–246.
5. Aiken JM, Williamson JL, Borchardt LM, Marsh RF. Presence of mitochondrial D-loop DNA in scrapie-infected brain preparations enriched for the prion protein. *J Virol* 1990;64:3265–3268.
6. Aiken JM, Williamson JL, Marsh RF. Evidence of mitochondrial involvement in scrapie infection. *J Virol* 1989;63:1686–1694.
7. Akowitz A, Sklaviadis T, Manuelidis EE, Manuelidis L. Nuclease-resistant polyadenylated RNAs of significant size are detected by PCR in highly purified Creutzfeldt-Jakob disease preparations. *Microb Pathog* 1990;9:33–45.
8. Akowitz A, Sklaviadis T, Manuelidis L. Endogenous viral complexes with long RNA cosediment with the agent of Creutzfeldt-Jakob disease. *Nucleic Acids Res* 1994;22:1101–1107.
9. Alper T, Cramp WA, Haig DA, Clarke MC. Does the agent of scrapie replicate without nucleic acid? *Nature* 1967;214:764–766.
10. Alper T, Haig DA, Clarke MC. The scrapie agent: evidence against its dependence for replication on intrinsic nucleic acid. *J Gen Virol* 1978; 41:503–516.
11. Alper T, Haig DA, Clarke MC. The exceptionally small size of the scrapie agent. *Biochem Biophys Res Commun* 1966;22:278–284.
12. Alpers M. Epidemiology and clinical aspects of kuru. In: Prusiner SB, McKinley MP, eds. *Prions—novel infectious pathogens causing scrapie and Creutzfeldt-Jakob disease*. Orlando: Academic Press, 1987;451–465.
13. Alpers M. Kuru: a clinical study. In: Mimeographed Manuscript, Reissued. Bethesda: USDHEW, NIH, NINCDS, 1964;1–38.
14. Alpers MP. Epidemiology and ecology of kuru. In: Prusiner SB, Hadlow WJ, eds. *Slow transmissible diseases of the nervous system*, Vol. 1. New York: Academic Press, 1979;67–90.
15. Alter M, Kahana E. Creutzfeldt-Jakob disease among Libyan Jews in Israel. *Science* 1976;192:428.
16. Amyx HL, Gibbs CJ Jr., Gajdusek DC, Greer WE. Absence of vertical transmission of subacute spongiform viral encephalopathies in experimental primates (41092). *Proc Soc Exp Biol Med* 1981;166:469–471.
17. Anderson JR, Allen CMC, Weller RO. Creutzfeldt-Jakob disease following human pituitary-derived growth hormone administration. *Br Neuropatholog Soc Proc* 1990;16:543.
18. Anderson RGW. Caveolae: where incoming and outgoing messengers meet. *Proc Natl Acad Sci USA* 1993;90:10909–10913.
19. Anonymous. The first victim of mad cow disease? *Daily Mail* 1993; Sept. 13, 1993:1.
20. Anonymous. Second farmer's death raises fear of "mad cow" cover-up. *Times* 1993;8/13/93:4.
21. Askanas V, Bilak M, Engel WK, Leclerc A, Tomé F. Prion protein is strongly immunolocalized at the postsynaptic domain of human normal neuromuscular junctions. *Neurosci Lett* 1993;159:111–114.
22. Azzarelli B, Muller J, Ghetti B, Dyken M, Conneally PM. Cerebellar plaques in familial Alzheimer's disease (Gerstmann-Sträussler-Scheinker variant?). *Acta Neuropathol* (Berl) 1985;65:235–246.
23. Baker HF, Poulter M, Crow TJ, Frith CD, Lofthouse R, Ridley RM. Amino acid polymorphism in human prion protein and age at death in inherited prion disease. *Lancet* 1991;337:1286.
24. Baker HF, Ridley RM, Wells GAH. Experimental transmission of BSE and scrapie to the common marmoset. *Vet Rec* 1993;132:403–406.
25. Baringer JR, Bowman KA, Prusiner SB. Replication of the scrapie agent in hamster brain precedes neuronal vacuolation. *J Neuropathol Exp Neurol* 1983;42:539–547.
26. Baringer JR, Prusiner SB. Experimental scrapie in mice: ultrastructural observations. *Ann Neurol* 1978;4:205–211.
27. Baringer JR, Prusiner SB, Wong J. Scrapie-associated particles in postsynaptic processes. Further ultrastructural studies. *J Neuropathol Exp Neurol* 1981;40:281–288.
28. Barry RA, Prusiner SB. Monoclonal antibodies to the cellular and scrapie prion proteins. *J Infect Dis* 1986;154:518–521.
29. Basler K, Oesch B, Scott M, et al. Scrapie and cellular PrP isoforms are encoded by the same chromosomal gene. *Cell* 1986;46:417–428.

30. Bastian FO. Bovine spongiform encephalopathy: relationship to human disease and nature of the agent. *ASM News* 1993;59:235–240.
31. Bastian FO. Spiroplasma-like inclusions in Creutzfeldt-Jakob disease. *Arch Pathol Lab Med* 1979;103:665–669.
32. Beck E, Daniel PM, Parry HB. Degeneration of the cerebellar and hypothalamo-neurohypophysial systems in sheep with scrapie; and its relationship to human system degenerations. *Brain* 1964;87:153–176.
33. Bellinger-Kawahara C, Cleaver JE, Diener TO, Prusiner SB. Purified scrapie prions resist inactivation by UV irradiation. *J Virol* 1987;61: 159–166.
34. Bellinger-Kawahara C, Diener TO, McKinley MP, Groth DF, Smith DR, Prusiner SB. Purified scrapie prions resist inactivation by procedures that hydrolyze, modify, or shear nucleic acids. *Virology* 1987; 160:271–274.
35. Bellinger-Kawahara CG, Kempner E, Groth DF, Gabizon R, Prusiner SB. Scrapie prion liposomes and rods exhibit target sizes of 55,000 Da. *Virology* 1988;164:537–541.
36. Bendheim PE, Barry RA, DeArmond SJ, Stites DP, Prusiner SB. Antibodies to a scrapie prion protein. *Nature* 1984;310:418–421.
37. Bendheim PE, Bockman JM, McKinley MP, Kingsbury DT, Prusiner SB. Scrapie and Creutzfeldt-Jakob disease prion proteins share physical properties and antigenic determinants. *Proc Natl Acad Sci USA* 1985;82:997–1001.
38. Berger JR, David NJ. Creutzfeldt-Jakob disease in a physician: a review of the disorder in health care workers. *Neurology* 1993;43: 205–206.
39. Bernouilli C, Siegfried J, Baumgartner G, et al. Danger of accidental person to person transmission of Creutzfeldt-Jakob disease by surgery. *Lancet* 1977;1:478–479.
40. Bernouilli CC, Masters CL, Gajdusek DC, Gibbs CJ Jr., Harris JO. Early clinical features of Creutzfeldt-Jakob disease (subacute spongiform encephalopathy). In: Prusiner SB, Hadlow WJ, eds. *Slow transmissible diseases of the nervous system*, vol. 1. New York: Academic Press, 1979;229–251.
41. Bertoni JM, Brown P, Goldfarb L, Gajdusek D, Omaha NE. Familial Creutzfeldt-Jakob disease with the PRNP codon 200^{lys} mutation and supranuclear palsy but without myoclonus or periodic EEG complexes. *Neurology* 1992;42(No. 4, Suppl. 3):350 [Abstr].
42. Bessen RA, Marsh RF. Biochemical and physical properties of the prion protein from two strains of the transmissible mink encephalopathy agent. *J Virol* 1992;66:2096–2101.
43. Bessen RA, Marsh RF. Identification of two biologically distinct strains of transmissible mink encephalopathy in hamsters. *J Gen Virol* 1992; 73:329–334.
44. Billette de Villemeur T, Beauvais P, Gourmelon M, Richardet JM. Creutzfeldt-Jakob disease in children treated with growth hormone. *Lancet* 1991;337:864–865.
45. Blum B, Bakalara N, Simpson L. A model for RNA editing in kinetoplastid mitochondria: "guide" RNA molecules transcribed from maxicircle DNA provide edited information. *Cell* 1990;60:189–198.
46. Bobowick AR, Brody JA, Matthews MR, Roos R, Gajdusek DC. Creutzfeldt-Jakob disease: a case-control study. *Am J Epidemiol* 1973; 98:381–394.
47. Bockman JM, Kingsbury DT, McKinley MP, Bendheim PE, Prusiner SB. Creutzfeldt-Jakob disease prion proteins in human brains. *N Engl J Med* 1985;312:73–78.
48. Bockman JM, Prusiner SB, Tateishi J, Kingsbury DT. Immunoblotting of Creutzfeldt-Jakob disease prion proteins: host species-specific epitopes. *Ann Neurol* 1987;21:589–595.
49. Bolton DC, McKinley MP, Prusiner SB. Identification of a protein that purifies with the scrapie prion. *Science* 1982;218:1309–1311.
50. Bolton DC, Meyer RK, Prusiner SB. Scrapie PrP 27-30 is a sialoglycoprotein. *J Virol* 1985;53:596–606.
51. Borchelt DR, Koliatsis VE, Guarnieri M, Pardo CA, Sisodia SS, Price DL. Rapid anterograde axonal transport of the cellular prion glycoprotein in the peripheral and central nervous systems. *J Biol Chem* 1994;269:14711–14714.
52. Borchelt DR, Rogers M, Stahl N, Telling G, Prusiner SB. Release of the cellular prion protein from cultured cells after loss of its glycoinositol phospholipid anchor. *Glycobiology* 1993;3:319–329.
53. Borchelt DR, Scott M, Taraboulos A, Stahl N, Prusiner SB. Scrapie and cellular prion proteins differ in their kinetics of synthesis and topology in cultured cells. *J Cell Biol* 1990;110:743–752.
54. Borchelt DR, Taraboulos A, Prusiner SB. Evidence for synthesis of scrapie prion proteins in the endocytic pathway. *J Biol Chem* 1992; 267:16188–16199.
55. Bouteille M, Fontaine C, Vedrenne CL, Delarue J. Sur un cas d'encephalite subaigue a inclusions. Étude anatomoclinque et ultrastructurale. *Rev Neurol* 1965;118:454–458.
56. Braham J. Jakob-Creutzfeldt disease: treatment by amantadine. *Br Med J* 1971;4:212–213.
57. Braig H, Diringer H. Scrapie: concept of a virus-induced amyloidosis of the brain. *EMBO J* 1985;4:2309–2312.
58. Brown P. Infectious cerebral amyloidosis: clinical spectrum, risks and remedies. In: Brown F, ed. *Developments in biological standardization*. Basel, Switzerland: Karger, 1993;91–101.
59. Brown P. The phenotypic expression of different mutations in transmissible human spongiform encephalopathy. *Rev Neurol* 1992;148: 317–327.
60. Brown P. The decline and fall of Creutzfeldt-Jakob disease associated with human growth hormone therapy. *Neurology* 1988;38:1135–1137.
61. Brown P. Virus sterility for human growth hormone. *Lancet* 1985;2: 729–730.
62. Brown P, Cathala F, Castaigne P, Gajdusek DC. Creutzfeldt-Jakob disease: clinical analysis of a consecutive series of 230 neuropathologically verified cases. *Ann Neurol* 1986;20:597–602.
63. Brown P, Cathala F, Raubertas RF, Gajdusek DC, Castaigne P. The epidemiology of Creutzfeldt-Jakob disease: conclusion of 15-year investigation in France and review of the world literature. *Neurology* 1987;37:895–904.
64. Brown P, Cervenáková L, Goldfarb LG, et al. Iatrogenic Creutzfeldt-Jakob disease: an example of the interplay between ancient genes and modern medicine. *Neurology* 1994;44:291–293.
65. Brown P, Coker-Vann M, Gajdusek DC. Immunological study of patients with Creutzfeldt-Jakob disease and other chronic neurological disorders: Western blot recognition of infection-specific proteins by scrapie virus antibody. In: Bignami A, Bolis CL, Gajdusek DC, eds. *Molecular mechanisms of pathogenesis of central nervous system disorders*. Geneva: Foundation for the Study of the Nervous System, 1986; 107–109.
66. Brown P, Coker-Vann M, Pomeroy K, et al. Diagnosis of Creutzfeldt-Jakob disease by Western blot identification of marker protein in human brain tissue. *N Engl J Med* 1986;314:547–551.
67. Brown P, Gajdusek DC, Gibbs CJ Jr., Asher DM. Potential epidemic of Creutzfeldt-Jakob disease from human growth hormone therapy. *N Engl J Med* 1985;313:728–731.
68. Brown P, Gibbs CJ Jr., Rodgers-Johnson P, et al. Human spongiform encephalopathy: the National Institutes of Health series of 300 cases of experimentally transmitted disease. *Ann Neurol* 1994;35:513–529.
69. Brown P, Goldfarb LG, Kovanen J, et al. Phenotypic characteristics of familial Creutzfeldt-Jakob disease associated with the codon 178^{Asn} PRNP mutation. *Ann Neurol* 1992;31:282–285.
70. Brown P, Goldfarb LG, McCombie WR, et al. Atypical Creutzfeldt-Jakob disease in an American family with an insert mutation in the PRNP amyloid precursor gene. *Neurology* 1992;42:422–427.
71. Brown P, Kaur P, Sulima MP, Goldfarb LG, Gibbs CJJ, Gajdusek DC. Real and imagined clincopathological limits of "prion dementia". *Lancet* 1993;341:127–129.
72. Brown P, Preece MA, Will RG. "Friendly fire" in medicine: hormones, homografts, and Creutzfeldt-Jakob disease. *Lancet* 1992;340:24–27.
73. Brown P, Rodgers-Johnson P, Cathala F, Gibbs CJ Jr., Gajdusek DC. Creutzfeldt-Jakob disease of long duration: clinicopathological characteristics, transmissibility, and differential diagnosis. *Ann Neurol* 1984; 16:295–304.
74. Bruce M, Chree A, McConnell I, Foster J, Fraser H. Transmissions of BSE, scrapie and related diseases to mice. In: *IXth International Congress of Virology*. Glasgow, Scotland, Aug. 8–13, 1993:1993; 93.
75. Bruce ME, Dickinson AG. Biological evidence that the scrapie agent has an independent genome. *J Gen Virol* 1987;68:79–89.
76. Bruce ME, Dickinson AG. Biological stability of different classes of scrapie agent. In: Prusiner SB, Hadlow WJ, eds. *Slow transmissible diseases of the nervous system*, vol. 2. New York: Academic Press, 1979;71–86.
77. Bruce ME, Dickinson AG, Fraser H. Cerebral amyloidosis in scrapie in the mouse: effect of agent strain and mouse genotype. *Neuropathol Appl Neurobiol* 1976;2:471–478.
78. Bruce ME, McConnell I, Fraser H, Dickinson AG. The disease characteristics of different strains of scrapie in Sinc congenic mouse lines:

implications for the nature of the agent and host control of pathogenesis. *J Gen Virol* 1991;72:595–603.
79. Buchanan CR, Preece MA, Milner RDG. Mortality, neoplasia and Creutzfeldt-Jakob disease in patients treated with pituitary growth hormone in the United Kingdom. *Br Med J* 1991;302:824–828.
80. Büeler H, Aguzzi A, Sailer A, et al. Mice devoid of PrP are resistant to scrapie. *Cell* 1993;73:1339–1347.
81. Büeler H, Fischcr M, Lang Y, et al. Normal development and behaviour of mice lacking the neuronal cell-surface PrP protein. *Nature* 1992; 356:577–582.
82. Butler DA, Scott MRD, Bockman JM, et al. Scrapie-infected murine neuroblastoma cells produce protease-resistant prion proteins. *J Virol* 1988;62:1558–1564.
83. Buyukmihci N, Rorvik M, Marsh RF. Replication of the scrapie agent in ocular neural tissues. *Proc Natl Acad Sci USA* 1980;77:1169–1171.
84. Carlson GA, Ebeling C, Torchia M, Westaway D, Prusiner SB. Delimiting the location of the scrapie prion incubation time gene on chromosome 2 of the mouse. *Genetics* 1993;133:979–988.
85. Carlson GA, Ebeling C, Yang S-L, et al. Prion isolate specified allotypic interactions between the cellular and scrapie prion proteins in congenic and transgenic mice. *Proc Natl Acad Sci USA* 1994;91:5690–5694.
86. Carlson GA, Goodman PA, Lovett M, et al. Genetics and polymorphism of the mouse prion gene complex: the control of scrapie incubation time. *Mol Cell Biol* 1988;8:5528–5540.
87. Carlson GA, Kingsbury DT, Goodman PA, et al. Linkage of prion protein and scrapie incubation time genes. *Cell* 1986;46:503–511.
88. Carlson GA, Westaway D, DeArmond SJ, Peterson-Torchia M, Prusiner SB. Primary structure of prion protein may modify scrapie isolate properties. *Proc Natl Acad Sci USA* 1989;86:7475–7479.
89. Carp RI, Callahan SM. Variation in the characteristics of 10 mouse-passaged scrapie lines derived from five scrapie-positive sheep. *J Gen Virol* 1991;72:293–298.
90. Carp RI, Moretz RC, Natelli M, Dickinson AG. Genetic control of scrapie: incubation period and plaque formation in mice. *J Gen Virol* 1987;68:401–407.
91. Casaccia P, Ladogana A, Xi YG, et al. Measurement of the concentration of amphotericin B in brain tissue of scrapie-infected hamsters with a simple and sensitive method. *Antimicrobial Agents & Chemotherapy* 1991;35:1486–1488.
92. Casaccia-Bonnefil P, Kascsak RJ, Fersko R, Callahan S, Carp RI. Brain regional distribution of prion protein PrP27-30 in mice stereotaxically microinjected with different strains of scrapie. *J Infect Dis* 1993; 167:7–12.
93. Cashman NR, Loertscher R, Nalbantoglu J, et al. Cellular isoform of the scrapie agent protein participates in lymphocyte activation. *Cell* 1990;61:185–192.
94. Cathala F, Baron H. Clinical aspects of Creutzfeldt-Jakob disease. In: Prusiner SB, McKinley MP, eds. *Prions—novel infectious pathogens causing scrapie and Creutzfeldt-Jakob disease*. Orlando: Academic Press, 1987;467–509.
95. Caughey B, Race RE. Potent inhibition of scrapie-associated PrP accumulation by Congo red. *J Neurochem* 1992;59:768–771.
96. Caughey B, Race RE, Ernst D, Buchmeier MJ, Chesebro B. Prion protein biosynthesis in scrapie-infected and uninfected neuroblastoma cells. *J Virol* 1989;63:175–181.
97. Caughey B, Raymond GJ. Sulfated polyanion inhibition of scrapie-associated PrP accumulation in cultured cells. *J Virol* 1993;67:643–650.
98. Caughey B, Raymond GJ. The scrapie-associated form of PrP is made from a cell surface precursor that is both protease-and phospholipase-sensitive. *J Biol Chem* 1991;266:18217–18223.
99. Caughey B, Raymond GJ, Ernst D, Race RE. N-terminal truncation of the scrapie-associated form of PrP by lysosomal protease(s): implications regarding the site of conversion of PrP to the protease-resistant state. *J Virol* 1991;65:6597–6603.
100. Caughey BW, Dong A, Bhat KS, Ernst D, Hayes SF, Caughey WS. Secondary structure analysis of the scrapie-associated protein PrP 27–30 in water by infrared spectroscopy. *Biochemistry* 1991;30:7672–7680.
101. Chandler RL. Experimental scrapie in the mouse. *Res Vet Sci* 1963;4: 276–285.
102. Chandler RL. Encephalopathy in mice produced by inoculation with scrapie brain material. *Lancet* 1961;1:1378–1379.
103. Chesebro B. PrP and the scrapie agent. *Nature* 1992;356:560.
104. Chesebro B, Caughey B. Scrapie agent replication without the prion protein? *Curr Biol* 1993;3:696–698.
105. Chesebro B, Race R, Wehrly K, et al. Identification of scrapie prion protein-specific mRNA in scrapie-infected and uninfected brain. *Nature* 1985;315:331–333.
106. Cho HJ. Inactivation of the scrapie agent by pronase. *Can J Comp Med* 1983;47:494–496.
107. Cho HJ. Requirement of a protein component for scrapie infectivity. *Intervirology* 1980;14:213–216.
108. Clousard C, Beaudry P, Elsen JM, et al. Different allelic effects of the codons 136 and 171 of the prion protein gene in sheep with natural scrapie. *J Gen Virol* 1995 (*in press*).
109. Cochius JI, Hyman N, Esiri MM. Creutzfeldt-Jakob disease in a recipient of human pituitary-derived gonadotrophin: a second case. *J Neurol Neurosurg Psychiatry* 1992;55:1094–1095.
110. Cochius JI, Mack K, Burns RJ, Alderman CP, Blumbergs PC. Creutzfeldt-Jakob disease in a recipient of human pituitary-derived gonadotrophin. *Aust N Z J Med* 1990;20:592–593.
111. Cohen AS, Shirahama T, Skinner M. Electron microscopy of amyloid. In: Harris JR, ed. *Electron microscopy of proteins,* vol. 3. New York: Academic Press, 1982;165–206.
112. Cohen FE, Pan K-M, Huang Z, Baldwin M, Fletterick RJ, Prusiner SB. Structural clues to prion replication. *Science* 1994;264:530–531.
113. Collinge J, Brown J, Hardy J, et al. Inherited prion disease with 144 base pair gene insertion. 2. Clinical and pathological features. *Brain* 1992;115:687–710.
114. Collinge J, Harding AE, Owen F, et al. Diagnosis of Gerstmann-Sträussler syndrome in familial dementia with prion protein gene analysis. *Lancet* 1989;2:15–17.
115. Collinge J, Owen F, Poulter H, et al. Prion dementia without characteristic pathology. *Lancet* 1990;336:7–9.
116. Collinge J, Palmer M. CJD discrepancy. *Nature* 1991;353:802.
117. Collinge J, Palmer MS, Dryden AJ. Genetic predisposition to iatrogenic Creutzfeldt-Jakob disease. *Lancet* 1991;337:1441–1442.
118. Collinge J, Whittington MA, Sidle KC, et al. Prion protein is necessary for normal synaptic function. *Nature* 1994;370:295–297.
119. Come JH, Fraser PE, Lansbury PT Jr. A kinetic model for amyloid formation in the prion diseases: importance of seeding. *Proc Natl Acad Sci USA* 1993;90:5959–5963.
120. Coschigano PW, Magasanik B. The URE2 gene product of Saccharomyces cerevisiae plays an important role in the cellular response to the nitrogen source and has homology to glutathione S-transferases. *Mol Cell Biol* 1991;11:822–832.
121. Cousens SN, Harries-Jones R, Knight R, Will RG, Smith PG, Matthews WB. Geographical distribution of cases of Creutzfeldt-Jakob disease in England and Wales 1970–84. *J Neurol Neurosurg Psychiatry* 1990; 53:459–465.
122. Cox BS, Tuite MF, McLaughlin CS. The psi factor of yeast: a problem in inheritance. *Yeast* 1988;4:159–178.
123. Crow TJ, Collinge J, Ridley RM, et al. Mutations in the prion gene in human transmissible dementia. *Seminar on Molecular Approaches to Research in Spongiform Encephalopathies in Man, Medical Research Council*, London, Dec 14, 1990 (Abstr) 1990.
124. Croxson M, Brown P, Synek B, et al. A new case of Creutzfeldt-Jakob disease associated with human growth hormone therapy in New Zealand. *Neurology* 1988;38:1128–1130.
125. Cuillé J, Chelle PL. Experimental transmission of trembling to the goat. *CR Seances Acad Sci* 1939;208:1058–1060.
126. Cunningham AA, Wells GAH, Scott AC, Kirkwood JK, Barnett JEF. Transmissible spongiform encephalopathy in greater kudu (Tragelaphus strepsiceros). *Vet Rec* 1993;132:68.
127. Czub M, Braig HR, Diringer H. Replication of the scrapie agent in hamsters infected intracerebrally confirms the pathogenesis of an amyloid-inducing virosis. *J Gen Virol* 1988;69:1753–1756.
128. Czub M, Braig HR, Diringer H. Pathogenesis of scrapie: study of the temporal development of clinical symptoms of infectivity titres and scrapie-associated fibrils in brains of hamsters infected intraperitoneally. *J Gen Virol* 1986;67:2005–2009.
129. Davanipour Z, Goodman L, Alter M, Sobel E, Asher D, Gajdusek DC. Possible modes of transmission of Creutzfeldt-Jakob disease. *N Engl J Med* 1984;311:1582–1583.
130. David-Ferreira JF, David-Ferreira KL, Gibbs CJ Jr., Morris JA. Scrapie in mice: ultrastructural observations in the cerebral cortex. *Proc Soc Exp Biol Med* 1968;127:313–320.

131. Davison C, Rabiner AM. Spastic pseudosclerosis (disseminated encephalomyelopathy; corticopal-lidospinal degeneration). Familial and nonfamilial incidence (a clinico-pathologic study). *Arch Neurol Psychiatry* 1940;44:578–598.
132. Dawson M, Wells GAH, Parker BNJ. Preliminary evidence of the experimental transmissibility of bovine spongiform encephalopathy to cattle. *Vet Rec* 1990;126:112–113.
133. Dawson M, Wells GAH, Parker BNJ, Scott AC. Primary parenteral transmission of bovine spongiform encephalopathy to the pig. *Vet Rec* 1990;Sept. 29:338.
134. De Fea KA, Nakahara DH, Calayag MC, et al. Determinants of carboxyl-terminal domain translocation during prion protein biogenesis. *J Biol Chem* 1994;269:16810–16820.
135. Dealler SF, Lacey RW. Transmissible spongiform encephalopathies: the threat of BSE to man. *Food Microbiol* 1990;7:253–279.
136. DeArmond SJ, McKinley MP, Barry RA, Braunfeld MB, McColloch JR, Prusiner SB. Identification of prion amyloid filaments in scrapie-infected brain. *Cell* 1985;41:221–235.
137. DeArmond SJ, Mobley WC, DeMott DL, Barry RA, Beckstead JH, Prusiner SB. Changes in the localization of brain prion proteins during scrapie infection. *Neurology* 1987;37:1271–1280.
138. DeArmond SJ, Yang S-L, Lee A, et al. Three scrapie prion isolates exhibit different accumulation patterns of the prion protein scrapie isoform. *Proc Natl Acad Sci USA* 1993;90:6449–6453.
139. Deleu C, Clavé C, Bégueret J. A single amino acid difference is sufficient to elicit vegetative incompatibility in the fungus Podospora anserina. *Genetics* 1993;135:45–52.
140. Deslys J-P, Lasmézas C, Dormont D. Selection of specific strains in iatrogenic Creutzfeldt-Jakob disease. *Lancet* 1994;343:848–849.
141. Deslys J-P, Marcé D, Dormont D. Similar genetic susceptibility in iatrogenic and sporadic Creutzfeldt-Jakob disease. *J Gen Virol* 1994;75: 23–27.
142. Dickinson AG, Bruce ME, Outram GW, Kimberlin RH. Scrapie strain differences: the implications of stability and mutation. In: Tateishi J, ed. *Proceedings of workshop on slow transmissible diseases*. Tokyo: Japanese Ministry of Health and Welfare, 1984;105–118.
143. Dickinson AG, Fraser H. An assessment of the genetics of scrapie in sheep and mice. In: Prusiner SB, Hadlow WJ, eds. *Slow transmissible diseases of the nervous system*, vol. 1. New York: Academic Press, 1979;367–386.
144. Dickinson AG, Fraser H. Scrapie: effect of Dh gene on the incubation period of extraneurally injected agent. *Heredity* 1972;29:91–93.
145. Dickinson AG, Fraser H. Genetical control of the concentration of ME7 scrapie agent in mouse spleen. *J Comp Pathol* 1969;79:363–366.
146. Dickinson AG, Fraser H, Outram GW. Scrapie incubation time can exceed natural lifespan. *Nature* 1975;256:732–733.
147. Dickinson AG, Meikle VM. A comparison of some biological characteristics of the mouse-passaged scrapie agents, 22A and ME7. *Genet Res* 1969;13:213–225.
148. Dickinson AG, Meikle VM, Fraser H. Genetical control of the concentration of ME7 scrapie agent in the brain of mice. *J Comp Pathol* 1969;79:15–22.
149. Dickinson AG, Meikle VMH. Host-genotype and agent effects in scrapie incubation: change in allelic interaction with different strains of agent. *Mol Gen Genet* 1971;112:73–79.
150. Dickinson AG, Meikle VMH, Fraser H. Identification of a gene which controls the incubation period of some strains of scrapie agent in mice. *J Comp Pathol* 1968;78:293–299.
151. Dickinson AG, Outram GW. Genetic aspects of unconventional virus infections: the basis of the virino hypothesis. In: Bock G, Marsh J, eds. *Novel infectious agents and the central nervous system*. Ciba Foundation Symposium 135. Chichester, UK: John Wiley and Sons, 1988; 63–83.
152. Dickinson AG, Outram GW. The scrapie replication-site hypothesis and its implications for pathogenesis. In: Prusiner SB, Hadlow WJ, eds. *Slow transmissible diseases of the nervous system,* vol. 2. New York: Academic Press, 1979;13–31.
153. Dickinson AG, Stamp JT. Experimental scrapie in Cheviot and Suffolk sheep. *J Comp Pathol* 1969;79:23–26.
154. Dickinson AG, Young GB, Stamp JT, Renwick CC. An analysis of natural scrapie in Suffolk sheep. *Heredity* 1965;20:485–503.
155. Diedrich J, Weitgrefe S, Zupancic M, et al. The molecular pathogenesis of astrogliosis in scrapie and Alzheimer's disease. *Microb Pathog* 1987;2:435–442.
156. Diener TO. PrP and the nature of the scrapie agent. *Cell* 1987;49: 719–721.
157. Diener TO. Is the scrapie agent a viroid? *Nature* 1972;235:218–219.
158. Diener TO, McKinley MP, Prusiner SB. Viroids and prions. *Proc Natl Acad Sci USA* 1982;79:5220–5224.
159. Diringer H. Hidden amyloidoses. *Exp Clin Immunogenet* 1992;9: 212–229.
160. Diringer H. Transmissible spongiform encephalopathies (TSE) virus-induced amyloidoses of the central nervous system (CNS). *Eur J Epidemiol* 1991;7:562–566.
161. Diringer H, Ehlers B. Chemoprophylaxis of scrapie in mice. *J Gen Virol* 1991;72:457–460.
162. Diringer H, Gelderblom H, Hilmert H, Ozel M, Edelbluth C, Kimberlin RH. Scrapie infectivity, fibrils and low molecular weight protein. *Nature* 1983;306:476–478.
163. Dlouhy SR, Hsiao K, Farlow MR, et al. Linkage of the Indiana kindred of Gerstmann-Sträussler-Scheinker disease to the prion protein gene. *Nat Genet* 1992;1:64–67.
164. Doh-ura K, Kitamoto T, Sakaki Y, Tateishi J. CJD discrepancy. *Nature* 1991;353:801–802.
165. Doh-ura K, Tateishi J, Sasaki H, Kitamoto T, Sakaki Y. Pro-:Leu change at position 102 of prion protein is the most common but not the sole mutation related to Gerstmann-Sträussler syndrome. *Biochem Biophys Res Commun* 1989;163:974–979.
166. Doms RW, Russ G, Yewdell JW. Brefeldin A redistributes resident and itinerant Golgi proteins to the endoplasmic reticulum. *J Cell Biol* 1989;109:61–72.
167. Duffy P, Wolf J, Collins G, Devoe A, Streeten B, Cowen D. Possible person to person transmission of Creutzfeldt-Jakob disease. *N Engl J Med* 1974;290:692–693.
168. Duguid JR, Rohwer RG, Seed B. Isolation of cDNAs of scrapie-modulated RNAs by subtractive hybridization of a cDNA library. *Proc Natl Acad Sci USA* 1988;85:5738–5742.
169. Ehlers B, Diringer H. Dextran sulphate 500 delays and prevents mouse scrapie by impairment of agent replication in spleen. *J Gen Virol* 1984; 65:1325–1330.
170. Eklund CM, Hadlow WJ, Kennedy RC. Some properties of the scrapie agent and its behavior in mice. *Proc Soc Exp Biol Med* 1963;112: 974–979.
171. Eklund CM, Kennedy RC, Hadlow WJ. Pathogenesis of scrapie virus infection in the mouse. *J Infect Dis* 1967;117:15–22.
172. Ellis CJ, Katifi H, Weller RO. A further British case of growth hormone induced Creutzfeldt-Jakob disease. *J Neurol Neurosurg Psychiatry* 1992;55:1200–1202.
173. Endo T, Groth D, Prusiner SB, Kobata A. Diversity of oligosaccharide structures linked to asparagines of the scrapie prion protein. *Biochemistry* 1989;28:8380–8388.
174. Falls DL, Rosen KM, Corfas G, Lane WS, Fischbach GD. ARIA, a protein that stimulates acetylcholine receptor synthesis, is a member of the neu ligand family. *Cell* 1993;72:801–815.
175. Farlow MR, Yee RD, Dlouhy SR, Conneally PM, Azzarelli B, Ghetti B. Gerstmann-Sträussler-Scheinker disease. I. Extending the clinical spectrum. *Neurology* 1989;39:1446–1452.
176. Field EJ. The significance of astroglial hypertrophy in scrapie, kuru, multiple sclerosis and old age together with a note on the possible nature of the scrapie agent. *Dtsch Z Nervenheilkd* 1967;192:265–274.
177. Field EJ, Joyce G, Keith A. Failure of interferon to modify scrapie in the mouse. *J Gen Virol* 1969;5:149–150.
178. Fink JK, Warren JT Jr., Drury I, Murman D, Peacock BA. Allele-specific sequencing confirms novel prion gene polymorphism in Creutzfeldt-Jakob disease. *Neurology* 1991;41:1647–1650.
179. Forloni G, Angeretti N, Chiesa R, et al. Neurotoxicity of a prion protein fragment. *Nature* 1993;362:543–546.
180. Fradkin JE, Schonberger LB, Mills JL, et al. Creutzfeldt-Jakob disease in pituitary growth hormone recipients in the United States. *JAMA* 1991;265:880–884.
181. Fraser H. Neuropathology of scrapie: the precision of the lesions and their diversity. In: Prusiner SB, Hadlow WJ, eds. *Slow transmissible diseases of the nervous system,* vol. 1. New York: Academic Press, 1979;387–406.
182. Fraser H, Dickinson AG. Targeting of scrapie lesions and spread of agent via the retino-tectal projection. *Brain Res* 1985;346:32–41.
183. Fraser H, Dickinson AG. Scrapie in mice. Agent-strain differences in the distribution and intensity of grey matter vacuolation. *J Comp Pathol* 1973;83:29–40.
184. Fraser H, Dickinson AG. The sequential development of the brain lesions of scrapie in three strains of mice. *J Comp Pathol* 1968;78: 301–311.

185. Fraser H, McConnell I, Wells GAH, Dawson M. Transmission of bovine spongiform encephalopathy to mice. *Vet Rec* 1988;123:472.
186. Friede RL, DeJong RN. Neuronal enzymatic failure in Creutzfeldt-Jakob disease. A familial study. *Arch Neurol* 1964;10:181–195.
187. Gabizon R, McKinley MP, Groth DF, Kenaga L, Prusiner SB. Properties of scrapie prion liposomes. *J Biol Chem* 1988;263:4950–4955.
188. Gabizon R, McKinley MP, Groth DF, Prusiner SB. Immunoaffinity purification and neutralization of scrapie prion infectivity. *Proc Natl Acad Sci USA* 1988;85:6617–6621.
189. Gabizon R, McKinley MP, Prusiner SB. Purified prion proteins and scrapie infectivity copartition into liposomes. *Proc Natl Acad Sci USA* 1987;84:4017–4021.
190. Gabizon R, Meiner Z, Cass C, et al. Prion protein gene mutation in Libyan Jews with Creutzfeldt-Jakob disease. *Neurology* 1991;41:160.
191. Gabizon R, Prusiner SB. Prion liposomes. *Biochem J* 1990;266:1–14.
192. Gabizon R, Rosenmann H, Meiner Z, et al. Mutation and polymorphism of the prion protein gene in Libyan Jews with Creutzfeldt-Jakob disease. *Am J Hum Genet* 1993;33:828–835.
193. Gabriel J-M, Oesch B, Kretzschmar H, Scott M, Prusiner SB. Molecular cloning of a candidate chicken prion protein. *Proc Natl Acad Sci USA* 1992;89:9097–9101.
194. Gajdusek DC. Subacute spongiform encephalopathies: transmissible cerebral amyloidoses caused by unconventional viruses. In: Fields BN, Knipe DM, Chanock RM, et al., eds. *Virology,* 2nd ed. New York: Raven Press, 1990;2289–2324.
195. Gajdusek DC. Transmissible and non-transmissible amyloidoses: autocatalytic post-translational conversion of host precursor proteins to β-pleated sheet configurations. *J Neuroimmunol* 1988;20:95–110.
196. Gajdusek DC. Subacute spongiform virus encephalopathies caused by unconventional viruses. In: Maramorosch K, McKelvey JJ Jr, eds. *Subviral pathogens of plants and animals: viroids and prions*. Orlando: Academic Press, 1985;483–544.
197. Gajdusek DC. Unconventional viruses and the origin and disappearance of kuru. *Science* 1977;197:943–960.
198. Gajdusek DC, Gibbs CJ Jr. Brain amyloidoses-precursor proteins and the amyloids of transmissible and nontransmissible dementias: scrapie-kuru-CJD viruses as infectious polypeptides or amyloid enhancing vector. In: Goldstein A, ed. *Biomedical advances in aging.* New York: Plenum Press, 1990;3–24.
199. Gajdusek DC, Gibbs CJ Jr., Alpers M. Experimental transmission of a kuru-like syndrome to chimpanzees. *Nature* 1966;209:794–796.
200. Gajdusek DC, Gibbs CJ Jr., Asher DM, et al. Precautions in medical care of and in handling materials from patients with transmissible virus dementia (CJD). *N Engl J Med* 1977;297:1253–1258.
201. Gajdusek DC, Zigas V. Clinical, pathological and epidemiological study of an acute progressive degenerative disease of the central nervous system among natives of the eastern highlands of New Guinea. *Am J Med* 1959;26:442–469.
202. Gajdusek DC, Zigas V. Degenerative disease of the central nervous system in New Guinea—The endemic occurrence of "kuru" in the native population. *N Engl J Med* 1957;257:974–978.
203. Gard S. Encephalomyelitis of mice. II. A method for the measurement of virus activity. *J Exp Med* 1940;72:69–77.
204. Gasset M, Baldwin MA, Fletterick RJ, Prusiner SB. Perturbation of the secondary structure of the scrapie prion protein under conditions associated with changes in infectivity. *Proc Natl Acad Sci USA* 1993;90:1–5.
205. Gasset M, Baldwin MA, Lloyd D, et al. Predicted a-helical regions of the prion protein when synthesized as peptides form amyloid. *Proc Natl Acad Sci USA* 1992;89:10940–10944.
206. Gerstmann J, Sträussler E, Scheinker I. Über eine eigenartige hereditär-familiäre Erkrankung des Zentralnervensystems zugleich ein Beitrag zur frage des vorzeitigen lokalen Alterns. *Z Neurol* 1936;154:736–762.
207. Ghetti B, Tagliavini F, Masters CL, et al. Gerstmann-Sträussler-Scheinker disease. II. Neurofibrillary tangles and plaques with PrP-amyloid coexist in an affected family. *Neurology* 1989;39:1453–1461.
208. Giaccone G, Tagliavini F, Verga L, et al. Neurofibrillary tangles of the Indiana kindred of Gerstmann-Sträussler-Scheinker disease share antigenic determinants with those of Alzheimer disease. *Brain Res* 1990;530:325–329.
209. Gibbons RA, Hunter GD. Nature of the scrapie agent. *Nature* 1967;215:1041–1043.
210. Gibbs CJ Jr., Amyx HL, Bacote A, Masters CL, Gajdusek DC. Oral transmission of kuru, Creutzfeldt-Jakob disease and scrapie to nonhuman primates. *J Infect Dis* 1980;142:205–208.
211. Gibbs CJ Jr., Asher DM, Brown PW, Fradkin JE, Gajdusek DC. Creutzfeldt-Jakob disease infectivity of growth hormone derived from human pituitary glands. *N Engl J Med* 1993;328:358–359.
212. Gibbs CJ Jr., Asher DM, Kobrine A, Amyx HL, Sulima MP, Gajdusek DC. Transmission of Creutzfeldt-Jakob disease to a chimpanzee by electrodes contaminated during neurosurgery. *J Neurol Neurosurg Psychiatry* 1994;57:757–758.
213. Gibbs CJ Jr., Gajdusek DC. Experimental subacute spongiform virus encephalopathies in primates and other laboratory animals. *Science* 1973;182:67–68.
214. Gibbs CJ Jr., Gajdusek DC. Infection as the etiology of spongiform encephalopathy. *Science* 1969;165:1023–1025.
215. Gibbs CJ Jr., Gajdusek DC, Amyx H. Strain variation in the viruses of Creutzfeldt-Jakob disease and kuru. In: Prusiner SB, Hadlow WJ, eds. *Slow transmissible diseases of the nervous system,* Vol. 2. New York: Academic Press, 1979;87–110.
216. Gibbs CJ Jr., Gajdusek DC, Asher DM, et al. Creutzfeldt-Jakob disease (spongiform encephalopathy): transmission to the chimpanzee. *Science* 1968;161:388–389.
217. Gibbs CJ Jr., Joy A, Heffner R, et al. Clinical and pathological features and laboratory confirmation of Creutzfeldt-Jakob disease in a recipient of pituitary-derived human growth hormone. *N Engl J Med* 1985;313:734–738.
218. Glenner GG. Amyloid deposits and amyloidosis. *N Engl J Med* 1980;302:1283–1292.
219. Glenner GG, Eanes ED, Bladen HA, Linke RP, Termine JD. Beta-pleated sheet fibrils—a comparison of native amyloid with synthetic protein fibrils. *J Histochem Cytochem* 1974;22:1141–1158.
220. Goate A, Chartier-Harlin M-C, Mullan M, et al. Segregation of a missense mutation in the amyloid precursor protein gene with familial Alzheimer's disease. *Nature* 1991;349:704–706.
221. Goldfarb L, Brown P, Goldgaber D, et al. Identical mutation in unrelated patients with Creutzfeldt-Jakob disease. *Lancet* 1990;336:174–175.
222. Goldfarb L, Korczyn A, Brown P, Chapman J, Gajdusek DC. Mutation in codon 200 of scrapie amyloid precursor gene linked to Creutzfeldt-Jakob disease in Sephardic Jews of Libyan and non-Libyan origin. *Lancet* 1990;336:637–638.
223. Goldfarb LG, Brown P, Gajdusek DC. The molecular genetics of human transmissible spongiform encephalopathy. In: Prusiner SB, Collinge J, Powell J, Anderton B, eds. *Prion diseases of humans and animals.* London: Ellis Horwood, 1992;139–153.
224. Goldfarb LG, Brown P, Goldgaber D, et al. Creutzfeldt-Jakob disease and kuru patients lack a mutation consistently found in the Gerstmann-Sträussler-Scheinker syndrome. *Exp Neurol* 1990;108:247–250.
225. Goldfarb LG, Brown P, Haltia M, et al. Creutzfeldt-Jakob disease cosegregates with the codon 178^{Asn} PRNP mutation in families of European origin. *Ann Neurol* 1992;31:274–281.
226. Goldfarb LG, Brown P, Haltia M, Ghiso J, Frangione B, Gajdusek DC. Synthetic peptides corresponding to different mutated regions of the amyloid gene in familial Creutzfeldt-Jakob disease show enhanced in vitro formation of morphologically different amyloid fibrils. *Proc Natl Acad Sci USA* 1993;90:4451–4454.
227. Goldfarb LG, Brown P, McCombie WR, et al. Transmissible familial Creutzfeldt-Jakob disease associated with five, seven, and eight extra octapeptide coding repeats in the PRNP gene. *Proc Natl Acad Sci USA* 1991;88:10926–10930.
228. Goldfarb LG, Brown P, Mitrová E, et al. Creutzfeldt-Jacob disease associated with the PRNP codon 200^{Lys} mutation: an analysis of 45 families. *Eur J Epidemiol* 1991;7:477–486.
229. Goldfarb LG, Brown P, Vrbovská A, et al. An insert mutation in the chromosome 20 amyloid precursor gene in a Gerstmann-Sträussler-Scheinker family. *J Neurol Sci* 1992;111:189–194.
230. Goldfarb LG, Haltia M, Brown P, et al. New mutation in scrapie amyloid precursor gene (at codon 178) in Finnish Creutzfeldt-Jakob kindred. *Lancet* 1991;337:425.
231. Goldfarb LG, Mitrová E, Brown P, Toh BH, Gajdusek DC. Mutation in codon 200 of scrapie amyloid protein gene in two clusters of Creutzfeldt-Jakob disease in Slovakia. *Lancet* 1990;336:514–515.
232. Goldfarb LG, Petersen RB, Tabaton M, et al. Fatal familial insomnia and familial Creutzfeldt-Jakob disease: disease phenotype determined by a DNA polymorphism. *Science* 1992;258:806–808.
233. Goldgaber D, Goldfarb LG, Brown P, et al. Mutations in familial Creutzfeldt-Jakob disease and Gerstmann-Sträussler-Scheinker's syndrome. *Exp Neurol* 1989;106:204–206.
234. Goldhammer Y, Gabizon R, Meiner Z, Sadeh M. An Israeli family with Gerstmann-Sträussler-Scheinker disease manifesting the codon 102 mutation in the prion protein gene. *Neurology* 1993;43:2718–2719.

235. Goldmann W, Hunter N, Benson G, Foster JD, Hope J. Different scrapie-associated fibril proteins (PrP) are encoded by lines of sheep selected for different alleles of the Sip gene. *J Gen Virol* 1991;72:2411–2417.
236. Goldmann W, Hunter N, Foster JD, Salbaum JM, Beyreuther K, Hope J. Two alleles of a neural protein gene linked to scrapie in sheep. *Proc Natl Acad Sci USA* 1990;87:2476–2480.
237. Goldmann W, Hunter N, Manson J, Hope J. The PrP gene of the sheep, a natural host of scrapie. *VIIIth International Congress of Virology,* Berlin, Aug 26–31 (Abstr) 1990;284.
238. Goldmann W, Hunter N, Martin T, Dawson M, Hope J. Different forms of the bovine PrP gene have five or six copies of a short, G-C-rich element within the protein-coding exon. *J Gen Virol* 1991;72:201–204.
239. Goldmann W, Hunter N, Multhaup G, et al. The PrP gene in natural scrapie. *Alzheimer Disease and Associated Disorders [Abstract supplement]* 1988;2:330.
240. Gordon WS. Variation in susceptibility of sheep to scrapie and genetic implications. In: *Report of scrapie seminar,* ARS 91–53. Washington, DC: U.S. Department of Agriculture, 1966;53–67.
241. Gordon WS. Advances in veterinary research. *Vet Res* 1946;58:516–520.
242. Griffith JS. Self-replication and scrapie. *Nature* 1967;215:1043–1044.
243. Hadlow WJ, Kennedy RC, Race RE. Natural infection of Suffolk sheep with scrapie virus. *J Infect Dis* 1982;146:657–664.
244. Hadlow WJ, Kennedy RC, Race RE, Eklund CM. Virologic and neurohistologic findings in dairy goats affected with natural scrapie. *Vet Pathol* 1980;17:187–199.
245. Haltia M, Kovanen J, Goldfarb LG, Brown P, Gajdusek DC. Familial Creutzfeldt-Jakob disease in Finland: epidemiological, clinical, pathological and molecular genetic studies. *Eur J Epidemiol* 1991;7:494–500.
246. Haraguchi T, Fisher S, Olofsson S, et al. Asparagine-linked glycosylation of the scrapie and cellular prion proteins. *Arch Biochem Biophys* 1989;274:1–13.
247. Hardy J. Prion dimers—a deadly duo. *Trends in Neurosciences* 1991; 14:423–424.
248. Harries-Jones R, Knight R, Will RG, Cousens S, Smith PG, Matthews WB. Creutzfeldt-Jakob disease in England and Wales, 1980-1984: a case-control study of potential risk factors. *J Neurol Neurosurg Psychiatry* 1988;51:1113–1119.
249. Harris DA, Falls DL, Johnson FA, Fischbach GD. A prion-like protein from chicken brain copurifies with an acetylcholine receptor-inducing activity. *Proc Natl Acad Sci USA* 1991;88:7664–7668.
250. Healy DL, Evans J. Creutzfeldt-Jakob disease after pituitary gonadotrophins. *Br J Med* 1993;307:517–518.
251. Hecker R, Taraboulos A, Scott M, et al. Replication of distinct prion isolates is region specific in brains of transgenic mice and hamsters. *Genes Dev* 1992;6:1213–1228.
252. Herishanu Y. Antiviral drugs in Jakob-Creutzfeldt disease. *J Am Geriatr Soc* 1973;21:229–231.
253. Herzberg L, Herzberg BN, Gibbs CJ Jr., Sullivan W, Amyx H, Gajdusek DC. Creutzfeldt-Jakob disease: hypothesis for high incidence in Libyan Jews in Israel. *Science* 1974;186:848.
254. Heston LL, Lowther DLW, Leventhal CM. Alzheimer's disease: a family study. *Arch Neurol* 1966;15:225–233.
255. Hope J, Morton LJD, Farquhar CF, Multhaup G, Beyreuther K, Kimberlin RH. The major polypeptide of scrapie-associated fibrils (SAF) has the same size, charge distribution and N-terminal protein sequence as predicted for the normal brain protein (PrP). *EMBO J* 1986;5:2591–2597.
256. Hope J, Reekie LJD, Hunter N, et al. Fibrils from brains of cows with new cattle disease contain scrapie-associated protein. *Nature* 1988;336: 390–392.
257. Hornabrook RW. Kuru and clinical neurology. In: Prusiner SB, Hadlow WJ, eds. *Slow transmissible diseases of the nervous system,* vol. 1. New York: Academic Press, 1979;37–66.
258. Hornabrook RW. Kuru—a subacute cerebellar degeneration: the natural history and clinical features. *Brain* 1968;91:53–74.
259. Hsiao K, Baker HF, Crow TJ, et al. Linkage of a prion protein missense variant to Gerstmann-Sträussler syndrome. *Nature* 1989;338: 342–345.
260. Hsiao K, Dlouhy S, Farlow MR, et al. Mutant prion proteins in Gerstmann-Sträussler-Scheinker disease with neurofibrillary tangles. *Nat Genet* 1992;1:68–71.
261. Hsiao K, Meiner Z, Kahana E, et al. Mutation of the prion protein in Libyan Jews with Creutzfeldt-Jakob disease. *N Engl J Med* 1991;324: 1091–1097.
262. Hsiao K, Prusiner SB. Inherited human prion diseases. *Neurology* 1990; 40:1820–1827.
263. Hsiao KK, Cass C, Schellenberg GD, et al. A prion protein variant in a family with the telencephalic form of Gerstmann-Sträussler-Scheinker syndrome. *Neurology* 1991;41:681–684.
264. Hsiao KK, Doh-ura K, Kitamoto T, Tateishi J, Prusiner SB. A prion protein amino acid substitution in ataxic Gerstmann-Sträussler syndrome. *Ann Neurol* 1989;26:137.
265. Hsiao KK, Groth D, Scott M, et al. Serial transmission in rodents of neurodegeneration from transgenic mice expressing mutant prion protein. *Proc Natl Acad Sci USA* 1994;91:9126–9130.
266. Hsiao KK, Groth D, Scott M, et al. Neurologic disease of transgenic mice which express GSS mutant prion protein is transmissible to inoculated recipient animals. In: *Prion diseases of humans and animals symposium*, London, Sept. 2–4, 1991 (abstract)
267. Hsiao KK, Scott M, Foster D, Groth DF, DeArmond SJ, Prusiner SB. Spontaneous neurodegeneration in transgenic mice with mutant prion protein. *Science* 1990;250:1587–1590.
268. Huang Z, Gabriel J-M, Baldwin MA, Fletterick RJ, Prusiner SB, Cohen FE. Proposed three-dimensional structure for the cellular prion protein. *Proc Natl Acad Sci USA* 1994;91:7139–7143.
269. Humphery-Smith I, Chastel C, Le Goff F. Spirosplasmas and spongiform encephalopathies. *Med J Aust* 1992;156:142.
270. Hunter GD. Scrapie: a prototype slow infection. *J Infect Dis* 1972;125: 427–440.
271. Hunter GD, Gibbons RA, Kimberlin RH, Millson GC. Further studies of the infectivity and stability of extracts and homogenates derived from scrapie affected mouse brains. *J Comp Pathol* 1969;79: 101–108.
272. Hunter GD, Kimberlin RH, Collis S, Millson GC. Viral and non-viral properties of the scrapie agent. *Ann Clin Res* 1973;5:262–267.
273. Hunter GD, Kimberlin RH, Gibbons RA. Scrapie: a modified membrane hypothesis. *J Theor Biol* 1968;20:355–357.
275. Hunter GD, Millson GC. Attempts to release the scrapie agent from tissue debris. *J Comp Pathol* 1967;77:301–307.
274. Hunter GD, Millson GC. Studies on the heat stability and chromatographic behavior of the scrapie agent. *J Gen Microbiol* 1964;37: 251–258.
276. Hunter N, Foster JD, Benson G, Hope J. Restriction fragment length polymorphisms of the scrapie-associated fibril protein (PrP) gene and their association with susceptiblity to natural scrapie in British sheep. *J Gen Virol* 1991;72:1287–1292.
277. Hunter N, Foster JD, Dickinson AG, Hope J. Linkage of the gene for the scrapie-associated fibril protein (PrP) to the Sip gene in Cheviot sheep. *Vet Rec* 1989;124:364–366.
278. Hunter N, Hope J, McConnell I, Dickinson AG. Linkage of the scrapie-associated fibril protein (PrP) gene and Sinc using congenic mice and restriction fragment length polymorphism analysis. *J Gen Virol* 1987; 68:2711–2716.
279. Ikeda S, Yanagisawa N, Allsop D, Glenner GG. A variant of Gerstmann-Sträussler-Scheinker disease with b-protein epitopes and dystrophic neurites in the peripheral regions of PrP—immunoreactive amyloid plaques. In: Natvig JB, Forre O, Husby G, et al., eds. *Amyloid and amyloidosis 1990.* Dordrecht: Kluwer Academic Publishers, 1991; 737–740.
280. Jacob H, Pyrkosch W, Strube H. Die erbliche Form der Creutzfeldt-Jakobschen Krankheit. *Arch Psychiatr Zeitsch Neurol* 1950;184: 653–674.
281. Jarrett JT, Lansbury PT Jr. Seeding "one-dimensional crystallization" of amyloid: a pathogenic mechanism in Alzheimer's disease and scrapie? *Cell* 1993;73:1055–1058.
282. Jeffrey M, Scott JR, Williams A, Fraser H. Ultrastructural features of spongiform encephalopathy transmitted to mice from three species of bovidae. *Acta Neuropathol* 1992;84:559–569.
283. Jendroska K, Heinzel FP, Torchia M, et al. Proteinase-resistant prion protein accumulation in Syrian hamster brain correlates with regional pathology and scrapie infectivity. *Neurology* 1991;41:1482–1490.
284. Johnson RT. Prion disease. *N Engl J Med* 1992;326:486–487.
285. Kahana E, Milton A, Braham J, Sofer D. Creutzfeldt-Jakob disease: focus among Libyan Jews in Israel. *Science* 1974;183:90–91.
286. Kahana E, Zilber N, Abraham M. Do Creutzfeldt-Jakob disease patients of Jewish Libyan origin have unique clinical features? *Neurology* 1991;41:1390–1392.
287. Kane PM, Yamashiro CT, Wolczyk DF, Neff N, Goebl M, Stevens TH. Protein splicing converts the yeast TFP1 gene product to the 69-kD

subunit of the vacuolar H+-adenosine triphosphatase. *Science* 1990;250: 651–657.

288. Kascsak RJ, Rubenstein R, Merz PA, et al. Mouse polyclonal and monoclonal antibody to scrapie-associated fibril proteins. *J Virol* 1987;61: 3688–3693.
289. Kato S, Hirano A, Umahara T, Llena JF, Herz F, Ohama E. Ultrastructural and immunohistochemical studies on ballooned cortical neurons in Creutzfeldt-Jakob disease: expression of aB-crystallin, ubiquitin and stress-response protein 27. *Acta Neuropathol* 1992;84:443–448.
290. Katz M, Koprowski H. Failure to demonstrate a relationship between scrapie and production of interferon in mice. *Nature* 1968;219:639–640.
291. Keller GA, Siegel MW, Caras IW. Endocytosis of glycophospholipid-anchored and transmembrane forms of CD4 by different endocytic pathways. *EMBO J* 1992;11:863–874.
292. Kellings K, Meyer N, Mirenda C, Prusiner SB, Riesner D. Further analysis of nucleic acids in purified scrapie prion preparations by improved return refocussing gel electrophoresis (RRGE). *J Gen Virol* 1992;73:1025–1029.
293. Kimberlin RH. Scrapie and possible relationships with viroids. *Semin Virol* 1990;1:153–162.
294. Kimberlin RH, Cole S, Walker CA. Temporary and permanent modifications to a single strain of mouse scrapie on transmission to rats and hamsters. *J Gen Virol* 1987;68:1875–1881.
295. Kimberlin RH, Field HJ, Walker CA. Pathogenesis of mouse scrapie: evidence for spread of infection from central to peripheral nervous system. *J Gen Virol* 1983;64:713–716.
296. Kimberlin RH, Hunter GD. DNA synthesis in scrapie-affected mouse brain. *J Gen Virol* 1967;1:115–124.
297. Kimberlin RH, Walker CA. The antiviral compound HPA-23 can prevent scrapie when administered at the time of infection. *Arch Virol* 1983;78:9–18.
298. Kimberlin RH, Walker CA. Pathogenesis of mouse scrapie: dynamics of agent replication in spleen, spinal cord and brain after infection by different routes. *J Comp Pathol* 1979;89:551–562.
299. Kimberlin RH, Walker CA. Pathogenesis of mouse scrapie: effect of route of inoculation on infectivity titres and dose-response curves. *J Comp Pathol* 1978;88:39–47.
300. Kimberlin RH, Walker CA. Evidence that the transmission of one source of scrapie agent to hamsters involves separation of agent strains from a mixture. *J Gen Virol* 1978;39:487–496.
301. Kimberlin RH, Walker CA, Fraser H. The genomic identity of different strains of mouse scrapie is expressed in hamsters and preserved on reisolation in mice. *J Gen Virol* 1989;70:2017–2025.
302. Kingsbury DT, Kasper KC, Stites DP, Watson JD, Hogan RN, Prusiner SB. Genetic control of scrapie and Creutzfeldt-Jakob disease in mice. *J Immunol* 1983;131:491–496.
303. Kirkwood JK, Cunningham AA, Wells GAH, Wilesmith JW, Barnett JEF. Spongiform encephalopathy in a herd of greater kudu (Tragelaphus strepsiceros): epidemiological observations. *Vet Rec* 1993;133:360–364.
304. Kirkwood JK, Wells GAH, Wilesmith JW, Cunningham AA, Jackson SI. Spongiform encephalopathy in an arabian oryx (Oryx leucoryx) and a greater kudu (Tragelaphus strepsiceros). *Vet Rec* 1990;127:418–420.
305. Kirschbaum WR. Zwei eigenartige Erkrankungen des Zentralnervensystems nach Art der spastischen Pseudosklerose (Jakob). *Z Ges Neurol Psychiatr* 1924;92:175–220.
306. Kitamoto T, Iizuka R, Tateishi J. An amber mutation of prion protein in Gerstmann-Sträussler syndrome with mutant PrP plaques. *Biochem Biophys Res Commun* 1993;192:525–531.
307. Kitamoto T, Muramoto T, Mohri S, Doh-ura K, Tateishi J. Abnormal isoform of prion protein accumulates in follicular dendritic cells in mice with Creutzfeldt-Jakob disease. *J Virol* 1991;65:6292–6295.
308. Kitamoto T, Ohta M, Doh-ura K, Hitoshi S, Terao Y, Tateishi J. Novel missense variants of prion protein in Creutzfeldt-Jakob disease or Gerstmann-Sträussler syndrome. *Biochem Biophys Res Commun* 1993; 191:709–714.
309. Kitamoto T, Shin R-W, Doh-ura K, et al. Abnormal isoform of prion proteins accumulates in the synaptic structures of the central nervous system in patients with Creutzfeldt-Jakob disease. *Am J Pathol* 1992; 140:1285–1294.
310. Kitamoto T, Tateishi J. Human prion diseases with variant prion protein. *Phil Trans R Soc Lond B* 1994;343:391–398.
311. Kitamoto T, Tateishi J, Sawa H, Doh-Ura K. Positive transmission of Creutzfeldt-Jakob disease verified by murine kuru plaques. *Lab Invest* 1989;60:507–512.
312. Kitamoto T, Tateishi J, Tashima I, et al. Amyloid plaques in Creutzfeldt-Jakob disease stain with prion protein antibodies. *Ann Neurol* 1986;20: 204–208.
313. Klatzo I, Gajdusek DC, Zigas V. Pathology of kuru. *Lab Invest* 1959; 8:799–847.
314. Klitzman RL, Alpers MP, Gajdusek DC. The natural incubation period of kuru and the episodes of transmission in three clusters of patients. *Neuroepidemiology* 1984;3:3–20.
315. Kneller DG, Cohen FE, Langridge R. Improvement in protein secondary structure prediction by an enhanced neural network. *J Mol Biol* 1990;214:171–182.
316. Koch TK, Berg BO, DeArmond SJ, Gravina RF. Creutzfeldt-Jakob disease in a young adult with idiopathic hypopituitarism. Possible relation to the administration of cadaveric human growth hormone. *N Engl J Med* 1985;313:731–733.
317. Kocisko DA, Come JH, Priola SA, et al. Cell-free formation of protease-resistant prion protein. *Nature* 1994;370:471–474.
318. Kondo K, Kuroina Y. A case control study of Creutzfeldt-Jakob disease: association with physical injuries. *Ann Neurol* 1981;11:377–381.
319. Kretzschmar HA, Honold G, Seitelberger F, et al. Prion protein mutation in family first reported by Gerstmann, Straussler, and Scheinker. *Lancet* 1991;337:1160.
320. Kretzschmar HA, Kufer P, Riethmüller G, DeArmond SJ, Prusiner SB, Schiffer D. Prion protein mutation at codon 102 in an Italian family with Gerstmann-Sträussler-Scheinker syndrome. *Neurology* 1992;42: 809–810.
321. Kretzschmar HA, Prusiner SB, Stowring LE, DeArmond SJ. Scrapie prion proteins are synthesized in neurons. *Am J Pathol* 1986;122:1–5.
322. Kretzschmar HA, Stowring LE, Westaway D, Stubblebine WH, Prusiner SB, DeArmond SJ. Molecular cloning of a human prion protein cDNA. *DNA* 1986;5:315–324.
323. Lacroute F. Non-Mendelian mutlation allowing ureidosuccinic acid uptake in yeast. *J Bacteriol* 1971;106:519–522.
324. Lamar CH, Gustafson DP, Krasovich M, Hinsman EJ. Ultrastructural studies of spleens, brains, and brain cell cultures of mice with scrapie. *Vet Pathol* 1974;11:13–19.
325. Lantos PL, McGill IS, Janota I, et al. Prion protein immunocytochemistry helps to establish the true incidence of prion diseases. *Neurosci Lett* 1992;147:67–71.
326. Laplanche J, Chatelain J, Beaudry P, Dussaucy M, Bounneau C, Launay J. French autochthonous scrapied sheep without the 136Val PrP polymorphism. *Mammalian Genome* 1993;4:463–464.
327. Laplanche JL, Chatelain J, Launay JM, Gazengel C, Vidaud M. Deletion in prion protein gene in a Moroccan family. *Nucleic Acids Res* 1990;18:6745.
328. Laplanche JL, Chatelain J, Westaway D, et al. PrP polymorphisms associated with natural scrapie discovered by denaturing gradient gel electrophoresis. *Genomics* 1993;15:30–37.
329. Latarjet R, Muel B, Haig DA, Clarke MC, Alper T. Inactivation of the scrapie agent by near monochromatic ultraviolet light. *Nature* 1970; 227:1341–1343.
330. Levy E, Carman MD, Fernandez-Madrid IJ, et al. Mutation of the Alzheimer's disease amyloid gene in hereditary cerebral hemorrhage, Dutch type. *Science* 1990;248:1124–1126.
331. Lewin P. Infectious peptides in slow virus infections: a hypothesis. *Can Med Assoc J* 1981;124:1436–1437.
332. Lewin P. Scrapie: an infective peptide? *Lancet* 1972;1:748.
333. Liberski PP, Budka H, Sluga E, Barcikowska M, Kwiecinski H. Tubulovesicular structures in Creutzfeldt-Jakob disease. *Acta Neuropathol* 1992;84:238–243.
334. Liberski PP, Yanagihara R, Gibbs CJ Jr., Gajdusek DC. Appearance of tubulovesicular structures in experimental Creutzfeldt-Jakob disease and scrapie preceeds the onset of clinical disease. *Acta Neuropathol* 1990;79:349–354.
335. Lifson S, Sander C. Composition, cooperativity and recognition in proteins. In: Jaenicke R, ed. *Protein folding*. Amsterdam: Elsevier/North-Holland Biomedical Press, 1980;289–314.
336. Little BW, Brown PW, Rodgers-Johnson P, Perl DP, Gajdusek DC. Familial myoclonic dementia masquerading as Creutzfeldt-Jakob disease. *Ann Neurol* 1986;20:231–239.
337. Locht C, Chesebro B, Race R, Keith JM. Molecular cloning and complete sequence of prion protein cDNA from mouse brain infected with the scrapie agent. *Proc Natl Acad Sci USA* 1986;83:6372–6376.
338. Lopez CD, Yost CS, Prusiner SB, Myers RM, Lingappa VR. Unusual topogenic sequence directs prion protein biogenesis. *Science* 1990;248: 226–229.

339. Lugaresi E, Medori R, Montagna P, et al. Fatal familial insomnia and dysautonomia with selective degeneration of thalamic nuclei. *N Engl J Med* 1986;315:997–1003.
340. Lumley Jones R, Benker G, Salacinski PR, Lloyd TJ, Lowry PJ. Large-scale preparation of highly purified pyrogen-free human growth hormone for clinical use. *Br J Endocrinol* 1979;82:77–86.
341. M'Fadyean J. Scrapie. J Comp Pathol 1918;31:102–131.
342. M'Gowan JP. *Investigation into the disease of sheep called "scrapie."* Edinburgh: William Blackwood and Sons, 1914:114.
343. Macario ME, Vaisman M, Buescu A, Neto VM, Araujo HMM, Chagas C. Pituitary growth hormone and Creutzfeldt-Jakob disease. *Br Med J* 1991;302:1149.
344. Malmgren R, Kurland L, Mokri B, Kurtzke J. The epidemiology of Creutzfeldt-Jakob disease. In: Prusiner SB, Hadlow WJ, eds. *Slow transmissible diseases of the nervous system,* vol. 1. New York: Academic Press, 1979;93–112.
345. Malone TG, Marsh RF, Hanson RP, Semancik JS. Evidence for the low molecular weight nature of the scrapie agent. *Nature* 1979;278: 575–576.
346. Malone TG, Marsh RF, Hanson RP, Semancik JS. Membrane-free scrapie activity. *J Virol* 1978;25:933–935.
347. Manetto V, Medori R, Cortelli P, et al. Fatal familial insomnia: clinical and pathological study of five new cases. *Neurology* 1992;42: 312–319.
348. Manuelidis E, Gorgacz EJ, Manuelidis L. Interspecies transmission of Creutzfeldt-Jakob disease to Syrian hamsters with reference to clinical syndromes and strains of agent. *Proc Natl Acad Sci USA* 1978;75: 3422–3436.
349. Manuelidis E, Kim J, Angelo J, Manuelidis L. Serial propagation of Creutzfeldt-Jakob disease in guinea pigs. *Proc Natl Acad Sci USA* 1976; 73:223–227.
350. Manuelidis EE, Manuelidis L. Experiments on maternal transmission of Creutzfeldt-Jakob disease in guinea pigs. *Proc Soc Biol Med* 1979; 160:233–236.
351. Manuelidis L, Manuelidis EE. Creutzfeldt-Jakob disease and dementias. *Microb Pathog* 1989;7:157–164.
352. Manuelidis L, Valley S, Manuelidis EE. Specific proteins associated with Creutzfeldt-Jakob disease and scrapie share antigenic and carbohydrate determinants. *Proc Natl Acad Sci USA* 1985;82:4263–4267.
353. Marsh RF, Bessen RA, Lehmann S, Hartsough GR. Epidemiological and experimental studies on a new incident of transmissible mink encephalopathy. *J Gen Virol* 1991;72:589–594.
354. Marsh RF, Kimberlin RH. Comparison of scrapie and transmissible mink encephalopathy in hamsters. II. Clinical signs, pathology and pathogenesis. *J Infect Dis* 1975;131:104–110.
355. Marsh RF, Malone TG, Semancik JS, Lancaster WD, Hanson RP. Evidence for an essential DNA component in the scrapie agent. *Nature* 1978;275:146–147.
356. Martinez-Lage JF, Sola J, Poza M, Esteban JA. Pediatric Creutzfeldt-Jakob disease: probable transmission by a dural graft. *Child's Nerv Syst* 1993;9:239–242.
357. Marzewski DJ, Towfighi J, Harrington MG, Merril CR, Brown P. Creutzfeldt-Jakob disease following pituitary-derived human growth hormone therapy: a new American case. *Neurology* 1988;38:1131–1133.
358. Masters CL, Gajdusek DC, Gibbs CJ Jr. Creutzfeldt-Jakob disease virus isolations from the Gerstmann-Sträussler syndrome. *Brain* 1981;104: 559–588.
359. Masters CL, Gajdusek DC, Gibbs CJ Jr. The familial occurrence of Creutzfeldt-Jakob disease and Alzheimer's disease. *Brain* 1981;104: 535–558.
360. Masters CL, Gajdusek DC, Gibbs CJ Jr., Bernouilli C, Asher DM. Familial Creutzfeldt-Jakob disease and other familial dementias: an inquiry into possible models of virus-induced familial diseases. In: Prusiner SB, Hadlow WJ, eds. *Slow transmissible diseases of the nervous system,* vol. 1. New York: Academic Press, 1979;143–194.
361. Masters CL, Harris JO, Gajdusek DC, Gibbs CJ Jr., Bernouilli C, Asher DM. Creutzfeldt-Jakob disease: patterns of worldwide occurrence and the significance of familal and sporadic clustering. *Ann Neurol* 1978; 5:177–188.
362. Masters CL, Richardson EP Jr. Subacute spongiform encephalopathy Creutzfeldt-Jakob disease—the nature and progression of spongiform change. *Brain* 1978;101:333–344.
363. Masullo C, Macchi G, Xi YG, Pocchiari M. Failure to ameliorate Creutzfeldt-Jakob disease with amphotericin B therapy. *J Infect Dis* 1992;165:784–785.
364. Masullo C, Pocchiari M, Macchi G, Alema G, Piazza G, Panzera MA. Transmission of Creutzfeldt-Jakob disease by dural cadaveric graft. *J Neurosurg* 1989;71:954.
365. McBride PA, Eikelenboom P, Kraal G, Fraser H, Bruce ME. PrP protein is associated with follicular dendritic cells of spleens and lymph nodes in uninfected and scrapie-infected mice. *J Pathol* 1992;168: 413–418.
366. McKinley MP, Bolton DC, Prusiner SB. A protease-resistant protein is a structural component of the scrapie prion. *Cell* 1983;35:57–62.
367. McKinley MP, Masiarz FR, Isaacs ST, Hearst JE, Prusiner SB. Resistance of the scrapie agent to inactivation by psoralens. *Photochem Photobiol* 1983;37:539–545.
368. McKinley MP, Meyer R, Kenaga L, et al. Scrapie prion rod formation in vitro requires both detergent extraction and limited proteolysis. *J Virol* 1991;65:1440–1449.
369. McKinley MP, Taraboulos A, Kenaga L, et al. Ultrastructural localization of scrapie prion proteins in cytoplasmic vesicles of infected cultured cells. *Lab Invest* 1991;65:622–630.
370. McKinley MP, Taraboulos A, Kenaga L, et al. Ultrastructural localization of scrapie prion proteins in secondary lysosomes of infected cultured cells. *J Cell Biol* 1990;111:316a.
371. McKnight S, Tjian R. Transcriptional selectivity of viral genes in mammalian cells. *Cell* 1986;46:795–805.
372. Medori R, Montagna P, Tritschler HJ, et al. Fatal familial insomnia: a second kindred with mutation of prion protein gene at codon 178. *Neurology* 1992;42:669–670.
373. Medori R, Tritschler H-J, LeBlanc A, et al. Fatal familial insomnia, a prion disease with a mutation at codon 178 of the prion protein gene. *N Engl J Med* 1992;326:444–449.
374. Meggendorfer F. Klinische und genealogische Beobachtungen bei einem Fall von spastischer Pseudosklerose Jakobs. *Z Ges Neurol Psychiatr* 1930;128:337–341.
375. Merz PA, Kascsak RJ, Rubenstein R, Carp RI, Wisniewski HM. Antisera to scrapie-associated fibril protein and prion protein decorate scrapie-associated fibrils. *J Virol* 1987;61:42–49.
376. Merz PA, Rohwer RG, Kascsak R, et al. Infection-specific particle from the unconventional slow virus diseases. *Science* 1984;225:437–440.
377. Merz PA, Somerville RA, Wisniewski HM, Iqbal K. Abnormal fibrils from scrapie-infected brain. *Acta Neuropathol (Berl)* 1981;54:63–74.
378. Merz PA, Somerville RA, Wisniewski HM, Manuelidis L, Manuelidis EE. Scrapie-associated fibrils in Creutzfeldt-Jakob disease. *Nature* 1983;306:474–476.
379. Merz PA, Wisniewski HM, Somerville RA, Bobin SA, Masters CL, Iqbal K. Ultrastructural morphology of amyloid fibrils from neuritic and amyloid plaques. *Acta Neuropathol (Berl)* 1983;60:113–124.
380. Meyer N, Rosenbaum V, Schmidt B, et al. Search for a putative scrapie genome in purified prion fractions reveals a paucity of nucleic acids. *J Gen Virol* 1991;72:37–49.
381. Meyer RK, McKinley MP, Bowman KA, Braunfeld MB, Barry RA, Prusiner SB. Separation and properties of cellular and scrapie prion proteins. *Proc Natl Acad Sci USA* 1986;83:2310–2314.
382. Millson G, Hunter GD, Kimberlin RH. An experimental examination of the scrapie agent in cell membrane mixtures. II. The association of scrapie infectivity with membrane fractions. *J Comp Pathol* 1971;81: 255–265.
383. Millson GC, Hunter GD, Kimberlin RH. The physico-chemical nature of the scrapie agent. In: Kimberlin RH, ed. *Slow virus diseases of animals and man.* New York: American Elsevier, 1976;243–266.
384. Miyashita K, Inuzuka T, Kondo H, et al. Creutzfeldt-Jakob disease in a patient with a cadaveric dural graft. *Neurology* 1991;41:940–941.
385. Miyazono M, Kitamoto T, Doh-ura K, Iwaki T, Tateishi J. Creutzfeldt-Jakob disease with codon 129 polymorphism (Valine): a comparative study of patients with codon 102 point mutation or without mutations. *Acta Neuropathol* 1992;84:349–354.
386. Mobley WC, Neve RL, Prusiner SB, McKinley MP. Nerve growth factor increases mRNA levels for the prion protein and the beta-amyloid protein precursor in developing hamster brain. *Proc Natl Acad Sci USA* 1988;85:9811–9815.
387. Monari L, Chen SG, Brown P, et al. Fatal familial insomnia and familial Creutzfeldt-Jakob disease: different prion proteins determined by a DNA polymorphism. *Proc Natl Acad Sci USA* 1994;91:2839–2842.
388. Morris JA, Gajdusek DC, Gibbs CJ Jr. Spread of scrapie from inoculated to uninoculated mice. *Proc Soc Exp Biol Med* 1965;120: 108–110.
389. Mullan M, Houlden H, Windelspecht M, et al. A locus for familial early-onset Alzheimer's disease on the long arm of chromosome 14, proximal to the a1-antichymotrypsin gene. *Nat Genet* 1992;2:340–342.

390. Muramoto T, Kitamoto T, Hoque MZ, Tateishi J, Goto I. Species barrier prevents an abnormal isoform of prion protein from accumulating in follicular dendritic cells of mice with Creutzfeldt-Jakob disease. *J Virol* 1993;67:6808–6810.
391. Muramoto T, Kitamoto T, Tateishi J, Goto I. Accumulation of abnormal prion protein in mice infected with Creutzfeldt-Jakob disease via intraperitoneal route: a sequential study. *Am J Pathol* 1993;143:1–10.
392. Muramoto T, Kitamoto T, Tateishi J, Goto I. Successful transmission of Creutzfeldt-Jakob disease from human to mouse verified by prion protein accumulation in mouse brains. *Brain Res* 1992;599:309–316.
393. Murdoch GH, Sklaviadis T, Manuelidis EE, Manuelidis L. Potential retroviral RNAs in Creutzfeldt-Jakob disease. *J Virol* 1990;64:1477–1486.
394. Narang HK. Relationship of protease-resistant protein, scrapie-associated fibrils and tubulofilamentous particles to the agent of spongiform encephalopathies. *Res Virol* 1992;143:381–386.
395. Narang HK. Scrapie-associated tubulofilamentous particles in scrapie hamsters. *Intervirology* 1992;34:105–111.
396. Narang HK. Scrapie-associated tubulofilamentous particles in human Creutzfeldt-Jakob disease. *Res Virol* 1992;143:387–395.
397. Narang HK. Ruthenium red and lanthanum nitrate a possible tracer and negative stain for scrapie "particles"? *Acta Neuropathol (Berl)* 1974; 29:37–43.
398. Narang HK, Asher DM, Gajdusek DC. Evidence that DNA is present in abnormal tubulofilamentous structures found in scrapie. *Proc Natl Acad Sci USA* 1988;85:3575–3579.
399. Narang HK, Asher DM, Gajdusek DC. Tubulofilaments in negatively stained scrapie-infected brains: relationship to scrapie-associated fibrils. *Proc Natl Acad Sci USA* 1987;84:7730–7734.
400. Narang HK, Asher DM, Pomeroy KL, Gajdusek DC. Abnormal tubulovesicular particles in brains of hamsters with scrapie. *Proc Soc Exp Biol Med* 1987;184:504–509.
401. Neary K, Caughey B, Ernst D, Race RE, Chesebro B. Protease sensitivity and nuclease resistance of the scrapie agent propagated in vitro in neuroblastoma-cells. *J Virol* 1991;65:1031–1034.
402. Neugut RH, Neugut AI, Kahana E, Stein Z, Alter M. Creutzfeldt-Jakob disease: familial clustering among Libyan-born Israelis. *Neurology* 1979;29:225–231.
403. New MI, Brown P, Temeck JW, Owens C, Hedley-Whyte ET, Richardson EP. Preclinical Creutzfeldt-Jakob disease discovered at autopsy in a human growth hormone recipient. *Neurology* 1988;38:1133–1134.
404. Nguyen J, Baldwin MA, Cohen FE, Prusiner SB. Prion protein peptides induce α-helix to β-sheet conformational transitions. *Biochemistry* 1995;34:4186–4192.
405. Nisbet TJ, MacDonaldson I, Bishara SN. Creutzfeldt-Jakob disease in a second patient who received a cadaveric dura mater graft.*J Am Med Assoc* 1989;261:1118.
406. Nochlin D, Sumi SM, Bird TD, et al. Familial dementia with PrP-positive amyloid plaques: a variant of Gerstmann-Sträussler syndrome. *Neurology* 1989;39:910–918.
407. O'Brien SJ. In: *Genetic maps—locus maps of complex genomes,* 6th ed. Cold Spring Harbor, NY: Cold Spring Harbor Laboratory Press, 1993;4.42–4.45.
408. Oesch B, Groth DF, Prusiner SB, Weissmann C. Search for a scrapie-specific nucleic acid: a progress report. In: Bock G, Marsh J, eds. *Novel infectious agents and the central nervous system, Ciba Foundation Symposium 135.* Chichester, UK: John Wiley and Sons, 1988;209–223.
409. Oesch B, Teplow DB, Stahl N, Serban D, Hood LE, Prusiner SB. Identification of cellular proteins binding to the scrapie prion protein. *Biochemistry* 1990;29:5848–5855.
410. Oesch B, Westaway D, Wälchli M, et al. A cellular gene encodes scrapie PrP 27–30 protein. *Cell* 1985;40:735–746.
411. Onodera T, Ikeda T, Muramatsu Y, Shinagawa M. Isolation of scrapie agent from the placenta of sheep with natural scrapie in Japan. *Microbiol Immunol* 1993;37:311–316.
412. Otto D. Jacob-Creutzfeldt disease associated with cadaveric dura. *J Neurosurg* 1987;67:149.
413. Outram G, Dickinson A, Fraser H. Reduced susceptibility of scrapie in mice after steroid administration. *Nature* 1974;249:855–856.
414. Owen F, Poulter M, Collinge J, Crow TJ. Codon 129 changes in the prion protein gene in Caucasians. *Am J Gen Health* 1990;46:1215–1216.
415. Owen F, Poulter M, Collinge J, et al. A dementing illness associated with a novel insertion in the prion protein gene. *Mol Brain Res* 1992; 13:155–157.
416. Owen F, Poulter M, Collinge J, et al. Insertions in the prion protein gene in atypical dementias. *Exp Neurol* 1991;112:240–242.
417. Owen F, Poulter M, Lofthouse R, et al. Insertion in prion protein gene in familial Creutzfeldt-Jakob disease. *Lancet* 1989;1:51–52.
418. Owen F, Poulter M, Shah T, et al. An in-frame insertion in the prion protein gene in familial Creutzfeldt-Jakob disease. *Mol Brain Res* 1990; 7:273–276.
419. Ozel M, Diringer H. Small virus-like structure in fraction from scrapie hamster brain. *Lancet* 1994;343:894–895.
420. Packer RJ, Cornblath DR, Gonatas NK, Bruno LA, Asbury AK. Creutzfeldt-Jakob disease in a 20-year-old woman. *Neurology* 1980; 30:492–496.
421. Palmer MS, Dryden AJ, Hughes JT, Collinge J. Homozygous prion protein genotype predisposes to sporadic Creutzfeldt-Jakob disease. *Nature* 1991;352:340–342.
422. Palmer MS, Mahal SP, Campbell TA, et al. Deletions in the prion protein gene are not associated with CJD. *Hum Molec Genet* 1993;2: 541–544.
423. Palsson PA. Rida (scrapie) in Iceland and its epidemiology. In: Prusiner SB, Hadlow WJ, eds. *Slow transmissible diseases of the nervous system,* vol. 1. New York: Academic Press, 1979;357–366.
424. Pan KM, Baldwin M, Nguyen J, et al. Conversion of a-helices into b-sheets features in the formation of the scrapie prion proteins. *Proc Natl Acad Sci USA* 1993;90:10962–10966.
425. Pan KM, Stahl N, Prusiner SB. Purification and properties of the cellular prion protein from Syrian hamster brain. *Protein Sci* 1992;1: 1343–1352.
426. Parry HB. Scrapie: a transmissible and hereditary disease of sheep. *Heredity* 1962;17:75–105.
427. Parry HB. Scrapie—natural and experimental. In: Whitty CWM, Hughes JT, MacCallum FO, eds. *Virus diseases and the nervous system.* Oxford: Blackwell Publishing, 1969;99–105.
428. Parry HB. *Scrapie disease in sheep*, Oppenheimer DR, ed. New York: Academic Press, 1983.
429. Pattison IH. Fifty years with scrapie: a personal reminiscence. *Vet Rec* 1988;123:661–666.
430. Pattison IH. The relative susceptibility of sheep, goats and mice to two types of the goat scrapie agent. *Res Vet Sci* 1966;7:207–212.
431. Pattison IH. Experiments with scrapie with special reference to the nature of the agent and the pathology of the disease. In: Gajdusek DC, Gibbs CJ Jr., Alpers MP, eds. *Slow, latent and temperate virus infections, NINDB Monograph 2.* Washington, DC: U.S. Government Printing, 1965;249–257.
432. Pattison IH. The spread of scrapie by contact between affected and healthy sheep, goats or mice. *Vet Rec* 1964;76:333–336.
433. Pattison IH, Hoare MN, Jebbett JN, Watson WA. Spread of scrapie to sheep and goats by oral dosing with foetal membranes from scrapie-affected sheep. *Vet Rec* 1972;90:465–468.
434. Pattison IH, Jones KM. The possible nature of the transmissible agent of scrapie. *Vet Rec* 1967;80:1–8.
435. Pattison IH, Millson GC. Experimental transmission of scrapie to goats and sheep by the oral route. *J Comp Pathol* 1961;71:171–176.
436. Pattison IH, Millson GC. Scrapie produced experimentally in goats with special reference to the clinical syndrome. *J Comp Pathol* 1961; 71:101–108.
437. Pearlman RL, Towfighi J, Pezeshkpour GH, Tenser RB, Turel AP. Clinical significance of types of cerebellar amyloid plaques in human spongiform encephalopathies. *Neurology* 1988;38:1249–1254.
438. Pearson GR, Wyatt JM, Gruffydd-Jones TJ, et al. Feline spongiform encephalopathy: fibril and PrP studies. *Vet Rec* 1992;131:307–310.
439. Petersen RB, Tabaton M, Berg L, et al. Analysis of the prion protein gene in thalamic dementia. *Neurology* 1992;42:1859–1863.
440. Pocchiari M, Peano S, Conz A, et al. Combination ultrafiltration and 6 M urea treatment of human growth hormone effectively minimizes risk from potential Creutzfeldt-Jakob disease virus contamination. *Horm Res* 1991;35:161–166.
441. Pocchiari M, Salvatore M, Cutruzzola F, et al. A new point mutation of the prion protein gene in familial and sporadic cases of Creutzfeldt-Jakob disease. *Ann Neurol* 1993;34:802–807.
442. Pocchiari M, Salvatore M, Ladogana A, et al. Experimental drug treatment of scrapie: a pathogenetic basis for rationale therapeutics. *Eur J Epidemiol* 1991;7:556–561.
443. Poidinger M, Kirkwood J, Almond W. Sequence analysis of the PrP protein from two species of antelope susceptible to transmissible spongiform encephalopathy. *Arch Virol* 1993;131:193–199.
444. Poulter M, Baker HF, Frith CD, et al. Inherited prion disease with 144 base pair gene insertion. 1. Genealogical and molecular studies. *Brain* 1992;115:675–685.

445. Powell-Jackson J, Weller RO, Kennedy P, Preece MA, Whitcombe EM, Newsome-Davis J. Creutzfeldt-Jakob disease after administration of human growth hormone. *Lancet* 1985;2:244–246.
446. Presnell SR, Cohen BI, Cohen FE. MacMatch: a tool for pattern-based protein secondary structure prediction. *Cabios* 1993;9:373–374.
447. Prusiner SB. The prion diseases. *Sci Am* 1995;272:48–57.
448. Prusiner SB. Inherited prion diseases. *Proc Natl Acad Sci USA* 1994; 91:4611–4614.
449. Prusiner SB. Transgenetics and cell biology of prion diseases: investigations of PrP^{Sc} synthesis and diversity. *Brit Med Bul* 1993;49:873–912.
450. Prusiner SB. Prion diseases. In: Scriver CR, Beaudet AL, Sly WS, Valle D, eds. *Metabolic basis of inherited disease,* 7th ed. New York: McGraw-Hill Publishing Co., 1993;
451. Prusiner SB. Molecular biology and genetics of neurodegenerative diseases caused by prions. *Adv Virus Res* 1992;41:241–280.
452. Prusiner SB. Molecular biology of prion diseases. *Science* 1991;252: 1515–1522.
453. Prusiner SB. Scrapie prions. *Annu Rev Microbiol* 1989;43:345–374.
454. Prusiner SB. Molecular structure, biology and genetics of prions. *Adv Virus Res* 1988;35:83–136.
455. Prusiner SB. The biology of prion transmission and replication. In: Prusiner SB, McKinley MP, eds. *Prions—novel infectious pathogens causing scrapie and Creutzfeldt-Jakob disease*. Orlando: Academic Press, 1987;83–112.
456. Prusiner SB. Novel proteinaceous infectious particles cause scrapie. *Science* 1982;216:136–144.
457. Prusiner SB, Bolton DC, Groth DF, Bowman KA, Cochran SP, McKinley MP. Further purification and characterization of scrapie prions. *Biochemistry* 1982;21:6942–6950.
458. Prusiner SB, Cochran SP, Alpers MP. Transmission of scrapie in hamsters. *J Infect Dis* 1985;152:971–978.
459. Prusiner SB, Cochran SP, Groth DF, Downey DE, Bowman KA, Martinez HM. Measurement of the scrapie agent using an incubation time interval assay. *Ann Neurol* 1982;11:353–358.
460. Prusiner SB, DeArmond SJ. Prion diseases and neurodegeneration. *Ann Rev Neurosci* 1994;17:311–339.
461. Prusiner SB, Fuzi M, Scott M, et al. Immunologic and molecular biological studies of prion proteins in bovine spongiform encephalopathy. *J Infect Dis* 1993;167:602–613.
462. Prusiner SB, Gajdusek DC, Alpers MP. Kuru with incubation periods exceeding two decades. *Ann Neurol* 1982;12:1–9.
463. Prusiner SB, Groth D, Serban A, et al. Ablation of the prion protein (PrP) gene in mice prevents scrapie and facilitates production of anti-PrP antibodies. *Proc Natl Acad Sci USA* 1993;90:10608–10612.
464. Prusiner SB, Groth D, Serban A, Stahl N, Gabizon R. Attempts to restore scrapie prion infectivity after exposure to protein denaturants. *Proc Natl Acad Sci USA* 1993;90:2793–2797.
465. Prusiner SB, Groth DF, Bildstein C, Masiarz FR, McKinley MP, Cochran SP. Electrophoretic properties of the scrapie agent in agarose gels. *Proc Natl Acad Sci USA* 1980;77:2984–2988.
466. Prusiner SB, Groth DF, Bolton DC, Kent SB, Hood LE. Purification and structural studies of a major scrapie prion protein. *Cell* 1984;38: 127–134.
467. Prusiner SB, Groth DF, Cochran SP, Masiarz FR, McKinley MP, Martinez HM. Molecular properties, partial purification, and assay by incubation period measurements of the hamster scrapie agent. *Biochemistry* 1980;19:4883–4891.
468. Prusiner SB, Groth DF, Cochran SP, McKinley MP, Masiarz FR. Gel electrophoresis and glass permeation chromatography of the hamster scrapie agent after enzymatic digestion and detergent extraction. *Biochemistry* 1980;19:4892–4898.
469. Prusiner SB, Hadlow WJ, Eklund CM, Race RE. Sedimentation properties of the scrapie agent. *Proc Natl Acad Sci USA* 1977;74:4656–4660.
470. Prusiner SB, Hadlow WJ, Eklund CM, Race RE, Cochran SP. Sedimentation characteristics of the scrapie agent from murine spleen and brain. *Biochemistry* 1978;17:4987–4992.
471. Prusiner SB, Hadlow WJ, Garfin DE, et al. Partial purification and evidence for multiple molecular forms of the scrapie agent. *Biochemistry* 1978;17:4993–4997.
472. Prusiner SB, Hsiao KK. Human prion diseases. *Ann Neurol* 1994;35: 385–395.
473. Prusiner SB, Hsiao KK, Bredesen DE, DeArmond SJ. Prion disease. In: Vinken PJ, Bruyn GW, Klawans HL, eds. *Handbook of clinical neurology,* vol. 12 (56): *Viral disease*. Amsterdam: Elsevier Science Publishers, 1989;543–580.
474. Prusiner SB, McKinley MP, Bolton DC, et al. Prions: methods for assay, purification and characterization. In: Maramorosch K, Koprowski H, eds. *Methods in virology*. New York: Academic Press, 1984;293–345.
475. Prusiner SB, McKinley MP, Bowman KA, et al. Scrapie prions aggregate to form amyloid-like birefringent rods. *Cell* 1983;35:349–358.
476. Prusiner SB, McKinley MP, Groth DF, et al. Scrapie agent contains a hydrophobic protein. *Proc Natl Acad Sci USA* 1981;78:6675–6679.
477. Prusiner SB, Scott M, Foster D, et al. Transgenetic studies implicate interactions between homologous PrP isoforms in scrapie prion replication. *Cell* 1990;63:673–686.
478. Puckett C, Concannon P, Casey C, Hood L. Genomic structure of the human prion protein gene. *Am J Hum Genet* 1991;49:320–329.
479. Race RE, Graham K, Ernst D, Caughey B, Chesebro B. Analysis of linkage between scrapie incubation period and the prion protein gene in mice. *J Gen Virol* 1990;71:493–497.
480. Raeber AJ, Borchelt DR, Scott M, Prusiner SB. Attempts to convert the cellular prion protein into the scrapie isoform in cell-free systems. *J Virol* 1992;66:6155–6163.
481. Ridley RM, Baker HF. Occupational risk of Creutzfeldt-Jakob disease. *Lancet* 1993;341:641–642.
482. Riesner D, Kellings K, Meyer N, Mirenda C, Prusiner SB. Nucleic acids and scrapie prions. In: Prusiner SB, Collinge J, Powell J, Anderton B, eds. *Prion diseases of humans and animals*. London: Ellis Horwood, 1992;341–358.
483. Ripoll L, Laplanche J-L, Salzmann M, et al. A new point mutation in the prion protein gene at codon 210 in Creutzfeldt-Jakob disease. *Neurology* 1993;43:1934–1938.
484. Roberts GW, Lofthouse R, Allsop D, et al. CNS amyloid proteins in neurodegenerative diseases. *Neurology* 1988;38:1534–1540.
485. Roberts GW, Lofthouse R, Brown R, Crow TJ, Barry RA, Prusiner SB. Prion-protein immunoreactivity in human transmissible dementias. *N Engl J Med* 1986;315:1231–1233.
486. Rogers M, Serban D, Gyuris T, Scott M, Torchia T, Prusiner SB. Epitope mapping of the Syrian hamster prion protein utilizing chimeric and mutant genes in a vaccinia virus expression system. *J Immunol* 1991;147:3568–3574.
487. Rogers M, Taraboulos A, Scott M, Groth D, Prusiner SB. Intracellular accumulation of the cellular prion protein after mutagenesis of its Asn-linked glycosylation sites. *Glycobiology* 1990;1:101–109.
488. Rogers M, Yehiely F, Scott M, Prusiner SB. Conversion of truncated and elongated prion proteins into the scrapie isoform in cultured cells. *Proc Natl Acad Sci USA* 1993;90:3182–3186.
489. Rohwer RG. The scrapie agent: "a virus by any other name." *Curr Top Microbiol Immunol* 1991;172:195–232.
490. Rohwer RG. Estimation of scrapie nucleic acid molecular weight from standard curves for virus sensitivity to ionizing radiation. *Nature* 1986; 320:381.
491. Rohwer RG. Scrapie infectious agent is virus-like in size and susceptibility to inactivation. *Nature* 1984;308:658–662.
492. Rohwer RG. Virus-like sensitivity of the scrapie agent to heat inactivation. *Science* 1984;223:600–602.
493. Rohwer RG, Gajdusek DC. Scrapie—virus or viroid, the case for a virus. In: Boese A, ed. *Search for the cause of multiple sclerosis and other chronic diseases of the central nervous system*. Weinheim: Verlag Chemie, 1980;333–355.
494. Roos R, Gajdusek DC, Gibbs CJ Jr. The clinical characteristics of transmissible Creutzfeldt-Jakob disease. *Brain* 1973;96:1–20.
495. Rosen DR, Siddique T, Patterson D, et al. Mutations in Cu/Zn superoxide dismutase gene are associated with familial amyotrophic lateral slerosis. *Nature* 1993;362:59–62.
496. Rosenthal NP, Keesey J, Crandall B, Brown WJ. Familial neurological disease associated with spongiform encephalopathy. *Arch Neurol* 1976;33:252–259.
497. Safar J, Ceroni M, Piccardo P, et al. Subcellular distribution and physicochemical properties of scrapie associated precursor protein and relationship with scrapie agent. *Neurology* 1990;40:503–508.
498. Safar J, Roller PP, Gajdusek DC, Gibbs CJ Jr. Scrapie amyloid (prion) protein has the conformational characteristics of an aggregated molten globule folding intermediate. *Biochemistry* 1994;33:8375–8383.
499. Safar J, Roller PP, Gajdusek DC, Gibbs CJ Jr. Conformational transitions, dissociation, and unfolding of scrapie amyloid (prion) protein. *J Biol Chem* 1993;268:20276–20284.
500. Safar J, Roller PP, Gajdusek DC, Gibbs CJJ. Thermal-stability and conformational transitions of scrapie amyloid (prion) protein correlate with infectivity. *Protein Sci* 1993;2:2206–2216.

501. Safar J, Wang W, Padgett MP, et al. Molecular mass, biochemical composition, and physicochemical behavior of the infectious form of the scrapie precursor protein monomer. *Proc Natl Acad Sci USA* 1990;87: 6373–6377.
502. Sakaguchi S, Katamine S, Yamanouchi K, et al. Kinetics of infectivity are dissociated from PrP accumulation in salivary glands of Creutzfeldt-Jakob disease agent-inoculated mice. *J Gen Virol* 1993;74:2117–2123.
503. Sawcer SJ, Yuill GM, Esmonde TFG, et al. Creutzfeldt-Jakob disease in an individual occupationally exposed to BSE. *Lancet* 1993;341:642.
504. Schätzl HM, Da Costa M, Taylor L, Cohen FE, Prusiner SB. Prion protein gene variation among primates. *J Mol Biol* 1995;245:362–374.
505. Schellenberg GD, Bird TD, Wijsman EM, et al. Genetic linkage evidence for a familial Alzheimer's disease locus on chromosome 14. *Science* 1992;258:668–671.
506. Scott M, Foster D, Mirenda C, et al. Transgenic mice expressing hamster prion protein produce species-specific scrapie infectivity and amyloid plaques. *Cell* 1989;59:847–857.
507. Scott M, Groth D, Foster D, et al. Propagation of prions with artificial properties in transgenic mice expressing chimeric PrP genes. *Cell* 1993; 73:979–988.
508. Scott MR, Köhler R, Foster D, Prusiner SB. Chimeric prion protein expression in cultured cells and transgenic mice. *Protein Sci* 1992;1: 986–997.
509. Selvaggini C, De Gioia L, Cantu L, et al. Molecular characteristics of a protease-resistant, amyloidogenic and neurotoxic peptide homologous to residues 106–126 of the prion protein. *Biochem Biophys Res Commun* 1993;194:1380–1386.
510. Serban D, Taraboulos A, DeArmond SJ, Prusiner SB. Rapid detection of Creutzfeldt-Jakob disease and scrapie prion proteins. *Neurology* 1990;40:110–117.
511. Shyng SL, Heuser JE, Harris DA. A glycolipid-anchored prion protein is endocytosed via clathrin-coated pits. *J Cell Biol* 1994;125:1239–1250.
512. Shyng SL, Huber MT, Harris DA. A prion protein cycles between the cell surface and an endocytic compartment in cultured neuroblastoma cells. *J Biol Chem* 1993;21:15922–15928.
513. Siakotos AN, Gajdusek DC, Gibbs CJ Jr., Traub RD, Bucana C. Partial purification of the scrapie agent from mouse brain by pressure disruption and zonal centrifugation in sucrose-sodium chloride gradients. *Virology* 1976;70:230–237.
514. Siakotos AN, Raveed D, Longa G. The discovery of a particle unique to brain and spleen subcellular fractions from scrapie-infected mice. *J Gen Virol* 1979;43:417–422.
515. Sigurdsson B. Rida, a chronic encephalitis of sheep with general remarks on infections which develop slowly and some of their special characteristics. *Br Vet J* 1954;110:341–354.
516. Simpson DA, Lander H, Robson HN. Observations on kuru. II. Clinical features. *Aust Ann Med* 1959;8:8–15.
517. Sklaviadis T, Akowitz A, Manuelidis EE, Manuelidis L. Nucleic acid binding proteins in highly purified Creutzfeldt-Jakob disease preparations. *Proc Natl Acad Sci USA* 1993;90:5713–5717.
518. Sklaviadis T, Akowitz A, Manuelidis EE, Manuelidis L. Nuclease treatment results in high specific purification of Creutzfeldt-Jakob disease infectivity with a density characteristic of nucleic acid-protein complexes. *Arch Virol* 1990;112:215–229.
519. Sklaviadis T, Dreyer R, Manuelidis L. Analysis of Creutzfeldt-Jakob disease infectious fractions by gel permeation chromatography and sedimentation field flow fractionation. *Virus Res* 1992;26:241–254.
520. Sklaviadis TK, Manuelidis L, Manuelidis EE. Physical properties of the Creutzfeldt-Jakob disease agent. *J Virol* 1989;63:1212–1222.
521. Somerville RA, Millson GC, Hunter GD. Changes in a protein-nucleic acid complex from synaptic plasma membrane of scrapie-infected mouse brain. *Biochem Soc Trans* 1976;4:1112–1114.
522. Somerville RA, Ritchie LA, Gibson PH. Structural and biochemical evidence that scrapie-associated fibrils assemble in vivo. *J Gen Virol* 1989;70:25–35.
523. Sparkes RS, Simon M, Cohn VH, et al. Assignment of the human and mouse prion protein genes to homologous chromosomes. *Proc Natl Acad Sci USA* 1986;83:7358–7362.
524. St. George-Hyslop P, Haines J, Rogaev E, et al. Genetic evidence for a novel familial Alzheimer's disease locus on chromosome 14. *Nat Genet* 1992;2:330–334.
525. Stahl N, Baldwin MA, Hecker R, Pan K-M, Burlingame AL, Prusiner SB. Glycosylinositol phospholipid anchors of the scrapie and cellular prion proteins contain sialic acid. *Biochemistry* 1992;31:5043–5053.
526. Stahl N, Baldwin MA, Teplow DB, et al. Structural analysis of the scrapie prion protein using mass spectrometry and amino acid sequencing. *Biochemistry* 1993;32:1991–2002.
527. Stahl N, Borchelt DR, Hsiao K, Prusiner SB. Scrapie prion protein contains a phosphatidylinositol glycolipid. *Cell* 1987;51:229–240.
528. Stender A. Weitere Beiträge zum Kapitel "Spastische Pseudosklerose Jakobs". *Z Neurol Psychiat* 1930;128:528–543.
529. Tagliavini F, Prelli F, Ghisto J, et al. Amyloid protein of Gerstmann-Sträussler-Scheinker disease (Indiana kindred) is an 11-kd fragment of prion protein with an N-terminal glycine at codon 58. *EMBO J* 1991; 10:513–519.
530. Tagliavini F, Prelli F, Verga L, et al. Synthetic peptides homologous to prion protein residues 106–147 form amyloid-like fibrils in vitro. *Proc Natl Acad Sci USA* 1993;90:9678–9682.
531. Taguchi F, Tamai Y, Miura S. Experiments on maternal and paternal transmission of Creutzfeldt-Jakob disease in mice. *Arch Virol* 1993; 130:219–224.
532. Tamai Y, Kojima H, Kitajima R, et al. Demonstration of the transmissible agent in tissue from a pregnant woman with Creutzfeldt-Jakob disease. *N Engl J Med* 1992;327:649.
533. Tange RA, Troost D, Limburg M. Progressive fatal dementia (Creutzfeldt-Jakob disease) in a patient who received homograft tissue for tympanic membrane closure. *Eur Arch Otorhinolaryngol* 1989;247: 199–201.
534. Taraboulos A, Jendroska K, Serban D, Yang S-L, DeArmond SJ, Prusiner SB. Regional mapping of prion proteins in brains. *Proc Natl Acad Sci USA* 1992;89:7620–7624.
535. Taraboulos A, Raeber AJ, Borchelt DR, Serban D, Prusiner SB. Synthesis and trafficking of prion proteins in cultured cells. *Mol Biol Cell* 1992;3:851–863.
536. Taraboulos A, Rogers M, Borchelt DR, et al. Acquisition of protease resistance by prion proteins in scrapie-infected cells does not require asparagine-linked glycosylation. *Proc Natl Acad Sci USA* 1990;87: 8262–8266.
537. Taraboulos A, Scott M, Semenov A, Avrahami D, Laszlo L, Prusiner SB. Cholesterol depletion and modification of C-terminal targeting sequence of the prion protein inhibit formation of the scrapie isoform. *J Cell Biol* 1995;129:121–132.
538. Taraboulos A, Serban D, Prusiner SB. Scrapie prion proteins accumulate in the cytoplasm of persistently infected cultured cells. *J Cell Biol* 1990;110:2117–2132.
539. Tateishi J, Doh-ura K, Kitamoto T, et al. Prion protein gene analysis and transmission studies of Creutzfeldt-Jakob disease. In: Prusiner SB, Collinge J, Powell J, Anderton B, eds. *Prion diseases of humans and animals*. London: Ellis Horwood, 1992;129–134.
540. Tateishi J, Kitamoto T. Developments in diagnosis for prion diseases. *Br Med Bull* 1993;49:971–979.
541. Tateishi J, Kitamoto T, Doh-ura K, et al. Immunochemical, molecular genetic, and transmission studies on a case of Gerstmann-Sträussler-Scheinker syndrome. *Neurology* 1990;40:1578–1581.
542. Tateishi J, Ohta M, Koga M, Sato Y, Kuroiwa Y. Transmission of chronic spongiform encephalopathy with kuru plaques from humans to small rodents. *Ann Neurol* 1979;5:581–584.
543. Taylor DM, Dickinson AG, Fraser H, Robertson PA, Salacinski PR, Lowry PJ. Preparation of growth hormone free from contamination with unconventional slow viruses. *Lancet* 1985;2:260–262.
544. Telling GC, Scott M, Hsiao KK, et al. Transmission of Creutzfeldt-Jakob disease from humans to transgenic mice expressing chimeric human-mouse prion protein. *Proc Natl Acad Sci USA* 1994;91:9936–9940.
545. Terzano MG, Montanari E, Calzetti S, Mancia D, Lechi A. The effect of amantadine on arousal and EEG patterns in Creutzfeldt-Jakob disease. *Arch Neurol* 1983;40:555–559.
546. Thadani V, Penar PL, Partington J, et al. Creutzfeldt-Jakob disease probably acquired from a cadaveric dura mater graft. Case report. *J Neurosurg* 1988;69:766–769.
547. Titner R, Brown P, Hedley-Whyte ET, Rappaport EB, Piccardo CP, Gajdusek DC. Neuropathologic verification of Creutzfeldt-Jakob disease in the exhumed American recipient of human pituitary growth hormone: epidemiologic and pathogenetic implications. *Neurology* 1986;36:932–936.
548. Tranchant C, Doh-ura K, Warter JM, et al. Gerstmann-Sträussler-Scheinker disease in an Alsatian family: clinical and genetic studies. *J Neurol Neurosurg Psychiatry* 1992;55:185–187.
549. Turk E, Teplow DB, Hood LE, Prusiner SB. Purification and proper-

ties of the cellular and scrapie hamster prion proteins. *Eur J Biochem* 1988;176:21–30.
550. Udovitch AL, Valensi L. *The last Arab Jews: the communities of Jerba, Tunisia*. London: Harwood Academic Publishers, 1984:178.
551. Van Broeckhoven C, Backhovens H, Cruts M, et al. Mapping of a gene predisposing to early-onset Alzheimer's disease to chromosome 14q24.3. *Nat Genet* 1992;2:335–339.
552. Van Broeckhoven C, Haan J, Bakker E, et al. Amyloid b protein precursor gene and hereditary cerebral hemorrhage with amyloidosis (Dutch). *Science* 1990;248:1120–1122.
553. Vernon ML, Horta-Barbosa L, Fuccillo DA, Sever JL, Baringer JR, Birnbaum G. Virus-like particles and nucleoprotein-type filaments in brain tissue from two patients with Creutzfeldt-Jakob disease. *Lancet* 1970;1:964–966.
554. Vnencak-Jones CL, Phillips JA. Identification of heterogeneous PrP gene deletions in controls by detection of allele-specific heteroduplexes (DASH). *Am J Hum Genet* 1992;50:871–872.
555. Weissmann C. A "unified theory" of prion propagation. *Nature* 1991; 352:679–683.
556. Weitgrefe S, Zupancic M, Haase A, et al. Cloning of a gene whose expression is increased in scrapie and in senile plaques. *Science* 1985; 230:1177–1181.
557. Wells GAH, Scott AC, Johnson CT, et al. A novel progressive spongiform encephalopathy in cattle. *Vet Rec* 1987;121:419–420.
558. Westaway D, Cooper C, Turner S, Da Costa M, Carlson GA, Prusiner SB. Structure and polymorphism of the mouse prion protein gene. *Proc Natl Acad Sci USA* 1994;91:6418–6422.
559. Westaway D, DeArmond SJ, Cayetano-Canlas J, et al. Degeneration of skeletal muscle, peripheral nerves, and the central nervous system in transgenic mice overexpressing wild-type prion proteins. *Cell* 1994; 76:117–129.
560. Westaway D, Goodman PA, Mirenda CA, McKinley MP, Carlson GA, Prusiner SB. Distinct prion proteins in short and long scrapie incubation period mice. *Cell* 1987;51:651–662.
561. Westaway D, Mirenda CA, Foster D, et al. Paradoxical shortening of scrapie incubation times by expression of prion protein transgenes derived from long incubation period mice. *Neuron* 1991;7:59–68.
562. Westaway D, Zuliani V, Cooper CM, et al. Homozygosity for prion protein alleles encoding glutamine-171 renders sheep susceptible to natural scrapie. *Genes Dev* 1994;8:959–969.
563. Wexler NS, Young AB, Tanzi RE, et al. Homozygotes for Huntington's disease. *Nature* 1987;326:194–197.
564. Wickner RB. Evidence for a prion analog in S. Qcerevisiae: the [URE3] non-Mendelian genetic element as an altered URE2 protein. *Science* 1994;264:566–569.
565. Wilesmith J, Wells GAH. Bovine spongiform encephalopathy. *Curr Top Microbiol Immunol* 1991;172:21–38.
566. Wilesmith JW. Bovine spongiform encephalopathy: a brief epidemiography, 1985–1991. In: Prusiner SB, Collinge J, Powell J, Anderton B, eds. *Prion diseases of humans and animals*. London: Ellis Horwood, 1992;243–255.
567. Wilesmith JW, Hoinville LJ, Ryan JBM, Sayers AR. Bovine spongiform encephalopathy: aspects of the clinical picture and analyses of possible changes 1986–1990. *Vet Rec* 1992;130:197–201.
568. Wilesmith JW, Ryan JBM, Hueston WD, Hoinville LJ. Bovine spongiform encephalopathy: epidemiological features 1985 to 1990. *Vet Rec* 1992;130:90–94.
569. Wilesmith JW, Wells GAH, Cranwell MP, Ryan JBM. Bovine spongiform encephalopathy: epidemiological studies. *Vet Rec* 1988;123: 638–644.
570. Will RG, Matthews WB. Evidence for case-to-case transmission of Creutzfeldt-Jakob disease. *J Neurol Neurosurg Psychiatry* 1982;45: 235–238.
572. Williams ES, Young S. Spongiform encephalopathy of Rocky Mountain Elk. *J Wildl Dis* 1982;18:465–471.
571. Williams ES, Young S. Chronic wasting disease of captive mule deer: a spongiform encephalopathy. *J Wildl Dis* 1980;16:89–98.
573. Williams RC. Electron microscopy of viruses. *Adv Virus Res* 1954;2: 183–239.
574. Willison HJ, Gale AN, McLaughlin JE. Creutzfeldt-Jakob disease following cadaveric dura mater graft. *J Neurol Neurosurg Psychiatry* 1991;54:940.
575. Willoughby K, Kelly DF, Lyon DG, Wells GAH. Spongiform encephalopathy in a captive puma (Felis concolor). *Vet Rec* 1992;131: 431–434.
576. Wilson DR, Anderson RD, Smith W. Studies in scrapie. *J Comp Pathol* 1950;60:267–282.
577. Worthington M. Interferon system in mice infected with the scrapie agent. *Infect Immun* 1972;6:643–645.
578. Xi YG, Ingrosso L, Ladogana A, Masullo C, Pocchiari M. Amphotericin B treatment dissociates in vivo replication of the scrapie agent from PrP accumulation. *Nature* 1992;356:598–601.
579. Ying Y-S, Anderson RGW, Rothberg KG. Each caveola contains multiple glycosyl-phosphatidylinositol-anchored membrane proteins. *Cold Spring Harb Symp Quant Biol* 1992;57:593–604.
580. Yost CS, Lopez CD, Prusiner SB, Myers RM, Lingappa VR. Non-hydrophobic extracytoplasmic determinant of stop transfer in the prion protein. *Nature* 1990;343:669–672.
581. Zigas V, Gajdusek DC. Kuru: clinical study of a new syndrome resembling paralysis agitans in natives of the Eastern Highlands of Australian New Guinea. *Med J Aust* 1957;2:745–754.
582. Zilber N, Kahana E, Abraham MPH. The Libyan Creutzfeldt-Jakob disease focus in Israel: an epidemiologic evaluation. *Neurology* 1991; 41:1385–1389.

Subject Index

Subject Index

C

O

W

X

ISBN 0-7817-0284-4

9 780781 702843